2019 IEEE 15th Brazilian Power Electronics Conference and 5th IEEE Southern Power Electronics Conference (COBEP/SPEC 2019)

Santos, Brazil
1-4 December 2019

Pages 1-723

IEEE Catalog Number:	CFP1977F-POD
ISBN:	978-1-7281-4181-7

Copyright © 2019 by the Institute of Electrical and Electronics Engineers, Inc.
All Rights Reserved

Copyright and Reprint Permissions: Abstracting is permitted with credit to the source. Libraries are permitted to photocopy beyond the limit of U.S. copyright law for private use of patrons those articles in this volume that carry a code at the bottom of the first page, provided the per-copy fee indicated in the code is paid through Copyright Clearance Center, 222 Rosewood Drive, Danvers, MA 01923.

For other copying, reprint or republication permission, write to IEEE Copyrights Manager, IEEE Service Center, 445 Hoes Lane, Piscataway, NJ 08854. All rights reserved.

*** This is a print representation of what appears in the IEEE Digital Library. Some format issues inherent in the e-media version may also appear in this print version.

IEEE Catalog Number: CFP1977F-POD
ISBN (Print-On-Demand):   978-1-7281-4181-7
ISBN (Online):   978-1-7281-4180-0
ISSN: 2165-0454

Additional Copies of This Publication Are Available From:

Curran Associates, Inc
57 Morehouse Lane
Red Hook, NY 12571 USA
Phone: (845) 758-0400
Fax: (845) 758-2633
E-mail: curran@proceedings.com
Web: www.proceedings.com

TABLE OF CONTENTS

SIMPLIFIED SINGLE-PHASE PV GENERATOR MODEL FOR DISTRIBUTION FEEDERS WITH HIGH PENETRATION OF POWER ELECTRONICS-BASED SYSTEMS 1

Mariana Altoé Mendes ; Murillo Cobe Vargas ; Oureste Elias Batista ; Yongheng Yang ; Frede Blaabjerg

MODULATED MODEL PREDICTIVE CURRENT CONTROL FOR PMSM OPERATING WITH THREE-LEVEL NPC INVERTER 8

Qi Wang ; Marco Rivera ; Jose A. Riveros ; Patrick Wheeler

MODEL PREDICTIVE CONTROLLER FOR TWO-PHASE THREE-WIRE GRID-CONNECTED CONVERTERS 13

Pablo C. De S. Furtado ; Pedro Gomes Barbosa

A NEW ZERO-SEQUENCE VOLTAGE COMPENSATION ALGORITHM FOR A DSTATCOM BASED ON CONSUMER UNBALANCE 19

Bruno C. Souza ; Samuel N. Duarte ; Pedro M. Almeida ; Pedro G. Barbosa ; Leandro R. Araújo

DESIGN OF RESONANT CONTROLLERS FOR COMPENSATION OF THIRD HARMONIC RIPPLE IN THE DC CAPACITORS VOLTAGES OF NPC CONVERTERS 25

Andrei De O. Almeida ; Pedro M. De Almeida ; Pedro G. Barbosa

SINGLE PHASE-SHIFT CONTROL OF DAB CONVERTER USING ROBUST PARAMETRIC APPROACH 31

Kevin E. Lucas ; Daniel J. Pagano ; Renan L. P. Medeiros

OPERATION BOUNDARIES OF A SINGLE PHASE THYRISTOR DRIVEN DC-MOTOR 37

Thomas M. Mertens ; Richard M. Stephan

INTEGRATED LOCAL CONTROL OF ACTIVE POWER AND VOLTAGE SUPPORT FOR THREE-PHASE THREE-WIRE CONVERTERS 43

Ya Zhang ; Gabriel Tibola ; Maurice G. L. Roe ; Jorge L. Duarte

IMPLEMENTATION OF A DIDACTIC PLATFORM FOR A GENERIC LOAD TORQUE EMULATOR USING INDUCTION MACHINES AND PWM INVERTERS 49

Luiz Otávio Campos De Medeiros ; José Carlos Grilo Rodrigues ; Angelo José Junqueira Rezek ; Nery De Oliveira Junior ; Rafael Di Lorenzo Corrêa ; Alexandre Viana Braga ; Christel Enock Ghislain Ogoulola ; Vinicius Zimmermann Silva ; Marcos Leonardo Ramos

ENERGY EFFICIENT CONTROL OF SYNCHRONOUS MACHINES IN DEEP FIELD-WEAKENING OPERATION INCLUDING SATURATION EFFECTS 55

Joao Bonifacio ; Ralph Kennel

PREDICTIVE VOLTAGE CONTROL OPERATING AT FIXED SWITCHING FREQUENCY OF A NEUTRAL-POINT CLAMPED CONVERTER 61

Felipe Herrera ; Roberto Cárdenas ; Marco Rivera ; José A. Riveros ; Patrick Wheeler

DEVELOPMENT OF CURRENT-SOURCE-INVERTER-BASED INTEGRATED MOTOR DRIVES USING WIDE-BANDGAP POWER SWITCHES 67

Renato A. Torres ; Hang Dai ; Woongkul Lee ; Thomas M. Jahns ; Bulent Sarlioglu

POWER CONTROL OF A DOUBLY FED INDUCTION WIND GENERATOR EMPLOYING A TAKAGI-SUGENO FUZZY LOGIC CONTROLLER 73

C. M. Rocha-Osorio ; J. S. Solís-Chaves ; Eliomar R. Conde D. ; J. L. Azcue Puma ; Fernando Lino ; A. J. Sguarezi Filho

NONISOLATED DC-DC QUADRATIC CUK CONVERTER FOR WIDE CONVERSION RANGE APPLICATIONS 79

Tatiane Martins Oliveira ; Lara Ana Rodarte Rios ; Fernando Lessa Tofoli ; Aniel Silva De Morais

IMPLEMENTATION OF AUTOMATIC BATTERY CHARGING TEMPERATURE COMPENSATION ON A PEAK-SHAVING ENERGY STORAGE EQUIPMENT 85

Wilson Cesar Sant'Ana ; Robson Bauwelz Gonzatti ; Germano Lambert-Torres ; Erik Leandro Bonaldi ; Pedro Andrade De Oliveira ; Bruno Silva Torres ; Joao Gabriel Luppi Foster ; Rondineli Rodrigues Pereira ; Luiz Eduardo Borges-Da-Silva ; Denis Mollica ; Jos

CASCADE CONTROL VS FULL-STATE FEEDBACK 92

Débora M. Soares ; Henrique A. M. Calil ; Richard M. Stephan

DESIGN AND PERFORMANCE ANALYSIS OF ISOLATED CUK CONVERTER EMPLOYED IN MULTIPLE PULSE RECTIFIER SYSTEMS 98

Ana L. Soares ; Antônio O. Costa Neto ; Gustavo B. Lima ; Luiz C. G. Freitas ; Ernane A. A. Coelho

A RESEARCH ON CONSTANT VOLTAGE OUTPUT CHARACTERISTICS OF WIRELESS POWER TRANSFER SYSTEM WITH A DC-DC CONVERTER 104

Zhimeng Liu ; Lifang Wang ; Chengliang Yin ; Yanjie Guo ; Chengxuan Tao

FINITE CONTROL SET MODEL BASED PREDICTIVE CONTROL OF GRID-TIED SIX-SWITCH CONVERTER APPLIED TO INDUCTION GENERATOR 108

Paulo R. U. Guazzelli ; Allan G. De Castro ; Stefan T. C. A. Dos Santos ; Carlos M. R. De Oliveira ; José R. B. A. Monteiro ; Manoel L. De Aguiar

APPLYING COUPLED INDUCTORS TO THE CLAMPED-RESONANT INTERLEAVED BOOST CONVERTER 114

Giorgio Spiazzi

LPV MODELING OF BOOST CONVERTER AND GAIN SCHEDULING MPC CONTROL 120

Rosana C. B. Rego

ZERO-CROSSING DETECTION FREQUENCY ESTIMATOR METHOD COMBINED WITH A KALMAN FILTER FOR NON-IDEAL POWER GRID 125

Tiago Davi Curi Busarello ; Sérgio Luiz Sambugari Junior ; Newton Da Silva

A REVIEW OF FCS-MPC IN MULTILEVEL CONVERTERS APPLIED TO ACTIVE POWER FILTERS 131

João G. L. Foster ; Rondineli R. Pereira ; Robson B. Gonzatti ; Wilson C. Sant'Ana ; Denis Mollica ; Germano Lambert-Torres

DISTANCE DETECTION SYSTEM FOR DIGITAL TRANSMITTER COIL ACHIEVING DISTANCE-VARIATION-TOLERANT WIRELESS POWER TRANSFER 137

Hao Qiu ; Yoshiaki Narusue ; Yoshihiro Kawahara ; Takayasu Sakurai ; Makoto Takamiya

DIGITAL CURRENT CONTROL FOR A BIDIRECTIONAL INTERLEAVED BOOST CONVERTER WITH COUPLED INDUCTORS 143

Francesco Toniolo ; Qing Liu ; Tommaso Caldognetto ; Simone Buso ; Giorgio Spiazzi

PEMSYN: A FREE SOFTWARE TO ASSIST THE DESIGN AND PERFORMANCE ASSESSMENT OF PERMANENT MAGNETS SYNCHRONOUS MACHINES 149

Khristian M. De Andrade ; Hugo E. Santos ; Wellington M. Vilela ; Thales E. P. De Almeida ; Geyverson T. De Paula

DEVELOPMENT OF LINEAR GENERATOR PROTOTYPE AS PART OF A POINT ABSORBER WAVE ENERGY CONVERTER 155

Eduardo Martins ; Wagner Marcilio ; Marcos Moreira ; João Santos

THE TRUE UNITY POWER FACTOR CONVERTER APPLIED TO PHOTOVOLTAIC APPLICATIONS 160

Marcos Henrique Da Silva Alves ; Thiago Morais Parreiras ; Braz De Jesus Cardoso Filho

HIGH-VOLTAGE STEP-UP DC-DC CONVERTER EMPLOYING THE FOUR STATE SWITCHING CELL AND VOLTAGE MULTIPLIER CELLS 166

Paulo Henrique Feretti ; Enio Roberto Ribeiro ; Fernando Lessa Tofoli

STABILIZATION OF DC MICROGRIDS WITH POINT-OF-LOAD CONVERTERS AS CONSTANT POWER LOADS 172

Isaías V. Bessa ; Renan L. P. Medeiros ; Iury V. Bessa ; Florindo A. C. Ayres Junior ; Kevin E. Lucas

INDUCTOR DESIGN METHODOLOGY FOR POWER ELECTRONICS APPLICATIONS 178

P. H. J. Vilkn ; L. M. F. Morais ; R. A. S. Santana ; P. C. Cortizo ; P. F. Seixas

REAL-TIME IMPLEMENTATION OF A DC CONVERTER USING MODIFIED NODAL ANALYSIS, SPARSITY HANDLING AND PARALLELISM ON A DSP PLATFORM 184

Luiz Felipe Corrêa De Sá Santos Ribeiro ; Felipe Novaes Francis Dicler ; Luis Guilherme Barbosa Rolim ; Mauricio Aredes

INVESTIGATION OF CONTROL STRATEGIES TO MITIGATE THE OSCILLATION EFFECTS CAUSED BY INTERCONNECTED BUCK CONVERTERS 190

Isaías Valente De Bessa ; Renan L. P. Medeiros ; Iury Valente De Bessa ; Florindo A. C. Ayres Junior ; Kevin E. Lucas

INTEGRATING A SINGLE Z-SOURCE NETWORK WITH A MODULAR MULTILEVEL CONVERTER FOR VOLTAGE BOOSTING 196

Fatma A. Khera ; Christian Klumpner ; Pat W Wheeler

OPTIMIZATION OF ROBUST PI CONTROLLERS FOR GRID-TIED INVERTERS 202

Caio R. D. Osório ; Lucas C. Borin ; Gustavo G. Koch ; Vinícius F. Montagner

INVESTIGATION OF VOLTAGE REGULATION WITH ACTIVE AND REACTIVE POWER WITH DISTRIBUTED LOADS ON A RADIAL DISTRIBUTION FEEDER 208

Felipe Joel Zimann ; Alessandro Luiz Batschauer ; Marcello Mezaroba ; Eduardo Vasconcelos Stangler ; Francisco De Assis Dos Santos Neves

DESIGN AND ASSEMBLY OF A BIPOLAR MARX GENERATOR BASED ON FULL-BRIDGE TOPOLOGY APPLIED TO ELECTROPORATION 214

Fernando Imai ; Yales Rômulo De Novaes

IMPLEMENTATION OF A IUPQC CONTROL SCHEME FOR ENSURING AN IMPROVED COMPENSATION PERFORMANCE 220

E. V. Stangler ; F. A. S. Neves ; F. Bradaschia ; M. Mezaroba ; F. J. Zimann ; A. L. Batschauer

ASPECTS OF TRAVELLING WAVE BASED PROTECTION PHILOSOPHY FOR CONSIDERATION IN DC GRIDS OF THE FUTURE 227

Monday Ikhide ; Sarath Tennakoon ; Alison Griffiths ; Hengxu Ha

AN UNIT-LESS MATHEMATICAL MODEL FOR ANALYSIS AND DESIGN OF CLASS-E RESONANT CONVERTERS 233

Lucas S. Mendonça ; Thiago C. Naidon ; Rafael Fernandes Raposo ; Fábio E. Bisogno

A CURVE TRACER FOR PHOTOVOLTAIC MODULES BASED ON THE CAPACITIVE LOAD METHOD 239

Emerson Abreu Bastos Junior ; Caio Meira Amaral Da Luz ; Tatiane Martins Oliveira ; Lara Ana Rodarte Rios ; Eduardo Moreira Vicente ; Fernando Lessa Tofoli

LED DRIVER WITH REDUCED REDUNDANT POWER PROCESSING AND DIMMING FOR STREET LIGHTING APPLICATIONS 245

Kleber Chan Bekoski ; Cassio Gobbato ; Cassiano Ferro Moraes ; Gustavo Weber Denardin ; Juliano De Pelegrini Lopes

ACTIVE-CAPACITOR FOR POWER DECOUPLING IN SINGLE-PHASE GRID-CONNECTED CONVERTERS 251

Thiago A. Pereira ; Denizar C. Martins ; Roberto F. Coelho

RELIABILITY ANALYSIS OF AIRCRAFT STARTER GENERATOR DRIVE CONVERTER 257

Jayakrishnan Harikumaran ; Giampaolo Buticchi ; Michael Galea ; Alessandro Costabeber ; Pat Wheeler

MULTIVARIABLE CONTROL OF A GRID FORMING SYSTEM BASED ON BACK-TO-BACK TOPOLOGY 263

Igor D. N. De Souza ; Gabriel A. Fogli ; Marcelo C. Fernandes ; Ademir S. T. Júnior ; Pedro G. Barbosa ; Pedro M. De Almeida

3-PHASE MULTI-FUNCTIONAL GRID-TIED INVERTER FOR COMPENSATION OF OSCILLATING INSTANTANEOUS POWER 269

José De Arimatéia Olímpio Filho ; Helmo Kelis Morales Paredes ; Augusto Matheus Dos Santos Alonso ; Jakson Paulo Bonaldo ; Fernando Pinhabel Marafão ; Marcelo Godoy Simões

COMPARATIVE STUDY OF DIFFERENT CORRECTION METHODS TO ANALYZE WIND TURBINE PERFORMANCE 275

Hércules Araújo Oliveira ; Luiz Antonio De Souza Ribeiro ; Jerson Rogério Pinheiro Vaz ; Osvaldo Ronald Saavedra ; José Gomes De Matos

NONISOLATED QUADRATIC SEPIC CONVERTER WITHOUT ELECTROLYTIC CAPACITORS FOR LED DRIVER APPLICATIONS 281

Douglas Rosa Corrêa ; Aniel Silva De Morais ; Fernando Lessa Tofoli

A PERFORMANCE ANALYSIS OF ACTIVE ANTI-ISLANDING METHODS BASED ON FREQUENCY DRIFT 288

Ênio C. Resende ; Henrique T. M. Carvalho ; Fernando C. Melo ; Ernane A. A. Coelho ; Gustavo B. De Lima ; Luiz C. G. De Freitas

SIMULATION OF THE MODEL, DESIGN AND CONTROL OF A CURRENT SOURCE INVERTER WITH UNIPOLAR SPWM MODULATION 294

Jeimy C. Sanabria Rojas ; Daniel M. Barrera Leguizamón ; Diego A. Bautista López ; Fabián R. Jiménez López

ANALYSIS AND COMPARISON OF THE DYNAMIC RESPONSE OF DIRECT AND INDIRECT ROTOR FLUX CONTROL APPLIED TO AN ASYMMETRICAL TWO-PHASE INDUCTION MOTOR 299

Rafael De Farias Campos ; José De Oliveira ; Ademir Nied

DYNAMIC ANALYSIS OF SELF-EXCITED SRG OPERATING IN OPEN LOOP 305

Lucas José Lemes ; Victor Regis Bernardeli ; Luciano Coutinho Gomes ; Darizon Alves De Andrade ; Ghunter Paulo Viajante ; Marcos Antonio Arantes De Freitas

THREE-PHASE, FOUR-WIRE PWM RECTIFIER APPLIED TO VARIABLE FREQUENCY AC SYSTEMS IN AIRPLANE ELECTRIC GRID UNDER FAULT CONDITIONS 311

Alexandre Galvão Bueno ; José Antenor Pomilio

MODEL, SIMULATION AND ANALYSIS OF BLDCM FOR A DIFFERENTIAL CONTROLLED ELECTRIC-POWERED WHEELCHAIR 317

Augusto Nery De Lima Neto ; José Antenor Pomilio

ANALYSIS AND OPTIMAL DESIGN OF MAGNETIC COMPONENTS IN DUAL-ACTIVE-BRIDGE CONVERTER FOR 1 MVA SOLID-STATE TRANSFORMER 323

Haonan Tian ; Sriram Vaisambhayana ; Anshuman Tripathi

CENTER-TAPPED π-TYPE SINGLE-PHASE CELL 329

Domingos S. L. Simonetti ; Xibo Yuan

NOVEL MTPA APPROACH FOR IPMSM WITH NON-SINUSOIDAL BACK-EMF 335

Allan Gregori De Castro ; Paulo R. U. Guazzelli ; Mateus M. Lumertz ; Carlos M. R. De Oliveira ; Geyverson T. De Paula ; José R. B. A. Monteiro

NON-ISOLATED HIGH CURRENT BATTERY CHARGER WITH PFC SEMI-BRIDGELESS RECTIFIER 341

Rodrigo Patrício Dacol ; Joselito Anastácio Heerdt ; Gierri Waltrich

COMPARATIVE ANALYSIS BASED ON THE SWITCHING FREQUENCY OF MODULATION TECHNIQUES FOR MMC APPLICATIONS 347

Juan C. Colque ; Ernesto Ruppert ; Rodrigo Z. Vargas ; José L. Azcue

DESIGNING OF THE TRANSMITTING COILS AND COMPENSATION NETWORK OF A SEGMENTED DWPT SYSTEM 353

Shufan Li ; Lifang Wang ; Chengxuan Tao ; Fang Li ; Liye Wang

PROPOSAL OF AN ISOLATED TWO-SWITCH DC-DC SEPIC CONVERTER 358

Marcos Vinícius Mosconi Ewerling ; Telles Brunelli Lazzarin ; Carlos Henrique Illa Font

COMPARISON AMONG TWO-PHASE THREE-WIRE AC OFF-GRID POWER SYSTEMS 364

Rafael M. Silva ; Danilo I. Brandao ; Gabriel A. Fogli ; Victor F. Mendes ; Clodualdo V. Sousa

EFFICIENCY ANALYSIS FOR INTERLEAVED BUCK CONVERTERS EMPLOYED AS EXTRA-HIGH CURRENT COB LED DRIVERS 370

Dênis De Castro Pereira ; Everton Bernard Figueiredo Rabelo ; Pedro Santos Almeida ; Guilherme Marcio Soares ; Fernando Lessa Tofoli ; Henrique Antônio Carvalho Braga

MODELING OF A THREE-LEVEL QUADRATIC BOOST CONVERTER 376

Mauricio Borchardt ; Mateus C. Orige ; Mateus De F. Bueno ; Gerardo Escobar ; Yales R. De Novaes

DISCRETE SPS CONTROL OF A DAB CONVERTER USING PARTIAL FEEDBACK LINEARIZATION 382

Eduardo Luiz Santos Da Silva ; André Luís Kirsten ; Daniel Juan Pagano

STEADY-STATE ANALYSIS OF A SINGLE-PHASE MODIFIED BRIDGELESS BOOST RECTIFIER IN DCM 388

Mateo D. Roig G. ; Caio G. Da S. Moraes ; Telles B. Lazzarin

A TWO-STAGE BATTERY CHARGER WITH ACTIVE POWER DECOUPLING CELL FOR SMALL ELECTRIC VEHICLES 394

Caio G. Da S. Moraes ; Mateo D. Roig G. ; Telles B. Lazzarin

MODELLING AND ANALYSIS OF THE ISOLATED MICROGRID INSTALLED AT THE LENÇÓIS ISLAND USING PSCAD/EMTDC 400

Silvangela L. Barcelos ; José Gomes De Matos ; Luiz Antonio De Souza Ribeiro

METHODOLOGY FOR EXPERIMENTAL DETERMINATION OF EQUIVALENT GRID IMPEDANCE BY USING EXTERNAL COMMANDS OF PV INVERTERS 406

Valesca Bettim Feltrin ; Thiago Brezolin Dalmolin ; Lucas Vizzotto Bellinaso ; Leandro Michels

DEVELOPMENT OF A FPGA-BASED CONTROL SYSTEM FOR MODULAR MULTILEVEL CONVERTER APPLICATIONS 411

Lucas Koleff ; Manoel Conde ; Pedro Hayashi ; Francesco Sacco ; Kelly Enomoto ; Eduardo Pellini ; Wilson Komatsu ; Lourenço Matakas

VARIABLE-STEP DFT ALGORITHM TO SPEED UP OPTIMIZATION ROUTINES APPLIED TO A THREE-PHASE INTERLEAVED VIENNA RECTIFIER 417

B. Bertoldi ; E. F. C. Grabovski ; A. B. Lange ; M. L. Heldwein

DESIGN METHOD TO REDUCE THE DC LINK VOLTAGE OF A THREE-WIRE THREE-PHASE HYBRID ACTIVE POWER FILTER 423

Mateus Freitas Braga ; Samuel Neves Duarte ; Guilherme Márcio Soares ; Pedro Gomes Barbosa

DEVELOPMENT OF A MODULAR OPEN SOURCE POWER ELECTRONICS DIDACTIC PLATFORM 429

Lucas Koleff ; Gustavo Valentim ; Victor Rael ; Luciana Marques ; Wilson Komatsu ; Eduardo Pellini ; Lourenço Matakas

FLEXIBLE DIDACTIC PLATFORM FOR THYRISTOR-BASED CIRCUITS 435

Lucas Koleff ; Lucas Araújo ; Mário Zambon ; Wilson Komatsu ; Eduardo Pellini ; Lourenco Matakas

CONTROL OF WIRELESS POWER TRANSFER SYSTEMS UNDER LARGE COIL MISALIGNMENTS 441

Yeran Liu ; Udaya K. Madawala ; Ruikun Mai ; Zhengyou He

CONSIDERATIONS ON COMMUNICATION INFRASTRUCTURES FOR COOPERATIVE OPERATION OF SMART INVERTERS 445

Augusto Matheus Dos Santos Alonso ; Leonardo Carlos Afonso ; Danilo Iglesias Brandao ; Elisabetta Tedeschi ; Fernando Pinhabel Marafão

250 W SINGLE STAGE STEP-UP INVERTER CONNECTED TO THE GRID 451

Jessika Melo De Andrade ; Roberto Francisco Coelho ; Telles Brunelli Lazzarin

EXPERIMENTAL ANALYSIS FOR LOW POWER SERIES-SERIES COMPENSATED INDUCTIVE POWER TRANSFER SYSTEM 455

Macklyster Lãnucy Scherre Stofel De Lacerda ; Tatiana Saviato Macedo ; Denizar Cruz Martins ; Walbermark Marques Dos Santos

METHOD TO TRACE THE PHOTOVOLTAIC CHARACTERISTIC CURVE WITH REVERSE VOLTAGE FOR SHADING CONDITIONS 461

Richard G. Cornelius ; Amanda C. Maia ; Matheos C. Wermuth ; Guilherme S. Da Silva

HARMONIC COMPENSATION STRATEGIES APPLIED TO MULTIFUNCTIONAL PHOTOVOLTAIC INVERTERS 466

Lucas S. Xavier ; Joice D. S. Zacarias ; Allan F. Cupertino ; Heverton A. Pereira ; Danilo I. Brandao ; Victor F. Mendes

EVENT MANAGER AND CONTROL STRUCTURE FOR HIGH PERFORMANCE THREE-PHASE GRID-TIED INVERTERS 472

Victor E. S. Barbosa ; Rodrigo A. S. Kraemer ; Emerson G. Carati ; Jean P. Da Costa ; Rafael Cardoso ; Carlos M. O. Stein

PARTIAL HARMONIC CURRENT COMPENSATION APPLIED TO MULTIPLE PHOTOVOLTAIC INVERTERS IN A RADIAL DISTRIBUTION LINE 478

André L. P. De Oliveira ; Lucas S. Xavier ; João M. S. Callegari ; Allan F. Cupertino ; Victor F. Mendes ; Heverton A. Pereira

ENHANCED SPACE-STATE REDUCED-ORDER MICROGRID MODEL IN COMMON DQ-REFERENCE FRAME 484

Sebastián De J. Manrique Machado ; Sérgio Augusto Oliveira Da Silva ; José R. B. A. Monteiro ; Azauri A. De Oliveira

OPTIMAL VOLTAGE COORDINATED CONTROL FOR GRID-CONNECTED PHOTOVOLTAIC SYSTEMS 490

Thiago William Pires Sousa ; Gustavo Kaefer Dill

COMPARATIVE STUDY OF RC SNUBBER CONFIGURATIONS IN SWITCHING CIRCUITS 495

Ana Carolina Moreira ; Daniel Cesar Piccoli ; Júlio Cesar Lopes De Oliveira ; Luiz Fernando Henning ; Rodrigo Jose Piontkewicz

COMPARATIVE STUDY OF CONTROL SYSTEMS FOR A PHOTOVOLTAIC INVERTER WITH LCL FILTER 501

Leandro T. Omine ; Moacyr A. G. Brito

DESIGN OF A LOW-COST PHASOR MEASUREMENT UNIT (PMU) FOR THREE-PHASE DISTRIBUTION POWER SYSTEMS ACCORDING IEEE C37.118.1 507

Alex Guamán ; Marcelo Pozo ; Isaac Pozo ; Mario Pacas ; Ana Cabrera ; Nataly Pozo

CONTROLLER COEFFICIENTS TUNING FOR A SINGLE-PHASE PHOTOVOLTAIC SYSTEM IN SYNCHRONOUS REFERENCE FRAME THROUGH GENETIC ALGORITHM 513

Marcos V. A. Vedovatte ; Moacyr A. G. De Brito ; Luigi G. Junior

ROBUST PID CONTROLLERS OPTIMIZED BY PSO ALGORITHM FOR POWER CONVERTERS 519

Lucas C. Borin ; Everson Mattos ; Caio R. D. Osorio ; Gustavo G. Koch ; Vinicius F. Montagner

IMPACTS OF THE MULTI-VARIABLES MODULATION ON TRANSFORMER AND SOFT-SWITCHING OF A DAB 525

Jeferson Fraytag ; André Luís Kirsten ; Marcelo Lobo Heldwein

DEVELOPEMENT OF A MULTILEVEL DVR WITH BATTERY CONTROL AND HARMONIC COMPENSATION 531

Bruno P. B. Guimarães ; Wilson Cesar Santana ; Paulo F. Ribeiro ; Robson Bauwelz Gonzatti ; Guilherme G. Pinheiro ; Rondineli R. Pereira ; Fernando Nunes Belchior ; Carlos Henrique Da Silva ; Luiz Eduardo Borges Da Silva

MODELING OF ELECTROCHEMICAL BATTERIES BEHAVIOR AND LIFETIME DEGRADATION FOR PV APPLICATIONS 537

Rafael César Nolasco ; Victor Flores Mendes

COMPARISON BETWEEN TWO MODULATION STRATEGIES FOR THE 3L-DC-SSI 543

Matheos C. Wermuth ; Amanda C. Maia ; Richard G. Cornelius ; Guilherme S. Da Silva

ROBUST CONTROL OF SWITCHED RELUCTANCE GENERATOR IN CONNECTION WITH A GRID-TIED INVERTER 549

Caio R. D. Osório ; Filipe P. Scalcon ; Rodrigo P. Vieira ; Vinícius F. Montagner ; Hilton A. Gründling

ANALYSIS OF PERFORMANCE AND OPPORTUNITY FOR IMPROVEMENTS IN THE MICROGRID OF ILHA GRANDE 555

Leonilson Dos Santos Veras ; Hércules A. Oliveira ; José G. De Matos ; Osvaldo Ronald Saavedra ; Luiz A. De Sousa Ribeiro ; Lucas De Paula Assunção Pinheiro

STEADY-STATE CHARACTERIZATION OF THE THREE-PHASE ISOLATED DC-DC BIDIRECTIONAL CONVERTER WITH LLC RESONANT TANK 561

Kristian Pessoa Dos Santos ; Hermínio Miguel Oliveira Filho ; Paulo Peixoto Praça ; Demercil De Souza Oliveira Júnior

GAS MICROTURBINES FOR DISTRIBUTED GENERATION SYSTEM 567

Walquíria Do N. Silva ; Janaína G. De Oliveira ; Bruno H. Dias ; Leonardo W. De Oliveira

INHERENT REDUNDANCY OF SDBC-MMCC BASED STATCOM IN THE OVERMODULATION REGION............573

D. C. Mendonça ; A. F. Cupertino ; H. A. Pereira ; S. I. Seleme ; R. Teodorescu

ACTIVE COOLING AND THERMAL SIMULATION APPLIED TO AN EXTRA-HIGH CURRENT COB LED............579

Dênis De Castro Pereira ; Rúbio Campos Marques ; Pedro Santos Almeida ; Guilherme Marcio Soares ; Fernando Lessa Tofoli ; Pedro Laguardia Tavares ; Henrique Antônio Carvalho Braga

VOLTAGE REGULATION OF A REMOTE BUS OF A DISTRIBUTION NETWORK BY STATIC SYNCHRONOUS COMPENSATOR............585

Samuel N. Duarte ; Bruno C. Souza ; Pedro M. Almeida ; Pedro G. Barbosa

SINGLE-STAGE SINGLE-PHASE AC/DC CONVERTER WITH HIGH FREQUENCY ISOLATION FEASIBLE TO MICROGENERATION............591

Samanta Gadelha Barbosa ; Bruno Ricardo De Almeida ; Debora Pereira Damasceno ; Demercil De S. Oliveira

SMARTBATTERY: AN ACTIVE-BATTERY SOLUTION FOR ENERGY STORAGE SYSTEM............597

Lucas S. Araujo ; Nicolas T. D. Fernandes ; Danilo I. Brandao ; Braz J. Cardoso Filho

FEEDFORWARD COMPENSATION OF THE ESS LOW-FREQUENCY CURRENT RIPPLE IN THE THREE-PORTS ANPC CONVERTER............603

Silvio Antonio Teston ; Kaio Vinicius Vilerá ; Marcello Mezaroba ; Cassiano Rech

SEPIC DC/DC CONVERTER CONTROL BY OBSERVED-STATE FEEDBACK............609

Silas M. Sousa ; Valter J. S. Leite ; Samir W. Fernandes ; Isabel R. H. Oliveira

THIRD HARMONIC INJECTION METHOD FOR RELIABILITY IMPROVEMENT OF SINGLE-PHASE PV INVERTERS............615

R. C. De Barros ; R. P. Silva ; D. B. Da Silveira ; W. C. Boaventura ; A. F. Cupertino ; H. A. Pereira

DIDACTIC SYSTEM FOR CONTROL OF ELECTRICAL MACHINES IN EDUCATION AND RESEARCH LABORATORIES............621

Allan V. S. Andrade ; Richard M. Stephan

ON THE INFLUENCE OF AREA VARIATIONS OF THE PHOTOVOLTAIC ;SURFACE IN SOLAR CELL ANTENNAS............627

Eduardo Vicente Valdés Cambero ; Humberto Pereira Da Paz ; Vinícius Santana Da Silva ; Humberto Xavier De Araújo ; Ivan Roberto Santana Casella ; Carlos Eduardo Capovilla

NEW SEMICONDUCTOR TECHNOLOGIES FOR POWER ELECTRONICS............633

Alisson Mengatto ; José Adriano Damacena Diesel ; Pedro Henrique Thiesen De França ; Joselito Anastácio Heerdt

SMALL SCALE COMPRESSED AIR ENERGY STORAGE (SS-CAES) STRATEGIES OVERVIEW............639

Luiz Fernando Martins Pastuch ; Roberto Francisco Coelho ; Telles Brunelli Lazzarin ; Marcos Antonio Salvador

SYNCHRONOUS REFERENCE FRAME PLL FREQUENCY ESTIMATION UNDER VOLTAGE VARIATIONS............645

Eliabe Duarte Queiroz ; José Antenor Pomilio

HIGH STEP-UP DC-DC CONVERTER WITH INPUT CURRENT SHARING BASED ON THE FORWARD CONVERTER............651

Víctor Ferreira Gruner ; Lucas Fiamoncini ; Lenon Schmitz ; Denizar Cruz Martins ; Roberto Francisco Coelho

SRF-PLL INFLUENCE ON THE STABILITY OF A CURRENT SOURCE CONVERTER IN DROOP MODE............657

Eliabe Duarte Queiroz ; José Antenor Pomilio

A LOW COST BI-DIRECTIONAL WIRELESS POWER TRANSFER SYSTEM............663

Lei Wang ; Udaya K. Madawala ; Man-Chung Wong

WIND TURBINE EMULATOR WITH DC MOTOR............668

Tadeu F. Dos Santos ; Igor V. Chacon ; Géssica C. De A. Souza ; Guilherme A. P. De C. A. Pessoa ; Felipe O. S. Gama ; Rodrigo De A. Teixeira ; Andrés O. Salazar

GENERALIZED MATHEMATICAL MODEL FOR AN N-CELL INTERLEAVED BOOST CONVERTER............674

Luana K. Melgaço Pereira ; Seleme I. Seleme ; João Lucas Da Silva

AC-DC CONVERTER WITH HIGH-FREQUENCY ISOLATION OPERATING UNDER ZVS............680

Bruno Alves Sousa Da Silva ; Dalton De Araújo Honório ; Demercil De Souza Oliveira Júnior ; Bruno Ricardo De Almeida ; Samanta Gadelha Barbosa ; Luiz Henrique Silva Colado Barreto ; Caio Kerson O. Veras

ENHANCED POWER MANAGEMENT SYSTEM FOR DROOP CONTROL IN A GRID CONNECTED DC MICROGRID............686

Pedro Jose Dos Santos Neto ; Joao Pedro C. Silveira ; Tárcio André Dos S. Barros ; Ernesto Ruppert Filho ; Juan Carlos Vasquez ; Josep M. Guerrero

MULTI-PORT SYSTEM FOR STORAGE AND MANAGEMENT OF REGENERATIVE BRAKING ENERGY IN DIESEL-ELECTRIC LOCOMOTIVES 692

Caio G. Da S. Moraes ; Sergio L. B. Junior ; Pedro P. Cavilha ; Antonio L. S. Pacheco ; Marcelo L. Heldwein ; Gierri Waltrich

SENSORLESS CONTROL OF NONSINUSOIDAL BACK-EMF PMSM BASED ON STATE OBSERVER 698

Thiago Lazzari ; Filipe Scalcon ; Cesar Volpato ; Thieli Gabbi ; Márcio Stefanelo ; Rodrigo P. Vieira

POWER DEMAND PREDICTION BASED ON MIXED DRIVING CYCLE APPLIED TO ELECTRIC VEHICLE HYBRID ENERGY STORAGE SYSTEM 704

Lucas F. R. Lago ; Silvana T. Faceroli ; Rodrigo A. F. Ferreira ; Marcio C. B. P. Rodrigues

SYMMETRICAL HYBRID MULTILEVEL VSI AND CSI INVERTERS DERIVED FROM DC-DC CONVERTERS 710

Domingo Ruiz-Caballero ; Luis Colque Miranda ; Carlos Paredes ; Javier Riedemann ; Werner Jara Montecinos ; Marcelo Lobo Heldwein ; Samir Ahmad Mussa

STRUCTURAL AND PERFORMANCE COMPARISON BETWEEN HARMONIC SELECTIVE REPETITIVE CONTROLLERS FOR SHUNT ACTIVE POWER FILTER 716

R. C. Neto ; F. A. S. Neves ; E. V. Stangler ; F. Bradaschia ; H. E. P. De Souza

IMPACT OF CAPACITOR DESIGN METHODOLOGY ON FC INVERTERS 724

Marcos Vinicius Bressan ; Cassiano Rech ; Alessandro L. Batschauer

MODELING AND CONTROL OF A BACK-TO-BACK SYSTEM FOR TURBOELECTRIC PROPULSION 730

Saulo O. Nascimento ; Vinicius M. De Albuquerque ; Marcelo De C. Fernandes ; Manuel A. Rendón ; Janaína G. Oliveira ; Pedro S. Almeida

AN ENHANCED THÉVENIN EQUIVALENT CIRCUIT OF A RESONANT-CONTROLLER-BASED UTILITY-INTERFACE 736

Joel Filipe Guerreiro ; Hildo Guillardi Júnior ; João Inácio Yutaka Ota ; José Antenor Pomilio

REAL TIME SIMULATION IN A DISTRIBUTION SYSTEM INCLUDING PV INVERTER AND VOLTAGE REGULATOR: VOLTAGE IMPACT ANALYSIS 742

João A. G. Archetti ; Lilian V. Pinheiro ; Mateus L. Lima ; Bernardo F. Musse ; Janaína. G. De Oliveira ; Leonardo. W. De Oliveira

USING SYNCHRONVERTER IN DISTRIBUTED GENERATION FOR FREQUENCY AND VOLTAGE GRID SUPPORT 748

Guilherme Penha Da Silva Junior ; Luciano Sales Barros

DESIGN PROCEDURES AND PROTOTYPING OF A FULL-BRIDGE HIGH FREQUENCY POWER INVERTER 754

Joel Filipe Guerreiro ; Hildo Guillardi Júnior ; José Antenor Pomilio

SVC OPERATING AS AN UNBALANCE COMPENSATOR WITH CONTROL SYSTEM BASED ON THE STEINMETZ METHOD AND THE INSTANTANEOUS POWER THEORY 760

Andressa De Melo Rodrigues ; Laylla Fernandes Silva ; Luciano De Souza Da Costa E Silva ; Lucas Carvalho Souza

DC CURRENT REDISTRIBUTOR FOR ELECTRIC AIRCRAFT SYSTEM 766

Mateus P. Dias ; Hildo Guillardi ; Joel F. Guerreiro ; João I. Y. Ota ; José A. Pomilio ; Paolo Mattavelli

TWO-STAGE STAND ALONE PHOTOVOLTAIC SYSTEM FOR WATER PUMPING SYSTEM 772

Renata C. Silva ; Gabriel G. Bacheti ; Rafael M. Silva ; Vitor G. Neves ; Clodualdo V. Sousa ; Seleme I. Seleme

AN IMPROVED IMPEDANCE ESTIMATION METHOD BASED ON POWER VARIATIONS IN GRID-CONNECTED INVERTERS 778

José H. Suárez ; Hugo M. C. Gomes ; Luan S. Santana ; Alfeu J. Sguarezi Filho ; Leandro L. O. Carralero ; Fabiano F. Costa

HYBRID SWITCHED CAPACITOR DC-DC CONVERTER BASED ON MMC 784

Marcus Vieira Soares ; Gustavo Lambert ; Yales Rômulo De Novaes

THERMAL STRESS EVALUATION OF A MULTIFUNCTIONAL MODULAR MULTILEVEL CONVERTER – STATCOM OPERATING AS ACTIVE FILTER 790

R. O. De Sousa ; W. C. S. Amorim ; D. C. Mendonça ; A. F. Cupertino ; L. M. F. Morais ; H. A. Pereira

TOTEM-POLE BRIDGELESS PFC CONVERTER IN DCM WITH SYNCHRONOUS RECTIFICATION 796

Leonardo S. Mai ; Alexsandra Rospirski ; Samir A. Mussa ; Telles B. Lazzarin

PASSIVE CAPACITOR VOLTAGE BALANCING IN MODULAR MULTILEVEL CONVERTER DURING ITS PRECHARGE: ANALYSIS AND DESIGN 802

Luiz H. T. Schmidt ; Gean J. M. De Sousa ; Marcelo L. Heldwein ; Daniel J. Pagano

DESIGN OF AN INTEGRATED CIRCUIT FOR LED DRIVING IN VISIBLE LIGHT COMMUNICATION APPLICATIONS 808

Nicolas Parma Rios ; Guilherme Márcio Soares ; Estêvão Coelho Teixeira

CONNECTION TIME IN MODBUS/TLS FOR SECURE COMMUNICATIONS ON PHOTOVOLTAIC SYSTEMS 814

Matheus K. Ferst ; Hugo F. M. De Figueiredo ; Gustavo W. Denardin

MODELING AND SIMULATION OF THE BATTERY ENERGY STORAGE SYSTEM FOR ANALYSIS IMPACT IN THE ELECTRICAL GRID. 820

Carolina A. Caldeira ; Henrique R. Schlickmann ; Antonio D. D. De Almeida ; Lucas V. Hartmann ; Camila S. Gehrke ; Fabiano Salvadori ; Gilielson F. Da Paz

COMPREHENSIVE AND DIDACTIC DC SERVOMOTOR CONTROL PLATFORM 826

Bruno De Almeida Regina ; Maria Júlia Rosa Aguiar ; André Augusto Ferreira

PERFORMANCE ANALYSIS OF ACTIVE ANTI-ISLANDING TECHNIQUES FOR PHOTOVOLTAIC APPLICATION 832

Marcos Roberto Oshiro ; Ruben Barros Godoy ; Moacyr A. G. De Brito ; Luigi Galotto

A HYBRID BIDIRECTIONAL PUSH-PULL DC-DC CONVERTER WITH A LADDER SWITCHED-CAPACITOR CELL 838

Marcos Dantas ; Fellipe Oliveira ; Lorena Albuquerque ; Isaac Freitas ; Romero Andersen

HIGH POWER FACTOR THREE-PHASE THREE-SWITCH STEP-DOWN CONVERTER 844

João Olímpio Caliman ; Walbermark Marques Dos Santos ; Tiara Rodrigues Smarssaro De Freitas ; Domingos Simonetti

SELECTION OF THE NUMBER OF LEVELS OF A MODULAR MULTILEVEL CONVERTER FOR AN ELECTRIC DRIVE 850

Paulo R. M. Júnior ; João. V. M. Farias ; Allan F. Cupertino ; Gabriel A. Mendonça ; Marcelo M. Stopa ; Heverton A. Pereira

HIGH VOLTAGE GAIN DC-DC CONVERTER BASED ON A SIMPLE CONFIGURATION OF SWITCHED CAPACITOR AND COUPLED INDUCTOR 856

Henrique J. Hoch ; Tiago M. K. Faistel ; Mauricio M. Da Silva ; António M. S. S. Andrade ; Mário L. Da S. Martins

WIND POWER SYSTEM CONNECTED TO THE GRID FROM SQUIRREL CAGE INDUCTION GENERATOR (SCIG) 862

Ângelo Marcílio M. Dos Santos ; Lucas Taylan P. Medeiros ; Leonardo P. S. Silva ; Ildenor Davi S. Junior ; Vanessa Siqueira De C. Teixeira ; Adson Bezerra Moreira

SINGLE-PHASE TO THREE-PHASE AC-DC-AC CONVERTER BASED ON CASCADED TRANSFORMERS RECTIFIER AND OPEN-END WINDING INDUCTION MOTOR 868

Antonio D. D. Almeida ; Nady Rocha ; Edgard L. L. Fabricio ; Carolina A. Caldeira ; Gleice M. S. Rodrigues ; Isaac S. Freitas

AN OVERVIEW ABOUT DETECTION OF CYBER-ATTACKS ON POWER SCADA SYSTEMS 874

Hugo F. M. De Figueiredo ; Matheus K. Ferst ; Gustavo W. Denardin

COMPARATIVE STUDY BETWEEN VIRTUAL SYNCHRONOUS MACHINE AND VIRTUAL IMPEDANCE TECHNIQUES FOR TWO PARALLELED INVERTERS SHARING A LOAD 880

Bruno Laurindo ; Fábio Alves ; Marcello Neves ; Jorge Caicedo ; Bruno França ; Maurício Aredes

ANALYSIS OF PARTIAL-POWER PROCESSING CONVERTERS FOR SMALL WIND TURBINES SYSTEMS 886

Anderson José Balbino ; Ronny Glauber De Almeida Cacau ; Telles Brunelli Lazzarin

PROPORTIONAL WAVELET SLIDING MODE CONTROLLER FOR TORQUE RIPPLE REDUCTION IN BLDC MOTOR 892

Mateus Moro Lumertz ; Carlos Matheus R. De Oliveira ; Allan Gregori De Castro ; Paulo Roberto U. Guazzelli ; Manuel Luis De Aguiar ; Jose Roberto B. A. Monteiro

VOLTAGE REGULATOR BEHAVIOR ON POWER DISTRIBUTION GRIDS WITH HIGH INTEGRATION OF PVDG 898

Lucas F. S. Azeredo ; Luiz G. R. Tonini ; Mariana A. Mendes ; Murillo C. Vargas ; Oureste E. Batista ; Carla J. Espindula

AN APPLICATION OF THE MULTI-PORT BIDIRECTIONAL THREE-PHASE AC-DC CONVERTER IN ELECTRIC VEHICLE CHARGING STATION MICROGRID 904

Raphael A. Da Câmara ; Luis M. Fernández-Ramírez ; Paulo P. Praça ; Demercil De S. Oliveira ; Pablo García-Triviño ; Raúl Sarrias-Mena

UNIFIED ROBUST CONTROL DESIGN FOR BTB-VSC SUBJECT TO UNCERTAINTIES IN GRID EQUIVALENT CIRCUIT 910

Marcelo De Castro ; Igor D. N. De Souza ; Gabriel A. Fogli ; Ademir Da S. T. Junior ; Pedro M. De Almeida ; Janaína G. De Oliveira ; Pedro G. Barbosa

A ISOP AC-AC HYBRID SWITCHED-CAPACITOR SRC FOR SOLID STATE TRANSFORMER APPLICATIONS 917

Victor Luiz Flor Borges ; Rogerio Luiz Da Silva ; Carlos Eduardo Possamai ; Arlan Luiz Bettiol ; Ivo Barbi

LOW COMPUTATIONAL COST TECHNIQUE FOR SPMSM SENSORLESS DRIVE USING ACTIVE FLUX CONCEPT..923

Camila R. S. Bartsch ; Luis F. F. De Campos ; Arthur G. Bartsch ; José De Oliveira ; Ademir Nied

A MODEL PREDICTIVE CONTROL APPLIED TO SINGLE-PHASE PACKED-U-CELLS CONVERTER..929

Jordan Zucuni ; Dimas Alã Schuetz ; Felipe Bovolini Grigoletto ; Fernanda Carnielutti De Morais ; Margarita Norambuena ; José Rodriguez ; Humberto Pinheiro

DESIGN OF ROBUST STRUCTURED CONTROL STRATEGY FOR SINGLE-PHASE DYNAMIC VOLTAGE RESTORER..934

João Amin Moor Neto ; Guilherme Giglio De Andrade ; Thiago Americano Do Brasil ; Mauro Sandro Dos Reis ; Julio Cesar Ferreira

REFERENCE GRID IMPEDANCE FOR TESTS OF GRID-CONNECTED POWER CONVERTERS FOR DISTRIBUTED ENERGY RESOURCES: THE BRAZILIAN CASE..940

Gabriel Avila Saccol ; Charles Schardong ; Leandro Michels ; Lucas Vizzotto Bellinaso ; Cassiano Rech

AN ELECTRONIC DRIVE FOR A SWITCHED RELUCTANCE MOTOR USING A DSC..946

Brayan Sobral Da Fonseca ; José Andrés Santisteban

DC MOTOR MODEL FOR WINDOWS PINCH PROTECTION APPLICATIONS..952

Gabriel F. Idalgo ; José A. Torrico Altuna ; Carlos E. Capovilla ; Ademir Pelizari ; Wolfgang Schulter

POWER CONTROL AND HARMONIC CURRENT MITIGATION FROM A WIND POWER SYSTEM WITH PMSG..958

Leonardo P. S. Silva ; Denisia De V. Mota ; Flávia P. Ruiz ; Levy R. Cavalcante ; Lucas Taylan P. Medeiros ; Vanessa S. C. Teixeira ; Adson B. Moreira

PV EMULATOR BASED ON A FOUR-SWITCH BUCK-BOOST DC-DC CONVERTER..964

Leandro L. O. Carralero ; Gabriel S. Barbara Da S. E Silva ; Fabiano F. Costa ; André P. N. Tahim

SINGLE-STAGE BRIDGELESS AC-DC PFC FLYBACK INTERLEAVED..969

G. S. Schowantz ; R. P. Barcelos ; T. B. Lazzarin

PREMAGNETIZED INDUCTORS IN SINGLE PHASE DC-AC AND AC-DC CONVERTERS..975

Jens Friebe ; Siqi Lin ; Leon Fauth ; Tobias Brinker

THREE-PHASE INDUCTION MOTORS EFFICIENCY ANALYSIS USING A PROGRAMMABLE POWER SUPPLY..981

Cássio Alves De Oliveira ; Josemar Alves Dos Santos Junior ; Marcos José De Moraes Filho ; Vinícius Marcos Pinheiro ; Augusto W. Fleury Veloso Da Siveira ; Luciano Coutinho Gomes

REACTIVE POWER CONTROL OF DISTRIBUTED PHOTOVOLTAIC GENERATION SYSTEM IN LOW VOLTAGE ELECTRICAL GRIDS..987

Vanessa Da Costa Marques ; Rogério Gaspar De Almeida ; Nady Rocha ; Darlan Alexandria Fernandes ; Gleice Mylena Da Silva Rodrigues

MICROINVERTER WITH REDUCED NUMBER OF SEMICONDUCTOR SWITCHES..993

Paulo R. Cagnini ; Luiz H. Meneghetti ; Victor E. S. Barbosa ; Emerson G. Carati ; Carlos M. Stein ; Zeno L. I. Nadal ; Jean M. S. Lafay ; Jean Patric Da Costa ; Rafael Cardoso

APPLICATION OF MODEL PREDICTIVE CONTROL IN A RESOLVER-TO-DIGITAL CONVERTER..998

Thyago Vasconcelos Estrabis ; Raymundo Cordero García ; Edson Antonio Batista ; Cristiano Quevedo Andrea ; Márcio Afonso Soleira Grassi

ON THE APPLICATION OF A POWER ELECTRONICS-BASED ARC-FLASH SUPPRESSOR..1004

Fernando Venâncio Amaral ; Matheus H. M. Zanchetta Oliveira ; Claudio Alvares Conceição ; Sidelmo Magalhães Silva ; Cleber Onofre Inácio ; Braz De J. Cardoso Filho

DERIVING STABILITY CONDITION FOR ONE-CYCLE CONTROL WITH TRIANGULAR CARRIER BY POINCARÉ MAPS..1010

Armando J. G. Abrantes-Ferreira ; Lucas Do N. Gomes ; Robson F. Da S. Dias ; Luís G. B. Rolim

THREE-PHASE ADAPTIVE FREQUENCY ESTIMATOR WITH A DELAYED SIGNAL CANCELLATION PRE-FILTER UNDER HEAVILY DISTORTED GRID CONDITIONS..1016

E. F. C. Grabovski ; M. L. Heldwein ; S. A. Mussa

ENHANCED PHASE-SHIFTED CARRIER PWM APPLIED TO 3-PHASE MULTILEVEL COUPLED INDUCTORS INVERTERS..1022

Emerson L. Soares ; Lucas Fabrício M. De Lucena ; Nady Rocha ; Cursino Brandão Jacobina ; Edison Roberto C. Da Silva

DIMENSIONING AND DEVELOPEMENT OF AN AC MICROGRID IN THE UFJF CAMPUS..1028

Paula Stael S. Barbosa ; Marina V. C. Monteiro ; Jessica S. Dohler ; Andre Ferreira ; Janaina G. De Oliveira

UNBALANCED VOLTAGE MITIGATION USING D2VC WITH PROPORTIONAL RESONANT CONTROLLER IN Aß-FRAME..1034

Elienai O. Macedo ; Robson F. S. Dias ; Silvangela L. S. L. Barcelos ; Edson H. Watanabe

DYNAMIC MODELING AND CONTROL OF A THREE-PORT ZVS-PWM THREE-PHASE PUSH PULL DC-DC CONVERTER..1040

Lorena Lorraine Oliveira Albuquerque ; Marcos Victor Dantas De Sá ; Fellipe André Lucena De Oliveira ; Isaac Soares De Freitas ; Romero Leandro Andersen

THERMAL MODELING OF CONVERTERS FOR WIND CONVERSION SYSTEMS EMPLOYING DFIG TECHNOLOGY..1046

Rodrigues De Oliveira ; Tofoli Lessa ; Mendes Flores

INTERLEAVED BIDIRECTIONAL DC-DC CONVERTER FOR APPLICATION IN HYBRID PROPULSION SYSTEM: MODELING AND CONTROL..1052

Vitor C. S. Torres ; Vinicius M. De Albuquerque ; Manuel A. Rendón ; Pedro S. Almeida ; Janaina G. Oliveira ; Márcio C. B. P. Rodrigues

ANALYSIS AND OPERATION OF A PV-BATTERY SYSTEM USING A MULTI-FUNCTIONAL CONVERTER..1058

Jessica S. Dohler ; Lorrana F. Da Rocha ; Dalmo C. Silva Junior ; Pedro M. De Almeida ; Andre A. Ferreira ; Janaina G. Oliveira

STATOR CURRENT CONTROLLER FOR HARMONIC AND UNBALANCE COMPENSATION APPLIED TO SEIG BASED SYSTEMS..1064

Gabriel Attuati ; Robinson Figueiredo De Camargo ; Lucas Giuliani Scherer

ENERGY-BALANCE BASED VOLTAGE REGULATION METHOD FOR MULTIPLE DC-LINKS IN ASYMMETRICAL CASCADED MULTILEVEL INVERTERS..1070

Andre Felicio De Sousa Silva ; Sergio Pires Pimentel ; Enes Goncalves Marra ; Bernardo Pinheiro De Alvarenga

COMPARISON OF THE PLL CONTROL TECHNIQUES APPLIED IN PHOTOVOLTAIC SYSTEM..1076

Marenice M. De Carvalho ; Renan L. P. Medeiros ; Iury V. Bessa ; Florindo A. C. Junior ; Kevin E. Lucas ; David A. Vaca

A CRITICAL ANALYSIS OF PSO AND ITS VARIATIONS APPLIED TO MPPT FOR PV SYSTEMS UNDER PARTIAL SHADING CONDITION...1082

Lucas Mendonça Andrade ; Paula Dos Santos Vicente ; Fernando Lessa Tofoli ; Eduardo Moreira Vicente

EVALUATION OF TECHNIQUES TO REDUCE THE EFFECTS OF PARTIAL SHADING ON PHOTOVOLTAIC ARRAYS..1088

André Augusto Rodrigues ; Paula Dos Santos Vicente ; Fernando Lessa Tofoli ; Eduardo Moreira Vicente

ANALYSIS OF NEUTRAL-POINT VOLTAGE BALANCING IN THREE-PORTS ACTIVE NEUTRAL-POINT-CLAMPED CONVERTER..1094

Kaio Vinicius Vilerá ; Cassiano Rech ; Silvio Antonio Teston

DIRECT TORQUE CONTROL SCHEME FOR A NINE-PHASE INDUCTION MOTOR WITH REDUCED CURRENT HARMONIC..1100

Gilielson F. Da Paz ; Isaac S. De Freitas ; Victor F. M. B. Melo ; Alexandre G. F. Da Silva

SINGLE-PHASE AC-DC-AC FIVE-LEVEL X-TYPE CURRENT SOURCE CONVERTER.............................1106

Louelson A. Costa ; Montiê A. Vitorino ; Maurício B. R. Corrêa

APPLICATION OF A SHE-PWM MODULATION FOR A LOW SWITCHING FREQUENCY MOTOR DRIVE WITH HARMONIC INVESTIGATION USING THE DTFT...1112

Marcos Paulo Brito Gomes ; Marina Hassen De Souza ; Gabriel Vilkn Ramos ; Alex-Sander Amável Luiz ; Marcelo Martins Stopa

A RESONANT-SWITCHED-CAPACITOR STEP-DOWN DC–DC CONVERTER IN CCM OPERATION AS AN LED DRIVER..1118

Fábio De P. Neres ; Antonia F. Da Rocha ; Rodrigo L. Dos Santos ; Pedro S. Almeida ; Fernando L. M. Antunes ; Edilson M. Sá

HIGH-GAIN BIDIRECTIONAL DC-DC CONVERTER FOR BATTERY CHARGING IN DC NANOGRID OF RESIDENTIAL PROSSUMER..1124

Fernando Queiroz ; Paulo Praça ; Alisson Freitas ; Fernando Antunes

CONCEPTS AND CASE STUDY OF MISMATCH LOSSES IN PHOTOVOLTAIC MODULES.......................1130

Elson Yoiti Sakô ; João Lucas De Souza Silva ; Daniel De Bastos Mesquita ; Rafael Espino Campos ; Hugo Soeiro Moreira ; Marcelo Gradella Villalva

NOVEL BUCK-BOOST PFC CONVERTER WITH THREE-STATE SWITCHING CELL.............................1136

Douglas Carvalho Morais ; Falcondes José Mendes De Seixas ; Luís De Oro Arenas ; Lucas Carvalho Souza ; Luciano De Souza Da Costa E Silva

CONTROL STRATEGY FOR MULTIFUNCTIONAL PV CONVERTER..1142

Luiz Henrique Meneghetti ; Edivan Laercio Carvalho ; Emerson Giovani Carati ; Jean Patric Costa ; Carlos Marcelo Oliveira Stein ; Zeno Luiz Iensen Nadal ; Rafael Cardoso

PHOTOVOLTAIC BOOST CONVERTER CONTROL OPERATING IN THE MPPT AND LPPT MODES..1148

Cássia C. C. Dos Santos ; Cassiano F. Moraes ; Jean P. Da Costa ; Carlos M. O. Stein ; Emerson G. Carati ; Rafael Cardoso

PREDICTIVE CONTROL FOR A HALF-CONTROLLED BOOST RECTIFIER 1154
Gleice Mylena Da S. Rodrigues ; Nady Rocha ; Edison Roberto C. Da Silva ; Filipe V. Rocha

LIFETIME INVESTIGATION OF DC-LINK CAPACITORS IN MULTIPLE SLIM DRIVES SYSTEM 1160
Shili Huang ; Haoran Wang ; Dinesh Kumar ; Guorong Zhu ; Huai Wang

DIRECT POWER CONTROL WITH SPACE VECTOR MODULATION APPLIED FOR THE BRUSHLESS DC MOTOR 1166
Henrique De Toledo ; José L. Azcue

MODELING AND CONTROL OF A FORWARD DC-DC CONVERTER FOR BATTERY VOLTAGE BALANCING 1172
Pábulo F. Ciarnoscki ; Daniel J. Pagano ; Gabriel M. Da Silva

ENERGY YIELD ASSESSMENT METHODOLOGY FOR PHOTOVOLTAIC MICROINVERTERS 1178
Andrii Chub ; Kosenko Roman ; Oleksandr Korkh ; Dmitri Vinnikov ; Samir Kouro

DIRECT POWER CONTROL STRATEGY TO ENHANCE THE DYNAMIC BEHAVIOR OF DFIG DURING VOLTAGE DIP 1183
Fernando Lino ; Rogério Vani Jacomini ; Alfeu J. Sguarezi Filho

ELECTRICAL SIMULATION OF TRACTION SUBWAY SYSTEM FOR ENERGY RECOVERY AND ENERGY SAVING STUDIES 1189
Marcio Annibal Pimenta ; Wilson Komatsu ; Lourenço Matakas Junior

MODELING AND SIMULATION OF A STIRLING-ENGINE-BASED GENERATOR CONNECTED TO THE GRID 1195
Augusto Hayashi ; Moacyr Brito ; João Onofre ; Lourenço Matakas ; Raymundo Cordero ; Luigi Galottojunior

COMPARATIVE STRATEGIES OF CONTROL FOR REGENERATIVE BRAKING IN ELECTRIC VEHICLES 1201
Marina G. S. P. Paredes ; José Antenor Pomilio

THEORETICAL SOLUTION OF THE OUTPUT VOLTAGE HARMONIC SPECTRA OF DUAL-INVERTER FED OPEN-END WINDING LOADS WITH DEAD TIME EFFECT 1206
Brunno Monteiro Guimarães Ribeiro ; Marcelo Martins Stopa ; Alex-Sander Amável Luiz

BRIDGELESS BUCK-BOOST PFC CONVERTER WITH THREE-STATE SWITCHING CELL 1211
Douglas Carvalho Morais ; Falcondes José Mendes De Seixas ; Luís De Oro Arenas ; Lucas Carvalho Souza ; Luciano De Souza Da Costa E Silva

EXPERIMENTAL RESULTS OF A BIDIRECTIONAL COUPLED INDUCTOR DC-DC CONVERTER 1217
Murilo Brunel Da Rosa ; Menaouar Berrehil El Kattel ; Robson Mayer ; Sérgio Vidal Garcia Oliveira

DESIGN AND TEST OF A SRF-PLL BASED ALGORITHM FOR POSITIVE-SEQUENCE SYNCHROPHASOR MEASUREMENTS 1223
Gabriel Ubirajara De Carvalho ; Gustavo Weber Denardin ; Rafael Cardoso ; Cassiano Ferro Moraes

ECONOMIC ANALYSIS OF A PEAK SHAVING SYSTEM WITH DIESEL GENERATOR 1229
Myrlena R. M. Ferreira ; Luiz A. De S. Ribeiro ; José G. De Matos

THYRISTOR TRIGGERING, STATIC AND DYNAMIC CHARACTERISTICS 1235
Ricardo Alves Do Prado ; Ulrich Nicolai

DEVELOPMENT OF A FPGA-BASED REAL-TIME SIMULATION SYSTEM 1241
Yago F. Oliveira ; Filipe A. La-Gatta ; Rodrigo A. F. Ferreira ; Márcio C. B. P. Rodrigues

POWER ELECTRONICS LAB: CONVERGING KNOWLEDGE AND TECHNOLOGIES 1247
José Antenor Pomilio

INTEGRATION OF SOLAR PHOTOVOLTAIC (PV) SYSTEMS WITH CCM INVERTERS INTO VCM DROOP-CONTROLLED ISLANDED AC MICROGRIDS 1253
Marcus E. T. Souza ; Fernando C. Melo ; Ernane A. A. Coelho ; Luiz C. G. De Freitas

CLOSED-FORM SOLUTIONS FOR CORE AND WINDING LOSSES CALCULATION IN SINGLE-PHASE BOOST PFC RECTIFIERS 1259
Marcos José Jacoboski ; André De Bastiani Lange ; Marcelo Lobo Heldwein

DEVELOPMENT OF A HYBRID PV-THERMOELECTRIC SYSTEM 1265
Rafael Magalhães Nóbrega De Araújo ; Hugo Álisson Alves Da Costa ; João T. De Carvalho Neto ; Alexandre Magnus F. Guimarães ; Andrés Ortiz Salazar

EXPERIMENTAL WORKBENCH: A TOOLS TO HELP TO TEACHING THE TECHNIQUES OF DRIVES OF ELECTRIC MACHINES 1271
Victor Camargo Reis ; Carlos Henrique Silva De Vasconcelos

PREDICTING THE LIFE TIME OF POWER SEMICONDUCTOR MODULES 1277
Clovis N. L. Gajo ; Arendt Wintrich ; Paul Drexhage

WORKBENCH FOR MONITORING AND OPERATION OF ELECTRIC MACHINES 1283
Guilherme Afonso Pillon De C. A. Pessoa ; Evandro Ailson De Freitas Nunes ; Werbet Luiz Almeida Da Silva ;
Ricardo Ferreira Pinheiro ; Andres Ortiz Salazar

**DC-LINK VOLTAGES BALANCE METHOD FOR SINGLE-PHASE NPC INVERTER
OPERATING WITH REACTIVE POWER COMPENSATION**... 1289
Marco Fajardo ; Julio Viola ; José Manuel Aller ; Flavio Quizhpi

**NEW FIVE-LEVEL FLYING CAPACITOR INVERTER FED BY A BOOST-FLYBACK DC-DC
VOLTAGE SOURCE**... 1295
Antonio Venâncio De M. Lacerda Filho ; André E. L. Da Costa ; Edison R. C. Da Silva ; Ronnan De B. Cardoso ;
Darlan A. Fernandes

**INTEGRATED ZETA INVERTER APPLIED IN A SINGLE-PHASE GRID-CONNECTED
PHOTOVOLTAIC SYSTEM**... 1301
Leonardo Poltronieri Sampaio ; Sérgio Augusto Oliveira Da Silva ; Paulo Júnior Silva Costa

**A NOVEL PHASE-LOCKED LOOP WITH POSITIVE AND NEGATIVE SEQUENCE
DETECTION CAPABILITY**.. 1307
Jean M. L. Fonseca ; Francisco Kleber A. Lima ; Carlos Gustavo C. Branco ; Renato G. Araujo

**PERFORMANCE ANALYSIS OF ALTERNATIVES SWITCHING COMMANDS FOR THREE-
LEVEL BOOST RECTIFIER WITH HYSTERESIS CURRENT CONTROL.**.................................... 1313
Gabriel Vilkn Ramos ; Alex-Sander Amável Luiz ; Marcelo Martins Stopa ; Marcos Paulo Brito Gomes ; Ramon
Henriques De Souza ; Daniel Franco Leal

HIGH POWER FACTOR RECTIFIER USING ONE CYCLE CONTROL STRATEGY 1319
Amanda Thayla S Monteiro ; Rafael Rocha Matias ; Jailson Leite Silva ; Jose Antonio Dos Santos Neto

**A COMPARISON AMONG DIFFERENT FINITE CONTROL SET APPROACHES AND CONVEX
CONTROL SET MODEL-BASED PREDICTIVE CONTROL APPLIED IN A THREE-PHASE
INVERTER WITH RL LOAD** 1325
Arthur G. Bartsch ; Dayse M. Cavalcanti ; Mariana S. M. Cavalca ; Ademir Nied

**DESIGN AND ANALYSIS OF OUTPUT FILTER WITH LONG LIFETIME E-CAP FOR AC-DC
LED DRIVER** 1331
Andressa Da S. Fernandes ; Felipe C. Do Nascimento ; Antonia F. Da Rocha ; Rodrigo L. Dos Santos ; Fernando
L. M. Antunes ; Edilson M. Sá

HYBRID MODEL OF ELECTRIC VEHICLE 1337
Débora De Souza Martins ; Danilo P. E Silva ; Jussara Farias Fardin

**STATIC TRANSFER SWITCH APPLIED TO SINGLE-PHASE UNINTERRUPTIBLE POWER
SUPPLY**.. 1343
Marcus Vinícius Maia Rodrigues ; Newton Da Silva ; Flávio Alessandro Serrão Gonçalves

**ANALYSIS OF RECTIFIERS FOR RF ENERGY HARVESTING AIMING LOW POWER
SENSING APPLICATIONS**... 1349
Humberto Pereira Da Paz ; Vinícius Santana Da Silva ; Eduardo Vicente Valdés Cambero ; Humberto Xavier De
Araújo ; Ivan Roberto Santana Casella ; Carlos Eduardo Capovilla

**A SIMPLIFIED ANALYSIS OF BUCK-TYPE INTERLEAVED DC-DC CONVERTER FOR
BATTERY CHARGERS APPLICATION**.. 1354
Menaouar Berrehil El Kattel ; Robson Mayer ; Maicon Douglas Possamai ; Sérgio Vidal Garcia Oliveira

**PERFORMANCE OF A MULTI-GROUNDED DISTRIBUTION NETWORK WITH A FOUR-
WIRE THREE-PHASE POWER CONDITIONER** 1360
Samuel N. Duarte ; Bruno C. Souza ; Pedro M. Almeida ; Leandro R. Araújo ; Pedro G. Barbosa

**A NEW STRATEGY OF MODULATION BASED ON SPACE VECTOR MODULATION AND
ANNOTATED PARACONSISTENT LOGIC FOR A THREE-PHASE CONVERTER**...................... 1366
Antonio Carlos Duarte Ricciotti ; João Inácio Da Silva Filho ; Raphael A. Bispo De Oliveira ; Viviane B. Da S.
Duarte Ricciotti ; Hyghor Miranda Côrtes ; André Miguel Nicolini

**TECHNIQUES OF SOLAR IRRADIANCE ESTIMATION FROM DATASHEET INFORMATION
OF PHOTOVOLTAIC PANELS** 1372
Itaiara F. Carvalho ; Maurício B. R. Correa

**EXTRA REACTIVE POWER ANALYSIS ON A DISTRIBUTION GRID WITH HIGH
INTEGRATION OF PV GENERATION** 1378
Daniel C. Pompermayer ; Caroline Marin ; Mariana A. Mendes ; Luann G. O. Queiroz ; Luiz G. R. Tonini ;
Murillo C. Vargas ; Oureste E. Batista

ALTERNATIVE FCS-MPC CONCEPTS FOR CASCADE FREE MOTOR SPEED CONTROL 1384
Filipe Fernandes ; José De Oliveira ; Ademir Nied ; Arthur G. Bartsch

**COMPARATIVE STUDY OF THE POWER SHARING TECHNIQUES FOR MICROGRIDS IN
AUTONOMOUS MODE**... 1390
Gustavo M. S. Azevedo ; Erik C. Oliveira ; Marcelo C. Cavalcanti ; Leonardo R. Limongi

HYBRID ONE-CYCLE CONTROL STRATEGY FOR BUCK+BOOST PFC BATTERY CHARGER 1396
Aluísio Alves De Melo Bento ; Raphael Perci Santiago

TRAPEZOIDAL CURRENT MODE FOR BIDIRECTIONAL HIGH STEP RATIO MODULAR MULTILEVEL DC-DC CONVERTER 1402
C. Pineda ; J. Pereda ; S. Neira ; P. Bravo ; J. Rodríguez ; C. García

HOMOTHETIC METHOD TO COMPUTE WINDING LOSSES IN THE DESIGN OF POWER INDUCTORS 1408
André Furlan ; Alvaro Morentin ; Guillaume Fontes ; Guillaume Delamare ; Marcelo L. Heldwein ; Thierry Meynard

EVALUATION OF POWER QUALITY IMPACTS DUE TO PHOTOVOLTAIC PENETRATION IN DISTRIBUTION GRIDS VIA TIME-DOMAIN SIMULATION 1414
Henrique Pires Corrêa ; Flávio Henrique Teles Vieira

CONTROL STRATEGY AND POWER MANAGEMENT FOR MULTIFUNCTIONAL INVERTERS WITH BESS AND REACTIVE POWER COMPENSATION 1420
Luiz Henrique Meneghetti ; Edivan Laercio Carvalho ; Gustavo B. K. Schmidt ; Emerson Giovani Carati ; Jean Patric Da Costa ; Carlos Marcelo De Oliveira Stein ; Zeno Luiz Iensen Nadal ; Rafael Cardoso

SIZING OF SUPERCAPACITOR AND BESS FOR PEAK SHAVING APPLICATIONS 1426
Kaique Ferreira ; Walbermark Marques Dos Santos ; Augusto César Rueda Medina

TWO-STAGE SEPIC-BUCK TOPOLOGY FOR NEIGHBORHOOD ELECTRIC VEHICLE CHARGER 1431
Rafael H. Eckstein ; Telles B. Lazzarin ; Gierri Waltrich

MODELING BATTERY ENERGY STORAGE SYSTEM OPERATING IN DC MICROGRID WITH DAB CONVERTER 1435
Rafael Dos Santos ; Flávio A. S. Gonçalves ; José A. Olimpio Filho ; Fernando P. Marafão ; Eric Gil

MULTI-MODULAR SCALABLE DC-AC POWER CONVERTER FOR CURRENT INJECTION TO THE GRID BASED ON PREDICTIVE VOLTAGE CONTROL 1441
S. Toledo ; M. Rivera ; E. Maqueda ; M. Ayala ; J. Pacher ; C. Romero ; R. Gregor ; T. Dragicevic ; P. Wheeler

FSC-MPC CURRENT CONTROL OF A 5-LEVEL HALF-BRIDGE/ANPC HYBRID THREE-PHASE INVERTER 1447
Jailson Leite Silva ; Rafael Rocha Matias ; Max Dannyel De Carvalho Alves ; Ranoyca Nayana Alencar Leão E Silva ; Amanda Thayla Silva Monteiro ; Kristian Pessoa Dos Santos

MODELING AND SIMULATION OF A STIRLING ENGINE IN SCILAB 1453
Raymundo Cordero ; Thyago Estrabis ; João Onofre ; Felipe Alexandre Monteiro ; Augusto Hayashi

A LOW COST PHOTOVOLTAIC PANEL EMULATOR 1458
Aluísio Alves De Melo Bento ; Raphael Perci Santiago

COMBINING MODEL-BASED AND EXTREMUM SEEKING CONTROL FOR FAST TRACKING THE MAXIMUM POWER POINT OF LUNDELL ALTERNATOR 1463
Gabriel Sales Lins Rodrigues ; Alexandre Cunha Oliveira ; Antonio Marcus Nogueira Lima

Z-SOURCE INVERTER FOR PHOTOVOLTAIC MICROGENERATION 1469
Felipe A. F. Almeida ; Felipe Guerra ; Flávio Alessandro Serrão Gonçalves

IFOC FOR A NINE-PHASE INDUCTION MOTOR DRIVE WITH CURRENT HARMONIC INJECTION 1475
Alexandre G. F. Da Silva ; Isaac S. De Freitas ; Simplicio A. Da Silva ; Victor F. M. B. Melo ; Gilielson F. Da Paz

A COMPARISON OF A DISCRETE-TIME PI AND AN INDIRECT MPC CURRENT CONTROLLERS FOR A SINGLE-PHASE GRID-CONNECTED INVERTER OPERATING WITH DISTORTED GRID AND SIGNIFICANT COMPUTATION FEEDBACK DELAY 1481
Sergio Pires Pimentel ; Oleksandr Husev ; Dmitri Vinnikov ; Serhii Stepenko ; Lauri Kutt ; Jose Rodriguez

A NON-ISOLATED DC-DC BOOST CONVERTER WITH THREE-STATE SWITCHING CELL 1487
Lucas Carvalho Souza ; Falcondes José Mendes De Seixas ; Luís De Oro Arenas ; Douglas Carvalho Morais ; Luciano De Souza Da Costa E Silva

INPUT VOLTAGE RANGE EXTENSION METHODS IN THE SERIES-RESONANT DC-DC CONVERTERS 1493
Andrii Chub ; Dmitri Vinnikov ; Jih-Sheng Lai

EXTENDED KALMAN FILTER BASED SPEED ESTIMATION FOR THE CONTROL OF PMSG 1499
Mukhtiar Singh

Author Index

Welcome Message

Welcome to Brazil. Welcome to Santos, one of the oldest cities in the Americas. Welcome to the **15th Brazilian and the 5th IEEE Southern Power Electronics Conferences**.

This joint Power Electronics conference is the largest one in the southern hemisphere. It was possible due to the effort of many people: organization and steering committees, financial supporters, SOBRAEP, IEEE-PELS, and above all, the authors and the reviewers. This event is therefore a collective and cooperative creation.

Much of the success of this SPEC-COBEP is due to the participation of the Guests responsible for the seven Tutorials and five Plenary Sessions. They bring to the discussion the most current and challenging topics in Power Electronics. In addition, the industrial sessions and the Rap session bring information and allow debates on trends and challenges of research and applications. All these activities together with the 256 papers, selected from 324 submissions, will be followed by more than 300 attendees, from five continents and 20 countries.

The COBEP-SPEC is also an opportunity to honor the pioneers of Power Electronics in Brazil, after half century of research, teaching and technological development. Thus, the plenary room is named "Edison da Silva", the first doctor in Power Electronics to work in Brazil. The technical session rooms have the names of "Ivo Barbi", and "Edson Watanabe", the greatest supervisors of masters and doctors in Power Electronics in Brazil. The room "Waldir Pó" honors the pioneer in Power Electronics at the University of São Paulo. The room "Celso Bottura" honors the author of the first doctoral thesis in Power Electronics in Brazil.

Thanks to the sponsors who made this event viable. Fapesp, São Paulo Research Foundation, and the companies: PHB-Solar, Ohmini, Typhoon-Hil, Supplier, Opal-RT, Semikron, Keysight, Wurth, WEG, Yokogawa, Tektronix, Precision Solutions, Hioki and Eldorado Institute. Thanks also to Master Class for the organization support.

Thanks to all students and professionals for making this COBEP-SPEC 2019 a memorable conference.

José Antenor Pomilio – General Chair

Organizing Committee

General Chair:
José Antenor Pomilio
University of Campinas/Brazil

Program Chair:
Fernando Pinhabel Marafão
São Paulo State University/Brazil

Program Co-chairs:
Paolo Mattavelli
University of Padova/Italy

Euzeli C. dos Santos Jr.
Purdue University/USA

Finance Chair:
Flávio A. Serrão Gonçalves
São Paulo State University/Brazil

Finance Co-chair:
Wilson Komatsu
University of São Paulo/Brazil

Tutorials Chair:
Alfeu J. Sguarezi Filho
Federal University of ABC/Brazil

Student Activities Chair:
Helmo K. Morales Paredes
São Paulo State University/Brazil

SPEC Steering Committee Chair
Udaya Madawala
The University of Auckland/New Zealand

COBEP Steering Committee Chair
Henrique A. C. Braga
Federal University of Juiz de Fora/Brazil

Steering Committee

J. Marcos Alonso - University of Oviedo/Spain

Fernando Antunes - Federal University of Ceará/Brazil

Luiz Henrique Barreto - Federal University of Ceará/Brazil

Jorge L. Duarte - Eindhoven University/Neederland

Josep M. Guerrero - Aalborg University/Denmark

Samir Kouro - Universidad Tecnica Federico Santa Maria/Chile

João Onofre P. Pinto - Federal University of Mato Grosso do Sul/Brazil

Marcos Rivera – Univ. de Talca/Chile

Edison da Silva – Federal University of Paraíba/Brazil

Elisabetta Tedeschi - Norwegian Univ. of Science and Technology, Norway

Paolo Tenti - University of Padova/Italy

Maria Ines Valla - National University of La Plata/Argentina

Edson Watanabe – Federal University of Rio de Janeiro/Brazil

Patrick Wheeler – The University of Nottingham/UK

Luiz Lopes – Concordia University, Canada

S. M. Muyeen - Curtin University, Australia

Jose Renes Pinheiro – Federal University of Bahia/Brazil

Track Editors

Alessandro Luiz Batschauer - UDESC
Alfeu Joãozinho Sguarezi Filho - UFABC
Cassiano Rech - UFSM
Danilo Iglesias Brandão - UFMG
Denizar Cruz Martins - UFSC
Elisabetta Tedeschi - NTNU
Euzeli Cipriano Santos Jr. – IUPUI
Fernando Pinhabel Marafão - UNESP
Fernando Luiz Marcelo Antunes   - UFC
Flávio A. Serrão Gonçalves - UNESP
Francisco A. dos Santos Neves - UPFE
Henrique Antônio Carvalho Braga - UFJF
Joao Onofre Pereira Pinto - UFMS
Jorge Duarte – TU/e
Lourenço Matakas Junior - USP
Luiz Henrique S. C. Barreto - UFC
Marcelo Lobo Heldwein - UFSC
Maurício B. de Rossiter Correa - UFCG
Paolo Mattavelli - UNIPD
Pedro Gomes Barbosa - UFJF
Renato de Oliveira Magalhães - INPE
Ricardo Quadros Machado - USP
Udaya Madawala – Auckland University

Technical Support Team

Augusto Nery - UNICAMP
Eliabe Queirós - UNICAMP
Eliomar Conde - UFABC
Fernando Lino - UFABC
Fernando Ortiz Martinz - USP
Joel Filipe Guerreiro - UNICAMP
Ligia Soster Ramos - UNESP
Lucas Kollef - USP
Mateus P. Dias - UNICAMP
Rafael Santos - UNESP
Victor Arruda - UNICAMP

Session Chairs

Alexandre Rocco
Alfeu Joãozinho Sguarezi Filho
Andrés Ortiz Salazar
Bulent Sarlioglu
Cassiano Rech
Dalton Honório
Danilo Iglesias Brandão
Demercil de Souza Oliveira Júnior
Denizar Cruz Martins
Dmitri Vinnikov
Domingos Simonetti
Edison R. Cabral da Silva
Edson H. Watanabe
Elisabetta Tedeschi
Enes Gonçalves Marra
Enio Ribeiro
Fabrício Bradaschia
Fernando L. Marcelo Antunes
Fernando Lessa Tofoli
Fernando Ortiz Martinz
Francisco A. dos Santos Neves
Gierri Waltrich
Giorgio Spiazzi
Grant Covic
Helmo Kelis Morales Paredes
Henrique Antônio Carvalho Braga
Janaína Gonçalves de Oliveira
João Ota
Joao Onofre Pinto

Jorge Duarte
Jose Bertuzzo
José Pomilio
José Puma
Joselito Heerdt
Kleber Lima
Leandro Michels
Leonardo Rodrigues Limongi
Lourenço Matakas Jr
Luiz Henrique S. C. Barreto
Marcelo Cavalcanti
Marcelo Heldwein
Maria Inés Valla
Mario Pacas
Maurício B. de Rossiter Correa
Paolo Mattavelli
Pat Wheeler
Prasad Enjeti
Richard M. Stephan
Sarath Tennakoon
Seleme Isaac Seleme Jr.
Sérgio Pires Pimentel
Sérgio Augusto Oliveira   da Silva
Sergio Vidal Oliveira
Sheldon Williamson
Telles Brynelli Lazzarin
Tommaso Caldognetto
Udaya Madawala
Wilson Komatsu

Reviewers List

Abilio Variz
Ademir Nied
Ademir Pelizari
Aditya Shekhar
Adriano Péres
Adson Bezerra Moreira
Agnaldo Dias
Alceu André Badin
Alef Silva
Alessandro Braatz
Alessandro Luz
Alessandro Batschauer
Alexandre Cunha Oliveira
Alexandre Candido Moreira
Alfeu Joãozinho Sguarezi Filho
Allan Fagner Cupertino
Aluísio Alves de Melo Bento
Alvaro Paladino
Alysson Seidel
Amilcar Gonçalves
Anderson da Silva Martins
Andre Cavalcante Do Nascimento
André A. Ferreira
André Nicolini
André Luís Kirsten
André Pires Nóbrega Tahim
Andrei de Oliveira Almeida
Andrés Ortiz Salazar
Angelo Lunardi
Angelo César Lourenço
Antonio José Bento Bottion
António M. S. Spencer   Andrade
Antonio Samuel Neto
Arnaldo J. Perin
Arthur Costa
Arthur Bartsch
Augusto M. dos Santos Alonso

Caio Ruviaro Dantas Osório
Camila Gehrke
Carlos Duque
Carlos Henrique Illa Font
Cassiano Rech
Cássio Luciano Baratieri
Cesar da Costa
Chrystian Remes
Cicero Dos Santos
Clodualdo Venicio Sousa
Cristiane Cauduro Gastaldini
Dalmo Cardoso
Dalton Honório
Daniel Pagano
Daniel Silveira
Daniel Soares Dos Santos
Andrade
Danilo Wollz
Danilo Caldas
Danilo Iglesias Brandão
David Campos Gaona
Demercil de Souza Oliveira Júnior
Denis de Castro Pereira
Denizar Cruz Martins
Diego Greff
Diego Madeira
Dmitri Vinnikov
Domingos Simonetti
Douglas de Assis Ferreira
Douglas Dotto de Oliveira
Edgar Maqueda
Edhuardo Grabovski
Edilson Sá
Edisio Alves
Edison Silva
Edivan Laercio Carvalho
Edson Acordi

Bruno França
Bruno Cortes
Bruno R. de Almeida
Eduardo Vasconcelos Stangler
Emerson Soares
Emerson R. Cabral da Silva
Emerson Giovani Carati
Enes Gonçalves Marra
Enio Ribeiro
Ernane A. Alves Coelho
Esio Eloi dos Santos Filho
Estêvão Teixeira
Everson Mattos
Fabiano F. Costa
Fábio Ecke Bisogno
Fabrício Bradaschia
Fabrício Campos
Fabrício Nogueira
Faete Filho
Falcondes José Mendes de Seixas
Felipe Almeida
Felipe Bovolini Grigoletto
Felipe Joel Zimann
Felipe Yoshimatsu Abe
Fernanda de Morais
Fernando Botterón
Fernando L. Marcelo Antunes
Fernando Amaral
Fernando Pinhabel Marafão
Fernando José da Costa Jr.
Fernando Lessa Tofoli
Fernando Ortiz Martinz
Fernando Soares Dos Reis
Flavio A. Serrão Gonçalves
Francieli Lima de Sá
Francisco Paz
Francisco Nunes
Francisco Neves
Frank Gonzatti
Frank Weiner Heerdt

Eduardo Verri Liberado
Eduardo Beline
Eduardo Barbosa
Eduardo Lorenzetti Pellini
Eduardo Moreira Vicente
Geovane Luciano Reis
Gideon Lobato
Gierri Waltrich
Gilberto Cunha
Gilberto Valentim Silva
Gilson Schiavon
Giorgio Spiazzi
Guilherme Marcio Soares
Guilherme Colnago
Guilherme da Silva Fischer
Guilherme Márcio Soares
Gustavo Koch
Gustavo Lambert
Haihao Jiang
Hans-Peter Schmidt
Helber de Souza
Hélio Marcos André Antunes
Helmo Kelis Morales Paredes
Henrique Antônio Carvalho Braga
Henrique Cabral
Henrique Chaves
Herminio Miguel de Oliveira Filho
Heverton Augusto Pereira
Igor A. Pires
Isaac Soares Freitas
Israel Divan
Israel Lopes
Italo R. F. M. Pinheiro da Silva
Ivan Chabu
Ivo Barbi
J. Alexis Andrade Romero
Jacson Oliveira
Jakson Bonaldo
Janaina Oliveira
Janaina Almada

Gabriel Macedo
Gabriel Tibola
Gabriel Ayres de Oliveira
Gabriel Negri
Gabriel Azevedo Fogli
Gaurav Kalra
Gean Jacques Maia de Sousa
Joel Melo
Jonatan Rafael Rakoski Zientarski
Jonatas Rodrigo Kinas
Jorge Duarte
Jorge Luiz Wattes
Jose Filho
Jose Junior
José Puma
José Oliveira
José Antenor Pomilio
Jose Alberto Torrico Altuna
José Andrés Santisteban
José R. B. de Almeida Monteiro
Joselito A. Heerdt
Josemar Quevedo
Juan Carlos Colque Carita
Juan Sebastian Solis Chaves
Julian Cezar Giacomini
Juliano Pacheco
Juliano de Pelegrini Lopes
Julio Teixeira
Julio Cesar Dias
Jurandir Soares
Ka-Hong Loo
Kelly Carvalho Mingorancia
Kleber Lima
Kleber Oliveira
Kleber Souza
Kristian Pessoa Santos
Laurinda Lúcia Nogueira Dos Reis
Leandro da Silva
Leandro Michels
Leandro Freitas

Jean Prigol
Jeferson Fraytag
Jens Friebe
Jessika Andrade
Joao Dias
João Ota
João Carlos de Oliveira
João Manoel Lenz
Joaquim Henrique   Reis
Lourenço Matakas Junior
Luan Carlos Dos Santos Mazza
Lucas Giuliani Scherer
Lucas Brighenti
Lucas Bellinaso
Lucas Menezes
Lucas Monogios Koleff
Lucas Pereira   Pires
Lucas Savoi Araujo
Luciano Barros
Luciano Schuch
Luis de Oro Arenas
Luis Fernando C. Monteiro
Luis J. Castelo Branco Camurca
Luiz A. Maccari Jr.
Luiz Antonio de Souza Ribeiro
Luiz Carlos Gili
Luiz Daniel Santos Bezerra
Luiz Henrique S. C. Barreto
Maicon Douglas Possamai
Maikel Menke
Makoto Takamiya
Marcel Hendrix
Marcelo Cavalcanti
Marcelo Heldwein
Marcelo Lima
Marcelo de Castro
Marcelo Santana
Marcelo Tirolli
Marcelo Gradella Villalva
Marcio Stefanello

Leandro Manso da Silva
Leandro Leysdian Oro Carralero
Lenin Martins Ferreira   Morais
Leo Lorenz
Leonardo Tabosa
Leonardo Bruno Garcia Campanhol
Leonardo Poltronieri Sampaio
Leonardo Rodrigues Limongi
Lisandra Kittel Ries
Loan Silva
Lorrana Rocha
Maria Dias Bellar
Mario Martins
Mario Oleskovicz
Mateus Braga
Matheus Garcia Soares
Matthew Pearce
Maurício Aredes
Maurício Beltrão de Rossiter Correa
Mauricio Mendes da Silva
Mauro Pagliosa
Milan Ilic
Milena Vargas Gil
Milton Evangelista
Moacyr Brito
Moises Lessa
Montie Vitorino
Murilo Lohn
Nady Rocha
Naji Rajai Nasri Ama
Nayara Brandao de Freitas
Neilor Dal Pont
Ni Lin
Nustenil Marinus
Olympio Cipriano
Osvaldo Ronald Saavedra
Pablo Fernando Soardi Costa
Paolo Tenti
Paolo Magnone
Paolo Mattavelli

Márcio Kimpara
Márcio Ortmann
Márcio Carvalho
Marcio Do Carmo Rodrigues
Marco Dalla Costa
Marco Aurelio Castro
Marcos Gutierrez Alves
Marcos Bressan
Marcos Alonso
Marcos Jacoboski
Marcos André Barros Galhardo
Marcus Gomes
Marcus Felipe Calori Jorgetto
Reginaldo Ferreira
Renato Magalhães
Renato Monaro
René Pastor Torrico Bascopé
Ricardo Quadros
Ricardo Benedito
Ricardo Lúcio   Ribeiro
Richard M. Stephan
Roberto Buerger
Roberto Francisco   Coelho
Robinson Figueiredo de Camargo
Robson Gonzatti
Rodnei Melo
Rodolfo Lacerda
Rodolfo Castanho
Rodolfo Valle
Rodolpho Neves
Rodrigo Linhares
Rodrigo Bento
Rodrigo Arruda Ferreira
Rodrigo Vieira
Rodrigo Binotto
Rodrigo Barriviera
Rodrigo Arruda Felício   Ferreira
Rodrigo de Souza Santos
Rogers Demonti
Romero Leandro Andersen

Paulo Peixoto Praça
Paulo Giacomo Milani
Pedro Machado de Almeida
Pedro Gomes Barbosa
Priscila Oliveira
Qing Liu
Qinhao Zhang
Rafael Cardoso
Rafael Cunha
Rafael Matias
Rafael Concatto Beltrame
Raimundo Nonato   de Oliveira
Raphael Amaral da Câmara
Raul Rabinovici
Raymundo Cordero
Simone Buso
Tárcio Barros
Telles Brunelli Lazzarin
Teresa Assunção
Thainan Theodoro
Thales Maia
Thiago Pereira
Thiago Parreiras
Thiago Fonseca Rech
Thiago de Oliveira
Tiago Dezuo
Tiago Henrique
Tiago Davi Curi Busarello
Tiago Miguel Klein Faistel
Tomas Correa
Tomasz Balkowiec
Tommaso Caldognetto
Victor Guzman
Victor Mendes Flores
Victor Régis Bernardeli
Victor Leonardo Yoshimura

Ronaldo Junior
Ronaldo Antonio Guisso
Ruben Godoy
Rubens Hock
Rudolf Riehl
Samanta Barbosa
Samir A. Mussa
Samuel Duarte
Samuel Figueiredo
Samuel Queiroz
Sarath B.Tennakoon
Sergio Daher
Sérgio Pires Pimentel
Sérgio Augusto Oliveira da Silva
Sérgio Vidal Garcia Oliveira
Shuang Xu
Sidelmo Silva
Vinicius Fuerback
Vinicius Maia
Vinícius Hoffmann
Vinicius Foletto Montagner
Walbermark Marques Dos Santos
Walter Kaiser
Walter L. Suemitsu
Waner Silva
Welflen Ricardo Nogueira Santos
Welton Lima
William Venturini
Wilson Komatsu
Ya Zhang
Yales Novaes
Yeran Liu
Ygo Batista
Yijie Wang
Yongheng Yang
Yonglu Liu
Zhongyi Quan

Program

Tutorial 1: Short-course on photovoltaic (PV) technology

Marcelo Gradella Villalva

Tutorial 3: Smart Power Electronics Battery Energy Management Solutions for Electric Transportation

Sheldon S. Williamson

Tutorial 4: Electric Machine Design and Applications and Wide Bandgap Power Electronics Drive

Bulent Sarlioglu

Tutorial 5: Advances in Wireless Power Transfer Technologies for EV charging

Udaya Madawala & Grant Covic

Tutorial 6: Power Electronics Opportunities and Challenges for Plugged and Wireless Fast Charging of Autonomous E-transport and Mobility

Sheldon S. Williamson

Tutorial 7: Conservative Power Theory and Cooperative Control of Distributed Power Converters

Elisabetta Tedeschi & Danilo I. Brandao & Tommaso Caldognetto

Opening Plenary: Power electronics in Brazil: Evolution and Challenges

Edson Roberto Cabral da Silva (Federal University of Paraíba)

Plenary 1: Grid Integration of EVs for V2G Applications: Wired or Wireless?

Udaya Madawala (The University of Auckland)

Plenary 2: Finite state Model Predictive Control of Multilevel Converters

Maria Ines Valla (Universidad Nacional de La Plata)

Plenary 3: On recent control oriented research issues in microgrids developed at the University of Padova

Paolo Mattavelli (University of Padova)

Plenary 4: Integrated Solid State Transformer Concepts for Utility Interface of Power Conversion Systems

Prasad Enjeti (Texas A&M University)

Rap Session: Transversality of research involving power electronics

Marcelo Heldwein *(Universidade Federal de Santa Catarina)*

Elisabetta Tedeschi *(Norwegian University of Science & Technology (NTNU))*

Jose Bertuzzo *(Eldorado)*

Pat Wheeler *(University of Nottingham)*

Industrial keynote 1: Power Test Solutions for Transportation Electrification and Renewable Energy - NH Research (NHR)

Industrial keynote 2: Typhoon-HIL

Industrial keynote 3: Magnetics in the GaN/SiC Power Electronics World - Würth Elektronik

Education 1

Edison R. Cabral da Silva

Maria Inés Valla

Experimental Workbench: a tools to help to teaching the techniques of drives of electric machines.

Victor Reis (Federal University of Juiz de Fora (UFJF)), Carlos Vasconcelos (Federal Center for Technological Education of Minas Gerais)

Modeling Battery Energy Storage System operating in DC microgrid with DAB converter

Rafael Santos (UNESP), Flavio A. S. Gonçalves (UNESP-São Paulo State University, Brazil), José de Arimatéia Olímpio Filho (São Paulo State University - UNESP), Fernando Pinhabel Marafão (UNESP-São Paulo State University, Brazil), Eric Gil (Emerson Process Management AB)

Implementation of a Didactic Platform for a Generic Load Torque Emulator Using Induction Machines and PWM Inverters

Christel Enock Ghislain Ogoulola (Federal University of Itajubá), Luiz Otavio Campos de Medeiros (Federal University of Itajubá), João Carlos Grilo (Federal University of Itajubá), Angelo José Junqueira Rezek (Federal University of Itajubá), Nery de Oliveira Junior (Nery Engenharia Ltda), Rafael Correa (Federal University of Itajubá), Alexandre Viana Braga (Federal University of Itajubá), Vinicius Zimmermann Silva (Federal University of Itajubá), Marcos Leonardo Ramos (Federal University of Itajubá)

A Low Cost Photovoltaic Panel Emulator

Aluísio Alves De Melo Bento (State University of Rio de Janeiro), RAPHAEL SANTIAGO (State University of Rio de Janeiro)

Power Quality 1

Elisabetta Tedeschi *(Norwegian University of Science & Technology (NTNU))*

João Ota *(University of Campinas)*

Zero-Crossing Detection Frequency Estimator Method Combined with a Kalman Filter for Non-ideal Power Grid

Tiago Davi Curi Busarello (Federal University of Santa Catarina), Sergio Luiz Sambugari Júnior (Universidade Estadual de Londrina), Newton da Silva (Universidade Estadual de Londrina)

Investigation of Voltage Regulation with Active and Reactive Power with Distributed Loads on a Radial Distribution Feeder

Felipe Joel Zimann (Universidade do Estado de Santa Catarina), Alessandro Batschauer (Universidade do Estado de Santa Catarina), Marcello Mezaroba (Universidade do Estado de Santa Catarina), Eduardo Stangler (Universidade Federal de Pernambuco), Francisco Neves (Universidade Federal de Pernambuco)

SVC OPERATING AS AN UNBALANCE COMPENSATOR WITH CONTROL SYSTEM BASED ON THE STEINMETZ METHOD AND THE INSTANTANEOUS POWER THEORY.

Andressa de Melo Rodrigues (Federal Institute of Education, Science and Technology of Goiás), Laylla Fernandes Silva (Federal Institute of Education, Science and Technology of Goiás), Luciano de Souza da Costa e Silva (Federal Institute of Education, Science and Technology of Goiás), Lucas Carvalho Souza (Federal Institute of Education, Science and Technology of Goiás)

Thermal Stress Evaluation of a Multifunctional Modular Multilevel Converter - STATCOM Operating as Active Filter

Renata de Sousa (Federal University of Minas Gerais (UFMG)), William Amorim (Federal Center for Technological Education of Minas Gerais), Dayane Mendonça (Federal Center for Technological Education of Minas Gerais), Allan Cupertino (Centro Federal de Educação Tecnológica de Minas Gerais), Lenin Morais (Federal University of Minas Gerais (UFMG)), Heverton Augusto Pereira (Universidade Federal de Viçosa)

Multilevel 1

Cassiano Rech *(Federal University of Santa Maria)*

Prasad Enjeti *(Texas A&M University)*

Integrating a Single Z-Source Network with a Modular Multilevel Converter for Voltage Boosting
Fatma Khera (University of Nottingham), Christian Klumpner (University of Nottingham), Patrick Wheeler (University of Nottingham)

Analysis of Neutral-Point Voltage Balancing in Three-Ports Active Neutral-Point-Clamped Converter
Kaio Vinicius Vilerá (Federal University of Santa Maria), Silvio Antonio Teston (Federal University of Fronteira Sul), Cassiano Rech (Federal University of Santa Maria)

Trapezoidal Current Mode for Bidirectional High Step Ratio Modular Multilevel dc-dc Converter
Cristian Pineda (Pontificia Universidad Católica de Chile), Javier Pereda (Pontificia Universidad Católica de Chile), Sebastián Neira (Pontificia Universidad Católica de Chile), Pablo Bravo (Pontificia Universidad Católica de Chile), Jose Rodriguez (Universidad Andres Bello), Cristian Garcia (Universidad Andres Bello)

Development of a FPGA-Based Control System for Modular Multilevel Converter Applications
Lucas Monogios Koleff (University of São Paulo), Manoel Conde (University of São Paulo), Pedro Henrique Itio Hayashi (University of São Paulo), Francesco Sacco (University of São Paulo), Kelly Carvalho Mingorancia (University of São Paulo), Eduardo Lorenzetti Pellini (University of São Paulo), Wilson Komatsu (University of São Paulo), Lourenço Matakas Junior (University of São Paulo)

Power Devices

Henrique Antônio Carvalho Braga *(Federal University of Juiz de Fora (UFJF))*

Giorgio Spiazzi *(University of Padova - Dept. of Information Engineering (DEI))*

Inductor Design Methodology for Power Electronics Applications
Pedro Vilkn (Universidade Federal de Minas Gerais), Lenin Martins Ferreira Morais (Universidade Federal de Minas Gerais), Renato Santana (Universidade Federal de Minas Gerais), Porfirio Cortizo (Universidade Federal de Minas Gerais), Paulo Fernando Seixas (universidade federal de minas)

Lifetime Investigation of DC-link Capacitors in Multiple Slim Drives System
Shili Huang (Wuhan University of Technology), Haoran Wang (Aalborg University), Dinesh Kumar (Danfoss Drives A/S), Guorong Zhu (Wuhan University of Technology), Huai Wang (Aalborg University)

Thyristor Triggering, Static and Dynamic Characteristics
Ricardo Prado (SEMIKRON Semicondutores Ltda), Ulrich Nicolai (SEMIKRON Elektronik GmbH & Co.)

Predicting the Life Time of Power Semiconductor Modules
Clovis Gajo (SEMIKRON Semicondutores Ltda), Arendt Wintrich (SEMIKRON International GmbH), Paul Drexhage (SEMIKRON, Inc.)

Modeling and Control 1

Richard M. Stephan *(Universidade Federal do Rio de Janeiro)*

Bulent Sarlioglu *(University of Wisconsin-Madison)*

Cascade Control vs Full-State Feedback
Debora Soares (Universidade Federal do Rio de Janeiro), Henrique Calil (Universidade Federal do Rio de Janeiro), Richard M. Stephan (Universidade Federal do Rio de Janeiro)

SEPIC DC/DC converter control by observed-state feedback
Silas Sousa (Federal Institute of Minas Gerais), Valter Leite (Federal Center of Minas Gerais), Samir Fernandes (Federal University of Santa Catarina), Isabel Oliveira (Federal University of São João del-Rei)

Variable-step DFT Algorithm to Improve Optimization Routines Performance for a Three-Phase Interleaved Vienna Rectifier

Bruno Bertoldi (Federal University of Santa Catarina), Edhuardo Grabovski (Federal University of Santa Catarina), Marcelo Heldwein (Federal University of Santa Catarina), André Lange (Federal University of Santa Catarina)

Comparative Study Between Virtual Synchronous Machine and Virtual Impedance Techniques for Two Paralleled Inverters Sharing a Load

Bruno Laurindo (Universidade Federal Fluminense), Fábio Alves (Universidade Federal do Rio de Janeiro), Marcello Neves (Universidade Federal do Rio de Janeiro), Jorge Caicedo (Universidade Federal do Rio de Janeiro), Bruno França (Universidade Federal Fluminense), Maurício Aredes (Universidade Federal do Rio de Janeiro)

Applications 1: Motors and drivers

Francisco Neves *(Universidade Federal de Pernambuco)*

Fernando Lessa Tofoli *(Universidade Federal de São João del-Rei)*

Development of current-source-inverter-based integrated motor drives using wide-bandgap power switches

Renato Amorim Torres (University of Wisconsin-Madison), Hang Dai (University of Wisconsin-Madison), Woongkul Lee (University of Wisconsin-Madison), Thomas Jahns (University of Wisconsin-Madison), Bulent Sarlioglu (University of Wisconsin-Madison)

Analysis and Comparison of the Dynamic Response of Direct and Indirect Rotor Flux Control Applied to an Asymmetrical Two-Phase Induction Motor

Rafael Campos (Universidade do Estado de Santa Catarina), José Oliveira (Universidade do Estado de Santa Catarina), Ademir Nied (Universidade do Estado de Santa Catarina)

Novel MTPA Approach for IPMSM with Non-Sinusoidal Back-EMF

Allan Gregori de Castro (University of São Paulo), Paulo Roberto Ubaldo Guazzelli (University of São Paulo), Mateus Moro Lumertz (University of São Paulo), Carlos Matheus Rodrigues de Oliveira (University of São Paulo), Geyverson Teixeira de Paula (Federal University of Goiás), José Roberto Boffino de Almeida Monteiro (University of São Paulo)

Third Harmonic Injection Method for Reliability Improvement of Single-Phase PV Inverters

Rodrigo de Barros (Federal University of Minas Gerais (UFMG)), Rafaela Silva (Universidade Federal de Viçosa), Diogo Silveira (Universidade Federal de Viçosa), Wallace do Couto Boaventura (Universidade Federal de Minas Gerais), Allan Cupertino (Federal Center for Technological Education of Minas Gerais), Heverton Augusto Pereira (Universidade Federal de Viçosa)

Sensorless Control of Nonsinusoidal Back-EMF PMSM Based on State Observer

Thiago Lazzari (Federal University of Santa Maria), Filipe Pinarello Scalcon (Federal University of Sa), Cesar Volpato (Federal University of Santa Maria), Thieli Smidt Gabbi (Federal University of Santa Maria), Marcio Stefanello (Federal University of Pampa), Rodrigo Vieira (Federal University of Santa Maria)

Renewables 1

Demercil de Souza Oliveira Júnior (Federal University of Ceará)

Sérgio Augusto Oliveira da Silva (Federal Technological University of Paraná)

Thermal Modeling of Converters for Wind Conversion Systems Employing DFIG Technology

Igor Rodrigues de Oliveira (Federal University of Minas Gerais (UFMG)), Fernando Lessa Tofoli (Universidade Federal de São João del-Rei), Victor Mendes Flores (Federal University of Minas Gerais (UFMG))

Comparative study of different correction methods to analyze wind turbine performance

Hércules Oliveira (Federal University of Maranhão), Luiz Antonio de Souza Ribeiro (Federal University of Maranhão), Jerson Vaz (Federal University of Pará), Osvaldo Ronald Saavedra (Federal University of Maranhão), José Gomes de Matos (Federal University of Maranhão)

Integrated Zeta Inverter Applied in a Single-Phase Grid-Connected Photovoltaic System

Leonardo Poltronieri Sampaio (Federal University of Techonology – UTFPR-CP), Sérgio Augusto Oliveira da Silva (Federal Technological University of Paraná), Paulo Júnior Silva Costa (Federal University of Techonology – UTFPR-CP)

Extra Reactive Power Analysis on a Distribution Grid with High Integration of PV Generation

Daniel Campos Pompermayer (Universidade Federal do Espírito Santo), Caroline Marim (Universidade Federal do Espírito Santo), Mariana Altoé Mendes (Federal University of Espírito Santo), Luann Georgy Oliveira Queiroz (Universidade Federal do Espírito Santo), Luiz Guilherme Riva Tonini (Universidade Federal do Espírito Santo), Murillo Cobe Vargas (Universidade Federal do Espírito Santo), Oureste Elias Batista (Federal University of Espírito Santo)

Analysis of Partial-Power Processing Converters for Small Wind Turbines Systems

Anderson José Balbino (Federal University of Santa Catarina), Ronny Glauber de Almeida Cacau (Federal University of Santa Catarina), Telles B. Lazzarin (Federal University of Santa Catarina)

Power Converters 1

Domingos Simonetti *(Universidade Federal do Espírito Santo)*

Telles B. Lazzarin *(Federal University of Santa Catarina)*

NOVEL BUCK-BOOST PFC CONVERTER WITH THREE-STATE SWITCHING CELL

Douglas Carvalho (UNESP-São Paulo State University), Falcondes Seixas (UNESP-São Paulo State University), Luis De Oro Arenas (UNESP-São Paulo State University), Lucas Carvalho Souza (UNESP-São Paulo State University, Brazil), Luciano de Souza da Costa e Silva (Federal Institute of Education, Science and Technology of Goiás)

Steady-State Characterization of the Three-Phase Isolated DC-DC Bidirectional Converter with LLC Resonant Tank

Kristian Santos (Federal Institute of Education, Science and Technology of Piauí), Paulo Peixoto Praça (Federal University of Ceará), Herminio Miguel De Oliveira Filho (University for the International Integration of the Afro-Brazilian), Demercil de Souza Oliveira Júnior (Federal University of Ceará)

Center-Tapped π-Type Single-Phase Cell

Domingos Simonetti (UFES), Xibo Yuan (University of Bristol)

Comparison Between Two Modulation Strategies for the 3L-DC-SSI

Matheos Coletto Wermuth (Unipampa), Guilherme Sebastião Da Silva (Unipampa), Amanda Costa Maia (Unipampa), Richard Gonçalves Cornelius (Unipampa)

New Five-Level Flying Capacitor Inverter Fed by a Boost-Flyback DC-DC Voltage Source

Antonio Lacerda Filho (Federal University of Paraíba), Edison Roberto Cabral da Silva (UFPB/UFCG), ANDRE COSTA (FEDERAL UNIVERSITY OF CAMPINA GRANDE), Ronnan Cardoso (FEDERAL UNIVERSITY OF PERNAMBUCO), Darlan Fernandes (Federal University of Paraíba)

Modeling and Control 2: Predictive Control

Janaína Gonçalves de Oliveira *(Universidade Federal de Juiz de Fora - UFJF)*

Joao Onofre Pinto *(Universidade Federal do Mato Grosso do Sul - UFMS)*

Modulated Model Predictive Current Control for a PMSM Operating with a Three-level NPC Inverter

Qi Wang (Southeast University, Nanjing), Marco Rivera (Universidad de Talca), Pat Wheeler (University of Nottingham, UK and China), Jose Riveros (Universidad de Talca)

Model Predictive Controller for Two-Phase Three-Wire Grid-Connected Converters

Pablo C. de S. Furtado (Federal Institute of Southeast of Minas Gerais (IF Sudeste MG)), Pedro Gomes Barbosa (Federal University of Juiz de Fora (UFJF))

A Model Predictive Control Applied To Single-Phase Packed-U-Cells Converter

Jordan Zucuni (Federal University of Santa Maria), Dimas Schuetz (Federal University of Santa Maria), Fernanda Carnielutti (Federal University of Santa Maria), Felipe Bovolini Grigoletto (Federal University of Pampa), Humberto Pinheiro (Federal University of Santa Maria), Margarita Valdivia (Universidad Técnica Federico Santa María), Jose Rodriguez (Universidad Técnica Federico Santa María)

Multi-modular scalable DC-AC power converter for current injection to the grid based on predictive voltage control

Sergio Toledo (Universidad de Talca), Edgar Maqueda (Universidad Nacional de

Asunción), Magno Ayala (Universidad Nacional de Asunción), Carlos Romero (Universidad Nacional de Asunción), Julio Pacher (Universidad Nacional de Asunción), Marco Rivera (Universidad de Talca), Pat Wheeler (University of Nottingham), Raul Gregor (Universidad Nacional de Asunción), Tomislav Dragicevic (Aalborg University)

Digital Current Control for a Bidirectional Interleaved Boost Converter with Coupled Inductors

Simone Buso (University of Padova - Dept. of Information Engineering (DEI)), Giorgio Spiazzi - University Of Padova (University of Padova - Dept. of Information Engineering (DEI)), Tommaso Caldognetto (University of Padova - Dept. of Management and Engineering (DTG))), Qing Liu (University of Padova - Dept. of Information Engineering (DEI)), Francesco Toniolo (University of Padova - Dept. of Management and Engineering (DTG)))

Smart grid applications 1

Paolo Mattavelli (University of Padova)

Luiz Henrique S. C. Barreto - UFC (Federal University of Ceará (UFC))

Aspects of travelling wave based protection philosophy for consideration in DC Grids of the future

Monday Ikhide (Coventry University), Sarath Tennakoon (Carnegie Mellon University; CMU-Africa), Alison Griffiths (Staffordshire University), Hengxu Ha (GE Renewable Energy - Stafford)

Stabilization of DC Microgrids with Point-of-Load Converters as Constant Power Loads

Isaías de Bessa (Federal University of Amazonas), Renan Landau Paiva de Medeiros (Universidade Federal do Amazonas), Iury Bessa (Federal University of Amazonas), Floryndo Antônio Ayres Junior (Federal University of Amazonas), Kevin Lucas (Federal University of Santa Catarina)

Implementation of Automatic Battery Charging Temperature Compensation on a Peak-Shaving Energy Storage Equipment

Wilson Santana (Instituto Gnarus), Robson Gonzatti (Universidade Federal de Itajuba), Germano Lambert-Torres (Instituto Gnarus), Erik Bonaldi (Instituto Gnarus), Pedro Oliveira (Universidade Federal de Itajuba), Bruno Torres (Universidade Federal de

Itajuba), João Foster (Universidade Federal de Itajuba), RONDINELI PEREIRA (Universidade Federal de Itajuba), Luiz Eduardo Borges-da-Silva (Universidade Federal de Itajuba), Denis Mollica (EDP Sao Paulo Distribuicao de Energia), Joselino Santana-Filho (EDP Sao Paulo Distribuicao de Energia)

Voltage Regulator Behavior on Power Distribution Grids with High Integration of PVDG

Lucas Azeredo (UFES), Jussara Fardin (Universidade Federal do Espírito Santo), Mariana Altoé Mendes (Federal University of Espírito Santo), Murillo Cobe Vargas (Universidade Federal do Espírito Santo), Oureste Elias Batista (Federal University of Espírito Santo), Carla Espindula (UFES), Luiz Guilherme Riva Tonini (Universidade Federal do Espírito Santo)

Modelling and Analysis of the Isolated Microgrid Installed at the Lençóis Island using PSCAD/EMTDC

Silvangela Lima Barcelos (Federal University of Maranhão), Jose Gomes de Matos (Federal University of Maranhão), Luiz Antonio de Souza Ribeiro (Federal University of Maranhão)

Vehicular technologies

Dalton Honório (Federal University of Ceará)

Fernando Antunes (Federal University of Ceará (UFC))

Reliability Analysis of aircraft starter generator drive converter

Jayakrishnan Harikumaran (University of Nottingham), Giampaolo Buticchi (University of Nottingham, UK and China), Michael Galea (University of Nottingham), Alessandro Costabeber (University of Nottingham), Pat Wheeler (University of Nottingham)

A Two-Stage Battery Charger with Active Power Decoupling Cell for Small Electric Vehicles

Mateo Roig (Universidade Federal de Santa Catarina), Caio Moraes (Universidade Federal de Santa Catarina), Telles B. Lazzarin (Federal University of Santa Catarina)

Application of Model Predictive Control in a Resolver-to-Digital Converter

Thyago Estrabis (Federal University of Mato Grosso do Sul), Raymundo Cordero

(Federal University of Mato Grosso do Sul), Edson Batista (Federal University of Mato), Cristiano Andrea (Federal University of Mato Grosso do Sul), Marcio Grassi (Federal University of Mato Grosso do Sul)

Combining Model-Based and Extremum Seeking Control for Fast Tracking the Maximum Power Point of Lundell Alternator

Gabriel Sales Lins Rodrigues (Universidade Federal de Campina Grande), Alexandre Cunha Oliveira (Universidade Federal de Campina Grande), Antonio Marcus Nogueira Lima (Universidade Federal de Campina Grande)

Multi-Port System for Storage and Management of Regenerative Braking Energy in Diesel-Electric Locomotives

Gierri Waltrich (Federal Univer), Caio Moraes (Universidade Federal de Santa Catarina), Sérgio Brockveld (Federal University of Santa Catarina), Pedro Cavilha (Federal University of Santa Catarina), Antonio Luiz Schalata Pacheco (Federal University of Santa Catarina), Marcelo Heldwein (Federal University of Santa Catarina)

Multilevel 2

Maurício Beltrão De Rossiter Correa (FEDERAL UNIVERSITY OF CAMPINA GRANDE)

Sergio Vidal Oliveira (Santa Catarina State University - UDESC)

Design of Resonant Controllers for Compensation of Third Harmonic Ripple in the DC Capacitors Voltages of NPC Converters

Andrei de Oliveira Almeida (Centro Federal de Educação Tecnológica de Minas Gerais), Pedro Machado de Almeida (Federal University of Juiz de Fora (UFJF)), Pedro Gomes Barbosa (Federal University of Juiz de Fora (UFJF))

Enhanced Phase-Shifted Carrier PWM Applied to 3-Phase Multilevel Coupled Inductors Inverters

Emerson Soares (UFCG), Lucas Lucena (UFPB), Nady Rocha (UFPB), Cursino Brandão Jacobina (UFCG), Edison Roberto Cabral da Silva (UFPB/UFCG)

Passive Capacitor Voltage Balancing in Modular Multilevel Converter During Precharge: Analysis and Design

Luiz Schmidt (Universidade Federal de Santa Catarina), Gean Jacques Maia de Sousa

(Universidade Federal de Santa Catarina), Marcelo Heldwein (Universidade Federal de Santa Catarina), Daniel Pagano (Federal University of Santa Catarina)

Hybrid Switched Capacitor DC-DC Converter Based on MMC

Marcus Soares (Santa Catarina State University), Gustavo Lambert (Santa Catarina State University), YALES NOVAES (Santa Catarina State University)

Predictive Voltage Control Operating at Fixed Switching Frequency of a Neutral-Point Clamped Converter

Felipe Herrera (Universidad de Chile), Marco Rivera (Universidad de Talca), Jose Riveros (Universidad de Talca), Roberto Cardenas (Universidad de Chile), Pat Wheeler (University of Nottingham)

Power Converters 2

Andrés Ortiz Salazar (Universidade Federal do Rio Grande do Norte)

Tommaso Caldognetto (University of Padova - Dept. of Management and Engineering (DTG)))

Design and Performance Analysis of Isolated Cuk Converter Employed in Multiple Pulse Rectifier Systems

Ana L. Soares (Federal University of Uberlândia), Antonio O. Costa Neto (Federal University of Uberlândia), Gustavo B. Lima (Federal University of Uberlândia), Luiz C. G. Freitas (Federal University of Uberlândia), Ernane A. Alves Coelho (Federal University of Uberlândia)

Steady-State Analysis of a Single-phase Modified Bridgeless Boost Rectifier in DCM

Mateo Roig (Universidade Federal de Santa Catarina), Caio Moraes (Universidade Federal de Santa Catarina), Telles Lazzarin (Universidade Federal de Santa Catarina)

Single-Stage Single-phase AC/DC Converter with High Frequency Isolation Feasible to Microgeneration

Samanta Barbosa (Federal University of Ceará), Bruno R. de Almeida (University of Fortaleza), Debora Damasceno (Federal University of Ceará), Demercil de Souza Oliveira Júnior (Federal University of Ceará)

Totem-Pole Bridgeless PFC Converter in DCM with Synchronous Rectification
Leonardo S. Mai (Federal University of Santa Catarina), Samir A. Mussa (Federal University of Santa Catarina), Alexandra Rospirski (Federal University of Santa Catarina), Telles B. Lazzarin (Federal University of Santa Catarina)

Single-Stage Bridgeless AC-DC PFC Flyback Interleaved in DCM
Guilherme Schowantz S. (UFSC/INEP), Renan P. Barcelos (UFSC/INEP), Telles B. Lazzarin (Federal University of Santa Catarina)

Modeling and Control 3: PLL

Sérgio Pires Pimentel (Federal University of G)

Lourenço Matakas Jr (University of São Paulo)

A Novel Phase-Locked Loop with Positive and Negative Sequence Detection Capability
Kleber Lima (Federal University of Cea), Jean M. L. Fonseca (Federal University of Ceará (UFC)), Carlos Gustavo Castelo Branco (Federal University of Ceará (UFC)), Renato Guerreiro Araujo (Federal University of Ceará (UFC))

Synchronous reference frame PLL frequency estimation under voltage variations
Eliabe Queiroz (University of Campinas), José Pomilio (University of Campinas)

SRF-PLL Influence on the Stability of a Current Source Converter in Droop Mode
Eliabe Queiroz (University of Campinas), José Pomilio (University of Campinas)

Design and Test of a SRF-PLL Based Algorithm for Positive-Sequence Synchrophasor Measurements
Gabriel Ubirajara de Carvalho (Universidade Tecnológica Federal do Paraná), Gustavo Weber Denardin (Universidade Tecnológica Federal do Paraná), Rafael Cardoso (Universidade Tecnológica Federal do Paraná), Cassiano Ferro Moraes (Universidade Tecnológica Federal do Paraná)

Comparison of the PLL Control techniques applied in Photovoltaic System
Marenice Melo de Carvalho (Federal University of Amazonas), Renan Landau Paiva de Medeiros (Universidade Federal do Amazonas), Iury Bessa (Federal University of

Amazonas), Floryndo Antônio Ayres Junior (Federal University of Amazonas), Kevin Lucas (Federal University of Santa Catarina), David Vaca Benavides (Escuela Superior Politécnica del Litoral)

Poster Session 1

Henrique Antônio Carvalho Braga *(Federal University of Juiz de Fora (UFJF))*

Denizar Cruz Martins *(Federal University of Santa Catarina)*

Alexandre Rocco *(UNISANTA)*

Three-Phase Induction Motors Efficiency Analysis Using a Programmable Power Supply

Cássio Alves de Oliveira (Federal University of Uberlândia), Josemar Alves dos Santos Junior (Federal Institute of Education, Science and Technology of Goias - IFG), Marcos José de Moraes Filho (Federal University of Uberlândia), Vinícius Marcos Pinheiro (Federal University of Uberlândia), Augusto Wohlgemuth Fleury Veloso da Silveira (Federal University of Uberlândia), Luciano Coutinho Gomes (Federal University of U)

Flexible Didactic Platform for Thyristor-Based Circuits

Lucas Monogios Koleff (University of São Paulo), Lucas Araújo (University of São Paulo), Mário Zambon (University of São Paulo), Wilson Komatsu (University of São Paulo), Eduardo Lorenzetti Pellini (University of São Paulo), Lourenço Matakas Junior (University of São Paulo)

Comparative study of RC snubber configurations in switching circuits

Ana Carolina Moreira (Instituto Federal de Santa Catarina), Daniel Piccoli (Instituto Federal de Santa Catarina), Julio Oliveira (Instituto Federal de Santa Catarina), Rodrigo Piontkewicz (Instituto Federal de Santa Catarina), Luiz Henning (Instituto Federal de Santa Catarina)

Comparative Strategies of Control for Regenerative Braking in Electric Vehicles

Marina Paredes (University of Campinas), José Antenor Pomilio (University of Campinas)

Hybrid Model of Electric Vehicle

Débora Martins (Universidade Federal do Espírito Santo), Danilo Silva (Universidade Federal do Espírito Santo), Jussara Fardin (Universidade Federal do Espírito Santo)

New semiconductor technologies for power electronics

Pedro França (Universidade do Estado de Santa Catarina), Joselito A. Heerdt (Universidade do Estado de Santa Catarina), José Diesel (Universidade do Estado de Santa Catarina), Alisson Mengatto (Universidade do Estado de Santa Catarina)

Design and Analysis of Output Filter with Long Lifetime E-Cap for AC-DC LED Driver

Andressa Fernandes (UNIVERSIDADE FEDERAL DO CEARÁ), Felipe Costa (Instituto Federal de Educação, Ciência e Tecnologia do Ceará), Antônia Rocha (UNIVERSIDADE FEDERAL DO CEARÁ), Rodrigo Santos (UNIVERSIDADE FEDERAL DO CEARÁ), Fernando Antunes (Federal University of Ceará (UFC)), Edilson Sá (Instituto Federal de Educação, Ciência e Tecnologia do Ceará)

Analysis of Rectifiers for RF Energy Harvesting Aiming Low Power Sensing Applications

Humberto Pereira da Paz (Universidade Federal do ABC), Vinícius Silva (Universidade Federal do ABC), Eduardo Vicente Valdés Cambero (Universidad Federal do ABC), Humberto Xavier de Araújo (Universidade Federal do Tocantins), Ivan Casella (Universidade Federal do ABC), Carlos Capovilla (Universidade Federal do ABC)

On the Influence of Area Variations of the Photovoltaic Surface in Solar Cell Antennas

Eduardo Vicente Valdés Cambero (Universidade Federal do ABC), Humberto Pereira da Paz (Universidade Federal do ABC), Vinícius Silva (Universidade Federal do ABC), Humberto Xavier de Araújo (Universidade Federal do Tocantins), Ivan Casella (Universidade Federal do ABC), Carlos Capovilla (Universidade Federal do ABC)

Using Synchronverter in Distributed Generation for Frequency and Voltage Grid Support

Guilherme Penha (Federal University of Rio Grande do Norte (UFRN)), Luciano Barros (Federal University of Rio Grande do Norte (UFRN))

Evaluation of Power Quality Impacts due to Photovoltaic Penetration in Distribution Grids via Time-Domain Simulation

Henrique Pires Corrêa (Universidade Federal de Goiás), Flávio Henrique Teles Vieira (Universidade Federal de Goiás)

Non-Isolated High Current Battery Charger with PFC Bridgeless Rectifier

Gierri Waltrich (Federal University of Santa Catarina), RODRIGO DACOL (Federal

University of Santa Catarina), Joselito Heerdt (Universidade do Estado de Santa Catarina)

Proposal of an Isolated Two-Switch DC-DC SEPIC Converter

Carlos Henrique Illa Font – UTFPR (Federal University of Technology - Paraná), Marcos Ewerling (Federal University of Santa Catarina), Telles B. Lazzarin (Federal University of Santa Catarina)

Impacts of the Multi-Variables Modulation on Transformer and Soft-Switching of a DAB

Jeferson Fraytag (Federal University of Santa Catarina), Andre Kirsten (Federal University of Santa Catarina), Marcelo Heldwein (Federal University of Santa Catarina)

A Hybrid Bidirectional Push-Pull DC-DC Converter with a Ladder Switched-Capacitor Cell

Marcos Dantas (Federal University of Paraíba), Fellipe Oliveira (Federal University of Paraíba), Lorena Albuquerque (Federal University of Paraíba), Isaac Soares Freitas (Federal University of Paraíba), Romero Leandro Andersen (Federal University of Paraíba)

Applying Coupled Inductors to the Clamped-Resonant Interleaved Boost Converter

Giorgio Spiazzi - University Of Padova (University of Padova - Dept. of Information Engineering (DEI))

Two-stage stand alone photovoltaic system for water pumping system

Renata Cristina da Silva (Federal University of Minas Gerais (UFMG)), Gabriel Gaburro Bacheti (Federal University of Itajubá), Rafael Mario da Silva (Federal University of Minas Gerais (UFMG)), Vitor Gomes Neves (Federal University of Itajubá), Clodualdo Venicio Sousa (Federal University of Itajubá), Seleme Isaac Seleme Jr. (Universidade Federal de Minas Gerais)

Performance Analysis of Active Anti-islanding Techniques for Photovoltaic Application

Marcos Roberto Oshiro (Federal University of Mato Grosso do Sul), Ruben Barros Godoy (Federal University of Mato Grosso do Sul), Moacyr A. G. de Brito (Federal University of Mato Grosso do Sul), Luigi Galotto Junior (Federal University of Mato Grosso do Sul)

Evaluation of Techniques to Reduce the Effects of Partial Shading on Photovoltaic Arrays

André Augusto Rodrigues (Universidade Federal de São João del-Rei), Paula dos Santos Vicente (Universidade Federal de São João del-Rei), Fernando Lessa Tofoli (Universidade Federal de São João del-Rei), Eduardo Moreira Vicente (Universidade Federal de São João del-Rei)

A critical analysis of PSO and its variations applied to MPPT for PV Systems under Partial Shading Condition

Lucas Mendonça Andrade (Universidade Federal de São João del-Rei), Paula dos Santos Vicente (Universidade Federal de São João del-Rei), Fernando Lessa Tofoli (Universidade Federal de São João del-Rei), Eduardo Moreira Vicente (Universidade Federal de São João del-Rei)

Gas Microturbines for Distributed Generation System

Walquíria Do Nascimento Silva (Universidade Federal de Juiz de Fora - UFJF), Janaína Gonçalves de Oliveira (Universidade Federal de Juiz de Fora - UFJF), Bruno Henrique Dias (Universidade Federal de Juiz de Fora - UFJF), Leonardo Willer de Oliveira (Universidade Federal de Juiz de Fora - UFJF)

Real-Time implementation of a DC converter using Modified Nodal Analysis, Sparsity Handling and Parallelism on a DSP platform

Luiz Felipe Ribeiro (Universidade Federal do Rio de Janeiro), Felipe Dicler (Universidade Federal do Rio de Janeiro), Luis Guilherme Rolim (Universidade Federal do Rio de Janeiro), Maurício Aredes (Universidade Federal do Rio de Janeiro)

Model, Simulation and Analysis of BLDCM for a Differential Controlled Electric-Powered Wheelchair

Augusto Nery de Lima Neto (University of Campinas), José Pomilio (University of Campinas)

Comparative Study of Control Systems for a Photovoltaic Inverter with LCL Filter

Moacyr Brito (Universidade Federal do Mato Grosso do Sul - UFMS), Leandro Takeshi (Universidade Federal do Mato Grosso do Sul - UFMS)

Generalized Mathematical Model for an N-cell Interleaved Boost Converter

Luana Pereira (Federal University of Minas Gerais (UFMG)), Seleme Isaac Seleme Jr.

(Universidade Federal de Minas Gerais), João Lucas da Silva (Federal University of Itajubá)

Interleaved Bidirectional DC-DC Converter for Application in Hybrid Propulsion System: Modeling and Control

Vitor Torres (Federal University of Juiz de Fora (UFJF)), Vinicius Albuquerque (Federal University of Juiz de Fora (UFJF)), Janaína Gonçalves de Oliveira (Universidade Federal de Juiz de Fora - UFJF), Manuel A. Rendón (Federal University of Juiz de Fora (UFJF)), Marcio do Carmo Rodrigues (Federal In), Pedro Santos Almeida (Federal University of Juiz de Fora (UFJF))

Unified Robust Control Design for BTB-VSC Subject to Uncertainties in Grid Equivalent Circuit

Marcelo de Castro Fernandes (Universidade Federal de Juiz de Fora- UFJF), Igor Souza (Universidade Federal de Ouro Preto), Gabriel Azevedo Fogli (Universidade Federal de Minas Gerais), Ademir Da Silva Toledo Junior (Federal Uni), Pedro Machado de Almeida (Federal Univer), Janaína Gonçalves de Oliveira (Universidade Federal de Juiz de Fora - UFJF), Pedro Gomes Barbosa (Federal University of Juiz de Fora (UFJF))

Deriving Stability Condition for One-Cycle Control with Triangular Carrier by Poincaré Maps

Armando Abrantes-Ferreira (COPPE/UFRJ), Lucas Gomes (COPPE/UFRJ), Robson Francisco Dias (COPPE/UFRJ), Luís Rolim (COPPE/UFRJ)

Three-phase, four-wire PWM rectifier applied to variable frequency AC systems in airplane electric grid under fault conditions

Alexandre G. Bueno (University of Campinas), José Pomilio (University of Campinas)

PERFORMANCE OF A MULTI-GROUNDED DISTRIBUTION NETWORK WITH A FOUR-WIRE THREE-PHASE POWER CONDITIONER

Samuel Duarte (Federal University of Juiz de Fora (UFJF)), Bruno Cortes (Federal University of Juiz de Fora (UFJF)), Pedro Machado de Almeida (Federal University of Juiz de Fora (UFJF)), Leandro Araújo (Federal University of Juiz de Fora (UFJF)), Pedro Gomes Barbosa (Federal University of Juiz de Fora (UFJF))

An Unit-Less Mathematical Model for Analysis and Design of Class-E Resonant Converters

Lucas Mendonça (Universidade Federal de Santa Maria), Thiago Cattani Naidon

(Universidade Federal de Santa Maria), Rafael Raposo (Universidade Tecnológica Federal do Paraná), Fábio Ecke Bisogno (Universidade Federal de Santa Maria)

A Research on Constant Voltage Output Characteristics of Wireless Power Transfer System with A DC-DC Converter

Zhimeng Liu (University of Chinese Academy of sciences), Lifang Wang (Institute of Electrical Engineering Chinese Academy of Sciences), Chengliang Yin (Shanghai Jiaotong University), Yanjie Guo (University of Chinese Academy of sciences), Chengxuan Tao (Institute of Electrical Engineering Chinese Academy of Sciences)

Power Demand Prediction Based on Mixed Driving Cycle Applied to Electric Vehicle Hybrid Energy Storage System

Lucas Lago (Universidade Federal de Juiz), Silvana Faceroli (Federal Institute of Education, Science and Technology of Southeast of Minas Gerais (IF Sudeste MG)), Rodrigo Arruda Ferreira (Federal Institute of Education, Science and Technology Southeast of Minas Gerais), Marcio do Carmo Rodrigues (Federal Institute of Education, Science and Technology of Southeast of Minas Gerais (IF Sudeste MG))

Application of a SHE-PWM modulation for a low switching frequency motor drive with harmonic investigation using the DTFT

Marcos Gomes (Centro Federal de Educação Tecnológica de Minas Gerais), Marina Hassen Souza (Centro Federal de Educação Tecnológica de Minas Gerais), Gabriel Vilkn Ramos (Centro Federal de Educação Tecnológica de Minas Gerais), Alex-Sander Amável Luiz (Centro Federal de Educação Tecnológica de Minas Gerais), Marcelo Martins Stopa (Centro Federal de Educação Tecnológica de Minas Gerais)

Power Quality 2

Fabrício Bradaschia (Universidade Federal de Pernambuco)

Sarath Tennakoon (Carnegie Mellon University; CMU-Africa)

Harmonic Compensation Strategies Applied to Multifunctional Photovoltaic Inverters

Lucas Xavier (Universidade Federal de Minas Gerais), Joice Zacarias (Universidade Federal de Viçosa), Allan Cupertino (Centro Federal de Educação Tecnológica de Minas Gerais), Heverton Augusto Pereira (Universidade Federal de Viçosa), Danilo Brandao

(Universidade Federal de Minas Gerais), Victor Mendes Flores (Universidade Federal de Minas Gerais)

Implementation of a iUPQC control scheme for ensuring an improved compensation performance

Eduardo Vasconcelos Stangler (Universidade Federal de Pernambuco), Francisco Neves (Universidade Federal de Pernambuco), Fabrício Bradaschia (Universidade Federal de Pernambuco), Marcello Mezaroba (Universidade do Estado de Santa Catarina), Alessandro Batschauer (Universidade do Estado de Santa Catarina), Felipe Joel Zimann (Universidade do Estado de Santa Catarina)

Homothetic Method to Compute Winding Losses in the Design of Power Inductors

André Furlan (Universidade Federal de Santa Catarina), Morentin Alvaro (Power Design Technologies), Guillaume Fontes (Power Design Technologies), Guillaume Delamare (Power Design Technologies), Marcelo Heldweing (Universidade Federal de Santa Catarina), Thierry Meynard (LAPLACE, Université de Toulouse, CNRS, INPT, UPS)

Inherent Redundancy of SDBC-MMCC based STATCOM in the Overmodulation Region

Dayane Mendonça (Centro Federal de Educação Tecnológica de Minas Gerais), Allan Cupertino (Centro Federal de Educação Tecnológica de Minas Gerais), Heverton Augusto Pereira (Universidade Federal de Viçosa), Seleme Isaac Seleme Jr. (Universidade Federal de Minas Gerais), Remus Teodorescu (Aalborg University)

Education 2

Fernando Antunes (Federal University of Ceará (UFC))

Grant Covic (University of Auckland)

PeMSyn: a Free Software to Assist the Design and Performance Assessment of Permanent Magnets Synchronous Machines

Geyverson Teixeira de Paula (Federal University of Goiás), Thales Eugenio Portes de Almeida (Federal Technological University of Paraná), Khristian Andrade Jr (Federal University of Goiás), Hugo Santos (Federal University of Goiás), Wellington Vilela (Federal University of Goiás)

Comprehensive and Didactic DC Servomotor Control Platform

Bruno Regina (Federal University of Juiz de Fora (UFJF)), Maria Júlia Rosa Aguiar (Federal University of Juiz de Fora (UFJF)), André Ferreira (Federal University of Juiz de Fora (UFJF))

An Electronic Drive for a Switched Reluctance Motor Using a DSC

Brayan Fonseca (Universidade Federal do Rio de Janeiro), José Santisteban (Universidade Federal Fluminense)

Development of a Modular Open Source Power Electronics Didactic Platform

Lucas Monogios Koleff (University of São Paulo), Gustavo Valentim (University of São Paulo), Victor Rael (University of São Paulo), Luciana Marques (University of São Paulo), Wilson Komatsu (University of São Paulo), Eduardo Lorenzetti Pellini (University of São Paulo), Lourenço Matakas Junior (University of São Paulo)

Applications 2

Giorgio Spiazzi *(University of Padova - Dept. of Information Engineering (DEI))*

Leandro Michels *(Federal University of Santa Maria)*

On the Application of a Power Electronics-based Arc-Flash Suppressor

Fernando Amaral (CEFET-MG), Matheus Oliveira (CEFET-MG), Claudio Conceicao (Petrobras), Sidelmo Silva (UFMG), Cleber Inacio (Petrobras), Braz de Jesus Cardoso Filho (Universidade Federal de Minas Gerais)

Static Transfer Switch Applied to Single-Phase Uninterruptible Power Supply

Vinicius Maia (Universidade Estadual Paulista (UNESP)), Newton da Silva (Universidade Estadual de Londrina), Flavio A. S. Gonçalves (UNESP-São Paulo State University, Brazil)

Design and Assembly of a Bipolar Marx Generator Based on Full-Bridge Topology Applied to Electroporation

Fernando Imai (Instituto Federal de Santa Catarina), YALES NOVAES (Universidade do Estado de Santa Catarina)

Proportional Wavelet Sliding Mode Controller for Torque Ripple Reduction in BLDC Motor

Mateus Moro Lumertz (University of São Paulo), Carlos Matheus Rodrigues de Oliveira

(University of São Paulo), Allan Gregori de Castro (University of São Paulo), José Roberto Boffino de Almeida Monteiro (University of São Paulo), Paulo Roberto Ubaldo Guazzelli (University of São Paulo), Manoel Luís de Aguiar (University of São Paulo)

Modeling and Control 4

João Ota *(University of Campinas)*

Fernando Ortiz Martinz *(University of São Paulo)*

Closed Form Solution for Ferrite Core and Winding Losses in Single Phase Boost PFC Converters
Marcos Jacoboski (Federal University of Santa Catarina), Marcelo Heldwein (Federal University of Santa Catarina), André Lange (Federal University of Santa Catarina)

Control Strategy for Multifunctional PV Converter
Luiz Henrique Meneghetti (Federal University of Technology - Paraná), Edivan Carvalho (Federal University of Santa Maria), Emerson Giovani Carati (Federal University of Technology - Paraná), Jean P. da Costa (Federal University of Technology - Paraná), Carlos Marcelo De Oliveira Stein (Federal University of Technology - Paraná), Zeno Luiz Iensen Nadal (Companhia Paranaense de Energia), Rafael Cardoso (Federal University of Technology - Paraná)

Energy-Balance Based Voltage Regulation Method for Multiple DC-links in Asymmetrical Cascaded Multilevel Inverters
André Felicio de Sousa Silva (Federal University of Goiás), Sérgio Pires Pimentel (Federal University of Goiás), Enes Gonçalves Marra (Federal University of Goiás), Bernardo Alvarengae (Federal University of Goiás)

An Enhanced Thévenin Equivalent Circuit of a Resonant-Controller-Based Utility-Interface
Joel Guerreiro (University of Campinas), Hildo Guillardi Júnior (University of Campinas), João Ota (University of Campinas), José Pomilio (University of Campinas)

Power Quality 3

Jorge Duarte (*Eindhoven University*)

Maurício Beltrão De Rossiter Correa (*FEDERAL UNIVERSITY OF CAMPINA GRANDE*)

Three-Phase Adaptive Frequency Estimator with a Delayed Signal Cancellation Pre-Filter Under Heavily Distorted Grid Conditions

Marcelo Heldwein (Federal University of Santa Catarina), Edhuardo Grabovski (Universidade Federal de Santa Catarina), Samir A. Mussa (Universidade Federal de Santa Catarina)

Developement of a Multilevel DVR with Battery Control and Harmonic Compensation

Bruno Guimaraes (Universidade Federal de Itajuba), Wilson Santana (Instituto Gnarus), Paulo Ribeiro (Universidade Federal de Itajuba), Robson Gonzatti (Universidade Federal de Itajuba), Guilherme Pinheiro (Universidade Federal de Itajuba), Rondineli Pereira (Universidade Federal de Itajuba), Fernando Belchior (Universidade Federal de Goiás), Carlos Henrique da Silva (Universidade Federal de Ouro Preto), Luiz Eduardo Borges-da-Silva (Universidade Federal de Itajuba)

3-Phase Multi-Functional Grid-Tied Inverter for Compensation of Oscillating Instantaneous Power

Helmo Kelis Morales Paredes (São Paulo State University - UNESP), José de Arimatéia Olímpio Filho (São Paulo State University - UNESP), Augusto Alonso (Norwegian University of Science & Technology), Jakson Paulo Bonaldo (Federal University of Mato Grosso), Fernando Pinhabel Marafão (São Paulo State University - UNESP), Marcelo Godoy Simões (Colorado School of Mines)

DC-link voltages balance method for single-phase NPC inverter operating with reactive power compensation

Marco Fajardo (Universidad Simón Bolívar), Julio Viola (Universidad Politécnica Salesiana), José Manuel Aller (Universidad Politécnica Salesiana), Flavio Quizhpi (Universidad Politécnica Salesiana)

Renewables 2

Gierri Waltrich *(Federal University of Santa Catarina)*

Edson H. Watanabe *(Universidade Federal do Rio de Janeiro)*

Power Control of a Doubly Fed Induction Wind Generator employing a Takagi-Sugeno Fuzzy Logic Controller

Carlos Rocha (Universidade Federal do ABC), Juan Sebastian Solís (Universidad Escuela Colombiana de Carreras Industriales), Eliomar Conde (Universidade Federal do ABC), José Puma (Universidade Federal do ABC), Fernando Lino (IFSP), Alfeu Sguarezi (Universidade Federal do ABC)

Robust Control of Switched Reluctance Generator In Connection With a Grid-Tied Inverter

Caio Ruviaro Dantas Osório (Federal University of Santa Maria), Filipe Pinarello Scalcon (Federal University of Santa Maria), Rodrigo Vieira (Federal University of Santa Maria), Vinicius Foletto Montagner (Federal University of Santa Maria), Hilton Abílio Gründling – UFSM (Federal University of Santa Maria)

Wind power system connected to the grid from Squirrel Cage Induction Generator (SCIG)

Ângelo Marcílio (Campus Sobral, Federal University of Ceará), Lucas Taylan Pontes Medeiros (Campus Sobral, Federal University of Ceará), Leonardo Pires de Sousa Silva (Campus Sobral, Federal University of Ceará), Ildenor Júnior (Campus Sobral, Federal University of Ceará), Vanessa Teixeira (Campus Sobral, Federal University of Ceará), Adson Bezerra Moreira (Campus Sobral, Federal University of Ceará)

Finite Control Set Model Based Predictive Control of Grid-Tied Six-Switch Converter Applied to Induction Generator

Paulo Roberto Ubaldo Guazzelli (University of São Paulo), Allan Gregori de Castro (University of São Paulo), Stefan Thiago Cury Alves dos Santos (University of São Paulo), Carlos Matheus Rodrigues de Oliveira (University of São Paulo), José Roberto Boffino de Almeida Monteiro (University of São Paulo), Manoel Luís de Aguiar (University of São Paulo)

Modeling and Simulation of a Stirling-Engine-based Generator Connected to the Grid

Augusto Hayashi (USP), Moacyr Brito (Universidade Federal do Mato Grosso do Sul - UFMS), Joao Onofre Pinto (Universidade Federal do Mato Grosso do Sul - UFMS), Lourenço Matakas Junior (University of São Paulo), Raymundo Cordero (Federal University of Mato Grosso do Sul), Luigi Galotto Junior (Universidade Federal do Mato Grosso do Sul - UFMS)

Power Converters 3

Luiz Henrique S. C. Barreto - UFC *(Federal University of Ceará (UFC))*

Pat Wheeler *(University of Nottingham)*

Design Procedures and Prototyping of a Full-Bridge High Frequency Power Inverter

Joel Guerreiro (University of cam), Hildo Guillardi Júnior (University of Campinas), José Pomilio (University of Campinas)

Premagnetized Inductors in Single Phase dc-ac and ac-dc Converters

Jens Friebe (Leibniz University Hannover), Siqi Lin (Leibniz University Hannover), Leon Fauth (Leibniz University Hannover), Tobias Brinker (Leibniz University Hannover)

Performance analysis of alternatives switching commands for Three-level Boost Rectifier with Hysteresis Current Control.

Gabriel Vilkn Ramos (CEFET/MG), Alex-Sander Amável Luiz (CEFET-MG), Marcelo Stopa (CEFET-MG), Marcos Paulo Gomes (CEMIG), Ramon Souza (CEFET/MG), Daniel Leal (CEFET/MG)

BRIDGELESS BUCK-BOOST PFC CONVERTER WITH THREE-STATE SWITCHING CELL

Douglas Carvalho (UNESP-São Paulo State University), Falcondes Seixas (UNESP-São Paulo State University), Luis De Oro Arenas (UNESP-São Paulo State University), Lucas Carvalho Souza (UNESP-São Paulo State University, Brazil), Luciano de Souza da Costa e Silva (Federal Institute of Education, Science and Technology of Goiás)

Single-Phase AC–DC–AC Five-level X-type Current Source Converter
Louelson Costa (FEDERAL UNIVERSITY OF CAMPINA GRANDE), Montie Vitorino (FEDERAL UNIVERSITY OF CAMPINA GRANDE), Maurício Beltrão De Rossiter Correa (FEDERAL UNIVERSITY OF CAMPINA GRANDE)

Modeling and Control 5: Motors

Janaína Gonçalves de Oliveira *(Universidade Federal de Juiz de Fora - UFJF)*

Seleme Isaac Seleme Jr. *(Universidade Federal de Minas Gerais)*

Energy Efficient Control of Synchronous Machines in Deep Field-Weakening Operation Including Saturation Effects
Joao Bonifacio (Technical University of Munich), Ralph Kennel (Technical University of Munich)

Low computational cost technique for SPMSM sensorless drive using active flux concept
Camila Bartsch (Santa Catarina State University), Luís Fernando Ferreira De Campos (South Santa Catarina University), Arthur Bartsch (Federal Institute of Education, Science and Technology of Santa Catarina), José Oliveira (Universidade do Estado de Santa Catarina), Ademir Nied (Universidade do Estado de Santa Catarina)

Alternative FCS-MPC concepts for cascade free motor speed control
Filipe Fernandes (Santa Catarina State University), José Oliveira (Universidade do Estado de Santa Catarina), Ademir Nied (Universidade do Estado de Santa Catarina), Arthur Bartsch (Federal Institute of Education, Science and Technology of Santa Catarina)

DC MOTOR MODEL FOR WINDOWS PINCH PROTECTION APPLICATIONS
Gabriel Fernandes Idalgo (Universidade Federal do ABC), Carlos Capovilla (Universidade Federal do ABC), Jose Alberto Torrico Altuna (Universidade Federal do ABC), Ademir Pelizari (Universidade Federal do ABC), Wolfgang Schulter (Hochschule Ravensburg Weingarten)

Modeling and Control of a Back-to-Back system for turboelectric propulsion
Saulo Oliveira Nascimento (Federal University of Juiz de Fora (UFJF)), Vinicius Albuquerque (Federal University of Juiz de Fora (UFJF)), Marcelo de Castro Fernandes

(Universidade Federal de Juiz de Fora- UFJF), Pedro Santos Almeida (Federal University of Juiz de Fora (UFJF)), Janaína Gonçalves de Oliveira (Universidade Federal de Juiz de Fora - UFJF), Manuel A. Rendón (Federal University of Juiz de Fora (UFJF))

Wireless

Udaya Madawala (University of Auckland)

Lourenço Matakas Jr (University of São Paulo)

Distance Detection System for Digital Transmitter Coil Achieving Distance-Variation-Tolerant Wireless Power Transfer

Hao Qiu (The University of Tokyo), Yoshiaki Narusue (The University of Tokyo), Yoshihiro Kawahara (The University of Tokyo), Takayasu Sakurai (The University of Tokyo), Makoto Takamiya (The University of Tokyo)

Control of Wireless Power Transfer Systems under Large Coil Misalignments

Yeran Liu (Southwest Jiaotong University), Udaya Madawala (University of Auckland), Ruikun Mai (Southwest Jiaotong University), Zhengyou He (Southwest Jiaotong University)

A Low Cost Bi-directional Wireless Power Transfer System

Lei Wang (Hunan University), Udaya Madawala (University of Auckland), Man Chung Wong (university of Macau)

Renewables 3: PV

Dmitri Vinnikov (TalTech University)

José Puma (Universidade Federal do ABC)

Simplified Single-phase PV Generator Model for Distribution Feeders With High Penetration of Power Electronics-based Systems

Mariana Altoé Mendes (Federal University of Espírito Santo), Murillo Cobe Vargas (Federal University of Espírito Santo), Oureste Elias Batista (Federal University of Espírito Santo), Yongheng Yang (Aalborg University), Frede Blaabjerg (Aalborg University)

Method to trace the photovoltaic characteristic curve with reverse voltage for shading conditions

Richard Gonçalves Cornelius (Unipampa), Amanda Costa Maia (Unipampa), Matheos Coletto Wermuth (UNIPA), Guilherme Sebastião da Silva (Unipampa)

ANALYSIS AND OPERATION OF A PV-BATTERY SYSTEM USING A MULTI-FUNCTIONAL CONVERTER

Jessica Dohler (Federal University of Juiz de Fora), Lorrana Rocha (Federal University of Juiz de Fora (UFJF)), Dalmo Silva (Federal University of Juiz de Fora (UFJF)), Janaína Gonçalves de Oliveira (Federal University of Juiz de Fora (UFJF)), Pedro Machado de Almeida (Federal University of Juiz de Fora (UFJF)), André Ferreira (Federal University of Juiz de Fora (UFJF))

Energy Yield Assessment Methodology for Photovoltaic Microinverters

Andrii Chub (Tallinn University of Technology), Roman Kosenko (Ubik Solutions OÜ), Oleksandr Korkh (Tallinn University of Technology), Dmitri Vinnikov (Tallinn University of Technology), Samir Kouro (Universidad Técnica Federico Santa María)

Methodology for experimental determination of equivalent grid impedance by using external commands of PV inverters

Valesca Bettim Feltrin (Federal University of Santa Maria), Thiago Brezolin Dalmolin (Federal University of Santa Maria), Lucas Bellinaso (Federal University of Santa Maria), Leandro Michels (Federal University of Santa Maria)

AN IMPROVED IMPEDANCE ESTIMATION METHOD BASED ON POWER VARIATIONS IN GRID-CONNECTED INVERTERS

Jose Suarez (Federal University of Bahia), Hugo Matheus Teixeira Cotrim Gomes (Federal University of Bahia), Luan Santana (Federal University of Bahia), Leandro Leysdian Oro Carralero (Federal University of Bahia), Alfeu Sguarezi Filho (Federal University of ABC), Fabiano Fragoso Costa (UFBA)

Smart grid applications 2

João Onofre Pereira Pinto - UFMS

Leonardo Rodrigues Limongi

Comparison among Two-Phase Three-Wire AC Off-Grid Power Systems

Rafael Mario da Silva (Federal University of Minas Gerais (UFMG)), Danilo Brandao (Federal University of Minas Gerais (UFMG)), Gabriel Fogli (Federal University of Minas Gerais (UFMG)), Victor Mendes Flores (Federal University of Minas Gerais (UFMG)), Clodualdo Venicio Sousa (Federal University of Minas Gerais (UFMG))

Enhanced Space-State Reduced-Order Microgrid Model in Common DQ-Reference Frame

Sebastian De Jesus Manrique Machado (Federal Technological University of Paraná), Sérgio Augusto Oliveira da Silva (Federal Technological University of Paraná), José Roberto Boffino de Almeida Monteiro (University of São Paulo), Azauri Albano de Oliveira Jr (University of São Paulo)

A ISOP AC-AC Hybrid Switched-Capacitor SRC for Solid State Transformer Applications

Victor Luiz Flor Borges (Brazilian Institute of Power Electronics and Renewable Energies - IBEPE), Rogerio Luiz Da Silva Júnior (Brazilian Institute of Power Electronics and Renewable Energies - IBEPE), Carlos Eduardo Possamai (Brazilian Institute of Power Electronics and Renewable Energies - IBEPE), Arlan Luiz Bettiol (Brazilian Institute of Power Electronics and Renewable Energies - IBEPE), Ivo Barbi (Brazilian Institute of Power Electronics and Renewable Energies - IBEPE)

Multivariable Control of a Grid Forming System Based on Back-To-Back Topology

Igor Souza (Federal University of Ouro Preto (UFOP)/ Federal University of Juiz de Fora (UFJF)), Gabriel Fogli (Federal University of Minas Gerais (UFMG)), Marcelo de Castro Fernandes (Universidade Federal de Juiz de Fora- UFJF), Ademir Da Silva Toledo Junior (Federal Uni), Pedro Gomes Barbosa (Federal University of Juiz de Fora (UFJF)), Pedro Machado de Almeida (Federal University of Juiz de Fora (UFJF))

Modeling and Control 6: Grid applications

Sérgio Pires Pimentel *(Federal University of G)*

Kleber Lima *(Federal University of Cea)*

A Review of FCS-MPC in Multilevel Converters Applied to Active Power Filters
João Foster (Universidade Federal de Itajuba), RONDINELI PEREIRA (Universidade Federal de Itajuba), Robson Gonzatti (Universidade Federal de Itajuba), Wilson Santana (Instituto Gnarus), Denis Mollica (EDP Sao Paulo Distribuicao de Energia), Germano Lambert-Torres (Instituto Gnarus)

Active-Capacitor for Power Decoupling in Single-Phase Grid-Connected Converters
Thiago Pereira (Federal University of Santa Catarina), Denizar Cruz Martins (Federal University of Santa Catarina), Roberto Francisco Coelho (Federal University of Santa Catarina)

Controller Coefficients Tuning for a Single-Phase Photovoltaic System in Synchronous Reference Frame Through Genetic Algorithm
Moacyr Brito (Universidade Federal do Mato Grosso do Sul - UFMS), Luigi Galotto Junior (Universidade Federal do Mato Grosso do Sul - UFMS), Marcos Vedovatte (Universidade Federal do Mato Grosso do Sul - UFMS)

Feedforward Compensation of the ESS Low-Frequency Current Ripple in the Three-Ports ANPC Converter
Silvio Antonio Teston (Federal University of Fronteira Sul), Kaio Vinicius Vilerá (Federal University of Santa Maria), Marcello Mezaroba (University of Santa Catarina State), Cassiano Rech (Federal University of Santa Maria)

A Comparison of a Discrete-Time PI and an Indirect MPC Current Controllers for a Single-Phase Grid-Connected Inverter Operating with Distorted Grid and Significant Computation Feedback Delay
Sérgio Pires Pimentel (Federal University of G), Oleksandr Husev (Tallinn University of Technology), Dmitri Vinnikov (Tallinn University of Technology), Serhii Stepenko (Tallinn University of Technology (TalTech)), Lauri Kutt (Tallinn University of Technology (TalTech)), Jose Rodriguez (Universidad Andres Bello)

Poster Session 2

João A. Moor Neto *(CEFET-RJ)*

Richard M. Stephan *(Universidade Federal do Rio de Janeiro)*

Enes Gonçalves Marra *(Federal University of Goias - UFG)*

Operation Boundaries of a Single Phase Thyristor Driven DC-Motor
Thomas Mertens (UFRJ), Richard Stephan (UFRJ)

Didactic System For Control Of Electrical Machines In Education And Research Laboratories
Allan Andrade (Universidade Federal do Rio de Janeiro), Richard Stephan (UFRJ)

Modeling and Simulation of a Stirling Engine in SCILAB
Raymundo Cordero (Federal University of Mato Grosso do Sul), Thyago Estrabis (Federal University of Mato), Joao Onofre Pinto (Universidade Federal do Mato Grosso do Sul - UFMS), Felipe Alexandre Monteiro (Universidade Federal do Mato Grosso do Sul - UFMS), Augusto Hayashi (USP)

Experimental Analysis for Low Power Series-Series Compensated Inductive Power Transfer System
Macklyster Lacerda (Federal University of Espírito Santo), Tatiana Macedo (Federal University of Espírito Santo), Denizar Cruz Martins (Federal University of Santa Catarina), Walbermark Marques dos Santos (Federal University of Espírito Santo)

Active Cooling and Thermal Simulation Applied to an Extra-High Current COB LED
Denis de Castro Pereira (Federal University of Juiz de Fora (UFJF)), Rubio Campos Marques (Federal University of Juiz de Fora (UFJF)), Pedro Santos Almeida (Federal University of Juiz de Fora (UFJF)), Guilherme Marcio Soares (Federal University of Juiz de Fora (UFJF)), Fernando Lessa Tofoli (Universidade Federal de São João del-Rei), Pedro Tavares (Federal University of Juiz de Fora (UFJF)), Henrique Antônio Carvalho Braga (Federal University of Juiz de Fora (UFJF))

Comparative analysis based on the switching frequency of modulation techniques for MMC applications
Juan Carlos Ccarita (University of Campinas), Ernesto Ruppert Filho (University of Campinas), Rodrigo Zambrana Vargas (Universidade Federal do ABC), José Puma (Universidade Federal do ABC)

High Power Factor Rectifier Using One Cycle Control Strategy

Amanda Thayla S Monteiro (Federal University of Piaui), Rafael Matias (Federal University of Piaui), Jailson Leite Silva (Federal University of Piaui), José Santos Neto (Federal University of Piaui)

Considerations on Communication Infrastructures for Cooperative Operation of Smart Inverters

Augusto Matheus dos Santos Alonso (UNESP-São Paulo State University), Leonardo Carlos Afonso (UNESP-São Paulo State University), Danilo Brandao (Federal University of Minas Gerais (UFMG)), Elisabetta Tedeschi (Norwegian University of Science & Technology (NTNU)), Fernando Pinhabel Marafão (UNESP-São Paulo State University, Brazil)

Integration of Solar Photovoltaic (PV) Systems with CCM Inverters into VCM Droop-Controlled Islanded AC Microgrids

Marcus Evandro Teixeira Souza Junior (Universidade Federal de Uberlândia), Fernando Cardoso Melo (Universidade Federal de Uberlândia), Ernane A. Alves Coelho (Federal University of Uberlândia), Luiz C. G. Freitas (Federal University of Uberlândia)

An Overview About Detection of Cyber-Attacks on Power SCADA Systems

Hugo Figueiredo (Federal University of Technology - Paraná), Matheus Ferst (Federal University of Technology - Paraná), Gustavo Weber Denardin (Universidade Tecnológica Federal do Paraná)

An Application of the Multi-Port Bidirectional Three-Phase AC-DC Converter in Electric Vehicle Charging Station Microgrid

Raphael Amaral Da Câmara (Federal University of Ceará (UFC)), Luis Fernández-Ramírez (University of Cadiz), Paulo Peixoto Praça (Federal University of Ceará), Demercil de Souza Oliveira Júnior (Federal University of Ceará), Pablo García-Triviño (University of Cadiz), Raúl Sarrias-Mena (University of Cadiz)

Hybrid One-Cycle Control Strategy for Buck+Boost PFC Battery Charger

Aluísio Alves De Melo Bento (State University of Rio de Janeiro), RAPHAEL SANTIAGO (State University of Rio de Janeiro)

Z-Source Inverter for Photovoltaic Microgeneration

Felipe Augusto Ferreira de Almeida (Federal Institute of Education, Science and Technology of São Paulo- IFSP), Felipe Guerra Soares (UNESP-São Paulo State University, Brazil), Flavio A. S. Gonçalves (UNESP-São Paulo State University, Brazil)

Input Voltage Range Extension Methods in the Series-Resonant DC-DC Converters

Andrii Chub (Tallinn University of Technology), Dmitri Vinnikov (Tallinn University of Technology), Jason Lai (Virginia Tech)

Symmetrical Hybrid Multilevel VSI and CSI Inverters Derived from Dc-Dc Converters

Samir A. Mussa (Universidade Federal de Santa Catarina), Marcelo Heldwein (Universidade Federal de Santa Catarina), Domingo Ruiz Caballero (Pontifical Catholic University of Valparaíso), Javier Riedemann (Pontifical Catholic University of Valparaíso), Werner Jara (Pontifical Catholic University of Valparaíso), Luis Colque (Pontifical Catholic University of Valparaíso), Carlos Paredes (Pontifical Catholic University of Valparaíso)

Experimental Results of a Bidirectional Coupled Inductor DC-DC Converter

Murilo brunel (Santa Catarina State University), Menaouar Berrehil El Kattel (Santa Catarina State University), Robson Mayer (Santa Catarina State University), Sérgio Vidal Garcia Oliveira (Santa Catarina State University)

Development of Linear Generator Prototype As Part of A Point Absorber Wave Energy Converter

Eduardo Beline (Instituto Federal Fluminense), Marcos Moreira (Instituto Federal Fluminense), João Santos (Instituto Federal Fluminense), Wagner Marcilio (Instituto Federal Fluminense)

Reference Grid Impedance for Tests of Grid-connected Power Converters for Distributed Energy Resources: The Brazilian Case

Gabriel Avila Saccol (Universidade Federal de Santa Maria), Charles Schardong (Universidade Federal de Santa Maria), Leandro Michels (Universidade Federal de Santa Maria), Lucas Bellinaso (Universidade Federal de Santa Maria), Cassiano Rech (Universidade Federal de Santa Maria)

Power Control and Harmonic Current Mitigation from a Wind Power System with PMSG

Leonardo Pires de Sousa Silva (Campus Sobral, Federal University of Ceará), Denisia de Vasconcelos Mota (Campus Sobral, Federal University of Ceará), Flávia Peroza Ruiz (Campus Sobral, Federal University of Ceará), Levy Rodrigues Cavalcante (Campus Sobral, Federal University of Ceará), Lucas Taylan Pontes Medeiros (Campus Sobral,

Federal University of Ceará), Vanessa Teixeira (camp), Adson Bezerra Moreira (Campus Sobral, Federal University of Ceará)

Concepts and Case Study of Mismatch Losses in Photovoltaic Modules

Elson Sakô (University of Campinas), João Silva (University of Campinas), Daniel Mesquita (University of Campinas), Rafael Campos (University of Campinas), Hugo Moreira (University of Campinas), Marcelo Gradella Villalva (University of Campinas)

A Performance Analysis of Active Anti-Islanding Methods Based on Frequency Drift

Enio Costa Resende (Universidade Federal de Uberlândia), Henrique Carvalho (Universidade Federal de Uberlândia), Fernando Cardoso Melo (Universidade Federal de Uberlândia), Ernane A. Alves Coelho (Federal University of Uberlândia), Gustavo B. Lima (Federal University of Uberlândia), Luiz C. G. Freitas (Federal University of Uberlândia)

Event Manager and Control Structure for High Performance Three-Phase Grid-tied Inverters

Victor Barbosa (Universidade Tecnológica Federal do Paraná), Rodrigo Antonio Sbardeloto Kraemer (Universidade Tecnológica Federal do Paraná), Emerson Giovani Carati (Universidade Tecnológica Federal do Paraná), Jean P. da Costa (Universidade Tecnológica Federal do Paraná), Rafael Cardoso (Universidade Tecnológica Federal do Paraná), Carlos Marcelo De Oliveira Stein (Universidade Tecnológica Federal do Paraná)

PV Emulator based on a Four-Switch Buck-Boost DC-DC Converter

Leandro Leysdian Oro Carralero (Federal University of Bahia), Gabriel Santa Barbara da Silva e Silva (Federal University of Bahia), Fabiano Fragoso Costa (Federal University of Bahia), André Pires Nóbrega Tahim (Federal University of Bahia)

Discrete SPS Control of a DAB converter using partial Feedback Linearization

Eduardo Luiz Santos da Silva (Federal University of Santa Catarina), André Luís Kirsten (Federal University of Santa Catarina), Daniel Pagano (Federal University of Santa Catarina)

Stator Current Controller for Harmonic and Unbalance Compensation Applied to SEIG Based Systems

Gabriel Attuati (UFSM), Robinson Figueiredo de Camargo (UFSM), Lucas Giuliani Scherer (UFSM)

Development of a FPGA-Based Real-Time Simulation System

Yago Oliveira (Federal Institute of Education, Science and Technology of Southeast of Minas Gerais (IF Sudeste MG)), Filipe La-Gatta (Federal Institute of Education, Science and Technology of Southeast of Minas Gerais (IF Sudeste MG)), Rodrigo Arruda Ferreira (Federal Institute of Education, Science and Technology Southeast of Minas Gerais), Marcio do Carmo Rodrigues (Federal In)

IFOC for a Nine-phase Induction Motor Drive with Current Harmonic Injection

Alexandre Silva (Federal University of Paraíba), Isaac Soares Freitas (Federal University of Paraíba), Simplício Silva (Federal University of Paraíba), Victor Melo (Federal University of Paraíba), Gilielson Paz (Federal University of Paraíba)

Single-phase to three-phase ac-dc-ac converter based on cascaded transformers rectifier and open-end winding induction motor

Antonio Almeida (Federal University of Paraíba), Nady Rocha (Federal University of Paraíba), Edgard Fabricio (Federal Institute of Paraíba), Carolina Caldeira (Federal University of Paraíba), Isaac Soares Freitas (Federal University of Paraíba), Gleice Rodrigues (Federal University of Paraíba)

Modeling of a three-level quadratic boost converter

Mauricio Borchardt (Universidade do Estado de Santa Catarina), Mateus Constantino Orige (Universidade do Estado de Santa Catarina), Mateus Bueno (Universidade do Estado de Santa Catarina), Gerardo Escobar (Tecnologico de Monterrey), YALES NOVAES (Universidade do Estado de Santa Catarina)

Workbench for Monitoring and Operation of Electric Machines

Andrés Ortiz Salazar (Universidade Federal do Rio Grande do Norte), Guilherme Afonso Pillon de Carvalho Alves Pessoa (Universidade Federal do Rio Grande do Norte), Evandro Ailson de Freitas Nunes (Universidade Federal do Rio Grande do Norte), Ricardo Ferreira Pineiro (Universidade Federal do Rio Grande do Norte), Werbet Silva (UFRN)

Designing of the transmitting coils and compensation network of a segmented DWPT system

Shufan Li (University of Chinese Academy of sciences), Lifang Wang (Institute of Electrical Engineering Chinese Academy of Sciences), Chengxuan Tao (Institute of Electrical Engineering Chinese Academy of Sciences), Fang Li (Institute of Electrical Engineering Chinese Academy of Sciences), Liye Wang (Institute of Electrical Engineering Chinese Academy of Sciences)

DIRECT POWER CONTROL STRATEGY TO ENHANCE THE DYNAMIC BEHAVIOR OF DFIG DURING VOLTAGE DIP

Fernando Lino (Universidade Federal do ABC- UFABC), Rogerio Jacomini (Federal Institute of Education, Science and Technology of São Paulo- IFSP), Alfeu Sguarezi (Universidade Federal do ABC)

Poster Session 3

Sérgio Augusto Oliveira da Silva *(Federal Technological University of Paraná)*

Cassiano Rech *(Federal University of Santa Maria)*

Analysis and Optimal Design of Magnetics Component in Dual-Actual-Bridge Converter for 1 MVA Solid-State Transformer

Sriram Vaisambhayana (Nanyang Technological University), Haonan Tian (Nanyang Technological University), Anshuman Tripathi (Nanyang Technological University)

Theoretical Solution of the Output Voltage Harmonic Spectra of Dual-Inverter Fed Open-End Winding Loads with Dead Time Effect

Brunno Ribeiro (CEFET-MG), Marcelo Stopa (CEFET-MG), Alex-Sander Amável Luiz (CEFET-MG)

Selection of the Number of Levels of a Modular Multilevel Converter for an Electric Drive

Paulo Júnior (Centro Federal de Educação Tecnológica de Minas Gerais), João Victor Matos Farias (Centro Federal de Educação Tecnológica de Minas Gerais), Allan Cupertino (Centro Federal de Educação Tecnológica de Minas Gerais), Gabriel Mendonça (Federal University of Minas Gerais (UFMG)), Marcelo Martins Stopa (Centro Federal de Educação Tecnológica de Minas Gerais), Heverton Augusto Pereira (Federal University of Viçosa)

A NEW ZERO-SEQUENCE VOLTAGE COMPENSATION ALGORITHM FOR A DSTATCOM BASED ON CONSUMER UNBALANCE

Bruno Cortes (Federal University of Juiz de Fora (UFJF)), Samuel Duarte (Federal University of Juiz de Fora (UFJF)), Pedro Machado de Almeida (Federal University of Juiz de Fora (UFJF)), Pedro Gomes Barbosa (Federal University of Juiz de Fora (UFJF)), Leandro Araújo (Federal University of Juiz de Fora (UFJF))

VOLTAGE REGULATION OF A REMOTE BUS OF A DISTRIBUTION NETWORK BY STATIC SYNCHRONOUS COMPENSATOR

Samuel Duarte (Federal University of Juiz de Fora (UFJF)), Bruno Cortes (Federal University of Juiz de Fora (UFJF)), Pedro Machado de Almeida (Federal University of Juiz de Fora (UFJF)), Pedro Gomes Barbosa (Federal University of Juiz de Fora (UFJF))

Design of a Low-Cost Phasor Measurement Unit (PMU) for Three-Phase Distribution Power Systems according IEEE C37.118.1

POZO MARCELO (ESCUELA POLITECNICA NACIONAL, QUITO), Alex Guaman (ESCUELA POLITECNICA NACIONAL, QUITO), Mario Pacas (University of Siegen), Nataly Pozo (Universidad San Francisco de Quito), Isaac Pozo (ESCUELA POLITECNICA NACIONAL, QUITO), Ana Cabrera (Universidad Técnica del Norte-Ibarra)

Analysis of Performance and Opportunity for Improvements in the Microgrid of Ilha Grande

Leonilson Veras (Federal University of Maranhão), Hércules Oliveira (Federal University of Maranhão), José Gomes de Matos (Federal University of Maranhão), Osvaldo Ronald Saavedra (Federal University of Maranhão), Luiz Antonio de Souza Ribeiro (Federal University of Maranhão)

Comparative Study of the Power Sharing Techniques for Microgrids in Autonomous Mode

Gustavo Medeiros de Souza Azevedo (UFPE), Erik Oliveira (UFPE), Marcelo Cavalcanti (UFPE), Leonardo Limongi (UFPE)

High-Gain Bidirectional DC-DC Converter for Battery Charging in DC Nanogrid of Residential Prossumer

Fernando Queiroz (Federal University of Ceará (UFC)), Paulo Peixoto Praça (Federal University of Ceará (UFC)), Alisson Freitas (Federal Rural University of the Semi-Arid), Fernando Antunes (Federal University of Ceará (UFC))

High Power Factor Three-Phase Three-Switch Step-Down Converter

Joao Caliman (Universidade Federal do Espírito Santo), Walbermark Marques dos Santos (Federal University of Espírito Santo), Tiara Freitas (UFES), Domingos Simonetti (UFES)

A Simplified Analysis of Buck-Type Interleaved DC-DC Converter for Battery Chargers Application

Menaouar Berrehil El Kattel (Santa Catarina State University), Robson Mayer (Santa

Catarina State University), Maicon Douglas Possamai (Santa Catarina State University), Sérgio Vidal Garcia Oliveira (Santa Catarina St)

High Voltage Gain DC-DC Converter based on a Simple Configuration of Switched Capacitor and Coupled Inductor

Henrique Hoch (Federal University of Sant), Tiago Miguel Klein Faistel (Federal University of Santa Maria), Mauricio Mendes (Universidad Tecnologica del Uruguay), Mario Martins (Federal University of Santa M), António Manuel Santos Spencer Andrade (Federal University of Santa Maria)

Micro inverter with reduced number of semiconductor switches

Paulo Cagnini (Universidade Tecnológica Federal do Paraná), Luiz Meneghetti (Universidade Tecnológica Federal do Paraná), Jean P. da Costa (Universidade Tecnológica Federal do Paraná), Emerson Giovani Carati (Universidade Tecnológica Federal do Paraná), Rafael Cardoso (Universidade Tecnológica Federal do Paraná), Carlos Marcelo De Oliveira Stein (Universidade Tecnológica Federal do Paraná), Jean Lafay (Federal University of Technology – Paraná (UTFPR), Pato Branco - PR), Victor Barbosa (Universidade Tecnológica Federal do Paraná), Zeno Luiz Iensen Nadal (Companhia Paranaense de Energia)

Dynamic Modeling and Control of a Three-Port ZVS-PWM Three-Phase Push Pull DC-DC Converter

Lorena Albuquerque (Federal University of Paraíba), Marcos Dantas (Federal University of Paraíba), Fellipe Oliveira (Federal University of Paraíba), Romero Leandro Andersen (Federal University of Paraíba), Isaac Soares Freitas (Federal University of Paraíba)

Development of a Hybrid PV-Thermoelectric System

Andrés Ortiz Salazar (Universidade Federal do Rio Grande do Norte), Rafael Magalhaes Nóbrega de Araújo (Universidade Federal do Rio Grande do Norte), Hugo Costa (Universidade Federal do Rio Grande do Norte), João Teixeira de Carvalho Neto (Federal Institute of Education, Science and Technology Rio Grande do Norte - IFRN), Alexandre Magnus Fernandes Guimarães (Universidade Federal do Rio Grande do Norte)

Techniques of Solar Irradiance Estimation from Datasheet Information of Photovoltaic Panels

Itaiara Felix Carvalho (FEDERAL UNIVERSITY OF CAMPINA GRANDE), Maurício Beltrão De Rossiter Correa (FEDERAL UNIVERSITY OF CAMPINA GRANDE)

Dynamic Analysis of Self-excited SRG Operating in Open Loop

Lucas José Lemes (Federal Institute of Education, Science and Technology of Goias - IFG), Victor Regis Bernardeli (Federal Institute of Education, Science and Technology of Goias - IFG), Luciano Coutinho Gomes (Federal University of Uberlândia - UFU), Darizon Alves de Andrade (Federal University of Uberlândia - UFU), Ghunter Paulo Viajante (Federal Institute of Education, Science and Technology of Goias - IFG), Marcos Freitas (Federal Institute of Education, Science and Technology of Goias - IFG)

Reactive Power Control of Distributed Photovoltaic Generation System in Low Voltage Electrical Grids

Vanessa Marques (Federal University of Paraíba), Rogerio Almeida (Federal University of Paraíba), Nady Rocha (Federal University of Paraíba), Darlan Fernandes (Federal University of Paraíba), Gleice Rodrigues (Federal University of Paraíba)

Direct Torque Control Scheme for a Nine-Phase Induction Motor With Reduced Current Harmonic

Gilielson Paz (Federal University of Paraíba), Isaac Soares Freitas (Federal University of Paraíba), Victor Melo (Federal University of Paraíba), Alexandre Silva (Federal University of Paraíba)

LPV Modeling of Boost Converter and Gain Scheduling MPC Control

Rosana Rego (Universidade Federal Rural do Semi-Árido)

Predictive Control for a Half-Controlled Boost Rectifier

Gleice Rodrigues (Federal University of Paraíba), Nady Rocha (Federal University of Paraíba), Edison Roberto Cabral da Silva (Federal University of Paraíba), Filipe Rocha (Federal University of Paraíba)

Simulation of the Model, Design and Control of a Current Source Inverter with Unipolar SPWM Modulation

Jeimy Carolina Sanabria Rojas (Universidad Pedagógica y Tecnológica de Colombia), Daniel Mauricio Barrera Leguizamón (Universidad Pedagógica y Tecnológica de Colombia), Diego Bautista (Universidad Pedagógica y Tecnológica de Colombia), Fabián Rolando Jiménez López (Universidad Pedagógica y Tecnológica de Colombia)

Photovoltaic Boost Converter Control Operating in the MPPT and LPPT Modes

Cássia C. C dos Santos (Universidade Tecnológica Federal do Paraná), Cassiano Ferro Moraes (Universidade Tecnológica Federal do Paraná), Jean P. da Costa (Universidade Tecnológica Federal do Paraná), Carlos Marcelo De Oliveira Stein (Universidade

Tecnológica Federal do Paraná), Emerson Giovani Carati (Universidade Tecnológica Federal do Paraná), Rafael Cardoso (Universidade Tecnológica Federal do Paraná)

Direct Power Control with Space Vector Modulation Applied for the Brushless DC Motor

Henrique de Toledo (Federal University of ABC), José Puma (Universidade Federal do ABC)

A new strategy of modulation based on Space Vector Modulation and Annotated Paraconsistent Logic for a three-phase converter

Antonio Carlos Ricciotti (Universidade Federal de Rondonia), João Inácio da Silva Filho (Universidade Santa Cecília dos Bandeirantes), Raphael Adamelk Bispo de Oliveira (Universidade Santa Cecília dos Bandeirantes), Viviane Barrozo da Silva Duarte Ricciotti (Universidade Federal de Rondonia), Hyghor Miranda Côrtes (Universidade Santa Cecília dos Bandeirantes), André Nicolini (Universidade Federal de Santa Maria)

FSC-MPC Current Control of a 5-level HalfBridge/ANPC Hybrid Three-phase Inverter

Jailson Leite Silva (Federal University of Piaui), Rafael Matias (Federal University of Piaui), Max Alves (Federal University of Piaui), Ranoyca Alencar (University of International Integration of the Afro-Brazilian Lusophony), Kristian Santos (Federal Institute of Education, Science and Technology of Piauí), Amanda Thayla S Monteiro (Federal University of Piaui)

Extended Kalman Filter based Speed Estimation for the Control of PMSG

Mukhtiar Singh (Delhi Technological University)

Integrated local control of active power and voltage support for three-phase three-wire converters

Ya Zhang (Eindhoven University of Technology), Gabriel Tibola (Eindhoven University of Technology), Maurice Roes (Eindhoven University of Technonolgy), Jorge Duarte (Eindhoven University of Technonolgy)

Electrical Simulation of Traction Subway System for Energy Recovery and Energy Saving Studies

Marcio Pimenta (University of São Paulo), Wilson Komatsu (University of São Paulo), Lourenço Matakas Jr (University of São Paulo)

Two-Stage Sepic-Buck Topology for Neighborhood Electric Vehicle Charger

Rafael Eckstein (UFSC), Telles B. Lazzarin (Federal University of Santa Catarina), Gierri Waltrich (Federal Univer)

Wind Turbine Emulator with DC Motor

Andrés Ortiz Salazar (Universidade Federal do Rio Grande do Norte), Tadeu Felix dos Santos dos Santos (Universidade Federal do Rio Grande do Norte), Felipe de Oliveira Simões Gama (Universidade Federal do Rio Grande do Norte), Rodrigo de Andrade Teixera (Universidade Federal do Rio Grande do Norte), Igor Vieira Chacon (Universidade Federal do Rio Grande do Norte), Gessica Costa de Araujo Souza (Universidade Federal do Rio Grande do Norte), Guilherme Afonso Pillon de Carvalho Alves Pessoa (Universidade Federal do Rio Grande do Norte)

Power Electronics Lab: Converging Knowledge and Technologies

José Pomilio (University of Campinas)

Impact of Capacitor Design Methodology on FC Inverters

Alessandro Batschauer (Universidade do Estado de Santa Catarina), Cassiano Rech (Federal University of Santa Maria), Marcos Bressan (Universidade do Estado de Santa Catarina)

Small Scale Compressed Air Energy Storage (SS-CAES) Strategies Overview

Luiz Fernando Martins Pastuch (Federal University of Santa Catarina), Marcos Antonio Salvador (Federal University of Santa Catarina), Telles B. Lazzarin (Federal University of Santa Catarina), Roberto Coelho (Federal University of Santa Catarina)

Lighting

Fernando Lessa Tofoli *(Universidade Federal de São João del-Rei)*

Wilson Komatsu *(University of São Paulo)*

LED Driver With Reduced Redundant Power Processing And Dimming For Street Lighting Applications

Kleber Chan Bekoski (Universidade Tecnológica Federal do Paraná), Cassio Gobbato (Universidade Tecnológica Federal do Paraná), Cassiano Ferro Moraes (Universidade Tecnológica Federal do Paraná), Gustavo Weber Denardin (Universidade Tecnológica

Federal do Paraná), Juliano de Pelegrini Lopes (Universidade Tecnológica Federal do Paraná)

Nonisolated Quadratic SEPIC Converter Without Electrolytic Capacitors for LED Driver Applications

Aniel Silva de Morais (Universidade Federal de Uberlândia), Douglas Rosa Corrêa (Federal University of Uberlândia), Fernando Lessa Tofoli (Universidade Federal de São João del-Rei)

Design of an Integrated Circuit for LED Driving in Visible Light Communication Applications

Nicolas Parma Rios (Universidade Federal de Juiz de Fora - UFJF), Guilherme Márcio Soares (Universidade Federal de Juiz de Fora - UFJF), Estêvão Teixeira (Universidade Federal de Juiz de Fora - UFJF)

Efficiency Analysis for Interleaved Buck Converters Employed as Extra-High Current COB LED Drivers

Denis de Castro Pereira (Federal University of Juiz de Fora (UFJF)), Everton Bernard Figueiredo Rabelo (Federal University of Juiz de Fora (UFJF)), Pedro Santos Almeida (Federal University of Juiz de Fora (UFJF)), Guilherme Marcio Soares (Federal University of Juiz de Fora (UFJF)), Fernando Lessa Tofoli (Universidade Federal de São João del-Rei), Henrique Antônio Carvalho Braga (Federal University of Juiz de Fora (UFJF))

A Resonant-Switched-Capacitor Step-Down DC–DC Converter in CCM Operation as an LED Driver

Fábio Paiva (Instituto Federal de Educação, Ciência e Tecnologia do Ceará), Antônia Rocha (UNIVERSIDADE FEDERAL DO CEARÁ), Rodrigo Santos (UNIVERSIDADE FEDERAL DO CEARÁ), Pedro Santos de Almeida (Federal University of Juiz de Fora (UFJF)), Fernando Antunes (Federal University of Ceará (UFC)), Edilson Sá (Instituto Federal de Educação, Ciência e Tecnologia do Ceará)

Energy storage

Telles B. Lazzarin *(Federal University of Santa Catarina)*

Sheldon Williamson

Modeling of Electrical Behavior and Lifetime Degradation of Electrochemical Batteries

Rafael Nolasco (Universidade Federal de Minas Gerais), Victor Mendes Flores (Federal University of Minas Gerais (UFMG))

SmartBattery: An Active-Battery Solution for Energy Storage System

Lucas Savoi Araujo (Federal University of Minas Gerais (UFMG)), Nicolas Fernandes (Federal University of Minas Gerais (UFMG)), Danilo Brandao (Federal University of Minas Gerais (UFMG)), Braz de Jesus Cardoso Filho (Federal University of Minas Gerais (UFMG))

Modeling and Simulation of the Battery Energy Storage System for Analysis Impact in the Electrical Grid.

Carolina Caldeira (Federal University of Paraíba), Henrique Schlickmann (Federal University of Paraíba), Antonio Almeida (UFPB), Fabiano Salvadori (Federal University of Paraíba), Camila Gehrke (Federal University of Paraíba), Lucas Hartmann (Federal University of Paraíba), Gilielson Paz (Federal University of Paraíba)

Sizing of Supercapacitor and BESS for peak shaving applications

Kaique Ferreira (Federal University of Espírito Santo), Walbermark Marques dos Santos (Federal University of Espírito Santo), Augusto Medina (Federal University of Espírito Santo)

Power Converters 4

Enio Ribeiro *(Universidade Federal de Itajuba)*

Leonardo Rodrigues Limongi

High Step-Up DC-DC Converter with Input Current Sharing Based on the Forward Converter

Victor Ferreira Gruner (Federal University of Santa), Lucas Fiamoncini (Federal

University of Santa Catarina), Lenon Schmitz (Federal University of Santa), Denizar Cruz Martins (Federal University of Santa Catarina), Roberto Coelho (U)

High-Voltage Step-Up DC-DC Converter Employing The Four State Switching Cell and Voltage Multiplier Cells

Paulo Feretti (Universidade Federal de São João del-Rei), Enio Ribeiro (Universidade Federal de Itajuba), Fernando Lessa Tofoli (Universidade Federal de São João del-Rei)

A NON-ISOLATED DC-DC BOOST CONVERTER WITH THREE-STATE SWITCHING CELL

Lucas Carvalho Souza (UNESP-São Paulo State University), Falcondes Seixas (UNESP-São Paulo State University), Luis De Oro Arenas (UNESP-São Paulo State University), Douglas Carvalho (UNESP-São Paulo State University), Luciano de Souza da Costa e Silva (Federal Institute of Education, Science and Technology of Goiás)

NonIsolated DC-DC Quadratic Ćuk Converter for Wide Conversion Range Applications

Tatiane Oliveira (Universidade Federal de São João del-Rei), Lara Rodarte (Universidade Federal de São João del-Rei), Aniel Silva de Morais (Universidade Federal de Uberlândia), Fernando Lessa Tofoli (Universidade Federal de São João del-Rei)

AC-DC Converter with High-Frequency Isolation Operating Under ZVS

Bruno Alves (Federal University of Ceará), Dalton Honório (Federal University of Ceará), Demercil de Souza Oliveira Júnior (Federal University of Ceará), Bruno R. de Almeida (University of Fortaleza), Samanta Barbosa (Federal University of C), Luiz Henrique S. C. Barreto - UFC (Federal University of Ceará (UFC)), CAIO VERAS (UNIVERSIDADE FEDERAL DO CEARÁ)

Modeling and Control 7

Marcelo Heldwein (Universidade Federal de Santa Catarina)

Mario Pacas (University of Siegen)

Investigation of Control Strategies to Mitigate the Oscillation Effects caused by Interconnected Buck Converters

Isaías de Bessa (Federal University of Amazonas), Renan Landau Paiva de Medeiros

(Universidade Federal do Amazonas), Iury Bessa (Federal University of Amazonas), Floryndo Antônio Ayres Junior (Federal University of Amazonas), Kevin Lucas (Federal University of Santa Catarina)

Cascaded Bidirectional Interleaved Buck-Boost DC-DC Converter Designed for Grid-To-Vehicle and Vehicle-To-Grid Technologies

Wagner Leal (University of São Paulo), Cassius Aguiar (Federal University of Technology - Paraná), Ricardo Quadros (University of São Paulo), Guilherme Fuzato (Federal Institute of Education, Science and Technology of Sao Paulo), Artur Piardi (Itaipu Technological Park Foundation), Almir Braggio (Itaipu Technological Park Foundation), Dabit Sonoda (Itaipu Technological Park Foundation), Rodrigo Bueno (Itaipu Technological Park Foundation), Zeno Luiz Iensen Nadal (Companhia Paranaense de Energia)

DC Current Redistributor for Electric Aircraft System

Mateus Dias (University of Campinas), Hildo Guillardi Júnior (University of Campinas), Joel Guerreiro (University of Campinas), João Ota (University of Campinas), José Pomilio (University of Campinas), Paolo Mattavelli (University of Padova - Dept. of Information Engineering (DEI))

Modeling and control of a Forward DC-DC converter for Battery Voltage Balancing

Pabulo Felipe Ciarnoscki (Federal University of Santa Catarina), Daniel Pagano (Federal University of Santa Catarina), Gabriel Manoel da Silva (Federal University of Santa Catarina)

A comparison among different Finite Control Set approaches and Convex Control Set Model-based Predictive Control applied in a Three-Phase Inverter with RL load

Arthur Bartsch (Federal Institute of Education, Science and Technology of Santa Catarina), Dayse Cavalcanti (Santa Catarina State University), Mariana Cavalca (Santa Catarina State University), Ademir Nied (Universidade do Estado de Santa Catarina)

Power Quality 4

Francisco Neves *(Universidade Federal de Pernambuco)*

Joselito Heerdt *(Universidade do Estado de Santa Catarina)*

Partial Harmonic Current Compensation Applied to Multiple Photovoltaic Inverters in a Radial Distribution Line

André Oliveira (Universidade Federal de Viçosa), Lucas Xavier (Universidade Federal de Minas Gerais), João Marcus Callegari (Universidade Federal de Viçosa), Allan Cupertino (Centro Federal de Educação Tecnológica de Minas Gerais), Victor Mendes Flores (Federal University of Minas Gerais (UFMG)), Heverton Augusto Pereira (Universidade Federal de Viçosa)

DESIGN METHOD TO REDUCE THE DC LINK VOLTAGE OF A THREE-WIRE THREE-PHASE HYBRID ACTIVE POWER FILTER

Mateus Braga (Federal University of Juiz de Fora (UFJF)), Samuel Duarte (Federal University of Juiz de Fora (UFJF)), Guilherme Marcio Soares (Federal University of Juiz de Fora (UFJF)), Pedro Gomes Barbosa (Federal University of Juiz de Fora (UFJF))

Structural and Performance Comparison Between Harmonic Selective Repetitive Controllers for Shunt Active Power Filter

Rafael Neto (Universidade Federal de Pernambuco), Helber de Souza (Instituto Federal de Educação, Ciência e Tecnologia de Pernambuco), Francisco Neves (Universidade Federal de Pernambuco), Eduardo Stangler (Universidade Federal de Pernambuco), Fabrício Bradaschia (UFPE)

Unbalanced Voltage Mitigation using D²VC with Proportional Resonant Controller in αβ-Frame

Elienai Macedo (Universidade Federal do Rio de Janeiro), Robson Francisco Dias (Universidade Federal do Rio de Janeiro), Silvangela Lima Barcelos (Federal University), Edson H. Watanabe (Universidade Federal do Rio de Janeiro)

Renewables 4

Gierri Waltrich *(Federal University of Santa Catarina)*

Demercil de Souza Oliveira Júnior *(Federal University of Ceará)*

Real Time Simulation in a Distribution System Including PV Inverter and Voltage Regulator: Voltage Impact Analysis

João Archetti (Universidade Federal de Juiz de Fora- UFJF), Lilian Venturi Pinheiro (Universidade Federal de Jui), Mateus Lopes (Universidade Federal de Juiz de Fora- UFJF), Bernardo Musse (Universidade Federal de Juiz de Fora- UFJF), Janaina Oliveira (Universidade Federal de Juiz de Fora- UFJF), Leonardo Willer (Universidade Federal de Juiz de Fora- UFJF)

A Curve Tracer for Photovoltaic Modules Based on The Capacitive Load Method

Emerson Abreu Bastos Junior (Universidade Federal de São João del-Rei), Caio Meira Amaral da Luz (Universidade Federal de São João del-Rei), Tatiane Oliveira (Universidade Federal de São João del-Rei), Lara Rodarte (Universidade Federal de São João del-Rei), Eduardo Moreira Vicente (Universidade Federal de São João del-Rei), Fernando Lessa Tofoli (Universidade Federal de São João del-Rei)

Optimal Voltage Coordinated Control for Grid-connected Photovoltaic Systems

Gustavo Dill (Celso Suckow da Fonseca Federal Center of Technology), Thiago William Pires Sousa (Celso Suckow da Fonseca Federal Center of Technology)

The True Unity Power Factor Converter Applied to Photovoltaic Applications

Marcos Alves (Universidade Federal de Minas Gerais), Thiago Parreiras (Universidade Federal de Minas Gerais), Braz de Jesus Cardoso Filho (Universidade Federal de Minas Gerais)

250 W Single Stage Step-up Inverter Connected to the Grid

Jéssika Andrade (Federal University of Santa Catarina), Roberto Francisco Coelho (Federal University of Santa Catarina), Telles B. Lazzarin (Federal University of Santa Catarina)

Smart grid applications 3

Denizar Cruz Martins *(Federal University of Santa Catarina)*

Helmo Kelis Morales Paredes *(São Paulo State University - UNESP)*

Economic Analysis of a Peak Shaving System with Diesel Generator
Myrlena Ferreira (Federal University of Maranhão), Luiz Antonio de Souza Ribeiro (Federal University of Maranhão), Jose Gomes de Matos (Federal University of Maranhão)

Enhanced Power Management System for Droop Control in a Grid Connected DC Microgrid
Pedro José dos Santos Neto (University of Campinas), João Pedro Carvalho Silveira (University of Campinas), Tárcio André dos Santos Barros (University of Campinas), Ernesto Ruppert Filho (University of Campinas), Juan Carlos Vasquez (Aalborg University), Josep Maria Guerrero (Aalborg University)

Connection Time in Modbus/TLS for Secure Communications on Photovoltaic Systems
Matheus Ferst (Federal University of Technology - Paraná), Hugo Figueiredo (Federal University of Technology - Paraná), Gustavo Weber Denardin (Universidade Tecnológica Federal do Paraná)

Dimensioning and Developement of a Hybrid Microgrid in the UFJF Campus
Paula Stael Silva Barbosa (Federal University of Juiz de Fora), Marina Monteiro (Federal University of Juiz de Fora), Jessica Dohler (Federal University of Juiz de Fora), André Ferreira (Federal University of Juiz de Fora), Janaína Gonçalves de Oliveira (Universidade Federal de Juiz de Fora - UFJF)

Modeling and Control 8

Kleber Lima *(Federal University of Ceará)*

Danilo Iglesias Brandão *(Federal University of Minas Gerais)*

Design of Robust Structured Control Strategy for Single-Phase Dynamic Voltage Restorer
João A. Moor Neto (CEFET-RJ), Guilherme Giglio (CEFET-RJ), Thiago Brasil (CEFET-RJ), Mauro dos Reis (CEFET-RJ), Julio Carvalho (CEFET-RJ)

Single Phase-Shift Control of DAB Converter using Robust Parametric Approach
Kevin Lucas (Federal University of Santa Catarina), Daniel Pagano (Federal University of Santa Catarina), Renan Landau Paiva de Medeiros (Universidade Federal do Amazonas)

Optimization of Robust PI Controllers for Grid-Tied Inverters
Caio Ruviaro Dantas Osório (Federal University of Santa Maria), Lucas Cielo Borin (Federal University of Santa Maria), Gustavo Koch (Federal University of Santa Maria), Vinicius Foletto Montagner (Federal University of Santa Maria)

Robust PID Controllers Optimized by PSO Algorithm for Power Converters
Lucas Cielo Borin (Federal University of Santa Maria), Everson Mattos (Federal University of Santa Maria), Caio Ruviaro Dantas Osório (Federal University of Santa Maria), Gustavo Koch (Federal University of Santa Maria), Vinicius Foletto Montagner (Federal University of Santa Maria)

Control Strategy and Power Management for Multifunctional Inverters with BESS and Reactive Power Compensation
Luiz Henrique Meneghetti (Federal University of Technology - Paraná), Edivan Carvalho (Federal University of Santa Maria), Gustavo Balduino Kruger Schmidt (Federal University of Technology - Paraná), Emerson Giovani Carati (Federal University of Technology - Paraná), Jean P. da Costa (Federal University of Technology - Paraná), Carlos Marcelo De Oliveira Stein (Federal University of Technology - Paraná), Zeno Luiz Iensen Nadal (Companhia Panaense de Energia), Rafael Cardoso (Federal University of Technology - Paraná)

Simplified Single-phase PV Generator Model for Distribution Feeders With High Penetration of Power Electronics-based Systems

Mariana Altoé Mendes[x], Murillo Cobe Vargas[x], Oureste Elias Batista[x], Yongheng Yang*, and Frede Blaabjerg*

[x]*Postgraduate Program in Electrical Engineering, PPGEE*
Federal University of Espírito Santo, UFES
Vitória, Brazil
murillo.vargas@aluno.ufes.br; mariana.a.mendes@aluno.ufes.br; oureste.batista@ufes.br
**Department of Energy Technology*
Aalborg University, Aalborg, Denmark
yoy@et.aau.dk; fbl@et.aau.dk

Abstract—Power systems simulations become more relevant to analyze the feeder behavior due to the increasing integration of inverter-based resources (IBR), like solar photovoltaic (PV). However, simulations based on dynamic and switching-level PV generator (DSLPVG) models are complicated to implement, due to the multi-variable dependence, and they are time-consuming. Thus, this paper introduces a simple single-phase grid-connected PV generator model. This simplified PV generator (SPVG) model is only dependent of the PV generator active output power, power factor, and the voltage at the point of common coupling (PCC). It is developed in Simulink/MATLAB with a single-phase feeder. Simulations are conducted to verify the proposed model by comparing SPVG with DSLPVG models in terms of feeder behavior, accuracy and time elapsed during simulations. The implementation of the SPVG model is easier than the DSLPVG, the feeder behavior is similar, and the simulations are almost 120 times faster. With the SPVG model, fast simulations with many PV generators may be done, like power system protection studies, while maintaining good accuracy. Furthermore, the model can be easily extended to other simulation software packages.

Index Terms—Distributed power generation; photovoltaic systems; renewable energy sources; system modeling.

I. INTRODUCTION

The integration of distributed energy resources (DER) into power systems plays a vital role in the new energy scenario due to the environmental impact of conventional fossil fuels. This concern, combined with recent initiatives to promote the distributed power generation (DG) and premium feed-in tariffs, foments the use of renewable energy in a large scale [1], [2].

Solar photovoltaic (PV) systems in the DG have great worldwide growth and gained much popularity because of several factors like: cost reduction, increase in efficiency, reduction of environmental impacts (compared to conventional fossil energy resources), and strong regulations and initiatives (e.g., premium feed-in tariffs). The PV panels do not emit pollutants during energy generation, and do not have moving or rotating parts, which reduces the maintenance

This study was financed in part by the Coordenação de Aperfeiçoamento de Pessoal de Nível Superior - Brazil (CAPES) - Finance Code 001. *(Mariana Altoé Mendes and Murillo Cobe Vargas are co-first authors.)*
(Corresponding author: Murillo Cobe Vargas.)

costs to some extent [1]–[3]. On the other hand, the weather dependence, change in the irradiance level, shading, and the fault current behaviour of PV generators, and others types of inverter-based resources (IBR), are examples of challenges when it comes to plan, implement, predict and control the integration of PV systems into the grid [4].

Large-scale PV plants are in general connected to the transmission grid through three-phase power converters. In small-scale PV systems (e.g., residential applications), single-phase power converters are common due to the low power ratings, and reduced cost [3]. Along with that, extensive studies are being developed in new converter topologies, control strategies and modulation methods [5], [6], to increase the reliability, efficiency and security of these devices.

However, the integration of PV systems in power distribution systems may change the conventional network design, where the main source of power and short-circuit capacity is the primary substation. This integration may cause voltage and protection issues, vulnerability and overvoltages due to the intentional or unintentional islanding [7]. Therefore, it is important to predict the network behavior in this new scenario using, for example, computational simulation tools.

Computational simulation tools are being used by power system engineers to model and analyze transmission and distribution grids, including representations of substations, equipment, loads and generators [8]. However, simulations that use PV generators modeled at the dynamic level [9] require many variables and are complicated to implement. Moreover, PV generators modeled at the switching-level have small steps, becoming time- and computational-consuming [10]. They are not practical for large network models, for a considerable quantity of PV systems, or for a high PV penetration level. As the number of PV generators into grid tends to grow, it is necessary to develop a model that can represent the entire PV generator in a simple way with the least time consumption and computational process.

Additionally, related simulation research usually represents the grid-connected PV generator as an ac current source. This, however, does not explain why it can be represented like this and the control of the angle of injected current into grid is often omitted [11], [13].

978-1-7281-4181-7/19 $31.00 © 2019 IEEE

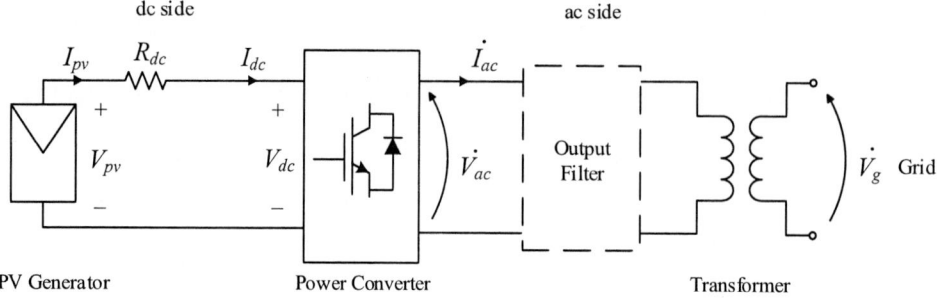

Fig. 1. Single-phase grid-connected photovoltaic generator diagram.

A steady-state analysis allows to study, for example, system power flow and grid voltage profile, but is also important to analyze the grid even under transient conditions, like fault scenarios. Although it does not fall in the scope of this paper, PV generators may require to support a faulty network even beyond the slowest time of the time-current curves from breakers, reclosers and fuses [10]. Moreover, the IEEE Std. 1547-2018 [14] recommends varying clearing time for different terminal voltage ranges of the DER is response to abnormal voltages. Hence, PV generators located at different points of a distribution feeder will experience distinct terminal voltages when a fault occurs. If a terminal voltage is in a range of protection, the PV generator will remain connected and injecting currents to support the grid recovery [11].

This work aims to propose a simple single-phase grid-connected PV generator model under the steady-state condition with the output active power and power factor control at the PCC. This simplified PV generator (SPVG) model may be applied to explore electrical power systems in terms of inverse time overcurrent protection, power grid operation planning, generation expansion, and feeders with many PV generators, and/or with a high PV penetration level.

II. SINGLE-PHASE STEADY-STATE MODEL OF GRID-CONNECTED PHOTOVOLTAIC GENERATOR

The schematic diagram of a common grid-connected single-phase PV system is illustrated in Fig. 1. The system consists of three main parts: the PV generator, the power electronic converter between the dc and ac sides, and the output filter and grid-coupling transformer.

For this study, some initial considerations were made: the grid-coupling transformer was neglected to simplify the model; the output filter was ignored considering that it acts only for frequencies near and above switching frequencies, and the value of the cable resistance (R_{dc}), which connects the PV arrays to the power converter, is small enough and can also be neglected [12]. Thus, the schematic diagram can be simplified, as illustrated in Fig. 2.

A. Simplified Photovoltaic Generator

Single-phase PV systems have in general small-scale power, a small number of PV modules connected in parallel (to meet the operational voltage range of the power converter), and are installed at the same location (e.g., house

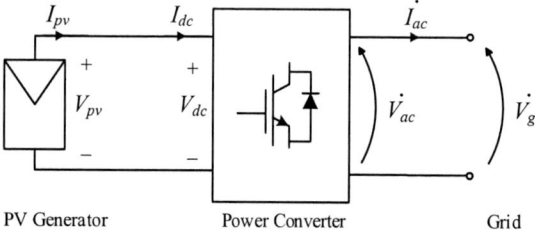

Fig. 2. Simplified grid-connected photovoltaic generator diagram.

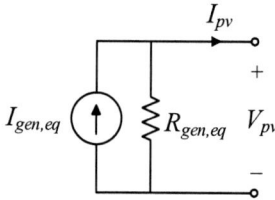

Fig. 3. Norton equivalent circuit of a PV generator.

rooftop). Therefore, it is reasonable that the effects of temperature difference, irradiance variation and shading – affect the current generated by each PV module, and the overall PV system power performance – can be neglected.

As presented in [15], if all PV modules are of the same type and under the same environmental conditions (e.g., irradiance and temperature), an overall PV generator model from the dc side can be represented as a Norton equivalent circuit, as shown in Fig. 3. Here, I_{pv} and V_{pv} represent the terminal current and voltage of the PV generator, respectively. The equivalent current generated by parallel modules is depicted by $I_{\mathrm{gen,eq}}$. Generally, $R_{\mathrm{gen,eq}}$, which represents the parallel equivalent resistance of the PV cells and resistance model of the leakage current of the *p-n* junction, is high and some authors neglect it [16]. Therefore, it is assumed that $I_{\mathrm{gen,eq}} = I_{\mathrm{pv}}$.

The total power generated by the PV generator (P_{pv}) can then be expressed by

$$P_{\mathrm{pv}} = V_{\mathrm{pv}} I_{\mathrm{pv}} \tag{1}$$

It is considered that the PV generator is always operating at the maximum power point (MPP). This operation point, P_{pv}, is assumed as a constant.

978-1-7281-4181-7/19 $31.00 © 2019 IEEE

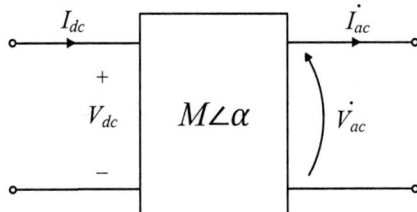

Fig. 4. Single-phase full-bridge power converter model.

B. Simplified Power Converter Model

With the above considerations, the single-phase full-bridge dc-ac converter with a sinusoidal pulse-width modulation (SPWM) can be modeled, as represented in Fig. 4. Here, I_{dc} and V_{dc} depict the input current and voltage at the dc side, and i_{ac} and v_{ac} are the output current and voltage at the ac side of the power converter. The voltage at the grid side, $v_g(t)$, can be represented as

$$v_g(t) = V_{\text{g-peak}} cos(wt + \theta) \qquad (2)$$

with the angular frequency being w and phase angle being θ. As the output filter and the coupling transformer were neglected, $v_g(t) = v_{ac}(t)$. If the output filter and the coupling transformer were considered, this assumption can not be made because the voltage magnitude and angle will be different due to the coupling transformer internal losses and magnetization current, and the phase shifting in the output filter.

It is assumed that the power converter will operate only by controlling the current injected at the PCC, i_{ac}, i.e., well-known as the grid-feeding or grid-following mode. This mode applies to the Current Source Inverter (CSI) [17], [18], and also to the current-controlled Voltage Source Inverter (VSI) [3], [5], [19], [20] topologies, which may be applied as a grid interface of the DER.

To perform this control, the power converter should be synchronized with the grid. It can be achieved by using a phase-locked loop (PLL) technique [21], [22], which estimates w and θ from v_{ac}. The power converter strategies use these values as input signals to control the system. In Fig. 4, M is the amplitude modulation ratio and α is the angle of output current [12], [17]. With the reference signals from the PLL, $i_{ac}(t)$ can be represented by

$$i_{ac}(t) = I_{\text{ac-peak}} cos(wt + \alpha) = \frac{M I_{dc}}{2} cos(wt + \alpha) \qquad (3)$$

which is synchronized with the grid voltage $v_{ac}(t)$.

C. Grid-connected Photovoltaic Generator Model

As show Fig. 2, $V_{pv} = V_{dc}$ and $I_{pv} = I_{dc}$, P_{pv} is equal to the power converter input power, P_{dc}. Neglecting the switching internal losses in the power converter and based on the principle of instantaneous power balance, the real power injected into grid, P_{ac}, is equal to P_{pv} under the steady-state operation. Thus, it is obtained that

$$P_{pv} = P_{ac} = \frac{V_{\text{ac-peak}} I_{\text{ac-peak}}}{2} cos(\theta - \alpha) \qquad (4)$$

Accordingly, $I_{\text{ac-peak}}$ can be defined as

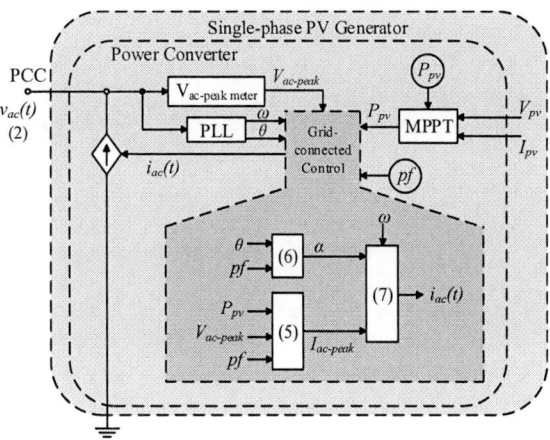

Fig. 5. Control block diagram of the simplified PV generator model.

$$I_{\text{ac-peak}} = \frac{2P_{pv}}{V_{\text{ac-peak}} cos(\theta - \alpha)} \qquad (5)$$

which shows that α enables the control of the PV generator power factor (p_f), $cos(\theta - \alpha)$. The p_f control is important because θ may vary depending on the PCC in the distribution feeder. For a specific value of p_f, the system can indirectly calculate the corresponding angle α,

$$\alpha = \theta - arccos(p_f) \qquad (6)$$

Thus, combining (5) and (3) gives

$$i_{ac}(t) = \frac{2P_{pv}}{p_f V_{\text{ac-peak}}} cos(wt + \alpha) \qquad (7)$$

Due to the solar intermittency, a maximum power point tracking (MPPT) is usually adopted in order to guarantee that P_{pv} will always be the maximum [23]. The value of P_{pv} may be set as a constant with the aim to model the MPPT control in a simple way for different p_f conditions. Fig. 5 shows the schematic block diagram of the simple single-phase PV generator model with the PLL, MPPT, input variables (P_{pv} and p_f, whose values are user-defined and may vary according to applications), as well as the current and α control. In Fig. 5, $V_{\text{ac-peak meter}}$ was chosen to match with the type of the simulation where the model will be implemented – and discussed in Section III. Using the RMS value, (7) can be represented as

$$i_{ac}(t) = \sqrt{2} \frac{P_{pv}}{p_f V_{\text{ac-RMS}}} cos(wt + \alpha) \qquad (8)$$

Since the system is synchronized only in one angular frequency, w, (2) and (7) can be represented as phasor notations,

$$\dot{V}_{ac} = V_{\text{ac-peak}} \angle \theta \qquad (9)$$

$$\dot{I}_{ac} = \frac{2P_{pv}}{p_f V_{\text{ac-peak}}} \angle \alpha \qquad (10)$$

This representation simplifies the model for a steady state analysis and may be more feasible for a software simulation environment.

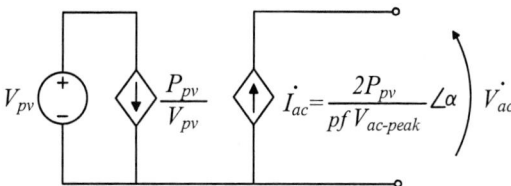

Fig. 6. Coupling circuit model between dc and ac sides.

The model can then be illustrated as a coupling circuit between dc and ac sides, as shown in Fig. 6. At the dc side, there is a dc voltage source V_{pv} and a dc-dependent current source in terms of P_{pv} and V_{pv}. At the ac side, the generator is represented by an ac-dependent current source according to (10).

D. Applications and Constraints

For a steady-state condition analysis, the model has no restriction in terms of the input power P_{pv}, and it is independent of the values of the PV side voltage and current, V_{pv} and I_{pv}. The PCC voltage may assume any nominal value. The value of α may be set or calculated to be equal or different from θ to perform a unit or leading/lagging power factor, respectively. For a simulation environment, the model can be summarized as just the portion of the ac dependent current source setting the desired rated power and power factor. The restrictions for the maximum current and power capacity of the power converter are not in the scope of this paper, and also the behavior under abnormal voltage conditions, like faults [24], [25]. These will be investigated in future research.

III. MODEL SIMULATION AND RESULTS

A. Simulation Methodology

The SPVG model, as demonstrated previously, was developed and simulated in Simulink/MATLAB, with the phasor simulation type, referring to Figs. 6 and 7. The inputs of the model are the desired active output power and power factor, P_{out} and p_f, i.e. the grey blocks shown in Fig. 7. Due to the simulation type, the magnitude of the controlled current source is a peak value, calculated according to (10) and the angle is calculated by (6).

To evaluate the behavior of the model, it was connected to the PCC of a system composed of a simple single-phase feeder (as shown in Fig. 8) with a 240-V_{RMS}/60-Hz voltage source, three parallel loads (with active and inductive reactive power) and four RL branches. Without the connection of the PV generator, the RMS value of the voltage at the PCC is equal to 209.5 V. Parameters of the feeder are shown in Table I. The negative signal in loads refers to a power absorption from the grid.

To validate the SPVG model, its performance was compared with a dynamic switching-level PV generator (DSLPVG) model: single-phase, 240-V_{RMS}, 60-Hz, 3500-W, transformerless grid-connected PV array available in Simulink/MATLAB with the discrete simulation type. The DSLPVG model is composed by a PV array, a dc-link, single-phase dc-ac power converter using pulse width modulation (PWM) control, single-phase full-bridge Insulated Gate Bipolar Transistor (IGBT) module (H-bridge), and a classical

TABLE I
SINGLE-PHASE FEEDER DATA

Symbol	Parameter	Value
V_s	Voltage Source	240∠0° V_{RMS}
f	Frequency	60 Hz
Load 1	Constant Z load	-1000 - j500 VA
Load 2	Constant Z load	-2000 - j1000 VA
Load 3	Constant Z load	-1500 - j500 VA
Branches (1, 2, 3, 4)	RL branches	0.5 + j0.823 Ω
$|V_{pcc}|$	PCC voltage magnitude	209.5 V_{RMS}

TABLE II
DYNAMIC SWITCHING-LEVEL PV GENERATOR MODEL DATA

Parameter	Specification
PV module model	Trina Solar TSM-250
Number of strings	1
Series PV modules	14
Parallel PV strings	1
PV Cell temperature	25 °C
PV rated power	3500 W (at 1000 W/m²)
Dc nominal voltage	400 V
Dc-link capacitor	3 mF
PWM method	Bipolar
Carrier frequency	3780 Hz
R series filter parameter	8.23 mΩ
L series filter parameter	2.183 mH
R parallel filter parameter	10.5 W
C parallel filter parameter	525 VAr
Ac voltage	240 V_{RMS}
Frequency	60 Hz

LCL output filter. The control system contains a MPPT controller based on the "Perturb and Observe" technique; a dc-link voltage regulator, which determines the reference of I_d (active current) for the current regulator; a current regulator based on the dq-frame; a PLL and PWM generator. Further information is shown in Table II.

The simulations were carried out by varying arbitrarily the irradiance level and the reactive current reference I_q, followed by the measurement of the power delivered from the PV array (P_{pv}), active power delivered to the grid (P_{out}), reactive power delivered or absorbed to/from the grid (Q_{out}), voltage at the PCC (V_{pcc}) in RMS, the output power factor (p_f) and the total time elapsed during each simulation. To perform a comparison, the values of P_{pv} and p_f from the DSLPVG model were inserted in blocks of P_{out} and p_f from Fig. 7, respectively, of the SPVG model. This assumption is made because the values of P_{pv} and P_{out} are different for the DSLPVG model, since its power converter has switching internal losses and an output filter. Once the power converter of the SPVG model was considered ideal and without the output filter, to make a fair comparison and analyze the impacts of those assumptions on other parameters (such as Q_{out} and V_{pcc}), the value of P_{out} from the SPVG model should then be equal to P_{pv} from the DSLPVG model. The values of Q_{out}, V_{pcc} and the total time elapsed during each simulation from the SPVG model were registered. The simulation time was set to 1.2 s for all cases for both models,

978-1-7281-4181-7/19 $31.00 © 2019 IEEE

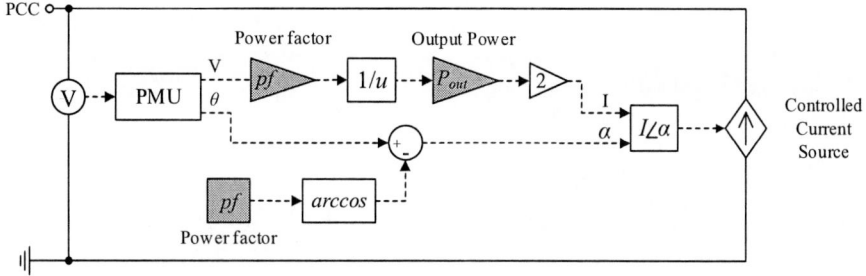

Fig. 7. Proposed simplified PV generator (SPVG) model circuit developed in Simulink/MATLAB.

TABLE III
SIMULATION RESULTS AND RELATIVE ERROR (%) FROM SPVG MODEL

		DSLPVG Model					SPVG Model				Relative error (%)	
Simulation ID	Irradiance (W/m^2)	P_{pv} (W)	I_q	P_{out} (W)	Q_{out} (VAr)	V_{pcc} (V)	P_{out} (W)	Q_{out} (VAr)	p_f	V_{pcc} (V)	ΔV_{pcc}	ΔQ_{out}
1	1,000	3,500.0	0.8	3,452.0	-2,281.0	196.6	3,500.0	-2,312.7	-0.8343	196.3	-0.17	1.39
2	1,000	3,500.0	0.4	3,476.0	-1,237.0	216.1	3,500.0	-1,245.5	-0.9421	216.5	0.19	0.69
3	1,000	3,500.0	0.0	3,451.0	36.2	235.2	3,500.0	36.8	0.9999	235.5	0.11	1.42
4	1,000	3,500.0	-0.3	3,388.0	1,121.0	248.0	3,500.0	1,158.1	0.9494	249.0	0.39	3.31
5	1,000	3,500.0	-0.8	3,436.0	3,167.0	268.8	3,500.0	3,226.0	0.7353	270.2	0.52	1.86
6	750	2,610.0	0.5	2,609.0	-1,488.0	207.1	2,610.0	-1,488.6	-0.8687	207.2	0.04	0.04
7	750	2,610.0	0.3	2,598.0	-919.1	216.3	2,610.0	-923.3	-0.9427	216.5	0.07	0.46
8	750	2,610.0	0.0	2,501.0	-16.5	229.5	2,610.0	17.2	0.9999	229.7	0.07	4.36
9	750	2,610.0	-0.2	2,548.0	728.4	238.3	2,610.0	746.1	0.9615	239.0	0.29	2.43
10	750	2,610.0	-0.3	2,560.0	1,111.0	242.8	2,610.0	1,132.7	0.9173	243.6	0.31	1.95
11	500	1,732.0	0.3	1,691.0	-904.5	210.5	1,732.0	-926.4	-0.8818	210.4	-0.04	2.42
12	500	1,732.0	0.2	1,686.0	-606.6	215.0	1,732.0	-623.2	-0.9410	215.0	0.01	2.73
13	500	1,732.0	0.0	1,689.0	31.0	223.6	1,732.0	31.8	0.9998	224.3	0.29	2.55
14	500	1,732.0	-0.2	1,682.0	718.0	232.3	1,732.0	739.3	0.9197	233.2	0.37	2.97
15	500	1,732.0	-0.3	1,677.0	1,079.0	236.8	1,732.0	1,114.4	0.8410	237.6	0.33	3.28
16	250	851.0	0.2	818.8	-591.2	208.2	851.0	-614.4	-0.8108	208.2	0.02	3.93
17	250	851.0	0.1	824.7	-290.7	212.3	851.0	-300.0	-0.9431	212.9	0.28	3.19
18	250	851.0	0.0	823.5	24.4	216.8	851.0	25.2	0.9996	217.4	0.27	3.34
19	250	851.0	-0.1	800.7	348.2	221.6	851.0	370.1	0.9170	221.9	0.15	6.28
20	250	851.0	-0.2	807.5	692.5	225.6	851.0	729.8	0.7591	226.5	0.38	5.39

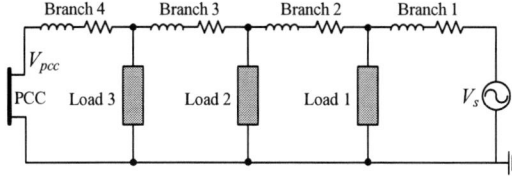

Fig. 8. Single-phase feeder developed in Simulink/MATLAB.

and each simulation was identified by an ID number.

The results of the comparison between the models, with the relative error (ΔV_{pcc} or ΔQ_{out}) from the SPVG model, as shown in (11), are presented in Table III. In terms of active or reactive power, a positive value represents the power injection into the grid, while a negative represents the power absorption. In terms of p_f, a positive value represents a leading power factor, while a negative represents a lagging power factor. The relative error is defined as

$$\Delta = \frac{|SPVG| - |DSLPVG|}{|DSLPVG|} \quad (11)$$

where $|SPVG|$ can be the voltage at the PCC in Fig. 8 with the proposed SPVG model connected at the PCC, or the reactive power that the proposed SPVG model injects/absorbs to/from the grid, and $|DSLPVG|$ assumes these values for the DSLPVG model connected at the PCC.

B. Comparison of the Behavior Between the Models

The behavior of both models was similar for each simulation ID, considering the power injection/absorption and the variation of the value of V_{pcc}. In the DSLPVG model, for a higher irradiance level, P_{pv} and P_{out} increase. As mentioned previously, these values are different, P_{out} is smaller than P_{pv} due to the switching internal losses of the power converter and the presence of the LCL filter. Comparing P_{pv} values at a unit power factor, the PCC voltage raises when P_{out}

increases (ID 3, 8, 13 and 18). For the same irradiance level, when p_f is lagging, the voltage at the PCC is smaller than the case when p_f is unit which is, in turn, smaller when p_f is leading. This behavior can be visualized by comparing the values of simulations ID 1, 3 and 5, or ID 7, 8 and 9. The SPVG model behaves likewise. Increasing P_{out}, the voltage at the PCC raises (ID 3, 8, 13 and 18). For the same values of P_{out}, the behavior of the voltage at the PCC when p_f varies from lagging to leading is equivalent to the behavior of the DSLPVG model (ID 12, 13, 14).

On average, the relative error in terms of V_{pcc} was 0.19%, with a maximum value of 0.52% for the simulation ID 5. The relative error of Q_{out} is larger than V_{pcc} because the power converter internal losses of the SPVG model were neglected, and also the control loop. However, this has insignificant influence on the behavior of the PCC voltage between the cases. In all, the average relative error of Q_{out} was 2.70% with a maximum value of 6.28% for the simulation ID 19.

For $I_q = 0$ (i.e., the unit power factor), the DSLPVG model still injects/absorbs reactive power to/from the grid due to the dynamics of the internal control process. Additionally, for different values of P_{out} with $I_q = 0$, the resultant p_f is not always equal (ID 3, 8, 13 and 18). For the proposed SPVG model, when $p_f = 1.0$, it will inject only active power to grid, i.e., $Q_{out} = 0$, being independent of the value of P_{out}, due to the ideal considerations for the modeling.

Table IV shows the simulation elapsed time for each simulation condition, DSLPVG model and the SPVG model, as well as the average, i.e., the elapsed time of each approach. The average elapsed time for the DSLPVG model was 35.59 s, against 0.3 s for the SPVG model. The results show that, for a steady-state analysis, the SPVG model is on average almost 120 times faster than the DSLPVG model. This fact is because of the simplicity of the model and the phasor simulation type.

C. Discussion and Applications of the SPVG Model

The proposed SPVG model presents three advantages compared to the DSLPVG model: simplicity, accuracy and speed. The implementation of the SPVG model in Simulink/MATLAB is simpler than the DSLPVG model. It was not necessary to model, for example, the MPPT control, dc-link capacitor, the dq reference frame control, and output filter. Thus, it can be easily implemented in other simulation software packages.

Despite the low relative errors, the applications of the proposed model must be taken into account. For voltage limits violation studies, for example, the proposed model may be suitable. For energy quality studies, depending on the variable studied, the error may be greater than the permissible desired range. Regarding the reactive power, for studies with low irradiance, the model presents larger errors than the high irradiance levels when it is injecting reactive power into grid. It may affect the results on networks with a high X/R ratio – which is not the case of distribution feeders.

Because of the low relative errors, the model can be used on power system protection studies, generation expansion planning, which the accuracy is not the main requirement, but is also relevant (e.g., impacts on inverse time overcurrent relays due to the change in load current at distribution feeders

TABLE IV
TOTAL SIMULATION TIME COMPARISON

Simulation ID	DSLPVG Model time (t_{dslpvg}) (s)	SPVG Model time (t_{spvg}) (s)	t_{dslpvg}/t_{spvg}
1	41.99	0.21	199.99
2	37.08	0.30	123.60
3	59.98	0.31	193.49
4	36.20	0.30	120.67
5	36.01	0.29	124.18
6	33.47	0.32	104.59
7	34.82	0.30	116.09
8	40.53	0.31	130.76
9	35.46	0.33	107.47
10	31.61	0.33	95.79
11	32.11	0.32	100.36
12	33.45	0.32	104.55
13	32.59	0.33	98.76
14	33.15	0.32	103.60
15	32.05	0.29	110.53
16	31.87	0.30	106.24
17	33.82	0.29	116.63
18	31.35	0.30	104.50
19	31.91	0.29	110.05
20	32.31	0.30	107.72
Average	35.59	0.30	118.98

with high PV penetration level), and on another IBR grid impacts studies.

Due to the fast speed, the SPVG model may enable extensive simulations in large distribution feeders (e.g., IEEE 13, 34 and 123-Node Test Feeders) with many PV generator units, with less time and less computational consumption, where the steady-state behavior of the system and its components are the focus of the research. Examples of the application of this proposed SPVG model on IEEE 13-Node Test Feeder considering a high penetration scenario of PV generation are presented:

- in [26], where the voltage profile of the distribution feeder was evaluated considering different p_f conditions for the PV generator;
- in [27], where the impacts on apparent impedance seen by the substation impedance relay of the distribution feeder were discussed (proposed SPVG model adapted for fault conditions); and,
- in [28], where the variation of the fault current for different fault types and different fault locations – downstream and upstream of overcurrent relays – was presented and discussed (proposed SPVG model adapted for fault conditions).

IV. CONCLUSION

The SPVG model demonstrated and proposed in this work was implemented in Simulink/MATLAB. Its behavior was compared with a DSLPVG model. The proposed model has a good accuracy, with the advantage of a simple implementation on a simulation environment, and simulation time is almost 120 times faster than the DSLPVG model. Moreover, a high or low demand and feeder nominal voltage are not

978-1-7281-4181-7/19 $31.00 © 2019 IEEE

restrictions for the model application, which can be easily adjusted for different scenarios. With the simple implementation, accuracy and speed advantages, the SPVG model can be used in a wide range of applications like power system protection studies, generation expansion planning, IBR grid impacts studies, and extensive simulations in large feeders with a high PV penetration level. Furthermore, the model can be easily adapted to other simulation software packages, since it has time and phasor-domain representations. The applied methodology can be extended to other types of the IBR to develop other simple models.

REFERENCES

[1] F. Blaabjerg, Y. Yang, D. Yang, and X. Wang, "Distributed Power-Generation Systems and Protection," *Proceedings of the IEEE*, vol. 105, no. 7, pp. 1311–1331, Jul. 2017.

[2] T. Adefarati and R. Bansal, "Integration of renewable distributed generators into the distribution system: a review," *IET Renewable Power Generation*, vol. 10, no. 7, pp. 873–884, Aug. 2016.

[3] S. Kouro, J. I. Leon, D. Vinnikov, and L. G. Franquelo, "Grid-Connected Photovoltaic Systems: An Overview of Recent Research and Emerging PV Converter Technology," *IEEE Industrial Electronics Magazine*, vol. 9, no. 1, pp. 47–61, Mar. 2015.

[4] D. Lew, M. Asano, J. Boemer, C. Ching, U. Focken, R. Hydzik, M. Lange, and A. Motley, "The Power of Small: The Effects of Distributed Energy Resources on System Reliability," *IEEE Power Energy Magazine*, vol. 15, no. 6, pp. 50–60, Nov. 2017.

[5] S. Chatterjee, P. Kumar, and S. Chatterjee, "A techno-commercial review on grid connected photovoltaic system," *Renewable and Sustainable Energy Reviews*, vol. 81, no. 1, pp. 2371–2397, Jan. 2018.

[6] H. Athari, M. Niroomand, and M. Ataei, "Review and Classification of Control Systems in Grid-tied Inverters," *Renewable and Sustainable Energy Reviews*, vol. 72, no. 1, pp. 1167–1176, May 2017.

[7] R. Walling, R. Saint, R. Dugan, J. Burke, and L. Kojovic, "Summary of Distributed Resources Impact on Power Delivery Systems," *IEEE Transactions on Power Delivery*, vol. 23, no. 3, pp. 1636–1644, Jul. 2008.

[8] A. Isaacs, "Simulation Technology: The Evolution of the Power System Network [History]," *IEEE Power and Energy Magazine*, vol. 15, no. 4, pp. 88–102, Jul. 2017.

[9] M. E. Ropp and S. Gonzalez, "Development of a MATLAB/Simulink Model of a Single-Phase Grid-Connected Photovoltaic System," *IEEE Transactions on Energy Conversion*, vol. 24, no. 1, pp. 195–202, Mar. 2009.

[10] J. Seuss, M. J. Reno, R. J. Broderick, and S. Grijalva, "Determining the Impact of Steady-State PV Fault Current Injections on Distribution Protection," Tech. Rep., Sandia National Laboratories (SNL), United States, 2017.

[11] H. Hooshyar and M. E. Baran, "Fault Analysis on Distribution Feeders With High Penetration of PV Systems," *IEEE Transactions on Power Systems*, vol. 28, no. 3, pp. 2890–2896, Aug. 2013.

[12] Y.-B. Wang, C.-S. Wu, H. Liao, and H.-H. Xu, "Steady-state model and power flow analysis of grid-connected photovoltaic power system," in *2008 IEEE International Conference on Industrial Technology*, pp. 1–6, Apr. 2008.

[13] A. Canova, L. Giaccone, F. Spertino, and M. Tartaglia, "Electrical Impact of Photovoltaic Plant in Distributed Network," *IEEE Transactions on Industry Applications*, vol. 45, no. 1, pp. 341–347, 2009.

[14] IEEE Std 1547-2018, "IEEE Standard for Interconnection and Interoperability of Distributed Energy Resources with Associated Electric Power Systems Interfaces," 2018.

[15] A. Yazdani, A. R. Di Fazio, H. Ghoddami, M. Russo, M. Kazerani, J. Jatskevich, K. Strunz, S. Leva, and J. A. Martinez, "Modeling Guidelines and a Benchmark for Power System Simulation Studies of Three-Phase Single-Stage Photovoltaic Systems," *IEEE Transactions on Power Delivery*, vol. 26, no. 2, pp. 1247–1264, Apr. 2011.

[16] M. Villalva, J. Gazoli, and E. Filho, "Comprehensive Approach to Modeling and Simulation of Photovoltaic Arrays," *IEEE Transactions on Power Electronics*, vol. 24, no. 5, pp. 1198–1208, May 2009.

[17] B. Sahan, S. V. Araújo, C. Nöding, and P. Zacharias, "Comparative Evaluation of Three-Phase Current Source Inverters for Grid Interfacing of Distributed and Renewable Energy Systems," *IEEE Transactions on Power Electronics*, vol. 26, no. 8, pp. 2304–2318, Aug. 2011.

[18] S.-H. Lee, S.-G. Song, S.-J. Park, C.-J. Moon, and M.-H. Lee, "Grid-connected photovoltaic system using current-source inverter," *Solar Energy*, vol. 82, no. 5, pp. 411–419, May 2008.

[19] B. Boukezata, J.-P. Gaubert, A. Chaoui, and M. Hachemi, "Predictive current control in multifunctional grid connected inverter interfaced by PV system," *Solar Energy*, vol. 139, no. 1, pp. 130–141, Dec. 2016.

[20] G. Zhang, Z. Li, B. Zhang, and W. A. Halang, "Power electronics converters: Past, present and future," *Renewable and Sustainable Energy Reviews*, vol. 81, no. 1, pp. 2028–2044, Jan. 2018.

[21] S. Chung, H. Shin, and H. Lee, "Precision control of single-phase PWM inverter using PLL compensation," *IEE Proceedings - Electric Power Applications*, vol. 152, no. 2, pp. 429–436, 2005.

[22] J.-F. Chen and C.-L. Chu, "Combination voltage-controlled and current-controlled PWM inverters for UPS parallel operation," *IEEE Transactions on Power Electronics*, vol. 10, no. 5, pp. 547–558, 1995.

[23] J. Enslin, M. Wolf, D. Snyman, and W. Swiegers, "Integrated photovoltaic maximum power point tracking converter," *IEEE Transactions on Industrial Electronics*, vol. 44, no. 6, pp. 769–773, 1997.

[24] A. Q. Al-Shetwi, M. Z. Sujod, and F. Blaabjerg, "Low voltage ride-through capability control for single-stage inverter-based grid-connected photovoltaic power plant," *Solar Energy*, vol. 159, no. 1, pp. 665–681, Jan. 2018.

[25] W. Kou and D. Wei, "Fault ride through strategy of inverter-interfaced microgrids embedded in distributed network considering fault current management," *Sustainable Energy, Grids and Networks*, vol. 15, no. 1, pp. 43–52, Sep. 2018.

[26] M. C. Vargas, M. A. Mendes, and O. E. Batista, "Impacts of High PV Penetration on Voltage Profile of Distribution Feeders Under Brazilian Electricity Regulation," in *2018 13th IEEE International Conference on Industry Applications (INDUSCON)*, pp. 38–44, Nov. 2018.

[27] M. C. Vargas, M. A. Mendes, and O. E. Batista, "Faults Location Variability in Power Distribution Networks with High PV Penetration Level," in *2018 13th IEEE International Conference on Industry Applications (INDUSCON)*, pp. 459–466, Nov. 2018.

[28] M. C. Vargas, M. A. Mendes, and O. E. Batista, "Fault Current Analysis on Distribution Feeders with High Integration of Small Scale PV Generation," in *2019 IEEE Power and Energy Society General Meeting (PESGM)*, pp. 1–5, Aug. 2019.

Modulated Model Predictive Current Control for PMSM Operating With Three-level NPC Inverter

Qi Wang
School of Electrical Engineering
Southeast University
Nanjing, CHINA
Email: qwang.seu@outlook.com

Marco Rivera, Jose A. Riveros
Department of Electrical Engineering
Universidad de Talca
Curico, CHILE
Email: marcoriv@utalca.cl

Patrick Wheeler
Department of Electrical and
Electronic Engineering
University of Nottingham
Nottingham, U.K
Email:Pat.Wheeler@nottingham.ac.uk

Abstract—In finite control set model predictive control (FCS-MPC) strategy only one basic voltage vector is to be selected in per periodic time, which causes big current ripple as well as the torque ripple of permanent magnet synchronous motor (PMSM). To solve this problem, an improved model predictive control method, named modulated model predictive control (M2PC) is proposed. The proposed control strategy can produce a modulated waveform, which can reduce torque ripple and improve power quality. Simulation results verify that the proposed current controller has a better control performance than the classical FCS-MPC strategy.

Keywords—*Finite control set model predictive control (FCS-MPC), inverter, neutral-point-clamped (NPC) inverter, predictive control, Modulated Model Predictive Control (M2PC).*

I. INTRODUCTION

Multi-level converters are widely used in higher voltage range than conventional two-level converter [1]. They can reduce common mode voltages and total harmonic distortion (THD). In multi-level inverters, the three-level inverter has the least number of switches. It can be implemented easily in the technical field of a high voltage. The neutral-point-clamped (NPC) converter is often known as the three-level diode clamped converter which can improve total harmonic distortion and has bigger bandwidth than conventional two-level converter[2]. It becomes more and more popular in many industrial application fields[3].

Finite control set model predictive control (FCS-MPC) has many advantages, such as simple structure, easy implementation and good multivariable control ability [4]. The FCS-MPC is easily extendible for different industrial applications. It has been widely concerned by academic and industrial communities [5]–[9]. More and more researchers apply MPC strategy for the multilevel converter. In [10], a finite control set model predictive control strategy was used in the five-level active neutral-point-clamped (ANPC) topology inverter for induction machine (IM). In [11], two MPC controllers are proposed for grid-side NPC inverter and generator-side converter respectively. An improved MPC controller for a high power wind energy conversion system using the three-level boost (TLB) converter and NPC inverter was proposed in [11]. In [12], a fast finite switching state MPC was proposed for T-type three-level NPC converter. However, there are still some disadvantages on this control method. The main drawback is that only one basic vector can be selected per periodic time, which causes big current ripple as well as the torque ripple of permanent magnet synchronous motor (PMSM). Another MPC control strategy named deadbeat current predictive control having a fixed switching frequency[13]–[17]. This control strategy only predictive the

reference voltage in static coordinate and use the conventional space vector pulse width modulation (SVPWM) algorithm to generate the firing pulses. However, it is very complex in the calculation of switching time[4]. In order to overcome above drawbacks, a novel model predictive control strategy, named modulated model predictive control (M2PC) was proposed [18]–[20]. The M2PC control strategy adds a modulator to generate the duty cycles by selecting two active voltage vectors and zero voltage vectors, and the modulation time of each vector is calculated by minimization of the cost function. In [19], [20], the M2PC strategy was used in a seven-level H-bridge converter. In [21], the M2PC was proposed for brushless doubly fed IM control. In [2], [22], the M2PC was proposed for balancing of the DC-link capacitor voltages and regulating the load currents with a NPC converter.

M2PC for PMSM current control operating with a NPC inverter is proposed in this paper. The M2PC strategy can produce a modulated waveform by operating a cost function minimization algorithm, which can reduce torque ripple of PMSM and improve power quality. A simulation is implemented comparing with the conventional MPC strategy. The simulation results prove the effectiveness of M2PC strategy.

II. MATHEMATICAL MODEL OF THE THREE-LEVEL NPC INVERTER

The three-level NPC inverter, which is widely used in many industrial application fields [1], [12], [23]–[26], include twelve switches and six clamping diodes.

The structure of three-level NPC inverter is shown in Fig.1. Switches S_{1x} and S_{3x}, S_{2x} and S_{4x} are complementary states. In the current loop of PMSM control system, the three-level NPC inverter can produce higher switching frequency to reduce the THD than conventional two-level inverter.

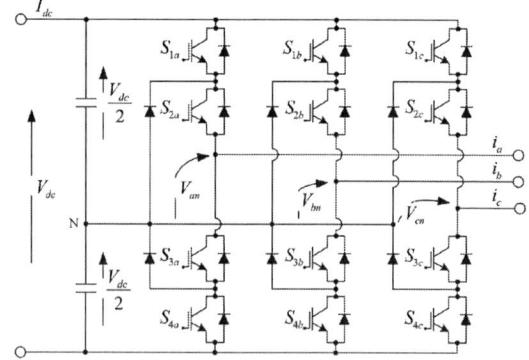

Fig. 1. Three-level neutral-point-clamped (NPC) inverter.

A total of 27 switching states are available in the three-level NPC inverter. These switching states can produce 27 voltage vectors in the stationary axis, including 8 redundant voltage vectors and 18 non-redundant vectors. The available voltage vectors of three-level NPC inverter are shown in Fig.2. Switches states and phase voltages of the three-level NPC inverter can be express in table I, where, the $x=a$, b, c.

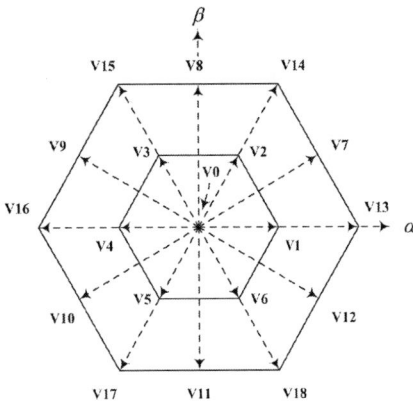

Fig. 2. The available voltage vectors of three-level NPC inverter

TABLE I
SWITCHING STATES AND VOLTAGE OF NPC INVERTER

S_x	S_{1x}	S_{2x}	S_{3x}	S_{4x}	V_{xn}
1	1	1	0	0	V_{dc}
0	0	1	1	0	$V_{dc}/2$
-1	0	0	1	1	0

III. CONVENTIONAL MODEL PREDICTIVE CURRENT CONTROL FOR PMSM

A. PMSM Mathematical Model

The d-q axis mathematical model of permanent magnet synchronous motor is shown as follows:

$$\begin{cases} u_d = R_s i_d + L_d \dfrac{di_d}{dt} - L_q \omega_e i_q \\ u_q = R_s i_q + L_q \dfrac{di_q}{dt} + L_d \omega_e i_d + \psi \omega_e \end{cases} \quad (1)$$

Where, u_d and u_q represent the d-q-axis voltages; ω_e is the electrical rotor speed of PMSM; i_d and i_q represent the d-q-axis currents; ψ is the flux linkage of permanent magnet; L_d and L_q are the d-q-axis inductances and R_s is the stator resistance.

The d-axis inductance and the q-axis inductance are approximately equal ($L_d = L_q$) in surface permanent magnet synchronous motor (SPMSM)[27].

B. Conventional FCS-MPC current control of PMSM

Assuming sampling time is T_s, the stator current derivatives can be discretized using the Euler approximation method, that is:

$$\frac{di}{dt} \approx \frac{i(k+1) - i(k)}{T_s} \quad (2)$$

Replacing (2) into (1), d-q axis predictive stator currents in the next sampling time can be obtained as:

$$\begin{cases} i_d^p(k+1) = \left(1 - \dfrac{R_s T_s}{L_d}\right) i_d(k) + \dfrac{T_s L_q}{L_d} \omega_e i_q(k) + \dfrac{T_s}{L_d} u_d(k) \\ i_q^p(k+1) = \left(1 - \dfrac{R_s T_s}{L_q}\right) i_q(k) - \dfrac{T_s L_d}{L_q} \omega_e i_d(k) - \psi_m \omega_e T_s + \dfrac{T_s}{L_q} u_q(k) \end{cases} \quad (3)$$

Where, $i_d^p(k+1)$ and $i_q^p(k+1)$ represent the d-q axis predictive stator currents in next sampling time; T_s is the sampling time. The field-oriented control (FOC) scheme of PMSM current control using FCS-MPC strategy is shown in Fig.3. Here, a PI speed controller is used to generate the q-axis reference current. The FCS-MPC current controller is used for tracking the d-q axis reference currents. The discrete-time model of PMSM is used to predict the stator current. During each sampling period, one voltage vector that minimizes the cost function is selected from the nineteen basic voltage vectors and applied to the three-level NPC inverter.

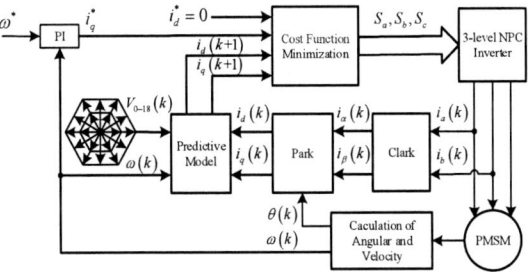

Fig. 3. Control diagram of classical MPC current control method with three-level NPC inverter.

The cost function can be shown as follow:

$$g = \left(i_d^p(k+1)\right)^2 + \left(i_q^* - i_q^p(k+1)\right)^2 + \hat{f}\left(i_d^p(k+1), i_q^p(k+1)\right) \quad (4)$$

Where, g is the cost function; i_q^* is the reference value of q-axis current, which is output by PI controller of speed loop; $\hat{f}\left(i_d^p(k+1), i_q^p(k+1)\right)$ is restriction of d-q axis currents, it can be shown as follow:

$$\hat{f}\left(i_{sd}^p(k+1), i_{sq}^p(k+1)\right) = \begin{cases} \infty & \text{if } \left|i_{sd}^p\right| > i_{max} \text{ or } \left|i_{sq}^p\right| > i_{max} \\ 0 & \text{if } \left|i_{sd}^p\right| \le i_{max} \text{ and } \left|i_{sq}^p\right| \le i_{max} \end{cases} \quad (5)$$

The task of the FCS-MPC strategy is selected the optimal switching state, which executes nineteen times (each for different basic voltage vector) to calculate the optimal cost function.

IV. MODULATED MODEL PREDICATIVE CURRENT CONTROL FOR PMSM

Same as the conventional MPC strategy, M2PC also has prediction and optimization sections. The cost function g is evaluated for each case. The only difference is the M2PC includes a suitable modulation scheme. The M2PC strategy select two adjacent active voltage vectors which minimize the cost function in each sector at every sampling time. The FOC control scheme of PMSM current control with three-level NPC inverter using M2PC strategy is shown in Fig.4. For example, two adjacent voltage vector $\mathbf{v}_1$ and $\mathbf{v}_2$ are selected in the first sector. Each prediction is calculated based on (3) and (4). g_1, g_2, g_0 are cost functions of voltage vectors $\mathbf{v}_1$, $\mathbf{v}_2$ and zero voltage vector, respectively. As shown in (6). The duty

978-1-7281-4181-7/19 $31.00 © 2019 IEEE

cycles of two adjacent active voltage vectors $\mathbf{v}_1, \mathbf{v}_2$ and zero voltage vector are calculated respectively.

$$
\begin{aligned}
d_0 &= K/g_0 \\
d_1 &= K/g_1 \\
d_2 &= K/g_2 \\
T_s &= d_0 + d_1 + d_2
\end{aligned} \tag{6}
$$

Where d_0, d_1, d_2 correspond to the duty cycles of zero voltage vector and two adjacent active voltage vectors $\mathbf{v}_1, \mathbf{v}_2$. From (6), duty cycles for each vector and the parameter K can be calculated as follows:

$$
K = \frac{T_s}{\dfrac{1}{g_1} + \dfrac{1}{g_2} + \dfrac{1}{g_o}} \tag{7}
$$

$$
\begin{aligned}
d_0 &= T_s g_1 g_2 / (g_0 g_1 + g_1 g_2 + g_0 g_2) \\
d_1 &= T_s g_0 g_2 / (g_0 g_1 + g_1 g_2 + g_0 g_2) \\
d_2 &= T_s g_0 g_1 / (g_0 g_1 + g_1 g_2 + g_0 g_2)
\end{aligned} \tag{8}
$$

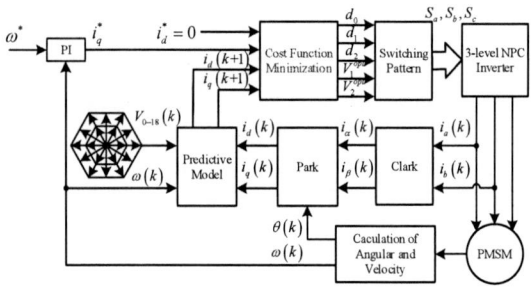

Fig. 4. Control diagram of M2PC current control method with three-level NPC inverter.

With these above equations, the total cost function g is defined as follow:

$$
g = d_1 g_1 + d_2 g_2 \tag{9}
$$

The minimum total cost function g, which is evaluated by two active voltage vectors, is applied to the three-level NPC inverter in the next sampling period.

V. SIMULATION RESULTS

To verify the performance of M2PC strategy, a simulation model is built in Matlab2018a. The parameters of the simulation are listed in table II. The simulation step is 1e-6, the current loop sampling time is 50e-6, the speed loop sampling time is same as the current loop. The speed controller is a PI controller, the proportion parameter is 0.009, and the integral parameter is 1.2. Both classical MPC strategy and proposed M2PC are evaluated in simulation.

The initial load of PMSM is 0.2N.m and target speed of PMSM is 1800rpm. In order to verify transient performance of the PMSM system, the load increases to 0.5N.m suddenly in 0.05s.

The waveforms of the d-q axis currents responses by classical MPC and M2PC are shown in Fig.5 and 6, respectively. It can be observed that the load increased suddenly at 0.05s, the q-axis current can response quickly. M2PC has a higher switching frequency than classical MPC in the same DSP interrupt time 50e-6, therefore, classical MPC strategy has a bigger ripple than M2PC strategy in the same sample time. The current THD of both methods will be discussed in details in Fig.13 and 14.

TABLE II
PARAMETER SETTING OF THE SIMULATION

Parameter	Symbol	Value
Rated Voltage	V	36 V
Rated Current	I	4.6 A
Maximum Current	I_{max}	13.8 A
Rated Power	P	100 W
Rated Torque	T	0.318 N.m
Stator Phase Resistance	R	0.375 Ohm
Motor Inertia	J	0.0588 kg.m^2.10^{-4}
Pole Pairs	P_n	4 Pair
q-axis Inductance	L_q	0.001 H
d-axis Inductance	L_d	0.001 H
Simulation Time	T_m	1e-6 s
Sampling Time	T_s	50e-6 s
Switching frequency	f_s	20kHz
Incremental Encoder Lines	N	2500 PPR

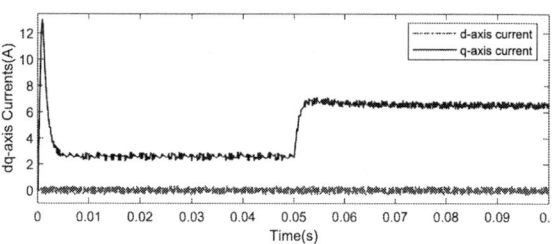

Fig. 5. d-q axis currents responses under classical MPC current control strategy in the presence of load torque disturbance at 1800rpm.

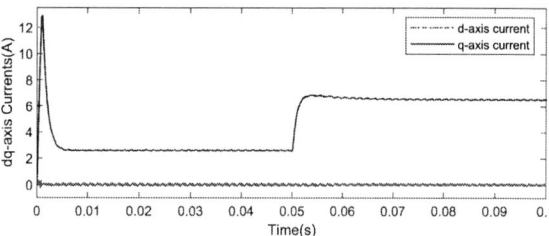

Fig. 6. d-q axis currents responses under M2PC current control strategy in the presence of load torque disturbance at 1800rpm.

Fig. 7 and Fig. 8 show the waveforms of the three phase currents for the classical MPC and M2PC, respectively. As evident, the M2PC shows lower ripple than the classical MPC. The step response of the PMSM for both methods is shown in Fig. 9 and Fig. 10, observing again, a better performance of the speed for the M2PC strategy. It can be observed that the speed of PMSM decrease as the load increased in 0.05s, then it can restore to target speed very quickly. However, the speed fluctuation of classical MPC strategy is bigger than M2PC strategy. Similarly, Fig. 11 and Fig. 12 show the line voltages for classical MPC and M2PC. FFT analysis for the A phase current under classical MPC strategy and M2PC strategy are shown in Fig.13 and 14. The fundamental frequency is 120Hz, A phase current is analyzed in two cycles. It can be seen from the Fig.13 and 14, classical MPC strategy has a higher THD value than M2PC strategy in the same sample time 50e-6. The M2PC has a higher switching frequency than classical MPC.

978-1-7281-4181-7/19 $31.00 © 2019 IEEE

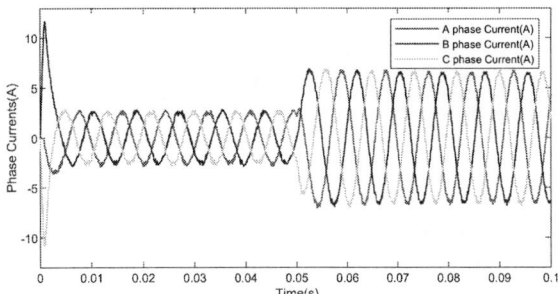

Fig. 7. Phase currents responses under classical MPC current control strategy in the presence of load torque disturbance at 1800rpm.

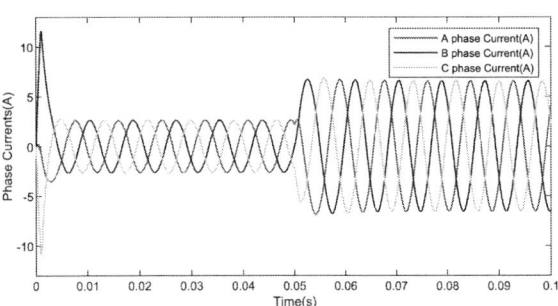

Fig. 8. Phase currents responses under M2PC current control strategy in the presence of load torque disturbance at 1800rpm.

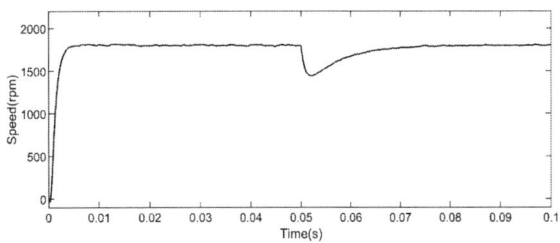

Fig. 9. Speed response under classical MPC current control strategy in the presence of load torque disturbance at 1800rpm.

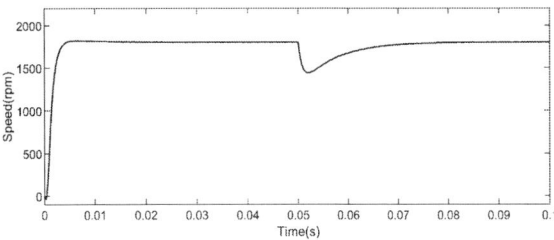

Fig. 10. Speed response under M2PC current control strategy in the presence of load torque disturbance at 1800rpm.

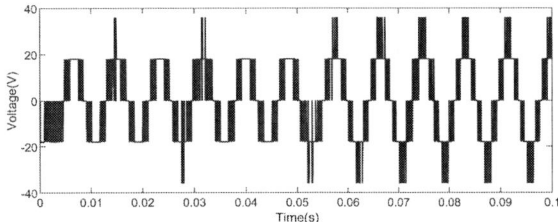

Fig. 11. Line voltage of inverter by classical MPC current control strategy.

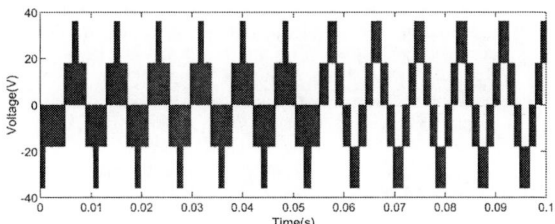

Fig. 12. Line voltage of inverter by M2PC current control strategy.

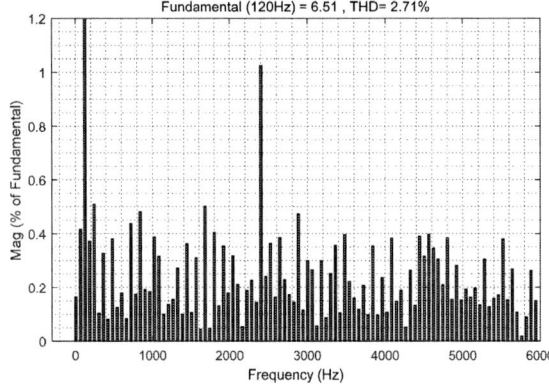

Fig. 13. FFT analysis for the A phase current by classical MPC strategy.

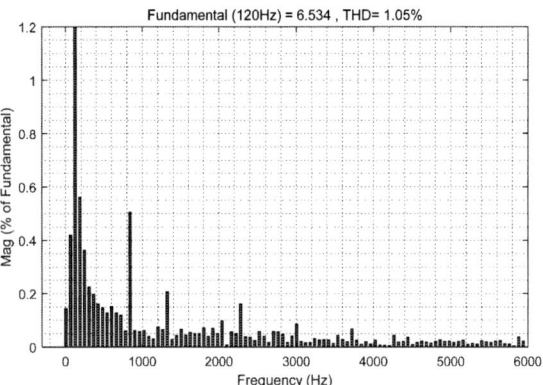

Fig. 14. FFT analysis for the A phase current by M2PC strategy.

VI. CONCLUSIONS

A M2PC strategy for a three-level NPC inverter feeding a PMSM was proposed in this paper. Compared with the classical MPC strategy, the proposed M2PC strategy produced a modulated waveform to reduce the THD value, and has a smaller current and torque ripple. The simulation was implemented in Matlab2018a, and the results show the feasibility and effectiveness of M2PC strategy.

ACKNOWLEDGMENT

The authors would like to thank the financial support of the Clean Sky 2 Joint Undertaking under Grant 807081 and the Fondecyt Regular Project 1160690.

REFERENCES

[1] S. Janous, D. Janik, T. Kosan, P. Kamenický, and Z. Peroutka, "Comparative study of vector PWM and FS-MPC for 3-level Neutral Point Clamped converter," in *Proceedings of the 16th International Conference on Mechatronics - Mechatronika 2014*, 2014, pp. 158–163.

[2] M. Rivera *et al.*, "Modulated Model Predictive Control (M2PC) with fixed switching frequency for an NPC converter," in *2015 IEEE 5th International Conference on Power Engineering, Energy and Electrical Drives (POWERENG)*, Riga, Latvia, 2015, pp. 623–628.

[3] J. Rodriguez, S. Bernet, P. K. Steimer, and I. E. Lizama, "A Survey on Neutral-Point-Clamped Inverters," *IEEE Transactions on Industrial Electronics*, vol. 57, no. 7, pp. 2219–2230, Jul. 2010.

[4] "Model Predictive Control," in *Predictive Control of Power Converters and Electrical Drives*, Chichester, UK: John Wiley & Sons, Ltd, 2012, pp. 31–39.

[5] S. Kouro, P. Cortes, R. Vargas, U. Ammann, and J. Rodriguez, "Model Predictive Control—A Simple and Powerful Method to Control Power Converters," *IEEE Transactions on Industrial Electronics*, vol. 56, no. 6, pp. 1826–1838, Jun. 2009.

[6] H. Miranda, P. Cortes, J. I. Yuz, and J. Rodriguez, "Predictive Torque Control of Induction Machines Based on State-Space Models," *IEEE Transactions on Industrial Electronics*, vol. 56, no. 6, pp. 1916–1924, Jun. 2009.

[7] J. Rodriguez *et al.*, "State of the Art of Finite Control Set Model Predictive Control in Power Electronics," *IEEE Transactions on Industrial Informatics*, vol. 9, no. 2, pp. 1003–1016, May 2013.

[8] S. Vazquez *et al.*, "Model Predictive Control: A Review of Its Applications in Power Electronics," *IEEE Industrial Electronics Magazine*, vol. 8, no. 1, pp. 16–31, Mar. 2014.

[9] M. Rivera *et al.*, "A Comparative Assessment of Model Predictive Current Control and Space Vector Modulation in a Direct Matrix Converter," *IEEE Transactions on Industrial Electronics*, vol. 60, no. 2, pp. 578–588, Feb. 2013.

[10] T. Geyer and S. Mastellone, "Model Predictive Direct Torque Control of a Five-Level ANPC Converter Drive System," *IEEE Transactions on Industry Applications*, vol. 48, no. 5, pp. 1565–1575, Sep. 2012.

[11] V. Yaramasu and B. Wu, "Predictive Control of a Three-Level Boost Converter and an NPC Inverter for High-Power PMSG-Based Medium Voltage Wind Energy Conversion Systems," *IEEE Transactions on Power Electronics*, vol. 29, no. 10, pp. 5308–5322, Oct. 2014.

[12] Y. Yang, H. Wen, M. Fan, M. Xie, and R. Chen, "Fast Finite-Switching-State Model Predictive Control Method Without Weighting Factors for T-Type Three-Level Three-Phase Inverters," *IEEE Transactions on Industrial Informatics*, vol. 15, no. 3, pp. 1298–1310, Mar. 2019.

[13] Y. Zhang and H. Yang, "Two-Vector-Based Model Predictive Torque Control Without Weighting Factors for Induction Motor Drives," *IEEE Transactions on Power Electronics*, vol. 31, no. 2, pp. 1381–1390, Feb. 2016.

[14] P. Kakosimos and H. Abu-Rub, "Deadbeat Predictive Control for PMSM Drives With 3-L NPC Inverter Accounting for Saturation Effects," *IEEE Journal of Emerging and Selected Topics in Power Electronics*, vol. 6, no. 4, pp. 1671–1680, Dec. 2018.

[15] A. D. Alexandrou, N. K. Adamopoulos, and A. G. Kladas, "Development of a Constant Switching Frequency Deadbeat Predictive Control Technique for Field-Oriented Synchronous Permanent-Magnet Motor Drive," *IEEE Transactions on Industrial Electronics*, vol. 63, no. 8, pp. 5167–5175, Aug. 2016.

[16] Y. Jiang, W. Xu, C. Mu, and Y. Liu, "Improved Deadbeat Predictive Current Control Combined Sliding Mode Strategy for PMSM Drive System," *IEEE Transactions on Vehicular Technology*, vol. 67, no. 1, pp. 251–263, Jan. 2018.

[17] X. Zhang, B. Hou, and Y. Mei, "Deadbeat Predictive Current Control of Permanent-Magnet Synchronous Motors with Stator Current and Disturbance Observer," *IEEE Transactions on Power Electronics*, vol. 32, no. 5, pp. 3818–3834, May 2017.

[18] L. Tarisciotti, P. Zanchetta, A. Watson, J. C. Clare, M. Degano, and S. Bifaretti, "Modulated Model Predictive Control for a Three-Phase Active Rectifier," *IEEE Transactions on Industry Applications*, vol. 51, no. 2, pp. 1610–1620, Mar. 2015.

[19] L. Tarisciotti, P. Zanchetta, A. Watson, S. Bifaretti, and J. C. Clare, "Modulated model predictive control for a seven-level cascaded h-bridge back-to-back converter," *IEEE Transactions on Industrial Electronics*, vol. 61, no. 10, pp. 5375–5383, 2014.

[20] L. Tarisciotti, P. Zanchetta, A. Watson, P. Wheeler, J. C. Clare, and S. Bifaretti, "Multiobjective Modulated Model Predictive Control for a Multilevel Solid-State Transformer," *IEEE Transactions on Industry Applications*, vol. 51, no. 5, pp. 4051–4060, Sep. 2015.

[21] X. Li *et al.*, "A Modulated Model Predictive Control Scheme for the Brushless Doubly Fed Induction Machine," *IEEE Journal of Emerging and Selected Topics in Power Electronics*, vol. 6, no. 4, pp. 1681–1691, Dec. 2018.

[22] F. Donoso, A. Mora, R. Cárdenas, A. Angulo, D. Sáez, and M. Rivera, "Finite-Set Model-Predictive Control Strategies for a 3L-NPC Inverter Operating With Fixed Switching Frequency," *IEEE Transactions on Industrial Electronics*, vol. 65, no. 5, pp. 3954–3965, May 2018.

[23] A. R. Beig, G. Narayanan, and V. T. Ranganathan, "Modified SVPWM Algorithm for Three Level VSI With Synchronized and Symmetrical Waveforms," *IEEE Transactions on Industrial Electronics*, vol. 54, no. 1, pp. 486–494, Feb. 2007.

[24] X. Gao, W. Tian, X. Liu, Z. Zhang, and R. Kennel, "Model Predictive Control of a Three-Level NPC Rectifier with a Sliding Manifold Term," in *2018 International Power Electronics Conference (IPEC-Niigata 2018 -ECCE Asia)*, Niigata, 2018, pp. 1661–1665.

[25] B. A. Welchko, M. B. de Rossiter Correa, and T. A. Lipo, "A Three-Level MOSFET Inverter for Low-Power Drives," *IEEE Transactions on Industrial Electronics*, vol. 51, no. 3, pp. 669–674, Jun. 2004.

[26] W.-S. Oh, S.-K. Han, S.-W. Choi, and G.-W. Moon, "Three Phase Three-Level PWM Switched Voltage Source Inverter With Zero Neutral Point Potential," *IEEE Transactions on Power Electronics*, vol. 21, no. 5, pp. 1320–1327, Sep. 2006.

[27] A. Al-Janabi, *Vector Control Of Permanent Magnet Synchronous Motor: LEARN Types of Motors and Application. The PMSM motor Applications. MATLAB simulation of PMSM with vector control idea.* LAP LAMBERT Academic Publishing, 2016.

Model Predictive Controller for Two-Phase Three-Wire Grid-Connected Converters

Pablo C. de S. Furtado[1,2]

[1]*Federal Institute of Education, Science and Technology of Southeast of Minas Gerais*
Santos Dumont, Brazil
pablo.furtado@ifsudestemg.edu.br

Pedro Gomes Barbosa[2]

[2]*Power Electronics and Automation Group (NAEP)*
Federal University of Juiz de Fora
Juiz de Fora, Brazil
pedro.gomes@ufjf.edu.br

Abstract—This paper presents an adaptation of a Model Predictive Controller for current control in two-phase three-wire grid connected converters. The discrete model of the system and the controller algorithm are described in detail. Fixed switching frequency is achieved by integrating the proposed predictive controller with Space Vector Modulation. This integration eases the implementation in digital signal controllers. Simulation studies were performed to test the suitability of this controller in tracking sinusoidal and nonsinusoidal current references. Simulation results are presented to confirm the effectiveness of the proposed controller.

Index Terms—Grid-connected converters, model predictive control, two-phase three-wire networks.

I. INTRODUCTION

Modern electric networks, commonly regarded in literature as Smart Grids (SGs), are the product of constant modernization aiming to improve efficiency, reliability, safety, cost-effectiveness and quality of the service provided to consumers [1]–[4]. Technical advances with these objectives have lead to the introduction of modern resources into electric networks. Examples of these resources are the distributed generation, renewable and alternative energy sources, smart meters for bidirectional and dynamic billing, electric vehicles, energy storage, active power filters, microgrids, as well as communication, data processing, automation and decision making systems [1]–[4].

The introduction of these technologies into SGs results in the massive presence of Grid-Connected Converters (GCCs), which are controlled for grid synchronization, power flow control and power quality improvement, for example. Most works in technical literature describe converter topologies and control schemes for single-phase and three-phase GCCs [5]–[7].

In practice, however, there are grid ramifications and consumer installations fed with two phases and one neutral conductor derived from three-phase four-wire grids. A simple $2\phi 3w$ system is illustrated in Figure 1, which shows a three-phase four-wire power grid feeding a $2\phi 3w$ load and a shunt GCC. The common point between grid, load and the GCC is called Point of Common Coupling (PCC). This kind of circuits are regarded in this work as two-phase three-wire ($2\phi 3w$) or simply two-phase. It is the case of residential and small commercial consumer installations, for example. In this context,

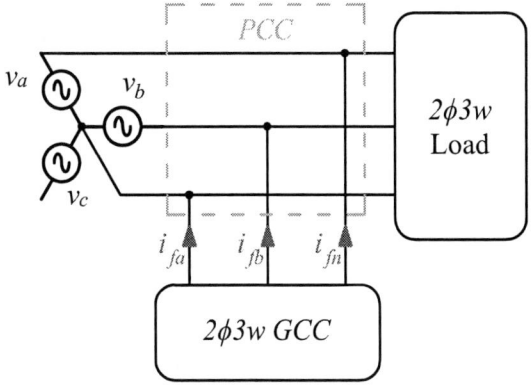

Figure 1: Two-phase three-wire.

the development of specific topologies and control schemes for $2\phi 3w$ power converters may contribute do the introduction of SG resources into small consumer installations. This work deals specifically with the control of the currents synthesized by a shunt $2\phi 3w$ GCC, represented in Figure 1 as i_{fa}, i_{fb} and i_{fn}. For simplicity, the two-phase converter is treated in this work as an Active Power Filter (APF). However, the current control method presented here can be applied whichever is the function performed by the $2\phi 3w$ GCC. It is possible to find in technical literature a wide variety of techniques for current control in shunt grid-connected converters. As examples of linear controllers, there are Proportional-Integral (PI) controllers in synchronous reference frame (PI-SRF), with multiple rotating integrators (PI-MRI) and with resonant integrators in stationary reference frame (PI-RES) [8], [9]. Non-linear techniques are also described for current control, including hysteresis controllers [10], Sliding-Mode Controllers (SMCs) [11], Model Predictive Controllers (MPCs) [12] and controllers based on artificial intelligence [13].

According to [12], MPCs present advantages like being based in simple and intuitive concepts, simple digital implementation, making no restrictions to the behaviour of reference signals and dealing naturally with the nonlinear nature of power converters. Due to these characteristics, MPCs showed to be an interesting option for current control in applications

978-1-7281-4181-7/19 $31.00 © 2019 IEEE

of $2\phi3w$ GCCs.

In this context, this work aims to describe a MPC to control the output currents of a shunt APF with two-phases and neutral conductors. The proposed controller also presents features of dead-time compensation and fixed switching frequency. Section II aims to describe the system considered in this study. The $2\phi3w$ MPC proposed in this work is described in detail in Section III. Simulation results demonstrating the effectiveness of the described $2\phi3w$ MPC are presented and discussed in Section IV. Finally, the conclusions achieved in this work are discussed in Section V.

II. SYSTEM MODEL

The design of a MPC begins by obtaining a discrete model of the system to be controlled [12], [14]. The topology of the system considered in this work is illustrated in Figure 2. The voltages v_a and v_b represent phase-to-neutral voltages of a three-phase four-wire power grid. The $2\phi3w$ APF is connected to phase a, phase b and neutral at PCC. The APF is composed by a three-leg Voltage Sourced Converter (VSC), dc-bus capacitance C_f, and inductive output filters represented by the inductances L_f and internal resistances r_f. The APF output voltages, measured in relation to the neutral point N, are represented by v_{fa}, v_{fb} and v_{fn}.

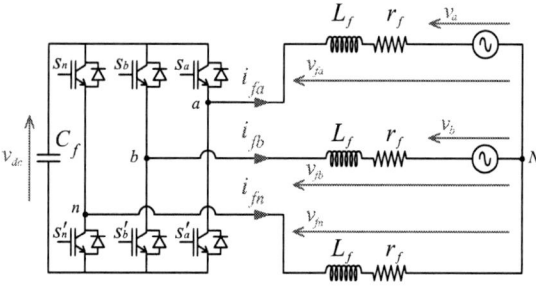

Figure 2: System topology used to design the predictive current controller.

The semiconductor switches in the VSC must be driven so that currents i_{fa}, i_{fb} and i_{fn} track the reference signals i_{fa}^*, i_{fb}^* and i_{fn}^*, respectively. Thus, the current controller described in this work must generate the binary pattern signals S_a, S_b and S_n, as well as their complementaries S_a', S_b' and S_n'. It is worth to mention that the method used to calculate reference currents for $2\phi3w$ grid-connected converters is not the scope of this work. Other works in technical literature [15], [16] deal with this subject in detail.

As shown in Figure 2, the VSC employed for grid-connection in $2\phi3w$ networks is composed by three semiconductor legs. The upper and lower switches in each leg are driven in a complementary manner. It follows that the VSC can be set in eight different states, also referred as switching states [14]. These states are denoted by x and listed from $x = 1$ to $x = 8$ in Table I. This table also presents the voltage values at VSC output terminals for each switching state, adopting the grid neutral conductor as voltage reference. Despite the

Table I
VSC output voltages (abc).

Switching state (x)	Switches States			VSC Output		
	S_a	S_b	S_n	$v_{fa}(x)$	$v_{fb}(x)$	$v_{fn}(x)$
1	0	0	0	0	0	0
2	1	0	0	$\frac{2}{3}v_{dc}$	$-\frac{1}{3}v_{dc}$	$-\frac{1}{3}v_{dc}$
3	1	1	0	$\frac{1}{3}v_{dc}$	$\frac{1}{3}v_{dc}$	$-\frac{2}{3}v_{dc}$
4	0	1	0	$-\frac{1}{3}v_{dc}$	$\frac{2}{3}v_{dc}$	$-\frac{1}{3}v_{dc}$
5	0	1	1	$-\frac{2}{3}v_{dc}$	$\frac{1}{3}v_{dc}$	$\frac{1}{3}v_{dc}$
6	0	0	1	$-\frac{1}{3}v_{dc}$	$-\frac{1}{3}v_{dc}$	$\frac{2}{3}v_{dc}$
7	1	0	1	$\frac{1}{3}v_{dc}$	$-\frac{2}{3}v_{dc}$	$\frac{1}{3}v_{dc}$
8	1	1	1	0	0	0

Table II
VSC output voltages (two-phase $\alpha\beta$).

Switching state (x)	VSC Output	
	$v_{f\alpha}(x)$	$v_{f\beta}(x)$
1	0	0
2	$\frac{2}{3}v_{dc}$	0
3	$\frac{1}{3}v_{dc}$	$\frac{\sqrt{3}}{3}v_{dc}$
4	$-\frac{1}{3}v_{dc}$	$\frac{\sqrt{3}}{3}v_{dc}$
5	$-\frac{2}{3}v_{dc}$	0
6	$-\frac{1}{3}v_{dc}$	$-\frac{\sqrt{3}}{3}v_{dc}$
7	$\frac{1}{3}v_{dc}$	$-\frac{\sqrt{3}}{3}v_{dc}$
8	0	0

different switching signals, it can be observed that the states 1 and 8 produce the same output voltages, resulting in seven different voltages that can be applied at the VSC output.

Aiming to deal with $2\phi3w$ circuits as an unity, a linear transformation was presented in [16] to take ab quantities into the orthogonal $\alpha\beta$ reference frame. This transformation is expressed as

$$\begin{bmatrix} u_\alpha \\ u_\beta \end{bmatrix} = \frac{1}{\sqrt{3}} \begin{bmatrix} \sqrt{3} & 0 \\ 1 & 2 \end{bmatrix} \begin{bmatrix} u_a \\ u_b \end{bmatrix}, \qquad (1)$$

where u_α and u_β represent the voltages or currents expressed in the $2\phi3w$ $\alpha\beta$ reference frame.

By applying this transformation to the voltages at Table I, one obtains the VSC output voltages represented in the two-phase $\alpha\beta$ reference frame, as presented in Table II. The voltages in this table equals the output voltages presented by [14], despite they were calculated by different methods. The eight possible VSC switching states are graphically represented as voltage vectors, in abn and $\alpha\beta$ coordinates, in Figure 3.

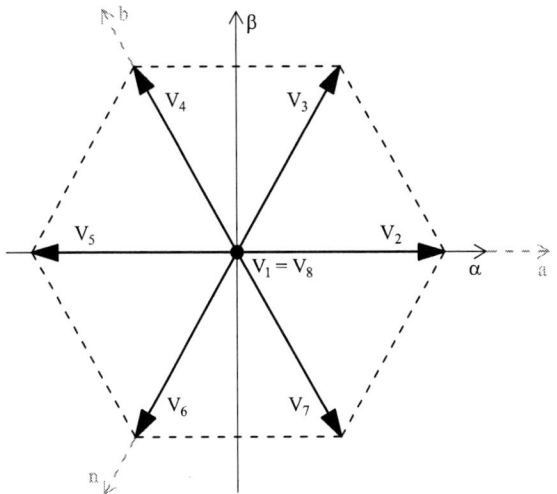

Figure 3: The eight voltage vectors which can be produced by a three-leg VSC.

The differential equations which describe mathematically the behaviour of the system illustrated in Figure 2 are

$$
\begin{cases}
v_{fa}(t) = & L_f \dfrac{d}{dt} i_{fa}(t) + r_f i_{fa}(t) + v_a(t) \\[2mm]
v_{fb}(t) = & L_f \dfrac{d}{dt} i_{fb}(t) + r_f i_{fb}(t) + v_b(t) \\[2mm]
v_{fn}(t) = & L_f \dfrac{d}{dt} i_{fn}(t) + r_f i_{fn}(t)
\end{cases} \quad (2)
$$

These model equations differ from a three-phase system because there is no grid voltage in the third leg, which in this case is referred to the neutral conductor.

By applying the two-phase $\alpha\beta$ transformation (1) to the system of equations in (2), one obtains the system model expressed in the orthogonal $\alpha\beta$ reference frame as

$$
\begin{cases}
v_{f\alpha}(t) = & L_f \dfrac{d}{dt} i_{f\alpha}(t) + r_f i_{f\alpha}(t) + v_\alpha(t) \\[2mm]
v_{f\beta}(t) = & L_f \dfrac{d}{dt} i_{f\beta}(t) + r_f i_{f\beta}(t) + v_\beta(t)
\end{cases} \quad (3)
$$

This step shows an advantage of modelling $2\phi3w$ systems using the linear transformation (1). The system of differential equations (2) is simplified to a two-equation system which does not depend on quantities associated to the neutral conductor. Only the voltages and currents of phases a and b are needed to implement the MPC described in this work, reducing the need of sensors and the number of calculations in digital implementations.

Aiming to obtain a discrete-time model, a forward Euler approximation is employed to calculate derivatives

$$
\frac{d}{dt} i(t) = \frac{i(k+1) - i(k)}{T_s}, \quad (4)
$$

where T_s and k are the the sampling period and the discrete time variable, respectively. Thus, the discrete model of system is obtained by changing the time variable and substituting (4) into (3).

III. TWO-PHASE MODEL PREDICTIVE CONTROLLER

The MPC presented in this work is a $2\phi3w$ adaptation of the controller reported by [14]. Basically, the controller employs the discrete mathematical model described in Section II to predict future values of controlled variables for each of the possible output states the VSC may assume [12]. These predictions are used to evaluate a predefined cost function for each possible VSC output state. The VSC output state to be applied at each sampling period is the one which implies the minimum value of the cost function.

It follows from the discrete model described in Section II that the predicted values of currents for each possible VSC output state x can be evaluated by

$$
\begin{cases}
i^p_{f\alpha}(k+1) = & \left(1 - \dfrac{r_f T_s}{L_f}\right) i_{f\alpha}(k) + \\[2mm]
& + \dfrac{T_s}{L_f} \left[v_{f\alpha}(x) - v_\alpha(k)\right] \\[4mm]
i^p_{f\beta}(k+1) = & \left(1 - \dfrac{r_f T_s}{L_f}\right) i_{f\beta}(k) + \\[2mm]
& + \dfrac{T_s}{L_f} \left[v_{f\beta}(x) - v_\beta(k)\right]
\end{cases} \quad (5)
$$

where $i^p_{f\alpha}(k+1)$ and $i^p_{f\beta}(k+1)$ are the predicted values of currents in the α and β axis, respectively.

Aiming to compensate the time delay between the data acquisition and the actual application of the selected VSC output, the MPC in this work also includes a dead-time compensation technique [12]. Aiming to implement this feature, currents must be predicted two sampling periods in the future:

$$
\begin{cases}
i^p_{f\alpha}(k+2) = & \left(1 - \dfrac{r_f T_s}{L_f}\right) i^p_{f\alpha}(k+1) + \\[2mm]
& + \dfrac{T_s}{L_f} \left[v_{f\alpha}(x) - v_\alpha(k)\right] \\[4mm]
i^p_{f\beta}(k+2) = & \left(1 - \dfrac{r_f T_s}{L_f}\right) i^p_{f\beta}(k+1) + \\[2mm]
& + \dfrac{T_s}{L_f} \left[v_{f\beta}(x) - v_\beta(k)\right]
\end{cases} \quad (6)
$$

where $i^p_{f\alpha}(k+2)$ and $i^p_{f\beta}(k+2)$ are the α and β values of predicted currents at the sampling period $k+2$.

After predicting currents for each VSC state x, a cost function is evaluated. The cost function $g(x)$ used in this work is a measurement of the future current error if each switching

978-1-7281-4181-7/19 $31.00 © 2019 IEEE

state x is applied at the present sampling period. This cost function is composed by the sum of absolute errors in the α and β reference frames, considering that reference signals do not change significantly in two sampling periods. The cost function is expressed by:

$$g\left(x\right) = \left| i_{f\alpha}^{*}\left(k\right) - i_{f\alpha}^{p}\left(k+2\right)\right| + \\ + \left| i_{f\beta}^{*}\left(k\right) - i_{f\beta}^{p}\left(k+2\right)\right| \tag{7}$$

After the cost function is evaluated for $x = 1 \ldots 8$, the value of x which returns the optimum (i.e., minimum) value of $g(x)$ is stored as x_{opt}.

According to the general MPC presented by [14], the switches states are set directly by setting s_a, s_b and s_n. This working mode leads to variable switching frequency, which causes the current spectrum to present a wide range of high frequency content due to switching [12], [17]. This characteristic can cause problems such as interferences and difficult in the design of passive filters [12], [17]. Aiming to overcome this issue, [18] proposed the use of a band-stop filter at the cost function. A different approach found in literature is the integration of MPC and the fixed switching frequency provided by the Space Vector Modulation (SVM) [17]. In this configuration, the MPC selects the voltage which minimizes the current error. This voltage reference is provided to the SVM algorithm, which generates the drive signals for VSC switches. SVM presents advantages such as the fixed switching frequency, optimization of converter hardware utilization [19] and it is so well established that optimal SVM algorithms can be found ready for use in digital signal controllers. Thus, this work adopts SVM to generate the VSC output voltages which minimizes the cost function (i.e., the current error) at each sampling period.

Finally, the complete algorithm for the implementation of the $2\phi3w$ MPC described in this work is illustrated in the block diagram of Figure 4.

IV. SIMULATION RESULTS

The system illustrated in Figure 1 and the $2\phi3w$-adapted MPC were implemented in a digital simulation software for validation. The parameters of converter output filters are $r_f = 50$ mΩ and $L_f = 2$ mH. The grid voltages at PCC v_a and v_b are sinusoidal with 180 V$_{pk}$ and phase displaced by $2\pi/3$ rad, whereas the DC voltage was set constant at 400 V. Despite sampling and switching frequencies can be set different in the proposed control scheme, they were both set equal to 32 kHz in these simulation studies.

The first simulation deals with sinusoidal current reference tracking. The current references i_{fa}^{*} and i_{fb}^{*} were provided to the controller with peak values of 20 A. At $t = 33$ ms these references were increased by 50 %. The results obtained in this simulation are shown for phase a, phase b and neutral in Figures 5(a), 5(b) and 5(c), respectively. These figures compare the GCC output currents with their reference signals. The reference for the neutral current, which is not used directly

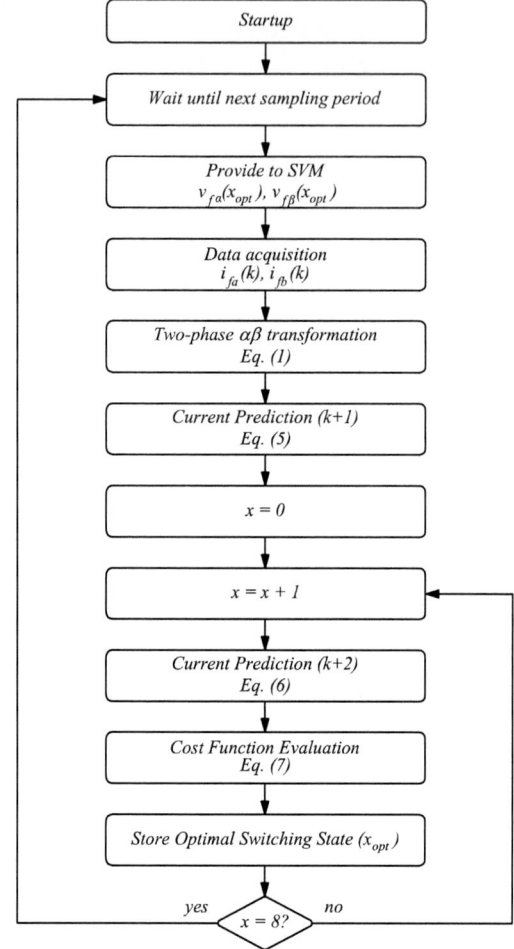

Figure 4: Algorithm of the $2\phi3w$ MPC.

in the control system, was calculated for comparison purpose as

$$i_{fn}^{*} = i_{fa}^{*} + i_{fb}^{*}. \tag{8}$$

The detail in Figure 5(d) shows the behaviour of the controller during the simulated transient in the current reference. It is possible to observe that VSC output currents follow adequately their reference signals.

The second simulation deals with non-sinusoidal current reference tracking. The current references i_{fa}^{*} and i_{fb}^{*} were provided to the controller with peak values of 20 A at the fundamental frequency. At $t = 33$ ms this component was increased by 50 %. Harmonic components of 3rd, 5th and 7th orders were included. The results obtained in this simulation are shown for phase a, phase b and neutral in Figures 6(a), 6(b) and 6(c), respectively. The detail in Figure 6(d) shows the behaviour of the controller during the simulated transient in the current reference. It is possible to observe that VSC output currents follow adequately their reference signals.

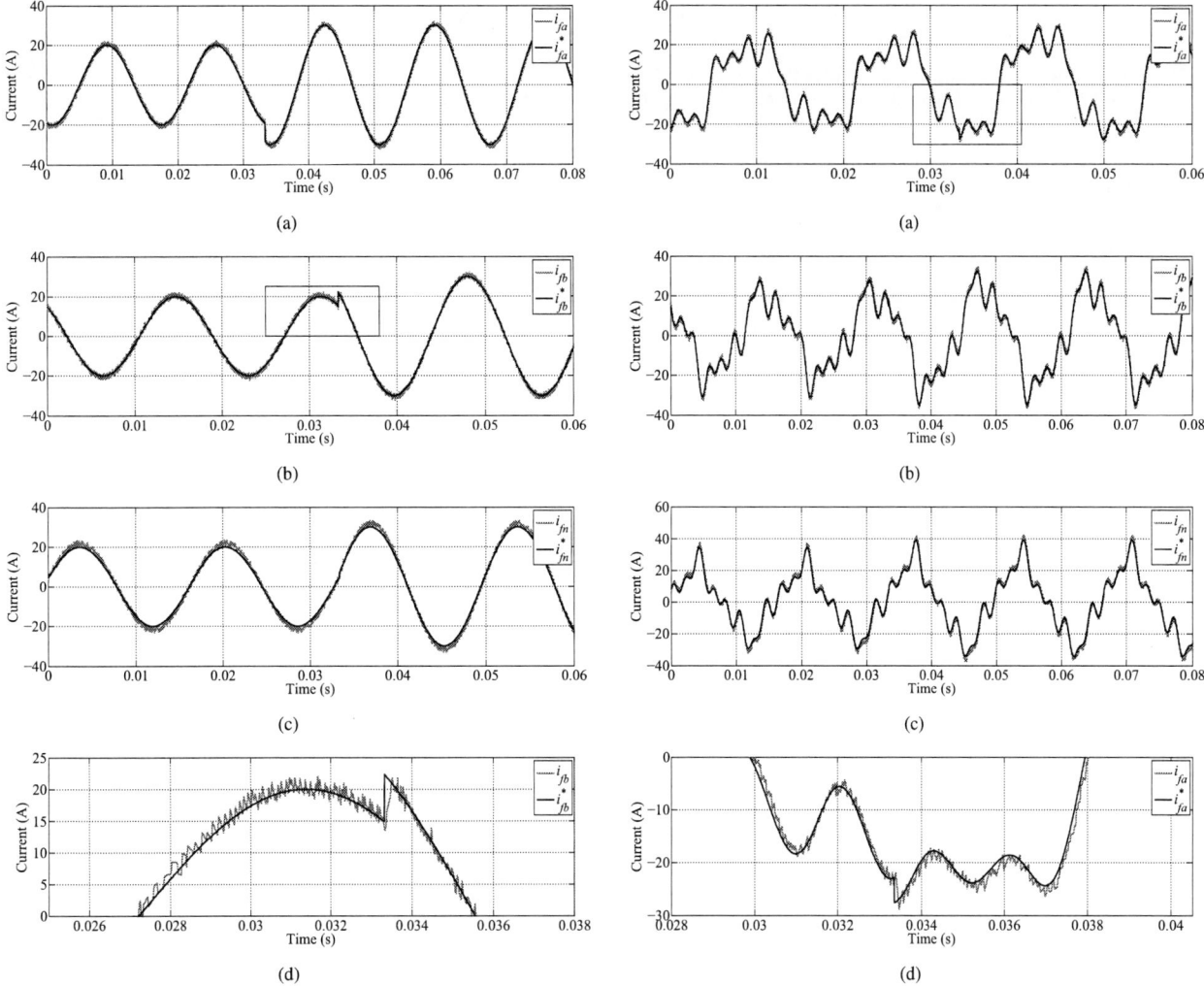

Figure 5: Sinusoidal reference current tracking: (a) phase a, (b) phase b, (c) neutral and (d) detail of phase b.

Figure 6: Non-sinusoidal reference current tracking: (a) phase a, (b) phase b, (c) neutral and (d) detail of phase b.

V. CONCLUSION

This work described an adaptation of a MPC for current control in $2\phi3w$ grid-connected converters. This approach provides a controller capable of naturally dealing with $2\phi3w$ quantities and systems, whereas other controllers described in literature typically deal with three-phase or single-phase systems. An advantage of the $2\phi3w$ MPC is that, in addition to v_{dc}, only 2 voltages (v_a, v_b) and 2 currents (i_{fa}, i_{fb}) must be measured. The current in the neutral leg of the converter is controlled indirectly. The integration of the MPC with SVM showed to be an interesting approach to guarantee fixed switching frequency. Simulation results showed the effectiveness of the described controller in tracking sinusoidal and non-sinusoidal current references. The controller also presented stable behaviour after step changes in reference signals.

ACKNOWLEDGEMENT

Authors would like to thank the Brazilian agencies FAPEMIG, CNPq and CAPES for the financial support of this work.

REFERENCES

[1] Math HJ Bollen, Jin Zhong, Francisc Zavoda, Jan Meyer, Alex McEachern, and Felipe Córcoles López. Power quality aspects of smart grids. In *International conference on renewable energies and power quality (ICREPQ'10)*, pages 1–6, March 2010.

[2] Hassan Farhangi. The path of the smart grid. *IEEE Power and Energy Magazine*, 8(1):18–28, 2010.

[3] Cédric Clastres. Smart grids: another step towards competition, energy security and climate change objectives. *Energy Policy*, 39(9):5399–5408, 2011.

[4] U.S. Department of Energy. The smart grid: an introduction. 2015.

[5] Amirnaser Yazdani and Reza Iravani. *Voltage-sourced converters in power systems*. John Wiley & Sons, 2010.

[6] Remus Teodorescu, Marco Liserre, and Pedro Rodriguez. *Grid converters for photovoltaic and wind power systems*, volume 29. John Wiley & Sons, 2011.

978-1-7281-4181-7/19 $31.00 © 2019 IEEE

[7] Sudipta Chakraborty, Marcelo G Simões, and William E Kramer. *Power electronics for renewable and distributed energy systems*. Springer, 2013.

[8] Marian P Kazmierkowski and Luigi Malesani. Current control techniques for three-phase voltage-source pwm converters: A survey. *IEEE Transactions on industrial electronics*, 45(5):691–703, 1998.

[9] Frederico T Ghetti, Pedro G Barbosa, Henrique AC Braga, and André A Ferreira. Estudo comparativo de tecnicas de controle de corrente aplicadas a filtros ativos shunt. In *XVIII Congresso Brasileiro de Automática*, pages 12–16, 2010.

[10] C. S. Lam, M. C. Wong, and Y. D. Han. Hysteresis current control of hybrid active power filters. *IET Power Electronics*, 5(7):1175–1187, August 2012.

[11] G. A. Fogli, P. M. de Almeida, V. M. L. Rodrigues, and P. G. Barbosa. Sliding mode control of a shunt active power filter with indirect current measurement. In *2015 IEEE 13th Brazilian Power Electronics Conference and 1st Southern Power Electronics Conference (COBEP/SPEC)*, pages 1–5, Nov 2015.

[12] Jose Rodriguez and Patricio Cortes. *Predictive control of power converters and electrical drives*. John Wiley & Sons, 2012.

[13] M. Qasim and V. Khadkikar. Application of artificial neural networks for shunt active power filter control. *IEEE Transactions on Industrial Informatics*, 10(3):1765–1774, Aug 2014.

[14] J. Rodriguez, J. Pontt, C. A. Silva, P. Correa, P. Lezana, P. Cortes, and U. Ammann. Predictive current control of a voltage source inverter. *IEEE Transactions on Industrial Electronics*, 54(1):495–503, Feb 2007.

[15] P. C. de S. Furtado, M. do C. B. P. Rodrigues, H. A. C. Braga, and P. G. Barbosa. Two-phase three-wire shunt active power filter control by using the single-phase p-q theory. *Eletrnica de Potncia*, 19(3):303–311, Aug 2014.

[16] P. C. de S. Furtado, M. do C. B. P. Rodrigues, and P. G. Barbosa. Adaptation of the instantaneous power theory for two-phase three-wire systems and its application in shunt active power filters. In *2015 IEEE 13th Brazilian Power Electronics Conference and 1st Southern Power Electronics Conference (COBEP/SPEC)*, pages 1–5, Nov 2015d.

[17] S Vazquez, JI Leon, LG Franquelo, JM Carrasco, O Martinez, José Rodriguez, P Cortes, and Samir Kouro. Model predictive control with constant switching frequency using a discrete space vector modulation with virtual state vectors. In *2009 IEEE International Conference on Industrial Technology*, pages 1–6. IEEE, 2009.

[18] Patricio Cortés, José Rodríguez, Daniel E Quevedo, and Cesar Silva. Predictive current control strategy with imposed load current spectrum. *IEEE Transactions on Power Electronics*, 23(2):612–618, 2008.

[19] Simone Buso and Paolo Mattavelli. *Digital control in power electronics*. Morgan & Claypool, 2006.

A NEW ZERO-SEQUENCE VOLTAGE COMPENSATION ALGORITHM FOR A DSTATCOM BASED ON CONSUMER UNBALANCE

Bruno C. Souza, Samuel N. Duarte, Pedro M. Almeida, Pedro G. Barbosa, Leandro R. Araújo

Federal University of Juiz de Fora, Graduate Program in Electrical Engineering, Juiz de Fora–MG, Brazil
Email: bruno.cortes@engenharia.ufjf.br, samuel.neves@engenharia.ufjf.br,
pedro.machado@ufjf.edu.br, pedro.gomes@ufjf.edu.br, leandro.araujo@ufjf.edu.br

Abstract - **This paper presents a methodology to identify in "real-time" the parcels of responsibility of the utility and the consumer for the zero-sequence unbalance at the point of common coupling. The output signal of the proposed algorithm is used to generate the reference signals for the voltage control-loop of a four-wire three-phase distribution static synchronous compensator. Digital simulation results are used to validate the presented methodology. In addition to the proposed algorithm to identify the responsibility of the consumer unbalance, the capacity of the static compensator can be designed to only compensate for the zero-sequence voltage in the PCC caused by the consumer.**

Keywords – **DSTATCOM, Unbalance Responsibility, Voltage Compensation, Zero-sequence.**

I. INTRODUCTION

Distribution networks commonly operate with unbalanced voltages due to the presence of single-phase and two-phase loads, asymmetries on feeder impedances and short-circuits in motor and transformer coils. These electric networks has become a very discussed topic due to the ascendancy of smart grids. The growth in the number of battery chargers for electric vehicles and the penetration of distributed generation (DG) systems, (*e.g.*, photovoltaic systems) may also increase the unbalance and losses in distribution networks, besides compromising the voltage security [1–3]. The voltage imbalances cause the appearance of negative and zero-sequence voltage components that are responsible for producing pulsating torques, increasing losses and causing speed variation in electrical motors [4].

In this way, several power quality indexes and connection guidelines were established by technical agencies to ensure the safe operation of electric networks [2]. For example, the International Electrotechnical Commission (IEC) [5] proposes that the voltage unbalance in distribution networks should not exceed 2 %, while the National Electrical Manufacturers Association (NEMA) [6] recommends that the voltage unbalance at the electrical motors' terminals should be less than 1 %. Concerned about these issues, the National Electric Energy Agency (ANEEL) [7] and the National Electric System Operator (ONS) [8, 9] established procedures and power quality indexes for the Brazilian electric network. It emphasizes that the compliance with the recommended limits must be followed to prevent malfunction and turn off of critical loads [10].

Although balancing the loads along the feeders of distribution networks would come up as a simple solution, its practical application is limited due to the load variation. On the other hand, specialists have been discussing the economic penalization for consumers that are responsible for compromising the power quality indexes of the grid [11–13]. For this achievement, it is important to quantify the individual responsibilities between the utility and consumer on the power quality indexes. Some papers have already been discussed the harmonic pollution responsibility [11] and most of these methodologies can also be adapted to identify the voltage and current unbalance responsibility in distribution networks [13,14]. However, none of them deals with zero-sequence components. Furthermore, these methodologies can not be used in real-time operation.

Therefore, this paper presents a new methodology to calculate, in real-time, the zero-sequence voltage unbalance, at the fundamental frequency, caused by a consumer connected to a distribution network. The output signal of the proposed algorithm is used to generate the reference for a four-wire three-phase DSTATCOM used to compensate for the zero-sequence voltage at PCC. Results of digital simulations are used to validate the proposed methodology.

II. UNBALANCE RESPONSIBILITY METHODOLOGY

The zero-sequence voltage unbalance responsibility methodology developed in this work is based on the adaptation of [11] proposed for harmonic distortion. However, the method presented here differs from the previous one because it does not require measurements of the consumer before its connection to the network. The proposed method needs only the previous knowledge of the values of the equivalent impedances of the consumer and network [14].

Fig. 1 (a) shows the Norton equivalent circuit for the utility and consumer, in the frequency domain. This circuit can be separated as shown in Fig. 1 (b) and (c), where the contributions of the utility and consumer for the zero-sequence current at PCC are highlighted.

According to the circuit of Fig. 1 (a), the following equation can be written:

$$I_{c0}(s) = \frac{V_{pcc0}(s)}{Z_{c0}(s)} - I_{pcc0}(s), \qquad (1)$$

where $I_{c0}(s)$ is the zero-sequence current of the consumer, $V_{pcc0}(s)$ and $I_{pcc0}(s)$ are the zero-sequence voltage and current at PCC, respectively, and $Z_{c0}(s) = (R_{c0} + sL_{c0})$ is the zero-sequence equivalent impedance of the consumer.

On the other hand, according to the circuit of Fig. 1 (c), the

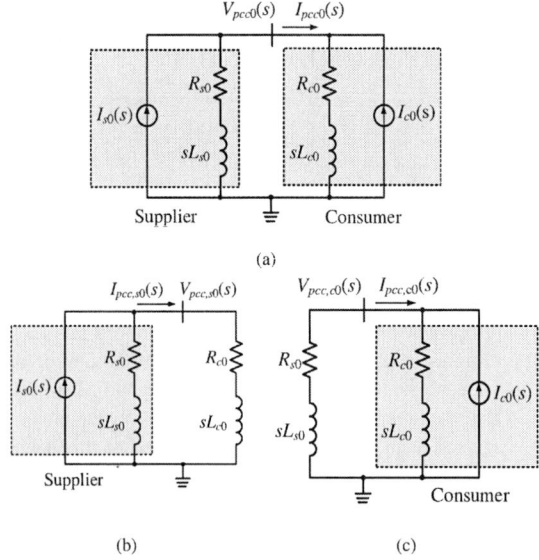

Fig. 1: Norton equivalent circuits: (a) zero-sequence equivalent circuit, (b) contribution of the utility for the current I_{pcc0} and (c) contribution of the consumer for the current I_{pcc0}.

following relations for the voltage and current at PCC can be written:

$$V_{pcc,c0}(s) = -Z_{s0}(s) I_{pcc,c0}(s) \qquad (2)$$

and

$$I_{c0}(s) = \frac{V_{pcc,c0}(s)}{Z_{c0}(s)} - I_{pcc,c0}(s), \qquad (3)$$

where $V_{pcc,c0}(s)$ and $I_{pcc,c0}(s)$ are the zero-sequence voltage and current at PCC due to the consumer contribution, respectively, and $Z_{s0}(s) = (R_{s0} + sL_{s0})$ is the zero-sequence equivalent impedance of the supplier seen from PCC.

The substitution of (1) and (2) in (3) provides the following relation:

$$\frac{V_{pcc0}(s)}{Z_{c0}(s)} - I_{pcc0}(s) = -\left[\frac{Z_{s0}(s)}{Z_{c0}(s)} + 1 \right] I_{pcc,c0}(s). \qquad (4)$$

By manipulating (4), the contribution of the consumer for the zero-sequence current is given by:

$$I_{pcc,c0}(s) = -\left(\frac{1}{Z_{s0}(s) + Z_{c0}(s)} \right) V_{pcc0}(s) +$$
$$+ \left(\frac{Z_{c0}(s)}{Z_{s0}(s) + Z_{c0}(s)} \right) I_{pcc0}(s). \qquad (5)$$

It can be noted in (5) that, knowing the values of the equivalent impedances Z_{s0} and Z_{c0}, the contribution of the consumer for the zero-sequence current at PCC can be determined by measuring the zero-sequence voltage and current at PCC. Although it is not the aim of this work, different methods can be used to estimate the grid and consumer impedances when these parameters are unknown [15].

A. Adaptation of the Methodology of Unbalance Responsibility to the Time Domain

Applying the Laplace inverse transform in (5) and rearranging the terms of the resulting expression, the following relation, in the time domain, can be written:

$$\frac{di_{pcc,c0}}{dt} = -\left(\frac{R_{t0}}{L_{t0}} \right) i_{pcc,c0} + \left(\frac{R_{c0}}{L_{t0}} \right) i_{pcc0} +$$
$$+ \left(\frac{L_{c0}}{L_{t0}} \right) \frac{di_{pcc0}}{dt} - \left(\frac{1}{L_{t0}} \right) v_{pcc0}, \qquad (6)$$

where $R_{t0} = (R_{s0} + R_{c0})$, $L_{t0} = (L_{s0} + L_{c0})$ and the lower case letters v and i are used to represent the voltages and currents in the time domain, respectively.

The expression (6) can be rewritten in the following way:

$$\frac{d\xi}{dt} = -\left(\frac{R_{t0}}{L_{t0}} \right) \xi - \left(\frac{1}{L_{t0}} \right) v_{pcc0} +$$
$$+ \left(\frac{R_{c0}L_{t0} - R_{t0}L_{c0}}{(L_{t0})^2} \right) i_{pcc0}, \qquad (7)$$

where $\xi = i_{pcc,c0} - (L_{c0}/L_{t0}) i_{pcc0}$.

The differential equation given by (7) can be rewritten in the state-space form as shown bellow:

$$\begin{cases} \dot{x} = Ax + Bu \\ y = Cx + Du \end{cases}, \qquad (8)$$

where $A = \left[\frac{-R_{t0}}{L_{t0}} \right]$, $B = \left[\frac{R_{c0}L_{t0} - R_{t0}L_{c0}}{(L_{t0})^2} \quad \frac{-1}{L_{t0}} \right]$, $C = 1$, $D = \left[\frac{L_{c0}}{L_{t0}} \quad 0 \right]$, $x = \xi$, $y = i_{pcc,c0}$, $u = \left[i_{pcc0} \quad v_{pcc0} \right]^t$, being the index t used to represent the transposed vector.

Determining $y = i_{pcc,c0}$ by the solution of (7)-(8) and rewritten (2) in the time domain, the consumer contribution for the zero-sequence voltage unbalance at PCC is given by:

$$v_{pcc,c0} = -L_{s0} \frac{di_{pcc,c0}}{dt} - R_{s0}i_{pcc,c0}. \qquad (9)$$

Fig. 2 shows the block diagram used to quantify, in real time, the consumer responsibility parcel on the voltage unbalance at PCC. It can be noted that the consumer contribution is calculated by using the measured values of voltage and current at PCC.

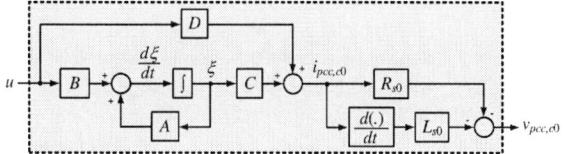

Fig. 2: State-space block diagram for the real-time computation of the voltage unbalance caused by the consumer.

The dynamic equations presented in this section can be easily incorporated in the controller of a DSTATCOM in order to allow the compensation of only the zero-sequence voltage

978-1-7281-4181-7/19 $31.00 © 2019 IEEE

unbalance at the point of connection caused by the consumer.

III. THE STATIC SYNCHRONOUS COMPENSATOR

Fig. 3 shows the topology of the four-wire three-phase DSTATCOM connected to a distribution network through a first-order passive filter [16, 17]. The DSTATCOM is used to compensate for the unbalanced voltages at PCC. The parameters R_f and L_f represent the resistance and inductance of the interface filter. The electric network is modeled by three ideal sinusoidal voltage sources ($v_{s,abc}$) and by the parameters R_s and L_s that represent its equivalent resistance and inductance, respectively. The equivalent DC capacitor of the DSTATCOM is modeled by the parameter C_{eq}.

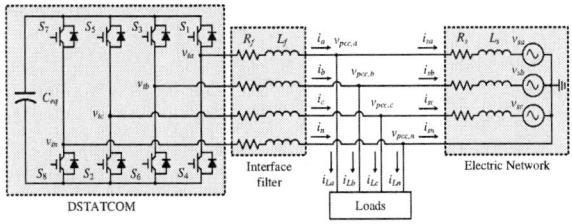

Fig. 3: Topology of the DSTATCOM connected to a distribution network.

In the next subsections the voltage and current control-loops based on the mathematical modelings proposed in [18] for a DSTATCOM will be presented, where the symbols ($\tilde{\ }$) and ($\bar{\ }$) are used to indicate the small signal and steady-state variables.

A. The Positive-Sequence Voltage and Current Control-Loops

Fig. 4 shows the simplified block diagram of the direct and quadrature positive-sequence current control-loop. In this figure, m_{d1} and m_{q1} are the modulation indexes used to control the currents I_{d1} and I_{q1} of the DSTATCOM, respectively, while V_{dc} is the DC bus voltage of the compensator and $V_{t,d1}$ and $V_{t,q1}$ are the direct and quadrature-axis terminal voltages of the DSTATCOM. The gains of the proportional-integral (PI) controller $K_{i1}(s)$ are designed to be equal to $k_{p,i1} = (L_f/\tau_{i1})$ and $k_{i,i1} = (R_{eq}/\tau_{i1})$, where $R_{eq} = (R_f + R_{igbt})$, R_{igbt} is the resistance of the IGBTs and τ_{i1} is the desired time constant.

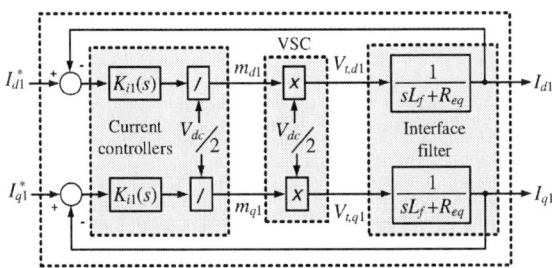

Fig. 4: Simplified block diagram for the direct and quadrature-axis positive-sequence current control-loops.

Fig. 5 shows the positive-sequence voltage control-loop of the DSTATCOM. The block $K_{v_1}(s)$ is the positive-sequence

voltage controller while $H_{i1}(s)$ represents the closed-loop transfer function of the positive-sequence current control-loop of Fig. 4. The block $G_{q1}(s) = \bar{\omega}L_{s1}$ represent the plant of the control-loop, where $\bar{\omega}$ is the steady state network angular frequency. The gain of the integral controller $K_{v_1}(s)$ can be designed by choosing $\tau_{v1} \geq (10\tau_{i1})$, where $\tau_{v1} = 1/(k_{v1}\bar{\omega}L_{s1})$ is the time constant of the voltage closed-loop; k_{v1} is the gain of the integral controller $K_{v1}(s) = k_{v1}/s$, $H_{i1}(s) = 1$ and L_{s1} is the positive-sequence inductance of the network.

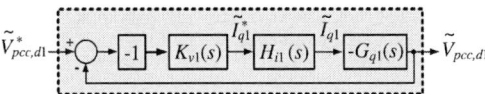

Fig. 5: Block diagram for the direct-axis positive-sequence voltage control loop.

B. The Zero-Sequence Current and Voltage Control-Loops

The use of the synchronous reference frame allows the implementation of PI controllers which makes the design of the controllers gains easier. However, for the case of zero-sequence quantities, it is not possible directly to use the synchronous reference frame transformation. In order to overcome this problem, in [18] it is presented a control strategy that allows the use of PI controllers to control the zero-sequence voltage and current. Thus, a second order generalized integrator (SOGI) and a transport delay buffer are used to generate 90-degrees shift-delay signals for the zero-sequence voltage and current, respectively.

Despite de good performance of the SOGI circuit, that will be used to obtain the zero-sequence voltage in the $\alpha\beta$ coordinates, it will not be used for the zero-sequence current since the DSTATCOM currents present a faster dynamic than the PCC voltages. Instead, a transport delay buffer will be used to generate the fictitious component $i_{\beta 0}$ for the zero-sequence current $i_0 = i_{\alpha 0}$. Fig. 6 shows the block diagram of the transport delay buffer, where $T = (1/f)$ is the fundamental period of the zero-sequence current. Thus, the synchronous reference frame transformation can then be applied in the currents $i_{\alpha 0}$ and $i_{\beta 0}$ in order to obtain and control the zero-sequence currents I_{d0} and I_{q0}, as shown in the block diagram for the zero-sequence current control-loop of Fig. 7. The gains of the PI controller $K_{i0}(s)$ can be chosen equal to $k_{p,i0} = (L_f/\tau_{i0})$ and $k_{i,i0} = (R_{eq}/\tau_{i0})$, where τ_{i0} is the desired zero-sequence time constant.

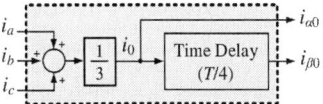

Fig. 6: Block diagram of the transport delay buffer used to generate the zero-sequence quadrature currents.

Fig. 8 shows the block diagram of the algorithm used to convert the zero-sequence voltage $v_{pcc,0}$ at PCC into a fictitious $\alpha\beta$ reference frame, providing the voltages $v_{pcc,\alpha 0}$ and $v_{pcc,\beta 0}$. Thus, the synchronous reference frame transformation can also be applied in these voltages and the zero-

978-1-7281-4181-7/19 $31.00 © 2019 IEEE

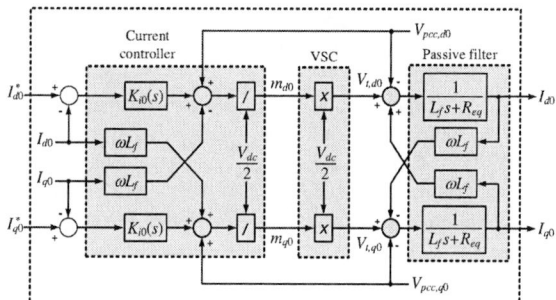

Fig. 7: Block diagram of the direct and quadrature zero-sequence current control-loops.

sequence voltage can be controlled in a similar way the positive-sequence voltage is controlled, as shown in the zero-sequence voltage control-loop of Fig. 9 [18]. However, two controllers are used here, one for the direct-axis and another for the quadrature-axis. The block $K_{v_0}(s)$ represents the integral controller used to control the zero-sequence voltage while $H_{i0}(s)$ represents the closed-loop transfer function of the zero-sequence current control-loop of Fig. 7. The gain k_{v0} of the zero-sequence voltage controller can be designed by choosing the time constant of the zero-sequence voltage control-loop $\tau_{v0} = 1/(k_{v0}\bar{\omega}L_{s0})$, where L_{s0} is the zero-sequence inductance of the network.

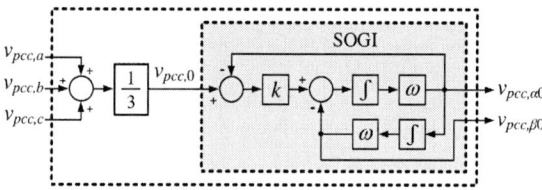

Fig. 8: Block diagram of the SOGI circuit.

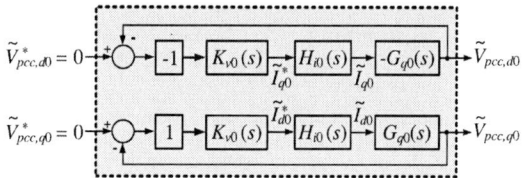

Fig. 9: Block diagram of the zero-sequence voltage control-loop.

C. The DC Voltage Control Loop

Fig. 10 shows the block diagram for the DC voltage control-loop of the DSTATCOM. The block $K_{vdc}(s)$ represents the PI controller while $\bar{V}_{pcc,d1}$ is the steady-state direct-axis positive-sequence voltage at PCC. Since the time constant of the DC voltage controller is designed to be greater than the positive-sequence current controller of the DSTATCOM, the current closed-loop transfer function H_{i1} can be considered as an unitary gain.

Since the DC bus voltage starts oscillating with twice the grid frequency due to the unbalances at PCC, a notch filter $F_n(s)$ tuned at $\omega_o = 2\omega$ and with an unitary quality factor is

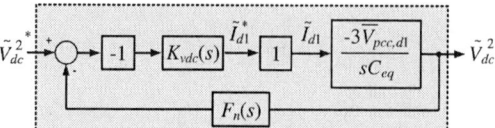

Fig. 10: Block diagram of the DC voltage control-loop.

used in the feed-back of the control-loop of Fig. 10 in order to ensure the correct operation of the DC voltage controller [18].

IV. RESULTS AND ANALYSIS

The circuit of Fig. 3 and all controllers of the DSTATCOM were modeled in an electromagnetic transient program in order to demonstrate and validate the proposed voltage unbalance responsibility strategy. Tables I, II, III and IV show the parameters of the electric network, load, DSTATCOM and controllers, respectively. The IGBTs of the DSTATCOM are switched with a sinusoidal PWM strategy. Two cases are investigated: (i) compensation of all the zero-sequence voltage at PCC and (ii) compensation of only the parcel of the zero-sequence voltage unbalance caused by the consumer (load).

TABLE I
Network parameters

Parameter	Value
Line RMS voltage (V_s)	4.16 kV
Fundamental frequency (f_s)	60 Hz
Phase "a" equivalent impedance	$0.4894 + j0.9176\ \Omega$
Phase "b" equivalent impedance	$0.4699 + j0.9757\ \Omega$
Phase "c" equivalent impedance	$0.4872 + j0.9395\ \Omega$
Mutual impedance - phases "a" and "b"	$j0.3664\ \Omega$
Mutual impedance - phases "a" and "c"	$j0.2854\ \Omega$
Mutual impedance - phases "b" and "c"	$j0.2552\ \Omega$

TABLE II
Load parameters

Parameter	Value
Phase "a" impedance	$5.74 + j3.25\ \Omega$
Phase "b" impedance	$6.04 + j4.40\ \Omega$
Phase "c" impedance	$5.10 + j2.92\ \Omega$

TABLE III
DSTATCOM parameters

Parameter	Value
Equivalent capacitance (C_{eq})	3000 μF
DC voltage (V_{dc}^*)	8.5 kV
Inductance of the passive filter (L_f)	24 mH
Resistance of the passive filter (R_f)	90 mΩ
Resistance of the IGBTs (R_{igbt})	13 mΩ
Switching frequency (f_{sw})	9 kHz

A. Case 1

In this case the DSTATCOM compensates for all the zero-sequence voltage unbalance at PCC. Fig. 11 (a), (b), (c), (d) and (f) show the waveforms of the consumer currents, the DSTATCOM currents, the RMS voltages at PCC, the zero-sequence voltage at PCC, the zero-sequence voltage unbalance factor (VUF) and the DC voltage of the DSTATCOM and the output signal of the notch filter, respectively. At t = 0.35 s

TABLE IV

Controller gains of the DSTATCOM

Controller	Gain	Value
K_{i1} and K_{i0}	Proportional	48 V/A
	Integral	206 V/(A s)
K_{v_1}	Integral	104 A/(V s)
K_{v_0}	Integral	43 A/(V s)
K_{vdc}	Proportional	39.2 μA/V^2
	Integral	1.3 mA/(V^2 s)

the DSTATCOM starts synthesizing zero-sequence currents to compensate for the voltages at PCC. Note that the zero-sequence voltage reaches the zero in 50 ms. It can also be noted that the static compensator synthesizes phase currents of 54 A (peak) in order to compensate for all the zero-sequence voltage at PCC. It is important to remember that the DSTATCOM should be designed to synthesize a neutral current three times greater than the phase currents.

B. Case 2

In this second case the voltages and currents measured at PCC are used to extract the parcel of the zero-sequence voltage unbalance caused by the consumer ($v_{pcc,c0}$), as shown in Fig. 2. After the voltage $v_{pcc,c0}$ be subtracted from the PCC voltage ($v_{pcc,0}$), the resulting signal ($v_{pcc,s0}$) is transformed to a fictitious $\alpha\beta$ reference frame by using the SOGI circuit, generating the signals $v_{pcc,\alpha0,s}$ and $v_{pcc,\beta0,s}$. These voltages are then transformed to the synchronous reference frame to generate the voltages $v_{pcc,d0,s}$ and $v_{pcc,q0,s}$ that are sent as reference signals for the zero-sequence voltage controllers of Fig. 9. Fig. 12 shows the same waveforms of Fig. 11. It can be noted that the peak value of the phase currents synthesized by the DSTATCOM is equal to 40 A, lower than the previous case. Another interesting conclusion is that the oscillation in the DC voltage of the DSTATCOM is also lower, since the zero-sequence current of the static compensator is lower.

V. CONCLUSIONS

This paper presented an algorithm to extract the responsibility parcel, between the utility and consumer, on the zero-sequence voltage unbalance at PCC. The proposed methodology was used to compute in real time the zero-sequence voltage at PCC caused by a consumer. The responsibility algorithm was used to generate the reference signals for the controllers of a DSTATCOM in order to compensate for the voltages at PCC. Results of digital simulations were also presented to validate the proposed algorithm. The results shown that, besides identifying the responsibility parcel of the consumer, the ratings of the DSTATCOM can be reduced when only the voltage unbalance caused by the consumer is compensated.

ACKNOWLEDGEMENT

This study was financed in part by the Coordenação de Aperfeiçoamento de Pessoal de Nível Superior - Brasil (CAPES) - Finance Code 001, the National Council for Scientific and Technological Development (CNPq), the State Funding Agency of Minas Gerais (FAPEMIG) and the National Institute for Electric Energy (INERGE).

REFERENCES

[1] A. Von Jouanne and B. Banerjee, "Assessment of voltage unbalance," *IEEE transactions on power delivery*, vol. 16, no. 4, pp. 782–790, 2001.

[2] L. R. de Araujo, D. R. R. Penido, J. L. R. Pereira, and S. Carneiro, "Voltage security assessment on unbalanced multiphase distribution systems," *IEEE Transactions on Power Systems*, vol. 30, no. 6, pp. 3201–3208, 2015.

[3] P. M. Almeida, K. M. Monteiro, P. G. Barbosa, J. L. Duarte, and P. F. Ribeiro, "Improvement of PV grid-tied inverters operation under asymmetrical fault conditions," *Solar Energy*, vol. 133, pp. 363–371, 2016.

[4] B. C. Souza, A. M. Variz, P. G. Barbosa, and L. R. Araujo, "Behavior of induction motors fed by distorted and unbalanced sources," in *11th IEEE/IAS International Conference on Industry Applications - INDUSCON*, 2014.

[5] IEC, "61000-3-13 - electromagnetic compatibility (EMC) - limits - assessment of emission limits for the connection of unbalanced installations to MV, HV and EHV power systems," 2008.

[6] N. S. P. M. 1-2016, "Motors and Generators," 2016.

[7] ANEEL, "Distribution procedures of Electric Energy in the National Power System - PRODIST: module 8 - Power Quality - (Rev. 10)," National Electric Power Agency, 2018. [Online]. Available: http://www.aneel.gov.br/modulo-8

[8] ONS, "Network procedure: submodule 2.8 - Managing of Performance Indicators of the Transmission Network and its Components," National Electrical System Operator, 2008. [Online]. Available: http://www.ons.org.br/pt/paginas/sobre-o-ons

[9] ONS', "Network procedure: submodule 3.6 - Minimum Technical Requirements for Connection to the Transmission Network." National Electrical System Operator, 2009. [Online]. Available: http://www.ons.org.br/pt/paginas/sobre-o-ons

[10] B. Singh, A. Chandra, and K. Al-Haddad, *Power quality: problems and mitigation techniques*. John Wiley & Sons, 2015.

[11] W. Xu and Y. Liu, "A method for determining customer and utility harmonic contributions at the point of common coupling," *IEEE Transactions on Power Delivery*, vol. 15, no. 2, pp. 804–811, 2000.

[12] M. Shojaie and H. Mokhtari, "A method for determination of harmonics responsibilities at the point of common coupling using data correlation analysis," *IET Generation, Transmission & Distribution*, vol. 8, no. 1, pp. 142–150, 2014.

[13] Y. Sun, X. Xie, and P. Li, "Unbalanced source identification at the point of evaluation in the distribution power systems," *International Transactions on Electrical Energy Systems*, vol. 28, no. 1, p. e2460, 2018.

[14] R. C. F. Gregory, T. M. Scotti, and J. C. Oliveira, "Evaluation of the methodologies for allocating the parcels of responsibility on imbalances," *Brazilian Symposium on Electrical Systems (SBSE)*, 2018.

[15] H. L. M. Monteiro, "Method of impedance estimation

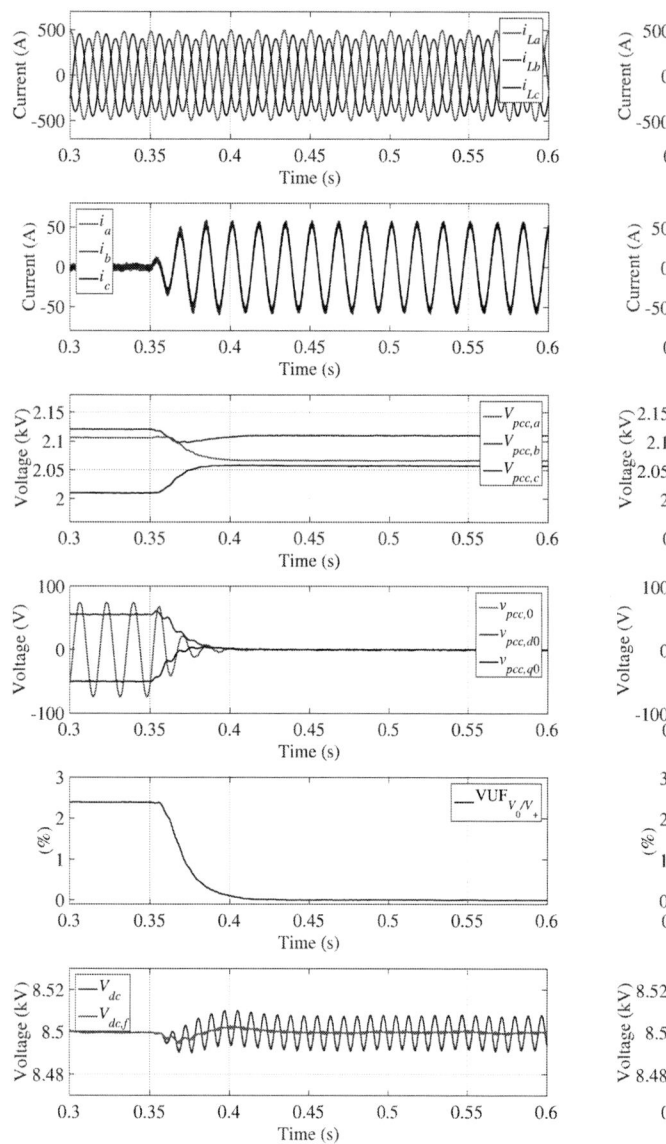

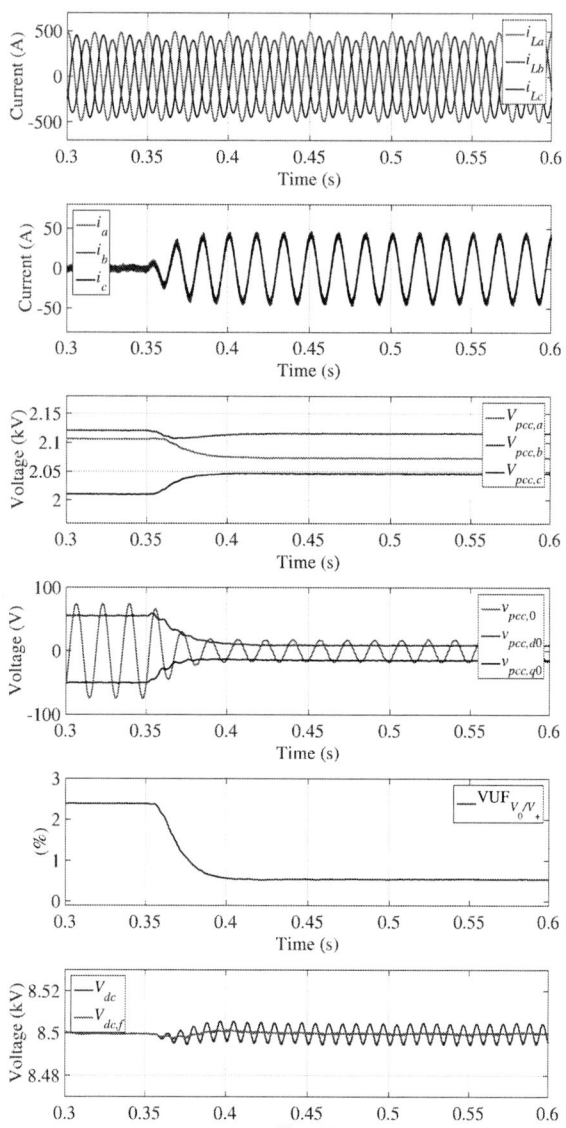

Fig. 11: Waveforms of the Case 1: (a) consumer currents, (b) AC currents of the DSTATCOM, (c) RMS voltages at PCC, (d) zero-sequence voltages at PCC, (e) zero-sequence voltage unbalanced factor and (f) DC voltage of the DSTATCOM.

Fig. 12: Waveforms of the Case 2: (a) consumer currents, (b) AC currents of the DSTATCOM, (c) RMS voltages at PCC, (d) zero-sequence voltages at PCC, (e) zero-sequence voltage unbalanced factor and (f) DC voltage of the DSTATCOM.

using the injection of controlled signals," Doctoral Thesis (Electric Engineering), Federal University of Juiz de Fora, Juiz de Fora, Brazil, 2018.

[16] J. C. da Cunha, W. de Oliveira Rossi, A. L. Batschauer, and M. Mezaroba, "A simple control scheme to a voltage regulator based in a current controlled STATCOM," in *2013 Brazilian Power Electronics Conference.* IEEE, 2013, pp. 1212–1218.

[17] J. V. M. Farias, A. F. Cupertino, V. N. Ferreira, S. I. Seleme, H. A. Pereira, and R. Teodorescu, "Design and lifetime analysis of a DSCC-MMC STATCOM," in

2017 Brazilian Power Electronics Conference (COBEP). IEEE, 2017, pp. 1–6.

[18] S. N. Duarte, F. T. Ghetti, P. M. de Almeida, and P. G. Barbosa, "Zero-sequence voltage compensation of a distribution network through a four-wire modular multilevel static synchronous compensator," *International Journal of Electrical Power & Energy Systems*, vol. 109, pp. 57–72, 2019.

978-1-7281-4181-7/19 $31.00 © 2019 IEEE

Design of Resonant Controllers for Compensation of Third Harmonic Ripple in the DC Capacitors Voltages of NPC Converters

Andrei de O. Almeida*†, Pedro M. de Almeida*, Pedro G. Barbosa*

*Graduate Program of Electrical Engeenering, Federal University of Juiz de Fora, Juiz de Fora, MG, Brazil
†Federal Center of Technological Education of Minas Gerais, Nepomuceno, MG, Brazil
Emails: andrei.almeida@engenharia.ufjf.br, pedro.machado@ufjf.edu.br, pedro.gomes@ufjf.edu.br

Abstract—This paper proposes a method to design an adaptive proportional-resonant (PR) controller for equalize and suppress the third harmonic in the capacitors voltages of NPC converters applied to a wind energy conversion system. This problem with NPC converters capacitors voltages occurs in all applications, but when the frequency of the AC system varies, as occurs in wind systems, a fixed frequency resonant controller could not be sufficient. Then, using a PR controller with adaptive resonant frequency is possible to improve the controller performance. A mathematical model for the capacitors voltages dynamic behavior is developed, as well as the control system used to equalize these voltages and suppress the third harmonic. A *quasi*-PR controller is used, which allows to increase the bandwidth, make it less sensitive to frequency variations. At least, is a method to design the controllers parameters. In order to verify the system operation and the controllers performance, results from simulations in PSCAD/EMTDC are used.

Index Terms—Wind Energy Conversion Systems, Neutral-Point Clamped Converter, Adaptive Resonant Controller

I. INTRODUCTION

In recent decades, due to rising costs and the ecological footprint related to the use of fossil fuels, some renewable energy sources such as wind and solar have been increasingly used [1]. The use of sophisticated aerodynamic designs, the development of new static converters and the use of modern control techniques have made it possible to generate electricity in a controlled, reliable and efficient way in high-power high-voltage wind turbines [2]. In the case of large-scale generation, offshore wind farms have been an attractive option [3], due to the limitations of onshore installations, such as required areas.

Figure 1 shows a schematic of a wind energy conversion system (WECS). The wind turbine is coupled to the generator by its mechanical shaft. In order to inject the generated energy into the grid, in a controlled and reliable way, two converters (rectifier and inverter) in back-to-back connection are used. A transformer can be used to connect the converter to the grid, in order to adjust the voltage level and ensure galvanic insulation. This configuration is called full-capacity converter since the rectifier and the inverter process all the generated energy [4]. These and other components are allocated inside the wind turbine nacelle and tower [5]. For offshore installations the connection to the grid is not direct, because the energy is transmitted through submarine cables to an onshore substation.

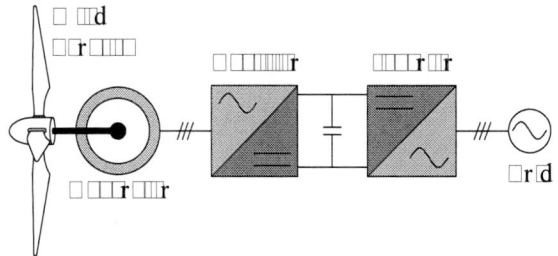

Figure 1. Schematic of the main electrical components of a wind energy conversion system.

In high-power wind turbines, different types of generators, with different voltage levels, can be used. Nevertheless, for power ratings greater than 3 MW, the medium voltage generators are the most suitable and economical option [6]. A widely studied converter for medium voltage WECS is the neutral point diode clamped (NPC) converter [4], [7]. In addition, the NPC converter is already offered by some manufacturers [8].

Figure 2 shows the structure of a three-phase NPC converter. Each leg of the converter has four Insulated Gate Bipolar Transistors (IGBT) and two clamped diodes connected to the midpoint between the DC bus capacitors. This structure allows the converter to handle higher voltages when its is compared to the two level voltage sourced converter (VSC) [9]. In addition, the three-phase phase voltages present lower harmonic content since the terminal voltages will have three levels.

A particularity of the NPC converter is the virtual neutral point "0" created by the DC link division, which allows zero sequence currents to flow through the converter. As a result, the capacitors voltages do not remain naturally equalized. Another problem is the third harmonic current, which causes oscillations of the same order in the capacitors voltages [9]. Some strategies to equalize the capacitors voltages and suppress the third harmonic were proposed in literature [10], [11]. In the strategy proposed in [12], a *quasi*-PR (proportional-resonant) controller is used to equalize the capacitors voltages and suppress the third harmonic. This strategy was tested in a system where the a NPC inverter feeds an AC load from a DC source, but nothing prevents it from being used for a

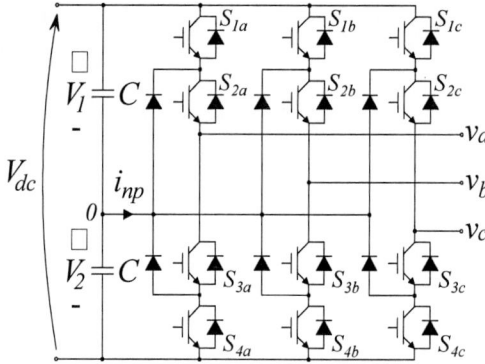

Figure 2. Structure of a three-phase neutral point diode clamped converter.

WECS. Although it works well for a fixed frequency system, for a WECS where the generator frequency is subject to wind speed variations, this strategy may not be as effective.

In this paper, is proposed to use an adaptive *quasi*-PR controller for equalize and suppress the third harmonic in the NPC's capacitors voltages. Furthermore, a method for design the controller gains based on the system transfer functions will be developed. Both the use of the adaptive controller and the design method of the controllers are the main contributions of this paper, since there is no mention in the literature. Simulations in the PSCAD/EMTDC software will be used to verify the proposed strategy and the system behavior with both the fixed and adaptive frequency *quasi*-PR controller.

II. WECS MODELING AND CONTROL

In Figure 4 an overall block diagram of the WECS with two NPC converters in back-to-back connection is shown. In this paper, a permanent magnet synchronous generator (PMSG) is used. In this system, NPC 1 is responsible for controlling the power supplied by the wind turbine, while NPC 2 controls the DC link voltage. Rectifier and inverter are connected by their DC terminals.

The generator is connected to the rectifier NPC 1 through an inductive filter (R_g, L_g) by its AC terminals. Controlling the three-phase currents supplied by the generator, is possible to control the power supplied by the wind turbine. The current control is made in the synchronous reference frame (SRF or *dq*-frame) [9]. The reference for *d-axis* current is zero, so that there is no reactive power flow at the generator terminals. The reference for *q-axis* current, which is related to the active power flow, is sent by the maximum power point tracking (MPPT), with optimum torque control (OTC) strategy [5]. The current control block generates three reference signals m_{1j}, where $j \in \{a, b, c\}$, which is added to a zero sequence signal m_0 and sent to the pulse width modulation (PWM). The signal m_0 is related to the equalization and third harmonic suppression in the capacitors voltages. The signals m_j, sent to the modulation, are the voltage references to be synthesized by the NPC converter on its AC terminals, and it is given by:

$$
\begin{cases}
m_a = m_0 + m_{1a}, \\
m_b = m_0 + m_{1b}, \\
m_c = m_0 + m_{1c}.
\end{cases} \tag{1}
$$

In Figure 3a the comparison between a triangular carrier and two reference signals, for one phase of the NPC converter, is shown. The carrier is compared with the modulation signal (m_j) and it multiplied by -1 ($-m_j$), where $j \in \{a, b, c\}$. According to the diagram of the Figure 3b, it is possible to define the switch functions, i.e. which switches will be triggered.

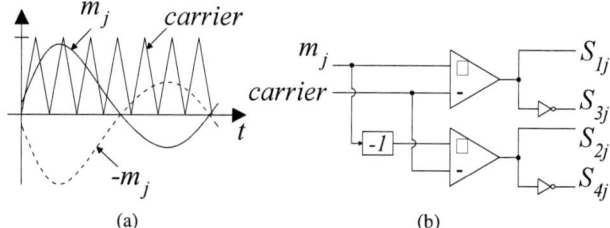

Figure 3. PWM for NPC converters: (a) comparison between carrier and reference signals and (b) implementation to turn on the switches.

The inverter NPC 2 is connected to the electrical grid by its AC terminals through a $Y - \Delta$ transformer, which is used to avoid the zero sequence harmonics to be injected into the grid and to ensure galvanic insulation between the WECS and the grid. This converter controls the DC voltage V_{dc}, which is done controlling the three-phase currents injected into the grid. As with NPC 1, the current control is made in *dq*-frame and the *d-axis* reference current is zero, so that there is no reactive power compensation. On the other hand, the *q-axis* current reference is generated by the DC voltage controller [13]. To synchronize the control in *dq*-frame with the grid voltages, a Phase Locked Loop (PLL) is used. The PWM reference signals are formed by the positive and the zero sequences parcels, just as it occurs in the rectifier.

In this paper, the three-phase currents control, the DC link voltage control and the blocks related to these controls will not be discussed, because it is not the focus. Despite this, it is possible to know more about these control strategies in [5], [9], [13]. The following discussions will focus on the capacitors voltages equalization and the third harmonic suppression.

A. Capacitors Voltages Mathematical Model

To develop a mathematical model for the capacitors voltages, the NPC converter switched model is considered, according to [9], where this model is shown with more details. For a symmetrical three-phase system, the midpoint current i_{np}, indicated in Figure 2, can be written as a function of the modulations signals m_j and the currents at the AC terminals i_j, for $j \in \{a, b, c\}$, as follows:

$$
i_{np} = -[f_a + f_b + f_c] \tag{2}
$$

978-1-7281-4181-7/19 $31.00 © 2019 IEEE

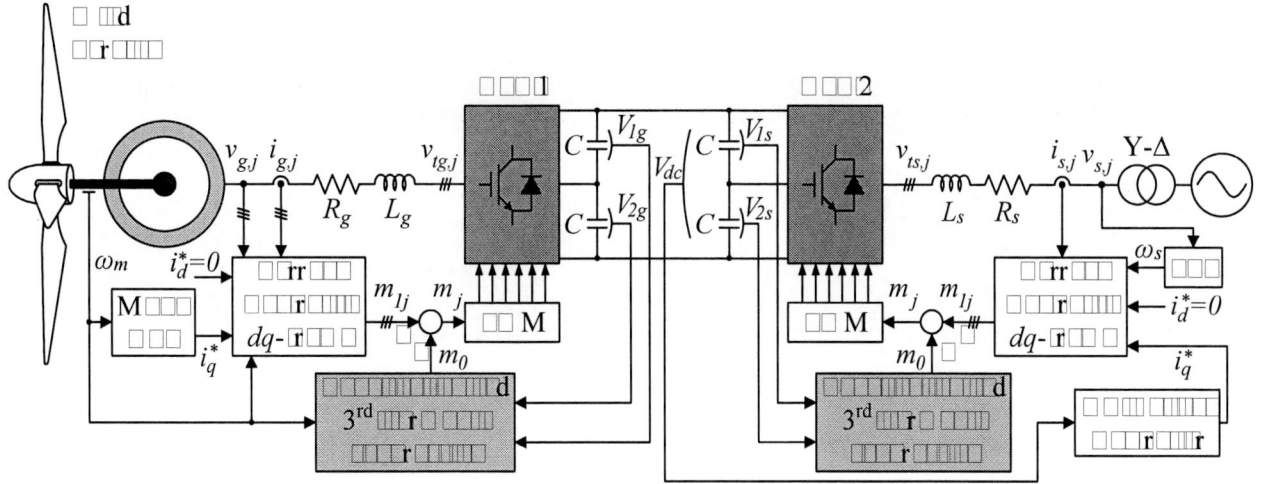

Figure 4. Block diagram of the controllers used in a WECS with two NPC converters in back-to-back connection.

where the functions f_j are given by:

$$\begin{cases} f_a = m_a i_a \left[\sigma(m_a) - \sigma(-m_a) \right] \\ f_b = m_b i_b \left[\sigma(m_b) - \sigma(-m_b) \right] \\ f_c = m_c i_c \left[\sigma(m_c) - \sigma(-m_c) \right] \end{cases} \quad (3)$$

and finally, the function $\sigma(m_j)$ is defined as:

$$\sigma(m_j) = \begin{cases} 1, & m_j \geq 0, \\ 0, & m_j < 0. \end{cases} \quad (4)$$

Replacing (1) in (3), (3) in (2) and solving the result to find the average value, is possible to relate i_{np} with the zero sequence modulation signal m_0, which is written as:

$$i_{np0} = -\frac{6\hat{i}\cos\delta}{\pi} m_0, \quad (5)$$

where $\hat{i}$ is the current peak value, δ is the phase angle between i_j and m_{1j} and the index "0" indicates an average value. As discussed previously, the positive sequence modulation signals m_{1j} are given by the three-phase current control.

Relation (5) allows to describe i_{np0} as a function of m_0, but is desired to control the capacitors voltages. Having in mind the circuit of the Figure 2 once again, the midpoint current can be described by the sum of the currents in node "0", as follows:

$$i_{np} = C \frac{d}{dt} \left(V_1 - V_2 \right). \quad (6)$$

Still based on [9], the capacitors voltages can be expressed as:

$$V_1 = \langle V_1 \rangle_0 + \hat{V}_3 \sin\left(3\omega t + \zeta\right), \quad (7)$$

$$V_2 = \langle V_2 \rangle_0 - \hat{V}_3 \sin\left(3\omega t + \zeta\right), \quad (8)$$

where $\hat{V}_3$ is the peak voltage of the third order component. Subtracting (7) and (8), the fallowing equation is obtained:

$$V_1 - V_2 = \langle V_1 \rangle_0 - \langle V_2 \rangle_0 + 2\hat{V}_3 \sin\left(3\omega t + \zeta\right). \quad (9)$$

B. Capacitors Voltages Control

The capacitors voltages difference given by (9), can be called "DC-partial voltage", and it is related to the midpoint current by (6). Applying the Laplace transform in (6) and (5) it is possible to define a closed loop control using a *quasi*-PR controller, as shown in Figure 5.

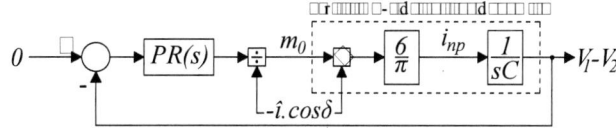

Figure 5. Closed loop control for the DC-partial voltage in the NPC converter.

The controller output is divided by $(-\hat{i}\cos\delta)$, according to (5), similar to a feed-forward compensation. In addition, if the current direction is not considered, the controller may not operate correctly. For the current control strategy adopted in this paper, $i_d = 0$ in both NPC converters. So, it is possible to say that $\hat{i}\cos\delta = i_q$, because i_q is the current peak value and there is no phase shift between voltage and current [5]. This statement is only true if the coordinate transformations are invariant in amplitude.

The *quasi*-PR controller transfer function is given by [12]:

$$PR(s) = k_p + \frac{k_r \omega_a s}{s^2 + \omega_a s + \omega_c^2}, \quad (10)$$

where k_p and k_r are the proportional and resonant gain, respectively, ω_a is the bandwidth and ω_c is the central (or resonant) frequency. Knowing that the DC-partial voltage has a DC component and a third harmonic, the proportional gain will be ensure the equalization, since the plant have an integrator, and the resonant portion of the controller will suppress the third harmonic.

As discussed previously, the main proposal of this paper is use an adaptive *quasi*-PR controller to equalize and suppress the third harmonic in NPC converter capacitors voltages, in a wind energy conversion system. To implement the adaptive *quasi*-PR controller, the block diagram shown in Figure 6, which represents the transfer function (10), can be used.

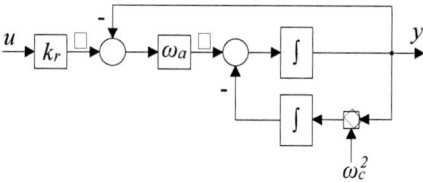

Figure 6. Block diagram of the adaptive *quasi*-PR controller.

C. Quasi-PR Controller Parameters Design

The design of the *quasi*-PR controller gains k_p and k_r and the band width ω_a can be done using the closed loop transfer function of the system shown in Figure 5, which is given by:

$$H(s) = \frac{\frac{k_p}{T}\left(s^2 + \left(1 + \frac{k_r}{k_p}\right)\omega_a s + \omega_c^2\right)}{s^3 + \left(\omega_a + \frac{k_p}{T}\right)s^2 + \left(\omega_c^2 + \frac{(k_p+k_r)\omega_a}{T}\right)s + \frac{k_p\omega_c^2}{T}},$$
(11)

where $T = \frac{\pi C}{6}$ is the plant time constant. Comparing the denominator of (11) with the canonical form $C(s) = s^3 + 2\omega_n s^2 + 2\omega_n^2 s + \omega_n^2$ [14], is possible to calculate the controller parameters as a function of the undamped natural frequency ω_n, defining the following relations:

$$\begin{cases} k_p = \dfrac{T\omega_n^3}{\omega_c^2}, \\ \omega_a = 2\omega_n - \dfrac{k_p}{T}, \\ k_r = \dfrac{T}{\omega_a}\left(2\omega_n^2 - \omega_c^2\right) - k_p \end{cases}$$
(12)

With the relations (12) is possible to design the controller parameters choosing an undamped natural frequency ω_n, or one of the three parameters (k_p, k_r or ω_a).

Note that the whole analysis done in this section is for one NPC converter, but it can be applied for both converters of the back-to-back structure of Figure 4.

III. SIMULATIONS RESULTS

To test the system behavior and verify the controllers operation, the WECS of Figure 4 was modeled in the PSCAD/EMTDC environment. The main parameters of the simulated system are shown in Table I. The other parameters of the wind turbine and generator was taken from [15].

In Table II the *quasi*-PR controllers parameters are shown. The bandwidth ω_a was chosen to ensure adequate frequency selectivity, while the other parameters were calculated related to it. For NPC 1, although the *quasi*-PR is adaptive, as is proposed in this paper, the design of this controller was done considering only the nominal frequency ($2\pi60$ rad/s). However, the frequency variation may change the controller stability region [16]. This did not occur in the situations analyzed, but is a topic to be addressed in future studies. On

Table I
SIMULATED SYSTEM PARAMETERS

Parameter	Value	Unit
WECS rated power	5	MW
Rated Wind Speed (v_w)	11.4	m/s
Generator Rated Voltage (V_g)	3	kV
Generator Rated Frequency (f_g)	20	Hz
Number of Polo Pairs	8	
Generator Winding Resistance	36	mΩ
Generator Synchronous Inductances	5.73	mH
RL Filter Resistance (R_g)	14	mΩ
RL Filter Inductance (L_g)	4.27	mH
Converters Switching Frequency	12	kHz
DC Rated Voltage (V_{dc})	10	kV
DC capacitors (C)	5	mF
Transformer Winding Voltages	3/34.5	kV
Transformer Leakage Inductance	0.48	mH
Grid Rated Voltage (V_s)	34.5	kV
Grid Rated Frequency (f_s)	60	Hz
Grid Filter Resistance (R_s)	50	mΩ
Grid Filter Inductance (L_s)	4.52	mH

the other hand, for NPC 2, which is grid connected, the *quasi*-PR is not adaptive. The grid frequency does not vary much from its nominal value, so choosing a not so small bandwidth, the controller is not so sensitive to frequency variations.

Table II
QUASI-PR CONTROLLERS PARAMETERS

Parameter	Value	Unit
NPC 1 (Generator Side)		
Center Frequency (ω_c)	variable	rad/s
Bandwidth (ω_a)	$2\pi3.5$	rad/s
Proportional Gain (k_p)	2.70	A/V
Resonant Gain (k_r)	46.64	A/V
Natural Undamped Frequency (ω_n)	$2\pi83.96$	rad/s
NPC 2 (Grid Side)		
Center Frequency (ω_c)	$2\pi180$	rad/s
Bandwidth (ω_a)	$2\pi20$	rad/s
Proportional Gain (k_p)	7.88	A/V
Resonant Gain (k_r)	67.79	A/V
Natural Undamped Frequency (ω_n)	$2\pi249.4$	rad/s

The system shown in Figure 4 and described above was simulated in PSCAD/EMTDC software and the simulations results are shown next. Most of the results were obtained using the adaptive frequency *quasi*-PR controller, except the one which is used to show the problem with the fixed frequency controller.

The system starts operating in rated conditions, without the *quasi*-PR, which is turned on in $t = 0.1$ s. As is shown in Figure 7a, from $t = 0.4$ s to $t = 0.5$ s the wind speed on the turbine decreases in ramp from 11.4 m/s to 7.4 m/s and the power generated decreases as well. In Figures 7b and 7c the line voltages, between phases a and b, on NPC 1 ($v_{tg,ab}$) and on generator ($v_{g,ab}$) terminals and the three-phase currents ($i_{g,abc}$) are shown, respectively. From $t = 0.4$ s, is possible to observe that the frequency reduces, in both the voltages and currents. The magnitude of the generator voltage and of the currents decreases as well, because the power supplied by the turbine reduces. In Figures 7d and 7e the line voltages on NPC 2 ($v_{ts,ab}$) and on grid ($v_{s,ab}$) terminals and the three-phase currents ($i_{s,abc}$) are shown, respectively. On the grid side, the frequency remains constant. Only the currents magnitudes decreases due to the power reduction.

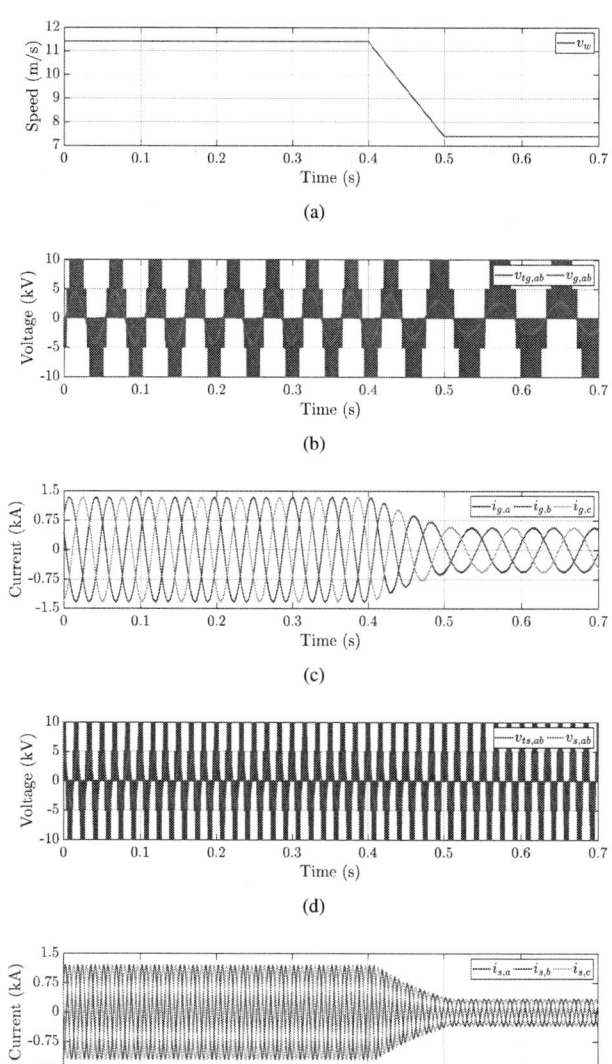

Figure 7. AC waveforms of the generator and grid sides: (a) wind speed, (b) line voltages of the NPC 1 and the generator, (c) three-phase currents, (d) line voltages of the NPC 2 and the grid and (e) three-phase currents.

In Figure 8a the DC-link voltage, measured in NPC 2 terminals, is shown. When the wind speed decreases, the voltage V_{dc} pass through a transient period and then return to track the reference value 10 kV again. In Figure 8b the capacitors voltages of the NPC 2 are shown. In $t = 0.1$ s the resonant parcel of the *quasi*-PR controller is activated and the third order oscillations are suppressed, remaining only the switching frequency harmonics. When the wind speed drops between $t = 0.4$ s and $t = 0.5$ s, is possible to observe the same transient oscillations of the DC link voltage (V_{dc}) in the

capacitors voltages (V_{1g} and V_{2g}).

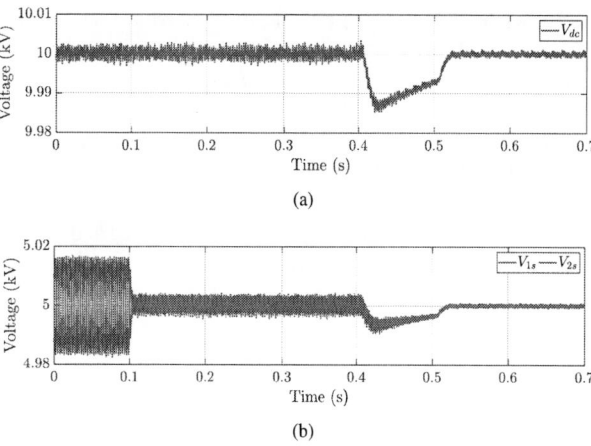

Figure 8. NPC 2 DC terminals: (a) DC link voltage and (b) capacitors voltages of the NPC 2.

In Figures 9a and 9b the capacitors voltages of the NPC 1 converter, connected to the generator, are shown, for the simulations with fixed and adaptive frequency *quasi*-PR controller, respectively. As said before, the system starts operating without the resonant controllers, which are turn on in $t = 0.1$ s. Regardless of whether the controller is adaptive or not, when it is activated the third-order oscillations are practically eliminated, with only switching harmonics remaining. Observing Figure 9a, when the wind speed and the frequency of the voltages and currents reduces, is possible to observe that the third order oscillations reappear, because the central frequency of the *quasi*-PR controller is not the generator frequency anymore. Otherwise, in Figure 9b the third order oscillations do not reappear, because the controller adapts to the new frequency. So, the comparison of these two graphics allows to verify the adaptive resonant controller effect.

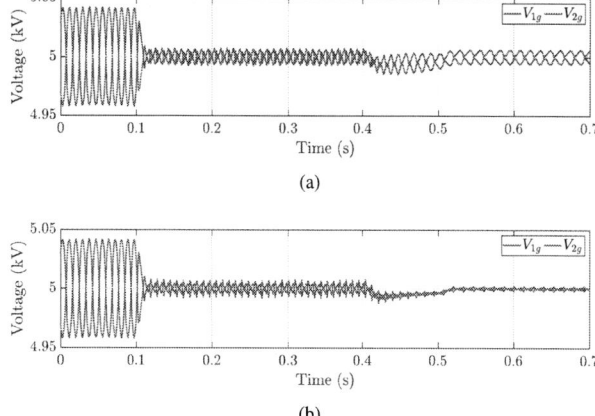

Figure 9. Capacitors voltages of the NPC 1 with (a) fixed and (b) adaptive *quasi*-PR controller.

IV. CONCLUSIONS

In this paper it was proposed the use of an adaptive *quasi*-PR controller for the equalization and third harmonic suppression in NPC converters used in a wind energy conversion system. The use of fixed frequency *quasi*-PR controller has been proposed in literature, but not for this application. In this sense, for a WECS, where the generator frequency varies with the wind speed, this controller could be improved using an adaptive frequency.

A mathematical model for the capacitors voltages dynamics was shown, which allowed to define a control strategy using a *quasi*-PR controller. Furthermore, another contribution of this paper was a method to design the controller parameters based in the closed loop transfer function.

The simulations results allowed to verify the performance of the designed controllers and to observe the problem in use a fixed frequency resonant controller for a NPC converter applied to a WECS. When the wind speed varies, the voltages and currents frequency varies in generator too. Thus, the comparison between simulations with fixed and adaptive frequency *quasi*-PR controllers allowed to observe the improvement in using the adaptive.

ACKNOWLEDGMENT

The authors would like to thank CEFET-MG, CAPES, CNPq, FAPEMIG and INERGE for the financial support and scholarships for the development of this research.

REFERENCES

[1] R. Teodorescu, M. Liserre, and P. Rodriguez, *Grid converters for photovoltaic and wind power systems*. John Wiley & Sons, 2011, vol. 29.

[2] T. Ackermann, *Wind power in power systems*. John Wiley & Sons, 2005.

[3] O. Anaya-Lara, D. Campos-Gaona, E. Moreno-Goytia, and G. Adam, *Offshore wind energy generation*. John Wiley & Sons, 2014.

[4] V. Yaramasu, B. Wu, P. C. Sen, S. Kouro, and M. Narimani, "High-power wind energy conversion systems: State-of-the-art and emerging technologies," *Proceedings of the IEEE*, vol. 103, no. 5, pp. 740–788, 2015.

[5] B. Wu, Y. Lang, N. Zargari, and S. Kouro, *Power conversion and control of wind energy systems*. John Wiley & Sons, 2011, vol. 76.

[6] V. Yaramasu and B. Wu, "Predictive control of a three-level boost converter and an npc inverter for high-power pmsg-based medium voltage wind energy conversion systems," *IEEE Transactions on Power Electronics*, vol. 29, no. 10, pp. 5308–5322, 2014.

[7] S. Alepuz, A. Calle, S. Busquets-Monge, S. Kouro, and B. Wu, "Use of stored energy in pmsg rotor inertia for low-voltage ride-through in back-to-back npc converter-based wind power systems," *IEEE Transactions on Industrial Electronics*, vol. 60, no. 5, pp. 1787–1796, 2013.

[8] J. Chivite-Zabalza, I. Larrazabal, I. Zubimendi, S. Aurtenetxea, and M. Zabaleta, "Multi-megawatt wind turbine converter configurations suitable for off-shore applications, combining 3-1 npc pebbs," in *2013 IEEE Energy Conversion Congress and Exposition*. IEEE, 2013, pp. 2635–2640.

[9] A. Yazdani and R. Iravani, *Voltage-sourced converters in power systems*. John Wiley & Sons, 2010.

[10] ——, "A generalized state-space averaged model of the three-level npc converter for systematic dc-voltage-balancer and current-controller design," *IEEE Transactions on Power Delivery*, vol. 20, no. 2, pp. 1105–1114, 2005.

[11] J. Pou, J. Zaragoza, P. Rodríguez, S. Ceballos, V. M. Sala, R. P. Burgos, and D. Boroyevich, "Fast-processing modulation strategy for the neutral-point-clamped converter with total elimination of low-frequency voltage oscillations in the neutral point," *IEEE Transactions on Industrial Electronics*, vol. 54, no. 4, pp. 2288–2294, 2007.

[12] Y. Zhang, J. Li, X. Li, Y. Cao, M. Sumner, and C. Xia, "A method for the suppression of fluctuations in the neutral-point potential of a three-level npc inverter with a capacitor-voltage loop," *IEEE Transactions on Power Electronics*, vol. 32, no. 1, pp. 825–836, 2017.

[13] A. O. Almeida, A. S. Ribeiro, F. T. Ghetti, P. M. de Almeida, and P. G. Barbosa, "State feedback control of a back-to-back converter for microgrids applications," in *Simpósio Brasileiro de Sistemas Elétricos (SBSE), Niterói*. IEEE, 2018.

[14] S. N. Duarte, F. T. Ghetti, P. M. de Almeida, and P. G. Barbosa, "Zero-sequence voltage compensation of a distribution network through a four-wire modular multilevel static synchronous compensator," *International Journal of Electrical Power & Energy Systems*, vol. 109, pp. 57–72, 2019.

[15] V. Yaramasu, B. Wu, M. Rivera, and J. Rodriguez, "A new power conversion system for megawatt pmsg wind turbines using four-level converters and a simple control scheme based on two-step model predictive strategy – part ii: Simulation and experimental analysis," *IEEE Journal of Emerging and Selected Topics in Power Electronics*, vol. 2, no. 1, pp. 14–25, 2014.

[16] A. V. Timbus, M. Ciobotaru, R. Teodorescu, and F. Blaabjerg, "Adaptive resonant controller for grid-connected converters in distributed power generation systems," in *Twenty-First Annual IEEE Applied Power Electronics Conference and Exposition, 2006. APEC'06*. IEEE, 2006, pp. 6–pp.

Single Phase-Shift Control of DAB Converter using Robust Parametric Approach

Kevin E. Lucas[1], Daniel J. Pagano[1], and Renan L. P. Medeiros[2]

[1]Federal University of Santa Catarina (UFSC), Department of Automation and Systems, Florianópolis 88040-900, Brazil
[2]Federal University of Amazonas (UFAM), Department of Electricity, Manaus 69080-900, Brazil
Email: kevin.lucas@posgrad.ufsc.br, daniel.pagano@ufsc.br, renanlandau@ufam.edu.br

Abstract—**This paper presents the design and evaluation of a robust controller, based on linear programming, Kharitonov's theorem and Chebyshev's theorem, in order to regulate the output voltage of the Dual Active Bridge (DAB) converter under parametric uncertainties. These robust control approach take into account all possible uncertainties in the system such as load and input voltage variations, improving the system performance. The converter is controlled by single phase-shift-modulation (PSM) with a fixed duty cycle. The proposed controller ensures robust control performance and stability with a minor performance degradation compared to an interval robust controller and a conventional controller. Simulation results are presented to validate the robustness and effectiveness of the proposed controller.**

Index Terms—**Dual active bridge (DAB), robust control, single phase-shift control, dc-dc converter, voltage regulation, parametric uncertainties.**

I. INTRODUCTION

In the last years, the penetration of renewable energy sources into the main grid power system is rising every time because of increasing load demand, crisis of conventional energy and the environmental issue [1], [2].

DC transmission and distribution systems, and Microgrid have been introduced to integrate the renewable energy sources, storage units, and DC loads, where power electronic converters are widely used as power interfaces in these applications to manage energy more efficiently and satisfy uninterruptedly the power demand of modern society [3].

Power electronic interfaces are responsible for controlling the DC-DC power conversion in the applications of electric vehicles, renewable energy, energy storage system, transmission and distribution systems and power electronics transformers [4], [5]. Among DC-DC converters, the Dual Active Bridge (DAB) DC-DC converter takes advantages into the mentioned applications due to its benefits of high power density, bidirectional power transfer capability, zero-voltage switching (ZVS), and symmetrical structure [5].

The DAB converter, which consists from two single-phase semi-conductor bridges connected through a high-frequency transformer, was firstly proposed in early 1990s [6]. In almost all DAB converter applications, high efficiency and superior dynamic performance of the converter is desired, particularly

This study was financed in part by the Personnel Improvement Coordination of Superior Level - Brazil (CAPES) - Finance Code 001.

under input voltage fluctuation and output load disturbance. Thus, the robust and fast dynamic response is an essential requirement for DAB converters in industry applications.

In order to improve the DAB converter efficiency, several advanced control schemes has been proposed by the scientific community. Many research focuses on modulation schemes to regulate the average power flow by the Phase-Shift (PS) control [5]. To analyze the dynamic characteristics, the state space averaging modeling and small-signal modeling schemes of DAB converters based on the single-phase-shift (SPS) control are proposed [7]–[9].

Based on the linearized dynamic model of DAB DC-DC converters, [9] and [10] present, a novel model-based phase-shift (MPS) control and model-based feed-forward control scheme, respectively, to improve the dynamic characteristics of DAB DC-DC converters under the load disturbance condition.

A boundary control scheme by using the natural switching surface, is reported in [11] for DAB DC-DC converters, which is an excellent control scheme for dealing with various dynamic change conditions. However, the proposed method belongs to variable frequency control schemes and the transformer saturation will occur. Transformer saturation is a critical issue in this control scheme.

In traditional SPS control, the transformer saturation may occur due to the power flow of the DAB converter is mainly dependent on the transformer leakage inductor [5].

The load current feed-forward (LCFF) control is an alternative solution [12] to improve dynamic performance for load changes. The use of LCFF control to achieve robust dynamic response of DAB DC-DC converters is successfully employed in [13] facing extreme conditions, such as start-up, load step-change, and the input voltage fluctuation.

In addition, Sliding-mode control (SMC) is reported in [14] to provide the robust control performance for the DC reference tracking applied to DAB converter with uncertain parameters.

On the other hand, in order to overcome the shortcomings of SPS control and improve the efficiency of DAB converters, various optimized phase-shift control strategies are reported [5], including dual phase-shift (DPS) control, extended phase-shift (EPS) control and triple phase-shift (TPS) control.

Although, these control approaches have more advantages such as smaller back-flow power and peak current than SPS control, the SPS is the main control strategy in real applications due to its easy implementation and convenience of control.

Recently, the studies developed in [1] and [15] addressed an outstanding contribution for the current state-of-the-art on the study of parametric uncertainties in DC-DC converters. Their approaches are focused on robust controller synthesis by using Robust Parametric Control (RPC) techniques [16].

To the best of the authors' knowledge, the application of the RPC method into DAB converter is not widely discussed in the literature.

Therefore, this paper proposes a robust control schemes based on SPS control to ensure robust performance and stability with a minor control effort when the DAB converter is operating outside its nominal operating point. To tune the parameters of the proposed controller, its approach is based on convex optimization solving the LMI optimization problem by using the Linear Programming approach with Chebyshev Theorem [1].

Aiming to evaluate the performance of the proposed robust controller under parametric uncertainties, the proposed robust controller is compared with a robust controller, proposed by Keel and S. P. Bhattacharyya [17], [18], and a classical controller based on pole-placement, carrying out several simulation tests. All the experiments are performed with simulations in Matlab/Simulink. The performance index (ISE) is computed to analyze the control methodologies compared in this work. The results show the proposed methodology outperforms the other approaches.

The remainder of this paper is organized as follows. Section II introduces the mathematical model of DAB converter related to the output-voltage control loop under phase-shift control. The small-signal control-to-output transfer function of DAB converter is also introduced. Section III presents the classical and robust controller design. Section IV presents an assessment of the simulation results. Finally, Section V presents the main conclusions.

II. MATHEMATICAL MODEL OF DAB CONVERTER UNDER SPS CONTROL

Fig. 1 shows the circuit diagram of the DAB converter with two full bridges where C_o is the DC output capacitor; L_r is the DAB storage inductor and the turns radio of the mediate-/high-frequency transformer is N_1/N_2; R_o is the load resistance; V_i is the DC input voltage while V_o is the DC output voltage; i_1 is the input current of the DAB input H-bridge; i_{L_r} is the DAB inductor current; i_2 is the output current of the DAB output H-bridge; i_o is the load current.

Control wise, the two full-bridges are usually square-wave modulated with an appropriate phase-shift inserted between them for fast regulation of power flow.

Power flow $\langle p(t) \rangle_{T_s}$ from the leading to lagging bridge can then be expressed as [6]:

$$\langle p(t) \rangle_{T_s} = \frac{N_1}{N_2} \frac{\langle v_i(t) \rangle_{T_s} \langle v_o(t) \rangle_{T_s}}{2\pi f_{sw} L_r} \varphi \left(1 - \frac{|\varphi|}{\pi} \right) \quad (1)$$

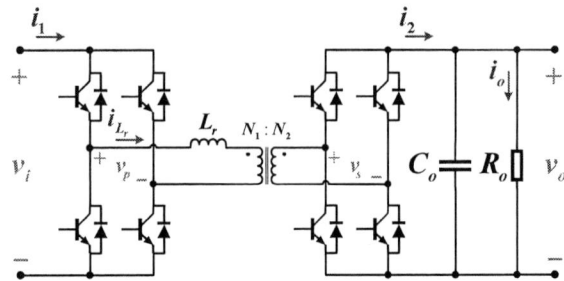

Fig. 1. Circuit diagram of the DAB converter.

where φ is the phase-shift, f_{sw} is the switching frequency, $T_s = \frac{1}{f_{sw}}$ is the switching period, $\langle v_i(t) \rangle_{T_s}$ is the average value of v_i in a switching period, and $\langle v_o(t) \rangle_{T_s}$ is the average value of v_o in a switching period. v_p and v_s are all the square waves whose duty cycle is 50 %, and the phase difference between them is φ $\left(-\frac{\pi}{2} \leq \varphi \leq \frac{\pi}{2} \right)$. If $\varphi > 0$, v_p gets ahead of v_s. Otherwise, v_p is lagging behind v_s.

To simplify analysis, let's define the following variable assuming that $\varphi > 0$,

$$\omega_{sw} = 2\pi f_{sw} \ , \quad g_m = \frac{N_1}{N_2} \frac{1}{\pi \omega_{sw} L_r} \quad (2)$$

then, substituting (2) into (1)

$$\langle p(t) \rangle_{T_s} = g_m \langle v_i(t) \rangle_{T_s} \langle v_o(t) \rangle_{T_s} \varphi \left(\pi - \varphi \right) \quad (3)$$

Assuming that the DAB converter has no losses (efficiency of 100%), the transmission power can be expressed as

$$\langle p(t) \rangle_{T_s} = \langle v_i(t) \rangle_{T_s} \langle i_1(t) \rangle_{T_s} = \langle v_o(t) \rangle_{T_s} \langle i_2(t) \rangle_{T_s} \quad (4)$$

where $\langle i_1(t) \rangle_{T_s}$ and $\langle i_2(t) \rangle_{T_s}$ are the average values of i_1 and i_2, respectively, during a switching period.

From (2) and (3), $\langle i_1(t) \rangle_{T_s}$ and $\langle i_2(t) \rangle_{T_s}$ are defined.

$$\langle i_1(t) \rangle_{T_s} = g_m \langle v_o(t) \rangle_{T_s} \varphi \left(\pi - \varphi \right) \quad (5)$$

$$\langle i_2(t) \rangle_{T_s} = g_m \langle v_i(t) \rangle_{T_s} \varphi \left(\pi - \varphi \right) \quad (6)$$

On the other hand, the average value of i_2 can be expressed as the sum of the average values of capacitor current and i_o.

$$\langle i_2(t) \rangle_{T_s} = C_o \frac{d \langle v_o(t) \rangle_{T_s}}{dt} + \frac{1}{R_o} \langle v_o(t) \rangle_{T_s} \quad (7)$$

Then, the equation that represents the variation of average value of v_o is defined as follows.

$$C_o \frac{d \langle v_o(t) \rangle_{T_s}}{dt} = g_m \langle v_i(t) \rangle_{T_s} \varphi \left(\pi - \varphi \right) - \frac{1}{R_o} \langle v_o(t) \rangle_{T_s} \quad (8)$$

In the above average variables, the switching ripple is filtered. Therefore, these average signals can be expressed with their corresponding DC values plus the superimposed small ac variations

$$\begin{cases} \langle i_2(t) \rangle_{T_s} = I_2 + \tilde{i}_2(t) \\ \varphi = D_\varphi + \tilde{\varphi} \end{cases} \text{and} \begin{cases} \langle v_i(t) \rangle_{T_s} = V_i + \tilde{v}_i(t) \\ \langle v_o(t) \rangle_{T_s} = V_o + \tilde{v}_o(t) \end{cases}$$

978-1-7281-4181-7/19 $31.00 © 2019 IEEE

where I_2, D_φ, V_i and V_o are the DC values, $\tilde{i}_2(t)$, $\tilde{\varphi}(t)$, $\tilde{v}_i(t)$ and $\tilde{v}_o(t)$ are the superimposed small ac variations.

Separate the disturbance and linearize the equations, then

$$\begin{cases} \tilde{i}_2(s) = G_{sd}\tilde{\varphi}(s) + G_{sv}(s)\tilde{v}_i(s) \\ \tilde{v}_o(s) = Z_{out}\tilde{i}_2(s) \end{cases} \quad (9)$$

with

$$G_{sd} = \frac{N_1}{N_2}\frac{1}{\pi\omega_{sw}L_r}(\pi - 2D_\varphi)V_i$$

$$G_{sv} = \frac{N_1}{N_2}\frac{1}{\pi\omega_{sw}L_r}D_\varphi\left(\pi - D_\varphi\right)$$

$$Z_{out} = \frac{R_o}{1 + sR_oC_o}$$

Hence, the small-signal control-to-output transfer function of the DAB converter is found.

$$G_{v\varphi}(s) = \left.\frac{\tilde{v}_o(s)}{\tilde{\varphi}(s)}\right|_{\tilde{v}_i(s)=0} = Z_{out}G_{sd} \quad (10)$$

$$G_{v\varphi}(s) = \frac{N_1}{N_2}\frac{(\pi - 2D_\varphi)}{\pi\omega_{sw}L_rC_o}\left(\frac{V_i}{s + \frac{1}{R_oC_o}}\right) \quad (11)$$

III. CONTROLLER DESIGN

The direction and the magnitude of the power flow can be adjustable and relative to phase-shift ratio φ. Under SPS control, the output voltage can be controlled and regulated by outer-phase shift ratio φ, as shown in the control-to-output transfer function of a DAB converter (11). The design procedure of the voltage controller is performed by using a voltage PI controller, which is adopted to adjust the output voltage.

In the following paper, the emphasis will be placed on the design procedure of the voltage controller.

Fig. 2 illustrates the control block diagram for DAB converter.

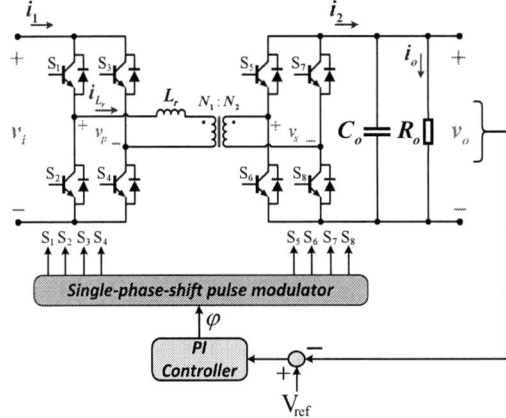

Fig. 2. Block diagram of the SPS control for DAB converter.

The nominal plant and the controller structure are defined in (12) and (13). Note that the nominal plant is the control-to-output transfer function of a DAB converter (11).

$$G_{v\varphi}(s) = \frac{n_p(s)}{d_p(s)} = K_{dc}\frac{V_i}{s + a_o} \quad (12)$$

where $K_{dc} = \frac{g_m(\pi - 2D_\varphi)}{C_o}$ and $a_o = \frac{1}{R_oC_o}$.

$$C(s) = \frac{n_c(s)}{d_c(s)} = \frac{K_ps + K_i}{s} \quad (13)$$

The nominal values of the parameters of DAB converter, operational point and the meaning of each symbol in (11) are presented in Table 1.

TABLE I
PARAMETERS OF THE DAB CONVERTER

Par.	Unit	Var (%)	Nom. Val.	Description
V_i	V	15	800	Source input voltage
V_o	V	–	400	Output voltage
L_r	mH	–	1.10	Auxiliary inductor
C_o	μF	–	104.17	Output Capacitor
R_o	Ω	25	80	Resistive load
D_φ	rad	–	$\pi/6$	Nominal phase-shift
f_{sw}	kHz	–	20	Switching frequency
N_2/N_1	–	–	0.50	Transformer turn ratio

A. Classical Pole-Placement Methodology

According to [16], the solution of the Diophantine equation (14) summarizes the classical pole-placement problem, as follows:

$$\Delta(s) = d_p(s)d_c(s) + n_p(s)n_c(s) \quad (14)$$

where, $\Delta(s)$ is the closed-loop characteristic polynomial.

Assuming that the desired dynamic of closed-loop system is represented by

$$\Delta_d(s) = s^2 + 2\xi\omega_n s + \omega_n^2 \quad (15)$$

In order to tune the controller, the closed-loop parameters obtained are compared with the parameters of the closed-loop desired polynomial, which represent the desired dynamics of the system as follows

$$\Delta(s) = \Delta_d(s) \quad (16)$$

This problem can be written in its matrix representation, presenting the following relationship:

$$\begin{bmatrix} K_{dc}V_i & 0 \\ 0 & K_{dc}V_i \end{bmatrix}\begin{bmatrix} K_p \\ K_i \end{bmatrix} = \begin{bmatrix} 2\xi\omega_n - a_o \\ \omega_n^2 \end{bmatrix} \quad (17)$$

B. Robust Controller Design

To design the controller, a region of uncertainty is previously defined, considering that the uncertainty is contained in the parameter variation of the plant-model (cf. Table 1). The controller is designed according to Keel and Bhattacharyya [17], [18], and K. Lucas *et al.* [1], associated with a linear goal programming formulation, which will lead to a set of linear inequality constraints.

The box region of uncertainties is built based on a previously specified uncertainty range according to Table 1. Thus, the interval plant is defined as

$$G_{v\varphi}(s) = \frac{[\,b_o\,]}{s + [\,a_o\,]} = \frac{[b_o^-, b_o^+]}{s + [a_o^-, a_o^+]} \quad (18)$$

where $[\,b_o\,] = K_{dc}[\,V_i\,]$ and $[\,a_o\,] = [\,1/(R_oC)\,]$

When the system is subject to parametric uncertainties, the controller performance may deteriorate. Therefore, the controller must ensure robust performance within an acceptable region of closed-loop parameters variation, so that the closed-loop poles are located in a certain region. Thereby, a desired region is defined as follows:

$$\Phi = [\Delta_d^-, \Delta_d^+] = s^2 + [\,\phi_1\,]s + [\,\phi_o\,] \quad (19)$$

where $[\,\phi_1\,] = [\phi_1^-, \phi_1^+]$ and $[\,\phi_o\,] = [\phi_o^-, \phi_o^+]$

The Closed-loop interval polynomial is obtained by replacing uncertainties into (14).

$$[\,\Delta(s)\,] = s^2 + (K_pK_{dc}[\,V_i\,] + [\,a_o\,])s + (K_iK_{dc}[\,V_i\,]) \quad (20)$$

In order to tune the robust controller, (17) can be rewritten by using (19) and (20).

$$\begin{bmatrix} [b_o] & 0 \\ 0 & [b_o] \end{bmatrix} \begin{bmatrix} K_p \\ K_i \end{bmatrix} = \begin{bmatrix} [\phi_1] - [a_o] \\ [\phi_1] \end{bmatrix} \quad (21)$$

By replacing interval parameters by their vertices and after simplification by eliminating redundant inequalities, we construct the following set of linear inequalities that the controller parameters should satisfy.

$$\begin{bmatrix} \phi_1^- - a_0^- \\ \phi_1^- - a_0^+ \\ \phi_1^- - a_0^- \\ \phi_1^- - a_0^+ \\ \phi_o^- \\ \phi_o^- \end{bmatrix} \leq \begin{bmatrix} b_o^- & 0 \\ b_o^- & 0 \\ b_o^+ & 0 \\ b_o^+ & 0 \\ 0 & b_o^- \\ 0 & b_o^+ \end{bmatrix} \begin{bmatrix} K_p \\ K_i \end{bmatrix} \leq \begin{bmatrix} \phi_1^+ - a_0^- \\ \phi_1^+ - a_0^+ \\ \phi_1^+ - a_0^- \\ \phi_1^+ - a_0^+ \\ \phi_o^+ \\ \phi_o^+ \end{bmatrix} \quad (22)$$

Thus,

$$B(\phi^-) \leq AX \leq B(\phi^+) \quad (23)$$

The robust controller design problem is summarized in the choice of $X = [\,K_p \quad K_i\,]^T$ (if possible) to satisfy the set of inequality (22).

According to Keel and Bhattacharyya [17], [18], the aforementioned robust performance control design problem for the pre-established conditions can be rewritten as the following optimization problem:

$$X = \arg(\min f(X))$$
$$s.t. \begin{bmatrix} A(p) \\ -A(p) \end{bmatrix} X \leq \begin{bmatrix} B(\phi^+) \\ -B(\phi^-) \end{bmatrix} \quad (24)$$

where, $f(X)$ is a linear cost function that must be built and minimized according to the control goals. In this study, the cost function $f(X)$ has been chosen to be the sum of the elements of vector of the controller parameter X, such as suggested by Keel and Bhattacharyya [17], [18].

The proposed controller is based on the work developed by K. Lucas *et al.* [1], thus, the optimization problem for the pre-established conditions can be rewritten as a problem of local minimization, subject to following restrictions

$$X' = \arg(\min f(X'))$$
$$s.t. \begin{bmatrix} A' \\ -A' \end{bmatrix} X' \leq \begin{bmatrix} B(\phi^+) \\ -B(\phi^-) \\ 0 \end{bmatrix} \quad (25)$$

with

$$X' = \begin{bmatrix} X \\ R \end{bmatrix}, A' = \begin{bmatrix} A & \|a\|_2 \\ -A & \|a\|_2 \\ 0_{1\times 2} & -1 \end{bmatrix} \quad (26)$$

where, $\|a\|_2$ is the euclidian norm of coefficients of A; the cost function is defined as the sum of controller gains within the radio R and the parameter vector X' contains the controller gains and the radio of the largest ball of Chebyshev Theorem.

Particularly, it was chosen for a maximum settling time of less than 30 ms and a damping factor (ξ) greater than 0.69, defining the desired performance region (15).

With the desired performance region defined, the parameters of the classical controller are tuned (27) using (17).

By replacing interval parameters values into the desired performance region and solving the linear goal programming problem given in (24) and (25), the robust controllers, (28) and (29), are tuned according to Keel and Bhattacharyya [39] and K. Lucas *et al* [39], respectively.

$$C(s)_{Classical} = \frac{0.00198s + 0.5039}{s} \quad (27)$$

$$C(s)_{Robust} = \frac{0.003066s + 0.5866}{s} \quad (28)$$

$$C(s)_{Proposed} = \frac{0.004513s + 0.6372}{s} \quad (29)$$

IV. SIMULATION RESULTS

The following experiments compare performance of controllers tuned under different operating condition using PI control structure to regulated the output voltage of DAB converter. Table 1 presents the simulation parameters of DAB converter.

These experiments aim to show that the proposed robust controller is able to better compensates for the oscillations caused by parametric uncertainties (load and input voltage variations). In addition to ensuring better reference tracking.

978-1-7281-4181-7/19 $31.00 © 2019 IEEE

A. Reference Tracking

The reference tracking is performed to check the closed-loop performance for different operating condition of voltage reference.

The system is set to its initial operating condition until the steady state is achieved (t = 0.05s). Then, the voltage reference is changed from 500 V to 600 V (t = 0.1s). Finally, the voltage reference returns to its original value again, 500 V (t = 0.2s). Fig. 3 shows the simulated results of closed-loop system performance for reference tracking.

Fig. 4 shows the inductor current and the phase-shift control for this experiment under the three control approaches.

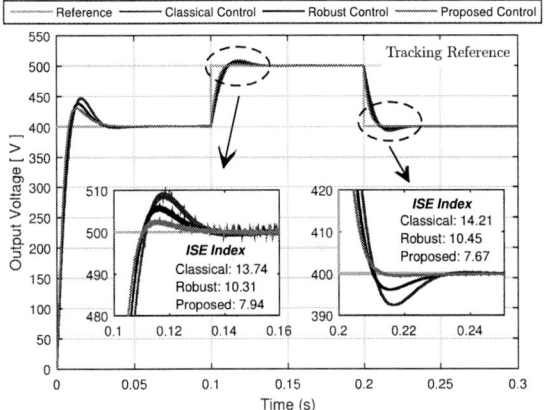

Fig. 3. Closed-loop system performance for reference tracking.

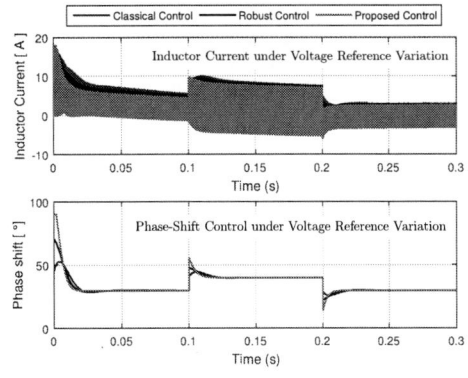

Fig. 4. Closed-loop system performance for reference tracking.

The proposed robust controller achieves better performance with minor overshoot (cf. Fig. 3) in comparison with other controllers. Note that all controllers achieve the desired performance requirement. The ISE index performance ratifies the effectiveness of the proposed controller for tracking reference.

B. Input Voltage Variation

This experiment evaluates the closed-loop performance under input voltage variation. The system starts to operate in the same way as the experiment of subsection IV(A) until the system achieves its steady state (t = 0.05s). After that, the system is subjected to input voltage variation from 800 V to 900 V (t = 0.1s). Then, the input voltage returns to its initial condition (t = 0.2s). Fig. 5 shows the simulated results of closed-loop system performance for input voltage variation.

Fig. 6 shows the inductor current and the phase-shift control for the three control approaches under input voltage variation.

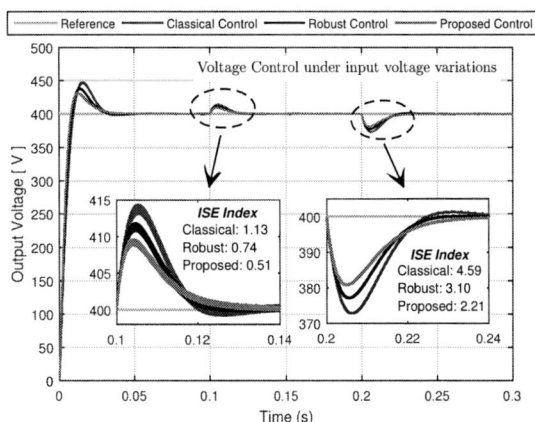

Fig. 5. Closed-loop system performance for input voltage variation.

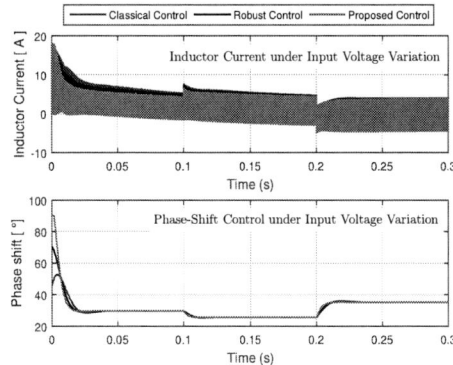

Fig. 6. Closed-loop system performance for input voltage variation.

The proposed controller more effectively compensates the oscillations caused by input voltage variation reducing the oscillation amplitude in comparison with other control approaches (cf. Fig. 5). Therefore, the impact of input voltage variation is lower for the proposed controller as shown the ISE performance index computed inside of Fig. 5.

C. Load Variation

The load variation test is performed to evaluate the closed-loop performance under load variation. After the system reaches its stable state condition with its nominal operating points (t = 0.05s), a load variation, in t = 0.1s is performed from 80 Ω to 100 Ω. Then, the load resistance returns to 80 Ω (t = 0.2s).

Fig. 7 shows the simulated results of closed-loop system performance for input voltage variation.

Fig. 8 shows the inductor current and the phase-shift control for the three control approaches under input voltage variation.

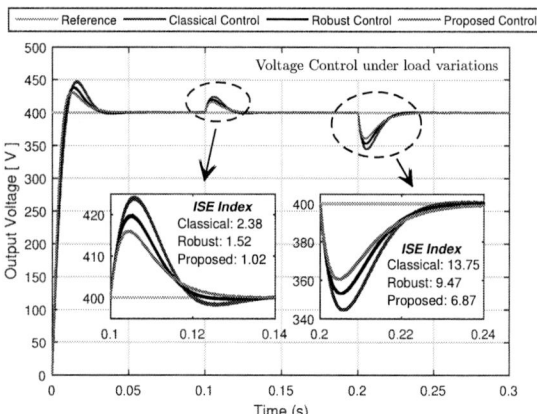

Fig. 7. Closed-loop system performance for load variation.

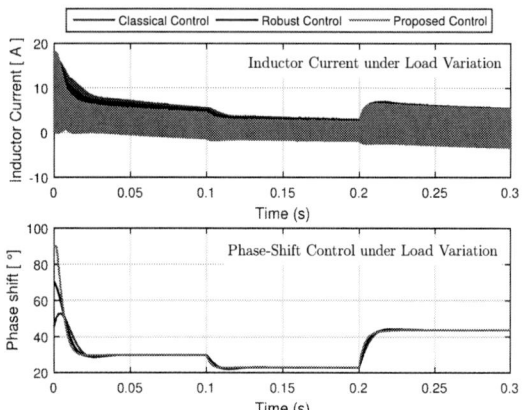

Fig. 8. Closed-loop system performance for load variation.

According to the ISE performance index, the proposed robust controller provides a better performance with reduced oscillation amplitude in comparison with other control approaches (cf. Fig. 7). Therefore, the impact of load variation is lower for the proposed controller.

V. CONCLUSION

This paper presents the design of a robust control strategy to ensure robust performance and stability for an entire predefined uncertainty region using PI control structure. For reference tracking, the proposed controller shows excellent dynamic behavior, such as minor overshoot and fast transient response. Moreover, the performance indicators comparison show that the proposed controller more effectively compensates the oscillations caused by load and input voltage variations ensuring the desired performance.

Therefore, the proposed controller ensures robustness. Hence, the results indicate that the proposed robust controller is justified and present relevant performance improvement.

ACKNOWLEDGMENT

Kevin E. Lucas acknowledges CAPES/BRAZIL for the financial support (Finance Code 001) given to this research and Daniel J. Pagano acknowledges CNPq/BRAZIL for partially funding its work under Project 302229/2018-3.

REFERENCES

[1] K. E. Lucas *et al.*, "Novel robust methodology for controller design aiming to ensure dc microgrid stability under CPL power variation," *IEEE Access*, vol. 7, pp. 64 206–64 222, May 2019.

[2] S. Singh, A. R. Gautam, and D. Fulwani, "Constant power loads and their effects in dc distributed power systems: A review," *Renewable and Sustainable Energy Reviews*, vol. 72, pp. 407–421, May 2017.

[3] M. Z. Hossain, N. A. Rahim, and J. a/l Selvaraj, "Recent progress and development on power dc-dc converter topology, control, design and applications: A review," *Renewable and Sustainable Energy Reviews*, vol. 81, no. 1, pp. 205–230, January 2018.

[4] J. A. Mueller and J. W. Kimball, "Modeling dual active bridge converters in dc distribution systems," *IEEE Transactions on Power Electronics*, vol. 34, no. 6, pp. 5867–5879, June 2019.

[5] B. Zhao, Q. Song, W. Liu, and Y. Sun, "Overview of dual-active-bridge isolated bidirectional dc-dc converter for high-frequency-link power-conversion system," *IEEE Transactions on Power Electronics*, vol. 29, no. 8, pp. 4091–4106, August 2014.

[6] R. W. A. A. Doncker, D. M. Divan, and M. H. Kheraluwala, "A three-phase soft-switched high-power-density dc/dc converter for high-power applications," *IEEE Transactions on Industry Applications*, vol. 27, no. 1, pp. 63–73, Jan/Feb 1991.

[7] C. Zhao, S. D. Round, and J. W. Kolar, "Full-order averaging modeling of zero-voltage-switching phase-shift bidirectional dc-dc converters," *IET Power Electronics*, vol. 3, no. 3, pp. 400–410, May 2010.

[8] H. Qin and J. W. Kimball, "Generalized average modeling of dual active bridge dc-dc converter," *IEEE Transactions on Power Electronics*, vol. 27, no. 4, pp. 2078–2084, April 2012.

[9] H. Bai, Z. Nie, and C. Mi, "Experimental comparison of traditional phase-shift, dual-phase-shift, and model-based control of isolated bidirectional dc-dc converters," *IEEE Transactions on Power Electronics*, vol. 25, no. 6, pp. 1444–1449, June 2010.

[10] H. Bai, C. Mi, C. Wang, and S. Gargies, "The dynamic model and hybrid phase-shift control of a dual-active-bridge converter," in *34th Annual Conference of IEEE Industrial Electronics*, Orlando, FL, USA, November 2008, pp. 2840–2845.

[11] G. Oggier, M. Ordonez, M. Galvez, and F. Luchino, "Fast transient boundary control and steady-state operation of the dual active bridge converter using the natural switching surface," *IEEE Transactions on Power Electronics*, vol. 29, no. 2, pp. 946–957, February 2014.

[12] X. Wang and H. Lin, "DC-link current estimation for load-side converter of brushless doubly-fed generator in the current feed-forward control," *IET Power Electronics*, vol. 9, no. 8, pp. 1703–1710, June 2016.

[13] W. Song, N. Hou, and M. Wu, "Virtual direct power control scheme of dual active bridge dcdc converters for fast dynamic response," *IEEE Transactions on Power Electronics*, vol. 33, no. 2, pp. 1750–1759, February 2018.

[14] Y.-C. Jeung and D.-C. Lee, "Voltage and current regulations of bidirectional isolated dual-active-bridge dc-dc converters based on a double-integral sliding mode control," *IEEE Transactions on Power Electronics*, vol. 34, no. 7, pp. 6937–6946, July 2019.

[15] K. E. Lucas, W. Barra, D. A. Plaza, R. L. P. Medeiros, E. M. Rocha, and D. A. Vaca, "Interval robust controller to minimize oscillations effects caused by constant power load in a dc multi-converter buck-buck system," *IEEE Access*, vol. 7, pp. 26 324–26 342, February 2019.

[16] K. E. Lucas, "Performance evaluation of robust parametric control strategies applied on suppression of oscillations effects due to constant power loads in multi-converter buck-buck systems," Master's thesis, Federal University of Pará, Instituto de Tecnologia, Belém, 2018. [Online]. Available: http://repositorio.ufpa.br/jspui/handle/2011/10257

[17] S. P. Bhattacharyya, H. Chapellat, and L. H. Keel, *Robust Control: The Parametric Approach.* New Jersey: Prentice Hall, January 1995.

[18] L. H. Keel and S. P. Bhattacharyya, "Robust stability and performance with fixed-order controllers," *Automatica*, vol. 35, no. 10, pp. 1717–1724, October 1999.

Operation Boundaries of a Single Phase Thyristor Driven DC-Motor

Thomas M. Mertens
UFRJ
Rio de Janeiro, Brazil
mmertens.thomas@poli.ufrj.br

Richard M. Stephan
UFRJ
Rio de Janeiro, Brazil
richard@dee.ufrj.br

Abstract—The focus of this work is the study of a DC motor driven by a full-bridge SCR thyristor rectifier. In order to highlight the nonlinear effects, a single-phase bridge with ideal components and instantaneous commutation was considered and the ohmic losses were neglected. The continuous and discontinuous modes of current conduction and their implications for static and dynamic behavior were explained. And, in addition to analytical expressions and the development of characteristics curves for steady-state operation, time-domain simulations and comparisons with the analytical results were made.

Index Terms—AC-DC Converter, Boundary conditions, DC-motor, Single Phase, Full Bridge, Continuous and discontinuous current modes.

I. INTRODUCTION

AC-DC converters, also known as Rectifiers, are composed of controlled, semi-controlled or uncontrolled switches with the objective of transforming the alternating waveform at the input of the circuit into direct voltage and current. Here "direct current" is understood as unidirectional current flow.

Based on analytical equations, steady state characteristic curves of Votage x Current are obtained. Simplifications of null Joule losses (zero resistance) and ideal semiconductor switches are considered.

This paper analyzes the behavior of a direct current motor driven by a full-bridge SCR thyristor rectifier powered by a sinusoidal voltage source, as shown in Fig. 1. Results from [1], [2] will be deepened, explained and simulation results will be presented.

The two modes of current conduction are analyzed:

- Continuous-Current conduction, or Continuous mode (in which the current never ceases);
- Discontinuous-Current conduction, or Discontinuous mode (in which there are periods in which the current remains zero).

The original contribution of this work is the analysis and simulations of firing angles greater than 147.5° for this circuit topology. The transition between the conduction modes shows discontinuities (a gap) revealed in the analytical study and proved by simulations. Angles in this region of operation are not commonly used due to the recovery time of the thyristors and the influence of AC-side impedance, but the nonlinear effect it presents is very interesting and justifies this research as a didactic approach to the non-linear effects present in power electronic circuits.

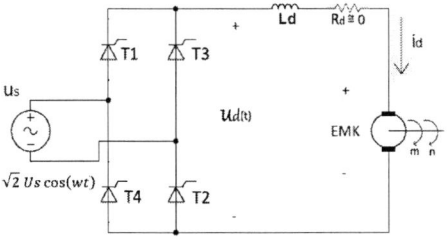

Fig. 1: Electrical equivalent of the studied circuit.

II. GENERAL EQUATION OF THE SYSTEM

From Fig. 1, the inductance of the voltage source u_s and the internal resistance of the motor are disregarded. Lowercase letters represent the instantaneous value of a variable and capital letters represent its mean value or its RMS value.

The zero resistance approximation implies a time constant $\tau = \frac{L_d}{R_d} \to \infty$, turning all transients extremely long. This does not affect the analytical study done for "Power Electronics Steady-State", in which the system always behaves in the same way every cycle, i.e., the initial and final currents of the inductor will be the same. Therefore, the mean voltage over the inductor in a period is zero, and the hatched areas A in Figs. 2 and 3 became equal at the instant the current returns to its initial value. These figures respectively represent continuous and discontinuous conduction modes, and are the basis for the following equations.

A. Voltage Equations

Analyzing Fig. 1, considering the voltage of each element it follows:

$$u_d = v_{L_d} + EMK = L_d \frac{di_d}{dt} + EMK \qquad (1)$$

Calculating the mean voltage U_d over the period T, the integration of the previous Eq. shows that the area of function $u_d(t) - EMK$, between initial instants must be null:

$$\frac{1}{T} \int_0^T u_d \, dt = \frac{L_d}{T} \int_0^T di_d + EMK$$

$$\therefore U_d = EMK \qquad (2)$$

978-1-7281-4181-7/19 $31.00 © 2019 IEEE

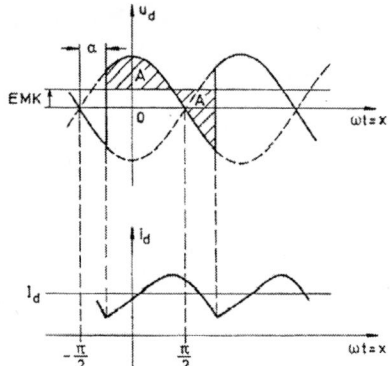

Fig. 2: Voltages and current of the circuit in continuous mode.

Considering U_s the RMS value of the voltage u_s and $\omega = 2\pi f$ the angular frequency for frequency f:

$$u_d(\omega t) = \begin{cases} \sqrt{2}U_s\cos(\omega t) & , \ -\frac{\pi}{2}+\alpha \leq \omega t \leq -\frac{\pi}{2}+\alpha+\tau_d \\ EMK & , \ -\frac{\pi}{2}+\alpha+\tau_d \leq \omega t \leq \frac{\pi}{2}+\alpha \end{cases}$$

$$(3)$$

The mean voltage for firing angle α ($U_{di\alpha}$), considering $x = \omega t$ and the conduction period τ_d, is:

$$U_{di\alpha} = EMK = \frac{1}{\tau_d}\int_{-\frac{\pi}{2}+\alpha}^{-\frac{\pi}{2}+\alpha+\tau d} u_d(x)dx$$

$$\therefore U_{di\alpha} = EMK = U_{dio}\frac{\pi}{2\tau_d}\left([cos\alpha - \cos(\alpha+\tau_d)]\right), \quad (4)$$

where $U_{dio} = \frac{2\sqrt{2}U_s}{\pi}$.

It is important to notice that Puschlowski's [8] abacus could be used to calculate some variables of this problem, but the computational facilities available nowadays allows the solution of analytical equations without bigger issues, giving a richer insight about the problem.

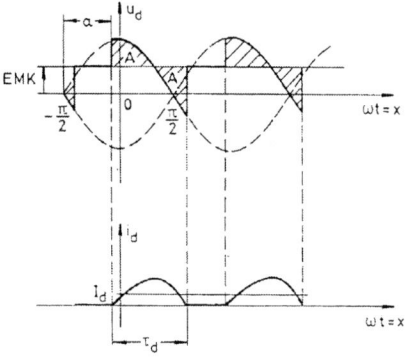

Fig. 3: Voltages and current in discontinuous mode.

B. Current Equations

Considering again the Eq. 1 and $X_d = \omega L_d$:

$$u_d(x) - EMK = X_d\frac{di}{dx}$$

$$i_d(x) = I_0 + \frac{1}{X_d}\int_{-\frac{\pi}{2}+\alpha}^{x} (u_d(x) - EMK)\,dx \quad (5)$$

Therefore, for $-\frac{\pi}{2}+\alpha \leq \omega t \leq -\frac{\pi}{2}+\alpha+\tau_d$:

$$i_d = \sqrt{2}U_s(\sin x + \cos\alpha) - \frac{EMK}{X_d}(x - \alpha + \frac{\pi}{2}) + I_0, \quad (6)$$

and for $-\frac{\pi}{2}+\alpha+\tau_d \leq \omega t \leq \frac{\pi}{2}+\alpha$:

$$i_d = 0,$$

where I_0 is the current value i_d at the firing. The mean value of the current i_d, I_d, will be:

$$I_d = \frac{1}{\pi}\int_{-\frac{\pi}{2}+\alpha}^{-\frac{\pi}{2}+\alpha+\tau_d} i_d dx = \frac{1}{2}\frac{U_{dio}}{X_d}[\sin\alpha -$$

$$\sin(\alpha+\tau_d) + \tau_d\cos\alpha] - \frac{EMK}{\pi X_d}\left(\frac{\tau_d^2}{2}\right) + I_0$$

Substituting EMK given by Eq. 4:

$$I_d = \frac{1}{2}\frac{U_{dio}}{X_d}\left\{\sin\alpha - \sin(\alpha+\tau_d) + \frac{\tau_d}{2}[\cos\alpha + \cos(\alpha+\tau_d)]\right\} + I_0$$

$$(7)$$

These are the mean voltage (Eq. 4) and current (Eq. 7) equations for the system, valid for the continuous mode, where $\tau_d = \pi$ and $I_0 > 0$, and for the discontinuous mode, where $\tau_d < \pi$ and $I_0 = 0$.

III. FRONTIER BETWEEN CONTINUOUS AND DISCONTINUOUS MODES

When the requested torque decreases, the mean current also decreases. At one point, the instantaneous current passes through zero, and since it can not reverse direction, it enters the discontinuous current conduction mode. Therefore, there is a boundary between current conduction modes, which will be analyzed in this item.

Since $\tau_d = \pi$, the current I_{dg} at the frontier of the continuous and discontinuous modes, Eq. 7 become:

$$I_{dg} = \frac{U_{di0}}{X_d}sin\alpha + I_0 \quad (8)$$

978-1-7281-4181-7/19 $31.00 © 2019 IEEE

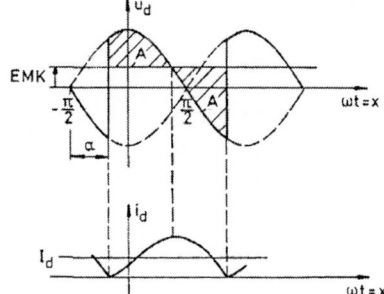

Fig. 4: Voltage and current at the frontier between modes.

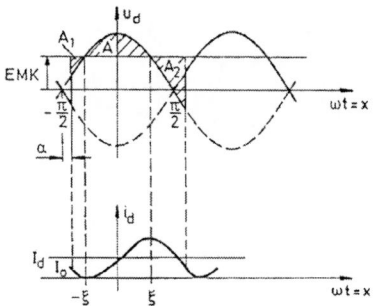

Fig. 5: Voltage and current at the frontier for $\alpha < 32.5°$.

A. Cases for $32.5° < \alpha < 147.5°$

The waveforms for the boundary between the driving modes are shown in Fig. 4. In this limiting case, $\tau_d = \pi$ and $I_0 = 0$, which replaced in Eq. 7 results in:

$$I_{dg(32.5°<\alpha<147.5°)} = \frac{U_{dio}}{X_d}sin\alpha \qquad (9)$$

Outside this region, the zero current condition does not occur at the instant of firing, as indicated in Figs. 5 and 6.

B. Minimum Firing Angle

If there is no current flowing to maintain the thyristor conduction, it is impossible to trigger for certain angles α, since the condition $u_{d(\alpha)} > EMK$ is not being satisfied. Eq. 4 shows that the smaller the angle α the greater will be EMK, hindering the correct polarization of the thyristor. This α_{limit} can be calculated by considering operation at the frontier ($\tau_d = \pi$ e $I_0 = 0$) for the case $u_{d(\alpha)} = EMK$.

$$u_d(t) = EMK \rightarrow \sqrt{2}U_s \cos(-\frac{\pi}{2} + \alpha_{limit}) = U_{dio} \cos \alpha_{limit}$$

$$\frac{U_{dio}}{\sqrt{2}U_s} = \frac{sin\alpha_{limit}}{cos\alpha_{limit}} = \tan(\alpha_{limit})$$

Since $U_{dio} = \frac{2\sqrt{2}U_s}{\pi}$:

$$\tan(\alpha_{limit}) = \frac{2}{\pi} \implies \alpha_{limit} = 32.5°$$

Therefore, it is impossible to have stable operation in the discontinuous current mode for angles less than α_{limit}.

C. Cases for $\alpha < 32.5°$

In this situation, area A needs A1 and A2 for compensation, as shown in Fig. 5.

When $v_{L_d} = 0$, $\frac{di_d}{dt} = 0$, which means that i_d has a maximum or minimum at $x = \pm\xi$. Fig. 5 shows that for $x = -\xi$, $i_d(-\xi) = 0$. Since it is at the frontier of the continuous mode, $EMK = U_{dio}cos(\alpha)$ is valid. From Eq. 6:

$$i_d = 0 = \frac{\pi}{2}\frac{U_{dio}}{X_d}\left(\sin\xi + \cos\alpha \right.$$
$$\left. - \frac{U_{dio}}{X_d}\left(-\xi - \alpha + \frac{\pi}{2}\right)\cos\alpha + I_0\right)$$

$$I_0(0 < \alpha < 32.5°) = \frac{\pi}{2}\frac{U_{dio}}{X_d}\left[\sin\xi + \frac{2}{\pi}(-\xi - \alpha)\cos\alpha\right]$$
$$(10)$$

In addition, the current becomes maximum only at the moment that $u_d(x) = EMK$, at $x = \omega t = \xi$. Therefore, from Eqs. 3 and 4:

$$u_d(\xi) = \sqrt{2}U_s \cos\xi = EMK = \frac{2\sqrt{2}U_s}{\pi\tau_d}[cos\alpha - cos(\alpha+\tau_d)]$$

At the frontier $\tau_d = \pi$, then:

$$cos(-\xi) = cos\xi = \frac{2}{\pi}cos\alpha$$

Substituting in Eq. 10:

$$I_0(0 < \alpha < 32.5°) = \frac{\pi}{2}\frac{U_{dio}}{X_d}[sin\xi - (\xi + \alpha)cos\xi] \quad (11)$$

Then there is a minimum current I_0 at firing so that these angles can be operated in continuous mode. Values for some angles are seen in the table I by applying this result in Eq. 8.

D. Cases for $\alpha > 147.5°$

It would be expected that there should be no discontinuous conduction above $180° - 32.5° = 147.5°$. However, in this case, as shown in Fig. 6, at the firing the switches are polarized, maintaining the conduction.

I_0 can be defined again by the Eq. 6, making $x = \pi + \xi$ which leads to $i_d(\pi + \xi) = 0$, and therefore:

$$i_d(\pi + \xi) = 0 = \frac{\pi}{2}\frac{U_{dio}}{X_d}[sin(\pi + \xi) + cos\alpha]$$
$$- \frac{U_{dio}}{X_d}(\xi - \alpha + \frac{3\pi}{2})cos\alpha + I_0$$

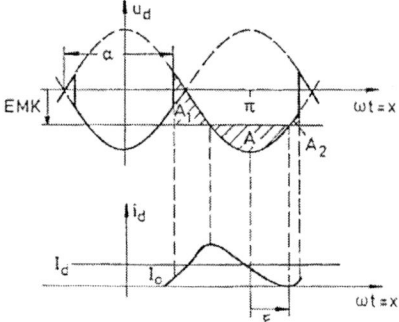

Fig. 6: Voltage and current at the frontier for $\alpha > 147.5°$.

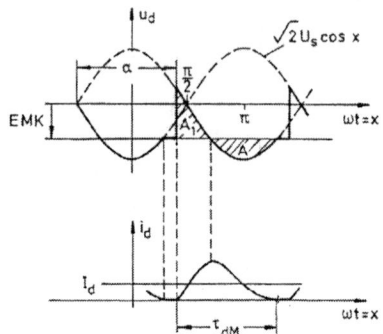

Fig. 7: Voltage and current in the discontinuous mode for τ_{dM}

$$I_0(\alpha > 147.5°) = \frac{\pi}{2}\frac{U_{dio}}{X_d}\left[sin\xi + \frac{2}{\pi}(\xi - \alpha + \pi)cos\alpha\right] \quad (12)$$

Once again, by Eqs. 3 and 4:

$$u_d(\pi + \xi) = \sqrt{2}U_s cos(\pi + \xi) = EMK$$
$$= 2\sqrt{2}\frac{U_s}{\pi}\frac{1}{\tau_d}(cos\alpha - cos(\alpha + \tau_d))$$

As $\tau_d = \pi$:

$$cos(\pi + \xi) = -cos\xi = \frac{2}{\pi}cos\alpha \implies cos\xi = -\frac{2}{\pi}cos\alpha$$

Replacing in 12:

$$I_0(\alpha > 147.5°) = \frac{\pi}{2}\frac{U_{dio}}{X_d}\left[sin\xi - (\xi - \alpha + \pi)cos\xi\right] \quad (13)$$

Unlike the previous case, there is both continuous and discontinuous conduction. Values for some angles are seen in Table I when applied to Eq. 9.

TABLE I: Minimum Currents on Continuous Mode

α	I_{dg}	α	I_{dg}	α	I_{dg}
0°	$0.33\frac{U_{dio}}{X_d}$	45°	$0.707\frac{U_{dio}}{X_d}$	147.5°	$0.537\frac{U_{dio}}{X_d}$
20°	$0.399\frac{U_{dio}}{X_d}$	90°	$\frac{U_{dio}}{X_d}$	165°	$0.367\frac{U_{dio}}{X_d}$
32.5°	$0.537\frac{U_{dio}}{X_d}$	120°	$0.866\frac{U_{dio}}{X_d}$	180°	$0.33\frac{U_{dio}}{X_d}$

E. Maximum Conduction Period

For $\alpha \leq 147.5°$, the maximum conduction period τ_{dM} is always π rad. But, in the discontinuous mode exists a $\tau_{dM} < \pi$ for $\alpha > 147.5°$. Fig. 7 shows such a limit situation where A compensates exactly for A1.

Considering Eq. 3, the voltage at the final conducting period is:

$$u_d(-\frac{\pi}{2}+\alpha+\tau_{dM}) = \sqrt{2}U_s cos(-\frac{\pi}{2} + \alpha + \tau_{dM})$$
$$= \sqrt{2}U_s sin(\alpha + \tau_{dM}).$$

At this point, u_d is also equal to EMK, given by Eq. 6:

$$u_d = \sqrt{2}\frac{U_s}{\tau_{dM}}(cos\alpha - cos(\alpha + \tau_{dM}))$$

Equating these equations, it follows:

$$\therefore \tau_{dM} = \frac{cos\alpha - cos(\alpha + \tau_{dM})}{sin(\alpha + \tau_{dM})} \quad (14)$$

The numerical solution for $\alpha = 150°$, $\alpha = 165°$ and $\alpha = 180°$ is presented in the second column of Table II.

In the same Table, the boundary conditions for both conduction modes are also shown. There exists a gap for steady state operating points for $\alpha = 147.5°$. For instance, for $\alpha = 165°$, the current jumps from 0.316 to 0.368 at the same time that the voltage jumps from -1.010 to -0.966. This situation will be explored in the section V.

IV. NORMALIZED LIMITS

Defining I_{dgM} as the maximum boundary current, from Table I:

$$I_{dgM} = \frac{U_{dio}}{X_d}. \quad (15)$$

Choosing U_{dio} and I_{dgM} as reference voltage and current, Eqs. 4 and 7 can be rewritten as:

$$\frac{EMK}{U_{dio}} = \frac{\pi}{2\tau_d}[cos\alpha - cos(\alpha + \tau_d)] \quad (16)$$

$$\frac{I_d}{I_{dgM}} = \frac{1}{2}\{sin\alpha - sin(\alpha + \tau_d)$$
$$+ \frac{\tau_d}{2}[cos\alpha + cos(\alpha + \tau_d)]\} + \frac{I_0}{I_{dgM}} \quad (17)$$

In the frontier between the two modes, $\tau_d = \pi$ and $I_0 = 0$ so the Eqs. 4 ans 7 can be rewritten:

$$cos\alpha = \frac{EMK}{U_{dio}} \quad sin\alpha = \frac{I_{dg}X_d}{U_{dio}}$$

From the trigonometric relation $(sinx)^2 + (cosx)^2 = 1$, the circle equation and Eq. 15:

$$\left(\frac{I_{dg}}{I_{dgM}}\right)^2 + \left(\frac{EMK}{U_{dio}}\right)^2 = 1.$$

TABLE II: Mean voltages and currents in the limits of both conduction modes $147,5° \leq \alpha \leq 180°$.

α	τ_{dM}	Discontinuous Mode - $\tau_d = \tau_{dM}$		Continuous Mode - $\tau_d = \pi$	
		$\dfrac{EMK}{U_{dio}}$	$\dfrac{I_d}{I_{dgM}}$	$\dfrac{EMK}{U_{dio}}$	$\dfrac{I_d}{I_{dgM}}$
147.5°	π rad	-0.844	0.537	-0.844	0.537
150°	3.08 rad	-0.867	0.500	-0.866	0.503
165°	2.71 rad	-1.010	0.316	-0.966	0.368
180°	2.33 rad	-1.139	0.181	-1.000	0.331

This means that the frontier between continuous and discontinuous modes describes a circle in the voltage-current curves, to the left is the discontinuous mode and to the right the continuous mode. This results are shown Fig. 9.

A. Maximum EMK

The minimum conduction period (τ_{dm}) is shown in Fig.8.

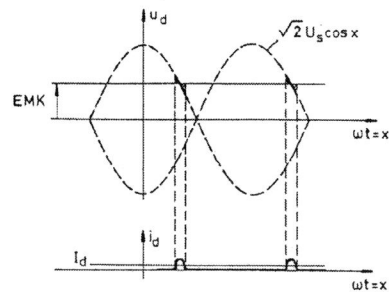

Fig. 8: Voltage and current for $I_d \to 0$.

From Eq. 16 the limit can be calculated.

$$\lim_{\tau_d \to 0} \frac{EMK}{U_{dio}} = \lim_{\tau_d \to 0} \frac{\pi}{2}[cos\alpha - cos(\alpha + \tau_d)] = \frac{\pi}{2}sin\alpha$$

These result is depicted in Fig. 9.

V. COMPUTATIONAL RESULTS

The DC machine model used is available in simulation software PSIM. The system shown in Fig. 10 required high simulation time to reach the steady state, due to the large time constant τ already discussed.

In order to reach a steady state operation point, in addition to the internal inductance $L_d = 30mH$, it was necessary to add an internal resistance of $R_d = 0,01\Omega$.

A. Cases for $32.5° < \alpha < 147.5°$

The simulations of these cases are always in agreement with the model discussed in item II, therefore the computational results of the continuous mode, discontinuous mode and frontier will be omitted, due to the space limitation. The behavior of Fig. 8, for current tending to zero, is repeated in the Fig. 11.

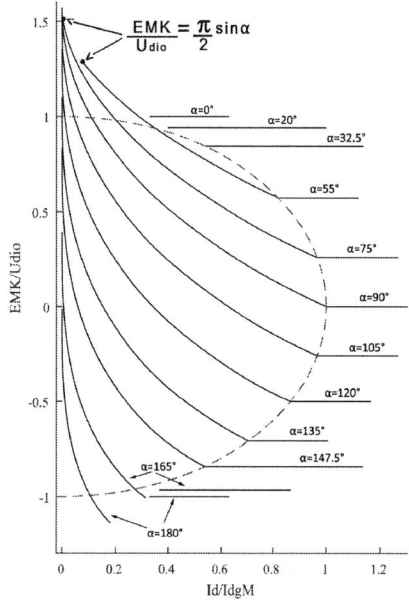

Fig. 9: Normalized analytical result $I_d/I_{dgM} \times EMK/U_{dio}$.

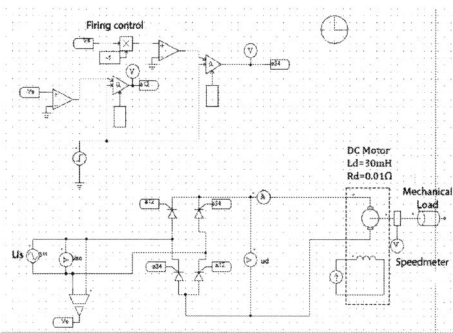

Fig. 10: Simulated System in PSIM.

B. Cases for $\alpha < 32.5°$

- Continuous Mode

    As already commented in the section III-B, this mode does not present major problems in steady state, since it has a minimum current. Then, the simulation results was omitted due to space limitations.
- Discontinuous Mode

978-1-7281-4181-7/19 $31.00 © 2019 IEEE

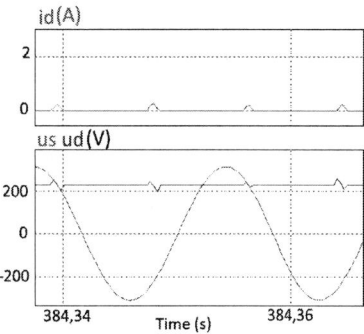

Fig. 11: Conduction limit with τ_{dmin} for $\alpha = 120°$.

The first point to be analyzed is the non-existence of steady state in discontinuous mode for $\alpha < 32.5°$. For example, Fig. 12 shows the dynamic behavior for $\alpha = 20°$.

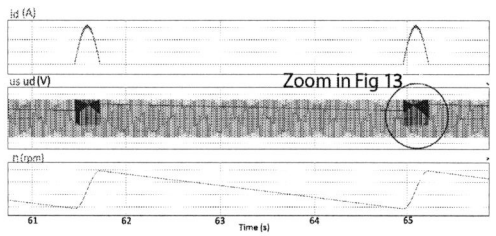

Fig. 12: Unstable operation in discontinuous mode for $\alpha = 20°$.

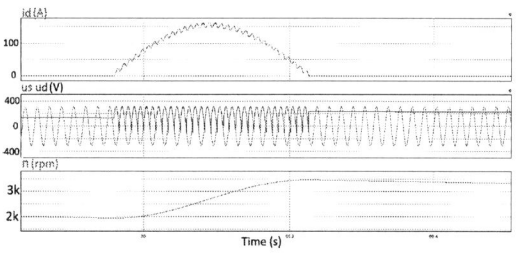

Fig. 13: Unstable operation in discontinuous mode, $\alpha = 20°$, zoom.

It is evident that it is not possible to operate in a discontinuous mode with low load. The firing is effective just when EMK (or speed) is low enough (Fig. 13).

C. Cases with $\alpha > 147.5°$

Simulating cases for $\alpha = 165°$.

- Continuous and Discontinuous Modes This cases was exactly as expected in section II. Then, the results was omitted due to space limitation.
- Gaps
  The load torque was adjusted to give a mean current that fall in the gaps mentioned before. Simulation are presented in Figs. 14 and 15. The dynamic changes from

continuous to discontinuous mode , since there is no stable operation point at the gap.

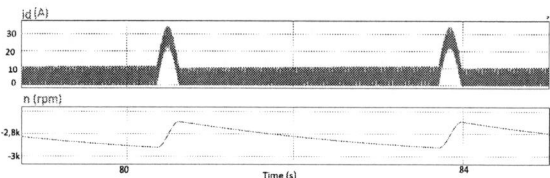

Fig. 14: Current and speed for discontinuities for $\alpha = 165°$.

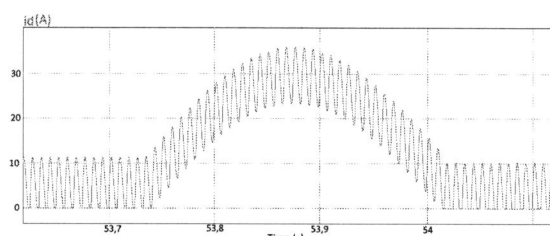

Fig. 15: Current behavior in the region of the discontinuities between the conduction modes for $\alpha = 165°$.

VI. CONCLUSIONS

This work established the operational limits of a DC motor powered by a full wave rectifier bridge, consisting of 4 SCR's.

The simplification of zero Joule losses was adopted to highlight the non-linear behavior of the circuit. The analysis of firing angles greater than $147.5°$ revealed a peculiar gap reinforced by simulations and not mentioned in the literature so far, at least to the author's knowledge. All results are summarized in a single figure (Fig. 9), keeping it as simple as possible.

Further investigation can be carried out considering the parameters neglected in this first study and pursuing experimental studies.

ACKNOWLEDGMENT

This study was financed in part by the Coordenação de Aperfeiçoamento de Pessoal de Nível Superior - Brasil (CAPES) - Finance Code 001.

REFERENCES

[1] R. M. Stephan, "A simple Model for a Thyristor Driven DC Motor Considering Continuous and Discontinuous Current Modes", IEEE Trans. on Education, v. 34, pp. 330-335, 1991.

[2] U. Keuchel and R. M. Stephan "Microcomputer Based Adaptive Control Applied to Thyristor-driven DC Motor", London: Springer Verlag, 1994.

[3] N. Mohan, T. M. Underland, W. P. Robbins, "Power Electronics. Converters, Applications and Design". John Wiley and Sons, 2003.

[4] A. Buxbaum, K. Schierau, "Berechnung von Regelkreisen der Antriebstechnik". AEG Telefunken, 1980.

[5] K. Bystron, "Leistungselektronik - Technische Elektronik Band II", Hanser 1979.

[6] J. J. Keljik, "Electricity 4: AC/DC Motors, Controls, and Maintenance", Cengage Learning 2013.

[7] L. N. Hulley and D. T. W. Liang, "Power Electronics and Motor Control", Cambridge University Press 1995.

[8] K. P. Puchlowski, "Voltage and Current Relations for Controlled Rectification with Inductive and Generative Loads", Trans. AIEE64 1945.

Integrated Local Control of Active Power and Voltage Support for Three-Phase Three-Wire Converters

Ya Zhang, Gabriel Tibola, Maurice G. L. Roes, Jorge L. Duarte

Department of Electrical Engineering
Eindhoven University of Technology
Eindhoven, The Netherlands
ya.zhang@tue.nl

Abstract—The derivation of a robust control algorithm is presented to provide decoupled active power regulation and local grid voltage support in three-phase three-wire grid-connected converters (GCCs). Unlike conventional control schemes, the proposed strategy is designed to be harmonic sequence asymmetric for the purpose of local voltage unbalance correction. A frequency-domain Norton equivalent model is derived to illustrate the working principle of the strategy. Accordingly, by following a frequency-domain decoupled method, the fundamental positive-sequence, the harmonic symmetrical sequences and the fundamental negative-sequence components are regulated independently. Consistent to the model analysis, simulation results validate reduction of local voltage unbalance and total harmonic distortion. Since no external sensors are required for the implementation of the strategy, it is a local approach, applicable to already-existing GCC systems. Moreover, in view of the higher switching frequencies as attainable by devices from the next SiC generation, the accuracy and dynamic behavior of the control algorithms can be much enhanced, improving therefore the quality of the processed energy.

Index Terms—Grid-interactive power converters, control, voltage unbalance, harmonics, compensation.

I. INTRODUCTION

Unbalanced voltages cause adverse effects on electrical loads and power distribution networks. Compensation for voltage unbalance is usually implemented using an active power filter acting as a voltage source in series with the power distribution line [1]. Another solution is to apply a shunt converter behaving as a current source to absorb or share the unbalanced currents [2], [3]. However, most local voltage support strategies are based on measuring the polluting load current, being therefore not always easy to implement, especially when multiple loads are present. Hence, the application of shunt converters to enhance the local voltage quality on the basis of local measurements, which are not taken beyond the converter's point of common coupling (PCC), has received increasing attention in recent years [4]–[10].

However, limited research has been carried out on how to correct the negative-sequence component and harmonics at the same time by means of three-phase grid-connected converters using only local measurements. Specifically, discussion on three-phase local voltage support is mostly carried out in the synchronous reference frame, making it difficult to build a natural analogy between single-phase and three-phase converters when talking about harmonic compensation, particularly in the context of Norton equivalent model derivation.

In this paper an integrated control method that decouples active power regulation and local voltage support for a three-phase three-wire GCC system is proposed on the basis of only stationary reference frames. The active power regulation is achieved by accurate control of the GCC output current fundamental positive-sequence component. Simultaneously, local voltage support is realized by providing an adjustable low-impedance path (through the GCC) for the unbalanced and harmonic components coming from the grid network. By doing so, the distorting current components, as introduced by local asymmetrical or non-linear loads, take the path through the GCC, instead of flowing towards the grid.

In Section II the GCC system architecture and control strategy are briefly presented. Then, in Section III analysis of the control approach through complex harmonic transfer functions is carried on. The accuracy of the control algorithms is assessed by simulation results, shown in Section IV, where the improvement on local voltage unbalance and total harmonic distortion is quantified by figures-of-merit. Section V points out design issues related to the switching frequency, and Section VI presents the main conclusions.

II. SYSTEM ARCHITECTURE AND CONTROL STRATEGY

A. System architecture

As can be seen in Fig. 1, only the local voltage at the PCC and the local converter output current are measured by the controller, characteristics which distinguish the proposed control strategy from conventional ones measuring the load current beyond the PCC for voltage support.

B. Stationary reference frames

The converter output line currents and the PCC line voltages are denoted by real vectors in the abc stationary frame as

$$\mathbf{i}_{abc} = \begin{bmatrix} i_a & i_b & i_c \end{bmatrix}^T$$
$$\mathbf{v}_{abc} = \begin{bmatrix} v_a & v_b & v_c \end{bmatrix}^T. \qquad (1)$$

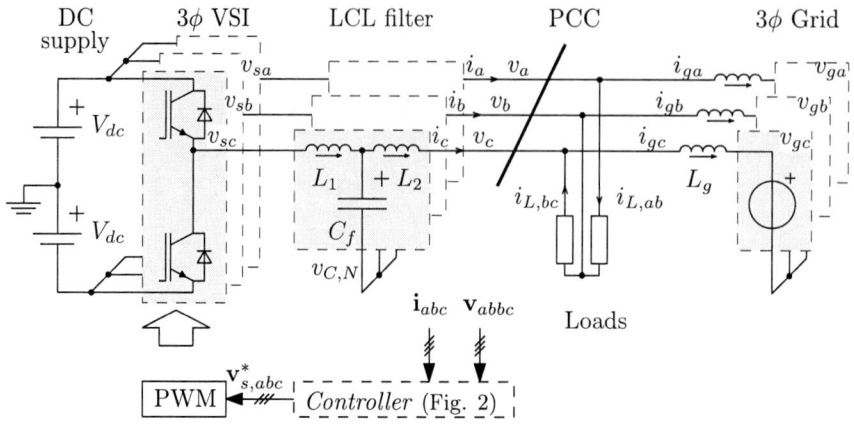

Fig. 1. Diagram of a grid-connected three-phase three-wire voltage-source converter system with local loads.

Because of practical convenience, stationary reference frames are applied in the rest of this paper for the control of three-phase GCC voltage and current quantities. The line currents and voltages in (1) are transformed to the $\alpha\beta$ quantities as

$$\mathbf{i}_{\alpha\beta} = \mathbf{T}_C \mathbf{i}_{abc}$$
$$\mathbf{v}_{\alpha\beta} = \mathbf{T}_C \mathbf{v}_{abc} \qquad (2)$$

where $\mathbf{T}_C$ is a Clarke transformation matrix,

$$\mathbf{T}_C = \frac{2}{3} \begin{bmatrix} 1 & -1/2 & -1/2 \\ 0 & \sqrt{3}/2 & -\sqrt{3}/2 \end{bmatrix}. \qquad (3)$$

In (2), the corresponding signals in the $\alpha\beta$ reference frame are denoted as

$$\mathbf{i}_{\alpha\beta} = \begin{bmatrix} i_\alpha & i_\beta \end{bmatrix}^T$$
$$\mathbf{v}_{\alpha\beta} = \begin{bmatrix} v_\alpha & v_\beta \end{bmatrix}^T. \qquad (4)$$

Furthermore, in order to take advantage of the circuit symmetry in Fig. 1, the real-valued $\alpha\beta$ quantities are grouped in complex numbers

$$\underline{i}_{\alpha\beta} = i_\alpha + j i_\beta$$
$$\underline{v}_{\alpha\beta} = v_\alpha + j v_\beta. \qquad (5)$$

In the following, it should be noted that the symbols fo real vectors and matrices are in bold, symbols with a underline, e.g. $\underline{(.)}$, represent complex quantities or transfer function polynomials with complex coefficients, and, unless mentioned otherwise, symbols without a underline bar denote real-valued quantities or transfer functions with only real-valued polynomial coefficients.

C. Control strategy

This section elaborates on the structure of the controller in Fig. 1 for add-on voltage support. Note that the two objectives of the control strategy are to regulate the active power injection and to support the local voltage simultaneously. Since the average (not the instantaneous) active power is dominantly determined by the fundamental positive-sequence component of the converter output current, the active (and reactive) power

injection can be controlled by regulating the fundamental positive-sequence component of the converter output current. At the same time, because the fundamental negative-sequence components and other harmonics hardly contribute to the average active power, they can be regulated to correct the local voltage for unbalanced and harmonic compensation. Therefore, the controller is composed of two loops: one for the control of the output current fundamental positive-sequence component and the other for the regulation of the PCC voltage fundamental negative-sequence component and other harmonics.

Accordingly, a parallel current-voltage control architecture is considered in this paper, because it can well generalize a current controller [2], [3], a hybrid current-voltage control scheme [5], [6] and a parallel scheme. Fig. 2 shows the block diagram of the current-voltage control architecture. The current reference consists of only fundamental positive-sequence component, whose magnitude is formulated by an active power regulation loop shown in Fig. 3.

The current and voltage controllers in Fig. 2 are decoupled in the frequency domain, following [11]:

$$\underline{C}_{i,\alpha\beta}(s) = K_p + K_i \frac{\omega_1}{s} + K_{res,1}^+ \underline{H}_{res,1}^+(s)$$
$$\underline{C}_{v,\alpha\beta}(s) = K_{res,1}^- \underline{H}_{res,1}^-(s) + K_{res,h} \sum_{h \in \mathbb{N}_h} H_{res,h}(s) \quad (6)$$

in which normalized complex transfer functions are given by

$$\underline{H}_{res,1}^+(s) = \frac{\omega_1}{s - j\omega_1 + \delta_1\omega_1}$$
$$\underline{H}_{res,1}^-(s) = \left(\frac{\omega_1}{s + j\omega_1 + \delta_2\omega_1} \right) \left(\frac{s - j\omega_1}{s - j\omega_1 + \delta_4\omega_1} \right)$$
$$H_{res,h}(s) = \frac{\omega_1 s}{s^2 + \delta_3(2h\omega_1)s + (h\omega_1)^2} \quad (7)$$

where ω_1 is the fundamental angular grid frequency, and δ_1, δ_2, δ_3 and δ_4 are damping factors to shape the bandwidth of the resonant filters in order to improve convergence to steady-state operation [12]. Further, K_p, K_i, $K_{res,1}^+$, $K_{res,1}^-$ and $K_{res,h}$ in (6) are real-valued adjustable gains that should be designed for accuracy and stability.

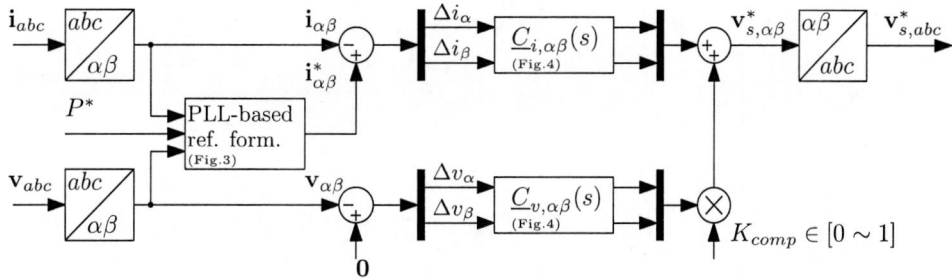

Fig. 2. Block diagram of decoupled (and, at last, integrated) current and voltage control in stationary reference frame.

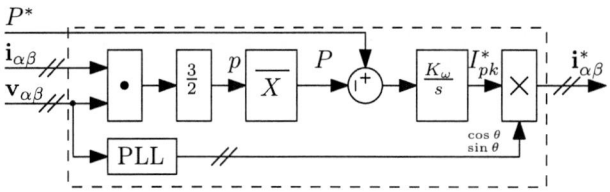

Fig. 3. Diagram of the control loop for (average) active power regulation.

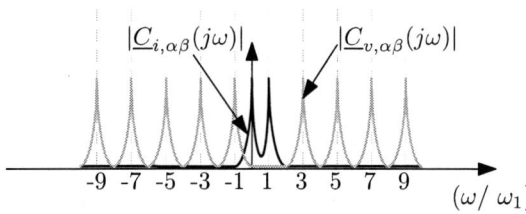

Fig. 4. Complementary operation of the current and voltage resonant controllers in Fig. 2, as function of frequency.

Fig. 4 illustrates the complementary operation of the resonant current and voltage controllers. As outlined, the current controller works actively at the zero and positive fundamental frequencies ($\{0, 1\}$), and, complementarily, the voltage controller works actively at the negative fundamental and harmonic frequencies ($\{-h, \cdots, -3, -1, 3, \cdots h\}$). Note in (6) that, in order not to interfere with the control of the fundamental positive-sequence component in the GCC system output current, the harmonic sequence component of order $h = 1$ is not included but attenuated by a notch filter in the voltage controller ($1 \notin N_h$).

III. MODELING AND ANALYSIS OF THE CLOSED-LOOP CONVERTER SYSTEM

In this section a frequency-domain equivalent model is derived for the closed-loop three-phase three-wire GCC system in Fig. 2.

A. Open-loop system modeling

Neglecting the PWM switching-frequency harmonics, the per-phase two-level VSI is modelled as a controlled average voltage source, as depicted in Fig. 5.

B. Equivalent sequence model

It is assumed that the devices of the 3ϕ LCL filters in Fig. 5 (inverter-side inductor Z_1, filter capacitor Z_c and grid-side inductor Z_2) are identical for each phase, and the mid-point voltage of the LCL filter, $v_{C,N}$, is referenced to the middle point of the dc supply. Therefore, the relation between per-phase currents and voltages can be described as

$$\frac{\mathbf{v}^*_{s,abc}}{Z_1(s)} - \frac{\mathbf{v}_{abc}}{Z_1(s)} - \frac{Z_2(s)}{Z_1(s)}\mathbf{i}_{abc} =$$

$$= \mathbf{i}_{abc} + \frac{\mathbf{v}_{abc}}{Z_c(s)} + \mathbf{i}_{abc}\frac{Z_2(s)}{Z_1(s)} - \frac{1}{Z_c(s)}\begin{bmatrix} v_{C,N}(s) \\ v_{C,N}(s) \\ v_{C,N}(s) \end{bmatrix}. \quad (8)$$

Applying the Clarke transformations of (2) to (8) results in

$$\frac{\mathbf{v}^*_{s,\alpha\beta}}{Z_1(s)} - \frac{\mathbf{v}_{\alpha\beta}}{Z_1(s)} - \frac{Z_2(s)}{Z_1(s)}\mathbf{i}_{\alpha\beta} = \mathbf{i}_{\alpha\beta} + \frac{\mathbf{v}_{\alpha\beta}}{Z_c(s)} + \mathbf{i}_{\alpha\beta}\frac{Z_2(s)}{Z_c(s)}. \quad (9)$$

After some rearrangements, (9) becomes

$$\mathbf{i}_{\alpha\beta} = -\frac{1}{Z_o(s)}\mathbf{v}_{\alpha\beta} + \frac{k(s)}{Z_o(s)}\mathbf{v}^*_{s,\alpha\beta} \quad (10)$$

in which

$$k(s) = \frac{Z_c(s)}{Z_c(s) + Z_1(s)} \quad (11)$$

$$Z_o(s) = \frac{Z_c(s)Z_1(s)}{Z_c(s) + Z_1(s)} + Z_2(s) \quad (12)$$

where the gain $k(s)$ and the equivalent open-loop impedance $Z_o(s)$ are determined from the LCL filter parameters, being

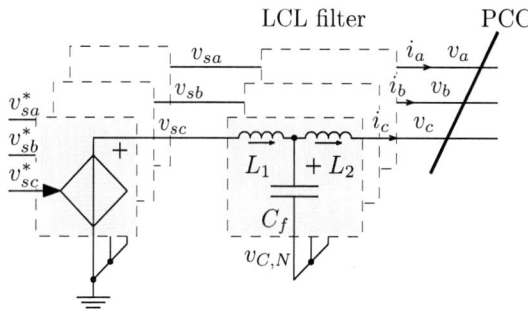

Fig. 5. Circuit representation of the 3-phase 3-wire GCC system in Fig. 1 in open loop control.

978-1-7281-4181-7/19 $31.00 © 2019 IEEE

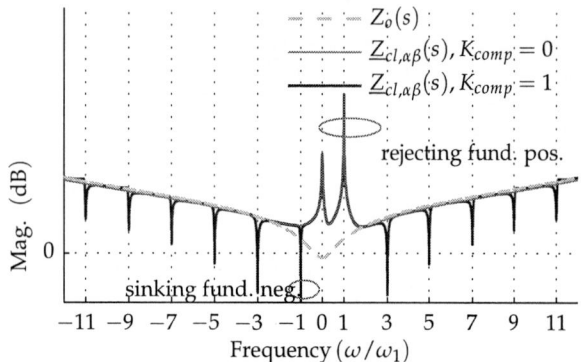

Fig. 6. Magnitude of the closed-loop GCC system equivalent output impedance as function of frequency, in comparison to the converter natural impedance.

therefore designated hereafter as the converter natural gain and impedance, respectively.

In view of the symmetric and decoupled impedances, the definitions in (5) are applied to (10), yielding a reduced open-loop model description in complex quantities as

$$\underline{i}_{\alpha\beta}(s) = \frac{k(s)}{Z_o(s)} \underline{v}_{s,\alpha\beta}^*(s) - \frac{1}{Z_o(s)} \underline{v}_{\alpha\beta}(s) \qquad (13)$$

which is a much more convenient representation for control design purposes of the three-phase GCC system.

C. Closed-loop system sequence model

In view of (13), the control law in Fig. 2 is rewritten as

$$\underline{v}_{s,\alpha\beta}^*(s) = (\underline{i}_{\alpha\beta}^*(s) - \underline{i}_{\alpha\beta}(s))\underline{C}_{i,\alpha\beta}(s) - K_{comp}\underline{v}_{\alpha\beta}(s)\underline{C}_{v,\alpha\beta}(s) \qquad (14)$$

where $\underline{i}_{\alpha\beta}^*(s)$ is the current reference. Combining the controller in (14) and the plant model in (10) yields

$$\underline{i}_{\alpha\beta}(s) = \underline{i}_{\alpha\beta}^*(s)\underline{k}_{cl,\alpha\beta}(s) - \frac{1}{\underline{Z}_{cl,\alpha\beta}(s)}\underline{v}_{\alpha\beta}(s) \qquad (15)$$

where $\underline{k}_{cl,\alpha\beta}(s)$ corresponds to an equivalent gain of a controllable current source with equivalent internal impedance $\underline{Z}_{cl,\alpha\beta}(s)$. It is found from (14) that

$$\underline{k}_{cl,\alpha\beta}(s) = \frac{\underline{C}_{i,\alpha\beta}(s)k(s)}{Z_o(s) + \underline{C}_{i,\alpha\beta}(s)k(s)}$$

$$\underline{Z}_{cl,\alpha\beta}(s) = \frac{Z_o(s) + \underline{C}_{i,\alpha\beta}(s)k(s)}{1 + K_{comp}\underline{C}_{v,\alpha\beta}(s)k(s)} \qquad (16)$$

where $K_{comp} \in [0 \sim 1]$ is a so-called compensating effort index. When $K_{comp} = 0$, no active voltage support is provided since only the current controller loop is active.

D. Closed-loop output impedance

Fig. 6 shows the magnitude of the complex output impedance $\underline{Z}_{cl,\alpha\beta}(s)$ in (16) as function of frequency.

The equivalent impedance is enhanced for the harmonic sequence components of order $\{0, 1\}$ and, compared to the converter natural impedance, lowered for the components of order $\{-11, \cdots, -3, -1, 3, \cdots 11\}$. As already illustrated [12], the grouped $\alpha\beta$-signal of order $h = 1$ corresponds

TABLE I
PARAMETERS OF THE THREE-PHASE THREE-WIRE GCC SYSTEM IN FIG. 1

Description	Symbol	Value
Supply voltage	V_{dc}	400V
PWM frequency	f_{sw}	10kHz
VSI-side inductor	L_1	3.6mH (0.4Ω)
Filtering capacitor	C_f	10μF (0.4Ω)
Grid-side inductor	L_2	2.0mH (0.4Ω)
Grid impedance	L_g	6mH(0.2Ω)
Grid per phase voltage	$V_{g,rms}$	220V
Grid frequency	f_g	50Hz

TABLE II
CONTROLLER PARAMETERS IN REFERENCE TO (6)

Component	Quantity	Value	Quantity	Value
Controller $\underline{C}_i(s)$	K_p	10	K_i	0.32
	$K_{res,1}^+$	4.78	δ_1	$2 \cdot 10^{-3}$
Controller $\underline{C}_v(s)$	$K_{res,1}^-$	0.25	δ_2	$4 \cdot 10^{-3}$
	$K_{res,h}$	0.20	$\mathbb{N}_h$	$\{3,5,\ldots 11\}$
	δ_3	10^{-3}	δ_4	0.10
Power loop	K_ω	$\omega_1/4000$	ω_1	$2\pi f_g$

to the fundamental positive-sequence components, $h = 0$ to the dc component, and $h = -1$ to the fundamental negative-sequence components. Therefore, under the proposed control strategy the sensitivity of the output current of the closed-loop converter to the fundamental positive-sequence and dc disturbance from the PCC voltage is lowered, and it is enhanced with respect to the fundamental negative-sequence and harmonic disturbances .

IV. SIMULATION RESULTS

The parameters of the three-phase three-wire GCC system (in reference to Fig. 1) in Table I are used for simulation tests. The controller parameters are listed in Table II.

A. Double-side harmonic spectrum

For the purpose of analysis of positive- and negative-sequence harmonic components, the double-side harmonic spectrum of the PCC voltage is represented as a complex Fourier series

$$\underline{v}_{\alpha\beta}(t) = \sum_{h \in \mathbb{Z}} V_h e^{jh\omega_1 t} e^{j\varphi_h} \qquad (17)$$

where $\mathbb{Z}$ is a set of integer numbers, and V_h and φ_h are the respective magnitude and initial phase of the synthesized harmonic sequence component of order h. As such, V_h denotes a positive- or negative-sequence harmonic component amplitudes when $h > 0$ and $h < 0$, respectively. Otherwise stated, the double-side harmonic spectrum in (17) allows for a compact representation of both harmonic and unbalance distortion of three-phase quantities, and it is applied to assess the local voltage quality in the sequence.

In order to quantify the quality of three-phase current/voltage signals like in (17), a figure-of-merit, so-called total harmonic distortion, is defined as

$$\mathrm{THD}_{ps} = \frac{\sqrt{\sum_{h \in \mathbb{Z}, h \neq 1} V_h^2}}{V_1^2} \qquad (18)$$

which is calculated in relation to the fundamental positive-sequence component of the signal.

B. Local voltage support

As mentioned earlier, the GCC output impedance for the harmonic and unbalanced components can be adjusted (by changing K_{comp}) for different degrees of local voltage support.

1) $K_{comp} = 0$: When the voltage support controller is disabled, the output impedance of the GCC system for the harmonic and unbalanced components is determined by the LCL filter. The simulation results in this case are shown in Fig. 7.

2) $K_{comp} = 0.1$: Fig. 8 shows the results when K_{comp} is set deliberately small in order to shape the output impedance of the GCC system to an intermediate low value (see Fig. 6 and (16)). It can be seen from Fig. 8 that compared to the results in Fig. 7, the total harmonic distortion of the PCC voltage and the grid current is reduced.

3) $K_{comp} = 1$: In Fig. 9 the compensation index is set to unity. In this case, the equivalent GCC output impedance for the targeting harmonic sequence signals of order $\{-11, \cdots, -3, -1, , 3, \cdots, 11\}$ is forced lower. It can be seen that, compared to the case in Fig. 8, the total harmonic distortion of the PCC voltage and the grid current is significantly reduced.

Summarizing, it can be seen from Figs.7 to 9 that increasing the compensation index K_{comp} helps to reduce the distortion of the local PCC voltage and grid current, which is in agreement with analysis of the harmonic transfer functions in Fig. 6. When K_{comp} is zero, the LCL filter is dominant in determining the GCC system output impedance for grid unbalanced and harmonic components. This makes the GCC control insensitive to unbalanced and harmonic disturbances from the grid. An increased K_{comp} enhances the control sensitivity, allowing the GCC system to compensate for a desired level of the local unbalanced or non-linear load currents.

C. Current reference step for active power regulation

Steering of active power flow is illustrated in Fig. 10. The current reference for active power regulation (I_{pk}^* in Fig. 3) steps from 0A to 6A. As a result, it is shown that, while also performing local voltage support as in Fig. 9, the GCC system achieves fast and stable tracking of the desired current components related to active power flow control.

V. IMPACT OF THE SWITCHING FREQUENCY

Due to its good performance regarding tracking of sinusoidal references, a first-order resonant filter at the grid fundamental frequency is selected for positive-sequence current control in (6), which sets the needed phase shift with respect

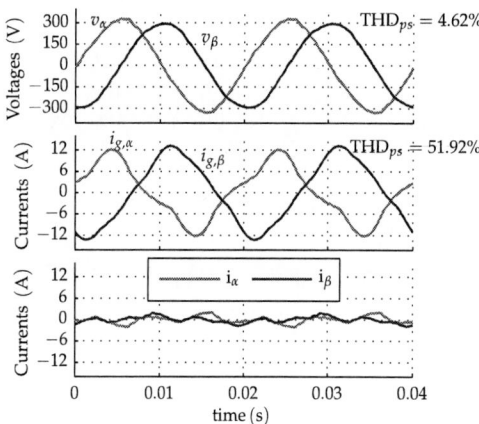

Fig. 7. Simulation results showing that the voltage harmonic distortion is slightly reduced when the GCC is operational and $K_{comp} = 0$.

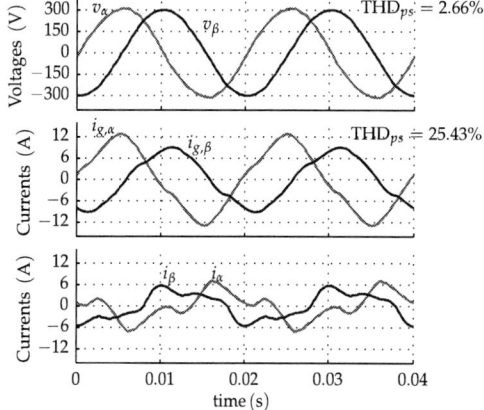

Fig. 8. Simulation results showing that the voltage harmonic distortion is further reduced when the GCC is operational and $K_{comp} = 0.1$.

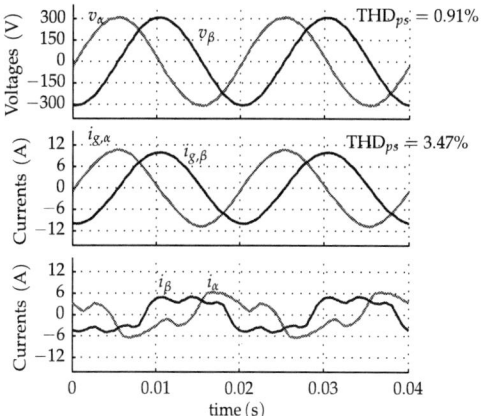

Fig. 9. Simulation results showing that the voltage harmonic distortion is significantly reduced when the GCC is operational and $K_{comp} = 1$.

to the fundamental grid voltage component (yielding therefore the desired charge/discharge active and reactive power flow). Voltage support is also achieved through first- and second-order resonant filters in (6) to attenuate a selection of harmonic components other than the fundamental one.

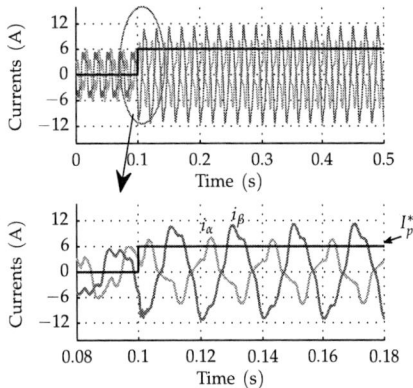

Fig. 10. Simulation results showing the GCC system's response when the current reference steps up aiming at an increase of active power transfer. The initial conditions (only voltage support before 0.1s) are the same as in Fig.9.

It is a common practical design recommendation to choose in (6) no more than a maximal number of harmonics to be attenuated such that $h < |h|_{\max}$, with

$$|h|_{\max} = 0.1\,\omega_s/\omega_1$$

where ω_s is the angular switching frequency and ω_1 the fundamental angular grid frequency. Furthermore, with regard to control stability and accuracy, the values of the adjustable gains for the first- and second-order filters in (6) should be high enough but limited, preferably chosen such that

$$K_{res,1}^+ = \omega_c L_2 \ \text{with} \ \omega_c \le 0.1\,\omega_s$$
$$K_{res,1}^-, K_{res,h} \ll \omega_c/\omega_1.$$

Therefore, the higher the switching frequency ω_s, the broader the design choices for achieving good filter performance. Due to superior material properties, wide band-gap semiconductors enable higher voltages, switching frequencies, and operating temperatures when compared to conventional Si technology. As a consequence, SiC semiconductors lead to the advancement of the proposed filtering methods aiming at improving power quality in distribution grids.

VI. CONCLUSION

The proposed control scheme for three-phase three-wire grid-connected converters consists of two decoupled loops, one for current control and another for voltage support. The local voltage at the PCC is measured and used in both loops: in one loop the voltage fundamental component is extracted via a PLL for converter output current synchronization with the grid; and in the other path the full voltage profile is used for unbalance and harmonic compensation. The resulting duty-cycle value that steers the pulse-width modulation of the converter switching legs is just the sum of the two previously calculated duty-cycles by each control loop. Since no external sensors beyond the PCC are required, the approach is readily applicable to already existing GCC systems as add-on local voltage support.

Results from numerical simulations validate the performance of the method, showing grid current harmonic distortion

being reduced from 51.92% to 3.47% and PCC voltage distortion from 4.62% to 0.91%. The superior material properties of the next SiC generation allows an advanced practical realization of the proposed algorithms.

ACKNOWLEDGMENT

This work has been conducted within HiPERFORM project and has received funding from the ECSEL Joint Undertaking (JU) under the Grant Agreement No. 783174. The JU receives support from the European Unions Horizon 2020 research and innovation programme and Austria, Spain, Belgium, Germany, Slovakia, Italy, Netherlands, and Slovenia.

REFERENCES

[1] H. Fujita and H. Akagi, "The unified power quality conditioner: the integration of series and shunt-active filters," *IEEE Transactions on Power Electronics*, vol. 13, no. 2, pp. 315–322, Mar. 1998.

[2] B. Singh and C. Jain, "A decoupled adaptive noise detection based control approach for a grid supportive SPV system," *IEEE Transactions on Industry Applications*, vol. 53, no. 5, pp. 4894–4902, Sep. 2017.

[3] R. S. R. Chilipi, N. A. Sayari, K. H. A. Hosani, and A. R. Beig, "Adaptive notch filter-based multipurpose control scheme for grid-interfaced three-phase four-wire DG inverter," *IEEE Transactions on Industry Applications*, vol. 53, no. 4, pp. 4015–4027, Jul. 2017.

[4] Z. Dai, H. Lin, H. Yin, and Y. Qiu, "A novel method for voltage support control under unbalanced grid faults and grid harmonic voltage disturbances," *IET Power Electronics*, vol. 8, no. 8, pp. 1377–1385, 2015.

[5] F. Nejabatkhah, Y. W. Li, and B. Wu, "Control strategies of three-phase distributed generation inverters for grid unbalanced voltage compensation," *IEEE Transactions on Power Electronics*, vol. 31, no. 7, pp. 5228–5241, Jul. 2016.

[6] C. Xu, K. Dai, X. Chen, and Y. Kang, "Unbalanced PCC voltage regulation with positive- and negative-sequence compensation tactics for MMC-DSTATCOM," *IET Power Electronics*, vol. 9, no. 15, pp. 2846–2858, 2016.

[7] F. H. M. Rafi, M. J. Hossain, and J. Lu, "Improved neutral current compensation with a four-leg PV smart VSI in a LV residential network," *IEEE Transactions on Power Delivery*, vol. 32, no. 5, pp. 2291–2302, Oct. 2017.

[8] S. D'Arco, M. Ochoa-Gimenez, L. Piegari, and P. Tricoli, "Harmonics and interharmonics compensation with active front-end converters based only on local voltage measurements," *IEEE Transactions on Industrial Electronics*, vol. 64, no. 1, pp. 796–805, Jan. 2017.

[9] X. Zhao, L. Meng, C. Xie, J. M. Guerrero, X. Wu, J. C. Vasquez, and M. Savaghebi, "A voltage feedback based harmonic compensation strategy for current-controlled converters," *IEEE Transactions on Industry Applications*, vol. PP, no. 99, pp. 1–1, 2017.

[10] M. M. Shabestary and Y. A. I. Mohamed, "Advanced voltage support and active power flow control in grid-connected converters under unbalanced conditions," *IEEE Transactions on Power Electronics*, vol. 33, no. 2, pp. 1855–1864, Feb. 2018.

[11] Y. Zhang, M. G. L. Roes, M. A. M. Hendrix, and J. L. Duarte, "Symmetric-component decoupled control of grid-connected inverters for voltage unbalance correction and harmonic compensation," *International Journal of Electrical Power & Energy Systems*, vol. 115, p. 105490, Feb. 2020.

[12] F. Wang, M. C. Benhabib, J. L. Duarte, and M. A. M. Hendrix, "Sequence-decoupled resonant controller for three-phase grid-connected inverters," in *Twenty-Fourth Annual IEEE Applied Power Electronics Conference and Exposition*, Feb. 2009, pp. 121–127.

978-1-7281-4181-7/19 $31.00 © 2019 IEEE

Implementation of a Didactic Platform for a Generic Load Torque Emulator Using Induction Machines and PWM Inverters

Luiz Otávio Campos de Medeiros
Energy and Electrical Systems Institute
Federal University of Itajubá (UNIFEI)
Itajubá, Brazil
lotavio@unifei.edu.br

José Carlos Grilo Rodrigues
Energy and Electrical Systems Institute
Federal University of Itajubá (UNIFEI)
Itajubá, Brazil
jcgrilo@unifei.edu.br

Angelo José Junqueira Rezek
Energy and Electrical Systems Institute
Federal University of Itajubá (UNIFEI)
Itajubá, Brazil
rezek@unifei.edu.br

Nery de Oliveira Junior
Director of the Nery Engenharia Ltda
Nery Engenharia Ltda
Delfim Moreira, Brazil
nery@nery.com.br

Rafael Di Lorenzo Corrêa
Energy and Electrical Systems Institute
Federal University of Itajubá (UNIFEI)
Itajubá, Brazil
rafaeldlcorrea@unifei.edu.br

Alexandre Viana Braga
Energy and Electrical Systems Institute
Federal University of Itajubá (UNIFEI)
Itajubá, Brazil
avbdsc@gmail.com

Christel Enock Ghislain Ogoulola
Energy and Electrical Systems Institute
Federal University of Itajubá (UNIFEI)
Itajubá, Brazil
christel@unifei.edu.br

Vinicius Zimmermann Silva
Energy and Electrical Systems Institute
Federal University of Itajubá (UNIFEI)
Itajubá, Brazil
vinicius.zimmermann@yahoo.com.br

Marcos Leonardo Ramos
Energy and Electrical Systems Institute
Federal University of Itajubá (UNIFEI)
Itajubá, Brazil
marcoslramos@hotmail.com

Abstract— **This work proposes a implementation of a generic load torque emulator using PWM inverters and induction machines, as a continuation of a previous work, whose implementation has used a load simulator, using scalar type control. This one will also be described in this paper and took place at the electrical drives laboratory of UNIFEI (Federal University of Itajubá – Brazil), to aid the understanding of this proposed alternative continuation work, using vector control. Different types of loads, as the most commonly found in industry, such as constant load torque, speed dependent linear load torque, speed dependent quadratic load torque and others, will be considered. It is intended that with the use of the proposed workbench, the determination of the tested motors performance will be available. The torque signal, to be registered, is originated from the motor pulse-width modulation inverter, torque transducer, isq current. Experimental results will be presented and discussed.**

Keywords— *AC drives, Torque emulator, AC machines, PWM inverter.*

I. Introduction

The induction motor drive systems [1–2] are the most used in industrial applications, because of their simplicity, easy maintenance, lower cost and robustness. The motors at the UNIFEI electrical drives laboratory, are operating with no load conditions and in order to propitiate a load torque to these ones, a generic load torque emulator [8-13] using pulse-width modulation (PWM) inverters [7], is being provided to improve the teaching of electrical drives inserted to the related courses. In the previous work, developed software has been employed using hysteresis control, to provide the load torque simulation. In this present work, it will be used the own software of the master drives, with vector control, type closed loop torque control (only for

induction motors) by means of the parametrization of the inverter, parameter P163=5.

II. Methodology and Overall Description of the Proposed Technique

Fig. 1 illustrates the curve torque versus slip, for induction motors, showing that if the slip s is negative (inferior part of the curve), the second machine runs at generator, being so the load for the coupled motor. This is obtained when the mechanical rotor speed is higher than the synchronous speed of the rotating field N_s, shown in the equation 2 [3-6].

$$N_s = 60f/p \, [rpm] \qquad (1)$$

$$s = (N_s - N_r)/N_s \qquad (2)$$

where N_s = synchronous speed of the rotating field[rpm], N_r= mechanical rotor speed, s = slip, f = frequency of the supply voltage of the machine, p = number of pair poles. Fig. 2 [18] shows the proposed experimental arrangement. In this one, there are two units. The first one is the PWM 1 (motor unit) and the second one is the PWM 2 (generator unit), both units coupled in the same shaft, by using an elastic device. This second unit operates as load to the motor. The capacitors of the DC links are connected in parallel, in order to avoid overvoltage in these ones, as described in [18]. The rectified output voltages of the AC/DC no controlled converters, 6-pulse rectifier converters, using diode semiconductors, feed the DC links.

978-1-7281-4181-7/19 $31.00 © 2019 IEEE

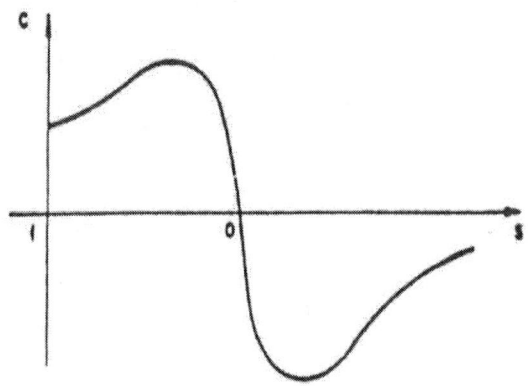

Fig.1. Curve torque versus slip for induction machines.

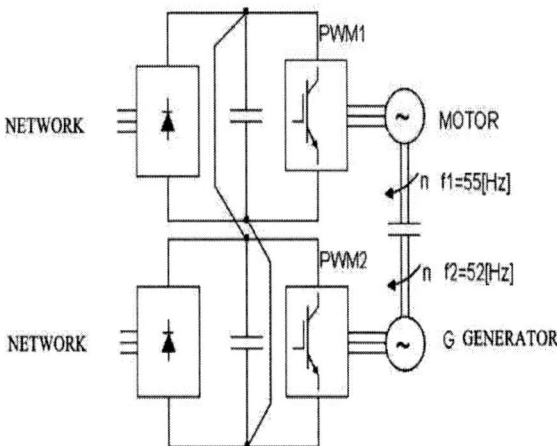

Fig. 2: Experiment arrangement proposed.

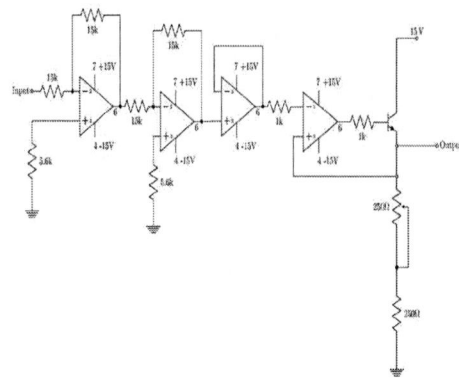

Fig.3. Driver circuit for the frequency adjustment of the inverter 2 .

Fig. 3 shows the used analogic driver circuit [18]. The input is the signal of D/A (Digital-to-Analogic converter) of the PCL 711 board, which has been used in both works. The output (0 – 10V) signal, supplies the PWM 2 analogic input, in order to change the frequency of this one. The use of the driver circuit is necessary in order to avoid undesirable voltage droop in the frequency reference signal, employed in the previous work [18]. The idea of the continuation work, is the using of the dedicated own software of the inverter to substitute the developed software of the previous work, using C++ language [14], which can be used in inverters which do not have the resource of torque control, justifying so this mentioned previous work. Therefore, the signal to change the inverter frequency, instead of the driver circuit output signal of Fig. 3, will be provided by the software of the inverter 2. Therefore, the PWM master drives of Siemens has the possibility of using a own software to enable the load torque control as a continuation and option to the previous work, in order to make possible an alternative load torque implementation. Therefore a comparison of both results related to the drive dynamic performance will be possible [17].

The torque transducer is obtained from the analogic output signal 1 of inverter 1, using isq motor current, as the motor torque transducer, parametrization inverter 1, motor torque, P 655.1=264. This analogic signal is the input 2 of the A/D PCL 711B board; the input 1 of the A/D of PCL 711B is the analogic signal of the tachogenerator speed transducer and both signals are used to be processed by the developed software in the previous work, to provide the output control signal D/A converter of the PCL 711B board, to modify the frequency of the inverter 2 with the parametrization P443 = 1003, analogic input 1 of this one (see fig. 4). This control can be explained in a way that if the torque is greater than the referenced torque, the control acts increasing the frequency of the inverter 2 and in the contrary, if the torque is smaller than the referenced torque, the control acts decreasing the frequency of the inverter 2. If the error torque is inside of a defined window, no action is required (hysteresis torque control). In figure 2, consider, for instance, the case in which the frequency of the inverter 1(motor unit) is 55 [Hz], corresponding to a mechanical speed n in the shaft, greater than the synchronous speed of the rotating field of unit 2, in which the frequency is 52 [Hz]. Therefore the machine of unit 2 operates as generator, as is desirable. If the frequency of the unit 1 is modified to a lower value, corresponding to a mechanical speed also lower than the previous one, the frequency of the unit 2 will be automatically changed by the control system, and the operation of unit 2 is maintained as generator. Fig. 4 illustrates the scheme of the implemented previous work prototype.

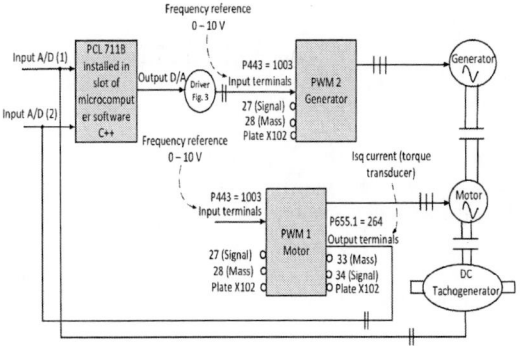

Fig. 4: Scheme of previous work prototype

Fig. 5 shows the circuit scheme of the alternative proposed work prototype.

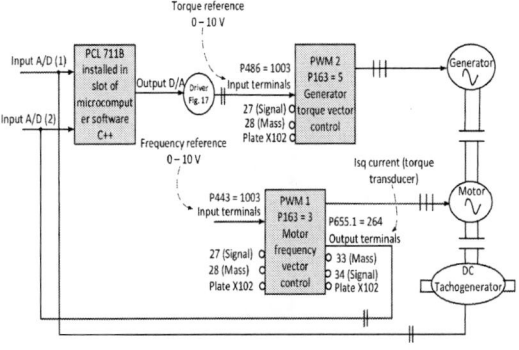

Fig. 5: Scheme of the alternative proposed work prototype

The basic difference of both works is that in the proposed continuation work, the adjustment of the frequency of the PWM 2 generator, for propitiating the desirable load torque, is made automatically by the own software of the PWM 2 inverter obtained by the option closed loop torque control, available in master drives of the referred workbench, parametrization P163 = 5 (vector control- closed loop control torque-only for induction motors). The reference torque is the analogic signal input 1, of PWM 2 inverter, parametrization P 486=1003, input terminals 27 (signal) and 28 (mass). Therefore in this alternative work of continuation, using the option torque control of the generator PWM 2 inverter, the parametrization should be P486 = 1003, instead of P443 = 1003, as in the previous work, in which there is a frequency control, instead of torque control, as shown in figures 4 (previous work) and 5 (proposed alternative work). The input signal for the reference torque is connected in plate X102 of the PWM 2 inverter, signal of the D/A PCL 711B [15], (see fig. 5), output control signal of the program written in C++ of the PCL 711B which is enabled by the MENU:

1) Loads with constant torque(constant load torque);

2) Loads with torque proportional to the speed (linear load torque);

3) Loads with torque proportional to the square of the speed(quadratic load torque);

4) Another generic type of load torque with suitable equation written in the C++ program.

The equipment data used in the workbench are as follow: PWM inverters - Siemens Master Drives VC (Manufactured in Germany), 380/460 V, 25.5 A; the voltage used in the implementation was three-phase 440 V.

Machines - (generator WEG, Jaraguá do Sul-SC, Brazil) and (motor EBERLE; Caxias do Sul [RS], Brazil), both three phase induction machines of 10 Hp, 380 V, and 16 A.

Microcomputer - Pentium 3, 600 MHz, card PCL 711 B – Advantech.

III. OBTAINED EXPERIMENTAL RESULTS

Figs. 6, 7, 8 and 9 show the results obtained in the previous work, loads torques in pu, type constant torque = kt, linear torque = kt*n, and quadratic torque = $kt*n^2$, with reduced load factor, fc (or torque constant Kt), equal to 0.6, with 1 p.u. rated torque, 22 Nm and 1 p.u. speed n 1800 rpm.

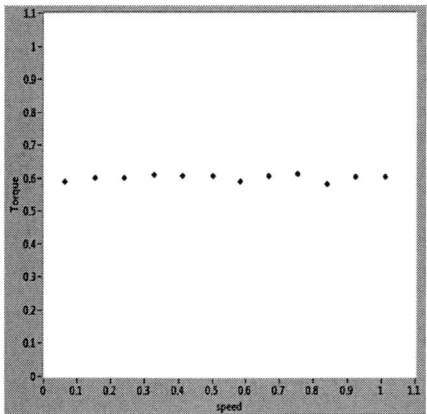

Fig.6: Constant load torque.

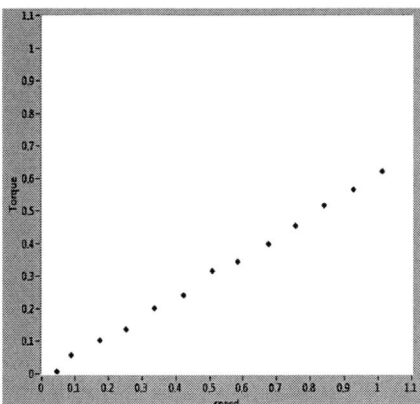

Fig.7: Linear load torque.

978-1-7281-4181-7/19 $31.00 © 2019 IEEE   51

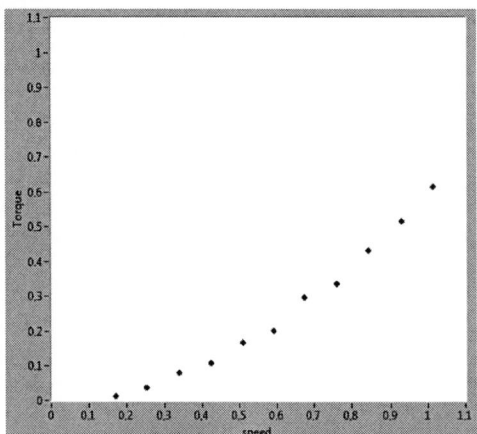

Fig.8: Quadratic load torque.

Fig. 9 shows the dynamic response for linear torque control, speed curve(red) and black curve (torque), both in pu, top curve, non-filtered signals and bottom curve, filtered signals, using the previous work [18]. Fig. 10 shows the workbench that has been used.

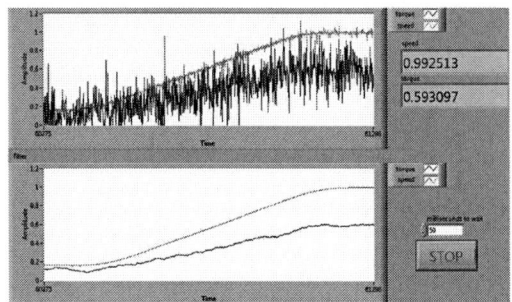

Fig. 9: Dynamic response for linear torque control speed curve (red) and black curve (torque).

Fig. 10 shows an overview of the workbench utilized[16]. The inverters are seen in the front, the computer on the right, and the machines in the rear. The inverters are Siemens Master Drives VC (Germany), 380/460 V, 25.5 A; the voltage used was 440V. The machines are a WEG (generator; Jaraguá do Sul [SC], Brazil) and EBERLE (motor; Caxias do Sul [RS], Brazil) and three-phase induction motors of 10 CV, 380 V, and 16 A.

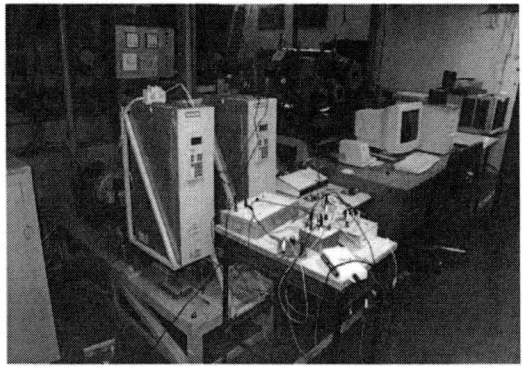

Fig.10: Overview of the used workbench.

The PWM 2 inverter was parametrized to enable the vector control, with torque control in the menu of parametrization P163= 5 (closed loop torque control), only for induction motors, resource available in this Siemens master drives inverter. Remembering that not all inverters have this feature, so the previous work of dedicated software for inverter torque control is in fact justified.

In the proposed work the frequency of the inverter will be controlled by the (closed loop torque control), parametrization of inverter PWM 2 P163 = 5, as already mentioned, analogic input 1 of PWM inverter 2, parametrization P 486 = 1003, for terminals 27 (signal) and 28 (mass) of the terminals plate X102 of the generator PWM 2 inverter.

IV. RESULTS FOR THE CLOSED LOOP TORQUE CONTROL

Fig. 11 shows the dynamic response of the closed loop control torque for linear torque with also the linear speed increasing (t = 40s), for kt = 0.6 (qualitative aspect).

Fig.12 shows the dynamic response of the closed loop control torque for quadratic torque with also the linear speed increasing, (t = 40s), for kt = 0.6 (qualitative aspect).

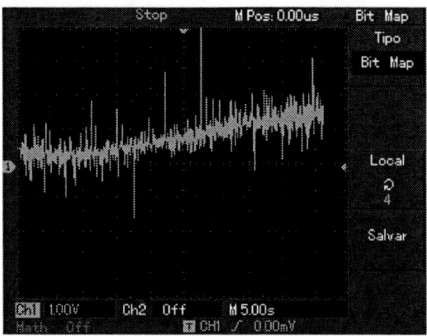

Fig.11: Dynamic response of the closed loop torque control for linear torque, with also the linear speed increasing, for kt=0.6.

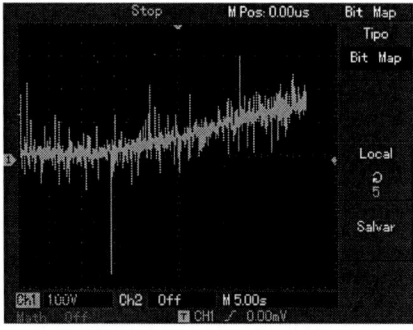

Fig.12: Dynamic response of the closed loop torque control for quadratic torque, with also the linear speed increasing, for kt=0.6.

Figs. 13 and 14 show the setpoint torque for linear and quadratic torque control with respect to the speed. The

978-1-7281-4181-7/19 $31.00 © 2019 IEEE

setpoint should be negative, because the torque of the generator emulates the load torque. The final speed is 1800 rpm. Fig. 15 shows the torque response for constant load torque equal to 0.6 pu, with speed variation from 300 rpm to 1800 rpm. Fig.16 shows steps (positive and negative) of applied load torque (maximum value equal to 0.6 pu), proving the dynamic stability of the proposed load torque application system.

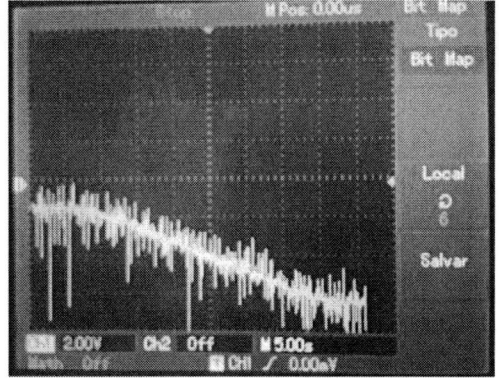

Fig.13: Setpoint signal for linear torque control.

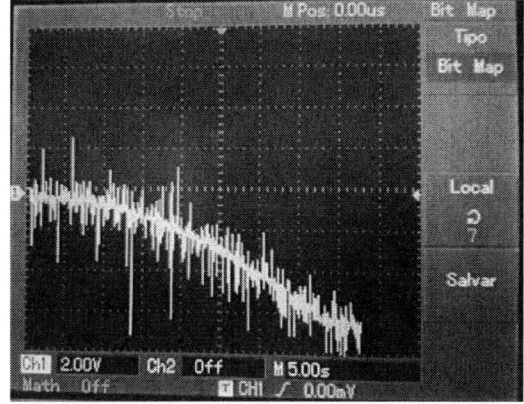

Fig.14: Setpoint signal for quadratic torque control.

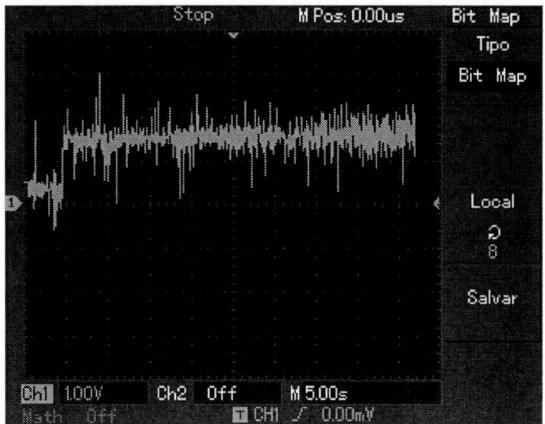

Fig. 15: Response for constant load torque equal to 0.6 pu.

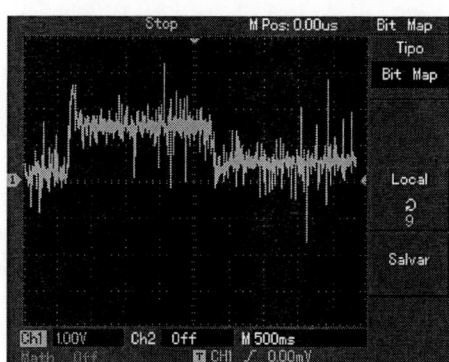

Fig. 16: Steps (positive and negative) of applied load torque.

The driver for the option of closed loop load torque should be modified. Therefore, instead of the driver presented in Fig. 3, the driver used is shown in Fig. 17, because the setpoint load torque should be negative. So, a PNP transistor PNP B601, has been used for this purpose.

Fig. 18 shows the implemented protection circuit, so when a fault occurs in the master (motor unit), the generator unit (slave) will be also turned off, avoiding the over speed of this unit [16]. When the terminal 6 and 7 are closed, this protection is active, by closing also the contacts 13 and 16. A rearguard protection circuit will be also implemeted, considering the case in which the speed of the motor generator group exceeds 2000 [rpm].

The proposed alternative continuation work using vector control will present a fast dynamic response with respect to the step changes in the load torque reference, comparatively with the results of the previous work, which has used scalar techniques of load torque control.

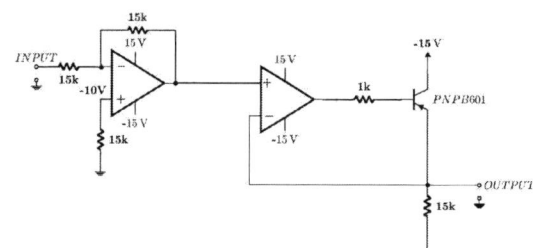

Fig. 17: Modified driver circuit.

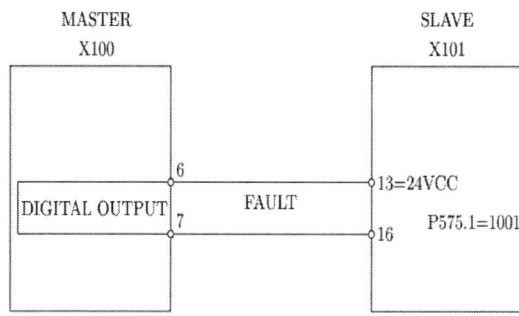

Fig. 18: Implemented protection circuit.

V. CONCLUSION

In this work an implementation of a simple load torque emulator using induction machines and PWM inverters is proposed, as a continuation of a previous work which has used a developed software to modify the frequency of the PWM generator inverter, for obtainment of the load torque. This resource can be used in inverters in general, because the most of industrial inverters do not have the option of closed loop torque control. However, some inverters have this option, as for instance the Siemens PWM master drives and so the modifying of the frequency inverter, to enable the desirable load torque, can be made by the own software of the inverter (vector control), with expected also of better dynamics response, because the previous work uses a technique of scalar type control. A comparison of the dynamic performance of the obtained load torque is possible, taking into account that in the proposed work the control using vector control (expert parametrization of the PWM generator inverter), provides a faster response with respect to modifications in load torque references, comparatively with those provided in the previous work, using scalar type control. The comparison of the dynamic response of the torque, by analyzing the results obtained in Fig. 9 (previous work) and in Fig. 11 (vector control) are similar, because the speed has been modified linearly, from zero to the final speed 1800 rpm, with a relatively long time (about 40 s). In the cases of fast variation in the speed reference, the vector control acts in fact with better faster dynamic. The load torque reference is negative, because the generator should actuate as a load torque for the shaft coupled motor. The load torque, which can be obtained by the choice in the written C++ program, is enabled by the output control signal of D/A converter of PCL 711B Advantech board. Therefore, the load torque can be as the type generic one.

Finally one conclude that a useful and didactic resource has been obtained, with this reseach work, by the implementation in laboratory of this workbench, important improvement contribution for teaching AC drives in our university. This work is also in the sense of promoting the interaction university-industry, very important aspect nowadays, by the possibility of using the proposed workbench in training courses offered to industry. The presented development is also subject of MSc dissertation of the first author of this paper.

REFERENCES

[1] J. Rodríguez, R. M. Kennel, and J. R. Espinoza, "High-performance control strategies for electrical drives: An experimental assessment," IEEE Trans. Ind. Electron., Vol. 59, No. 2, pp. 1208–1216, February 2012.

[2] B. K. Bose, "Modern Power Electronics and Variable FrequencyAC Motor Drives", IEEE Press, 1997.

[3] A. E. Fitzgerald, C. Kingsley Jr., and . Kusko, "Electrical Machines", McGraw Hill do Brasil, 1977 (in Portuguese).

[4] I. L. Kosow, "Electrical Machines and Transformers,", Brazil, Globo S.A., 1997.

[5] A. J. J. Rezek, "Electrical Machines Basic Fundamentals: Theory and Tests", Brazil: Acta and Synergia, 2011 (in Portuguese).

[6] M. G. Simoes, and F. A., and Farret, "Alternative Energy Systems: Design and Analysis with Induction Generators", (2nd ed.), Boca Raton, FL: Taylor & Francis, CRC Press, 2008.

[7] Simovert, Master Drives, Siemens: Operation Manual.

[8] Z. HakanAkpolat, G. M. Asher, and J. C. Clare, "Experimental dynamometer emulation ofnon-linear mechanical load" IEEE Trans. Ind. Electron., Vol. 46, No. 2, pp. 532–539, 1998.

[9] O. Vodyakho, M. Steurer, C. S. Edrington, and F. Fleming, "An induction machine emulator for high-power applications utilizing advanced simulation tools with graphical user interfaces," IEEE Trans. Energy Conversion, Vol. 27, No. 1, pp. 160–172, March 2012.

[10] Z. HakanAkpolat, G. M. Asher, and J. C. Clare, "Experimental dynamometer emulationof nonlinear mechanical loads," IEEE Trans. Ind. Appl., Vol. 35, No. 6, pp. 1367–1373, November/December 1999.

[11] Hassania, B., Sicard, P., and Ba-razzouk, A., "Solutions to typical motor load emulation control problems," Electrimacs, 18–21 August 2002.

[12] V. Fernão Pires, J. F. Martins, and T. G. Amaral, "Web based teaching of electrical drive susing a mechanical load simulator," Proc. 34th IEEE IECON, pp. 3545–3550, 10–13 November 2008.

[13] M. A. A. Pedrasa, and V. L. S. Delfin, "Low cost mechanical load emulator," IEEE Region 10. Conference, pp. 1–3, 14–17 November 2006.

[14] Pappas, C. H., and Murray, W. H., Turbo C Total and Complete, Macron Books do Brasil, 1991 (in Portuguese).

[15] Advantech Co., Ltda., PCL-711B—PC-MultiLab User's Manual, Taiwan: Advantech Co., Ltd., August 1993.

[16] Nery Trade and Representations Engineering Ltd, available at: www.nery.com.br.

[17] Davari, S. A., Khaburi, D. A., Wang, F., and Kennel, R. M., "Using full order and reduced order observers for robust sensorless predictive torque control of induction motors," IEEE Trans. Power Electron., Vol. 27, No. 7, pp. 3424–3432, July 2012.

[18] Rodrigues, J. C. G.; Rezek A. J. J ;Martinez, M. L. B.; Bernardes, D. F. ; Oliveira Junior N. "Implementation of a Simulator for the Most Commonly Found Industrial Motor Loads Based on Pulse- width Modulation Inverters and Torque Estimator". Electric Power Components and Systems, v. 41, p. 345-364, 2013.

Energy Efficient Control of Synchronous Machines in Deep Field-Weakening Operation Including Saturation Effects

Joao Bonifacio[1,2]
[1]System-House E-Mobility
ZF Friedrichshafen AG
Friedrichshafen, Germany

Ralph Kennel[2]
[2]Institute for Electrical Drive Systems and Power
Electronics
Technical University of Munich
Munich, Germany

Abstract— **In this paper, a generic numerical framework for considering saturation in the reference current generation regarding energy optimization in deep field weakening for synchronous machines with strong saliencies is proposed. It is based on the geometrical interpretation of the optimization problem in the flux plane where the saturation effect is considered by using differential inductances. The algorithm is validated through Finite Element Analyses in a PMa-SynRM and a RSM.**

Keywords—Synchronous machines; deep field weakening; loss minimization; Maximum torque per flux.

I. INTRODUCTION

Recent trends on electrification and stringent emission regulations, especially related to automotive applications, have brought the energy efficiency of electrical drives to the focus of academic research and industrial development. The core of these developments has taken place in the study of more efficient machine designs and control strategies [1]. In this context, synchronous machines, especially the Permanent Magnet Assisted Synchronous Machine (PMa-SynRM) and the Reluctance Synchronous Machine (RSM), are good candidates for high performance industry applications due to their inherent advantages like high power density, high efficiency, cost, flux weakening capability, and so on.

Especially for traction applications, the machines have usually to be able to be operated in deep field weakening range. In order to take full advantage of the intrinsic characteristics of synchronous machines in these situations, it is necessary to use control algorithms that can maximally exploit the available voltage even at higher speeds. This can reached through the use of Maximum Torque per Flux (MTPF), since the core losses, which are flux dependent, are dominant in this speed range.

There are several approaches available in literature for solving the issue of generating reference currents for synchronous machines in deep field weakening operation. These approaches can be classified into (a) Look-up table (LUT) based, (b) voltage controller (c) analytical and numerical methods. In [2, 3] several LUTs with the optimal points for all operating regions of the electrical machine, including deep field weakening, are used for reference current generation. Although being a straightforward solution for the problem, the use of LUTs has the disadvantage of requiring a huge commissioning effort for generating them and great amount of memory for their storage. Moreover, the compensation of temperature related effects can be very challenging.

A combination of a voltage controller and feed-forward LUTs is proposed in [4]-[6]. It has the advantage of reaching a certain parameter independency in field weakening operation, while keeping a simple structure for generating the reference currents. Nevertheless this approach still has many disadvantages of the LUT-based algorithms as well as the difficulty of tuning the dynamics of the voltage control loop. In [7] a strategy for coping with oscillations generated by the voltage controller in deep field weakening is proposed.

Authors in [8]-[11] propose solving the optimal conditions of MTPF analytically. It is performed either by directly solving the resulting fourth order polynomials using the Ferrari method or by reducing them into second order ones. These methods have the advantage that parameter variations can be compensated online and the measuring effort related to the LUTs can be avoided in some extent. Nevertheless, the real-time implementation of some of these methods remains challenging for industrial microcontrollers. Moreover, the effect of saturation and the stator resistance are neglected in all these works. Recently, an analytical solution considering saturation as well as the stator resistance was proposed in [12].

In this work a generic numerical framework for solving the MTPF condition considering saturation effects is proposed. It is based on the MTPF formulation in the flux plane and uses a geometrical argument to redefine the optimization problem. Newton Method is used for solving the iterations and a convergence proof is derived. The proposed strategy is validated both for a PMa-SynRM and a RSM using Finite Element Analysis.

II. MACHINE MODELING AND ENERGY EFFICIENT CONTROL

A. Mathematical Model of PMSM

The set of stator voltages u_d and u_q in the synchronous reference frame of a permanent magnet synchronous machine is shown in Eqs. (1) and (2).

$$u_d = R_s i_d + \frac{d\Psi_d}{dt} - \omega_s \Psi_q \qquad (1)$$

$$u_q = R_s i_q + \frac{d\Psi_q}{dt} + \omega_s \Psi_d \qquad (2)$$

where i_d and i_q, Ψ_d and Ψ_q are the d- and q-axes stator currents and fluxes respectively, R_s is the stator resistance and ω_s is the electrical speed.

Generally, the flux in d- and q-axis are functions of the currents in both axes, i.e. $\Psi_d = \Psi_d(i_d, i_q)$ and $\Psi_q = \Psi_q(i_d, i_q)$. Eq. (3) shows the definition of the flux linkage vector, where $\mathbf{L}$ is the inductance matrix and Ψ_{pm} is the magnetic flux linkage due to the permanent magnets.

$$\begin{bmatrix} \Psi_d \\ \Psi_q \end{bmatrix} = L \begin{bmatrix} i_d \\ i_q \end{bmatrix} + \begin{bmatrix} \Psi_{pm} \\ 0 \end{bmatrix} \qquad (3)$$

In order to model the nonlinear effects, like saturation and cross-coupling, the matrix $\mathbf{L}$ can be defined as in Eq. (4) [13].

$$L = \begin{bmatrix} L_d(i_d, i_q) & 0 \\ 0 & L_q(i_d, i_q) \end{bmatrix} \qquad (4)$$

The electromagnetic torque of a synchronous machine is given by Eq. (5), where z_p is the number of pole pairs.

$$T = \frac{3}{2} z_p \left(\Psi_d i_q - \Psi_q i_d \right) \qquad (5)$$

By replacing the flux definition of Eq. (3) in (5) it is possible to write the torque equation as a function of the machine inductances and the permanent magnet flux linkage, as shown in Eq. (6)

$$T = \frac{3}{2} z_p \left(\Psi_{pm} + \Delta L i_d \right) i_q \qquad (6)$$

where ΔL is defined as the difference between the d- and q-axis' inductances respectively.

Although Equations (1) to (6) are valid for both machines, it is necessary to add Equation (7) when modeling the RSM, since this machine does not have permanent magnets in its rotor.

$$\Psi_{pm} = 0 \qquad (7)$$

B. Absolute and differential inductances

It is necessary to distinguish between the concepts of absolute and differential inductances. Absolute inductances represent the gradient of the flux vs. current curve in relation to the origin, while the differential inductances represent the local derivative of the flux. In Fig. 1 it is possible to see the difference between these two concepts.

The inductances in Equation (3) are absolute, since they are defined as [13]:

$$L_d(i_d, i_q) = \frac{\Psi_d(i_d, i_q) - \Psi_d(0, i_q)}{i_d} \qquad (8)$$

$$L_q(i_d, i_q) = \frac{\Psi_q(i_d, i_q)}{i_q} \qquad (9)$$

Mathematically the differential inductances are defined as

$$L_{\text{diff}} = \frac{\partial \Psi}{\partial i} \qquad (10)$$

Obviously, if the flux is a linear function of the current, i.e. without or with negligible levels of saturation, the differential and absolute inductances will be equal.

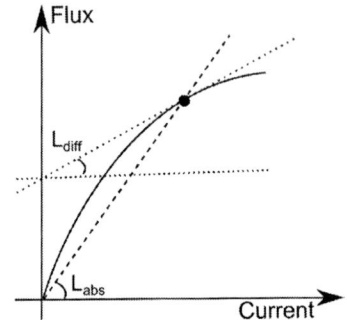

Fig.1. Relationship between absolute and differential inductances.

C. Analytical development of the MTPF control

The MTPF strategy consists on minimizing the total stator flux for a given reference torque. This optimization problem has been solved in [14] using the Lagrange method. The implicit relation of the optimal fluxes in d- and q-directions is given by:

$$\lambda_d = \frac{\Psi_{pm}}{L_d} \pm \frac{L_d L_q}{2(L_q - L_d)} \sqrt{\frac{\Psi_{pm}^2}{L_d^2} + 8\lambda_s^2 \left(\frac{1}{L_q} - \frac{1}{L_d} \right)^2} \qquad (11)$$

It will be shown in that this relationship inherently neglects saturation and cross-coupling effects and, therefore, cannot represent the optimal solution regarding the total flux for a highly anisotropic machine.

III. REFERENCE CURRENT GENERATION FOR DEEP FIELD-WEAKENING OPERATION

In order to solve the reference current generation problem for deep field weakening operation, the torque, current and voltage equations are, in a first step, mapped from the current plane into the flux plane. The mapping is carried out by replacing Equation (12) into (6).

$$\begin{cases} i_d = \frac{\lambda_d - \Psi_{pm}}{L_d} \\ i_q = \frac{\lambda_q}{L_q} \end{cases} \qquad (12)$$

Then, the torque (assuming linear flux distribution), flux and current limitations can be written as Equations (13), (14), and (15). The torque equation (5) will be still valid in the flux plane. It is possible to see that the torque will still be a hyperbole in this new plane, while the current and flux limitations will be ellipses and circumferences respectively.

$$T = \frac{\frac{3}{2} z_p (\Delta L \lambda_d \lambda_q + \Psi_{pm} \lambda_q L_q)}{L_d L_q} \qquad (13)$$

$$I^2 = \left(\frac{\lambda_d - \Psi_{pm}}{L_d} \right)^2 + \left(\frac{\lambda_q}{L_q} \right)^2 \qquad (14)$$

$$\lambda_0^2 = \lambda_d^2 + \lambda_q^2 \qquad (15)$$

Analogously to [15], a geometrical argument can then be developed in order to find the optimal flux for a given torque, since the fluxes limitations are now represented by circumferences. The gradient of the torque on the $\lambda_d \times \lambda_q$ plane, which defines a vector perpendicular to the curve at each point, can be defined as

$$\overrightarrow{\nabla T} = \frac{\partial T}{\partial \lambda_d}\vec{i} + \frac{\partial T}{\partial \lambda_q}\vec{j} \qquad (16)$$

Where $\vec{i}$ and $\vec{j}$ are unitary vectors in the direction of the d- and q-axes respectively.

Let $\vec{k} = \alpha\vec{i} + \beta\vec{j}, \alpha, \beta \in \mathbb{R}$, be a vector which is tangent to the torque curve at each point, i.e. it satisfies $\overrightarrow{\nabla T} \bullet \vec{k} = 0$. We will have, simplifying constant terms which change only the magnitude and not the direction of the vector, that

$$\alpha = -\frac{\partial T}{\partial \lambda_q} = -\lambda_d \frac{\partial i_q}{\partial \lambda_q} + i_d + \lambda_q \frac{\partial i_d}{\partial \lambda_q} \qquad (17)$$

$$\beta = \frac{\partial T}{\partial \lambda_d} = i_q + \frac{\partial i_q}{\partial \lambda_d}\lambda_d - \lambda_q \frac{\partial i_d}{\partial \lambda_d} \qquad (18)$$

The cross-coupling terms can be neglected by setting $\frac{\partial i_d}{\partial \lambda_q} \approx \frac{\partial i_q}{\partial \lambda_d} \approx 0$. Using the definition of the absolute and differential inductances, the vector $\vec{k}$ can then be written as

$$\vec{k} = \left(-\frac{\lambda_d}{L_{qq}} + \frac{\lambda_d - \Psi_{pm}}{L_d}\right)\vec{i} + \left(\frac{\lambda_q}{L_q} - \frac{\lambda_q}{L_{dd}}\right)\vec{j} \quad (19)$$

Where L_{dd} and L_{qq} are the differential inductances in d- and q-axis respectively.

By defining a function $f : \mathbb{R} \to \mathbb{R} | f = \vec{k} \bullet \nabla(\lambda_0^2)$, which will be proportional to the cosine of the angle between $\vec{k}$ and the gradient of the flux $\nabla(\lambda_0^2)$, it is possible to write:

$$f = \lambda_d^2\left(\frac{1}{L_d} - \frac{1}{L_{qq}}\right) + \lambda_q^2\left(\frac{1}{L_q} - \frac{1}{L_{dd}}\right) - \Psi_{pm}\left(\frac{\lambda_d}{L_d}\right) \qquad (20)$$

with

$$\lambda_q = \frac{L_d L_q T}{1.5 z_p(\Psi_{pm}L_q + \Delta L \lambda_d)} \qquad (21)$$

The searched optimal point is the root of the function f. The derivative of f regarding the flux λ_d is:

$$f' = 2\lambda_d\left(\frac{1}{L_d} - \frac{1}{L_{qq}}\right) - 2\frac{\lambda_q^2 \Delta L}{(\Delta L \lambda_d + \Psi_{pm}L_q)}\left(\frac{1}{L_q} - \frac{1}{L_{dd}}\right) - \frac{\Psi_{pm}}{L_d}(22)$$

It is then possible to define Newton Method's iterations for finding the desired solution

$$\lambda_{d(n+1)} = \lambda_{d(n)} - \frac{f(n)}{df(n)/d\lambda_{d(n)}} \qquad (23)$$

With λ_q given through Eq. (21).

A. Special case: Synchronous Reluctance Machine

For the synchronous reluctance machine, since $\Psi_{pm} = 0$, the function f and its derivative are given by Eqs. (24) and (25).

$$f = \lambda_d^2\left(\frac{1}{L_d} - \frac{1}{L_{qq}}\right) + \lambda_q^2\left(\frac{1}{L_q} - \frac{1}{L_{dd}}\right) \qquad (24)$$

$$f' = \frac{df}{d\lambda_d} = 2\lambda_d\left(\frac{1}{L_d} - \frac{1}{L_{qq}}\right) - 2\frac{\lambda_q^2}{\lambda_d}\left(\frac{1}{L_q} - \frac{1}{L_{dd}}\right) \qquad (25)$$

B. Convergence analysis

In order to prove ensure the convergence of the algorithm, the derivative of f must be nonzero and its second derivative must be bounded, as described in Eq.(26).

$$\begin{cases} f'(i_d) \neq 0 \\ \exists c \in \mathbb{R} \mid -c \leq f''(i_d) \leq c \end{cases} \qquad (26)$$

For the PMa-SynRM, considering the fact that the term $-\frac{\Psi_{pm}}{L_d}$ will always be negative, the derivative of f will be different from zero if $\frac{1}{L_d} - \frac{1}{L_{qq}} > 0$ and $\frac{1}{L_q} - \frac{1}{L_{dd}} < 0$.

The second derivative of f is given by Eq. (27)

$$f'' = 2\left(\frac{1}{L_d} - \frac{1}{L_{qq}}\right) + 6\frac{\lambda_q^2(\Delta L)^2}{(\Delta L \lambda_d + \Psi_{pm}L_q)^2}\left(\frac{1}{L_q} - \frac{1}{L_{dd}}\right) \quad (27)$$

The second derivative will be bounded if Eq. (28) holds. This implies that λ_d should be always different from a given positive real number, because ΔL is assumed negative. Given the fact that λ_d will always be negative for deep field weakening operation, because the MTPV curve is located on the left side of the point $i_d = -\frac{\Psi_{pm}}{L_d}$ it can be concluded that this equation holds.

$$\lambda_d \neq -\frac{\Psi_{pm}L_q}{\Delta L} > 0 \qquad (28)$$

For the Reluctance Synchronous Machine, the following convergence condition is found:

$$\frac{\lambda_d^2}{\lambda_q^2} \neq \frac{\left(\frac{1}{L_q} - \frac{1}{L_{dd}}\right)}{\left(\frac{1}{L_d} - \frac{1}{L_{qq}}\right)} \qquad (29)$$

The RSM is constructively determined for $\frac{1}{L_q} - \frac{1}{L_{dd}} > 0$ and $\frac{1}{L_d} - \frac{1}{L_{qq}} < 0$, which implies that Eq. (29) will always hold. The second derivative will always be bounded because $\lambda_d \neq 0$.

C. Operation under voltage limitation

Under field-weakening operation, the requested torque can be reached, but not by using the MTPC. This is caused by the fact that the voltage limitation does not allow the MTPC points to be settled. In this case, the energy-optimal operation point will be at the intersection between the torque curve and the voltage limiting ellipse. Since this optimization is also a highly nonlinear problem, a simple numerical solution is analyzed in this work. We supposed that the MPTC and MTPF points are known. While the MTPF point can be found by using the proposed algorithm, the MTPC can be found by using the algorithm proposed in [16]. By having both points, the operation point considering the voltage limitation can be found by using a bisection algorithm iterating over the torque curve between both points. The point on the left side must be always chosen to be inside the voltage limitation and the point on the right side to be outside it. The stator resistance is considered in the definition of the available voltage. Fig. 2 illustrates this procedure.

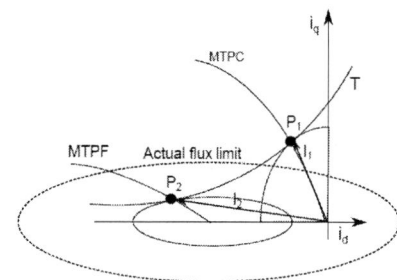

Fig.2. Illustration of the proposed field-weakening strategy

IV. RESULTS AND DISCUSSION

In order to validate the proposed approach, simulations using Finite Element models of a 400V PMa-SynRM and a 12V RSM, whose parameters are shown in Table I, were carried out.

TABLE I.   MACHINE PARAMETERS

Parameter	Machine	
	PMa-SynRM	**RSM**
Rated power	10.8kW	2kW
Rated current	40A	60A
Stator resistance (@20°C)	95.6mΩ	0.2Ω
d-axis inductance	17μH	5.6mH
q-axis inductance	62.2μH	4.4mH
PM flux linkage	0.0477Vs	0
Pole pairs	2	2

The following simulation procedure was used in this work: For each given torque request, the optimal operation points are calculated using the proposed method and are then applied to the machine model. The resulting torque is compared to the request and the fluxes are used for performing parameter calculations for the next iteration. For comparison purposes, the state-of-the-art method that neglects saturation [14] is also implemented. The requested torque is varied stepwise from 0.5 Nm to around 26 Nm for the PMa-SynRM and around 3 Nm for the RSM. The reason for these limitations is that the currents needed for setting higher torques exceed the current limitation of the drive system.

Fig.3 shows the locus in the id − iq plane of the reference currents calculated by the proposed approach and the method without the differential inductances. It is possible to see that with the consideration of the differential inductances, the calculated MTPF curve approaches the real MTPF curve. Although these curves approach, there is still a gap between the calculated and the real MTPF curves. The reason for this relies on the neglection of the cross-coupling differential inductances.

In Fig. 4 it is possible to see the inductance's variations as a function of the requested torque. For all requested torques, the convergence conditions discussed in Section III-b remain valid.

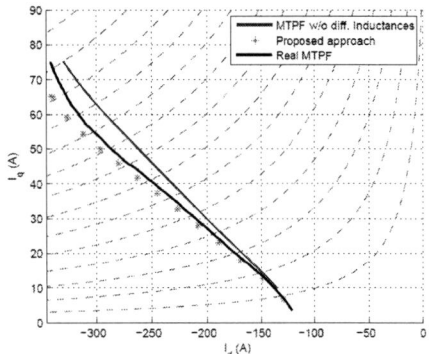

Fig.3. FE Results of the MTPF optimization in the current plane for the PMa-SynRM

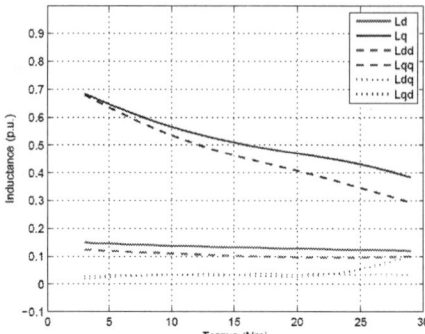

Fig.4. Inductances evolution of the PMa-SynRM under MTPF operation

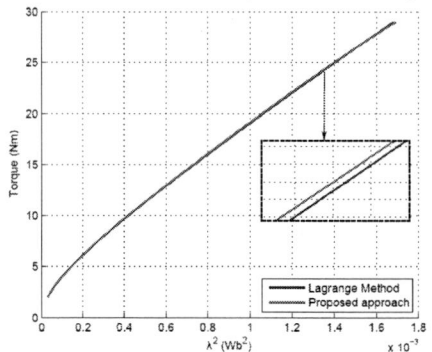

Fig.5. PMa-SynRM: Produced torque as a function of the squared flux for MTPF operation

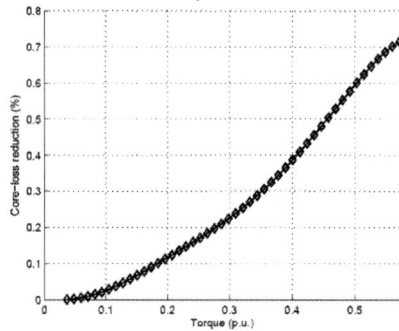

Fig.6. PMa-SynRM: Core loss reduction as a function of the torque

In Fig. 5 the produced torque as a function of the squared total flux is displayed. The squared total flux is used because it is proportional to the total core losses. The proposed method shows a better performance when compared to the method without differential inductances, because it produces more torque with the same stator flux. Fig.6 shows the core loss reduction brought by the consideration of the differential inductances for the PMa-SynRM. The performance improvement reaches up to 0.7%. This relatively low improvement can be explained by the fact that the saturation is not very prominent in this operating range, as it can be seen in Fig. 4.

For the RSM, the reference algorithm was chosen to be the $\lambda_d = \lambda_q$ control, since this strategy optimizes the flux if no saturation is present. Fig.7 shows similar results as in the PMa-SynRM case. It is possible to see that with the consideration of saturation, the calculated MTPF approaches the real MTPF curve.

In Fig. 8 it is possible to see that, as in the PMa-SynRM case, the saturation does not play a major role for the MTPF optimization for the RSM. The fact that the absolute and differential inductances are almost equal, explain the fact that the core-loss reduction shown in Fig.10 is only around 1.2%.

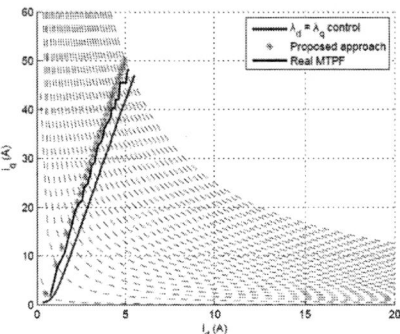

Fig.7. FE Results of the MTPF optimization in the current plane for the RSM

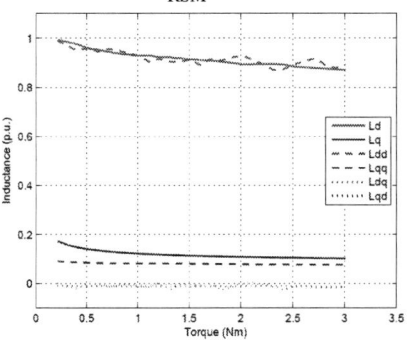

Fig.8. Inductances evolution of the RSM under MTPF operation

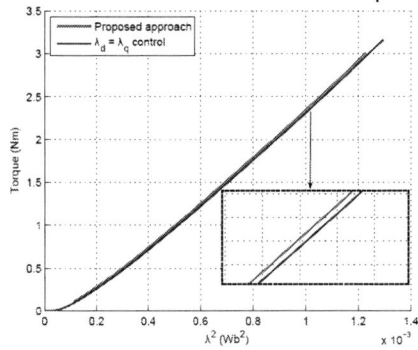

Fig.9. RSM: Produced torque as a function of the squared flux for MTPF operation

For performing field weakening operation, the same algorithm discussed in Section III-c, is also applied to the PMa-SynRM and the RSM. As described before, by having the two optimal points minimizing the current and the flux, field weakening operation can be achieved by iterating over the torque curve between these two extrema. Figures 11 and 12 show respectively the results for the PMa-SynRM and the RSM. The algorithm is able to find the intersection point between the voltage ellipse and the torque curve without explicitly modeling saturation and cross-coupling through differential inductances because no derivation of the flux is needed for constructing this field-weakening algorithm.

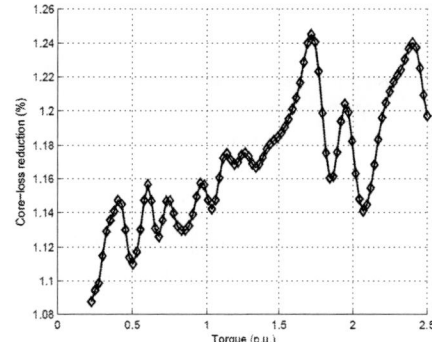

Fig.10. RSM: Core loss reduction as a function of the torque

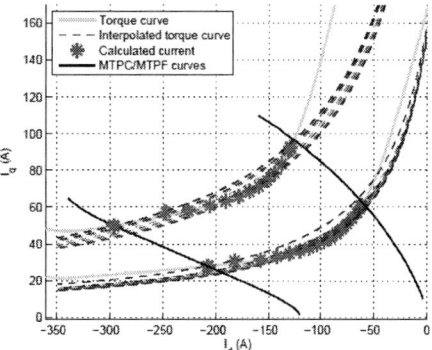

Fig.11. Field weakening operation of the PMa-SynRM

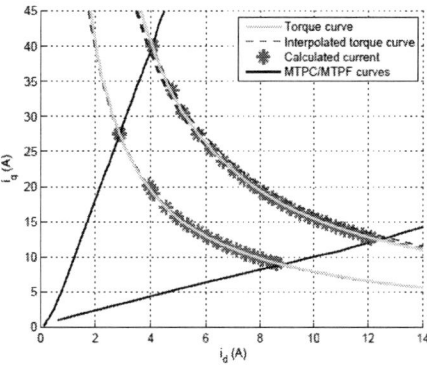

Fig.12. Field weakening operation of the RSM

It is interesting to note that in both figures, the interpolated torque curves are obtained through the use of the parameter values for calculated current. It is remarkable that these curves deviate heavily from the real torque curve for points that are not near the reference points used for obtaining the parameters. This effect is more prominent in Fig. 11 due to the presence of heavier saturation and cross-coupling on this machine. This shows clearly the importance of adapting the parameter values between iterations of the algorithm in a real implementation. Failing to update the parameters in an acceptable rate can lead to oscillations and wrong convergence.

V. CONCLUSIONS

An algorithm for considering saturation in the reference current generation in deep field weakening was introduced in this work. The optimization problem is solved numerically by using a geometrical interpretation of it in the flux plane. The convergence can be guaranteed by constructive aspects of the electrical machines. Saturation phenomena are taken into account by the use of differential inductances. Simulations using Finite Element Analysis show a positive effect regarding the core loss reduction for both the PMa-SynRM and RSM when considering saturation for the optimization. Moreover it has also been shown that it is possible to generate the reference currents for field weakening operation if the MTPF and MTPC points are known. Further research shall include numerical optimizations for allowing the real-time implementation of this strategy.

REFERENCES

[1] Rajashekara, K, "Present Status and Future Trends in Electric Vehicle Propulsion Technologies" *IEEE Journal of Emerging and Selected Topics in Power Electronics*, Vol.1, No.1, pp.3-10, 2013.

[2] Meyer, M.; Böcker, J.; "Optimum Control for Interior Permanent Magnet Synchronous Motors (IPMSM) in Constant Torque and Flux Weakening Range", *12th International Power Electronics and Motion Control Conference*, 2006.

[3] Windisch, T.; Hofmann, W.; "Loss-minimization of an IPMSM drive using pre-calculated optimized current references"; *37th Annual Conference of the IEEE Industrial Electronics Society*, IECON 2011.

[4] Bae, B.; Patel, N.; Schulz, S.; Sul, S.; "New field weakening techniue for high saliency interior permanent magnet motor"; *38th IAS Annual Meeting on Conference Record of the Industry Applications Conference*, 2003.

[5] Huber, T.; Peters, W.; Böcker, J.;"Voltage controller for flux weakening operation of interior permanent magnet synchronous motor in automotive traction applications"; *IEEE International Electric Machines and Drives Conference*, IEMDC 2015.

[6] Sepulchre, L.; Fadel, M.; Pietrzak-David, M.; Porte, G.; "MTPV Flux-Weakening Strategy for PMSM High Speed Drive"; *IEEE Transactions on Industry Applications*, Vol. 54, No 6, Nov. 2018.

[7] Hu, D.; Zhu, L.; Xu, L.; "Maximum Torque per Volt operation and stability improvement of PMSM in deep flux-weakening region"; *IEEE Energy Conversion Congress and Exposition*, ECCE 2012.

[8] Ekanayake, S.; Dutta, R.; Rahman, M.; Xiao D.; "Deep flux weakening control of a segmented interior permanent magnet synchronous motor with maximum torque per voltage control"; *41st Annual Conference of the IEEE Industrial Electronics Society*, IECON 2015.

[9] Morimoto, S.; Sanada, M. & Takeda, Y.; "Wide-Speed Operation of Interior Permanent Magnet Synchronous Motors with High-Performance Current Regulator"; *IEEE Transactions on Industry Applications*, 1994, 30, 920-92

[10] Inoue, Y.; Morimoto, S.; Sanada, M.; "A novel control scheme for maximum power operation of synchronous reluctance motors including maximum torque per flux control", *IEEE Transactions on Industry Applications*, 2011, Vol. 47, No. 1.

[11] Preindl, M.; Bolognani, S.; "Optimal State Reference Computation With Constrained MTPA Criterion for PM Motor Drives"; *IEEE Transactions on Power Electronics*, 2015, Vol. 30, No. 8.

[12] Hackl, C.; Kullick, J.;Eldeeb, H.; Horlbeck, L.; "Analytical computation of the optimal reference currents for MTPC/MTPA, MTPV and MTPF operation of anisotropic synchronous machines considering stator resistance and mutual inductance"; *19th European Conference on Power Electronics and Applications*, 2017.

[13] Qi G.,Chen, J.T.; Zhu Z. Q.; Howe, D.; Zhou, L. B.;Gu, C.L.; "Influence of skew and cross-coupling on flux-weakening performance of permanent-magnet brushless AC machines" *IEEE Trans. on Magnetics*, Vol. 45, No. 5, 2009

[14] Schröder, D. „Elektrische Antriebe – Regelung von Antriebssystemen" *Springer Verlag*, 2009

[15] Bonifacio, J.; Kennel, R.;" Online copper-loss minimization of interior permanent magnet synchronous machines for automotive applications: A geometrical approach". *4th International Conference on Electric Power and Energy Conversion Systems* (EPECS), 2015

[16] Bonifacio, J.; Kennel, R.; "On considerig saturation and cross-coupling effects for copper loss minimization on highly anisotropic synchronous machines" *IEEE Transactions on Industry Applications*, Vol. 54, No 5, 2018

Predictive Voltage Control Operating at Fixed Switching Frequency of a Neutral-Point Clamped Converter

Felipe Herrera
Department of Electrical Engineering
Universidad de Chile
Santiago, Chile
fherreraibanez@gmail.com

Roberto Cárdenas
Department of Electrical Engineering
Universidad de Chile
Santiago, Chile
rcardenas@ing.uchile.cl

Marco Rivera
Facultad de Ingeniería
Universidad de Talca
Curicó, Chile
marcoriv@utalca.cl

José A. Riveros
Facultad de Ingeniería
Universidad de Talca
Curicó, Chile
jriverosv@utalca.cl

Patrick Wheeler
PEMC Research Group
The University of Nottingham
Nottingham, UK, and Ningbo, China
Pat.Wheeler@nottingham.ac.uk

Abstract—**Uninterruptible power supply units are system formed by power electronics converters to supply sinusoidal voltages to feed critical loads. In this paper, a fixed switching frequency model predictive control strategy is presented for the control of the output voltage in an *LC* filter connected to a three-level NPC converter. The control objectives of the system are the tracking of the voltage reference and balance of the voltages of the dc-link capacitors. The mathematical model of the converter and the *LC* filter is developed and the control strategy is explained. Simulation results obtained in the Matlab/Simulink enviroment are presented to validate the control strategy**

Index Terms—**Predictive control, DC-AC power converters, voltage control.**

I. Introduction

When regulated sinusoidal voltages are required by a load, the solution is to use inverters with an *LC* output filter to generate ac output voltages with very low harmonic content [1]. One application of this is as the main inverter of an uninterruptible power supply (UPS) system. UPS units are composed of power electronics converters designed to feed critical linear and non-linear loads such as medical and industrial equipment [2].

For the operation of the inverter the most common approach is to use linear control because of its well-known design and simple implementation. These linear control strategies are PI-based linear cascaded control loops with coordinate transformation (such as the Clarke transform and the Park transform) and modulation techniques such as carrier-based PWM and space vector modulation [3]. The modulation technique is used to linearize the converter and generate the commutation signals based on a time-average principle. Advantages of linear control strategies are the fixed switching frequency and easiness to extend the method to different converter topologies by changing the modulation technique used, etc. but it has some disadvantages such as the difficulty to include contraints and nonlinearities and the necessity of

a modulator resulting in slower dynamics. The development of semiconductor technology has increased the processing capability of digital microprocessors and reduced their price allowing the exploration and implementation of new and more complex control schemes. These techniques such as fuzzy control, robust control, sliding mode control and model preditive control (MPC) are more advanced than standard PID control and thus are denominated as advanced control [4].

Predictive control refers to a wide class of controllers such as deadbeat control, hysteresis-based, trayectory-based and MPC [5]. These control techniques shares the same common characteristic which is the use of the mathematical model of the system to predict the future behaviour of the controlled variables over a prediction horizon N to select the appropiate control action based on an optimization criterion [6]. Advantages of predictive control are: applicability to a variety of systems, nonlinearities can be included in the model avoiding the need of linearizing for a given operating point, constraints can be included in the optimization criterion, the multivariable case can be easily included and the possibility to avoid the cascaded structure of linear control schemes resulting in fast transient response. The disadvantages of predictive control strategies are the high computational burden and the need of very good mathematical models of the system under control [4].

FCS-MPC is a predictive control strategy which use the discrete nature of the converter. In every sampling instant the controlled variables are measured and fed to the discrete-time model of the system to predict their future behavior for all possible switching states of the converter and compute a cost function. The cost function depends on the control objectives and can be the absolute error, quadratic error or time-average error between the reference and measured variables [7]. The switching state who minimizes the cost function is stored and applied in the next sampling instant. The strategy does not need a modulator resulting in a variable switching frequency. A FCS-MPC algorithm with constant switching frequency is preferred because it allows easy filter design in applications such as grid-connected converters and inverters with output LC filter [4, 8]. A solution to the switching frequency problem is a FCS-MPC algorithm called Modulated Model Predictive Control (M^2PC). In M^2PC, a modulation scheme is included in the cost function minimization by selecting and applying, in every sampling instant, two or more switching states with their corresponding application times [9]. This approach has been applied to many power converter topologies including the NPC converter [8-14] .

In this paper, a M^2PC strategy is proposed for the output voltage control and dc-link capacitor voltage balance of a NPC converter connected to a LC filter feeding a linear resistive load. In Section II the mathematical model of the converter and load is developed, in section III the M^2PC strategy is explained and in section IV the control strategy is validated with a Matlab/Simulink simulation.

II. Topology and Mathematical Model of the Converter

In Fig. 1 a three-level neutral-point clamped (NPC) converter connected to a resistive load through a LC filter is shown.

The NPC converter is a multilevel inverter which means that the converter transforms a fixed dc voltage to a ac voltage with variable magnitude and frequency. The dc-link is formed by two cascaded dc capacitors who provide a floating neutral point (O). The converter topology consist of three phases (or legs) connected in parallel with four high-power switching devices per phase. The high-power switching devices can be either IGBT or GCT [15]. Two series connected diodes (denominated as clamping diodes) are connected to the node

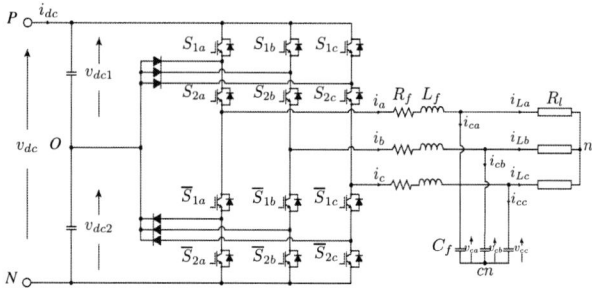

Fig. 1. NPC converter connected to a resistive load through an LC filter.

Table I
Switching States and Inverter Terminal Voltage.

S_x	S_{1x}	S_{2x}	$\overline{S_{1x}}$	$\overline{S_{2x}}$	v_{xN}
1(P)	1	1	0	0	$v_{dc1} + v_{dc2}$
0(O)	0	1	1	0	v_{dc2}
-1(N)	0	0	1	1	0

between the upper switching devices (with switching signals S_{1x} and S_{2x}, with $x \in \{a, b, c\}$) and the node between the lower switching devices (with switching signals $\overline{S_{1x}}$ and $\overline{S_{2x}}$). Since the converter operation don't allow all the switching devices to be in the ON state at the same time, the lower switching devices work in a complementary manner with the upper switching devices. The node between the clamping diodes is connected to the floating neutral point of the dc-link. The switching state, S_x, summarise the switching devices state (ON or OFF) of the four switches in one of the three phases. Table I shows the relationship between the switching state, the switching signals of the switching devices in one phase and the inverter terminal voltage, $v_{xN}(t)$. The voltage $v_{xN}(t)$ can be expressed as a function of the switching signals of the upper switching devices and the voltages of the dc-link capacitors as follows:

$$v_{xN}(t) = S_{1x}v_{dc1}(t) + S_{2x}v_{dc2}(t) \qquad (1)$$

The voltages of the dc-link capacitors can be expressed with the differential equations in (2), where $i_{dc1}(t)$ and $i_{dc2}(t)$ are the current through the upper capacitor and the current through the lower capacitor, respectively.

$$\begin{aligned} \frac{d\,v_{dc1}(t)}{dt} &= \frac{1}{C_1}i_{dc1}(t) \\ \frac{d\,v_{dc2}(t)}{dt} &= \frac{1}{C_2}i_{dc2}(t) \end{aligned} \qquad (2)$$

The currents through the dc-link capacitors, $i_{dc1}(t)$ and $i_{dc2}(t)$, are a function of the output currents $i_x(t)$ and the switching states of the converter and can be determined by the equation presented in [16] as follows:

$$\begin{aligned} i_{dc1}(k) &= i_{dc}(k) - H_{1a}i_a(k) - H_{1b}i_b(k) - H_{1c}i_c(k) \\ i_{dc2}(k) &= i_{dc}(k) + H_{2a}i_a(k) + H_{2b}i_b(k) + H_{2c}i_c(k) \end{aligned} \qquad (3)$$

H_{1x} and H_{2x} are piecewise-defined functions whose values dependes on the switching states of the converter. The function $H_{1x} = 1$ if and only if $S_x = 1$; $H_{1x} = 0$ in any other case. The function $H_{2x} = 1$ if and only if $S_x = -1$; $H_{2x} = 0$ in any other case.

This converter has 3 switching states per phase, therefore there are $3^3 = 27$ possible combinations of the switching states. Considering a three-phase balanced load the ac side analysis can be simplified using the Clarke transform. Applying the Clarke transform to the three-phase inverter terminal voltages, 19 different voltage space vectors are generated from the 27 combinations of the switching states. Those combinations who generate the same voltage vector are called redundant. The vectors are classified based on their length in zero vector ($\mathbf{v_0}$), small vectors ($\mathbf{v_1}$-$\mathbf{v_6}$), medium vectors ($\mathbf{v_7}$-$\mathbf{v_{12}}$) and large vectors ($\mathbf{v_{13}}$-$\mathbf{v_{18}}$). The projection of the vectors in the $\alpha - \beta$ plane is shown in Fig. 2. The plane is divided in six triangular sectors and each sector is divided in four regions. The following current and voltage space vectors are defined for the LC filter and load variables:

978-1-7281-4181-7/19 $31.00 © 2019 IEEE

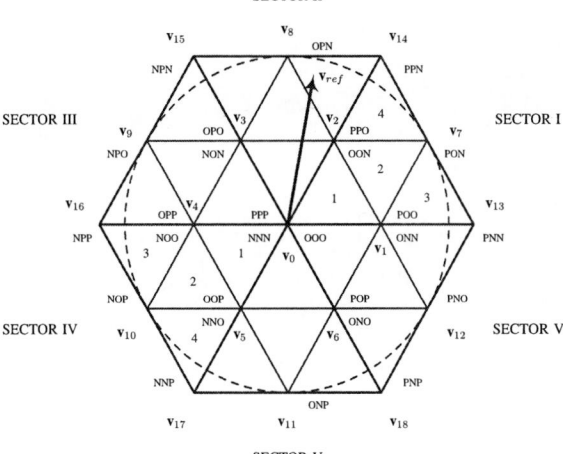

Fig. 2. Space vector diagram of the NPC converter.

$$\begin{aligned}
\mathbf{v}(t) &= (2/3)\left(v_{aN}(t) + \mathbf{a}v_{bN}(t) + \mathbf{a}^2 v_{cN}(t)\right) \\
\mathbf{i}(t) &= (2/3)\left(i_a(t) + \mathbf{a}i_b(t) + \mathbf{a}^2 i_c(t)\right) \\
\mathbf{v}_c(t) &= (2/3)\left(v_{ca}(t) + \mathbf{a}v_{cb}(t) + \mathbf{a}^2 v_{cc}(t)\right) \\
\mathbf{i}_L(t) &= (2/3)\left(i_{La}(t) + \mathbf{a}i_{Lb}(t) + \mathbf{a}^2 i_{Lc}(t)\right)
\end{aligned} \quad (4)$$

where $\mathbf{a} = e^{j(2\pi/3)}$, $\mathbf{v}(t)$ is the space vector of the inverter terminal voltages, $\mathbf{i}(t)$ is the space vector of the inductors current in the LC filter, $\mathbf{v}_c(t)$ is the space vector of the capacitors voltages in the LC and $\mathbf{i}_L(t)$ is the space vector of the load currents. The differential equations who describe the dynamics of the ac side are:

$$\frac{d\mathbf{v}_c}{dt} = \frac{1}{C_f}\mathbf{i} - \frac{1}{C_f}\mathbf{i}_L \quad (5)$$

$$\frac{d\mathbf{i}}{dt} = \frac{1}{L_f}\mathbf{v} - \frac{1}{L_f}\mathbf{v}_c - \frac{R_f}{L_f}\mathbf{i} \quad (6)$$

where C_f, L_f y R_f are the filter capacitance, filter inductance and filter resistance, respectively. To implement the control algorithm in a digital platform the continuous-time equations need to be discretized. The discrete-time equations are obtained applying the Euler forward method.

$$\frac{d\mathbf{x}(t)}{dt} \approx \frac{\mathbf{x}(k+1) - \mathbf{x}(k)}{T_s} \quad (7)$$

Replacing (7) in (2), (5) and (6):

$$v_{dcz} = v_{dcz}(k) + \frac{T_s}{C_z}i_{dcz}(k), \quad z \in \{1, 2\} \quad (8)$$

$$\mathbf{i}(k+1) = \left[1 - \frac{R_f T_s}{L_f}\right]\mathbf{i}(k) + \frac{T_s}{L_f}\left[\mathbf{v}(k) - \mathbf{v}_c(k)\right] \quad (9)$$

$$\mathbf{v}_c(k+1) = \mathbf{v}_c(k) + \frac{T_s}{C_f}\left[\mathbf{i}(k+1) - \mathbf{i}_L(k)\right] \quad (10)$$

III. FIXED SWITCHING FREQUENCY PREDICTIVE VOLTAGE CONTROL

M^2PC is a combination of the operation principles of SVM and FCS-MPC. In every switching instant T_s a switching sequence is applied to the converter, the switching sequence consist in the controlled application of the switching states of the voltage vectors that form the region where $\mathbf{v}_{ref}$ is located. The transition between one switching state and the next should follow a criterion. The criteria are: (a) at every transition only one leg of the converter can change his switching state and (b) in one sampling interval only two switches per leg can switch states, one for turn-ON and then for turn-OFF. Following these criteria the NPC space vector diagram is analized finding 36 possible switching sequences. Consider region 1 in sector I which is formed by the vectors $\mathbf{v}_0$, $\mathbf{v}_1$ and $\mathbf{v}_2$. These are redundant vectors enabling the possibility to form two switching sequences with them: $\mathbf{v}_0(\text{OOO})\text{-}\mathbf{v}_{1P}(\text{POO})\text{-}\mathbf{v}_{2P}(\text{PPO})\text{-}\mathbf{v}_0(\text{PPP})$ and $\mathbf{v}_0(\text{OOO})\text{-}\mathbf{v}_{2N}(\text{OON})\text{-}\mathbf{v}_{1N}(\text{ONN})\text{-}\mathbf{v}_0(\text{NNN})$. The difference between both switching sequences is the use of the dc-link. The switching sequence of region 1 in sector I is shown in Fig. 3 and the switching sequence of the optimal vectors is shown in Fig. 4, the vector $\mathbf{v}_0$ is the vector with most redundant states in the sequence. The control objetives are: (a) track of the reference voltage and (b) voltage balance in the dc-link capacitors. The following cost funcion is defined:

$$g_j = (v_{c\alpha}^* - v_{c\alpha}^p)^2 + \left(v_{c\beta}^* - v_{c\beta}^p\right)^2 + \lambda_{dc}\left(v_{dc1}^p - v_{dc2}^p\right)^2 \quad (11)$$

where λ_{dc} is a weighting factor and $j \in \{0, 1, 2\}$. The discrete equations of the load are used for the prediction of the state variables with a prediction horizon $N = 1$. The sequence that minimize the global cost function g is applied:

$$g = d_0 g_0 + d_1 g_1 + d_2 g_2 \quad (12)$$

The variables d_j are the duty cycles of the voltage vectors and depends on the cost function of every vector in the sequence.

$$\begin{aligned}
d_0 &= 1 - d_1 - d_2 \\
d_1 &= g_1 g_3 / (g_1 g_2 + g_2 g_3 + g_1 g_3) \\
d_2 &= g_1 g_2 / (g_1 g_2 + g_2 g_3 + g_1 g_3)
\end{aligned} \quad (13)$$

The application time of every vector depends on the duty cycles and the sampling time.

$$\begin{aligned}
T_0 &= T_s d_0 \\
T_1 &= T_s d_1 \\
T_2 &= T_s d_2
\end{aligned} \quad (14)$$

In Fig. 5 the block diagram of M^2PC is shown.

IV. SIMULATION RESULTS

In this section the simulation results of the proposed control strategy are presented. In Table II the simulation parameters are shown. The results are presented for steady state and transient state with a step change of the reference voltage.

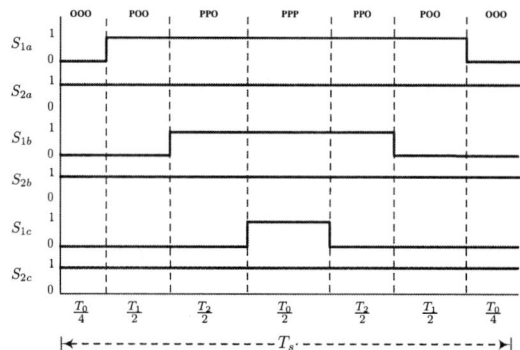

Fig. 3. Switching sequence formed by the vectors $\mathbf{v}_0(OOO)$ - $\mathbf{v}_{1P}(POO)$ - $\mathbf{v}_{2P}(PPO)$ - $\mathbf{v}_0(PPP)$

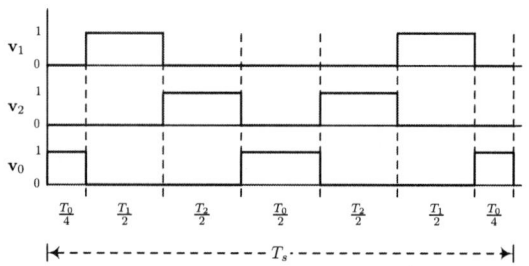

Fig. 4. Switching sequence for the optimal vectors.

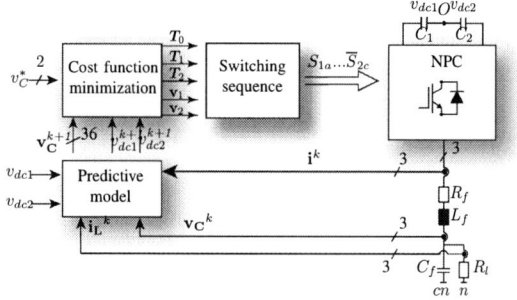

Fig. 5. Block diagram of M^{2}PC applied to a NPC converter connected to a LC filter.

A. Results in steady state

In Fig. 6 the reference voltage and the voltage in the capacitors of the LC filter is shown in the upper graph and the current through the LC filter inductors is shown in the lower graph. In steady state, the system is capable of following the reference voltage fulfilling the first control objective.

In Fig. 7 the voltage in the capacitors of the dc-link is shown in the upper graph and the current to the load are shown in the lower graph. The voltages of the dc-link capacitors are kept at half the voltage applied to the dc-link each and the voltage error between v_{dc1} and v_{dc2} oscillates between ± 0.1 [V]. The M^{2}PC is capable of balance the dc-link capacitors in steady state.

Table II
SIMULATION PARAMETERS

Variables	Description	Value
V_f	dc-source voltage	300 [V]
C_{dc}	dc-link capacitors	4700 [μF]
R_f	LC filter resistance	0.05 [Ω]
L_f	LC filter inductance	10 [mH]
C_f	LC filter capacitance	80 [μF]
R_L	Load resistance	10 [Ω]
f_r	Reference signal frequency	50 [Hz]
T_s	Sampling time	50 [μs]
	Simulation step	1 [μs]
λ_{dc}	Weighting factor	0.4

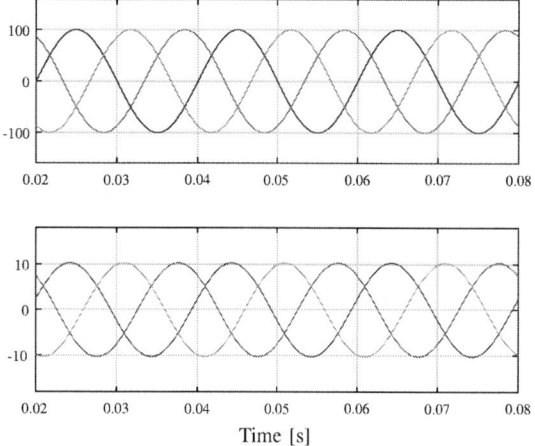

Fig. 6. Reference voltage $\mathbf{v}_c^*(t)$ and voltage in the LC filter capacitors, $\mathbf{v}_c(t)$ [V] (upper). Currents in the inductors of the LC filter, $\mathbf{i}(t)$ [A] (lower.)

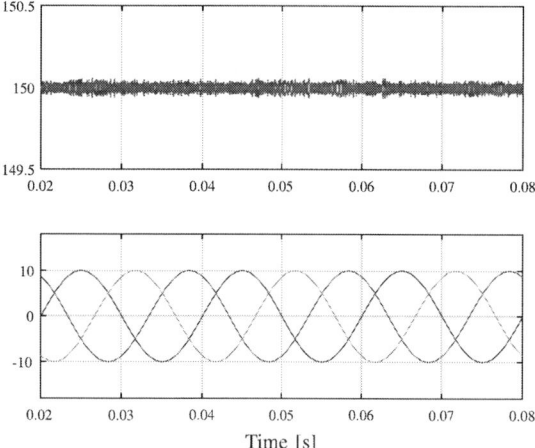

Fig. 7. dc-link capacitors voltages, $v_{dc1}(t)$ [V] and $v_{dc2}(t)$ [V] (upper). Load currents $\mathbf{i_L}(t)$ [A] (lower).

In Fig. 8 the harmonic spectrum of the voltage and current

978-1-7281-4181-7/19 $31.00 © 2019 IEEE

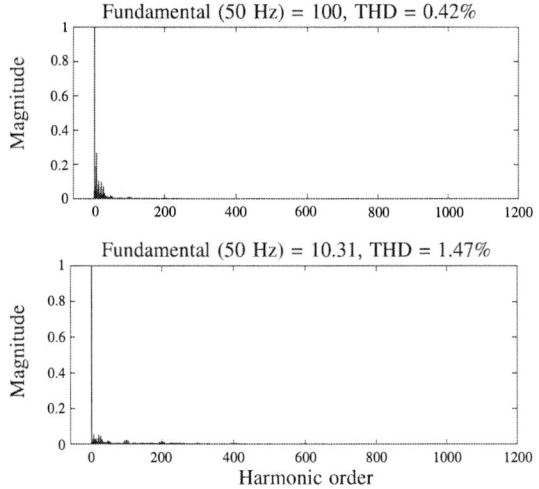

Fig. 8. Harmonic spectrum of the LC filter capacitor in phase a, $v_{ca}(t)$ (upper). Harmonic spectrum of the current through the LC filter inductor in phase a $i_a(t)$ (lower).

in the capacitors and inductors of the LC filter is shown. The harmonic distorsion of the current and the voltages is very low with a THD of 0.42% for the voltage of the LC filter capacitors and a THD of 1.47% for the current through the inductors of the filter. This values are obtained when the system is working with a device switching frequency of 10 [kHz]. The device switching frequency is half the sampling frequency of 20 [kHz].

B. Results in transient state

In Fig. 9 the reference voltages and voltages in the capacitors of the LC filter are shown in the upper graph for a step change in the reference from 100 [V] to 150 [V] and the currents in the LC filter inductors is shown in the lower graph. For a step load in the voltage refence, the system is capable of tracking the reference with a fast transient response fulfilling the tracking objective.

In Fig. 10 the voltage in the capacitors of the dc-link is shown in the upper graph and the currents to the load are shown in the lower graph for a step change in the voltage reference of 100 [V] to 150 [V]. The voltage error between the capacitors increase from $\pm$ 0.1 [V] to $\pm$ 0.4 [V] at the step time. This is explained because the weigthing factor of the voltage balance objetive in the cost function is not adjusted to the new conditions of the system. Since the difference is small, the control strategy is capable of fulfill the voltage balance in the dc-link capacitors objective.

V. CONCLUSIONS

In this paper, a fixed switching frequency model predictive control strategy was developed for the voltage control of an LC filter connected to a three-level NPC converter. The mathematical model of the NPC converter and the LC filter is developed and used for the prediction of the states variables

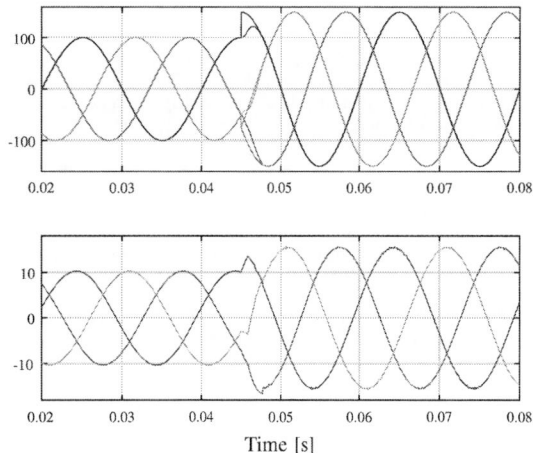

Fig. 9. Reference voltage $\mathbf{v}_c^*(t)$ and voltage in the LC filter capacitors, $\mathbf{v}_c(t)$ [V] (upper). Currents in the inductors of the LC filter, $\mathbf{i}(t)$ [A] (lower.)

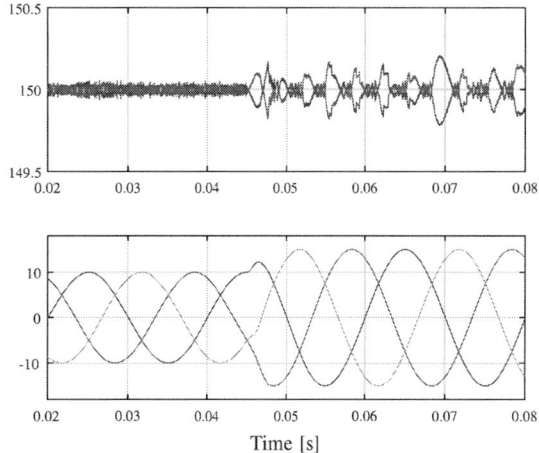

Fig. 10. dc-link capacitors voltages, $v_{dc1}(t)$ [V] and $v_{dc2}(t)$ [V] (upper). Load currents $\mathbf{i_L}(t)$ [A] (lower).

and the M^{2}PC control strategy is explained. Simulation results of the control strategy were presented for steady state and transient state with a step change of the reference voltage from 100 [V] to 150 [V]. The simulations results shows that the control strategy is capable of tracking the reference voltage and mantain balanced the voltage in the capacitors of the dc-link under both simulation conditions presented. The control strategy is based on the SVM operation principle and is restricted by the same operational limits of the modulation such as the maximum modulation index of $\left(\sqrt{(3)}/3\right)v_{dc}$.

ACKNOWLEDGMENT

The authors are thankful for the funding to the program FONDECYT Regular through the project 1160690, Postdoctoral 3170014 and the Universidad de Talca through the fund for scientific research of undergraduate students.

References

[1] V. K. Singh, R. N. Tripathi, and T. Hanamoto, "Model-based design approach for implementation of finite state MPC," in *2018 13th IEEE Conference on Industrial Electronics and Applications (ICIEA)*, May 2018, pp. 881–886.

[2] R. E. Carballo, F. Bottem, G. G. Oggier, and G. O. Garca, "Multiple resonant controllers strategy to achieve fault ride-through and high performance output voltage in UPS applications," *IET Power Electronics*, vol. 11, no. 15, pp. 2415–2426, 2018.

[3] S. Kouro, M. A. Perez, J. Rodriguez, A. M. Llor, and H. A. Young, "Model Predictive Control: MPC's Role in the Evolution of Power Electronics," *IEEE Industrial Electronics Magazine*, vol. 9, no. 4, pp. 8–21, Dec. 2015.

[4] P. Cortes, M. P. Kazmierkowski, R. M. Kennel, D. E. Quevedo, and J. Rodriguez, "Predictive Control in Power Electronics and Drives," *IEEE Transactions on Industrial Electronics*, vol. 55, no. 12, pp. 4312–4324, Dec. 2008.

[5] J. Rodriguez and P. Cortes, "Model Predictive Control," in *Predictive Control of Power Converters and Electrical Drives*. Wiley-IEEE Press, 2012, pp. 248–. [Online]. Available: http://ieeexplore.ieee.org/xpl/articleDetails.jsp?arnumber=6199000

[6] S. Vazquez, J. I. Leon, L. G. Franquelo, J. Rodriguez, H. A. Young, A. Marquez, and P. Zanchetta, "Model Predictive Control: A Review of Its Applications in Power Electronics," *IEEE Industrial Electronics Magazine*, vol. 8, no. 1, pp. 16–31, Mar. 2014.

[7] S. Kouro, P. Cortes, R. Vargas, U. Ammann, and J. Rodriguez, "Model Predictive ControlA Simple and Powerful Method to Control Power Converters," *IEEE Transactions on Industrial Electronics*, vol. 56, no. 6, pp. 1826–1838, Jun. 2009.

[8] F. Donoso, A. Mora, R. Crdenas, A. Angulo, D. Sez, and M. Rivera, "Finite-Set Model-Predictive Control Strategies for a 3l-NPC Inverter Operating With Fixed Switching Frequency," *IEEE Transactions on Industrial Electronics*, vol. 65, no. 5, pp. 3954–3965, May 2018.

[9] L. Tarisciotti, P. Zanchetta, A. Watson, J. C. Clare, M. Degano, and S. Bifaretti, "Modulated Model Predictive Control for a Three-Phase Active Rectifier," *IEEE Transactions on Industry Applications*, vol. 51, no. 2, pp. 1610–1620, Mar. 2015.

[10] L. Tarisciotti, A. Formentini, A. Gaeta, M. Degano, P. Zanchetta, R. Rabbeni, and M. Pucci, "Model Predictive Control for Shunt Active Filters With Fixed Switching Frequency," *IEEE Transactions on Industry Applications*, vol. 53, no. 1, pp. 296–304, Jan. 2017.

[11] M. Rivera, M. Perez, V. Yaramasu, B. Wu, L. Tarisciotti, P. Zanchetta, and P. Wheeler, "Modulated model predictive control (M2pc) with fixed switching frequency for an NPC converter," in *2015 IEEE 5th International Conference on Power Engineering, Energy and Electrical Drives (POWERENG)*, May 2015, pp. 623–628.

[12] M. Vijayagopal, P. Zanchetta, L. Empringham, L. D. Lillo, L. Tarisciotti, and P. Wheeler, "Modulated model predictive current control for direct matrix converter with fixed switching frequency," in *2015 17th European Conference on Power Electronics and Applications (EPE'15 ECCE-Europe)*, Sep. 2015, pp. 1–10.

[13] L. Tarisciotti, J. Lei, A. Formentini, A. Trentin, P. Zanchetta, P. Wheeler, and M. Rivera, "Modulated Predictive Control for Indirect Matrix Converter," *IEEE Transactions on Industry Applications*, vol. 53, no. 5, pp. 4644–4654, Sep. 2017.

[14] H. Mahmoudi, M. Aleenejad, and R. Ahmadi, "Modulated Model Predictive Control of Modular Multilevel Converters in VSC-HVDC Systems," *IEEE Transactions on Power Delivery*, vol. 33, no. 5, pp. 2115–2124, Oct. 2018.

[15] "Diode-Clamped Multilevel Inverters," in *High-Power Converters and AC Drives*. John Wiley & Sons, Ltd, 2016, pp. 143–183. [Online]. Available: https://onlinelibrary.wiley.com/doi/abs/10.1002/9781119156079.ch8

[16] J. Rodriguez and P. Cortes, "Predictive Control of a Three-Phase Neutral-Point Clamped Inverter," in *Predictive Control of Power Converters and Electrical Drives*. IEEE, 2012. [Online]. Available: http://ieeexplore.ieee.org/xpl/articleDetails.jsp?arnumber=6199002

978-1-7281-4181-7/19 $31.00 © 2019 IEEE

Development of Current-Source-Inverter-based Integrated Motor Drives using Wide-Bandgap Power Switches

Renato A. Torres   Hang Dai   Woongkul Lee   Thomas M. Jahns   Bulent Sarlioglu

Wisconsin Electric Machines and Power Electronics Consortium (WEMPEC)
University of Wisconsin - Madison
Madison, WI USA
sarlioglu@wisc.edu

Abstract— The miniaturization of power electronics is made possible by using wide-bandgap (WBG) power switches that can operate at high switching frequencies and high temperatures. Thus, WBG devices open the door for promising opportunities for the development of integrated motor drives (IMDs) that places the power electronics and the electric machine in the same enclosure. The concept of IMDs can replace traditional adjustable speed drives consisting of motor and inverter connected by a cable. However, WBG-based IMDs still faces many challenges regarding electromagnetic interference, cooling, reliability, and packaging. Nowadays, the majority of IMDs use voltage-source-based inverters since this topology is naturally compatible with conventional silicon-based power devices. Nevertheless, major advances in the development of new types of power semiconductor switches made from WBG semiconductor materials such as gallium nitride create new competitive alternatives for the development of reverse-voltage blocking (RVB) switches that are well-suited for use in current-source inverters (CSIs). This paper investigates the use of WBG power devices in CSIs for use in IMD applications. A 3-kW concept demonstrator CSI is successfully built and tested exciting an RL load that emulates a surface permanent-magnet (PM) synchronous motor. This CSI-IMD achieves appealing efficiency values >97.6% at both half- and full-load conditions with a PWM switching frequency of 125 kHz and an inverter power density of 9.84 kW/L.

Keywords— current-source inverter, integrated motor drives, wide-bandgap devices.

I. INTRODUCTION

More than half of the consumed global electricity is used in electric motor systems, and it is estimated that this consumption will almost double by 2040 [1]. Therefore, mandatory energy efficiency standards cover nearly 90% of the industrial electric motors sold nowadays [1]. Nevertheless, in most applications, the losses in the electric motor itself are a minor share of the total system loss. Thus, system-level energy efficiency improvement approaches are necessary to achieve larger energy savings. An effective strategy to increase the overall system efficiency is to replace fixed-speed drives by adjustable speed drives (ASD). According to [1], installing an ASD can increase system efficiency by 15-35% in a standard motor system with a variable load.

The choice for fixed-speed motors generally lies in low cost, small footprint, fast commissioning, and low maintenance. Therefore, when replacing fixed-speed motors with ASD, it is vital to avoid the addition of bulky drives and

to minimize the cost. In these situations, integrated motor drives (IMDs) can play a crucial role since they combine the power electronics and the machine into a single structure designed to minimize mass, volume and cost while maximizing efficiency and reliability. Fig. 1 shows the concept of the proposed IMD implemented with a current-source inverter (CSI).

Emergent wide-bandgap (WBG) power devices can efficiently operate at higher switching frequencies and higher junction temperatures than conventional silicon (Si)-based devices. Thus, these new devices can substantially benefit IMDs by shirking differential filters, energy storage elements, and cooling system [2]. Nevertheless, the replacement of Si-based switches by WBG switches raises new system challenges. The extremely fast switching speed (high dv/dt) allowed by WBG devices generates significant electromagnetic interference (EMI) [3], particularly in a frequency range where the standards are stricter. Besides, the presence of high dv/dt at the terminals of the machine can reduce its lifetime due to faster degradation of the winding isolation and the bearings [2][4]. Another critical point is the operating temperature of all the components of the motor drive system. Even though the WBG power switches can operate at higher temperatures, it is fundamental that the surrounding components should also withstand such temperatures.

Most of the motor drives nowadays use voltage-source inverters (VSIs) due to their natural compatibility with Si-based power devices such as the traditional insulated-gate bipolar transistor (IGBT), which does not block reverse voltage. However, the VSI topology imposes some critical limitations associated with high dv/dt at the output terminals, high EMI, temperature-sensitive dc-link capacitors and poor fault tolerance when exciting permanent magnet (PM)

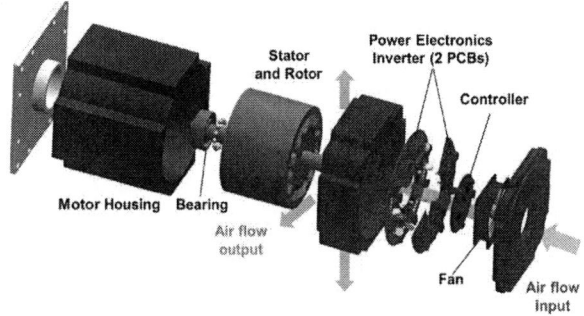

Fig. 1. 3D exploded model of WBG-based CSI-IMD.

Financial support for this work was provided by the US Advanced Research Projects Agency – Energy (ARPA-E) under Grant No. DE-AR0000893.

synchronous machines [5] [6]. In contrast, the CSI topology presents promising advantages for operations at high frequency and high temperatures. For example, the CSI dc-link inductor can be designed to withstand temperatures up to 200 °C, the dc-link inductor intrinsically reduces the EMI by providing a higher impedance path for high-frequency common-mode currents, and the combination of its dc-link inductor and output filter capacitors allow the CSI to deliver sinusoidal output voltages to the machine terminals.

During recent years, some attention has been devoted to the use of CSIs in motor drive applications. In [7], the paralleling of CSIs to increase the system power range and improve the output quality is investigated. The use of dual-gate monolithic bidirectional Gallium Nitride (GaN) switch and a new dc-link current control strategy is investigated in [8] to increase the efficiency of a buck-boost CSI. In [9], an optimal space vector modulation that allows a reduction of the dc-link inductance is presented. In [10], the uncontrolled generator operating characteristics of a CSI implemented with normally-on devices are investigated when driving PM machines. Additionally, CSIs have been recently investigated for a wide range of applications such as electric vehicles, wind turbines, and photovoltaic [11] [12] [13]. Nevertheless, very few studies have evaluated the use of CSI using WBG devices and their application to IMDs [4][5].

Therefore, the main contributions of this paper are to present essential design aspects of CSI-IMDs using WBG power switches, to provide simulation results showing that CSIs can achieve higher efficiency than VSIs when using WBG devices, and to demonstrate experimental results of a 3-kW concept demonstrator CSI exciting an RL load that emulates a surface PM synchronous motor.

The paper is structured as follows. In Section II, an overview of the investigated system is provided, including information on different design choices. Section III provides detailed numerical simulations that evaluated and compared the efficiency of the CSI and VSI topologies for the desired system. Section IV presents information on the control strategy and the controller hardware. Finally, the experimental results of the 3-kW concept demonstrator CSI are presented in Section IV.

II. SYSTEM OVERVIEW

Fig. 2 shows the schematic of the investigated system. It consists of two current source inverters in a back-to-back configuration where the first inverter interfaces with the power grid while the second is connected to a PM machine.

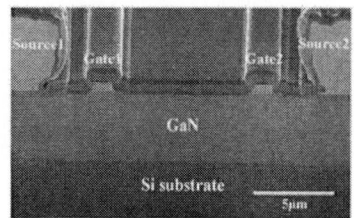
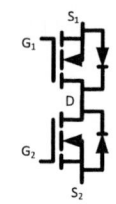

Fig. 3. Cross-sectional image of the AlGaN/GaN monolithic FQ switch on a Si substrate [16].

Fig. 4. Equivalent representation of common-source FQ switch.

A three-phase system is preferred since it is commonly used for most of the commercial heating, ventilation, air-conditioning (HVAC), and pumping systems, which are the target applications. The input voltage level of 240V is defined as complying with the standard line-line voltage rms values of commercial building electric service in the U.S. [14]. The traditional CSI converter topology (H6-CSI) is used due to its simplicity, which requires a minimal number of switches for a three-phase inverter system. Note that, since the aim is to replace fixed-speed drives, the cost of the system must be low enough to be competitive. Finally, the PM machine is selected due to its increased efficiency and power density, which are the primary goals of the IMD. Table I summarizes the system parameters and ratings.

A. Reverse-Voltage Blocking Hybrid Switch Realizations

In CSIs, the power switches must have reverse-voltage blocking (RVB) capability to avoid short circuit between output phases (due to the presence of output capacitive filters) [15]. In traditional CSIs, such capability is provided by connecting a power diode in series with the switching device. Unfortunately, this approach increases the conduction loss of the inverter and represents a major drawback of the CSI topology, which is often discarded due to its low efficiency.

The symmetric lateral structure of some new WBG power switches opens opportunities for the development of monolithic four-quadrant (FQ) switches that have both reverse-current flowing and RVB capabilities. Fig. 3 shows an example of a monolithic device developed by *Panasonic* [16]. The device has two gate terminals allowing for an FQ operation (the equivalent model of such FQ switch is shown in Fig. 4). Note that the monolithic FQ device has a common-drain structure that shares the drift region with both switches, thus reducing the total on-resistance of the FQ switch to a value almost equivalent to that of a standard non-FQ switch of similar ratings. Therefore, the use of these new monolithic

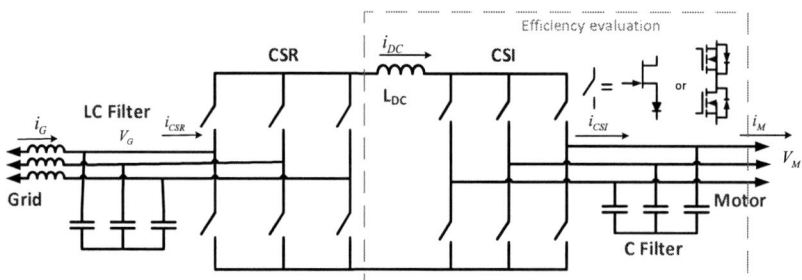

Fig. 2. – Schematic of investigated CSI-CSR system.

TABLE I – SYSTEM RATINGS AND MACHINE PARAMETERS

Metric	Value
Rated Power	3 kW
Rated voltage (l-l)	240 Vrms
Rated Phase Current	≈ 8 A
Maximum dc-link current	12 A
Motor # of Poles	10
Motor inductance	2.95 mH
Stator Resistance	0.19 Ω
Back EMF constant	137 V/kRPM

978-1-7281-4181-7/19 $31.00 © 2019 IEEE   68

FQ switches in CSIs allows an efficiency boost of the topology by eliminating the in-series diode.

Note that the introduction of FQ switches in CSIs requires an adaptation of the gate signals to avoid switching faults caused by interphase short-circuit or dc-link open-circuit. Such adaptation requires minimal software and hardware overhead and is thoroughly explained in [17].

In Section III, two switch configurations for the CSI are investigated. The first corresponds to the traditional configuration of a switch in series with a diode (*Cree* C3M0065090J + *Infineon* IDH16G65C6). The second configuration corresponds to a virtual common-drain FQ switch of the same on-state resistance as the standard switch (C3M0065090J). Note that the experimental setup presented in this paper adopts the configuration of a switch in series with a diode because of the lack of commercially available monolithic FQ switches at the required ratings.

B. Motor selection

Motor voltage: The CSI topology requires that there are always two power switches gated on at every time instant to provide a current path for the dc-link inductor current [17]. Therefore, even if the modulation index is zero and no current is being delivered to the load, the inverter still generates conduction losses. In other words, the conduction losses of the CSI are only dependent on the dc-link current level regardless of the state of the switches and load conditions. However, the dc-link current level is ultimately related to the modulation index by (1) where i_{ph} is the peak value of the required load phase current.

$$i_{ph} = m\, i_{DC} \qquad (1)$$

From (1), it is clear that, in order to minimize the dc-link current (conduction losses), the modulation index must be maximized. Therefore, for the most efficient operation of the CSI, the CSI modulation index is set to its maximum level ($m = 1$) without entering the overmodulation region ($m > 1$). From (1), if m is constant, it means that any change in the dc-link current level will be directly reflected at the load current. Therefore, it important to have a controlled front-end rectifier that continuously adjusts the dc-link current amplitude to correspond to the exact required output current at every time instant, i.e., $m = 1$ for the CSI.

The same idea applies to the CSR. Therefore, to maximize the whole CSR-CSI system efficiency at rated steady-state operation, it is important that both CSR and CSI are operating at a modulation index close to unity ($m \approx 1$). In this case, no voltage boosting or reduction occurs from CSR input to CSI output. Therefore, the rated voltage of the motor should be similar to the power line input voltage (240V).

Speed and Power: The target speed range is set to 1200 to 2000 rpm and the power rating is set to 3 kW to maximize the compatibility of the drive with general HVAC applications.

Frontal end-bell area: Since the power electronics converters should be integrated into one of the motor end-bells at the end of the stator stack, it is beneficial to choose a motor that has sufficient area for mounting the power electronics. Such consideration favors the selection of machines with a higher diameter-to-length aspect ratio.

Power factor: Maximizing the machine terminal power factor is extremely important for reducing the current rating of the inverter and, consequently, its conduction losses. Note that, while VSIs require machines with high series inductance to minimize motor current ripple, CSIs favor low series inductance to maintain a high-power factor operation. It should be noted that the current ripple in the CSI is low even if the machine inductance is small because the CSI output filter capacitors already provide an additional filtering stage between the inverter and the terminals of the machine.

Herein, machines from several different manufacturers were evaluated and the most promising models were simulated using their datasheet parameters. Based on this investigation, the motor AKM73M-ACGN2-00 manufactured by *Kollmorgen* is selected.

C. DC-link inductance selection and inductor design

The adoption of higher switching frequencies allowed by the WBG devices, unfortunately, cannot relieve the requirement for stored energy for the low voltage ride-through prerequisite. Therefore, to achieve the maximum possible reduction of the mass and volume of the motor drive, the ride-through capability is disregarded when designing the CSI. It should be noted that for many applications, particularly commercial HVAC, ride-through protection is not a requirement. The dc-link inductance value is calculated using (2), provided on [4]:

$$L_{dc} = \frac{\sqrt{3}V_m}{4\Delta I_{dc-max}F_s} \qquad (2)$$

where $F_S = 125kHz$ is the PWM switching frequency, $V_m = 196\,V$ is the peak fundamental frequency phase voltage at the CSI's ac output terminals, and $\Delta I_{dc-max} = 2A$ is the maximum dc-link current ripple. Note that (2) ignores some important aspects, such as the PWM states sequence. Hence, the final value adopted for the dc-link inductance ($320\,\mu H$) is set based on numerical simulations. The goal is to limit the maximum dc-link current ripple to approximately 2A peak-to-peak during steady-state full-load (3 kW) operation.

Detailed information on the dc-link inductor design procedure is available in [18]. The design procedure tries to optimize three major variables that are key to IMD: loss, mass, and volume. High-quality-factor powder core material that can withstand temperatures up to 200 °C is used. A non-gapped toroidal geometry is selected to retain the flux inside the core and to reduce radiated EMI. Note that, since this is an IMD, the inductor will be very close to the controller, which is susceptible to EMI radiation.

D. Motor Filter Capacitance Selection

The motor filter capacitance is defined according to (3) [4]:

$$C_f = \frac{I_{dc}}{4\,f_{sw}\Delta V_{cf}} \qquad (3)$$

where ΔV_{cf} is the maximum allowable voltage ripple during each switching event across the capacitor. The ΔV_{cf} value is initially defined to be 5% of the machine's peak line-to-line voltage, which yields a filter capacitance of $C_f = 1.88\,\mu F$. Therefore, a capacitor of $2\mu F$ is used. For this capacitance value, the simulated total harmonic distortion (THD) of the

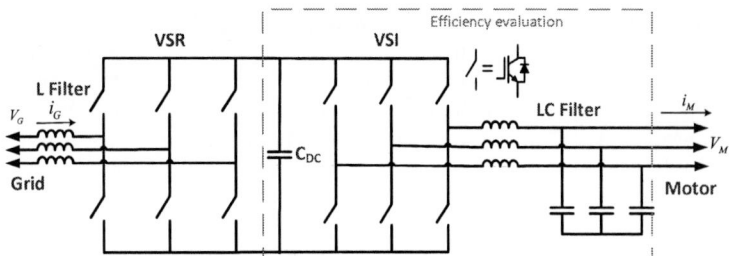

Fig. 5. Schematic of investigated VSI-VSR system.

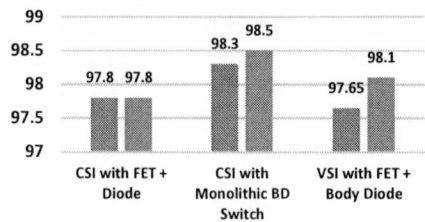

Fig. 6. Predicted efficiencies of the two CSI versions with different switch configurations and the VSI (with output LC filter). In blue, half load operation. In orange, full-load operation.

motor phase current is almost negligible due to the significant phase inductance of the motor. The voltage ripple is approximately 10 V, which is nearly equal to the 5% value initially selected as the target ripple amplitude.

III. EFFICIENCY AND LOSS EVALUATION

This section describes the technical approach used to carry out the numerical simulations that evaluated the efficiency performance of the CSI. In parallel, a simulation of a VSI is performed for comparison. Both the CSI and VSI are designed to deliver high-quality sinusoidal voltage and current waveforms at their output terminals to ensure a fair comparison.

The simulations included not only the inverter stage but also the front-end since the front-end can have a significant impact on the system efficiency. The inclusion of the rectifier stage is particularly important for the CSR-CSI system, in which the dc-link current amplitude is regulated according to the required output current with the aim of reducing the conduction loss in the dc-link inductor and inverter switches. Nevertheless, the efficiency results presented here do not account for the loss in the rectifier stage. A red dashed line in Fig. 2 and Fig. 5 shows the CSI and VSI circuit sections in which the efficiency is evaluated for each topology.

In the CSI (Fig. 2), the three-phase capacitive filter (labeled as "C Filter") interacts with the dc-link inductor to deliver sinusoidal output voltages. In the VSI (Fig. 5), three inductors in addition to three capacitors (labeled as "LC Filter") are necessary to provide the filtering action required to deliver comparable sinusoidal three-phase voltages at the inverter output.

Two versions of the CSI with different types of RVB switches are simulated. The first CSI uses a discrete hybrid RB switch with the FET + Diode configuration, as described in Section II.A. The second CSI uses a virtual monolithic FQ switch that consists of two SiC MOSFET devices connected in anti-series with a common-drain connection (see Fig. 4). Noteworthy, the total drift region resistance in this simulated monolithic FQ switch is adjusted to be equivalent to the resistance of a single MOSFET device (C3M0065090J) to emulate a common-drain monolithic device that contains only one drift region.

For the VSI, a single SiC MOSFET is used as the switch, which is the same used for CSI. No additional anti-parallel discrete diode is required in this model as the quality of the inherent body diode of the C3M0065090J MOSFET is sufficient to provide reverse conduction with very low losses.

The software PLECS is used for efficiency simulation of both topologies (VSI and CSI). The complete systems were modeled including power electronics, modulation strategy, and controls. Note that the PLECS software estimates conduction and switching losses by interpolating the values of stored loss look-up tables. The first look-up table corresponds to a map of the switching losses for switching transients at different voltage and current levels. The second look-up table provides conduction losses for different current levels. These look-up tables are obtained based on experimental results from double pulse tests (DPTs). Detailed information of the DPTs can be found in [19].

Fig. 6 shows the predicted half- and full-load efficiencies values for both CSI switch configurations and for the VSI at a PWM switching frequency of 125 kHz. The junction temperature of 90 °C for full-load and 60 °C for half-load is defined according to a combination of simulation and experimental results (see section V). The following losses are considered: switching, conduction, dc-link inductor (for the CSI case) and output filters.

Closer examination shows that the CSI topology using the virtual monolithic FQ switches enjoys a loss reduction of over 32% compared to the traditional CSI topology (switch + diode). This major loss reduction can be attributed to the much lower conduction loss that is achieved by using a monolithic FQ switch as compared to the FET + Diode configuration.

Additionally, it can be seen that the VSI efficiency is lower than the efficiency of the CSI implemented with monolithic FQ switches. The main reason for that is the lossy inductors of the VSI three-phase LC filter that has a high core loss component due to the ripple current at high PWM frequency. Increasing the inductance of the LC filter while maintaining its cutoff frequency allows the reduction of core loss by reducing the current ripple. However, it ends-up increasing the inductor conduction loss component and its size. Increasing the capacitance of the LC filter, on the other hand, may seem a reasonable approach to reducing the filter size since capacitors are generally more energy-dense than inductors. Nevertheless, even though this approach may deliver the same output waveforms to the motor, it increases the current ripple in the inductor, increasing the loss of the filter. Thus, it was verified that, for the same total mass of passive components and the same quality of voltage and current waveforms at the output of the simulated system, the CSI topology delivers higher efficiency than the VSI with an output LC filter.

978-1-7281-4181-7/19 $31.00 © 2019 IEEE

It is worth to point out that conventional VSIs without LC output filter is likely to be more efficient than both CSI configurations; however, this basic VSI topology has high *dv/dt* at the output terminals that results in many undesirable effects for the motor drive system including significant motor winding overvoltage, high bearing current, reduced motor life and high EMI.

IV. CONTROLLER

Fig. 7 shows the equivalent circuit of an average switching model for a CSI. Note that the CSR portion is not shown in the figure, but it corresponds to a reflection of the CSI side. The voltages v_{CSR} and v_{CSI}, at the terminals of the dc-link inductor, are a function of the modulation index vectors $m = (m_a, m_b, m_c)$ of the CSI and CSR, respectively.

As mentioned in Section II.B, it is desirable to operate the CSI with a modulation index close to unity. Therefore, a simplified control strategy of the CSI-CSR system is used where the CSI modulation index vector (m_{CSI}) is set to a constant unitary amplitude and only the modulation index vector angle with respect to the electrical angle of the motor is controlled. The motor electrical angle is provided by an encoder.

Note that, since the $|m_{CSI}|$ is constant, any disturbance in the motor terminal voltage is directly reflected as a disturbance in the v_{CSI} voltage. Thus, to keep the dc-link current constant, the modulation index of the CSR must be properly commanded to assure that the average voltage of v_{CSR} equals the average voltage of v_{CSI}. For this purpose, a PI regulator is used where the dc-link current is controlled by commanding the CSR modulation index. The reference for the dc-link current reference must be set to the required load peak phase current since the modulation index of the CSI is fixed (see (1)).

The control algorithm is implemented in a digital signal processor (DSP) TMS320F28377D from *Texas Instruments*. Instead of industrial controller boards, preference has been given to small chip candidates that facilitate the controller hardware integration into the PCBs, reducing the system size and cost. The selected DSP has two cores, and each core is programmed to control a single converter (CSI or CSR). Finally, the selected DSP has a total of 24 PWM channels available, which will be all required in case dual gates are used for both the CSI and CSR [17].

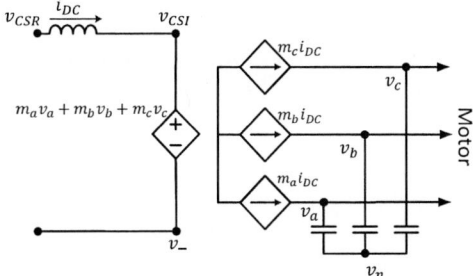

Fig. 7. Equivalent circuit of an average switching model for CSI.

V. EXPERIMENTAL RESULTS

Fig. 8 shows the final configuration of a preliminary CSI-IMD demonstrator. At the moment, the power electronics system is a bench-top prototype. However, its dimensions are defined to fit the enclosure of the selected motor. It can be seen that there are three types of PCB: 1) the motherboard, which includes sensors, and power connections; 2) six half-bridge daughter boards required for both CSI and CSR, which contains the WBG switching devices, and 3) a controller board, where the DSP is mounted on. Note that, in future design, the controller board will be integrated into the motherboard to reduce the overall height of the system. Finally, it can also be observed the presence of the dc-link inductor that occupies a small section of the motherboard.

Fig. 9 shows the CSI inside a polycarbonate enclosure that emulates the motor enclosure geometry. A cooling fan is placed at the top of the enclosure to generate a constant airflow that cools down the switching devices. Note that the switching devices are soldered to PCB pads that are exposed to the air, behaving as heatsinks (see Fig. 8). In future versions of the prototype, the objective is to integrate the fan to the shaft of the machine.

Fig. 10 shows the measured voltage and current waveforms at full-load operation at 125 kHz PWM switching frequency when exciting an RL load that emulates a surface PM synchronous motor. It can be observed that the harmonic distortions of both the voltage and current waveforms are very small (< 5%). It should be emphasized that no additional filters at the CSI output terminals are required to deliver these sinusoidal waveforms for both the voltage and current, contrasting the CSI sharply with the VSI that needs an

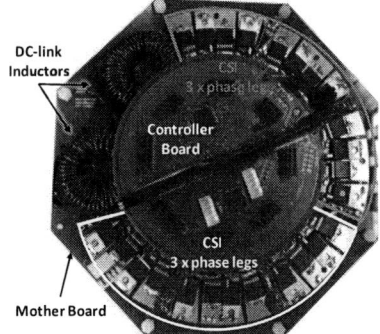

Fig. 8. View of the CSI-CSR-IMD including mother board, half-bridfes and controller board.

Fig. 9. View of the preliminary demonstrator CSI machine drive.

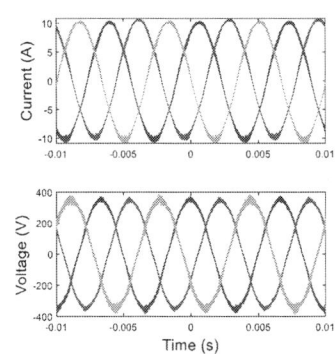

Fig. 10. Line-line voltage and phase current waveforms of CSI-IMD when driving an RL load.

978-1-7281-4181-7/19 $31.00 © 2019 IEEE

additional three-phase LC filter to achieve similarly sinusoidal waveforms. The high quality of the machine excitation waveforms reflects the strong advantages offered by the CSI over a conventional VSI when using high-frequency WBG switching devices.

The efficiency measurements are obtained using the *Yokogawa* Precision Power Analyzer WT1800. The measured CSI efficiencies are 97.6% and 97.7% at half-load and full-load operation, respectively. Finally, Fig. 11 shows the maximum temperature of the power devices under full load operation. Based on the estimated losses via computational simulation, on the thermal resistance of the devices provided in their datasheet and on the measured case temperatures, a junction temperature of 92 °C and 91 °C is estimated for the SiC MOSFET (C3M0065090J) and SiC Schottky diode (IDH16G65C6), which are well within their maximum junction operating temperatures of 150 °C and 175 °C, respectively.

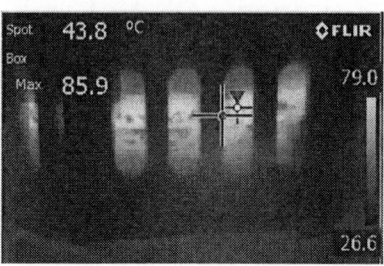

Fig. 11. Measured temperature of power devices under full load operation (3kW) using thermal camera. Maximum temperature of 85.9 °C.

VI. CONCLUSIONS

This paper presents an ongoing project of a 3-kW CSI-IMD aimed for HVAC applications. The benefits of using the CSI topology against the conventional VSI topology are discussed. Detailed information on different design choices such as motor selection, passive components sizing, and controller hardware implementation is provided. Numerical simulations evaluated for both the CSI and VSI topologies demonstrated that the CSI using monolithic FQ switches can achieve higher efficiency than the VSI for the desired system. Finally, experimental results have shown that the CSI prototype is able to achieve appealing efficiency values >97.6% at both half- and full-load conditions with a PWM switching frequency of 125 kHz and an inverter power density of 9.84 kW/L. These results are in accordance with the simulation data and demonstrate the feasibility of the system and the high competence of the CSI topology.

ACKNOWLEDGMENT

The information, data, or work presented herein was funded in part by the Advanced Research Projects Agency-Energy (ARPA-E), U.S. Department of Energy, under Award Number DE-AR0000893. The views and opinions of authors expressed herein do not necessarily state or reflect those of the United States Government or any agency thereof. The authors also gratefully acknowledge the support of the Wisconsin Electric Machines and Power Electronics Consortium (WEMPEC) including access to its laboratory facilities.

REFERENCES

[1] International Energy Agency, "Energy Efficiency 2018 - Analysis and outlooks to 2040," 2018.

[2] A. K. Morya et al., "Wide bandgap devices in AC electric drives: opportunities and challenges," in IEEE Transactions on Transportation Electrification, vol. 5, no. 1, pp. 3-20, March 2019.

[3] N. Oswald, P. Anthony, N. McNeill and B. H. Stark, "An experimental investigation of the tradeoff between switching losses and EMI generation with hard-switched all-Si, Si-SiC, and all-SiC device combinations," *IEEE Transactions on Power Electronics*, May 2014.

[4] H. Dai and T. M. Jahns, "Comparative investigation of PWM current-source inverters for future machine drives using high-frequency wide-bandgap

power switches," *IEEE Applied Power Electronics Conference and Exposition*, San Antonio, TX, 2018, pp. 2601-2608.

[5] R. A. Torres, H. Dai, W. Lee, T. M. Jahns and B. Sarlioglu, "Current-source inverters for integrated motor drives using wide-bandgap power switches," *IEEE Transportation Electrification Conference and Expo*, Long Beach, CA, 2018, pp. 1002-1008.

[6] H. Wang and F. Blaabjerg, "Reliability of capacitors for DC-Link applications in power electronic converters—an overview," *IEEE Trans. Ind. Appl.*, vol. 50, no. 5, pp. 3569–3578, Sep. 2014.

[7] L. Ding, Y. Li and Y. W. Li, "Interleaved SPWM of Parallel CSC System with Low Common-mode Voltage," *IEEE Applied Power Electronics Conference and Exposition*, Anaheim, CA, USA, 2019, pp. 2505-2510.

[8] M. Guacci, M. Tatic, D. Bortis, J. W. Kolar, Y. Kinoshita and H. Ishida, "Novel Three-Phase Two-Third-Modulated Buck-Boost Current Source Inverter System Employing Dual-Gate Monolithic Bidirectional GaN e-FETs," *International Symposium on Power Electronics for Distributed Generation Systems*, Xi'an, China, 2019, pp. 674-683.

[9] X. Guo, Y. Yang and X. Wang, "Optimal Space Vector Modulation of Current-Source Converter for DC-Link Current Ripple Reduction," in *IEEE Transactions on Industrial Electronics*, vol. 66, no. 3, pp. 1671-1680, 2019.

[10] Y. Zhang and T. M. Jahns, "Uncontrolled generator operation of PM synchronous machine drive with current-source inverter using normally on switches," *IEEE Trans. Ind. Appl.*, Jan. 2017.

[11] G.-J. Su, L. Tang, and Z. Wu, "Extended constant-torque and constant-power speed range control of permanent magnet machine using a current source inverter," *IEEE Veh. Power Propuls. Conf.*, 2009.

[12] Q. Wei, B. Wu, D. Xu and N. R. Zargari, "Further Study on a PWM Current-Source-Converter-Based Wind Energy Conversion System Considering the DC-Link Voltage," in *IEEE Transactions on Power Electronics*, vol. 34, no. 6, pp. 5378-5387, June 2019.

[13] E. Lorenzani, F. Immovilli, G. Migliazza, M. Frigieri, C. Bianchini and M. Davoli, "CSI7: A Modified Three-Phase Current-Source Inverter for Modular Photovoltaic Applications," in IEEE Transactions on Industrial Electronics, vol. 64, no. 7, pp. 5449-5459, July 2017.

[14] National Electrical Manufacturers Association (NEMA), "American National Standard For Electric Power Systems and Equipment— Voltage Ratings (60 Hertz)," American National Standards Institute, ANSI C84.1, 2016.

[15] G. Ledwich, "Current source inverter modulation," *IEEE Trans. Power Electron.*, vol. 6, no. 4, pp. 618–623, 1991.

[16] T. Morita, M. Yanagihara, H. Ishida, M. Hikita, K. Kaibara, H. Matsuo, Y. Uemoto, T. Ueda, T. Tanaka, D. Ueda, "650 V 3.1mΩcm^2 GaN-based monolithic bidirectional switch using normally-off gate injection transistor", *in Proc. IEEE Int. Electron Devices Meeting*, pp. 865-868, 2007.

[17] R. A. Torres, H. Dai, T. M. Jahns and B. Sarlioglu, "Operation and analysis of current-source inverters using dual-gate four-quadrant wide-bandgap power switches," *IEEE Energy Conversion Congress and Exposition*, Baltimore, MD, 2019.

[18] R. A. Torres, H. Dai, T. M. Jahns and B. Sarlioglu, "Design of high-performance toroidal dc-link inductor for current-source inverters," *IEEE Applied Power Electronics Conference and Exposition*, Anaheim, CA, USA, 2019, pp. 2694-2701.

[19] H. Dai, R. A. Torres, T. M. Jahns and B. Sarlioglu, "Characterization and implementation of hybrid reverse-voltage-blocking and bidirectional switches using WBG devices in emerging motor drive applications," *IEEE Applied Power Electronics Conference and Exposition*, Anaheim, CA, USA, 2019, pp. 297-304.

Power Control of a Doubly Fed Induction Wind Generator employing a Takagi-Sugeno Fuzzy Logic Controller

C. M. Rocha-Osorio*, J.S. Solís-Chaves†, Eliomar R. Conde D.*, J.L. Azcue Puma*, Fernando Lino‡,
A.J. Sguarezi Filho*

*Federal University of ABC (UFABC)

Avenida dos Estados, 5001 - Bairro Santa Terezinha - Santo André, SP/Brazil

{rocha.carlos@ufabc.edu.br, eliomar.conde@ufabc.edu.br, jose.azcue@ufabc.edu.br, alfeu.sguarezi@ufabc.edu.br}

†Universidad Escuela Colombiana de Carreras Industriales (UECCI)

jsolisc@ecci.edu.co

‡Federal Institute Sao Paulo (IFSP)

flino@ifsp.edu.br

Abstract—This paper presents the active and reactive power control of the Doubly Fed Induction Generator employing Fuzzy Logic for a wind power generation system. The proposed controller is a Fuzzy Logic Controller Takagi-Sugeno type which determines the rotor voltage by a set of rules base defined by a linear combination of the rotor current's error membership sets, reducing the structure complexity of the control system and consequently a lower computational cost is achieved. Additionally, it is easy to implement on a digital signal processing board when it's compared with other controllers such as the Fuzzy Mamdani type. The control system is accomplished using the Stator Flux Oriented Control with Space Vector Modulation technique and a computer simulation is implemented in Matlab/Simulink. The results indicates that the TS FLC could be an interesting alternative to conventional controllers in the power control of the DFIG during normal operation conditions.

Index Terms—Doubly Fed Induction Generator, vector control, Stator Flux Field Oriented Control, Fuzzy Logic control, Takagi Sugeno, voltage sag, wind power, intelligent control.

I. INTRODUCTION

From a technological point of view, the control of wind energy is difficult, particularly due to the non-linear nature of the wind. That is, the energy produced changes as a function of its speed, which in addition to increasing the complexity in the control, increases the costs in the operation of the system [1]. With the evolution of power electronics, the control of large amounts of energy (between 6 and 8 MW) had been possible, allowing the operation of wind turbines at variable speed, which increases the efficiency of these systems [2]. One important evolution in this type of renewable energy generation occurred with the inclusion of the Doubly Fed Induction Generator (DFIG), mainly because it is a device with access to the rotor terminals, which results in a greater controllability; its connection to the mains by means of a bidirectional electronic converter, enables a bidirectional power flow; besides operating as a generator in sub-synchronous speed, it can be considered a supplier of reactive power under

some operating conditions; finally, as an additional advantage, the power of the electronic converter employed to control the generator, processes a maximum of 30% of its nominal power, which reduces the implementation costs of the entire system [3], [4]. However, although the DFIG is considered one of the suitable devices for wind energy applications, it is necessary to apply on it some control strategies to guarantee the optimal power supply to the network, both in normal operating conditions and during disturbances.

On the other hand, of all the possible control strategies that can be applied to the DFIG, this work focuses on the power vector control, by means of a controller based on Fuzzy Logic, also called FLC (Fuzzy Logic Controller). The control of electrical machines using FLCs had been reported from nineties to today [5], [6], the main reason for that is because the Fuzzy Logic is a powerful tool to resolve typical control problems through the emulation of human thoughts when a well-known behavior plant is present, and this is the typical case of a wind energy generator, such as in [7] and [8], where a kind of FLC called Fuzzy-PI Controller was proposed to control the active and the reactive powers. Some Direct Torque Control (DTC) strategies were described in [9] and [10], using different types of generators for testing a variety of FLC algorithms. In 2017, Kalaivani et al. [11] presents a FLC for a DFIG powered by a Back–to–Back converter and operating under fault conditions. The error and the change in the error of the rotor voltage signals are the inputs for the FLC based on the Mamdani inference. In 2018, Elkhadiri et al. [12] propose a FLC to control a DFIG wind turbine by means of the Rotor Flux Oriented Vector (RFOC) technique. The FLC uses the rotor current error and the evolution of this error to estimate the control signal.

In [13] a Direct Torque Control technique for a DFIG was proposed to test the reactive power and the electromagnetic torque step changes according to the grid conditions using wireless references signals. A Fuzzy-tuned PI controller was included to optimize the calculation of the proportional and the

Identify applicable funding agency here. If none, delete this.

integral gains. At the beginning of 2019, a FLC for improvement of the steady-state response of a Doubly Fed Induction Generator used in a wind energy system, and governed by means of a Deadbeat Power Controller was proposed. Control responses follows the power references imposed, despite the fact that the generator parameters were varied in a 30%. A lower steady-state error is also achieved when compared with a Deadbeat and a classical PI controller [14].

In this context, this paper presents the control of the DFIG's active and reactive powers using a FLC in a wind power generation system. The controller presented here is a variation of the control scheme proposed in [15] but adapted for the DFIG. The control system is performed using a Takagi-Sugeno fuzzy controller (TS FLC) based on Stator Flux Oriented Vector Control (SFOC) with Space Vector Modulation (SVM) by determining the rotor voltage vector "$\vec{v}_{r,dq}$" functions of "$e_{i_{rd}}$" and "$e_{i_{rq}}$". The proposed controller has a simplified structure because it has a lower number of rules compared to other FLCs such as the Mamdani type, which implies a lower computational cost [15]. The studies are carried out using computer simulation and using mathematical models implemented in Matlab/Simulink under different operating conditions, such as variable power factor and fixed and variable mechanical speed.

To demonstrate the performance of this TS FLC algorithm the following sections are written: Section II describes the DFIG power control, Section III depicts the TS FLC, rules and membership functions. Section IV addresses the simulation results focusing on two different tests, first one at constant speed and second one at variable rotor speed. Finally, in Section V some conclusions are summarized.

II. POWER CONTROL OF DOUBLY FED INDUCTION GENERATOR

The Power Control of the DFIG is done by means of a Vector Control technique, which is perhaps the most known one. It is very similar to the well known classical vector control of a squirrel cage machine, in which the main purpose is to achieve a decoupled control over two variables. In a similar way, the Vector Control technique applied to control the DFIG powers entails that a decoupled control of active power and reactive power can be achieved.

It consists in the use of the Park transformation to reference the system magnitudes to a rotating reference frame called 'dq' which will rotate at the synchronous speed of the system. Once the this transformation is used, the model of the DFIG in the 'dq' reference frame is given by [16]:

$$\vec{v}_{sdq} = R_s \vec{i}_{sdq} + \frac{d\vec{\psi}_{sdq}}{dt} + j\omega_s \vec{\psi}_{sdq} \tag{1}$$

$$\vec{v}_{rdq} = R_r \vec{i}_{rdq} + \frac{d\vec{\psi}_{rdq}}{dt} + j\left(\omega_s - PP\omega_{mec}\right)\vec{\psi}_{rdq} \tag{2}$$

The expressions corresponding to active and reactive power in the stator are:

$$P_s = \frac{3}{2}\left(v_{sd}i_{sd} + v_{sq}i_{sq}\right) \tag{3}$$

$$Q_s = \frac{3}{2}\left(v_{sq}i_{sd} - v_{sd}i_{sq}\right) \tag{4}$$

The fluxes in the stator and the rotor

$$\vec{\psi}_{sdq} = L_s \vec{i}_{sdq} + L_m \vec{i}_{rdq} \tag{5}$$

$$\vec{\psi}_{rdq} = L_m \vec{i}_{sdq} + L_r \vec{i}_{rdq} \tag{6}$$

The subscripts "s" and "r" correspond to the stator and the rotor respectively, v and i represents the voltage and current variables respectively (with the corresponding subscript "s" or "r"), R the resistance, L the inductance, L_m the mutual inductance, ω_{mec} is the mechanical speed and PP the pair of poles of the machine.

Aligning the reference frame with the stator flux (5) can be written as the following two equations:

$$i_{sd} = \frac{\psi_s}{L_s} + \frac{L_m}{L_s}i_{rd} \tag{7}$$

$$i_{sq} = -\frac{L_m}{L_s}i_{rq} \tag{8}$$

Substituting (7) and (8) into the expressions (3) and (4) results:

$$P_s = -\frac{3}{2}\left(v_s\frac{L_m}{L_s}i_{rq}\right) \tag{9}$$

$$Q_s = \frac{3}{2}v_s\left(\frac{\psi_s}{L_s} - \frac{L_m}{L_s}i_{rd}\right) \tag{10}$$

Where is observed that now in this new synchronous reference frame the decoupled control of the active and reactive power is possible through the 'dq' currents.

A. Power control using fuzzy Takagi-Sugeno controller

In this section is presented the proposed control scheme, which consists of PI controllers and a FLC TS as shown in Fig. 1.

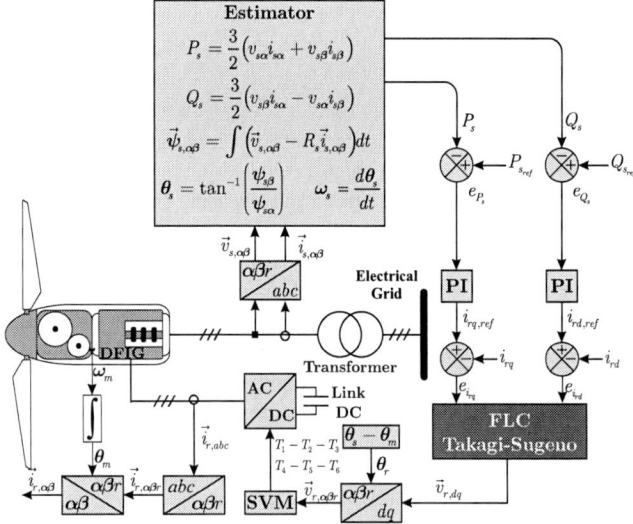

Fig. 1: Power control using Takagi-Sugeno FLC

The control scheme previously presented in Fig. 1 has additionally two stages, estimation and modulation.

1) Estimation: In this step the active power (P_s) and reactive power (Q_s) are estimated by means of the stator voltage ($v_{s,\alpha\beta}$) and stator current ($i_{s,\alpha\beta}$), where:

$$P_s = \frac{3}{2}\left(v_{s\alpha}i_{s\alpha} + v_{s\beta}i_{s\beta}\right) \qquad (11)$$

$$Q_s = \frac{3}{2}\left(v_{s\beta}i_{s\alpha} - v_{s\alpha}i_{s\beta}\right) \qquad (12)$$

In addition, the stator angular position and speed are estimated respectively in (13) and (14) by means of the stator magnetic flux components represented in (5).

$$\theta_s = \tan^{-1}\left(\frac{\psi_{s\beta}}{\psi_{s\alpha}}\right) \qquad (13)$$

$$\omega_s = \frac{d\theta_s}{dt} \qquad (14)$$

The estimation of the stator magnetic flux is given by:

$$\vec{\psi}_{s,\alpha\beta} = \int \left(\vec{v}_{s,\alpha\beta} - R_s\vec{i}_{s,\alpha\beta}\right)dt \qquad (15)$$

The angle of the rotor θ_m, it can be obtained from:

$$\theta_m = \int \frac{p}{2}\omega_m dt \qquad (16)$$

And the slip angle or, the also called, rotor flux angle θ_r, can be obtained from equations (13) and (16):

$$\theta_r = \theta_s - \theta_m \qquad (17)$$

2) Space Vector Modulation: From the rotor voltages obtained through the TS FLC, the SVM strategy is used to generate the switching times (T_1, T_2, T_3, T_4, T_5 and T_6) that provided the rotor side converter.

III. TAKAGI-SUGENO FUZZY LOGIC CONTROLLER

The Takagi Sugeno type Fuzzy Controller shown in Fig. 2 and used in the control scheme of Fig. 1 was studied. An important difference with respect to other control schemes that use Mamdani type fuzzy controllers is that the defuzzification phase is not necessary, since each rule of the controller already provides a numerical value and the total result is determined by the average of the weighted sum of each rule [17].

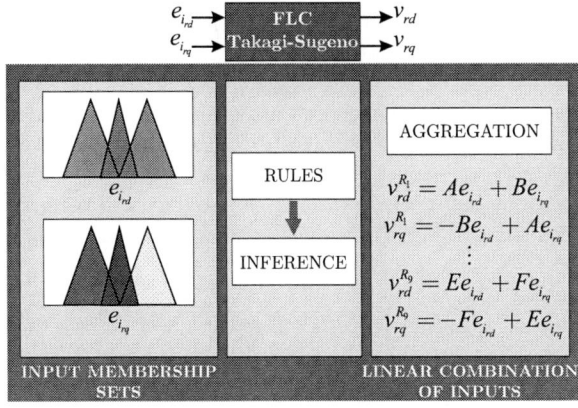

Fig. 2: Fuzzy Logic Controller Takagi-Sugeno Type

The TS FLC controller has the following considerations:

The rotor reference currents are calculated by PI controllers that have as inputs the stator active and reactive power errors.

The rotor voltage components v_{rd} and v_{rq} are calculated by means of equations (18) and (19), in this way:

$$v_{rd} = \frac{\sum_{i=1}^{n}\mu^{R_i}v_{rd}^{R_i}}{\sum_{i=1}^{n}v_{rd}^{R_i}} \qquad (18)$$

$$v_{rq} = \frac{\sum_{i=1}^{n}\mu^{R_i}v_{rq}^{R_i}}{\sum_{i=1}^{n}v_{rq}^{R_i}} \qquad (19)$$

1) Membership Functions: The main characteristics of the Membership Functions employed in this FLC are shown in Table I. The membership functions were constructed using a heuristic method of trial and error, since one of the problems of the use of FLC is that there is no systematic procedure for the adjustment of the controller [18].

TABLE I: Membership Functions Characteristics for $e_{i_{rd}}$ and $e_{i_{rq}}$

Variable linguistic	Function	Points			
		A	B	C	D
NE	Trapezoidal	-161.9	-100	-60	0
ZE	Triangular	-0.77	0	0.77	NA
PO	Trapezoidal	0	60	100	161,9

$e_{i_{rd}}$ (top table header)

Variable linguistic	Function	Points			
		A	B	C	D
NE	Trapezoidal	-172	-100	-75	0
ZE	Triangular	-0,77	0	0,77	NA
PO	Trapezoidal	0	75	100	172

$e_{i_{rq}}$ (bottom table header)

2) Base Rules: The FLC TS type proposed here, takes into account the rules presented in Table II:

Where, each linguistic term has the following meaning: **NE:** Negative - **ZE:** Zero - **PO:** Positive, and the constants A, B, C, D, E, and F are the coefficients of a first order polynomial function that is typically present in the consequent part of the fuzzy first-order Takagi-Sugeno controllers [15].

The following data were used as constants: A=20, B=200, C=100, D=80, E=120 and F=0, and were obtained by the trial and error heuristic method.

Hence, some examples for the base rules are:

If $e_{i_{rd}}$ is NE and $e_{i_{rq}}$ is NE then $v_{rd}^{R_1} = 20e_{i_{rd}} + 200e_{i_{rq}}$ and $v_{rq}^{R_1} = -200e_{i_{rd}} + 20e_{i_{rq}}$

If $e_{i_{rd}}$ is NE and $e_{i_{rq}}$ is PO then $v_{rd}^{R_3} = 100e_{i_{rd}} + 80e_{i_{rq}}$ and $v_{rq}^{R_3} = -80e_{i_{rd}} + 100e_{i_{rq}}$

If $e_{i_{rd}}$ is PO and $e_{i_{rq}}$ is ZE then $v_{rd}^{R_8} = 120e_{i_{rd}} + 0e_{i_{rq}}$ and $v_{rq}^{R_8} = -0e_{i_{rd}} + 120e_{i_{rq}}$

TABLE II: Base Rules for the outputs estimation (v_{rd} and v_{rq})

$e_{i_{rd}}/e_{i_{rq}}$	NE	ZE	PO
NE	$v_{rd}^{R_1} = Ae_{i_{rd}} + Be_{i_{rq}}$ $v_{rq}^{R_1} = -Be_{i_{rd}} + Ae_{i_{rq}}$	$v_{rd}^{R_2} = Ae_{i_{rd}} + Be_{i_{rq}}$ $v_{rq}^{R_2} = -Be_{i_{rd}} + Ae_{i_{rq}}$	$v_{rd}^{R_3} = Ce_{i_{rd}} + De_{i_{rq}}$ $v_{rq}^{R_3} = -De_{i_{rd}} + Ce_{i_{rq}}$
ZE	$v_{rd}^{R_4} = Ae_{i_{rd}} + Be_{i_{rq}}$ $v_{rq}^{R_4} = -Be_{i_{rd}} + Ae_{i_{rq}}$	$v_{rd}^{R_5} = Ce_{i_{rd}} + De_{i_{rq}}$ $v_{rq}^{R_5} = -De_{i_{rd}} + Ce_{i_{rq}}$	$v_{rd}^{R_6} = Ee_{i_{rd}} + Fe_{i_{rq}}$ $v_{rq}^{R_6} = -Fe_{i_{rd}} + Ee_{i_{rq}}$
PO	$v_{rd}^{R_7} = Ce_{i_{rd}} + De_{i_{rq}}$ $v_{rq}^{R_7} = -De_{i_{rd}} + Ce_{i_{rq}}$	$v_{rd}^{R_8} = Ee_{i_{rd}} + Fe_{i_{rq}}$ $v_{rq}^{R_8} = -Fe_{i_{rd}} + Ee_{i_{rq}}$	$v_{rd}^{R_9} = Ee_{i_{rd}} + FE_{i_{rq}}$ $v_{rq}^{R_9} = -Fe_{i_{rd}} + Ee_{i_{rq}}$

IV. SIMULATION RESULTS

The implemented system is shown in Fig. 1. This configuration was tested under different power conditions according to Table III:

TABLE III: Network conditions

Condition	Powers		Power Factor	Time Interval [s]
	P [W]	**Q [var]**		
1	-1000	-620	1177	0.85 leading	3 - 4
2	-1500	930	1765	0.85 lagging	4 - 5
3	-2000	0	2000	1	5 - 6

In condition 1, the active power reference is equal to -1000 W and the reactive power reference is -620 var, which means that the generator is supplying active and reactive power to the grid with a power factor in advance of 0.85. In condition 2, the power factor is 0.85 in delay, which means that the DFIG is only supplying active power (-1500 W) and that the reactive power is being consumed (930 var). Finally, in condition 3, the DFIG is tested under unit power factor i.e. without generating or consuming reactive power. The above conditions were tested at 1 second time intervals.

In addition to the previously proposed conditions, the system was tested under fixed rotor mechanical speed at 200 rad/s and with variable rotor mechanical speed.

A. Test 1: Constant rotor mechanical speed, fixed in 200 rad/s

The control scheme shown in Fig. 1, was tested under different active and reactive power conditions according to the table III, and in the super-synchronous DFIG's operation with the mechanical speed of the rotor at 200 rad/s to evaluate the dynamic response of controllers under different power factor conditions.

In Fig. 3a, the results of the DFIG active and reactive power control are shown, it can be seen that the powers follow the references appropriately, in Fig. 3b, the rotor currents are shown in the direct and quadrature axis.

In Fig. 3c, the variation of the power factor when it changes from condition 1 (leading power factor) to condition 2 (lagging power factor) is shown, the Fig. 3d shows the variation of the power factor from condition 2 to condition 3 (unit power factor), in both figures the current was multiplied by a factor of 10 to visualize the differences between the voltage and the current adequately.

1) Controllers Comparison during test 1: The response of the proposed TS FLC was compared with both PI controller and Mamdani FLC responses (see 3e and 3f). The PI control was adjusted using the methodology proposed in [19] and the Mamdani FLC was adjusted using the trial and error heuristic method. The TS FLC has a similar response to the Mamdani FLC and a faster response than the PI controller in the rotor currents control. However, when comparing the two fuzzy controllers, it can be established the following advantages of the proposed TS FLC:

- The proposed control system employs a smaller number of controllers: a single FLC TS controller with two inputs and two outputs. The conventional system with PI and/or FLC Mamdani controllers employs two controllers (two inputs, one output).
- The proposed FLC TS employs 18 base rules (9 rules for each output). The Mamdani FLC employs 98 base rules (49 base rules for each output).

B. Test 2 : Variable mechanical rotor speed

Now, testing the control scheme of Fig. 1 under different power factor conditions and considering variable mechanical speed according with Fig. 4. Fig. 5a shows that the DFIG active and reactive powers follow its references, Fig. 5b shows the rotor currents on the direct and quadrature axes.

Fig. 5c shows the power factor (PF) variations when the condition 1 (leading PF) changes to condition 2 (lagging PF), and in Fig. 5d, presents the power factor variation when the condition 2 changes to condition 3 (unit PF), in both Figs. rotor current was multiplied by a 10x factor for a better visualizing between currents and voltages differences.

Fig. 5e shows the alternating rotor currents. Here, in time 3.48 s it can be perceived a change in the machine performance, first generating under sub-synchronous speed and then a new change in the machine performance is shown, generating energy at super-synchronous speed in 5.51 s, and then generating under super-synchronous speed to sub-synchronous speed. At last, Fig. 5f shows alternating stator currents.

V. CONCLUSION

In this work the active and reactive power control for the Doubly Fed Induction Generator (DFIG) was presented using Fuzzy Logic for a wind power generation system. The control of the system was carried out using a FLC Takagi Sugeno type based on vector control by stator flux orientation with SVM,

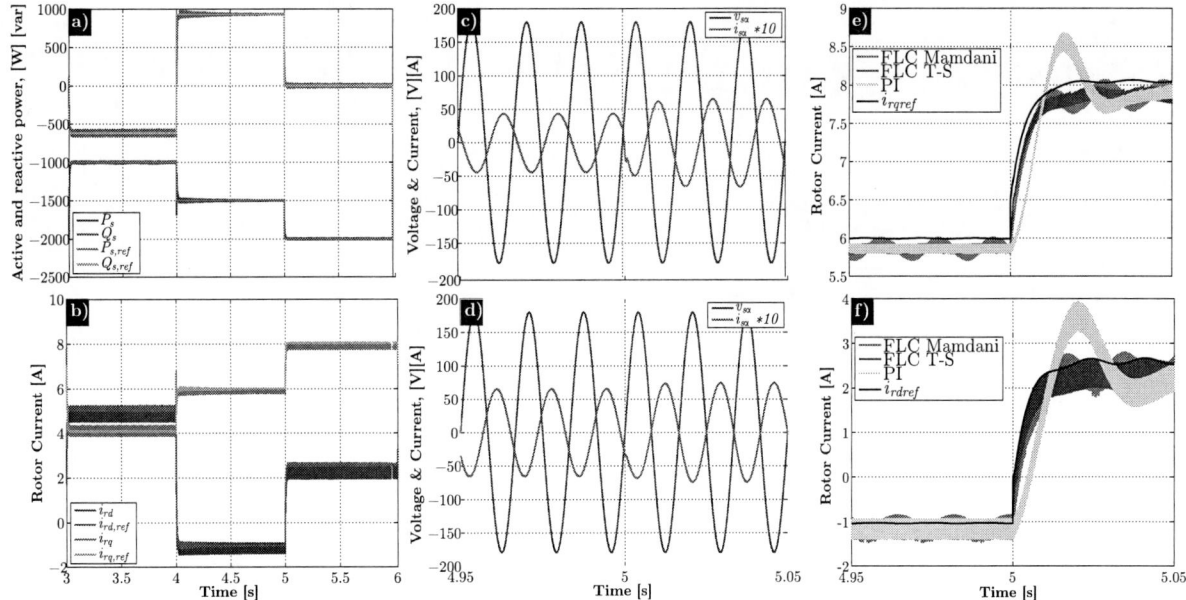

Fig. 3: TS FLC response. a) Active and reactive powers, b) rotor current, c) power factor (conditions 1-2). d) power factor (conditions 2-3), e) rotor current - quadrature component, i_{rq}, f) rotor current - direct component, i_{rd}.

the studies were carried out through computer simulation using mathematical models implemented in Matlab/Simulink. The system was tested under different operating conditions, varying the power factor and varying the mechanical speed of the rotor.

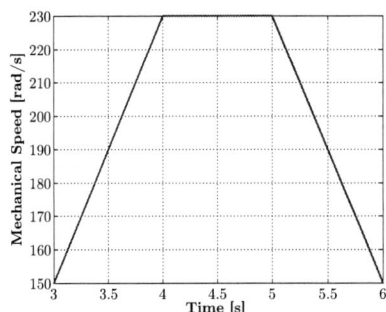

Fig. 4: Mechanical speed.

As a contribution, it can be highlighted the implementation of the FLC TS type in the DFIG, based mainly on the experience and knowledge of the system to perform the basic rules of control. The proposed controller has two inputs (e_{ird}, e_{irq}) and a single rule base to obtain the two components of the rotor voltage vector $\vec{v}_{r,dq}$, the former allowed to reduce the structure and complexity of the controller, resulting in a lower computational cost and ease of implementation in the experimental bench in relation to the Mamdani type fuzzy controller. Therefore, the Takagi Sugeno fuzzy controller can be an interesting alternative to the conventional controllers (PI, Fuzzy Mamdani, Neural...), applied in the DFIG control.

ACKNOWLEDGMENT

The authors would like to thank to CAPES and FAPESP for their financial support.

TABLE IV: Parameters of the DFIG.

Parameter	Value
Rated Stator Voltage $V_{s,n}$	220/380 Δ–Y
Rated Stator Current $I_{s,n}$	12 A
Rated Power P_n	3 kW
Rated Speed	1800 rpm
Rated Frequency f_s	60 Hz
Stator Resistance R_s	1 Ω
Rotor Resistance R_r	3.1322 Ω
Mutual Inductance L_m	0.1917 H
Stator Inductance L_s	0,2010 H
Rotor Inductance L_r	0.2010 H
Number of Poles p	4
Lumped Inertia Constant J	0.05 kgm^2
Frequency Modulation, f_m	10000 Hz

REFERENCES

[1] R. Peña, R. Cárdenas, and G. Asher, "Overview of control systems for the operation of DFIGs in wind energy applications," pp. 88–95, Nov 2013.

[2] F. Blaabjerg and K. Ma, "Future on power electronics for wind turbine systems," *Emerging and Selected Topics in Power Electronics, IEEE Journal of*, vol. 1, no. 3, pp. 139–152, Sept 2013.

[3] M. T. Lamchich and N. Lachguer, "Matlab simulink as simulation tool for wind generation systems based on doubly fed induction machines," *Edited by Vasilios N. Katsikis*, p. 139, 2012.

[4] E. Bim, "Máquinas elétricas e acionamento," *Campinas: Campus*, 2009.

[5] Y.-H. Song and A. T. Johns, "Application of fuzzy logic in power systems. ii. comparison and integration with expert systems, neural networks and genetic algorithms," *Power Engineering Journal*, vol. 12, no. 4, pp. 185–190, Aug 1998.

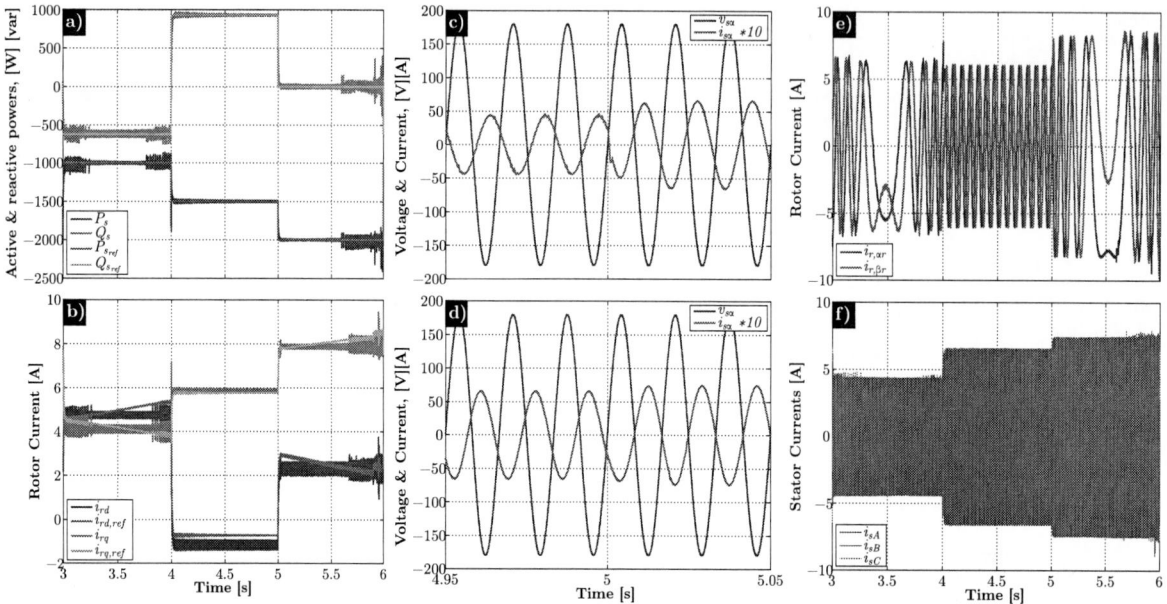

Fig. 5: System response with TS FLC under variable mechanical speed. a) Active and reactive powers, b) rotor current, c) power factor during network conditions 1-2. d) power factor during network conditions 2-3), e) rotor current $i_{r,\alpha\beta r}$, f) stator currents $i_{s_{ABC}}$.

[6] L. Suganthi, S. Iniyan, and A. A. Samuel, "Applications of fuzzy logic in renewable energy systems a review," *Renewable and Sustainable Energy Reviews*, vol. 48, pp. 585 – 607, 2015. [Online]. Available: http://www.sciencedirect.com/science/article/pii/S136403211500307X

[7] Y. Xing-jia, L. Zhong-liang, and C. Guo-sheng, "Decoupling control of doubly-fed induction generator based on fuzzy-PI controller," *International Conference on Mechanical and Electrical Technology (ICMET 2010)*, pp. 226 – 230, 2010.

[8] S. Louarem, S. Belkhiat, and D. Belkhiat, "A control method using PI/fuzzy controllers based DFIG in wind energy conversion system," in *2013 IEEE Grenoble Conference*, June 2013, pp. 1–6.

[9] X. Yao, Y. Jing, and Z. Xing, "Direct torque control of a doubly-fed wind generator based on grey-fuzzy logic," in *Proceedings of the 2007 IEEE International Conference on Mechatronics and Automation*, 2007.

[10] S. Tamalouzt, T. Rekioua, and R. Abdessemed, "Direct torque and reactive power control of grid connected doubly fed induction generator for the wind energy conversion," in *2014 International Conference on Electrical Sciences and Technologies in Maghreb (CISTEM)*, Nov 2014, pp. 1–7.

[11] S. Kalaivani, T. Karthick, S. C. Raja, and P. Venkatesh, "Mitigation of voltage disturbances using fuzzy logic controller in a grid connected dfig for different types of fault," in *2017 Innovations in Power and Advanced Computing Technologies*, April 2017, pp. 1–7.

[12] S. Elkhadiri, P. L. Elmenzhi, and P. A. Lyhyaoui, "Fuzzy logic control of DFIG-based wind turbine," in *2018 International Conference on Intelligent Systems and Computer Vision*, April 2018, pp. 1–5.

[13] C. M. Rocha-Osorio, J. S. Solís-Chaves, I. R. Casella, C. Capovilla, J. A. Puma, and A. S. Filho, "Gprs/egprs standards applied to dtc of a dfig using fuzzy pi controllers," *International Journal of Electrical Power Energy Systems*, vol. 93, pp. 365 – 373, 2017. [Online]. Available: http://www.sciencedirect.com/science/article/pii/S0142061517303514

[14] C. M. Rocha-Osorio, J. S. Solís-Chaves, L. L. Rodrigues, J. L. A. Puma, and A. S. Filho, "Deadbeatfuzzy controller for the power control of a doubly fed induction generator based wind power system," *ISA Transactions*, vol. 88, pp. 258 – 267, 2019. [Online]. Available: http://www.sciencedirect.com/science/article/pii/S0019057818304804

[15] J. L. Azcue P., A. J. Sguarezi Filho, and E. Ruppert, "Ts fuzzy controller applied to the dtc-svm scheme for three-phase induction motor," in *XI Brazilian Power Electronics Conference*, Sep. 2011, pp. 201–206.

[16] G. Abad, J. Lopez, M. Rodriguez, L. Marroyo, and G. Iwanski, *Doubly fed induction machine: modeling and control for wind energy generation.* John Wiley & Sons, 2011, vol. 85.

[17] R. Leonid, *Fuzzy controllers*, 1997, vol. 1.

[18] M. Santos, J. de la Cruz, S. Dormido, and A. de Madrid, "Between fuzzy-pid and pid-conventional controllers: a good choice," in *Fuzzy Information Processing Society, 1996. NAFIPS., 1996 Biennial Conference of the North American*, Jun 1996, pp. 123–127.

[19] A. Luiz Lacerda Ferreira Murari, J. Alberto Torrico Altuna, R. Vani Jacomini, C. M. Rocha Osorio, J. S. Solís-Chaves, and A. Joaozinho Sgaurezi Filho, "A proposal of project of PI controller gains used on the control of doubly-fed induction generators," *IEEE Latin America Transactions*, vol. 15, no. 2, pp. 173–180, Feb 2017.

978-1-7281-4181-7/19 $31.00 © 2019 IEEE

NonIsolated DC-DC Quadratic Ćuk Converter for Wide Conversion Range Applications

Tatiane Martins Oliveira, Lara Ana Rodarte Rios, and Fernando Lessa Tofoli
Department of Eletrical Engeneering
Federal University of São João del-Rei
São João del-Rei, Brazil
tatiane.martins@oi.com.br, laara.rodarte@gmail.com, fernandolessa@ufsj.edu.br

Aniel Silva de Morais
Faculty of Eletrical Engeneering
Federal University of Uberlândia
Uberlândia, Brazil
aniel@ufu.br

Abstract – **A nonisolated dc-dc converter for applications that demand wide conversion range is proposed in this work. Using the integration technique known as graft scheme, it is possible to design a single-switch quadratic Ćuk converter, whose input current and output stage current present reduced ripple, resulting in minimized electromagnetic interference (EMI) levels. The topology is able to operate in step-up or step-down mode depending on the value assumed by the duty cycle. The qualitative and quantitative analyses in continuous conduction mode (CCM) are presented in detail, from which it is possible to design and analyze the converter. Simulation results are presented and discussed in to validate the theoretical assumptions.**

Keywords – Ćuk converter, dc-dc converters, nonisolated converters with wide conversion range.

I. INTRODUCTION

The development of wide voltage conversion range dc-dc converters has attracted great interest recently, whose applications include renewable energy systems [1], light-emitting diode (LED) drivers [2], and telecommunications equipment [3]. Although isolated dc-dc converters are the usual solution to provide high-voltage step-up/step-down from the adjustment of the duty cycle and/or the transformer turns ratio of a high-frequency transformer, nonisolated dc-dc converters can be used instead if galvanic isolation is not mandatory, also leading to reduced dimensions and increased efficiency [4].

The classical nonisolated dc-dc buck, boost, buck-boost, Ćuk, SEPIC (Single-Ended Primary Inductance Converter), and Zeta converters are suitable only for applications where the conversion ratio involving the input-output is not high. In practice, wide voltage conversion range cannot be typically achieved with very low or high duty ratios in the aforementioned converters, otherwise very fast and costly drive circuits would be required [4-6].

Several techniques and topologies of nonisolated wide conversion ratio converters were presented in the literature with inherent advantages and disadvantages in terms of component count, efficiency, and stresses regarding the semiconductor elements [7]. Most approaches are based on the use of a basic structure associated with an additional technique, e.g., cascaded conversion stages [4], voltage multiplier cells [8], coupled inductors [9], and switched capacitors/switched inductors [10].

The graft scheme was proposed in [11] and allows integrating multiple converters to obtain a single-switch converter as long as the active switches of cascade-connected stages are connected to each other through a common point. Thus it is possible to aggregate the advantages of distinct converters in a single-stage structure, while also extending the conversion ratio as a consequence. In this context, this paper introduces a nonisolated converter with wide conversion range, which is obtained from the connection of two cascaded Ćuk converters. Analogously to its classical counterpart, it can be used for step-up and step-down applications; both the input current and the current through the output stage can be continuous with low ripple, thus implying minimized electromagnetic interference (EMI) levels; however, the output voltage has the polarity as the input voltage [12]. The qualitative and quantitative analysis of the converter in continuous conduction mode (CCM) are presented in detail to provide a consistent design procedure. Besides, simulation results are presented and discussed to support the theoretical assumptions.

II. PROPOSED CONVERTER

The cascaded structure consisting of two Ćuk converters is shown in Fig. 1 (a), which employs two active switches. According to [11], since there is a common node, switches S_1 and S_2 can be replaced by two diodes D_3 and D_4 and only one switch S in Fig. 1 (b). The latter structure can be optimized in order to reduce component count, resulting in the topology represented in Fig. 2, while also preserving the very same characteristics in terms of wide conversion ratio. The circuit is composed of the following elements: input voltage source V_i; inductors L_1, L_2, and L_3; capacitors C_1, C_2, and C_3; switch S; diodes D_1, D_2, D_3, D_4 (considering that operation may occur in both step-up and step-down modes); and the load R.

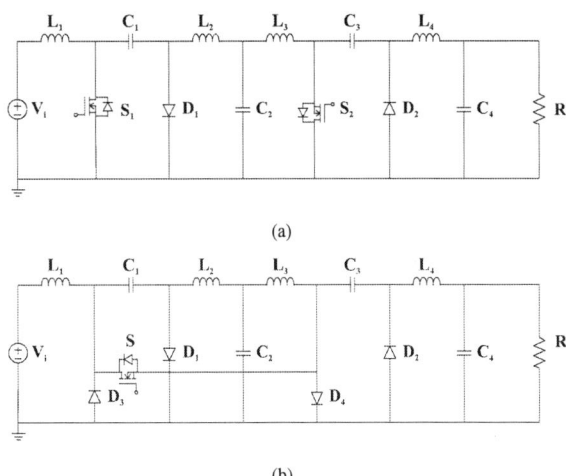

(a)

(b)

Fig. 1. Conception of the dc-dc quadratic Ćuk converter (a) cascaded structure with two switches; (b) resulting single-switch converter.

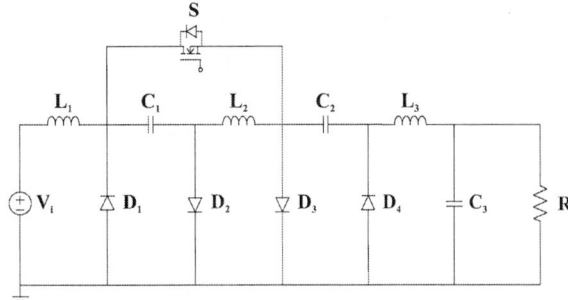

Fig. 2. Optimized single-switch quadratic Ćuk converter.

A. Qualitative Analysis

If the converter is supposed to operate only in step-down mode, the circuit can be simplified by removing diode D_3. On the other hand, if operation occurs in step-up mode, the simplification consists in removing diode D_1. This is possible because there is no current flowing through the such diodes in any operating stage considering the aforementioned conditions. The forthcoming analysis considers only the operation in step-up mode and steady-state condition, while the currents through all inductors do not become null over the switching period, thus characterizing the CCM. The operating stages corresponding to the step-up condition are shown in Fig. 3, while the respective theoretical waveforms are represented in Fig. 4.

<u>First stage $[t_0, t_1]$</u> (Fig. 3 (a)): Switch S is turned on. The currents through L_1, L_2, and L_3 increase linearly. Capacitor C_1 is discharged and capacitor C_2 is charged, while C_3 provides energy to the load R. Only diode D_3 is forward biased. This stage ends when S is turned off. Analyzing the corresponding equivalent circuit, the following expressions can be obtained:

$$V_i - L_1 \cdot \frac{di_{L1}(t)}{dt} = 0 \tag{1}$$

$$V_{C1} - L_2 \cdot \frac{di_{L2}(t)}{dt} = 0 \tag{2}$$

$$V_o - V_{C2} - L_3 \cdot \frac{di_{L3}(t)}{dt} = 0 \tag{3}$$

where $i_{L1}(t)$, $i_{L2}(t)$, and $i_{L3}(t)$ are the instantaneous currents through inductors L_1, L_2, and L_3, respectively; V_{C1} and V_{C2} are the average voltages across capacitors C_1 and C_2, respectively; V_i is the input average voltage; and V_o is the average voltage.

Furthermore, the time interval that defines this stage is:

$$t_1 - t_0 = D \cdot T_s \tag{4}$$

where D is the duty cycle and T_s is the switching period.

<u>Second Stage $[t_1, t_2]$</u> (Fig. 3 (b)): When S is turned off, all inductors are discharged linearly as diodes D_2 and D_4 are forward biased. Capacitor C_1 is charged while C_2 is

discharged. Besides, capacitor C_3 provides energy to the output stage composed of C_3 and R. Expressions (5), (6) and (7) represent the circuit behavior during this stage.

$$V_i - V_{C1} - L_1 \cdot \frac{di_{L1}(t)}{dt} = 0 \tag{5}$$

$$L_2 \cdot \frac{di_{L2}(t)}{dt} + V_{C2} = 0 \tag{6}$$

$$L_3 \cdot \frac{di_{L3}(t)}{dt} + V_o = 0 \tag{7}$$

The expression that defines the time interval of this stage is:

$$t_2 - t_1 = (1 - D) \cdot T_s \tag{8}$$

B. Quantitative Analysis

Using the volt-second balance, the average voltages across the inductors represented V_{L1}, V_{L2} and V_{L3} are null over the switching cycle, i.e.:

$$\frac{1}{T_s} \cdot \left[\int_{t_0}^{t_1} V_i \cdot dt + \int_{t_1}^{t_2} (V_i - V_{C1}) \cdot dt \right] = 0 \tag{9}$$

$$\frac{1}{T_s} \cdot \left[\int_{t_0}^{t_1} V_{C1} \cdot dt + \int_{t_1}^{t_2} (-V_{C2}) \cdot dt \right] = 0 \tag{10}$$

$$\frac{1}{T_s} \cdot \left[\int_{t_0}^{t_1} (V_o - V_{C2}) \cdot dt + \int_{t_1}^{t_2} (V_o) \cdot dt \right] = 0 \tag{11}$$

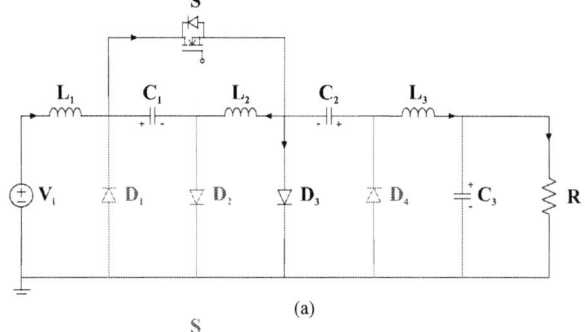

(a)

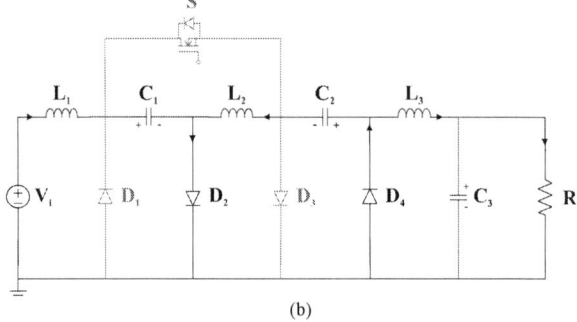

(b)

Fig. 3. Operating stages in step-up condition: (a) 1st stage; (b) 2nd stage.

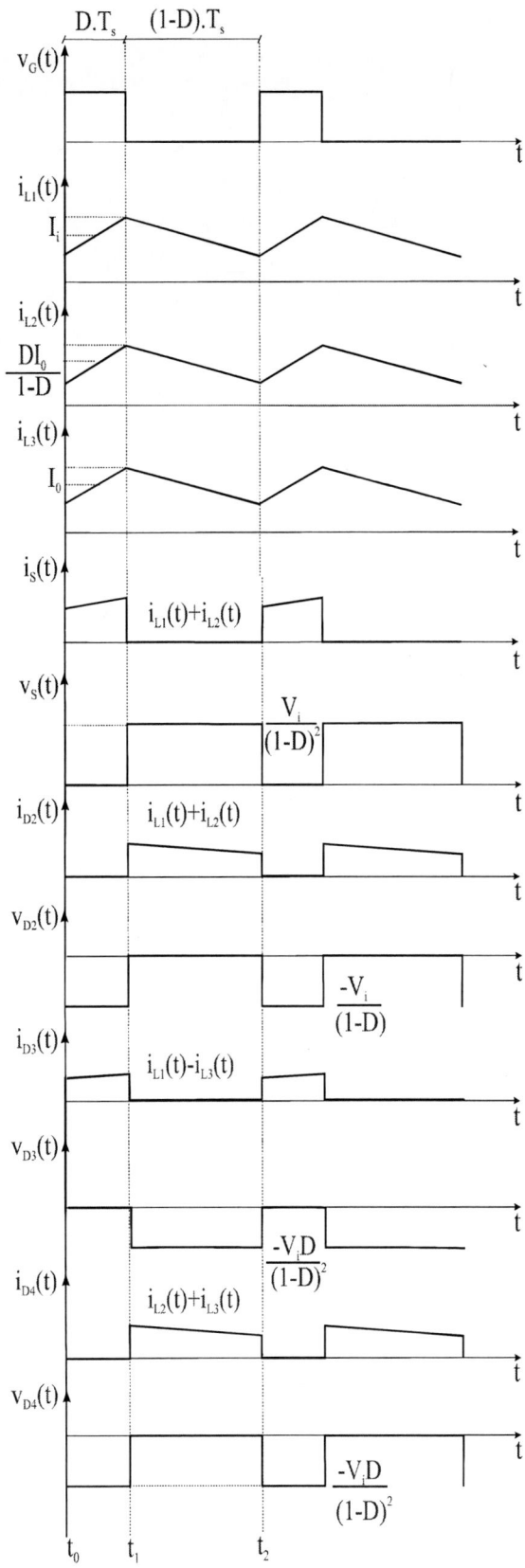

Fig. 4. Theoretical waveforms in step-up condition: currents and voltages.

Substituting (4) and (8) in (9) gives:

$$V_{C1} = \frac{V_i}{1-D} \quad (12)$$

Analogously, substituting (4), (8) and (12) in (10) gives:

$$V_{C2} = \frac{D \cdot V_i}{(1-D)^2} \quad (13)$$

Finally, the static gain is obtained after substituting (4), (8), and (13) in (11):

$$G = \frac{V_o}{V_i} = \left(\frac{D}{1-D}\right)^2 \quad (14)$$

According to (14), it can be stated that the proposed converter is able to provide voltage step-up and step-down for $D>0.5$ and $D<0.5$, respectively, while also achieving wide conversion range. Besides, it presents significant advantages over other quadratic-type converters. For instance, the quadratic buck-boost converter is only able to operate with $D<0.5$ and provide voltage step-down [6]. Another example lies in the SEPIC-buck-boost converter presented in [13], whose static gain is the same as that of the proposed topology. However, the current through the output stage is pulsating, thus implying higher EMI levels [13].

From the operating stages, it is possible to obtain expressions (15), (16) and (17) to calculate inductances L_1, L_2, and L_3, respectively.

$$L_1 = \frac{D \cdot V_i}{f_s \cdot \Delta I_{L1}} \quad (15)$$

$$L_2 = \frac{D \cdot V_i}{f_s \cdot \Delta I_{L2} \cdot (1-D)} \quad (16)$$

$$L_3 = \frac{D^2 \cdot V_i}{f_s \cdot \Delta I_{L3} \cdot (1-D)} \quad (17)$$

where $f_s = 1/T_s$ is the switching frequency; and ΔI_{L1}, ΔI_{L2}, ΔI_{L3} are the peak-to-peak current ripples through inductors L_1, L_2, L_3, respectively.

Capacitors C_1, C_2, and C_3 can be determined as follows:

$$C_1 = \frac{D^2 \cdot I_o}{f_s \cdot \Delta V_{C1} \cdot (1-D)} \quad (18)$$

$$C_2 = \frac{D \cdot I_o}{f_s \cdot \Delta V_{C2}} \quad (19)$$

$$C_3 = \frac{D^2 \cdot V_i}{8 \cdot L_3 \cdot f_s^2 \cdot \Delta V_{C3} \cdot (1-D)} \qquad (20)$$

where I_o is the average output current; ΔV_{C1}, ΔV_{C2}, ΔV_{C3} are the peak-to-peak voltage ripples across capacitors C_1, C_2, C_3, respectively.

The current and voltage stresses on switch S can be determined from the analysis of the theoretical waveforms shown in Fig. 4, resulting in:

$$I_{S(avg)} = \frac{D^2 \cdot I_o}{(1-D)^2} \qquad (21)$$

$$I_{S(rms)} = I_o \cdot \frac{D \cdot \sqrt{D}}{(1-D)^2} \qquad (22)$$

$$V_{S(max)} = \frac{V_i}{(1-D)^2} + \frac{\Delta V_{C1} + \Delta V_{C2}}{2} \qquad (23)$$

where $I_{S(avg)}$ is the average current through S; $I_{S(rms)}$ is the rms current through S; and $V_{S(max)}$ is the maximum voltage across S.

The average and rms currents through the diodes can also be determined from Fig. 4 as:

$$I_{D2(avg)} = \frac{D \cdot I_o}{(1-D)} \qquad (24)$$

$$I_{D2(rms)} = \frac{D \cdot I_o \cdot \left(\sqrt{1-D}\right)}{(1-D)^2} \qquad (25)$$

$$I_{D3(avg)} = \frac{D \cdot I_o \cdot (2 \cdot D - 1)}{(1-D)^2} \qquad (26)$$

$$I_{D3(rms)} = \frac{\sqrt{D} \cdot I_o \cdot (2 \cdot D - 1)}{(1-D)^2} \qquad (27)$$

$$I_{D4(avg)} = I_o \qquad (28)$$

$$I_{D4(rms)} = \frac{\left(\sqrt{1-D}\right) \cdot I_o}{(1-D)} \qquad (29)$$

The maximum reverse voltages across the diodes are given by:

$$V_{D2(max)} = -\left(\frac{V_i}{(1-D)} + \frac{\Delta V_{C1}}{2} \right) \qquad (30)$$

$$V_{D3(max)} = -\left(\frac{D \cdot V_i}{(1-D)^2} + \frac{\Delta V_{C2}}{2} \right) \qquad (31)$$

$$V_{D4(max)} = -\left(\frac{D \cdot V_i}{(1-D)^2} + \frac{\Delta V_{C2}}{2} \right) \qquad (32)$$

III. SIMULATION RESULTS

In order to validate the theoretical analysis, the quadratic Ćuk converter operating in CCM and step-up mode has been properly designed and thoroughly evaluated according to the specifications given in Table I. Simulation results obtained in PSIM® software are discussed as follows.

Fig. 5 shows the voltages across capacitors C_1, C_2, C_3, whose respective ripples are in accordance with the design specifications. Besides, it can be seen that the voltage across C_3, i.e., the output voltage has the same polarity as the input voltage. In this case, the rated duty cycle is $D=0.691$, while the same conversion ratio could be obtained with a boost converter operating with $D=0.8$. Therefore, operation with extremely high duty cycle is avoided, also contributing to the reduction of conduction losses and reverse recovery issues.

The inductor currents are shown in Fig. 6, thus denoting the operation in CCM. In addition, the current ripples are strictly in agreement with the specifications given in Table I. Besides, the currents through L_1 and L_3 are nonpulsating, with consequent reduction of EMI levels.

The current and voltage waveforms of the active switch and diodes are represented in Fig. 7 and Fig. 8, respectively. A comparison between the calculation performed with the expressions derived in the quantitative analysis and values obtained by simulation is given in Table II, thus validating the theoretical analysis properly.

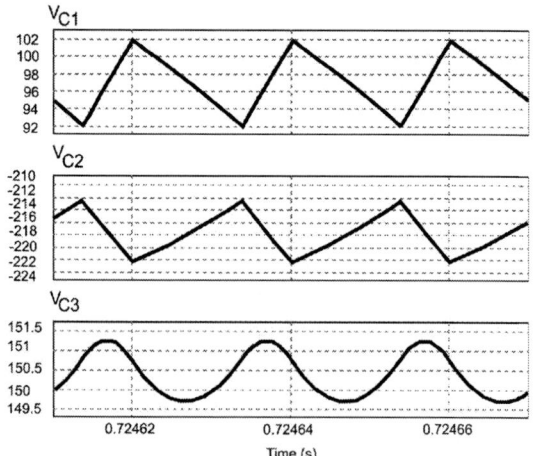

Fig. 5. Voltages across the filter capacitors.

TABLE I
DESIGN SPECIFICATIONS OF THE QUADRATIC ĆUK CONVERTER.

Parameters	Values
Input voltage	V_i=30 V
Output voltage	V_o=150 V
Output power	P_o=200 W
Switching frequency	f_s=50 kHz
Voltage ripple across the capacitors	ΔV_{C1}= 6.5%·V_o ΔV_{C2}=6.5%·V_o ΔV_{C3}=1%·V_o
Current ripples through the inductors	ΔI_{L1}=0.5 A ΔI_{L2}=0.8 A ΔI_{L3}=0.7 A
Duty cycle	D=0.691
Inductors	L_1=829.18 µH L_2=1.677 mH L_3=1.324 mH
Capacitors	$C_1$4.226 µF C_2=1.89 µF C_3=1.167 µF
Load resistance	R = 112.5 Ω

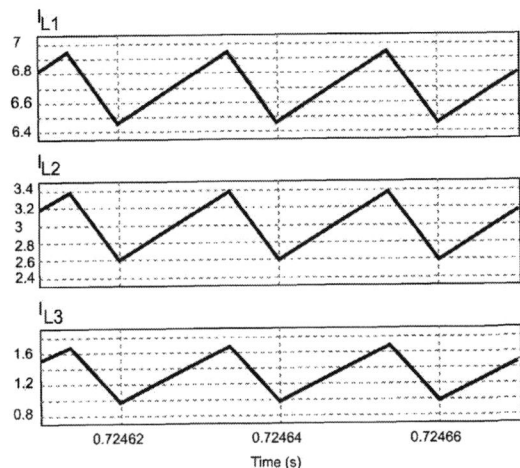

Fig. 6. Currents through the filter inductors.

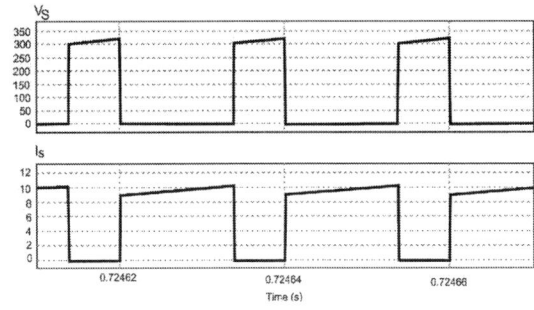

Fig. 7. Current and voltage waveforms of switch S.

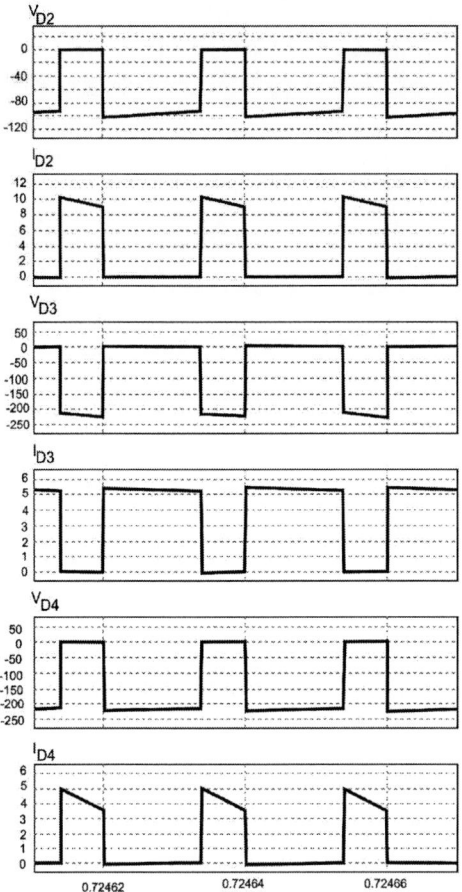

Fig. 8. Current and voltage waveforms of diodes D_1, D_2, and D_4.

TABLE II
STRESSES ON THE SEMICONDUCTORS OBTAINED FROM THEORETICAL CALCULATION AND SIMULATION.

Parameters	Simulation	Calculation
$I_{S(avg)}$	6.693 A	6.667 A
$I_{S(rms)}$	8.065 A	8.02 A
$V_{S(max)}$	324.067 V	323.914 V
$I_{D2(avg)}$	3.009 A	2.981 A
$I_{D2(rms)}$	5.405 A	5.363 A
$V_{D2(max)}$	101.854 V	101.957 V
$I_{D3(avg)}$	3.697 A	3.685 A
$I_{D3(rms)}$	4.453 A	4.433 A
$V_{D3(max)}$	222.135 V	221.957 V
$I_{D4(avg)}$	1.339 A	1.333 A
$I_{D4(rms)}$	2.416 A	2.399 A
$V_{D4(max)}$	222.056 V	221.957 V

IV. CONCLUSION

Considering that nonisolated converters with wide conversion range are a modern research topic in the field of power electronics, this work has presented a quadratic Ćuk converter able to provide both high-voltage step-up and step-down without requiring the use of extreme duty ratios. The introduced topology is adequate not only for renewable energy systems where there is the need to step low voltages up to supply a cascaded inverter, but also LED drivers and telecommunications equipment where the input voltage is much higher than the output voltage.

The converter employs a single active switch due to the use of the well-known graft scheme, while both the input and output stage currents are nonpulsating and present reduced ripple considering CCM operation, thus contributing to the minimization of EMI issues. The disadvantage is the large number of passive elements in the circuit, that demands a careful design to avoid problems in the dynamic model operation. The theoretical analysis was thoroughly validated through simulation tests performed in PSIM® software.

ACKNOWLEDGEMENT

The authors acknowledge CAPES, CNPq, FAPEMIG, INERGE, and PPGEL for the support to this work.

REFERENCES

[1] P. K. Maroti, S. Padmanaban, P. Wheeler, F. Blaabjerg, and M. Rivera, "Modified high voltage conversion inverting cuk DC-DC converter for renewable energy application," in *2017 IEEE Southern Power Electronics Conference (SPEC)*, 2017, pp. 1-5.

[2] Y. G. Kim and N. Terschak, "Development of a programmable DC-DC converter module for driving a scalable LED array," in *2015 IEEE Electrical Power and Energy Conference (EPEC)*, 2015, pp. 381-384.

[3] P. Saravana Prakash and R. Kalpana, "Configurations of modular push-pull buck dc-dc converters for 12KW telecom SMPS and its design," in *2016 Biennial International Conference on Power and Energy Systems: Towards Sustainable Energy (PESTSE)*, 2016, pp. 1-7.

[4] F. L. Tofoli, I. Rodrigues de Oliveira, and A. S. Morais, "Single-Switch, Integrated DC-DC SEPIC-Buck Converter for High-Voltage Step-Down Applications," *IET Power Electronics*, 2019.

[5] A. Mostaan, S. A. Gorji, M. N. Soltani, and M. Ektesabi, "A novel single switch transformerless quadratic DC/DC buck-boost converter," in *2017 19th European Conference on Power Electronics and Applications (EPE'17 ECCE Europe)*, 2017, pp. P.1-P.6.

[6] D. Maksimovic and S. Cuk, "Switching converters with wide DC conversion range," *IEEE Transactions on Power Electronics*, vol. 6, no. 1, pp. 151-157, 1991.

[7] J. P. d. Souza, P. d. Oliveira, R. Gules, E. F. R. Romaneli, and A. A. Badin, "A high static gain CUK DC-DC converter," in *2015 IEEE 13th Brazilian Power Electronics Conference and 1st Southern Power Electronics Conference (COBEP/SPEC)*, 2015, pp. 1-6.

[8] B. P. Baddipadiga and M. Ferdowsi, "A high-voltage-gain dc-dc converter based on modified dickson charge pump voltage multiplier," *IEEE Transactions on Power Electronics*, vol. 32, no. 10, pp. 7707-7715, 2017.

[9] H. Liu, H. Hu, H. Wu, Y. Xing, and I. Batarseh, "Overview of High-Step-Up Coupled-Inductor Boost Converters," *IEEE Journal of Emerging and Selected Topics in Power Electronics*, vol. 4, no. 2, pp. 689-704, 2016.

[10] X. Zhu, B. Zhang, Z. Li, H. Li, and L. Ran, "Extended Switched-Boost DC-DC Converters Adopting Switched-Capacitor/Switched-Inductor Cells for High Step-up Conversion," *IEEE Journal of Emerging and Selected Topics in Power Electronics*, vol. 5, no. 3, pp. 1020-1030, 2017.

[11] W. Tsai-Fu and C. Yu-Kai, "A systematic and unified approach to modeling PWM DC/DC converters based on the graft scheme," *IEEE Transactions on Industrial Electronics*, vol. 45, no. 1, pp. 88-98, 1998.

[12] M. H. Rashid, *Power Electronics - Circuits, Devices and Applications* Second Edition ed.: Prentice Hall, 1993.

[13] P. S. Almeida, G. M. Soares, D. P. Pinto, and H. A. Braga, "Integrated SEPIC buck-boost converter as an off-line LED driver without electrolytic capacitors," in *IECON 2012-38th Annual Conference on IEEE Industrial Electronics Society*, 2012, pp. 4551-4556.

Implementation of Automatic Battery Charging Temperature Compensation on a Peak-Shaving Energy Storage Equipment

Wilson Cesar Sant'Ana*[†], Robson Bauwelz Gonzatti[†], Germano Lambert-Torres*, *Fellow, IEEE*,
Erik Leandro Bonaldi*, Pedro Andrade de Oliveira[†], Bruno Silva Torres[†], Joao Gabriel Luppi Foster[†]
Rondineli Rodrigues Pereira[†], Luiz Eduardo Borges-da-Silva[†], Denis Mollica[‡] and Joselino Santana Filho[‡]
*Instituto Gnarus, Itajuba, MG, Brasil
[†]Universidade Federal de Itajuba - UNIFEI, Itajuba, MG, Brasil
[‡]EDP Sao Paulo Distribuicao de Energia, Sao Paulo, SP, Brasil
Emails: wilson_santana@ieee.org, bauwelz@gmail.com, germanoltorres@gmail.com
erik@InstitutoGnarus.com.br, p_andrade100@hotmail.com, bs_torres@hotmail.com, fosterelt@yahoo.com.br,
rondinelirp@gmail.com, leborgess@gmail.com, denis.mollica@edpbr.com.br, joselino.filho@edpbr.com.br

Abstract—This paper presents the implementation of an automatic temperature compensation for the charging of Lead-Acid batteries on a peak-shaving equipment. The equipment is composed by a multilevel converter, controlled by DSP, in a cascaded H-bridge topology and injects active power from the batteries into the grid in order to provide support to the system during peak times. When the energy price is lower, the batteries are charged from the grid, preparing for the next cycle. During charge times, the batteries floating voltages must be compensated as a function of temperature, in order to preserve their lifetime. The temperature information is provided by an RTD amplifier board, that communicates with the DSP via SPI bus. The paper presents the details of SPI communication on the TMS320F28335 DSP and the implementation of its control algorithm that sets the reference floating voltage for the batteries. Experimental results are presented.

Index Terms—batteries, DSP programming, energy storage, multilevel converters, peak-shaving

I. INTRODUCTION

Energy storage systems have been installed in order to optimize the grid usage. A review on such systems is presented in [1], where the energy may be stored under different forms: electrochemical energy (stored in batteries), magnetic field energy (stored in SMES), kinetic energy (stored in flywheels), potential energy (stored in hydro dams) or as compressed air (stored in CAES). Among the listed solutions, [1] highlights that the storage on batteries is the most cost-efficient technology.

In a recent paper [2], the authors have presented a detailed implementation of a peak-shaving equipment using storage on VRLA (Valve Regulated Lead Acid - also called sealed) batteries, including all necessary control loops and control algorithms. The equipment is composed by a multilevel converter (three cascaded H-bridges) controlled by a TMS320F28335 DSP. During the peak-times, the equipment injects active power on the grid from the battery banks. During the off-peak-times, the equipment charges its batteries from the grid, with

lower cost energy. The precise time of the day is obtained from a GPS module, that communicates with the DSP via serial UART. A guide on the TMS320F28335 UART usage is provided in [2], as well as a decoding procedure to extract the time of the day information.

An important issue concerning operation of VRLA batteries, and not taken into consideration in [2], is the loss of lifetime when operating at high temperatures. According with [3], the optimum temperature for sealed lead-acid batteries is 25°C and every 8°C rise from this temperature results in a reduction of its lifetime by half. Thus, ideally, batteries should operate in a controlled environment. As this may not be practical for some applications, extra care should be taken while charging the batteries at temperatures above 25°C. According with [4], an adjustment of charging float voltage of VRLA batteries is required in order to prevent thermal runaway and is imperative when the ambient conditions change significantly. Also according with [4], in case of vented/flooded batteries, the temperature compensation is rarely used in practice - as their operating environments are already controlled. It is important to notice that the operating temperature is not the only cause of loss of lifetime in batteries - [5] lists as well the number of charge/discharge cycles and the levels of discharge currents. However, according with [6], the charging technique is the most important factor that affects the life-time of the batteries.

This paper is an addendum to [2], presenting an automatic compensation on the charging float voltage based on the temperature. The temperature information is obtained from an RTD module, which communicates with the DSP via SPI bus. Section II presents the developed peak-shaving equipment with a succinct explanation on its control algorithms. Section III presents the algorithm for temperature compensation of the batteries floating voltage. Section IV presents a guide on the use of the SPI interface on the TMS320F28335 DSP and its communication with the RTD module. Section V presents the experimental results.

978-1-7281-4181-7/19 $31.00 © 2019 IEEE

II. PEAK-SHAVING AND BATTERY CONTROL

Fig. 1 presents the electrical circuit of the Peak-Shaving equipment developed in [2]. This equipment is composed by three cascaded H-bridges, aiming for seven levels at the output v_{AN}. Banks of lead-acid batteries supply each DC link. The multilevel output v_{AN} is filtered and coupled to the grid via transformer.

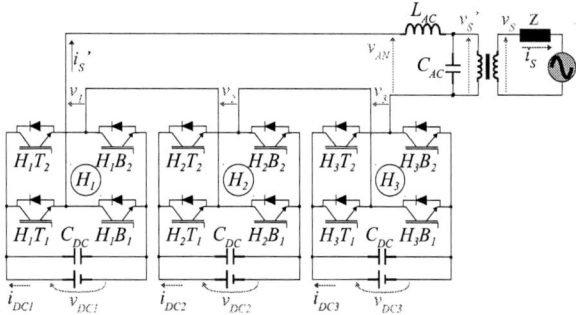

Fig. 1. Electrical circuit of the Peak-Shaving equipment developed in [2].

Fig. 2 presents an overview of the control of the Peak-Shaving equipment. The implementation details of the pink colored blocks have already been discussed in references [2] and [7] and will be reviewed superficially here. The battery control is centralized, based on the average value of the measured DC currents (i_{DCx}, with x representing each of the H-bridges) and voltages (v_{DCx}). The reference value for the DC currents is obtained based on the information of the time of the day (from a GPS module) in order to control the amount of AC power to be injected on the grid. The output of the battery control block is the reference amplitude of the AC current i_S'. This amplitude is multiplied by the output of a PLL, in order to obtain an AC current in phase with the installation site voltage v_S. In order to perfectly track the reference AC current, a PR (Proportional + Resonant) controller is used. The output of the PR controller (v_{pwm}^*) represents the reference voltage at the output of each of the H-bridges (v_x), which are obtained through PWM modulation. A detailed procedure for the implementation of multilevel PWM on the TMS320F28335 DSP is presented in [7].

The original control implemented in [2] considered a fixed reference for the batteries voltage (v_{DC}^*). When considering the automatic temperature compensation of battery floating voltage (discussed in Section III), the whole control implemented in [2] is maintained, with the exception that v_{DC}^* is determined based on the temperature of the batteries. Section IV presents a procedure for extraction of the temperature information based on a low cost RTD module. As this module communicates with the DSP via SPI, details on the configuration of the SPI on the TMS320F28335 DSP are also presented in Section IV.

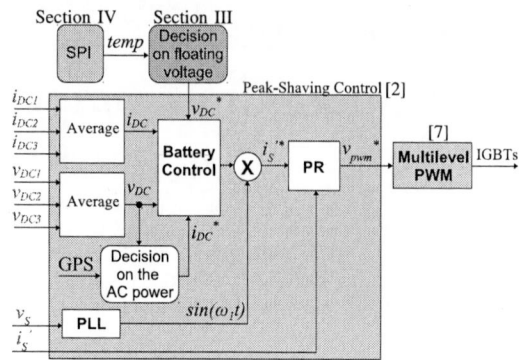

Fig. 2. Overview of the control of the Peak-Shaving equipment developed in [2] with the addition of battery voltage reference compensation.

III. TEMPERATURE COMPENSATION OF BATTERY FLOATING VOLTAGE

In order to be able to operate the batteries, even at high temperatures, [4] discusses the adjustment of the floating voltage as a function of the temperature. Although According to the research performed in [4], battery manufacturers determine a compensating factor from 2.0mV to 5.5mV (per 2V cell) for each degree Celsius of variation from 25°C, resulting in (1).

$$v_{DC}^* = v_{float}\big|_{25°C} - n_{cells} \cdot \alpha \cdot (temp - 25°C) \ , \quad (1)$$

where $v_{float}\big|_{25°C}$ is the manufacturer recommended floating voltage at 25°C, n_{cells} is the number of 2V cells inside the battery ($n_{cells} = 3$ for 6V batteries or $n_{cells} = 6$ for 12V batteries), α is the manufacturer recommended compensating factor (2.0mV$\leq \alpha \leq$5.5mV) and $temp$ is the measured temperature (in degrees Celsius).

Considering the prototype developed in [2], three 12V batteries are connected in series on each of the three DC links. Consulting a catalog specific for the batteries used, the recommended floating voltage for each of the batteries at 25°C is $v_{float}\big|_{25°C}$ =13.5V. Also, for these batteries, the recommended compensating factor is $\alpha = 4mV/cell$ (with 6 cells per battery). Hence, considering a bank of three series connected batteries, (1) can be rewritten as (2).

$$v_{DC}^* = 3 \cdot [13.5 - 6 \cdot 0.004 \cdot (temp - 25°C)] \ . \quad (2)$$

The batteries temperature ($temp$) must be constantly monitored, in order to automatically adjust the reference floating voltage (v_{DC}^*). A review on the temperature sensors commonly used on Industrial IOT devices is presented in [8]. Among them, the RTDs are the most stable, precise and linear. For a given variation of temperature, the RTD sensor proportionally varies its resistance. Thus an analog circuitry is necessary to convert the resistance into a proportional voltage, that any Analog-to-Digital (A/D) converter is capable to read. Section IV presents the use of an Adafruit RTD board [9] (based on the MAX31865 chip [10]), that already includes all required circuitry and also includes an integrated A/D converter. The board, then, sends the analog readings to a microcontroller/DSP/FPGA via SPI communication interface. Section IV also presents the setup procedure of the SPI interface on the TMS320F28335 DSP, used in the developed prototype.

It is important to notice that the ideal location of the temperature sensor would be inside the battery pack, closer to the cells. However, as the installation of the sensor inside the battery is impossible, it can be installed directly on the plastic enclosure of the batteries - and, actually, the compensating factors given by the manufacturers already consider ambient temperature.

IV. SPI COMMUNICATION BETWEEN THE TMS320F28335 DSP AND THE MAX31865 BOARD

A. Basics of SPI Communication

In SPI communication, there is always an unique master and one or more slaves. Usually, a microcontroller or a DSP has the option to be configured either as master or slave and the devices to be connected (A/D converters, sensors, etc) are always slaves. The master is responsible for the control of the clock signal that is sent to the slaves and to select the desired slave (only one slave can communicate at a time). Fig. 3 presents the SPI signals between the TMS320F28335 (configured as master) and the MAX31865 (slave). There are variations in the nomenclature of the signals from one manufacturer to the other, but their function is always the same:

• Clock line: always from master to the slaves. Called SPI-CLKA on the TMS320F28335 and CLK on the MAX31865 (and on the majority of other devices).

• Serial data line from master to slave: in case of the TMS320F28335, it is called SPISIMOA (SPI Slave In Master Out channel A). On other devices, usually, it is called SDI (Slave Data In) or MOSI (Master Out Slave In).

• Serial data line from slave to master: in case of the TMS320F28335, it is called SPISOMIA (SPI Slave Out Master In channel A). On other devices, usually, it is called SDO (Slave Data Out) or MISO (Master In Slave Out).

• Control line to enable one slave at a time: in case of the TMS320F28335, it is called SPISTEA (SPI Slave Transmit Enable channel A). On other devices, usually, it is called CS (Chip Select) or SS (Slave Select).

```
            SPICLKA ────→ CLK
Master      SPISOMIA ←──── SDO   Slave
TMS320F28335 SPISIMOA ────→ SDI  MAX31865
            SPISTEA ────→ CS
```

Fig. 3. SPI connection between the TMS320F28335 and MAX31865 chips.

Both the master and the connected slaves must agree on when, in relation to the clock signal, the data on the serial lines are valid. This is performed based on two parameters, one related to the clock polarity and the other related to the clock phase.

The clock polarity, known as CPOL on most devices, determines the clock logic level when the device is idle:

• For CPOL=0, the clock line remains at logic level 0 when no data is being transfered and alternates between 0 and 1 when data are being transfered. This implies that the clock pulses always have leading edges from 0 to 1 and trailing edges from 1 to 0.

• For CPOL=1, the clock line remains at logic level 1 when no data is being transfered and alternates between 0 and 1 when data are being transfered. This implies that the clock pulses always have leading edges from 1 to 0 and trailing edges from 0 to 1.

The clock phase, known as CPHA on most devices, determines in what transition of the clock signal, leading or trailing, the serial data is shifted through the line.

• For CPHA=0, the serial data are always shifted on the trailing transitions. This implies that, at the leading transitions (the opposite transition) the data bit is always valid and can be sampled from the reading device.

• For CPHA=1, the serial data are always shifted on the leading transitions. This implies that, at the trailing transitions (the opposite transition) the data bit is always valid and can be sampled from the reading device.

B. SPI Configuration on the TMS320F28335

The TMS320F28335 has all of its hardware peripherals (including the SPI) mapped in memory with configuration and access through registers. These peripherals have all their I/Os multiplexed into some GPIO pins. Fig. 4 presents the procedure to configure the SPI peripheral.

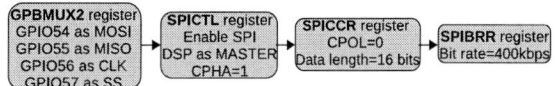

Fig. 4. Setup procedure for the SPI.

The first setup step presented in Fig. 4 is the configuration of the MUX registers, in order to select in which GPIO pins the SPI I/Os will be available. Then, the register SPICTL is configured in order to set the DSP as the Master with $CPHA=1$ (which is the only option available for the MAX31865 slave). Then, the SPICCR register configures $CPOL=0$ and the number of bits in each transfer to 16 bits. Finally the bit rate is configured in register SPIBRR, as (3). Thus, for an internal clock $LSPCLK=37.5$MHz, the SPIBRR register must be configured with the integer value $SPIBRR=93$, in order to obtain a bit rate of $bps=400$k bits/second.

$$SPIBRR = (LSPCLK/bps) - 1 . \qquad (3)$$

The configuration procedure of Fig. 4 is performed only once, at the initialization of the DSP. Then, the code enters a real-time infinite loop, where the whole algorithm of Fig. 2 is performed at each sampling interruption (at each T_s seconds). Inside this infinite loop, the DSP (configured as SPI master) sends specific commands (discussed in sub-section IV-C) to the slave and reads the data in return. All commands sent from the DSP must be written at the 16 bit register $SPITXBUF$.

978-1-7281-4181-7/19 $31.00 © 2019 IEEE

Also, the DSP reads the received data through another 16 bit register: *SPIRXBUF*. The data is transferred serially. After the last bit is transferred, a flag called *SPI INT_FLAG* (bit 6 of register *SPIST*) becomes active. The next sub-section presents an usage example of these registers.

C. SPI Communication with the MAX31865 board

The RTD PT100 sensor presents a resistance of 100Ω at the temperature $0°C$ and have an approximately linear variation of this resistance with temperature. In order to precisely measure this resistance (hence, an indirect measure of the temperature), the MAX31865 chip offers a complete solution: with integrated analog circuitry, analog to digital conversion (ADC) and a SPI slave controller. The internal ADC has 15 bits resolution, however its SPI has a length of only 8 bits. Thus, two 8-bit transfers are required for each ADC conversion.

From a list of SPI commands that the MAX31865 is able to respond (see [10]), the two most important are:

• 0x01: chip MAX31865 sends back bits 14 to 7 of the conversion;

• 0x02: chip MAX31865 sends back bits 6 to 0 of the conversion, justified to the left.

Fig. 5 presents the algorithm of the SPI communication between the DSP and the MAX31865 and the decoding of the received temperature information. This algorithm is executed at every DSP sampling time T_s, inside an infinite loop. The SPI transfers are executed by a hardware module internal to the DSP chip, thus no processing resources are wasted. However, due to the fact that two 8-bit transfers are required to get the correct reading, some coordination is necessary among the tasks. The algorithm of Fig. 5 can be divided into three main tasks (which are coordinated by the logic state of the flags *ReadMSB*, *WaitingMSB*, *ReadLSB*, *WaitingLSB* and *DecodeRTD*):

• The first main task sends the 8-bit command 0x01 to the MAX31865 chip and waits for the reception of the Most Significant Bits (MSB) of the conversion. It is important to note that, for an 8-bit transfer over a 16-bit frame, the 8 bits are justified to the left (with the operator $<<8$). When transfer is completed, the *INT_FLAG* becomes active and the 8MSBs of the measured RTD resistance are stored in a variable *R_MSB*.

• The second main task sends the 8-bit command 0x02 and waits for the reception of the Least Significant Bits (LSB) of the conversion (also using the operator $<<8$). When transfer is completed, the 8LSBs of the measured RTD resistance are stored in a variable *R_LSB*.

• The third main task decodes the RTD resistance and converts its value to the temperature information. A variable *R_16* is created with the concatenation of the 8 bits of *R_MSB* (at position 15 to 8) with the 8 bits of *R_LSB* (at position 7 to 0). As the ADC of the MAX31865 has a 15-bit resolution (and *R_LSB* was 1 bit justified to the left), the valid 15-bit digital representation of the RTD resistance (*R_15*) is obtained with a 1 bit shift to the right ($>>1$). Its analog decimal value (*Res*) can be obtained using Eq. (4). Finally, a conversion is made from the measured RTD resistance to the temperature, using Eq. (5).

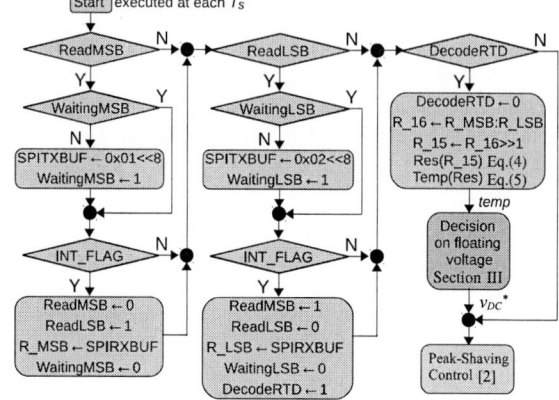

Fig. 5. SPI communication and decoding of temperature.

$$Res = R\_15 \cdot R_{REF}/(2^{15} - 1) , \qquad (4)$$

where R_{REF} is the value of a reference resistor. In case of the Adafruit board [9], $R_{REF} = 430\Omega$.

$$Temp = (Res - 100)/\alpha_{sensor} , \qquad (5)$$

where α_{sensor}=$0.385\Omega/°C$ is the linear coefficient of the PT100 sensor.

V. Experimental Results

Fig. 6 presents a photo of the experimental setup developed in [2] with the addition of the RTD module (shown in the lower left zoom). Its elements are identified in Table I. Fig. 7 presents a block diagram indicating the connections between each of the elements. The H bridges H_1, H_2 and H_3 are manufactured by a Brazilian vendor called *Supplier* and communicate with the DSP through optical fibers.

TABLE I
IDENTIFICATION OF THE ELEMENTS IN THE TEST SETUP OF FIGURE 6.

1	TMS320F28335 DSP
2	Adafruit GPS board
3	Adafruit RTD board
4,6,8,10,12,14,16,18	Hall effect sensors
5,7,9,11,13,15,17,19	Signal conditioning with OpAmps
20,21,22	Conversion 3.3V⇔15V
23,24,25,26,27,28	Fiber Optic Transceivers
29,30,31	Gate drivers
32,33,34	IGBT power blocks

Fig. 8 presents a log of 7 hour operation of the setup of Fig. 6, obtained with a Fluke power and energy analyzer. It is shown the DC voltage and DC current at the bridge H_1 - the results for the other two bridges are supposed to be similar. In order to speed-up the test and avoid an overnight work

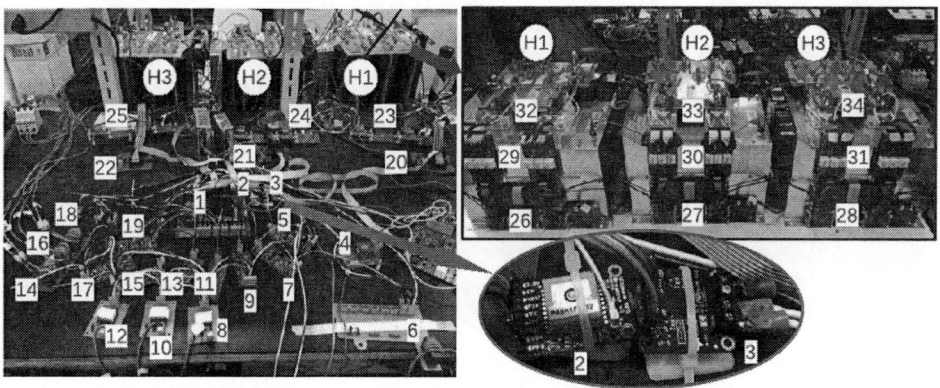

Fig. 6. Experimental setup used in [2] with the addition of the RTD module (identified as number 3).

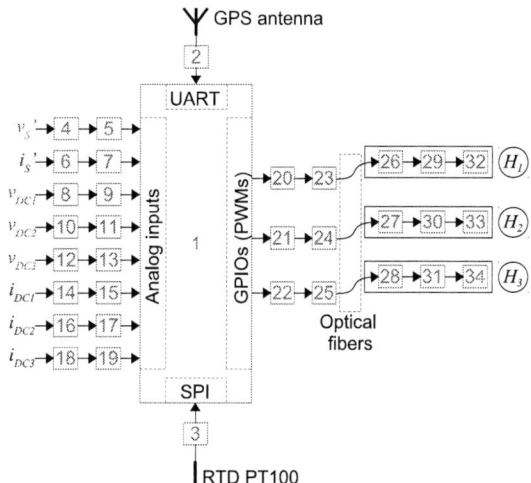

Fig. 7. Connections between each of the elements of Figure 6.

shift, the peak-times were defined as Table II. At some specific times, concerning different battery operation stages, marked in the figure, a Tektronix oscilloscope has been employed to capture the waveforms (from Fig. 9 to Fig. 12) of the AC voltage (v_S), AC current (i_S'), battery voltage at bridge H_1 (v_{DC1}) and battery current at bridge H_1 (i_{DC1}).

TABLE II
PEAK TIMES DEFINED FOR THE EXPERIMENT.

Start time of ramp-up	t_1	10:15
Start time of maximum injection	t_2	10:30
End time of maximum injection	t_3	11:30
End time of ramp-down	t_4	11:45

During the whole test, the room temperature (hence, the temperature at the batteries) is kept controlled by air conditioner at near 25°C, which is the nominal temperature for the batteries. However, in order to test the floating voltage com-

pensation, the temperature sensor (PT100, which is connected to the board number 3 of Figure 6 by a cable) is kept near a soldering iron. The proximity of the sensor to the hot soldering iron is, then, adjusted in order to simulate an increase/decrease at the temperature of the batteries.

From the start of the test (9:08) until t_1=10:15, the batteries are fully charged, hence only "floating", consuming a minimum current, near zero. At the start, the temperature sensor is far from the soldering iron and this distance is slowly being decreased, in order to increase its readings. At 9:14 the temperature reading is around 25°C, as shown at the first detail of Fig. 8. This detail is a screen capture of the watch-window of the DSP programing environment (*Code Composer Studio*), that shows some selected variables in a sort of real-time (the algorithms are performed in real-time, only the data displayed at the watch-window are shown with some delay). The information of the variables *hours*, *minutes* and *seconds* is obtained from the GPS (as discussed in [2]). For a measured temperature near the nominal, the reference floating voltage v_{DC}^* is calculated near 40.5V, according with (2).

Then, the distance between the temperature sensor and the soldering iron is reduced. At 9:40 the temperature reading is around 29°C, as shown at the second detail of Fig. 8. At this temperature, the reference floating voltage v_{DC}^* drops to around 40.2V. The figure shows a drop at the DC voltage and the DC current becomes less negative (as less power is needed from the grid).

At 9:56, with the positioning of the temperature sensor very close to the soldering iron, the temperature is further increased to around 35°C and the reference floating voltage v_{DC}^* is further reduced to around 39.8V. It is important to note that the sensor has a time constant in order to heat-up or cool-down and the positioning of the soldering iron was performed by hand, thus some peaks can be seen in Fig. 8 between 9:40 and 9:56.

Then, the sensor is removed away from the soldering iron. At 10:07 the temperature has already been stabilized around the nominal 25°C and, also, the floating voltage at around

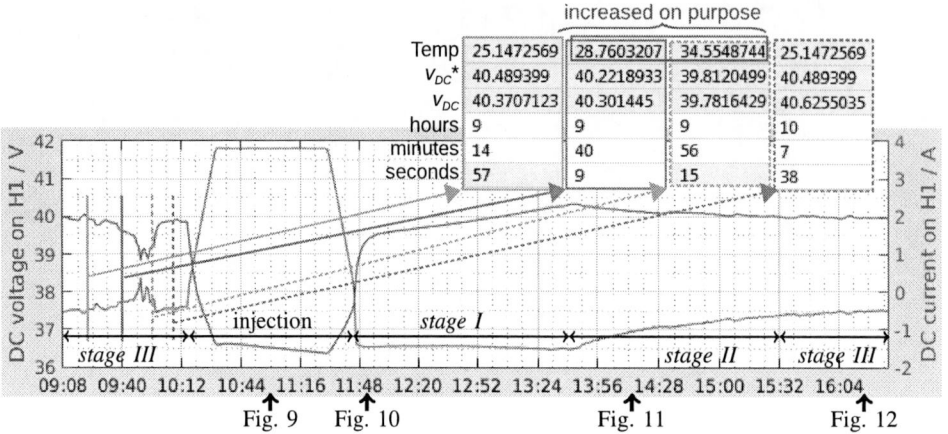

	increased on purpose		
Temp	25.1472569	28.7603207	34.5548744	25.1472569
$v_{DC}{}^*$	40.489399	40.2218933	39.8120499	40.489399
v_{DC}	40.3707123	40.301445	39.7816429	40.6255035
hours	9	9	9	10
minutes	14	40	56	7
seconds	57	9	15	38

Fig. 8. Logs of DC voltage and DC current, sampled at each minute.

40.5V.

At t_1=10:15, the beginning of the peak-time, the converter starts to inject power into the grid with an ascending ramp, as defined in [2]. From t_2=10:30 until t_3=11:30, the converter is injecting its maximum constant power into the grid. During this interval, at 11:00, Fig. 9 is obtained with the oscilloscope. The AC current $i_S{}'$ is shown in blue color and in phase with the AC voltage v_S (yellow color) - hence, power flows from the batteries to the grid, according with the signal convention adopted in Fig. 1. The voltage at the batteries of bridge H_1 (magenta color) is shown with almost null ripple. The current at the batteries of bridge H_1 (green color) has a second harmonic ripple and a positive average value - also meaning power flow from the batteries to the grid according with the adopted signal convention.

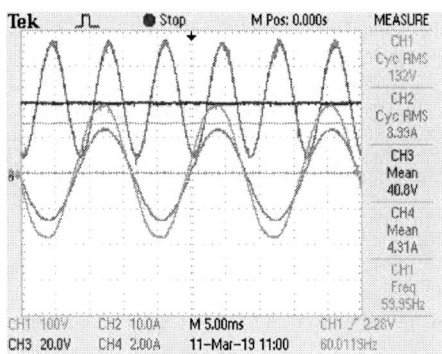

Fig. 9. Measurements at 11:00 - Converter injecting maximum active power on the grid.

From t_3=11:30 until t_4=11:45, the equipment is reducing (as a descending ramp) the power injected into the grid. At t_4=11:45, the end of the the peak time, the converter begins to charge its batteries (with three different stages of charging), according with the algorithm discussed in [2].

Fig. 10 presents the oscillography taken at 11:52, during charging at *stage I*. At this stage, the DC current at the batteries is kept constant while their DC voltages $v_{DC1,2,3}$ slowly increases until the reference $v_{DC}{}^*$. It can be noticed a negative average value on the current i_{DC1} (green) and an AC current $i_S{}'$ (blue) 180° out of phase in relation to v_S (yellow) - meaning power flow from the grid to the batteries according with the adopted signal convention.

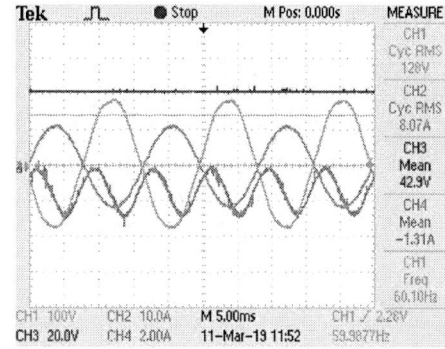

Fig. 10. Measurements at 11:52 - Converter charging the batteries (*stage I*).

When the DC voltage reaches its reference $v_{DC}{}^*$ (at around 13:45, as it can be seen in Fig. 8), the charging algorithm switches to *stage II*, where the batteries voltage is kept constant (at around $v_{DC}{}^*$) and the negative DC current average value starts to decrease (becoming less negative). Fig. 11 presents the oscillography taken at 14:14, during charging at *stage II*. Comparing Fig. 11 with Fig. 10 (*stage I*), it can be noticed an increase at the DC voltage (magenta) and a decrease in amplitude of the AC current (blue) and in the negative average DC current (green).

At *stage II* the negative DC current decreases until near zero, when stage *III* starts. At both stages *II* and *III* the DC voltage is kept around $v_{DC}{}^*$. There is not a clear boundary

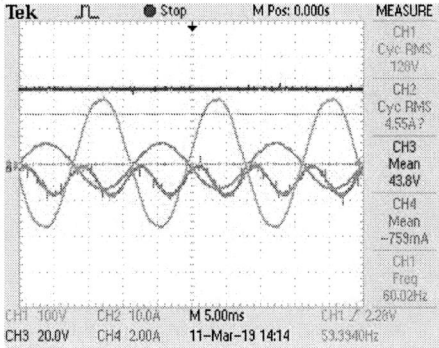

Fig. 11. Measurements at 14:14 - Converter charging the batteries (stage II).

between stages *II* and *III*. Considering Fig. 8, it can be estimated that after 15:30 the batteries were already at *stage III*. At this stage a minimum current is required to keep the batteries "floating" around $v_{DC}{}^*$. Fig. 12 presents the oscillography taken at 16:16, during charging at *stage III*. Comparing Fig. 12 with Fig. 11 (*stage II*), it can be noticed that the DC voltage (magenta) has been kept constant and that the amplitude of the AC current (blue) and the negative average DC current (green) are minimum.

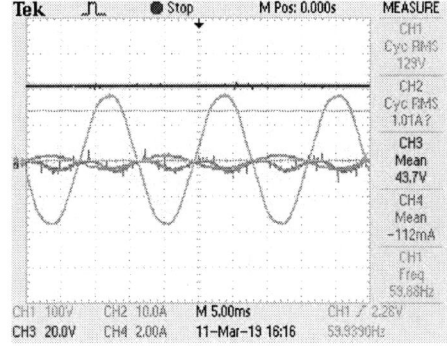

Fig. 12. Measurements at 16:16 - Converter charging the batteries (stage III).

During the entire *stage II* and *stage III*, the algorithm for the compensation of the batteries floating voltage as a function of the temperature is active. At the conditions of the performed experiment, since 10:07 the temperature sensor has been placed far away from the heat source and near the batteries (thus at a temperature controlled by air conditioner at around 25°C). Thus, in Fig. 8, since the start of *stage II* until the end of the experiment, only minor variations at the DC voltage (and, as a consequence, at the DC current) are present.

VI. CONCLUSION

This paper presented an automatic battery charging floating voltage compensation based on temperature. This compensation had been suggested as "future work" of a recent paper [2] and has been implemented on the same peak-shaving energy storage equipment of the aforementioned paper, with new experimental results. The added functionality is important in order to prevent loss of lifetime of the lead-acid battery banks used as energy storage. Also, this paper has presented a procedure of SPI communication between the TMS320F28335 DSP and the MAX31865 RTD board.

The experimental results show, besides the expected operational conditions of a peak-shaver (injection of power on the grid from the battery banks during peak-times and charging of the batteries with lower cost energy during off-peak-times), that, the floating voltage is properly adjusted according with the temperature readings. It is important to notice, however, that there are other operating factors, besides charging at elevated temperatures, that cause reduction on the lifetime of VRLA batteries and this paper has dealt with only one measure in order to improve this lifetime.

ACKNOWLEDGMENT

The authors would like to thank the following Brazilian Research Agencies: CNPq, CAPES, FAPEMIG and ANEEL R&D for the financial support of this project.

REFERENCES

[1] V. A. Boicea, "Energy storage technologies: The past and the present," *Proceedings of the IEEE*, vol. 102, no. 11, pp. 1777–1794, Nov 2014.

[2] W. C. Sant'Ana, R. B. Gonzatti, G. Lambert-Torres, E. L. Bonaldi, B. S. Torres, P. A. de Oliveira, R. R. Pereira, L. E. Borges-da Silva, D. Mollica, and J. Santana Filho, "Development and 24 hour behavior analysis of a peak-shaving equipment with battery storage," *Energies*, vol. 12, no. 11, 2019.

[3] R. Hutchinson, "Temperature effects on sealed lead acid batteries and charging techniques to prolong cycle life." Sandia National Laboratories, Tech. Rep., 2004.

[4] S. S. Misra and A. J. Williamson, "On temperature compensation for lead acid batteries in float service: its impact on performance and life," in *Proceedings of Intelec'96 - International Telecommunications Energy Conference*, Oct 1996, pp. 25–32.

[5] A. T. Elsayed, C. R. Lashway, and O. A. Mohammed, "Advanced battery management and diagnostic system for smart grid infrastructure," *IEEE Transactions on Smart Grid*, vol. 7, no. 2, pp. 897–905, March 2016.

[6] H. A. Serhan and E. M. Ahmed, "Effect of the different charging techniques on battery life-time: Review," in *2018 International Conference on Innovative Trends in Computer Engineering (ITCE)*, Feb 2018, pp. 421–426.

[7] W. Sant'Ana, R. Gonzatti, B. Guimaraes, G. Lambert-Torres, E. Bonaldi, R. Pereira, L. E. B. da Silva, C. Ferreira, L. de Oliveira, G. Pinheiro, C. H. da Silva, C. Salomon, D. Mollica, and J. S. Filho, "Development of a multilevel converter for power systems applications based on DSP," in *Procedings of the VII Simposio Brasileiro de Sistemas Eletricos (SBSE)*, Niteroi-RJ, Brazil, may 2018.

[8] D. Tranca, D. Rosner, R. Tataroiu, S. C. Stegaru, A. Surpateanu, and M. Peisic, "Precision and linearity of analog temperature sensors for industrial IoT devices," in *2018 17th RoEduNet Conference: Networking in Education and Research (RoEduNet)*, Sep. 2018, pp. 1–6.

[9] Adafruit, "Adafruit MAX31865 RTD PT100 or PT1000 Amplifier," mar 2018.

[10] Maxim Integrated, "MAX31865 RTD-to-Digital Converter, Rev 3," jul 2015.

Cascade Control vs Full-State Feedback

Débora M. Soares
COPPE - Universidade Federal do Rio
de Janeiro
Rio de Janeiro – RJ, Brasil
deboramicroni@gmail.com

Henrique A. M. Calil
COPPE - Universidade Federal do Rio
de Janeiro
Rio de Janeiro – RJ, Brasil
hamcalil@gmail.com

Richard M. Stephan
COPPE - Universidade Federal do Rio
de Janeiro
Rio de Janeiro – RJ, Brasil
rms@ufrj.br

Abstract—**Cascade control systems find a wide range of applications and acceptance in the industry. They, however, are not as featured in the academy. Taking the case of a DC motor position control, this paper points out the advantages and limitations of the cascade control system. The current, speed and position controllers are tuned following the Root Locus and the Magnitude and Symmetric Optimum criteria. These last two methods are commonly found in German literature but not in America. The methods and the adjustment procedure for the nested loops are explained. Up next, the importance of a feedforward compensator, in order to mitigate the natural delay in the system time response due to the nested structure, is proved. Finally, the cascade control system is compared to state-space feedback control, highlighting limitations of a full state feedback design in terms of disturbance rejection, adjustability of control parameters and implementation challenges.**

Keywords—*Cascade Control, DC Motor, Feedforward Control, Full State Feedback.*

NOTATION

t_s Settling time.
τ Electrical time constant
T_H Mechanical time constant
m_L Load torque.
u_c Rectifier input signal
$e*$ Electromotive force
$n*$ Rotor angular speed.
$\Psi*$ Magnetic field excitation.

I. INTRODUCTION

The industrial sector demands automated processes, envisioning increasingly efficiency, standardization, and reliability [1]. Automation viability is directly linked to the implementation costs of controllers, sensors and actuators. PID-controllers are widely deployed in cascade control [1].

Cascade control systems have, as an inherent advantage, the ability to compensate for disturbances in its internal loops without influencing the external loops control law. The acquisition of adequate variables (state variables) from intermediate processes is necessary, although this is not always an easy task [2]. Furthermore, the system response dynamics also depends on the actuation time of the inner loops, which may further slow it down.

On the other hand, state-space representations have the advantage of being able to represent linear and non-linear systems, as well as time-variant and invariant systems. State equations provide a general mathematical model, which is adequate for systems with non-zero initial conditions [3] and multiple-input, multiple-output (MIMO) systems. This model also has the advantage of utilizing compact matrix notation, which makes complex manipulations easier [4]. State feedback control offers the designer the elegant solution of freely positioning the closed-loop system poles. However, for the design of a controller with zero-error in steady-state response, integrators must be added at the plant input [4]. Another disadvantage of state-space representations is that it is less intuitive for not being able to break down the plant into smaller loops [3].

In this paper, a comparison between these two different control systems applied to the position control of a thyristor driven DC motor is presented.

Initially, current, speed and position regulators are designed and combined in a nested cascade structure.

Next, predictive signals from a pre-defined reference system are fed to each individual controller. These are later referred to as feedforward (FF) signals [5]. The intent of selecting a reference model is to obtain a system response which is quicker than that of the pure cascade system.

The two systems – with and without the reference model – are set-up and their output position to a step reference input are compared.

A third study consists in designing the plant for a state-space feedback controller.

The three systems are compared for variations in reference input and disturbance rejection. The disturbance is represented by the load torque, m_L.

II. DC MOTOR MODEL

The DC motor is supplied by a full bridge thyristor rectifier. The thyristor driven DC motor used as a background for the comparative studies has been extracted from [6]. For simplification, only the continuous conduction mode was considered, and the field current was kept at a constant value. The plant model and its parameters are shown in Fig. 1 and Table I, respectively. The M block in Fig. 1 is a signal multiplier.

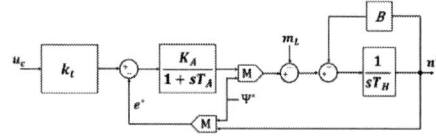

Fig. 1. DC Motor Plant.

TABLE I.   DC MOTOR - PLANT DATA

Parameter	Description	Value	Unit
k_t	Rectifier gain	1.0	p.u.
K_A	Armature gain	2.0	p.u.
T_A	Armature time constant	0.03	s
T_H	Mechanical time constant	0.50	s
B	Viscous friction coefficient	1.0	p.u.
$\Psi*$	Magnetic field excitation	0.5	p.u.

In the following, the DC motor position control will be studied considering as input signals the motor current, speed and angular position.

III. DESIGN OF REGULATORS FOR NESTED CASCADE SYSTEM

A total of three control loops were designed, starting from the faster (current loop) to the slowest (position loop). Whenever designing an outer loop, the loop that is nested inside is replaced by an equivalent system of the lowest possible order, for the sake of simplicity.

Cascade Control can be found in the literature [7][8], but these references are oriented to a Power Electronics audience and are not the natural books that will be searched by someone interested in control issues.

A. Current regulator design

The current loop is characterized by the motor armature dynamics and a PI regulator, as shown in Fig. 2.

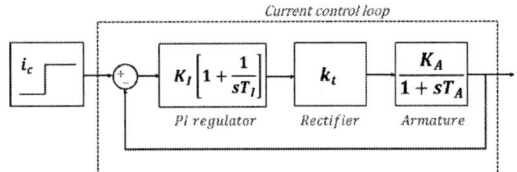

Fig. 2. Current control loop with a PI regulator.

For the current loop control design, the root-locus method was used [3].

The regulator gains have been tuned for a settling time lower than 0.05 seconds, within a 1% error margin. Another design requirement was a zero-overshoot condition for the controller output signal. By means of computational tools, a range of admissible gains for the proposed criteria was obtained and the values selected as expressed in (1).

$$C_i(s) = K_I[1 + \frac{1}{T_I s}] = 1{,}67[1 + \frac{1}{0{,}03\,s}] \tag{1}$$

B. Speed regulator design

The speed loop is characterized by the dynamics of the inner current loop, the rotor inertia, its viscous friction in rotation and the speed PI controller itself, as shown in Fig. 3.

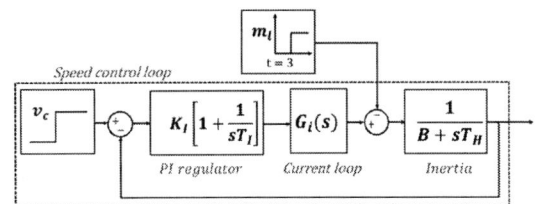

Fig. 3. Speed control loop with the PI regulator.

As mentioned before, to design cascade control systems, the inner loop is replaced by an equivalent system of the lowest possible order, that represents the internal loop dynamics. This significantly reduces complexity in the design of the outer loop.

It was possible to calculate the time constant of an equivalent first-order system by inspecting the current control response to a step reference input. The setting time,

t_s, for a 1% output error, is 0.04s. The relation between t_s and the constant time τ, for a first order system, is given by (2):

$$t_s = 4.6 * \tau \qquad \tau = 0.009s \tag{2}$$

Thus, the transfer function of the current loop equivalent system is (3):

$$G_i(s) = \frac{1}{1 + 0.009s} \tag{3}$$

This way, the plant is then characterized by two first-order dynamic functions, being one of fast dynamics (τ), which consists in the system that is equivalent to the current loop, and the other, the slower function, characterizes the rotor inertia and friction response, with a time constant T_H. This structure often appears in multiple electro-mechanical systems and it is Kessler's object of study in [9], summarized in Table II.

TABLE II. KESSLER'S PI OPTIMAL REGULATORS

Kessler's PI	Gain	Integral Time	Objective
Magnitude	$\left[\frac{T_H}{2K\tau}\right]$	T_H	Reference tracking
Symmetric	$\left[\frac{T_H}{2K\tau}\right]$	4τ	Disturbance rejection

Given that the speed loop is subject to a disturbance, here represented by the load torque (m_L), the speed regulator design was carried out following the Symmetric Optimum criterion, since that kind of control provides faster plant disturbance rejection. In that sense, the final design follows the equation (4):

$$C_v(s) = K_I\left[1 + \frac{1}{T_I s}\right] = \frac{s + 27.78}{0.036s} \tag{4}$$

Where,

K_I $K_I = \left[\frac{T}{2K_s\tau}\right]$

T_I $T_I = 4\tau = 0.036$ s

τ   Small time constant (0.009 seconds)

K_s Open loop gain (1.0)

For comparison purposes, the speed loop controller was also designed following the Magnitude Optimum criterion, which advantage is the faster response in reference tracking.

The speed controller transfer function, $C_v(s)$, is expressed by (5):

$$C_V(s) = K_I\left[1 + \frac{1}{T_I s}\right] = \frac{13.89s + 27.78}{0.5s} \tag{5}$$

Where,

K_I $K_I = \left[\frac{T}{2K_s\tau}\right]$

T_I $T_I = T$

T   Large time constant (0.500 seconds)

τ   Small time constant (0.009 seconds)

K_s Open loop Gain (1.0)

The speed loop response for both controller designs were observed for a given disturbance step signal of 4 p.u at t=3s, while keeping the same reference input of a unitary step at t=0s.

As expected and observed in Fig. 4, the speed loop whose controller is designed by the Magnitude Optimum criterion, takes longer to reject the disturbance at t=3s when compared to the Symmetric Optimum designed one.

978-1-7281-4181-7/19 $31.00 © 2019 IEEE

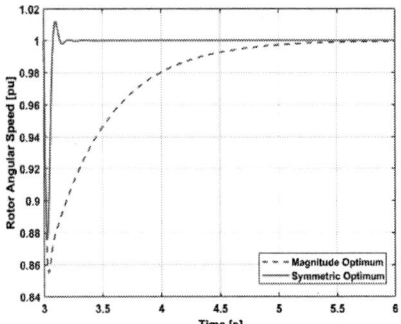

Fig. 4. Speed loop controller time-response to a disturbance step signal at t=3s.

Therefore, the speed controller designed by the Symmetric Optimum criterion was chosen.

C. Position regulator design

The position loop is characterized by the speed loop dynamics, an integrator, and the position controller, as schematically represented in Fig. 5.

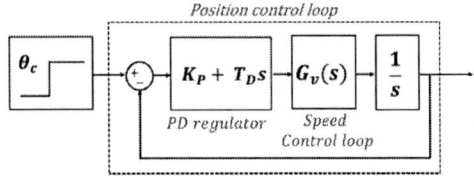

Fig. 5. Position control loop with a PD regulator.

Similarly, as in the design of the speed controller, the speed loop will be approximated by a 1st order system. In Fig. 6, the speed loop response to a step input is shown together with its equivalent first-order function, that will be used in the design of the position regulator. Both take the same time to settle but sharply differ when looking at the overshoot magnitude. Still, for the sake of simplicity, this first-order representation will be used for designing the position-loop controller. The results will show that the equivalence assumption is valid in the present study.

By inspection of Fig. 6, the settling time (t_s), for an error of 1% in the output signal, equals to 0.17s. From (2) the equivalent time constant can be calculated and the equation yields τ=0.04s. The equivalent system's transfer function is represented by (6):

$$G_v(s) = \frac{1}{1+0,04s} \qquad (6)$$

In order to achieve the desired output for a step input at the position loop, a Proportional-Derivative (PD) compensator was introduced. The controller was designed by positioning poles and zeros in the function root locus [3], to make sure that the settling time was less than 0.7s for a 1% error of the output signal.

With the assistance of computational tools, the gains for the position loop controller were defined as: K_p = 15 and TD=0.09s and its time response to a step input can be seen in Fig. 7.

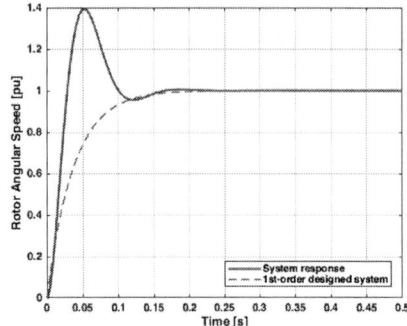

Fig. 6. Time response to a step input for a speed-loop equivalent 1st-order system, compared to the output of the Symmetric Optimum designed speed controller.

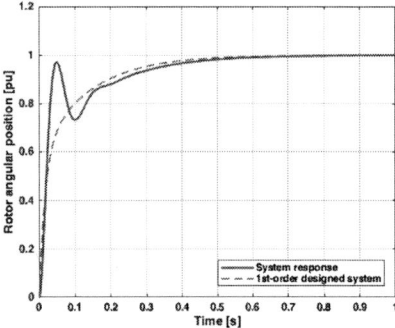

Fig. 7. Position loop response to a step input after the introduction of a PD controller.

D. Feedforward Reference Model

For testing purposes and potential improvement in the system response, especially in reducing its delay, a reference model was introduced to provide current, speed and position feedforward signals to its respective comparators. This is a third order equivalent system, which dynamics is mainly characterized by a damping constant of 0.7 and settling time of 0.3s. The reference model third pole was positioned so far that it should not affect the system behavior (positioned at $30s^{-1}$).

It should be noticed in Fig. 8 that the position reference signal fed to the control system is the position output signal of the reference system.

Thus, the characteristic equation of the reference system, considering a settling time of 0.3s, is given by:

$$D_{(s)} = (s^2 + 2\xi\omega_n s + \omega_n^2)(s + p_3) \qquad (7)$$
$$D_{(s)} = [s^2 + (2 \times 0,7 \times 21,9)s + 21,9^2](s + 30)$$
$$D_{(s)} = s^3 + K_2 s^2 + K_1 s + K_0$$

The system constant values are specified in Table III.

TABLE III.   FEEDFORWARD REFERENCE SIGNAL CALCULATED PARAMETERS

Parameter	Calculated Value
K_0	14.4
K_1	1.4
K_2	61.0

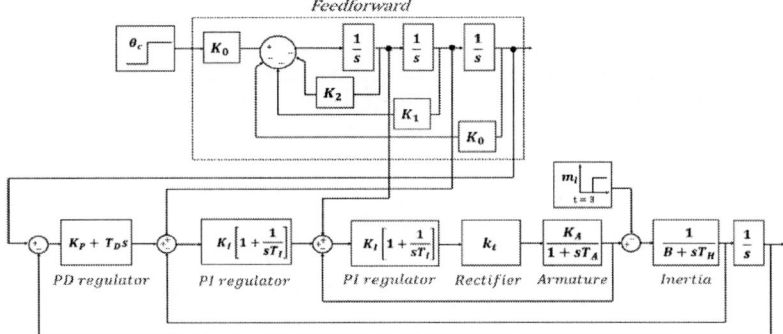

Fig. 8. System with all controllers and feedforward.

The output for the cascade system with feedforward tracks the reference signal and, therefore, yields to a shorter settling time when compared to the cascade system without the feedforward model.

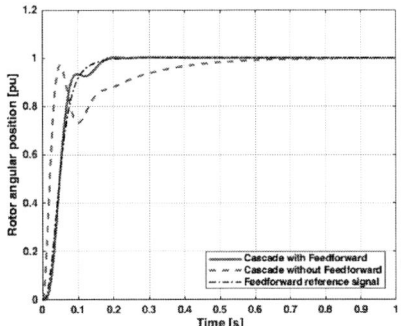

Fig. 9. Third-order system representation for ts=0.1s.

An additional analysis of each cascade system was carried out considering a settling time of 0.1 seconds for the reference model. The position output signals are shown in Fig. 9 for each one of the three control methods.

Clearly, the cascade system with feedforward has not entirely tracked the reference input. This occurs because the chosen settling time is too close to the system dynamic limitation.

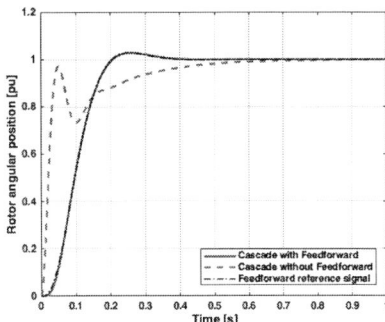

Fig. 10. Position loop time response for a step input with and without feedforward reference (ts=0.3s).

When compared to the 0.3s reference model in Fig. 10, it can be noticed a slight improvement in the system reference tracking delay. However, the reference model with a 0.3s settling time was kept for the cascade system to be compared with a system designed following the state-feedback

criterion. The selection of the system with a longer settling time is explained by the comparison of that time with the dynamics of the inner loops. It would not be real a model in which its outer loop settles faster than the inner one. The speed loop settling time is 0.2s, as shown in Fig. 6 for instance.

Obviously, with different controllers the time response will change. It is important to recognize that with cascade control the performance can be as that good as for state feedback control. Moreover, cascade control presents some implementation advantages that will be pointed out at the conclusion.

IV. STATE-FEEDBACK CONTROL

The transfer function which represents the DC Motor [6], in which Y(s) is the position and U(s) the control signal is given by (7):

$$\frac{Y(s)}{U(s)} = \frac{\psi K_A k_T}{(T_A T_H)s^3 + (BT_A + T_H)s^2 + (B+\psi)^2 K_A s} \tag{8}$$

The matrices A, B, C, and D for the system representation in state-space variables are defined in (8):

$$A = \begin{bmatrix} -33.33 & -33.33 & 0 \\ 1 & -2 & 0 \\ 0 & 1 & 0 \end{bmatrix} \quad B = \begin{bmatrix} 66.67 \\ 0 \\ 0 \end{bmatrix} \quad \begin{matrix} C = [0\ 0\ 1] \\ \\ D = [0] \end{matrix} \tag{9}$$

And state-feedback control law is given by (9):

$$\begin{matrix} u(t) = -kx(t) \\ \boldsymbol{u} = -K\boldsymbol{x} \end{matrix} \qquad \boldsymbol{x} = \begin{bmatrix} i \\ v \\ \Theta \end{bmatrix} \tag{10}$$

Where K is an m x n dimension matrix with the corresponding gains for each feedback state. And **x** is the vector carrying the state variables for current, angular speed, and angular position (i, v, and Θ, respectively).

It is worth noticing that matrices A, B, C, and D are built in such way that the system is described as a function of current, speed and position state variables.

From (9), it is self-evident that the gain matrix K defines the system control law. For proper comparison with the cascade control system, the feedforward reference model is again used here as a reference for the state-space system representation.

In order to find out the corresponding gain matrix, which will give the state feedback system the same response of the cascade system reference model, both systems characteristic polynomials must have identical poles.

By means of computational tools, state feedback system poles were placed at the roots of the reference model characteristic polynomial, and the corresponding gain k_i, k_v k_Θ are 0.38, 18.737, 215.92, respectively.

V. DYNAMICS COMPARISON

A. Reference signal tracking

For the three presented control system variants (cascade system with and without feedforward, and full-state feedback) current, speed, and position response plots were generated for a step reference input.

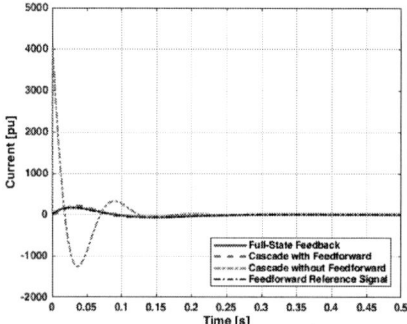

Fig. 11. Current signal comparison for the three control systems response and the reference model for a position step input.

From Fig. 11 to Fig. 13, one can verify that the introduction of a reference signal makes the response of all current, speed, and position loops faster. Besides, an additional positive characteristic from the introduction of a reference model is the decrease in the magnitude of the overshoot in all signals.

Furthermore, it should be noticed that the feedforward in a cascade control system provides the fastest time response. Added to that, it is possible with that control structure to individually tune each control loop.

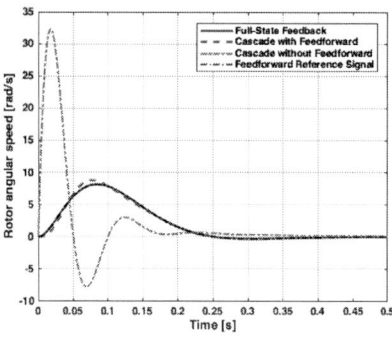

Fig. 12. Speed signal comparison for the three control systems response and the reference model for a position step input.

B. Disturbance Rejection

When evaluating disturbance rejection capability between the two cascade systems, with and without feedforward, it is seen in Fig. 14, that both systems respond identically. That is self-evident since the reference signals that are fed to the cascade system can't model the disturbances the plant may be subject to. Hence the

feedforward only affects the cascade system reference tracking capability.

For better visualization of the position output in Fig. 14, a disturbance of magnitude 4 p.u. was considered at a shifted time t=3s.

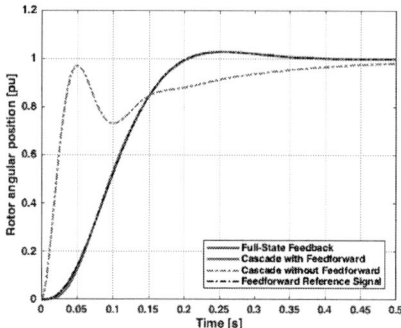

Fig. 13. Position signal comparison for the three control systems response and the reference model for a position step input.

It is worth highlighting that the cascade system quickly rejects disturbances because of its Optimal-Symmetric designed speed controller [9].

It can be seen from Fig. 15 that the state feedback system loses track of the reference input after the disturbance input.

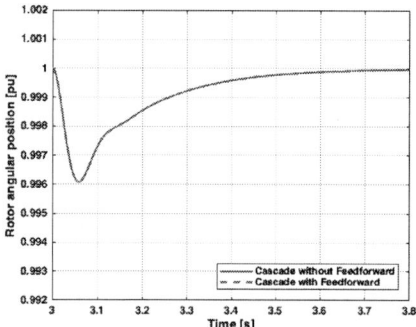

Fig. 14. Position time response to a disturbance input at t=3s.

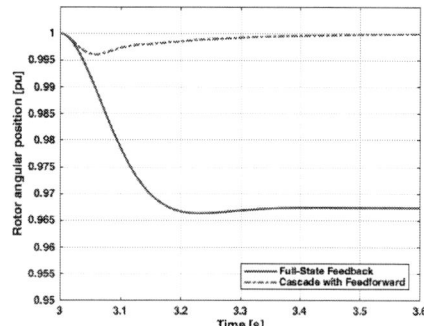

Fig. 15. Comparison of position signal for a step disturbance at t=3s. Cascade control with feedforward vs state feedback control.

VI. ZERO-ERROR FULL-STATE FEEDBACK

As just shown in Fig. 15, the state feedback control system does not reject disturbance inputs. The system maintains a steady state error when a step in load torque (m_L) is applied. One way to handle disturbances is to introduce an integrator associated with a gain at the plant input [11], thereby increasing the system transfer function order as indicated in Fig. 16.

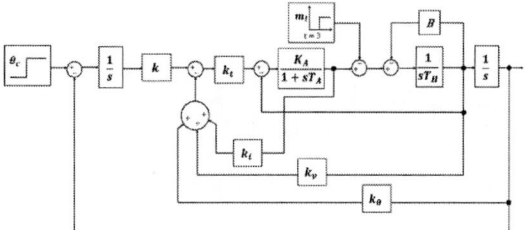

Fig. 16. Block-diagram representation of a state feedback system considering the implementation of an additional integrator for disturbance rejection.

The k-gains, shown in Fig. 16, were calculated by pole placement taking the reference model as paradigm.

For evaluation of the integrator block addition, Fig. 17 presents the position output signals for a step disturbance input of 4 p.u. magnitude at t=3s, in cascade and state-feedback systems.

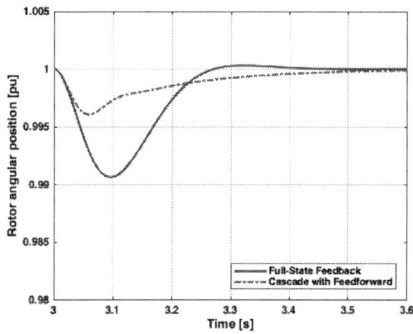

Fig. 17. Position output signal for a step disturbance signal at t=3s of a feedforward cascade control system and state feedback control with an additional integrator.

As it can be seen, the state feedback control system with the additional integrator rejects the disturbance signal practically at the same instant as the feedforward cascade control variant. The state feedback position output signal oscillates higher than in the cascade system.

The performance of "state feedback + observer" control is always worse than that without the observer, since the observer introduces a dynamic effect in the feedback loop. On the other hand, FF improves the dynamic, since it looks forward, as the name says.

VII.   CONCLUSION

With the information gathered in this study, it is possible to list the advantages and disadvantages of cascade control systems in comparison with the characteristics of state feedback controls.

Among the advantages of the cascade control system, we have:

- Control structure easy to understand and implement

- The analysis of the system/plant is intuitive since the physical problem is directly represented.

- Actuation limiters can be separately added to each individual loop, although this was not implemented in this work.

- Integral reset wind-up [10] can be easily implemented.

- The control system can be gradually commissioned, with subsequent tests of the internal loops

- Disturbances and non-linear effects are easily compensated.

- The nested cascade system uses standard PID regulators, widely known and deployed both in the industry and academy.

The main disadvantage of a nested cascade structure, however, is its slower response because of the successive computations at each feedback loop. This effect, however, can be mitigated with the introduction of feedforward signals by a reference linearized model, as presented in this study.

ACKNOWLEDGMENT

This study was financed in part by the Coordenação de Aperfeiçoamento de Pessoal de Nível Superior - Brasil (CAPES) - Finance Code 001.

REFERENCES

[1] A. B. Lugli, G. H. F. Floriano, H. J. Gonzaga, L. V. Carvalho, R. M. Volpato, Y. M. C. Masselli, "Sistema aplicado a controle de posição utilizando algoritmo PID", II Seminário de Automação Industrial e Sistemas Eletro-Eletrônicos – SAISEE (ISSN 2319-0280), Mar. 2015.

[2] Isermann R., "Cascade Control Systems", Digital Control Systems. Springer, Berlin, Heidelberg (ISBN 978-3-642-86422-3), 1991.

[3]   K. Ogata, "Engenharia de controle moderno". 5. ed. Pearson Prentice Hall (ISBN 978-85-4301-375-6), 2010.

[4] F. Leonardi, P. A. Maya, "Controle Essencial" 1.ed. Pearson (ISBN: 857605700x) 2011.

[5] W. Leonhard, "Control of Electrical Drives" Springer, Berlin, 2001.

[6] R. M. Stephan, "A simple model for a thyristor driven dc motor considering continuous and discontinuous current modes", *IEEE Trans.Edu.*, vol. 34, no. 4, November 1991.

[7] Simone Buso, Paolo Mattavelli, "Digital Contol in Power Electronics" Morgan & Claypool Publishers, 2006.

[8] Donald Grahame Holmes, Thomas Lipo, "Pulse Width Modulation for Power Converters" IEEE Press 2003.

[9] J. Umland, M. Safuddin, "Magnitude and symmetric optimum criterion for the design of linear control systems: what is it and how does it compare with others?", *IEEE Trans. Ind. Appl.*, v. 26, p. 489–497, 1990.

[10] R. M. Stephan, "Acionamento, Comando e Controle de Máquinas Elétricas", Editora Ciência Moderna, 2013.

[11] G. F. Franklin, J. D. Powell, A. Emami-Naeini "Feedback control for dynamic system" 4th edition. Ed. Prentice Hall.

Design and Performance Analysis of Isolated Cuk Converter Employed in Multiple Pulse Rectifier Systems

Ana L. Soares, Antônio O. Costa Neto, Gustavo B. Lima, Luiz C. G. Freitas, Ernane A. A. Coelho

Núcleo de Pesquisa em Eletrônica de Potência (NUPEP)
Faculdade de Engenharia Elétrica (FEELT) - Universidade Federal de Uberlândia (UFU)
Uberlândia, Brazil 38400-902
e-mails: anajeali@yahoo.com.br, antonio.costaneto@hotmail.com, lcgfreitas@yahoo.com.br

Abstract— This work presents a 12-pulse rectifier in order to reduce the harmonic distortion of line currents. The topology is based on a Delta autotransformer connection and isolated Cuk converters operating in continuous conduction mode. This structure provides higher power factor and a regulated cc output. Simulation and experimental results are presented in order to verify the reduction of the harmonic content, improving the quality of the energy.

Keywords— *autotransformer, Cuk converters, harmonic distortion, multipulse rectifier, power factor.*

I. INTRODUCTION

The increasing in harmonic distortions caused by the growing use of non-linear characteristic loads in the electrical system is one of the problems that affect the quality of the electric energy, since these disturbances causing extra losses in conductors and equipment, a reduction in the life of electric machines and transformers and others. There are international standards that limit the Total Harmonic Distortion rate of current (THDi) and the power factor [1], [2], making it necessary to mitigate the harmonic content of the input currents. There are several ways to solve this problem, such as the use of passive and active alternatives. or active power factor correction, hybrid and multipulse.

There are several ways to solve this problem, such as the use of passive and active filters, power factor correction, hybrid and multipulse converters.

Several topologies of multipulse converters have been developed, operating in conjunction with passive auxiliary circuits, maintaining the correct power processing and correcting the unbalance of the secondary voltages caused by the phase shifting transformer, such as the Interphase (IPR) and the Interphase Transformer (IPT) or active as static converters.

Following are some of the main topologies involving multi-pulse rectifiers with autotransformer connection.

In [3] there is an 18-pulse autotransformer with Delta differential connection connected to IPTs. It has low harmonic distortion input currents and high power factor. However, these equipments grow the weight and the volume of the converter.

In the literature there are several differential connections of autotransformers that use DC-DC converters instead of IPTs, correcting the power factor and mitigating the harmonic components in the input current. Some topologies [4], [5], [6] use the Full-Bridge converters in conjunction with autotransformers, since this switched converter has high frequency insulation, nevertheless it requires an LC filter at the output and has a high number of semiconductors. There are also structures that use the autotransformers and Boost converters [7] that have current source characteristics at the input, but which not isolated galvanically.

An 18-pulse autotransformer with Sepic converters can be seen in [8]. The Sepic converter also has the current source characteristic at the input and allows the voltage regulation of the output,and can act as voltage step-down or step-up.

In [9] a 12-pulse rectifier with autotransformer and Sepic converters was presented. It has the same characteristics shown in [8], but with the control used, it obtained better values of power factor and THDi.

This work presents a 12-pulse rectifier with a differential Delta autotransformer and two isolated Cuk converters. The Cuk converter has a current source input, uses capacitive transfer between input and output, has galvanic isolation and has a high frequency transformer, which does not influence weight and volume since it is very low.

II. PROPOSED STRUCTURE

Figure 1 shows a 12-pulse rectifier with a delta differential autotransformer and two isolated Cuk converters operating in continuous conduction mode (MCC), connected in parallel. This rectifier gives the system a low harmonic distortion in the current, a high power factor and a flexibility in the regulation of the output voltage. Each Cuk converter works independently, offering 50% of the total power in the load

A. Autotransformer with Generalized Delta Differential Connection

The autotransformer used was designed according to [3] - [5]. It is 12 pulses, with three primary windings and twelve auxiliary windings, generating two three-phase systems shifted in 30 degrees with each other, which are connected to two six-pulse rectifier bridges each.

The authors would like to acknowledge UFU, CAPES and CNPq (Under processes: 304489/2017-4 and 420602/2016-0) for the financial support.

978-1-7281-4181-7/19 $31.00 © 2019 IEEE

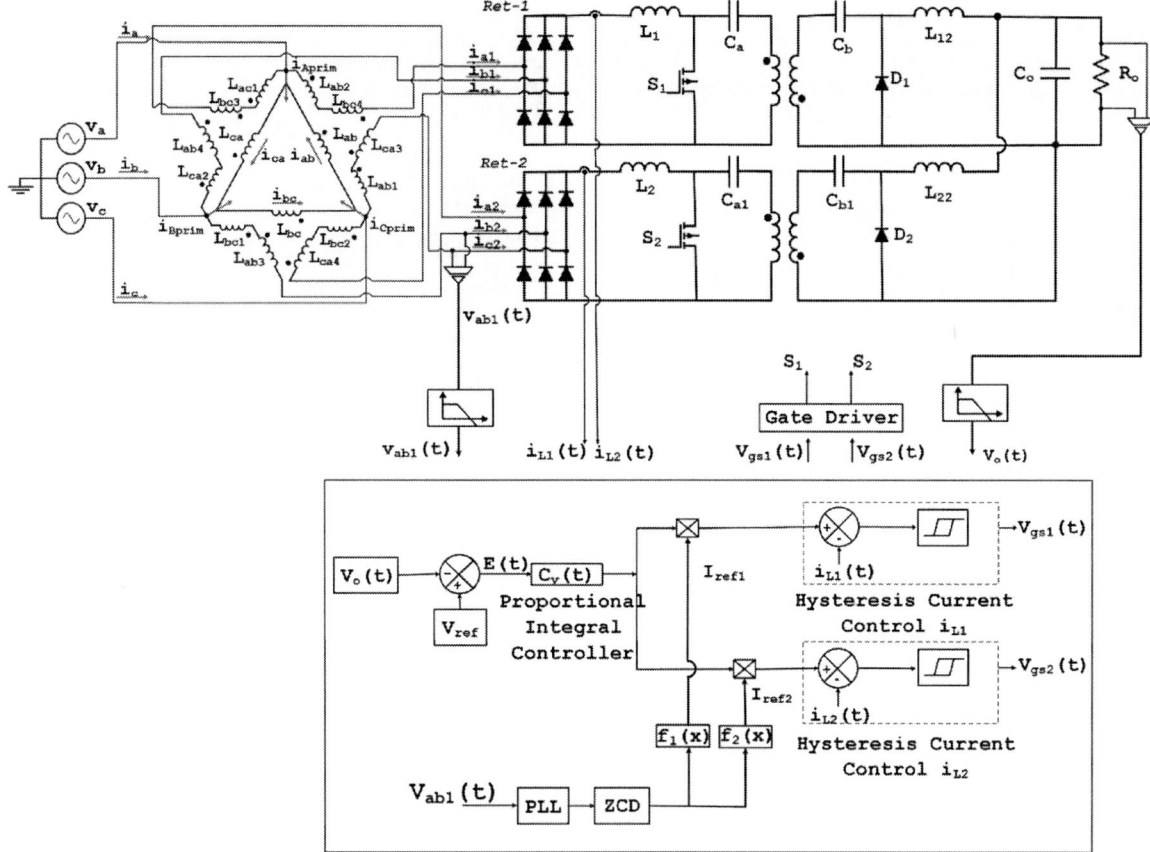

Fig. 1 – Differential Delta Autotransformer with Isolated Cuk Converters. (Adapted from [8])

This autotransformer does not have galvanic isolation, processing only a portion of the power, about 18%, through the magnetic coupling, which is responsible for the phase shifts and voltage adjustments. This results in a core with reduced weight and volume.

B. Isolated Cuk Converter

The isolated Cuk converter has two capacitors (C_a and C_b) and a high-frequency transformer between them. The Cuk is a DC-DC converter that has the following advantages: the input with current source characteristic, besides being a fundamental characteristic in the current imposition, it contributes to the absorption of abrupt variations of the input voltage, it uses capacitive energy transfer between input and output allowing the isolation by means of transformers without the need of gap as in toroidal cores with lower values of dispersion inductance contributing in the decrease of unwanted voltage peaks in the semiconductors [10] [11], in addition, it has a reduced number , low input current ripple and wide output voltage range [12].

1) Operation Steps

The Cuk converter in continuous conduction mode has two steps of operation, as observed in figure 2:

- 1ª: The capacitor C_a stores the energy of the source and when the switch S_1 turns on, this energy is transferred through the transformer forming a current outside the point (negative) in the primary winding, which in turn forms a current at the point (positive) in the secondary, passing through the capacitor C_b

and discharging into the inductor L_{12}, the capacitor C_o and the charge R_o.

- 2ª: When the switch S_1 is switched off, both the source and the energy stored in the inductor L_1 transfer their energy through C_a creating a positive current in the primary of the transformer and, consequently, a negative current in the secondary, polarizing the diode D_1 and finally discharging in the inductor L_{12}, the capacitor C_o and the load R_o.

C. Control Strategy

The control strategy is done by imposing triangular currents with frequency around 360Hz in the input inductors of each Cuk converter. For this, it is necessary that these currents are in phase with any line voltage of the secondary of the autotransformer of the same rectifier group in which the line current is sensed, as in Fig. 3 which shows the line voltage V_{ab1} and the input currents (i_{L1} and i_{L2}) from each rectifier group (Ret-1 and Ret-2). With this methodology it is possible to obtain a sinusoidal composition at the entrance of the structure from the sum of the areas between the secondary currents of the autotransformer and the current flowing inside the delta connection (i_{Lab} - i_{Lca}), as seen in Fig. 4.

The triangular imposition increases the average current of the autotransformer while maintaining the same level of power required by the load. This imposition, although beneficial to the quality of the input currents and the power factor, generates an increase in the processed power of the autotransformer's magnetic core, which was initially from 18.34% to 20.5%, generating a decrease in the output power at which the autotransformer was originally designed.

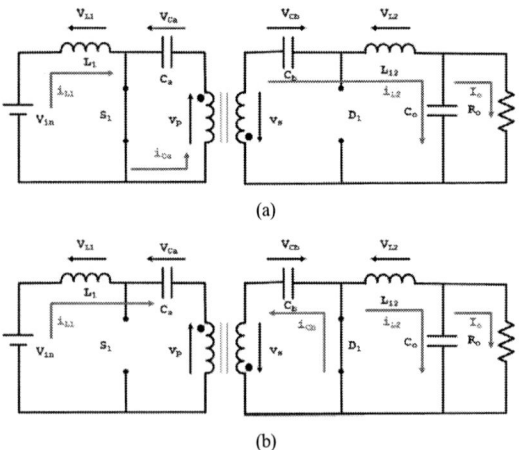

(a)

(b)

Fig. 2 – Operation steps of the Cuk converter: (a) S_1 closed and (b) S_1 open.

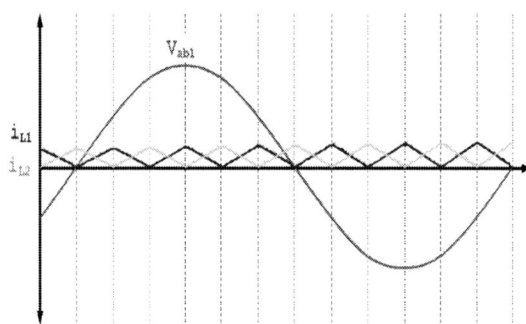

Fig. 3 – Secondary line voltage waveforms of the autotransformer and current imposed at the input of each DC-DC converter.

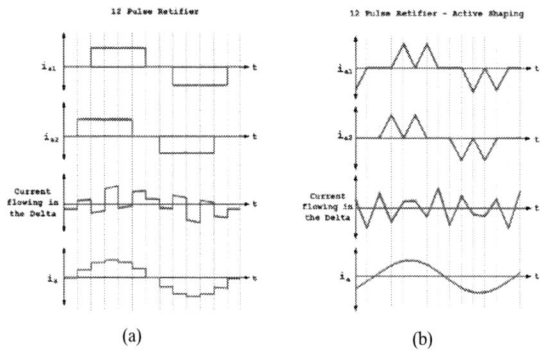

(a)   (b)

Fig. 4 – Theoretical waveforms: (a) without and (b) with imposition..

As shown in Fig. 1, the adopted control strategy employs a Phase Locked Loop (PLL) is technique implemented to ensure phase and frequency synchronization of the autotransformer secondary line voltage with the reference voltage used in the Zero Crossing Detector (ZCD) to define the point of generation of the triangular reference $f_1(x)$ and $f_2(x)$ imposed on each inductor. The blocks $f_1(x)$ and $f_2(x)$ are

also responsible for generating the phase shift of 180° between each triangular reference of the rectifier groups (I_{ref1} and I_{ref2}), which is a necessary characteristic for the correct imposition of the input currents.

The output voltage signal (V_o) is compared to a desired voltage reference signal (V_{ref}). This error is sent to an integral proportional compensator (C_v) with a frequency attenuation filter of 360 Hz, which is the observed oscillation frequency at the output voltage of each Cuk converter. Thus, the signal $f_1(x)$ and $f_2(x)$ - synchronized triangular reference signal - is multiplied by the control voltage from the voltage PI controller, creating the current reference signal (I_{ref1} and I_{ref2}) with magnitude and shape of desired waves.

Finally, the current imposed on the input inductors of each Cuk converter will follow the current reference signal (I_{ref1} and I_{ref2}) using a hysteresis controller. The choice of current controller for hysteresis is due to the fact that it has optimum performance for wide range of load variation and as well as in transient conditions.

This technique stands out for the simplicity in obtaining the voltage compensation using only a PI compensator. In addition, constant hysteresis modulation provides a fast and accurate response to the input current induction [13].

To guarantee the natural cancellation of the characteristic harmonics of 12-pulse rectifiers, the system needs to have currents in constant equilibrium [3], [8]. Hysteresis modulation operates with identical reference current values obtained by the PI controller response. From this, the converters operate with current balance, ensuring the power processing in 50% each group (Ret-1 and Ret-2), eliminating the use of Interfase Transformers that in parallel connections would do this balancing by acting in the absorption of the instantaneous differences of the output voltages of the rectifiers [4], [5], [7], [8].

D. Determination of the Transfer Functions of the Plant and the Voltage Controller

By means of the analysis of the average state space, we can determine the transfer function of the Cuk converter, seen in (1).

The output voltage compensator, seen in (2), was obtained in the SISOTOOL of the MATLAB software, using the modeled function of the plant in (1). Applying the Tustin method, we obtain the discrete controller transfer function in (3).

$$G_{Voil}(s) = \frac{1.36383 \cdot 10^{20} \cdot s^2 - 1.09655 \cdot 10^{24} \cdot s + 1.05476 \cdot 10^{28}}{10^{17} \cdot s^3 + 1.01305 \cdot 10^{21} \cdot s^2 + 1.015204 \cdot 10^{25} \cdot s + 4.30181 \cdot 10^{26}} \quad (1)$$

$$\frac{Kv(s)}{Ev(s)} = \frac{2 \cdot (s + 0.03)}{s^2 + 0.00442 \cdot s} \quad (2)$$

$$\frac{Kv(z)}{Ev(z)} = \frac{10^{-5} + 3 \cdot 10^{-12} \cdot z^{-1} - 10^{-5} \cdot z^{-2}}{1 - 2 \cdot z^{-1} + z^{-2}} \quad (3)$$

The complete closed-loop system was simulated using the MATLAB Simulink tool, as shown in Figure 5, where it is verified that the step response exceeds around 20% and enters steady state at about 182 ms. This demonstrates that the voltage compensator has a response fast enough to control the output voltage. Note that the voltage controller has been designed with the objective of reducing output voltage oscillation, having an additional 360Hz pole and a sampling frequency of 100kHz. The stability of the system can be

978-1-7281-4181-7/19 $31.00 © 2019 IEEE

proved by Fig. 6 which shows the location of the roots with the insertion of the voltage controller PI, where the gain margin is 40dB and the phase angle is 48.6°.

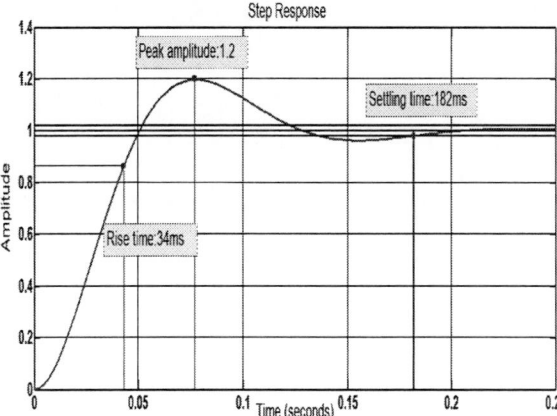

Fig. 5 – Step response of the internal voltage loop.

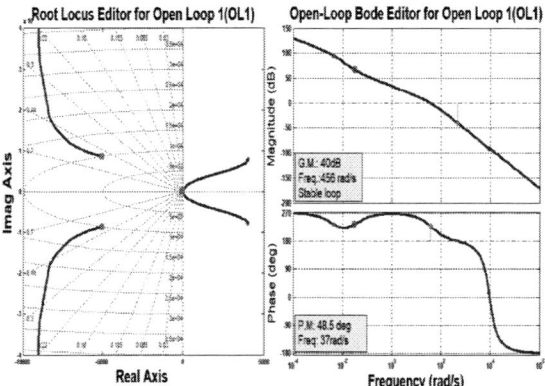

Fig. 6 – Root locus of the internal voltage loop.

E. Simulation Results

To validate the proposal, computational simulations were performed using the Psim® software as shown in Fig. 7, which shows the waveform of the line currents of the autotransformer input, indicating that with the control used these currents are practically sinusoidal. Fig. 8 shows the response of the output voltage controllers when subjected to a step of descent and rise of ± 50%, where the accomodation time is 160ms with a overshoot of 1.28, demonstrating a satisfactory behavior in relation.

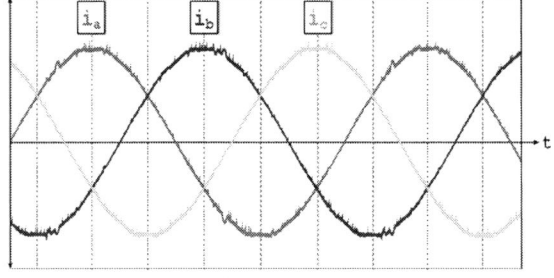

Fig. 7 – Autotransformer input line currents.

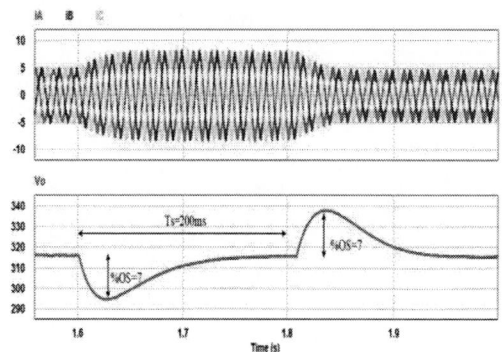

Fig. 8 – Response to the step up and step down of the converter.

F. Experimental Results

To verify the simulation results, corroborating the presented topology, a prototype of 2kW, as in Fig. 9, was developed and analyzed in laboratory and the specifications of this rectifier are in table I.

The waveforms of the currents obtained and their respective harmonic spectra for the nominal condition are shown in Figure 10 and table II, showing the topology compared to the IEC61000 3-2 standard and that it operates with a harmonic distortion of 3.44%. Figures 11 and 12 show the dynamic response to the rise and fall steps of ± 50%, proving the satisfactory behavior in relation to the output. Figure 13 shows a comparison of the harmonic spectrum of the proposed topology, a conventional 12-pulse rectifier alone and the IEC 61000-3-2 standard. It can be seen that the conventional 12-pulse autotransformer exceeds the norm in the individual harmonics characteristic of the expression $12 \cdot k \pm 1$; while the proposed structure does not, due to the imposition made on the input inductor currents of each Cuk.

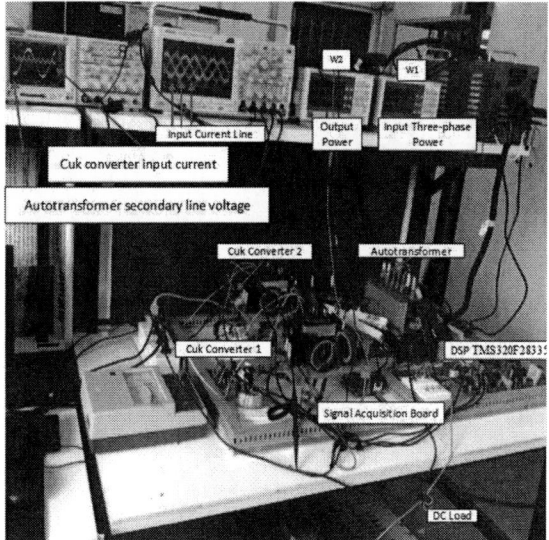

Fig. 9 – Prototype of the rectifier.

TABLE I
Project specifications

Input Line Voltage	220Vrms
DC Bus Voltage	315Vcc
THD$_I$	in accordance with IEC 61000 3-2
Cuk Converters	
Power output for each converter	2000W
Maximum switching frequency	50kHz
Inductances Lin1, Lin2, Lout1, Lout2	5mH
Transformer Magnetizing Inductance	1mH
Capacitances Ca1, Ca2, Cb1, Cb2	564nF
Output capacitance Cout	900µF
Switches S1 e S2	MOSFET SCT20N120 (20A/1200V)
Diodes D1 e D2	RHRG30120 (30A / 1200V)

TABLE II
Harmonic distortion for each phase compared to IEC in percent (%)

Order	iA	IEC$_{iA}$	iB	IEC$_{iB}$	iC	IEC$_{iC}$
3	0.3867	84.152	0.2131	77.670	0.0297	23.116
5	0.4127	41.710	0.4174	38.497	0.4404	12.009
7	0.2927	28.173	0.4468	26.003	0.1438	9.9071
9	0.0516	14.636	0.102	13.508	0.0561	6.3045
11	0.5806	12.074	0.9962	11.144	1.5365	4.5032
13	1.2257	7.6834	1.3214	7.0916	2.3990	3.9028
15	0.1379	5.4882	0.0679	5.0654	0.0108	3.3024
17	0.4720	4.7564	0.3401	4.3901	0.2484	3.0021
19	0.3670	4.0247	0.1969	3.7147	0.2038	2.7019
21	0.1013	3.6588	0.1347	3.377	0.0270	2.7019
23	0.6275	3.2929	0.2327	3.0393	0.7592	2.4017
25	0.5514	3.2929	0.3211	3.0393	1.3459	2.1015
27	0.1431	2.9270	0.0807	2.70166	0.0862	2.1015
29	0.1346	2.5611	0.1560	2.36399	0.0603	1.8013
31	0.2792	2.5611	0.1496	2.36399	0.1651	1.8013
33	0.1167	2.1953	0.21588	2.02622	0.1186	1.8013
35	0.4833	2.1953	0.23848	2.02622	0.1216	23.116
37	0.6209	2.1953	0.26078	2.02622	0.2726	12.009

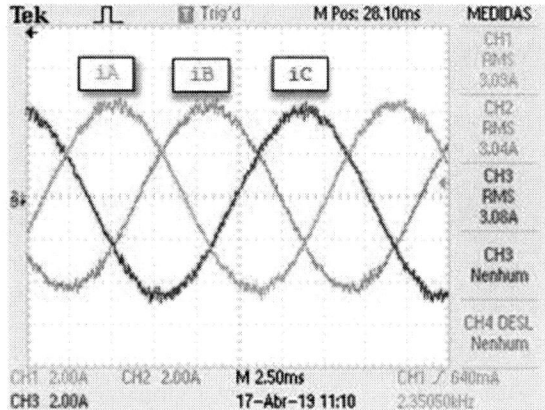

Fig. 10– Experimental result for the 2kW prototype. (Vo = 315V, Po = 2kW)

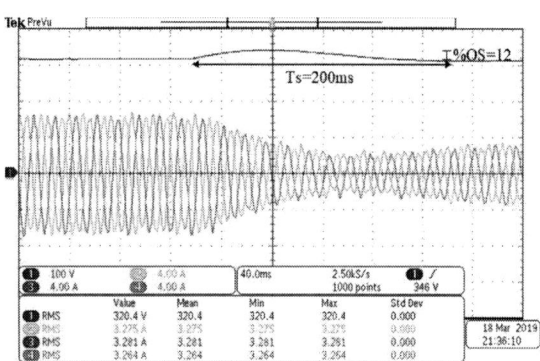

Fig. 11 – Dynamic response during a load step-down of ±50%.

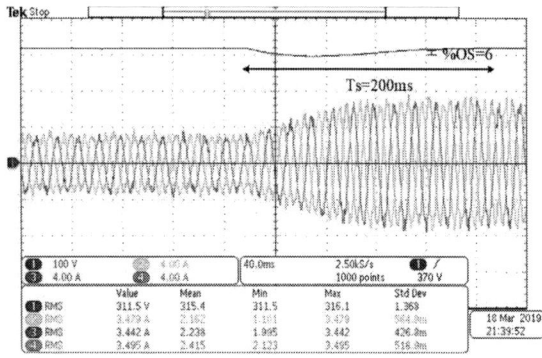

Fig. 12 – Dynamic response during a load step-up of ±50%.

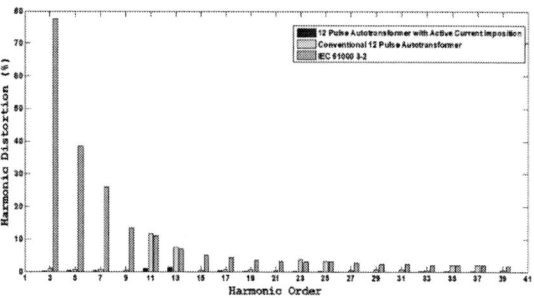

Fig. 13 – Harmonic spectrum of the proposed topology compared to a conventional isolated 12-pulse autotransformer and IEC 61000-3-2.

III. CONCLUSIONS

In this work a generalized connection autotransformer operating with two isolated Cuk converters was presented. It showed the operation of the Cuk converter and brought a description of the control used, demonstrated through the simulation results and experimental results of the high performance of the structure in relation to the reduction of THD when compared to conventional rectifiers of 12 pulses and also showed that with a simple control by constant hysteresis it is possible to obtain excellent results with regard to the voltage regulation and imposition of the topology line currents.

The proposed topology has proven operationally flexible and can be used in a variety of telecommunications applications, electric machine drives, battery chargers and more electric aircraft (MEA).

IV. REFERENCES

[1] I. E. Commission et al., Electromagnetic Compatibility (EMC). Part 3: Limits—Section 6: Assessment of emission limits for distorting loads in MV and HV *power systems*. [S.l.], 1996.

[2] R. Langella, A. Testa, E. Alii, IEEE Recommended practice and requirements for harmonic control in electric power systems. IEEE, 2014.

[3] R.C. Fernandes, P S. Oliveira and F. J. M. Seixas, "A Family of Autoconnected Transformers for 12-and 18-Pulse Converters – Generalization for Delta and Wye Topologies," IEEE Trans. Power Electron, vol.26, no. 7, pp. 2065-2078, Jul. 2011.

[4] F. J. M. Seixas and I. Barbi, "A New Three-phase Low THD Power Supply with High-Frequency Isolation and 60 V/200 A Regulated DC Output," Proc. IEEE32nd Annu. Power Electron. Spe. Conf., pp. 1629-1634, 2001.

[5] F. J. M. Seixas and I. Barbi, "A New Three-phase Low THD Rectifier with High-Frequency Isolation Regulated DC Output," IEEE Trans. Power Electron., vol. 19, no. 2, pp. 371-377, Mar. 2004.

[6] S. Choi, "A Three-phase Unity-power-factor Diode Rectifier with Active Input Current Shaping," IEEE Trans. Ind. Electron.,vol. 52, no. 6, pp. 1711-1714, Dec. 2005.

[7] R.C. Fernandes and F. J. M. Seixas, "AC-DC Three-phase Multipulse Converter with Boost DC-DC Stage and Constant hysteresis Control," Proc. IEEE Brazilian Power Electron. Conf. pp. 403-408, 2011.

[8] A. C. Lourenço, F. J. M. Seixas, J. C. Pelicer and P S. Oliveira, "18-pulse Autotransformer Rectifier Unit Using Sepic Converters for Regulated DC-bus abd High Frequency Isolation," Proc. IEEE 13th Brazilian Power Electron. Conf. 1st Southern Power Electron. Conf. COBEP/SPEC), pp. 1-6, 2015.

[9] A. O. C. Neto, A. L. Soares, G. B. Lima, D. B. Rodrigues, E. A. A. Coelho, L. C. G. Freitas, "Optimized 12-pulse Rectifier with Generalized Delta Connection Autotransformer and Isolated Sepic

Converters for Sinusoidal Input Line Current Imposition," IEEE Trans. Power Electron., vol. 34. pp. 3204-3213, Jul. 2018.

[10] J. R. de Britto, A. E. Demian, L. C. de Freitas, V. J. Farias, E. A. A. Coelho, J. B. Vieira, "A Proposal of Led Lamp Driver for Universal Input Using Cuk Converter," Proc. IEEE Power Electron. Special Conf., pp. 2640-2644, June 2008.

[11] A. Newton, T. C. Green, D. Andrew, " AC/DC Power Factor Correction Using Interleaved Boost and Cuk Converters," Proc. Inst. Electr. Eng. PEVCD Conf.., pp. 293-298, 2000.

[12] N. Mohan, T. M. Undeland, "Power Electronics Converters Applications and Design," John Wiley & Sons, 2007.

[13] A. V. Costa, D. B. Rodrigues, G. B. Lima, L. C. Freitas, E. A. A. Coelho, V. J. Farias, L. C. G. Freitas, "New Hybrid High-power Rectifier with Reduced THDi and Voltage-sag Ride-through Capability using Boost Converter," IEEE Trans. Industry Applications, vol. 49, no.6, pp. 2421-2436, 2013.

A Research on Constant Voltage Output Characteristics of Wireless Power Transfer System with A DC-DC Converter

Zhimeng Liu[1,2]
1. Key Laboratory of Power Electronics
and Electric Drives
Institute of Electrical Engineering
Chinese Academy of Sciences
2. University of Chinese Academy of
Sciences
Beijing, China
liuzhimeng@mail.iee.ac.cn

Lifang Wang[1,2]
1. Key Laboratory of Power Electronics
and Electric Drives
Institute of Electrical Engineering
Chinese Academy of Sciences
2. University of Chinese Academy of
Sciences
Beijing, China
wlf@mail.iee.ac.cn

Chengliang Yin[1,]
1. Shanghai Jiaotong University
Shanghai, China
yjguo@mail.iee.ac.cn

Yanjie Guo[1,2]
1. Key Laboratory of Power Electronics
and Electric Drives
Institute of Electrical Engineering
Chinese Academy of Sciences
2. University of Chinese Academy of Sciences
Beijing, China
yjguo@mail.iee.ac.cn

Chengxuan Tao[1]
1. Key Laboratory of Power Electronics
and Electric Drives
Institute of Electrical Engineering
Chinese Academy of Sciences
Beijing, China
taochengxuan@mail.iee.ac.cn

Abstract—**This paper presents the analysis of constant voltage (CV) output wireless power transfer (WPT) system with a DC-DC converter load, including characteristics of system efficiency and output voltage. Firstly, the WPT system model with LCC/S compensation network and a DC-DC converter is established. Then, system transmission power and efficiency are deduced on the CV condition, and the boost circuit's influence on the WPT system is analyzed. Finally, a 1kW WPT prototype is built and simulations are conducted. Experimental results have proved that a proper DC-DC converter can improve the WPT system's efficiency and the WPT system can keep a constant output voltage.**

Keywords—*wireless power transfer (WPT) system, DC/DC converter, constant voltage (CV) output, system efficiency*

I. INTRODUCTION

In recent years, wireless power transfer (WPT) technology has drawn more and more attention [1, 2]. In order to improve the carrying capacity of the system, it is necessary to the study constant voltage (CV) output WPT system [3]. Compensation networks [4-6] or control strategies [7] are designed to keep WPT system's output voltage constant. Some kinds of compensation networks can realize output voltage constant, such as LCC/C and LCL/C [4-6]. Also, for the sake of improving the WPT system's stability, primary-side controller and load identification approach are used to adjust voltage [7], but due to limitation of estimation accuracy, it cannot match more precise reference goal.

Based on the above research results, the WPT system model with a DC-DC converter load is established in this paper firstly. Secondly, the method of keeping system's output voltage constant is deduced and the equivalent two-port network is built to analyze system characteristics. Finally, experimental results have verified that the WPT

This work was supported by National Key R&D Program of China (2018YFB0106300).

system can realize constant voltage output and improve efficiency with an accurate DC-DC converter.

II. MODELING AND ANALYSIS

The WPT system model with a DC/DC converter load is shown in Fig. 1. U_d is the input DC voltage of the inverter, the inverter is composed of $G_1\sim G_4$; the primary compensation network is composed of L_{11}, C_{p1} and C_{p2}; C_{s1} is the secondary compensation capacitor. L_1 and L_2 are the self-inductances of the coupled coil; M is the mutual inductance; U_O is the input voltage of the rectifier; R_e is the input resistance of the rectifier, R_{DC} is the boost converter input resistance; $D_1\sim D_4$ constitute the rectifier, and C_O is the output filter capacitor of the rectifier. The DC/DC converter is a boost converter, and R_L is the load of the WPT system.

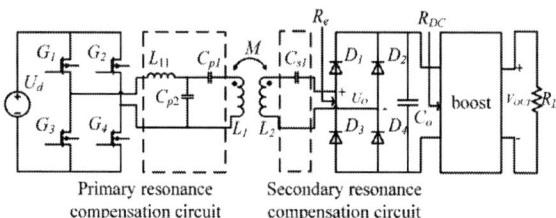

Primary resonance   Secondary resonance
compensation circuit   compensation circuit

Fig. 1. The WPT system model

The boost converter is used as the DC-DC converter to analyze the WPT system in this paper, for the following reasons: when the transmitting coil current is constant, and the secondary circuit is designed to be a constant voltage source, the amplitude of U_O is constant, which makes the constant WPT output voltage constant. However, on this condition, the real load is larger than optimal load, and results in lower efficiency. So, in order to decrease the input impedance of the rectifier to improve the WPT system efficiency, boost converter is adopted in this paper.

To analyze the WPT system transmission efficiency and power, two-port network of the coupling coils and

978-1-7281-4181-7/19 $31.00 © 2019 IEEE

compensation capacitors is shown in Fig. 2. Where, Z_{IN} is the inverter output impedance.

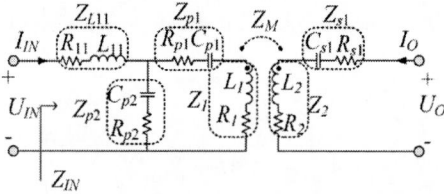

Fig. 2. Two-port network equivalent circuit of the coupling coils and compensation capacitors

Firstly, based on Fig. 2, Z-parameter equation of the equivalent circuit is deduced as shown in equation (1):

$$\begin{bmatrix} U_{IN} \\ U_O \end{bmatrix} = \begin{bmatrix} Z_{11} & Z_{12} \\ Z_{21} & Z_{22} \end{bmatrix} \begin{bmatrix} I_{IN} \\ I_O \end{bmatrix} \quad (1)$$

Where, $Z_{11}, Z_{12}, Z_{21}, Z_{22}$ are given by (2):

$$\begin{cases} Z_{11} = Z_{L11} + \dfrac{Z_{p2}(Z_{p1} + Z_1)}{Z_{p2} + Z_{p1} + Z_1} \\ Z_{12} = Z_M \dfrac{Z_{p2}}{Z_{p1} + Z_{p2} + Z_1} \\ Z_{21} = Z_{12} \\ Z_{22} = Z_{s1} + Z_2 - \dfrac{(Z_M)^2}{Z_{p1} + Z_{p2} + Z_1} \end{cases} \quad (2)$$

When performing the CV analysis, the parasitic resistances R_{11}, R_{p1}, R_{p2}, R_1, R_2, R_{s1} can be neglected [8]. Assuming L_{11} and C_{p2}, L_2 and C_{s1} are resonant as given by (3), U_O will be constant with the change of R_e. Where, ω_0 is the resonance angle frequency.

$$\begin{cases} \omega_0 L_{11} = \dfrac{1}{\omega_0 C_{p2}} \\ \omega_0 L_2 = \dfrac{1}{\omega_0 C_{s1}} \end{cases} \quad (3)$$

Then, to analyze system's characteristics, the parasitic parameters will be considered. On the basis of (3), (2) can be simplified as:

$$\begin{cases} Z_{11} = \dfrac{(Z_{L11} + 1)(Z_{p1} + Z_1) + Z_{p1}(R_{11} + R_{p2})}{Z_{p2} + Z_{p1} + Z_1} \\ Z_{12} = Z_M \dfrac{Z_{p2}}{Z_{p1} + Z_{p2} + Z_1} \\ Z_{21} = Z_{12} \\ Z_{22} = R_{s1} + R_2 - \dfrac{(Z_M)^2}{Z_{p1} + Z_{p2} + Z_1} \end{cases} \quad (4)$$

Thus, the relationship between output voltage of boost converter V_{OUT} and U_{IN} can be obtained and given by:

$$G_{vv} = \frac{V_{OUT}}{U_{IN}} = \frac{\omega_0^2 M Z_{p2}}{1 - D} \quad (5)$$

Furthermore, for the rectifier load, assuming the input current of the rectifier is continuous, and the input resistance of the rectifier bridge is $R_e = 8 R_{DC}/(\pi^2)$ [9]. The relationship between R_e and R_L is derived as:

$$R_e = \frac{8(1-D)^2 R_L}{\pi^2} \quad (6)$$

Equations (5) and (6) suggest that equivalent input resistance of the rectifier R_e change with the duty cycle of the boost converter. Once the system parameters are determined, the output of different voltage values and adjustment of compensation network parameters can be realized by changing the duty cycle D.

Finally, the transmission efficiency η_{trans} and power p_{trans} of the WPT system are deduced as equations (7) and (8), respectively. Where $r(Z_{IN})$ is the real part of the inverter's output impedance.

$$\begin{cases} \eta_{trans} = \dfrac{|I_O|^2 R_e}{|I_{IN}|^2 r(Z_{IN})} \\ \qquad = \dfrac{8*(1-D)^2 R_L}{\pi^2 * r\left(Z_{11} - \dfrac{Z_{12}^2}{Z_{Re} + Z_{22}}\right)} \left| \dfrac{Z_{12}}{Z_{Re} + Z_{22}} \right|^2 \\ Z_{IN} = \dfrac{U_{IN}}{I_{IN}} \\ \qquad = Z_{11} - \dfrac{Z_{12}^2}{Z_{Re} + Z_{22}} \end{cases} \quad (7)$$

$$\begin{aligned} p_{trans} &= |I_O|^2 R_e \\ &= \left| \frac{V_{OUT} Z_{21}}{G_{vv}(Z_{12}^2 - R_e Z_{11} - Z_{11} Z_{22})} \right|^2 * \frac{8(1-D)^2 R_L}{\pi^2} \end{aligned} \quad (8)$$

Equations (6)、(7) and (8) show that R_e will change according to the change of the duty cycle D, which affects system efficiency η_{trans} and power p_{trans}. Therefore, when the WPT system is designed, the final output voltage is determined, an appropriate duty cycle D can be adjusted to improve the system efficiency, while satisfying output power requirement.

III. SIMULATIONS AND EXPERIMENTS

The parameters of the WPT system and the required performance indexes of the load are shown in Table I.

Table I. The WPT system parameters

System parameters	Values
Operating Frequency f/kHz	85
DC Voltage of Inverter U_d/V	400
Self-inductance of Primary Coil L_1/µH	232.19
Internal Resistance of Primary Coil R_1/Ω	0.207
Self-inductance of Secondary Coil L_2/µH	211
Internal Resistance of Secondary Coil R_2/Ω	0.171
Mutual Inductance M/µH	29.73
Rated Power P_{OUT}/W	1000
Load Voltage Required V_{OUT}/V	320
Duty Cycle	0.35

978-1-7281-4181-7/19 $31.00 © 2019 IEEE

To verify performance of the WPT system with adding boost converter, two groups of parameters of the compensation networks are designed as shown in Table II.

Table II. Parameters of compensation networks

Resonance Compensation Networks	No Boost Converter	With Boost Converter
C_{p1}/nF	17.5	19.5
C_{p2}/nF	104.8	63.87
C_{s1}/nF	14.9	15.8
L_{11}/μH	33.4	54.86

Based on the system prototype parameters, simulations are conducted to analyze the system output power and transmission efficiency when the system outputs the same voltage 320V. Table III shows the efficiency of the WPT system before and after adding the boost converter.

Table III. Comparison of parameters before and after system optimization

Characteristics of System Parameters	No Boost Converter	With Boost Converter	Rate of Change
Efficiency of the WPT System	90.2%	93.76%	Rise 3.56%

Table III shows that adding boost converter can improve the efficiency of the WPT system. In order to further verify the constant voltage output characteristics of the WPT system, an experimental prototype is built as shown in Fig. 3.

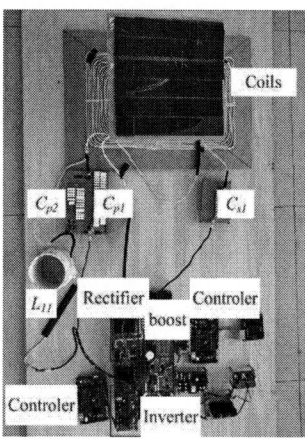

Fig. 3. Photograph of the developed WPT system prototype

In the experimental prototype, the air gap between coils is 18cm. The transmitting coil is constructed by 20 turns of 4mm diameter Litz wire, with a winding width of 60mm. Its size is 500mm × 300mm. The receiving coil is constructed by 20 turns of 4mm diameter Litz wire, with a winding width of 40mm. Its size is 300mm × 300mm. SiC MOSFET C3M0065090D is selected as the switching device of the single-phase full-bridge inverter. SiC Schottky diode C3D16060D is selected as the switching device of single-phase uncontrolled rectifier bridge.

Considering the disturbance of misalignment in practical applications, closed-loop control with PI controller is used for boost converter. Setting the electronic load to work on constant resistance mode and changing different resistance values, experimental results of output voltage efficiency (from DC power supply to load side) are shown in Fig. 4.

Where, Urms3, Irms3 and P3 are output voltage, output current and output power, respectively; Urms4, Irms4 and P4 represent input voltage, input current and input power, respectively; $\eta1$ means system efficiency, the channel of $\eta2$ does not work.

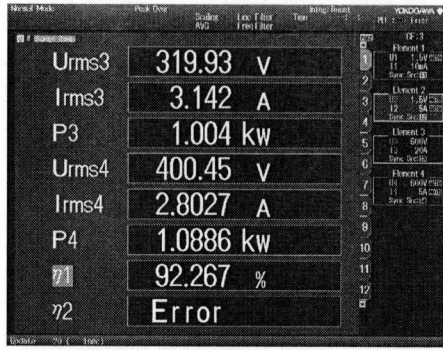

(a)

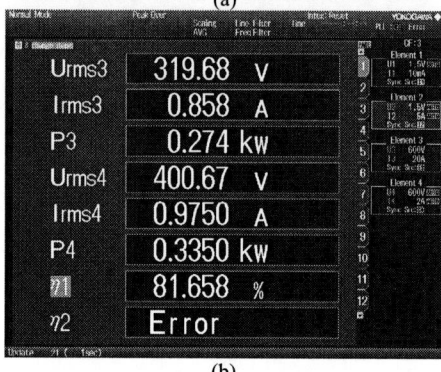

(b)

Fig. 4. Experimental results of system output power and efficiency on the conditions of different load resistances. (a) rated load of100Ω (b) light load of 350Ω

Fig. 4(a) shows that system efficiency achieves 92.267% and output voltage is 319.93V when the system works on the condition of rated load. Fig. 4(b) indicates that system efficiency achieves 81.658% and output voltage is 319.68V when the system works on the condition of light load. These results have proved that the designed WPT system with boost converter can work well on the conditions of both rated and light loads.

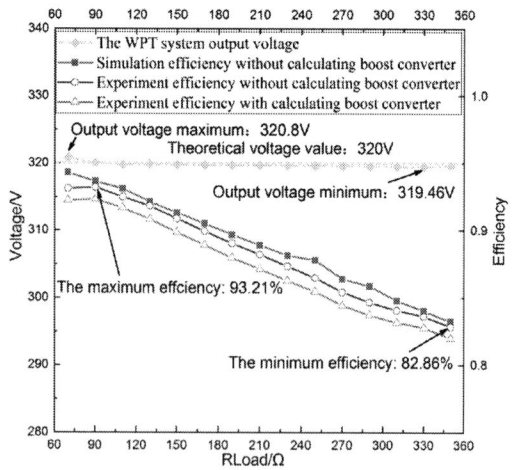

Fig. 5. Output voltage and efficiency of the system, changing with

different load resistance values.

In order to further verify characteristics of the overall system, Fig. 5 shows that output voltage of the whole WPT system ranges from 320.8V to 319.46V with a change rate of 0.41%. Output voltage of the WPT system can keep constant on the condition of wide load range. Maximum efficiency of the WPT system without calculating boost converter reaches 93.21%, and the maximum efficiency of the whole system with calculating boost converter is 92.267%. The results indicates that adding boost converter can improve efficiency of the WPT system.

IV. CONCLUSION

In this paper, the WPT system model with boost converter on the secondary side is established. Based on the model, the principle of realizing constant voltage output and boost converter's impact on the WPT system are analyzed. Efficiency of the WPT system is raised from 90.2% to 93.76%. The efficiency is obviously improved. Experiments show that output voltage of the system ranges from 320.8V to 319.46V, the change rate is 0.41%, and the maximum efficiency of the overall system is 92.32%. These works will be helpful for the characteristics optimization of WPT system with a DC-DC converter on the secondary side.

REFERENCES

[1] S. Li and C. C. Mi, "Wireless Power Transfer for Electric Vehicle Applications," *IEEE J. Emerg. Sel. Topics Power Electron.*, vol. 3, no. 1, pp. 4-17, Mar. 2015

[2] Z. Zhang, H. L. Pang, A. Georgiadis and Cecati. C, "Wireless Power Transfer-An Overview," *IEEE* Trans. *Ind. Electron.*, vol. 66, no. 2, pp. 1044-1058, Feb. 2019.

[3] X. Qu, H. Han, S. Wong, C. K. Tse, and W. Chen, "Hybrid IPT Topologies With Constant Current or Constant Voltage Output for Battery Charging *Applications*," *IEEE Trans. Power Electron.*, vol. 30, no. 11, pp. 6329-6337, Nov. 2015.

[4] X. H. Qu, Y. Y. Jing, H. D. Han, S. C. Wong, and C. K. Tse, "Higher Order Compensation for Inductive-Power-Transfer Converters With Constant-Voltage or Constant-Current Output Combating Transformer Parameter Constraints," *IEEE Trans. Power Electron.*, vol. 32, no. 1, pp. 394-405, Jan. 2017.

[5] Z. Li, G. Wei, S. Dong and C. Zhu, "Constant current/voltage charging for the inductor–capacitor–inductor-series compensated wireless power transfer systems using primary-side electrical information," *IET Power Electron.*, vol. 11, no. 14, pp. 2302-2310, Nov. 2018

[6] K. Song, Z. Li, J. Jiang and C. Zhu, "Constant Current/Voltage Charging Operation for Series–Series and Series–Parallel Compensated Wireless *Power* Transfer Systems Employing Primary-Side Controller," *IEEE Trans. Power Electron.*, vol. 33, no. 9, pp. 8065-8080, Sept. 2018.

[7] V. Vu, D. Tran and W. Choi, "Implementation of the Constant Current and Constant Voltage Charge of Inductive Power Transfer Systems With the Double-Sided LCC Compensation Topology for Electric Vehicle Battery Charge Applications," *IEEE Trans. Power Electron.*, vol. 33, no. 9, pp. 7398-7410, Sept. 2018.

[8] Y. Wang, Y. Yao, X. Liu, D. Xu and L. Cai, "An LC/S Compensation Topology and Coil Design Technique for Wireless Power Transfer," *IEEE Trans. Power Electron.*, vol. 33, no. 3, pp. 2007-2025, Mar. 2018.

[9] Y. Guo, L. Wang, Y. Zhang, S. Li and C. Liao, "Rectifier Load Analysis for *Electric* Vehicle Wireless Charging System," *IEEE Trans. Ind. Electron.*, vol. 65, no. 9, pp. 6970-6982, Sept. 2018.

Finite Control Set Model Based Predictive Control of Grid-Tied Six-Switch Converter Applied to Induction Generator

Paulo R. U. Guazzelli
School of Engineering of São Carlos
University of São Paulo
São Carlos, SP, Brazil
paulo.ubaldo@usp.br

Allan G. de Castro
School of Engineering of São Carlos
University of São Paulo
São Carlos, SP, Brazil
allangregori@usp.br

Stefan T. C. A. dos Santos
School of Engineering of São Carlos
University of São Paulo
São Carlos, SP, Brazil
stefan.santos@usp.br

Carlos M. R. de Oliveira
School of Engineering of São Carlos
University of São Paulo
São Carlos, SP, Brazil
carlosmro@usp.br

José R. B. A. Monteiro
School of Engineering of São Carlos
University of São Paulo
São Carlos, SP, Brazil
jrm@sc.usp.br

Manoel L. de Aguiar
School of Engineering of São Carlos
University of São Paulo
São Carlos, SP, Brazil
aguiar@sc.usp.br

Abstract—**Finite Control Set Model Based Predictive Control (FCS-MPC) established itself as an effective technique for the control of power electronic converters, with fast dynamics and the absence of voltage modulators. However, there are still converters in the literature which did not exploit the advantages of FCS-MPC, as the Six-Switch Converter (SSC), a multiport converter which can replace the back-to-back converter in power generation with 50% of the switch count. As a result, this paper investigates the use of FCS-MPC in the SSC for grid-tied power generation with the squirrel-cage induction generator. Two Predictive Current Control (PCC) approaches are presented: a Concentrated PCC (PCC-C), in which a centralized cost function is employed for control decision evaluation; and a Decoupled PCC (PCC-D), with independent cost functions for the grid and generator control. Simulation results show PCC-D provided better torque ripple factor and better grid current distortion, when compared to the concentrated approach. Also, PCC-D eliminated the diverging bus capacitor voltage offset of PCC-C, and the superior performance was obtained with a lower computational cost.**

Index Terms—**six-switch converter, predictive current control, squirrel cage induction generator**

I. INTRODUCTION

The performance requirements for control systems in power electronics become more and more critical. Fast dynamics and easy incorporation of restrictions without voltage modulators are advantages which encouraged the investigation of Finite Control Set Model Based Predictive Control (FCS-MPC) in power electronics [1]. FCS-MPC deals with the minimization of a control cost function within a limited universe of voltage vectors, according to the power converter of the system. A model-based prediction feeds the cost function evaluation, and

This work is supported in part by Conselho Nacional de Desenvolvimento Científico e Tecnológico (CNPq) and in part by Coordenação de Aperfeiçoamento de Pessoal de Nível Superior (CAPES) - Finance Code 001.

the converter switch state that leads to the lower cost is applied in the system [2].

FCS-MPC application to power converters is wide in range and type. A voltage and current control for an active filter was developed in [3] for DC bus voltage balance and reactive power compensation. Authors in [4] developed an FCS-MPC current control for HVDC multilevel converters, and two independent FCS-MPC were applied in [5] to the back-to-back converter for power generation systems. A predictive torque control was developed for the SCIG, and a Predictive Current Control (PCC) for the grid connection.

One important front for the development of FCS-MPC is the improvement of the steady state performance [1]. Some efforts for overcoming this issue are the adjustment of the weighting factors, the addition of new control objectives and the feasible implementation of longer prediction horizons [6], [7].

Advances are proposed not only in the controllers, but also in the power converters. In fact, the reduction on the switch count of converters has also been a topic of interest for power electronics, aiming cost and size reduction [8]. The B4 inverter was proposed in [9] as an alternative for the six-switch inverter. The nine-switch converter reduced 25% of the switches of the back-to-back converter, for AC/AC conversion [10]. In this segment, the Six-Switch Converter (SSC) [11] goes further and uses only half of the back-to-back 12 switches, as showed in Fig. 1 [12].

The fewer switches impose advantages and difficulties for MPC implementation. Fewer switches imply on the number of voltage vectors available for enumeration. If there are less voltage vectors, a lower computational burden is achieved but also there is a limited universe of possibilities at each time step. Authors in [13] proposed an FCS-MPC for the B4 inverter with four voltage vectors, and the voltage offset

978-1-7281-4181-7/19 $31.00 © 2019 IEEE

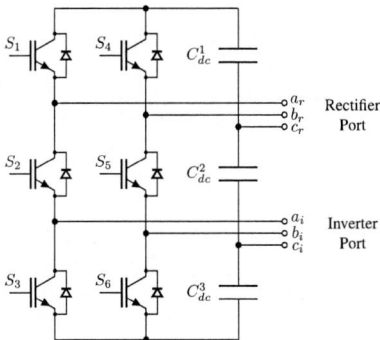

Fig. 1. Six-switch AC/AC converter.

problem was solved by an additional term in the cost function. The opposite happen in [14], where the authors proposed an FCS-MPC for the nine-switch converter connected to the grid with 27 vectors, due to the centralized cost function for two once separated ports.

The separation of MPC into smaller subsystems has been pointed as a way for improvement of MPC [15]. In fact, [16] showed the PCC applied to a nine-switch converter benefits from the decoupling of the generator and grid cost functions.

The SSC also can benefit from FCS-MPC. Based on above discussions, this paper proposes two PCC structures for the SSC applied for power generation: 1) a conventional PCC with a single concentrated cost function, named PCC-C, and 2) a decoupled PCC, named PCC-D, where the cost functions for the generator and for the grid control remain separates. One PCC loop controls the currents of the SCIG, for torque and flux control, while the other one controls the grid currents, for DC bus voltage control. Both approaches control the desired objectives, however PCC-C presents a voltage offset which is solved by PCC-D. The correction of the voltage offset between the upper and lower capacitors helps the improvement of torque ripple and DC voltage ripple.

II. SYSTEM MODELLING

A. Six-Switch Converter

The SSC of Fig. 1 has only two active legs. In each one, two and only two switches must be turned on. As a result, there are three combinations per leg, and nine different combinations in total. These combinations are listed in Table I. Each one applies a different voltage vector in each port. Let $\vec{v}_1$ to $\vec{v}_4$ be the vectors of the rectifier side, and $\vec{v}_5$ to $\vec{v}_8$ be the vector of the inverter side. These vectors are represented in an $\alpha\beta$ reference frame according to Fig. 2, for the rectifier port and the inverter port. As can be seen, there are no null vectors in the SSC. When the two top switches are turned on, there is still a minimal vector $\vec{v}_1$ applied to the rectifier port, and when both bottom switches are tuned on there is a minimal vector $\vec{v}_5$ applied to the inverter port of the SSC.

B. MPC Control

Figure 3 shows the control system applied to the SSC. The SCIG and the grid connect to the rectifier and inverter to ports

TABLE I
SSC SWITCHING STATES AND THE RESPECTIVE VOLTAGE VECTORS
APPLIED TO EACH SIDE.

Switching States	Applied Vectors	
$[S_1S_2S_3S_4S_5S_6]$	Rectifier Side	Inverter Side
[110110]	$\vec{v}_1$	$\vec{v}_7$
[110011]	$\vec{v}_4$	$\vec{v}_6$
[110101]	$\vec{v}_1$	$\vec{v}_6$
[011110]	$\vec{v}_2$	$\vec{v}_8$
[011011]	$\vec{v}_3$	$\vec{v}_5$
[011101]	$\vec{v}_2$	$\vec{v}_5$
[101110]	$\vec{v}_1$	$\vec{v}_8$
[101011]	$\vec{v}_4$	$\vec{v}_5$
[101101]	$\vec{v}_1$	$\vec{v}_5$

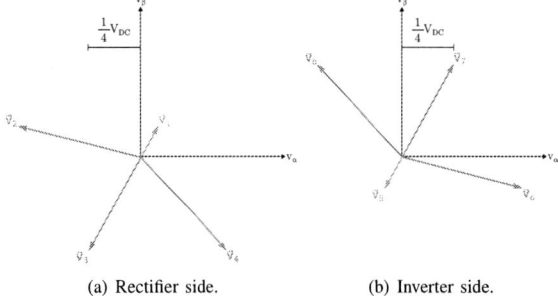

(a) Rectifier side.   (b) Inverter side.

Fig. 2. Voltage vectors in $\alpha\beta$ reference frame, for the rectifier (a) and inverter (b) sides of the SSC.

of the SSC, respectively. Measurements of SCIG currents and speed are used for estimation of rotor flux and the conversion of the currents to the dq reference frame. A PI flux controller determines the reference current i_d, and a PI speed controller determines the torque reference of the SCIG, converted to a i_q reference. On the grid side, a dq control is oriented on the voltage grid angle, and a PI controller for the DC bus voltage determines the i_d reference current, while the reference of i_q is set to zero. An L filter connects the SSC to the grid.

The basic algorithm of a PCC is depicted in the pseudo code of Algorithm 1, for one step of execution. After the measurements and estimations, PCC is effectively evaluated by the iterative process. For each possible voltage vector, the system predicts the SCIG and grid currents, and these predictions are used with their respective references to calculate the associated cost of each voltage vector. After the extensive evaluation, the system chooses the voltage vector with the lowest cost to be applied to the SSC.

Algorithm 1 Pseudo code for one step of control.

Measurements of system variables
Estimation of system variables
for each voltage vector of SSC **do**
    Current prediction for each voltage vector
    Cost function calculation for each voltage vector
    Update of the optimal voltage vector
end for
return Optimal voltage vector

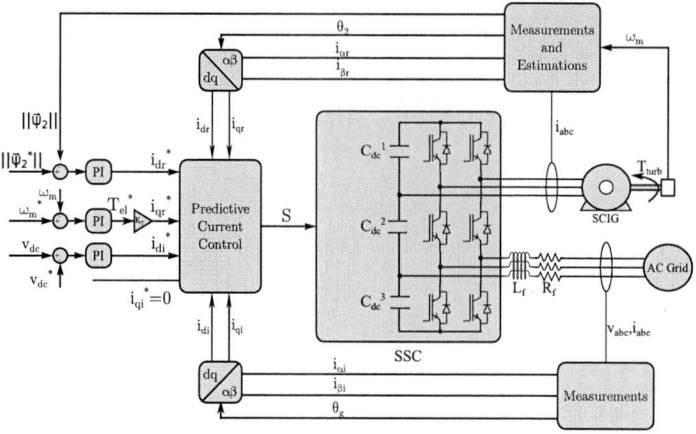

Fig. 3. Diagram of the SSC connected to the SCIG and to the grid.

III. PREDICTIVE CURRENT CONTROL

A. Grid-Side Control

The grid currents i_{dqi} are predicted in time step $k+1$ according to (1),

$$\hat{i}_{dqi}^{k+1} = i_{dqi}^k + \frac{t_D}{L_f}\left(v_{dqj} - v_{dqg}^k - R_f \cdot i_{dqi}^k\right) \qquad (1)$$

where k is the current time step, t_D is the discretization time, and L_f and R_f are the filter inductance and resistance, respectively. The grid voltage is given by v_{dqg} and the SSC voltage vectors are given by v_{dqj}. The prediction is executed for each j voltage vector.

From the predicted currents and the reference currents, the cost of each voltage vector is calculated according (2), which uses the squared L2-norm [17] and I_I is the dimensionless overcurrent penalty factor [18]:

$$g_{inv_j} = \left(i_{di}^* - \hat{i}_{di_j}^{k+1}\right)^2 + \left(i_{qi}^* - \hat{i}_{qi_j}^{k+1}\right)^2 + I_I. \qquad (2)$$

B. Generator-Side Control

The prediction of SCIG currents i_{dqr} is obtained from the machine model [19], as below:

$$\hat{i}_{dqr}^{k+1} = \hat{i}_{dqr}^k + t_D\left(\frac{L_H}{\sigma L_1 L_2}\left(\frac{R_2}{L_2} - jp\omega_m^k\right)\|\hat{\vec{\psi}}_2^k\| + \frac{v_{dqj}}{\sigma L_1} + \right.$$
$$\left. -\left(\frac{R_2 L_H^2}{\sigma L_1 L_2^2} + \frac{R_1}{\sigma L_1} + j\omega_2^k\right)\hat{i}_{dqr}^k\right) \qquad (3)$$

where the stator and rotor resistances are given by R_1 and R_2, the stator and rotor inductances are given by L_1 and L_2, L_H denotes the mutual inductance and the number of pole pairs is given by p. Also, $\sigma := 1 - \frac{L_H^2}{L_1 L_2}$. The flux rotor $\vec{\psi}_2$ was estimated according to the observer from [20].

The cost function evaluation is similar to the inverter one, and the i_q is a function of the torque reference:

$$g_{rect_j} = \left(i_{dr}^* - \hat{i}_{dr_j}^{k+1}\right)^2 + \left(i_{qr}^* - \hat{i}_{qr_j}^{k+1}\right)^2 + I_R. \qquad (4)$$

$$i_{qr}^* = K_T \cdot T_{el}^* = \frac{2L_2}{3pL_H\|\hat{\vec{\psi}}_2\|} \cdot T_{el}^* \qquad (5)$$

C. Control Implementation

This paper considers two approaches for PCC implementation. The first one is the conventional PCC, named PCC-C, depicted in Fig. 4 (a). The cost functions of (2) and (4) are added in a single cost function, which determines the combination of switches to be applied to the SSC, according Table I. Therefore, the enumeration of PCC-C extends to a set of eight voltage vectors.

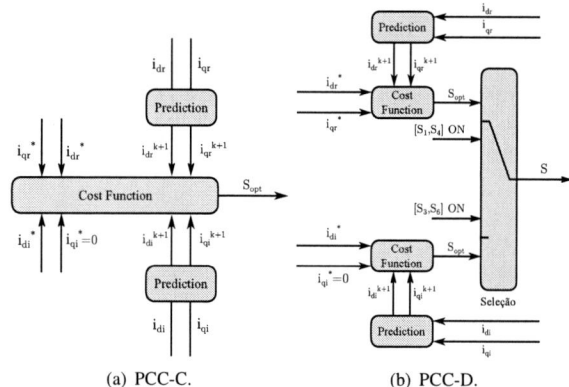

(a) PCC-C.   (b) PCC-D.

Fig. 4. Diagram of PCC-C (a) and PCC-D (b) for the SSC.

The second approach is the decoupled PCC, named PCC-D, as can be seen in Fig. 4 (b). Instead of combining the two cost functions, PCC-D executes the controls loops alternately. In one control period, switches S_1 and S_4 (the two top one of the SSC) are turned on, and the grid side predictive control loop determines the optimal vector to be applied at the grid port. In the following time step, the predictive SCIG control is evaluated, and it determines the optimal voltage vector to be applied at the rectifier port, while switches S_3 and S_6 (the two bottom one of the SSC) are turned on.

978-1-7281-4181-7/19 $31.00 © 2019 IEEE

This approach leads to the scheme of Fig. 5. At each time instant, only one of the controls is being evaluated. On the other port, there is the application of a minimal vector, $\vec{v}_1$ for the rectifier port, and $\vec{v}_5$ for the inverter one. This changes the vector to be evaluated in each port. The effective voltage vector applied are the average between the vectors of Fig. 2 and the minimal vectors. Therefore, PCC-D calculates the cost of applying the vectors of Fig. 6. The average vectors are the same for both ports, and have the coordinates showed in Table II. It is also important to point that the evaluation set of PCC-D is limited to four voltage vectors and that only one of the controls is executed at each time step. Therefore, PCC-D exhibits a lower computational burden than PCC-C.

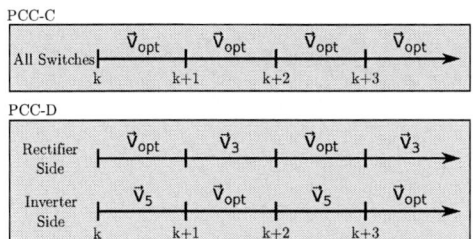

Fig. 5. Comparison between the discrete execution of PCC-C and PCC-D, for the SSC.

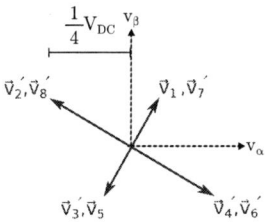

Fig. 6. Average voltage vectors applied by SSC, in $\alpha\beta$ reference frame.

TABLE II
$\alpha\beta$ COORDINATES OF THE AVERAGE VOLTAGE VECTORS APPLIED BY SSC.

Vector	v_α	v_β	Amplitude
$\vec{v}_1', \vec{v}_7'$	$\frac{1}{12}V_{DC}$	$\frac{\sqrt{3}}{12}V_{DC}$	$\frac{1}{6}V_{DC}$
$\vec{v}_2', \vec{v}_8'$	$-\frac{1}{4}V_{DC}$	$\frac{\sqrt{3}}{12}V_{DC}$	$\frac{\sqrt{3}}{6}V_{DC}$
$\vec{v}_3', \vec{v}_5'$	$-\frac{1}{12}V_{DC}$	$-\frac{\sqrt{3}}{12}V_{DC}$	$\frac{1}{6}V_{DC}$
$\vec{v}_4', \vec{v}_6'$	$\frac{1}{4}V_{DC}$	$-\frac{\sqrt{3}}{12}V_{DC}$	$\frac{\sqrt{3}}{6}V_{DC}$

IV. RESULTS AND DISCUSSIONS

Both PCC strategies were simulated in MATLAB®, from MathWorks®, with a current sampling frequency of 40 Hz. The control frequency of PCC-D is 20 kHz, which is half of the one from PCC-C, 40 kHz. The simulations utilized a 220/380 V 1 hp SCIG and a 142/246 V grid with L filter whose data appear in Table III. The PI controller gains were set to: $K_P = 1$ and $K_I = 6$ for the speed control; $K_P = 0.7$

TABLE III
SCIG AND GRID PARAMETERS USED IN SIMULATIONS.

Parameter	Value	Unit
Nominal torque	2	Nm
Nominal speed	3420	rpm
Nominal flux	0.7	V·s
R_1	7.5022	Ω
R_2	4.8319	Ω
L_1 and L_2	718.5	mH
L_m	694.1	mH
p	1	—
J	0.0517	kg·m^2
K_D	0.001	N·m·s
$C_{dc}^1/C_{dc}^2/C_{dc}^3$	16000/8000/16000	μF
L_f	80	mH
R_f	0.05	Ω
Line-to-line grid voltage	246	V
Grid frequency	60	Hz

and $K_I = 1$ for the voltage control; and $K_P = 2$ and $K_I = 14$ for the flux control.

The steady state at 3500 rpm was simulated with a DC voltage reference of 2156 V, for 2 Nm of turbine torque. The graphs of the SCIG speed, torque, dq currents and phase a current can be seen in Figs. 7 (a) and (b), for the PCC-C and PCC-D, respectively. Both strategies feature i_d and i_q current ripples, leading to the SCIG torque ripple.

Figs. 7 (c) and (d) contain the graphs of DC bus voltage, dq currents and phase a current for the PCC-C and PCC-D, respectively, for the same simulated condition. PCC-D was able to improve the quality of the grid current, compared to PCC-C. This can be specially seen in the controlled dq grid current, where the ripple in PCC-D is visually lower than the one of PCC-C. This can be explained by the two independent cost functions. All current errors are in Amperes, however the operating points of SCIG and grid ports are different. The separation of the cost function overcome this matter.

The difference between the grid i_d performance on both PCCs is explained by the graphs of Fig. 7, which depict the DC voltages on each capacitor of SSC for PCC-C (a) and PCC-D (b), respectively. It shows how even though PCC-C controlled the total DC bus voltage to 2155 V, it was unable to meet each voltage on the desired levels of 1/4, 1/2 and 1/4 of the total DC bus voltage, instead presenting an increasing offset on the DC bus. The PCC-D, however, indeed maintained the capacitor voltages in their desired levels, according to the 1:2:1 ratio.

The controls were simulated for a range of SCIG speeds, from 2000 to 4000 rpm. The different steady states produced the graphs depicted in Figs. 9 (a) and (b), for SCIG torque and DC bus voltage ripple factors, respectively. The results show how PCC-D even reduced both ripples for the whole range of speeds, with a lower computational burden.

V. CONCLUSIONS

The MPC strategy for current control was investigated for use in the SSC, a reduced switch count converter. The voltage vectors of the SSC were mapped, and two approaches for MPC were implemented: a concentrated and a decoupled predictive

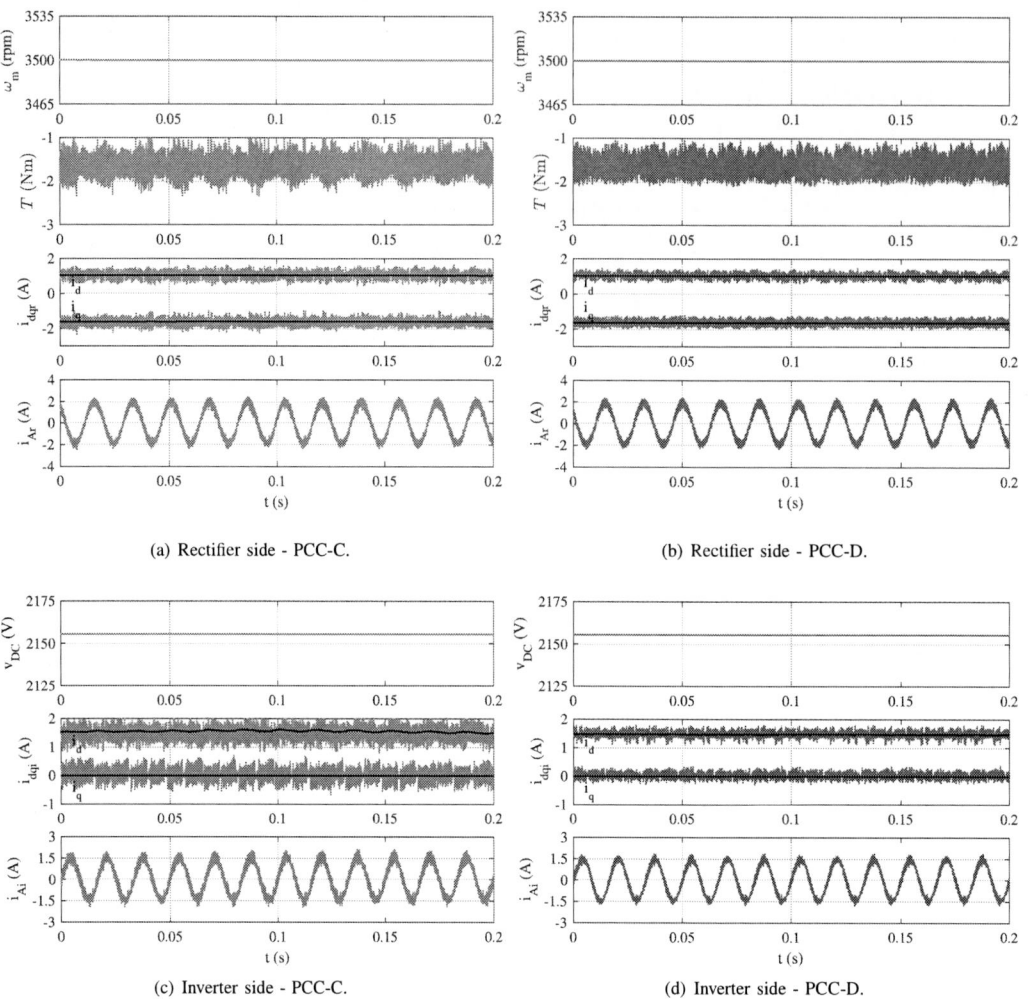

(a) Rectifier side - PCC-C.

(b) Rectifier side - PCC-D.

(c) Inverter side - PCC-C.

(d) Inverter side - PCC-D.

Fig. 7. Steady state simulation at 3500 rpm, under 2 Nm turbine torque - SCIG speed, torque and currents on the rectifier side for the PCC-C (a) and PCC-D (b); grid currents, voltages and DC bus voltage for the PCC-C (c) and PCC-D (d) on the inverter side.

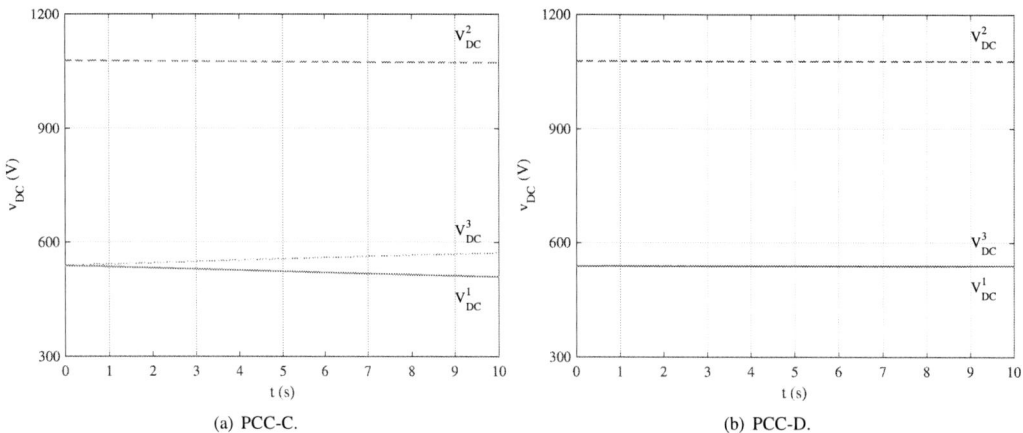

(a) PCC-C.

(b) PCC-D.

Fig. 8. DC bus voltages on each capacitor for the PCC-C (a) and PCC-D (b) after 10 s of simulation.

978-1-7281-4181-7/19 $31.00 © 2019 IEEE

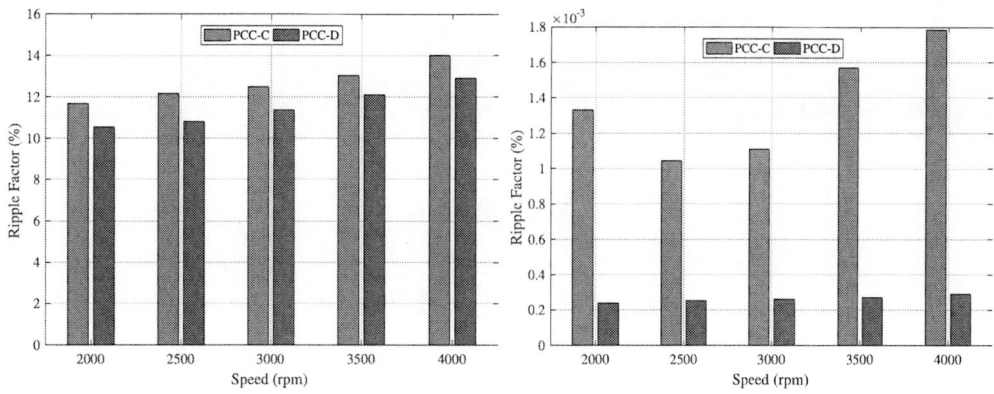

(a) Torque ripple factor.   (b) DC bus voltage ripple factor.

Fig. 9. Torque (a) and DC bus voltage (b) ripple factors for a range of speeds, under 2 Nm turbine torque, for the PCC-C and PCC-D.

current control, for a grid-tied generation system with the SCIG. The generator PCC loop controls the SCIG currents for torque and flux control, and the grid PCC loop controls the grid currents for DC bus voltage of control.

The results showed the decoupled strategy provided better steady state performance in the SCIG torque and specially on the DC bus voltage in each capacitor of the SSC. Also, PCC-D reduced the number of enumeration at each time step as well the size of the control, decreasing the computational burden of PCC for the SSC. Therefore, the advantages brought by the decoupling improves the interest of applying MPC in the control of not only SSC but also other multiport converters in power electronics.

REFERENCES

[1] S. Vazquez, J. Rodriguez, M. Rivera, L. G. Franquelo, and M. No-rambuena, "Model Predictive Control for Power Converters and Drives: Advances and Trends," *IEEE Trans. Ind. Electron.*, vol. 64, no. 2, pp. 935–947, feb 2017.

[2] S. Kouro, P. Cortes, R. Vargas, U. Ammann, and J. Rodriguez, "Model Predictive Control - A Simple and Powerful Method to Control Power Converters," *IEEE Trans. Ind. Electron.*, vol. 56, no. 6, pp. 1826–1838, jun 2009.

[3] P. Acuna, L. Moran, M. Rivera, R. Aguilera, R. Burgos, and V. G. Age-lidis, "A Single-Objective Predictive Control Method for a Multivariable Single-Phase Three-Level NPC Converter-Based Active Power Filter," *IEEE Trans. Ind. Electron.*, vol. 62, no. 7, pp. 4598–4607, jul 2015.

[4] J.-W. Moon, J.-S. Gwon, J.-W. Park, D.-W. Kang, and J.-M. Kim, "Model Predictive Control With a Reduced Number of Considered States in a Modular Multilevel Converter for HVDC System," *IEEE Trans. Power Deliv.*, vol. 30, no. 2, pp. 608–617, apr 2015.

[5] A. S. Lunardi, A. J. Sguarezi, and Filho, "Experimental results for predictive direct torque control for a squirrel cage induction motor," in *2017 Brazil. Power Electron. Conf. (COBEP)*, Nov 2017, pp. 1–5.

[6] J. Rodriguez, M. P. Kazmierkowski, J. R. Espinoza, P. Zanchetta, H. Abu-Rub, H. A. Young, and C. A. Rojas, "State of the Art of Finite Control Set Model Predictive Control in Power Electronics," *IEEE Trans. Ind. Informat.*, vol. 9, no. 2, pp. 1003–1016, may 2013.

[7] F. Wang, X. Mei, J. Rodriguez, and R. Kennel, "Model predictive control for electrical drive systems - an overview," *CES Trans. Electr. Mach. Syst.*, vol. 1, no. 3, pp. 219–230, 2017.

[8] C. Brandao Jacobina, M. de Rossiter Correa, A. Nogueira Lima, and E. Cabral da Silva, "AC motor drive systems with a reduced-switch-count converter," *IEEE Trans. Ind. Appl.*, vol. 39, no. 5, pp. 1333–1342, sep 2003.

[9] H. W. Van Der Broeck and J. D. Van Wyk, "A comparative investigation of a three-phase induction machine drive with a component minimized voltage-fed inverter under different control options," *IEEE Trans. Ind. Appl.*, vol. IA-20, no. 2, pp. 309–320, March 1984.

[10] C. Liu, B. Wu, N. Zargari, D. Xu, and J. Wang, "A Novel Three-Phase Three-Leg AC/AC Converter Using Nine IGBTs," *IEEE Trans. Power. Electron.*, vol. 24, no. 5, pp. 1151–1160, may 2009.

[11] M. Heydari, A. Y. Varjani, M. Mohamadian, and H. Zahedi, "A novel variable-speed wind energy system using permanent-magnet syn-chronous generator and nine-switch ac/ac converter," in *2011 2nd Power Electron., Drive Syst. Technol. Conf.*, Feb 2011, pp. 5–9.

[12] M. Heydari, A. Fatemi, and A. Y. Varjani, "A reduced switch count three-phase ac/ac converter with six power switches: Modeling, analysis, and control," *IEEE J. Emerg. Sel. Topics Power Electron.*, vol. 5, no. 4, pp. 1720–1738, Dec 2017.

[13] D. Zhou, J. Zhao, and Y. Liu, "Predictive torque control scheme for three-phase four-switch inverter-fed induction motor drives with dc-link voltages offset suppression," *IEEE Trans. Power. Electron.*, vol. 30, no. 6, pp. 3309–3318, June 2015.

[14] S. S. Lee, Y. E. Heng, and M. A. Roslan, "Finite control set model predictive control of nine-switch AC/DC/AC converter," in *2016 IEEE Int. Conf. Power and Energy (PECon)*, no. 3. IEEE, nov 2016, pp. 746–751.

[15] M. G. Forbes, R. S. Patwardhan, H. Hamadah, and R. B. Gopaluni, "Model Predictive Control in Industry: Challenges and Opportunities," *IFAC-PapersOnLine*, vol. 48, no. 8, pp. 531–538, 2015, 9th IFAC Symp. Adv. Control of Chem. Proc. ADCHEM 2015.

[16] P. R. U. Guazzelli, A. G. de Castro, S. T. C. A. dos Santos, C. M. R. Oliveira, W. C. A. Pereira, J. R. B. A. Monteiro, and M. L. de Aguiar, "Dual predictive current control of grid connected nine-switch converter applied to induction generator," in *2018 13th IEEE Int. Conf. Ind. Appl. (INDUSCON)*, Nov 2018, pp. 1038–1044.

[17] P. Karamanakos, T. Geyer, and R. Kennel, "On the Choice of Norm in Finite Control Set Model Predictive Control," *IEEE Trans. Power. Electron.*, vol. 8993, no. 1, pp. 1–1, 2017.

[18] F. Wang, S. Li, X. Mei, W. Xie, J. Rodriguez, and R. M. Kennel, "Model-Based Predictive Direct Control Strategies for Electrical Drives: An Experimental Evaluation of PTC and PCC Methods," *IEEE Trans. Ind. Informat.*, vol. 11, no. 3, pp. 671–681, jun 2015.

[19] P. Vas, *Vector control of AC machines*, ser. Monographs in electrical and electronic engineering. Clarendon Press, 1990.

[20] Z. Yongchang and Z. Zhengming, "Speed sensorless control for three-level inverter-fed induction motors using an Extended Luenberger Ob-server," in *2008 IEEE Veh. Power Prop. Conf.* IEEE, sep 2008, pp. 1–5.

Applying Coupled Inductors to the Clamped-Resonant Interleaved Boost Converter

Giorgio Spiazzi

Dept. of Information Engineering (DEI) – University of Padova – Padova, ITALY
e-mail: giorgio.spiazzi@dei.unipd.it

Abstract—**This paper investigates the possibility to use coupled inductors in the DC-DC converter topology called Clamped-Resonant Interleaved Boost converter (CRIB). Such topology employs a resonant L-C tank connected between the switch drain terminals to achieve zero-voltage turn-on and turn-off of both switches, as well as zero-current turn-off for the rectifier diodes, independently of the load current. Moreover, no-load operation is demonstrated providing the voltage gain is higher than a minimum value. The converter analysis and proposed design criteria were verified by an experimental prototype rated at $42-54\,\mathrm{V}$ to $400\,\mathrm{V} - 300\,\mathrm{W}$, showing a very good efficiency in the whole input voltage range. Performance comparison with the same topology adopting separated input inductors is included as well.**

Index Terms—**Boost converter, soft-switching, interleaving, coupled inductors**

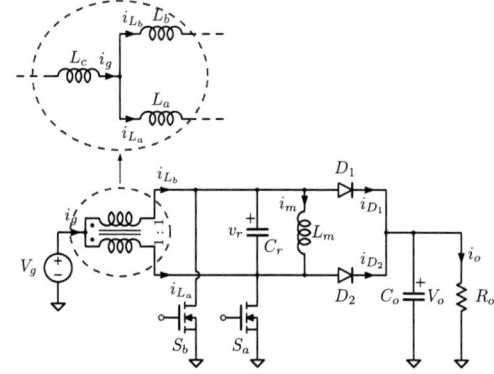

Fig. 1. Clamped-Resonant Interleaved Boost topology with coupled inductors.

I. INTRODUCTION

Interleaving arrangements of basic switching cells is an effective way to increase the output power of DC-DC converters while reducing devices current rating and the total input current ripple. This is especially advantageous in applications where a low-voltage high-current source, like photovoltaic panels and fuel-cells, has to be interfaced with a high voltage DC bus or load. The basic interleaved boost converter is the simplest topology that can be employed for such applications, but its performance are limited by the hard-switching characteristic, which hinders the increase of the switching frequency and is a source of electromagnetic noise.

Passive auxiliary snubbers are widely employed in literature to achieve soft switching, like in [1], and [2], where coupled inductors are used to achieve higher voltage gains with ZCS switches turn-on. Alternatively, soft-switching cells using auxiliary switches can be used, like in [3]–[10]. In some of these examples, like in [4], [9], and [10], the auxiliary switch drain-source voltage is not clamped, thus giving space for high frequency ringing and possible overvoltage across it.

The solution proposed in [11], explores the transformer magnetizing inductance as resonant inductor connected between the switch drain nodes to achieve ZVS at both turn on and turn off through a proper resonance with the switch drain-source capacitances (an external capacitor may be used as well). Despite being proposed for the push-pull topology, the same approach can also be used in an interleaved boost converter, as presented in [12]–[14]. In particular, [14] reports a detailed analysis of the converter working principles, in-

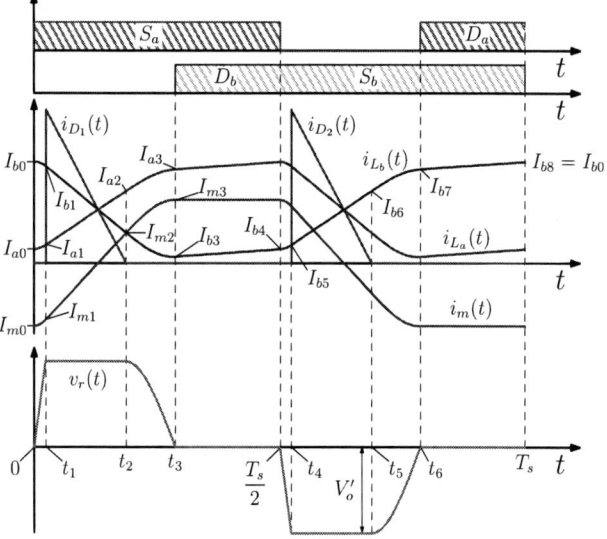

Fig. 2. Main waveforms in a switching period.

cluding suitable design guidelines. This paper proposes the application of coupled inductors instead of the two uncoupled input inductors of the original topology, thus saving one magnetic core, leading to a more compact implementation. The studied topology of the Clamped-Resonant Interleaved Boost converter (CRIB) with coupled inductors is shown in Fig. 1. The coupled-inductor equivalent model, shown in the same figure, adds an input inductor L_c to the original topology, thus modifying the inductor current waveforms. Consequently, the analysis needs to be modified compared to what has been presented in [14], and is reported in section II. Section III reviews some design considerations, including the no-load operation, leading to the modified design steps reported in section IV. Finally, experimental results, taken on a $42 - 54\,\text{V}$ to $400\,\text{V}$, $300\,\text{W}$ prototype, are included in section V showing the performance of the proposed topology that confirm the theoretical expectations.

II. Converter operation

Similarly to what has been done in [14], to account for the possibility of adding a voltage doubler rectifier, the considered scheme uses an equivalent output voltage $V_o' = \alpha V_o$, with the parameter α equal to 1, in the case of the basic topology of Fig. 1, or equal to 0.5, for the case with the voltage doubler rectifier. The circuit is assumed symmetrical, i.e. $L_a = L_b = L_\ell$. The switching period is divided into eight sub-intervals, described in Fig. 2. However, being the converter operation similar in each half-switching period, we limit the analysis to interval $0 \leq t \leq \frac{T_s}{2}$, that starts from the instant the switch S_b is turned off. During this first half switching period, the equivalent topology is as shown in Fig. 3a, which can be further simplified, by applying Thevenin, into the scheme of Fig. 3b, where $\gamma = \frac{L_\ell}{L_\ell + L_c}$ and $L = L_\ell + L_\ell \parallel L_c = L_\ell(2 - \gamma)$. Doing this, we can explore the steady-state analysis provided in [14], with only few adjustments, starting from the definition of the intrinsic voltage gain $M_i = \frac{V_o'}{V_g'} = \frac{V_o'}{\gamma V_g} = \frac{M}{\gamma}$.

The analysis proceeds in normalized form using the following base quantities and parameters ($u = v/V_N, j = i/I_N$):

$$V_N = V_o', \quad I_N = \frac{V_N}{Z_N}, \quad Z_N = Z_r\frac{1+\lambda}{\lambda}, \quad \lambda = \frac{L_m}{L}, \quad \theta = \omega_r t,$$

$$\tag{1}$$

where

$$Z_r = \sqrt{\frac{L_r}{C_r}}, \quad \omega_r = \frac{1}{\sqrt{L_r C_r}}, \quad L_r = \frac{1}{\frac{1}{L} + \frac{1}{L_m}}. \tag{2}$$

– *Interval* $0 \leq t \leq t_1$ (see Fig. 3c). When S_b turns off, a resonant tank is formed involving C_r, L, and L_m, with initial conditions $i_{L_b}(0) = I_{b0}$, $i_m(0) = I_{m0}$, $v_r(0) = 0$, (see Fig. 2). Currents $i_m(t)$, and $i_{L_b}(t)$, and the resonant capacitor voltage $v_r(t)$ in normalized form are, respectively:

$$j_m(\theta) = J_{m0} + \frac{J_{C0}}{1+\lambda}\left(1 - \cos(\theta)\right)$$
$$+ \frac{1}{M_i(1+\lambda)}\left(\theta - \sin(\theta)\right), \tag{3}$$

$$j_{L_b}(\theta) = J_{C0}\cos(\theta) + \frac{1}{M_i}\sin(\theta) + j_m(\theta), \tag{4}$$

$$u_r(\theta) = \frac{\lambda}{1+\lambda}\left[J_{C0}\sin(\theta) + \frac{1}{M_i}\left(1 - \cos(\theta)\right)\right], \tag{5}$$

where $J_{C0} = J_{b0} - J_{m0}$ is the normalized initial resonant capacitor current. For the steady-state analysis we must know

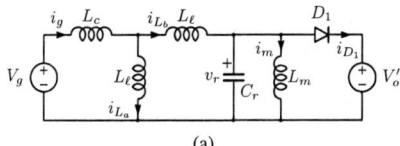

(a)

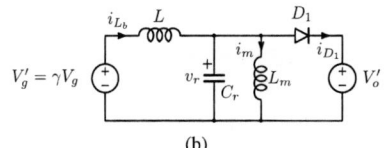

(b)

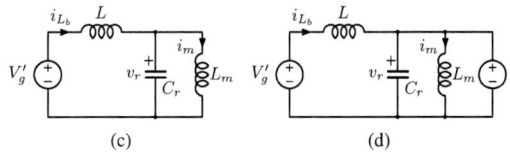

(c)   (d)

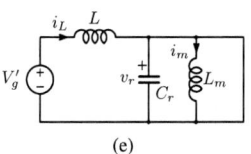

(e)

Fig. 3. Sub-topologies in a switching period: (a) equivalent topology in a switching half period; (b) simplified topology in a switching half period; (c) intervals $0 \leq t \leq t_1$ and $t_2 \leq t \leq t_3$; (d) interval $t_1 \leq t \leq t_2$; (e) intervals $t_3 \leq t \leq \frac{T_s}{2}$.

the evolution of current $i_{L_a}(t)$ (see Fig. 3a), which is the same of $i_{L_b}(t)$ in the second switching half period. Observing that $\frac{di_{L_a}(t)}{dt} = \frac{\gamma V_g + (1-\gamma)v_r(t)}{L}$, we can write:

$$j_a(\theta) = J_{a0} + (1 - \gamma)\frac{\lambda}{1+\lambda}\left[J_{C0}\left(1 - \cos(\theta)\right) + \frac{\theta}{M_i}\right.$$
$$\left. - \frac{1}{M_i}\sin(\theta)\right] + \frac{\theta}{M_i}. \tag{6}$$

When, at instant t_1, $v_r(t_1) = V_o'$ ($u_r(\theta_1) = 1$), D_1 turns on. In the converter steady-state analysis, this short interval will be approximated with a linear behavior, so that the interval duration can be estimated as:

$$\theta_{01} \approx \frac{1+\lambda}{\lambda}\frac{1}{J_{C0}}. \tag{7}$$

– *Interval* $t_1 \leq t \leq t_2$ (see Fig. 3d). During the clamping phase, currents i_m and i_{L_a} rise, while current i_{L_b} decreases, all in a linear manner, i.e.

$$j_m(\theta) = J_{m1} + \frac{1}{\lambda}(\theta - \theta_1), \tag{8}$$

$$j_{L_b}(\theta) = J_{b1} - \left(1 - \frac{1}{M_i}\right)(\theta - \theta_1), \tag{9}$$

$$j_{L_a}(\theta) = J_{a1} + \left(1 - \gamma + \frac{1}{M_i}\right)(\theta - \theta_1), \tag{10}$$

where J_{b1}, J_{a1}, and J_{m1} are the normalized current values in the corresponding inductor at the end of the previous sub-interval. At instant t_2, the diode current $i_{D_1}(t) = i_{L_b}(t) -$

978-1-7281-4181-7/19 $31.00 © 2019 IEEE

$i_m(t)$ goes to zero, causing the turn off of D_1. The interval duration is calculated as follows:

$$\theta_{12} = \theta_2 - \theta_1 = \lambda \frac{J_{C1}}{1 + \lambda \left(1 - \frac{1}{M_i}\right)}. \quad (11)$$

– *Interval* $t_2 \leq t \leq t_3$ (see Fig. 3c). This is the same topology of the first sub-interval, with different initial conditions, $i_{L_a}(t_2) = I_{a2}$, $i_{L_b}(t_2) = i_m(t_2) = I_{m2}$, $v_r(t_2) = V_o'$. Accordingly, we have:

$$j_m(\theta) = J_{m2} + \frac{\theta - \theta_2}{M_i(1 + \lambda)} + \frac{1}{\lambda}\left(1 - \frac{\lambda}{1 + \lambda}\frac{1}{M_i}\right)\sin(\theta - \theta_2), \quad (12)$$

$$j_{L_b}(\theta) = J_{b2} + \frac{\theta - \theta_2}{M_i(1 + \lambda)} - \left(1 - \frac{\lambda}{1 + \lambda}\frac{1}{M_i}\right)\sin(\theta - \theta_2), \quad (13)$$

$$j_{L_a}(\theta) = J_{a2} + \frac{\theta - \theta_2}{M_i}\left(1 + \frac{\lambda}{1 + \lambda}(1 - \gamma)\right) - (1 - \gamma)\left(\frac{\lambda}{1 + \lambda}\frac{1}{M_i} - 1\right)\sin(\theta - \theta_2), \quad (14)$$

$$u_r(\theta) = \left(1 - \frac{\lambda}{1 + \lambda}\frac{1}{M_i}\right)\cos(\theta - \theta_2) + \frac{\lambda}{1 + \lambda}\frac{1}{M_i}. \quad (15)$$

At t_3 voltage $v_r(t)$ tends to reverse polarity, thus forward polarizing S_b body diode D_b. From (15), we can derive:

$$\theta_{23} = \theta_3 - \theta_2 = \arccos\left(\frac{1}{1 - M_i\left(\frac{1 + \lambda}{\lambda}\right)}\right). \quad (16)$$

– *Interval* $t_3 \leq t \leq \frac{T_s}{2}$ (see Fig. 3e). In this phase, the two switches are conducting, thus keeping $v_r(t)$ to zero and current $i_m(t)$ constant at its peak value. Both currents i_{L_a} and i_{L_b} are increasing, i.e.:

$$j_{L_{a,b}}(t) = J_{a,b3} + \frac{1}{M_i}(\theta - \theta_3), \quad (17)$$

III. DESIGN CONSIDERATIONS

The analysis reported in the previous section is formally identical to the one reported in [14], except for the added $i_{L_a}(t)$ behavior. Similarly, the design considerations here reported are based on the previous analysis, with only few modifications to account for the different input inductor current waveforms.

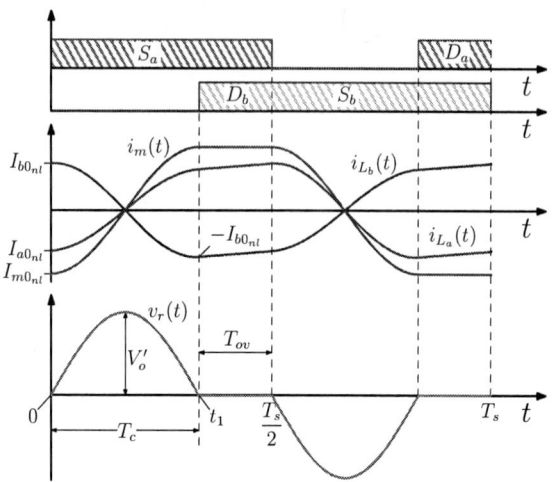

Fig. 4. Main waveforms in a switching period at no-load (boundary condition).

A. No-load condition

As already demonstrated in [14], the converter can operate at no load whenever the resonant voltage $v_r(t)$ peak remains below the equivalent output voltage. Corresponding waveforms, at the boundary condition, are shown in Fig. 4. Conduction angle θ_c comes from (5):

$$u_r(\theta_c) = 0 \quad \Rightarrow \quad J_{C0_{nl}}\cos\left(\frac{\theta_c}{2}\right) + \frac{1}{M_i}\sin\left(\frac{\theta_c}{2}\right) = 0, \quad (18)$$

yielding

$$\theta_c = 2\arctan\left(-M_i J_{C0_{nl}}\right). \quad (19)$$

The condition for zero energy transfer is $u_r\left(\frac{\theta_c}{2}\right) \leq 1$, that, together with (18) and (5), gives:

$$\cos\left(\frac{\theta_c}{2}\right) \leq \frac{1}{1 - \left(\frac{1 + \lambda}{\lambda}\right)M_i}. \quad (20)$$

From (20), assuming a known parameter λ, angle θ_c is found, and from (18) the initial condition $J_{C0_{nl}}$ is calculated as

$$J_{C0_{nl}} = -\frac{1}{M_i}\tan\left(\frac{\theta_c}{2}\right). \quad (21)$$

Using (3) and (18) together with the condition $j_m(\theta_c/2) = 0$, the initial condition $J_{m0_{nl}}$ results:

$$J_{m0_{nl}} = -\frac{1}{1 + \lambda}\left(J_{C0_{nl}} + \frac{\theta_c}{2M_i}\right). \quad (22)$$

Then, the input inductor current initial value is derived from

$$J_{b0_{nl}} = J_{C0_{nl}} + J_{m0_{nl}} = \frac{1}{1 + \lambda}\left(\lambda J_{C0_{nl}} - \frac{\theta_c}{2M_i}\right). \quad (23)$$

Similarly, using (6) to impose the condition $j_{L_a}\left(\frac{\theta_c}{2}\right) = 0$, yields the initial condition

$$J_{a0_{nl}} = -\frac{\lambda}{1 + \lambda}(1 - \gamma)J_{C0_{nl}} - \frac{\theta_c}{2M_i}\left(1 + \frac{\lambda}{1 + \lambda}(1 - \gamma)\right). \quad (24)$$

978-1-7281-4181-7/19 $31.00 © 2019 IEEE

The maximum switching frequency occurs at no load, and its value can be calculated as

$$f_{s_{max}} = \frac{1}{2(T_c + T_{ov})} = \frac{\pi f_r}{\theta_c + \theta_{ov}}, \tag{25}$$

where the overlapping angle $\theta_{ov} = \omega_r T_{ov}$ is calculated imposing a steady-state condition for the inductor current i_{L_b}. In fact, from Fig. 4, it must be $i_{L_b}\left(\frac{T_s}{2}\right) = I_{a0_{nl}}$, i.e.

$$j_{L_b}\left(\frac{\theta_{sw}}{2}\right) = -J_{b0_{nl}} + \frac{\theta_{ov}}{M_i} = J_{a0_{nl}}, \tag{26}$$

which gives:

$$\theta_{ov} = -\frac{1}{1+\lambda}\left[\gamma\lambda\tan\left(\frac{\theta_c}{2}\right) + \frac{\theta_c}{2}\left(2 + \lambda(2-\gamma)\right)\right]. \tag{27}$$

From (27), the overlapping angle θ_{ov} shows a weak dependency on λ, and also reveals a minimum value for the voltage gain that guarantees a correct no-load operation, i.e. a positive θ_{ov} value. By imposing the condition $\theta_{ov} = 0$ from (27), the minimum value of the voltage gain $M_{\text{No-Load}}^{\min} = \gamma M_{i_{\text{No-Load}}}^{\min}$ for no-load operation is in the range $3.15 - 3.25$ for any λ value.

B. ZVS condition

The considerations for the ZVS conditions are the same as the ones in [14]. In particular, guaranteeing a correct no load operation, automatically ensures the ZVS condition.

C. Average output current

The expression for the normalized output current, exploring the output capacitor *charge-balance*, is:

$$J_o' = \frac{\lambda}{\theta_{sw}} \frac{J_{C1}^2}{1 + \lambda\left(1 - \frac{1}{M_i}\right)}, \tag{28}$$

which is the same expression of the original topology.

D. Output power transfer

The calculation of the power transferred to the load at a given switching frequency needs to account for the different inductor current waveforms, compared with the original topology. Everything starts from the steady-state condition applied to current i_{L_b}. In this calculation, sub-interval θ_{01} is considered small enough to assume $\sin(\theta_{01}) \approx \theta_{01}$ and $\cos(\theta_{01}) \approx 1 - \theta_{01}^2/2$. In this way, expressions (3), (4), and (6) simplify into linear equations, so that

$$J_{m1} \approx J_{m0} + \frac{\theta_{01}}{2\lambda}, \tag{29}$$

$$J_{b1} \approx J_{b0} + \left(\frac{1}{M_i} - \frac{1}{2}\right)\theta_{01}, \tag{30}$$

$$J_{a1} \approx J_{a0} + \left(\frac{1}{M_i} + \frac{1-\gamma}{2}\right)\theta_{01}. \tag{31}$$

Combining (29) and (30), the normalized capacitor current $J_{C1} = J_{b1} - J_{m1}$ at the beginning of the energy transfer sub-interval is

$$J_{C1} \approx J_{C0} + \left(\frac{1}{M_i} - \frac{1}{2} - \frac{1}{2\lambda}\right)\theta_{01}. \tag{32}$$

Using (7) into (32) we obtain

$$J_{C1} = J_{C0} + \frac{B_4}{J_{C0}}, \tag{33}$$

with

$$B_4 = \frac{1+\lambda}{\lambda}\left(\frac{1}{M_i} - \frac{1}{2} - \frac{1}{2\lambda}\right). \tag{34}$$

This represents the first equation in the two unknowns J_{C0} and J_{C1}, which is identical to the original topology. A second equation is derived imposing the steady-state condition for i_{L_b}. To this purpose, let's observe, from Fig. 2, that $I_{an} = I_{b(n+4)}, n = 0, 1, 2, 3$. Thus, using (9), (13), (17), together with (10), (14), (11), and (16) and imposing the condition $I_{b8} = I_{b0}$ we obtain:

$$\frac{B_1}{J_{C0}} + B_2 J_{C1} + B_3 = 0, \tag{35}$$

where

$$B_1 = \gamma \frac{1+\lambda}{2\lambda}, \tag{36}$$

$$B_2 = \gamma \frac{\lambda}{1 + \lambda\left(1 - \frac{1}{M_i}\right)}, \tag{37}$$

$$B_3 = \gamma\left[\frac{\theta_{23}}{M_i}\frac{\lambda}{1+\lambda} + \left(1 - \frac{\lambda}{1+\lambda}\frac{1}{M_i}\right)\sin(\theta_{23})\right] - \frac{\theta_{sw}}{M_i}, \tag{38}$$

Note how parameter γ modifies the original coefficients, valid for the topology with separated input inductors. Combining (33) and (35), a simple second order equation is found that allows to determine the normalized capacitor current values J_{C0} and J_{C1}. Using J_{C1} into (28) we obtain the average output current that, thanks to the selected base voltage $V_N = V_o'$, represents also the normalized output power, since $P_o = J_o' I_N V_o' = J_o' P_N$ (this is true also for the voltage doubler configuration).

E. Input current ripple

From Fig. 2, the total instantaneous input current increases, approximately, during interval T_{34}, where both currents i_{L_a} and i_{L_b} are increasing. Using (17) and neglecting interval T_{01} at nominal power we can write:

$$\Delta J_{g_{PP}} \approx \frac{2}{M_i}\left(\frac{\theta_{sw}}{2} - \theta_{12} - \theta_{23}\right). \tag{39}$$

Considering the average input current, under unity efficiency, is expressed as $I_g = \gamma M_i I_o'$, the relative input current ripple is (the normalized output current is used):

$$r_{ig} \approx \frac{2}{\gamma M_i^2 J_o'}\left(\frac{\theta_{sw}}{2} - \theta_{12} - \theta_{23}\right). \tag{40}$$

TABLE I
CONVERTER SPECIFICATIONS

Parameter	Symbol	Value	
Minimum input voltage	$V_{g_{min}}$	42	V
Nominal input voltage	$V_{g_{nom}}$	48	V
Maximum input voltage	$V_{g_{max}}$	54	V
Nominal output voltage	V_o'	200	V
Nominal output power	P_o	300	W
Maximum Switching frequency	$f_{s_{max}}$	400	kHz
Minimum Switching frequency	$f_{s_{min}}$	150	kHz

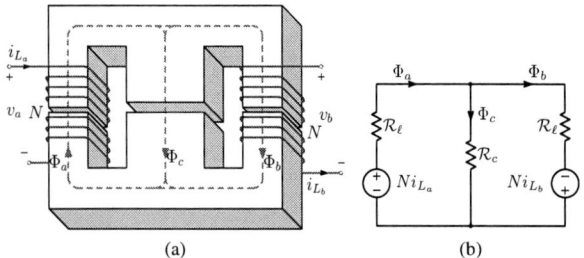

Fig. 5. Coupled-inductors structure: (a) windings in the lateral legs of an E-core; (b) equivalent electric model.

IV. DESIGN STEPS

The design procedure follows the same guidelines outlined in [14], and starts from the same specifications reported in Table I. However, with the coupled inductors, an additional degree of freedom is added with parameter γ, which is related to the magnetic structure. For example, if we use an E-core, as shown in Fig. 5a, the equivalent electric model is illustrated in Fig. 5b, where a symmetrical construction is considered. Reluctances $\mathcal{R}_\ell$ and $\mathcal{R}_c$ account for the gaps in the magnetic paths, neglecting the core contribution. From Fig. 5b, the flux linked to leg a is given by:

$$N\Phi_a = \frac{N^2}{\mathcal{R}_T}i_{L_a} + \frac{N^2}{\mathcal{R}_T}\frac{\mathcal{R}_c}{\mathcal{R}_c + \mathcal{R}_\ell}i_{L_b}$$
$$= (L_\ell + L_c)\,i_{L_a} + L_c i_{L_b}. \quad (41)$$

A similar equation holds for leg b simply swapping i_{L_a} with i_{L_b}. In the above equation $\mathcal{R}_T = \mathcal{R}_\ell + \mathcal{R}_\ell \parallel \mathcal{R}_c$ is the total reluctance seen by each magneto-motive force. from the parameter γ definition and (41), we get:

$$\gamma = \frac{L_\ell}{L_\ell + L_c} = 1 - \frac{L_c}{L_\ell + L_c} = \frac{\mathcal{R}_\ell}{\mathcal{R}_c + \mathcal{R}_\ell}. \quad (42)$$

The simplest magnetic coupling realization employs the same gap thickness in all lateral and central legs. In this way, the central leg reluctance results one half of the lateral legs reluctance, i.e. $\mathcal{R}_c = 0.5\mathcal{R}_\ell$, yielding $\gamma = 2/3$. We set this value in the design process.

A. Resonance frequency

The converter resonance frequency is calculated so as to guarantee no-load operation at the maximum input voltage and at the desired maximum switching frequency. In such conditions, the converter behavior is hardly affected by parameter

λ. Thus, combining (25) with (27) and setting $\lambda = 1$, the resonance frequency is calculated as follows:

$$f_r \approx \gamma\frac{f_{s_{max}}}{\pi}\left(\sqrt{M_{i_{min}}^2 - M_{i_{min}}} + \frac{1}{2}\arccos\left(\frac{1}{1-2M_{i_{min}}}\right)\right). \quad (43)$$

With the given specifications, the above equation yields $f_r = 513\,\text{kHz}$.

B. Equivalent input inductors

The value of the equivalent input inductors L affects the input current ripple. For a better comparison with the case with separate input inductors, we selected the same value, i.e. $L = 33\,\mu\text{H}$.

C. Resonant components

From the desired nominal output power P_o and the minimum switching frequency that occurs at the minimum input voltage, a constraint on the normalized output current J_o' is found as

$$J_{O_{design}}' = \frac{P_o}{V_o'I_N} = P_o\frac{\omega_r L}{V_o'^2}. \quad (44)$$

Using (28) at the minimum input voltage and switching frequency, the value λ_{design}, needed to obtain (44), turns out to be 1.35. If we use the same inductor $L_m = 34.5\,\mu\text{H}$ of the original topology, the λ value is $\lambda = 1.05$, lower than the calculated one. This will imply a minimum switching frequency slightly higher than the specified value. Using the resonance frequency definition (2), the needed resonant capacitor turned out to be $C_r = 5.9\,\text{nF}$.

D. Input current ripple

Using (40) at the operating point corresponding to the minimum input voltage and switching frequency and nominal power the input current ripple is estimated. The outcome of this calculation is $r_{ig} = 0.39$, value much lower than the original topology, as also predicted by (40). In fact, parameter γ appears also into the definition of M_i, making the overall relative ripple roughly proportional to γ. This means that, comparing with the original topology, the relative input current ripple reduces by a factor $\approx \gamma = 2/3$.

V. EXPERIMENTAL RESULTS

In order to verify the theoretical analysis, a converter prototype based on the CRIB topology with a voltage doubler rectifier (see [14]) was built. The coupled inductors have been realized with an ETD34 magnetic core (material N87), each winding made by 21 turns of Litz wire (bundle of 400 strands, having each a diameter of $50\,\mu\text{m}$). The values obtained were $L_\ell = 28.5\,\mu\text{H}$, $L_c = 11\,\mu\text{H}$, measured at $200\,\text{kHz}$ with the Agilent 4294A precision impedance analyzer. The measured value differs slightly from the calculated ones, while all the other components are the same of the original topology of [14].

The main converter waveforms, recorded at nominal input voltage ($V_g = 48\,\text{V}$) and output power ($P_o = 300\,\text{W}$), are shown in Fig. 6. The switching frequency was $f_s = 191\,\text{kHz}$,

978-1-7281-4181-7/19 $31.00 © 2019 IEEE

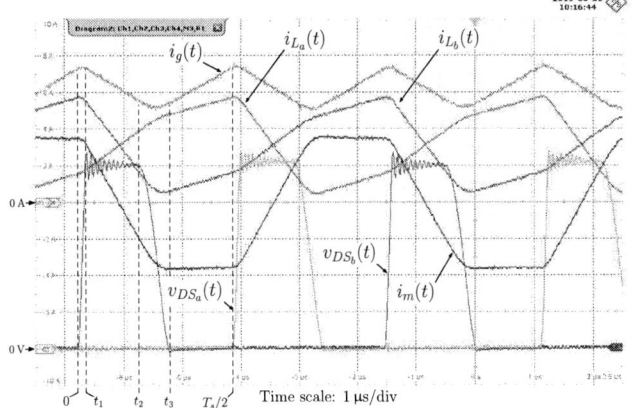

Fig. 6. Main waveforms in few switching periods taken at nominal input voltage and output power ($f_s = 191\,\text{kHz}$). Voltage scale: 40 V/div; Current scale: 2 A/div.

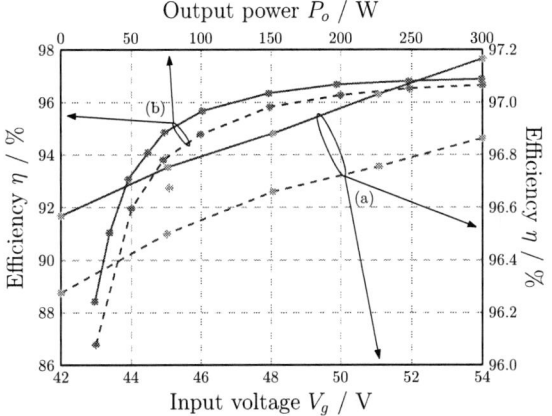

Fig. 7. Measured converter efficiency (dashed line = original topology; continuous line: with coupled inductors): (a) as a function of input voltage at nominal load (axes on the bottom and to the right); (b) as a function of output power at nominal input voltage (axes on the top and to the left).

very close to the theoretical value of $190\,\text{kHz}$. On the same figure are reported the time instants corresponding to the theoretical analysis, for an immediate comparison with the expected waveforms in Fig. 2. No-load operation, at the maximum input voltage, is achieved at $f_s = 427\,\text{kHz}$, while the minimum switching frequency was $f_{s_{min}} = 157\,\text{kHz}$ (minimum input voltage and nominal power).

The measured conversion efficiency (power stage only) was compared with the one taken with the original topology: the results are shown in Fig. 7, where curves (a) are taken varying the input voltage at nominal output power, while curves (b) show the efficiency variation with the output power at nominal input voltage. A slight improvement was found, especially at reduced output power, mainly thanks to the reduced core losses (lower volume compared with two separated cores, and lower flux swing in the central leg due to flux cancellation). The input current ripple in Fig. 6 is roughly 70% of the original value, i.e. reduced by a factor γ, as predicted.

VI. Conclusions

This paper investigates the use of coupled inductors in the Clamped-Resonant Interleaved Boost converter, a step-up topology featuring soft-switching for all devices independently of the load current. Similarly to the original topology that employs two separated input inductors, no-load operation is possible, provided that the minimum voltage gain is higher than a given limit value. A simple design procedure is suggested in order to satisfy the given specifications.

The original experimental prototype, rated at $42 - 54\,\text{V}$ to $400\,\text{V}$- $300\,\text{W}$ was modified substituting the input inductors with a coupled inductor structure, improving the conversion efficiency, especially at light load, and reducing the input current ripple.

References

[1] W. Li and X. He, "An interleaved winding-coupled boost converter with passive lossless clamp circuits," *IEEE Transactions on Power Electronics*, vol. 22, no. 4, pp. 1499–1507, July 2007.

[2] D. Wang, Y. Deng, X. He, and F. Cao, "Designing and analysis of an interleaved boost converter with passive lossless clamp circuits," in *2008 34th Annual Conference of IEEE Industrial Electronics*, Nov 2008, pp. 2149–2155.

[3] N. Park and D. Hyun, "N interleaved boost converter with a novel zvt cell using a single resonant inductor for high power applications," in *Conference Record of the 2006 IEEE Industry Applications Conference Forty-First IAS Annual Meeting*, vol. 5, Oct 2006, pp. 2157–2161.

[4] C. Wang, C. Lin, C. Wu, and C. Chuang, "A soft-switching interleaved boost dc/dc converter," in *2016 19th International Conference on Electrical Machines and Systems (ICEMS)*, Nov 2016, pp. 1–5.

[5] C. Wang, C. Lin, and S. Hsu, "A zvs-pwm interleaved transformer-isolated boost dc/dc converter with a simple zvs-pwm auxiliary circuit," in *2012 IEEE Third International Conference on Sustainable Energy Technologies (ICSET)*, Sept 2012, pp. 299–304.

[6] G. Yao, A. Chen, and X. He, "Soft switching circuit for interleaved boost converters," *IEEE Transactions on Power Electronics*, vol. 22, no. 1, pp. 80–86, Jan 2007.

[7] C. Wang, C. Lin, C. Lu, and J. Li, "Analysis, design, and realisation of a zvt interleaved boost dc/dc converter with single zvt auxiliary circuit," *IET Power Electronics*, vol. 10, no. 14, pp. 1789–1799, 2017.

[8] W. Li, Y. Deng, R. Xie, j. Shi, and X. He, "Interleaved zvt boost converters with winding-coupled inductors and built-in lc low pass output filter suitable for distributed fuel cell generation system," in *2007 IEEE Power Electronics Specialists Conference*, June 2007, pp. 697–701.

[9] H. Bahrami, E. Adib, S. Farhangi, H. Iman-Eini, and R. Golmohammadi, "Zcs-pwm interleaved boost converter using resonance-clamp auxiliary circuit," *IET Power Electronics*, vol. 10, no. 3, pp. 405–412, 2017.

[10] R. N. A. L. e Silva Aquino, F. L. Tofoli, P. P. Praca, D. d. S. Oliveira, and L. H. S. C. Barreto, "Soft switching high-voltage gain dc-dc interleaved boost converter," *IET Power Electronics*, vol. 8, no. 1, pp. 120–129, 2015.

[11] M. Shoyama and K. Harada, "Zero-voltage-switching realized by magnetizing current of transformer in push-pull current-fed dc-dc converter," in *Power Electronics Specialists Conference, 1993. PESC '93 Record., 24th Annual IEEE*, Jun 1993, pp. 178–184.

[12] M. Veerachary and J. Prakash, "Low source current ripple soft-switching boost converter," in *2016 Biennial International Conference on Power and Energy Systems: Towards Sustainable Energy (PESTSE)*, Jan 2016, pp. 1–6.

[13] Q. Luo, H. Yan, S. Chen, and L. Zhou, "Interleaved high step-up zero-voltage-switching boost converter with variable inductor control," *IET Power Electronics*, vol. 7, no. 12, pp. 3083–3089, 2014.

[14] G. Spiazzi, "Clamped-resonant interleaved boost converter: Analysis and design," in *2019 IEEE 10th International Symposium on Power Electronics for Distributed Generation Systems (PEDG)*, June 2019, pp. 986–992.

978-1-7281-4181-7/19 $31.00 © 2019 IEEE

LPV Modeling of Boost Converter and Gain Scheduling MPC Control

1st Rosana C. B. Rego
Department of Engineering and Technology
Federal Rural University of the Semi-Arid
Mossoró, Brazil
rosana.rego@ufersa.edu.br

Abstract—This paper proposes a modeling time-varying dc-dc converters in linear parameter variant (LPV) form. The LPV model is able to describe almost any nonlinear system since the system matrix is time-varying. The proposed modeling allows taking into account the uncertainties of the system and common specifications of power and load. The main contribution of this work is to demonstrate how to reformulate the Continuous Conduction Mode (CCM) model into an LPV one. The system is controlled using the model predictive control (MPC) with gain scheduling. The LPV formulation allows for obtaining improved robustness and performance properties. The results are verified by DC-DC boost converter numerical simulations.

Index Terms—boost converter, linear parameter varying, model predictive control, MPC-LPV

I. INTRODUCTION

The common need for accurate and efficient control of current industrial applications is driving the field of system identification to face the constant challenge of providing better models of physical phenomena [1]. One of the classes of systems that is being used is the class of linear parameter varying (LPV) systems due to the ability of embedding nonlinearities of the system into the variable parameters [2]. LPV model is able to describe almost any nonlinear system since the system matrix is time varying, thus the linear control theory is able to solve the nonlinear control problems [3].

In this way, LPV modeling can be used to take parameter uncertainty into consideration. In power electronics, a switching converter system can be stable around the operating point, however, it may be unstable when the system meets parameters varying [4]. In order to take parameter uncertainty into consideration a LPV representation could be used.

In recent works [4]–[6] advantages are shown in using parametric system identification for DC-DC converter. Thus, in this paper is proposed the LPV modeling for the DC-DC boost converter. The proposed LPV modeling allows taking into account the uncertainties of the system and common specifications of power and load.

A boost converter has the ability to raise a given DC voltage. It has a simplified topology but presents some singularities in its modeling, such as the variations of load resistance, input voltage and the effects of the non-minimum phase as presented in [7]–[10]. Performing the control of these converters is considered a complicated task because the model is non-minimum phase [11], [12]. Therefore, it was decided to control

the converter LPV modeling with the MPC control technique because the MPC proved to be a very robust control type in most applications, such as in static converters and in electric drive devices. The main causes of this control are that it can be applied either to linear or non-linear multivariate [10], [12]–[14].

The paper is organized as follows. First, in Section 2, the mathematical LPV modeling of the boost converter is presented. In Section 3, the control strategy is presented and some basic concepts concerning model predictive control are recalled. In Section 4, the numerical example is presented. Finally, in Section 5, is discussed the conclusions of the study.

Notation. The symbol * is used in some matrix expressions to induce a symmetric structure. For example

$$\begin{bmatrix} Q & * \\ S & P \end{bmatrix} = \begin{bmatrix} Q & S^T \\ S & P \end{bmatrix}.$$

II. LPV MODELING OF BOOST CONVERTER

Figure 1 shows the boost converter used [10], [12], [15].

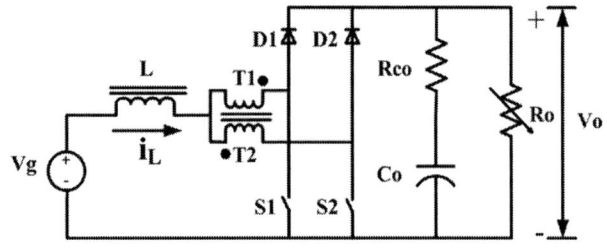

Fig. 1. Boost converter.

The expressions in the state space A_t, B_t, C_t and D_t operating in Continuous Conduction Mode (CCM) [10], [16] are:

$$\dot{x} = A_t(t)x + B_t(t)u,$$
$$y = C_t(t)x + D_t(t)u,$$
(1)

where,

$$A_t = \begin{bmatrix} -\frac{(1-D_{cycle})(R_{co}\|R_o)}{L} & -\frac{(1-D_{cycle})R_o}{L(R_{co}+R_o)} \\ \frac{(1-D_{cycle})R_o}{C_o(R_{co}+R_o)} & -\frac{1}{C_o(R_{co}+Ro)} \end{bmatrix}, \quad (2)$$

$$B_t = \frac{V_g}{R'} \begin{bmatrix} \frac{R_o}{L} & \frac{(1-D_{cycle})R_o+R_{co}}{R_o+R_{co}} \\ & -\frac{R_o}{R_o+R_{co}} \end{bmatrix}, \quad (3)$$

978-1-7281-4181-7/19 $31.00 © 2019 IEEE

$$C_t = \begin{bmatrix} (1 - D_{cycle})(R_{co}\|R_o) & \frac{R_o}{R_{co}+R_o} \end{bmatrix}, \quad (4)$$

$$D_t = -V_g \frac{R_{co}\|R_o}{R'}. \quad (5)$$

such that $R' = (1-D_{cycle})^2 R_o + D_{cycle}(1-D_{cycle})(R_{co}\|R_o)$, $x = [i_L \ V_c]^T$ where i_L is the inductor current, V_c is capacitor voltage, u is the control signal, D_{cycle} is the duty cycle and $y = V_o$, V_o is the the output voltage.

Analyzing the matrices (6), (7), (8) and (9) it is seen that the model is directly influenced by the variations of the input voltage V_g and the load R_o, since the variation of the input voltage changes the duty cycle (D_{cycle}) permanent regime and the load variation influences the load resistance. In this way if the duty cycle is a function of the input voltage and the load is a function of the power, the system (1) could be LPV and can be written as,

$$A(\alpha_1) = \begin{bmatrix} -\frac{\alpha_1(R_{co}\|R_o)}{L} & -\frac{\alpha_1 R_o}{L(R_{co}+R_o)} \\ \frac{\alpha_1 R_o}{C_o(R_{co}+R_o)} & -\frac{1}{C_o(R_{co}+Ro)} \end{bmatrix}, \quad (6)$$

$$B(\alpha_2, \alpha_3) = \begin{bmatrix} \frac{V_g R_o}{L} \frac{\alpha_3 R_o + \alpha_2 R_{co}}{R_o + R_{co}} \\ -\frac{V_g \alpha_2 R_o}{R_o + R_{co}} \end{bmatrix}, \quad (7)$$

$$C(\alpha_1) = \begin{bmatrix} \alpha_1(R_{co}\|R_o) & \frac{R_o}{R_{co}+R_o} \end{bmatrix}, \quad (8)$$

$$D(\alpha_2) = -V_g \alpha_2(R_{co}\|R_o). \quad (9)$$

where $\alpha_2 = (1-D_{cycle})^2 R_o + D_{cycle}(1-D_{cycle})(R_{co}\|R_o)^{-1}$, $\alpha_3 = (1 - D_{cycle})R_o + D_{cycle}(R_{co}\|R_o)^{-1}$ and $\alpha_1 = (1 - D_{cycle})$. Whereas, $R_o \in [\underline{R_o}\ \overline{R_o}]$, $V_g \in [\underline{V_g}\ \overline{V_g}]$ and $\alpha \in [\alpha_1\ \alpha_2\ \alpha_3]$.

The load and duty cycle of the model can be defined by [10], [12]:

$$R_o = f(Pot) = \frac{V_o^2}{Pot} \quad Pot \in [\underline{Pot}\ \overline{Pot}], \quad (10)$$

$$D_{cycle} = f(V_g) = 1 - \frac{V_g}{V_o} \quad V_g \in [\underline{V_g}\ \overline{V_g}]. \quad (11)$$

With this variation in input voltage $V_g \in [\underline{V_g}\ \overline{V_g}]$ and load $R_o \in [\underline{R_o}\ \overline{R_o}]$, the system becomes variant with uncertain parameters. Thus, one way of representing this system is by using polytopic modeling. Therefore, the system (1) belongs to the following polytope formed by the four local models,

$$[A(\alpha))|B(\alpha)|C(\alpha)|D(\alpha)] \in Co\{[A_0, B_0, C_0, D_0],$$
$$[A_1, B_1, C_1, D_1], [A_2, B_2, C_2, D_2], \ [A_3, B_3, C_3, D_3]\}, \quad (12)$$

where, $Co\{\cdot\}$ denotes the convex hull of the polytope and $[A_j, B_j, C_j, D_j]$ are vertices of the polytopic.

So, the matrices can be rewritten as,

$$A(\alpha) = A_0 + \sum_{i=1}^{n} \alpha_i A_i \Rightarrow A_0 + \alpha_1 A_1 \quad (13)$$

where,

$$A_0 = \begin{bmatrix} 0 & 0 \\ 0 & -\frac{1}{C_o(R_{co}+R_o)} \end{bmatrix},$$
$$A_1 = \begin{bmatrix} \frac{-R_{co}\|R_o}{L} & -\frac{R_o}{L(R_{co}+R_o)} \\ \frac{R_o}{C_o(R_{co}+R_o)} & 0 \end{bmatrix}. \quad (14)$$

$$B(\alpha) = B_0 + \sum_{i=1}^{n} \alpha_i B_i \Rightarrow B_0 + \alpha_1 B_1 + \alpha_2 B_2 + \alpha_3 B_3, \quad (15)$$

such that,

$$B_0 = B_1 = \begin{bmatrix} 0 \\ 0 \end{bmatrix},$$
$$B_2 = \begin{bmatrix} \frac{V_g R_o}{L}\left(\frac{R_{co}}{R_o+R_{co}}\right) \\ -\frac{V_g R_o}{R_o+R_{co}} \end{bmatrix}, \quad (16)$$
$$B_3 = \begin{bmatrix} \frac{V_g R_o}{L}\left(\frac{R_o}{R_o+R_{co}}\right) \\ 0 \end{bmatrix},$$

$$C(\alpha) = C_0 + \sum_{i=1}^{n} \alpha_i C_i \Rightarrow C_0 + \alpha_1 C_1, \quad (17)$$

where,

$$C_0 = \begin{bmatrix} 0 & \frac{R_o}{R_o+R_{co}} \end{bmatrix}, \quad C_1 = \begin{bmatrix} (R_{co})\|R_o & 0 \end{bmatrix}. \quad (18)$$

$$D(\alpha) = D_0 + \sum_{i=1}^{n} \alpha_i D_i \Rightarrow D_0 + \alpha_1 D_1 + \alpha_2 D_2, \quad (19)$$

with,

$$D_0 = 1, \ D_1 = 0, \ D_2 = -V_g(R_{co}\|R_o). \quad (20)$$

A discrete-time LPV model can be obtained using Euler approximation with a sample time T_s to allow a digital implemantation of the overral scheme:

$$\begin{bmatrix} i_L(k+1) \\ V_c(k+1) \end{bmatrix} = (I + T_s A(\alpha)) \begin{bmatrix} i_L(k) \\ V_c(k) \end{bmatrix} + T_s B(\alpha)u(k),$$

$$y(k) = C(\alpha) \begin{bmatrix} i_L(k) \\ V_c(k) \end{bmatrix} + D(\alpha)u(k) \quad (21)$$

Thus, the LPV model allows to consider the synthesis of the boost converter as a linear system.

A. Block diagram - Model with integral action

In Figure 2, g, h are the matrices that correspond to the degree freedom of the integral action block diagram. K_i and K are respectively the integral action gain and MPC controller gain. ref is the input reference of the system, y is the output of the system. The expressions of the model based on the block diagram are given by,

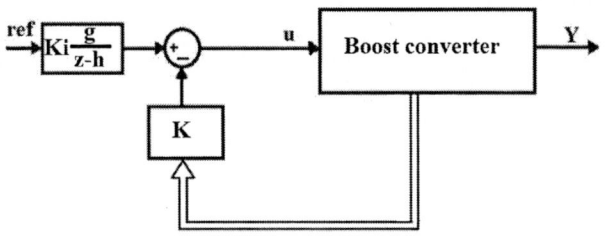

Fig. 2. Block diagram.

$$\hat{\mathcal{A}}(\alpha) = \begin{bmatrix} A(\alpha) & \cdots & 0 \\ \vdots & \ddots & \vdots \\ -hC(\alpha) & \cdots & g \end{bmatrix}, \tag{22}$$

$$\hat{\mathcal{B}}(\alpha) = \begin{bmatrix} B(\alpha) \\ \vdots \\ -hD(\alpha) \end{bmatrix}, \tag{23}$$

$$\hat{\mathcal{C}}(\alpha) = \begin{bmatrix} C(\alpha) & \cdots & 0 \end{bmatrix}, \tag{24}$$

whose closed-loop expressions are given by,

$$\bar{\mathcal{A}}(\alpha) = \begin{bmatrix} A(\alpha) - BK & B_j K_I \\ -h(C_j - D_j K) & g - h D_j K_I \end{bmatrix}, \tag{25}$$

$$\bar{\mathcal{B}} = \begin{bmatrix} 0 \\ h \end{bmatrix}, \tag{26}$$

$$\bar{\mathcal{C}} = \begin{bmatrix} (C_j - D_j K) & D_j K_I \end{bmatrix}, \tag{27}$$

$$\bar{\mathcal{D}} = 0. \tag{28}$$

where $\bar{\mathcal{A}}, \bar{\mathcal{B}}, \bar{\mathcal{C}}$ and $\bar{\mathcal{D}}$ are the closed-loop matrices whose state is defined by,

$$\hat{x} = \begin{bmatrix} x(k) \\ v(k) \end{bmatrix}, \tag{29}$$

where $v(k)$ is the integral action.

III. MPC CONTROL

The formulation of the offline MPC used are given by the following inequalities, as proposed in [17]–[20],

$$\begin{bmatrix} \hat{\mathcal{A}}(k+i) & \hat{\mathcal{B}}(k+i) \end{bmatrix}_{\in \Omega, i \geq 0}^{\max} J_\infty(k) \leq \\ \leq V(k+i|k) \leq \gamma(k) \tag{30}$$

$$\begin{bmatrix} 1 & \hat{x}(k|k) \\ \hat{x}(k|k) & Q \end{bmatrix} \geq 0, Q_i > 0, \tag{31}$$

$$\begin{bmatrix} Q_i & * & * & * \\ \hat{\mathcal{A}}_i Q_i + \hat{\mathcal{B}}_i Y_i & Q_i & * & * \\ \delta^{1/2} Q_i & 0 & \gamma(k)I & * \\ \mathcal{R}^{1/2} Y & 0 & 0 & \gamma(k)I \end{bmatrix} \geq 0, i = 1, \ldots, L \tag{32}$$

where, $F_i = Y_i Q_i^{-1}$.

$$\begin{bmatrix} X & Y \\ Y^T & Q_i \end{bmatrix} \geq 0, X_{rr} \leq u_{r,\max}^2, r = 1, 2, \ldots, n_u \tag{33}$$

$$\begin{bmatrix} Z & \hat{\mathcal{C}}\left(\hat{\mathcal{A}}_i Q_i + \hat{\mathcal{B}}_i Y_i\right) \\ * & Q_i \end{bmatrix} \geq 0, Z_{rr} \leq y_{r,\max}^2,$$

$$r = 1, 2, \ldots, n_y \tag{34}$$

$$\gamma(k) < \gamma(k-1) \tag{35}$$

A. Synthesis MPC Algorithm

To implement the algorithm, the following procedure is performed: for an offline system, given an initial condition x_1 it is generated a sequence of minimizers γ_j, Q_j, X_j, Y_j based in (30), (31), (32), (33), and (34). Do $j := 1$

1. compute the minimizers γ_j, Q_j, X_j, Y_j. Save Q_j^{-1}, F_j, and Y_j.
2. if $j < N$ choice the state $\hat{x}_{j+1}$ satisfying $\|\hat{x}_{j+1}\|_{Q^{-1}}^2 \leq 1$. Do $j := j + 1$ and return to step 1.
3. Find α_i satisfying $\|\hat{x}_{j+1}\|_{Q^{-1}}^2 \leq 1$.
4. calculate $F_j = Y_j Q_j^{-1}$.
5. Save F_j
6. apply the law of control $u(k) = -[K \ K_I]\hat{x}(k)$.

IV. NUMERICAL SIMULATION

The circuit implementation considered non-linear continuous modeling using the Runge Kutta 4th order method. For simulation and implementation, it was used MATLAB software with the free toolbox YALMIP and solver SEDUMI.

The initial states of the system (1) is assumed as $x = [38.4615 \ 26]^T$. The set reference voltage was $V_o = 48V$. The maximum value of the control signal was $u_{max} = 0.5$ and the operating points of the converter is $380W - 1000W$ for sample time $T_s = 1ms$ and simulation step of the $1\mu s$, is used $g = h = 1$.

The boost converter parameters used are in table I.

TABLE I
BOOST CONVERTER PARAMETERS

Parameters	Values
Input Voltage (V_g)	26-36 [V]
Output Voltage (V_o)	48 [V]
Duty Cycle (D_{cycle})	0.25-0.46
Frequency (f_s)	22 [kHz]
Inductor (L)	36[μH]
Inductive Resistance (R_L)	0 [Ω]
Capacitor (C_o)	4400[μF]
Resistance (R_{co})	26.7[$m\Omega$]
Resistance (R_o)	2.034-6.06 [Ω]
Output power (Pot)	380-1000 [W]

The weighting matrices are

$$\delta = \begin{bmatrix} 1 & 0 & 0 \\ 0 & 1 & 0 \\ 0 & 0 & 1 \end{bmatrix}, \text{ and } \mathcal{R} = 0.1. \tag{36}$$

A step variation of the input voltage of $26V - 36V$ was made, at the time of $0.125s$ and $0.375s$, whose analysis interval was between $0s$ and $0.5s$, as shown in the Figure 3. The Figure 4 shows the variation of the power (Pot) applied in the system simulation.

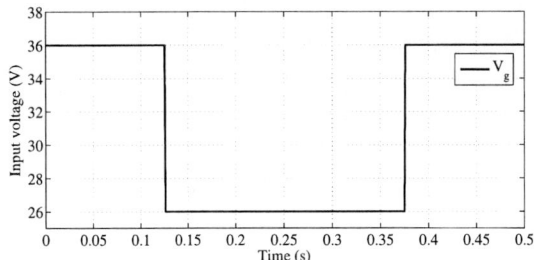

Fig. 3. Input voltage V_g

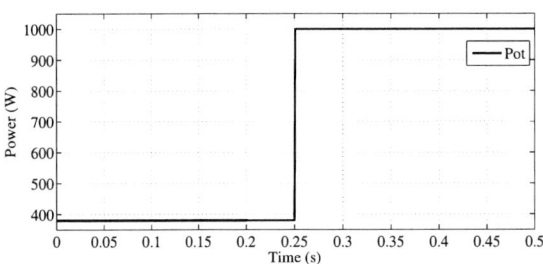

Fig. 4. Variation of the power Pot

The controller gain is given by,

$$K_{MPC} = [K \,|K_I] = F_0 + \alpha_1 F_1 + \alpha_2 F_2 + \alpha_3 F_3 \quad (37)$$

where, in the simulation the values obtained $\alpha_1 = 0.6458$, $\alpha_2 = 0.6644$, $\alpha_3 = 0.4399$. And

$$F_0 = [0.03 \ -0.11 \ -0.04] \times 10^{-2}, \quad (38)$$

$$F_1 = [0.0746 \ -0.2861 \ -0.0927] \times 10^{-3}, \quad (39)$$

$$F_2 = [0.1067 \ -0.3720 \ -0.1360] \times 10^{-3}, \quad (40)$$

$$F_3 = [0.04 \ -0.15 \ -0.05] \times 10^{-2}. \quad (41)$$

Therefore,

$$K_{MPC} = [0.06 \ -0.22 \ -0.08] \times 10^{-2}. \quad (42)$$

The uncertain or time-varying parameters are available in simulation time. The Figures 5, 6 and 7 show the variation of the uncertain parameters.

Figure 8 shows the output voltage $y(k)$. The controller followed the preset reference voltage. In Figure 9 is shown control signal $u(k)$. It is possible to see that the restriction (u_{max}) was satisfied.

Figure 10 shows the inductor current (i_L). The current oscillates when a change in load applies. When the input voltage is increased back to $36V$ at $0.375s$ the inductor current has a

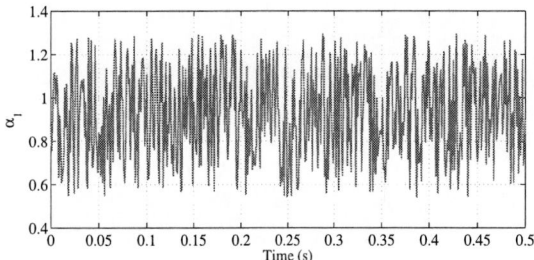

Fig. 5. Variation of the uncertain parameter $\alpha_1(t)$

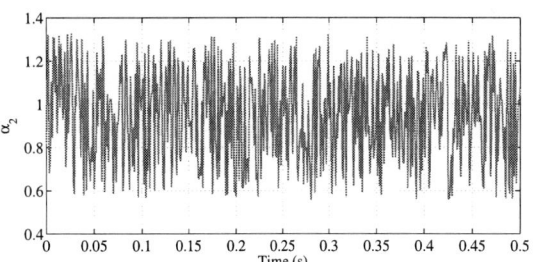

Fig. 6. Variation of the uncertain parameter $\alpha_2(t)$

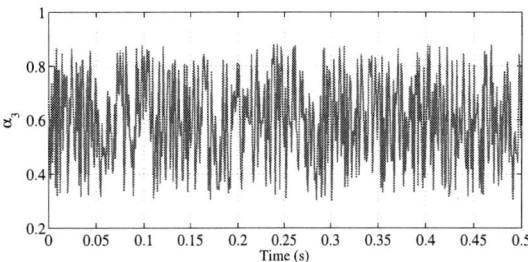

Fig. 7. Variation of the uncertain parameter $\alpha_3(t)$

peak of $105A$. This will certainly be unfeasible in practice and may lead to permanent damage to the converter. So, to avoid this overshoot is necessary to apply some technique known as anti-windup [21].

However, the simulation results show that the LPV approach can achieve the robust stabilization and tight transient performances.

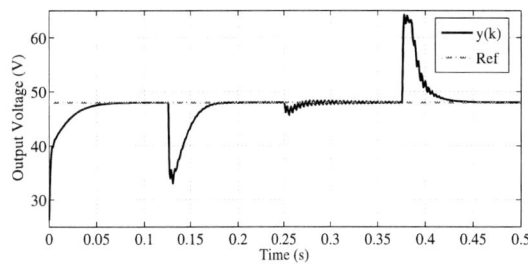

Fig. 8. Output voltage $y(k)$

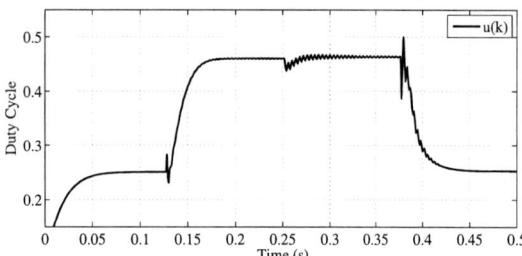

Fig. 9. Control signal $u(k)$

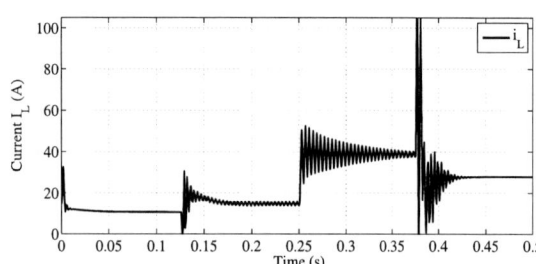

Fig. 10. Inductor current i_L

V. CONCLUSION

In this work, a LPV formulation for the boost converter was presented. This model allows to consider the synthesis of the boost converter as a linear system. The LPV system was controlled using the MPC control. A contribution present in the paper consists in showing that with the parameter varying modeling is possible embedding the nonlinearities of the system into the variable parameters. And this modeling is closer to the real system.

As future work, the author intends to perform the experimental simulation in the converter and perform the LMI relaxation of the controller to reduced conservativeness of polytopes.

ACKNOWLEDGMENT

This study was financed in part by the Coordenação de Aperfeiçoamento de Pessoal de Nível Superior - Brazil (CAPES) - Finance Code 001.

REFERENCES

[1] R. Tóth, *Modeling and identification of linear parameter-varying systems.* Springer, 2010, vol. 403.
[2] W. Yan, Y. Zhu, L. Zhu, and X. Liu, "Identification of systems with slowly sampled outputs using lpv model," *Computers & Chemical Engineering*, vol. 112, pp. 316–330, 2018.
[3] H. Zhang and L. Chen, "Robust variable gain hinf control for multi-propeller aerostat based on lpv." Computer Simulation, 2014.
[4] Z. Liu, L. Xie, A. Bemporad, and S. Lu, "Fast linear parameter varying model predictive control of buck dc-dc converters based on fpga," *IEEE Access*, vol. 6, pp. 52 434–52 446, 2018.
[5] C.-C. Chen, C.-L. Chen, J.-X. Chang, and C.-F. Yang, "Lpv gain-scheduling control for a phase-shifted pwm full-bridge soft switched converter," *IFAC Proceedings Volumes*, vol. 47, no. 3, pp. 6135–6140, 2014.

[6] S. Padhee, U. C. Pati, and K. Mahapatra, "Closed-loop parametric identification of dc-dc converter," *Proceedings of the Institution of Mechanical Engineers, Part I: Journal of Systems and Control Engineering*, vol. 232, no. 10, pp. 1429–1438, 2018.
[7] J. Linares-Flores, A. H. Mendez, C. Garcia-Rodriguez, and H. Sira-Ramirez, "Robust nonlinear adaptive control of a boost converter via algebraic parameter identification," *IEEE Transactions on Industrial Electronics*, vol. 61, no. 8, pp. 4105–4114, 2014.
[8] R. Ortega, J. A. L. Perez, P. J. Nicklasson, and H. J. Sira-Ramirez, *Passivity-based control of Euler-Lagrange systems: mechanical, electrical and electromechanical applications.* Springer Science & Business Media, 2013.
[9] L. Cheng, P. Acuna, R. P. Aguilera, J. Jiang, S. Wei, J. Fletcher, and D. D.-C. Lu, "Model predictive control for dc-dc boost converters with reduced-prediction horizon and constant switching frequency," *IEEE Transactions on Power Electronics*, 2017.
[10] M. V. Costa, F. E. Reis, J. C. Campos *et al.*, "Controlador robusto mpc-lmi aplicado ao conversor boost com célula de comutação de três estados," *Eletrônica de Potência, Campo Grande*, pp. 81–90, 2017.
[11] R. Amirifar, "Extended dynamic matrix control design for a dc-dc power converter," in *System Theory, 2005. SSST'05. Proceedings of the Thirty-Seventh Southeastern Symposium on.* IEEE, 2005, pp. 191–195.
[12] M. V. S. Costa, "Controle mpc robusto aplicado ao conversor *boost* ccte otimizado por inequações matriciais lineares," Engenharia Elétrica, Universidade Federal do Ceará, Fortaleza, 2017.
[13] E. F. Camacho and C. B. Alba, *Model predictive control.* Springer Science & Business Media, 2013.
[14] L. A. Aguirre, A. H. Bruciapaglia, P. E. Miyagi, and J. R. C. Piqueira, *Enciclopédia de automática: controle e automação.* Blucher, 2007.
[15] G. T. Bascopé and I. Barbi, "Generation of a family of non-isolated dc-dc pwm converters using new three-state switching cells," in *Power Electronics Specialists Conference, 2000. PESC 00. 2000 IEEE 31st Annual*, vol. 2. IEEE, 2000, pp. 858–863.
[16] R. Middlebrook and S. Cuk, "A general unified approach to modelling switching-converter power stages," in *Power Electronics Specialists Conference, 1976 IEEE.* IEEE, 1976, pp. 18–34.
[17] R. C. B. Rego, M. V. S. Costa, F. E. U. Reis, and R. P. T. Bascopé, "Análise e simulação do controlador mpc-aw-lmi aplicado ao conversor ccte operando em condições de saturação no sinal de controle." *XXII Congresso Brasileiro de Automática*, 2018.
[18] M. V. Kothare, V. Balakrishnan, and M. Morari, "Robust constrained model predictive control using linear matrix inequalities," *Automatica*, vol. 32, no. 10, pp. 1361–1379, 1996.
[19] J.-H. Park, T.-H. Kim, and T. Sugie, "Output feedback model predictive control for lpv systems based on quasi-min–max algorithm," *Automatica*, vol. 47, no. 9, pp. 2052–2058, 2011.
[20] Z. Wan and M. V. Kothare, "An efficient off-line formulation of robust model predictive control using linear matrix inequalities," *Automatica*, vol. 39, no. 5, pp. 837–846, 2003.
[21] R. C. B. Rego, "Controle MPC robusto com anti-windup aplicado a sistemas LPV e LTV baseado no algoritmo quasi-min-max com relaxao em LMIS," Master's thesis, Federal Rural University of Semi-Arid Region, Brazil, 2019.

978-1-7281-4181-7/19 $31.00 © 2019 IEEE

Zero-Crossing Detection Frequency Estimator Method Combined with a Kalman Filter for Non-ideal Power Grid

Tiago Davi Curi Busarello
Department of Engineering
Federal University of Santa Catarina
Blumenau - Brazil
tiago.busarello@ufsc.br

Sérgio Luiz Sambugari Junior
Departamento de Engenharia Elétrica
Universidade Estadual de Londrina
Londrina, Brazil
slsj08@gmail.com

Newton da Silva
Departamento de Engenharia Elétrica
Universidade Estadual de Londrina
Londrina, Brazil
newton.silva@uel.br

Abstract—**This paper proposes a Zero-Crossing Detection Frequency Estimator Method combined with a Kalman Filter for Non-ideal Power Grid. The Kalman filter generates the in-phase and in-quadrature signals from the voltage grid. Due to the adaptive feature of the Kalman figure, the in-phase and in-quadrature signals are free of noise and harmonics and it guarantees an accurate frequency estimation for a long range of grid frequency. With both clean signals in-phase and in-quadrature, the frequency estimator computes the arc tangent of the relationship between them. The result is a ramp signal varying from -pi to +pi. Such a signal is used to estimate the frequency. Considering zero-crossing detection, the frequency estimator counts the number of samples within a fundamental period. Experimental results show the efficacy of the proposed method. The frequency estimator was implemented C2000 Delfino MCU F28379D LaunchPad development kit.**

Keywords—Frequency estimator, Kalman Filter, F28379D LaunchPad, Power grid, PLL, distorted voltage.

I. INTRODUCTION

Measurements of electrical variables in a power grid is being obtained since the beginning of first installations of the electric sector. For a long time, such measurements were easy and most of them were clean signals. With the advancement of electronic technology, several disturbances that were once unworried are now causing problems for an accurate measurement. The grid frequency which hardly deviated from its nominal value is nowadays suffering vast variations due to occurrences like connection of distributed generators and nonlinear loads with intermittent behavior [1].

Measuring the grid frequency has an important and indispensable role for system operators. The value of the frequency may be used in algorithms designated to manage the grid, to decide the best power dispatch of generators, to stabilize the grid and so on. Therefore, there is a need for an accurate, fast and reliable measurement of the grid frequency. Deviation from the normal condition such as the presence of noise and harmonic, sudden changes in phase and amplitude cannot compromise the frequency measurements.

The grid frequency value is not directly measured. It is necessary an indirect method. Therefore, the frequency is estimated. For this reason, it is common to say frequency estimation instead of frequency measurement. Frequency estimation methods are constantly being reported in the literature [2]–[13]. In [2] the authors proposed an approach using a novel circular limit cycle oscillator (CLO) coupled

with frequency-locked loop (FLL). Due to the nonlinear structure of the CLO, the proposed frequency adaptive CLO technique is robust against various perturbations faced in the practical settings like discontinuous jump of phase, frequency and amplitude. In [3] the proposal is about an adaptive sliding mode observer for frequency and phase estimation. The observer is simple, easy to tune and suitable for real-time implementation.

A Fourier Transform-Based Frequency Estimation Algorithm is proposed in [4] where the algorithm is a modified synchronous clock generator that together with a modified frequency interpolation method provides an accurate measurement of the input signal frequency when the only available information about the sampling clock is a given integer multiple of the input signal fundamental frequency. In [10], similar Fourier-Based transform algorithm shows a wide frequency range applicability.

Frequency estimation methods are also found in solution based on Phase-Locked Loop (PLL) [14]–[18]. Knowing the frequency value of the grid is also essential in power electronics application like grid-connected converters [19]–[22].

All the above-mentioned researches have their efficacy and legitimacy. However, new issues can be addressed. This paper proposes a Zero-Crossing Detection Frequency Estimator Method using Kalman Filter for Non-ideal Power Grid. The Kalman filter generates the in-phase and in-quadrature signals from the voltage grid. Due to the adaptive feature of the Kalman figure, the in-phase and in-quadrature signals are free of noise and harmonics and it guarantees an accurate frequency estimation for a long range of grid frequency. With both clean signals in-phase and in-quadrature, the frequency estimator computes the arc tangent of the relationship between them. The result is a ramp signal varying from $-\pi$ to $+\pi$. Such a signal is used to estimate the frequency. Considering zero-crossing detection, the frequency estimator counts the number of samples within a fundamental period.

II. THE FREQUENCY ESTIMATOR METHOD

Fig. 1 presents a simplified diagram of the proposed Zero-Crossing Detection Frequency Estimator Method using Kalman Filter. The grid voltage is the input signal (z). This signal passes through the Kalman filter, which in turn, produces two filtered signals ($\hat{x}_1 and \hat{x}_2$). These signals are the estimated fundamental frequency component of the input signal and its orthogonal component. Later, these signals are

---

The authors thank to National Council for Scientific and Technological Development – CNPq (grant 421281/2016- 2).

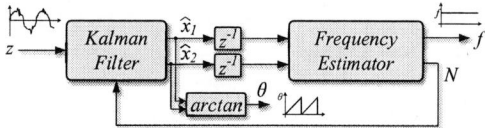

Fig. 1. Simplified diagram of the proposed Zero-Crossing Detection Frequency Estimator Method using Kalman Filter.

sent to the frequency estimator block. The unit delay blocks are necessary to avoid algebraic loop. The frequency is estimated with a method based on the zero-crossing detection. The arc tangent of the filtered signals is computed and a ramp signal varying from $-\pi$ to $+\pi$ is obtained. The number of samples is then computed (N) in one period of the input signal and then sent it back to the Kalman filter. This assures a satisfactory performance over a large input signal frequency range. The value of N depends on the frequency of the input signal. The value of theta θ can be computed by taking the arctan of signals $\hat{x}_1$ and $\hat{x}_2$. Actually, such computation is done within the block of Frequency Estimator. The arctan is shown outside for the sake of simplicity. The result of this operation is the phase angle of the input signal. Therefore, the proposed zero-crossing detection frequency estimator method using Kalman Filter may also be used as PLL.

III. BACKGROUND ON KALMAN FILTERING

The Kalman filter estimates a state of a linear system in the presence of uncertainties, noise and inaccurate measurement. This ability of estimating a state is due to a recursive method where the goal is to minimize the sum of squares of the difference between the real and estimated values. Such a method has two processes: prediction and estimation.

The prediction is also known as prior estimate. It estimates the current state taking into account the estimate and error covariance of the previous step. The prediction process does not consider the current input data of the Kalman filter. Later, the estimation process uses the prediction values to update the current state.

A linear system can be modeled in state-space equations such as:

$$\begin{cases} x_k = A_k x_{k-1} + w_k \\ z_k = H_k x_k + v_k \end{cases} \tag{1}$$

where:

    k is the current state

    x_k the state at $t = k$

    w_k is the noise of the process

    z_k is the state observation at $t = k$ (2)

    v_k is the noise of the measurments

    A_k and H_k are the inputs of the system

The noises w_k and v_k are white band-limited Gaussian noises with zero average. Moreover, the noises have covariance Q_k and R_k. Therefore, the noises are described by:

$$\begin{cases} w_k \sim N(0, Q_k) \\ v_k \sim N(0, R_k) \end{cases} \tag{3}$$

In order to compute the state at $t = k$, the Kalman filter computes the following set of equations. The first is the definition of the initial values for the estimate and the error covariance, given by:

$$\hat{x}_0, P_0 \tag{4}$$

The next step of the algorithm consist of two parts. The first is the computation of the state prediction, given by:

$$\hat{x}_k^- = A\hat{x}_{k-1} \tag{5}$$

The superscipt $^-$ means prediction. The second part is the computation of the covariance error prediction, given by:

$$P_k^- = AP_{k-1}A^T + Q \tag{6}$$

Where P_k^- is the covariance error prediction.

The next step is the computation of the Kalman gain (K), given by:

$$K_k = P_k^- H^T \left(HP_k^- H^T + R \right)^{-1} \tag{7}$$

The estimate of the state is the next computation, given by:

$$\hat{x}_k = \hat{x}_k^- + K_k \left(z_k - H\hat{x}_k^- \right) \tag{8}$$

Notice that (8) is the only equation where the state observation is used.

After computing the estimate, the error covariance must be computed in order to be used in the next prediction. The error covariance is given by:

$$P_k = P_k^- - K_k H P_k^- \tag{9}$$

IV. SYSTEM MODELING

In order to estimate the fundamental frequency in a non-ideal power grid, the power grid must be modeled in state-space equations, containing dynamic behavior as well as possible noise in the grid.

Considering initially a sinusoidal voltage given by:

$$x_k = M_k \sin\left(\omega_k t_k + \theta_k \right) \tag{10}$$

where M_k in amplitude, ω_k is angular frequency, θ_k is phase angle and t_k is time at instant k.

Considering also two orthogonal signals, x_{1k} and x_{2k}, given by:

$$\begin{cases} x_{1k} = M_k \sin\left(\omega_k t_k + \theta_k \right) \\ x_{2k} = M_k \cos\left(\omega_k t_k + \theta_k \right) \end{cases} \tag{11}$$

If $M_k \approx M_{k+1}$, $\omega_k \approx \omega_{k+1}$, $\theta_k \approx \theta_{k+1}$ and $t_{k+1} \approx t_k + T_s$ with T_s the sampling period, one may find:

$$\begin{cases} x_{1_{k+1}} = M_k \sin\left(\omega_k t_k + \omega_k T_s + \theta_k \right) \\ x_{2_{k+1}} = M_k \cos\left(\omega_k t_k + \omega_k T_s + \theta_k \right) \end{cases} \tag{12}$$

Resulting in:

978-1-7281-4181-7/19 $31.00 © 2019 IEEE

$$\begin{cases} x_{1_{k+1}} = x_{1k}\cos(\omega_k T_s) + x_{2k}\sin(\omega_k T_s) \\ x_{2_{k+1}} = -x_{1k}\sin(\omega_k T_s) + x_{2k}\cos(\omega_k T_s) \end{cases} \quad (13)$$

Writing (13) using matrix, it results:

$$\begin{bmatrix} x_1 \\ x_2 \end{bmatrix}_{k+1} = \begin{bmatrix} \cos(\omega_k T_s) & \sin(\omega_k T_s) \\ -\sin(\omega_k T_s) & \cos(\omega_k T_s) \end{bmatrix} \begin{bmatrix} x_1 \\ x_2 \end{bmatrix}_k \quad (14)$$

Taking into account the noise of the measurements and process w_k and v_k, respectively, the non-ideal power grid is modeled in state-space equations as:

$$\begin{bmatrix} x_1 \\ x_2 \end{bmatrix}_{k+1} = \begin{bmatrix} \cos(\omega_k T_s) & \sin(\omega_k T_s) \\ -\sin(\omega_k T_s) & \cos(\omega_k T_s) \end{bmatrix} \begin{bmatrix} x_1 \\ x_2 \end{bmatrix}_k + \begin{bmatrix} w_1 \\ w_2 \end{bmatrix}_k \quad (15)$$

$$z_k = \begin{bmatrix} 0 & 1 \end{bmatrix} \begin{bmatrix} x_1 \\ x_2 \end{bmatrix}_k + v_k \quad (16)$$

With (15) and (16) it is possible to apply the Kalman filter and estimate the fundamental component of voltage of the non-ideal power grid, even though the voltage has noise and harmonic distortion.

V. FREQUENCY ESTIMATION

The frequency estimator block of Fig. 1 estimates the frequency based on a zero-crossing detection and the relationship between the frequency of the input signal and the number of samples.

The number of samples is calculated according to:

$$N = \frac{f_s}{f_{signal}} = \frac{T_{signal}}{T_s} \quad (17)$$

where f_s is the sampling frequency and f_{signal} is the frequency of the input signal.

In order to detect the number of samples, it would be enough to use the input signal. However, the presence of uncertainties, noise and disturbance in the input signal, the frequency is better estimated with the filtered signal from the Kalman filter.

The number of samples plays an important role in the Kalman filter. This value is used in the matrix A of the Kalman filter, which is updated in every sampling period. Depending on the frequency of the input signal, a value of N is obtained. For an input signal at 60 Hz, the value of N is 200. It means that the 360 degree of one cycle of the input signal is divided into 200 samples. As a result, each sample corresponds to 1.8 degree. In case the input signal changes its frequency from 60 Hz to 50 Hz, the value of N is updated to 240. Consequently, each sample corresponds to 1.5 degree. The updated value of N is feedback to the Kalman filter. Therefore, matrix A is updated every time a change happens in the frequency of the input signal. This shows the efficacy of the proposed frequency estimation method for non-ideal power grid.

The instantaneous angle is given by:

$$\theta_k = \arctan\left(-\frac{x_{1k}}{x_{2k}}\right) \quad (18)$$

Eq. (18) is variable in time. The theta is the ramp signal mentioned previously. From this signal, the number of samples is calculated. The value of N begins to be counted in a zero-cross of the theta signal. The counting ends in the next zero-crossing. Notice that since the in-phase and in-quadrature signal are clean signals due to the Kalman filter, the computation of N has accuracy on is value.

VI. EXPERIMENTAL RESULTS

The proposed frequency estimator method of Fig. 1 was experimentally verified in the Digital Signal Controller (DSC) C2000 Delfino MCU F28379D LaunchPad development kit. Tab. I presents the parameters of the prototype. The results were collected through the Digital-Analog Converter (DAC) of the DSC.

Fig. 2 presents the input signal (z), which is also known as measurement, and the output signals of the Kalman filter ($\hat{x}_1$ and $\hat{x}_2$). The input signal is a sinusoidal waveform with amplitude equals to 1.5 V and the frequency is 60 Hz. The ($\hat{x}_1$) is in-phase with the input signal while the ($\hat{x}_2$) is in-quadrature. The Kalman filter produces the in-phase and in-quadrature signals orthogonal related to each other.

TABLE I. PARAMETERS OF THE PROTOTYPE

Parameter	Value
Matrix A	$A = \begin{bmatrix} \cos(\pi/N) & -\sin(\pi/N) \\ \sin(\pi/N) & \cos(\pi/N) \end{bmatrix}$
Matrix H	$H = \begin{bmatrix} 1 & 0 \end{bmatrix}$
Matrix Q	$Q = \begin{bmatrix} 0.01 & 0 \\ 0 & 0.01 \end{bmatrix}$
Matrix R	$R = 25$
Matrix x_0	$x_0 = \begin{bmatrix} 0 & 0 \end{bmatrix}$
Matrix P_0	$P_0 = \begin{bmatrix} 1 & 0 \\ 0 & 1 \end{bmatrix}$
Sampling Frequency	12 kHz

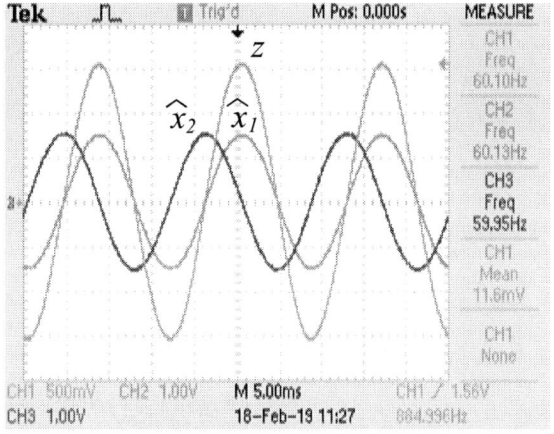

Fig. 2. The input signal (z) and the output signals of the Kalman filter ($\hat{x}_1$ and $\hat{x}_2$).

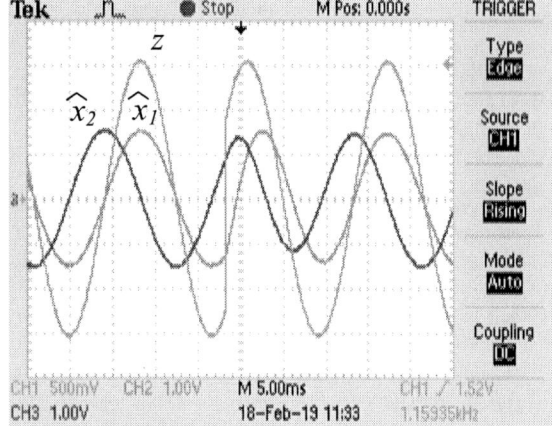

Fig. 3. The input signal (z) and the output signals of the Kalman filter $(\hat{x}_1 \, and \, \hat{x}_2)$ when the input signal suffers a 90 degrees phase shift.

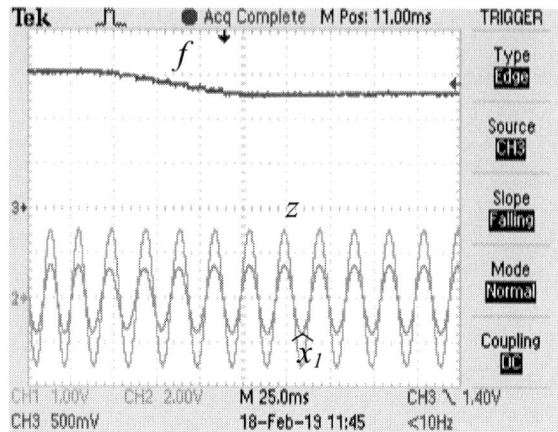

Fig. 5. The estimated frequency (f), the input signal (z) and the in-phase signal when the input signal $(\hat{x}_1)$ changes from 60 Hz to 50 Hz (CH3: 500 mV/ 20 Hz).

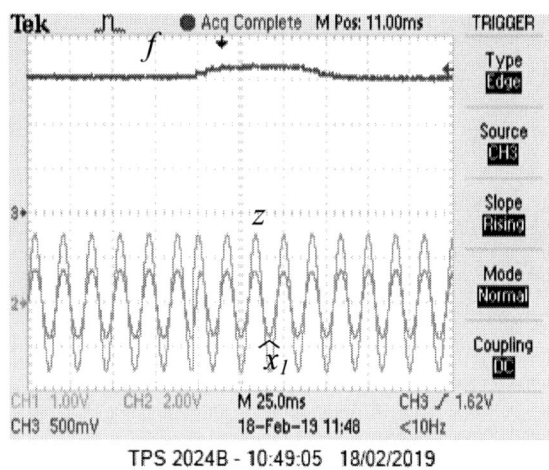

Fig. 4. The estimated frequency (f), the input signal (z) and the in-phase signal when the input signal $(\hat{x}_1)$ suffers a 90 degrees phase shif (CH3: 500 mV/ 20 Hz).

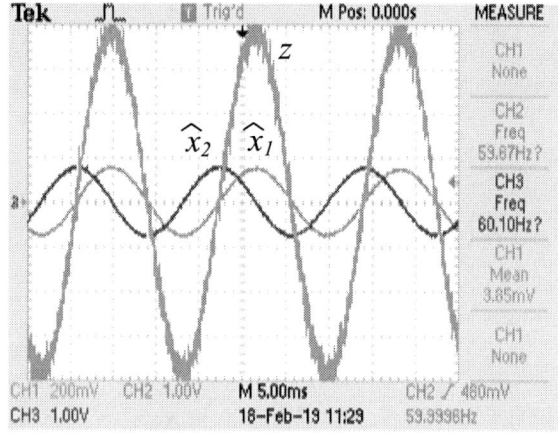

Fig. 6. The input signal (z) and the output signals of the Kalman filter $(\hat{x}_1 \, and \, \hat{x}_2)$ when the input signal is polluted with noise.

Fig. 3 presents input signal (z) and the output signals of the Kalman filter $(\hat{x}_1 \, and \, \hat{x}_2)$ when the input signal suffers a 90 degrees phase shift. Before the phase shift, the signals are in accordance to that presented in the previous figure. After the phase shift, the in-phase and in-quadrature signals of the Kalman filter reach steady-state in less than one fundamental cycle of the input signal. In other words, the in-phase signal in again in-phase with the input signal in less than one cycle.

Fig. 4 presents the estimated frequency (f), the input signal (z) and the in-phase signal when the input signal $(\hat{x}_1)$ suffers a 90 degrees phase shift. The estimated frequency returns to its steady-state value in approximately 75 ms after the phase shift.

Fig. 5 presents the estimated frequency (f), the input signal (z) and the in-phase signal when the input signal $(\hat{x}_1)$ changes

from 50 Hz to 60 Hz. The estimated frequency reaches the steady-state condition after approximately 60 ms. The estimated frequency signal did not present neither oscillatory nor unpredictable behavior.

Fig. 6 presents the input signal (z) and the output signals of the Kalman filter $(\hat{x}_1 \, and \, \hat{x}_2)$ when the input signal is polluted with noise. The in-phase and in-quadrature signal $(\hat{x}_1 \, and \, \hat{x}_2)$ are free of noise and they are in-phase and in-quadrature related to the input signal. This result shows the filtering efficacy of the Kalman.

Fig. 7 presents the estimated frequency (f), the input signal (z) and the in-phase signal $(\hat{x}_1)$ at the moment the input signal is polluted with noise. The amplitude of the input signal is also reduced to half. The estimated frequency keeps unchanged on

978-1-7281-4181-7/19 $31.00 © 2019 IEEE

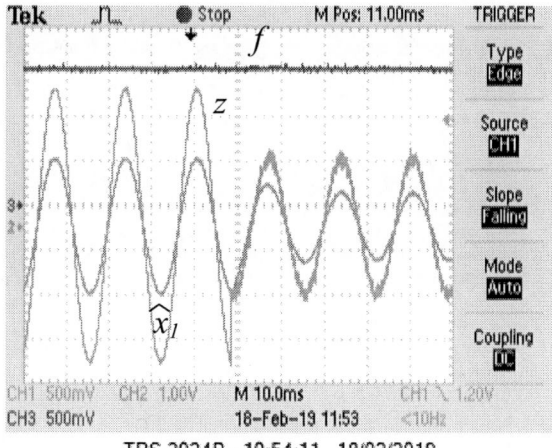

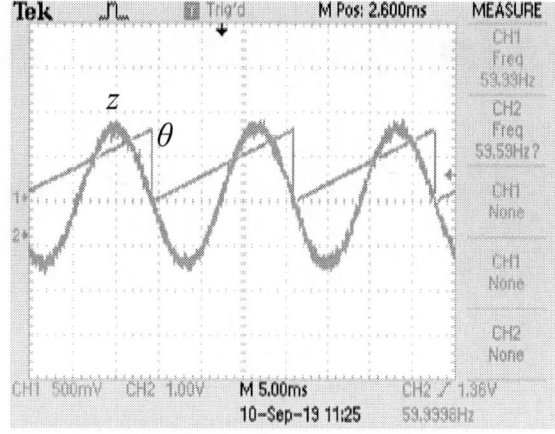

Fig. 7. The estimated frequency (f), the input signal (z) and the in-phase signal ($\hat{x}_1$) at the moment the input signal is polluted with noise (CH3: 500 mV/ 20 Hz).

Fig. 9. The input signal (z) and the theta signal (θ) when the input signal is polluted with noise.

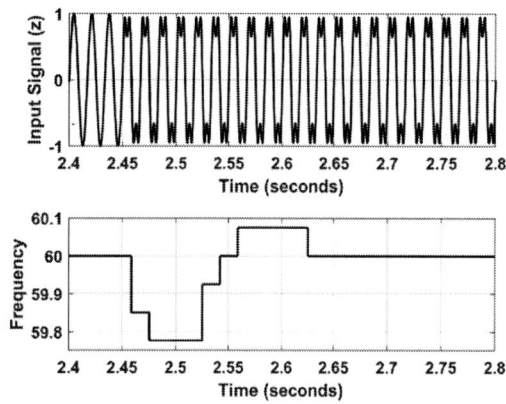

Fig. 10. Simulated result for he input signal (z) (at the top) and the estimated frequency (f) during the inclusion of harmonic distortion in the input signal.

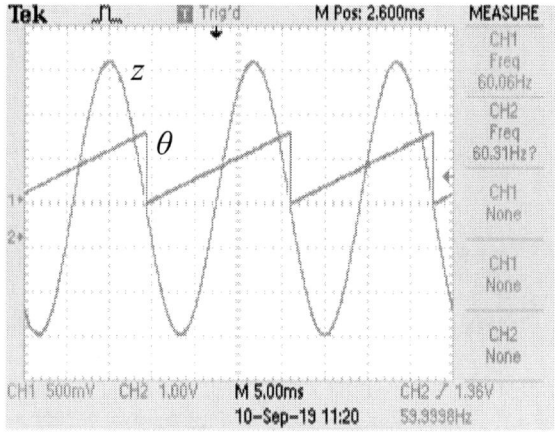

Fig. 8. The input signal (z) and the theta signal (θ).

its steady-state value, showing the efficacy of the proposed estimation method.

Fig. 8 presents the input signal (z) and the theta signal (θ). The theta signal is synchronized with the input signal. This result shows that the proposed frequency estimator method can also be used as PLL. The values varies from $-\pi$ to $+\pi$ but the result shows a signal with offset due to the range of the DAC.

Fig. 9 presents the input signal (z) and the theta signal (θ) when the input signal is polluted with noise. Even in this case, the theta signal is synchronized with the input signal.

Fig. 10 presents a simulated result for the input signal (z) and the estimated frequency (f) during the inclusion of harmonic distortion in the input signal. Initially, the input signal is sinusoidal. At t = 2.45, a third harmonic component (180 Hz) with amplitude equals to 0.35 is added to the input signal. The estimated frequency varies and returns to 60 Hz after a short transient time. This result was presented through simulation due to the incapability of the signal generator to produce such a signal. The value of 0.35 corresponds to 35% of the fundamental component. This value is not practical in voltage power grid, but it was set in this study only to show the efficacy of the proposed frequency estimator.

Fig. 11 presents a picture of the DSP, signal generator and the oscilloscope used to verify experimentally the proposed frequency estimator method.

978-1-7281-4181-7/19 $31.00 © 2019 IEEE

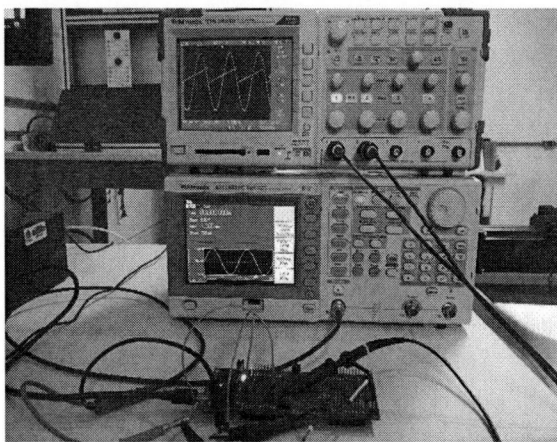

Fig. 11. picture of the DSP, signal generator and the oscilloscope used to verify experimentally the proposed frequency estimator method.

The computational effort and cost of implementing the proposed zero-crossing detection frequency estimator method using Kalman Filter for non-ideal power grid is beyond the scope of this paper, but may be presented by these authors in a future paper.

VII. Conclusion

This paper presented a Zero-Crossing Detection Frequency Estimator Method using Kalman Filter for Non-ideal Power Grid. The Kalman filter generated the in-phase and in-quadrature signals from the voltage grid. Due to the adaptive feature of the Kalman figure, the in-phase and in-quadrature signals were free of noise and harmonics.

Experimental results collected in C2000 Delfino MCU F28379D LaunchPad development kit showed the efficacy of the proposed frequency estimator method. The proposed method was verified under the application of phase displacement, frequency variation and inclusion of noise in the input signal. For all cases, the estimated frequency did not present neither oscillatory nor unpredictable behavior. Therefore, the proposed frequency estimator method based on Kalman filter is an attractive solution to estimating the frequency value of a non-ideal power grid. The simulation files used in this paper is freely available on the author's webpage http://busarello.prof.ufsc.br.

References

[1] M. H. Bollen, *Integration of Distributed Generation in the Power System*. John Wiley & Sons, 2011.

[2] H. Ahmed, S. Amamra, e M. H. Bierhoff, "Frequency-Locked Loop Based Estimation of Single-Phase Grid Voltage Parameters", *IEEE Transactions on Industrial Electronics*, p. 1–1, 2018.

[3] H. Ahmed, S. Amamra, e I. Salgado, "Fast Estimation of Phase and Frequency for Single Phase Grid Signal", *IEEE Transactions on Industrial Electronics*, p. 1–1, 2018.

[4] A. Carboni e A. Ferrero, "A Fourier Transform-Based Frequency Estimation Algorithm", *IEEE Transactions on Instrumentation and Measurement*, vol. 67, nº 7, p. 1722–1728, jul. 2018.

[5] V. Choqueuse, A. Belouchrani, F. Auger, e M. Benbouzid, "Frequency and Phasor Estimations in Three-Phase Systems: Maximum Likelihood Algorithms and Theoretical Performance", *IEEE Transactions on Smart Grid*, p. 1–1, 2018.

[6] Z. Dai, Z. Zhang, Y. Yang, F. Blaabjerg, Y. Huangfu, e J. Zhang, "A Fixed-Length Transfer Delay-based Adaptive Frequency Locked Loop for Single-Phase Systems", *IEEE Transactions on Power Electronics*, p. 1–1, 2018.

[7] J. Li, Z. Teng, Y. Wang, S. You, Y. Liu, e W. Yao, "A Fast Power Grid Frequency Estimation Approach Using Frequency-Shift Filtering", *IEEE Transactions on Power Systems*, p. 1–1, 2019.

[8] S. Tomar e P. Sumathi, "Amplitude and Frequency Estimation of Exponentially Decaying Sinusoids", *IEEE Transactions on Instrumentation and Measurement*, vol. 67, nº 1, p. 229–237, jan. 2018.

[9] H. Wen, C. Li, e W. Yao, "Power System Frequency Estimation of Sine-Wave Corrupted With Noise by Windowed Three-Point Interpolated DFT", *IEEE Transactions on Smart Grid*, vol. 9, nº 5, p. 5163–5172, set. 2018.

[10] L. Zhan, Y. Liu, e Y. Liu, "A Clarke Transformation-Based DFT Phasor and Frequency Algorithm for Wide Frequency Range", *IEEE Transactions on Smart Grid*, vol. 9, nº 1, p. 67–77, jan. 2018.

[11] S. Golestan, J. M. Guerrero, J. C. Vasquez, A. M. Abusorrah, e Y. Al-Turki, "A Study on Three-Phase FLLs", *IEEE Transactions on Power Electronics*, vol. 34, nº 1, p. 213–224, jan. 2019.

[12] M. S. de Padua, "Tecnicas digitais para sincronização com a rede eletrica, com aplicação em geração distribuida", *Dissertação de Metrado - Universidade Estadual de Campinas*, p. 165, 2006.

[13] S. L. S. Junior, "APLICAÇÃO DE FILTRO DE KALMAN PARA FILTRAGEM DE SINAIS DA REDE ELÉTRICA", *Trabalho de Conclusão de Curso - Universidade Estudual de Londrina*, p. 77, 2016.

[14] F. K. A. Lima, R. G. Araujo, F. L. Tofoli, e C. G. C. Branco, "A Phase-Locked Loop Algorithm for Single-Phase Systems with Inherent Disturbance Rejection", *IEEE Transactions on Industrial Electronics*, p. 1–1, 2019.

[15] M. S. Reza, F. Sadeque, M. M. Hossain, A. M. Y. M. Ghias, e V. Agelidis, "Three-Phase PLL for Grid-Connected Power Converters under Both Amplitude and Phase Unbalanced Conditions", *IEEE Transactions on Industrial Electronics*, p. 1–1, 2019.

[16] F. K. de Araújo Lima, R. Guerreiro Araújo, F. L. Tofoli, e C. G. C. Branco, "A three-phase phase-locked loop algorithm with immunity to distorted signals employing an adaptive filter", *Electric Power Systems Research*, vol. 170, p. 116–127, maio 2019.

[17] J. F. Guerreiro, J. A. Pomilio, e T. D. C. Busarello, "Design of a multilevel Active Power Filter for More Electrical Airplane variable frequency systems", in *2013 IEEE Aerospace Conference*, 2013, p. 1–12.

[18] J. F. Guerreiro, J. A. Pomilio, e T. D. C. Busarello, "Design and implementation of a multilevel active power filter for more electric aircraft variable frequency systems", in *2013 Brazilian Power Electronics Conference*, 2013, p. 1001–1007.

[19] M. Babakmehr, F. Harirchi, A. A. Durra, S. M. Muyeen, e M. G. Simões, "Exploiting Compressive System Identification for Multiple Line Outage Detection in Smart Grids", in *2018 IEEE Industry Applications Society Annual Meeting (IAS)*, 2018, p. 1–8.

[20] H. Sartipizadeh e F. Harirchi, "Robust Model Predictive Control of DC-DC Floating Interleaved Boost Converter under Uncertainty", in *2017 Ninth Annual IEEE Green Technologies Conference (GreenTech)*, 2017, p. 320–327.

[21] M. Babakmehr, F. Harirchi, A. Alsaleem, A. Bubshait, e M. G. Simões, "Designing an intelligent low power residential PV-based Microgrid", in *2016 IEEE Industry Applications Society Annual Meeting*, 2016, p. 1–8.

[22] F. Harirchi, M. G. Simões, M. Babakmehr, A. AlDurra, S. M. Muyeen, e A. Bubshait, "Multi-functional double mode inverter for power quality enhancement in smart-grid applications", in *2016 IEEE Industry Applications Society Annual Meeting*, 2016, p. 1–8.

A Review of FCS-MPC in Multilevel Converters Applied to Active Power Filters

João G. L. Foster
IESTI
Universidade Federal de Itajubá
Itajubá, Brasil
fosterelt@yahoo.com.br

Rondineli R. Pereira
IESTI
Universidade Federal de Itajubá
Itajubá, Brasil
rondinelirp@gmail.com

Robson B. Gonzatti
IESTI
Universidade Federal de Itajubá
Itajubá, Brasil
bauwelz@gmail.com

Wilson C. Sant'Ana
Instituto Gnarus
Itajubá, Brasil
wilson_santana@ieee.org

Denis Mollica
EDP São Paulo
São Paulo, Brasil
denis.mollica@edpbr.com.br

Germano Lambert-Torres
Instituto Gnarus
Itajubá, Brasil
germanoltorres@gmail.com

Abstract—In the last decade finite control set model predictive control (FCS-MPC) has shown satisfying results when applied to active power filters (APF). Nevertheless, when multilevel topologies are used some specific topics need to be properly addressed, such as computational burden, DC-link balance and redundancies. This paper brings a review of several proposes found in literature to deal with these issues. To verify the discussed strategies a simulation of a 5-level cascaded H-bridge (CHB) hybrid active power filter (HAPF) topology using an LCL-filter is also presented.

Keywords—FCS-MPC, HAPF, APF, Multilevel Converter.

I. INTRODUCTION

Non-linear loads such as adjustable-speed motor drives, arc-furnaces, uninterruptible power supplies (UPS), consumer electronics etc. drain current harmonics and reactive power from the grid. This causes the deterioration of the provided power quality, higher losses in transmission and distribution systems and can be harmful to sensible loads [1]–[3].

The proliferation of these kind of loads has been one of the main motivators in industry and in the academy for the continuous research of new techniques to mitigate the issues that may arouse from those components.

Those issues were first addressed in the 40's with the utilization of passive power filters (PPF) that are able to filter harmonic content and compensate reactive power. The major drawback of PPF is that they are tuned for specific harmonics and reactive compensation in addition to the risk of resonance with the power system. In order to solve this problems active power filters began to be developed in the 70's with the utilization of power converters to work around the resonance problems, compensate for any desired harmonics and deliver dynamic reactive power compensation [4], [5]. In the late 80's topologies of hybrid active power filters were proposed, with the intent to reduce the ratings of the power electronics in the filter [6], [7].

The application of APF's to medium and high voltages is limited by the power switches rating, so in order to overcome this issue, the utilization of multilevel topologies is preferred. Besides that, multilevel topologies also provide an improved output voltage waveform and, in some cases, makes possible the omission of interface transformers [8], [9].

The main multilevel topologies of power converters are the three-level neutral point clamped (3L-NPC), the three-level flying capacitor (FC), the modular multilevel converter (MMC) and the cascaded H-bridge (CHB). Among these structures, the CHB has the advantage of having a modular structure, which simplifies and reduce the cost of design for more levels [9].

Although the same control techniques applied to traditional converters can also be used for multilevel topologies, a new trend has emerged in the last decade for the utilization of model predictive control (MPC) in multilevel converters [10]. One of the factors that contribute to this tendency is the growing capabilities of modern microprocessors, since MPC has a high increase in its computational cost when working with multilevel converters. The main advantages of applying MPC to multilevel converters are the fast dynamic response achieved, the simplicity to work with multivariable systems and to deal with nonlinearities and constraints of the system [7], [10]–[12].

Finite control set MPC (FCS-MPC) has already proven to achieve good results when applied to APF's [7], [13], [14], even when integrated with renewable energy sources [15].

There are three main aspects that needs to be properly addressed when designing an MPC based controller. The system model responsible for the prediction of the outputs, the cost function, which defines the desired dynamics of the system, and the optimization algorithm used to compute the next state to be applied [11].

This paper discusses those issues for FCS-MPC applied to a multilevel HAPF topology with LCL-filter, focusing on the relevance of the cost function design, and is organized as follows: Section II presents an explanation of FCS-MPC operation and main advantages for multilevel converters. Section III discusses the design of a proper cost function and the enhancements provided by the optimization algorithm. Section IV shows the modelling and discretization of the system. Section V presents simulation results of the HAPF used to compensate reactive power as an application example and section VI presents the conclusions.

II. FCS-MPC

MPC is a control technique that uses a model of the system to calculate the resulting output for each of the possible control inputs. The control action that will be applied is then chosen to be the one that minimizes a cost function that describes the desired dynamics of the system [11].

978-1-7281-4181-7/19 $31.00 © 2019 IEEE

The FCS-MPC is a class of MPC controllers that takes advantage of the discrete nature of the power converters in order to reduce the computational burden inherent to MPC strategies. This is possible because the system response needs to be evaluated only for the limited switching states that can be provided by those devices [16].

The operation principle of FCS-MPC is straightforward and is summarized below:

1. Measure the state variables $x_n(k)$;

2. Apply $S_{opt}(k)$ – Calculated in the previous period;

3. Predict $x_n(k+1)$ for $S_{opt}(k)$;

4. Predict $x_n(k+2)$ for each possible switching state;

5. Select as $S_{opt}(k+1)$ the state that minimizes the cost function;

Where S_{opt} is the optimal switching state. Since FCS-MPC works with the direct states of the switches, once the optimal switching state is selected the IGBT gates are directly turned on or off depending on the chosen state.

The predictions of the state variables are calculated through the dynamic model of the system and will be discussed in Section IV.

One of the drawbacks of FCS-MPC is that it produces a variable switching frequency, which can create resonance problems when used with higher order filters. This issue has already been approached in literature by using a modulation stage after the selection of the optimal switching states [17]–[19]. This modulation is done by selecting more than one optimal state and switching them in equal intervals inside the period. In [19] the authors made a comparison between the frequency spectrum of the variable and the fixed switching frequency methods, which makes clear the increased quality in grid current.

When working with multilevel converters one of the main issues that need to be addressed is the computational burden related to the rise in the number of possible switching states [20]. For example, [10] shows that for a three-phase CHB converter the number of possible switching states can be calculated as:

$$K_s = 2^{6C} \tag{1}$$

Where C is the number of cells per phase, which gives a total of 4096 possible switching states for a 5-level converter. Although there is a high number of possible switching states most of them are redundant, since they generate the same output voltage level. It is important to notice that these redundant states do not affect the current control [21].

The difference between these redundancies is the behavior of the power switches, and this is well exploited in literature for several control purposes, as shown in Table 1.

TABLE 1. CONTROL PURPOSES ACHIEVED WITH THE REDUNDANCIES

Control purpose	Reference
Reduce computational cost	[10], [22], [23]
Reduce common-mode voltages	[10]
Regulate inductor currents	[18]
Regulate capacitor voltages	[21], [24]
Thermal redistribution	[25]

The approaches used to work with these redundancies are related to the cost function and will be further discussed in Section III.

III. COST FUNCTION

The design of a proper cost function is the most critical and complex step of the MPC strategy, since it will be responsible for the dynamic response of the system. When designing the cost function it is necessary to take into account not only the main control objective, which for APF's is usually the current control, but also the inclusion of system non-linearities and constraints [12].

The cost function can be described in general as:

$$g = \sum_{l=k+1}^{k+N_p} \sum_{j=0}^{m-1} \lambda_j \left(x_{j,l}^* - x_{j,l}^p \right)^2 + \sum_{r=k}^{k+N_c-1} \sum_{i=0}^{n-1} \lambda_i \left(u_{i,r} \right)^2 \tag{2}$$

Where $x_{j,l}^p$ is the predicted value and $x_{j,l}^*$ is the reference value for the state variable x_j in the moment l, $u_{i,r}$ is the control input u_i in the moment r, N_p is the prediction horizon and N_c is the control horizon. The weighting factors λ_j and λ_i influence the system's performance and stability [11].

It is always important to remember that the switching state selected as optimal for the next sampling period is the one that minimizes a cost function g.

The simplest cost functions used for APF's are:

$$g = \left(i_f^* - i_f^p \right)^2 \tag{3}$$

$$g = \left(i_\alpha^* - i_\alpha^p \right)^2 + \left(i_\beta^* - i_\beta^p \right)^2 \tag{4}$$

Both of them are responsible for the current control in the output of the filter (i_f), being (3) for single-phase and (4) for three-phase systems [11], [12]. When working with three-phase systems the utilization of the α-β coordinates (i_α, i_β) simplifies the system, reduces the computational cost and does not affect the three-phase currents [21].

As already discussed in Section II the flexibility of the cost function can be used to achieve several different control objectives. Some of those will be discussed below.

A. Multivariable Control

The use of multivariable control with FCS-MPC is simple and based on the design of a proper cost function to guarantee the tracking of both references.

In [7], the authors use multivariable control in order to track both inverter output current I_{inv}, responsible for the stability, and filter capacitor voltage V_f, responsible for active resonance damping and source voltage harmonic blocking. To achieve this simultaneous control, the authors designed the cost function as (5):

$$g = \lambda_{inv} \left(i_{inv}^* - i_{inv}^p \right)^2 + \lambda_v \left(V_f^* - V_f^p \right)^2 \tag{5}$$

In order to achieve an appropriate control of both variables, the weighting factor choice is the most important step, and is not trivial, being normally defined through an heuristic approach [11]. In [7] the weighting factors λ_{inv} and λ_v were chosen based on simulation results.

B. Reduce Computational Cost

Some ways to reduce computational cost while using FCS-MPC for multilevel converters can be found in literature.

In [10] the authors take a more direct approach to reduce the computational cost, while at the same time reducing the common-mode voltages. First the redundant switching states are removed, by choosing to keep the ones that minimizes the common-mode voltages, then, for each voltage vector the authors calculated offline a subset of adjacent voltage vectors, so that only the 7 voltage vectors closest to the last one applied are considered in the prediction calculations.

Another way to deal with the redundancies of multilevel converters is by using what was first described in [26] as hierarchical FCS-MPC. This method is based on the utilization of more than one cost function layer. For example, for the control of the output current of an APF it is possible to use a cost function g_1 in the form of (3), in order to find the voltage level that minimizes the filter current error and then use another cost function g_2, as in (6):

$$g_2 = n_c \qquad (6)$$

Where n_c is the number of commutations needed to achieve a certain switching state, evaluated only for the redundant states related to the voltage level already selected by g_1. This helps to reduce the computational burden, while at the same time reducing the losses caused by several commutations of the power switches [11].

A very similar method was used in [22] and [23] for two different multilevel topologies, proving to provide a significantly faster response for both cases, while slightly improving the output current THD. The difference from the previous example is that for g_2 the authors used a cost function that has the inner voltages as state variables and the redundant states as control inputs, thus finding the switching state redundancy that keeps the inner voltages balanced. Another advantage of this technique is that it has no need for weighting factors, simplifying the overall control design.

C. Regulate Inductor Currents and Capacitor Voltages

In [24] a single cost function is used in a way that both control objectives, current control and DC capacitor voltages balancing, are achieved. For each phase the designed cost function is:

$$g = \left(i_o^* - i_o^p\right)^2 + \lambda_{cap}\left(\sum_{i=1}^{2}\left(v_c^* - v_{ci}^p\right)^2\right) \qquad (7)$$

Where v_c^* is the reference value for the DC-link capacitor voltage and the weighting factor λ_{cap} is calculated as:

$$\lambda_{cap} = \frac{i_o}{v_c^*} \qquad (8)$$

Where i_o is the nominal output current.

A similar approach is used in [18] to maintain the inductor current in a multilevel current-source converter.

D. Thermal Redistribution

To solve the problems related to the thermal stress and the unequal losses in 3L-NPC converters, it was proposed in [25] an algorithm that, without any thermal modeling of the devices, is able to lower the power switches junctions temperature and also relieve the clamping diodes.

The algorithm is based on the fact that, if a specific switching state generates a high current through the power switch, then that state should be avoided when possible.

The designed cost function is composed by three different parts, one related to the output voltage tracking, one for the DC-link balancing and one responsible for the thermal redistribution:

$$g = \left(v_{c\alpha}^* - v_{c\alpha}^p\right)^2 + \left(v_{c\beta}^* - v_{c\beta}^p\right)^2 + \lambda_{dc}.g_{dc} + \lambda_t.g_t \qquad (9)$$

Where g_{dc} and λ_{dc} are related to the DC-link balancing and g_t and λ_t are related to the thermal redistribution.

E. Optimization Algorithm

Until this point all the enhancements described were provided solely by the design of a proper cost function. But to avoid problems related to the computational burden, it is also possible to improve the algorithm used to perform the calculations of the cost function.

An algorithm that drastically reduces the execution time of FCS-MPC for a multilevel CHB STATCOM is shown in [21]. First the authors use the same principle of the hierarchical FCS-MPC by creating two cost functions, the first one with the aim to track the current error while keeping the voltage vector close to the last one applied, the second one is responsible for DC-link balancing while reducing the number of commutations of the power switches. The cost functions used are described as:

$$g_1 = \left|I_{\alpha\beta}^* - I_{\alpha\beta}^p\right| + \lambda_1\left|S_{\alpha\beta}(k+1) - S_{\alpha\beta}(k)\right| \qquad (10)$$

$$g_2 = \left|V_{c(k+2)}^* - V_{cxi(k+1)}^p\right| + \lambda_2\left|s_{xi}(k+1) - s_{xi}(k)\right| \qquad (11)$$

Where $S_{\alpha\beta}$ is related to the voltage vectors, V_{cxi} is the capacitor voltage for phase x and cell i and s_{xi} relates to the possible commutations for phase x and cell i, including redundancies.

Each cost function is evaluated through a different algorithm. First g_1 is evaluated with a dynamic programming algorithm and then, using the result provided by the minimization of g_1, g_2 is evaluated by a mixed optimization approach, consisting of the same dynamic programming algorithm and also a greed algorithm. Details of the operation principle each algorithm can be found in [21].

IV. SYSTEM MODEL ANALYSIS

When working with MPC methodologies, the calculation of an accurate but, at the same time, simple model is of upmost importance. Since it reflects directly on the system response quality and the time needed to calculate the several predictions [27].

In literature it is common to find modeling and applications of simple L-filters when working with FCS-MPC in APF's [14], [15], [17]. The reasons are the dynamic model simplicity, the satisfactory response and the lack of resonance problems even when working with a variable switching frequency.

In order to present the modeling and operation of a more sophisticated filter, this paper will consider the single phase

HAPF topology described in [7] which uses an LCL-filter, as shown in Fig. 1.

Although this filter introduces more complex resonance problems the improved dynamics and higher attenuation factor justify its use for APF's [28], [29].

The detailed modeling and operating principle of this system can be found in [7] and will be briefly discussed below.

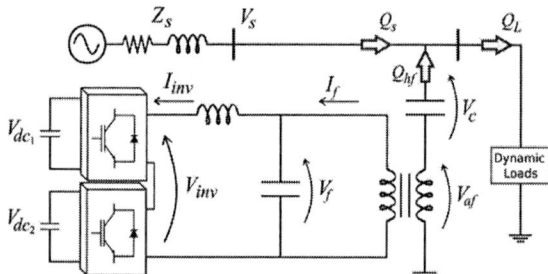

Fig. 1 - Topology of multilevel HAPF with LCL-filter

The equivalent circuit to be modeled is depicted in Fig. 2. Where L_f and R_f are the inverter side inductance and its inner resistance; C_f and R_{cf} are the filter capacitor capacitance and its series resistance; L_t and R_t are the transformer short circuit resistance and inductance; V_c is the capacitor bank voltage and is only included in the reference calculations. The state variables of the system are the inverter current (I_{inv}), the filter capacitor voltage (V_f) and the hybrid filter current (I_f). The control inputs are the inverter voltage (V_{inv}) and the transformer voltage (V_{af}).

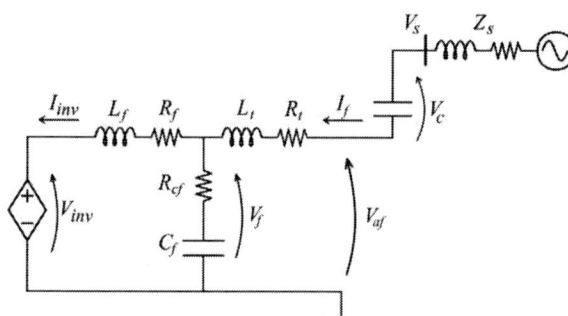

Fig. 2 - System equivalent circuit

The inner resistances are needed in order to obtain a more precise model.

The system will be represented by its state-space equations:

$$\dot{x} = Ax(t) + Bu(t) \tag{12}$$

Where $x(t)$ is the state variables vector, A is the state matrix, $u(t)$ is the control variables vector, and B is the input matrix. Using the state variables and the control inputs described above (13) is obtained:

$$\frac{d}{dt}\begin{bmatrix} I_{inv}(t) \\ I_f(t) \\ V_f(t) \end{bmatrix} = A\begin{bmatrix} I_{inv}(t) \\ I_f(t) \\ V_f(t) \end{bmatrix} + B\begin{bmatrix} V_{inv}(t) \\ V_{af}(t) \end{bmatrix} \tag{13}$$

From [7] and [29] it is possible to achieve the following formulation for the state and input matrix:

$$A = \begin{bmatrix} -\dfrac{R_f}{L_f} & 0 & \dfrac{1}{L_f} \\ 0 & -\dfrac{R_t}{L_t} & -\dfrac{1}{L_t} \\ \dfrac{R_f \cdot R_{cf}}{L_f} - \dfrac{1}{C_f} & \dfrac{1}{C_f} - \dfrac{R_t \cdot R_{cf}}{L_t} & \dfrac{R_{cf}}{L_f} + \dfrac{R_{cf}}{L_t} \end{bmatrix} \tag{14}$$

$$B = \begin{bmatrix} -\dfrac{1}{L_f} & 0 \\ 0 & -\dfrac{1}{L_t} \\ \dfrac{R_{cf}}{L_f} & \dfrac{R_{cf}}{L_t} \end{bmatrix} \tag{15}$$

In order to work with continuous systems in a digital environment i.e. microcontrollers, microprocessors, DSP's, FPGA's etc. the system needs to be discretized.

The most commonly applied discretization method is the Euler approximation, due to its simplicity, as can be seen below:

$$x(k + 1) = A_d x(k) + B_d u(k) \tag{16}$$

$$A_d = e^{AT_s} \approx I + AT_s \tag{17}$$

$$B_d \approx BT_s \tag{18}$$

Where T_s is the sampling period and k is a multiple of the sampling period ($t = kT_s$). This was the discretization method used in [7] and proved to provide a good dynamic response, while keeping a simple structure.

However, as the authors described in [27], the utilization of the truncated Taylor series expansion can improve the accuracy of the model while assuring a one sample period delay for higher order systems, such as output LCL-filters.

V. SIMULATION RESULTS AND DISCUSSION

The operating principle of this HAPF topology is fully discussed in [7] and will be resumed below.

In order to compensate for the dynamic reactive power of the load, the HAPF controls I_inv, and V_f to indirectly control the current flowing through the filter branch.

The voltage V_{af} has two components, one in phase with the reactive part of the filter branch current (V_{af}^q), responsible for DC-link charging, and one in phase with the active part of the filter branch current (V_{af}^d), responsible for the reactive power control. To dynamically compensate for reactive power variations on the load the HAPF can operate in two different regions, as described below:

1) Overcompensation: When the load reactive power surpasses the capacitor bank nominal power, the applied voltage V_{af}^d is in opposite phase with V_s, which increases the provided reactive power Q_{hf}.

2) Undercompensation: When the reactive power of the load is lower than the capacitor bank nominal power, the applied voltage V_{af}^d is in phase with V_s, decreasing the provided Q_{hf}.

This way what causes the variation in the supplied reactive power is the adjustment provided to the voltage over the capacitor bank.

The LCL-filter parameters used in the simulation are shown in Table 2.

TABLE 2. LCL-FILTER PARAMETERS

L_f	R_f	L_t	R_t	C_f	R_{cf}
5.64 mH	0.2 Ω	1.06 mH	0.17 Ω	11.4 μF	1 Ω

In the simulation a 246.69 μF capacitor bank was used, which, for 127 V/60 Hz, provides a nominal 1500 Var. The system parameters used for the simulation are presented in Table 3.

TABLE 3. SYSTEM PARAMETERS

Sampling	f_s = 40080 Hz
DC Link	V_{dc} = 220 V ; C_{dc} = 10000 μF
DC Link PI gain	k_p = 5 ; k_i = 11
Transformer	S = 7.5 kVA (127/440 V)
Load 1	2000 W/1900 Var
Load 2	2000 W/1100 Var

The hierarchical FCS-MPC technique was used to achieve the multivariable control and DC-link capacitor balancing while reducing the computational burden, and the cost functions used were:

$$g_1 = \lambda_{inv}\left|i_{inv}^* - i_{inv}^p\right| + \lambda_v\left|V_f^* - V_f^p\right| \quad (19)$$

$$g_2 = \left|V_{dc}^* - V_{dc1}^p\right| + \left|V_{dc}^* - V_{dc2}^p\right| \quad (20)$$

The cost function g_1 selects the voltage level that minimizes the error in the inverter current and in the capacitor filter voltage, then, using this voltage level, g_2 selects the redundancy that provides the best DC-link capacitor voltage balancing. The weighting factors $\lambda_{inv} = 1$ and $\lambda_v = 80$ were used based on the results provided in [7].

The simulation results for the reactive power are shown in Fig. 3. The HAPF is connected in the beginning of the simulation together with Load 1, this makes the HAPF work in the overcompensation region. After 0.1 seconds Load 1 is disconnected and Load 2 is connected, this makes the HAPF transition from the overcompensation to the undercompensation region. In both operating regions the current source THD is kept around 3.6% in relation to the fundamental, and it is possible to notice that the source current is kept in phase with the source voltage for both cases, and the time to achieve stability after the load is around 2 fundamental cycles.

The DC-link voltage reference used was 220 V and in Fig. it can be seen that the capacitor voltages are kept close to the reference for both compensating regions. The ripple was around 0.2 V.

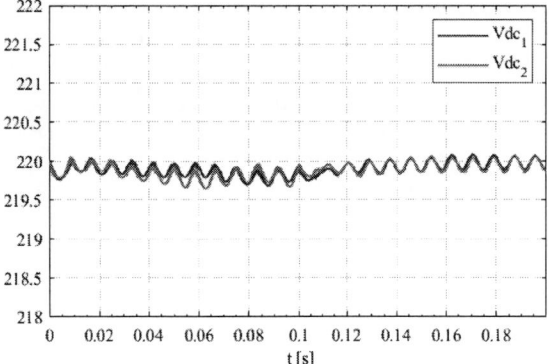

Fig. 4 – DC-link voltages

VI. CONCLUSION

This paper presents as main contribution an assemble of techniques found in literature to deal with the most common issues when applying FCS-MPC to multilevel APF topologies. It is possible to note that more than one approach can be used in order to solve the same problem, having each one different advantages and drawbacks.

The presented simulation shows the results obtained by FCS-MPC applied to a multilevel HAPF topology that uses an LCL-filter for dynamic reactive power compensation. It was shown that through the utilization of two cost functions it is possible to perform a multivariable control while, at the same time, balancing the DC-link voltages of the two cells of the 5-level CHB inverter utilized.

ACKNOWLEDGMENT

The authors would like to thank the following Brazilian Research Agencies: CNPq, CAPES, FAPEMIG and ANEEL R&D for the financial support of this project.

REFERENCES

[1] H. Rudnick, J. Dixon and L. Moran, "Delivering clean and pure power," in *IEEE Power and Energy Magazine*, vol. 1, no. 5, pp. 32-40, Sept.-Oct. 2003.

[2] V. Gali, N. Gupta and R. A. Gupta, "Mitigation of power quality problems using shunt active power filters: A comprehensive review," *2017 12th IEEE Conference on Industrial Electronics and Applications (ICIEA)*, Siem Reap, 2017, pp. 1100-1105.

[3] S. R. Das, P. K. Ray and A. Mohanty, "Enhancement of power quality disturbances using hybrid power filters," *2017 International*

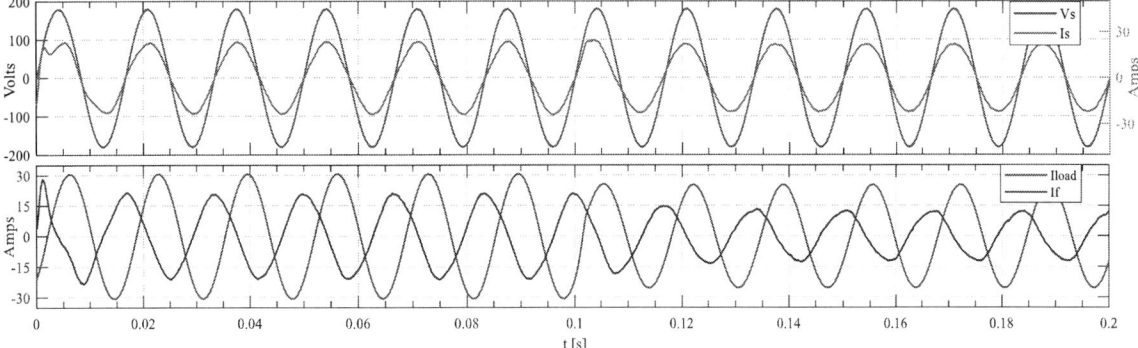

Fig. 3 - Simulation results for reactive power compensation

Conference on Circuit ,Power and Computing Technologies (ICCPCT), Kollam, 2017, pp. 1-6.

[4] Yang-Wen Wang, Man-Chung Wong and Chi-Seng Lam, "Historical review of parallel hybrid active power filter for power quality improvement," *TENCON 2015 - 2015 IEEE Region 10 Conference*, Macao, 2015, pp. 1-6.

[5] A. Cleary-Balderas, A. M. Senior and O. Cruz-Hernéndez, "Hybrid active power filter based on the IRP theory for harmonic current mitigation," *2016 IEEE International Autumn Meeting on Power, Electronics and Computing (ROPEC)*, Ixtapa, 2016, pp. 1-5.

[6] H. Akagi, "Active Harmonic Filters," in *Proceedings of the IEEE*, vol. 93, no. 12, pp. 2128-2141, Dec. 2005.

[7] S. C. Ferreira, R. B. Gonzatti, R. R. Pereira, C. H. da Silva, L. E. B. da Silva and G. Lambert-Torres, " ," in *IEEE Transactions on Industrial Electronics*, vol. 65, no. 3, pp. 2608-2617, March 2018.

[8] B. Geethalakshmi, M. Kavitha and K. Delhibabu, "Harmonic compensation using multilevel inverter based shunt active power filter," *2010 Joint International Conference on Power Electronics, Drives and Energy Systems & 2010 Power India*, New Delhi, 2010, pp. 1-6.

[9] S. Kouro *et al.*, "Recent Advances and Industrial Applications of Multilevel Converters," in *IEEE Transactions on Industrial Electronics*, vol. 57, no. 8, pp. 2553-2580, Aug. 2010.

[10] P. Cortes, A. Wilson, S. Kouro, J. Rodriguez and H. Abu-Rub, "Model Predictive Control of Multilevel Cascaded H-Bridge Inverters," in *IEEE Transactions on Industrial Electronics*, vol. 57, no. 8, pp. 2691-2699, Aug. 2010.

[11] S. Vazquez, J. Rodriguez, M. Rivera, L. G. Franquelo and M. Norambuena, "Model Predictive Control for Power Converters and Drives: Advances and Trends," in *IEEE Transactions on Industrial Electronics*, vol. 64, no. 2, pp. 935-947, Feb. 2017.

[12] J. Rodriguez *et al.*, "State of the Art of Finite Control Set Model Predictive Control in Power Electronics," in *IEEE Transactions on Industrial Informatics*, vol. 9, no. 2, pp. 1003-1016, May 2013.

[13] B. Gutierrez and S. -. Kwak, "Finite set model predictive control method of shunt hybrid power filter," *2015 9th International Conference on Power Electronics and ECCE Asia (ICPE-ECCE Asia)*, Seoul, 2015, pp. 2849-2852.

[14] V. K. Gonuguntala, A. Fröbel and R. Vick, "Performance analysis of finite control set model predictive controlled active harmonic filter," *2018 18th International Conference on Harmonics and Quality of Power (ICHQP)*, Ljubljana, 2018, pp. 1-6.

[15] S. A. Taher, M. H. Alaee and Z. Dehghani Arani, "Model predictive control of PV-based shunt active power filter in single phase low voltage grid using conservative power theory," *2017 8th Power Electronics, Drive Systems & Technologies Conference (PEDSTC)*, Mashhad, 2017, pp. 253-258.

[16] S. Vazquez *et al.*, "Model Predictive Control: A Review of Its Applications in Power Electronics," in *IEEE Industrial Electronics Magazine*, vol. 8, no. 1, pp. 16-31, March 2014.

[17] L. Tarisciotti *et al.*, "Model Predictive Control for Shunt Active Filters With Fixed Switching Frequency," in *IEEE Transactions on Industry Applications*, vol. 53, no. 1, pp. 296-304, Jan.-Feb. 2017.

[18] S. A. Verne, M. Rivera and M. I. Valla, "Current-source multilevel inverter operated at constant switching frequency with a hybrid FCS-MPC strategy," *2017 IEEE Southern Power Electronics Conference (SPEC)*, Puerto Varas, 2017, pp. 1-6.

[19] L. Comparatore, R. Gregor, J. Rodas, J. Pacher, A. Renault and M. Rivera, "Model based predictive current control for a three-phase cascade H-bridge multilevel STATCOM operating at fixed switching frequency," *2017 IEEE 8th International Symposium on Power Electronics for Distributed Generation Systems (PEDG)*, Florianopolis, 2017, pp. 1-6.

[20] M. Norambuena, C. Garcia and J. Rodriguez, "The challenges of predictive control to reach acceptance in the power electronics industry," *2016 7th Power Electronics and Drive Systems Technologies Conference (PEDSTC)*, Tehran, 2016, pp. 636-640.

[21] Y. Zhang, X. Wu, X. Yuan, Y. Wang and P. Dai, "Fast Model Predictive Control for Multilevel Cascaded H-Bridge STATCOM With Polynomial Computation Time," in *IEEE Transactions on Industrial Electronics*, vol. 63, no. 8, pp. 5231-5243, Aug. 2016.

[22] M. Norambuena, S. Dieckerhoff, S. Kouro and J. Rodriguez, "Finite control set model predictive control of a stacked multicell converter with reduced computational cost," *IECON 2015 - 41st Annual Conference of the IEEE Industrial Electronics Society*, Yokohama, 2015, pp. 001819-001824.

[23] M. Norambuena, C. Garcia, J. Rodriguez and P. Lezana, "Finite Control Set Model Predictive Control reduced computational cost applied to a Flying Capacitor converter," *IECON 2017 - 43rd Annual Conference of the IEEE Industrial Electronics Society*, Beijing, 2017, pp. 4903-4907.

[24] M. Narimani, Bin Wu, V. Yaramasu, Zhongyuan Cheng and N. R. Zargari, "Finite Control-Set Model Predictive Control (FCS-MPC) of Nested Neutral Point-Clamped (NNPC) Converter," in *IEEE Transactions on Power Electronics*, vol. 30, no. 12, pp. 7262-7269, Dec. 2015.

[25] M. Novak, T. Dragicevic and F. Blaabjerg, "Finite Set MPC Algorithm for Achieving Thermal Redistribution in a Neutral-Point-Clamped Converter," *IECON 2018 - 44th Annual Conference of the IEEE Industrial Electronics Society*, Washington, DC, 2018, pp. 5290-5296.

[26] F. Kieferndorf, P. Karamanakos, P. Bader, N. Oikonomou and T. Geyer, "Model predictive control of the internal voltages of a five-level active neutral point clamped converter," *2012 IEEE Energy Conversion Congress and Exposition (ECCE)*, Raleigh, NC, 2012, pp. 1676-1683.

[27] C. A. Silva and J. I. Yuz, "On sampled-data models for model predictive control," *IECON 2010 - 36th Annual Conference on IEEE Industrial Electronics Society*, Glendale, AZ, 2010, pp. 2966-2971.

[28] Guohong Zeng, T. W. Rasmussen, Lin Ma and R. Teodorescu, "Design and control of LCL-filter with active damping for Active Power Filter," *2010 IEEE International Symposium on Industrial Electronics*, Bari, 2010, pp. 2557-2562.

[29] N. Panten, N. Hoffmann and F. W. Fuchs, "Finite Control Set Model Predictive Current Control for Grid-Connected Voltage-Source Converters With LCL Filters: A Study Based on Different State Feedbacks," in *IEEE Transactions on Power Electronics*, vol. 31, no. 7, pp. 5189-5200, July 2016.

978-1-7281-4181-7/19 $31.00 © 2019 IEEE

Distance Detection System for Digital Transmitter Coil Achieving Distance-Variation-Tolerant Wireless Power Transfer

Hao Qiu
The University of Tokyo
Tokyo, Japan
hqiu@iis.u-tokyo.ac.jp

Yoshiaki Narusue
The University of Tokyo
Tokyo, Japan
narusue@akg.t.u-tokyo.ac.jp

Yoshihiro Kawahara
The University of Tokyo
Tokyo, Japan
kawahara@akg.t.u-tokyo.ac.jp

Takayasu Sakurai
The University of Tokyo
Tokyo, Japan
tsakurai@iis.u-tokyo.ac.jp

Makoto Takamiya
The University of Tokyo
Tokyo, Japan
mtaka@iis.u-tokyo.ac.jp

Abstract— **In this paper, a distance detection system that only processes the information in the transmitter (TX) side is proposed. By using the distance detection system, the radius of the digital TX coil can change automatically depending on the distance and thus achieve the maximum efficiency tracking. A wireless power transfer (WPT) prototype is fabricated and experimental results demonstrate that, compared with using the conventional TX coil with a constant radius, the efficiency is improved from 12% to 20% at the distance of 4 times the RX coil radius.**

Keywords—Coil-to-coil efficiency, coil design, distance variation, distance detection, impedance measurement, magnetic resonance coupling, mutual inductance, optimal radius, wireless power transfer.

I. INTRODUCTION

Wireless power transfer (WPT) using magnetic resonance coupling has been applied in a wide range of applications such as electric vehicles (EVs), portable electronics and biomedical implanted devices [1]–[4]. The distance (d) variation between the transmitter (TX) and receiver (RX) coils is a common problem. For example, EVs with different ground clearances [2] can have a different d. In portable electronics [3], it is preferred that these devices are not put on the wireless charging pad but can be held in hand and operated during wireless charging. Thus, d can change during operations. In biomedical implanted devices [4], the different body postures of the patients can also change d and affect the system's efficiency. Thus, a d-variation-tolerant WPT method is required for practical applications.

The efficiency of a WPT system is determined by both the coil-to-coil efficiency (η) and the impedance matching

condition [1]. When d changes, the coupling between the TX and RX coils varies and then affects η. In addition, the coupling variation changes the input impedance of the TX coil and there will be an impedance mismatch with the source impedance. Therefore, the maximum system efficiency cannot be achieved when d varies.

Several adaptive impedance matching methods with respect to d have been proposed. The first method uses the adaptive frequency tracking [5]. The optimal frequency is tracked to overcome the frequency-splitting phenomenon even when the coupling between TX and RX coils enters the over-coupling region. However, it can be problematic since the industrial, scientific, and medical (ISM) bands are very narrow. The second method has the same operating frequency but uses the adaptive impedance matching network [1], [6]–[7], which requires complex control and affects the system's reliability. The third method uses coil repeaters between the TX and RX coils [8]–[12], in which a large number of repeaters adds the system's complexity. In addition, the impedance matching condition in [8]–[9] is adjusted manually, which makes the system impractical.

Though the above methods can achieve the impedance matching condition with respect to d, η cannot be improved since the coil design is not discussed. In [13], the optimal design of the RX coil was discussed. From a practical point of view, however, a TX-only tuning is preferred since the RX coil should be as simple as possible. On the other hand, [14] shows the optimal layout of the TX coil, but how to adaptively change the coil's radius is not discussed.

In [15], the topology of digital TX coil that consists of several sub-coils connected in parallel is proposed and its radius can be changed by selectively turning on one of these

---

This work was partially supported by JST ERATO Grant Number JPMJER1501, Japan.

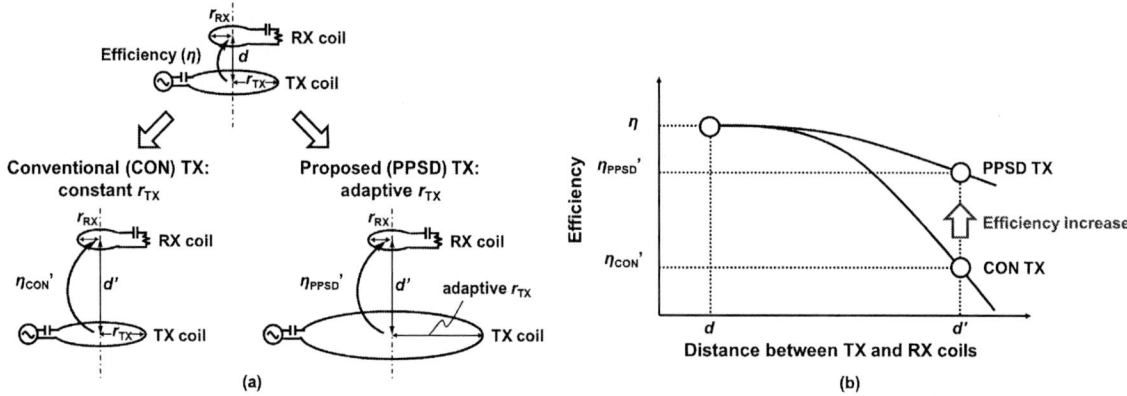

Fig. 1. (a) Conventional TX coil with constant r_{TX} and the proposed TX coil with adaptive r_{TX} depending on d. (b) Schematic of d dependence of η.

sub-coils. However, its radius is changed manually, which is far from practical applications. In this paper, we propose a d detection system that only processes the information in the TX side. Based on the d detection system, the digital TX coil can change its radius automatically and realize the tracking of the maximum η at different d. The rest of the paper is organized as follows. Section II discusses the optimal radius of the TX coil for the maximum η. In Section III, the digital TX coil integrated with the d detection system is proposed. Experimental results are presented in Section IV. Finally, conclusions are given in Section V.

II. OPTIMAL TRANSMITTER COIL RADIUS FOR MAXIMUM EFFICIENCY

In this paper, as shown in Fig. 1, it is assumed that the RX coil is perfectly aligned with the TX coil and changes its position along the central axis. The radii of the TX and RX coils are r_{TX} and r_{RX}, respectively. In a conventional (CON) design, the TX coil has a constant radius, whereas in the proposed (PPSD) design, the TX coil has an adaptive r_{TX} that depends on d. As d increases to d', η_{PPSD}' using the PPSD TX coil is higher than η_{CON}' using the CON TX coil.

In a WPT system, η is defined as the ratio of the power delivered to the load (R_L) to the power input into the TX coil. Under the condition that R_L is the optimal value, η can be expressed as [16]

$$\eta = \frac{k^2 Q_{TX} Q_{RX}}{\left(1 + \sqrt{1 + k^2 Q_{TX} Q_{RX}}\right)^2},\qquad(1)$$

where

$$k^2 Q_{TX} Q_{RX} = \frac{(2\pi f_0 M)^2}{R_{TX} R_{RX}}.\qquad(2)$$

Here, f_0 is the resonance frequency. M is the mutual inductance between the TX and RX coils. R_{TX}, R_{RX} are the parasitic resistances of the TX and RX coils, respectively. k is the coupling coefficient between the TX and RX coils. Q_{TX} and Q_{RX} are the quality factors of the TX and RX coils, respectively.

Thus, η can be optimized by maximizing $k^2 Q_{TX} Q_{RX}$.

$k^2 Q_{TX} Q_{RX}$, on the other hand, is correlated with the physical parameters of the coils. Firstly, we discuss the calculation of the coils' resistances. For loosely wounded coils, the proximity effect is negligible, so R_{TX} and R_{RX} can be simplified to [14]

$$R_{TX} = \frac{2mr_{TX}}{\sigma\varpi\delta}, \quad R_{RX} = \frac{2nr_{RX}}{\sigma\varpi\delta}.\qquad(3)$$

Here, σ is the conductivity of copper (5.96×10^7 S/m), ϖ is the diameter of the copper wire (1 mm). The skin depth (δ) can be calculated as $(\pi\mu_0 f_0 \sigma)^{-1/2}$, where μ_0 is the vacuum permeability ($4\pi \times 10^{-7}$ H/m). m and n are the number of turns of the TX and RX coils, respectively.

Secondly, under the condition that $r_{RX} \ll r_{TX}$, M can be obtained as [17]

$$M = \frac{\mu_0 \pi}{2} mn \frac{r_{TX}^2 r_{RX}^2}{\left(d^2 + r_{TX}^2\right)^{3/2}}.\qquad(4)$$

By substituting (3) and (4) into (2), $k^2 Q_{TX} Q_{RX}$ is expressed as

$$k^2 Q_{TX} Q_{RX} = \left(\frac{\pi^2 \mu_0 \sigma \varpi \delta f_0}{2}\right)^2 mn \frac{r_{RX}^3}{\left(r_{TX} + d^2 / r_{TX}\right)^3}.\qquad(5)$$

978-1-7281-4181-7/19 $31.00 © 2019 IEEE

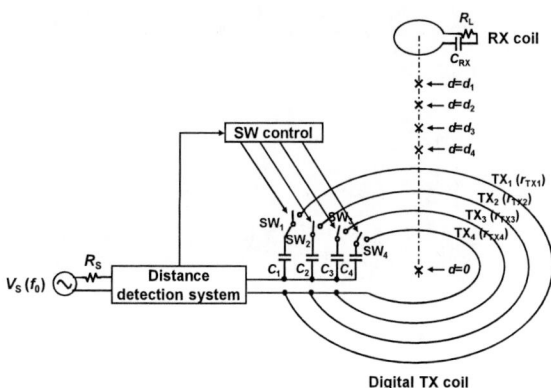

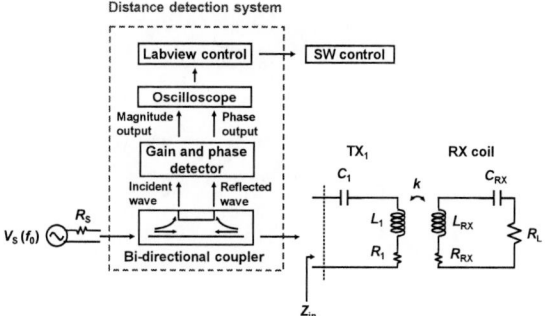

Fig. 2. A WPT system consisting of the proposed digital TX coil, the RX coil, and the d detection system.

Fig. 3. Block diagram of d detection system in Fig. 2. TX_1 in the digital TX coil is used. d can be detected by measuring Z_{in}.

It can be derived that when f_0, m, n, and r_{RX} are constants, $k^2 Q_{\text{TX}} Q_{\text{RX}}$ reaches the maximum when r_{TX} equals its optimal value ($r_{\text{TX, OPT}}$) which is obtained as

$$r_{\text{TX, OPT}} = d . \qquad (6)$$

Equation (6) means, under the condition that $r_{\text{RX}} \ll r_{\text{TX}}$, using the TX coil with r_{TX} that equals d, we can achieve the maximum η.

III. DIGITAL TRANSMITTER COIL WITH PROPOSED DISTANCE DETETION SYSTEM

A. Digital Transmitter Coil Topology

Fig. 2 shows a WPT system consisting of a voltage source (V_S) with a source resistance (R_S), the digital TX coil, the RX coil, and the proposed d detection system followed by a switch (SW) control unit. In this paper, we assume the RX coil has four positions, namely, d_1, d_2, d_3, and d_4. The digital TX coil consists of four concentric sub-coils (TX_1, TX_2, TX_3, and TX_4). For TX_p ($p = 1, 2, 3$, and 4), its radius ($r_{\text{TX}p}$) is determined according to (6). Its number of turns, inductance and parasitic resistance are represented by m_p, L_p, and R_p, respectively. The compensation capacitor has the capacitance of C_p. For the RX coil, the inductance and compensation capacitance are represented by L_{RX} and C_{RX}, respectively. f_0 of each sub-coil in the digital TX coil is the same as that of the RX coil. With respect to d_p detected by the d detection system, the corresponding TX_p is selected.

B. Proposed Distance Detection System

Fig. 3 shows the block diagram of the proposed d detection system in Fig. 2. According to (4), d is correlated with M. At the same time, M can be known by measuring the input impedance (Z_{in}) of the TX coil [16]. Thus, d can be detected by measuring Z_{in}. Different from [10] and [11], where the

communication channel between the TX and RX coils is necessary, the d detection system here only processes the information of the TX side and makes the whole system easier and more reliable. In this paper, TX_1 in the digital TX coil is used for d detection. The details are given as follows.

Using a bidirectional coupler and a gain phase detector, Z_{in} can be measured from the ratio of the amplitude of the reflected and incident waves, namely, the reflection coefficient (Γ).

$$Z_{\text{in}} = Z_0 \frac{1+\Gamma}{1-\Gamma}, \qquad (7)$$

where Z_0 is the characteristic impedance of the transmission line and equals 50 Ω.

M is correlated with Z_{in} and expressed by [16]

$$M = \frac{\sqrt{(Z_{\text{in}} - R_1)(R_L + R_{\text{RX}})}}{2\pi f_0}. \qquad (8)$$

Thus, by combining (4), (7), and (8), d can be detected as

$$d = \sqrt{\left(\frac{\left(\mu_0 \pi^2 f_0 m_1 n r_{\text{TX1}}^2 r_{\text{RX}}^2\right)^2}{(Z_{\text{in}} - R_1)(R_L + R_{\text{RX}})}\right)^{1/3} - r_{\text{TX1}}^2}. \qquad (9)$$

With the developed d detection system, the algorithm of selecting the corresponding sub-coil in the digital TX coil is given. Firstly, SW_1 is turned on and d_p is detected, followed by turning off SW_1 and then turning on the corresponding SW_p. Here, $p = 1, 2, 3$, and 4. Thus, in an electronical way, the radius of the digital TX coil can be automatically controlled to the optimal value for the maximum η at each d.

978-1-7281-4181-7/19 $31.00 © 2019 IEEE

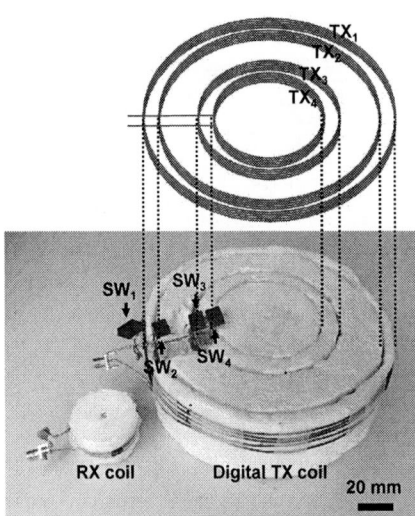

Fig. 4. Photograph of fabricated digital TX coil and RX coil.

TABLE I
MEASURED PARAMETERS OF DIGITAL TX COIL AND RX COIL

	Digital TX Coil				RX Coil
	TX_1	TX_2	TX_3	TX_4	
f_0	13.6 MHz				
Radius	r_{TX1} = 80 mm	r_{TX2} = 70 mm	r_{TX3} = 45 mm	r_{TX4} = 35 mm	r_{RX} = 20 mm
Inductance	L_1 = 15.3 µH	L_2 = 12.4 µH	L_3 = 4.77 µH	L_4 = 2.73 µH	L_{RX} = 0.554 µH
Parasitic resistance	R_1 = 2.6 Ω	R_2 = 2.3 Ω	R_3 = 1.0 Ω	R_4 = 0.70 Ω	R_{RX} = 0.18 Ω
Capacitance	C_1 = 11 pF	C_2 = 13 pF	C_3 = 32 pF	C_4 = 52 pF	C_{RX} = 247 pF
Quality factor	503	461	408	333	263

Note: Number of turns of the digital TX and RX coils are 5 and 3, respectively. Thicknesses of the digital TX and RX coils are 15 mm and 7 mm, respectively.

IV. MEASUREMENT RESULTS

In order to verify the effectiveness of the proposed d detection system in the digital TX coil, a WPT prototype is fabricated, and experimental results are presented.

A. WPT Prototype Implementation

In this paper, it is assumed that the RX coil has 4 known positions, namely, d_1 = 80 mm, d_2 = 65 mm, d_3 = 50 mm, and d_4 = 35 mm. The RX coil and digital TX coil consisting of four concentric sub-coils were fabricated, as shown in Fig.4. Four relays (TQ2-L2-4.5, Panasonic Electric Works) with an on resistance of less than 50 mΩ were used as SWs. r_{RX} is 20 mm, and all radii of the sub-coils in the digital TX coil are determined by the electromagnetic simulation results in [15]. The difference from (6) is analyzed in [15] and mainly ascribed to the approximation made in (4). Table I lists the measured parameters of each coil. Considering the coils' thicknesses, d is defined as the distance between the

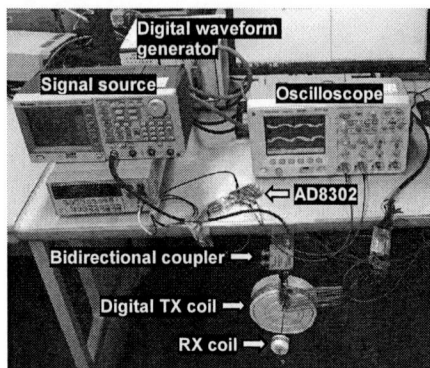

Fig. 5. Photograph of the WPT prototype consisting of digital TX coil with the developed d detection system and the RX coil. R_L = 50 Ω.

geometrical center points of the digital TX coil and the RX coil.

Fig. 5 shows the photograph of the WPT prototype consisting of the digital TX coil with the proposed d detection system and the RX coil. A signal source (AFG3252, Tektronix) is used to generate a sinusoidal wave signal with a peak-to-peak value of 5 V at 13.6 MHz. R_S is 50 Ω. A bidirectional coupler (ZFBDC20-61HP+, Mini-Circuits) and a gain phase detector (AD8302, Analog Devices) perform Z_{in} measurement. The output waveforms are measured by an oscilloscope (D5054A, Agilent Technologies). A labview control program is run to perform the d detection and then generate the control voltages for the SWs through a digital waveform generator (PXIe-6555, National Instruments).

B. Evaluation of WPT Prototype

Firstly, the d detection system was evaluated, in which measured M according to (8) and calculated M according to (4) are compared in Fig. 6. In addition, the calculated M by including the coils' thicknesses is also shown for comparison. It is shown that the difference between measured and calculated M becomes smaller when the coils' thicknesses are included. The percentage error is estimated as 14% at d = 35 mm and decreases to 2.7% at d = 80 mm. The remaining difference is mainly ascribed to the fact that no impedance calibration is performed in Z_{in} measurement. To guarantee the correct d detection, the measured M is calibrated using the calculated M at d = 65 mm as a reference. After M calibration, the detected d (d_{det}) is calculated according to (4). Fig. 7 shows the relationship between d_{det} and d. Their strong correlation guarantees the correct operation of SW control unit.

In addition, a network analyzer (E5061B, Keysight Technologies) was used to measure the S parameters between the TX and RX coils, and η is calculated as $|S_{21}|^2/(1-|S_{11}|^2)$. Fig. 8 shows the measured d dependence of η. Compared with the conventional TX coil with a constant radius, the proposed

978-1-7281-4181-7/19 $31.00 © 2019 IEEE

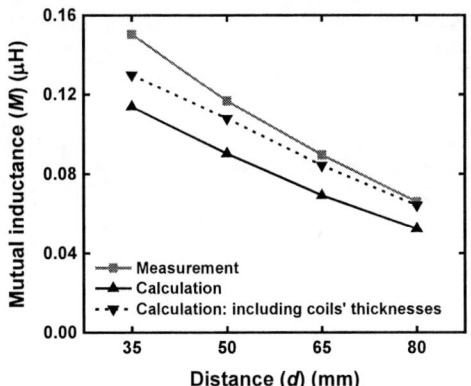

Fig. 6. *d* dependence of measured and calculated *M*.

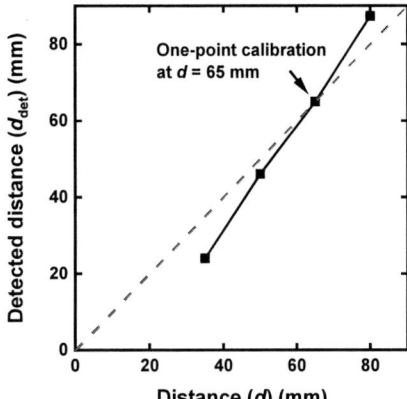

Fig. 7. Correlation between d_{det} and *d*.

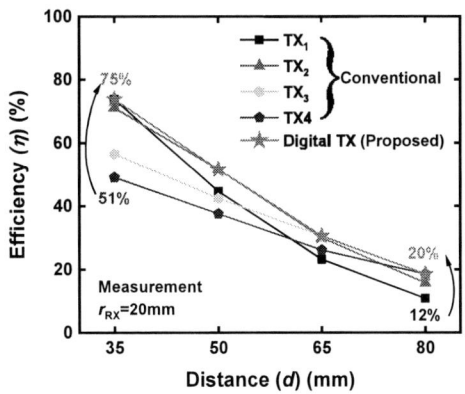

Fig. 8. Measured *d* dependence of η using the proposed digital TX coil, compared with the conventional TX coil with a constant r_{TX}.

digital TX coil with an adaptive radius achieves the maximum η at different *d*. For example, at *d* = 80 mm, η is improved from 12% to 20% (almost doubled) by turning on TX_1 rather than TX_4. Similarly, for *d* = 35 mm, η is improved from 51% to 75% by turning on TX_4 rather than TX_1.

V. CONCLUSIONS

A distance detection system that only processes information in the TX side is proposed and implemented in the digital TX coil for wireless power transfer robust against distance variation. Experimental results demonstrate that, based on the proposed distance detection system, the radius of the digital TX coil can adjust automatically depending on the distance. Compared with the conventional TX coil with a constant radius, the WPT efficiency is improved from 12% to 20% (almost doubled) at the distance of 4 times the RX coil radius.

REFERENCES

[1] T. C. Beh, M. Kato, T. Imura, O. Sehoon, and Y. Hori, "Automated impedance matching system for robust wireless power transfer via magnetic resonance coupling," *IEEE Trans. Ind. Electron.*, vol. 60, no. 9, pp. 3689–3698, Sep. 2013.

[2] J. Schneider *et al.*, "Bench testing validation of wireless power transfer up to 7.7kW based on SAE J2954," *SAE Int. J. Passeng. Cars – Electron. Electr. Syst.*, vol. 11, no.2, pp. 89–108, Apr. 2018.

[3] C. Kim, D. Seo, J. You, J. Park, and B. Cho, "Design of a contactless battery charger for cellular phone," *IEEE Trans. Ind. Electron.*, vol. 48, no. 6, pp. 1238–1247, Jun. 2001.

[4] A. K. RamRakhyani, S. Mirabbasi, and M. Chiao, "Design and optimization of resonance-based efficient wireless power delivery systems for biomedical implants," *IEEE Trans. Biomed. Circuits. Syst.*, vol. 5, no. 1, pp. 48–63, Feb. 2011.

[5] J. Park, Y. Tak, Y. Kim, Y. Kim, and S. Nam, "Investigation of adaptive matching methods for near-field wireless power transfer," *IEEE Trans. Antennas Propag.*, vol. 59, no. 5, pp. 1769–1773, May 2011.

[6] J. Lee, Y.-S. Lim, W.-J. Yang, and S.-O. Lim, "Wireless power transfer system adaptive to change in coil separation," *IEEE Trans. Antennas Propag.*, vol. 62, no. 2, pp. 889–897, Feb. 2014.

[7] W.-S. Lee, K.-S. Oh, and J.-W. Yu, "Distance-insensitive wireless power transfer and near-field communication using a current-controlled loop with a loaded capacitance," *IEEE Trans. Antennas Propag.*, vol. 62, no. 2, pp. 936–940, Feb. 2014.

[8] T. P. Duong and J.-W. Lee, "Experimental results of high-efficiency resonant coupling wireless power transfer using a variable coupling method," *IEEE Microwave Wireless Compon. Lett.*, vol. 21, no. 8, pp. 442–444, Aug. 2011.

[9] J. Lee, Y. Lim, H. Ahn, J.-D. Yu, and S.-O. Lim, "Impedance-matched wireless power transfer systems using an arbitrary number of coils with flexible coil positioning," *IEEE Antennas Wireless Propag. Lett.*, vol. 13, pp. 1207–1210, Jun. 2014.

[10] B.-C. Park, and J.-H. Lee, "Adaptive impedance matching of wireless power transmission using multi-loop feed with single operating frequency," *IEEE Trans. Antennas Propag.*, vol. 62, no. 5, pp. 2851–2856, May 2014.

[11] J. Kim and J. Jeong, "Range-adaptive wireless power transfer using multiloop and tunable matching techniques," *IEEE Trans. Ind. Electron.*, vol. 62, no. 10, pp. 6233–6241, Oct. 2015.

[12] G. Lee *et al.*, "A reconfigurable resonant coil for range adaption wireless power transfer," *IEEE Trans. Microw. Theory Techn.*, vol. 64, no. 2, pp. 624–632, Feb. 2016.

[13] S. B. Lee and I. G. Jang, "Layout optimization of the secondary coils for wireless power transfer systems," in *IEEE Wireless Power Transfer Conf.*, Jun. 2015, pp. 1–4.

[14] B. Waters, B. Mahoney, G. Lee, and J. Smith, "Optimal coil size ratios for wireless power transfer applications," in *Proc. IEEE Int. Symp. Circuits Syst.*, Jun. 2014, pp. 2045–2048.

[15] H. Qiu, Y. Narusue, Y. Kawahara, T. Sakurai, and M. Takamiya, "Digital coil: transmitter coil with programmable radius for wireless powering against distance variation," in *IEEE Wireless Power Transfer Conf.*, Jun. 2018, pp. 1–4.

[16] K. Hata, T. Imura, and Y. Hori, "Simplified measuring method of kQ product for wireless power transfer via magnetic resonance coupling based on input impedance measurement," in *Proc. 43rd Annu. Conf. IEEE Ind. Electron. Soc.*, Oct. 2017, pp. 6974–6979.

[17] S. Ramo, J. R. Whinnery, and T. Van Duzer, *Fields and Waves in Communication Electronics*, 3rd ed., New York: Wiley, 2007.

Digital Current Control for a Bidirectional Interleaved Boost Converter with Coupled Inductors

Francesco Toniolo[†], Qing Liu[*], Tommaso Caldognetto[†], Simone Buso[*], Giorgio Spiazzi[*]

[*]Dept. of Information Engineering (DEI) – University of Padova – Padova, ITALY
e-mail: [name].[surname]@dei.unipd.it
[†]Dept. of Management and Engineering (DTG) – University of Padova – Vicenza, ITALY
e-mail: [name].[surname]@unipd.it

Abstract—This paper describes the implementation of a digital current controller for a Bidirectional Interleaved Boost converter with Coupled Inductors (BIBCI). The final goal of the project is to develop a complete, high efficiency, battery management system (BMS). In order to do that, an optimized, triple phase shift (TPS) digital PWM strategy is used to maximize the BIBCI conversion efficiency. On top of that, a battery current controller is implemented, whose dynamic modeling is analyzed. The controller design procedure is then explained. Experimental results, taken on a $1.5\,\text{kW}$ converter prototype, validate the theoretical analysis and control design procedure.

Index Terms—Interleaved converters, soft-switching, battery charger, digital control.

I. INTRODUCTION

THE Bidirectional Interleaved Boost converter with Coupled Inductors (BIBCI), shown in Fig. 1a, is an interesting solution for interfacing low-voltage batteries with a high voltage DC link. The topology has numerous merits, the main ones being the continuous current on the low voltage side, with reduced ripple at twice the switching frequency, the flexible step-up ratio and the Zero-Voltage-Switching (ZVS) capability for all switches [1]–[4]. The latter, however, calls for a dedicated and optimized modulation strategy to be fully exploited.

This paper describes the modeling and design of the BIBCI current controller, that represents the first

Francesco Toniolo's research activity is sponsored by Fondazione *Cariverona.*

and essential building block of any battery management system (BMS). After a brief recap of the converter's energy transfer mechanism and of its operating modes, the adopted modulator properties are explained. Based on that, the modeling approach is discussed that allowed us to derive a small-signal, open-loop, dynamic model of both the converter and the modulator. Then, we analyze the current controller design and present a set of experimental results validating the modeling and design procedures.

II. BIBCI PRINCIPLE OF OPERATION

The energy transfer mechanism of the BIBCI is identical to that of the well known Dual Active Bridge (DAB) topology. Indeed, the converter operation is such that the power transferred from one side to the opposite depends on the voltage applied to an energy transfer inductor, as shown in Fig. 1b. Notably, this inductor results from the secondary side leakage inductance of the two only converter magnetic components (i.e. the coupled inductors). Voltage sources $v_A(t)$ and $v_B(t)$ derive from the two full bridge circuits, with a remarkable property: the amplitude V_A of the three-level voltage $v_A(t)$ is not constant (as it would be in a DAB converter), but, rather, is proportional to the clamp capacitor voltage v_{CL} and, as such, depends on the duty-cycle of the interleaved boost-type cells that form the low voltage bridge. This feature is usually exploited to compensate for the input voltage variations and keep the amplitudes V_A and V_B of the two voltages

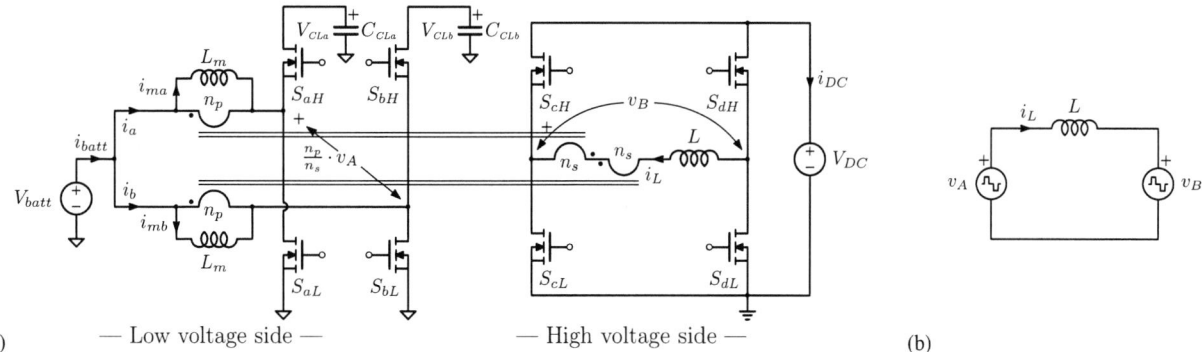

(a) — Low voltage side —   — High voltage side —   (b)

Fig. 1. Bidirectional Interleaved Boost with Coupled Inductors (BIBCI) converter: (a) basic topology schematic; (b) equivalent circuit.

$v_A(t)$ and $v_B(t)$ equal, implementing the so-called *Plus-Phase-Shift* (PPS) modulation [1], [2]. The power flow between input and output ports is then regulated by adjusting the phase shift between $v_A(t)$ and $v_B(t)$, while keeping the latter a pure square-wave. However, with such a modulation strategy, the good efficiency obtained at nominal load rapidly degrades at light load.

On the other hand, the two controllable full bridges allow more sophisticated modulation strategies, that can be exploited to optimize the efficiency. An effective one is detailed in [5], whose features are briefly outlined in the next section. Based on that, it is possible to set-up any kind of higher level control organization. In the case considered in this paper, the low voltage side input current (i.e. the battery current) is closed-loop regulated by a digital controller.

III. TRIPLE PHASE SHIFT (TPS) MODULATION

As is discussed in detail in [5], the BIBCI converter exhibits a large set of different operating modes, 12 in number. These result from the particular values of the three modulation parameters that govern the switch commutations, namely: the duty-cycle of the interleaved boost at the low voltage side, D_B, the duty-cycle of the three-level voltage generated by the controlled bridge at the high voltage side, D_H, and the phase-shift, φ, between the two three-level voltages v_A and v_B (please, see Fig. 2). The modulation strategy that employs the three parameters and adjusts them simultaneously, indicated as *Triple-Phase-Shift* modulation (TPS) in the following, has been shown to offer a significantly higher efficiency at low power levels with respect to the conventional PPS modulation. The typical waveforms with PPS/TPS modulation are shown in Fig. 2, highlighting the capability of three-level operation on the high voltage bridge. The TPS modulator is built around a pre-defined, off-line computed, look-up table (LUT), where the optimal combinations of the three parameters are stored. Optimality is achieved by selecting the (φ, D_B, D_H) triplet that, for any given battery power and voltage set-points, that represent the entry points of the LUT, achieves ZVS on all switches (whenever possible) and the minimum RMS current on the energy transfer inductor. The ZVS region obtainable for the converter design whose parameters are listed in Tab. I is shown in Fig. 3a. Clearly, the TPS modulation allows to extend the ZVS condition at light load with respect to PPS modulation, improving the overall conversion efficiency. The effect is well visible in Fig. 3b, where the light load efficiency is improved by almost 2% at $V_{batt} = 60\,\mathrm{V}$ with respect to PPS modulation. The efficiency is therefore higher than 96% over the whole applicable power range.

TABLE I
CONVERTER SPECIFICATIONS AND PARAMETER VALUES

Parameter	Symbol	Value	
Battery voltage	V_{batt}	$40 \div 60$	V
Nominal DC link voltage	V_{DC}	400	V
Nominal power	P	± 1500	W
Switching frequency	f_{sw}	60	kHz
Energy transfer inductance	L	150	μH
Magnetizing inductance	L_m	30	μH
Clamp capacitors	C_{CL}	6.8	μF
Turns ratio	n_p/n_s	6/20	

Once the modulator is implemented, any control loop can be built on top of it, maintaining the optimal conversion efficiency and achieving reference tracking capability. To do so, the controller just needs to provide the automatic adjustment of the LUT entry points. For a BMS, the power set-point is used as the control input. At the same time, the battery voltage entry is updated in a simple feed-forward manner, based on the assumption that the battery voltage will change very slowly and, therefore, will have a negligible impact on the converter dynamic response.

IV. BATTERY CURRENT CONTROL

The BIBCI converter is used to interface a Li-ion battery system operating at 48 V (nominal) with a 400 V DC link, regulated by a dedicated converter (a grid-tied inverter). The purpose of the circuit is to allow the implementation of a BMS, so as to integrate the battery in a more complex power system (i.e. a hybrid nano-grid). In order to do that, the battery current needs to be closed loop regulated in the first place. As far as the regulation dynamics are concerned, the only constraint is to limit the battery stress by setting a current slew-rate limitation. In the case of interest, this is set to $100\,\mathrm{A/s}$ and it is assumed that the full current swing, from $-30\,\mathrm{A}$ to $+30\,\mathrm{A}$ and vice-versa, cannot be repeated earlier than every 6 s. This is equivalent to track a trapezoidal current reference with a 12 s period and 60 A peak to peak amplitude. Clearly, these specifications pose only mild requirements to the control system, so that a very low (e.g., below 10 Hz) regulation bandwidth is applicable. However, in order to properly design the battery current digital controller, the dynamic response of both the TPS modulator and the power converter must be suitably characterized.

A. TPS modulator model

The small-signal model of the TPS modulator is needed to describe how its three outputs, φ, D_B and D_H respond to input perturbations. Because the modulation frequency, f_{sw}, is so much larger than the target

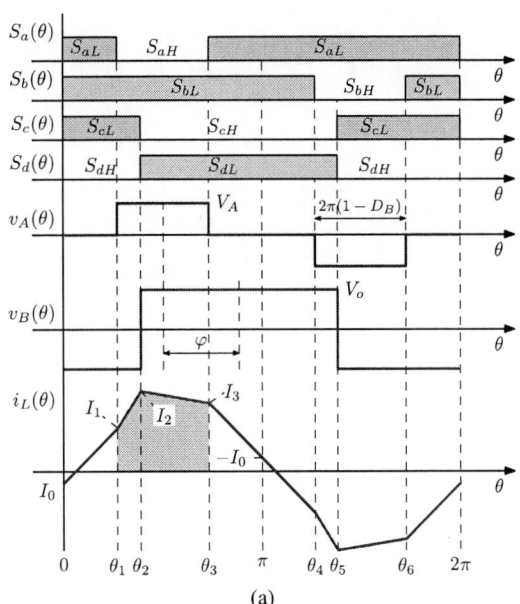
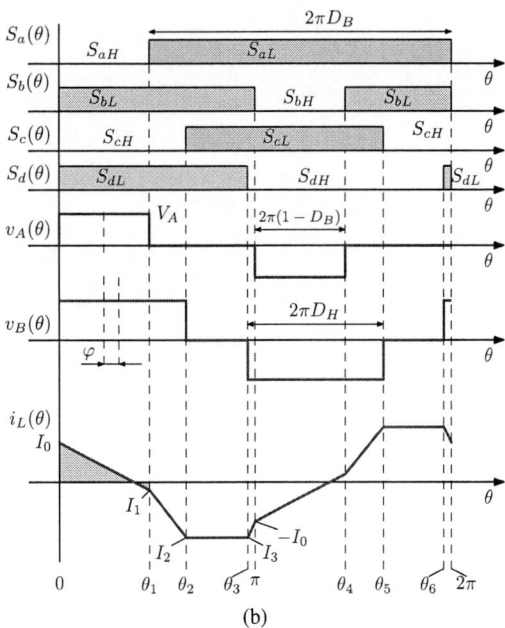

Fig. 2. Example of main converter waveforms for $D_B > 0.5$: (a) PPS using only D_B and φ; (b) TPS, using D_B, D_H and φ. By reducing D_H the RMS current can be reduced and ZVS achieved for all switches. The shaded areas are relevant for the average model derivation.

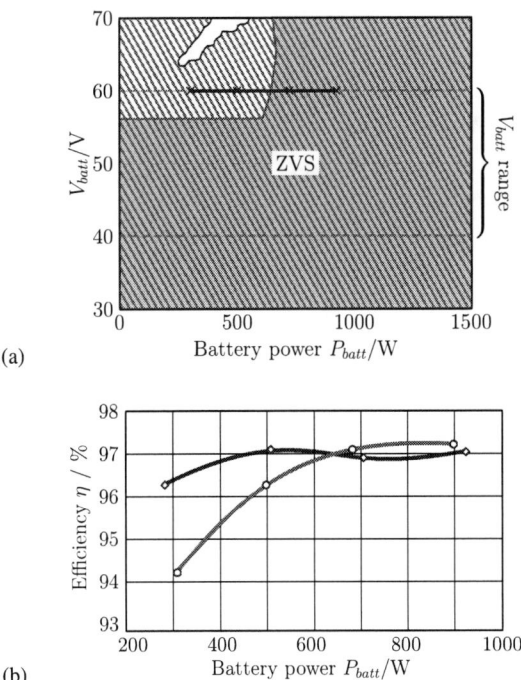

Fig. 3. TPS modulation effect. (a) Region where ZVS is achieved for all switches with TPS/PPS modulation. The darker area is the ZVS region for PPS modulation. The line connects the points where efficiency has been measured. (b) Measured efficiency at $V_{batt} = 60\,\text{V}$: TPS modulation (blue trace) and PPS modulation (red trace).

controller bandwidth, there is virtually no delay in its response. Therefore, it will be modeled as an instantaneous system. The static characteristics of the modulator are shown in Fig. 4, where φ, D_B and D_H are plotted against

P_{batt} for three different V_{batt} voltage levels, including the expected minimum and maximum. It is evident that the optimal TPS modulation mainly exploits the phase-shift φ to adjust the transferred power, while both the boost duty-cycle D_B and the high voltage bridge duty-cycle D_H are adjusted in response to battery voltage variations, to optimize the converter efficiency. Based on the plots of Fig. 4, we can calculate the small-signal gains of the modulator. As an example, we can express the modulator gain that relates power and phase-shift variations as

$$g_{\varphi p} \triangleq \frac{\tilde{\varphi}}{\tilde{p}_{batt}} = \left.\frac{\partial \varphi}{\partial p_{batt}}\right|_{p_{batt}=P_{batt}}, \qquad (1)$$

where the ~ operator indicates a small-signal perturbation of the steady-state operating condition. The other gains can be calculated in the same way. Considering the plots of Fig. 4, one also sees that $g_{d_H p}$ and $g_{d_B p}$ will be zero anywhere the respective plots are flat.

B. BIBCI small-signal model

The small-signal model for the BIBCI converter can be derived from its averaged non-linear dynamic model. In order to keep it simple, only the low voltage side dynamics are taken into account. This is shown in Fig. 5, where the typical structure of a boost converter can be recognized. The model is given just for one phase, since the interleaved operation of the two input boost cells is equivalent, in average terms, to a current doubling effect, i.e. the average battery current is just given by

$$I_{batt} = I_{ma} + I_{mb} = 2 \cdot I_m \qquad (2)$$

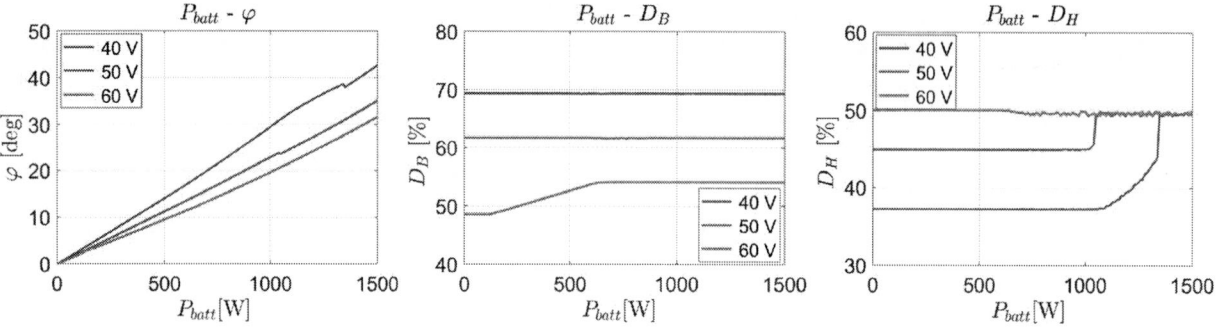

Fig. 4. Plots of LUT outputs as functions of the entries for three battery voltages. From left to right: phase shift φ, boost duty-cycle D_B and high voltage bridge duty-cycle D_H. The plots for negative power values (battery charging mode) are identical (only φ has opposite sign).

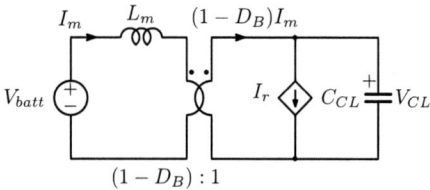

Fig. 5. Non-linear, average, dynamic model of the low voltage converter side.

since, for symmetry, $I_{ma} = I_{mb} \triangleq I_m$. Observing the boost switching cell, we can set the following constraint for the average current flowing into the clamp capacitor,

$$I_{CL} = \overline{\left(i_m - \frac{i_L}{n}\right)}\Bigg|_{S_{aH}=on} = I_m(1 - D_B) - \overline{i_r} \quad (3)$$

where the bar operator indicates the time average and $n = n_p/n_s$. While the first term of (3) is easily determined, the calculation of $\overline{i_r}$ is a little more involved, because the averaging integral

$$\overline{i_r} = I_r = \frac{1}{2\pi} \int_{S_{aH}=on} \frac{i_L}{n}(\theta)\, d\theta, \quad (4)$$

depends on the converter operating region. For example, in Fig. 2, the two shaded areas correspond both to (4), but are obviously different. However, with the considered design, the converter *only operates in a limited number of regions,* indicated as #5, #6 and #8 in [5], as is shown in Fig. 6. Therefore, three different expressions of (4) need to be computed. After a few cumbersome calculations, the expressions, valid for $D_B > 0.5$ (that is the typical case for our design), are the ones shown in Tab. II, where a normalization current $I_N = V_{DC}/(2\pi f_{sw}L)$ has been considered. As can be seen, in each region, I_r is a different non linear function of the circuit inputs φ, D_B, D_H, V_{DC}, but, interestingly, does not depend on state variable v_{CL}. In order to derive a linear model, the functions in Tab. II and the circuit of Fig. 5 have to be linearized around any selected operating point

TABLE II
EXPRESSIONS OF $I_{rN} = I_r/I_N$, $\quad I_N = \dfrac{V_{DC}}{2\pi f_{sw}L}$

R. #5	$\frac{1}{4}(1 - D_B)(2\varphi + 2\pi D_H - (1 - D_B)\pi) - \frac{\pi}{4}\left(\frac{\varphi}{\pi} - D_H\right)^2$
R. #6	$\frac{\pi}{2}(D_H(1 - D_H) + D_B(1 - D_B)) - \frac{\varphi^2}{4\pi} - \frac{\pi}{4}\left(\frac{\varphi}{\pi} - 1\right)^2$
R. #8	$\varphi(1 - D_B)$

$OP = [P_{batt}, V_{batt}]$. That requires the determination of all the small-signal gains in

$$\tilde{i}_r = \tilde{\varphi} \cdot \frac{\partial I_r}{\partial \varphi}\bigg|_{OP} + \tilde{d}_B \cdot \frac{\partial I_r}{\partial d_B}\bigg|_{OP} + \tilde{d}_H \cdot \frac{\partial I_r}{\partial d_H}\bigg|_{OP} =$$
$$= \tilde{\varphi} \cdot h_{r\varphi} + \tilde{d}_B \cdot h_{rB} + \tilde{d}_H \cdot h_{rH}. \quad (5)$$

The effect of V_{DC} has been neglected on purpose, as this is considered constant and independent from the

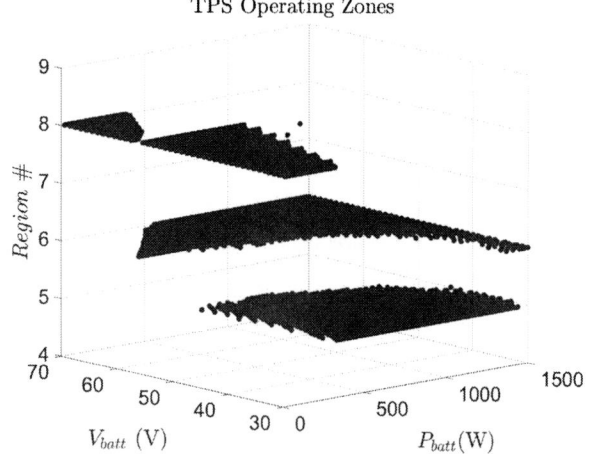

Fig. 6. Operating regions of the BIBCI converter over the whole control space (defined by voltage and power ranges) for the considered application design. Out of the 12 operating regions, only #5, #6 and, marginally, #8 are relevant.

978-1-7281-4181-7/19 $31.00 © 2019 IEEE

power flowing through the BIBCI. The calculation of all the gains is a little involved, but can be easily done numerically. The results are given in Tab. III for the nominal operating point $[1.5\,\text{kW}, 48\,\text{V}]$, that is located in region #6.

TABLE III
PARAMETERS OF THE SMALL-SIGNAL MODEL

$g_{\varphi p}$	0.023	$[deg]\text{W}^{-1}$	$g_{d_H p}$	0	W^{-1}
$g_{d_B p}$	0	W^{-1}			
$h_{r\varphi}$	2.161	$\text{A}/[deg]$	h_{rH}	0	A
h_{rB}	-2.22	A	I_m	15	A
G_{iP0}	0.252	AW^{-1}	ω_z	155	s^{-1}
ω_p	$28\cdot 10^3$	s^{-1}			

Finally, the small-signal, linear circuit shown in Fig. 7 is determined, based on which the transfer function to be stabilized can be found. By superposition and taking into account the appropriate modulator gains for each converter control input, the transfer function is given by

$$G_{i_{batt}P^\star_{batt}}(s) = G_{iP0}\frac{1+s/\omega_z}{1+(s/\omega_p)^2} \tag{6}$$

where

$$G_{iP0} = 2\cdot\frac{g_{\varphi p}h_{r\varphi} + g_{d_H p}h_{rH} + g_{d_B p}(h_{rB}+I_m)}{1-D_B},$$

$$\omega_z = \frac{G_{i0}(1-D_B)}{2\,C_{CL}V_{batt}}, \tag{7}$$

$$\omega_p = \frac{1-D_B}{\sqrt{L_m C_{CL}}}.$$

V. CONTROLLER DESIGN

The controller for the BMS discussed in this paper is described by the block diagram of Fig. 8. A simple integral controller is considered, on behalf of the very low cross-over frequency required by the application. Indeed, keeping the closed loop system time constant τ_{CL} below $0.1\,\text{s}$ is more than adequate for a good trapezoidal reference tracking, yielding a delay limited

to 0.8% of the period of the fastest applicable reference signal. Referring to the values in Tab. III, it is possible to verify that, at the frequency of interest, neither the transfer function's (7) complex poles nor the zero have any significant effect, so (7) can be well approximated by its DC gain G_{iP0}. Likewise irrelevant is the sampling and processing delay, provided that at least a few kHz sampling frequency is adopted for the battery current signal. Therefore, the controller can be designed in the continuous time domain and directly translated into a control algorithm without incurring into any phase margin penalization. In conclusion, the controller gain can be easily sized as:

$$K_I = \frac{2\pi f_{CR}}{G_{iP0}K_T} \cong 1\cdot 10^3 \tag{8}$$

having chosen $f_{CR} = 4\,\text{Hz}$ (i.e. $\tau_{CL} \cong 0.04\,\text{s}$) and being $K_T = 0.1\,\text{VA}^{-1}$ the current transducer gain.

VI. DIGITAL CONTROLLER IMPLEMENTATION

The battery current controller is implemented on a general purpose converter control board [6], currently used at our laboratory. It comprises three main parts: the integral regulator, the LUT and the PWM modulator. The integral regulator, together with the LUT, including the access logic and a bilinear interpolator, are realized on the software programmable part of the control board, namely a PowerPC. This allows to store a large number of operating points and to easily implement the interpolator filter and the controller in C language. The PWM logic, including counters, binary comparators and protections are instead synthesized on the FPGA chip of the control board, namely a Xilinx Spartan-6 LX45, that also performs the A/D conversions. For safety reasons, the eight gate signals are sent to the power board through a galvanic isolation barrier, that is applied to the sampled signals as well. A possible limitation of this architecture is represented by the communication delay between the PowerPC and the FPGA, that is about $2\,\text{ms}$. For the target application, however, such delay has negligible consequence. If higher performance were needed, with some further effort, the LUT and the controller could as well be synthesized within the FPGA chip, thus practically eliminating any control delay.

VII. EXPERIMENTAL VERIFICATION

In order to verify the theoretical analysis and the controller design, several experimental tests have been performed on a converter with the parameters listed in Tab. I. An example of the system functionality is shown in Fig. 9, where the battery current reference is stepping between $6\,\text{A}$ and $13\,\text{A}$ (left column) and

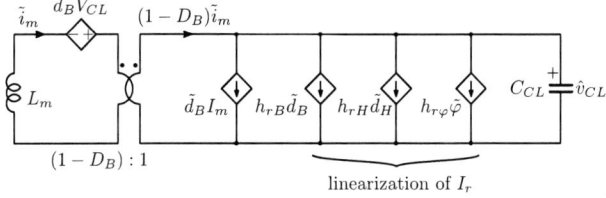

Fig. 7. Linear, small-signal model of the low voltage side of the BIBCI converter.

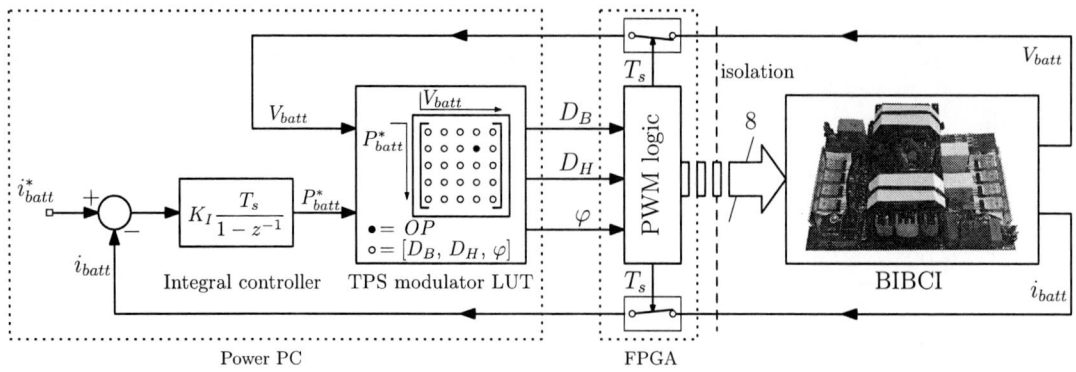

Fig. 8. Block diagram of the adopted control architecture. T_s is the selected sampling period.

ramping between 7 A and 14 A (right column). The estimated time constants (32 ms - 44 ms) are consistent with the design specifications and, in any case, well below the maximum acceptable (100 ms). At 80 A/s, the tracking delay matches the measured time constants, as expected. To make the delay more visible, the period of the trapezoidal reference has been substantially shortened with respect to the specifications. Finally, the sensitivity to the battery voltage level is very limited, as anticipated. Some minor quantization effect is visible on the step responses, that has been solved in the ramp test by better scaling the current feedback signal and increasing the resolution of the digital integrator.

VIII. Conclusions

This paper discusses the use of a BIBCI converter as a BMS for Li-ion batteries. In particular, the design and implementation of a digital battery current controller is presented. Leveraging on the TPS optimal modulation strategy, the controller achieves the expected dynamic performance, while maintaining the maximum conversion efficiency compatible with the considered design. The converter and modulator's small-signal models are derived and used to design the regulator. To verify the functionality of the controller, measurements are taken on a 1.5 kW BMS prototype, showing encouraging results.

References

[1] H. Xiao and S. Xie, "A ZVS Bidirectional DC-DC Converter with Phase-Shift Plus PWM Control Scheme," *Power Electronics, IEEE Transactions on*, vol. 23, no. 2, pp. 813–823, March 2008.

[2] W. Li, H. Wu, H. Yu, and X. He, "Isolated Winding-Coupled Bidirectional ZVS Converter with PWM Plus Phase-Shift (PPS) Control Strategy," *Power Electronics, IEEE Transactions on*, vol. 26, no. 12, pp. 3560–3570, Dec 2011.

[3] H. Bahrami, S. Farhangi, H. Iman-Eini, and E. Adib, "A New Interleaved Coupled-Inductor Nonisolated Soft-Switching Bidirectional DC-DC Converter with High Voltage Gain Ratio," *IEEE Transactions on Industrial Electronics*, vol. 65, no. 7, pp. 5529–5538, July 2018.

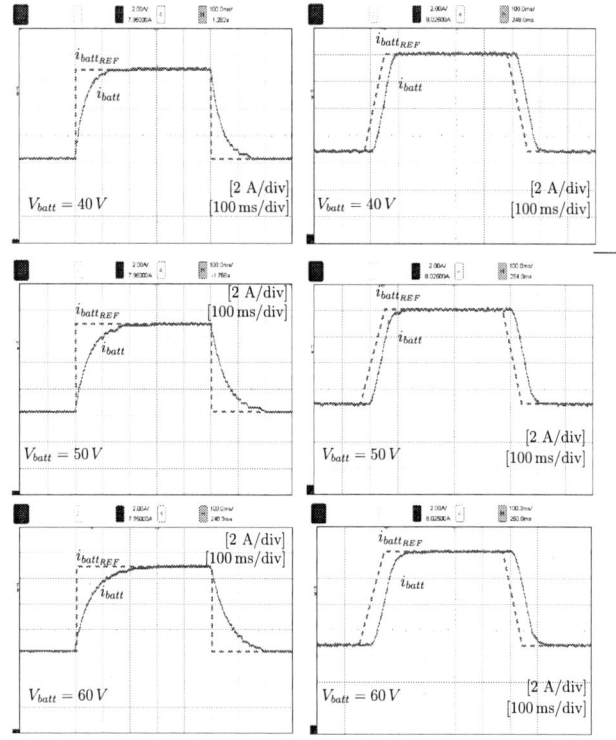

Fig. 9. Experimental step and ramp response test for the battery current controller at three different battery voltages. The measured time constants and transport delays are respectively 32 ms at $V_{batt} = 40$ V (top), 40 ms at $V_{batt} = 50$ V (middle), 44 ms at $V_{batt} = 60$ V (bottom).

[4] G. Spiazzi, S. Buso, D. Biadene, G. Rossetto, and F. Mela, "High Efficiency Battery Charger for Photovoltaic Inverters," in *2017 IEEE Southern Power Electronics Conference (SPEC)*, Dec 2017, pp. 1–6.

[5] F. Toniolo, P. Mattavelli, and G. Spiazzi, "Design Criteria and Modulation Strategies for Complete ZVS Operation of the Bidirectional Interleaved Boost Converter with Coupled Inductors," in *Twentieth IEEE Workshop on Control and Modeling for Power Electronics, IEEE COMPEL 2019*, 2019.

[6] "Single Board RIO - General Purpose Inverter Controller," National Instruments, Tech. Rep., 2015.

PeMSyn: a Free Software to Assist the Design and Performance Assessment of Permanent Magnets Synchronous Machines

Khristian M. de Andrade Jr[1], Hugo E. Santos[1], Wellington M. Vilela[1],
Thales E. P. de Almeida[2], and Geyverson T. de Paula[1]

Abstract—This paper presents PeMSyn, a free and open source graphical user interface based in Matlab/FEMM, for the design and simulation of Permanent Magnet Synchronous Machines. PeMSyn aggregates several functionalities normally found separated in other similar softwares, such as: automation of the drawing and sizing of Permanent Magnet Machines, winding design, Finite Element Analysis, pole segmentation analysis and performance assessment of Permanent Magnet Machines. PeMSyn is also modular, allowing the user to employ each one of these features individually. Moreover, it can be easily translated to any language by means of a language file and allows adding other machine topologies.

This paper discusses each one of the modules in detail, the equations used in the design process, input parameters and output results, as well as an application example.

Index Terms—Graphical User Interface, Permanent Magnet Synchronous Machine, Design and simulation of electrical machines.

I. Introduction

Permanent magnet machines are compact, have high torque density and low maintenance [1]. Due to these characteristics, this type of machine has been employed in numerous applications and some of them have been catching attention of several researchers around the world, such as in wind turbines [2], electric vehicles [3], boat and submarine propulsion [4]. Nowadays, during the machine design process, it is a common practice to use design and finite element analysis softwares, such as FEMM (Finite Element Method Magnets software)[5], SyR-e (Synchronous Reluctance (machines)-evolution) [6] and QuickField [7], usually integrated with a simulation and analysis software like Matlab [8] for numerical, data and machine's performance assessment. Although a large number of softwares for machine design exist, a feature not commonly found in them is the analysis of pole segmentation, a cogging torque reduction technique [9].

In order to facilitate the study of electrical machines, the authors have developed the PeMSyn program. It is a graphical user interface (GUI) based on Matlab script that interacts with FEMM to design and evaluate permanent magnet machines parameters and solve electromagnetic problems. PeMSyn is composed of six independent modules, which can be run separately.

Two of these modules are focused on the design, dimensioning and finite element analysis of surface mounted and inset mounted PMSM, both with inner rotor. Other two modules

perform a similar design procedure but focused on outer rotor machines. The user must input some machine parameters, such as pole number, stator slot number, line voltage, rated speed, etc. Then, the other parameters will be calculated by PeMSyn and the results (a CAD created in FEMM and sizing data) can be saved.

The winding design module, similar to Koil [10], allows the design and distribution of three-phase windings. The user should input rotor pole number, stator slots and the coil pitch. An option for double-layer or single-layer winding is also available. The module outputs the three-phase winding distribution as a vector as well as it displays a figure to illustrate the assessed distribution. Both, vector and image can be exported.

The last module simulates the performance of the machine at no-load and on-load conditions. At no-load, the outputs are: PM flux-linkage, cogging torque, phase and line back-EMF and their harmonic content. At on-load, the outputs are: the current waveform (sinusoidal or square, as chosen by the user) and the electromagnetic torque. The results can be saved as .mat or .xlsx formats, as well as all figures are automatically exported as .png.

It is important to note that the performance of the machine is approximated by a sequence of static simulations. FEMM is used to solve magnetic problems for rotor position. The rotor is rotated and the process is repeated in a succession of small steps.

It is intended that PeMSyn can be used as an educational and research tool. Therefore, in order to make the experience of non-native speakers of portuguese or english more pleasant, a language file in .m extension with all variables and field values is available. So users are able to translate PeMSyn to their own languages. PeMSyn is available for download on https://pemsyn.sourceforge.io/. Feedback from users is encouraged to improve the software, making it more comprehensive and detailed.

II. Machine Design

The equations used in the machine design modules are presented in the following subsections. For more details, see [11]-[15]. First, the user must specify the supply voltage, V, the machine power output, P, or the output torque, T_D, and its nominal/maximum speed, ω and ω_{max}, respectively. Since three-phase machines are considered, the supply voltage must be the line voltage.

The aforementioned parameters define basic machine characteristics. The supply voltage, alongside maximum speed,

---

[1]The authors are with School of Electrical, Mechanical and Computer Engineering, Federal University of Goiás, Goiás 74605-010, Brazil
[2] The author is with Federal Technological University of Paraná, Paraná, 86812-460, Brazil

978-1-7281-4181-7/19 $31.00 © 2019 IEEE

determine the back-emf constant, k_E (V-s/rad). The air gap size, g, is related to power, whereas rotor diameter and stack length, L_{STK}, are related to the torque per unit volume, i.e., the machine volume is directly related to the developed torque [11]-[13].

The design procedures are analyzed in three parts: rotor, stator and winding design. Nevertheless, these parts are inter-related but their explanation is divided for sake of clarity.

A. Rotor

All dimensions mentioned throughout this section can be found in Fig.1, for both outer and inner rotor. D_{sh} is the shaft diameter.

At first, the user must select the rotor type to be employed: inner or outer, with surface mounted or inset magnets. The choice of rotor type will depend on its application, speed and available space. Inner rotors are suitable for applications where high torque and low speed are required, whereas outer rotors are suitable for applications where high speeds and inertia are required [13].

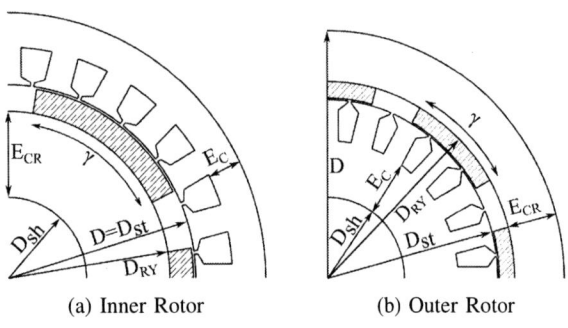

(a) Inner Rotor   (b) Outer Rotor

Figure 1: Machine Dimensions

Next, the magnet shape should be selected, block or arc, as well as the magnet type, e.g. SmCo, NdFeB or Alnico. The air gap length and the permeance coefficient, CP, define the magnet thickness, L_M, as shown in (1). Furthermore, with this coefficient and the magnet demagnetization curve supplied by the manufacturer, it is possible to determine the magnet flux density, B_M, during operation.

$$L_M = CP \cdot g \qquad (1)$$

If multiple magnet blocks (segments) are desired to build a pole, the machine may perform similarly to one whose magnets are arc-shaped. The number of segments, however, is limited since too small dimensions would make the machine construction unfeasible. To illustrate the pole segmentation see Fig. 2. Studies on the effects of pole segmentation can be appreciated in [16]-[18].

Following the previous step, the machine pole number, $2p$, or the pole pairs, p, must be chosen. This number influences a considerable number of parameters, e.g. rotor and stator yoke thickness. An interesting discussion about the selection of this number can be seen in [19]. Regarding the aforementioned thicknesses, the maximum flux density value, B_T, desired

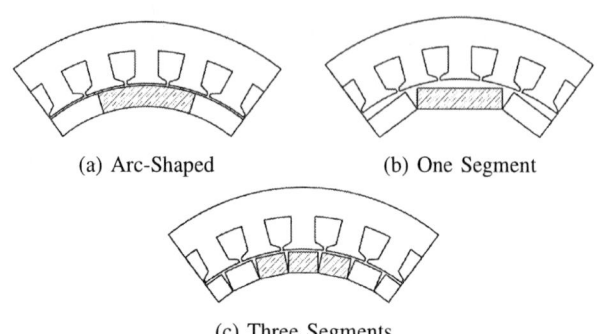

(a) Arc-Shaped   (b) One Segment

(c) Three Segments

Figure 2: Poles Segmentation

in the rotor, stator and teeth should be chosen in order to minimize the material and machine volumes but avoiding electromagnetic saturation.

The computation of the polar area, A_P, is the goal of this step. Its value depends on the pole pitch, γ, the magnets inner diameter, D_{RY}, the stack length and the pole number, as can be seen in (2). Thus the flux per pole, Φ_P, can be calculated using (3).

$$A_P = \frac{\frac{\gamma}{\pi} D_{RY} L_{STK}}{2p} \qquad (2)$$

$$\Phi_P = B_M A_P \qquad (3)$$

For the outer rotor, its yoke thickness , $E_{C_R}^{OR}$, can be obtained by (4), where D corresponds to the rotor outer diameter. Now, for the inner rotor, its yoke thickness, $E_{C_R}^{IR}$, is given by (5), with D being the diameter over slots. In the case of inner rotor, if there is sufficient space, part of this yoke may be replaced by a lighter non-magnetic material, reducing its inertia, weight and cost.

$$E_{C_R}^{OR} = \frac{\pi B_M (D - 2L_M)}{4p B_T + 2\pi B_M} \qquad (4)$$

$$E_{C_R}^{IR} = \frac{\frac{1}{2} B_M A_P}{B_T L_{STK}} \qquad (5)$$

B. Stator

First, the slot number, N, is selected. The user should know that in order to accommodate a three-phase winding, this number must be a multiple of three. In addition, the feasibility of the winding must be verified for the chosen number of slots and poles.

Afterwards, the slot opening, w_o, must allow the accommodation of the coils in the stator, facilitating the manufacturing process. The choice of slot number and slot opening influences the cogging torque value [13].

Leakage factor, f_{lkg}, takes into account the part of the flux that is leaked through the air gap. This value is chosen by the user, between 0 (irrelevant leakage) and 1 (no flux links the stator), and impacts the sizing of stator quantities.

978-1-7281-4181-7/19 $31.00 © 2019 IEEE

The stator yoke thickness, E_C, is obtained by means of (6). In the case of outer rotor, if there is sufficient space, part of this yoke can be replaced by non-magnetic material, reducing the machine weight and cost. Observe again Fig. 1.

$$E_C = \frac{\frac{1}{2}(1 - f_{lkg})B_M A_P}{B_T L_{STK}} \quad (6)$$

The teeth width, w_t, is calculated by (7). The total shoe width, w_s, is given by (8), and its height, h_s, by (9). The ratio between shoe height and tooth height, h_t, is recommended to be between 0.25 and 0.5 [13]. In PeMSyn, the default ratio is 0.25. The last value to be defined is the tooth-tip, tt, which will be half the slot opening, as shown in (10).

$$w_t = (1 - f_{lkg}) \frac{(2p)\Phi_P}{N B_T L_{STK}} \quad (7)$$

$$w_s = \frac{\pi}{N} D_{st} - w_o \quad (8)$$

$$h_s = \frac{w_s - w_t}{2} \quad (9)$$

$$tt = \frac{w_o}{2} \quad (10)$$

The aforementioned teeth dimensions can be seen in Fig. 3, as well as teeth types available in PeMSyn (equal, unequal and straight). In case of unequal teeth, the slot number must be even, and a scale must be chosen to reduce one set and increase the remainder, proportionally.

$$t = \gcd(N, p) \quad (11)$$

The angle between adjacent slots is calculated by (12). See Fig. 4.

$$\alpha = \frac{2\pi}{N} \quad (12)$$

The angle between the phasors is the conversion of the angle between adjacent slots from mechanical to electrical radians.

By taking a slot as reference and using the angle obtained with (12) (in electrical radians), it is possible to determine the "star of slots", as shown in Fig. 5a. Next, the phasors are divided into sectors. For the three-phase winding considered here, the division is done in six sectors of 60°. Each phase winding occupies two sectors, placed 180° from each other, one will be considered "positive" and the other "negative" [20]. See Fig. 5b.

The "go" slots of the positive coil sector are considered to come out of the page. Those of the negative sector are directed into the page. Therefore, to obtain a balanced winding, the sequence of the sectors must be a+, c-, b+, a-, c+ and b-.

Finally, the "return" slots of the coils must be determined. For this, the coil pitch, ρ, is either defined by the user or calculated by the software, using (13). It is worth mentioning that the value obtained with (13) comprises the maximum coil pitch.

$$\rho = \text{int}\left(\frac{N}{2p}\right) \quad (13)$$

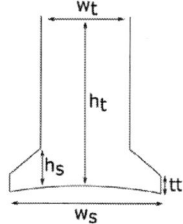 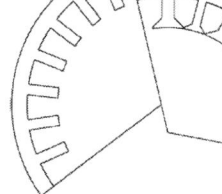

(a) Teeth Dimensions   (b) Straight and Unequal Teeth

Figure 3: Teeth Types and Dimensions

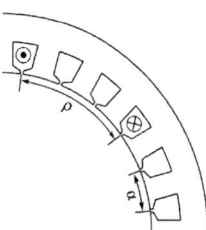

Figure 4: Stator with One Coil Drawn

C. Winding

The module for winding design can be initialized independently or within one of the design modules. If it is the case, the designed winding data may be exported to the design module.

The "star of slots" method is used, [10], [20]-[21]. This method considers the induced voltages in the wires of each slot represented by phasors. The result is a symmetrical and balanced winding [10].

First, one must verify the winding feasibility. The periodicity of the machine, t, is defined in (11) by means of the greatest common divisor (gcd). The Winding is feasible if $N/3t$ is an integer.

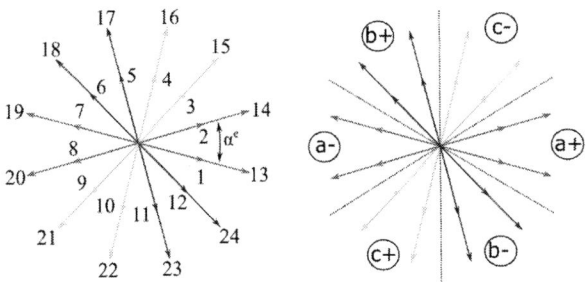

(a) Phasor Distribution   (b) Phasors divided in sectors

Figure 5: Star of slots for $N = 24$ and $2p = 4$

The back-emf constant of the machine is obtained by means of (14).

$$k_E = 0.9 \frac{V}{\omega_{max}} \tag{14}$$

With the back-emf constant, the number of turns per coil, N_t, is obtained using (15). If the torque constant, k_T, is considered equal to the back-emf constant, the peak current, I_{max}, can be obtained by (16).

$$N_t = \text{int}\left(\pi \frac{k_E}{4p(1-f_{lkg})N_c\Phi_P}\right) \tag{15}$$

$$I_{max} = \frac{T_D}{k_T} = \frac{T_D}{k_E} \tag{16}$$

If the straight teeth has been chosen as the teeth type, the value obtained in (15) should be updated. This type of teeth has no shoes, therefore the flux-linkaged value is lower. The updated number of turns, N_ts, is obtained with (17).

$$N_{ts} = \text{int}\left(N_t \frac{w_s}{w_t}\right) \tag{17}$$

Finally, the wire gauge must be determined. The slot area, A_{slt}^{OR} for outer rotors or A_{slt}^{IR} for inner rotors, is a key parameter for this. For outer rotors it can be computed by (18), and (19) for inner ones. In these equations, D_{st} is the diameter over slots (see Fig. 1).

$$A_{slt}^{OR} = \frac{(D_{st}-2h_s)^2 - (D_{st}-2h_t)^2}{4N} - w_t(h_t - h_s) \tag{18}$$

$$A_{slt}^{IR} = \frac{(D_{st}+2h_t)^2 - (D_{st}+2h_s)^2}{4N} - w_t(h_t - h_s) \tag{19}$$

A slot fill factor, f_{slt}, is taken into account since the slot area is not completely occupied by wires. Its value varies from 0.3 to 0.35 for double layer windings and from 0.65 to 0.7 for single layer ones [13]. Therefore, the maximum wire diameter is obtained by (20). The next smaller wire gauge is selected, otherwise less turns per coil can be accommodated in the slot.

$$D_t = \sqrt{f_{slt} \frac{A_{slt}}{N_t}} \tag{20}$$

III. Machine Design and Performance Assessment Using PeMSyn: Example

This section describes how PeMSyn interface works. The example comprises the design of a machine and the assessment of its performance.

Table I presents the input data machine parameters considered in this example. The materials are 1020 steel for both the rotor and stator yoke and 316 stainless steel for the shaft. Magnet Arcs made of NdFeB 40 MGOe are used, mounted on the surface of an inner rotor.

If the number of turns per coil and wire gauge are unknown, these fields must be kept as "0" and "-", respectively. Their values will be computed by PeMSyn.

Table I: Machine Input Data

Parameter	Value
Pole Pairs	8
Number of Slots	24
Rated Power (kW)	10
Air gap (mm)	1
Diameter over Slots (mm)	150
Stack Length (mm)	100
Slot opening (mm)	2
Shaft Diameter (mm)	20
Line Voltage (V)	380
Rated Speed (RPM)	1500
Maximum Speed (RPM)	2000
Permeance Coefficient	6
Magnet Flux Density (T)	1.011
Allowance (T)	1.5
Leakage Factor	0.05

Once the basic data are known, the main module must be initialized, followed by SM-PMSM (Surface Mounted-Permanent Magnet Synchronous Machine) module, as shown in figs. 6-7. Next, the user must fill in the data required by the design module and, for this example, described in Table I, as mentioned.

The panel named "Mesh Size" relates to the finite elements mesh density created by FEMM. The user can enter a fixed value for this. Otherwise, the "Auto" option must be selected, then FEMM will determine the density.

To design the winding the user can use the winding module available by pressing the "Wizard" button within the design module. After the data is filled in and the "Design" button pressed, the winding designed can be visualized, as shown in Fig. 8. Its data can be exported to the design module through the "Export" button.

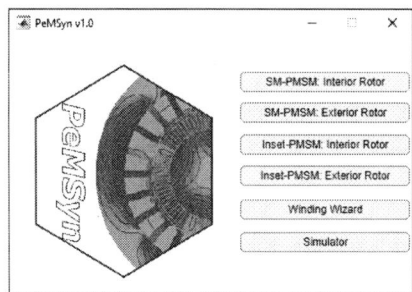

Figure 6: Main Module

With the data from Table I, the materials considered and the winding data, the user must now press "Calculate" and then "Draw". At this point, PeMSyn computes the remaining machine parameters, draws its CAD and assigns all materials in each part of the machine model.

The model created in FEMM is shown in Fig. 9. The input data, the design results and the model can be saved by pressing the "Save" button. The results are shown in Table II .

It is worth mentioning that PeMSyn automatically assigns some conditions to the model for the minimization of numer-

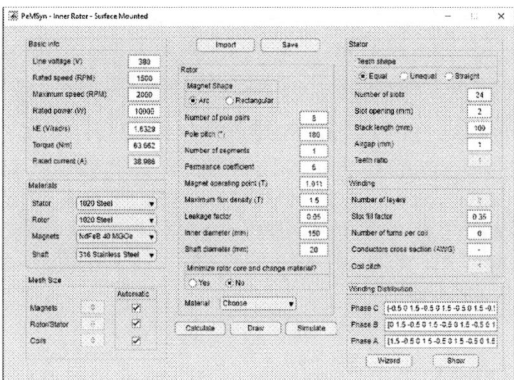

Figure 7: Inner Rotor with Surface Mounted Magnets Machine Design Module

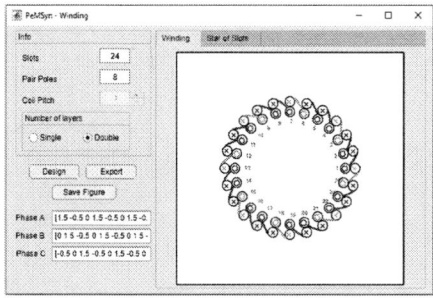

Figure 8: Winding Design Module

ical errors, since FEMM works with first order elements.

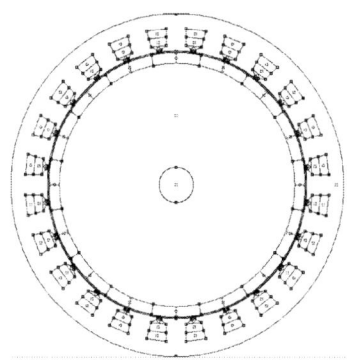

Figure 9: Model Created in FEMM

The performance parameters can be assessed through the simulation of the previously created model. For this, the "Simulate" button within the design module must be pressed. The simulation module is opened, as shown in Fig. 10. The user then must fill in some simulation data, choose the simulation type (no-load, on-load or both) and the current waveform (sinusoidal or square).

Once the simulation is finished, the results are saved automatically. Also, the user can navigate through the tabs available in the module to analyze the desired results. The no-

Table II: Design Results

Parameter	Result
Tooth Width (mm)	11.40
Total Shoes Width (mm)	17.63
Tooth-Tip (mm)	1.00
Shoes Height (mm)	3.12
Tooth Height (mm)	12.47
Stator Yoke Thickness (mm)	12.47
Rotor Yoke Thickness (mm)	8.55
Magnet Thickness (mm)	9.00
Wire Gauge (AWG)	13
Turns per Coil	13

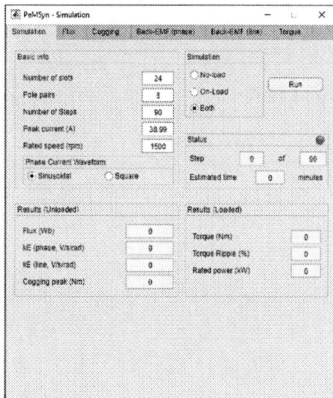

Figure 10: Simulation Module

load simulation comprises the PM flux-linkage, the back-emf and the cogging torque. Such data are necessary for the on-load simulation, being the developed electromagnetic torque its main output.

For this example, the waveforms of the simulation results can be seen in Figs. 11-12. The values are detailed in Table III.

IV. CONCLUSIONS

This paper presents PeMSyn, a free and open source graphical user interface for drawing, sizing and simulating the performance of Permanent Magnets Synchronous Machines.

PeMSyn puts together functionalities that in other softwares are usually found separated. For instance, Koil only designs machine windings. SyR-e does the drawing and sizing of a machine. QuickField is specific only for finite element analysis. All these features are also present in PeMSyn. Moreover, PeMSyn also integrates Matlab/FEMM, analyzes pole segmentation, allows the inclusion of additional machine topologies and is capable of evaluating the performance of any PM machine.

Another advantage of PeMSyn is its modularity, allowing flexibility in program expansion and addition of other features. Due to the modularity, the user is also able to access each functionality independently. For example: once a CAD for the machine is already available, the user can go directly to

978-1-7281-4181-7/19 $31.00 © 2019 IEEE   153

the simulations. Furthermore, PeMSyn has a simple and user-friendly interface, so great computing skills are not required. It can also be easily translated to any language through an available language file.

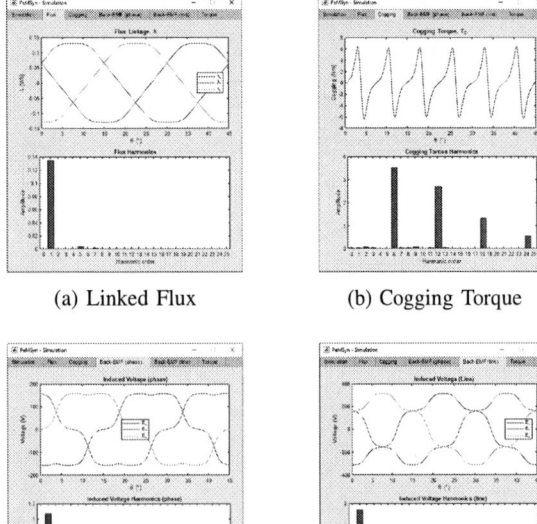

(a) Linked Flux   (b) Cogging Torque

(c) Back-EMF (phase)   (d) Back-EMF (line)

Figure 11: No-load Simulation Results

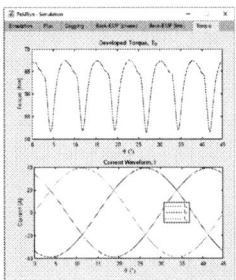

Figure 12: On-load Simulation Results

Table III: Performance Parameters

Parameter	Result
Average Torque (Nm)	62.64
Torque Ripple (%)	12.76
Power (kW)	9.84
Cogging Peak (Nm)	6.40
Currente Peak (A)	38.99
Line Voltage (V)	313.01
Back-EMF Constant (V-rad/s)	1.9927

The example in this paper proves that PeMSyn is a powerful and reliable tool, being nowadays an interesting alternative for design and simulation of electrical machines.

Authors hope PeMSyn will be useful as an educational and research tool, aiding future research projects and helping students understanding machine performance.

V. ACKNOWLEDGEMENTS

All the authors would like to thank CNPq, Coordenação de Aperfeiçoamento de Pessoal de Nível Superior – Brasil (CAPES) (Finance Code 001) and FAPEG (Fundação de Amparo à Pesquisa do Estado de Goiás) for funding guarantee and support.

REFERENCES

[1] Arkkio, N. Bianchi, S. Bolognani, T. Jokinen, F. Luise, and M. Rosu, "Design of synchronous PM motor for submersed marine propulsion systems," inICEM 2002 Proc., Brugge, Belgium, Aug. 2002, pp. 25 – 28.

[2] K. T. Chau, C. C. Chan, Chunhua Liu, "Overview of Permanent-Magnet Brushless Drives for Electric and Hybrid Electric Vehicles" IEEE Transactions on Industrial Electronics, May. 2008, pp. 2246 - 2257

[3] Gieras JF. Advancements in Electric Machines. London: Springer, 2008.

[4] Kane, D. M., & Warburton, M. R. (n.d.). "Integration of permanent magnet motor technology.", IEEE Power Engineering Society Summer Meeting, July 2002.

[5] FEMM - Finite Element Method Magnetics, site http://www.femm.info/wiki/HomePage, acessed 20/05/2019.

[6] Syre, site https://sourceforge.net/projects/syr-e/, acessed 02/06/2019.

[7] QuickField, site https://quickfield.com/, acessed 02/06/2019.

[8] MathWorks, site https://www.mathworks.com/products/matlab.html, acessed 22/05/2019.

[9] T. M. Jahns and W. L. Soong," Pulsating torque minimization techniques for permanent magnet AC motors drives—A review," IEEE Trans. Ind. Electron., vol. 43, no. 2, pp. 321– 330, 1996.

[10] L. Alberti, "Koil: A Tool to Design the Winding of Rotating Electric Machinery," 2018 XIII International Conference on Electrical Machines (ICEM), Alexandroupoli, 2018, pp. 805-811.

[11] T.J.E Miller, *Brushless Permanent-Magnet and Reluctance Motor Drives*. Oxford: Clarendon Press, 1989.

[12] A. Hebala, W. A. M. Ghoneim and H. A. Ashour, "Detailed Design Procedures for Low-Speed, Small-Scale, PMSG Direct-Driven by Wind Turbines," 2018 XIII International Conference on Electrical Machines (ICEM), Alexandroupoli, 2018, pp. 697-703.

[13] J.R. Hendershot Jr and T.J.E. Miller, *Design of Brushless Permanent-Magnet Motors*. Oxford: Magna Physics Publications, 1994.

[14] Duane Hanselman, *Brushless Permanent Magnet Motor Design*. Lebanon: Magna Physics Publishing, 2006.

[15] S. A. Nasar, I. Boldea, L. E. Unnewehr, *Permanent Magnet, Reluctance and Self-Synchronous Motors*. Boca Raton: CRC Press, 1993.

[16] A. Mansouri and H. Trabelsi, "Effect of the number magnet-segments on the output torque and the iron losses of a SMPM," 10th International Multi-Conferences on Systems, Signals & Devices 2013 (SSD13), Hammamet, 2013, pp. 1-5.

[17] R. Lateb, N. Takorabet and F. Meibody-Tabar, "Effect of magnet segmentation on the cogging torque in surface-mounted permanent-magnet motors," in IEEE Transactions on Magnetics, vol. 42, no. 3, pp. 442-445, March 2006.

[18] Sang-Moon Hwang and Kyung-Tae Kim, "Effects of segmented poles on motor performances," in IEEE Transactions on Magnetics, vol. 35, no. 5, pp. 3712-3714, Sept. 1999.

[19] Xiaolong Zhang and Ronghai Qu, "Pole number selection strategy of low-speed multiple-pole permanent magnet synchronous machines," 2013 International Electric Machines & Drives Conference, Chicago, IL, 2013, pp. 1267-1274.

[20] I. Abdennadher and A. Masmoudi, "Star of slots-based graphical assessment of the back-EMF of fractional-slot PM synchronous machines," 10th International Multi-Conferences on Systems, Signals & Devices (SSD13), Hammamet, 2013, pp. 1-8.

[21] N. Bianchi and M. Dai Pre, "Use of the star of slots in designing fractional-slot single-layer synchronous motors," in IEE Proceedings - Electric Power Applications, vol. 153, no. 3, pp. 459-466, 1 May 2006.

Development of Linear Generator Prototype As Part of A Point Absorber Wave Energy Converter

Eduardo Martins
Instituto Federal Fluminense (IFF)
Macaé, Brazil
eduardo.beline@iff.edu.br

Wagner Marcilio
Instituto Federal Fluminense (IFF)
Macaé, Brazil
wagnerm289@hotmail.com

Marcos Moreira
Instituto Federal Fluminense (IFF)
Macaé, Brazil
macruz@iff.edu.br

João Santos
Instituto Federal Fluminense (IFF)
Macaé, Brazil
j.j.james@hotmail.com

Abstract— **The need to produce energy from renewable sources has gained importance in recent years. The ocean has an energetic potential that justifies studies in search of energy conversion. One such source is the sea waves. In this work the behavior of waves in the Região dos Lagos - Brazil for the year 2017 was verified, where the month of December was the most energetic. The permanent magnets used in the assembly were tested and evaluated before installed. A preliminary Transverse Flux Generator prototype was designed, simulated (by Finite Element Analysis) and assembled to understand the dynamics of a linear generator. The final prototype passed through the same steps and in the end it was possible to verify its behavior in different frequencies of movement.**

Keywords—linear generator, sea wave, energy, sustainability

I. INTRODUCTION

The ocean is a gigantic source of renewable energy that spans about three quarters of the Earth's surface and energy can be drawn from waves in many ways. Ocean wave has high power density compared to solar or wind power and it is available and environment friendly [1]. Generating energy from renewable sources is not such a simple task because it requires many types of technologies, which can add a lot of cost to the system. For this reason, investments in research may be an alternative for economic viability [2].

Nowadays, with global attention being attracted by climate change and increasing CO2 levels, the focus on electricity generation from renewable sources is certainly an important area of research. Among emerging options for generating electricity from renewables, the energy present in the oceans is one of the most promising [3]. Globally, ocean energy sources have a significantly higher total potential than global electricity demand of 16,000 TWh per year [4].

The generator proposed in this work is a Transverse Flux Permanent Magnet Linear Generator, and according to [5], due to the direct coupling between wave movement and the generating part, there are several studies focused on projects of this type of generator for Wave Energy Converters - WECs. Reference [6] raise the question that mechanical interfaces such as gearboxes, hydraulic or pneumatic systems increase the complexity of the system and generate more maintenance procedures, lower efficiency and lower reliability. The converter using a linear generator solves this problem by simplifying the system's power take-off (PTO). In a direct drive power take-off, the linear generator is coupled directly to the buoy or via a simple mechanical connection.

Reference [7], after completing his work, pointed out that the energy extraction of the waves by means of a linear generator coupled to a Point Absorber type converter is feasible. Reference [8] verified some topologies similar to the one presented in this work in the attempt to find improvements, based on the work of [9] that investigated a project where the magnets were fixed in the translator. After testing its prototype, [9] concluded that this kind of generator has its construction easier than other types of linear generators and its efficiency reached acceptable values.

Based on the work of [9] and the work performed by [7], a prototype of linear generator was constructed to be applied to a Point Absorber wave energy converter.

II. WAVE BEHAVIOR IN REGIÃO DOS LAGOS - RJ (BRAZIL)

In order to understand the wave pattern in the region where the work was performed, data were obtained from the sensored buoy located in Cabo Frio (Fig. 1), located approximately 70km from the coast. The data are provided by the National Buoys Program. The buoy works with hourly measurements 24 hours a day and can measure variables such as wave height, wave period and others. In one month there can be up to 744 measurements available, if we consider that all the measurements have been validated. It was the database used in this analysis for the year 2017.

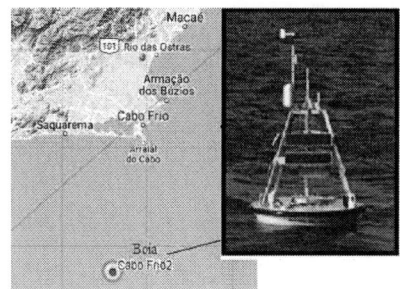

Fig. 1. Buoy location

The total energy in a wave, according to [10], is the combination of the potential energy, due to the vertical displacement of the surface of the water, with the kinetic energy, due to the oscillatory movement of the wave. It is directly proportional to the square of its height.

For the year 2017 the average monthly value of wave height (Hs) and wave period (Tp) were found and can be seen in TABLE I.

978-1-7281-4181-7/19 $31.00 © 2019 IEEE

TABLE I.   AVERAGE VALUES FOR THE YEAR 2017

Months of 2017	Height Hs (m)	Period Tp (s)
January	1.69	7.65
February	1.55	7.85
March	1.89	9.72
April	2.03	9.84
May	2.28	10.86
June	1.84	9.68
July	1.88	9.38
August	2.18	9.70
September	2.26	9.04
October	3.06	9.25
November	3.59	9.76
December	4.13	10.26
YEARLY	2.37	9.42

It can be noticed that the average height of the wave for 2017 had variations between 1.55 and 4.13 meters, while the period varied between 7.65 and 10.86 seconds. The month of December was the most energetic because it had the highest average height, followed by November and October. The mean annual value of Hs was 2.37m and Tp 9.42s.

III. REVIEW OF THE LAST TWO YEARS

A literature search was carried out in the Institute of Electrical and Electronic Engineers IEEE-Xplore database for the term "linear generator" in publications of the years 2018 and 2019. There were found 53 articles whose most common keywords were: linear machines; generation of wave energy; permanent magnet generators; finite element analysis; and electric generators. Most of the articles (64%) were directly related to the conversion of wave energy, which reinforces the applicability of the linear generator in the generation of electricity from the sea waves.

For the 53 works verified after bibliographic research, 29 were related to the tubular structure and 21 to the flat. Some works are highlighted by their innovations.

IV. CHARACTERISTICS OF THE MAGNETS USED

The magnets used in this project were Nd2Fe14B grade N35 and 20 mm x 20 mm x 10 mm dimensions. With a mass of 30g, nickel plated coating, maximum working temperature of 80 ºC and a tensile force of 12.5Kg according to the manufacturer, they have an excellent ratio induction / mass. In order to obtain a reference about its magnetic field at a given distance, measurements were made with a 20 Gauss scale magnetometer (1 Gauss = 10^{-4} Tesla).

Nine points were tested and a Magnetic Field x Distance curve was created (Fig. 2).

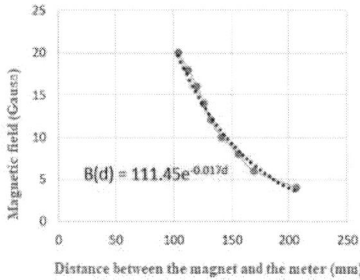

Fig. 2. Magnetic Field x Distance curve

The dotted curve is the exponential approximation whose function B(d) is presented in this same Fig. 2. If we want to infer the magnetic field on the magnet surface, we will find a magnetic field of 111.45 Gauss or 11.14 mTesla. Due to meter limitations, this inference was required.

V. PROTOTYPES - SIMULATIONS AND TESTS

A. Preliminary Prototype

To verify and analyze the dynamics of electric power generation from the translation movements of a linear generator, a prototype of simpler construction was thought. Through Finite Element Analysis (FEA) with the Maxwell software it was possible to simulate this device and, even before its construction, it was possible to modify and test it virtually. These modifications were important because we were able to arrange geometry and available materials with induced voltage values on the coil.

Fig. 3 represents the two metallic parts designed for this prototype, in the part that corresponds to the translator there are two magnets installed with alternating polarities. Point A is coil location.

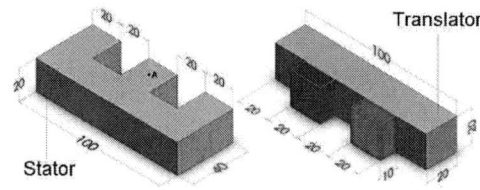

Fig. 3. Stator and translator parts

The design was submitted to simulation and in Fig. 4 the expected behavior of the magnetic field can be seen when the exposed faces of the magnets are 5mm apart from the stator.

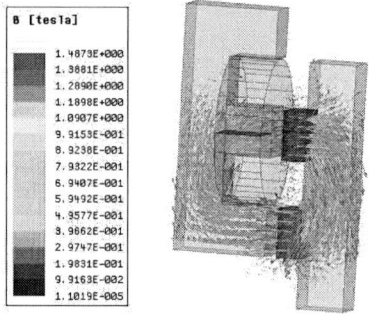

Fig. 4. Behavior of the magnetic field

After simulations the prototype of Fig. 5 was manufactured.

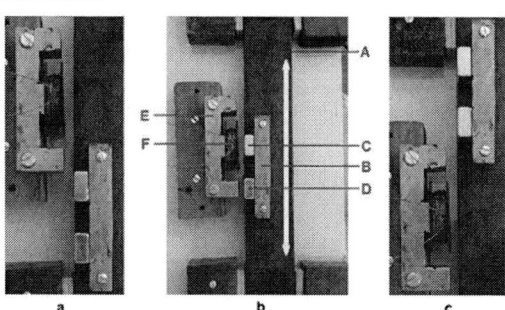

Fig. 5. Preliminary prototype

The Fig. 5b shows the structure of the first prototype where the translator is represented by A. Parts B, C and D belong to this translator and the other parts to the stator. Through Fig. 5a and 5c it is possible to identify the two extremes of the vertical translation movement. Parts B and E were made with overlapping blade cuts taken from low power transformers. F is a coil of enameled copper wire wound on a plastic spool, C and D are N35 grade NdFeB magnets mounted with alternating polarities. For reasons of assembly, the smallest distance between the face of the magnet and the part E is 5mm. A complete translation cycle corresponds to a total displacement of 400 mm, 200 mm down and 200 mm up. Much of the structure of this generator was built in wood to avoid interaction with the magnetic circuit.

The Fig. 6 shows the waveforms for the induced voltage in a coil of 200 turns of 26 AWG wire and for an approximate frequency of 1.2 Hz (0.48 m/s). The left figure (Simulation), copy of the software screen, represents a translation cycle with an approximate amplitude of 4.00 V peak to peak (Vpp). The figure of the right (Real), copy of the screen of the oscilloscope, presents several cycles generated with the vertical translation of the prototype, where a amplitude of 4.40 Vpp was reached. By making a comparison between the real and the simulated, a coefficient of variation of 10% was found.

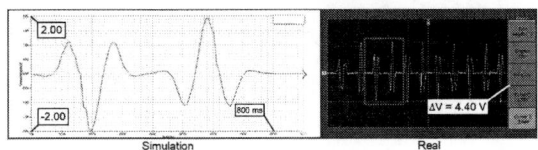

Fig. 6. Waveforms of induced voltage, simulated and real

Some tests were performed with this prototype, both in simulation and in practice. Different amounts of turns were tested for a frequency of approximately 1.2 Hz. The results can be seen in TABLE II.

TABLE II. VOLTAGES INDUCED AT DIFFERENT TURNS FOR F = 1.2 Hz

Number of turns (26 AWG)	Induced voltage, Vpp (Simulation)	Induced voltage, Vpp (Real)	Coefficient of Variation
20	0.38	0.46	21%
40	0.80	0.92	15%
80	1.60	1.74	9%
200	4.00	4.40	10%

By placing the prototype to translate at an approximate frequency of 2.2 Hz (0.88 m/s) with coils of 20, 40, 80 and 200 turns of 26 AWG wire, we found 0.66 Vpp, 1.42 Vpp, 2.96 Vpp and 7.12 Vpp respectively. When the distance between the magnet and the stator was reduced from 5mm to 3mm it was possible to measure in the coil of 200 turns a voltage of 6.40Vpp for the frequency of 1.2 Hz and 11.00Vpp for 2.2 Hz.

B. Final Prototype

This model includes some modifications in relation to the models proposed and realized by [9] and [7]. The main modification is the distance from one magnet to another that is vertically positioned. The spacing was 20 mm, it would fit exactly another magnet. A complete translation cycle corresponds to a total displacement of 160 mm, 80 mm down and 80 mm up. In Fig. 7 it is possible to visualize the upper and lower limits of the translation movement. The simulation through finite element analysis allowed a predictability of results even before the assembly of this final prototype.

The Fig. 7 represents the stator/translator set with some information, thought and developed after simulation conclusions. In the part that corresponds to the translator there are eight magnets installed with alternating polarities and stainless steel plates on the sides of each pair of magnets. Both the stator and the translator were made with laminated iron-silicon alloy steel. This single-phase permanent magnet linear generator features four 26 AWG copper wire coils with 200 turns each connected in series. Plastic parts and Epoxy resin were used for fixing the parts.

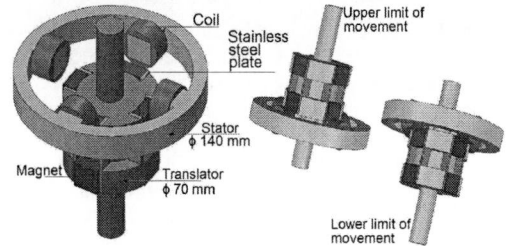

Fig. 7. Stator/translator set

The design was subjected to simulation and in Fig. 8 the magnetic flux density B [Tesla] in the machine can be seen. The most intense points are the coil installation places. The software used does not allow transient simulation for cylindrical bodies. The parts that should have been cylindrical were simulated based on regular polyhedra of 24 segments.

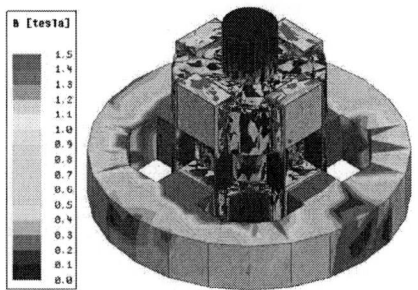

Fig. 8. Magnetic flux density

In Fig. 9 it is possible to visualize the magnetic flux lines A [Wb/m] for a top view of the prototype. The figure corresponds to the exact moment when the upper magnets are aligned with the stator.

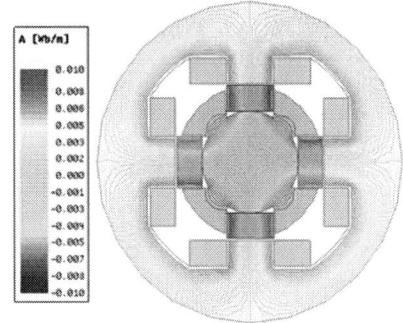

Fig. 9. Magnetic flux lines

Transient simulation requires a very high processing load. Depending on the machine that is running the software, the simulation can take hours or even crashes can happen. Therefore, in complex simulations, it is necessary to use machines with excellent working capacity.

The Fig. 10 shows different views of the prototype already constructed.

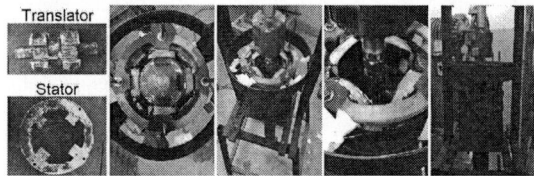

Fig. 10. Final prototype

The Fig. 11 shows the induced voltage waveforms at an approximate frequency of 3.5 Hz. The left figure (Simulation), copy of the software screen, represents a translation cycle with an approximate amplitude of 52 Vpp. The right figure (Real), copy of the oscilloscope screen, presents several cycles generated where a amplitude of 66 Vpp was reached. Making a comparison between the simulated and real, there is a coefficient of variation of 27% for this situation.

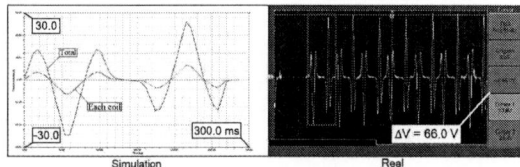

Fig. 11. Voltage generated, simulated and real

The linear generator was tested with three different translation frequencies and the results can be verified in TABLE III. As a load, a resistor with Beryllium Oxide substrate was used due to its low reactance characteristic.

TABLE III. VOLTAGES INDUCED AT DIFFERENT FREQUENCIES

Frequency (Hz)	Speed (m/s)	No Load		50 Ohms Load	
		Total voltage (Vpp)	Total voltage (VRMS)	Total voltage (Vpp)	Total voltage (VRMS)
1.5	0.24	38	8.1	26	6.1
2.5	0.40	44	9.4	35	7.7
3.5	0.56	66	13.1	54	10.4

With data from TABLE III it is possible to verify that, as the flow variation increases (frequency increases), the induced voltage in the coils also increases. The induced voltage in each coil corresponds to ¼ of the total voltage generated. When a load is used, it is already expected that the voltage generated by the equipment would tend to decrease.

When comparing the values found during simulation to the values measured with the aid of the oscilloscope, we verified significant differences based on the coefficients of variation, TABLE IV. These coefficients are relatively higher than those found in the first prototype due to the amount of variables involved in each project . In both cases, the voltage generated was higher than the simulated voltage.

TABLE IV. COMPARISON BETWEEN SIMULATION AND REAL

Frequency of translation (Hz)	Induced total voltage, Vpp (Simulation)	Induced total voltage, Vpp (Real)	Coefficient of Variation
1.5	26	38	46%
2.5	35	44	26%
3.5	52	66	27%

One of the difficulties found in the construction of this prototype is the perfect alignment between translator and stator. In Fig. 12 the waveform of the generated voltage can be seen when the misalignment was present, in this case the translator touched the stator both in the down and up movements generating a noisy waveform when compared to Fig. 11. Another fact is the displacement of the translator less than 80 mm where the magnets do not completely exit the stator direction. An asymmetrical movement can also be verified, condition that can be seen in Fig. 12 due to the lack of symmetry of the waveform. This test was performed at a frequency of 2.5 Hz (no load) and reached a voltage of 32 Vpp (7.4 VRMS). When compared to the same translation frequency as in TABLE III, there is a loss of 21.3% in the RMS voltage generated due to the existing problem.

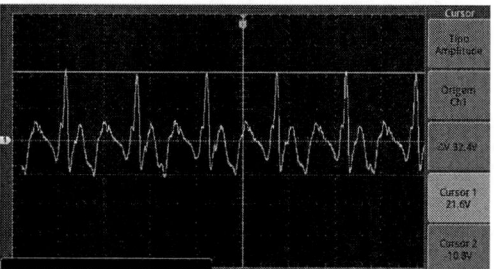

Fig. 12. Waveform with misaligned generator

VI. CONCLUSION

The Linear Generator has been a key part of recent Wave Energy Converters (WEC) projects and it was proven by the survey conducted at the IEEE base. The tubular structure has been shown to be more present in the work in different countries. The simulations through Finite Element Analysis, both in the preliminary prototype and in the final prototype, were of great importance for the learning and assembly of the equipments. The first prototype, because it was simpler and had a smaller number of variables involved, had a low coefficient of variation for the generated voltages when comparing values of simulations and tests.

The final low-cost prototype was designed and assembled with reuse of recyclable material, so it was not possible to make a full comparison with the works of [9] and [7]. [9] used eight NdFeB N40 magnets six times larger volume and performed its test with a translation frequency of 10 Hz, frequency not supported by the final prototype of this work. [7] carried out a similar assembly with eight magnets of the same model, and concluded that for a rotation of 160 rpm generated a peak voltage of 48.5 V. This rotation corresponds to a frequency of 5.3 Hz of the generated voltage waveform. These data demonstrate a significant gain of energy produced in the current generator. From the qualitative point of view, [7] perceived a greater presence of harmonics, as well as lack of symmetry due to its topology (something close to Fig. 12). This characteristic was not so present in the generated voltage of Fig. 11.

The linear generator of the final prototype was shown suitable for use in a Point Absorber WEC. A perceived challenge during assembly would be the mechanical encapsulation designed to keep the generator intact at sea. The sealing of the equipment to avoid contact with the water or even the saline effect will be of great importance in a continuous and uninterrupted generation of energy by this machine.

The waveform generated by the final prototype described here, from the qualitative point of view, has harmonic content. This characteristic is related to the topology used. For continuity of the work, it is suggested a rectification of the voltage generated and its storage in batteries. Tests with different loads can be performed to verify the power delivered by the generator, respecting its constructive limits.

ACKNOWLEDGMENT

This work was carried out in the laboratories of the Instituto Federal Fluminense in the Macaé city. Thanks to IFF per research grant for undergraduate student, and FAPERJ by software license.

REFERENCES

[1] O. Farrok, Md. R. Islam, Md. R. I. Sheikh, Y. Guo, J. Zhu, and W. Xu, "A Novel Superconducting Magnet Excited Linear Generator for Wave Energy Conversion System", *IEEE Transactions on Applied Superconductivity*, vol. 26, no. 7, 2016.

[2] F. Weschenfelder, P. Pauletti, S. D. Bittencourt, L. Pelegrini, D. K. Ito, and L. Schaeffer, "Situação atual e perspectivas da produção de ímãs permanentes e reservas de terras raras: brasil × mundo", *Tecnologia em Metalurgia, Materiais e Mineração*, vol. 9, no. 4, 2012 [In Portuguese].

[3] Y. Gao, S. Shao, H. Zou, M. Tang, H. Xu, and C. Tian, "A fully floating system for a wave energy converter with direct-driven linear generator", *Energy*, vol. 95, 2016.

[4] N. Khan, A. Kalair, N. Abas, and A. Haider, "Review of ocean tidal, wave and thermal energy technologies", *Renewable and Sustainable Energy Reviews*, vol. 72, 2017.

[5] L. Huang, M. Chen, L. Wang, F. Yue, R. Guo, and X. Fu, "Analysis of a Hybrid Field-Modulated Linear Generator For Wave Energy Conversion", *IEEE Transactions on Applied Superconductivity*, vol. 28, no. 3, 2018.

[6] J. Faiz, and A. Nematsaberi, "Linear electrical generator topologies for direct-drive marine wave energy conversion - an overview", *IET Renewable Power Generation*, vol. 11, no. 9, 2017.

[7] I. P. C. Cunha, M. A. C. Moreira, R. C. Santos, and J. A. M. Santos, "Estudo da Viabilidade de Utilização de uma Máquina de Fluxo Transversal Linear como parte da Estrutura do Conversor de Energia das Ondas Point Absorber", *Revista Vértices*, vol. 19, no. 2, 2017 [In Portuguese].

[8] L. H. Joubert, J. Schutte, J. M. Strauss, and R. T. Dobson, "Design Optimisation of a Transverse Flux, Short Stroke, Linear Generator", *XXth International Conference on Electrical Machines*, Marseille, France, 2012.

[9] J. Schutte, "Optimisation of a transverse flux linear PM generator using 3D Finite Element Analysis", Master's thesis, University of Stellenbosch, 2011.

[10] R. G. Dean, and R. A. Dalrymple, *Water wave mechanics for engineers and scientists*, World Scientific: USA, 1991.

The True Unity Power Factor Converter Applied to Photovoltaic Applications

Marcos Henrique da Silva Alves
Graduate Program in Elect. Engineering
Universidade Federal de Minas Gerais
Belo Horizonte - MG, Brazil
marcoshenrique@ieee.org

Thiago Morais Parreiras
Graduate Program in Elect. Engineering
Universidade Federal de Minas Gerais
Belo Horizonte - MG, Brazil
thiago.m.parreiras@ieee.org

Braz de Jesus Cardoso Filho
Department of Elect. Engineering
Universidade Federal de Minas Gerais
Belo Horizonte - MG, Brazil
braz.cardoso@ieee.org

Abstract—**The True Unity Power Factor (TUPF) converter is a grid-connected active rectifier capable of delivering a sinusoidal current through a combination of Selective Harmonic Elimination (SHE) technique in two power electronics converters and harmonic cancellation at a three-winding transformer. This configuration provides a cost-effective solution with low switching frequency and simple converter and transformer designs without using bulk sinusoidal filters. The application of the TUPF converter in Photovoltaic (PV) systems demands dc link voltage control in response to a proper selected and adjusted Maximum Power Point Tracker (MPPT). This work proposes a control scheme for this operation and verifies its results with respect to power quality and power tracking. The results are validated though simulation and hardware-in-the-loop (HIL) tests.**

Index Terms—**power quality, harmonics, photovoltaic, voltage control, power factor**

I. INTRODUCTION

At the end of 2016, the Photovoltaic (PV) global installed capacity surpassed the 300 GW mark and Solar Power Europe prospects as a medium scenario an increase to over 700 GW by 2021 [1] which demonstrates the importance of searching for grid interface solutions that provides an adequate connection to the grid in terms of power management, ancillary services, reliability, efficiency and power quality.

Although much research is performed to construct high power grid-tied converters composed of several smaller power modules connected to smaller strings in order to improve the extraction of the energy available in the Solar Panels [2], [3], central inverters still constitute the solution more used in Solar Power Plants, mainly due to its low cost, complexity and therefore high reliability [4].

The majority of such converters uses a bulk sinusoidal filter at the grid side in order to comply with restrict power quality requirements of countries grid codes [5], [6] and international standards recommendations [7], [8]. These filters contains a combination of inductors and capacitors that can be a source

This work was supported in part by the Brazilian federal government agencies Coordenação de Aperfeiçoamento de Pessoal de Nível Superior (CAPES), and Conselho Nacional de Desenvolvimento Científico e Tecnológico (CNPq), in part by the Minas Gerais state government agency Fundação de Amparo à Pesquisa do Estado de Minas Gerais (FAPEMIG), and in part by CELPE through the ANEEL R&D Program "Chamada ANEEL: 021/2016" under Research Grant PD-00043-0516/2016.

of systems resonances and therefore contributing to the very problem they were installed to prevent [9].

In this context, a True Unity Power Factor (TUPF) converter is proposed as a high power grid-tied solution that provides sinusoidal currents, from the normative perspective, without capacitive elements at the grid filter, with low switching frequencies and conventional transformer design and converters topologies [10].

This technology has proven to be able of providing better power quality results than other inverters technologies in large wind farms [11] by shifting the resonance frequencies to higher orders [12]. It was also presented as an alternative for ultra fast charging of electrical buses [13] and high-capacity belt conveyors in comparison to conventional configurations [14].

The main contribution of this work is to present the TUPF converter as a suitable central inverter for solar PV applications in a situation where the dc voltage, and therefore the modulation index, needs to be continuously modified according to a proper Maximum Power Point Tracking (MPPT) algorithm, which differs from the previous presented applications where dc voltage reference is a constant. Furthermore, it presents for the first time complete closed-loop results in a HIL test bench.

II. TUPF CONVERTER

A. System Description

Fig. 1 shows the basic structure of the TUPF converter applied as central inverter in a PV application. It consists of two Voltage Source Converters (VSCs) with their dc links connected in parallel. Each VSC is connected to a secondary of a three-winding (3-w) transformer. As there is a phase shift of 30° between secondaries, due to delta and star connections, only characteristic harmonics of order according to (1) can be present at the primary winding [15].

$$h = 12k \pm 1, \quad k = 1, 2, 3, 4, \ldots \tag{1}$$

The VSCs in the TUPF converter are modulated with precalculated phase voltage waveforms to eliminate harmonic voltages of the same order indicated by (1) until $h = 49$ through a proper implementation of Selective Harmonic Elimination (SHE) PWM technique [7]. The combined effect of

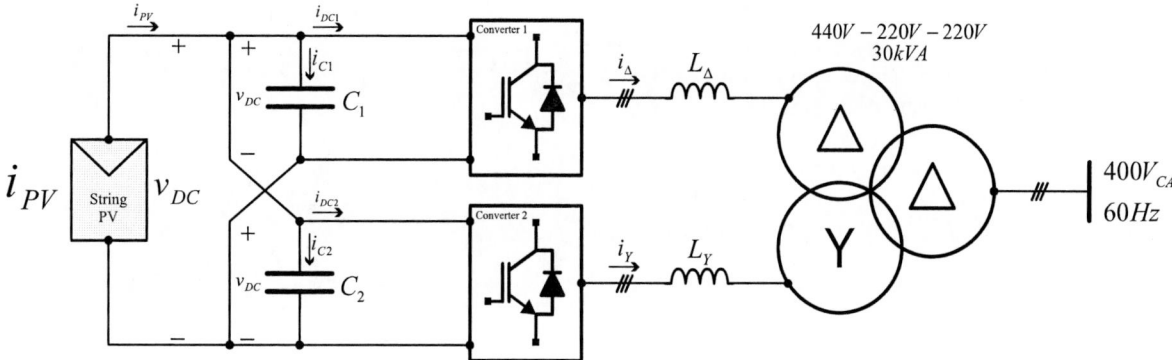

Fig. 1. Basic structure of the TUPF applied to PV plants.

the harmonics eliminated by the VSCs and cancelled by the transformer, summarized in Table I, is the absence of harmonics until the 50th order in the primary winding.

TABLE I
DISTRIBUTION OF HARMONIC SUPPRESSION

Element	Harmonics eliminated or cancelled
3-w transformer	5, 7, 17, 19, 29, 31, 41, 43, ...
VSCs	11, 13, 23, 25, 35, 37, 47, 49

It is worthwhile to mention that even order harmonics and third order harmonics are non-characteristic due to the quarter-wave symmetry and to the three-phase three-wire system, respectively.

As most grid codes and international standards specifies individual harmonic and quality index limits until the 50th order only, it can be said that the current at the transformer primary is pure sinusoidal from the normative perspective, therefore resulting in an unity distortion power factor (2). That is the reason for the converter name: meaning that true unity power factor (3) can be achieved.

It is important to note, however, that this is not a limitation of the converter concerning the flux of reactive power since it can be commanded to work with non-unity displacement power factor (4) as will become evident in the next section.

$$PF_{dist.} = \frac{1}{\sqrt{1 + THD_v^2}\sqrt{1 + THD_i^2}} \qquad (2)$$

$$PF_{true} = PF_{dist.} \cdot PF_{disp.} \qquad (3)$$

$$PF_{disp.} = \frac{P_1}{V_1 I_1} \qquad (4)$$

where P_1, V_1 and I_1 are the fundamental average active power, rms voltage and current, respectively. THD_v and THD_i stand for voltage and current Total Harmonic Distorion (THD), respectively, and can be calculated by (5).

$$THD_x = \frac{\sum_{h=2}^{50} \sqrt{X_h^2}}{X_1} \qquad (5)$$

B. TUPF Control

The TUPF SHE PWM efficient implementation along with VSC current control is first presented in [16] for the context of three-level (3L) VSCs and extended in [10] to two-level (2L) VSCs. Therefore this section focuses on presenting the adopted current and dc voltage control strategy along with its adjustment technique.

Fig. 2 presents the current control block diagram for the converter in the d,q rotating frame aligned with the grid voltage space phasor. The active power reference is generated by the dc voltage control presented in Fig. 3, while the reactive power reference is free for the user ($Q^{ref} = 0$ gives unity displacement power factor).

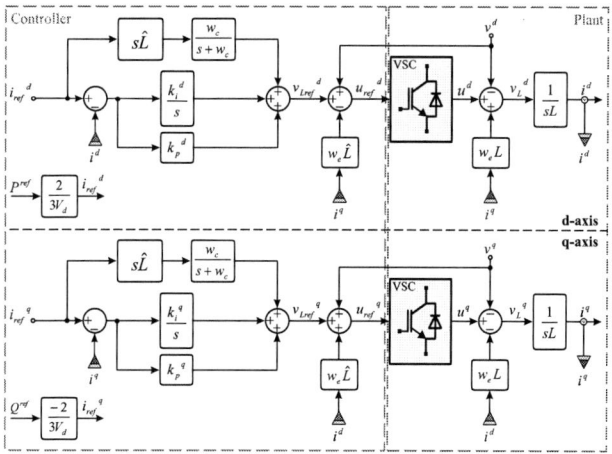

Fig. 2. Current control in a d,q rotating frame.

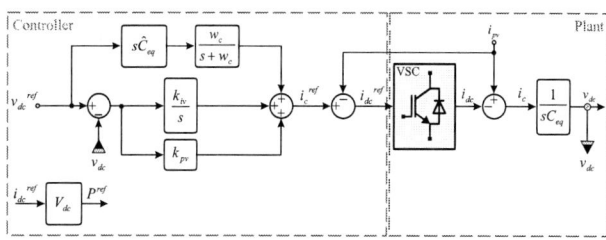

Fig. 3. dc voltage control.

978-1-7281-4181-7/19 $31.00 © 2019 IEEE

The equations describing the disturbance-output relationships of both controls are (6) and (7), which can be represented by the curves of Fig. 4. The control is designed to obtain a desired disturbance-output relationship [17] by calculating integral and proportional gains in such a way that the poles of (6) and (7) are located in the maximum frequencies possible, but adequately spaced among them to avoid mutual interference and to keep the SHE PWM technique working properly.

$$\frac{V^{dq}(s)}{I^{dq}(s)} = sL + k_p^{dq} + \frac{k_i^{dq}}{s} \qquad (6)$$

$$\frac{I_{pv}(s)}{V_{dc}(s)} = sC_{eq} + k_{pv} + \frac{k_{iv}}{s} \qquad (7)$$

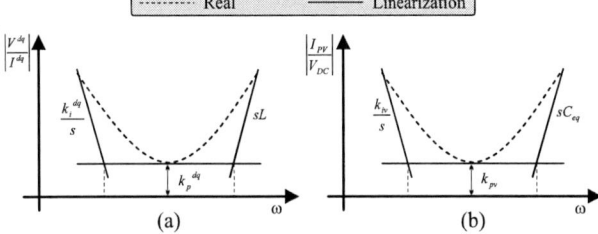

Fig. 4. Disturbance-output relationship for: (a) current control and (b) DC voltage control.

From the previous description, Table II illustrates briefly the controller gains used in this work. Each secondary has its own current controller and there is just one dc voltage controller since dc links are parallel connected.

TABLE II
PARAMETERS OF CURRENT AND VOLTAGE CONTROLLERS

Parameter	Voltage mesh	Current mesh
Proportional gain	$k_p = 0.9$	$k_p^d = 0.8,$ $k_p^q = 0.8$
Integral gain	$k_i = 7.3$	$k_i^d = 0,$ $k_i^q = 33$
Converter 1 capacitance	4500 μF	
Converter 2 capacitance	4500 μF	

III. PV SYSTEM UNDER STUDY

A. Plant Description

The performance of the TUPF converter will be evaluated considering a case study based on the solar PV plant Tesla, located at the Universidade Federal de Minas Gerais (UFMG) in Belo Horizonte, Brazil. The plant consists of 144 Yingli 245-32b modules distributed in three arrays, each string consisting of 10 modules in series, being 14 strings in parallel as described in Fig. 5.

Since the open circuit and maximum power voltages for each module are $V_{oc} = 40.8\ V$ and $V_{mp} = 32.2\ V$, respectively, the arrangement offers the dc bus $322\ V$ at full power and $408\ V$ at open circuit.

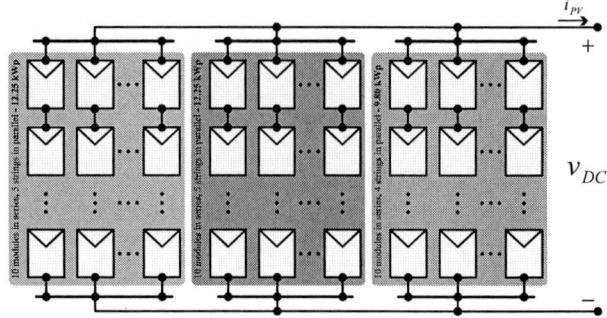

Fig. 5. Layout of the 35.28 kWp solar power plant.

In Fig. 6, the I-V and P-V curves are shown as characteristics of the arrangement considered. According to the illustration, the maximum power point is obtained around 330 V.

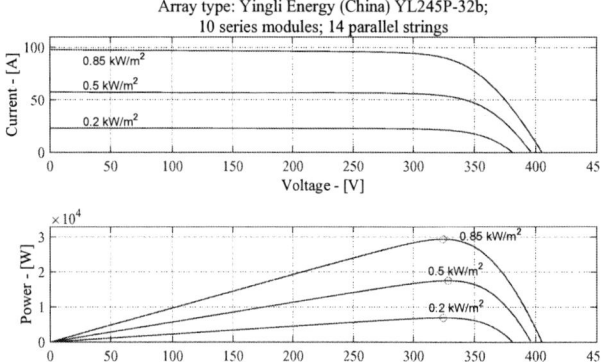

Fig. 6. I-V and P-V curves for the selected module.

B. MPPT Algorithm

Several algorithms for maximum power point tracking are presented in the literature [18], among them the techniques Disturb and Observe (D & O), and Incremental Conductance (IC) are the most widely referenced in the literature due to their simplicity and no need for prior knowledge of the PV arrangement characteristics.

These algorithms are compared in [19] and [20], regarding their capabilities of tracking of the maximum power point under transient and steady state conditions. In [20] their performance are also evaluated in the presence of noise in the measurement signals.

In all cases, the IC algorithm presented by [21] and represented in Fig. 7 presented better response and therefore, was considered for the application in this work.

The PV solar plant presented in Fig. 5 was simulated in a Matlab/Simulink environment to validate the use of TUPF in solar PV systems. Complementarily, the energy quality is evaluated through the comparison with the limits recommended in [7] and [8].

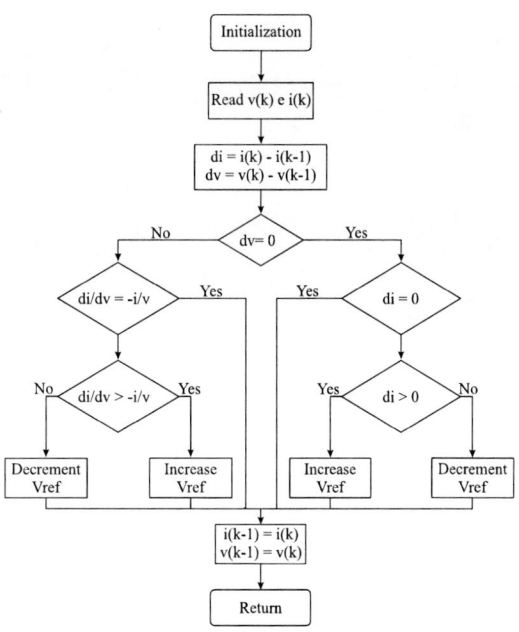

Fig. 7. MPPT algorithm for IC.

IV. SIMULATION RESULTS

The irradiance and temperature profile shown in Fig. 8 was chosen to test the two most widely used scenario in the literature: ramp and step variation.

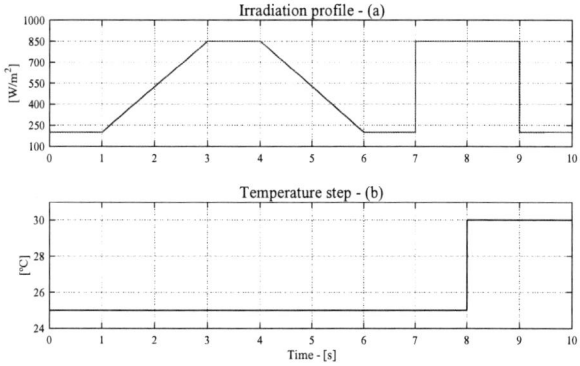

Fig. 8. Irradiation and temperature profile chosen.

The ramp variation is an approximation of the behavior verified in practice, and seeks to validate the tuning of the MPPT. On the other hand, the variation in step has the objective to test the control under severe operating conditions, causing fluctuations in the dc bus.

Fig. 9 - (a) shows that the MPPT follows the irradiation variations smoothly, providing the voltage reference needed to control the extracted power. It also shows that the voltage measured in the capacitor of the dc bus also follows suitably close to the reference voltage in steady state. Fig. 9 - (b) shows the currents in the transformer primary to the same irradiation

profile. This demonstrates the proper working of the MPPT combined with the voltage and current controls.

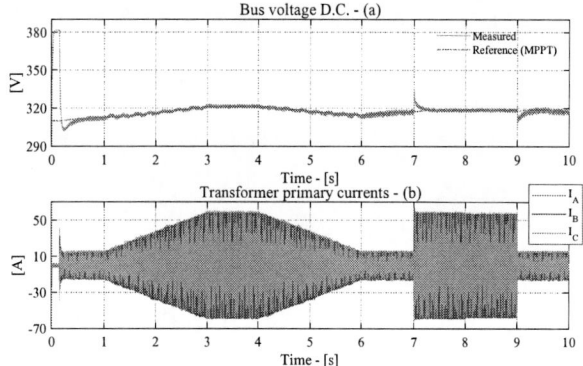

Fig. 9. dc voltage variation with irradiation.

For a severe transient irradiation, Fig. 10 shows the stability of the response in the currents and voltages of the transformer secondary, at the point of connection to the grid. The synchrony of the voltages and currents testify the unit power factor with load and without load.

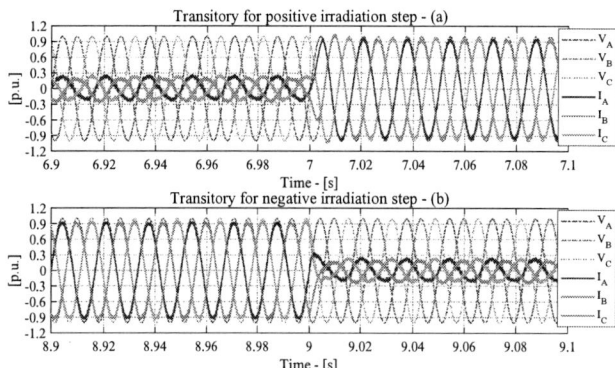

Fig. 10. Effect of the positive and negative irradiation step on the secondary current.

The compliance with the power quality indices is verified considering the Table III, which presents the limits of even and odd harmonics, according to the IEEE standards [7], [8]. Here, I_L is maximum demand load current (fundamental frequency component) at the PCC under normal load operating conditions.

TABLE III
MAXIMUM HARMONIC CURRENT DISTORTION IN PERCENT OF I_L, [7], [8]

Individual harmonic order	Even	Odd
$2 \leq h < 11$	1	4
$11 \leq h < 17$	0.5	2
$17 \leq h < 23$	0.375	1.5
$23 \leq h < 35$	0.15	0.6
$35 \leq h \leq 50$	0.075	0.3
TDD = 5 %		

Fig. 11 presents the harmonic spectrum of the current in phase A at the grid connection, comparing it with the respective percentage indices and total demand distortion (TDD). It is possible to observe that all residual harmonics ($2 \leq h \leq 50$) are below the respective recommended percentage. Allied to this, the TDD is less than half of the recommended value.

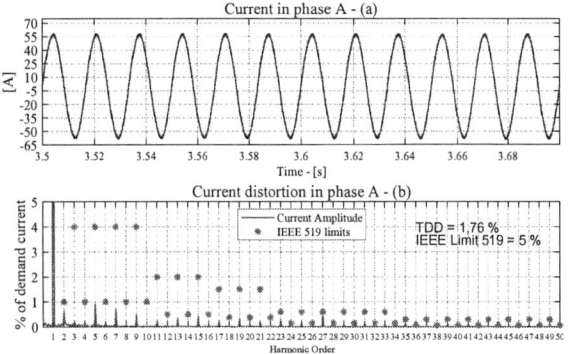

Fig. 11. Harmonic spectrum of the secondary current in phase A.

Considering a balanced three-phase system, the currents of phases B and C, follow the same principle and have similar power quality indices.

V. HIL RESULTS

Finally, the proposed control strategies are validated in the HIL test bench of Fig. 12, in which the closed control loop output signals, modulation index and voltage phase angle, are sent through parallel communication from the digital signal processor (DSP) to the FPGA board [22], where the SHE PWM was implemented.

The plant under analysis, the TUPF converter applied to a PV arrangement, was implemented in a Typhoon HIL 600 [23] real-time simulator to obtain complete and reliable results.

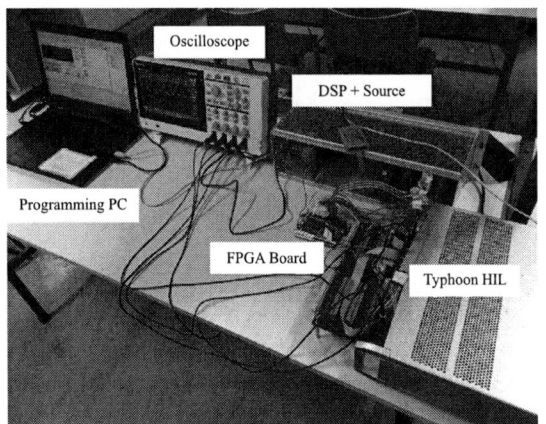

Fig. 12. Structure of tests and validation.

Fig. 13 shows the results of the current at the output of the array (CH1), the current at the secondary delta connection (CH2), the current at the primary (CH3), as well as the voltage at the dc bus (CH4) to a negative irradiation step. On the other hand, Fig. 14 shows the same measurements, however a positive irradiation step is applied.

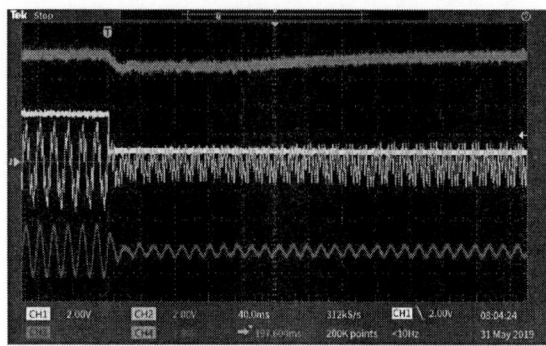

Fig. 13. Delta secondary current and voltage on the dc bus for a negative irradiation step, $850 \frac{W}{m^2}$ to $200 \frac{W}{m^2}$.

In both cases one can see the control was capable of reponding to the abrupt change in the irradiation while keeping the DC voltage at the MPPT generated reference.

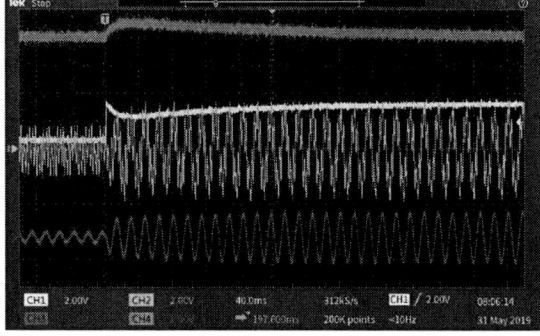

Fig. 14. Delta secondary current and voltage on the dc bus for a positive irradiation step, $200 \frac{W}{m^2}$ to $850 \frac{W}{m^2}$.

In Fig. 15, the current at the delta (CH2) and star (CH4) secondaries of the transformer, together with the voltage (CH1) and current (CH3) in phase A of the primary are highlighted. It is evident that despite the rather distorted waveform present in the secondaries, the current delivered to the network is practically sinusoidal and maintains the unit power factor.

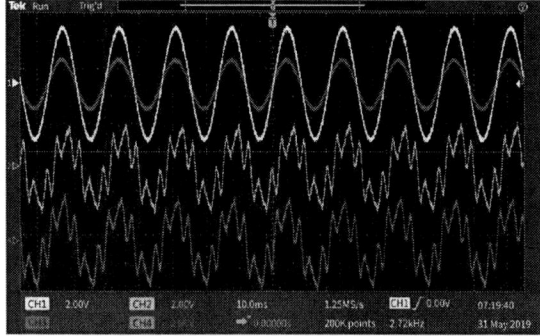

Fig. 15. Voltage and current in the grid, and delta and star secondary currents, for irradiation of $850 \frac{W}{m^2}$.

Fig. 16 shows that the TUPF converter is is supplying three phase balanced sinusoidal currents (CH2, CH3, CH4) to the grid.

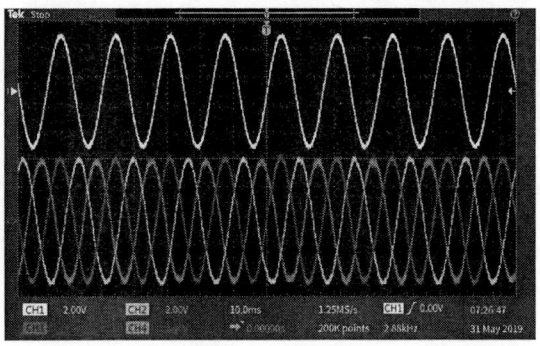

Fig. 16. Voltage and current in the grid for irradiation of 850 $\frac{W}{m^2}$.

Finally, Fig. 17 highlights the harmonic spectrum of the current in phase A at the grid connection, comparing it with the respective percentage indices and total demand distortion (TDD).

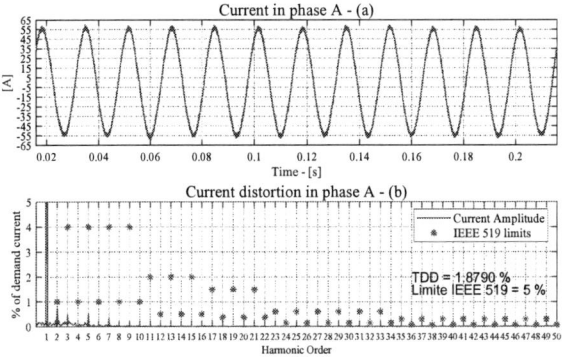

Fig. 17. Harmonic spectrum of the secondary current in phase A.

The results obtained with HIL validate the TUPF's ability to improve power quality, especially TDD.

VI. CONCLUSIONS AND FUTURE WORK

The global use of renewable energy as an alternative to fossil fuels is already a reality. According to [1], photovoltaic solar energy has been highlighted by the constant growth, besides enabling distrubuted generation and microgrids. The main solutions for the interface of generation with the grid, usually present low efficiency and above all problems of power quality.

Thus, this paper proposed a new application niche for the *True Unit Power Factor* (TUPF) converter, as a suitable central inverter for PV applications. In this context, it was demonstrated in Fig. 9 the satisfactory performance of the MPPT algorithm for this topology, as well as the robustness of the current and voltage controls for rapid irradiation variations, in Fig. 10.

Therefore, the results illustrated in Fig. 11 and Fig. 17 as well as Fig. 15 and Fig. 16, lead to the conclusion that the TUPF is able to function properly under dc voltage variation, promoting better power quality at low cost and high efficiency for high power PV applications.

A partial scale prototype (30 kVA) is under construction at the TESLA Power Lab to allow complete tests at the experimental PV plant.

REFERENCES

[1] SolarPower Europe, "Global Market Outlook For Solar Power / 2017 - 2021," 2017.

[2] N. Foureaux, A. Machado, E. Silva, I. Pires, J. Brito, and F. B. Cardoso, "Central inverter topology issues in large-scale photovoltaic power plants: Shading and system losses," 2015 IEEE 42nd Photovolt. Spec. Conf. PVSC 2015, pp. 49, 2015.

[3] F. V. Amaral, T. M. Parreiras, G. C. Lobato, A. A. P. Machado, I. A. Pires, and B. de Jesus Cardoso Filho, "Operation of a Grid-Tied Cascaded Multilevel Converter Based on a Forward Solid-State Transformer Under Unbalanced PV Power Generation," IEEE Trans. Ind. Appl., vol. 54, no. 5, pp. 5493–5503, Sep. 2018.

[4] S. Kouro, J.I.Leon, D. Vinnikov, and L.G. Franquelo, "Grid-Connected Photovoltaic Systems: An Overview of Recent Research and Emerging PV Converter Technology",Ind. Electron.Mag. IEEE, vol.9, no.1, pp.47-61,2015.

[5] ANEEL, "Mdulo 8   Qualidade da Energia Eltrica," Procedimentos Distrib. Energ. Eltrica no Sist. Eltrico Nac. PRODIST, 2017.

[6] ONS, "Submdulo 2.8 - Gerenciamento dos indicadores de qualidade da energia eltrica da Rede Bsica," Procedimentos Rede, 2017.

[7] IEEE STANDARDS, "IEEE Std 519-2014. Recommended Practice and Requirements for Harmonic Control in Electric Power Systems," IEEE Std 519-2014 (Revision IEEE Std 519-1992), 2014.

[8] IEEE, IEEE Standard for Interconnection and Interoperability of Distributed Energy Resources with Associated Electric Power Systems Interfaces. New York, NY: IEEE Std 1547-2018, 2018.

[9] M. Bradt et al., "Harmonics and resonance issues in wind power plants," in PES T&D 2012, 2012, pp. 1-8.

[10] T. M. Parreiras, J. C. G. Justino, and B. de J. Cardoso Filho, "The True Unity Power Factor converter - A practical filterless solution for sinusoidal currents," in 2015 9th International Conference on Power Electronics and ECCE Asia (ICPE-ECCE Asia), 2015, pp. 2557-2565.

[11] C. E. Almeida and B. de J. C. Filho, "Impact of active front end topology on wind farm resonance," in 2017 IEEE Power & Energy Society General Meeting, 2017, pp. 1-5.

[12] C. E. Almeida and B. de J. Cardoso Filho, "Shifting Resonances in Wind Farms to Higher Frequencies due to TUPF Converters," J. Control. Autom. Electr. Syst., vol. 29, no. 6, pp. 805-815, Dec. 2018.

[13] J. Justino, T. Parreiras, and B. Filho, "Hundreds kW Charging Stations for e-Buses Operating Under Regular Ultra-Fast Charging," IEEE Trans. Ind. Appl., vol. 52, no. 2, pp. 1766–1774, 2015.

[14] T. M. Parreiras,J. C. G. Justino,A. V. Rocha, and B. de J. C. Filho,"True Unit Power Factor Active Front End for High-Capacity Belt-Conveyor Systems,"IEEE Trans.Ind.Appl.,vol.52,no.3,pp. 2737-2746, May 2016.

[15] J. Arrillaga and N. Watson, "Harmonic Sources," in Power System Harmonics, Chichester, UK: JohnWiley & Sons, Ltd,2003, pp. 61-142.

[16] T. M. Parreiras and B. J. C. Filho, "Current control of three level neutral point clamped voltage source rectifiers using selective harmonic elimination," in IECON 2014 - 40th Annual Conference of the IEEE Industrial Electronics Society, 2014, pp. 4608-4614.

[17] J. G. Bollinger and N. A. Duffie, "Discrete Controller Design," in Computer Control of Machines and Processes, Reading, MA: Addison-Wesley Publishing Company, Inc., 1988, pp. 111–172.

[18] B. Subudhi and R. Pradhan, "A Comparative Study on Maximum Power Point Tracking Techniques for Photovoltaic Power Systems," IEEE Trans. Sustain. Energy, vol. 4, no. 1, pp. 89-98, Jan. 2013.

[19] H. N. Zainudin and S. Mekhilef, "Comparison Study of Maximum Power Point Tracker Techniques for PV Systems," in Proceedings of the 14th International Middle East Power Systems Conference (MEPCON10), Cairo University, Egypt, 2010, no. 1, pp. 750-755.

[20] M. Calavia, J. Peri, et al. "Comparison of MPPT strategies for solar modules," Int. Conf. Renew. Energies Power Qual., 2010.

[21] K. H. Hussein, I. Muta, T. Hoshino, and M.Osakada,"Maximum Photovoltaic Power Tracking: an Algorithm for Rapidly Changing Atmospheric Conditions," IEEProc.- Gener. Transm. Distrib., vol.142,no.1, pp.59-64, 1995.

[22] Terasic Inc., "DE10-Lite Board," 2019. [Online]. Available: http://de10-lite.terasic.com/. [Accessed: 30-Apr-2019].

[23] Typhoon HIL, "Hardware in the LoopSoftware e Hardware," 2018. [Online]. Available: https://www.typhoon-hil.com/. [Accessed: 18-Dec-2018].

High-Voltage Step-Up DC-DC Converter Employing The Four State Switching Cell and Voltage Multiplier Cells

Paulo Henrique Feretti, Enio Roberto Ribeiro
Federal University of Itajubá
Itajubá, Brazil
pauloferetti@gmail.com, enio.k@unifei.edu.br

Fernando Lessa Tofoli
Federal University of São João del-Rey
São João del-Rey, Brazil
fernandolessa@ufsj.edu.br

Abstract — **Distinct modern applications require wide voltage conversion range, e.g., renewable energy systems, uninterruptible power supplies (UPS), among others. However, this requirement brings some problems associated with the power converters used for this purpose, mainly related to the system efficiency, stresses on the semiconductor devices, and component count. This work presents a high-voltage step-up non-isolated dc-dc converter based on the four-state switching cell (4SSC) and voltage multiplier cells (VMC), whose voltage and current stresses on the semiconductors are reduced. Besides, the reactive components are smaller due to the increased operating frequency, which is three times higher than the switching frequency. The operating principle, theoretical analysis, and simulation results are presented in detail. The introduced topology is also compared with other similar approaches that exist in the literature.**

Keywords— dc-dc converters, four-state switching cell, step-up converter, voltage multiplier cells.

I. Introduction

Renewable energy systems often require wide conversion range because the primary voltage levels are generally much lower than the nominal values employed in the ac grid [1]. For instance, in solar photovoltaic (PV) and wind energy conversion systems, the dc voltages must be stepped up from 48 Vdc to a range between 200 Vdc and 800 Vdc [2]. Other potential applications include uninterruptible power systems (UPS) and transportation [3] [4].

In order to achieve step-up, boost-based converters are preferred over other topologies such as the Ćuk, SEPIC (Single-Ended Primary Inductance Converter) and Zeta ones due to simplicity, reduced stresses on the components, and lower component count. However, in order to achieve high-voltage step-up in the classical boost converter, the use of high duty ratios is necessary, which brings some problems related to the parasitic resistance of the filter inductor and reverse recovery losses that compromise overall efficiency. Nowadays, several solutions are presented in the literature in order to solve the aforementioned problems, e.g., cascaded, coupled-inductor-based, and switched-capacitor/switched-inductor-based, and voltage multiplier cell- (VMC) based converters [2] [5] [6].

Interleaving is a good solution in order to increase efficiency and achieve high current levels while sharing the processed power among many phases [7]. Besides, it can be associated with distinct strategies to extend the voltage conversion gain. The work proposed in [8] describes an interleaved boost converter associated with a voltage doubler circuit in order to achieve high voltage gain. However, efficiency is significantly affected, thus restricting this

topology to lower power applications. Other drawbacks are the need of large-size capacitors; existence of unbalanced voltages across the capacitors; and duty cycle limitation to 50%.

The integration of voltage multiplier cells (VMC) to dc-dc converters has proven to be an interesting alternative in order to solve the problem of high voltage stresses on the active switches [9]. A comprehensive review on voltage boosting techniques is presented in [10], as well the use of VMCs in order to achieve wide conversion range. An approach for generating VMC structures is addressed in [11], where VMCs are specifically applied to generate novel step-up dc-dc converters in [12]. A considerable number of topologies based on VMC structures has been proposed in the current literature. The use of a VMC associated with a small inductor is suggested in [12] in order to achieve the zero-current-switching (ZCS) when the active switch is turned on and minimize the negative effects of the reverse recovery current of all diodes, in addition to the high voltage gain. Besides, the Dickson multiplier is used in [13] and [14] to increase the voltage gain and reduce the voltage stresses on the switches. A three-state switching cell- (3SSC) based dc-dc converter using VMCs is presented in [15], whose claimed advantages are: reduction of the voltage stress on the switches, smaller reactive components and reduction of conduction losses. Even though high voltage gain was obtained, a large number of VMCs was employed, while the prototype was designed for a low rated power of only 250 W.

In this context, this work proposes the association of the four-state switching cell (4SSC) and VMCs to achieve high voltage gain, reduced voltage and current stresses on the semiconductors, and smaller filter elements, thus leading to improved efficiency and high-power density for high-power applications.

II. 4SSC-Based DC-DC Step-Up Converter Using VMCs

The proposed dc-dc step-up converter is presented in Fig. 1, which employs the 4SSC and VMCs. The 4SSC is composed of three active switches S_1, S_2, and S_3; three diodes D_1, D_2, and D_3; and one autotransformer with unity turns ratio. The VMCs are composed of a set of three capacitors C_m and three diodes D_m for each stage. Even though the converter is capable of operating over the complete duty cycle range $0 \leq D \leq 1$, operation will be analyzed for $1/3 < D < 2/3$ only, also considering the continuous conduction mode (CCM).

978-1-7281-4181-7/19 $31.00 © 2019 IEEE

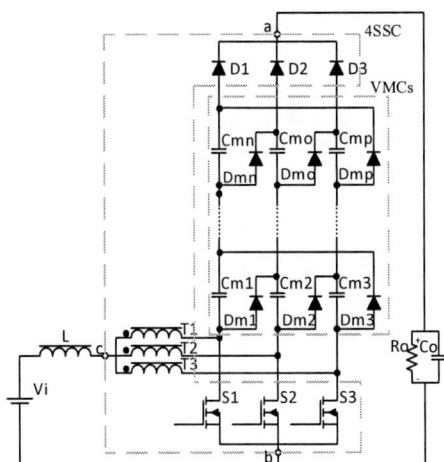

Fig. 1. 4SSC VMC dc-dc step-up converter.

A. Operation at CCM and 1/3<D<2/3

The converter operation can be defined according to six operating stages as shown in Fig. 2, while the main theoretical waveforms are represented in Fig. 3.

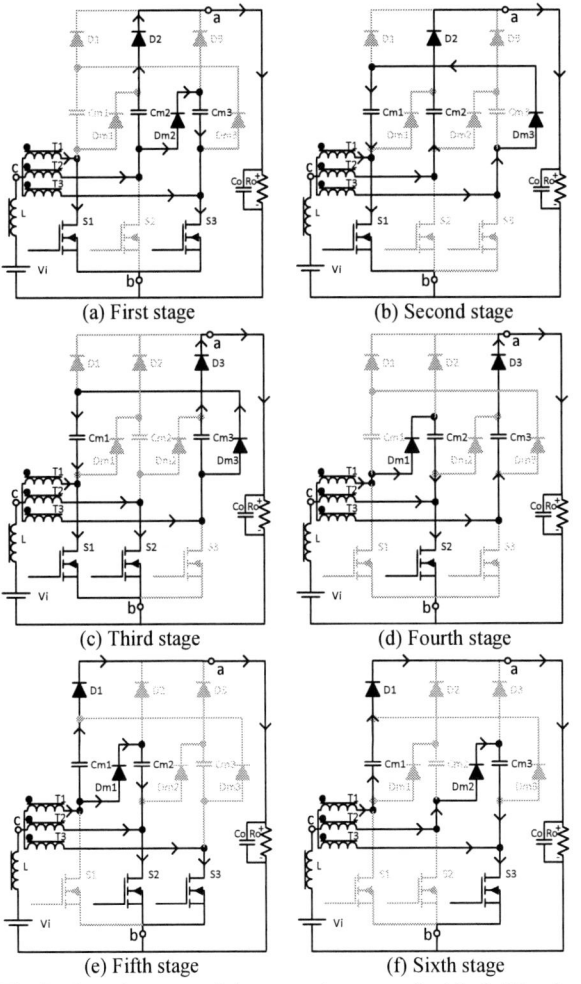

(a) First stage

(b) Second stage

(c) Third stage

(d) Fourth stage

(e) Fifth stage

(f) Sixth stage

Fig. 2. Operating stages of the proposed converter for 1/3<D<2/3 and CCM.

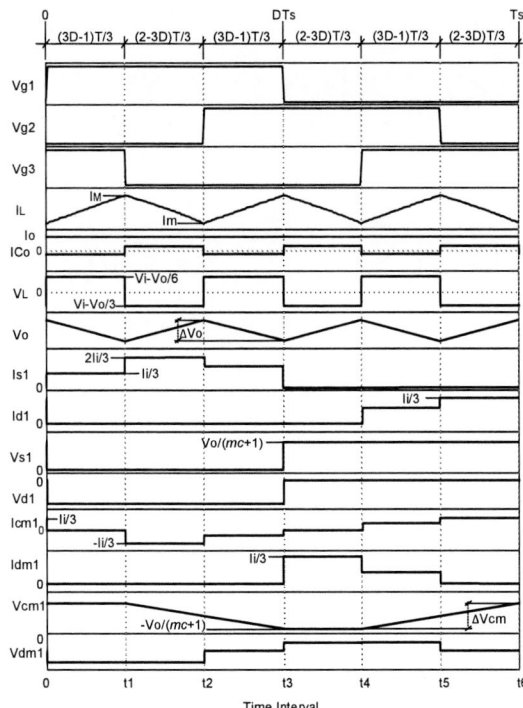

Fig. 3. Main theoretical waveforms of the proposed converter for 1/3<D<2/3 and CCM.

First stage [t_0, t_1] (Fig. 2 (a)): Switch S_1 is turned on, but switches S_2 and S_3 remain off and on, respectively. The current through filter inductor L increases linearly and is equally shared among the autotransformer windings due to unity turns ratio. The first part flows through T_1 and S_1. The second part flows through T_2, being shared between two branches, i.e., through D_{m2}-C_{m3} and through C_{m2}-D_2, as energy is supplied to the load. This stage finishes when S_3 is turned off.

Second stage [t_1, t_2] (Fig. 2 (b)): Switches S_1 and S_2 remain on and off, respectively. Switch S_3 is turned off. The voltage across inductor L is inverted. The inductor current decreases linearly, flowing through three different paths: T_1-S_1; T_3-D_{m3}-C_{m1}; and T_2-C_{m2}-D_2, and the output stage. This stage finishes when S_2 is turned on.

Third stage [t_2, t_3] (Fig. 2 (c)): Due to the existing symmetry in the circuit, this stage is analogous to the first one, but switch S_2 is turned on instead, while S_3 is off. Analogously to the first stage, S_1 is turned on.

Fourth stage [t_3, t_4] (Fig. 2 (d)): The behavior of the converter in this stage is similar to the second stage, but switch S_1 is turned off instead, while S_2 remains on and S_3 is turned off.

Fifth stage [t_4, t_5] (Fig. 2 (e)) and sixth stage [t_5, t_6] (Fig. 2 (f)): Such stages are identical to the first and second ones, respectively.

Applying the volt-second balance to the converter and considering a generic number of VMCs defined as mc, the static gain for 1/3<D<2/3 and CCM is obtained as:

$$G_V = \frac{V_o}{V_i} = \frac{mc+1}{1-D} \qquad (1)$$

where V_o is the average output voltage and V_i is the average input voltage. Fig.4 presents the behavior of the voltage gain G_v as a function of the duty cycle considering the use of several VMCs.

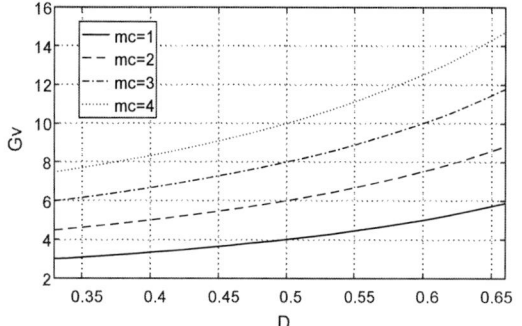

Fig. 4. Voltage gain versus duty cycle for $1/3 < D < 2/3$ and CCM.

Analyzing Fig. 2 and Fig. 3, it is possible to obtain the following expressions for the filter inductance L, the output filter capacitor C_o, and the multiplier capacitances C_m for $m = 1 \ldots mc$:

$$L = \frac{(2-3D)(3D-1)V_o}{9(mc+1)f_s \Delta I_L} \quad (2)$$

$$C_o = \frac{(2-3D)(mc-2+3D)I_o}{9 f_s \Delta V_o (1-D)\eta} \quad (3)$$

$$C_m = \frac{(mc+1)(3D-1)I_o}{6\Delta V_{Cm} f_s (1-D)} \quad (4)$$

where ΔI_L is the inductor current ripple, f_s is the switching frequency, I_o is the average output current, ΔV_o is the output voltage ripple; and ΔV_{Cm} is the voltage ripple across the capacitors used by the VMC.

The average inductor current I_L and its respective maximum and minimum values I_M and I_m are obtained from expressions (5) to (7).

$$I_L = \frac{(mc+1)I_o}{(1-D)\eta} \quad (5)$$

$$I_M = \frac{(mc+1)I_o}{(1-D)\eta} + \frac{(2-3D)(3D-1)V_o}{9(mc+1)Lf_S} \quad (6)$$

$$I_m = \frac{(mc+1)I_o}{(1-D)\eta} - \frac{(2-3D)(3D-1)V_o}{9(mc+1)Lf_S} \quad (7)$$

In order to verify the behavior of the normalized inductor current ripple $\overline{\Delta I_L}$ as a function of the duty cycle, expression (8) can be used. Fig. 5 shows that the normalized current ripple is maximum and equal to 0.125 and 0.083 at $D = 0.5$ for $mc = 1$ and $mc = 2$, respectively.

$$\overline{\Delta I_L} = \frac{9\Delta I_L L f_s}{V_o} = \frac{(2-3D)(3D-1)}{(mc+1)} \quad (8)$$

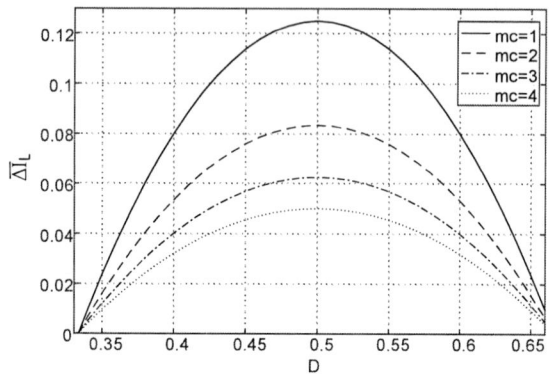

Fig. 5. Normalized inductor current ripple versus duty cycle for $1/3 < D < 2/3$ and CCM.

The average current $I_{S(avg)}$, the rms current $I_{S(rms)}$, and the peak current $I_{S(pk)}$ through the actives switches are given by (9), (10), and (11), respectively.

$$I_{S(avg)} = \frac{I_o}{18}\left(\frac{5(1+mc)+3(D-1)}{(1-D)}\right) \quad (9)$$

$$I_{S(rms)} = \frac{I_o}{3}\sqrt{\begin{array}{l} 3(3D-1)+ \\ \dfrac{4(mc+1)(3D-1)}{(1-D)}+ \\ \dfrac{(mc+1)^2(D+1)}{(1-D)^2} \end{array}} \quad (10)$$

$$I_{S(pk)} = \frac{2}{3}\frac{(mc+1)I_o}{(1-D)} \quad (11)$$

The average current $I_{D(avg)}$, the rms current $I_{D(rms)}$, and the peak current $I_{D(pk)}$ through the diodes of the 4SSC are given by (12), (13), and (14), respectively.

$$I_{D(avg)} = \frac{I_o}{3} \quad (12)$$

$$I_{D(rms)} = \frac{1}{3}\sqrt{\begin{array}{l}\left(C_o^2 \Delta V_o^2 + 3C_o I_o \Delta V_o + 3I_o^2\right) \\ \times\left(\dfrac{(2-3D)}{f_S}+6D-2\right)\end{array}} \quad (13)$$

$$I_{D(pk)} = \frac{(mc+1)I_o}{3(1-D)} \quad (14)$$

Finally, the maximum voltage stress on the switches is:

$$V_{S(\max)} = \frac{1}{(mc+1)}\left(V_o + \Delta V_{Cm} mc\right) \quad (15)$$

The normalized voltage stress on the switches with respect to the output voltage is a function of parameter mc and can be approximated by (16). As expected, the stress is further reduced as more VMCs are added, being this relationship nonlinear as demonstrated in Fig. 6.

$$\frac{V_{S(\max)}}{V_o} \cong 0.5 mc^{-0.5} \quad (16)$$

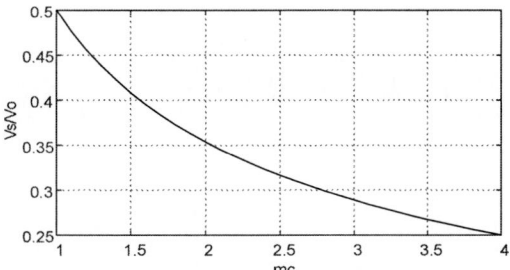

Fig. 6. Normalized voltage stress on the active switches as a function of mc for $1/3 < D < 2/3$ and CCM.

The maximum reverse voltage across the diodes used by the 4SSC and multiplier diodes are given by (17) and (18), respectively.

$$V_{D(\max)} = -\frac{1}{(mc+1)}\left(V_o + \Delta V_{Cm}mc\right) \qquad (17)$$

$$V_{Dm(\max)} = -\frac{2}{(mc+1)}V_o \qquad (18)$$

The average voltage across the multiplier capacitors is:

$$V_{Cm} = -\frac{V_o}{(mc+1)} \qquad (19)$$

III. SIMULATION RESULTS

In order to validate theoretical analysis, the proposed converter was designed according to the specifications given in Table I. In this case, two VMCs are adopted so that it is possible to achieve the same operating conditions given in [2], which are adopted for comparison purposes. The resulting circuit is then represented in Fig. 7. Some simulation results are presented in Fig. 8. In order to validate the quantitative analysis and the derived design procedure, Table II presents a comparison between the calculated and simulated quantities.

Based on Fig. 8, it can be stated that the operating frequency of the filter inductor current is three times higher than f_s, thus leading to reduced dimensions. Besides, all required capacitors present low capacitance ratings, as it is possible to use film components that allow extending the useful life of the converter. The currents through the branches that constitute the 4SSC remain properly balanced due to the transformer with unity turns ratio without the use of special and complex control strategies. Since the input current is shared among them, reduced current stresses result as a consequence, thus contributing to the minimization of switching losses.

TABLE I   OPERATING POINT OF THE 4SSC BOOST CONVERTER

Parameter	Value
Input voltage	V_i=48 V
Output voltage	V_o=400 V
Output Power	3 kW
Number of VMCs	mc=2
Rated duty cycle	D=0.64
Output voltage ripple	ΔV_o=1% V_o
VMC capacitor voltage ripple	ΔV_m=10% V_o
Inductor current ripple	ΔI_L=10% I_L
Switching frequency	f_s=33 kHz
Transformer turns ratio	1
Boost inductor	L=5.3 µH
Output filter capacitor	C_o=2.7 µF
VMC capacitors	C_m=7.2 µF

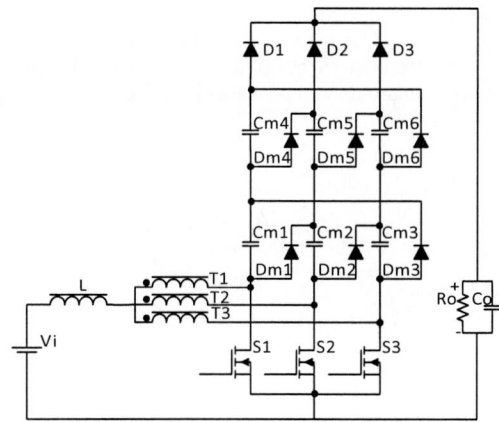

Fig. 7. 4SSC boost converter using two VMCs.

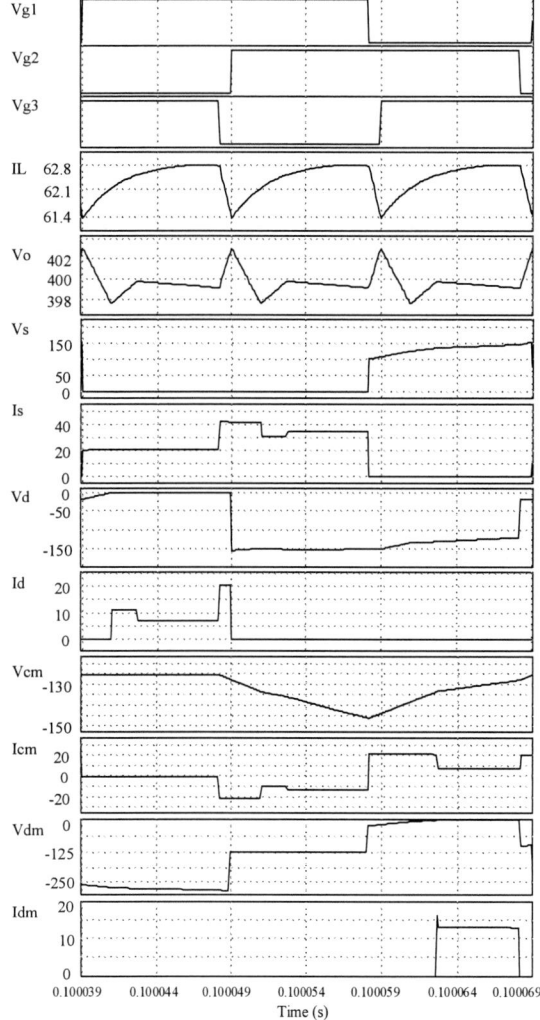

Fig. 8. Simulation results of the 4SSC dc-dc converter with mc=2 in steady-state condition.

The voltage across the diodes of the 4SSCM is decreased as the number of VMCs increases, being equal to two thirds of the output voltage for mc=2. According to Table II, the maximum voltage stress on the switch is approximately 38% of the output voltage, which is close to the calculated value (around 40%) and the one estimated in the graph shown in Fig. 6, i.e., 35%. Due to the multiplier capacitors, the voltage stress on the active switches is also reduced, as it is possible to employ metal-oxide-semiconductor field effect transistors (MOSFETs) with reduced drain-source on resistance, implying minimized conduction losses and increased efficiency at high power levels.

TABLE II   CALCULATED AND SIMULATED QUANTITIES OF THE PROPOSED CONVERTER

Parameter	Calculated Value	Simulated Value
Average output voltage	V_o=400 V	V_o=399.7 V
Output voltage ripple	ΔV_o=4 V	ΔV_o=5 V
Average inductor current	I_L=62.5 A	I_L=62.6 A
Inductor current ripple	ΔI_L=6.25 A	ΔI_L=5.74 A
Maximum switch voltage	$V_{S(max)}$=160 V	$V_{S(max)}$=154.88 V
Average switch current	$I_{S(avg)}$=16.11 A	$I_{S(avg)}$=18.14 A
Rms switch current	$I_{S(rms)}$=23.18 A	$I_{S(rms)}$=23.68 A
Maximum switch current	$I_{S(max)}$=41.66 A	$I_{S(max)}$=41.88 A
Diode reverse voltage	$V_{D(max)}$=-160 V	$V_{D(max)}$=-154.7 V
Average diode current	$I_{D(avg)}$=2.5 A	$I_{D(avg)}$=2.44 A
Rms diode current	$I_{D(rms)}$=5.87 A	$I_{D(rms)}$=5.2 A
Peak diode current	$I_{D(max)}$=20.83 A	$I_{D(max)}$=20.54 A
Reverse voltage across the multiplier diodes	$V_{Dm(max)}$=-267 V	$V_{Dm(max)}$=-273 V
Average voltage across the multiplier capacitors	V_{Cm}=-133 V	V_{Cm}=-132.3 V

IV. COMPARISON WITH OTHER DC-DC BOOST TOPOLOGIES

As it was previously mentioned, the operating point described in [2] was employed in the design of the proposed converter for comparison purposes. Table III can also be employed in this case, where the 4SSC-based topology is compared with other similar approaches that exist in the literature in terms of voltage gain, voltage stresses on the active switches, and component count.

TABLE III   COMPARISON AMONG THE PROPOSED CONVERTER AND OTHER TOPOLOGIES

Parameter	Topology			
	[2]	[16]	[17]	4SSC (mc=2)
Static gain	$\dfrac{n+1}{1-D}$	$\dfrac{1}{1-2D}$	$\dfrac{n}{1-D}$	$\dfrac{mc+1}{1-D}$
Voltage stress on the switches	$\dfrac{V_o}{n+1}$	$\dfrac{(1-D)V_o}{1-2D}$	$\dfrac{V_o}{n}$	$\dfrac{V_o}{2\sqrt{mc}}$
Active switches	3	2	2	3
Diodes	9	2	2	9
Magnetic cores	2	1	2	2
Capacitors	3	1	1	7
Duty cycle range	$0<D<1$	$D<0.5$	$0<D<1$	$0<D<1$

Parameter n is the transformer turns ratio in Table III. Even though the proposed converter requires higher component count when compared with its similar counterparts, it presents the widest conversion range and lowest normalized voltage stresses on the switches among the analyzed solutions considering n=1, mc=2, and 0.33<D<0.66 as seen in Fig. 9 and Fig. 10. It is also worth mentioning that such topology is capable of operating within the whole range of the duty cycle, although the theoretical analysis carried out in this work is only valid for the aforementioned interval.

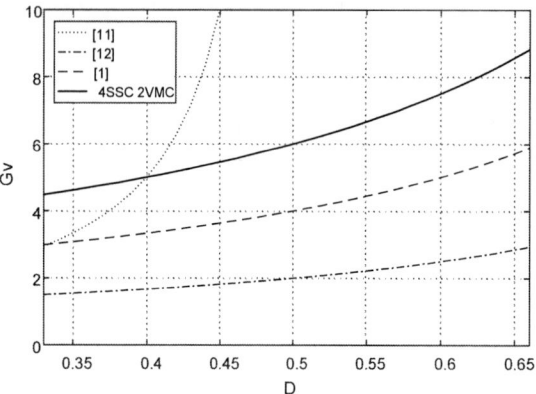

Fig. 9. Voltage gain versus duty cycle for 1/3<D<2/3 in CCM considering n=1 and mc=2.

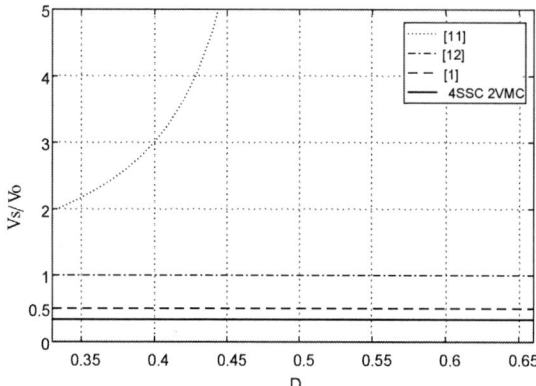

Fig. 10. Voltage stress on the active switches as a function of the duty cycle for 1/3<D<2/3 in CCM considering n=1, mc=2, and 0.33<D<0.66.

V. CONCLUSION

This paper has presented the development of a nonisolated dc-dc boost converter based on the 4SSC and VMCs. The main advantages of the proposed topology include the fact that the filter elements present reduced dimensions since the ripple frequency is three times higher than the switching frequency; the input current is continuous and equally shared among the semiconductors due to the autotransformer with unity turns ration; reduced voltage stresses on the active switches, which can be further minimized as more VMCs are added to the circuit. Being this a modular approach, the main disadvantage lies in the high number of required components when compared with other similar topologies.

A thorough theoretical analysis has been carried out, being properly validated through simulation tests. Considering the obtained results, it is reasonable to state that this is a viable solution for high-power, high-current applications where high-voltage step-up is required, e.g. renewable energy conversion systems. Future work includes the development of an experimental prototype, analysis of other VMC configurations, and the development of a family of nonisolated dc-dc converters.

ACKNOWLEDGMENT

The authors acknowledge CAPES, CNPq, FAPEMIG, INERGE, UFSJ and UNIFEI for the support to this work.

REFERENCES

[1] H. Gholizadeh, R. Babazadeh-Dizaji, and M. Hamzeh, "High-Gain Buck-Boost Converter Suitable for Renewable Applications," in *2019 27th Iranian Conference on Electrical Engineering (ICEE)*, 2019, pp. 777-781: IEEE.

[2] G. H. de Alcântara Bastos, L. F. Costa, F. L. Tofoli, G. V. T. Bascopé, and R. P. T. Bascopé, "Nonisolated DC-DC Converters with Wide Conversion Range for High-Power Applications," *IEEE Journal of Emerging and Selected Topics in Power Electronics,* 2019.

[3] F. Shang, G. Niu, and M. Krishnamurthy, "Design and analysis of a high-voltage-gain step-up resonant DC–DC converter for transportation applications," *IEEE Transactions on Transportation Electrification,* vol. 3, no. 1, pp. 157-167, 2017.

[4] S. Chakraborty, H.-N. Vu, M. M. Hasan, D.-D. Tran, M. E. Baghdadi, and O. Hegazy, "DC-DC Converter Topologies for Electric Vehicles, Plug-in Hybrid Electric Vehicles and Fast Charging Stations: State of the Art and Future Trends," *Energies,* vol. 12, no. 8, p. 1569, 2019.

[5] M. S. Bhaskar, M. Meraj, A. Iqbal, S. Padmanaban, P. K. Maroti, and R. Alammari, "High gain transformer-less double-duty-triple-mode DC/DC converter for DC microgrid," *Ieee Access,* vol. 7, pp. 36353-36370, 2019.

[6] X. Liu, X. Hu, H. Chen, L. Chen, and Y. Zhang, "An Interleaved Boost Converter with Parallel Input and Output Series for Renewable energy system," in *2019 IEEE 10th International Symposium on Power Electronics for Distributed Generation Systems (PEDG),* 2019, pp. 993-998: IEEE.

[7] K. Guépratte, P.-O. Jeannin, D. Frey, and H. Stephan, "High efficiency interleaved power electronics converter for wide operating power range," in *2009 Twenty-Fourth Annual IEEE Applied Power Electronics Conference and Exposition,* 2009, pp. 413-419: IEEE.

[8] G. A. Henn, R. Silva, P. P. Praça, L. H. Barreto, and D. S. Oliveira, "Interleaved-boost converter with high voltage gain," *IEEE*

transactions on power electronics, vol. 25, no. 11, pp. 2753-2761, 2010.

[9] S. Chen, S. Yang, C. Huang, and Y. Chen, "High Step-Up Interleaved Converter With Three-Winding Coupled Inductors and Voltage Multiplier Cells," in *2019 IEEE International Conference on Industrial Technology (ICIT),* 2019, pp. 458-463.

[10] M. Forouzesh, Y. P. Siwakoti, S. A. Gorji, F. Blaabjerg, and B. Lehman, "Step-up DC–DC converters: a comprehensive review of voltage-boosting techniques, topologies, and applications," *IEEE Transactions on Power Electronics,* vol. 32, no. 12, pp. 9143-9178, 2017.

[11] P. Lin and L. Chua, "Topological generation and analysis of voltage multiplier circuits," *IEEE Transactions on Circuits Systems,* vol. 24, no. 10, pp. 517-530, 1977.

[12] M. Prudente, L. L. Pfitscher, G. Emmendoerfer, E. F. Romaneli, and R. Gules, "Voltage multiplier cells applied to non-isolated DC–DC converters," *IEEE Transactions on Power Electronics,* vol. 23, no. 2, pp. 871-887, 2008.

[13] A. Alzahrani, P. Shamsi, and M. Ferdowsi, "Boost converter with bipolar Dickson voltage multiplier cells," in *2017 IEEE 6th International Conference on Renewable Energy Research and Applications (ICRERA),* 2017, pp. 228-233: IEEE.

[14] A. Alzahrani, M. Ferdowsi, and P. Shamsi, "High-voltage-gain DC-DC Step-up Converter with Bi-fold Dickson Voltage Multiplier Cells," *IEEE Transactions on Power Electronics,* 2019.

[15] S. Araujo, R. Bascope, G. Bascope, and L. Menezes, "Step-up converter with high voltage gain employing three-state switching cell and voltage multiplier," in *2008 IEEE Power Electronics Specialists Conference,* 2008, pp. 2271-2277: IEEE.

[16] P.-W. Lee, Y.-S. Lee, D. K. Cheng, and X.-C. Liu, "Steady-state analysis of an interleaved boost converter with coupled inductors," *IEEE Transactions on Industrial Electronics,* vol. 47, no. 4, pp. 787-795, 2000.

[17] W. Li and X. He, "High step-up soft switching interleaved boost converters with cross-winding-coupled inductors and reduced auxiliary switch number," *IET Power Electronics,* vol. 2, no. 2, pp. 125-133, 2009.

Stabilization of DC Microgrids with Point-of-Load Converters as Constant Power Loads

Isaías V. Bessa[1], Renan L. P. Medeiros[1], Iury V. Bessa[1], Florindo A.C. Ayres Junior[1], and Kevin E. Lucas[2]

[1]Federal University of Amazonas (UFAM), Department of Electricity, Manaus 69080-900, Brazil
[2]Federal University of Santa Catarina (UFSC), Department of Automation and Systems, Florianópolis 88040-900, Brazil
Emails: isaias.97.ib@gmail.com, renanlandau@ufam.edu.br, iurybessa@ufam.edu.br,
florindoayres@ufam.edu.br, kevin.lucas@posgrad.ufsc.br

Abstract—DC-DC switching power converters exhibit the important characteristic of almost-perfect regulation at the output terminals independent of the input perturbations. However, such characteristic reflects at the input terminal as a constant power load (CPL). Microgrids (MGs) consists of multiple converters to satisfy the different voltage levels that loads need and stability problems are caused due to the interaction of Point-of-Load (POL) converters. In this context, this paper focuses on the stability analysis of POL converters as CPLs. In particular, POL converters are modeled using boost and buck converter topologies and the stabilization of POL converters is performed with different classic and modern control strategies, such as output feedback and state feedback. The performance analysis of the controllers is carried out using the Integral Squared Error (ISE), the Integral Time-weighted Squared Error (ITSE) and the Integral Squared of Control Signal (ISSC) performance indices.

Index Terms—Constant power load, Microgrid Stabilization, buck converter, boost converter.

I. INTRODUCTION

DC-DC converters are important electronic devices used for power transfer and voltage regulation in electrical systems. Multi-converter electronic systems are used to satisfy the voltage level requirements in different industrial application. However, the interaction between the power electronics converter may cause undesired oscillations in the system due to the point-of-load (POL) converters, which behave as a constant power load (CPL) [1].

DC-DC converters are elements commonly used in DC microgrids (MG) due to their simplicity in structure and control [2], [3]. The DC-DC converters are also used to emulate the behavior of CPL which when connected to microgrid causes instability in the DC power link [1], [2], [4]. Several control techniques for power electronics converter have been proposed by researchers, for instance, in [5], linear-quadratic regulator (LQR) is used to regulate the output voltage of a buck converter, satisfying reference tracking, settling time and disturbance rejection requirements. In [6], the authors use fractional order control techniques to reduce the output voltage oscillations due to loads and parametric variation.

The CPL effects on an MG are addressed in [2], where the POL converter tends to destabilize the MG due to the CPL negative incremental impedance. Some techniques for CPL stabilization are addressed in [1], [4]. A passive damping method is used in [1] by adding a input LC filter due to its simplicity of implementation, however, the use of an input LC filter adds system dynamics, increases system cost and can lead to impractical capacitive and inductive element sizes, as seen in [2]. In [4], a robust controller based on the Kharitonov

theorem is employed to reduce the CPL oscillation effects in a multi-converter system where the use of a controller external to the system allows for the reduction of the passive elements of the system and the regulation of the output voltage without changing the input power, however this structure is more susceptible to output noise due to the feedback loop, as seen in [2].

This work aims to perform the stability analysis of POL converters using boost and buck converter topologies to model a CPL. Different control techniques are performed to regulate the CPL power, such as: output feedback control solving the Diophantine equation and the graphical root locus tool, and state feedback control through linear-quadratic regulator (LQR) controller and solving the Lyapunov equation. The dynamics of CPL is depicted by means of phase-portraits. The performance of these different controllers is investigated by means of ISE, ITSE and ISSC performance indices.

The remainder of this work is divided as follows: Section II presents the dynamic behavior of converters and CPLs; Section III summarizes the control theory of controllers addressed in this work; Section IV presents the design procedure of the control strategies and the experimental procedures; Section V presents the experimental results; and finally, Section VI draws the final considerations.

II. OPERATION OF CONVERTERS SUCH AS CPL

Fig.1a shows a CPL which is an element in the DC system responsible for a constant power consumption [1], [2], [4]. Thus, according to (1), input current I_{CPL} increases when input voltage U_{CPL} decreases, and vice versa in order to maintain a constant power level P_{CPL}. This behavior is illustrated in Fig.1b.

$$P_{CPL} = U_{CPL}I_{CPL} \qquad (1)$$

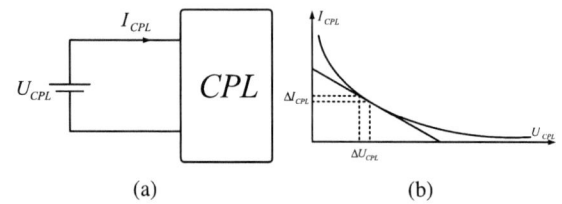

(a)   (b)

Fig. 1: (a) CPL fed by source U_{CPL} consuming current I_{CPL} (b) CPL behavior.

Thus, the rate of change in voltage and current is described as follows:

$$\frac{\Delta U_{CPL}}{\Delta I_{CPL}} \cong \frac{\partial U_{CPL}}{\partial I_{CPL}} = -\frac{P_{CPL}}{I_{CPL}^2} \qquad (2)$$

where $R = P_{CPL}/I_{CPL}^2$ represents a resistance. Hence, the system may become unstable due to destabilizing effects of CPL, which cause significant oscillations leading to instability or voltage collapse [1], [3]. DC-DC converters operating with tight closed-loop regulation present a behavior, at the input terminals, similar to a CPL in a certain range of frequencies and input voltages [3], [4]. Therefore, buck and boost converters are modeled as a CPL in the next subsections.

A. Modeling of Buck Converter as CPL

The design of DC-DC converters is related to determinate the converter parameters, i.e., resistance, inductance and capacitance, to ensure steady state conditions. Given the buck converter shown in Fig.2a with input voltage V_{i1}, operating with duty cycle D_1, the output voltage V_{o1} of this converter is given in (3).

$$V_{o1} = D_1 V_{i1} \qquad (3)$$

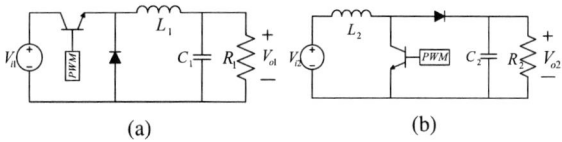

(a)   (b)

Fig. 2: (a) Buck converter (b) Boost converter.

The power consumed by load R_1 of the buck converter in p.u. and being the highest voltage value at the converter output occurs for $D_1 = 1$, then

$$P_{o1} = D_1^2 \qquad (4)$$

The choice of buck converter parameter values must satisfy the set of equations, defined in (5), to ensure continuous conduction mode (CCM) operation, i.e., $I_{L_1} > 0$, in addition to satisfying a specify a ripple current ΔI_{L_1} in the inductor L_1 and ripple output voltage ΔV_{o1} for a given frequency of switch f_1 set by the designer.

$$\begin{cases} L_1 > \frac{(1-D_1)R_1}{2f_1} \\ C_1 > \frac{1-D_1}{8L_1(\Delta V_{o1}/V_{o1})f_1^2} \\ \Delta I_{L_1} < \frac{V_{i1}D_1(1-D_1)}{L_1 f_1} \end{cases} \qquad (5)$$

The equations shown in (3)-(5) indicate steady-state ratios of the converter. The dynamic behavior of the buck converter is performed by modeling the system using the averaged state-space model. According to the switching circuit shown in Fig.2a, the averaged state-space model is given in (6) [4].

$$\dot{x} = D_1 \dot{x}_{on} + (1-D_1)\dot{x}_{off} \qquad (6)$$

Solving (6), the averaged state-space model of the buck converter is found (7).

$$\begin{bmatrix} \dot{I}_{L_1} \\ \dot{V}_{C_1} \end{bmatrix} = \begin{bmatrix} 0 & -\frac{1}{L_1} \\ \frac{1}{C_1} & \frac{1}{R_1 C_1} \end{bmatrix} \begin{bmatrix} I_{L_1} \\ V_{C_1} \end{bmatrix} + \begin{bmatrix} \frac{D_1}{L_1} \\ 0 \end{bmatrix} V_{i1} \qquad (7)$$

The construction of the CPL is related to ensuring constant power consumption at the converter output. Note that equation (7) is a nonlinear continuous-time equation. Thus, it can be linearized by small-signal perturbation with $I_{L_1} = I_{L_1}^o + \Delta I_{L_1}$, $V_{C_1} = V_{C_1}^o + \Delta V_{C_1}$, and $D_1 = D_1^o + \Delta D_1$, where: ΔI_{L_1}, ΔV_{C_1}, and ΔD_1 represent a small-signal values, and $I_{L_1}^o$, $V_{C_1}^o$, and D_1^o represent the dc value, i.e., the operating points. It is important to note that $I_{L_1}^o >> \Delta I_{L_1}$, $V_{C_1}^o >> \Delta V_{C_1}$, and $D_1^o >> \Delta D_1$. The perturbation yields the linear small-signal state-space model in (8).

$$\begin{bmatrix} \Delta \dot{I}_{L_1} \\ \Delta \dot{V}_{C_1} \end{bmatrix} = \begin{bmatrix} 0 & -\frac{1}{L_1} \\ \frac{1}{C_1} & \frac{1}{R_1 C_1} \end{bmatrix} \begin{bmatrix} \Delta I_{L_1} \\ \Delta V_{C_1} \end{bmatrix} + \begin{bmatrix} \frac{V_{i1}}{L_1} \\ 0 \end{bmatrix} \Delta D_1 \qquad (8)$$

The output of the converter is the power in the load R_1, hence, the power output is given by $P_{o1} = \frac{V_{C_1}^2}{R_1}$. The power output can be linearized by small-signal perturbation around the operating point given the expression below (9).

$$\Delta P_{o1} = \frac{2 V_{C_1}^o \Delta V_{C_1}}{R_1} \qquad (9)$$

B. Modeling of boost converter as CPL

Fig.2b shows the boost converter topology. The boost converter aims to step up from its input (V_{i2}) to its output (V_{o2}), i.e., the output voltage is greater than the input voltage. Equation (10) shows that the output voltage is always higher than the input voltage where D_2 is the duty cycle that goes from 0 to 1.

$$V_{o2} = \frac{V_{i2}}{1-D_2} \qquad (10)$$

The power consumed by load R_2 of the boost converter, in p.u. and as high duty cycle values condition the output to mathematically high voltage values, the base $D_2 = 0.5$ was chosen to normalize the system, is

$$P_{o2} = \frac{1}{4(1-D_2)^2} \qquad (11)$$

The boost converter is designed based on the same criteria presented in the Section II-A. Thus, the choice of boost converter parameter values, i.e., R_2, L_2 and C_2, are given by the following set of equations

$$\begin{cases} L_2 > \frac{D_2(1-D_2)^2 R_2}{2f_2} \\ C_2 > \frac{D_2}{R_2(\Delta V_{o2}/V_{o2})f_2} \\ \Delta I_{L_2} < \frac{V_{i2}D_2}{L_2 f_2} \end{cases} \qquad (12)$$

For the dynamic modeling of the boost converter, the same methodology presented by the buck converter is used. Thus, the averaged state-space model of the boost converter is shown in (13).

$$\begin{bmatrix} \dot{i}_{L_2} \\ \dot{V}_{C_2} \end{bmatrix} = \begin{bmatrix} 0 & -\frac{1-D_2}{L_2} \\ \frac{1-D_2}{C_2} & \frac{1}{R_2 C_2} \end{bmatrix} \begin{bmatrix} I_{L_2} \\ V_{C_2} \end{bmatrix} + \begin{bmatrix} \frac{1}{L_2} \\ 0 \end{bmatrix} V_{i2} \qquad (13)$$

The linear small-signal state-space model of the boost converter (14) can be found linearizing the model (13) by small-signal perturbation. where: ΔI_{L_2}, ΔV_{C_2}, and ΔD_2 represent a small-signal values, and $I_{L_2}^o$, $V_{C_2}^o$, and D_2^o represent the dc value, i.e., the operating points.

$$\begin{bmatrix} \Delta \dot{i}_{L_2} \\ \Delta \dot{V}_{C_2} \end{bmatrix} = \begin{bmatrix} 0 & -\frac{1-D_2^o}{L_2} \\ \frac{1-D_2^o}{C_2} & \frac{1}{R_2 C_2} \end{bmatrix} \begin{bmatrix} \Delta I_{L_2} \\ \Delta V_{C_2} \end{bmatrix} + \begin{bmatrix} \frac{V_{C_2}^o}{L_2} \\ -\frac{I_{L_2}^o}{C_2} \end{bmatrix} \Delta D_2 \quad (14)$$

The output of the boost converter is the power in the load R_2, given by $P_{o2} = \frac{V_{C_2}^2}{R_2}$. The power output can be linearized by small-signal perturbation around the operating point given the expression below (15).

$$\Delta P_{o2} = \frac{2 V_{C_2}^o \Delta V_{C_2}}{R_2} \quad (15)$$

C. Stability of converters

Depending on the chosen operating point, the stability region changes. Suppose that λ_1 and λ_2 are the eigenvalues of the linearized buck converter, and λ_3 and λ_4 are the eigenvalues of the boost converter, if real parts of all eigenvalues are negative, then the equilibrium point is stable. If at least one eigenvalue has a positive real part, then the equilibrium point is unstable. The phase-portrait of buck and boost converter is simulated using the linearized models with $D_1^o = 0.6$, $D_2^o = 0.7$, $R_1 = R_2 = 4\ \Omega$, $L_1 = L_2 = 1$ mH, and $C_1 = C_2 = 2000\ \mu$F as shown in Figs.3a and 3b, respectively.

Since the eigenvalues are stable, the trajectory of both systems converges to the point of operation. If the system is damped oscillation, this trajectory occurs in the form of a spiral around the equilibrium point.

III. Control strategies for CPL stabilization

When a power electronic converter tightly regulates its output, it behaves as a CPL [3]. Hence, a power electronic converter can emulate a CPL by controlling the power output with appropriate control strategies. This section presents a brief review about control strategies to design a CPL from DC-DC power electronics converter using output feedback structures, Fig. 4a and state feedback structures, Fig. 4b.

A. The Diophantine equation

The open-loop transfer function (16) can be determined from the output feedback structure shown in Fig.4a.

$$G(s) = \frac{B(s)}{A(s)} = C(sI - A)^{-1} B \quad (16)$$

The solution of the Diophantine equation (17) summarizes the classical pole-placement problem, as follows:

$$\Delta_C(s) = A(s)M(s) + B(s)N(s) \quad (17)$$

where, $\Delta_C(s)$ is the closed-loop characteristic polynomial, and $C(s) = \frac{N(s)}{M(s)}$ is the controller structure. Assuming that the desired dynamic of closed-loop system is represented by (18).

$$\Delta_D(s) = s^n + a_1 s^{n-1} + \cdots + a_{n-1} s + a_n \quad (18)$$

If $\Delta_C(s)$ and $\Delta_D(s)$ are polynomials of the same degree and if $A(s)$ and $B(s)$ are known polynomials that describe the behavior of the plant, then equating (17) and (18) yields

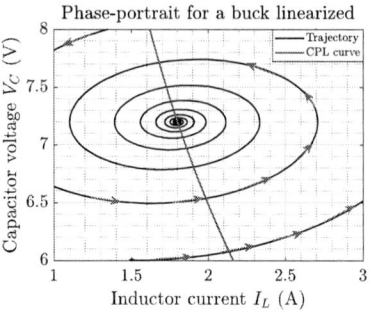

(a)

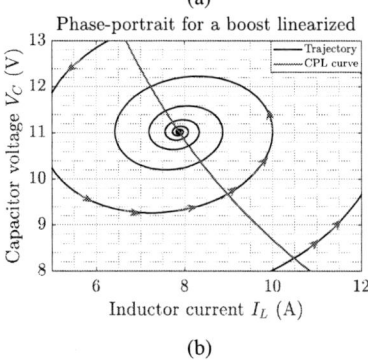

(b)

Fig. 3: Phase-portrait of the linearized systems (a) Buck converter. (b) Boost converter.

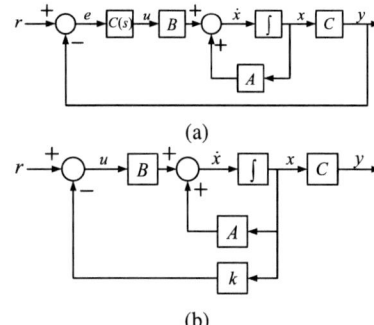

Fig. 4: Control loops: (a) output feedback. (b) state feedback.

the Diophantine equation (DE) (19) that allows to find the coefficients $M(s)$ and $N(s)$ of the controller $C(s)$ [7].

$$\Delta_D(s) = A(s)M(s) + B(s)N(s) \quad (19)$$

B. The root locus graphic tool

The closed-loop desired polynomial $\Delta_D(s)$ (18) has dominant poles $s_{1,2} = \omega_n(-\zeta \pm j\sqrt{1 - \zeta^2})$, these poles then determine on the s-plane a desired performance region, as shown in Fig.5. Where ω_n is the natural frequency of the closed-loop system, represented by the distance from the origin to the dominant pole [7]. Moreover the angle of this line with the real axis is related to the value of the damping coefficient ζ [7]. The real part $\zeta \omega_n$ of the dominant pole is the projection of segment ω_n on the real axis. The damping coefficient is specified for a given maximum overshoot O_{vs} of the system

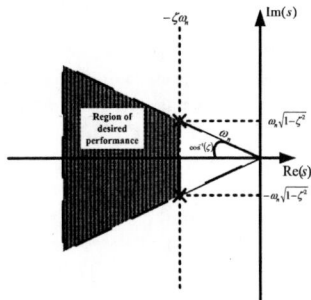

Fig. 5: S-plane root locus.

while the natural frequency is related to the settling time t_{ss} of the system [7]. Eq. (20) shows the above relationships, where e_{ss} is the permanent steady state error, and determine the performance region where the closed-loop poles will be located depending on the chosen controller structure $C(s)$ (cf. Fig.5).

By determining the root locus curve and the performance region specified in Fig.5, the controller gains of $C(s)$ are determined in order to allocate the closed-loop poles in the desired region.

$$O_{vs} = e^{-\frac{\zeta \omega_n}{\sqrt{1-\zeta^2}}}, \quad t_{ss} = -\frac{\ln(e_{ss})}{\zeta \omega_n} \quad (20)$$

C. The linear-quadratic regulator

The LQR control is a technique based on the state feedback structure considering an observable system as shown in Fig.4b. This methodology is performed from the development of the cost function (21) [5].

$$J = \int_0^\infty (x^T Q x + u^T R u) dt \quad (21)$$

where Q is a positive semi-definite matrix responsible for weighing the states of the system while R is a positive definite matrix responsible for limiting the control action.

By developing the equation (21), the Riccati equation (22) is obtained [5].

$$PA + A^T P - PBR^{-1}B^T P = -Q \quad (22)$$

Note that the matrix P is obtained from the choice of Q and R. The the feedback gain K is determined from the obtained matrix P.

$$K = R^{-1}B^T P \quad (23)$$

D. The Lyapunov equation

Consider the controllable system with state feedback structure shown in Fig.4b, the Lyapunov equation (LE) presented (24) allows to determine the matrix T given a desired performance matrix F for a given $\bar{K}$ [7].

$$AT - TF = B\bar{K} \quad (24)$$

The choice of $\bar{K}$ is such that $(F, \bar{K})$ is observable. From T, the feedback gain K is determinable using equation (25).

$$K = \bar{K}T^{-1} \quad (25)$$

IV. METHODOLOGY

This section presents the methodology for design of POL converters as a CPL from buck and boost converter topologies with the control strategies addressed in this study.

A. Design of CPL controllers

For the buck converter, a current ripple $\Delta I_{L_1} < 1$ A is specified for a switching frequency $f_1 = 5$ kHz limited by a $L_1 = 1$ mH inductor and a voltage ripple $\Delta V_{o1}/V_{o1} < 10\%$ is also defined. For the boost converter, the same above ripple current and voltage specifications are chosen, limited by the same inductor value ($L_2 = 1$ mH) operating at a switching frequency of $f_2 = 10$ kHz. Therefore, the parameters for the buck and boost converter are obtained using (5) and (12), respectively. Table I summarizes the parameters for both converters.

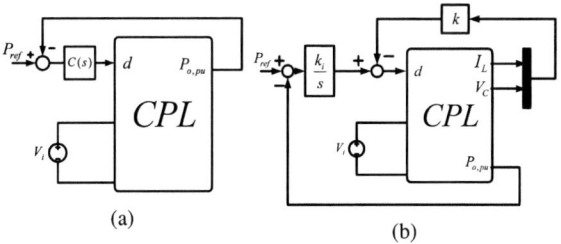

Fig. 6: (a) CPL with output feedback (b) CPL with state feedback and integrator.

TABLE I: CPL parameters

Buck CPL			
Parameter	**Symbol**	**Value**	**Unit**
Input voltage	V_{i1}	12.0	V
Duty cycle	D_1	0.6	-
Switching frequency	f_1	5.0	kHz
Load resistance	R_1	4.0	Ω
Inductor	L_1	1.0	mH
Capacitor	C_1	2.2	mF
Output power	P_{o1}	0.36	p.u.
Boost CPL			
Parameter	**Symbol**	**Value**	**Unit**
Input voltage	V_{i2}	18.0	V
Duty cycle	D_2	0.7	-
Switching frequency	f_2	10.0	kHz
Load resistance	R_2	4.0	Ω
Inductor	L_2	1.0	mH
Capacitor	C_2	2.2	mF
Output power	P_{o2}	0.51	p.u.

The controllers are designed to meet the following specifications: settling time $t_{ss} \leq 0.02$ s; maximum overshoot $O_{vs} \leq 10\%$; and steady state error $e_{ss} = 0$. From (20), the constraints for ω_n and ζ are computed resulting $\omega_n \geq 389.07$ rad/s, $\zeta \leq 0.59$.

The controllers with output feedback structures, shown in Fig.6a, used PID structures, where for the buck converter a classic $C_1(s)$ model is used and for the boost a PID with filter $C_2(s)$ is used. The structures are indicated in (26).

$$C_1(s) = \frac{k_d s^2 + k_p s + k_i}{s}, \quad C_2(s) = \frac{k_d s^2 + k_p s + k_i}{s(s+N)} \quad (26)$$

In state feedback structure (cf. Fig.6b), an external loop with an integrator is used to ensure the step reference tracking. The desired polynomial for the buck and boost converter are shown in (27). Note that non-dominant auxiliary poles are added in the closed-loop desired polynomial in order to be feasible the solution of DE (19).

$$\Delta_{D1} = (s^2 + 2\zeta\omega_n s + \omega_n^2)(s + p_1)$$
$$\Delta_{D2} = (s^2 + 2\zeta\omega_n s + \omega_n^2)(s + p_2)(s + p_3) \quad (27)$$

On the other hand, for controller design by RL, the desired performance region is defined, as shown in Fig.5 and with the help of the `rltool()` tool of MATLAB, the gains of the buck and boost controllers are determined.

The matrices R_{lqr1} and Q_{lqr1} for the buck converter and R_{lqr2} and Q_{lqr2} for the boost converter are shown in (28).

$$R_{lqr1} = 0.01, \ R_{lqr2} = 50$$
$$Q_{lqr1} = \begin{bmatrix} 10 & 0 \\ 0 & 1 \cdot 10^{-6} \end{bmatrix}, \ Q_{lqr2} = \begin{bmatrix} 1 & 0 \\ 0 & 1 \cdot 10^{-10} \end{bmatrix} \quad (28)$$

To find the value gains of the feedback vector K, we use the MATLAB command `lqr()`.

Finally, to solve the LE, we use the desired performance matrix F shown in (29).

$$\begin{bmatrix} \zeta\omega_n & \omega_n\sqrt{1-\zeta^2} \\ -\omega_n\sqrt{1-\zeta^2} & \zeta\omega_n \end{bmatrix} \quad (29)$$

For the buck converter the vector $\bar{K}_1$ is determined and to boost the vector $\bar{K}_2$ indicated in (30).

$$\bar{K}_1 = [1 \ 10], \quad \bar{K}_2 = [10^{-3} \ 10^{-3}] \quad (30)$$

The LE solution was solved using the MATLAB `lyap()` command. With the matrix T determined, the state realignment gain K is calculated with the equation (25). Table II summarizes the gains of the proposed controller.

TABLE II: Controller parameters

Buck CPL				
Gains	DE	RL	LQR	LE
k_p	0.38	1.81	—	—
k_i	284.1	287.6	$20 \cdot 10^3$	400
k_d	0.003	0.003	—	—
N	—	—	—	—
K	—	—	[0.49 1.94]	[0.02 − 0.006]
Boost CPL				
Gains	DE	RL	LQR	LE
k_p	740.2	$1.06 \cdot 10^4$	—	—
k_i	$6.2 \cdot 10^5$	$3.2 \cdot 10^6$	$4.0 \cdot 10^3$	150
k_d	6.2	33.82	—	—
N	$3.7 \cdot 10^3$	$3.4 \cdot 10^4$	—	—
K	—	—	[0.13 0.21]	[0.02 − 0.006]

B. Test description

To verify the performance of the CPL controllers, the POL converters that act as CPL are tested under variation of power reference. Initially the converters start operating in open-loop, i.e., the converters do not act as a CPL. At time $t = 1.0$ s, the converters achieve their steady state condition. At time $t = 2.0$ s, the converters operate in closed-loop and start to act as a CPL.

The power variation starts at time $t = 5.0$ s and $t = 5.5$ s with variations of 0.1 p.u. of power returning to the reference in the $t = 6.0$ s. At $t = 6.5$ s and $t = 7.0$ s, there are variations of -0.1 p.u.. At $t = 7.5$ s and $= 8.0$ s, there are variation of 0.1 p.u., after that, the system returns to the original reference.

The simulated test shows that the system moves away from the designed operating point when the system is subject to a variation of power reference. The system performance is analyzed by ISE, ITSE and ISSC performance indices. In particular, ISE allows to evaluate which strategy accumulate the least error, ITSE indicates which strategy provides the fastest reference tracking, and ISSC verifies which strategy requires the greatest control effort [8].

Furthermore, the stability analysis of the system is performed in $t = 6$ s using the phase-portrait of the CPL with the proposed controllers.

V. DISCUSSION OF RESULTS

This section analyzes the results obtained during the simulated test. The simulated test are performed according to Section IV.

Fig.7a and Fig.7b shows the performance of the controllers for the buck and boost converter, respectively.

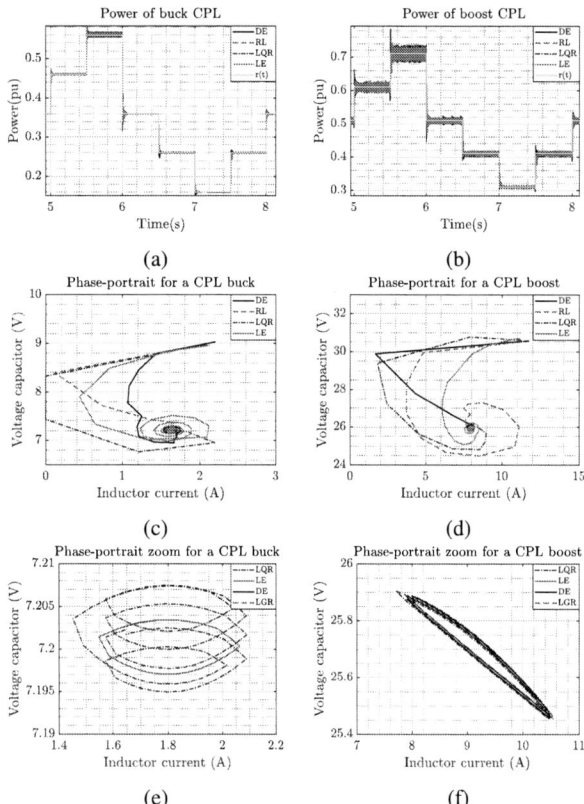

Fig. 7: (a) Power variation test on CPL buck (b) Power variation test on CPL buck (c) Phase portrait of the CPL buck (d) Phase portrait of the CPL boost (e) Zoom in the phase-portrait around the operating point of the buck (f) Zoom in the phase-portrait around the operating point of the boost.

The CPL is stable due to the states converge to the equilibrium point forming a spiral in the phase-portrait, characteristic of a damped oscillatory system as shown in Fig. 7c-7d.

Voltage and current ripple are caused by switching of the power electronic converter leading to voltage oscillations. Due to these oscillations, a limit-cycle is formed around the equilibrium point as shown in Fig.7e-7f. Because of the existence of a limit-cycle, the system is purely oscillatory.

Fig.8a-8c show the performance indices of the controllers for buck converter as CPL under variation of power reference.

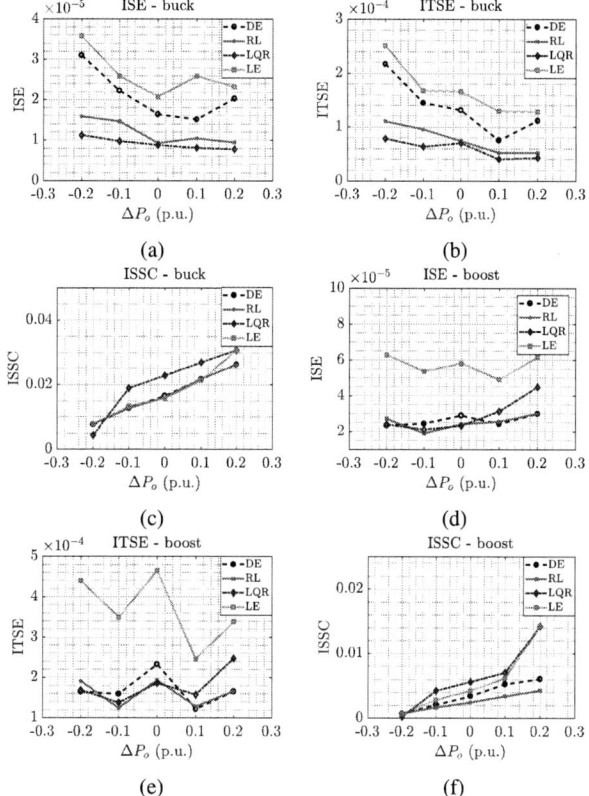

Fig. 8: (a) ISE for a CPL buck (b) ITSE for a CPL buck (c) ISSC for a CPL buck (d) ISE for a CPL boost (e) ITSE for a CPL boost (f) ISSC for a CPL boost.

LQR approach provides a better performance with reduced oscillation amplitude in the minor time as shown the ISE and ITSE performance indices in Fig.8a and Fig.8b, respectively. However, LQR approach presents greater control effort according to the ISSC performance index shown in Fig.8c. In contrast, the other approaches presents a similar control effort according to ISSC values (cf. Fig.8c). Therefore, LQR presents less accumulation of error in a shorter time resulting in a greater control effort compared to the other approaches. However, RL approach presents better performance-to-effort ratio, having a close performance than the LQR approach for the accumulation of error (ISE) and time response (ITSE). The best LQR controller performance is due to the design model that weighted the output performance against the control effort. While the controllers performed by the classical methodolo-

gies, DE and RL, were specifically designed to meet the design criteria presented in the previous section.

Fig.8d-8f show the performance indices of the controllers for boost converter as CPL under variation of power reference. Fig.8d and Fig.8f show the ISE and ITSE for each methodology. Note that the LE approach presents the worst performance in comparison with other approaches. Moreover, the other approaches presents a similar performance, however, the LQR approach presents a increase of ISE and ITSE as the power variation increases. On the other hand, LQR and LE approaches presents greater control effort according to the ISSC performance index shown in Fig.8f. Thus, LQR approach presents greater performance degradation as the power variation increases. Therefore, for boost converter topology, the RL approach provides better performance in a shorter time with a less control effort.

VI. CONCLUSION

In this work, the design of a POL converter as CPL is performed using buck and boost converter topologies and different control techniques are employed to ensure the point-of-load regulation of converters.

The stability of a damped oscillatory system caused by a CPL is checked by the phase-portrait. Both system are stable because the trajectory have a form of a spiral around the equilibrium point. However, the system present significant and undesirable limit-cycle oscillations due to the switchings. In addition, the simulation results indicate that the LE approach presents the worst performance in comparison with the other approaches for both, buck and boost topologies. On the other hand, the LQR approach presents a better performance minimizing the error and settling time, altought it required a greater control effort.

ACKNOWLEDGMENT

This work was supported in part by CNPq under grant 432341/2018-8, in part by CAPES, FAPEAM, and in part by the PROPG-CAPES/FAPEAM Scholarship Program.

REFERENCES

[1] H. Mosskull, "Constant power load stabilization," *Control Engineering Practice*, vol. 72, pp. 114 – 124, 2018.
[2] S. Singh, A. R. Gautam, and D. Fulwani, "Constant power loads and their effects in DC distributed power systems: A review," *Renewable and Sustainable Energy Reviews*, vol. 72, pp. 407 – 421, 2017.
[3] K. E. Lucas, D. A. Plaza, W. B. Jr, R. L. P. Medeiros, E. M. Rocha, D. Benavides, S. J. Ríos, and E. V. Herrera, "Novel robust methodology for controller design aiming to ensure DC microgrid stability under CPL power variation," *IEEE Access*, vol. 1, no. 7, pp. 1–16, May 2019.
[4] K. E. Lucas, D. A. Plaza, W. Barra, R. L. P. Medeiros, E. M. Rocha, D. Vaca, and F. G. Nogueira, "Interval robust controller to minimize oscillations effects caused by constant power load in a DC multi-converter buck-buck system," *IEEE Access*, vol. 7, pp. 26 324–26 342, February 2019.
[5] L. A. Maccari, R. L. Valle, A. Ferreira, P. G. Barbosa, and V. Montagner, "A LQR design with rejection of disturbances and robustness to load variations applied to a buck converter," *Eletrônica de Potência*, vol. 21, pp. 7–15, 02 2016.
[6] Z. Yichen, X. Hejin, and L. Deming, "Feedback control of fractional PIλDμ for DC/DC buck converters," in *2017 International Conference on Industrial Informatics - Computing Technology, Intelligent Technology, Industrial Information Integration (ICIICII)*, Dec 2017, pp. 219–222.
[7] C.-T. Chen, *Linear system theory and design*, 3rd ed. Oxford university press, 1999.
[8] M. A. Duarte-Mermoud and R. A. Prieto, "Performance index for quality response of dynamical systems," *ISA Transactions*, vol. 43, no. 1, pp. 133 – 151, 2004.

Inductor Design Methodology for Power Electronics Applications

P. H. J. Vilkn, L. M. F. Morais, R. A. S. Santana, P. C. Cortizo, P. F. Seixas

Graduate Program in Electrical Engineering Universidade Federal de Minas Gerais

Av. Antônio Carlos 6627, 31270-901, Belo Horizonte, MG, Brazil

pedrovilkn@gmail.com, lenin@cpdee.ufmg.br, rass.eletrica@gmail.com, porfirio@cpdee.ufmg.br, paulos@cpdee.ufmg.br

Abstract—In this work, a methodology is presented that allows the automatization of the inductor design, estimating copper losses, considering skin and proximity effects and also core losses through improved generalized Steinmetz equation. The inductors designed with this methodology were tested in a Buck converter, comparing the calculation of losses with measurements through a calorimeter.

Index Terms—Inductor Design, Core Losses, IGSE, Calorimeter, Loss measurement

I. INTRODUCTION

With the increment of the switching frequency, made possible by the development of the wide band gap semiconductors, the balance between power density, eficiency and cost of static converters was severely altered due to the reduction of volume and change in the loss profile of the devices.

The amount of losses represented by the inductances becomes more significant due to the reduction of the losses in the wide band gap semiconductors and also to the increase of the losses by skin and proximity effects in the inductor.

This work presents a methodology for the design of inductors used in switched static converters that allows the obtaining of projects with losses, volume or minimum costs.

The proposed methodology performs projects for the combination of provided parameters for the cores, frequency, current density and inductance, identifying the possible projects and comparing the results to obtain the most suitable inductor.

The inductor design is independent of the type of converter used, as long as it is possible to obtain the voltage and current vectors at the terminals of the inductor and to identify the harmonic components of the current, it is then possible to perform all calculations present in the design proposed.

Section II explains the design of the inductor, section III describes the system created for loss measurement and section IV describes the experiments performed and presents the results obtained.

This work has been supported by the Brazilian agency CAPES and Engetron Engenharia Eletrônica Ind e Com LTDA.

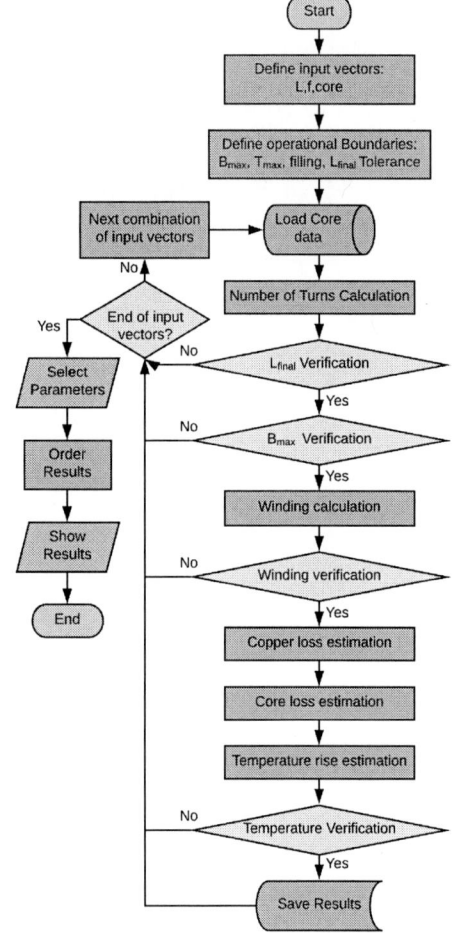

Fig. 1. Flowchart of the inductor design methodology

the generation of a large number of projects, verifying the possibility of realization and then identifying the best results.

A design attempt is made for each combination of inductance, frequency, current density and core inserted in the input vectors.

Boundaries for magnetic flux density B, temperature T, core window fill and inductance tolerance are set to ensure that the design is achievable and meets minimum operating limits.

II. INDUCTOR PROJECT IN NON SINUSOIDAL CONDITIONS

The flowchart in Fig. 1 demonstrates the project methodology proposed in this work, which is based on

When one of the boundaries is not respected in a project attempt, it is then discarded and the next attempt is initiated until all combinations are performed.

The core data is loaded from a library that was created to standardize information of the cores from different manufacturers.

At the end of the calculation of all combinations the saved projects are sorted according to the priority of parameters (volume, losses, temperature, cost) defined by the user.

A. calculation of the number of turns

The calculation of the number of turns, modified from the proposed one in [1] and [2], is an iterative process since the magnetic permeability μ is affected by the frequency f, temperature T, density of flux B_{AC} and magnetic field H_{DC}, which depends on the number of turns used.

These relationships between frequency, temperature, flux density *a.c.*, magnetic field *d.c.* and magnetic permeability are given in curves or equations by manufacturers, however it is unusual to find all curves available.

The loop of each iteration starts with (1) obtaining the new number of turns, where A_L is the specific inductance given by the manufacturer or obtained from the reluctance of the magnetic path $\mathcal{R}$ per (2), with A_c the cross-sectional area and l_e the average length of the magnetic path.

$$N = \sqrt{\frac{L}{A_L}} \qquad (1)$$

$$A_L = \frac{1}{\mathcal{R}} = \frac{\mu_e A_c}{l_e} \qquad (2)$$

Using air gap adds a term in the reluctance that is independent of the permeability of the material, which facilitates the definition of the number of turns but generates greater emission of electromagnetic radiation and additional terms of losses. These effects are intensified in applications of higher power and frequency, and are generally avoided in switched sources [1].

While f is known and can be used directly H_{DC}, B_{AC} and T require additional calculations.

To obtain H_{DC} is used (3), where I_{DC} in a switched inverter context is the peak current of the low frequency sinusoidal component, in this case is calculated H_{DC} maximum and its instantaneous value will vary, as will the final inductance.

$$H_{DC} = \frac{N I_{DC}}{l_e} \qquad (3)$$

It is possible to obtain B_{AC} using (3) for the maximum and minimum values of the ripple current and using the $B \times H$ curve of the material when available.

However, an independent method of the $B \times H$ curve requires only the voltage vector in the inductor $V_L(t)$, N and A_c, obtaining the vector $B(t)$ from (4) [3].

$$B(t) = \frac{\int V_L(t)\, dt}{N\, A_c} \qquad (4)$$

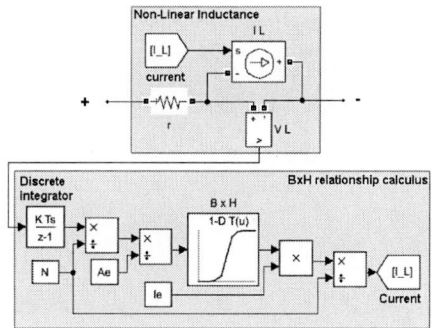

Fig. 2. Saturated inductance model for simulation

The temperature value can be estimated using basic loss calculations with (17), (27) and (31), but in general this effect is not significant except in projects specific for high temperatures.

One can then use the relations $\mu \times H_{DC}$, $\mu \times B_{AC}$, $\mu \times T$, and $\mu \times f$ to obtain $\mu_\%$ which is the reduction of the initial permeability given by (5).

$$\mu_\% = \frac{\mu_{initial} - \mu_{effective}}{\mu_{initial}} \qquad (5)$$

With this we calculate the effective $A_{L,ef}$ and L_{ef} in (6) and (7) respectively.

$$A_{L,ef} = A_L\, \mu_\% \qquad (6)$$

$$L_{ef} = N^2\, A_{L,ef} \qquad (7)$$

If L_{ef} is not within the tolerance limits we return to (1) using $A_L = A_{L,ef}$, until L_{ef} is within tolerance or the maximum number of iterations is reached.

Finding the number of turns the vector obtained in (4) is checked for saturation limits. Alternatively a better result is obtained by simulating the converter using a saturable inductance model which includes the $B \times H$ curve of the material as in Fig. 2, best explained in [3].

B. Winding design

In the design of the winding the wires used and the way they are coiled are defined, obtaining the filling of the window and the resulting number of layers, as well as the length and d.c. resistance of the conductor. For this it is necessary to calculate the conductor cross-sectional area A_{copper}, obtained by (8).

$$A_{copper} = \frac{I_{ef}}{j} \qquad (8)$$

The conductor is a set of N_{wires} parallel insulated conductors of diameter d_{wire} twisted to reduce skin and proximity effect. Using $d_{wire} \leq 2\delta$ being δ the depth of penetration, we reduce the skin effect considerably. On copper $\delta = \sqrt{\rho/(\pi\mu f)}$ can be approximated by $\delta = 65/\sqrt{f}$ at a temperature of 25^oC [1].

The equations (9) and (10) relate the diameter of the wires and the number of parallel wires required. High frequencies require a large number of very thin wires in parallel, which leads to problems in the winding construction. In these cases the use of Litz wires is recommended [4].

$$d_{wire} = 2\sqrt{\frac{A_{copper}}{N_{wires}\,\pi}} \qquad (9)$$

$$N_{wires} = \mathbb{N}\left\{\frac{A_{copper}}{\pi\left(\frac{d_{wire}}{2}\right)^2}\right\} \qquad (10)$$

The fill of the core window area A_W is calculated by (11). If the fill is above the operational boundary defined the project is discarded. A reference for the filling value expected for each core format can be found in [1].

$$filling = \frac{N\,A_{copper}}{A_w} \qquad (11)$$

To estimate the number of layers N_C it is necessary to also estimate the diameter of the loop by $D_e = 2\sqrt{A_{copper}/\pi}$ and consider a coil spacing d_e which will vary with the technique used to braid the wires and to make the winding, being an estimated value between approximately $0.5mm$ and $3mm$.

Considering the spacing between adjacent turns and between the turns of the first layer and the core equal to d_e, for toroidal cores can be calculated the internal diameter of each layer with (12), where ID is the inner core diameter and k_c is the layer number.

$$D_c(k_c) = ID - (2(k_c-1)+1)D_e - 2\,k_c\,d_e \qquad (12)$$

The inner perimeter of each layer is calculated directly by $P_c(k) = \pi D_c(k_c)$ and the number of wires in the layer is obtained in (13).

$$n_{wires}(k) = \mathbb{N}\left\{\frac{P_c(k)}{D_e+d_e}\right\} \qquad (13)$$

The number of layers N_C is found when satisfying (14) by comparing the number of turns with the number of wires accumulated in each layer

$$\sum_{k_c=0}^{N_C-1} n_{wires}(k_c) \le N < \sum_{k_c=0}^{N_C} n_{wires}(k_c) \qquad (14)$$

This process is simplified in non-toroidal cores, like E and U cores, since the diameter is replaced by the length of the winding region which, like the number of wires per layer, is constant.

The length of each wire C_{wire} is calculated in (15) using the mean length per turn MLT generally given by the core manufacturers, but can also be found by simple geometrical calculations.

$$C_{wire} = MLT * N \qquad (15)$$

The DC resistance of the wire is calculated from (16), where ρ is the resistivity of the conductor used.

$$R_{wire} = C_{wire}\frac{\rho}{n_{wire}\,\pi\,(d_{wire}/2)^2} \qquad (16)$$

C. Copper losses estimation

The power dissipated in the winding due to the low frequencies can be calculated using (17).

$$P = R_{wire}\,I_{ef}^2 \qquad (17)$$

Pelicular and proximity effects can be calculated by obtaining the effective conductor cross-sectional area for each frequency component of the current signal.

Considering only the film effect, one can calculate the penetration depth δ_k by (18), where k is the index of the harmonic component.

$$\delta_k = \frac{65}{\sqrt{k\,f}} \qquad (18)$$

The equations of (19) to (22) [5] are approximations to obtain the effective area for each δ.

$$p'_k = \delta_k\left(1 - e^{-\frac{d_{wire}}{2\,\delta_k}}\right) \qquad (19)$$

$$z_k = 0.62006\,\frac{d_{wire}}{2\,\delta_k} \qquad (20)$$

$$y_k = \frac{0.189774}{(1+0.272481\,(z_k^{1.82938} - z_k^{-0.99457})^2)^{1.0941}} \qquad (21)$$

$$A_{ef,k} = \pi\left(d_{wire}\,p'_k - (p'_k)^2\right)(1+y_k) \qquad (22)$$

The winding resistance is then calculated for each harmonic component with (23).

$$R_k = \frac{C_{wire}\,\rho}{n_{wires}\,A_{ef,k}} \qquad (23)$$

Alternatively the skin and proximity effects can be calculated collectively by using (24) and (25) [6].

$$A_k = \left(\frac{\pi}{4}\right)^{\frac{3}{4}}\frac{d_{wire}^{\frac{3}{2}}}{\delta_k\,\sqrt{d_{wire}+d_e}} \qquad (24)$$

$$R_k = R_{wire}\,A_k\left(\frac{e^{2A_k}-e^{-2A_k}+2sen(2A_k)}{e^{2A_k}+e^{-2A_k}-2cos(2A_k)}+\,...\right.$$
$$\left....\,\frac{2\,N_C^2-1}{3}\frac{e^{2A_k}-e^{-2A_k}-2sen(2A_k)}{e^{2A_k}+e^{-2A_k}+2cos(2A_k)}\right) \qquad (25)$$

With the final winding resistance R_k for each frequency, found in (25), the power is obtained by summing the product of each harmonic component of current I_k squared by the corresponding resistance R_k, as in (26):

$$P = \sum_0^k R_k\,I_k^2 \qquad (26)$$

D. Core loss estimation

The equation (27) presents the Steinmetz, the basic loss estimation formula for the core. The Steinmetz equation which is a modified form of the one proposed originally in [7], whose parameters are always supplied by the manufacturers of magnetic materials, is a function of the frequency f and the amplitude of the oscillation of the magnetic flux density B_{pk}.

$$P = K \, f^{\alpha} \, B_{pk}^{\beta} \qquad (27)$$

This equation is valid only for sinusoidal conditions. Different methods were created to obtain equations for non-sinusoidal estimations using the same parameters, K, α and β of the Steinmetz equation.

The improved generalized steinmetz equation (IGSE) developed in [8], an improved form of the GSE method [9], enables losses estimation using the parameters of the Steinmetz equation and the density vector of time flow. Considering smaller loops within the hysteresis loop, identifying, separating and arranging the magnetic flow vector as explained in [10]. Calculating the volumetric density of energy dissipated between each point of vector $B(t)$ for each of the smaller and larger cycles separately by (28) with Bpk being the amplitude of the *loop* in question.

$$E = K_i \left| \frac{dB}{dt} \right|^{\alpha} B_{pk}^{\beta - \alpha} \, \Delta t \qquad (28)$$

Where the K constant must be changed according to (29) to get K_i:

$$K_i = \frac{K}{(2\pi)^{\alpha+1} \int_0^{2\pi} 2^{\beta - \alpha} |cos(\theta)|^{\alpha} d\theta} \qquad (29)$$

The power is then calculated by the total energy dissipated in a period T:

$$P = \frac{V_c}{T} \sum_{t=0}^{T} E(t) \qquad (30)$$

The use of (28) in place of the Steinmetz equation makes the estimation less dependent on the switching frequency, but since faithful representation of the flux density is required for at least an entire fundamental cycle, it is important to note that with increasing of the switching frequency the calculation step must be decresaed to obtain a good representation of the instantaneous flux density, which progressively increases the time of simulation and size of the vectors.

E. Temperature rise estimation

The expected temperature rise ΔT can be calculated using (31), obtaining the inductor temperature by $T = T_{amb} + \Delta T$ [1] [2]. $A_{surface}$ refers to the surface area of the inductor including the coil, wich is usualy an aproximation.

$$\Delta T = 450 \left(\frac{P_{total}}{A_{surface}} \right)^{0.826} \qquad (31)$$

Fig. 3. Front view of the calorimeter built

III. Loss measurement System

A calorimeter developed for the purpose of loss measurement in switched inverters and their components, shown in Fig. 3, was used to obtain the loss results of the next section. Figure 4 show the supervisory diagram used with the calorimeter. The functioning of the calorimeter is explained in detail in [11].

The calorimeter has a topology similar to that developed in [11]. It has two thermally insulated chambers, an internal chamber where the test object is placed. And an external chamber that has its temperature controlled by resistive loads to be equal to that of the inner chamber, reducing the flow of heat through the wall of the inner chamber. Ventilation systems are present in both chambers to homogenize the temperature.

The main temperature change of the internal chamber is made by a closed thermal system of water, being possible to reach an equilibrium where all internally generated power is transferred to water and the internal chamber

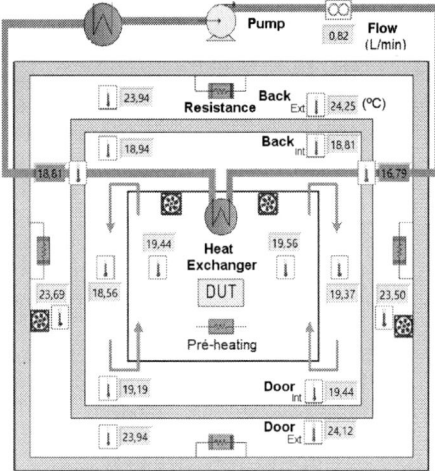

Fig. 4. LabView supervisory diagram of the calorimeter

temperature is constant. The thermal system exchanges heat with the inner chamber and the environment by radiators and also with a commercial refrigeration system for conventional refrigerators. The water flow is regulated by a hydraulic pump controlled by PWM.

At the inlet and outlet of water from the inner chamber are measured the inlet temperature T_{in} and outlet temperature T_{out} of the water and the flow $\dot{V}$. The equation (32) calculates the power, being the first term referring to the one transferred to the water and the second term being the sum of the power that passes through the insulating walls. Where c_p is the thermal capacity of the water, ρ_{H_2O} the volumetric density of the water, T_{int} and T_{ext} the temperatures on the inner and outer chambers and R_{th} is the thermal resistance of the insulation wall.

$$P = c_p\, \rho_{H_2O}\, \dot{V}\, (T_{out} - T_{in}) + \sum \frac{T_{int} - T_{ext}}{R_{th}} \quad (32)$$

The temperature and flow sensors were individually calibrated and the power measure was also calibrated to remove the error caused by the internal ventilation system of the chambers required to homogenize the temperature. but slow oscillations in the power dissipated by the ventilation system will cause an error in the power measurement of the device under test.

IV. RESULTS

In order to evaluate the proposed methodology, four inductors, identified in table I by A, B, C and D, were designed using different materials and different inductance values to be used in a buck converter, shown in Fig. 5.

The converter parameters are present in the table II and the design data of the designed inductors in the table I.

A. Inductance measurement

Using a constant duty cycle $d = 0.5$, one can estimate the inductance, when in saturation, by the slope of the current in the inductor ΔI by (33).

$$L = \frac{V_{dc}}{2\,\Delta I\, f} \quad (33)$$

The table III shows the inductance values obtained by the measurement through an RLC bridge L_{RLC}, by estimating (33) through the voltage and current measurement L_{est}, by the ideal inductance value $L_{ideal} = AL * N^2$.

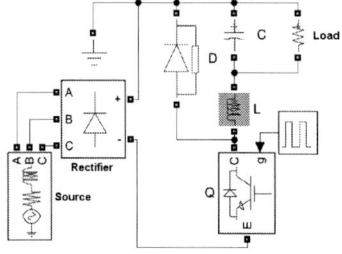

Fig. 5. Diagram of the buck converter used to test the inductances

TABLE I
INDUCTOR DESIGN DATA

Inductor	A	B	C	D	
Core	T520	T400	58090	78090	
Manufacturer	Micrometals		Magnetics		
Material	34D	14D	High Flux	Xflux	
L_{proj}	1,06	0,230	0,155	0,097	mH
$L_{initial}$	1,077	0,236	0,162	0,098	mH
L_{final}	1,038	0,231	0,141	0,096	mH
L_{leak}	0,128	0,065	0,021	0,014	mH
L_{total}	1,165	0,296	0,163	0,110	mH
N	102	80	49	34	
winding	24,81	28,82	40,16	27,87	%
NC	2	2	3	2	
$P_{copper,dc}$	14,70	12,00	6,872	4,440	W
$P_{copper,ac}$	0,0281	0,462	0,292	0,484	W
P_{copper}	14,72	12,46	7,164	4,923	W
B_{pk}	0,0228	0,058	0,372	0,536	T
P_{core}	0,586	1,961	2,659	18,46	W
P_{total}	15,31	14,43	9,823	23,38	W
ΔT	17,10	24,11	67,31	137,1	C
R_{dc}	23,5	19,2	11,0	7,1	mΩ
R_{ac}	216,1	157,3	44,9	29,0	mΩ
Volume	1336	637	108	88	cm^3

TABLE II
CONVERTER PARAMETERS

V_{DC}	150	V
I_{med}	25	A
f	15360	Hz
d	0.5	
V_{load}	75	V

and the values projected for inductance, $L_{project}$ and its expected final value in the project, L_{end}.

Looking at the table IV, which shows the inductance error L_{est} in relation to L_{proj}, we realize that the projected value for the inductance lies within the expected error limits , considering the tolerance of 8% in the permeability μ of the materials.

B. Loss Masurement

The measurement of losses by the calorimeter was carried out to obtain the total losses P_{total}, the loss in the copper P_{copper} approximated by the d.c. was obtained by measuring the average voltage and current in the inductor. For a switching frequency of $15360 Hz$, the expected high-frequency effects are significantly smaller than the measurement errors encountered. And the loss in the core P_{core} was estimated by the difference between P_{total} and P_{copper}. The values of losses can be observed in the table V.

The calorimeter developed has two error components, one relative to the measured power of $\pm 1\%$ and another

TABLE III
COMPARISON OF INDUCTANCE VALUES OBTAINED

Inductor	A	B	C	D	
L_{proj}	1,06	0,23	0,155	0,097	mH
L_{final}	1,038	0,231	0,141	0,096	mH
L_{RLC}	1,434	0,269	0,211	0,108	mH
L_{est}	1,092	0,249	0,153	0,104	mH
L_{ideal}	1,379	0,299	0,214	0,109	mH

TABLE IV
INDUCTANCE ERROR $L_{est} - L_{proj}$

Inductor	A	B	C	D	
Error L	0,032	0,019	-0,002	0,007	mH
	3,02	8,26	-1,29	7,22	%

TABLE V
MEASURED AND CALCULATED LOSSES IN THE INDUCTORS

Inductor	A	B	C	D	
$P_{core,calc}$	0,56	2,0	2,7	18,5	W
$P_{core,est}$	3,5	2,5	3,7	16,3	W
Error P_{core}	2,86	0,51	1,06	-2,14	W
$P_{copper,calc}$	14,7	12,5	7,2	4,9	W
$P_{copper,meas}$	13,3	13,3	10,1	6,5	W
Error P_{copper}	-1,38	0,86	2,91	1,55	W
$P_{total,calc}$	15,3	14,4	9,8	23,4	W
$P_{total,meas}$	16,8	15,8	13,8	22,8	W
Error P_{total}	1,48	1,36	3,97	-0,59	W

independent of the measured device load of absolute $\pm 3W$, coming from the calorimeter system itself, making measurements of small powers inaccurate.

Figures 6 to 8 compares the values: Calculated, wich are the values obtained using the equations proposed in this text; Traditional, wich are the values obtained using only the Steinmetz equation for P_{core} and effective current for P_{copper}; Measured, wich are the experimental values obtained, by calorimeter for P_{total}, by $I \times V$ obtained by osciloscope for P_{copper} and by the energy on the $B \times H$ loops as in [12] for P_{core}; And the estimated is the diference between the measured total and core loss $P_{core,est} = P_{total,meas} - P_{copper,meas}$.

V. CONCLUSIONS

In order to demonstrate the applicability of the proposed methodology, four different inductors were designed, achieving adequate inductance values when in operation.

The values of losses estimated by the proposed project methodology can be confirmed by being consistent with the power values obtained by the traditional methodology. However, the great uncertainty found in the loss measurement system prevents a better analysis of losses.

A calorimeter better suited to the inductor power levels should be used, or another method to measure losses together with the calorimeter should be used to assure the accuracy of the proposed losses and temperature rise estimates in the methodology.

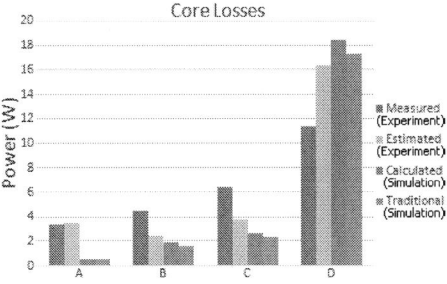

Fig. 6. Comparison of losses in the core

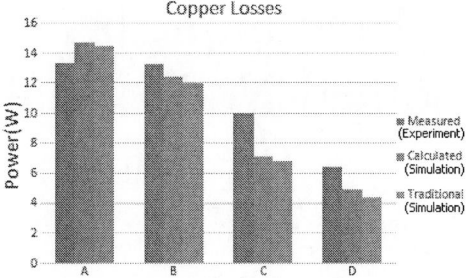

Fig. 7. Comparison of losses in the copper

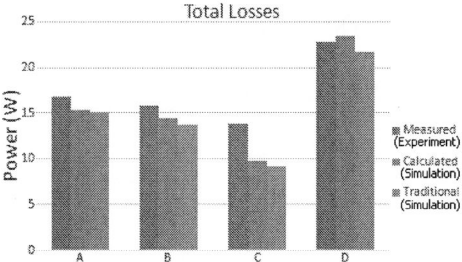

Fig. 8. Comparison of total losses

REFERENCES

[1] Colonel Wm. T. McLyman, Transformer and Inductor Design Handbook, 3rd ed., Marcel Dekker, Inc., 2004.

[2] J. R. Zientarski, "Análise modelagem e validação experimental de uma metodologia de projeto do indutor em conversores Boost PFC," Universidade Federal de Santa Catarina, 2009.

[3] J. Koscelnik, J. Sedo and B. Dobrucky, "Modeling of resonant converter with nonlinear inductance," 2014 International Conference on Applied Electronics, pp. 153–156 september 2014.

[4] S. Hiruma, Y. Otomo and H. Igarashi, "Eddy current analysis of Litz wire using homogenization-based FEM in conjunction with integral equation," IEEE Transactions on Magnetics, vol. 54, pp. 1–4, march 2018.

[5] Knight, David, "Practical continuous functions for the internal impedance of solid cylindrical conductors." 2016.

[6] M. Bartoli, A. Reatti and M. K. Kazimierczuk, "Modelling iron-powder inductors at high frequencies," Proceedings of 1994 IEEE Industry Applications Society Annual Meeting, vol. 2. pp. 1225–1232, 1994.

[7] C. P. Steinmetz, "On the law of hysteresis," Transactions of the American Institute of Electrical Engineers, pp. 1–64, vol. IX. January 1892.

[8] K. Venkatachalam, C. R. Sullivan, T. Abdallah and H. Tacca, "Accurate prediction of ferrite core loss with nonsinusoidal waveforms using only Steinmetz parameters," 2002 IEEE Workshop on Computers in Power Electronics, 2002. Proceedings. pp. 36–41, June 2002.

[9] Jieli Li, T. Abdallah and C. R. Sullivan, "Improved calculation of core loss with nonsinusoidal waveforms," Conference Record of the 2001 IEEE Industry Applications Conference. 36th IAS Annual Meeting (Cat. No.01CH37248), vol. 4, pp. 2203–2210, september 2001.

[10] M. J. Jacoboski, A. de Bastiani Lange and M. L. Heldwein, "Closed-form solution for core loss calculation in single-phase bridgeless PFC rectifiers Based on the iGSE method," IEEE Transactions on Power Electronics, vol. 33, pp. 4599–4604, 2018.

[11] D. Christen, U. Badstuebner, J. Biela and J. W. Kolar, "Calorimetric power loss measurement for highly efficient converters," The 2010 International Power Electronics Conference - ECCE ASIA -, pp. 1438–1445, june 2010.

[12] IEC 60404-6, Magnetic materials - Part 6: Methods of measurement of the magnetic properties of magnetically soft metallic and powder materials at frequencies in the range 20 Hz to 200 kHz by the use of ring specimens. [S.l.], 2018.

Real-Time implementation of a DC converter using Modified Nodal Analysis, Sparsity Handling and Parallelism on a DSP platform

Luiz Felipe Corrêa de Sá Santos Ribeiro
Laboratório Eletrônica de Potência e Média Tensão
Federal University of Rio de Janeiro
Rio de Janeiro, Brazil
luizfelipecss@poli.ufrj.br

Felipe Novaes Francis Dicler
Gerência de Engenharia de Instalações
Operador Nacional do Sistema
Rio de Janeiro, Brazil
dicler@poli.ufrj.br

Luis Guilherme Barbosa Rolim
Laboratório de Fontes Alternativas de Energia
Federal University of Rio de Janeiro
Rio de Janeiro, Brazil
rolim@poli.ufrj.br

Mauricio Aredes
Laboratório Eletrônica de Potência e Média Tensão
Federal University of Rio de Janeiro
Rio de Janeiro, Brazil
aredes@lemt.ufrj.br

Abstract—**Modified Nodal Analysis(MNA) is a generalized method of discrete linear circuit analysis as a matrix equation, in which the matrix order is the sum of the number of nodes and the number of voltage sources, it also have additional lines for dynamic elements, such as capacitors and inductors, but for this work these lines were kept implicit using Nodal Analysis' equations for historic sources. As this order grows, the computational effort for solving the equation grows to the square of its size. Therefore, optimizations are in place to better utilize the processing power of a hardware, so that more complex circuits can fit in less powerful simulators, avoiding problems associated with high values of time step, such as mathematical inaccuracy and instability. This paper presents a boost converter and a PI controller, each implemented in a separate Texas Instruments' F28377S, assembled in a Hardware in the Loop configuration. The MNA is used for the plant discretization and a matrix multiplier is developed to solve it. This multiplier is then optimized with two methods: sparse matrix handling and a parallel multiplier. The sparsity is dealt with by storing the matrix in a Compressed Sparse Row(CSR) format and adjusting the multiplier. The parallel matrix multiplier is implemented using both the Control Law Accelerator(CLA) and the main CPU to reduce the processing time. The results presented by the comparison of each program shows that the The results are drawn from each multiplier and comparison is made between them, leading to a conclusion of the benefits of such implementation.**

Index Terms—**real-time, simulation, sparsity, parallelism, boost**

I. INTRODUCTION

The advance of technology, evident by the decrease of size and increase of computational power of modern day electronics, has made possible for increasingly complex systems to be simulated at faster rates. As a result, some simulations processed each iteration faster than the step assigned to it, allowing, with an insertion of a deterministic frequency, for a simulation to be a time-precise mimic of its real system. This kind of simulation was called Digital Real-Time Simulation, or just Real-Time Simulation.

This time-precise mimic capability of simulations evolved to some applications, among which Hardware in the Loop (HIL) is one of the most important. This particular usage of real-time simulation was developed to test responses of real systems in a laboratory environment, to validate its results before applying it in the field.

As for the real-time simulation modeling, describing a specific circuit as a discrete Transfer Function can be the best way of achieving a low computational effort. For generic implementations, this method generates a problem because of the enlargement of the circuit, as this would make the choosing of state variables a complex matter. This complexity come from cases that can be very commonly found, as capacitors and voltage sources loop in the transmission line model, or hard to identify, as inductors and current sources in a linear combination of one another, due to the network around them [1].

To achieve a generalized solution for a circuit, the Modified Nodal Analysis(MNA) discretization method can be used instead, which can supply a model based solely on the circuit design. As a result of such generalization, the discretization is written in a matrix equation with the number of variables equal to the number of nodes plus the number of voltage sources, increasing the necessary computational effort.

The difficulties of this method are that matrix becomes sparser the larger the circuit grows, making for a waste of space and processing time to handle a large quantity of zeroes in a multiplication algorithm. Another drawback is the fact that for switching devices, the admittance matrix has to be saved for each possible combination of switches states, e.g. allocating 64 matrices for a six switch inverter. The matrix formulation is a problem in itself as it calculates all of the variables, even the ones that have no use as a result.

To apply this concepts, this work presents an implementation of a boost using Modified Nodal Analysis in a Texas Instruments' F28377S DSP controlled by a PI controller in a Hardware in the Loop configuration. Afterwards, two optimization methods for the matrix multiplier will be presented, the first being a method of storage and multiplication for sparse matrices and the second being the introduction of parallelism to this system. The data extracted from the different implementations is shown, compared and discussed. Then, a conclusion of the benefits of this kind of implementation is presented.

This paper is organized with Section II explaining the Modified Nodal Analysis method and Section III showing the implementation of both the plant and the controller. Section IV, V and VI show the implementation of the programs with no optimization, sparse optimization and parallel and sparse optimization, respectively. Section VII is the Results section, where they are presented and discussed, leading to the conclusion in Section VIII.

II. MODIFIED NODAL ANALYSIS

Nodal analysis is a type of circuit analysis used to determine the voltage of the nodes in a circuit, which is widely used for studying circuits analytically. It is based on the Kirchhoff Current Law that states that for every node in a circuit, the sum of the current flowing inwards is equal to the sum of the currents flowing outwards.

Therefore, for a given linear circuit, arranging the equations of each node and using its voltage as variables creates a linear system that may include differential equations. In a discrete, iterational and generic study, this analytical system must be converted into a series of equations compatible with a discrete analysis. The arrangement of said equations constitutes a matrix equation for the circuit [2].

This Discrete Nodal Analysis was first developed for the purpose of network simulation, and one of its characteristics, as a initial solution, was the necessity of all voltage sources to be set between a node and the ground. In a later development of this method, another formulation for the voltage sources made possible the creation of such elements between any two chosen nodes. It also introduced the possibility of current variables for some elements, enlarging the matrix but giving direct access to variables that had the necessity of being calculated after the solution of the matrix equation [3]. The Modified Nodal Analysis, as it was called, has the possibility of using, for each element, either the Nodal Analysis' algorithm or its own, depending in the needs of the project.

This paper proposes to achieve the lowest time step possible, so it will be used only the ungrounded source algorithm of the Modified Nodal Analysis and all the other elements will be written using Nodal Analysis(NA) to ensure a small matrix.

III. IMPLEMENTATION

The boost implemented in this work is shown in figure 1.

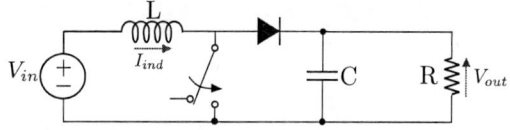

Fig. 1. Boost converter used in this work.

The matrix equation of the MNA uses the formula **Ax=b** that is solved in this paper by inverting A. As the boost converter is not a linear circuit, there are four possible A^{-1}, one for each state combination of the diode and controlled switch. These states are:

1) Both opened;
2) Diode closed, Switch opened;
3) Diode opened, Switch closed;
4) Both closed.

For this paper, a trapezoidal discretization was used for the inductor and capacitor. Using NA approach, the result is the dynamic element represented as a companion model of a conductance in parallel with a current source, as show in figure 2.

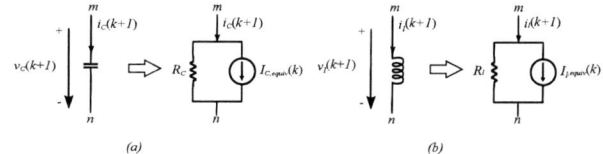

Fig. 2. Companion model for: (a) Capacitor; (b) Inductor.

The nonlinear devices were written as $1\ G\Omega^{-1}$ or $1\ n\Omega^{-1}$ conductance for a closed or opened state, respectively. So the matrices follow the pattern bellow:

$$\begin{bmatrix} \frac{\Delta t}{2L} & -\frac{\Delta t}{2L} & 0 & 1 \\ -\frac{\Delta t}{2L} & \frac{\Delta t}{2L} + G_s + G_d & -G_d & 0 \\ 0 & -G_d & \frac{1}{R} + \frac{2C}{\Delta t} + G_d & 0 \\ 1 & 0 & 0 & 0 \end{bmatrix} \quad (1)$$

As Δt represents the time step, G_d, the diode conductance, G_s, the switch conductance. The matrix equation in which A for the correct state is used in each step of the simulation is:

$$\begin{bmatrix} A \end{bmatrix} \times \begin{bmatrix} v_1 \\ v_2 \\ v_3 \\ i_v \end{bmatrix} = \begin{bmatrix} -I_{h_L} \\ +I_{h_L} \\ -I_{h_C} \\ V_{in} \end{bmatrix} \quad (2)$$

For the above, I_{h_L} is the historic current source of the companion model for the inductor, while I_{h_C} refers to the historic current source for the capacitor [2].

The selection of states comes from the reading of the voltage source's current and the input from the PWM, this is done by

978-1-7281-4181-7/19 $31.00 © 2019 IEEE

assuming that if the switch is closed the diode has to be open, this means that the model doesn't have a state where both devices are closed. However, if the switch is open, a positive current in the source forces a open diode as a negative forces a closed one, this accounts for discontinuous mode.

The PI discretization was done by applying Tustin Transform to the PI Transfer Function. This formula was used in the controller DSP, where it receives the voltage and exports a PWM signal comparing its output to a 3 V carrier.

Using the notation for the controller bellow, K_p and K_i were established as 1×10^{-2} and 60, respectively.

$$\frac{U(s)}{E(s)} = \frac{K_i}{s} + K_p \qquad (3)$$

IV. PROGRAM I - C28X MULTIPLICATION

In this section will be discussed the program developed for the first step, the straight forward matrix multiplier using C28X.

The names of the important variables used in the programs presented from now on: A1, A2, A3 and A4 are the four inverted matrices, the **b** vector is named just Vector, the variable vector calculated in the previous iteration is named Solved, the state of the switches is S. The historic current sources (I_h) are called jc and jl, for the capacitor and inductor respectively.

A. Matrix Formualtion

The matrices were created first using a C++ program, also developed by the authors, that reads a netlist and stamps the matrix with the element in the place assigned for it, for the passive elements. Then, it was copied to a MATLAB routine that stamps the matrix with open and closed switches. After all matrices are finished, the routine then inverts them and stores in a .h and .c files in the DSP's project folder, using C syntax, for it to be included automatically.

B. DSP Code

The code for this program uses a periodic interruption with a 200 kHz frequency. This interrupt is measured with a GPIO toggle at its start and end. It has four main steps:

- Change the DAC output to the value calculated in the previous iteration;
- Calculate Vector based on the Solved array;
- Calculate S, based on the GPIO entry and the circuit;
- The correct matrix is used, along with the Vector, in the multiplication algorithm and the result is the updated Solved for the next timestep.

The multiplication algorithm is as shown in algorithm 1:

V. PROGRAM II - C28X SPARSE MATRIX MULTIPLICATION

The inverse of the admittance matrix for the boost has five zeroes within it, 31.25% of the total of terms. The optimization made for this program ignores these zeroes inside the multiplication algorithm.

Algorithm 1: Matrix Multiplier

```
i=0;
k=0;
for i < 4 do
    Solved[i] = 0;
    for k < 4 do
        Solved[i]+=Vector[k]*A[i][k];
        k++;
    end
    i++;
end
```

A. Matrix Formulation

The Compressed Sparse Row(CSR) format is used to handle only the nonzero elements of the matrix. It splits a matrix in 3 vectors, A is the vector containing all the nonzero terms, I and J are auxiliary vectors containing information in regard of the positioning of each entry of A.

The four inverse matrices stored have the same shape, described in the generic matrix bellow:

$$\begin{bmatrix} 0 & 0 & 0 & A_0 \\ 0 & A_1 & A_2 & A_3 \\ 0 & A_4 & A_5 & A_6 \\ A_7 & A_8 & A_9 & A_{10} \end{bmatrix}$$

In CSR format, the previous matrix is written first taking all nonzero entries and fitting into A, as bellow:

$$A = \begin{bmatrix} A_0 & A_1 & A_2 & A_3 & A_4 & A_5 & \dots & A_{10} \end{bmatrix}$$

Then, I is written starting with a 0, and then, the ith is the sum of the previous term and the number of nonzero entries in the ith row, with the length equal to the numbers of rows plus one. The first row has 1 entry and as the vector starts with a 0, the second item is a 1. Then as the second row has 3 items, the number to follow the 1 is 4(1 + 3) and so on. So, the I vector is written as:

$$I = \begin{bmatrix} 0 & 1 & 4 & 7 & 11 \end{bmatrix}$$

The J vector is simply made by inputting the column index of each one of the nonzero entries in order, as this is written in C, the index starts at 0, For example, the following pairs are the A and J of the same index: (A_0,3), (A_1,1), (A_2,2), (A_3,3) and so forth. This means:

$$J = \begin{bmatrix} 3 & 1 & 2 & 3 & 1 & 2 & \dots & 3 \end{bmatrix}$$

Altogether, the matrix writen in CSR format has the following 3 vectors:

$$A = \begin{bmatrix} A_0 & A_1 & A_2 & A_3 & A_4 & A_5 & \dots & A_{10} \end{bmatrix}$$

$$I = \begin{bmatrix} 0 & 1 & 4 & 7 & 11 \end{bmatrix}$$

$$J = \begin{bmatrix} 3 & 1 & 2 & 3 & 1 & 2 & \dots & 3 \end{bmatrix}$$

This matrix is built using the same matrix created in the C++ program as before, but the MATLAB routine is different, the stamped switches and the inversion are the same, the .h and .c file storage that differ. The new routine dissects the inverted matrices and stores each one with its correspondent three vectors.

B. DSP Code

The code for this example is the same as before, except for the multiplication algorithm. The new multiplications, written bellow, has no zero multiplication, given that the Vector is written as in (2) and there is very low probability that either I_{h_L} or I_{h_C} are going to be zero after the first iteration.

Algorithm 2: Sparse Matrix Multiplier

```
i=1;
for i < 5 do
    Solved[i-1] = 0;
    k=I[i-1];
    for k < I[i] do
        Solved[i-1]+=Vector[J[k]]*A[k];
        k++;
    end
    i++;
end
```

The Sparse Matrix Multiplier algorithm is not as easy to read as the simple Matrix Multiplier. It uses the variable i to get the row which will be multiplied(i-1) and sets its Solved value to zero, then it creates a *for loop* for that row from I[i-1] to I[i], this *for loop* gives the indexes of A that are from that row and the indexes of J for which element of Vector should the term from A be multiplied for. The accumulation is done until it hits the end of the row and another row is done until it finishes the matrix.

VI. PROGRAM III - C28X-CLA PARALLEL SPARSE MATRIX MULTIPLICATION

The parallelism is one of the most efficient ways of cutting down a processing time. And as the Control Law Accelerator(CLA), available in the hardware used in this work, can function independently from the main CPU, the program can be split in two parallel threads. This secondary core is a 32-bit floating-point math accelerator, clocked in the same frequency.

A. Matrix Formulation

The matrices are built in the same CSR format, with the same MATLAB routine. Also, this program uses floats and not doubles, which have been used in the previous codes, as the CLA can only handle 32-bit datatype, this being a minor drawback for the accuracy.

B. DSP Code

The code for this implementation is changed in the multiplication algorithm, where the Sparse Multiplication Algorithm is divided in the first two rows for the CLA and the last two for the C28X. The pseudo-algorithms bellow show the C28X and the CLA multiplication of this implementation.

Algorithm 3: C28X Parallel Sparse Matrix Multiplier

```
Trigger CLA Task;
i=3;
for i < 5 do
    Solved[i-1] = 0;
    k=I[i-1];
    for k < I[i] do
        Solved[i-1]+=Vector[J[k]]*A[k];
        k++;
    end
    i++;
end
Wait CLA Task;
```

Algorithm 4: CLA Task

```
i=1;
for i < 3 do
    Solved[i-1] = 0;
    k=I[i-1];
    for k < I[i] do
        Solved[i-1]+=Vector[J[k]]*A[k];
        k++;
    end
    i++;
end
```

VII. RESULT

Three sets of results were taken from the programs talked about in the latter sections:

- Processing time;
- Steady State regime;
- Step response.

The first set of results tells the improvement of the optimization method, the second shows if the program correlates to a valid simulation, the third makes a comparison of how accurate the program was according to a well established commercial simulator.

A. Interruption Time

The graphs in this test show the interruption period, that is the same as the simulation's time step, the interruption processing time and the multiplication algorithm measured.

The multiplication algorithm measured is a case structure where the switch state defines what case is chosen and the actual multiplication algorithm, showed in the section of each program.

The figures 3 through 5 shows the total interruption time, in blue, and the multiplication algorithm processing time, in orange, for each of the programs. The interruption period can be inferred as the time between two rises in the interruption signal.

Program I

Figure 3 shows the results obtained in the first implementation.

The period in this program was 5 μs, being 4.4 μs the total interrupt time and 3.8 μs the multiplication algorithm.

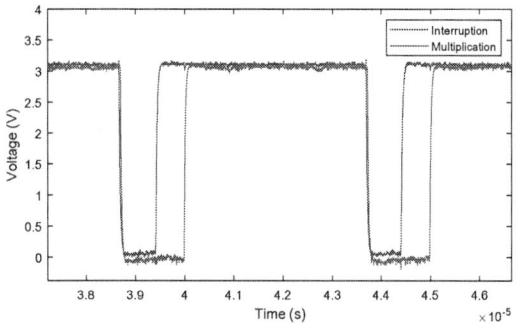

Fig. 3. Processing times of Program I.

Program II

Figure 4 consists on the results collected from the sparsity handling implementation

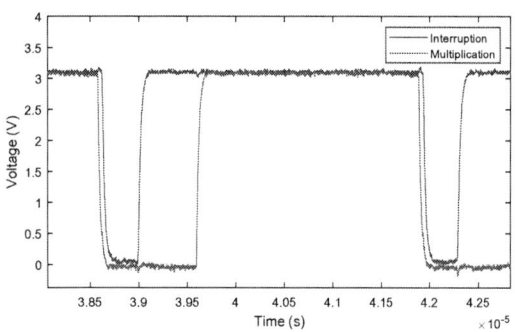

Fig. 4. Processing times of Program II.

The period was reduced to 4.5 μs as the total interruption and the algorithm were both reduced by 0.4 μs to 4.0 μs and the 3.4 μs, respectively.

Program III

Figure 5 shows the times for the program with sparsity handling and parallel computation.

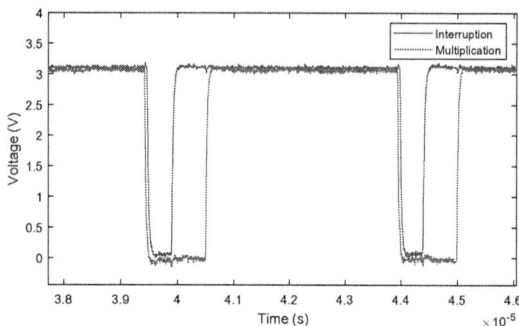

Fig. 5. Processing times of Program III.

The period was reduced to 3.3 μs as the total interruption was reduced to 3.0 μs and the algorithm, to 2.3 μs.

Each optimization showed an improvement for the times in the program. The improvement for the sparse matrix was not very relevant, but the sparsity of this matrix was 31.25% while the sparsity of larger circuits is higher, for example IEEE 39-bus test system has a 95.92% sparsity in its Y^{-1} [4].

The total improvement of the interruption period was of 27%, making for a valid option if parallelism is achievable. Even with such lower times, the algorithm for the parallelism was not optimal, as it was done based on rows. The CLA had 4 elements and the CPU had 7 elements to handle. As they are clocked equally, the CPU time was the limiting factor, so the improvement was from 11 to 7 element-wise, making a 36% expected improvement of the multiplication algorithm, which shows validity by the 32% improvement measured. Even more, taking into account that it has a case structure to select the correct matrix for the state inside the measured time.

B. Steady State test

For this test, the voltage reference of the PI controller was set to 2.5 V and the steady state regime was captured in an oscilloscope. The data was compared in MATLAB and is presented in figure 6, bellow.

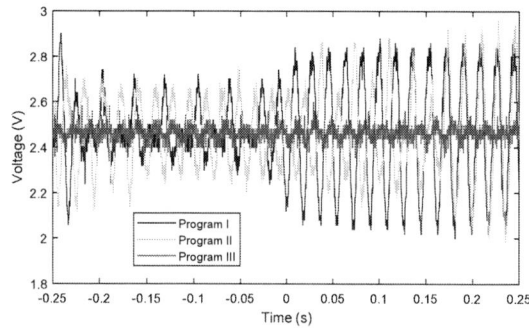

Fig. 6. Comparison of steady states for all program.

It is notable the instability of Program I and II, this is caused by inter-step switching [5], which is the problem associated with a switch changing state in mid step, this makes for a delay of a fraction of a time step for the plant to respond and this delay can cause the system to lose stability. The problem was expected for time steps of 5 μs and higher, but it also appeared in Program II, with a 4.5 μs time step, invalidating the simulation without a correction algorithm. The parallelism, however, was able to pass the threshold of this problem and the simulation was able to run.

C. Step Response

As it was shown in figure 6 the programs I and II were not able to function as an accurate simulation, and for this reason this test was done solely to program III.

This test is a measurement of the response of the controlled plant with a square wave as reference. This wave has a DC value of 2 V and a 1 $V_{peak-peak}$. As in the test before, the data was taken by an oscilloscope and imported to MATLAB. A PSIM simulation of the circuit of figure 7 was done and compared with the measured data. This comparsion is show in figure 8.

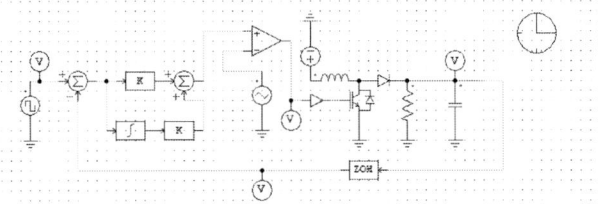

Fig. 7. Comparison of steady states for all program.

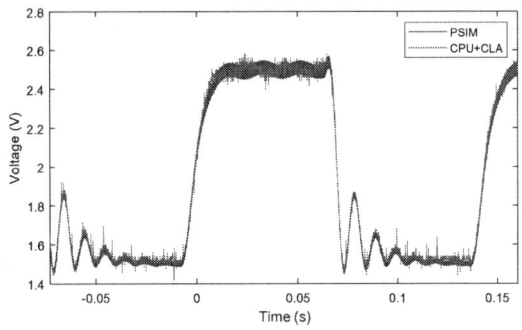

Fig. 8. Comparison of step response of the program III and a PSIM simulation.

The comparison shows accordance between both curves, the rise and fall times are the same, and the oscillation after the fall are similar. This result validates the plant and the controller. There is a small oscillation at the top of both curves, they appear to be the same but phased from one another, and this matter will be studied further.

These test showed not only that the optimization worked as intend, but also that, without the parallelism, this boost is above the hardware capabilities to run in Nodal Analysis. So a valid simulation was possible because of these improvements.

VIII. CONCLUSION

The Modified Nodal Analysis is a method of discretization and generalization of circuit analysis to calculate variables within this circuit. This method requires a great computational power, which makes it difficult for a complex circuit to be simulated in a hardware not as powerful.

The MNA writes a matrix equation of the type $Ax = b$. For the usage of this analysis, it is required both a matrix inversion and for the system to be linear. This paper dealt with the necessity of linearization of switching devices by storing the matrices of every possible switch state, determining what is the current state in mid simulation.

Three programs were made to optimize the implementation of the boost shown in figure 1 to avoid problems of low sampling frequency. The programs II and III accomplished a better performance from the original case, even with not optimal implementations, but resulting simulation only was viable at the final optimization, overcoming the instability caused by inter-step switching delay.

The treatment of sparse matrices as CSR format revealed to be not very powerful for this circuit, but for circuits with much more nodes it should be a defining factor for the simulation to run at a viable time step.

The parallelism showed a powerful improvement in the simulation times. The 27% improvement off of the sparse handling program, means a total of 34% from the first implementation. The multiplier only had a 32% improvement in performance.

Overall, this paper shows the importance of parallelism and sparsity in a MNA simulation. The results of this paper could be improved if the circuit was larger and there were more cores to use as parallel threads. There was a discrepancy between the oscillation in the top of the measured and simulated step response, which can be a small contribution of the inter-step switching delay.

REFERENCES

[1] N. Watson and J. Arrillaga, *Power Systems Electromagnetic Transients Simulation.*

[2] H. W. Dommel, "Digital computer solution of electromagnetic transients in single-and multiphase networks," *IEEE Transactions on Power Apparatus and Systems*, vol. PAS-88, no. 4, pp. 388–399, 04 1969.

[3] C.-W. Ho, A. Ruehli, and P. Brennan, "The modified nodal approach to network analysis," *IEEE Transactions on Circuits and Systems*, vol. 22, no. 6, pp. 504–509, 6 1975.

[4] Y. Chen, "Large-scale real-time electromagnetic transient simulation of power systems using hardware emulation on fpgas."

[5] F. Dicler, G. G. Matheus Soares, T. Tricarico, M. Neves, L. G. B. Rolim, and M. Aredes, "Dsp implementation of a real-time boost converter simulator handling inter-step switching delay," in *Proceedings XXII Congresso Brasileiro de Automática*, 10 2018.

978-1-7281-4181-7/19 $31.00 © 2019 IEEE

Investigation of Control Strategies to Mitigate the Oscillation Effects caused by Interconnected Buck Converters

Isaías Valente de Bessa[1], Renan L. P. Medeiros[1], Iury Valente de Bessa[1],
Florindo A.C. Ayres Junior[1], and Kevin E. Lucas[2]

[1]Federal University of Amazonas, UFAM, Department of Electricity, Av. Rodrigo Octávio, 6200,
Coroado I, CEP: 69080-900, Brazil
[2]Federal University of Santa Catarina, UFSC, Department of Automation and Systems,
Florianópolis 88040-900, SC, Brazil
Email: isaias.97.ib@gmail.com[1], renanlandau@ufam.edu.br[1], iury.bessa@gmail.com[1],
florindoayres@ufam.edu.br[1], kevin.lucas@posgrad.ufsc.br[2]

Abstract—The use of renewable energy requires the integratation between alternative energies sources and the main grid. These sources, such as solar and wind energy, are increasingly used due to their simplicity in structure, high power efficiency, low cost and high reliability. In these systems, DC-DC converters are used to ensure regulation and voltage stability. However, the dynamic behavior of the converters is a big concern due to nonlinear characteristic caused by switching operation. Thus, The linearization around the operating point is necessary to regulate the output of converters in order to preserve stability and dynamic performance of the system. On the other hand, when the converters are in series connection create another problem which, depending on the operating conditions, can destabilize the whole system. Instability in cascaded systems may occur due to the constant power load, which exhibit incremental negative resistance behavior causing a high risk of instability in interconnected converters. Aiming to effectively mitigate oscillations effects caused by the interconnection of DC-DC buck converts, this papers presents the studies and comparison of difference control techniques. It is investigated state and output feedback structures through pole-placement methods using classical tools such as solving the Diophantine equation, root locus design, in addition to other modern control tools as the use of the Riccati equation to solve the optimization control problem by using LQR. Through performance index graphs, the controllers ability, speed and effort to mitigate oscillations are observed.

Index Terms—DC-DC converters, buck converters, pole-placement, Diophantine equation, root locus design, Riccati equation, LQR.

I. Introduction

The use of renewable energy, such as solar and wind energy, is growing rapidly because of increasing load demand, crisis of conventional energy and the environmental issues. However, increased penetration of distributed energy resources into the main grid increases control challenges [1]. An efficient option to integrate emerging renewable energy sources are DC microgrids (DCMGs) due to DC nature of these sources [1], [2]. Apart from a reduction in AC-DC conversions losses, DCMGs

can supply continuous high-quality power when voltage sags or blackouts occur in utility grids [3].

In DCMGs, the DC-DC converters play an important role in the control of DC bus voltage and in maintaining the system stability [1], [4].

The operation of a DC system consists of one or more elements responsible for supplying a DC bus with constant voltage to transfer power from a power source to loads having a constant power characteristic, which are responsible for current demand [2]. Among these loads are the constant power load (CPL), which consumes a constant level of power of the system [5]. CPLs are common in DC distribution systems due to point-of-load (POL) converters which act as a CPL when they tightly regulate their outputs [1], [2], [3]. CPL introduces a destabilizing effect in the system, which causes significant oscillations in the intermediate DC bus voltage between the feeder and load (CPL) subsystem [1], [2], [3].

Power electronic converters are widely used to implement the CPL effects in DCMGs through the interconnection of POL converters [6], [7]. In [8]–[11], the CPL behavior is performed through buck converter. These works represent the DCMG using two subsystems, the first one is a feeder that regulates the intermediate DC bus voltage, and the second one is a load subsystem that is represented by a POL buck converter. They focus on ensuring DC bus voltage stability, minimizing the oscillation effects caused by a CPL, thus, control techniques are applied on the feeder subsystem to overcome the CPL negative incremental impedance instability problem [1], [3], [12].

To cope with the destabilizing effect of CPL in a DCMG, several methods have been proposed in the literature [13]. An LQR control scheme is reported in [14] to regulate DCMG distribution bus voltage with CPL. Robust control techniques are introduced [1] and [2] to mitigate the oscillations effects due to CPL.

This paper presents a comparative study between different

978-1-7281-4181-7/19 $31.00 © 2019 IEEE

control techniques applied on a DCMG to overcome the destabilization effect of CPL. DCMG consists of two POL buck converters in series connection. The control strategies are applied on the feeder buck converter, which regulates the DC bus voltage, in order to mitigate oscillations effects caused by a CPL power variation.

The remaining of this paper is outlined as follows: Section II presents the dynamic model of buck converters that constitute the proposed DCMG. Section III presents a brief review about control strategies discussed in this work. Section IV describes the experiments to be performed in this paper. Section V presents an assessment of the simulation results. Finally, Section VI presents the main conclusions.

II. BUCK CONVERTER MODELING

This section presents a brief review on dynamic and static behavior of DC-DC buck converters.

A. Dynamic analysis of buck converter

Fig. 1a shows a typical topology of DC-DC buck converter which provides an output voltage level V_o less than the input voltage level V_i [15] that is controlled by a variation of duty cycle D of a PWM modulation. When buck converter operates in continuous conduction mode (CCM) and due to the nonlinearity introduced by static switching, buck converter assumes two states per switching cycle [15], on and off state as shown in Fig. 1b and Fig. 1c.

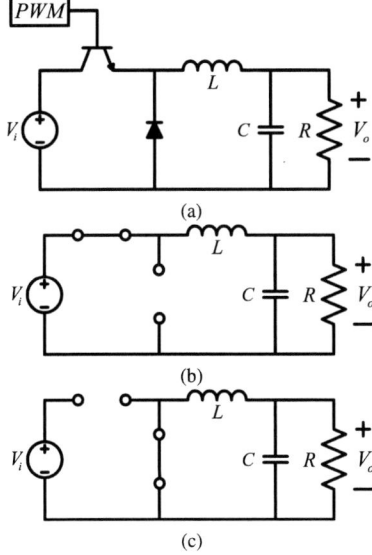

Fig. 1: (a) Buck converter (b) to static switch on and (c) to static switch off

Applying the Kirchhoff's voltage law and Kirchhoff's current law in the circuit shown in Fig. 1b, the dynamic behavior for on and off state can be defined by equations (1) and (2).

$$ON \Rightarrow \begin{cases} \dot{I}_L = -\frac{V_C}{L} + \frac{V_i}{L} \\ \dot{V}_C = \frac{I_L}{C} - \frac{V_C}{RC} \end{cases} \quad (1)$$

$$OFF \Rightarrow \begin{cases} \dot{I}_L = -\frac{V_C}{L} \\ \dot{V}_C = \frac{I_L}{C} - \frac{V_C}{RC} \end{cases} \quad (2)$$

Thus, by using the linear small-signal averaged state-space [15], the state-space equations of buck converter, described in (1) and (2), can be rewritten as follows

$$\begin{bmatrix} \Delta \dot{I}_L \\ \Delta \dot{V}_C \end{bmatrix} = \begin{bmatrix} 0 & -\frac{1}{L} \\ \frac{1}{C} & -\frac{1}{RC} \end{bmatrix} \begin{bmatrix} \Delta I_L \\ \Delta V_C \end{bmatrix} + \begin{bmatrix} \frac{V_i}{L} \\ 0 \end{bmatrix} [\Delta D] \quad (3)$$

For the small-signal averaged model of buck converter in CCM, the variation of output voltage, V_o, around the operating point is value of the variation of the steady-state capacitor voltage. In addition, for a buck converter in power mode control operating as a CPL, the power variation around its operating point is given by (4).

$$\begin{cases} \Delta V_o = \Delta V_C \\ \Delta P_o = \frac{2V_C}{R} \Delta V_C \end{cases} \quad (4)$$

B. Static analysis of the buck converter

The buck converter reduces the input voltage by duty cycle-control, thus, the output voltage V_o can be represented as follows

$$V_o = DV_i \quad (5)$$

To ensure the operation of the converter in the CCM region, the inductor must be big enough such that its energy will not be depleted until the next charging cycle occurs [15]. When the inductor current is continuous, it will not be null at any switching cycle f_{sw}. The conditions for ensuring the CCM operation mode, as well as the equations to compute inductor ripple current and output ripple current are provided below [15].

$$\begin{cases} L > \frac{R(1-D)}{2f_{sw}} \\ \Delta I_L = \frac{D(1-D)V_i}{Lf_{sw}} \\ \frac{\Delta V_o}{V_o} = \frac{1-D}{8f_{sw}^2 RC} \end{cases} \quad (6)$$

In addition, (7) shows the power that is consumed by the system load (R). Furthermore, an equivalent expression, in pu, is described in (8), adopting a base power $P_b = \frac{V_i^2}{R}$ with $D = 1$, eq. (7) is rewritten for a power value in as represented.

$$P_o = \frac{(DV_i)^2}{R} \quad (7)$$

$$P_{o,pu} = D^2 \quad (8)$$

III. CONTROL STRATEGIES APPLIED TO THE DC MICROGRID

In this paper, the controller design uses three different procedures to control the feeder and load subsystem (CPL). First, a classical approach by solving the Diophantine Equation (DE) is presented. The next control approach is based on root locus (RL) design by using graphical tool. And finally, the linear-quadratic regulator (LQR) control approach is introduced. These approaches are based on the open-loop and closed-loop structures shown in Fig. 2.

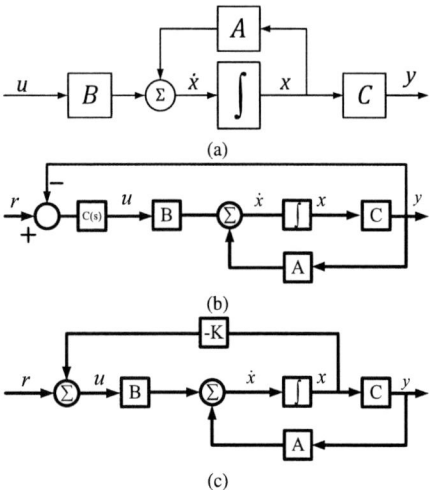

(a)

(b)

(c)

Fig. 2: (a) Block diagram of the system (b) System with output feedback (c) System with state feedback.

A. The Diophantine equation

Considering the transfer function of the open-loop system $G(s) = \frac{Y(s)}{U(s)}$ and the output feedback structure presented in Fig. 2a and Fig. 2b, respectively, and with a controller $C(s) = \frac{N(s)}{M(s)}$, the closed-loop characteristic polynomial $\Delta_c(s)$ is

$$\Delta_c(s) = U(s)M(s) + Y(s)N(s) \qquad (9)$$

Assuming that the desired dynamic of closed-loop system is represented by

$$\Delta_d(s) = s^n + \alpha_1 s^{n-1} + \dots + \alpha_{n-1}s + \alpha_n \qquad (10)$$

In order to tune the controller, the closed-loop characteristic polynomial $\Delta_c(s)$ is compared with closed-loop desired polynomial $\Delta_d(s)$, obtaining the Diophantine equation (11) [16].

$$\Delta_d(s) = U(s)M(s) + Y(s)N(s) \qquad (11)$$

By solving (11), the parameters of numerator $N(s)$ and denominator $M(s)$ of the controller $C(s)$ are tuned.

B. Root locus graph tool

Another control approach to guarantee the desired closed-loop performance is the root locus (RL) tool. The root locus

plots the poles of the closed-loop transfer function in the complex s-plane as a function of the parameters gain of a given controller. In addition to determining the stability of the system, the root locus can be used to design the damping ratio (ζ) and natural frequency (ω_n) of a feedback system [16].

By selecting a point along the root locus that coincides with a desired damping ratio and natural frequency, the parameters gain of a given controller can be calculated.

Therefore, the closed-loop desired polynomial is formed by the desired performance conditions such as settling time t_{ss} and damping coefficient ζ.

The settling time is related to the time constant $\tau \simeq \frac{t_{ss}}{5}$ of the open-loop system. Where the damping σ of the system is related to the time constant as indicated in (12).

$$\sigma = \frac{1}{\tau}, \quad \zeta = \cos\theta \qquad (12)$$

The damping coefficient ζ is related to the angle θ that the line segment from the origin to the dominant pole realizes with the real axis, as indicated in (12). Thus, the desired performance region in the plane of the complex is determined, as shown in Fig. 3, where the desired performance region is hatched.

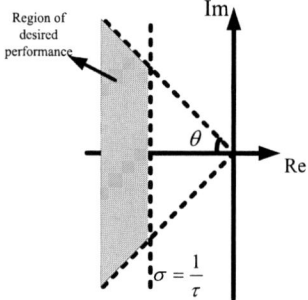

Fig. 3: Desired performance region on root locus

C. Linear-Quadratic Regulator

The LQR is an optimal control methodology capable of allowing the designer to choose between minimizing energy or minimizing time, by developing the quadratic performance index shown in (13).

$$I = \int_t (x^T Q x + u^T Z u)dt \qquad (13)$$

For this, consider an observable system represented for the block diagram of the system is showing in Fig. 2a.

To use the LQR approach, it is necessary a state feedback structure, as shown in Fig. 2c. The solution of the optimal control problem is to determine the gain value K for the state feedback, by means of the solution of the Riccati equation, see [16], shown in (14) resulting from the development of (13).

$$PA + A^T P - PBZ^{-1}B^T P = -Q \qquad (14)$$

Where Q is symmetric semi-definite positive matrix and Z is a scalar defined by the designer. The solution of (14) results

978-1-7281-4181-7/19 $31.00 © 2019 IEEE

in the definition of the vector P. The values of the gains of the vector K are given by (15).

$$K = Z^{-1}B^T P \tag{15}$$

The weighting of Q and Z are related to the project where need to prioritize better performance or less control effort.

IV. METHODOLOGY

The aim of this work is to evaluate the performance of the proposed controllers aiming to regulate the outputs of the feeder and load (CPL) subsystem. Fig. 4 represents the proposed DCMG with two decoupled outputs, V_o DC Bus voltage) and P_o (CPL output), and the topology employed to control the system.

A. Determination of the parameters of the DC Microgrid

The DC-DC buck converter of the feeder subsystem is set with a fixed switching frequency $f_1 = 1$ KHz, Limited by a inductor, $L_1 = 1$ mH, and an input voltage $V_{i1} = 15$ V. In order to satisfy the CCM operation, the parameter values of resistance, R_1, capacitor, C_1, Duty Cycle, D, and output voltage, V_{o1}, are calculated according to the set of equation defined in (6) with a small inductor ripple current, $\Delta I_{L_1} < 1$ A.

On the other hand, the POL converter that acts as a CPL is set with a switching frequency $f_2 = 5$ KHz, limited by a inductor, $L_2 = 1$ mH and an input voltage $V_{i2} = 12$ V. The small inductor ripple current for this converter is $\Delta I_{L_2} < 1$ A in order to ensure the CCM operation. Table I shows the computed parameters of the DCMG.

TABLE I: Microgrid parameters

Feeder			
Parameter	Symbol	Value	Unit
Input voltage	V_{i1}	15.0	V
Duty cycle	D_1	0.8	-
Switching frequency	f_1	1.0	kHz
Load resistance	R_1	1.0	Ω
Inductor	L_1	1.0	mH
Capacitor	C_1	2.0	mF
Output voltage	V_{o1}	12.0	V
CPL			
Parameter	Symbol	Value	Unit
Input voltage	V_{i2}	12.0	V
Duty cycle	D_2	0.6	-
Switching frequency	f_2	5.0	kHz
Load resistance	R_2	4.0	Ω
Inductor	L_2	1.0	mH
Capacitor	C_2	2.0	mF
Output power	P_{o2}	0.36	p.u.

B. Controller design

According to the DCMG parameters previously calculated, the small-signal averaged model (cf. Fig. 2a) is found for each subsystem according to (3). Fig. 5 presents the controller structures used in this paper.

In order to design the controllers, the requirements (settling time, t_{ss}, maximum overshoot, ovs, and steady-state error, e_{ss})

to regulate output 1 (V_o) and output 2 (P_o) are presented in (16).

$$t_{ss} < 0.02\text{s}, \quad ovs < 10\%, \quad e_{ss} = 0 \tag{16}$$

Both controllers are tuned by classical pole-placement approach by solving the DE presented in Section 3. By using the small-signal averaged models of each subsystem and the proposed controller structures (cf. Fig. 5a), the parameters of controllers are tuned by solving (11). the closed-loop desired polynomial is calculated by the damping coefficient ζ and the natural frequency ω_n given in (17).

$$ovs = e^{-\frac{\zeta\pi}{\sqrt{1-\zeta^2}}}, \quad t_{ss} = \frac{4.6}{\zeta\omega_n} \tag{17}$$

An auxiliary pole, p_1 is added to the closed-loop desired polynomial to ensure the solution of the Diophantine equation is feasible with the proposed controller structure to regulate output 1 (V_o). The same procedure is performed for the proposed controller structure to regulate output 2 (P_o), thus an auxiliary pole, p_2 is added. These auxiliary poles must be non-dominant poles so that so that their characteristic do not influence in the system behavior. The location of these non-dominant poles is at the discretion of designer and their relationships are shown in (18).

$$p_1 = 3\zeta\omega_n, \quad p_2 = 5\zeta\omega_n \tag{18}$$

From (17) and (18), the desired polynomials for the feeder, $\Delta_{df}(s)$ and CPL, $\Delta_{dc}(s)$ are calculated in (19).

$$\begin{cases} \Delta_{df}(s) = (s^2 + 2\zeta\omega_n s + \omega_n^2)(s + p_1) \\ \Delta_{dc}(s) = (s^2 + 2\zeta\omega_n s + \omega_n^2)(s + p_2) \end{cases} \tag{19}$$

The controller for output 1, V_o is also designed by RL. With the expected performance of the system presented in the system and the desired pole placement performance region is determined with the aid of the `rltool()`. The performance region is similar to that shown in Fig. 3.

Moreover, the controller for output 1, V_o is also designed by the LQR method using the structure shown in Fig. 5b. For the problem solving of the Riccati equation the matrices Q and Z were chosen as indicated in (20). Where the weighting of a value elevated to Z causes a penalty in the control action. The value of the integral gain was determined from the increased matrix of the system, in order to guarantee zero error to the steady state step. The gain vector K is calculated using the MATLAB `lqr()` command. Table II presents the controllers' gains.

$$Q = \begin{bmatrix} 1 & 0 \\ 0 & 10^{-7} \end{bmatrix}, Z = 0.2 \tag{20}$$

C. Test environment

The experiment described below is performed using MATLAB/Simulink. The aim is to check the oscillations effects in the DC bus voltage caused by small variations of CPL

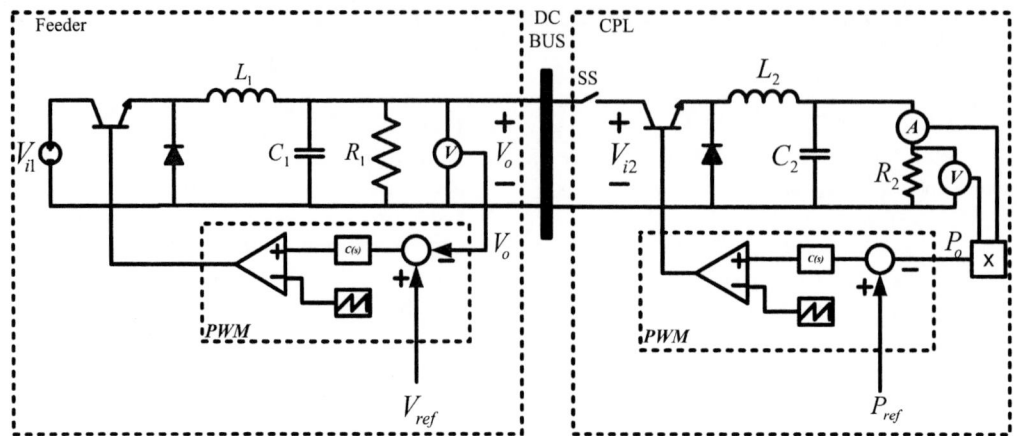

Fig. 4: Proposed DC microgrid.

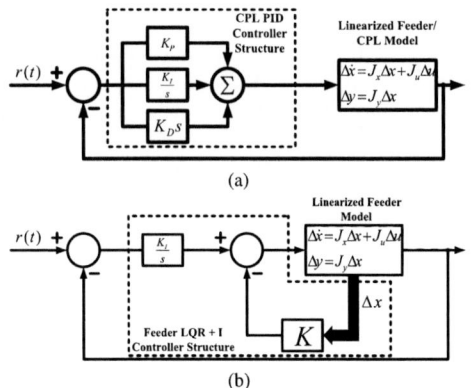

Fig. 5: Structure proposed for the design of the buck converter controller (a) with output feedback (b) with state feedback

TABLE II: Projected controller gains

Gain	Feeder			CPL
	DE	RL	LQR	DE
K_p	-0.0042	0.0244	-	0.2729
K_i	13.9265	15.7153	10	258.2576
K_d	$1.2 \cdot 10^{-4}$	$6.3 \cdot 10^{-5}$	-	0.0023
K	-	-	$[0.0385 \quad -0.0099]$	-

power when the feeder subsystem is regulated for the control approaches mentioned in Table II.

The DCMG (cf. Fig. 4) is started with the feeder subsystem operating in open-loop and with the load subsystem (CPL) disconnected. The feeder subsystem achieves its nominal operating points (see Table I) at $t = 1.0$ s. After that, the feeder subsystem stars to operate in closed-loop. CPL is connected at $t = 3.0$ s operating in open-loop. CPL achieves its steady state at $t = 4.0$ s. Then, CPL starts to operate in closed-loop. Once whole system reaches its stable state condition (t = 4.5s), a variation of CPL power is performed ($t = 5.0$ s) within amplitude range from 0.1 to 0.2 pu. It is worth to note that all analysis are performed from time $t = 5.0$ s when the variation of CPL power occurs.

V. RESULTS ANALYSIS

The simulated test are performed as mentioned in Subsection IV-C. The simulated results show that all controllers that regulate the DC bus voltage can deal with oscillations caused by CPL power variation, ensuring system stability. In order to improve the analysis of system performance, the integral of square error (ISE) is used to evaluate the performance of the control strategies addressed in this study, more detail about performance index in [1]. In addition, the integral of square error weighted by time (ISET) is introduced to verify which controller of feeder subsystem obtained the faster transient response when the system is subjected to a variation of CPL power. And, finally, the integral of square of signal control (ISSC) is introduced to analyze the control effort of mentioned controllers [1]. Fig. 6 shows the all simulated results.

A. Performance analysis of the proposed feeder controllers

As mentioned above, when the DCMG reaches its stable state condition ($t = 4.5$ s), variations of CPL power are performed as shown in Fig. 6b.

The simulated results show that all control approaches that regulates the DC bus voltage to ensure system stability under CPL power variation as shown in Fig. 6a.

Fig. 6c shows the ISE performance indices of all control approaches. The impact of CPL power variations is reduced when the DC bus voltage is regulated by the LQR control approach. On the other hand, the DC bus voltage show more oscillation amplitude when the feeder subsystem is controlled by RL control approach. Therefore, the LQR control approach more effectively compensates the oscillations caused by CPL power variation, reducing the oscillation amplitude in comparison with other control approaches as shown the ISE performance indices (cf. Fig. 6c).

Fig. 6d shows the ISET performance indices of all control approaches. According with the ISE results, LQR control approach outperforms the other control approaches. Therefore, the LQR control approach achieves better performance with faster transient response.

978-1-7281-4181-7/19 $31.00 © 2019 IEEE

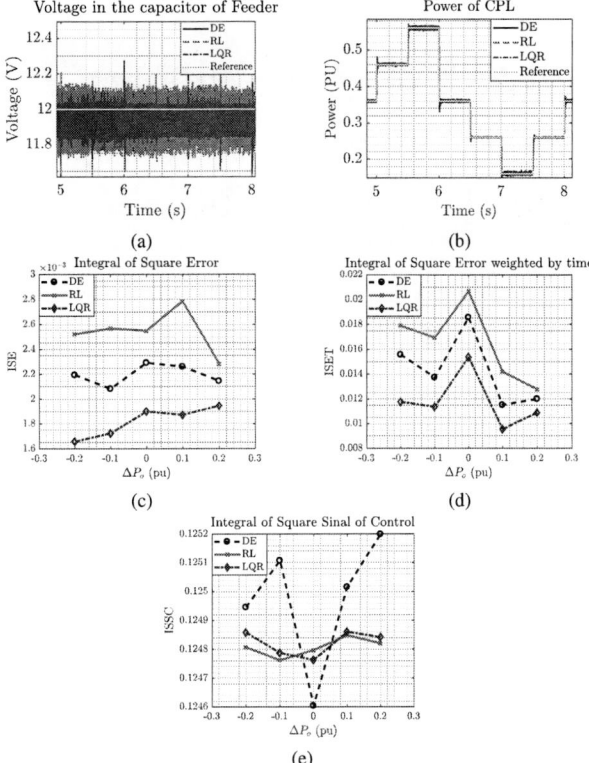

Fig. 6: (a) Voltage at the output of the Feeder for (b) power variation in the CPL and perfomance index for design controller in the feeder (c) integral of square error (ISE) (d) integral of square error weighted by time (ISET) (e) integral of square signal of control (ISSC).

Finally, the control effort of all control approaches is analyzed by the ISSC performance indices as shown in Fig. 6e. In the operating point, the DE control approach shows a minor control effort, but when the system is subjected to variation of CPL power, it presents more degradation in the control system performance. On the other hand, RL and LQR control approaches show less degradation in the control system performance when the system system is subjected to variation of CPL power. Note that RL and LQR present a similar control effort (cf. Fig. 6e).

VI. CONCLUSION

The aim of this paper is to evaluate control methodologies (LQR, RL and DE) applied to ensure stability of a DC microgrid. Test results show that all controllers are able to compensate the impact of CPL in the DC bus voltage. However, the LQR control approach provides a better performance with reduced oscillation amplitude in comparison with other control approaches. The performance indicators comparison show that the LQR control approach more effectively compensates for disturbances offering better performance and stability with a minor control effort.

ACKNOWLEDGMENT

This work was supported in part by CNPq under grant 432341/2018-8, in part by CAPES, FAPEAM, and in part by the PROPG-CAPES/FAPEAM Scholarship Program.

REFERENCES

[1] K. E. Lucas, D. A. Plaza, W. B. Jr, R. L. P. Medeiros, E. M. Rocha, D. Benavides, S. J. Ríos, and E. V. Herrera, "Novel robust methodology for controller design aiming to ensure DC microgrid stability under CPL power variation," *IEEE Access*, vol. 1, no. 7, pp. 1–16, May 2019.

[2] K. E. Lucas, W. Barra, D. A. Plaza, R. L. P. Medeiros, E. M. Rocha, and D. A. Vaca, "Interval robust controller to minimize oscillations effects caused by constant power load in a DC multi-converter buck-buck system," *IEEE Access*, vol. 7, pp. 26 324–26 342, February 2019.

[3] S. Singh, A. R. Gautam, and D. Fulwani, "Constant power loads and their effects in DC distributed power systems: A review," *Renewable and Sustainable Energy Reviews*, vol. 72, pp. 407–421, May 2017.

[4] A. Jusoh, H. Baamodi, and S. Mekhilef, "Active damping network in DC distributed power system driven by photovoltaic system," *Solar Energy*, vol. 87, no. 1, pp. 254–267, January 2013.

[5] A. Riccobono and E. Santi, "Comprehensive review of stability criteria for DC power distribution systems," *IEEE Transactions on Industry Applications*, vol. 50, no. 5, pp. 3525–3535, Sept.-Oct. 2014.

[6] W. Du, J. Zhang, Y. Zhang, and Z. Qian, "Stability criterion for cascade system with constant power load," *IEEE Transactions on Power Electronics*, vol. 28, no. 4, pp. 1843–1851, April 2013.

[7] H. Mosskull, "Constant power load stabilization," *Control Engineering Practice*, vol. 72, pp. 114–124, December 2017.

[8] Q. Xu, C. Zhang, C. Wen, and P. Wang, "A novel composite nonlinear controller for stabilization of constant power load in DC microgrid," *IEEE Transactions on Smart Grid*, vol. 10, no. 1, pp. 752–761, January 2019.

[9] A. Kwasinski and C. N. Onwuchekwa, "Dynamic behavior and stabilization of DC microgrids with instantaneous constant-power loads," *IEEE Transactions on Power Electronics*, vol. 26, no. 3, pp. 822–834, March 2011.

[10] J. You, M. Vilathgamuwa, and N. Ghasemi, "DC bus voltage stability improvement using disturbance observer feedforward control," *Control Engineering Practice*, vol. 75, pp. 118–125, June 2018.

[11] S. Sumsurooah, M. Odavic, S. Bozhko, and D. Boroyevich, "Robust stability analysis of a DC/DC buck converter under multiple parametric uncertainties," *IEEE Transactions on Power Electronics*, vol. 33, no. 6, pp. 5426–5441, June 2018.

[12] A. T. Elsayed, A. A. Mohamed, and O. A. Mohamed, "DC microgrids and distribution systems: An overview," *Electric Power Systems Research*, vol. 119, pp. 407–417, February 2015.

[13] M. K. Al-Nussairi, R. Bayindir, S. Padmanaban, L. Mihet-Popa, and P. Siano, "Constant power loads (CPL) with microgrids: Problem definition, stability analysis and compensation techniques," *Energies*, vol. 10, no. 10, pp. 1656–1676, October 2017.

[14] A. J. Mills and R. W. Ashton, "Parameter optimization of an interval robust controller of a buck converter subject to parametric uncertainties," in *2017 IEEE International Conference on Industrial Technology (ICIT)*, Toronto, Canada, March 2017, pp. 498–503.

[15] R. W. Erickson and D. Maksimovic, *Fundamentals of Power Electronics*, 2nd ed. New York: Springer, January 2004.

[16] J. J. D'Azzo and C. H. Houpius, *Linear Control System Analysis and Design*. Lisle, Illinois, USA: McGraw-Hill Inc, October 1981.

Integrating a Single Z-Source Network with a Modular Multilevel Converter for Voltage Boosting

Fatma A. Khera*†   Christian Klumpner*   Pat W Wheeler*

* Power Electronics, Machines and Control Research Group,
University of Nottingham, Nottingham, NG7 2RD, UK
† Electrical Power and Machines Engineering Department, Tanta University, Egypt
E-mail: Fatma.Khera@nottingham.ac.uk; klumpner@ieee.org; pat.wheeler@nottingham.ac.uk

Abstract— **This paper proposes the integration of a Z-source network with the modular multilevel converter (MMC) to add voltage boosting capability to a voltage step down converter. To limit the increase in complexity, the proposed Z-source modular multilevel converter uses a single Z-source network that is interconnected between the corresponding terminals of the DC-input source and the DC-link terminals of the MMC. The authors previously presented a modulation technique for quasi Z-source MMC (qZS-MMC) referred to as the reduced inserted cells (RICs) PWM but a large size inductor for the two quasi Z-source networks were needed. This paper shows that utilising the RICs scheme with the-source MMC is more advantageous compared with the qZS-MMC. The operation principle of the Z-source MMC using RICs scheme and the derivation of key design parameters is presented in this paper by analysing the relevant current and voltage waveforms. The simulation results verify the operation and showcase the excellent waveform performance of the proposed topology.**

Keywords—Z-source modular multilevel converter, quasi Z-source MMC, reduced inserted cells.

I. INTRODUCTION

The modular multilevel converter (MMC) proposed in [1] is a relatively new competitive concept in both medium and high voltage applications. It provides several features such as modularity, voltage and power scalability and failure management capability [2-6].

The basic building block of the MMC phase-leg is the sub-module (SM). There are different configurations for implementing the SMs such as half bridge [7], full bridge [8] and three level neutral point clamped [6]. The half-bridge SMs based MMC has the lowest power losses due to lower number of semiconductor devices in the current path compared to other configurations. However, MMC with half bridge SMs is constrained by its inability to generate output voltages greater than half the DC-source voltage which may be needed in some applications.

The integration of an impedance network to MMC has been proposed in [9, 10] where a quasi Z-source modular multilevel converter (qZS-MMC) based on half bridge SMs has been used. The operating principle of voltage boosting by using the impedance network relies on producing a short circuit (shoot-through) at the DC-link terminals in order to increase the

stored energy in the inductors. This energy is then transferred to the capacitors leading to an extra voltage that adds up to the DC source voltage providing a voltage boosting capability.

In [9], two modulation techniques have been proposed: the reduced inserted cells (RICs) and simultaneously shorted (SS) techniques. Using the SS technique limits the output voltage to the average value of the DC-link voltage which leads to a high stress voltage on the qZS-network switching devices. On the other hand, the RICs technique allows the output voltage to be equal to the peak value of the DC-link voltage and results in a lower stress voltage on the qZS-network switching devices. However; the RICs technique depends on using the partial shoot-though concept, where the upper DC-link switches can perform a shooting through in negative half of the fundamental frequency cycle while the lower DC-link switches can perform in the positive half. This means that the inductors will de-energize during one half of output frequency cycle leading to a high inductor current ripple at the fundamental frequency which increases the inductor size.

This paper proposes the integration of a single Z-source network with the MMC to realise a Z-source modular multilevel converter (ZS-MMC). Applying the RICs modulation technique to the ZS-MMC will reduce the inductors size as the largest inductor current ripple will be at switching frequency.

The paper is organized as follow: the ZS-MMC equivalent circuit, operation modes and capacitor voltage balancing mechanism are presented. The suitable PWM scheme investigating by referring to voltage and current waveforms. Finally, the analysis of the proposed converter is validated by simulation results using a MATLAB/PLECS model.

II. PROPOSED Z-SOURCE MODULAR MULTILEVEL CONVERTER

The proposed Z-source modular multilevel converters (ZS-MMCs) topology is shown in Fig. 1, where a single Z-source impedance network is connected between the DC-source terminals and DC-link terminals of the MMC phase-leg. The Z-source network consists of two inductors and two capacitors, connected between the input split DC-source and MMC-leg. The DC-source can be split using a single DC source and two series-connected capacitors where the mid-

978-1-7281-4181-7/19 $31.00 © 2019 IEEE

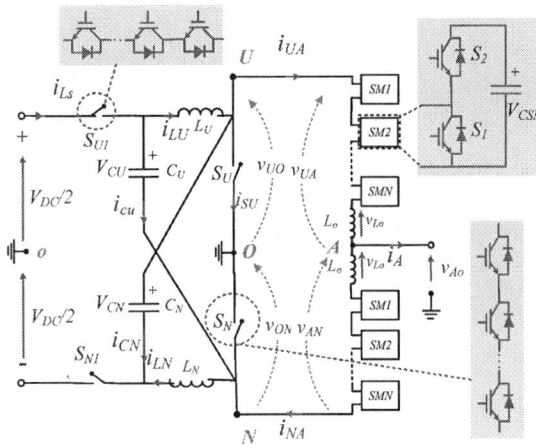

Fig. 1: Structure of a Z-source modular multilevel converter ZS-MMC

point "o" can be used as a reference point for the output voltage v_{AO} as indicated in Fig. 1. The operating principle for the Z-source network depends on introducing shoot-through (ST) at its DC-link terminals [11].

The MMC leg contains an upper and lower arm each formed by -connecting in series N sub-modules (SMs), and an arm inductor (L_o). Each SM has a half-bridge configuration with a floating capacitor C_{SM} as depicted in Fig. 1. The two switches (S_1, and S_2) in the SM are controlled by complementary gating signals, when S_1 is on, SM capacitor is bypassed, and SM terminal voltage is zero. If S_1 is off, S_2 is on, therefore SM terminals voltage is equal to SM capacitor voltage V_{CSM} which is inserted into the arm circuit. The upper and lower arms voltages are represented by v_{UA} and v_{UA}, respectively. The upper and lower DC-link voltages at the terminals of MMC-leg are defined by v_{UO} and v_{ON} respectively. The output voltage equation is given by:

$$v_{AO} = (v_{AN} - v_{UA})/2 + (v_{UO} - v_{ON})/2 \qquad (1)$$

The MMC arm requires a voltage balancing strategy to balance and keep the sub-modules capacitor voltages at their desired average values. The capacitor voltages balancing can be achieved by using different strategies [2]. The most widely used balancing strategy is based on the sorting method [7] which is summarized in the following four steps:

1. Measure and sort the upper and lower capacitor voltages.
2. The modulation scheme will determine the number of inserted SM capacitors (N_U and N_N) from upper and lower arms respectively.
3. If the upper (lower) arm current is positive (same as the current direction shown in Fig. 1), choose the N_U (N_N) cells with lower voltage to be inserted. Therefore, the corresponding cell capacitor is charged and its voltage increases;
4. If the upper (lower) arm current is negative, choose the N_U (N_N) cell with higher voltage to be inserted. Therefore, the corresponding cell capacitor is discharged and its voltage decreases.

As reported in [12], two chain-link of series connected switches S_U and S_N are linked at the DC-link terminals with a mid-point "o" to provide a shoot-through current path. This is attributed to difficulty to use the MMC leg to produce the shoot-through by bypassing all the SMs in both the upper and lower arms which would lead to a drop in upper and lower arm voltage levels to zero, which would cause a high distortion in the output voltage and the benefit of having a multilevel functionality will be compromised.

The basic configuration of the ZS-network depends on using input diodes which are mandatory for the voltage boost mechanism and cannot be removed [11]. As derived in [12] for single phase converter, the active switches should be connected in antiparallel to the input diodes to avoid any undesired operation modes. The two switches S_{UI} and S_{NI} have to be gated by complementary gating signals to S_N and S_U respectively.

The ZS network can be fully shorted by turning on both the upper and lower DC-link switches or partially shorted by turning on the upper or the lower switches separately [13]. The operation modes for ZS-network are described in the subsequent subsections.

A. Upper or/and lower shoot-through (ST) mode:

In this mode, if the upper (lower) DC-link switch S_U (S_N) is turned-on, the S_{NI} (S_{UI}) should be turned-off. Fig. 2a, Fig, 2b and Fig. 2c show the upper, lower and full shoot-through equivalent circuit respectively. For any case of the upper, lower, and full shoot-through modes, both Z-source inductors are energized and therefore both inductor currents increase. The inductor voltages for any case are defined by:

$$v_{LU} = v_{LN} = V_{DC}/2 \qquad (2)$$

Assume the ZS-capacitor voltages are $V_{CU} = V_{CN} = V_C$, the voltage expressions for the upper shoot-through (UST), lower shoot-through (LST) and full shoot-through (FST) modes are:

1. The upper shoot-through (UST)

$$v_{UO} = 0 \quad , \quad v_{ON} = -V_{DC}/2 + V_C \qquad (3)$$

2. The lower shoot-through (LST)

$$v_{UO} = -V_{DC}/2 + V_C \quad , \quad v_{ON} = 0 \qquad (4)$$

3. The full shoot-through (FST)

$$v_{UO} = 0 \quad , \quad v_{ON} = 0 \qquad (5)$$

B. Non-shoot-through mode (NST):

In this mode, the upper and lower DC-link switches (S_U, and S_N) are turned-off and consequently, the series switches are turned-on as shown in Fig. 2d. The Z-source inductors are

978-1-7281-4181-7/19 $31.00 © 2019 IEEE

discharging and therefore the inductor currents decrease. The expression for NST mode is as follows:

$$v_{L1} = v_{L2} = V_{DC} - V_C$$
$$v_{UO} = v_{ON} = V_{UN} / 2 = -V_{DC}/2 + V_C \tag{6}$$

where V_{UN} is the peak value of the DC-link voltage. In NST mode, the capacitors charging/discharging states depend on their current direction. If the upper (lower) arm current i_{UA} (i_{NA}) is higher than upper (lower) inductor current i_{LU} (i_{LN}), the capacitor charges and the voltage increases otherwise the capacitor voltage decreases. The capacitor currents are:

$$i_{CU} = i_{LU} - i_{UA}$$
$$i_{CN} = i_{LN} - i_{NA} \tag{7}$$

With defining the switching period T, non-shoot-through NST period T_{NST}, the upper ST period T_U and lower ST periods T_N, therefore:

$$T_{NST} + T_U + T_N = T$$
$$T_U = T_N = T_{ST} / 2$$
$$D_{ST} = T_{ST} / 2T \tag{8}$$

where D_{sh} is the average ST duty ratio. T_U and T_N are equal to ensure symmetrical operation. At steady state, the average value of the inductor voltage over one switching period is zero. Therefore, the average capacitor voltages are given by:

$$V_C = \frac{1 - D_{sh}}{1 - 2D_{sh}} V_{DC} \tag{9}$$

Substituting from (9) in to (6), the expression of peak value of the upper and lower DC-link voltage V_{UO} and V_{ON} can be expressed by:

$$V_{uo} = V_{on} = V_{un} / 2 = \frac{1}{1 - 2D_{sh}} V_{DC} / 2 \tag{10}$$

III. MODULATION TECHNIQUE

In this work, phase disposition SPWM (PD-SPWM) is used as illustrated in Fig. 3. Two complementary reference signals v_{UA-m} and v_{ON-m} are used to modulate the upper arm and the lower arm respectively. Each carrier is responsible for producing the gating signals of two cells (one from upper and one from lower arm). This technique produces $(2N_{SM} - 1)$ output voltage level where N_{SM} is the number of SMs per arm.

If one of the upper or the lower DC-link switches are turned on, the output voltage will be highly distorted, and the output voltage level will drop/rise by $N_{SM}/2$. From (1), if the upper DC-link switch S_U is turned on ($v_{UO} = 0$), the upper arm voltage v_{UA} should be decreased by an amount equal to half the DC-link voltage to keep unchanged the output voltage level. To attain that, if the upper (lower) DC-link switches are doing a shooting-through, $N_{SM}/2$ SMs that are initially on should be selected from the upper (lower) arm to be bypassed.

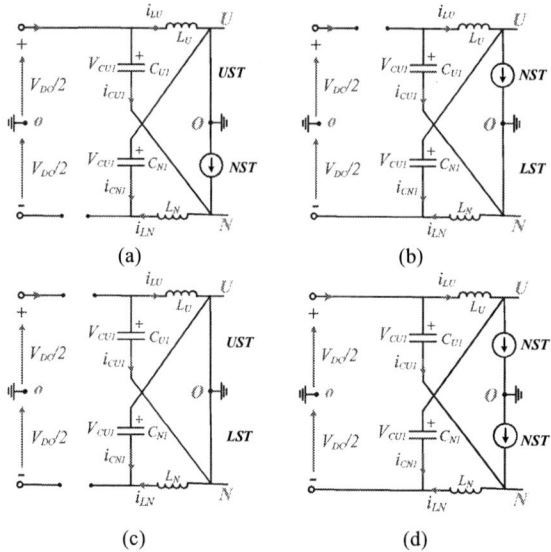

Fig. 2: Operation modes: a) Upper shoot-through UST, b) Lower shoot-through LST, c) Full shoot-through, and d) Non shoot-through modes

This scenario is named as "reduced inserted cells" (RICs) which has been proposed in [9] and applied for quasi Z-source modular multilevel converter (qZS-MMC). During the upper (lower) ST intervals shown in Fig. 3, the number of upper (lower) arm inserted cells greater than or equal to $N_{SM}/2$ is realized during the second (first) half-cycle of the upper (lower) arm modulating signal as illustrated in Fig. 4. Hence, the number of inserted cells in the upper arm can be reduced by $N_{SM}/2$. As a result, ST reference signals for the upper and the lower arms v_{sh-U} and v_{sh-U} at a unity ST carrier height shown in Fig. 3a can be defined by:

$$v_{sh-U} = \begin{cases} 0 & \rightarrow & 0 : \pi \\ T_{sh} = 2D_{sh} & \rightarrow & \pi : 2\pi \end{cases}$$
$$v_{sh-N} = \begin{cases} T_{sh} = 2D_{sh} & \rightarrow & 0 : \pi \\ 0 & \rightarrow & \pi : 2\pi \end{cases} \tag{11}$$

The upper and lower ST modulating signals and gated signals of the upper and lower DC-link switches are shown in Fig. 3a and Fig. 3b respectively. The upper and lower arm modulation signals need to be modified by controlling the reference signal with the actual shoot-though pulses as shown in Fig. 3c where during the ST intervals, the amplitude of original modulating signals is level-shifted by $N_{SM}/2$ units of SMs carrier. During the UST or LST intervals, the arm inductor voltage is defined by:

$$v_{Lo} = V_{UN} / 2 - (N_U + N_N - N_{SM} / 2)V_{CSM} \tag{12}$$

where $N_U + N_N$ has average value N_{SM} over one switching period. During the NST interval, the arm inductor voltage is defined by:

$$v_{Lo} = V_{UN} - (N_U + N_N)V_{CSM} \tag{13}$$

Using (12) and (13), at steady state, the average value of the arm inductor voltage is zero.

$$\left(\frac{V_{UN}}{2} - \frac{N}{2}V_{CSM}\right)D_{sh} + (V_{UN} - N_{SM}V_{CSM})(1-D_{sh}) = 0 \quad (14)$$

The SMs average capacitor voltages are given by:

$$V_{CSM} = \frac{1}{N(1-2D_{sh})}V_{DC} = \frac{V_{UN}}{N} \quad (15)$$

From (15), the SMs capacitor are charged according to the peak value of the DC-link voltage. As a result, the peak value of the fundamental output phase voltage can be calculated by:

$$V_m = \frac{m}{2}V_{UN} = \frac{m}{2}N_{SM}V_{CSM} = G\frac{E}{2} \quad (16)$$

where G in the converter gain:

$$G = \frac{m}{(1-2D_{sh})} \quad (17)$$

IV. EVALUATING THE INDUCTOR SIZE REDUCTION FOR ZS-MMC

The RICs technique has an advantage of requiring significantly smaller ZS inductor size when using in ZS-MMC compared to qZS-MMC [9] and the size reduction potential will be evaluated next. This technique depends on using partial shoot-though concept, where the upper DC-link switches can perform a shooting through in half of the fundamental frequency cycle while the lower DC-link switches can perform in the other half. In ZS-MMC, the two ZS inductors are charging all together in both LST and UST modes and discharging in NST mode, leading to inductor current ripple at the switching frequency. Fig. 3d shows the ZS inductor voltages and currents waveforms. Generally, the inductance has been obtained by:

$$L = \frac{v_L \Delta t}{\Delta i_L} \quad (18)$$

Δt, Δi_L and v_L can be defined during shoot-through mode by:

$$\Delta t = 2D_{sh}/f_s \quad , \quad v_L = V_{DC}/2 \quad , \quad \Delta i_L = k_i I_L \quad (19)$$

where f_s is the switching frequency, and k_i is the inductor current ripple ratio. Substituting from (19) into (18), the ZS inductance can be expressed by:

$$L_U = L_N = \frac{D_{sh}V_{DC}^2}{K_i f_s P} \quad (20)$$

where P is the converter input power which is equal to $V_{DC}I_L$. On the other hand, in qZS-MMC, the upper (lower) qZS-inductor is only energized during UST (LST). Consequently, the inductors will be de-energized along one half of output frequency cycle leading to a high inductor current ripple at the fundamental frequency. Fig. 3e shows the voltages and currents waveforms of the qZS-MMC inductors L_U and L_N. The parameters given in (19) are re-defined for qZS-MMC during NST mode by:

$$\Delta t = 1/2f_o \quad , \quad \Delta i_L = k_i I_L$$
$$v_L = -V_{C2} = \frac{-D_{sh}}{1-2D_{sh}}V_{DC}/2 \quad (21)$$

Using (18) and (21), the qZS-network inductances (L_U, and L_N) are calculated by:

$$L_U = L_N = \frac{D_{sh}V_{DC}^2}{4k_i(1-2D_{sh})f_o P} \quad (22)$$

As shown in Fig. 3f, the qZS-MMC source inductor is charging in both LST and UST modes and discharging in NST mode (which is similar to ZS-MMC inductors), leading to inductor current ripple at the switching frequency. The parameters Δt, Δi_L and v_L are re-defined for qZS-MMC during the half of the output frequency cycle by:

$$\Delta t = 2D_{sh}/f_s \quad , \quad \Delta i_L = k_i I_L$$
$$v_L = -2V_{C2} = \frac{-D_{sh}}{1-2D_{sh}}V_{DC} \quad (23)$$

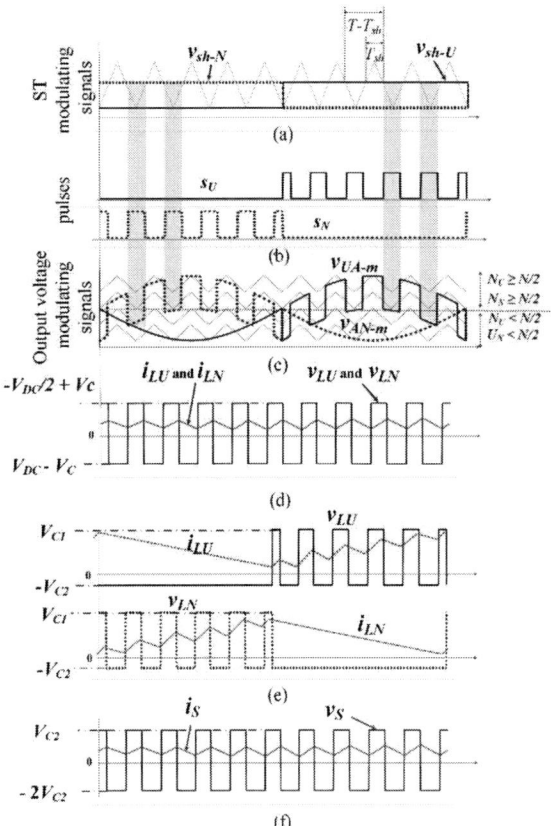

Fig. 3: The circuit waveforms including: a) ST reference signals, b) pulses for S_U and S_N, c) Output voltage reference signals, d) qZS-MMC inductor (L_U, L_N) voltages and currents waveforms, e) ZS-MMC inductor (L_U, L_N) voltages and currents waveforms, and f) ZS-MMC source inductor (L_S) voltage and current waveforms

From (23) and (18), the qZS-network source inductance (L_S) is calculated by:

$$L_s = \frac{2D_{sh}^2 V_{DC}^2}{k_i(1-2D_{sh})f_s P} \qquad (24)$$

The inductances L_U and L_N depends on switching frequency f_s (20) and output frequency f_o (22) for ZS-MMC and qZS-MMC respectively. It is noted that the inductor current ripple for the ZS-MMC in much lower compared to qZS-MMC, at a ratio of 1/20 for f_s = 2 kHz, f_o = 50 Hz and D_{sh} = 0.25.

V. SIMULATION RESULTS

To verify the theoretical analysis of the proposed ZS-MMC topology and compare it with qZS-MMC, single phase simulation models are implemented for both using MATLAB/PLECS. The parameters used in the simulation models are given in Table I. The simulation study has been carried out using the same passive RL load for both. The modulation index M was set at "1" and the gain factor G was commanded by setting the shoot-through ratio D_{sh} 0.25. From (17), this yields to a gain factor equal to 1/ (1-2*0.25) = 2.

For qZS-MMC, the upper and lower DC-link voltages (v_{UO} and v_{ON}), qZS-network inductor currents (i_{LU} and i_{LN}) and source current i_{LS} are shown in Fig. 4. The inductor currents i_{LU} and i_{LN} have fundamental frequency ripples with 2.4 times of the average current as shown in Fig. 4. However, the source current i_s has a switching frequency ripples with a ratio 7 % (17A) of the average value.

Fig. 5a shows the ZS-MMC simulation results including, the upper and lower DC-link voltages (v_{UO} and v_{ON}) and ZS-network inductor currents (i_{LU} and i_{LN}) and their zooming are shown in Fig. 5b. The DC-link voltage has a peak value of 5.3 kV at Gain factor of "2". The inductor currents are fairly constant and have a very small switching frequency ripples with a ratio 7% (17A) of the average value "250 A" which proves their size can be significantly reduced if needed.

The upper and the lower SMs capacitor voltages are well balanced and their average value equals 2.65 kV as shown in Fig. 6. It is noted that the SMs capacitor voltages have been charged according to the peak value of the DC-link voltage. It is well known that, the peak value of fundamental phase voltage that can be attained theoretically from MMC with half bridge SMs is half of the supply voltage which is 5.5kV/2. For ZS-MMC/qZS-MMC, the peak value of fundamental phase voltage is around 5.25 kV as expected from (16) at D_{sh} equals 0.25 where an FFT spectrum clear of low order harmonics and having a fundamental of 5.25 kV is shown in Fig. 7. The difference between the actual output voltage value (5.25 kV) and the expected value (5.5 kV) is as a result of the converter losses. Finally, the output voltage of ZS-MMC is doubled to full DC supply voltage compared to a traditional MMC which can only output a peak voltage of half the DC-source. The output voltage has 9-level as a result of using 4 SMs per arm and using PD-SPWM technique as shown in Fig. 7.

It can be concluded that using the RICs technique for ZS-MMC is more advantageous compared to qZS-MMC in terms of inductor size. However, ZS-MMC has discontinuous source current compared to qZS-MMC which would need an additional filter that may require an additional inductance, an aspect usually omitted in most ZS converter papers. It may be worthwhile to investigate if the reduction in ZS inductance for the ZS-MMC can cover for the added filter extra inductance needed to provide same DC-source current ripple as produced by the qZS-MMC.

VI. CONCLUSIONS

This paper proposed a novel Z-source modular multilevel converter (ZS-MMC) topology that can achieve buck and boost voltage capabilities. The reduced inserted cells (RICs) technique has been used. Compared to quasi Z-source modular multilevel converter (qZS-MMC), RICs technique is more convenient with ZS-MMC because the inductor currents have much lower ripple allowing for smaller inductor size. The theoretical analysis has been verified by simulation.

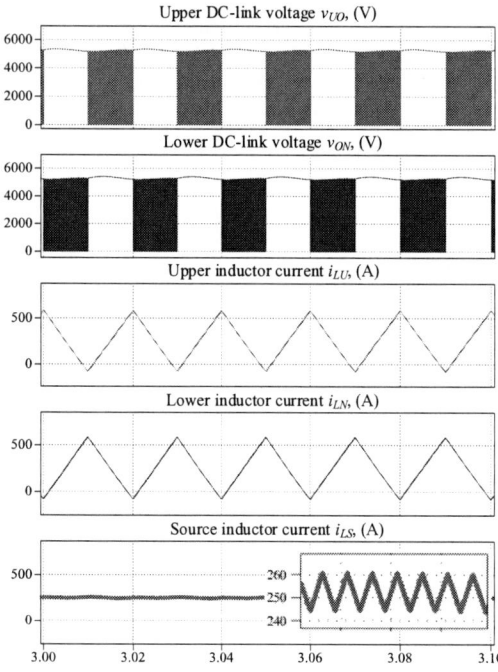

Fig. 4: The simulation results of qZS-MMC: The upper and lower DC-link voltages and inductor currents (top to bottom)

TABLE I. ZS-MMC SIMULATION MODEL PARAMETERS

Parameter	Value
Source voltage E	5.5 kV
No. of Submodules	4
MMC-arm inductance	2.5 mH
MMC-arm capacitances	3.3 mF
ZS inductances	20 mH
ZS capacitances	3mF
Switching frequency	4 kHz
RL load	10 mH, 10 Ω

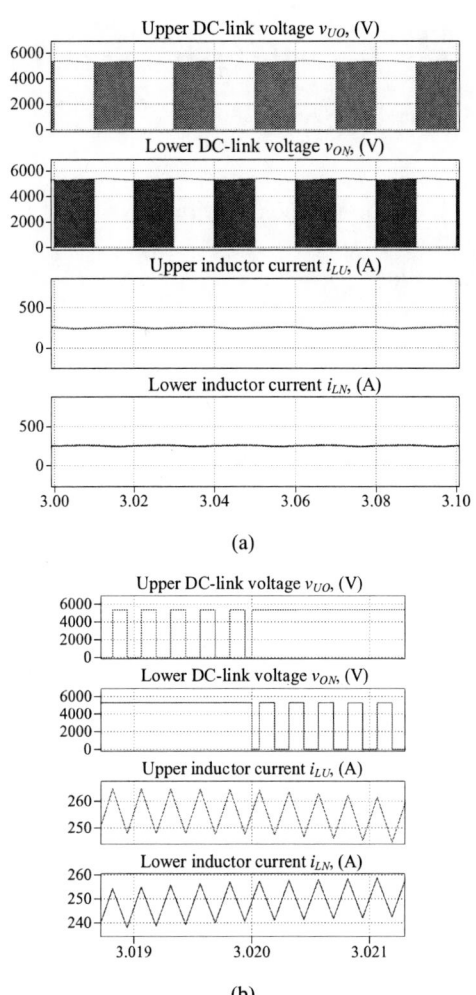

(a)

(b)

Fig. 5: The simulation results of ZS-MMC: a) The upper and lower DC-link voltages and inductor currents (top to bottom), b) their Zoom-in

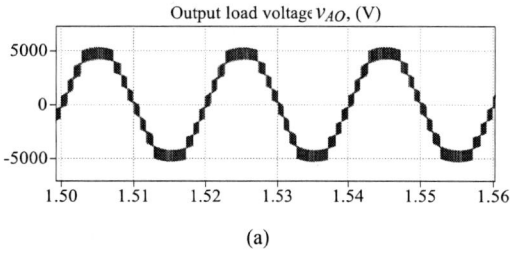

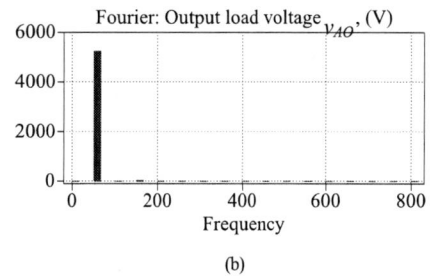

Fig. 6: The simulation results of ZS-MMC: The upper and lower sub-modules capacitor voltages.

(a)

(b)

Fig. 7: The simulation results of ZS-MMC: a) The output voltage waveform v_{ao}. b) FFT spectrum

REFERENCES

[1] A. Lesnicar and R. Marquardt, "An innovative modular multilevel converter topology suitable for a wide power range," in *2003 IEEE Bologna Power Tech Conference Proceedings*, 2003, p. 6 pp. Vol.3.

[2] M. A. Perez, S. Bernet, J. Rodriguez, S. Kouro, and R. Lizana, "Circuit Topologies, Modeling, Control Schemes, and Applications of Modular Multilevel Converters," *IEEE Transactions on Power Electronics*, vol. 30, pp. 4-17, 2015.

[3] A. Dekka, B. Wu, R. L. Fuentes, M. Perez, and N. R. Zargari, "Evolution of Topologies, Modeling, Control Schemes, and Applications of Modular Multilevel Converters," *IEEE Journal of Emerging and Selected Topics in Power Electronics*, vol. 5, pp. 1631-1656, 2017.

[4] R. Zeng, L. Xu, and L. Yao, "An improved modular multilevel converter with DC fault blocking capability," in *2014 IEEE PES General Meeting | Conference & Exposition*, 2014.

[5] S. Debnath, J. Qin, B. Bahrani, M. Saeedifard, and P. Barbosa, "Operation, Control, and Applications of the Modular Multilevel Converter: A Review," *IEEE Transactions on Power Electronics*, vol. 30, pp. 37-53, 2015.

[6] A. Nami, J. Liang, F. Dijkhuizen, and G. D. Demetriades, "Modular Multilevel Converters for HVDC Applications: Review on Converter Cells and Functionalities," *IEEE Transactions on Power Electronics*, vol. 30, pp. 18-36, 2015.

[7] M. Saeedifard and R. Iravani, "Dynamic performance of a modular multilevel back-to-back HVDC system," in *2011 IEEE Power and Energy Society General Meeting*, 2011, pp. 1-1.

[8] N. Thitichaiworakorn, M. Hagiwara, and H. Akagi, "Experimental Verification of a Modular Multilevel Cascade Inverter Based on Double-Star Bridge Cells," *IEEE Transactions on Industry Applications*, vol. 50, pp. 509-519, 2014.

[9] F. A. Khera, C. Klumpner, and P. W. Wheeler, "New modulation scheme for bidirectional qZS modular multi-level converters," *The Journal of Engineering*, pp. 3836-3841, 2019.

[10] F. A. Khera, C. Klumpner, and P. W. Wheeler, "A Comparison of Modulation Techniques for Three-phase quasi Z-Source Modular Multilevel Converter Able to Provide DC-link Fault Blocking Capability," in *2018 20th European Conference on Power Electronics and Applications (EPE'18 ECCE Europe)*, 2018, pp. P.1-P.10.

[11] F. Z. Peng, "Z-source inverter," in *Conference Record of the 2002 IEEE Industry Applications Conference. 37th IAS Annual Meeting (Cat. No.02CH37344)*, 2002, pp. 775-781 vol.2.

[12] F. A. Khera, C. Klumpner, and P. W. Wheeler, "Operation principles of quasi Z-source modular multilevel converters," in *2017 IEEE Southern Power Electronics Conference (SPEC)*, 2017, pp. 1-6.

[13] F. Gao, P. C. Loh, F. Blaabjerg, R. Teodorescu, and D. M. Vilathgamuwa, "Five-level Z-source diode-clamped inverter," *IET Power Electronics*, vol. 3, pp. 500-510, 2010.

Optimization of Robust PI Controllers for Grid-Tied Inverters

Caio R. D. Osório, Lucas C. Borin, Gustavo G. Koch, Vinícius F. Montagner

Power Electronics and Control Research Group
Federal University of Santa Maria
Santa Maria, Brazil
caio.osorio@gmail.com

Abstract—This paper contributes with a procedure that can provide PI current regulators for grid-tied inverters, with robustness against uncertain grid parameters ensured in the control design stage, without the need of *a posteriori* exhaustive tests of robust performance. The procedure is suitable for inverters with LCL output filters, taking into account the control delay and not using any approximation of lower order for the filter. The gains of fixed PI controllers are obtained offline, in an automated synthesis relying on particle swarm optimization with an objective function based on commonly used frequency domain design criteria, which is also different from similar works, based predominantly on time domain criteria. The results obtained are suitable, respecting the requirements of the IEEE 1547 Standard for grid currents, which makes the proposed procedure an alternative for providing robust PI controllers for grid-tied inverters, avoiding time consuming designs based on human-machine iterations.

Index Terms—Automated design, Grid-tied inverter, Particle swarm optimization, PI controller, Robust control.

I. INTRODUCTION

Grid-tied inverters (GTIs) are key elements in the context of renewable energy generation, allowing to properly control the power flow between the sources and the electrical grid. The current injected into the grid must comply with stringent limits of harmonic distortion, such as the requirements of the IEEE 1547 Std. Since GTIs usually comprise PWM voltage source inverters, the connection to the grid requires the inclusion of passive filters in order mitigate the high frequency harmonics. Among the commonly used filter topologies, one can highlight the LCL filters, usually preferable due to the better harmonic attenuation and size reduction of the magnetic components, when compared to the L filters [1]–[3].

The GTIs must operate properly even when subject to uncertain grid impedance at the point of common coupling, ensuring performance and stability for a set of uncertain parameters [4]. Control systems for LCL filters can be implemented in synchronous reference frame, which allows to employ proportional-integral (PI) controllers to track fundamental components with zero steady-state error [5]. Even though PI controllers are widely used for their simplicity,

This study was financed in part by the Coordenação de Aperfeiçoamento de Pessoal de Nível Superior - Brasil (CAPES/PROEX) - Finance Code 001. The authors would also like to thank the INCT-GD and the finance agencies (CNPq 465640/2014-1, CNPq 309536/2018-9, CAPES 23038.000776/2017-54 e FAPERGS 17/2551-0000517-1).

conventional designs are usually sufficient to ensure good performance under stiff grid conditions and when parametric uncertainties are not significant. However, when dealing with uncertain grid impedance, to overcome the issues related to the LCL filter inherent resonance, the low frequency gain and the crossover frequency have to be reduced to ensure stability of the system in the entire range of the parameters. Thus, the design becomes challenging, dependent on heuristic choices, usually demanding a large amount of time in a design based on trial and error [6], [7].

To overcome this, artificial intelligence can be employed in automated control tuning for power electronics applications. Among these techniques, the particle swarm optimization (PSO) deserves attention, since it has low numerical complexity and less sensitivity to the complexity of the system [8]. PSO is a population based stochastic search algorithm, which evolves guided by the minimization of an objective function, not requiring the knowledge of its derivative, and with good ability to escape local minima [9], [10]. In the specific context of PSO applied to tune PI controllers for GTI applications, one has, for instance, [11]–[14]. In [11] and [12] online optimizations are used to adapt the control gains, while [13] and [14] offline optimizations are used to tune fixed controllers. A common point in these works is not including frequency domain specifications in the objective functions, which are often employed in control design of power converters.

In this context, the key contribution of this paper is to provide an offline design procedure for robust PI controllers, with fixed gains, applied to LCL-filtered grid-tied inverters under uncertain grid impedance. The control gains are automatically tuned, overcoming the usual spent of time from manual heuristic procedures. The proposed procedure is based on the optimization of an objective function that takes into account reference values for phase margin and crossover frequency, also avoiding choices of control gains that lead to low gain margins or instability. The objective function is evaluated for the extreme values of grid impedance, and the worst case scenario is used to guide the algorithm evolution in each epoch. The search for the gains is limited to a region obtained based on the Routh condition applied to the characteristic polynomial of the closed-loop transfer functions, taking into account the parametric uncertainty. Hardware-in-the-loop based real-time results with PI controllers designed by means of the proposed procedure were obtained, showing the effectiveness and compliance with the IEEE 1547 Standard.

II. MODELING AND PROBLEM DEFINITION

Consider the three-phase inverter connected to the grid through an LCL filter, as shown in Figure 1. L_c and L_{g1} are the converter-side and grid-side filter inductances, respectively. r_c and r_{g1} are the filter parasitic resistances. C_f is the filter capacitance and R_f is the damping resistance. The grid is modeled with a background voltage v_g, grid impedance L_{g2} and grid resistance r_{g2}.

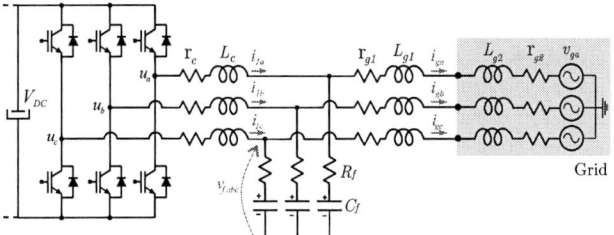

Fig. 1. LCL-filtered three-phase GTI

Based on the Kirchhoff's laws and considering the Park transformation from abc to the synchronous reference frame, it is possible to write the following equations [5], [15], [16]

$$
\begin{aligned}
u_d - v_{fd} &= L_c \frac{d}{dt} i_{1d} + r_c i_{1d} - \omega L_c i_{1q} \\
u_q - v_{fq} &= L_c \frac{d}{dt} i_{1q} + r_c i_{1q} + \omega L_c i_{1d} \\
i_{1d} - i_{gd} &= i_{fd} = C_f \frac{d}{dt}(v_{fd} - R_f i_{fd}) - \omega C_f v_{Cfq} \\
i_{1q} - i_{gq} &= i_{fq} = C_f \frac{d}{dt}(v_{fd} - R_f i_{fd}) + \omega C_f v_{Cfd} \\
v_{fd} - v_{gd} &= L_g \frac{d}{dt} i_{gd} + r_g i_{gd} - \omega L_c i_{gq} \\
v_{fq} - v_{gq} &= L_g \frac{d}{dt} i_{gq} + r_g i_{gq} + \omega L_c i_{gd}
\end{aligned}
\tag{1}
$$

where

$$
L_g = L_{g1} + L_{g2}, \quad r_g = r_{g1} + r_{g2} \tag{2}
$$

From (1), using the Laplace transform, it is possible to represent the GTI presented in Figure 1, in d-axis, by the model in Figure 2. Similar model is valid for the q-axis.

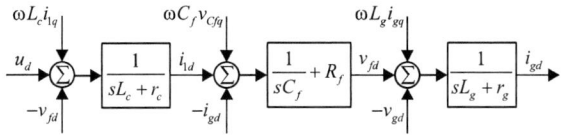

Fig. 2. MIMO model for LCL filters in synchronous reference frame [17].

Notice that the block diagram of Figure 2 represents a multiple-input multiple-output (MIMO) system with coupling between d-axis and q-axis. For modeling purposes, it is assumed that the three-phase grid voltages are sinusoidal and balanced. Thus, they can be neglected at this point ($v_{gd} = v_{gq} = 0$), being treated as external disturbances [18].

In this condition, the coupled MIMO transfer matrix of the system is given by

$$
\begin{bmatrix} i_{gd} \\ i_{gq} \end{bmatrix} = \begin{bmatrix} G_{dd} & G_{dq} \\ -G_{dq} & G_{qq} \end{bmatrix} \begin{bmatrix} u_d \\ u_q \end{bmatrix} \tag{3}
$$

For the purpose of control design, a usual way to derive the transfer function of the LCL system is to neglect the coupling terms between d-axis and q-axis, also considering them as external disturbance signals, which greatly simplify the modeling [16], [17].

As an important point, due to the uncertainty on the grid inductance, the parameter L_g is assumed from now on as uncertain, lying in bounded interval whose limits are provided, using the notation for real interval parameter $[L_g] \in [L_{gmin}, L_{gmax}]$. Differently from most of the works in the literature, here the interval of uncertainty in the grid parameter will be carried out in the control design stage, to obtain a controller robust also against parameter uncertainties.

In this context, the single-input single-output transfer function, valid for both d-axis and q-axis, is given by

$$
G(s) = \frac{i_g(s)}{u(s)} = \frac{a_1 s + 1}{[b_3]s^3 + [b_2]s^2 + [b_1]s + b_0} \tag{4}
$$

where

$$
\begin{aligned}
a_1 &= C_f R_f \\
b_3 &= C_f L_c [L_g] \\
b_2 &= C_f (L_c + [L_g]) R_f + C_f L_c r_g + C_f [L_g] r_c \\
b_1 &= L_c + [L_g] + R_f r_g C_f + C_f r_c (R_f + r_g) \\
b_0 &= r_g + r_c
\end{aligned}
\tag{5}
$$

and $[b_3]$, $[b_2]$ and $[b_1]$ are interval coefficients that depend on the grid inductance L_g.

Notice that neglecting the coupling terms allows to simplify the modeling and also to control the system with a single current loop control, which requires a small number of sensors. However, the actual systems can loose performance when the interaction between the axes appear [17], [19]. A feasible approach in this regard is to feedforward compensation decoupling terms (dec) aiming to mitigate the dynamic effect of the inherent coupling. Moreover, since in practical conditions grid voltages are usually distorted and may be unbalanced, the grid voltages can also be feedforwarded within the control structure to improve the dynamics of the closed-loop system [18].

A single-loop grid current feedback control is shown in Figure 3, for d-axis, including the feedforward terms. A PI controller is employed and a Padé approximation for the delay effect is also included, where $\alpha = T_s/4$, being T_s the sampling period. The same block diagram is valid for the q-axis grid current control.

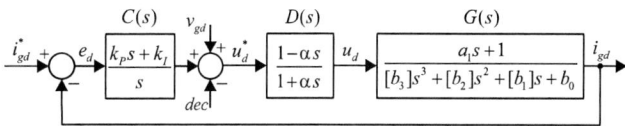

Fig. 3. Block diagram of the closed-loop system, including PI controller, Padé approximation for the delay and plant with uncertain parameters.

978-1-7281-4181-7/19 $31.00 © 2019 IEEE

Considering the block diagram of Figure 3, it should be noted that the single-loop control method for LCL filters is very susceptible to the resonance peak, being performance and stability highly dependent on the damping of the system. Moreover, the stability and performance are also highly dependent on parameter uncertainties, since the low frequency gain and control bandwidth must be reduced to ensure stability of the system in the entire range of the parameters [7], [17]. In this context, the system performance and stability margins are usually specified by the crossover frequency (ω_c), phase margin (PM) and gain margin (GM) of the system [20].

Control Problem Definition

The control problem to be solved here is to synthesize, with an offline procedure, fixed gains for PI controllers (in d and q axes) that ensure, for both extreme situations of grid inductance, phase margin and crossover frequency as close as possible of reference values, defined by PM^* and ω_c^*, and gain margin not less than a prescribed value, defined by GM^*.

III. PROPOSED DESIGN SOLUTION

This section proposes a solution, based on a PSO algorithm, for the control problem stated in the previous section. For the application of this algorithm in the design of a fixed gain PI controller, one has a search space of dimension 2 (gains k_P and k_I). Each possible solution of the optimization problem can be seen as a particle ℓ, with position x_ℓ defined by

$$x_\ell = [k_{I\ell} \quad k_{P\ell}] \tag{6}$$

The set of possible solutions is called a swarm of particles and, based on the evaluation of an objective function and the computation of the velocity for each particle, the swarm moves in the search space as the algorithm evolves.

The velocity of each particle is given by

$$v_\ell(k+1) = \lambda v_\ell(k) + \phi_1 \text{rand1}()(P_{\ell.best} - x_\ell(k)) \\ + \phi_2 \text{rand2}()(G_{best} - x_\ell(k)) \tag{7}$$

and then, the position is updated as

$$x_\ell(k+1) = x_\ell(k) + v_\ell(k+1) \tag{8}$$

where k represents a given iteration (epoch) of the algorithm.

The scalar λ represents the inertia, which is the tendency to keep the velocity previously calculated. ϕ_1 is the cognitive factor, which determines how the particle movement is influenced by $P_{\ell.best}$, its best position considering all epochs until the present calculation. In a similar way, ϕ_2 is the social factor, which determines how the particle movement is influenced by G_{best}, the global best position of the swarm considering all epochs. The inclusion of the random functions rand1 and rand2 has the objective of maintaining a certain randomness in the behavior of the particle with respect to itself and with respect to the swarm, respectively.

For the PI controller, in each epoch, each particle is evaluated based on its current position, which is the pair of gains in (6). For that, the open loop transfer function

$$T(s) = C(s)D(s)G(s) \tag{9}$$

of the system in Figure 3 is updated with those control gains, and then the following objective function is computed

$$OF_j = \{|PM^* - PM_j| + \gamma_1 |\omega_c^* - \omega_{cj}|\} \gamma_2, \quad j = 1, 2.$$
$$OF = \max(OF_1, OF_2) \tag{10}$$

In this objective function, PM^* and ω_c^* are the reference values for the phase margin and the crossover frequency, while PM and ω_c are the values obtained for the position under evaluation, in the open-loop transfer function $T(s)$, in (9). Assuming L_g as an uncertain parameter, $j = 1$ and $j = 2$ represent the minimum and maximum values, respectively. The weight γ_1 is included to make the order of magnitude of the summed terms compatible, while γ_2 is included as a penalty for the following undesired conditions: if the closed-loop system has unstable poles or if the gain margin (GM) is lower than a prescribed limit.

In order to increase the algorithm efficiency, the search for the gains is limited to a region obtained based on the Routh condition applied to the characteristic polynomial of the closed-loop transfer function. For the system shown in Figure 3, the characteristic polynomial is given by

$$p(s) = [p_5]s^5 + [p_4]s^4 + [p_3]s^3 + [p_2]s^2 + p_1 s + p_0 \tag{11}$$

where

$$[p_5] = [b_3]\alpha, \quad [p_4] = [b_3] + [b_2]\alpha, \quad [p_3] = [b_2] + [b_1]\alpha - a_1\alpha k_P,$$
$$[p_2] = [b_1] + b_0\alpha + a_1 k_P - \alpha a_1 k_I - \alpha k_P,$$
$$p_1 = b_0 + a_1 k_I + k_P - \alpha k_I, \quad p_0 = k_I \tag{12}$$

Considering the positivity of the coefficients, which is a necessary condition for Hurwitz stability, the following inequalities can be obtained

$$(a_1\alpha)k_P < [b_2][b_1]\alpha$$
$$(\alpha - a_1)k_P + (\alpha a_1)k_I < [b_1] + b_0\alpha \tag{13}$$
$$(-a_0)k_P + (a_0\alpha - a_1)k_I < b_0$$

This inequalities are used to define the search space of the control gains, as will be shown in the case study.

The execution of the proposed procedure can be summarized as:

a. Define the system parameters, given by nominal values or intervals;
b. Choose reference values PM^*, ω_c^* and the minimum value for GM, and the weights for the objective function;
c. Define the search space limits, given by $k_{Pmin} \leq k_P \leq k_{Pmax}$ and $k_{Imin} \leq k_I \leq k_{Imax}$;
d. Define the PSO configuration parameters;
e. Run algorithm ;
f. Test the obtained control gains in simulation. If the control gains are suitable for practical application, then stop. Else, if the PSO has not converged, return to step d and redefine the PSO parameters. If the PSO has converged but with unsuitable results, return to step b and redefine specifications.

978-1-7281-4181-7/19 $31.00 © 2019 IEEE

IV. CASE STUDY

In order to show the effectiveness of the proposed procedure, consider the system shown in Figure 1, with the parameters presented in Table I.

TABLE I
SYSTEM PARAMETERS

System description	
Switching frequency f_{sw}	10020 Hz
Sampling frequency f_s	20040 Hz
DC-link V_{dc}	400 V
Grid voltage V_g	220 Vrms, 60 Hz
Converter inductance L_c	1 mH
Grid-side inductance L_{g1}	0.3 mH
Filter capacitor C_f	62 μF
Max. grid inductance L_{g2max}	1.5 mH
Min. grid inductance L_{g2min}	0.1 mH
Grid resistance r_g	0.1 Ω
Damping resistance R_f	1 Ω

The parasitic resistances r_c and r_{g1} are neglected. The grid inductance is an uncertain parameter lying the interval from 0.1 mH to 1.5 mH. Thus, in this case study, the ratio between the resonance and sampling frequencies is lower than 1/6 for the entire range of parameters. Therefore, the LCL filter resonance needs to be damped to achieve a stable operation with current feedback PI control [18], [21], [22].

The reference values for system performance and stability margins, in this case study, are specified as $\omega_c{}^* = 6000$ rad/s and $PM^* = 60^o$. The weight γ_1 in the objective function is choses as 10^{-2}. The weight γ_2 is chosen as 10^6 to penalize unstable systems or systems with $GM < 5$ (14 dB).

The search space for the gains k_P and k_I is defined based on the inequalities in (13), which lead to the regions presented in Figure 4, for both L_{gmin} and L_{gmax}. In order to inform to the algorithm the limits of the search space as simple upper and lower values, a rectangular region with edges parallel to the axes is considered. It should be pointed out that belonging to this search space is a necessary (but no sufficient) condition to the system stability over the entire range of parameters.

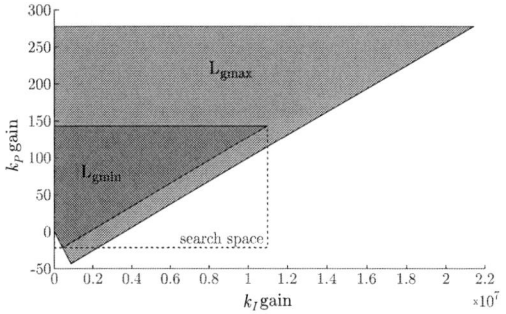

Fig. 4. Definition of the search space for the PI control gains.

The configuration parameters chosen for the PSO algorithm are: swam size = 500 particles, number of iterations = 30

epochs, $\phi_1 = 0.5$ and $\phi_2 = 0.5$. The swarm is randomly initialized, and evolves with the fitness curve presented in Figure 5. As the search space contains also particles in positions that represent unstable control gains, they are penalized in the objective function with higher values. After 7 epochs, there are stable particles in the swarm, and the algorithm keeps running until converge to a minimum objective function value.

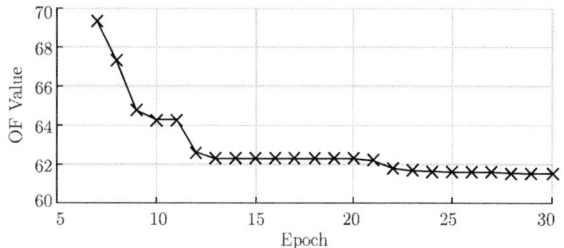

Fig. 5. Evolution of the objective function over the epochs.

To verify the repeatability of the results, the algorithm was executed 10 times, having always converged in about 3 minutes to the solution

$$k_P = 0.8759638104404, \quad k_I = 301.6784610842813$$
$$GM_{min} = 5.01, \quad PM_{min} = 54.34 \text{ deg}, \quad f_{cmin} = 54.3 \text{Hz} \quad (14)$$

with a standard deviation limited to 1%.

The system in Figure 3 was simulated with the control gains in (14). The frequency response of the open-loop transfer function (9) is shown in Figure 6, for L_{gmin} and L_{gmax}, which confirms system stability for the entire range.

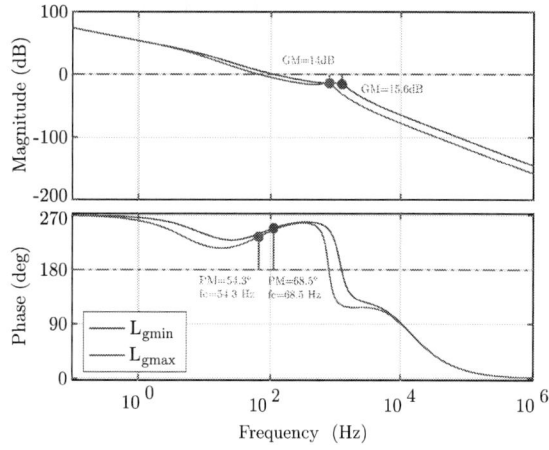

Fig. 6. Bode diagram from i_{ref} to i_g, for the closed-loop system with control gains in (14).

The step responses of the closed loop system for both conditions are shown in Figure 7.

Finally, the closed-loop poles in Figure 8 confirms the stability of the closed-loop system over the entire range of grid parameters, from L_{gmin} to L_{gmax}. For better visualization, the high frequency real pole it is omitted in this figure.

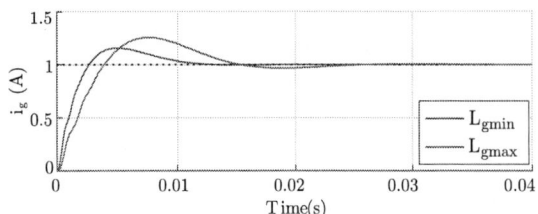

Fig. 7. Step responses for L_{gmin} and L_{gmax}, with control gains in (14)

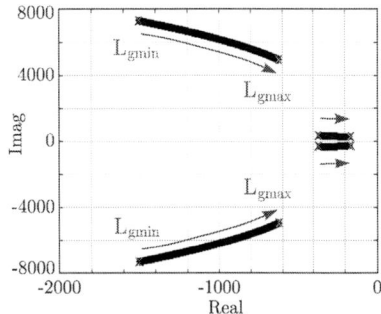

Fig. 8. Closed-loop poles with control gains in (14), for a sweep from L_{gmin} to L_{gmax}.

It should be noted that it would be possible to improve the dynamic performance of the system using the same algorithm if the design was carried out considering only a nominal condition for the grid inductance. For instance, if only L_{gmin} was considered, it would be possible to further optimize the objective function, achieving the reference value for the phase margin with a significant increase in the crossover frequency. However, it would not be possible to guarantee, in the design stage, stability for the entire range of grid inductances, as is the case here.

V. Experimental Results based on HIL

In this section, real-time tests based on hardware-in-the-loop (HIL) are presented to validate the effectiveness of the design procedure proposed in Section III. The LCL-filtered GTI in Figure 1 was emulated using a Typhoon HIL, model 402, and the parameters in Table I. The dq-axes controllers are implemented using a digital signal processor (DSP) TMS32F28335, from Texas Instruments. The interface between DSP, HIL and osciloscope is carried out by a μ-grid interface board.

For the digital implementation, consider the discretization of the PI controller in (14) using the Tustin method, with the sampling frequency given in Table I. The discrete control gains are given by

$$k_{Pk} = 0.8834907181521, \quad k_{Ik} = -0.8684369027287 \quad (15)$$

and are implemented recursively, including the one sample delay, by the following equation

$$u(k+1) = u(k) + k_{Pk}\, e(k) + k_{Ik}\, e(k-1) \quad (16)$$

For the results presented here, the controller (16) is implemented without the feedforward terms shown in Figure 3,

without significant loss of performance. In order to validate the system robustness and performance, sudden variations in the grid current references (d and q axes) are performed, considering the extremes of the uncertain parameter L_g.

The references and closed-loop system responses are presented in Figure 9, for L_{gmin}, and in Figure 10, for L_{gmin}. In these figures, the first variation represents the start-up of the system, injecting active power into the grid, while the second variation represents a transient from active to reactive power.

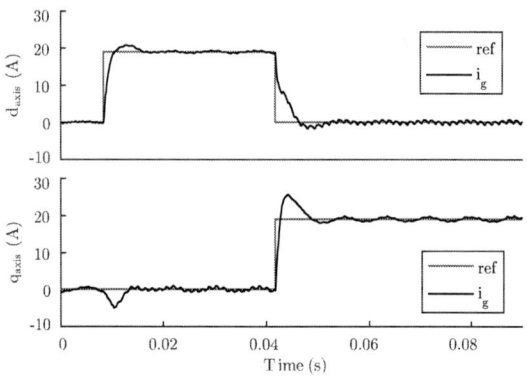

Fig. 9. Step responses in dq-axis (active and reactive power) for L_{gmin}.

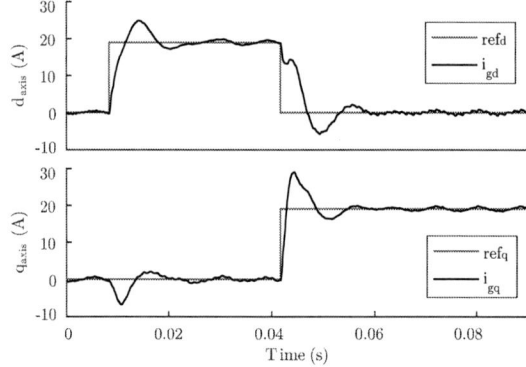

Fig. 10. Step responses in dq-axes (active and reactive power) for L_{gmax}.

Figure 11 presents the three-phase grid currents with respect to the dq-axes signals presented in Figure 9. From Figures 9 to 11, it is possible to verify that the closed-loop system is able to track the references with suitable settling times, for both grid conditions. The transient responses for L_{gmax} are slower, representing a small loss of performance due to the need of ensuring robustness over the entire range of parameters with a simple fixed gain controller.

The grid currents in steady-state are highlighted in Figure 12(a), for operation with L_{gmin}. To verify that the proposed controller is able to properly synthesize grid currents with low harmonic distortion, Figure 12(b) shows the harmonic spectra for the current in one of the phases in this condition (THD=2.25%). The THD and the individual harmonics comply with the requirements of the IEEE 1547 Standard. Similar results are verified for operation with L_{gmax}.

978-1-7281-4181-7/19 $31.00 © 2019 IEEE

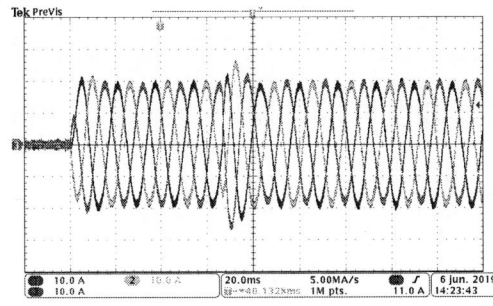

Fig. 11. Three-phase grid currents with respect to the dq-axes signals presented in Figure 9.

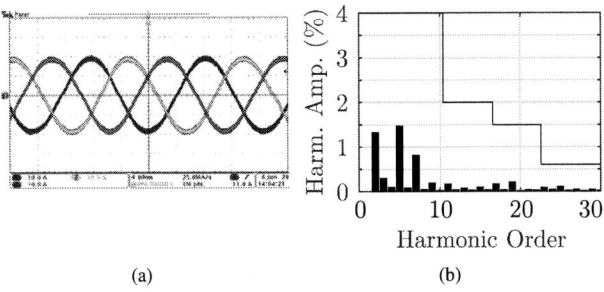

(a)   (b)

Fig. 12. Steady state responses for L_{gmin}: (a) three-phase grid currents; (b) current harmonic spectrum and limits from IEEE 1547.

VI. CONCLUSION

This paper presented an automated design procedure for robust PI controllers applied to GTIs under uncertain grid impedance. The proposed procedure can be seen as an alternative for systematic design of fixed gains PI controllers for this application, avoiding time consuming processes for control engineers, which is an issue specially considering designs under uncertain parameters. The control gains were tuned with a PSO algorithm, that optimizes an objective function comprised by important frequency domain criteria, given by crossover frequency, phase margin and gain margin. A case study was presented, showing that the procedure can provide control gains in reasonable computation time and with good repeatability for different executions. Real-time results based on HIL were given, showing that the proposed procedure leads to controllers that ensure, for a set of plant parameters, stable operation, suitable dynamic performance and also compliance with the harmonic distortion requirements of the IEEE 1547 Standard.

REFERENCES

[1] IEEE:1547 standard for interconnecting distributed resources with electric power systems, 2011.

[2] R. Teodorescu, M. Liserre, and P. Rodríguez. *Grid Converters for Photovoltaic and Wind Power Systems*. Wiley - IEEE. John Wiley & Sons, 2011.

[3] M. Ben Saïd-Romdhane, M. W. Naouar, I. Slama-Belkhodja, and E. Monmasson. Robust active damping methods for lcl filter-based grid-connected converters. *IEEE Transactions on Power Electronics*, 32(9):6739–6750, Sep. 2017.

[4] M. Lu, A. Al-Durra, S. M. Muyeen, S. Leng, P. C. Loh, and F. Blaabjerg. Benchmarking of stability and robustness against grid impedance variation forlcl-filtered grid-interfacing inverters. *IEEE Transactions on Power Electronics*, 33(10):9033–9046, Oct 2018.

[5] Xianwen Bao, Fang Zhuo, Yuan Tian, and Peixuan Tan. Simplified feedback linearization control of three-phase photovoltaic inverter with an lcl filter. *Power Electronics, IEEE Transactions on*, 28(6):2739–2752, June 2013.

[6] X. Chen, C. Y. Gong, H. Z. Wang, and L. Cheng. Stability analysis of lcl-type grid-connected inverter in weak grid systems. In *2012 International Conference on Renewable Energy Research and Applications (ICRERA)*, pages 1–6, Nov 2012.

[7] D. Pan, X. Ruan, C. Bao, W. Li, and X. Wang. Optimized controller design for LCL-type grid-connected inverter to achieve high robustness against grid-impedance variation. *IEEE Transactions on Industrial Electronics*, 62(3):1537–1547, March 2015.

[8] Y. A. I. Mohamed and E. F. El Saadany. Hybrid variable-structure control with evolutionary optimum-tuning algorithm for fast grid-voltage regulation using inverter-based distributed generation. *IEEE Transactions on Power Electronics*, 23(3):1334–1341, May 2008.

[9] Russell Eberhart and James Kennedy. A new optimizer using particle swarm theory. In *In Proceedings of the Sixth International Symposium on Micro Machine and Human Science.*, pages 39–43. IEEE, 1995.

[10] I. Chung, W. Liu, D. A. Cartes, E. G. Collins, and S. Moon. Control methods of inverter-interfaced distributed generators in a microgrid system. *IEEE Transactions on Industry Applications*, 46(3):1078–1088, May 2010.

[11] A. Althobaiti, M. Armstrong, and M. A. Elgendy. Control parameters optimization of a three-phase grid-connected inverter using particle swarm optimisation. In *8th IET International Conference on Power Electronics, Machines and Drives (PEMD 2016)*, pages 1–6, April 2016.

[12] W. Al-Saedi, S. W. Lachowicz, and D. Habibi. An optimal current control strategy for a three-phase grid-connected photovoltaic system using particle swarm optimization. In *2011 IEEE Power Engineering and Automation Conference*, volume 1, pages 286–290, Sep. 2011.

[13] M. A. Hassan and M. A. Abido. Optimal design of microgrids in autonomous and grid-connected modes using particle swarm optimization. *IEEE Transactions on Power Electronics*, 26(3):755–769, March 2011.

[14] F. M. de Oliveira, S. A. Oliveira da Silva, F. R. Durand, L. P. Sampaio, V. D. Bacon, and L. B. G. Campanhol. Grid-tied photovoltaic system based on pso mppt technique with active power line conditioning. *IET Power Electronics*, 9(6):1180–1191, 2016.

[15] Ramu Krishnan. *Permanent Magnet Synchronous and Brushless DC Motor Drives*. Mechanical Engineering (Marcel Dekker). CRC Press, 1 edition, September 2009.

[16] P. Xuetao, Y. Tiankai, Q. Keqing, Z. Jinbin, L. Wenqi, and C. Xuhui. Analysis and evaluation of the decoupling control strategies for the design of grid-connected inverter with lcl filter. In *International Conference on Renewable Power Generation (RPG 2015)*, pages 1–6, Oct 2015.

[17] D. Sivadas and K. Vasudevan. Stability analysis of three-loop control for three-phase voltage source inverter interfaced to the grid based on state variable estimation. *IEEE Transactions on Industry Applications*, 54(6):6508–6518, Nov 2018.

[18] M. Hanif, V. Khadkikar, W. Xiao, and J. L. Kirtley. Two degrees of freedom active damping technique for lcl filter-based grid connected pv systems. *IEEE Transactions on Industrial Electronics*, 61(6):2795–2803, June 2014.

[19] J. Dannehl, C. Wessels, and F.W. Fuchs. Limitations of voltage-oriented PI current control of grid-connected PWM rectifiers with filters. *Industrial Electronics, IEEE Transactions on*, 56(2):380 –388, feb. 2009.

[20] C. Bao, X. Ruan, X. Wang, W. Li, D. Pan, and K. Weng. Step-by-step controller design for lcl-type grid-connected inverter with capacitor–current-feedback active-damping. *IEEE Transactions on Power Electronics*, 29(3):1239–1253, March 2014.

[21] J. Dannehl, F. W. Fuchs, S. Hansen, and P. B. Thogersen. Investigation of active damping approaches for PI-based current control of grid-connected pulse width modulation converters with LCL filters. *IEEE Transactions on Industry Applications*, 46(4):1509–1517, July 2010.

[22] R. Errouissi and A. Al-Durra. Design of pi controller together with active damping for grid-tiedlcl-filter systems using disturbance-observer-based control approach. *IEEE Transactions on Industry Applications*, 54(4):3820–3831, July 2018.

978-1-7281-4181-7/19 $31.00 © 2019 IEEE

Investigation of Voltage Regulation with Active and Reactive Power with Distributed Loads on a Radial Distribution Feeder

Felipe Joel Zimann
Alessandro Luiz Batschauer, Marcello Mezaroba
Department of Electrical Engineering
Santa Catarina State University
Joinville – SC, Brazil
felipezimann@ieee.org
{alessandro.batschauer ; marcello.mezaroba}@udesc.br

Eduardo Vasconcelos Stangler
Francisco de Assis dos Santos Neves
Power Electronics and Electric Drives Group
Federal University of Pernambuco
Recife – PE, Brazil
eduardo.vasconcelosstangler@ufpe.br
fneves@ufpe.br

Abstract—**This paper addresses the voltage-drop problem in a low-voltage network with distributed loads. A radial distribution feeder is used to show how voltage drop occurs in such systems. Theoretical analysis shows why traditional reactive-power approaches are not fully effective in resistive grids. Power management is used to coordinate between active and reactive power for proper voltage regulation. The control strategy ensures voltage regulation with reactive and active power under a high R/X ratio and load change conditions. The control method is evaluated in a low-voltage distribution network, and the simulation results validate such method's effectiveness.**

Index Terms—**Active and reactive power, low-voltage network, power management, voltage regulation, weak grid.**

I. INTRODUCTION

Reliability and quality are both fundamental requirements in electric distribution systems. This concerns not only sensitive loads, but also the end users [1]. Low-quality power effects are well documented and include voltage sags and swells, voltage unbalance, momentary interruptions, distorted waveform, and under and over voltages [2]–[5]. Voltage delivered to end users and utilities is one of many parameters used to measure power quality. Consumers located at the end of distribution lines are more likely to experience voltage-reduction issues [6].

Furthermore, reactive control strategies, such as a distribution static synchronous compensator (DSTATCOM), are not fully effective to maintain a regulated voltage profile in a low–voltage (LV) distribution network [7]. Excessive reactive power leads to higher losses in the feeder and a lower power factor in the LV network [6]. Active power is a more suitable approach for voltage regulation in grids with a high R/X ratio [8], [9]. Many solutions have been proposed to solve voltage-drop/rise issues [9]–[13]. In [9],

This work was developed with support from the *Programa Nacional de Cooperação Acadêmica da Coordenação de Aperfeiçoamento de Pessoal de Nível Superior* – CAPES/Brazil. The authors would like to thank CNPq (465640/2014-1), CAPES (23038.000776/2017-54), FAPERGS (17/2551-0000517-1), PROCAD (88881.068414/2014-01), INCT–GD, FITEJ, FAPESC, FACEPE, UFPE and UDESC for their financial support.

a hybrid power solution for voltage regulation is used with active and reactive power, but there is no priority for active power usage. In [10], stored energy is used for voltage power-quality improvement, although excessive charge and discharge cycles may reduce battery lifetime. In [11], stored photovoltaic generation for energy shifting and solar smoothing strategies is used, although it does not use system inactivity, i.e. nighttime or cloudy days, to coordinate between reactive and active power.

This paper uses a coordinated control between reactive and stored active energy for voltage regulation in a LV distribution feeder as presented in [13] and evaluates the control strategy in another scenario. A LV distribution feeder with typical resistive behavior and loads connected at different locations is used. The rms voltage value is evaluated at different line points according to the position chosen for the voltage regulator. Moreover, additional connection points are discussed based on voltage regulation within the standard limits.

This paper is organized as follows: In Section II, the description and analysis of the voltage- drop issue are presented. In Section III, the voltage-regulation strategy is presented and discussed. In Section IV, the control strategy and the power management approach are introduced. Section V studies the power converter positioning based on rms voltage regulation with a typical voltage-drop event reproduced through simulations, with problem discussion and analysis. Finally, Section VI concludes the paper.

II. VOLTAGE PROFILE IN DISTRIBUTION FEEDER

The rms voltage at the end of a distribution line is calculated based on an equivalent circuit, as shown in Fig. 1. The equivalent impedance represents both the distribution lines' and the transformer's impedances, and the loads are connected at the point of common connection (PCC), represented by Z_{load} at the end of the line. The power converter is connected to the same PCC location for proper voltage regulation.

978-1-7281-4181-7/19 $31.00 © 2019 IEEE

The equations which represent the equivalent circuit are:

$$\vec{v}_g = \vec{v}_{pcc} + \vec{v}_l = \vec{v}_{pcc} + \vec{Z}_g \cdot \left(\vec{i}_{load} - \vec{i}_F \right) \qquad (1)$$

$$\vec{v}_g \vec{v}_{pcc}^* = \vec{v}_{pcc} \vec{v}_{pcc}^* + \vec{Z}_g \cdot \left(\vec{S}_{load}^* - \vec{S}_F^* \right) \qquad (2)$$

where $\vec{v}_g$ represents the feeder voltage, $\vec{v}_{pcc}$, the PCC voltage at the end of the line, $\vec{v}_l$ the drop voltage at the impedance line, $\vec{i}_{load}$ the load current, $\vec{i}_F$ the converter current, $\vec{Z}_g$ the line impedance, $\vec{S}_{load}$ the apparent power consumed by the load and $\vec{S}_F$ the apparent power supplied by the converter. After some algebraic manipulations, an expression for the PCC voltage can be obtained, represented by:

$$V_{pcc}^4 + V_{pcc}^2 \left(2A - V_g^2 \right) + B^2 + A^2 = 0 \qquad (3)$$

where:

$$\begin{cases} A = R_g \cdot \left(P_{load} - P_F \right) + X_g \cdot \left(Q_{load} - Q_F \right) & (4) \\ B = R_g \cdot \left(Q_{load} - Q_F \right) - X_g \cdot \left(P_{load} - P_F \right) & (5) \end{cases}$$

The condition (3) is satisfied by the function's roots (6):

$$V_{pcc}' = \sqrt{ \frac{ V_g^2 - 2A \pm \sqrt{ \left(2A - V_g^2 \right)^2 - 4(A^2 + B^2) } }{2} } \qquad (6)$$

Among the four solutions of (6), only the two positive ones are suitable, since a negative rms voltage value does not have a physical meaning. Among those, one is unstable and the other is the stable rms voltage value at the PCC. Distributed lines have a maximum load point, called the critical point or the nose point [14]. This point indicates the limit for maximum power transfer under adequate and secure operation and, for clear reasons, the system should work above such limit, as shown in Fig. 2.

Fig. 3 (a) shows the feeder equivalent circuit used to obtain the final impedance value at the end of a distribution line. Fig. 3 (b) shows the impedance versus feeder distance along the line. Initially, only the transformer's impedance can be seen with a lower R/X ratio. The resistance increases with the distance and the R/X ratio gets higher, mainly due to the feeder cables' resistive characteristic. Therefore, the distribution lines are more resistive with greater distances. From such analysis, it can be concluded that the technique proposed in [13] uses more active power when located at the end of long lines and, probably, less reactive power in points near the feeder.

In this analysis, a 30 kVA/380 V power transformer feeder with a 3.5 % base impedance and 4 AWG distribution cables with $R_g = 1.7179\,\Omega/\text{km}$ and $X_g = 0.4495\,\Omega/\text{km}$ is considered based on a real network. The impedance values for the transformer and the distribution cables are shown in Table I.

TABLE I. Total impedance values.

Parameter	Resistance	Inductance	Ratio
Cables	687.2 mΩ	476.9 μH	3.822
Transformer	87.4 mΩ	381.96 μH	0.607
Total	774.6 mΩ	858.9 μH	2.392

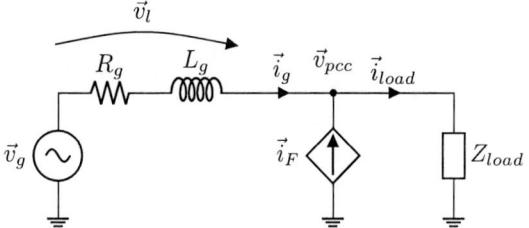

Fig. 1. Single–phase equivalent model.

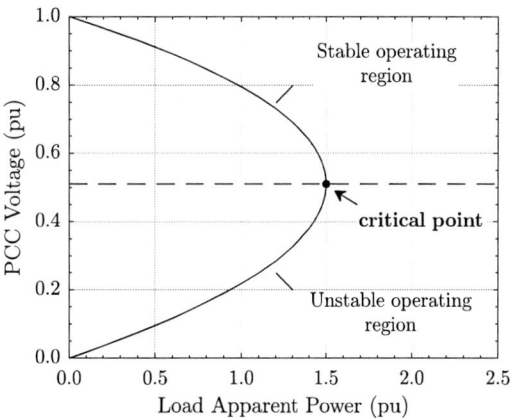

Fig. 2. Relationship between voltage and power load.

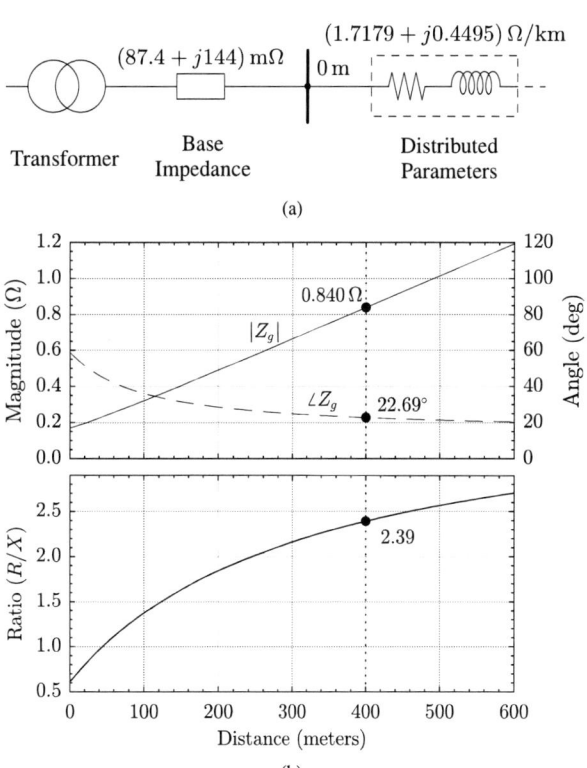

Fig. 3. Impedance along the feeder: (a) Equivalent circuit; (b) Magnitude, angle and ratio of impedance along the feeder.

TABLE II. Voltage limit ratings.

Rating	Voltage Limits
Adequate	$0.921\,\mathrm{pu} \le v_o \le 1.047\,\mathrm{pu}$
Precarious	$0.866\,\mathrm{pu} \le v_o < 0.921\,\mathrm{pu}$ $1.047\,\mathrm{pu} < v_o \le 1.063\,\mathrm{pu}$
Critical	$v_o < 0.866\,\mathrm{pu}\;;\;v_o > 1.063\,\mathrm{pu}$

III. Voltage Regulation Strategy

The voltage regulation strategy is based on active and reactive power injection at the point of common connection. Traditionally, only reactive power is used to raise the rms voltage the way DSTATCOMs work. However, due to the high R/X ratio of the line feeder, reactive power will not be effective, leading to a poor quality and undesirable voltage level. To overcome this issue, active power is used along with the reactive power, since it is more effective and provides better results for voltage regulation in such networks. The operation is summarized below:

- Initially, reactive power is used until it is no longer possible to increase the rms voltage to the desired level;
- From that moment on, active power is used until an acceptable voltage regulation is achieved;
- The reactive power is reduced as the active power increases due to the converter's power limit.

Under these conditions, active power will only be used when the reactive power is not enough to establish an adequate rms voltage level. Active power needs some type of energy-storage system (ESS), such as batteries (BESS), which are the most widely used storage device for power system applications. For that reason, this strategy prioritizes reactive power over active power, which increases battery lifetime due to the reduced number of cycles and lower discharge rates [15].

Table II shows Brazilian standard voltage ratings [16], which are based on international standards [17]–[19]. There are three ratings for rms voltage: Adequate levels, which means no losses for consumers and no active corrections required by the distributor, Precarious and Critical, which mean corrective actions are required and affected consumers must be refunded.

In Fig. 4, voltage rms curves at the PCC with no correction are shown. The rms voltage value is lower with a higher load power, due to the higher losses in the feeder cables. Therefore, in this system, the voltage is dependent on feeder impedance and load current. In order to correct the voltage, at least one of the variables must be changed. Since the load characteristic or the distribution network can't be changed, a power converter can be allocated at the PCC.

In the described case, the power converter uses reactive power and stored active energy, i.e. from batteries, to raise voltage levels. This process can be seen in Fig. 5 for three different voltage settings. First, on Fig. 5 (a), the minimum acceptable voltage is the lowest value in the adequate rating. In this case, only loads with a power factor (FP) higher than 0.5 need stored active energy to raise the rms voltage level. In Fig. 5 (b), a case with a higher voltage level $v_{min} = 0.95\,\mathrm{pu}$

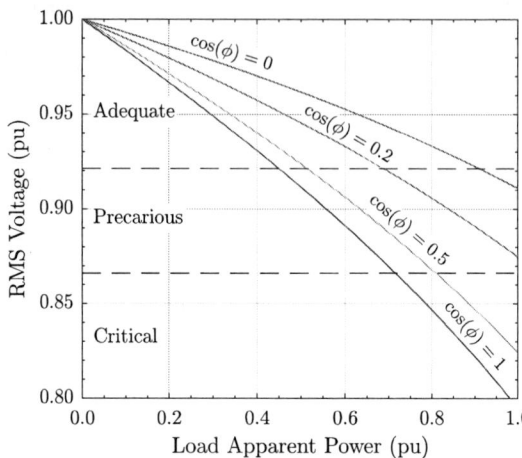

Fig. 4. Relationship between voltage and load apparent power.

is shown where active power is used even before FP = 0.5. Finally, cases where the voltage is very close to the nominal value are shown in Fig. 5 (c). In this case, active power is most demanded. All three cases consider a 400-meter-long feeder to calculate rms voltage levels. Voltage drop is even worse in longer lines with the predominantly resistive characteristic of the cables, which needs more active power for actual voltage raise.

IV. Control Strategy and Power Management

In order to mitigate the voltage drop, the proposed system controls the converter's output current and provides reactive and active power to the grid. The control strategy is based on two different control loops, inner and outer. The inner loop provides a sinusoidal reference current with low steady–state error at the fundamental frequency and high rejection to harmonic distortion. The outer loop controls the rms voltage regulation and provides a current reference for the inner loop. It is a low-bandwidth control to provide decoupling with the inner loop, which has a high dynamic bandwidth. Power management is performed inside the outer loop, where the references of in–phase and quadrature currents are changed to meet the requirements of the described strategy.

A. Inner control loop

The inner loop is a current control loop, which uses the inductor's output current for a proper control design. An equivalent single–phase circuit as shown in Fig. 6 (a) is used to obtain the current-transfer function (7):

$$G_i(s) = \frac{i_{L2}(s)}{d(s)} = \frac{E}{L_1 L_2 C_1 \cdot s^3 + (L_1 + L_2) \cdot s} \quad (7)$$

The current control uses natural reference (abc) to control individual per-phase currents as three individual single–phase systems. A practical proportional–resonant (PR) controller is used to improve reference tracking and disturbance rejection

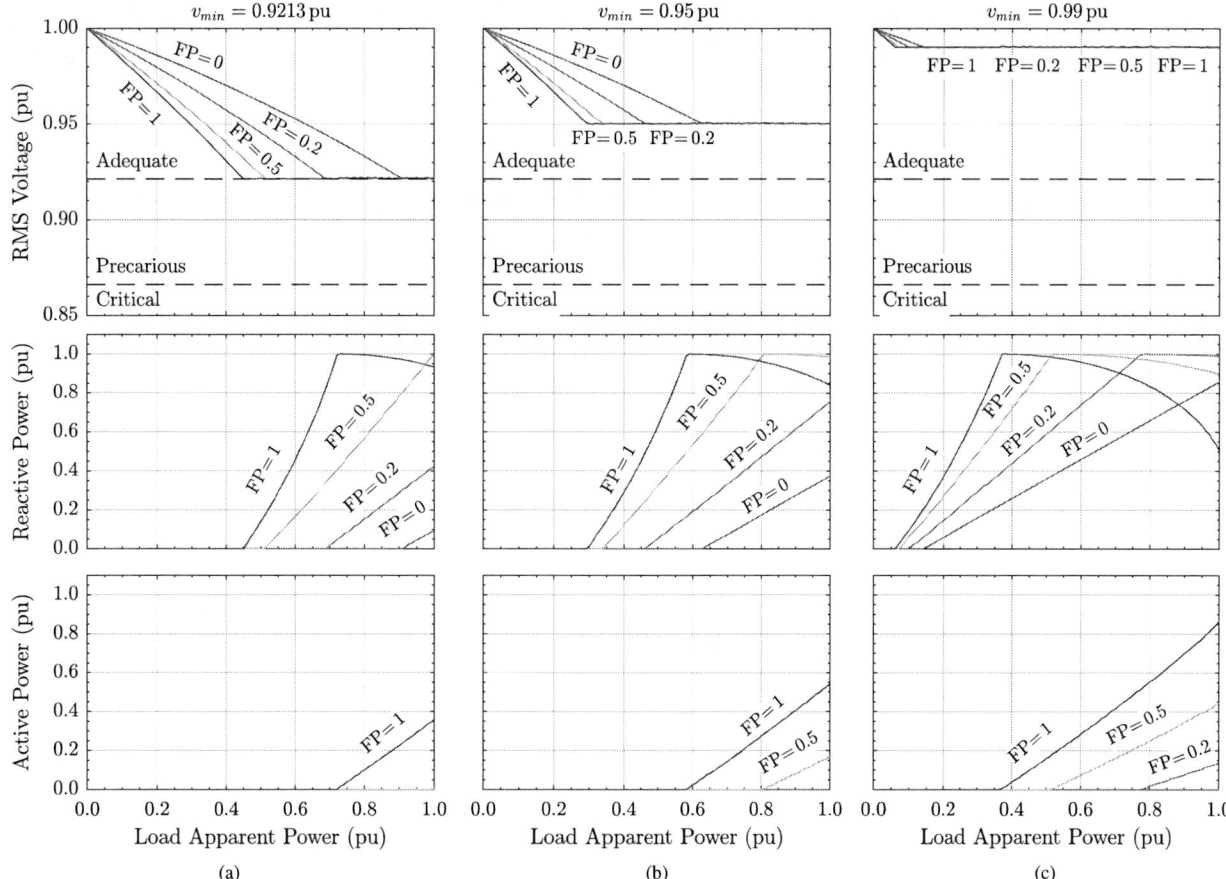

Fig. 5. Three cases of voltage-regulation settings: (a) Minimum acceptable rms voltage; (b) Intermediate rms voltage value; (c) Almost nominal rms voltage value.

[20]–[22]. The practical PR transfer function for multiple frequencies (8) is given by:

$$C_i(s) = k_p + \sum_{h=1,3,5,7,9} k_{ih} \cdot \left(\frac{2\zeta_h \omega_h \cdot s}{s^2 + 2\zeta_h \omega_h \cdot s + \omega_h^2} \right) \quad (8)$$

The current model to be controlled has a resonance peak, which can lead to system instability [23]. Therefore, a PR controller was designed with active damping and harmonic components $\{1, 3, 5, 7, 9\}$, leading to a final open–loop transfer function with a 41.5° phase margin, a 645 Hz zero–crossing frequency, $\{3, 1, 0.75, 0.5, 0.25\}$ gains with a 0.001 damping factor and the proportional gain of 0.0105.

B. Outer control loop

The outer loop is a low-bandwidth voltage control which uses the rms voltage at the PCC for voltage regulation. An equivalent single–phase circuit as shown in Fig. 6 (b) is used to obtain the voltage-transfer function (9):

$$\frac{v_{pcc}(s)}{i_{L2}(s)} = \frac{s^2 L_g R_l L_l + s R_g R_l L_l}{s^2 L_g L_l + s\left(L_g R_l + R_g L_l + R_l L_l\right) + R_g R_l} \quad (9)$$

Knowing the load power is necessary for the design of the voltage loop. In this case it was modeled as an equivalent impedance (R_l, L_l). A proportional–integral (PI) controller is used in the voltage-control loop. This controller provides null error in reference tracking for a constant input [24]. Two distinct controllers are used, one for active $(C_{0°})$ reference and another one for reactive $(C_{90°})$. Both controllers are designed with the same specification, a crossing frequency of 6 Hz and a phase margin of 63°. The outer-loop sampling frequency is adjusted to be one tenth of the inner loop's.

The coordination between active and reactive power is performed through a logic algorithm that works within the outer control loop and is shown in Fig. 6 (c). The scheme coordinates the current references in–phase and in quadrature to maintain the rms voltage regulated. The control algorithm works in two steps: Initially, the switch Sw_1 remains open and the voltage regulation occurs only with reactive power injection. Therefore, active power is not used while the reactive power is enough to raise the voltage to the set point.

The second step begins when the quadrature reference reaches its maximum value, indicating that the reactive power is no longer enough to raise the voltage. Then, a signal triggers

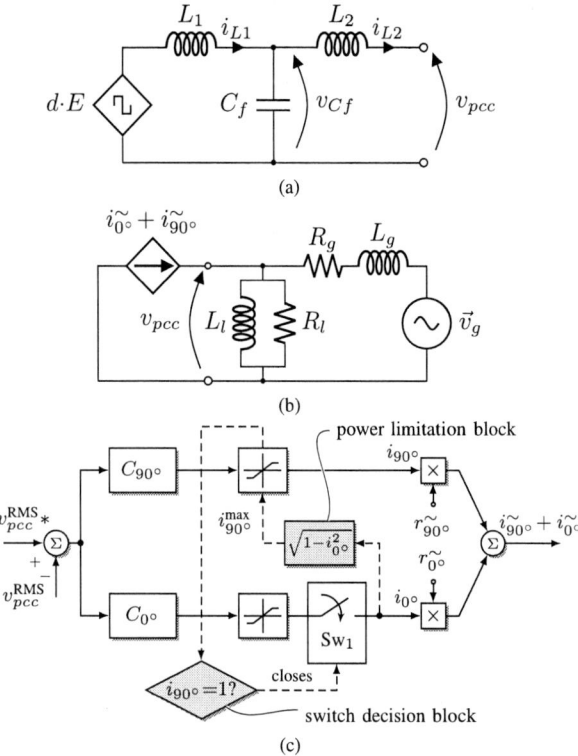

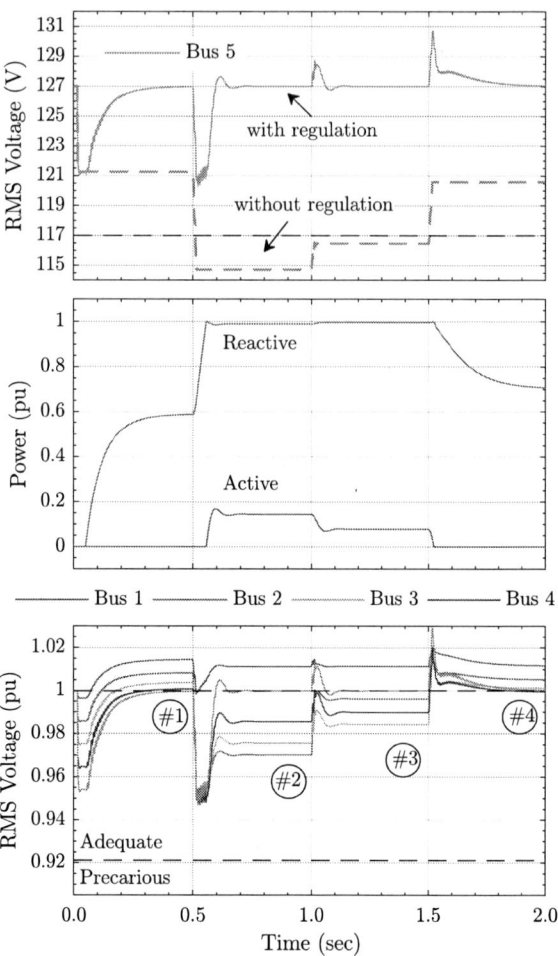

Fig. 6. Equivalent circuits and power management: (a) Equivalent circuit for current loop; (b) Equivalent circuit for voltage loop; (c) Logic algorithm used within outer control loop.

the switch Sw_1 closed ($=1$) and the control increases the in-phase current reference $i_{0°}$, injecting active power into the PCC. Finally, a power limitation block imposes a saturation value ($i_{90°}^{max}$) for the quadrature current to hold the apparent power within the converter's limits.

V. PROBLEM ANALYSIS AND RESULTS

A radial distribution feeder with multiple connection points is used to analyze voltage regulation issues, as shown in Fig. 8. Each connection point has local loads, representing a group of individual customers. Feeder voltage is 127 V with a 600 m length, divided down to three sections: first and second ones are 150 m long, and the last is 300 m long. A 3 kW load with $1\,s < t < 2\,s$ switching times is placed on the first bus, 8 kW with $0.5\,s < t < 1\,s$ on the second, $(2 + j2)$ kVA load with $1\,s < t < 1.5\,s$ on the third, while a 2 kW load is always connected to the last connection point. There is no load at bus 4 and the voltage regulator of 12 kVA is placed at the last bus. A dc source is placed on the converter's dc link to provide active power.

Fig. 7 shows the voltage at bus 5, where the voltage regulator is placed. The voltage violates adequate ratings when there is no regulation, otherwise it is is regulated to 127 V with the presented strategy. In this case, active and reactive power is used depending on line loading. For example, at 0.5 s, a load requires more power from the feeder and, consequently, a

Fig. 7. Simulation results for voltage at bus 5, processed power by converter and voltage across buses.

voltage drop is observed, followed by a raise caused by the converter's power injection. In such case, active power is used for proper voltage regulation according to the power management scheme. The opposite occurs at 1.5 s, where the active power is no longer needed and the reactive power is enough to raise the voltage.

Fig. 9 shows the voltage deviation from the reference value. Voltage across buses have different values depending on the load in each connection point, meaning that the converter can not ensure voltage regulation at different points from its connection. At the other points, the voltage floats around the reference value, while it is kept regulated on the bus 5.

VI. CONCLUSION

This paper analyzed voltage regulation with active and reactive power with distributed loads on a radial feeder. A circuit analysis shows that reactive power is not enough for proper voltage raise in distribution lines with poor grids. A detailed analysis with active and reactive voltage regulation is validated to overcome voltage-drop issues. The voltage raise capability of the selected approaches was tested in a

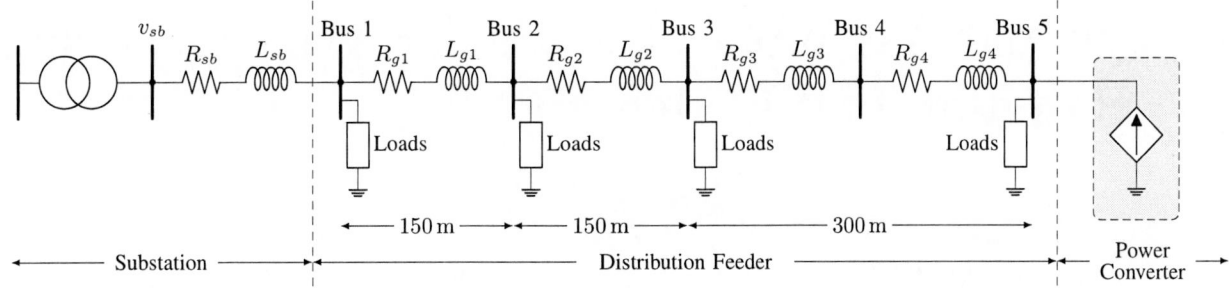

Fig. 8. Typical radial distribution feeder with multiple connection points.

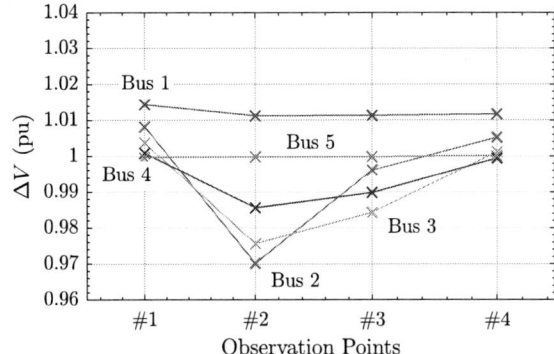

Fig. 9. Simulation results for voltage deviation.

more realistic scenario. A typical radial distribution feeder was covered in the analysis. Results can be applied to other cases, with other line parameters. The selected approach behaves similarly for different R/X ratios, using more active power with higher ratios. A simulation was used to validate the theoretical results for voltage raise.

In future works, the influence of the converter's position on the voltage profile along the feeder can be explored. In a distributed case, bus voltages seem to be influenced by the connected loads and other loads. A different position than at the final line connection point may show better results for global voltages along the feeder. Estimations of line impedance, voltage drop and local loads could improve the positioning decision.

REFERENCES

[1] R. Dugan, S. Santoso, M. McGranaghan, and H. Beaty, *Electrical Power Systems Quality*, ser. McGraw-Hill professional engineering. Mcgraw-hill, 2002.

[2] A. McEachern, "Designing electronic devices to survive power-quality events," *IEEE Ind. Appl. Mag.*, vol. 6, no. 6, pp. 66–69, Nov 2000.

[3] M. H. Bollen, *Overview of Power Quality and Power Quality Standards*. IEEE, 2000.

[4] H. Rudnick, J. Dixon, and L. Moran, "Delivering clean and pure power," *IEEE Power Energy Mag.*, vol. 1, no. 5, pp. 32–40, Sep. 2003.

[5] F. J. Zimann, R. C. Neto, F. A. S. Neves, H. E. P. de Souza, A. L. Batschauer, and C. Rech, "A complex repetitive controller based on the generalized delayed signal cancelation method," *IEEE Trans. Ind. Electron.*, vol. 66, no. 4, pp. 2857–2867, April 2019.

[6] Y. Wang, K. T. Tan, X. Y. Peng, and P. L. So, "Coordinated control of distributed energy-storage systems for voltage regulation in distribution networks," *IEEE Trans. Power Del.*, vol. 31, no. 3, pp. 1132–1141, June 2016.

[7] R. T. Hock, Y. R. de Novaes, and A. L. Batschauer, "A voltage regulator for power quality improvement in low–voltage distribution grids," *IEEE Trans. Power Electron.*, vol. PP, no. 99, pp. 1–1, 2017.

[8] P. M. S. Carvalho, P. F. Correia, and L. A. F. M. Ferreira, "Distributed reactive power generation control for voltage rise mitigation in distribution networks," *IEEE Trans. Power Syst.*, vol. 23, no. 2, pp. 766–772, May 2008.

[9] M. N. Kabir, Y. Mishra, G. Ledwich, Z. Y. Dong, and K. P. Wong, "Coordinated control of grid–connected photovoltaic reactive power and battery energy storage systems to improve the voltage profile of a residential distribution feeder," *IEEE Trans. Ind. Informat.*, vol. 10, no. 2, pp. 967–977, May 2014.

[10] S. J. Lee, J. H. Kim, C. H. Kim, S. K. Kim, E. S. Kim, D. U. Kim, K. K. Mehmood, and S. U. Khan, "Coordinated control algorithm for distributed battery energy storage systems for mitigating voltage and frequency deviations," *IEEE Trans. Smart Grid*, vol. 7, no. 3, pp. 1713–1722, May 2016.

[11] A. Nagarajan and R. Ayyanar, "Design and strategy for the deployment of energy storage systems in a distribution feeder with penetration of renewable resources," *IEEE Trans. Sustain. Energy*, vol. 6, no. 3, pp. 1085–1092, July 2015.

[12] T. Stetz, F. Marten, and M. Braun, "Improved low voltage grid-integration of photovoltaic systems in germany," *IEEE Trans. Sustain. Energy*, vol. 4, no. 2, pp. 534–542, April 2013.

[13] F. J. Zimann, A. L. Batschauer, M. Mezaroba, and F. A. S. Neves, "Energy storage system control algorithm for voltage regulation with active and reactive power injection in low-voltage distribution network," *Electr. Power Syst. Res.*, vol. 174, p. 105825, 2019.

[14] P. Acharjee, "Identification of maximum loadability limit and weak buses using security constraint genetic algorithm," *International Journal of Electrical Power & Energy Systems*, vol. 36, no. 1, pp. 40 – 50, mar 2012.

[15] K. Divya and J. Østergaard, "Battery energy storage technology for power systems – an overview," *Electr. Power Syst. Res.*, vol. 79, no. 4, pp. 511 – 520, 2009.

[16] *PRODIST: Module 8 – Power Quality*, ANEEL, 2018.

[17] *50160: Voltage characteristics of electricity supplied by public electricity networks*, ES/EN, 2010.

[18] *ANSI/C84.1: Electric Power Systems And Equipment – Voltage Ratings (60 Hz)*, NEMA, Jan. 2011.

[19] *Recommended Practice for Monitoring Electric Power Quality*, IEEE, June 2009.

[20] L. Herman, I. Papic, and B. Blazic, "A proportional-resonant current controller for selective harmonic compensation in a hybrid active power filter," *IEEE Trans. Power Del.*, vol. 29, no. 5, pp. 2055–2065, Oct 2014.

[21] F. Blaabjerg, R. Teodorescu, M. Liserre, and A. V. Timbus, "Overview of control and grid synchronization for distributed power generation systems," *IEEE Trans. Ind. Electron.*, vol. 53, no. 5, pp. 1398–1409, Oct 2006.

[22] L. R. Limongi, R. Bojoi, G. Griva, and A. Tenconi, "Digital current–control schemes," *IEEE Ind. Electron. Mag.*, vol. 3, no. 1, pp. 20–31, March 2009.

[23] J. Dannehl, F. W. Fuchs, S. Hansen, and P. B. Thogersen, "Investigation of active damping approaches for pi-based current control of grid-connected pulse width modulation converters with LCL filters," *IEEE Trans. Ind. Appl.*, vol. 46, no. 4, pp. 1509–1517, July 2010.

[24] K. Ogata, *Modern Control Engineering*, 5th ed. Prentice Hall, 2010.

Design and Assembly of a Bipolar Marx Generator Based on Full-Bridge Topology Applied to Electroporation

Fernando Imai
Instituto Federal de Santa Catarina (IFSC)
Jaraguá do Sul, Brasil
fernando.imai@ifsc.edu.br

Yales Rômulo de Novaes
Universidade do Estado de Santa Catarina (UDESC)
Joinville, Brasil
yales.novaes@udesc.br

Abstract—In this work is presented a bipolar Marx Generator (MG) to create pulsed electric fields in electroporation, technique consisting of creating pores on plasma membranes in different medical treatments. The Marx Generator is a voltage multiplier that charges capacitors in parallel and discharges them in series on the output. A new position for the power switches is proposed in order to allow a better control of capacitor individual usage. The main electrical parameters are mentioned and explained how they affect the design of the electroporator. At the end, experimetal tests are performed using vegetable samples as load, to analyze the current response and visual effects of the pulses application.

Index Terms—Capacitors, electrical parameters, electroporation, Marx Generator, pulsed electric field.

I. Introduction

Electroporation is a procedure consisting of applying pulsed electric fields (PEF) in cells, live tissues, organs, bacterial cultures, or other types of samples, aiming the creation of temporary or permanent pores on their plasma membranes [1]. Processes like gene and molecules insertion, non-heat tumor treatments, and even destruction of membrane are optimized by this technique.

Electrochemotherapy is an *in vivo* process benefited by electroporation. It consists of enhancing drug insertion in neoplastic cells through the pores generated. *In vitro* procedures can also be aided, as in gene transfer.

Electroporators are the equipments responsible by the pulse generation, as they differ from each other according to their electrical parameters: voltage level, width, frequency, quantity, shape and polarity.

The plasma membrane is the structure responsible to separate internal and external enviroments of cells. Composed by a phospholipid double layer [1], it has the function of selecting the materials to penetrate or leave the cell. Having a thickness of nanometers [2] this double layer has one border attracted to water molecules and other with the inverted behaviour, not absorbing water.

The resting transmembrane electric potential (V_m) is approximately 75 *mV* [3]. When an electric field is applied, electrically charged microparticles flow to internal and external borders, increasing their potential. If this value reaches

approximately 200 *mV* (depending on the kind of cell) the electropermeabilization occurs [4]. It consists of breaking the membrane barrier and easing the traffic of substances.

Diverse applications of electroporation have been recorded and published in the past two decades. It was used to analyze cell migration in the healing process of a wounded intestinal tissue [5]; sterilize samples of blueberry juice [6]; treatment of hepatic and pancreatic cancer by means of irreversible electroporation [7]; water desinfection to human consumption [8]; DNA insertion in animal fetus [9]; destruction of a *Candida Albicans* fungal colony [10]; i.a.

Some types of circuits were proposed aiming to obtain pulsed voltage and current. Solutions using a high voltage power supply (same level of pulses) with an inverter were presented in [9]-[12]. Employing this high voltage power supply can represent an inconvenience due to pricing and space required.

Another possibility is the pulsed forming network (PFN) of Blumlein. It also employs a power supply with same level as the output, at least one switch and transmission lines, normally composed by coaxial cables, as presented in [13] and [14]. In [15] this system earns functionality with parallel lines, allowing voltage boost.

As the response time of PFN (tens of nanoseconds) is usually lower than the ones related to electroporation, this solution is aplied to other kinds of pulsed applications. Besides, the cost and volume of the coaxial cables are also a relevant problem.

Another category of circuits capable of generating high voltage pulses is the multilevel inverter (MLI). They can be divided by the power supply level (symmetric, asymmetric or hybrid), topologies and components (Diode Clamped, Flying Capacitor and H-Cascaded) [16].

Examples of asymmetric MLI were presented in [17] and [18], while symmetric ones are shown in [19] and [20]. The need for several independent sources tends to increase size and price of the system.

DC-DC converters can also be used to generate PEF. Proposals with boost circuits were presented in [21] and [22], a buck-boost topology was presented in [23] and a push-pull

978-1-7281-4181-7/19 $31.00 © 2019 IEEE

was shown in [24]. The main negative issue pointed out by the authors of the boost circuits is that the fixed values of the passive components (especially the inductors) limit the variation of parameters.

Given the presence of negative points in the systems mentioned, a widespread circuit (especially in the last two decades) is the Marx Generator (MG). It is a set composed generally of switches (IGBTs or MOSFETs), diodes and capacitors. It charges the capacitors in parallel and discharges them in series, in order to multiply the value on its output.

MG proposals were presented in [25]-[32]. The circuit presented in [32] based the system topology developed in this work. The choice was to employ only one low voltage source, to have redundant gate command signals (repeated, regardless of the number of cells used) and to enable the generation of pulses with good adjustments of the parameters quantity, width, frequency and intensity. The Full-Bridge topology of capacitive cells allows the generation of bipolar pulses, characterized by reducing muscle spasms of *in vivo* applications, presenting lower toxicity, increasing electrode lifespan and distributing a more homogeneous electric field [33] and [34].

II. BIPOLAR MARX GENERATOR BASED ON FULL-BRIDGE TOPOLOGY

The topology adopted in this paper is presented in Fig. 1. The main difference of the proposed circuit in relation to that presented in [32], is with respect to the position of the switches responsible for the capacitor charging, S_{ch}. Arranged in parallel, not in series, this new configuration allows alternation of capacitors to be charged during electroporation protocols. If only two capacitive cells were used in a given electroporation procedure, by the original topology, the first two capacitors would necessarily act in the pulse generation.

With the arrangement in parallel, it is possible to divide in a more balanced way the operation of each capacitor, not being a cell overloaded in detriment to the others. In addition, this topology avoids current peaks when two consecutive protocols employing different amounts of capacitive cells occur, which would cause significant losses to the system.

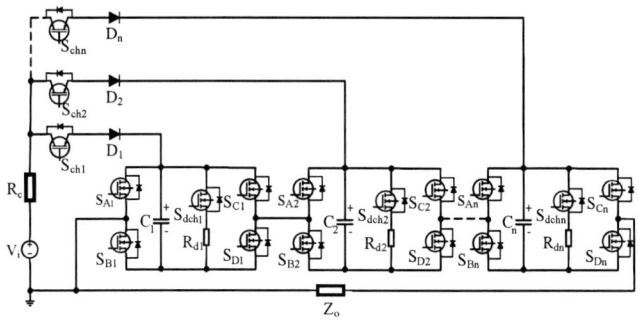

Fig. 1. Adopted Marx Generator topology.

The generator operation initiates by performing the charge of capacitors that will contribute to the formation of the pulses. The resistor R_C is used to limit the peak current of this step, but

so as not to raise this period too long and make the operation of the electroporator ineffective. All switches S_B and S_D are driven, in addition to the S_{ch} of the cells involved in the pulse protocol. This way, a common branch is created at the lower terminal of capacitors, connected straight to the negative pole of the source, while the positive terminal is connected to the upper side of the capacitors, according to Fig. 2 (a).

After charging, it is possible to apply energy to the load. During positive pulses, switches S_B and S_C are turned on, in the negative, switches S_A and S_D are enabled, as shown in Fig. 2 (b) and (c).

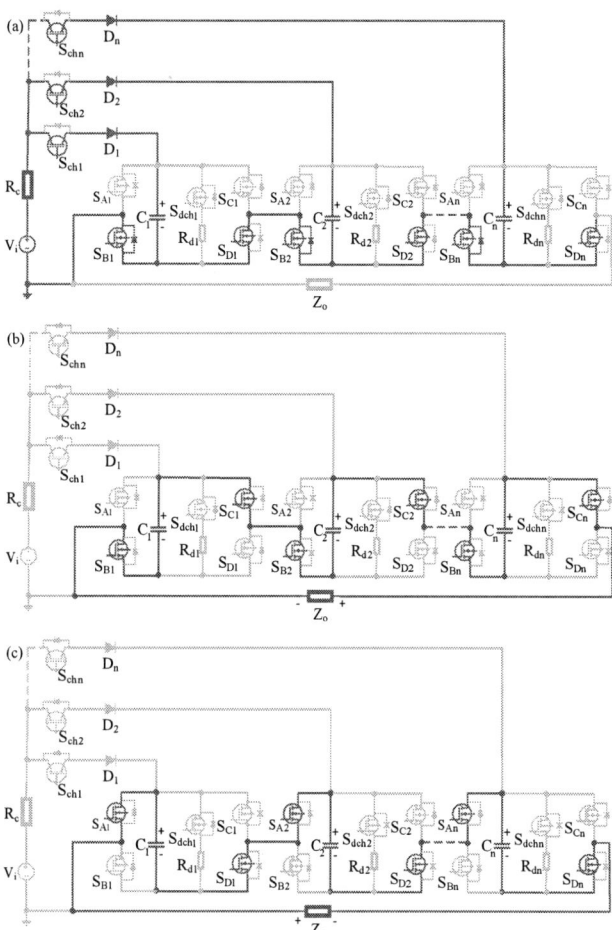

Fig. 2. Operating stages of the MG. (a) Capacitor charge, (b) Positive pulse e (c) Negative pulse.

Once the parameters of pulses have been configured, it is possible to transfer the energy from the capacitors to the load. The procedure always consists of a positive pulse, an interval between pulses, a negative application and another interval. After the pulse application protocols are finished, capacitors must be discharged in order to provide safety to the operators. It is only necessary activating the switches S_{dch} to discharge the energy stored in the capacitors to specific resistors.

The prototype assembled in this work has its parameters designed to electrochemotherapy, process consisting of in-

978-1-7281-4181-7/19 $31.00 © 2019 IEEE

creasing drug introduction in neoplastic cells along with cellular electropermeabilization. Tab. I presents the configurable specifications of the electroporator proposed in this paper.

TABLE I
ELECTRICAL PARAMETERS OF PULSES

Parameter	Minimum value	Maximum value
Quantity (k)	2	20
Width (t_p)	50 μs	500 μs
Voltage level (V_o)	100 V	1200 V
Frequency (f)	80 Hz	8000 Hz
Current (I_o)	–	15 A
Voltage drop (ΔV_{max})	–	12 %

The generator gain, ratio of output (V_o) and input voltage (V_i), is given by the amount of charged capacitors (n) and is shown in (1). The prototype mounted in this paper has four cells, thus, the peak voltage of 1200 V can be obtained with a 300 V maximum power supply.

$$V_o = \pm n V_i \qquad (1)$$

Once the pulses are rectangular, capacitance is calculated to keep voltage above maximum drop (ΔV_{max}) stipulated in Tab. I. The most critical situation happens with highest amount of charged cells (n), pulse time (t_p), number of pulses (k) and lowest load impedance (Z_{min}), according to (2).

$$C_{min} = \frac{-t_p n k}{ln(1 - \Delta V_{max}) Z_{min}} \qquad (2)$$

Z_{min} can be obtained according to (3).

$$Z_{min} = \frac{V_{o_{max}}}{I_{o_{max}}} \qquad (3)$$

The minimum capacitance of each cell is 3,91 mF. Six 680 μF/400 V capacitors were used in parallel, totaling 4,08 mF.

About the switches, it is necessary to analyze their voltage and current stresses, noting that some only act in charging phase, others only in the pulses and some in both. During charge step, the switches S_A and S_C register capacitor voltage, increasing exponentially from zero to the input level. In the phase of pulses, the cell switches divide the voltage of the capacitor in the intervals and are subjected to maximum level when they are not in operation. Related to the currents, the switches of the cells are subject to the peak of pulses. The root mean square (RMS) value of the switch current is obtained according to (4), being (I_p) the pulsed current, (Z_o) the impedance load and (T_P) the total time of protocol.

$$I_{S_{B_{rms}}} = \sqrt{\frac{-I_p^2 Z_o C}{2n T_P} \sum_{i=1,3,5...}^{k-1} e^{-2nt/Z_o C} \Big|_{(i-1)t_p}^{it_p}} \qquad (4)$$

The charging semiconductors are then subjected to the RMS current stress according to (5), where (I_c) is the current peak and (t_c) the charging period.

$$I_{S_{ch_{rms}}} = \sqrt{\frac{-n R_c C I_c^2}{2t_c} e^{-2t/n R_c C} \Big|_0^{t_c}} \qquad (5)$$

Using a 55 Ω resistor to limit the input current the charging period will last approximately 6 seconds. The maximum stresses of each semiconductor are presented in Tab. II.

TABLE II
MAXIMUM VOLTAGE AND CURRENT STRESS ON SEMICONDUCTORS

Semiconductor	Pulsed current [A]	RMS current [A]	Blocking voltage [V]
Load switch	1,36	0,37	1200
Cell switch	15	9,48	300
Diode	1,36	0,37	900

Based on these values MOSFETs APT5010B2VR (500 V/47 A) are used as cell switches, IGBTs G4PH40UD (1200 V/41 A) as S_{ch3} and S_{ch4}, IGBTs GP50B60PD (600 V/75 A) as S_{ch1} and S_{ch2} and FR307 (1000 V/3 A) as diodes.

In addition to the MG, the electroporator needs auxiliary circuits to drive MOSFETs and IGBTs. Their arrangement is shown in Fig. 3.

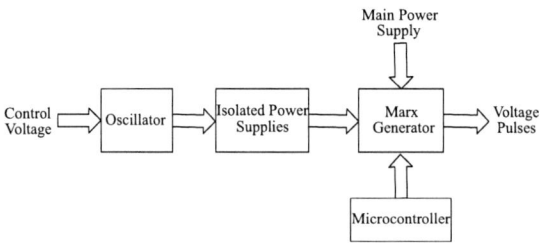

Fig. 3. Auxiliary circuits arrangement.

A microcontroller model PIC18F4550 is used to process the information required by the operator and generate the vectors with pulse configurations. A C programming language was developed to command the electroporator, using Microchip MPLAB IDE v5.00.

The switches are driven by HCPL3120 optocouplers, the voltage from isolated sources feed the IC, and the LEDs excitation signals are provided by the microcontroller. An overview of the prototype is shown in Fig.4.

Another point to be analyzed in the circuit is the use of electrolytic capacitors as main elements for energy storage. These components have wide application in the industrial and electronic power processing due to their high capacitances and rated voltages. However, they have a low lifetime and become one of the main causes of replacement needs in electronic circuits [35].

Like any real component, electrolytic capacitors have parasitic resistances that generate heat, increasing the risks of damaging them.

Several physical parameters contribute to shorten the lifetime of these components, such as humidity, atmospheric

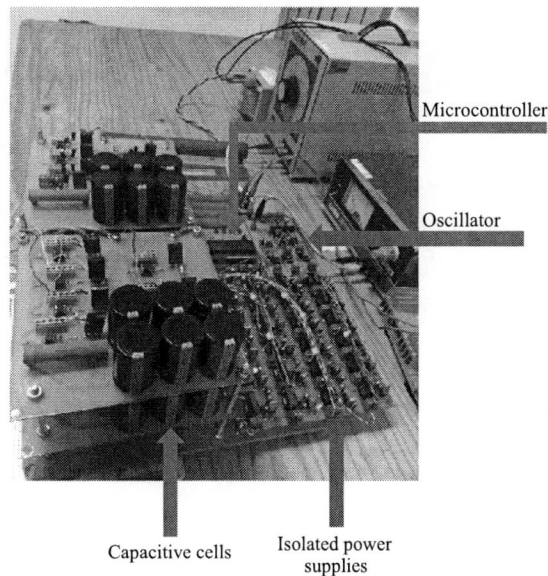

Fig. 4. Prototype overview.

pressure, site vibration index and number of charging and discharging cycles. But the main ones are operating temperature, current ripple and supply voltage.

The amount of charge and discharge cycles may be considered a minor factor because the electroporator usage does not occur for long consecutive periods of time (range of minutes or hours), but rather during specific treatments or applications. Also, since capacitors are not fully discharged at the end of protocols, it is not necessary to perform a full charge at the beginning of another.

As for the operating temperature of the capacitors it may be proposed to use forced ventilation elements to keep it within permissible limits. In addition, it is important paying attention to the thermal specifications informed by the manufacturer and ensure these values are not extrapolated, since a 10 °C decrease in capacitor tends to double its lifetime [35].

Furthermore it can be suggested to use smaller capacitors in parallel instead of applying only one with full capacitance (as was done in this paper) in order to decrease the equivalent series resistance (ESR) of the set and also the individual current ripple in the components.

About the voltage applied in the capacitors, this factor is often not feasible to be modified because using an element with rated voltage much greater than the operating one generates relevant increases in its dimensions.

III. EXPERIMENTAL RESULTS

Measurements of electrical quantities were performed using a Tektronix MDO3014 oscilloscope, voltage probes THDP0100 and a current probe TCP0030A.

It is possible to point out that there is no undesired voltage rise in the gates of the switches that must remain open, as seen in Fig. 5. Even so, negative voltage is an important safety

feature to this type of topology (switches in an inverter leg) aiming to avoid short circuits.

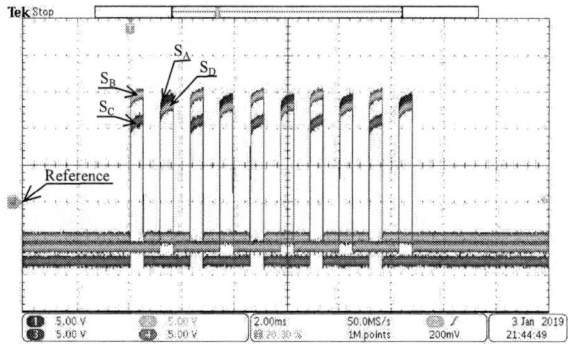

Fig. 5. Gate voltage of cell switches in protocol with ten pulses

Laboratory tests were performed using potato (*Solanum tuberosum*) pieces as load to verify responses to different pulse parameters. Authors [36] and [37] emphasized that potato tissue can be used instead of animal materials to validate concepts of electroporation. While electrochemotherapy applications may take weeks or months to analyze the results, the effects of electroporation on the potato start appearing 6 hours after the application [36]. Parallel needles 2 mm apart each other, with 0.5 mm of diameter and 1 cm high were used as electrode.

Ten distinct protocols were performed for tests, as Tab. III. Protocols 1 to 4 compare different voltage values, with the other parameters being identical. From fifth protocol on, they were divided in pairs to compare high and low levels of pulse width, frequency, and quantity.

TABLE III
ELECTROPORATION PROTOCOLS APPLIED ON VEGETABLE SAMPLES

Protocol	Voltage level [V]	Quantity	Width [μs]	Frequency [Hz]
1	60	10	200	1100
2	100	10	200	1100
3	200	10	200	1100
4	370	10	200	1100
5	300	10	500	440
6	300	10	50	4400
7	300	10	200	200
8	300	10	200	2000
9	300	4	200	1100
10	300	20	200	1100

The analysis of pulse applications were performed visually and also through current load. An example is shown in Fig. 6 which refers to protocol 3.

Analyzing the behavior of the load current it is possible to notice an increasing tendency during first two pulses. As proposed in [4], it confirms the occurrence of electropermeabilization, since pores in cell membranes decrease their impedances gradually.

However, the increase of conductance is limited and the rate tends to reduce, because the membrane reaches a saturation

978-1-7281-4181-7/19 $31.00 © 2019 IEEE

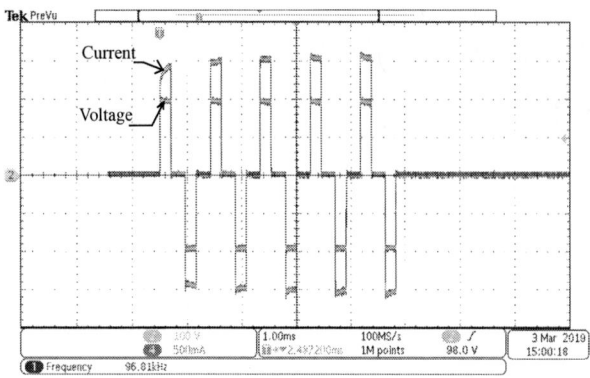

Fig. 6. Voltage and current in protocol 3.

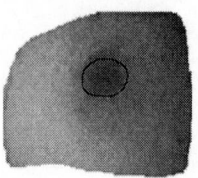

Fig. 8. Result of protocol 5 in the vegetable sample.

Fig. 9. Hues of non electroporated (a) and electroporated (b) areas.

point due to its high number of pores. Thus, after the initial behavior the current tends to follow the voltage shape. Hence, the biological load can be modeled as a series RL circuit at the first pulses and as a resistance at following ones.

Protocol 5 was the only that presented a different current, seen in Fig. 7.

The most visible electroporation results were recorded in protocols 4 and 5, which have higher voltage level and pulse width. Disparities in the number of pulses and frequency were not so relevant in this case.

IV. CONCLUSIONS

In this paper a bipolar Marx Generator was proposed and tested. The results were positive for the experiments performed, the circuit proved to be an interesting option for applications employing rectangular voltage pulses, as electroporation. Modifying the power switches position improved the control of capacitos involved in the protocols, without causing overcurrents during their charges. On the other hand, the voltage stress increased. However, it did not represent a relevant issue, since it was still possible employing discrete package IGBTs. Yet on the use of IGBTs and MOSFETs, it showed to be an adequate and efficiente choice, due to their wide ranges of electrical parameters.

The experiments proceeded in laboratory using live vegetable tissues showed how electrical parameters of pulses are relevant in protocols involving electroporation. As an example the application of high voltage levels and width produced much more evident visual effects in the samples.

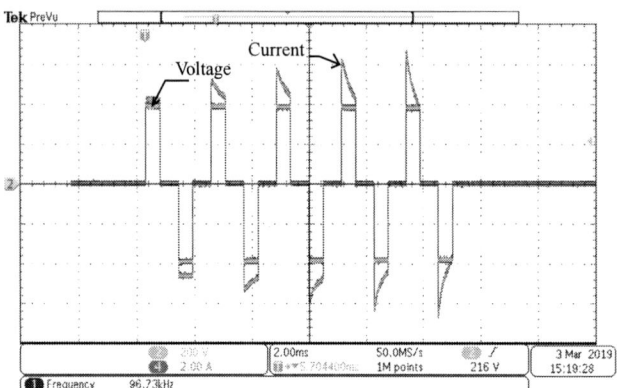

Fig. 7. Voltage and current in protocol 5.

It can be assumed that an irreversible electroporation process occurred due to the high pulse widths and voltage level, which resulted in a high energy transfer. The distinct behavior of the current can be justified by a possible destruction of the plasma membrane, which would cause the tissue to have a different electrical response.

In addition a visual analysis of the effects on the samples was performed. They were photographed and processed through the software ImageJ v. 1.52a, to measure the contrast between areas that received discharges.

The analysis of contrasts occurred through the RGB (Red - Green - Blue) color system, where each hue is composed of an index from 0 to 255 for the basic colors of this system.

The result of protocol 5 is presented to exemplify the methodology adopted. Fig. 8 shows applications results of pulses in the sample. The tonalities of the region that did not receive the pulses and the electroporated area are presented in Fig. 9.

ACKNOWLEDGMENT

The present work had the support of the Instituto Federal de Santa Catarina.

REFERENCES

[1] M. D. da Silva, "Desenvolvimento e avaliação de um gerador programável de pulsos monofásicos de campo elétrico para eletroporação." MSc. Thesis. UFRJ, 2011.
[2] B. Alberts, "Molecular Biology of the Cell".
[3] J. Shi et al, "A review on electroporation-based intracellular delivery" Molecules, vol. 23, nº 11, pp. 3044 1–19.
[4] R. L. Weinert, "Estudo experimental e computacional de eletropermeabilização de tecidos biológicos," MSc. Thesis. UDESC, 2017.
[5] H. B. Mamman, A. A. Sadiq, M. N. Adon and M. M. A. Jamil, "Study of electroporation effect on ht29 cell migration properties."IEEE International Conference on Control System, Computing and Engineering, pp. 342–346, November 2015, Malaysia.

[6] J. Chen et al, "Influence of pulsed electric field and thermal treatments on the quality of blueberry juice." International Journal of Food Properties, vol. 17, pp. 1419–1427, 2014.

[7] L. M. Wu, L. L. Zhang, X. H. Chen and S. S. Zheng, "Is irreversible electroporation safe and effective in the treatment of hepatobiliary and pancreatic cancers?". Hepatobiliary & Pancreatic Diseases International, 8 p, 2019.

[8] A. A. Elserougi, M. Faiter, A. M. Massoud and S. Ahmed, "A transformerless bipolar/unipolar high-voltage pulse generator with low-voltage components for water treatment applications". IEEE TRANSACTIONS ON INDUSTRY APPLICATIONS, vol. 53, n° 3, pp. 2307–2319. 1

[9] T. Bullmann, T. Arendt, U. Frey and C. Hanashima, "A transportable, inexpensive electroporator for in utero electroporation". Japanese Society of Developmental Biologists, vol. 57, pp. 369–377, 2015.

[10] V. Novickij et al, "High-frequency submicrosecond electroporator". Biotechnology & Biotechnological Equipment, vol. 30, n° 3, pp. 606–613.

[11] A. Grainys, V. Novickij and J. Novickij, "High-power bipolar multilevel pulsed electroporator". INSTRUMENTATION SCIENCE AND TECHNOLOGY, vol. 44, n° 1, pp. 65–75, 2016.

[12] M. A. Elgenedy, A. Darwish, S. Ahmed and B. W. Williams, "A transition arm modular multilevel universal pulse-waveform generator for electroporation applications". IEEE TRANSACTIONS ON POWER ELECTRONICS, vol. 32, n° 12, pp. 8979–8991, 2017.

[13] A. de Angelis, J. F. Kolb, L. Zeni and K. H. Schoenbach, "Kilovolt blumlein pulse generator with variable pulse duration and polarity". Review of Scientific Instruments, vol. 79, n° 4, 5 p., 2008.

[14] J. P. M. Mendes, L. M. Redondo, H. Canacsinh, M. Vieira and J. Rossi, "Modelling of n-stage blumlein stacked lines for bipolar pulse generation". IFIP Advances in Information and Communication Technology, pp. 395–402, 2012.

[15] S. Romeo, C. D. Avino, O. Zeni and L. Zeni, "A blumlein-type, nanosecond pulse generator with interchangeable transmission lines for bioelectrical applications". IEEE TRANSACTIONS ON DIELECTRICS AND ELECTRICAL INSULATION, vol. 20, n° 4, pp. 1224–1230, 2013.

[16] J. Venkataramanaiah, Y. Suresh and A. K. Panda, "A review on symmetric, asymmetric, hybrid and single dc sources based multilevel inverter topologies". Renewable and Sustainable Energy Reviews, vol. 76, pp. 788–812, 2017.

[17] E. Samadaei, S. A. Gholamian, A. Sheikholeslami and J. Adabi, "An envelope type (e-type) module: asymmetric multilevel inverters with reduced components". IEEE TRANSACTIONS INDUSTRIAL ELECTRONICS, vol. 63, n° 11, pp. 7148–7156, 2016.

[18] E. Samadaei, A. Sheikholeslami, S. A. Gholamian and J. Adabi, "A square t-type (st-type) module for asymmetrical multilevel inverters". IEEE TRANSACTIONS ON POWER ELECTRONICS, vol. 33, n° 2, pp. 987–996, 2018.

[19] I. Abdelsalam, M. A. Elgenedy, S. Ahmed and B. W. Williams, "Full-bridge modular multilevel submodule-based high-voltage bipolar pulse generator with low-voltage dc, input for pulsed electric field applications". IEEE TRANSACTIONS ON PLASMA SCIENCE, vol. 45, n° 10, pp. 2857–2864, 2017.

[20] J. S. Choi and F. S. Kang, "Seven-level pwm inverter employing series-connected capacitors paralleled to a single dc voltage source". IEEE TRANSACTIONS ON INDUSTRIAL ELECTRONICS, vol. 62, n° 6, pp. 3448–3459, 2015.

[21] A. Elserougi, A. M. Massoud and S. Ahmed, "A boost-inverter-based bipolar high-voltage pulse generator". IEEE TRANSACTIONS ON POWER ELECTRONICS, vol. 32, n° 4, pp. 2846–2855, 2017.

[22] R. Bondade, Y. Wang and D. Ma, "Design of integrated bipolar symmetric output dc–dc power converter for digital pulse generators in ultrasound medical imaging systems". IEEE TRANSACTIONS ON POWER ELECTRONICS, vol. 29, n° 4, pp. 1821–1829, 2014.

[23] A. A. Elserougi, A. M. Massoud and S. Ahmed, "A unipolar/bipolar high-voltage pulse generator based on positive and negative buck–boost dc–dc converters operating in discontinuous conduction mode". IEEE TRANSACTIONS ON INDUSTRIAL ELECTRONICS, vol. 64, n° 7, pp. 5368–5379, 2017.

[24] M. Ilic, L. Laskai, J. L. Reynolds and R. Encallaz, "An isolated high-voltage dc-to-dc converter with fast turn-off capability for x-ray tube gridding". IEEE TRANSACTIONS ON INDUSTRY APPLICATIONS, vol. 38, n° 4, pp. 1139–1146, 2002.

[25] S. Zabihi, Z. Zabihi and F. Zare, "A solid state marx generator with a novel configuration". 19th Iranian Conference on Electrical Engineering, 6 p., 2011.

[26] M. Rezanejad, A. Sheikholeslami and J. Adabi, "High-voltage modular switched capacitor pulsed power generator". IEEE TRANSACTIONS ON PLASMA SCIENCE, vol. 42, n° 5, pp. 1373–1379, 2014.

[27] M. R. Delshad, M. Rezanejad and A. Sheikholeslami, "A new modular bipolar high voltage pulse generator". IEEE TRANSACTIONS ON INDUSTRIAL ELECTRONICS, vol. 64, n° 2, pp. 1195–1203, 2017.

[28] L. M. Redondo, H. Canacsinh and J. F. Silva, "Generalized solid-state marx modulator topology". IEEE TRANSACTIONS ON DIELECTRICS AND ELECTRICAL INSULATION, vol. 16, n° 4, pp. 1037–1042, 2009.

[29] H. Canacsinh, L. M. Redondo and J. F. Silva, "Marx-type solid-state bipolar modulator topologies: performance comparison". IEEE TRANSACTIONS ON PLASMA SCIENCE, vol. 40, n° 10, pp. 2603–2610, 2012.

[30] C. Yao, S. Dong, Y. Zhao, Y. Mi and C. Li, "A novel configuration of modular bipolar pulse generator topology based on marx generator with double power charging". IEEE TRANSACTIONS ON PLASMA SCIENCE, vol. 44, n° 10, pp. 1872–1878, 2016.

[31] H. Shi, Y. Lu, T. Gu, J. Qiu and K. Liu, "High-voltage pulse waveform modulator based on solid-state marx generator". IEEE TRANSACTIONS ON DIELECTRICS AND ELECTRICAL INSULATION, vol. 22, n° 4, pp. 1983–1990, 2015.

[32] T. Sakamoto, A. Nami, M. Akiyama and H. Akiyama, "A repetitive solid state marx-type pulsed power generator using multistage switch-capacitor cells". IEEE TRANSACTIONS ON PLASMA SCIENCE, vol. 40, n° 10, pp. 2316–2321, 2012.

[33] C. B. Arena et al, "High-frequency irreversible electroporation (h-fire) for non-thermal ablation without muscle contraction". BioMedical Engineering OnLine, vol. 10, 20 p., 2011.

[34] M. B. Sano, C. B. Arena, M. R. DeWitt, D. Saur and R. V. Davalos, "In-vitro bipolar nano- and microsecond electro-pulse bursts for irreversible electroporation therapies". Bioelectrochemistry, vol. 100, pp. 69–79, 2014.

[35] M. Frivaldsky, J. Cuntala, P. Spanik and A. Kanovsky, "Investigation of thermal effects and lifetime estimation of electrolytic double layer capacitores during repeated charge and discharge cycles in dedicated application". Electrical Engineering, vol. 100, n° 1, pp. 11–25, 2018.

[36] J. Berkenbrock, G. Pintarelli, A. Antônio and D. Suzuki, "In vitro simulation of electroporation using potato model". Conference of The Canadian Medical and Biological Engineering, n° 40, 5 p., 2017.

[37] N. Boussetta, N. Grimi, N. I. Lebovka and E. Vorobiev, "Cold electroporation in potato tissue induced by pulsed electric field". Journal of food engineering, vol. 115, n° 2, pp. 232–236, 2013.

Implementation of a iUPQC control scheme for ensuring an improved compensation performance

E. V. Stangler, F. A. S. Neves, F. Bradaschia
Federal University of Pernambuco
Power Electronics and Eletric Drives Group
Recife - PE, Brazil
eduardo.vasconcelosstangler@ufpe.br,
fneves@ufpe.br,
fabricio.bradaschia@ufpe.br

M. Mezaroba, F. J. Zimann, A. L. Batschauer
Santa Catarina State University
Electric Power Processing Group
Joinville - SC, Brazil
marcello.mezaroba@udesc.br,
felipezimann@ieee.org,
alessandro.batschauer@udesc.br

Abstract—The unified power quality conditioner has been considered one of the most complete solutions to power quality related issues. Thus, the development of control strategies that improve its performance has gained increasing research interest. This paper presents an enhanced control scheme for ensuring an improved compensation performance for the dual unified power quality conditioner (iUPQC). The basics of the control implementation, the reference generators and the control loops specifications of both converters are discussed in detail. Besides, the design step of the controllers is carefully explained in order to allow its reproduction. Furthermore, the complete control scheme is validated through experimental results using a three-phase three-wire iUPQC prototype.

Index Terms—unified conditioners, control schemes, reference generators

I. INTRODUCTION

In recent decades, the increasing use of non linear loads has raised concerns about the power quality (PQ) of the electrical systems, especially at the distribution grids. Such scenario has made the active power conditioning solutions to become one of the major research subject of the scientific community. The development of active power filters (APF) has taken an increasing attention, since such devices may provide a suitable compensation for several PQ related problems.

The unified power quality conditioner (UPQC) was developed in order to compensate current and voltage related issues simultaneously. In its most usual topology, the UPQC is formed by a shunt and a series APF connected to the same DC-bus in a back-to-back configuration [1]–[4]. Besides, both converters are controlled as the conventional series (srAPF) and shunt (shAPF) APFs. Alternatively, the UPQC can be controlled by a dual or inverted control strategy, the iUPQC [5], [6]. In such control strategy, the converters control plots are inverted. Thus, the srAPF is current controlled aiming to impose sinusoidal grid currents which must be in-phase with the fundamental-frequency positive-sequence (FFPS) grid voltages. On the other side, the shAPF is voltage controlled

The authors would like to thank *Conselho Nacional de Desenvolvimento Científico e Tecnológico - CNPq, Fundação de Amparo à Ciência e Tecnologia do Estado de Pernambuco - FACEPE* and *Fundação de Amparo à Pesquisa e Inovação do Estado de Santa Catarina - FAPESC*, for the financial support.

and must provide a sinusoidal three-phase voltage with low harmonic distortion for the load bus.

Several iUPQC control strategies with different purposes have been developed in recent years. Even with purely sinusoidal references quatities, for both APFs, the controllers must be able to reject harmonic disturbances caused by distorted or unbalanced grid currents and load currents. For achieving these goals, it is possible to use a few different control configurations, such as proportional-integral in the dq reference frame, proportional-resonant or repetitive-based controllers in natural and $\alpha\beta$ reference frames, or predictive-type controllers [7], [8]. Along with the reference generators, the well-known strategies employ p-q-theory-based operations or synchronous reference frame (SRF) compensators [9], [10].

The iUPQC control scheme outlined in this paper ensures a desirable compensation performance and is suitable for low cost implementation. A description of the iUPQC operation characteristics and the control requirements for both APFs is performed in section II, along with the characterization of the implemented system. Each part of the developed control scheme is precisely described in sections III and IV. Futhermore, an experimental validation in order to verify its effectiveness is presented in section V.

II. IUPQC BASICS AND IMPLEMENTED SYSTEM CHARACTERIZATION

Due to the iUPQC's operation principles, only a balanced three-phase current is demanded from the grid. Thus, the srAPF must synthesize purely sinusoidal currents equivalent to the active part of load current FFPS added to the current component demanded by the DC voltage control loop for providing its regulation. On the other hand, the shAPF acts for ensuring a low distorted balanced three-phase voltage at the load bus. Therefore, the load currents harmonic components and reactive power demand are provided by the shAPF.

According to the above-mentioned statements, the reference signals of the iUPQC's APFs are both purely sinusoidal, what should improve the waveform quality of the supplied load voltages and the demanded grid currents, once these quantities are the main control variables. Besides the sinusoidal quantities control, the DC voltage regulation must be carried out carefully, since it is a high demanding condition for a suitable sinusoidal signal supplying.

As can be seen, good performance indicators depend on both the accuracy of the reference generation algorithm adopted and the proper design of the control system.

The iUPQC system implemented is based on a three-phase three-wire (3P3W) hardware digitally controlled. Therefore, the sinusoidal quantities control schemes are implemented in the $\alpha\beta$, given that this adoption requires only two control loops for 3P3W systems. Nevertheless, it is sufficient to consider an additional zero sequence control loop for ensuring safe operation for a four wire system.

The complete control scheme developed and the circuit parameters are presented in Fig. 1 and TABLE I, respectively. The control scheme characteristics of each APF, such as reference generation algorithms, control loops and adopted controllers, are widely discussed in following sections.

TABLE I: Experimental prototype parameters.

Trafos	srAPF		shAPF				DC-bus
L_{ct} [1]	L_{sr} [1]	$R_{L_{sr}}$ [1]	L_{sh} [1]	$R_{L_{sh}}$ [1]	C_{sh} [1]	$R_{C_{sh}}$ [2]	C_b
(mH)	(mH)	(mΩ)	(mH)	(mΩ)	(μF)	(Ω)	(mF)
1,48	2,56	307,5	2,87	408,7	25,2	2,2	7

[1] Parameters obtained through eletric tests.
[2] Physical resistor inserted to ensure passive damping.

Regarding the digital control implementation, a sampling frequency of $f_s = 19,2$kHz has been selected. Besides, all the voltage and current sensors have been calibrated for ensuring unitary gain to the measured variables. Moreover, the anti-alising filters implemented have all been first-order filters with unitary gain and cutoff frequency of $f_{aa} = 5,1$kHz.

III. SRAPF CONTROL

As can be seen in Fig. 1, the srAPF is controlled by a double-loop control scheme. The internal control loop is responsible for imposing the sinusoidal reference current with low steady-state error. Thus, it needs to provide a fast dynamic response and to have high gain in a few frequencies of interest. On the other hand, the external loop task is to control the DC bus voltage and must have a low bandwidth, ensuring the decoupling between the control loop responses. Therefore, this loop control action, added to the reference generator output, governs the internal control loop.

A. Current Reference Generator

The developed reference generation algorithm is presented in Fig. 2. Such algorithm is based on the p-q theory [11], where the srAPF reference currents are derived from the active power demanded from the grid ($\overline{p}_s^*$), which is equal to the sum of the DC bus control action ($p_{3\phi_{dc}}$) with the calculated load average active power ($\overline{p}_l$).

Once the $\vec{i}_{sr,\alpha\beta}^*$ must be in-phase with the fundamental-frequency positive-sequence grid voltage ($\vec{v}_{s,\alpha\beta}^{+1}$), a GDSC-PLL algorithm is implemented to ensure the desired conditions of synchronization [12].

B. Input Current Control

Considering the circuit depicted in Fig 3, the input current space vector in the frequency domain, obtained through Kirchoff's law, is presented in (1).

$$\vec{I}_{sr}(s) = G_{i_{sr}}(s) \left(\vec{V}_i(s) - \vec{V}_{sr}(s) \right), \qquad (1)$$

where $G_{i_{sr}}(s)$ is given by

$$G_{i_{sr}}(s) = \frac{1}{R_{L_{sr}} + sL_{eq}} \qquad (2)$$

and $L_{eq} = L_{ct} + L_{sr}$.

It can be observed through (1), that the srAPF voltage (v_{sr}) acts as a disturbance in the input current model. Thus, this control loop must be able not even to ensure a sinusoidal reference tracking, as to reject harmonic disturbance, once, harmonic components may appear in the v_{sr} voltage. Therefore, a proportional-resonant controller, with multiple resonant control actions, has been chosen to reach a suitable control performance.

The resonant terms have been implemented at 60 Hz and at typical harmonic frequencies, which may appear sometimes at the grid voltage (i.e., 3rd, 5th, 7th, 11th and 13th). Besides, resonant terms with computation delay compensation ($R_{i,comp}^h$), whose transfer function is presented in (3), have been chosen in order to improve system stability [13].

$$R_{i,comp}^h(s) = k_{ri}^h \cdot \frac{s\cos(\phi_i^h) - h\omega_1\,sen(\phi_i^h)}{s^2 + (h\omega_1)^2}. \qquad (3)$$

The input current controller was designed based on the frequency response analysis in the discrete-time domain considering the control loop depicted in Fig. 4. Besides, it can be seen that the v_{sr} was used as a feed-forward action aiming to improve the dynamic response. In addiction to the gain and phase stability margins, the minimum distance of the Nyquist diagram to the critical point (η) [14] was also considered. The selected crossover frequency was $f_{c_{i_{sr}}} = 1500$Hz. Therefore, the Bode and Nyquist plots of the input current open loop transfer function ($OLTF_{i_{sr}}(z)$) are presented in Fig. 5.

As can be seen, the controlled system reached $GM = 5,4$ dB at $2,6$ kHz, $PM = 36,6°$ at $1,5$ kHz and $\eta = 0,39$, which should be considered suitable for sinusoidal current control purposes [14].

The resulting proportional gain has been $k_{p_{i_{sr}}} = 43,08$ and the parameters of each $R_{i,comp}^h$ term implemented are presented in TABLE II.

C. DC-bus Voltage Control

The DC-bus voltage is governed through the active power balance between DC and AC sides, what means that the total average power required to regulate the DC voltage and to

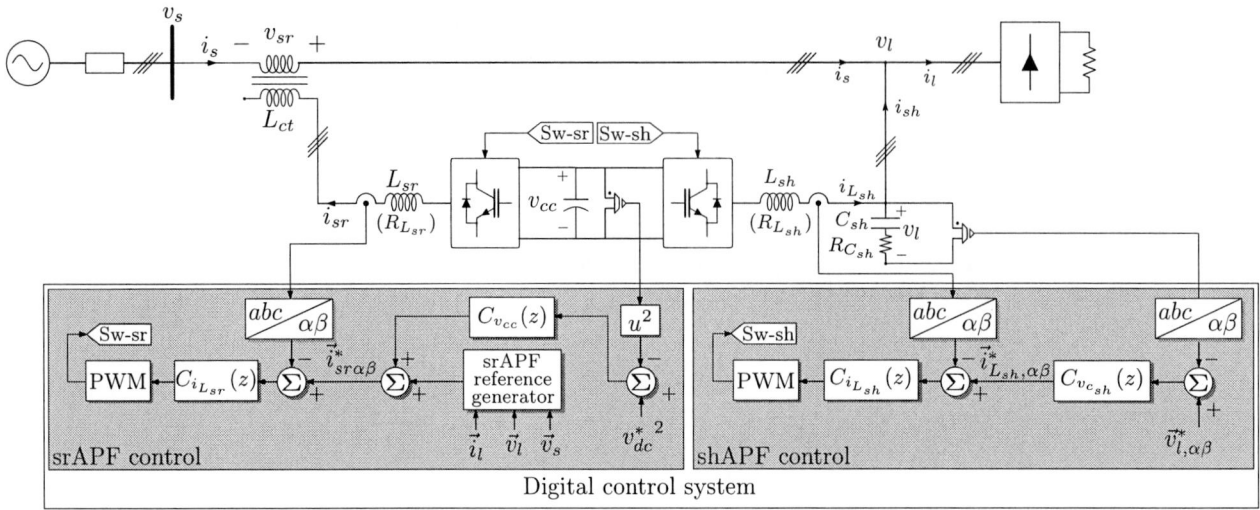

Fig. 1: Developed iUPQC's control scheme.

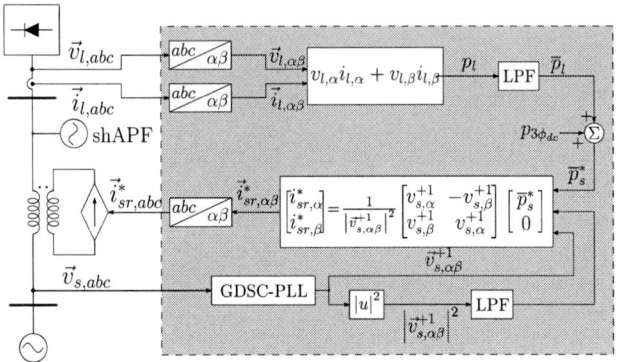

Fig. 2: srAPF sinusoidal current reference generator.

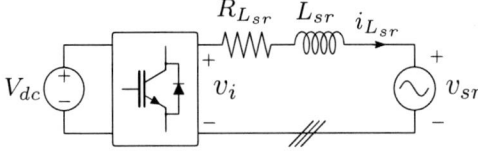

Fig. 3: Equivalent circuit for srAPF internal loop modeling.

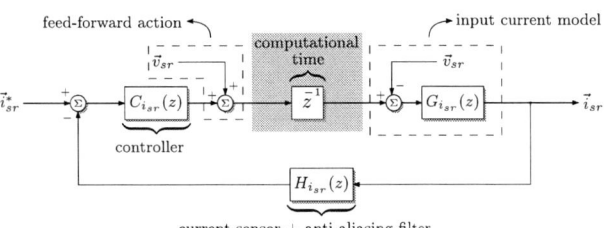

Fig. 4: srAPF input current control loop in discrete-time domain.

supply converters losses is equal to the amount of active power drained from the grid at the AC side. Therefore, a modeling, which related the DC voltage square value (v_{dc}^2) and the grid

TABLE II: Parameters of the srAPF controller resonant terms with computational delay compensation capability.

k_{ri}^h						
k_{ri}^1	k_{ri}^3	k_{ri}^5	k_{ri}^7	k_{ri}^9	k_{ri}^{11}	k_{ri}^{13}
4000	4000	4000	2000	2000	2000	2000

ϕ_i^h (rad)						
ϕ_i^1	ϕ_i^3	ϕ_i^5	ϕ_i^7	ϕ_i^9	ϕ_i^9	ϕ_i^{13}
$\pi/10$	$\pi/7$	$\pi/5$	$\pi/4$	$\pi/4$	$\pi/4$	$\pi/4$

active power demanded by the DC-bus control ($p_{3\phi}$), was developed, resulting in the transfer function presented in (4)

$$G_{v_{dc}}(s) = \frac{2}{sC_b},\qquad(4)$$

where C_b is the DC-bus capacitance.

The choice for this form of modeling was due to the srAPF reference generator, at which the total active power demanded from the grid to regulate the DC-voltage is added to the load average active power. Thus, the DC voltage controller action is equivalent to $p_{3\phi_{dc}}$, which has a linear dynamic relation with v_{dc}^2.

Considering above presented model, given by (4), the DC voltage controller can be designed based on the control loop shown in Fig 6.

The controller design was developed based on the frequency response method in the auxiliar domain w [15]. As previously described, the DC-bus control loop must have a low bandwidth for ensuring the decoupling between the srAPF control loops. Thus, a 6 Hz crossover frequency ($f_{c_{v_{cc}}}$) was set to this control loop. Besides, a linear proportional-integral controller has been chosen, since it is capable to ensure zero steady-state error for DC signals. Therefore, the frequency response of the controlled open loop transfer function in w-domain ($OLTF_{v_{dc}}(w)$) is shown in Fig. 7.

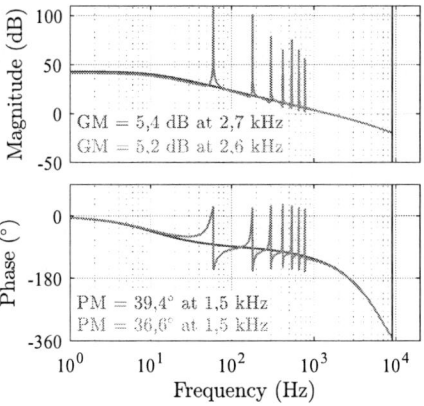

(a) Bode plot. Response without the resonant terms in blue and with resonant terms in green.

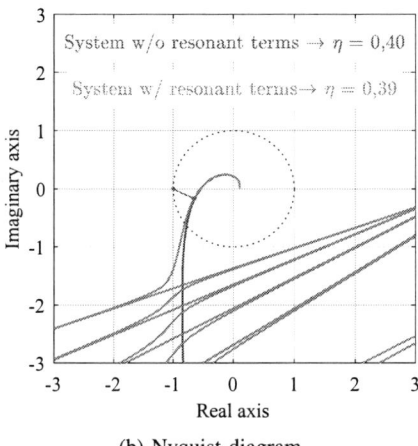

(b) Nyquist diagram.

Fig. 5: Bode and Nyquist diagrams for the input current open loop transfer function.

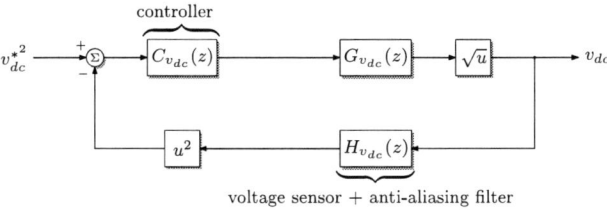

Fig. 6: DC-bus voltage control loop

As can observed, the controlled system reached a gain margin (GM) of $65,9\,\text{dB}$ at $16,2\,\text{kHz}$ and a phase margin (PM) of $71,5°$ at a crossover frequency of $6\,\text{Hz}$. Therefore, not only the GM and PM were acceptable, but also the crossover frequency matches with its desired value. The proportional and integral resulting gains were $k_{p_{v_{dc}}} = 0,12$ and $k_{i_{v_{dc}}} = 1,57$, respectively.

IV. SHAPF CONTROL

According to the iUPQC operation principles, the shAPF must impose a low distorted sinusoidal voltage, which is in-phase with the grid voltage, at the load bus. Thus, a fast

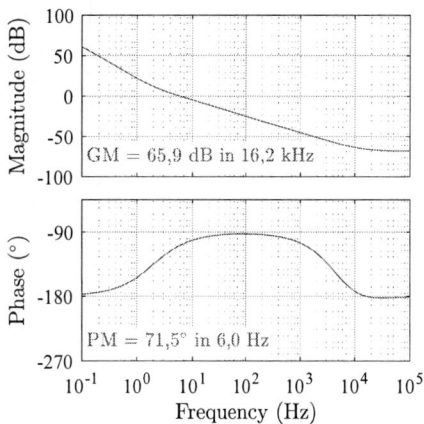

Fig. 7: Frequency response of the DC-bus voltage controlled open loop transfer function.

voltage control loop is demanded, what is normally reached by implementing a single-loop voltage control scheme [16]. However, when the single-loop control scheme is selected, the protection the would be provided by the internal current control loop is lost. Therefore, a double-loop voltage control has been selected for the shAPF control loop, as presented in Fig. 1.

A. Voltage Reference Generator

The developed reference generation algorithm for the shAPF is presented in Fig. 8. As can be seen, the GDSC-PLL provides the grid voltage unitary space vector which is multiplied by the rated grid voltage amplitude. Therefore, the resulting reference space vector is always at rated value and is in-phase with the instantaneous grid voltage space vector.

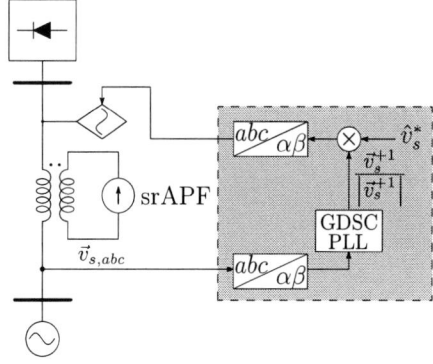

Fig. 8: shAPF sinusoidal voltage reference generator

The proposed reference generator acts as a SRF-compensator [5] but without demanding any transformation for dq reference frame, what reduces the computational effort and is desirable for control implementation in $\alpha\beta$ reference frame.

B. Double-Loop Load Voltage Control

As previously described, the double-loop voltage control is suitable for controlling the inverter's output current enabling its saturation for switches protection. Nevertheless, the implementation of this voltage control scheme becomes particularly difficult for sinusoidal reference tracking, mainly when high distorted currents are supplied by the inverter. This difficulty is due to the need of high bandwidth for both external and internal control loops, once the internal loop bandwidth is limited by the switching frequency. Thus, the decoupling between loops becomes quite compromised.

This issue may be solved by adopting a dead-beat controller in the internal control loop, which ensures a very fast closed loop response with a delay of only two sampling periods [17]. Therefore, it is possible to increase the external loop bandwidth, without causing loops interaction.

1) Internal Control Loop: The internal control loop model has been developed the same of the input current model of the srAPF, presented in Section III-B, thus, its model is given by

$$G_{i_{sh}}(s) = \frac{1}{R_{L_{sh}} + sL_{sh}}. \tag{5}$$

In order to reduce the effect of the difference between the modeled and real system in the controller design, the transfer function of the anti-aliasing filters was considered in the plant. Thus, the conventional dead-beat controller design procedure, which considers the dynamic equation of the inductor's current, may not be applied. Therefore, a digital controller with dead-beat response has been designed on the basis of the procedure described in [18], considering the control loop depicted in Fig. 9.

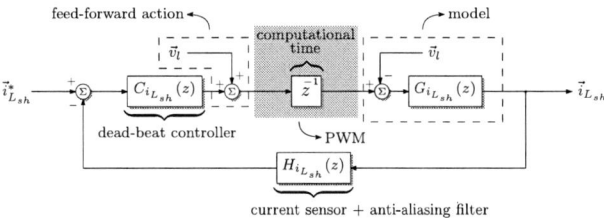

Fig. 9: shAPF internal control loop in discrete-time domain.

The resulting dead-beat response controller transfer function is presented in (6).

$$Ci_{L_{sh}}(z) = \frac{z^3 - 1,037z^2 + 0,04294z}{0,01937z^3 - 0,01409z - 0,005284} \tag{6}$$

2) External Control Loop: Considering the circuit depicted in Fig 10, the load voltage space vector in the frequency domain, obtained through Kirchoff's law, is presented in (7).

$$\vec{V}_l = G_{v_l}(s)\left(\vec{I}_{L_{sh}} - \vec{I}_{sh}\right), \tag{7}$$

where $G_{v_l}(s)$ is given by

$$G_{v_l}(s) = \frac{sR_{C_{sh}}C_{sh} + 1}{sC_{sh}}. \tag{8}$$

It can be seen through (7), that the current provided by the shAPF to the load ($\vec{i}_{sh}$) acts as a disturbance in the load voltage model. Therefore, besides ensuring a sinusoidal

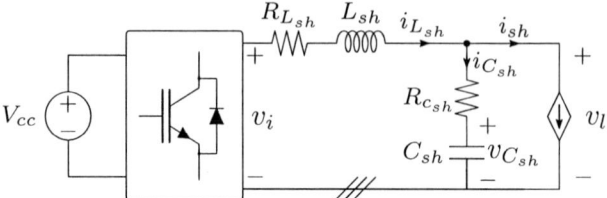

Fig. 10: Equivalent circuit for shAPF external loop modeling.

reference tracking, the shAPF external control loop must be able to reject the harmonic disturbance due to non-linear load currents. Thus, a proportional-resonant controller, with multiple resonant terms, has been also chosen to reach a suitable control performance.

In this case, besides the action with 60 Hz of resonant frequency, additional terms with resonant frequency that match the typical low order harmonics for a three-phase rectifier (i.e., $6k \pm 1$ for $k = 1, 2, 3$ and 4) were also implemented.

The load voltage controller was designed based on the frequency response analysis in the discrete-time domain, as the same way as the input current controller design presented in Section III-B. The control loop considered for this design is depicted in Fig. 11. It can be noticed that the internal loop, controlled by a dead-beat control action, has been represented as a delay of two samples [17].

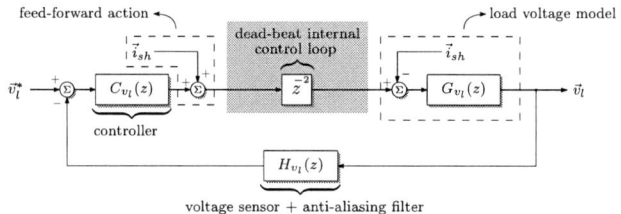

Fig. 11: shAPF external control loop in discrete-time domain.

The selected crossover frequency was $f_{c_{v_l}} = 1600\,\text{Hz}$, once the last resonant term has been implemented at the 25th order harmonic (1,5kHz). Therefore, the Bode and Nyquist plots of the load voltage open loop transfer function are presented in Fig. 12.

As can be seen, the controlled system reached $GM = 2,1\,\text{dB}$ at 2,4 kHz, $PM = 26,5°$ at 1,7 kHz and $\eta = 0,21$, which has been considered acceptable, in order to avoid compromising the dynamic performance. The resulting proportional gain has been $k_{p_{i_{sr}}} = 0,20$ and the parameters of each $R_{i,comp}^h$ term implemented are presented in TABLE III.

V. EXPERIMENTAL RESULTS

A three-phase iUPQC prototype was used in order to experimentally validate the developed control scheme. The system has been digitally controlled by a dSPACE microprocessor using its real-time-interface along with a computer. Regarding the experimental tests, the conditioner has been connected to a 127 V_{RMS} three-phase grid with a 380 V_{dc} DC-bus voltage. The load setup used was formed by a three-phase diode

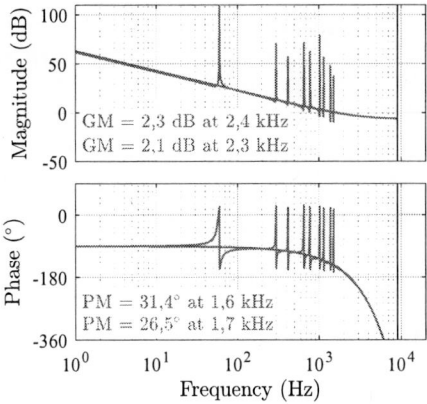

(a) Bode plot. Response without the resonant terms in blue and with resonant terms in orange.

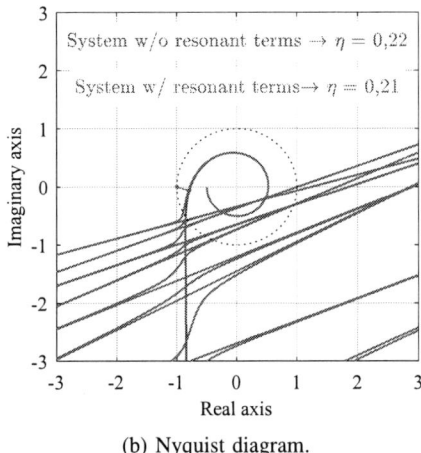

(b) Nyquist diagram.

Fig. 12: Bode and Nyquist diagrams for the load voltage open loop transfer function.

TABLE III: Parameters of the shAPF controller resonant terms for compensating the two-sample delay of the internal-loop dead-beat control.

k_{rv}^h								
k_{rv}^1	k_{rv}^5	k_{rv}^7	k_{rv}^{11}	k_{rv}^{13}	k_{rv}^{17}	k_{rv}^{19}	k_{rv}^{23}	k_{rv}^{25}
10	8	4	20	20	10	10	5	5
ϕ_v^h (°)								
ϕ_v^1	ϕ_v^5	ϕ_v^7	ϕ_v^{11}	ϕ_v^{13}	ϕ_v^{17}	ϕ_v^{19}	ϕ_v^{23}	ϕ_v^{25}
25	36	36	50	50	60	60	70	70

rectifier - with commutation inductors of $L_{com} = 1,8mH$ and a resistive load of $R_{dc} = 36\,\Omega$ at the DC side - along with a linear load of $R_{ac} = 48\,\Omega$ and $L_{ac} = 128mH$. The complete experimental setup is depicted in Fig. 13.

The measured sinusoidal quantities are depicted in Fig. 14. Regarding the current compensation capability, it can be verified, in Fig. 14a, that the grid current has a sinusoidal waveform shape with a low THD_{i_s} of 1,14%, while the load current is highly-distorted with $THD_{i_l} = 20,13\%$. On

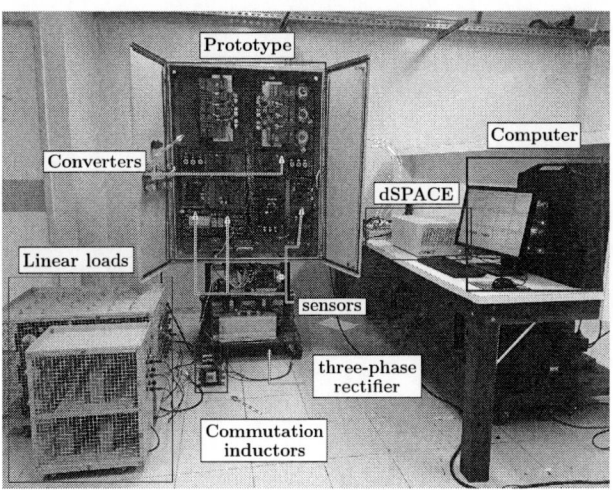

Fig. 13: Experimental setup.

the other hand, the voltages of the system are depicted in Fig. 14b. Although the THD of the grid voltage is low ($THD_{v_s} = 1,96\%$), it can be observed that even with a highly distorted current provided by the shAPF, the conditioner was able to provide a very low distorted load voltage with only 0,45% of THD_{v_l}.

Lastly, the regulated DC-bus voltage is depicted in Fig. 15. As can be seen, such voltage matches its reference value except for the 60 Hz ripple, which was chosen not to be controlled by the DC-bus voltage control loop.

VI. CONCLUSION

This paper presents a iUPQC control scheme for ensuring an improved compensation performance. The reference generators and control schemes of both converters, as well as the steps of designing each controller are presented. The experimental results, obtained through a three-phase prototype, demonstrate the effectiveness of the developed scheme in suitably compensating grid voltage and load current related issues.

REFERENCES

[1] H. Akagi and H. Fujita, "A new power line conditioner for harmonic compensation in power systems," *IEEE Transactions on Power Delivery*, vol. 10, no. 3, pp. 1570–1575, July 1995.

[2] H. Fujita and H. Akagi, "The unified power quality conditioner: the integration of series and shunt-active filters," *IEEE Transactions on Power Electronics*, vol. 13, no. 2, pp. 315–322, Mar 1998.

[3] M. Aredes, K. Heumann, and E. H. Watanabe, "An universal active power line conditioner," *IEEE Transactions on Power Delivery*, vol. 13, no. 2, pp. 545–551, April 1998.

[4] L. F. C. Monteiro, M. Aredes, and J. A. M. Neto, "A control strategy for unified power quality conditioner," in *2003 IEEE International Symposium on Industrial Electronics (Cat. No.03TH8692)*, vol. 1, June 2003, pp. 391–396.

[5] S. A. Oliveira da Silva, P. Donoso-Garcia, P. C. Cortizo, and P. F. Seixas, "A three-phase line-interactive ups system implementation with series-parallel active power-line conditioning capabilities," in *Conference Record of the 2001 IEEE Industry Applications Conference. 36th IAS Annual Meeting (Cat. No.01CH37248)*, vol. 4, Sep. 2001, pp. 2389–2396 vol.4.

978-1-7281-4181-7/19 $31.00 © 2019 IEEE

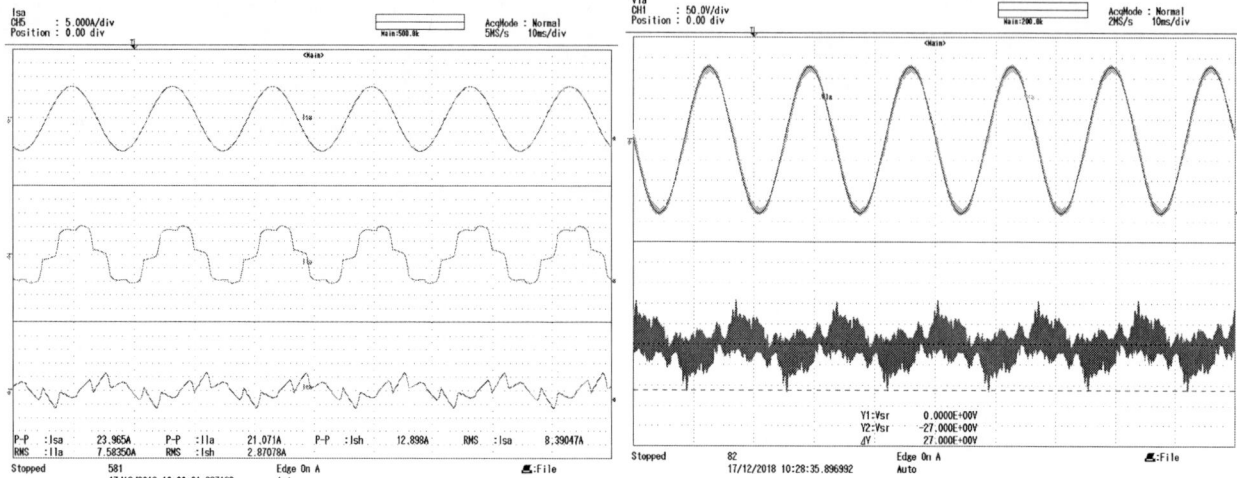

(a) Current signals. grid current in red; load current in blue; shAPF current in puple. Scale: current - 5A/div; time - 10ms/div.

(b) Voltage signals. grid voltage in yellow; load voltage in blue; coupling transformers voltage in red. Scales: voltages - 50V/div (upper scope); 12,5V/div (bottom scope); time - 10ms/div.

Fig. 14: Experimentally measured sinusoidal quantities.

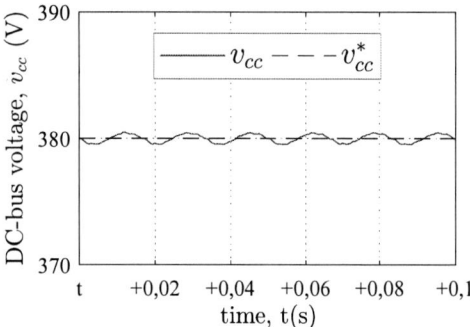

Fig. 15: DC-link voltage

[6] M. Aredes and R. M. Fernandes, "A dual topology of unified power quality conditioner: The iupqc," in *2009 13th European Conference on Power Electronics and Applications*, Sept 2009, pp. 1–10.

[7] S. A. Oliveira da Silva, P. Donoso-Garcia, P. C. Cortizo, and P. F. Seixas, "Performance analysis of three-phase line-interactive ups system with active power-line conditioning," in *IECON'03. 29th Annual Conference of the IEEE Industrial Electronics Society (IEEE Cat. No.03CH37468)*, vol. 1, Nov 2003, pp. 353–360 vol.1.

[8] B. W. Frana, L. F. da Silva, and M. Aredes, "Comparison between alpha-beta and dq-pi controller applied to iupqc operation," in *XI Brazilian Power Electronics Conference*, Sep. 2011, pp. 306–311.

[9] B. W. Frana, L. F. da Silva, M. A. Aredes, and M. Aredes, "An improved iupqc controller to provide additional grid-voltage regulation as a statcom," *IEEE Transactions on Industrial Electronics*, vol. 62, no. 3, pp. 1345–1352, March 2015.

[10] R. A. Modesto, S. A. Oliveira da Silva, and A. A. de Oliveira Jnior, "Power quality improvement using a dual unified power quality conditioner/uninterruptible power supply in three-phase four-wire systems," *IET Power Electronics*, vol. 8, no. 9, pp. 1595–1605, 2015.

[11] H. Akagi, Y. Kanazawa, and A. Nabae, "Generalized theory of the instantaneous reactive power in three-phase circuits," in *IPEC'83 - Int. Power Electronics Conf.*, 1983, pp. 1375–1386.

[12] F. A. S. Neves, M. C. Cavalcanti, H. E. P. de Souza, F. Bradaschia, E. J. Bueno, and M. Rizo, "A generalized delayed signal cancellation method for detecting fundamental-frequency positive-sequence three-

phase signals," *IEEE Transactions on Power Delivery*, vol. 25, no. 3, pp. 1816–1825, Jul. 2010.

[13] A. G. Yepes, F. D. Freijedo, . Lopez, and J. Doval-Gandoy, "High-performance digital resonant controllers implemented with two integrators," *IEEE Transactions on Power Electronics*, vol. 26, no. 2, pp. 563–576, Feb 2011.

[14] ——, "Analysis and design of resonant current controllers for voltage-source converters by means of nyquist diagrams and sensitivity function," *IEEE Transactions on Industrial Electronics*, vol. 58, no. 11, pp. 5231–5250, Nov 2011.

[15] K. Ogata, *Discrete-time Control Systems (2nd Ed.)*. Upper Saddle River, NJ, USA: Prentice-Hall, 1995.

[16] X. Li, P. Lin, Y. Tang, and K. Wang, "Stability design of single-loop voltage control with enhanced dynamic for voltage-source converters with a low lc-resonant-frequency," *IEEE Transactions on Power Electronics*, vol. 33, no. 11, pp. 9937–9951, Nov 2018.

[17] S. Buso and P. Mattavelli, *Digital Control in Power Electronics*, 2nd ed. Morgan & Claypool Publishers, 2015.

[18] F. Golnaraghi and B. C. Kuo, *Automatic Control Systems*, 9th ed. Wiley Publishing, 2009.

978-1-7281-4181-7/19 $31.00 © 2019 IEEE

Aspects of travelling wave based protection philosophy for consideration in DC Grids of the future

Monday Ikhide
Faculty of Engineering,
Environment and Computing,
Coventry University, UK
monday.ikhide@coventry.ac.uk

Sarath Tennakoon
College of Engineering
Carnegie Mellon University,
CMU-Africa Kigali, Rwanda
stennako@andrew.cmu.edu

Alison Griffiths
School of Creative Arts & Engineering
Staffordshire University,
Stoke-on-Trent, UK
a.l.Griffiths@staffs.ac.uk

Hengxu Ha
Grid Solutions,
GE Renewable Energy Stafford – UK
hengxu.ha@ge.com

Abstract—This paper presents a transient based protection principle utilising the magnitude of the travelling wave power derivative at only local terminal to formulate a non-unit protection scheme, where the directional transient relay and boundary relay have been involved. In terms of boundary relay, for an internal fault in the forward direction with respect to a local relay, the traveling wave power derivative will exceed a pre-determined setting, otherwise the fault is external to the relay, due to the discontinuity of surge impedance at the boundary point. Directional detection is achieved by the comparison of the forward travelling wave power to the backward travelling wave power. The ratio of forward traveling wave to backward traveling wave is less than unity for all forward directional faults and much greater than unity for reverse directional faults. Simulation studies carried out in PSCAD and the results presented shows the suitability of the proposed protection principle as all fault scenario indicated were detected within $250\mu s$ following the application of the faults.

Keywords—HVDC grids; DC Line protection; travelling wave power derivative, forward internal and forward external faults

I. INTRODUCTION

Grid protection remains a major issue in the development of DC interconnections, leading to DC grids [1][2]. Generally, DC grids are considered to be the best option for utilising the full potential of offshore wind energy in the north sea and future exploitation of the solar resources in North Africa; whereas the realisation would require reliable protection schemes. This is because protection algorithms for conventional HVAC systems are not suitable for the protection of DC grids due to the characteristic differences in their fault current profiles. In general, DC grids are susceptible to overcurrent which is largely due to the high rate of propagation of fault currents resulting from the low inductance compared to AC interconnections. Therefore selective and non-unit protection algorithms are required to detect and isolate these faults to prevent the power electronic converters from damage. Generally, converters for future DC grids will be based on Voltage Source Converters (VSC) technology, where the Modular Multilevel Converters (MMCs) has been identified as a viable option. This is due to their numerous advantages over the two-level voltage VSCs such as low switching frequency resulting in reduced losses , scalability, flexibility in control of voltage level, and the ability to control the submodule voltage to synthesize the desired waveform[2][3]. MMCs can either be of *half-bridge* or *full-bridge* sub-module arrangements and are generally referred to as *non-blocking* and *blocking* converters respectively (Fig. 1). The term "*blocking*" implies the ability to *block* or *oppose* the fault current. Details have been well documented by Whitehouse [1]. The types of fault that could occur on a DC grid are (*i*) DC side faults such as pole-to-ground faults (P-G), pole-to-pole (P-P) faults (*ii*) AC Side faults such as three phase short-circuits, phase to ground and phase to phase faults (*iii*) converter internal faults such as sub-

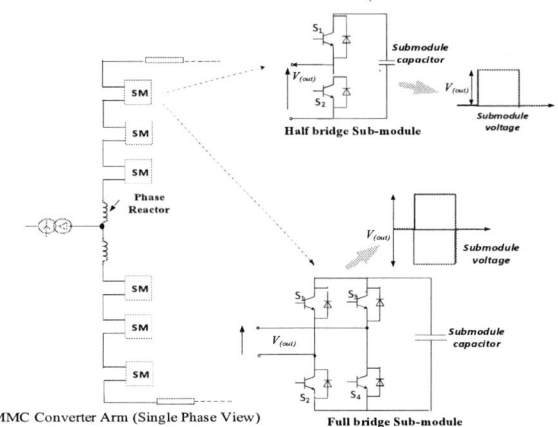

Fig. 1 MMC Terminology and configurations[1]

module internal faults (*iv*) Faults at busbar terminal or a short circuit in the DC inductor [3]. This study focuses on the DC side faults, and hence consider the P-P and P-G faults.

II. OPTIONS AND STRATEGIES FOR DC GRID PROTECTION

Three possible protection strategies have been proposed in literature [2][4][5]. The first option is to open all AC side breakers to isolate the entire DC grid from the AC system once a DC side fault is detected (with the *MMC* being a *half-bridge* sub-module arrangement). Thereafter a mechanical isolator is used to isolate the faulty section. Following this, the entire system is restored back. The major drawback of this strategy is that in the event of a fault, the entire grid is de-energised thereby disrupting the continuity of service in the healthy sections of the grid and other associated sub-grids that may be connected to the network. The second option is the use of *blocking converters* (as per full bridge MMC) [1][5], to oppose the fault current following a DC short circuit; and there after using a fast mechanical isolator to isolate the fault section. This strategy doesn't need the use of HVDC breakers which ultimately will reduce cost. However, similarly as in option 1, this strategy will also de-energise the entire DC grid. In option 3, a DC side circuit breaker is proposed to isolate only the faulty section, while maintaining service delivery in the healthy section. The advantage of this strategy is that only the faulty section is isolated in the event of fault while maintaining service delivery on the healthy sections. This study adopts option 3, and therefore the protection strategies and algorithms discussed in the following sections assumes DC breakers located at both ends of a DC transmission line. It is anticipated that DC breakers will be commercially available in the near future considering the recent breakthrough where prototypes DC breakers have been developed [6][7], following successive attempts dated back to 1980's [8].

978-1-7281-4181-7/19 $31.00 © 2019 IEEE   227

III. FAULT CHARACTERISATION IN HVDC GRIDS

In developing protection algorithms for DC grids, the starting point is to understand the characteristics of the different types of fault with a view to predicting their magnitude and the rate of rise. To achieve this, simulations were carried out in PSCAD in the first instance based on CICRE benchmark for four terminal MMC-DC grid [9]. However, the network was modified to reflect a 4- terminal meshed network, as per the scenario considered in this paper (Fig. 2). As shown, converters 1, 2 and 3 are connected to AC sources respectively, whilst converter 4 is connected to a load. AC sources represent generation such as wind farms either an onshore or offshore and could represent an AC grid. Each converter station is capable of operating either as a *rectifier* or an *inverter,* and as such power can flow in either direction, that is, from the AC side into the DC grid or vice versa. The MMCs are of HB submodules and therefore this configuration would require HVDC breakers located at the cable ends. Each cable section is 200km in length. Details of the converter parameters and cable configurations are given in Appendix 1. The inductors located at the cable ends are representative of the inductive effect of the HVDC breaker [2]. The reference direction for current for a relay is that from the bus into the cable, just as shown in Fig. 2. Still considering Fig. 2 with respect to relay R_{12} in the first instance, F_{12} (close-up fault) and F_{21} (remote fault) are internal faults whereas all other faults indicated are external faults. In addition, F_{14} is a reverse fault whilst F_{12}, F_{21} and F_{23} forward faults. Therefore, the relay must distinguish between (*i*) forward and reverse faults (*ii*) forward internal and forward external faults (FIF and FEF); thus operating for only faults along cable section 1 as shown. The same scenario holds for all other faults indicated with respect to other relays shown Generally, the worst case scenario for relay R_{12} is to discriminate between a high resistance remote *FIF* (F_{21}=500Ω) and a low resistance FEF (e.g. F_{23}=0.01Ω as shown) on an adjacent feeder close to the busbar.

A. Pole-Pole versus Pole-ground fault

To investigate the characteristic differences between the voltage and current characteristics following a *P-P* or a *P-G* faults, the voltages and current signals recorded at the relay terminal for both the positive pole and negative pole were measured and recorded as per Fig. 2. All faults were applied along cable section 1; at 2s following the start of the simulation. As shown in Fig. 3, a *P-P* fault results in a high magnitude of fault currents which is driven by the converter. Both the positive and negative pole voltages collapse to zero following the application of the fault. However in Fig. 4, a *P-G* fault results in a significant shift in the healthy pole voltage, typically up

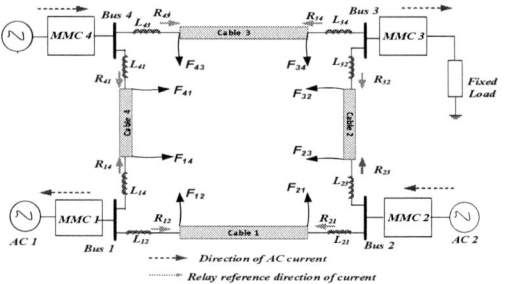

Fig. 2 Four Terminal MMC-HVDC Test Grid.

to 2*pu*; whilst the faulty pole voltage collapse suddenly. The converter will experience a sudden high transient current owing to the discharge of the energy store in the transmission system as shown. The excess voltage on the healthy cable can results in excessive stress on the DC cables if the fault is not cleared during the first few milliseconds following fault inception.

B. Effect of fault distance

Studies were also carried out to establish the effect of fault distance on the resulting fault current profile following the occurrence of fault. The fault distance was varied as 50km, 100km, 150km, 200km, and 400km; with respect to relay R_{12}. As shown in Fig. 5, the magnitude of the fault current and voltage following the occurrence of the fault varies with the fault distance.

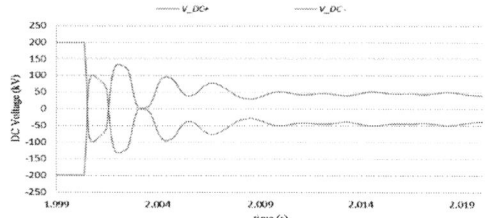

(a) Positive and negative pole voltages; v_{DC}^+ and v_{DC}^-

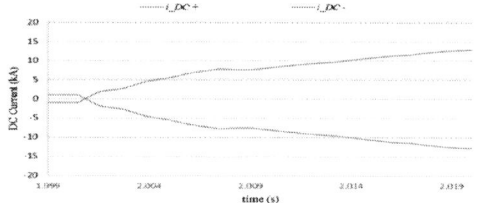

(a) Positive and negative pole currents ; i_{DC}^+ and i_{DC}^-

Fig. 3. Voltages and Current for a Pole-Pole faults

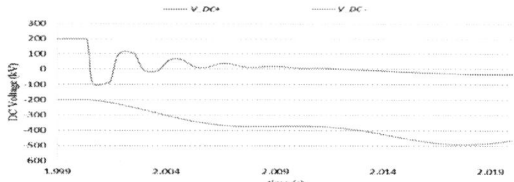

(a) Positive and negative pole voltages; v_{DC}^+ and v_{DC}^-

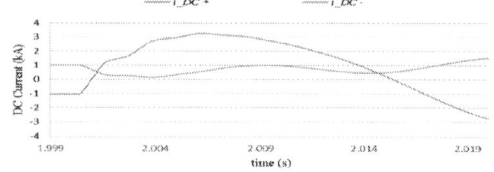

(a) Positive and negative pole currents ; i_{DC}^+ and i_{DC}^-

Fig. 4. Voltages and Current for a Pole-Ground faults

978-1-7281-4181-7/19 $31.00 © 2019 IEEE

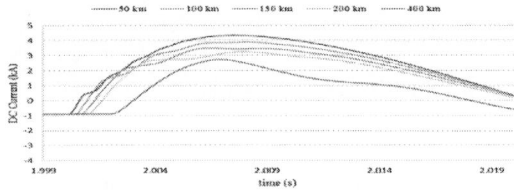

(a) Positive Pole current, i_{DC}^+

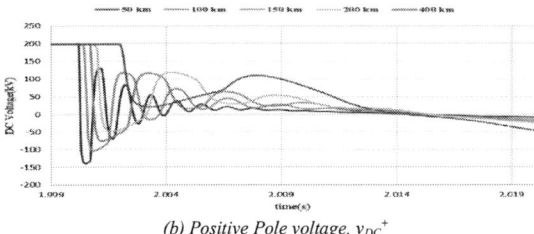

(b) Positive Pole voltage, v_{DC}^+

Fig. 5 Effect of fault distance

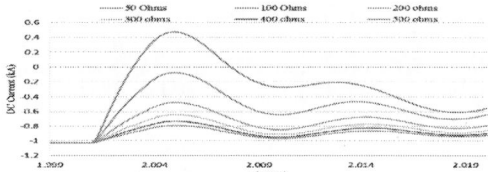

(a) Positive Pole current, i_{DC}^+

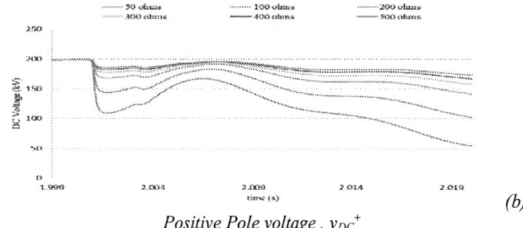

Positive Pole voltage , v_{DC}^+

(b)

Fig. 6 Effect of fault resistance

Fig. 7 Boundary conditions

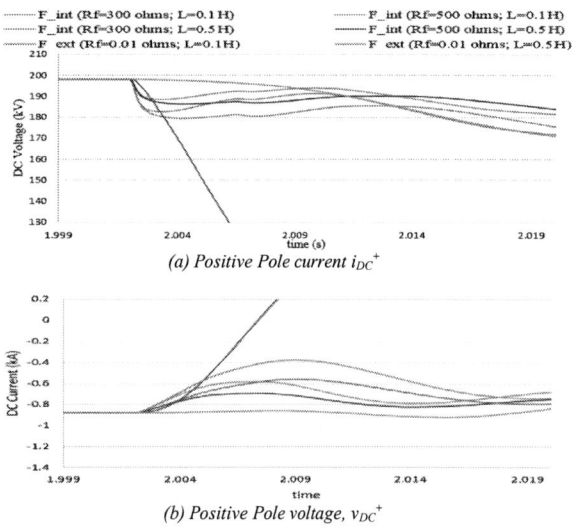

C. Effect of fault resistance

The effect of fault resistance on the voltage and current following the occurrence of fault was also investigated by simulations. For this purpose, the fault distance was kept fixed at 200km and the fault resistances were varied. The simulation results presented in Fig. 6 also revealed that the magnitude and the rate of change (*RoC*) of the DC pole current and voltage following the occurrence of fault also depends largely on the fault resistance. The higher the fault resistance, the more the waveforms are damped thus reducing their magnitude and *RoC*.

D. Effect of the boundary

In this study, the boundary refers to the busbar and the DC link inductors at the cable ends. In practice, inductors are representative of the inductive effect of the HVDC breaker and fault current limiters as per Fig. 2. The effect of variation in the DC side inductors was also considered in the simulations. For this purpose, high fault resistances (R_f = 300Ω and 500Ω) were considered for the internal faults whilst a low resistance (R_f = 0.01Ω) external fault, where the subscript *"int"* and *"ext"* depicts *internal* and *external* respectively. These scenarios (Fig.7) were assumed to represent the most critical conditions for relay R_{12}. The plots of the currents and voltage considering F_{int} and F_{ext} for varying inductor sizes at the boundaries are presented in Fig. 8. As shown, the characteristic differences in the magnitude and *RoC* of the voltages and currents shown is largely due to the differences in the inductors at the boundaries. The larger the inductors, the more the current and voltage are attenuated. For an internal fault with low fault resistance (F_{ext} = 0.01Ω), the magnitude of the current exceeds that for high resistance internal fault (F_{int} = 500Ω). Generally, this scenario represent the most critical conditions for the relay R_{12} shown in Figure 1. However, by increasing the size the DC inductor from 0.1H to 0.5H, the magnitude was significantly reduced. This shows that the DC link inductor has a significant effect on the fault current following fault. This is also the case for the voltage. Cleary, the magnitude and *RoC* of the associated current and voltage following a fault depends on several factors such as the fault resistance, fault distance and the boundary condition

(a) Positive Pole current i_{DC}^+

(b) Positive Pole voltage, v_{DC}^+

Fig. 8 Effect of varying boundary conditions

IV. PROTECTION ALGORITHMS FOR DC GRIDS

The general requirements for DC network protection with the third strategy presented in section II are the same as those for AC systems and includes selectivity, speed, sensitivity and security. However as indicated in [10], the above four attributes are in contradiction with each other. For example, if the protection scheme has higher sensitivity, then it must somehow be less secure; if the protection has higher selectivity, then it must lose some speed. Therefore a compromise must be reached when designing or developing protection algorithms for DC grids but should be without prejudice to the reliability and security of service delivery. However, researchers and system developers have long worked on the compromise of these four attributes to provide a secure and reliable protection for power system. Generally, studies have shown that transient based protection algorithms are ideal candidates for DC grid if the protection system must be reliable. Transient based protection algorithm can be broadly classified into two: *unit* and *non-unit* schemes. Those relying on information from the local terminal only are referred to as non-*unit* protection scheme while those relying on information from both the local and remote terminals are referred to as *unit* protection scheme. The information refers to the current and voltage superimposed components recorded at the relaying terminal following fault inception. Example of unit protection schemes are techniques comparing the polarity of the incremental quantities following fault inception at both local and remote end relays; proposed in [11] in the 80's. Notable examples of non-unit schemes are current derivative, and voltage derivative. Distance protection schemes developed for AC transmission can either be unit or non-unit based depending on the requirement. Considering the need for fast and intelligent DC line fault detection algorithm, this study adopts the transient and non-unit protection scheme. As earlier stated, two criteria must be established to provide full discrimination on a particular branch or section of a DC grid with respect to a local relay. These are *(i)* criterion for directional comparison *(ii)* criterion for forward internal versus forward external fault.

A. Directional comparison technique

Considering cable section *AB* of Fig. 9; and with respect to relays R_{12} and R_{21}, F_1 is an internal fault whilst F_2 and F_3 are external fault. For an internal fault, the polarity of the superimposed transient voltage and current (Δi and Δv respectively) at both terminals of the protected line are of opposite signs [11]. However, for external faults, the polarity is the same at one terminal and opposite at the other terminal. This is illustrated in Table 1. Directional comparison techniques have also been achieved using the forward and backward travelling wave components' ratio at the respective relay terminals [12]. Generally, a fault on the DC link will results in an abrupt injection and travelling waves are generated which propagates towards the line terminations or busbar where the relays are located. These waves are captured at the relay terminals for fault identification. The fundamental principle is explained as given in Fig. 10. The DC inductors have been omitted for the sake of convenience; and the superscripts "*f*", "*b*" represents the forward and backward travelling wave respectively; and subscripts "*A*", "*B*" represents the bus bar terminal where the relays are located.

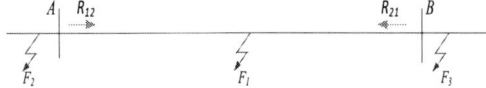

Fig. 9 Directional comparison utilising superimposed transient components

TABLE 1 POLARITY IDENTIFICATION

Fault	Terminal A		Terminal B	
	Δi	Δv	Δi	Δv
F_1	Positive	Negative	Positive	Negative
F_2	Negative	Negative	Positive	Negative
F_3	Positive	Negative	Negative	Negative

For a forward directional fault (*FDF*), the ratio of the forward travelling wave components to the backward travelling wave components is always less than unity during the travelling wave period; the converse is the case for reverse directional faults - *RDF* (less than unity). Details have been well documented in [12] and also adapted in [13] for directional comparison.

(a) F_1 is Forward Internal fault with respect to R_{12} and R_{21}

(b) F_2 is Reverse fault to R_{12} and Forward external with respect to R_{21}

(c) F_3 is Reverse fault to R_{21} and Forward external with respect to R_{12}

- - - - →   *Direction of travelling wave*

———→   *Relay reference direction of current*

Fig. 10 Travelling wave on a transmission

TABLE 2 DIRECTIONAL COMPARISON UTILISING TRAVELLING WAVE COMPONENTS RATIO

Fault	Terminal A	Terminal B
	$\dfrac{v^f}{v^b}$	$\dfrac{v^f}{v^b}$
F_1	< 1	< 1
F_2	> 1	< 1
F_3	< 1	> 1

Considering fault F_1 shown in Figure 10a, and following an abrupt injection, the first incident wave seen by R_{12} at terminal A is v^b_A and thereafter v^f_A is reflected. Also, R_{21} at terminal B sees v^b_B in the first instance, thereafter v^f_B is reflected. Therefore at the instance of the fault, $v^f_B < v^b_B$, implying that F_1 is an internal fault with respect to terminals "A" and "B". This is also the case for terminal "B" *(fault F_2)* and terminal "A"*(fault F_3)* in figures 10b and 10c respectively. However in terminal "A" of figure 10b and terminal "B" of figure 10c, the relays (R_{12} & R_{21} respectively) sees a forward travelling wave in the first instance, and thereafter a backward travelling wave. Therefore $v^f_A > v^b_A$ and $v^f_B > v^b_B$ in terminals A & B of figures 10b & 10c respectively.

B. Forward Internal versus forward external fault

In this study, the discrimination between a FIF and FEF was achieved by making use of the DC inductor which helps to block the high frequency contents in the fault induced transient components resulting from an external fault. Generally techniques relying on the attenuation produced at the boundary to distinguish between FIF and FEF can be classified into two: those utilising the magnitudes of the fault induced components (such as incremental changes in voltages and currents and power) and those utilising the *RoC* of these components such as current derivative, voltage derivatives and among others. Generally, the techniques utilizing the *RoC* would present the highest level of sensitivity and therefore was adopted in this paper. In general, the discriminative characteristic between FIF and FEF faults actually resides in the wave front of the traveling waves, which makes the *RoC* of current or voltage for an internal fault significantly different from that of an external fault, and hence the *RoC* of the travelling wave power.

V. PROPOSALS FOR DC GRID PROTECTION

In this paper, the character of the power developed by the forward and backward travelling wave, p^f and p^b was used to provide discrimination between a FIF and FEF. Thus from **[13]**,

$$p^f = \frac{\Delta v_{DC}(t)^2 + 2\Delta v_{DC}(t) \times \Delta i_{DC}(t) + \Delta i_{DC}(t) Z_C^2}{4 Z_C} \quad (1)$$

$$p^b = -\frac{\Delta v_{DC}(t)^2 - 2\Delta v_{DC}(t) \times \Delta i_{DC}(t) + \Delta i_{DC}(t) Z_C^2}{4 Z_C} \quad (2)$$

Δv_{DC}, Δi_{DC} are the incremental changes or superimposed components of the DC voltages and current recorded at the relay terminals following a fault; Z_C is the surge impedance of the DC cable.

$$\Delta v_{DC} = v_{DC(post-fault)} - v_{DC(pre-fault)}$$
$$\Delta i_{DC} = i_{DC(post-fault)} - v_{DC(pre-fault)}$$

For FDF: $\frac{p^f}{p^b} < 1$; For RDF: $\frac{p^f}{p^b} > 1$

Directional discrimination has been established and well documented in literature [13][14], however the contribution of this paper is to provide discrimination between FIF and FEF using a *non-unit* based approach with an improved sensitivity.

A. *Proposal utilising the magnitude of the power developed by the forward and backward travelling waves, p^f, p^b*

The ratio of p^f and p^b provides directional comparison whilst their magnitudes provides discrimination between FIF and FEF. Studies have been carried out in [11] regarding this approach and simulation studies carried out in PSCAD based on Fig. 2 is presented in Fig. 11. For simplicity, only cable section 1 is considered, and with R_{12} being the reference relay. Generally, this scenario is a representative of other cable sections on the grid. Two fault scenarios critical to the relays were considered - F_{21}=500Ω, F_{23}=0.01Ω; corresponding to a high resistance remote *FIF* and a low resistance *FEF*. It can be seen from Fig. 11 that discrimination between a FIF and FEF was achieved during the travelling wave period. Therefore measurements must be taken during this period. (typically less than *1ms* following the inception of fault). A proof-of-concept implementation on a COMPACT-RIO FPGA running a LABVieW interface is shown in Fig. 12. The simulation results from PSCAD was exported to an Excel file; and thereafter the data was imported on to the LabVIEW work space using the "read from spreadsheet" function. These data are the voltage (v_{DC}) and current (i_{DC}), recorded at the relay terminals as per the HVDC grid text model shown in Fig. 2. As shown in Fig. 12, the hardware comprises two Compact RIOs (CRIO 1 and CRIO2 respectively), two Personal Computers, oscilloscope, connecting leads and other accessories as shown. CRIO 1 serves as the signal generator that simulate an analogue signal based on the data store in the real time host (PC1). The resulting analogue output from CRIO1 is then sampled and the resulting signal fed into CRIO2 onto which the algorithm has been implemented. A digital output which is representative of a relay trip signal is displayed via an LED connected to the digital output of CRIO2 as shown. The resulting wave forms (Fig. 13) shows that the travelling wave components can be capture during the measurement period.

B. *Proposal utilising the derivative of the power developed by the forward and backward travelling waves, $\frac{dp^f}{dt}, \frac{dp^b}{dt}$*

The contribution of this paper is to improve the sensitivity and hence the *RoC* of p^f, p^b is adopted in this paper for fault identification; where the subscripts "*cal*" and "*set*" represents

the calculated and setting components respectively. Thus, for FIF

$$\left|\frac{dp^f}{dt}\right|_{cal} > \left|\frac{dp^f}{dt}\right|_{set} \quad \left|\frac{dp^b}{dt}\right|_{cal} > \left|\frac{dp^b}{dt}\right|_{set};$$

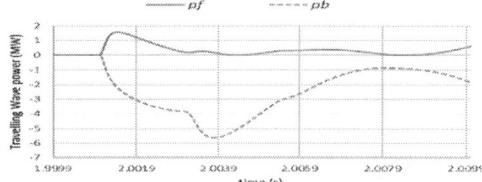

(a) Forward Internal Fault (R_f=500Ω)

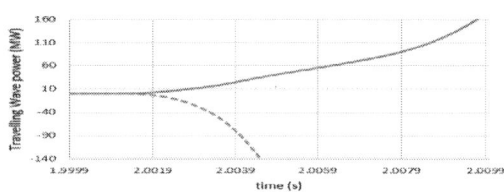

(b) Forward External Fault (R_f=0.01Ω)

Fig. 11 Simulation results showing forward and backward travelling wave power.

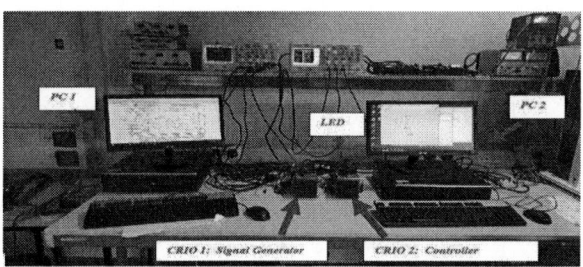

Fig. 12 LabView Compact-RIO Experimental Set up

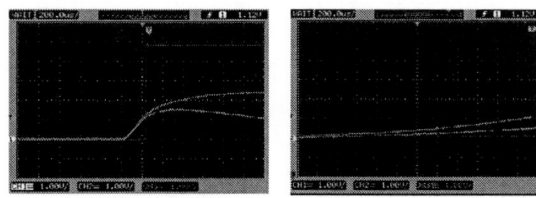

Fig. 13 Travelling wave components as per Fig. 12
(a)F_{21}=500Ω   (b)F_{23}=0.01Ω

The protection settings (PS) was calculated; thus

$$P_{set}^f = \frac{1}{N}\sum_{k=1}^{N}\frac{\left|\frac{dp^f}{dt}\right|_{cal}}{\left|\frac{dp^f}{dt}\right|_{set}}, \qquad P_{set}^b = \frac{1}{N}\sum_{k=1}^{N}\frac{\left|\frac{dp^b}{dt}\right|_{cal}}{\left|\frac{dp^b}{dt}\right|_{set}};$$

For a FIF; P_{set}^f **OR** $P_{set}^f > 1$

For a FEF; P_{set}^b **AND** $P_{set}^b < 1$

Due to space limitations, considerations was given to critical conditions for the relay. As shown in Fig. 14, discrimination between FIF and FEF was archived at less than *250μs* following the application of the fault. As shown, t_{inc} = arrival time of the first incident; t_d = fault (travelling wave) detection time. In this study, time required to detect the fault is t_d - t_{inc} = *2.00112 – 2.00102=0.11ms.*

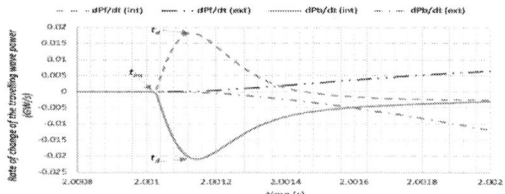

Fig. 14 Calculated *RoC* of the travelling wave power for FIF and FEF

Generally, utilising the second derivative of the travelling wave power, $\frac{d^2p^f}{dt^2}$, $\frac{d^2p^b}{dt^2}$ (Fig. 15) would presents a higher sensitivity; however this will depend on the level of sensitivity required as well as the resources available. Generally, as the complexity of the algorithm increases, the computation time increases. There this will be a matter of compromise.

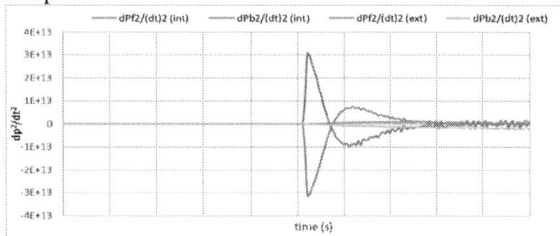

Fig. 15 Calculated second derivative of the travelling wave power for FIF and FEF

VI. CONCLUSIONS

A travelling-wave-based protection strategy for DC grid utilising the rate of change of the power developed by the forward and backward travelling wave is presented. The simulation results presented shows the suitability of the proposed technique. The protection principle therefore meets the requirements of DC grid protection as the circuit breaker trip signal can be generated within *250μs* including the time delay in the hardware. Attempts made on the proof-of-concept implementation is also presented.

ACKNOWLEDGMENT

Special thanks are due to GE Grid solutions, Stafford UK for supporting this work. Our unreserved gratitude also goes to David Martin, a technical skills specialist at Staffordshire University for his input in the proof-of-concept implementation.

APPENDIX 1

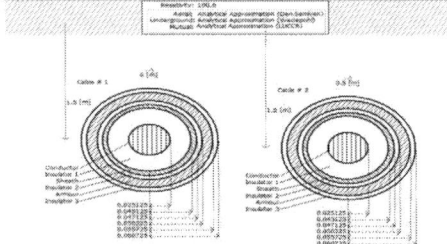

Fig. 16 Cable Configuration

TABLE 3 CONVERTERS AND AC SYSTEMS PARAMETERS

Item	Ratings
Rated Power of Converter	800MVA
Rated DC Voltage of Converter	400kV
Converter arm inductance	29mH
Cell DC Capacitor	10000μF
Nominal Frequency	50Hz
Transformer nominal voltage (L-L) RMS	380kV
Nominal voltage at VSC side (L-L) RMS	220kV
Leakage reactance of transformer	0.18pu
Rated real power per phase of Load	33MW
Rated reactive power per phase of Load	0.0MW
Rated load voltage(L-G) RMS	83.72kV

TABLE 4 CONDUCOR AND INSULATION PARAMETER

Item	Ratings
Resistivity of core conductor	2.2×10^{-8} Ωm
Resistivity of 1st conducting layer (sheath)	27.4×10^{-8} Ωm
Resistivity of 2nd conducting layer	18.15×10^{-8} Ωm
Outer radius of core conductor	2.51×10^{-2} m
Thickness of 1st conducting layer	2×10^{-3} m
Thickness of 2nd conducting layer	5.5×10^{-3} m
Thickness of 1st insulation layer	2×10^{-2} m
Thickness of 2nd insulation layer	3.1×10^{-3} m
Thickness of 3rd insulation layer	5×10^{-3} m
Relative permittivity of all insulation layer	2.3
All relative permeability	1
Ground resistivity	100 Ωm
Length of Cable	200km

REFERENCES

[1] R. Whitehouse, "Technical challenges of realising multi-terminal networking with VSC," 14th IEEE Int. Conf.. (EPE2011), 2011, pp 1-12,

[2] J. Sneath and A. D. Rajapakse, "Fault Detection and Interruption in an earthed HVDC Grid using ROCOV and Hybrid DC Breakers," *IEEE Trans. Power Deliv.*, vol. 31 (3), 2016 pp. 973-981,

[3] H. Liu, P. C. Loh, F. Blaabjerg, "Review of Fault Diagnosis and Fault-tolerant Control for Modular Multilevel Converter of HVDC" IEEE Conf (IECON2013),pp1242-1247, 2013

[4] J. Wang; B. Berggren; K. Linden. 'Multi-Terminal DC System line Protection Requirement and High Speed Protection Solutions', CIGRE, *Lund, Sweden, 2015, pp 1-9,*

[5] A. Adamczyk, C. D. Barker, and H. Ha, "Fault Detection and Branch Identification for HVDC Grids," 12th IET International Conference, (DPSP2014), 2014, pp. 1–6

[6] C. C. Davidson; R. S. Whitehouse; C. D. Barker; J. P. Dupraz; W. Grieshaber "A new ultra-fast HVDC Circuit breaker for meshed DC networks, 11th IET Int. Conference (ACDC), 2015, pp 1-7,

[7] M. Callavik and A. Blomberg, "The Hybrid HVDC Breaker An innovation breakthrough enabling reliable HVDC grids," ABB grid Systems, Technical Paper, 2012, pp 1-6,

[8] B. Bachmann, G. Mauthe, E. Ruoss, H.P. Lips, J. porter, J. Vithayathil, Development of a 500KV Air blast HVDC Circuit Breaker, ,IEEE Transactions,," Vol. PAS 104 (9), 1985p, p. 2460–2466,

[9]]Manitoba HVDC, https://hvdc.ca/

[10] Implementing the protection and control of future HVDC grids, http://www.think-grid.org/implementing-protection-and-control-future-hvdc-grids

[11] M. Chamia S. Liberman, "Ultra High Speed Relay for EHV/UHV Transmission Lines-Development, Design and Applications", IEEE Transc.,PAS97 no.4, 1978,

[12] A. T. Johns, "New ultra-high-speed directional comparison technique for the protection of EHV transmission lines," *Gener. Transm. Distrib. IEE Proc. C*, vol. 127, no. 4, 1980, pp. 228–239,

[13] M. Ikhide; S. Tennakoon; A. Griffiths; S. Subramanian; H. Ha, A. Adamczyk, "A transient based protection technique for future DC grids utilising travelling wave power"; *IET Journal of Engineering*, vol. (15)10, 2018, pp. 1267 - 1273

An Unit-less Mathematical Model for Analysis and Design of Class-E Resonant Converters

Lucas S. Mendonça
Power Processing Engineering Department
Universidade Federal de Santa Maria
Santa Maria, Brazil
lucassangoi1993@gmail.com

Thiago C. Naidon
Power Processing Engineering Department
Universidade Federal de Santa Maria
Santa Maria, Brazil
thiagonaidon@gmail.com

Rafael Fernandes Raposo
Universidade Tecnológica Federal do Paraná
Apucarana, Brazil
rafaelfernandes.raposo@gmail.com

Fábio E. Bisogno
Power Processing Engineering Department
Universidade Federal de Santa Maria
Santa Maria, Brazil
fbisogno@gepoc.ufsm.br

Abstract—Considering sustainable policies for energy systems, resonant converters are suitable systems to realize the desirable flow of energy in several applications, such as: renewable sources, energy harvesting systems, electric and hybrid vehicles. However, analysis and design of resonant converters are more complex than their pulse-width modulation counterparts, which leads to simplifying assumptions that result in limitations in the size of the designed components. This paper proposes a result-oriented methodology for analysis and design Class-E converters by means of an unit-less mathematical model that is independent of circuit parameters, such as, inductances and capacitances. The system is modeled in a generalized representation and normalized gain curves can be obtained to design the converter. A 500 kHz Class-E resonant converter was simulated and implemented to validate the theoretical approach.

Index Terms—modeling, resonant inverters, state-space methods

I. INTRODUCTION

Resonant converters reveal their excellence by means of shaping the circuits waveforms in a sinusoidal fashion, which leads to advantages such as, soft-switching, reduce electromagnetic interference, increase operating frequency, which are important requirements for high efficiency converters. Furthermore, resonant converters have been used in several applications, like as, renewable sources, energy harvesting systems, electric and hybrid vehicles, induction cookers and so on, which are trends considering sustainable policies for energy systems. These applications have been extensively explored in literature [1]-[12]. Notwithstanding, resonant converters analysis is more complex than their PWM counterparts due to multi resonant procedure in the converter, so the analytical solutions are hard to be obtained. Conventional methodologies make use of simplifying assumptions, such as: small-ripple

approximation, high loaded Q-factor, resonant circuit model as an ideal sinusoidal source, which result in limitations in the size of the designed components. Furthermore, analysis and design procedures of power converters are performed in an iterative manner that depends on circuit-level simulations. This type of procedure is useful and can quickly solve engineer's tasks, however, it can be time-consuming and inaccurate for higher complexity systems, as in the case of resonant converters. This paper proposes a result-oriented methodology for analysis and design Class-E resonant converters by means of an unit-less mathematical model that the system is represented by state-space matrices, which their terms are resonant unit-less parameters. The main advantages of the proposed method are: to model the converter in a generalized representation that do not depend of real circuit parameters, such as inductances, capacitances and resistances; to obtain component stress curves, gain curves and normalized waveforms for any operating point; and design by normalized parameters. These advantages allow analysis and design Class-E converters without prior specifications, which is an all-purpose tool to improve energy systems. In this work, a $500\,kHz$ Class-E resonant converter was simulated and implemented to validate the theoretical approach.

II. THEORETICAL APPROACH

The core of the proposed method is the use of resonant frequencies and quality factor to represent the system in an unit-less model. First, the Class-E converter is represented in a state-space model considering circuit parameters, such as, inductances, capacitances and resistances. Subsequently, the terms of the state-space matrices are represented by resonant unit-less parameters, which allows a generalized model that is valid for any operating point.

This study was financed in part by the Coordenação de Aperfeiçoamento de Pessoal de Nível Superior - Brasil (CAPES/PROEX) - Finance Code 001. The authors would like to thank the national council for scientific and technological development (CNPq) for all financial, technical and scientific support.

978-1-7281-4181-7/19 $31.00 © 2019 IEEE

A. Class-E Resonant Converter

The topology of the Class-E resonant converter is shown in Fig. 1. This converter is one of the most efficient and can operate in zero-voltage switching or zero-current switching.

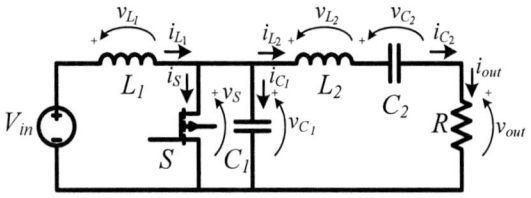

Fig. 1. Class-E resonant converter

It is considered: input voltage, V_{in}; inductors, L_1 and L_2; capacitors, C_1 and C_2; load, R; switch S and circuit variables, i_{L_1}, i_{L_2}, i_{C_1}, i_{C_2}, i_S, i_{out}, v_{L_1}, v_{L_2}, v_{C_1}, v_{C_2}, v_S and v_{out}. The converter has two operation modes: Mode I ($0 < t \leq D_c 2\pi$): Switch S is *on* and C_1 is short-circuited by the switch; Mode II ($D_c 2\pi < t \leq 2\pi$): Switch S is *off* and the resonant circuit consists of C_1, L_2, C_2 and R connected in series. The zero-voltage switching is approached in this work, thereby, the energy stored in C_1 is zero when S turns on, ensuring soft-switching.

B. Circuit-based Model

First, the system is represented in a state-space model based on circuit parameters. By modeling the system in an appropriate state vector, an unit-less equivalent representation can be obtained afterward. The state-space model for the Class-E converter is expressed as follows:

$$\frac{d\mathbf{x}(\omega t)}{d\omega t} = \mathbf{A}_I \mathbf{x}(\omega t) + \mathbf{B}_I \frac{u}{V_{in}} \tag{1}$$

$$\frac{d\mathbf{x}(\omega t)}{d\omega t} = \mathbf{A}_{II} \mathbf{x}(\omega t) + \mathbf{B}_{II} \frac{u}{V_{in}} \tag{2}$$

$$\mathbf{y}(\omega t) = \mathbf{C}_I \mathbf{x}(\omega t) + \mathbf{D}_I \frac{u}{V_{in}} \tag{3}$$

$$\mathbf{y}(\omega t) = \mathbf{C}_{II} \mathbf{x}(\omega t) + \mathbf{D}_{II} \frac{u}{V_{in}} \tag{4}$$

where $\mathbf{x}(\omega t)$ is the state vector, $\mathbf{A}_I$ and $\mathbf{A}_{II}$ are the state matrices, $\mathbf{B}_I$ and $\mathbf{B}_{II}$ are the input matrices, $\mathbf{C}_I$ and $\mathbf{C}_{II}$ are the output matrices, $\mathbf{D}_I$ and $\mathbf{D}_{II}$ are the transmission matrices, where I and II represent the indexes for the converter mode, $\mathbf{y}(\omega t)$ is the output vector and u is the input source. Differently of conventional state vectors, which are represented by inductor currents and capacitor voltages, a slight modification in state and output vectors is going to be performed. In the proposed method, the state vector for the Class-E converter is expressed as: $\mathbf{x}(\omega t) = \left[\frac{i_{L_1} L_1 \omega}{V_{in}} \quad \frac{i_{L_2} \sqrt{L_1}\sqrt{L_2}\omega}{V_{in}} \quad \frac{v_{C_1}\sqrt{C_1}\sqrt{L_1}\omega}{V_{in}} \quad \frac{v_{C_2}\sqrt{C_2}\sqrt{L_1}\omega}{V_{in}} \right]^T$ and the output vector:

$$\mathbf{y}(\omega t) = \left[\frac{I_{in}}{I_{in}} \quad \frac{i_{L_2}}{I_{in}} \quad \frac{i_{C_1}}{I_{in}} \quad \frac{i_{C_2}}{I_{in}} \quad \frac{i_S}{I_{in}} \quad \frac{i_{out}}{I_{in}} \quad \frac{v_{C_1}}{V_{in}} \quad \frac{v_{C_2}}{V_{in}} \quad \frac{v_S}{V_{in}} \quad \frac{v_{out}}{V_{in}} \right]^T.$$

In this case, the state-space matrices are represented as.

$$\mathbf{A}_I = \begin{pmatrix} 0 & 0 & 0 & 0 \\ 0 & -\frac{R}{L_2\omega} & 0 & -\frac{1}{\sqrt{C_2 L_2}\omega} \\ 0 & 0 & 0 & 0 \\ 0 & \frac{1}{\sqrt{C_2 L_2}\omega} & 0 & 0 \end{pmatrix}, \mathbf{B}_I = \begin{pmatrix} 1 \\ 0 \\ 0 \\ 0 \end{pmatrix}, \mathbf{A}_{II} =$$

$$\begin{pmatrix} 0 & 0 & -\frac{1}{\sqrt{C_1 L_1}\omega} & 0 \\ 0 & -\frac{R}{L_2\omega} & \frac{1}{\sqrt{C_1 L_2}\omega} & -\frac{1}{\sqrt{C_2 L_2}\omega} \\ \frac{1}{\sqrt{C_1 L_1}\omega} & -\frac{1}{\sqrt{C_1 L_2}\omega} & 0 & 0 \\ 0 & \frac{1}{\sqrt{C_2 L_2}\omega} & 0 & 0 \end{pmatrix} \mathrm{e} \; \mathbf{B}_{II} =$$

$$\begin{pmatrix} 1 \\ 0 \\ 0 \\ 0 \end{pmatrix}.$$

$$\mathbf{C}_I = \begin{pmatrix} \frac{V_{in}}{I_{in} L_1 \omega} & 0 & 0 & 0 \\ 0 & \frac{V_{in}}{I_{in}\sqrt{L_1 L_2}\omega} & 0 & 0 \\ 0 & 0 & 0 & 0 \\ 0 & \frac{V_{in}}{I_{in}\sqrt{L_1 L_2}\omega} & 0 & 0 \\ \frac{V_{in}}{I_{in} L_1 \omega} & -\frac{V_{in}}{I_{in}\sqrt{L_1 L_2}\omega} & 0 & 0 \\ 0 & \frac{V_{in}}{I_{in}\sqrt{L_1 L_2}\omega} & 0 & 0 \\ 0 & 0 & \frac{1}{\sqrt{C_1 L_1}\omega} & 0 \\ 0 & 0 & 0 & \frac{1}{\sqrt{C_2 L_1}\omega} \\ 0 & 0 & \frac{1}{\sqrt{C_1 L_1}\omega} & 0 \\ 0 & \frac{R}{\sqrt{L_1 L_2}\omega} & 0 & 0 \end{pmatrix} \quad \text{and}$$

$$\mathbf{C}_{II} = \begin{pmatrix} \frac{V_{in}}{I_{in} L_1 \omega} & 0 & 0 & 0 \\ 0 & \frac{V_{in}}{I_{in}\sqrt{L_1 L_2}\omega} & 0 & 0 \\ \frac{V_{in}}{I_{in} L_1 \omega} & -\frac{V_{in}}{I_{in}\sqrt{L_1 L_2}\omega} & 0 & 0 \\ 0 & \frac{V_{in}}{I_{in}\sqrt{L_1 L_2}\omega} & 0 & 0 \\ 0 & 0 & 0 & 0 \\ 0 & \frac{V_{in}}{I_{in}\sqrt{L_1 L_2}\omega} & 0 & 0 \\ 0 & 0 & \frac{1}{\sqrt{C_1 L_1}\omega} & 0 \\ 0 & 0 & 0 & \frac{1}{\sqrt{C_2 L_1}\omega} \\ 0 & 0 & \frac{1}{\sqrt{C_1 L_1}\omega} & 0 \\ 0 & \frac{R}{\sqrt{L_1 L_2}\omega} & 0 & 0 \end{pmatrix}.$$

$$\mathbf{D}_I = \mathbf{D}_{II} = 0.$$

Being, I_{in} the input current. Considering the circuit-based model, some points are discussed. The independent variable from the differential operator of the state-space model is ωt. Considering $u = V_{in}$, the system is normalized by the input source. Furthermore, this representation allows to develop an unit-less model considering resonant parameters, which is going to be performed in the next section.

C. Resonance-based Model

The relations of reactive components define resonant parameters that are going to be used for the unit-less model. The Class-E converter has four reactive elements, then, three resonant frequencies can be obtained as follows:

$$\omega_1 = \frac{1}{\sqrt{L_2 C_2}}; \; \omega_2 = \frac{1}{\sqrt{L_2 C_1}}; \; \omega_3 = \frac{1}{\sqrt{L_1 C_1}} \tag{5}$$

978-1-7281-4181-7/19 $31.00 © 2019 IEEE

$$A_1 = \frac{\omega_1}{\omega}; \; A_2 = \frac{\omega_2}{\omega}; \; A_3 = \frac{\omega_3}{\omega} \qquad (6)$$

Furthermore, the quality factor Q and the inverse transfer power ratio, a, must be defined as follows [15]:

$$Q = \frac{L_2 \omega_1}{R} = \frac{1}{C_2 \omega_1 R} \qquad (7)$$

$$a = \frac{V_{in}}{I_{in} R} \qquad (8)$$

The relations (5), (6), (7) and (8) are used to obtain the unit-less model. The terms of the state-space matrices must be equated as function of resonant parameters as follows:

- $\frac{1}{\sqrt{C_2 L_2} \omega} = \frac{\omega_1}{\omega} = A_1$;
- $\frac{1}{\sqrt{C_1 L_1} \omega} = \frac{\omega_3}{\omega} = A_3$;
- $\frac{1}{\sqrt{C_1 L_2} \omega} = \frac{\omega_2}{\omega} = A_2$;
- $\frac{R}{L_2 \omega} = \frac{R \omega_2}{L_2 \omega_2 \omega} = \frac{A_2}{Q} a$;
- $\frac{V_{in}}{I_{in} L_1 \omega} = \frac{Ra}{L_1 \omega} = \frac{Ra L_2 \omega_2}{L_1 \omega L_2 \omega_2} = \frac{a A_2 L_2}{Q L_1} = \frac{a A_2 L_2 C_1}{Q L_1 C_1} = \frac{a A_2 \omega_3^2}{Q \omega_2^2} = \frac{a A_3^2}{Q A_2}$;
- $\frac{V_{in}}{I_{in} \sqrt{L_1 L_2} \omega} = \frac{Ra C_1}{\sqrt{L_1 C_1} \sqrt{L_2 C_1} \omega} = \frac{\omega_2 \omega_3 Ra C_1}{\omega} = \frac{\omega_2 \omega_3 Ra C_1 L_2 \omega_2}{\omega L_2 \omega_2} = \frac{a \omega_2 \omega_3 \omega_2^2}{Q \omega \omega_2^2} = \frac{a A_3}{Q}$;
- $\frac{R}{\sqrt{L_1 L_2} \omega} = \frac{A_3}{Q}$;
- $\frac{1}{\sqrt{C_2 L_1} \omega} = \frac{\sqrt{L_2}}{\sqrt{C_2 L_2} \sqrt{L_1} \omega} = \frac{A_1 \sqrt{L_2 C_1}}{\sqrt{L_1 C_1}} = \frac{A_1 A_3}{A_2}$.

The unit-less mathematical model for the Class-E resonant converter can be represented in state-space model described as:

$$\dot{\mathbf{x}}(\omega t) = \mathbf{E}_{I[A_i, Q_i, a]} \mathbf{x}(\omega t) + \mathbf{F}_I \qquad (9)$$

$$\dot{\mathbf{x}}(\omega t) = \mathbf{E}_{II[A_i, Q_i, a]} \mathbf{x}(\omega t) + \mathbf{F}_{II} \qquad (10)$$

$$\mathbf{y}(\omega t) = \mathbf{G}_{I[A_i, Q_i, a]} \mathbf{x}_e(\omega t) + \mathbf{H}_I \qquad (11)$$

$$\mathbf{y}(\omega t) = \mathbf{G}_{II[A_i, Q_i, a]} \mathbf{x}_e(\omega t) + \mathbf{H}_{II} \qquad (12)$$

Being, $\mathbf{E}_I = \begin{pmatrix} 0 & 0 & 0 & 0 \\ 0 & -\frac{a A_2}{Q} & 0 & -A_1 \\ 0 & 0 & 0 & 0 \\ 0 & A_1 & 0 & 0 \end{pmatrix}$, $\mathbf{F}_I = \begin{pmatrix} 1 \\ 0 \\ 0 \\ 0 \end{pmatrix}$, $\mathbf{E}_{II} = \begin{pmatrix} 0 & 0 & -A_1 & 0 \\ 0 & -\frac{a A_2}{Q} & A_2 & -A_1 \\ A_3 & -A_2 & 0 & 0 \\ 0 & A_1 & 0 & 0 \end{pmatrix}$, $\mathbf{F}_{II} = \begin{pmatrix} 1 \\ 0 \\ 0 \\ 0 \end{pmatrix}$,

$$\mathbf{G}_I = \begin{pmatrix} \frac{a A_3^2}{Q A_2} & 0 & 0 & 0 \\ 0 & \frac{a A_3}{Q} & 0 & 0 \\ 0 & 0 & 0 & 0 \\ 0 & \frac{a A_3}{Q} & 0 & 0 \\ \frac{a A_3^2}{Q A_2} & -\frac{a A_3}{Q} & 0 & 0 \\ 0 & \frac{a A_3}{Q} & 0 & 0 \\ 0 & 0 & A_3 & 0 \\ 0 & 0 & 0 & A_2 \\ 0 & 0 & A_3 & 0 \\ 0 & \frac{a A_3}{Q} & 0 & 0 \end{pmatrix}, \quad \mathbf{H}_I = \begin{pmatrix} 0 \\ 0 \\ 0 \\ 0 \\ 0 \\ 0 \\ 0 \\ 0 \\ 0 \\ 0 \end{pmatrix}, \quad \mathbf{G}_{II} =$$

$$\begin{pmatrix} \frac{a A_3^2}{Q A_2} & 0 & 0 & 0 \\ 0 & \frac{a A_3}{Q} & 0 & 0 \\ \frac{a A_3^2}{Q A_2} & -\frac{a A_3}{Q} & 0 & 0 \\ 0 & \frac{a A_3}{Q} & 0 & 0 \\ 0 & 0 & 0 & 0 \\ 0 & \frac{a A_3}{Q} & 0 & 0 \\ 0 & 0 & A_3 & 0 \\ 0 & 0 & 0 & A_2 \\ 0 & 0 & A_3 & 0 \\ 0 & \frac{a A_3}{Q} & 0 & 0 \end{pmatrix} \quad \text{and } \mathbf{H}_{II} = \begin{pmatrix} 0 \\ 0 \\ 0 \\ 0 \\ 0 \\ 0 \\ 0 \\ 0 \\ 0 \\ 0 \end{pmatrix}.$$

where $\mathbf{E}_I$ and $\mathbf{E}_{II}$ are the new state matrices, $\mathbf{F}_I$ and $\mathbf{F}_{II}$ are the new input matrices, $\mathbf{G}_I$ and $\mathbf{G}_{II}$ are the new output matrices, $\mathbf{H}_I$ and $\mathbf{H}_{II}$ are the new transmission matrices. In this form, the Class-E resonant converter is generically represented, and its waveforms can be obtained for any operating point independently of circuit parameters. As long as the behavior of power converters is ruled by duty cycle D_c, it will be indicated as a parameter in the system solution. Given the initial conditions, the system can be solved for all operating points by sweeping A_1, A_2, A_3, Q and D_c, however, a needs a numerical value, which can be calculated as follows:

$$\frac{1}{a} = \frac{1}{2\pi} \int_0^{D_c 2\pi} \left[\frac{v_{out}}{V_{in}} \right]_I^2 d\omega t + \frac{1}{2\pi} \int_{D_c 2\pi}^{2\pi} \left[\frac{v_{out}}{V_{in}} \right]_{II}^2 d\omega t \qquad (13)$$

The solution of the system returns normalized waveforms for any operating point. Furthermore, component stress and gain curves can be obtained independently of specifications and circuit parameters, such as: input voltage, output power, frequency, inductances, capacitances and so on. An algorithm was developed in Wolfram Mathematica to solve the system. Normalized waveforms for the Class-E converter for $D_c = 0.5$ are shown in Fig. 2. As it can be seen, the converter normalized waveforms are represented by the relation of a circuit variable and the input voltage/current, shown in Fig. 2(a)-(i). Switch S turns *on* with zero voltage, yielding zero-voltage switching, as shown in Fig. 2(h). By sweeping D_c from 0.1 to 0.9, the Class-E gain curve can be obtained for different values of A_1. The relation of the transfer power ratio $\frac{1}{a}$ for the duty cycle D_c, considering quality factor $Q = 1$, for a set of $A_1 = \{0.1, 0.5, 0.8, 1.0, 1.2, 1.5, 2.0\}$, is shown in Fig. 3. The normalized peak switch voltage and normalized peak switch current as function of duty cycle for $Q = 1$ are

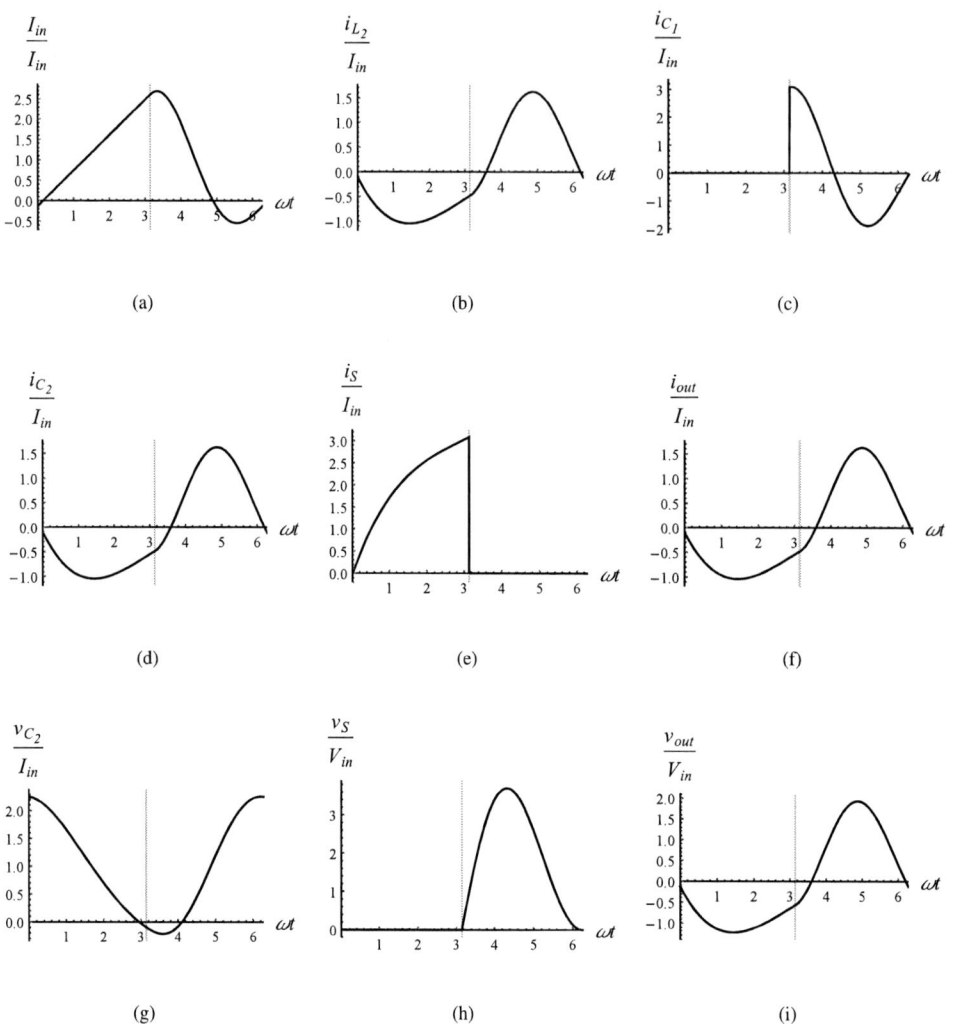

Fig. 2. Normalized waveforms for Class-E resonant converter considering $D_c = 0.5$

shown in Fig. 4 and Fig. 5, respectively. As far as duty cycle increases, transfer power ratio increases too, however, peak switch current decreases. In Fig. 4, peak switch voltage is higher for $A_1 = 2.0$, and increases as long as duty cycle increases. These results are valid for any operating point independently of specifications and can be used to design the Class-E resonant converter.

III. DESIGN BY NORMALIZED PARAMETERS

The unit-less model allows the normalized design of the converter, in other words, the resonant parameters can be used to design the circuit. The design equations are derived from (5) and (7). The resonant parameters are extracted from the gain curves. The steps for the design by normalized parameters can be given by:

1) Normalized results and operating point chosen: From the theoretical approach, the system is solved for any operating point, then, a desirable normalized result should be chosen

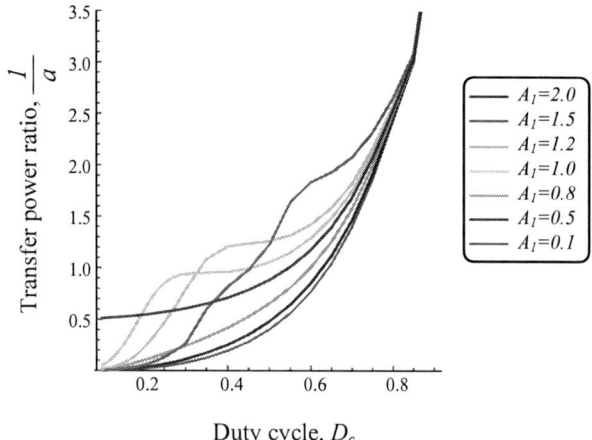

Fig. 3. Transfer power ratio as function of duty cycle for the Class-E resonant converter for $Q = 1$

978-1-7281-4181-7/19 $31.00 © 2019 IEEE   236

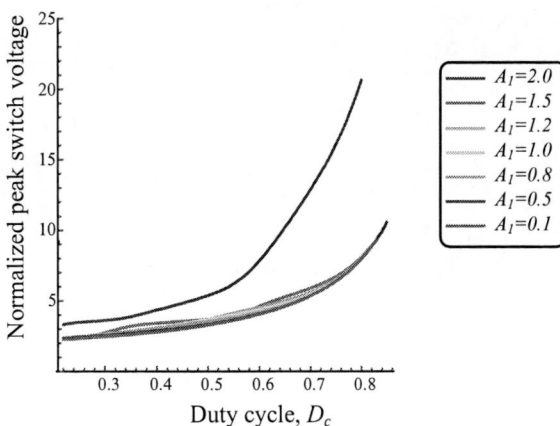

Fig. 4. Normalized peak switch voltage as function of duty cycle for the Class-E resonant converter for $Q = 1$

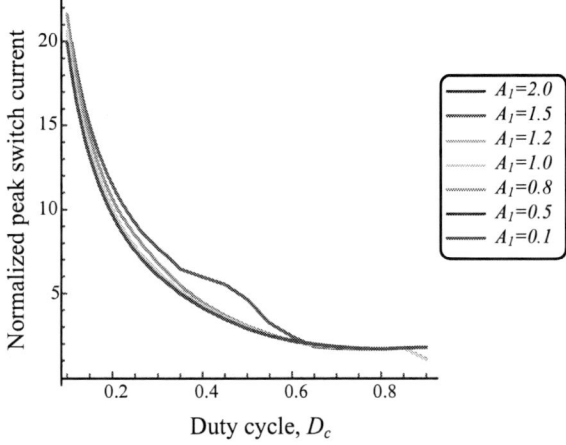

Fig. 5. Normalized peak switch current as function of duty cycle for the Class-E resonant converter for $Q = 1$

and the resonant parameters are obtained.

2) *Resonant parameters based on operating point:* As an example, by choosing $D_c = 0.5$, the resonant parameters are: $A_1 = 0.8$, $A_2 = 1.02$, $A_3 = 1.18$, $Q = 1$ and $a = 0.821$.

3) *Specifications:* Input voltage, $V_{in} = 10\,V$; operating frequency, $f_s = 500\,kHz$ and output power, $P_{out} = 5\,W$.

4) Calculate output voltage V_o and angular operating frequency ω according to:

$$V_o = V_{in}\sqrt{\frac{1}{a}} \tag{14}$$

$$\omega = 2\pi f_s \tag{15}$$

5) Design of load R and reactive elements, L_1, L_2, C_1 and C_2, according to:

$$R = \frac{V_o^2}{P_{out}} \tag{16}$$

$$L_1 = \frac{RA_2^2}{QA_1A_3^2\omega} \tag{17}$$

$$L_2 = \frac{R}{QA_1\omega} \tag{18}$$

$$C_1 = \frac{QA_1}{A_2^2R\omega} \tag{19}$$

$$C_2 = \frac{Q}{A_1R\omega} \tag{20}$$

From (16), (17), (18), (19) and (20) the circuit components are obtained as: $R = 22\,\Omega$, $L_1 = 7.31\,\mu H$, $L_2 = 9.68\,\mu H$, $C_1 = 10\,nF$ and $C_2 = 15\,nF$. This design is going to be used for the simulation and experimental results in the next section.

IV. RESULTS

The Class-E resonant converter was simulated and implemented to compare the theoretical, simulation and experimental results. Specifications and components for the simulation and experimental results are summarized in TABLE I.

TABLE I
SIMULATION AND EXPERIMENTAL PARAMETERS

Parameter	Value	Model
Input voltage, V_{in}	$10\,V$	
Operating frequency, f_s	$500\,kHz$	
Duty cycle, D_c	0.5	
Resistor, R	$22\,\Omega$	
Inductor, L_1	$7.31\,\mu H$	Air core AWG18
Inductor, L_2	$9.68\,\mu H$	Ring core AWG28
Capacitor, C_1	$10\,nF$	
Capacitor, C_2	$15\,nF$	
Switch MOSFET, S		K03H1202 - Infineon

A comparison of theoretical, simulation and experimental results for the output voltage is shown in Fig. 6. The normalized output voltage $\frac{v_{out}}{V_{in}}$, output voltage from simulation v_{out} and output voltage from experimental are shown in Fig. 6(a), Fig. 6(b) and Fig. 6(c) respectively. A comparison of theoretical, simulation and experimental results for the switch S voltage is depicted in Fig. 7(a). The normalized switch voltage $\frac{v_S}{V_{in}}$, switch voltage from simulation v_S and switch voltage from experimental are shown in Fig. 7(a), Fig. 7(b) and Fig. 7(c) respectively.

V. CONCLUSION

In this work, an unit-less mathematical model was developed for Class-E resonant converters. This methodology is based on system modeling considering resonant parameters, which allows a generalized representation that is independent of specifications and circuit parameters, such as inductances, capacitances and resistances. The main advantages of the proposed method are: to model the converter in a generalized

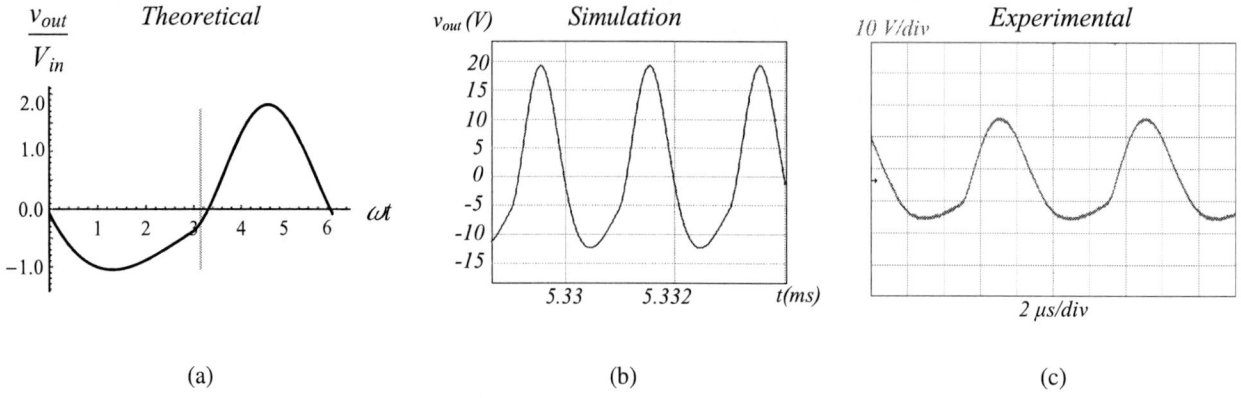

Fig. 6. Comparison of theoretical, simulation and experimental results for output voltage

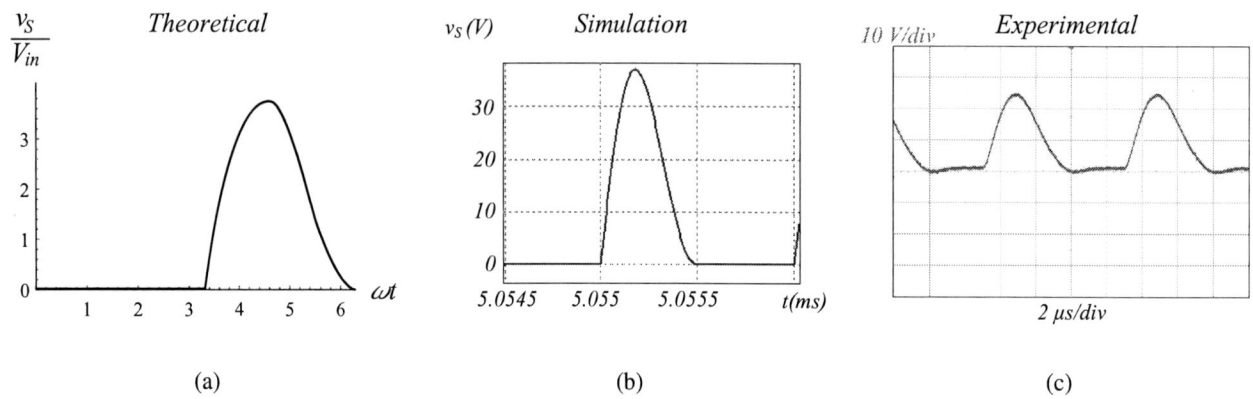

Fig. 7. Comparison of theoretical, simulation and experimental results for switch voltage

representation that do not depend of real circuit parameters, to obtain component stress curves, gain curves and normalized waveforms for any operating point, and design by normalized parameters. The Class-E converter was simulated in implemented to validate the theoretical results.

REFERENCES

[1] W. Chen, X. Wu, L. Yao, W. Jiang and R. Hu, "A Step-up Resonant Converter for Grid-Connected Renewable Energy Sources," IEEE Transactions on Power Electronics, vol. 30, pp. 3017–3029, 2015.

[2] C. Cecati, H. A. Khalid, M. Tinari, G. Adinolfi and G. Graditi, "DC Nanogrid for Renewable Sources with Modular DC/DC LLC Converter Building Block," IET Power Electronics, vol. 10, pp. 536–544, 2017.

[3] S. I. Annie, K. M. Salim, Z. Tasneem and M. R. Uddin, "Design and Performance Analysis of a ZVS Parallel Quasi Resonant Converter for a Solar Based Induction Cooking System," IEEE Region 10 Conference (TENCON), February 2015.

[4] S. M. Mousavi, R. Beiranvand, S. Goodarzi and M. Mohamadian, "Designing a 48 V to 24 V DC-DC Converter for Vehicle Application Using Resonant Switched Capacitor Converter Topology," 2015 6th Power Electronics, Drives Systems and Technologies Conference, February 2015.

[5] T. Mishima, Y. Nakawaga and M. Nakaoka, "A Bridgeless BHB ZVS-PWM AC-AC Converter for High-frequency Induction Heating Applications," IEEE Transactions on Industry Applications, vol. 51, pp. 3304–3315, 2015.

[6] H. Sarnago, O. Lucia, A. Mediano and J. M. Burdio, "Design and Implementation of a High-efficiency Multiple-output Resonant Converter for Induction Heating Applications," IEEE Transactions on Power Electronics, vol. 29, pp. 2539–2549, May 2015.

[7] M. T. Outeiro and A. Carvalho, "Design, Implementation and Experimental Validation of a DC-DC Resonant Converter for PEM Fuel Cell Applications," IEEE Ind. Electron. Soc. Annu. Conference (IECON), pp. 619–624, 2013.

[8] K. Colak, E. Asa and M. Bojarski, "Hybrid Control Approach of CLL Resonant Converter for EV Battery Chargers," IEEE Ind. Electron. Soc. Annu. Conference (IECON), pp. 5041–5046, 2014.

[9] J. Deng, S. Hum C. C. Mi and R. Ma, "Design Methodology of LLC Resonant Converters for Electric Vehicle Battery Chargers," IEEE Trans. Veh. Technol., vol. 63, pp. 1581–1592, May 2014.

[10] K. Rakhi, K. Ilango, H. V. Manjunath and M. G. Nair, "Simulation Analysis of Half Bridge Series Parallel Resonant Converter based Batery Charger for Photovoltaic System," in Proc. Power and Energy Syst. Conf.: Towards Sustainable Energy, pp. 1–5, 2014.

[11] S. Xu, K. D. T. Ngo, T. Nishida, G. Chung and A. Sharma, "Low Frequency Pulsed Resonant Converter for Energy Harvesting," IEEE Transactions on Power Electronics, vol. 22, pp. 63–68, January 2007.

[12] Y. Tang and A. Khaligh, "A Multiinput Bridgeless Resonant AC-DC Converter for Electromagnetic Energy Harvesting," IEEE Transactions on Power Electronics, vol. 31, pp. 2254–2263, April 2016.

[13] F. E. Bisogno, "Energy-related System Normalization and Decomposition Targeting Sensitivity Consideration," PhD Dissertation, 2006.

A Curve Tracer for Photovoltaic Modules Based on The Capacitive Load Method

Emerson Abreu Bastos Junior, Caio Meira Amaral da Luz, Tatiane Martins Oliveira, Lara Ana Rodarte Rios,
Eduardo Moreira Vicente, and Fernando Lessa Tofoli
Department of Eletrical Engeneering
Federal University of São João del-Rei
São João del-Rei, Brazil
emersonabjr1@gmail.com, caiomeiramaral@hotmail.com, tatiane.martins@oi.com.br, laara.rodarte@gmail.com,
eduardovicente@ufsj.edu.br, fernandolessa@ufsj.edu.br

Abstract – This work presents the development of a current versus voltage (*I-V*) curve tracer of photovoltaic (PV) modules based on the capacitive load method, being this an important approach for checking the integrity of PV devices and also the extraction of intrinsic parameters for modeling purposes. The tracer employs a microcontroller to manipulate the data acquired from current and voltage sensors associated with the module output quantities. For comparison purposes, three different values of capacitance are used to trace *I-V* characteristic of a 5-Wp module, as it is possible to verify that the capacitor charging time has direct influence on the number of acquired points that constitute the curve.

Keywords – curve tracer, capacitive load method, photovoltaic solar energy, photovoltaic system modeling.

I. INTRODUCTION

With the advancement of technology and growth of population, the energy demand has increased significantly. In this context, renewable energy sources have drawn significant attention from both industry and academia in an attempt to promote a more sustainable environment in a near future [1]. Among the existing solutions available for the generation of electricity, photovoltaic (PV) solar energy is regarded as an abundant, clean, and sustainable source that is often associated with low operation and maintenance costs [2].

Even though the cost of kilowatt hour generated by PV modules is still high when compared with other primary sources, it is worth mentioning that manufacturing costs have been drastically reduced over the last few years [3]. Besides, although the sun is considered to be a virtually inexhaustible source, the generation of electricity through this process has the disadvantage of having low conversion efficiency. Being intrinsically nonlinear devices, it is necessary to analyze the behavior of PV modules through their respective current versus voltage (*I-V*) and power versus voltage (*P-V*) characteristics. It is also worth mentioning that such curves vary according to the environmental weather conditions, i.e. temperature and irradiance, as it is necessary to determine the maximum power point (MPP) accurately [4], [5].

The aforementioned characteristics can be measured using commercial curve tracers, which often present high cost. Other possible solutions include the use of variable resistors, capacitive loads, electronic loads, bipolar power amplifiers, four-quadrant power supplies, and dc-dc converters [6, 7]. Each approach has proper characteristics is terms of flexibility, modularity, accuracy, response, display of results, and cost.

According to [7], capacitive loads are capable of providing *I-V* with reasonable flexibility and accuracy. Furthermore, compared with others methods available in the literature, the capacitive load method proves to be simple, cheap, and scalable from module level to array level [8]. A detailed sizing of the capacitive load is needed to improve the accuracy of measurements. A single-pole, double-throw switch can be used instead of a MOSFET (metal oxide semiconductor field effect transistor) is employed to charge and discharge the capacitor, thus eliminating the need for a driver and the aforementioned calculation. This approach is adopted in the forthcoming study.

In this context, this work presents a curve tracer employing a capacitive load connected to the output of the PV module. The *I-V* curve can be properly measured while controlling the charging and discharging of the capacitor, considering that the operating point varies from the short-circuit current to the open-circuit voltage [9]. Simulation and experimental results are presented and discussed in detail to validate the theoretical assumptions.

II. CAPACITIVE LOAD METHOD

In order to obtain the *I-V* characteristic of a given module through a capacitive load, let us consider the circuit shown in Fig. 1, where *C* represents an ideal capacitor and *R* is the discharge resistor. Considering the well-known single-diode representation of the module [10], the output current *I* can be described as:

$$I(V) = I_{ph} - I_d - I_{sh} \qquad (1)$$

where I_{ph} is the photogenerated current, I_d is the current that circulates through the diode, I_{sh} is the leakage current, and *V* is the output voltage of the module.

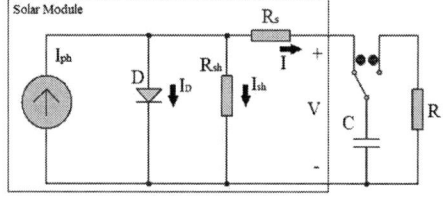

Fig. 1. Single-diode model of a PV module supplying a capacitive load.

The tracer must be capable of controlling the current and voltage from the short-circuit current (I_{SC}) to the open-circuit voltage (V_{OC}) condition. If such output quantities are properly monitored, the characteristic can be adequately plotted in the form on current and voltage points. From Fig. 2, it can be stated that it behaves initially as a current source

978-1-7281-4181-7/19 $31.00 © 2019 IEEE

at I_{SC}, but as a voltage source at V_{OC} [10]. Thus, it is necessary to calculate the ideal capacitance required by the circuit while also taking into account the transient behavior of the circuit in both aforementioned conditions.

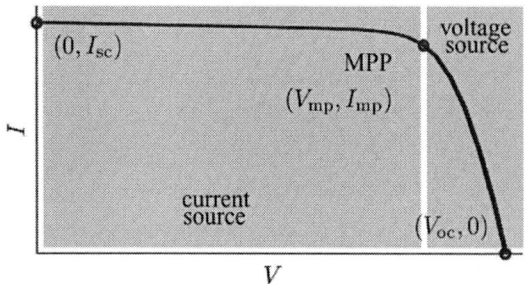

Fig. 2. *I-V characteristic curve* [10].

In short circuit condition, the PV module behaves as an ideal current source since the short-circuit current in most crystalline silicon devices is close to the current at the maximum power point (MPP). Thus, the leakage current that flows through the shunt resistor can be disregarded, while the diode behaves an open circuit [8]. Therefore, Fig. 1 can be simplified, resulting in the photogenerated current flowing through the series resistor R_s as shown in Fig. 3 (a). When the module is considered as a voltage source, the diode D and R_s are associated in series, resulting in a Thévenin equivalent representation in terms of R_{eq} as in Fig. 3 (b). In both transient conditions, the capacitor C is considered to be the only element connected to the module and the input impedance is infinite.

When the switch is closed at t=0 in Fig. 3 (a), the dc source provides some current to the series resistor and the capacitor, resulting in:

$$i(t) = I_{SC}, \; 0 < t < t_0 \qquad (2)$$

At the beginning of the transient, the capacitor behaves as a short circuit and the voltage increases linearly until the transitory finishes at t=t_0, as the following expression is valid:

$$v(t) = \frac{1}{C}\int_0^{t_0} I_{SC} \; dt = \frac{I_{SC}}{C} \cdot t, \;\; 0 < t < t_0 \qquad (3)$$

The first stage finishes at t_0 when the capacitor is charged with voltage V_0, which is obtained from (4).

$$v(t = t_0) = V_0 = \frac{I_{SC} \cdot t_0}{C} \cong V_{MPP} \qquad (4)$$

where V_{MPP} is the voltage at the MPP.

From instant t_0, the PV module starts operating as a dc voltage source. The circuit behavior is described by a second-order differential equation, whose unknown variable is the capacitor voltage $v(t)$:

$$\frac{dv(t)}{dt} + \frac{v(t)}{R_{eq} \cdot C} = \frac{V_{OC}}{R_{eq} \cdot C} \qquad (5)$$

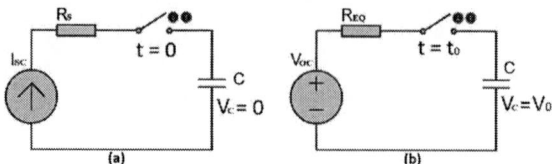

Fig. 3. Equivalent circuits during the transient conditions when the module behaves as: (a) an ideal current source and (b) an ideal voltage transitory.

From the initial condition described by (4), the capacitor voltage as a function of time can be obtained solving (5):

$$v(t) = V_{OC} + \left(\frac{I_{SC} \cdot t_0}{C} - V_{OC} \right) \cdot e^{\frac{-(t-t_0)}{R_{eq} \cdot C}}, \; t_0 < t < \infty \qquad (6)$$

The instantaneous current that flows through the capacitor is represented as:

$$i(t) = C \cdot \frac{dv(t)}{dt} = \left(\frac{V_{OC}}{R_{eq}} - \frac{I_{SC} \cdot t_0}{C} \right) \cdot e^{\frac{-(t-t_0)}{R_{eq} \cdot C}}, \; t_0 < t < \infty \qquad (7)$$

The expression that represents the capacitance during t_0, which corresponds to the time interval comprising the two transients, is given by:

$$C = \frac{I_{SC}}{V_{OC} - R_{eq} \cdot I_{SC}} \cdot t_0 \qquad (8)$$

The final instant t_f is the time interval for which the switch can be closed after the whole *I-V* curve is swept. This parameter corresponds to the sum of the time intervals required for the capacitor to be charged with V_0 and the operation of the RC circuit in steady state-condition for $5 \cdot \tau$, where τ is the time constant.

$$t_f = t_0 + 5 \cdot \tau = t_0 + 5 \cdot R_{eq} \cdot C \qquad (9)$$

Considering the whole time interval t_f, the capacitance required by the tracer can be obtained substituting (8) in (9):

$$C = \frac{t_f \cdot I_{SC}}{V_{OC} + 4 \cdot R_{eq} \cdot I_{SC}} \qquad (10)$$

According to (10), first it is necessary to determine R_{eq}, which can be obtained from the analysis of Fig. 3 (b) using Ohm's law:

$$R_{eq} = \frac{V_{OV} - V_{MPP}}{I_{SC}} \qquad (11)$$

III. Simulation and Experimental Results

The circuit presented in Fig. 4 was employed to simulate the *I-V* curve tracer in PSIM® software. Fig. 5 (a) shows the behavior of the voltage and current of the module in standard test condition (STC), i.e., 1,000 W/m² and 25 °C. Plotting the current as a function of the voltage in Fig. 6

978-1-7281-4181-7/19 $31.00 © 2019 IEEE

shows that the values of I_{SC} and V_{OC} are quite similar to the ones obtained from the datasheet of module KM(P)5 listed in Table I, which are measured in the simulation as I_{SC}=0.3099 A and V_{OC}=21.557 V in steady state.

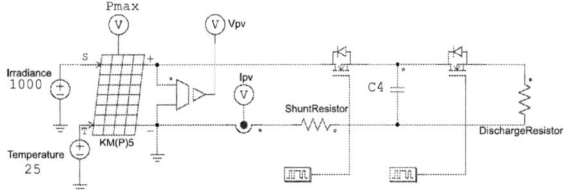

Fig. 4. Simulated circuit in PSIM®.

TABLE I
ELECTRICAL CHARACTERISTICS OF MODULE KM(P)5 IN STC.

Parameter	Value
Maximum power	$P_{MPP(ref)}$=5 W
Maximum power voltage at $P_{MPP(ref)}$	$V_{MPP(ref)}$=17.56 V
Maximum power current at $P_{MPP(ref)}$	$I_{MPP(ref)}$=0.286 A
Open-circuit voltage	$V_{OC(ref)}$=21.52 V
Short-circuit current	$I_{SC(ref)}$=0.31 A

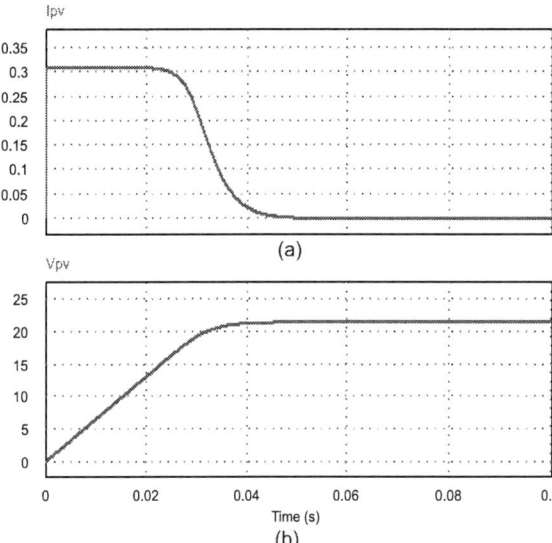

Fig. 5. Behavior of the (a) current and (b) voltage of module KM(P)5 in STC.

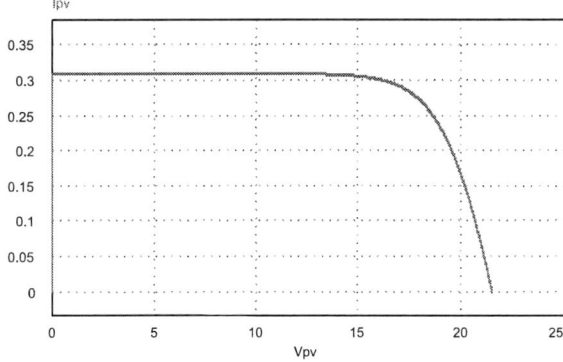

Fig. 6. *I-V* curve obtained for module KM(P)5 in STC.

The work developed in [8] suggests that parameter t_f must be between 20 ms and 100 ms for plotting the curves accurately. From this assumption and also considering the substitution of the values listed in Table I in (10), it can be stated that the rated capacitance ranges between 165.97 µF and 829.87 µF. Three distinct capacitance values were considered in this work comparison purposes, i.e., 470 µF, 4.7 mF and 22 mF.

Experimental tests were then performed using an adjustable dc voltage source for supplying an operational amplifier with 15 V, while microcontroller Arduino was connected to a personal computer (PC) via serial port. The discharge resistance used was chosen as 1.8 Ω to test module KM(P)5, while the experimental setup is shown in Fig. 7. A detailed view of the printed circuit board (PCB) that corresponds to the tracer is also depicted in Fig. 8.

Fig. 9, Fig. 10 and Fig. 11 present the *I-V* curves obtained for module KM(P)5 in a single measurement session for an irradiance of 980 W/m². Considering that the tests were performed in an uncontrolled external environment, the module temperature could not be maintained constant during the measurements. From the mathematical model of the module, whose parameters were obtained following the methodology described in [10], the corresponding curves could also be obtained by simulation for comparison purposes using the circuit shown in Fig. 4. The main results are summarized in Table II in terms of a comparison among the values obtained by simulation and experimentally for the short-circuit current, open-circuit voltage, and maximum power. According to Table III, it can be stated that the capacitive load method provides accurate results with low percent errors even though this is a low-power module.

Fig. 7. Experimental setup.

Fig. 8. Curve tracer prototype.

978-1-7281-4181-7/19 $31.00 © 2019 IEEE

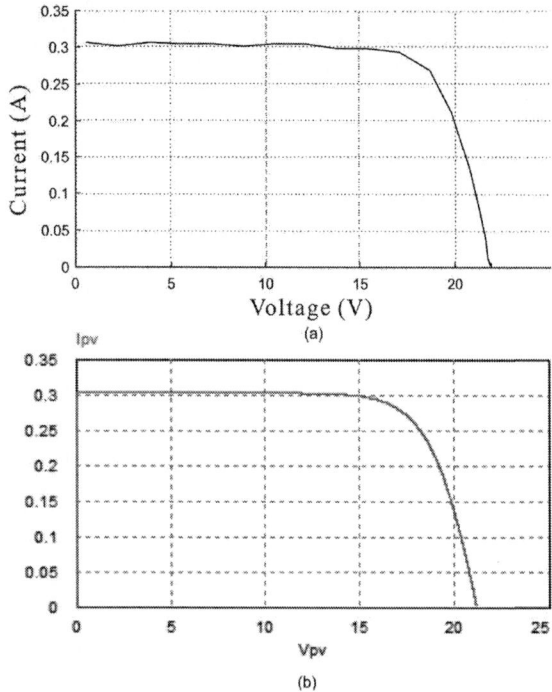

Fig. 9. *I-V* curves of module KM(P)5 obtained (a) experimentally and (b) by simulation for *C*=470 µF.

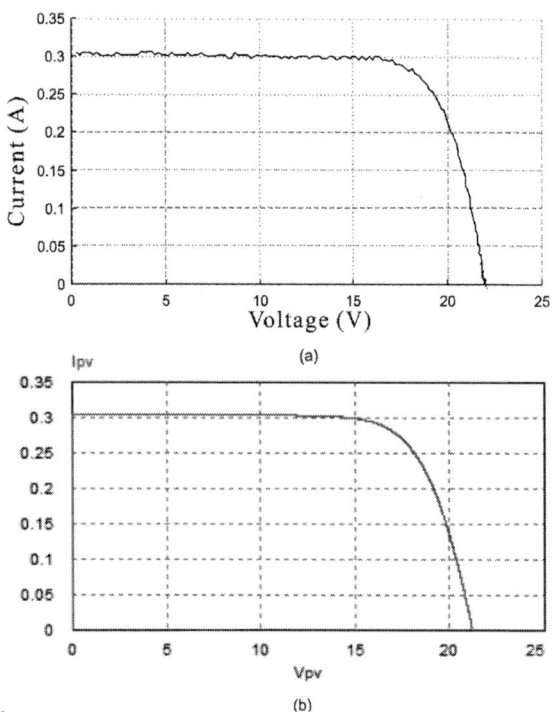

Fig. 10. *I-V* curves of module KM(P)5 obtained (a) experimentally and (b) by simulation for *C*=4.7 mF.

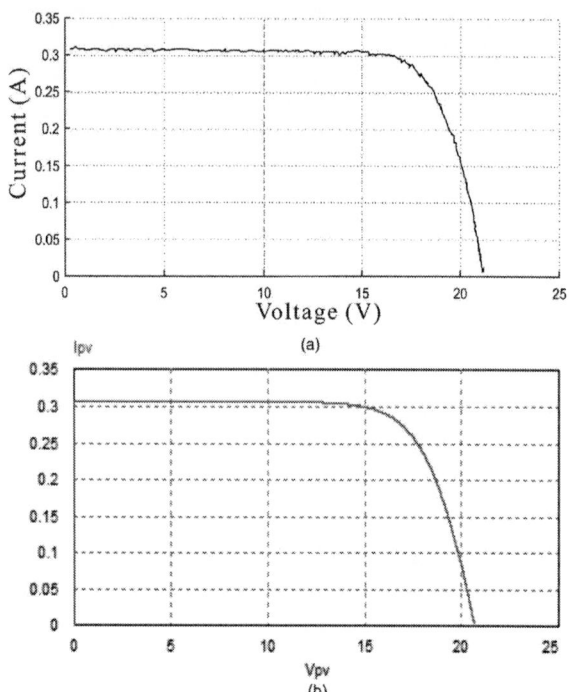

Fig. 11. *I-V* curves of module KM(P)5 obtained (a) experimentally and by simulation for *C*=22mF.

TABLE II
COMPARISON OF RESULTS OBTAINED BY SIMULATION AND
EXPERIMENTALLY FOR DISTINCT CAPACITANCE RATINGS.

C=470 µF	Temp. (°C)	Irradiance (W/m²)	V_{OC} (V)	I_{SC} (A)	P_{MPP} (W)
Simulation	29	980	21.22	0.3051	4.82
Experiment	29	980	21.9	0.3065	5.009

C=4.7 mF	Temp. (°C)	Irradiance (W/m²)	V_{OC} (V)	I_{SC} (A)	P_{MPP} (W)
Simulation	27	980	21.37	0.3044	4.857
Experiment	27	980	22.05	0.3043	5.149

C=22 mF	Temp. (°C)	Irradiance (W/m²)	V_{OC} (V)	I_{SC} (A)	P_{MPP} (W)
Simulation	36	980	20.72	0.3074	4.687
Experiment	36	980	21.25	0.301	5.045

TABLE III
PERCENT ERROR BETWEEN THE SIMULATION AND EXPERIMENTAL RESULTS.

Capacitor	Temp. (°C)	Irradiance (W/m²)	V_{OC} (%)	I_{SC} (%)	Max Power (%)
470 µF	29	980	3.2	0.46	3.92
4,700 µF	27	980	3.18	0.03	6.01
22,000 µF	36	980	2.56	2.08	7.64

In order to provide a better comparison among the aforementioned *I-V* curves, let us consider Fig. 12, where all characteristics are displayed in a same grid. Besides, the black, red, and green traces correspond to *C*=470 µF, *C*=4.7 mF, and *C*=22 mF, respectively. Fig. 13, Fig. 14, and Fig. 15 represent the output voltage and output current of module KM(P)5 measured during the experimental tests. A total of 24 points, 186 points, and 786 points could be acquired when the capacitance ratings are *C*=470 µF, *C*=4.7 mF, and *C*=22 mF, respectively, thus showing that the larger the capacitance is, the more accurate the plotted curves will be.

However, it is worth mentioning that large-size capacitors lead to increased cost and longer time to obtain the desired results.

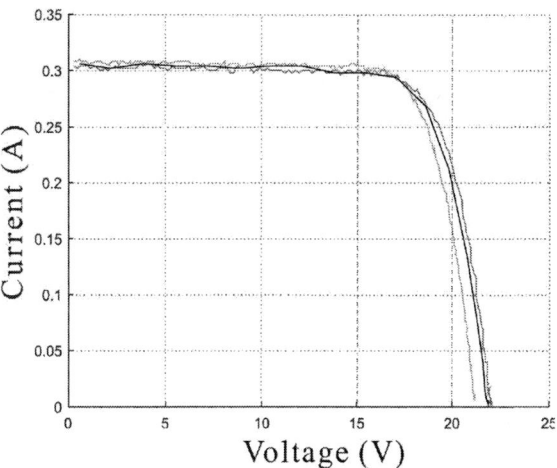

Fig. 12. *I-V* curves obtained experimentally for three distinct capacitances.

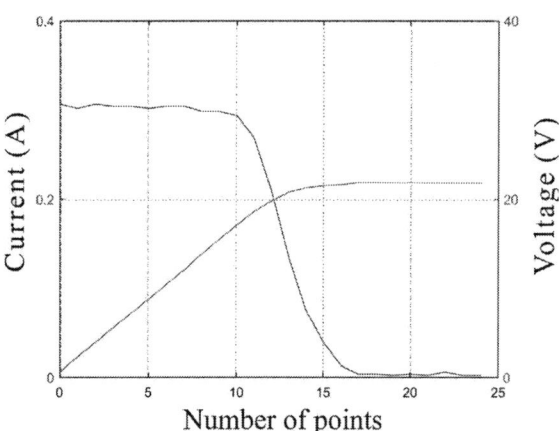

Fig. 13. Current and voltage of module KM(P)5 as a function of the number of plotted points in the *I-V* curve obtained with C=470 μF.

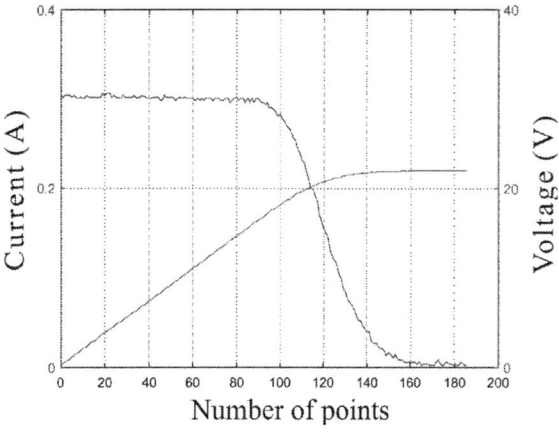

Fig. 14. Current and voltage of module KM(P)5 as a function of the number of plotted points in the *I-V* curve obtained with C=4.7 mF.

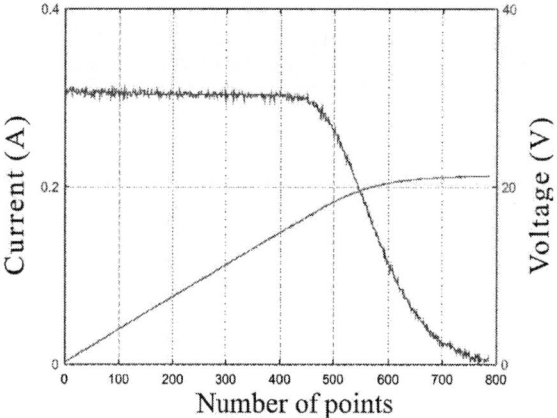

Fig. 15. Current and voltage of module KM(P)5 as a function of the number of plotted points in the *I-V* curve obtained with C=22 mF.

VI. CONCLUSION

This paper has presented a curve tracer based on the capacitive load method to obtain the *I-V* characteristics of PV modules. From the acquired results, it is observed that the exact points corresponding to the open-circuit voltage and short-circuit current could not be measured, which is possibly due to errors associated with the conversion of the analog quantities into bits performed by the controller. However, considering that the first and last points of the plotted curves are very close to the aforementioned values, this limitation does not lead to significant errors. For instance, the short-circuit current varies from 0.304 A to 0.308 A at 980 W/m^2 while the expected theoretical value is 0.31 A, also considering that the temperature varying between 27 °C and 36 °C does not influence this parameter significantly. Besides, the values obtained for the open-circuit voltage and maximum power are quite close to the theoretical ones.

ACKNOWLEDGEMENT

The authors acknowledge CAPES, CNPq, FAPEMIG, INERGE, and PPGEL for the support to this work.

REFERENCES

[1] E. Kabir, P. Kumar, S. Kumar, A. A. Adelodun, and K.-H. Kim, "Solar energy: Potential and future prospects," *Renewable and Sustainable Energy Reviews*, vol. 82, pp. 894-900, 2018.

[2] A. K. Pathak, K. Chopra, H. M. Singh, V. Tyagi, R. Kothari, S. Anand, *et al.*, "Role of Solar Energy Applications for Environmental Sustainability," in *Environmental Biotechnology: For Sustainable Future*, ed: Springer, 2019, pp. 341-374.

[3] M. Brito, L. Sampaio, G. Melo, and C. Canesin, "Contribuição ao estudo dos principais algoritmos de extração da máxima potência dos painéis fotovoltaicos," *Eletrônica de Potência*, vol. 17, pp. 592-600, 2012.

[4] K. Harini and S. Syama, "Simulation and analysis of incremental conductance and Perturb and Observe MPPT with DC-DC converter topology for PV array," in *2015 IEEE International Conference on Electrical, Computer and Communication Technologies (ICECCT)*, 2015, pp. 1-5.

[5] E. M. Vicente, R. L. Moreno, and E. R. Ribeiro, "MPPT technique based on current and temperature measurements," *International Journal of Photoenergy*, vol. 2015, 2015.

[6] E. Durán, M. Ferrera, J. Andújar, and M. Mesa, "I–V and P–V curves measuring system for PV modules based on DC-DC converters and portable graphical environment," in *2010 IEEE International Symposium on Industrial Electronics*, 2010, pp. 3323-3328.

[7] E. Durán, J. Andújar, J. Enrique, and J. Pérez-Oria, "Determination of PV generator I-V/P-V characteristic curves using a DC-DC converter controlled by a virtual instrument," *International Journal of Photoenergy,* vol. 2012, 2012.

[8] F. Spertino, J. Ahmad, A. Ciocia, P. Di Leo, A. F. Murtaza, and M. Chiaberge, "Capacitor charging method for I–V curve tracer and MPPT in photovoltaic systems," *Solar Energy,* vol. 119, pp. 461-473, 2015.

[9] Í. F. Silva, P. S. Vicente, F. L. Tofoli, and E. M. Vicente, "Portable and low cost photovoltaic curve tracer," in *2017 Brazilian Power Electronics Conference (COBEP),* 2017, pp. 1-6.

[10] M. G. Villalva, J. R. Gazoli, and E. Ruppert Filho, "Comprehensive approach to modeling and simulation of photovoltaic arrays," *IEEE Transactions on power electronics,* vol. 24, pp. 1198-1208, 2009.

LED Driver With Reduced Redundant Power Processing And Dimming For Street Lighting Applications

Kleber Chan Bekoski, Cassio Gobbato, Cassiano Ferro Moraes, Gustavo Weber Denardin
and Juliano de Pelegrini Lopes

PPGEE - Postgraduate Program in Electrical Engineering
UTFPR - Federal University of Technology - Paraná
Pato Branco, Brazil
kleberchan@gmail.com, cassiogobbato@gmail.com, cassianofmoraes@gmail.com, gustavo@utfpr.edu.br,
julianolopes@utfpr.edu.br

Abstract—This paper presents a light-emitting diode (LED) driver with dimming capability for street lighting application. The power reprocessing is reduced by the use of a partial cascade structure, which requires an isolated converter at the reprocessing stage. A SEPIC converter is used in discontinuous conduction mode of current for power factor correction. At the power control stage, a flyback converter is used in discontinuous current mode, reprocessing only 20 % of the total output power. A proportional integral (PI) controller maintains the output current regulated. Through the methodology employed it was possible to reduce the capacitances of the circuit and substitutes the electrolytic technology for capacitances with longer lifespan. Experimental results, obtained with a 70 W prototype shows that the circuit complies with the IEEE Std. 1789-2015 over the entire operating range. In addition, due the low total harmonic distortion of the input current, the prototype complies with the IEC 61000-3-2 Class C. The measured results for efficiency and power factor were above 90 % and 0.99, respectively.

Index Terms—Dimming, Flicker-free, Light-emitting diode drivers, Partial cascade, Power factor correction.

I. INTRODUCTION

Light Emitting Diode (LED) are becoming increasingly popular as lighting source due to its features such as high efficacy, reliability, long lifetime, low power consumption, lack of a warm-up period and high-power density compared with other traditional lighting sources [1]–[3]. Others advantages attributed to LED lamps, such the ability to control both light intensity and color, universal input voltage, visual light communication (VLC), and even detect people in indoor environments or cars in highway, are actually achieved thanks to the LED driver. [4]–[6].

Because of that, LED drivers has been the topic of many research papers, chasing ways to make the driver increasingly efficient and reliable. Therefore, several approaches have been proposed in the literature over the past few years [1], [7]. Additionally, being compliant with the standards and recommendations such as IEC 61000-3-2 Class C [8] and IEEE Std. 1789-2015 [9] are essential to make a quality driver.

In order to achieve a LED-compliant driver lifespan, the driver circuitry must use components with a long lifespan, which makes the use of electrolytic capacitors not recommended [4], [7].

Due to its performance, active drivers are more used than the passive ones. The simplest active driver is the single-stage. Although it is possible to reduce low-frequency ripple in the LEDs current using a control loop, it is difficult to incorporate many functions for such drivers, such as power factor correction (PFC), constant output voltage, dimming and universal voltage input simultaneously. Because of that, single-stage drivers are best suited for low power, in which size and cost are more relevant than PFC and other functions [4], [10].

Incorporating several functions to a single converter is not a simple task. Thus, multiple stages can be employed, each stage being responsible for part of the driver assignments. Typically, one stage is used for PFC and the other one for power control (PC) [10]. The multiple stages can be in cascade, integrated or in partial cascade connection. The cascade connection allows design flexibility, reduced capacitances, and the addition of advanced functions to the driver. However, in this topology, the total power is processed twice, increasing the power losses. By integrating both stages in only one switch, the number of active switches and their command circuits are reduced, but the control complexity and the switch voltage and/or current stresses are increased. Moreover, the integrated converters must operate at the same switching frequency and duty cycle, making it difficult to add advanced functions, such as dimming or universal voltage input [11], [12].

The partial cascade connection, also reported in the literature as reduced redundant power processing (R^2P^2) or optimized cascade, is based in two stages as in cascade connection, but in this configuration the second stage does not reprocess all the power demanded by the load. Such an approach reduces the power losses in one stage and increases the overall efficiency of the driver. Among the works that employ R^2P^2 for LED drivers, mostly use 35 W or less, and few of them have dimming capability. [2], [4], [11], [13], [14].

In this context, this paper proposes a dimming capable

978-1-7281-4181-7/19 $31.00 © 2019 IEEE   245

LED driver for street lighting application with partial cascade connection. The partial cascade will be detailed in Section II and the stages design is realized in Section III. Section IV presents the modeling and control design, followed by the dimming technique in Section V. The experimental results are shown in Section VI and the conclusion in Section VII.

II. PARTIAL CASCADE

The R^2P^2 can be achieved in different configurations, as shown in [14], but the partial cascade used in this work refers to the type I-IIIB or optimized cascade as presented in [13]. This type of connection between stages allows that one converter connects to the grid and the input current quality will be as good as this converter can perform. Besides, the second stage is not in full cascade and reprocesses only a fraction of the total output power, reducing the losses and increasing the overall efficiency of the driver. The Fig. 1 presents the partial cascade in a LED driver application.

The load voltage (V_{leds}) is the sum of the both stages output voltage (V_{pfc} and V_{pc}). In V_{pfc} it is allowed a high ripple to reduce the capacitance of the bus capacitor, which makes it possible to replace the electrolytic technology. The AC coupling of V_{pc} is in opposite phase with V_{pfc}, which reduces the ripple on the LEDs. The fraction of reprocessed power is denominated k factor, given by

$$k = \frac{P_{pc}}{P_{leds}} = \frac{I_{leds}V_{pc}}{I_{leds}V_{leds}} = \frac{V_{pc}}{V_{leds}} \quad (1)$$

being:

P_{pc} - Power processed at PC stage;

P_{leds} - Power demanded by the load;

I_{leds} - LEDs current.

However, as reported in [15], the implementation of partial cascade requires an isolated converter at reprocessing stage, since the non-isolated converter does not represent a correct implementation of the concept. Indeed, in [15] is demonstrated that such approach using a non-isolated converter at the reprocessing stage is equivalent to solutions that reprocess the entire load power.

III. CONVERTERS DESIGN

In this section, the parameters of the load and the design of the converters are presented. As previously stated, the LED driver has two stages based on DC-DC converters, one of them in the PFC and the other one in the PC. In this work, a SEPIC converter is used for the PFC stage and a flyback converter is used for the PC stage, both of them in discontinuous-current mode (DCM). The schematic of the proposed circuit is presented in Fig. 2, and the source $|V_{in}|$ represents the sinusoidal mains voltage after a full wave rectifier.

A. Design Parameters

In Fig. 2, the electric model of LEDs can be approximated by an ideal diode D, a forward diode voltage drop V_f and the internal series resistance R_{led} [16]. Assuming 600 mA the average current on the LEDs, to reach 70 W output power 37

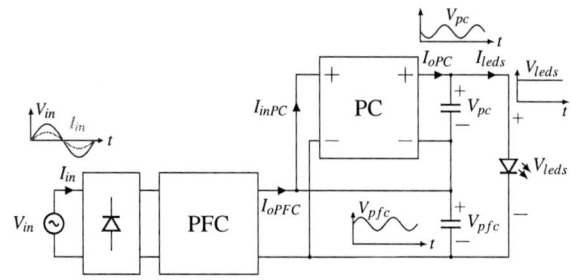

Fig. 1. Partial cascade in a LED driver application.

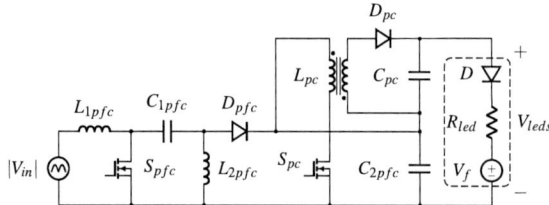

Fig. 2. Partial cascade configuration using SEPIC converter as PFC and flyback converter as PC stage.

LEDs BRIDGELUX model PEANUT 3W are used in series connection, whose features are presented in Table I.

The other parameters required for the design of the converters are also presented in Table I. The k factor determines the output voltage of each stage regarding the output voltage, and in this work is chosen 0.2 as utilized in [13], which means that just 20 % of the total power is processed twice.

B. SEPIC Converter Design

The PFC stage is used to correct the power factor (PF) and limit the harmonics of the input current in order to comply with IEC 61000-3-2 Class C standard. The SEPIC converter is appropriate to use as PFC stage, due to its characteristics, for example, operate as a voltage step-up or step-down and voltage output in the same polarity. Moreover, the SEPIC

TABLE I
DESIGN PARAMETERS

Symbol	Specification	Value
V_{in}	Input voltage (RMS)	127 V / 60 Hz
ΔI_{in}	maximum ripple input current	30 %
V_{leds}	LEDs voltage (average)	117.77 V
P_{leds}	LEDs power	70 W
I_{leds}	LEDs current (average)	0.6 A
V_{pfc}	PFC output voltage (average)	94.22 V
ΔV_{pfc}	PFC output voltage ripple	30 %
η_{pfc}	PFC efficiency	95 %
V_{pc}	PC output voltage (average)	23.55 V
ΔV_{pc}	PC output voltage ripple	1 %
η_{pc}	PC efficiency	90 %
f_s	Switching frequency	60 kHz
$D_{1pfc,max}$	maximum PFC duty cycle	0.3
$D_{1pc,max}$	maximum PC duty cycle	0.5
k	k factor	0.2
n_{leds}	Number of LEDs	37
R_{led}	LED resistance	0.725 Ω
V_f	LED forward voltage	2.7475 V

converter has only one active switch in the same ground of the input and output voltage, with simplifies the gate driver circuit. Furthermore, due to the input inductance, the circuit requires a small Electromagnetic Interference (EMI) filter, and in some cases, it may be eliminated [17]. The DCM is widely used in the PFC stage because when operating at fixed frequency and duty cycle, the converter emulates a resistance and theoretically has unitary PF, without a control loop [18].

The design of the SEPIC converter in DCM is developed according to the methodology presented in [17] and [18]. The power transfer of the circuit is determined by the equivalent inductance L_{eq}, as given by

$$L_{eq} = \frac{V_{pk}^2 D_{1pfc}^2 T_s}{4 P_{pfc}} \eta_{pfc} \qquad (2)$$

being:

V_{pk} - Peak line voltage;
T_s - Switching period;
P_{pfc} - Power processed at PFC stage;

The inductance L_{eq} is equivalent to the parallel association of the inductances L_{1pfc} and L_{2pfc}, which are respectively determined by

$$L_{1pfc} = \frac{V_{pk} D_{1pfc} T_s}{\Delta I_{in}} \qquad (3)$$

$$L_{2pfc} = \frac{L_{1pfc} L_{eq}}{L_{1pfc} - L_{eq}} \qquad (4)$$

where ΔI_{in} corresponds to the maximum ripple input current, which occurs at the peak line voltage.

The capacitor C_{1pfc} is designed to present a nearly constant voltage within a switching period, and, at the same time, to follow the line-voltage low-frequency variation [17], [19], being calculated by

$$C_{1pfc} = \frac{1}{(2\pi f_r)^2 (L_{1pfc} + L_{2pfc})} \qquad (5)$$

where the resonant frequency f_r is above the line frequency and below the switching frequency. To design C_{2pfc} the following expression is used

$$C_{2pfc} = \frac{V_{pk}^2 D_{1pfc}^2 T_s}{4 L_{eq} V_{pfc} \omega_e \Delta V_{pfc}} \qquad (6)$$

being ω_e the grid angular frequency.

The value of the used components are shown in Table II. The capacitor C_{2pfc} is obtained by the parallel association of the capacitors $C_{2pfc(1)}$ and $C_{2pfc(2)}$.

C. Flyback Converter Design

As mentioned in Section II, the partial cascade connection requires an isolated converter. Among the options of isolated converters, the most common are flyback, forward, push-pull, half-bridge and full-bridge. The flyback converter uses a coupled inductor as the transformer and inductance of the converter, reducing the number of components. However, it is only intended for applications ranging from tens to a few hundreds of watts. Although the other converters are able to handle a larger power transfer, they also use more

components and are more complex than the flyback converter [20]. Therefore, the flyback converter is used for the PC stage.

Regarding the conduction mode, the continuous current mode (CCM) has smaller inductance, lower current ripple and peak current at the switch than DCM. However, the DCM has advantages such as zero current turn-on of the active switch and zero current turn-off of the diode. Furthermore, the transfer function of the flyback converter operating in DCM is of order one and the control loop is less demanding than in CCM [21], [22]. Thus, the DCM is chosen for this work.

The design of the flyback converter in DCM is developed as followed by [23]. When the switch is closed, the inductor stores energy from the input voltage, which in this case is the PFC output voltage and the inductance is determined by

$$L_{pc} = \frac{V_{pfc,min}^2 D_{1pc}^2}{2 P_{pc} f_s} \eta_{pc} \qquad (7)$$

where P_{pc} is the amount of power processed by the flyback converter, which is equal to the product of the total output power by the k factor.

The design of the capacitor C_{pc} is developed from the first subinterval when the active switch is on and the diode is off. Thus, the capacitor voltage is the same as the output of the converter, and the capacitance is given by

$$C_{pc} = \frac{I_{leds} D_{1pc,max}}{\Delta V_{pc} f_s} \qquad (8)$$

In the flyback converter using MOSFET, when it turns-off, a high-voltage spike occurs on the drain pin because of a resonance between the leakage inductance of the main transformer and the output capacitor of the MOSFET. An additional circuit is utilized to clamp the voltage in order to prevent the failure of the semiconductor, as presented in [24]. The designed components of such circuit are shown in Table II.

TABLE II
DESIGNED COMPONENTS

Symbol	Specification	Value
Full bridge rectifier		
$D_{fb(1-4)}$	1N4007	700 V / 1 A
PFC - SEPIC		
L_{1pfc}	Thorton E30/15/7	3.79 mH
L_{2pfc}	Thorton E30/15/7	179.95 μH
C_{1pfc}	B32529C3154J	150 nF / 250 V
$C_{2pfc(1)}$	B32678G3476K	47 μF / 300 V
$C_{2pfc(2)}$	B32676G3206K	20 μF / 300 V
S_{pfc}	IRFP360	400 V / 23 A / $R_{ds,on}$: 0.20 Ω
D_{pfc}	15ETH06FP	600 V / 15 A
PC - Flyback		
L_{pc}	Thorton E25/10/6	416.81 μH
$n1$	Primary turns	57
$n2$	Secondary turns	18
C_{pc}	B32676G3206K	20 μF / 300 V
S_{pc}	IRFP460	500 V / 20 A / $R_{ds,on}$: 0.27 Ω
D_{pc}	MUR1560	600 V / 15 A
Flyback snubber		
D_{sn}	MUR1560	600 V / 15 A
R_{sn}	Film resistor	68 kΩ / 3 W
C_{sn}	Ceramic capacitor	4.7 nF / 1 kV

IV. MODELING AND CONTROL

In order to prevent lighting variation that may pose possible health risks to users, since the luminous flux is directly proportional to the current through the LED and due to LED unique characteristics, a current control loop is used in the PC stage to obtain a constant output current [1], [9].

The control is designed based on the transfer function of the converter. To obtain the transfer function of flyback converter in DCM it is adopted the small-signal AC model that derives from the averaged switch model according to the methodology presented in [22]. By the use of such method, the transfer function that relates the output current and the duty cycle is

$$G_{id}(s) = \cfrac{\cfrac{V_{pfc}^2 D_{1pc} T_s}{V_{pc} L_{pc} C_{pc} n_{leds} R_{led}}}{s + \left(\cfrac{1}{n_{leds} R_{led} C_{pc}} + \cfrac{V_{pfc}^2 D_{1pc}^2 T_s}{2 L_{pc} V_{pc}^2 C_{pc}} \right)} \quad (9)$$

A Proportional Integral (PI) controller is used to control the current through the LEDs. This controller has features like satisfactory input noise immunity, reduced complexity and has been widely employed in recent literature on LED drivers [11], [13]. The block diagram of the LEDs current control system is shown in Fig. 3. Where:

  I_{ref} - Current reference;
  $e(s)$ - Error;
  $C(s)$ - Controller transfer function;
  $u(s)$ - Control action;
  $H_i(s)$ - Feedback transfer function.

The controller has been designed using the SISOTOOL from MATLAB. It is admitted that the controller must be fast enough to compensate the low-frequency ripple (120 Hz), but its bandwidth must be limited to at least one decade below the switching frequency to filter the switching noise. Also, the phase margin has been chosen near the 60° to avoid overshoot. However, the control does not respond fast enough to mitigate the low-frequency ripple with a phase margin of 60° and low bandwidth. Thus, the bandwidth has been increased to 1.1 kHz, which increases the overshoot as well. A high overshoot increases the output current oscillation and can raise the flicker level. So, the overshoot limit is its impact at the flicker level and, therefore, care must be taken with the overshoot in control design. However, as shown in Fig. 4, the step response of such controller is in tenths of milliseconds and it is enough to compensate the low-frequency oscillations. The overshoot of 13.8 % is not critical for the circuit because, as will be presented in the experimental results, the topology complies with the standards. The transfer function of the designed PI controller is

$$C(s) = 0.4003 \frac{(s + 11631)}{s} \quad (10)$$

To implement the control loop a TIVA TM4C123G microcontroller was used. Therefore, a discrete-time approximation of this controller transfer function was implemented based on

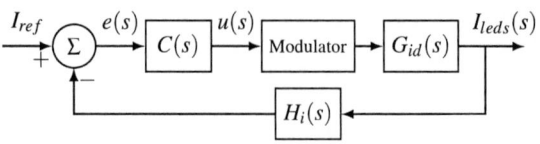

Fig. 3. Block diagram of the LEDs current control system.

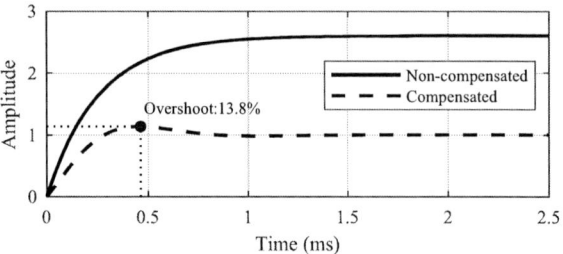

Fig. 4. Step response of the control system.

the Zero Order Hold (ZOH) method, with sample time equal to T_s.

V. DIMMING TECHNIQUE

The LED brightness is proportional to its average current. Methods such as amplitude modulation (AM) and pulse width modulation (PWM) can be used for dimming control. The AM method is accomplished by adjusting the output current reference and it does not require hardware modification. However, a current variation in the LEDs may cause a color shift. The PWM method can prevent color shift but require an additional circuit to control the output current, increasing cost and complexity to the driver. Also, the PWM method may cause flicker, which is not desired. Since that street lighting does not require a highly accurate color reproduction, the AM method can be employed without major disadvantages. Also, due to its simplicity of implementation, the AM method is commonly used in street lighting applications. Thus, the AM method is employed in this work [25], [26].

Due to the partial cascade connection, the output voltage is the sum of the PFC and PC stages. Therefore, for dimming based on the AM method, both the PFC output voltage and the output current reference must be changed. According to the LEDs model (Fig. 2), V_f remains constant independently to the LEDs current, but the voltage drop in R_{leds} does not. So, the output voltage as a function of dimming is given by

$$V_o(dim) = n_{leds} V_f + n_{leds} R_{led} I_{leds} dim \quad (11)$$

and the PFC output voltage is determined by

$$V_{pfc}(dim) = (1 - k) V_o(dim) \quad (12)$$

To determines the D_{1pfc} according to the dimming level, the SEPIC DCM voltage ratio is used, which is given by

$$M = D_{1pfc} \sqrt{\frac{T_s V_{pfc}^2}{2 L_{eq} P_{pfc}}} \quad (13)$$

As the power processed by the PFC stage also changes according to the dimming and rearranging the terms of (13), the expression

$$D_{1pfc}(dim) = \frac{\sqrt{2L_{eq}P_{pfc}dimf_s}}{V_{in}} \qquad (14)$$

returns the duty cycle to operate the SEPIC converter according to the dimming.

Therefore, to implement dimming in this work, the dimming reference should change the current reference of the control in the PC stage and change the duty cycle in the PFC stage as well.

VI. RESULTS

In this section, the main results achieved with the prototype of the proposed LED driver are presented. The prototype was built with the components present in Table II. The auxiliary circuits were powered by an external power supply and to measure the output current, the hall effect current sensor WCS 2702 was used. The results are shown in 20 % dimming steps from one to another operating point. Once the results are consistent in these points, it is expected that the circuit is capable to operate consistently between these points as well.

Fig. 5 shows the input power, output power, and overall efficiency. The rated output power is 70 W as designed and the efficiency is above 90 % over the entire operating range, with a maximum of 91.63 %, measured by the Yokogawa WT1800 high performance power analyzer.

The values of average current and ripple current in LEDs are presented in Fig. 6, with 600 mA at nominal power and decreasing linearly for lower output power. The current ripple through the LEDs results in flicker modulation, which is presented in Fig. 7 in comparison with the IEEE Std. 1789-2015. It is noted that all points are complying with the standard.

The measurements of PF and total harmonic distortion (THD) are shown in Fig. 8. In the whole range of operation

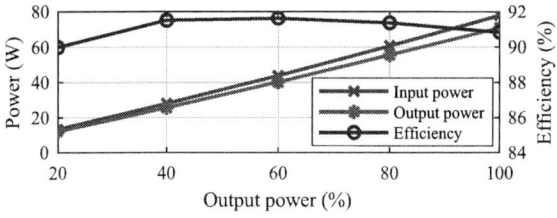

Fig. 5. Input power, output power and overall efficiency of the proposed LED driver.

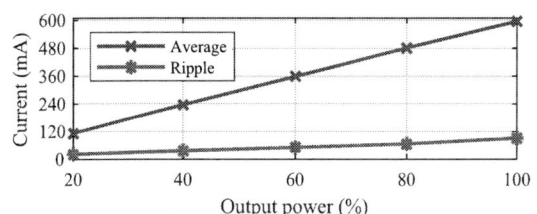

Fig. 6. Output current, average and ripple of the proposed LED driver.

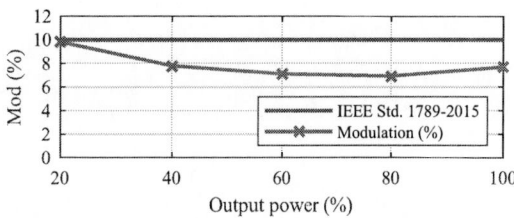

Fig. 7. LED current modulation (%) of the proposed LED driver and recommendation by IEEE Std. 1789-2015.

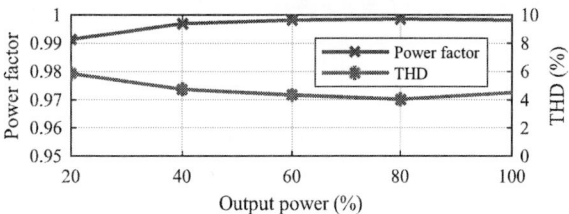

Fig. 8. Power factor and total harmonic distortion of the input current of the proposed LED driver.

the PF is above 0.99, with THD lower than 6 %. In Fig. 9 is presented the harmonic content of the circuit to rated power and minimum power in comparison with the limits imposed by the standard. However, the harmonic content of the input current has been verified over the entire operating range and complies with IEC 61000-3-2 Class C standard. The waveforms of input and output current and voltage are presented in Fig. 10(a), Fig. 10(b) and Fig. 10(c) to rated output power, 60 % load and 20 % load, respectively.

VII. CONCLUSION

This paper proposed a LED driver with dimming capability, using the partial cascade connection and no electrolytic capacitor. As required by the connection, an isolated converter was used at the reprocessing stage, and according to the presented experimental results, the driver was able to achieve all the designed specifications. The dimming has been properly implemented and the flicker complies with the IEEE Std. 1789-2015 to the whole operation range as well as the THD complies with the IEC 61000-3-2 Class C. All this with more than 90 % efficiency and PF higher than 0.99. So, the authors believe that the presented work is a good alternative to implement a street lighting LED driver with dimming capability.

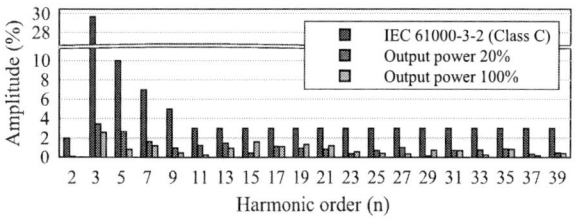

Fig. 9. Harmonic spectrum input current for 100 % and 20 % output power in comparison with IEC 61000-3-2 Class C.

978-1-7281-4181-7/19 $31.00 © 2019 IEEE

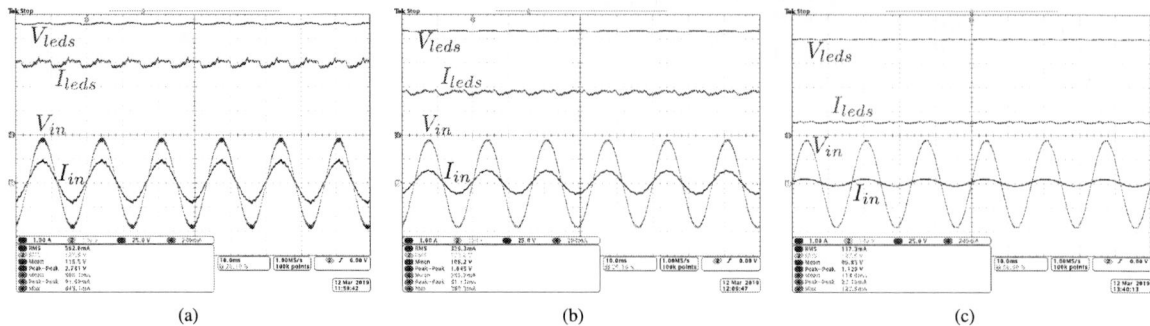

(a)   (b)   (c)

Fig. 10. Experimental results. (a) Rated output power. (b) 60 % output power. (c) 20 % output power; Input current (CH1 - 1 A/div); Input voltage (CH2 - 100 V/div); LEDs voltage (CH3 - 25 V/div); LEDs current (CH4 - 200 mA/div); Time scale 10 ms/div.

ACKNOWLEDGMENT

This study was financed in part by the Coordenação de Aperfeiçoamento de Pessoal de Nível Superior - Brasil (CAPES) - Finance Code 001, Conselho Nacional de Desenvolvimento Científico e Tecnológico (CNPq), Fundação Araucária (FA) and Financiadora de Estudos e Projetos (FINEP).

REFERENCES

[1] Y. Wang, J. M. Alonso, and X. Ruan, "A Review of LED Drivers and Related Technologies," *IEEE Transactions on Industrial Electronics*, vol. 64, no. 7, pp. 5754–5765, 2017.

[2] Z. Shan, X. Chen, J. Jatskevich, and C. K. Tse, "AC–DC LED Driver With an Additional Active Rectifier and a Unidirectional Auxiliary Circuit for AC Power Ripple Isolation," *IEEE Transactions on Power Electronics*, vol. 34, no. 1, pp. 685–699, jan 2019.

[3] Peng Fang, Yan-Fei Liu, and P. C. Sen, "A Flicker-Free Single-Stage Offline LED Driver With High Power Factor," *IEEE Journal of Emerging and Selected Topics in Power Electronics*, vol. 3, no. 3, pp. 654–665, sep 2015.

[4] I. Castro, A. Vazquez, M. Arias, D. G. Lamar, M. M. Hernando, and J. Sebastian, "A review on flicker-free ac-dc LED drivers for single-phase and three-phase ac power grids," *IEEE Transactions on Power Electronics*, 2019.

[5] J. Wang, C. Jiang, H. Zhang, X. Zhang, V. C. M. Leung, and L. Hanzo, "Learning-Aided Network Association for Hybrid Indoor LiFi-WiFi Systems," *IEEE Transactions on Vehicular Technology*, vol. 67, no. 4, pp. 3561–3574, apr 2018.

[6] S. A. E. Mohamed, "Smart Street Lighting Control and Monitoring System for Electrical Power Saving by Using VANET," *International Journal of Communications, Network and System Sciences*, vol. 06, no. 08, pp. 351–360, 2013.

[7] P. S. Almeida, D. Camponogara, M. Dalla Costa, H. Braga, and J. M. Alonso, "Matching LED and Driver Life Spans: A Review of Different Techniques," *IEEE Industrial Electronics Magazine*, vol. 9, no. 2, pp. 36–47, jun 2015.

[8] IEC61000-3-2, "Electromagnetic compatibility (emc) - part 3-2: Limits for harmonic current emissions (equipment input current $\leq$ 16 a per phase)," *International Standard*, 2018.

[9] "IEEE Recommended Practices for Modulating Current in High-Brightness LEDs for Mitigating Health Risks to Viewers," *IEEE Std 1789-2015*, pp. 1–80, June 2015.

[10] S. Li, S.-C. Tan, C. K. Lee, E. Waffenschmidt, S. Y. R. Hui, and C. K. Tse, "A Survey, Classification, and Critical Review of Light-Emitting Diode Drivers," *IEEE Transactions on Power Electronics*, vol. 31, no. 2, pp. 1503–1516, feb 2016.

[11] C. Gobbato, S. V. Kohler, I. H. de Souza, G. W. Denardin, and J. d. P. Lopes, "Integrated Topology of DC–DC Converter for LED Street Lighting System Based on Modular Drivers," *IEEE Transactions on Industry Applications*, vol. 54, no. 4, pp. 3881–3889, jul 2018.

[12] Tsai-Fu Wu and Te-Hung Yu, "Unified approach to developing single-stage power converters," *IEEE Transactions on Aerospace and Electronic Systems*, vol. 34, no. 1, pp. 211–223, 1998.

[13] D. Camponogara, G. F. Ferreira, A. Campos, M. A. D. Costa, and J. Garcia, "Offline LED Driver for Street Lighting With an Optimized Cascade Structure," *IEEE Transactions on Industry Applications*, vol. 49, no. 6, pp. 2437–2443, nov 2013.

[14] M. Chow and C. Tse, "An efficient PFC voltage regulator with reduced redundant power processing," *30th Annual IEEE Power Electronics Specialists Conference. Record. (Cat. No.99CH36321)*, vol. 1, pp. 87–92, 1999.

[15] G. Spiazzi, "Reduced redundant power processing concept: A reexamination," in *2016 IEEE 17th Workshop on Control and Modeling for Power Electronics (COMPEL)*. IEEE, jun 2016, pp. 1–8.

[16] R.-L. Lin, S.-Y. Liu, C.-C. Lee, and Y.-C. Chang, "Taylor-Series-Expression-Based Equivalent Circuit Models of LED for Analysis of LED Driver System," *IEEE Transactions on Industry Applications*, vol. 49, no. 4, pp. 1854–1862, jul 2013.

[17] M. F. da Silva, J. Fraytag, M. E. Schlittler, T. B. Marchesan, M. A. Dalla Costa, J. Alonso, and R. N. do Prado, "Analysis and Design of a Single-Stage High-Power-Factor Dimmable Electronic Ballast for Electrodeless Fluorescent Lamp," *IEEE Transactions on Industrial Electronics*, vol. 60, no. 8, pp. 3081–3091, aug 2013.

[18] J. D. P. Lopes, S. V. Kohler, M. F. Menke, and A. R. Seidel, "Tubular Fluorescent Lamps Detection Based on the Self-Oscillating Electronic Ballast," *IEEE Journal of Emerging and Selected Topics in Power Electronics*, vol. 6, no. 3, pp. 1259–1272, sep 2018.

[19] D. Simonetti, J. Sebastian, and J. Uceda, "The discontinuous conduction mode Sepic and Cuk power factor preregulators: analysis and design," *IEEE Transactions on Industrial Electronics*, vol. 44, no. 5, pp. 630–637, 1997.

[20] Art Pini Arthur, "Design a Switch Mode Power Supply Using an Isolated Flyback Topology," *Digi-Key's North American Editors*, 2018.

[21] M. Ferdowsi, A. Emadi, M. Telefus, and C. Davis, "Pulse Regulation Control Technique for Flyback Converter," *IEEE Transactions on Power Electronics*, vol. 20, no. 4, pp. 798–805, jul 2005.

[22] R. W. Erickson and D. Maksimović, *Fundamentals of Power Electronics*, 2nd ed. Boston: Kluwer Academic, 2001.

[23] P. C. V. Luz, M. R. Cosetin, P. E. Bolzan, T. Maboni, M. F. da Silva, and R. N. do Prado, "An integrated insulated buck-Flyback converter to feed LED's lamps to street lighting with reduced capacitances," in *2015 IEEE International Conference on Industrial Technology (ICIT)*, vol. 5. IEEE, mar 2015, pp. 908–913.

[24] G.-B. Koo, "Application Note AN-4147 Design Guidelines for RCD Snubber of Flyback Converters," Fairchild Semiconductor Corporation, 2006.

[25] C.-S. Moo, Y.-J. Chen, and W.-C. Yang, "An Efficient Driver for Dimmable LED Lighting," *IEEE Transactions on Power Electronics*, vol. 27, no. 11, pp. 4613–4618, nov 2012.

[26] M. F. Menke, R. V. Tambara, F. E. Bisogno, M. F. Da Silva, and A. R. Seidel, "Universal input voltage LED driver with dimming capability and reduced DC-link capacitance," *IECON Proceedings (Industrial Electronics Conference)*, pp. 3629–3634, 2016.

978-1-7281-4181-7/19 $31.00 © 2019 IEEE

Active-Capacitor for Power Decoupling in Single-Phase Grid-Connected Converters

Thiago A. Pereira, Denizar C. Martins and Roberto F. Coelho

Power Electronics Institute (INEP)
Federal University of Santa Catarina (UFSC)
Florianópolis, Santa Catarina, Brasil
thiago.pereira@inep.ufsc.br, denizar@inep.ufsc.br and roberto@inep.ufsc.br

Abstract—**In power conversion systems connected to the single-phase power grid it is usual the utilization of electrolytic capacitors to handle with low-frequency power ripple. However, such capacitors have a limited lifetime. Thereby, this paper proposes the application of an active power decoupling cell (PDC) connected to the dc bus of a two-stage photovoltaic microinverter, operating as an active-capacitor. The purpose of the PDC is to shift the power ripple from the dc bus to a decoupling capacitor, whose voltage ripple may oscillate more than the voltage of the main bus, allowing the use of low capacitance and high lifetime capacitors. Thus, the paper presents an analysis of the active-capacitor based on PDC and a new control structure to meet all the aforementioned objectives. Experimental results obtained from the prototype of a photovoltaic microinverter with PDC validate the performance of the active-capacitor.**

Index Terms—**Active-Capacitor, dc Bus voltage, second order frequency, Power decoupling cell, power ripple**

I. INTRODUCTION

In a single-phase grid-connected PV application, as shown in Figure 1, the power flow is given from the PV module to the grid. For simplicity, the effects of power losses and output filters are neglected. Thus, the power extracted from the PV module, $p_{pv}(t)$, is considered constant, as illustrated in Figure 1 (a), whereas the instantaneous power provided to the grid is time varying, as illustrated in Figure 1 (c). As can be seen, the instantaneous input power, $p_{pv}(t)$, is not equal to the instantaneous output power, $p_{grid}(t)$. This difference creates a voltage ripple containing *nth*-order harmonics on the dc bus voltage (cf. Figure 1 (b)), which reduces the system efficiency and reliability when it is not decoupled from the PV module output. Thereby, this power mismatch must be handled by an energy storage element able to decouple this unbalance between the input $p_{pv}(t)$ and output $p_{grid}(t)$ power [1].

The energy storage might be performed by means an element other than a capacitor, such as an inductor. However, the use of an inductor may be a poor option, because of its high weight and cost. Hence, capacitors are usually selected as energy storage component at dc bus voltage.

As mentioned, the power ripple flows through the bus capacitor, C_{bus} as illustrated in Figure 1 (b). In effect the voltage across the bus capacitor $v_{cb}(t)$ varies accordingly to the amount of power that is processed, as is shown in Figure 1 (c) and (d). In this way, when $p_{grid}(t) < p_{pv}(t)$ the excess of energy is stored in C_{bus}, consequently the voltage $v_{cb}(t)$

increases. Conversely, when $p_{grid}t > p_{pv}(t)$, the capacitor deliveries the energy to the grid, such that the voltage $v_{cb}(t)$ decreases. Therefore, the capacitor voltage must be allowed to increase and decrease its voltage as necessary to store and release the required energy.

This swing implies in a pulsating voltage Δv_{Cb} that oscillates on same frequency of $p_{grid}(t)$. Thus, depending on applications, this voltage ripple may negatively affect the system performance. In the particular case of a PV system, it reduces the PV conversion efficiency and degrades the MPPT performance if not decoupled from the PV module output, since the propagation of this low-frequency ripple to the PV side causes a swing of the operating point around the true maximum power point (MPP).

As shown above, there are some situations that it is desirable to compensate the voltage ripple on the dc bus. In response to these issues, the use of a capacitor in parallel with the dc bus is the only solution to be considered on the passive power decoupling method. This technique is simple to implement and the value of the bus capacitance is designed using [2]:

$$C_{bus} = \frac{P_{pv}}{\omega_0 V_{Cb} \Delta v_{Cb}} \tag{1}$$

Equation (1) shows that high capacitances (C_{bus}) are required to ensure a low-voltage ripple (Δv_{Cb}), thus in order to increase the stored energy density (i.e. joules/cm^3), electrolytic capacitors are still the most useful technology in these applications. However, bulky electrolytic capacitors have the drawback of a short lifetime, especially under high temperature, and are not compatible with the lifetime of the others elements that compose the PV Microinverter. Thus, these capacitors are the limiting components that determine the lifetime and the volume of the PV Microinverter [3], [4].

Based on this fact, the literature has proposed several active decoupling strategies to shift the power ripple to another specific energy storage component with a relatively small size and a long lifetime by means a bidirectional power converter. These strategies have been extensively studied in the literature in recent years and a comprehensive overview is given in the following papers [5]–[12].

In this paper, a solution of power decoupling cell to be integrated into a two-stage PV Microinverter is proposed, in order to solve the issues regarding the volume, lifetime and

978-1-7281-4181-7/19 $31.00 © 2019 IEEE

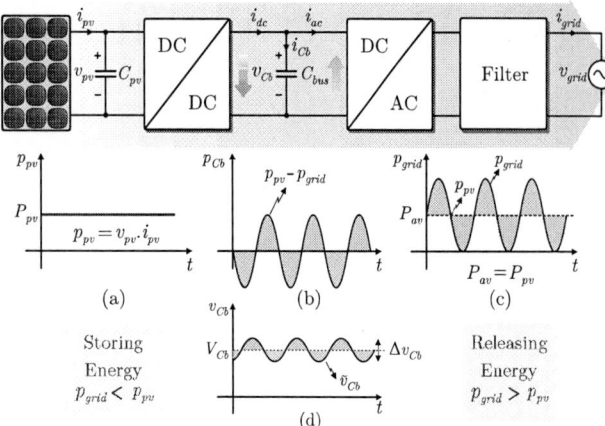

Fig. 1. Power flow in a PV Microinverter connected to the grid, where: (a) is the power generated by the PV module; (b) is the power decoupled in energy storage element and (c) is the power delivered to the grid.

reliability of the dc bus. For this reason, it is proposed a power decoupling stage controlled to reduce the voltage ripple. As it will be presented, this power decoupling stage can be understood as an active-capacitor, so that it must alternately store and release energy within a quarter of the grid period T, where $T = 2\pi/\omega_0$. Hence, the main contribution of this paper is to present the power decoupling stage under another point of view presented in the literature. In that case, as an active-capacitor tuned on the double-line-frequency $(2\omega_0)$.

II. Active Power Decoupling Cell

In general, the active power decoupling methods are conveniently divided as independent and dependent decoupling technique. In the first strategy, the Power Decoupling Cell (PDC) works independently, which can be connected in parallel or in series to the dc bus. In the second case, the PDC shares partially and even completely the dc-ac converter switches. As a result, the shared switches work coordinately to achieve simultaneously the power decoupling and the power conversion.

Among both, the parallel power decoupling techniques are the favorite, so that most of the active topologies proposed in the literature belong to the parallel methods. This popularity is due to the low capacitance value when compared with series structure, since only the power ripple is processed by the decoupling system. In addition, the parallel power decoupling topologies allows flexibility in employed control and modulation strategies, so that the PDC operates independently of the PV Microinverter. Thereby, the dc-ac converter is responsible to regulate the dc bus voltage while the PDC deal with the power ripple, shifting it away from dc bus to the decoupling capacitor located at PDC output.

Once the focus of this work is oriented to the two-stage PV Microinverter shown in Figure 2, the most convenient active power decoupling strategies is the independent PDC, which can be connected in parallel to the dc bus. Thus, a set of active power decoupling topologies can be applied, such as: Buck, Boost, Buck-Boost and H-bridge [1], [2], [5]–[12].

The chosen topology of the PDC to be integrated to the PV Microinverter consists of a bidirectional dc-dc buck converter which is also known as a synchronous-buck, as shown in Figure 2 (a).

The selected buck-type PDC is connected in parallel to the dc bus capacitor C_{bus} at the input of the single-phase PV Microinverter. This capacitor is still needed at the input of the inverter to filter the switching ripple energy and the residual second-order harmonic ripple energy, not fully decoupling by the PDC. A low voltage decoupling capacitor, C_f, placed at the output filter of the buck converter, is used as an energy storage element; while the inductor L_f is used as an energy transfer component. The switches of the PDC must be properly actioned to allow the storage of energy into C_f and to release it back to C_{bus}. Thus, the switch S_5 controls the PDC as a buck converter, establishing the power flow from the dc bus voltage $v_{Cb}(t)$ to output filter capacitor voltage $v_{Cf}(t)$ (Storage Mode), whereas the switch S_6 controls the PDC as a boost converter, so that the previously storage energy is released to the main dc bus capacitor (Releasing Mode). In other words, the PV module provides constant energy during each switching cycle, while the decoupling capacitor stores the excessive energy when $P_{pv} > p_{grid}(t)$ and supplement the PV module with additional energy by releasing the buffered energy when $P_{pv} < p_{grid}(t)$,

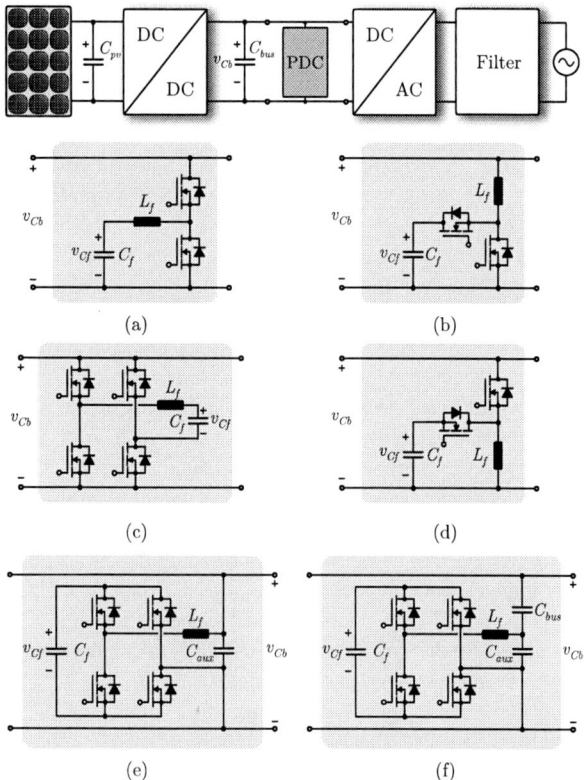

Fig. 2. Overview of common Active Power Decoupling Cell (PDC) connected in parallel to the dc bus voltage: (a) Buck; (b) Boost; (c) Buck-Boost; (d) and (e) Half-bridge and (f) Half-bridge connected in series with the bus capacitor.

978-1-7281-4181-7/19 $31.00 © 2019 IEEE

so that this process allows to mitigate the voltage ripple on the bus voltage [2], [12]–[14].

Under operation with unity power factor, the instantaneous power delivered to grid $p_{grid}(t)$ has a double-line-frequency oscillation, which peak value is exactly the double of the PV module power, $2P_{pv}$. The energy stored or released from the decoupling capacitor during a quarter of cycle can be calculated by integrating one of the shadowed area of Figure 1 (c), according to:

$$E_{Cf_{(in)}} = \frac{2}{\omega_0} \int_{\pi/4}^{3\pi/4} P_{pv} cos(2\omega_0 t) d\omega_0 t = -\frac{P_{pv}}{\omega_0} \quad (2)$$

Since the conventional positive flow of energy in the PV Microinverter is given from the PV module to the grid, the positive flow of energy in the PDC is given from the decoupling capacitor to the grid. Thus, the negative signal in (2) represents the reverse energy flow into the PDC, which means that the excess of energy flows from PV module to the decoupling capacitor C_f.

III. ACTIVE-CAPACITOR CONCEPT

The switched electrical circuit of the Buck-type PDC converter is illustrated in Figure 3 (a). The aim of the *LC* filter with damping branch (C_{fd} and R_{fd}) is to minimize the magnitude of the output quantity (current or voltage) around the resonance frequency, avoiding oscillations and instabilities [1].

The resulting small-signal model of the buck converter is illustrated in Figure 3 (b); this model can be solved using conventional circuit analysis techniques, to find the small-signal transfer functions, output impedance, and other frequency-dependent properties.

Since the small-signal ac model is a linear and time-invariant system, the set of linear small-signal ac equations obtained by means of perturbation and linearization is given by:

$$\tilde{i}_{S5}(s) = I_{Lf}\tilde{d}(s) + \tilde{i}_{Lf}(s)D$$
$$\tilde{v}_{S6}(s) = V_{Cb}\tilde{d}(s) + \tilde{v}_{Cb}(s)D \quad (3)$$

Given the aforementioned analysis, one can understand the PDC as a capacitor with high capacitance value, since the behaviour of both on reducing the voltage ripple on the dc bus is quite similar. Indeed, in this section, its shown that the frequency response of the decoupling cell is equivalent with that obtained by a physical capacitor, allowing represent it whereby an equivalent capacitor $C_{pdc(eq)}$.

For this purpose, it is necessary to determine the input impedance $Z_i(s)$ of the PDC using the small-signal ac circuit model of the developed buck converter, cf. Figure 3 (b).

Since the objective is to obtain an expression for $Z_i(s)$, a test voltage source $\tilde{v}_{test}(s)$ is inserted at the converter input port, as shown in the Figure 3 (c). Thus, the input impedance $Z_i(s)$ can be seen as the transfer function that relates to $\tilde{v}_{test}(s)$ with $\tilde{i}_{test}(s)$, in according to:

$$Z_i(s) = \frac{\tilde{v}_{test}(s)}{\tilde{i}_{test}(s)} \quad (4)$$

In addition, the equivalent impedance of the passive elements at the PDC output, defined by $Z_{o(eq)}$, is expressed as:

$$Z_{o(eq)} = \frac{\left(1 + sC_{fd}R_{fd}\right)}{s^2 C_f C_{fd} R_{fd} + s\left(C_f + C_{fd}\right)} \quad (5)$$

Therefore, the current that flows through decoupling inductor is given by:

$$\tilde{i}_{Lf} = \frac{D\tilde{v}_{test} + \tilde{d}V_{Cb}}{sL_f + Z_{o(eq)}} \quad (6)$$

As consequence, applying (6) into (3), yields:

$$\tilde{i}_{test} = \left[\frac{D^2\tilde{v}_{test} + \tilde{d}\left(V_{Cb}D + sL_f I_{Lf} + Z_{o(eq)}I_{Lf}\right)}{sL_f + Z_{o(eq)}}\right] \quad (7)$$

The PDC impedance in open loop $Z_{i,(OL)}$, i.e. when $\tilde{d} = 0$ in (7), is described by (8), as follows:

$$Z_{i,(OL)}(s)\big|_{\tilde{d}=0} = \frac{\tilde{v}_{test}}{\tilde{i}_{test}} = \frac{a_1 s^3 + a_2 s^2 + a_3 s + 1}{b_1 s^2 + b_2 s} \quad (8)$$

wherein, the coefficients are listed below:

$$a_1 = 2C_{fd}C_f L_f R_d \quad a_2 = 2L_f(C_{fd} + C_f) \quad a_3 = C_d R_d$$
$$b_1 = C_{fd}C_f R_d D^2 \quad b_2 = D^2(C_{fd} + C_f) \quad a_4 = D^2$$

Once the decoupling cell is designed to operate at low-frequency ($s = j\omega \to 0$), it is possible to simplify (8):

$$Z_{i,(eq)}(s) \approx \frac{1}{b_2 s} = \frac{1}{D^2}\frac{1}{(C_{fd} + C_f)s} \quad (9)$$

The magnitude of the power decoupling input impedance in open-loop is dominated by the capacitor impedance at low-frequency, and by the inductor at high-frequency. The capacitor and inductor asymptotes intersect at the resonance frequency. So, the magnitude at low-frequency is approximately equal to:

$$\left\|Z_{i,(eq)}(j\omega)\right\| \approx \frac{1}{D^2}\frac{1}{\omega(C_{fd} + C_f)} \quad (10)$$

Equation reveals that the PDC behaves as equivalent capacitance $C_{pdc(eq)}$ defined in function of C_f, C_{fd} and D, in which the asymptote is illustrated in Figure 4. The Bode plot of Figure 4 (a) represents the input impedance of the power decoupling $Z_{i,(OL)}$, when it operates in the open-loop at the operating point determined by $D_{buck} = V_{Cf}/V_{Cb}$. As can be seen, the behaviour of the active-capacitor is verified, since the asymptote capacitive $Z_{i,(eq)}$ and the Bode plot of $Z_{i,(OL)}$ are equals during the low-frequency ($\omega < \omega_r$). From (9), one can be observed that the equivalent capacitance is given by:

$$C_{pdc(eq)} = D^2\left(C_f + C_{fd}\right) \quad (11)$$

By extending the analysis to other operating points, it is possible to attain the Bode plot depicted in Figure 4 (b), wherein the duty cycle varies between 0.1 and 1.0. As can be seen, the additional capacitor can only be enlarged if the duty cycle is equal to unity, which means to connected the decoupling capacitor directly to the dc bus voltage, since the effective bus capacitance C_{bus} is defined by $C_{bus} + D^2\left(C_f + C_{fd}\right)$.

978-1-7281-4181-7/19 $31.00 © 2019 IEEE   253

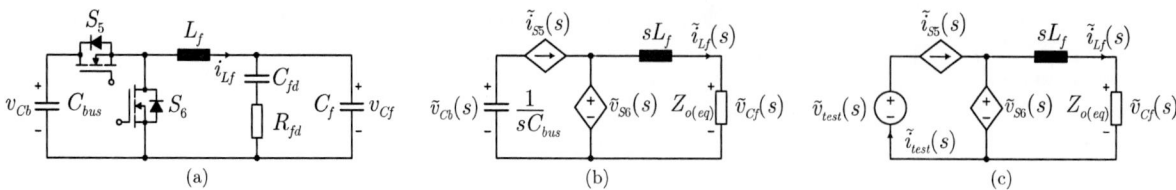

Fig. 3. (a) Equivalent electrical circuit of the Buck-type PDC; (b) Small-signal ac equivalent circuit model of the buck converter after replacing the output elements by the output equivalent impedance $Z_{o(eq)}$ in the frequency-domain; (c) Addition of a test current source at the small-signal ac equivalent circuit model to prescribe the input impedance $Z_i(s)$.

Therefore, the Bode plot shows that the equivalent capacitance in open loop is ineffective to decouple the power ripple, so that its effect on the dc bus may be neglected. In practic, the reduction of the voltage ripple when the active-capacitor behaves as shown in Figure 4 (a) is similar to a reduction caused by controlling the average voltage V_{Cf} on the decoupling capacitor, where the duty cycle is kept equals D_{buck}.

Once that it has been analyzed the equivalent capacitance operating in open-loop, the next step is to verify how the active-capacitor behaviour is influenced by the closed-loop that will be presented in the next section. For this purpose, it is necessary to define the closed-loop input impedance. Thus, the quantity $\tilde{d}$ must be calculated and applied into (7).

IV. CONTROL STRATEGY

In order to compensate the ac power ripple, the employed decoupling capacitor must alternately store and release the energy $E_{Cf} = P_{pv}/\omega_0$ within a quarter of the mains period $T = 1/f_{grid}$, where E_{Cf} is the amount of energy storage/released by the decoupling capacitor C_f. For this purpose, it is employed a low-bandwidth control loop to ensure a regulated mean voltage value V_{Cf} on the decoupling capacitor, aiding the stability of the system under transients. Additionally, a second feedback control loop is added to shift the voltage

ripple away from the dc bus voltage to the decoupling capacitor. Thereby, the proposed active power decoupling control can be represented in the diagram block illustrated in Figure 5.

The average value V_{Cf}, is controlled by the compensator $C_{vcf}(s)$ to follow the reference $V_{Cf}^*(t)$, in order to keep the decoupling capacitor with a mean value voltage V_{Cf}. In addition, to proper regulate the dc bus voltage ripple, a second control loop is applied. Since the average value of the dc bus voltage V_{Cb} is controlled by the dc-ac converter of the PV Microinverter, the PDC has as function to control only the pulsating component ($\tilde{v}_{Cb}$). Therefore, in order to split the instantaneous bus voltage $v_{Cb}(t)$ in its ac and dc components, a band-stop filter $BSF(s)$ has been employed. This filter is tuned at second order grid harmonic ($2\omega_0$), as depicted in Figure 5 [1].

Thereby, the control action $\tilde{v}_{C\tilde{v}cb}$ should operate on the voltage ripple, yielding a voltage ripple on the decoupling capacitor ($\tilde{v}_{Cf}$) proportionally to the voltage ripple ($\tilde{v}_{Cb}$) on the dc bus, such that this ripple is transferred to the decoupling capacitor instead of swing on the bus capacitor C_{bus}. Simultaneously, the mean value across the decoupling capacitor is held a dc bias voltage V_{Cf}. In other words, this control loop tries to null de voltage ripple on the dc bus ($\tilde{v}_{Cb} \rightarrow 0$). For this purpose, there are two control strategies to accomplish this task: Conventional Feedforward Controller and Feedforward employing a Proportional-Resonant (PR) Controller.

Fig. 4. Bode plot of magnitude and phase of the: (a) power decoupling input impedance for the operating point designed by D_{buck} and (b) power decoupling input impedance for several operating points with duty cycle ranging between 0.1 and 1.0.

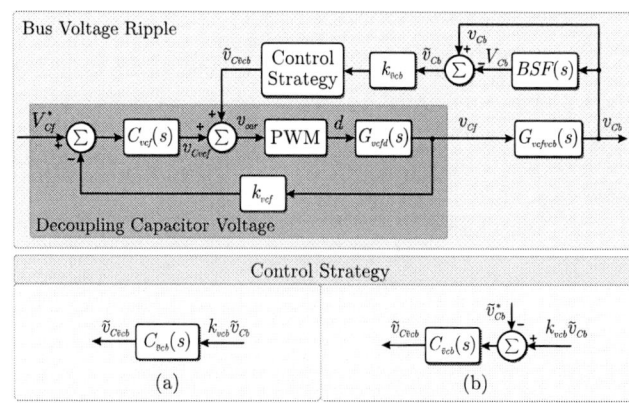

Fig. 5. Proposed control structure applied to the buck-type PDC applied to regulate the decoupling capacitor voltage V_{Cf} and to shift the power ripple away from the dc bus voltage employing a feedforward control based on: (a) Proportional Compensator and (b) Proportional-Resonant Compensator.

978-1-7281-4181-7/19 $31.00 © 2019 IEEE

The first strategy consists on adding the pulsating component of $v_{Cb}(t)$ into the feedback loop which controls the mean voltage V_{Cf}, in accordance with the control system shown in Figure 5 (a). This strategy may understood as a feedforward controller $C_{\tilde{v}cb}(s) = k_{ff(\tilde{v}cb)}$ which is responsible for amplifying the dc bus pulsating component (i.e., the ripple portion of $v_{Cb}(t)$) and then insert it to the output of the feedback controller $v_{car} = v_{Cvcf} + v_{C\tilde{v}cb}$.

The gain of the feedforward compensator $k_{ff(\tilde{v}cb)}$ can be calculated by assuming that the dynamic of the decoupling capacitor voltage control loop is much slower than the feedforward control loop, since the compensator $C_{vcf}(s)$ acts only to maintain the mean value of the V_{Cf}. So, neglecting the dynamics of $C_{vcf}(s)$, the gain is given by:

$$k_{ff(\tilde{v}cb)} = \frac{V_{Cf}}{k_{\tilde{v}cb}k_{PWM}\hat{V}_{Cb}V_{Cb}} \quad (12)$$

Moreover, in order to provide a more meaningful reduction on the voltage ripple, a PR controller tunned at $2\omega_0$ (resonance frequency) may be employed to replace the proportional feedforward controller.

$$C_{\tilde{v}cb}(s) = k_p + \frac{2k_r s}{s^2 + 4(\omega_0)^2} \quad (13)$$

Hence, the voltage ripple $\tilde{v}_{Cb}$ is amplified in the loop control by the high-gain provided by the resonant compensator. Figure 5 (b) illustrates the second control strategy by adopting a PR controller to minimizes the ripple of the dc bus voltage to nearby of zero. Note that the choice of $\tilde{v}_{Cb}^* = 0$ is justified as a tentative to null the voltage ripple of the $v_{Cb}(t)$. Thus, by the inspection of Figure 5, one can defines:

$$\tilde{d} = \frac{k_{\tilde{v}cb}C_{\tilde{v}cb}(s)\big(1 - BSF(s)\big)\tilde{v}_{test}(s)}{k_{vcf}C_{vcf}(s)G_{vcfd}(s) + k_{PWM}} = \Upsilon(s)\tilde{v}_{test}(s) \quad (14)$$

wherein $G_{vcfd}(s)$ is the control-to-input transfer function and $C_{vcf}(s)$ is the proportional-integral controller employed to the decoupling capacitor voltage control loop. Thereby, substituting $\Upsilon(s)$ into (7), one finds the closed-loop input impedance:

$$Z_{i,(CL)} = \frac{sL_f + Z_{o(eq)}}{D^2 + \Upsilon(s)\big[sL_f I_{Lf} + V_{Cb}D + Z_{o(eq)}I_{Lf}\big]} \quad (15)$$

Now, applying both described control strategies into (15), one has the closed loop input impedance of the PDC for both controllers. Figure 6 (a) shows the input impedance by regarding the feedforward control loop $k_{ff(\tilde{v}cb)}$ while the Figure 6 (b) depicts the input impedance considering the use of the PR controller in closed-loop $C_{\tilde{v}cb}(s)$.

As has been noted, the control strategy increases the effective capacitance around the double-line frequency, such that the impedance is meaningfully reduced. For this reason, the PDC can be understood as an active-capacitor inserted to the dc bus.

V. EXPERIMENTAL RESULTS

The hardware prototype has been built to evaluate the active-capacitor proposed in this paper, as shown in Figure 7. The

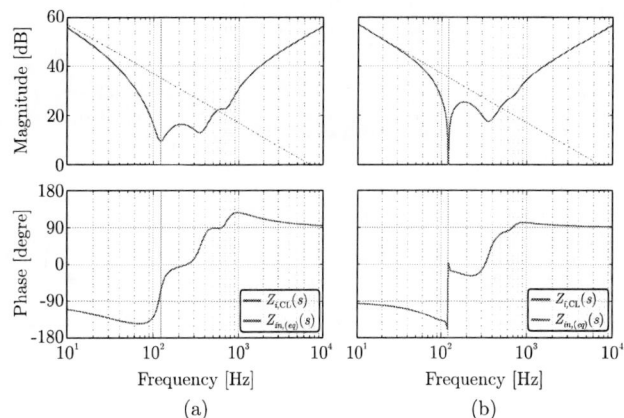

Fig. 6. Bode plot of the magnitude and phase of the: (a) closed-loop input impedance employing the feedforward controller $C_{ff(\tilde{v}cb)}(s)$ and (b) closed-loop input impedance employing the PR controller tunned at ω_0, $C_{\tilde{v}cb}(s)$.

main specifications and components designed are summarized in Table I and II, respectively [1].

TABLE I
SYSTEM PARAMETERS & ELECTRICAL SPECIFICATION OF THE PDC.

Rated Average Power ($P_{pv} = P_{av}$)	250 W
Decoupling Capacitor Voltage (V_{Cf})	250 V
Bus Voltage (V_{Cb})	420 V
Grid Frequency (f_{grid})	60 Hz
PDC Switching Frequency (f_S)	50 kHz
Modulator Gain ($k_{PWM} = 1$)	1.0
Decoupling Capacitor Voltage Sensor Gain (k_{vcf})	1.0
Bus Voltage Sensor Gain ($k_{\tilde{v}cb}$)	1.0

TABLE II
MAIN POWER COMPONENTS.

C_{bus}	50 μF/500 V	C3D2H506KMCAC00
C_f and C_{fd}	30 μF/500 V	MKP1848S63050JY5C
R_{fd}	15 Ω/3 W	ERG-3SJ150V
Switches	900 V	C3M0280090J-TR
L_f	2 x 1.0 mF	MMTF75T2711 (Toroidal)

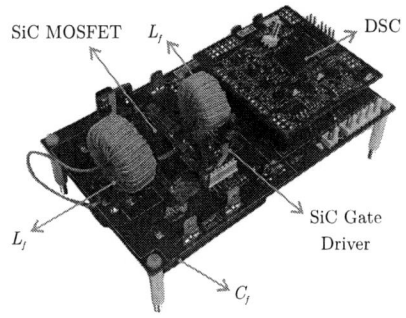

Fig. 7. Hardware implementation of the PDC. The dimensions of the shown power decoupling cell prototype are 80 mm x 135 mm x 30 mm (power density of 1.45 W/cm³ without Digital Signal Controller - DSC).

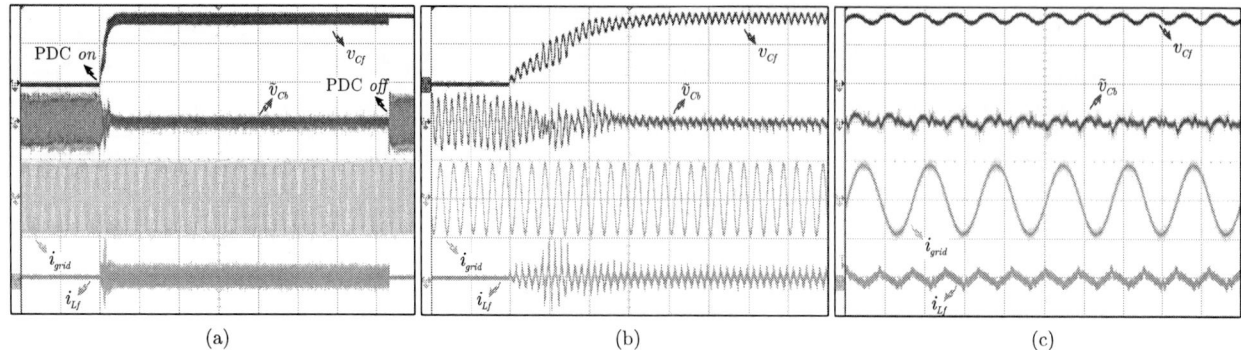

(a)   (b)   (c)

Fig. 8. Experimental waveforms of the decoupling capacitor voltage v_{Cf} (Ch 1), drain-source voltage of S_5 (high-side switch) $v_{DS(\text{HS})}$ (Ch 2), and decoupling inductor current i_{Lf} (Ch 3): (a) during the PDC startup and disconnection; (b) zoom on the startup and (c) steady-state behaviour. Timebase of the measurement is: (a) 1.0 s/div and (b) 10 μs/div. Scales: v_{Cf} (100 V/div); $v_{DS(\text{HS})}$ (200 V/div); (a) i_{Lf} (2.0 A/div) and (b) i_{Lf} (1.0 A/div).

Figure 8 (a) displays the connection of the PDC (PDC *on*) on the dc bus at the instant $t = 4.25$ s, whereas the Figure 8 (b) presents a zoom in the instant that occurs this connection, emphasizing the startup of the PDC.

Immediately afterward the insertion of the PDC (*PDC on*), the voltage ripple is compensated, so that the ripple across the bus capacitor is significantly reduced from $\Delta V_{Cb} = 30.2$ V (7.2 % of ripple) to $\Delta V_{Cb} = 5.0$ V (1.12 % of ripple). As consequence, the ripple decreases to 17 % of its initial value.

The steady-state performance at the rated power of the implemented PV Microinverter without and with the PDC is also illustrated in Figure 8 (a) and (b), respectively. As can been noted from the measured dc bus voltage, v_{Cb} and the inductor current, i_{Lf}, the voltage ripple was successfully shifted from the dc bus to the decoupling capacitor, C_f, which presents a 120 Hz voltage ripple equal to $\Delta V_{Cf} = 54$V superimposed to mean value voltage of 258 V. Further, even with the insertion of the PDC to the dc bus, it has been observed that the THD$_i$ of the grid current $i_{grid}(t)$ presents a low value, around 2.808 %. Before insertion of the PDC, the THD$_i$ was around 3.13 %.

VI. SUMMARY AND CONCLUSIONS

This paper has presented a power decoupling cell for a photovoltaic microinverter which can be understood as an electronic capacitor with high-capacitance able to mitigate the voltage ripple on the bus, suppressing its undesired double-line-frequency component $2\omega_0$. The theoretical analysis demonstrated that ripple suppression is only effective when a suitable strategy is applied to control the decoupling cell and then emulate a high-capacitance.

Finally, experimental results showed that the voltage ripple decreases about 17 % of its initial value after the power decoupling cell be inserted on the bus of the photovoltaic microinverter. Despite the decoupling cell was here used to voltage ripple suppressing, understanding it as an active-capacitor is the first step to expand its usage for new and still more relevant applications.

ACKNOWLEDGMENT

The authors would like to thank the CNPq.

REFERENCES

[1] T. A. Pereira, "Compensation of the double-line frequency voltage ripple on single-phase two-stage photovoltaic microinverter," Master's thesis, Universidade Federal de Santa Catarina, 2018.

[2] D. Neumayr, D. Bortis, and J. W. Kolar, "Ultra-compact Power Pulsation Buffer for single-phase DC/AC converter systems," *2016 IEEE 8th International Power Electronics and Motion Control Conference, IPEMC-ECCE Asia 2016*, pp. 2732–2741, 2016.

[3] M. Keshani, E. Adib, and H. Farzanehfard, "Micro-inverter based on single-ended primary-inductance converter topology with an active clamp power decoupling," *IET Power Electronics*, vol. 11, no. 1, pp. 73–81, 2018.

[4] C. Y. Liao, W. S. Lin, Y. M. Chen, and C. Y. Chou, "A PV Micro-inverter with PV Current Decoupling Strategy," *IEEE Transactions on Power Electronics*, vol. 32, no. 8, pp. 6544–6557, 2017.

[5] S. Xu, L. Chang, R. Shao, and Shao, "Evolution of single-phase power converter topologies underlining power decoupling," *Chinese Journal of Electrical Engineering*, vol. 2, no. June, 2016.

[6] H. Hu, S. Harb, N. Kutkut, I. Batarseh, and Z. J. Shen, "A review of power decoupling techniques for microinverters with three different decoupling capacitor locations in PV systems," *IEEE Transactions on Power Electronics*, vol. 28, no. 6, pp. 2711–2726, 2013.

[7] R. S. Balog, "Capacitance, dc Voltage Utilization, and Current Stress," no. September, pp. 37–49, 2017.

[8] A. S. Morsy and P. N. Enjeti, "Comparison of Active Power Decoupling Methods for High-Power-Density Single-Phase Inverters Using Wide-Bandgap FETs for Google Little Box Challenge," *IEEE Journal of Emerging and Selected Topics in Power Electronics*, vol. 4, no. 3, pp. 790–798, 2016.

[9] B. Gu, J. Dominic, J. Zhang, L. Zhang, B. Chen, and J. S. Lai, "Control of electrolyte-free microinverter with improved MPPT performance and grid current quality," *Conference Proceedings - IEEE Applied Power Electronics Conference and Exposition - APEC*, pp. 1788–1792, 2014.

[10] M. A. Vitorino, L. F. S. Alves, R. Wang, and M. B. De Rossiter Correa, "Low-frequency power decoupling in single-phase applications: a comprehensive overview," *IEEE Transactions on Power Electronics*, vol. 32, no. 4, pp. 2892–2912, 2017.

[11] Y. Sun, Y. Liu, M. Su, W. Xiong, and J. Yang, "Review of Active Power Decoupling Topologies in Single-Phase Systems," *IEEE Transactions on Power Electronics*, vol. 31, no. 7, pp. 4778–4794, 2016.

[12] Z. Qin, Y. Tang, P. C. Loh, and F. Blaabjerg, "Benchmark of AC and DC active power decoupling circuits for second-order harmonic mitigation in kW-scale single-phase inverters," *2015 IEEE Energy Conversion Congress and Exposition, ECCE 2015*, vol. 6777, no. c, pp. 2514–2521, 2015.

[13] K. Mozaffari, M. Amirabadi, and Y. Deshpande, "A Single-Phase Inverter/Rectifier Topology with Suppressed Double-Frequency Ripple," *IEEE Transactions on Power Electronics*, vol. 8993, no. c, 2018.

[14] S. Komeda and H. Fujita, "A Power Decoupling Control Method for an Isolated Single-Phase AC-to-DC Converter Based on Direct AC-to-AC Converter Topology," *IEEE Transactions on Power Electronics*, vol. XX, no. XX, 2018.

978-1-7281-4181-7/19 $31.00 © 2019 IEEE

Reliability Analysis of aircraft starter generator drive converter

Jayakrishnan Harikumaran, Giampaolo Buticchi, Michael Galea, Alessandro Costabeber, Pat Wheeler

Power Electronics, Machines and Control Group

University of Nottingham, Nottingham, UK, Jayakrishnan.Harikumaran@nottingham.ac.uk

Abstract—Under the more electric aircraft (MEA) theme, many aircraft systems are being electrified. In a previous research work, an electric starter generator system for aircrafts was implemented. This paper presents the reliability of the drive converter – a 3 level neutral point clamped converter (3L-NPC) under an expected mission profile of a short haul aircraft. Wear-out failure based reliability of semiconductors and capacitors are considered along with reliability of gate drivers to estimate the overall reliability of the converter.

Index Terms—More electric aircraft (MEA), Power Electronics, Reliability, 3L-NPC

I. INTRODUCTION

Aviation industry is exploring means to cap emissions level at 2020 level and achieve net reduction of 50% in emissions by 2050. To achieve these goals, electrification of aircraft systems is being pursued [1]. Such electrified systems would include power electronic converters and their reliability must meet aerospace safety requirements. Prior works have estimated the reliability of power converters in various applications - railway traction converters [2], electric vehicle traction [3], wind power converters [4], PV inverters [5].

In this work, the system under study is an electric starter generator system designed under the Aircraft Electrical Starter-Generation System with Active Rectification Technology (AE-GART) project [6]. A 3L-NPC converter is interfacing the electric motor/generator to the DC power grid.

The main research problem addressed in this work is estimation of reliability of power converters under aerospace operating conditions. Simulations are carried out considering the torque speed profile, DC link voltage and other parameters applicable to the AEGART system. Wear out failure mechanism of power converters are usually studied in detail. But random failures are also to be considered for applications in safety critical and high reliability applications. In a prior work on cosmic ray induced random failures [7], the effect of random failures and early failures are shown in relation to the overall system reliability. The importance of considering random failures are shown as it may have higher influence on the system reliability. Gate driver failure due to random failures are considered in this work to bring out the effect of random failures on the reliability of the converter.

This work was supported by the INNOVATIVE Doctoral Program. The INNOVATIVE program is supported in part by the Marie Curie Initial Training Networks action under Grant 665468, in part by the Institute for Aerospace Technology, University of Nottingham, and in part by the University of Nottingham Propulsion Futures Beacon.

II. AEGART SYSTEM DESCRIPTION

The operation of the starter generator system is briefly described. At startup the AEGART system acts as a starter to the main turbine. Hence the drive converter functioning as inverter is supplying power at maximum torque to spin-up the main turbine upto start-up speed. Once the startup speed is achieved AEGART system stops powering the motor. The main turbine continues to speed up and the AEGART system starts generating power and supplies power to the DC grid once minimum generation speed is reached. During the entire flight phase after startup, the AEGART system continues to operate at rated operating speed and provides power to the electrical network. The torque speed profile of the system is shown in Fig. 1.

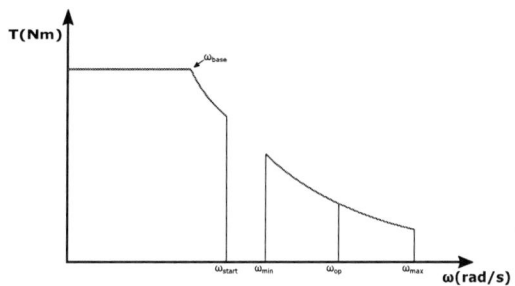

Fig. 1. Torque Speed profile of AEGART system

The torque speed details are summarized in Table I.

TABLE I
AEGART MOTOR SPEED-TORQUE SUMMARY

Start-up Torque	17.9 Nm
ω_{base}	8000 rpm
ω_{start}	12000 rpm
ω_{genmin}	19200 rpm
ω_{genop}	24000 rpm
ω_{genmax}	32000 rpm

It is assumed that the system achieves base speed in 40 seconds and the same rate of motor speed change is assumed in the rest of the operation. A time delay of 30 seconds is considered from main turbine start to turbine achieving minimum generator speed. Based on the torque-speed profile as well as the flight phase a mission profile is generated for reliability calculatoins. The schematic of the 3L-NPC

978-1-7281-4181-7/19 $31.00 © 2019 IEEE

converter is shown in Fig.2. The design parameters of the 3L-NPC converter is summarized in Table II.

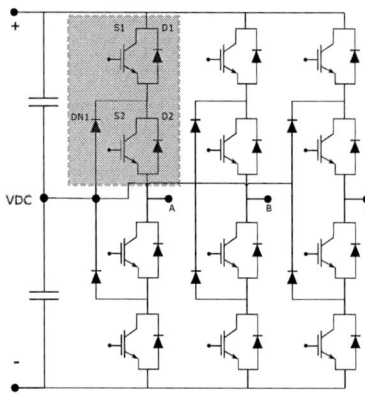

Fig. 2. 3L-NPC Bidirectional Rectifier

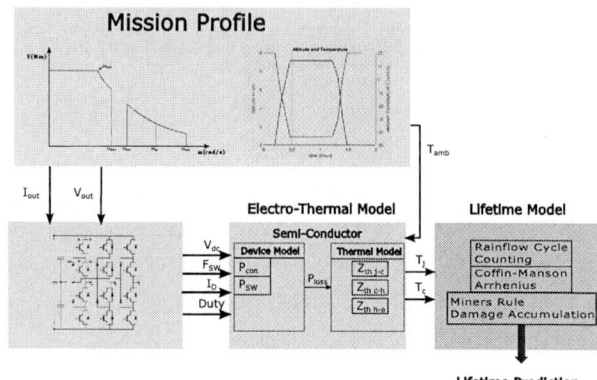

Fig. 3. Mission Profile based reliability estimation methodology for IGBTs

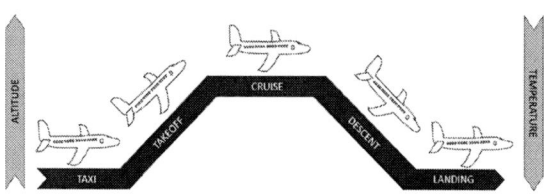

Fig. 4. Flight Stages of an Aircraft

TABLE II
3L-NPC PARAMETERS

Vdc	270
SW frequency	16 kHz
Rated Power	45kW

The rest of the paper reports detailed reliability calculations of the 3L-NPC converter. Mission profile for reliability calculations is derived based on the output power of the converter and ambient temperature. Loss estimation and thermal simulation is carried out to estimate the junction temperature of IGBTs and hot spot temperature of the capacitors. The simulated stress values (temperature and voltage) due to the expected mission profile is applied in Physics of Failure (PoF) models of lifetime estimation for IGBTs and DC link capacitors. For reference, a block diagram of lifetime estimation of IGBTs is shown in Fig. 3. Capacitor wear out based lifetime estimation follows similar steps. A simulation study on various failure rates for gate drivers is included to bring out the effect of random failures. Finally, Reliability Block Diagram (RBD) method is applied to derive system reliability for the converter from individual component reliability profiles.

III. MISSION PROFILE OF AEGART SYSTEM IN A SHORT HAUL AIRCRAFT

Airlines aim to maximize flight time of aircrafts, typically of short haul flights. A typical short haul flight is around 1 hour flying time. The cruising altitude is assumed to be 30000 feet. Accounting for flight preparation, taxi, take-off, landing etc. a typical flight requires roughly 2 hours time from gate to gate. Assuming only day time operations, 6 flights per day are assumed. Summary of flight stages are shown in Fig.4.

The duration of different flight phases assumed in this study is provided in Table III.

The following assumptions are made on the temperature gradient of the atmosphere. In the lower atmosphere upto

5000m a gradient of -6.5 degC/km is assumed. The cruising altitude temperature is assumed to be -50 degC at 10 km. Three different average ground level ambient temperature is assumed – 5, 15, 20 degC. Hence simulations are carried out 3 times to estimate losses and consequent thermal stresses in devices during operation in a year. Climb rates assumed in building the ambient temperature profile in summarized in Table IV.

The ambient temperature profile for one simulation run along with altitude variation is shown in Fig. 5. The ambient temperature profile based on the flight data along with the torque speed requirements of the AEGART system, as de-

TABLE III
FLIGHT PHASE DURATION

Flight Phase	Duration (mins)
Taxiing time	10
Cruise time	45
Climb/Descent time	Based on climb rate

TABLE IV
CLIMB/DESCENT RATE OF AIRCRAFT

Altitude Range (feet)	Climb Rate (ft/min)
0-15,000	2,400
15,000-30,000	1,500
30,000-24,000	-1,000
24,000-10,000	-3,000
10,000-0	-2,000

scribed in [6], is applied to generate mission profile - Power and Ambient temperature of the 3L-NPC converter.

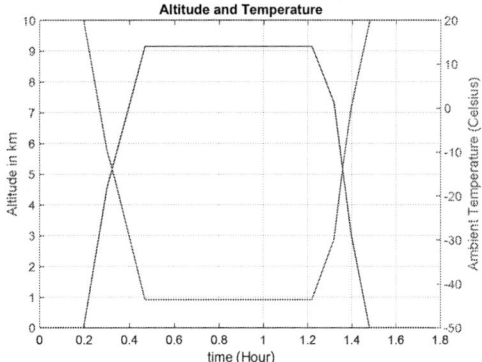

Fig. 5. Altitude and Temperature variation during Flight

IV. POWER LOSS ESTIMATION AND THERMAL MODELLING OF 3L NPC CONVERTER

The main lifetime limiting factor of power electronic components is temperature stress. Thermal simulation is carried out to estimate temperature stress imposed on semiconductors and capacitors. Thermal simulation is carried out using Foster thermal impedance networks. It is graphically represented in Fig.6 for IGBTs.

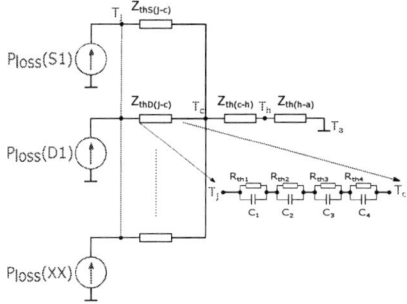

Fig. 6. Thermal Simulation using Foster Network

A. Junction temperature estimation of IGBTs

The IGBT module used in AEGART 3L-NPC is the module - F3L400R07ME4_B22 from Infineon. The module contains the semiconductors in the red highlighted portion of Fig. 2 - 2 IGBTs with free wheeling diodes as well the neutral point clamp diode. The datasheet values for switching losses and on-state voltage drop is used to estimate the losses in the IGBTs.

$$P_{SW\_IGBT} = fsw((E_{on} + E_{off}).(\frac{I}{I_{ref}})^{K_i}.(\frac{V_{off}}{V_{ref}})^{K_v}) \quad (1)$$

$$P_{cond} = Vce_{on}.I \quad (2)$$

Ki is assumed to be 1 and Kv is taken as 1.3 in (1). Junction temperature can be obtained as shown in Fig. 6 by applying the power losses to the Foster network.

$$T_{j(IGBT)}(t) = Ploss_{tot-IGBT}(t).Z_{th-IGBT(j-c)}(t) \quad (3)$$
$$+T_c(t)$$

$$T_c(t) = \left(\sum Ploss_{IGBTx}(t) + \sum Ploss_{Diodex}(t) \right). \quad (4)$$
$$[Z_{th(c-h)}(t) + Z_{th(h-a)}(t)] + T_a(t)$$

Tj, Tc and Ta are junction temperature, case temperature and ambient temperature respectively. As can be seen in (4), the junction temperature is dependent on losses from other semiconductors in the same module. In the case of Infineon module, the outer and inner IGBT experiences different losses. The modulation of IGBTs in 3L NPC follows the output voltage requirement and to ensure no unsafe operating modes. The modulation details are described in the Semikron application note [8]. The simulated junction temperature of IGBTs are used in reliability calculations.

B. Hot spot temperature estimation of Capacitors

The main source of loss in capacitors can be modelled using the equivalent series resistor (ESR) of the capacitor. Hence the rms current through the capacitor would yiled the loss figure that can be fed into the thermal network to yield the capacitor hot spot temperature.

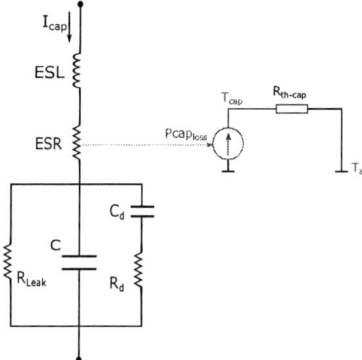

Fig. 7. Capacitor hot spot estimation

An equation to derive rms current through DC link capacitors for a 3L-NPC converter is derived in [9].

$$I_{Crms}^2 = \frac{3Im^2M}{4\pi}\left(\sqrt{3} + \frac{2}{\sqrt{3}}.\cos(2\phi)\right)$$
$$-\frac{9}{16}(I_mM)^2\cos^2(\phi) \quad (5)$$

Based on the voltage rating of 270V and potential for higher voltage operation, B32778G4107K000 from TDK Epcos (100 uF, 400V) is considered for reliability analysis. A bank of 4 pairs of capacitors make up the DC link. Along with capacitance, the temperature rise in capacitor is also a design input for number of capacitor selection as per TDK Epcos

978-1-7281-4181-7/19 $31.00 © 2019 IEEE   259

application note. Hence the number of capacitors is chosen such that the capacitor temperature rise is limited to datasheet specification. Using (5) and the datasheet value for thermal impedance, the capacitor hotspot temperature is calculated during the flight.

V. Reliability of Inverter System

Based on temperature values and DC link voltage from previous section, the following components are separately treated to estimate reliability of the converter system – Outer IGBT switch, Inner IGBT switch and DC link capacitor.

A. Lifetime model and Reliability estimation of IGBTs

The well-known extended Coffin-Manson model with Arrhenius term is chosen as the lifetime model in this study. The process of extracting lifetime data using junction temperature and lifetime models is described in [10]. Rainflow algorithm is used to extract thermal cycles to be applied in the lifetime model. Thermal cycling causes mechanical stresses in the semiconductor module. Miners rule of linear damage accumulation is used to calculate the lifetime of the module based on thermal cycles experienced by the power module. As per miners rule, once the accumulated damage equals 1, the useful lifetime of a device is consumed.

$$ Nf = A.dT_j^\alpha .e^{\frac{E_a}{kB*T_{jmean}}} \tag{6} $$

$A = 3.025*10^5; \alpha = -5.04; E_a = 9.891*10^{-20}$

Application of the miner's rule on damage accumulation caused by thermal cycles gives a lifetime of 2027 years for outer IGBT (S1) and 451 years for the inner IGBT (S2) based on the mission profile considered. The inverter is operating in generating mode during most of the flight and hence the lifetime predictions are not surprising. The outer IGBTs hardly experience any losses during generation mode at high power factor while the inner IGBTs would be conducting current during both generation and motoring mode. As with any lifetime model, statistical variations need to be considered to get reliability figures from lifetime data. Hence the following process as described in [11] is followed to perform a Monte-Carlo simulation assuming a 5% variation in device parameters. Weibull distribution is used to fit lifetime distribution data of failure modes caused by repetitve mechanical stress. As thermal stresses leads to repetitive mechanical stress, Weibull distribution is used to fit lifetime distribution from Monte-Carlo simulation.

The reliability figures thus obtained for the IGBTs and the β and η factors of the Weibull distribution are shown in Fig. 8. The loss figures and predicted lifetime values are highly dependent on the switching losses and conductions losses of the devices as well as the expected mission profile of the converter.

B. Lifetime model and Reliability estimation of Capacitors

The major factors affecting the lifetime of film capacitors are temperature, voltage stress and humidity [12]. A lifetime model considering the effect of all three parameters are

provided in [13]. In this work the effect of humidity is not considered.

$$ L = L_0.e^{\frac{E_a}{kB}\left(\frac{1}{T}-\frac{1}{T_0}\right)}.e^{-\beta\left(\frac{U-U_0}{U_0}\right)} \tag{7} $$

$L_0 = 20000 hours; T_0 = 25C; \beta = 3.5; U_0 = 400V$

As with IGBTs, application of the Miner's rule of linear damage accumulation on the capacitor damage gives a lifetime of 254 years based on the mission profile considered. Statistical effects is to be considered for the capacitor as well. The method to derive reliability figure by considering uncertainties in capacitor lifetime model is described in [14]. The unreliability curve of capacitor based on 5% variation of parameters is shown in Fig. 9.

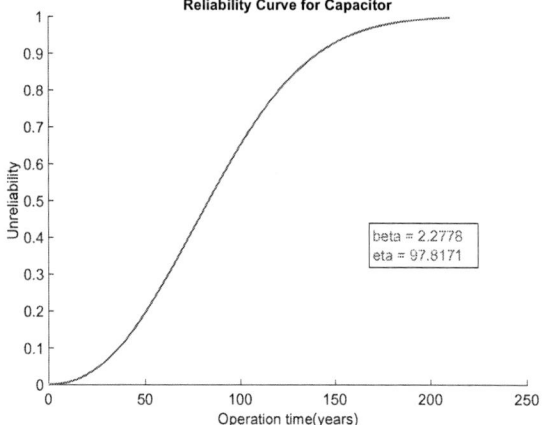

Fig. 9. Unreliability of Capacitors in NPC

C. Lifetime model of Gate Drivers

The effect of the reliability of gate drivers is usually not included in many reliability estimations. For high reliability applications like aerospace, every component that can cause a failure needs to be considered while estimating system reliability. Failure mechanisms of gatedrivers would be a combination of semiconductor and capacitor failure modes. Gate driver lifetime models are not readily available in literature. Due to lack of sufficient prior work, gate driver reliability is modelled using a constant failure rate. There are 12 gate drivers in a 3L-NPC and the effect of gate driver reliability when considering the whole converter is shown considering three failure rates - 0.01 (1 failure every 100 years), 0.001 (1 failure every 1000 years) and 0.0001 (1 failure every 10000 years).

D. Reliability Block Diagram method for full converter reliability

RBD method is used to combine the reliability figures of individual components to generate the overall reliability of the inverter system. In RBD each component that is necessary for operation of the system is connected in series and the reliability curves are multiplied together. Redundancy as well as m out of n systems can also be handled by RBD method.

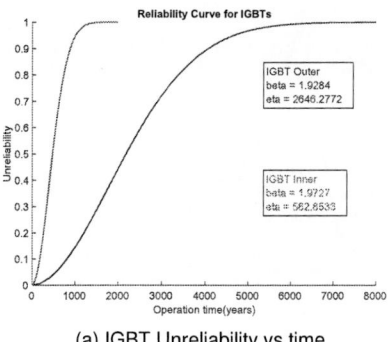

(a) IGBT Unreliability vs time

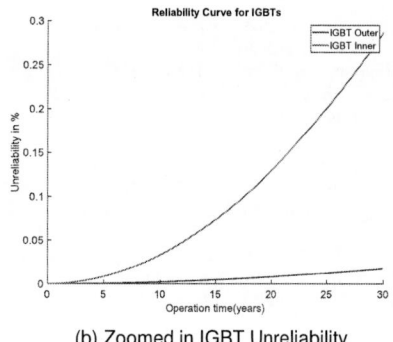

(b) Zoomed in IGBT Unreliability

Fig. 8. Unreliability of IGBTs in NPC

The reliability block diagram for the 3L NPC converter is shown in the Fig. 10. In the 3L-NPC case, the reliability of the 6 outer IGBTs, 6 inner IGBTs, 8 DC link capacitors as well as 12 Gate Drivers are combined to obtain the inverter system reliability.

Fig. 10. Reliability block diagram of 3L NPC converter

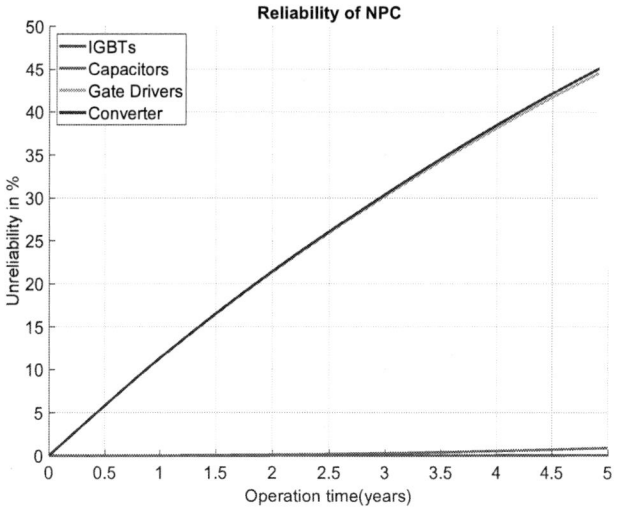

Fig. 11. Reliability of NPC with failure rate of 0.01 for gate drivers

Figure 11 shows the reliability of the whole converter when a gate driver failure rate of once every 100 years (1% per year) is modelled. It can be clearly seen that the reliability of the whole system is dominated by the reliability of gate drivers. Hence the importance of focussing attention on random failures along with wearout failures is clear. The importance of gate driver reliability is further clarified in Fig.

12 where different gate driver failure rates are modelled. As can be seen, to achieve a system reliability dominated by wearout failure modes, a failure rate of once every 10000 years per gate driver (0.01% per year) should be achieved by design.

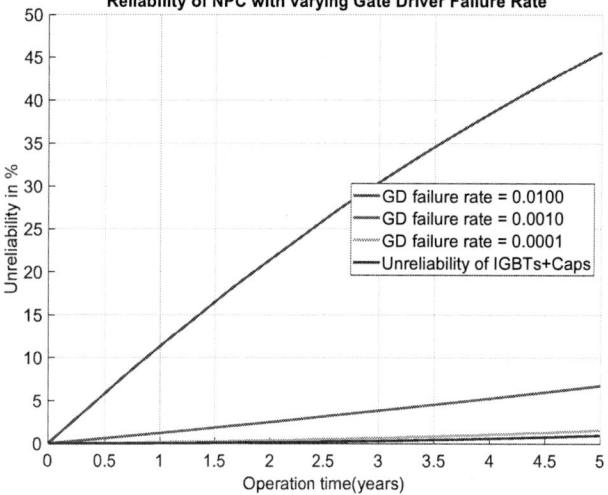

Fig. 12. Reliability of NPC with varying gate driver failure rate

VI. RESULTS AND DISCUSSION

In this work, the reliability calculations of a starter generator drive converter for MEA is presented. Relaibility of IGBTs and capacitors are derived using PoF models. As expected, outer and inner IGBTs in the 3L-NPC converter has different reliabilities with the inner IGBT having a lower lifetime than the outer IGBT. The effect of gate driver reliability in the analysis highlights the importance of considering the whole system reliability in case of multi-level converters. Effect of random failures on system reliability is clearly shown in the reliability of the 3L-NPC converter. Random failure modes require more careful analysis and particular attention must be placed on them in case of high reliability and mission critical applications.

978-1-7281-4181-7/19 $31.00 © 2019 IEEE

REFERENCES

[1] IATA, DLR, and G. I. of Technology, *IATA TECHNOLOGY ROADMAP 2013*, International Air Transport Association, 2013.

[2] X. Perpiñà, J. Serviere, L. Navarro, M. Mermet-Guyennet, M. Vellvehi, and X. Jordà, *Reliability and Lifetime Prediction for IGBT Modules in Railway Traction Chains*. INTECH Open Access Publisher, 2012. [Online]. Available: https://books.google.co.uk/books?id=LkLYoAEACAAJ

[3] D. Hirschmann, D. Tissen, S. Schroder, and R. W. D. Doncker, "Reliability prediction for inverters in hybrid electrical vehicles," *IEEE Transactions on Power Electronics*, vol. 22, no. 6, pp. 2511–2517, Nov 2007.

[4] K. Ma, M. Liserre, F. Blaabjerg, and T. Kerekes, "Thermal loading and lifetime estimation for power device considering mission profiles in wind power converter," *IEEE Transactions on Power Electronics*, vol. 30, no. 2, pp. 590–602, Feb 2015.

[5] Y. Yang, H. Wang, F. Blaabjerg, and K. Ma, "Mission profile based multi-disciplinary analysis of power modules in single-phase transformerless photovoltaic inverters," in *Proceedings of the 15th European Conference on Power Electronics and Applications, EPE 2013*. IEEE Press, 2013, pp. 1–10.

[6] S. Bozhko, T. Yang, J. Le Peuvedic, P. Arumugam, M. Degano, A. La Rocca, Z. Xu, M. Rashed, W. Fernando, C. I. Hill, C. Eastwick, S. Pickering, C. Gerada, and P. Wheeler, "Development of aircraft electric starter–generator system based on active rectification technology," *IEEE Transactions on Transportation Electrification*, vol. 4, no. 4, pp. 985–996, Dec 2018.

[7] U. Scheuermann and U. Schilling, "Impact of device technology on cosmic ray failures in power modules," *IET Power Electronics*, 05 2016.

[8] I. Staudt, *Application Note AN-11001 3L NPC & TNPC Topology*, SEMIKRON International GmbH, 2015.

[9] K. GOPALAKRISHNAN, S. JANAKIRAMAN, S. Das, and G. Narayanan, "Analytical evaluation of dc capacitor rms current and voltage ripple in neutral-point clamped inverters," *Sadhana - Academy Proceedings in Engineering Sciences*, vol. 42, pp. 1–13, 05 2017.

[10] Y. Yang, H. Wang, F. Blaabjerg, and K. Ma, "Mission profile based multi-disciplinary analysis of power modules in single-phase transformerless photovoltaic inverters," in *2013 15th European Conference on Power Electronics and Applications (EPE)*, Sep. 2013, pp. 1–10.

[11] P. D. Reigosa, H. Wang, Y. Yang, and F. Blaabjerg, "Prediction of bond wire fatigue of igbts in a pv inverter under a long-term operation," *IEEE Transactions on Power Electronics*, vol. 31, no. 10, pp. 7171–7182, Oct 2016.

[12] H. Wang and F. Blaabjerg, "Reliability of capacitors for dc-link applications in power electronic converters 2014;an overview," *IEEE Transactions on Industry Applications*, vol. 50, no. 5, pp. 3569–3578, Sept 2014.

[13] R. Gallay, "Metallized film capacitor lifetime evaluation and failure mode analysis," 05 2014.

[14] H. Wang, P. Davari, H. Wang, D. Kumar, F. Zare, and F. Blaabjerg, "Lifetime estimation of dc-link capacitors in adjustable speed drives under grid voltage unbalances," *IEEE Transactions on Power Electronics*, pp. 1–1, 2018.

Multivariable Control of a Grid Forming System Based on Back-To-Back Topology

1st Igor D. N. de Souza
Department of Electrical Engineering
Federal University of Ouro Preto
João Monlevade, MG, Brazil
Electrical Engineering Program
Federal University of Juiz de Fora
Juiz de Fora, MG, Brazil
igor.souza@ufop.edu.br

2nd Gabriel A. Fogli
Department of Electronic Engineering
Federal University of Minas Gerais
Belo Horizonte, MG, Brazil
gabrielfogli@ufmg.br

3rd Marcelo C. Fernandes,
4th Ademir S. T. Júnior,
5th Pedro G. Barbosa,
6th Pedro M. de Almeida
Electrical Engineering Program,
Federal University of Juiz de Fora
Juiz de Fora, MG, Brazil
pedro.machado@ufjf.edu.br

Abstract—This paper presents steps of an unified modeling and controller design for a back-to-back converter applied to a grid forming system. The proposed procedure captures in a single state-space model a grid-interface converter with an output L filter, a grid-forming converter with an output LC filter and DC capacitor. This approach deals with a modern multivariable linear quadratic regulator control strategy for taking full advantage of the unified model. A full state-feedback controller has been designed for control the grid-interface converter input currents, grid-forming converter output voltages and DC-bus voltage behavior. Resonant controllers are also proposed to reject the current harmonic disturbance drawn by nonlinear loads. The controllers are designed and tested in PSIM, then simulation results are show to verify the proposed control strategy.

Index Terms—Unified modeling, multivariable Control, linear quadratic regulator, back-to-back converter, grid-forming converter.

I. INTRODUCTION

The Back-to-Back converter (BTB) has been commonly used in different applications, such as, electrical machines, to improve the motor performance [1]; High Voltage Direct Current transmission (HVDC) for asynchronous interconnection and power flow control [2]; Unified Power Quality Conditioners (UPQC) to improve the power quality of the power supply and the load [3]; and Wind Power Generators, to achieve high performance control of the electromechanical power conversion with minimum impact on the grid [4].

An important property of the BTB lies in its capacity to permit isolating two AC systems, working as a firewall in terms of power quality disturbances. This idea enables to manage an AC microgrid as an invisible load for the upstream system, avoiding the propagation of any disturbance to and from the microgrid [5], [6]. In this implementation the BTB should be capable of generate a high quality sinusoidal output voltage, regardless of the power and current waveform demanded by local load. Therefore, the voltage source inverters (VSI) with an LC output filter topology should be part of the BTB structure.

Usually, the conventional control approach and structure used in BTB converters assumes an inner controller for one primary variable (the current injected in both AC sides) and

an outer controller for other variable with relative slower dynamics (the DC-bus voltage) [7]. In its basic configuration several control strategies have been developed and applied. It can be cited, for instance, a conventional PI (proportional-integrator), proportional resonant, etc [5]. However, PI-based controllers are not capable of perfect tracking of sinusoidal references and they present a poor disturbance rejection. On the other hand, in the case of grid-forming application, a set of resonant controllers must be used to guarantee reference tracking and rejection of harmonics disturbances at specific frequencies according with the load [8]. Multiple resonant controllers design is not an easy task [9]. Mainly, when classical design methodologies are used. Although the easiness provided by the parallel connection of several resonant modes, every time a resonant compensator is included, the stability and its margins must be checked, and sometimes the whole control system must be redesigned.

The state-space (SS) control techniques have been extensively used in several fields. This approach allows, theoretically, free choice of the system dynamics based on arbitrarily pole placement, as long as the system is reachable. In the context of SS control, the design and implementation of linear quadratic regulators (LQRs) for BTB is a type of strategy that should be investigated. The classic LQR deals with the optimization of performance index or a cost function. Therefore, the designer can weight which states and inputs are more important. The great advantage of using state-space LQR technique is that the design is systematic, regardless of the number of resonant modes included, and there are few parameters to be chosen. Furthermore, good stability margins are guaranteed.

In this paper an unified modeling and control procedure for BTB converters is presented. Based on [10] a linear state-space model of the BTB converter will be presented. A linear multivariable control algorithm is then used to control inverter input current of the grid-interface converter (GIC), output voltage at the LC filter of the grid-forming converter (GFC) and the DC-capacitor voltage. In order to ensure low output voltage total harmonic distortion (THD), resonant controllers

978-1-7281-4181-7/19 $31.00 © 2019 IEEE

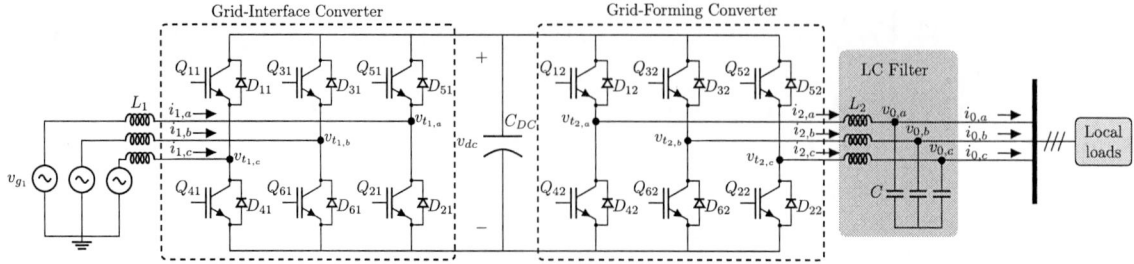

Fig. 1. Back-to-back converter with a LC output filter.

are added at the load side. The simulation verification of the proposed procedure is performed, then, results are used to validate the proposed control law.

II. SYSTEM DESCRIPTION AND MODELING

The system under study is depicted in Fig. 1, which is composed of a BTB converter, a second order LC output filter and local loads, which can be linear, nonlinear, balanced or unbalanced. There are two stages of power processing. The grid-interface converter, connected to the utility-grid via L filter, draws the energy demanded by the local loads from the grid.

A. Grid-interface Converter

Based on Fig. 1 three equivalent circuit per phase can be obtained. Neglecting the current and voltage switching harmonics the grid-connected system can be presented in Fig. 2.

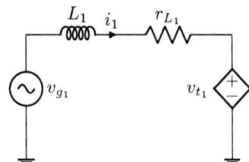

Fig. 2. Grid-interface converter equivalent circuit.

where L_1 represents the output filter and r_{L_1} is its series equivalent resistance; the converter's output instantaneous averaged voltage is represented by the controlled voltage source v_{t_1}. The subscript "1" refers to the grid-side variable.

The aforementioned circuits, in natural frame, may be reduced in two circuits for each axis, d and q, if a Park's transformation is used. Therefore, v_{t_1} can be written as a function of the voltage v_{dc} in the DC bus capacitor, and the direct and quadrature axis modulation index m_1^d and m_1^q, respectively.

By applying the Kirchhoff's laws on the equivalent circuit depicted in Fig. 2, the dynamic model of the inverter in dq reference frame can be derived as follows.

$$
\begin{cases}
\dfrac{di_1^d}{dt} = -\dfrac{r_{L_1}}{L_1}i_1^d + \omega i_1^q - \dfrac{m_1^d v_{dc}}{2L_1} + \dfrac{v_{g_1}^d}{L_1} \\[2ex]
\dfrac{di_1^q}{dt} = -\omega i_1^d - \dfrac{r_{L_1}}{L_1}i_1^q - \dfrac{m_1^q v_{dc}}{2L_1} + \dfrac{v_{g_1}^q}{L_1}
\end{cases} \quad , \quad (1)
$$

where ω represents the grid angular fundamental frequency.

B. Grid-forming Converter

Fig. 3 shows the simplified equivalent instantaneous averaged circuit per phase.

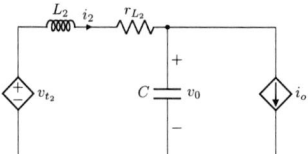

Fig. 3. Grid-forming converter equivalent circuit.

where L_2 and C represent the output filter inductor and capacitor, respectively; r_{L_2} is the series equivalent resistance of the inductor; the converter's output instantaneous averaged voltage is represented by the controlled voltage source v_{t_2}. Which also can be written as a function of the voltage v_{dc}, the direct and quadrature axis modulation index m_2^d and m_2^q, respectively. The load current is considered as disturbance and because of that, it is modeled as an unknown current source i_0. Therefore, the grid former dynamics, in dq frame, can be expressed as:

$$
\begin{cases}
\dfrac{di_2^d}{dt} = -\dfrac{r_{L_2}}{L_2}i_2^d + \omega i_2^q + \dfrac{m_2^d v_{dc}}{2L_2} - \dfrac{v_0^d}{L_2} \\[2ex]
\dfrac{di_2^q}{dt} = -\omega i_2^d - \dfrac{r_{L_2}}{L_2}i_2^q + \dfrac{m_2^q v_{dc}}{2L_2} - \dfrac{v_0^q}{L_2} \\[2ex]
\dfrac{dv_0^d}{dt} = +\dfrac{i_2^d}{C} + \omega v_0^q - \dfrac{i_0^d}{C} \\[2ex]
\dfrac{dv_0^q}{dt} = +\dfrac{i_2^q}{C} - \omega v_0^d - \dfrac{i_0^q}{C}.
\end{cases} \quad (2)
$$

C. DC-Link Modeling

In order to complete the model, the dynamic behavior of the voltage on DC-link should be included. According to Fig. 4, both inverters are linked via the DC-bus voltage, sharing the DC-capacitor dynamics.

By assuming that there are no power losses in inverters, the power balance can be written as,

$$
p_{t1} = p_{cap} + p_{t2}, \quad (3)
$$

978-1-7281-4181-7/19 $31.00 © 2019 IEEE   264

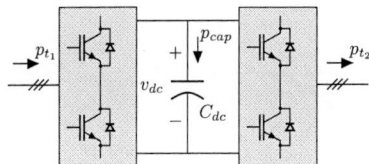

Fig. 4. DC-Link terminals power flow representation.

where p_{t1} and p_{t2} are the instantaneous active powers at the grid-interface converter and grid-forming converter terminals, respectively; p_{cap} is the power flowing into the DC-capacitor. Therefore, voltage dynamics over the capacitor placed between the converters can be expressed as:

$$p_{cap} = C_{dc} \left(\frac{d}{dt} v_{dc} \right) v_{dc}. \tag{4}$$

Furthermore, using instantaneous power theory [11], p_{t1} and p_{t2} can be expressed in terms of the related current (i_1, i_2) and voltage (v_{t1}, v_{t2}) dq components. However, as discussed before, v_{t1} can be written as a function of v_{dc}, m_{d_1}, m_{q_1} and v_{t2} as a function of v_{dc}, m_{d_2}, m_{q_2}. Consequently, (3) can be rewritten as:

$$\frac{dv_{dc}}{dt} = \frac{3m_1^d}{4C_{dc}}i_1^d + \frac{3m_1^q}{4C_{dc}}i_1^q - \left(\frac{3m_2^d}{4C_{dc}}i_2^d + \frac{3m_2^q}{4C_{dc}}i_2^q \right). \tag{5}$$

Therefore, the dynamic of the whole BTB system is given by the combination of (1), (2) and (5)

$$
\begin{cases}
\dfrac{di_1^d}{dt} = -\dfrac{r_{L_1}}{L_1}i_1^d + \omega i_1^q - \dfrac{m_1^d v_{dc}}{2L_1} + \dfrac{v_{g_1}^d}{L_1} \\[2mm]
\dfrac{di_1^q}{dt} = -\omega i_1^d - \dfrac{r_{L_1}}{L_1}i_1^q - \dfrac{m_1^q v_{dc}}{2L_1} + \dfrac{v_{g_1}^q}{L_1} \\[2mm]
\dfrac{di_2^d}{dt} = -\dfrac{r_{L_2}}{L_2}i_2^d + \omega i_2^q + \dfrac{m_2^d v_{dc}}{2L_2} - \dfrac{v_0^d}{L_2} \\[2mm]
\dfrac{di_2^q}{dt} = -\omega i_2^d - \dfrac{r_{L_2}}{L_2}i_2^q + \dfrac{m_2^q v_{dc}}{2L_2} - \dfrac{v_0^q}{L_2} \\[2mm]
\dfrac{dv_0^d}{dt} = +\dfrac{i_2^d}{C} + \omega v_0^q - \dfrac{i_0^d}{C} \\[2mm]
\dfrac{dv_0^q}{dt} = +\dfrac{i_2^q}{C} - \omega v_0^d - \dfrac{i_0^q}{C} \\[2mm]
\dfrac{dv_{dc}}{dt} = \dfrac{3m_1^d}{4C_{dc}}i_1^d + \dfrac{3m_1^q}{4C_{dc}}i_1^q - \left(\dfrac{3m_2^d}{4C_{dc}}i_2^d + \dfrac{3m_2^q}{4C_{dc}}i_2^q \right).
\end{cases}
\tag{6}
$$

Linearizing (6) around an equilibrium point, the BTB state-space representation is denoted in (7). Where $\tilde{\mathbf{x}} = \begin{bmatrix} \tilde{i}_1^d & \tilde{i}_1^q & \tilde{i}_2^d & \tilde{i}_2^q & \tilde{v}_0^d & \tilde{v}_0^q & \tilde{v}_{dc} \end{bmatrix}^T$ is the state vector, $\tilde{\mathbf{u}} = \begin{bmatrix} \tilde{m}_1^d & \tilde{m}_1^q & \tilde{m}_2^d & \tilde{m}_2^q \end{bmatrix}^T$ is the input vector and $\tilde{\mathbf{w}} = \begin{bmatrix} \tilde{v}_{g_1}^d & \tilde{v}_{g_1}^q & \tilde{i}_0^d & \tilde{i}_0^q \end{bmatrix}^T$ is the disturbance vector. It is important to emphasize that the subscript ($\sim$) denotes the small-signal around the operation point. While voltages, currents and modulation indexes with uppercase letters represent the variables at the operation point of steady-state. Furthermore, the matrix $\mathbf{A}$ is called the dynamics matrix, the matrix $\mathbf{B_u}$ is called the control matrix and the matrix $\mathbf{B_w}$ is called disturbance matrix.

Under nominal operating conditions, the grid voltages are approximately constant or vary slowly due to the estimation angle obtained by the Phase-Locked-Loop (PLL) [12]. Therefore, $v_{g_1}^d$ and $v_{g_1}^q$ can be approximated as the phase peak grid-voltage (V_{gp}) and zero voltage, respectively.

III. UNIFIED CONTROL SCHEME

A multivariable system has been proposed and designed for the simultaneous control of both sides of BTB. As mentioned, this approach has been chosen to overcome the conventional control strategies issues.

In order to ensure zero steady-state tracking error in the system, integral actions should be added to the space-state model. Thus, four integral actions have been added.

Therefore, the closed-loop controlled variables are the quadrature current of the grid-interface converter ($\tilde{i}_1^q$), the direct and quadrature output voltages of the output LC-filter ($\tilde{v}_0^d$, $\tilde{v}_0^q$) and the DC-bus voltage ($\tilde{v}_{dc}$). The controller can be represented in the following form:

$$\dot{\mathbf{x}}_c = \mathbf{A_c}\mathbf{x_c} + \mathbf{B_c}\mathbf{e}, \tag{8}$$

where $\mathbf{x_c}$ and $\mathbf{e} \in \mathbb{R}^{4 \times 7}$ are the states related to the additional dynamics and the controller's error input, respectively. Moreover, $\mathbf{x_c}$ is called the vector of integral actions,

$$\mathbf{A_c} = \begin{bmatrix} \mathbf{0}_{4 \times 4} \end{bmatrix} \quad \text{and} \quad \mathbf{B_c} = \begin{bmatrix} \mathbf{I}_{4 \times 4} \end{bmatrix}, \tag{9}$$

where $\mathbf{0}_{4 \times 4}$ represents block matrices of zeros with dimensions 4×4 and $\mathbf{I}_{4 \times 4}$ is a 4-dimensional identity matrix.

From the aforementioned discussion, the augment system can be expressed in Fig. 5. It is important highlight that the matrix $\mathbf{C} \in \mathbb{R}^{4 \times 7}$ is called the sensor matrix, and $\mathbf{r} \in \mathbb{R}^{4 \times 7}$ is the reference vector.

A. Resonant Controller

The feedback of the plant states are not sufficient to reject disturbances caused by the nonlinear and unbalanced loads. Therefore, according to the internal model principle, resonant controllers tunned on the frequency of the signal to be followed or reject must be included. In the case under study, the aforementioned additional dynamics are included in a dq-frame through resonant controller tunned to 6th and 12th harmonic frequencies, to ensure the disturbances mitigation of 5th, 7th, 11th and 13th harmonics frequencies in abc-frame [13]. A space-state representation of such controller for an angular frequency tunned to 6th harmonics frequencies ω_{r_6} in a direct-axis is shown in (10).

$$\frac{d}{dt}\begin{bmatrix} x_{r_6,1}^d \\ x_{r_6,2}^d \end{bmatrix} = \overbrace{\begin{bmatrix} 0 & 1 \\ -\omega_{r_6}^2 & 0 \end{bmatrix}}^{\mathbf{R}_6^{\mathbf{d}}}\begin{bmatrix} x_{r_6,1}^d \\ x_{r_6,2}^d \end{bmatrix} + \overbrace{\begin{bmatrix} 0 \\ 1 \end{bmatrix}}^{\mathbf{S}_6^{\mathbf{d}}}\tilde{v}_0^d. \tag{10}$$

The system described in (10) can be extended for quadrature axis. Therefore, a general dq-frame state-space representation for resonant controllers is:

$$\dot{\mathbf{x}}_{\mathbf{r_6}} = \overbrace{\begin{bmatrix} \mathbf{R}_6^{\mathbf{d}} & \mathbf{0}_{2 \times 2} \\ \mathbf{0}_{2 \times 2} & \mathbf{R}_6^{\mathbf{q}} \end{bmatrix}}^{\mathbf{R_6}}\mathbf{x_{r_6}} + \overbrace{\begin{bmatrix} \mathbf{S}_6^{\mathbf{d}} & \mathbf{0}_{2 \times 1} \\ \mathbf{0}_{2 \times 1} & \mathbf{S}_6^{\mathbf{q}} \end{bmatrix}}^{\mathbf{S_6}}\tilde{\mathbf{v}}_{\mathbf{0}}. \tag{11}$$

978-1-7281-4181-7/19 \$31.00 © 2019 IEEE

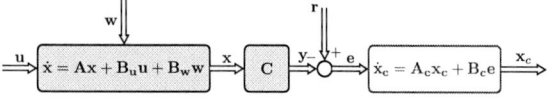

$$\dot{\tilde{\mathbf{x}}} = \overbrace{\begin{bmatrix} \frac{-r_{L_1}}{L_1} & \omega & 0 & 0 & 0 & 0 & \frac{-M_1^d}{2L_1} \\ -\omega & \frac{-r_{L_1}}{L_1} & 0 & 0 & 0 & 0 & \frac{-M_1^q}{2L_1} \\ 0 & 0 & \frac{-r_{L_2}}{L_2} & \omega & \frac{-1}{L_2} & 0 & \frac{+M_2^d}{2L_2} \\ 0 & 0 & \omega & \frac{-r_{L_2}}{L_2} & 0 & \frac{-1}{L_2} & \frac{-M_2^q}{2L_2} \\ 0 & 0 & \frac{1}{C} & 0 & 0 & \omega & 0 \\ 0 & 0 & 0 & \frac{1}{C} & -\omega & 0 & 0 \\ \frac{3M_1^d}{4C_{dc}} & \frac{3M_1^q}{4C_{dc}} & \frac{-3M_2^d}{4C_{dc}} & \frac{-3M_2^q}{4C_{dc}} & 0 & 0 & 0 \end{bmatrix}}^{\mathbf{A}} \tilde{\mathbf{x}} + \overbrace{\begin{bmatrix} \frac{-V_{dc}}{2L_1} & 0 & 0 & 0 \\ 0 & \frac{-V_{dc}}{2L_1} & 0 & 0 \\ 0 & 0 & \frac{V_{dc}}{2L_2} & 0 \\ 0 & 0 & 0 & \frac{V_{dc}}{2L_2} \\ 0 & 0 & 0 & 0 \\ 0 & 0 & 0 & 0 \\ \frac{3I_1^d}{4C_{dc}} & \frac{3I_1^q}{4C_{dc}} & \frac{-3I_2^d}{4C_{dc}} & \frac{-3I_2^q}{4C_{dc}} \end{bmatrix}}^{\mathbf{B_u}} \tilde{\mathbf{u}} + \overbrace{\begin{bmatrix} \frac{1}{L_1} & 0 & 0 & 0 \\ 0 & \frac{1}{L_1} & 0 & 0 \\ 0 & 0 & 0 & 0 \\ 0 & 0 & 0 & 0 \\ 0 & 0 & \frac{-1}{C} & 0 \\ 0 & 0 & 0 & \frac{-1}{C} \\ 0 & 0 & 0 & 0 \end{bmatrix}}^{\mathbf{B_w}} \tilde{\mathbf{w}} \quad (7)$$

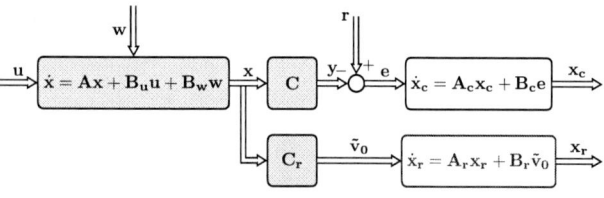

Fig. 5. Augmented system.

Consequently, a general state-space representation for two resonant controllers tunned to 6th and 12th frequencies is:

$$\dot{\mathbf{x}}_{\mathbf{r}} = \overbrace{\begin{bmatrix} \mathbf{R_6} & \mathbf{0}_{4\times4} \\ \mathbf{0}_{4\times4} & \mathbf{R_{12}} \end{bmatrix}}^{\mathbf{A_r}} \mathbf{x}_{\mathbf{r}} + \overbrace{\begin{bmatrix} \mathbf{S_6} \\ \mathbf{S_{12}} \end{bmatrix}}^{\mathbf{B_r}} \tilde{\mathbf{v}}_0. \quad (12)$$

where $\mathbf{x}_{\mathbf{r}} = \begin{bmatrix} \mathbf{x}_{\mathbf{r_6}}^T & \mathbf{x}_{\mathbf{r_{12}}}^T \end{bmatrix}^T$ and $\tilde{\mathbf{v}}_0 = \begin{bmatrix} \tilde{v}_0^d & \tilde{v}_0^q \end{bmatrix}^T$.

Fig. 6 depicts the whole augment system, including the resonant modules. It should be noted that $\mathbf{C_r} \in \mathbb{R}^{2\times7}$.

Fig. 6. Augmented system with resonant module.

By analysing the aforementioned figure, the whole augment system can be expressed as:

$$\dot{\mathbf{x}}_{\mathbf{a}} = \begin{bmatrix} \mathbf{A} & \mathbf{0}_{7\times4} & \mathbf{0}_{7\times8} \\ -\mathbf{B_cC} & \mathbf{A_c} & \mathbf{0}_{7\times8} \\ \mathbf{B_rC_r} & \mathbf{0}_{8\times7} & \mathbf{A_r} \end{bmatrix} \mathbf{x}_{\mathbf{a}} + \begin{bmatrix} \mathbf{B_u} \\ \mathbf{0}_{4\times4} \\ \mathbf{0}_{8\times4} \end{bmatrix} \tilde{\mathbf{u}} + \begin{bmatrix} \mathbf{B_w} \\ \mathbf{0}_{4\times4} \\ \mathbf{0}_{8\times4} \end{bmatrix} \tilde{\mathbf{w}} + \begin{bmatrix} \mathbf{0}_{7\times4} \\ \mathbf{B_c} \\ \mathbf{0}_{8\times4} \end{bmatrix} \mathbf{r}$$
$$(13)$$

where $\mathbf{x}_{\mathbf{a}} = \begin{bmatrix} \tilde{\mathbf{x}}^T & \mathbf{x}_{\mathbf{c}}^T & \mathbf{x}_{\mathbf{r}}^T \end{bmatrix}^T$.

IV. LQR CONTROLLER

The state-feedback controller for system (13) can be designed as a Linear Quadratic Regulator. Therefore, control input $\tilde{\mathbf{u}}$ is determined using the following equation [14]:

$$\tilde{\mathbf{u}} = -\mathbf{Kx}_{\mathbf{a}} \quad (14)$$

Gain matrix $\mathbf{K}$ can be determined by using relation (14) and minimizing a cost function J described in (15).

$$J = \int_0^\infty \left(\mathbf{x}_{\mathbf{a}}^T \mathbf{Q} \mathbf{x}_{\mathbf{a}} + \tilde{\mathbf{u}}^T \mathbf{R} \tilde{\mathbf{u}} \right) dt. \quad (15)$$

The gains of the controller matrix $\mathbf{K}$ can be tuned by choosing appropriate values for the matrices $\mathbf{Q}$ and $\mathbf{R}$, which weight the effects of the minimization on the state variables and on the control variables, respectively [15]. In the system under study, the $\mathbf{Q} \in \mathbb{R}^{19\times19}$ and $\mathbf{R} \in \mathbb{R}^{4\times4}$ matrices are determined so that the control system presented the desired response. Both matrices are presented in (16).

$$\mathbf{Q} = \text{diag}([1\ 1\ 10^{-9} 10^{-9} 1\ 1\ 1\ 10^{13} 10^{15} 10^{15} 10^{13} 10^3 \dots 10^3])$$

$$\mathbf{R} = \text{diag}([10^{12}\ 10^{12}\ 10^{12}\ 10^{12}]). \quad (16)$$

From the previous analysis, it can be concluded that a steady-state operation point must be chosen before setting the values of matrix $\mathbf{K}$. In addition, this point must satisfy a equilibrium point, in which the system (6) is linearized. For this analyze, the simulation system parameters are listed in Table I.

Based on the aforementioned attributes the gain matrix $\mathbf{K}_{\mathbf{4\times19}}$ can be calculated using the values in (16) and minimizing the cost function in (15).

A comparison between the open-loop (Z_0^d) and closed-loop ($Z_{0_{cl}}^d$) output impedance shows that proposed control improves its frequency response and rejection capability. Lower values of $Z_{0_{cl}}^d$ implies that higher current levels should be drawn from the converter to result in a considerable output voltage drop. Notice, in Fig. 7, that the resonant controllers acts as band reject filters.

V. SIMULATION RESULTS

The system from Fig. 1 is modeled in the PSIM simulation package in order to investigate the effectiveness of the proposed controller. The grid-interface converter its synchronized with the power grid.

Firstly, the proposed control is evaluated at nonlinear load conditions. Fig. 8 shows the load current waveforms, the three-phase voltage and current provided by the grid-interface

978-1-7281-4181-7/19 $31.00 © 2019 IEEE

TABLE I
PARAMETERS USED IN SIMULATION.

System parameters	
Description	Value
Switching and Sampling frequency (f_{sw}, f_s)	20 kHz
Fundamental frequency (f)	60 Hz
Phase *peak* grid-voltage (V_{gp})	180 V
DC link voltage (V_{dc})	500 V
Grid-interface filter inductance (r_{L_1})	75 $m\Omega$
Grid-interface filter resistance (L_1)	2 mH
Output filter resistance (r_{L_2})	75 $m\Omega$
Output filter inductance (L_2)	175 μH
Output filter capacitance (C)	100 μF
DC-link capacitance (C_{dc})	2 mF
Operation point parameters	
Description	Value
Output voltage d-axis (V_0^d)	180 V
Output voltage q-axis (V_0^q)	0 V
Output current d-axis (I_0^d)	30 A
Output current q-axis (I_0^q)	0 A
Modulation index d-axis (M_1^d)	0.71076
Modulation index q-axis (M_1^q)	-0.09285
Modulation index d-axis (M_2^d)	0.72720
Modulation index q-axis (M_2^q)	0.00995

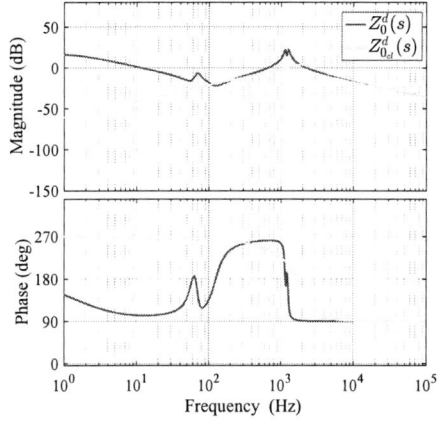

Fig. 7. Output impedance bode diagram: open-loop Z_0^d and closed-loop $Z_{0_{cl}}^d$.

converter. It is clear that the unified control system is capable of providing a high quality voltage waveform. This statement is verified by the low THD (0.58%). Moreover, it's noticed in the Fig. 8 (c) that the current is not affected by nonlinear load, maintaining sinusoidal shaped waveforms.

Despite the fact that the previous results presented good output voltage power quality for balanced nonlinear load, a more demanding scenario should be investigated. In this sense, a single phase bridge rectifier with a capacitive output filter is connected at the BTB's output terminals. In Fig. 9 (a) and (b) are depicted the output currents and voltages waveforms, and in Fig. 9(c) the grid-interface current for unbalanced nonlinear load connected between phases "b" and "c" are shown. In this case, the phase "a" acts as an open-circuit. It is worth noting that even in the case that one phase is open, the output voltage THD is low, keeping a good power quality. When there is imbalance load , a third order harmonic appears in the output voltage. It is important to note that, for this load, the DC-bus

voltage and the grid-interface current is affected. This occurs because a third order harmonic will also appear, inevitably. Although the THD has increased to about 1.06%, it is still reasonably low.

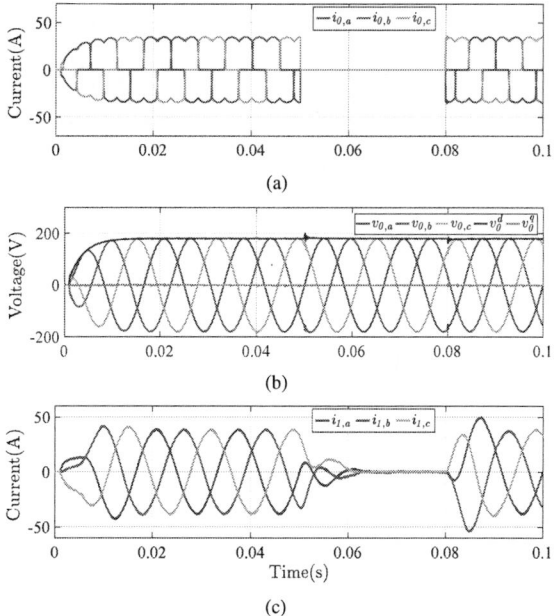

Fig. 8. Simulation results for nonlinear load: (a) Three-phase output currents; (b) Three-phase output voltages; (c) Three-phase grid-interface inverter currents.

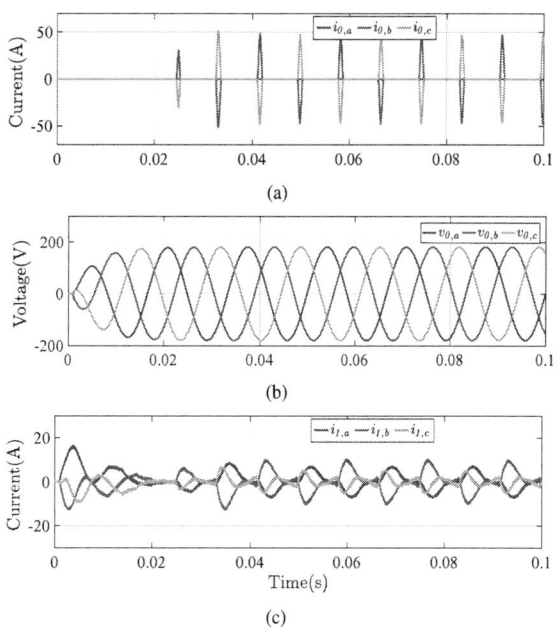

Fig. 9. Simulation results for the single-phase bridge rectifier:(a) Three-phase output currents; (b) Three-phase output voltages; (c) Three-phase grid-interface inverter currents.

Finally, to analyze the behavior of the closed-loop system. It is tested under load step changes. Therefore, between 0 s and

978-1-7281-4181-7/19 $31.00 © 2019 IEEE

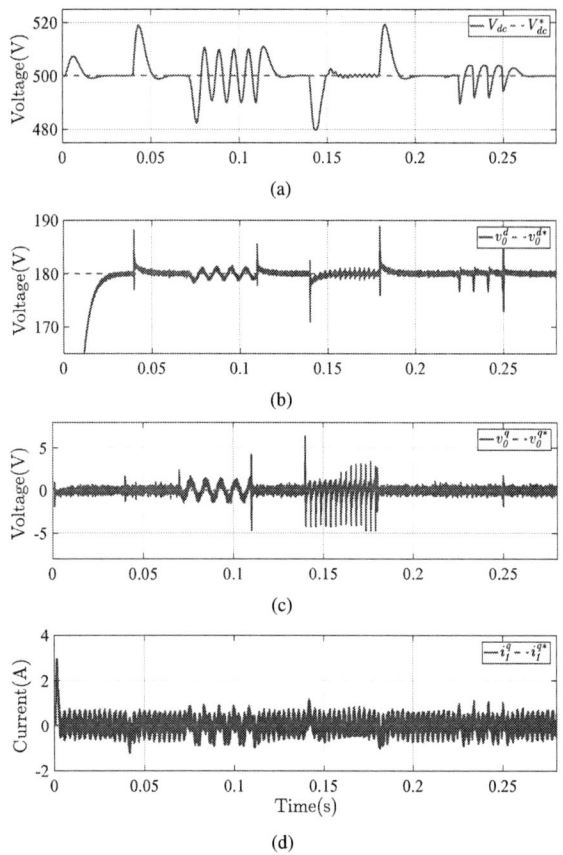

Fig. 10. Waveforms under different load conditions:(a) DC-Link voltage; (b) Output d-axis voltage; (c) Output q-axis voltage; (d) Input q-axis current.

0.25 s, linear, nonlinear, balanced and unbalanced loads are connected and disconnected of BTB . The sequence of load changes is: at 0 s a three-phase linear load is connected, in 0.04 s is disconnected; a unbalance resistive load is turned-on in 0.07 s and turned-off in 0.11 s; further, between 0.14 s and 0.18 s a nonlinear load is working. Finally, an unbalanced nonlinear load is connected load is connected in 0.21s and disconnected in 0.25s. Fig. 10 shows the behavior of DC voltage (a), dq-frame output voltage (b,c) and quadrature axis grid-interface inverter current (d) during the 0.25 seconds of simulation. Note that during load transitions, all controlled variables vary, but they still stabilize and maintain its value bounded by the reference value.

As discussed before, a load imbalance makes it appear a third order harmonic in the output voltage. Consequently, it is possible to note in all the results of Fig. 10 that, at these unbalance moments, a second order harmonic appears at the DC voltage, the dq-frame output voltage and quadrature axis grid-interface inverter current.

VI. CONCLUSION

This paper presents a BTB converter using LQR control. Design steps of voltage controller and LQR are detailed. Besides that a multivariable control for the BTB with LC filter is proposed, which is based on a state-space representation.

This multivariable controller can take advantage of all dynamic interactions between both inverters and the DC-capacitor included in the proposed modeling procedure.

The simulation results have demonstrated the effectiveness of the proposed technique under different load conditions, therefore, a satisfactory quality of output voltage is guaranteed.

ACKNOWLEDGMENT

This study was financed in part by the Coordenação de Aperfeiçoamento de Pessoal de Nível Superior - Brasil (CAPES). The authors also gratefully acknowledge the financial support in part of CNPq, INERGE, and FAPEMIG (Project 01666/15), and express gratitude for the educational support of UFOP, UFMG and UFJF.

REFERENCES

[1] M. De Santis, S. Agnelli, F. Patanè, O. Giannini, and G. Bella, "Experimental study for the assessment of the measurement uncertainty associated with electric powertrain efficiency using the back-to-back direct method," *Energies*, vol. 11, no. 12, 2018.

[2] W. Xiang, R. Yang, C. Lin, J. Zhou, J. Wen, and W. Lin, "A cascaded converter interfacing long distance hvdc and back-to-back hvdc systems," *IEEE Journal of Emerging and Selected Topics in Power Electronics*, pp. 1–1, 2019.

[3] A. M. Rauf, A. V. Sant, V. Khadkikar, and H. H. Zeineldin, "A novel ten-switch topology for unified power quality conditioner," *IEEE Transactions on Power Electronics*, vol. 31, no. 10, pp. 6937–6946, 2016.

[4] F. Taveiros, L. Barros, and F. Costa, "Back-to-back converter state-feedback control of dfig (doubly-fed induction generator)-based wind turbines," *Energy*, vol. 89, pp. 896 – 906, 2015.

[5] R. Majumder, A. Ghosh, G. Ledwich, and F. Zare, "Power management and power flow control with back-to-back converters in a utility connected microgrid," *IEEE Transactions on Power Systems*, vol. 25, no. 2, pp. 821–834, May 2010.

[6] A. Ortiz, S. Klyapovskiy, T. Østrem, and W. Sulkowski, "Radial microgrid operation based on a btb converter," in *IECON 2011 - 37th Annual Conference of the IEEE Industrial Electronics Society*, Nov 2011.

[7] N. Zhou, C. Yuan, and Q. Wang, "Control strategies for microgrid power quality enhancement with back-to-back converters connected to a distribution network," in *2012 IEEE 15th International Conference on Harmonics and Quality of Power*, June 2012, pp. 384–389.

[8] I. D. de Souza, P. M. de Almeida, P. G. Barbosa, C. A. Duque, and P. F. Ribeiro, "Digital single voltage loop control of a vsi with lc output filter," *Sustainable Energy, Grids and Networks*, vol. 16, pp. 145 – 155, 2018. [Online]. Available: http://www.sciencedirect.com/science/article/pii/S2352467718300092

[9] P. M. Almeida, P. G. Barbosa, J. G. Oliveira, J. L. Duarte, and P. F. Ribeiro, "Digital proportional multi-resonant current controller for improving grid-connected photovoltaic systems," *Renewable Energy*, vol. 76, pp. 662 – 669, 2015.

[10] A. Rodríguez-Cabero, F. H. Sánchez, and M. Prodanovic, "A unified control of back-to-back converter," in *2016 IEEE Energy Conversion Congress and Exposition (ECCE)*, Sep. 2016, pp. 1–8.

[11] H. Akagi, Y. Kanazawa, and A. Nabae, "Instantaneous reactive power compensators comprising switching devices without energy storage components," *IEEE Transactions on Industry Applications*, vol. IA-20, no. 3, pp. 625–630, May 1984.

[12] P. Rodríguez, R. Teodorescu, I. Candela, A. V. Timbus, M. Liserre, and F. Blaabjerg, "New positive-sequence voltage detector for grid synchronization of power converters under faulty grid conditions," in *37th IEEE Power Electronics Specialists Conference*, June 2006.

[13] R. Teodorescu, M. Liserre, and P. Rodriguez, *Grid Converters for Photovoltaic and Wind Power Systems*, ser. Wiley - IEEE. Wiley, 2011.

[14] E. Sontag, *Mathematical Control Theory: Deterministic Finite Dimensional Systems*. Springer New York, 2013.

[15] E. V. Kumar, J. Jerome, and K. Srikanth, "Algebraic approach for selecting the weighting matrices of linear quadratic regulator," in *2014 International Conference on Green Computing Communication and Electrical Engineering (ICGCCEE)*, March 2014, pp. 1–6.

3-Phase Multi-Functional Grid-Tied Inverter for Compensation of Oscillating Instantaneous Power

José de Arimatéia Olímpio Filho
Group of Automation and Integrated Systems (GASI)
São Paulo State University (Unesp)
Sorocaba, Brazil
jose.olimpio@unesp.br

Helmo Kelis Morales Paredes
Group of Automation and Integrated Systems (GASI)
São Paulo State University (Unesp)
Sorocaba, Brazil
helmo.paredes@unesp.br

Augusto Matheus dos Santos Alonso
Department of Electric Power Engineering
Norwegian University of Science & Technology (NTNU)
Trondheim, Norway
augusto.alonso@ntnu.no

Jakson Paulo Bonaldo
Department of Electrical Engineering
Federal University of Mato Grosso (UFMT)
Cuiaba, Brazil
jaksonpaulo@ufmt.br

Fernando Pinhabel Marafão
Group of Automation and Integrated Systems (GASI)
São Paulo State University (Unesp)
Sorocaba, Brazil
fernando.marafao@unesp.br

Marcelo Godoy Simões
Electrical Engineering Department
Colorado School of Mines
Golden, Colorado, USA
msimoes@mines.edu

Abstract— **The primary objective of this paper is to present the basic terms of Conservative Power Theory (CPT) and its applications for compensating instantaneous power oscillations in three-phase three-wire systems. The reference signals for a three-phase multi-functional grid-tied inverter are derived using the instantaneous power and the instantaneous reactive energy terms defined by CPT, having the oscillating power components defined directly in *a-b-c* coordinates. The compensation method proposed in this study is validated through simulations based on practical power system modeling. The results presented here show the feasibility and singularities of the proposed method.**

Keywords—Conservative Power Theory, Instantaneous Power Oscillations, Multifunctional Grid-Tied Inverters, Power Quality, Renewable Energy Sources.

I. INTRODUCTION

The electrical power generation from Renewable Energy Sources (RES) is a topic that has been extensively researched all over the world. Such a trend has been motivated by sustainable energy policies such as credit incentive and tax subsidy, among other reasons [1], [2]. Accordingly, RES recent technologies have been presenting promising solutions for integration of Distributed Energy Resources (DER) [3], [4], such as hybrid/electric vehicles, energy storage and microgrids. In particular, microgrids have been considered an appealing and efficient solution for industrial parks, commercial, rural and residential areas, in addition to several other alternatives considering renewable based DERs [5], [6].

A microgrid is comprised of a set of loads and DER elements that may or may not have energy storage, knowing that such elements are connected to a distribution grid through a Point of Common Coupling (PCC). In general, microgrids are connected through a Power Electronics Interface (PEI), which is usually a voltage source inverter operated by an Energy Management System (EMS) [4]-[7]. Microgrids can operate in either on-grid or off-grid mode, imposing that its infrastructure should be robust and able to switch between these two operation modes smoothly [5], [8].

Also, a microgrid could be implemented as a low voltage distribution system [9]. Therefore, it must be able to supply energy to single- and two-phase (linear or non-linear) loads in three-phase systems. For instance, possible loads that may be connected to a three-phase three-wire microgrid are computers, lighting reactors, home appliances, battery chargers, among others. Although, the majority of such loads are non-linear, draining unbalanced and distorted currents.

RES connected to microgrids by PEIs usually consist of a DC-DC converter (Buck-Boost), which can track the Maximum Power Point, as well as a DC-AC converter. The DC-AC converter is responsible for maintaining a constant voltage on the DC link and controlling the active power flow at its AC side, in addition to providing power quality ancillary services [10]-[14].

However, some RES, especially PV and wind systems, are highly affected by weather conditions and their physical location within the microgrid. Hence, the energy generated and consumed may vary significantly through time, creating a dynamic and complex power system with high level of interaction between sources and loads. In this context, the main challenges for operating microgrids are the RES intermittency, as well as the variable load consumption, which consequently causes instantaneous power oscillations [15]-[17]. Such variations can cause braking torque oscillations on rotating machines, trigger tie-line power fluctuations [18], also affecting the energy dispatch on wind power generation [19]-[21].

As consequence, it is of importance to mention that power quality issues in a three-wire three-phase microgrid are associated with harmonics, unbalance and instantaneous power oscillations. For example, the nature of the instantaneous power and average power of different load configurations, considering the same rated power, are shown in Fig. 1. Note that the energy oscillation between the load and the source depends on the load configuration. Therefore, different from classic compensation strategies that aim to mitigate current harmonics and unbalance [11]-[14], [22], this paper focuses on the instantaneous power oscillation problem to be suppressed by a three-phase Multi-Functional Grid-Tied Inverter (MFGTI).

To achieve the such goal, this study proposes an innovative use of the instantaneous power and reactive energy terms defined by the Conservative Power Theory (CPT) [23], as a new alternative for the design and control of a three-wire three-phase MFGTI. The proposed control algorithm for

---

This work was supported by São Paulo Research Foundation (FAPESP) under grants, 2017/22629-9, 2017/20987-5, 2018/22172-1, 2016/08645-9, and National Council for Scientific and Technological Development (CNPq) under grand 311332/2018-8.

978-1-7281-4181-7/19 $31.00 © 2019 IEEE

generating the reference signals for the operation of the inverter is obtained directly in *a-b-c* coordinates. Thus, the main advantages of the proposed approach are the enhanced efficiency and fast dynamic response of the MFGTI control architecture. Herein, the method is presented and discussed on the basis of simulation results.

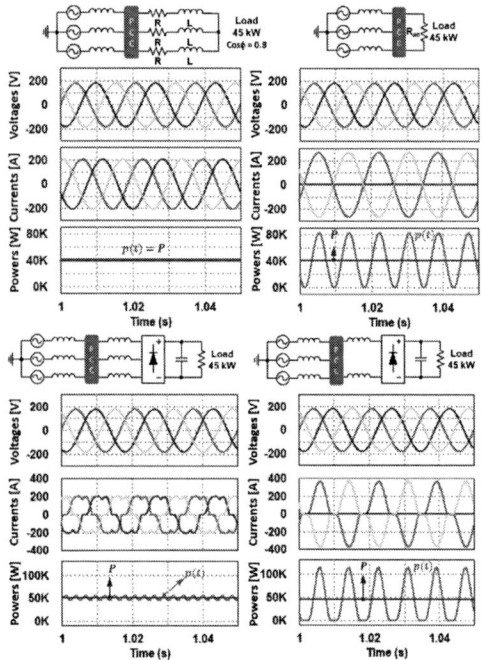

Fig.1 – Time evolution of instantaneous power, $p(t)$ and average power, P to different load configuration with the same nominal power (45 kW).

II. CPT CONCEPTS AND BASIC DEFINITIONS

Consider a set of real electrical quantities, continuous and periodic of period T, with fundamental frequency $f = 1/T$ and angular frequency $\omega = 2\pi f$. As detailed in [23], the CPT proposes an approach in time domain developed entirely based on a-b-c coordinates, on which two instantaneous main terms are defined. First, the instantaneous power term is defined by the scalar product between the voltage (v) and current (i) vectors:

$$p(t) = v \circ i = \begin{bmatrix} v_a & v_b & v_c \end{bmatrix} \circ \begin{bmatrix} i_a \\ i_b \\ i_c \end{bmatrix} \quad (1)$$

Secondly, instantaneous reactive energy is defined as:

$$w(t) = \hat{v} \circ i = \begin{bmatrix} \hat{v}_a & \hat{v}_b & \hat{v}_c \end{bmatrix} \circ \begin{bmatrix} i_a \\ i_b \\ i_c \end{bmatrix} \quad (2)$$

where, $\hat{v}$ is a vector that contains the unbiased integrals of phase voltages. This quantity is calculated by the difference between the time integral and its mean value, as shown in (3).

$$\hat{v}_m = \int_0^t v_m(\tau)d\tau - \frac{1}{T}\int_0^T \left[\int_0^t v_m(\tau)d\tau \right] dt \quad (3)$$

The "m" index represents the variables for each phase. Also, according to the CPT, in three-wire three-phase circuits, the phase voltages are measured using a virtual reference point [24].

The mean values corresponding to (1) and (2) are given by (4) and (5):

$$\bar{p} = \frac{1}{T}\int_0^T p(t)dt = \frac{1}{T}\int_0^T (v_a i_a + v_b i_b + v_c i_c)dt = P \quad (4)$$

$$\bar{w} = \frac{1}{T}\int_0^T w(t)dt = \frac{1}{T}\int_0^T (\hat{v}_a i_a + \hat{v}_b i_b + \hat{v}_c i_c)dt = W \quad (5)$$

where P is the active power in Watts e W is the reactive energy in Joules.

Based on the definitions of (4) and (5), the CPT decomposes the load phase currents as the sum of five subcomponents that are: balanced active currents, balanced reactive currents, unbalanced active currents, unbalanced reactive currents, and residual currents [23], [24]. For a sinusoidal voltage operation, independently of balanced or unbalanced characteristics, the active and reactive current components, irrespectively of being balanced or unbalanced related, correspond to a portion of the fundamental current. On the other hand, the residual portion defined by the CPT corresponds to the harmonic components generated by the non-linear loads. These current subcomponents have been satisfactorily applied to generate reference signals for single-phase and three-phase active power filters (APF) [25], [26], as well as for multifunctional grid-tied inverters [11], [12]. Such decompositions provided by the CPT allow selective and oriented identification of different disturbances on a generic load (non-linearities, unbalances and reactive power [24]). Hence, it is possible to set compensation strategies for undesirable currents with high flexibility level.

Since the application of the CPT focusing on the compensation of oscillating powers in three-wire three-phase systems has not yet been explored in literature, the main contribution of this work is settled. It is important to highlight that, although other previously proposed strategies approach the goal of providing constant instantaneous power in electric circuits [15]-[17], their mathematical and physical interpretations, as well as their implementation for the control of MFGTIs, differ from the CPT's definitions discussed herein. In this context, the following section is presented aiming to analyze the defined power terms within (1), (2), (4) and (5) and their relationship with the generation of control signals for a MFGTI in three-phase three-wire networks.

III. COMPENSATION STRATEGY AND PROPOSED CONTROL ALGORITHM

The strategy to compensate the instantaneous power oscillations by means of the CPT can be obtained through the decomposition of (1) and (2), devising mean and oscillating terms. Therefore, the instantaneous power and reactive energy terms defined in a-b-c coordinates result in (6) and (7), respectively:

$$p(t) = \bar{p} + \tilde{p} \quad (6)$$

$$w(t) = \bar{w} + \tilde{w} \quad (7)$$

where "~" represents the oscillating components of each instantaneous term. The mean components, which are represented by "-" can be derived by (4) and (5) and are valid independently of voltage and current waveforms. This means that such approach is valid to be applied for both sinusoidal and non-sinusoidal voltage conditions. Yet, the mean terms comprised in (6) and (7) can also be derived by low-pass filters, as it is accomplished in alternative methods like the p-q theory [27] or other control algorithms for APFs [15].

The instantaneous power, $p(t)$, represents the useful energy per unit of time that flows from the source to the load (or from the load to the source, if negative). The mean component of $p(t)$, if positive, comprises the energy per unit of time that is transferred from the source to the load. While the oscillating component ($\tilde{p}$) corresponds to the energy per unit of time which is exchanged between the source and the load. Certainly, the mean value of the oscillating component is zero, but at each instant, it represents a quantity of energy that flows on the electric circuit due to an undesirable current.

In general, the mean component ($\bar{p}$) can be calculated considering the grid frequency period as in (4). The oscillating component ($\tilde{p}$), on the other hand, correspond to the components with higher frequencies than the grid's, or it is given by the components of the negative sequence. Furthermore, upon the existence of RES with intermittent and unpredictable behavior, $\tilde{p}$ can also be found within harmonic frequencies, unbalanced loads, or due to resonances triggered by single- or two-phase converters connected to three-wire three-phase circuits. Particularly, in three-phase electric circuits, with or without neutral wire, where voltages and currents consist only of positive sequence and fundamental components, the energy transfer is unidirectional, usually flowing from source to load. In this case, the instantaneous power contains just the mean component ($p(t) = \bar{p} = P$).

Additionally, there are other specific scenarios on which energy is transferred unidirectionally from source to load. For instance, it occurs when voltages and currents are in phase, balanced, and present the same harmonics (common harmonics). In any other situation, if voltages and currents are distorted (i.e., presenting uncommon harmonics) or with unbalanced components, the instantaneous power will present mean and oscillating components with the bidirectional flow.

A. Reference Signal Generation for Compensation of Instantaneous Power Oscillations

With the active power and reactive energy portion defined in (6) and (7), two instantaneous current components, $\boldsymbol{i}_p$ and $\boldsymbol{i}_w$, can be defined as:

$$\boldsymbol{i}_p = \frac{\bar{p}}{v_{abc}^2}\begin{bmatrix} v_a \\ v_b \\ v_c \end{bmatrix} + \frac{\tilde{p}}{v_{abc}^2}\begin{bmatrix} v_a \\ v_b \\ v_c \end{bmatrix} \tag{8}$$

$$\boldsymbol{i}_w = \frac{\bar{w}}{\hat{v}_{abc}^2}\begin{bmatrix} \hat{v}_a \\ \hat{v}_b \\ \hat{v}_c \end{bmatrix} + \frac{\tilde{w}}{\hat{v}_{abc}^2}\begin{bmatrix} \hat{v}_a \\ \hat{v}_b \\ \hat{v}_c \end{bmatrix} \tag{9}$$

where $v_{abc}^2 = v_a^2 + v_b^2 + v_c^2$ and $\hat{v}_{abc}^2 = \hat{v}_a^2 + \hat{v}_b^2 + \hat{v}_c^2$ are the collective instantaneous values of voltage and unbiased integrals of voltage, respectively. The bold variables represent vectors.

Therefore, the current vectors associated with instantaneous terms $[p(t)$ e $w(t)]$ can be decomposed into two subcomponents:

$$\boldsymbol{i}_p = \boldsymbol{i}_{\bar{p}} + \boldsymbol{i}_{\tilde{p}} \tag{10}$$

$$\boldsymbol{i}_w = \boldsymbol{i}_{\bar{w}} + \boldsymbol{i}_{\tilde{w}} \tag{11}$$

The oscillating components ($\boldsymbol{i}_{\tilde{p}}$ e $\boldsymbol{i}_{\tilde{w}}$) represent the harmonic components and/or the unbalanced current resulting from voltage and current uncommon harmonics and unbalanced voltages. These oscillating portions do not contribute to the active power (P) or to the reactive energy (W). Hence, $\boldsymbol{i}_{\tilde{p}}$ e $\boldsymbol{i}_{\tilde{w}}$ are responsible for the instantaneous power oscillations in the electric circuit, as well as for any additional losses.

From the load point of view, the decomposed portions shown in (10) and (11) do not represent any specific load behavior or characteristic; only on the condition of sinusoidal and balanced voltages, $\boldsymbol{i}_{\bar{p}}$ e $\boldsymbol{i}_{\bar{w}}$ coincide with active and reactive balanced currents defined by CPT.

Since oscillations on reactive energy and instantaneous power can be caused by voltage and current distortions, as well as by unbalance on voltage and current, the reference signals for compensation can be represented by the sum of $\boldsymbol{i}_{\tilde{w}}$, $\boldsymbol{i}_{\tilde{p}}$ e $\boldsymbol{i}_{\bar{w}}$ as presented in (12).

$$\begin{bmatrix} i_{a-APF} \\ i_{b-APF} \\ i_{c-APF} \end{bmatrix} = \begin{bmatrix} i_{a-\tilde{w}} \\ i_{b-\tilde{w}} \\ i_{c-\tilde{w}} \end{bmatrix} + \begin{bmatrix} i_{a-\tilde{p}} \\ i_{b-\tilde{p}} \\ i_{c-\tilde{p}} \end{bmatrix} + \begin{bmatrix} i_{a-\bar{w}} \\ i_{b-\bar{w}} \\ i_{c-\bar{w}} \end{bmatrix} \tag{12}$$

It is noted in (12) that such references for compensation purposes can be applied in a selective manner, where the sum of its components $\boldsymbol{i}_{\tilde{p}}$ e $\boldsymbol{i}_{\tilde{w}}$ is related to the oscillations and $\boldsymbol{i}_{\bar{w}}$ is associated with the reactive energy (power) flow in the system. Hence, the components ($\boldsymbol{i}_{\tilde{p}} + \boldsymbol{i}_{\tilde{w}}$) and ($\boldsymbol{i}_{\bar{w}}$) can be compensated independently by the MFGTI, or even by the combination of capacitor banks, which would compensate $\boldsymbol{i}_{\bar{w}}$, and the MFGTI ensuring compensation of $\boldsymbol{i}_{\tilde{p}} + \boldsymbol{i}_{\tilde{w}}$. Therefore, since the main goal of this study is to compensate the oscillations, the sum of the oscillating components will be taken as reference, in other words, $\boldsymbol{i}_{APF} = \boldsymbol{i}_{\tilde{p}} + \boldsymbol{i}_{\tilde{w}}$.

B. Reference Signal Generation to Inject Active Power

The current reference signal for the injection of the energy generated by a Local Energy Source (LES) is created based on the synthesis of sinusoidal currents. According to this strategy, the injected current waveform should match the waveform of the positive sequence fundamental component of the PCC voltages (v_{m1}^+). As discussed in [11], such strategy guarantees a smaller distortion level in the current. Thus, the active current reference $\boldsymbol{i}_{LES}$ of the MFGTI is determined by (13).

$$\boldsymbol{i}_{LES} = \frac{P_{LES}}{V_1^2}\begin{bmatrix} v_{a1}^+ \\ v_{b1}^+ \\ v_{c1}^+ \end{bmatrix} = G_{LES}\begin{bmatrix} v_{a1}^+ \\ v_{b1}^+ \\ v_{c1}^+ \end{bmatrix}, \tag{13}$$

where G_{LES} is the equivalent conductance of the multifunctional converter and P_{LES} is the liquid power generated from the LES, which should be injected into the grid:

$$P_{LES} = \frac{1}{T}\int_0^T v_{DC}(t)i_{DC}(t)dt. \tag{14}$$

The fundamental voltage component at the converter point of coupling (v_{m-1}) can be obtained using a Phase Locked Loop (PLL) or through a low-pass filter with a narrow bandwidth tuned into the grid frequency. Considering that the grid frequency variation is relatively small, a low-pass filter was chosen to attain v_{m-1}, followed by the RMS (V_{m-1}) value calculation.

C. Reference Signal Generation for the Three-Phase Multifunctional Converter

Considering the reference signal shown in (12) and (13), the final current reference ($\boldsymbol{i}_{ref}^*$) that is effectively synthesized by the MFGTI is given by:

$$i_{ref}^* = i_{LES}^* - i_{APF}^*. \tag{15}$$

If the multifunctional converter was to operate only as a PEI, without doing the compensation of the instantaneous power oscillations, the signal i_{APF}^* would be zero and the converter would only inject the energy generated by the LES into the grid. On the other hand, as the component $i_{\tilde{w}}$ does not contribute to the instantaneous power oscillation, to completely compensate the instantaneous power oscillations it is enough to assign the terms $i_{\tilde{p}} + i_{\tilde{w}}$ to the compensation reference, as:

$$i_{ref}^* = i_{LES}^* - i_{\tilde{p}}^* - i_{\tilde{w}}^*. \tag{16}$$

IV. CONTROL SYSTEM MODELLING

A modelling method and control design for three-phase converters using an APF is presented in reference [28] and the same approach is used for the MFGTI presented in this study. The system presented in Fig. 2 summarizes the topology of the adopted electrical circuit and converter. Also, the main electrical system parameters are shown in Table I. Because the system presented in Fig. 2 is a three-wire circuit with no neutral, the mesh control can be achieved considering the voltage and current values in only two phases of the three-phase system. For instance, for the converter, it is sufficient to control currents i_{Fa} and i_{Fb}, since the current i_{Fc} is related to the previous currents, being $i_{Fc} = -i_{Fa} - i_{Fb}$. Regarding the virtual point voltages, they can be obtained from the line voltage [24], according with (17). Thus, to v_{ab} e v_{bc} could be measured and the third line voltage component is derived applying Kirchhoff's voltage law, having $v_{ca} = -(v_{ab} + v_{bc})$, for instance.

TABLE I. THREE-PHASE CONVERTER AND SYSTEM PARAMETERS.

Parameter	Value	Parameter	Value
Grid, V_{Line}	220 V / 60 Hz	L_G; R_G	0.25 mH; 0.1 Ω
P_{LES}	2.4 kW	L_1; R_1	0.5 mH; 10 mΩ
C_0	3,3 μF	L_2; R_2	0.5 mH; 10 mΩ
R_x, L_x	5 Ω; 40 mH	L_z	1 mH
L_y, R_y	70 mH; 30 Ω	R_z; C_z	50 Ω; 470 μF

$$v_a = \frac{1}{3}(v_{ab} + v_{ac})$$
$$v_b = \frac{1}{3}(v_{ba} + v_{bc}) \tag{17}$$
$$v_c = \frac{1}{3}(v_{cb} + v_{ca})$$

The converter control system comprises two main control loops. The first is a fast loop used to control the converter output current, whereas the second is a voltage control loop that presents a slow dynamic response. The DC bus control keeps the energy balance between the power delivered to the system at the converter's output side and the power at its DC link [29]. Fig. 3 shows the control strategy adopted with the control loops proposed for the MFGTI.

The current control is based on a proportional resonant controller and another resonant harmonic controller (PR+HC) [30], [31], as:

$$G_C(s) = K_C + \sum_{h=1,3,5,\dots,11} \frac{2K_{IPR}\omega_{cPR}s}{s^2 + 2\omega_{cPR}s + (h\omega_o)^2} \tag{18}$$

In (18), h is the harmonic order, ω_o is the grid fundamental frequency and K_C, K_{IPR}, ω_{cPR} are the proportional gain, integral gain and a resonant frequency band pass controller, respectively. The K_{IPR} value is chosen to produce a high gain on harmonic frequencies and ω_{cPR} should be designed to compensate harmonic frequencies and is supposed to be small to amplify the compensator selectivity.

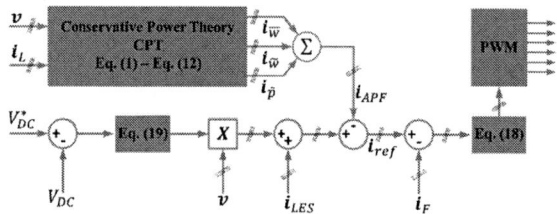

Fig.3 – Proposed multifunctional structure block diagram.

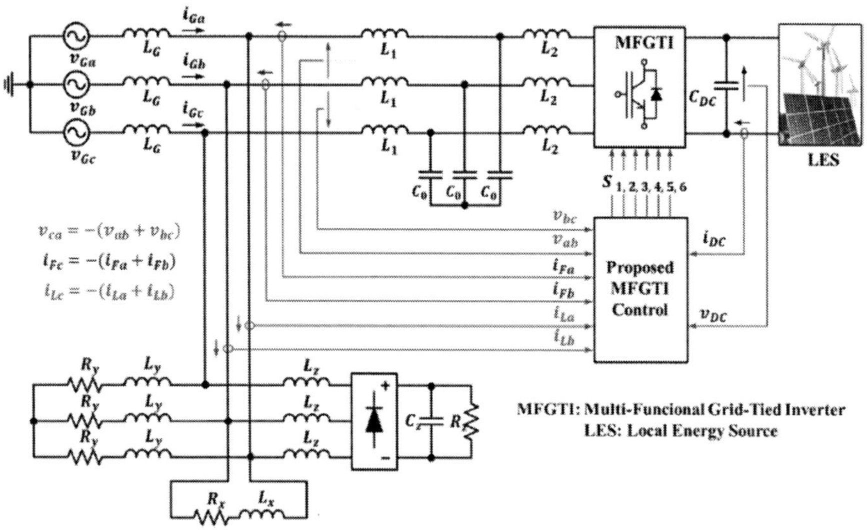

Fig.2 – Three-phase MFGTI connected to the electric grid.

The voltage controller used to maintain constant the DC link voltage is a proportional-integral (PI) regulator and it is given by (19):

$$PI_{DC}(s) = K_{P_{DC}} + \frac{K_{I_{DC}}}{s} \qquad (19)$$

where $K_{P_{DC}}$ is the controller proportional gain, while $K_{I_{DC}}$ is the integral gain. The MFGTI was designed with a switching frequency of 20 kHz. Table II shows the adopted values for the voltage and current controllers. The converter DC link voltage is adjusted to 400 V.

TABLE II. CONTROLLER PARAMETERS

$K_c = 2$	$K_{I_{PR}} = 100$
$\omega_{c_{PR}} = 6{,}28$ [rad/s]	$\omega_o = 377$ [rad/s]
$K_{P_{DC}} = 5$	$K_{I_{DC}} = 50$

V. SIMULATION RESULTS

In order to validate the compensation strategy and the proposed control algorithm for the three-phase MFGTI, the circuit shown in Fig. 2 was simulated using PSIM®. The circuit comprises an electronic converter (DC-AC) with a LCL output filter, a three-phase non-linear unbalanced load and the electric grid.

The main simulation goal is to validate the proposed method to achieve a constant instantaneous power, and at the same time inject, active power into the grid. The system operation dynamic is described by voltages, currents, power terms and instantaneous reactive energy in different measuring points at the grid, load and converter, respectively. Following, the simulation results are presented for two distinct scenarios.

A. Scenario 1: MFGTI injecting only active power

In this first scenario, the MFGTI is operated to function only as a power electronic interface (PEI). This simulation result is shown in Fig. 4. It can be seen that initially at $t < 0{,}6\ s$, the converter currents (i_{Fa}, i_{Fb} e i_{Fc}) are zero and the grid currents (i_{Ga}, i_{Gb} e i_{Gc}) are highly distorted and unbalanced. With the actioning of the PEI function at $t > 0{,}6\ s$, the converter currents are sinusoidal and synchronized with the PCC voltages. Hence, the mean grid component, $\overline{p}_G$ (active power), decreases while the grid currents continue to be distorted and unbalanced. In addition, consequently, the oscillating components $\tilde{p}_G$ e $\tilde{w}_G$ present a non-sinusoidal oscillation and continue to flow through the grid.

B. Scenario 2: MFGTI injecting active power and compensating power oscillations

In this second scenario, the active power filter (APF) ancillary service is added to the MFGTI. The simulation result can be seen in Fig. 5. Before the APF function is activated ($t < 1{,}2\ s$) it can be observed that the harmonic distortion and current unbalance create a non-sinusoidal oscillation at the $\tilde{p}_G$ e $\tilde{w}_G$ components. In addition, it can be seen a slightly distortion and voltage unbalance at the PCC, caused by the unbalanced non-linear load. After starting the APF service in the MFGTI at $t > 1{,}2\ s$, the current unbalance as well as the harmonics generated by the non-linear load (residual component) that are present in the grid are compensated, which results in the elimination of all oscillating components ($\tilde{p}_G$ e $\tilde{w}_G$); therefore, improving the voltage waveforms at the PCC. The compensation results can also be observed at the grid current waveform that becomes sinusoidal and balanced.

However, the grid current is still lagging the voltage because of the average reactive energy ($\overline{w}_G$), which is not being compensated.

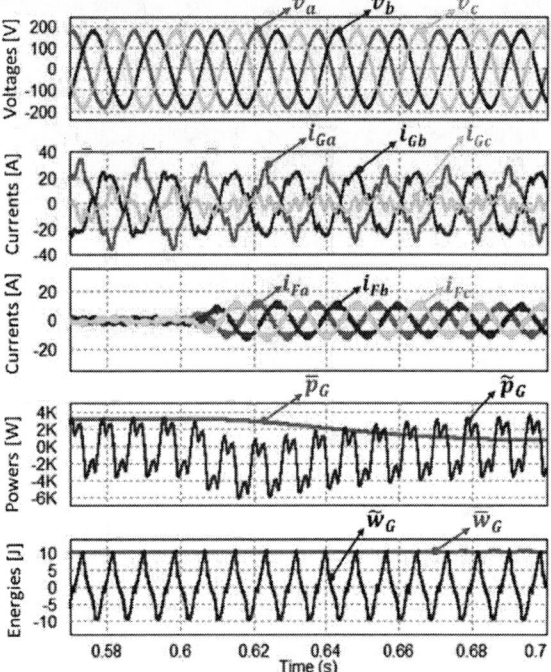

Fig. 4. Simulation results when the MFGTI is injecting only active power into the grid (Scenario 1).

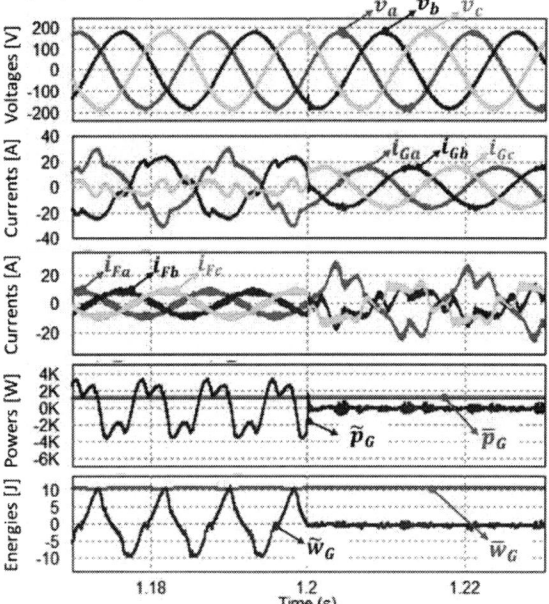

Fig. 5. Simulation results when the MFGTI is injecting active power and compensating the oscillations (Scenario 2).

VI. CONCLUSIONS

It was presented in this paper a strategy of using grid-connected multifunctional power electronic converters for injecting active power into the grid and compensating instantaneous power oscillations. The proposed control is based on instantaneous and average terms ($p(t)$, $w(t)$, $\overline{p}(t)$ and $\overline{w}(t)$) of the Conservative Power Theory (CPT).

As shown in the simulation results, the proposed control strategy allows not only injection of active power, but also supports power system operation at constant instantaneous power, $p(t) = \bar{p}(t)$, even under different linear/non-linear unbalanced load configurations. This feature is extremely important since the instantaneous power oscillations are undesirable in any power system due to rotating electric machines deterioration, and operation issues regarding power quality, especially on scenarios of weak electric grids such as microgrids.

Furthermore, considering the compensation results presented, future work is needed to verify the feasibility of the proposed strategy under distorted and unbalanced conditions, not only at PCC, but also at the voltage source (grid). Thus, power system operation at constant instantaneous power is expected even under adverse conditions. Yet, extensions of this work intend to further evaluate the singularities of the methodology proposed, exploring its advantages and disadvantages, when compared to other methods in the literature.

REFERENCES

[1] Y. Yang, E. Koutroules, A. Sangwongwanich, F. Blaabjerg, "Pursuing Photovoltaic Cost-Effectiveness," *IEEE Ind. Appl. Mag.*, vol. 23, pp. 40-49, June 2017.

[2] H. F. Camilo et al., "Assessment of Photovoltaic Distributed Generation – Issues of Grid-connected Systems Through the Consumer Side Applied to a Case Study of Brazil," *Renew. Sust. Energy Reviews*, vol. 71, pp. 712-719, May 2017.

[3] G. Spanuolo et al, "Renewable Energy Operation and Conversion Schemes: A Summary of Discussions During the Seminar on Renewable Energy Systems," *IEEE Ind. Electron. Mag.*, vol. 4, pp. 38-51, March 2010.

[4] R. Palma-Behnke, C. Benavides, F. Lanas, B. Severino, L. Reyes, J. Llanos and D. Sáez, "A Microgrid Energy Management System Based on the Rolling Horizon Strategy," *IEEE Trans. Smart Grid*, vol. 4, no. 2, pp. 996-1006, 2013.

[5] D. E. Olivares, A. Mehrizi-Sani, A. H. Etemadi, C. A. Cañizares, R. Iravani, M. Kazerani, A. H. Hajimiragha, O. Gomis-Bellmunt, M. Saeedifard, R. Palma-Behnke, G. A. Jiménez-Estévez and N. D. Hatziargyriou, "Trends in Microgrid Control," *IEEE Trans. Smart Grid*, vol. 5, pp. 1905-1919, July 2014.

[6] R. Bhoyar and S. Bharatkar, "Potential of MicroSources, Renewable Energy sources and Application of Microgrids in Rural areas of Maharashtra State India," *Energy Procedia*, vol. 14, pp. 2012-2018, 2012.

[7] J. M. Carrasco, L. G. Franquelo, J. T. Bialasiewicz, E. Galvan, R. C. P. Guisado, M. A. M. Prats, J. I. Leon, and N. Moreno-Alfonso, "Power Electronic Systems for the Grid Integration of Renewable Energy Sources: A Survey," *IEEE Trans. Ind. Electron.*, vol. 53, pp. 1002–1016, June 2006.

[8] D. I. Brandao, P. Tenti, T. Caldognetto, and S. Buso, "Control of Utility Interfaces in Low-voltage Microgrids," *Brazilian J. Power Electron. – SOBRAEP*, vol. 20, no. 4, pp. 373-382, Nov. 2015.

[9] G.-H. Kim, C. Hwang, J.-H. Jeon, J.-B. Ahn and E.-S. Kim, "A Novel Three-phase Four-leg Inverter Based Load Unbalance Compensator for Stand-alone Microgrid," *Elec. Power and Energy Syst.*, vol. 65, p. 70–75, February 2015.

[10] M. Singh, V. Khadkikar, A. Chandra, R.K. Varma. "Grid Interconnection of Renewable Energy Sources at the Distribution Level with Power-Quality Improvement Features". *IEEE Trans. Power Del.*, vol.26, no.1, pp.307-315, January 2011.

[11] F. Marafão, D. Brandão, A. Costabeber, H. K. M. Paredes, "Multi-task Control Strategy for Grid-tied Inverters Based on Conservative Power Theory," *IET Renew. Power Gen.*, vol. 9, pp. 154-165, February 2015.

[12] J. P. Bonaldo, H. K. M. Paredes, J. A. Pomilio, "Control of Single-Phase Power Converters Connected to Low-Voltage Distorted Power Systems with Variable Compensation Objectives", *IEEE Trans. Power Electron.*, vol. 31, pp. 2039-2052, March 2016.

[13] J. He, Y. W. Li, F. Blaabjerd, X. Wang, "Active Harmonic Filtering Using Current-controlled, Grid-connected DG Units with Closed-loop Power Control," *IEEE Trans. Power Electron.*, vol. 29, pp. 642-653, February 2014.

[14] J. He, Y. W. Li, M. S. Munir, "A Flexible Harmonic Control Approach Through Voltage Controlled DG-grid Interfacing Converters," *IEEE Trans. Ind. Electron.*, vol. 59, pp. 444-455, January 2012.

[15] F. Z. Peng, G. O. Jr., D. J. Adams, "Harmonic and Reactive Power Compensation Based on the Generalized Instantaneous Reactive Power Theory for Three-Phase Four-Wire Systems," *IEEE Trans. Power Electron.*, vol. 13, November 1998.

[16] H. Akagi, Y. Kanazawa, A. Nabae, "Instantaneous Reactive Power Compensator Comprising Switching Devices Without Energy Storage Components," *IEEE Trans. Ind. Appl.*, vol. IA-20, pp. 625-630, May 1984.

[17] E. H. Watanabe, M. Aredes, J. L. Afonso, J. G. Pinto, L. F. C. Monteiro, H. Akagi, "Instantaneous p–q Power Theory for Control of Compensators in Micro-grids," *in Proc.* 2010 International School on Nonsinusoidal Currents and Compensation, Lagow, 2010, pp. 17-26.

[18] Y. Sun et al, "Microgrid Tie-line Power Fluctuation Mitigation with Virtual Energy Storage," *The Journal of Engineering*, vol. 2019, no 16, pp. 1001-1004, April 2019.

[19] Z. Miao, L. Fan, D. Osborn, S. Yuvarajan, "Control of DFIG-Based Wind Generation to Improve Interarea Oscillation Damping," *IEEE Trans. Energy Conv.*, vol. 24, pp. 415-422, June 2009.

[20] L. Fan, Z. Miao, "An Explanation of Oscillations Due to Wind Power Plants Weak Grid Interconnection," *IEEE Trans. Sust. Energy*, vol. 9, pp. 488-490, January 2018.

[21] D. Gautam, V. Vittal, R. Ayyanar, T. Harbour, "Supplementary Control for Damping Power Oscillations Due to Increased Penetration of Doubly fed Induction Generators in Large Power Systems", *in Proc.* IEEE/PES Power Sys. Conf. Exp., PSCE 2011.

[22] F. Z. Peng, "Harmonic Sources and Filtering Approaches," *IEEE Ind. Appl. Mag.*, vol. 7, pp. 18-25, July 2001.

[23] P. Tenti, H. K. Morales-Paredes, P. Mattavelli, "Conservative Power Theory, a Framework to Approach Control and Accountability Issues in Smart Microgrids," *IEEE Trans. Power Electron.*, vol. 26, pp. 664-673, March 2011.

[24] P. Tenti, P. Mattavelli and H. K. Morales, "Conservative Power Theory, Sequence Components and Accountability in Smart Grids," *Prz. Elektrotech*, vol. 6, pp. 30-37, 2010.

[25] F. P. Marafão, D. I. Brandão, F. A. S. Gonçalves, and H. K. Morales Paredes, "Decoupled Reference Generator for Shunt Active Filters Using the Conservative Power Theory," *J. Control Autom. Electr. Syst*, pp. 522–534, August 2013.

[26] D. I. Brandão, H. K. Morales-Paredes, A. Costabeber, F. P. Marafão, "Flexible Active Compensation Based on Load Conformity Factors Applied to Non-Sinusoidal and Asymmetrical Voltage Conditions", *IET Power Electron.*, vol. pp. 1-9, February 2015.

[27] H. Akagi, E. H. Watanabe and M. Aredes, Instantaneous Power Theory and Applications to Power Conditioning, New Jersey, 2007.

[28] Yi Tang, P.C. Loh, Peng Wang, F.H. Choo, F. Gao. "Generalized Design of High Performance Shunt Active Power Filter with Output LCL Filter", *IEEE Trans. Ind. Electron.*, vol.59, March 2012.

[29] M. G. Villalva, J. R. Gazoli, E. R. Filho, "Modeling and Control of a Three-phase Isolated Grid-connected Converter fed by a Photovoltaic Array," *in Proc.* Brazilian Power Electronics Conference, Dec. 2009.

[30] S. Buso, P. Mattavelli, "Digital Control in Power Electronics", First edition, Morgan & Claypoo, United States of America, 2006.

[31] R. Teodorescu, F. Blaabjerg, M. Liserre, P. C. Loh, "Proportional-Resonant Controllers and Filters for Grid-connected Voltage-source Converters," *Electric Power Appl., IEE Proceedings*, vol. 153, no.5, pp.750-762, September 2006.

Comparative study of different correction methods to analyze wind turbine performance

Hércules Araújo Oliveira
Institute of Electrical Engineering
Federal University of Maranhão
São Luís, Brazil
hercules.oli@hotmail.com

Luiz Antonio de Souza Ribeiro
Institute of Electrical Engineering
Federal University of Maranhão
São Luís, Brazil
l.a.desouzaribeiro@ieee.org

Jerson Rogério Pinheiro Vaz
Faculty of Mechanical Engineering
Federal University of Pará
Belém, Brazil
jerson@ufpa.br

Osvaldo Ronald Saavedra
Institute of Electrical Engineering
Federal University of Maranhão
São Luís, Brazil
o.saavedra@ieee.org

José Gomes de Matos
Institute of Electrical Engineering
Federal University of Maranhão
São Luís, Brazil
gomesdematos@ieee.org

Abstract — **In this work, a comparative study of different correction models applied to the Blade Element Moment Theory (BEM) is carried out, in order to analyze the performance of horizontal-axis wind turbines. Although, BEM is the best-known method for this type of analysis, it still needs corrections regarding the effects of tip and root vortices, high induction factors and aerodynamic stall on the turbine blades. Thus, to evaluate different correction models, experimental data of MOD-0 turbine are used. A critical and comparative analysis is performed between such corrections, identifying the most appropriate approach for predicting the performance of horizontal-axis wind turbines.**

Keywords — *wind turbine, BEM method, correction factors, turbine performance.*

I. INTRODUCTION

Identification and analysis of factors that influence the performance of wind turbines are very important to estimate energy extraction considering turbine modeling under different environmental conditions [1].

The generation of electric energy from wind turbines occurs due to the air flow passing through the wind rotor, transforming kinetic energy transported by the flow into shaft mechanical energy. An electric generator, in which the turbine shaft is coupled, transforms mechanical energy into electric. The kinetic energy of the wind is not completely converted into mechanical energy. This energy conversion is penalized by a power coefficient (C_p), which can be understood as the mechanical efficiency of the turbine rotor. Similarly, it is considered an efficiency η for the electric set (electric generator + power converter) in the conversion of mechanical energy into electric [2]. These conversion steps are illustrated in Fig. 1.

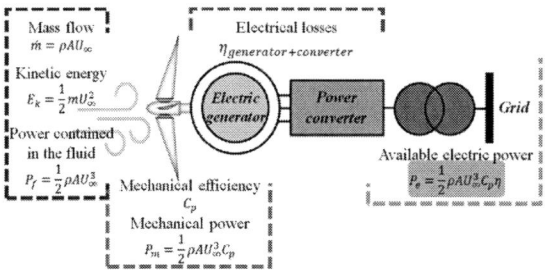

Fig. 1. Illustration of the complete wind turbine system.

According to Albert Betz, the maximum theoretical C_p is 59.26 % (usually called Betz Limit), a condition in which the axial induction factor is $a = 1/3$, [3]. However, in practice, high performance turbines operate with a C_p of approximately 0.45, due to losses along the system [1] [2]. Hence, studies on turbine rotor designs are still conducted, being the main objective improvements on the mechanical performance.

This paper analyzes the performance of horizontal-axis wind turbines based on Blade Element Moment Theory (BEM), by comparing different correction methods available in the literature [2] [3]. BEM theory is the most frequently used model for designing and assessment of wind rotors. This method is essentially an integral method, with semi-empirical information from aerodynamic forces in airfoil sections issued from two-dimensional flow model or experimental data, widely used by designers and engineers.

The aim of this paper is to evaluate the most appropriate approach to be used with BEM. Through a case study, critical and comparative analysis are carried out, in order to identify the best correction model to be used within BEM. This is important because the classical BEM method is not valid for all types of analysis, requiring complementary mathematical models capable of representing correctly the performance of a wind turbine. Thus, here mathematical models are combined with the classical BEM method. The results are compared and validated using the experimental data of the MOD-0 turbine data provided by NASA [3].

This work is part of a larger project, aiming an integrated modeling (hydraulic-mechanical-electrical-conversion) with rotor, electric generator and power converter, allowing maximum power point tracker (MPPT).

II. MATHEMATICAL MODELING

A. BEM method

BEM is the result of a combination of two theories, axial momentum theory and blade element theory. In these analyzes, a free, one-dimensional and incompressible flow is considered. In one-dimensional analysis, the energy transportation occurs on both directions, axial and rotational, based on studies of the conservation laws (mass, moment and energy quantities) [1] [2]. The result of this analysis is the simple momentum theory, in which the turbine thrust dT is defined through (1), while the torque dQ is given by (2). These equations are obtained applying Bernoulli equation in sections 1 to 4 of Fig. 2.

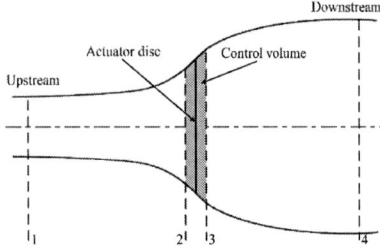

Fig. 2. Control volume of an ideal wind turbine.

$$dT = 4\rho U_\infty^2 a(1-a)\pi r dr \tag{1}$$

$$dQ = 4\rho U_\infty \Omega a'(1-a)\pi r^3 dr \tag{2}$$

where ρ is the density of the fluid, U_∞ the velocity of the free upstream fluid, a e a' the axial and rotational induction factors, respectively, r the local radius of the ideal turbine and Ω is the rotational speed of the rotor.

For the blade element theory, the energy transportation is based on studies of fluid mechanics and aerodynamic. The resulting aerodynamic forces are calculated mainly by the blade geometry and the angle of attack at each blade section, taking into account the Reynolds number [1] [2].

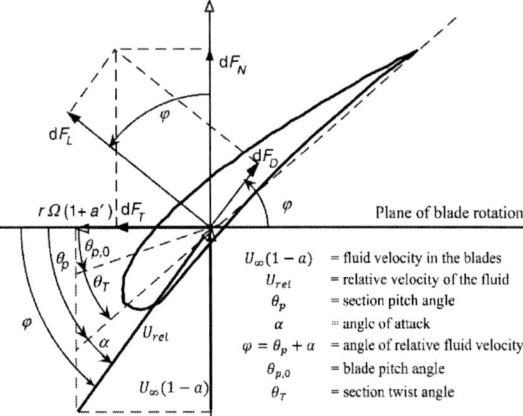

Fig. 3. Wind velocity diagram based on the blade element theory [1].

From Fig. 3, we define φ in (3) and the force components in (4) and (5).

$$tan(\varphi) = \frac{U_\infty(1-a)}{\Omega r(1+a')} \tag{3}$$

$$dF_L = \frac{1}{2}C_l \rho U_{rel}^2 c dr \tag{4}$$

$$dF_D = \frac{1}{2}C_d \rho U_{rel}^2 c dr \tag{5}$$

where φ is the flow angle formed between the plane of rotation and the flow relative velocity, dF_L and dF_D are drag and lift forces acting on each blade section, where C_l and C_d are lift and drag coefficients, respectively, and c is the sectional chord.

In (6) and (7) the resulting thrust and torque are defined, which are derived from equations (3) to (5), as well as from velocity and force components shown in Fig. 3 [1].

$$dF_N = \sigma'\pi \left[\rho \left(\frac{U_\infty(1-a)}{sen(\varphi)}\right)^2 C_n r dr\right] \tag{6}$$

$$dQ = \sigma'\pi \left[\rho \left(\frac{U_\infty(1-a)}{sen(\varphi)}\right)^2 C_t r^2 dr\right] \tag{7}$$

$$\sigma' = \frac{Bc}{2\pi r} \tag{8}$$

$$C_n = C_l cos(\varphi) + C_d sen(\varphi) \tag{9}$$

$$C_t = C_l sen(\varphi) - C_d cos(\varphi) \tag{10}$$

where σ' is the local solidity, defined in (8), B is the number of blades; C_n and C_t are the coefficients of normal and tangential forces, respectively shown in (9) and (10).

Prandtl's tip-loss factor

The easiest extension of BEM to consider finite number of blades is through the use of a tip-loss factor, F. Okulov and Sørensen [4] extended the Betz limit to finite number of blades using Prandtl's approximation [5]. Thus, Prandtl's tip-loss factor F is formally defined as the ratio between the actual bound circulation and that of a rotor with infinite number of blades. Prandtl's tip-loss factor is the most commonly-used form of F [6], [7]. Blade element calculations made with Prandtl's model show good agreement with test data [8] at high tip speed ratio, however, Wood et al. [9] showed that it is inaccurate at low tip speed ratio (less than 1.0) and does not reproduce the increase in F above unity near the axis for any tip speed ratio. In this work, all simulations are done for tip speed ratios higher than 1.0. The Prandtl's correction factor, for the tip region (11) and for the root region (12), [3] [10],where R is the nominal radius of the rotor and r_{hub} the radius of the root of the blades.

$$F_{tip} = \left(\frac{2}{\pi}\right)cos^{-1}\left[exp^{-\left(\frac{B}{2}\frac{(R-r)}{(rsen(\varphi))}\right)}\right] \tag{11}$$

$$F_{hub} = \left(\frac{2}{\pi}\right)cos^{-1}\left[exp^{-\left(\frac{B}{2}\frac{(r-r_{hub})}{(r_{hub}sen(\varphi))}\right)}\right] \tag{12}$$

The total losses from vortices are represented by (13). In some blades, the geometry is uniform up to the rotor body, so F_{hub} can be disregarded [1] [2] [3] [10].

$$F_{loss} = F_{tip}F_{hub} \tag{13}$$

The combination of the one-dimensional and two-dimensional analysis is based on the equality of the thrust axial forces, in (1) and (6), and of the torque, in (2) and (7). From this equality the equations of the axial (14) and rotational (15) induction factors are derived, considering the losses F_{loss}, [1] [2] [3] [10].

$$a = \frac{1}{1 + \frac{F_{loss}4sen^2(\varphi)}{\sigma'C_l cos(\varphi)}} \tag{14}$$

$$a' = \frac{1}{\frac{F_{loss}4cos(\varphi)}{\sigma'C_l} - 1} \tag{15}$$

Glauert's Correction Method

According to the empirical relation of Glauert, under turbulent wake conditions ($a > 0.5$), the thrust coefficient C_T

increases until approximately 2.0 for $a = 1$, [1]. This correction is rather important because the classical BEM model fails for $a > 0.4$. In this case, this correction is given by (16) and (17), valid for $a > 0.4$ or $C_T > 0.96$. Otherwise, if $C_T < 0.96$, the calculation of a and a' should be determined using (14) and (15), [1].

$$C_T = \frac{\sigma'(1-a)^2 C_n}{sin^2(\varphi)} \quad (16)$$

$$a = \left(\frac{1}{F_{loss}}\right)(0.143 + \sqrt{0.0203 - 0.6427(0.889 - C_T)}) \quad (17)$$

Classical structure of the BEM method

The structure of classical BEM method is further detailed in [4]. Here, it can be summarized by the following sequence:

I   Defines the input parameters: tolerance and characteristics of the fluid, rotor (RPM, internal and external diameters) and blade (profile, chord and twist).

II Divide the blade into x parts for analysis in each region (consistent with the geometry of the blade).

III For each section of the blade:

Defines an arbitrary value for a and a';

Calculate the fluid velocity components;

Calculates φ and then α;

Calculates the tip loss correction (Prandtl method);

Calculates C_n and C_t;

Calculates the correction for high values of a (Glauert's methods);

Calculates the error. Only if the error is less than the tolerance, the calculations in the next section will be executed.

IV Calculates the integral of forces acting on the area of each section.

V   Calculates the total values of thrust, torque and power.

B. Mathematical models complementary to the classical BEM method

Corrections to the tip-loss

In addition to the method applied to the tip and root, Prandtl proposed together with Betz two other methods (18) and (19), [11]. Where λ and λ_r are the speed ratio of the nominal and local tip respectively.

$$F_{loss} = \frac{2}{\pi} acos\left(exp\left(-\frac{B}{2}\left(1 - \frac{\lambda_r}{\lambda}\right)\sqrt{1 + \lambda^2\left(\frac{1 + a'(R)}{1 - a(R)}\right)^2}\right)\right) \quad (18)$$

$$F_{loss} = \frac{2}{\pi} acos\left(exp\left(-\frac{B}{2}\left(1 - \frac{\lambda_r}{\lambda}\right)\sqrt{1 + \left(\frac{\lambda}{1 - a(R)}\right)^2}\right)\right) \quad (19)$$

Burton [2] proposed a method similar to that of Prandtl which is exposed in (20), [11].

$$F_{loss} = \frac{2}{\pi} acos\left(exp\left(-\frac{B}{2}\left(\frac{\lambda}{\lambda_r} - 1\right)\sqrt{1 + \left(\frac{\lambda_r}{1 - a}\right)^2}\right)\right) \quad (20)$$

The simplest method to compensate for tip-loss is by reducing the Effective Radius of the blade [3]. This reduction is generally about 3 %, and it can be implemented by (21) and (22), where $r_e = 0.97R$.

$$F = 1 \quad if \ 0 < r < r_e \quad (21)$$
$$F = 0 \quad if \ r_e \leq r \leq R \quad (22)$$

Goldstein proposed a method that promises to be more accurate than the Prandtl approximation, but it is little used because of the complexity, in which parameter information characteristic of the vortex region is required. Due to the absence of this information, the Goldstein method was not implemented in this work, but it can be found in [12].

Correction for high values of induction factor

The mathematical models for correction of BEM, when working with high values of the axial induction factor, are usually based on empirical relations based on the preliminary study done by Glauert [1] [2] [3] [10]. These methods are described below.

Method described by Peter Smith [13] [14], for $a > 0.5$, can be implemented by (23) to (25).

$$a = 1 - \sqrt{\frac{f_t}{F_t}} \quad (23)$$

$$f_t = \frac{1}{\left[0.11\left(\frac{1}{F_t}\right)^3 - 0.70\left(\frac{1}{F_t}\right)^2 + 2.15\left(\frac{1}{F_t}\right) + 2.15\right]} \quad (24)$$

$$F_t = \frac{a}{(1 - a)} \quad (25)$$

Method described by Robert Wilson [15] [12], for $a > 0.4$, which can be implemented by (26) and (27).

$$S = \frac{\sigma' C_l cos(\varphi)}{8 sin^2(\varphi)} \quad (26)$$

$$a = \frac{2S + F_{loss} - \sqrt{F_{loss}^2 + 4SF_{loss}(1 - F_{loss})}}{2(S + F_{loss}^2)} \quad (27)$$

Method described by Robert Wilson and Spera [10] [3], for $a > a_c$, where $a_c = 0,2$, use (28) and (29).

$$K = \frac{4F_{loss} sin^2(\varphi)}{\sigma' C_n} \quad (28)$$

$$a = \frac{1}{2}\left[2 + K(1 - 2a_c) - \sqrt{(K(1 - 2a_c) + 2)^2 + 4(Ka_c^2 - 1)}\right] \quad (29)$$

Method described by Marshall Buhl [16] [17], for $a > 0.4$ or equivalent $C_T > 0.96$, use (30) and (31).

$$C_T = \frac{8}{9} + \left(4F_{loss} - \frac{40}{9}\right)a + \left(\frac{50}{9} - 4F_{loss}\right)a^2 \quad (30)$$

$$a = \frac{18F_{loss} - 20 - 3\sqrt{C_T(50 - 36F_{los}) + 12F_{loss}(3F_{loss} - 4)}}{36F_{loss} - 50} \quad (31)$$

Method described by Madsen [11], for $C_T < C$, which uses (32), and for $C_T > C$, (33) is used, where $C = 2.5$.

$$a = k_0 + k_1 C_T + k_2 C_T^2 + k_3 C_T^3 \qquad (32)$$

$$a = (k_1 + 2Ck_2 + 3Ck_2^2)(C_T - C) + k_0 + 2{,}5k_1 C + k_2 C^2 \\ + k_3 C^3 \qquad (33)$$

The coefficients of (32) and (33) are given in (34).

$$k_0 = -0.001701 \quad k_1 = 0.251163 \\ k_2 = 0.054496 \quad k_3 = 0.089207 \qquad (34)$$

Correction of high angle of attack

The blades of the turbine rotor can be subjected to high angle of attack when the fluid velocity is also high. This is because the angle φ is directly proportional to the velocity of the fluid (see equation (3)), as well as the angle α, since a constant pitch angle is assumed (see Fig. 3). In this condition, the blade is subjected to the stall phenomenon, characterized by a decrease in aerodynamic efficiency (decrease in lift and increase in drag). The classical BEM is not able to properly represent the resulting aerodynamic forces in the post-stall region and, in general, the polar data (aerodynamic coefficients) are available for a small range of angle of attack, $-15^o < \alpha < 15^o$, approximately. Regarding stall on turbine blades, the best known correction method is the one proposed by Viterna and Corrigan in (35) and (36), [3] [13] [11].

$$C_l = A_1 \sin(2\alpha) + A_2 \left(\frac{\cos^2(\alpha)}{\sin(\alpha)} \right) \qquad (35)$$

$$C_d = B_1 \sin^2(\alpha) + B_2 \cos(\alpha) \qquad (36)$$

where A_1, A_2, B_1 and B_2 are defined in (37) to (40).

$$A_1 = \frac{B_1}{2} \qquad (37)$$

$$A_2 = C_{ls} - C_{dmax} \sin(\alpha_s) \cos(\alpha_s) \left(\frac{\sin(\alpha_s)}{\cos^2(\alpha_s)} \right) \qquad (38)$$

$$B_1 = C_{dmax} \qquad (39)$$

$$B_2 = \frac{C_{ds} - C_{dmax}\sin^2(\alpha_s)}{\cos(\alpha_s)} \qquad (40)$$

where C_{ls} and C_{ds} are lift and drag coefficients given in the stall angle, α_s. The maximum drag coefficient C_{dmax} is calculated for $\alpha = 90^o$, where it is assumed that the flow of one side of the profile is totally separated. Viterna and Corrigan, and Montgomerie and Radkey proposed equations for the calculation of this coefficient in (41), (42) and (43), respectively [18].

$$C_{dmax} = 1.11 + 0.018\mu \qquad (41)$$

$$C_{dmax} = 1.98 - 0.81 \left(1 - \exp\left(-\frac{20}{\mu} \right) \right) \qquad (42)$$

$$C_{dmax} = 1.98 - 0.81 \tanh\left(\frac{12.22}{\mu} \right) \qquad (43)$$

where μ is the aspect ratio between the radius and the blade chord, defined in (44).

$$\mu = \frac{R - r_{hub}}{c} \qquad (44)$$

III. Results and Discussions

The mathematical models are analyzed separately for each type of correction previously discussed: tip-loss, axial induction factor and angle of attack. The MOD-0 turbine data are used to validate the combinations of the complementary methods to the classic BEM. The characteristics of the MOD-0 turbine are: 2 blades, rated power of 100 kW, $C_p = 0.37$ approximately, NACA 23000 series, primary rotor nominal speed of 27 rpm, pitch angle of 0°, rotor and hub diameters are 38 m and 7.84 m, respectively. The distribution of the chord is 1.96 m for $r/R = 0.234$, varying linearly up to 0.67 m for $r/R = 1$, [11] [19].

A. Correction for the tip-loss

Fig. 4 shows the performance of MOD-0 turbine, which is compared using different tip-loss corrections. It is observed that, in this case, the methods proposed by Burton and the two methods of Prandtl and Betz are not able to represent tip-losses for wind speeds below 9 m/s. In this velocity region the power is higher than the MOD-0 data, which leads to the conclusion that torque and lift forces on the blades are larger than the actual data, overestimating the turbine power. Prandtl and Effective Radius methods are similar. However, in the velocity range of 8 to 9 m/s, the Effective Radius method presents a greater error than that of Prandtl. Thus, Prandtl approximation is shown to be closer to the actual data power curve.

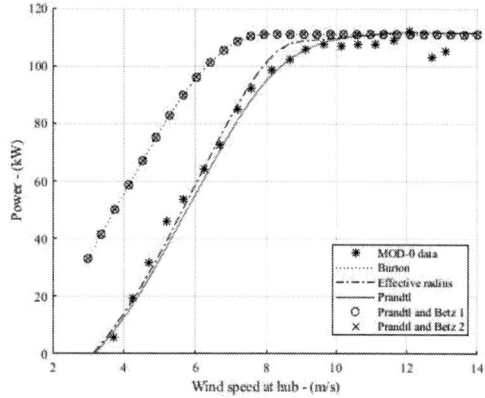

Fig. 4. Power curve as a function of tip-loss correction models.

In Fig. 5, it can be seen that the methods proposed by Burton and the two Prandtl and Betz methods are also not able to properly represent C_p. For these methods, Betz limit is exceeded, demonstrating unphysical behavior. Although the Effective Radius and Prandtl methods present similar curves of C_p, the Prandtl curve represents the best tip correction because it is closer to the real value of C_p, considering the real optimum point of the curve [11] and [19].

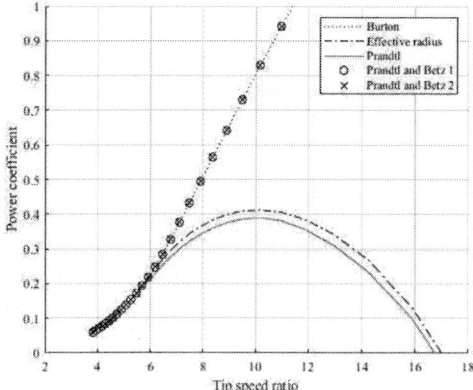

Fig. 5. C_p curve as a function of tip-loss correction models.

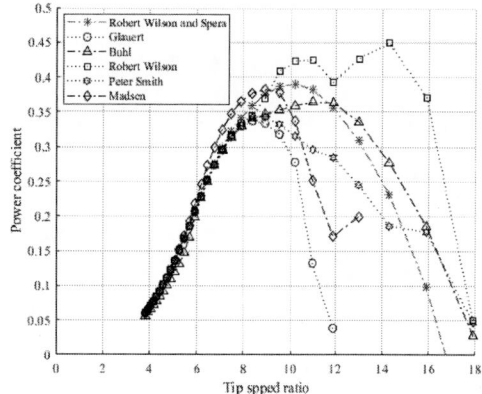

Fig. 7. C_p curve as a function of induction factor correction models.

B. Correction for high values of induction factor

Fig. 6 depicts the MOD-0 turbine performance, which is compared using the correction methods for high axial induction factors. It is observed that the methods proposed by Glauert, Peter Smith and Madsen present bigger errors than the other ones for the wind speed range of 2 to 6 m/s. Madsen errors are also higher in the wind speed range from 8 to 9 m/s. Buhl method presents a greater error than the other methods for wind speed range of 9 to 14 m/s. The methods of Robert Wilson and Robert Wilson with Spera are the ones closest to the actual data power curve.

C. Correction for high angle of attack

In Fig. 8, the turbine performance is compared for different angle of attack corrections. It is shown that the approaches proposed by Viterna and Corrigan, Hibbs and Radkey, and Bjorn Montgomerie are practically equal and close to the real curve. This is due to the fact that all these angle of attack corrections are based on the post-stall correction proposed by Viterna and Corrigan. Montgomerie and Radkey proposed the correction only for the C_{dmax}, which takes into account the drag of the nacelle and the blade as a function of the geometry. Therefore, any of these three methods can be used to correct the angle of attack during post-stall, since the drag effect does not differ significantly between these methods. It is also observed that only the use of an interpolation for C_l and C_d as a function of α, introduces great errors in the power curve of the turbine, being it not sufficient for the performance calculation.

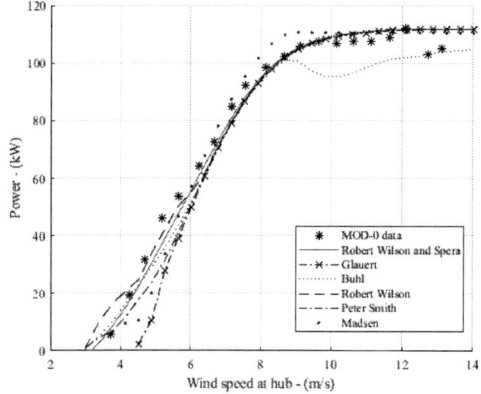

Fig. 6. Power curve as a function of induction factor correction models.

In Fig. 7, it can be seen that the methods proposed by Glauert, Robert Wilson, Peter Smith and Madsen are not able to correctly represent C_p. This occurs because the curve of C_p must be smooth (no steps), since it is the ratio between the calculation of two powers for the same speed range, considering the same wind properties (see Fig. 1). The methods of Robert Wilson and Spera, and of Buhl are those that best represent the curves of C_p. However, that of Robert Wilson and Spera becomes the one that best represents the correction of the axial induction factor, considering the real optimum point of C_p.

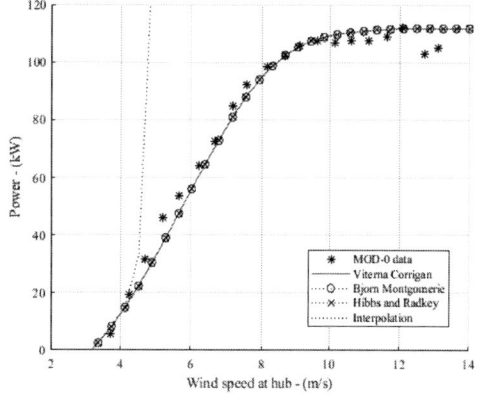

Fig. 8. Power curve as a function of angle of attack correction models.

Through Fig. 9, it can be seen that the methods proposed by Viterna and Corrigan, Hibbs and Radkey, and Bjorn Montgomerie maintain the good agreement for the C_p curve.

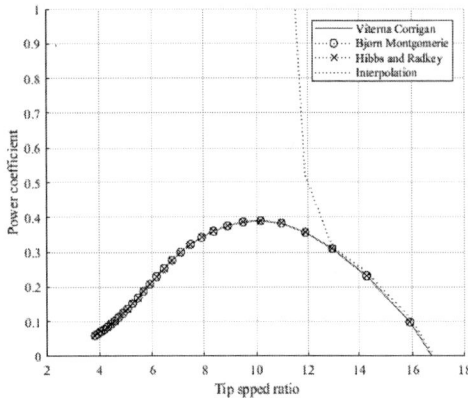

Fig. 9. C_p curve as a function of angle of attack correction models.

IV. CONCLUSIONS

An assessment of different correction methods applied to the classical BEM theory is presented. The results confirm that BEM is not enough to analyze the performance of a turbine as it is, requiring corrections. These corrections are implemented to analyze tip and root loss effects, high values of the axial induction factor and high angle of attack during post-stall.

Based on the comparisons, the BEM method combined with Prandtl (tip and root loss), Robert Wilson and Spera corrections (high values of the axial induction factor), and Viterna and Corrigan (high angle of attack), presents a good prediction of the performance of horizontal-axis turbines when compared to other methods available. The correction of C_{dmax} can be done by the methods of Viterna and Corrigan, Hibbs and Radkey, and Bjorn Montgomerie, because the results are approximately equal.

ACKNOWLEDGMENT

The authors would like to thank the support and motivation provided by the Federal University of Maranhão (UFMA), CEMAR, FAPEMA, CNPq – Brazil, as well as MEIHAPER CYTED, PROCAD/CAPES (Agreement: 88881.200549/2018-01), and PROPESP/UFPA.

REFERÊNCIAS

[1] J. F. a. M. J. G. Manwell, Wind Energy Explained: Theory, Design and Application, 2nd ed., 2th Edition ed., Chippenham, Wiltshire: John Wiley & Sons Ltd, 2009.

[2] T. Burton, D. Sharpe and N. a. B. E. Jenkins, Wind Energy Handbook, 1th Edition ed., Chichester, West Sussex: John Wiley & Sons, Ltd, 2001.

[3] D. A. Spera, Wind Turbine Technology Fundamental Concepts of Wind Turbine Engineering, New York: ASME Press, 2009.

[4] V. Okulov and J. Sørensen, "Refined Betz limit for rotors with a finite number of blades," *Wind Energy*, pp. 415-26, 2008.

[5] B. A. Prandtl L, *Four essays on the hydrodynamics and aerodybamics*, Göttinger Nachr, Göttingen, 1927.

[6] Q. Wald, "The aerodynamics of propellers," *Prog Aerosp Sci*, vol. 42, pp. 85-128, 2006.

[7] J. Vaz and W. DH, "Performance analysis of wind turbines at low tip-speed ration using the Betz-Goldstein model," *Energy Conversion and Management*, no. 126, pp. 662-672, 2016.

[8] R. Wilson, Aerodynamic behavior of wind turbines, New York, USA: Spera DA. Wind turbine techonology. ASM Press, 2009.

[9] D. Wood, V. Okulov and D. Bhattacharjee, "Direct calculatio od wind turbine tip loss," *Renew Energy*, pp. 95:269-76, 2016.

[10] M. O. L. Hansen, Aerodynamics of Wind Turbines, 3th Edition ed., London and New York: Routledge, 2015.

[11] E. Branlard, Wind Turbine Aerodynamics and Vorticity-Based Methods: Fundamentals and Recent Applications, vol. 7, Cham, Switzerland: Springer International Publishing, 2017.

[12] R. E. Wilson and P. B. S. a. W. S. N. Lissaman, Aerodynamic Performance of Wind Turbines, Oregon: National Renewable Energy Laboratory, 1976.

[13] C. Bak, P. Fuglsang, N. N. Sørensen and H. A. Madsen, *Airfoil Characteristics for Wind Turbines*, Roskilde: Risø National Laboratory, 1999.

[14] P. Smith, *Aerodynamic Theory: A General Review of Progress – Vol. IV. Applied*, Peter Smith Publisher, Inc., 1976.

[15] E. R. Wilson and S. B. P. Lissaman, Applied Aerodynamics of Wind Power Machines, Oregon: Oregon State University and Science National Foundation, 1974.

[16] P. a. H. A. Moriarty, AeroDyn Theory Manual, Golden, Colorado: National Renewable Energy Laboratory, 2005.

[17] J. Marshall L. Buhl, *A New Empirical Relationship between Thrust Coefficient and Induction Factor for the Turbulent Windmill State*, Golden: National Renewable Energy Laboratory, 2005.

[18] C. Lindenburg, *Stall Coefficients - Aerodynamic airfoil coefficients at large angeles of attrak*, Colorado, 2001.

[19] J. R. P. Vaz, J. T. Pinho and A. L. A. Mesquita, "An extension of BEM method applied to horizontal-axis wind turbine design," *Renewable Energy*, pp. 1734-1740, 8 January 2011.

978-1-7281-4181-7/19 $31.00 © 2019 IEEE

Nonisolated Quadratic SEPIC Converter Without Electrolytic Capacitors for LED Driver Applications

Douglas Rosa Corrêa,
Faculty of Eletrical Engeneering
Federal University of Uberlândia
Uberlândia, Brazil
drcorrea@ufu.br

Aniel Silva de Morais
Faculty of Eletrical Engeneering
Federal University of Uberlândia
Uberlândia, Brazil
aniel@ufu.br

Fernando Lessa Tofoli
Department of Eletrical Engeneering
Federal University of São João del-Rei
São João del-Rei, Brazil
fernandolessa@ufsj.edu.br

Abstract—**This work presents an ac-dc quadratic Single-Ended Primary Inductance Converter (SEPIC) to provide input power factor correction (PFC) for light-emitting diode (LED) driver applications. This structure allows obtaining wide conversion range and operation in either step-up or step-down. Considering that the lifetime of LEDs is much longer than that of electrolytic capacitors, a prominent advantage on the proposed converter lies in the use of only film counterparts. The integration of two SEPIC stages using the graft technique allows independent PFC and power control (PC) while using a single active switch. A detailed mathematical study is presented in terms of a consistent qualitative and quantitative analysis. Experimental results are also discussed to demonstrate the claimed advantages and validate the theoretical assumptions.**

Keywords — **dc-dc converters, LED drivers, power factor correction, SEPIC converter.**

I. INTRODUCTION

The increasing demand for electricity has led to the need to improve conversion efficiency and many other related aspects. It is worth mention that a considerable amount of the generated electricity around the world is used for lighting purposes. In this context, light-emitting diodes (LEDs) have drawn significant attention due to increased efficiency and long service life. According to [1], phosphorus-based LEDs can produce up to 160 lm/W and normal life of 50,000 hours.

A good lighting system should be concerned with both the power quality in terms of the current drawn from the grid and luminous efficacy. LEDs often require a specific voltage and current range for its operation, which can be obtained using power electronic converters as drivers. Besides, according to [2-4], power factor correction is an essential feature associated with LED drivers in order to keep reduced total harmonic distortion (THD) of the input current. A general representation of a LED driver is presented in Fig. 1, which must meet the load requirements while maintain good power quality indices with long useful life. One of the obstacles found in this context lies in the use of electrolytic capacitors, which are found to be the main component associated with failures [5]. According to [6], electrolytic capacitors are sensitive to large variations in temperature, current and operating voltage, thus causing incompatibility between the driver and LED lifetime. In order to avoid this problem, several alternatives have been studied, while one the main solutions is the replacement of electrolytic capacitors by film counterparts [7]. However, such component has lower energy density, lower capacitance ratings for a same volume and also higher acquisition cost, as it is necessary to reduce the capacitance used in the driver design.

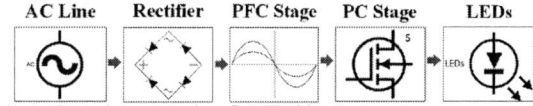

Fig. 1. General representation of a LED driver.

Quadratic converters can be obtained from the cascade association of two topologies while sharing the same switch. The main reason for integration is to reduce costs and simplify the control circuit [8]. In the case of LED drivers, it is possible to perform PFC in the first stage, while power control (PC) is associated with the second one. Besides, it is also possible to reduce the filter capacitances as consequence. In this context, this paper proposes a quadratic Single-Ended Primary Inductance Converter (SEPIC), whose qualitative and quantitative analysis is presented in detail to obtain a consistent design procedure. A control circuit is also proposed to reduce the output current ripple. Experimental results are also obtained and thoroughly discussed in order to support the theoretical assumptions.

II. DESCRIPTION OF PROPOSED TOPOLOGY

Among the various types of high-frequency switched power converters, the SEPIC topology is adequate for LED-related applications because it is possible to achieve high-voltage step-up or step-down depending on the value of the duty cycle. Other prominent advantages include the continuous nature of the input current in continuous conduction mode (CCM) and the fact that the input source and load are connected to a same reference with the a same polarity associated with the respective voltages. Let us consider the general structure shown in Fig. 1, where PFC and PC (power control) circuit are performed by two distinct converters. Two SEPIC converters can be used for such purpose, resulting in the structure represented in Fig. 2.

The graft scheme was introduced in [9], which allows obtaining an alternative single-switch representation as long as there is a common connection node for two active switches. In this case, the T-type configuration exists and only one switch with two additional diodes D_{n1} and D_{n2} can be used as in Fig. 3. Thus, the quadratic SEPIC converter results as shown in Fig. 4.

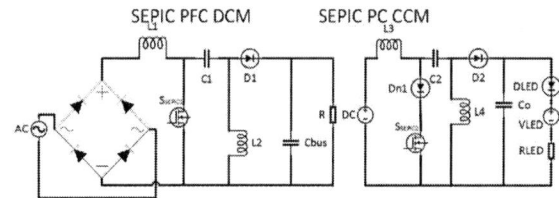

Fig. 2. Cascade association of two SEPIC converters.

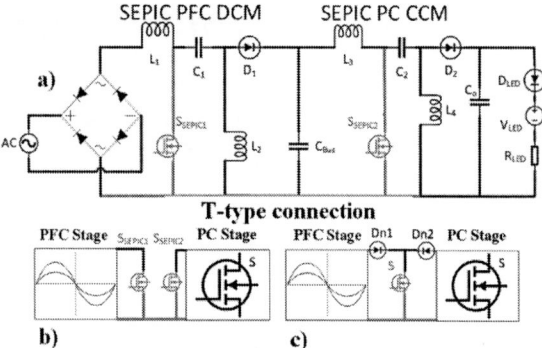

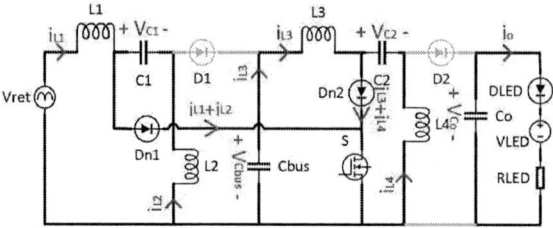

Fig. 3. (a) Common node between the two active switches, (b) T-type connection, and (c) resulting single-switch configuration.

Fig. 5. First operating stage.

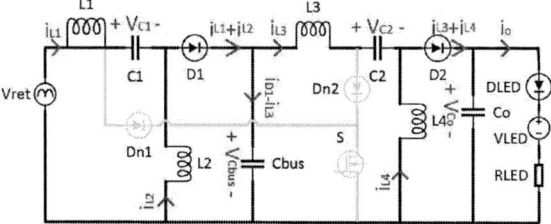

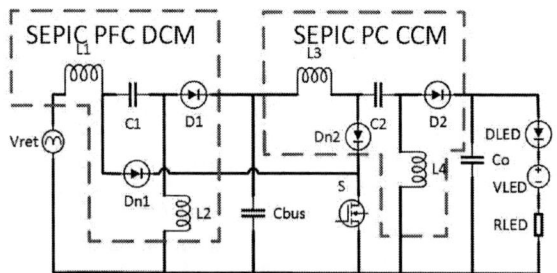

Fig. 6. Second operating stage.

Fig. 4. Proposed quadratic SEPIC converter.

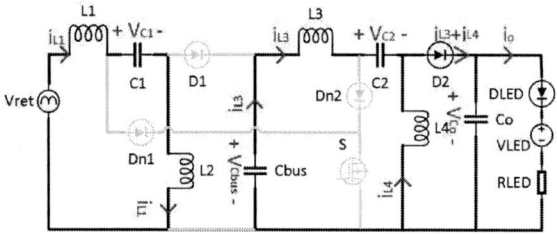

Considering that the PFC stage operates in discontinuous conduction mode (DCM) while the PC one operates in CCM, three operating stages will exist over one switching period T_s.

Fig. 7. Third operating stage.

In the first stage shown in Fig. 5 ($t_o \le t < t_1$), the active switch S is turned on. Diodes D_{n1} and D_{n2} forward biased, while diodes D_1 and D_2 are reverse biased. The current flowing through the switch is the sum of the currents in the four inductors. The currents through the inductors increase linearly while capacitors C_1, C_2, and C_{Bus} are discharged. Since diode D_1 is off, capacitor C_{Bus} is responsible for supplying the second stage. Diode D_2 is also reverse biased and the output capacitor C_o is responsible for supplying the load current.

In the second stage shown in Fig. 6 ($t_1 \le t < t_2$), switch S is turned off, diodes D_{n1} and D_{n2} are reverse biased, but diodes D_1 and D_2 are forward biased. The capacitors are charged and all inductors are discharged. The current through D_1 is the sum of the currents through L_1 and L_2. Capacitor C_{bus} is charged and inductor L_2 is discharged until the direction of the current in L_2 is opposite to that in L_1. From this moment on, diode D_1 is reverse biased, characterizing the end of the second stage. The current flows through D_2 to supply the load and charge the capacitor C_o.

In the third step shown in Fig. 7 ($t_2 \le t < T_s$), the switch S is still off and diodes D_{n1} and D_{n2} remain reverse biased, as well as D_1, thus characterizing the DCM operation of the PFC stage. Capacitor C_{Bus} is responsible for supplying the second stage and diode D_2 is still forward biased, while current is supplied to C_o and the output stage.

Based on the waveforms shown in Fig. 8 and the aforementioned operating stages, it is possible to conclude that the main characteristics of the SEPIC operating in DCM and CCM are preserved, as it is possible to design each stage separately as in [10].

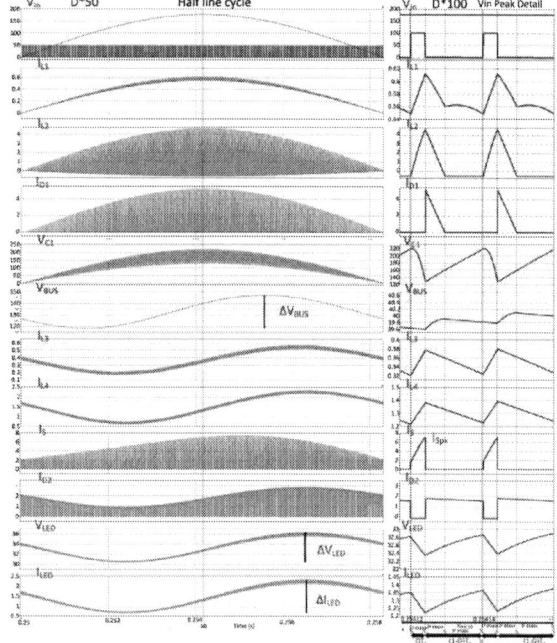

Fig. 8. Main theoretical waveforms in terms of low-frequency (left) and high-frequency behavior (right).

978-1-7281-4181-7/19 $31.00 © 2019 IEEE

III. DESIGN PROCEDURE

The first operating stage allows determining the voltage across capacitor C_{bus}, which is represented by V_{Bus}. In order to determine the best value of V_{Bus} that best fits the design, it is necessary to plot it as a function of the duty cycle D. In this case, the curves define the boundary conditions for each stage of the converter. The operating region must be between the two curves shown in Fig. 9.

The critical duty cycles of the PFC and PFC stages are given by D_{crit1} and D_{crit2} in the form:

$$D_{crit1} = \frac{V_{Bus}}{V_{Bus} + V_{in}} \tag{1}$$

$$D_{crit2} = \frac{V_o}{V_o + V_{Bus}} \tag{2}$$

where V_o is the average output voltage.

From Fig. 9, it is possible to observe the intersection point between the two curves and choose a value approximately 40% larger for V_{Bus}. This factor aims to avoid high voltages and provide a safety margin for the duty cycle variation so that the PFC stage does not operate in CCM mode. The duty cycle must be less than D_{crit1} as calculated in (1) and greater than or equal to D_{crit2} as calculated in (2), thus ensuring that the PFC stage operates in DCM. The peak value of the input voltage V_{inpk} and the output voltage V_o required for this purpose.

The chosen values of $D_{crit1}=D_{max}$, $D_{crit2}=D_{min}$, and V_{Bus} are highlighted in the graph. The next step is to calculate inductances L_1 and L_2, which can be found through an equivalent inductance L_{eq} as in equation (3). The input power P_{in} includes the estimated efficiency for each stage (η_1 and η_2) according to . The switching period (T_s) is the inverse of the switching frequency (f_s) defined in the design.

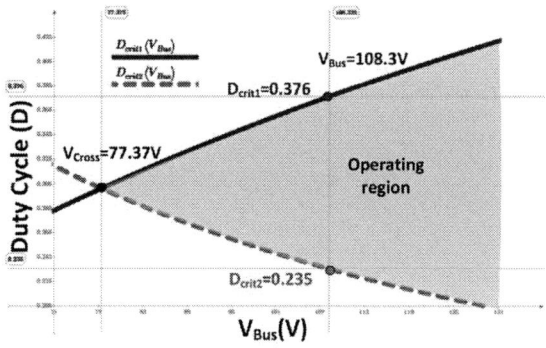

Fig. 9. Duty cycle as a function of V_{bus}.

The chosen values of $D_{crit1}=D_{max}$, $D_{crit2}=D_{min}$ and V_{Bus} are highlighted in the graph. The next step is to calculate the value of the inductors L_1 and L_2 that can be found through an equivalent inductance value (L_{eq}) as in equation . The input power (P_{in}) includes the estimated efficiency for each stage (η_1 and η_2) according to (4). Besides, $T_s=1/f_s$, where f_s is the switching frequency.

$$L_{eq} = \frac{V_{inpk}^2 \cdot D^2 \cdot T_s}{4 \cdot P_{in}} \tag{3}$$

$$P_{in} = \frac{P_o}{\eta_1 \cdot \eta_2} = \frac{P_o}{\eta} \tag{4}$$

Inductance L_1 can be obtained from (5) which considers the peak input current I_{pk} calculated in (6) [11]. Thus, inductance L_2 can be obtained as in (7).

$$L_1 = \frac{V_g \cdot D \cdot T_s}{I_{pk} \cdot \%\Delta I_L} \tag{5}$$

$$I_{pk} = \frac{V_g \cdot D^2 \cdot T_s}{2 \cdot L_{eq} \cdot \eta} \tag{6}$$

$$L_2 = \frac{L_1 \cdot L_{eq}}{L_1 - L_{eq}} \tag{7}$$

The voltage across capacitor C_1 (V_{C1}) is considered constant over the switching period, but at the same time it follows the grid voltage [11]. The first stage operates in PFC mode and has a resonance frequency f_{Res} defined by C_1, L_1, and L_2, which must be much higher than the line frequency f_L and much lower than the switching frequency f_s. Ideally, the resonance frequency should be at least 10 times lower than f_s. Besides, the resonance frequency associated with L_2 and C_1 should be lower than f_s so that the voltage does not oscillate within a switching period. From such conditions, an initial value for capacitor C_1 can be calculated according to (8). This value has great influence on the waveform of the input current and can be adjusted by simulation [10], obtaining a better response.

$$C_1 = \frac{1}{\left(2 \cdot \pi \cdot f_{res}\right)^2 \left(L_1 + L_2\right)} \tag{8}$$

The output current of the PFC stage is the current that flows through diode D_1, which can be divided into a DC component $\bar{I}_{D1}$ and an ac component $\tilde{i}_{D1}$ with twice the line frequency as represented by (9) and (10), respectively.

$$\bar{I}_{D1} = \frac{V_{inpk}^2 \cdot D^2}{4 \cdot V_{Bus} \cdot L_{eq} \cdot f_s} \tag{9}$$

$$\tilde{i}_{D1} = -\frac{V_{inpk}^2 D^2}{4 \cdot V_{Bus} L_{eq} f_s} \left[\cos\left(4\pi f_L t\right)\right] \tag{10}$$

Assuming that the current flowing through C_{Bus} is equal to $\tilde{i}_{D1}$, the voltage variation ΔV_{Bus} can be calculated in (11) according to [12].

$$\Delta V_{Bus} = \frac{V_{inpk}^2 \cdot D^2}{8 \cdot \pi \cdot V_{Bus} \cdot L_{eq} \cdot f_s \cdot f_L C_{Bus}} \tag{11}$$

It is worth mentioning that capacitor C_{Bus} must be present a low capacitance value so that film capacitors can be used due to their high reliability and service life of up to 200,000 hours [13].

The second stage can be designed as a SEPIC converter operating in CCM. Considering it is supplied by a dc voltage, it can be designed according to the expressions given in [14], i.e.:

$$\Delta I_{L3} = \frac{D}{1-D} \cdot I_{LED} \cdot \%\Delta i_L \qquad (12)$$

$$L_3 = \frac{V_{Bus} \cdot D \cdot T_s}{\Delta I_{L3}} \qquad (13)$$

$$\Delta I_{L4} \cong I_{LED} \cdot \%\Delta I_L \qquad (14)$$

$$L_4 = \frac{V_o \cdot (1-D) \cdot T_s}{\Delta I_{L4}} \qquad (15)$$

where $\%\Delta i_L$ is the percent current ripple defined in terms of the average inductor current, and I_{LED} is the average current through the LED array.

Capacitor C_2 is dimensioned in (16) based on its respective percent ripple $\%\Delta V_{C2HF}$, being $\%\Delta V_{C2HF} < 5\% * V_{C2}$, where V_{C2} is the average voltage across C_2.

$$C_2 = \frac{I_{L4} \cdot D}{\%\Delta V_{C2HF} \cdot V_{Bus} \cdot f_s} \qquad (16)$$

where I_{L4} is the average current through L_4.

The average output current of the PC stage can be calculated in (17), being equal to the current through D_2 (I_{D2}), which is responsible for supplying the load and charging capacitor C_o during the time interval $t_{off}=(1-D)T_s$. Diode D_2 remains forward biased and the capacitor is responsible for supplying the load during the time interval $t_{on}=T_s$.

$$I_{D2} = (I_{L3} + I_{L4}) \cdot (1-D) \qquad (17)$$

Since the second stage operates in CCM, the low-frequency ripple of the output voltage is given by (18). The output capacitor C_o is defined a percent high-frequency ripple $\%\Delta V_{o\_HF} < 2\% \cdot V_o$ according to (19). Besides, the output voltage ripple defined in (18) depends exclusively on ΔV_{Bus}.

$$\Delta V_{o\_LF} = \frac{D}{1-D} \cdot \Delta V_{Bus} \qquad (18)$$

$$C_{o\_HF} = \frac{I_{LED} \cdot D \cdot T_s}{\Delta V_{o\_HF}} \qquad (19)$$

Another important aspect associated with the quadratic SEPIC converter is the control circuit, which must be designed to provide low current ripple in the LEDs through an effective control of the duty cycle, also considering the value calculated for C_{Bus}. For this purpose, the transfer function of the duty cycle to the LED current was derived using the well-known average state space technique [15, 16], which allows representing the converter in terms of a small-signal model. The Bode plots of the model and the converter were obtained through the ac Sweep tool in PSIM® software, as shown in Fig. 10, thus denoting that the propose model can describe the converter accurately.

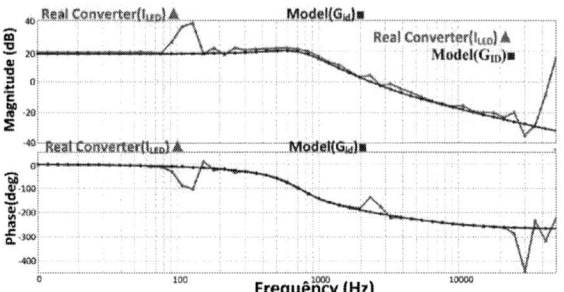

Fig. 10. Comparison of the Bode plots representing transfer function of the duty cycle to the LED current for the converter and small-signal model.

The behavior of the plant differs from that of the mathematical model for frequencies close to the f_s and twice the line frequency f_L. The bandwidth used for the control loop design is 25% of the switching frequency, while the model matches the circuit behavior at f_L. A proportional-integral-derivative (PID) controller was designed in this case using MATLAB®.

IV. PROTOTYPE DESIGN

The specifications in Table I was used to design the prototype elements. The first step consists in plotting the graph using equations (1) and (2) to choose V_{Bus} and the minimum duty cycle D_{min} according to Fig. 9. Then, $V_{Bus}=108.32$ V is chosen which is about 40% above the intersection point of the curves $V_{cross}=77.375$ V. From V_{Bus}, one can find the duty cycle using (2), resulting in $D=0.235$, which must be lower than $D_{crit1}=0.366$ as calculated from (1), thus ensuring that the first stage operates in DCM.

Using equations (3) and (4), it is possible to calculate $L_{eq}=144.677$ µH and $P_{in}=61.728$ W. From (5) and (6), $L_1=11$ mH is obtained for $\Delta I_{L1}=10\%$ and $I_{pk}=0.764$ A. Then, expression (7) is used to obtain $L_2=146.6$ µH. With the adopted resonance frequency $f_{res}=5$ kHz and using (8), capacitor $C_1=90.35$ nF is calculated, while the closest commercial value $C_1=100$ nF is adopted. Capacitor $C_{Bus}=40$ µF is determined from (11) for a ripple of approximately $\Delta V_{Bus}=31.4\% \cdot V_{Bus}=34.01$ V.

Using equations (13) to (15) gives $L_3=9.94$ mH and $L_4=3.399$ mH considering $\Delta I_{L3}=\Delta I_{L4}=10\%$. According to (16), $C_2=2.172$ µF is obtained, while expression (19) gives $C_o=10.589$ µF. From (18), the resulting low-frequency ripple is $\%\Delta V_o=31.4\%=10.589$ V. It is also worth mentioning that the closes commercially available capacitance values were used in the design.

The input power factor (PF) and THD of the input current are established by IEC 61000-3-2 for class C equipment [3]. The minimum requirements for the converter are PF=0.92 and $THD_I=30\% * PF$.

978-1-7281-4181-7/19 $31.00 © 2019 IEEE

TABLE I
DESIGN SPECIFICATIONS OF THE QUADRATIC SEPIC CONVERTER.

Symbol	Description	Value
V_{inpk}	Peak input voltage	179.6 V
f_l	Line frequency	60 Hz
f_s	Switching frequency	50 kHz
I_o	Output current	1.5 A
V_o	Output voltage	33.33 V
V_f	LED forward voltage	28 V
r_d	LED dynamic resistance	3.556 Ω
P_o	Output power	50 W
$\eta_1\eta_2$	Overall efficiency	90%
$\%\Delta I_L$	Current ripple through the inductors	10%

Through the transfer function of the duty cycle to the LED current represented numerically in (20) as obtained from the small-signal modeling, the transfer function of the PID controller can be obtained as in,

$$G_{io}(s) = \frac{-7986s + 2.098 \cdot 10^8}{s^2 + 4002s + 2.247 \cdot 10^7} \quad (20)$$

$$C_{pid}(s) = 2 \cdot \frac{s^2 + 11500s + 3.3 \cdot 10^7}{s^2 + 12000s} \quad (21)$$

V. EXPERIMENTAL RESULTS

The components used in the prototype are described in Table II.

TABLE II
COMPONENTS USED IN THE PROTOTYPE.

Component	Description	Model	Value
D_{Rect}	Rectifier diodes	RS-606	6 A, 1000 V
D_1, D_2	Diode Hyperfast	HFA15TB60	15 A, 600 V
D_{n1}/D_{n2}	Diode Ultrafast	HFA30TA60C	15 A, 600 V
S	Power MOSFET	IRFP460A	20 A, 500 V
L_1, L_2	Inductor	E-42/20 E-30/14	11 mH 146.6 µH
L_2, L_3	Inductor	E-42/15 E-42/20	9.94 mH 3.4 mH
C_1, C_2	Capacitor		100 nF/650 V 2.2 µF/400 V
C_{Bus}, C_o	Film Capacitor		40 µF/1k1V 11 µF/450V
LED	LED Chip		50 W, 1.5 A, V_f=28 V, r_d=3.556 Ω

Experimental results on a 50-W LED were obtained, being discussed in detail as follows. Fig. 11 presents the voltage and current waveforms for the active switch, whose maximum voltage is 280 V. The currents through diodes D_1 and D_2 clearly evidence that the PFC and PC stages operate in DCM and CCM, respectively, as predicted in the theoretical analysis.

The input voltage and input current are strictly in phase in Fig. 12, resulting in FP=0.985 and THD_F=8.37%. Besides, the harmonic content of the input current waveform is shown in Fig. 13 as measured by the oscilloscope. According to Fig. 12, the output current ripple is 7.2% of the average value. A detailed view of the LED current and voltage can also be seen in Fig. 14, where the low-frequency current ripple is measured as 108 mA.

The analysis of the main parameters associated with the converter operation is presented in Table III, where simulation and experimental results are properly compared. It is reasonable to state that good agreement exists between the theoretical assumptions and practice.

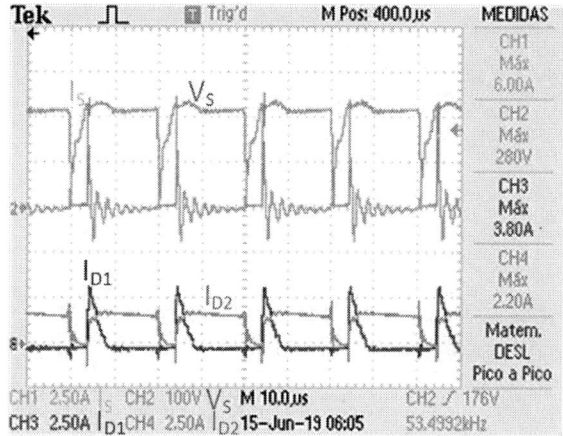

Fig. 11. Waveforms representing the commutation of the semiconductors: switch current I_S (Ch. 1, 2.5A/div), switch voltage V_S (Ch. 2 100V/div), diode current I_{D1} (Ch. 3, 2.5A/div), diode current I_{D2}I (Ch. 4, 2.5A/div); time scale: 10 µs/div.

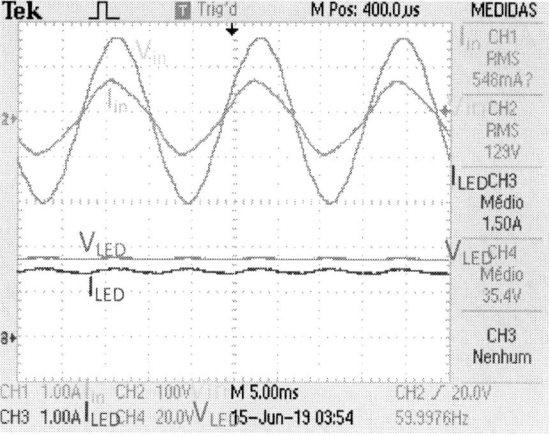

Fig. 12. Waveforms of the input current I_{in} (Ch. 1, 1 A/div), input voltage V_{in} (Ch. 2, 100 V/div), LED current I_{LED} (Ch. 3, 1 A/div), and LED voltage V_{LED} (Ch. 4, 20 V/div); time scale: 5 ms/div.

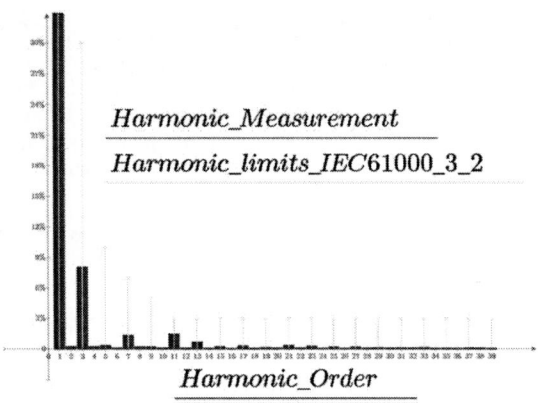

Fig. 13. Harmonic spectrum of the input current and limits stated in the IEC 61000-3-2 Class C Regulations.

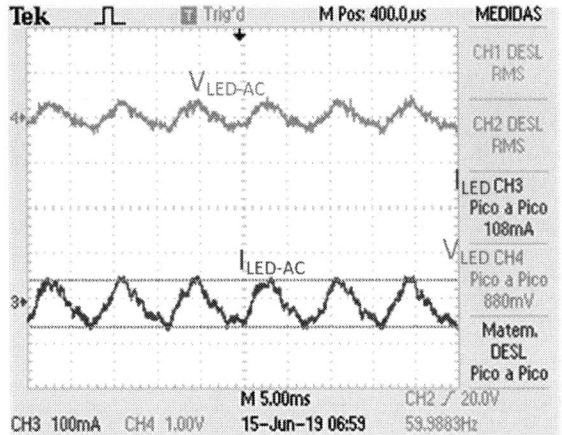

Fig. 14. Detailed view of the ac components associated with the LED current $I_{LED\text{-}ac}$ (Ch. 3, 100 mA/div) and LED voltage $V_{LED\text{-}ac}$ (Ch. 4, 1 V/div); time scale: 5 ms/div.

TABLE III
COMPARISON BETWEEN SIMULATION AND EXPERIMENTAL RESULTS.

| Symb. | Project | Simulation | | Closed-Loop Experimental Results |
		Fixed D	PID Control	
V_o	33.33 V	33.30 V	33.31 V	35.4 V
ΔV_o	3.33 V	6.1 V	842 mV	880 mV
I_o	1.5 A	1.493 A	1.496 A	1.5 A
ΔI_o	150 mA	1.446 A	0.199 A	108 mA
I_{Inpk}	764 mA	613 mA	628m A	774 mA
P_{in}	61.73 W	62.27 W	61.32 W	69.9 W
PF	0.92	0.999	0.993	0.985
THD_I	29.31%	3.031%	8.64%	8.37%

VI. CONCLUSION

This work has presented a quadratic SEPIC converter as a LED driver. The first stage of the converter operates in DCM to provide PFC, while the second stage is responsible for controlling the load current and operates in CCM. The proposed design methodology aims at the use of only film capacitors to extend the driver lifetime, which must match that of the LEDs.

A small-signal model has been derived to control the output current ripple and avoid the use of electrolytic capacitors. It has been effectively demonstrated that operation with high input power factor and low THD_I can be obtained as stipulated by IEC 61000-3-2 for class C equipment. Besides, experimental results are close to the theoretical values, reaching a load current ripple of less than 10% without significantly impact on the input power factor and harmonic content of the input current.

VII. REFERENCES

[1] (2017, June 1st, 2019.). *U. S. Deparment of Energy - Solid-State Lighting 2017 Suggested Research Topics* Available: https://www.energy.gov/sites/prod/files/2017/09/f37/ssl_suggested-research-topics_sep2017.pdf

[2] L. Oliveira, M. Fortes, D. C. R. Tomas, A. Fragoso, and A. Queiroz, "A Power Quality Analysis and Thermal Properties of the System associated with the change of Fluorescent Lamps for LED Lamps," *Order*, vol. 2, p. 2, 2017.

[3] "Electromagnetic compatibility (EMC) - Part 3-2: Limits - Limits for harmonic current emissions (equipment input current ≤16 A per phase)," *IEC 61000-3-2*, 2014.

[4] "Electromagnetic compatibility (EMC) - Part 3-4: Limits - Limitation of emission of harmonic currents in low-voltage power supply systems for equipment with rated current greater than 16 A," *IEC 61000-3-4*, 2014.

[5] Y. Zhou, X. Li, X. Ye, and G. Zhai, "A remaining useful life prediction method based on condition monitoring for LED driver," in *Proceedings of the IEEE 2012 Prognostics and System Health Management Conference (PHM-2012 Beijing)*, 2012, pp. 1-5.

[6] A. Arora, N. K. Medora, and J. Swart, "Failures of electrical/electronic components: Selected case studies," in *2007 IEEE Symposium on Product Compliance Engineering*, 2007, pp. 1-6.

[7] L. Gu, X. Ruan, M. Xu, and K. Yao, "Means of eliminating electrolytic capacitor in AC/DC power supplies for LED lightings," *IEEE Transactions on Power Electronics*, vol. 24, no. 5, pp. 1399-1408, 2009.

[8] A. Álvarez, J. Marcos, D. Gacio Vaquero, A. J. Calleja Rodríguez, F. Sichirollo, M. F. d. Silva, *et al.*, "Reducing storage capacitance in off-line LED power supplies by using integrated converters," in *Conference Record-IAS Annual Meeting (IEEE Industry Applications Society)*, 2012.

[9] W. Tsai-Fu and C. Yu-Kai, "A systematic and unified approach to modeling PWM DC/DC converters based on the graft scheme," *IEEE Transactions on Industrial Electronics*, vol. 45, no. 1, pp. 88-98, 1998.

[10] M. Cosetin, T. Bolzan, P. Luz, M. da Silva, J. M. Alonso, and R. N. do Prado, "Dimmable single-stage SEPIC-Ćuk converter for LED lighting with reduced storage capacitor," in *2014 IEEE Industry Application Society Annual Meeting*, 2014, pp. 1-7.

[11] D. S. L. Simonetti, J. Sebastian, and J. Uceda, "The discontinuous conduction mode Sepic and Cuk power factor preregulators: analysis and design," *IEEE Transactions on Industrial Electronics*, vol. 44, no. 5, pp. 630-637, 1997.

[12] P. S. Almeida, G. M. Soares, D. P. Pinto, and H. A. Braga, "Integrated SEPIC buck-boost converter as an off-line LED driver without electrolytic capacitors," in *IECON 2012-38th Annual Conference on IEEE Industrial Electronics Society*, 2012, pp. 4551-4556.

[13] EPCOS. (June 1st, 2019). *Metalized Polyester Film Capacitors*. Available: https://www.infineon.com/dgdl/ir1150.pdf?fileId=5546d462533600a4015355c41aa21642

[14] M. K. Kazimierczuk, *Pulse-width modulated DC-DC power converters* vol. 1: Wiley Online Library, 2008.

[15] S. Cuk and R. D. Middlebrook, "A general unified approach to modelling switching DC-tO-DC converters in discontinuous conduction mode," in *1977 IEEE Power Electronics Specialists Conference*, 1977, pp. 36-57.

[16] R. D. Middlebrook and S. Cuk, "A general unified approach to modelling switching-converter power stages," in *1976 IEEE Power Electronics Specialists Conference*, 1976, pp. 18-34.

A Performance Analysis of Active Anti-Islanding Methods Based on Frequency Drift

Ênio C. Resende, Henrique T. M. Carvalho, Fernando C. Melo, Ernane A. A. Coelho, Gustavo B. de Lima, Luiz C. G. de Freitas

Faculdade de Engenharia Elétrica (FEELT)
Universidade Federal de Uberlândia (UFU)
Uberlândia, Brasil
eniocostaresende@gmail.com, fernando_cmelo@outlook.com, lcgfreitas@yahoo.com.br

Abstract — **In Grid-Tied Distributed Generation systems (GTDG), the islanding phenomena is a fundamental subject, since it can possibly lead to safety hazards for the line workers during maintenance and damage to equipment. Therefore, it is required that GTDG have anti-islanding techniques, which can be passive and/or active. This paper focuses on the computational implementation of three active anti-islanding techniques based on the frequency drift of the injected current of the inverter. These methods are the Classic Active Frequency Drift (AFD), AFD with Positive Feedback (AFDPF) and AFD with Pulsating Chopping Factor (AFDPCF). For the tests carried out, was considered a PV system of 1 kWp using a VSI inverter, evaluating the time for detecting the islanding, the NDZ size and the THD_i caused by the anti-islanding method. The AFDPCF proved to present reduced NDZ, the SFS performed the faster islanding detection and both presented reduced THD_i, for all the analyzed cases.**

Keywords —**Active Frequency Drift, AFD with Positive Feedback, AFD with Pulsating Chopping Factor, Anti-Islanding, Non-Detection Zone.**

I. INTRODUCTION

The growing concern about the environmental problems such as air pollution, acid rains and global warming are the great catalyst to the searching of renewable energy solutions. However, it is important to mention that large electrical energy production centers, even those that use clean energy sources such as hydroelectric plants, for instance, are intrinsically associated to a significant human interference in the ecosystem where they are implanted, leading to impacts on the local biodiversity, as the extinction of fauna and flora species and the expropriation of local populations. Thus, it is necessary to diversify the global energetic matrix, introducing on it, clean alternatives sources and gradually replacing the large power plants in remote areas by distributed ones, closer to the loads. This goal is achieved by Distributed Generation (DG) [1].

For Grid-Tied Distributed Generation systems (GTDG), the operation is only allowed through the

compliance with the requirements imposed by several standards, like IEEE 1547.1 [2], IEEE 929-2000 [3] and IEC 62116 [4]. These documents have the objective of defining the thresholds of voltage, frequency, total harmonic distortion of the injected current and power factor in which the DG inverter should operate to avoid power quality degradation and the prerequisites to avert electrical accidents and equipment damage. Among these requirements are listed a series of protection systems that must be implemented to avoid electrical accidents. One of those protections is related to the unintentional islanding and it will be the focus of this work.

According to [3], islanding is an operating condition in which a GTDG system remains supplying power to the local loads and/or a portion of the electrical grid after the grid unavailability. This means a serious hazard to line workers safety, because this phenomenon can lead to electrical accidents during maintenance procedures and to electrical devices damage caused by variations of frequency and voltage magnitude. Therefore, it is important to analyze the anti-islanding protection schemes proposed by the technical literature to know their effectiveness and the trade-off relations associated to their implementation. Indeed, several anti-islanding protection (AIP) strategies have been proposed in the past years and consequently is important to divide them in three different groups: passive [5], active [6] and those ones that use communication between the utility and the inverter that will not be considered in this paper [7].

The passive solutions are based on the pure monitoring of electrical variables of the inverter operation (phase, frequency and voltage magnitude) and the islanding occurrence is detected by changes that occur after the grid disconnection. The main examples of passive methods are Phase Jump Detection, Over/Under Frequency and Over/Under Voltage and Current and Harmonic Detection [6]. The main advantage of these methods lies on the fact that they don't insert any disturbance in the inverter output, not impacting the power quality. However, their detection capability is compromised if there is a balance between the power produced by the GTDG and the power consumed by the load [5]-[7].

The active solutions overwhelm these weaknesses by inserting some distortions in the system, such as harmonic

---

This study was financially supported by the Brazilian agencies: Coordenação de Aperfeiçoamento de Pessoal de Nível Superior (CAPES), Conselho Nacional de Desenvolvimento Científico e Tecnológico (CNPq) and Fundação de Amparo à Pesquisa do Estado de Minas Gerais (FAPEMIG).

978-1-7281-4181-7/19 $31.00 © 2019 IEEE

currents into the PCC, for instance, in order to make some variable extrapolate the range allowed by the standards after the utility interruption. Therefore, these techniques present better performance when it comes to islanding detection if compared with the passive ones. Nonetheless, its functionality depends on some degradation of power quality that must not compromise the attendance of the Standards aforementioned. Some examples of active AIP are: Frequency Drift Based, Voltage Drift Based and Harmonic Insertion Based [7].

In [8] was proposed an analysis performance of the Sandia Frequency Shift (SFS) anti-islanding method for a Single-Phase Photovoltaic DG, pointing the advantages related to its implementation if compared to the Classic Active Frequency Drift (AFD). However, the paper does not approach the concept of the Non-Detection Zone (NDZ) that is an important criterion for the determination of the AIP efficiency and the anti-islanding test is executed for only one load condition.

In [9] was proposed a novel active AIP scheme based on frequency drift with variable chopping factor, called Active Frequency Drift with Pulsating Chopping Factor (AFDPCF). In this work, is also presented a performance evaluation between the new strategy and the classic AFD for different load conditions. The experimental results demonstrate the effectiveness of this technique, nevertheless, it was not compared to the SFS method.

The principal objective of this work is to perform a comparative analysis among the classic active frequency drift (AFD) anti-islanding algorithm and its variation, the Sandia Frequency Shift (SFS) [8] and the AFD with Pulsating Chopping Factor (AFDPCF) [9]. The comparison criteria are: the time for detecting the islanding formation, the total harmonic distortion of the current (THD_i) inserted in the grid and the NDZ area, defined as the set of cases that each method fails to detect the utility disconnection. The tests will be performed for 4 different types of load and inverter output conditions.

This paper is structured as follows: in Section II, is presented the general theory about the classic AFD method and the variations it has suffered over the past years and, beyond that, will be presented the NDZ concept and the procedures for its determination; in Section III, is pointed the computational implementation and the tests methodology for the worst case scenario (RLC load in resonance with the nominal frequency) and in Section IV are shown the tests results and a brief discussion about the effectiveness of each strategy, considering their strengths and weaknesses.

II. ACTIVE ANTI-ISLANDING TECHNIQUES

A. Islanding Strategies and Non-Detection Zone (NDZ)

One of the most important criteria for evaluating the performance of the anti-islanding techniques is the NDZ. It is defined as the set of GTDG operating situations in which the AIP is not able to detect the grid disconnection [3]. Considering the grid-tie condition, the active and the reactive power transmitted to the load is given by (1) and (2):

$$P_{load} = P_{DG} + \Delta P \quad (1)$$

$$Q_{load} = Q_{DG} + \Delta Q \quad (2)$$

One can see that, if there is a considerable power flow from the utility grid to the load, during the islanding operation, the values of frequency and voltage of the loads will suffer abrupt variations that will facilitate the detection of the grid disconnection [5]. For this case, the inverter will avoid the islanding formation. However, if ΔP and ΔQ are close to zero, the variation of frequency or voltage will be very small, and the system will operate indeterminately [7]. For the passive methods, the NDZ can be mapped in the ΔP x ΔQ space, by (3) and (4) [5]-[7]:

$$\left(\frac{V}{V_{max}}\right)^2 - 1 \le \frac{\Delta P}{P_{DG}} \le \left(\frac{V}{V_{min}}\right)^2 - 1 \quad (3)$$

$$Q_f\left(1 - \left(\frac{f}{f_{min}}\right)^2\right) \le \frac{\Delta Q}{P_{DG}} \le Q_f\left(1 - \left(\frac{f}{f_{max}}\right)^2\right) \quad (4)$$

Where Q_f is the load quality factor defined by (5).

$$Q_f = R\sqrt{\frac{L}{C}} \quad (5)$$

Where:

- R is the load Resistance;

- L is the load Inductance;

- C is the load Capacitance;

However, as stated by [10] the ΔP x ΔQ space is not accurate to represent the NDZ of the active AIP. Consequently, there are several proposals of different methodologies to mapping the set cases scenarios in which a given AIP is incapable of detecting the grid disconnection. In this work, will be adopted the Q_f x C_{norm} plan, defined by (6) [11]:

$$\tan^{-1}\left[Q_f\left(C_{norm} - 1 + \frac{2\Delta\omega}{\omega_0}\right)\right] = \theta_{inv} \quad (6)$$

Where:
- $\Delta\omega$ is the oscillation frequency range allowed by the standards;

- ω_0 is the nominal angular frequency [rad/s];

- C_{norm} is the ratio between the Load Capacitance and the Capacitance that resonate with the Load Inductance at the nominal frequency;

- θ_{inv} is the phase difference between the Voltage and Current of the Inverter Output;

B. Classic AFD

The classic AFD strategy implementation is based on the insertion of a zero-conduction time (t_z) at the end of each half-cycle of the current's reference waveform and impose a frequency drift in the inverter operation. The dead time is responsible for producing disturbances in the PCC voltage frequency [11]. In grid tie operation, the synchronization between the grid and the inverter voltage promoted by the phase-locked loop (PLL) is responsible for correcting this small perturbation and keep the system stability. Nevertheless, after the islanding occurrence is observed a natural tendency of the frequency to drift out from the nominal value. Fig. 1 compares the current waveform with and without the AFD implementation.

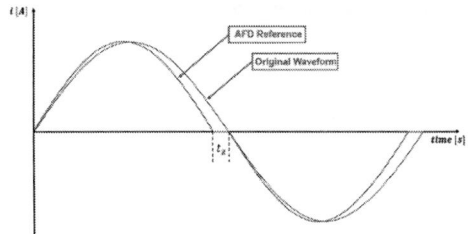

Fig. 1.   Classic AFD Reference

Regarding the operating principle of the Classic AFD, one important parameter necessary for its implementation is the *chopping factor (cf)*. It is defined as the ratio between the amount of dead time inserted at the end of each half-cycle and the whole period as expressed by (7). It is mandatory to state that the cf value is linked to the total of current harmonic distortion inserted by the AIP strategy and to its effectiveness, once it determines the location of the NDZ as will be shown forward [12]:

$$c_f = \frac{2t_z}{T} \tag{7}$$

Thus, the reference current waveform can be expressed by (8):

$$
\begin{aligned}
&i_{afd}(t) \\
&= \begin{cases}
I\sin(2\pi f't), & 0 < \omega t \leq \pi - t_z \\
0, & \pi - t_z < \omega t \leq \pi \\
I\sin(2\pi f't), & \pi < \omega t \leq 2\pi - t_z \\
0, & 2\pi - t_z < \omega t \leq 2\pi
\end{cases}
\end{aligned} \tag{8}
$$

Where:

$$f' = \frac{f}{1 - c_f} \tag{9}$$

The phase of the fundamental component of the inverter output current under Classical AFD implementation is given by (10):

$$\tan(\theta_{inv}) = \frac{\pi c_f}{2} \tag{10}$$

The total harmonic distortion presented in the waveform from Fig.1 (in red) can be approximated by [12]:

$$THD_i \approx c_f \tag{11}$$

The biggest advantage for the Classical AFD application lies on its simplicity. Indeed, it can be easily implemented in the microprocessor where the inverter controlling code is embedded. Though, its effectiveness is committed in the multiple inverter case and for RLC load configurations [10]-[12].

C. Sandia Frequency Shift (SFS)

The Sandia Frequency Shift anti-islanding method was proposed by the National Laboratories to overcome the

disadvantages related to the classical AFD operation previously mentioned. This algorithm works with the same waveform of the output current, but the fixed c_f from the traditional strategy is replaced by a chopping factor that varies with the error between the operation frequency and the nominal frequency at the end of each cycle as stated by (12) [9], [11]:

$$c_f = c_{f_0} + K(f_{op} - 60) \tag{12}$$

Where:

- c_{f_0} is the initial value for the chopping factor;
- K is the accelerating gain;
- f_{op} is the frequency measured by the PLL;

During the normal operation, the frequency error is corrected by the PLL. However, after the disconnection, this error increases substantially what forces the c_f to increases as well what accelerate the islanding detection. The phase of the fundamental component of the inverter output is obtained by replacing (12) in (10), as represented by (13) [8], [11].

$$\tan(\phi_{inv}) = \frac{\pi c_{f_0} + \pi K(f_{op} - 60)}{2} \tag{13}$$

The THD_i imposed by this AIP is also expressed by (11) and the NDZ area is determined by (6). Nevertheless, once the c_f value is variable, the harmonic distortion and the NDZ size are highly dependent from the values of c_{f_0} and K [10]. In [14], for instance, is proposed an inequation for the determination of the gain K that eliminates completely the NDZ for a given quality factor. This inequation is expressed by (14).

$$K > \frac{4Q_{f_{SFS}}}{\pi f_0} \tag{14}$$

Where:

- f_0 is the resonance frequency;
- $Q_{f_{SFS}}$ is the quality factor for the SFS implementation;

Undeniably, the choosing of a gain K that respects the mathematical relation defined by (14), guarantees that the SFS AIP will not present NDZ for load configurations in which the quality factor is lower than $Q_{f_{SFS}}$. Therefore, as shown by [10], [14] it is possible to consider $c_{f_0} = 0$ in (12), what decreases considerably the amount of harmonic distortion injected into the PCC.

D. AFD with pulsating chopping factor (AFDPCF)

In [9] was proposed a variation of the classical AFD in which the fixed c_f is replaced by a pulsating signal that is composed by positive and negative values and zero values as expressed by Fig 2. The c_f waveform illustrated in Fig. 3, is also represented by (15):

$$
c_f = \begin{cases}
c_{f_{max}}, & if\ T_{cmax\ on} \\
c_{f_{min}}, & if\ T_{cmin\ on} \\
0, & if\ T_{coff\ on}
\end{cases} \tag{15}
$$

The AFDPCF has a very clear advantage over the conventional AFD strategy considering the total harmonic distortion. From (11) it is possible to infer a relationship between the chopping factor and the THD_i. Observing Fig. 2 and (15) it is conceivable to note that during the $T_{coff\,on}$ the AFDPCF does not insert any kind of perturbance into the PCC, thus, the average level of distortion over the normal operation is lower if compared with the ones caused by the classical AFD [15].

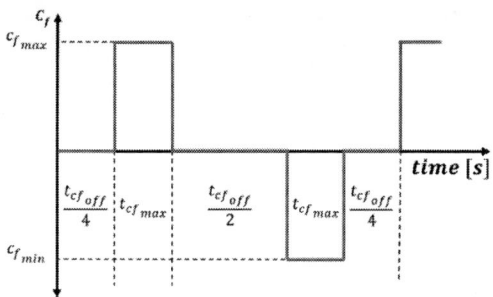

Figure 2. c_f behavior over time for AFDPCF.

III. METHODOLOGY

A. DG system parameters

After the study of the general theory of the three AIP based on the frequency drift, computational simulations (PSIM® software) were implemented, in order to make a quantitative analysis of their performance. For these simulations, it was considered a GTDG system with rated active power of 1kW connected to a utility grid of 127 Volts RMS and frequency of 60 Hz. The DC/AC converter, which is connected to the output of the step-up converter, is composed of the conventional Voltage Source Inverter (VSI) topology modulated by unipolar PWM (switching frequency at 15 kHz). The output filter is composed by LCL and the RLC load is determined by the recommendations of the IEEE 1547, as depicted in Fig. 3.

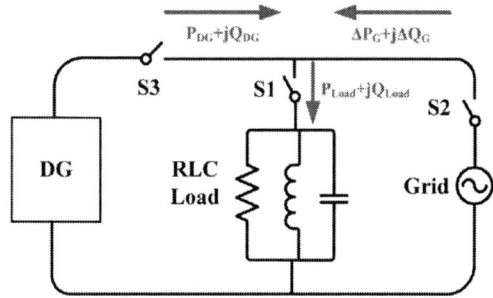

Figure 3. Simulated System.

The inverter controlling system is composed by two loops. The first one is the external loop responsible for maintaining the DC-link voltage at a constant value, even when the power generated by the GTDG varies along the time. The internal loop is responsible for controlling the injected current into the grid, tracking the reference generated by the control of the respective AIP used guaranteeing a low THD_i. A Phase-Locked Loop (SOGI-PLL [16]) was used for maintaining the internal references of the inverter in synchronism with the grid voltage. The grid current control is based on the Proportional-Resonant

(PR). Moreover, Table I summarizes all the parameters of the DG inverter used in the tests [17].

TABLE I. ELECTRICAL PARAMETERS OF THE INVERTER FOR THE IMPLEMENTATION OF COMPUTATIONAL TESTS

Electrical Parameters	
Parameters	**Values**
P_{rated}	1000 W
V_{grid}	127 V
$I_{grid\ rated}$	7.87 A
Frequency	60 Hz
Power Factor	> 0.99
LCL Filter	1.3 mH - 20 uF - 15 uH
Current THD	< 5%
Grid Impedance	0.2 +j0.06 Ω

B. Standards and tests methodology

The comparative analysis proposed by this paper is based on the recommendations from and IEEE 1547.1 [2] and IEEE 929 2000 [3] that, beyond to establish the correct procedure to the realization to the anti-islanding test, they define the voltage, frequency and THD_i limits to the normal conditional operation. These thresholds are shown in Table II.

TABLE II. STANDARDS THRESHOLDS

Parameter	Limits	TripTime Limit	Limits	Trip Time Limit
Voltage	V < 50 %	6 cycles	V < 50 %	6 cycles
	50 ≤ V< 80 [%]	120 cycles	50 ≤ V< 80 [%]	120 cycles
	88 ≤ V < 110 [%]	∞	88 ≤ V < 110 [%]	∞
	110≤V<137 [%]	120 cycles	V > 115 %	2.0 s
	V ≥ 137 %	2 cycles	V ≥ 144 %	0.16 s
Frequency	59.3 ≤ V < 60.5 [Hz]	∞	59.3 ≤ V < 60.5 [Hz]	∞
			Other	0.16 s
Current THD%	< 5 %	∞	< 5 %	∞
Islanding Detection	-	-	-	2s

The RLC load parameters calculations is based on the recommendations from IEEE 1547. The resistor R is dimensioned to match the active power presented in the inverter output and the LC combination must resonate at the nominal frequency. Thus, the load does not demand active neither reactive power from the utility grid. Equations (16), (17) and (18) represents the load components and Tab. III presents the values of load and inverter output power that should be respected during the anti-islanding test [2].

$$R = \frac{V^2}{P} \tag{16}$$

$$L = \frac{V^2}{2\pi f_0 P Q_f} \tag{17}$$

$$C = \frac{Q_f P}{2\pi f_0 V^2} \tag{18}$$

TABLE III. Anti-islanding tests

Tests	Local Load	Inverter Output
1	25%	25%
2	50%	50%
3	100%	100%
4	125%	100%

IV. COMPUTATIONAL RESULTS

As was previously mentioned, all the electrical Standards impose a very stringent threshold to the current harmonic distortion. Consequently, to determinate the performance of each one of the AIP studied by this paper, it is important to determinate the amount of theoretical and real THD_i expected after the method implementation. Beyond this, it is very important to determine the set of operational cases in which each method is not capable of detecting the grid disconnection. Finally, each one of the implemented AIP must disable the inverter respecting the time of two seconds defined by IEEE 1547.1. Thus, the evaluation criteria are: THD_i, size of the NDZ and the time detection.

The THD_i is calculated over 2 seconds of grid tie operation and its value is defined by the PSIM (software in which the simulation was executed) calculation. It is necessary to state that, without the AIP implementation, the measured THD_i is 0.9%. The NDZ, in its turn, is a region from the $C_{norm} x Q_f$ plan limited by two curves generated by (6). The first curve is related to the maximum frequency threshold allowed by the standards ($\Delta\omega_{max} = 2\pi.0.5\ rad/s$) and the second to the minimum frequency threshold ($\Delta\omega_{min} = -2\pi.0.7\ rad/s$) as expressed by (19) [14]. It is necessary to state that the position of this area in the plan depends on the anti-islanding parameters highlighted by Table IV.

$$1 - \frac{2(2\pi.0.5)}{\omega_0} + \frac{tg(\theta_{inv})}{Q_f} < C_{norm}$$
$$< 1 + \frac{2(2\pi.0.7)}{\omega_0} + \frac{tg(\theta_{inv})}{Q_f} \quad (19)$$

TABLE IV. Anti-Islanding Parameters

AIP	Parameters
Classic AFD	$c_f = 0.03$
SFS	$c_{f_0} = 0$
	$k = 0.07$
AFDPCF	$c_{f_{max}} = 0.035$
	$c_{f_{min}} = -0.035$
	$t_{cf_{max}} = 0.2\ s$
	$t_{cf_{min}} = 0.2\ s$
	$t_{cf_{off}} = 0.6\ s$

The Fig. 4 is the NDZ representation of each AIP. From Fig 4. is possible to observe that the method that present the larger NDZ is the Classic AFD. Indeed, the SFS and the AFDPCF represent a remarkable evolution in the anti-islanding theory, since they eliminate the NDZ for a range of Q_f values that, in this case, goes from 0 to 2.62 to the SFS and from 0 to 2.76 to the AFDPCF. Lastly, it is necessary to calculate the time each AIP takes to identify the grid disconnection. Figs. 5, 6 and 7 shows the behavior

of the grid current, the frequency and the inverter state for each strategy during Test 3, before and after the grid disconnection that occurs at 2s.

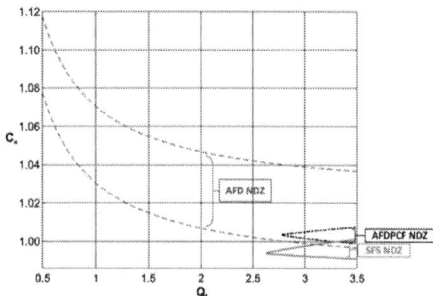

Figure 4. NDZ for each method

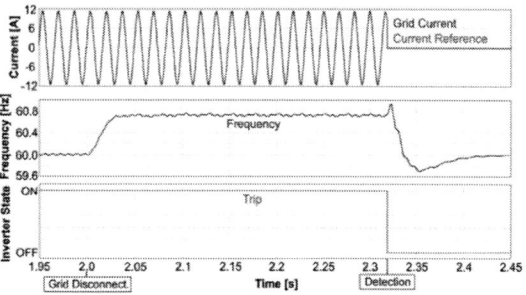

Figure 5. Test 3 for Classic AFD

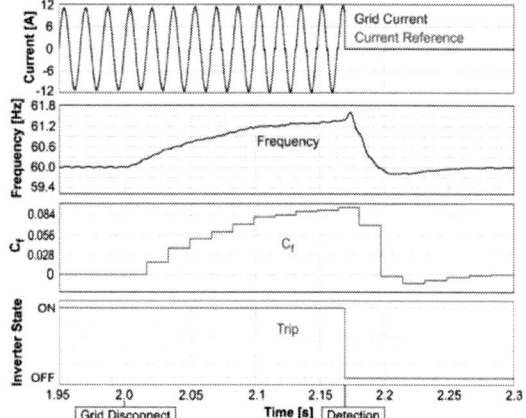

Figure 6. Test 3 for SFS

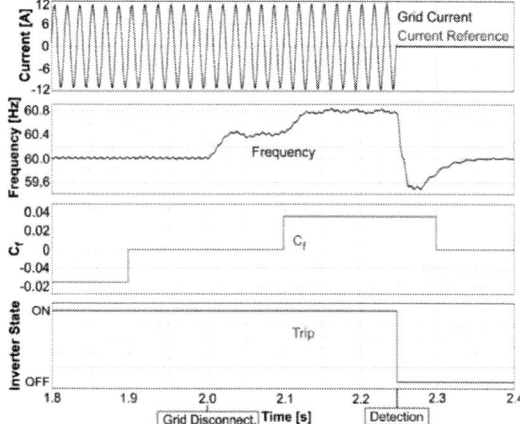

Figure 7. Test 3 for AFDPCF

TABLE V. TESTS RESULTS

Tests	AFD Method	Detection Time (s)	Reference THD (%)	Measured THD (%)
1	Classic	0.321	3	5.63
	SFS	0.168	1	3.99
	AFDPCF	0.259	1.10	3.99
2	Classic	0.319	3	4.62
	SFS	0.168	1	2.28
	AFDPCF	0.253	1.10	2.07
3	Classic	0.315	3	3.42
	SFS	0.167	1	1.24
	AFDPCF	0.249	1.10	1.23
4	Classic	0.311	3	3.4
	SFS	0.166	1	1.23
	AFDPCF	0.244	1.10	1.23

From Table V it is possible to infer that both SFS Frequency Drift and AFDPCF introduce lower rate of harmonic distortion in the grid current, respecting the harmonic threshold from Table III for all tested scenarios. By the other hand, the classic strategy was responsible for inserting a considerably high THD_i at the PCC, extrapolating the Standards recommendations for Test 1, what means that this method can affect the inverter operation when its output power is significantly lower if compared to the nominal one.

Beyond that, SFS and AFDPCF present a much-reduced NDZ size as shown in Fig. 5, eliminating the cases in which the AIP fails for a range of Q_f values. Finally, it is fundamental to state that the method that reached the fastest islanding detection for all the performed tests was the SFS, what occurred due to the connection between its c_f value and the frequency error exposed by (12).

V. CONCLUSIONS

In this paper was proposed a computational analysis among three AIP based on frequency drift through which a performance comparison was carried out. From the computational results it is possible to notice that the THD_i associated to SFS and AFDPCF implementation is considerably lower than the classic AFD. All the methods respected the Standards requisites with exception of the Classic AFD in the Test 1. Beyond that, both are responsible to a significant reduction in the NDZ area compared to the classic strategy. Finally, the SFS presented the lowest time detection followed by the AFDPCF and the classic. Moreover, all of them accomplished the time detection lower than 2s.

VI. REFERENCES

[1] S. K. Gill, K. Vu and C. Aimone, "Quantifying fossil fuel savings from investment in renewables and energy storage," 2017 Saudi Arabia Smart Grid (SASG), Jeddah, 2017, pp. 1-6.

[2] IEEE, "Standard for Interconnection and Interoperability of Distributed Energy Resources with Associated Electric Power Systems Interfaces" IEEE Std 1547-2018 (Revision of IEEE Std 1547-2003), April 2018.

[3] IEEE, "Recommended Practice for Utility Interface of Photovoltaic (PV) Systems" IEEE Std 929-2000, vol., no., pp.i-, 2000

[4] I.E.C., "Test procedure of islanding prevention measures for utility-interconnected photovoltaic inverters," IEC 62116:2014, Ed.2.0, September 2014.

[5] F. De Mango, M. Liserre, A. Dell'Aquila and A. Pigazo, "Overview of Anti-Islanding Algorithms for PV Systems. Part I: Passive Methods," 2006 12th International Power Electronics and Motion Control Conference, Portoroz, 2006, pp. 1878-1883.

[6] F. De Mango, M. Liserre and A. Dell'Aquila, "Overview of Anti-Islanding Algorithms for PV Systems. Part II: Active Methods," 2006 12th International Power Electronics and Motion Control Conference, Portoroz, 2006, pp. 1884-1889.

[7] R. Teodorescu, M. Liserre, P. Rodriguez, "Grid Synchronization in Single-Phase Power Converters," in *Grid Converters for Photovoltaic and Wind Power Systems*, New York: Wiley-IEEE Press, 2007, pp.106-122

[8] M. V. G. Reis, T. A. S. Barros, A. B. Moreira, P. S. Nascimento F., E. Ruppert F. and M. G. Villalva, "Analysis of the Sandia Frequency Shift (SFS) islanding detection method with a single-phase photovoltaic distributed generation system" in 2015 IEEE PES Innovative Smart Grid Technologies Latin America (ISGT LATAM), Montevideo, pp. 125-129, 2015.

[9] Y. Jung, J. Choi, B. Yu, J. So, J. Choi, "A Novel Active Frequency Drift Method of Islanding Prevention for the grid-connected Photovoltaic Inverter", Proc. IEEE 36thPower Electronics Specialists Conference, pp. 1915-1921, 2005.

[10] H. T. da. SILVA, Estudo sobre a interação de métodos anti-ilhamento para sistemas fotovoltaicos conectados à rede de distribuição de baixa tensão com múltiplos inversores. 2016. Dissertação de Mestrado - Escola Politécnica, Universidade de São Paulo, São Paulo, 2016.

[11] F. Liu, Y. Kang and S. Duan, "Analysis and optimization of active frequency drift islanding detection method," APEC 07 - Twenty-Second Annual IEEE Applied Power Electronics Conference and Exposition, Anaheim, CA, USA, 2007, pp. 1379-1384.

[12] A. Yafaoui, B. Wu and S. Kouro, "Improved Active Frequency Drift Anti-Islanding Detection Method for Grid Connected Photovoltaic Systems," in IEEE Transactions on Power Electronics, vol. 27, no. 5, pp. 2367-2375, May 2012.

[13] F. Liu, Y. Kang and S. Duan, "Analysis and optimization of active frequency drift islanding detection method," APEC 07 - Twenty-Second Annual IEEE Applied Power Electronics Conference and Exposition, Anaheim, CA, USA, 2007, pp. 1379-1384.

[14] H. H. Zeineldin and S. Kennedy, "Sandia Frequency-Shift Parameter Selection to Eliminate Nondetection Zones," in IEEE Transactions on Power Delivery, vol. 24, no. 1, pp. 486-487, Jan. 2009.

[15] Yuan Ling, Zhang Xian-fei, Zheng Jian-yong, Mei Jun and Gao Ming, "An improved islanding detection method for grid-connected photovoltaic inverters," 2007 International Power Engineering Conference (IPEC 2007), Singapore, 2007, pp. 538-543.

[16] M. Ciobotaru, V. Agelidis and R. Teodorescu, "Accurate and less-disturbing active anti-islanding method based on PLL for grid-connected PV Inverters," 2008 IEEE Power Electronics Specialists Conference, Rhodes, 2008, pp. 4569-4576.

[17] F. C. Melo, L. S. Garcia, L. C. de Freitas, E. A. A. Coelho, V. J. Farias and L. C. G. de Freitas, "Proposal of a Photovoltaic AC-Module With a Single-Stage Transformerless Grid-Connected Boost Microinverter," in IEEE Transactions on Industrial Electronics, vol. 65, no. 3, pp. 2289-2301, March 2018.

Simulation of the Model, Design and Control of a Current Source Inverter with Unipolar SPWM Modulation

Jeimy C. Sanabria Rojas
Engineering Faculty, Departament of Electronic Engineering
Universidad Pedagógica y Tecnológica de Colombia
Tunja, Colombia
jeimy.sanabria@uptc.edu.co

Daniel M. Barrera Leguizamón
Engineering Faculty, Departament of Electronic Engineering
Universidad Pedagógica y Tecnológica de Colombia
Tunja, Colombia
daniel.barrera02@uptc.edu.co

Diego A. Bautista López
Engineering Faculty, Departament of Electronic Engineering
Universidad Pedagógica y Tecnológica de Colombia
Tunja, Colombia
diego.bautista03@uptc.edu.co

Fabián R. Jiménez López
Engineering Faculty, Departament of Electronic Engineering
Universidad Pedagógica y Tecnológica de Colombia
Tunja, Colombia
fabian.jimenez02@uptc.edu.co

Abstract— **This article presents the topology, large and small signal analysis of the mathematical model, sizing of passive components and design of a proportional resonant controller for a Current Source Inverter. The performance of the inverter controller was verified by simulation, obtaining satisfactory results in terms of the harmonic distortion factor in the voltage output and the regulation indices for the load with values close to the operating point. The response of the control system in static considerations and operation dynamic demonstrated its viability for electronic implementation in the future.**

Keywords—*CSI, control, modeling, unipolar, simulation, inverter.*

I. INTRODUCTION

To supply energy home appliances using photovoltaic system, to DC to AC converter is needed [1]. The device used for converting direct current to alternating current is called an inverter. Some inverters topologies are: Impedance Source Inverters ZSI, Voltage Source VSI Inverters, Multi-level Inverters [2], [3], [4]. The topology discussed in this article is the Current Source Inverter or CSI. This topology is characterized by having a constant current source at the input and also by increasing the output voltage, which allows behave how a low voltage input source such as a solar panel. In this work the modulation used for the devices switching was the Sinusoidal Pulse Width Modulation or SPWM.

Initially is presented the bridge type topology of the Current Source Inverter CSI, then is described the comparison between bipolar and unipolar sinusoidal pulse width modulation implemented on the CSI, in order to evaluate which modulation has lower harmonic content. Subsequently, the CSI inverter mathematical model, is obtained by the large and small signal analysis of its switching states using the modulation that presented the best results. Finally, the design of a proportional resonance voltage controller based on transfer function is described and the controller's performance is evaluated by means of simulation of the open-loop inverter and the closed-loop controlled system.

II. CURRENT SOURCE INVERTER

The monophasic inverter CSI in bridge type configuration is composed of four switching devices, an inductance in series with the input source and a capacitor at the bridge output. The input inductance acts as a constant current source that is subsequently modulated to obtain the sinusoidal signal at the output [5], [6]. The topology of the CSI inverter is shown in Fig. 1. And has like switching devices S1, S2, S3 and S4.

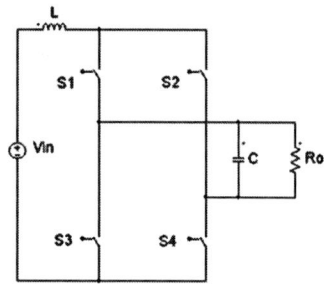

Fig. 1. Current Source Inverter CSI in bridge type topology. .

III. SINUSOIDAL PULSE WIDTH MODULATION

The most commonly used technique to drive power electronics switching devices is the sinusoidal pulse width modulation or SPWM. This modulation is generated by comparison of the amplitude between a sinusoidal reference signal with frequency f_r and a triangular carrier signal with frequency f_C [7]. The reference signal with frequency f_r determines the output frequency of the inverter. The modulation index or duty cycle M_a is determined by (1) where V_{ref} is the amplitude of the reference signal and V_{car} the amplitude of the carrier signal.

$$M_a = V_{ref}/V_{car} \qquad (1)$$

The SPWM modulation is classified into two types, bipolar and unipolar. Bipolar modulation has a single reference signal [7], where the comparison between these generates a control signal D, that acts on the switches S1 and S4 and $\overline{D}$ over S2 and S3. Unipolar modulation has two reference signals [7], generates two control signals: D y D1 which active S1 and S3 respectively and $\overline{D}$ and $\overline{D1}$ to S2 and S4.

Fig. 2 shows the spectral components of the unipolar modulation (red) and the bipolar modulation (blue), having values for f_r=60 Hz and for f_C=50 kHz in the topology of the inverter without using passive components. In bipolar modulation, a higher harmonic content is observed, unlike unipolar modulation, where only the even multiples components of the carrier frequency are presented with less amplitude.

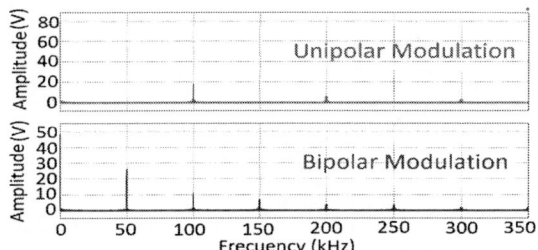

Fig. 2. Harmonic content of unipolar and bipolar SPWM modulation.

For the CSI inverter, the unipolar modulation was selected because it presented a lower harmonic distortion factor in the output, with similar results presented in [8].

IV. CURRENT SOURCE INVERTER WITH UNIPOLAR SPWM MODULATION

Switching devices of the CSI inverter must handle bidirectional voltages and positive currents. Therefore, it is necessary to use three-quadrant switches. The device chosen to meet this requirement is the MOSFET connected in series with a diode. Fig. 3 shows the bridge type CSI inverter with switching elements.

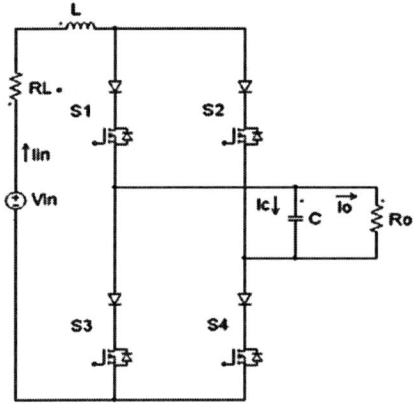

Fig. 3. CSI Topology with switching elements.

The model with losses of the inverter is determined given that the MOSFETS have intrinsically an activation resistance of the drain source R_{dson}, the diodes have a resistance of diode R_d and a voltage direct V_d. In addition, the input inductance has a resistance R_L series generated by the losses that the winding has.

In the first state the switches S1-S4 are on, while S2-S3 remain off, Fig. 4 shows the behavior of the circuit in the first switching state, allowing the passage of current in positive direction to the load.

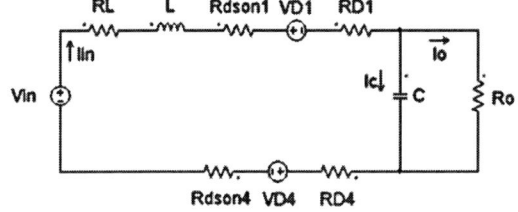

Fig. 4. Current Source Inverter when S1-S4 are on and S2-S3 are off.

In the second switching state, the switches S1-S3 are on and S2-S4 are off. Fig. 5 shows the operation of the CSI inverter in the second switching state. It is observed that the input source is in short circuit, therefore, the circuit load and the power supply are disconnected.

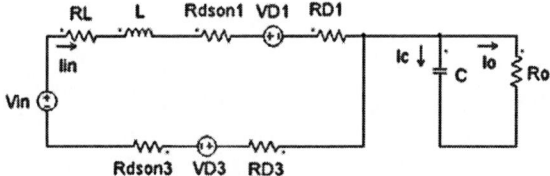

Fig. 5. Current Source Inverter when S1-S3 are on and S2-S4 are off.

In the third state of commutation, switches S2-S3 are on and S1-S3 are off. Fig. 6 shows that the current supplied by the source goes in the opposite direction to the load, generating a negative voltage at the output.

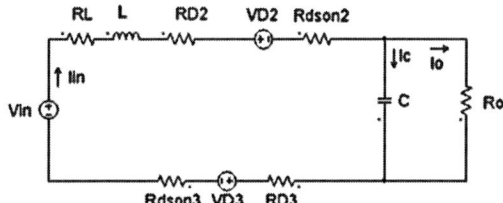

Fig. 6. Current Source Inverter when S2-S3 are on and S1-S4 are off.

In the fourth and last state of commutation, switches S2-S4 are on and S1-S3 are off. Fig. 7 represents the behavior of the inverter, presenting the same situation of the second state, where the input source is in short circuit.

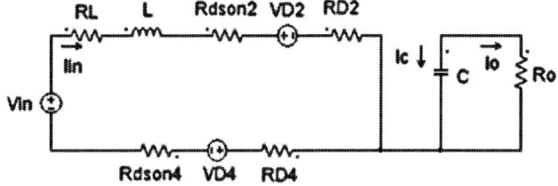

Fig. 7. Current Source Inverter when S2-S4 are on and S1-S3 are off.

V. INVERTER MODELING

To model the CSI inverter, the large and small signal analysis was used, because it has a non-linear dynamic. The large and small signal analysis considers the average value and the variation that the different current and voltage signals have within the circuit [9]. The equations that define the behavior of the first, second, third and fourth switching states, are represented by (2) and (3), (4) and (5), (6) and (7), (8) and (9) respectively. R_X corresponds to the sum of the resistive losses generated by the MOSFET, the diode and the series resistance of the inductance shown in (10).

These equations consider the voltage drop in the inductance and the current through the capacitor. It is assumed that the switching devices have the same reference and therefore the same nominal value.

$$<V_L>=<V_{in}>-R_x^*<I_{in}>-2V_D-<V_c> \qquad (2)$$

$$<I_C>=<I_{in}>-<I_O> \qquad (3)$$

$$<V_L>=<V_{in}>-R_x^*<I_{in}>-2V_D \qquad (4)$$

$$\langle I_C \rangle = -\langle I_0 \rangle \quad (5)$$

$$\langle V_L \rangle = \langle V_{in} \rangle - R_x{}^*\langle I_{in} \rangle - 2V_D + \langle V_C \rangle \quad (6)$$

$$\langle I_C \rangle = -\langle I_{in} \rangle - \langle I_0 \rangle \quad (7)$$

$$\langle V_L \rangle = \langle V_{in} \rangle - R_x{}^*\langle I_{in} \rangle - 2V_D \quad (8)$$

$$\langle I_C \rangle = -\langle I_0 \rangle \quad (9)$$

$$R_x = 2{}^*R_{dson} + 2{}^*R_d + R_L \quad (10)$$

The control signals are determined by the unipolar SPWM modulation, these signals must have a large and small signal component so that the circuit analysis is effective, as well as the voltages and currents of the passive elements. Thus, the values of the control signals are defined as $\langle D \rangle$, $\langle \overline{D} \rangle$, $\langle D1 \rangle$ y $\langle \overline{D1} \rangle$. Equations (11) and (12), represent the complete model of the CSI inverter by unipolar modulation having the behavior of the four switching states with their respective control signals.

$$\langle V_L \rangle = \langle V_{in} \rangle - R_x{}^*\langle I_{in} \rangle - 2V_D - \langle V_C \rangle(\langle D \rangle - \langle D1 \rangle) \quad (11)$$

$$\langle I_C \rangle = \langle I_{in} \rangle(\langle D \rangle - \langle D1 \rangle) - \langle I_0 \rangle \quad (12)$$

The components classification of the previous equations was performed at their large and small signal values. Small signal values were considered because the large signal values do not include the commutation of the carrier signal in the circuit waveforms, in order to maintain accuracy in the analysis [10]. The control signal is reduced to (13). Equations (14) and (15) represent the control signal at large signal U and the control signal at small signal û, respectively.

$$\langle U \rangle = (\langle D \rangle - \langle D1 \rangle) \quad (13)$$

$$U = D - D1 \quad (14)$$

$$\hat{u} = \hat{d} - \widehat{d1} \quad (15)$$

Normalizing and separating the small signal component of (11) and (12) in (16) and (17) the differential equations of the CSI inverter with unipolar modulation was obtained considering the switching variations of the carrier signal.

$$\frac{d\widehat{i_{in}}}{dt} = \frac{\widehat{v_{in}}}{L} - \frac{R_x{}^*\widehat{i_{in}}}{L} - \frac{V_C{}^*\hat{u}}{L} - \frac{\widehat{v_C}{}^*U}{L} \quad (16)$$

$$\frac{d\widehat{v_C}}{dt} = \frac{I_{in}{}^*\hat{u}}{C} + \frac{\widehat{i_{in}}{}^*U}{C} - \frac{\widehat{i_o}}{C} \quad (17)$$

In order to obtain the transfer function of the CSI inverter, the representation in state space was obtained, only using small signal values. The representation by state space was made from (16) and (17). Equation (18) represents the matrices of state, input, output and direct transmission equivalents.

$$\frac{d}{dt}\begin{bmatrix} \widehat{i_{in}} \\ \widehat{v_C} \end{bmatrix} = \begin{bmatrix} -\frac{R_x}{L} & -\frac{U}{L} \\ \frac{U}{C} & -\frac{1}{RC} \end{bmatrix} * \begin{bmatrix} \widehat{i_{in}} \\ \widehat{v_C} \end{bmatrix} + \begin{bmatrix} -\frac{V_C}{L} \\ \frac{I_{in}}{C} \end{bmatrix} * [\hat{u}] + \begin{bmatrix} \frac{1}{L} & 0 \end{bmatrix}[\widehat{v_{in}}] + [0]$$

$$(18)$$

From the representation in the state space defined in (18) the transfer function of the system was obtained by (19), with the purpose of subsequently designing the control loop. The transfer function of the CSI inverter is shown in (20) having as an output function the voltage in the capacitor and as an input function the control signal related to the modulation index defined in (1).

$$G(s) = \frac{Y(s)}{U(s)} = C(sI-A)^{-1}B \quad (19)$$

$$G(s) = \frac{\frac{I_{in}}{C}s + \frac{I_{in}R_x}{LC}}{s^2 + \frac{RCR_x + L}{RLC}s + \frac{R_x}{RLC}} \quad (20)$$

VI. CALCULATION OF PASSIVE COMPONENTS

The minimum value of the input inductance was calculated using (11), having the ripple current at the inverter input behaves like a triangular waveform with a slope determined by the input voltage [11]. Equation (21) represents the value of the inductance in relation to the parameters previously mentioned, where ΔI_{in} is the maximum desired ripple in the inverter's input current.

$$L = \frac{V_{in}}{\omega \Delta I_{in}} \quad (21)$$

Based on (12) the minimum value of the output capacitor is determined using the voltage ripple and the average value considering the control signals D and D1, that depend on a triangular waveform [11]. The capacitor value is represented in (22), where ΔV_c is the maximum desired ripple in the voltage signal at the output generated by the switching frequency of the carrier signal.

$$C = \frac{D(1-D1)(I_{in} - I_o)}{2\Delta V_c fc} \quad (22)$$

Table I shows the established design parameters, where it was assumed an ideal behavior of the inverter establishing that the input power is the same as the output.

TABLE I. SPECIFICATION DEISGN

Specification	Value
Output power (P_o)	200 W
Input power (P_{in})	200 W
Output voltage (V_o)	120 V_{rms}
Input voltage (V_{in})	60 V
Modulation index (D)	0.82
Modulation index (D1)	0.82
Carrier frecuency (f_C)	18 kHz
Angular frecuency (ω)	120π rad/s
Current ripple on input (ΔI_{in})	0.25*I_{in} A
Voltage ripple on output (ΔV_c)	0.0025*V_o V
Losses in resistors (R_x)	0.2 Ω

VII. DESIGN OF CONTROL LOOP

Due to the inverter behavior, the controllers generally use proportional integral PI control action when the model of the plant is unknown and no demanding design specifications are required [12], but it has the disadvantage of not following sinusoidal references due to its integral term [13]. Thus, the selected controller is the proportional resonant control, that allows a better tracking of sinusoidal references and guarantees a low steady state error in the system [14]. This controller is represented as the function in the Laplace domain by (23), where K_P is the proportional gain, K_i is the integral term, ω_a the allowable bandwidth of the frequency ω_o and ω_o the fundamental angular resonance frequency of the inverter output.

978-1-7281-4181-7/19 $31.00 © 2019 IEEE

$$C_{PR}(s)=K_P+\frac{K_i s}{s^2+\omega_a s+\omega_o{}^2} \qquad (23)$$

To determine the values of the constants K_P and K_i of the controller, from the transfer function described in (20), the Tune Tool of the parameter block of the PID controller available in Simulink - MATLAB® was used. The following design parameters will be selected: stabilization time $t_s<1.2$ ms, percentage of maximum peak % MP≤ 10%, and ω_a as 30% of ω_o, being equal to 120π rad/s. The tuning results gave gain values $K_P=0.07$ and $K_i=258$, however, a manual adjustment was made in the simulation of the gain $K_P=0.0007$ to reduce the error in steady state. Because the plant only had the analysis of the small signal, it was necessary to add the value of the large signal (13), to have the controller in its complete and correct operation. The value of U average, is represented in (24) where V_{op} and I_{op} the peak output voltage and current at the inverter output, considering the phase shift generated by the capacitor and the sinusoidal waveform in the current.

$$U = \frac{C*\omega*V_{op}Cos(\omega t)+I_{op}Sin(\omega t)}{I_{in}+\Delta I_{in}\sin(240\pi*t)} \qquad (24)$$

Fig. 8 shows the desired temporal response for the tuning of resonant proportional controller in closed loop, having as input a unit step.

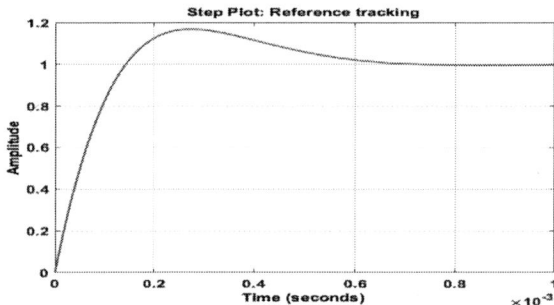

Fig. 8. System step response in closed loop.

VIII. SIMULATIONS

The behavior of the differential equations obtained to model the CSI inverter represented in (11) and (12) were simulated by functional blocks on the Simulink® platform. Fig. 9 illustrates the block diagram made to check the operation of the mathematical model, where is composed of the control signal <U> on the voltage in the capacitor and the input current.

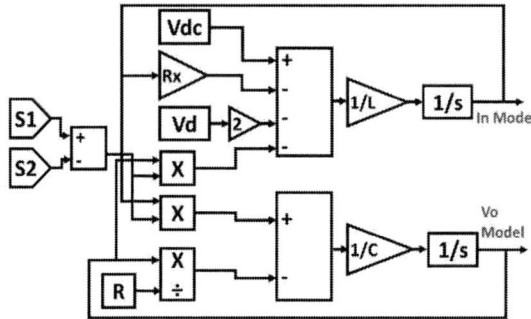

Fig. 9. Block diagram of the mathematical model of the inverter in Simulink®.

Fig. 10 shows the behavior of the voltage waveform at the output of the mathematical model (blue) and the simulated ideal circuit (red) and Fig. 11 shows the behavior current waveform at the input inductance of the mathematical model (blue) and the simulated circuit (red) in open loop.

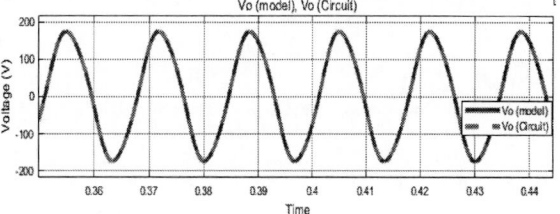

Fig. 10. Output voltage waveforms of the mathematical model and the simulated circuit.

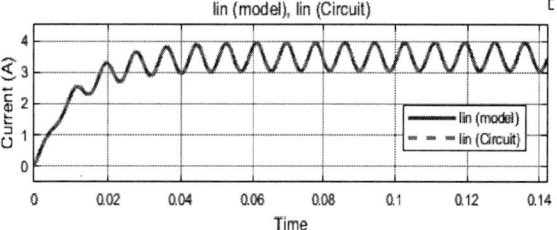

Fig. 11. Input current waveforms of the mathematical model and the simulated circuit.

Fig. 12 shows the behavior of the voltage at the output against variations in the nominal value of the load. In steady state the load has a value of $R_o=72$ Ω and the output voltage is 120.5 V_{rms} at a frequency of 60 Hz.

In a time of 1.5 seconds a change is made in the value of the load that passes to a value of $R_o=80$ Ω, where it is observed that the system stabilizes in the next negative semi cycle of the reference (blue) around 1.52 s, coinciding in phase and frequency of 60 Hz, and with a voltage value at the output 123.4 V_{rms}. The error signal (green) oscillates around ±9V, less than $10\%V_P$.

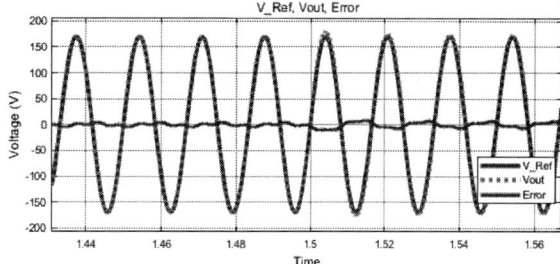

Fig. 12. Voltage waveforms at the circuit output with variation in the load.

Fig. 13 shows the behavior of the current injected by the input inductor (blue), the output current (red) and the output current in DC (brown). In contrast to the variation of the load, a change in the injected current is observed in order to compensate the voltage at the output, going from 3.37 A_{rms} to 3.24 A_{rms}.

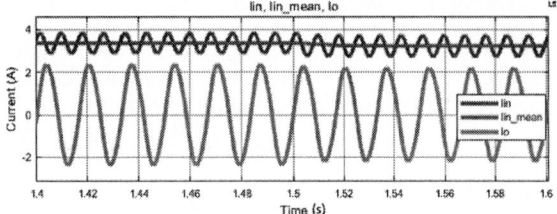

Fig. 13. Current waveforms at the input and output with variation in the load.

Table II shows the results obtained in terms of the performance parameters to evaluate the operation of the inverter as: RMS and DC values of output voltage and input current; and total harmonic distortion THD. The performance indices are reviewed for the answers of the theoretical model obtained, the response of the simulated circuit in real operating conditions and simulated circuit in closed loop, where similar values were reached in the measurements based on a nominal power of operation close to 200 W.

TABLE II.   RESULTS OF SIMULATION

Type of simulation	Parameter			
	Output voltage (V_o)		Input current (I_{in})	
	RMS (V)	% THD	RMS (A)	DC (A)
Model	120.5	3.81	3.41	3.4
Circuit	120.5	3.81	3.41	3.4
Circuit with control	119.6	3.88	3.37	3.36

IX. CONCLUSION

The mathematical model obtained from the CSI inverter by switching techniques of the control signals, for the four possible switching states, exhibits the same behavior as the ideal CSI inverter circuit using unipolar SPWM modulation. From obtaining the mathematical model of the inverter, the transfer function used for the PR Proportional Resonant controller design was obtained.

Due to the non-linear behavior of the inverter, the PR controller responds to values close to the designed operating point. The voltage THD obtained at the output of the CSI inverter complies with the standards recommended in the IEEE Std 519 ™-2014 Standard. being appropriate for their further implementation.

REFERENCES

[1] O. Perpiñan, Energía Solar Fotovoltaica. Spain: Creative Comons, 2018.

[2] A. K. Gupta, P. Samuel and D. Kumar, "A state of art review and challenges with impedance networks topologies," 2016 IEEE 7th Power India International Conference (PIICON), Bikaner, 2016, pp. 1-6.

[3] A. L. Julian and G. Oriti, "A Comparison of Redundant Inverter Topologies to Improve Voltage Source Inverter Reliability," Conference Record of the 2006 IEEE Industry Applications Conference Forty-First IAS Annual Meeting, Tampa, FL, 2006, pp. 1674-1678.

[4] M. Anzari, J. Meenakshi and V. T. Sreedevi, "Simulation of a transistor clamped H-bridge multilevel inverter and its comparison with a conventional H-bridge multilevel inverter," 2014 International Conference on Circuits, Power and Computing Technologies [ICCPCT-2014], Nagercoil, 2014, pp. 958-963.

[5] A. Pérez, "Inversor para sistema fotovoltaico aislado," M.S. thesis, Pontificia Universidad Javeriana, Bogotá, CUND, Colombia, 2016.

[6] F. Lin Luo and H. Yo, Advanced DC/AC Inverters. New York: CRC Press, 2012.

[7] S. Maheshri and P. Khampariya." Simulation of single phase SPWM (Unipolar) inverter," 2014 International Journal of Innovative Research in Advanced Engineering [IJIRAE-2014], Tamil Nadu, 2014, pp. 12-18.

[8] E. H. E. Aboadla, S. Khan, M. H. Habaebi, T. Gunawan, B. A. Hamidah and M. B. Yaacob, "Effect of modulation index of pulse width modulation inverter on Total Harmonic Distortion for Sinusoidal," 2016 International Conference on Intelligent Systems Engineering (ICISE), Islamabad, 2016, pp. 192-196.

[9] S. Bacha, I. Munteanu and A. I. Bratcu, Power Electronic Converters Modeling and Control with Case Studies. London: Springer, 2014.

[10] A. Yazdani and R. Iravani, Voltage-Sourced Converters in power Systems Modeling, Control and Applications. New Jersey: John Wiley & Sons, 2010.

[11] R. W. Erickson and D. Maksimovic, Fundamentals of Power Electronics. New York: Kluwer Academic Publishers, 2004.

[12] E. Ballester and P. Robert, Electrónica de Potencia. Ciudad de México: Alfa Omega Grupo Editor, 2012.

[13] R. Borque, "Diseño de un controlador para una mini-turbina eólica," Degree work, Universitat Politécnica de Catalunya, Terrasa, CAT, Spain, 2015.

[14] R. Alzate, "Diseño e Implementación de un Controlador Resonante para Sistemas de Conversión DC/AC Bidireccionales," Degree work, Universidad Autónoma de Occidente, Santiago de Cali, VALLE, Colombia, 2017.

978-1-7281-4181-7/19 $31.00 © 2019 IEEE

Analysis and Comparison of the Dynamic Response of Direct and Indirect Rotor Flux Control Applied to an Asymmetrical Two-Phase Induction Motor.

Rafael de Farias Campos
Department of Electrical Engineering
Santa Catarina State University-Udesc
Joinville, Brazil
rafacfar@gmail.com

José de Oliveira
Department of Electrical Engineering
Santa Catarina State University-Udesc
Joinville, Brazil
jose.oliveira@udesc.br

Ademir Nied
Department of Electrical Engineering
Santa Catarina State University-Udesc
Joinville, Brazil
ademir.nied@udesc.br

Abstract— This paper discusses two methods for asymmetrical two-phase induction motor control. The direct rotor field oriented control (DRFOC) and the indirect rotor field control (IRFOC) methods are investigated. Both control strategies were implemented in the same hardware with little modifications in the control algorithm. However, the intrinsic model asymmetry causes extra coupling between the stator windings. To use the field-orientated method, the asymmetry must be eliminated by using a transformation based on the mutual inductances. Experimental investigation was carried out and a comparative analysis of the performance of the two methods of control is presented. Based on experimental results of the DRFOC and IRFOC control systems, it is shown that the speed control and the flux control present a fast response even under load disturbance. In addition, the complexity for implementing both systems is identical. Although some questions were raised concerning the use of such transformation, it is still a viable solution as is demonstrated by the results.

Keywords— *Asymmetrical two-phase induction motor, direct field oriented control, indirect field oriented control.*

I. INTRODUCTION

Asymmetrical two-phase induction motors can be found in households and commercial applications. The bulk of these applications are air conditioning systems, washers, dryers, fans, refrigerators, pumps, etc. In terms of energy saving, the impact of these types of motors is related to their efficiency to convert electrical energy in mechanical energy. These types of motors have a main and an auxiliary winding 90 degrees apart and make use of constant speed drives yielding low efficiency and an increase in energy consume. However, the requirement for energy saving in electrical machines has led to the use of adjustable speed drives (ASDs) in all rages of motor power. In that sense, different ASDs topologies were presented to increase the performance of asymmetrical two-phase induction motors: two-leg, three-leg and four-leg two-phase inverters [2]-[4] and [14]. That way a combination adjustable speed drives and control strategies can be used to achieve higher performance by increasing speed rage and lowering the rotor inertia. Suitable control strategies applied to ASDs systems can be employed to guarantee the correct speed regulation improving performance and efficiency.

Different control strategies have been presented for the control of three-phase induction motors, each one with its own degree of complexity. These strategies can be adapted to control of two-phase induction motors, as presented in [5]-[7]. The use of a particular strategy depends on the nature of the application. Generally, simple methods as volt/Hertz and current-slip frequency control fail to provide good transient performance, presenting oscillation during transients which is not adequate for processes that requires fast and precise torque response.

To improve performance, field orientation method can be applied so the model is valid for transient conditions. The advantage is that the highly coupled nonlinear dynamics of the induction motor becomes decoupled and linearized. The most used method of field oriented control is called quadrature control, which can be divided in two modes: the direct mode and the indirect mode. The indirect mode considers constant flux amplitude and the desired spatial flux position is computed using known motor parameters to obtain the motor slip frequency. The direct mode of field orientation uses a control loop for the flux and requires flux acquisition, which is realized by flux estimator using motor terminal quantities.

The orientation of the field can be obtained using stator flux, rotor flux or air-gap flux. However, the rotor-flux control strategy delivers a natural decoupling of the stator current. This paper presents an ASD applied to asymmetrical two-phase induction motor with fractional horsepower and controlled by the direct rotor field oriented control (DRFOC) and indirect rotor filed oriented control (IRFOC). The use of field-oriented control systems relies on the elimination of the asymmetry between the stator windings of the motor, which is useful for vector applications. The asymmetry is a result of different d and q axes parameters, which is very common in off -the-shelve standard machines. So, to create a symmetrical model a relation between the mutual inductances is employed to define a transformation for the stator variables, as it was done in [6]. However, in [13] this transformation is questioned due to the fact that the winding factor can affect the value of such relation and, consequently, the implementation of a vector controlled system cannot maintain balanced operation. The solution presented in the paper was simply to change the asymmetrical two-phase motor for a symmetrical two-phase induction motor. The main problem with this solution is the need to manufacture a new motor, which brings an extra cost. Since most applications that require improvement in performance make use of already existing machines, it is natural to use a solution the can be implemented without any change in the structure of the motor.

The objective of this paper is to demonstrate that the transformation presented in [6] is effective for driving asymmetrical two-phase induction motors. To verify the performance of a these types of motors under symmetrical transformation, the direct and indirect methods of vector control are examined and compared. The similarities and differences of both methods are investigated from mathematical point of view, and are shown that the control structure of these two methods is almost identical. Experimental results are presented to demonstrate the dynamic characteristics of both control strategies. The dynamic operation will be analyzed under load and no load condition.

II. ASYMMETRICAL TWO-PHASE INDUCTION MOTOR MODEL

Considering linearity of the magnetic circuit, constant air-gap length and the motor windings producing a sinusoidal distribution of magnetic field in the air-gap, the motor equations can be in the stationary reference frame can be obtained as presented in [1]:

$$\begin{bmatrix} v_{ds}^s \\ v_{qs}^s \end{bmatrix} = \begin{bmatrix} r_{ds} & 0 \\ 0 & r_{qs} \end{bmatrix} \begin{bmatrix} i_{ds}^s \\ i_{qs}^s \end{bmatrix} + \frac{d}{dt}\begin{bmatrix} \lambda_{ds}^s \\ \lambda_{qs}^s \end{bmatrix} \tag{1}$$

$$\begin{bmatrix} 0 \\ 0 \end{bmatrix} = \begin{bmatrix} r_r & 0 \\ 0 & r_r \end{bmatrix} \begin{bmatrix} i_{dr}^s \\ i_{qr}^s \end{bmatrix} + \frac{d}{dt}\begin{bmatrix} \lambda_{dr}^s \\ \lambda_{qr}^s \end{bmatrix} + \omega_r \begin{bmatrix} 0 & 1 \\ -1 & 0 \end{bmatrix}\begin{bmatrix} \lambda_{dr}^s \\ \lambda_{qr}^s \end{bmatrix} \tag{2}$$

$$\begin{bmatrix} \lambda_{ds}^s \\ \lambda_{qs}^s \end{bmatrix} = \begin{bmatrix} L_{ds} & 0 \\ 0 & L_{qs} \end{bmatrix} \begin{bmatrix} i_{ds}^s \\ i_{qs}^s \end{bmatrix} + \begin{bmatrix} L_{dsr} & 0 \\ 0 & L_{qsr} \end{bmatrix} \begin{bmatrix} i_{dr}^s \\ i_{qr}^s \end{bmatrix} \tag{3}$$

$$\begin{bmatrix} \lambda_{dr}^s \\ \lambda_{qr}^s \end{bmatrix} = \begin{bmatrix} L_r & 0 \\ 0 & L_r \end{bmatrix} \begin{bmatrix} i_{dr}^s \\ i_{qr}^s \end{bmatrix} + \begin{bmatrix} L_{dsr} & 0 \\ 0 & L_{qsr} \end{bmatrix} \begin{bmatrix} i_{ds}^s \\ i_{qs}^s \end{bmatrix} \tag{4}$$

The torque equation and the mechanical equation can be written as

$$T_e = P[L_{qsr} i_{qs}^s i_{dr}^s - L_{dsr} i_{ds}^s i_{qr}^s] \tag{5}$$

$$P(T_e - T_L) = J\frac{d\omega_r}{dt} + F.\omega_r \tag{6}$$

This set of equations represents the single-phase induction motor as an asymmetric two-phase motor. In [5] and [6] a transformation of the stator variables employing the mutual inductances is given by $k = L_{dsr}/L_{qsr}$. The variables transformed can be written as

$$i_{ds}^s = i_{ds1}^s \; ; \; i_{qs}^s = k i_{qs1}^s , \tag{7}$$

$$\lambda_{ds}^s = \lambda_{ds1}^s \; ; \; \lambda_{qs}^s = k^{-1}\lambda_{qs1}^s , \tag{8}$$

$$v_{ds}^s = v_{ds1}^s \; ; \; v_{qs}^s = k^{-1}v_{qs1}^s , \tag{9}$$

For rotor flux orientation, the motor equations can be rewritten using the rotor flux vector as the frame of reference (denoted by superscript er). Thus, taking into account the relations (7)-(9)

$$\begin{bmatrix} v_{ds1}^{er} \\ v_{qs1}^{er} \end{bmatrix} = \begin{bmatrix} r_{dds} & r_{dqs} \\ r_{qds} & r_{qqs} \end{bmatrix}\begin{bmatrix} i_{ds1}^{er} \\ i_{qs1}^{er} \end{bmatrix} + \frac{d}{dt}\begin{bmatrix} \lambda_{ds1}^{er} \\ \lambda_{qs1}^{er} \end{bmatrix} + \omega_{er}\begin{bmatrix} 0 & -1 \\ 1 & 0 \end{bmatrix}\begin{bmatrix} \lambda_{ds1}^{er} \\ \lambda_{qs1}^{er} \end{bmatrix} \tag{10}$$

$$\begin{bmatrix} v_{dr}^{er} \\ v_{qr}^{er} \end{bmatrix} = \begin{bmatrix} r_r & 0 \\ 0 & r_r \end{bmatrix}\begin{bmatrix} i_{dr}^{er} \\ i_{qr}^{er} \end{bmatrix} + \frac{d}{dt}\begin{bmatrix} \lambda_{dr}^{er} \\ \lambda_{qr}^{er} \end{bmatrix} + (\omega_{er} - \omega_r)\begin{bmatrix} 0 & -1 \\ 1 & 0 \end{bmatrix}\begin{bmatrix} \lambda_{dr}^{er} \\ \lambda_{qr}^{er} \end{bmatrix} \tag{11}$$

$$\begin{bmatrix} \lambda_{ds1}^{er} \\ \lambda_{qs1}^{er} \end{bmatrix} = \begin{bmatrix} L_{ds} & 0 \\ 0 & L'_{qs} \end{bmatrix}\begin{bmatrix} i_{ds1}^{er} \\ i_{qs1}^{er} \end{bmatrix} + \begin{bmatrix} L_{dsr} & 0 \\ 0 & L_{dsr} \end{bmatrix}\begin{bmatrix} i_{dr}^{er} \\ i_{qr}^{er} \end{bmatrix} \tag{12}$$

$$\begin{bmatrix} \lambda_{dr}^{er} \\ \lambda_{qr}^{er} \end{bmatrix} = \begin{bmatrix} L_r & 0 \\ 0 & L_r \end{bmatrix}\begin{bmatrix} i_{dr}^{er} \\ i_{qr}^{er} \end{bmatrix} + \begin{bmatrix} L_{dsr} & 0 \\ 0 & L_{dsr} \end{bmatrix}\begin{bmatrix} i_{ds1}^{er} \\ i_{qs1}^{er} \end{bmatrix} \tag{13}$$

The torque equation now has the following form:

$$T_e = PL_{dsr}[i_{qs1}^{er} i_{dr}^{er} - i_{ds1}^{er} i_{qr}^{er}] , \tag{14}$$

where $L'_{qs} = k^2 L_{qs}$ and ω_{er} is the instantaneous angular velocity. The asymmetry presented in the stator winding appears in the resistance matrix in (10) and is represented by

$$\begin{cases} r_{dds} = \dfrac{r_{ds} + r'_{qs}}{2} + \dfrac{r_{ds} - r'_{qs}}{2}.\cos(2\theta_{er}) \\[2mm] r_{dqs} = \dfrac{r_{ds} - r'_{qs}}{2}.sen(2\theta_{er}) \\[2mm] r_{qds} = \dfrac{r_{ds} - r'_{qs}}{2}.sen(2\theta_{er}) \\[2mm] r_{qqs} = \dfrac{r_{ds} + r'_{qs}}{2} - \dfrac{r_{ds} - r'_{qs}}{2}.\cos(2\theta_{er}) \end{cases} \tag{15}$$

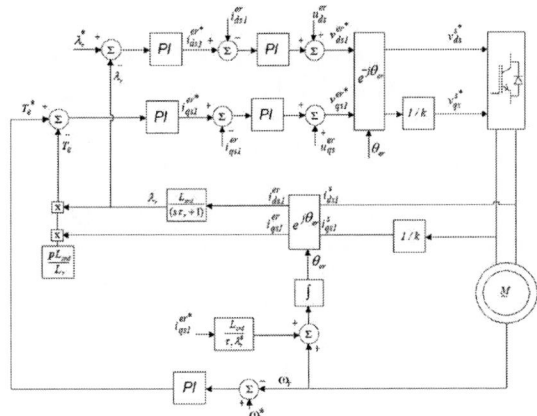

Fig. 1: Block diagram of the indirect rotor flux control

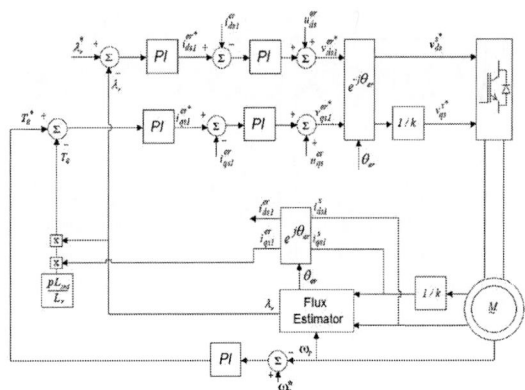

Fig. 2: Block diagram of the direct rotor flux control

The instantaneous electrical angle between the rotor flux vector and the stationary reference frame is represented by θ_{er} and $r'_{qs} = k^2 r_{qs}$, respectively.

III. ROTOR FLUX CONTROL

The control schemes IRFOC and DRFOC are presented in Figure 1 and Figure 2, respectively. As can be observed, the decoupling of torque and flux control occurs naturally when the rotor flux reference frame is used. The $abc{\rightarrow}dq$ transformation and inverse transformation that are needed in vector control of three-phase induction motor are omitted in the implementation of the direct vector control and the indirect vector control of asymmetrical two-phase induction motor. The flux controller uses the difference between the actual rotor flux magnitude and the rotor flux reference as input and the output is the reference i^{er}_{ds1} current. The same way, the i^{er}_{qs1} reference current is the output of the torque error controller. Since a voltage inverter is used, these current references are converted to voltage references. This stage uses a voltage decoupler that properly positions the voltages in the rotor-flux reference frame. The knowledge of the desired position of the rotor flux differs in both modes of control. The direct vector control needs a flux control loop based on the estimation of the rotor flux to establish the flux position, while the indirect control is obtained by integrating the sum of the rotor speed ω_r and the slip frequency ω_{err}.

Based on the mathematical model of the two-phase asymmetric induction motor, the field orientation can be applied to the control of the electromagnetic torque and the rotor flux. To achieve vector control of induction motor, the d-axis of the reference frame must be oriented along the rotor flux vector [8, 9],

$$\lambda^{er}_{dr} = \lambda_r \; ; \; \lambda^{er}_{qr} = 0, \tag{17}$$

where $\lambda_r = \lambda^{er}_{dr} + j\lambda^{er}_{qr}$. The control system can be established by (11) and (13) and rewritten (14) in terms of stator current and rotor flux, yielding

$$\begin{bmatrix} \dfrac{d\lambda^{er}_{dr}}{dt} \\ \dfrac{d\lambda^{er}_{qr}}{dt} \end{bmatrix} = \begin{bmatrix} -\dfrac{1}{\tau_r} & (\omega_{er} - \omega_r) \\ -(\omega_{er} - \omega_r) & -\dfrac{1}{\tau_r} \end{bmatrix} \begin{bmatrix} \lambda^{er}_{dr} \\ \lambda^{er}_{qr} \end{bmatrix}$$
$$+ \begin{bmatrix} \dfrac{L_{dsr}}{\tau_r} & 0 \\ 0 & \dfrac{L_{dsr}}{\tau_r} \end{bmatrix} \begin{bmatrix} i^{er}_{ds1} \\ i^{er}_{qs1} \end{bmatrix} \tag{18}$$

$$T_e = P \frac{L_{dsr}}{L_r} [i^{er}_{qs1} \lambda^{er}_{dr} - i^{er}_{ds1} \lambda^{er}_{qr}] \tag{19}$$

where $\tau_r = L_r / r_r$ is the rotor time constant. Appling the conditions (17) to (18) and (19), the control equations can be written as

$$\frac{d\lambda_r}{dt} + \frac{\lambda_r}{\tau_r} = \frac{1}{\tau_r} L_{dsr} i^{er}_{ds1} \tag{20}$$

$$\omega_{err} \lambda_r = \frac{1}{\tau_r} L_{dsr} i^{er}_{qs1} \tag{21}$$

$$T_e = p \frac{L_{dsr}}{L_r} i^{er}_{qs1} \lambda_r \tag{22}$$

where $\omega_{err} = \omega_{er} - \omega_r$ is the slip frequency. According to Equation (20) the flux is controlled by i^{er}_{ds1} and Equation (22) shows that the electromagnetic torque is proportional to i^{er}_{qs1}. A three-leg inverter is used to drive a single-phase induction motor in a two-phase configuration, presented in Figure 3. To diminish the harmonic distortion in the output voltages, the ripples in stator current, a space vector PWM (SVPWM) is employed. In [9,11-12], a SVPWM applied to a single-phase induction motor drive system making use of a three-leg inverter is presented and is used in this paper.

978-1-7281-4181-7/19 $31.00 © 2019 IEEE

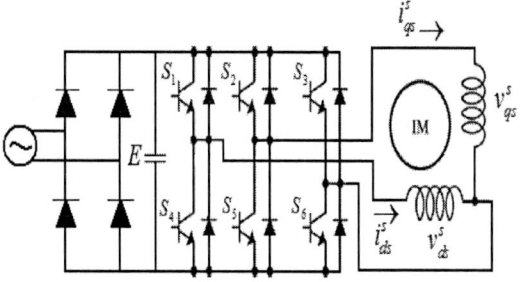

Fig. 3: Thee-leg inverter topology

A. Indirect Rotor Field Oriented Control (IRFOC)

The indirect strategy uses the slip frequency to estimate the position of the rotor flux to obtain the field orientation. Then rewriting (21),

$$\omega_{er} - \omega_r = \frac{1}{\tau_r \lambda_r^*} L_{dsr} i_{qs1}^{er*} \qquad (23)$$

The rotor flux angle can be calculated by

$$\theta_{er} = \int (\omega_{er}) = \int (\omega_{err} + \omega_r) \qquad (24)$$

Using Laplace transformation in (20), the rotor flux is estimated by

$$\lambda_r = \frac{L_{dsr}}{(s\tau_r + 1)} i_{ds1}^{er} \qquad (25)$$

The slip frequency ω_{er} can be adjusted to a specific value that can guarantee the decoupling of the stator currents in components related to the flux (i_{ds1}^{er}) and related to the torque (i_{qs1}^{er}). However, observing (23) one can notice the dependency on the rotor resistance, which can brings an orientation error of rotor flux vector due to parametric variations. As presented in Figure 1, T_e^* and λ_r^* represent the reference electromagnetic torque and the amplitude of the rotor flux, respectively. The coordinate transformation is represented by $e^{j\theta_{er}}$.

B. Direct Rotor Field Oriented Control (DRFOC)

To establish a direct vector control system, the estimation of the rotor flux is made by the direct measure of the stator currents is stationary frame. This can be accomplished making use of (3) and (4) and relations (7) and (8) as followed

$$\lambda_{qr}^s = \frac{L_r}{L_{dsr}} (\lambda_{qs1}^s - \sigma_{qs} L_{qs}' i_{qs1}^s) \qquad (26)$$

$$\lambda_{dr}^s = \frac{L_r}{L_{dsr}} (\lambda_{ds1}^s - \sigma_{ds} L_{ds} i_{ds1}^s) \qquad (27)$$

where $\sigma_{ds} = 1 - \dfrac{L_{dsr}^2}{L_{ds}.L_r}$ and $\sigma_{qs} = 1 - \dfrac{L_{qsr}^2}{L_{sq}.L_r}$.

From (1), (7) and (8) is possible to estimate the stator flux as

$$\lambda_{qs1}^s = \int (v_{qs1}^s - r_{qs}' i_{qs1}^s) \qquad (28)$$

$$\lambda_{ds1}^s = \int (v_{ds1}^s - r_{ds} i_{ds1}^s) \qquad (29)$$

Observing (28) and (29), the flux estimation requires an ideal integration which can be a problem when working at low frequencies, but it is not dependent of the rotor time constant which varies with the temperature of rotor resistance [9]. The block diagram that represents the system is shown in Figure 2.

C. Decoupling Compensation

Since a voltage command PWM is used, the stator dq currents need an independent control. That way, a dynamic equation that relates the stator currents and the stator voltages must be derived. Using (10), (12) and (13) the decoupling equations can be written as

$$v_{ds1}^{er} = \left(\frac{r_{ds} + r_{qs}'}{2} + \frac{L_{dsr}^2}{L_r \tau_r} \right) i_{ds1}^{er} + \sigma_{ds} L_{ds} \frac{di_{ds1}^{er}}{dt} + u_{ds}^{er} \quad (30)$$

$$v_{qs1}^{er} = \left(\frac{r_{ds} + r_{qs}'}{2} + \frac{L_{dsr}^2}{L_r \tau_r} \right) i_{qs1}^{er} + \sigma_{ds} L_{ds} \frac{di_{qs1}^{er}}{dt} + u_{qs}^{er} \quad (31)$$

where u_{ds}^{er} and u_{qs}^{er} represent the decoupling terms associated with the asymmetry of the stator windings, as mentioned before. The decoupling terms are given by

$$u_{ds}^{er} = -\frac{L_{dsr}\lambda_r}{L_r \tau_r} - \omega_{er} i_{qs1}^{er} \sigma_{ds} L_{ds} + \frac{r_{ds} - r_{qs}'}{2} \cos(2\theta_{er}) i_{ds1}^{er}$$

$$+ \frac{r_{ds} - r_{qs}'}{2} sen(2\theta_{er}) i_{qs1}^{er}$$

$$(32)$$

$$u_{qs}^{er} = -\omega_r \frac{L_{srd}\lambda_r}{L_r \tau_r} - \omega_{er} \sigma_{ds} L_{ds} i_{ds1}^{er} - \frac{r_{ds} - r_{qs}'}{2} \cos(2\theta_{er}) i_{qs1}^{er}$$

$$+ \frac{r_{ds} - r_{qs}'}{2} sen(2\theta_{er}) i_{ds1}^{er}$$

$$(33).$$

978-1-7281-4181-7/19 $31.00 © 2019 IEEE

IV. EXPERIMENTAL RESULTS

In order to verify the dynamic behavior of both control schemes, IRFOC and DRFOC, an experimental set up was implemented according to the block diagrams shown in Figure 1 and Figure 2. The two-phase asymmetric induction motor has the following parameters: $r_{ds} = 1.7\Omega$, $r_{qs} = 5.3\Omega$, $r_r = 2.42\Omega$, $L_{ds} = 67.2mH$, $L_{qs} = 168.3mH$, $L_r = .21mH$, $L_{qsr} = 100.3mH$, $L_{dsr} = 63mH$. The motor is 4 poles, 60Hz and 1/2 Hp, the rated speed is 180 rad/s and the rated rotor flux is 0.4 Wb. The systems were tested using a TMS320F2812 DSP connected to a three-leg inverter. The switching frequency was set to 5 kHz.

The motor performance was verified under load conditions. The shaft speed is indicated by ω_r and tracks the reference input ω_r^*. The rotor speed reference is kept at 180 rad/s and a load of approximately $2Nm$ was imposed at two instants. For the test it was used as load a 0.8 HP DC motor connected to a load resistor that serves as the dynamometer. Figure 4(a) presents the speed response under DRFOC showing that the system senses the load but keeps tracking the reference speed. For the IRFOC, shown in Figure 4(b), the load does not have a significant impact and the rotor speed response is satisfactory. The corresponding magnitude of the rotor flux is shown in Figure 5(a) and Figure 5(b) for DRFOC and IRFOC, respectively. The starting transient of the speed reflects directly in the flux waveform, which increases simultaneously.

Under rated load the flux control command acts properly and after removing the load the flux keeps tracking the reference. This demonstrates the decoupling between the torque and flux control. It can be seen that for the DRFOC the flux estimation exhibits noise problems. It occurs due to the open-loop integrator which is acting on the back emf voltage terms. Figure 6(a) and Figure 6(b) show the stator currents waveforms observed when the load is applied. The difference in amplitude of the stator currents shows the asymmetry of the single-phase induction motor windings. Experimental results show that the dynamic performance of both control strategies is satisfactory.

V. CONCLUSIONS

This paper discussed two vector control strategies applied to a two-phase asymmetric induction motor. The main objective was to compare DRFOC and IRFOC based on their dynamic behavior and show that the symmetrical transformation used is a viable solution. Observing their control structure, it is possible to implement both methods in the same hardware system with little modifications, since the main difference is the determination of the orientation angle. The dynamic modeling proposed makes use of decoupling terms that acts as input for the PWM modulator when using a voltage source inverter.

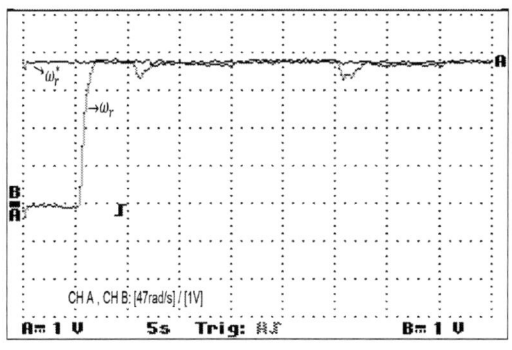

(a)

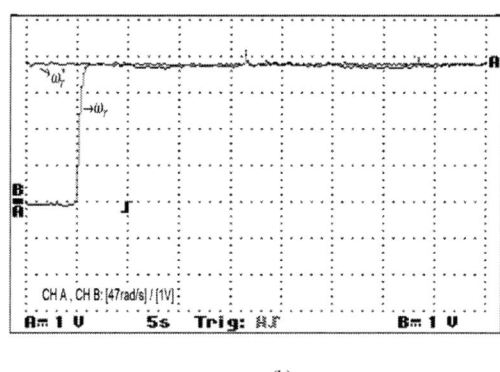

(b)

Fig. 4. Convergence of the shaft speed at rated load for (a) DRFOC and (b) IRFOC

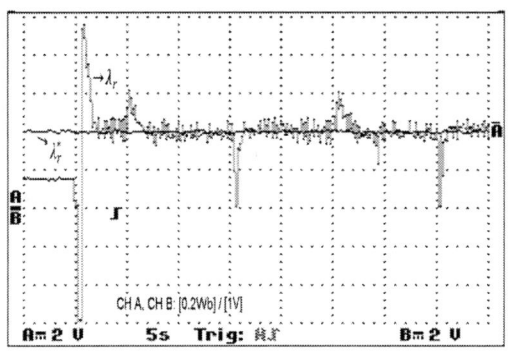

(a)

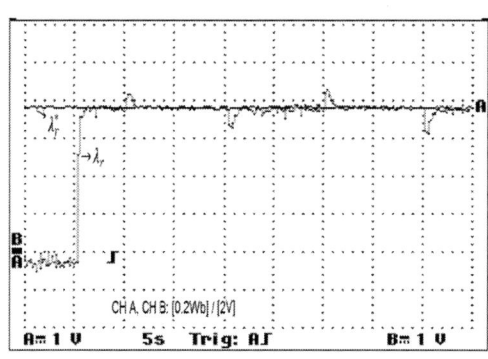

(b)

Fig. 5. Rotor flux waveform, reference and actual, for (a) DRFOC and (b) IRFOC

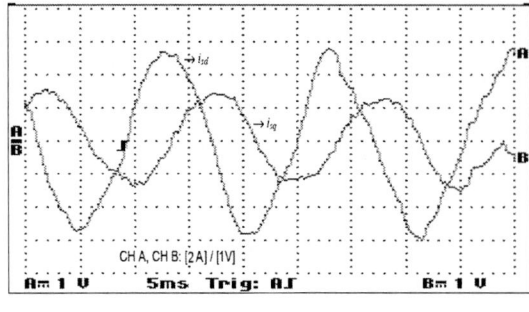 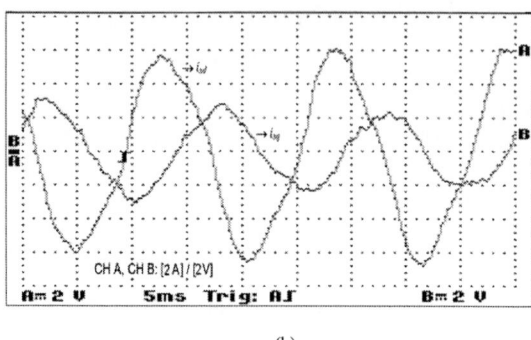

(a)   (b)

Fig. 6. Stator current waveform at rated load, for (a) DRFOC and (b) IRFOC.

Experimental results were considered satisfactory and shown that field oriented control can enable two-phase asymmetric induction motor to produce a better performance which is very useful in low power applications.

In order to show the effectiveness of the control systems two conditions were tested: no load and loaded. At no load, the motor speed was reversed with the shaft speed tracking the reference input during transients. Both control methods presented good dynamic performance. For the test with load disturbance, the induction motor was driven with a constant speed profile. The control algorithms gave satisfactory results in terms of speed tracking and load rejection. Also, the flux response for the DRFOC algorithm presented some noise, which was caused by the ideal integrator used in the rotor flux loop. In addition, in IRFOC algorithm the flux response presented a smooth convergence. It is important to notice that vector control systems may not increase the final cost of the system, since the capability of processing of the integrated circuits has increased over the years

REFERENCES

[1] P. C. Krause, O. Wasynczuk, and S. D. Sudhoff, *Analysis of Electric Machinery.* Piscataway, NJ: IEEE Press, 1995.

[2] E. R. Collins Jr., H. B. Puttgen, W. E. Sayle, "Single-phase induction motor adjustable speed drive: direct phase angle control of the auxiliary winding supply," *Conf. Rec. of the IEEE Industry Applications Society Annual Meeting,* 1988 vol. 1, pp. 246-252.

[3] D. G. Holmes, A. Kotsopoulos, "Variable speed control of single and two phase induction motors using a three phase voltage source inverter," *Conf. Rec. of the IEEE Industry Applications Society Annual Meeting,* 1993, vol. 1, pp. 613-620.

[4] S. S. Wekhande, B. N. Chaudhari, S. V. Dhopte and R. K. Sharma, "A Low Cost Inverter Drive for 2 Phase Induction Motor" *in IEEE 1999 International Conference on Power Electronics and Drive Systems,* PEDS'99, July 1999, Hong Kong.

[5] M. B. R. Corrêa, C. B. Jacobina, A. M. N. Lima and E. R. C. da Silva "Rotor-flux-oriented control of a single-phase induction motor drive," *IEEE Trans. Ind. Electron.,* vol. 47, no. 4, pp. 832-841, Aug. 2000.

[6] M. B. R. Corrêa, C. B. Jacobina, A. M. N. Lima, and E. R. C. da Silva,"Vector Control Strategies for Single-Phase Induction Motor Drive Systems," *IEEE Trans. Ind. Electron.,* vol. 51, no. 5, pp. 1073–1080, Oct. 2004.

[7] M. Popescu, D.M. Ionel and D.G. Dorrell, "Vector control of unsymmetrical two-phase induction machines," *in Conf. Rec. IEEE IEMDC'01,* 2001, pp. 95-101.

[8] W. Leonard, *Control of Electrical Drives.* Springer, Berlin: Heidelberg, 1985.

[9] B. K. Bose, *Power Electronics and AC Drives.* Prentice Hall, 1986.

[10] M. B. R. Corrêa, C. B. Jacobina, A. M. N. Lima and E. R. C. da Silva, "A Three-Leg Voltage Source Inverter for Two-Phase AC Motor Drive Systems," *Transactions on Power Electronics,* vol. 17, no. 4, pp. 517- 523, July 2002.

[11] J. Do-Hyun and Y. Duck-Yong, "Space-Vector PWM Technique for Two-Phase Inverter-Fed Two-Phase Induction Motors," *IEEE Trans. On Industry Applications,* vol. 39, no. 2, pp. 542-549, Mar/Apr 2003.

[12] M. A. Jabbar, A. M. Khambadkone and Z. Yanfeg, "Space-Vector Modulation in a Two Phase Induction Motor Drive for Constant-Power Operation," *Transactions Ind. Electron.,* vol. 51, no. 5, pp. 1081-1088, Oct. 2004.

[13] D. H. Jang, "Problems Incurred in a Vector-Controlled Single-Phase Induction Motor, and a Proposal for a Vector-Controlled Two-Phase Induction Motor as a Replacement," in *IEEE Transactions on Power Electronics,* vol. 28, no. 1, pp. 526-536, Jan. 2013.

[14] F. Blaabjerg, F. Lungeanu, K. Skaug and M. Tonnes, "Two-phase induction motor drives," in IEEE Industry Applications Magazine, vol. 10, no. 4, pp. 24-32, July-Aug. 2004.

978-1-7281-4181-7/19 $31.00 © 2019 IEEE

Dynamic Analysis of Self-excited SRG Operating in Open Loop

Lucas José Lemes
Academical Area - IFG
Federal Institute of Education, Science
and Technology of Goias - IFG
Itumbiara, Brazil.
ljlemes1@gmail.com

Victor Regis Bernardeli
Academical Area - IFG
Federal Institute of Education, Science
and Technology of Goias - IFG
Itumbiara, Brazil.
victor.bernadeli@ifg.edu.br

Luciano Coutinho Gomes
Faculty of Electrical Engineering
Federal University of Uberlândia -
UFU
Uberlândia, Brazil.
lcgomes2005@gmail.com

Darizon Alves de Andrade
Faculty of Electrical Engineering
Federal University of Uberlândia -
UFU
Uberlândia, Brazil
darizon.andrade@gmail.com

Ghunter Paulo Viajante
Academical Area - IFG
Federal Institute of Education, Science
and Technology of Goias - IFG
Itumbiara, Brazil.
ghunter.viajante@ifg.edu.br

Marcos Antonio Arantes de Freitas
Academical Area - IFG
Federal Institute of Education, Science
and Technology of Goias - IFG
Itumbiara, Brazil
marcos.freitas@ifg.edu.br

Abstract—**This paper presents an analysis of the SRG operating in self-excited mode, with Half Bridge converter and a parallel capacitor. The simulations are performed based on a nonlinear model considering the effect of magnetic saturation, the model equations are described on this paper. The model allows static simulations to build the inductance, flux and torque profiles, as well as the dynamic simulations. Analyses about the effect of rotor speed and triggers angles over a resistive load are realized.**

Keywords— Inductance, magnetic saturation, modeling, motor drives, nonlinear systems, reluctance machine, self-excited.

I. INTRODUCTION

The Switched reluctance machine (SRM), has been a frequent target of study, mostly on the last twenty years, due to the advances in control methods and sensors that allowed the operation of the machine in different topologies. Its robustness, applicability to variable speeds and reduced cost due to the absence of windings on rotor, make the SRM an interesting option for some applications, such as automotive [1], aerospace [2] and renewable energy [3,4].

The modeling and representation of the machine is a challenge due to the high nonlinearity characteristics of the intrinsic relationship between inductance, current and rotor position, which if ignored, the model may fail in simulate the dynamic of the SRM operating as a self-excited generator (SRG). A strategy to deal with the nonlinear aspects of the machine is use Fourier series to consider magnetic saturation, this kind of models were developed by Andrade [5] and Hannoun [6]. Other strategy is to use look up tables builded with measured or FEA data in association with dynamic nonlinear models as presented by Sotelo [7]. There are also analytical models used for simulation and real-time implementation, such as those presented by Hossain [8] and Ding [9].

Accurate models are used in important investigations about the SRM dynamic, a model considering asymmetries in SRM structure and allow simulation under fault conditions was presented by Weiss [10]. A study about the impact of voltage fluctuation on phase current done by Wang [11]. Those investigations provide crucial information for determination of better controls methods and improve project phase.

This paper presents a study about the dynamic operation of the SRG in self-excited mode based on a nonlinear model; a test of a linear model in same conditions; the voltage behavior over the same load with different speeds; the effect of a variation of on/off angles on the output voltage. The modeling will consider the nonlinearity of the system using Fourier series.

Additionally, for self-excited operation mode, in open loop, it is indispensable that the model includes magnetic saturation. If the model neglects the magnetic circuit saturation, the voltage generated at terminals shall not stabilize, therefore, it will not represent a real operation of SRG. This paper brings as new contribution a study about the operation with different loads and different on/off angles, thus bringing optimal point of operation for the generator in an operation range with different load values.

II. STATIC MODEL

A. Inductance as function of rotor position

The inductance x rotor position curve, shows one of the SRG nonlinearities characteristic, a strategy to consider that, is fractionate the curve in line segments, see Fig 1.

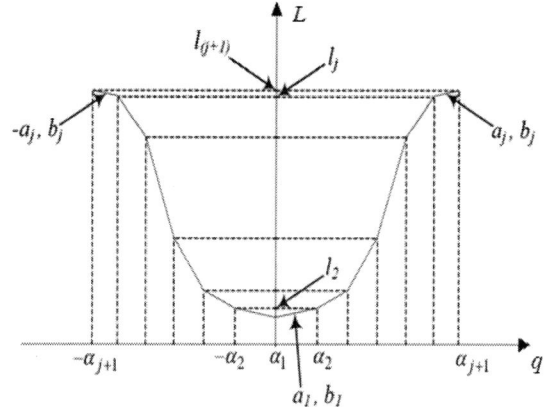

Fig. 1. Inductance x rotor position

Thus the inductance can be expressed as function of rotor position, through the straight line equation:

$$L(q) = \begin{cases} -a_j q + b_j & para\ q < 0 \\ a_j q + b_j & para\ q > 0 \end{cases} \qquad (1)$$

Where a_j is the angular coefficient and b_j the linear coefficient, expressed as:

$$a_j = \frac{l_{j+1} - l_j}{\alpha_{j+1} - \alpha_j} \qquad (2)$$

$$b_j = l_j - a_j \alpha_j \qquad (3)$$

Starting from these assumptions, the inductance as function of rotor position can be expressed with Fourier series:

$$L(q) = \frac{1}{G} \sum_{j=1}^{m} \left[(l_{j+1} + l_j)(\alpha_{j+1} - \alpha_j) \right]$$
$$+ \sum_{n=1}^{\infty} \left\{ \sum_{j=1}^{m} \left[(l_{j+1} A_{(j,n)} \right. \right.$$
$$\left. \left. - l_j B_{(j,n)}) \right] \right\} \cos(n P_r q) \qquad (4)$$

Where $A_{(j,n)}$ and $B_{(j,n)}$ are:

$$A_{(j,n)} = \frac{2}{np} \left(sen(n P_r \alpha_{j+1}) + \frac{\cos(n P_r \alpha_{j+1}) - \cos(n P_r \alpha_j)}{(\alpha_{j+1} - \alpha_j) n P_r} \right)$$
$$B_{(j,n)} = \frac{2}{np} \left(sen(n P_r \alpha_j) + \frac{\cos(n P_r \alpha_{j+1}) - \cos(n P_r \alpha_j)}{(\alpha_{j+1} - \alpha_j) n P_r} \right) \qquad (5)$$

This model already has been published by Bernardeli, for further information about the modeling progress presented in this entire section see [12].

B. Incremental Inductance

The equation of the inductance as function of rotor position was introduced, next step is to adjust the model with the current influence, a strategy is to make an interpolation, thus, a third order polynomial was used:

$$l_j = l_j(i, \alpha_j) = C_{3j} i^3 + C_{2j} i^2 + C_{1j} i + C_{0j}$$

The inductance as function of current and rotor position may be written:

$$L(i,q) = \frac{1}{G} \sum_{j=1}^{m} \left\{ \left[(C_{3j+1} + C_{3j}) i^3 + (C_{2j+1} + C_{2j}) i^2 + (C_{1j+1} + C_{1j}) i \right. \right.$$
$$\left. + (C_{0j+1} + C_{0j}) \right] (\alpha_{j+1} - \alpha_j) \right\}$$
$$+ \sum_{n=1}^{\infty} \left\{ \sum_{j=1}^{m} \left[\left((C_{3j+1} i^3 + C_{2j+1} i^2 + C_{1j+1} i + C_{0j+1}) A_{(j,n)} \right. \right. \right.$$
$$\left. \left. \left. - (C_{3j} i^3 + C_{2j} i^2 + C_{1j} i + C_{0j}) B_{(j,n)} \right) \right] \right\} \cos(n P_r q) \qquad (6)$$

For the representation of inductance as function of current and rotor position is given the name Incremental Inductance. This paper uses initial data from FEA simulation due to the current desired values close to 50A, which are difficult to obtain through experimental methods. Using the initial data and (6) is possible to perform a static simulation to create an Inductance profile of the SRG:

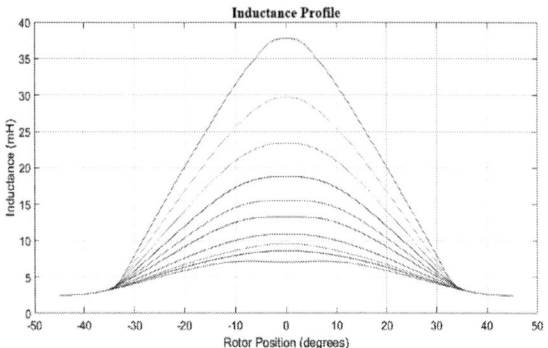

Fig. 2. Inductance profile 2d.

Note that in Fig.2 the top curve is the lowest current and the bottom is the highest, the increasing of current level results on a drop of inductance, this effect is caused by the magnetic saturation which limit the flux, the inductance and current are two variables inversely proportional.

C. Flux, Coenergy and Torque

The flux equation can be deducted as the product of inductance and current:

$$\lambda(i,q) = \frac{1}{G} \sum_{j=1}^{m} \left\{ \left[(C_{3j+1} + C_{3j}) i^4 + (C_{2j+1} + C_{2j}) i^3 + (C_{1j+1} + C_{1j}) i^2 \right. \right.$$
$$\left. + (C_{0j+1} + C_{0j}) i \right] (\alpha_{j+1} - \alpha_j) \right\}$$
$$+ \sum_{n=1}^{\infty} \left\{ \sum_{j=1}^{m} \left[\left((C_{3j+1} i^4 + C_{2j+1} i^3 + C_{1j+1} i^2 + C_{0j+1} i) A_{(j,n)} \right. \right. \right.$$
$$\left. \left. \left. - (C_{3j} i^4 + C_{2j} i^3 + C_{1j} i^2 + C_{0j} i) B_{(j,n)} \right) \right] \right\} \cos(n P_r q) \qquad (7)$$

From (7), the flux profile of SRG may be obtained through static simulation:

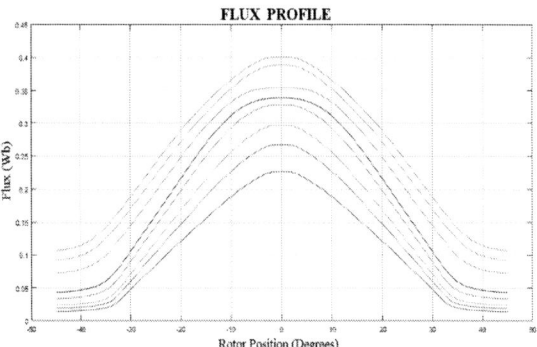

Fig. 3. Flux profile 2d.

Fig. 3 is the Flux profile of the SRG, Flux x Rotor Position, the top curves corresponds to the highest currents, and the bottom curves to the lowest. If the magnetic saturation were neglect the relation between flux and current could be direct proportional.

Integrating flux in relation of current, the coenergy is given as:

978-1-7281-4181-7/19 $31.00 © 2019 IEEE

$$W_j^{coe}(i,q) = \frac{1}{G}\sum_{j=1}^{m}\left\{\left[(C_{3j+1}+C_{3j})\frac{i^5}{5}+(C_{2j+1}+C_{2j})\frac{i^4}{4}+(C_{1j+1}\right.\right.$$
$$+C_{1j})\frac{i^3}{3}+(C_{0j+1}+C_{0j})\frac{i^2}{2}\bigg](\alpha_{j+1}-\alpha_j)\bigg\}$$
$$+\sum_{n=1}^{\infty}\left\{\sum_{j=1}^{m}\left[\left(\left(C_{3j+1}\frac{i^5}{5}+C_{2j+1}\frac{i^4}{4}+C_{1j+1}\frac{i^3}{3}\right.\right.\right.$$
$$\left.+C_{0j+1}\frac{i^2}{2}\right)A_{(j,n)}\right.$$
$$\left.\left.\left.-\left(C_{3j}\frac{i^5}{5}+C_{2j}\frac{i^4}{4}+C_{1j}\frac{i^3}{3}+C_{0j}\frac{i^2}{2}\right)B_{(j,n)}\right)\right]\right\}\cos(nP_r q) \tag{8}$$

The coenergy can be used to find the electromagnetic torque, deriving the equation in relation of rotor position:

$$T_{emag} = -nP_r\sum_{n=1}^{\infty}\left\{\sum_{j=1}^{m}\left[\left(\left(C_{3j+1}\frac{i^5}{5}+C_{2j+1}\frac{i^4}{4}+C_{1j+1}\frac{i^3}{3}\right.\right.\right.\right.$$
$$\left.+C_{0j+1}\frac{i^2}{2}\right)A_{(j,n)}$$
$$\left.\left.\left.-\left(C_{3j}\frac{i^5}{5}+C_{2j}\frac{i^4}{4}+C_{1j}\frac{i^3}{3}+C_{0j}\frac{i^2}{2}\right)B_{(j,n)}\right)\right]\right\}\operatorname{sen}(nP_r q) \tag{9}$$

Using (9) is possible to build the torque profile:

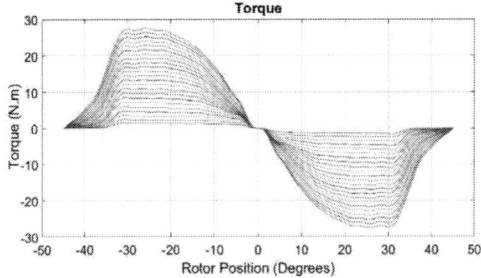

Fig. 4. Torque profile 2d

Fig. 4 show the periodic form of the electromagnetic torque, the increasing of current causes also an increase of electromagnetic torque, the switches conduct angles must consider this profile, determine if the machine will provide or consume torque, operating in motoring or generating mode.

III. Dynamic Model

A. Electrical Equation

Considering that inductance between phases may be neglected due to its levels are lower than 1%, the electric equation that describes one phase of machine for constant speed is given as:

$$v = Ri + L(i,q)\frac{di}{dt} + e \tag{10}$$

Where e is the back electromotive force:

$$e = i\omega\frac{\partial L(i,q)}{\partial q} \tag{11}$$

Where ω is the angular speed, substituting (11) in (10):

$$v = Ri + L(i,q)\frac{di}{dt} + i\frac{dq}{dt}\frac{\partial L(i,q)}{\partial q} \tag{12}$$

This equation represents one phase of the machine.

B. Mechanical Equation

The SRG must receive torque enough to balance with rotor inertia, friction and the electromagnetic torque, thus the equation that define the mechanic torque is expressed as:

$$T_m = T_{emag} + D\omega + J\frac{d\omega}{dt} \tag{13}$$

Where T_{emag} is the sum of electromagnetic torque of all phases:

$$T_m = \left(\frac{\partial W_1^{coe}}{\partial q} + \frac{\partial W_2^{coe}}{\partial q} + \cdots + \frac{\partial W_F^{coe}}{\partial q}\right) + D\omega + J\frac{d\omega}{dt} \tag{14}$$

The equations can be grouped in a matrix, this will allow the application of computational tools:

$$[V] = [R][I] + [L][\dot{I}] \tag{15}$$

$$\begin{bmatrix} v_1 \\ v_2 \\ v_3 \\ T_m \\ 0 \end{bmatrix} = \begin{bmatrix} R_1 & 0 & 0 & 0 & 0 \\ 0 & R_2 & 0 & 0 & 0 \\ 0 & 0 & R_3 & 0 & 0 \\ T_{e1} & T_{e2} & T_{e3} & D & 0 \\ 0 & 0 & 0 & -1 & 0 \end{bmatrix}\begin{bmatrix} i_1 \\ i_2 \\ i_3 \\ \omega \\ q \end{bmatrix} + \begin{bmatrix} L_1 & 0 & 0 & 0 & i_1\frac{\partial L_1(i,q)}{\partial q} \\ 0 & L_2 & 0 & 0 & i_2\frac{\partial L_2(i,q)}{\partial q} \\ 0 & 0 & L_3 & 0 & i_3\frac{\partial L_3(i,q)}{\partial q} \\ 0 & 0 & 0 & J & 0 \\ 0 & 0 & 0 & 0 & 1 \end{bmatrix}\begin{bmatrix} \dot{i}_1 \\ \dot{i}_2 \\ \dot{i}_3 \\ \dot{\omega} \\ \dot{q} \end{bmatrix} \tag{16}$$

C. Preliminary results

The study object of this paper is the SRM showed in Fig.5. Using the matrix (16), the inductance profile and data of the SRM from table I:

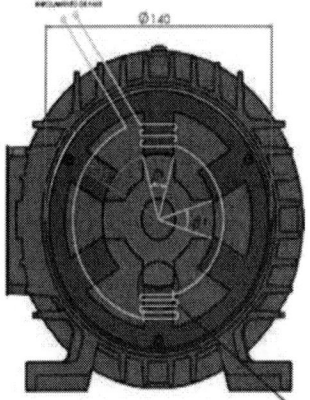

Fig. 5. Prototype in study

TABLE I. Machine Characteristics

Item	Value
Power	1 HP
Nominal Voltage	220 V
Nominal Current	10V
Nominal Speed	1200 rpm
Stator Poles	6
Rotor Poles	4
Friction Coefficient	0.026 N.m.s
Inertia	0.0028 kg.m²

The simulation of the SRG operating in self-excited mode is possible. The converter used was the Half bridge with a parallel capacitor, using open loop control on switches.

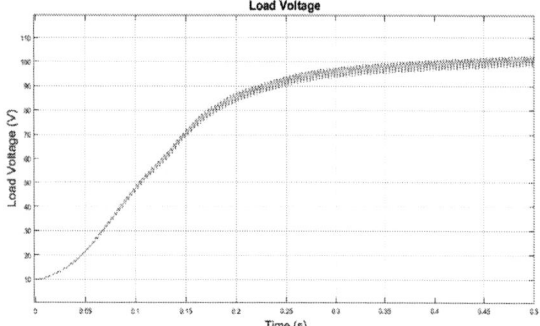

Fig. 6.   Voltage x Time

The Fig. 6 shows the waveform of voltage on load during the transient. Fig. 7 shows the current and inductance on phase A during the steady state. It's possible to see the inductance decreasing faster with the increase of current.

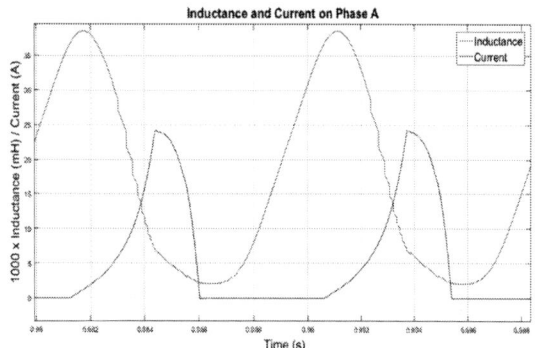

Fig. 7.   Inductance and Phase Current x Time

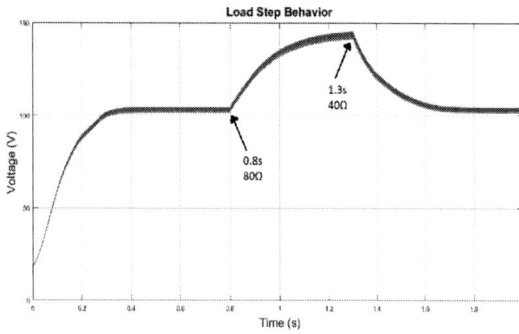

Fig. 8.   Load Step, Voltage x Time.

Fig. 8 shows SRM voltage behavior during load steps, the simulation starts with 40 Ω load, at 0,8s load is changed to 80 Ω, and finally at 1,3s the load is changed back to 40 Ω. The open loop operation presents no fast answers to the step, as may be observed in the softly behavior transition between loads.

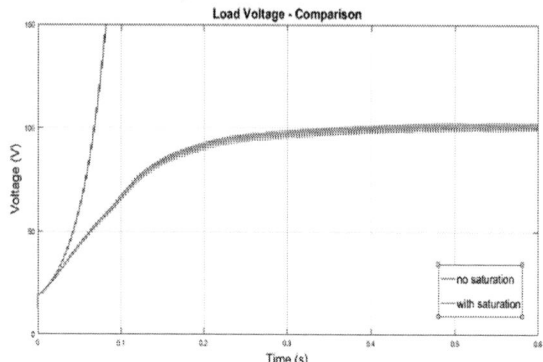

Fig. 9.   Voltage x Time

Fig. 9 shows a comparison between the previous simulation and a simulation neglecting the magnetic saturation, the instability of system causes the rise of voltage on load to infinite values. Thus, It proves the impossibility of simulate SRG in self-excited mode without considering the saturation.

IV.   SIMULATION ANALYSIS

For speed effect analyses, the system was simulated with the speeds, 800rpm, 1200rpm, 1600rpm e 2000rpm, and the loads, 10Ω, 20Ω, 40Ω e 80Ω. Fig. 10 shows the results.

The influence of the speed on load voltage is almost linear, and very expressive, for 10Ω load (Fig. 10 A) the variation of speed from 800rpm to 2000rpm causes a rise of 48,27% on voltage, with the same speed range for a load of 80Ω (Fig. 10 D) the voltage variation was 145,58%.

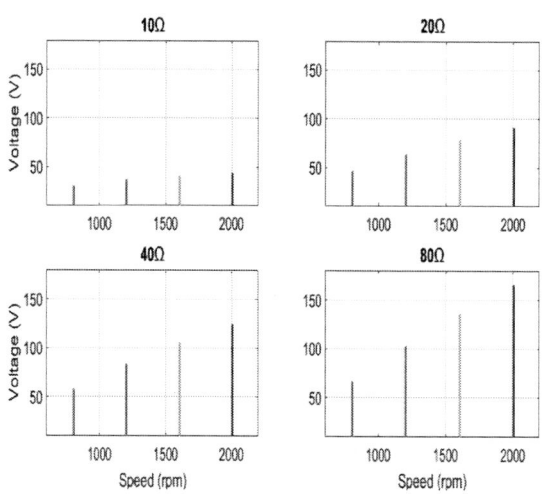

Fig. 10. Voltage x Speed

The next analyses are the effect of delaying or advancing on/off angles on load voltage. Fig. 11 shows results for 10Ω, delaying on/off angles the transient period becomes faster (Fig. 11 A), however, voltage on steady state is smaller (Fig. 11 B). Advance in on/off increases the transient period (Fig. 11 C) also (-4°) results on failure of operation, it's interesting to note that -1° causes a rise of voltage on steady state (Fig. 11 D).

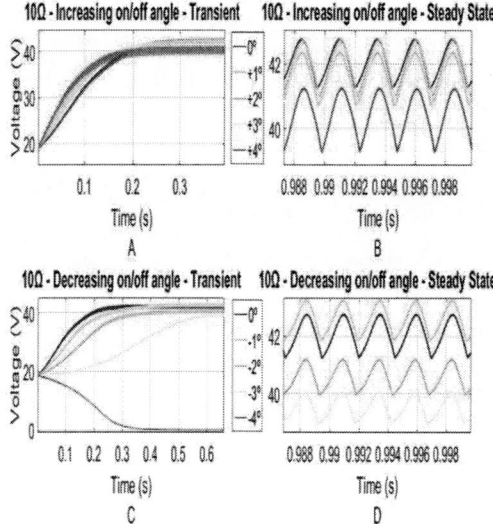

Fig. 11. 10Ω Voltage x Time, changing angles.

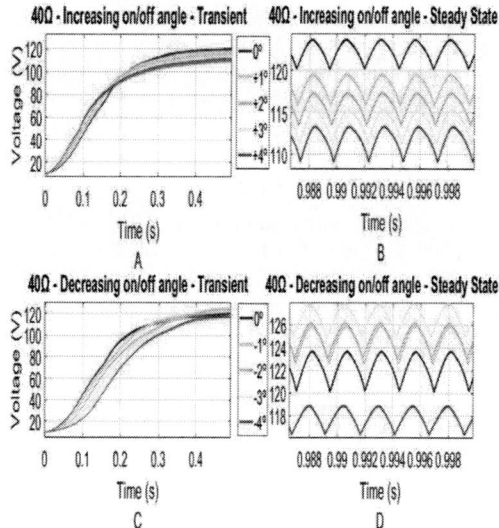

Fig. 13. 40Ω Voltage x Time, changing angles.

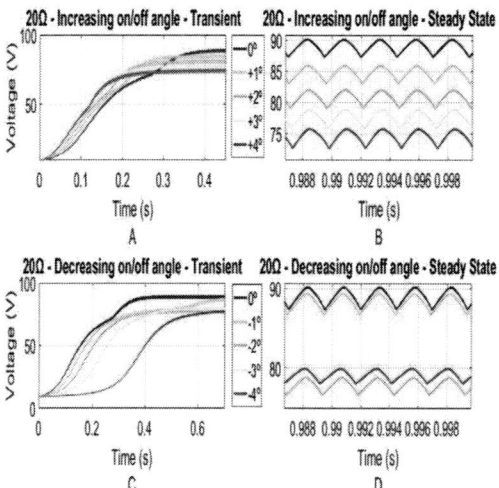

Fig. 12. 20Ω Voltage x Time, changing angles.

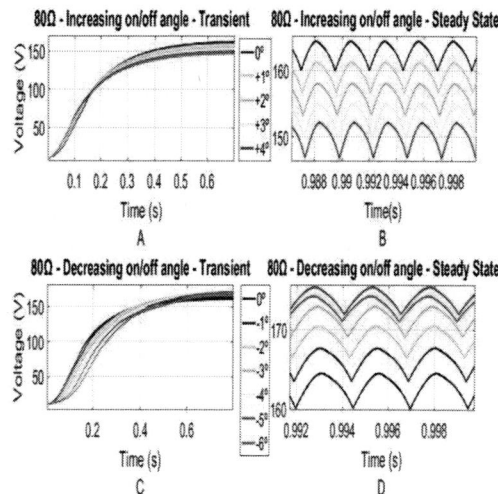

Fig. 14. 80Ω Voltage x Time, changing angles.

For 20Ω load, the increase of on/off angles has a similar effect to the previous load (Fig. 12 A and B). This time, the decrease of angles presented no operation failures, however, all angles resulted on lower voltage levels than the reference.

For 40Ω load, similar effects on transients can be noted (Fig. 13 A and C), increased angles keep giving lower voltage levels on steady state (Fig. 13 B), however, from -1° to -3° the voltage levels became higher than reference, -4° resulted on a drop (Fig. 13 D).

For 80Ω load, delayed angles presented the same effect of previous simulations (Fig. 14 A and B). The reduced angles keep increasing the transient period (Fig. 14 C), however, voltage level increased during steady state for negative angles, simulations were realized until the voltage drop (Fig. 14 D), from -1° to -5° the voltage progressively rises.

V. CONCLUSIONS

The representation of the SRM is challenge due to its highly nonlinear characteristics, the model used in this paper include the magnetic saturation through Fourier series, it allows the simulation of the SRG operating in self excited mode.

The analyses of speed effect on load voltage, shows an almost linear behavior, speed rising may imply in high increasing of voltage level, in the simulation the highest variation was of 145,58%.

The study about on/off angle has presented some findings, an increase on angles speeds up the transient period and decrease the voltage level during steady state, a decrease on angles slow the transient period, and depending of load, increase voltage level during steady state, this effect increase with load.

It's important to note that in the simulations were only used resistive loads, for future researches the study may be expanded using different loads and speed ranges, also it's possible to evaluate proficiency of machine in the specified circumstances by studying the power flow.

ACKNOWLEDGMENT

The authors thank the Goiás Research Support Foundation (FAPEG), through the so-called nr. 03/2015 of the First Projects Program (PPP) in partnership with CNPq.

REFERENCES

[1] UDDIN W., HUSAIN T., SOZER Y., HUSAIN I. "Design Methodology of a Switched Reluctance Machine for Off-Road Vehicle Applications" IEEE Transactions on Industry Applications, vol. 52, Issue 3, pag. 2138-2147, May-June 2016.

[2] BARTOLO J. B., DEGANO M., ESPINA J., GERADA C. "Design and Initial Testing of a High-Speed 45-kW Switched Reluctance Drive for Aerospace Application" IEEE Transactions on Industrial Electronics, vol. 64, Issue 2, pag. 988-997, Feb. 2017.

[3] BARROS T. A. S., SANTOS NETO P. J., NASCIMENTO FILHO P. S., MOREIRA A. B., RUPPERT FILHO E. "An Approach for Switched Reluctance Generator in a Wind Generation System With a Wide Range of Operation Speed", IEEE Transactions on Power Electronics, IEEE, Vol. 32, Issue 11, pag. 8277-8292, Nov. 2017.

[4] SANTOS NETO P. J., BARROS T. A. S., NASCIMENTO FILHO P. S., PAULA M. V., SOUZA R. R., RUPPERT FILHO E. "Design of Computational Experiment for Performance Optimization of a Switched Reluctance Generator in Wind Systems", Transactions on Energy Conversion, IEEE Vol. 33, Issue: 1, pag. 406-419, March 2018.

[5] ANDRADE D. A., HRISHNAN R. "Characterization of Switched Reluctance Machines Using Fourier Series Approach", Conference Record of the 2001 IEEE Industry Applications Conference. 36th IAS Annual Meeting, September 2001.

[6] HANNOUN H., HILAIRET M., MARCHAND C. "Analytical Modeling of Switched Reluctance Machines Including Saturation", IEEE International Electric Machines & Drives Conference, May 2007.

[7] SOTELO G. G., RIBEIRO M. R., EL-MANN M., ROLIM L. G. B., SILVA NETO J. L., ANDRADE JR R., FERREIRA A. C., STEPHAN R. M., SUEMITSU W. I. "Dynamic Non-Linear Model of a Switched Reluctance Machine for Operation as Motor/Generator" Sobraep, Eletronica de Potência, Vol. 15, no. 1, pag. 21-30, Feb. 2010.

[8] HOSSAIN S. A., HUSAIN I. "A Geometry Based Simplified Analytical Model of Switched Reluctance Machines for Real-Time Controller Implementation", 2002 IEEE 33rd Annual IEEE Power Electronics Specialists Conference. Proceedings , Cat. No.02CH37289, June 2002.

[9] DING W., LIANG D. "A fast Analytical Model for an Integrated Switched Reluctance Starter/Generator", IEEE Transactions on Energy Conversion, Vol. 25, Issue: 4, pag. 948 – 956, Dec. 2010.

[10] WEISS C. P., HUEBNER M., HENNEN M. D., DONCKER R. W. "Switched Reluctance Machine Model Considering Asymmetries and Enabling Dynamic Fault Simulation", IEEE, 2013 International Electric Machines & Drives Conference, Chicago, IL, USA, May 2013.

[11] WANG W., HUO Y., YAN S., DOU Y. "The impact of Excitation Voltage Fluctuation on Phase Current of Switched Reluctance Generator", IEEE Southern Power Electronics (SPEC), Dec 2017.

[12] BERNARDELI, V. R. ; Andrade ; FLEURY, A. W. ; Gomes, L. C. ; VIAJANTE, G. P. ; CABRAL, L. G. . Self-excited switched reluctance generator. In: COBEP 211, 2011, Natal. XI Brazilian Power Electronics Conference, 2011. v. 11.

Three-phase, four-wire PWM rectifier applied to variable frequency AC systems in airplane electric grid under fault conditions

Alexandre Galvão Bueno
School of Electrical and Computer Engeneering
Universsity of Campinas
Campinas, Brazil
amgbueno@gmail.com

José Antenor Pomilio
School of Electrical and Computer Engeneering
Universsity of Campinas
Campinas, Brazil
antenor@fee.unicamp.br

Abstract — **This paper analyses the use of PWM rectifier in variable frequency, three-phase, four-wire AC system as present in modern airplanes. The converter is analysed in terms of reliability for maintaining the correct operation of the symmetrical high-voltage DC bus and the power distribution among the phases of the AC source in case of fault. The paper discusses the standards for voltage and current quality and testing in aircraft environment. The PWM three-phase four-wire rectifier with double DC outputs is considered. Based on simulation, the topology is analysed regarding the stress on the circuits and the impact on the AC grid. It is shown the necessity of maintaining balanced DC load during the normal operation and also under fault in order to help preserving the AC power quality.**

Keywords—component, formatting, style, styling, insert (key words)

I. Introduction

The traditional aeronautical electric system operates at fixed frequency (400 Hz). To maintain regulated frequency while the turbine speed changes, an automatic gear box regulates the electric generator speed. The elimination of this heavy device, contributes to reduce the airplane weight, and operational costs, increasing the overall efficiency. This kind of improvement occurs in the context of the so called "More Electric Aircraft" – MEA - concept [1]. Thus, in modern airplanes, like the Boeing 787 and the Airbus A380 and A350, the electric generator operates at variable frequency (360 Hz to 800 Hz), transferring to the power electronics converters the role of correctly maintaining the electric system operation [2]. Notice that this situation is very different of the commercial electric grid, not only due to the higher frequency, but mainly due to the variable frequency.

Besides this, the aeronautical standards establish as a normal operation a frequency change as fast as 250 Hz/s. Additionally defines an upper limit of 500 Hz/s in which electronic equipment may disconnect but must automatically resume normal operations as soon as the frequency stabilizes [3,4], that represents an additional and challenging restriction.

This work was supported by the São Paulo Research Foundations (FAPESP - grant 2017/05565-7), and National Council for Scientific and Technological Development (CNPq – grants 302257/2015-2 and 401216/2016-0).

Due to the increasing electric power demand, the systems tend to use higher voltage levels. For example, the DC bus migrates from 270 V to 540 V (± 270 V), so minimizing the cabling mass and volume [5] for a given power demand.

Concerning the standards, the *MIL-STD-704F* [3] defines the conditions the embedded power grid must present. Its focus is the voltage (AC and DC) quality. Like in the commercial grid, there are limits for RMS levels, for voltage distortion and unbalance, etc. For example, Fig. 1 and Fig. 2 show the limits for the RMS AC voltage and the DC distortion, respectively.

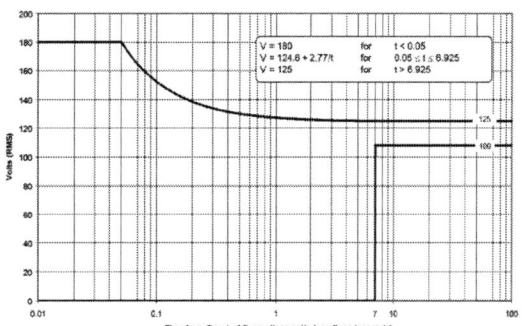

Fig. 1 AC over and under voltage limits, according to [3].

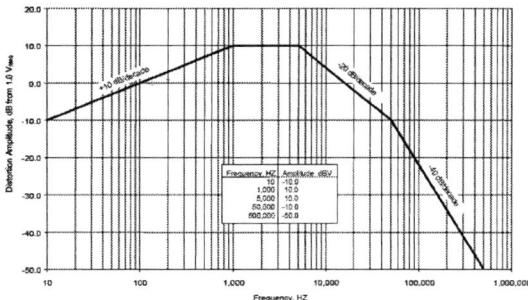

Fig. 2 DC voltage distortion limits, according to [3].

On the other hand, the standard *RTCA DO-160G* [4] determines limits and test procedures for the equipment that will be installed in the aircraft grid. This standard determines, among other aspects, the current distortion and unbalance, the ride-through capacity, etc. For example, Tab. 1 shows the current harmonic limits for a three-phase equipment.

TABLE 1 CURRENT HARMONIC LIMITS FOR BALANCED THREE-PHASE ELECTRICAL EQUIPMENT [4]

Harmonic order	Limits
3rd, 5th, 7th	$I_3 = I_5 = I_7 = 0.02I_1$
Odd triplen harmonics (9, 15, 21, . . . , 39)	$I_h = 0.1I_1/h$
11th	$I_{11} = 0.1I_1$
13th	$I_{13} = 0.08I_1$
Odd non triplen harmonics 17, 19	$I_{17} = I_{19} = 0.04I_1$
Odd non triplen harmonics 23, 25	$I_{23} = I_{25} = 0.03I_1$
Odd non triplen harmonics 29, 31, 35, 37	$I_h = 0.3I_1/h$
Even harmonics 2 and 4	$I_h = 0.01I_1/h$
Even harmonics > 4 (h = 6, 8, 10, . . . , 40)	$I_h = 0.0025I_1$

I_1 = maximum fundamental current of the equipment that is measured during the maximum steady-state power demand operating mode condition, at a single test frequency
h = order of harmonic
I_h = maximum harmonic current of order h obtained for all normal steady state modes of operation.

A. Rectifiers

There are some rectifier topologies for aircraft applications, most of them regarding single DC output. In fact, the AC and DC levels as defined in the standard, are adequate for a passive (diode) rectifier, with 12 or more pulses (due to the current harmonic limitation, see Tab. 1). For 115 V (phase voltage), equivalent to 200 V (line voltage), which peak value is 282 V, results ideally, the same 282 V at the DC side. Considering the diodes and other circuit element losses, the DC filter behavior, and the load current, the average DC voltage converges to the desired level (270 V). Such topology is shown in Fig. 3. This solution is used in most of the conventional airplanes, whose generation system operates at constant frequency (400 Hz), usually with an auto-transformer arrangement. For a constant frequency it is possible to design input (AC) and output (DC) filters to comply with the standards. However, for variable frequency systems, such solution isn't able to comply with all the standards requirements [6,7] for single and double DC output.

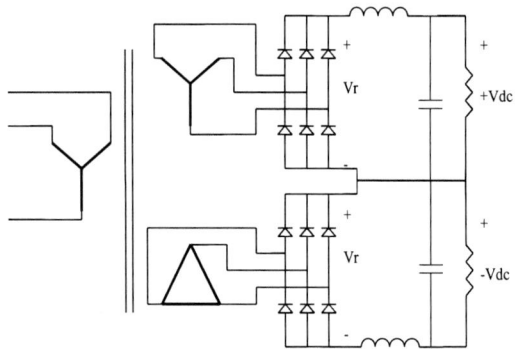

Fig. 3 Twelve-pulse passive rectifier with double DC output.

For the best operation of the generator, the rectifier unit must provide unitary power factor. Consequently, the use of controlled rectifiers with thyristors can be preliminarily discarded. Single-switch power factor correctors [8], usually employed in switched mode power supplies, also can be cast off due to the power level necessary in airplane applications. The remaining alternatives are the PWM rectifiers, from which one can highlight the transistor bridge and the Vienna topology [9]. This paper focuses on the use of bridge PWM rectifiers.

II. PWM RECTIFIER WITH DOUBLE DC OUTPUT

The converter is a three-phase bridge, as in Fig. 4. The control strategy must be able to simultaneously regulate the individual DC outputs. This can be done by regulating the total voltage to 540 V and the differential voltage to 0. The input current must result sinusoidal and in phase with the respective voltage. Notice that the rectifier input voltage is the source phase-to-phase voltage, what means the converter has a voltage step-up behavior.

For getting a symmetrical DC output (+/- 270 V) it is necessary to access the DC bus intermediate point. Such point is connected to the airplane reference ("ground"), as well as the neutral point of the AC generator, creating a common mode (zero sequence) current path.

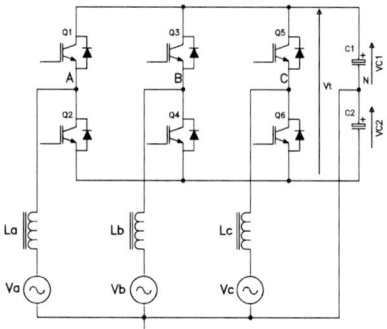

Fig. 4. Three-phase, four-wire transistor bridge PWM rectifier with double DC bus.

The control strategy can be developed considering each phase independently. Fig. 5 shows the single-phase model. Fig. 6 shows the control block diagram, considering a digital implementation [10]. The voltage waveform used in the multiplier is the respective phase voltage. In this case it's necessary to control all the three currents, since it's a four-wire system.

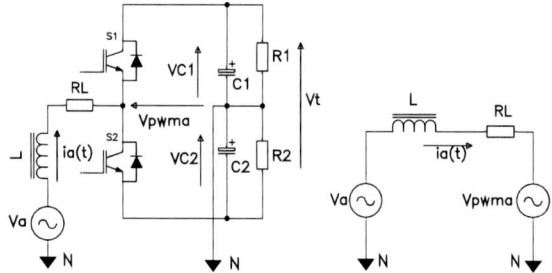

Fig. 5. Single-phase circuit and equivalent circuit for control purposes.

The total voltage is regulated by a PI controller that adjusts the amplitude of the AC current reference. Such reference is achieved by multiplying the PI output signal to the AC voltage waveform, thus producing a current reference with the same waveform, assuring unitary power factor. The control strategy doesn't use a PLL since the synchronization with the AC voltage is guaranteed by the multiplication. Even fast frequency variations, which in normal operation can reach 250 Hz/s, according to the standards, immediately reflect on the current reference. A fast current loop defines the switching pattern in order to minimize the current error. Design procedures are presented in [10].

978-1-7281-4181-7/19 $31.00 © 2019 IEEE

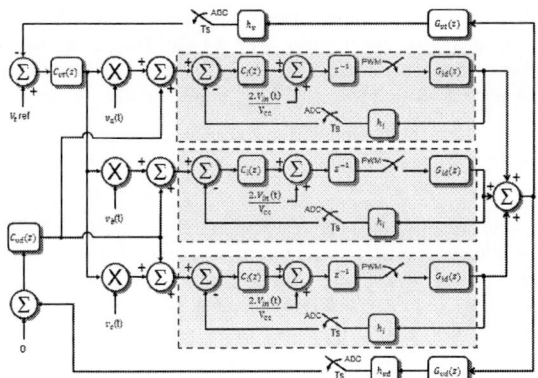

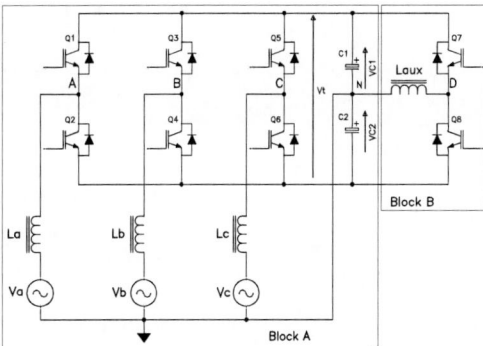

Fig. 6. Block diagram of the digital control structure.

Fig. 8. Regenerative DC voltage balancing circuit [13].

For balanced load, the DC voltages difference remains zero and no correction is required. However, in case of unbalanced DC load the voltages will change (the PI controller adjusts the total DC voltage), accordingly to the respective load. This means that the positive bus current is different of the negative DC bus. For equalizing the voltages it's necessary to compensate such difference and consequently to inject a DC current in the AC side, as shown in Fig. 7. All the phase currents assume a positive DC level. The common mode current returns via the neutral, in which it's possible to see also the high frequency (switching) component). Obviously, this fact is not allowed, due to the resulting unbalance (zero sequence component) on the generator magnetic flux, increasing the saturation and distorting the magnetizing current [10,11].

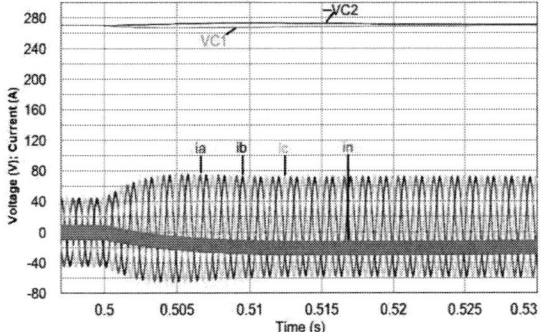

Fig. 7. Above, the DC voltages, bellow, the DC current injection in the AC side due to load unbalance compensation, grid frequency 800 Hz.

It is necessary a procedure for maintaining the DC voltages balanced in case of unbalanced loads, avoiding the DC current injection. There are some alternatives, as a current balancer [12] or the neutral current compensator [13].

The current balancer [12] needs 6 power switches and at least 3 PWM commands. Its advantage is the action on the three DC current branches, what can be useful for dynamically regulating the DC busses.

The solution presented in [13], shown in Fig. 8, uses only two additional switches (Block B) and a circulating inductor to provide a current path that zeroes the neutral current, and guarantees the voltage balance without losses (except for the switches and inductor losses).

III. POWER CONVERTER OPERATION UNDER FAULT

Considering that any load unbalance effect has been solved by the DC load equalizer, no DC current will be injected into the AC grid. The next point is to analyze the operation of the topology in case of fault in the converter or in the AC grid.

A. Aeronautic three-phase AC grid

The standard MIL-STD-704F [3] determines that, at abnormal situations, the AC voltage can be zero up to 7 seconds (see Fig. 1). Such restriction is valid for the whole three-phase system but also in case of fail in one or two phases. According to the procedure described in the standard, the voltage must decrease from the steady state value to 0 within ½ cycle, remain at 0 for the duration of the test condition, and return from 0 to the steady state voltage within ½ cycle. The test uses a programmable source and not a real AC generator, what means it's easy to make the voltage variations.

Summarizing, [14] defines that "the utilization equipment must maintain the specified performance during one and two phase power failures". More than this, it states that the equipment "must automatically return to the performance specified for normal aircraft electrical conditions when the power returns to within normal limits. The utilization equipment must not suffer damage or cause an unsafe condition".

To minimize the impact of these severe incidents, it's used to provide the airplane electric system with a back-up source (like an Uninterruptable Power Source – UPS) to substitute the feeder under fault, maintaining the normal operation of the electric system [15].

On the other hand, the standard *RTCA DO-160G* [4] that defines the test procedures for the equipment, establishes that under abnormal situation: "all three phase AC loads (in this case, the PWM rectifier) are to be designed such that no damage or unsafe condition will occur during and following removal of one or more input phase connections". This means that such situation must be considered for designing the equipment. The test procedure indicates the equipment under test (EUT) must operate at least 30 min after the removal of one or two phases. The test procedure indicates that the connections must be removed.

B. Simulations of faults

The simulations show results according to both procedures, i.e. the input voltage at the phase under fault goes to zero or the feeder is disconnected. Table 2 shows the converter parameters used in simulations.

978-1-7281-4181-7/19 $31.00 © 2019 IEEE   313

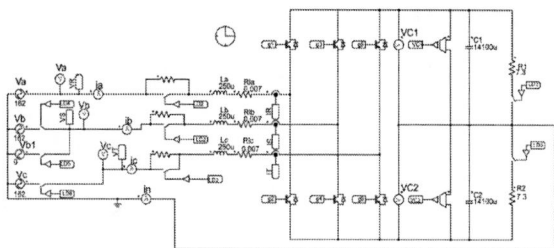

Fig. 9. PSIM circuit for fault simulation.

TABLE 2. THREE PHASE PWM RECTIFIER PARAMETERS

Parameter	Symbol	Value
Output Power	P_o	20 kW
Phase voltage RMS	V_a, V_b, V_c	115 V
Phase peak voltage	V_p	162 V
Total bus voltage	V_t	540 V
Partial bus voltages	V_{C1} and V_{C2}	270 V
Input choke	$L = L_a, L_b, L_c$	250 µH
Input choke resistance	$R_1 = R_{1a}, R_{1b}, R_{1c}$	7 mΩ
Bus capacitors	$C = C_1, C_2$	14.1 mF
PWM frequency	f_s	40 kHz
Sampling frequency	f_a	40 kHz

In the simulations the AC grid operates at 800 Hz, in spite of the same behavior is gotten in the full frequency range. The DC load is 20 kW @ 540 V (+/- 270 V).

Initially all the phases work properly, as shown in Fig. 10. The currents are balanced and the power factor is unitary. The DC ripple is negligible. At 0.42 s phase *b* voltage goes to zero (for example, due to a short circuit to the ground). According to the control structure, the voltage measurement is used in the multiplier. As it goes to zero, the respective current reference also goes to zero, and the PWM modulator continues working. The currents no more are balanced. The remaining phases increase the respective currents, maintaining the regulation of the DC voltage and the unity power factor. The DC voltage ripple increases and current circulates through the neutral conductor.

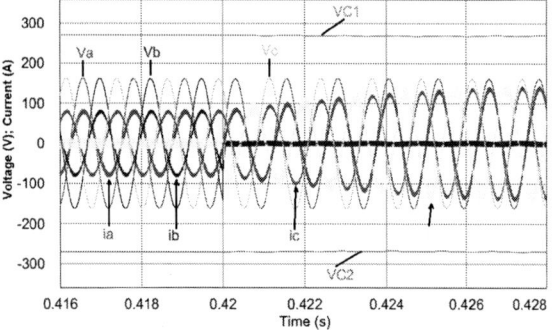

Fig, 10. Input voltages and current and DC voltages after single-phase fault in phase *b*. Top and bottom: DC regulated voltages [V]. Inner: Voltage [V] - phase *a* (red), phase *b* (blue), and phase *c* (green); Currents [A] - phase *a* (red), phase *b* (blue), phase c (green), neutral (yellow). Time: 2 ms/div.

Following the procedure of the *RTCA DO-160G*, the results are shown in Fig. 11. The branch connected to phase *c* opens indicating, for example, that a protection device interrupts the current on that phase. This protection action can be due to a fail in the feeder or in the converter (for

example, a short-circuit in the bridge arm). The DC voltage ripple increases. After a fast transient the voltage regulator is able to restore the correct set point. The currents no more are balanced. Phases *a* and *b* have to deliver the full power.

In both fault simulations, the phases that continue working increase the RMS current in 50% and the power factor remains unitary.

Fig. 12 shows the result with fault in two phases. Clearly, a single phase must be able to furnish the full load, still under unity power factor. The current in the remaining phase is 3 times the initial one, flowing back through the neutral connection. The DC ripple increases, but it's under control.

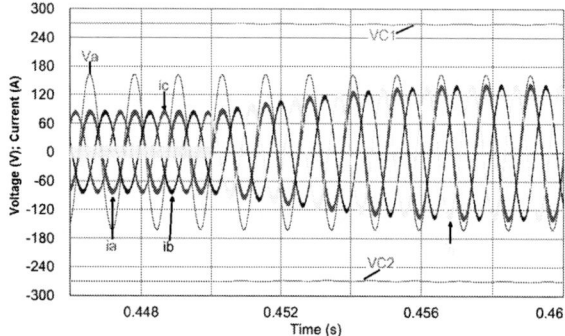

Fig. 11. Three-phase converter waveforms. Top and bottom: DC regulated voltages [V]. Inner: Voltage [V] - phase *a* (gray); Currents [A] - phase *a* (red), phase *b* (blue), phase c (green), neutral (yellow). Time: 2 ms/div.

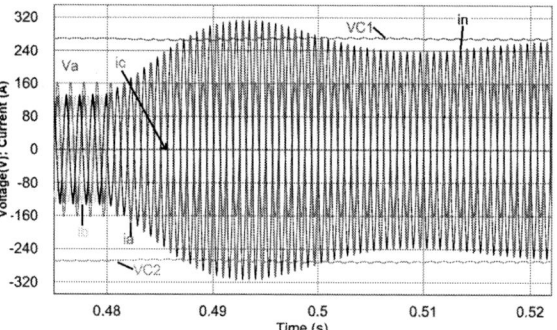

Fig. 12. Three-phase converter waveforms. Top and bottom: DC regulated voltages [V]. Inner: Voltage [V] - phase *a* (red); Currents [A] - phase *a* (blue), phase *b* (green), phase *c* (pink), neutral (black). Time: 10 ms/div.

It is known that the AC synchronous generator operating under flux and torque fluctuation presents many drawbacks [16]: the torque oscillation caused by the negative sequence current; the consequent additional heating, acoustic noise, vibration and bearing wear, etc. Such operation under fault must be considered in the design stage. The same is valid for the protective devices.

The sequence components are determined considering the Fortescue theorem according to (1), (2), and (3).

$$\begin{bmatrix} I^0 \\ I^+ \\ I^- \end{bmatrix} = A.\begin{bmatrix} I_a \\ I_b \\ I_c \end{bmatrix} \tag{1}$$

$$A = \frac{1}{3}\begin{bmatrix} 1 & 1 & 1 \\ 1 & \alpha & \alpha^2 \\ 1 & \alpha^2 & \alpha \end{bmatrix} \tag{2}$$

$$\alpha = e^{-j\frac{2\pi}{3}} \tag{3}$$

978-1-7281-4181-7/19 $31.00 © 2019 IEEE

The unbalance index is useful for evaluating the impact on the AC generator. Negative sequence and zero sequence components represent serious operational disturbances. Zero sequence current disturbs the magnetization, while the negative sequence current is responsible for negative torque.

According to the standard [17], the AC generator must operate continuously delivering power if the difference between any two phases is 15% of the rated current. This means that for reduced load, the unbalance can be proportionally high. The same standard determines the generator must be able to deliver 125% of the rated current for 5 minutes and 150% for 5 seconds.

However, the AC generator feeds a set of loads, not only the rectifier [18, 19]. Approximately 15% of the power load is exclusively DC, while 25% are fully AC, The remaining 60% can be feed by AC or DC power. In spite of the distribution between AC and DC loads is continuously changing due to MEA concept, the purely dissipative loads (heating and anti-icing systems) probably will not migrate from AC to DC. These systems represent an important part of the total power, and are typically balanced loads. Consequently, the effective current unbalance applied to the AC generator will depend on the power share between AC balanced and unbalanced loads.

TABLE 3. AC AND SEQUENCE COMPONENT CURRENTS @20 KW; DC VOLTAGE, +/- 270 V, INPUT VOLTAGE 115 V (PHASE-NEUTRAL)

	Three-phase bridge
Normal operation	
Source currents [A]	$I_a=I_b=I_c=59$
Converter current [A]	$I_a=I_b=I_c=59$
One phase fault	
Sequence components [A]	$I^+ = 59$, $I^- = 29.5$, $I^0 = 29.5$
Source currents [A]	$I_a = 88.4$, $I_b = 0$, $I_c = 88.4$
Current unbalance index	$\dfrac{I^-}{I^+} = 0.5; \dfrac{I^0}{I^+} = 0.5$
Two phases fault	
Sequence components[A]	$I^+ = 59$, $I^- = 59$, $I^0 = 59$
Current unbalance index	$\dfrac{I^-}{I^+} = 1; \dfrac{I^0}{I^+} = 1$
Source currents [A]	$I_a = 177$, $I_b = 0$, $I_c = 0$
Power factor (on remaining phases)	1
Maximum current stress	300%

Fig. 13 shows the response to a frequency variation., from 300 to 400 Hz at a rate of 500 Hz/s. As can be seen the control structure is able to maintain the DC voltage regulation (only positive voltage shown in the figure) as well as the synchronization between the current and the voltage.

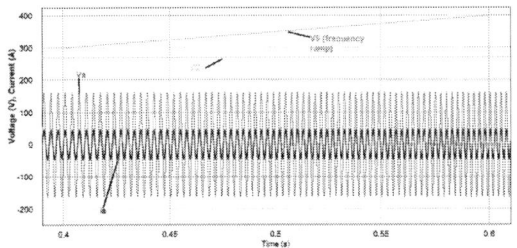

Fig. 13. Frequency variation test. Top: DC voltage [V] (green) and AC frequency [Hz] (orange). Bottom: Phase *a* voltage [V] (red) and current [A] -(blue),Time: 50 ms/div.

IV. CONCLUSIONS

The compliance with the standards that regulate the power quality in an aircraft electrical system puts severe restrictions to the power converters that convert the AC power to feed the DC busses. For the variable frequency system, the use of passive (diode) rectifiers is quite complicated due to the variable AC frequency. Consequently active solutions, like PWM rectifiers must be applied.

The paper has considered the three-phase, four wire PWM bridge operating as rectifier. The analyses carried out consider the occurrence of faults in the converter itself, in the AC feeder or in the generator.

The rectifier is able to maintain the regulated DC voltage after the loss of one and even two phases. In the first case the current increases 50% in the remaining phases. After losing two phases, the increase is 200% in the lasting phase. The rectifier capability to hold up these situations must be established as a design requirement.

Note that this limit fault situation (zero AC voltage) must be supported up to seven seconds, and not in steady state. This means that the design rule shall be the semiconductor devices current capability, and not the converter thermal behavior.

If the rectifier processes 15% of the generator power capability, a 100% current unbalance is acceptable according to the AC generation specification. If the DC power demand is higher, it's needed to analyze the total load composition in order to avoid damage the generator.

ACKNOWLEDGMENT

Thanks to Eng. Joao Viniccius G. Alves for the valuable comments about the use of the converters in modern aircrafts.

REFERENCES

[1] K. Emadi and M. Ehsani, "Aircraft power systems: technology, state of the art, and future trends", *IEEE Aerospace and Electronic Systems Magazine*, v. 15, n. 1, p. 28–32, Jan 2000.

[2] B. Sarlioglu and C. T. Morris, "More electric aircraft: Review, challenges, and opportunities for commercial transport aircraft". *IEEE Transactions on Transportation Electrification*, v. 1, n. 1, p. 54–64, June 2015.

[3] U. States, *Aircraft Electric Power Characteristics - MIL-STD-704F.* Department of Defense - United States of America, 2008. 1-40 p.

[4] U. States, *RTCA DO-160G – Environmental conditions and test procedures for airborne equipment.* U.S. Dept. of Transportation, Federal Aviation Administration, Dec. 2010.

[5] P. Wheeler, and S. Bozhko, "The more electric aircraft: Technology and challenges", *IEEE Electrification Magazine,* v. 2, n. 4, p. 6–12, Dec 2014.

[6] L. A. Vitoi, J. A. Pomilio, and D. I. Brandao, "Analysis of 12-pulse diode rectifier operating in aircraft systems with constant frequency". *Proc. of the 14th Brazilian Power Electronics Conference*, COBEP 2017, Juiz de Fora, Brazil.

[7] L. A. Vitoi, J. A. Pomilio, and D. I. Brandao, "Analysis of 12-pulse diode rectifier operating in aircraft systems with variable frequency". *Proc. of the 3rd IEEE Southern Power Electronics Conference*, SPEC 2017, Puerto Varas, Chile.

[8] J. A. Pomilio and G. Spiazzi, "High-precision current source using low-loss, single-switch, three-phase AC/DC converter", *IEEE Transactions on Power Electronics*, v. 11, no. 4, p. 561-566, 1996.

[9] G. Gong, M. L. Heldwein, U. Drofenik, J. Minibock, K. Mino, and J. W. Kolar, "Comparative evaluation of three-phase high-power-factor ac-dc converter concepts for application in future more electric

aircraft". *IEEE Transactions on Industrial Electronics,* v. 52, n. 3, p. 727–737, June 2005.

[10] A. G. Bueno, and J. A. Pomilio, "Balancing Voltage in the DC Bus with Split Capacitors in Three-Phase Four-Wire PWM Boost Rectifier", *13th INDUSCON - IEEE/IAS International Conference on Industry Applications*, São Paulo, Brasil 11-14 december 2018.

[11] A. G. Bueno, *Three-phase, four-wire, PWM Rectifier with High Power Factor* (in Portuguese), Master dissertation, University of Campinas, Brazil, 2018.

[12] J. Lago, and M. L. Heldwein, "Operation and Control-Oriented Modeling of a Power Converter for Current Balancing and Stability Improvement of DC Active Distribution Networks", *IEEE Transactions on Power Electronics*, Vol. 26, Issue 3, Pages: 877 – 885, 2011.

[13] H. Jank, *"Auxiliary circuit to voltage balancing and current reduction of the DC bus center point applied to a transformerless UPS"* (in Portuguese), Master dissertation, Federal University of Santa Maria, Santa Maria, Brazil, 2016.

[14] U. States, *MIL-HDBK-704-5 Handbook, Guidance For Test Procedures For Demonstration Of Utilization Equipment Compliance To Aircraft Electrical Power Characteristics Three Phase, Variable Frequency, 115 Volt,* U.S. Dept. of Defense, 9 April 2004.

[15] J. V. G. Alves, P. A. Figueiredo, I. F. Malizia, and J. A Pomilio, "Analysis and tests of Power Quality in aviation environment", *17th IEEE International Conference on Harmonics and Quality of Power (ICHQP)*, Pages 272-277, Belo Horizonte, Brazil, 2016.

[16] R. D. Yulisetiawan, E. S. Koenhardono, and S. Sarwito, "Effect Analysis of Unbalanced Electric Load in Ship at Three Phase Synchronous Generator on Laboratory Scale", *Jurnal Teknik Its*, Vol. 5, No. 2, p. G-389, G-395, 2016.

[17] U. States, MIL-E-85583A, *General Specification for Electric Power Generating Channel, Variable input Speed, Alternating Current, 400 Hz, Aircraft;* 1987.

[18] J. Brombach, T. Schröter, A. Lücken, D. Schulz, "Optimizing the Weight of an Aircraft Power Supply System through a +/- 270 VDC Main Voltage", *Przeglad Elektrotechniczny*, pages 47-50, January 2012.

[19] V. Madonna, P. Giangrande, and M. Galea, "Electrical Power Generation in Aircraft: review, challenges and opportunities", *IEEE Trans. on Transportation Electrification*, May 2018.

Model, Simulation and Analysis of BLDCM for a Differential Controlled Electric-Powered Wheelchair

Augusto Nery de Lima Neto
Power Conditioning Laboratory
University of Campinas
Campinas, Brazil
augustoneryln@gmail.com

José Antenor Pomilio
Power Conditioning Laboratory
University of Campinas
Campinas, Brazil
antenor@fee.unicamp.br

Abstract—One application of the permanent magnetic motor is in the field of electric vehicles. The electric-powered wheelchair is one of its example.

This paper will present a model, a simulation and analysis of a differential controlled electric-powered wheelchair using brushless DC motors. Bringing the theory of the electronics, mechanics and kinematcs, and the steps to simulate the wheelchair, including the driving and passive caster wheels. This article analyses the trajectory of the wheelchair in distinct situations as different slip surfaces, friction coefficient and gravity center over the wheelchair.

Index Terms—Electric-Powered Wheelchair, Differential Motors, BLDC, Wheel Slip, Passive Caster

I. INTRODUCTION

It is estimated that more than one billion people, approximately 15% of the world population, are living with some kind of desability [1]. Within this percentage, it is believed that more than 190 million people worldwide have severe disabilities and over 70 million people need wheelchairs to assist in their lives [2].

Electric-powered wheelchair (EPW) is one of the alternatives for those people who have severe disabilities or do not have strength to move the wheels with their own arms. Unfortunatelly, depending on the person's difficulty in moving certain parts of the body, it is impossible or inconvenient to control the wheelchair using a joystick. Conventional EPW on the market usually have their own control logic already embedded in the command of the joystick (right, left, front and back directions), being impossible to rewrite and remodel the logic control of the engines for their convenience according to their needs. In this case, it is possible to build an EPW using mechatronics and robotics [3] to have a semi-autonomous or autonomous mechanism and, subsequently, focus on the control of the two differential motor drive with normal, shared or autonomous control, depending on the person's disability and dexterity to control it by themselves, without external help [4].

Brushless DC motor (BLDCM) is typically used for high-performance and high-efficiency motor drives. It is characterized by controlled rotation over the entire speed range, full

This study was financed in part by the Coordenação de Aperfeiçoamento de Pessoal de Nível Superior - Brasil (CAPES) - Finance Code 001 (132984/2018-0 and 302257/2015-2).

torque control at zero speed and fast acceleration and deceleration [5]. These features make it suitable for certain high-performance applications as in electric vehicles and robotics.

BLDCM has lower inertia when compared to other machines because of the absence rotor cage [6], which means it has a faster response for given electric torque. Strictly speaking, the torque to inertia ratio is higher.

BLDCM has a higher efficiency than other machines, primarily because of the inexistence Joule losses in the rotor [7]. The energy transfer due the rotor losses at the BLDCM will not affect the machine operation, considering the motor will not transfer the heat while working.

The use of the permanent magnet in the rotor makes it unnecessary to supply magnetizing current. As being smaller on size comparing to some other machines, BLDCM weight less, which means the power density is higher. If size or weight are reduced and the efficiency is high, permanent magnets consistently show superiority [8]. The BLDCM can be applied in many frame designs, once space is a serious limitation when building a EPW.

This paper presents a new simulation model of electric-powered wheelchair, including the two driving wheels and two passive caster wheels in its composition. The analysis of trajectories with particular influences of slip, different friction coefficients and gravity center in distinct types of surfaces are the main purpose of this study.

II. EPW MODELING

The electric-powered wheelchair model will be developed in three parts: the electronics, the mechanics and the kinematics.

A. Electronics

Equations of armature winding and torque are used to modeling the BLDC motor. The equation of each armature is represented by (1). Taking into consideration that the mutual inductance is constant while the motor is turning [9].

$$V_x = RI_x + l\frac{di_x}{dt} + E_x \qquad (1)$$

Where x implies the phase a, b or c, l is the armature inductance [H], R is the armature resistence [Ω], V is the terminal voltage [V], i is the phase current [A] and E is the back EMF [V].

The torque equation is represented by (2).

$$T_x = K_{T_x} \cdot I_x \qquad (2)$$

Where K_T is the torque constant.

Assuming the reluctance torque is small, the total torque output (T [N.m]) can be represent by summing the torque of each phase (3).

$$T = T_a + T_b + T_c \qquad (3)$$

The conservation of moments in a rigid body allows to relate the torque to the speed (4).

$$T - \tau_c - \tau_{sl} = J\frac{d\omega}{dt} + B\omega \qquad (4)$$

Where τ_c represents the torque against the rolling movement of the tires and the force that resist to the movement by the misalignment of caster wheels; τ_{sl} is the torque needed to overcome a slope surface; J is the inertia of the wheelchair, B is the damping constant and ω is the angular speed.

B. Mechanics

The forces acting on the EPW are the motor force, the friction force, the aerodynamic force and the gravitacional force, as shown in Fig. 1.

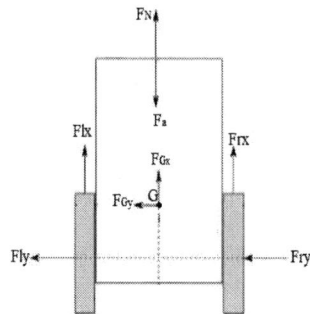

Fig. 1. Forces acting on the wheelchair.

F_N is the normal force, F_a is the aerodynamic force, $F_{G_{x,y}}$ are the gravitacional forces and $F_{x,y}$ are the friction and rolling forces of the motor, acting on the left and right wheels.

C. Kinematics

The kinematic model is used to describe the movement of the EPW. It determines the speed of each driving wheel (right and left), position and geometry of the system.

The structure of the electrical-powered wheelchair is a set of two driving wheels, with independent controls and two caster wheels. The caster wheels are able to rotate completely and do not have the ground contact point aligned with the vertical rotation [10].

These characteristics make the EPW kinematics equal to the differential robot kinematics [11]. The equations (5), (6), (7) and (8) represent the kinematics of the wheelchair.

$$v = \frac{v_r + v_l}{2} \qquad (5)$$

$$w = \frac{v_r - v_l}{L} \qquad (6)$$

$$w_r = v_r \cdot R \qquad (7)$$

$$w_l = v_l \cdot R \qquad (8)$$

v is the linear speed of the wheelchair, expressed by the individual velocities v_r and v_l (linear velocity right and left), w is the angular speed of the wheelchair, w_r is the right and w_l is the left angular speeds of the wheels, L is the distance between the left and right wheels and R is the radius of the wheel. Fig 2 represents the structure of the wheelchair and the speeds components.

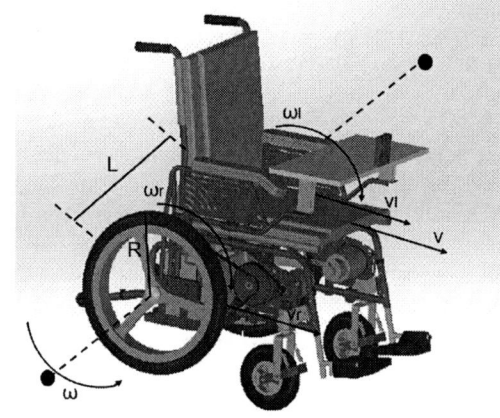

Fig. 2. Wheelchair indicating the various speeds components and its frame.

As stated in [10], caster wheels would not have any influence on the kinematics of the wheelchair, but according to [12], [13] this statement is not true, caster wheels do influencing on the moving of the EPW when not aligned with the driving wheels, making the behavior not similar to a differential robot.

Therefore, the driving wheels will be the only wheels directly influencing on the EPW, considering that the caster wheels are passive and aligned, at the beginning.

The trajectory of the wheelchair will exclusively depends on the linear speed of each wheel. There are 8 possibilities of outcomes from these speeds:

1) If v_l and v_r are equal to zero, the EPW will stay still.
2) If v_l is equal to v_r, the EPW will have a linear motion in a straight line.
3) If v_l is greater than v_r, the EPW moves towards clockwise direction, having a negative angular speed $(-\omega)$.

978-1-7281-4181-7/19 $31.00 © 2019 IEEE   318

4) If v_l is lower than v_r, the EPW moves towards anti-clockwise direction, having a positive angular speed (ω).

5) If v_l is zero and v_r is positive, the EPW will rotate in the anti-clockwise direction with a L radius rotation.

6) If v_l is positive and v_r is zero, the EPW will rotate in the clockwise direction with a L radius rotation.

7) If v_l is equal to $-v_r$, the EPW will rotate clockwise about its own central axis (midpoint of the wheel axis).

8) If $-v_l$ is equal to v_r, the EPW will rotate anti-clockwise about its own central axis.

Notice that while the angular speed of the wheels can be measured precisely, the same is not true to the linear movement, since it depends on the interaction between the tires and the floor. This aspect will be discued in section V.

III. EPW SIMULATION

The model was divided in four blocks created in the software Simulink:

- Electronics;
- Mechanics;
- Tires and Body;
- Kinematics.

A. Electronic block

The electronic block of the EPW simulation contains the brushless motors and respective drivers. The schematic is shown in Fig. 3.

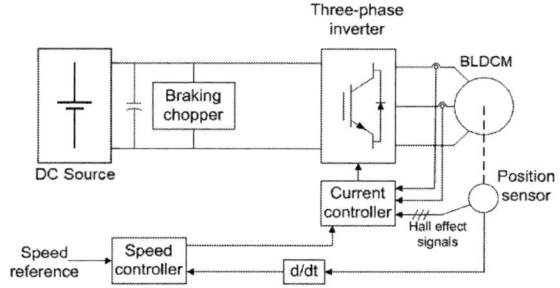

Fig. 3. Schematic of the BLDCM.

There are two BLDCMs, one for each wheel. The motor follows the reference speed, which develops a torque reference that controls the stator currents using a hysteresis controller that determines the switching pattern. The block representation of the speed controller and the current controller with hysteresis modulation are shown in Fig. 4 and 5, respectively.

B. Mechanical block

The mechanical block represents the necessary signal conversions to obtain the input torque to make the wheel turn in subsection III-C.

According to Simulink procedures, the output speed signal from the motor must be converted to a speed physical signal in order to connect the "electronic" block to the "mechanical" block. With the combination of an angular speed source, a torque sensor and a rotational motion sensor, it is possible to

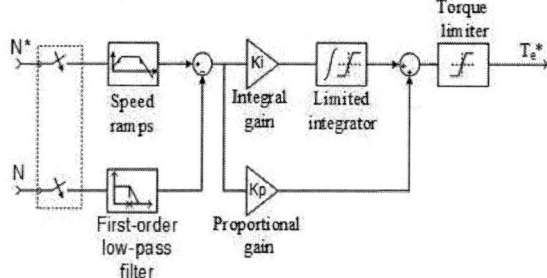

Fig. 4. Speed controller of the BLDCM.

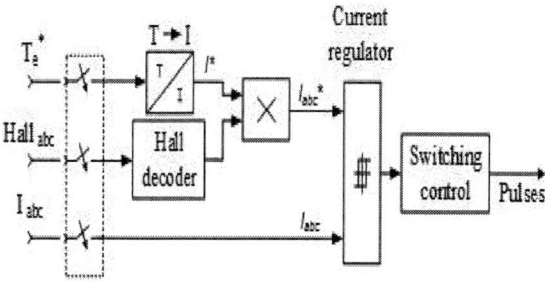

Fig. 5. Current controller with hysteresis modulation of the BLDCM.

get the mechanical torque signal that will supply the wheels to spin, Fig. 6.

This mechanical torque causes the wheels to rotate. A gearbox reduces the angular speed while multiplies the torque, adapting the BLDC output to the EPW needs. The reduction ratio used was $\frac{1}{32}$, which means that every 32 full revolutions of the rotor the wheel completes one turn. This angular speed was established to generate an appropriate and comfortable linear speed for the wheelchair.

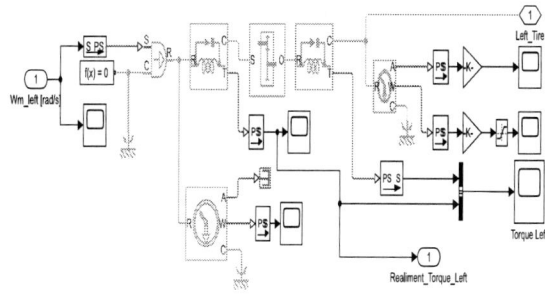

Fig. 6. Mechanics representation of the left side on Simulink of the EPW.

C. Wheelchair model: Tires and Vehicle body

The vehicle body was parameterized to have the same features as a wheelchair. The friction block, Fig. 7, represents the slip that each wheel may or may not suffer because of the contact with the floor. It uses de Pacejka model [14]. There are different types of surfaces in this model as, for example dry and wet.

Depending on the type of floor the wheelchair is moving, the wheel will have a different slip [15].

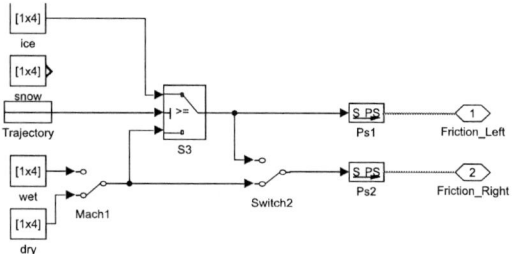

Fig. 7. Friction representation on Simulink of the EPW.

This slip calculation is done using the so called "magic formula", present in the Simulink model, stablishing the rolling radius, longitudinal stiffness, longitudinal damping and inertia of the tire [16]. It is also possible to include slope surface, as shown in Fig. 8. The response from this block is the linear speed of each wheel.

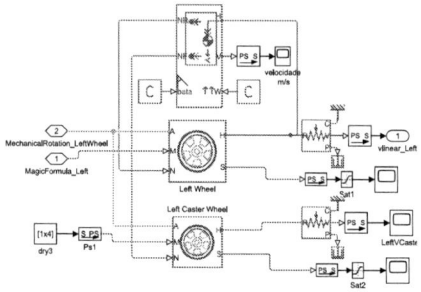

Fig. 8. Left tires and vehicle body representation on Simulink of the EPW.

D. Kinematic block

The kinematic block of the EPW is represented in Fig. 9. It was used the equations of the linear and angular speeds of the model in II-C.

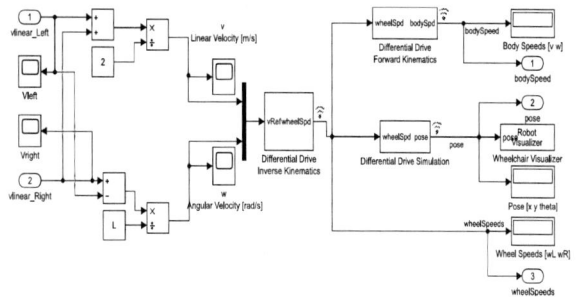

Fig. 9. Kinematics representation on Simulink of the EPW.

With the linear speeds (v_l and v_r), it was used the differential drive inverse and differential drive simulation blocks [17] to get the angular wheel speeds (ω_r and ω_l) and the location of the EPW in the cartesian plan in x, y, θ.

IV. SIMULATIONS

To represent different traveled paths for the EPW, it was simulated three cases: 2, 6 and 7 of the model described in II-C. The outcome of the trajectory of the wheelchair is shown in Fig. 10.

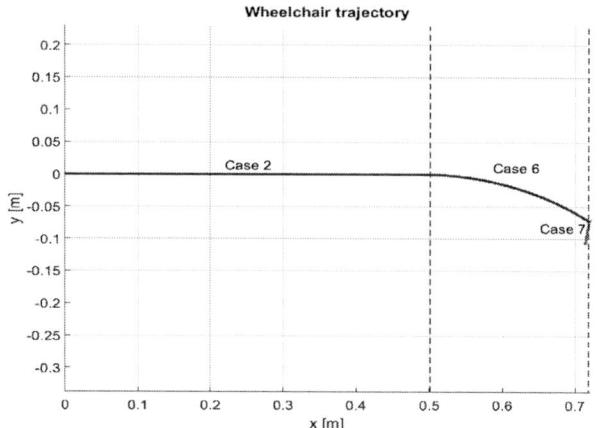

Fig. 10. Trajectory traveled by the EPW.

In the first 3 seconds, the wheelchair has a path with equal linear speeds on both wheels, $v_l = v_r$, having a forward straight linear motion in a straight line, moving only in the x direction in the cartesian plan.

In the second case, from 3 to 6 seconds, the wheelchair has a curvilinear path in which the left wheel maintains the same speed, but the right wheel decelerates to 0 m/s, the EPW travels towards clockwise direction with L radius rotation, moving in the x, y and θ directions, in this case it is possible to see the negative torque due to deceleration of the wheelchair.

And finally, in the third case, from 6 to 9 seconds, the left wheel maintains the same speed, and the right wheel will have the same speed but with negative module, $v_l = -v_r$, causing the wheelchair to rotate clockwise around its own axis, moving only θ in the cartesian plane.

Fig. 11 and 12 represent the current of the BLDC motors and torque before and after the reducer.

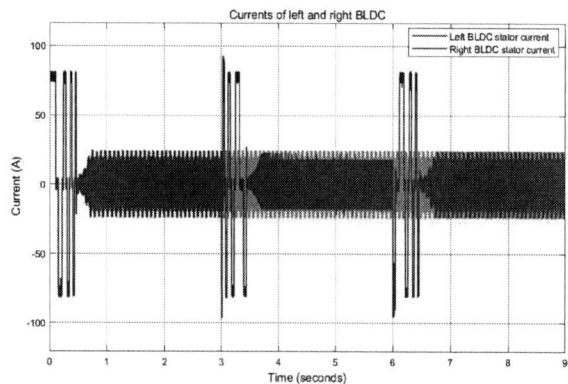

Fig. 11. Current response of the BLDCM for the trajectory of the EPW.

978-1-7281-4181-7/19 $31.00 © 2019 IEEE

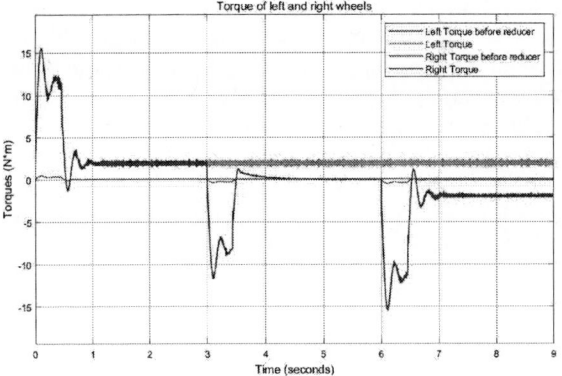

Fig. 12. Torque response of the wheels for the trajectory of the EPW.

As a result of the different path traveled by the wheelchair, the current and the torque behave in different ways. It is possible to see that the current has its peak due to acceleration and deceleration. The same happens in relation to the torque (0, 3 and 6 seconds).

When the speed in the wheels reaches the reference, the current and torque reduce to its cruise value, compensating only the travel losses.

Defiant situations were simulated to verify the model. To analyse the torque and position of the wheelchair, it was created an ideal trajectory, Fig. 13, representing the torque of the wheels over an ideal surface, equal friction coefficient in both wheels and central gravity center. In this case the wheelchair travels in a straight line, increasing in the x direction, while y and θ remain at zero.

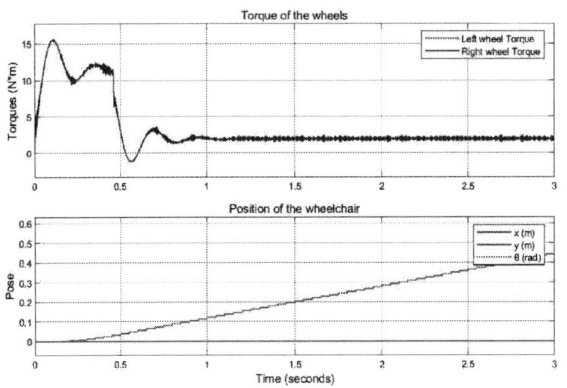

Fig. 13. Torque of the wheels over an ideal surface, equal friction coefficient in both wheels and central gravity center on the top; Position x, y, θ of the wheelchair on the bottom.

In the first event, Fig. 14, the left driving and caster wheels were put in a slippery surface, while the right driving and caster wheels kept the same with no influence of the slippery surface. The response to this event in comparison to the ideal trajectory is that the left wheels kept turning in the same position, not beeing able to escape of the place and the right wheels carried on normally, consequently the wheelchair

changes its direction turning to the left, when y and θ kept increasing while the wheelchair travels.

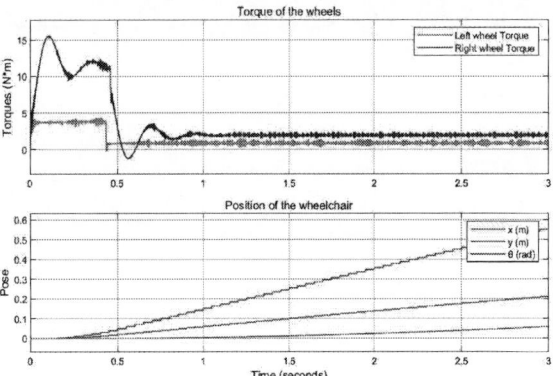

Fig. 14. Torque over the wheels when there is a slippery surface and position x, y, θ of the wheelchair.

In Fig. 15 the friction coefficient of the wheels are different. The right tire is flat in comparison to the left one, which means its contact with the floor is greater. The outcome to this event is that the torque damping on the left wheel is bigger when compared to the ideal trajectory. However, it does not significantly affect the trajectory once the speed control compensates the torque value of the wheels. The average torque of both wheels are balanced and the position y and θ kept at zero, maintaining the trajectory of the wheelchair.

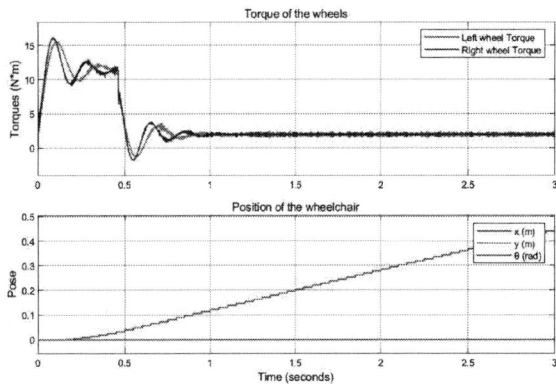

Fig. 15. Torque over the wheels when the friction coefficient are different and position x, y, θ of the wheelchair.

In the last event developed, Fig. 16, the gravity center of the wheelchair was not in the center of vehicle. The mass distribution over the EPW was made to represent a wheelchair user that supports his body to the right. This distribution makes the gravity center of the wheelchair to move from the center to the right. The outcoming of this situation is a higher torque over the right wheel in comparison to the left to compensate the speed control of both BLDCM. This situation makes the angular speed to decrease, slightly changing the direction of the wheelchair to the right, .

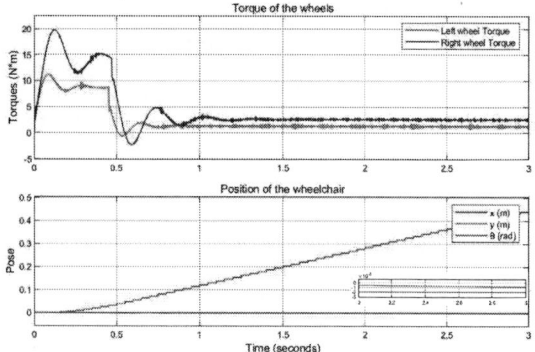

Fig. 16. Torque of the wheels when the gravity center is placed on the right side of the wheelchair and position x, y, θ of the wheelchair.

Even with precise speed control over the wheels, in some situations it was not possible to prevent the wheelchair from leaving its ideal path slightly changing y and θ directions, while still maintaining speed during the trajectory.

V. EXPERIMENTAL RESULTS

A prototype of the EPW is being made, Fig. 17. Experimental results of the study will be than compared with the simulation data to certify the veracity of simulation. The characteristics of the BLDC motor are shown in Table I.

TABLE I
MIDWEST MMP BL86-425E-24V BLDC MOTOR

Rated DC Voltage	24 V
Peak Torque	4,6324 N.m (Max)
Peak Torque Current	80 A (±10%)
Torque Constant (Kt)	$57,9.10^{-3}$ N.m/A (±10%)
Armature Resistance	0.2 Ω (±15%)
Armature Inductance	0.3 mH (±15%)

Fig. 17. Prototype construction of the EPW.

VI. CONCLUSION

This paper has presented the model, the simulation and the analysis of a differential controlled electric-powered wheelchair using brushless DC motors.

This model can be used to predict the location of the wheelchair traveling on different types of surfaces, such as dry, wet, etc.

With the simulation of challenging cases, when the wheelchair and its user are suffering slippage on the wheels, different friction, and center of gravity displaced from the central point, it is possible to obtain the necessary torque to feedback the motor and thus maintain more precise speed control. And, with the accuracy of electronic outcome signals, apply in different types of navigation control (autonomous or semi-autonomous) to maintain the desired speed of the wheelchair while preserving the comfort of the user, regardless of their disabilities.

ACKNOWLEDGMENT

The authors want to thank SEW-EURODRIVE for supporting the design and construction of the EPW prototype.

REFERENCES

[1] World Health Organization, "World Report on Disability". 2011, Geneva, Switzerland, HV 1553.
[2] World Health Organization, "Wheelchair Service Training Package". 2012, Geneva, Switzerland, WB 320.
[3] R. Velázquez and C. A. Gutiérrez, "Modeling and Control Techniques for Electric Powered Wheelchairs: An Overview", *CONCAPAN XXXIV*, 2014.
[4] H. Yanco, J. Gips, "Preliminary investigation of a semi-autonomous robotic wheelchair directed through electrodes", *In Proceedings of the Rehabilition Engineering Society of North America Annual Conference*, Pittsburgh, PA, 20-24 June 1997.
[5] S. Derammelaere, M. Haemers, J. De Viaene, F. Verbelen and K. Stockman, "A quantitative comparison between BLDC, PMSM, brushed DC and stepping motor technologies", *2016 19th International Conference on Electrical Machines and Systems (ICEMS)*, Chiba, 2016, pp. 1-5.
[6] P. Pillay and R. Krishnan, "Application Characteristics of Permanent Magnet Synchronous and Brushless dc Motors for Servo Drives", *IEEE TRANSACTIONS ON INDUSTRY APPLICATIONS*, VOL. 21, NO. 5, SEPTEMBER/OCTOBER 1991.
[7] P. Pillay and R. Krishnan, "Modeling, Simulation, and Analysis of Permanent-Magnet Motor Drives, Part I: The Permanent-Magnet Synchronous Motor Drive", *IEEE TRANSACTIONS ON INDUSTRY APPLICATIONS*, VOL. 25, NO. 2, MARCH/APRIL 1989.
[8] H.R. Kirchmayr, "Permanent magnets and hard magnetic materials", *J. Phys. D. Appl. Phys.*, 29 (1996), pp. 2763-2778.
[9] Y. S. Jeon, H. S. Mok, G. H. Choe, D. K. Kim and J. S. Ryu, "A new simulation model of BLDC motor with real back EMF waveform", *COMPEL 2000. 7th Workshop on Computers in Power Electronics. Proceedings*, Blacksburg, VA, USA, 2000, pp. 217-220.
[10] SIEGWART, R.; NOURBAKHSH, I.; SCARAMUZZA, D. *Introduction to Autonomous Mobile Robots*. 2nd Ed. MIT Press, 2011, 455 p. ISBN 9780262015356.
[11] Cox, I. J.; Wilfon, G. T. *Autonomous Robot Vehicle*. Ed. Springer-Verlag, 1990, 4-24p. ISBN 139781461389996.
[12] LEE, D.-A.; JUNG, D.-G.; WOO, K.-S.; KIM, L.-K.; MOK, H.; HAN, S. Orientation compensation for initially misaligned caster wheels. *International Journal of Control, Automation and Systems*, v. 11, n. 5, p. 1071–1074, 2013.
[13] GERSDORF, B.; SHI, H. A castor wheel controller for differential drive wheelchairs. In: FILIPE, J. A. C. J.; FERRIER, J.-L. (Ed.). *ICINCO* 2010. [S.l.: s.n.], 2010. v. 2, p. 174–179. ISBN 978-989-8425-01-0
[14] L. C. A. Silva; F. C. Corrêa; J. J. Eckert; F. M. Santciolli; F. G. Dedini (2017) "A lateral dynamics of a wheelchair: identification and analysis of tire parameters", *Computer Methods in Biomechanics and Biomedic Engineering*, vol. 20, no. 3, pp. 332–341, 2017. PMID: 28095721.
[15] H. B. Pacejka and E. Bakker, "The magic formula tyre model," *Vehicle System Dynamics*, vol. 21, no. sup001, pp. 1–18, 1992.
[16] MathWorks®, SimMechanics User's Guide. *"A SimMechanics motorcycle tyre model for real time purposes"*, March 2008.
[17] MathWorks®, *Simulink Example Differential Drive*, 2019. (www.mathworks.com/examples/simulink/community/37564-differential-drive?s_cid=rlcnt_ME).

Analysis and Optimal Design of Magnetic Components in Dual-Active-Bridge Converter for 1 MVA Solid-State Transformer

Haonan Tian[1]*, Sriram Vaisambhayana[1], Anshuman Tripathi[1]

[1] Energy Research Institute @ NTU, Nanyang Technological University, Singapore
* TIANHN@ntu.edu.sg

Abstract— The high frequency (HF) isolation transformer is the core element of a dual-active-bridge (DAB) bi-directional high-power DC-DC converter. At high frequency high power conversion, providing compact and efficient transformer design solutions need to consider the magnetic, electric as well as thermal aspect thoroughly and carefully to ensure the reliability and efficiency of the whole system. Detailed design considerations for a HF isolation transformer in a DAB conversion system are presented in this paper taking into consideration the insulation, high frequency magnetics characteristics, skin-proximity effects at high switching frequency as well as the thermal-fluid interaction analysis. Special attention is paid to the leakage inductance since improper value leads to an undesirable overshoot on the device voltage which may cause the SiC failure. This paper also presents the a multi-objective optimization design methodology applied to a 25kVA/25kHz, 1500V/750V transformer intended for DC-DC stage in 1MVA SiC-based solid state transformer (SST) and the Pareto optimal sets under different cooling methods are derived and analyzed. Comparison between different magnetic materials and different operation conditions are also carried out to find the characteristics variation of the designed transformer. The merit of the optimal design are validated through finite element modeling (FEM) and computational fluid dynamics (CFD) simulations.

I. INTRODUCTION

Nowadays, the concept of solid state transformer is significantly gaining attention in overall power conversion system for higher power density and better power controllability. It is expected to play an essential role and replace their line frequency counterparts due to its novel merits in terms of the superior controllability, space advantage and improved power quality. SST has been utilized especially in the applications where space and weight restrictions are critical, like traction power system [1]. Rapid development of the offshore wind energy power generation [2], the energy storage systems as well as the electric vehicles also led to the popularity of the SST building block to contribute to the next-generation power networks [3].

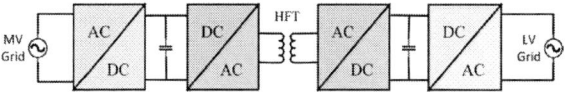

Figure 1：Three-stage SST structure

Figure 1 shows the typical three-stage SST system structure, which is composed of the AC-DC rectifier connected to the MV grid, the DC-DC converter controlling the power

flow and the DC-AC inverter regulating the total active power and LV-side power factor [4]. One of the potential architecture of the three phase SST is shown in Figure 2, where the AC-DC converter modules are cascaded in series to achieve a higher voltage and higher power requirement. In the MV-22kV side, the line-to-neutral voltage is used as the input voltage for cascaded H-bridge (CHB) converter system. DC-DC converter enables the bi-directional power flow and integrate with high frequency transformer providing the isolation between MV and LV side grid. Among the construction of the DC-DC converter module, the design of the HFT is a major challenge consisting of comprehensive and complicated considerations. High operating frequency leads to the significant reduction in magnetics size but increases the loss density, especially for the copper loss considering the enhanced skin and proximity effects in the conductors. Besides, the hysteresis losses and eddy current losses are also enhanced in the magnetic core. The significant size reduction and the increase of the loss density results in higher thermal challenge as well as potential insulation damage. It is reported that a lifespan of 20 years can be expected under temperatures near 353 K, while temperatures near 373 K should be avoided [3]. Therefore, an efficient, reliable and compact transformer design is highly concerned considering higher efficiency, stronger insulation, smaller size and better heat dissipation simultaneously.

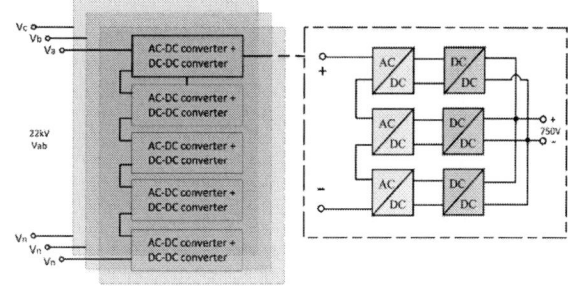

Figure 2: System architecture of three phase 22kV 1MW SST

This paper presents detailed design considerations for high-frequency high-power isolation transformers for the DC-DC stage which takes the insulation, leakage inductance, skin-proximity effects at high switching frequency as well as the thermal-fluid interaction analysis into consideration. In section III, the design framework of the high frequency transformer is presented with special attention given to the high frequency transformer efficiency, power density and heat dissipation. The optimal design and pareto front of a 25kVA, 25kHz,

---

*Resrach supported by National Research Foundation.

978-1-7281-4181-7/19 $31.00 © 2019 IEEE

1500V/750V high frequency transformer used in the 22kV 1MVA SST topology is derived and validated with FEM-CFD simulation studies, which substantiates the effectiveness of the presented design methodology and models. Different operation conditions and comparison between different magnetic materials are also carried out to find the characteristics variation of the designed transformer.

II. DESIGN CONSIDERATIONS FOR HIGH FREQUENCY TRANSFORMER

One of the most typical configuration of the bi-directional isolated high-power DC-DC converter is the Dual Active Bridge, which is illustrated in Figure 3. Basically, the DAB DC-DC converter integrates a core element high-frequency isolation transformer and two active bridges located on the primary and secondary side of the high-frequency transformer, respectively. The high frequency transformer (HFT) ensures the required galvanic isolation and voltage matching between the high-voltage and low-voltage buses.

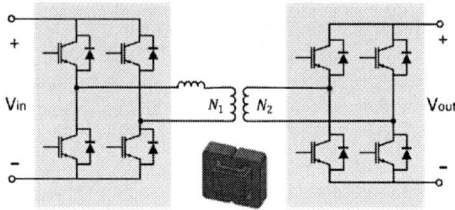

Figure 3: Bidirectional isolated dc–dc converter with dual active full bridges and high frequency transformer integrated.

The properties and behavior of a high frequency transformer is dramatically different from a line frequency transformer. The size has been significantly reduced compared to a conventional transformer, however, at high frequency and high power, the losses occurred in the magnetic core and the windings are increased. Smaller size and higher loss in other hand will give the burden for heat dissipation of a transformer. The design of the HFT is a major challenge consisting of comprehensive and complicated considerations, which will be discussed in the following.

A. Insulation design

Under high power and high voltage operation conditions, strong voltage insulation capacity between layers and windings is required to ensure the reliability and safety of a transformer. In the meanwhile, it should also assist in the dissipation of produced heat from the heated up elements in a transformer. Epoxy was chosen as the insulation material exhibiting high thermal conductivity and high inherent dielectric strength and mechanical toughness. The thickness of the insulation is determined with respect to the dielectric strength of the insulation material as well as the maximum withstand voltage.

Z-winding (flyback winding) structure can be employed where all the winding layers are wound in the same direction to have the same voltage difference between two adjacent winding layers, as shown in Figure 4. The Z-winding structure is slightly more complicated than the U-type, but can mitigate the possibility of insulation breakdown. On the other hand, it also reduce the winding capacitance compared to the U-type.

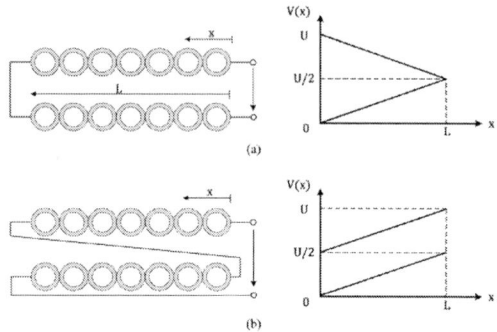

Figure 4: Layer voltage distribution for (a) U-winding structure (b) Z-winding structure

B. Core characteristics

Several materials are the favorable soft type magnetic materials used in a higher range of frequency, such as ferrite (FN.), Amorphous alloy (Amorp.) and Nano-crystalline (Nano.c.) [5] [6], whose performances are compared in Table I. There are several different approach for core loss estimation and the most widely used model is the empirical model based on the original Steinmetz equation. Due to the square-waves excitation, iGSE has been shown to be the most accurate model dealing with non-sinusoidal excitation waveforms without using additional parameters other than the k, α, β. The mean core loss density can be described as below [7]

$$P_c = \frac{1}{T} \int_0^T k_i \left| \frac{dB(t)}{dt} \right|^\alpha (\Delta B)^{\beta - \alpha} dt \qquad (1)$$

With

$$k_i \approx \frac{K}{2^{\beta+1} \pi^{\alpha-1} \left(0.2761 + \frac{1.7061}{\alpha + 1.354} \right)}$$

Table I: Characteristics of magnetic materials for HFT

	Amorp.	*Nano.c.*	*FN.*
Bsat (T)	1.56	1.23	0.39
Specific losses(kW/kg) @0.1T 100kHz	0.2	0.01	0.009
Continuous operating temperature (°C)	150	120	140
Initial Permeability	6.5-8k	20-200k	1.5-15k
k[8] W/m³	1.3773	0.01137	0.03
α	1.51	1.8	1.78
β	1.74	2.3	2.62

C. High-frequency winding losses

With rising frequency, the effective winding resistances increase as a consequence of the more conspicuous skin and proximity effect. To take the eddy current effect into consideration, two classical analytical models are used in this paper to accurately evaluate the AC resistance, which are applicable to foil-type and litz-type conductors respectively.

The AC-to-DC resistance factor of the jth layer (Fr_j) of the

978-1-7281-4181-7/19 $31.00 © 2019 IEEE

foil conductor are expressed by the classical Dowell's equation [9]:

$$Fr_j = \Delta \left[\xi_1 + \frac{2}{3}(j^2 - 1)\xi_2 \right]$$

$$\Delta = \frac{d_{foil}}{\delta}$$

$$\delta = \sqrt{\frac{1}{\pi f \varepsilon \sigma}} \qquad (2)$$

$$\xi_1 = \frac{\sinh(2\Delta) + \sin(2\Delta)}{\cosh(2\Delta) - \cos(2\Delta)}$$

$$\xi_2 = \frac{\sinh(\Delta) - \sin(\Delta)}{\cosh(\Delta) + \cos(\Delta)}$$

where Δ denotes the penetration ratio, d_{foil} is the foil thickness, δ is the skin depth, ξ_1 and ξ_2 represent the skin and proximity effect respectively.

For litz-type conductor, Fr_j can be calculated by (3), where pf is the packing factor, $\varphi_1(\xi)$ and $\varphi_2(\xi)$ represents the skin effect losses and proximity effect losses in round conductors [10].

$$Fr_j = \frac{\Delta}{\sqrt{2}} \left[\varphi_1(\Delta) - \frac{\pi^2 n_{st}.p_f}{24} (16j^2 - 1 + \frac{24}{\pi^2}) \varphi_2(\Delta) \right]$$

$$\Delta = \frac{d_{st}}{\sqrt{2}\delta}$$

$$p_f = n_{st}.(\frac{d_{st.}}{d_{outer}})^2 \qquad (3)$$

$$\varphi_1(\Delta) = 2\sqrt{2} \left(\frac{1}{\Delta} + \frac{1}{3 \times 2^8}\Delta^3 - \frac{1}{3 \times 2^{14}}\Delta^5 \right)$$

$$\varphi_2(\Delta) = \frac{1}{\sqrt{2}} (-\frac{1}{2^5}\Delta^3 + \frac{1}{2^{12}}\Delta^7)$$

The general losses occurred in each winding can be calculated by summing the copper loss component of each harmonic (h is the order of the harmonic) as below.

$$P_w = \sum_{h=1}^{H} \left(\sum_{i=1}^{n_1} R_{DC(i)} \cdot Fr_{(i)} \right) \cdot I_h^2 \qquad (4)$$

D. Leakage inductance

Leakage inductance in the HF transformer serves as the energy transfer device from the primary to the secondary side in the operation of the DC-DC converter. Since it produces substantial effect on the global behavior of both the transformer and the converter, the value of this parameter needs to be carefully selected. The core and winding arrangements, the isolation distance between windings and the number of turns shall be adjusted to obtain the desired value.

The leakage energy stored at the nth layer in the transformer winding can be obtained by volume integration of the magnetic field intensity distribution $H(x)$ as given in (5) [4].

$$E_{winding,n} = \frac{1}{2}\mu_0 \, MLT_n h_w \int_0^t H(x)^2 \cdot dx \qquad (5)$$

$$= \frac{\mu_0 \, MLT_n \, h_w \, H_0^2}{8\gamma \, \sinh^2(\gamma t)} [(2n^2 - 2n + 1)k_1 + 4n(n-1)k_2]$$

Where μ_0 is the permeability of the vacuum, h_w is the height of the magnetic core window, MLT_n is the mean turn length of the layer, x is the distance from the inner surface of the winding, t is the conductor thickness of the primary/secondary layer

$$k_1 = \sinh(2\gamma t) - 2\gamma t$$

$$k_2 = \gamma t \cosh(\gamma t) - \sinh(\gamma t)$$

The leakage energy stored at the nth insulation layer can be obtained by the formula (6), where $H(x)$ remains constant and equal to the value at the outer point of the nth winding layer.

$$E_{insu,n} = \frac{1}{2}\mu_0(MLT_n)h_w \int_0^{t_{insu}} H_n^2 \, dx \qquad (6)$$

$$= \frac{1}{2}n^2 \cdot (MLT_n) \cdot \mu_0 h_w H_0^2 t_{insu}$$

The value of it can be accurately estimated based on the total magnetic energy ($E_{leakage}$), which is the sum of the stored energies in winding and insulation sections.

$$E_{leakage} = E_p + E_{insu,p} + E_s + E_{insu,s} \qquad (7)$$

$$= \sum_{n=1}^{n_p} E_{p,n} + \sum_{n=1}^{n_p-1} E_{insu,n} + \sum_{n=1}^{n_s} E_{s,n} + \sum_{n=1}^{n_s-1} E_{insu,n} = \frac{1}{2}L_{\sigma(pri)}I_P^2$$

Where $E_{p/s}$ is the leakage energy stored inside primary /secondary winding, $E_{insu,p/s}$ is the leakage energy stored between interlayer insulations of primary/secondary section, $n_{p/s}$ is the number of layers in primary/secondary winding.

E. Thermal modeling

The heat generated from transformer during power conversion comes from energy losses including the iron loss in the core and the ohmic loss in the windings that causes the increase of overall temperature inside the transformer. In light of this, predicting the thermal behaviour for the HF transformer accurately is insightful for the design and practical operation of the overall system to reduce the overheat risk inside a HFT. The thermal model employed in this paper is based on the thermal nodal network presented in [11], which is computationally easy to simulate the dynamic thermal behaviors of different transformer components. The resulting equivalent thermal nodal network of the considered EE-type concentric-winding transformer is illustrated in Figure 5. The core is described by two thermal branches due to different heat transfer mechanisms involved in central core (cc) and external core (ce). For external core limb, convection and radiation are dominant due to its contact with the ambient. In terms of the winding, each winding section is characterized by a source of copper loss and the equivalent thermal resistance. The central core limb and the primary winding are thermally coupled via the bobbin by conduction.

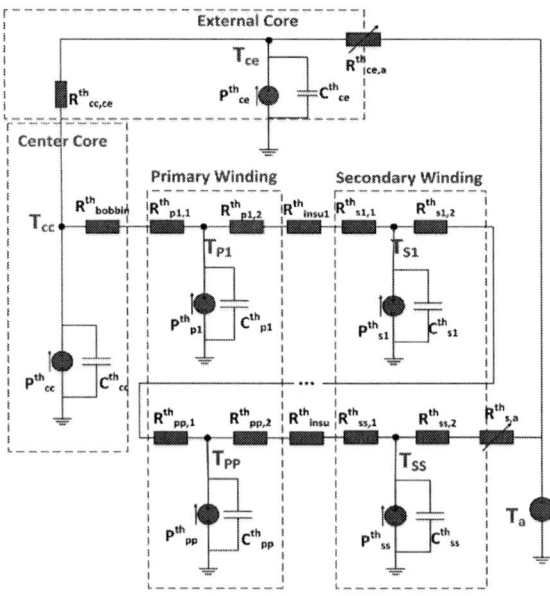

Figure 5. Equivalent thermal network for a shell-type multi-layer interleaved-winding transformer structure (P_i represents the corresponding heat source, R_{ij}^{th} represents the equivalent thermal resistance and C_{ij}^{th} represents the thermal capacitance)

With the geometry of the transformer model, thermal properties of the material and the heat transfer mechanism included, the equivalent thermal resistance of each thermal branch can be obtained and consequently, the governing equations describing energy balance can be easily obtained at each temperature nodes. The Heun's method can be used to estimate the equilibrium core and winding temperatures numerically.

III. HIGH FREQUENCY TRANSFORMER OPTIMIZATION

Based on the analytical models presented above, the proposed HFT design methodology follows the framework described in Figure 6, in which the weighted-sum approach and the constraint multi-objective genetic algorithm are combined to obtain the Pareto optimal set providing multiple optimal solutions. Individual weights (α_i) are assigned to the core volume (V_{core}), total loss ($P_c + P_w$) and maximum temperature rise (ΔT_{max}), which are set to be the optimization objectives. Weighting coefficients are varied within the range of (0,1) to offer multiple weight combinations. For one set of particular weighting coefficients, the cost function will be formulated based on the transformer size model, loss model and thermal model with respect to the design variables [x] (the length (E) and depth (T) of the core central limb, the height of the window (D) and flux density (B_{mag})) at high switching frequency.

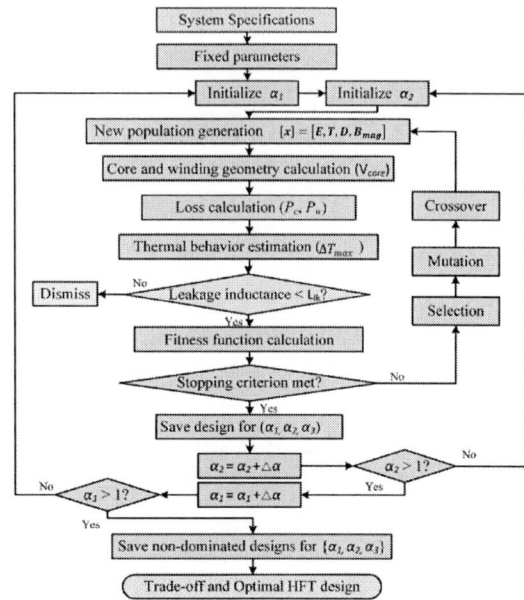

Figure 6: Design methodology flowchart

Table II: Design specifications of the high frequency transformer

Power rating (P)	25	[kVA]
Switching frequency (f)	25	[kHz]
Primary/ Secondary side voltage (V_1/V_2)	1500/750	[V]
Isolation voltage level (V_{iso})	80	[kV]

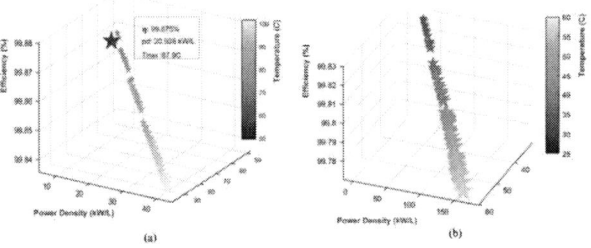

Figure 7: Optimization outcomes with respect to efficiency, power density and maximum temperature (a) Natural air cool (b) Oil natural cool.

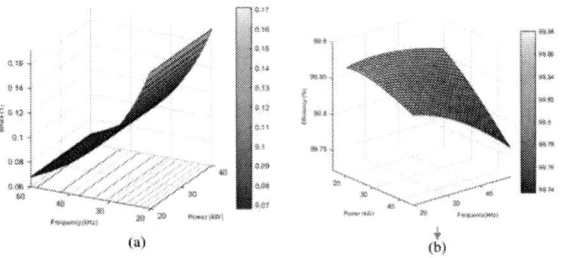

Figure 8: (a) Flux density (b) efficiency comparison for various operation conditions

Figure 7 (a) describes the Pareto front of the high frequency transformer design with the electrical specifications summarized in Table II under natural-air cool, while (b) constitutes the Pareto optimal front under natural-oil cool. It is seen that the power density increases as the efficiency decreases, which is attributed to the enhanced core loss and copper loss occurred at higher operation frequency. In the first cooling condition, maximum achievable power density is 41.57kW/L with a 99.84% efficiency but the maximum temperature inside the transformer rises beyond 100°C. The maximum temperature tends to rise with higher power density. At the same power rating and frequency level, the mineral oil is a better coolant than the air due to its superior convective properties and it enhances the insulation. And from Figure 7, the oil natural cool helps achieve better power density and thermal behavior.

Among these non-dominated solutions, 3 designs are chosen and the characteristics are listed in Table III with first design put the thermal performance at the priority and the second emphasize on the power density. The third design is using oil natural cool, the transformer itself exhibits higher power density and lower temperature rise, but considering the oil might cause environmental, maintenance issues, oil-free transformer is highly recommended for SST application. The solution with 99.875% efficiency at 25kW and a power density of 14.77 kW/L is selected and starred based on trade-off among these three objectives, whose maximum temperature rise in the transformer is limited within 50°C. The transformer is using the ferrite N87 core and 2 pairs of UU core shaped an EE core configuration, the winding is wound on the central limb of the magnetic core with primary winding inside and secondary winding outside. Litz conductors of type 1050_AWG36 are chosen for the high-frequency windings. Primary has 22 turns, arranged in 2 layers, each having 11 turns per layer while secondary side has 11 turns in total arranged in one layer.

Table III: Optimal Transformer Characteristics

Parameters	Transformer 1	Transformer 2	Transformer 3
Core	186_152_80	144_98_100	104_102_75
Wire	Litz_1050 _AWG36	Litz_1050 _AWG36	Litz_660 _AWG36
Pri. winding	11 turns, 2 layer	7 turns, 4 layers	13 turns, 2 layers
Sec. winding	11 turns, 1 layer	7 turns, 2 layers	13 turns, 1 layer
$L_{\sigma(pri)}$	2.5uH	7.4uH	11.5uH
Core volume	1693.44cm^3	679.94cm^3	585.6 cm^3
Coolant	Natural-air	Natural-air	Natural-oil
Core loss	12.93W	15.11W	23.39W
Copper loss	18.32W	22.96W	22.68W
Efficiency	99.875%	99.85%	99.827%
Power density	14.77kW/L	36.7kW/L	42.69kW/L
Maximum ΔT	47.9°C	72.9°C	17.65°C

Table IV shows the transformer characteristics comparison of the same magnetic size but composed from different

materials. The Nanocrystalline alloy core has the lowest core loss and minimum temperature rise among these three materials, being the best candidate among all for promising both the efficiency and heat dissipation. On the one hand, it is superior at higher frequency which can guarantee high power density operating requirement.

Figure 8 shows the efficiency variation under different operation conditions. Maximum achievable efficiency of 99.882% can be achieved at 22.22kW with the operating frequency at 50kHz, the corresponding maximum temperature rise is within 30C.

Table IV: Comparison of core materials characteristics for the HFT

Core material	Ferrite	Amorphous	Nanocrystalline
Core loss (W)	18.63	319.8	12.93
η (%)	99.85%	98.66%	99.875%
ΔT_{max} (°C)	50.93	410.01	47.9

IV. DESIGN VALIDATION WITH FEM-CFD SIMULATION

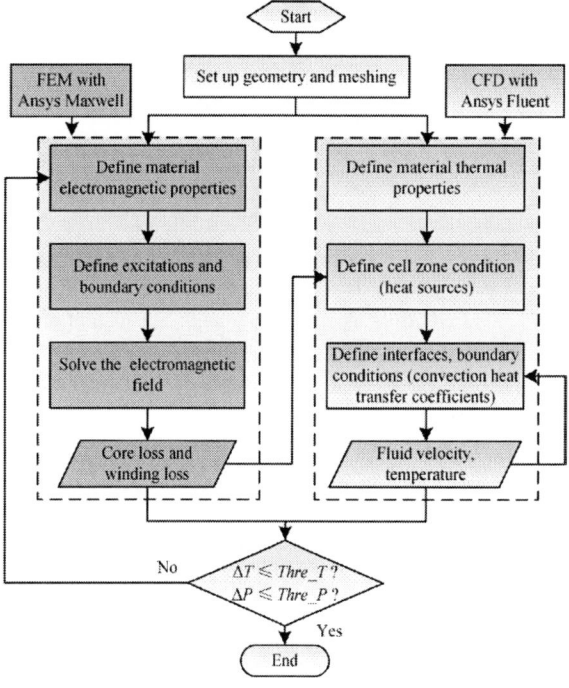

Figure 9: Framework with FEM and CFD for the electromagnetic and thermal-fluid coupled analysis of HF transformer

For validating the performance of the design choices, the transformer design is modeled and simulated in ANSYS Maxwell 3D software. The FEM-based electromagnetic analysis and the CFD-based thermal-fluid analysis are coupled together as shown in Figure 9. The losses and temperature calculation are solved iteratively with both FEM and CFD until the changes of them during two adjacent iterations (ΔT1 and ΔP) are no more than their respective thresholds. This iterative process formulates the macro close loop as shown in the completed framework in Figure 9.

Table V: Comparison Between Analytical Calculation and Maxwell Simulation

	P_{cc}	P_{ce}	P_p	P_s
Analytical	3.28	9.65	6.25	12.07
Maxwell	3.01	8.85	6.57	12.95
Deviation (%)	8.97	9.04	4.87	6.8

Table VI: Comparison Between Simulated And Analytical Temperature At Different Location

ΔT_j	ΔT_{cc}	ΔT_{ce}	ΔT_p	ΔT_s
Analytical	38.31	31.15	45.77	47.90
FEM-CFD	31.85	26.85	48.10	51.23
Deviation (%)	11.36	8.29	3.19	4.37

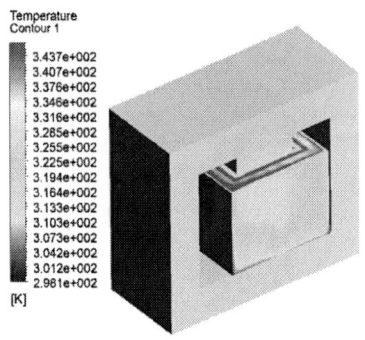

Figure 10: Temperature distribution in transformer active components and oil for the three investigated prototypes

Based on the coupled framework, the overall temperature distribution can be obtained as illustrated in Figure 10 and the simulated temperatures at different thermal zones are compared with the analytical calculations to evaluate the deviation. The relative deviations between simulation and measurement calculated by $(T_{simu} - T_{anal}) / T_{simu}$ are summarized in Table VI. Minimal relative deviation substantiates that the estimation of the core and winding loss as well as the temperatures are fairly accurate and reasonable.

V. CONCLUSION

In this paper, a multi-objective optimization methodology and design considerations are explored for the high frequency transformer in the dual-active-bridge DC-DC converter system. The multi-objective optimization design methodology combining the weighted-sum approach and genetic algorithm is applied on the 25kVA/25kHz, 1500V/750V transformer intended for DC-DC stage in 1MVA solid state transformer and the optimal design is derived under both air-natural cool and oil-natural cool. The methodology is computational effective and reliable to provide non-dominated solutions for further choices and offers absolute freedom without limitation on the manufacturers' standard library. Trade-off between size, efficiency and thermal behavior has been carefully considered targeting on compact, efficient and reliable design solutions. Core and copper losses are accurately calculated while non-linearities observed at higher switching frequency

are taken into account. And accurate estimation of the leakage inductance is also addressed considering the influences of both frequency and geometrical dimensions. An easy-to-implement equivalent thermal model for the HFT has also been included, which is more computationally effective and yields fair accuracy. Optimal design exhibits an efficiency larger than 99.8% with a power density of 14.77kW/L and maximum temperature rise limited to 50°C under air-natural cool. Close agreement among analytical and FEM-CFD results are observed, which corroborates accuracy and aptness of proposed method. Comparative study of HFT design with different magnetic materials and under different operation conditions is also performed. Further experimental validation will be carried out in the near future.

REFERENCES

[1] N. B. Kadandani, M. Dahidah, S. Ethni, and J. Yu, "Solid state transformer: an overview of circuit configurations and applications," 2019.

[2] S. Meier, T. Kjellqvist, S. Norrga, and H.-P. Nee, "Design considerations for medium-frequency power transformers in offshore wind farms," in *13th European Conference on Power Electronics and Applications (EPE 2009) Barcelona, SPAIN, SEP 08-10, 2009*, 2009, pp. 757-768: IEEE.

[3] X. She, A. Q. Huang, and R. Burgos, "Review of solid-state transformer technologies and their application in power distribution systems," *IEEE journal of emerging and selected topics in power electronics,* vol. 1, no. 3, pp. 186-198, 2013.

[4] H. Tian, Z. Wei, S. Vaisambhayana, M. P. Thevar, A. Tripathi, and P. C. Kjær, "Calculation and Experimental Validation on Leakage Inductance of a Medium Frequency Transformer," in *2018 IEEE 4th Southern Power Electronics Conference (SPEC)*, 2018, pp. 1-6: IEEE.

[5] X. She, R. Burgos, G. Wang, F. Wang, and A. Q. Huang, "Review of solid state transformer in the distribution system: From components to field application," in *2012 IEEE Energy Conversion Congress and Exposition (ECCE)*, 2012, pp. 4077-4084: IEEE.

[6] E. Agheb and H. K. Høidalen, "Medium frequency high power transformers, state of art and challenges," in *2012 International conference on renewable energy research and applications (ICRERA)*, 2012, pp. 1-6: IEEE.

[7] M. Leibl, G. Ortiz, and J. W. Kolar, "Design and experimental analysis of a medium-frequency transformer for solid-state transformer applications," *IEEE Journal of Emerging and Selected Topics in Power Electronics,* vol. 5, no. 1, pp. 110-123, 2017.

[8] M. A. Bahmani, "Design and optimization of hf transformers for high power dc-dc applications," 2014.

[9] P. Dowell, "Effects of eddy currents in transformer windings," in *Proceedings of the Institution of Electrical Engineers*, 1966, vol. 113, no. 8, pp. 1387-1394: IET.

[10] F. Tourkhani and P. J. I. T. o. m. Viarouge, "Accurate analytical model of winding losses in round Litz wire windings," vol. 37, no. 1, pp. 538-543, 2001.

[11] H. Tian, Z. Wei, M. P. Thevar, S. Vaisambhayana, A. Tripathi, and P. C. Kjaer, "Experimental Verification on Thermal Modeling of Medium Frequency Transformers," in *IECON 2018-44th Annual Conference of the IEEE Industrial Electronics Society*, 2018, pp. 5527-5534: IEEE.

978-1-7281-4181-7/19 $31.00 © 2019 IEEE

Center-Tapped π-Type Single-Phase Cell

Domingos S. L. Simonetti
Power Electronics and Drives Laboratory
Department of Electrical Engineering
Universidade Federal do Espírito Santo
Vitória, Brazil
domingos.simonetti@ufes.br

Xibo Yuan
Electrical Energy Management Group
Department of Electrical and Electronic Engineering
University of Bristol
Bristol, United Kingdom
xibo.yuan@bristol.ac.uk

Abstract— **A π-type derived four-level single-phase inverter is analyzed in this paper. It is demonstrated that the topology can be obtained both by inserting a neutral connection point in a four-level π-type cell, or removing the switches connected to ground in a five-level T-type cell. Five different pattern approaches to drive converter switches are presented, showing its influence on the high-frequency waveform of the output voltage. Simulation results for scenarios of feeding a single-phase load and a solar array connection to grid are given, making possible to demonstrate its functionality.**

Keywords—multilevel, single-phase, inverter component.

I. INTRODUCTION

Switching power converters are being used for several decades to perform ac-dc, dc-dc, dc-ac or ac-ac power conversion [1]. The advance of semiconductor technology has lead to increase switching frequency in order to reduce weight and size of passive components, although some penalty on semiconductor losses arise, reflecting on heatsink dimensions. By using multilevel converters [2,3], a trade-off between semiconductor losses and passive weight and size can be achieved. Multilevel inverters can be obtained by specific topologies, like neutral point clamped converter - NPC [4], capacitor-clamped multilevel inverter (flying capacitor inverter [5]), or even cascade association of converters, mainly half-bridge or full-bridge single-phase inverters [6]. In this last approach, converters operating at a not so high switching frequency are usually series associated to generate a waveform with a switching frequency Fs given by n*fs, being fs the switching frequency of each converter and n the number of converters. As a consequence, by appropriate control, the synthesized output voltage has an appearance of stairs, reducing harmonic content. Fig. 1 shows typical ac voltage waveforms for (a) two-level converter; (b) a five-step multilevel converter.

The multilevel approach is not a so new approach, as can be seen in [7], which employs multiple transformers and inverters in order to generate a multilevel voltage.

Initially multilevel inverters where thought to be applied at high voltage levels, allowing the use of standard 600V or 1200V switching devices in topologies operating at higher voltages, as industrial 3.3kV or 4.16kV, or higher. More recently, due to the advantages of lower switching losses as well as harmonic content reduction, multilevel converters are being applied also to low-voltage applications, up to around 400V phase-neutral voltage [8,9].

A classic half-bridge three-phase inverter, Fig. 2(a), produces a multilevel phase-neutral voltage at load. However, its single-phase counterpart (half-bridge, Fig. 2.b) is unable to produce such waveform, being its output voltage a two-level one.

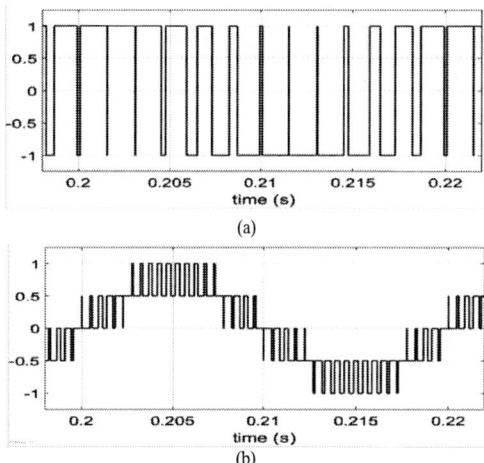

Fig. 1 - Typical ac voltage waveforms, pu: (a) two-level converter; (b) a five-step multilevel converter.

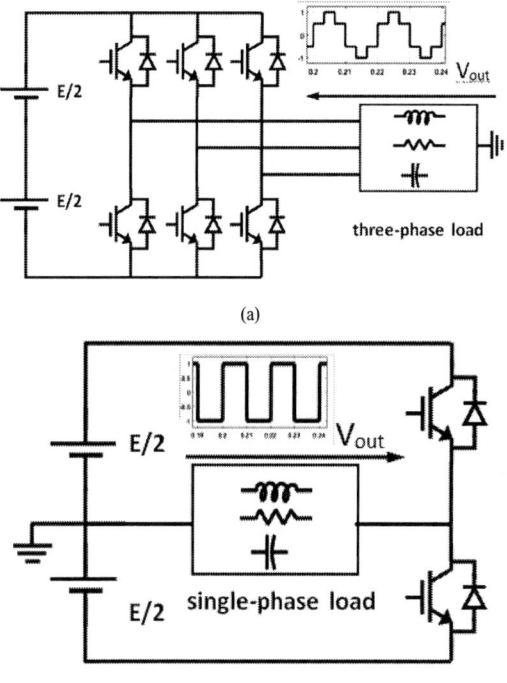

Fig. 2 – a) Half-bridge three-phase inverter; b) Half-bridge single-phase inverter (voltages using 180° modulation).

The single-phase T-cell (Fig. 3) [2] is able to produce a three-level output voltage, the same as can be generated using a full-bridge converter, or a three-level NPC single-phase cell.

978-1-7281-4181-7/19 $31.00 © 2019 IEEE   329

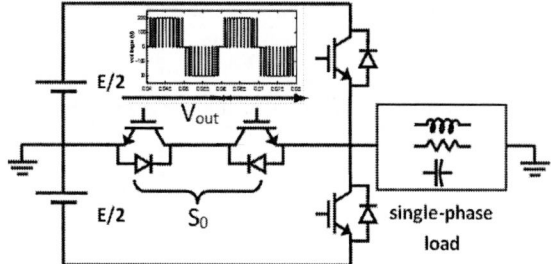

Fig. 3 – The single-phase T-cell and its three-level output voltage.

The basic π-type cell (Fig. 4) is a four-level converter, presented in prior works [10] applied to a three-phase converter. S_3 and S_4 are bidirectional switches with reverse blocking capacity. In this analysis they are considered as the back-to-back connection of two IGBT/diode devices, respectively S_{3a}, S_{3b}, S_{4a}, S_{4b}. Due to its topology, the cell isn´t able to produce positive and negative voltages, because multilevel converters with even levels need to have a source with a middle point to get the symmetrical condition [11]. To build a single-phase converter, two π-cells are necessary, in double star chopper-cells (DSCC) connection [12].

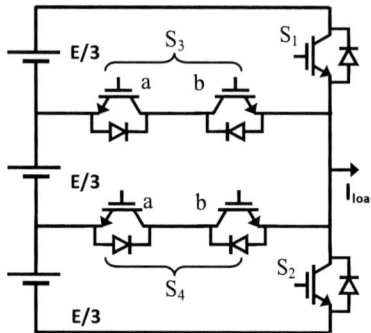

Fig. 4 - The single-phase π-type cell.

S1 and S2, as in a conventional two-level converter, are required to block the whole dc-link voltage. Hence, for low-voltage applications, e.g., dc-link voltage E = 600 V, S1 and S2 will usually be 900V or 1200-V devices in order to leave enough voltage margin. S3b and S4a need to withstand ⅔E, while S3a and S4b only need to withstand ⅓E. Therefore, they can be implemented with 600-V devices [13].

This paper analyzes a slight modification in the topology, introducing a neutral-point in the middle of the input sources, forming a Center-Tapped π-Type Single-Phase Cell.

II. THE CENTER-TAPPED Π-TYPE SINGLE-PHASE CELL

A. Derived from the four-level π-Type Single-Phase Cell

Several works in the literature present how, from one family of multilevel converters, other cells can be derived [14,15]. By proper simplifications in the generalized topology of a N-level T-type cell, the π-type cell can be obtained [16]. The minimum cell employs three dc voltage sources (see Fig. 4), therefore naturally does not have a medium point of connection. This is similar that occurs to the half-bridge converter (single or three-phase), that has one dc voltage source. In order to obtain a medium point it is necessary to halve the dc source and create a medium point. Doing so, the resulting center-tapped π-type single phase cell that is obtained is shown in Fig. 5.

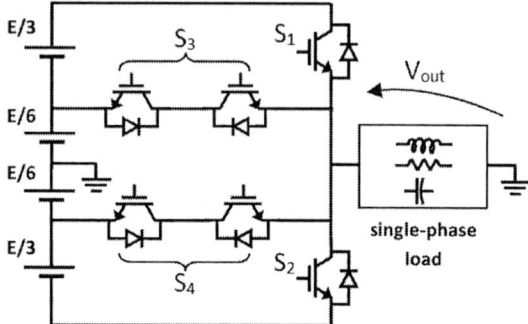

Fig. 5 – The proposed center-tapped π-type single phase cell.

Keeping voltage sources as in the original π-cell, the upper and lower voltage sources are E/3; and intermediate voltage sources are E/6. The output voltage is Vout=E/2 if S_1 is connected (-E/2 if S_2 is connected), whereas Vout=E/6 connecting S_3 (-E/6 connecting S_4). (This cell and respective voltage distribution can also be found in [11] in a generic topology of a multi-voltage cascaded multilevel inverter using fundamental switching frequency.)

1) Low switching frequency operation

It can be easily calculated that third harmonic voltage is null if S_3S_4 conducts for 40 degrees, and S_1S_2 for 100 degrees, in a sequence $S_3S_1S_3S_4S_2S_4$. The generated output voltage is shown in Fig. 6, presenting harmonics h5=7%, h7=7.5%, h9=13.2%, h11=4.8% and THD=21.8%.

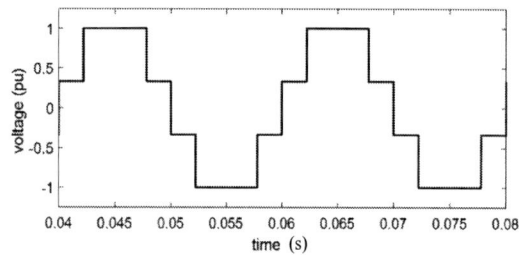

Fig. 6 – Output voltage for low switching frequency operation.

2) High switching frequency operation

By appropriate modulation control the center-tapped four–level π-type single-phase cell can operate at higher switching frequency and sinusoidal PWM, displacing harmonics to higher frequencies. Switching is done between adjacent switches, as follows in (1):

$$V_{ref}(pu) \begin{cases} -0.333 \leq V_{ref}(pu) \leq 0.333: & S_3S_4 \\ 0.333 \leq V_{ref}(pu): & S_1S_3 \\ -0.333 \geq V_{ref}(pu): & S_2S_4 \end{cases} \quad (1)$$

Figure 7 shows a) the reference voltage (pu) and carrier; b) Gate signals for this situation; c) PWM output voltage.

B. Derived from a five-level T-type converter

The same topology can be obtained from a five-level T-type converter as shown in Fig. 8, taking out the central switch pair S0. Compared to Fig. 5, the difference is on the voltage source values, all being E/4. The output voltage Vout=E/2 if S_1 is connected (-E/2 if S_2 is connected), and Vout=E/4 connecting S_3 (-E/4 connecting S_4).

978-1-7281-4181-7/19 $31.00 © 2019 IEEE

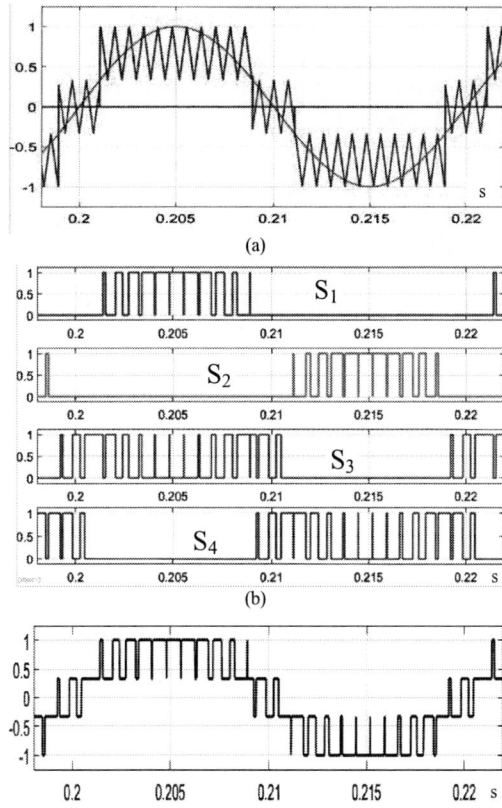

Fig. 7 - a) carrier signals and reference voltage; (b) gate signals; (c) the output voltage, pu.

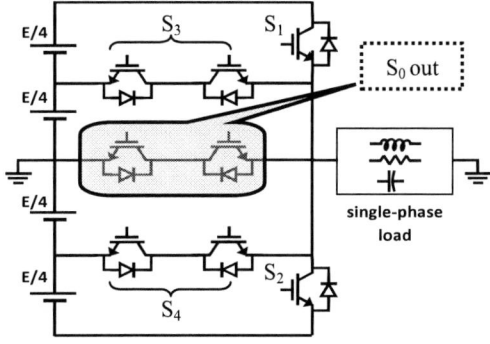

Fig. 8 – Obtaining the cell from a five-level T-type converter.

1) Low switching frequency operation.

A null third harmonic is obtained if we have $S_3S_1S_3S_4S_2S_4$ conducting sequentially for 60 degrees. The same phase-neutral voltage waveform of a three-phase half-bridge inverter, see Fig. 2.a, is obtained. Harmonics 5th and 7th are 20% and 14%, respectively, and THD=31%.

2) High switching frequency operation.

This converter can also operate at higher switching frequency under sinusoidal modulation. Switching is done between adjacent switches, as presented in (1), but voltage comparisons are 0.5 and -0.5. Carrier signals are similar to those in Fig. 7.a, adjusting limits (0.5 in place of 0.333). Figure 9 shows the generated output voltage.

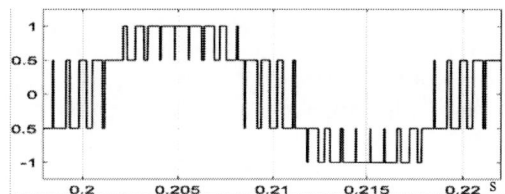

Fig. 9 - PWM output voltage from T-type derived.cell, pu.

Due the fact the comparison levels are different for π-type and T-type derived cell, conduction interval (degrees) as well as switching voltage also changes. A summary is presented in Table I for the PWM patterns of this section.

TABLE I. SWITCHING VOLTAGE COMPARISON

	π-type		T-type	
	S_1S_2	S_3S_4	S_1S_2	S_3S_4
voltage	E/3	E/3	E/4	E/4	E/2
interval	141°	180°	120°	120°	60°

III. OTHER MODULATION PAIRS

Three other PWM modulations are able to synthesize the necessary output voltage, for both π-type and T-type derived topology. The first one switches between S_3 and S_4 for reference voltage between -0.5 and 0.5 (or -0.333pu and 0.333pu); otherwise the output voltage is obtained switching S_1 and S_2. Fig. 10 shows (a) the reference voltage and carrier signals; (b) gate signals; and (c) the output voltage that is obtained. Modulations are shown for a T-type derived cell.

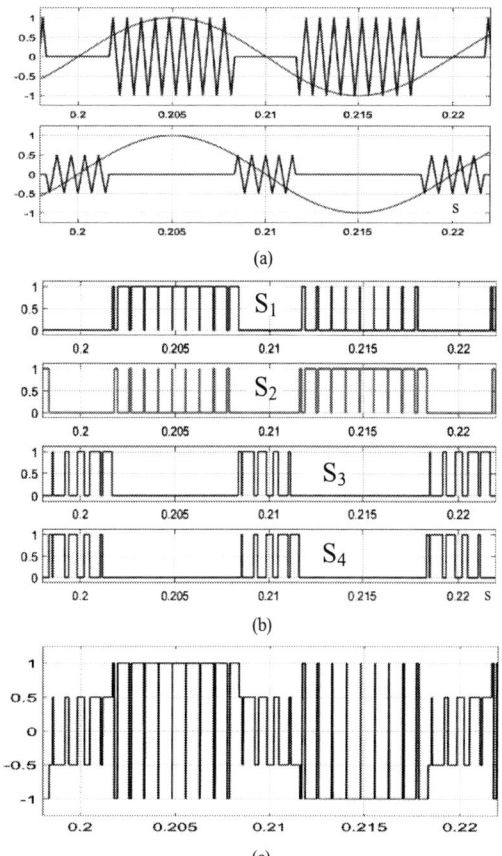

Fig. 10 – First alternative modulation: a) carrier signals and reference voltage; (b) gate signals; (c) the output voltage.

978-1-7281-4181-7/19 $31.00 © 2019 IEEE

In the second approach, switching occurs between S_1 and S_3 for v_{ref} above 0.5pu ($S_4 S_2$ if less than -0.5pu). And S_3 and S_4 commutates if the reference is between -0.5 and 0.5 pu (or ±0.333pu for the π-type derived). Triangular carriers and reference voltage for such approach are shown in Fig. 11.a; Fig. 11.b shows switching patterns, whereas 11.c shows the output voltage (pu).

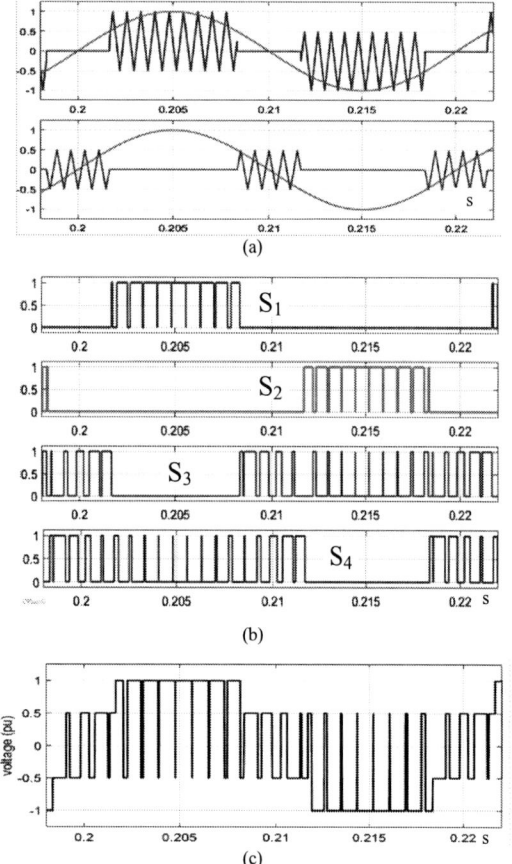

Fig. 11 - Second alternative modulation: a) Triangular carriers and reference voltage; b) switching patterns; c) output voltage (pu).

The third strategy synthesizes the output voltage by driving S_1/S_4 whenever the reference voltage is >0, otherwise driving S_2/S_3. The approach is shown in Fig. 12 (a) carriers and reference voltage; b) pattern switching; c) output voltage).

Table II shows, for all three cases, the value of switching voltages and the duration (in degrees) that switches are commutating. Comparing the results with those showed in Table I, it is easy to conclude that the switching patterns of this section leads to a higher commutation voltage on switches, consequently higher switching losses.

TABLE II. SWITCHING VOLTAGE COMPARISON

voltage and interval	π-type		T-type	
	S_1S_2	S_3S_4	S_1S_2	S_3S_4
Case 1	E - 282°	E/3 - 78°	E - 240°	E/2 - 120°
Case 2	E/3- 120°	2E/3- 120° E/3- 120°	3E/4- 120°	3E/4 - 120° E/2 - 120°
Case 3	2E/3- 180°	2E/3- 180°	3E/4- 180°	3E/4- 180°

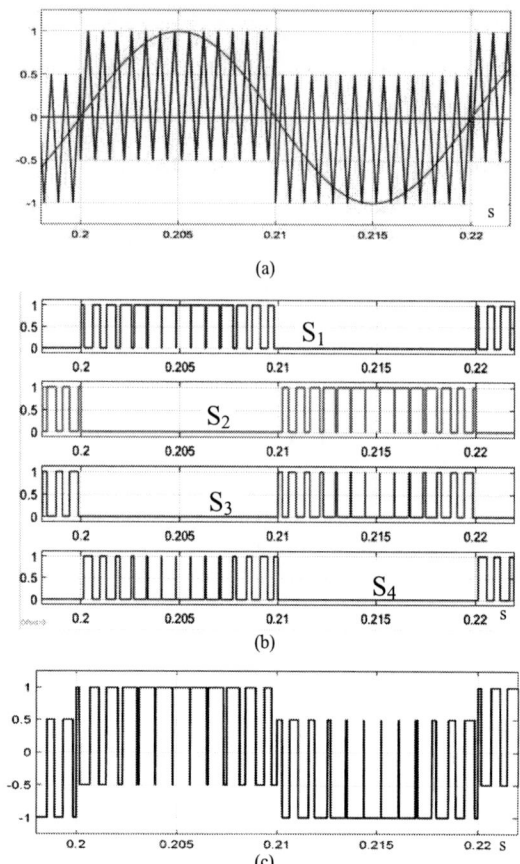

Fig. 12 – Third alternative modulation: a) Triangular carriers and reference voltage; b) switching patterns; c) output voltage (pu).

IV. SIMULATION ANALYSIS

Some situations were simulated to verify the behavior of the derived converter. The switching pattern presented in sections II.A.2 and II.B.2 were employed.

A. Feeding an RL load

From a total dc-bus of 600V and a switching frequency of 1350Hz, a 50-Hz 212V-rms is generated, and directly applied to a 2kW/500var load. A low switching frequency is used in order to better observe voltage and current waveforms.

1) For a T-type derived cell

The applied voltage is similar to that shown in Fig. 9, and the load current is shown in Fig. 13. The load current is delayed 14° respect the voltage.

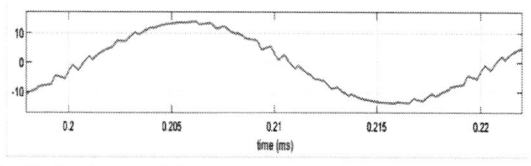

Fig. 13 – Load current (A)

Voltage and current in S_1 and S_2 can be seen in Fig. 14, and in composed switch S_3 can be seen in Fig. 15.

978-1-7281-4181-7/19 $31.00 © 2019 IEEE

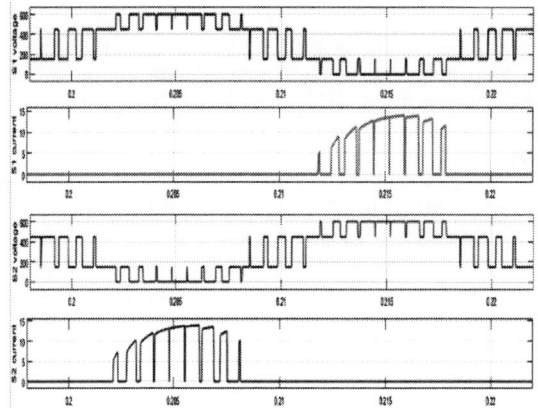

Fig. 14 – From up to down: S_1 voltage, S_1 current, S_2 voltage, S_2 current. (units: voltage: V, current: A, horizontal axis: ms)

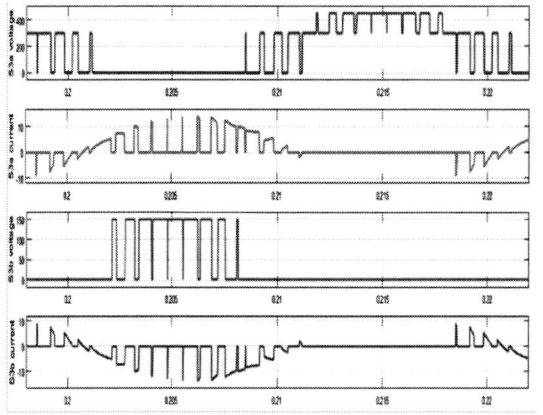

Fig. 15 – From up to down: S_{3a} voltage, S_{3a} current, S_{3b} voltage, S_{3b} current. (units: voltage: V, current: A, horizontal axis: ms)

2) For a π-type derived cell

The simulation was repeated considering the voltage sources as derived from the π-type converter. The applied voltage is similar to that shown in Fig. 7.c, and the load current similar to the one shown in Fig. 13.

Voltage and current in S_1 and S_2 can be seen in Fig. 16, and in composed switch S_3 is showed in Fig. 17.

3) Comparing results

By comparing voltage and current values in switches, some details can be observed. Basically the switches commutate the same current, differing in duration and voltage, as already summarized in Table I.

As most of time derived T-type is switching under a lower voltage, will produce lower commutation losses. Harmonics for both approaches are very similar, as is shown in Fig. 18 for the generated voltage. THD of the output voltage is about 39% for T-type derived and 35% for π-type derived.

B. Feeding power to a grid

It is considered a scenario of a 3kW renewable PV source connected to a single-phase 127V/60Hz bus. The converter is tied to grid by an equivalent 0.1 pu reactance (LCL filter), having a switching frequency of 19950 Hz. It is assumed that a MPPT algorithm along with a dedicated converter keeps all voltages at their nominal values.

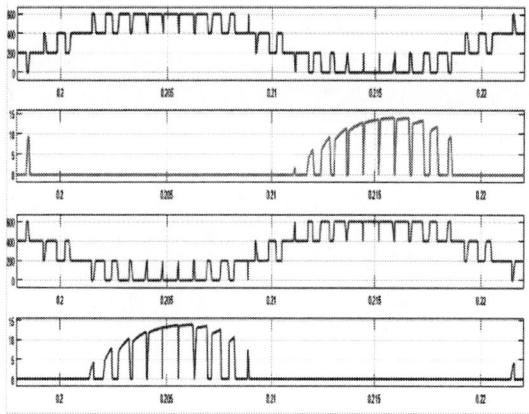

Fig. 16 – From up to down: S_1 voltage, S_1 current, S_2 voltage, S_2 current (units: voltage: V, current: A, horizontal axis: ms).

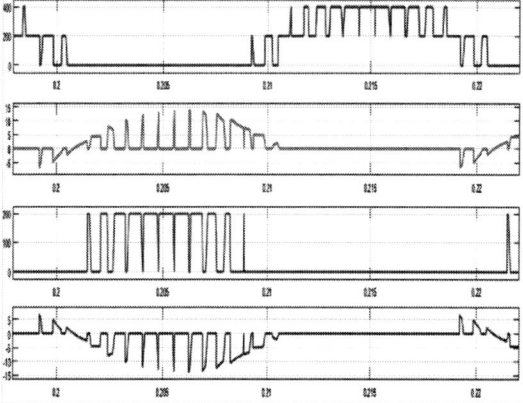

Fig. 17 – From up to down: S_{3a} voltage, S_{3a} current, S_{3b} voltage, S_{3b} current (units: voltage: V, current: A, horizontal axis: ms).

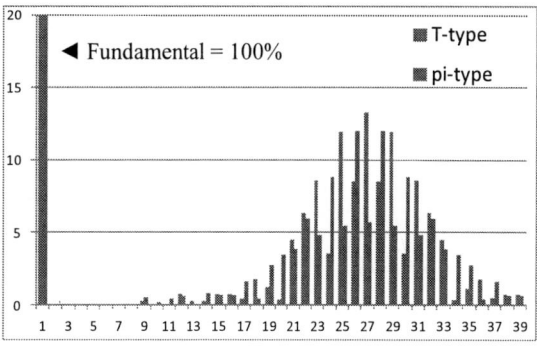

Fig. 18 – Load voltage harmonic spectrum.

Therefore, the solar array is always delivering the maximum available power along the day, and this information is given to the converter control. The simulation will be timescaled so that 1 hour will be represented by 1 second. The irradiance is considered as shown in Fig. 19 (top). Figure 19 also shows the angle displacement imposed between the converter and the grid (middle), and the rms reference voltage (bottom), calculated by the converter control. Fig. 20 shows the active (top) and reactive power (bottom) delivered to grid. Reactive power is set to zero. Between 0h-6h and 18h-24h both active and reactive references are zero, and omitted from the figure.

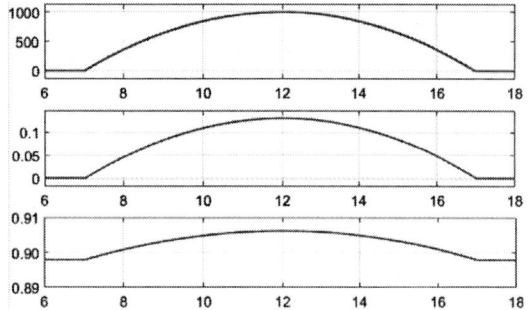

Fig. 19 – From top to bottom: Irradiance (W/m²), Phase angle between converter voltage and grid voltage (rad), rms reference voltage (pu). horizontal: hour.

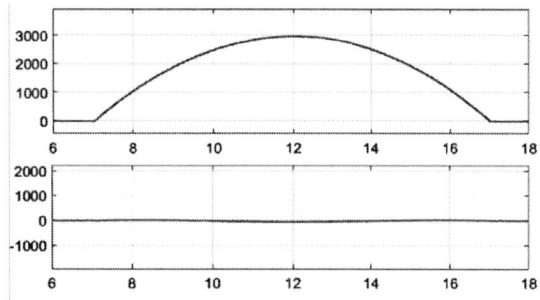

Fig. 20 – Output active (top, W) and reactive power (bottom, var); horizontal: hour.

Figure 21 shows, at maximum power, from top to bottom: PWM voltage (pu), the current injected to grid and current in S_2. It can be seen that the power factor can be considered unitary. It was employed the π-type derived converter.

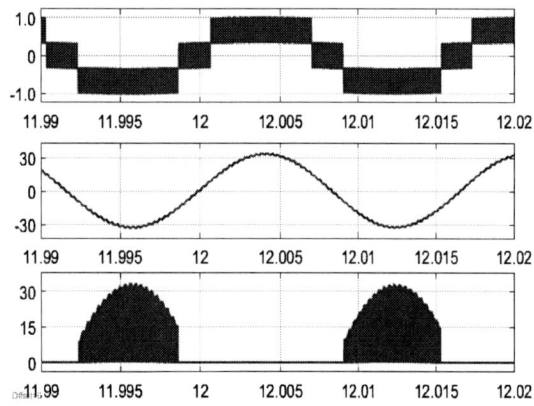

Fig. 21 – From top to bottom: PWM voltage (pu), injected current (A), and S_2 current (S). Horizontal: hour.

V. CONCLUSIONS

A four-level single-phase multilevel cell was presented in this paper. It can be derived from a π-type four-level cell or from a T-type five-level cell. Due to its diagram be similar to the π-type cell, it was named center-tapped π-type single-phase cell.

Related to its parent, two voltage source distribution where considered: four equal voltage sources, to the T-type derived or 2/3 and 1/3 relative voltage sources, to the π-type derived. Several alternatives to generate switching patterns, for low and high switching frequency under sinusoidal modulation strategy, were obtainable. Switching between

adjacent voltage levels allows lower switching losses, because turn-on and turn-off voltage transition is lower.

Some simulation studies were presented. First, a 50Hz single-phase source feeding an RL load for both equally distributed dc sources and 2/3-1/3 dc sources was useful to observe that using same values for the voltage sources allow lower losses than the unequal distribution for high power factor operation. Finally, the cell delivering power from a PV source to a 127V/60Hz grid was simulated, and again the operation was adequate. The analysis plus simulations presented have demonstrated the functionality of the cell. It is able to operate as a 4-level converter in single-phase applications providing low switching losses.

REFERENCES

[1] R. L Bright. "Junction transistors used as switches," in Trans. of the American Institute of Electrical Engineers, Part I: Communication and Electronics, vol. 74, no. 1, pp. 111-121, March 1955.

[2] P. M. Bhagwat and V. R. Stefanovic, "Generalized Structure of a Multilevel PWM Inverter," in IEEE Trans. on Industry Applications, vol. IA-19, no. 6, pp. 1057-1069, Nov. 1983.

[3] N. S. Choi, J. G. Cho and G. H. Cho, "A general circuit topology of multilevel inverter," PESC '91 Record 22nd Annual IEEE Power Electronics Specialists Conference, Cambridge, USA, 1991, pp. 96-103.

[4] A. Nabae, I. Takahashi and H. Akagi, "A New Neutral-Point-Clamped PWM Inverter," in IEEE Trans. on Industry Applications, vol. IA-17, no. 5, pp. 518-523, Sept. 1981.

[5] T. A. Meynard and H. Foch, "Multi-level conversion: high voltage choppers and voltage-source inverters," PESC '92 Record. 23rd Annual IEEE Power Electronics Specialists Conference, Toledo, Spain, 1992, pp. 397-403 vol.1.

[6] J. Lai and F. Z. Peng, "Multilevel converters-a new breed of power converters," IAS '95. Conference Record of the 1995 IEEE Industry Applications Conference Thirtieth IAS Annual Meeting, Orlando, FL, USA, 1995, pp. 2348-2356 vol.3.

[7] C. W. Flairty, "A 50-KVA Adjustable-Frequency 24-Phase Controlled Rectifier Inverter," in IRE Trans. on Industrial Electronics, vol. IE-9, no. 1, pp. 56-60, May 1962.

[8] R. Teichmann and S. Bernet, "A comparison of three-level converters versus two-level converters for low-voltage drives, traction, and utility applications," in IEEE Trans. on Industry Applications, vol. 41, no. 3, pp. 855-865, May-June 2005.

[9] T. B. Soeiro and J. W. Kolar, "The New High-Efficiency Hybrid Neutral-Point-Clamped Converter," in IEEE Trans. on Industrial Electronics, vol. 60, no. 5, pp. 1919-1935, May 2013.

[10] X. Yuan, "A four-level π-type converter for low-voltage applications," 2015 17th European Conference on Power Electronics and Applications (EPE'15 ECCE-Europe), Geneva, 2015, pp. 1-10.

[11] J. Dixon and L. Moran, "High-level multistep inverter optimization using a minimum number of power transistors," in IEEE Trans. on Power Electronics, vol.21-no. 2, pp. 330-337, March 2006.

[12] H. Akagi, "Classification, Terminology, and Application of the Modular Multilevel Cascade Converter (MMCC)," in IEEE Trans. on Power Electronics, vol. 26, no. 11, pp. 3119-3130, Nov. 2011.

[13] B. Jin, and X. Yuan. "Topology, efficiency analysis and control of a four-level π-type converter", in IEEE Journal of Emerging and Selected Topics in Power Electronics, vol. 7, no. 2, pp. 1044-1059, June 2019

[14] J. Rodriguez, J. Lai and F. Z. Peng, "Multilevel inverters: a survey of topologies, controls, and applications," in IEEE Trans. on Industrial Electronics, vol. 49, no. 4, pp. 724-738, Aug. 2002.

[15] M. N. Raju, J. Sreedevi, R. P Mandi and K. S. Meera, "Modular multilevel converters technology: a comprehensive study on its topologies, modelling, control and applications," in IET Power Electronics, vol. 12, no. 2, pp. 149-169, Feb. 2019.

[16] X. Yuan, "Derivation of Voltage Source Multilevel Converter Topologies," in IEEE Trans. on Industrial Electronics, vol. 64, no. 2, pp. 966-976, Feb. 2017.

978-1-7281-4181-7/19 $31.00 © 2019 IEEE

Novel MTPA Approach for IPMSM with Non-Sinusoidal Back-EMF

Allan Gregori de Castro
School of Engineering of São Carlos
University of São Paulo
São Carlos, Brazil
allangregori@usp.br

Paulo R. U. Guazzelli
School of Engineering of São Carlos
University of São Paulo
São Carlos, Brazil
paulo.ubaldo@usp.br

Mateus M. Lumertz
School of Engineering of São Carlos
University of São Paulo
São Carlos, Brazil
mateuslumertz@ieee.org

Carlos M. R. de Oliveira
School of Engineering of São Carlos
University of São Paulo
São Carlos, Brazil
carlosmro@usp.br

Geyverson T. de Paula
School of Electrical, Mechanical
and Computing Engineering
Federal University of Goiás
Goiânia, Brazil
geyverson@gmail.com

José R. B. A. Monteiro
School of Engineering of São Carlos
University of São Paulo
São Carlos, Brazil
jrm@sc.usp.br

Abstract—Interior permanent magnet synchronous machines (IPMSMs) are widely used due to features like high efficiency and high power density. For high performance drives, maximum torque per Ampère (MTPA) is employed to optimally balance the contribution of mutual and reluctance torque. However, conventionally the MTPA algorithm considers IPMSM with sinusoidal back-electromotive forces (back-EMFs). In practice, IPMSM often presents back-EMF waveforms that deviate from ideal sinusoid, exhibiting harmonic components. Thus, this paper presents a new algorithm for MTPA operation of IPMSM with non-sinusoidal back-EMF that includes the contribution of harmonic back-EMF components in average torque production. For the test machine the proposed optimal stator current provides approximately 2% higher average torque per Ampère ratio compared to the conventional MTPA. Due to the increase of torque undulations, this method is more suitable for applications non sensitive to torque ripple.

Index Terms—MTPA, IPMSM, non sinusoidal back-EMF, optimal stator current

I. INTRODUCTION

Permanent magnet synchronous machines (PMSMs) are a popular machine topology in motor drives and generation systems due to some special features like high efficiency and high power density. Due to rotor and stator saliencies, interior permanent magnet synchronous machines (IPMSMs) counts on the contribution of magnetic flux and reluctance in the electromagnetic torque production. The contribution of each source of torque depends on the injected current in the direct-quadrature-zero ($dq0$) reference system. Usually, the most favorable feeding strategy is the one which balances these components so the lowest current magnitude for a required torque output is achieved. This operation condition can be

This work is supported in part by Conselho Nacional de Desenvolvimento Científico e Tecnológico (CNPq), in part by Coordenação de Aperfeiçoamento de Pessoal de Nível Superior (CAPES) - Finance Code 001 -, in part by FAPEG (Fundação de Amparo à Pesquisa do Estado de Goiás) and in part by São Paulo Research Foundation (FAPESP), grant #2006/04226-0.

achieved by control strategies referred as maximum torque per Ampère (MTPA).

The typical MTPA approach considers IPMSM with sinusoidal flux distribution and, therefore, sinusoidal back electromotive forces (back-EMF) [1]–[3]. However, due to machine design, manufacturing limitations and mechanical tolerances, a perfect sinusoidal back-EMF is not feasible, presenting deviations in the form of harmonic components. In fact, non sinusoidal back-EMF may be composed of a wide range of harmonics and some components can present significant magnitude [4]–[7].

Because the interaction of back-EMF and stator currents is the major source of torque, the harmonic components of the non sinusoidal back-EMF also have their contribution. Efficient control algorithms for IPMSMs with non sinusoidal back-EMF have been addressed in several publications [5], [6], [8]. However, many techniques are interested in achieve MTPA operation combined to torque ripple reduction, but the invested energy to reduce torque ripple may decrease the torque per Ampère ratio [9], [10]. In a different approach presented in [11], the authors have proposed an optimum current waveform for a IPMSM with non sinusoidal back-EMF in order to increase the efficiency without the consideration of the torque ripple reduction. However, the presented optimization process require offline adjustment for each machine operation point.

Accordingly, the purpose of this paper is to develop an MTPA algorithm that accounts for the non sinusoidal back-EMF harmonic components and is feasible for real-time implementation. The method is based on the series expansion of the average torque production and optimization using Lagrangian approach.

In Section II, the average electromagnetic torque is modeled considering the harmonic components of back-EMF and stator current. Further, the MTPA optimum current reference is

978-1-7281-4181-7/19 $31.00 © 2019 IEEE

derived and illustrated for a test motor. In Section III a IPMSM drive is presented to implement the proposed MTPA and the results are compared to conventional IPMSM control strategies. Finally, in Sections IV conclusions are drawn.

II. IPMSM Average Torque Modeling and MTPA Approach

According to the magnetic co-energy model, the torque T_e produced by a three phase IPMSM, neglecting the cogging component, is [12]

$$T_e = n_p \frac{1}{2} i_{abc}^T \frac{d\mathbf{L}_{abc}}{d\theta_e} i_{abc} + i_{abc}^T e_{abc} \tag{1}$$

where $i = [i_a \ i_b \ i_c]^T$ is the three-phase stator current vector, $\mathbf{L}_{abc}$ is the matrix of self and mutual inductances [13], $e_{abc} = [e_a \ e_b \ e_c]^T$ corresponds to the back-EMF vector normalized to the mechanical speed ω_m, n_p is the number of pole pairs and θ_e is the electrical rotor position. Applying the rotor oriented dq0-axis transformation with power invariance property [13], the electromagnetic torque equation (1) becomes

$$T_e = L_\Delta i_d i_q + i_{dq0}^T e_{dq0} \tag{2}$$

where $L_\Delta = n_p(L_d - L_q)$ is the difference between dq-axis inductances multiplied by n_p, $i_{dq0} = [i_d \ i_q \ i_0]^T$ are the dq0-axis current vector and $e_{dq0} = [e_d \ e_q \ e_0]^T$ is the normalized back-EMF vector.

In order to model the IPMSM torque harmonics as a function of the harmonic content in back-EMF and the stator currents, the non-sinusoidal normalized back-EMF and stator current are represented in the form of Fourier series such as, respectively,

$$e_{dq0} = \begin{bmatrix} 0 \\ E_{q0} \\ 0 \end{bmatrix} + \begin{bmatrix} \sum\limits_{h=6,12,...}^{\infty} E_{dh,r}\cos(h\theta_e) + E_{dh,j}\sin(h\theta_e) \\ \sum\limits_{h=6,12,...}^{\infty} E_{qh,r}\cos(h\theta_e) + E_{qh,j}\sin(h\theta_e) \\ \sum\limits_{h=3,9,...}^{\infty} E_{0h,r}\cos(h\theta_e) + E_{0h,j}\sin(h\theta_e) \end{bmatrix} \tag{3}$$

and

$$i_{dq0} = \begin{bmatrix} I_{d0} \\ I_{q0} \\ 0 \end{bmatrix} + \begin{bmatrix} \sum\limits_{h=6,12,...}^{\infty} I_{dh,r}\cos(h\theta_e) + I_{dh,j}\sin(h\theta_e) \\ \sum\limits_{h=6,12,...}^{\infty} I_{qh,r}\cos(h\theta_e) + I_{qh,j}\sin(h\theta_e) \\ \sum\limits_{h=3,9,...}^{\infty} I_{0h,r}\cos(h\theta_e) + I_{0h,j}\sin(h\theta_e) \end{bmatrix}, \tag{4}$$

where the definitions of above normalized back-EMF and current components are as follows:

- E_{q0} is the q-axis average component of normalized back-EMF; $E_{dh,r}$, $E_{dh,j}$, $E_{qh,r}$, $E_{qh,j}$, $E_{0h,r}$ and $E_{0h,j}$ are the magnitudes of normalized dq0-axis back-EMF harmonics;
- I_{d0} and I_{q0} are dq-axis average values of dq-axis current; $I_{dh,r}$, $I_{dh,j}$, $I_{qh,r}$, $I_{qh,j}$, $I_{0h,r}$ and $I_{0h,j}$ are the magnitudes of stator current harmonics;
- h is the harmonic order. For PMSM dq-axis quantities, the usual dominant harmonic components are multiples of the sixth harmonic [14], [15]. The 0-axis components comprises the odd multiple of the third harmonic.

At first, to avoid the non linear stator current relationship in the reluctance torque portion in (2), in the proposed approach we linearize (2) regarding the stator current. The linearization is performed around a given value of stator current $\bar{i}_{dq0}$ considering constant insaturated L_Δ. This process allows to rewrite (2) as an approximated linearized torque expression such as

$$T_e = (L_\Delta \bar{i}_q + e_d)i_d + (L_\Delta \bar{i}_d + e_q)i_q + i_0 e_0 - L_\Delta \bar{i}_d \bar{i}_q \tag{5}$$

The substitution of (3) and (4) in (5) obtains the detailed model of the total torque produced by a IPMSM with non sinusoidal back-EMF and stator current. The resulting torque model is composed of an average component T_0 and of the harmonic components. The torque harmonics consist usually of 6th, 12th, 18th, 24th,... harmonics. Other harmonics may appear, for instance, in machines with unbalanced phase back-EMF [15]. However, only a small number of these harmonics are dominant [14]. In the case of this paper, the considered IPMSM presents the 6th and 12th torque harmonics as the dominant ones. The torque undulations may not be an issue in applications like wind turbines or other high inertia systems [11]. Furthermore, torque ripple mitigation strategies may reduce the torque per Ampère ratio [9], [10]. Thus, this paper focus in the maximization ratio of average torque per Ampère, not aiming torque ripple mitigation.

From the expansion of (5) through application of (3) and (4), the average torque component T_0 results in

$$T_0 = \Gamma_{d0}I_{d0} + \Gamma_{q0}I_{q0} - \Upsilon + \left(\frac{1}{2}\sum\limits_{h=6,12,...}^{\infty} \Gamma_{dh,r}I_{dh,r} + \Gamma_{dh,j}I_{dh,j} \right) + \left(\frac{1}{2}\sum\limits_{h=6,12,...}^{\infty} \Gamma_{qh,r}I_{qh,r} + \Gamma_{qh,r}I_{qh,r} \right) + \left(\frac{1}{2}\sum\limits_{h=3,9,...}^{\infty} I_{0h,r}E_{0h,r} + I_{0h,j}E_{0h,j} \right) \tag{6}$$

where

$$\Gamma_{d0} = L_\Delta \bar{I}_{q0}, \tag{7}$$

$$\Gamma_{q0} = L_\Delta \bar{I}_{d0} + E_{q0}, \tag{8}$$

$$\Gamma_{dh,r} = L_\Delta \bar{I}_{qh,r}\cos(h(\theta_e - \bar{\theta}_e)) - L_\Delta \bar{I}_{qh,j}\sin(h(\theta_e - \bar{\theta}_e)) + E_{dh,r}, \tag{9}$$

$$\Gamma_{dh,j} = L_\Delta \overline{I}_{qh,r} \sin(h(\theta_e - \overline{\theta}_e))$$
$$+ L_\Delta \overline{I}_{qh,j} \cos(h(\theta_e - \overline{\theta}_e)) + E_{dh,j}, \quad (10)$$

$$\Gamma_{qh,r} = L_\Delta \overline{I}_{dh,r} \cos(h(\theta_e - \overline{\theta}_e))$$
$$- L_\Delta \overline{I}_{dh,j} \sin(h(\theta_e - \overline{\theta}_e)) + E_{qh,r}, \quad (11)$$

$$\Gamma_{qh,j} = L_\Delta \overline{I}_{dh,r} \sin(h(\theta_e - \overline{\theta}_e))$$
$$+ L_\Delta \overline{I}_{dh,j} \cos(h(\theta_e - \overline{\theta}_e) + E_{dh,r} \quad (12)$$

and

$$\Upsilon = L_\Delta \left(\overline{I}_{d0}\overline{I}_{q0} + \frac{1}{2} \sum_{h=6,12,...}^{\infty} \overline{I}_{dh,r}\overline{I}_{qh,r} + \overline{I}_{dh,j}\overline{I}_{qh,j} \right).$$
$$(13)$$

To reach an MTPA operation, we aim to find the optimal current harmonic magnitudes

$$\{I_{d0}, \ I_{q0}, \ I_{dh,r}, \ I_{dh,j}, \ I_{qh,r}, \ I_{qh,j}\} \quad (14)$$

to produce T_0 and minimize the stator ohmic losses, which is to minimize the square of RMS of phase stator current, that is,

$$\text{minimize } \{i_{a,RMS}^2\}. \quad (15)$$

By expressing $i_{a,RMS}^2$ in terms of dq0-axis currents and using the Fourier expansion in (4), (15) can be rewritten as

$$\text{minimize } \left\{ \begin{array}{l} I_{d0}^2 + I_{q0}^2 + \\[4pt] \sum\limits_{h=6,12,...}^{\infty} I_{dh,r}^2 + I_{dh,j}^2 + \\[4pt] \sum\limits_{h=6,12,...}^{\infty} I_{qh,r}^2 + I_{qh,j}^2 + \\[4pt] \sum\limits_{h=3,9,...}^{\infty} I_{0h,r}^2 + I_{0h,j}^2 \end{array} \right\}. \quad (16)$$

Considering (6)-(13) and (16), the optimal stator current design for the production of average electromagnetic torque with minimal machine cooper losses leads to an optimization problem as

$$\text{minimize } (16)$$
$$\text{subject to } T_0 = T_e^{ref}. \quad (17)$$

where T_e^{ref} is the reference torque command.

In order to solve (17), it is established the following Lagrange cost function, which is based on (17) by the introduction of a Lagrange coefficient δ, such as

$$f(\lambda, I_{d0}, I_{q0}, I_{dh,r}, I_{dh,j}, I_{qh,r}, I_{qh,j}, I_{0h,r}, I_{0h,j}) =$$

$$i_{a,RMS}^2 + \delta(T_0 - T_e^{ref}). \quad (18)$$

In order to solve (18), we first compute the partial derivatives of (18) as

$$\frac{\partial f}{\partial \delta} = 0, \qquad \frac{\partial f}{\partial I_{d0}} = 0, \qquad \frac{\partial f}{\partial I_{q0}} = 0,$$
$$\frac{\partial f}{\partial I_{dh,r}} = 0, \quad \frac{\partial f}{\partial I_{dh,j}} = 0, \quad \frac{\partial f}{\partial I_{qh,r}} = 0, \quad (19)$$
$$\frac{\partial f}{\partial I_{qh,j}} = 0, \quad \frac{\partial f}{\partial I_{0h,r}} = 0, \quad \frac{\partial f}{\partial I_{0h,j}} = 0.$$

Then, by solving the set of linear equations (19) and applying the results in (4), the optimum stator currents are obtained and represented as

$$\boldsymbol{i}_{dq0}^{opt} = \Lambda \left[\begin{array}{c} \Gamma_{d0} + 2 \sum\limits_{h=6,12,...}^{\infty} \Gamma_{dh,r}\cos(h\theta_e) + \Gamma_{dh,j}\sin(h\theta_e) \\[8pt] \Gamma_{q0} + 2 \sum\limits_{h=6,12,...}^{\infty} \Gamma_{qh,r}\cos(h\theta_e) + \Gamma_{qh,j}\sin(h\theta_e) \\[8pt] 2 \sum\limits_{h=3,9,...}^{\infty} \Gamma_{0h,r}\cos(h\theta_e) + \Gamma_{0h,j}\sin(h\theta_e) \end{array} \right]$$
$$(20)$$

where

$$\Lambda = \frac{T_e^{ref} + \Upsilon}{\mathcal{K}} \quad (21)$$

and

$$\mathcal{K} = \Gamma_{d0}^2 + \Gamma_{q0}^2 + 2 \sum_{h=6,12,...}^{\infty} \left(\Gamma_{dh,r}^2 + \Gamma_{dh,j}^2 + \Gamma_{qh,r}^2 + \Gamma_{qh,j}^2 \right)$$
$$+ 2 \sum_{h=3,9,...}^{\infty} \left(\Gamma_{0h,r}^2 + \Gamma_{0h,j}^2 \right). \quad (22)$$

The resulting optimal current definition (20) specially depends on three factors: inductances L_d and L_q; harmonic back-EMF components; and linearization points. Concerning the linearizaton points, it is considered $\overline{I}_{d0}$, $\overline{I}_{q0}$, $\overline{I}_{dh,r}$, $\overline{I}_{dh,j}$, $\overline{I}_{qh,r}$, $\overline{I}_{qh,j}$ and $\overline{\theta}_e$ as the last values of reference harmonic current (20) and rotor position, calculated in the last discrete control period. Besides, by assuming constant speed during one control period, the terms $\theta_e - \overline{\theta}_e$ in (9)-(12) become simply $\omega_e T_s$, where ω_e is the electrical rotor speed and T_s is the control period.

To verify the proposed MTPA technique, it is considered a three phase spoke type IPMSM with the non sinusoidal phase back-EMF at 500 rpm depicted in Fig. 1(a) and the machine parameters presented in Appendix I. Although the formulated optimum current (20) is presented as an infinite series of harmonics, the test machine only presents significant phase back-EMF harmonic magnitudes below the 11th one, as shown in Fig. 1(b).

After evaluating (20) for one electric period, Fig. 1(a) also presents the resulting optimum waveform of stator phase current for the test machine under the 6 Nm constant load condition. Fig. 1(b) shows the harmonic distribution of the obtained current. It confirms that optimum current comprises all the harmonics of phase back-EMF. Fig. 1(c) shows the dq0-axis components of the optimum current in one electric cycle. It should be noticed that i_d and i_q components have average values, which means a phase angle is incorporated

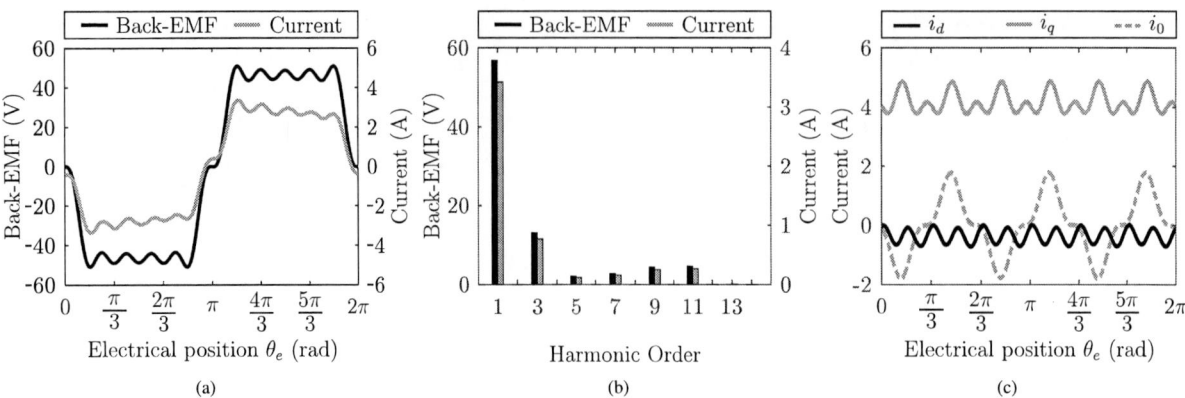

Fig. 1. Phase back-EMF and theoretical optimum stator current characteristics for 500 rpm and 6 Nm load condition of test machine.

in the fundamental component of phase current. This fact points out that the proposed method takes advantage from the reluctance torque contribution ($L_\Delta i_d i_q$).

III. SIMULATION RESULTS AND DISCUSSIONS

In order to inject the optimum current waveform in the stator winding, the converter and control scheme shown in Fig. 2 is implemented. Since the 0-axis current component is considered in this paper, the neutral point of the star connected stator winding is assumed to be available. The machine terminals are connected to the three phase four-leg two level inverter, with the neutral point connected to the forth inverter leg. Regarding the control part, an outer speed control loop based on a Proportional-Integrative (PI) controller produces the reference torque T_e^{ref}, which is forwarded to the optimum current reference generator block, based in (20). A hysteresis current controller is used to generate gate drive signals for the voltage source inverter and to make the stator phase currents follow the optimal current references.

As a basis of comparison, this paper considers two conventional approaches in the literature for IPMSM drives: the zero d-axis current method ($i_d = 0$); and the conventional MTPA method [1]–[3]. The former method only considers the sinusoidal portion of the back-EMF to generate average torque and does not include the reluctance torque. The latter also considers only the fundamental component of the back-EMF but benefits the drive with the reluctance torque contribution. Additionally, both of these methods considers null zero sequence current injection. Further details of their implementation are presented in Appendix II.

Figure 3(a) presents the steady state results for the three compared methods for $\omega_r^{ref} = 500$ rpm under 6 Nm load condition. From top to bottom, Fig. 3(a) presents the behaviour of: estimated electromagnetic instantaneous torque; dq0-axis currents with the reference values; and currents of phases 'b' (i_b) and 'c' (i_c). Regarding the current behaviour, the provided tracking of references ensures the machine operation under the proposed and the compared methods. About the instantaneous torque, both zero d-axis current and conventional

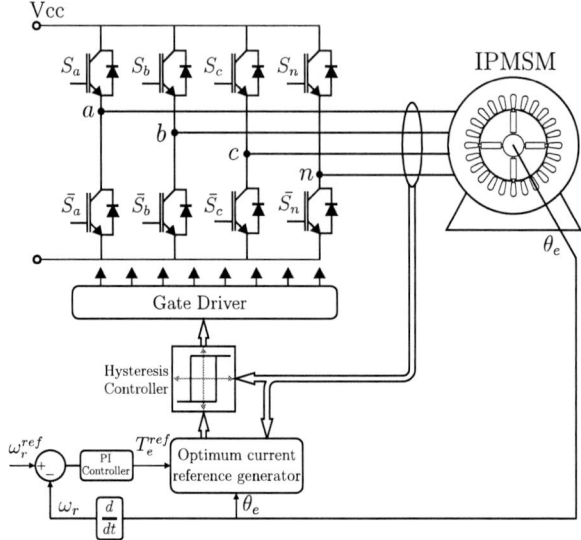

Fig. 2. Block diagram for proposed feeding strategy.

MTPA methods develop lower torque undulation compared to the proposed one. This fact is closer analysed in Fig. 3(b), which shows the harmonic torque components for the three methods. It can be seen that the proposed method increases the torque undulation: 215% higher in the 6th harmonic and 140% higher in the 12th for the experimented operation condition. Concerning to current requirement, Fig. 3(c) compares the RMS current required by the three methods in the experiment of Fig. 3(a), normalized to the value of the zero d-axis method. Despite the increase of torque undulations, it can be seen that the proposed method requires the lowest RMS value to deliver the same average torque. By dividing the average torque production by the RMS values shows the proposed method delivers approximately 2% higher torque per Ampère ratio than the conventional MTPA.

978-1-7281-4181-7/19 $31.00 © 2019 IEEE 338

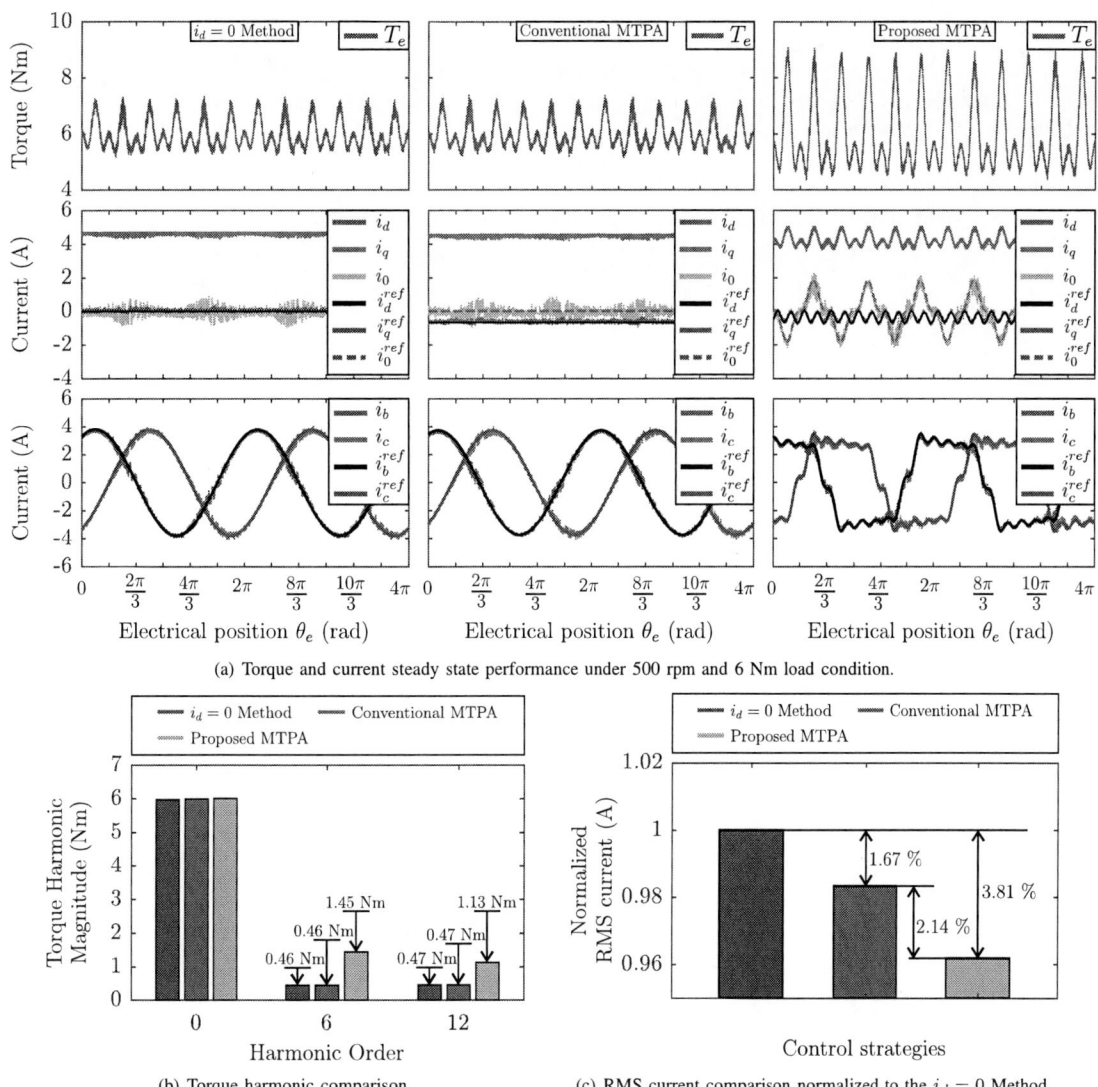

(a) Torque and current steady state performance under 500 rpm and 6 Nm load condition.

(b) Torque harmonic comparison.

(c) RMS current comparison normalized to the $i_d = 0$ Method.

Fig. 3. (Simulation results) Comparison of steady state results for the three methods.

IV. CONCLUSIONS

A new optimal current waveform calculation technique was presented in this paper. The new technique aims at maximizing the average torque production per Ampère for IPMSM including the participation of reluctance torque and harmonics of the non sinusoidal back-EMF. In contrast to existing techniques in the literature, no previous optimization process is necessary and the proposed method can be implemented in real time. Besides, the optimum current reference has a generic series form, suitable for the adaptation to machines with different back-EMF harmonic content. Further improvements for the performance of the proposed method are the incorporation of saturation and back-EMF harmonic variations to the calculation of optimum currents. Apart from improving torque per Ampère ratio, the proposed method increases the torque undulations and, therefore, it is more appropriate to applications not sensitive to torque ripples such as wind turbines due to high inertia.

APPENDIX I
TABLE I
MACHINE PARAMETERS

Description	Symbol	Value	
Number of pole pairs	n_p	2	
D-axis inductance	L_d	28.9	mH
Q-axis inductance	L_q	55.9	mH
Winding resistance		1.09	Ω
Back-EMF constant		0.89	V/(rad·s^{-1})
Rated stator current		4.2	A
Rated rotor speed		1200	rpm
Rated torque		6	Nm
Rated power		750	W

APPENDIX II

It is considered as conventional cases, for the sake of comparison, the well-known zero d-axis current reference method and the MTPA method for sinusoidal PMSM, as illustrated in Fig. 4 [1]–[3]. In the $i_d = 0$ method a speed control loop produces a torque reference that is converted in constant q-axis current in the synchronous reference frame such that

$$i_q^{ref} = \frac{T_e^{ref}}{n_p \Psi_m}, \tag{23}$$

where n_p is the number of pole pairs and Ψ_m is the magnitude of fundamental harmonic of rotor flux. The 0-axis current reference is also set to zero in this conventional approach. The resulting dq0-axis current references are forwarded to the hysteresis current control loop in Fig. 2.

For the conventional MTPA method, the speed control loop produces the sinusoidal stator phase current magnitude reference I_s, which is converted in dq0-axis reference current by

$$\beta = \sin^{-1}\left(\frac{-\Psi_m + \sqrt{\Psi_m^2 + 8(L_d - L_q)^2 I_s^2}}{4(L_d - L_q)I_s}\right)$$

$$i_d^{ref} = \frac{\sqrt{6}}{2} I_s \sin(\beta) \tag{24}$$

$$i_q^{ref} = \frac{\sqrt{6}}{2} I_s \cos(\beta)$$

$$i_0^{ref} = 0.$$

REFERENCES

[1] A. Rabiei, T. Thiringer, M. Alatalo, and E. A. Grunditz, "Improved Maximum-Torque-Per-Ampere Algorithm Accounting for Core Saturation, Cross-Coupling Effect, and Temperature for a PMSM Intended for Vehicular Applications," *IEEE Transactions on Transportation Electrification*, vol. 2, no. 2, pp. 150–159, 2016.

[2] Ching-Tsai Pan and Shinn-Ming Sue, "A linear maximum torque per ampere control for IPMSM drives considering magnetic saturation," in *30th Annual Conference of IEEE Industrial Electronics Society, 2004. IECON 2004*, vol. 3, no. 2. IEEE, 2005, pp. 2712–2717.

[3] S. Morimoto, Y. Takeda, T. Hirasa, and K. Taniguchi, "Expansion of Operating Limits for Permanent Magnet Motor by Current Vector Control Considering Inverter Capacity," *IEEE Transactions on Industry Applications*, vol. 26, no. 5, pp. 866–871, 1990.

[4] K. Wang, Z. Q. Zhu, G. Ombach, and W. Chlebosz, "Average Torque Improvement of Interior Permanent-Magnet Machine Using Third Harmonic in Rotor Shape," *IEEE Transactions on Industrial Electronics*, vol. 61, no. 9, pp. 5047–5057, sep 2014.

[5] G. Feng, C. Lai, and N. C. Kar, "An Analytical Solution to Optimal Stator Current Design for PMSM Torque Ripple Minimization With Minimal Machine Losses," *IEEE Transactions on Industrial Electronics*, vol. 64, no. 10, pp. 7655–7665, oct 2017.

[6] H. Zahr, E. Semail, B. Aslan, and F. Scuiller, "Maximum Torque per Ampere strategy for a biharmonic five-phase synchronous machine," *2016 International Symposium on Power Electronics, Electrical Drives, Automation and Motion, SPEEDAM 2016*, pp. 91–97, 2016.

[7] A. G. De Castro, W. C. A. Pereira, T. E. P. De Almeida, C. M. R. De Oliveira, J. Roberto Boffino De Almeida Monteiro, and A. A. De Oliveira, "Improved Finite Control-Set Model-Based Direct Power Control of BLDC Motor with Reduced Torque Ripple," *IEEE Transactions on Industry Applications*, vol. 54, no. 5, pp. 4476–4484, 2018.

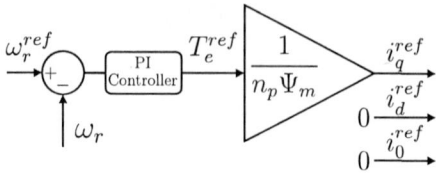

(a) Zero d-axis stator current method.

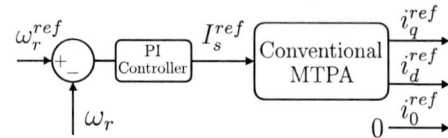

(b) Conventional MTPA method.

Fig. 4. Compared feeding strategy methods.

[8] Z. Li, J. Wang, L. Zhou, X. Liu, and F. Jiang, "Enhanced Generalized Vector Control Strategy for Torque Ripple Mitigation of IPM-type Brushless DC Motors," *IEEE Transactions on Power Electronics*, vol. PP, no. c, pp. 1–1, 2019.

[9] N. Nakao and K. Akatsu, "Torque ripple suppression of permanent magnet synchronous motors considering total loss reduction," *2013 IEEE Energy Conversion Congress and Exposition, ECCE 2013*, pp. 3880–3887, 2013.

[10] A. G. de Castro, P. R. U. Guazzelli, S. T. C. A. dos Santos, C. R. de Oliveira, W. C. A. Pereira, and J. R. B. A. Monteiro, "Zero Sequence Power Contribution on BLDC Motor Drives. Part I: A Theoretical Investigation," in *2018 13th IEEE International Conference on Industry Applications (INDUSCON)*. IEEE, nov 2018, pp. 1016–1023.

[11] J. D. De Kooning, J. Van De Vyver, B. Meersman, and L. Vandevelde, "Maximum Efficiency Current Waveforms for a PMSM Including Iron Losses and Armature Reaction," *IEEE Transactions on Industry Applications*, vol. 53, no. 4, pp. 3336–3344, 2017.

[12] N. Nakao and K. Akatsu, "Suppressing pulsating torques: Torque ripple control for synchronous motors," *IEEE Industry Applications Magazine*, vol. 20, no. 6, pp. 33–44, 2014.

[13] P. C. Krause, O. Wasynczuk, and S. D. Sudhoff, *Analysis of Electric Machinery and Drive Systems*. Wiley-IEEE Press, 2002.

[14] P. Chapman, S. Sudhoff, and C. Whitcomb, "Optimal current control strategies for surface-mounted permanent-magnet synchronous machine drives," *IEEE Transactions on Energy Conversion*, vol. 14, no. 4, pp. 1043–1050, 1999.

[15] A. H. Abosh, Z. Q. Zhu, and Y. Ren, "Reduction of torque and flux ripples in space vector modulation-based direct torque control of asymmetric permanent magnet synchronous machine," *IEEE Transactions on Power Electronics*, vol. 32, no. 4, pp. 2976–2986, 2017.

Non-Isolated High Current Battery Charger with PFC Semi-Bridgeless Rectifier

Rodrigo Patrício Dacol
Mobility Engineering Department
Federal University of Santa Catarina
Joinville, Brazil
rodrigo.dacol@gmail.com

Joselito Anastácio Heerdt
Electrical Engineering Department
University of State of Santa Catarina
Joinville, Brazil
joselito.heerdt@udesc.br

Gierri Waltrich
Mechanical Engineering Department
Federal University of Santa Catarina
Florianopolis, Brazil
gierri.waltrich@ufsc.br

Abstract— **In this paper a unidirectional non-isolated high current battery charger with power factor correction (PFC) is proposed. The structure of the battery charger is divided in two parts: a bridgeless PFC at the input stage and a interleaved Buck converter at the output stage. The structure uses multi-interfase transformer, to allow the magnetic volume reduction, and a bridgeless PFC rectifier to reduce the number of components. This paper presents the converter operational stage analysis, control and simulation results.**

For a 48V battery with 32A constant current of charging, the proposed converter reached a power factor of 0.9938.

Keywords — Battery charger, interleaved buck, couple inductor, PFC bridgeless, lithium-ion battery.

I. INTRODUCTION

With the increasing demand for portable high-power devices, high-capacity energy storage devices are becoming very popular, in especially the technologies involving lithium-ion batteries due to their higher power/energy density.

Lithium-ion battery packs have being used in many applications so far, for instance, in telecommunication, uninterrupted power supplies, computer, cameras, drones, and electric vehicles.

In the last years, many companies start to commercialize electric vehicles with large battery banks, with high current charge/discharge capability, to provide fast battery charging/discharging. However, in fast battery chargers the current can reach higher levels (>500A), and therefore, a new infrastructure to recharge this batteries need to be created [1] [2]. Not only fast charger infrastructures need to be built but also smaller charger, with high current capability, should be design for use mainly in residential applications.

The battery chargers are classified by the International Electrotechnical Commission (IEC) standard IEC61851 [3], into three categories. The first two categories are for a power range between 1 to 14.4 kW, used in small and onboard charger applications. The third category represents public charging stations, with power levels up to 240 kW. The study presented in this paper is focus on the first two categories.

Normally, the battery chargers, for the first categories, are composed by EMI filters, PFC rectifiers and dc-dc converters. The EMI filters are not the focus of this paper; therefore, they will not be study. The studies are related to single phase PFC and dc-dc converter at the output stage.

Single phase PFC topologies used in battery charger for categories 1 and 2 are usually classified as phase-shifted semi-bridgeless rectifier, bridgeless rectifier and controllable full-bridge transistor rectifier [2]. However, in commercial applications, normally it is used the simplest solution, i.e., using the conventional fully-bridge single-phase boost PFC with a step-down converter [4] (Buck, forward or flyback) at the output stage [5] [6].

Thus, with the goal of reducing the number of components and increase the charger current output, in this paper is proposed a converter with a bridgeless rectifier to reduce number of components, and an interleaved Buck converter to allow a higher charger current output. Furthermore, a CC/CV controller (Constant Current-Constant Voltage controller) [7] for the charging process is implemented.

II. PROPOSED TOPOLOGY

A battery charger, with an interleaved buck converter [8] and a bridgeless PFC [9] is proposed in this paper and shown Fig. 1. The PFC is used the obtain a high power factor (close to unity) and the interleaved Buck converter is used to control the battery charging current due to its output current source characteristic. The interleaved Buck converter has a multi-interfase transformer to reduce its magnetic volume, as described in [10] [11].

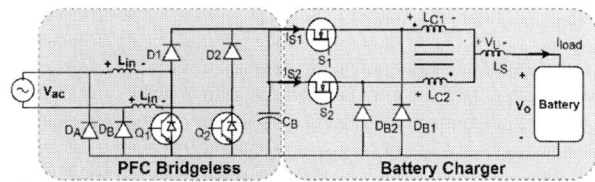

Fig. 1 - Battery charger proposed.

In the following subsections are presented the operational stages and main characteristics of both input and output stages of the proposed converter.

A. Semi-Bridgelessrectifier

The operational stages of the Semi-BridgelessPFC rectifier [9] are shown in Fig. 2. It is divided in four steps, two steps for the positive half-cycle and two steps for the negative half-cycle of utility grid.

In the stage I, Q_1 and D_B diode are conducting. In this situation the input current increases and the current in the load is supplied by the dc bus capacitor.

In the stage II, Q_1 is turned-off and diode D_B is still conducting. The input current flows through the diode D_1 to the load and to the dc bus capacitor. In this situation, the input current derivative is negative.

The input current follows the input voltage reference (in a closed lopping control) guarantying high power factor.

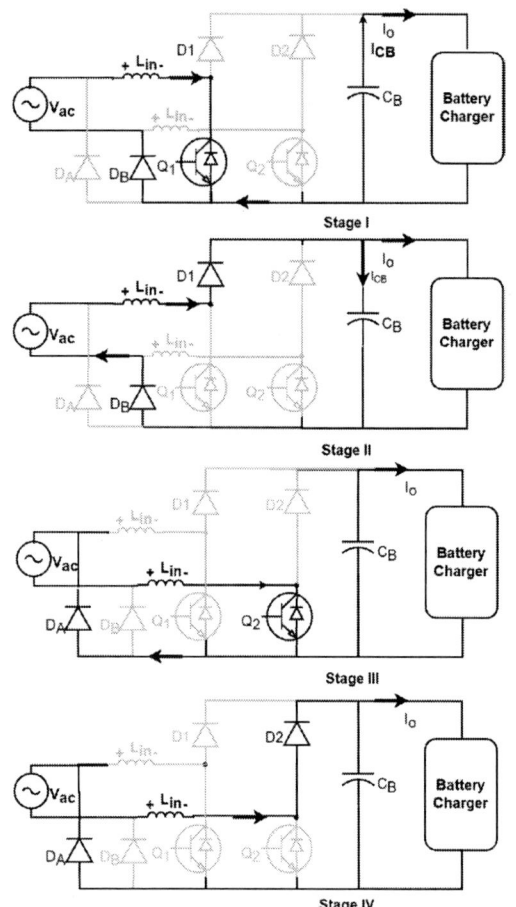

Fig. 2 - Semi-BridgelessPFC rectifier.

When the input current is negative, the rectifier works similarly, but the current is controlled by Q_2.

Fig. 3 shows the main PFC Semi-Bridgelessrectifier waveforms, including, input current, switches currents and dc bus capacitor voltage.

B. Interleaved Buck

Fig. 4 shows the operational stages of the Buck interleavead converter with a duty cycle lower than 50%. Fig. 5 shows the couple inductor currents (i_{C1}, i_{C2}), switches currents (i_{S1}, i_{S2}) and diode currents (i_{D1}, i_{D2}).

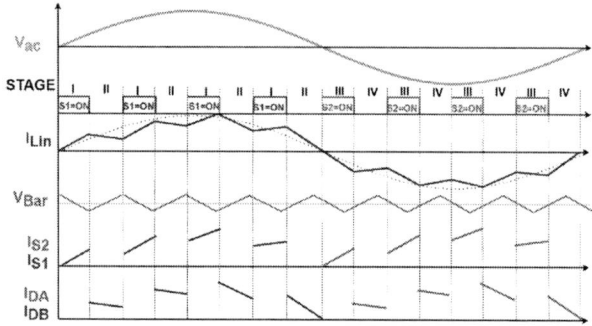

Fig. 3 - Semi-BridgelessPFC waveforms.

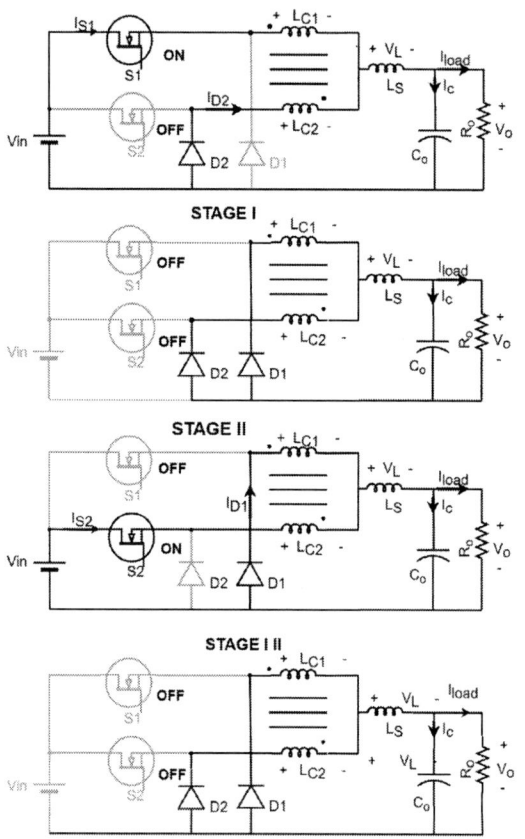

Fig. 4 - Buck interleaved operational stages.

In the stage I, S_1 is on, and S_2 is off. The input current flows through L_{C1} and L_{C2} and the current in the inductor, L_S, increases, as shown in Fig. 6.

In the Stage II, S_1 and S_2 are turned-off. The current flows through D_1 and D_2 and the current in L_S decreases. This stage is similar to Stage IV.

978-1-7281-4181-7/19 $31.00 © 2019 IEEE

The Stage III, occurs when the S_1 is off, and S_2 is on. The input current flows through L_{C1} and L_{C2} and the current in the inductor, L_S, increases.

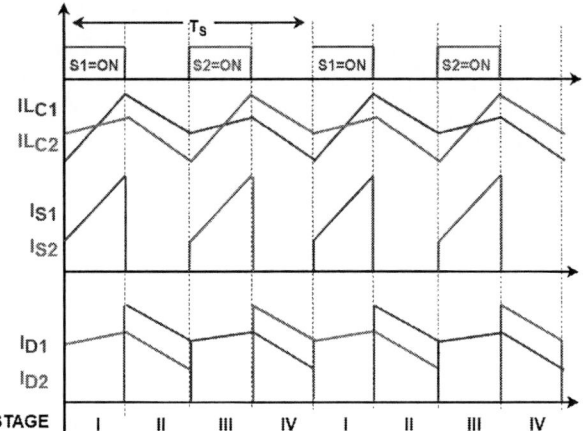

Fig. 5 - Buck interleaved main waveforms.

Using a interleaved buck at the output stage the current in L_S has a frequency twice higher compared with the switching frequency, thus, this magnetic has its volume reduced [10]. Furthermore, because the multi-interfase transformer s, L_{C1} and L_{C2} has its average flux canceled with each other, the induced magnetic field peak value is very low [10] [12]. In Fig. 6 is observed that ΔI_C has a null average value.

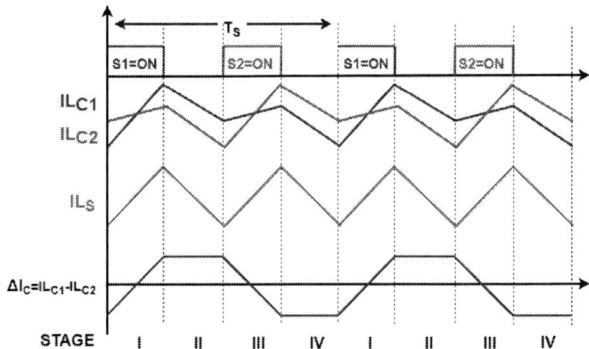

Fig. 6 - Typical waveforms of current on inductors.

III. Battery Charger Control

The battery charger control system [11] is very simple, as showed in Fig. 7. It only uses a proportional-integral (PI) controller for the converter current, with a duty cycle limiter, to protect against battery overvoltage.

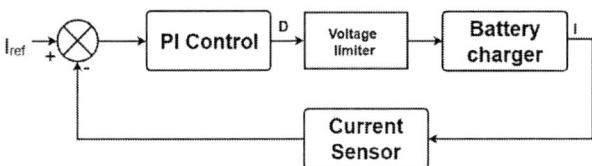

Fig. 7- Proposed battery charger control.

Some mechanisms were added to the DSP (used in the prototype) to disconnect the charger in case of overvoltage.

The PFC control system is more complex. It has two control loops; one for the dc bus voltage and another one for the input current, as shown in Fig. 8.

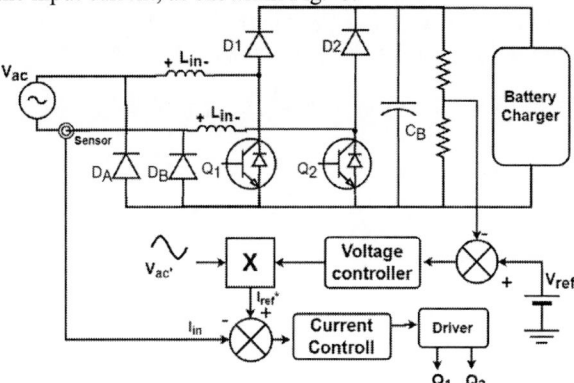

Fig. 8 – Semi-BridgelessPFC control.

The voltage control is the external loop, therefore, has a slower dynamic. This voltage loop controls the dc bus using a PI controller. The current reference is generated by multiplying the utility grid alternating voltage with the output of the voltage loop control, thus, it is guaranteed the high power factor control. The current control loop also uses a PI controller, however, with a faster dynamic.

IV. Prototype and Simulations

In order to verify the operation of the proposed battery charger, a 1.5 kW prototype was built and it is shown in Fig. 9. The battery charger specifications and parameters are shown in Table 1. For tests purposes it has been configured the battery charge with 48V, to charge a battery bank with 13-cell lithium-ion batteries in series with 3 cells in parallel, with capacity of 6600mAh, as shown in Fig. 10. This battery pack was built in laboratory with a spot welding machine and a commercial BMS (battery management system), for protection against over-discharge, over-voltage charge and short circuits. This battery pack was built using recycling battery cells from laptops.

TABLE 1 - Prototype specifications and parameters.

Input voltage	$V_{ac} = 220V$
Input frequency	60Hz
Switching frequency	f_s =50kHz
Output power	$P_{out} = 1.5kW$
Maximum output voltage	$V_{o\_max} = 55V$
Minimum output voltage	$V_{o_{min}}=38V$
Maximum output current	$I_{o_{max}} = 32A$
Input inductor	L_{in} =250uH
Dc bus capacitor	C_B =900uF
Output inductor	L_S =150uH
Battery configuration	13 cells in series and 3 in parallel (13S3P)

The multi-interfase transformer were built in accordance with the methodology described in [10]. The total volume of the inductors was 0.165L (0.041L for the multi-interfase transformer $L_{C1,2}$ and 0.125L for the output inductor L_S). Compared with the inductor volume of a Buck converter, built with the same specifications results in 0.427L, in which represents a reduction of almost 60%.

Fig. 9 - Prototype of the proposed battery charger.

Fig. 10 - 13S3P lithium-ion battery pack.

A. Semi-BridgelessPFC simulations

The simulations for the Semi-BridgelessPFC controlled rectifier, were carried out using the data from Table 1 and the results are presented in Fig. 11. Interleaved As shown in this figure, the input stage has a high power factor. The power factor for the Fig. 11 was calculated and the resulted obtained was equal to 0.9938. The Fast Fourier Transformer was also measured, and it is presented in Fig. 12. The first harmonics appears around twice the switching frequency, showing that the high frequency harmonics in fact are twice the switching frequency.

B. Interleaved Buck simulation

The simulation for the Buck interleaved were also carried out and the results are shown in Fig. 13.

Because the multi-interfase transformer were winding in opposite direction, the dc component of the induced magnetic field is cancelled. The dc flux cancellation, allows only the ac flows through the core with an small peak value.

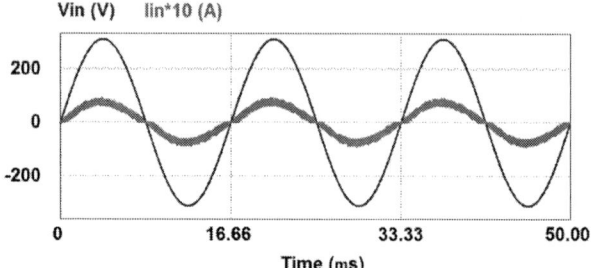

Fig. 11 – Proposed input voltage and input current battery charger simulation.

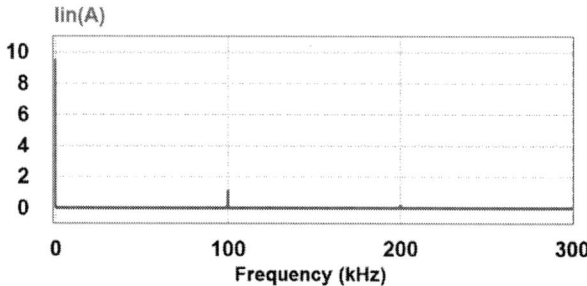

Fig. 12 – Fast Fourier Transformer of the input current.

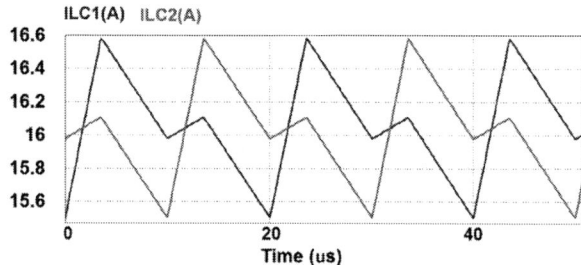

Fig. 13 – Multi-interfase transformer currents in I_{c1} and I_{c2}

In Fig. 14 is shown the current through the inductor L_S. As expected, the current frequency through this inductor is twice the switching frequency, allowing to reduce size in the inductor design. Actually, because this inductor has lower current ripple and high dc value (around 30.5A in this case), this inductor might be designed using iron powder core, which has higher induced magnetic flux saturation value. However, for the multi-interfase transformer , they were designed with ferrite.

The simulation of the lithium-ion battery charging process [7] is shown in Fig. 15. It is divided into two steps. In the first step, the battery pack was charged with a constant current, supplied by the interleaved Buck converter until the battery pack reaches its maximum value. For this case the maximum value is equal to 54.6V, because it has 13 cells in series and each cell cannot be charged over 4.2V. Thus, in the second step, when it reaches 54.6V, the current controller is designed to saturated, and therefore, the voltage controller actuates maintain the voltage at its maximum value and the

978-1-7281-4181-7/19 $31.00 © 2019 IEEE

charging current starts to reducing in a nonlinear way, until nearly zero.

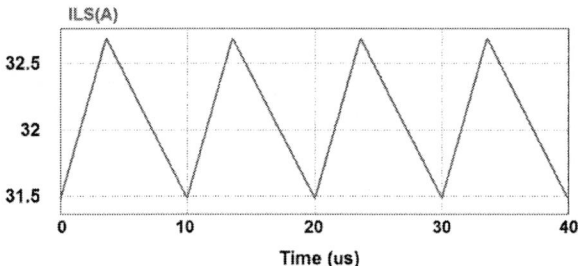

Fig. 14 – Output inductor current I_S.

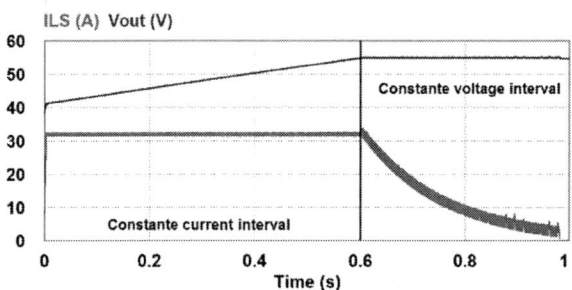

Fig. 15 - Current and voltage on battery during charging process.

V. CONCLUSIONS

This paper presents a battery charger structure composed by a Semi-BridgelessPFC controlled rectifier and a interleaved Buck converter with multi-interfase transformer to be used as a battery charger for electric vehicle applications. The proposed charger has high current in the output in combination with a high power factor at the input stage. It was presented the operating principles, analysis, and simulation for a prototype of 1.5kW power (already built) and a 48V lithium-ion battery. In the simulation results, the proposed converter presented a power factor of 0.9938.

In this paper the output current was limited to 32A, however, if a higher current is necessary more Buck converter can be interleaved at the output stage.

Because the prototype is already built, for the final paper version, the authors will show experimental results for the 1.5kW prototype, and more details about the components used in the construction will also be presented.

REFERENCES

[1]   M. Yilmaz and T. K. Philip, "Review of Battery Charger Topologies, Charging Power Levels, and Infrastructure for Plug-In Electric and Hybrid Vehicles," *IEEE TRANSACTIONS ON POWER ELECTRONICS,* vol. 28, no. 5, pp. 2151-2170, 2013.

[2]   T. Jalakas, I. Roasto and D. Vinnikov, "Analysis of Battery Charger Topologies for an Electric Vehicle," in *Biennial Baltic Electronics Conference,* Tallinn, 2012.

[3]   G. Joos, M. d. Freige and M. Dubois, "Design and Simulation of a Fast Charging Station for PHEV/EV batteries," *Electric Power and Energy Conference(EPEC),* pp. 58-59, april 2010.

[4]   T. Jalakas, I. Roasto and D. Vinnikov, "Analysis of Battery Charger Topologies for an Electric Vehicle," in *Biennial Baltic Electronics Conference,* Tallinn, Estonia,, 2012.

[5]   F. Pöttker de souza and I. Barbi, "A Unity Power Factor Buck Pre-Regulator with Feedforward of the Output Inductor Current," *Fourteenth Annual Applied Power Electronics Conference and Exposition,* vol. 1, no. 1, pp. 1130-1136, 1999.

[6]   o. Yoshiya and I. Jun-Ichi, "A Novel Single-Phase Buck PFC AC–DC Converter With Power Decoupling Capability Using an Active Buffer," *IEEE TRANSACTIONS ON INDUSTRY APPLICATIONS,,* vol. 50, no. 3, pp. 1905-1915, 2014.

[7]   S. Abinaya, Sivaranjani and S. Suja, "Methods of Battery Charging with Buck Converter Using Soft-Switching Techniques," *Bonfring International Journal of Power Systems and Integrated Circuits,,* vol. 1, no. 1, pp. 20-26, 2011.

[8]   S. Vijayalakshmi , E. Arthika and G. P. Shanmuga, "Modeling and Simulation of Interleaved Buck-Converter with PID Controller," *Sponsored 9th International Conference on Intelligent Systems and Control,* vol. 1, no. 1, pp. 978-985, 2015.

[9]   M. Fariborz , W. Eberle and W. G. Dunford, "A Phase-Shifted Gating Technique With Simplified Current Sensing for the Semi-Bridgeless AC–DC Converter," *IEEE TRANSACTIONS ON VEHICULAR TECHNOLOGY,* vol. 62, no. 4, pp. 1568-1577, 2013.

[10]   M. Hirakawa, Y. Watanabe, M. Nagano, K, Andoh, "High Power DC/DC Converter using Extreme Close-Multi-interfase transformer s aimed for Electric Vehicles," *The 2010 International Power Electronics Conference,* vol. 1, 2010.

[11]   O. Yoshiya and I. Jun-Ichi , "Buffer, A Novel Single-Phase Buck PFC AC–DC Converter With Power Decoupling Capability Using an Active,"

IEEE TRANSACTIONS ON INDUSTRY APPLICATIONS, vol. 3, no. 4, 2013.

[12]   M. hirakawa, M. nagano, Y. Watanabe and K. Ando, "High Power Density Interleaved DC/DC Converter using a 3-phase Integrated Close-Multi-interfase transformer   Set aimed for Electric Vehicles," *Energy Conversion Congress and Exposition,* pp. 2452-2457, 2010.

[13]   C. Ying-Chun , "High-Efficiency ZCS Buck Converter for Rechargeable Batteries," *TRANSACTIONS ON INDUSTRIAL ELECTRONICS,* vol. 57, no. 7, pp. 2463-2473, 2010.

[14]   M. Brandl, H. Gall, M. Wenger, V. Lorentz and M. Giegerich, "Batteries and Battery Management Systems for Electric Vehicles," *Design, Automation & Test in Europe Conference & Exhibition,* vol. 1, no. 1, pp. 971-977, 2012.

Comparative analysis based on the switching frequency of modulation techniques for MMC applications

Juan C. Colque
DSE
UNICAMP
Campinas, SP, Brasil
juanca.colque@outlook.com

Ernesto Ruppert
DSE
UNICAMP
Campinas, SP, Brasil
ruppert@fee.unicamp.br

Rodrigo Z. Vargas
CECS
UFABC
Santo André, SP, Brasil
r.zambrana@ufabc.edu.br

José L. Azcue
CECS
UFABC
Santo André, SP, Brasil
jose.azcue@ufabc.edu.br

Abstract—The Modular Multilevel Converter (MMC) is a consolidated attractive technology taking great importance in medium / high power applications compared to other topologies of multilevel converters. One of the main lines of research is on voltage modulation techniques and its influence on the converter performance. Specifically for a relatively high number of submodules (SMs) that may depend on the specific application type. In this paper, a brief overview of the MMC basic operation is presented and the modulation techniques based on switching frequency as the main parameter are studied. Also, three different modulation techniques are studied to later be performed in Matlab/Simulink, in simulation is considering two cases of MMC systems with 5 and 20 SMs/arm, in order to comparing the impact of modulation technique on MMC performance.

Index Terms—Modular Multilevel Converter, modulation techniques, harmonic analysis, CPU time

I. INTRODUCTION

The Multilevel Converters have well-marked advantages over the conventional 2-level converters, such as: transformless system, the voltage stress on the devices is reduced, the output filters are smaller, switching losses are lower because is not necessary a high switching frequency, among others [1], [2], receiving wide acceptance in the industry and energy systems in medium or high voltage applications.

In the literature, different modulation techniques for multilevel converters were proposed mainly for: Neutral Point Clamped (NPC), Flying Capacitor (FC), Cascaded H-Bridge (CHB) and, lately for the Modular Multilevel Converter (MMC) [2], [3]. The MMC has revolutionized the market of power electronic converters based on voltage source converters, and various application niches, such as: energy storage [2], [3], drive of medium and high-power motors [2], [3], photovoltaic solar energy conversion [4], offshore wind farms [5], electric vehicles [6], Static Synchronous Compensator in medium-voltage applications [7], and High Voltage Direct Current transmission systems [8], etc.

Advances in the converters technology and semiconductor devices enable these systems to be developed and tested in laboratories, for later it will be implemented in real scale depending on the application, operating voltage and classi-

fication of devices, in this case the IGBT. However, due to the reduced capacity of semiconductor devices, it is necessary to have several submodules (SMs) connected in series, e.g., for medium voltage motors with a rated voltage of 3.3-13.8 kV are required 5-20 SMs/arm, for STATCOM applications are required 15-200 SMs/arm to achieve an operating voltage of 13.9kV to 220 kV and for HVDC transmission systems are required 200-400 SMs/arm to achieve an operating voltage of $\pm$ 320 kV (DC line) [9], [10].

To date, several studies have been reported in order to improve the reliability and performance of MMC, which mainly include: circulating current minimization [3], [8], voltage balancing of capacitors [3], [11], control of the output current [2], types of SMs topologies [2], dynamic modeling in continuous time and discrete time [12], control strategies [2], modulation strategies [1], [13] and fault-tolerance operation in DC-bus and internal devices of the SMs [7], [14].

One of the main research lines regarding the performance of the MMC refers to its voltage modulation strategy. According to the type of application of the MMC, as shown above, a relative amount of submodules is required for proper operation, for this, it is necessary to study the methods of voltage modulation to understand the impact that modulation causes over the waveform in the output voltage of the MMC inverter (THD analysis and fundamental magnitude), and the advantages of one method over others.

This paper provides a comprehensive review of modulation techniques applied to the MMC, e.g. in Fig. 1 is shown a overview diagram of various modulation techniques applied to MMC, three main categories are considered, which are: High, Low and Fundamental frequency modulation. One method of each main technique is simulated and the results are compared. The CPU time of each method with same processing conditions is also analysed in two cases:

- **Case I**: MMC with 5 SM/arm
- **Case II**: MMC with 20 SM/arm

The methods are adapted to be compatible with a classification and selection algorithm to guarantee the adequate balance of each arm submodule voltage of the MMC. This paper

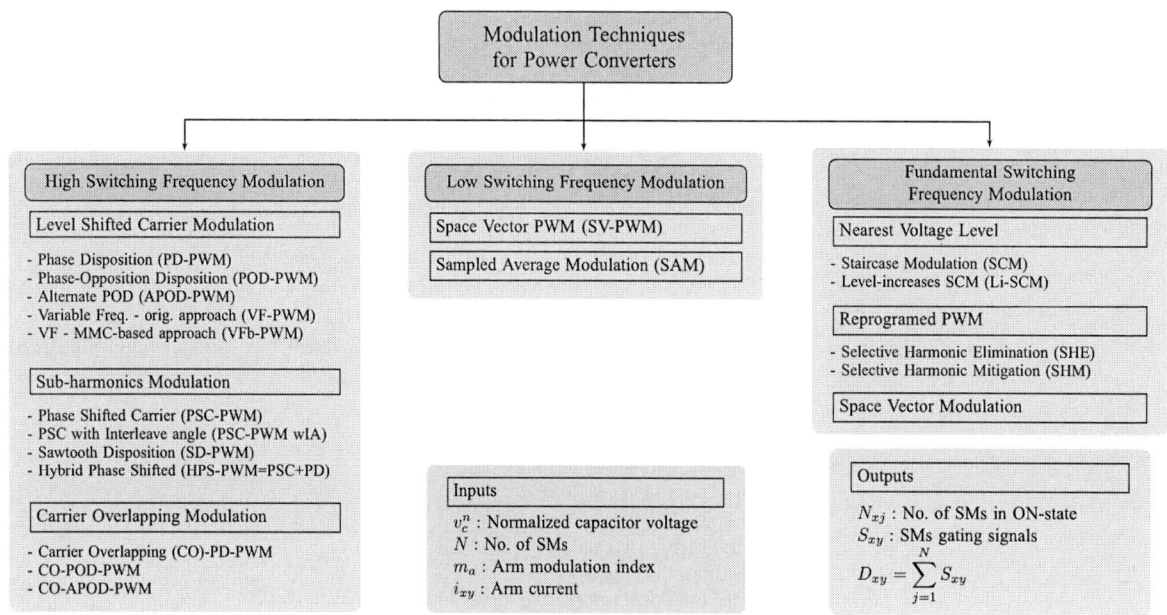

Fig. 1. Overview of Pulse Width Modulation techniques for the MMC.

is organized as follows: in Section II a brief mathematical analysis is presented, in Section III is briefly shown three techniques of voltage modulation, in Section IV are presented the simulation results and the conclusions are detailed in Section V.

II. THE MMC PRINCIPLES

The upper and lower arm submodules are modeled as controlled AC voltage sources. The DC system is modeled as two DC voltage sources. The single-phase equivalent model of MMC under ideal conditions is shown in Fig. 2 [15].

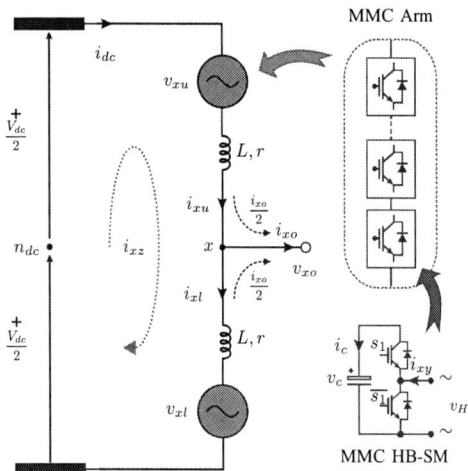

Fig. 2. Single-phase equivalent circuit and main parts of MMC

The upper and lower arm currents are given by:

$$i_{xu} = \kappa\, i_{dc} + i_{xz} + \frac{1}{2} i_{xo}$$
$$i_{xl} = \kappa\, i_{dc} + i_{xz} - \frac{1}{2} i_{xo} \qquad (1)$$

Where: i_{dc}, i_{xz}, i_{xo} are the DC-bus current, the AC circulating current and the AC output current, respectively. κ is the number phase index e.g. $\kappa = 1/3$ for three-phase and $\kappa = 1$ for single-phase. The upper and lower arm currents are summed, which gives a common-mode current component flowing through each leg.

$$i_{xz,f} = \frac{1}{2}(i_{xu} + i_{xl}) = \kappa\, i_{dc} + i_{xz} \qquad (2)$$

The upper and lower arm voltages are given by:

$$v_{xu} = \frac{V_{dc}}{2} - v_{xo} - L\frac{di_{xu}}{dt} - r\, i_{xu}$$
$$v_{xl} = \frac{V_{dc}}{2} + v_{xo} - L\frac{di_{xl}}{dt} - r\, i_{xl} \qquad (3)$$

The arm modulation signals at steady state are given by:

$$v_{xu} = \frac{V_{dc}}{2} - v_{xo} - v_{xz}$$
$$v_{xl} = \frac{V_{dc}}{2} + v_{xo} - v_{xz} \qquad (4)$$

Where v_{xz} represents the voltage drop across arm inductor.

This model simplifies the analysis and its possible to obtain the decoupled currents, which are used to control the arm current components. The simplified model is valid under the following assumptions [9]:

978-1-7281-4181-7/19 $31.00 © 2019 IEEE

1) Large SM capacitance, enough to model as a ideal battery sources.
2) Each MMC arm must be equipped with a large number of SMs, so that the harmonic components corresponding to the DC-source voltage can be neglected.

III. MODULATION TECHNIQUES FOR MMC

The main objective of a voltage modulation method is to control the voltage and current at its terminals. The AC output voltage waveform of the power converter is obtained from the variation of the duty cycle of the switching devices (IGBTs). Based on switching frequency, it is classified in:

1) High switching frequency: $f_{sw} > 2000$ Hz,
2) Low switching frequency: 200 Hz $< f_{sw} < 2000$ Hz,
3) Fundamental switching frequency: $f_{sw} = 50/60$ Hz.

• **Phase reference signals.** For MMC modulation is necessary to generate the phase reference signals for open-loop / closed-loop converter control, and its given by:

$$V_{xi} = m_a \times \frac{V_{dc}}{2} sin(\omega t + \theta_x) \tag{5}$$

Where ω is the fundamental angular frequency, and $\theta_x \in \{0, -\frac{2\pi}{3}, \frac{2\pi}{3}\}$ for three-phase system.

Thus, the total each arm voltage is given by:

$$V_{dc} = NV_c \tag{6}$$

The reference phase voltage in terms of number of submodules (N) and its capacitor voltage (V_c) is given by:

$$V_{xi} = \frac{NV_c}{2} \times m_a sin(\omega t + \theta_x) \tag{7}$$

This voltage is normalized by dividing it with the rated voltage of the submodule capacitor.

$$v_{xi}^n = \frac{N}{2} \times m_a sin(\omega t + \theta_x) \tag{8}$$

A. PSC–PWM with an Interleave Angle (PSC-PWM wIA)

In this method of modulation, its arranged carrier signals of identical triangles horizontally, all triangles have the same frequency and peak-to-peak amplitude, with an angle offset between them. This method has natural advantages over other modulation techniques [15]–[17]:

○ It provides a natural balance of SM capacitors voltage.
○ It distributes in a similar way the voltage stress on the semiconductors and the power handled by each SM.
○ It minimizes the DC-bus current ripple.

In Equation (8) is given the reference phase voltage. This method requires a number of triangle carriers similar to the number of submodules in each arm arranged horizontally.

$$\phi = \frac{360°}{N} \tag{9}$$

Hence, the interleave angle is given by:

$$\phi_i = \frac{360°}{2N} \tag{10}$$

With an interleave angle which shifts the carrier signals of upper and lower arms, the MMC can generate $2N+1$ voltage levels. The Fig. 3 shows the arrangement of triangular carriers.

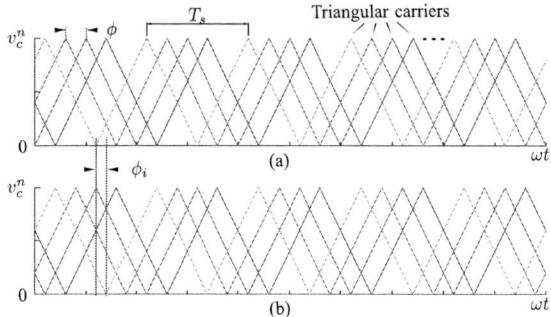

Fig. 3. PSC-PWM wIA carriers arrangement (a) carriers for upper arm, (b) carriers for lower arm.

The switching frequency for this method is given by:

$$f_{psc} = 2N \times f_c \tag{11}$$

B. Sampled Average Modulation (SAM)

In this method, the voltage reference in each phase is generated from the average of two nearest voltage levels in each sampling interval [18].

To the normalized reference phase voltage value from equation (8) an offset value of $\frac{N}{2}$ is added.

$$v_{xi,r}^n = \frac{N}{2} \times [m_a sin(\omega t + \theta_x) + 1] \tag{12}$$

The lower and upper voltage levels is given by:

$$
\begin{aligned}
v_{x1} &= \text{floor}(v_{xi,r}^n) \\
v_{x2} &= \text{floor}(v_{xi,r}^n) + 1
\end{aligned} \tag{13}
$$

The volt-sec balance theory of the reference phase voltage is expressed in two nearest voltage levels and its dwell times.

$$
\begin{aligned}
v_{xi,r}^* T_s &= V_{x1}T_{x1} + V_{x2}T_{x2} \\
T_s &= T_{x1} + T_{x2}
\end{aligned} \tag{14}
$$

Applying it during a sampling interval T_s, it is:

$$
\begin{aligned}
T_{x2} &= T_s \times \frac{v_{xi,r}^n - v_{x1}}{v_{x2} - v_{x1}} \\
T_{x1} &= T_s \times \left[1 - \frac{v_{xi,r}^n - v_{x1}}{v_{x2} - v_{x1}}\right]
\end{aligned} \tag{15}
$$

Where g_{t1} and g_{t2} are the pulses to be applied in the upper and lower arm drivers, and $y \in \{u, l\}$ is the arm variable.

Them, the arm voltage levels are given by:

$$
\begin{aligned}
N &= v_{xu,m} + v_{xl,m} \\
v_{xl,m} &= v_{xm} \\
v_{xu,m} &= N - v_{xl,m}
\end{aligned} \tag{16}
$$

Where $m \in \{1, 2\}$ are the voltage levels.

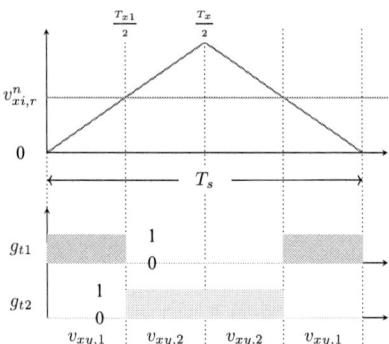

Fig. 4. Dwell times of voltage levels.

C. Level-increases Staircase Modulation (Li-SCM)

The staircase modulation approach is mainly applied for the high-voltage applications, where the SMs number is very large (100-200 SMs) [15]. This type of modulation proposed in [19] is the improved version of the conventional Staircase mod.

In order to obtain the normalized lower arm voltage an offset value of $\frac{N}{2}$ is created, to the offset value a normalized voltage from equation (8) is added, and for the upper arm voltage is subtracted, its given by:

$$
\begin{aligned}
v_{xu}^n &= \frac{N}{2} \times [1 - m_a sin(\omega t + \theta_x)] \\
v_{xl}^n &= \frac{N}{2} \times [1 + m_a sin(\omega t + \theta_x)]
\end{aligned}
\tag{17}
$$

Where v_{xu}^n and v_{xl}^n represents the normalized upper and lower arm voltages, this voltages has steps in range of 0 to $N+1$, and represents the number of ON-state SMs in an arm.

$$
\begin{aligned}
e_{xu}^n &= \text{round}_{1/4}(v_{xu}^n) \\
e_{xl}^n &= \text{round}_{1/4}(v_{xl}^n)
\end{aligned}
\tag{18}
$$

Where e_{xu}^n and e_{xu}^n are the upper and lower nearest arm voltage level (instantaneous).

D. The capacitor voltage balancing

The PWM techniques are adapted to be compatible with a sorting algorithm [17], in order to guarantee the adequate balance of each arm submodule voltage of the MMC. The Fig. 5 shows the overall scheme of PWM techniques plus the capacitor voltage balancing method.

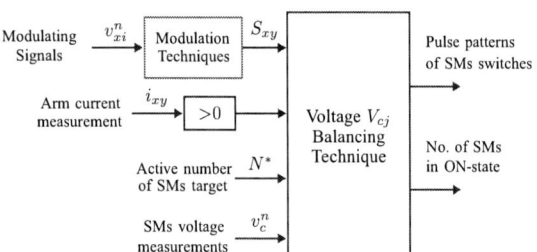

Fig. 5. Overall scheme of capacitor voltage balancing method.

The capacitors voltage balance of submodules is of fundamental importance to complement a modulation method, and so able to maintain the arms voltage in equilibrium, obtaining the control of charging and discharging of submodule capacitors , this balance is directly influenced by the current circulating between the phases.

IV. Simulation Results

The modulation techniques performance together with the voltage balancing algorithm is validated in a single-phase MMC system using the MATLAB/Simulink software with the system parameters shown in Table I. To verify waveforms, the system consists of a single-phase MMC system with a passive load connected between the midpoint of the DC-bus voltage and the inverter output. Is also analyzed the CPU time, that is the real time interval that the computer takes to perform the simulation, in these simulation are 1 sec. for each case.

TABLE I
SIMULATION PARAMETERS

Item	Variable	Values	
		Case I	Case II
Rated power	P_s	10 kW	50 kW
DC-bus voltage	V_{dc}	1000 V	1000 V
Arm inductor	L	4.3 mH	1.5 mH
SM capacitor	C_{SM}	1.4 mF	27.9 mF
No. of SMs per arm	N	5	20
SM voltage	V_c	200 V	50 V
Load resistance	R_L	125 Ω	125 Ω
Modulation index	m_o	1	1
Fundam. frequency	f	60 Hz	60 Hz
SAM frequency	f_{sam}	1200 Hz	1200 Hz
PSC-PWM wIA frequency	f_{psc}	6000 Hz	24000 Hz
Sample time	T_s	50e-6	
Simulink solver		Variable-Step Auto	
Processor		AMD A8-5500B APU	3.20 GHz

A. Simulation Case I

In this case of simulation is considered a single-phase MMC system with 5 SMs/arm. In Fig. 6(a1), 6(b1) and 6(c1) are shown the waveform of the upper and lower arm voltage levels, its clearly observed a difference in each waveform obtained for PSC-PWM wIA, SAM and Li-SCM technique, respectively.

The inverter output voltage is analyzed in Table II, the results indicate that SAM and PSC-PWM wIA maintain the output fundamental voltage amplitude close to the theory, whereas PSC-PWM wIA guarantees a low THD followed by Li-SCM, Although PSC-PWM wIA has the best average performance, this method requires much higher computational cost due to the large number of interactions, and its waveform are shown in Fig. 6(a1), 6(d1) and 6(g1).

The submodule capacitor voltage of the three techniques are shown in Fig. 6(g1), 6(h1) and 6(i1), these waveforms indicates a ripple factor: PSC-PWM wIA has 5%, SAM has 3% and Li-SCM has 7%, these values are calculated considering $\frac{\text{Fundamental}}{N}$.

978-1-7281-4181-7/19 $31.00 © 2019 IEEE

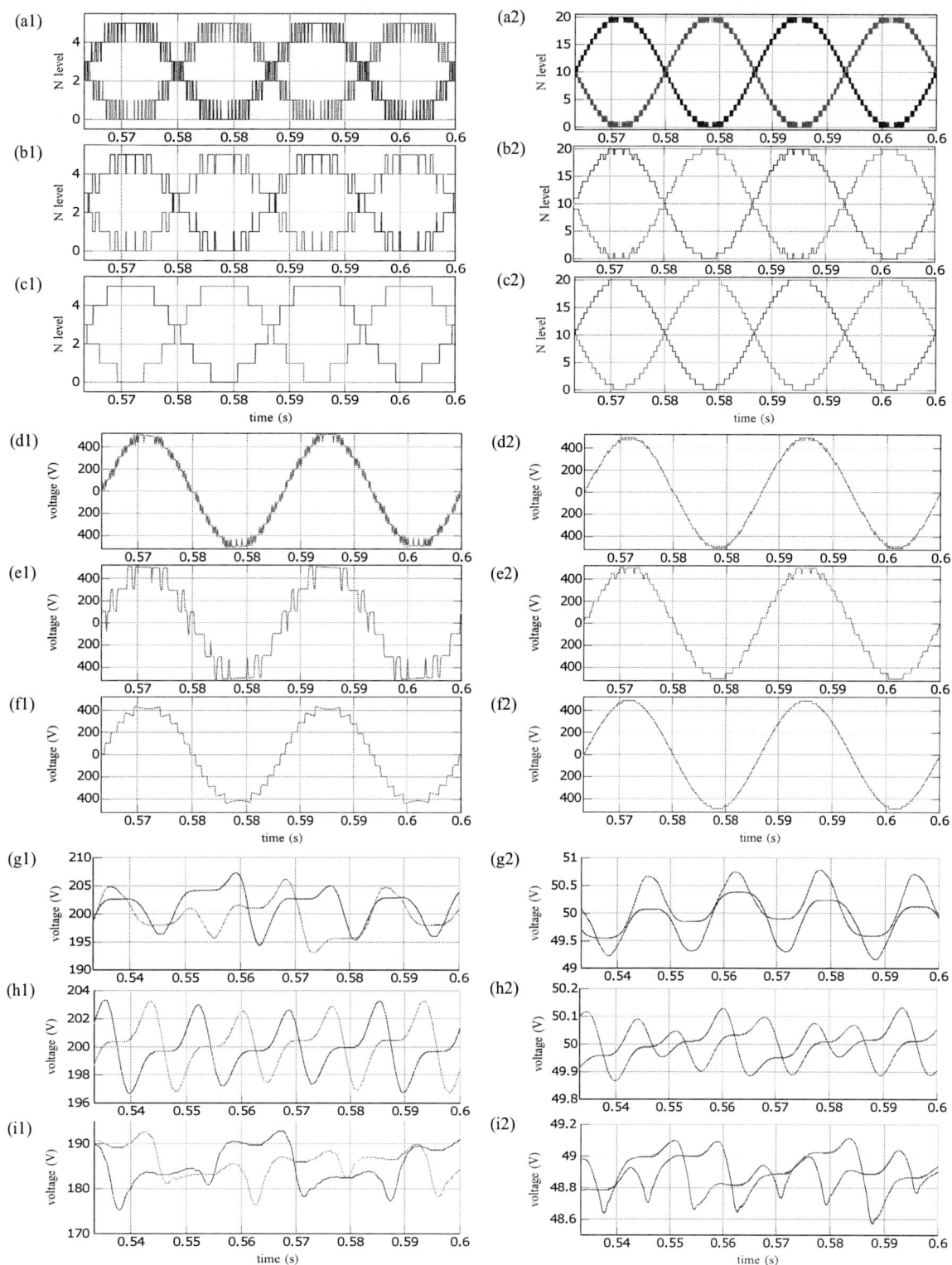

Fig. 6. Arm voltage levels, Inverter output voltage and SM capacitor voltage for **MMC with 5 SMs/arm (left)** and **20 SMs/arm (right)**. (a, d, g) PSC-PWM wIA; (b, e, h) SAM and (c, f, i) Li-SCM.

978-1-7281-4181-7/19 $31.00 © 2019 IEEE

A. PSC-PWM wIA: Considering the parameters in Table I, it is calculated:

$\phi = 72^o$, $\phi_i = 36^o$, with $f_c = 600$ Hz and $f_{psc} = 6000$ Hz.

B. Simulation Case II

Similarly to Case I, but with 20 SMs/arm. In Fig. 6(a2), 6(b2) and 6(c2) it is observed that, when the number of submodules is increased the frequency of operation increases progressively, therefore, also the CPU time increases.

In Table II the output fundamental voltage amplitude values are shown, and the theoretical value of 500 V is reached by PSC-PWM wIA and SAM, Also it is observed that Li-SCM improves greatly when compared with the previous case, this is due to the fact that as more submodules there is, the dischanging-time of submodules capacitor is lower, this waveforms are shown in Fig. 6(d2), 6(e2) and 6(f2).

In Fig. 6(g2), 6(h2) and 6(i2), the submodule capacitor voltage are shown and its ripple factor of: 2%, 0.5% and 1% for PSC-PWM wIA, SAM and Li-SCM, respectively.

A. PSC-PWM wIA: it is calculated.

$\phi = 18^o$, $\phi_i = 9^o$, with $f_c = 600$ Hz and $f_{psc} = 24000$ Hz.

TABLE II
INVERTER OUTPUT VOLTAGE COMPARISON

Case I - MMC with 5 SMs/arm			
	Fundamental	THD	CPU time
PSC-PWM wIA	496.6 V	7,72 %	26.34s
SAM	501.4 V	19.93 %	10.84s
Li-SCM	441.1 V	8.32 %	09.71s
Case II - MMC with 20 SMs/arm			
	Fundamental	THD	CPU time
PSC-PWM wIA	499.8 V	2,76 %	1m 41.12s
SAM	500.7 V	5.40 %	21.18s
Li-SCM	488.6 V	1.75 %	19.45s

V. CONCLUSION

In this paper, the PSC-PWM wIA, SAM and Li-SCM techniques were compared considering the output fundamental voltage amplitude, the THD and the CPU time.

The results indicate that: PSC-PWM wIA and SAM techniques maintain an adequate amplitude of fundamental in the inversor output voltage, while in Li-SCM technique this voltage tends to decrease, this is because when fewer submodules are used, the capacitors have more discharge time, however, it progressively improves if as more submodules are added.

The THD of the inverter output voltage is a parameter to be considered, it is observed that PSC-PWM and Li-SCM techniques maintain a low THD due to generating 2N+1 voltage levels at the output, and this ratio is maintained as more submodules are added, in comparison to SAM that shows a relatively high THD.

The SAM and Li-SCM techniques use a relatively low CPU time due to their simplicity of operation, whereas the PSC-PWMwIA technique uses a higher CPU time due to a large number of interactions and tends to increase exponentially if more submodules are added. All the results were obtained considering the same conditions, settings and simulation parameters.

REFERENCES

[1] K. K. Gupta, A. Ranjan, P. Bhatnagar, L. K. Sahu, and S. Jain, "Multilevel inverter topologies with reduced device count: a review," *IEEE transactions on Power Electronics*, vol. 31, no. 1, pp. 135–151, 2015.

[2] M. A. Perez, S. Bernet, J. Rodriguez, S. Kouro, and R. Lizana, "Circuit topologies, modeling, control schemes, and applications of modular multilevel converters," *IEEE transactions on power electronics*, vol. 30, no. 1, pp. 4–17, 2014.

[3] S. Debnath, J. Qin, B. Bahrani, M. Saeedifard, and P. Barbosa, "Operation, control, and applications of the modular multilevel converter: A review," *IEEE Transactions on Power Electronics*, vol. 30, no. 1, pp. 37–53, Jan. 2015.

[4] F. Rong, X. Gong, and S. Huang, "A novel grid-connected PV system based on mmc to get the maximum power under partial shading conditions," *IEEE Transactions on Power Electronics*, vol. 32, no. 6, pp. 4320–4333, Jun. 2017.

[5] M. Raza, E. Prieto-Araujo, and O. Gomis-Bellmunt, "Small-signal stability analysis of offshore AC network having multiple vsc-HVDC systems," *IEEE Transactions on Power Delivery*, vol. 33, no. 2, pp. 830–839, Apr. 2018.

[6] D. Ronanki and S. S. Williamson, "Modular Multilevel Converters for Transportation Electrification: Challenges and Opportunities," *IEEE Transactions on Transportation Electrification*, vol. 4, no. 2, pp. 399–407, Jun. 2018.

[7] T. H. Nguyen, K. A. Hosani, M. S. E. Moursi, and F. Blaabjerg, "An overview of modular multilevel converters in HVDC transmission systems with statcom operation during pole-to-pole DC short circuits," *IEEE Transactions on Power Electronics*, vol. 34, no. 5, pp. 4137–4160, May 2019.

[8] J. Li, G. Konstantinou, H. R. Wickramasinghe, J. Pou, X. Wu, and X. Jin, "Investigation of mmc-hvdc operating region by circulating current control under grid imbalances," *Electric Power Systems Research*, vol. 152, pp. 211–222, 2017.

[9] A. Dekka, B. Wu, R. L. Fuentes, M. Perez, and N. R. Zargari, "Evolution of topologies, modeling, control schemes, and applications of modular multilevel converters," *IEEE Journal of Emerging and Selected Topics in Power Electronics*, vol. 5, no. 4, pp. 1631–1656, 2017.

[10] M. Davies, M. Dommaschk, J. Dorn, J. Lang, D. Retzmann, and D. Soerangr, "HVDC plusBasics and principle of operation. Technical report," *Special Edition for Cigr Exposition*, 2008.

[11] Y. Luo, Z. Li, L. Xu, X. Xiong, Y. Li, and C. Zhao, "An adaptive voltage-balancing method for high-power modular multilevel converters," *IEEE Transactions on Power Electronics*, vol. 33, no. 4, pp. 2901–2912, 2017.

[12] M. H. Nguyen and S. Kwak, "Simplified indirect model predictive control method for a modular multilevel converter," *IEEE Access*, vol. 6, pp. 62 405–62 418, 2018.

[13] A. Antonio-Ferreira, C. Collados-Rodriguez, and O. Gomis-Bellmunt, "Modulation techniques applied to medium voltage modular multilevel converters for renewable energy integration: A review," *Electric Power Systems Research*, vol. 155, pp. 21–39, 2018.

[14] F. Deng, Y. Tian, R. Zhu, and Z. Chen, "Fault-tolerant approach for modular multilevel converters under submodule faults," *IEEE Transactions on Industrial Electronics*, vol. 63, no. 11, pp. 7253–7263, 2016.

[15] Du, Sixing and Dekka, Apparao and Wu, Bin and Zargari, Navid, *Modular multilevel converters*. John Wiley & Sons, 2017.

[16] F. Deng and Z. Chen, "Elimination of dc-link current ripple for modular multilevel converters with capacitor voltage-balancing pulse-shifted carrier pwm," *IEEE Transactions on Power Electronics*, vol. 30, no. 1, pp. 284–296, 2014.

[17] ——, "Voltage-balancing method for modular multilevel converters under phase-shifted carrier-based pulsewidth modulation," *IEEE Transactions on Industrial Electronics*, vol. 62, no. 7, pp. 4158–4169, 2015.

[18] A. Dekka, B. Wu, and N. R. Zargari, "A Novel Modulation Scheme and Voltage Balancing Algorithm for Modular Multilevel Converter," *IEEE Transactions on Industry Applications*, vol. 52, no. 1, pp. 432–443, Jan. 2016.

[19] P. Hu and D. Jiang, "A level-increased nearest level modulation method for modular multilevel converters," *IEEE Transactions on Power Electronics*, vol. 30, no. 4, pp. 1836–1842, 2014.

Designing of the transmitting coils and compensation network of a segmented DWPT system

Shufan Li[1,2]
1. Key Laboratory of Power Electronics and Electric Drives
Institute of Electrical Engineering
Chinese Academy of Sciences
2. University of Chinese Academy of Sciences
Beijing, China
lishufan@mail.iee.ac.cn

Lifang Wang[*]
Key Laboratory of Power Electronics and Electric Drives
Institute of Electrical Engineering
Chinese Academy of Sciences
Beijing, China
wlf@mail.iee.ac.cn

Chengxuan Tao
Key Laboratory of Power Electronics and Electric Drives
Institute of Electrical Engineering
Chinese Academy of Sciences
Beijing, China
taochengxuan@mail.iee.ac.cn

Fang Li
Key Laboratory of Power Electronics and Electric Drives
Institute of Electrical Engineering
Chinese Academy of Sciences
Beijing, China
lifang@mail.iee.ac.cn

Liye Wang
Key Laboratory of Power Electronics and Electric Drives
Institute of Electrical Engineering
Chinese Academy of Sciences
Beijing, China
wangliye@mail.iee.ac.cn

Abstract—**A multi-objective designing method of the dynamic wireless power transfer(DWPT) system considering both the coupling coils and compensation networks is proposed in this paper. The operating process of the DWPT system is analyzed based on LCC compensation network. The power transfer performance, electrical stress, ZVS condition and cost of the DWPT system are analyzed as the objectives or constraints of the optimization. Finally, a 4-coil segmented prototype is built to verify the proposed method.**

Keywords—*DWPT, transmitting coils, compensation network, designing method*

I. Introduction

Dynamic wireless power transfer(DWPT) technology[1-3] for electric vehicles(EVs) can help to extend the driven range of EVs and reduce the weight of the battery capacity, and many studies have been focused on the designing method[4-8] of it. The structure and size of the transmitting coils are optimized in [4-5] to charge for different kinds of vehicles. Also, LCC network[6-7] is studied as a compensation network for the DWPT system and designed to realize constant output current and ZVS condition. Although the coupling coils and compensation networks can be designed separately, the requirement for one of them may change when another is designed alone, which makes it difficult to obtain a good performance of the system.

In this paper, a multi-objective designing method of the DWPT system considering both the coupling coils and compensation networks is proposed. The power transfer performance, electrical stress, ZVS condition and cost of the DWPT system are analyzed as the objectives or constraints of the optimization.

This work is supported by the National Key Research and Development Program of China(No. 2018YFB0106300).

II. Overview of a Segmented DWPT System

A. Configuration of a DWPT system

A segmented DWPT system is shown in Fig.1. There are N transmitting coils in the system, the length of a transmitting coil is l_t, and its width is w_t. the spacing between the transmitting coils is s, and the total length of the system is l. For a single transmitting coil, the LCC compensation network is used to compensate for the reactive components. N switches($S_1, S_2...S_N$) are connected with the LCC networks and the high-frequency inverters to change the power-supplying coils. The detecting devices are mounted beside every coil to detect the position of the receiving coils.

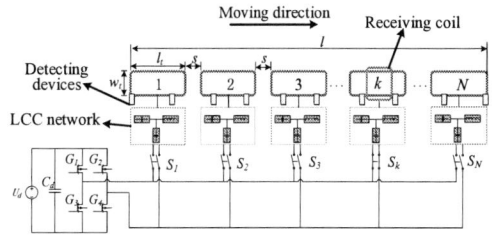

Fig.1 A segmented DWPT system

The LCC network is used in the presented DWPT system for its constant-current output characteristics. The LCC network is shown in Fig.2, L_{p_k} is the power-supplying transmitting coil, and Z_{r_k} is the reflected impedance from the receiving side. The values of the components in the LCC network can be calculated by

$$\begin{cases} L_{fp\_k} = X_p / \omega \\ C_{pp\_k} = 1 / \omega / X_p \\ C_{ps\_k} = 1 / \omega / (\omega L_{p\_k} - X_p) \end{cases} \quad (1)$$

In (1), ω is the angular frequency of the system, and X_p can be calculated by

$$X_p = \frac{U_i}{I_p} \quad (2)$$

where I_p is the current of the transmitting coil determined by the power demand.

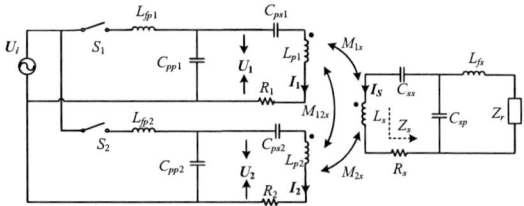

Fig.2 Circuit of the DWPT system

As for the receiving side, a similar LCC network is used as the compensation network, and the impedance of the three arms of the network is X_s.

B. Operating process of a DWPT system

The circuit of the DWPT system is shown in Fig.2. One or two transmitting coils are energized according to the coupling between the transceiving coils. There will be three stages in the operating process of a segmented DWPT system:

Stage 1: $S_1=1$, $S_2=0$

"1" and "0" represents the on and off state of the switches. In this stage, the KVL function of the system is

$$\begin{bmatrix} Z_1 & j\omega M_{12} & j\omega M_1 \\ j\omega M_{12} & Z_2 - jX_p & j\omega M_2 \\ j\omega M_1 & j\omega M_2 & Z_s \end{bmatrix} \begin{bmatrix} I_1 \\ I_2 \\ I_s \end{bmatrix} = \begin{bmatrix} U_i \\ 0 \\ 0 \end{bmatrix} \quad (3)$$

where Z_s is the equivalent input impedance of the secondary side, and

$$\begin{cases} Z_s = R_s + X_s^2 / Z_r \\ Z_1 = R_1 + j\beta X_p \\ Z_2 = R_2 + j\beta X_p \end{cases} \quad (4)$$

The current in the transmitting and receiving coils are

$$\begin{cases} I_1 = U_i / jX_p \\ I_2 = -I_1 \cdot \dfrac{\omega^2 M_1 M_2 + j\omega M_{12} Z_s}{\omega^2 M_2^2 + (Z_2 - jX_p) Z_s} \\ I_s = -j\omega (M_1 I_1 + M_2 I_2) / Z_s \end{cases} \quad (5)$$

The induced voltage in the receiving coil is

$$U_s = \omega M_1 I_1 + \omega M_2 I_2 \quad (6)$$

Stage 2: $S_1=1$, $S_2=1$, $\mu_{1x}>0$, $\mu_{2x}>0$

In this stage, the KVL function of the system is

$$\begin{bmatrix} Z_1 & j\omega M_{12} & j\omega M_1 \\ j\omega M_{12} & Z_2 & j\omega M_2 \\ j\omega M_1 & j\omega M_2 & Z_s \end{bmatrix} \begin{bmatrix} I_1 \\ I_2 \\ I_s \end{bmatrix} = \begin{bmatrix} U_1 \\ U_2 \\ 0 \end{bmatrix} \quad (7)$$

The current in the transmitting and receiving coils are

$$\begin{cases} I_1 = U_i / jX_p \\ I_2 = U_i / jX_p \\ I_s = -j\omega (M_1 I_1 + M_2 I_2) / Z_s \end{cases} \quad (8)$$

The induced voltage in the receiving coil can be calculated with (6).

Stage 3: $S_1=0$, $S_2=1$

This stage is similar with stage 1, and the analysis can be done by exchanging the subscript "1" and "2" in stage 1.

III. ANALYSIS OF THE DWPT SYSTEM

A. Performance analysis

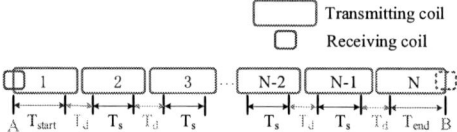

Fig.3 A segment of the DWPT system

Fig.3 shows a segment of the DWPT system containing N transmitting coils. Coil 1 is the starting coil and coil N is the ending coil. According to the analysis about the operating process of the system, the segment in Fig.3 can be described by three conditions: (1) power supplied by the starting coil or ending coil alone(T_{start} and T_{end}). Considering symmetry, T_{start} can be seen the same as T_{end}. (2) power supplied by two adjacent transmitting coils(T_d). (3) power supplied by non-start-non-end coils(T_s). There are one T_{start}, one T_{end}, (N-1) T_d and (N-2) T_s in a segment.

The output power and power losses at an arbitrary position of the segment are calculated by (9) and (10),

$$P_t = \frac{1}{2} |U_s| \cdot |I_s| \quad (9)$$

$$P_{loss\_t} = |I_1|^2 R_1 + |I_2|^2 R_2 + |I_s|^2 R_s \quad (10)$$

The average output and input power for an arbitrary part of the segment of length x are

$$P_{avr} = \frac{\int_0^x P_t dl}{x} \quad (11)$$

$$P_{in\_avr} = \frac{\int_0^x (P_t + P_{loss\_t}) dl}{x} \quad (12)$$

978-1-7281-4181-7/19 $31.00 © 2019 IEEE

The average output and input power of the whole segment can be described with (13) and (14)

$$P_{avr} = \left[2P_{T\_start}.l_{T\_start} + (N-1)P_{T\_d}.l_{T\_d} + (N-2)P_{T\_s}.l_{T\_s} \right]/l \tag{13}$$

$$P_{avr\_in} = \left[2P_{T\_start\_in}.l_{T\_start\_in} + (N-1)P_{T\_d\_in}l_{T\_d\_in} + (N-2)P_{T\_s\_in}l_{T\_s\_in} \right]/l \tag{14}$$

where P_{T_start}, P_{T_d} and P_{T_s} are the output power for T_{start}, T_d and T_s, while $P_{T\_start\_in}$, $P_{T\_d\_in}$ and $P_{T\_s\_in}$ are the input power for T_{start}, T_d and T_s.

The average transfer efficiency is

$$\eta_{av} = \frac{P_{avr}}{P_{avr\_in}} \tag{15}$$

Assuming the velocity of the vehicle is v, the energy transferred to the vehicle by an segment is

$$E_{seg} = \frac{P_{avr}.l}{v} \tag{16}$$

B. Electrical stress analysis

To ensure the DWPT system operating in an safe and steady condition, the electrical stress of the components in the system should be considered. The electrical stress includes the current stress in the transceiving coils, inductors in the compensation network and the switching devices, the voltage stress of the capacitors in the compensation network. The constraints about the electrical stress are

$$\begin{cases} \max(|\mathbf{I}_{L\_pk}|, |\mathbf{I}_{L\_fpk}|, |\mathbf{I}_{Ls}|, |\mathbf{I}_{Lfs}|) \le I_{liz\_\lim} \\ \max(|\mathbf{I}_i|) \le I_{mos\_\lim} \\ \max(|\mathbf{U}_{C\_ppk}|, |\mathbf{U}_{C\_psk}|, |\mathbf{U}_{C\_ss}|, |\mathbf{U}_{C\_sp}|) \le U_{C\_\lim} \end{cases} \tag{17}$$

where $|\mathbf{I}_{L\_pk}|$, $|\mathbf{I}_{L\_fpk}|$, $|\mathbf{I}_{Ls}|$ and $|\mathbf{I}_{Lfs}|$ are the amplitudes of the current in the inductors in the compensation network and the transceiving coils, $|\mathbf{I}_i|$ is the amplitude of the output current of the inverter, $|\mathbf{U}_{C\_ppk}|$、$|\mathbf{U}_{C\_psk}|$、$|\mathbf{U}_{C\_ss}|$、$|\mathbf{U}_{C\_sp}|$ are the amplitudes of the voltage in the capacitors in the compensation network. It should be mentioned that the three different operating stages should be considered when analyzing the above amplitudes. I_{Litz_lim} is the current limit of the Litz wire, I_{mos_lim} is the current limit of the switching device, and U_{C_lim} is the voltage limit of the capacitors.

C. ZVS analysis

The turn-off current of the switching devices of the inverter should be large enough to discharge the junction capacitor and guarantee a ZVS condition of the inverter. The turn-off current at x point can be calculated by

$$I_{off\_x} = \frac{|U_i| X_{ix}}{|Z_{ix}|^2} + \frac{|U_i|}{4X_p} \tag{18}$$

where Z_{ix} is the equivalent input impedance of the inverter at x point, and X_{ix} is the imaginary part of Z_{ix}. The value of I_{off_x} should be larger than that of the minimum turn-off current of the switching devices I_{off_min}.

D. Cost analysis

The cost of a segmented DWPT system can be divided into four parts: the transmitting coils, the magnetic shielding materials, the LCC compensation networks and detecting and switching devices. However, the cost of the transmitting coils and the magnetic shielding materials accounts for the majority of the total cost, therefore only these two parts are considered in this work.

The transmitting coils are made of Litz wire to avoid the skin effect. The total length of the transmitting coils l_{Litz} are determined by

$$l_{Liz} = N. \sum_{x=0}^{N_t-1} \left[2(l_t + w_t) - 8x.s_t \right] \tag{19}$$

where N_t is the turns of a transmitting coil, s_t is the turn-to-turn spacing of a transmitting coil.

The cost of the transmitting coils can be calculated by

$$C(Liz) = P_{Liz}.l_{Liz} \tag{20}$$

where P_{Litz} is the cost of the Litz wire per meter.

The ferrite magnetic materials are used to shield the leakage of the magnetics. The transmitting coils should be fully covered by ferrite, and the total area of the ferrite is

$$A_{fe} = \begin{cases} N.(l_t + \alpha)(w_t + \alpha) & s > \alpha \\ (l + \alpha).(w_t + \alpha) & s \le \alpha \end{cases} \tag{21}$$

where α is an extra length to ensure the coils to be fully covered.

The cost of the ferrite magnetic materials can be calculated by

$$C(fe) = P_{fe}.A_{fe} \tag{22}$$

where P_{fe} is the cost of the ferrite per square meter.

IV. Designing procedure of a Segmented DWPT System

Based on the analysis of III, a segmented DWPT system can be designed according to the procedure below:

Step 1: Design a series of schemes of the segments according to the size, spacing and number of turns of the transmitting coils, and then number them(Num=1, 2, 3… Every number is correlated with the a value of the length of the transmitting coil l_t, the spacing of the adjacent coils s and the number of turns of the transmitting coils N_r). FEM software will be used to analyze the schemes and calculate the inductance and mutual inductance of the transceiving coils. It should be mentioned that the receiving coil based on SAE J2954 VA WPT1/Z2 is used in this work to be compatible with the stationary WPT system.

Step 2, an optimization objective function and constraint function can be achieved with the Num in step 1, the parameters of the compensation network X_p, X_s and β, the amount of the transmitting coils in a segment Nc and the amount of the segments in a DWPT road as the independent variables, the average power transfer efficiency η_{av}, the total transferred energy E_{sum} and the total cost C_{sum} as the optimization objectives, and the turn-off current of the switching devices in the inverter I_{off}, the current stress in the transceiving coils, inductors in the compensation network and the switching devices, the voltage stress of the capacitors in the compensation network, the total length of the DWPT road l_{sum} and the average output power P_{avr} as the constraints.

The objective function is

$$\min \begin{cases} f(1) = k_1 / \eta(Num, X_p, X_s, \beta, N_s, N_c) + k_2 / E_{sum}(Num, X_p, X_s, \beta, N_s, N_c) \\ f(2) = C_{sum}(Num, X_p, X_s, \beta, N_s, N_c) \end{cases}$$

(23)

where k_1 and k_2 is the weight coefficients which are determined by the requirements and characteristics of the system.

The constraint function is

$$\begin{cases} I_{off} \geq I_{off\_min} \\ \max(|\mathbf{I}_{L\_pk}|, |\mathbf{I}_{L\_fpk}|, |\mathbf{I}_{Ls}|, |\mathbf{I}_{Lfs}|) \leq I_{liz\_lim} \\ \max(|\mathbf{I}_i|) \leq I_{mos\_lim} \\ \max(|\mathbf{U}_{C\_ppk}|, |\mathbf{U}_{C\_psk}|, |\mathbf{U}_{C\_ss}|, |\mathbf{U}_{C\_sp}|) \leq U_{C\_lim} \\ l_{sum} \leq l_{lim} \\ P_{avr} \leq P_{avr\_lim} \end{cases}$$

(24)

Step 3, According to the objective function and constraint function in step 2, an multi-objective particle swarm optimization method is used to obtain the optimum designing scheme of the DWPT segment and compensation network.

V. CASE STUDY

The above designing method is used to study a case of the segmented DWPT system. 308 schemes are designed according to values of l_f[500,750,1000...2750,3000](mm)), N_r([4,6,8...14,16]) and s([0,100,200,300](mm)), which means the range of Num is [1,2,3...307,308]. The other parameters of the system are shown in Table I.

TABLE 1 PARAMTERS OF THE SYSTEM

parameter	value	parameter	value
X_p/Ω	[1,100]	U_{c_lim}/V	2500
X_s/Ω	[1,100]	l_{lim}/m	100
b	[0,2]	P_{avr_lim}/W	4000
N_s	[5,200]	k_1	1000
N_c	[1,40]	k_2	10
I_{off_min}/A	3.5	P_{Litz}/\$	1.88
I_{Litz_lim}/A	32.2	α/mm	20
I_{mos_lim}/A	25	P_{fe}/\$	327.8
v/(km/h)	20		

The multi-objective particle swarm optimization method is used to design the parameters of the DWPT system. Fig.4 shows the pareto fronts of the optimization when the iterations are set as 10, 50 and 100. It can be seen that the method converges quickly.

One point of the pareto front after 100 iterations of optimization is chosen as the optimum solution: l_f=500mm, N_r=10, s=200mm, X_p=18.17Ω, X_s=13.98Ω, β=0.717, N_s=5, N_c=4. A prototype of one segment of the proposed DWPT system is shown in Fig.5. The mutual inductance between the transmitting coil 1 and the receiving coil is shown in Fig.6(a), and the measured values are slightly larger than the theoretical values. Simulation results of the system supplying by one transmitting coil is compared with the theoretical analysis in Fig.6(b-d). The efficiency of the simulation in Fig.6(c) is slightly lower than the theoretical value, which is mainly

because some power losses such as that of the parasitic resistance are not considered in the calculation. As for the turn-off current, the simulation results are affected by the harmonics, making it slightly lower than the theoretical value. It is shown that the output power is in positive correlation with the coupling, and ZVS condition is achieved though the mutual inductance changes with the movement of the receiver. In the future works, the detection and switching of the transmitting coils will be studied.

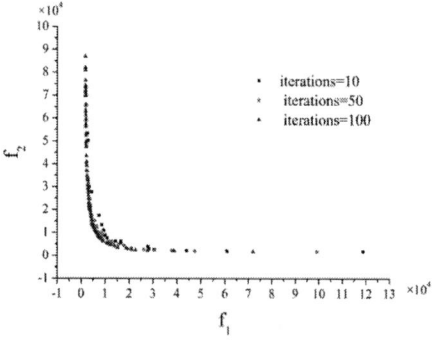

Fig.4 Pareto fronts of the optimization

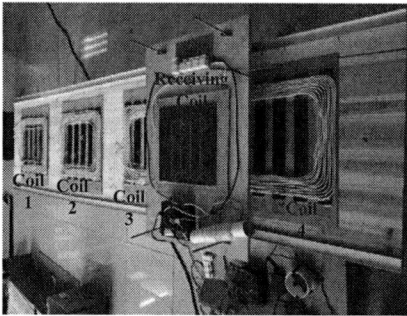

Fig.5 A prototype of one segment of the DPWT system

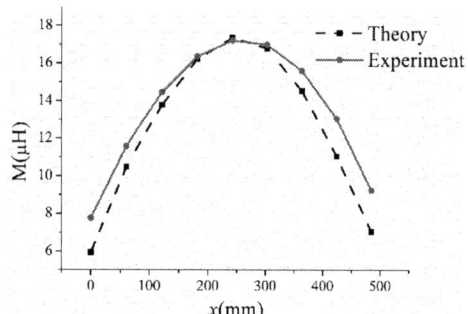

(a) mutual inductance

(b) output power

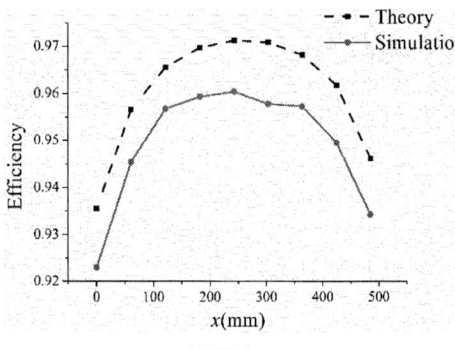

(c) Efficiency

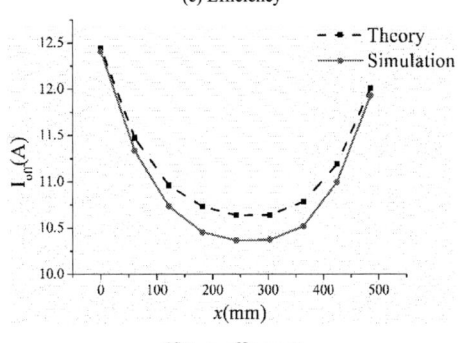

(d) turn-off current

Fig.6 Measured mutual inductance and simulation results

VI. CONCLUSION

A designing method of the DWPT system considering both the transmitting coils and the compensation network is proposed in this paper. The operating process of a segmented DWPT system is analyzed the LCC network is chosen as the compensation network. A multi-objective optimization method is proposed based on the analysis of the performance, electrical stress, ZVS condition and cost of the DWPT system.

REFERENCES

[1] J. Miller et al., "Demonstrating Dynamic Wireless Charging of an Electric Vehicle: The Benefit of Electrochemical Capacitor Smoothing," Power Electronics Magazine, IEEE, vol. 1, no. 1, pp. 12-24, 2014.

[2] S. Y. Choi, B. W. Gu, S. Y. Jeong, and C. T. Rim, "Advances in Wireless Power Transfer Systems for Roadway-Powered Electric Vehicles," Emerging and Selected Topics in Power Electronics, IEEE Journal of, vol. 3, no. 1, pp. 18-36, 2015.

[3] C. C. Mi, G. Buja, S. Y. Choi, and C. T. Rim, "Modern Advances in Wireless Power Transfer Systems for Roadway Powered Electric Vehicles," IEEE Transactions on Industrial Electronics, vol. 63, no. 10, pp. 6533-6545, 2016.

[4] G. R. Nagendra, G. A. Covic, and J. T. Boys, "Determining the Physical Size of Inductive Couplers for IPT EV Systems," IEEE Journal of Emerging and Selected Topics in Power Electronics, vol. 2, no. 3, pp. 571-583, 2014.

[5] G. R. Nagendra, G. A. Covic, and J. T. Boys, "Sizing of Inductive Power Pads for Dynamic Charging of EVs on IPT Highways," IEEE Transactions on Transportation Electrification, vol. 3, no. 2, pp. 405-417, 2017.

[6] Q. Zhu, L. Wang, Y. Guo, C. Liao, and F. Li, "Applying LCC Compensation Network to Dynamic Wireless EV Charging System," IEEE Transactions on Industrial Electronics, vol. 63, no. 10, pp. 6557-6567, 2016.

[7] S. Zhou and C. C. Mi, "Multi-Paralleled LCC Reactive Power Compensation Networks and Their Tuning Method for Electric Vehicle Dynamic Wireless Charging," IEEE Transactions on Industrial Electronics, vol. 63, no. 10, pp. 6546-6556, 2016.

[8] R. Tavakoli and Z. Pantic, "Analysis, Design, and Demonstration of a 25-kW Dynamic Wireless Charging System for Roadway Electric Vehicles," IEEE Journal of Emerging and Selected Topics in Power Electronics, vol. 6, no. 3, pp. 1378-1393, 2018.

Proposal of an Isolated Two-Switch DC-DC SEPIC Converter

Marcos Vinícius Mosconi Ewerling
Power Electronics Institute - INEP
Federal University of Santa Catarina
Florianópolis, Brazil
marcos.ewerling@inep.ufsc.br

Telles Brunelli Lazzarin
Power Electronics Institute - INEP
Federal University of Santa Catarina
Florianópolis, Brazil
telles@inep.ufsc.br

Carlos Henrique Illa Font
Department of Electronics Engineering
Federal University of Technology - Paraná
Ponta Grossa, Brazil
illafont@utfpr.edu.br

Abstract—The theoretical analysis and numerical simulations of a novel dc-dc converter conceived from SEPIC topology are presented in this paper. In this proposed converter, the voltage stress on switches and diodes is lower than a conventional SEPIC converter, when the same input and output voltage levels are considered. The proposed structure can be employed in applications with higher input voltage and high output current, once its output is connected in parallel. The paper approaches the topological stages, static analysis from DCM, design equations, and a comparative analysis. Furthermore, to evaluate the performance of the proposed topology is realized an open-loop numerical simulation for a 500 W rated power, with 120 V output voltage, 400 V input voltage and switching frequency of 50 kHz.

Index Terms—SEPIC, dc-dc converter, isolated-converter, DCM

I. Introduction

Recently, the dc-dc converters have been applying in a wide range of places, especially in the renewable energy field. [1], [2]. A classic converter widely used in renewable energy is the Single-Ended Primary-Inductance Converter (SEPIC) [3]–[5]. An advantage of the SEPIC-type converters is its feature of stepping-up or stepping-down voltages without inverting the output voltage, unlike buck-boost converters.

In addition, the SEPIC-type converter presents low ripple of input current due to a front inductor [6]–[8], different from the buck converter, which has an input pulsed current. Furthermore, the SEPIC converters can provide high frequency galvanic isolation employing coupled inductors [8]–[10].

However, a drawback of the conventional SEPIC converter is the voltage stress on switches, which is equal to the input voltage plus the output voltage [11], [12]. To solve this problem, some topologies derived from the SEPIC converter are proposed [13], [14], which has the characteristic of reducing the voltage across the switch.

In this paper is proposed a dc-dc converter conceived from SEPIC topology with reduced voltage stress on the semiconductors, which is the main advantage of the proposed structure. Furthermore, this proposed topology supplies high frequency galvanic isolation through two coupled inductors.

The theoretical analysis of the isolated dc-dc converter based on SEPIC topology operating DCM (discontinuous conduction mode) is presented in Section II. Section III

presents the static analysis for the power stage. The numerical simulation results of the converter operating in DCM are presented in Section IV. In Section V is presented a comparative analysis between conventional SEPIC converter and the proposed converter. Finally, in Section VI, the conclusions are stated.

II. Proposed Topology

Fig. 1 presents the proposed topology conceived from SEPIC topology. The proposed converter employs two input inductors (L_{i1} and L_{i2}), two active switches (S_1 and S_2), two input capacitors (C_{i1} and C_{i2}), two coupled inductors (L_{o1} and L_{o2}), two diodes (D_1 and D_2) and an output capacitor (C_o).

It should be noted in Fig 1 that the proposed structure can supply multiple outputs, since it has high-frequency galvanic isolation between the input and the output voltages.

A. Operating Stages

The ideal proposed converter operating in the DCM presents three operating stages in a switching period. Fig. 2 depicts the operating stages.

On the first operational stage, the switches are turned-on and the diodes are reverse biased. The inductors L_{i1}, L_{i2}, L_{o1} and L_{o2} obtain energy from the source V_{in} and the capacitors C_{i1} and C_{i2}, respectively. The currents in inductors L_{i1} and L_{i2} increase linearly according to the relation $V_{in}/2L_i$ and the

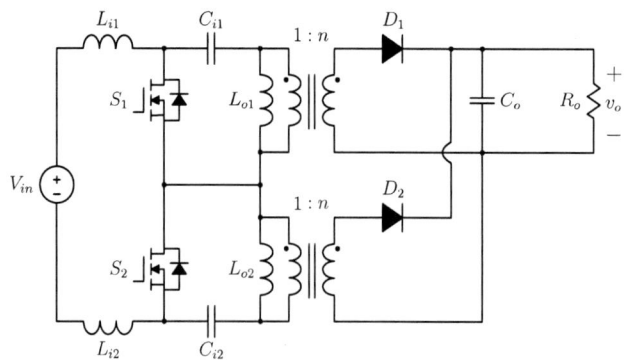

Fig. 1. Proposed structure of the dc-dc converter.

currents in inductors L_{o1} and L_{o2} increase linearly according the relation $V_{in}/2L_o$, where $L_i = L_{i1} = L_{i2}$ and $L_o = L_{o1} = L_{o2}$. The capacitor C_o supplies the equivalent load R_o.

The active switches S_1 and S_2 are turned-off in the second operating stage. Meanwhile, the diodes D_1 and D_2 are conducted. The energy stocked in inductors L_{i1}, L_{i2}, L_{o1} and L_{o2} is transferred to capacitors C_{i1}, C_{i2} and C_o and, to equivalent load R_o. The inductor currents L_{i1} and L_{i2} decrease linearly following the relation $-V_o'/L_i$ and the currrents in L_{o1} and L_{o2} decrease linearly as well following the relation $-V_o'/L_o$, where V_o' is the output voltage seen by the primary side.

The third operating stage starts when the currents in inductors reach the same absolute value, blocking the diodes D_1 and D_2. The active switches remain blocked. The capacitor C_o supplies the equivalent load R_o.

B. Ideal Waveforms from the Proposed Converter

The ideal waveforms of voltages and currents for inductors L_{i1}, L_{i2}, L_{o1} and L_{o2}, switches S_1 and S_2, diodes D_1 and D_2 and, capacitors C_{i1}, C_{i2} and C_o are depicted in Fig. 3 and Fig. 4, respectively for a switching period.

It should be noted that the voltages on the inductors L_{i1}, L_{i2}, L_{o1} and L_{o2} have equal maximum and minimum values. The voltage on the inductors is equal to the half of the input source during the first topological stage, while it is equal to the output voltage seen by the primary side during the second topological stage. In the third topological stage, when the inductor currents are the same, the voltage on the inductors is zero.

Regarding the voltage on the diodes D_1 and D_2, it is equal to the sum of the half of reflected input voltage to the secondary side and the output voltage, during the first operating stage.

Regarding the voltage on the switches S_1 and S_2, it is equal to the sum of the half of input source and the output voltage

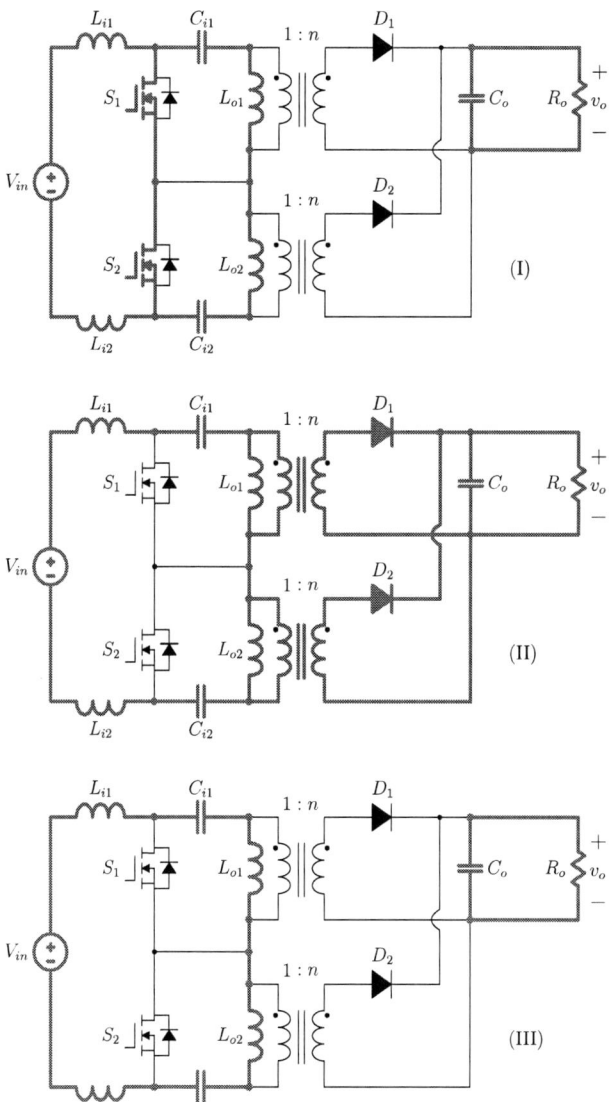

Fig. 2. Operation stages: I) first operational stage; II) second operational stage; III) third operational stage.

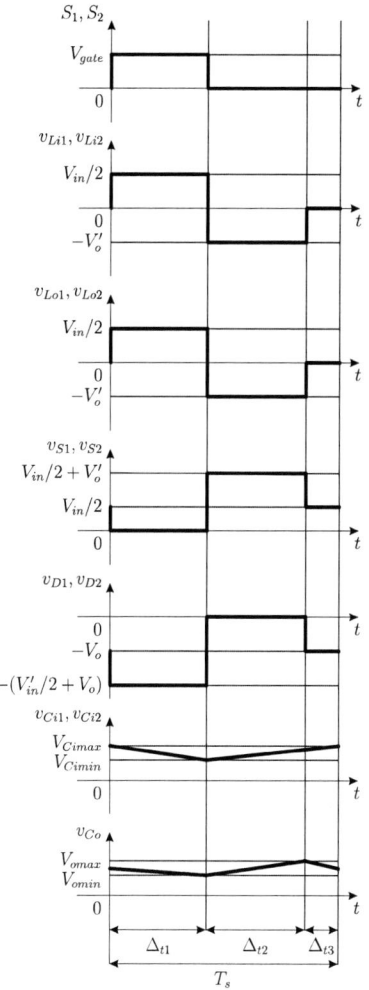

Fig. 3. Main voltage waveforms.

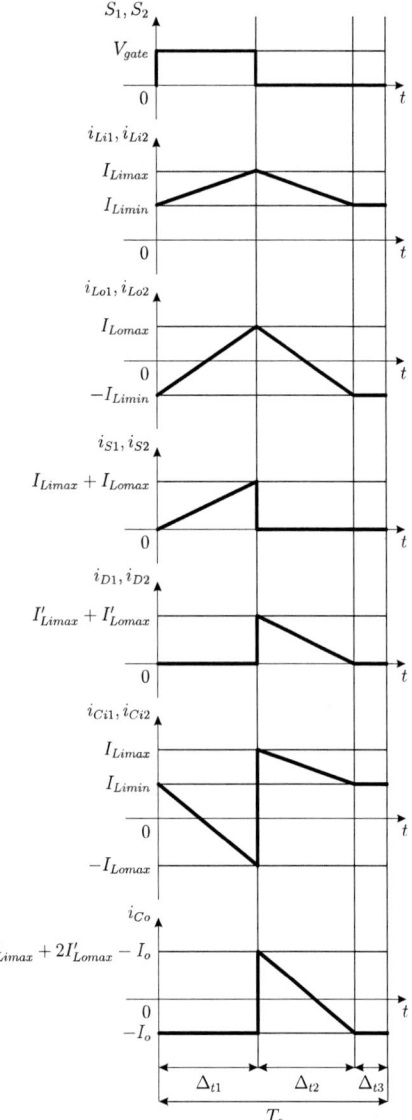

Fig. 4. Main current waveforms.

currents raise linearly until reaching their maximum value. Then, the current in switches S_1 and S_2 is the sum of the inductor currents in this stage. Since the capacitors C_{i1} and C_{i2} supplies energy to the inductors L_{o1} and L_{o2}, its current is equal to the current of these inductors in the first topological stage. In relation to the capacitor C_o, it supplies energy to the load during the all first topological stage, thus its current is equal to the output current.

For the second topological stage, the inductor currents decline linearly until their minimum value. During this period, the current in diodes D_1 and D_2 are the sum of the inductor currents replicated to the secondary side. The capacitors C_{i1} and C_{i2} receives energy from the inductors L_{i1} and L_{i2}, thus its currents is equal to the inductor currents. For capacitor C_o, the current is equal to the sum of the diodes D_1 and D_2 currents subtracted from the output current. Since the diodes currents are considered equal, the C_o current is equal to twice of the diode currents subtracted from the output current.

During the third topological stage, the inductor currents are equal. As the switches and diodes are blocked, their current are zero. The capacitors C_{i1} and C_{i2} remain receiving energy from the input inductors, the current is constant at the minimum value of the inductors during this step. Since the diodes are reverse biased, the capacitor C_o returns to supplying energy to the equivalent load and its current is the load current.

III. STATIC ANALYSIS

This section defines the main equations for the design of passive components and the static gain of the proposed converter.

A. Static Gain of Proposed Converter

The static gain is calculated using the ratio between output voltage (V_o) and the input voltage (V_{in}), as can be expressed in (1).

$$G_{DCM} = \frac{V_o}{V_{in}} = Dk_a \tag{1}$$

Where:

$$k_a = \sqrt{\frac{R_o\left(L_i + L_o\right)}{4L_iL_of_s}} \tag{2}$$

Fig 5 depicts the static gain of the converter as a function of some values of the parameter k_a and the duty cycle. It can be observed that for each value of the parameter k_a there is a maximum value of duty cycle, which limits the operation of the converter between DCM and CCM (continuous conduction mode).

B. Restriction of Operation

The proposed converter presents two restrictions of operation regarding the maximum equivalent load and maximum duty cycle, which allow the DCM operation.

The third operating stage of the converter is initialized when the current of the diodes D_1 and D_2 reaches zero. Based on this fact, it is possible to state that the operating discontinuity

seen by the primary side, during the first topological stage. This feature is an advantage of this topology when compared to the conventional SEPIC converter.

In the third operating stage, the voltage on the switches S_1 and S_2 is equal to the half of input voltage, while in the diodes D_1 and D_2 is equal to the output voltage.

The voltage stress on the capacitors C_{i1} and C_{i2} is equal to the half of input voltage, and the voltage stress on the capacitor C_o is equal to the output voltage (neglecting the high-frequency ripple). These voltage stress values on the capacitors C_{i1}, C_{i2} and C_o are calculated for steady-state, considering the average voltage values across the inductors L_{i1}, L_{i2}, L_{o1} and L_{o2} equal to zero.

On analyzing the currents waveforms in the components, it can be observed in the first topological stage that the inductor

978-1-7281-4181-7/19 $31.00 © 2019 IEEE

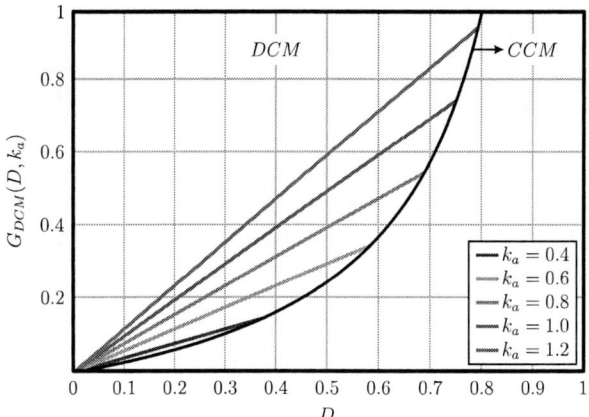

Fig. 5. Static gain curves.

limit of the converter occurs when the time interval Δ_{t3} tends to zero. Assuming that all energy stored in the first operating stage is transferred to the output during the second operating state, it yields to (3) and (4).

Equation (3) ensures that: using an equivalent load equal to or greater than the maximum value, the converter will present the operating discontinuity. Equation (4) ensures that: using a duty cycle value less than the maximum value, the converter will operate in DCM for the given equivalent load condition.

$$R_{o\,\min} = \frac{n^2 L_i L_o f_s}{(1-D)^2 (L_i + L_o)} \qquad (3)$$

$$D_{\max} = 1 - \sqrt{\frac{n^2 L_i L_o f_s}{R_o (L_i + L_o)}} \qquad (4)$$

C. Design Equations

The inductances of L_i and L_o are determined using (5) and (6), respectively. The inductance L_i is calculated from the criteria of maximum current ripple (peak to peak) and the inductance L_o is obtained from (1).

$$L_i = \frac{V_{in} D}{2\Delta_{iLi} f_s} \qquad (5)$$

$$L_o = \frac{V_{in}^2 D^2 R_o L_i}{4V_o^2 L_i f_s - V_{in}^2 D^2 R_o} \qquad (6)$$

The capacitors C_i and C_o are calculated from the criteria of maximum voltage ripple (peak to peak). From this, it yields to (7) and (8).

$$C_i = \frac{V_{in} D^2 [D (2V_o L_i + V_{in} n L_o) - 4V_o L_i]^2}{64V_o^2 L_i^2 f_s^2 L_o \Delta_{VCi}} \qquad (7)$$

$$C_o = \frac{V_{in}^2 D^2 (L_i + L_o)(V_{in} n D - 8V_o)^2}{256V_o^3 f_s^2 L_i L_o \Delta_{VCo}} \qquad (8)$$

IV. NUMERICAL SIMULATION RESULTS

The design specifications of the proposed converter are seen in Table I. Table II shows the values of passive components after application the specifications in design equations.

Considering the design specifications seen in Table I and the values of components depicted in Table II, an open-loop numerical simulation was made to verify the proposed design equations.

Fig. 6 presents the waveforms of output voltage and output current. It should be noted that the output voltage presents a value close to 123 V, whereas the average current is next to 4.3 A, considering a load power around of 529 W.

The voltage on the capacitors C_{i1} and C_{i2} can be seen in Fig. 7. They are close to the half of input voltage plus the voltage ripple, with an average value of approximately 200 V.

The voltages on the active switches S_1 and S_2 and the diodes D_1 and D_2 are illustrated in Fig. 8. It should be noted that the maximum voltage on the switches is the sum of the half of input voltage and the output voltage seen by the primary side, which their values is close to to 450 V. For the diodes, it is possible to note that the maximum voltage is equal to the half of reflected input voltage to the secondary side plus the output voltage, which is approximately -230 V.

The inductor currents L_{i1}, L_{i2}, L_{o1} and L_{o2} are shown in Fig. 9. As can be observed, the current of the inductors rise linearly during the first stage, reaching their maximum value, which are close to 1.47 A and 4.24 A, respectively. Moreover, it is also possible to observe in Fig. 9 three operating stages, which allows verifying the operation in DCM.

The currents waveforms in the active switches S_1 and S_2 and the diodes D_1 and D_2 are shown in Fig. 10. The maximum current in the switches is the sum of the inductor currents, thus its value is close to 5.71 A. In the diodes, the maximum

TABLE I
DESIGN SPECIFICATIONS

Specification	Value
Output Power (P_o)	500 W
Output voltage (V_o)	120 V
Input voltage (V_{in})	400 V
Switching frequency (f_s)	50 kHz
Duty cycle (D)	0.45
Turns ratio (n)	0.5
Ripple voltage in capacitors C_{i1} and C_{i2} (Δ_{VCi})	10%
Ripple voltage in capacitor C_o (Δ_{VCo})	1%
Ripple current in inductor L_i (Δ_{iLi})	20%

TABLE II
VALUES OF COMPONENTS

Parameter	Value
Equivalent load R_o	28.8 Ω
Inductor L_i	7.2 mH
Inductor L_o	339.267 μH
Capacitor C_i	366.718 nF
Capacitor C_o	45.844 μF
Maximum equivalent load R_{omin}	13.388 Ω
Maximum duty cycle D_{max}	0.625

978-1-7281-4181-7/19 $31.00 © 2019 IEEE

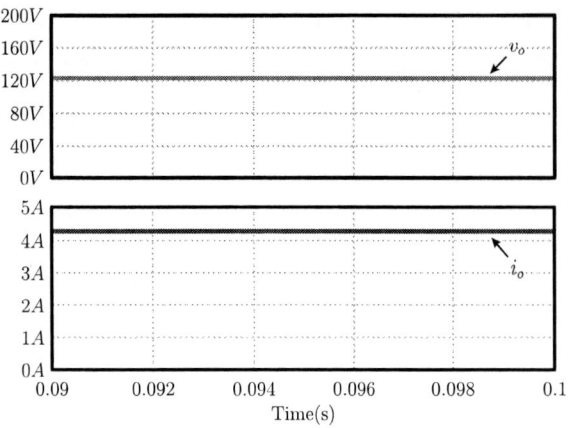

Fig. 6. Output voltage and output current waveforms.

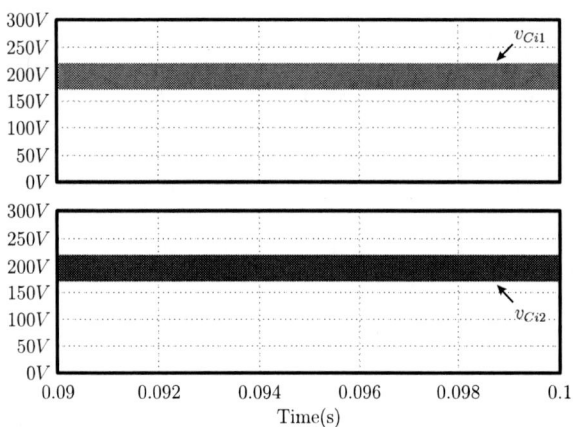

Fig. 7. Voltage waveforms across input capacitors.

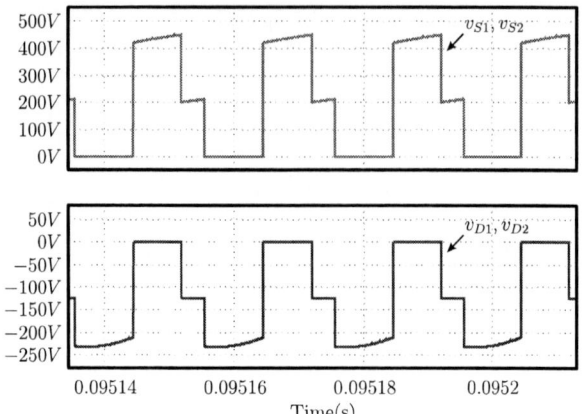

Fig. 8. Voltage waveforms across active switches and diodes.

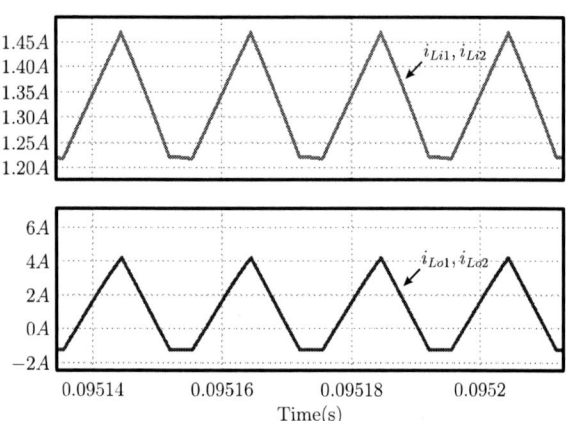

Fig. 9. Current waveforms in inductors L_{i1}, L_{i2}, L_{o1} and L_{o2}.

current is equal to the sum of the inductor current reflected to the secondary side, and its value is equal to approximately 11.4 A.

V. COMPARATIVE ANALYSIS

A comparative analysis of the theoretical efficiency curves between proposed converter and conventional SEPIC converter is presented in Fig. 11. It should be observed that the efficiency curve of the proposed converter is higher than the conventional SEPIC converter in all power range. The efficiency of the proposed converter is equal to 94.828% at rated power, while the conventional SEPIC converter reaches 93.604%.

In addition, the distribution of theoretical losses analysis at rated power between the elements that composes the conventional SEPIC converter and the proposed converter are presented in Fig. 12. It is noted that the largest amount of power losses in the converter herein proposed are concentrated in the switches, while in conventional SEPIC converter are placed in the diode.

VI. CONCLUSIONS

The paper proposes a dc-dc converter conceived from the SEPIC topology and it is analyzed in details on the paper. This topology presents input current with low ripple, requiring no starting auxiliary circuit to reduce current peaks. In addition, presents high frequency galvanic isolation through two coupled inductors.

The stead state analysis and the main design equations of the power components were presented for operation in DCM. Using numerical simulation, the topological stages and all equations were verified.

It should be noted that the proposed topology employs more components when compared to the conventional SEPIC converter, however the proposed converter presents reduced voltage stresses across the switches and diodes for a same output voltage level, being the main contribution of this topology.

The theoretical efficiency was analyzed and it can be observed that despite increasing the number of components, in all power range, the obtained efficiency is higher than a conventional SEPIC converter. Due to the reduction of

978-1-7281-4181-7/19 $31.00 © 2019 IEEE

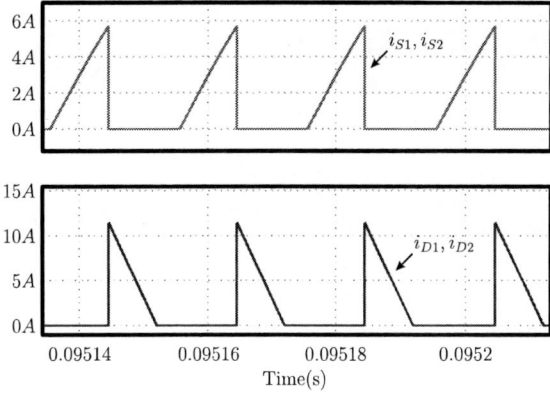

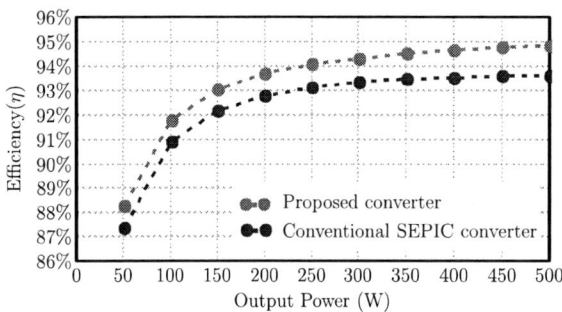

Fig. 10. Current waveforms in active switches and diodes.

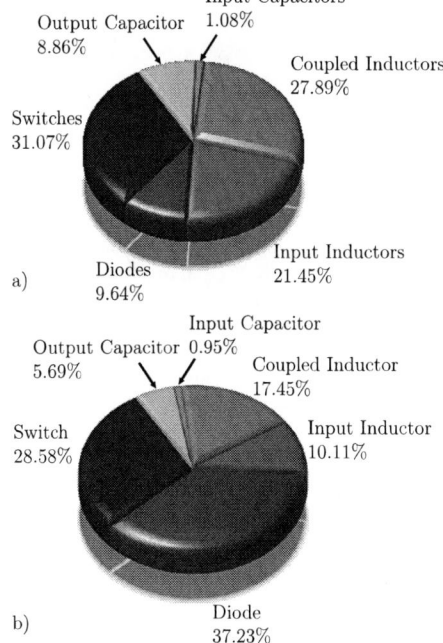

Fig. 11. Theoretical efficiency curves of the proposed converter and the conventional SEPIC converter.

voltage stresses on the semiconductors, it is possible to select devices with lower losses, which improves the efficiency of the proposed converter.

Finally, this dc-dc converter presents high robustness and high frequency galvanic isolation.

ACKNOWLEDGMENT

The authors would like to thank CAPES for their contribution to this work in the form of a grant provided to Marcos Vinícius Mosconi Ewerling.

REFERENCES

[1] R. Wai, R. Y. Duan, Uthra, R. Saravanakumar. "Analysis design and implementation of VSI fed high gain full bridge isolated DC-DC converter for renewable energy applications," IEEE Transactions on Power Electronics, vol. 20, no. 4, pp. 847-856, Jul. 2005.

[2] M. S. K. Reddy, D. Elangovan. "High-efficiency power conversion for low power fuel cell generation system," 2013 International Conference on Renewable Energy and Sustainable Energy (ICRESE), pp. 45-50, Oct. 2014.

[3] S.J. Chiang, H. J. Shieh, M. C. Chen. "Modeling and Control of PV Charger System With SEPIC Converter," IEEE Transactions on Industrial Electronics, vol. 56, no. 11, pp. 4344-4353, Sep. 2008.

[4] A. C. C. Hua, B. C. Tsai. "Design of a wide input range DC/DC converter based on SEPIC topology for fuel cell power conversion," 2010 International Power Electronics Conference - (ECCE ASIA), pp. 311-316, Jun. 2010.

[5] R. Gules, W. M. dos Santos, R. C. Annunziato, E. F. R. Romaneli, C. Q. Andrea. "A modified SEPIC converter with high static gain for renewable applications," XI Brazilian Power Electronics Conference (COBEP), pp. 162-167, Nov. 2011.

[6] H. -I Do. "Soft-Switching SEPIC Converter With Ripple-Free Input Current," IEEE Transactions on Power Electronics, vol. 27, no. 6, pp. 2879-2887, Nov. 2011.

[7] J. W. Yang, H. -L Do. "Bridgeless SEPIC Converter With a Ripple-Free Input Current," IEEE Transactions on Power Electronics, vol. 28, no. 7, pp. 3388-3394, Nov. 2012.

[8] S. W. Lee, H. -L Do. "Isolated SEPIC DC–DC Converter With Ripple-Free Input Current and Lossless Snubber," IEEE Transactions on Industrial Electronics, vol. 65, no. 2, pp. 1254-1262, Jul. 2017.

[9] A. Ghasemi, E. Adib, M. R. Mohammadi. "A new isolated SEPIC converter with coupled inductors for photovoltaic applications," 19th Iranian Conference on Electrical Engineering, Jul. 2011.

[10] M. V. M. Ewerling, C. H. Illa Font. "Single-stage AC/DC converter based on sepic topology operating in continuous conduction mode," 2017 Brazilian Power Electronics Conference (COBEP), Nov. 2017.

[11] M. Verma, S. S. Kumar. "Hardware Design of SEPIC Converter and its Analysis," 2018 International Conference on Current Trends towards Converging Technologies (ICCTCT), pp. 1-4, Nov. 2018.

[12] G. L. K. R. Tadi, P. Ramamurthyraju, V. R. Kumar. "Analysis of SEPIC for PV-Applications using PI Controller and Current Mode Control ," International Journal for Scientific Research e Development (IJSRD), vol. 1, no. 9, pp. 1821-1824, 2013.

[13] K. Mohanraj, S. Bharathnarayanan. "Three Level SEPIC For Hybrid Wind-Solar Energy Systems," 1st International Conference on Power Engineering, Computing and Control (PECCON), pp. 120-127, Mar. 2017.

[14] Z. Zhou, L. Li. "Isolated Sepic Three-Level DC-DC converter," 2011 6th IEEE Conference on Industrial Electronics and Applications (ICIEA), pp. 2162-2165, Aug. 2011.

Fig. 12. Theoretical distribution of losses: a) proposed converter; b) conventional SEPIC converter.

Comparison among Two-Phase Three-Wire AC Off-Grid Power Systems

Rafael M. Silva[1], Danilo I. Brandao[1], Gabriel A. Fogli[1], Victor F. Mendes[1], Clodualdo V. Sousa[2].

[1]*Graduate Program in Electrical Engineering (PPGEE)*
Federal University of Minas Gerais (UFMG), Belo Horizonte, MG, Brazil.
[2]*Institute of technological sciences (ICT)*
Federal University of Itajubá (Unifei), Itabira, MG, Brazil

rafaellmario@ufmg.br, dibrandao@ufmg.br, gabrielfogli@ufmg.br, victormendes@cpdee.ufmg.br,
clodualdosousa@unifei.edu.br

Abstract – **The update of the power system moves over to connection of small distributed energy resource through grid-interactive converters. The electrifying process of the distribution power system allows the implementation of different grid configurations, such as those based on two-phase three-wire systems that are appealing for residential zones, commercial areas and off-grid systems. The off-grid two-phase three-wire topology can be implemented considering voltages as an extension of the conventional three-phase system (*abn*), voltages in quadrature (*αβn*), or voltages with opposite phases (*xyn*). Then, this paper aims at comparing the three off-grid power system topologies in terms of power quality characteristics, sizing of DC capacitor of grid-connected inverters, sizing of neutral conductor, and integration with the conventional mains. To accomplish those goals, a mathematical approach considering the Conservative Power Theory and computational simulations are shown and discussed.**

Keywords – **off-grid power systems, two-phase three-wire systems, utility interface converter, voltages in quadrature.**

I. INTRODUCTION

The development of new power electronics technologies has been providing the use of small distributed resources for the electrical energy generation. This scenario allows the interconnection and interoperability of several kind of distributed generators (DGs), loads and storage systems into one single entity capable of isolated operation.

Due to the different characteristics of the loads and DGs, several off-grid power system topologies can be proposed, the present work focuses on AC ones, which can be classified according to the number of live conductors and voltage levels (line-to-neutral or line-to-line), as follows:

- **Single or two-phases two-wire system**: it presents one live conductor, with neutral current and one voltage level (line-to-neutral) [1], [2];

- **Two-phase three-wire system**: it presents two live conductors, with neutral current and two different voltage levels (line-to-neutral and line-to-line) [3]–[5];

- **Three-phase three-wire system**: that has three live conductors, no neutral current and only one voltage level (line-to-line) [6];

- **Three-phase four-wire system**: that shows three live conductors, with neutral current and two different voltage levels [7].

The best ratio among number of available voltage levels and conductors is the two-phase three-wire (*2ϕ3w*) system. Such feature classifies this configuration suitable for residential/commercial application and implementation of off-grid power systems in remote areas. Usually, the AC *2ϕ3w*

system are implemented using *abn* coordinates, i.e., the voltages v_a and v_b are displaced 120º from each other, considering the possibility of further connection with the mains. However, in remote applications, like rural areas or riverside communities, a grid-forming converter [8] must be used to supply the voltage and frequency references. Then, several island power system topologies, regarding the voltage magnitudes, phases and frequency can be approached.

The authors of [3]–[5] present a new concept of power system, in which the *2ϕ3w* topology has phase voltages in quadrature (*αβn*), thus, under sinusoidal and balanced conditions there is no instantaneous active power oscillation. This feature can also be founded in three-phase systems, but, considering lower number of available voltage levels, as in *3φ3w* systems or higher number of live conductors, as in *3φ3w*.

Another power system configuration is obtained by the symmetrical components concept [9], in which according to Fortescue's a set of *n*-phasors unbalanced is branched into *n* sets of symmetrical and balanced system, which are shifted from $k2\pi/n$ to each other, where *k* ranges from 0 to (*n-1*). Therefore, for a system with two-phases (i.e. *n = 2*), the phase voltage displacement is 180º, and can be decomposed into zero and positive sequence components. It means that the negative sequence does not exist in this configuration, that is called herein as *xyn* system.

According to [10]–[13], the instantaneous active power oscillation causes energy fluctuation in the DC link of grid-tied converters, it in turns boosts the losses in the DC link capacitors, reducing the converter life time. Thus, heavy instantaneous active power oscillations requires a higher capacitance value, which in turn means more volume and higher converter overall cost[11]. Therefore, the minimization of the instantaneous active power oscillation is a required feature from the converter design point of view. The *αβn* show this desired feature, however, it also presents some disadvantages that need to be considered, such as higher neutral currents magnitudes.

Thus, this paper aims at comparing the three AC off-grid power topologies (i.e. *abn*, *αβn* and *xyn*) in terms of power quality characteristics, integration with the mains and sizing of DC-link capacitors and neutral conductor. To fairly compare those topologies, the mathematical approach used herein is based on the Conservative Power Theory (CPT) [14]–[17], and the figures of merit are based on the instantaneous power values, voltage levels and neutral current magnitude.

This work is outlined as follows: Section II reviews the CPT, since it is used as the mathematical tool herein. Section III presents the mathematical development that describes the

978-1-7281-4181-7/19 $31.00 © 2019 IEEE

two-phase three-wire systems. Section IV shows the simulation results. Finally, Section V concludes the paper.

II. CONSERVATIVE POWER THEORY: CONCEPTS REVIEW

The discussions around the power quality characteristics of the $2\phi 3w$ power systems approached in this paper are based on the CPT, which the power and current terms are introduced in this section. Since the CPT uses the collective root mean square (rms) values of voltage and current, it allows a fair comparison among the three analyzed AC power systems.

A. Active power/current components

For a polyphase grid with M-phases, the instantaneous active power is calculated by (1) [14], [16].

$$p = \sum_{m=1}^{M} v_m \cdot i_m, \qquad (1)$$

where p is the instantaneous active power, v_m and i_m are the instantaneous voltage and current from the m-phase. Then, the average active power value is given by:

$$P = \frac{1}{T} \int_0^T p(\tau) \cdot d\tau, \qquad (2)$$

where T is the fundamental grid voltage period. The balanced active current of the m-phase is the minimum current needed to convey the active power P to the M-phases, given by:

$$i_{a\,m}^b = \frac{P}{||v||^2} \cdot v_m = G^b \cdot v_m, \qquad (3)$$

$|| v ||$ is the collective root mean square (rms) voltage and G^b is the equivalent balanced conductance [14].

B. Reactive power/current components

In a similar way to (1), the instantaneous reactive power is the sum of products from the instantaneous voltage unbiased time integral of m-phase, $\hat{v}_m$, and instantaneous current i_m, of the M-phases [14], [16]:

$$q = \omega \cdot \sum_{m=1}^{M} \hat{v}_m \cdot i_m, \qquad (4)$$

where ω is the fundamental frequency of the grid voltage and the unbiased time integral of m-phase voltage, $\hat{v}_m$, is defined by [17]:

$$\hat{v}_m = v_{m\int} - \bar{v}_m = \cdots$$
$$\int_0^t v_m(\varsigma) \cdot d\varsigma - \frac{1}{T} \int_0^T v_{m\int}(\tau) \cdot d\tau. \qquad (5)$$

The reactive power is the average value of (4), as follows:

$$Q = \frac{1}{T} \int_0^T q(\tau) \cdot d\tau. \qquad (6)$$

The balanced reactive current of the m-phase is the minimum value of rms currents needed to convey the reactive power Q to the M-phases, and it is defined by [14]:

$$i_{r\,m}^b = \frac{Q}{\omega \cdot ||\hat{v}_m||^2} \cdot \hat{v}_m = B^b \cdot \hat{v}_m, \qquad (7)$$

where B^b is the equivalent balanced reactivity, and $|| \hat{v}_m ||$ is collective rms value of the voltage unbiased time integral.

C. Unbalanced and void components

The unbalanced active current from the m-phase is defined according to (8) [14], [15]:

$$i_{a\,m}^u = \left(\frac{P_m}{||v_m||} - \frac{P}{||v||} \right) \cdot v_m = (G_m - G^b) \cdot v_m, \qquad (8)$$

where P_m is the active power and G_m is the equivalent conductance of the m-phase, respectively. Likely, the reactive unbalanced current is defined by (9) [14], [15]:

$$i_{r\,m}^u = \left(\frac{Q_m}{||\hat{v}_m||} - \frac{Q}{||\hat{v}||} \right) \cdot \frac{\hat{v}_m}{\omega} = (B_m - B^b) \cdot \hat{v}_m, \qquad (9)$$

such that Q_m is the reactive power, and B_m is the reactivity of the m-phase, respectively. Therefore, the unbalanced current term is the sum of the unbalanced active and reactive current components, i.e.:

$$i_m^u = i_{a\,m}^u + i_{r\,m}^u. \qquad (10)$$

The remaining term is the void current and it does not convey neither active nor reactive power. It reflects the presence of harmonic currents generated by the load[15]. Hence the void current is defined by:

$$i_{v\,m} = i_m - i_{a\,m}^b - i_{r\,m}^b - i_m^u. \qquad (11)$$

Using the collective rms unbalanced current $|| i^u ||$ and voltage $|| v ||$, it is possible to compute the unbalanced power term, as follows [15]:

$$N = ||v|| \cdot ||i^u|| = \sqrt{\sum_{m=1}^{M} (v_m)^2} \cdot \sqrt{\sum_{m=1}^{M} (i_m^u)^2}. \qquad (12)$$

The void power term is defined in (13) [16]:

$$D = ||v|| \cdot ||i_v|| = \sqrt{\sum_{m=1}^{M} (v_m)^2} \cdot \sqrt{\sum_{m=1}^{M} (i_{v\,m})^2}. \qquad (13)$$

Since the current components are orthogonal to each other, as well as the power terms, the apparent power is calculated according to:

$$A = ||v|| \cdot ||i|| = \sqrt{P^2 + Q^2 + N^2 + D^2}. \qquad (14)$$

III. TWO-PHASE THREE-WIRE OFF-GRID POWER SYSTEMS

Fig. 1 shows a generic configuration of a $2\phi 3w$ grid with different loads connected to. For the analysis carried out in this paper, the utility grid interface converter is assumed as an ideal source, since the converter topology or control aspects are not the focus herein.

A. abn power system

In the abn topology, the line-to-neutral voltages are described by the following set of equations:

$$v_a = V_a \cos(\omega t + \phi_a),$$
$$v_b = V_b \cos(\omega t - 2\pi/3 + \phi_b). \qquad (15)$$

where V_a, V_b are magnitudes and φ_a, φ_b are phases from the utility voltages v_a and v_b, respectively. Hence, under balanced voltage conditions, i.e., $V_a = V_b = V$ and $\varphi_a = \varphi_b = \varphi$, the line-to-line voltage is defined by (16).

$$v_{ab} = v_a - v_b = \sqrt{3} \cdot V \cdot \cos(\omega t + \phi - \pi/6). \qquad (16)$$

If the utility feeds generic linear loads, the line currents can be described as follows:

978-1-7281-4181-7/19 $31.00 © 2019 IEEE

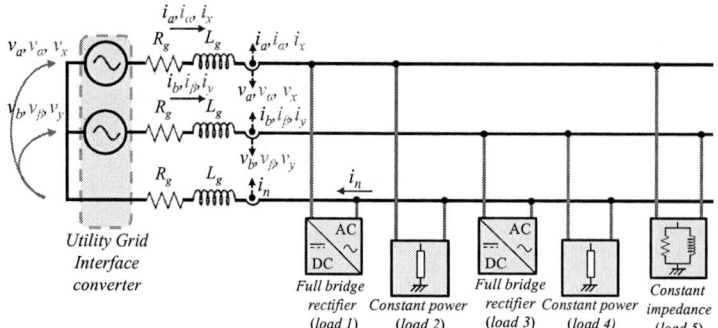

Fig. 1: Two-phase three-wire off-grid power system.

$$i_a = I_a \cos(\omega t + \theta_a),$$
$$i_b = I_b \cos(\omega t - 2\pi/3 + \theta_b), \qquad (17)$$

where I_a, I_b are the magnitudes and θ_a, θ_b are the phases of the *abn* line currents i_a and i_b, respectively. Then, operating with balanced voltages and loads, i.e., $I_a = I_b = I$ and $\theta_a = \theta_b = \theta$, the neutral current is defined by:

$$i_n = i_a + i_b = I \cdot \cos(\omega t + \theta + \pi/3). \qquad (18)$$

Replacing (17) and (15) in (1), the *abn* instantaneous active power is given by:

$$p = \frac{V_a I_a}{2}\{\cos(2\omega t + \phi_a + \theta_a) + \cos(\phi_a - \theta_a)\} \dots$$
$$+ \frac{V_b I_b}{2}\{\cos(2\omega t + \phi_b + \theta_b) + \cos(\phi_b - \theta_b)\}. \qquad (19)$$

The (19) are composed of a continuous component and oscillating one with twice the fundamental frequency and even for balanced load and voltages, i.e. $V_a = V_b = V$, $\varphi_a = \varphi_b = \varphi$, $I_a = I_b = I$ and $\theta_a = \theta_b = \theta$, the instantaneous active power oscillations are not cancelled, i.e.:

$$p = V I \cos(\phi - \theta) \dots$$
$$+ \frac{V I}{2}\cos(2\omega t + \phi + \theta - \pi/3). \qquad (20)$$

B. αβn power system

For the *αβn* topology, the line-to-neutral voltages are expressed as follows:

$$v_\alpha = V_\alpha \cos(\omega t + \phi_\alpha),$$
$$v_\beta = V_\beta \sin(\omega t + \phi_\beta), \qquad (21)$$

where V_α, V_β are magnitudes and φ_α, φ_β the phases from the voltages v_α and v_β, respectively. Thus, under balanced voltages operation, i.e., $V_\alpha = V_\beta = V$ and $\varphi_\alpha = \varphi_\beta = \varphi$, the line-to-line voltage is defined by the following equation:

$$v_{\alpha\beta} = v_\alpha - v_\beta = \sqrt{2} \cdot V \cdot \cos(\omega t + \phi + \pi/4). \qquad (22)$$

Taking into account only linear loads, the *αβn* line currents are defined as:

$$i_\alpha = I_\alpha \cos(\omega t + \theta_\alpha),$$
$$i_\beta = I_\beta \sin(\omega t + \theta_\beta), \qquad (23)$$

where I_α, I_β are the magnitudes and θ_α, θ_β are the phases of the *αβn* line currents. Then, operating with balanced voltages and loads, i.e., $I_\alpha = I_\beta = I$ and $\theta_\alpha = \theta_\beta = \theta$, the neutral current is defined by:

$$i_n = i_\alpha + i_\beta = \sqrt{2} \cdot I \cdot \cos(\omega t + \theta - \pi/4). \qquad (24)$$

Replacing (23) and (21) in (1) the *αβn* power system instantaneous active power is:

$$p = \frac{V_\alpha I_\alpha}{2}\{\cos(\omega t + \phi_\alpha + \theta_\alpha) + \cos(\phi_\alpha - \theta_\alpha)\} \dots$$
$$- \frac{V_\beta I_\beta}{2}\{\cos(\omega t + \phi_\beta + \theta_\beta) + \cos(\phi_\beta - \theta_\beta)\}. \qquad (25)$$

In a similar way to (19), (25) is composed of continuous and oscillating terms. Nevertheless, under balanced loads and voltage conditions, i.e. $V_\alpha = V_\beta = V$, $\varphi_\alpha = \varphi_\beta = \varphi$, $I_\alpha = I_\beta = I$ and $\theta_\alpha = \theta_\beta = \theta$, the instantaneous active power becomes:

$$p = V \cdot I \cdot \cos(\phi - \theta). \qquad (26)$$

C. xyn power system

For the *xyn* topology, the line-to-neutral voltages and line currents are generically described by the following equations:

$$v_x = V_x \cos(\omega t + \phi_x),$$
$$v_y = V_y \cos(\omega t + \pi + \phi_y), \qquad (27)$$

$$i_x = I_x \cos(\omega t + \theta_x),$$
$$i_y = I_y \cos(\omega t + \pi + \theta_y), \qquad (28)$$

where V_x, V_y, I_x and I_y are the magnitudes and φ_x, φ_y, θ_x and θ_y are the phase angles of *xyn* voltages and currents, respectively. Under balanced voltages and loads, i.e., $V_x = V_y = V$, $\varphi_x = \varphi_y = \varphi$, $I_x = I_y = I$ and $\theta_x = \theta_y = \theta$, the line-to-line voltage and neutral current are defined by:

$$v_{xy} = 2 \cdot V \cos(\omega t + \phi), \qquad (29)$$
$$i_n = 0. \qquad (30)$$

The instantaneous active power in the *xyn* power system is obtained replacing (27) and (28) in (1), as follows:

$$p = \frac{V_x I_x}{2}\{\cos(2\omega t + \phi_x + \theta_x) + \cos(\phi_x - \theta_x)\} \dots$$
$$+ \frac{V_y I_y}{2}\{\cos(2\omega t + \phi_y + \theta_y) + \cos(\phi_y - \theta_y)\}. \qquad (31)$$

Therefore, under balanced conditions, the instantaneous power is defined as bellow:

$$p = V \cdot I \cdot \{\cos(2\omega t + \phi + \theta) + \cos(\phi - \theta)\}. \qquad (32)$$

D. Instantaneous power oscillation analysis

Using only the oscillating terms of (19), it is possible to obtain the planes in Fig. 2(a) that show the instantaneous active power oscillation magnitude in the *abn* system, $\|\tilde{p}_{abn}\|$, with the *a*-phase parameters fixed in $V_a = 1$ pu, $\varphi_a = \theta_a = 0$ rad,

978-1-7281-4181-7/19 $31.00 © 2019 IEEE

while changing the voltage, current and power factor values in *b*-phase. Similarly, by means of the oscillating terms of (25), it is obtained the planes in Fig. 2(b) that show the magnitude of the instantaneous active power oscillation in the *αβn* system, $\|\tilde{p}_{\alpha\beta n}\|$, where it is kept the *α*-phase parameters fixed in $V\alpha = 1$ pu, $\varphi_\alpha = \theta_\alpha = 0$ rad, while changing the voltage, current and power factor values in the *β*-phase. Finally, the instantaneous active power oscillation in the *xyn* system $\|\tilde{p}_{xyn}\|$ is shown in Fig. 2(c) using the oscillating components of (31), where it is kept the *x*-phase parameters fixed in $Vx = 1$ pu, $\varphi_x = \theta_x = 0$ rad, while changing the voltage, current and power factor values in the *y*-phase.

As earlier discussed, and proven through Fig. 2(a) and (19), the instantaneous active power oscillation cannot be cancelled in the *abn* system, where the difference between the voltage and current magnitudes and load power factor from different phases increase the value of $\|\tilde{p}_{\alpha bn}\|$. The minimum magnitude of power oscillation is 0.433 pu that occurring with $V_b/V_a = 0.7$, $I_b/I_a = 0.7$ and $\varphi_b = 0$ rad.

In the *αβn* power system, as shown in Fig. 2(b), the difference between the voltage and current magnitudes and load power factor in different phases, can also increase the value of $\|\tilde{p}_{\alpha\beta n}\|$. However, for this topology, it is possible to obtain values of power oscillation lower than 0.433 pu as the system becomes balanced, $\|\tilde{p}_{\alpha\beta n}\|$ is zero when voltages and currents are balanced ($V\alpha/V\beta = I\alpha/I\beta = 1$ and $\varphi_\beta = 1$).

For the *xyn* system, the instantaneous power oscillation magnitude is higher than the others analyzed topologies, as shown in Fig. 2(c). The maximum value reaches 2 pu, and the minimum is 1 pu. It characterizes a significant disadvantage of this configuration over the others analyzed ones.

IV. CASE STUDY

To evaluates the topologies, computational simulations of the off-grid power systems shown in Fig. 1 are performed, using MATLAB/Simulink software. The line-to-neutral rms voltage is 127 V, frequency $f = 60$ Hz and the loads characteristics are described in Table I. The systems are evaluated under five different load conditions, described as follows:

TABLE I: TWO-PHASE LOADS.

	Connection	Characteristic	P (W)	Q (Var)
Load 1	line-to-neutral	Non-sinusoidal	1000	-
Load 2	line-to-neutral	Sinusoidal	746	442.65
Load 3	line-to-neutral	Non-sinusoidal	1000	-
Load 4	line-to-neutral	Sinusoidal	1492	442.65
			R (Ω)	L (mH)
Load 5	line-to-line	Sinusoidal	7.029	38.5
Rg, Lg	-	Sinusoidal	0.001	0.00265

- *A* - $0.0 \leq t < 0.05$: loads 2 and 4 are connected to the system, in which load 2 operates with 50% of its rated active power, representing a balanced system;

- *B* - $0.05 \leq t < 0.1$: the active power of load 4 is increased to its rated value and the systems operates with unbalanced loads;

- *C* - $0.1 \leq t < 0.15$: loads 2 and 4 are removed from the system, and loads 1 and 3 are connected to. At this

condition, the system operates with balanced and non-linear loads;

- *D* - $0.15 \leq t \leq 0.2$: loads 1 and 3 are removed, and load 5 is line-to-line connected to.

- *E* - $0.2 \leq t \leq 0.25$: all the loads are connected simultaneously.

A. Simulation results

Fig. 3 to Fig. 5 show the line and neutral currents profile of the *abn*, *αβn* and *xyn* power system, where the current collective rms values are shown in Table II. At the time intervals *(A)*, *(B)* and *(C)* the neutral current has higher values than the line one for the *αβn* system, where as in the *abn* system, it is observed only in *(C)* condition, because of the non-linear load. In the *xyn* the neutral current is always lower than the line ones for the analyzed cases. Since the load 5 has a linear voltage-to-current relation, the system that presents the highest line-to-line voltage also has the higher line current value, as can be seen at the time intervals *(D)* and *(E)*.

The CPT power components are shown in Table III for the five load conditions. For the cases in which the loads are subjected to the same line-to-neutral voltage magnitudes (cases *(A)*, *(B)* and *(C)*), the CPT power components are very similar, despite the difference among the instantaneous power oscillations shown in Fig. 6. At the time intervals *(D)* and *(E)*, the analyzed power systems present different CPT components due to two factors: *1)* the difference in the line-to-line voltage magnitudes, which reflects in the active and reactive power values; *2)* the *xyn* topology corresponds to the two-phase system of Fortescue's Theorem, then under balanced condition, there is not zero sequence components. In the other circuit topologies, even under balanced voltage and current condition, there are non-positive sequence component, because they are not based on the two-phase system of Fortescue's Theorem. Finally, the distortion power, *D*, are quite similar to all the AC power systems

Fig. 6 compares the instantaneous active power in the *abn*, *αβn* and *xyn*. For balanced conditions (*interval (A)*) the instantaneous active power in the *αβn* system is equivalent to the average value in *abn* and *xyn*. Moreover, as previously discussed in Fig. 2, when the systems are subjected to the same load unbalance level (*interval (B)*) for linear loads, the *αβn* system presents the lowest value of instantaneous power oscillation, while the *xyn* system shows the highest value among all the analyzed cases.

B. Two-phase Three-wire AC microgrid comparison

Table IV presents a comparison of the AC *2ϕ3w* off-grid power systems based on the mathematical developments and simulation results, allowing some discussions about the analyzed power systems.. The *xyn* system has shown the highest instantaneous active power oscillations values among the analyzed off-grid systems, thus, the grid-tied converter requires a greater storage energy capacity and consequently a larger dc-link capacitor value in relation to the conventional *abn* power system. Moreover, this system does not present negative sequence, since it is not possible connect line-to-line unbalanced loads, thus, the loads unbalanced is exclusively related to zero sequence components and, therefore, reflecting in the neutral current.

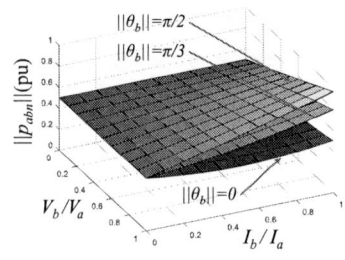 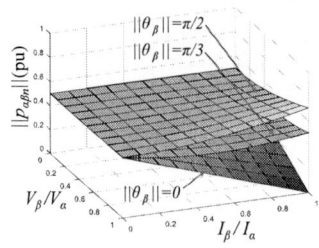 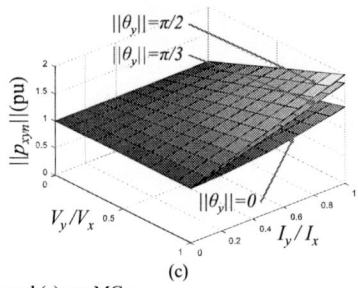

Fig. 2: Magnitude of instantaneous active power oscillation in (a) *abn*, (b) *αβn* and (c) *xyn* MGs.

TABLE II: LINE AND NEUTRAL COLLECTIVE RMS CURRENTS.

	A	B	C	D	E
Ia [A]	6.83	6.83	12.75	13.64	28.69
Ib [A]	6.83	12.25	12.75	13.64	31.59
In(abn) [A]	6.83	12.29	18.03	0.00	24.08
Iα [A]	6.83	6.83	12.75	11.13	23.45
Iβ [A]	6.83	12.25	12.75	11.13	31.31
In(αβn) [A]	9.66	12.48	18.03	0.00	27.07
Ix [A]	6.83	6.83	12.71	15.75	30.53
Iy [A]	6.83	12.25	12.71	15.75	35.87
In(xyn) [A]	0.00	5.87	0.00	0.00	5.87
In(αβn)/In(abn)	1.41	1.02	1.00	-	1.12
In(xyn)/In(abn)	0.00	0.48	0.00	-	0.24

TABLE III: CPT POWER COMPONENTS.

		A	B	C	D	E
abn	P [W]	1492.0	2238.0	1838.0	2700.0	6770.0
	Q [VA]	884.8	884.8	-336.2	1307.0	1855.0
	N [VA]	0.0	745.7	0.0	1732.0	1559.0
	D [VA]	0.0	0.0	2646.0	0.0	2646.0
αβn	P [VA]	1492.0	2238.0	1839.0	1800.0	5873.0
	Q [VA]	884.8	884.8	-337.9	871.2	1418.0
	N [VA]	0.0	745.7	0.0	2000.0	2418.0
	D [VA]	0.0	0.0	2644.0	0.0	2644.0
xyn	P [VA]	1492.0	2238.0	1839.0	3599.0	7671.0
	Q [VA]	884.9	884.9	-343.6	1742.0	2283.0
	N [VA]	0.0	745.8	0.0	0.0	745.9
	D [VA]	0.0	0.0	2632.0	0.0	2632.0

TABLE IV: FIGURES OF MERITS TO OFF-GRID POWER SYSTEM.

	abn	αβn	xyn
Power oscillation	Middle	Lower	Higher
Neutral Current	Middle	Higher	Lower
Line-to-lineVoltage	Middle	Lower	Higher

In opposite way in relation to *xyn* off-grid, the *αβn* system has shown the lowest power oscillation values, then, the instantaneous energy storage capacity not need to be as high as for the *abn* and a lower DC link capacitance value is required. This has higher neutral current values than the others analyzed AC power systems.

Finally, the *abn* configuration shown intermediate characteristics. It is worth noting that the *αβn* and *xyn* cannot

be connected to the mains without a grid interface converter, however, this advantage becomes less burdensome when it is taken into account the ability of island operating, and the greater range of possibilities, regarding modern power system scenarios considering so-called microgrids.

V. CONCLUSIONS

The aim of this paper was to compare the two-phase three-wire microgrids in *abn*, *αβn* and *xyn* coordinates, take into account the line-to-line voltage levels, neutral current magnitude, instantaneous active power oscillation and the conservative power theory components.

The mathematical developments and simulation results show that, for the analyzed systems feeding sinusoidal loads, the instantaneous power oscillations are inversely related to the neutral current magnitude, as can be observed in the *αβn* and *xyn* power systems. Then, the power systems with higher instantaneous power oscillations require grid-tied converters with higher capacitance values in the DC-link. However, in grid topologies with lower instantaneous power oscillations, the neutral current magnitudes should be considered in the grid-tied converter static switches design. Thus, despite the disadvantages, the *αβn* configurations are a very interesting solution for the off-grid power systems implementation based on power electronic devices (e.g., hybrid PV plus battery systems). Therefore, the *αβn* and *xyn* system show opposite characteristics that should be considered in the implementation of the off-grid systems. The off-grid power system, not based on power electronic devices, such as those based on diesel genset generator may be devised on conventional *abc* system, which is an intermediary solution.

ACKNOWLEDGMENT

The authors are grateful to CAPES (Finance Code 001), CNPQ (grant 420850/2016-3) and FAPEMIG (grant APQ-02518-16).

REFERENCES

[1] G. Kulia, M. Molinas, L. M. Lundheim, and O. B. Fosso, "Simple model for understanding harmonics propagation in single-phase microgrids," in *2017 6th International Conference on Clean Electrical Power (ICCEP)*, Santa Margherita Ligure, Italy, 2017, pp. 354–358.

[2] Q. Sun, J. Zhou, J. M. Guerrero, and H. Zhang, "Hybrid Three-Phase/Single-Phase Microgrid Architecture With Power Management Capabilities," *IEEE Trans. Power Electron.*, vol. 30, no. 10, pp. 5964–5977, Oct. 2015

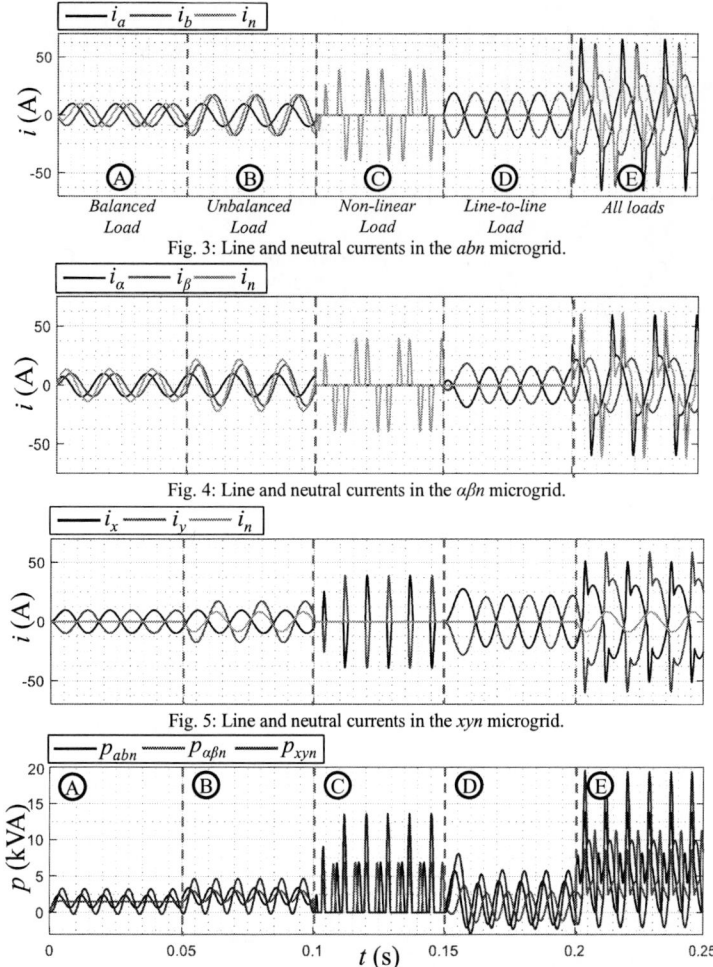

Fig. 3: Line and neutral currents in the *abn* microgrid.

Fig. 4: Line and neutral currents in the *αβn* microgrid.

Fig. 5: Line and neutral currents in the *xyn* microgrid.

Fig. 6: Instantaneous active power from off-grid topologies: *abn* (black curves), *αβn* (red curves) and *xyn* (*blue lines*)

[3] M. Alibeik, E. C. dos Santos, Y. Yang, X. Wang, and F. Blaabjerg, "Harmonic analysis and practical implementation of a two-phase microgrid system," in *2015 IEEE Applied Power Electronics Conference and Exposition (APEC)*, Charlotte, NC, USA, 2015, pp. 1830–1837.

[4] E. C. dos Santos and M. Alibeik, "Microgrid system with voltages in quadrature," in *2013 IEEE Energy Conversion Congress and Exposition*, Denver, CO, USA, 2013, pp. 1344–1349.

[5] M. Alibeik, E. C. dos Santos, and F. Blaabjerg, "Symmetrical components and power analysis for a two-phase microgrid system," in *2014 Power and Energy Conference at Illinois (PECI)*, Champaign, IL, USA, 2014, pp. 1–8.

[6] Y. W. Li, D. M. Vilathgamuwa, and P. C. Loh, "A grid-interfacing power quality compensator for three-phase three-wire microgrid applications," in *2004 IEEE 35th Annual Power Electronics Specialists Conference (IEEE Cat. No.04CH37551)*, 2004, vol. 3, pp. 2011-2017 Vol.3.

[7] D. I. Brandao, T. Caldognetto, F. P. Marafao, M. G. Simoes, J. A. Pomilio, and P. Tenti, "Centralized Control of Distributed Single-Phase Inverters Arbitrarily Connected to Three-Phase Four-Wire Microgrids," *IEEE Trans. Smart Grid*, vol. 8, no. 1, pp. 437–446, Jan. 2017.

[8] A. A. P. Machado, D. I. Brandao, I. A. Pires, and B. de J. C. Filho, "Fault-tolerant Utility Interface power converter for low-voltage microgrids," in *2017 IEEE 8th International Symposium on Power Electronics for Distributed Generation Systems (PEDG)*, Florianopolis, Brazil, 2017, pp. 1–5.

[9] G. Chicco and A. Mazza, "100 Years of Symmetrical Components," *Energies*, vol. 12, no. 3, p. 450, Jan. 2019.

[10] H. Wang and F. Blaabjerg, "Reliability of Capacitors for DC-Link Applications in Power Electronic Converters—An Overview," *IEEE Trans. Ind. Appl.*, vol. 50, no. 5, pp. 3569–3578, Sep. 2014.

[11] J. M. Lenz, J. R. Pinheiro, and H. C. Sartori, "DC-link electrolyte capacitor lifetime analysis for a PV boost converter," in *2017 IEEE 8th International Symposium on Power Electronics for Distributed Generation Systems (PEDG)*, Florianopolis, Brazil, 2017, pp. 1–6.

[12] Y. Yang, K. Ma, H. Wang, and F. Blaabjerg, "Instantaneous thermal modeling of the DC-link capacitor in PhotoVoltaic systems," in 2015 IEEE Applied Power Electronics Conference and Exposition (APEC), 2015, pp. 2733–2739.

[13] A. F. Cupertino, L. S. Xavier, E. M. S. Brito, V. F. Mendes, and H. A. Pereira, "Benchmarking of power control strategies for photovoltaic systems under unbalanced conditions," *Int. J. Electr. Power Energy Syst.*, vol. 106, pp. 335–345, Mar. 2019.

[14] P. Tenti, H. K. M. Paredes, and P. Mattavelli, "Conservative Power Theory, a Framework to Approach Control and Accountability Issues in Smart Microgrids," *IEEE Trans. Power Electron.*, vol. 26, no. 3, pp. 664–673, Mar. 2011.

[15] P. Tenti, H. K. M. Paredes, F. P. Marafao, and P. Mattavelli, "Accountability in Smart Microgrids Based on Conservative Power Theory," *IEEE Trans. Instrum. Meas.*, vol. 60, no. 9, pp. 3058–3069, Sep. 2011.

[16] H. K. M. Paredes, F. P. Marafão, D. I. Brandão, and A. Costabeber, "Multi-task control strategy for grid-tied inverters based on conservative power theory," *IET Renew. Power Gener.*, vol. 9, no. 2, pp. 154–165, Mar. 2015.

[17] T. D. C. Busarello, A. Mortezaei, A. Peres, and M. G. Simoes, "Application of the Conservative Power Theory Current Decomposition in a Load Power-Sharing Strategy Among Distributed Energy Resources," *IEEE Trans. Ind. Appl.*, vol. 54, no. 4, pp. 3771–3781, Jul. 2018.

978-1-7281-4181-7/19 $31.00 © 2019 IEEE

Efficiency Analysis for Interleaved Buck Converters Employed as Extra-High Current COB LED Drivers

Dênis de Castro Pereira
Federal University of Juiz de Fora
Juiz de Fora, Brazil
denis.castro@engenharia.ufjf.br

Everton Bernard Figueiredo Rabelo
Federal University of Juiz de Fora
Juiz de Fora, Brazil
everton.rabelo@engenharia.ufjf.br

Pedro Santos Almeida
Federal University of Juiz de Fora
Juiz de Fora, Brazil
pedro.almeida@engenharia.ufjf.br

Guilherme Marcio Soares
Federal University of Juiz de Fora
Juiz de Fora, Brazil
guilherme.marcio@engenharia.ufjf.br

Fernando Lessa Tofoli
Federal University of São João del-Rei
São João del-Rei, Brazil
fernandolessa@ufsj.edu.br

Henrique Antônio Carvalho Braga
Federal University of Juiz de Fora
Juiz de Fora, Brazil
henrique.braga@ufjf.edu.br

Abstract – **This work presents a performance evaluation concerning interleaved buck structures used to drive extra-high current (up to 12 A) chip-on-board (COB) light-emitting diodes (LEDs). The analysis is carried out employing nonideal elements in simulations so that a better and reliable approach can be obtained as close to real results as possible. Aiming at a thorough analysis of conduction losses at very high current levels in floodlighting applications, interleaved buck converters (IBCs) are employed to establish a comparative approach concerning global efficiency and component count for conventional and synchronous topologies. Furthermore, the 3-cell IBC is experimentally implemented to evaluate and compare the total losses for this very-high current lighting application.**

Keywords – **Extra-high current COB LEDs, Interleaved buck converter, Loss estimation, Synchronous converter.**

I. INTRODUCTION

According to [1], the electrical energy demand applied in artificial lighting has decreased from 30% to 15% in ten years. This aspect can be easily explained by the recent improvement of solid-state lighting (SSL) sources and their increased lumens per watt consumption, which results in great lighting efficiency. Thus, the light-emitting diode (LED) technology is considered a great solution for many lighting issues addressed in both outdoor and indoor environments due to photo-electrical characteristics like efficiency, different power spectral distributions, high luminous efficacy, long lifespan, high reliability and environmental friendliness [2]-[3].

In general, most practical applications are based on the discrete association of LEDs associated with high-voltage buses as the main alternative to achieve high power levels [3]. More recently, the very-high-current and low-voltage approach has been investigated in order to spot advantages and also the challenges of this method. In this case, LED chips are arranged in a matrix compact structure called COB (chip on-board). Such arrangements can provide light to large areas when they require extra-high luminous flux levels. Nowadays, COB LEDs are gradually being employed in stadiums, airport runways, international borders, maritime, and mining applications.

Flip Chip Opto (FCOpto) is a company that produces COB LEDs with high luminous flux (up to 230,000 lm) and extra-high current (EHC) levels (up to 48 A) [4]. Among FCOpto devices, the Apollo 600 has been chosen in this work [5]. The EHC COB LED Apollo 600 requires very high current levels (up to 12 A), which leads to driving efficiency issues carefully pointed out and

discussed throughout this paper. More information about this EHC COB LED model is properly described in [6]. Apollo 600 mounted on a 600-W heatsink [7] is presented in Fig. 1 (a), while its dimensions are described Fig. 1 (b).

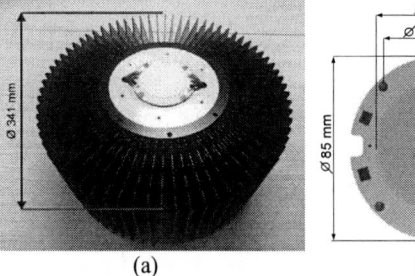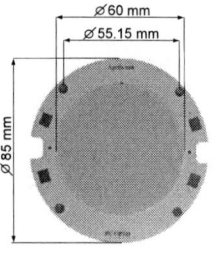

| (a) | (b) |

Fig. 1. Apollo 600 real picture (a) and main dimensions (b).

TABLE I. APOLLO 600 MAIN PARAMETERS FROM DATASHEET [5]

Parameter (Symbol)	Value
Maximum output power (P_o)	608.4 W
Maximum dc forward current (I_F)	12 A
Threshold voltage (V_t)	40.5 V
Dynamic resistance (r_d)	0.95 Ω
Maximum luminous flux	60840 lm

The luminaires for EHC COB LEDs must be conceived to handle high power levels. The driving circuits described in this work comprise interleaved buck structures supplied from a high-voltage controlled dc-link, which is often employed in such floodlighting applications. Moreover, an efficiency analysis is carried out to establish a comparison among different interleaved buck alternatives for very high current levels. An efficiency analysis based on the conduction losses for the buck structures applied to EHC COB LEDs is the main subject of this paper, whereas nonideal simulations are used considering intrinsic parameters of the main circuit components. Thus, Section II introduces the interleaved buck converters (IBCs) employed in this work, while Section III describes the main real components employed in the analysis. Section IV presents simulation results for nonideal conditions, and Section V performs an experimental validation for the 3-cell IBC, while a comparative losses analysis is also carried out. Finally, the conclusions are discussed in Section VI, where the main contributions are highlighted.

II. INTERLEAVED BUCK CONVERTERS SUPPLIED BY A HIGH-VOLTAGE DC LINK

Several requirements are necessary in LED drivers, such as output current control, reduced output current ripple, long lifespan, and high efficiency. Mainly when dealing with high current levels, efficiency is compromised in basic structures, while they are somehow restricted in terms of robustness, lifespan, and stresses regarding the semiconductors. Additionally, from recent works in technical literature, electrolytic capacitors are the main components responsible for compromising the useful life of LED drivers, as they should be avoided [8], [9].

A reliable solution that has been studied is the use of two-stage converters to process the energy from the grid to the LED [10]. The advantages lie on associating power factor correction (PFC) and power control (PC) functions to two distinct converters, whereas a high voltage dc-link may be used to employ only long-life film capacitors. In this work, a front-end PFC rectifier is responsible to generate a nearly constant dc-link, which represents the input voltage source for the analyzed PC stages, as shown in Fig. 2. Typically, when dealing with high power levels, both circuits (PFC and PC) operate in continuous conduction mode (CCM). This mode is widely used in high power levels and offers high efficiency due to minimal switching and conduction losses.

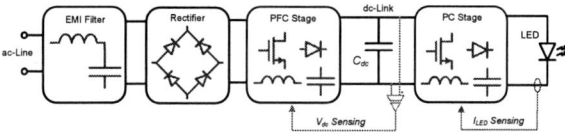

Fig. 2. General two-stage EHC COB LED driver with an intermediate high voltage dc-link.

A two-stage topology allows obtaining independent operation of the involved converters. In these cases, the intermediate dc-link voltage is often high, which decreases the dc-link capacitance (C_{dc}). It is important to notice that increasing the dc-link voltage has direct impact on the MOSFETs (Metal Oxide Semiconductor Field Effect Transistors) and diodes used in PC stage, which means high values of drain-source on-resistance $R_{ds(on)}$ and higher conduction losses. In this paper, conventional and interleaved buck topologies are employed to quantify the existing conduction losses, efficiency, and component count. The conduction losses correspond to the dominant portion in CCM structures, while the switching losses do not present significant impact due to the minimal current variation in inductors and consequent lower rms currents through active and passive semiconductors [11].

The conventional buck converter (CBC) has been widely explored in lighting technical literature as an LED driver. It is the most usual type of basic SMPS (switched-mode power supply), while used in industry to adapt high input voltage levels to low output voltage ones. This topology presents an output filter inductor as well as an output capacitor in the back-end side, providing dc output currents with reduced high-frequency ripple. Additional information on this basic step-down topology will be suppressed for the sake of simplicity, although designing issues of the CBC can be found in [11] and [12].

In order to increase efficiency in the CBC, the interleaved buck converter (IBC) is a reliable alternative when dealing with high current levels. The general n-cell structure is then shown in Fig. 3. The IBC is formed by n controlled switches, n diodes, and n inductors, which can

be coupled for volume reduction. Consequently, the load is fed by the sum of the currents through each cell. This connection leads to lower rms current levels through switches, which reduce significantly the conduction and switching losses [13].

Aiming at an in-depth analysis of the IBC topology, the synchronous rectification is also addressed in this work [14]. This connection replaces the diodes for synchronous MOSFETs, which are used as complementary switches in the synchronous interleaved buck converter (SIBC). The n-cell SIBC is depicted in Fig. 4. For this topology, n inductors are also employed, as they can be also coupled for decreased physical volume.

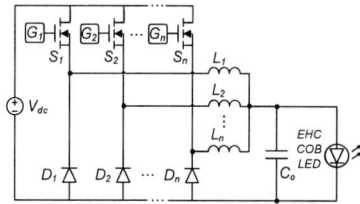

Fig. 3. N-cell IBC employed as an EHC COB LED driver.

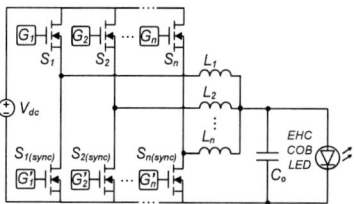

Fig. 4. N-cell SIBC employed as an EHC COB LED driver.

To provide proper operating conditions to the EHC COB LED when fed from a high dc-link voltage, the duty cycle of the MOSFETs must be very small. For the operating point in this work (V_{COB}=50 V and I_{COB}=10 A), and also considering V_{dc}=400 V, the duty cycle in PC stage (D_{PC}) would be 12.5% (for idealized condition) in the CBC and IBCs. In this case, some applications should also consider the PWM limitations that exist in step-down operation. In extreme conditions, a step-down structure with extended duty ratio may be feasible to ensure that this conversion will be properly achieved [15].

III. Intrinsic Parameters and Graphical Analysis of Conduction Losses in the IBC

In order to evaluate the theoretical performance regarding conduction losses in the aforementioned IBC topologies, the main parasitic elements are considered in nonideal simulations. This analysis aims to obtain the total amount of conduction losses, which are related to each one of the semiconductor elements, i.e., MOSFETs and diodes, as well as the magnetic elements, i.e., interleaved inductors. For lighting applications, capacitors based on metalized film technology are preferable, which present minimal equivalent series resistance (ESR) and lead to reduced I^2R losses [16]. This issue is generally required in LED drivers to improve the global luminaire lifespan [17], [18]. Therefore, this particular resistance can be neglected in the analysis.

Software PSIM 11.1® allows including the main parasitic elements in a general converter by selecting the required component nonideal level. Since the conduction losses will be mainly evaluated, the parameters to be included are the on resistance of switches, voltage drops of diodes, and also the copper resistances associated with the inductor windings. Table II and Table III describe the main parameters related to the analyzed MOSFETs and diodes, respectively.

The parameters described for MOSFETs are the drain-source breakdown voltage (V_{ds}), the average drain current (I_d), the drain-source on resistance ($R_{ds(on)}$), the rise time (t_r), the fall time (t_f), and the reverse recovery time (t_{rr}) of the body diode. On the other hand, the parameters for diodes are the peak reverse voltage (V_D), the average forward current (I_F), the forward voltage drop (V_F), and the reverse recovery time (t_{rr}), which is zero for silicon carbide (SiC) diodes.

TABLE II. PARAMETERS OF THE ANALYZED POWER MOSFETS [19]-[23]

Param.	IRFP 460	IPI50 R199CP	IPP50 R299CP	IPW50R 140CP	IPW50R 280CE
V_{ds}	500 V	550 V	550 V	550 V	550 V
I_d	20 A	17 A	12 A	23 A	18 A
$R_{ds(on)}$	270 mΩ	199 mΩ	299 mΩ	140 mΩ	280 mΩ
t_r	55 ns	14 ns	14 ns	14 ns	6.4 ns
t_f	39 ns	10 ns	12 ns	8 ns	7.6 ns
t_{rr}	480 ns	340 ns	260 ns	400 ns	230 ns

TABLE III. PARAMETERS OF THE ANALYZED POWER DIODES [24]-[28]

Param.	MUR860	IDW15G 120C5B	IDH12SG 60C	IDW12G 65C5	IDW16G 65C5
V_D	600 V	1200 V	600 V	650 V	650 V
I_F	8 A	23 A	12 A	12 A	16 A
R_B	40 mΩ	39 mΩ	37 mΩ	35 mΩ	32 mΩ
V_F	1.5 V	1.4 V	1.8 V	1.5 V	1.5 V
t_{rr}	70 ns	≈ 0	≈ 0	≈ 0	≈ 0

To define the best relationship between efficiency and complexity, the curves in Fig. 5, Fig. 6, and Fig. 7 are plotted to point out the optimal range for the operation of IBC-based drivers. Such curves have been plotted considering a duty cycle of 0.125 for the nominal output current point of 10 A, i.e., a step-down conversion rate from 400 V to 50 V. From Fig. 5, the best range is found to be between two and three cells, where the driver can achieve low conduction losses with acceptable driving complexity due to the number of cells. If more than four cells are employed, tradeoffs between efficiency and complexity are not satisfactory due to the high number of cells. Besides, it leads to high component count in terms of inductors, MOSFETs and diodes, with direct impact on the drive circuitry. The conduction losses in MOSFETs can be computed as in (1).

$$P_{S(cond.)} = R_{ds(on)} \cdot I_{d(rms)}^2 \qquad (1)$$

In Fig. 6, it has been verified that the body resistances for diodes (normally lower than 40 mΩ) can be neglected in the analysis. In this case, the curves become straight lines when considering only the significant portion due to the forward voltage drop V_F, being described in (2).

$$P_{D(cond.)} = V_F \cdot I_{F(avg)} \qquad (2)$$

The conduction losses associated with the synchronous rectification switches are also addressed in Fig. 7. In this case, the same conclusions as that for the active switches are valid. The range between two and three cells is quite attractive for the required application, which presents

significant loss reduction while also simplifying the number of gating drivers for synchronous operation. Expression (1) also can be applied in this case to compute the conduction losses in synchronous switches.

Aiming at the best parameter acquisition, the magnetic components have been designed according to [29]. Thus, Table IV describes the physical characteristics of the inductors, while their respective copper resistances are employed in the simulations for each specific measured case. It is important to highlight that such magnetic elements were physically implemented in the laboratory to quantify the inductances and copper resistance, while they were also employed in the experimental validation results. Furthermore, the respective measurements for each particular inductor have been obtained from a LCR meter according to Table IV.

To obtain the best tradeoff between the maximum inductor current and operating frequency, some important designing assumptions must be highlighted. In this work, the switching frequency has been chosen at 40 kHz. AWG (American Wire Gauge) wire number 22 is adequate for frequencies up to 41 kHz. This frequency rate is in accordance with the design specifications and also ensures proper depth of penetration to minimize the skin effect [30]. A significant drawback when dealing with basic single-cell converters is the need of many parallel-connected wires, since the rms current through the inductor is high. This issue is mitigated by using the interleaved structures, where the current is divided among many cells, thus resulting in magnetic elements with fewer paralleled wires.

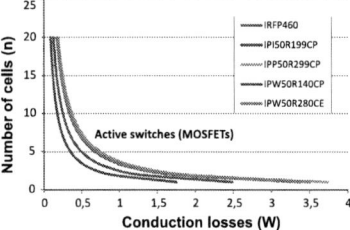

Fig. 5. Theoretical variation of conduction losses in the active switches (MOSFETs) of the IBC.

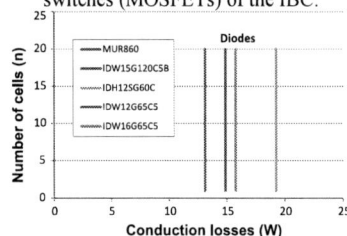

Fig. 6. Theoretical variation for conduction losses in the diodes of the IBC.

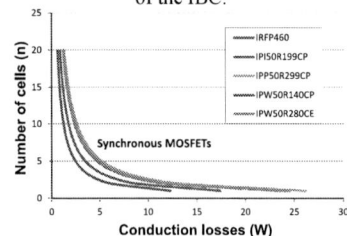

Fig. 7. Theoretical variation for conduction losses in the complementary switches (synchronous MOSFETs) of the SIBC.

978-1-7281-4181-7/19 $31.00 © 2019 IEEE

TABLE IV. PHYSICAL PARAMETERS OF THE INTERLEAVED INDUCTORS

Parameter	1-Cell	2-Cell	3-Cell
Ferrite core (Thornton)	NEE 42/21/20	NEE 42/21/20	NEE 42/21/15
Number of turns	42	40	45
AWG wire	22	22	22
Paralleled conductors	6	3	2
Measured inductance (μH)	502.4	501.3	499.2
Measured copper resistance in LCR meter (Ω)	0.45	0.44	0.42

IV. SIMULATION RESULTS FOR THE IBC DRIVERS

In order to evaluate the feasibility of the proposed drivers, the power circuits have been designed for the rated operating point, i.e., 10 A. The main key parameters are summarized in Table V, where the interleaved inductances have been suppressed due to their minimal variations already presented in Table IV. After designing the converters, simulations were performed in software PSIM 11.1®. As previously mentioned, the main intrinsic parameters mandatory to compute the total conduction losses were employed in the simulations, while IRFP460 (active and synchronous switches) and MUR 860 (diodes) models were chosen due to availability in laboratory for experimental validations. In the simulated waveforms, the voltage stresses on the switches are suppressed for sake of simplicity, as the peak values in these cases are equal to the dc-link voltage V_{dc}, i.e., 400 V. Besides, also for simplicity, the waveforms for CBC (i.e., the 1-cell IBC) are suppressed from this section, while its lowest simulated efficiency has been measured as 89.1%.

TABLE V. OPERATING POINT AND PARAMETERS OF THE IBC DRIVERS

Parameter (Symbol)	Value	Unit
Dc-link average voltage (V_{dc})	400	V
Output power (P_o)	500	W
Average output current (I_{COB})	10	A
Average output voltage (V_{COB})	50	V
Switching frequency (f_s)	40	kHz
Output capacitor (C_o)	40	μF

Fig. 8 depicts the waveforms for switches (S_1 and S_2) and diodes (D_1 and D_2) in the 2-cell IBC. The currents through the switches are approximately equal, although phase shifted by 180°. Besides, their peak values are approximately equal to half of the average forward current of the EHC COB LED, i.e., 5 A. From the diode waveforms, the long on time is due to the complementary duty cycle ($D_{PC}'≈0.875$). This feature impacts significantly on the conduction losses in this case, whereas synchronous MOSFETs may decrease conduction losses and improve the efficiency.

Fig. 9 shows the output current, output voltage, and currents through inductors L_1 and L_2. The output voltage and output current waveforms denote the EHC COB LED operation at the designed point (10 A). The output quantities present a ripple whose frequency is twice higher than that in the CBC. The inductors equally share the output current, whose average values are 5 A. Therefore, the IBC is an interesting alternative, since the output current can be reduced as more cells are added. However, current sharing issues due to intrinsic differences in the passive and active elements used in the different cells are of major concern, which must be considered in

experimental prototypes. The simulated efficiency for 2-cell IBC has been measured as 93.5% and 93.7%, for diodes and synchronous MOSFETs, respectively.

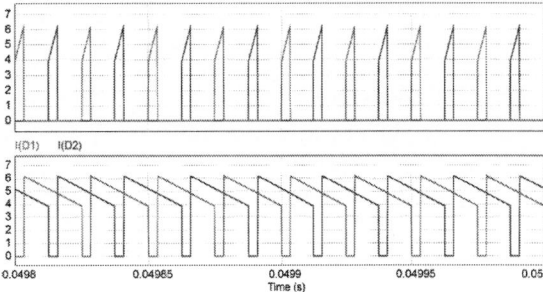

Fig. 8. Current waveforms in MOSFETs S_1 and S_2 (A), and diodes D_1 and D_2 (A) in the 2-cell IBC.

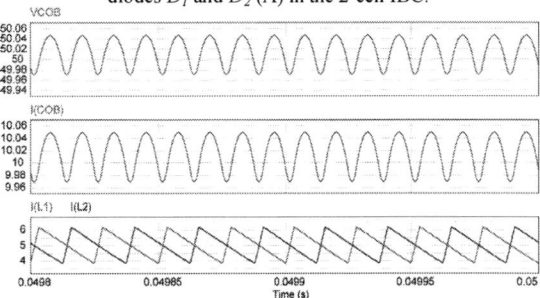

Fig. 9. Waveforms of the output voltage (V), output current (A), and inductors (A) in the 2-cell IBC.

Fig. 10 shows the waveforms for the switches (S_1, S_2, and S_3) and diodes (D_1, D_2, and D_3) in the 3-cell IBC. The currents through the switches are approximately the same, although phase shifted by 120°. Besides, their peak values are approximately equal to one third of the average forward current of EHC COB LED, i.e., 3.3 A. In addition, Fig. 11 presents the output current, output voltage, and inductor currents waveforms in the 3-cell IBC. The ripple frequency associated with the output quantities is three times higher when compared to the CBC, which minimizes their high-frequency output ripple for the same C_o. Still in Fig. 11, the inductor currents are satisfactorily balanced with an average value of 3.3 A for each cell. In this case, for the 3-cell IBC, the simulated efficiency has been measured as 94.7% and 95.5%, for diodes and synchronous MOSFETs, respectively. Therefore, an improved efficiency performance has been found for the 3-cell IBC, as this structure has been chosen to validate the losses values found by simulation results.

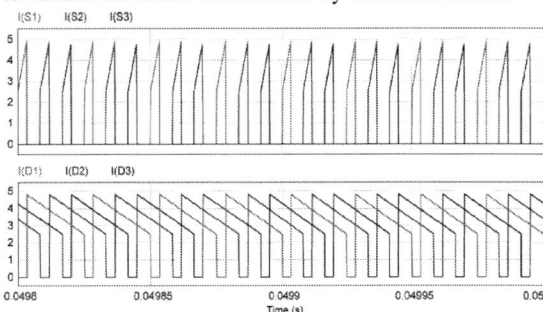

Fig. 10. Current waveforms in MOSFETs S_1, S_2, and S_3 (A), and diodes D_1, D_2, and D_3 (A) in the 3-cell IBC.

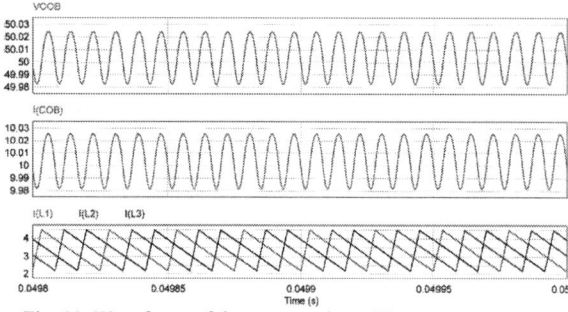

Fig. 11. Waveforms of the output voltage (V), output current (A), and inductors L_1, L_2, and L_3 (A) in the 3-cell IBC.

V. EXPERIMENTAL RESULTS FOR THE 3-CELL IBC DRIVER AND LOSSES COMPARATIVE EVALUATION

The 3-cell IBC structure has presented improved performance regarding conduction losses and efficiency, whereas the best tradeoffs could be achieved with this topology. Owing to achieve experimental validation, the 3-cell IBC prototype has been implemented in this section using diodes (MUR860) as complementary switches. The same operating point in Table V has been adopted, while the prototype picture is presented in Fig. 12.

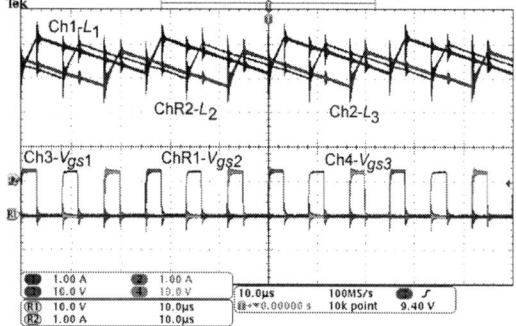

Fig. 12. 3-cell IBC prototype used to drive the EHC COB LED.

The waveforms for the interleaved inductors (L_1, L_2 and L_3) and gate-source voltages in the active switches are presented in Fig. 13. From the waveforms, the inductors charging periods correspond to the pulse width associated with the gate-source voltages (V_{gs}). Moreover, some minimal unbalance exists in such experimental waveforms, as the rms currents in inductors have been measured as 3.52 A, 3.29 A, and 3.19 A, respectively.

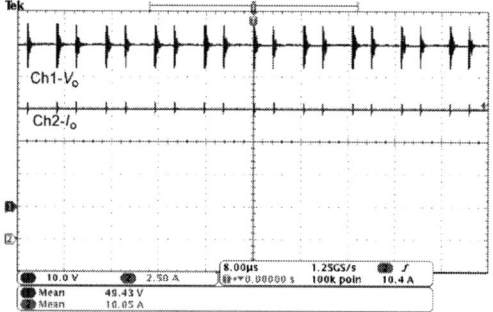

Fig. 13. Waveforms for interleaved inductors currents I_{L1}, I_{L2}, and I_{L3}; and for gate-source voltages in active switches (MOSFETs) V_{gs1}, V_{gs2}, and V_{gs3}.

The waveforms for the employed MOSFETs (IRFP460) are presented in Fig. 14, where approximately the same characteristics obtained in the simulations can also be seen in this case. Their respective current peak values are equally the same ($\approx$3.3 A), while the voltage stresses are equal to the dc-link voltage, i.e., 400 V.

Fig. 15 shows the output waveforms for the EHC COB LED. One can see that the average output voltage and output current are 50 V and 10 A, respectively. The high-frequency oscillations are due to switching transitions from on to off and vice-versa, which are transmitted to the output without major concerns to the load.

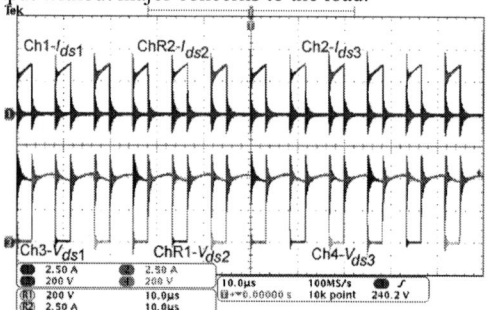

Fig. 14. Waveforms for MOSFETs currents I_{ds1}, I_{ds2} and I_{ds3}; and for drain-source voltages in MOSFETs (V_{ds1}, V_{ds2} and V_{ds3}).

Fig. 15. Waveforms for output voltage (V_o) and output current (I_o) through the EHC COB LED Apollo 600.

In order to perform a fair comparison of losses among the aforementioned structures, Table VI summarizes the obtained results, where the evaluated parameters are the measured conduction losses, component count (S_n, D_n, L_n, and C_o), and the overall efficiency for each converter. For the simulations in Section IV, diodes and synchronous MOSFETs (IRFP460) were employed as complementary switches to evaluate their respective performance in both cases. In simulations, the conduction losses in all components were directly measured in the software environment, and confirmed by expressions (1) and (2).

According to Table VI, the CBC (1-cell IBC) has the lowest efficiency, which is due to appreciable rms currents in the semiconductors, thus implying increased conduction losses. Still in CBC, the copper losses are the main portion, reaching 44.8 W in the 1-cell converter. The 2-cell IBC has presented intermediate results between total efficiency (93.5%) and component count (7). From 2-cells ahead, the synchronous alternative becomes compensatory, where the efficiency is increased to 93.7% in 2-cell IBC. Moreover, the 3-cell IBC has presented the highest simulated efficiency (94.7% and 95.5% for diodes and synchronous MOSFETs, respectively) with the lowest conduction losses through the main circuit elements, however, with the highest component count (10).

The losses distribution in the 3-cell IBC prototype is also shown in Table VI, while one can see that the values are very close to the ones obtained in nonideal simulation (an efficiency of 94.4% measured by a Yokogawa Power Meter). This experimental test validates the quantities acquired by simulation, while also ensuring that only the minimal difference, i.e., 0.3% of the total experimental losses are due to switching transitions or core losses in inductors. Therefore, the main losses in the 3-cell IBC driver are due to conduction losses in active switches and diodes, and due to copper losses in interleaved inductors.

TABLE VI. EFFICIENCY COMPARISON AMONG IBC TOPOLOGIES AS EXTRA-HIGH CURRENT COB LED DRIVERS

Conduction Losses	CBC (1-cell)	2-cell IBC	3-cell IBC	3-cell IBC (prototype)
MOSFETs (W)	3.38	1.68	1.12	1.25
Diodes (W)	13.2	13.2	13.2	13.9
Sync. MOSFETs (W)	23.62	11.81	7.9	-
Inductors (Copper) (W)	44.8	22.3	13.9	14.2
Overall efficiency	89.1%	93.5%	94.7%	94.4%
Overall efficiency (with sync. MOSFETs)	87.4%	93.7%	95.5%	-
Component count	4	7	10	10

VI. CONCLUSION

This work has proposed a comparative efficiency evaluation for interleaved buck topologies employed as EHC COB LED drivers. The analysis has been performed considering real intrinsic parameters for the main power circuit elements in order to achieve more reliable results. The parameters of EHC COB LED Apollo 600 as provided by the manufacturer have been also employed.

To overcome efficiency limitations in basic converters, interleaved topologies have been employed instead in order to improve the total efficiency. The comparative analysis has been performed for the CBC, 2-cell IBC, 3-cell IBC, and their respective synchronous alternatives. The 3-cell IBC has the highest efficiency (i.e., 94.7% and 95.5% for 3-cell SIBC) with the highest component count, while it has been chosen to experimental validation (94.4%). The results have shown that the experimental and simulated losses quantities are very close to each other, while also ensuring that they are mainly due to conduction losses in semiconductors and to copper losses in inductors.

ACKNOWLEDGMENT

The authors would like to acknowledge CAPES, CNPq, FAPEMIG, and INERGE for the financial support; and FCOpto Company for the product support.

REFERENCES

[1] G. Dreyfus, C. Gallinat, "Rise and shine: lighting the world with 10 billion LED bulbs", 2015. Available in: https://bit.ly/2YJayXc.

[2] T. Novak, K. Pollhammer, "Traffic-adaptive control of LED-based streetlights", *IEEE Industrial Electronics Magazine*, vol. 9, nº 2, pp. 48-50, 2015.

[3] Osram Opto Semiconductors, "Street lighting with LED light sources", *Application Note*, 2014. Available in: https://goo.gl/OSJ8AQ.

[4] Flip Chip Opto, "FCOpto – Starlite LED product catalogue", 2016. Available in: https://goo.gl/mxWZtx.

[5] Flip Chip Opto, "Apollo 600 datasheet", 2016. Available in: https://goo.gl/CxkdGi.

[6] D. C. Pereira, P. L. Tavares, P. S. Almeida, G. M. Soares, F. L. Tofoli, H. A. Braga. "Improved Photoelectrothermal Model with Thermal Parameters Variation Applied to an Extra-High Current COB LED." *Revista Eletrônica de Potência*, vol. 24, nº 2, pp. 147-156, 2019.

[7] Starlite LED, "600 W cold forged heatsink", 2016. Available in: https://goo.gl/mVZrpI.

[8] Philips, "LED driver lifetime and reliability", 2011. Available in: https://goo.gl/iQrkUU.

[9] B. Singh, A. Shrivastava, "Buck converter-based power supply design for low power light emitting diode lamp lighting", *IET PELs*, vol. 7, nº 4, pp. 946-956, 2014.

[10] X. Xie, M. Ye, Y. Cai, J. Wu, "An optocouplerless two-stage HPF LED driver", *IEEE Applied Power Electronics Conference and Exposition*, pp. 2078-2083, 2011.

[11] D. W. Hart, "Power Electronics", McGraw-Hill Ed., 2011.

[12] M. H. Rashid, "Power Electronics Handbook", AP, 2001.

[13] A. C. Schittler, D. Pappis, A. Campos, M. A. D. Costa, and J. M. Alonso. "Interleaved buck converter applied to high-power HID lamps supply: design, modeling and control", *IEEE Transactions on Industry Applications*, vol. 49, no. 4, pp. 1844-1853, Apr. 2013.

[14] F. Marvi, E. Adib, H. Farzanehfard, "Zero voltage switching interleaved coupled inductor synchronous buck converter operating at boundary condition", *IET Power Electronics*, vol. 9, nº 1, pp.126-131, 2016.

[15] I. O. Lee, S. Y. Cho, and G. W. Moon. "Interleaved Buck Converter Having Low Switching Losses and Improved Step-Down Conversion Ratio." *IEEE Transactions on Power Electronics*, vol. 27, no. 8, pp. 3664-3675, 2012.

[16] EPCOS – TDK, "Aluminum electrolytic capacitors series B41456", 2016. Available in: https://goo.gl/6JFYVM.

[17] G. M. Soares, P. S. Almeida, J. M. Alonso, H. A. C. Braga, "Capacitance minimization in offline LED drivers using an active-ripple-compensation technique", *IEEE Transactions on Power Electronics*, vol. 32, nº 4, pp. 3022-3033, 2017.

[18] U. R. Reddy, B. L. Narasimharaju, "Single-stage electrolytic capacitor less non-inverting buck-boost PFC based AC-DC ripple-free LED driver", *IET PELs*, vol. 10, nº 1, pp. 38-46, 2017.

[19] Infineon, "IRFP460 power MOSFET datasheet", 2012. Available in: https://bit.ly/2EmBLa7.

[20] Infineon, "IPI50R199CP power MOSFET datasheet", 2012. Available in: https://bit.ly/2VWWAUE.

[21] Infineon, "IPP50R299CP power MOSFET datasheet", 2012. Available in: https://bit.ly/2VTMIuE.

[22] Infineon, "IPW50R140CP power MOSFET datasheet", 2012. Available in: https://bit.ly/2WnbRxa.

[23] Infineon, "IPW50R280CE power MOSFET datasheet", 2012. Available in: https://bit.ly/2WXhyip.

[24] Infineon, "MUR860 fast switching power diode datasheet", 2013. Available in: https://bit.ly/2VWXS1W.

[25] Infineon, "IDW15G120C5B fast switching power diode datasheet", 2013. Available in: https://bit.ly/2ErCCXr.

[26] Infineon, "IDH12SG60C fast switching power diode datasheet", 2013. Available in: https://bit.ly/2VSipER.

[27] Infineon, "IDW12G65C5 fast switching power diode datasheet", 2013. Available in: https://bit.ly/2Qdl8T2.

[28] Infineon, "IDW16G65C5 fast switching power diode datasheet", 2013. Available in: https://bit.ly/30zjQH3.

[29] I. Barbi, C. H. Font, R. L. Alves, "Projeto físico de indutores e transformadores", Instituto de eletrônica de potência (INEP-UFSC), 2002. Available in: https://goo.gl/OM2Gk5.

[30] Solaris, "American wire gauge conductor evaluation", 2015. Available in: https://goo.gl/Y59azW.

Modeling of a three-level quadratic boost converter

Mauricio Borchardt[1], Mateus C. Orige[1], Mateus de F. Bueno[1], Gerardo Escobar[2], Yales R. de Novaes[1]

[1]Santa Catarina State University (UDESC), Joinville – Santa Catarina, Brazil

[2]Tecnológico de Monterrey, Monterrey – San Luis Potosi, Mexico

mauricioborchardt13@gmail.com, mateus.orige@edu.udesc.br, mateusbueno96@gmail.com, gescobar@ieee.org, yales.novaes@udesc.br

Abstract—**This work presents the modelling of a three-level quadratic boost converter. The dynamical model obtained for the converter turns out to be of non-minimal phase, which limits to a slow transient response, requiring the application of a cascade control. Hence, the present work is aimed to search and identify the model parts associated to each of the control loops involved in such a cascade controller. In the modelling process, the series resistors of the capacitors of the topology were considered with the aim to obtain a theoretical model as close as possible to the real system. Numerical results have been included for the model validation.**

Index Terms—**Three-level quadratic boost converter, modeling, validation.**

I. INTRODUCTION

The boost converter is a topology widely used to increase the DC voltage level out of an input voltage source. However, when a high gain is required, as in the case of photovoltaic (PV) systems, the conventional boost topology may be insufficient due to losses associated with the parasitic components in inductor, capacitive filter, rectifier diodes, and perhaps the switching frequency [1].

The cascade boost arises as a solution to increase the gain. This converter offers a quadratic voltage gain, however, both the volume of the structure increases as well as the voltage stress in second stage semiconductors [2]. The two switches of the cascade boost topology can be replaced by a single one, deriving in this way the so called quadratic boost converter, which has the same static gain as in the cascade boost converter [3]. Once again, the main disadvantage is the high voltage stress on the switch, which affects the overall efficiency.

Another solution consists in appling a three-level switching cell in the quadratic boost converter. In this case the blocking voltage is fractionated at the output of the converter [1]. Therefore, the three-level quadratic converter shown in Figure 1 offers the square gain characteristic, while the switches are under a reduced voltage stress since the total bus voltage has been fractioned between the two output capacitors. Although this converter may operate in open loop, a more precise adjustment of the output voltage requires the use of a controller, which will guarantee fast correction of any deviations from feed transients or load changes.

The small signal models of these quadratic converters usually present zeros in the right half-plane and, therefore, they belong to the family of non-minimal phase systems. This limits the speed of the transient response, deteriorates

the dynamical performance, and may cause unstability in the closed loop response [4]. To achieve a desired performance, a usual solution is the implementation of a cascading control based on multi loop controllers.

The design of such a controller design necessarily requires a representative mathematical model descriving the dynamics of the converter. In this case, the average state space model proposed by Midlebrook and Cuk and the PWM switch concept proposed by Vorperian [5] are the existing techniques to obtaining sufficiently accurate mathematical models for converters. Such models must represent the performance of a system as close as possible to reality, for this, pertinent non-idealities are considered, which sometimes contribute with additional complexity to the model.

In this work a model is presented for the three-level quadratic converter shown in Figure 1. The proposed model is first expressed in terms of its state-space average equations, which are later transformed to transfer functions, which are the base for a multi-loop control design.

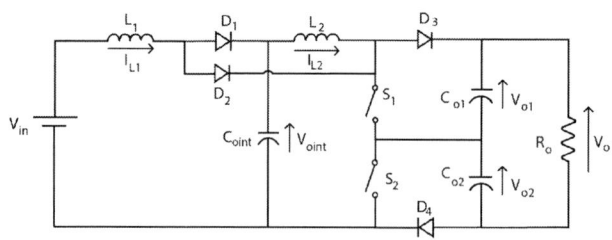

Fig. 1. Three-level quadratic boost converter circuit.

II. PRINCIPLES OF OPERATION

To better understand the operation of the proposed converter, the analysis starts with the Continuous Current Mode (CCM) operation, especially in the case where the duty cycle $D > 0.5$. The complete analyzes of all modes of operation of this converter together with their variations defined by the cyclic ratio values is explained in more detail in [1].

A. Continuous-Conduction Mode (CCM)

The operation steps in CCM for $D > 0.5$ are described below. The topological states of the converter operating steps are shown in Figure 2. A detailed analysis of this mode of operation together with the waveforms can be found in [1].

978-1-7281-4181-7/19 $31.00 © 2019 IEEE

Notice that, initially, that equivalent series resistors (ESR) in the capacitors are neglected for an ideal analysis.

- Stage 1 $(t_0 - t_1)$ – (Figure 2 (a)): S_1 and S_2 are conducting and the current of the inductors increases linearly. Capacitors C_{o1} and C_{o2} deliver energy to the load.
- Stage 2 $(t_1 - t_2)$ – (Figure 2 (b)): S_2 is blocked and the diodes D_1 and D_4 start conducting. The C_{oint} capacitor is charged with the energy of V_{in} and L_1 and the inductor L_2 transfers energy to the C_{o2} capacitor.
- Stage 3 $(t_2 - t_3)$ – (Figure 2 (a)): S_2 is turned on again. The topological state and the principle of operation of this stage are equal to the operation of the first stage.
- Stage 4 $(t_3 - t_4)$ – (Figure 2 (c)): S_1 is blocked. During this stage V_{in} and $L_1 1$ carry C_{oint} and C_{o1} receives the energy of inductor L_2.

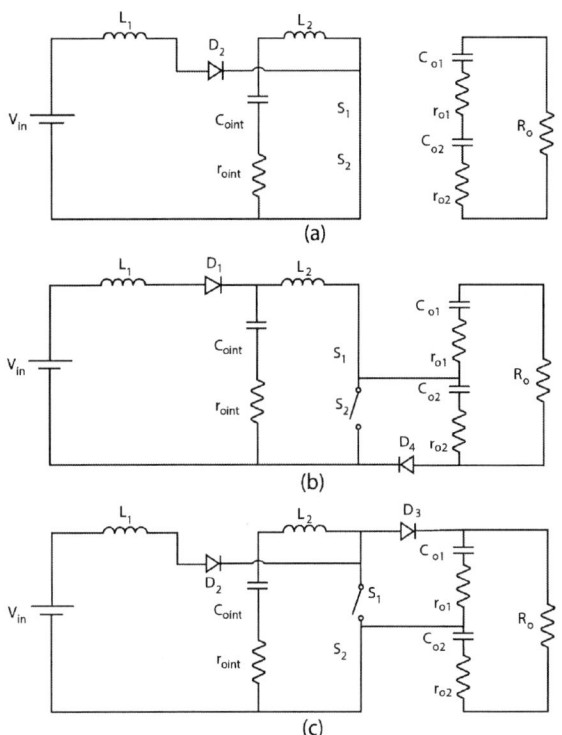

Fig. 2. Topological states of the T-LQ Boost converter in CCM for $D > 0.5$.

The modulation strategy adopted for the proposed converter is a phase-shifted PWM modulation, with a delay of 180°. For this modulation technique, and based on the value of D, two different operating conditions arise for the converter:

- For $D < 0.5$, a switching period occurs in which the two switches are blocked simultaneously.
- For $D > 0.5$, the switches conduct twice during a cycle.

III. Modelling of the T-LQ Boost Converter

In this work, the control design technique presented in [6] is considered. This technique comprises three control loops,

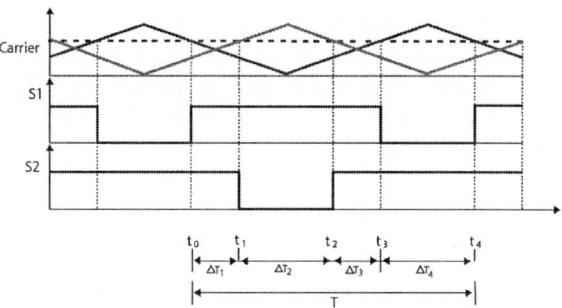

Fig. 3. Modulation applied to the proposed converter

namely, a voltage loop used to generate the current reference, a current loop to control the input inductor current, and a voltage difference loop to balance the output capacitors voltages.

The design of each of the three control loops, requires the mathematical model of the converter expressed in terms of the particular variable to be controlled. These models, in the form of transfer functions, are described next. The details on their structure is the main goal of the present work.

- $G_{id}(s) = \frac{i_{L1}(s)}{d(s)}$. Is the transfer function of the input inductor current increment with respect to the duty cycle.
- $G_{vi}(s) = \frac{v_o(s)}{i_{L1}(s)}$. Is the transfer function of the output voltage increment with respect to the increment of the input inductor current.
- $G_{vdif}(s) = \frac{\Delta V(s)}{\Delta d(s)}$. Is the transfer function of the increment of the capacitors voltages difference with respect to the duty cycle increment applied to each of the switches.

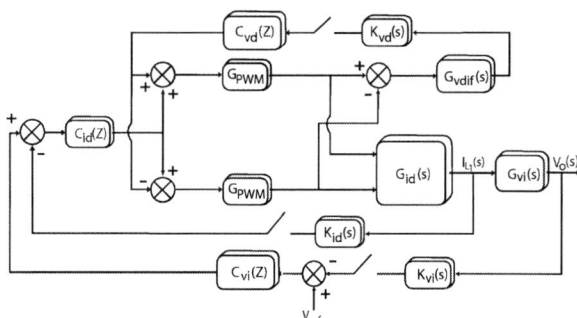

Fig. 4. Control strategy for T-LQ Boost.

The expressions for $G_{id}(s)$, $G_{vi}(s)$ and $G_{vdif}(s)$ are obtained using the small signal modelling approach, which has been widely used to model power converters. The rationale behind this method consist in checking the effect caused by a disturbance of a variable during a switching period.

1) $G_{id}(s)$ model: Recall that $G_{id}(s)$ represents the relationship between the current passing through the inductor L_1 and the duty cycle. As previously shown, in CCM, the converter has different operation stages depending on the duty cycle value. However, the modelling process was performed for the nominal operation of the converter, i.e., for $D > 0.5$. Moreover, the modelling process considers the equivalent series resistance ($ESRs$), with the aim to get a model that accurately approximates the real converter behaviour.

The expression of $G_{id}(s)$ involves the three circuits that correspond to the three stages of operation, since two of the four steps are equivalent to each other. For all steps, the inductors current equations and capacitors voltage equations were obtained through the Kirchoff's laws. The first and third, second and fourth stages to be considered are shown in Figures 2(a), (b) and (c), respectively.

The obtained equations can be represented in state-space form as $K\frac{\mathrm{d}x(t))}{\mathrm{d}t} = Ax(t) + Bu(t)$. Notice that in this representation the state vector $x(t)$ has the following composition $x(t) = \begin{bmatrix} i_{L1}(t) & i_{L2}(t) & v_{oint}(t) & v_{o1}(t) & v_{o2}(t) \end{bmatrix}^T$, while K is a matrix with its main diagonal showing the inductance and capacitance values, and the input vector is given by $u(t) = \begin{bmatrix} V_{in}(t) \end{bmatrix}$.

The corresponding representations of matrices $A_1 = A_3$ (for the first and third stages), A_2 (for the second stage) and A_4 (for the fourth stage) are shown in (1), (2) and (3).

$$A_1 = A_3 = \begin{bmatrix} 0 & 0 & 0 & 0 & 0 \\ 0 & -r_{oint} & 1 & 0 & 0 \\ 0 & -1 & 0 & 0 & 0 \\ 0 & 0 & 0 & -\frac{1}{r_{o1}+r_{o2}+R_o} & -\frac{1}{r_{o1}+r_{o2}+R_o} \\ 0 & 0 & 0 & -\frac{1}{r_{o1}+r_{o2}+R_o} & -\frac{1}{r_{o1}+r_{o2}+R_o} \end{bmatrix} \tag{1}$$

Matrices B_1, B_2, B_3 and B_4 are all equivalent for all operation stages, and are represented by (4).

$$B_1 = B_2 = B_3 = B_4 = \begin{bmatrix} 1 & 0 & 0 & 0 & 0 \end{bmatrix}^T \tag{4}$$

To remove the discontinuity of the model and to obtain the complete average model, a weighting of the stages according to periods $\Delta T_1, \Delta T_2, \Delta T_3$ and ΔT_4 (shown in Figure 3) is carried out as presented in (5).

$$\langle A \rangle = \overbrace{\frac{2d(t)-1}{2}}^{\Delta T_1} A_1 + \overbrace{(1-d(t))}^{\Delta T_2} A_2$$
$$+ \underbrace{\frac{2d(t)-1}{2}}_{\Delta T_3} A_3 + \underbrace{(1-d(t))}_{\Delta T_4} A_4 \tag{5}$$

To simplify the calculations, consider that
- $C_{o1} = C_{o2} = C_o$
- The voltage in both output capacitors are equal, i.e., $v_{o1} + v_{o2} = V_o$
- $ESRs$ of the output capacitors have the same value, i.e., $r_{o1} = r_{o2} = r_o$.

Application of the simplifications and the weights to matrix A yields the expression (6).

It is important to emphasize that the matrix K and vectors $x(t)$ and B also undergo the simplifications above described. Besides, it is necessary to obtain the equilibrium point, that

is, the expressions of the states in the steady state, which is a requirement in the modeling process. These expressions are obtained from (7). As these expressions are relatively long, they are omitted here, and can be consulted in [7].

$$X(t) = -A^{-1}Bu(t) \tag{7}$$

The above described system is not linear, therefore, it is proposed to perform a linearization following the small signal approach. For this, replace variables $v_{oint}(t), v_o(t), i_{L_1}(t), i_{L_2}(t)$ and $d(t)$ by a constant plus a small variation.

The perturbation in the input voltage $\hat{v_{in}}(t)$ has been neglected, hence, it is assumed that, for the validation of the models, the perturbation only occurs in the duty cycle. The other perturbations, current in the inductors and voltage in the capacitors, are consequences of the disturbance in the cyclic ratio.

Applying the Laplace transform to the linearized equations yields the following expressions in the frequency domain.

$$sL_1 I_{L_1}(s) = \left[(2I_{L_1} - 2I_{L_2})r_{oint} + 2V_{oint}\right]d(s)$$
$$+ 2(D-1)(V_{oint}(s) + r_{oint}(I_{L_1}(s) - I_{L_2}(s))) \tag{8}$$

$$sL_2 I_{L_2}(s) = \frac{1}{2r_o + R_o} [((2D-2)r_o - r_{oint})R_o$$
$$+ 2r_o((D-1)r_o - r_{oint})I_{L_2}(s)$$
$$+ (-2I_{L_1}r_{oint} + 2I_{L_2}r_o + V_o)R_o d(s)$$
$$+ (2r_o{}^2 I_{L_2} - 4I_{L_1}r_{oint}r_o)d(s)$$
$$- 2r_{oint}(2r_o + R_o)(D-1)I_{L_1}(s)$$
$$+ (2r_o + R_o)V_{oint}(s) + V_o(s)R_o(D-1)] \tag{9}$$

$$sC_{oint}V_{oint}(s) = 2(1-D)I_{L_1}(s) - 2I_{L_1}d(s) - I_{L_2}(s) \tag{10}$$

$$sC_o V_o(s) = \frac{-2}{2r_o + R_o} [(D-1)I_{L_2}(s)R_o]$$
$$- \frac{2}{2r_o + R_o} [I_{L_2}R_o d(s) + V_o(s)] \tag{11}$$

The previous equations involve terms I_{L1}, I_{L2}, V_{oint} and V_o, which represent expressions of the currents and voltages in the permanent regime. These terms were obtained through (7), and their expressions can be substituted in (8) to (11).

Out of the system (8) to (11), the transfer function $G_{id}(s)$ is derived. The symbolic final expression of $G_{id}(s)$ is extremely large and is not displayed in this article. Instead, a numerical evaluation of $G_{id}(s)$ was performed considering the operation at the nominal condition of the converter. The values adopted are the same as the parameters used for the construction of the prototype, which are collected in the Table I. This numerical evaluation of $G_{id}(s)$ is shown in (12).

$$A_4 = \begin{bmatrix} -r_{oint} & r_{oint} & -1 & 0 & 0 \\ r_{oint} & \frac{-r_{oint}(r_{o1}+r_{o2}+R_o)-r_{o1}(r_{o2}+R_o)}{r_{o1}+r_{o2}+R_o} & 1 & \frac{-r_{o2}-R_o}{r_{o1}+r_{o2}+R_o} & \frac{r_{o1}}{r_{o1}+r_{o2}+R_o} \\ 1 & -1 & 0 & 0 & 0 \\ 0 & \frac{r_{o2}+R_o}{r_{o1}+r_{o2}+R_o} & 0 & \frac{-1}{r_{o1}+r_{o2}+R_o} & \frac{-1}{r_{o1}+r_{o2}+R_o} \\ 0 & \frac{-r_{o1}}{r_{o1}+r_{o2}+R_o} & 0 & \frac{-1}{r_{o1}+r_{o2}+R_o} & \frac{-1}{r_{o1}+r_{o2}+R_o} \end{bmatrix} \tag{2}$$

$$A_2 = \begin{bmatrix} -r_{oint} & r_{oint} & -1 & 0 & 0 \\ r_{oint} & \frac{-r_{oint}(r_{o1}+r_{o2}+R_o)-r_{o2}(r_{o2}+R_o)}{r_{o1}+r_{o2}+R_o} & 1 & \frac{r_{o2}}{r_{o1}+r_{o2}+R_o} & \frac{-r_{o1}-R_o}{r_{o1}+r_{o2}+R_o} \\ 1 & -1 & 0 & 0 & 0 \\ 0 & \frac{-r_{o2}}{r_{o1}+r_{o2}+R_o} & 0 & \frac{-1}{r_{o1}+r_{o2}+R_o} & \frac{-1}{r_{o1}+r_{o2}+R_o} \\ 0 & \frac{r_{o1}+R_o}{r_{o1}+r_{o2}+R_o} & 0 & \frac{-1}{r_{o1}+r_{o2}+R_o} & \frac{-1}{r_{o1}+r_{o2}+R_o} \end{bmatrix} \tag{3}$$

$$\langle A \rangle = \begin{bmatrix} (-2+2d(t))r_{oint} & (2-2d(t))r_{oint} & -2+2d(t) & 0 \\ (2-2d(t))r_{oint} & \frac{((-2+2d(t))r_{o2}+(-1+d(t))R_o-r_{oint})r_{o1}+((-1+d(t))R_o-r_{oint})r_{o2}-R_o r_{oint}}{r_{o1}+r_{o2}+R_o} & 1 & \frac{(-1+d(t))R_o}{r_{o1}+r_{o2}+R_o} \\ 2-2d(t) & -1 & 0 & 0 \\ 0 & \frac{(2-2d(t))R_o}{r_{o1}+r_{o2}+R_o} & 0 & \frac{-2}{r_{o1}+r_{o2}+R_o} \end{bmatrix} \tag{6}$$

$$G_{id}(s) = \frac{0.000178016791 \cdot s^3 + 0.0873986998 \cdot s^2 + 1156.281114 \cdot s + 44457.68751}{7.3086364 \cdot 10^{-11} \cdot s^4 + 4.3291532 \cdot 10^{-8} \cdot s^3 + 4.8053911 \cdot 10^{-4} \cdot s^2 + 0.089578545 \cdot s + 156.01825} \tag{12}$$

2) $G_{vi}(s)$ model: Recall that $G_{vi}(s)$ represents the relationship between the output voltage variation and the current input on inductor L_1. The model of $G_{vi}(s)$ is calculated based on (13).

$$G_{vi}(s) = \frac{V_{out}(s)}{i_{L1}(s)} = \frac{V_{out}(s)}{d(s)} \cdot \frac{d(s)}{i_{L1}(s)} = \frac{G_{vd}(s)}{G_{id}(s)} \tag{13}$$

The following additional equation is used to obtain $G_{vd}(s)$, in addition to (8), (9), (10) and (11).

$$\langle v_{out}(t) \rangle_{T_s} = \langle v_o(t) \rangle_{T_s} + r_o C_o \frac{\mathrm{d}\langle v_o(t) \rangle_{T_s}}{\mathrm{d}t} \tag{14}$$

In (8) to (11), $V_o(s)$ represents the sum of the voltage over the output capacitors. Notice that the model must involve the sum of the voltages over the ERSs because the frequency components of the voltage on the $ERSs$ are not zero, although the average value is zero.

A small signal analysis is performed in (14) substituting expressions (15) and (16). The result obtained together with the realization of a linearization can be visualized in (17). In (18) shows the application of the Laplace transform to (17).

$$\langle v_o(t) \rangle_{T_s} = V_o + \hat{v}_o(t) \tag{15}$$

$$\langle v_{out}(t) \rangle_{T_s} = V_{out} + \hat{v_{out}}(t) \tag{16}$$

$$\hat{v_{out}}(t) = \hat{v}_o(t) + r_o C_o \frac{\mathrm{d}\hat{v}_o(t)}{\mathrm{d}t} \tag{17}$$

$$V_{out}(s) = V_o(s)(1 + r_o C_o s) \tag{18}$$

Notice that (8) to (11) and (18) form a system of equations, out of which expression $\frac{V_{out}(s)}{d(s)}$ can be isolated. The numerical transfer function for the conditions established in Table I is given by (19). Once $G_{vd}(s)$ is obtained, the expression (13) is used to obtain $G_{vi}(s)$. A numerical expression for this transfer function is shown in (20).

3) $G_{vdif}(s)$ model: Transfer function $G_{vdif}(s)$ establishes a relationship between the voltage difference in the output capacitors with respect to the increment in the duty cycle. This model is obtained based on the expressions during the conduction times of each of the switches. As shown in Figure 3, the ratios below the driving times can be obtained, where D_1 and D_2 are the duty cycle ratios of switches S_1 and S_2, respectively.

$$\Delta T_1 = \Delta T_3 = \frac{(D_1 + D_2 - 1)}{2} \tag{21}$$

$$\Delta T_2 = 1 - D_2 \tag{22}$$

$$\Delta T_4 = 1 - D_1 \tag{23}$$

Next, the operation steps (based on Figure 2) from the point of view of the output capacitors are considered. The equations for the capacitor voltages for the first and third stages (Figure 2(a)) are presented in (24).

$$C_{o1} \frac{\mathrm{d}v_{o1}(t)}{\mathrm{d}x} = C_{o2} \frac{\mathrm{d}v_{o2}(t)}{\mathrm{d}x} = \frac{-v_{o2}(t) - v_{o1}(t)}{r_{o2} + r_{o1} + R_o} \tag{24}$$

$$G_{vd}(s) = \frac{V_{out}(s)}{d(s)} = \frac{-9.36472469 \cdot 10^{-7} \cdot s^4 + 0.00260527 \cdot s^3 + 945.14 \cdot s^2 + 3.70884359 \cdot 10^{-5} \cdot s + 5.495639067 \cdot 10^9}{7.3086365 \cdot 10^{-7} \cdot s^4 + 0.000432915 \cdot s^3 + 4.80539113 \cdot s^2 + 895.78545 \cdot s + 1.5601826 \cdot 10^6} \quad (19)$$

$$G_{vi}(s) = \frac{V_{out}(s)}{i_{L_1}(s)} = \frac{-6.875614 \cdot 10^{-9} \cdot s^4 + 1.9112798 \cdot 10^{-5} \cdot s^3 + 6.9392517 \cdot s^2 + 2723.0462 \cdot s + 4.0349178 \cdot 10^7}{0.01307005649 \cdot s^3 + 6.416843792 \cdot s^2 + 84894.57281 \cdot s + 3.264099314 \cdot 10^6} \quad (20)$$

For the second step (illustrated in Figure 2(b)) the following expressions are obtained.

$$C_{o1}\frac{dv_{o1}(t)}{dt} = \frac{-i_{L2}(t)r_{o2} - v_{o2}(t) - v_{o1}(t)}{r_{o2} + r_{o1} + R_o} \quad (25)$$

$$C_{o2}\frac{dv_{o2}(t)}{dt} = \frac{i_{L2}(t)(r_{o1} + R_o) - v_{o2}(t) - v_{o1}(t)}{r_{o2} + r_{o1} + R_o} \quad (26)$$

Figure 2(c) is used to solve the fourth step, which yields

$$C_{o1}\frac{dv_{o1}(t)}{dt} = \frac{i_{L2}(t)(r_{o2} + R_o) - v_{o2}(t) - v_{o1}(t)}{r_{o2} + r_{o1} + R_o} \quad (27)$$

$$C_{o2}\frac{dv_{o2}(t)}{dt} = \frac{-i_{L2}(t)r_{o1} - v_{o2}(t) - v_{o1}(t)}{r_{o2} + r_{o1} + R_o} \quad (28)$$

Once the expressions of the steps are obtained, the average in a switching period can be obtained by a weighted sum of the stages expressions associated with the corresponding periods ($\Delta T_{1,2,3,4}$) for each of the output capacitors voltages.

To obtain a differential weighted sum of voltage $\langle v_{cdif}(t)\rangle_{T_s}$, the weighted expression C_{o1} is subtracted out of C_{o2}. Once this difference is performed, and considering that $C_{o1} = C_{o2} = C_o$, the result is given by (29).

$$C_o\frac{d\langle v_{cdif}(t)\rangle_{T_s}}{dt} = \langle i_{L2}(t)\rangle_{T_s}(d_2(t) - d_1(t)) \quad (29)$$

However, there is still a portion of the $ESRs$ to be added.

$$\langle v_{dif}(t)\rangle_{T_s} = \langle v_{cdif}(t)\rangle_{T_s} + r_oC_o\frac{d\langle v_{cdif}(t)\rangle_{T_s}}{dt} \quad (30)$$

Therefore, $\langle v_{dif}(t)\rangle_{T_s}$ is the difference of the output voltages taking into account the voltages on the $ESRs$.

Now, the small signal analysis for variables $v_{cdif}(t), v_{dif}(t), i_{L_2}(t), d_1(t)$ and $d_2(t)$ is performed. Then a linearization process is followed to get the following expressions.

$$C_o\frac{dv_{\hat{c}dif}(t)}{dt} = I_{L_2}(\hat{d}_2(t) - \hat{d}_1(t)) + \hat{i_{L_2}}(t)(D_2 - D_1) \quad (31)$$

$$v_{\hat{d}if}(t) = v_{\hat{c}dif}(t) + r_oC_o\frac{dv_{\hat{c}dif}(t)}{dt} \quad (32)$$

Moreover, (31) can be further reduced by considering $D_2 - D_1 = 0$ since the mean values for the two duty cycles are equal in steady state. Applying the Laplace transform to (31) and (32) gives (33) and (34).

TABLE I
SYSTEM PARAMETERS CONSIDERED FOR THE NUMERICAL VALIDATION

Input Voltage	V_{in}	34 V
Output Voltage	V_o	380 V
Frequency	f_s	50 kHz
Intermediate filter Capacitor	C_{oint}	940 μF
Output filter Capacitors	C_{o1}, C_{o2}	330 μF
Inductance	L_1, L_2	66 μH, 369 μH
Load	R_o	278 Ω
ESRs	r_{oint}, r_{o1}, r_{o2}	0.05 Ω, 0.1 Ω, 0.1 Ω

$$sC_oV_{cdif} = -I_{L_2}(d_1(s) - d_2(s)) \quad (33)$$

$$V_{dif}(s) = V_{cdif}(s)(1 + r_oC_os) \quad (34)$$

Solving this system of equations gives the following transfer function, where I_{L2} represents an expression in steady state. In the control method to be used there is a sum of the control action d_1 and subtraction of the control action d_2. However, since $d_1 = d_2 = d$, then $d_1 - d_2 = 2d$, so (35) must be multiplied by two, after the considered control method.

$$G_{vdif}(s) = \frac{V_{dif}(s)}{\Delta d(s))} = \frac{-I_{L_2}(1 + r_oC_os)}{sC_o} \quad (35)$$

$$G_{vdif}(s) = \frac{-2I_{L_2}(1 + r_oC_os)}{sC_o} \quad (36)$$

IV. VALIDATION OF MODELS THROUGH NUMERICAL SIMULATION

The validation of the mathematical models consists in tracing the bode graphs of the simulated circuit and the transfer function, and making the comparison between the curves.

The transfer functions are verified in the simulation by the addition of a disturbance in the duty cycle with sinusoids of different frequencies. The simulation was performed in the PSIM software.

A. Validation of $G_{id}(s)$

To facilitate the validation process, the simulation was divided in two parts, the first part consists of a simulation for the frequencies between 10 Hz and 500 Hz. The second simulation comprises the frequencies between 500 Hz and 10 kHz. As the converter has the characteristic of attenuation in high frequencies, it is necessary to apply a perturbation of greater amplitude and to execute the simulation with smaller calculation step due to the larger frequency of the perturbation.

Figure 5 shows the magnitude and phase bode plots. Notice that it is possible to verify that the performance of the theoretical model is compatible with the simulation in the entire frequency range analysed.

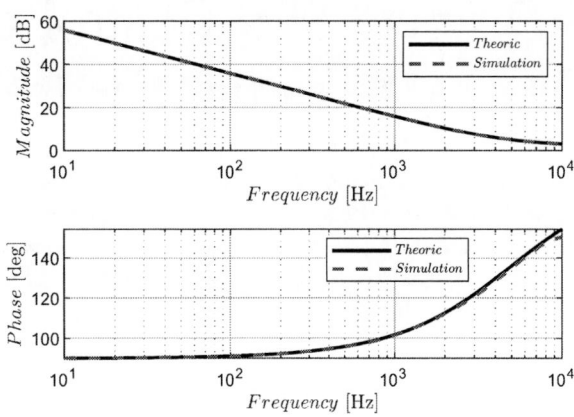

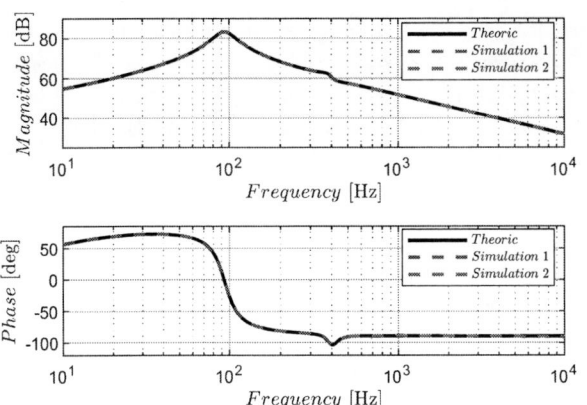

Fig. 5. Bode diagram for plant validation $G_{id}(s)$

B. Validation of $G_{vi}(s)$

The simulation to validate $G_{vi}(s)$ was divided in three parts. To verify the hypothesis that the plant has more zeros than the number of poles, a simulation was performed for a larger frequency range, going from 10 Hz to 500 kHz.

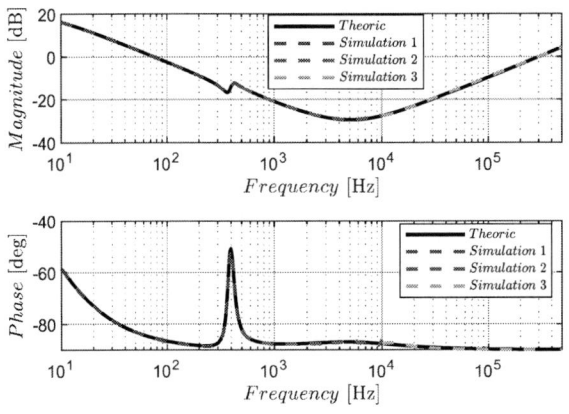

Fig. 6. Bode diagram for plant validation $G_{vi}(s)$.

The graphs verify that the behaviour of the theoretical model is compatible with the simulation in the analysed frequency range. As shown in Figure 6, there is a magnitude growth in the response of the transfer function at high frequency. All these are strong indications that it is due to the fact that the plant has a greater number of zeros than poles.

C. Validation of $G_{vdif}(s)$

Figure 7 shows the magnitude and phase bode diagram of $G_{vdif}(s)$. Notice that the simulation followed the theoretical behaviour in all analysed frequencies.

Fig. 7. Bode diagram for the validation of $G_{vdif}(s)$.

V. CONCLUSIONS

This work presented the three-level quadratic boost converter model considering the capacitors equivalent series resistors. The consideration of the capacitors ESRs favours the damping of the resonant peak, leaving the model closer to what is observed in practice. All the models were tested through simulation. The tests showed that the equations obtained from the analytical study correspond to the simulation results. It should be noted that the test example was designed for a nominal power of 520 W, which has been presented with greater detail in [8].

ACKNOWLEDGMENT

The authors would like to thank UDESC and FAPESC for supporting the project.

REFERENCES

[1] J. B. R. Cabral, T. Lemes da Silva, S. V. G. Oliveira and Y. R. de Novaes, "A new high gain non-isolated DC-DC boost converter for photovoltaic application," 2013 Brazilian Power Electronics Conference, Gramado, 2013, pp. 569-574.

[2] L. Huber and M. M. Jovanovic, "A design approach for server power supplies for networking applications," APEC 2000. Fifteenth Annual IEEE Applied Power Electronics Conference and Exposition, New Orleans, LA, USA, 2000, pp. 1163-1169 vol.2.

[3] Tsai-Fu Wu and Te-Hung Yu, "Unified approach to developing single-stage power converters," in IEEE Transactions on Aerospace and Electronic Systems, vol. 34, no. 1, pp. 211-223, Jan. 1998.

[4] J. Morales Saldana, R. Galarza-Quirino, J. Leyva Ramos, E. Carbajal Gutierrez e M. Ortiz Lopez, "Multiloop Controller Design for A Quadratic Boost Converter," Electric Power Applications, IET, vol. 1, n. 3, pp. 362-367, may 2007.

[5] V. Vorperian, "Simplified analysis of PWM converters using model of PWM switch. Continuous conduction mode," in IEEE Transactions on Aerospace and Electronic Systems, vol. 26, no. 3, pp. 490-496, May 1990.

[6] J. B. R. Cabral, "DC-DC Converter Not Isolated from High Gain For Application In Photovoltaic Solar Energy Processing," M.S. thesis, Santa Catarina State University (UDESC), Santa Catarina, Brazil, 2013.

[7] M. Borchardt, "Controle Digital de um conversor quadrático," Santa Catarina State University (UDESC), Santa Catarina, Brazil, 2018.

[8] M. C. Orige, "Modelagem de topologias Boost para estudo de chaveamento dependente do estado," Santa Catarina State University (UDESC), Santa Catarina, Brazil, 2018.

Discrete SPS Control of a DAB converter using partial Feedback Linearization

Eduardo Luiz Santos da Silva
Department of Automation and Systems
Federal University of Santa Catarina
Florianópolis - SC, Brazil
eduardoluizss@gmail.com

André Luís Kirsten
Institute of Power Electronics
Federal University of Santa Catarina
Florianópolis - SC, Brazil
kirsten.andre@gmail.com

Daniel Juan Pagano
Department of Automation and Systems
Federal University of Santa Catarina
Florianópolis - SC, Brazil
daniel.pagano@ufsc.br

Abstract—**This article presents the design and evaluation of a discrete non-linear controller with bilinear approximation to regulate the output voltage of a Dual Active Bridge (DAB) converter under parametric uncertainties. In this paper we consider the possible uncertainties in the system, such as variations in load and input voltage. The drive is controlled by single phase shift modulation (SPS) with a fixed duty cycle. The discrete controller developed ensures disturbance rejection, reference tracking and zero overshoot, with a small transient regime. The results obtained through the Typhoon HIL are presented to validate the effectiveness of the developed controller.**

Index Terms—**DAB converter, Discrete Modeling, Nonlinear control, Feedback Linearization Control, Hardware in the loop.**

I. Introduction

Dc transmission and distribution systems (dc microgrids), are rising due to the crisis of conventional energy, the environmental issue and the increasing load demanding. The power processing systems developed for renewable energies generation are encouraging the development of smarter electronic power converters that have high density of power and bidirectionality of energy.

Even if the basic converters like buck and boost could be used, the achieved power for these converters would be lower because they use only one switch when compared with other converters that have more switches. For this reason, when it comes to high power, other topologies of static converters must be used.

One of this topologies is the so called Dual Active Bridge (DAB), developed in the 1990s by [1]. The DAB was thought to reduce the weight of existing DC-DC converters at the time. Currently its applications are as diverse as smart grids [2], solid state transformers [3], and in photovoltaic systems [4].

Regarding the control, non-linear control techniques, such as: feedback linearization control; sliding mode; hysteresis control; when applied to static converters provide a more natural response. [5]–[9]

The purpose of this paper is to present the discrete modeling of the DAB converter and the design of a nonlinear control based on this discrete model. In order to do that, a partial feedback linearization is applied the model equations and a

This study was financed in part by the Personnel Improvement Coordination of Superior Level - Brazil (CAPES) - Finance Code 001.

PI discrete control is synthetized from the linearized system model. Experimental results of the DAB converter modeled in a Typhoon HIL system and controlled by a non-linear control implemented in a DSP are obtained and comparing with the results of a classic PI controller.

The remainder of this paper is organized as follows. Section II introduces the discrete mathematical model of DAB converter under phase-shift control. Section III, presents the reduction of the discrete model, by means of a bilinear approximation, to be used in the control design. The nonlinear control design based on a partial feedback linearization is presented in Section IV. Section IV-D, presents the control design of a classical discrete PI controller used to compare the performance of the proposed controller. Finally, experimental results obtaining from Typhoon-HIL emulator and a DSP-based controller are shown in Section V.

II. Dual Active Bridge

Fig. 1 shows the circuit diagram of the DAB converter with two full bridges where C_o is the DC output capacitor; L_t is the transformer inductor; r_t is the inductor resistance and the turns ratio of transformer is $N_p{:}N_s$; R_o is the load resistance; R_c is the capacitor resistance; V_i is the DC input voltage while V_o is the DC output voltage; i_{Lt} is the inductor current of the DAB input H-bridge; $\frac{i_{Lt}}{n}$ is the output current of the DAB output H-bridge; i_o is the load current.

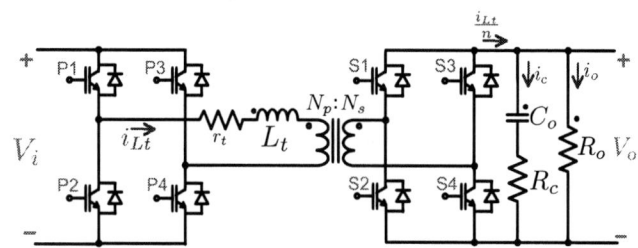

Fig. 1: Dual Active Bridge converter

According to [10]–[12], there are currently four popular modulation modes: Single Phase Shift (SPS), Extended Phase Shift(EPS), Dual Phase-Shift (DPS), Triple Phase-Shift (TPS).

978-1-7281-4181-7/19 $31.00 © 2019 IEEE

The most common type of modulation is the SPS because it is easier to control, the other modulations, called three levels, are more complex [10], [11], [13].

A. SPS modulation modeling

We consider in this work the SPS modulation where there are four steps of operation for the DAB converter [10], [14]. Table I shows where the current leads in each of the four steps. For these steps it is considered that the converter is in steady state, and the energy flow is from the primary to the secondary, but it is also possible to establish the reverse flow since it is equivalent.

TABLE I: DAB converter steps.

Steps	Conducting current
1	P1, P4, S2, S3
2	P1, P4, S1, S4
3	P2, P3, S1, S4
4	P2, P3, S2, S3

Based on the Kirchoff laws and transformer ratios it is possible to determine the capacitor voltage (V_c) and the inductance current (i_{Lt}) for each step. By placing these equations in the matrix mode according to the model of (1), we obtain the matrices as shown in (2) and (3).

$$\begin{bmatrix} \frac{di_L}{dt} \\ \frac{dV_c}{dt} \end{bmatrix} = A_i \begin{bmatrix} i_L \\ V_c \end{bmatrix} + B_i Vi \tag{1}$$

$$A_1 = A_4 = \begin{bmatrix} \dfrac{-n^2 R_t - \dfrac{R_o R_c}{(R_o + R_c)}}{n^2 L} & \dfrac{Ro}{nL(R_o + R_c)} \\ \dfrac{-R_o}{nC_o(R_o + R_c)} & \dfrac{-1}{C_o(R_o + R_c)} \end{bmatrix} \tag{2}$$

$$A_2 = A_3 = \begin{bmatrix} \dfrac{-n^2 R_t - \dfrac{R_o R_c}{(R_o + R_c)}}{n^2 L} & \dfrac{-Ro}{nL(R_o + R_c)} \\ \dfrac{R_o}{nC_o(R_o + R_c)} & \dfrac{-1}{C_o(R_o + R_c)} \end{bmatrix} \tag{3}$$

$$B_1 = B_2 = \begin{bmatrix} \frac{1}{L} & 0 \end{bmatrix}^T, \quad B_3 = B_4 = \begin{bmatrix} -\frac{1}{L} & 0 \end{bmatrix}^T \tag{3}$$

In this work, we will assume that C_o is an ideal capacitor, it means $R_c = 0$, so the matrices A of each step become (4), and $Vc = Vo$.

$$A_1 = \quad A_4 = \begin{bmatrix} -\dfrac{R_t}{L} & \dfrac{1}{nL} \\ -\dfrac{1}{nC_o} & -\dfrac{1}{C_o R_o} \end{bmatrix}$$

$$\tag{4}$$

$$A_2 = \quad A_3 = \begin{bmatrix} -\dfrac{R_t}{L} & \dfrac{1}{nL} \\ \dfrac{1}{nC_o} & -\dfrac{1}{C_o R_o} \end{bmatrix}$$

B. Time for each step

The duration time of each step of the DAB converter depends on the value of φ, in the SPS modulation steps 1 and 3 have the same time, likewise 2 and 4 as well. And steps 1 and 2 are complementary, as are 3 and 4. From this information we can arrive at the following time values as

$$t_1 = t_3 = \frac{\varphi}{\pi} \frac{Ts}{2}, \tag{5}$$

$$t_2 = t_4 = \left(1 - \frac{\varphi}{\pi}\right) \frac{Ts}{2}. \tag{6}$$

C. Calculation of DAB parameters

To simulate a DAB converter it is necessary to define the output power, the input and output voltages and the switching frequency. From this information it is possible to determine the other parameters necessary for the simulation.

Considering the equation of the output power of the DAB converter presented in (7), it is possible to find the value of the inductor as given by

$$P_o = \frac{N_p Vi Vo \varphi(\pi - \varphi)}{N_s \pi^2 2 f_s L} \tag{7}$$

$$L = \frac{N_p Vi Vo \varphi(\pi - \varphi)}{2 f_s N_s \pi^2 P_o} \tag{8}$$

The value of the bus capacitor to be used in the DAB can be calculated through

$$C_o = \frac{\varphi P_o}{\omega_s Vo^2 \Delta V_{(\%)}}. \tag{9}$$

To find the value of the resistance of the converter is used

$$R = \frac{Vo^2}{P_o}. \tag{10}$$

Using the following values shown in table II it is possible to find the values of the inductor (1.1 mH), capacitor (104,17μH) and resistance (80 Ω).

TABLE II: Converter values for the simulations.

Parameters	Values	Parameters	Values
P_o	2 kW	N_p	2
n	0.5	N_s	1
Vi	800 V	ΔV	5 %
Vo	400 V	φ_n	30°
f_s	20 kHz		

III. DISCRETE SYSTEM MODELING

In this section we develop the discrete system model of DAB. First let us considering (1) and the matrices (3), (4), it is possible to discretize each of the four operation steps of the SPS modulation, which we will call A_{di} and B_{di}, where i is equal to 1, 2, 3 or 4 according to the step operation. In (11) it is possible to see how was discretize each matrix.

$$A_{di} = e^{A_i t_i} \quad B_{di} = \int_0^{t_i} e^{A_i t_i} B_i dt = A_i^{-1}(e^{A_i t_i} - I)B_i \tag{11}$$

And as the operation steps are repeated at each period, in the discrete system it is possible to couple the four steps in only one equation as can be seen below [15].

$$\begin{bmatrix} i_L[n+1] \\ V_o[n+1] \end{bmatrix} = (A_{d4} \cdot A_{d3} \cdot A_{d2} \cdot A_{d1}) \begin{bmatrix} i_L \\ V_o \end{bmatrix}_n +$$
$$+ (A_{d4} \cdot A_{d3} \cdot A_{d2}B_{d1} + A_{d4} \cdot A_{d3} \cdot B_{d2}$$
$$+ A_{d4} \cdot B_{d3} + B_{d4})V_i \tag{12}$$

A. Bilinear Approximation

The model as presented previously in 12, although it works very well, makes the task, to perform a control of the system, arduous because it uses exponential matrices.

Therefore, a bilinear approximation approach is used. The approximations for the exponential matrices are seen in 13, where i is for steps 1 and 3, and j for steps 2 and 4

$$e^{A_i t_i} = e^{A_i \frac{\Phi}{2\pi f_s}} e^{A_i \left(t_i - \frac{\Phi}{2\pi f_s}\right)}$$
$$\approx e^{A_i \frac{\Phi}{2\pi f_s}} \left(I + A_i \left(t_i - \frac{\Phi}{2\pi f_s}\right)\right)$$

$$e^{A_j t_j} = e^{A_j \left(T_s/2 - \frac{\Phi}{2\pi f_s}\right)} e^{A_j \left(t_j - T_s/2 + \frac{\Phi}{2\pi f_s}\right)}$$
$$\approx e^{A_j \left(T_s/2 - \frac{\Phi}{2\pi f_s}\right)} \left(I + A_j \left(t_j - T_s/2 + \frac{\Phi}{2\pi f_s}\right)\right) \tag{13}$$

As you can see, the bilinear approximation determines a fixed φ to calculate the exponential matrix. This fixed value of φ is called Φ. Thus, when the value of φ equals the value set in Φ the exponential matrix will be exact, and as the value of φ is distancing from Φ the correction will be bilinear approach.

B. Study of the discrete bilinear model

For the circuit being studied, the input voltage, the phase shift angle and the resistive load are considered variables. However the resistive load is present in matrix A_i, and due to

the fact that to perform control it is not good to have variable values raised to an exponential it is necessary to fix a value for the load variable that is in the exponential matrix.

Applying (13) in (12) and made algebraic manipulations, we obtain a matrix M_{2x1}, but because of the page limit we don't show in this article, see more details in [15]. Because of the big size of the matrix is hard to be able to carry out any type of evaluation of the model for the control. Therefore, in order to evaluate the model, the values of frequency, inductor, resistors, capacitor and the transformation ratio that were presented in the section II-C, and Φ equals to the nominal value of φ_n ($=\frac{\pi}{6}$) were used to have an smaller equation.

1) Reducing model order: Nevertheless, the four-step system has two fourth order equations, with input and output voltage, resistive load and leakage current variables.

In order to reduce the order of the model and to facilitate the development of a control, the mean error between the switched model and the discrete model was calculated by eliminating the fourth, third and second order of the model. The table III displays the average errors of these models.

TABLE III: Average error of V_o and i_L between models for different bilinear model orders

Order	Average error V_o	Average error i_L
4th	0.2242	0.0031
3rd	0.1780	0.0037
2nd	0.3147	0.7602
1st	3.4100	3.5820

As can be seen in the table III, the mean error of the second order model remains small and therefore the second order reduction of the two system equations will be made.

2) Inductor current: In order to evaluate the influence of the current on the output voltage, an experiment was carried out comparing the bilinear 4th order model with and without the inductor current to the nominal values defined in section II-C and calculated the mean error between the models. The test result can be in the table IV.

TABLE IV: Average error of V_o between switched model and discrete bilinear model of 4th order with and without i_L

		Models with current	
		Switched	Bilinear
Bilinear	With current	0.224	0.000
Models	Without current	0.124	0.100

As can be seen in the table IV, the mean difference in the final value between the model with and without the current is only 0.1 volts and the mean difference between the switched model and the 4th order bilinear model without the current is 0.124 volts, which is a smaller difference than the model with the current. Therefore, the values multiplied by the current will be disregarded.

3) Resistive load: The value of the resistive load R is contained within the matrix A that is present in the exponential matrix as well as in the correction of the bilinear approximation.

978-1-7281-4181-7/19 $31.00 © 2019 IEEE

In order to verify if the bilinear model is sufficient to follow load changes an experiment was carried out by changing the value of the load for different situations of V_i and φ. For the accomplishment of this test the bilinear model of 4th was used and the switched for comparison.

The conclusion of this test was that the bilinear model can not keep up with the load variations of the system, and therefore the entire value of the resistive load will be constant in the nominal value, but as will be seen below, the controller can control these variations of resistive load.

C. Simplified discrete model

After carrying out the study on the discrete bilinear model, it was possible to eliminate the 3rd and 4th order terms, and consider the average of $i_L = 0$ and $R = 80$. In this way the discrete model for the DAB converter presented in (14).

$$
\begin{aligned}
V_o[n+1] = {} & 3.48445E^{-6}\, V_i[n] + 0.995233\, V_o[n] \\
& + (0.00690383\, V_i[n] - 0.00461305\, V_o[n])\varphi[n] \\
& + (-0.00218989\, V_i[n] + 0.00440413\, V_o[n])\varphi[n]^2
\end{aligned}
\tag{14}
$$

IV. CONTROL DESIGN

We propose a partial feedback linearization to linearize (14) and a discrete PI control.

A. Partial FLC

Considering the control signal $u_{[n]}$ equal to the non-linear term of (14) we have (15).

$$
\begin{aligned}
u_{[n]} = {} & (0.00690383\, Vi_{[n]} - 0.00461305\, Vo_{[n]})\varphi_{[n]} \\
& + (-0.00218989\, Vi_{[n]} + 0.00440413\, Vo_{[n]})\varphi_{[n]}^2
\end{aligned}
\tag{15}
$$

To control the DAB converter, $\varphi_{[n]}$ will be used as the control variable for the FLC proposed in this article. Therefore, from the (15) two solutions can be obtained in (16).

$$
\varphi_{[n]} = \frac{-0.00690383\, Vi_{[n]} + 0.00461305\, Vo_{[n]} \pm \sqrt{\Delta}}{2(-0.00218989\, Vi_{[n]} + 0.00440413\, Vo_{[n]})}
\tag{16}
$$

where Δ is:

$$
\begin{aligned}
\Delta = {} & (0.00690383\, Vi_{[n]} - 0.00461305\, Vo_{[n]})^2 \\
& + 4(-0.00218989\, Vi_{[n]} + 0.00440413\, Vo_{[n]})u_{[n]}
\end{aligned}
$$

Since $\varphi_{[n]}$ is restricted from 0° to 90° only the solution with $+\sqrt{\Delta}$ can be used for the development of the control.

B. PI control design

Considering the system, now linearized with the partial FLC, it is possible to write the transfer function as

$$
\frac{Vo_{[z]}}{U_{[z]}} = \frac{1}{(z - 0.995233)}.
\tag{17}
$$

With this presented transfer function was developed a PI controller using the pole-placement methodology with $\xi = 0.7$

and a settling time of 10 miliseconds. The controller is given by

$$
C(z) = \frac{0.02524(z - 0.9822)}{(z - 1)}.
\tag{18}
$$

C. Dynamic Saturation

The constraints on the φ variable $\left(0 \le \varphi_{[n]} \le \frac{\pi}{2}\right)$ can be defined for the control variable $u_{[n]}$ from (15) in such a way that for $\varphi_{[n]} = 0$ we have $u_{[n]min} = 0$ and for $\varphi_{[n]} = \frac{\pi}{2}$,

$$
u_{[n]max} = 0.00544117\, Vi_{[n]} + 0.00362059\, Vo_{[n]}.
\tag{19}
$$

Moreover for saturation conditions $u_{[n]min}$ and $u_{[n]max}$, an anti-windup *Back Calculation* mechanism was implemented.

D. Linear control

To compare the proposed FLC controller results, a classical discrete PI controller was designed. The mathematical model used for the design of this control is given by

$$
\frac{dVo}{dt} = \frac{N_p Vi\varphi(\pi - \varphi)}{N_s 2\pi^2 f_s \pi^2 f_s LC} - \frac{Vo}{RC},
\tag{20}
$$

and since (20) is a non-linear equation, we linearize (20) around an operating point, which was chosen from the nominal values of table II. Now, by applying Laplace transformation we obtain

$$
\frac{\Delta Vo(s)}{\Delta \varphi(s)} = \frac{74084.13}{(s + 120)}.
\tag{21}
$$

The PI controller was designed using the pole-placement methodology, and the same specifications design of the non-linear control. After discretization with Tustin method, the discrete PI controller is given by

$$
C(z) = \frac{0.0065411(z - 0.981)}{(z - 1)}.
\tag{22}
$$

V. HIL EXPERIMENTAL RESULTS

Experimental results for the proposed and classical controller were obtained using a Typhoon 402 hardware in the loop (HIL) setup, as shown in Fig. 2, and the control algorithms were implemented in a Texas Instruments DSP F28379D.

Fig. 2: Experimental HIL setup.

A. Reference tracking

The first experiment was carried out to verify if the control is able to track the output voltage reference (V_{ref}). The nominal parameters given in section II-C were used. The V_{ref} was changed from $400V$ to $500V$ at $t = 0.1s$, and at $t = 0.2s$ the V_{ref} was setting back to $400V$, as shown in Figs. 3-4.

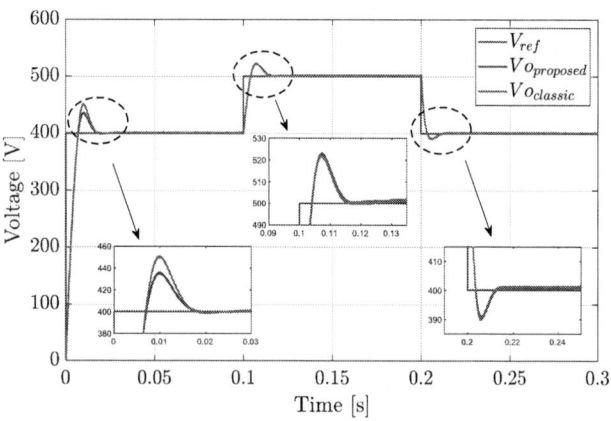

Fig. 3: Voltage in closed-loop system performance for reference tracking

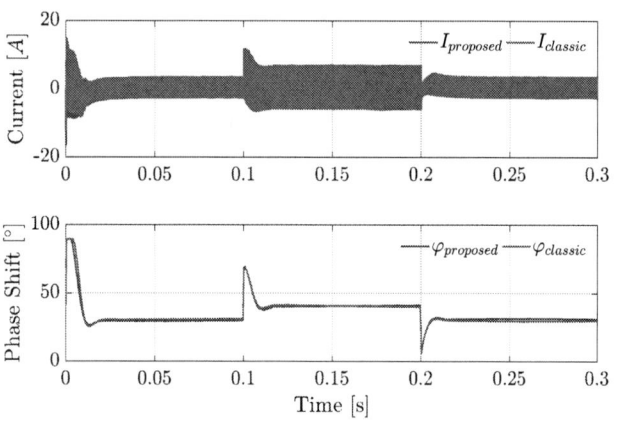

Fig. 4: Inductor current and phase shift in closed-loop system performance for reference tracking

Figures 3 and 4 show the tracking reference performance of both controllers with similar time-responses. The only difference between controllers is in the first step from 0 to 400 V, that the classic controller has a larger overshoot. We can confirm this affirmations with the table V that shows the settling time and the overshoot for the three reference variations.

TABLE V: Settling time and overshoot for experimental 1

Settling time		Overshoot	
Proposed	Classic	Proposed	Classic
0.0130	0.0139	8.9722	12.7242
0.1027	0.1027	4.6004	4.4225
0.2021	0.2022	2.4069	2.5539

B. Input Voltage Variation

The second experiment was performed to verify if the control is able to reject input voltage disturbances. The input voltage was chaged from $800V$ to $900V$ at $t = 0.1s$, and at $t = 0.2s$ was setting back to $800V$, as shown in Figs. 5-6.

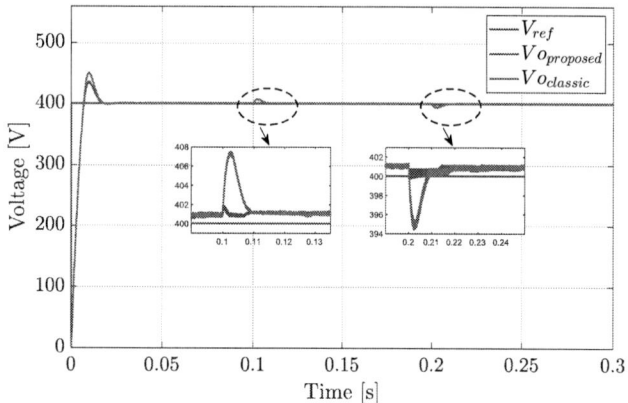

Fig. 5: Voltage in closed-loop system performance for input voltage variation

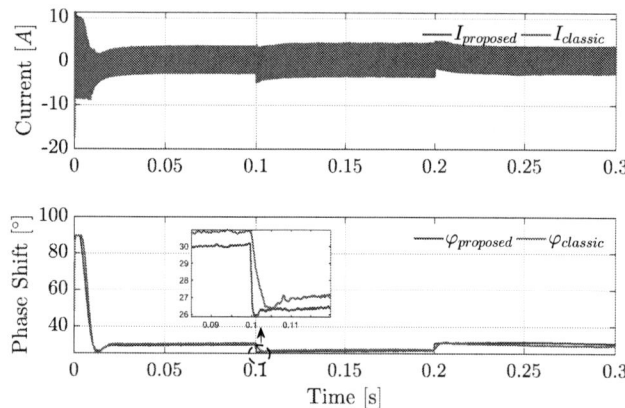

Fig. 6: Inductor current and phase shift in closed-loop system performance for input voltage variation

As expected the FLC controller reject the input voltage disturbances since it act as a feedforward controller for this signal. The response of the classical PI controller present a larger overshoot, as we can confirm in table VI. The control signal (φ) for both controllers is also shown in Fig.6. The FLC φ practically responds instantly when input voltage variations were applied, as shown in the zoom of the figure, and in table VI.

TABLE VI: Settling time and overshoot for experimental 2

Settling time		Overshoot	
Proposed	Classic	Proposed	Classic
0.1	0.1	0.4918	1.8708
0.2	0.2	0.0588	1.3996

978-1-7281-4181-7/19 $31.00 © 2019 IEEE

C. Load Variation

The third experiment was carried out to verify if the control should reject perturbations in the resistive load. The resistive load was changed from 80Ω to 90Ω at $t = 0.1s$, and at $t = 0.2s$ the R_o was setting back to 80Ω, as shown in Figs.7-8.

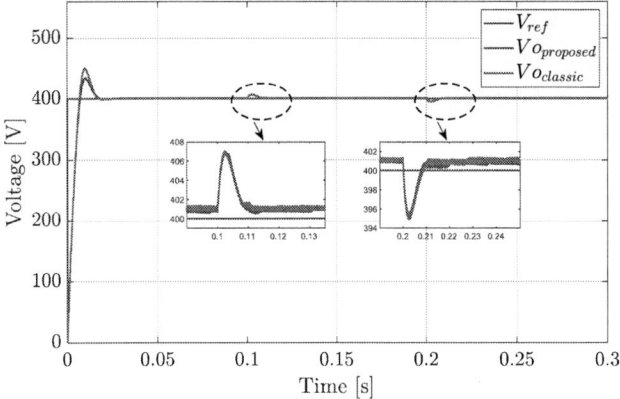

Fig. 7: Voltage in closed-loop system performance for load variation

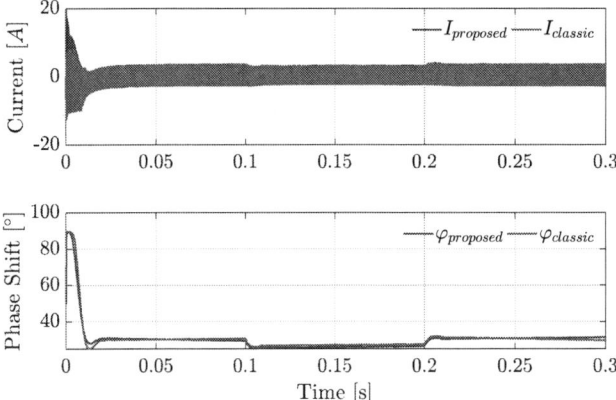

Fig. 8: Inductor current and phase shift in closed-loop system performance for load variation

As can be observed, with the figure 7 and the table VII, for the applied load variations both controllers have a similar response for the perturbation rejection of a resistive load. This happens because we have fixed the value of the resistor in the feedback linearization control.

TABLE VII: Settling time and overshoot for experimental 3

Settling time		Overshoot	
Proposed	Classic	Proposed	Classic
0.1	0.1	1.7559	1.7437
0.2	0.2	1.2616	1.2667

VI. CONCLUSION

This paper address the discrete modeling of the DAB converter and the design of a nonlinear controller. This control law is based on a partial feedback linearization of the DAB

equations and in a discrete PI control. Experimental results were obtained using a Typhoon-HIL module and a DSP-based controller. The proposed controller, when compared with a classical discrete PI controller, shows better performance for voltage input perturbations and similar performance for load disturbances. As we can see the non-linear part of the DAB converter is not very expressive. Further research will conducted to estimate the i_o current load to use in FLC and in this way to reject load disturbances.

REFERENCES

[1] R. W. A. A. D. Doncker, D. M. Divan, and M. H. Kheraluwala, "A three-phase soft-switched high-power-density dc/dc converter for high-power applications," *IEEE Transactions on Industry Applications*, vol. 27, pp. 63–73, Jan 1991.

[2] S. Inoue and H. Akagi, "A bi-directional isolated dc/dc converter as a core circuit of the next-generation medium-voltage power conversion system," in *37th IEEE Power Electronics Specialists Conference*, pp. 1–7, June 2006.

[3] J. Shi, W. Gou, H. Yuan, T. Zhao, and A. Q. Huang, "Research on voltage and power balance control for cascaded modular solid-state transformer," *IEEE Transactions on Power Electronics*, vol. 26, pp. 1154–1166, April 2011.

[4] P. Joebges, J. Hu, and R. W. D. Doncker, "Design method and efficiency analysis of a dab converter for pv integration in dc grids," in *2016 IEEE 2nd Annual Southern Power Electronics Conference (SPEC)*, pp. 1–6, Dec 2016.

[5] A. H. R. Rosa, L. M. F. Morais, and S. I. S. Junior, "Estudo e comparação de técnicas de controle não lineares aplicadas a conversores estáticos de potência.," in *XXI Congresso Brasileiro de Automática*, (Vitória, ES), pp. 1590 – 1595, Outubro 2016.

[6] V. M. Rao, A. K. Jain, K. K. Reddy, and A. Behal, "Experimental comparison of digital implementations of single-phase pfc controllers," *IEEE Transactions on Industrial Electronics*, vol. 55, pp. 67–78, Jan 2008.

[7] H. Rodriguez, R. Ortega, and G. Escobar, "A new family of energy-based non-linear controllers for switched power converters," in *ISIE 2001. 2001 IEEE International Symposium on Industrial Electronics Proceedings (Cat. No.01TH8570)*, vol. 2, pp. 723–727 vol.2, 2001.

[8] G. Escobar, R. Ortega, H. Sira-Ramirez, J. P. Vilain, and I. Zein, "An experimental comparison of several non linear controllers for power converters," in *Industrial Electronics, 1997. ISIE '97., Proceedings of the IEEE International Symposium on*, vol. 1, pp. SS89–SS94 vol.1, Jul 1997.

[9] C. Zhou, R. B. Ridley, and F. C. Lee, "Design and analysis of a hysteretic boost power factor correction circuit," in *21st Annual IEEE Conference on Power Electronics Specialists*, pp. 800–807, 1990.

[10] A. Tong, L. Hang, G. Li, and J. Xu, "Equivalent circuit model of dual active bridge converter," in *IECON 2017 - 43rd Annual Conference of the IEEE Industrial Electronics Society*, pp. 4677–4682, Oct 2017.

[11] M. Jafari, Z. Malekjamshidi, and J. G. Zhu, "Analysis of operation modes and limitations of dual active bridge phase shift converter," in *2015 IEEE 11th International Conference on Power Electronics and Drive Systems*, pp. 393–398, June 2015.

[12] B. Zhao, Q. Song, W. Liu, and Y. Sun, "Overview of dual-active-bridge isolated bidirectional dc-dc converter for high-frequency-link power-conversion system," *IEEE Transactions on Power Electronics*, vol. 29, pp. 4091–4106, Aug 2014.

[13] J. A. Mueller and J. Kimball, "An improved generalized average model of dc-dc dual active bridge converters," *IEEE Transactions on Power Electronics*, vol. PP, no. 99, pp. 1–1, 2018.

[14] Q. Tian and K. Bai, "Widen the zero-voltage-switching range and secure grid power quality for an ev charger using variable-switching-frequency single-dual-phase-shift control," *Chinese Journal of Electrical Engineering*, vol. 4, pp. 11–19, March 2018.

[15] L. Shi, W. Lei, Z. Li, J. Huang, Y. Cui, and Y. Wang, "Bilinear discrete-time modeling and stability analysis of the digitally controlled dual active bridge converter," *IEEE Transactions on Power Electronics*, vol. 32, pp. 8787–8799, Nov 2017.

Steady-State Analysis of a Single-phase Modified Bridgeless Boost Rectifier in DCM

Mateo D. Roig G., Caio G. da S. Moraes and Telles B. Lazzarin

Departament of Electronics and Electrical Engineering
Federal University of Santa Catarina UFSC
Power Electronics Institute INEP
Florianopolis, BRAZIL
mateodaniel.roigg@gmail.com, caio.guimoraes@gmail.com, telles@inep.ufsc.br

Abstract—This article presents an analysis and design of an AC-DC Boost single-phase converter operating in discontinuous conduction mode (DCM) with output voltage regulation and power factor correction (PFC). The topology dispenses the rectifier bridge reducing the conduction losses and, however, it is already known in the literature, there is no previously steady-state modeling presented for this operation mode. Furthermore, the article presents a comparison between the converter with the other two common used topologies for the same kind of application. All results were validated with a 500 W prototype, with a universal input $100 - 220$ V_{AC} and 450 V_{DC} voltage output. The results presented nearby 96% of efficiency.

Index Terms—Boost, modified, bridgeless, rectifier, modeling, discontinuous, comparison

I. INTRODUCTION

THE advancement in power electronics technology and the proposal of new topologies have increased the number of even more compact and efficient solutions. However, steady-state power converters mostly introduce a non-linear distortion characteristic in the circuit, especially for rectifiers with a capacitive filter in the input stage. To mitigate such problems, meet the standards and increase the efficiency of the rectifiers, Power Factor Correction (PFC) techniques are employed in several topologies.

With this purpose, the Boost family rectifiers draw attention to the simplicity of their design in addition to the input current source characteristic. In this context, the Boost converter cascaded with a bridge rectifier is the most used for active power factor correction. [1], [2]. However, due to the presence of the bridge rectifier, there are always three semiconductors in the current circulation path, which considerably increases the conduction losses in this topology.

In order to reduce the semiconductor losses due to the bridge rectifier and also to increase the efficiency, several alternatives were proposed for the Boost rectifier [3]. The conventional bridgeless topology originally proposed in [4], as well as some of its variations, reduces conduction losses by eliminating a diode from the current path. However, this structure exhibits significantly greater common-mode current over the conventional Boost rectifier [3], [5].

The common-mode current, though, can be reduced by modifying the bridgeless Boost rectifier circuit. This might be

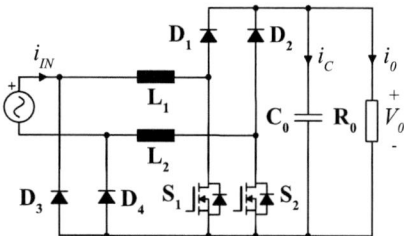

Fig. 1. Modified Bridgeless Boost rectifier circuit

done providing a low-frequency path between the AC power source and the circuit reference ground. One of the alternatives was originally proposed by [6] and is represented in Fig. 1. Beyond the inherent advantages of the bridgeless topology, the studied converter does not require isolated sensors to sample the current and the input voltage. It can be noted, in Fig. 1, that the switches S1 and S2 can be driven with the same switching signal, which simplifies the implementation of the control circuit.

Regarding the operation mode, the rectifier topology can operate both continuous (CCM) and discontinuous conduction mode (DCM). Still, the DCM operation simplifies the control strategy, besides avoiding reverse recovery of the output diode [7]. As a disadvantage, in this mode of operation, the input current shows some distortion due to the modulation of the discharge time of the inductors [8]. However, there is no detailed analysis in the literature about the steady-state modeling of the modified bridgeless Boost rectifier (MBBR) operating in DCM. On this, the article hand over an ideal steady-state analysis and design of the presented bridgeless rectifier, with output voltage regulation and power factor correction. Also, some comparison within the commonly used rectifiers of the Boost family, the Boost rectifier with diode bridge (BR) and the Conventional bridge-less Boost rectifier (CBBR), will be presented, highlighting the advantages that motivate the use of the topology.

II. STEADY-STATE ANALYSIS OF THE CONVERTER

The Modified Bridgeless Boost converter is not much different from the other rectifiers of the Boost family essentially for the analysis on the switching period. Even though, concerning

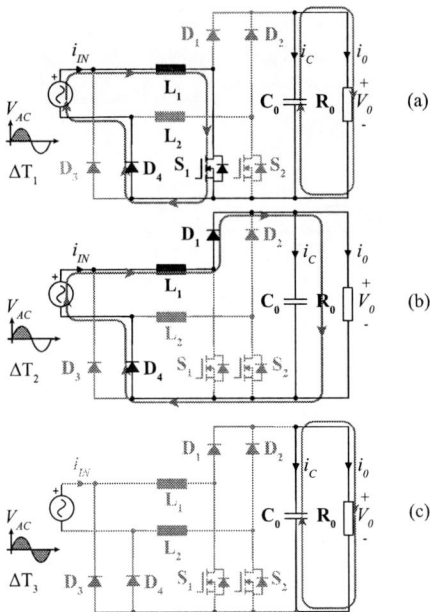

Fig. 2. Rectifier operating stages: during the grid positive half-cycle a) e b); during the negative half-cycle of the grid; c) in the discontinuos operating mode of the inductor.

the period of the input voltage, some substantial distinctions justify the use of the topology. This section may provide the theoretical steady-state background for further comparison.

A. Operation Stages

The bridgeless Boost rectifier works in five operation different stages. Two of them operates only in the positive half-cycle of the electric grid input while the other two stages can be just verified in the negative half-cycle of the input source. As both half-cycles are similar, however using complementary elements of the circuit, the analysis of only the first half cycle is sufficient to verify the complete operating dynamic of the rectifier as it is illustrated in Fig. 2. An additional stage is verified due to the discontinuity of the current over the inductor. This last stage can be verified over the complete input source cycle.

While the grid voltage source has a positive value, the first Boost converter ($L_1 - S_1 - D_1$) is activated by the D_4 diode polarization, which connects the CA source to the common reference of the circuit. The three operating stages (the two continuous stage plus the discontinuous one) are represented in Fig. 2-a, 2-b and 2-c. In the last one, which characterizes the discontinuous operating mode, the load is fed exclusively by the output capacitor, once that the energy accumulated in the inductor was completely over during the second stage, leading the diodes to block.

During the negative half-cycle of the grid voltage, the second Boost converter ($L_2 - S_2 - D_2$) is activated by the D_3 diode conduction. The behavior of these stages is similar to the analysis of the positive half-cycle of the grid voltage, due to the symmetry of the circuit. The third operating stage, in this case, is the same as in Fig. 2-c. 3 presents some of

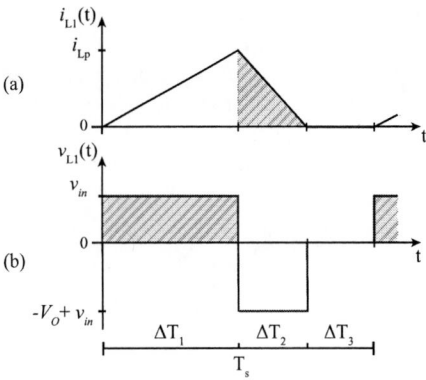

Fig. 3. Waveforms of the modified bridgeless Boost rectifier.

the main waveforms of the DCM operating mode, considering only one switching period. Given that the inductors conduct the input current in different periods of the source input cycle, by inference, it is possible to conclude that the current function expression is the same in both half-cycles, once that $L1 = L2 = L$. Therefore, from the analysis of waveforms from Fig. 3, it is possible to obtain the equations that describe the behavior of the current over the inductors, as described in

$$
i_L = \begin{cases} \frac{v_{in}}{L}t, & \text{for } t \in \Delta T_1 \\ i_{L_p} + \frac{(v_{in}-V_o)}{L}t, & \text{for } t \in \Delta T_2 \\ 0, & \text{for } t \in \Delta T_3 \end{cases} \quad (1)
$$

The peak value of the current from (1), i_{L_p}, is determined after the first operation stage, that is, when $t = \Delta T_1 = DT_s$. Knowing that the input voltage varies in sine form, the current maximum value can be written as

$$
i_{L_p} = \frac{v_{in}\Delta T_1}{L} = \frac{V_p D T_s}{L}\sin(\omega t), \quad (2)
$$

where V_p represents the peak value of the input voltage source and D is the duty cycle.

By the analysis of (2), it is verified that if D is maintained constant during a whole period of the electric grid source, i_{L_p} will present a sinusoidal envelope, just like the input voltage. This is one of the main advantages of operating in discontinuous conduction mode, once there is no necessity of current control.

The peak value of the inductors current can also be obtained based on the second operating stage, resulting in

$$
i_{L_p} = \frac{V_o - V_p\sin(\omega t)}{L}\Delta T_2. \quad (3)
$$

Equating (2) and (3), the second operation period (ΔT_2) is determined as

$$
\Delta T_2 = \frac{V_p\sin(\omega t)\Delta T_1}{V_o - V_p\sin(\omega t)} \xrightarrow{\alpha = v_p/v_o} \frac{\alpha\sin(\omega t)DT_s}{1 - \alpha\sin(\omega t)}, \quad (4)
$$

where $\alpha = V_p/V_o$ is the constant relationship between the output voltage and the maximum input value of the input voltage.

To ensure the operation of the converter at the discontinuous conduction mode throughout the entire cycle period is necessary to know the moment when the converter starts

978-1-7281-4181-7/19 $31.00 © 2019 IEEE

to operate in critical conduction. It is already known that this instant happens when the interval ΔT_3 becomes null, and consequently, ΔT_2 reaches its maximum value, that is, $\Delta T_2 = (1 - D) \cdot T_s$. By inspection, it is verified that this condition is reached at the instant of greater value of the grid voltage. For this reason, the maximum duty cycle that the converter can assume is

$$D_{max} = 1 - \alpha. \tag{5}$$

B. Output Characteristic

The output characteristic depends on the converter operating condition. This condition, in DCM, is directly related to the current absorbed by the load during the source cycle. Therefore, it is necessary to calculate the average value of this current component which, due to the nature of the circuit, corresponds to the average current in the diode in a switching period. By analyzing the waveform and based on the equation (4), for a switching period, it is possible to obtain the relationship expressed by

$$\langle i_o \rangle (\omega t) = \frac{1}{T_s} \cdot \frac{i_{L_p} \Delta T_2}{2} \Rightarrow \frac{V_p D^2}{2 f_s L} \cdot \frac{\alpha \sin(\omega t)^2}{1 - \alpha \sin(\omega t)}. \tag{6}$$

From the equation (6), in a grid period, the average value of the output current results in

$$\overline{I}_o = \frac{1}{\pi} \left(\int_0^\pi \overline{I}_o(\omega t) d\omega t \right) \Rightarrow \frac{V_p D^2}{2\pi f_s L} \cdot Y(\alpha), \tag{7}$$

where

$$
\begin{aligned}
Y(\alpha) &= \int_0^\pi \frac{\alpha \sin^2(\omega t)}{(1 - \alpha \sin(\omega t))} d\omega t \\
&= -2 - \frac{\pi}{\alpha} + \frac{2}{\alpha\sqrt{1 - \alpha^2}} \left(\frac{\pi}{2} + \tan^{-1}\left(\frac{\alpha}{\sqrt{1 - \alpha^2}} \right) \right).
\end{aligned} \tag{8}
$$

By the principle of energy conservation, considering an ideal converter, it is possible to state that the input power on the converter is equivalent to total power transferred to the load, that is

$$\frac{V_p^2 \cdot D T_s}{2L \cdot \overline{I}_o} = V_o \cdot \overline{I}_o. \tag{9}$$

Also, based on equation (7), it is possible to define a normalized current for analytical reference:

$$\overline{I}_{0_{norm}} = \frac{\overline{I}_o \cdot 2\pi f_s L}{V_p} = D^2 \cdot Y(\alpha). \tag{10}$$

To obtain the steady-state gain of the converter, which is a direct ratio of the output voltage to the input voltage source, the load current must be isolated from (10) and replaced at (9) as

$$M(D) = \frac{V_o}{V_p} = \frac{D \cdot \pi}{\overline{I}_{0_{par}}}. \tag{11}$$

A function of the output characteristic is interesting to be visualized and, for this, the critical limit to operate in MDC is necessary. The boundary is reached when the steady-state gain is equivalent to gain in the continuous condition mode.

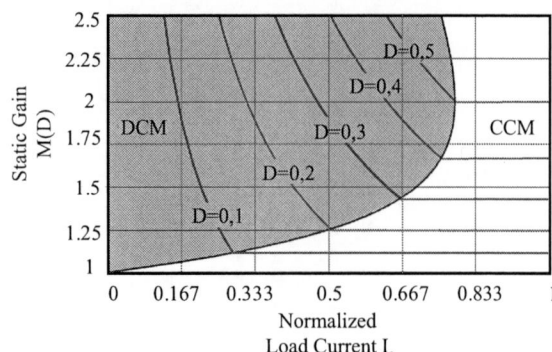

Fig. 4. Output characteristic of the steady-state gain according to the operating region for the converter

In this last circumstance the following equivalence is true: $D = 1 - \frac{1}{M(D)}$ By making the right substitutions in (11), the normalized limiting current is achieved as

$$\overline{I}_{0_{norm}} = \frac{\pi(M(D) - 1)}{M(D)^2} \tag{12}$$

Based on the last equation and from the steady-state gain relation achieved before, the graph from Fig. 4 illustrates the operating region of the converter in MCD.

C. Passive components design

At first, the design of the Boost inductor can be established by isolating the inductance term from the equation (7). The final expression is achieved by considering the processed power in the circuit $P_o = V_o \overline{I}_o$ along with the maximum duty cycle that ensures critical conduction, as shown in (18). It is important to note that the designed inductance value is the same for the two circuit inductors ($L_1 = L_2 = L_{in}$).

$$L_{in} = \frac{V_p^2}{2\pi f_s P_o} \frac{(1 - \alpha)^2}{\alpha} Y(\alpha) \tag{13}$$

The output capacitor, on the other hand, can be designed according to the total current energy delivered to the capacitor during the second stage of operation at the critical conduction instant $(\omega t) = \frac{\pi}{2}$ that guarantees the maximum specified variation over the output voltage. From the peak-to-peak current on the capacitor expression in (14) the maximum capacitance value expression is obtained, as presented in (15).

$$I_{cpp} = \frac{V_p D_{\max}}{L_{in} f_s} \frac{(1 - D_{\max})}{2} \tag{14}$$

$$\hookrightarrow C_o = \frac{I_{cpp}}{2\pi 2 f_r V_o \Delta V_{o\%}} \tag{15}$$

D. Current Efforts

The current in the semiconductors and other components of the converter circuit is directly related to the current ripple in the inductors. The average value of the current in the inductor in a switching period is obtained by the average current in the semiconductors as shown in

$$\langle i_L(t) \rangle = \frac{1}{T} \left[\underbrace{\int_0^{T_1} i_{s(1,2)}(\tau) d\tau}_{\frac{I_L(p) \cdot \Delta T_1}{2}} + \underbrace{\int_{T_1}^{T_2} i_{d(1,2)}(\tau) d\tau}_{\frac{I_L(p) \cdot \Delta T_2}{2}} \right]. \tag{16}$$

TABLE I
CURRENT EFFORTS

Current Item	Average current in grid cycle $\bar{i}_{(r)} = \frac{1}{2\cdot\pi}\int_0^\pi \bar{i}_{(s)}(\omega t)d\omega t$	Rms current in switching cycle $i_{rms(s)} = \sqrt{\frac{1}{T_s}\left(\int_0^{\Delta T_{1,2}} i^2(t,\omega t)dt\right)}$	Rms current in grid cycle $i_{rms(r)} = \sqrt{\frac{1}{2\pi}\left(\int_0^\pi i^2(\omega t)d\omega t\right)}$
$i_s(1,2)(t)$	$\frac{D^2 V_p}{2\pi f_s \cdot L_{in}}$	$\frac{V_{in}\sqrt{D^3}}{\sqrt{3}L_{in}f_s}$	$\frac{V_{in}\sqrt{D^3}}{2\sqrt{3}L_{in}f_s}$
$i_D(1,2)(t)$	$\frac{D^2 V_p}{4\pi f_s \cdot L_{in}}\int_0^\pi \frac{\alpha \sin^2(\omega t)}{1-\alpha\sin(\omega t)}d\omega t$	$\frac{V_{in}}{L_{in}f_s}\sqrt{\frac{V_{in}D^3}{3\cdot(V_o-V_{in})}}$	$\frac{V_p\sqrt{D^3}}{2\cdot\sqrt{3}L_{in}f_s}\sqrt{\frac{2}{\pi}\int_0^\pi \frac{\alpha\sin(\omega t)^3}{1-\alpha\sin(\omega t)}d\omega t}$
$i_L(1,2)(t)$	$\frac{D^2 V_o}{4\pi f_s \cdot L_{in}}\int_0^\pi \frac{\alpha \sin(\omega t)}{1-\alpha\sin(\omega t)}d\omega t$	$\sqrt{\left(\frac{V_{in}\sqrt{D^3}}{\sqrt{3}L_{in}f_s}\right)^2\cdot\left(1+\frac{V_{in}}{V_o-V_{in}}\right)}$	$\frac{V_p\sqrt{D^3}}{2\cdot\sqrt{3}L_{in}f_s}\sqrt{1+\frac{2}{\pi}\int_0^\pi \frac{\alpha\sin(\omega t)^3}{1-\alpha\sin(\omega t)}d\omega t}$

However, once that each inductor (L_1 and L_2) conduct the input current by only half-cycle of the grid, the average of the current on this element, in one grid period, equals half the current in the conventional Boost rectifier inductor. This equivalence is presented by

$$\overline{I}_L = \frac{1}{2\cdot\pi}\int_0^\pi \bar{i}_{L(s)}(\omega t)d\omega t \,. \tag{17}$$

Based on the inductor current analysis, it is possible to obtain the necessary equations for the calculation of the current efforts over some elements of the rectifier. These equations are summarized in Table I.

III. COMPARATIVE ANALYSIS OF THE BOOST RECTIFIER TOPOLOGIES

In order to verify the advantages and some disadvantages of the bridgeless Boost rectifier topology described in this work, this section brings a brief comparative analysis of the circuit with other two commonly used topologies of the Boost rectifier family. Fig. 5 present the topologies to be compared.

A good comparative analysis for the operation of DCM converters should deal with current stresses in the circuit components. Indeed, it can be verified that the effective current of the modified bridgeless Boost rectifier is smaller or equal small in each semiconductor component of the circuit in comparison to both other common used topologies from Fig. 5. This statement can be seen in Fig. 6 in which the effective current parameters were normalized as

$$\langle i_{\{S,D\},rms}(\alpha)\rangle_{norm} = \frac{i_{\{S,D\},rms}(\alpha)}{\overline{I}_o}\,. \tag{18}$$

Usually operating with lower levels of power, the conduction losses on the semiconductor have to be determinant in the global efficiency in DCM Boost rectifiers. This is most evident by the fact the operation in discontinuous mode reduces the losses by switching since the switches are conducting with zero current and the diodes are also blocked with this initial state. On this, Table II summarize the number of components that can result in conduction losses in each topology.

In contrast to the conventional Boost rectifier with a diode bridge, both bridgeless topologies eliminate a diode in the current path. As a result, the input current flows simultaneously through only two semiconductors, thus reducing the conduction losses. On the other hand, among the bridgeless topologies, the CBBR does not have any low-frequency connection between the common reference from the circuit and the ground of the source input. By this reason, in this last topology, the high-frequency pulsed charges and discharges an equivalent parasitic capacitance between these terminals, generating and considerable common-mode current noise in the structure [3], [5].

Meanwhile, the MBBR topology introduces two slow re-

TABLE II
COMPONENTS IN THE PATH OF THE CURRENT IN EACH TOPOLOGY

Parameters	Boost Rectifier Topology		
	BR	MBBR	CBBR
Number of Boost inductors	1	2	1
Maximum number of semiconductors in the current path	3	2	2
Number of fast diodes	1	2	2
Number of active switches	1	2	2
Number of slow diodes	4	2	0

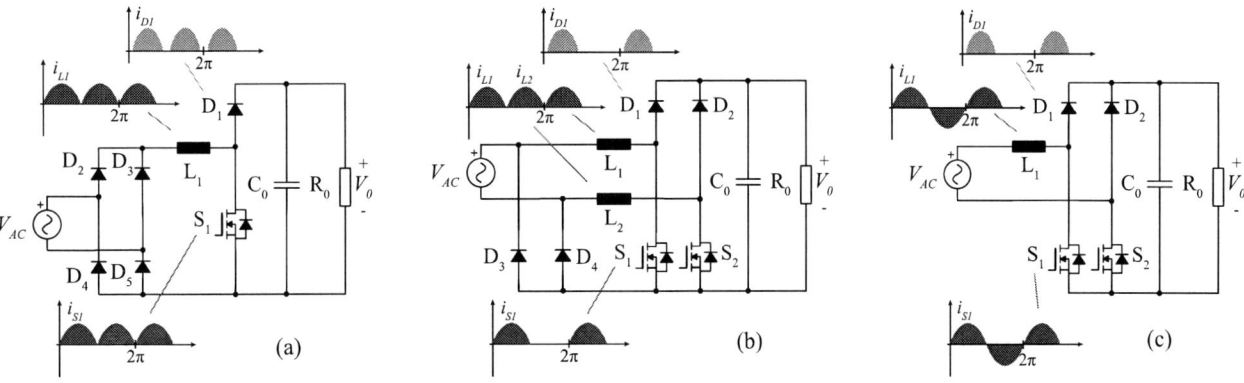

Fig. 5. a) Boost rectifier with diode bridge – BR; b) Modified bridgeless Boost rectifier – MBBR; c) Conventional bridgeless Boost rectifier – CBBR;

978-1-7281-4181-7/19 $31.00 © 2019 IEEE

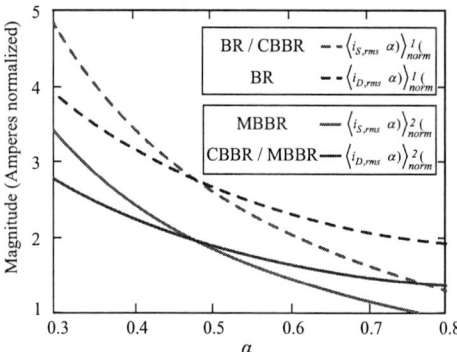

Fig. 6. Comparison between the root-mean-square value of the semiconductors in this project topology and the conventional Bridgeless Boost Rectifier

covery diodes (D_3 and D_4), which maintains a low-frequency path between the grid source and circuit reference terminals. This two recovery diodes eliminate common mode current just like happens in the conventional Boost rectifier. besides the addition of two recovery diodes, two inductors are needed instead of one as are in both other topologies compared in this section. Although two inductors are sharing the path of the current, this does not mean an increase in losses since each one operates in half cycle of the source of the network. Therefore, the disadvantages of the topology concern about the cost of adding more components and prototype size.

Since the differences between topologies in terms of efficiency are mostly related to losses in semiconductors, Fig. 7 compares the three circuit, by losses estimative. For that, the chart uses the same components from the prototype characteristic presented further in the results section of this article.

IV. CONTROL STRATEGY

As well known that the discontinuous Boost rectifier presents an inherent non-linear characteristic on its behavior, making it necessary to compensate the output with a control strategy. Indeed, some dynamic modeling realization is known in the literature and are based on the linearization of the circuit to a specific operation point, as were discussed by [9]. However, the non-linear characteristic is not fully compen-

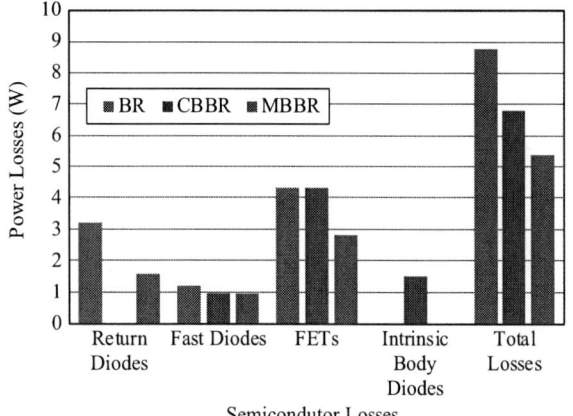

Fig. 7. Power losses estimative on the semiconductors of the three common used Boost rectifiers.

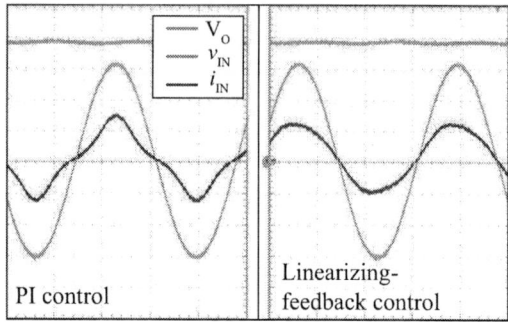

Fig. 8. Experimental results with different techniques for power factor correction and current THD reduction [7].

sated unless non-linear control techniques are applied to the converter.

The authors of this article published and linearizing-feedback strategy applied to the converter in [10] to solve the issue and reduce the input current THD. The methodology was earlier presented in [7] and extensively discussed by this article authors previous work. Although it will not be widely examined, to achieve a regulated output voltage, the feedback was applied to the output voltage and compared to the desired reference. The error generated by this difference is compensated by a PI-type controller.

To illustrate the difference between control techniques, concerning the THD reduction in the rectifier input current, the PI controller is compared to the nonlinear control strategy in Fig. 8.

V. EXPERIMENTAL RESULTS

To validate the converter design and the equations presented in the above sections, a prototype was developed according to Table III. All results and variables were validated with respect component values.

As the main focus of this paper is to verify the steady-state response of the rectifier operating in DCM, Fig. 9 present the current and voltage response of the converter in the grid frequency and in the switching frequency. It can be noted from the figure that the input current has an expressive third harmonic component. As mentioned before, this can be compensated with appropriate control. From Fig. 9 it is also possible to verify that the converter is operating in full-time in DCM. Thus, the viability of the DCM operation and

TABLE III
DESIGN ESPECIFICATIONS

Design component	Value	Design component	Value
Effective source voltage ($V_{in_{RMS}}$)	220 V_{AC}	Input Inductance (L_1 L_2)	180 μH
Grid frequency (f_r)	60 Hz	Filter Capacitance (C_f)	470 nF
Output voltage (V_o)	450 V	Filter Inductance (L_{f1} L_{f2})	850 μH
Output power (P_o)	500 W	Power Switches (S_1 S_2)	IRFP460
Switching Frequency (f_s)	58, 6 kHz	Fast Diodes (D_1 D_2)	MUR1660
Output capacitance (C_o)	560 μF	Return Diodes (D_3 D_4)	MUR460

978-1-7281-4181-7/19 $31.00 © 2019 IEEE

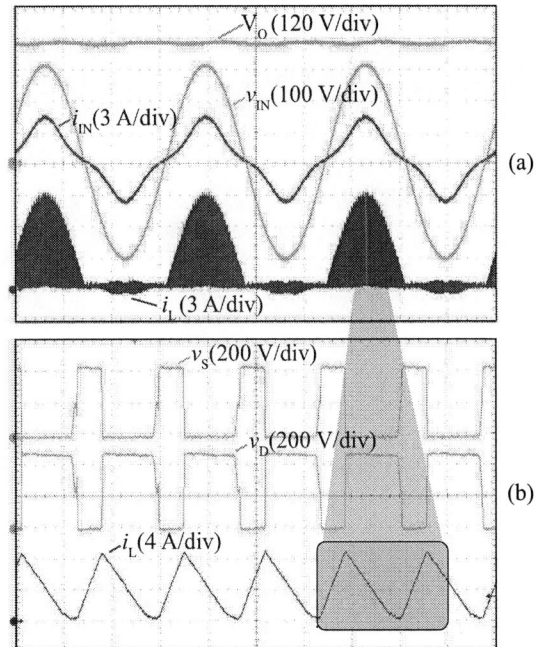

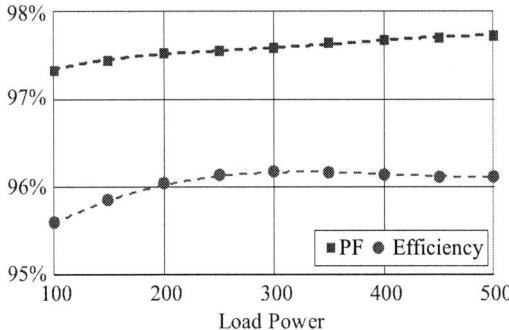

Fig. 9. Steady state performance of the modified b ridgeless B oost rectifier: a) in the input voltage frequency; b) in switching frequency.

the attendance of the limits designed for this application are assured, provided that appropriate control is designed for the converter.

Besides, the efficiency and power factor curves were obtained and are presented in Fig. 10. With nominal power, the rectifier achieved a total harmonic distortion (THD) of 22,17%, with a power factor of 0,977 and with an efficiency of 96,10%. The reached values are relevant for DCM operation.

By way of comparison, Figure 11 presents a chart comparing the harmonic content in the MBBR input current with the IEC61000-3-2 Class A Standard. As can be seen, there is an enhance in total current harmonic using a more refined control strategy, as presented in [10]. However, even operating with a constant duty cycle, i.e. without any different modulation, compliance with the standard is verified without further complications.

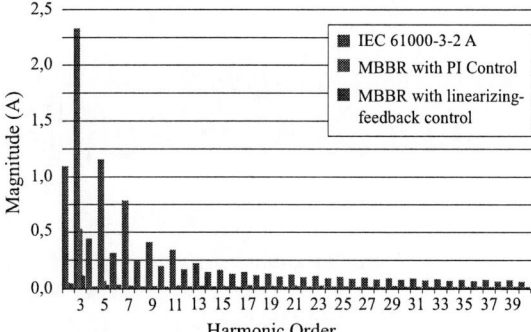

Fig. 11. Magnitude comparison of the harmonic spectrum between control techiniques applied to MMBR and the IEC61000-3-2 Class A Standard.

CONCLUSIONS

This article aimed to present the steady-state modeling of the modified bridgeless Boost rectifier and its advantages related to other common single-phase Boost-type rectifiers. The steady-state modeling of this converter in DCM was not previously presented in literature and is, therefore, a contribution of this paper. From the converter comparison, it was verified that the topology can present itself as a good alternative when switching losses are taken into account for the application. Although the volume of the converter may increase, as it needs two inductors instead of one, a reduction of conduction losses may also mean that it can be designed for a higher power.

ACKNOWLEDGMENT

The authors would like to thank CAPES for their contribution to this work in the form of a fellowship awarded to Telles Brunelli Lazzarin, Dr.[program: POS-DOC – Pesquisa Pós-doutoral no Exterior / Process n 88881.119841/2016-01].

REFERENCES

[1] W. Zhang, L. Ruan, and P. Ye, "The design and analysis of power factor pre-regulator based on Boost circuit," in Third International Power Electronics and Motion Control Conference, Aug. 2000
[2] H. Wei and I. Batarseh, "Comparison of basic converter topologies for power factor correction," Proceedings IEEE Southeastcon 98 Engineering for a New Era, April 1998.
[3] L. Huber, Y. Jang and M. M. Jovanovic, "Performance Evaluation of Bridgeless PFC Boost Rectifiers," IEEE Transactions on Power Electronics, v. 23, n. 3, p. 1381-1390, May 2008.
[4] R. Martinez and P. N. Enjeti, "A high-performance single-phase rectifier with input power factor correction," IEEE Transactions on Power Electronics, v. 11, n. 2, p. 311-317, Mar 1996.
[5] H. Ye, et al, "Common mode noise modeling and analysis of dual Boost PFC circuit," INTELEC 2004. 26th Annual International Telecommunications Energy Conference, Sept. 2004.
[6] A. F. Souza and I. Barbi, "High power factor rectifier with reduced conduction and commutation losses," 21st International Telecommunications Energy Conference. INTELEC 99, June 1999.
[7] H. S. Athab,"A duty cycle control technique for elimination of line current harmonics in single-stage DCM Boost PFC circuit,"TENCON 2008 - 2008 IEEE Region 10 Conference, Nov. 2008.
[8] K. H. Liu and Y. L. Lin, "Current waveform distortion in power factor correction circuits employing discontinuous-mode boost converters," 20th Annual IEEE Power Electronics Specialists Conference, June 1989.
[9] D. S. L. Simonetti, J. L. F. Vieira and G. C. D. Sousa,"Modeling of the High-Power-Factor Discontinuous Boost Rectifiers,"IEEE Transactions on Industrial Electronics, v. 46, n. 4, p. 788-795, Aug 1999.
[10] C. G. da S. Moraes, M. D. Roig G. and T. B. Lazzarin,"Técnica de Modulação para Redução da DHT de Corrente Aplicada a Retificador Boost Bridgeless em MCD," (in portuguese) Brazilian Journal of Power Electronics, July 2019.

Fig. 10. Efficiency and power factor curves from the prototype tests, from an operation range of 100 W to 500 W (nominal power).

A Two-Stage Battery Charger with Active Power Decoupling Cell for Small Electric Vehicles

Caio G. da S. Moraes, Mateo D. Roig G. and Telles B. Lazzarin

Power Electronics Institute (INEP) - Departament of Electronics and Electrical Engineering
Federal University of Santa Catarina (UFSC), Florianópolis, Brasil
caio.guimoraes@gmail.com, mateodaniel.roigg@gmail.com, telles@inep.ufsc.br

Abstract—This paper describes the design and control of a two-stage on-board battery charger for small electric vehicles, with slow and fast charging capacity. Although little discussed in the literature, this type of application requires greater attention due to the low nominal battery voltage, which results in high output current during the fast charging operation. To ensure a good quality of the grid current for both power conditions, a modified bridgeless boost rectifier is used as front-end converter. At the second stage a phase-shifted full-bridge with a current doubler rectifier is selected to ensure zero voltage switching and reduce the transformer secondary winding current. Moreover, an active power decoupling cell is employed in order to reduce dc-link capacitance and, consequently, increase the converter reliability and power density. Finally, the operation of the proposed battery charger under steady-state and dynamic conditions is verified by means of simulations.

Index Terms—on-board battery charger, small electric vehicles, active decoupling cell

I. INTRODUCTION

Unlike conventional electric cars, which have a battery package with a nominal voltage around 350 V, the smaller and cheaper alternatives, such as Renault Twizy [1], as well as electric forklifts, are driven by low voltage batteries usually using a range from 58 V to 72 V. In such cases, for a fast charge application, the battery charger design must withstand high current at the converter output. Besides, the converter must meet the requirements of high power density, high efficiency, low weight, and long lifetime.

A variety of circuit topologies has been proposed in the literature for electric vehicle (EV) on-board battery chargers [2]–[5]. Generally, they can be classified into two categories: single stage types and two stage types. Single-stage converters mainly have simpler power and control circuits, but require higher-rated components and may exhibit lower efficiency compared to two-stage converters when used in high-power applications [6], [7]. In addition, these structures have a low-frequency ripple in the output current, which can lead to an immoderate chemical reaction while charging the battery, causing the temperature to rise [8], [9]. Therefore, the two-stage approach is preferred for EV battery chargers, since the power rating of this type of application is relatively high and this topology provides inherent low-frequency ripple rejection.

The most widely used approach for two stage type consists of a single phase rectifier, with power factor correction (PFC), followed by an isolated DC-DC converter. To decouple both

systems, due to the existence of a second-order power ripple, an electrolytic capacitor (E-cap) bank is usually employed in the DC link. However, the bulky size and short lifetime of E-caps make this solution unfeasible for applications requiring high power density and reliability [10], [11]. Hence, to overcome this problem, some active power decoupling strategies have been investigated in literature [12]. The main idea of these circuits is to deviate ripple power to another energy storage element, which can be much smaller since it allows a heavy fluctuation of voltage/current. As a result, film-type capacitors – that have a longer lifetime than electrolytic capacitors – can be used as the main capacitor of the DC link.

In this paper the design and the control strategy of a two stage converter employing active power decoupling for small EV battery chargers are described in Section II and Section III, respectively. The circuit was designed to operate in both slow charging (1.5 kW) and fast charging (4.5 kW) conditions. In Section IV an application of the proposed topology is verified by simulation results. Finally, this paper is concluded in Section V.

II. SYSTEM OVERVIEW AND DESIGN CONSIDERATIONS

A circuit diagram of the on-board battery charger proposed in this work is given in Fig.1. The system consists of an EMI filter, a bridgeless rectifier with PFC, an active power decoupling cell and a dc-dc converter. Details of each power stage are presented in the following subsections.

A. PFC Stage

The front-end converter chosen in this work is a modified bridgeless boost rectifier operating in continuous conduction mode (CCM). This topology was first proposed in [13]. In comparison to the conventional bridgeless boost rectifier, it shows a reduced common mode noise due to the presence of diodes D_3 and D_4, providing a low-frequency path between the ac source and the dc circuit reference ground. As reported in [14], it tends to be more efficient than other conventional boost rectifiers, including interleaved topology, while the number of components remains attractive.

Due to the circuit symmetry, it is possible to analyze the converter operation considering only one half-cycle and an equivalent circuit of a conventional boost converter [13].

978-1-7281-4181-7/19 $31.00 © 2019 IEEE

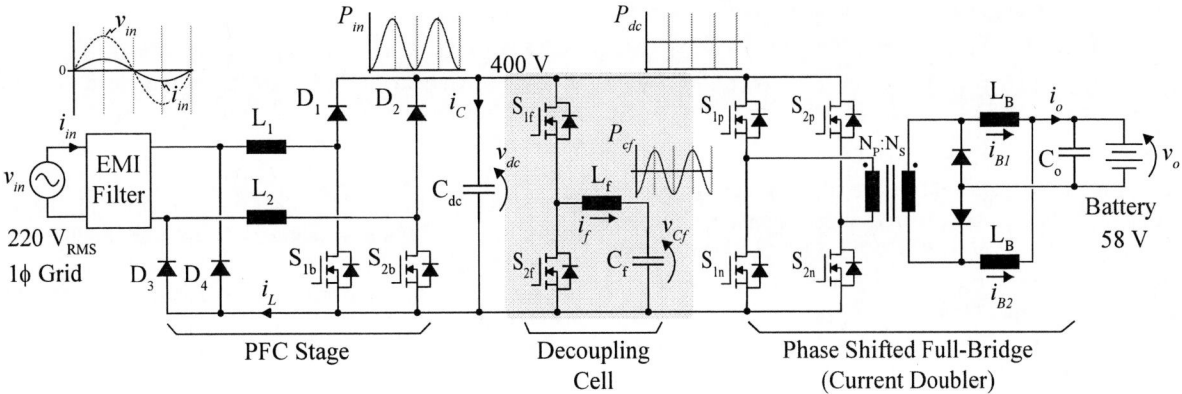

Fig. 1: Proposed two-stage on-board battery charger with active decoupling cell.

Taking this into account, the inductances L_1 and L_2 can be determined as

$$L_i \geq \begin{cases} \dfrac{V_{dc}}{4f_s\Delta I_L}, \text{ if } \alpha > 0.5 \\ \dfrac{V_p}{f_s\Delta I_L}\left(1 - \dfrac{V_p}{V_{dc}}\right), \text{ otherwise} \end{cases} \quad i \in [1,2], \quad (1)$$

where ΔI_L is the desired inductor current ripple, V_p is the peak value of the grid voltage, V_{dc} is the output voltage (i.e. dc-link voltage), f_s is the switching frequency and $\alpha = V_p/V_{dc}$.

If the power decoupling cell is not implemented, the dc-link capacitor (C_{DC}) will be sized to meet the low frequency voltage ripple requirement, according to

$$C_{DC} \geq \frac{P_{max}}{2\pi f_g \Delta V_{dc} V_{dc}} \quad (2)$$

where P_{max} is the maximum power of the system, f_g is the grid frequency and ΔV_{dc} is the desired low frequency voltage ripple.

B. DC-DC Stage

In order to achieve high efficiency on the second stage, a phase-shift full-bridge with a current doubler rectifier was selected (c.f Fig. 1). As a result, the secondary winding of the transformer is rated to half of the load current and its size is also reduced. Additionally, the ripple currents of both inductors partially cancel each other, resulting in a low output current ripple [15]

Another advantage of this topology is the possibility of achieving soft switching without additional circuits. This is done by using the transformer leakage inductance in addition to the MOSFET capacitance (C_{oss}) to achieve zero voltage switching (ZVS). For high frequency operation, however, the transformer primary winding parasitic capacitance ($C_{p,trf}$) should also be considered in the design of the resonant capacitance, which is given by

$$C_r \approx C_{p,trf} + 2C_{oss}. \quad (3)$$

To ensure ZVS operation, the energy stored in the primary leakage inductance must be greater than the energy stored in the resonant capacitance, even at light load [15]. Then, the

resonant inductance can be expressed as (4), where $i_{p,min}$ is the minimum primary current.

$$L_r \geq \frac{C_r V_{dc}^2}{i_{p,min}^2} \quad (4)$$

As a result, the resonant frequency is given by

$$f_r = \frac{1}{T_r} = \frac{1}{2\pi\sqrt{L_r(C_{p,trf} + 2C_{oss})}}. \quad (5)$$

Besides having sufficient energy for achieving ZVS, the deadtime must also be enough to guarantee the drain-source voltage (V_{ds}) transition. In practice, V_{ds} requires at least one quarter of the resonant period ($T_r/4$) to complete the transition [15], therefore the deadtime is defined as $t_d = T_r/4$.

Finally, the output filter inductances and the output filter capacitance can be obtained as (6) and (7), respectively.

$$L_B = \frac{(1 - D_{eff})V_o}{f_s\Delta I_B} \quad (6)$$

$$C_o = \frac{(1 - 2D_{eff})V_o}{16f_s^2 L_B \Delta V_o} \quad (7)$$

where $D_{eff} = (V_o/V_{dc})(N_p/N_s)$ is the effective duty cycle, ΔI_B is the desired inductor current ripple and ΔV_o is the desired output voltage ripple.

C. Active Power Decoupling Cell

Based on the rectifier structure employed in this work, it is more convenient to integrate the power decoupling cell (PDC) in parallel with the dc-link. Thus, there is no need for modification to the original converter. Under these circumstances, the rectifier works independently to regulate the average value of the dc-link voltage, while the PDC deals with the power ripple.

Among the various possibilities, the bidirectional buck type converter [16] was selected, since it uses only two switches and presents lower voltage stress over them, compared to boost and buck-boost topologies [17], [18]. To reduce the current efforts over the components, the continuous conduction mode (CCM) was adopted. Thereby, as in the conventional buck converter, the inductance L_f can be calculated according to

$$L_f = \frac{V_{cf}(V_{dc} - V_{cf})}{\Delta I_{Lf,hf}V_{dc}f_s}, \qquad (8)$$

where V_{cf} is the average voltage over the auxiliary capacitor (C_f) and $\Delta I_{Lf,hf}$ is the desired high-frequency inductor current ripple, which can be chosen as a fraction of the peak inductor current $I_{Lp} = P_o/V_{cf}$.

The minimum required capacitance can be determined by making the input power ripple equal to the auxiliary capacitor power [19], resulting in

$$C_f \geq \frac{P_o}{(2\pi f_g)V_{cf}^2}. \qquad (9)$$

To avoid the resonance problem of pure LC filters, one practical solution is to add a passive dumping branch in parallel to the capacitor C_f, as shown in Fig. 2. The optimum damping resistor (R_d) is determined by (10), where n is the ratio of the blocking capacitance (C_d) to the filter capacitance (C_f) [20]. Then, considering $C_b = C_f$ (i.e. $n = 1$), the optimum damping resistor is simply obtained as $R_d = 1.45\sqrt{L_f/C_f}$. In Fig. 2, it is possible to see the influence of the damping branch on the filter output impedance.

$$R_d = \sqrt{\frac{L_f}{C_f}}\sqrt{\frac{(2+n)(4+3n)}{2n^2(4+n)}} \qquad (10)$$

III. CONTROL STRATEGY

All controllers in this work were designed employing the frequency response technique, considering the digital implementation effects, and then discretized by Backward Euler's method. A 15 kHz sampling frequency was used for the PFC and decoupling cell control loops, and a 20 kHz sampling frequency was used for the battery charging control loop. Furthermore, the feedback signals considered a passive first-order low-pass filter with the cutoff frequency tuned a decade below the respective switching frequency.

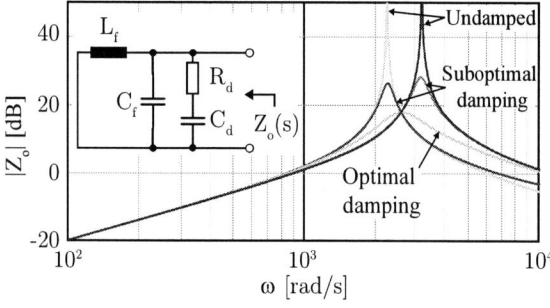

Fig. 2: Output impedance analysis of the decoupling power cell with passive damping branch. The first undamped peak occurs for $R_d = 0$ and the second one for $R_d = \infty$.

A. Power Factor Correction

Fig. 3a shows the control block diagram of the modified bridgeless rectifier. In the diagram, the reference of the current control loop (i_L^*) depends on equation (13), which will be discussed later. A proportional-integral (PI) controller is responsible for regulating the dc-link voltage according to its reference (V_{DC}^*). Typically, there is a twice mains frequency ripple on the dc-link voltage. For this reason, the voltage controller bandwidth must be much smaller than the line frequency to maintain low input current distortion. As a result, the dc-link voltage exhibits slow dynamics and may have long-term and high-amplitude overshoots. An alternative to minimize this effect is to employ a digital moving average filter [21]. It provides great attenuation for each harmonic order less than or equal to $N/2$, where N is the number of samples in a fundamental period [22]. Therefore, it is possible to increase the bandwidth of the voltage loop without affecting the current waveform.

The digital moving average filter can be represented by

$$y[k] = y[k-1] + \frac{1}{N}\left(x[k] - x[k-N]\right). \qquad (11)$$

where $y[k]$ is the output signal and $x[k]$ is the input signal of the filter.

To design the digital moving average filter, it is only necessary to define the number of samples N. However, it should be noted in (11) that the implementation of this kind of filter requires an N-word vector to store the last N samples.

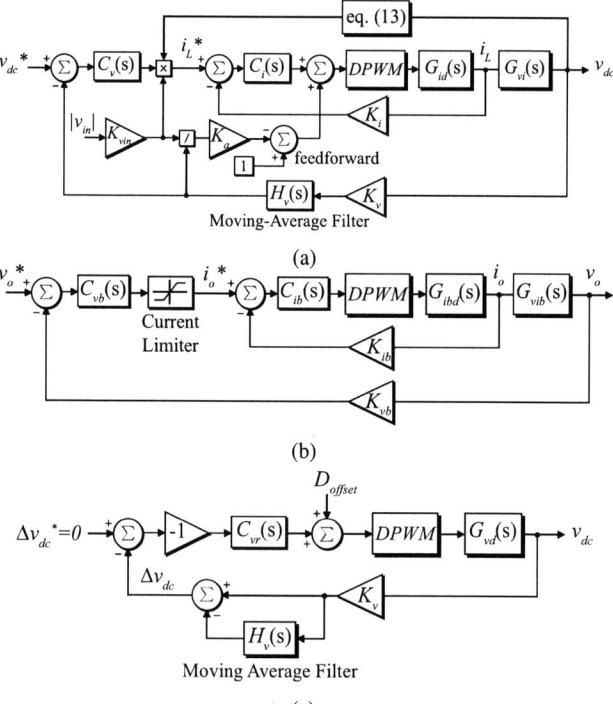

Fig. 3: (a) power factor correction control loop, where $K_a = K_v/K_{vin}$; (b) battery charger control loop; and (c) active power decoupling control loop.

978-1-7281-4181-7/19 $31.00 © 2019 IEEE

As concerns the current control loop, it was used a simple PI controller in addition to a duty cycle feedforward, as shown in Fig. 3a. This results in a smaller current error and a better tracking of the reference inductor current, with no need to use complex multi-resonant or repetitive controllers [23]. The feedforward duty cycle (d_{ff}) is given by

$$d_{ff} = 1 - \frac{V_p}{V_{dc}}|\sin(\omega t)| = 1 - \frac{|v_{in}|}{V_{dc}}. \quad (12)$$

B. Battery Charging Control

The control strategy adopted to charge the battery is shown in Fig. 3b, in which a multiloop control was employed to emulate a constant voltage source with current limiter. In this case, the charger initially sources a limited current into the battery in an attempt to force the battery voltage up to the desired reference. Once this voltage is reached, the charger will source only the necessary current to hold the voltage at this constant value. Therefore, the battery reaches full charge without risk of overvoltage.

C. Active Power Decoupling Control

A control scheme based on [24] was adopted to control the power decoupling cell, as depicted in Fig. 3c. Unlike other schemes proposed in literature, the only information required is the dc-link voltage measurement. This is because the dc-link voltage ripple alone should have enough information to predict the unbalanced power ripple [24].

As shown in Fig. 3c, the voltage ripple Δv_{dc} is obtained through the moving average filter – already existing in PFC stage control loop – by subtracting the filtered signal from the original one. Then, a proportional resonant (PR) controller, with the poles placed at the double-line frequency (120 Hz), tightly regulates the voltage ripple following a reference $\Delta v_{dc}^* = 0$. As a consequence, all the ac component presents in the dc-link voltage is transferred to the auxiliary capacitor (C_f). Finally, a 0.75 offset is added to the modulation signal such that the voltage across C_f will vary around $0.75 \cdot 400$ V $= 300$ V. This offset can also be obtained through a closed-loop system that controls the average value of v_{Cf}, as in [18], [19].

The impact of the PDC into the system can be verified in Fig. 4. At low frequencies (including dc) the power decoupling cell exhibits a high impedance, which means that it behaves like an open-circuit, no influencing on the dc voltage regulation in PFC stage. On the other hand, the very low impedance at double-line frequency implies that any 120 Hz ripple will be directly shorted to ground without going to the dc-dc stage. Then, it can be stated that the introduction of the decoupling cell into the system will only mitigate the double-line frequency ripple, and will not interfere with the other converters.

D. DC-Link Protection

In this paper, the dc-link voltage is controlled even if the battery is sudden disconnected from the system, and there is no significant increase in the dc-link voltage during this transient.

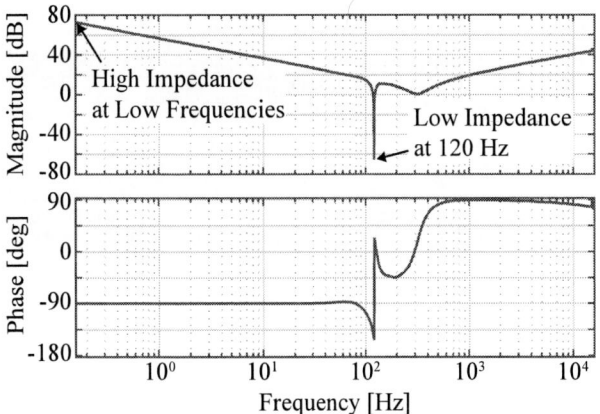

Fig. 4: Frequency response of the power decoupling cell input impedance with proportional-resonant controller.

The proposed dc-link protection strategy is based on [25] and acts in the input current reference, according to

$$i_{L_p}^* = \left(\frac{V_{dc,max} - v_{dc}}{V_{dc,max} - V_{dc,min}} \right) i_{L_p,Ref}, \quad (13)$$

where $V_{dc,max}$ is the maximum tolerable dc-link voltage and $V_{dc,min}$ is the minimum voltage to initiate the dc-link protection. In agreement with Fig. 3a, $i_{L_p,Ref}$ is the voltage loop output, and $i_{L_p}^*$ is the resulting peak value of current reference.

As can be verified in (13), $i_{L_p}^*$ decreases proportionally to the dc-link voltage (v_{dc}) variation, once $v_{dc} > V_{dc,min}$. Otherwise, the current reference is not change, since the result of the calculation in parentheses is saturated between 0 and 1. Therefore, this strategy makes the power injected into the dc-link, by the front-end converter, to cancel out when the dc-link voltage reaches $V_{dc,max}$, thus improving the dynamic response of the voltage control loop.

IV. SIMULATION RESULTS

To verify the effectiveness of the proposed on-board battery charger, a numerical simulation was carried out using PSIM® software. The system parameters considered in this study are presented in Table I.

Fig. 5 shows the converter startup, in which the dc-link capacitor is pre-charged from the grid until it reaches 400 V.

TABLE I: System Parameters.

Battery Pack Parameters	Nominal Voltage = 58 V
	Ah Capacity = 116 Ah
Utility Grid and DC Link Parameters	$V_{in} = 220$ V $\pm 10\%$
	$f_g = 60$ Hz, $V_{dc} = 400$ V
AC-DC Converter Parameters	$L_{1,2} = 2$ mH, $C_{DC} = 300$ μF,
	$f_{sw} = 30$ kHz
DC-DC Converter Parameters	$L_B = 170$ μH, $C_o = 2.2$ μF,
	$L_r = 10$ μH, $C_r = 1.74$ nF,
	$N_p = 8$, $N_s = 4$, $L_M = 1.44$ mH
	$t_d = 217$ ns, $f_{sw} = 100$ kHz
Decoupling Cell Parameters	$L_f = 1$ mH, $C_f = 200$ μF
	$C_d = 200$ μF, $R_d = 3.3$ Ω
	$f_{sw} = 30$ kHz

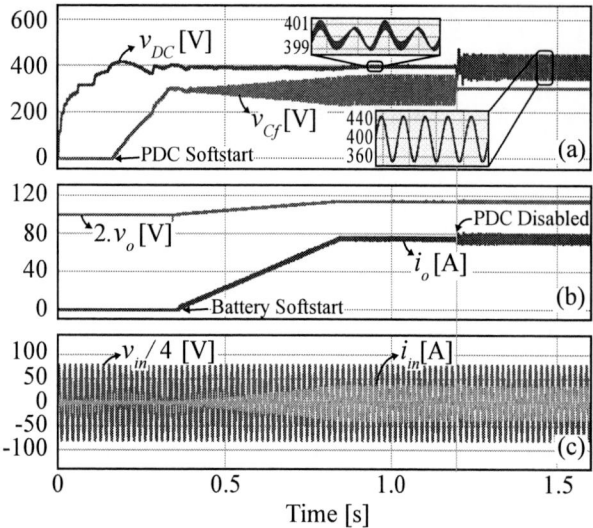

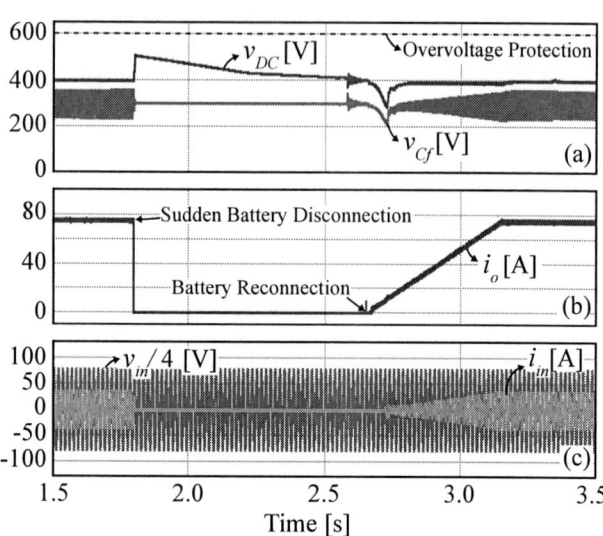

Fig. 5: Converter startup waveforms, considering fast charge condition (4.5 kW).

Fig. 6: Converter dynamic performance under a sudden disconnection of the battery, considering the proposed dc-link protection strategy.

After this, the power decoupling cell is activated through a soft-start routine, as can be seen in Fig. 5(a). In this case, the offset added to the modulation signal was gradually increased from 0 to 0.75 within a time interval of 170 ms. The battery charging process initiates after v_{Cf} achieves steady-state condition and, as reported in Fig. 5(b), the current i_o increases linearly to avoid battery stress. It is also important to note the influence of the decoupling cell on the dc-link voltage quality. During normal cell operation, the ripple on the dc-link voltage is about 2.47 V, which corresponds to only 0.618% of its average value. In order to achieve this same result without the active power decoupling technique, a 12072 μF capacitor would be required on the dc-link, according to (2). Nevertheless, when the PDC is disabled (c.f. Fig. 5(a)) the voltage ripple rises up to 98.7 V, causing a low-frequency ripple in the battery current.

Dynamic results during a sudden battery disconnection are shown in Fig. 6. In this test, the proposed protection strategy was considered and, although the dc-link voltage presents an overshoot of 25%, it does not reach the overvoltage protection limit. On the other hand, Fig. 7 shows the same test without the protection strategy. As reported, the dc-link voltage transient reaches the overvoltage limit, thus, disconnecting the charger from the grid.

In order to verify the quality of the input current in the proposed converter, Fig. 8 shows the steady-state results of the front-end stage for slow and fast charging conditions. In the first case the input current presented a total harmonic distortion (THD) of 2.06%, while in the second this measure was about 3.71%. Furthermore, in both conditions the power factor remained close to unit and the dc-link voltage was well regulated at the desired reference value (400 V).

Finally, to confirm soft switching operation, Fig. 9 shows the voltage and current waveforms of the switches S_{2p} and S_{2n}. As can be seen in figure, even for slow charging condition,

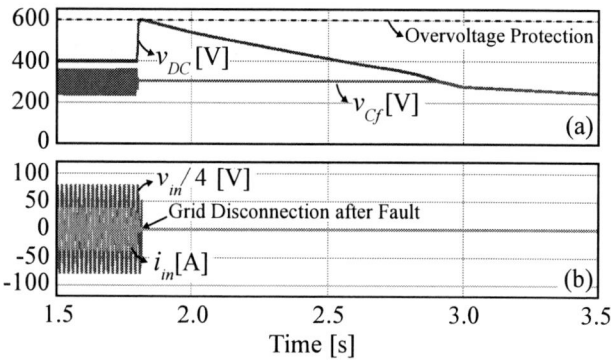

Fig. 7: Converter dynamic performance under a sudden disconnection of the battery, without the proposed dc-link protection.

the drain-to-source voltage (v_{DC}) reaches zero voltage before the drain current (i_D) becomes positive, meaning that these switches are turned on at ZVS.

V. CONCLUSION

This paper proposed a new solution for small EV on-board battery chargers. A two-stage topology was chosen due to the relatively high power. However, unlike conventional structures that use bulky size E-caps, the proposed converter employs an active power decoupling cell to reduce dc-link capacitance. Although the efficiency may be compromised by the increase of switches, the topology allows the use of film capacitors on dc-link, resulting in a more reliable and compact converter, which are important features for EV application. Simulation results for fast and slow charging requirements corroborate the stability and performance of the converter, presenting a nearly unitary power factor and a current THD of less than 5% for both power conditions. Moreover, ZVS operation was verified in all full-bridge switches for both power levels as well.

978-1-7281-4181-7/19 $31.00 © 2019 IEEE

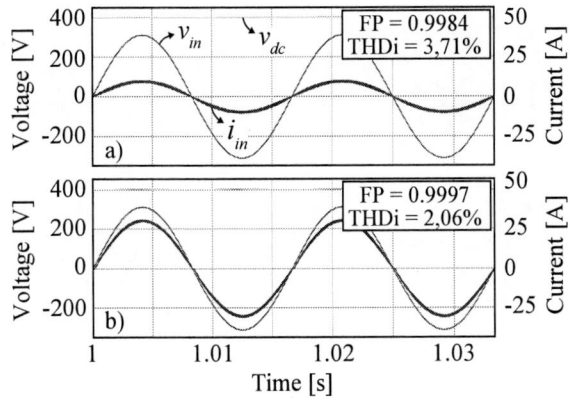

Fig. 8: Steady-state results of the input voltage (v_{in}), input current (i_{in}) and dc-link voltage (v_{dc}) for: (a) slow charging at 1.5 kW and (b) fast charging at 4.5 kW.

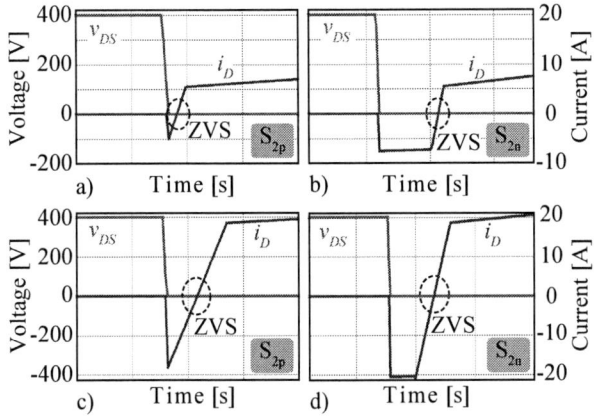

Fig. 9: Drain-to-source voltage (v_{DS}) and drain current (i_D) waveforms of the switches S_{2p} and S_{2n}. The upper figures were obtained for 1.5 kW and the lower figures for 4.5 kW.

Although the proposed dc-link protection strategy worked properly, as future work, it is suggested the implementation of different control techniques to improve the dynamic behavior, mainly to reduce the overvoltage/undervoltage during sudden variations of load. Also, loss modeling and practical implementation are fundamental to validate the topology advantages and its performance under real conditions.

REFERENCES

[1] "Renault twizy." [online]. https://www.renault.co.uk/vehicles/new-vehicles/twizy.html.

[2] J.-S. Kim, G.-Y. Choe, H.-M. Jung, B.-K. Lee, Y.-J. Cho, and K.-B. Han, "Design and implementation of a high-efficiency on- board battery charger for electric vehicles with frequency control strategy," *IEEE Vehicle Power and Propulsion Conference*, 2010.

[3] D. Gautam, F. Musavi, M. Edington, W. Eberle, and W. G. Dunford, "An automotive on-board 3.3 kw battery charger for phev application," *IEEE Vehicle Power and Propulsion Conference*, 2011.

[4] F. Musavi, M. Craciun, D. S. Gautam, W. Eberle, and W. G. Dunford, "An llc resonant dc-dc converter for wide output voltage range battery charging applications," *IEEE Transactions on Power Electronics*, vol. 28, pp. 5437–5445, Dec 2013.

[5] B. Whitaker, A. Barkley, Z. Cole, B. Passmore, D. Martin, T. R. McNutt, A. B. Lostetter, J. S. Lee, and K. Shiozaki, "A high-density, high-efficiency, isolated on-board vehicle battery charger utilizing silicon

[6] Jindong Zhang, M. M. Jovanovic, and F. C. Lee, "Comparison between ccm single-stage and two-stage boost pfc converters," in *Fourteenth Annual Applied Power Electronics Conference and Exposition (APEC)*, vol. 1, pp. 335–341 vol.1, March 1999.

[7] L. Petersen and M. Andersen, "Two-stage power factor corrected power supplies: the low component-stress approach," in *Seventeenth Annual IEEE Applied Power Electronics Conference and Exposition (APEC)*, vol. 2, pp. 1195–1201 vol.2, March 2002.

[8] J. Jiang and C. Zhang, *Fundamentals and Applications of Lithium-Ion Batteries in Electric Drive Vehicles*. Singapore: John Wiley and Sons Ltd., 1 ed., 2015.

[9] I. Puranik, L. Zhang, and J. Qin, "Impact of low-frequency ripple on lifetime of battery in mmc-based battery storage systems," in *2018 IEEE Energy Conversion Congress and Exposition (ECCE)*, pp. 2748–2752, Sep. 2018.

[10] H. Wang, M. Liserre, and F. Blaabjerg, "Toward reliable power electronics: Challenges, design tools, and opportunities," *IEEE Industrial Electronics Magazine*, vol. 7, pp. 17–26, June 2013.

[11] H. Wang and F. Blaabjerg, "Reliability of capacitors for dc-link applications in power electronic converters - an overview," *IEEE Transactions on Industry Applications*, vol. 50, pp. 3569–3578, Sep. 2014.

[12] Y. Sun, Y. Liu, M. Su, W. Xiong, and J. Yang, "Review of active power decoupling topologies in single-phase systems," *IEEE Transactions on Power Electronics*, vol. 31, pp. 4778–4794, July 2016.

[13] A. F. de Souza and I. Barbi, "High power factor rectifier with reduced conduction and commutation losses," in *21st International Telecommunications Energy Conference (INTELEC)*, pp. 158–, June 1999.

[14] F. Musavi, W. Eberle, and W. G. Dunford, "A Phase Shifted Semi-Bridgeless Boost Power Factor Corrected Converter for Plug in Hybrid Electric Vehicle Battery Chargers," *Twenty-Sixth Annual IEEE Applied Power Electronics Conference and Exposition (APEC)*, pp. 821–828, 2011.

[15] B.-R. Lin, K. Huang, and D. Wang, "Analysis and implementation of full-bridge converter with current doubler rectifier," *IEE Proceedings - Electric Power Applications*, vol. 152, no. 5, pp. 1193 – 1202, Sept. 2005.

[16] R. Wang, F. Wang, D. Boroyevich, R. Burgos, R. Lai, P. Ning, and K. Rajashekara, "A high power density single-phase pwm rectifier with active ripple energy storage," *IEEE Transactions on Power Electronics*, vol. 26, pp. 1430–1443, May 2011.

[17] A. C. Kyritsis, N. P. Papanikolaou, and E. C. Tatakis, "A novel parallel active filter for current pulsation smoothing on single stage grid-connected ac-pv modules," in *European Conference on Power Electronics and Applications*, pp. 1–10, Sep. 2007.

[18] X. Cao, Q. Zhong, and W. Ming, "Ripple eliminator to smooth dc-bus voltage and reduce the total capacitance required," *IEEE Transactions on Industrial Electronics*, vol. 62, pp. 2224–2235, April 2015.

[19] T. A. Pereira, "Compensation of the double-line frequency voltage ripple on single-phase two-stage photovoltaic microinverter," *(Master's thesis)*, Federal University of Santa Catarina, 2018.

[20] R. W. Erickson and D. Maksimović, *Fundamentals of Power Electronics*. New York: Kluwer Academic/ Plenum Publishers, 2nd ed., 2001.

[21] R. Ghosh and G. Narayanan, "A simple method to improve the dynamic response of single-phase pwm rectifiers," *IEEE Transactions on Industrial Electronics*, vol. 55, pp. 3627–3634, Oct 2008.

[22] L. M. Nodari, M. Mezaroba, L. Michels, and C. Rech, "A new digital control system for a single-phase half-bridge rectifier with fast dynamic response," in *IEEE Energy Conversion Congress and Exposition*, pp. 1204–1211, Sep. 2010.

[23] D. M. Van de Sype, Koen De Gusseme, A. P. M. Van den Bossche, and J. A. Melkebeek, "Duty-ratio feedforward for digitally controlled boost pfc converters," *IEEE Transactions on Industrial Electronics*, vol. 52, pp. 108–115, Feb 2005.

[24] S. Li, A. T. L. Lee, Siew-Chong-Tan, and S. Y. Hui, "A plug-and-play ripple mitigation approach for dc-links in hybrid systems," in *IEEE Applied Power Electronics Conference and Exposition (APEC)*, pp. 169–176, March 2016.

[25] G. A. Finamor, D. L. S. Solano, M. S. Ortmann, A. Ruseler, L. Munaretto, D. D. da Silva, R. F. Coelho, and M. L. Heldwein, "Solar photovoltaic static conversion system applied to a smart microgrid," in *IEEE 8th International Symposium on Power Electronics for Distributed Generation Systems (PEDG)*, pp. 1–6, April 2017.

Modelling and Analysis of the Isolated Microgrid Installed at the Lençóis Island using PSCAD/EMTDC

Silvangela L. Barcelos
Electrical Engineering Dept.
Federal University of Maranhão
São Luís-MA, Brazil
silvangela.barcelos@gmail.com

José Gomes de Matos
Electrical Engineering Dept.
Federal University of Maranhão
São Luís-MA, Brazil
gomesdematos@ieee.org

Luiz Antonio de Souza Ribeiro
Electrical Engineering Dept.
Federal University of Maranhão
São Luís-MA, Brazil
l.a.desouzaribeiro@ieee.org

Abstract—**A power system served by alternative energy sources to be deployed separately from the conventional power grid must be able to form a local power grid (also called a Microgrids), in which the load of this grid must be served in such a way that the concern with the degradation of the Electric Power Quality (PQ) indices is evidenced through studies that aim to characterize the state of operation of such electrical grid. In this sense, the proposition of solutions to improve the quality and continuity of service of the isolated Microgrids with intermittent generation integrated are extremely important. In this work, the results of the development of a simulation modelling using the PSCAD / EMTDC software to analyze and evaluate the impacts of several integration scenarios of intermittent generators in the Microgrid installed at the Lençóis Island community are presented.**

Keywords— *Simulation modelling, Alternative Energy Sources, Microgrids, Distribution Grids.*

I. INTRODUCTION

Many studies have been developed to evaluate the shared operation of conventional distribution grids with Distribution Generation (DG), such as solar energy, wind energy or tidal energy[1][2][3]. The basic idea of each of these studies is to evaluate the operating limits and the capacity of the current electrical systems to withstand the integration of such energy sources without degrading the quality of services, especially the voltage quality, as discussed by Sahan et. al. in [1]. Calderaro et al. In [2], for example, had studied the violation of the voltage limits during the peak load and the peak of intermittent generation periods and concluded that the load or the power generation leads to a poor voltage quality or even to equipment failure. Kruschel et al. in [3] have shown that it is possible to significantly increase a low voltage distribution network capacity for DG by implementing power-electronic-based voltage management strategies.

In the case of isolated systems, the electricity supply is usually made precariously, using diesel generators, which operate on average 3 to 4 hours a day [4][5]. This practice is due to the high operating cost to service the loads of these systems (usually isolated communities) 24 hours a day from diesel generation. This cost can be considered relatively high when one takes into account the per capita income of families living in these places, which is generally low. An alternative to service isolated systems with electric power may be the use of local renewable energy resources [6][7]. Although there are different types of renewable energy sources that can be used for this purpose, solar energy and wind energy have emerged as renewable sources with a natural vocation to be used in the composition of energy generation systems to enable the electric service of loads isolated.

In Brazil, this characteristic is particularly accentuated, because solar radiation is abundant during almost all the year and in almost all the regions of the country and the wind exists with good energetic density, mainly in the coastal regions, exactly where the existence of isolated communities prevails residents on islands. These types of systems have proved to be suitable for isolated applications in difficult to reach areas and are responsible for the reduction or even elimination of diesel oil consumption in these locations. An example of this type of system has been working since July 2008 in the Lençóis Island Community, belonging to the municipal district of Cururupu-MA, northeast of Brazil [8][9].

The project of the Microgrid installed at the Lençóis Island was executed by the IEE/UFMA and financed by the MME "Light for All" Program [10], aiming to develop solutions for the supply of electricity to isolated communities, with potential for use of wind and solar energy. The isolated Microgrid of Lençóis Island is composed by photovoltaic and wind systems, battery bank and diesel generator, which supply electric energy 24 hours a day for a community formed by approximately 200 consumers through two inverters working in a parallelism regime at a load of approximately 15kW [11].

According to module 8 of the Procedures for Distribution of Electric Energy in the National Electric System (PRODIST) [12], a microgrid such as that of Lençóis must form a grid meeting the criteria of electric power quality established for conventional distribution systems. In this sense, it is necessary to understand the different operating scenarios that occurs in the Microgrid installed at the Lençóis Island to actions to improve the quality of power supply of this system, as well as to predict changes in the operating conditions of the generation/distribution system.

In this paper, a simulation modelling using PSCAD/EMTDC of the isolated microgrid installed at the Lençóis Island is proposed. This simulation model will allow to analyze the impact of the insertion of the existing generations in this Microgrid, characteristics of parallelism of inverters as well as the load dynamics.

The paper is organized as follows: section III presents the main characteristics of the isolated microgrid of Lençóis; the distribution grid modeled in the PSCAD/EMTDC are presented in section III; The PSCAD/EMTDC modeling of Lençóis Microgrid including the representation of the inverters is presented in the section IV. Results of the dynamics analysis of the Microgrid modeled for different integration levels of DG and different conditions for the load are presented in Section V. Finally, the conclusions are presented in Section VI.

II. THE PSCAD/EMTDC SIMULATION TOOL

PSCAD/EMTDC is an industry standard simulation tool for studying and analysis of the transient behavior of electrical grids. Its graphical user interface enables all aspects of the simulation to be conducted within a single integrated environment including circuit assembly, run-time control, analysis of results and reporting. Its comprehensive library of models supports most ac and dc of power plant components and controls, in such a way that equipment FACTS (Flexible ac Transmission Systems), custom power, HVDC (High-Voltage Direct Current) systems as well as special grids, for example microgrids, can be modeled with speed and precision by PSCAD/EMTDC.

The main characteristics of the PSCAD/EMTDC including simplicity of use, many modelling capabilities and highly complex algorithms and method, that are transparent to the user, allowing him to concentrate his effort on the analysis results rather than on mathematical modeling. For more updating to the use of PSCAD/EMTDC, it is recommended the generic examples and characteristics available in [13], that allow a better understanding about the resources of this simulation tool.

III. THE ISOLATED MICROGRID OF LENÇÓIS ISLAND

The general diagram of the Lençóis Island generation system is shown in Fig. 1. The system is basically composed by the following modules [8][9][11]: 1 photovoltaic (PV) panel of 21 kWp, with load controller and MPPT (Maximum Power Point Tracker) algorithm to extrat the maximum power from the solar source; 3 wind turbines of 10 kW each at 12.4m/s (TE1, TE2 and TE3), each turbine with its respective controller (Wind Turbine Rectifier – RTE and MPPT algorithm to extrat the maximum power from the wind source; 2 diesel generator sets (GMG1 and GMG2), one of 48 kVA and another of 73 kVA, in "Prime Power"; 1 battery bank with 2 strings (BAT1 and BAT2), each string with 120 batteries in series, each lead-acid battery of 600Ah and 2V, composing a 240 V nominal, 1200 Ah direct current (DC) busbar; 1 rectifier with 3-phase nominal input voltage of 380 V_{AC}, nominal output voltage 240V_{DC} and rated output current equal to 150A and 2 inverters of 30 kVA each (INV 1 and INV 2), with nominal input voltage at 240V_{DC} and three-phase star output with accessible and grounded neutral, with 380V line to line and 220V phase to neutral, 60 Hz. The main operating data of the Lençoes Island system are shown in Table I.

When wind and sun are available, renewable sources (wind turbines and PV) produce power to supply the load (consumers). In the case of excess (generation greater than consumption), the energy excess is stored in the battery bank, for use in the night period or during the periods of low incidence of wind or solar radiation. In the event of a deficit of renewable sources, the battery bank provides the power complement required to attend the load. There are no converters to control the charging of the battery bank.

In periods of low solar radiation and without wind with electric power consumption for a long time, the battery bank may become discharged. In these situations, one of the GMGs needs to be turned on, through the RECTIFIER. Thus, the batteries life cycle is prolonged and the discontinuity of the electric power supply to the community is avoided. Operation Modes of the Inverters and the GMGs

According to Fig. 1, the consumers are supplied by alternating current (AC) power, which can be supplied by the inverters or by one of the GMGs. The Automatic Transfer

Board (QTA) hardware was assembled with electrical interlock between the KINV and KG contactors, so that the inverters never operate in parallel with the GMG, that is, when KINV is switched on KG is off and vice versa.

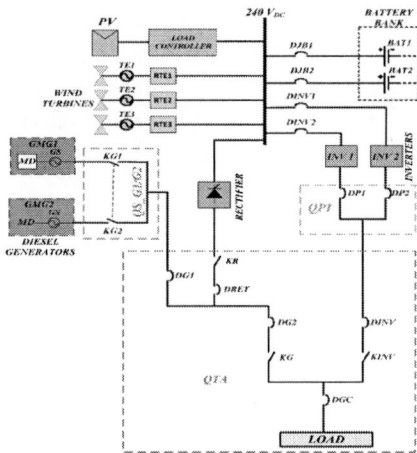

Fig. 1 Block diagram of the Lençóis Island.

There is also the possibility of choosing which is the main source. If inverters are choosing as the main source, KINV is turn on and KG is turn off. On the other hand, if the GMG is selected as the main source, it will assume the load (KG is turned on and KINV is turned off), regardless of the inverters are actives or not. If the selector switch is in the neutral or off position, it means that GMG and inverters were not selected as the priority source. The load (homes/power consumers) is supplied by the source that first generates voltage and is connected to the QTA. The generators GMG1 and GMG2 only run one at a time, that is, it is not allowed the two operating in parallel.

TABLE I. OPERATING DATA OF LENÇOES ISLAND

PCC Voltage	DC Voltage	Power demand	THDv / THDi
380V_{LLrms} (60Hz)	240V	15kW (PF≈0.85)	≈4% / ≈70%

A. Consumers characteristics

The hybrid generation system from the Island of Lençóis, provides the local community with 24-hour electric power per day, in houses and on public roads. The distribution grid of the Lençóis Island microgrid is made at low voltage, according to the standards established by ABNT and the local concessionaire [12]. The distribution lines are made with isolated multiplexed cables, 3 phases and 1 neutral, with 380V (line to line) and 220V (phase to neutral) [14].

IV. DISTRIBUTION GRID MODELING

A survey of the distribution grid of Lençóis Island through mapping with GPS (Global Position System - Model GPSMap 76S) was done by Oliveira in [14]. From its results, it was determined the length of each distribution line. All consumer units were visited with the purpose of collecting the geographical position and the average monthly consumption of each of them. From the software MapSource Version 6.16.3 the distances between generation, distribution, posts and consumer units were determined, as shown in Fig. 2.

A. The estimated eletric power demand

The measurements of the maximum and average demands of the consumer units of the Lençóis island are known from

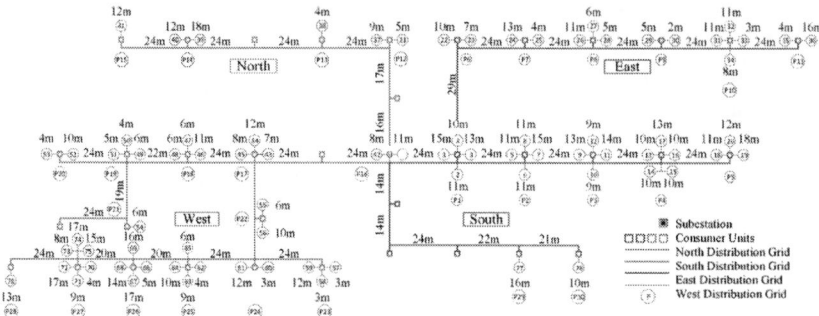

Fig. 2. Distribution system of the Microgrid installed in Lençóis Island.

the database created by IEE/UFMA during the monitoring of the microgrid operation from 2008 to 2016. Through the Maximum and Average Power Demands, it was possible to find the load factor (LF), according to:

$$LF = \frac{\text{Average Demand}}{\text{Peak Demand}} \qquad (1)$$

The survey of the average electric power demand of each consumer unit was estimated from the average of the monthly electric power consumption during the period of one year, according to Oliveira in [11]. From these data, it was possible to estimate the average electric power demand of the microgrid. The maximum individual electric power demand of each consumer unit was determined by LF using (1).

The knowledge of data obtained from the database created by IEE/UFMA during the monitoring period were compared with the data obtained from the survey of the average electric power demand of each consumer unit [11], determining the electricity consumption pattern in Lençóis Island. Thus, it was possible to make daily, weekly or even monthly estimates to study the behavior of the Lençóis Island Load. Finally, the standard electric power consumption was determined considering a correction factor of approximately 1.8 to make the values of individual standard electric power consumption and the individual demands compatible. According to results showing in [11], the actual maximum electric power demand of the Microgrid of Lençóis is approximately 15kW. For purposes of a complete distribution system representation, the

loads were distributed according to the demand requested per area (North, South, East and West) of the Lençóis Island, totaling the maximum demand raised.

B. The Distribution Lines

The distribution lines of the Microgrid installed at the Lençóis Island are short. In this case, the shunt capacitance of these lines must be disregard. Only the equivalent resistance and inductance are considered. Initially, the distribution grids are considered balanced to facilitate the system modelling analysis.

Fig. 3 shows the test system implemented in PSCAD/EMTDC to carry out simulations for the distribution system of the Lençóis Island. As can be seen, all four areas mentioned before are represented in this modelling. This test system comprises a $380V_{RMS}$ (line-to-line) distribution system with loads distributed throughout all areas of the Microgrid. A set of breakers shown in Fig. 3 were used to assist different integration scenarios or different loading condition being simulated with ease. All parameters of the distribution lines are available in [11].

To show the effectiveness of this distribution system modelling, simulations were carried out with different integration scenarios for the load system. For this, a voltage source to supply the same conditions of generation of the Lençóis Island microgrid was connected to the system. The set

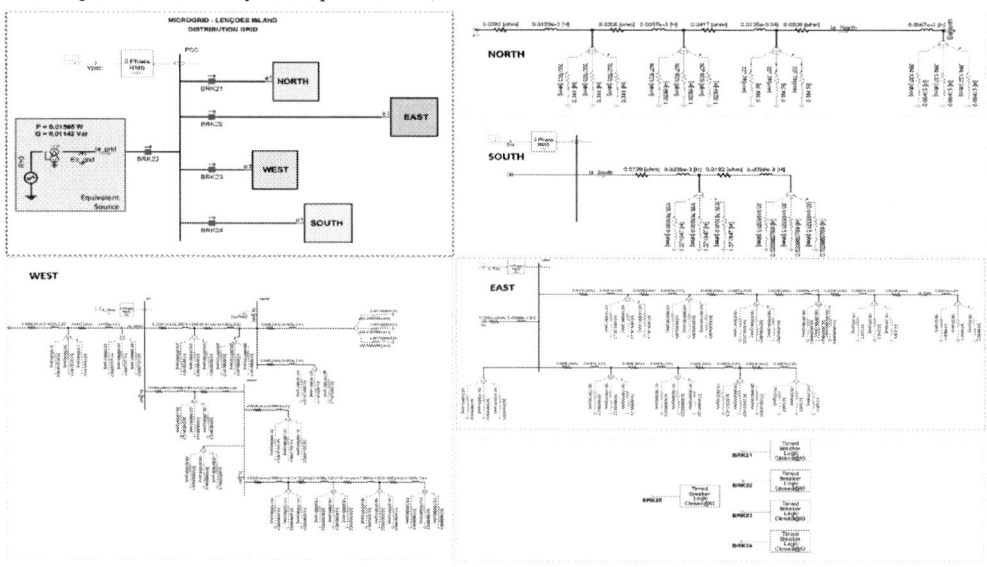

Fig. 3. Test system implemented in PSCAD/EMTDC to carry out simulations for the distribution system of the Lençóis Island.

978-1-7281-4181-7/19 $31.00 © 2019 IEEE

of simulation was carried for the test system shown in Fig. 3, relating three operating condition: (1) Loading changing on North area from t = 1.5 to t = 2.0s and (2)Loading changing on West area from t = 2.5 to t = 4.5s and (3) Connection of the East area at t =4.5s.

Fig. 4(a) shows the PCC voltage drops with respect the reference value due to load changing. The instantaneous values of these voltage are shown in Fig. 4(b). As can be shown in Fig. 4(c) and Fig. 5, after the steady state be stablished, in the simulation period (1) has occurred a disconnection/connection of the North area, decreasing the requested demand from 15kW to 12kW approximately, for 0.5 s. In this case, the load decreased by opening of *Breaker* BRK21. In the simulation period (2) has occurred a disconnection/connection of the West area during 1.5s by opening of *Breaking* BRK23. In this situation, the requested demand decreased to almost 10kW. In the third simulation event (3) the connection of the East area by opening of *Breaker* BRK22 has occurred at t=4.0s, resulting on a requested demand of 15kW.

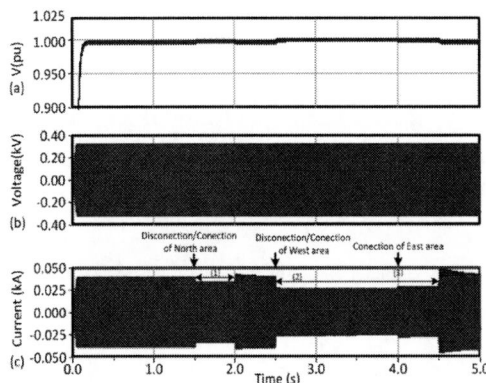

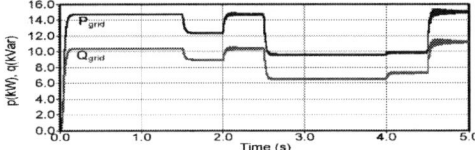

Fig. 4. (a) Voltage at PCC; (b) Instantaneous voltages at PCC and (c) Instantaneous currents demanded by the grid.

Fig. 5. active and reactive power measured on Source terminals.

V. THE GENETATION MODELING

The simplified block diagram of the generation system of the Microgrid installed at Lençóis Island is presented in Fig. 6. As can be seen, all energy sources share a dc bus, that is the input of the inverter sub-system. It is formed by two three-phase Voltage Source Inverters (VSI), each of 30kVA, configured to work in parallel, sharing equally the load without any communication interface between these units of data interchange. There is a supervisory control done by Programmable Logic Control (PLC) responsible to coordinate the parallel operation of all sources giving special attention to system efficiency; the charging control of the battery bank; the load control of the diesel generator (when its operation is required) and the measurement and transmission of all variable.

A. Parallel Connected Inverters

In case of the Microgrid of Lençóis Island, the VSIs are parallel connected without communication [9]. In the project

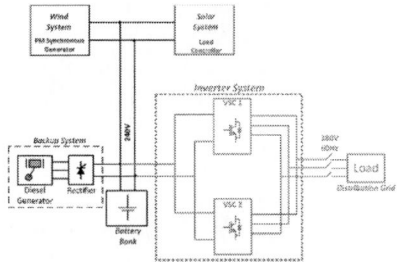

Fig. 6. Block diagram of the generation system.

of these inverters was developed a system with high reliability and had capacity for expansion as the load increase, without the need of alterations on equipment already installed. As mentioned for Lorençato in [14], the Mean Time Between Failure (MTBF) of inverters connected in parallel, in the N+1 and N+2 configurations, is higher than own useful life of the equipments, demonstrating the reliability that this type of operation provides. However, this reliability is only for system without communication, with units totally autonomous.

B. The Drooping Method

The inverter system of Lençóis has an operation totally autonomous through the implementation of a parallelism method that doesn't use any communication interface among the units, called *"Drooping Method"* [15]. In the Drooping Method the inverters present an electric behavior similar to that generators operating in parallel [15]. As described in [8] and [9], whenever an increase in the load active power occurs, the generators tend to reduce their rotation, reducing their frequency proportionally. In the developed system this idea was implemented. Thus, each inverter has its own circuit to accomplish the Drooping Method, where the frequency decreases to compensate for the variations in the active power, as shown in Fig. 7. According to Fig. 7 (a), the active power sharing between the parallel-connected inverters is based on the adjustment of the coefficient m, which define the slope of the decay curve as:

$$m_{máx} = 2\pi \frac{\Delta f_{máx}}{P_{Nom}} \qquad (2)$$

where, $\Delta f_{máx}$ is the maximum variável of frequency and P_{Nom} is the rated power of the grid. In the case of the Microgrd of Lençóis $\Delta f_{máx} = 1,8Hz$, resulting $m_{máx} = 0.69$. Fig 7(b) shows in a simply way two single-phase inverters parallel-connected as close as possible to each other so that the resistive components of the impedances are negligible. As a result, the output voltage controller systems acts to have zero steady state error. Z_1 and Z_2 are used to voltage unbalances compensation being much smaller than the load impedance Z_{Load}. The system balance is small if compared with Z_1 and Z_2, for this reason can be neglected on drooping

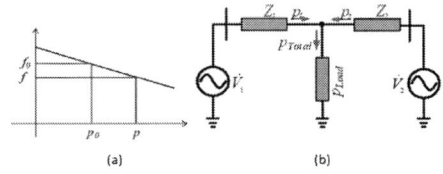

Fig. 7. (a) Drooping Method representation; (b) Equivalent circuit of two inverters in parallel.

978-1-7281-4181-7/19 $31.00 © 2019 IEEE   403

method equation. Approachs: $\sin \alpha \approx \alpha; \cos \alpha \approx 1$; inductive lines gives p e q expressed by:

$$p_{V_1} = -p_{V_2} = \frac{V_1 V_2}{X} \alpha \qquad (3)$$

$$q_{V_1} = \frac{V_1}{X} (V_1 - V_2) \qquad (4)$$

where, p_{V_1} and p_{V_2} are the active power supplied by inverters 1 and 2, respectively; q_{V_1} is the reactive power supplied by inverter 1; V_1 and V_2 are output voltage of inverter 1 and 2, respectively and α is the phase angle between V_1 and V_2.

The block diagram of the *Inverter System* implemented in PSCAD/EMTDC is presented in Figure 8. As can be seen, the inverter system is based on a 30kVA, 3-phase IGBT inverter. The output voltage (60Hz) was regulated based on Sinusoidal Pulse Width Modulation (SPWM) with 4kHz switching frequency. An output low pass filter was used in each phase to eliminate high frequency harmonic content due to inverter switch action. Transformers was used to provide galvanic isolation and change the output voltage level. There is an impedance used to share equally the load between the inverters. Others functions for the isolation transformer are load protection against fail or isolation loss, improving reliability and robustness.

Fig. 9 shows the control scheme implemented using the PSCAD/EMTDC to determine the power order and the reference angle for each inverter (the drooping).

In Figure 10 is shown the voltage control scheme of just one inverter implemented in PSCAD/EMTDC. The other inverter that works in parallel has the same dq control strategy. As shown in Figure 10, the voltage synthesized at ac side of the inverter depend on the reference angle defined by the Control scheme of drooping. The control scheme (implemented using PSCAD/EMTDC) to synchronize both inverter giving conditions for parallel operation is presented in Figure 11. This control scheme is based on Synchronous Reference Frame–Phase Locked Loop (SRF-PLL) to track the fundamental component of the ac side voltage from INV 1, giving condition for the VSI 2 working parallel connected.

VI. SIMULATION RESULTS

Fig. 12 shows a set of the main events simulated for the Microgrid of Lençóis Island implemented using the PSCAD/EMTDC. For all analysis, the dc bus was represented by a controlled dc source (Battery Bank) and a load profile of 12kW with a power factor equal to 0.8 inductive was considered.

Fig. 13 shows the active power measured at the ac side of the INV 1 and INV 2 when only INV 1 worked suppling the

load. The ac voltage control performance can be seen though the voltage and current at the ac side of INV 1 on dq synchronous reference frame are shown in Figure 14.

A simulation was carried out considering the inverters working in parallel. In this simulation the INV 2, after

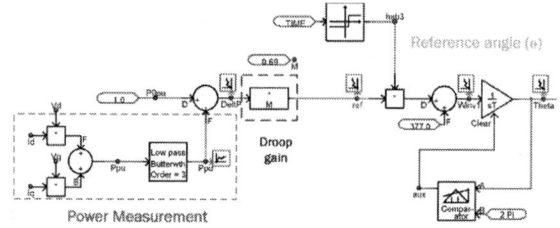

Fig. 9. Control scheme of drooping implemented in PSCAD/EMTDC.

synchronization, share the load with the INV 1. The load sharing at t=2s is shown in Figure 15 through the voltage and currents (phase a) of each inverter. It is observed that after the parallelism both inverters share approximately equally the load (50% of the load current in each inverter). The current in phase a of each inverter before and after the disconnection

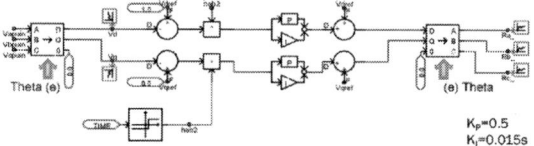

Fig. 10. Voltage control strategy implemented in PSCAD/EMTDC.

of INV 2 are presented in this figure. It is observed that after the disconnection, INV 1 assumes the load totally. Other simulation result is shown in Figure 16. In this case both inverters sharing the load while load changes occur due to the

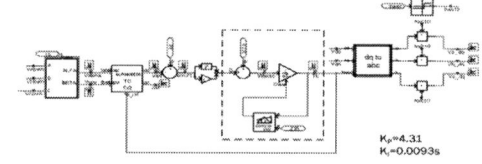

Fig. 11. Synchronization strategy implemented in PSCAD/EMTDC.

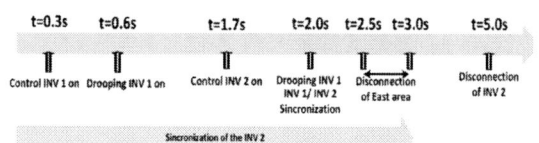

Fig. 12 Events simulated for the Microgrid modeling of Lençóis.

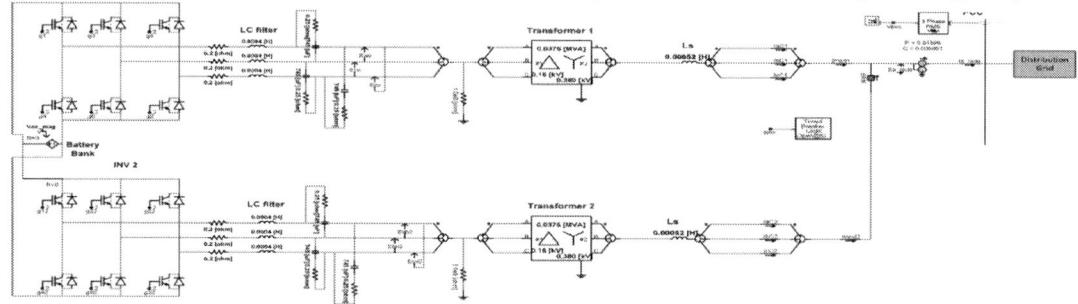

Fig. 8. Control Scheme of the Inverter System implemented in PSCAD/EMTDC.

disconnection of the *East Area*, resulting on reduction of the current provided by each inverter.

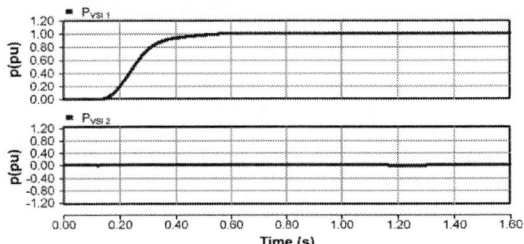

Fig. 13. Active power measured at the ac side of the INV 1 and INV 2 when only INV 1 works suppling the load.

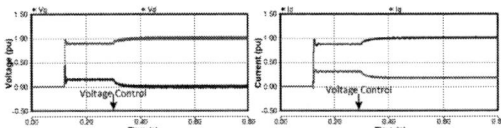

Fig. 14. dq voltages and currents at the ac side of INV 1.

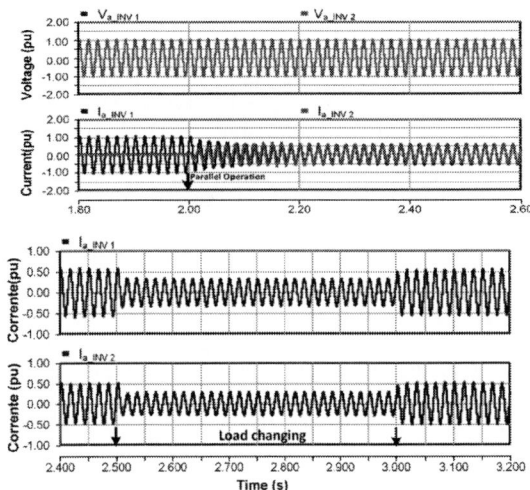

Fig. 16. Currents in phase a of each inverter during the load changing.

VII. CONCLUSIONS

This paper has presented the results obtained with the development of a simulation modelling for the isolated microgrid installed at the Lençóis Island using PSCAD/EMTDC, expressing its dynamic behavior in the face of different operating scenarios. This simulation model has allowed to analyze the impact of the insertion of the existing generations in this Microgrid, the characteristics of parallelism considered for the *Inverter System* as well as of changes in the dynamics of loads. The set of simulation results presented shown that is possible to represent operating scenarios similar to those that occur in the real microgrid of Lençóis, contributing to futures studies as expansion of the distribution grid, new technologies of converters, news control techniques and the insertion of others renewable sources in different points of the Microgrid. As improvement for future versions, the authors intend to represent the photovoltaic and wind systems, battery bank and diesel generator making the modeling more robust.

ACKNOWLEDGMENT

The authors would like to thank the support and motivation provided by the Federal University of Maranhão (UFMA), CEMAR, FAPEMA, CNPq – Brazil, and MEIHAPER CYTED.

REFERENCES

[1] Sahan, W., da Costa, J. P., Kruschel, W., Zacharias, P., 'Power electronics for voltage control in distribution networks'. In 16th Kassel Symposium on Energy Systems Technology, Kassel, Germany, 2011, pp. 1-15.

[2] Calderaro, V., Galdi, V., Lamberti, F., Piccolo, A.: 'A smart strategy

[3] for voltage control ancillary service in distribution networks'. IEEE Transactions on Power Systems, 2015, 30, (1), pp. 494–502, doi: 10.1109/TPWRS.2014.2326957, ISSN:0885-8950.

[4] Kruschel, W., da Costa J. P., Dombert B., Mendel, D., Blo, T., Zacharias, P.: 'Power electronics voltage regulator for increasing the distributed generation capacity in low voltage networks'. In 15th European Conference on Power Electronics and Applications (EPE), 2013, Lille, France, pp. 1-10.

[5] G. Bonan, A. S. Martins, L. A. de S. Ribeiro, O. R. Saavedra and J. G. de Matos, "Parallel-connected inverters applied in renewable energy systems," 2009 Brazilian Power Electronics Conference, Bonito-Mato Grosso do Sul, 2009, pp. 993-999. doi: 10.1109/COBEP.2009.5347645.

[6] H. A. Oliveira, J. G. de Matos, L. A. de Souza Ribeiros, A. S. Martins and G. C. Flores, "Operation of three-phase rectifiers with Diesel Generator Sets of similar power: Practical solution based on passive filters," 2017 IEEE 8th International Symposium on Power Electronics for Distributed Generation Systems (PEDG), Florianopolis, 2017, pp. 1-5. doi: 10.1109/PEDG.2017.7972533

[7] Duncan G. Labay, Thomas C. Kinnear, Exploring the Consumer Decision Process in the Adoption of Solar Energy Systems, Journal of Consumer Research, Volume 8, Issue 3, December 1981, Pages 271–278, https://doi.org/10.1086/208865

[8] Yanfeng Liu, Sisi Yu, Ying Zhu, Dengjia Wang, Jiaping Liu, "Modeling, planning, application and management of energy systems for isolated areas: A review," Renewable and Sustainable Energy Reviews Volume 82, Part 1, February 2018, Pages 460-470.

[9] L. A. de Souza Ribeiro, O. R. Saavedra, S. L. de Lima and J. G. de Matos, "Isolated Micro-Grids With Renewable Hybrid Generation: The Case of Lençóis Island," in IEEE Transactions on Sustainable Energy, vol. 2, no. 1, pp. 1-11, Jan. 2011. doi: 10.1109/TSTE.2010.2073723.

[10] Luiz A. de S. Ribeiro, Osvaldo R. Saavedra, Shigeaki. L. Lima, José G. de Matos, Guilherme Bonan, Making isolated renewable energy systems more reliable, Renewable Energy, Volume 45, 2012, Pages 221-231,ISSN0960-1481, ttps://doi.org/10.1016/j.renene.2012.02.014.

[11] Ministry of Mines and Energy (MME), "Brazil launches Distributed Generation Program with emphasis on solar energy,"Ministério de Minas e Energia, "Brasil lança Programa de Geração Distribuída com destaque para energia solar," 16 12 2015. [Online]. Available: http://www.mme.gov.br/web/guest/pagina-inicial/outras-noticas/-/asset_publisher/32hLrOzMKwWb/content/programa-de-geracao-distribuida-preve-movimentar-r-100-bi-em-investimentos-ate-2030 [Access in Juny 24, 2019].

[12] Oliveira H. A, *Rede Híbrida de Distribuição de Energia em CC e CA como uma Solução Alternativa para Microrredes Isoladas*, Dissertation, UFMA, São Luis-MA, 2017.

[13] Aneel, *Procedures for Distribution of Electric Energy in the National Electric System (PRODIST)*, last modification: 10/01/2017 15:35

[14] User's Guide, "EMTDC: Transient Analysis for PSCAD Power System Simulation,", 2005.

[15] A. A. Lorençato, A. S. Martins, G. Bonan and G. Gabiatti, "Single, Series and Parallel Redundant UPS", in Proc. of COBEP, vol. 01, pp. 1010 – 1013, 2007.

[16] S. J. Chapman, Electric Machinery Fundamentals, New York: McGraw-Hill, 5th, Ed., 1991.

Methodology for experimental determination of equivalent grid impedance by using external commands of PV inverters

Valesca Bettim Feltrin, Thiago Brezolin Dalmolin, Lucas Vizzotto Bellinaso, Leandro Michels
Federal University of Santa Maria (UFSM)
Power Electronics and Control Research Group (GEPOC)
Santa Maria, Rio Grande do Sul, Brasil
valesca.bfeltrin@gmail.com, thiago.dalmolin13@gmail.com, lucas@gepoc.ufsm.br, michels@gepoc.ufsm.br

Abstract — **Equivalent grid impedance is an important information for testing control stability of grid-connected inverters. Furthermore, grid impedance parameters impact on local grid voltage when the local load or generator changes its active and reactive power. In grid connection standards such as the Brazilian Association of Technical Standards (ABNT) Brazilian Regulatory Standard (NBR) 16149, Photovoltaic (PV) inverters higher than 6 kW shall have external control of active and reactive power. This manuscript proposes a methodology for experimental determination of grid impedance by using the external commands of PV inverters. Experimental results were obtained with a 100 kW inverter of a photovoltaic power plant, and then the local grid impedance was calculated. Finally, the impedance calculation was validated in simulation.**

Keywords—component; grid impedance; PV inverters external commands

I. INTRODUCTION

Distributed generation of grid-connected PV systems has started in Brazil in 2012 [8]. Since then, the number of PV generation units has increased exponentially [8]. In 2013, the Brazilian Association of Technical Standards published the NBR 16149 and the NBR 16150, which deal with the characteristics of the connection interface with the distribution grid and the conformity test procedure, respectively. Therefore, it is necessary to have minimum operation standards of the inverters connected to the low voltage grid such as: active and reactive power injection; injection limitation; current harmonics; flicker; voltage variation; anti-islanding, among others [7]. In addition, the photovoltaic system must identify the abnormal conditions of the power grid and disconnect if necessary [4].

Photovoltaic inverters operate in parallel with the distribution grid. The purpose of standardization of inverters is to guarantee the quality of the electrical grid, maintaining reliability and stability of the system for the connected loads [7]. The value of the grid impedance is important information for power system management, current control design, harmonic estimation, fault detection and grid unbalance detection [9].

A reference value of grid impedance is necessary for testing low frequency electromagnetic compatibility of PV inverters and other converters applied to distributed generation [1-5]. IEC TR 60725/2012 [6] (technical report) defines reference impedance for tests with currents equal to or less than 75A. The standard provides a methodology for field study and measurement of grid impedance. However, most collected data are from Europe, Asia and North America. There are no studies for the Brazilian grid, so the Brazilian grid code ABNT NBR 16149 [4] considers grid impedances defined by IEC standards.

Considering the importance of grid impedance determination, this manuscript proposes a methodology to estimate the grid impedance by using the external control capability of PV inverters higher than 6 kW [4]. Differently from IEC TR 60725/2012, no measurement setup is necessary. Using the communication of the PV inverter, the controlled variables are the inverter active and reactive power [7], and the measured value is the local rms voltage. The data is collected and the grid impedance is estimated by analyzing the power flow as an optimization problem. This methodology was applied to estimate impedance of the grid where a 100 kW PV power plant is installed. The calculated impedance was confirmed in simulation.

II. EXTERNAL COMMANDS OF PV INVERTERS THAT COMPLY WITH NBR 16149

NBR 16149 determines three power ranges of PV inverters. As shown in Table I, inverters higher than 6 kW shall have the feature of external control of active and reactive power. Also, these inverters shall be able to perform power factor between 0.9 (under excited) and 0.9 (over excited). The external control can usually be performed by TCP/IP communication.

TABLE I. POWER RANGE, PF AND INVERTER COMMANDS

Power range	Power factor range	External command		
		On-off	Active power control	Reactive power control
< 3 kW	1	yes	-	-
3 kW ≤ P < 6 kW	-0.95 ~ 0.95	yes	yes	yes
P ≥ 6 kW	-0.90 ~ 0.90	yes	yes	yes

Adaptation of NBR 16149 [4].

The active power command limits the value injected into the grid and must be executed within the maximum time of 1 min after sending the information [4]. The PV inverter cannot supply active power higher than the PV power, or higher than its maximum power. So, the active power is usually set as a percentage of the nominal active power, with a tolerance of ± 2,5% [4].

The reactive power command shall be carried out within the maximum time of 10s after sending the information, and with a tolerance of ± 2.5% [4]. It is worth mentioning that the inverter can operate with constant power factor, or with constant reactive power, but not both modes [4].

978-1-7281-4181-7/19 $31.00 © 2019 IEEE

III. DETERMINATION OF EQUIVALENT GRID IMPEDANCE ACCORDING IEC TR 60725

IEC TR 60725 contains references impedance for public utility grid equal to or less than 75A per phase, and in the annex are presented methodologies to determinate the impedance associated with equipment with more than 100A per phase [6]. This Technical Report presents recommended reference impedances, obtained based on the data acquired over the years, which differ according to the operating voltage. The grid impedance for residential consumers has been measured in several countries, but most data presented for low voltage 50Hz grids [6]. Table II shows the recommended reference impedances for different low voltage systems.

TABLE II. REFERENCE IMPEDANCES OF IEC TR 60725

System	Voltage	Reference impedance		
		Phase	Neutral	Total
Three-phase, four-wire Current < 100 A	230V/400V	0.24 + j0.15	0.16 + j0.10	0.40 + j0.25
Single-phase, two-wire	230V	0.4 + j0,25	-	0.4 + j0.25
Single-phase, three-wire	100V/200V 120V/240V	0.209 + j0.103	0.143 + j0.025	0.350 + j0.13
Single-phase, three-wire	200V/240V	0.209 + j0.103	0.209 + j0.103	0.42 +j0.21
Three-phase, four-wire Current ⩾ 100 A	230V/400V	0.15 + j0.15	0.10 + j0.10	0.25 + j0.25

IEC TR 60725 defines a method to determine the grid impedance in services of more than 100 A per phase, at 60Hz [6]. The basic model is a single phase system shown in Fig. 1. Therefore, some variables are need to be defined the impedance, such as: declared voltage U_{dec}; percentage of declared voltage variation range R_{ange}; upper voltage limit $+R_{up}$ is a percentage of U_{dec}; lower voltage limit $-R_{down}$ is a percentage of U_{dec}; transformer with electromotive force $(e.m.f)$ has a maximum voltage like $U_{dec}(1 + R_{up}/100)$; voltage regulation from no-load to full-load U_{reg} it's a percentage of U_{dec}; transformer impedance Z_T; full-load rated current $I_{full load}$; grid load L; impedance of distribution cable Z_{cable}; and service current I_{load} [6].

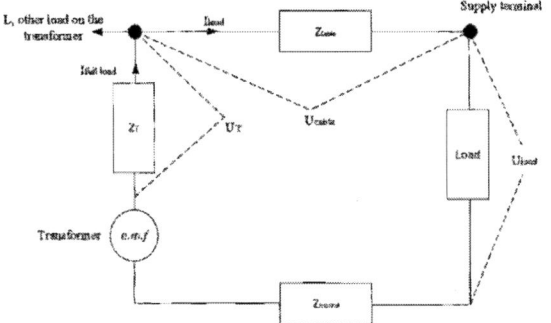

Fig. 1. Single phase basic model [6].

Before calculating the impedance, it is necessary to know what the value of the voltage U_{cable} across the cable impedance Z_{cable} [6]. The following equations are presented in IEC TR 60725 to determine the impedance:

$$U_{cable} = e.m.f - U_T - U_{load} \tag{1}$$

$$U_{cable} = U_{dec}\left(1 + \frac{R_{up}}{100}\right) - \frac{U_{reg}}{100}.U_{dec}\left(1 + \frac{R_{up}}{100}\right) - U_{dec}\left(1 - \frac{R_{down}}{100}\right) \tag{2}$$

$$U_{cable} = \frac{R_{ange}}{100}.U_{dec} - \frac{U_{reg}}{100}.\left(1 + \frac{R_{up}}{100}\right).U_{dec} \tag{3}$$

$$U_{cable} = \frac{U_{dec}}{100}\left[R_{ange} - U_{reg}\left(1 + \frac{R_{up}}{100}\right)\right] \tag{4}$$

The cable impedance Z_{cable} is given by dividing I_{load}, that is the service cut-out phase-current capacity, which is also U_{cable} [6]. On the other hand, is not possible development a representative model for 60 Hz single-phase and two-phase systems, for the lack of many configurations and plant parameters [6]. IEC TR 60725 recommends the calculus of cable impedance for cases using the equation defined before, and then adding the transformer impedance.

IV. MODEL FOR DETERMINATION OF EQUIVALENT GRID IMPEDANCE BASED ON STEADY-STATE MEASUREMENTS

PV inverters complying to NBR 16149 and NBR 16150 provide precise measurements of AC voltage, current, active and reactive power. In order to obtain parameters of the electrical grid, a model was developed which considers line and inverter voltages, grid and local load impedances. This model is presented in Fig. 2.

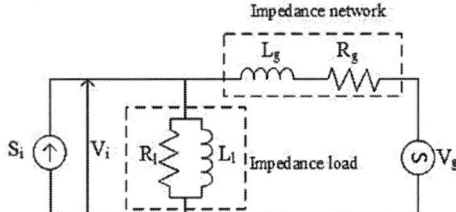

Fig. 2. Developed model.

In this model some variables are known as: the local rms voltage measured by the inverter (V_i), defined angle of the primary source voltage $(0°)$, inverter active power P_i, and reactive power Q_i. Moreover, the unknown variables are: local load impedance; grid impedance; primary source rms voltage V_g; and voltage angle of the local voltage.

Based on Fig. 2, we can determine the apparent power produced by the inverter that was injected into the grid (g) or consumed at the local load (l).

$$S_i = S_l + S_g \tag{5}$$

$$P_i = P_l + P_g \tag{6}$$

$$Q_i = Q_l + Q_g \tag{7}$$

The power dissipated by the local load can be determined by the following equation, considering a parallel load:

$$S_l = \frac{V_i^2}{Z_l^*} = P_l + jQ_l \qquad (8)$$

The active and reactive powers will be related to succeptance (B) and conductance (G) for the calculations:

$$P_l = GV_i^2 \quad , \quad G = 1/R_l \qquad (9)$$

$$Q_l = BV_i^2 \quad , \quad B = 1/X_l \qquad (10)$$

After the equation of the power dissipated in the load is necessary to know the active power that is being injected into the grid. Let us consider the local rms voltage as a complex variable $V = V_x + jV_y = V_i(\cos\theta + j\sin\theta)$.

$$S_g = VI_g^* = \left|I_g\right|^2 Z_g + V_g I_g^* = P_g + jQ_g \qquad (11)$$

$$S_g^* = I_g V^* = P_g - jQ_g \qquad (12)$$

Isolating the grid current (13) and replacing in equation (12), the grid power S_g^* can be obtained as a function of inverter complex rms voltage V, grid rms voltage V_g, and grid impedance Z_g:

$$I_g = \frac{V - V_g}{Z_g} \qquad (13)$$

$$P_g - jQ_g = \frac{VV^* - V_g V^*}{Z_g} = \frac{V_i^2 - V_g V^*}{Z_g} \qquad (14)$$

As the grid voltage source (V_g) is considered $0°$, we can determine the active and reactive power of the grid:

$$\frac{V_i^2 - V_g V_x + jV_g V_y}{\left|Z_g\right|^2}\left(R - jX\right) = P_g - jQ_g \qquad (15)$$

$$P_g = \frac{R}{\left|Z_g\right|^2}\left(V_i^2 - V_s V_x\right) + \frac{X}{\left|Z_g\right|^2}V_s V_y \qquad (16)$$

$$Q_g = \frac{X}{\left|Z_g\right|^2}\left(V_i^2 - V V_x\right) - \frac{R}{\left|Z_g\right|^2}V_g V_y \qquad (17)$$

In this way, with the intention of simplifying the equations (16) and (17) with only known variables, we can substitute equations (18), (19) and (20) in them.

$$V_x = V_i \cos(\theta) \qquad (18)$$

$$V_y = V_i \sin(\theta) \qquad (19)$$

$$\left|Z_g\right| = \sqrt{R_g^2 + X_g^2} \qquad (20)$$

Therefore, the active and reactive power of the grid are:

$$P_g = \frac{R}{R_g^2 + X_g^2}\left(V_i^2 - V_g V_i \cos(\theta)\right) + \frac{X}{R_g^2 + X_g^2}V_g V_i \sin(\theta) \qquad (21)$$

$$Q_g = \frac{X}{R_g^2 + X_g^2}\left(V_i^2 - V_g V_i \cos(\theta)\right) - \frac{R}{R_g^2 + X_g^2}V_g V_i \sin(\theta) \qquad (22)$$

The final active power of the inverter is the addition of equations (9) and (21), and similarly the final reactive power of the inverter is the addition of equations (10) and (22).

$$P_i = GV_i^2 + \frac{R_g}{R_g^2 + X_g^2}\left(V_i^2 - V_g V_i \cos(\theta)\right) + \frac{X_g}{R_g^2 + X_g^2}V_g V_i \sin(\theta) \qquad (23)$$

$$Q_i = BV_i^2 + \frac{X_g}{R_g^2 + X_g^2}\left(V_i^2 - V_g V_i \cos(\theta)\right) - \frac{R_g}{R_g^2 + X_g^2}V_g V_i \sin(\theta) \qquad (24)$$

In a similar way it is possible to do an analysis of the relation between a parallel load and a series load. But to simplify this analysis we use the succeptance (B) and conductance (G) from equation (16) and (17). The equations obtained are:

$$G_g = \frac{R_g}{R_g^2 + X_g^2} \rightarrow G_g\left(R_g^2 + X_g^2\right) = R_g \qquad (25)$$

$$G_g R_g^2 - R_g + G_g X_g^2 = 0 \qquad (26)$$

$$B_g = \frac{X_g}{R_g^2 + X_g^2} \rightarrow B_g R_g^2 - X_g + B_g X_g^2 = 0 \qquad (27)$$

$$R_g^2 - \frac{R_g}{G_g} + X_g^2 = 0 \qquad (28)$$

$$R_g^2 - \frac{X_g}{B_g} + X_g^2 = 0 \qquad (29)$$

$$-\frac{R_g}{G_g} + \frac{X_g}{B_g} = 0 \rightarrow R_g = \frac{G_g}{B_g}X_g \qquad (30)$$

$$\frac{G_g^2}{B_g^2}X_g^2 - \frac{X_g}{B_g} + X_g^2 = 0 \rightarrow X_g\left(\frac{G_g^2}{B_g^2} + 1\right) = \frac{1}{B_g} \qquad (31)$$

Uniting and simplifying equations (30) and (31) we obtain:

$$X_g = \frac{B_g}{G_g^2 + B_g^2} \qquad (32)$$

$$R_g = \frac{G_g}{G_g^2 + B_g^2} \qquad (33)$$

Substituting equations (32) and (33) into equations (21) and (22), we have a simplification of the expression.

$$\begin{cases} V_i^2 G_l + V_i^2 G_g - V_i G_g V_g \cos(\theta) + V_i B_g V_g \sin(\theta) - P_i = 0 \\ V_i^2 B_l + V_i^2 B_g - V_i B_g V_g \cos(\theta) - V_i G_g V_g \sin(\theta) - Q_i = 0 \end{cases} \qquad (34)$$

It is necessary to elaborate cases to solve the problem, in this way, at least five analysis points are being used. The known variables are $P_{i,1-5}$, $Q_{i,1-5}$, $V_{i,1-5}$. The calculated variables are G_g, B_g, G_l, B_l, V_g, and θ_{1-5}. The active and reactive power can be represented in 10 equations where $F_i = 0$:

$$V_{i,1}^2 G_l + V_{i,1}^2 G_g - V_{i,1} G_g V_g \cos(\theta_1) + V_{i,1} B_g V_g \sin(\theta_1) - P_{i,1} = F_1$$

$$V_{i,1}^2 B_l + V_{i,1}^2 B_g - V_{i,1} B_g V_g \cos(\theta_1) - V_{i,1} G_g V_g \sin(\theta_1) - Q_{i,1} = F_2$$

$$V_{i,2}^2 G_l + V_{i,2}^2 G_g - V_{i,2} G_g V_g \cos(\theta_2) + V_{i,2} B_g V_g \sin(\theta_2) - P_{i,2} = F_3$$

$$V_{i,2}^2 B_l + V_{i,2}^2 B_g - V_{i,2} B_g V_g \cos(\theta_2) - V_{i,2} G_g V_g \sin(\theta_2) - Q_{i,2} = F_4$$

$$V_{i,3}^2 G_l + V_{i,3}^2 G_g - V_{i,3} G_g V_g \cos(\theta_3) + V_{i,3} B_g V_g \sin(\theta_3) - P_{i,3} = F_5$$

$$V_{i,3}^2 B_l + V_{i,3}^2 B_g - V_{i,3} B_g V_g \cos(\theta_3) - V_{i,3} G_g V_g \sin(\theta_3) - Q_{i,3} = F_6$$

$$V_{i,4}^2 G_l + V_{i,4}^2 G_g - V_{i,4} G_g V_g \cos(\theta_4) + V_{i,4} B_g V_g \sin(\theta_4) - P_{i,4} = F_7$$

$$V_{i,4}^2 B_l + V_{i,4}^2 B_g - V_{i,4} B_g V_g \cos(\theta_4) - V_{i,4} G_g V_g \sin(\theta_4) - Q_{i,4} = F_8$$

$$V_{i,5}^2 G_l + V_{i,5}^2 G_g - V_{i,5} G_g V_g \cos(\theta_5) + V_{i,5} B_g V_g \sin(\theta_5) - P_{i,5} = F_9$$

$$V_{i,5}^2 B_l + V_{i,5}^2 B_g - V_{i,5} B_g V_g \cos(\theta_5) - V_{i,5} G_g V_g \sin(\theta_5) - Q_{i,5} = F_{10}$$

$$(35)$$

Finding the susceptances and conductances it is possible to find the impedance of the grid, through equations (32) and (33). The impedance of the load is needed to use the equations (9) and (10).

Hence, the mathematical solution of this problem was given by the creation of an algorithm. This algorithm works with the problem as an optimization with minimization of quadratic error. All the equations of (35) are a function F in the algorithm. The optimization problem is given by:

$$\text{minimize} \sum_{i=1}^{10} F_i^2 \tag{36}$$

V. EXPERIMENTAL DETERMINATION OF EQUIVALENT GRID IMPEDANCE BY USING EXTERNAL COMMANDS

The experimental results have been carried out with a Ingecon Sun 3Play 100 kW PV inverter installed in the PV power plant of Federal University of Santa Maria. By Modbus TCP communication, the following data are available: date, time, active power, AC voltage per phase, AC current per phase, frequency, reactive power, apparent power, phase angle, daytime power, active power generated, total DC power, DC current, DC current, string current, among others. Also, it is possible to externally control active and reactive power.

Five tests have been performed to find the initial parameters to the algorithm. In these tests active and reactive power are changed and the local voltage was measured. All tests have been carried out within the period of six minutes in order not to compromise energy generation. The tests and values obtain are shown in Table III.

The results of $P_{i,1-5}$, $Q_{i,1-5}$, and $V_{i,1-5}$ presented in Table III have been applied to (35), and then the minimization algorithm has been processed. The results per phase are demonstrated in Table IV. The resulted grid impedance for phase 1 was $Z_g = 0.010 + j0.046$. The other phases had similar results.

TABLE III. TESTS USING EXTERNAL COMMANDS

Test 1: set P=0% and Q=0%							
I_1 (A)	I_2 (A)	I_3 (A)	V_1 (V)	V_2 (V)	V_3 (V)	P_i (kW)	Q_i (VAr)
3.3	3.5	3.36	216.9	217.1	216.6	0.1	0
Test 2: set P=100% and Q=0%							
I_1 (A)	I_2 (A)	I_3 (A)	V_1 (V)	V_2 (V)	V_3 (V)	P_i (kW)	Q_i (VAr)
55.68	55.45	56.47	218.9	219	218.6	37.58	0.1
Test 3: set P=0% and Q=100%							
I_1 (A)	I_2 (A)	I_3 (A)	V_1 (V)	V_2 (V)	V_3 (V)	P_i (kW)	Q_i (VAr)
139.3	138.88	139.52	223.6	224	223.3	0	95.7
Test 4: set P=100% and Q=100%							
I_1 (A)	I_2 (A)	I_3 (A)	V_1 (V)	V_2 (V)	V_3 (V)	P_i (kW)	Q_i (VAr)
141.7	141.28	142.13	224.9	224.7	224.2	97.9	95.7
Test 5: set P=0% and Q=50%							
I_1 (A)	I_2 (A)	I_3 (A)	V_1 (V)	V_2 (V)	V_3 (V)	P_i (kW)	Q_i (VAr)
69.14	68.69	69.27	220.6	220.8	220.8	0	47.85

TABLE IV. RESULTS USING THE ALGORITHM

Phase 1		Phase 2		Phase 3	
R_l	3.567386	R_l	3.776471	R_l	3.414623
X_l	3.599081	X_l	3.727983	X_l	3.239586
R_g	0.010288	R_g	0.008314	R_g	0.009257
X_g	0.046276	X_g	0.045872	X_g	0.047602
V_g	220.5358	V_g	220.55942	V_g	220.4824
θ_1	-0.570278	θ_1	-0.559856	θ_1	-0.623937
θ_2	1.488221	θ_2	1.479045	θ_2	1.494398
θ_3	-0.968841	θ_3	-0.883917	θ_3	-0.987135
θ_4	0.774947	θ_4	0.8463243	θ_4	0.814318
θ_5	-0.772178	θ_5	-0.72435	θ_5	-0.807001

VI. VALIDATION AND DISCUSSION

In order to validate the results, an experimental model was created in PSIM to simulate the electrical grid. In this circuit are inserted the values of grid impedance and primary voltage, found with the algorithm, and the values of active and reactive power are set for each case. Five simulations per phase were performed to compare with the experimental results. The values obtained for phase 1, phase 2, and phase 3 are shown in Table V, Table VI, and Table VII, respectively. The results are plot in Fig. 3 (a-c).

TABLE V. SIMULATION RESULTS FOR PHASE 1

Test	Real Inverter			Simulation	Error (range)
	P_i	Q_i	V_i	V_{L_sim}	
0	3.3	0	216.9	217.1	2.50%
1	37580.3	3.3	218.9	218.77	1.62%
2	0	31900	223.6	223.59	0.12%
3	32633.3	31900	224.9	225.03	1.62%
4	0	15950	220.6	220.39	2.63%

TABLE VI. SIMULATION RESULTS FOR PHASE 2

Test	Real Inverter			Simulation	Error (range)
	P_i	Q_i	V_i	V_{L_sim}	
0	3.3	0	217.1	217.3	2.53%
1	37580.3	3.3	219	218.8	2.53%
2	0	31900	224	223.9	1.27%
3	32633.33	31900	225	225.1	1.27%
4	0	15950	220.8	220.7	1.27%

TABLE VII. SIMULATION RESULTS FOR PHASE 3

Test	Real Inverter			Simulation	Error (range)
	P_i	Q_i	V_i	V_{i_sim}	
0	3.3	0	216.6	**216.69**	1.18%
1	37580.3	3.3	218.6	**218.305**	3.88%
2	0	31900	223.3	**223.45**	1.97%
3	32633.33	31900	224.2	**224.504**	4.00%
4	0	15950	220.8	**220.5**	3.95%

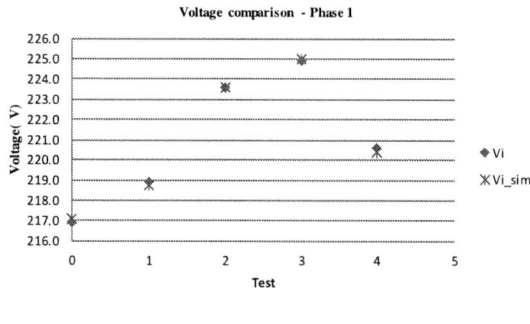

(a)

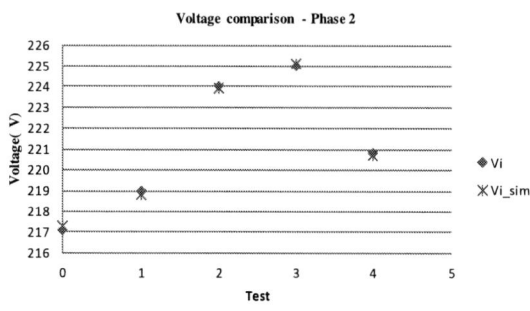

(b)

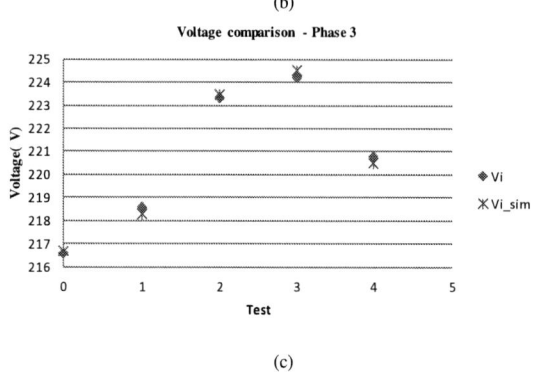

(c)

Fig. 3. Comparison of Experimental Results with Simulation results with calculated impedance: (a) phase 1; (b) phase 2; (c) phase 3;

Comparing of the results for the same conditions of power and impedance, the simulated voltages were similar to the experimentally obtained voltages. Possible causes for the error are: model does not represent all system parameters such as non-linear loads; inverter has measurement errors. In case the application needs low error, the circuit model can be improved to more accurately obtain the grid impedance.

VII. CONCLUSION

This paper presented a methodology for experimental determination of equivalent grid impedance for PV inverters by using the external commands. Experimental results were obtained in a 100 kW power plant. The grid impedance was calculated and validated by simulation results.

The main contribution of this methodology is the possibility to estimate the grid impedance without any special measurement setup. The grid impedance is estimated using the external control features provided by PV inverters with more than 6 kW active power. Possible applications of this methodology are: diagnose problems in the distribution grid; tune grid current controllers; and contribute for the definition of a Brazilian reference impedance for grid standards.

ACKNOWLEDGMENT

The present work was carried out with the support of the CPFL Energia Power Utility to the project "Soluções Inovadoras de Eficiência Energética e Minigeração em Instituição Pública Federal de Ensino Superior: Uma abordagem na UFSM" (P&D/ANEEL). The present work was carried out with the support of the INCTGD and the financing agencies (CNPq process 465640/2014-1, CAPES process No. 23038.000776/2017-54 and FAPERGS 17/2551-0000517-1) and CAPES-PROEX. This study was financed in part by the Coordenação de Aperfeiçoamento de Pessoal de Nível Superior - Brasil (CAPES) - Finance Code 001.

REFERENCES

[1] IEC 62786 2017 - Distributed energy resources connection with the grid

[2] IEC 62920 2017 - PV power generating systems - EMC requirements and test methods for power conversion equipment,

[3] EN 50549-1 2019 - Requirements for generating plants to be connected in parallel with distribution networks. Connection to a LV distribution network.

[4] ABNT NBR 16149:2013 Sistemas fotovoltaicos (FV) – Características da interface de conexão com a rede elétrica de distribuição

[5] IEC 61000-3-15 2011 - EMC - Part 3-15 Limits - Assessment of low frequency EMC requirements for dispersed generation systems in LV networkk

[6] IEC TR 60725 2012 Consideration of reference impedances and public supply network impedances for use in determining the disturbance (less 75A per phase)

[7] FIGUEIRA, H.H ; HEY, H. L. ; SCHUCH, L. ; RECH, C. ; MICHELS, L. Brazilian grid-connected photovoltaic inverters standards: A comparison with IEC and IEEE. IEEE 24th International Symposium on Industrial Electronics (ISIE), p. 1104, 2015.

[8] ANEEL. Geração Distribuída: Unidades Consumidoras com Geração Distribuída. http://www2.aneel.gov.br/scg/gd/GD_Fonte.asp

[9] TIMBUS, A.V. ; RODRIGUIZ, P. ; TEODORESCU, R. ; CIOBATARU, M. Line Impedance Estimation Using Active and Reactive Power Variations. IEEE Power Electronics Specialists Conference, p. 1273, 2007.

[10] RAGHAMI, A. ; LEDWICH, G. ; MISHRA, Y. Improved Reactive Power Sharing Among Photovoltaic Inverters Using Thévenin's Impedance Based Approach. IEEE Power & Energy Society General Meeting, p. 1, 2017.

[11] ABNT NBR 16150:2013 Sistemas fotovoltaicos (FV) – Características da interface de conexão com a rede elétrica de distribuição – Procedimento de ensaio de conformidade.

Development of a FPGA-Based Control System for Modular Multilevel Converter Applications

Lucas Koleff
Power Electronics Laboratory
University of São Paulo
São Paulo, Brazil
koleff.lucas@gmail.com

Manoel Conde
Power Electronics Laboratory
University of São Paulo
São Paulo, Brazil
maconde@terra.com.br

Pedro Hayashi
Power Electronics Laboratory
University of São Paulo
São Paulo, Brazil
phi_hayashi@outlook.com

Francesco Sacco
Power Electronics Laboratory
University of São Paulo
São Paulo, Brazil
francesco.sacco@hotmail.com

Kelly Enomoto
Power Electronics Laboratory
University of São Paulo
São Paulo, Brazil
kellymingorancia@gmail.com

Eduardo Pellini
Power Electronics Laboratory
University of São Paulo
São Paulo, Brazil
elpellini@usp.br

Wilson Komatsu
Power Electronics Laboratory
University of São Paulo
São Paulo, Brazil
wilsonk@usp.br

Lourenço Matakas Jr.
Power Electronics Laboratory
University of São Paulo
São Paulo, Brazil
matakas@pea.usp.br

Abstract—**Modular Multilevel Converters (MMCs) are convenient solutions for a variety of mid/high power energy conversion systems found in renewable energy plants, DC transmission networks and others. Nevertheless, such convenience, as the number of power modules grows, brings along an increased complexity in the hardware and control strategies used. The utilization of Field Programmable Gate Arrays (FPGAs), offering higher Input/Output availability, complex circuitry synthetization and parallel processing capabilities becomes a suitable alternative to Digital Signal Processor (DSP) based control circuits. This article explores the implementation of a system that integrates DSPs and FPGAs benefiting from their properties to come up with a flexible and scalable solution to control an MMC used for academic research purposes.**

Keywords—*power electronics, modular multilevel converters, FPGA, DSP*

I. Introduction

Modular Multilevel Converters (MMCs) are fundamental elements for the development of modern energy conversion applications such as connection of wind power plants, multiterminal HVDC networks, medium-voltage drives, among others [1]. Therefore, the study of the implementation of control systems for MMCs is a relevant subject.

One important advantage of MMCs is the ability to increase the operating voltage by adding more submodules (SMs) in series. The drawback, however, resides on the new challenges brought by a more complex hardware and the associated control algorithms.

In the field of power electronics, Digital Signal Processors (DSPs) are historically used to implement controllers; therefore, there is vast availability of control libraries for immediate use. However, the number of submodules that can be controlled directly by a single commercial DSP is limited. Field Programmable Gate Arrays (FPGAs) are a more scalable alternative as they offer much higher input/output (I/O) pin availability and flexibility and can be completely reprogrammed depending on the application.

Therefore, the growth in the number of submodules being controlled is easily supported. The inherent capacity for parallel processing of such devices also allows the execution of more complex control loops and tasks (such as signal filtering and data manipulation) in adequate time intervals. Except for latest high-end versions including floating point capabilities, FPGAs are better suited for fixed-point calculations. If floating-point, especially double precision, calculations are required, DSPs still demonstrate better performance.

Towards this scenario, the motivation and main goal of this paper is to present the development of a control system using a FPGA combined with a DSP, communicating through a high-speed (XINTF) memory bus and benefiting from the advantages of both devices. The presented control system has the advantage of allowing changes in the number of controlled submodules with strongly reduced programming efforts. Experimental results in a prototype are presented in order to validate the proposed architecture. It is important to note that comparisons with other topologies (FPGA-only or DSP-only implementations) are beyond the scope of this work.

II. System Architecture

A. Modular Multilevel Platform Overview

The MMC platform presented in this paper is developed for use in academic research and its block diagram can be found in Fig. 1. The Power Submodules feature a full-bridge converter, a local DC capacitor and the sensing circuitry for the capacitor voltage measurements. The Measurement Submodules feature sensing circuitry in order to measure the phase voltages, the arm currents and the overall DC link voltage.

The experimental implementation has 8 power submodules and 1 measurement submodule for each phase of the three-phase system. The processing unit is connected to all submodules through differential low voltage digital signaling (LVDS) transceivers, receiving measurement data and transmitting switching signals.

978-1-7281-4181-7/19 $31.00 © 2019 IEEE

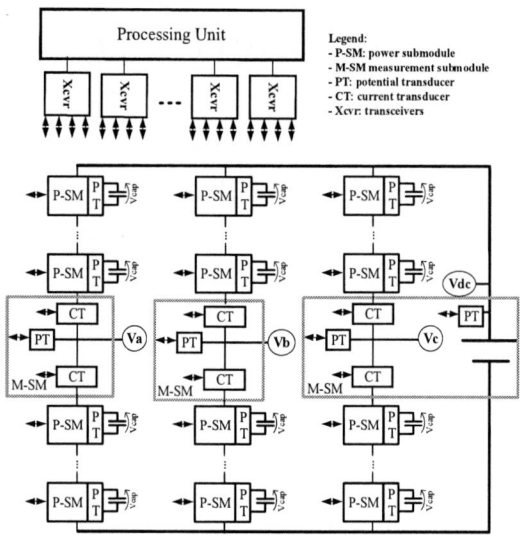

Fig. 1.   Block diagram of the MMC system.

B. Processing Board

The Processing Board features the transceivers needed for communication with all submodules, the FPGA (Xilinx Artix 200T), the DSP (TMS320F28335) and an ARM (STM32F429) microcontroller that handles communication with the user and coordinates the startup process of the converter. The synthetization of a microcontroller core into the FPGA or utilization of SoCs (System-on-a-Chip) has not been pursued to save logic and, specially, I/O resources and avoid the need for more expensive devices. The I2S protocol [2] is used in the communication link with the submodules. It is a simple and efficient solution given its synchronous frame serial communication and no need for data handshake protocol. In order to ensure the maximum possible speed for the communication interface between the FPGA and the DSP, a 32-bit wide external memory interface (XINTF) bus clocked at 150MHz is used. Considering all wait states (lead, active and trail) in the XINTF bus, a data exchange rate of 800 Mbit/s is achieved between the DSP and FPGA. It is a fact that not all this bandwidth is occupied with the current implementation but the XINTF bus, besides being the only available option for the DSP model used in this work, will provide the scalability to support more complex systems. The block diagram of the Processing Board is shown in Fig. 2.

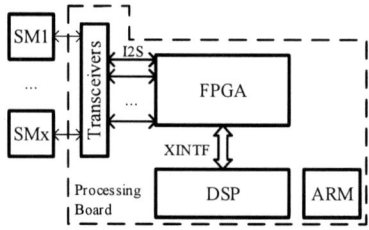

Fig. 2.   Block diagram of the Processing Board.

III. FPGA SYNTHESIS

A. Integration

The FPGA synthesis block diagram is shown in Fig. 3. The main part of the synthesized FPGA circuit is the Control Module. It has interfaces for the PWM modules, the acquisition modules and the DSP. Each Power Submodule

features local switching logic and transmission of the acquired local DC voltage. On the other hand, the Measurement Submodules do not feature switching logic, handling only acquisition and transmission of current and voltage. The **sync16** signal is used for synchronization of sampling with the peak of the PWM carriers and the **dsp_finished** and **fpga_finished** signals are used for memory bus access sequencing. The PWM modules feature the **enable** and **pwm** signals while the acquisition modules feature the **clk6M146** bit clock signal, the **sync192k** sampling synchronization and the serial data **sd** signals. The number of submodules can be adjusted depending on the application, being limited to the layout of the processing board and the number of available I/O pins of the FPGA.

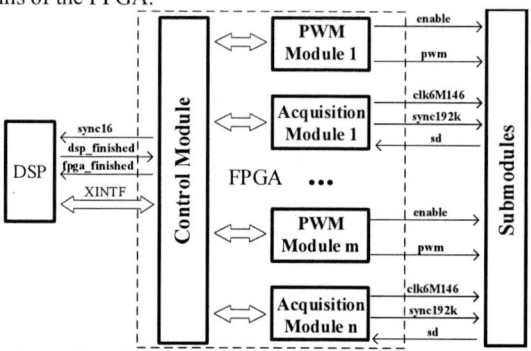

Fig. 3.   Block diagram of the FPGA synthesis.

B. Control Module

The Control Module orchestrates the execution of the three main steps of the MMC's control cycle: data acquisition, data processing and PWM update. The design combines the use of custom VHDL and Xilinx Intellectual Property (IP) blocks. It has been conceived to be scalable, requiring minimum re-programming efforts to work with MMC architectures with more levels. Figure 4 presents its main functional blocks.

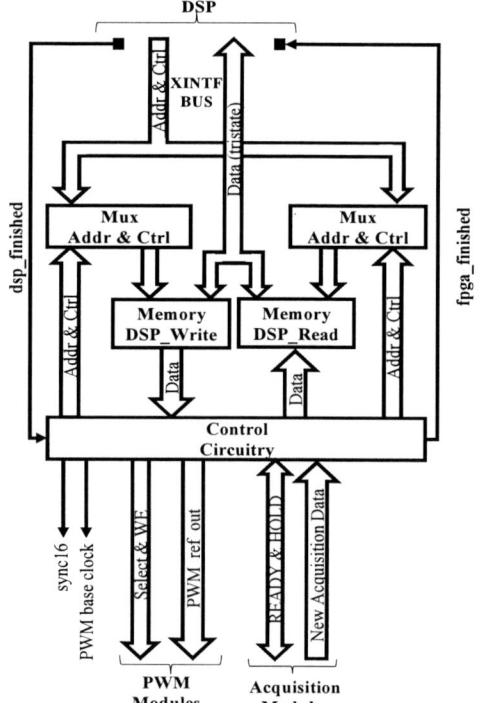

Fig. 4.   Block diagram of the Control Module.

978-1-7281-4181-7/19 $31.00 © 2019 IEEE   412

The Control Circuitry shown in Fig. 4 is a custom block written in VHDL. Concisely, it coordinates the operation of the Control Module by synchronizing the FPGA and DSP activities along the MMC control cycle, arbitrating the access to the FPGA memory blocks, reading new sensor data coming from Acquisition Modules and making them available for DSP manipulation as well as transmitting new modulation data coming from the DSP to the PWM Modules and generating synchronization signals and base clocks used in the MMC's control cycle and PWM carrier generation.

Two independent memory blocks (Xilinx IP) are used (Fig. 4). The **DSP_Read** memory block stores the data provided by the Acquisition Modules, which are 34 voltage and current measurements coming from the submodules to be processed by the DSP. The **DSP_Write** memory block stores the data provided by the DSP comprising 24 modulation levels to be transmitted to the PWM modules. Both memory blocks are 32-bit wide and 128 positions deep. As the MMC works with 16-bit data, each position stores a data pair. Consequently, not all memory positions are used in this three-phase implementation with 8 power submodules per phase. Non-used positions may be used to carry additional information between the DSP and the FPGA and/or support more complex implementations with a larger number of modules.

The data between the DSP and memory blocks flow through the tristate data bus and the memory address and control lines are multiplexed. The address and control lines multiplexers, **Mux Addr & Ctrl**, are custom blocks written in VHDL.

Figure 5 shows a state diagram describing the sequence of events along the three main steps of the control cycle.

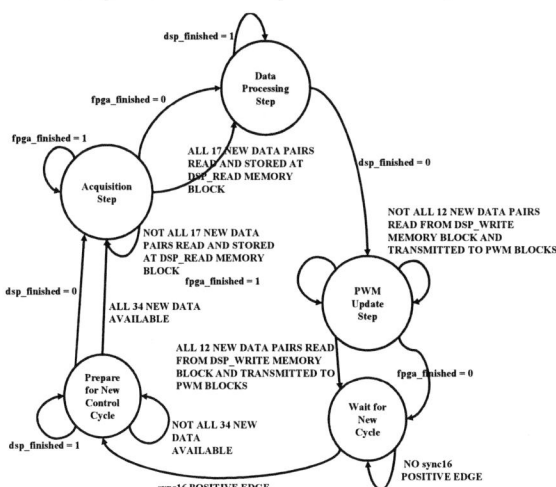

Fig. 5.   State diagram of the Control Module.

At every rising edge of the **sync16** signal the Control Circuitry checks if the DSP is idle (**dsp_finished** is low) and all Acquisition Modules have new data available (verification of the **READY** indication lines of the **READY & HOLD** bus). When both conditions are met, the execution of a new Data Acquisition step is allowed. The Control Circuitry takes control of the memory blocks setting the **dsp_finished** signal high. The **HOLD** signal is also set high to tell the Acquisition Modules to stop updating their outputs.

It will, then, sequentially read the Acquisition Modules and store the new data in the **DSP_Read** memory block. Proper selection of each acquisition module pair is done by an internal 17x32-bit multiplexer (part of the Control Circuitry).

When all the acquisition module pairs are read and the new data are stored in the **DSP_Read** block, the Data Acquisition step is finished. The Control Circuitry resets the **fpga_finished** and **HOLD** lines releasing the Memory blocks for the DSP and freeing up the Acquisition Modules to restart updating their outputs. A negative transition in the **fpga_finished** line indicates to the DSP that a new Data Processing step is allowed.

The DSP sets the **dsp_finished** line high, takes control of the memory blocks and starts the Data Processing step by reading the new acquisition data stored in the **DSP_Read** memory block. It, then, processes the new data, produces the new modulation data for the PWM modules and writes them into the **DSP_Write** memory block.

Once all the 12 PWM modulation data pairs are stored in the **DSP_Write** memory block, the Data Processing step is finished. The DSP resets the **dsp_finished** line releasing the memory blocks for the FPGA. The negative transition (falling edge) of the **dsp_finished** line shows that the Data Processing step has been completed and a PWM Update step is allowed. The control circuitry sets the **fpga_finished** line high again, takes control of the memory blocks and starts the PWM Update step by sequentially reading the **DSP_Write** memory block and loading the modulation data into the shadow register from each pair of PWM modules. Proper selection and loading of PWM pairs are carried out by setting the respective **Select** signal and pulsing the **WE** (write enable) lines of the **Select & WE** bus.

Once all the 12 PWM module pairs are updated, the PWM Update step is finished. The control circuitry resets the **fpga_finished** line, releases control of the memory blocks and waits for a new positive transition of the **sync16** signal.

C. Acquisition Module

The Acquisition Module, shown in Fig. 6, takes the serial data **sd** and the respective synchronization **sync192k** and clock **clk6M146** signals as inputs to the I2S decoder, which parallelizes the serial data using a shift register and outputs the final 16-bit word. A **DATA_GOOD** pulse is generated to indicate the completion of a Reception cycle, which is followed by the Median Filter step. The 16-bit word is then split into 14 bits of raw acquisition data and 2 status bits.

The acquisition data is then processed by a 7th order Median Filter implemented in C language using the Xilinx' High Level Synthesis (HLS) development-tool. This filter can be bypassed if the implementation does not require it. The status bits are handled accordingly in order to tell whether there are communication or internal submodule errors. The Median Filter pulses the **DONE** signal when processing is finished. Next, the Output Latch uses this pulse to register the filtered data. The final 16-bit word output of the Acquisition Module concatenates the 14-bit filtered data and the 2-bit decoded error signals. The Output Latch sets the **READY** signal high when there is a new data available. The **HOLD** signal input, when set, disables data updates (see Control Module item). The I2S Decoder, Input Latch, Output Latch and Status Handler are custom VHDL blocks.

978-1-7281-4181-7/19 $31.00 © 2019 IEEE   413

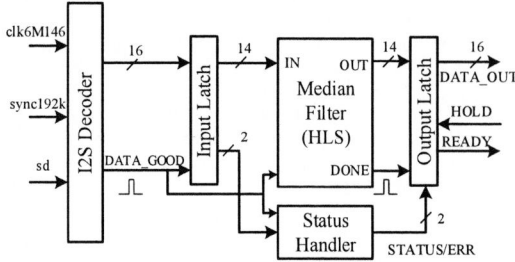

Fig. 6.   Block diagram of the Acquisition Module.

D. PWM Generation Module

The PWM generation is divided in the modulation itself and the carrier signal generation. Here, only custom VHDL blocks are used. The PWM Module is presented in Fig. 7. The Shadow Latch takes the 16-bit reference word, the **WE** and **Select** signals as inputs. The new reference provided by the Control Module is stored when **Select** is set high and **WE** is pulsed. The data transfer from the Shadow Latch to the Internal Latch occurs in the positive transition of **sync16**. This guarantees proper synchronization of the PWM signal with the MMC control cycle, so that the internal reference is updated in the right instant.

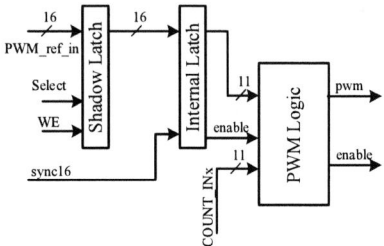

Fig. 7.   Block diagram of the PWM Module.

The PWM Logic then compares the internal reference with the carrier input in order to generate the modulated signal. Only 11 bits of the 16-bit word received are actually used for modulation. The two most significant bits are reserved for control functions. For instance, the most significant bit is used to generate the **enable** signal responsible for activating the gate drivers in the power submodules.

In Fig. 8, the PWM carrier generation is shown. There are eight triangular wave generators in this implementation, as eight power submodules are used per phase. Again, this circuit can easily accommodate the generation of a different number of carriers. In order to achieve proper harmonic cancellation in the MMC, they must be equally spaced as shown in Fig. 9 [3] [4]. All triangle waves have a frequency of 2kHz, and the counter is updated in the range between 0 and 2048 with a frequency of 8.146MHz. At each cycle of the **sync16** signal, one of the eight carriers reach its maximum value. At this instant the PWM references are updated and new acquisition samples are recorded for the processing of the control task.

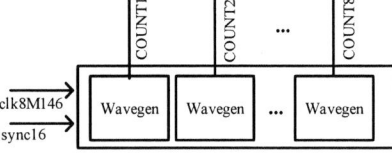

Fig. 8.   Block diagram of the PWM carrier generation.

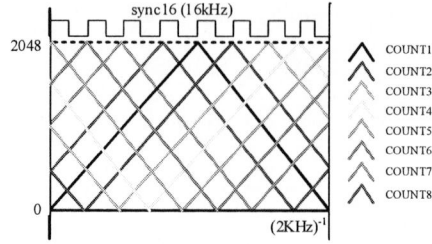

Fig. 9.   Generated carrier waveforms.

IV. DSP PROGRAM

A. Execution Steps

As discussed in the previous sections, the DSP is responsible for the execution of the high-level control task, using the measurement data to calculate new references for the PWM modulators. In Fig. 10, the flowchart of the DSP routine is presented. The rising edge transition of **sync16** triggers an internal interrupt in the DSP starting the process. At this moment, the FPGA updates the internal PWM latches and copies the acquisition data into the **DSP_Read** memory, lowering the **fpga_finished** signal when done. The DSP waits up to 50 internal clock cycles for the conclusion of the FPGA task. If the FPGA is finished within this interval, the DSP continues the execution of the program and raises the **dsp_finished** signal. Otherwise, a timeout error occurs.

The acquisition data stored in the **DSP_Read** memory is copied into the internal RAM and the Preprocessing step is executed when the 32-bit words are split and converted to floating-point values and the scaling factors and offsets are applied. The control loop calculations are done and new PWM references are generated. The Postprocessing steps are then performed converting and adjusting the references to fixed-point values and concatenating them into 32-bit words. The new references are then written into the **DSP_Write** memory and the DSP lowers the **dsp_finished** signal so that the FPGA can copy the references in the corresponding shadow registers.

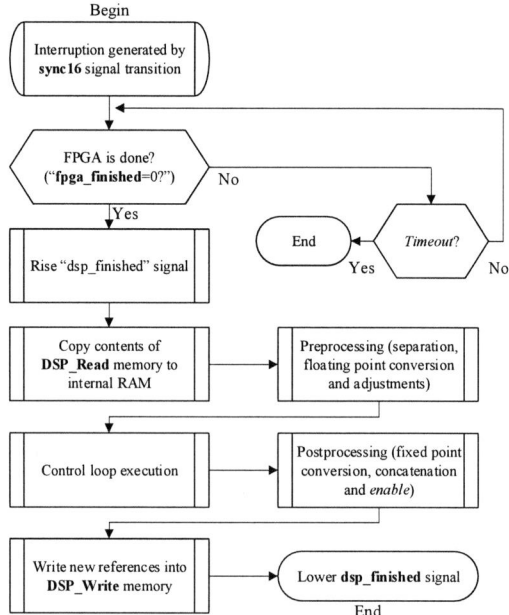

Fig. 10   Flowchart of the DSP program.

B. Control Loop

The MMC control loop itself is developed and explained in [5] [6] and this article focuses in the control system that allows the execution of those loops. For reference, a block diagram of the control loop for one phase is presented in Fig. 11. The submodule references are represented as V_{SMk}, where k is the index representing the position of the submodule in the phase. The DC voltages of the submodules are represented as V_{Ck} and the corresponding individual balancing reference ΔV_k. The upper arm current is i_p and the lower arm current is i_n. The measured circulating current is i_z and the DC link voltage is V_{DC}. The circulating current references $\bar{i}_{z,ref}$ and $\tilde{i}_{z,ref}$ are generated by the arm total capacitor voltage sum and difference loops. The upper arm total capacitor voltage is V_{Cp} and the lower arm total capacitor voltage is V_{Cn}. The MMC output current reference is i_{ref}, the measured output current is i and the AC grid (or load feedforward) voltage is v. The output current loop generates the v_s voltage reference and the circulating current loop generates the v_z voltage reference.

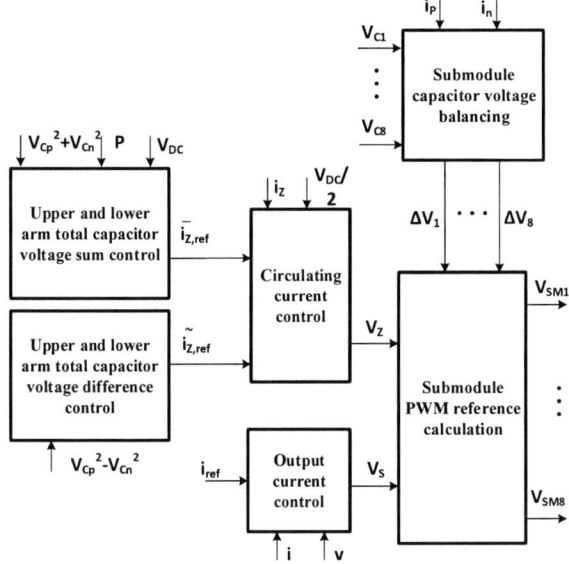

Fig. 11   Block diagram of the control loop executed by the DSP.

V. EXPERIMENTAL RESULTS

A. Overview

In order to validate the proposed control system, the three-phase MMC prototype shown in Fig. 12 is used. It features 24 power submodules, 3 measurement submodules, the processing board and all other auxiliary components necessary for the correct functioning of the platform. It is possible to notice the Ethernet cables pairs that carry the LVDS signals between the processing board and the submodules. The level of utilization of the FPGA resources is shown in Table I. It is possible to see that there is room for further extension of the system from the point of view of the FPGA.

TABLE I.   SUMMARY OF FPGA UTILIZATION

FPGA I/O Pin Usage	139/285 (48.77%)	FPGA Slice Register Usage	7737/269200 (2.87%)
FPGA Look Up Table Usage	6189/134600 (4.6%)	FPGA Block RAM Tile Usage	1/365 (0.27%)

The timing of all control cycle steps is measured and presented in Table II. It is possible to see that 38% of the cycle is used, giving enough room for more complex algorithms.

TABLE II.   SUMMARY OF TIMING OF THE CONTROL CYCLE

Acquisition Step	504ns	PWM Update Step	644ns
Processing Step	22750ns	Total Time Available per Cycle	62500ns

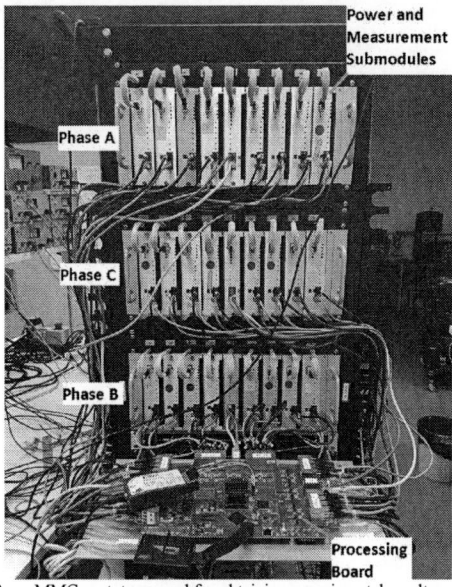

Fig. 12   MMC prototype used for obtaining experimental results.

A detailed view of the processing board highlighting the main parts is presented in Fig. 13. The FPGA is the most important component; therefore, it is positioned in the center of the board in order to equalize signal propagation delays. The LVDS transceivers and RJ45 connectors are distributed along the borders of the PCB. The DSP and ARM microcontroller are placed as close as possible to the FPGA to minimize signal propagation delays.

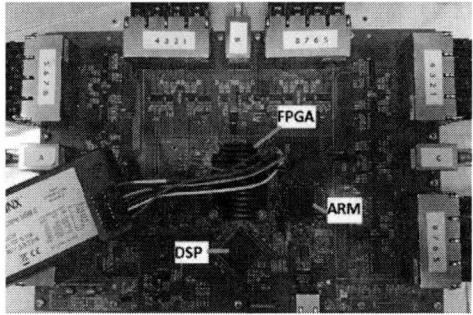

Fig. 13   Detail of the processing board.

B. Open Loop Tests

The open-loop tests are performed by energizing the local DC buses of the submodules using independent voltage sources and setting the modulation index to reproduce a sinusoidal waveform with 60Hz at the output. The upper arm always has a positive voltage and the lower arm a negative voltage, with the output voltage being the subtraction of them. In Fig. 14, it is possible to see the obtained results for phase B.

978-1-7281-4181-7/19 $31.00 © 2019 IEEE   415

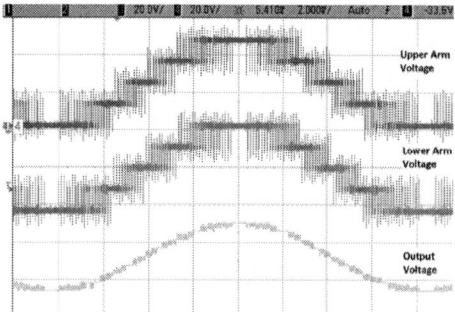

Fig. 14 Experimental waveforms of MMC arm voltages and output voltage during open-loop modulation tests in phase B.

In Fig. 15 the spectrum of both upper and lower arm voltages is presented. It is possible to see that with the carriers having a frequency of 2kHz, the switching spectral lines are concentrated at 8kHz and 16kHz, four and eight times the carrier frequency, respectively.

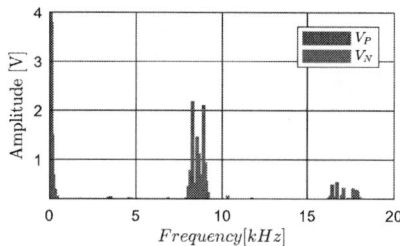

Fig. 15 FFT of the phase B arm voltages.

The output voltage spectrum is shown in Fig. 16 and is possible to see that the 8kHz switching spectral lines have been cancelled as expected, only the 16kHz components are remaining.

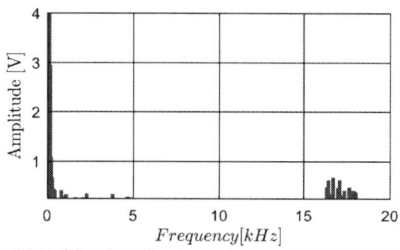

Fig. 16 FFT of the phase B output voltage.

C. Closed Loop Tests

The closed loop tests are performed with the MMC connected to a resistive load of 33Ω and the control loop is set to track 1A 60Hz sinusoidal output current reference. The obtained results are presented in Fig. 17. It is possible to see that the arm currents are composed of circulating current and output current components. The circulating current does not appear in the phase output current. The voltage balancing control is also successfully validated.

VI. CONCLUSIONS AND FUTURE WORK

The development of a circuit integrating FPGAs and DSPs has been described and its utilization in the control of an MMC verified. The chosen circuit architecture concentrates the control loop execution thus most floating-point mathematic calculation burden to the DSP.

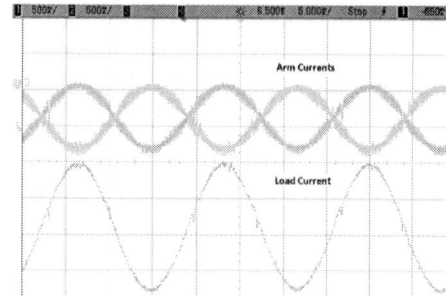

Fig. 17 Experimental waveforms of MMC arm currents and load current showing closed-loop current tracking in phase B.

This leaves to the FPGA the complex peripheral logic required for overall circuit synchronization, high speed data storage and transfer, multiple and parallel input sensor data acquisition and PWM signals output. This is proven to be effective and a convenient design that ensures scalability, a key and strategic feature to guarantee flexibility in MMC implementations. The experimental results show the ability of the circuit to achieve the functional and performance expectations of the MMC on both open and closed loop tests. The resources utilization level and measurements of the time consumed to execute control cycle activities show that both, DSP and FPGA, have bandwidth to support MMC implementations with higher number of modules and/or faster and more complex control loops. Exploring such scalability potential and the utilization of SoCs (System On Chip) components combining Processor and FPGA functionality are areas for future studies and developments.

ACKNOWLEDGMENTS

The authors would like to thank Texas Instruments, STMicroelectronics and TDK/EPCOS for their support. This work was financially supported by the Sao Paulo Research Foundation (FAPESP) research grants 2016/16542-5 and 2016/01930-0 and by the Brazilian National Council for Scientific and Technological Development (CNPq) grants 306970/2015-5 and 311789/2014-5. This study was financed in part by the Coordination for the Improvement of Higher Education Personnel (CAPES) - Finance Code 001.

REFERENCES

[1] R. Marquardt, "Modular Multilevel Converter: An universal concept for HVDC-Networks and extended DC-Bus-applications," *The 2010 International Power Electronics Conference - ECCE ASIA -*, Sapporo, 2010, pp. 502-507.

[2] Philips Semiconductors, I2S Bus Specifications, pp. 1-7.

[3] D. G. Holmes and B. P. McGrath, "Opportunities for harmonic cancellation with carrier based PWM for two-level and multi-level cascaded inverters," *Conference Record of the 1999 IEEE Industry Applications Conference. Thirty-Forth IAS Annual Meeting (Cat. No.99CH36370)*, Phoenix, AZ, USA, 1999, pp. 781-788 vol.2.

[4] B. Li, R. Yang, D. Xu, G. Wang, W. Wang and D. Xu, "Analysis of the Phase-Shifted Carrier Modulation for Modular Multilevel Converters," in *IEEE Transactions on Power Electronics*, vol. 30, no. 1, pp. 297-310, Jan. 2015.

[5] K. C. Mingorancia de Carvalho, W. Komatsu and L. Matakas Junior, "A Novel Fixed Switching Frequency Control Strategy for the Modular Multilevel Converter in the ABC Reference Frame," *2018 13th IEEE International Conference on Industry Applications (INDUSCON)*, São Paulo, Brazil, 2018, pp. 12-16.

[6] K. C. Mingorancia de Carvalho Enomoto, "Proposal for modelling and control of MMCs with fixed switching frequency," Ph.D. thesis, University of São Paulo, São Paulo, 2019.

Variable-step DFT Algorithm to Speed Up Optimization Routines Applied to a Three-Phase Interleaved Vienna Rectifier

B. Bertoldi, E. F. C. Grabovski, A. B. Lange, M. L. Heldwein
Federal University of Santa Catarina (UFSC)
Department of Electronics and Electrical Engineering (EEL)
Power Electronics Institute (INEP)
88040-970 PO box:5119 Florianópolis, SC, Brazil
e-mail: bertoldi.br@gmail.com; edhuado.celli@grad.ufsc.br; heldwein@inep.ufsc.br

Abstract—**A variable-step Discrete Fourier Transform with reduced complexity for the electrical quantities of power converters is introduced, as to reduce the time consumption of optimization routines which depend on voltage and current frequency spectrum derived from a complex time domain waveform typically generated by a numeric simulation of a power converter. A variable-step decimation algorithm is also described, reducing the number of points by detecting level transitions and/or derivative discontinuities. The algorithms are validated for a Three-Phase Vienna Interleaved Rectifier voltage and current waveforms employing Discontinuous Pulse Width Modulation.**

Index Terms—**Frequency spectrum, Vienna rectifier, Signal processing, Discrete Fourier Transform.**

I. INTRODUCTION

The optimized design of power converters is a topic that has been gaining attention over the last years, as to maximize performance criteria of the interface between different energy resources. Typical objective include minimizing volume, weight and cost associated to magnetic, semiconductor and filter devices. As the optimization for these problems typically leads to nonlinear programming, such cost functions results in time consuming tasks [1]–[4].

Various optimization challenges depend on frequency spectra of the electrical quantities, such as magnetic core and copper losses, semiconductor losses and differential and common mode filter design. Hence, the development of faster algorithms aims to minimize the optimization time consumption without altering the optimum point, i.e, the final optimization result.

Time domain waveforms are typically used to derive the behavior of the quantities of interest since power converters are modelled as non linear systems. In the frequency spectrum problem, a first solution would be to obtain analytic expressions for the spectrum through the Fourier Series, employing for instance a two-dimensional Fourier Transformation to describe the electrical quantities (voltages and/or currents) [5]. However, the mathematical description for some types of Pulse-Width Modulated (PWM) Power Converters might turn out to be a challenging and complex task.

Another solution is to obtain the spectrum through the Discrete Fourier Transform (DFT) and its Fast Fourier Transform (FFT) counterpart. However, PWM power converters usually present some inherent characteristics which might simplify the spectrum, as the energy of the electrical quantities are concentrated around fixed frequencies integer multiples of the switching frequency and a fundamental ac-side frequency.

Reference [6] also proposes some alternative methods to describe the time instants when the voltage transitions occur in a power converter, as to eliminate the need for circuit simulation. However, there is also a dependency for variable-step algorithms to obtain the frequency spectrum using this approach. Fig. 1 shows a diagram with different methods to obtain the frequency spectrum of a given electrical quantity.

This work proposes the use of a variable-step Discrete Fourier Transform based on power converter voltage (or current, flux, charge, power, etc) transitions as to reduce the complexity of traditionally employed algorithms for power

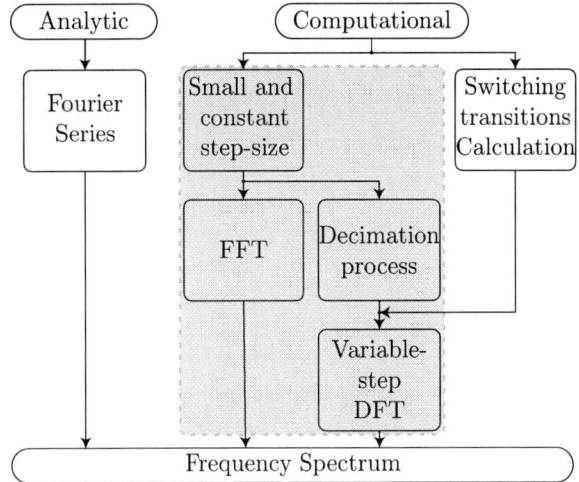

Fig. 1. Methods to obtain the spectrum of electrical quantities for power converters.

978-1-7281-4181-7/19 $31.00 © 2019 IEEE

converter applications. The algorithm is validated for a Vienna Interleaved Rectifier employing a Discontinuous PWM strategy, which presents a complex analytic description. In addition, a pre-processing algorithm to decimate a fixed-step waveform as to obtain the level transitions is also described. The variable-step DFT is validated through a spectrum comparison with MATLAB® *fft* function.

II. INTERLEAVED VIENNA RECTIFIER

The Vienna rectifier is a rectifier topology that enables Power Factor Correction (PFC) applications which do not require a bidirectional power flux [7]. Some of its inherent characteristics is the robustness, as there is no direct possibility of a dc-bus short circuit due to a determined switching state; high power density, as the voltage stress on the semiconductor devices is reduced if compared to a conventional 2-level Voltage-Source Rectifier (VSR); and a relatively simple structure [6]–[8].

This work considers a Vienna rectifier employing a multi-state switching cell, as introduced in [6], with a two channels switching cell per phase. The implemented topology and modulation strategy are described in the following.

A. Rectifier Topology

The Vienna rectifier with two-channel multi-state switching cells is shown in Fig. 2. The multi-state switching cells are connected through interphase transformers, which aims to reduce circulating currents, enabling high frequency currents and voltages balancing to promote even power sharing between the channels.

The dc-bus voltage is divided among the two output capacitors (C_o) and the bidirectional turn-off switches (S_{ij} with $i = \{1, 2\}$ and $j = \{a, b, c\}$) of each phase are connected to a central point. Hence, the voltage stress over these semiconductor devices is half of the dc-bus voltage. A strategy to balance the dc capacitor voltages is also needed. In addition, four-quadrant switches must be employed, although there is no need for multi-step transition schemes.

The rectifier is also composed of three boost inductors that provide an interface between rectifier and the grid (or an EMC input filter) and two three-phase diode bridges.

B. Modulation strategy

Let the input phase voltage of the rectifier be defined by a sinusoidal voltage $v_{i,\kappa}$ with amplitude $\hat{V}_i$ for phase $\kappa \in \{a, b, c\}$. Also, let the output voltage of the rectifier be defined by v_o with average value V_o. The modulation signals can, thus, be defined according to

$$\begin{cases} m_a = MV_o \sin(\omega_g t) \\ m_b = MV_o \sin\left(\omega_g t - \frac{2}{3}\pi\right), \\ m_c = MV_o \sin\left(\omega_g t + \frac{2}{3}\pi\right) \end{cases} \quad (1)$$

assuming sinusoidal voltage references, with a modulation index defined as $M = 2\hat{V}_i/V_o$.

However, modulation schemes in three-phase systems are often implemented by inserting a zero-axis component into the modulation signals, which opens a wide range of schemes, each presenting its own set of advantages. An alternative is the discontinuous pulse width modulation (DPWM) strategy, which presents a reduced number of switching transitions, leading to lower switching losses [9] and when used in the interleaved Vienna rectifier promotes low distortion zero crossings for the ac-side currents.

The DPWM strategy can be implemented by different methods. Throughout this work, the common-mode signal is defined as

$$m_0 = \tfrac{1}{2}\text{sign}\left(m'_{max}\right) - m'_{max}, \quad (2)$$

with the new modulation references defined by

$$m'_\kappa = \text{mod}\left(m_\kappa + 1, 1\right) - \tfrac{1}{2} \quad (3)$$

and

$$m'_{max} = \begin{cases} m'_a, & \text{if } |m'_a| = \max(|m'_a|, |m'_b|, |m'_c|) \\ m'_b, & \text{if } |m'_b| = \max(|m'_a|, |m'_b|, |m'_c|) , \\ m'_c, & \text{if } |m'_c| = \max(|m'_a|, |m'_b|, |m'_c|) \end{cases} \quad (4)$$

with $\kappa \in \{a, b, c\}$ and where $\text{mod}(x, y)$ returns the remainder after division of x by y.

The modulation signal for phase a ($m_a + M_0$), its common-mode component (m_0) and the normalized switched voltage $V_{i,a}$ with respect to the dc-bus mid point are depicted in Fig. 3, demonstrating the modulation discontinuous characteristic.

III. WAVEFORM DESCRIPTION

As the analytic expressions for modulated signals are typically highly complex, there is a tendency to generate such waveforms computationally through numeric simulations and/or algorithms. However, a large number of points might be necessary to properly reproduce a discrete signal meeting Shannon's theorem requirements while ensuring a large bandwidth and small aliasing.

The number of points depends on the application and the working frequency range. For instance, in a Electromagnetic Compatibility problem, it is necessary to evaluate the frequency spectrum to frequencies up to 30 MHz to properly model the conducted electromagnetic emissions. Hence, assuming a 60 Hz grid frequency and a sampling frequency of 60 MHz, a minimum of 10^6 points is required, which might imply into time consuming post-processing algorithms due to the large number of points.

However, the minimum necessary information to reconstruct or represent a time discontinuous waveform without information loss is to evaluate the function at its level transitions. An automated algorithm to generate these waveform level transitions might require previous knowledge of functions that describe the voltage transition times and the respective levels during the transitions, which can also be quite complex for multilevel converters with advanced modulation schemes.

A simpler approach is through the use of a variable-step decimation, as described in Alg. 1. In this case, the algorithm evaluates the boost inductance voltage of phase κ, obtaining a

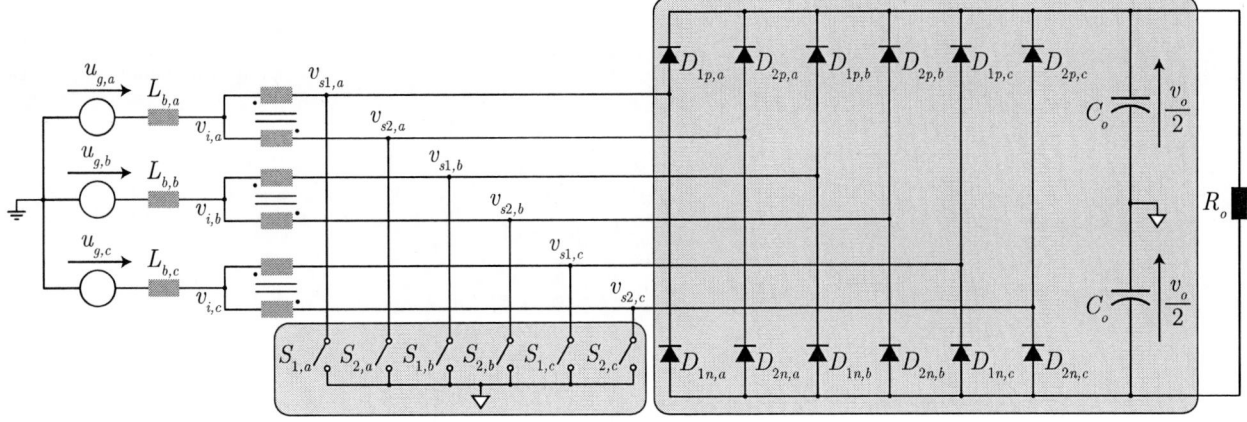

Fig. 2. Five-level Vienna Interleaved Rectifier topology, composed by two three-phase diode bridge, six four-quadrant switches and three interphase transformers.

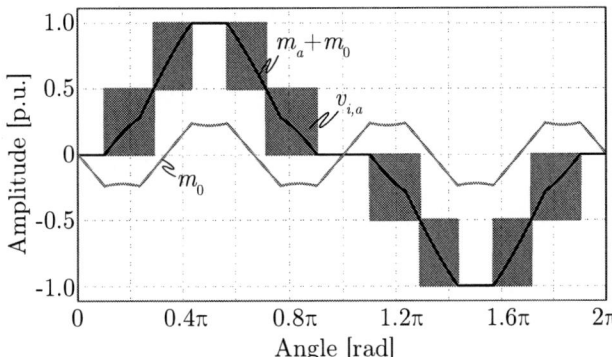

Fig. 3. Five-level Vienna rectifier phase voltage with Discontinuous PWM, demonstrating the modulation signal and the injected common-mode voltage.

Algorithm 1 Inductor voltage Variable-step decimation

1: **Input:** Converter Phase Voltage $\vec{v}_i = \{v_{i,a}, v_{i,b}, v_{i,c}\}$, Grid Voltage u_κ, Time Vector t
2: **Output:** Decimated Voltage Vector v_{dec}, Decimated Time Vector t_{dec}
3: $j \leftarrow 1$
4: Tol $\leftarrow 10^{-5}$
5: **for** $n \leftarrow 1$ **to** length(t) **do**
6: $\quad v_{cm}(n) \leftarrow \frac{1}{3} \sum_{\kappa=\{a,b,c\}} v_{i,\kappa}(n)$
7: $\quad v_{L,\kappa}(n) \leftarrow v_{i,\kappa}(n) - u_\kappa(n) - v_{cm}(n)[1\ 1\ 1]^T$
8: $\quad$ **if** $|v_{L,\kappa}(n+1) - v_{L,\kappa}(n)| >$ Tol **then**
9: $\quad\quad p(j) \leftarrow k$
10: $\quad\quad j \leftarrow j + 1$
11: $\quad$ **end if**
12: **end for**
13: **for** $i \leftarrow 1$ **to** length(p) **do**
14: $\quad t_{dec}(i) \leftarrow t(p(i))$
15: $\quad v_{dec}(i) \leftarrow v_{L,\kappa}(p(i))$
16: **end for**

---

vector of indexes p with the instants when voltage transitions occur. A tolerance is inserted as to minimize errors due to small signal deviations, which are commonly present in waveforms generated through analytic equations and numeric simulations with dc-bus voltage variations.

After the identification of voltage level transitions and decimation, the information stored in vectors t_{dec} and v_{dec} corresponds to the instants where transitions have occurred. Thus, the original signal can be recomposed through interpolation. For instance, the inductor current, which is given by the voltage integral, can be obtained through a first order interpolation between points. This interpolation is basis for the variable-step DFT presented in the following section.

IV. DFT METHOD

Traditional DFT and FFT algorithms typically employ fixed step sizes, which in turn compute the contribution of each sample, according to

$$X_k = \sum_{n=0}^{N-1} x(n) e^{\iota 2\pi \frac{k}{N} n}, \qquad (5)$$

with $\iota^2 \triangleq -1$, k as the harmonic order and N as the total number of samples (or vector components).

However, the power converters switched voltages energy is typically concentrated in small frequency ranges around multiples of the switching frequency, hence the definition of central frequencies (integer multiples of the switching frequency) and side-band frequencies [5]. Consequently, an approximation to the DFT can be computed by interpolating the variable-step voltage signal r, which can be obtained through the previously described algorithm and calculating the sum of the contribution of each interpolation segment. A simple approach is to interpolate each point by line segments, with its derivative set by s [10], as in

$$s_n = \frac{r_{n+1} - r_n}{t_{n+1} - t_n} \qquad (6)$$

where t is the discretized time vector. Fig. 4 illustrates the

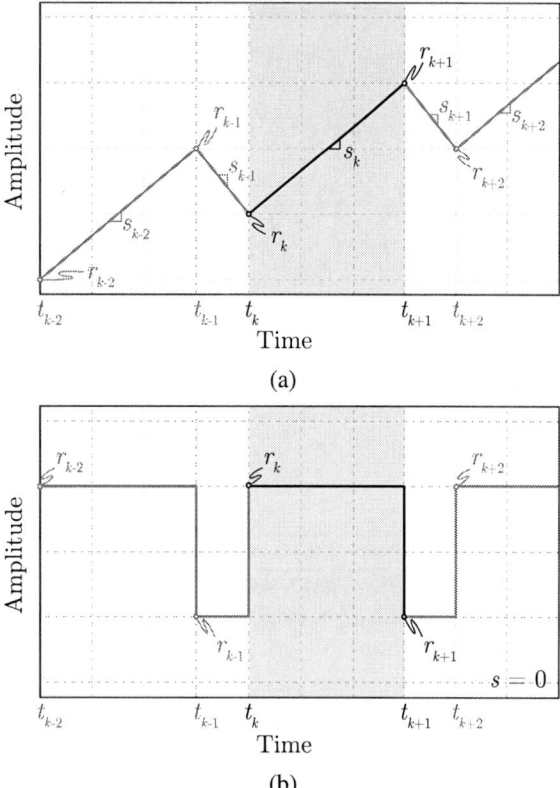

Fig. 4. Interpolation process for: (a) converter current; and (b) discontinuous converter voltages.

Algorithm 2 Variable-step DFT

1: **Input:** Variable-step Time Vector t, Decimated Electrical Quantity Vector r, Fundamental Frequency f_g, Carrier Frequency f_s, Central Band Set of Harmonics N_c, Number of Sideband Harmonics N_{sb}
2: **Output:** Fourier Series Coefficients Vectors a and b
3: **for** $i \leftarrow 1$ **to** length$(t) - 1$ **do**
4: $\quad s(i) \leftarrow \dfrac{r(i+1) - r(i)}{t(i+1) - t(i)}$
5: **end for**
6: **for** $i \leftarrow 1$ **to** length(N_c) **do**
7: $\quad k \leftarrow N_c(i)$
8: $\quad$ **for** $j \leftarrow 1$ **to** $2N_{sb} + 1$ **do**
9: $\quad\quad n \leftarrow k \dfrac{f_s}{f_g} - N_{sb} + j$
10: $\quad\quad$ **for** $m \leftarrow 1$ **to** length$(t) - 1$ **do**
11: $\quad\quad\quad A \leftarrow \sin\left(\pi f_g n t(m)\right)$
12: $\quad\quad\quad B \leftarrow \cos\left(\pi f_g n t(m)\right)$
13: $\quad\quad\quad C \leftarrow \cos\left(2\pi f_g n t(m+1)\right)$
14: $\quad\quad\quad D \leftarrow \sin\left(2\pi f_g n t(m+1)\right)$
15: $\quad\quad\quad a(i,j) \leftarrow a(i,j) - \dfrac{1}{n\pi}\left(2AB - D\right) r(m)$
16: $\quad\quad\quad\quad + \dfrac{1}{2n^2\pi^2 f_g}\left(2B^2 - C - 1\right) s(m)$
17: $\quad\quad\quad\quad + \dfrac{D}{n\pi}\left(t(m+1) - t(m)\right) s(m)$
18: $\quad\quad\quad b(i,j) \leftarrow b(i,j) + \dfrac{1}{n\pi}\left(2B^2 - C - 1\right) r(m)$
19: $\quad\quad\quad\quad - \dfrac{1}{2n^2\pi^2 f_g}\left(2AB - D\right) s(m)$
20: $\quad\quad\quad\quad - \dfrac{C}{n\pi}\left(t(m+1) - t(m)\right) s(m)$
21: $\quad\quad$ **end for**
22: $\quad$ **end for**
23: **end for**

interpolation procedure for two different processes, namely: discontinuous converter voltages (vanishing s) and boost inductance currents, with s defined as a linear interpolation between two defined samples. This example for inductor current and converter voltage applications with a variable-step sampling (see Fig. 4) illustrates some of the variables employed in the algorithm as well as the resulting interpolated line segments.

Hence, the Fourier series coefficients are obtained for each line segment and added together by solving

$$a_k + \iota b_k = \sum_{n=1}^{N} \frac{2}{T_g} \int_{t_n}^{t_{n+1}} r_n(\tau) e^{-\iota 2\pi k f_g \tau} d\tau, \quad (7)$$

which in turn results in

$$a_k = \sum_{n=1}^{N} - \frac{1}{k\pi}\left(2AB - D\right) r_n$$
$$+ \frac{1}{2k^2\pi^2 f_g}\left(2B^2 - C - 1\right) s_n \quad, \quad (8)$$
$$+ \frac{D}{k\pi}\left(t_{n+1} - t_n\right) s_n$$

and

$$b_k = \sum_{n=1}^{N} \frac{1}{k\pi}\left(2B^2 - C - 1\right) r_n$$
$$- \frac{1}{2k^2\pi^2 f_g}\left(2AB - D\right) s_n, \quad (9)$$
$$- \frac{C}{k\pi}\left(t_{n+1} - t_n\right) s_n$$

where

$$\begin{aligned} A &= \sin\left(\pi k f_g t_n\right) \\ B &= \cos\left(\pi k f_g t_n\right) \\ C &= \cos\left(2\pi k f_g t_{n+1}\right) \\ D &= \sin\left(2\pi k f_g t_{n+1}\right) \end{aligned} \quad (10)$$

f_g and T_g are the grid frequency and period, respectively.

The algorithm is described in Alg. 2 and its execution can be further optimized through a vectored implementation and parallel processing.

Considering the complexity of the DFT as $O(N^2)$ and the complexity of the FFT as $O(N\log(N))$, the proposed strategy might present a lower complexity since the number of points that defines the sums in (8) and (9) is equal to the number of

978-1-7281-4181-7/19 $31.00 © 2019 IEEE

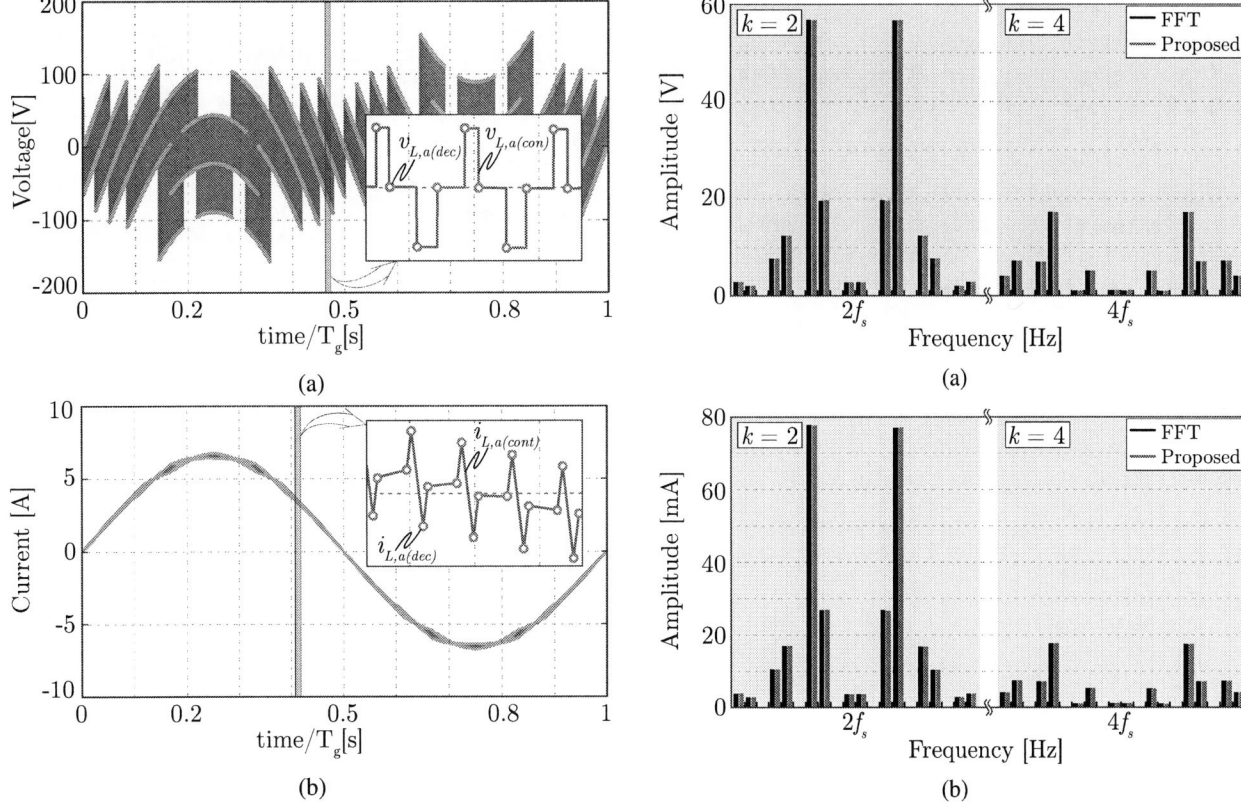

Fig. 5. Decimation process results based on the converter simulation for: (a) phase a inductor voltage waveform; and (b) phase a inductor current waveform.

Fig. 6. Comparison between the proposed Variable-step DFT method and MATLAB® fft function (10^6 points) for: (a) phase a inductor voltage waveform; and (b) phase a inductor current waveform.

voltage transitions and, thus, depends almost exclusively on the switching frequency, i.e, $O\left(\left(f_s/f_g\right)^2\right)$.

V. SIMULATION RESULTS

The proposed methodology validation was performed through numeric simulations for the Vienna Interleaved Rectifier. The converter parameters were calculated assuming a 3-kW system, with 220 V input phase voltage rms value, ac-side frequency of $f_g = 60$ Hz and $V_o = 800$ V total dc output voltage. The interphase transformers where modelled with near perfect magnetic coupling coefficient, i.e., no stray fields were considered. An integer multiple of the grid frequency must be chosen to guarantee an even number of transitions per grid period. Additionally, a quarter-wave symmetry is achieved by using a multiple of six times the grid period. Therefore, a $f_s = 50.04$ kHz switching frequency (834 switching periods in a grid period) was chosen.

The analysis was performed by evaluating the voltage and current waveforms of the boost inductor. As the rectifiers are operating in parallel, the interphase transformer causes a harmonic cancelling effect of the central band and sideband harmonics for odd multiples of the switching frequency, hence the first harmonic components appears at frequencies around $2f_s = 100.08$ kHz.

An initial simulation generated a 10^6 points vector for, both, boost inductor voltage and current, and the decimation algorithm described in Alg. 1 was then used, which resulted in a 6649 points voltage vector and a 13261 points current vector, which represents a reduction of approximately 150 and 75 times, respectively.

The method is illustrated in Fig. 5(a) for the phase a inductor voltage waveform and in Fig. 5(b) for its current waveform. It should be highlighted that the algorithm obtained satisfactory results also for the voltage waveform, which ignored the small derivatives by altering the algorithm tolerance, which in this case was set to 10^{-1}.

The harmonic spectrum was obtained exemplarily for the first two existing harmonic multiples of the switching frequency, which results in $2f_s$ and $4f_s$. MATLAB® fft function was applied to the vectors with the full number of points, while the proposed DFT method was applied only to the decimated vectors. A comparison is observed in Fig. 6(a) for the voltage and in Fig. 6(b) for the current. The achieved results demonstrate the efficacy of the variable-step DFT since similar spectra are found.

978-1-7281-4181-7/19 $31.00 © 2019 IEEE

VI. CONCLUSIONS

This work presented a methodology to obtain the harmonic spectra of a given power converter electrical quantities while reducing the typically involved computational burden by taking into account inherent characteristic of converter voltage and current waveforms.

The proposed variable-step DFT was validated for a 3-kW Vienna Interleaved Rectifier employing Discontinuous pulse-width modulation due to the inherent complexity of its analytic equations. The method is generic and can be employed in other power converters and modulation strategies. It might be very useful for precise EMC modeling of power converters based on time domain simulations and the design of magnetic components.

REFERENCES

[1] A. De Bastiani Lange and M. L. Heldwein, "Optimal inductor design for single-phase three-level bridgeless pfc rectifiers," in *2017 Brazilian Power Electronics Conference (COBEP)*, Nov 2017, pp. 1–6.

[2] M. Mogorovic and D. Dujic, "Medium frequency transformer design and optimization," in *PCIM Europe 2017; International Exhibition and Conference for Power Electronics, Intelligent Motion, Renewable Energy and Energy Management*, May 2017, pp. 1–8.

[3] ——, "100 kw, 10 khz medium-frequency transformer design optimization and experimental verification," *IEEE Transactions on Power Electronics*, vol. 34, no. 2, pp. 1696–1708, Feb 2019.

[4] J. Muhlethaler, M. Schweizer, R. Blattmann, J. W. Kolar, and A. Ecklebe, "Optimal design of lcl harmonic filters for three-phase pfc rectifiers," *IEEE Transactions on Power Electronics*, vol. 28, no. 7, pp. 3114–3125, July 2013.

[5] D. G. Holmes and T. A. Lipo, *Pulse Width Modulation for Power Converters: Principles and Practice*. IEEE, 2003. [Online]. Available: https://ieeexplore.ieee.org/document/5311996

[6] M. S. Ortmann, S. A. Mussa, and M. L. Heldwein, "Three-phase multilevel pfc rectifier based on multistate switching cells," *IEEE Transactions on Power Electronics*, vol. 30, no. 4, pp. 1843–1854, April 2015.

[7] J. W. Kolar, U. Drofenik, and F. C. Zach, "Vienna rectifier ii-a novel single-stage high-frequency isolated three-phase pwm rectifier system," *IEEE Transactions on Industrial Electronics*, vol. 46, no. 4, pp. 674–691, Aug 1999.

[8] T. Friedli, M. Hartmann, and J. W. Kolar, "The essence of three-phase pfc rectifier systemspart ii," *IEEE Transactions on Power Electronics*, vol. 29, no. 2, pp. 543–560, Feb 2014.

[9] M. S. Ortmann, S. A. Mussa, and M. L. Heldwein, "Evaluation of carrier-based pwm strategies for multi-state switching cells-based multilevel three-phase rectifiers," in *XI Brazilian Power Electronics Conference*, Sep. 2011, pp. 903–910.

[10] A. de Bastiani Lange, "Retificador pfc monofásico pwm bridgeless três-nveis de alto desempenho," Master's thesis, UFSC, Brazil, 2012.

DESIGN METHOD TO REDUCE THE DC LINK VOLTAGE OF A THREE-WIRE THREE-PHASE HYBRID ACTIVE POWER FILTER

Mateus Freitas Braga, Samuel Neves Duarte, Guilherme Márcio Soares, Pedro Gomes Barbosa

Federal University of Juiz de Fora, Graduate Program in Electrical Engineering, Juiz de Fora–MG, Brazil

Email: mateus.braga@engenharia.ufjf.br, samuel.neves@engenharia.ufjf.br,
guilherme.marcio@ufjf.edu.br, pedro.gomes@ufjf.edu.br

Abstract - **The hybrid active power filter is one of the topics most discussed by researchers concerning power quality issues. However, most of the works presented in literature do not deal with the design of the hybrid active power filter's components. Thus, this paper proposes a design method of a shunt hybrid active power filter used to compensate for a non-linear load connected to a three-wire three-phase network. The methodology is based on the superposition theorem and the instantaneous power theory. This method is performed in such a way to ensure a low voltage at the DC bus, compared to the DC voltage value commonly adopted in voltage-sourced inverters. Therefore, the ratings of the active power filter are reduced. Results of digital simulations are presented in order to validate the design method proposed in this paper.**

Keywords - **DC Capacitor, Hybrid Active Power Filter, Instantaneous Power Theory, Power Quality.**

I. INTRODUCTION

The increase in the number of electric non-linear and unbalanced loads connected to utility has been responsible for various power quality issues in modern electrical systems. The harmonic and reactive currents drained by these loads may cause overheating in equipment, increase in power losses, capacitor fuses blowing, low power factor, among others [1–3].

Several studies have been developed over the last years where passive and/or active solutions are proposed to limit the harmonic currents and voltage distortions in electric power system in order to meet international standards and recommendations [4, 5].

Passive filters, comprising inductors and capacitors, are a simple option to filter a harmonic or a set of harmonic components as well as to compensate for reactive power. Despite the lower cost, they have large volumes and do not allow dynamic compensation of the load, and may also cause problems of undesirable resonances with the electrical system [6].

On the other hand, shunt active power filters have been proposed as an alternative solution to mitigate the mentioned issues. They usually employ a voltage source inverter (VSI) comprising six semiconductor switches, a DC capacitor bank and three inductors connected to the AC terminals of the converter. Its main advantage compared to passive filters is the possibility of implementing an active control. On the other hand, active filters have a higher cost because they require switches operating with frequencies of the order of some kilohertz and blocking voltages that exceed at least twice the peak

value of the voltage at the point of common coupling (PCC) [7].

In this context, hybrid active power filters (HAPFs) have been emerged as solution to overcome some of the drawbacks of the passive and active power filters. They combine characteristics of active and passive filters, incorporating the controllability of the first and allowing to work with reduced voltages due to the second. This feature is one of the main advantages of HAPFs, allowing also to reduce the stress and rating of the HAPF semiconductor switches. [1, 8–12].

Although there are several papers dealing with hybrid filters [13–18], the authors do not explore all the characteristics of the HAPFs, specially, how to determine the operation value of the DC bus voltage. Thus, this paper presents a design method of a three-wire three-phase hybrid active power filter. The design method assure the HAPF will operate with the minimum DC bus voltage. Results of digital simulations of a HAPF correcting the power factor (PF) and compensating the total harmonic distortion (THD) of a non-linear load are used to validate the design methodology of the HAPF.

II. SYSTEM DESCRIPTION

The advantage of reducing the DC bus voltage of the hybrid filters occurs only if the load or electric network requires inductive reactive current. This is due to the fact that, when the HAPF synthesizes currents in the fundamental frequency, part of the AC voltage is across the capacitor of the passive portion of the hybrid filter. Thus, the voltage across the active part (voltage-sourced inverter) is reduced.

Therefore, to support the study presented in this paper, an inductive non-linear load is parallel-connected to a HAPF and a three-wire three-phase electric network, as shown in Fig. 1. The grid is modeled by the voltage v_s, with frequency f_s, and by the equivalent inductance L_s. The non-linear load is comprised by the inductance L_{AC}, the three-phase diode bridge $(D_1 - D_6)$, the capacitor C_d and the resistance R_L. The passive portion of the hybrid filter is modeled by the inductance L_f, the capacitance C_f and the resistance R_f. On the other hand, the active portion of the HAPF is comprised by six switches $(S_1 - S_6)$ and the DC capacitor C.

III. HYBRID ACTIVE POWER FILTER DESIGN

This section will be subdivided into two parts. The first one presents the design method of the passive portion, while the second one focuses on the active part of the HAPF.

The passive filter will be designed to compensate for the

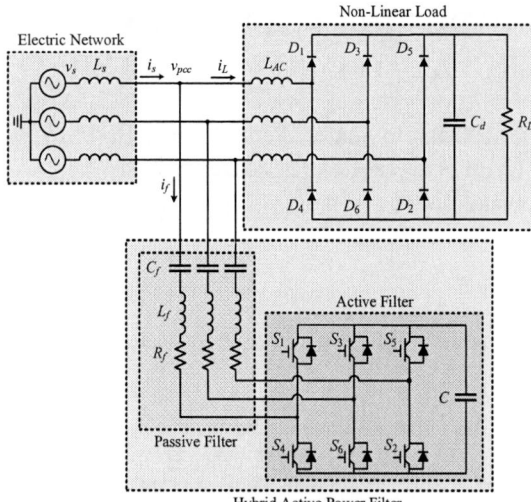

Fig. 1: Topology of the HAPF parallel-connected to a non-linear load and the electric network.

average reactive power (due to the fundamental components) and to be resonant at the higher amplitude among all harmonic currents drained by the load. On the other hand, the active portion of the HAPF will be designed to behave as a short-circuit at both fundamental component and the resonant frequency of the passive filter. Thus, the HAPF DC bus voltage will be much smaller compared to the DC voltage of active power filters that usually needs to be at least twice the PCC peak voltage. Moreover, the active part of the HAPF will compensate for the others harmonic currents drained by the load.

A. Design of the HAPF Passive Portion

Fig. 2 shows the single-line diagram for the harmonic component h of the circuit of Fig. 1.

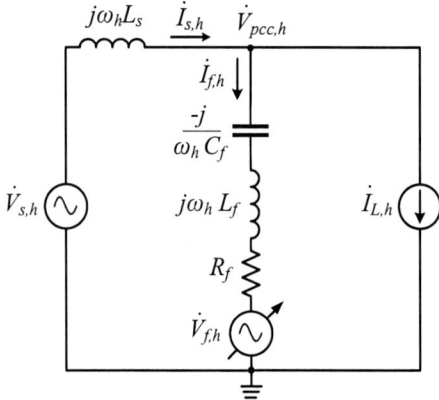

Fig. 2: Single-line diagram of the circuit.

Considering that, at the fundamental frequency, the active part of the HAPF behave as a short-circuit (which means a lower DC bus voltage), the passive part of the HAPF will com-

pensate for all the average reactive power consumed by the load. Thus, the following expression can be written for the reactance of the passive filter:

$$\frac{V_{pcc}^2}{\bar{q}_L} = X_{pf} = \omega_1 L_f - \frac{1}{\omega_1 C_f}, \quad (1)$$

where ω_1 is the fundamental angular frequency of the grid voltage, X_{pf} is the reactance of the passive filter, V_{pcc} is the RMS voltage at PCC and $\bar{q}_L$ is the average reactive power of the load.

At the highest amplitude among the current harmonic components drained by the load, the active portion of the HAPF is designed to behave as a short-circuit while the passive filter is in resonance, being the resonant frequency given by:

$$\omega_r = \sqrt{\frac{1}{L_f C_f}}. \quad (2)$$

Since the active portion of the HAPF behave as a short-circuit at frequency ω_r, the DC bus voltage of the HAPF can be further reduced.

The system of equations given by (1) and (2) can then be solved in order to design the inductance and capacitance values of the passive filter. The design parameters are the amount of reactive power ($\bar{q}_L$) to be compensated and the resonance frequency (ω_r) of the passive filter.

B. Design of the HAPF Active Portion

The reference current for the HAPF is basically a sum of two signals, each one subdivided into two others. The first one is composed by a reference current to compensate for the average reactive power of the load and the harmonic reference currents to minimize the THD. The second one contain reference currents to charge the DC capacitor and to regulate the DC bus voltage of the HAPF. The computation of all these reference currents for the HAPF will be presented in this section.

As the current control will be performed in the $\alpha\beta$ coordinates, the voltages and currents are transformed to this reference frame through the Clark transformation, as follows:

$$\begin{bmatrix} x_\alpha \\ x_\beta \end{bmatrix} = \frac{2}{3} \begin{bmatrix} 1 & -\frac{1}{2} & -\frac{1}{2} \\ 0 & \frac{\sqrt{3}}{2} & -\frac{\sqrt{3}}{2} \end{bmatrix} \begin{bmatrix} x_a \\ x_b \\ x_c \end{bmatrix}, \quad (3)$$

where x represents the voltages or currents of the system and the constant $\frac{2}{3}$ is used to ensure amplitude invariance.

The HAPF can then compensate for the average reactive power consumed by the load and the load harmonics by means of the instantaneous power theory. Thus, the instantaneous active (p_L) and reactive (q_L) powers demanded by the load are calculated as follows:

$$\begin{bmatrix} p_L \\ q_L \end{bmatrix} = \begin{bmatrix} v_{pcc,\alpha} & v_{pcc,\beta} \\ v_{pcc,\beta} & -v_{pcc,\alpha} \end{bmatrix} \begin{bmatrix} i_{L,\alpha} \\ i_{L,\beta} \end{bmatrix}, \quad (4)$$

where $i_{L,\alpha}$ and $i_{L,\beta}$ are the load currents and $v_{pcc,\alpha}$ and $v_{pcc,\beta}$ are the PCC voltages in the $\alpha\beta$ reference frame.

The load instantaneous powers given by (4) can be separated into average parcels ($\bar{p}_L$ and $\bar{q}_L$) and oscillating parcels

978-1-7281-4181-7/19 $31.00 © 2019 IEEE

$\tilde{p}_L$ and $\tilde{q}_L$). The average reactive load power ($\bar{q}_L$) impacts on the grid power factor. On the other hand, the oscillating instantaneous powers $\tilde{p}_L$ and $\tilde{q}_L$ are related to the harmonic content. This means that in order to guarantee low THD and unitary power factor the HAPF needs to compensate for those oscillating and average instantaneous reactive powers, respectively. A first order low-pass filter is used to separate the powers p_L and q_L into its average and oscillating parcels. Fig. 3 depicts the algorithm used to calculate the reference currents for the HAPF, where the block LPF represents the low-pass filter.

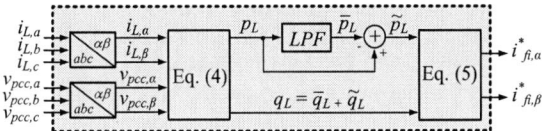

Fig. 3: Block diagram for the calculation of the reference currents for the power quality control.

The harmonic reference currents for the HAPF are then calculated by:

$$\begin{bmatrix} i_{fi,\alpha}^* \\ i_{fi,\beta}^* \end{bmatrix} = \frac{1}{\Delta} \begin{bmatrix} v_{pcc,\alpha} & v_{pcc,\beta} \\ v_{pcc,\beta} & -v_{pcc,\alpha} \end{bmatrix} \begin{bmatrix} \tilde{p}_L \\ \bar{q}_L + \tilde{q}_L \end{bmatrix}, \quad (5)$$

where $\Delta = v_{pcc,\alpha}^2 + v_{pcc,\beta}^2$ and the subscript ($_i$) indicates the reference currents $i_{fi,\alpha}^*$ and $i_{fi,\beta}^*$ are responsible to compensate for the load currents.

It is important to highlight that, ideally, the HAPF AC terminal voltage at the fundamental frequency will be null, since the passive filter was designed to compensate for all the reactive power consumed by the load.

The strategy to charge (energize) the DC capacitor and to regulate the DC bus voltage of the HAPF is presented in [19, 20] and has already been explained in [21]. Fig. 4 shows the block diagram for the DC voltage controllers.

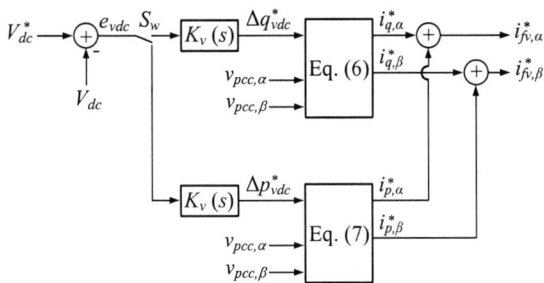

Fig. 4: Block diagram for the DC voltage controllers.

The DC voltage error $e_{vdc} = (V_{dc}^* - V_{dc})$, determined by the difference between the reference and measured DC voltages, feeds two Proportional-Integral (PI) controllers $K_v(s) = (k_{v,p} + k_{v,i}/s)$ in order to produce the active and reactive power control signals Δp_{vdc}^* and Δq_{vdc}^*, respectively. These signals should be converted into the reference currents $i_{q,\alpha}^*$ and $i_{q,\beta}^*$ to charge the HAPF DC capacitor and the reference cur-

rents $i_{p,\alpha}^*$ and $i_{p,\beta}^*$ to regulate the DC bus voltage, as follows:

$$\begin{bmatrix} i_{q,\alpha}^* \\ i_{q,\beta}^* \end{bmatrix} = \frac{1}{\Delta} \begin{bmatrix} v_{pcc,\alpha} & v_{pcc,\beta} \\ v_{pcc,\beta} & -v_{pcc,\alpha} \end{bmatrix} \begin{bmatrix} 0 \\ \Delta q_{vdc}^* \end{bmatrix} \quad (6)$$

and

$$\begin{bmatrix} i_{p,\alpha}^* \\ i_{p,\beta}^* \end{bmatrix} = \frac{1}{\Delta} \begin{bmatrix} v_{pcc,\alpha} & v_{pcc,\beta} \\ v_{pcc,\beta} & -v_{pcc,\alpha} \end{bmatrix} \begin{bmatrix} \Delta p_{vdc}^* \\ 0 \end{bmatrix}. \quad (7)$$

The switch S_w is used to select the power Δq_{vdc}^* to charge the DC capacitor or the power Δp_{vdc}^* to regulate the DC bus voltage. The subscript ($_v$) in the reference currents $i_{fv,\alpha}^*$ and $i_{fv,\beta}^*$ was used to distinguish these variables from the reference current signals given by (5).

Although both powers Δp_{vdc}^* and Δq_{vdc}^* can be used to regulate the DC bus voltage [19], in this paper, only the power Δp_{vdc}^* will be used to perform this task.

Fig. 5 shows the block diagram for the current control of the three-wire three-phase HAPF.

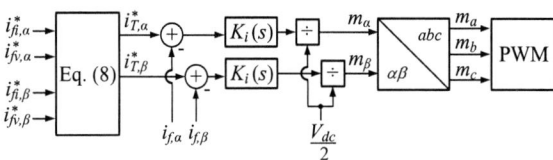

Fig. 5: Block diagram for the current control of the HAPF.

All the previous reference currents should be summed to obtain the total reference current in the $\alpha\beta$ coordinates, as follows:

$$\begin{cases} i_{T,\alpha}^* = i_{fi,\alpha}^* + i_{fv,\alpha}^* \\ i_{T,\beta}^* = i_{fi,\beta}^* + i_{fv,\beta}^* \end{cases}. \quad (8)$$

These total reference currents must be compared to the measured filter currents ($i_{f,\alpha}$ and $i_{f,\beta}$), producing the current errors that feed two Proportional-Resonant (PR) controllers ($K_i(s)$). These controllers are responsible to give high gain in specific harmonic frequencies in order to ensure the HAPF currents track its references.

The transfer function of the PR controller is given by:

$$K_i(s) = k_{i,p} + \sum_{n=1,7,11,13\ldots}^{h} k_{i,rh} \frac{2\omega_{ch}s}{s^2 + 2\omega_{ch}s + \omega_h^2}, \quad (9)$$

where the subscript ($_h$) indicates the harmonic order; $k_{i,p}$ and $k_{i,rh}$ are the proportional and resonant gains, respectively; ω_h and ω_{ch} are the resonant frequency and the band-pass of the resonant controller, respectively.

Note that, in (9) the 5th harmonic is omitted, although the load current has this harmonic component. This occurs because in the proposed design methodology the passive filter is in resonance, already providing a high gain at this frequency.

Finally, the modulation indexes m_α and m_β should be transformed to the abc coordinates (m_a, m_b and m_c), in order to be compared to the carrier and to generate the switching signals for the HAPF, as follows:

$$
\begin{bmatrix} m_a \\ m_b \\ m_c \end{bmatrix} = \begin{bmatrix} 1 & 0 \\ -\frac{1}{2} & \frac{\sqrt{3}}{2} \\ -\frac{1}{2} & -\frac{\sqrt{3}}{2} \end{bmatrix} \begin{bmatrix} m_\alpha \\ m_\beta \end{bmatrix}. \tag{10}
$$

IV. SIMULATION RESULTS

The circuit of Fig. 1 and the controllers of the HAPF were modeled and simulated on PSCAD software in order to validate the design method proposed in this paper. Tables I, II, III and IV present the parameters of the electric network, nonlinear load, HAPF and controllers, respectively. The gains of the PI controllers for the DC capacitor charging and the DC bus voltage regulation are the same. The PR controller of the current control-loop acts only until the 19th harmonic (1st, 7th, 9th, 11th, 13th, 17th and 19th), since the load current does not present expressive amplitudes above this harmonic order.

TABLE I
Grid parameters

Parameter	Value
Line RMS voltage (V_s)	3.3 kV
Fundamental frequency (f_s)	50 Hz
Equivalent inductance (L_s)	0.58 mH

TABLE II
Load parameters

Parameter	Value
Inductance (L_{AC})	11.6 mH
Capacitance (C_d)	150 μF
Resistance (R_L)	60 Ω

TABLE III
HAPF parameters

Parameter	Value
Filter capacitance (C_f)	29.29 μF
Filter inductance (L_f)	13.84 mH
Filter resistance (R_f)	49 mΩ
Resonant frequency (f_r)	250 Hz
DC bus voltage (V_{dc})	830 V
DC bus capacitance (C)	5000 μF
Switching frequency (f_{sw})	10 kHz

TABLE IV
Parameters of the HAPF controllers

Controller	Parameter	Value
K_v	$k_{v,p}$	100 A/V
	$k_{v,i}$	300 A/(Vs)
K_i	$k_{i,p}$	40 V/A
	$k_{i,r1}$	60 V/A
	$k_{i,rh}$	60 rad/s
	ω_{ch}	$0.01\omega_h$ rad/s

The reference value for the DC bus voltage was determined by:

$$
V_{dc}^* = \frac{2}{\hat{m}} \left(\sum_{h=7,11,13\ldots}^{19} |Z_{pf,h}| \hat{I}_{f,h} \right), \tag{11}
$$

where h is the harmonic order, Z_{pf} is the impedance of the passive filter, $\hat{I}_f$ is the peak current of the HAPF and $\hat{m}$ is the modulation index.

In (11) the fundamental and fifth harmonic components were not computed, since their respective active filter's terminal voltages are equal to zero, as discussed in Section III. Moreover, the reference DC voltage was calculated considering the worst case, in which the peak voltages at different frequencies coincide. The result of (11) (see Table III) was also rounded up to ensure a safe margin for the HAPF operation.

Fig. 6 shows the waveform of the DC bus voltage and its reference value. The HAPF starts operating at t = 0.2 s in order to compensate for the PF and harmonic currents, besides to charge the HAPF DC capacitor through the signal Δq_{dc}^*. At t = 1.2 s, the switch S_w of the DC voltage controller of Fig. 4 disables the DC capacitor charging control and enables the second PI controller in order to regulate the DC bus voltage.

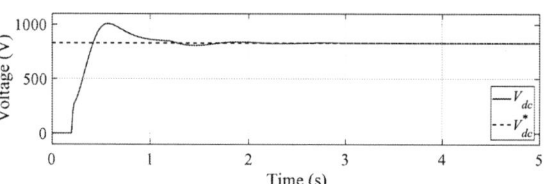

Fig. 6: Waveform of the DC bus voltage during the HAPF energization and DC bus voltage regulation.

Fig. 7 shows a detail of the waveform shown in Fig. 6. It is possible to note that the DC bus voltage is regulated at its reference value during the operation of the HAPF.

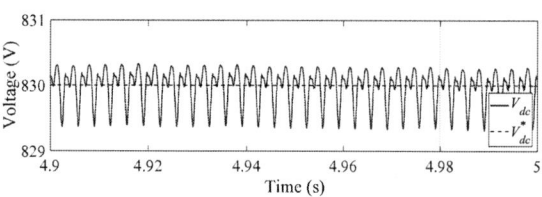

Fig. 7: Waveform detail of the DC bus voltage during its regulation.

Fig. 8 (a), (b), (c) and (d) show a detail of phase "a" waveforms of the voltage at PCC, load current, source current and HAPF current, respectively. It can be noted that the source current presents a low harmonic content, since the HAPF synthesizes the harmonic currents drained by the non-linear load. Also, it is possible to observe that the source current is in phase with the PCC voltage. Thus, it is worth mentioning that the designed HAPF was able to follow its references correctly.

Table V summarizes a comparison of the harmonic content and PF on source current before filtering, with only passive filter and with the appliance of the HAPF proposed to compensate for the non-linear load. It is possible to note a significant reduction in the THD and a PF almost unitary when the HAPF is operating. The passive filter by itself can also reduce the

978-1-7281-4181-7/19 $31.00 © 2019 IEEE

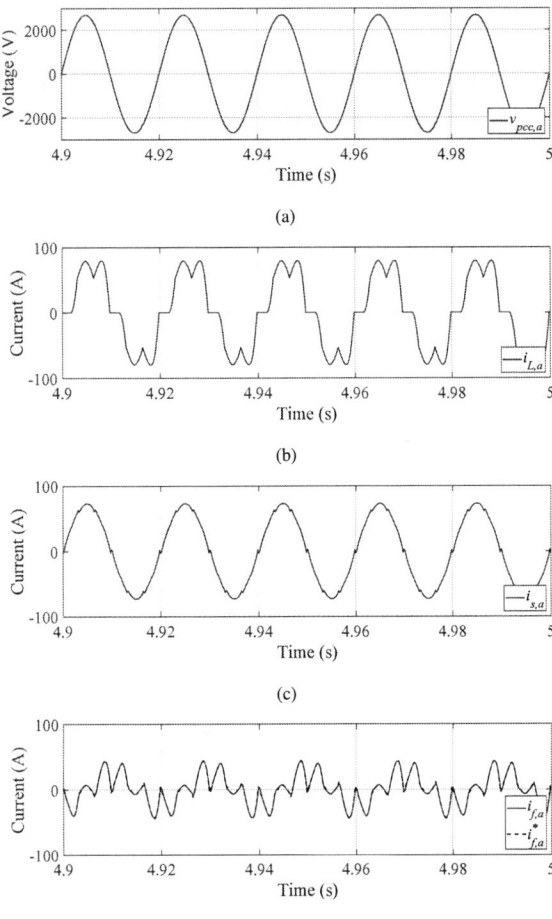

Fig. 8: Detail of phase "a" waveforms: (a) PCC voltage (b) load current (c) source current and (d) HAPF current and its reference.

harmonic content and correct the PF. However, if the load demands more or less reactive power and/or harmonic currents, the passive filter is limited to compensate for the same values, since it does not comprise an active parcel.

TABLE V
Results of THD and PF with and without filter

Parameter	Without filter	Passive filter	Hybrid filter
I_{s5} [A]	18.05	1.58	0.03
I_{s7} [A]	5.66	5.03	0.82
I_{s11} [A]	2.80	2.60	0.96
I_{s13} [A]	2.17	2.03	0.94
I_{s17} [A]	1.02	0.96	0.56
I_{s19} [A]	0.81	0.76	0.51
THD [%]	25.18	8.94	2.87
PF	0.912	0.996	0.999

V. CONCLUSIONS

This paper presented a method to design a hybrid active power filter used to compensate for a non-linear load con-

nected to a three-wire three-phase 3.3 kV electric network. The main objective of the method is to minimize the DC bus voltage of the HAPF. The study was divided into two parts. The first one was related to its passive portion, where the superposition theorem was used to design the inductance and capacitance values of the passive filter. The second one focused on the active portion of the hybrid filter, where two different signals were calculated: (i) reference current to increase the PF and to mitigate the THD and (ii) reference current to charge the DC capacitor and to regulate the DC bus voltage of the hybrid active power filter. In this section, a proportional resonant controller was adopted so that the filter could track its reference currents. Also, it was presented an equation to evaluate the minimum DC bus voltage for the hybrid filter. The simulation results shown a reduction in the source current THD from 25.18 % to 2.87 % and an increase in the PF from 0.912 to 0.999. This results shown that the proposed method to design and to reduce the DC bus voltage of the hybrid active power filter ensure its correct operation. It is also important to mention that the reduction in the DC bus voltage also reduces the total cost of the HAPF.

ACKNOWLEDGEMENT

This study was financed in part by the Coordenação de Aperfeiçoamento de Pessoal de Nível Superior - Brasil (CAPES) - Finance Code 001, the National Council for Scientific and Technological Development (CNPq), the State Funding Agency of Minas Gerais (FAPEMIG) and the National Institute for Electric Energy (INERGE).

REFERENCES

[1] P. Salmeron and S. P. Litran, "A control strategy for hybrid power filter to compensate four-wires three-phase systems," *IEEE Transactions on Power Electronics*, vol. 25, no. 7, pp. 1923–1931, 2010.

[2] H.-K. Chiang, B.-R. Lin, K.-T. Yang, and K.-W. Wu, "Hybrid active power filter for power quality compensation," in *2005 International Conference on Power Electronics and Drives Systems*, vol. 2. IEEE, 2005, pp. 949–954.

[3] V. F. Corasaniti, M. B. Barbieri, P. L. Arnera, and M. I. Valla, "Hybrid active filter for reactive and harmonics compensation in a distribution network," *IEEE Transactions on industrial electronics*, vol. 56, no. 3, pp. 670–677, 2008.

[4] IEC 61000, "Limits for harmonic current emissions (equipment input current up to and including 16 A per phase)," may 2014.

[5] IEEE 519, "IEEE recommended practice and requirements for harmonic control in electric power systems," june 2014.

[6] B. Singh, V. Verma, A. Chandra, and K. Al-Haddad, "Hybrid filters for power quality improvement," *IEE Proceedings-Generation, Transmission and Distribution*, vol. 152, no. 3, pp. 365–378, 2005.

[7] M.-C. Wong, J. Tang, and Y.-D. Han, "Cylindrical coordinate control of three-dimensional PWM technique in three-phase four-wired trilevel inverter," *IEEE Transac-*

978-1-7281-4181-7/19 $31.00 © 2019 IEEE

tions on Power Electronics, vol. 18, no. 1, pp. 208–220, 2003.

[8] S. Park, J.-h. Sung, and K. Nam, "A new parallel hybrid filter configuration minimizing active filter size," in *30th Annual IEEE Power Electronics Specialists Conference. Record.(Cat. No. 99CH36321)*, vol. 1. IEEE, 1999, pp. 400–405.

[9] H. Fujita, T. Yamasaki, and H. Akagi, "A hybrid active filter for damping of harmonic resonance in industrial power systems," *IEEJ Transactions on Industry Applications*, vol. 118, no. 10, pp. 1193–1200, 1998.

[10] W. Tangtheerajaroonwong, T. Hatada, K. Wada, and H. Akagi, "Design and performance of a transformerless shunt hybrid filter integrated into a three-phase diode rectifier," *IEEE Transactions on Power Electronics*, vol. 22, no. 5, pp. 1882–1889, 2007.

[11] R. Inzunza and H. Akagi, "A 6.6-kV transformerless shunt hybrid active filter for installation on a power distribution system," *IEEE Transactions on power electronics*, vol. 20, no. 4, pp. 893–900, 2005.

[12] S. Rahmani, A. Hamadi, N. Mendalek, and K. Al-Haddad, "A new control technique for three-phase shunt hybrid power filter," *IEEE Transactions on industrial electronics*, vol. 56, no. 8, pp. 2904–2915, 2009.

[13] A. Cleary-Balderas, A. M.-R. Senior, and O. Cruz-Hernéndez, "Hybrid active power filter based on the IRP theory for harmonic current mitigation," in *2016 IEEE International Autumn Meeting on Power, Electronics and Computing (ROPEC)*. IEEE, 2016, pp. 1–5.

[14] S. C. Behera, G. Choudhary, and R. Mandal, "Fuzzy based vector PI controller to mitigate the harmonics issue of a distribution system using three-phase hybrid power filter," in *2016 2nd International Conference on Advances in Electrical, Electronics, Information, Communication and Bio-Informatics (AEEICB)*. IEEE, 2016, pp. 610–616.

[15] S. Rahmani, A. Hamadi, K. Al-Haddad, and L. A. Dessaint, "A combination of shunt hybrid power filter and thyristor-controlled reactor for power quality," *IEEE Transactions on Industrial Electronics*, vol. 61, no. 5, pp. 2152–2164, 2013.

[16] S. Jamali, M. Masoum, and S. Mousavi, "Influence of controller high pass filter on the performance of shunt hybrid power filter," in *2008 Australasian Universities Power Engineering Conference*. IEEE, 2008, pp. 1–6.

[17] S. Suresh, N. Devarajan, and V. Rajasekaran, "Design and analysis of shunt hybrid filter control topology for mitigation of load current harmonics," in *2012 Annual IEEE India Conference (INDICON)*. IEEE, 2012, pp. 1189–1193.

[18] B. Kedra, "Reducing inverter power rating in active power filters using proposed hybrid power filter topology," in *2015 IEEE 15th International Conference on Environment and Electrical Engineering (EEEIC)*. IEEE, 2015, pp. 443–448.

[19] W.-H. Choi, C.-S. Lam, M.-C. Wong, and Y.-D. Han, "Analysis of DC-link voltage controls in three-phase four-wire hybrid active power filters," *IEEE Transactions*

on power electronics, vol. 28, no. 5, pp. 2180–2191, 2012.

[20] P. C. d. S. Furtado, G. A. Fogli, P. M. de Almeida, P. G. Barbosa, and J. G. de Oliveira, "Topology and control of a two-phase residential pv system with load compensation capability," in *2015 IEEE 24th International Symposium on Industrial Electronics (ISIE)*. IEEE, 2015, pp. 1127–1132.

[21] M. Braga, S. Duarte, P. Almeida, and P. Barbosa, "DC capacitor energization and voltage regulation of a single-phase hybrid filter," in *2017 Brazilian Power Electronics Conference (COBEP)*. IEEE, 2017, pp. 1–6.

Development of a Modular Open Source Power Electronics Didactic Platform

Lucas Koleff
Power Electronics Laboratory
University of São Paulo
São Paulo, Brazil
koleff.lucas@gmail.com

Gustavo Valentim
Power Electronics Laboratory
University of São Paulo
São Paulo, Brazil
gustavo.soares.valentim@usp.br

Victor Rael
Power Electronics Laboratory
University of São Paulo
São Paulo, Brazil
victor_praxedes@usp.br

Luciana Marques
Power Electronics Laboratory
University of São Paulo
São Paulo, Brazil
lucianadacostamarques@gmail.com

Wilson Komatsu
Power Electronics Laboratory
University of São Paulo
São Paulo, Brazil
wilsonk@usp.br

Eduardo Pellini
Protection and Automation Laboratory
University of São Paulo
São Paulo, Brazil
elpellini@usp.br

Lourenço Matakas Jr.
Power Electronics Laboratory
University of São Paulo
São Paulo, Brazil
matakas@pea.usp.br

Abstract— **This paper introduces a modular open source power electronics didactic platform developed for use in undergraduate and graduate courses, student projects and interdisciplinary applications. The five modules (inverter, current and voltage sensor, processing and filter) can be interconnected to study several different topics in power electronics converter control. The platform has reduced material costs and all hardware and software developed are available under an open source hardware and software licenses. Experimental results are presented in order to validate the operation of the system with the inverter working in open-loop and closed-loop, together with a graphical user interface.**

Keywords— power electronics, inverter control, education

I. INTRODUCTION

In the field of power electronics, there is a constant demand for experimental verification of control strategies, modulation schemes, among other aspects. Design and test of power converters are difficult tasks which can discourage students from working with power electronics because the learning curve is very steep. Modern approaches for education in power electronics use both simulations [1] and experiments [2] to facilitate the learning process. Therefore, the concept of a modular open source power electronics didactic platform is created. It consists of five modules (inverter, current sensor, voltage sensor, processing and L/LC/LCL filters) that can be interconnected as specified for a particular application. Important constraints in this project are that all modules should be affordable, of easy construction and also ensure safe operation. Therefore, expensive components [3] requiring complex soldering techniques are to be avoided [4] when possible. Following other open source initiatives [5,6], all design files (schematics, layouts), firmware and documentation are available online under an open source license at [7], allowing students (either in undergraduate or graduate courses), research assistants and even hobbyists to develop their own projects. The main contribution of this work is to stimulate innovation and entrepreneurship in the power electronics community by giving students the means necessary to develop hands-on experience in this field.

II. MODULAR PLATFORM

A. Overview

The interconnection between the modules is shown in Fig. 1, where the three-phase four wire inverter is found. It is connected to the filter boards, which consist in L (inductor) and C (capacitor) modules which can be connected to form a L, LC or LCL filter. There are also boards for current and voltage sensing. The processing board generates the gating signals for the inverter, receives measurements from the sensor boards and communicates with the user PC. There is a complete galvanic isolation barrier between the power stage, the control system and the PC, that guarantees the integrity and safety of the system.

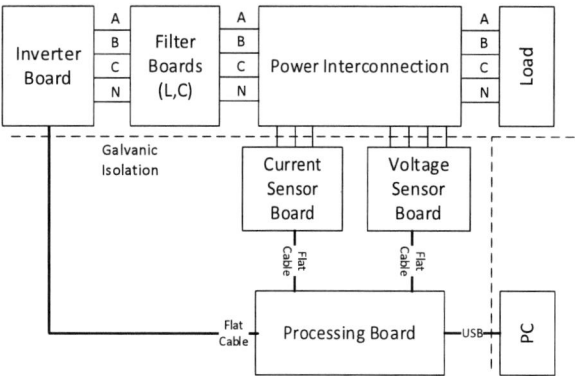

Fig. 1.   Interconnection diagram.

The system is currently connected to passive loads but can be adapted to be connected to the grid through a coupling transformer. The guideline for the development of the platform is to consider experiments with a 50Ω wye-connected resistive load with a $25V$ DC link voltage, resulting in a nominal current of $0.5A$ (peak), while still having a reasonable overload capacity and assuring safe current and voltage levels according to the low voltage directives in [8]. Even though many power electronics courses start with basic topologies (e.g., buck or step-down converters), the DC-AC converter is chosen in this didactic platform as it has more

978-1-7281-4181-7/19 $31.00 © 2019 IEEE   429

applications in the field of energy conversion such as integration of renewable sources and three-phase motor drives.

B. Inverter Board

The inverter board features a three-phase four-wire topology using two L6203 full bridge devices [9]. The specifications are shown in Table I. The maximum voltage is limited in order to comply with safety requirements and to keep costs low.

TABLE I.   INVERTER PARAMETERS

Parameter	Value	Parameter	Value
Maximum DC Link Voltage	48V	Rated Output Current	4A(RMS)
Maximum Switching Frequency	100kHz	Typical Drain-Source Resistance	0.3Ω

The block diagram of the inverter is shown in Fig. 2. It is possible to see the DC link capacitor C_{DC}. The switching devices feature integrated gate drivers and switching logic, including dead time insertion. In order to ensure galvanic isolation between the power stage and the rest of system, high speed optocouplers are used. The gate driver side of the optocouplers is supplied by an isolated DC/DC converter.

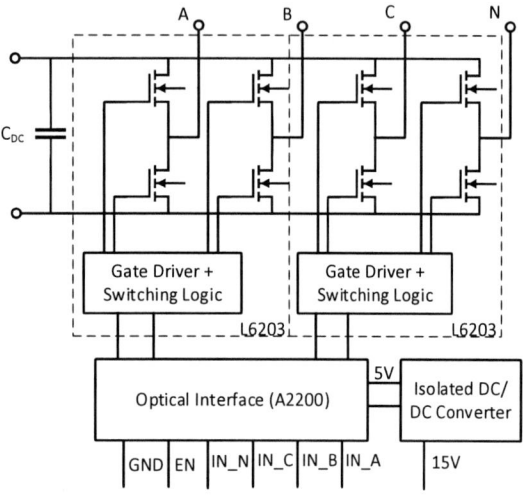

Fig. 2.   Block diagram of the inverter.

The first prototype is shown in Fig. 3. The final version of the platform uses commercially made prototype circuit boards.

Fig. 3.   Inverter prototype.

C. Current Sensor Board

The current sensor board is based on three LEM LTS6NP [10] current transducers. It can be used, for instance, to measure the three-phase load current, while the neutral wire current can be determined by applying the first Kirchoff's law. Even though the isolated Hall effect transducers are expensive components, they offer accuracy, which is important when studying certain aspects such as observation of switching ripple waveforms. The specifications for each channel of this board are shown in Table II. The nominal measurement range of the transducer is $6At$ (RMS). Therefore, 6 turns are used in the primary winding to reduce the measurement range to $1A$ (RMS), which is more adequate to didactic applications.

TABLE II.   CURRENT SENSOR PARAMETERS

Parameter	Value	Parameter	Value
Maximum Input Current	1A(RMS)	Supply Voltage	+5V
Bandwidth	200kHz	Rated Insulation Voltage	600V(RMS)

The block diagram of the current sensor board is presented in Fig. 4, highlighting the transducer and the signal conditioning stages, which is required to reduce the output voltage of the transducer in order to allow connection with the A/D converter of the microcontroller (which has an input voltage range of 3.3V). The first prototype of the current sensor board is shown in Fig. 5.

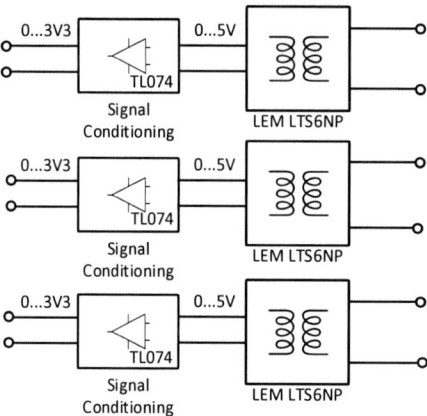

Fig. 4.   Block diagram of current sensor board.

Fig. 5.   Current sensor board prototype.

D. Voltage Sensor Board

Four LEM LV20NP [11] voltage transducers are used in the voltage sensor board. Again, they are used due to their accuracy, even though they are expensive. These sensors are based on a measuring resistance R_m, through which a small current, proportional to the input voltage, flows into the transducer, as shown in the block diagram of Fig. 6. The output voltage of the transducer is measured in R_1 and ranges between -15V and 15V. The measuring resistor is adjusted for a $60V$ peak voltage range, which is adequate to this didactic platform. The parameters of the voltage sensor board are listed

978-1-7281-4181-7/19 $31.00 © 2019 IEEE   430

in Table III. As in the current sensor board, a signal conditioning stage is required to make the measurement compatible with input of the A/D converter of the microcontroller. The first prototype is shown in Fig. 7.

TABLE III. VOLTAGE SENSOR PARAMETERS

Parameter	Value	Parameter	Value
Maximum Input Voltage	60V(peak)	Supply Voltage	±15V
Bandwidth	10kHz	Rated Insulation Voltage	1600V

Fig. 6. Block diagram of voltage sensor board.

Fig. 7. Voltage sensor board prototype.

E. Processing Board

The processing board hosts a low-cost embedded development kit, which has a 32-bit ARM Cortex-M4 STM32F407 microcontroller [12] with a single precision floating point unit, as shown in Fig. 8. It features three flat-cable connectors for communication with inverter and sensor boards. The measurements are redirected to the microcontroller pins corresponding to the A/D converter inputs and the PWM outputs are redirected to inverter.

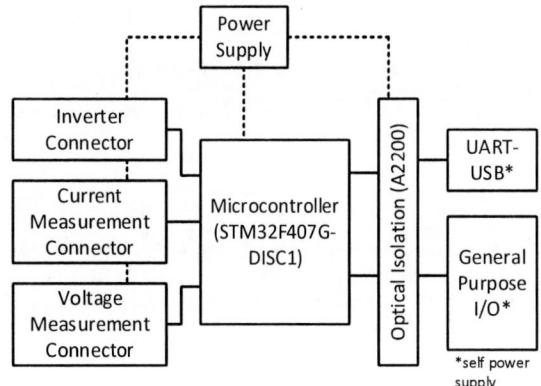

Fig. 8. Processing board block diagram.

The communication with the host computer is done using a USB-UART adapter, which is connected to the microcontroller through high-speed optocouplers. The maximum transmission rate has been measured to 189 KB/s. A general purpose isolated I/O header is also featured in the board. The processing board is also responsible for redirecting the DC voltages from the power supply (+5V and ± 15V) to the other boards. The first prototype is shown in Fig. 8.

Fig. 8. Processing board prototype.

F. Filter Boards

In order to allow different filter topologies, two independent modules were designed: the inductor board and the capacitor board. They can be combined in order to build L, LC and LCL filters [13, 14, 15], with the parameters shown in Table IV. If additional inductance is needed, two inductor boards can be connected in series. The system is expected to use a switching frequency up to 4.8 kHz. The filter cutoff frequency is designed to be below the switching frequency. For a simple L filter, the cutoff frequency is calculated as in (1), considering the load resistance to be 50Ω and one single 3.3mH inductor.

$$f_{c,L} = R/(2\pi L) = 2.4\ kHz \tag{1}$$

On the other hand, for the LC filter with the same inductance and a capacitance of 9.4μF, the cutoff frequency can be calculated as in (2).

$$f_{c,LC} = 1/(2\pi\sqrt{LC}) = 904\ Hz \tag{2}$$

Finally, for the LCL filter, the cutoff frequency is calculated in (3), where $L_1 = L_2 = L$ and $C_0 = 2C$. In this case, a dampening resistor is added to ensure stability [16]. The value of the resistance is calculated in (4).

$$f_{c,LCL} = \frac{1}{2\pi}\sqrt{(L_1 + L_2)/L_1 L_2 C_0} = 1277 Hz \qquad (3)$$

$$R_d = \sqrt{(L_1 + L_2)/C_0} = 27\Omega \qquad (4)$$

The goal of the filter board design is to enable changing the filter topology with minimal reconnection effort in order to demonstrate the difference in the filtering capability of L, LC and LCL filters.

TABLE IV.   L/LC/LCL FILTER PARAMETERS

Parameter	Value	Parameter	Value
Inductance (L)	3.3mH	LCL Filter Resonant Frequency	1277Hz
Capacitance (2C)	2x4.7uF	LC Filter Resonant Frequency	904Hz
Inductor Saturation Current	2.5A(peak)	Dampening Resistance	26Ω

The inductor is designed using a MMTS60T5715 metal powder toroidal core with a saturation flux density of $1.05T$ and relative permeability of 60. A software that analyzes various combinations of cores and wires was developed to design the inductor used in the filter [17]. The algorithm calculates the number of winding turns and the quality factor and discards unfeasible or inefficient designs. The windings are done using 159 turns of AWG 18 copper wire.

Fig. 9 illustrates the circuit diagram of the inductor board, which considers a three-phase four-wire circuit. The neutral wire inductor can be short-circuited by a switch depending on the topology being studied. The first prototype is presented in Fig. 10.

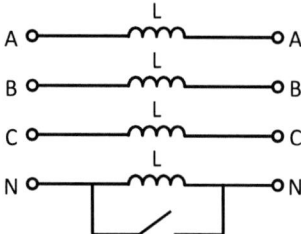

Fig. 9.   Inductor board circuit diagram.

Fig. 10.   Inductor board prototype.

The circuit diagram of the capacitor board is depicted in Fig. 11. It features two capacitors in parallel for each phase, while one of them features the dampening resistor, which can be short-circuited for comparison purposes. Fig. 12 shows the first prototype of this board.

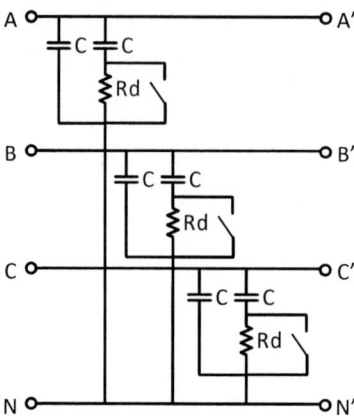

Fig. 11. Capacitor board circuit diagram.

Fig. 12. Capacitor board prototype.

III. DIDACTIC APPLICATIONS

A. Overview

The didactic platform is designed to be used by graduate and undergraduate students at the power electronics laboratory classes. The active learning educational methodology consists in theory lessons combined with experimental verification at the laboratory. First, basic concepts are explained in the classroom. Then, using the components (hardware and software) of this platform, the students are going to perform experiments and obtain data (such as waveforms) to verify the learned theories. As previously stated, the modules can be interconnected according to the application being studied.

Two initial demonstrations are presented next. The first one shows the study of PWM modulation techniques, enabling the students to verify the impact of the switching frequency and the modulation type on the final inverter output waveform. The graphical user interface that eases interaction of the students with the platform is also presented. The second one shows to study of a current control loop with a simple proportional controller, feeding a sinusoidal current in a RL load. The students can verify the impact of the proportional gain in the tracking performance.

Other experiments are planned, including other control loops (voltage and current), different controllers (such as proportional integral or proportional resonant), study of LC and LCL filters, reference frame transformations, phase locked loops and grid connection.

B. Communication with target PC with variation of PWM parameters

In order to facilitate the interaction of the user with the system, the graphical user interface shown in Fig. 13 has been developed in C# using Microsoft Visual Studio. It handles the communication with the processing board and allows experiment selection. The user can also easily change parameters such as the switching frequency, modulation index, reference frequency, among others.

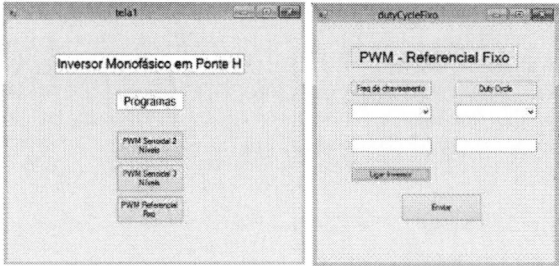

Fig. 13.   Graphical user interface.

Fig. 14 illustrates the interaction sequence designed for the user with the PC application. After initiating the program, the user must choose which Communication (COM) port will be used in the experiment. Once the choice is made, a verification step is initialized, sending test packages between the PC and the microcontroller in order to check the integrity of the communication channel. If the verification is well succeeded, then the user is directed to a screen in which is possible to choose which type of experiment will be performed.

Once the experiment is defined, another screen opens with specific parameters to be adjusted during the experiment by the user. The implementation shown in Fig. 14 allows experiments of fixed duty cycle, two level and three level sinusoidal modulation, but future implementation of other experiments is straightforward using the same code structure.

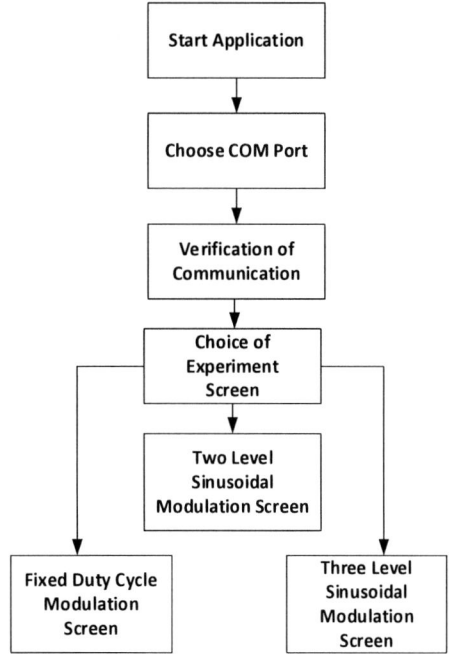

Fig.14.   Flowchart of interaction with the user interface application.

In Fig. 15, it is possible to see the study of the two-level PWM modulation with the inverter connected to the load with a simple inductive filter. One can observe and measure several important aspects, such as the amplitude of the switching

ripple and the frequency of the resulting waveforms. Fig. 16 shows the output signals of the microcontroller demonstrating the concept of the three-level PWM modulation in a full-bridge inverter.

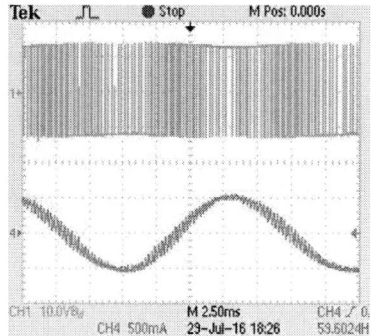

Fig. 15.   Two-level modulation (top: phase A-B inverter output, bottom: load current).

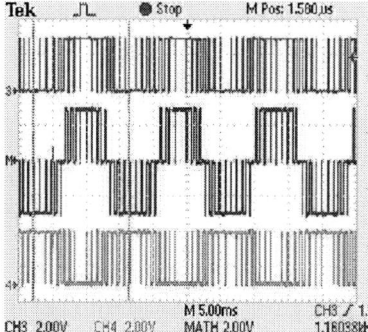

Fig. 16.   Three-level modulation (top: phase A, bottom: phase B, middle: phase A-B).

C. Current control loop

Another important application is the current control loop. In this application, the inverter board, the current sensor board, the L filter board and the processing board are used in order to track a reference current at the load. The reference voltage is normalized by the inverter DC link voltage, as shown in the block diagram in Fig. 17, where i_{ref} is the reference current, i_{meas} is the measured current, e is the control error signal, V_{DC} is the DC link voltage, K is the controller proportional gain, u_{inv} is the inverter output voltage and m is the modulation index.

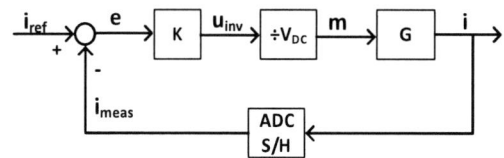

Fig. 17.   Block diagram of current control feedback loop.

The control system is set to track a 0.5A (peak) 60Hz current waveform at the 50Ω resistive load. The tracking performance is evaluated for two different proportional gains, $K = 100$ and 300. The obtained waveforms are shown in Fig. 18 and Fig. 19, where the reference signals are outputted by the digital-analog converters (DACs) of the microcontroller. It is possible to see that the increased proportional gain enables significantly smaller tracking errors.

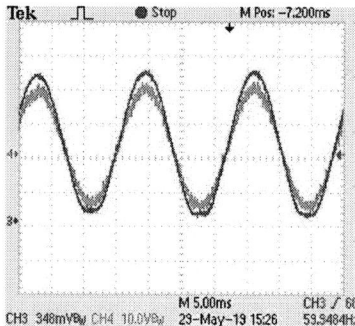

Fig. 18.   Experimental results for current tracking with proportional gain K=100 (green: measured current, purple: reference current).

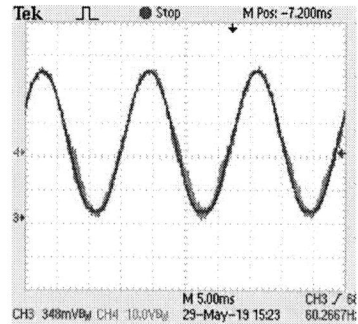

Fig. 19.   Experimental results for current tracking with proportional gain K=300 (green: measured current, purple: reference current).

IV. CONCLUSIONS

This article presents a modular open source power electronics didactic platform, that is composed of several modules that demonstrate fundamental aspects of power electronics. Those modules consist in the inverter, current sensor, voltage sensor, filter and processing boards. The total material costs (in US dollars) are calculated (including printed circuit boards) in Table V, which are 56% lower than the price of an equivalent commercial platform [18].

TABLE V.   MATERIAL COSTS OVERVIEW (IN US DOLLARS)

Item	Costs	Item	Costs
Inverter Board	$60.16	Processing Board	$57.62
Current Sensor Board	$71.85	Filter Boards	$63.92
Voltage Sensor Board	$199.80	Total	$453.35

The modules can be connected depending on the application being studied. The platform has been conceived for usage in undergraduate and graduate power electronics courses but can also be used in student projects and even non-academic maker projects. An experimental demonstration of the fundamental aspects of the PWM modulation and current control loop is presented and validates the proposed didactic platform. Concerning the continuity of this work, for manufacturing and reproducing the final version of the platform, professional circuit board manufacturing services are going to be used. Afterwards, the didactic platform is going to be extensively used in laboratory classes by undergraduate and graduate students. The impact of the platform should be then evaluated by distributing evaluation forms and measuring the impact on the grades of the students.

ACKNOWLEDGMENTS

The authors would like to thank Felipe Abe, Henrique Hokama, Brando Moraes, Kelly Enomoto and Moreno Oliveira for their contributions and Texas Instruments and STMicroelectronics for their support. This work was financially supported by the Sao Paulo Research Foundation (FAPESP) grants 2016/16542-5 and 2016/01930-0, by the National Council for Scientific and Technological Development (CNPq) grants 306970/2015-5, 311789/2014-5, by the Coordination for the Improvement of Higher Education Personnel (CAPES) Finance Code 001 and others.

REFERENCES

[1] U. Drofenik and J. W. Kolar, "Survey of modern approaches of education in power electronics," *APEC. Seventeenth Annual IEEE Applied Power Electronics Conference and Exposition (Cat. No.02CH37335)*, Dallas, TX, USA, 2002, pp. 749-755 vol.2..

[2] R. S. Balog, Z. Sorchini, J. W. Kimball, P. L. Chapman and P. T. Krein, "Modern laboratory-based education for power electronics and electric machines," in *IEEE Transactions on Power Systems*, vol. 20, no. 2, pp. 538-547, May 2005.

[3] L. P. D. Silva *et al.*, "Low-cost didactic module for single-phase inverter teaching," *2017 Brazilian Power Electronics Conference (COBEP)*, Juiz de Fora, 2017, pp. 1-6.

[4] M. B. Dias, F. A. S. Goncalves, F. P. Marafao and H. K. M. Paredes, "Low cost digital module for demonstration of modulation strategies in DC-to-AC converters," *2017 Brazilian Power Electronics Conference (COBEP)*, Juiz de Fora, 2017, pp. 1-6.

[5] A. Müsing and J. W. Kolar, "Successful online education - GeckoCIRCUITS as open-source simulation platform," *2014 International Power Electronics Conference (IPEC-Hiroshima 2014 - ECCE ASIA)*, Hiroshima, 2014, pp. 821-828.

[6] H. Guillardi, E. V. Liberado, J. A. Pomilio, and F. P. Marafão, "General-compensation-purpose Static var Compensator prototype," *HardwareX*, vol. 5, 2019.

[7] LEP-PEA-EPUSP, "Open Source Power Electronics". [Online]. Available: https://github.com/LEP-PEA-EPUSP.[Accessed Jun. 03, 2019].

[8] European Parliament and Council, *Directive 2014/35/EU*. Strasbourg, France: 2014.

[9] SGS Thomson Microeletronics, L6201-L6202 L6203 DMOS Full Bridge Driver, pp. 1-20.

[10] Liaisons Electroniques-Mecaniques LEM S.A., Current Transducer LTS 6-NP, pp. 1-4.

[11] Liaisons Electroniques-Mecaniques LEM S.A., Voltage Transducer LV 20-P, pp. 1-4.

[12] SGS Thomson Microeletronics, STM32F405xx/ STM32F407xx Datasheet - Production Data , pp 13-40.

[13] U. P. Yagnik and M. D. Solanki, "Comparison of L, LC & LCL filter for grid connected converters," *2017 International Conference on Trends in Electronics and Informatics (ICEI)*, Tirunelveli, 2017, pp. 455-458.

[14] J. Kim, J. Choi and H. Hong, "Output LC filter design of voltage source inverter considering the performance of controller," *PowerCon 2000*, Perth, WA, Australia, 2000, pp. 1659-1664 vol.3.

[15] Frede Blaabjerg,, ed. *Control of Power Electronic Converters and Systems*. Cambridge, MA: Academic Press, 2018.

[16] P. Channegowda and V. John, "Filter Optimization for Grid Interactive Voltage Source Inverters," in *IEEE Transactions on Industrial Electronics*, vol. 57, no. 12, pp. 4106-4114, Dec. 2010.

[17] G. S. Valentim, "Didactic platform for execution of experiments with self-commutated converters" (in Portuguese), Polytechnic School of the University of São Paulo, Brazil, 2017.

[18] Festo Didactic, "Materials and Equipment for Educational Technology and Occupational Training," *SKU 579718*. [Online]. Available: https://online.ogs.ny.gov/purchase/spg/pdfdocs/3822423077PL_Festo.pdf. [Accessed: 16-Sep-2019].

Flexible Didactic Platform for Thyristor-Based Circuits

Lucas Koleff
Power Electronics Laboratory
University of São Paulo
São Paulo, Brazil
koleff.lucas@gmail.com

Lucas Araújo
Power Electronics Laboratory
University of São Paulo
São Paulo, Brazil
lga261@gmail.com

Mário Zambon
Power Electronics Laboratory
. University of São Paulo
São Paulo, Brazil
mfelipe.zambon@gmail.com

Wilson Komatsu
Power Electronics Laboratory
University of São Paulo
São Paulo, Brazil
wilsonk@usp.br

Eduardo Pellini
Protection and Automation Laboratory
University of São Paulo
São Paulo, Brazil
elpellini@usp.br

Lourenco Matakas Jr.
Power Electronics Laboratory
University of São Paulo
São Paulo, Brazil
matakas@pea.usp.br

Abstract—**The main goal of this article is to present a didactic platform for thyristor-based circuits. The proposed platform has the advantage of being more flexible than existing academic and commercial teaching platforms. Using a basic thyristor element board, it is possible to mount different topologies, such as solid-state relays, thyristor bridge rectifiers, among others. The development steps are presented in detail. Application examples such as a switched capacitor bank are presented together with experimental results in order to demonstrate the functionality of the platform and its didactic applications.**

Keywords—power electronics, education, thyristors, reactive power, capacitor banks, rectifiers

I. INTRODUCTION

In power electronics courses, lectures about line-commutated converters and thyristor-based circuits are often found in the syllabus. However most didactic platforms are designed for voltage source inverters [1], electric drives [2] [3] or photovoltaic systems [4]. Therefore, a demand for didactic experimental platforms for power electronics circuits using thyristors exists. Some academic [5] and commercial didactic platforms [6] for this purpose are currently found, but they are not flexible regarding the topology, which is the main feature of the didactic platform presented in this work. Historically, a similar development is shown in [7], but the platform presented in this article features modern components.

The proposed platform is flexible, as its basic thyristor element is a circuit board featuring a single thyristor and the isolated gating circuit, as shown in Fig. 1. It is possible to interconnect the basic thyristor element on several different topologies.

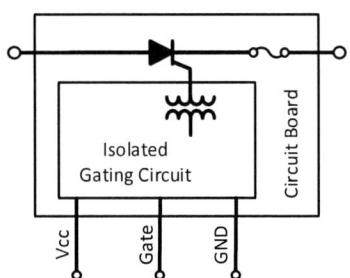

Fig.1. Basic thyristor element of the proposed platform.

In the next sections, this paper presents the development of the basic thyristor element, the most useful interconnection possibilities and some experimental application results.

II. DEVELOPMENT OF THE CIRCUIT

A. Specifications

The platform is intended to operate connected to AC mains, so the thyristor must withstand the peak voltage of the grid with a reasonable safety margin. The RMS current rating is specified as 7A, which is suitable for most didactic applications. The TIC126 thyristor [8] is chosen, as it supports up to 12A continuous RMS current and 800V peak reverse voltage, while having a low holding current (10mA). This component is also readily available at low cost in the market. Using the IPC-2221 standard [9], the width and clearance spacing of the printed circuit board (PCB) tracks are dimensioned for a 7A current and a 600V voltage (both RMS values). At the specified current rating, the power losses are 7.7W. An aluminum heatsink is installed at the thyristor to provide enough power dissipation and keep the junction temperature in acceptable limits. A fuse provides protection against overcurrents.

B. Gating Circuit

The gating circuit [10] shown in Fig. 2 is based on a MTPT17 pulse transformer [11], which provides galvanic isolation for the control circuit and has a measured voltage-time integral of 250 $\mu V \cdot s$.

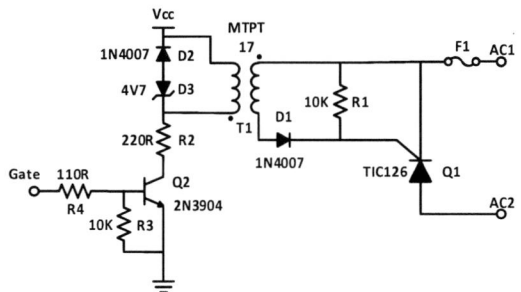

Fig. 2. Circuit diagram of the gating circuit.

On the secondary side of pulse transformer, the diode D1 ensures the correct direction of the gate current, while the resistor R1 avoids spurious thyristor firing due the anode-cathode voltage transients. On the primary side, the transistor Q2 is responsible for exciting the pulse transformer when the

978-1-7281-4181-7/19 $31.00 © 2019 IEEE

gate input signal is triggered. The resistor R2 limits the primary current when the transistor Q2 is in active conduction, while the resistor R4 limits the gate current and the resistor R3 is the base-emitter pull-down. The pulse transformer needs to be demagnetized after a gate pulse is generated. The demagnetizing current flows through the diodes D2 and D3. The nominal gate trigger current for the TIC126 is 5mA, and the peak current provided by this circuit has been measured to 30mA, so the thyristor is fired with a reasonable margin. The supply voltage V_{cc} is of $5V$ and a pulse train with a maximum frequency of 10 KHz (with a duty cycle of 50%) can be used as gating signal under these conditions.

The final PCB with populated components is shown in Fig. 3, which highlights the key elements. It is possible to see the up-mounting control and power headers, which are designed in order to allow mounting one board on top of the other, so two thyristor switches are connected in anti-parallel to form a solid state relay (SSR), as detailed in the next item.

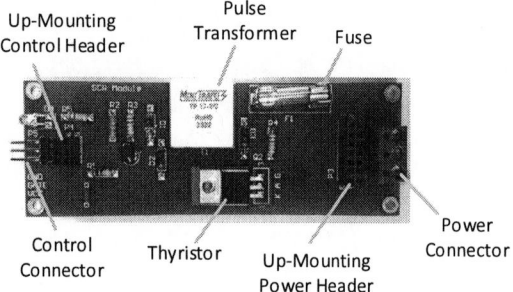

Fig. 3. Picture of the finished thyristor circuit element without dissipator.

C. Interconnection Flexibility

The first connection possibility is the stand-alone thyristor, also known as silicon controlled rectifier (SCR), shown in Fig. 4. In this case, only one board is used, which can be applied mainly for measurement and determination of the characteristics of the thyristor for didactic purposes, such as forward voltage drop and the holding current. It is also possible to obtain waveforms for the turn-on and turn-off transients.

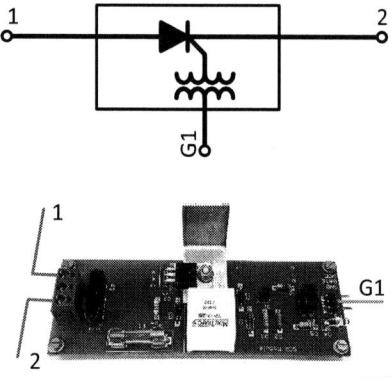

Fig. 4. Silicon controlled rectifier.

Two boards can be mounted on top of each other in order to form a bidirectional switch (or SSR), as shown in Fig. 5. In this case, the anode of the thyristor from the board under is going to be connected to the cathode of the thyristor of the board above and vice-versa through the up-mounting power

header. The gating circuit power supply (V_{cc} and GND) and the gating signal $G1$ are shared between the two boards through the up-mounting control headers in this case. The SSR can be used to control the connection and disconnection of several types of loads electronically, therefore, several didactic applications can be explored.

The SSR can also switch a capacitor bank for reactive power management purposes. For this application, the switching instant can be precisely controlled, which is an advantage over traditional mechanical switches. Further applications are in the control of static var compensators (SVCs), where the firing angle of thyristors is changed in order to adjust the amount of reactive power provided by a reactor [12].

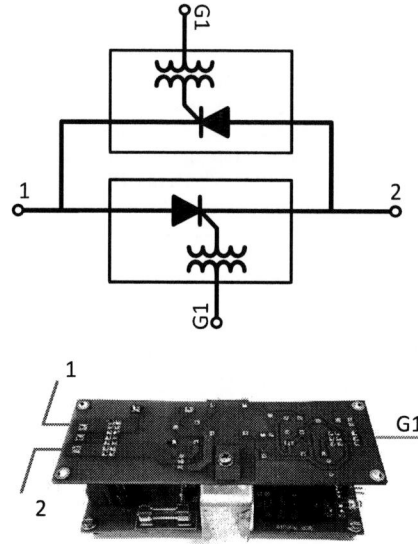

Fig. 5. Solid state relay.

In a different configuration, two boards can be connected using the up-mounting power header in order to form a controlled thyristor phase-leg, as shown in Fig. 6. This arrangement can be further extended by connecting more legs in parallel in order to form for example a three-phase controlled thyristor rectifier. The didactic applications are wide, ranging from demonstrating the commutation process in the controlled rectifier to machine drive control.

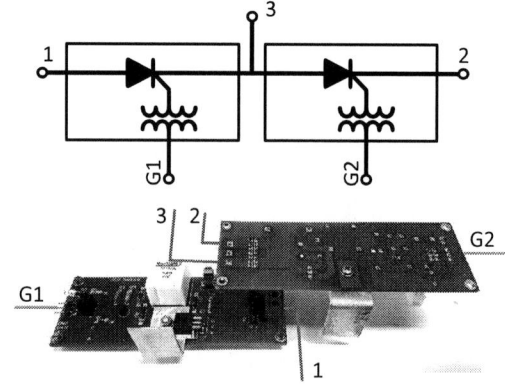

Fig. 6. Controlled thyristor phase-leg.

III. APPLICATIONS

A. Switched Capacitor Bank

The solid-state relay can be used to control the amount of reactive power injected by a capacitor bank. This is done by switching several capacitor branches in parallel. In the literature, this is also known as the thyristor switched capacitor (TSC) [13]. The didactic applications in this application are mainly the study of reactive power compensation strategies, determination and control of the switching transients of the capacitors. In Fig. 7, it is possible to see the variation of the amount of reactive power injected in a capacitor bank. The reactive power compensation capacitor C. The inductor L together with the resistances R are in series with the capacitor. The amount of reactive power injected in the grid is defined as Q and the amount of reactive power provided by a single capacitor branch is Q_C.Here, the switches are represented as ideal bidirectional switches.

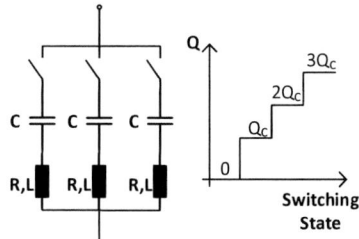

Fig. 7. Principle of operation of a capacitor bank.

The equivalent circuit diagram of a single switched branch is shown in Fig. 8, where it is possible to see the thyristor switches S. The capacitor voltage is represented as $v_C(t)$, the grid voltage as $v(t)$ and the branch current as $i(t)$.

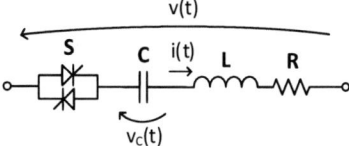

Fig. 8. Equivalent circuit diagram of a switched branch.

According to [12] and [13], the series inductor L is needed to limit switching transients. However, in [14], it is shown that if the switching instant is controlled, the transient overcurrents are greatly reduced allowing operation without the series inductor. On the other hand, if the switching instant is not controlled, the series inductor L must be used to limit overcurrents [14]. It is dimensioned to have the factor n shown in (1) between 3 and 5 [15]. The grid frequency is represented by ω_0.

$$n = \sqrt{\frac{X_C}{X_L}} = \sqrt{\frac{1}{\omega_0^2 LC}} \tag{1}$$

Moreover, starting from the model in [7], the optimal switching instant t_S is determined to be when the initial capacitor voltage is equal to the grid voltage, as shown in (2).

$$v(t_S) = v_C(t_S) \tag{2}$$

The optimal firing angle α is then determined in (3), where V_{C0} is the voltage stored at the capacitor and V is the peak voltage of the grid.

$$\alpha = \sin^{-1} \frac{V_{c0}}{V} \tag{3}$$

The parameters of the components in the equivalent circuit diagram are listed in Table I. In the experimental platform, the inductor can be short-circuited (bypassed) to study its impact on the switching transients.

TABLE I.   CIRCUIT PARAMETERS

Parameter	Value
L	$9.7mH$
C	$25\mu F$
R	0.5Ω
ω_0	$377rad/s$

The switching transients are simulated under four scenarios, considering inclusion (or not) of the series inductor and optimal (or not) switching. In Fig. 9, the resulting current waveforms are presented. Here, a grid voltage with a $127V$ RMS fundamental frequency component is considered. A voltage distortion considering individual harmonic distortion levels of 7.5% at the 3rd order, 1.25% at the 5th order and 0.625% at the 7th order of the fundamental frequency is used in order to simulate the distorted voltage waveforms of the laboratory grid. The simulations indicate that if the series inductor is not inserted, optimal switching must be performed in order to avoid overcurrents. The harmonic distortion of the grid voltage is causing the distortion of the capacitor current. The inductors available at the laboratory, when series connected with the capacitors of the bank, resonate around the 6th order harmonic of the grid frequency, thus the current distortion is changed with insertion of the inductor.

However, if the series inductor is inserted, the transients are limited for non-optimal switching. Experimental results are presented next in order to validate the simulation results.

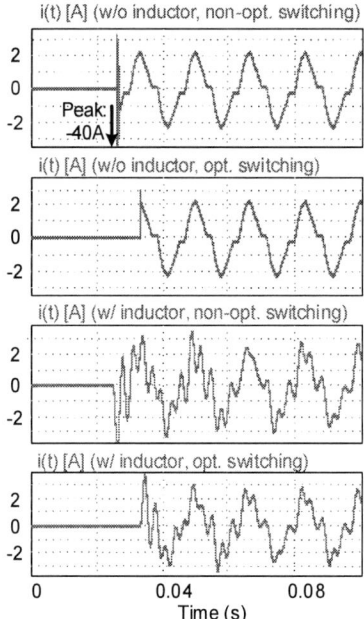

Fig. 9. Simulated current waveforms for capacitor bank switching.

978-1-7281-4181-7/19 $31.00 © 2019 IEEE

The experimental setup is depicted in Fig. 10, where the most important components are identified. In the current setup, reactive power variation is performed only in the single-phase configuration. In the three-phase configuration, the switching transients can be studied. The grid voltage is reduced to 1/10 of the full (127V RMS) voltage using a variable transformer in order to reduce transient overcurrents (which reached up to 40A peak in the simulations with full grid voltage) and avoid damage to the components of the platform.

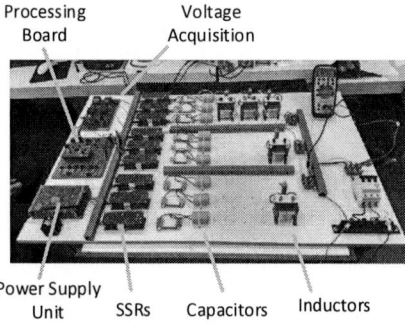

Fig. 10. Overview of the experimental platform.

In Fig.11, the filter current (in dark green) and the grid voltage (in dark purple) are shown for optimal switching without inductor. When the switching command is received (in dark yellow), the system waits until the grid voltage is equal to the capacitor voltage. For comparison, the switching without the inductor and outside the optimal switching instant is evaluated and the resulting waveforms are shown in Fig. 12. It is possible to see that the overcurrent, which is limited in Fig. 11, is now significantly higher, causing a voltage dip on the grid voltage during the transient.

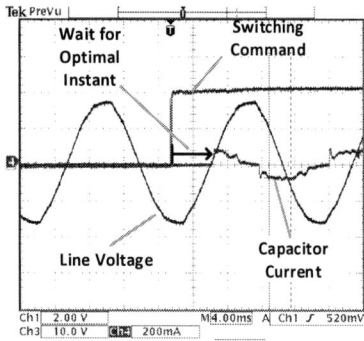

Fig. 11. Optimal switching without inductor.

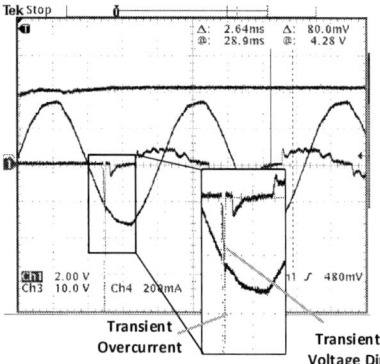

Fig. 12. Non-optimal switching without inductor.

The inductor is now inserted back in the system, and optimal switching is performed. The resulting waveform are shown in Fig. 13. It is possible to see that the association of the inductor with the capacitor changes the current waveform distortion, as it is acting similarly to a passive filter. The grid voltage harmonic distortion is exciting the resonance of the LC series circuit, which has low dampening, therefore the distorted current waveforms result. Switching is performed outside the optimal instant with the inductor and the resulting waveforms are presented in Fig. 14. It is possible to see that the transient overcurrents are mitigated through the inductor.

From the obtained results, it can be concluded that if the inductor is used, control of the switching instant is optional. On the other hand, if optimal switching is performed, usage of the inductor can be avoided.

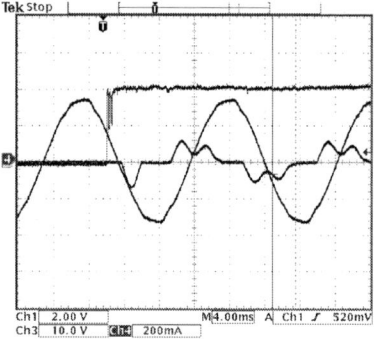

Fig. 13. Optimal switching with inductor.

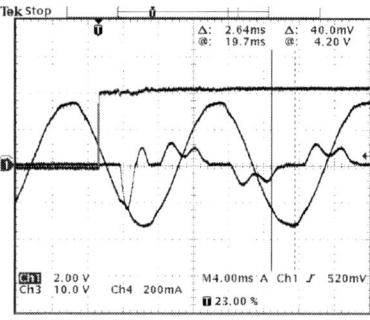

Fig. 14. Non-optimal switching with inductor.

In Fig. 15 the current waveform for step variation of the reactive power is shown for a single-phase system. This demonstrates how the capacitor bank can be used to compensate fundamental frequency displacement reactive power for improvement of the power factor of an installation.

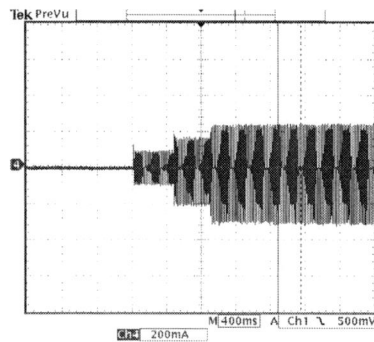

Fig. 15. Current waveform step variation of the reactive power.

978-1-7281-4181-7/19 $31.00 © 2019 IEEE   438

B. Single-Phase Controlled Rectifier

The equivalent circuit diagram of this application is found in Fig. 16, where four thyristor elements are connected to form a single-phase controlled rectifier. On the AC side of the rectifier, the grid voltage is represented by $v(t)$, the grid current by $i_{AC}(t)$ and the grid inductance by L_{AC}. On the DC side of the rectifier, the load inductance is represented by L_0, the load resistance by R_0, the load voltage by $v_0(t)$ and the load current by $i_0(t)$. The parameters of the circuit are listed in Table II.

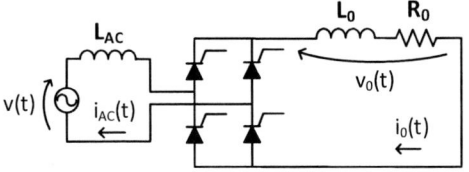

Fig. 16. Circuit diagram of the controlled rectifier.

TABLE II.   RECTIFIER CIRCUIT PARAMETERS

Parameter	Value
L_{AC}	$50\mu H$
V_{RMS}	$110V$
R_0	10Ω
L_0	$500mH$

The circuit is simulated using PSIM and the resulting waveforms are presented in Fig. 17. Here, a phase angle delay of 60° is used as an example. The average output voltage and current of the rectifier under these conditions is calculated in (4) according to [16] (without consideration of the grid inductance).

$$\bar{V}_0 = \frac{2\sqrt{2}}{\pi} V_{RMS} \cos \alpha = 49.5V; \bar{I}_0 = \frac{\bar{V}_0}{R} = 4.95A \quad (4)$$

Fig. 17. Simulated waveforms of the single-phase rectifier operating with a phase angle delay of 60°.

In order to validate the simulations presented in Fig. 17, experimental results are obtained using the setup shown in Fig. 18, where the four thyristor elements are interconnected to form a single-phase rectifier.

Fig. 18. Overview of the single-phase rectifier.

The obtained waveforms for the grid voltage and current are shown in Fig. 19 and for the output voltage and current are shown in Fig. 20.

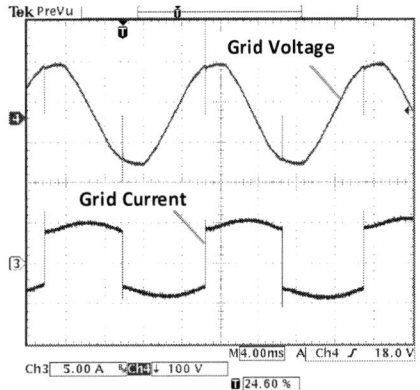

Fig. 19. Grid voltage and current waveforms.

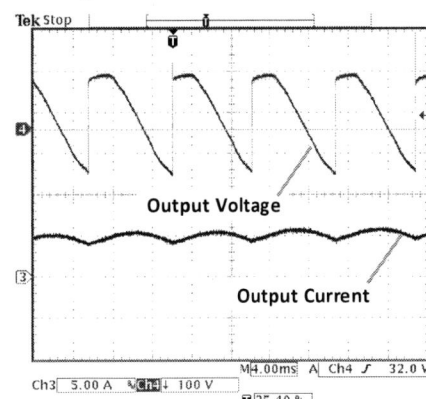

Fig. 20. Output voltage and current waveforms.

IV. CONCLUSIONS

A didactic platform for teaching thyristor-based power electronics circuits is proposed in this article, and the development steps are presented in detail. The main advantage of the proposed platform is that it is flexible, enabling study of different applications using the same basic thyristor element board. Some possible topologies are the SCR, the SSR, the single-phase and three-phase thyristor bridges.

Examples of didactic applications demonstrating a thyristor switched capacitor bank and a single-phase rectifier are presented. In the laboratory, through experimental verification of the concepts learned in theorical classes using the presented didactic platform, the students can improve their understanding of fundamental aspects in power electronics circuits based on thyristors.

The platform has reduced material costs, which are currently of US$8,75 for each module (including printed circuit board manufacturing. It is also available online at [17] under an open hardware license. Therefore, this platform can be used by undergraduate and graduate students which are developing their own research projects in the field of power electronics.

ACKNOWLEDGMENTS

The authors would like to thank Texas Instruments, STMicroelectronics and TDK/EPCOS for their support. This work was financially supported by the Sao Paulo Research Foundation (FAPESP) research grants 2016/16542-5 and 2016/01930-0 and by the Brazilian National Council for Scientific and Technological Development (CNPq) grants 306970/2015-5 and 311789/2014-5. This study was financed in part by the Coordination for the Improvement of Higher Education Personnel (CAPES) - Finance Code 001.

REFERENCES

[1] E. A Vendrusculo, A. A Ferreira, J. A. Pomilio, "Didactic platform for quick and experimental evaluation of control strategies in power electronics," (in Portuguese), *Eletrônica de Potência*, vol. 13, no. 2, pp. 99-108, May 2008.

[2] L. G. Rolim, W. Suemitsu, R. M. Stephan, M. B. Medeiros, "An experimental setup for the study of oriented control of AC machines," in *Brazilian Cong. on Automation*, 2000.

[3] M.V. Lazarini, E. R. Filho, "Induction motor control didactic set-up using sensorless and sliding mode DTC strategy," *Eletrônica de Potência*, vol. 13, no. 4, pp. 291-299, Nov. 2008.

[4] S. A. O. da Silva, D. H. Wollz, L. P. Sampaio, "Development of a didactic workbench using real-time monitoring system for teaching of photovoltaic systems," *Eletrônica de Potência*, vol. 23, no. 3, pp. 371-381, Jul. 2018.

[5] Brito, Marcelo R. S., et al. "Didactic Platform for Teaching of Three-Phase Rectifier Circuits in Power Electronics." International Journal of Electrical Engineering & Education, vol. 51, no. 4, Oct. 2014, pp. 279–291, doi:10.7227/ijeee.0001.

[6] Festo Didactic , "Thyristors and Power Control Circuits Module," Labvolt Series Datasheet, 2019.

[7] L. G. Rolim, R. M. Stephan, W. Suemitsu, "Power electronics teaching laboratory at UFRJ," in *Brazilian Power Electronics Conf.*, 1993.

[8] Power Innovations , "Silicon Controlled Rectifiers," TIC126 Series Datasheet, 1997.

[9] IPC Association, "Generic Standard on Printed Board Design," IPC-2221, Febr. 1998.

[10] W. Komatsu. PEA2487. Lecture Notes, Topic: "Power Electronics" Polytechnic School, University of São Paulo, São Paulo, Brazil, 2016.

[11] Minitrafo, "Pulse transformer specification," [Online]. Avaiable: http://minitrafos.com.br/clientes/u_228898_8ad467f2f3/imgRoot/Tab ela%20Site%20Pulso.htm. [Accessed June 25, 2019].

[12] R. M. Mathur and R. K. Varma, *Thyristor-based facts controllers for electrical transmission systems*. New York, NY: Wiley-Interscience, 2002.

[13] N. G. Hingorani and L. Gyugyi, *Understanding facts: concepts and technology of flexible AC transmission systems*. New York: Institute of Electrical and Electronics Engineers, 2000.

[14] L. Araújo and M. Zambon, "Static reactive power generation with capacitor switching - Gerador de reativos estático com chaveamento de capacitores," Undergraduate thesis, 2018.

[15] T. J. E. Miller, *Reactive power control in electric systems*. New York: Wiley, 1982.

[16] J. Specovius, A *First Course on Power Electronics Components, Circuits and Systems*, Wiesbaden: Vieweg Teubner (in German), 2011.

[17] LEP-PEA-EPUSP, "Open Source Power Electronics". [Online]. Available: https://github.com/LEP-PEA-EPUSP.[Accessed Jun. 03, 2019].

Control of Wireless Power Transfer Systems under Large Coil Misalignments

Yeran Liu
Department of Electrical Engineering
Southwest Jiaotong University
Chengdu, China
yeranliu@my.swjtu.edu.cn

Udaya K. Madawala
Department of Electrical and Computer
Engineering
University of Auckland
Auckland, New Zealand
u.madawala@auckland.ac.nz

Ruikun Mai
Department of Electrical Engineering
Southwest Jiaotong University
Chengdu, China
mairk@swjtu.edu.cn

Zhengyou He
Department of Electrical Engineering
Southwest Jiaotong University
Chengdu, China
hezy@swjtu.edu.cn

Abstract—Coil misalignment in wireless power transfer systems causes variation in self- and mutual inductances, resulting in reduction in power throughput at increased losses. This paper proposes a P-Q based primary side controller that regulates output voltage while minimizing losses under coil misalignment or load variations. The proposed controller varies the frequency of the converter to restore the tuned operation under coil misalignment while avoiding bifurcation and simultaneously regulating the output voltage through primary phase-shift modulation. Both theoretical analysis and simulation results are presented to show the validity of the proposed control method, which does not require a secondary controller or a dedicated communication link.

Keywords—*Wireless power transfer, coil misalignment, primary control, variable frequency, zero phase angle*

I. INTRODUCTION

Coil misalignment is an unavoidable challenge in wireless power transfer (WPT) systems. Both self- and mutual inductances inevitably vary with coil misalignment as shown in Fig. 1. As a result, not only the output power is significantly affected, but also the WPT system operates under de-tuned condition, drawing reactive power and thereby increasing losses.

A number of control concepts have been proposed to address this problem but with varying degrees of success [1]-[6]. Among these methods, the frequency tracking algorithms based on "perturb and observe" techniques were used to find an optimal operating frequency against load variation [5] and misalignment [1][6]. Minimum reflection power [1], maximum input power [5], and minimum input current [6] have been used as criteria to estimate a tuned frequency. However, these methods either required direct feedback from the secondary side to control power or failed to consider variations in both load and circuit components caused by coil misalignments. Moreover, no attempt has been made to avoid bifurcation frequencies to ensure stability during the frequency sweep. According to literature, a direct control strategy that ensures precisely tuned operation through frequency control, avoiding bifurcation and regulating output

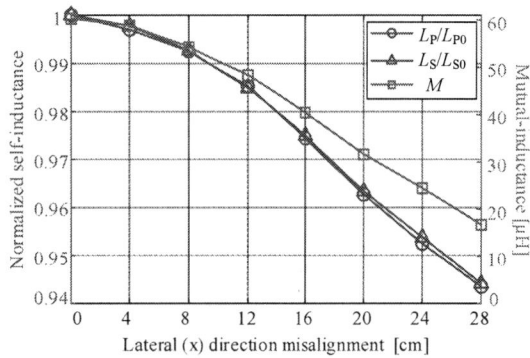

Fig. 1 Measured self- and mutual inductance of DD pads.

voltage under both large coil misalignments or load variations, is yet to be reported.

Therefore, this paper proposes a primary-side P-Q controller for WPT systems that regulates either output voltage while tracking the tuned frequency, which corresponds to zero-phase-angle (ZPA), under large coil misalignment and without communication link. Only the primary side information is used to measure reactive power and estimate output power. The tuned condition at (ZPA) frequency is achieved by minimizing the reactive power, but avoiding bifurcation. With the proposed method, the output power is maintained at the rated level despite the variations in self- and mutual inductances due to coil misalignment.

II. CIRCUIT ANALYSIS

A. Circuit Model

The schematic of a series-series WPT system is shown in Fig. 2(a). The full-bridge converter, used in the primary side, is operated as an inverter. To improve power transfer capability, a series-series (S-S) resonant network is chosen, but other compensation networks can also be employed. L_P and L_S are self-inductances of the charging coils, which may change under misalignment condition, and M is the mutual inductance between the primary and secondary coils, and R_L

978-1-7281-4181-7/19 $31.00 © 2019 IEEE   441

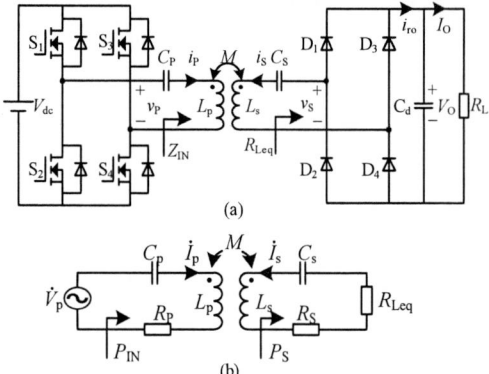

(a)

(b)

Fig. 2 (a) A series-series WPT system. (b) Equivalent circuit model.

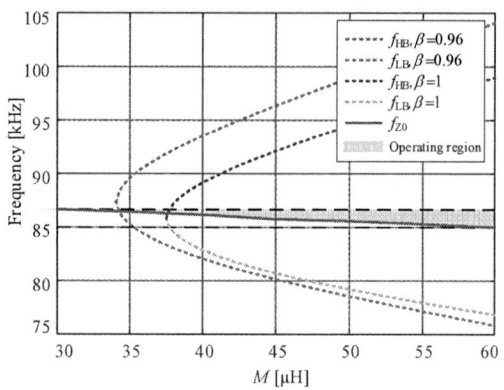

Fig. 3 ZPA and bifurcation frequencies with varied mutual inductance.

represents the dc load. C_P and C_S are respective compensation capacitors and are designed to tune the system under the aligned condition as:

$$C_P = \frac{1}{\omega_0^2 L_{P0}} \tag{1}$$

$$C_S = \frac{1}{\omega_0^2 L_{S0}} \tag{2}$$

where L_{P0} and L_{S0} are the self-inductances of the aligned coils, and ω_0 is the resonant frequency under the aligned condition.

The equivalent circuit model at the fundamental frequency is shown in Fig. 2(b), where a sinusoidal voltage source $\dot{V}_P$ represents the fundamental harmonic of the inverter voltage at angular frequency ω. The dc load and the diode rectifier are represented by an ac equivalent load resistor R_{Leq}. R_P and R_S are the equivalent-series resistors (ESRs) of primary and secondary side charging coils respectively.

The circuit model in Fig. 2(b) can be expressed by:

$$\dot{V}_P = Z_P \dot{I}_P + j\omega M \dot{I}_S \tag{3}$$

$$0 = j\omega M \dot{I}_P + Z_S \dot{I}_S + R_{Leq} \dot{I}_S \tag{4}$$

where Z_P and Z_S are the impedance in primary and secondary side:

$$Z_P = j\omega L_P + (j\omega C_P)^{-1} + R_P = jX_P + R_P \tag{5}$$

$$Z_S = j\omega L_S + (j\omega C_S)^{-1} + R_S = jX_S + R_S \tag{6}$$

where X_P and X_S represent the reactance in primary and secondary side.

Seen by the primary voltage, the input impedance can be expressed by:

$$Z_{IN} = R_P + \frac{\omega^2 M^2 (R_S + R_{Leq})}{(R_S + R_{Leq})^2 + X_S^2} + j\left(X_P - \frac{\omega^2 M^2 X_S}{(R_S + R_{Leq})^2 + X_S^2} \right) \tag{7}$$

B. ZPA & Bifurcation Frequencies

The frequency for ZPA condition is intended to be tracked by adjusting the operating frequency. The ZPA frequency ω_Z can be found by solving for ω in

$$\mathrm{Im}(Z_{IN}) = X_P - \frac{\omega^2 M^2 X_S}{X_S^2 + (R_{Leq} + R_S)^2} = 0 \tag{8}$$

Of the three positive roots in (8), one root is represented as

$$\omega_{Z0} = \omega_0 / \sqrt{\beta} \tag{9}$$

where β is the normalized self-inductance as shown in Fig. 1:

$$\beta = \frac{L_P}{L_{P0}} = \frac{L_S}{L_{S0}} \tag{10}$$

Because of the similarity of the primary and the secondary side DD coil, β can be represented as the normalized variation of both L_P and L_S, which is approximately constant.

The remaining two roots of (8) are the bifurcation frequencies f_{HB} (ω_{HB}) and f_{LB} (ω_{LB}) as shown in Fig. 3 and can be represented as:

$$\omega_{Z1} = \omega_0 \sqrt{\frac{\sqrt{K_1} + L_P R_{Leq}^2 - 2\omega_0^2 \beta L_P L_S^2}{2\omega_0^2 L_S M^2 - 2\omega_0^2 \beta^2 L_P L_S^2}} \tag{11}$$

$$\omega_{Z2} = \omega_0 \sqrt{\frac{-\sqrt{K_1} + L_P R_{Leq}^2 - 2\omega_0^2 \beta L_P L_S^2}{2\omega_0^2 L_S M^2 - 2\omega_0^2 \beta^2 L_P L_S^2}} \tag{12}$$

where

$$K_1 = 4\omega_0^4 L_P L_S^3 M^2 - 4\beta \omega_0^2 L_P^2 L_S^2 R_{Leq}^2 + L_P^2 R_{Leq}^2$$

f_{Z0} (ω_{Z0}) is the desired tuned frequency of operation. It can be seen from Fig. 3 that the operating region of the controller only intersects with the bifurcation frequency over a short range of M. However, the controller takes into account the leading and lagging nature of the phase angle between the voltage and current produced by the converter to ensure that the system will operate outside the bifurcation region when varying the frequency.

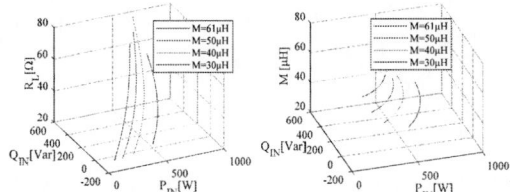

Fig. 4 R_L and M as a function of P_{IN} and Q_{IN}.

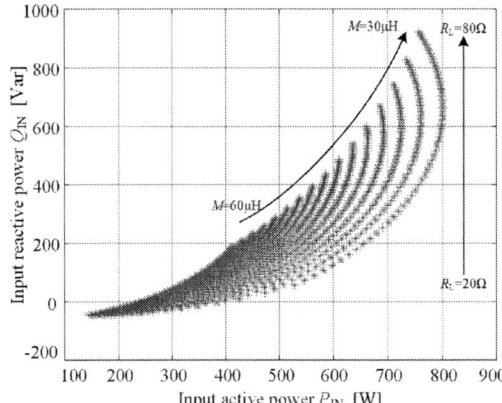

Fig. 5 Points in the PQ table.

C. Load and Mutual Inductance Estimation

Based on the observation that, in S-S compensation network, the output power may increase greatly with weak coupling or large load, the variation of P_{IN} and Q_{IN} against R_L and M is considered. The purpose is to see whether R_L and M can be estimated according to P_{IN} and Q_{IN}. Once M is known, L_P and L_S can be estimated according to the data in Fig. 1. It is complex to derive the explicit expressions of R_L and M as functions of P_{IN} and Q_{IN}. But the relationship among R_L (M), P_{IN} and Q_{IN} can be plotted and transformed into a lookup table.

With fixed input voltage (100V) and fixed operating frequency (86kHz), R_L and M as functions of P_{IN} and Q_{IN} are shown in Fig. 4. It is evident that the curves with various M do not intersect each other, which means a unique combination of (P_{IN}, Q_{IN}) corresponds to a unique load resistor R_L and a unique M. That provides an approach to identify R_L and M only according to the PQ look-up table.

The points in the lookup table with f=86kHz is shown in Fig. 5. Each cluster of points has the same mutual inductance from 60μH to 30μH by a step of 2μH. Among each cluster of points, the higher point in this figure corresponds to larger R_L. R_L increases from 20Ω to 80Ω by a step of 0.5Ω. Therefore, each two of the adjacent points have the same difference on R_L or M.

D. Output Voltage Estimation

The purpose of the control strategy is to regulate the output voltage from the primary side. Because the magnitude of the primary current is easy to be measured, the output voltage can be estimated according to the current- voltage gain G_{IV}, which is defined and solved based on (3) and (4) as:

$$G_{IV} = \frac{V_S}{I_P} = \frac{\omega M R_{Leq}}{\sqrt{(R_{Leq} + R_S)^2}} \quad (13)$$

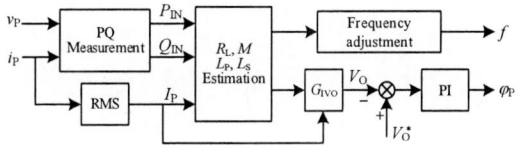

Fig. 6 Control block of the proposed method.

Because the dc voltage is $\pi^2/8$ times larger than the ac voltage of the rectifier, the estimated dc output voltage is given as:

$$V_O = I_P G_{IVO} = \frac{\pi^2}{8} I_P G_{IV} = \frac{\pi^2 \omega M R_{Leq}}{8\sqrt{(R_{Leq} + R_S)^2}} I_P \quad (14)$$

III. CONTROL METHOD

Using the estimation method proposed in Section II, the inductance and load resistor of the WPT system can be known only according to the primary-side measurement. That provides enough information for the primary-side controller to regulate the output voltage or output power without communication.

The control block of the proposed strategy is shown in Fig. 6. The instantaneous value of the inverter voltage v_P and the primary current i_P are sampled to calculate the input active power P_{IN} and reactive power Q_{IN} based on the PQ measurement in [8]. Then P_{IN}, Q_{IN}, and the rms value I_P are used to estimate M and R_L according to the PQ look-up table. Based on the estimation results, the inverter frequency is adjusted to achieve ZPA operation. The ZPA frequency ω_Z can be found by (9).

A PI controller is used to generate the phase-shift signal φ_P of the inverter to restore the output voltage, according to the error between the estimated voltage V_O and the reference value $V_O{}^*$.

IV. SIMULATION RESULTS

The simulation was carried out in SIMULINK/PLECS with the mutual inductance ranging from 30μH to 60μH. The compensation capacitors were designed to be resonant at 85kHz when the coils are aligned. Parameters of the simulated system are shown in Table I.

Table I Parameters of simulation

Symbol	Parameter	Value	Unit
f	inverter frequency	[80, 90]	kHz
V_{dc}	input voltage	250	V
P_{out}	output power	1000	W
L_P	inductance of primary coil	[192.9, 185.6]	μH
C_P	compensation capacitor of L_P	18.2	nF
R_P	ESR of primary coil	0.3	Ω
L_S	inductance of secondary coil	[191.5, 184.5]	μH
C_S	compensation capacitor of L_S	18.3	nF
R_S	ESR of secondary coil	0.3	Ω
M	mutual inductance	[60.3, 30]	μH
R_L	load resistor	25	Ω

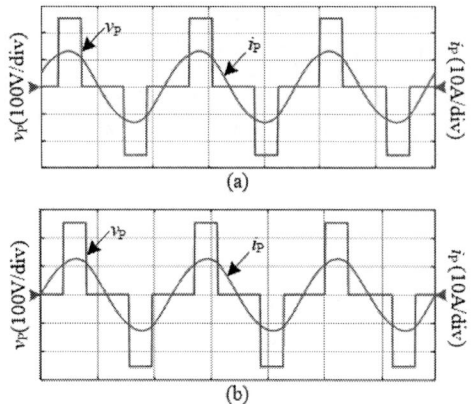

(a)

(b)

Fig. 7 Steady waveforms of primary voltage v_P and current i_P. (a) detuned at f=85kHz (b) tuned at f=86.6kHz.

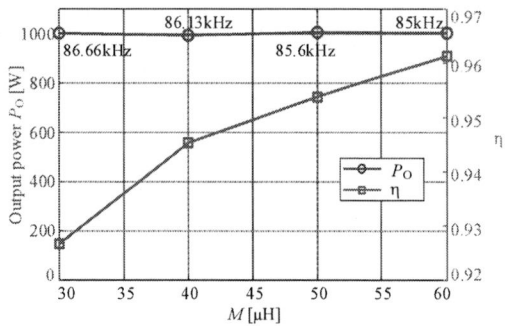

Fig. 8 Output power and system efficiency with various mutual inductance M.

Fig. 7 shows the steady waveforms of the primary voltage and current with M=30uH. When the operating frequency is same as the resonant frequency (85kHz) at the aligned position, there is a phase difference between i_P and v_P. When the operating is changed to 86.6kHz, i_P and v_P are in phase and the system is tuned.

With the proposed method, the system can regulate output power at 1kW and achieve high efficiency with large coil misalignment as shown in Fig. 8.

V. CONCLUSION

A P-Q controller has been proposed for WPT systems to restore its tuned operation as well as to regulate the output voltage throughput despite large coil misalignments and load variations. Theoretical analysis and simulation results have been presented to demonstrate the validity of the proposed method.

REFERENCES

[1] N. Y. Kim, K. Y. Kim, J. Choi and C. -. Kim, "Adaptive frequency with power-level tracking system for efficient magnetic resonance wireless power transfer," in Electronics Letters, vol. 48, no. 8, pp. 452-454, 12 April 2012.

[2] T. Diekhans and R. W. De Doncker, "A Dual-Side Controlled Inductive Power Transfer System Optimized for Large Coupling Factor Variations and Partial Load," IEEE Transactions on Power Electronics, vol. 30, no. 11, pp. 6320-6328, Nov. 2015.

[3] U. K. Madawala and D. J. Thrimawithana, "New technique for inductive power transfer using a single controller," in *IET Power Electronics*, vol. 5, no. 2, pp. 248-256, Feb. 2012.

[4] K. Song, Z. Li, J. Jiang and C. Zhu, "Constant Current/Voltage Charging Operation for Series–Series and Series–Parallel Compensated Wireless Power Transfer Systems Employing Primary-Side Controller," in IEEE Transactions on Power Electronics, vol. 33, no. 9, pp. 8065-8080, Sept. 2018.

[5] D. Patil, M. Sirico, L. Gu and B. Fahimi, "Maximum efficiency tracking in wireless power transfer for battery charger: Phase shift and frequency control," 2016 IEEE Energy Conversion Congress and Exposition (ECCE), Milwaukee, WI, 2016, pp. 1-8.

[6] Y. Gao, C. Zhou, J. Zhou, X. Huang and D. Yu, "Automatic Frequency Tuning with Power-Level Tracking System for Wireless Charging of Electric Vehicles," 2016 IEEE Vehicle Power and Propulsion Conference (VPPC), Hangzhou, 2016, pp. 1-5.

[7] Chwei-Sen Wang, G. A. Covic and O. H. Stielau, "Power transfer capability and bifurcation phenomena of loosely coupled inductive power transfer systems," in IEEE Transactions on Industrial Electronics, vol. 51, no. 1, pp. 148-157, Feb. 2004.

[8] Y. Tang, Y. Chen, U. K. Madawala, D. J. Thrimawithana and H. Ma, "A New Controller for Bidirectional Wireless Power Transfer Systems," IEEE Transactions on Power Electronics, vol. 33, no. 10, pp. 9076-9087, Oct. 2018.

Considerations on Communication Infrastructures for Cooperative Operation of Smart Inverters

Augusto Matheus dos Santos Alonso
G. of Automation and Int. Systems (GASI)
São Paulo State University (UNESP)
Sorocaba, Brazil
augusto.alonso@unesp.br

Leonardo Carlos Afonso
G. of Automation and Int. Systems (GASI)
São Paulo State University (UNESP)
Sorocaba, Brazil
leocafonso@gmail.com

Danilo Iglesias Brandao
Graduate Program in Electrical Engineering
Federal University of Minas Gerais (UFMG)
Belo Horizonte, Brazil
dibrandao@ufmg.br

Elisabetta Tedeschi
Department of Electric Power Engineering
Norwegian University of Science & Technology (NTNU)
Trondheim, Norway
elisabetta.tedeschi@ntnu.no

Fernando Pinhabel Marafão
G. of Automation and Int. Systems (GASI)
São Paulo State University (UNESP)
Sorocaba, Brazil
fernando.marafao@unesp.br

Abstract—The presence of distributed generation systems spread over low-voltage electrical networks is boosting the development of control methodologies aiming at coordinating and cooperatively managing the existing smart inverters. Although low-bandwidth data transmission links are constantly described to be required for a considerable number of centralized and decentralized control methodologies, there is a gap in literature concerning the plain understanding of the features of the related communication protocols available for such application. Thus, this paper brings considerations on some of the most relevant communication protocols that can be applied to the cooperative control of multiple smart inverters, taking into account the recent updates on interoperability requirements recommended by the IEEE 1547-2018 standard. The communication infrastructure, topology and features of a low-bandwidth data transmission link are discussed in this paper focusing on the SunSpec, DNP3 and SEP2 protocols. Yet, some critical comments are made regarding the practical interoperability of commercial inverters, also bordering cyber security matters.

Keywords—Communication protocol, cooperative control, IEEE 1547 standard, microgrid, interoperability, smart inverter.

I. INTRODUCTION

Since the past decades, the dense presence of distributed energy sources (DERs) in low-voltage distribution systems has been playing a key role in the decentralization of energy generation, allowing renewables to be inserted into the new digitized paradigm of electrical networks [1], which brings the Smart Grid (SG) concept to reality. As electrical grids move towards SG implementation, the adoption of intelligent mechanisms and provision of higher interactivity among electronic devices is inevitable [2]. Consequently, the employment of communication technologies within electrical networks, especially in dynamic and more interactive systems such as microgrids (MGs) [3], is being required for many related applications [4].

A particular application of communications, which is gaining significant attention in literature, is related to power electronic interfaces existing within DERs, in order to enable their cooperative operation. By driving smart inverters under an integrated approach, their provision of ancillary services can be coordinated to enhance the overall performance of MGs, especially by enabling power/current sharing and compensation functionalities such as reactive, unbalance and harmonic compensation [6]. Hence, several methodologies for cooperative operation of inverters, which are generally based

on centralized or decentralized control approaches [7], are being proposed to provide integration of DERs [8]-[10].

Regardless of their control architecture, most methods for cooperative steering of smart inverters rely on low-bandwidth data transmission links to exchange information among agents, even though communication may not be required for the overall operation and stability maintenance of the MG. By doing so, local information can be exchanged with neighbors (e.g., as done by consensual approaches [11]) or with a central management system/controller (e.g., as done by centralized control [10], [12]). Therefore, through communication means, the cooperation of agents in MGs can be achieved accounting for the status of several nodes and striving for enhancing the overall performance of the electrical system.

Nonetheless, although a significant amount of research has been done [6]-[12] aiming at developing cooperative control methodologies which take advantage of low-bandwidth data transmission links, there is a gap in literature in regard to the description of such communication infrastructure and the related protocols that are available for the fulfillment of this particular purpose. To the best knowledge of the authors, most of previous works related to SGs focus, majorly, on information and communication technologies (ICTs) required for applications in metering (AMI) [13] and data exchange in the utility level [14]. Yet, when addressing communication in MGs [3,4], focus is seldom given to the application of smart inverters. Consequently, literature lacks discussions about communication infrastructures and protocols embedded in smart inverters for cooperative operation. Since recent updates on the IEEE 1547 − 2018 standard [15] consider such ICT matters for the compliance of smart inverters, it is of importance to discuss the requirements of the interoperability protocols for cooperative operation and networked control [16], laying the groundwork of this study.

Thus, this paper has the main goal of discussing the communication protocols highlighted within the IEEE 1547 − 2018, since they are the likely candidates to be implemented in real applications of cooperation among inverters. The paper is organized as follows. Firstly, in Section II, cooperative control is defined and a brief explanation is presented to highlight the communication architectures that fall within the scope of analysis. Later, an overview of the three protocols comprised within [15] (i.e., SunSpec, DNP3, and SEP2) is presented in Section III. Such discussion aims at showing their technical aspects and their applicability to different approaches of cooperative control. At last, a brief discussion

978-1-7281-4181-7/19 $31.00 © 2019 IEEE

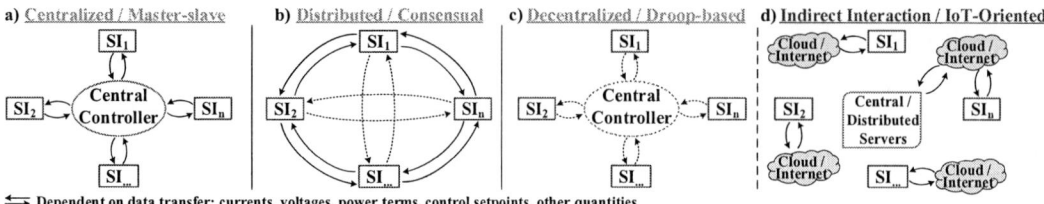

Fig. 1. Principle of operation of cooperative control strategies on which communication technologies are implemented to control "*n*" smart inverters.

oriented to communications is included in Section IV, concerning cyber security and the practical interoperability of commercial inverters.

II. COOPERATIVE CONTROL STRATEGIES AND THEIR USE OF DATA TRANSMISSION LINKS

The principle of coordinated operation of smart inverters consists on steering them toward a common goal that, in general, corresponds to extracting the maximum power from the MG, additionally offering support to the grid under abnormal conditions and improving power quality. Depending on how such inverters are driven, different control strategies are required [5]-[12], and their cooperative operation is usually designed to follow a centralized, distributed or decentralized architecture, as summarized in Fig. 1. Cooperation through indirect methods, as shown in Fig. 1(d), is essentially based on one of the three previous designs, being later discussed. The most significant contrast among those architectures concerns how communication infrastructures are implemented, or if they are not used at all. Apart from that, cooperation of inverters is formulated based on exchanging information among agents (e.g., voltages, currents, power, control setpoints, so forth).

Regarding the communication framework, cooperative operation can be split into two main approaches that focus on the centralization or decentralization of the related data processing and consequently supervised control algorithms [6]-[8], as depicted in Fig. 2. For this first classification, communication is basically required to allow local agents (i.e., inverters) to exchange information with a central controller (CC), demanding control setpoints to steer their operations as a feedback from this interaction [10,12,17]. Moreover, the CC is mostly considered as a master unit, whereas the inverters are driven as slave agents [10]. This occurs based on point-to-point (P2P) low-bandwidth data transmission [18], which is a basic communication method characterized by punctual interactions of the CC (i.e., server) with each of all other distributed inverters (i.e., clients). It is highlighted that P2P communication differs from peer-to-peer networking. The latter does not require a central server for data transmission among agents, allowing each communicating unit to operate both as a client and a server [18].

Other strategies of cooperative control avoid the extensive use of P2P communication to minimize latency, since significant time is expended if a considerable number of inverters have to communicate with the CC sequentially. As a consequence, communication alternatives, such as additionally allowing the CC to broadcast generalized operational references as control feedback to all inverters, are proposed on some approaches [10,12,19]. On the other hand, another possible communication approach is the use of the point-to-multipoint (PMP) technology [20], which is characterized by the centralization of the data processing in the CC through a shared network. In such way, all inverters

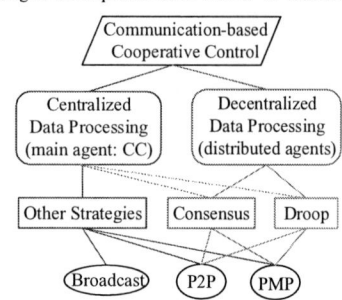

Fig. 2. Basic use of communication technologies for cooperative control.

have concomitant access to the same communication link carrying the operational control references. Briefly, by relying on communication with a central agent as a fundamental part of the overall coordinated operation of inverters, the strategy can be defined as centralized.

For the second approach (i.e., decentralized strategies), they are mostly based on droop control [8,9,21], which allows inverters to operate in parallel, not requiring an immediate communication link for the overall operation of the MG. However, several droop-based strategies present limitations in regard to, for instance, balancing the trade-off between accurate power sharing and deviations of frequency and/or voltages [22]. Therefore, alternative droop-based methods take advantage of communication links to implement control loops in the secondary and/or tertiary hierarchical layers to overcome such limitations [16], [21]. Note that, in this work, the definition of decentralized control strategy lies on the independence of communication links to perform basic cooperative control functionalities, such as active and reactive power sharing. Communication might be required to extend few functionalities of the above-mentioned decentralized methods by interacting with a CC, such as done in [17]; or yet, for inverters to exchange data among themselves, aiming at optimizing the overall operation of the MG. Thus, an infrastructure based on P2P or PMP communication may be used in such cases. Also, data exchange among inverters can occur based on novel methods of indirect interaction among agents (i.e., internet of things (IoT)-oriented), such as through the access to cloud servers [23] or by communicating using the concept of Energy Internet [24], as seen in Fig. 1(d). Note that such indirect concepts may be characterized as centralized, distributed or decentralized depending on how inverters interact with cloud/internet servers.

Finally, some consensus-based control methods [11], [25], lie in between centralized and decentralized operation due to the means required for the consensual convergence among inverters. Such strategies are decentralized since they do not rely on a centralized controller to perform local control of inverters, but they concomitantly depend on low latency communication channels to perform cooperative control. Usually, communication is simply set up for the consensus algorithm by just having distributed agents interacting with a

978-1-7281-4181-7/19 $31.00 © 2019 IEEE

few of their neighbors (i.e., adjacent inverters) through P2P, PMP or peer-to-peer data links.

Among all the aforementioned cooperative control strategies, there is a common agreement in relation to the requirement and use of the ICT infrastructure. Since inverters within MGs are usually distant from each other, or from the CC, the control strategy ought to stand operation following the use of low-bandwidth communication channels. Then, by low-bandwidth communication, it is meant a transmission mean with a low data transfer speed (i.e., up to a few hundreds of Kbps) between the communicating entities [18]. This holds independently of the physical layer employed for the exchange of data among agents (e.g., wireless, optical fiber, Ethernet). The consequence of this is that, if a strategy, either centralized or decentralized, requires an excessive amount of data to be transferred among agents, it may become unfeasible for real life applications. Such lack of feasibility may particularly occur due to the inherent delays and data transmission latency existing for all mentioned communication technologies [20]. In addition, although vaguely described in research literature, the mentioned low bit-rate communication interface embedded on inverters has to comply with interoperability features recently incorporated within standards, such as the IEEE 1547-2018 [15].

As a final remark, most of the previous works [6]-[14], superficially describe the communication link as a low-bandwidth channel and do not specify which of the three protocols proposed in [15] (i.e., SunSpec, DNP3, SEP2) should be used according to the type of communication required by each control layer of the method. Thus, the next section describes the features of each of these three protocols and initiates some discussion on their employment on different layers of application focusing on cooperative operation of inverters.

III. COMMUNICATION INFRASTRUCTURE AND PROTOCOLS FOR COOPERATIVE SMART INVERTERS

The recent updates comprised within the IEEE 1547-2018 focus on the interconnection and interoperability related to grid-connected inverters, particularly on their interaction among themselves and with other devices. The most significant changes on this regulation brings to reality the concept of smart inverters, which occurs through the provision of ancillary services, now recommending such converters to ride through abnormal voltage/frequency conditions, actively regulate voltage by adjusting reactive power, and many others [15]. Inverters must also comprise a communication interface. Thus, from the IEEE 1547-2018 recommendation that DERs must perform grid-support functions based on local measurements, while also comprising interoperability interfaces, the distributed generation sector steps forward to likely consider communication-based services, as expected for the ideal SG future [26]. Such communication-based functions enable more controllable and accurate power regulation in MGs, which in turns contributes to increase the system hosting capacity [26].

Particularly focusing on communication features, for any of the three protocols (i.e., SunSpec, DNP3 and SEP2), inverters must be able to communicate with external agents (e.g., other inverters, the CC, or metering devices) through the exchange of information under a certain infrastructure of data packet format as depicted in Fig. 3. Note that the following structure is independent on the data transmission technology

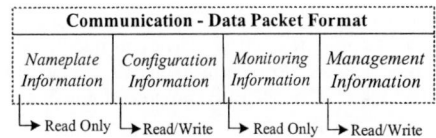

Fig. 3. Data packet configuration proposed in [15].

used (i.e., P2P, PMP, broadcast) and should be adequately considered on protocol level. Such communication should contain data classified within one of the four following categories, which are: nameplate, configuration data, monitoring measurement, and management information. This means that, upon the beginning of a new communication interaction between two agents, a data packet must be gathered by each inverter being grouped based on the following information:

- *Nameplate Information:* This piece of data presents a read-only feature, and consists of built-in information of the device. It comprises commercial information about the inverter, such as its model, manufacturer, serial number, as well as technical data given by its nominal and maximum active, reactive and apparent powers, its AC voltage ratings, along with many other defined inside [15]. This category presents read-only data, which allows the written information to be used, for instance, by a centralized controller for the calculation of operational setpoints as used in [10,12,19]. Nevertheless, this field of the data packet cannot be used for writing purposes. From this piece of the communication infrastructure it is also possible to interpret different types of devices exchanging data since, for example, data headers identify if the agent is an inverter (i.e., presenting ID = 1XX), a storage system (i.e., ID = 8XX), or many others [27]. Such feature may be used to facilitate implementation of control algorithms for cooperative strategies that require prior knowledge of the communication agents;

- *Configuration Information*: This category allows to modify the settings of the actual operational capacities of the inverter, which are by default based on the values presented in the nameplate information. This means that if, in any case, the MG operator intends to change the features of the converter by limiting its nominal ratings, it can be done by writing different data to this section of the communication packet. As consequence, by detecting nominal ratings different from the ones previously existing in the nameplate, the inverter must adjust its operation. This feature might be interesting while considering a coordinated operation in order to limit the output of inverters that are placed in critical nodes where resonances or other power quality issues may be triggered. In strategies on which specific quantities (i.e., beyond the basic ones already comprised within [15]) need be transmitted among the participating agents, such as values of peak currents [19], calculations of harmonic powers [8], etc), such information can be inserted in specific fields on this section of the data packet being exchanged;

- *Monitoring Information*: This section of the data packet mainly comprises the latest measurements performed by the inverter, not being accessible for writing data. In general, such information consists of active and reactive power, voltage and frequency, the state of charge of the possible energy storage system, and a few other quantities for supervision purposes

978-1-7281-4181-7/19 $31.00 © 2019 IEEE   447

[15]. Of course, the data transmitted in this part of the packet present the basic electrical quantities required to be read for most of the cooperative control strategies that take advantage of communication [6]-[12];

- *Management Information*: Finally, the last category allows to read and write some functional and mode settings of the inverter. Here, functionalities such as constant power factor mode, active/reactive power curve points, frequency droop parameters, and many others can be adjusted to steer the inverter to operate as desired. Therefore, cooperative strategies based on droop control like [9,16,21] can, for instance, directly act on writing control setpoints on this portion of the communication packet.

The most beneficial reason for adopting inverters compliant with the IEEE 1547-2018 on cooperative control strategies is given by this basic standardization on the format of data packets being exchanged during communications. By following similar patterns of data structure, interpreting the transmitted information is facilitated. Thus, the SunSpec, DNP3 and SEP2 communication protocols are presented in the following to demonstrate their features and applicability to the distributed operation of inverters.

A. SunSpec Modbus Protocol

The SunSpec protocol was developed by the SunSpec Alliance, which incorporates manufacturers, technology developers and commercial providers, aiming at specifying an information and communication model focused on the interoperability of inverters and other devices comprised within the scope of SGs [27], [28]. Thus, it can be upfront mentioned that the SunSpec protocol mostly focuses on "device level" communications (i.e., not specialized on clustering or "utility level" data). Consequently, being a suitable alternative for the local exchange of information occurring among inverters, as well as for their interactions with a CC, if required by the strategy.

Such protocol is mostly incorporated in the application layer, being at the top of the OSI model [20]. Since it is also based on the well-established modbus protocol, its implementation allows integration with most of the commercialized DERs technologies presenting communication interfaces. Yet, being on top of the OSI model, it supports different communication means (wireless, wired, optical fiber, etc), and it provides access to different data transmission approaches (P2P or PMP). For what concerns smart inverters, the SunSpec protocol defines a chained data model for the mapping of registers that follows the data transmission (i.e., different categories) defined in [15] and previously mentioned in Section II (i.e., placement of different data categories). This means that the protocol specifies which registers should be accessed for the reading/writing of: identification data, control variables, monitoring measurements and other operational data [28].

As an example, when an electrical quantity is required to be shared among inverters in a consensual strategy through SunSpec Protocol, each inverter must, as depicted in Fig. 4: *i*) identify its neighbors by reading and interpreting their ID headers (i.e., information comprised within the nameplate category); later, *ii*) know which specific register it should particularly read in order to attain the desired control variable (e.g., shared power [8]-[10],); and then, *iii*) adjust its local control references. Since the mapping proposed for the

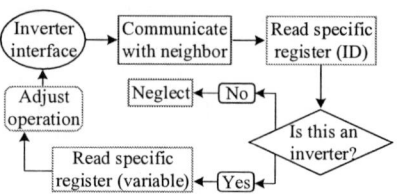

Fig. 4. Example of communication through SunSpec protocol in a general consensus-based cooperative strategy.

protocol is thoroughly described in [27], [28], it is not further discussed herein.

In [15], some additional technical features are defined, which demand, as compliance requirement, that an inverter using the SunSpec protocol should support, at least: *i*) TCP/IP implementation at the transport layer of the OSI model; as well as *ii*) RS-485 or Ethernet connectivity for the physical layer. This protocol also allows inverters to use low data-rate communication, comprising narrowband transmission with baud rates from 9600 bps to 115200 bps [27], which should be enough to fulfill the data transfer requirements of most of cooperative control strategies proposed for application in low-voltage MGs of small to medium size [29].

B. DNP3 Protocol

The DNP3 protocol [30], also known as IEEE 1815 − 2012 standard, is a widely used open protocol considered on higher level applications, mostly focused on data transmission within the power utility scale due to its particularly designed features and high immunity against noise. Yet, it has been widely used for the interface of DER devices with utility's supervisory systems (e.g., SCADA). Among many [30], some of the most significant features for the application of this protocol, particularly targeting cooperative control of inverters, are:

- *Broadcasting*: this is the most suitable protocol among the three comprised within [15] to transmit a single message to multiple dispersed inverters. This feature is highly notable for strategies like the ones within [10], [12] and [19];
- *Time-stamped data*: Accurate time-stamp can be performed regardless of the type of data being transmitted. Although other protocols, such as the SunSpec, are able to register time-stamps, they might not be as precise as DNP3 depending on the format of the data or the transmission rate of communication. This feature also is interesting for [10] and [19];
- *Accurate time synchronization*: Depending on the properties of the control strategy, synchronization techniques are required to timely align the communication among inverters to improve their coordinated operation. Hence, this feature may be suitable for approaches like the one in [31].

The infrastructure of the DNP3 protocol is set up mainly on the application and data-link layers of the OSI model. Even though it follows the categories defined within [15], it focuses on the formulation of groups of data according to their type (e.g., binary, analog, counter, etc), as well as to their feature of being part of the communication packet as an input (e.g., reading of a control variable) or output (e.g., a written variable to command the inverter). Also, data transmission is formulated by events, which is characterized by the occurrence of a significant change in the related system, or by an intended trigger on communication channels [30].

Compliance with this protocol is considered within [15] only comprising TCP/IP implementation and through Ethernet means. In addition, as evaluated in [32], DNP3 can perform data transmission with delays from 3 to 100 ms, depending on the type of message to be sent and the distances of the communicating nodes. This also encompasses expected latencies for data transmissions within MGs, of up to 100 ms, as discussed in [33]. Thus, in brief, this protocol should be a suitable solution for implementation of cooperative strategies, especially for approaches requiring a centralized controller, which could monitor and manipulate inverter's data just as already done by punctual utility related applications.

C. SEP2 Protocol

This last protocol, also named IEEE 2030.5 standard [34], is one of the most promising approaches employed for adequacy of data transmission means for DER-based systems. Beyond its presence in [15], it has been already incorporated as the default protocol within leading real applications of smart inverters and regulations such as the Rule 21 [35]. Such protocol focuses on the procedures and communication infrastructure for transferring data related to the monitoring and control of inverters. Its fundamental particularities are related to the application layer of the OSI model, taking advantage of TCP/IP to interact with the transport and internet layers, enabling the utilities to manage distributed devices in an integrated way [34]. One important feature to highlight is its flexibility to arrange data transfer links for supporting individual or grouped (i.e., clustered) inverters existing within the communication network. Additionally, other relevant features for MG control through cooperative inverters are:

- *Price data*: if desired, this protocol incorporates real time electricity pricing on the communicated data, which is an important matter to account while steering inverters to provide more economical gains, regardless if the focus is on the overall goal of MG or for single owners of DERs, as done in strategies [16], [36];

- *Communication to aggregators*: it is also possible to manage communication among several hubs or central units, allowing to coordinate clusters of inverters in different networks, aiming at providing higher dispatchability of MGs [12];

- *Cloud support*: since this technology is built to interact with the internet layer, it provides a singular capability to easily support communication interactions with cloud-based servers, which gives reinforcement for the implementation of strategies like [23].

The basic principle of interoperability behind the SEP2 protocol is the adoption of the RESTful architecture [34]. Hence, this communication infrastructure relies on a server-client interaction on which communication packets are formulated following the hypertext transfer protocol (HTTP) structure [20]. Although this protocol facilitates the provision of plug & play features for inverters, for such application, it may introduce latencies in communications on the order of 10's of seconds as stated in [37], which may not be suitable for cooperative strategies that totally rely on a centralized controller to provide fast frequency response.

A summarized comparison among these three protocols included in [15] is presented in Table I to demonstrate the main particularities among them. The adopted features within Table I are defined as follows. Regarding communication distances, the term "short" stands for maximum lengths of very few kilometers, "medium" and "long" refer up to tenths

TABLE I. SUMMARIZED COMPARISON OF THE PROTOCOLS IN [15].

Feature	Protocol		
	SunSpec	**DNP3**	**SEP2**
Application Range (Max. Distance)	Short to Medium	Medium to Long	Short to Long
Technology Readiness	Fair	Good	Good
Amount of Inverters Supported in a Communication Cluster	Small to Medium	Medium to Large	Large
Complexity of Implementation	Low	Medium	High

and hundreds of kilometers, respectively. Technology readiness indicates how consolidated the protocol is and how easily it can be found in commercial solutions in the electric sector. Finally, complexity of implementation is related to the required amount of code instructions. A brief discussion on constraints is then presented in relation to these protocols, focusing on their application on coordinating inverters.

IV. DISCUSSION ON CONSTRAINTS

Although standards and regulations like [15] and [35] encompass the main required features of the SunSpec, DNP3 and SEP2 protocols, some concerns still need to be further addressed when it comes to their employment as means for coordinating distributed inverters. Firstly, taking into consideration the merit of interoperability, regardless of the scope of application of these protocols, inverter manufacturers must provide support to at least one of these technologies, in order to fulfill the compliance requirements within [15]. From the commercial point of view, such requirement is beneficial since it facilitates for manufacturers to choose whichever protocol fits better their designs. Nevertheless, by having different protocol technologies embedded in inverters, which are not directly compatible with each other, interoperability becomes likely constrained. For instance, in real applications of cooperative control a MG management system would certainly consider distributed inverters from different manufacturers. In [35], a similar limitation occurs since communication interfaces of inverters are not required to support compatibility with more than one protocol. However, on the opposite side, by imposing SEP2 as the default protocol, standardization is facilitated.

Cyber security is another very relevant issue related to the employment of communication in DERs and SGs [38]. Likewise, this matter should also be accounted on the establishing of data networks for transmitting data among inverters. Each of the three previous protocols within the scope of this work presents particular features in respect to encryption and how data can be securely transmitted. Beyond the already mentioned interoperability issues, [15] does not determine any requirement related to this matter and states that data security should be defined based on the deployment of the inverters, also proposing it to occur following a mutual agreement among the participating DER owners. Such undefined condition makes more difficult to form clusters of inverters in MGs, since there is not default operational settings to securely access an inverter's data (i.e., even among devices using the same protocols, different manufacturers may adopt particular cyber security settings that could possibility limit access to control information).

CONCLUSION

Communication infrastructures are addressed in this paper focusing on their applicability for cooperative control strategies of smart inverters in MGs. The communication requirements comprised within the recently updated IEEE

978-1-7281-4181-7/19 $31.00 © 2019 IEEE

1457-2018 standard were taken as reference, leading to the highlighting of specifications from the SunSpec, DNP3 and SEP2 protocols. From discussions, it can be inferred that each of these technologies presents advantages and disadvantages upon different aspects, such as their main scope of application, the capability to accommodate different DER devices in a network, and how they formulate data transmission. Moreover, some critical comments remark that, although standards for the electrical sector are incorporating communication requirements to achieve higher interactivity among inverters, significant improvements are required in terms of interoperability. Standardization on cyber security is also critical and should be further evaluated.

As final remark, the adoption of the SunSpec protocol is seen, by the authors' point of view, as the most prominent technology to be employed on applications related to cooperative control of inverters in MGs. This is due to the flexibility of the approach to support low latency data transmission, which is structured on the device level. Additionally, through the mapping of registers provided by its data model, standardization of communication among inverters from different manufacturers might be facilitated, supporting higher interactivity and plug-&-play features.

ACKNOWLEDGMENT

The authors are grateful to FAPESP (Grants 2018/22172-1, 2017/24652-8, 2016/08645-9), NFR (Grant f261735/H30), and the agencies CAPES and CNPq for their financial support.

REFERENCES

[1] X. Fang, S. Misra, G. Xue, D. Yang, "Smart Grid – The New and Improved Power Grid: A Survey," *IEEE Commum. Surveys Tut.*, vol. 14, no 4, Dec. 2011.

[2] T. Legenthiran et al, "Intelligent Control System for Microgrids Using Multiagent System," *IEEE J. Emerg. Selec. Topics Power Electron.*, vol. 3, no 4, pp. 1036-1045, Jun. 2015.

[3] A. B. Ahmed, L. Weber, A. Nasiri, H. Hosseini, "Microgrid Communications: State of the Art and Future Trends," *in Proc.* Int. Conf. Renew. Energy Research Appl., Oct. 2014.

[4] S. Marzal, R. Sala, R. G. Medina, G. Garcera, E. Figures, "Current challenges and future trends in the field of communication architectures for microgrids," *Renew. Sust. Energy Reviews*, vol. 82, pp. 3610-3622, Nov. 2017.

[5] B. A. Zavar, E. J. P. Garcia, J. C. Vasquez, J. M. Guerrero, "Smart Inverters for Microgrid Applications: A Review," *MDPI Energies*, vol. 12, pp. 1-22, Mar. 2019.

[6] H. Han et al, "Review of Power Sharing Control Strategies for Islanding Operation of AC Microgrids," *IEEE Trans. Smart Grid*, vol. 7, no. 1, pp. 200-215, Jan. 2016.

[7] Z. Cheng et al, "To Centralize or to Distribute: That Is the Question: A Comparison of Advanced Microgrid Management Systems," *IEEE Ind. Electron. Mag.*, vol. 12, no. 1, pp. 6-24, Mar. 2018.

[8] Y. Han *et al*, "Review of Active and Reactive Power Sharing Strategies in Hierarchical Controlled Microgrids," *IEEE Trans. Power Electron.*, vol. 32, no. 3, pp. 2427-2451, Mar. 2017.

[9] U. B. Tayab, M. A. B. Roslan, J. J. Hwai, M. Kashif, "A review of droop control techniques for microgrid," *Renew. Sust. Energy Reviews*, vol. 76, pp. 717-727, Sep. 2017.

[10] T. Caldognetto, S. Buso, P. Tenti, and D. I. Brandao, "Power-Based Control of Low-Voltage Microgrids," *IEEE J. Emerg. Sel. Topics Power Electron.*, vol. 3, no. 4, pp. 1056-1066, Dec. 2015.

[11] F. Guo et al, "Distributed Secondary Voltage and Frequency Restoration Control of Droop-Controlled Inverter-Based Microgrids," *IEEE Trans. Ind. Electron.*, vol. 62, no 7, pp. 4355-4364, July 2015.

[12] Brandao et al, "Centralized Control of Distributed Single-Phase Inverters Arbitrarily Connected to Three-Phase Four-Wire Microgrids," *IEEE Trans. Smart Grid*, vol. 8, no 1, Jan. 2017.

[13] Y. Kabalci, "A survey on smart metering and smart grid communication," *Renew. Sust. Energy Reviews*, vol. 57, Jan. 2016.

[14] V. C. Gungor et al, "Smart Grid Technologies: Communication Technologies and Standards," *IEEE Trans. Ind. Inf.*, vol. 7, no 4, pp. 529-540, Nov. 2011.

[15] IEEE Standard 1547 for Interconnection and Interoperability of Distributed Energy Resources with Associated Electric Power Systems Interfaces, IEEE 1547 Std., 2018.

[16] W. Zhang, H. Ma, "Theoretical and Experimental Investigation of Networked Control for Parallel Operation of Inverters," *IEEE Trans. Ind. Electron.*, vol. 59, no 4, pp. 1961-1971, April 2012.

[17] Y. Han et al, "MAS-Based Distributed Coordinated Control and Optimization in Microgrid and Microgrid Clusters: A Comprehensive Overview," *IEEE Trans. Power Electron.*, vol. 33, no 8, Aug. 2018.

[18] G. Held, Understanding Data Communications: From Fundamentals to Networking, Joh Wiley & Sons, 3rd Ed, 2000.

[19] A. M. S. Alonso et al, "A Selective Harmonic Compensation and Power Control Approach Exploiting Distributed Electronic Converters in Microgrids," *Int. J. Elec. Power Energy Sys.*, vol 115, Aug. 2019.

[20] G. Held, Handbook of Communication Systems Management, CRC Press, 1999.

[21] J. M. Guerrero *et al*, "Hierarchical Control of Droop-Controlled AC and DC Microgrids - A General Approach Toward Standardization," *IEEE Trans. Ind. Electron.*, vol. 58, no. 1, pp. 158–172, Jan. 2011.

[22] T. D. C. Busarello, A. Mortezaei, A. Peres, M. G. Simoes, "Application of the Conservative Power Theory Current Decomposition in a Load Power-Sharing Strategy Among Distributed Energy Resources," *IEEE Trans. Ind. Appl.*, vol. 54, no 4, pp. 3771-3781 Aug. 2018.

[23] E. Harmon et al, "The Internet of Microgrids: A Cloud-Based Framework for Wide Area Networked Microgrids," *IEEE Trans. Ind. Inf.*, vol. 14, no 3, pp. 1262-1275, Mar. 2018.

[24] Q. Sun et al, "A Multi-Agent-based Consensus Algorithm for Distributed Coordinated Control of Distributed Generators in the Energy Internet," *IEEE Trans. Smart Grid*, vol. 6, no 6, Nov. 2015.

[25] J. Qin, Q. Ma, Y. Shi, L. Wang, "Recent Advances in Consensus of Multi-Agent Systems: A Brief Survey," *IEEE Trans. Ind. Electron.*, vol. 64, no 6, pp. 4972-4983, Jun. 2017.

[26] A. Hoke et al, "Setting the Smart Solar Standard: Collaborations Between Hawaiian Electric and the National Renewable Energy Laboratory," *IEEE Power Energy Mag.*, vol. 16, pp. 18-29, Dec. 2018.

[27] SunSpec Logging in SolarEdge Inverters, *SolarEdge SunSpec Alliance – Tech. Note*, pp. 1-29, 2019.

[28] SunSpec Technology Overview - Spec Alliance Interoperability Specification, *SunSpec Alliance - Technical Note*, pp. 1-8, Mar. 2015.

[29] CIGRE Working Group C6.04, Benchmark Systems for Network Integration of Renewable and Distributed Energy Resources. *CIGRE Task Force C6.04 – Tech. Note 575*, pp. 1-119, 2014.

[30] *IEEE Standard for Electric Power Systems Communications — Distributed Network Protocol (DNP3)*, IEEE 1815 Std., 2012.

[31] M. S. Golsorkhi, D. D. Lu, J. M. Guerrero, "A GPS-Based Decentralized Control Method for Islanded Microgrids," *IEEE Trans. Power Electron.*, vol. 32, no. 2, pp. 1615-1625, Feb. 2017.

[32] A. Orega et al, "Performance Evaluation of the DNP3 Protocol for Smart Grid Applications over IEEE 802.3/802.11 Networks and Heterogeneous Traffic," *in Proc.* Int. Conf. Commun., 2015.

[33] A. Angioni et al, "Coordinated Voltage Control in Distribution Grids with LTE Based Communication Infrastructure," *in Proc.* 15th Int. Conf. Environ. Electr. Eng., 2015.

[34] *IEEE Standard for Smart Energy Profile Application Protocol*, IEEE 2030.5 Std., 2018.

[35] Common Smart Inverter Profile: IEEE 2030.5 Implementation Guide for Smart Inverters, *SunSpec Alliance – Tech Note*, pp.1-63, Mar. 2018.

[36] I. U. Nutkani, P. C. Loh, F. Blaabjerg, "Droop Scheme With Consideration of Operating Costs," *IEEE Trans. Power Electron.*, vol. 29, no 3, Mar. 2014.

[37] F. Goodman et al, Module 2: Pre-Commercial Demonstration of Communication Standards for DER, San Diego Gas & Electric Company - Technical Note, Dec. 2017.

[38] S. Sridhar, A. Hahn, M. Govindarasu, "Cyber–Physical System Security for the Electric Power Grid," *Proceedings of the IEEE*, vol. 100, no 1, pp. 210-224, Jan. 2012.

250 W Single Stage Step-up Inverter Connected to the Grid

Jessika Melo de Andrade
Departament of Electrical Engineering
Federal University of Santa Catarina
Florianopolis, Brazil
jessika.melo@inep.ufsc.br

Roberto Francisco Coelho
Departament of Electrical Engineering
Federal University of Santa Catarina
Florianopolis, Brazil
roberto@inep.ufsc.br

Telles Brunelli Lazzarin
Departament of Electrical Engineering
Federal University of Santa Catarina
Florianopolis, Brazil
telles@inep.ufsc.br

Abstract—**This paper addresses the study of the switched-capacitor differential boost inverter (SCDBI) for the connection to the electrical grid. The corroboration of the theoretical analysis is made through experimental results obtained in a 250 W prototype connected to a 127 V RMS voltage grid, considering a variation rate in input voltage from 50 to 70 V. The topology efficiency peak was 90% and grid current presented THD less than 3%.**

Keywords— Boost Inverter; Connected to the grid; Control and Modelling; Switched Capacitor; Linearization.

I. INTRODUCTION

Renewable sources can be connected to the grid through a single or multi-stage system. In photovoltaic systems, when power level is above 200 W, energy processing is usually performed from two stages. Usually, these systems are made of a boost-type dc-dc converter connected to a buck-type inverter [1]. However, this approach usually results in weight gain, volume, and reduced efficiency and reliability [2].

On the other hand, aiming for improvements over double-stage topologies, there are non-isolated single-stage inverters (transformerless) [3], as inverters boost [2], buck-boost [4], zeta [5], among others. Such inverters are derived from the differential connection of two dc-dc converters [4]-[7]. In order to increase the static gain of the non-isolated inverters, in [8]-[9] is proposed the integration between switched capacitor cells (SC) and conventional differential boost inverter (DBI) [2]. The resulting topology is named as switched capacitor differential boost inverter (SCDBI). The step-up gain characteristic and the capability to add more SC cells to improve the gain makes it attractive for application in systems connected to the grid. In this context, this paper addresses the use of the SCDBI in applications that involve grid-connection.

II. INVERTER CONNECTED TO THE GRID

The SCDBI connected to the grid is depicted in Fig. 1 in which is highlighted the SC cell (shaded trace) and the conventional boost inverter (black trace). The proposed inverter consists of two hybrid bidirectional boost dc-dc converters, named as sub-converters *A* and *B*.

The output voltage of each converter (also named herein of sub-converter) is always positive and composed of two components: one continuous and one sinusoidal. The continuous portion is equal in both sub-converters, and the alternating components are 180° phase-shifted from each other. In this configuration, after the differential connection, the dc components are canceled, and the alternating components are added.

It should be highlight that, the rated power of this inverter, should be the range up to 1 kW, due the switched-capacitor

high peak current and the high current on the input inductors and boost switches (S_{1a} and S_{2a}).

The theoretical static gain of the SCDBI in continuous conduction mode (CCM) is given as a function of SC cell gains (k), as is described by:

$$\frac{v_o}{V_i} = \frac{v_a - v_b}{V_i} = \frac{k(2d-1)}{d(1-d)}.$$
(1)

A. Design of Passive Elements

The input inductors are defined from the current ripple specification (Δ_{iL}) and they are expressed by:

$$L_a = L_b = \frac{V_i D}{f_s I_{Lpk} \Delta i_L},$$
(2)

where V_i is the input voltage, D is the maximum duty-cycle, f_s is the switching frequency and I_{Lpk} is the peak value of the current in the inductors.

The output filter inductance is found from the resonance frequency (f_{LC}) between the inductor (L_o) and the equivalent capacitance of the multiplier cell (C_{eq}). It is given by:

$$L_o = \frac{2k^2}{\left(2\pi f_{LC}\right)^2 C_{eq}}.$$
(3)

The capacitances of SC cells are specified from the operating mode of the multiplier cell (Fig. 2). According to [10], the mode that provides the best cost/benefit between efficiency and volume is the partial charge mode, Fig. 2 (c). The operation of the converter in the partial charge mode is guaranteed when:

$$C = C_1 = C_2 = C_3 \geq \frac{0,1}{f_s\left(R_{SE} + R_{ds(on)}\right)},$$
(4)

where R_{SE} is the equivalent series of resistance of the capacitor and $R_{ds(on)}$ is the conduction resistance of the switch.

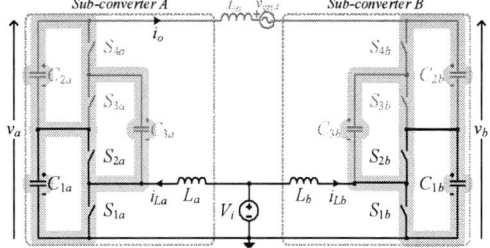

Fig. 1 – SCDBI inverter (boost inverter plus SC cells) connected to the grid.

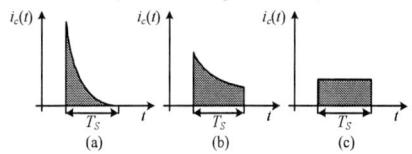

Fig. 2 – Modes of operation of the switched capacitor cell: (a) complete charge, (b) partial charge and (c) no charge.

B. Linearization Technique

The output voltage and / or the output current of boost type inverters are susceptible to distortions caused by the nonlinear behavior of these converters, which implies additional control efforts [11]-[12]. As an alternative, this paper linearizes the static gain of each sub-converter. As result, both partial voltages v_a and v_b are linearized, reducing the THD of the partial voltages and consequently of the differential output voltage.

The linearization function is given by:

$$d_{boost} = \frac{\alpha d + \beta - 1}{\alpha d + \beta}, \quad (5)$$

where α and β are the angular and linear coefficients of the linearization function and d represents the duty cycle signal provided from the control loop and applied to the linearization block.

III. MODELLING AND CONTROL

A. Control-oriented modelling

According to [13], n-order hybrid boost converters can be modeled considering only the dominant dynamics, as shown in Fig. 3. The differential connection of two of this equivalent converter, as seen in Fig. 4, represents the proposed simplified model to the SCDBI (referenced to the low voltage side).

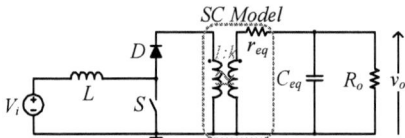

Fig. 3 – Hybrid boost converter equivalent circuit.

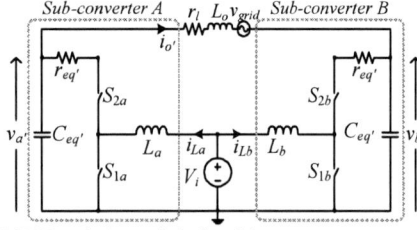

Fig. 4 – SCDBI equivalent switched model.

The voltages and currents on the switches can be described as a function of their near-instantaneous average values [14]. On replacing the switches by sources that described such values, it is obtained the mean large-signal model, which represents the SCDBI. In addition, on applying a small perturbation on these variables, it is found: i) a "dc model", which describes the converter at the operation point (Fig. 5 (a)), and; ii) a small signals "ac model", which represents its dynamics (Fig. 5 (b)). On the analyzing of "ac model", the transfer function that relates i_o (variable to be controlled) and d (control variable) follow the format given by:

$$G_{id}(\text{s}) = \frac{\hat{i}_o}{\hat{d}} = \frac{b_3 s^3 + b_2 s^2 + b_1 s + b_0}{a_5 s^5 + a_4 s^4 + a_3 s^3 + a_2 s^2 + a_1 s + a_0}. \quad (6)$$

In which the coefficients are summarized in Table I.

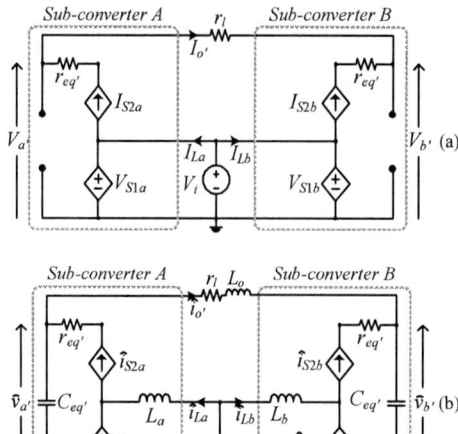

Fig. 5 – SCDBI equivalents model: (a) dc model and (b) ac model.

The obtained transfer function was validated by simulation using the software PSIM®. A small perturbation of 1% was applied on the duty cycle, in order to verify the grid current dynamic. Fig. 5 shows a comparison between the dynamic responses of both, switched (*Io_switched*) and the small signal average (*Io_average*) models, considering the step response (Fig. 6 (a)) and the frequency response (Fig. 6 (b)). It should be noticed that the average model satisfactorily represents the switched model.

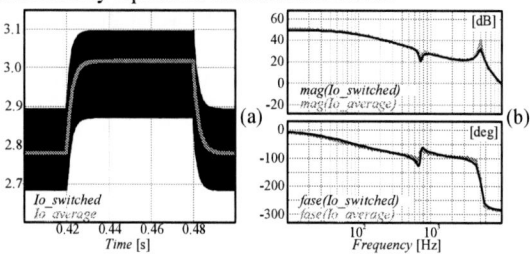

Fig. 6 – Validation of the transfer function: (a) time domain and (b) frequency domain.

B. Control

A control loop was used to regulate the current injected into the electric grid, as shown in Fig. 7. A feedforward loop is suggested as well to re-feed the grid voltage, seen as disturbance by the main control loop. The controller chosen is a proportional integral (PI), with addition of an extra pole, written as:

$$C(s) = \frac{K_c (s + \omega_z)}{s (s + \omega_p)}. \quad (7)$$

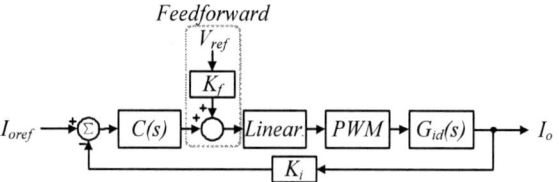

Fig. 7 – Block diagram of the strategy employed for the control of SCDBI connected to the electric grid.

978-1-7281-4181-7/19 $31.00 © 2019 IEEE

TABLE I. TRANSFER FUNCTION COEFFICIENTS

b_3	$-C_{eq}'L_a^2 k\left(I_{La}+I_{Lb}\right)$
b_2	$C_{eq}'L_aV_a'k\left(1-D\right) - C_{eq}'I_{Lb}L_a kr_{eq}'\left(1-D\right) + ...$ $C_{eq}'DL_a k\left(V_b' - I_{La}r_{eq}'\right)$
b_1	$-D^2 L_a k\left(I_{La}+I_{Lb}\right) + I_{Lb}L_a k\left(2D-1\right) + ...$ $C_{eq}'Dkr_{eq}'\left(V_a'+V_b'\right) - C_{eq}'D^2 kr_{eq}'\left(V_a'+V_b'\right)$
b_0	$DV_b'k + D^2 k\left(V_a'-2V_b'\right) + D^3 k\left(V_b-V_a\right)$
a_5	$C_{eq}'^2 L_a^2 L_o$
a_4	$C_{eq}'^2 L_a L_o r_{eq}' + C_{eq}'^2 L_a^2 r_l$
a_3	$L_o C_{eq}'^2 r_{eq}' D\left(1-D\right) - 2L_o C_{eq}'L_a D\left(1-D\right) + 2k^2 C_{eq}'L_a^2 + ...$ $L_o C_{eq}'L_a + C_{eq}'^2 L_a r_l r_{eq}'$
a_2	$L_o C_{eq}'Dr_{eq}'\left(1-D\right) + 2L_a k^2 C_{eq}'r_{eq}' + C_{eq}'^2 Dr_{eq}'r_l\left(1-D\right) - ...$ $2C_{eq}'L_a Dr_l\left(1-D\right) + C_{eq}'L_a r_l$
a_1	$L_a k^2 + D^2 L_o\left(1-D\right)^2 - 2DL_a k^2\left(1-D\right) + ...$ $2C_{eq}'Dk^2 r_{eq}'^2\left(1-D\right) + C_{eq}'Dr_{eq}'r_l\left(1-D\right)$
a_0	$k^2 r_{eq}'D\left(1-D\right) + r_l D^4 - 2r_l D^3 + r_l D^2$

IV. EXPERIMENTAL RESULTS

A 250 W prototype was designed to verify the proposed study. The specifications are described in Tables II, and a prototype photograph is shown Fig. 8.

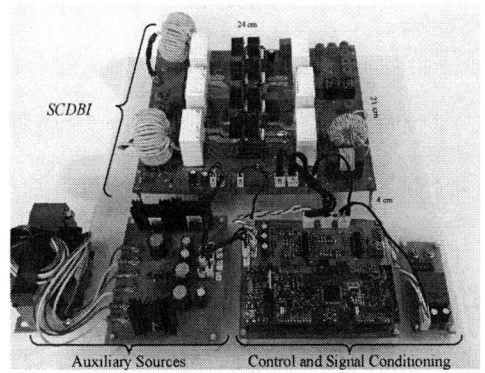

Fig. 8 – Prototype: boost inverter (250 W), auxiliary sources, control and signal conditioning.

TABLE II. PROTOTYPE DESIGN SPECIFICATION

Parameters	Values
Input voltage (V_i)	50 – 70 V
Output voltage RMS value (V_{orms})	127 V
Output power (P_o)	250 W
Switching frequency (f_s)	50 kHz
Gain of the multiplier cell (k)	2
Maximum duty cycle (D)	0.65

The SCDBI in closed loop with linearization block enabled was verified connected to the electrical grid. The grid voltage (v_{grid}), the injected current (i_o), at rated power (250 W) and the input voltage V_i equal 70, 60 and 50 V are shown in Figures 9, 10 and 11, respectively. The injected current THD for V_i equal 70 V, 60 V and 50 V was 1.61%, 2% and 2.43%, in this order, considering grid voltage THD of 3.36%.

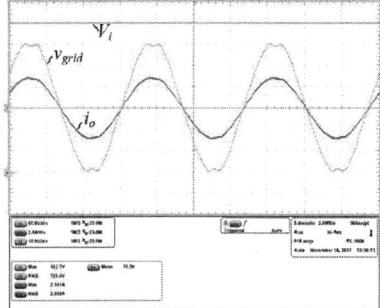

Fig. 9 – Experimental results for V_i = 70 V: grid voltage v_{grid} (60 V/div), current injected into the grid i_o (2 A/div) and input voltage V_i (10 V/div), time base (5 ms/div).

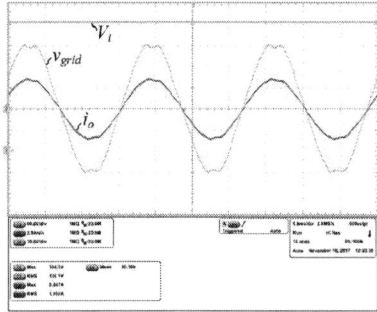

Fig. 10 – Experimental results for V_i = 60 V: grid voltage v_{grid} (60 V/div), current injected into the grid i_o (2 A/div) and input voltage V_i (10 V/div), time base (5 ms/div).

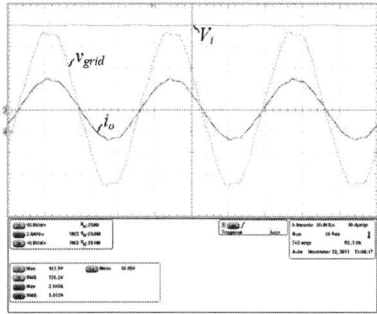

Fig. 11 – Experimental results for V_i = 50 V: grid voltage v_{grid} (50 V/div), current injected into the grid i_o (2 A/div) and input voltage V_i (10 V/div), time base (5 ms/div).

This study analyzed the current injected harmonic spectrum into the power grid at nominal power with input voltage equal to 50, 60 and 70 V. Figure 12 shows the worst case result (V_i = 50V). It is possible to verify that current injected into the electric grid complies with the limits established by the international norms IEC61727 and IEEE1547 [15], which dictate the distortion maximum limits per harmonic component.

The inverter was subjected to a ± 50% current reference step, as seen in Fig. 13 (a) and (b). Figure 13 (a) details the grid voltage and current injected during the reference reduction, while Fig. 13 (b) illustrates the behavior of the inverter during the increase of the reference. In both cases the controller response is satisfactory.

The efficiency curve of the topology is shown in Fig. 14. The maximum efficiency was 90%, and it occurred at around 100 W.

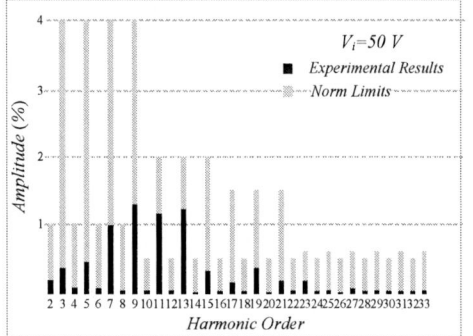

Fig. 12 – Harmonic analysis at nominal power and input voltage at 50 V.

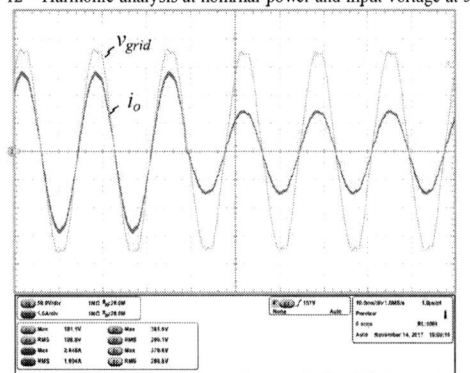

(a)

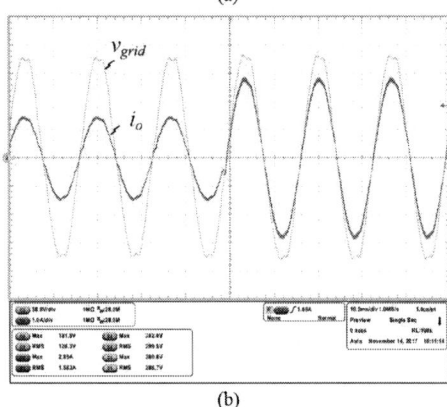

(b)

Fig. 13 – Experimental results: ± 50% grid current reference step: (a) details of grid voltage and grid current i_g - time base (10 ms/div) - during the negative reference step; (b) details of grid voltage and grid current i_g - time base (10 ms/div) - during the positive reference step.

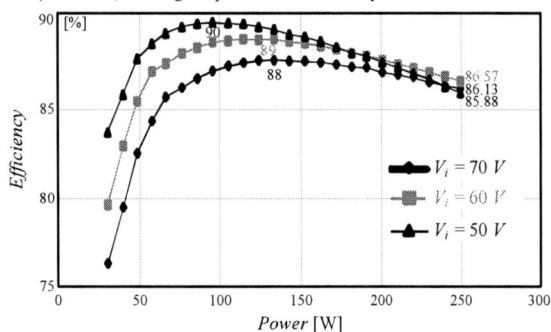

Fig. 14 – Efficiency curve of the prototype connected to the grid.

V. CONCLUSIONS

The proposed step-up inverter was verified in extensive experimental results for 127 V RMS. The theoretical study was validated and the inverter presented a good characteristic regarding efficiency, operation, and stability, which make it as an adequate solution for step-up inverters until 250 W.

The results from the converter operating connected to the grid in closed loop were also evaluated, confirming the inverter is capable of providing THD lower than 3% and efficiency peak of about 90%.

ACKNOWLEDGMENT

The authors would like to thank the CAPES - PROEX and CNPq (Process No. 422276/2016-2) for financial support.

REFERENCES

[1] G. R. Walker and P. C. Sernia, "Cascaded DC-DC converter connection of photovoltaic modules," *IEEE Transactions on Power Electronics*, vol. 19, no. 4, pp. 1130-1139, July 2004.

[2] R. O. Caceres; I. Barbi, "A boost DC-AC converter: analysis, design, and experimentation," *IEEE Transactions on Power Electronics*, vol.14, no.1, pp.134,141, Jan 1999

[3] B. N. Alajmi, K. H. Ahmed, G. P. Adam, B. W. Williams, "Single-Phase Single-Stage Transformer less Grid-Connected PV System," *IEEE Transactions on Power Electronics*, vol. 28, no. 6, pp. 2664-2676, June 2013.

[4] S. Xu, R. Shao, L. Chang and M. Mao, "Single-Phase Differential Buck-Boost Inverter with Pulse Energy Modulation and Power Decoupling Control," *IEEE Journal of Emerging and Selected Topics in Power Electronics.*

[5] G. L. Piazza , I. Barbi, "New Step-Up/Step-Down DC–AC Converter," IEEE Transactions on Power Electronics, vol. 29, pp. 4512-4520, 2014.

[6] S. Huang, F. Tang, Z. Xin, Q. Xiao and P. C. Loh, "Grid-Current Control of Differential Boost Inverter with Hidden LCL Filters," *IEEE Transactions on Power Electronics*, 2018.

[7] W. Yao, X. Wang, P. C. Loh, X. Zhang and F. Blaabjerg, "Improved Power Decoupling Scheme for a Single-Phase Grid-Connected Differential Inverter With Realistic Mismatch in Storage Capacitances," *IEEE Transactions on Power Electronics*, vol. 32, no. 1, pp. 186-199, Jan. 2017.

[8] G. V. Silva, R. F. Coelho, T. B. Lazzarin, "Switched capacitor boost inverter," in *IEEE 25th International Symposium on Industrial Electronics*, 2016, pp. 528-533.

[9] G. V. Silva, R. F. Coelho, T. B. Lazzarin, "Switched-capacitor differential boost inverter: Static gain and generalized structure," in *12th IEEE International Conference on Industry Applications*, 2016, pp. 1-8.

[10] P. J. S. Costa, C. H. Illa Font and T. B. Lazzarin, "Single-Phase Hybrid Switched-Capacitor Voltage-Doubler SEPIC PFC Rectifiers," *IEEE Transactions on Power Electronics*, vol. 33, no. 6, pp. 5118-5130, June 2018.

[11] D. Cortes, N. Vazquez, J. Alvarez-Gallegos, "Dynamical Sliding-Mode Control of the Boost Inverter," *IEEE Transactions on Industrial Electronics*, vol. 56, pp. 3467-3476, 2009.

[12] P. Sanchis, A. Ursaea, E. Gubia, L. Marroyo, "Boost DC-AC inverter: a new control strategy," *IEEE Transactions on Power Electronics*, vol. 20, pp. 343-353, 2005.

[13] G. V. Silva, R. F. Coelho and T. B. Lazzarin, "State space modeling of a hybrid Switched-Capacitor boost converter," in *2015 IEEE 13th Brazilian Power Electronics Conference and 1st Southern Power Electronics Conference*, Fortaleza, 2015, pp. 1-6.

[14] R. Middlebrook, S. Cuk, "A general unified approach to modelling switching-converter power stages," in *Power Electronics Specialists Conference*, 1976, pp. 18-34.

[15] H. H. Figueira, H. L. Hey, L. Schuch, C. Rech, L. Michels, "Brazilian grid-connected photovoltaic inverters standards: A comparison with IEC and IEEE," in Proc. of *IEEE 24th International Symposium on Industrial Electronics*, Buzios, 2015, pp. 1104-1109.

Experimental Analysis for Low Power Series-Series Compensated Inductive Power Transfer System

Macklyster Lãnucy Scherre Stofel de Lacerda
Electrical Engineering Department
Federal University of Espírito Santo
Vitória, Brazil
macklysterstofel@hotmail.com

Tatiana Saviato Macedo
Electrical Engineering Department
Federal University of Espírito Santo
Vitória, Brazil
saviato@gmail.com

Denizar Cruz Martins
Electrical Engineering Department
Federal University of Santa Catarina
Santa Catarina, Brazil
denizar@inep.ufsc.br

Walbermark Marques dos Santos
Electrical Engineering Department
Federal University of Espírito Santo
Vitória, Brazil
walbermark.santos@ufes.br

Abstract— **The research in wireless power transfer has grown in the last years, motivated mainly by the convenience and the security that technology offers, including applications where the feeding of equipment by cables is technically complicated or dangerous. Among the wireless transfer technologies, the Inductive Power Transfer (IPT) has been highlighted in several works in the literature. However, because it is an emerging technology, the subject still has an opportunity for analysis. In this paper, experimental results of a 20 W IPT system were used to validate a simplified Thevenin equivalent circuit (which directly relates the input and output DC voltages) and a coupling factor k equation as a function of the circuit measurable electrical parameters. Power has been limited due to available Litz wire specifications. The equations are supported with computational simulations and experimental results.**

Keywords— *Inductive Power Transfer, Thevenin Equivalent Circuit, Experimental Results.*

I. INTRODUCTION

Many scientists contributed to the development of Wireless Power Transfer (WPT) technology for a large number of applications. In 1864, scientist James C. Maxwell assembled twenty equations capable of describing all magnetic and electrical phenomena. Maxwell's Equations boosted the study of electromagnetic waves [1]. In 1884 John H. Poynting formulated a theorem that resulted in the Poynting vector. This vector represents the density and direction of the power flow [2]. His studies gained historical importance due to his contributions to the theory on energy and electromagnetic waves. Subsequently, many other scientists have studied data transfer and energy. In 1891, scientist Nikola Tesla studied and experimented in the field of wireless energy transfer, providing various contributions for the subject [3].

The emergence of new technologies, such as mobile communication devices [4], electric vehicles [5], medical implants [6] and others, as well as the demand for WPT surveys has increased. Inductive Power Transfer (IPT) is currently the most studied and applied method of WPT, presenting advantages such as the use for various distances and powers [7].

To mathematically represent the inductive energy transfer system, the articles [8]–[12] present equations and models that use parameters such as coupling factor and operating frequency to relate the input and output voltages and currents. Through these equations it is possible to estimate the physical behavior of the IPT according to its configuration.

Among the models representing the IPT, Thevenin equivalent circuit may relate input and output voltages by calculating Thevenin equivalent voltage and impedance using the IPT parameters.

Some articles [13]–[17] use Thevenin equivalent models to relate transmission and reception. However, they present complex equations for the calculation of voltage and equivalent impedance. In [13] a model is presented where only Thevenin equivalent impedance is found, and the effects of coil resistances are neglected. In a different approach, [14], [16], [17] present a model in which equivalent voltage and resistance are calculated for maximum power and efficiency frequency tracking. However, it uses capacitive and inductive impedance ratios, which may be suppressed if the IPT is operating at the resonant frequency. In [18] Thevenin's voltage and resistance are calculated considering that the system is operating at the resonant frequency, with appropriate simplifications. However, the proposed model does not relate the input DC voltage and load resistance values directly to Thevenin voltage and resistance.

In order to simplify the mathematical model that represents the IPT, this article presents a Thevenin equivalent circuit between the input DC source and the output DC voltage at the load, and an equation for determining the coupling factor as a function of the electrical circuit parameters. The IPT configuration analyzed is showed in Fig. 1. Experimental results of an IPT prototype is confronted with theoretical analyses.

II. THEORETICAL DEVELOPMENT

A. The IPT system

In Inductive Power Transfer (IPT) systems, power is transferred through the inductive coupling between the primary or transmission coil and the secondary or receiving coil. The amount of magnetic flux enclosed in the secondary coil is given by the mutual inductance M [19], whose well-known relationship is presented in (1).

$$M = k\sqrt{L_1 L_2} \qquad (1)$$

Where k, L_1 e L_2 are respectively, the coupling factor and the self-inductances of the primary and secondary coils.

Figure 1 shows the IPT system diagram developed in this article. A full bridge inverter operating at high frequency imposes a symmetric square wave voltage V_P at the inductive link input. A full-wave rectifier together with an RC filter

978-1-7281-4181-7/19 $31.00 © 2019 IEEE

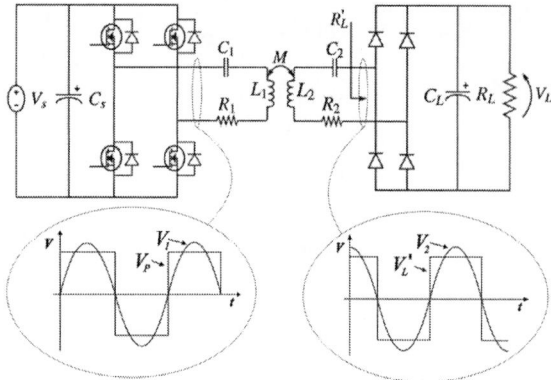

Fig. 1. IPT System Diagram

form the transmission system load. V_1 and V_2 are the fundamental voltage components at the inductive link input V_P and output $V_L{}'$, respectively.

B. The IPT Circuit Model

Considering only the inverter fundamental voltage component (since only the fundamental component transmits active power) and all the components reflected to their respective inductive link sides, it may be achieved the circuit shown in Fig. 2 [9], which results in the equations (2) and (3) [8], [10], [20].

$$V_1 - I_1\left(R_1 + j\omega(L_{lk1} + M) - \frac{j}{\omega C_1}\right) + I_2 j\omega M = 0 \quad (2)$$

$$I_1 j\omega M - I_2\left(R_2 + R_L{}' + j\omega(L_{lk2} + M) - \frac{j}{\omega C_2}\right) = 0 \quad (3)$$

Where V_1 represents the fundamental voltage component imposed on the inductive link, I_1 the transmission circuit current, I_2 the receiving circuit current, ω the inverter switching angular frequency, M the mutual coupling inductance, R_1 and R_2, L_{lk1} and L_{lk2}, C_1 and C_2 the resistances, dispersion inductances and the transmission and reception coils compensation capacitances, respectively.

In the resonance frequency the capacitive reactance and the inductive reactance cancel out, resulting in the circuit shown in Fig. 3 and the simplified equations presented in (4) and (5) [21].

$$V_1 - I_1 R_1 + I_2 j\omega M = 0 \quad (4)$$

$$I_1 j\omega M - I_2\left(R_2 + R_L{}'\right) = 0 \quad (5)$$

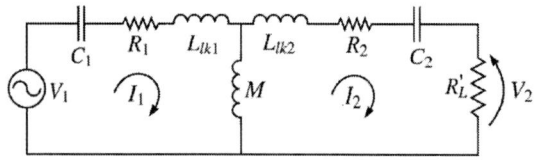

Fig. 2. IPT system equivalent circuit

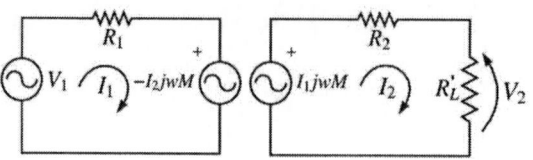

Fig. 3. Simplified equivalent circuit

Substituting (5) into (4), the relationship between the output voltage and the input voltage is given in (6) [22].

$$\frac{V_2}{V_1} = \frac{j\omega M R_L{}'}{R_1\left(R_2 + R_L{}'\right) + \left(\omega M\right)^2} \quad (6)$$

Where $V_2 = I_2 R_L{}'$.

Equation (6) shows that the inductive link alternating voltages are out of phase by $\pi/2$ rad. Thus, the relationship between the voltage effective values is given by (7).

$$\frac{V_{2rms}}{V_{1rms}} = \frac{\omega M R_L{}'}{R_1\left(R_2 + R_L{}'\right) + \left(\omega M\right)^2} \quad (7)$$

The reflected load resistance $R_L{}'$ is determined by (8) [12], [23].

$$\frac{R_L{}'}{R_L} = \frac{8}{\pi^2} \quad (8)$$

C. Thevenin Equivalent Circuit

According to Fig. 1, the inductive link input and output voltages are square waves. The ratio between the square wave effective value and its fundamental effective value is constant and equal to $2\sqrt{2}/\pi$ [12], [24]–[26]. Moreover, the inverter DC input voltage V_S and rectifier output voltage V_L are constant and equal to the effective values of their respective square waves. Therefore, (7) is used to directly relate the DC voltages of the source V_S and load V_L.

Substituting (8) into (7) and rearranging the terms yields (9).

$$\frac{V_L}{V_T} = \frac{R_L}{R_T + R_L} \quad (9)$$

Where V_T is the Thevenin voltage given by (10) and R_T is the Thevenin resistance given by (11).

$$V_T = V_S \frac{\omega M}{R_1} \quad (10)$$

$$R_T = \frac{\pi^2}{8}\left(R_2 + \frac{\left(\omega M\right)^2}{R_1}\right) \quad (11)$$

Equation (9) may be interpreted as a Thevenin equivalent circuit, shown in Fig. 4, which relates the DC load voltage V_L with the Thevenin equivalent voltage V_T.

978-1-7281-4181-7/19 $31.00 © 2019 IEEE

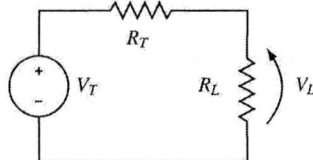

Fig. 4. Thevenin equivalent circuit

D. Equation of the Coupling Factor Using Circuit Variables

The direct measurement of the inductive link coupling factor of an IPT prototype is a work that requires expensive equipment capable to measure the magnetic flux intensity. An alternative is estimating it through the assembled or simulated circuit parameters. The approximation may be obtained through an easily acquired mathematical equation, substituting (1) into (7) and considering only the resulting equation positive root, it is concluded that k may be determined by (12).

$$k = \frac{4R_L V_S}{\pi^2 \omega \sqrt{L_1 L_2} V_L} + \sqrt{\left(\frac{4R_L V_S}{\pi^2 \omega \sqrt{L_1 L_2} V_L}\right)^2 - \frac{R_1\left(R_2 + 8R_L/\pi^2\right)}{\omega^2 L_1 L_2}} \quad (12)$$

III. COMPUTATIONAL SIMULATIONS

In order to validate the presented theoretical equation, an IPT Series-Series computer simulation was implemented in *Simulink MATLAB®* software, whose parameters are presented in Table 1. The resistance, inductance and capacitance values are considered equal for both transmitter circuit and receiver circuit.

TABLE I.   EXPERIMENTAL PARAMETERS

Frequency	Prototype Parameters			
	Coil Resistance	Inductance	Capacitance	Load Resistances
125 kHz	1.03 Ω	34.6 µH	46.85 nF	15.7 Ω 46.4 Ω
250 kHz	1.03 Ω	32.2 µH	12.59 nF	69.4 Ω 92.7 Ω

In the simulations, the load resistance is varied according to the presented values in Table I, it is observed the load voltage, the coupling factor and the transmission frequency variation.

Fig. 5 shows the load voltage V_L for the circuit in Fig. 1, when it is varied: load resistance R_L, coupling factor k and transmission frequency f. In the analyzed cases, the largest relative error between the results of (9) and the simulation, in relationship to the simulated values, was 5.89 %, which is equivalent to 2.09 V.

Fig. 6 shows the load voltage $V_L(pu)$ versus the load resistance $R_L(pu)$ (both in pu, using the source voltage and the coil resistance as the base) and coupling factor k.

From Fig. 6 (a) and Fig. 6 (b) it is noted that the voltage is directly proportional to the load resistance $R_L(pu)$, i.e. for the same coupling factor k and transmission frequency f, the increase/decrease of one is followed by the other's increase/decrease. This behavior is predicted by (9), where for a fixed value of R_T, V_L is directly proportional to R_L, so when

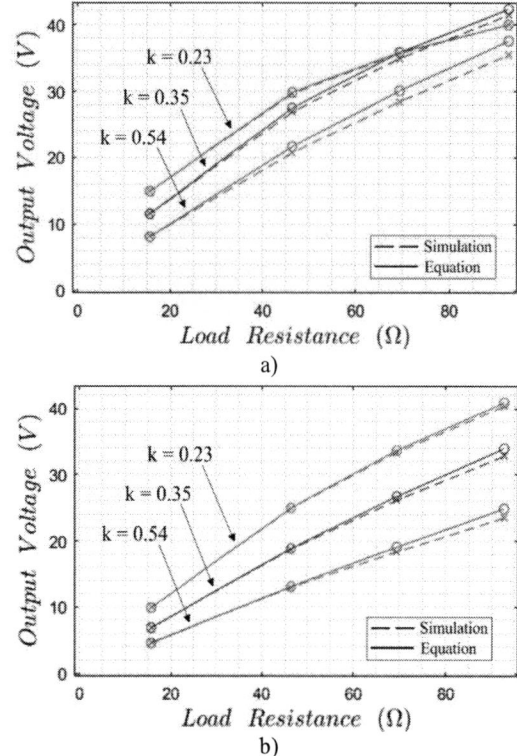

Fig. 5. Results comparison between (9) and computational simulation for k = 0.23 (distance of 3 cm), k = 0.35 (distance of 2 cm), and k = 0.54 (distance of 1 cm) at a) 125 kHz and b) 250 kHz.

$R_L = R_T$, the voltage that provides the load maximum power is transferred.

When the load resistance $R_L(pu)$ is fixed, in Fig. 6 (a) and Fig 6 (b) it is noted that a coupling factor variation Δk causes a voltage variation $\Delta V_L(pu)$. Therefore, for $R_L(pu)$ and f fixed, there is a k value that provides a load maximum voltage and, consequently, the maximum power is observed. The maximum load power coupling factor k_{mPo} is obtained from the denominator of (9) and may be calculated by (13).

$$k_{mPo} = f^{-1}\left(\frac{1}{2\pi}\sqrt{\frac{R_1\left(R_2 + 8R_L/\pi^2\right)}{L_1 L_2}}\right) \quad (13)$$

According to (13) and comparing the Fig. 6 (a) to Fig. 6 (b) it may be observed that, for fixed $R_L(pu)$, an increase of f causes the decrease in the same proportion of the value of k_{mPo}, not leading to changes in the load maximum voltage, retaining the maximum amount of power that may be transmitted. It is concluded that, according to the analyzed data, the increasing of operation frequency allows the power transmission at greater distances, since the same voltage value is transmitted with a lower coupling.

IV. EXPERIMENTAL RESULTS

A. The Prototype

Fig. 8 shows the developed prototype. The DC source used was the Skill-Tec model SKFA-05S and the data was acquired

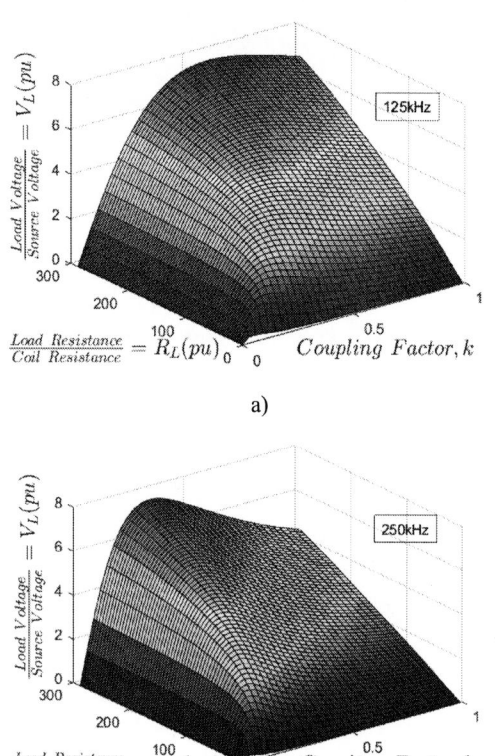

a)

b)

Fig. 6. Load voltage behavior under load resistance and coupling factor variation at a) 125 kHz and b) 250 kHz.

through a Tectronix TPS2024B oscilloscope connected via USB cable to a laptop.

For the complete bridge inverter assembly it was used MOSFETS BUK9507-30B, IR2110 drivers and electrolytic capacitors in the DC bus. The control signals are produced by ESP-32 microcontroller. Table II summarizes the main components specifications used in the prototype fabrication.

TABLE II. PROTOTYPE COMPONENTS

Element	Features	
Capacitive compesations	125 kHz	8 polypropylene capacitors B32692 4x8.2 nF, 1x6.8 nF, 2x2.7 nF and 1x1.8 nF
	250 kHz	5 polypropylene capacitors B32692 4x2.7 nF and 1x1.8 nF
Inductors	12 turns of Lits wire 41 x 20 AWG and internal radius of 10 cm	
Rectifier	4 UF5408 diodes and 100 µF electrolytic capacitor	
Load	30 resistors of 3 W and 4.7 Ω	

Fig. 7 shows the IPT prototype waveforms. As shown previously, in Fig. 7 (a) and Fig. 7 (e) the inductive link input voltage V_1 and the inductive link output voltage V_2 are out of phase by $\pi/2$ rad. However, this delay is observed in all inductive link magnitudes. Fig. 7 (b) and Fig. 7 (f) shows the capacitive compensation voltages. These voltages are sinusoidal because they are proportional to the transmit and receive current's integrals. The others square wave harmonic components of V_P appear on the inductors, as shown in Fig. 7 (c) and Fig. 7 (g).

The power transmission occurs only at the fundamental component V_1 of the primary voltage V_p, thereby the transmission and reception currents are sinusoids at the transmission frequency, as shown in Fig. 7 (d) and Fig. 7 (h).

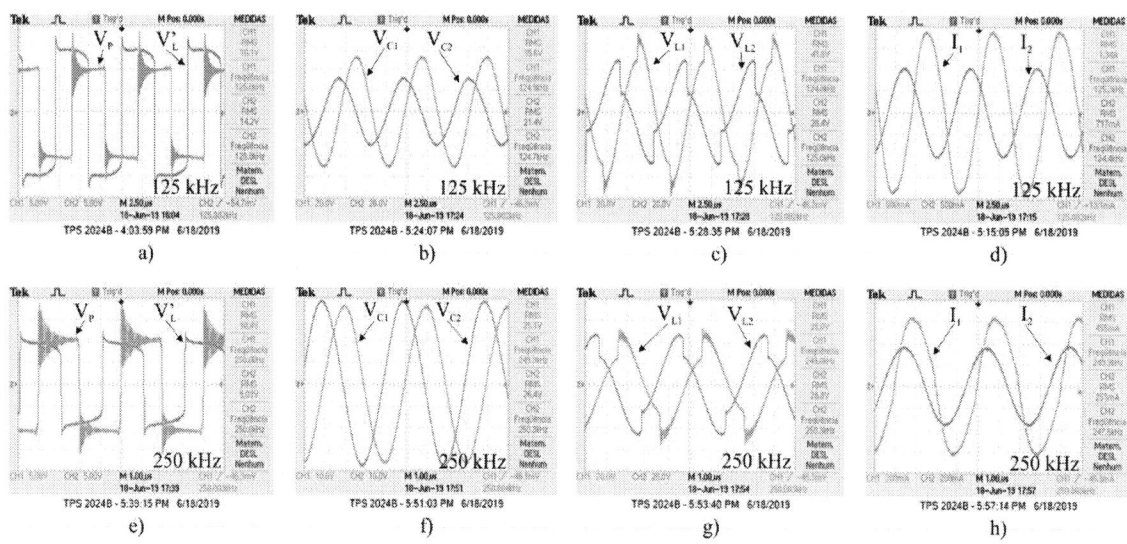

Fig. 7. Prototype waveforms with $V_S = 10$ V, $R_L = 15.7$ Ω, $f = 125$ kHz from (a) to (d) and $f = 250$ kHz from (e) to (h). In orange is showed the transmission waveforms and in blue, the reception waveforms. a) Input V_1 and output V_2 inductive link voltage at 125 kHz; b) Capacitive compensations voltage at 125 kHz; c) Inductors voltage at 125 kHz; d) Transmission and reception currents at 125 kHz; e) Input V_1 and output V_2 inductive link voltage at 250 kHz; f) Capacitive compensations voltage at 250 kHz; g) Inductors voltage at 250 kHz; and h) Transmission and reception currents at 250 kHz.

978-1-7281-4181-7/19 $31.00 © 2019 IEEE

Fig. 8. IPT prototype

B. Thevenin Circuit Validation

In order to validate the Thevenin equivalent circuit shown in Fig. 4, it was simulated in *Simulink* the circuit shown in Fig. 9. The simulation is composed of a DC voltage source V_T, calculated at (10); a resistor R_T, calculated at (11) and a variable resistor representing the load resistance R_L. The other components are simulation tool requirements and they are present exclusively for the simulator needs. A block, which outputs the simulation time, was used as the load resistance R_L control signal. The voltage behavior at the load resistor is similar to the observed in an IPT system.

Fig. 10 shows a comparison between simulation and experimental results of the load voltage behavior V_L when it varies: load resistance R_L, operating frequency f and coupling factor k. In the analyzed cases, the largest relative error, in relationship to the experimental values, was 7.51 %, which is equivalent to 0.69 V. In addition, the experimental load power and efficiency were calculated. Fig. 11 shows the resulting graphs. The largest load power relative error, in relationship to the experimental power values, was 15.58 %, which equals 0.84 W.

V. CONCLUSIONS

An IPT prototype Series-Series system was used to validate a DC gain equation, a coupling factor equation with circuits parameters, and a simplified Thevenin equivalent circuit presented in this article.

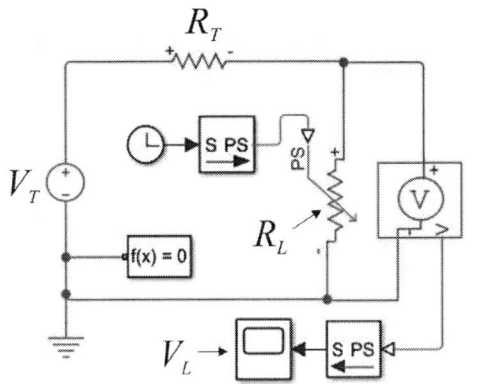

Fig. 9. Simulated Thevenin equivalent circuit.

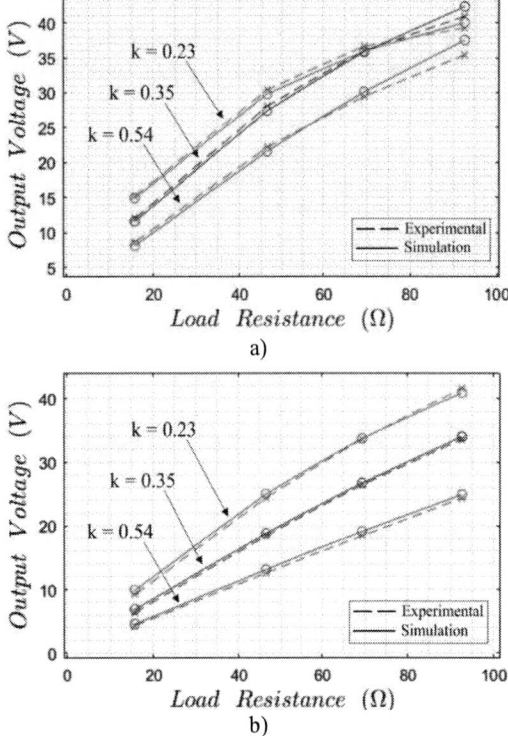

Fig. 10. Comparison between the load voltage provided by Fig. 9 simulation and experimental data for k = 0.23 (distance of 3 cm), k = 0.35 (distance of 2 cm), and k = 0.54 (distance of 1 cm) at a) 125 kHz and b) 250 kHz.

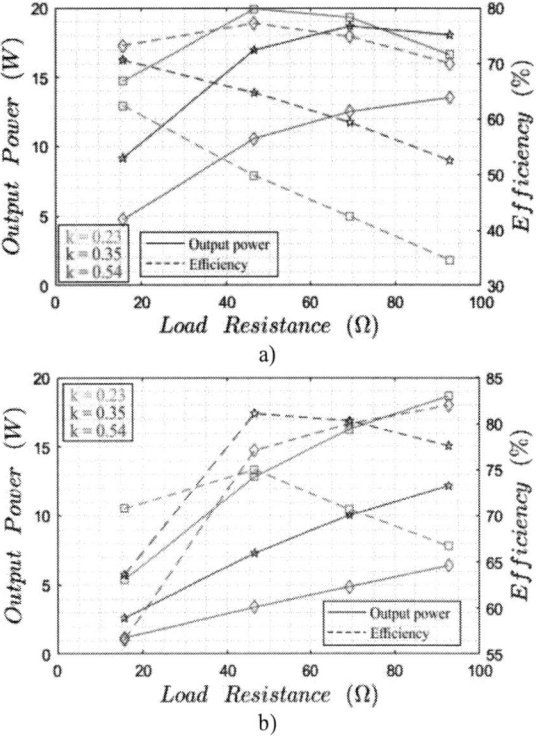

Fig. 11. Experimental output power and efficiency for k = 0.23 (distance of 3 cm), k = 0.35 (distance of 2 cm), and k = 0.54 (distance of 1 cm) at a) 125 kHz and b) 250 kHz.

In order to validate the proposals, twenty-four different operating points were compared between simulation and experimental results varying the follow parameters: load, frequency and coupling factor. The average deviation between simulation and experimental voltage and power values was 1.14 % and 2.29 %, respectively.

It must be emphasized that, despite being applied in linear circuits, the Thevenin's theory had an acceptable error of at most 7.51 % within the analyzed range.

The coupling factor equation facilitates his determination, since it depends only on the system measurable elements. More tests are needed to evaluate the acceptable error region of the proposed equations.

ACKNOWLEDGMENT

The authors thanks the Federal University of Espírito Santo (UFES), the Electronic Power and Electrical Drives Laboratory (Lepac) for infrastructure and devices for prototype construction and testing, and the National Council for Scientific and Technological Development (CNPQ) for the financial support.

REFERENCES

[1] J. C. Rautio, "The shoulders of the giants on which we stand," *IEEE MTT-S Int. Microw. Symp. Dig.*, pp. 1–3, 2012.

[2] G. Pelosi and S. Selleri, "Energy in electromagnetism: The poynting vector," *IEEE Antennas Propag. Mag.*, vol. 59, no. 6, pp. 148–153, 2017.

[3] H. Wiki, "Nikola Tesla," *IEEE Eng. Manag. Rev.*, vol. 45, no. 3, pp. 9–10, 2017.

[4] S. M. Khan, "Analysis of wireless power transfer (WPT) scheme with connected ground planes," *2017 Usn. Radio Sci. Meet. (Joint with AP-S Symp. Usn. 2017*, pp. 21–22, 2017.

[5] Y. Hori, "Application of electric motor, supercapacitor, and wireless power transfer to enhance operation of future vehicles," *IEEE Int. Symp. Ind. Electron.*, pp. 3633–3635, 2010.

[6] K. . Keerthi., K. Ilango., and G. N. Manjula., "Study of Midfield Wireless Power Transfer for Implantable Medical Devices," *2018 2nd Int. Conf. Biomed. Eng.*, pp. 44–47, 2018.

[7] J. Dai and D. C. Ludois, "A Survey of Wireless Power Transfer and a Critical Comparison of Inductive and Capacitive Coupling for Small Gap Applications," *IEEE Trans. Power Electron.*, vol. 30, no. 11, pp. 6017–6029, 2015.

[8] D. Patil, M. Sirico, L. Gu, and B. Fahimi, "Maximum efficiency tracking in wireless power transfer for battery charger: Phase shift and frequency control," *ECCE 2016 - IEEE Energy Convers. Congr. Expo. Proc.*, pp. 1–8, 2016.

[9] Ö. Cenk and M. A. Naci, "A frequency-tracking algorithm for inductively coupled wireless power transfer systems," *10th Int. Conf. Electr. Electron. Eng.*, pp. 1461–1465, 2017.

[10] J. Li and D. Wang, "Perturb and Observe method of Impedance Matching for Magnetically Coupled Wireless power transfer System," *2018 Chinese Autom. Congr.*, pp. 2513–2517, 2018.

[11] V. V. Burlaka, S. K. Podnebennaya, and S. V. Gulakov, "Analysis of Approaches to the Efficiency Improvement of Wireless Power Transmission Systems Using Low-Frequency Magnetic Fields," *2018 IEEE 38th Int. Conf. Electron. Nanotechnology, ELNANO 2018 - Proc.*, pp. 572–575, 2018.

[12] X. Li, X. Dai, Y. Li, Y. Sun, Z. Ye, and Z. Wang, "Coupling coefficient identification for maximum power transfer in WPT system via impedance matching," *IEEE PELS Work. Emerg. Technol. Wirel. Power, WoW 2016*, pp. 27–30, 2016.

[13] F. Nasr, M. Madani, and M. Niroomand, "Precise analysis of frequency splitting phenomenon of magnetically coupled wireless power transfer system," *Asia-Pacific Microw. Conf. Proceedings, APMC*, pp. 219–224, 2018.

[14] M. Stănculescu, M. Iordache, L. I. Bobaru, D. Niculae, and V. Bucată, "Analysis of wireless power transfer systems using s parameters to

generate the power gains," *Proc. - 2017 Int. Conf. Optim. Electr. Electron. Equipment, OPTIM 2017 2017 Intl Aegean Conf. Electr. Mach. Power Electron. ACEMP 2017*, pp. 72–77, 2017.

[15] R. Bansal, "Did maxwell pull a fast one? [AP-S Turnstile]," *IEEE Antennas Propag. Mag.*, vol. 52, no. 5, p. 186, 2010.

[16] D. H. Tran, V. B. Vu, and W. Choi, "Design of a High-Efficiency Wireless Power Transfer System with Intermediate Coils for the On-Board Chargers of Electric Vehicles," *IEEE Trans. Power Electron.*, vol. 33, no. 1, pp. 175–187, 2018.

[17] Y. Zhang, T. Kan, Z. Yan, and C. C. Mi, "Frequency and Voltage Tuning of Series-Series Compensated Wireless Power Transfer System to Sustain Rated Power under Various Conditions," *IEEE J. Emerg. Sel. Top. Power Electron.*, vol. 7, no. 2, pp. 1311–1317, 2019.

[18] Y. Zhang, T. Kan, Z. Yan, and C. Mi, "Analytical Models of Wireless Power Transfer Systems with a Constant-Power Load," *2018 IEEE Transp. Electrif. Conf. Expo, ITEC 2018*, pp. 720–724, 2018.

[19] M. Lu and K. D. T. Ngo, "Systematic Design of Coils in Series-Series Inductive Power Transfer for Power Transferability and Efficiency," *IEEE Trans. Power Electron.*, vol. 33, no. 4, pp. 3333–3345, 2018.

[20] W. Hu, H. Zhou, Q. Deng, and X. Gao, "Optimization algorithm and practical implementation for 2-coil wireless power transfer systems," in *Proceedings of the American Control Conference*, 2014, pp. 4330–4335.

[21] X. Tang, J. Zeng, K. P. Pun, S. Mai, C. Zhang, and Z. Wang, "Low-Cost Maximum Efficiency Tracking Method for Wireless Power Transfer Systems," *IEEE Trans. Power Electron.*, vol. 33, no. 6, pp. 5317–5329, 2018.

[22] W. Zhang, S. C. Wong, C. K. Tse, and Q. Chen, "Design for efficiency optimization and voltage controllability of series-series compensated inductive power transfer systems," *IEEE Trans. Power Electron.*, vol. 29, no. 1, pp. 191–200, 2014.

[23] H. Hu *et al.*, "Constant maximum power control for dynamic wireless power transmission system," *2017 IEEE PELS Work. Emerg. Technol. Wirel. Power Transf. WoW 2017*, no. 51361130150, pp. 295–299, 2017.

[24] Y. Yang, W. Zhong, S. Kiratipongvoot, S. C. Tan, and S. Y. R. Hui, "Dynamic Improvement of Series-Series Compensated Wireless Power Transfer Systems Using Discrete Sliding Mode Control," *IEEE Trans. Power Electron.*, vol. 33, no. 7, pp. 6351–6360, 2018.

[25] R. Mai, Y. Liu, Y. Li, P. Yue, G. Cao, and Z. He, "An Active-Rectifier-Based Maximum Efficiency Tracking Method Using an Additional Measurement Coil for Wireless Power Transfer," *IEEE Trans. Power Electron.*, vol. 33, no. 1, pp. 716–728, 2018.

[26] H. Li, K. Wang, J. Fang, and Y. Tang, "Pulse Density Modulated ZVS Full-Bridge Converters for Wireless Power Transfer Systems," *IEEE Trans. Power Electron.*, vol. 34, no. 1, pp. 369–377, 2018.

978-1-7281-4181-7/19 $31.00 © 2019 IEEE

Method to trace the photovoltaic characteristic curve with reverse voltage for shading conditions

Richard G. Cornelius[1], Amanda C. Maia[1], Matheos C. Wermuth[2], Guilherme S. da Silva[1]

[1]Group of Integrated Exploration of Energy Resources – EIRE[1]
[2]Eletronic Systems Research Group – GEPSEl[2]
Federal University of the Pampa – UNIPAMPA
Alegrete, Rio Grande do Sul, Brazil
richardgc98@gmail.com, amanda_maia15@hotmail.com, conetto197@gmail.com, guilhermesilva@unipampa.edu.br

Abstract—**The I-V curve is an important parameter, because it contains the PV module characteristics. It shows what happens in the module due to external variations. This paper presents a method to trace the photovoltaic curves with reverse voltage under partial shading conditions. In addition, it mentions the effects of the bypass diode. Lastly, it makes comparisons between curves of shaded and unshaded modules and the shunt resistance effect.**

Keywords—*I-V curve, photovoltaic module, reverse voltage, shading conditions.*

I. Introduction

The distributed generation is increasing around the world [1], where the power generation is mostly centralized systems. In other words, the energy comes from large power plants far away from consumer centers. However, with the development of renewable energy sources, such, as photovoltaic systems, wind generation and biomass, the worldwide power generation is changing and becoming more decentralized [2].

In Brazil, the photovoltaic (PV) distributed generation is growing since 2014. The reason for this is related to the government policy incentives. Besides, the country has a great radiation levels and the characteristics of photovoltaic modules contributes for its growth. According to [3] the increase of this generation is 125% in 2019. Thus, all these factors are influencing on the development of photovoltaic distributed generation.

For the correct application of the PV modules, it is necessary to know their technical features. The PV module power production depends on the parameters and other variables like the module irradiance and temperature. In addition, they are represented on the module characteristic curve (I-V curve), which consists of the module current and voltage [4]. Moreover, the I-V curve represents the behavior of the PV module, and any alteration on its parameters reflects on the curve.

Irregular irradiances modify the I-V curve. The irradiation on a PV cell influences directly on the output current. Then, when a cell is partially shaded, its output drops proportionally. However, the voltage stays the same. As the other cells are not shaded, it creates an inflection voltage [5]. Consequently, the power generated by the cell drops and the P-V curve acquiers more than one maximum power point.

The hotspots phenomenon causes irreversible damage on PV modules [6]. Hotspots can happen on a series of situations, such as damages in the PV cell, dirt residues accumulation on the panel's surface or from any kind of barrier capable of blocking the sunrays to the photovoltaic module. Moreover, when a cell is shaded, it becomes a load: the shunt resistance drops, causing more power dissipation, which creates a hotspot. If a hotspot causes a reverse voltage beyond of the breakdown voltage, the cell stops working.

This paper presents the difference between PV array models for uniform irradiation and for mismatching conditions. It also present the I-V and P-V curve for shading conditions, considering and not considering the shunt resistance. Also, the consequence on the breakdown voltage is analyzed. Finally, the I-V curve is analyzed and parameterized for partial shading conditions.

II. Photovoltaic Cell Modeling for Regular Irradiance Condition with Reverse Voltage

An equivalent circuit can represent the photovoltaic cell model. This enables the model analysis through equations for evaluating its physics behavior. Among the cell equivalents circuits, a succint representation is in the model with one diode.

Besides that, the reverse voltage model is also relevant, because it shows the rupture voltage when the cell is operating like a load. That way, the completed model depends on the direct and reverse voltages and is shown in Fig. 1 [7].

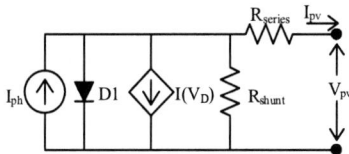

Fig. 1. One diode equivalent circuit of PV cell.

The I_{pv} is the current generated by the PV cell at the output terminal. The diode (D1) represents the *pn* semiconductor junction. Besides that, the shunt resistance (R_{shunt}) express the impurities and the *pn* structure defect. The series resistance (R_{series}) represents the metal joints and the $I(V_D)$ is the controlled current source by reverse voltage [7]-[8].

From the Kirchhoff Currents Law, the transcendental equation (1) is obtained that represents the PV output current I_{pv}.

$$I_{pv} = I_{ph} - I_0 \cdot \left(e^{q \cdot \frac{V_{pv} + I_{pv} \cdot R_{serie}}{a \cdot k \cdot T}} - 1 \right) - \frac{V_{pv} + I_{pv} \cdot R_{series}}{R_{shunt}} - b \cdot \left(V_{pv} + I_{pv} \cdot R_{serie} \right) \cdot \left(1 - \frac{V_{pv} + I_{pv} \cdot R_{serie}}{V_{br}} \right)^{-m} \quad (1)$$

This equation is described according to the following parameters:

- I_0 is the reverse diode saturation current in A;
- I_{ph} is the photo generated current in A;
- q is the electron charge in C;

978-1-7281-4181-7/19 $31.00 © 2019 IEEE

- k is the *Boltzmann* constant in m^2kg/s^2K;
- a is the ideality factor;
- T is the temperature in K;
- V_{pv} is voltage in V;
- b is the breakdown correction factor of the PV cell;
- m is the breakdown coefficient of the PV cell;
- V_{br} is the breakdown voltage of the PV cell.

Solving (1) is possible to obtain the PV module characteristic I-V curve, shown on Fig. 2. This curve contains the main characteristics of a PV module which consist of the short-circuit current (I_{sc}), the open circuit voltage (V_{oc}), maximum power point current (I_{mpp}), the maximum power point voltage (V_{mpp}) and the breakdown voltage characteristic.

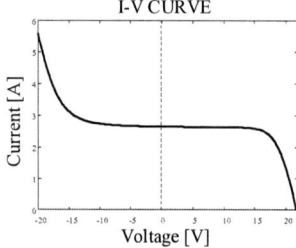

Fig. 2. PV module I-V curve.

The parameters shown on Fig. 2 can be found on the PV module datasheet. However, the parameters such as I_0, R_{series}, and R_{shunt} cannot be obtained in the datasheet. For this application, the *Newton-Raphson* method is used to obtain these parameters, which were used in [9].

III. PHOTOVOLTAIC MODULE MODEL FOR MISMATCHING CONDITIONS

The photovoltaic module model for mismatching conditions needs special attention. The cells on the PV module are connected in series, which sum its voltages and keep the current constant. However, if a shading occurs on the module, its current is limited by current of the shaded cell. This creates an inflection point in the I-V curve and a hotspot. Therefore, mismatching conditions modify the I-V curve [5]-[10].

A. Entirely Shaded PV Cells

An entirely shaded PV cell becomes a load, dissipating the module power and creating hotspots. Besides that, for an elevated reverse voltage applied in the cell, it suffers serious damage [11]. On the other hand, PV modules usually contain bypass diodes, connected in antiparallel at every amount of cells. Thus, they form an alternative path for the flow current.

Thereby, for entirely shaded PV cells, the bypass diode starts to conduct, and the respective cells group is disabled, as shown in Fig. 3. Consequently, the PV module current continues to be the same; however, the voltage goes down. Due to this , the R_{shunt} and R_{series} change as well, because the numbers of activate cells on the module goes lower [12].

Due to the parameters change, the I-V curve for shaded conditions is different when compared to the unshaded one, as shown on Fig. 4.

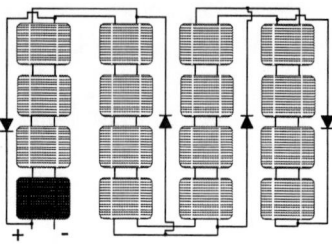

Fig. 3. PV module for entirely shaded cell.

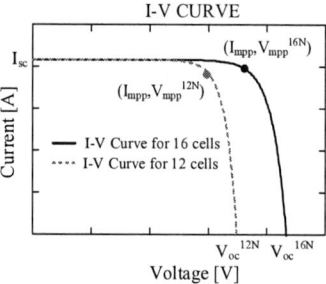

Fig. 4. Comparison between the I-V curve for regular irradiance and for an enterily shaded cell.

Therefore, if shaded conditions happen on different cells of each group that contains the bypass diode, all the diodes start to conduct and the module stops to generate power. This is represented on Fig. 5.

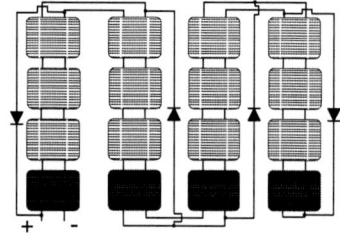

Fig. 5. PV module with different shaded cells.

B. Partially Shaded Photovoltaic Cells

In the partially shaded PV cell situations, shown on Fig. 6, different consequences occurs. The PV module current is limited by cell with mismatching conditions in the inflection point. Thus, there is power dissipation in the shaded part [13].

This way, the R_{series} remains the same, because none of the physical characteristics changes, while R_{shunt} suffers an alteration in its value [5].

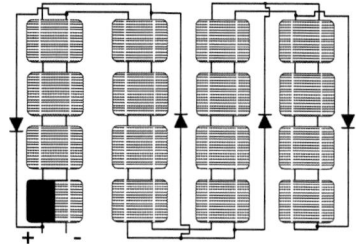

Fig. 6. PV cell with partially shaded conditions.

978-1-7281-4181-7/19 $31.00 © 2019 IEEE

The I-V curve for irregular irradiance changes if it is compared with the curve for normal conditions. The I-V curve characteristic for shaded PV module is shown on Fig. 7.

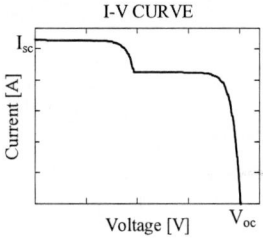

Fig. 7. Partially shaded PV cell.

1) Mathematic Model for the Inflection Voltages

The mathematic model is based on the inflection voltages [14]. The partially shaded PV cell imposes a current limit in the modules with regular irradiance. This is shown on Fig. 8 and Fig 9.

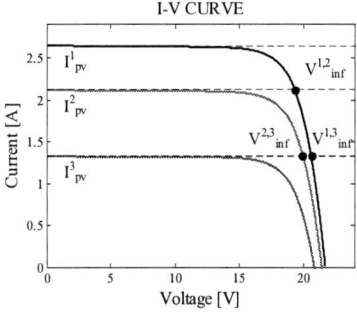

Fig. 8. I-V curve for partially shaded cells with the representation of inflection voltages.

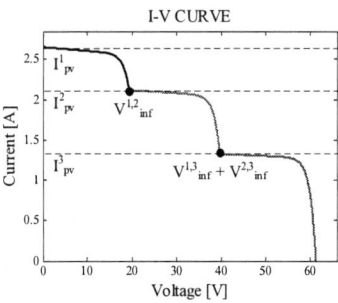

Fig. 9. I-V curve of a partially shaded module.

Fig. 9 presents the inflection points related with three PV modules connected in series. Each module has different shading percentage. In the $V^{1,2}_{inf}$ point, the current I^1_{pv} becomes equal to I^2_{pv}, and at this voltage it generates an inflection point. In addition to that, the same happen for the $V^{k,\,k+1}_{inf}$. Thereby, the inflection voltage for a string with k modules is given by (2).

$$V^{k,k+1}_{inf} = -a \cdot N \cdot V_t \cdot lambert_W \left(0, I^k_0 \cdot C^k \cdot e^{D^k} \right) - ...$$
$$...I^k_{ph} \cdot \left(R^k_{shunt} + R_{series} \right) + R^k_{shunt} \cdot \left(I^k_{pv} + I^k_0 \right) \tag{2}$$

The lambert function is used because the voltage and current are implicit variables. This function make them explicit [14].

The variables C and D are given by (3) and (4), and the current I^k_{pv} of each iteration is given by (5). Initially, R_{shunt} is considered as a constant given by (6). Moreover, the reverse saturation current for each iteration is given by (7).

$$C^k = \frac{R^k_{shunt}}{a \cdot N \cdot V_T} \tag{3}$$

$$D^k = C^k \cdot \left(I^k_{pv} + I^k_0 - I^k_{ph} \right) \tag{4}$$

$$I^k_{pv} = \left(1 + \left(\frac{R_{series}}{R^k_{shunt}} \right) \right) \cdot I^k_{sc} \tag{5}$$

$$R^k_{shunt} = R_{shunt} \tag{6}$$

$$I^k_0 = \frac{\left(I^k_{ph} - \frac{V^k_{pv}}{R^k_{shunt}} \right)}{e^{\frac{V^k_{pv}}{a \cdot N \cdot V_T}} - 1} \tag{7}$$

The V_T is the PV cell thermal voltage represented by (8), and the I^k_{ph} is the shaded module short circuit current given by (9).

$$V_T = \frac{k \cdot T}{q} \tag{8}$$

$$I^k_{ph} = \left(\frac{S}{S_{REF}} \right) \cdot \left(I^{k-1}_{ph} + k_i \cdot \left(T - T_{REF} \right) \right) \tag{9}$$

The S is the module irradiance, S_{ref} is the reference irradiance and k_i is the temperature coefficient, which changes the current.

C. Tracing the I-V curve for partially shaded PV modules

Aiming to trace the I-V curve for partially shaded PV modules, is necessary to calculate the inflection point. This is the point when the PV module current is limited by current of the shaded cells. Each shading implicates different inflection points.

Thus, it is defined the standard datasheet parameters and the parameters k and module irradiances are stipulated. Then, R_{series}, R_{shunt} and I_0 are estimated from *Newton-Raphson* method.

Afterwards, V^k_{oc}, I^k_{sc}, I^k_0 and R^k_{shunt} are calculated for each module irradiance and temperature. Therefore, the inflection voltages are calculated from (2).

Thus, I_{pv} is calculated by *Newton-Raphson* method and P from (10), then the I-V curve is traced.

$$P = I_{pv} \cdot V_{pv} \tag{10}$$

The Fig. 10 presents the flowchart with the following procedures to trace the I-V curve with partial shading conditions.

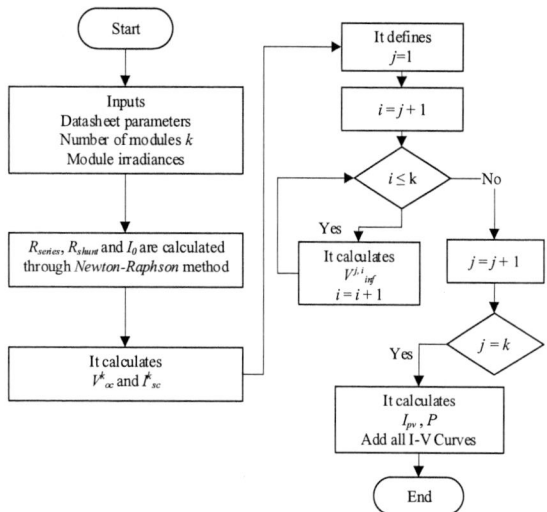

Fig. 10. Flowchart to trace the partially shaded I-V Curve.

IV. RESULTS

The simulation results was obtained with the PV module *Kyocera KC40T*. Table I shows the module parameters available on the datasheet and calculated through the *Newton-Raphson* method. For this simulation, it was used an irradiance of 1000 W/m² and temperature of 298 K.

TABLE I.   VALUES USED FOR THE *NEWTON-RAPHSON* METHOD

Description	Value
Maximum Power	$P_{max} = 43W$
Maximum Power Voltage	$V_{mpp} = 17.4V$
Maximum Power Current	$I_{mpp} = 2.48A$
Open Circuit Voltage	$V_{oc} = 21.7V$
Short Circuit Current	$I_{sc} = 2.65A$
Number of Cells	$N = 36$
Ideally Factor	$a = 1.3$
Boltzmann Constant	$k = 1.38 \times 10^{-23}\ m^2kg/s^2K$
Electron Charge	$q = 1.6021 \times 10^{-19}\ C$
Temperature	$T = 298\ K$
Series Resistance	$R_{series} = 0.122\ \Omega$
Shunt Resistance	$R_{shunt} = 224.4268\ \Omega$

According to Fig. 10, an algorithm was made to trace the I-V curve with shading conditions. First, it established a string containing three modules, which parameters are shown on Table I. In addition, different types of shadow were considered. The used parameters of irradiance and the global maximum power point (GMPP) are on the Table II, Table III and Table IV.

TABLE II.   RED I-V CURVE AND P-V CURVE

k module	Irradiance [W/m²]	GMPP [W]
1	1000	
2	400	43.35
3	200	

TABLE III.   BLACK I-V CURVE AND P-V CURVE

k module	Irradiance [W/m²]	GMPP [W]
1	1000	
2	700	63.37
3	400	

TABLE IV.   BLACK I-V CURVE AND P-V CURVE

k module	Irradiance [W/m²]	GMPP [W]
1	1000	
2	800	97.79
3	700	

The Fig. 11 (a), (b) e (c) shows the results for each shadow condition. The I-V curve has many inflection points that reflect on the P-V curve, generating local maximum power points (LMPP). Moreover, the curves show that the maximum power changes for different irregular irradiances.

When a string is under mismatching conditions, the module resistances change. However, the model represented by the Fig. 10 does not consider these effects on R_{shunt} [15]. When a module is partially shaded, its shunt resistance drops proportionally with the irradiance.

The Fig. 12 shows the difference between the shaded module with the decrease of R_{shunt} and without this effect.

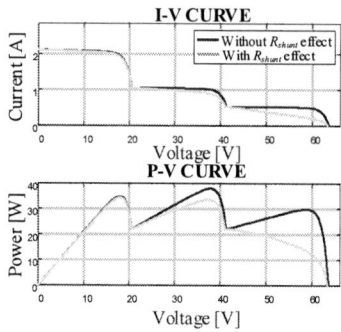

Fig. 12. I-V and P-V curve showing the difference considering or not the R_{shunt} decrease.

The shunt resistance causes power reduction. For the simulation, the same mentioned PV modules were used, and their stipulated irradiances were 800 W/m², 400 W/m² and 200 W/m². These shaded conditions increase the power dissipation due to the drop of R_{shunt}, causing a significant reduction on the local maximum power point and also in the global maximum power point.

In addition, the shunt resistance interferes on the breakdown voltage. When there is a power dissipation on R_{shunt} resistance, it creates a hotspot [16]. Moreover, it reduces the breakdown voltage, and the cell does not work anymore [7].

The Fig. 13 shows this effect for different types of irradiance.

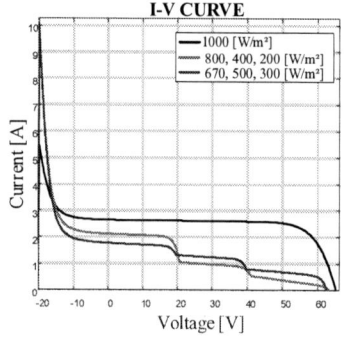

Fig. 13. I-V curve for diferrent breakdown voltages.

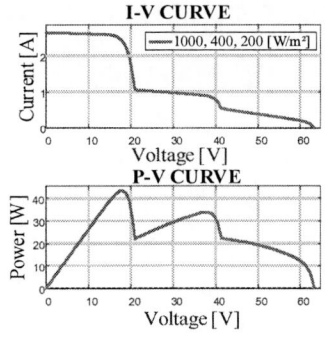 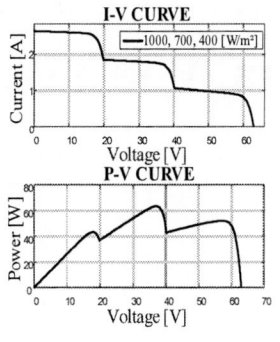 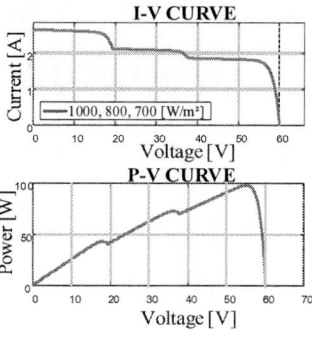

(a) I-V curve for 1000, 400 e 200 W/m²   (b) I-V curve for 1000, 700 e 400 W/m²   (c) I-V curve for 1000, 800 e 700 W/m²

Fig. 11. I-V curve for different shading conditions.

V. CONCLUSIONS

When a shading occurs, the PV module current is limited by shaded cell. This effect happens due to the cell connection and, therefore, the current cells with regular irradiance dissipates in the R_{shunt} of shaded cell. This way, partially shading creates hotspots and inflection points.

Given that, the shaded cell turns into a load, creating hotspots. In long-term can damage the PV cell and decrease the module voltage. Besides that, it tends to decrease the rupture voltage like in the presented results.

The breakdown voltage is the higher reverse voltage value that a cell supports without being damaged. Thus, if the cell does not have the bypass diode and it is shaded, a reverse voltage will be submitted on this cell. In addition, the shading effect causes a decrease of the breakdown voltage value increasing the chance of the cell to stop working.

The paper presented graphically that the inflection point modifies the I-V curve. When a shading occurs, an inflection point is formed by the shaded cell. This causes reduction of the shunt resistance and the power generated by string. Moreover, it causes a bad operation of the inverter devices.

In addition, it is also observed a great decrease on the local maximum power points of a shaded string. Which offers the current MPPT devices a challenge when trying to find the global maximum power point. In a practical context, the PV cells inserted on an electric vehicle are exposed to irregular irradiances, making, necessary the application of fast MPPT methods and knowledge about the characteristic curve behavior.

REFERENCES

[1] G. A. Xavier *et al.*, "Simulation of Distributed Generation with Photovoltaic Microgrids—Case Study in Brazil," pp. 4003–4023, 2015.

[2] O. Moraes, D. Oliveira, and A. Diniz, "Distributed photovoltaic generation and energy storage systems : A review," vol. 14, pp. 506–511, 2010.

[3] M. Brazil, "Solar Energy is Expected to Grow 44% in Brazil at 2019," 2019. [Online]. Available: http://www.brazilmonitor.com/index.php/2019/01/18/solar-energy-is-expected-to-grow-44-in-brazil-at-2019/.

[4] F. M. González-longatt, "Model of Photovoltaic Module in Matlab ™," pp. 1–5, 2005.

[5] M. C. Alonso-García, J. M. Ruiz, and F. Chenlo, "Experimental study of mismatch and shading effects in the I – V characteristic of a photovoltaic module," vol. 90, pp. 329–340, 2006.

[6] M. Simon and E. Meyer, "Detection and analysis of hot-spot formation in solar cells," *Sol. Energy Mater. Sol. Cells*, vol. 94, no. 2, pp. 106–113, 2010.

[7] E. I. Batzelis, I. A. Routsolias, and S. A. Papathanassiou, "An Explicit PV String Model Based on the Lambert W Function and Simplified MPP Expressions for Operation Under Partial Shading," vol. 5, no. 1, pp. 301–312, 2014.

[8] A. L. C. de Carvalho, "Metodologia para análise , caracterização e simulação de células fotovoltaicas," Universidade Federal de Minas Gerais, 2014.

[9] R. G. Cornelius, A. C. Maia, A. P. C. De Mello, J. W. M. Kaehler, and G. S. Silva, "Desenvolvimento De Um Dispositivo Eletrônico Para Caracterização De Painéis Fotovoltaicos," *Sepoc*, p. 6, 2018.

[10] H. Assunção, "Degradação De Módulos Fotovoltaicos De Silicio Cristalino Instalados No Dee - Ufc," 2014.

[11] J. Peroza and G. A. Rampinelli, "Análise De Desempenho E Atuação De Diodos De Bypass Em Um Módulo Fotovoltaico Comercial," no. 1988, 2018.

[12] M. C. M. Pedro, "Modelling of Shading Effects in Photovoltaic Optimization," Universidade Nova de Lisboa, 2016.

[13] S. Silvestre and A. Chouder, "Effects of Shadowing on Photovoltaic Module Performance," no. September 2007, pp. 141–149, 2008.

[14] J. Rodríguez, C. Ramos-Paja, and E. Mejía, "Modeling and parameter calculation of photovoltaic fields in irregular weather conditions," vol. 17, no. 1, pp. 37–48, 2012.

[15] H. Kawamura, K. Naka, N. Yonekura, S. Yamanaka, H. Kawamura, and H. Ohno, "Simulation of I-V characteristics of a PV module with shaded PV cells," vol. 75, pp. 613–621, 2003.

[16] O. Breitenstein *et al.*, "Understanding junction breakdown in multicrystalline solar cells," vol. 071101, no. September 2010, 2011.

978-1-7281-4181-7/19 $31.00 © 2019 IEEE

Harmonic Compensation Strategies Applied to Multifunctional Photovoltaic Inverters

Lucas S. Xavier[1], Joice D. S. Zacarias[2], Allan F. Cupertino[1,3], Heverton A. Pereira[2], Danilo I. Brandao[1] and Victor F. Mendes[1]

[1]Graduate Program in Electrical Engineering
Universidade Federal de Minas Gerais
Av. Antônio Carlos 6627, 31270-901
Belo Horizonte, MG, Brazil
lsx@ufmg.br, dibrandao@ufmg.br
victormendes@cpdee.ufmg.br

[2]Department of Electrical Engineering
Federal University of Viçosa
Av. P. H. Rolfs s/n°, 36570-000
Viçosa, MG, Brazil
joice.zacarias@ufv.br
heverton.pereira@ufv.br

[3]Department of Materials Engineering
Federal Center for Technological
Education of Minas Gerais
Belo Horizonte 30421-169, Brazil
afcupertino@ieee.org

Abstract— **The multifunctional photovoltaic inverter may not be able to fully compensate the harmonic currents of a nonlinear load, either because the various services that it may be performing in a certain moment, or because the increase of the load harmonic content. Therefore, it is found in the literature some compensation strategies that can be applied in those situations, such as selective and partial compensation. Despite the diversity of harmonic compensation strategies, it is important to compare them to understand their advantages and disadvantages under limited operating capability of the inverter. Therefore, this paper proposes to compare partial and selective harmonic current compensation strategies. In addition, a variant of these compensation approaches is introduced and compared to, which is the partial-selective compensation strategy. A case study is simulated to verify the impact of each compensation strategy on the grid power quality and the photovoltaic inverter effort in terms of thermal stress.**

Keywords— Distributed Generation, Harmonic Compensation, Multifunctional Inverter, Photovoltaic, Power Quality.

I. INTRODUCTION

The basic power electronics of a grid-connected photovoltaic (PV) system is the voltage source inverter, which interfaces the PV modules to the electrical grid. Due to the variation of the solar irradiance, in a few hours of the day the PV inverter operates in its nominal current capacity. Thus, other services can be performed by this inverter, characterizing its multifunctional operation. For example, a PV inverter can perform reactive power injection to grid support and harmonic current compensation of nonlinear loads [1]–[3].

The growth of nonlinear load in the electrical power system raises concerns about power quality. Electronic loads like digital devices, computers, equipment with rectifiers and many others cause high level of harmonic currents circulation over the power system. Such undesirable harmonic currents circulation causes part of the power losses in the grid [4]. Therefore, the PV inverters can be used to compensate these harmonics over the grid, acting as an active power filter [5].

However, the multifunctional PV inverter may not be able to perform the total harmonic current compensation of a load current, due to the other more prioritized services that may be performed concomitantly, due to the high active power generation or due to the increase of the harmonic content of the load current. These events can exceed the PV inverter rated current during the harmonic current compensation [6].

There are two main limits of a PV inverter when it performs harmonic compensation. The first one is the minimum dc-link voltage required to compensate a certain amount of harmonic current, which is associated with its output filter [7]. If the harmonic compensation requires a minimum voltage higher than exists in the dc-link, the modulator runs into overmodulation region. The second one is the inverter maximum current constraint that is associated with the temperature of its semiconductor switches. Some works in the literature have demonstrated a correlation between the increase of the average operating temperature and the high thermal cycling, under semiconductor switches, with the reduction of inverter lifetime expectancy [8], [9].

Therefore, in addition to the total harmonic current compensation strategy, it is important to have more flexible compensation strategies to ensure that the inverter operates within its dc-link voltage and current capacity during limited operating conditions. In [6] and [10], the partial harmonic compensation is proposed through current dynamic saturation techniques. The detected load harmonic currents are scaled by a factor ranging from 0 to 1 to scale the compensation in order to not exceed the rated current of the PV inverter. In this way, the PV inverter contributes as much as possible to the grid power quality enhancement.

Another compensation strategy is the selective harmonic current compensation. In this case, only some harmonic orders are compensated. For this reason, the selectivity can be achieved through the algorithms that previously extract the information only of the selected harmonic orders of the load [11], [12], [13]. Despite the diversity of harmonic compensation strategies, it is important to compare them to understand their advantages and disadvantages under tight operating conditions of the inverter, in which the total harmonic current compensation may not be possible due to limited inverter capability.

In view of the points aforementioned, this paper proposes to contribute by comparing the harmonic current compensation strategies under critical operating conditions of the multifunctional PV inverter. The compared strategies are the

978-1-7281-4181-7/19 $31.00 © 2019 IEEE

partial, selective and partial-selective harmonic current compensation. The compensation approaches are simulated to verify the impact of each strategy on the grid power quality and the PV inverter effort in terms of thermal stress.

This paper is outlined as follows. Section II presents the control strategy of the PV system analyzed herein. Section III describes the harmonic compensation strategies compared in this paper. Section IV shows the current control design and parameters of the PV inverter used in the case studies. Section V describes the case studies, while Section VI shows the simulation results. Finally, Section VII concludes.

II. POWER AND CONTROL STRUCTURE

The grid-connected single-phase PV system considered herein is shown in Figure 1. A boost converter is used to control the output voltage of the PV modules. For this purpose, the maximum power point tracking (MPPT) algorithm based on incremental conductance is used to maximize the energy extraction of the PV array [14].

Generally, the control of a PV inverter is composed of an outer voltage loop, responsible for controlling the dc-link voltage and compute the injection of active power, and an inner current loop responsible for controlling the inverter output current. Figure 2 shows the control scheme used in this paper.

The method based on v_{dc}^2 control is used to control the external loop of the dc-link voltage [15]. Considering the references of active power (P^*) and the reactive power (Q^*), calculated by the outer loop, the fundamental current reference is computed by using the PQ theory for single-phase systems:

$$i_\alpha = \frac{2}{v_\alpha^2 + v_\beta^2}\left(v_\alpha P^* + v_\beta Q^*\right), \qquad (1)$$

where v_α and v_β are the direct and quadrature components of the grid voltage (v_G), respectively. These components are calculated by a second order generalized integrator (SOGI) along with a phase locked-loop (PLL) [16].

The load current (i_L) is estimated by measuring the inverter (i_S) and grid current (i_G). This approach ensures that harmonic currents of all nonlinear loads connected to the point of common coupling (PCC) are computed. In the harmonic current detection (HCD) algorithm, the harmonic load currents are identified and processed according to the desired compensation strategy, generating the harmonic current reference (i_h^*) for the current control loop. In this paper, the proportional multi-resonant controller (PMR) is used [17].

During the harmonic compensation, the PV inverter has the limitation of the current that circulates through its switches and the available voltage in the dc-link [6], [7], [18]. If the compensated harmonic current requires a voltage beyond the one available on the dc-link, the pulse width modulation (PWM) runs into nonlinear region (i.e., overmodulation). In this operating condition, the inverter injects undesired low order harmonics. This region is normally avoided and not considered a normal operating condition in this paper.

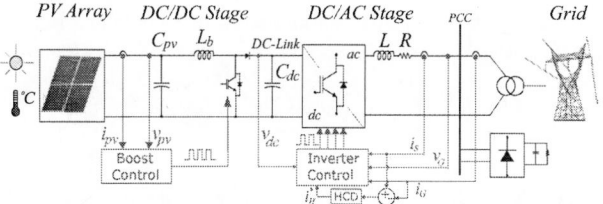

Figure 1. Grid-connected single-phase PV power system.

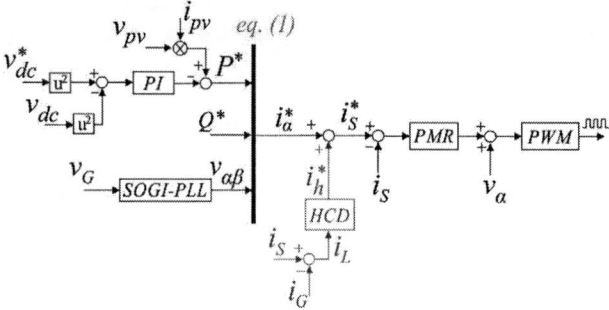

Figure 2. Control loop scheme of the PV inverter.

III. HARMONIC COMPENSATION STRATEGIES

In this section, the harmonic current compensation strategies of nonlinear loads are defined: *a*) total, *b*) partial, *c*) selective and *d*) partial-selective harmonic compensation.

A. Total Harmonic Compensation

Generally, all harmonic current content of the load current is detected and added to the inverter current control loop [5]. This can be classified as total harmonic current compensation. It is important to note that odd harmonics are more common in the electric power system than the even ones, and therefore, only they are considered. Thereby, the harmonic current reference is equal to the harmonic load current, given by:

$$i_h^* = \sum_{n=2}^{\infty} A_{2n-1} \cos\left[(2n-1)\omega_f + \varphi_{2n-1}\right], \quad (2)$$

where n defines the harmonic order presented in the load current. A_{2n-1} is the amplitude of each harmonic component, ω_f is the fundamental frequency and $\varphi_{(2n-1)}$ is the phase of each harmonic component.

The load current can present high harmonic levels and, in this case, the inverter cannot be able to fully compensate the harmonic content. For this reason, more flexible compensation strategies are required to be applied to PV inverter for maximize its harmonic current compensation, complying with its current rating and dc-link voltage capacity.

B. Partial Harmonic Compensation

A way to ensure that the inverter does not exceed its current and voltage constraints during harmonic compensation is using partial compensation. For this compensation strategy, the harmonic content of the load current is identified, similar to the total compensation. However, the harmonic currents are all equally scaled by a factor K ($0 < K < 1$). This scaling is performed if the PV inverter does not have enough capacity to compensate the total load harmonic current content. Thereby, the harmonic current reference is given by:

$$i_h^* = K \cdot \sum_{n=2}^{\infty} A_{2n-1} \cos\left[(2n-1)\omega_f + \varphi_{2n-1}\right]$$

$$\{K \in \mathbb{R} \mid 0 < K < 1\}, \qquad (3)$$

The K factor can be settled by an operator through a supervisory system or calculated by a technique based on an open-loop algorithm used to limit the inverter current, as presented in [6] and [19]. In [20], the K factor is calculated using a distortion factor desired for the grid, settled through a supervisory control center.

Note that, applying this strategy, the inverter current reference may present harmonic frequency components with higher frequencies than the designed controller bandwidth (CBW). Therefore, the current control loop, trying to track these terms outside the CBW, may degrade the injected inverter current in relation to its reference, not contributing to the power quality improvement.

C. Selective Harmonic Compensation

In this strategy, only some selected harmonic components are used to generate the inverter current reference (i_h^*). It is performed in order to comply with the PV inverter limits and designed controller bandwidth. For this purpose of selecting the harmonic components, frequency domain algorithms can be used such as bandpass filters (BPF) and methods based on Discrete Fourier Transform(DFT) [21], [22]. In this case, the harmonic current reference is given by:

$$i_h^* = \sum_{m=2}^{M} A_{2n-1} \cos\left[(2n-1)\omega_f + \varphi_{2n-1}\right]$$

$$\{m \in \mathbb{Z} \mid 2 \le m \le M\}, \qquad (4)$$

where m defines the selected harmonic orders from the load current and it can assume integer values between 2 and M.

Using this strategy, the inverter current reference is comprised only of harmonic frequency components with lower frequencies than the CBW, enhancing the current tracking accuracy, and not degrading the injected inverter current.

D. Partial-Selective Harmonic Compensation

The partial and selective compensation strategies are combined in the partial-selective harmonic current compensation. For this strategy, some load current harmonic components are selectively detected and added to the current control, similar to the selective compensation. However, the selected harmonic components are scaled by a K factor ($0 < K < 1$), similar to the partial compensation. Therefore, the harmonic current reference is given by:

$$i_h^* = K \cdot \sum_{m=2}^{M} A_{2n-1} \cos\left[(2n-1)\omega_f + \varphi_{2n-1}\right]$$

$$\{(m \in \mathbb{Z} \text{ and } K \in \mathbb{R} \mid 2 \le m \le M \text{ and } 0 \le K < 1\}, \quad (5)$$

Therefore, using the partial-selective harmonic compensation, the inverter current reference presents only harmonic frequency components with lower frequencies than

the CBW, enhancing the current tracking accuracy, and the selected harmonic components are scaled improving the system flexibility to adapt to critical operating conditions under tight PV inverter available capability to harmonic mitigation.

The main features of the harmonic compensation strategies discussed in this paper are summarized in Figure 3, where $HFS\{i_h^*\}$ means the harmonic frequency spectrum of the inverter harmonic current reference.

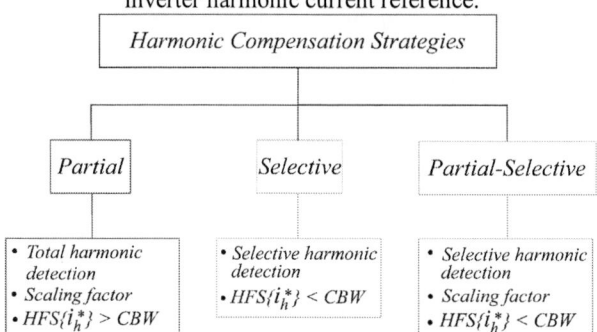

Figure 3. Overview of harmonic current compensation strategies.

IV. INVERTER CURRENT CONTROL DESIGN

The behavior of the injected inverter current is modelled by:

$$P(s) = \frac{i_s(s)}{v_s(s)} = \frac{1}{sL + R}, \qquad (6)$$

where L and R are the equivalent inductance and equivalent series resistance of the inverter output filter, respectively. v_s is the inverter output voltage. A proportional multi-resonant controller (PMR) is used in this paper, in which the transfer function is given by:

$$G_c = K_P + \sum_{h}^{n_h} K_{Ih} \overbrace{\frac{\frac{R(s)}{s}}{s^2 + h^2 \omega_f^2}}, \qquad (7)$$

where K_P is the proportional gain, h is the harmonic order ($h = 1,2,3 \dots, n_h$), ω_f is the fundamental frequency and K_{Ih} is the resonant gain for each harmonic frequency. $R(s)$ is the resonant parcel of the controller. The design of the gains K_P and K_{Ih} is based on the desired cut-off frequency and phase margin for the open-loop transfer function $[P(s)G_c(s)]$ [23]. The designed cut-off frequency for the open-loop transfer function is 800 Hz. Therefore, it is designed the resonant gains for odd harmonics until the 11th harmonic order. Thus, $h \in \{1,3,5,7,9,11\}$ in (7). The Bode diagram of $[P(s)G_c(s)]$ is shown in Figure 4, considering $K_P = 13$, $K_{Ih} = \frac{26616}{h+1}$, $L = 3\,mH$ and $R = 0.028$.

V. CASE STUDY

The grid-connected single-phase PV system with harmonic current compensation capability is simulated using the PLECS software in order to compare the harmonic compensation strategies. The PV array model proposed in [24] is used. The specifications are of a traditional commercial PV module of 250 W. The PV modules are arranged to achieve 6 kWp. For this case of study, it is considered 0.3 pu of generated active power

978-1-7281-4181-7/19 $31.00 © 2019 IEEE

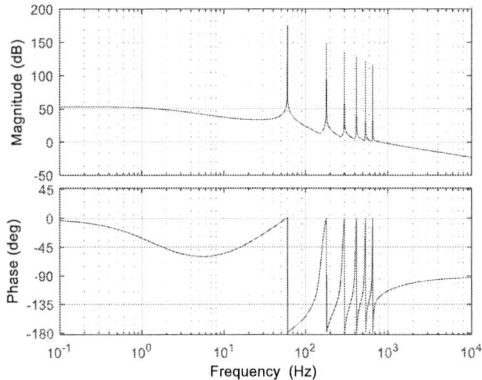

Figure 4. Bode diagram of $[P(s)G_c(s)]$ with 800 Hz cut-off frequency.

In this case study, the inverter dc-link voltage and rated current are not enough to perform the total harmonic compensation of the load current. Therefore, the other compensation strategies are applied and compared.

The grid power quality is quantified through the total demand distortion (TDD) of the grid current, as recommended by IEEE 519-2014. The load current TDD is 10.55 %, as illustrated in Figure 5. It is worth emphasizing that, without the harmonic compensation, the grid current TDD is also 10.55 %.

The case studies are set in order the guarantee a fair comparison among the strategies, in which a baseline based on the inverter current is defines as 37 A of maximum value, including the fundamental frequency component corresponding to active power generation. Therefore, the K factor for partial compensation is set to 0.3875. The $3^{rd}, 5^{th}$ and 7th harmonic components are compensated using the selective compensation. For partial-selective compensation, the $3^{rd}, 5^{th}, 7^{th}, 9^{th}$ and 11th are compensated for $K = 0.8727$. Therefore, the harmonic current reference spectra for the harmonic compensation strategies are shown in Figure 6.

In Figure 6, the sum of the blue areas of every bar in a graphic is identical to every compensation strategy, corresponding to the same amount of current processed through the PV inverter. However, one can see that the partial-selective compensation shows the slowest TDD value among the strategies.

The thermal mode is used to estimate the junction temperature of the inverter semiconductor switches for each harmonic compensation strategy. For this purpose, the losses and thermal data of the Infineon power module FF50R12RT4 are used considering a switch frequency of 18 kHz. Therefore, the average junction temperature (T_j) and junction temperature variation (ΔT_j) of the inverter IGBTs and diodes are indicators of the stress on the inverter′s semiconductors for each harmonic compensation strategy.The controller tracking accuracy for each harmonic compensation strategy is quantified by the maximum error between the inverter current reference (i_S^*) and the measured current (i_S).

VI. SIMULATION RESULTS

The comparison data of the harmonic compensation strategies are shown in Table I. Figure 7 shows the current processed through the inverter considering each compensation

strategy. Using partial harmonic compensation with $K = 0.3875$, the grid current TDD is improved for 8.21 %. The controller error is 4.45 A. This high value is explained due to the presence of harmonic frequencies higher than CBW. The IGBTs T_j and ΔT_j are 112.71 °C and 9.89 °C, respectively. The diodes T_j and ΔT_j are 87.53 °C and 2.17 °C, respectively.

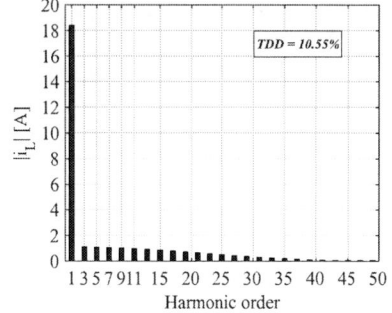

Figure 5. Load current spectrum.

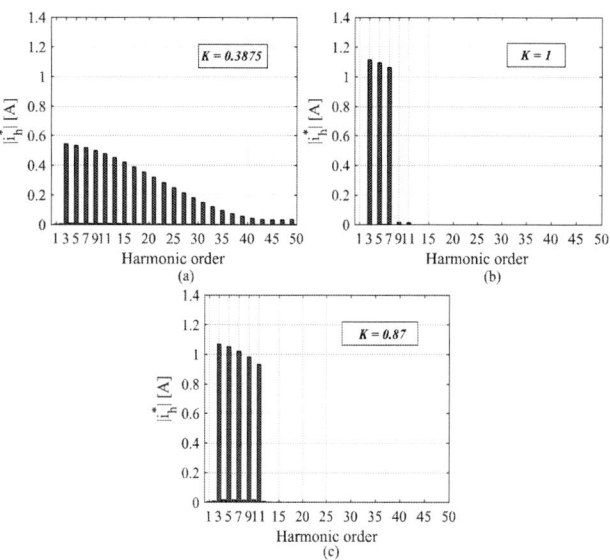

Figure 6. Harmonic current reference (i_h^*) spectra for the harmonic current compensation strategies. (a) partial compensation. (b) selective compensation. (c) partial-selective compensation.

Through the selective compensation of the $3^{rd}, 5^{th}$ and 7th, the grid current TDD is 8.32 %. The controller error is reduced to 0.40 A due to the presence of only harmonic frequencies lower than CBW. This shows that by ensuring only harmonic components lower than CBW in the inverter harmonic reference, the grid current TDD is not improved in relation to the partial compensation. The IGBT′s T_j and ΔT_j are 112.22 °C and 9.33 °C, respectively. The diodes T_j and ΔT_j are 87.23 °C and 2.15 °C, respectively. Note that, ΔT_j is approximately similar in relation to the partial compensation. However, T_j is slightly higher for partial compensation because this compensation strategy provides more harmonic components in the inverter current.

978-1-7281-4181-7/19 $31.00 © 2019 IEEE

Table I. Comparison between harmonic compensation strategies.

Harmonic Compensation Strategy	K factor	Peak value of Inverter Current [A]	DC-link voltage [V]	Grid Current TDD [%]	Controller Maximum Error [A]	IGBT T_J [°C]	IGBT ΔT_J [°C]	Diode T_J [°C]	Diode ΔT_J [°C]
Partial	0.3875	37	400	8.21	4.45	112.71	9.89	87.53	2.17
Selective: 3rd + 5th + 7th harm.	-	37	400	8.32	0.40	112.22	9.33	87.23	2.15
Partial-selective: (3rd + 5th + 7th + 9th + 11th)	0.8727	37	400	6.90	0.32	111.91	9.32	86.71	2.12

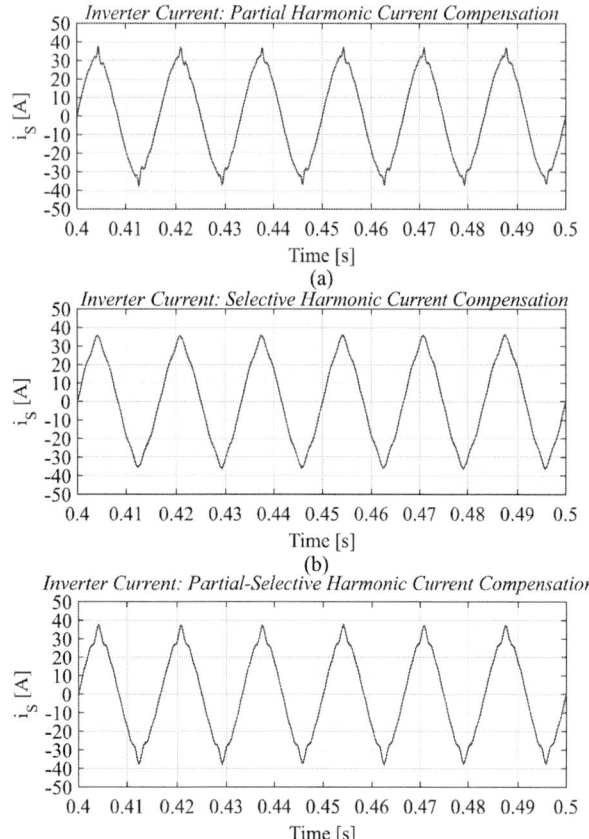

Figure 7. Injected inverter current waveform for each harmonic compensation strategy. (a) Partial harmonic compensation. (b) Selective harmonic compensation. (c) Partial-selective harmonic compensation.

Finally, using the partial-selective grid current TDD is reduced to 6.90%, the lowest value among the strategies. The control error is 0.32 A since only harmonic components below the CBW are compensated. The IGBTs T_J and ΔT_J are 111.91 °C and 9.32 °C, respectively. The diodes T_J and ΔT_J are 86.71 °C and 2.12 °C, respectively. Note that, in this case, with approximately the same inverter effort than the selective compensation, this strategy contributes more for grid power quality enhancement.

Figure 8 shows the load and grid current waveform for each harmonic current compensation strategy. The grid current with the partial-selective compensation presents a

current waveform closer to a sinusoidal form. However, it is possible to note the improvement in the grid current waveform for each strategy.

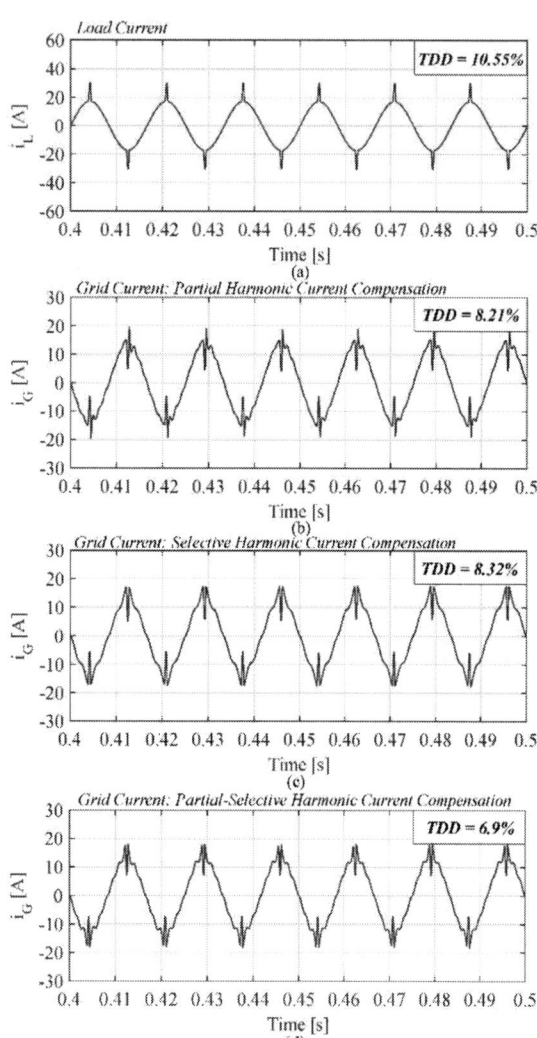

Figure 8. Current waveforms when applying each harmonic compensation strategy. (a) Load current. (b) Grid current for partial compensation. (c) Grid current for selective compensation. (d) Grid current for partial-selective compensation.

VII. Conclusion

This paper proposed a comparison between harmonic current compensation strategies applied to the grid-connected single-phase PV system. The partial, selective and partial-selective harmonic current compensation are compared in terms of grid energy quality and the inverter effort to perform this service. The comparisons are made under extreme nonlinear load conditions, and tight inverter capability. Therefore, the strategies are compared in terms of thermal stress, control performance, and power quality improvement.

The case studies show that the partial and partial-selective compensation have similar thermal impact on the inverter's semiconductors; however, the partial-selective compensation contribute more for the grid power quality improvement than the partial compensation due to selected harmonics with lower frequency than the inverter's control bandwidth, which improves the current control performance. Thus, the partial-selective compensation strategy is recommended to be applied to multifunctional PV inverters.

Acknowledgment

This work has been supported by the Brazilian agencies CAPES, CNPq and FAPEMIG.

References

[1] R. A. Mastromauro, M. Liserre, T. Kerekes, and A. Dell'Aquila, "A Single-Phase Voltage-Controlled Grid-Connected Photovoltaic System With Power Quality Conditioner Functionality," *IEEE Trans. Ind. Electron.*, vol. 56, no. 11, pp. 4436–4444, 2009.

[2] P. Acuña, L. Morán, M. Rivera, J. Dixon, and J. Rodriguez, "Improved Active Power Filter Performance for Renewable Power Generation Systems," *IEEE Trans. Power Electron.*, vol. 29, no. 2, pp. 687–694, 2014.

[3] Y. Yang, F. Blaabjerg, H. Wang, and M. G. Simões, "Power control flexibilities for grid-connected multi-functional photovoltaic inverters," *IET Renew. Power Gener.*, vol. 10, no. 4, pp. 504–513, 2016.

[4] W. Abbas and M. A. Saqib, "Effect of nonlinear load distributions on total harmonic distortion in a power system," *2007 Int. Conf. Electr. Eng. ICEE*, pp. 3–8, 2007.

[5] H. K. M. Paredes, F. P. Marafão, D. I. Brandão, and A. Costabeber, "Multi-task control strategy for grid-tied inverters based on conservative power theory," *IET Renew. Power Gener.*, vol. 9, no. 2, pp. 154–165, 2015.

[6] L. S. Xavier, A. F. Cupertino, H. A. Pereira, and V. F. Mendes, "Partial Harmonic Current Compensation for Multifunctional Photovoltaic Inverters," *IEEE Trans. Power Electron.*, no. In press, 2019.

[7] V. M. R. De Jesus, L. S. Xavier, A. F. Cupertino, V. F. Mendes, and H. A. Pereira, "Operating limits of three-phase multifunctional photovoltaic converters applied for harmonic current compensation," *IEEE 8th Int. Symp. Power Electron. Distrib. Gener. Syst. PEDG 2017*, pp. 1–8, 2017.

[8] R. C. d. Barros, E. M. S. Brito, G. G. Rodrigues, V. F. Mendes, A. F. Cupertino, and H. A. Pereira, "Lifetime evaluation of a multifunctional PV single-phase inverter during harmonic current compensation," *Microelectron. Reliab.*, vol. 88–90, no. July, pp. 1071–1076, 2018.

[9] A. Sangwongwanich, Y. Yang, D. Sera, and F. Blaabjerg, "Lifetime Evaluation of Grid-Connected PV Inverters Considering Panel Degradation Rates and Installation Sites," *IEEE Trans. Power Electron.*, vol. 33, no. 2, pp. 1125–1236, 2018.

[10] J. P. Bonaldo, H. K. M. Paredes, A. Costabeber, and J. A. Pomilio, "Adaptive saturation system for grid-tied inverters in low voltage residential micro-grids," *2015 IEEE 15th Int. Conf. Environ. Electr. Eng. EEEIC 2015 - Conf. Proc.*, pp. 784–789, 2015.

[11] P. Mattavelli and F. P. Marafao, "Repetitive-based control for selective harmonic compensation in active power filters," *IEEE Trans. Ind.*

Electron., vol. 51, no. 5, pp. 1018–1024, 2004.

[12] M. Sonnenschein and M. Weinhold, "Comparison of time-domain and frequency-domain control schemes for shunt active filters," in *Euro. Trans. Electr. Power*, 1999.

[13] R. Teodorescu, F. Blaabjerg, M. Liserre, and P. C. Loh, "Proportional-resonant controllers and filters for grid-connected voltage-source converters," in *IEE Proceedings - Electric Power Applications*, 2006.

[14] D. Sera, L. Mathe, T. Kerekes, S. V Spataru, and R. Teodorescu, "On the Perturb-and-Observe and Incremental Conductance MPPT Methods for PV Systems," *IEEE J. Photovoltaics*, vol. 3, no. 3, pp. 1070–1078, 2013.

[15] L. S. Xavier, A. F. Cupertino, and H. A. Pereira, "Ancillary services provided by photovoltaic inverters: Single and three phase control strategies," *Comput. Electr. Eng.*, vol. 70, pp. 102–121, 2018.

[16] M. Ciobotaru, R. Teodorescu, and F. Blaabjerg, "A new single-phase PLL structure based on second order generalized integrator," *PESC Rec. - IEEE Annu. Power Electron. Spec. Conf.*, 2006.

[17] A. G. Yepes, F. D. Freijedo, O. Lopez, J. Doval-Gandoy, Ó. López, and J. Doval-Gandoy, "Analysis and Design of Resonant Current Controllers for Voltage-Source Converters by Means of Nyquist Diagrams and Sensitivity Function," *IEEE Trans. Ind. Electron.*, vol. 58, no. 11, pp. 5231–5250, 2011.

[18] L. Harnefors, A. G. Yepes, A. Vidal, and J. Doval-Gandoy, "Multifrequency Current Control with Distortion-Free Saturation," *IEEE J. Emerg. Sel. Top. Power Electron.*, vol. 4, no. 1, pp. 37–43, 2016.

[19] D. I. Brandao, H. Guillardi, H. K. Morales-Paredes, F. P. Marafao, and J. A. Pomilio, "Optimized Compensation of Unwanted Current Terms by AC Power Converters under Generic Voltage Conditions," *IEEE Trans. Ind. Electron.*, vol. 63, no. 12, pp. 7743–7753, 2016.

[20] J. P. Bonaldo, H. K. M. Paredes, and J. A. Pomilio, "Control of Single-Phase Power Converters Connected to Low Voltage Distorted Power Systems with Variable Compensation Objectives," *IEEE Trans. Power Electron.*, vol. 31, no. 3, pp. 2039–2052, 2016.

[21] R. I. Bojoi *et al.*, "Current control strategy for power conditioners using sinusoidal signal integrators in synchronous reference frame," *IEEE Trans. Power Electron.*, vol. 20, no. 6, pp. 1402–1412, 2005.

[22] E. Jacobsen and R. Lyons, "The sliding DFT," *IEEE Signal Process. Mag.*, vol. 20, no. 2, pp. 74–80, 2013.

[23] S. Buso and P. Mattavelli, *Digital Control in Power Electronics*. Morgan & Claypool, 2006.

[24] M. G. Villalva, J. R. Gazoli, and E. R. Filho, "Comprehensive Approach to Modeling and Simulation of Photovoltaic Arrays," *Power Electron. IEEE Trans.*, vol. 24, no. 5, pp. 1198–1208, 2009.

Event Manager and Control Structure for High Performance Three-Phase Grid-tied Inverters

Victor E. S. Barbosa, Rodrigo A. S. Kraemer, Emerson G. Carati, Jean P. da Costa,
Rafael Cardoso, Carlos M. O. Stein
PPGEE - Graduate Program in Electrical Engineering
UTFPR - Federal University of Technology - Parana
Pato Branco - PR, Brazil
victorbarbosa@alunos.utfpr.edu.br, rodrigorask@gmail.com, emerson@utfpr.edu.br, jpcosta@utfpr.edu.br,
rcardoso@utfpr.edu.br, cmstein@utfpr.edu.br

Abstract—This paper presents a time and event based control strategy that guarantees the connection and operation of the grid-tied inverter to the IEEE 1547-2018 standard. The control structure is composed by a synchronization algorithm and current controller with active damping based on capacitor current estimation. This structure guarantees the high performance of the real time control presented in this paper. Moreover, the proposed control structure also includes the distributed generation (DG) manager, which is responsible for determining the operation mode of DG system: Connection; Active power limitation (Normal operation); Reactive power limitation; Dispatch interruption; and, Disconnection. Besides, flow charts are presented in order to describe the connection and disconnection procedures, as well the modes of operation. The proposed strategy is implemented in a DSP (Digital Signal Processor) connected to a Hardware in the Loop (HIL) system. Real time domain emulations are performed considering a 15 kW three-phase grid-tied inverter and the results are discussed.

Index Terms—Distributed generation, Inverter control, Microgrids, Smart inverter.

I. INTRODUCTION

The electrical power generation in the customer side of network is defined as distributed generation. From 2010 and 2015 the DG photovoltaic systems (PV) has experienced an increase of approximately 40% of global capacity [1]. This increase is leading to a high insertion of DG in the main grid. To deal with the DG insertion the regulation standards are becoming more rigid [2].

In order to ensure the properly operation of the DG systems the unity power factor and THD must be kept in standard limits. To achieve these specifications the main control structures that need to be observed are the PLL (Phase Locked Loop) algorithm [3–6] and the current controller [7–9]. For photovoltaic systems the DG can also be refered as grid-tied inverters. Moreover, a LCL filter is commonly used in order to connect the three-phase voltage source inverter (VSI) to the grid. The LCL filter provides a better harmonic attenuation at the inverter switching frequency and results in smaller passive components than the L filter [10–13]. However, the LCL filter introduces resonances, which can be affected by

This work was funded by the Research and Development project PD 2866-0468/2017, granted by the Agencia Nacional de Energia Eletrica (ANEEL) and Companhia Paranaense de Energia (COPEL).

grid impedances and can lead the current control loop to instability [11], [12]. Therefore, an active damping technique is implemented in order to attenuate the resonance frequency and to increase the stability of the system. This procedure allows to use higher gains in the current controller [10], [13].

Many problems related to the connection and operation of DG systems are addressed in literature, such as the coordination between modes of operation, references configuration of the DG system required to obtain reference output power and maintaining the voltage level of the DC bus [14].

In this way, the main contribution of this paper is to propose a control structure for three-phase grid-tied inverter ensuring the IEEE 1547-2018 [15] standard limits on the total harmonic distortion (THD), range of voltage and frequency to connection, and others parameters that affect the power quality. Also, an inverter system manager was designed and implemented in order to coordinate the operation modes: Connection; Active power limitation (Normal operation); Reactive power limitation; Dispatch interruption; and, Disconnection. Additional protection modes are also implemented to obtain the correct performance of the grid-tied inverter [16].

This paper is arranged as follows: section II presents and describe the proposed system and its control structure; section III describes the operations modes of inverter manager; section IV analyzes the results obtained from real time domain emulation; and, section V concludes the paper.

II. SYSTEM DESCRIPTION

The grid-tied inverter system proposed in this paper is shown in Fig. 1 and includes the electrical circuit and the control structure implemented in a Digital Signal Processor. The AC stage is composed of a three-phase inverter, an LCL filter and the main grid. The grid parameters includes grid impedances (L_g and r_g) and LCL filter internal resistances (r_{L1}, r_{Cf} and r_{L2}). The DC-link voltage (v_{dc}), the PCC (Point of Common Coupling - v_{pcc}) voltage and the inverter current (i_{inv}) are measured in order to implement the DC-link control and the output current control synchronized with AC grid.

Additionally, the diagram also presents the photovoltaic array and the DC stage composed by a Boost converter. Though the DC stage is not the focus, it is important make it clear

978-1-7281-4181-7/19 $31.00 © 2019 IEEE

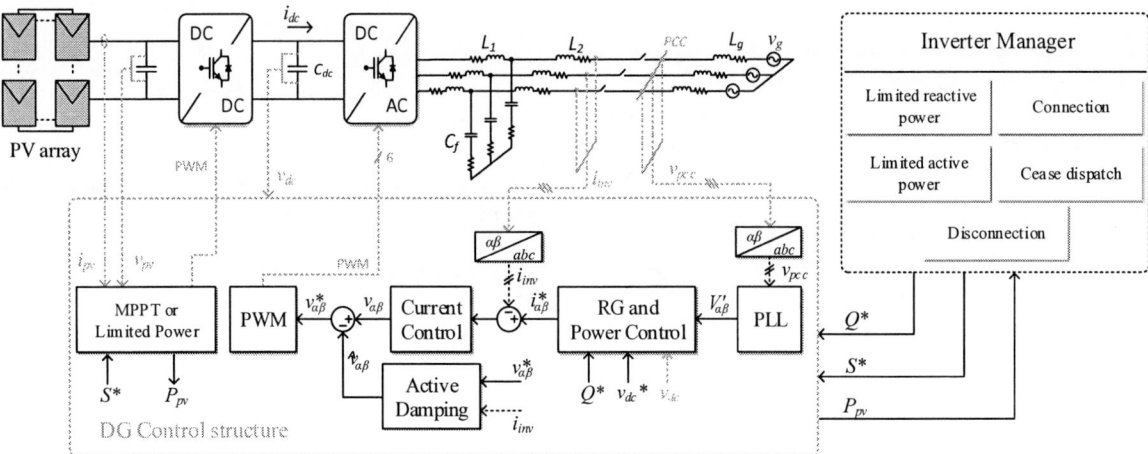

Fig. 1. Control structure for three-phase grid-tied inverter.

that the proposal presented in this paper is applied in grid-tied photovoltaic systems and the DC stage will provide the active power to the DC-link, operating in MPPT (Maximum Power Point Tracking) or limited power mode, that will be described in section III. The maximum power of the photovoltaic array (P_{mppt}) connected to the DC stage is 15 kW at 1000 W/m² of solar irradiation with $T_{env} = 25\ ^oC$ (environment temperature). The systems parameters is shown in Table I. The inverter is designed for 15 kW of nominal power, and connected to the grid at 220 V_{rms} and 60 Hz.

The system transfer function that relates the inverter grid side current (i_{inv}) to the inverter PWM voltage ($v^*_{\alpha\beta}$) is presented in (1). In addition, since no local loads are considered, the grid parameters is included in the system transfer function, hence, $L_{2g} = L_2 + L_g$ and $r_{L2g} = r_{L2} + r_g$.

$$G_{LCL}(s) = \frac{i_{inv}(s)}{v^*_{\alpha\beta}(s)} = \frac{Z_2}{Z_1Z_2 + Z_1Z_3 + Z_2Z_3} \quad (1)$$

In (1), $v^*_{\alpha\beta}$ is the inverter voltage, i_{inv} the current in L_2, $Z_1 = (L_1s + r_{L1})$, $Z_2 = \left(\frac{1}{sC_f} + r_{C_f}\right)$ and $Z_3 = (L_{2g}s + r_{L2g})$.

The control structure implemented in DSP is detailed in Fig. 1 and it contains the grid synchronization algorithm (PLL), current reference generator (RG) and power control, current controller and the active damping stabilizer. At right side, it is shown the inverter manager, which is responsible for the definition of the inverter operation mode. These control structures are described in following subsections. The inverter manager will be described in section III.

A. Synchronization algorithm

A PLL (Phase Locked Loop) algorithm is used to synchronize the converter voltage with the grid, generating the reference voltages ($V'_{\alpha\beta}$). The DSOGI (Double Second Order Generalized Integrator) with FLL (Frequency Locked Loop) is chosen here [3], [8], due to the smooth response during transient faults. While most PLLs estimates the grid phase-angle (see DDSRF-PLL - Decoupled Double Synchronous Reference [3]), the DSOGI-FLL estimates the grid frequency, which is a much more stable variable.

TABLE I
SYSTEM PARAMETERS

Parameters	Description	Value
L_1	Inverter side LCL filter inductance	232.3 μ[H]
r_{L1}	L_1 series resistance	41.5 m[Ω]
L_2	Grid side LCL filter inductance	163.3 μ[H]
r_{L2}	L_2 series resistance	22.1 m[Ω]
C_f	LCL filter capacitance	10 μ[F]
r_{C_f}	C_f series resistance	5.2 m[Ω]
L_r	Grid equivalent inductance	1.56 m[H]
r_g	Grid equivalent resistance	0.8 [Ω]
V_g	Grid voltage	220 [V]
S_{dg}	Nominal apparent power	15 k[VA]
f_g	Grid frequency	60 [Hz]
f_{sw}	Switching frequency	18 k[Hz]
v_{dc}	DC bus voltage	400 [V]
C_{dc}	DC bus capacitance	3 m[F]
P_{mppt}	Maximum power of photovoltaic array	15 k[W]

The DSOGI-FLL equations in stationary reference of α axis and the FLL grid frequency estimation are given by

$$\frac{V'_\alpha(s)}{v_\alpha(s)} = \frac{K_e\hat{\omega}s}{s^2 + K_e\hat{\omega}s + \hat{\omega}^2} \quad (2)$$

$$\frac{qV'_\alpha(s)}{v_\alpha(s)} = \frac{K_e\hat{\omega}^2}{s^2 + K_e\hat{\omega}s + \hat{\omega}^2}, \quad (3)$$

$$\hat{\omega} = \frac{1}{s}\frac{-K_e\hat{\omega}\gamma}{(V'^+_\alpha)^2 + (V'^+_\beta)^2}\frac{\varepsilon_f}{2} + \omega_g, \quad (4)$$

where $\varepsilon_f = (v_\alpha - V'_\alpha)qV'_\alpha + (v_\beta - V'_\beta)qV'_\beta$. The β axis variables are implemented by the equivalent equations. In addition, the resulting voltages V'_α, qV'_α, V'_β and qV'_β are the input signals of PNSC (Positive-Negative Sequence Calculation) structure, which computes the positive sequence (5) to be used as the reference for inverter to grid synchronization.

$$V^{+'}_\alpha = \frac{V'_\alpha - qV'_\beta}{2}; \quad V^{+'}_\beta = \frac{V'_\beta + qV'_\alpha}{2}. \quad (5)$$

In (2) - (4) the gains used are $K_e = 1$ and $\gamma = 30.667$.

B. Power control and reference generator

The power control block is responsible for the voltage regulation in DC-link (v_{dc}) through the power dispatch to the grid. The control of v_{dc} is obtained using a PI controller to generates the power to the grid reference such it becomes the same the one produced by the DC stage. Therefore, the power control block indirectly applies the active power reference (P^*), through the DC stage, and the reactive power reference (Q^*) by the inverter manager.

From the power references and voltages synchronized with the grid, generated by DSOGI-FLL, the reference generator provides the current references ($i^*_{\alpha\beta}$) trough the PQ Open-Loop Control in stationary reference frame [17], according to

$$
\begin{bmatrix} i^*_\alpha \\ i^*_\beta \end{bmatrix} = \frac{1}{(V^{+'}_\alpha)^2 + (V^{+'}_\beta)^2} \begin{bmatrix} V^{+'}_\alpha & -V^{+'}_\beta \\ V^{+'}_\beta & V^{+'}_\alpha \end{bmatrix} \begin{bmatrix} P^* \\ Q^* \end{bmatrix} \quad (6)
$$

C. Current control

Considering that stationary reference frame is used by DSOGI-FLL and reference generator, the PR (Proportional Resonant) controller is applied [7], [17] in this work. The PR approach allows to obtain selective harmonic compensation (HC) by cascading several resonant blocks tuned to attenuate the desired order harmonics, in this case, 5th, 7th and 11th harmonics. The PR controller with harmonic compensation (HC) are given by

$$
G_{PR+HC}(s) = \frac{v_{\alpha\beta}(s)}{i_{\alpha\beta}(s)} = K_p + \sum_{h=1,5,7,11} \frac{2K_{hi}s}{s^2 + (h\hat{\omega})^2}, \quad (7)
$$

where $\hat{\omega}$ is the grid frequency estimated by the FLL in (4).

D. Active damping

The active damping approach usually applies the feedback of current capacitor of the LCL filter to attenuate the resonant component caused by the LCL filter resonance frequency. To avoid additional measurements, the capacitor current can be determined using robust estimators [10]. Moreover, the estimation technique presented in [10] demonstrates improved robustness compared to measured technique. The PR with active damping control structure allows higher gains for the current controller and improves the performance and robustness of the system in a wide range of grid parameters.

The current estimation of the capacitor is given by (8), and it only depends on the inverter voltage reference ($v^*_{\alpha\beta}$) and inverter current (i_{inv}) signals.

$$
\hat{i}_{cf}(s) = \frac{C_f s v^*_{\alpha\beta}(s) - C_f L_1 s^2 i_{inv}(s) - C_f K_{ic} s\, i_{inv}(s)}{C_f L_1 s^2 + C_f K_{ic} s + 1} \quad (8)
$$

In (8), K_{ic} is the virtual resistance introduced in the LCL filter model, according to [10]. The active damping voltage is obtained as $\hat{v}_{AD} = K_m \hat{i}_{cf}$, where K_m is the resistance equivalent gain. The inverter reference voltage is obtained as $v^*_{\alpha\beta} = v_{\alpha\beta} - \hat{v}_{AD}$.

E. Stability Analysis

In order to ensure the robustness of the control structure against grid parameters variations, the design procedure follow the methodology presented in [8], [10]. Using the transfer functions of open-loop ($G_{io}(s)$) in continuous-time and closed-loop ($G_{ic}(z)$) in discrete time, see [8], the frequency response and the root locus of the system for grid parameters variation are presented in Fig. 2 and Fig. 3 respectively. The gains of the controllers are shown in Table II.

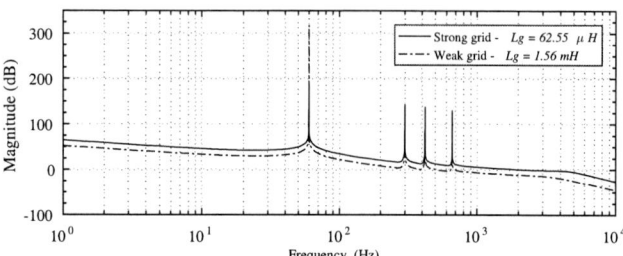

Fig. 2. Frequency response of $G_{io}(s)$ for grid impedance variation: Strong grid: $L_g = 65\ \mu H$, $r_g = 32.3$ mΩ. Weak grid :$L_g = 1.56 mH$, $r_g = 0.8\Omega$.

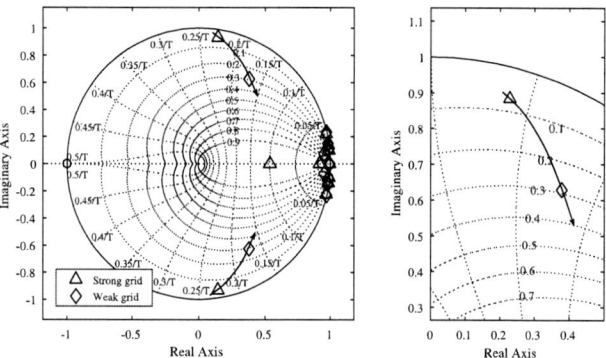

Fig. 3. Root locus of $G_{ic}(z)$ for grid impedance variation: Strong grid: $L_g = 65\ \mu H$, $r_g = 32.3$ mΩ. Weak grid :$L_g = 1.56 mH$, $r_g = 0.8\Omega$.

In Fig. 2 it is possible to observe that the frequency response decreases by an offset for weak grids. Moreover, as shown in Fig. 3, weaker grid moves the system poles toward inside the unit circle. This behavior makes it possible to choose higher proportional gains (K_p) of the PR+HC controller [10].

TABLE II
CONTROLLER PARAMETERS

Parameters	Description	Value
K_p	Proportional gain of PR+HC controller	5
K_i	Resonant gain of PR+HC controller	3000
K_{r5}	5th Harmonic resonant gain	300
K_{r7}	7th Harmonic resonant gain	300
K_{r11}	11th Harmonic resonant gain	300
K_m	Active damping gain	4

III. INVERTER MANAGER

The main priority of the DG control is to extract the maximum available energy from the renewable energy resource, in this case, photovoltaic panels. However, the distribution

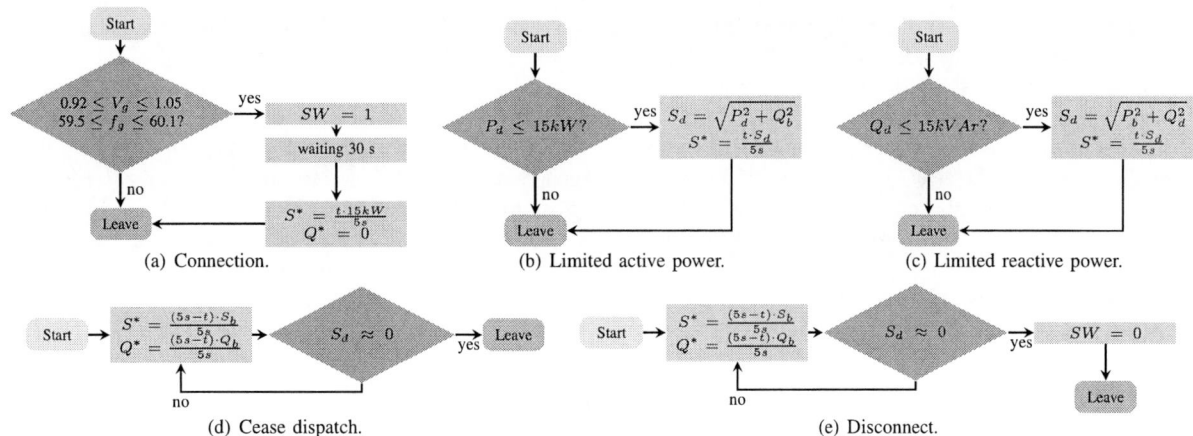

(a) Connection. (b) Limited active power. (c) Limited reactive power.

(d) Cease dispatch. (e) Disconnect.

Fig. 4. Modes decisions flowcharts.

and transmission network operators (DNO and TNO) are increasingly imposing specific technical requirements, through the standards such IEEE Std 1547-2018, to connect the DG systems in the main grid. Furthermore, ancillary services are required such as Low Voltage Ride Through, active and reactive power control, modes of operation and protection functions to increase the resilience, stability and efficiency of distributed generation [1], [19].

In order to achieve the DNO and TNO requirements, a new component is proposed for the grid-tied inverter system, which is refered as Inverter Manager in this paper. The Inverter Manager is designed using modular modeling. The main inverter operation modes focused in this paper are:

1) Connection;
2) Limited Active Power;
3) Limited Reactive Power;
4) Dispatch Interruption;
5) Disconnection.

The Inverter Manager must be constantly monitoring the network parameters in order to perform the specified functions when necessary, such commutation beetween operation modes, protection modes and exceptions. In addition, it needs to maintain communication with the DNO to attend dispatch commands. The Connection Mode procedures are shown in decision tree of Fig. 4(a), where is primarily verified if the grid parameters are inside the acceptable parameters. If it is right, the connection switch is closed and a counter is applied for system stabilization. After the counting, a ramp for power dispatch is applied. The S^* is the apparent power reference to be dispatched to the grid, Q^* is the reference reactive power, and t is the time in seconds, and SW is the On/Off signal to open and close the inverter connection switch.

The procedures of the Limited Active Power Mode are shown in the decision tree of Fig. 4(b), where S_d is the apparent power definition calculated from the active power P_d defined by the DNO. The reactive power Q_b is the actual reactive power reference. The apparent active power reference P^* is reached in a ramp of 5 seconds.

Fig. 4(c) shows the decision tree of the Limited Reactive Power Mode. Where Q_d is the DNO required reactive power, and P_b is the actual active power reference. In the same way as in Limited active power mode, in this mode S^* is reached in a ramp of 5 seconds.

In the Dispatch Interruption Mode, the procedures are the same that is shown in the decision tree of Fig. 4(d), where S_b is the actual apparent power reference that are decreased in a ramp of 5 seconds, and the SW continues closed. The procedures of the Disconnect Mode are shown in the Fig. 4(e), where the SW are opened after the S^* reaches close to zero.

IV. REAL-TIME DOMAIN EMULATION

The DG system described in the previous sections was implemented in a real-time domain emulation (RTDE) system using the Typhoon HIL 602+. The Inverter Manager and Control Structure were implemented in a DSP-TMS320F28335. The grid parameters are set to a weak grid, where the PCC voltage is more susceptible to distortions and present a greater challenge to the controller [11], [12], [20]. Also, the irradiation and temperature that affect the photovoltaic panel are constant, with values of 1000 W/m^2 and 25 ^{o}C. In this way, the maximum power produced by the DC stage is equal to the maximum power of the photovoltaic array (15 kW).

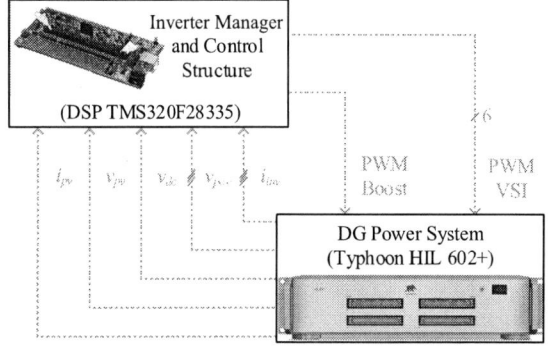

Fig. 5. Real-time domain emulation setup.

In order to analyze all the operation modes addressed in the section III, the following transitions are made during the emulation of the system:

- Connection of the DG in the main grid with the limited active power, where the reference power (P^*) is equal to the nominal power of the DG (P_{dg}). Therefore, the Boost controller is set to MPPT mode;
- Change of the reference power in Limited Active Power mode, from P_{dg} to 10 kW, at the instant of 32 s;
- Change from Limited Active Power mode to Limited Reactive Power mode, where $Q^* = -5$ kVAr, at the instant of 55 s;
- Change of the reference power in Limited Reactive Power mode, from $Q^* = -5$ kVAr to $Q^* = +5$ kVAr, at the instant of 82 s;
- Interrupt dispatch and disconnect the DG from the grid, at the instant of 106 s.

The results of the apparent (S_{gd}), active (P_{gd}) and reactive (Q_{gd}) power of the DG dispatched to the grid are presented in Fig. 6. In the same figure are shown the power produced by the DC stage (P_{pv}). The 5s reference ramp generated by DG Management in the transitions of the power references assisted the PV inverter dispatch power to the main grid in a smooth way, avoiding abrupt changes in the power references that could compromise the quality of the injected power.

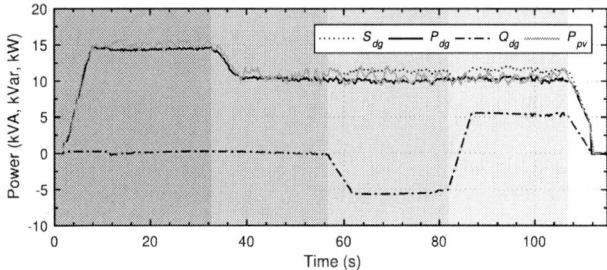

Fig. 6. DG power dispatch during the operation modes.

In the Limited Active Power mode, where the boost controller is set to MPPT mode, the current injected in the grid with the PCC voltage are show in Fig. 7. Furthermore, the total harmonic distortion (THD) of the inverter current (i_{inv}) is 0.79 %, lower than the 5.0 % limit of the IEEE 1547-2018. The spectral analysis is presented in Fig. 8, where the individual harmonics of i_{inv} are compared with the limits of IEEE 1547-2018.

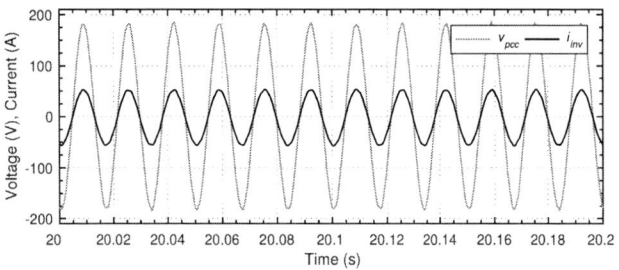

Fig. 7. DG operation in limited active power, with $P^* = P_{mppt}$ and $Q^* = 0$ kVAr.

The operation of the DG in the Limited Reactive Power mode is also evaluated under capacitive and inductive reactive power variation, as it can be seen in Fig. 9 and in Fig. 10,

respectively. The power factor in both cases was 0.88. In both cases the THD remained below 1.0 %.

The last mode of operation analyzed is Dispatch Interruption and then the Disconnection of the DG from the grid. The procedure is presented in Fig. 11. The power references generated by the DG Management are decreased until to be close to zero, so the DG can disconnect from the grid.

Additionally, with the purpose to evaluate the performance of the proposed control strategy closer to practical conditions, a real measured irradiation is applied to the photovoltaic array, while the temperature remains constant at 25 °C. The power of the DG (DC and AC stage) and the irradiation data is presented in Fig. 12 (a) and (b), respectively. The transition between the operation modes was performed in the same way as Fig. 6, but in different instants.

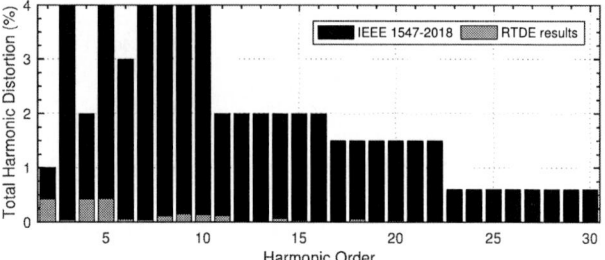

Fig. 8. DG inverter current spectral analysis.

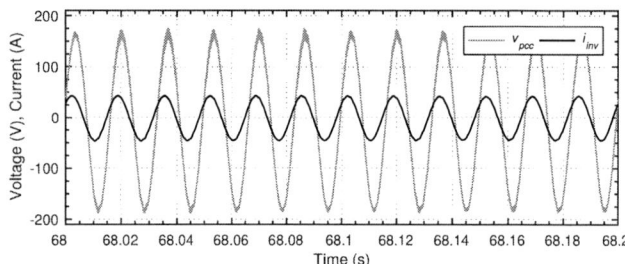

Fig. 9. DG operation in limited reactive power mode, with $Q^* = -5$ kVAr and $P^* = 10$ kW.

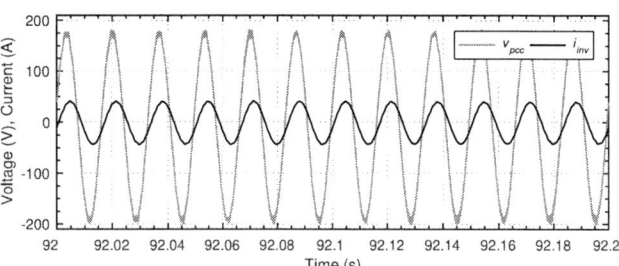

Fig. 10. DG operation in limited reactive power mode, with $Q^* = +5$ kVAr and $P^* = 10$ kW.

V. CONCLUSION

A PV inverter system compliant with IEEE 1547-2018 was presented in this paper and the main control aspects were pointed out. Furthermore, the issues related to connection and operation of the DG was discussed and solutions were proposed. The main purpose of the paper is to provide a complete structure for connection and different operation modes

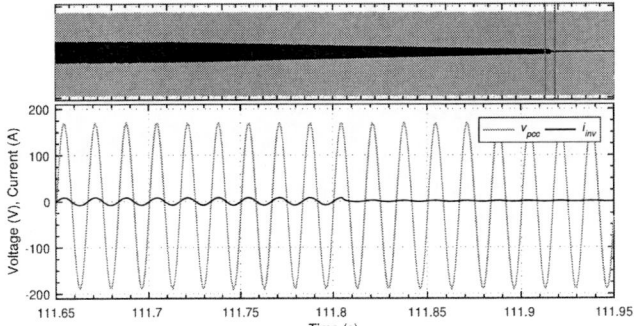

Fig. 11. DG dispatch interruption and disconnection.

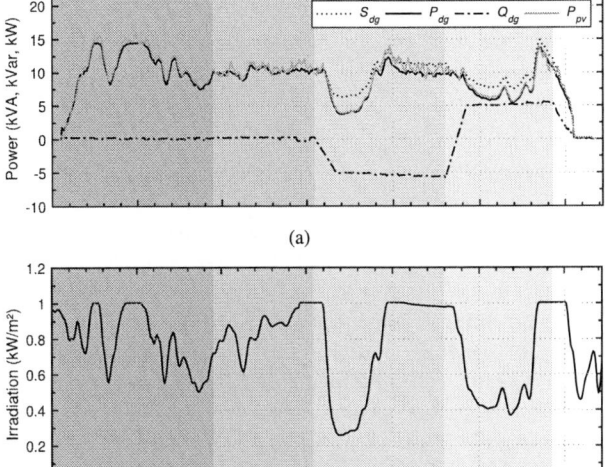

(a)

(b)

Fig. 12. DG operation with a real irradiation data. (a) DG power dispatch. (b) Irradiation data.

of the grid-tied inverter taking in account the IEEE 1547-2018 definitions. The real time evaluation of the proposed control strategy demonstrates the feasibility of the grid-tied inverter operation under several operation conditions, as variant solar irradiation and reactive power dispatch. Results shown the system is stable for strong and weak grids and presents very low voltage distortion (THD$<$ 1%) in all cases.

ACKNOWLEDGMENT

This study was financed in part by the Coordenacao de Aperfeicoamento de Pessoal de Nivel Superior - Brasil (**CAPES**) - Finance Code 001, of Conselho Nacional de Desenvolvimento Cientifico e Tecnologico (**CNPq**), from Fundacao Araucaria (**FA**) and from Financiadora de Estudos e Projetos (**FINEP**).

REFERENCES

[1] F. Blaabjerg, Y. Yang, D. Yang, X. Wang, "Distributed Power-Generation Systems and Protection", *Proceedings of the IEEE*, vol. 105, no. 7, pp. 1311–1331, July 2017, doi:10.1109/JPROC.2017.2696878.

[2] Y.-K. Wu, J.-H. Lin, H.-J. Lin, "Standards and guidelines for grid-connected photovoltaic generation systems: A review and comparison", *IEEE Transactions on Industry Applications*, vol. 53, no. 4, pp. 3205–3216, 2017.

[3] P. Rodrguez, A. Luna, R. S. Muoz-Aguilar, I. Etxeberria-Otadui, R. Teodorescu, F. Blaabjerg, "A Stationary Reference Frame Grid Synchronization System for Three-Phase Grid-Connected Power Converters Under Adverse Grid Conditions", *IEEE Transactions on Power Electronics*, vol. 27, no. 1, pp. 99–112, Jan 2012.

[4] Y. Han, M. Luo, X. Zhao, J. M. Guerrero, L. Xu, "Comparative Performance Evaluation of Orthogonal-Signal-Generators-Based Single-Phase PLL AlgorithmsA Survey", *IEEE Transactions on Power Electronics*, vol. 31, no. 5, pp. 3932–3944, May 2016.

[5] T. Tran, T. Chun, H. Lee, H. Kim, E. Nho, "PLL-Based Seamless Transfer Control Between Grid-Connected and Islanding Modes in Grid-Connected Inverters", *IEEE Transactions on Power Electronics*, vol. 29, no. 10, pp. 5218–5228, Oct 2014.

[6] R. Cardoso, R. F. D. Camargo, H. Pinheiro, H. A. Grundling, "Kalman filter based synchronisation methods", *IET Generation, Transmission Distribution*, vol. 2, no. 4, pp. 542–555, July 2008.

[7] R. Teodorescu, F. Blaabjerg, U. Borup, M. Liserre, "A new control structure for grid-connected LCL PV inverters with zero steady-state error and selective harmonic compensation", *in Applied Power Electronics Conference and Exposition, 2004. APEC '04. Nineteenth Annual IEEE*, vol. 1, pp. 580–586 Vol.1, 2004, doi:10.1109/APEC.2004.1295865.

[8] A. S. Rodrigo Kraemer, E. G. Carati, J. P. d. Costa, R. Cardoso, M. O. Carlos Stein, "Robust Design of Control Structure for Three-Phase Grid-Tied Inverters", *in 2018 13th IEEE International Conference on Industry Applications (INDUSCON)*, pp. 636–643, Nov 2018, doi: 10.1109/INDUSCON.2018.8627068.

[9] C. H. Van Der Broeck, S. A. Richter, J. V. Bloh, R. W. De Doncker, "Methodology for analysis and design of discrete time current controllers for three-phase PWM converters", *CPSS Transactions on Power Electronics and Applications*, vol. 3, no. 3, pp. 254–264, Sep. 2018, doi: 10.24295/CPSSTPEA.2018.00025.

[10] R. A. L. Junior, E. G. Carati, J. P. da Costa, R. Cardoso, C. M. O. Stein, "Robust Design of Active Damping with Current Estimator for Single-Phase Grid-Tied Inverters", *IEEE Transactions on Industry Applications*, pp. 1–1, 2018, doi:10.1109/TIA.2018.2838074.

[11] M. Lu, A. Al-Durra, S. M. Muyeen, S. Leng, P. C. Loh, F. Blaabjerg, "Benchmarking of Stability and Robustness Against Grid Impedance Variation forLCL-Filtered Grid-Interfacing Inverters", *IEEE Transactions on Power Electronics*, vol. 33, no. 10, pp. 9033–9046, Oct 2018, doi:10.1109/TPEL.2017.2784685.

[12] S. G. Parker, B. P. McGrath, D. G. Holmes, "Regions of Active Damping Control for LCL Filters", *IEEE Transactions on Industry Applications*, vol. 50, no. 1, pp. 424–432, Jan 2014, doi:10.1109/TIA.2013.2266892.

[13] Y. He, X. Wang, X. Ruan, D. Pan, X. Xu, F. Liu, "Capacitor-Current Proportional-Integral Positive Feedback Active Damping for LCL-Type Grid-Connected Inverter to Achieve High Robustness Against Grid Impedance Variation", *IEEE Transactions on Power Electronics*, pp. 1–1, 2019, doi:10.1109/TPEL.2019.2906021.

[14] Y. Atwa, E. F. El-Saadany, M. Salama, R. Seethapathy, M. Assam, S. Conti, "Adequacy evaluation of distribution system including wind/solar DG during different modes of operation", *IEEE Transactions on Power systems*, vol. 26, no. 4, pp. 1945–1952, 2011.

[15] "IEEE Standard for Interconnection and Interoperability of Distributed Energy Resources with Associated Electric Power Systems Interfaces", *IEEE Std 1547-2018 (Revision of IEEE Std 1547-2003)*, pp. 1–138, April 2018, doi:10.1109/IEEESTD.2018.8332112.

[16] Y. Han, H. Li, P. Shen, E. A. A. Coelho, J. M. Guerrero, "Review of active and reactive power sharing strategies in hierarchical controlled microgrids", *IEEE Transactions on Power Electronics*, vol. 32, no. 3, pp. 2427–2451, 2016.

[17] P. Rodriguez, R. E. Teodorescu, M. Liserre, *Grid Converters for Photovoltaic and Wind Power Systems*, 1 ed., Wiley-IEEE Press, 2011.

[18] M. Liserre, R. Teodorescu, F. Blaabjerg, "Stability of photovoltaic and wind turbine grid-connected inverters for a large set of grid impedance values", *IEEE Transactions on Power Electronics*, vol. 21, no. 1, pp. 263–272, Jan 2006, doi:10.1109/TPEL.2005.861185.

[19] R. H. Lasseter, "Smart distribution: Coupled microgrids", *Proceedings of the IEEE*, vol. 99, no. 6, pp. 1074–1082, 2011.

[20] W. Wu, Y. Liu, Y. He, H. S. Chung, M. Liserre, F. Blaabjerg, "Damping Methods for Resonances Caused by LCL-Filter-Based Current-Controlled Grid-Tied Power Inverters: An Overview", *IEEE Transactions on Industrial Electronics*, vol. 64, no. 9, pp. 7402–7413, Sep. 2017, doi:10.1109/TIE.2017.2714143.

Partial Harmonic Current Compensation Applied to Multiple Photovoltaic Inverters in a Radial Distribution Line

André L. P. de Oliveira[1], Lucas S. Xavier[2], João M. S. Callegari[3], Allan F. Cupertino[2,4],
Victor F. Mendes[2], Heverton A. Pereira[1]

[1]Electric Engineering Department, Federal University of Viçosa, 36570-900, Viçosa, Brazil

[2] Graduate Program in Electrical Engineering - Federal University of Minas Gerais -
Av. Antônio Carlos 6627, 31270-901, Belo Horizonte, MG, Brazil

[3] Graduate Program in Electrical Engineering - Federal Center for Technological Education of Minas Gerais
- Av. Amazonas 5253, 30421-169, Belo Horizonte, MG, Brazil

[4]Department of Materials Engineering - Federal Center for Technological Education of Minas Gerais
- Av. Amazonas 5253, 30421-169, Belo Horizonte - MG, Brazil

e-mail: andre.pires@ufv.br, lsantx@gmail.com, jmcallegari@hotmail.com, allan.cupertino@yahoo.com.br,
victormendes@cpdee.ufmg.br, heverton.pereira@ufv.br

Abstract—**The increase of the current harmonic content due to nonlinear loads is one of the main concerns in power systems. Therefore, the possibility to use the current margin of photovoltaic (PV) systems to provide harmonic current compensation have been addressed in several works. In this context, an important point is to ensure that the PV inverter current does not extrapolate its rated current peak. For this purpose, dynamic current saturation techniques are adopted, incorporating the partial harmonic current compensation capability in PV inverters. Thereby, this work proposes to implement the partial harmonic current compensation applied to multiple PV inverters using a dynamic saturation technique based on an open-loop algorithm. Furthermore, a radial distribution line is used to analyze the effects of harmonic compensation sharing among PV inverters under different solar irradiance conditions. The current margin of PV systems are combined to compensate nonlinear load currents in the line. Simulation results show the improvements in the grid power quality, ensuring a total demand distortion (TDD) lower than 5%, as recommended by the IEEE Std 519-2014.**

Index Terms—**partial harmonic current compensation, photovoltaic inverters, dynamic saturation, radial distribution line**

I. INTRODUCTION

The Distributed Generation (DG) have been used to diversify the energy matrix in several countries around the world. This generation model allows a greater proximity between the energy generation and the consumer, avoiding the adversities caused by distant transmission systems. In this context, the photovoltaic (PV) systems stand out due to the decentralized generation, ease of installation and less impact on the environment if compared to other generation systems such as hydroelectric, thermoelectric and others [1].

With the remarkable growth of grid-connected photovoltaic power plants in the last decade, the possibility of integrating

The authors would like to thank the Brazilian agencies CNPq, CAPES and FAPEMIG by funding.

extra functionalities into this system has been widely addressed. For low solar irradiance conditions, the PV inverter has a current margin that can be used to provide ancillary services to the grid, such as reactive power support and harmonic current compensation [2]–[4].

One of the main concerns in the several stages of eletric power systems is the harmonic distortion. The growth of the nonlinear loads connection causes the reduction of the grid power quality. Passive and active power filters can be used to filter the harmonic current of nonlinear loads [5]. The active filters are capable of operating over a wide frequency range and are designed to compensate harmonic current content in the grid by means of its injected current. Due to the similarity between the grid interface of PV systems and the active filters, references [6]–[8] suggest to use the PV inverters for this purpose.

However, some details must be observed for PV inverters to be able to provide the harmonic current compensation (HCC). Firstly, the current controller must be tuned to different frequencies, since the controlled signal is composed of several frequencies. For this purpose, references [9] and [10] address resonant and repetitive controllers with the capability to adjust high gains in the interested frequencies. Secondly, an important point is the detection of the load harmonic current content. The harmonic detection method is responsible to extract the load harmonic current information. In literature, several methods are proposed based on instantaneous power theory [11], [12], conservative power theory [8], second order generalized integrator (SOGI) [13], and discrete Fourier transform [14]. Finally, another point to be observed is the inverter current capacity. PV inverters have a maximum current limitation and the device may be damaged if this value is exceeded. Thus, techniques are required to limit the inverter current when the HCC is performed [15].

978-1-7281-4181-7/19 $31.00 © 2019 IEEE

In fact, there are techniques proposed in literature to limit the current peak injected by PV inverters during HCC. Thereby, the dynamic saturation (DS) can be adopted. This technique decreases the amplitude of the compensated harmonic components according to the current availability of PV inverter, leading to a partial HCC. Reference [16] proposes a DS technique based on an open-loop algorithm to limit the peak value of the inverter current during HCC with low computational burden and memory requirements. The DS technique, proposed by this reference, inserts the partial HCC capability on the PV inverter. Therefore, the PV inverter contributes for the maximum HCC according to its rated current. However, this same reference made an analysis of the partial HCC for one PV inverter and indicates, for future works, the partial HCC sharing analysis between multiple PV inverters, connected in the line.

Taking into consideration the previously mentioned points, this work proposes the implementation and the study of the partial HCC to multiple PV systems using a DS technique to limit the current peak injected by the PV inverter. Furthermore, this work analyze the effects of partial HCC in a radial distribution line with several PV systems, according to the current margin of each one.

This paper is outlined as follows: Section II presents the power and control structure of the proposed system used to investigate the partial HCC to multiple PV systems. Section III presents the DS technique adopted and its application for partial HCC to multiple PV inverters. The case studies to evaluate the effects of partial HCC to multiple PV systems in radial distribution line are presented in Section IV. The simulation results are presented in Section V. Finally, the conclusions about the work are carried out in Section VI.

II. POWER AND CONTROL STRUCTURE

The proposed system to investigate the effects of partial HCC in a radial distribution line is shown in Fig. 1. Each PV system is connected to the grid, by a dc/ac inverter, and the local load at the own point of common coupling. The connection of each single-phase PV system is shown in Fig. 2. A boost converter is used to increase the operation range of the converter and, as shown in Fig. 3 (a), its control strategy is based on proportional-integral (PI) controller. A maximum power point tracking (MPPT) algorithm calculates the PV arrays voltage reference (v_{pv}^*) in order to extract the maximum performance from the PV arrays [17]. A LCL filter, designed in [18], is used to attenuate high frequencies due to the switching frequencies of the dc/ac stage.

Fig. 3 (b) shows the PV inverter control block diagram. The outer loop is responsible for controlling the dc-link voltage and to provide the active power reference (P^*) [16]. Moreover, the reactive power reference (Q^*) is also informed by the outer loop, as addressed in references [11] and [19]. Using the single-phase PQ-theory, the fundamental current component i_α^* can be calculated as:

$$i_\alpha^* = \frac{2}{v_\alpha^2 + v_\beta^2}(v_\alpha P^* + v_\beta Q^*), \tag{1}$$

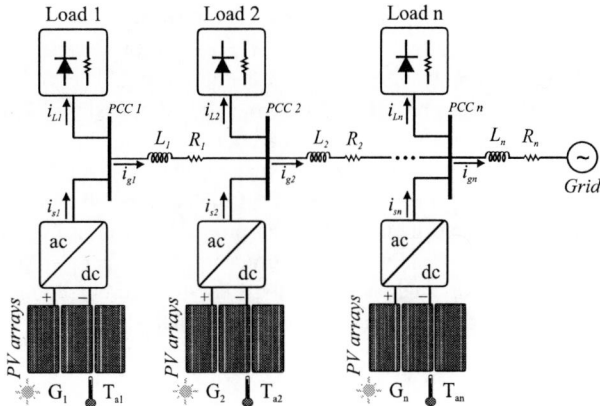

Fig. 1. PV inverters and loads in the radial power system studied in this paper.

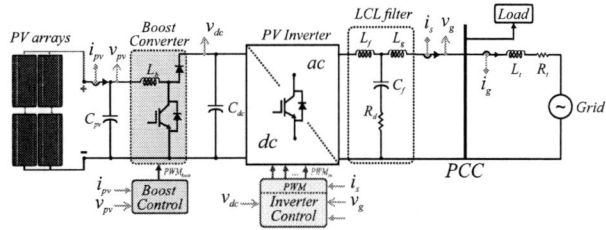

Fig. 2. Grid-connected photovoltaic system for single-phase applications.

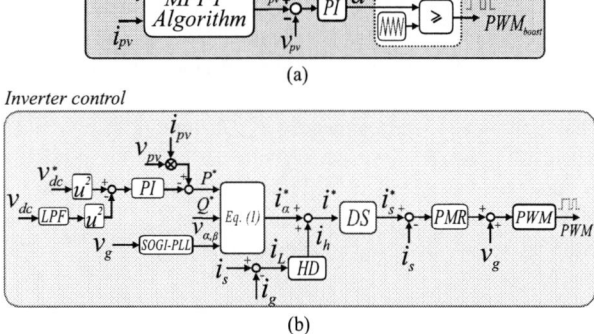

Fig. 3. PV system control strategy. (a) Boost converter control diagram. (b) Inverter control diagram with ancillary service capability.

where v_α and v_β are the grid voltages at fundamental frequency in stationary reference. A phase-locked loop (PLL) with orthogonal signal generator based on SOGI structure is used to calculate these components [20].

Measurements of inverter current (i_s) and PCC_x current (i_{gx}) are made to obtain the load current (i_L) information for HCC. The current can be measured in PCC and the harmonics amplitude and phase sent to the PV inverters, that would make the communication more reliable as presented in [21]. Depending on the distance between the PCC and the PV inverter, current i_g measurement delays may occur, which should be considered in the control design. However, this approach is beyond the scope of this work.

Then, the harmonic detection (HD) block provides the load harmonic current component (i_h) to the control. The sliding discrete Fourier transform (SDFT) method addressed in [22] is used in this work to detect the load current harmonic content. The DS block receives the reference current i^*, given by the sum between i_α^* and i_h, and verify if the HCC can be total or partial. The inner loop receives the processed inverter current reference (i_s^*). Finally, a proportional multi-resonant (PMR) controller, addressed in [9], is used for the current control, tuned at the frequencies of the current signal.

III. Partial HCC Applied To Multiple PV systems

The DS technique is important to avoid damaging the PV inverter components. In this work, the DS technique based on an open-loop algorithm, proposed and described by [16], is adopted. Fig. 4 shows the DS technique diagram.

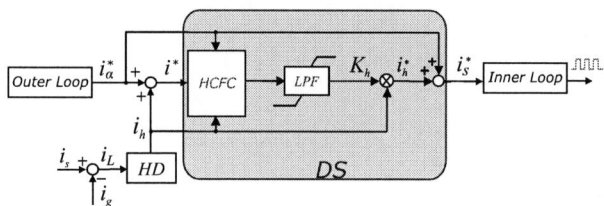

Fig. 4. DS technique based on a open-loop alorithm diagram.

The harmonic compensation factor calculation (HCFC) block is an algorithm which has as inputs the load harmonic current component (i_h), the fundamental current reference (i_α^*) and the total inverter current reference (i^*). Firstly, the algorithm establishes the contribution value of i_h and i_α^* for the peak value of i^*, represented by I_h, I_α and I_m^*, respectively. Then, a factor (K_h) is calculated to designate the portion of i_h that can be added to the converter current reference i_s^* without exceeding the inverter rated current I_m^*. The following equation shows how K_h is calculated:

$$K_h = \frac{I_m^* - I_\alpha}{I_h}. \qquad (2)$$

The complete description of the algorithm for K_h factor calculation can be find in [16]. This work has as objective propose a strategy that can be applied in the distribution of HCC between PV inverters.

Based on these concepts, this paper evaluates the applications of this DS technique in a radial distribution line. Each PV inverter calculates a compensation factor (K_h) and the harmonic current references are given by:

$$\begin{cases} i_{h1}^* = K_{h1} \cdot i_{Lh1} \\ i_{h2}^* = K_{h2} \cdot (i_{Lh2} - i_{gh1}) \\ \qquad \vdots \\ i_{hn}^* = K_{hn} \cdot (i_{Lhn} - i_{gh(n-1)}) \end{cases} \qquad (3)$$

where i_{Lh} is the local load harmonic current components and i_{gh} is the harmonic current components of i_g. It is important

to highlight that each inverter makes its own current measurements without communicating with the others. Thus, the inverters K_h factor calculation is exclusive for each converter and an K_h optimization between them is not addressed in this work.

Starting the analysis from the second inverter, if the previous inverter does not have current margin to compensate all the local load harmonic content, then the next inverter connected to the electrical system can help in the HCC, as long as it has a current margin. For example, if K_{h1} is lower than 1, the load 1 remaining harmonic content is provided by the grid, by means of the current i_{g1}. In this way, the harmonic current reference i_{h2}^* has the local load harmonic current components (i_{Lh2}) and part of the harmonic current components of i_{g1} (i_{gh1}). The latter one is due to the lack of compensation margin from previous converter. This idea can be extended for the next PV inverters, so each subsequent PV inverter can auxiliary the previous ones in the line.

IV. Case Study

A system with $n = 3$ converters is developed and simulated in PLECS and MATLAB softwares. The parameters of the PV inverters are shown in Tab. I. Among the converters, the inverter-3 is oversized to be able to offer ancillary services to the grid.

TABLE I
PARAMETERS OF THE MULTIPLE PV INVERTERS AND LOADS CONNECTED
INTO THE RADIAL LINE.

Parameters	Inverter 1	Inverter 2	Inverter 3
Grid voltage - line to neutral (V_g)	220 V	220 V	220 V
Inverter: switching frequency (f_{sw})	12 kHz	12 kHz	12 kHz
PV array rated power [1] (P_{mpp})	4 kW	2 kW	6 kW
PV array configuration (N_s,N_p)	8x2	8x1	12x2
Inverter rated power (S_{max})	4 kVA	2 kVA	8 kVA
Peak inverter current (I_{max})	25.7 A	12.9 A	51.4 A
LCL filter inductance ($L_f = L_g$)	1.5 mH	1.5 mH	1.5 mH
LCL filter capacitance (C_f)	2 μF	2 μF	2 μF
LCL filter damping resistor (r_d)	2 Ω	2 Ω	2 Ω
LCL filter inductance X/R ratio	20	20	20
dc-link capacitance (C_{dc})	3 mF	3 mF	3 mF
Boost: inductance (L_b)	0.8 mH	0.8 mH	0.8 mH
Boost: inductor resistance (r_b)	10 mΩ	10 mΩ	10 mΩ
Boost: capacitance (C_{pv})	0.5 mF	0.5 mF	0.5 mF
Boost: switching frequency (f_{swb})	12 kHz	12 kHz	12 kHz
Load rectifier inductor	1.5-3.5 mH	3.5 mH	-
Load rectifier capacitor	3 mF	3 mF	-
Load active power	4 kW	2 kW	6 kW

The inverters control is developed with sampling frequency equal to the switching frequency. Moreover, all PI controllers are discretized by the Tustin method, while the resonant controllers are discretized by the Tustin method with prewarping. The controllers parameters are shown in Tab. II.

It is assumed that distributed generators with the HCC capability are separated by impedance distribution lines (see Tab. III) and different conditions of solar irradiance may occur on the PV arrays surface connected to inverters 1, 2 and 3.

The nonlinear loads connected to the PCCs are represented by rectifiers. Besides, a resistive load is connected to the PCC-3.

978-1-7281-4181-7/19 $31.00 © 2019 IEEE

TABLE II
PARAMETERS OF THE CONTROLLERS.

Parameter	Inverter 1, 2 and 3
v_{dc}^2: Proportional gain	0.207
v_{dc}^2: Integral gain	2.369
i_g: Proportional gain	22.249
i_g: Resonant gain	4000
Boost input voltage: Proportional gain	0.754
Boost input voltage: Integral gain	66.139
PLL: Proportional gain	76.454
PLL: Integral gain	5684.892

TABLE III
LINE IMPEDANCES.

Parameter	Label	Value
Inductance PCC1 $\rightarrow$ PCC 2	L_1	0.2 mH
Resistance PCC1 $\rightarrow$ PCC 2	R_1	0.267 Ω
Inductance PCC2 $\rightarrow$ PCC 3	L_2	0.2 mH
Resistance PCC2 $\rightarrow$ PCC 3	R_2	0.267 Ω
Inductance PCC3 $\rightarrow$ grid	L_g	0.212 mH
Resistance PCC3 $\rightarrow$ grid	R_g	0.267 Ω

Two case studies are evaluated in this work. Firstly, the solar irradiance profiles shown in Figs. 5(a), (b) and (c) are applied into the PV arrays connected to the converters 1, 2 and 3, respectively. The inverter input power variation allows a distribution of harmonic current compensation between the converters according to their available current margins.

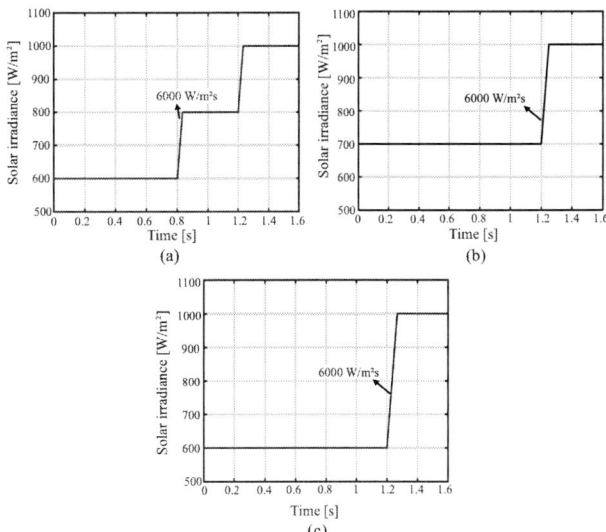

Fig. 5. Solar irradiance profile on the PV array connected to the inverter (a) 1, (b) 2, (c) 3.

The second case study is based on a 3-day solar irradiance profile shown in Fig. 6, applied to all inverters. It is expected that the margin for the current HCC is reduced near midday, especially on the third day. Moreover, outside the hours of 8 a.m. to 5 p.m., the converters presents maximum margin to perform harmonic current compensation.

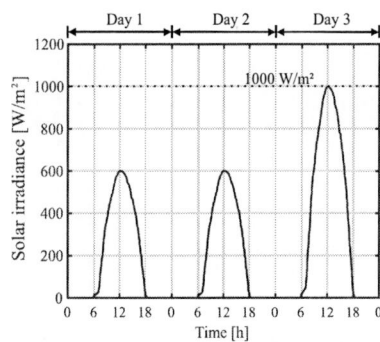

Fig. 6. 3-day solar irradiance profile applied in the inverters.

V. RESULTS

The operation of the three PV systems is simulated to check the partial HCC applied to multiple PV systems. The solar irradiance profiles for the three systems are shown in Fig. 5. The HCC of the PV systems is started in 0.4 s with the aim of reducing the harmonic components in the grid current i_{g3}. In 0.8 s the harmonic content of load 1 is varied and in 1.2 s the solar irradiance is increased in all PV systems to verify the behavior of the compensation strategy under load and solar irradiance variation condition. However, the PV inverters have current limitation and the total harmonic compensation cannot be possible at all times. The K_h factors for the three inverters are shown in Fig. 7.

In $0.4\ \mathrm{s} < t < 0.8\ \mathrm{s}$ the PV inverter-1 has current margin to compensate $K_h = 0.47$ of the nonlinear load-1. The inverter-2 has current margin to compensate $K_h = 0.2$ of the load-2 harmonic current and the remaining harmonic components of the load-1 current. The inverter-3 compensates all the remaining harmonic components ($K_h = 1$), reducing the propagation of the harmonic currents of load-1 and 2 to the grid. The current waveforms during the beginning of the HCC in 0.4 s are shown in Fig. 8. Before 0.4 s, the current of the inverters are sinusoidal, since the HCC is disabled. In this case, the TDD of the PCC-3 grid current (i_{g3}) is 34.08 %. When the HCC is enabled, the i_{g3} TDD reduces to 1.40 %. It is important to note that the inverters do not exceed their current limit I_{nm}.

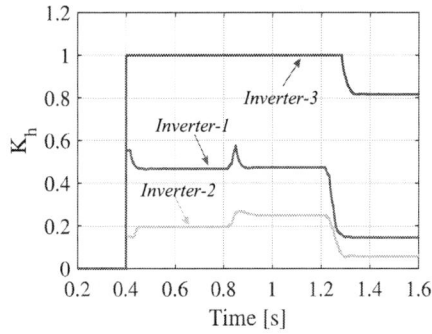

Fig. 7. K_h factor for the three PV systems.

978-1-7281-4181-7/19 $31.00 © 2019 IEEE

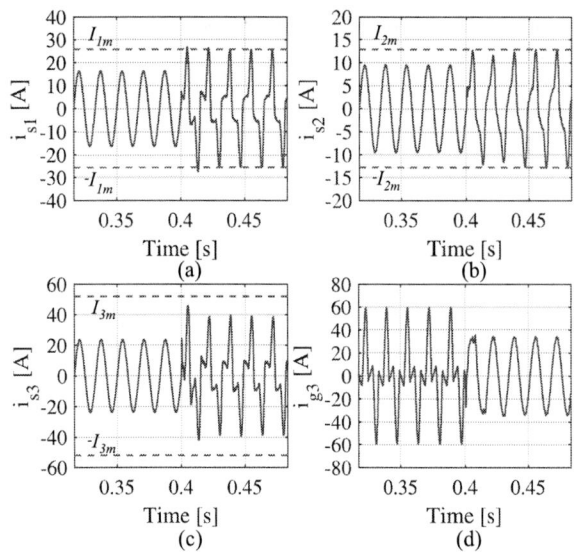

Fig. 8. Current waveforms during the beginning of the HCC in 0.4 s. (a) Inverter-1 current (i_{s1}). (b) Inverter-2 current (i_{s2}). (c) Inverter-3 current (i_{s3}). (d) PCC-3 grid current (i_{g3}).

In 0.8 s, the load-1 TDD current is reduced and the solar irradiance in the PV system-1 increases to 800 W/m². The K_h factors of the inverters-1 and 2 are increased to 0.47 and 0.25, respectively. The inverter-3 continues compensating all the remaining harmonic components ($K_h = 1$), as shown in Fig. 7. The current waveforms in the PCCs, around 0.8 s, are shown in Fig. 9. Note that, the currents of the PCC-1 and 2 are distorted and the PV inverter-3 prevents these distortions from propagating to grid current in PCC-3. After 0.8 s the TDD of the PCC-3 grid current (i_{g3}) reduces to 0.75 %.

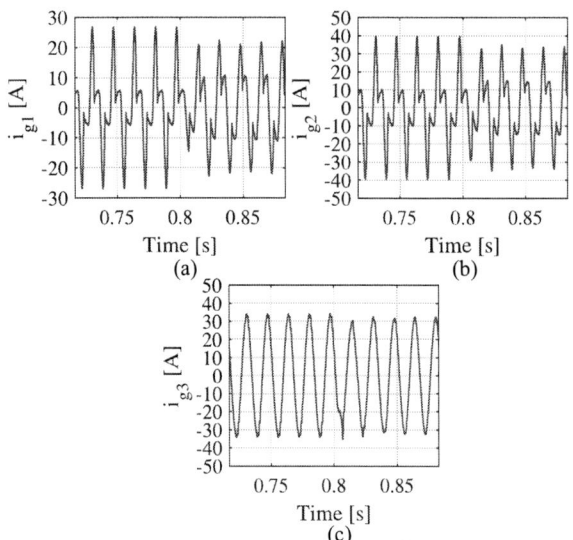

Fig. 9. Current waveforms around 0.8 s. (a) PCC-1 current (i_{g1}). (b) PCC-2 current (i_{g2}). (c) PCC-3 current (i_{g3}).

In 1.2 s < t < 1.6 s the solar irradiance in all PV systems

increase to 1000 W/m². Therefore, the K_h factor of the inverters-1, 2 and 3 reduces to 0.05, 0.14 and 0.8, as shown in Fig. 7. After 1.2 s, with high solar irradiance conditions in all PV systems, the TDD of the PCC-3 grid current (i_{g3}) is 5 %. Even with rated solar irradiance, the PV inverter-3 has current margin to compensate part of the harmonic content, keeping the i_{g3} TDD respecting the recommendations in the IEEE Std 519-2014. The current waveforms around 1.2 s are shown in Fig. 10. It is possible to observe that the currents injected by the inverters exceed the maximum limits during some cycles in the transients, when the dynamic saturation technique of each inverter is calculating the amount of harmonics that each converter can compensate. In steady-state, the currents of the inverters do not exceed the limits of the power switches.

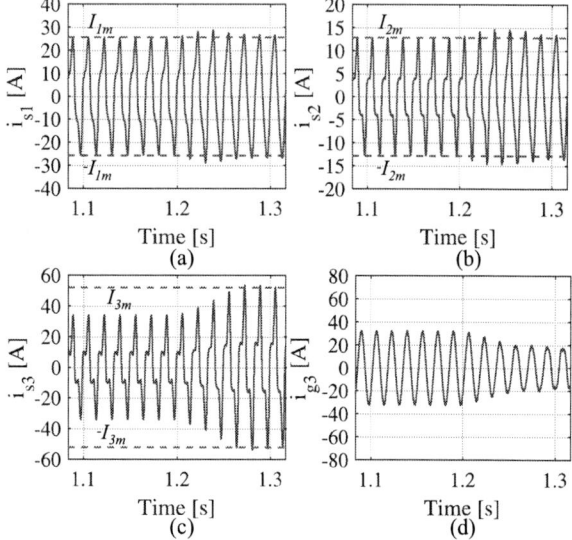

Fig. 10. Current waveforms around 1.2 s. (a) Inverter-1 current (i_{s1}). (b) Inverter-2 current (i_{s2}). (c) Inverter-3 current (i_{s3}). (d) PCC-3 grid current (i_{g3}).

Three days of operation of these three PV systems are simulated to verify the partial compensation applied to multiple PV systems. The solar irradiance profile is shown in Fig. 6. The PV power generation by the three inverters is shown in Fig. 11(a). The PV power generation is smaller than the rated power (P_{nm}^*) of the inverters on days 1 and 2. Afterwards, the PV power generation is equal to the rated power of the inverter 1 and 2 on day 3. The inverter-3 has a power design above the rated PV power of the system 3 to perform ancillary services.

The TDD profile of PCC-3 grid current is shown in Fig. 11(b). Without the HCC, the i_{g3} TDD is 29.08 % on days 1 and 3 and 39.08 % on day 2. With HCC, the i_{g3} TDD is below the 5 %, recommended by the IEEE Std 519-2014. The K_h factors of the three systems are shown in Fig. 11(c). The K_h factor are minimum around 12 h, when the current margin of the PV inverters to perform harmonic compensation is minimum due to the PV power generation. Note that the inverters 1 and 2 partially compensate the load current 1 and

978-1-7281-4181-7/19 $31.00 © 2019 IEEE

2 during the day and the inverter-3 has current margin to compensate all remaining harmonic currents on day 1 and 2. On day 3, the inverter-3 partially compensate these harmonic currents, but enough to keep the TDD of i_{g3} below 5 %. The peak current (I_m) injected by each inverter during these three daily operation is shown in Fig. 11(d). Note that in any moment the inverters operate above their current peak limit (I_n^*m) due to the dynamic current saturation.

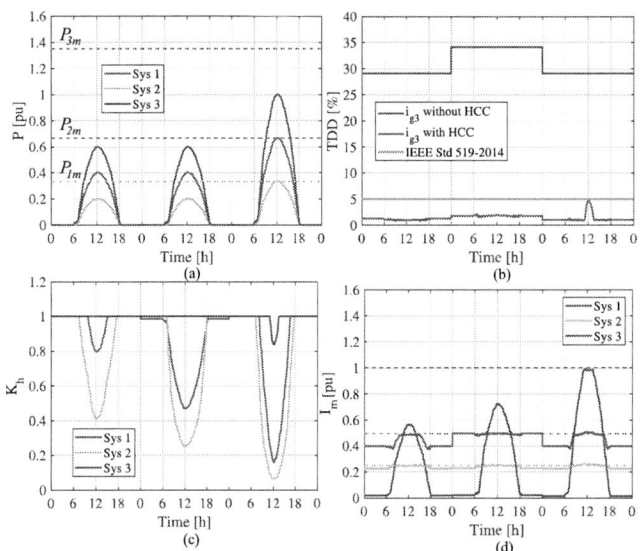

Fig. 11. Three days of operation of the three PV systems with partial HCC capability. (a) PV power generation. (b) TDD of the PCC-3 grid current (i_{g3}). (c) K_h factors of the three PV inverters. (d) Peak current (I_m) injected by each PV inverter. PU base is the converter power.

VI. CONCLUSIONS

This work proposed to analyze the partial harmonic current compensation applied to multiple photovoltaic systems in radial distribution lines. This harmonic compensation strategy is achieved through dynamic current saturation algorithm. The algorithm calculates the among of load harmonic current can be compensated in order to respect the current limitation of the PV inverters. Therefore, the PV systems can be configured to combine their current margins to compensate harmonic currents generated by nonlinear loads connected in the system.

Simulation results showed the performance of the harmonic compensation strategy applied in three PV systems and the improvements in the grid current quality. It can be noticed that the current margin of the inverters were combined to perform harmonic current compensation. Under such conditions, if several photovoltaic systems are already installed into the power system or active power filters, each one can contribute for the current compensation and, in the overall result, have a positive impact on the grid power quality.

REFERENCES

[1] F. Blaabjerg, Y. Yang, D. Yang, and X. Wang, "Distributed power-generation systems and protection," *Proceedings of the IEEE*, vol. 105, no. 7, pp. 1311–1331, July 2017.

[2] P. Acuna, L. Moran, M. Rivera, J. Dixon, and J. Rodriguez, "Improved active power filter performance for renewable power generation systems," *IEEE Trans. Power Electron.*, vol. 29, no. 2, pp. 687–694, Feb 2014.

[3] Y. Yang, F. Blaabjerg, H. Wang, and M. G. Simões, "Power control flexibilities for grid-connected multi-functional photovoltaic inverters," *IET Renewable Power Generation*, vol. 10, no. 4, pp. 504–513, 2016.

[4] H. K. Morales-Paredes, J. P. Bonaldo, and J. A. Pomilio, "Centralized control center implementation for synergistic operation of distributed multifunctional single-phase grid-tie inverters in a microgrid," *IEEE Trans. Ind. Electron.*, vol. 65, no. 10, pp. 8018–8029, Oct 2018.

[5] F. C. De la Rosa, *Harmonics and Power Systems*. Boca Raton: CRC Press, 2006.

[6] J. P. Bonaldo, H. K. M. Paredes, and J. A. Pomilio, "Control of Single-Phase Power Converters Connected to Low Voltage Distorted Power Systems with Variable Compensation Objectives," *IEEE Trans. Power Electron.*, vol. 31, no. 3, pp. 2039–2052, 2016.

[7] S. Munir and Y. W. Li, "Residential distribution system harmonic compensation using pv interfacing inverter," *IEEE Trans. Smart Grid*, vol. 4, no. 2, pp. 816–827, June 2013.

[8] H. K. M. Paredes, F. P. Marafão, D. I. Brandão, and A. Costabeber, "Multi-task control strategy for grid-tied inverters based on conservative power theory," *IET Renewable Power Generation*, vol. 9, no. 2, pp. 154–165, 2015.

[9] A. G. Yepes, F. D. Freijedo, O. Lopez, and J. Doval-Gandoy, "Analysis and Design of Resonant Current Controllers for Voltage-Source Converters by Means of Nyquist Diagrams and Sensitivity Function," *IEEE Trans. Ind. Electron.*, vol. 58, no. 11, pp. 5231–5250, 2011.

[10] X. H. Wu, S. K. Panda, and J. X. Xu, "Design of a plug-in repetitive control scheme for eliminating supply-side current harmonics of three-phase pwm boost rectifiers under generalized supply voltage conditions," *IEEE Trans. Power Electron*, vol. 25, no. 7, pp. 1800–1810, July 2010.

[11] H. Akagi, Y. Kanazawa, and A. Nabae, "Instantaneous reactive power compensators comprising switching devices without energy storage components," *IEEE Trans. Ind. Appl.*, vol. IA-20, no. 3, pp. 625–630, May 1984.

[12] F. Z. Peng, G. W. Ott, and D. J. Adams, "Harmonic and reactive power compensation based on the generalized instantaneous reactive power theory for three-phase four-wire systems," *IEEE Trans. Power Electron.*, vol. 13, no. 6, pp. 1174–1181, Nov 1998.

[13] P. Rodriguez, A. Luna, I. Candela, R. Mujal, R. Teodorescu, and F. Blaabjerg, "Multiresonant frequency-locked loop for grid synchronization of power converters under distorted grid conditions," *IEEE Trans. Ind. Electron.*, vol. 58, no. 1, pp. 127–138, Jan 2011.

[14] H. Chen, H. Liu, Y. Xing, H. Hu, and K. Sun, "Analysis and design of enhanced dft-based controller for selective harmonic compensation in active power filters," in *IEEE Applied Power Electronics Conference and Exposition (APEC)*, 2018.

[15] L. Xavier, A. Cupertino, and H. Pereira, "Ancillary services provided by photovoltaic inverters: Single and three phase control strategies," *Computers and Electrical Engineering*, vol. 70, no. 7, pp. 102–121, Aug. 2018.

[16] L. S. Xavier, A. F. Cupertino, H. A. Pereira, and V. F. Mendes, "Partial harmonic current compensation for multifunctional photovoltaic inverters," *IEEE Trans. Power Electron.*, pp. 1–1, 2019.

[17] D. Sera, L. Mathe, T. Kerekes, S. V. Spataru, and R. Teodorescu, "On the perturb-and-observe and incremental conductance mppt methods for pv systems," *IEEE Journal of Photovoltaics*, vol. 3, no. 3, pp. 1070–1078, July 2013.

[18] R. Peña-Alzola, M. Liserre, F. Blaabjerg, M. Ordonez, and Y. Yang, "Lcl-filter design for robust active damping in grid-connected converters," *IEEE Trans. Ind. Inf.*, vol. 10, no. 4, pp. 2192–2203, Nov 2014.

[19] Y. Yang, F. Blaabjerg, and H. Wang, "Low-voltage ride-through of single-phase transformerless photovoltaic inverters," *IEEE Trans. Ind. Appl.*, vol. 50, no. 3, pp. 1942–1952, May 2014.

[20] M. Ciobotaru, R. Teodorescu, and F. Blaabjerg, "A new single-phase PLL structure based on second order generalized integrator," *IEEE Annual Power Electronics Specialists Conf.*, 2006.

[21] M. Savaghebi, A. Jalilian, J. C. Vasquez, and J. M. Guerrero, "Secondary control for voltage quality enhancement in microgrids," *IEEE Trans. Smart Grid*, vol. 3, no. 4, pp. 1893–1902, Dec 2012.

[22] E. Jacobsen and R. Lyons, "The sliding dft," *IEEE Signal Processing Magazine*, vol. 20, no. 2, pp. 74–80, March 2003.

Enhanced Space-State Reduced-Order Microgrid Model in Common DQ-Reference Frame

Sebastián de J. Manrique Machado
Department of Electrical Engineering
Federal University of Technology of
Paraná
Apucarana, Brazil
sebastiand@utfpr.edu.br

Sérgio Augusto Oliveira da Silva
Department of Electrical Engineering
Federal University of Technology of
Paraná
Cornélio Procópio, Brazil
augus@utfpr.edu.br

José R. B. A. Monteiro[1] and Azauri A.
de Oliveira Jr[2]
Department of Electrical and
Computational Engineering
University of Sao Paulo, School of
Engineering of Sao Carlos
São Carlos, Brazil
jrm@sc.usp.br[1] azauri@sc.usp.br[2].

Abstract— Inverter-based droop-controlled microgrids systems are becoming attractive solutions to meet future energy demand in a sustainable way. The development of accurate models is crucial for control loops design and performance assessment of the system. However, extracting essential information about the overall microgrid behavior from complete models can be a hard task, as well as can involve a high computational burden depending on the microgrid size. Therefore, the development of a reduced-order model that properly represents the dynamic performance of the microgrid in a computationally efficient way is mandatory. In this paper, an enhanced reduced-order model able to represent the influence of network dynamics on the performance of the droop-controlled microgrid is presented. By means of simulation results the fidelity improvement of the proposed microgrid model is confirmed.

Keywords— *Droop control, microgrids, small-signal model, stability analysis.*

I. INTRODUCTION

Microgrid concept is expected to enable reliable large-scale integration of distributed generation (DG) based on renewable energy sources in distribution networks, in order to meet future energy demand in a sustainable way. For this purpose, droop control has been largely adopted as a solution for parallel operation of several converters and load sharing among them.[1], [2]. In order to properly design the involved control loops and assess the performance of the system, the development of suitable models that represent the dominant dynamics of the microgrid is required.

Microgrids are considered complex systems since the control loops of several inverters interact with each other, as well as with the distribution network and loads. Hence, the mentioned interactions can result in complex models in such a way that could be difficult to figure out the cause-effect relationships between the model parameters and the microgrid dynamic behavior. For this reason, several reduced-order models have been recently proposed, in order to accurately account for dominant microgrid dynamics, while the model is kept simple enough to easily identify the parameters that cause certain dynamic behaviors [3]–[9].

A third-order model based on DQ-reference frame has been proposed in [10]. This model neglects the internal control loops, the output filter of the inverters, as well as the dynamics of the network and the load. Thus, the model addressed in [10] could be regarded as a reduced-order model, however, it does not predict the instability caused by the adoption of large droop coefficients. In [11]–[13] a complete droop-controlled

microgrid models presenting a high accuracy associated to high complexity and computational burden have been developed. Theoretical and experimental evidences have been provided in [4]–[6], [8] demonstrating the importance of network and load dynamics on the region of stability predicted by the microgrid model. Particularly, dynamic phasors modeling (DPM) technique has been used in [4], which was previously introduced in stability studies of bulk power systems in [14]. In [5], the network and load dynamics were represented introducing two new states for each line or load. As a result, the accuracy was improved but at the cost of increasing the complexity of the model. Different from the aforementioned works, in [6] only the currents flowing through both loads and transmission lines have been considered as state variables, while both voltage amplitudes and relative phase-angles at the bus bars have been considered as inputs. Then, by means of singular perturbations technique [15], the order of the developed model was reduced, deriving new models that are valid for particular microgrid conditions.

The singular perturbations technique seems an interesting choice to be used in microgrids modeling since it enables to derive models that represent the influence of fast state variables on the slow ones. Indeed, this technique has been recently used in [7]–[9]. In [7], the first derivative of the fast state variables associated with the distribution lines were considered null, which is equivalent to regard only the influence of the steady-state solution of the fast state variables. Instead, in [8], the first derivative of fast state variables was approximated using Taylor series. Hence, a better representation related to the influence of the dynamics of both distribution network and loads on the slow states associated with droop controllers was achieved.

In this paper, the main insights provided in [8] are taken into account to include both network and loads dynamics into the model presented in [10]. Although the limitations of small signal model addressed in [10] are well known, the proposed modeling procedure has been proved to be very scalable to easily account for an arbitrary number of inverters and higher hierarchical control layers [5], [16]. Thus, based on the modeling procedure proposed in [10], the main contribution of this paper is to enhance its fidelity, without increasing its complexity.

This paper is organized as follows: In section II an overview of the model developed in [10] is presented, as well as the proposed procedures for including into the model the network dynamics without increasing the order of the microgrid model. In section III, the simulation setup is presented in detail. Then, in section IV, the effectiveness of

978-1-7281-4181-7/19 $31.00 © 2019 IEEE

the proposed model is verified through simulations results and comparisons between both models. Finally, in section V the conclusions are presented.

II. MICROGRID MODEL

A. Third-order model overview

In this paper, the model developed in [10] is referred as "third-order model" since each inverter adds three states to the closed-loop microgrid model. In this section an overview of this model is presented.

The expressions of conventional frequency and voltage droop control are, respectively, given by [1]:

$$\omega_i = \omega_n - k_{p_i} P_{AV_i}. \tag{1}$$

$$E_i = E_n - k_{q_i} Q_{AV_i}, \tag{2}$$

Where the quantities ω_i and E_i represent the voltage angular frequency and amplitude references of the "i" inverter, respectively; ω_n and E_n are the values of ω and V at no load; k_{p_i} and k_{q_i} are the respective $P - \omega$ and $Q - V$ droop coefficients of the "i" inverter; and P_{AV_i} and Q_{AV_i} are the average active and reactive powers.

The quantities P_{AV_i} and Q_{AV_i} are computed as follows:

$$P_{AV_i} = \frac{\omega_{f_i}}{s + \omega_{f_i}} p_i. \tag{3}$$

$$Q_{AV_i} = \frac{\omega_{f_i}}{s + \omega_{f_i}} q_i. \tag{4}$$

where ω_{f_i} is the low-pass filter (LPF) cutoff frequency.

In accordance with [10], [16], each inverter voltage (E_i) can be expressed in the direct-quadrature synchronous reference frame (DQ-frame) as given by:

$$E_i = e_{d_i} + j e_{q_i} = E_i cos(\delta_i) + j E_i sin(\delta_i). \tag{5}$$

$$\delta_i = \arctan\left(\frac{e_{q_i}}{e_{d_i}}\right). \tag{6}$$

$$|E_i| = \sqrt{e_{d_i}^2 + e_{q_i}^2}. \tag{7}$$

As detailed in [10], [16], using (1)-(7) it is possible to obtain the third-order model for each microgrid inverter, as follows:

$$\frac{d}{dt}\underbrace{\begin{bmatrix}\Delta\omega_i\\\Delta e_{d_i}\\\Delta e_{q_i}\end{bmatrix}}_{\Delta\dot{X}_{si}} = \mathbf{M_i}\underbrace{\begin{bmatrix}\Delta\omega_i\\\Delta e_{d_i}\\\Delta e_{q_i}\end{bmatrix}}_{\Delta\dot{X}_{si}} + \mathbf{B_{si}}\underbrace{\begin{bmatrix}\Delta p_i\\\Delta q_i\end{bmatrix}}_{\Delta s_i}. \tag{8}$$

$$\mathbf{M_i} = \begin{bmatrix} -\omega_{f_i} & 0 & 0 \\ \dfrac{n_{q_i}}{d_{en_i}} & \dfrac{n_{d_i}m_{q_i}\omega_{f_i}}{d_{en_i}} & \dfrac{n_{q_i}m_{q_i}\omega_{f_i}}{d_{en_i}} \\ -\dfrac{n_{d_i}}{d_{en_i}} & -\dfrac{n_{d_i}m_{d_i}\omega_{f_i}}{d_{en_i}} & -\dfrac{n_{q_i}m_{d_i}\omega_{f_i}}{d_{en_i}} \end{bmatrix}. \tag{9}$$

$$\mathbf{B_{si}} = \begin{bmatrix} -k_{p_i}\omega_{f_i} & 0 \\ 0 & \dfrac{k_{q_i}m_{q_i}\omega_{f_i}}{d_{en_i}} \\ 0 & -\dfrac{k_{q_i}m_{d_i}\omega_{f_i}}{d_{en_i}} \end{bmatrix}, \tag{10}$$

where the parameters d_{en_i}, m_{d_i}, m_{q_i}, n_{d_i} and n_{q_i} are given by:

$$d_{en_i} = m_{d_i}n_{q_i} - m_{q_i}n_{d_i}. \tag{11}$$

$$m_{d_i} = \frac{\partial\delta_i}{\partial e_{d_i}} = \frac{-e_{q_i}}{e_{d_i}^2 + e_{q_i}^2}. \tag{12}$$

$$m_{q_i} = \frac{\partial\delta_i}{\partial e_{q_i}} = \frac{e_{d_i}}{e_{d_i}^2 + e_{q_i}^2}. \tag{13}$$

$$n_{d_i} = \frac{\partial E_i}{\partial e_{d_i}} = \frac{e_{d_i}}{\sqrt{e_{d_i}^2 + e_{q_i}^2}}. \tag{14}$$

$$n_{q_i} = \frac{\partial E_i}{\partial e_{q_i}} = \frac{e_{q_i}}{\sqrt{e_{d_i}^2 + e_{q_i}^2}}. \tag{15}$$

Considering a microgrid with n inverters, the individual inverter models in the form of (8) can be grouped leading to:

$$\underbrace{\begin{bmatrix}\Delta\dot{X}_{s1}\\\Delta\dot{X}_{s2}\\\vdots\\\Delta\dot{X}_{sn}\end{bmatrix}}_{\Delta\dot{X}_s} = \underbrace{\begin{bmatrix}\mathbf{M_1} & 0 & \cdots & 0\\0 & \mathbf{M_2} & \cdots & 0\\\vdots & \vdots & \ddots & \vdots\\0 & 0 & \cdots & \mathbf{M_n}\end{bmatrix}}_{\mathbf{M_s}}\begin{bmatrix}\Delta X_{s1}\\\Delta X_{s2}\\\vdots\\\Delta X_{sn}\end{bmatrix}$$
$$+ \underbrace{\begin{bmatrix}\mathbf{B_{s1}} & 0 & \cdots & 0\\0 & \mathbf{B_{s2}} & \cdots & 0\\\vdots & \vdots & \vdots & \vdots\\0 & 0 & \cdots & \mathbf{B_{sn}}\end{bmatrix}}_{\mathbf{B_{ss}}}\underbrace{\begin{bmatrix}\Delta s_1\\\Delta s_2\\\vdots\\\Delta s_n\end{bmatrix}}_{\Delta s}. \tag{16}$$

The active and reactive instantaneous powers can be, respectively, defined in DQ-frame for each inverter as follows:

$$p_i = e_{d_i}i_{d_i} + e_{q_i}i_{q_i}. \tag{17}$$
$$q_i = e_{d_i}i_{q_i} - e_{q_i}i_{d_i} \tag{18}$$

Linearizing the instantaneous powers leads to:

$$\mathbf{\Delta s} = \mathbf{E_s}\mathbf{\Delta i_{dq}} + \mathbf{I_s}\mathbf{\Delta e_{dq}}, \tag{19}$$

where $\mathbf{E_s}$ and $\mathbf{I_s}$ are the matrices containing, the steady-state values of inverter voltages and currents, respectively.

The steady-state currents through the lines can be written in terms of the bus admittance matrix in rectangular coordinates ($\mathbf{Y_s}$) and the bus voltages ($\mathbf{\Delta e_{dq}}$) as:

$$\mathbf{\Delta i_{dq}} = \mathbf{Y_s}\mathbf{\Delta e_{dq}} \tag{20}$$

From (16), (19), (20) and defining the matrix $\mathbf{K_e}$ like $\mathbf{\Delta e_{dq}} = \mathbf{K_e}\mathbf{\Delta X_s}$, the microgrid model can be represented by:

978-1-7281-4181-7/19 $31.00 © 2019 IEEE

$$\Delta \dot{X}_s = [M_s + B_{ss}(I_s + E_s Y_s)K_e]\Delta X_s \quad (21)$$

Thus, as can be noted, in the third-order-model each inverter adds into the entire microgrid model the state variables $\Delta\omega_i$, Δe_{d_i} and Δe_{q_i}.

B. Proposed reduced-order model

As can be observed, (20) does not consider the electromagnetic transient of the RL equivalent circuit of the network lines and/or loads. Neglecting this dynamic behavior leads to optimistic stability boundaries. Thus, in this section a method that takes into account the dynamics of the network lines and loads is proposed.

In order to include into the model the network and load dynamics, the circuit presented in Fig. 1 is considered. Then, in the Laplace domain, the respective direct and quadrature currents are given by:

$$L_{ij}s\, I_{dij} = (E_{di} - E_{dj}) - R_{ij}I_{dij} + L_{ij}\omega_n I_{qi}. \quad (22)$$

$$L_{ij}s\, I_{qij} = (E_{qi} - E_{qj}) - R_{ij}I_{qij} - L_{ij}\omega_n I_{di}. \quad (23)$$

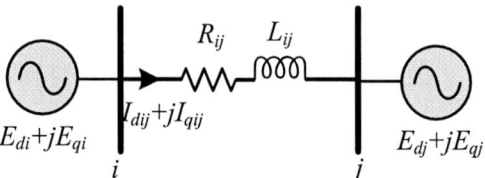

Fig. 1. Single line diagram of a RL equivalent circuit of a line.

From (22) and (23), it is possible to isolate the DQ-currents, as follows:

$$I_{dij}(s) = \frac{(L_{ij}s + R_{ij})(E_{di} - E_{dj}) + L_{ij}\omega_n(E_{qi} - E_{qj})}{(L_{ij}s + R_{ij})^2 + (L_{ij}\omega_n)^2} \quad (24)$$

$$I_{qij}(s) = \frac{(L_{ij}s + R_{ij})(E_{qi} - E_{qj}) - L_{ij}\omega_n(E_{di} - E_{dj})}{(L_{ij}s + R_{ij})^2 + (L_{ij}\omega_n)^2} \quad (25)$$

This set of equations can be approximated using Taylor series such that:

$$I_{dqij}(s) \approx I_{dqij}(0) + \frac{d}{ds}I_{dqij}(0)(s - 0). \quad (26)$$

According to the second term of (26) it is necessary to compute the derivative of (23) and (24) in relation to "s". Both derivatives can be expressed in two terms that are called here as $G_{dij}(s)$ and $B_{dij}(s)$. These derivatives are identical for both direct and quadrature axis currents, but with opposing signs. Hence, $G_{dij}(s)$ and $B_{dij}(s)$ are obtained as:

$$G_{dij}(s) = \frac{L_{ij}\left((L_{ij}s + R_{ij})^2 - (L_{ij}\omega_n)^2\right)}{\left((L_{ij}s + R_{ij})^2 + (L_{ij}\omega_n)^2\right)^2}. \quad (27)$$

$$B_{dij}(s) = \frac{2L_{ij}^2\omega_n(L_{ij}s + R_{ij})}{\left((L_{ij}s + R_{ij})^2 + (L_{ij}\omega_n)^2\right)^2}. \quad (28)$$

Thus, considering (26)-(28) in the time domain, the DQ-currents can be approximated as follows:

$$I_{dij}(t) \approx I_{dij}(0) - G_{dij}(\dot{e}_{di} - \dot{e}_{dj}) - B_{dij}(\dot{e}_{qi} - \dot{e}_{qj}) \quad (29)$$

$$I_{qij}(t) \approx I_{qij}(0) + B_{dij}(\dot{e}_{di} - \dot{e}_{dj}) - G_{dij}(\dot{e}_{qi} - \dot{e}_{qj}) \quad (30)$$

where $I_{dqi}(0)$ are the steady-state currents, and G_{dij} and B_{dij} are, respectively, given by:

$$G_{dij} = \frac{L_{ij}\left(R_{ij}^2 - (L_{ij}\omega_n)^2\right)}{\left(R_{ij}^2 + (L_{ij}\omega_n)^2\right)^2}. \quad (31)$$

$$B_{dij} = \frac{2R_{ij}L_{ij}^2\omega_n}{\left(R_{ij}^2 + (L_{ij}\omega_n)^2\right)^2}. \quad (32)$$

It is noticed that to capture the influence of the network dynamics on the dominant poles of the system (i.e. the slow ones), the Taylor series approximation is done around $s = 0$ as it is expressed in (26). Furthermore, the steady-state currents $I_{dqij}(0)$ are obtained replacing $s = 0$ into (29) and (30) so that it becomes equivalent to the expression obtained in (20).

In addition, it can be highlighted that the current quantities represented in (29) and (30) also depend on the bus voltages derivatives. Thus, these terms represent the dynamics of distribution network and loads. It can be noticed that G_{dij} and B_{dij} exist only when the inductances are different from zero. This occurs because when the inductances are null, there are no electromagnetic transients ($\tau = L/R$).

Generically, the currents through the lines can be computed by adding both the steady-state and transient currents. Thereby, the calculation involving the transient currents requires the construction of a new transient admittance matrix ($\mathbf{Y_{dr}}$) according to (29)-(32), such that the currents and, consequently, the instantaneous powers are given by:

$$\Delta i_{dq} = Y_s \Delta e_{dq} + Y_{dr}\Delta \dot{e}_{dq} \quad (33)$$

$$\Delta s = (I_s + E_s Y_s)K_e \Delta X_s + E_s Y_{dr}K_e \Delta \dot{X}_s \quad (34)$$

Replacing (34) into (16) derives in the proposed microgrid model, as follows:

$$\Delta \dot{X}_s = T_{id}[M_s + B_{ss}(I_s + E_s Y_s)K_e]\Delta X_s \quad (35)$$

where $T_{id} = (I - B_{ss}E_s Y_{dr}K_e)^{-1}$ and I is the identity matrix of suitable size.

Comparing (21) and (35) to each other, it is possible to observe that T_{id} involves the dynamics of the network and load. The effect of this matrix on the overall model fidelity is studied in the next Section.

978-1-7281-4181-7/19 $31.00 © 2019 IEEE

III. SIMULATION SETUP

In this section the microgrid topology and the control structure of the DG system regarded in this study are presented.

A. Microgrid topology

In Fig. 2 illustrates the microgrid topology composed of four dispatchable DG system, three distributions lines and three loads. This topology is adopted since it has been regarded in several other studies and have become a benchmark in microgrid studies. The microgrid parameters and the equilibrium point considered in the simulations results presented in next section are listed in Table I.

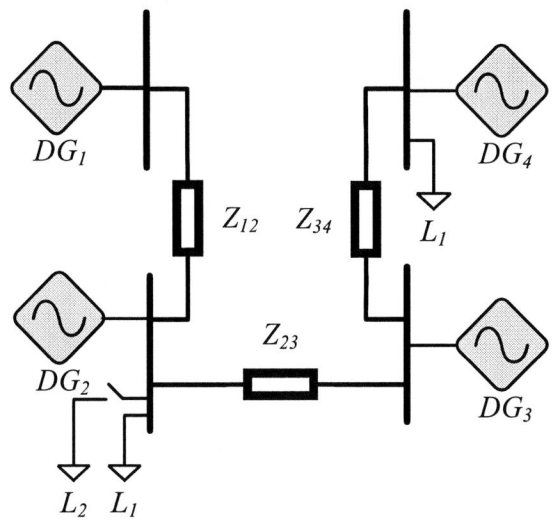

Fig. 2. Single-line diagram of Microgrid topology.

B. Inverter Control

Each controlled voltage source shown in Fig. 2 represents a dispatchable inverter operating in grid-forming mode according to Fig. 3. From the power calculation and the droop control the amplitude (E^*) and frequency (ω^*) references are obtained. These quantities are used by the reference generator to compute the DQ-voltage references for the voltage loop. The internal control loops of the grid-forming converters are implemented by using two cascaded synchronous controllers in DQ-frame. Both the outer voltage controller ($G_{cv}(s)$) and the inner current controller ($G_{ci}(s)$) consist of the conventional proportional- integral (PI) controllers. A current feedforward term (k_{ff}) is added to improve the dynamic behavior of the inverter. The parameters of the inverters and the outer and inner control loops gains are listed in Table II.

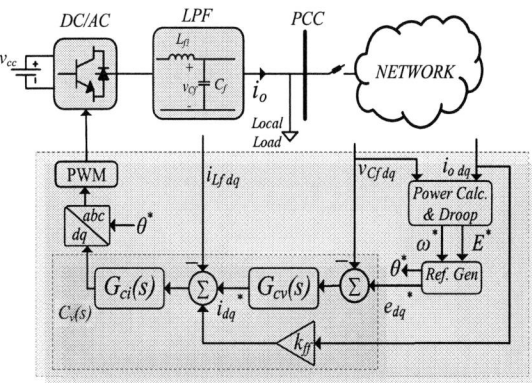

Fig. 3. Schematic of internal control loops of DG operating as grid-forming inverters.

TABLE I
MICROGRID PARAMETERS AND EQUILIBRIUM POINT

Parameter	Symbol	Value	Unit
Nominal frequency	ω_n	376.99	rad/s
Nominal phase to phase voltage	V_{pp}	220	V
Load 1	Z_{L1}	$1000 + j0$	Ω
Load 2	Z_{L2}	$75 + j15.08$	Ω
Line impedance (1,2)	Z_{12}	$0.2 + j1.2$	Ω
Line impedance (2,3) and (3,4)	Z_{23}, Z_{34}	$0.1 + j0.6$	Ω
Measuring filter cutoff frequency	ω_{f_i}	30	rad/s
Frequency-droop coefficient	k_{p_i}	0.001	rad/s/W
Voltage-droop coefficient	k_{q_i}	0.0005	V/Var
Inverter 1 output power	S_{AV_1}	$76.60 + j16.95$	Va
Inverter 2 output power	S_{AV_2}	$76.60 - j29.44$	Va
Inverter 3 output power	S_{AV_3}	$76.60 + j35.40$	Va
Inverter 4 output power	S_{AV_4}	$76.60 - j23.14$	Va
Inverter 1 output voltage	E_1	$220.44 + j0$	V
Inverter 2 output voltage	E_2	$220.47 - j0.43$	V
Inverter 3 output voltage	E_3	$220.44 - j0.42$	V
Inverter 4 output voltage	E_4	$220.46 - j0.63$	V

TABLE II
INVERTER PARAMETERS

Parameter	Symbol	Value	Unit
Inductor of inverter LC filter	L_{f_i}	1.5	mH
Capacitor of inverter LC filter	C_{fp_i}	60	μF
Proportional gain voltage controller	$k_{p_{v_i}}$	0.15	$1/\Omega$
Integral gain voltage controller	$k_{i_{v_i}}$	2075	$1/\Omega s$
Current Feedforward gain	k_{ff_i}	0.7	-
Proportional gain current controller	$k_{p_{c_i}}$	38.66	Ω
Integral gain current controller	$k_{i_{c_i}}$	55134	Ω/s
DC Bus voltage	V_{cc_i}	550	V
Sampling Frequency	f_{s_i}	20	kHz

IV. SIMULATION RESULTS

In this section, simulation results of the islanded microgrid illustrated in Fig. 2 are presented. Comparisons between the third-order model and the proposed reduced-order model are performed. The tests were carried out by means of MATLAB/Simulink software considering the parameters defined in Tables I. and II.

A. Eigenvalue analysis

Firstly, an eigenvalue analysis of both models is presented. Fig. 4 presents the eigenvalues found by each model for different values of k_{p_i}. In both cases, frequency-droop coefficient mainly affects the low-frequency complex-conjugated eigenvalues. However, in the conventional third-order model, the stability boundary does not depend on the selection of k_{p_i}. Instead, the eigenvalues of the proposed model tend to the right half of the complex plane when k_{p_i} increases. This behavior is consistent with those found from complete models [11], [12]. Thus, the approximation of the network and load dynamics enables the proposed reduced-order model to predict microgrid instability due to the adoption of large droop coefficients. In this specific case, the microgrid instability is found when the frequency droop coefficient is chosen close to $k_{p_i} = 3.8 \times 10^3$ rad/s/W. Furthermore, different from the model presented in [8], the proposed model have a pole on the origin, which is a characteristic of the modeling technique used in this work. However, this pole is not taken into account for dynamical analysis according to [10], [11].

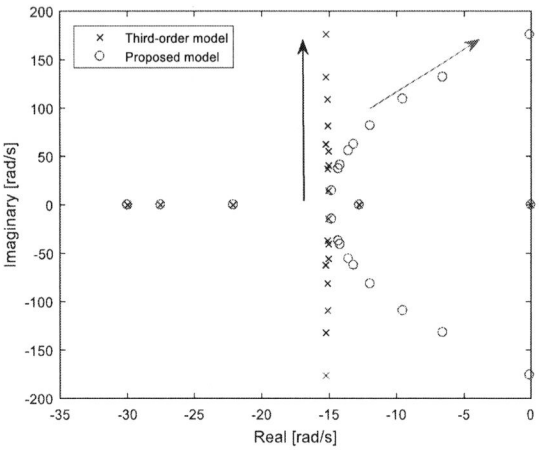

Fig. 4. Eigenvalue sensitivity for frequency-droop coefficient variations ($0.5 \leq k_{p_i} \leq 3.8$).

B. Time-domain simulations

The following results correspond to the simulation considering a step of local load 2 (Z_{L2}) occurring in the distributed generator 2 (see Fig. 2).

Fig. 5 presents the comparison of the frequency at different DG obtained by numerical simulation using MATLAB/Simulink software, as well as by means of the state transition matrix of the models. The results presented in Fig. 5 (a) correspond to the conventional third-order model. Such results clearly show that the responses obtained from the third-order model have higher damping than those obtained by numerical simulation. Instead, in Fig. 5 (b) is observed a better correlation between the simulated results and the expected time-domain response calculated by means of the proposed model.

Similarly, in Fig. 6 the results obtained for active and reactive power are presented. The behavior of the active powers in Fig. 6 (a) and Fig. 6 (c) is almost the same observed in Fig. 5 for the frequency. However, a comparison between the results obtained in Fig. 6 (b) and Fig. 6 (d) reveal that

conventional third-order model is especially inaccurate to represent the transient reactive power oscillations.

Although the proposed model improves the accuracy, it is important to highlight that both models disregard the parametric variations of the network reactances due to frequency deviations imposed by droop control. Hence, the accuracy of both models is affected as the value of k_{p_i} is increased.

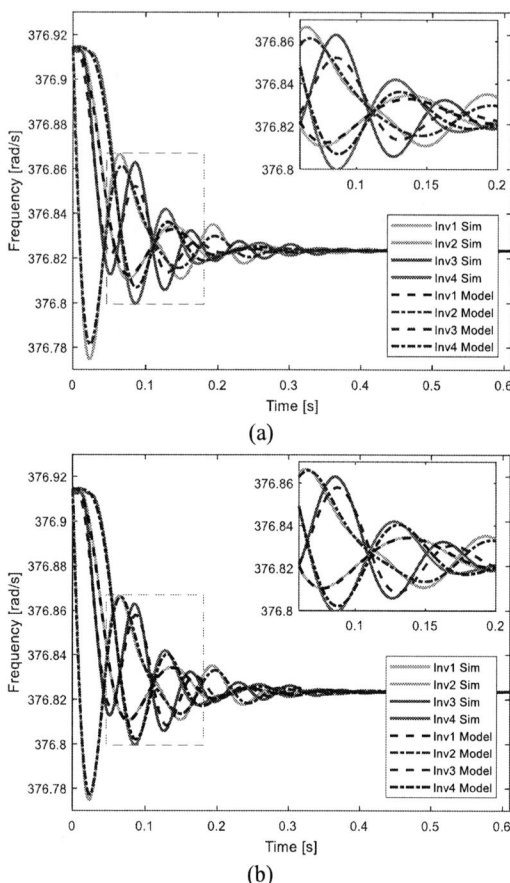

Fig. 5. Time-domain simulation and comparison with the expected response of the model: (a) ω_i (3rd-order model); (b) ω_i (Proposed model).

V. CONCLUSIONS

In this paper, a reduced-order model in common DQ-frame that includes the influence of network and load dynamics on microgrid stability was proposed. Hence, more accurate stability boundaries and time-domain responses were achieved.

In addition, the order of the microgrid model was not increased and the scalability of the conventional third-order model was maintained. Furthermore, through simulation results, the improvement of the proposed model fidelity was verified and confirmed.

The proposed model also neglects the parametric variation of the network reactances caused by frequency deviations. Hence, the representativeness of the proposed model is affected as frequency-droop coefficient increases since at this condition higher frequency deviations are expected.

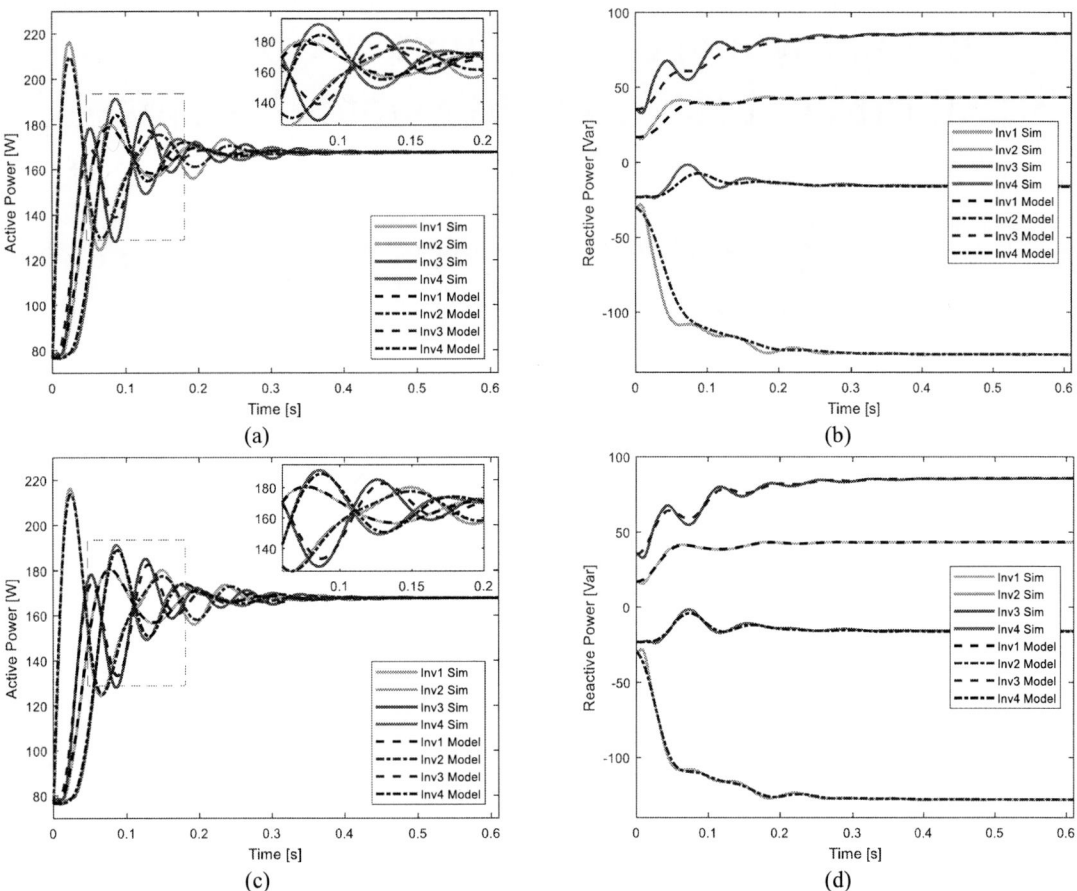

Fig. 6. Time-domain simulation and comparison with the expected response of the model. (a) P_{AV_i} (3rd-order model); (b) Q_{AV_i} (3rd-order model); (c) P_{AV_i} (Proposed model); (d) Q_{AV_i} (Proposed model).

REFERENCES

[1] M. C. Chandorkar, D. M. Divan, and R. Adapa, "Control of parallel connected inverters in standalone ac supply systems," IEEE Trans. Ind. Appl., vol. 29, no. 1, pp. 136–143, 1993.

[2] Y. Sun, X. Hou, J. Yang, H. Han, M. Su, and J. M. Guerrero, "New Perspectives on Droop Control in AC Microgrid," IEEE Trans. Ind. Electron., vol. 64, no. 7, pp. 5741–5745, 2017.

[3] P. Vorobev, P. H. Huang, M. Al Hosani, J. L. Kirtley, and K. Turitsyn, "A framework for development of universal rules for microgrids stability and control," in 2017 IEEE 56th Annual Conference on Decision and Control, CDC 2017, 2018, pp. 5125–5130.

[4] X. Guo, Z. Lu, B. Wang, X. Sun, L. Wang, and J. M. Guerrero, "Dynamic phasors-based modeling and stability analysis of droop-controlled inverters for microgrid applications," IEEE Trans. Smart Grid, vol. 5, no. 6, pp. 2980–2987, 2014.

[5] E. P. Correa, M. Mezaroba, and G. M. S. Azevedo, "Reduced-order model of AC microgrid for stability analysis and adjustment of droop control," 14th Brazilian Power Electron. Conf. COBEP 2017, vol. 2018-January, pp. 1–6, 2018.

[6] V. Mariani, F. Vasca, J. C. Vásquez, and J. M. Guerrero, "Model Order Reductions for Stability Analysis of Islanded Microgrids With Droop Control," IEEE Trans. Ind. Electron., vol. 62, no. 7, pp. 4344–4354, 2015.

[7] I. P. Nikolakakos, H. H. Zeineldin, E. I. Moursi, and J. L. Kirtley, "Reduced-Order Model for Inter-Inverter Oscillations in Islanded Droop-Controlled Microgrids," IEEE Trans. Smart Grid, vol. 9, no. 5, pp. 4953–4963, 2018.

[8] P. Vorobev, P. Huang, M. Al Hosani, J. L. Kirtley, and K. Turitsyn, "High-Fidelity Model Order Reduction for Microgrids Stability Assessment," IEEE Trans. Power Syst., vol. 33, no. 1, pp. 874–887, 2018.

[9] L. Luo and S. V. Dhople, "Spatiotemporal model reduction of inverter-based islanded microgrids," IEEE Trans. Energy Convers., vol. 29, no. 4, pp. 823–832, 2014.

[10] E. Antônio, A. Coelho, P. C. Cortizo, P. Francisco, and D. Garcia, "Small-Signal Stability for Parallel-Connected Inverters in Stand-Alone AC Supply Systems," IEEE Trans. Ind. Appl., vol. 38, no. 2, pp. 533–542, 2002.

[11] N. Pogaku, M. Prodanović, and T. C. Green, "Modeling, analysis and testing of an inverter-based microgrid," IEEE Trans. Power Electron., vol. 22, no. 2, pp. 613–625, 2007.

[12] M. Zhu, H. Li, and X. Li, "Improved state-space model and analysis of islanding inverter-based microgrid," in 2013 IEEE International Symposium on Industrial Electronics, 2013, pp. 1–5.

[13] M. Rasheduzzaman, J. A. Mueller, and J. W. Kimball, "An accurate small-signal model of inverter-dominated islanded microgrids using (dq) reference frame," IEEE J. Emerg. Sel. Top. Power Electron., vol. 2, no. 4, pp. 1070–1080, 2014.

[14] E. H. Allen and M. D. Ilić, "Interaction of transmission network and load phasor dynamics in electric power systems," IEEE Trans. Circuits Syst. I Fundam. Theory Appl., vol. 47, no. 11, pp. 1613–1620, 2000.

[15] H. K. Khalil, Nonlinear Systems, Third. New Jersey: Prentice Hall, 2002.

[16] E. A. A. Coelho et al., "Small-Signal Analysis of the Microgrid Secondary Control Considering a Communication Time Delay," IEEE Trans. Ind. Electron., vol. 63, no. 10, pp. 6257–6269, 2016.

Optimal Voltage Coordinated Control for Grid-connected Photovoltaic Systems

Thiago William Pires Sousa
School of Electrical Engineering
Celso Suckow da Fonseca Federal Center of Technology
Rio de Janeiro, 20271-110, Brasil
thiagowpsousa@gmail.com

Gustavo Kaefer Dill
School of Electrical Engineering
Celso Suckow da Fonseca Federal Center of Technology
Rio de Janeiro, 20271-110, Brasil
gkdill@hotmail.com

Abstract—The intermittent penetration of Grid-connected photovoltaic systems and load changing in distribution feeders can cause voltage fluctuation. This paper aims to analyze the performance of different power control strategies to avoid overvoltage in MV feeders. The strategies use Volt/Var control and Volt/Watt control techniques to find the optimal active or reactive power to be injected/absorbed by the PV inverters with the aim to minimize the voltage deviation and power losses and also coordinate the grid-connected photovoltaic system with the voltage regulation devices of distribution feeders. The strategies are tested in the IEEE 13 node test feeder.

Keywords - Voltage regulation, grid-connected photovoltaic systems, active and reactive power control.

I. INTRODUCTION

Grid-connected photovoltaic systems (GCPVS) are distributed generation sources connected to distribution feeders through power inverters. GCPVS have grown greatly due to environmental problems with carbon sources, government incentives, power loss reduction, feeders loading reduction and electricity costs. GCPVS come as a quick way to generate electricity, directly to the consumer, in line with the power load increases, islanding and emergency loads or contingencies. Consequently GCPVS can reduce the feeders operational costs and postpone investments in the long term for distribution companies. On the other hand, this generation alters some relations already established due to constraints and particularities of the feeders to which they are associated. The penetration of GCPVS in medium or low-voltage feeders can cause power quality problems like frequency and power fluctuation, harmonic distortion and undervoltage/overvoltage [1]. The presence of GCPVS can interfere with the on-load tap changer (OLTC) and capacitor banks (CB) and change the operation and control of feeders [2]. The new technical challenge for power distribution engineers is to propose solutions for those new paradigms. The operational voltage control of feeders in the presence of GCPVS needs control actions with the already established equipments to ensure the voltage limits. In the literature, the voltage control of distribution feeders takes place using Volt/Var control and power droop control [3], [4], [5]. The Volt/Var control is based on injecting or absorbing some reactive power by the inverters to achieve the voltage limits [3], [5], [6]. The Volt/Var control can also be coordinated with controllable devices like OLTC and CB to regulate the terminal voltage and minimize the power losses of distribution feeders when optimization is taken into account. In this case the efficiency of the feeders can be increased [2], [7], [8]. The power droop

control (PDC) is based on constraining the active power generation to prevent voltage rise along the feeders [4], [7]. In the PDC technique, the inverter output active power is curtailed by moving the maximum power point tracker (MPPT) to a lower level when the voltage reaches its maximum limit. The power droop is defined as a function of the inverters voltage [7]. This technique does not take into account the reactive power injection/absorption in the feeders.

The rise of GCPVS and the concept of smart-grids in distribution feeders demand more and more cutting-edge technology. In this paper the Volt/Var control and the Volt/Watt control based on power droop control technique will be investigated through different strategies which aim to optimize the active or reactive power to be injected/absorbed by GCPVS inverters and control overvoltages in MV feeders. The proposal aims to coordinate GCPVS with a smart power electronic on-load tap changer (EOLTC) and minimize the voltage deviation and power losses through optimal active or reactive power injection/absorption.

This paper is summarized in VI chapters. Apart from the abstract and the introduction, the rest of the paper is organized as follow. In section II the coordinated voltage control review is introduced. In section III the voltage control strategies based on Volt/Var and Volt/Watt are presented. Section IV presents brief description of the optimization methods and techniques applied to solve the proposed strategies. In section V comparative results obtained with different strategies when applied to an unbalanced distribution test feeder are shown. Section VI presents the conclusions of the proposed paper and, finally, the references on which this proposal is based are listed.

II. COORDINATED VOLTAGE CONTROL STATEMENTS

The high penetration of GCPVS in distribution feeders can increase the terminal voltages specially at noon times when the irradiation level is high. The regulations in Brazil require that GCPVS have to maintain a terminal voltage within 0.93 to 1.05 p.u. and operate with a minimum power factor (FP) of 90% capacitive/inductive [9]. In this paper the voltage regulation aims to control the terminal voltages of feeders through different optimal power strategies and improve the robustness by coordinating EOLTC, CB and GCPVS.

The coordinated voltage control aims to inject/absorb the optimal power from GCPVS inverters to maintain the voltage in the boundary limit. If the terminal voltage is under the limit, the capacitor banks, for the specified nodes, are switched one at a time and the load flow is performed to check the terminal voltages of each node. If the feeder has no CB the EOLTC is set to increase its tap. The EOLTC can change its tap up to 9 times to increase the terminal voltages within a 10% maximum increase in the slack bus of each phase. If the terminal voltages are over the limit, the EOLTC is set to decrease its tap up to 9 times to decrease the terminal voltages within 10% maximum decrease in the slack bus. The EOLTC is only set to satisfy the terminal voltages when GCPVS penetration is higher and even using the strategies defined here can not achieve the voltage limits. This proposal aims to reduce the operation of the EOLTC and CB using the optimal power from GCPVS inverters.

The EOLTC is a single phase distribution transformer with a smart power electronics on-load tap changer. The electronic control is based on silicon controlled rectifiers (SCR's) which are set to perform zero current crossing detection. The EOLTC presents fast response, high reliability and better performance when compared with electromechanical on-load tap changer [10] and is suited for renewable power sources applied to distribution feeders.

III. ACTIVE AND REACTIVE POWER STRATEGIES

In previous works that apply Volt/Var control to regulate the voltage, power loss and voltage deviation were the most considered indices in the optimization problem [7]. The application of power droop control to regulate the voltage was based on the critical voltage, the terminal voltage and the inverter MPPT power [7]. In this section, the Volt/Var control is represented by three different strategies that consider single and multi-objective functions and the Volt/Watt control is defined as variants of those strategies.

A. Volt/Var strategies

The Volt/Var control take into account the reactive power strategies to inject/absorb reactive power by GCPVS inverters into the feeders to regulate the terminal voltages. The strategies consider a fixed active power P_{mppt} and an optimal reactive power.

1) Voltage Deviation (VD): This strategy aims to minimize the terminal voltage of each node, at any time, considering the quadratic deviation of the terminal voltage and a reference voltage. The optimal reactive power is achieved for the minimum voltage deviation to ensure a flat voltage profile along the feeder. The VD strategy is defined by

$$VD(t) = \sum_{i=1}^{N} |V_i(t) - V_{ref}|^2 \qquad (1)$$

where $VD(t)$ is the total voltage deviation which is represented by the sum of all terminal voltages of each node, $V_i(t)$ is the terminal voltage of each ith node and V_{ref} is the reference voltage which is set to 1.0 p.u.

The objective function and the optimization problem of this strategy is defined by:

$$\begin{aligned} \min_{t} \quad & F_1 = VD(t) \\ \text{s.t.} \quad & V_{min} \leq V_i(t) \leq V_{max} \\ & P_{opt}(t) = P_{mppt} \\ & Q_{min} \leq Q_{opt}(t) \leq Q_{max} \end{aligned} \qquad (2)$$

which is equivalent to approximate the terminal voltage of each node to the reference voltage. In (2), V_{min} and V_{max} represent the minimum and maximum voltage for MV feeders in Brazil. P_{opt} is the GCPVS optimal active

power which is set to the maximum PV power P_{mppt} and Q_{opt} is the GCPVS optimal reactive power which is limited by $Q_{min} = -\sqrt{(S_{inv}^2 + P_{mppt}^2)}.F_{lim}$ and $Q_{max} = \sqrt{(S_{inv}^2 + P_{mppt}^2)}.F_{lim}$, where F_{lim} represent the limit of reactive power for the maximum active power produced by the GCPVS [9].

2) Power Loss (PL): The reduction of power losses imply in a greater availability of power flow and increase the stability margin of feeders. This strategy aims to minimize the feeder's power loss by injecting or absorbing a minimum reactive power to regulate the terminal voltages. The PL strategy is defined by

$$PL(t) = \sum_{i=1}^{N} g_{ij}[V_i(t)^2 + V_j(t)^2 - 2.V_i(t).V_j(t)cos(\sigma_i(t) - \sigma_j(t))] \qquad (3)$$

where $PL(t)$ is the total power loss dissipated by the feeder which is represented by the sum of all power losses of each branch taking into account the terminal voltage $V_i(t)$ and its angle $\sigma_i(t)$ of each ith node and the terminal voltage $V_j(t)$ and its angle $\sigma_j(t)$ of each jth node, g_{ij} is the conductance of the branch $i - j$.

The objective function and the optimization problem of this strategy is defined by:

$$\begin{aligned} \min_{t} \quad & F_2 = PL(t) \\ \text{s.t.} \quad & V_{min} \leq V_i(t) \leq V_{max} \\ & P_{opt}(t) = P_{mppt} \\ & Q_{min} \leq Q_{opt}(t) \leq Q_{max} \end{aligned} \qquad (4)$$

which is equivalent to minimize the power loss of a feeder considering the voltage and reactive limits as restrictions.

3) Voltage Deviation and Power Loss (VD+PL): This strategy aims to minimize the power loss and voltage deviation as a multi-objective problem satisfying both objectives, choosing one optimal reactive power to be injected or absorbed by the inverters.

The objective function and the optimization problem of this strategy is defined by:

$$\begin{aligned} \min_{t} \quad & F_3 = [VD(t), PL(t)] \\ \text{s.t.} \quad & V_{min} \leq V_i(t) \leq V_{max} \\ & P_{opt}(t) = P_{mppt} \\ & Q_{min} \leq Q_{opt}(t) \leq Q_{max} \end{aligned} \qquad (5)$$

which is equivalent to approximate the terminal voltage of each node to the reference voltage and also considering the minimum power loss.

B. Volt/Watt strategies

The Volt/Watt control takes into account the active power strategies to inject power by GCPVS inverters into feeders and regulate the terminal voltages. The strategies consider no reactive power to be injected/absorbed by inverters and an optimal active power limited by P_{mppt}. The active power control strategies proposed here consider the optimal active power to minimize the voltage deviation and power losses.

1) Voltage Deviation (VD): Similarly to the reactive power VD strategy, in this strategy the optimal active power is achieved for the minimum voltage deviation to ensure a flat voltage profile along the feeder.

The objective function and the optimization problem of this strategy is defined by:

$$\min_{t} \quad F_4 = VD(t)$$
$$\text{s.t.} \quad V_{min} \leq V_i(t) \leq V_{max}$$
$$0 \leq P_{opt}(t) \leq P_{mppt} \quad (6)$$
$$Q_{opt}(t) = 0$$

which is equivalent to approximate the terminal voltage of each node to the reference voltage without injecting reactive power.

2) Power Loss (PL): This strategy aims to minimize the feeder's power loss by injecting an adequate active power to regulate the voltages of the system. The objective function and the optimization problem of this strategy is defined by:

$$\min_{t} \quad F_5 = PL(t)$$
$$\text{s.t.} \quad V_{min} \leq V_i(t) \leq V_{max}$$
$$0 \leq P_{opt}(t) \leq P_{mppt} \quad (7)$$
$$Q_{opt}(t) = 0$$

which is equivalent to minimize the feeders power loss considering the voltage limits as a restriction and the active power droop without injecting any reactive power.

3) Voltage Deviation and Power Loss (VD+PL): Power loss and voltage deviation are also minimized, considering the optimal active power injection. The objective function and the optimization problem of this strategy is defined by:

$$\min_{t} \quad F_6 = [VD(t), PL(t)]$$
$$\text{s.t.} \quad V_{min} \leq V_i(t) \leq V_{max}$$
$$0 \leq P_{opt}(t) \leq P_{mppt} \quad (8)$$
$$Q_{opt}(t) = 0$$

which is equivalent to approximate the terminal voltage of each node to the reference voltage considering the minimum power loss without injecting any reactive power.

IV. SOLUTION OF THE OPTIMIZATION STRATEGIES

The solutions of the Volt/Var and Volt/Watt strategies presented here require optimization methods to handle with single and multi-objective functions and nonconvex problems as the management of reactive power optimization is nonconvex [5]. The single objective functions are solved by a heuristic method to deal with nonconvex problems. The multi-objective solutions are obtained applying an evolutionary algorithm which presents a good performance for nonconvex problems [11] and uses the concept of Pareto front [12] to select nondominated solutions, which are, solutions that cannot be improved for one of the objective function without degrading any other objective function. The multi-objective functions are solved by the Nondominated Sorting in Genetic Algorithm (NSGA) [12] and the single objective functions are solved by the Particle Swarm Optimization (PSO) [13]. The optimal load flow (OPF) is performed using a modified linear load flow [14]. The modified linear load flow takes into account a linear approximation on the complex plane to solve MV and LV feeders load flow with low X/R rate and unbalanced loads and branches with accuracy.

The PSO is a heuristic algorithm [13] inspired by the choreography of a bird flock, where the members tend to follow the leader or the member that presents the best performance. The algorithm is performed based on velocity, that represents the search direction and the position, that represents the solution of the problem. At each iteration the best position of each individual and the best position of the whole swarm are determined. The best local and global position are selected to determine the new velocity and then, update the new positions. The process is repeated until the number of generations is reached.

The NSGA is an evolutionary genetic algorithm [11] based on genes mutation and nondominated chromosomes of a population. At each iteration, the chromosomes which dominate the rest of the individuals, are evaluated. Then, two indicators are created for each chromosome, one with the number of individuals it dominates and another based on the crowding distance, which measures how close a chromosome is to its neighbors. Based on these indicators, the best 50% individuals are selected to perform the crossover and mutation. As two offspring are generated by crossing, the new population has the same size of the initial population. For the new population, the dominance and crowding distance are evaluated and the process is repeated until the number of generations defined by the user is reached.

V. RESULTS

In this section, the proposed approach is performed using the IEEE 13 node test feeder which is an unbalanced small test system widely known for its test characteristics [15]. The system studied here with GCPVS and the smart interface of the central control management (CCM) is shown in Figure 1.

The results present studies where the effectiveness of the coordinated voltage control, with different strategies are compared. In this paper, the inverter reactive power injected/absorbed is limited to 43.58% ($F_{lim} = 0.4358$) of the available reactive power according to Brazilian standards [9] and the population size and the number of generation for both optimization methods were set to 50 and 10 respectively.

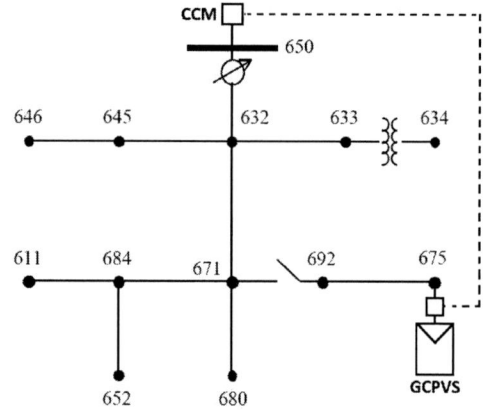

Fig. 1. IEEE 13 node test feeder with GCPVS

The load characteristics were represented by a real-power data time series from a substation in Rio de Janeiro. The solar irradiation, temperature and wind data were taken from PVGIS [16], considering 22.1° south latitude and 43.9° west longitude, which represent the location of the substation in Rio de Janeiro. The solar irradiation power, temperature and wind data were represented in per-unit considering the base values of the standard test condition (STC) [17]. The load and weather data were expressed by one specific day ($24h$) of each season. The active and reactive power load, the solar

978-1-7281-4181-7/19 $31.00 © 2019 IEEE

power and the variable temperature are presented in Figure 2.

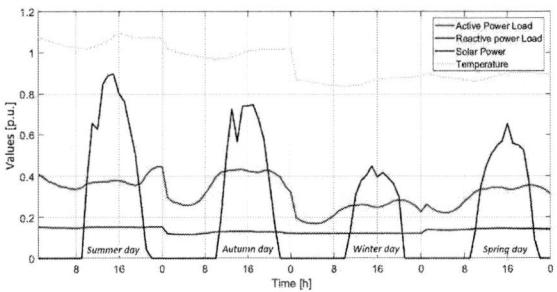

Fig. 2. Load and Solar power

The IEEE 13 node test feeder presents a satisfactory load flow for the load characteristics of Figure 2 without the presence of the GCPVS, EOLTC and CB for all nodes. The GCPVS penetration is limited to 24% of the feeders capacity. Any active power penetration above 24% of the feeders capacity can cause overvoltage in some periods of the day. As an example, the voltage profiles of phase B at 675 node with an active power penetration of 24% and 75% of the feeders capacity without EOLTC is compared with the 75% GCPVS penetration with the EOLTC in Figure 3. The inverter reactive power was set to zero.

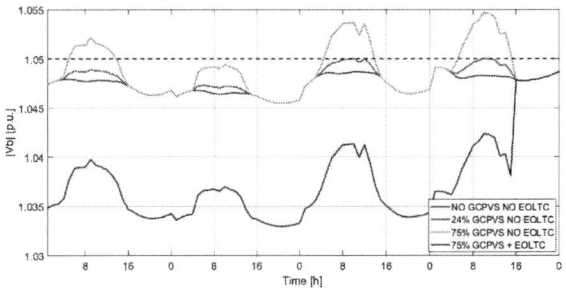

Fig. 3. Voltages at 675 node (phase B) with and without the GCPVS/EOLTC

As exposed in Figure 3, the voltages of phase B are above the limit for some periods of the day, when the GCPVS penetration is above 24%. The problem of GCPVS penetration in distribution feeders is evidenced and the presence of the EOLTC mitigated the problem. The voltages at phases A and C present satisfactory load flow for this load characteristics even for 100% GCPVS penetration without EOLTC and this study focus on phase B because phase B is the most critical.

A. Optimal power strategies

Once the GCPVS penetration problem is evidenced and voltage coordination control with the EOLTC is accomplished the active and reactive power strategies are taken into account to improve the performance of the feeder. This study evaluates the coordinated voltage control of the EOLTC and CB with 75% and 100% GCPVS penetration. The P_{mppt} was fixed to the maximum PV power for the reactive power strategies and the optimal reactive power was obtained. For the active power strategies the reactive power was set to zero and the inverters active power was optimized. The voltages of phase B at 675 node, the power losses and the active and reactive power injected/absorbed by the inverters are presented in Figures 4 to 7.

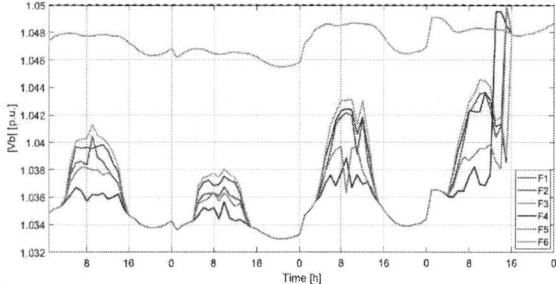

Fig. 4. Voltages at 675 node (phase B) with 75% GCPVS penetration

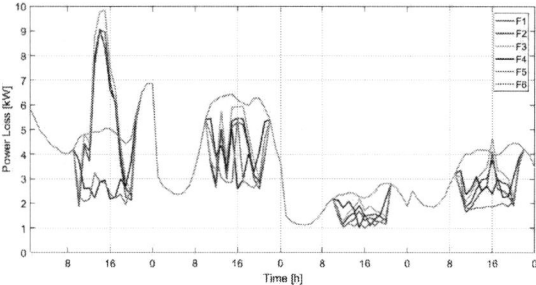

Fig. 5. Power Losses with 75% GCPVS penetration

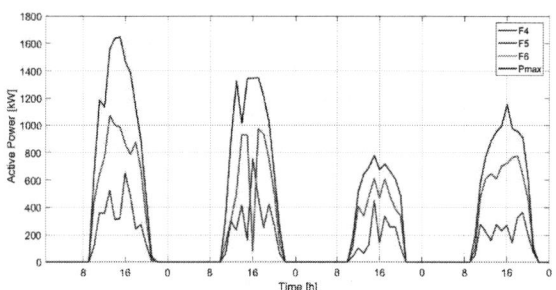

Fig. 6. Active Power injected with 75% GCPVS penetration

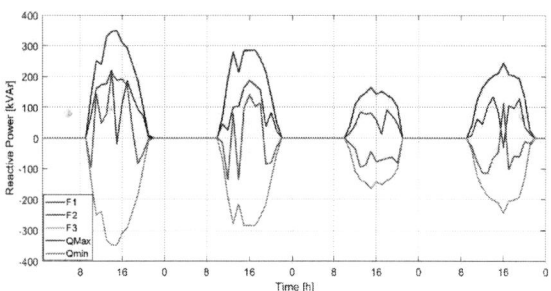

Fig. 7. Reactive Power injected/absored with 75% GCPVS penetration

The voltage profiles of the solutions obtained with active power strategies present lower voltage deviation and power losses when compared with reactive power strategies results. The results obtained with F_4 present the lowest voltage deviation, as shown in Figure 4, and the results obtained with F_5 present the lowest power losses, as illustrated in Figure 5. The numerical solutions obtained with single objective functions solved by PSO achieved the objectives with more accuracy but in the solutions obtained with F_6 the

978-1-7281-4181-7/19 $31.00 © 2019 IEEE   493

active power was curtailed to zero and the voltage profile is similar to the results obtained without GCPVS and EOLTC. For this load characteristics the active power was curtailed because no extra power is required. The strategy F_5 leads to inject active power to reduce the power losses of the feeder, as presented in Figure 6, while the strategy F_2 aims to inject or absorb reactive power to reduce the power losses, as shown in Figure 7. The optimal reactive power obtained when the strategy F_3 was performed was set to the maximum available reactive power (magenta line) to compensate the maximum active power injected by the inverters. In general, the Volt/Watt strategies where the active power was optimized presented better performance. To corroborate the results, the voltage profile of phase B at 675 node is presented at Figure 8 with 100% GCPVS penetration. All the results were similar to the 75% GCPVS penetration when 100% GCPVS penetration were taken into account and also with a better performance for the Volt/Watt strategies.

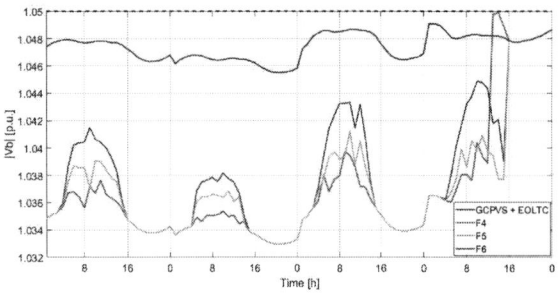

Fig. 8. Voltage at 675 node with 100% GCPVS penetration

TABLE I
AVERAGE POWER LOSS AND EOLTC OPERATION

| | 100% Penetration | |
Strategy	Power Loss (kW)	EOLTC set
No GCPVS No EOLTC	3.78	0
GCPVS + EOLTC	4.33	1
GCPVS + EOLTC + F1	4.49	1
GCPVS + EOLTC + F2	4.47	1
GCPVS + EOLTC + F3	4.77	1
GCPVS + ROLTC + F4	2.92	1
GCPVS + EOLTC + F5	2.88	1
GCPVS + EOLTC + F6	3.78	0

In Table I a comparison in terms of power losses and EOLCT switches is performed. The average power losses of the results obtained with the Volt/Watt strategies were lower than the Volt/Var strategies. The EOLTC was set down once for all strategies apart from F_6.

According to the results presented here it is clear the improvements in terms of power quality for a feeder managed by a CCM. The Volt/Var strategies can solve the problem but increase the power losses while the Volt/Watt strategies presented better performance in terms of voltage deviation, power losses and also in the number of times the EOLTC tap was changed taking into account the load characteristics represented in this study. The computing burden of the proposed design methods were around 1 $minute$ for both optimization methods using the strategies F_1 to F_6 taking into account a Intel(R) Core(TM) i5-5200U CPU @ 2.20GHz computer.

VI. CONCLUSION

GCPVS penetration in distribution feeders can harm the voltages profile of its nodes according to the penetration and the consumption. The Volt/Var and the Volt/Watt coordinated control applied to unbalanced MV distribution power feeders with low relation X/R can be used to ensure the voltages profile.

The strategies and the EOLTC coordination control were implemented on Matlab as a computational tool, and the results show that the voltage coordination with EOLTC ensures that GCPVS can be easily inserted in the feeders without violating the voltages terminal. Different strategies to inject/absorb power by the inverters were tested and the strategies based on optimizing the active power has presented better results in terms of voltage deviation, power losses and also in terms of EOLTC switches.

The control strategies proposed here can be implemented in the central control of distribution feeders where the power generated by distributed generation sources can be managed as a smart-grid system.

REFERENCES

[1] Shareef H. Zayandehroodi H. Farhoodnea M., Mohamed A. Power quality impact of renewable energy based generators and electric vehicles on distribution systems. *The 4th International Conference on Electrical Engineering and Informatics (ICEEI 2013)*, 11:11–17, 2013.

[2] Yee J. Zeinalzadeh A., Ghorbani R. Stochastic model of voltage variations in the presence of photovoltaic systems. *American Control Conference(ACC)*, page 50325037, July 2016.

[3] Regassa R. Kim I., Harley R. G. The investigation of the maximum effect of the volt/var control of distributed generation on voltage regulation. *Photovoltaic Specialist Conference (PVSC)*, June 2015.

[4] El-Fouly T. H. M. Tonkoski R., Lopes L. A. C. Coordinated active power curtailment of grid connected pv inverters for overvoltage prevention. *IEEE Transactions on Sustainable Energy*, 2(2):139 – 147, April 2011.

[5] Conejo A. J.-Giannakis G. B. Kekatos V., Wang G. Stochastic reactive power management in microgrids with renewables. *IEEE Transactions on Power Systems*, 30(6):3386–3395, November 2017.

[6] Na Li. Qu G. An optimal and distributed feedback voltage control under limited reactive power. *Power Systems Computation Conference (PSCC)*, June 2018.

[7] McGrath B. P. Kabiri R., Holmes D. G. The influence of pv inverter reactive power injection on grid voltage regulation. *IEEE 5th International Symposium on Power Electronics for Distributed Generation Systems (PEDG)*, June 2014.

[8] Palizban A. Arzanpour S. Manbachi M., Farhangi H. Smart grid adaptive volt-var optimization: Challenges for sustainable future grids. *Sustainable Cities and Society*, 28(3):242–255, january 2017.

[9] Associaç ao Brasileira de Normas Técnicas ABNT. Nbr 16149 sistemas fotovoltaicos (fv) características de interface de conexão com a rede elétrica de distribuição. *Norma Técnica*, 2014.

[10] Rivas D. Betancourt E., Mendes O. Distribution transformer with electronic tap changer featuring robust low current zero switching. *IEEE PES T&D Conference and Exposition*, April 2014.

[11] K. Y. Lee and M. A. El-Sharkawi. *Modern Heuristic Optimization Techniques: Theory and Applications on Power Systems*. Wiley - IEEE Press, 2008.

[12] A. Seshadri. Multi-objective optimization using evolutionary algorithms (MOEA). *IEEE Trans. on Evolutionary Computation*, pages 1–20, 2002.

[13] Kennedy J. and Eberhat R. C. Particle swarm optimization. *Proceedings of IEEE International Conference on Neural Networks*, IV:1942–1948, 1995.

[14] Shareef H. Zayandehroodi H. Farhoodnea M., Mohamed A. A linear three-phase load flow for power distribution systems. *IEEE Transactions on Power Systems*, 31(1):827 – 828, January 2016.

[15] Kersting W. H. Radial distribution test feeders. *PES Summer Meeting*, 2000.

[16] The European Commission's science and knowledge service. Photovoltaic geographical information system (pvgis). *https://ec.europa.eu/jrc/en/pvgis*.

[17] Deambi S. *Photovoltaic System Design: Procedures, tools and applications*. CRC Press, 2016.

Comparative study of RC snubber configurations in switching circuits

Ana Carolina Moreira
Instituto Federal de Educação,
Ciência e Tecnologia (IFSC)
Jaraguá do Sul – Brazil
moreira.ana.1996@gmail.com

Daniel Cesar Piccoli
Instituto Federal de Educação,
Ciência e Tecnologia (IFSC)
Jaraguá do Sul – Brazil
danicpiccoli@hotmail.com

Júlio Cesar Lopes de Oliveira
Instituto Federal de Educação,
Ciência e Tecnologia (IFSC)
Jaraguá do Sul – Brazil
julio.oliveira@ifsc.edu.br

Luiz Fernando Henning
Instituto Federal de Educação,
Ciência e Tecnologia (IFSC)
Jaraguá do Sul – Brazil
luizh@ifsc.edu.br

Rodrigo Jose Piontkewicz
Instituto Federal de Educação,
Ciência e Tecnologia (IFSC)
Jaraguá do Sul – Brazil
rodrigo.piontkewicz@ifsc.edu.br

Abstract— **This article shows a study of the dissipative passive RC snubber configurations, by using theoretical and experimental data. For that, the buck converter is use to analyse the impact of these devices in the circuit. Three projects of the RC snubber are analyse and compare, with that the device that shows a better performance in the circuit will be obtained.**

Index Terms—**Snubber, RC Snubber, Buck Converter, Switching Circuits.**

I. Introduction

With the advance of power electronic, the utilization of semiconductor switches for the operation of these devices are very important. This switching process results in undesirable disturbances in the power process.

"DC/DC converters switching has high frequency oscillations that can occur due to parasitic inductances and capacitances in the transistors, diodes and printed circuit board that occur with the high voltage and current variations." [1]

All switches has limitations such as peak voltage, peak and average current, power dissipation, switching speed, etc. Snubbers are use to improve the performance and reliability of switches imbedded in power circuits but to properly apply snubber techniques it is important to understand how switches themselves behave. [2]

The objective of this article is to compare the RC snubbers project methods, analysing their effects in the oscillation damping of a switching circuit. For that, their experimental and simulate values will be compare with a situation without snubber to be able to observe the differences in their waveforms.

With this data the values of voltage overshoot and settling time will be obtain in the simulation and experimentally. The RC snubber that can better reduce both of this parameters will be the most efficient device for the load situation.

In the scientific publications do not has a research about what is the best method for damping a circuit, because of that, the objective of this article is to compare different mathematical methods and show which has more impact with the RC snubber.

II. Theoretical Foundation

The snubber has the function of act like a damping for a switch. They can be passive, formed with resistors, diodes, capacitors and inductors; and actives, formed basically of transistors and other active elements. In this article, only the RC snubbers will be analyse. For these analyses, the buck converter will be use, as is represented at Fig. 1.

The input voltage (V_{cc}) used in this circuit is 12 V with a resistance (R_L) of 1 Ω. The control circuit has a voltage peak (V_{Peak}) of 20 V, a frequency of 200 kHz and a Duty Cycle of 50%.

The buck converter is a DC/DC voltage converter, that always produces a lower voltage output than the input or, in the theoretical limit, equals to the input. Due to the fact that this converter has the characteristics of a voltage source in the input, the current drained by a buck converter is naturally pulsed, implying high harmonic frequencies and destructive voltages to the switch during it is opening, caused by the parasitic inductance. [3]

In this article, the methods will be referred to the main author that wrote it, being then: The Boylestad's method [4],

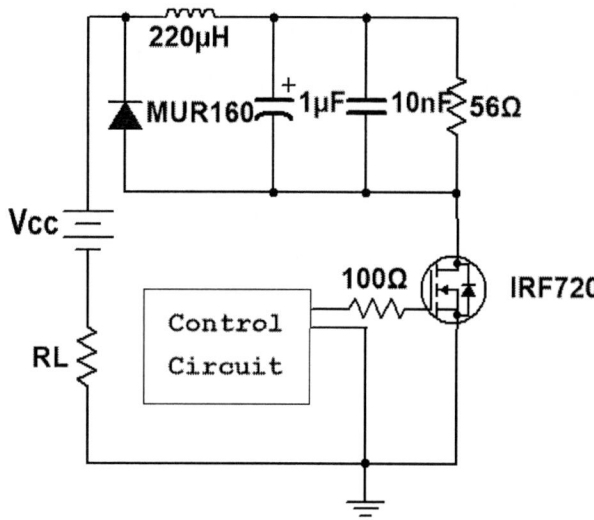

Fig. 1. Buck converter.

the Albuquerque's [5] method and the Vaculick's method [6].

To better describe this RC snubbers projects, the mathematical data is obtained by the bibliographies, where the values of resistance and capacitance are obtain. Then, all the methods are compare with simulation an practical values, being that the best method will be determinate.

III. PROJECT

In this section the RC snubbers parameters are obtain with a mathematical analysis, using the buck converter as a reference. The RC snubber will be represent by the addition of a capacitor C_s and resistor R_s in parallel with the buck converter MOSFET switch.

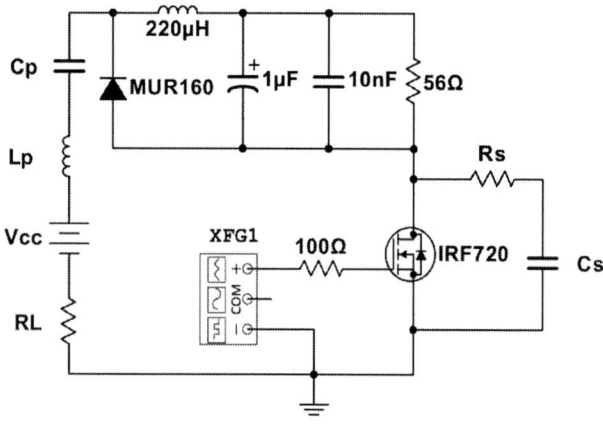

Fig. 2. RC snubber.

The voltage peak in the MOSFET drain can cause faults

and damage the switch. To overcome this problem, the RC snubber can be use. The damping circuit is use to limit overvoltage and overcurrent. [7]

The snubber resistor improves the efficiency in oscillation frequency, while the capacitor reduce the switching frequency dissipation. To design an RC snubber, the mathematical data present in the references of this article are use.

A. Vaculik's RC snubber project

The first project of RC snubber present is the Vaculick's method. In this mathematical process, the circuit's parasitic inductance and capacitance are use as basis to obtain the resistor and capacitor values.

To design an RC snubber circuit mathematically the equations (1) and (2) can be use. [6]

$$\zeta = (\sqrt{L_p/C_p})/(2 \cdot R_S) \tag{1}$$

$$f_{osc} = 1/(2\pi \cdot C_S \cdot R_S) \tag{2}$$

Where ζ is the damping coefficient, L_p is the total parasitic inductance, C_p the parasitic capacitance, R_s and C_s are the RC snubber values that are required, and f_{osc} is the circuit oscillation frequency.

Rewriting the equations (1) and (2) in order to obtain the resistance and capacitance values, the equations (3) and (4) are obtained:

$$R_S = (\sqrt{L_p/C_p})/(2 \cdot \zeta) \tag{3}$$

$$C_S = 1/(2\pi \cdot R_S \cdot f_{osc}) \tag{4}$$

For the resistor value calculation it is necessary to know the ζ, L_p and C_p values. The chosen value of ζ is 1, in order to achieve a critical damp. [8]

The C_p is the parasitic capacitance that can be approach to the MOSFET output capacitance (C_{oss}). This capacitance can be obtain from the component datasheet, that in this case is the IRF720. [9]

In the datasheet, this tabulated value has a drained voltage to the source of 25 V, and in the circuit this voltage has a maximum value of 12 V. In this case the C_{oss} is 120 pF and the oscillation frequency (f_{osc}) is show at the Fig. 3, and it is 7.692 MHz, but for mathematical effects it will be use 8 MHz.

The buck converter equivalent circuit with the parasitic inductances (L_p) and with the RC snubber can be represent as follows in Fig. 2. A capacitor and a inductor in series in a alternate current circuit has a angular frequency (ω) according to equation (5).

$$\omega = 1/\sqrt{C_p \cdot L_p} \tag{5}$$

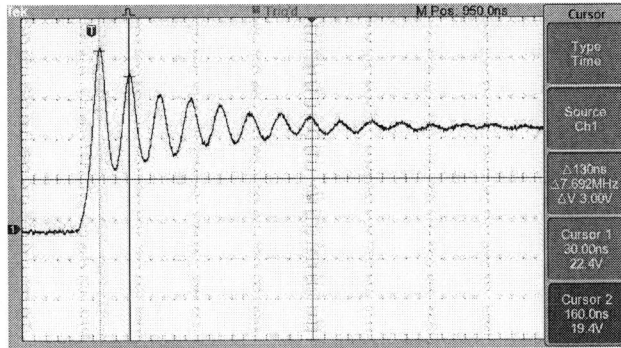

Fig. 3. Oscillation frequency.

With that the angular frequency is $2\pi \cdot f_{osc}$ an isolating L_p, the equation (6) is obtain:

$$L_p = (1/(2\pi \cdot f_{osc}))^2/C_p \qquad (6)$$

Replacing the frequency and capacitance values the equation (7) is obtain.

$$L_p = (1/2\pi \cdot 8M)^2/120p \approx 3.3 \ \mu H \qquad (7)$$

Replacing the frequency, damping coefficient, parasitic capacitance and inductance values in equations (3) and (4), the RC snubber capacitor and resistor values will be reach. As equations (8) and (9) shows.

$$R_S = \sqrt{3.3\mu/120p}/2 \cdot 1 \approx 82.89 \ \Omega \qquad (8)$$

The closest resistor comercial value is 91 Ω, with that $R_s = 91 \ \Omega$.

$$C_S = 1/(2\pi \cdot 51 \cdot 8M) \approx 218.61 \ pF \qquad (9)$$

To capacitor comercial chosen is 220 pF.

B. Albuquerque's RC snubber project

In the Albuquerque's method, to obtain the values of the RC snubber the current and voltage of the circuit are necessary to use.

A RC snubber, place in parallel with the switch, can be use to reduce the voltage on it during switch-off, and damp the ripples. In most cases, a simplify analysis can be used to determine the values of the R_S and C_S components. In general, for a good damping, $C_S > C_p$, for example, C_S may be double of C_p which represents the sum of the mounting capacitance. For R_S, a good approximation is $R_S = V_0/I_0$, being that the voltage in the resistance can be estimate by the energy stored in C_S. [5]

The parameters of capacitance and resistance of this snubber will be obtain through mathematical calculations, by the equations (10) and (11).

$$C_S = 2C_p \qquad (10)$$

$$R_S = V_0/I_0 \qquad (11)$$

Being that I_0 is the current of the circuit and V_0 the voltage of the circuit, that has the values of 110 mA and 12 V respectively, and the parasitic capacitance C_p have a value of 120 pF. With this values the parameters of the snubber can be calculate by the equations (12) and (13).

$$C_S = 2 \times (120\mu) = 240pF \qquad (12)$$

$$R_S = 12/110m = 112\Omega \qquad (13)$$

These components values must be adapt to the commercial values of them, being that the resistance and capacitance use in the obtain of the experimental data are 100Ω and $240pF$. Another important parameter of the circuit is the resistor power that can be obtain by equation (14), being that the circuit frequency is $f_s = 200kHz$.

$$P_{Rs} = C_S V_0{}^2 f_s$$

$$P_{Rs} = (240p) \times (12)^2 \times (200k) = 6.91 \ mW \qquad (14)$$

So the resistor use in the circuit must has a minimum power of 6.91 mW to the correct action of the RC snubber in the circuit.

C. Boylestad's RC snubber project

The Boylestad's method is a more theoretical than mathematical process. The author recommend a resistor of 100 Ω and a capacitor that have a maximum value of 10 nF.

The capacitor reactance is determined by $X_C = 1/(2\pi f C)$, so as higher is the frequency, lower is the resistance. Due to high voltages, the ceramic capacitors are use, with values around 10 nF. High value capacitors are not use, because in this devices the voltage rise slowly, what causes a decrease in the system operation velocity. The 100 Ω resistor in series with the capacitor are use to limit the current outbreaks, that are result of the system state change. [4]

The Boylestad's RC snubbers will be analyse varying the capacitors values within the limits present by the author. There will be use three capacitance values: 1 nF, 4.7 nF and 10 nF.

IV. RESULTS

In this section the circuits simulated and experimental data, with and without snubber will be compared. With that, the waveforms can be analysed to see the difference between the circuits.

978-1-7281-4181-7/19 $31.00 © 2019 IEEE

A. Practice Value X Computer Simulation

The comparison is make between the simulate data, with a circuit software, and the real data, with practical values. The buck converter is assemble and simulate, measuring the voltage and the current values in the MOSFET, before and after the addition of the RC snubbers.

In the simulation, a series inductance value is add to the circuit to simulate the parasitic inductance of the circuit. This inductance, in the experimental circuit, is low, so the authors is not be able to measure it. With that this parameter is varied until a close result is obtain between simulate and practical experiences, that results in 700 nH.

1) Without snubber: Before start the results comparison, obtained with the RC snubbers, the measurement is performe without any snubber, to be able to check it is efficiency as shows the Fig. 4. The experimental and computational charts are use as a basis of comparison with the RC snubber.

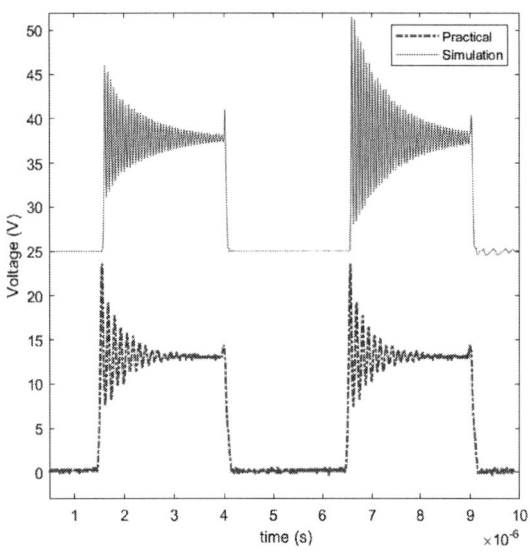

Fig. 4. Voltage without snubber

In this graphic some oscillations, as the result of semiconductor switching, can be seen, so, this circuit does not have a effective damping without the snubber device. This graphics and results will be use as a base of comparation between the RC snubbers study in this article.
To be able to get a better comparison between the circuits, some values of the graphics are analyse, being then the settling time, the voltage overshoot and the output voltage.

The criterion of the settling time use is the time that the voltage takes to enter in the interval of 3% of variation of it is steady state voltage value ($t_{s(97\%)}$). [10]

Without the snubber the output voltage peak was 13 V, the voltage overshoot 8.75 V and the settling time 2.15 μs.

2) Vaculick's RC snubber: The voltage graphic, in the buck converter circuit, with the Vaculick's RC snubber performe experimentally and simulate as shown in Fig. 5.

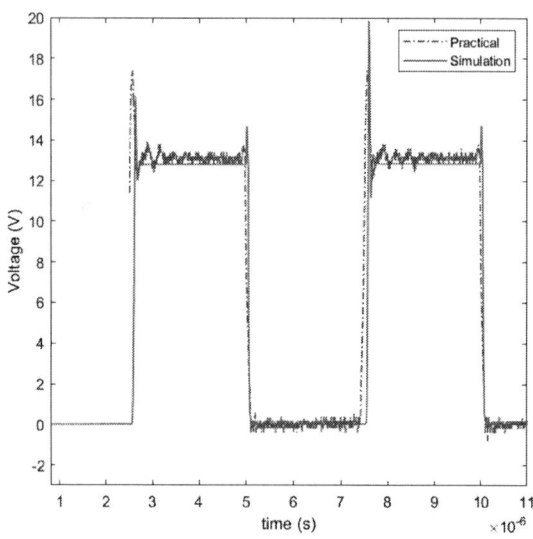

Fig. 5. Vaculick RC – Experimentally and Computer Simulation.

So, with the graphical analysis, the RC snubber has a output voltage peak of 13 V, a voltage overshoot of 6 V and a settling time of 1,18 μs.

The values of overshoot and settling time is better than the the values without the snubber, due to the better damping that the RC snubber provides.

3) Albuquerque's RC snubber: The simulate and experiment graphic with Albuquerque's RC snubber is shown in Fig. 6.

So, with the graphical analysis, the RC snubber has a output voltage peak of 13 V, a voltage overshoot of 3.73 V and a settling time of 0.4 μs. This values is similar to the first RC snubber analyse, with a better impact in the settling time of the circuit.The voltage overshoot was practically the same in both cases, due to the similarity of their component values.

4) Boylestad's RC snubber: This snubber is analyse with three different values of capacitor, being the of 1 nF, 4.7 nF and 10 nF. The capacitor of 1 nF had a low impact in the damping of the oscillations of the circuit, due to it is capacitance level. The 10 nF has a big impact in the damping, but it generate a distortion in the output signal,

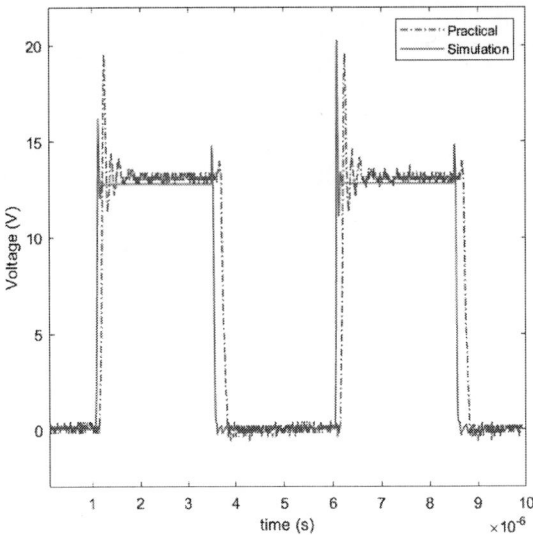

Fig. 6. Albuquerque's RC – Computer Simulation.

what is not great in electronic circuits.

So the capacitor of 4.7 nF is utilize, because it has a impact in the oscillation and no distortions in the signal. All the graphics of this section are base in this capacitor value. The simulate and the experimental graphics, of this snubber is shown in Fig. 7.

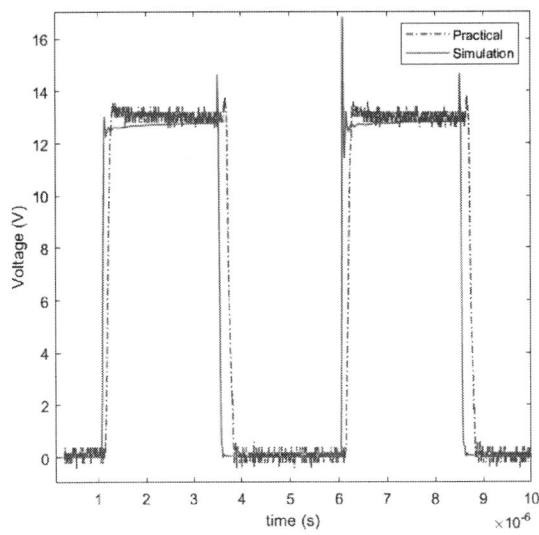

Fig. 7. Boylestad's RC – Computer Simulation.

This RC snubber has a output voltage peak of 13 V, a voltage overshoot of 0.1 V and a settling time of 0.05 μs.

As can be seen by the comparison of the three methods, higher values of capacitance had a bigger impact in the damping of the system, however, the capacitance value can't be higher then 10 nF, to avoid distortions in the output signal.

The Table (1) shows the comparison between the values obtained without snubber and with the RC snubber methods.

TABLE I
VALUES COMPARATION

	Without	Vaculick	Albuquerque	Boylestad
Output Voltage (V)	13	13	13	13
Overshoot (V)	8.75	3.73	3,05	0.1
Settling time (us)	2.15	0.40	0.15	0.05

This comparison shows that both snubbers had a great impact in the parameters, when compared with a non-snubber situation, and does not influenced in the output voltage. Being that, there is not a significant voltage drop in this devices.

The Boylestad RC snubber has a better performance in all the parameters, having the shortest oscillation time and voltage overshoot. With that it is the more recommended passive snubber to damp the voltage oscillations in a switching circuit.

V. CONCLUSION

With the mathematical and graphical analyses, the influence of the parameters of the RC snubbers in the circuit is obtaine. All the three methods has a considerable impact in the voltage damping of the buck convert.

This analyses also shows a great proximity in between the practical and experimental data that is study. This factor has a important value to validate the research done, being that the simulation has to predict the influence of external factors in the behavior of the circuit.

The Boylestad's method is the one who has the better performance in the damping, having the lowers settling time and voltage overshoot. The other methods were close in performance, the Albuquerque's RC has a higher impact in the settling time than the Vaculik's, and both has a similar voltage overshoot.

In this process is note that the capacitor has a high influence as high is it is value. But this value must be lower than 10 nF to avoid distortions on the output signal due to the charging time of the capacitor.

After calculate and compare the charts, the conclusion is witch the Boylestad's RC snubber method is most effective, what makes it the better device to damp the oscillations in a switching circuit.

This research can be continue with the analyse of other type of passive snubber, as the snubber C and the RCD, as well the active snubbers, with the same methodology adopted in this article.

REFERENCES

[1] J. C. L. de Oliveira, "Projeto e estudo de um conversor ca/cc de alta potência, 14,4 v e 300 a para aplicações automotivas." mathesis, Universidade Estadual de Londrina (UEL), Paraná, 2014.

[2] R. Severns, *Snubber Circuits for Power Electronics*, 1st ed. n.a: Rudolf Severns, 2008.

[3] R. R. Coelho, "Estudo dos conversores buck e boost aplicados ao rastreamento de máxima potência de sistemas solares fotovoltaicos." Master's thesis, Universidade Federal de Santa Catarina (UFSC), 2008.

[4] R. L. Boylestad and L. Nashelsky, *Dispositivos Eletrônicos e Teoria de Circuitos.*, 11th ed. São Paulo: Person Education, 2013.

[5] R. O. Albuquerque and A. C. Seabra, *Utilizando Eletrônica com AO, SCR, TRIAC, UJT, PUT, CI 555, LDR, LED, IGBT e FET de potência*, 12th ed. São Paulo: Érica Ltda., 2013.

[6] P. Vaculik, "The experience with sic mosfet and buck converter snubber design." in *World Academy of Science, Engineering and Technology*, London, 2014.

[7] A. Algaddafi and K. Elnaddab., "Modelling and designing the RC snubber circuit for a buck converter and testing its effectiveness," in *2016 International Renewable and Sustainable Energy Conference (IRSEC)*, Marrakech, 2016, pp. 554–559.

[8] S. Maniktala, *Switching Power Suplies A to Z*, 1st ed. n.a: Elsivier/Newnes, 2006.

[9] *Power MOSFET IRF720*, Vishay Siliconix, Jun. 2008.

[10] R. C. Dorf and R. H. Bishop, *Sistemas de Controle Modernos*, 2nd ed. São Paulo: LTD, 2013.

Comparative Study of Control Systems for a Photovoltaic Inverter with LCL Filter

Leandro T. Omine
Faculdade de Engenharias, Arquitetura e Urbanismo e Geografia
Universidade Federal do Mato Grosso do Sul
Campo Grande, Brasil
leandro9029@gmail.com

Moacyr A. G. Brito
Faculdade de Engenharias, Arquitetura e Urbanismo e Geografia
Universidade Federal do Mato Grosso do Sul
Campo Grande, Brasil
moa.brito@gmail.com

Abstract—**The LCL filter for on-grid inverters are a cheaper and smaller substitute for the L filter. Many researches have proposed solutions to the control system of LCL plus inverter plant, which is challenging due to the resonance effects. This paper presents a review of some linear and nonlinear control methods frequently cited in literature through simulation comparisons under three scenarios: ideal case with no grid inductance, with a small value of inductance and finally with an increase in the line frequency.**

Keywords—LCL filter, Nonlinear control, Distributed generation.

I. INTRODUCTION

In recent years the advances in photovoltaic (PV) panels manufacturing technologies have increased the viability of this generation system by reducing the cost of the related devices [1]. Moreover, many countries have been funding programs for the integration of renewable energy sources into the electrical grid [2].

In any of these renewable or so-called distributed generation (DG) systems, the grid-tie inverter makes the connection between the generator, local loads and grid [3]. Due to the high frequency harmonics caused by the switching, a passive filter needs to be used at the grid side. Initially a simple L filter was adopted due to its simplicity [4], however the constraints on total harmonic distortion (THD) and on the size and cost of the product have been encouraging the industry to research for better choices.

The LCL filter has been a popular candidate to replace the L filter as it has a -60 dB/dec attenuation rate at high frequencies, and the total inductance required is smaller [3-5], making its size and cost decrease, even if the number of components increases.

There are a few challenges in using an LCL filter [3], with one being the peak of resonance introduced by the inductors-capacitor interactions, which increases the difficulty of stabilization by conventional proportional-integral (PI) based controllers. Passive or active damping methods [3] may be used to reduce this peak and support the control system [6].

Many controllers have been proposed individually for the current control loop of the LCL inverter. In [7] a linear quadratic regulator (LQR) obtained by the linear matrix inequalities (LMI) [8] approach is used to achieve 3.4% of THD with a single-phase inverter. In [6] the LMI is used to solve a H_2/H_∞ mixed constraint and create a robust controller, that is applied in a three-phase inverter under simulation scenarios. Some authors applied nonlinear methods as the sliding-mode control (SMC) [9-10], that inserts a nonlinear term in the feedback control loop; the finite control set model predictive controller (FCS-MPC) [11-12], that predicts the state variables for all the possible switching states and then

apply the best one based on a cost function; the passivity based control (PBC) [13-14], that incorporates an active damping similar of passive system concept, among others.

This paper presents a brief introduction about the project of some of these linear and nonlinear controllers, as well as comparative performance analysis among them. The paper presents the project of an LCL filter and its modelling in state space form by using small signal analysis. The basic theory behind each of the controllers and the procedures to project and verify the stability are also presented. Finally, simulation results are used to track the inverter side current in a single-phase LCL inverter.

II. THE LCL FILTER

An overall representation of the PV system is illustrated in Fig. 1. The scenario considered has four PV modules of 265 W each, sending a total of 1060 W into a local grid of 127 V, through a single-phase inverter. Considering that a voltage controller embedded into a common boost converter keeps the dc side voltage at 250 V, the circuit may be simplified as in Fig. 2; remaining the inverter plus the LCL filter as the main part under analysis.

The LCL filter design follows the one presented by [15], and the inductances L_1, L_2, and capacitance C_1 are calculated using (1), (2) and (3), with the parameters given in Table I.

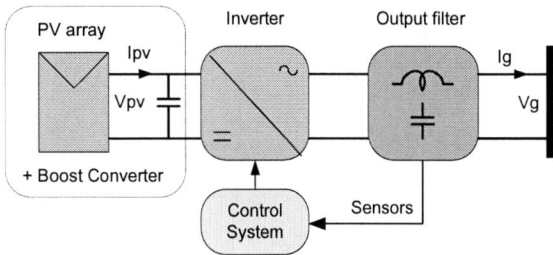

Fig. 1. Block diagram of an on-grid PV generation system.

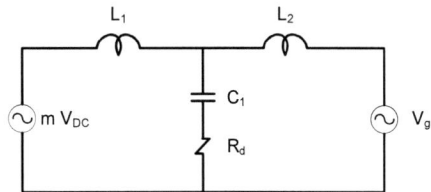

Fig. 2. Simplified circuit of the inverter AC side.

$$L_1 = \frac{V_{dc}}{6f_{sw}\Delta I_L} \qquad (1)$$

$$C_1 = 0.03\left(\frac{P_{dc}}{2\pi f_g V_g^2}\right) \qquad (2)$$

978-1-7281-4181-7/19 $31.00 © 2019 IEEE

$$L_2 = \frac{1+\frac{1}{k_a}}{C_1(2\pi f_{sw})^2} \tag{3}$$

In addition, the damping resistance value is obtained by (4), using the filter's resonance frequency ω_{res} given by (5). All the calculated results were inserted in Table II.

$$R_d = \frac{1}{3C_1\omega_{res}} \tag{4}$$

$$\omega_{res} = \sqrt{\frac{L_1+L_2}{L_1 L_2 C_1}} \tag{5}$$

For control purposes, a state space model of the LCL filter might be obtained by a small signal analysis of the plant in Fig. 2. The state equation of each element is described in (6).

$$L_1 \frac{di_{L_1}}{dt} = mV_{dc} - v_{C_1} - R_d(i_{L_1} - i_{L_2});$$
$$L_2 \frac{di_{L_2}}{dt} = v_{C_1} + R_d(i_{L_1} - i_{L_2}) - V_g; \tag{6}$$
$$C_1 \frac{dv_{C_1}}{dt} = i_{L_1} - i_{L_2};$$

Where m is the inverter modulation index.

Applying small perturbations in (6) and removing the constant and second-order terms results in:

$$\begin{bmatrix} \dot{\hat{\imath}}_{L_1} \\ \dot{\hat{\imath}}_{L_2} \\ \dot{\hat{v}}_{C_1} \end{bmatrix} = \overbrace{\begin{bmatrix} -R_d & R_d & -1 \\ R_d & -R_d & 1 \\ 1 & -1 & 0 \end{bmatrix}}^{A_d} \overbrace{\begin{bmatrix} \hat{\imath}_{L_1} \\ \hat{\imath}_{L_2} \\ \hat{v}_{C_1} \end{bmatrix}}^{x} + \overbrace{\begin{bmatrix} V_{dc} \\ 0 \\ 0 \end{bmatrix}}^{B_d} \overbrace{\hat{m}}^{u} \tag{7}$$

Where hats indicates small perturbations around some nominal condition.

TABLE I.   PV SYSTEM'S PARAMETERS

	Description	Value
V_{dc}	Dc side voltage	250 V
V_g	Grid side RMS voltage	127 V
P_{dc}	Nominal power	1060 W
f_{sw}	Switching frequency	20 kHz
f_g	Grid frequency	60 Hz
ΔI_L	Inductor current ripple	1.18 A
k_a	Attenuation factor	0.1

TABLE II.   LCL FILTER'S PARAMETERS

	Description	Value
L_1	Inverter side inductance	1.80 mH
L_2	Grid side inductance	133.19 μH
C_1	Filter capacitance	5.23 μF
ω_{res}	Resonance frequency of the filter	39.29 krad/sec
R_d	Damping resistance	1.62 Ω

III. MODERN CONTROL TECHNIQUES

The current controller of an inverter must be able to track a sinusoidal reference. While a simple PI controller is not sufficient, some modification based on the internal model principle leads to the PR controller, which amplifies the Bode gain of the open loop system around a pre-defined frequency. Equation (8) shows the equation of a PR controller in frequency domain.

$$C_{PR} = K_p + \frac{K_r s}{s^2 + \omega^2} \tag{8}$$

Where K_p and K_r are the respective proportional and resonant gains, s is the Laplace operator, and ω is the required resonant frequency.

Another approach using the classical control theory is by substituting the proportional term in (8) with a PI term while keeping the resonant one (PI+RES). Equation (9) shows its representation:

$$C_{PI+RES} = K_p + \frac{K_i}{s} + \frac{K_r s}{s^2 + \omega^2} \tag{9}$$

Where K_i is the integrator gain.

Modern control theories mostly use state space models and state feedback to achieve the goals, which are defined by cost functions instead of classical phase margin and bandwidth. The LQR is part of this modern theory, being based on the cost function (10) [7]:

$$J = \int_0^t x^T Q x + u^T R u \, d\tau \tag{10}$$

Where x and u are the state and input vectors respectively, and $Q \geq 0$ and $R > 0$ are square matrices of appropriate size that specify the importance of each state and the input in the optimization.

By choosing a state feedback in the form of (11):

$$u = -K_{LQR} x \tag{11}$$

the LQR gives a gain vector K_{LQR} that minimizes (10).

Considering an upper limit function $V(x)$ for the integral in (10), such as (12):

$$V(x) = x^T P x \tag{12}$$

Where P is a square, positive definite matrix, solving the LMI problem in (13) also gives the solution for K_{LQR}, as in (14):

$$\min tr(-S)$$
$$s.t. \begin{bmatrix} SA^T + AS - G^T B^T - BG & S & G^T \\ S & -Q^{-1} & 0 \\ G & 0 & -R^{-1} \end{bmatrix} \leq 0 \tag{13}$$

$$K_{LQR} = GS^{-1} \tag{14}$$

Where $S = P^{-1} = S^T$ and $G = K_{LQR}P^{-1}$.

As one of nonlinear control techniques, the sliding-mode control inserts a discontinuous signal function in the control loop. It creates a surface in the state space, which contains the desired steady-state conditions and makes the space coordinates to move toward and stay nearby this surface [9]. As it is impossible for the states to be exactly on the surface, they will be "sliding" around it in steady-state conditions, as illustrated in Fig 3.

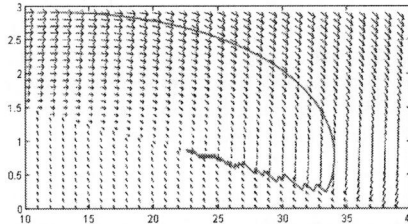

Fig. 3. State-space path of a system under the sliding-mode control.

Because of the forced movement that depends on the measured values, the SMC is robust against parameters uncertainties of the model when operating in sliding mode [16].

To bring the states to the surface, denoted by $s(x) = 0$, the control law must contain a term δu that obeys the relation (15).

$$\begin{aligned} s(x) > 0 \rightarrow \delta u < 0 \\ s(x) < 0 \rightarrow \delta u > 0 \end{aligned} \tag{15}$$

Equation (16) shows an example of δμ that satisfies (15):

$$\delta u = -\rho\, sgn\big(s(x)\big) - k\, s(x),\ \rho > 0 \tag{16}$$

The control law of SMC is then represented by (17), where the steady-state equivalent control law u_{eq} is obtained by taking the time derivative of $s(x)$.

$$u = u_{eq} + \delta u \tag{17}$$

The SMC's stability is verified by Lyapunov's method. Taking a Lyapunov function in the format of (18), with **P** as an identity matrix, its time derivative results in (19).

$$V = \frac{1}{2} s(x)^T s(x) \tag{18}$$

$$\dot{V} = s(x)s(\dot{x}) \tag{19}$$

Asymptotic stability will be guaranteed as long as (19) is a negative value [9].

The last controller uses concepts of passivity. The concept of passivity originates in electrical circuit analysis, where the term passive indicates a circuit component that dissipates or stores, but cannot generate energy.

In analogy, a system is said to be passive if there exist a state dependent storage function that is always smaller than the energy introduced by the inputs [17]. Equation (20) as well as its time derivative, (21), express this relation by inequalities.

$$\int_0^t u(t)y(t)d\tau \geq V\big(x(t)\big) - V(x(0)) \tag{20}$$

$$u(t)y(t) \geq \dot{V}\big(x(t)\big) \tag{21}$$

If the stored function $V(x)$ is positive definite, then the system is also stable in the sense of Lyapunov. An important consequence of passivity is that the state-space origin of a passive system is asymptotically stable.

A system in Euler-Lagrange formulation, as in (22), can be made passive by shaping its state variables.

$$D\dot{x} = \big(J_0 + J_1(u)\big)x + F(u) \tag{22}$$

Where D is a diagonal matrix, J_0 and J_1 are skew-symmetric matrices and F is the input vector. The state vector x can be substituted by the sum of the desired value and the error:

$$D(\dot{x}_e + \dot{x}^*) = \big(J_0 + J_1(u)\big)(x_e + x^*) + F(u) \tag{23}$$

Rearranging (23) and adding a dissipative term in both sides result in (24):

$$D\dot{x}_e - \big(J_0 + J_1(u)\big)x_e + Rx_e = -D\dot{x}^* + \big(J_0 + J_1(u)\big)x^* + F(u) + Rx_e \tag{24}$$

Where R is a positive diagonal matrix. For instance assuming that the right hand side of (24) is zero, the error dynamics are described by (25):

$$D\dot{x}_e = \big(J_0 + J_1(u)\big)x_e - Rx_e \tag{25}$$

By choosing a storage function as in (26), its time derivative results in (27):

$$V(x) = \frac{1}{2}x_e^T D x_e \tag{26}$$

$$\dot{V}(x) = x_e^T D \dot{x}_e \tag{27}$$

Substituting (25) in (27), it is obtained:

$$\dot{V}(x) = -x_e^T R x_e \tag{28}$$

Equation (28) alongside (26) proves the asymptotic stability of the error dynamics in the sense of Lyapunov, as long as the condition in (29) is attended [18].

$$D\dot{x}^* = \big(J_0 + J_1(u)\big)x^* + F(u) + Rx_e \tag{29}$$

The control laws for each controller and the parameters adopted for simulations were inserted in Table III and Table IV, respectively. The PR, PI+RES and PBC sensors only the grid side current; the LQR demands all state variables to be sensored; the SMS needs the capacitor voltage and grid side current. All controllers, in addition has a grid voltage sensor to synchronize the reference.

TABLE III. CONTROL LAWS

Controller	Control Law
PR	$u = (K_{PR1} + \frac{K_{PR2}\, s}{s^2 + \omega^2})(i_{L1}^* - i_{L1})$
PI+RES	$u = (K_{PI1} + \frac{K_{PI2}}{s} + \frac{K_{PI3}\, s}{s^2 + K_{PI4}s + \omega^2})(i_{L1}^* - i_{L1})$
LQR	$u = K_{LQR1}(i_{L1}^* - i_{L1}) - K_{LQR2}i_{L2} - K_{LQR3}v_{C1}$
SMC	$u = \frac{v_c}{V_{dc}} - \rho\, sgn(i_{L1} - i_{L1}^*) - \sigma(i_{L1} - i_{L1}^*)$
PBC	$u = \frac{v_g}{V_{dc}} - R_1(i_{L1} - i_{L1}^*)$

TABLE IV. CONTROL PARAMETERS

Description	Value
K_{PR1}, K_{PR2}	0.05, 15.10
$K_{PI1}, K_{PI2}, K_{PI3}, K_{PI4}$	0.047, 80.84, 300, 75.40
$K_{LQR1}, K_{LQR2}, K_{LQR3}$	0.3056, 0.0107, -0.0017
ρ, σ	0.012, 0.217
R_1	40.5

IV. SIMULATIONS

The controllers were tested under three situations: case a with a grid inductance $L_g = 0$, case b with an $L_g = 500\mu H$, and case c with an increase of 5% in the grid frequency. The results are illustrated in Fig. 4 to Fig.8. Fig. 4 presents the results for the PR, Fig. 5 for PI+RES, Fig 6. for LQR, Fig. 7 for SMC and finally Fig. 8 presents the results for PBC.

All simulations start with the nominal current as the reference, depicted in blue, and at 0.06 seconds it is reduced to 50%.

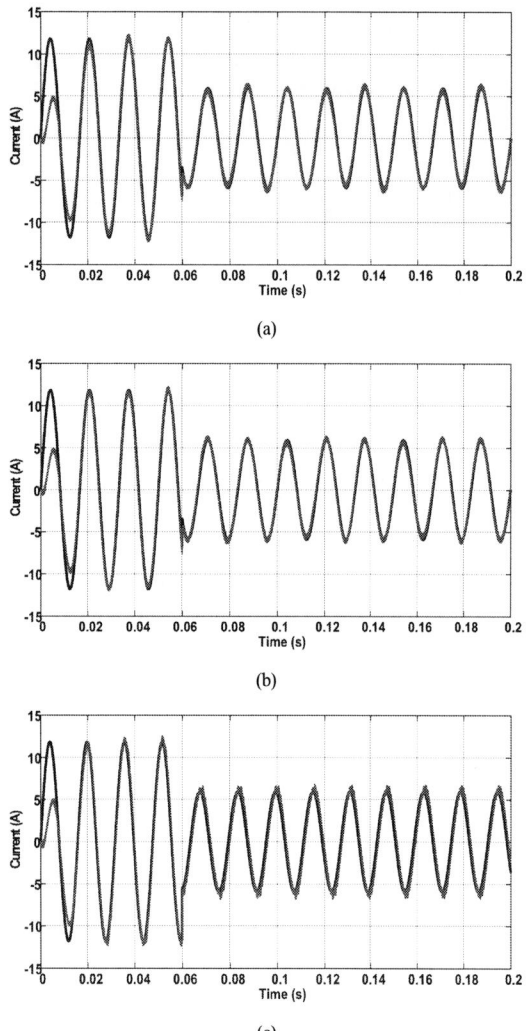

Fig. 4. Simulation results for PR controler .

The PR shows similar results for both cases a and b, however the frequency change in case c creates a phase delay in the output waveform. This tracking error was expected as the resonant term was specifically projected for 60Hz. The same applies to the PI+RES controller, Fig.5, besides it is faster in achieving the steady-state amplitude.

Although the LQR keeps the same performance despite the scenario and its tracking capacity is considerably high, it generates a great amount of harmonics, as seen in Fig. 6 a), b), and c), and in Table V.

The SMC, Fig. 7, presented a good performance for case a, but the grid inductance affects its ripple significantly. Other than this, the steady-state error was kept almost the same even with the frequency shift.

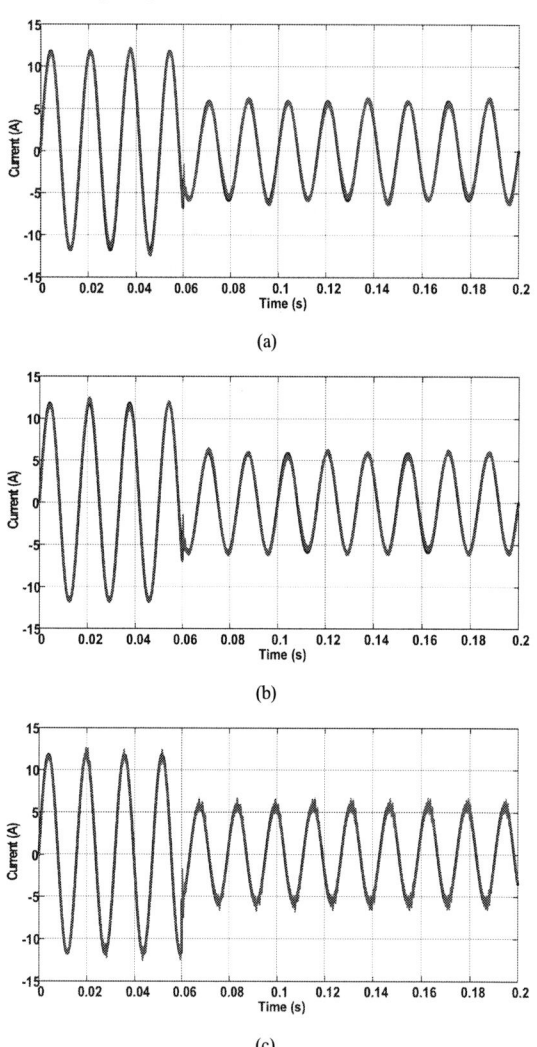

Fig. 5. Simulation results for the PI+RES controller.

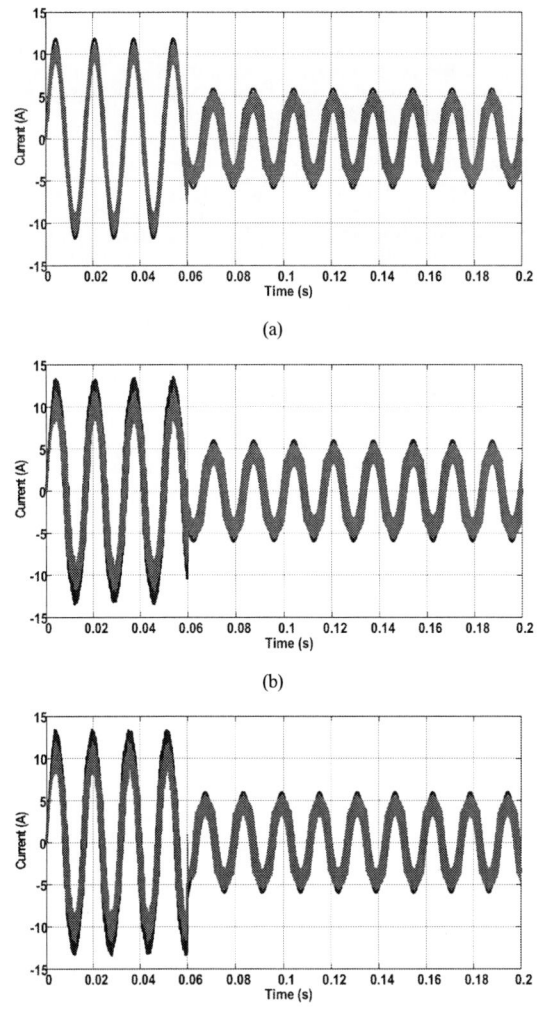

Fig. 6. Simulation results for the LQR control.

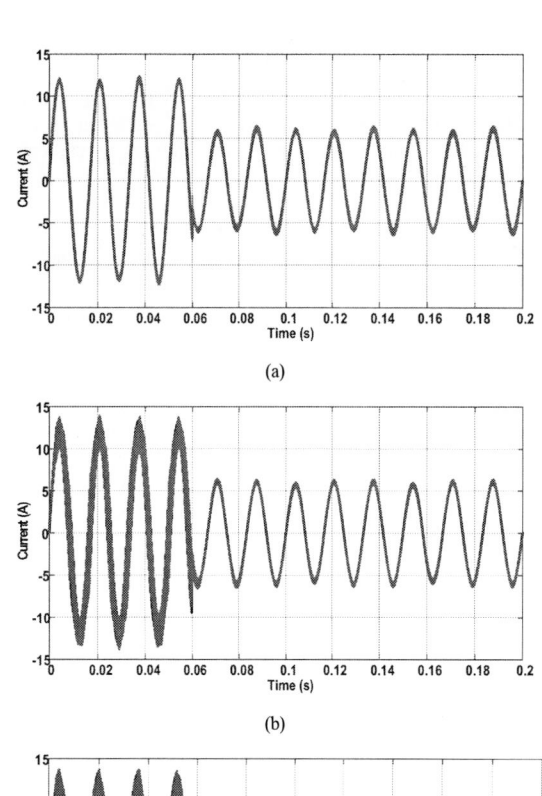

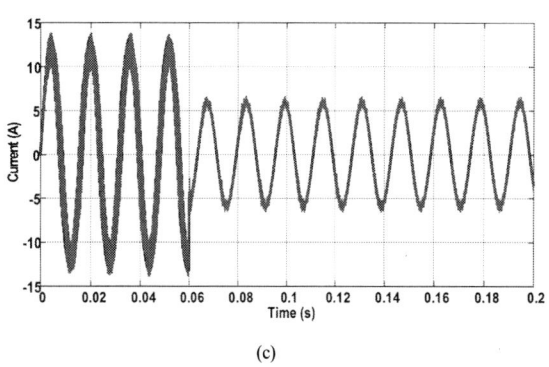

Fig. 7. Simulation results for the SMC control.

The PBC, Fig. 8, had similar responses for all the scenarios with a slight increase in THD in case c. It is not capable however, to eliminate the steady-state error since the control law is based on constant gains.

TABLE V.   THD OF GRID-SIDE CURRENT

-	Case		
Controller	a	b	c
PR	3.22%	1.36%	7.99%
PI+RES	3.23%	1.44%	10.72%
LQR	20.21%	11.90%	15.52%
SMC	3.44%	1.68%	12.69%
PBC	3.21%	1.31%	11.01%

V. CONCLUSIONS

PV generation systems can be installed in a vast of environments, each of them with different characteristics. The inverter connects them into the grid and must support slight variations on the parameters, i.e., the inductance, voltage amplitude and the line frequency.

The resonant type controllers were able to track the reference with similar transient characteristics even with the inductance variation, showing a certain robustness against this kind of parameter variation. However, a slight increase in the line frequency led to a steady-state error in tracking, which was expected since the controllers were projected to resonate on a specific frequency. It should be noted that these controllers need only one variable, the inverter side current, to be measured besides the grid voltage.

Both nonlinear controllers presented almost instantaneous transient responses, although the PBC could keep a small steady-state ripple even for the variations in the grid inductance and line frequency. It is noted that both of them still are not capable of eliminate completely the error between the output and reference.

In an overall point of view, the PI+RES controller showed the best results since it is fast, has a high tracking capability, and requires minimum quantity of sensors. The PBC has similar characteristics, moreover it is immune to frequency variations and its control law could be implemented easily without the necessity of a microprocessor, being an excellent alternative to be considered.

From the internal model principle the tracking capacity of the SMC and PBC could be improved by inserting a resonant term into the control loop. This interaction has already been explored by some researches and could be inserted into the comparative list as well as other methods such as the finite control set model predictive control, in future works.

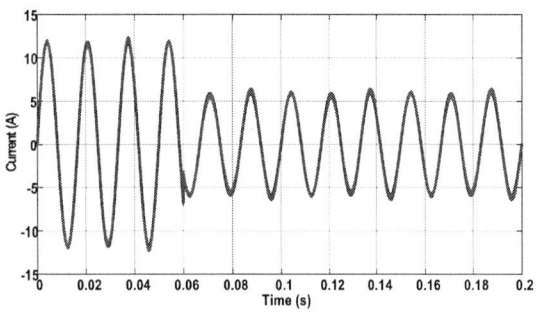

(a)

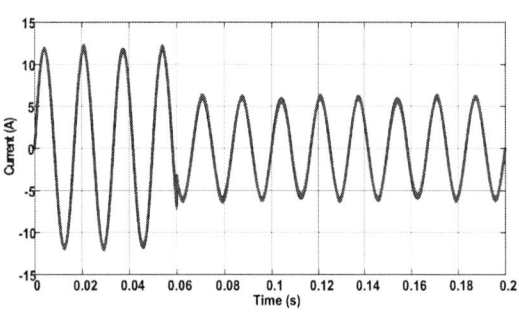

(b)

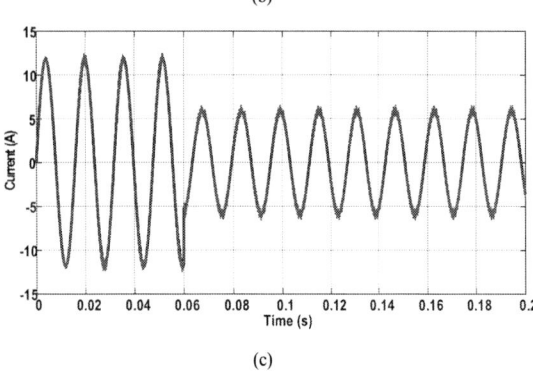

(c)

Fig. 8. Simulation results for the PBC control.

REFERENCES

[1] N. Kumar, T. K. Saha, and J. Dey, "Sliding-Mode Control of PWM Dual Inverter-Based Grid-Connected PV System: Modeling and Performance Analysis," IEEE J. Emerg. Sel. Topics Power Electron., vol. 4, pp. 435-444, June 2016.

[2] V. N. Lal, S. N. Singh, "Control and Performance Analysis of a Single-Stage Utility-Scale Grid-Connected PV System", IEEE Syst. J., vol. 11, pp 1601-1611, September 2017.

[3] W. Wu, et al., "Damping Methods for Resonances Caused by LCL-Filter-Based Current-Controlled Grid-Tied Power Inverters: An Overview," IEEE Trans. Ind. Electron., vol. 64, pp 7402-7413, Semptember 2017.

[4] S. Jayalath, M. Hanif, "Generalized LCL-Filter Design Algorithm for Grid-Connected Voltage-Source Inverter," IEEE Trans. Ind. Electron., vol. 64, pp 1905-1915, Semptember 2017.

[5] R. N. Beres, X. Wang, M. Liserre, F. Blaabjerg, and C. L. Bak, "A Review of Passive Power Filters for Three-Phase Grid-Connected Voltage-Source Converters," IEEE J. Emerg. Sel. Topics Power Electron., vol. 4, pp. 54-69, March 2016.

[6] R. K. Sharma, S. Mishra, and S. Mudliyar, "Robust state feedback current controller with harmonic compensation for single stage grid connected PV inverter with LCL filter," IEEMA Engineer Infinite Conference, India, March 2018.

[7] P. Buduma, and G. Panda, "Robust nested loop control scheme for LCL-filtered inverter-based DG unit in grid-connected and islanded modes," IET Renewable Power Generation, vol. 12, pp 1269-1285, August 2018.

[8] G. Chesi, "LMI Techniques for Optimization Over Polynomials in Control: A Survey," IEEE Trans. Autom. Control, vol. 55, pp 2500-2510, November 2010.

[9] R. P. Vieira, L. T. Martins, J. R. Massing, and M. Stefanello, "Sliding Mode Controller in a Multiloop Framework for a Grid-Connected VSI With LCL Filter," IEEE Trans. Ind. Electron., vol. 65, pp 4714-4723, June 2018.

[10] M. Rezkallah, S. K. Sharma, A. Chandra, B. Singh, and D. R. Rousse, "Lyapunov Function and Sliding Mode Control Approach for the Solar-PV Grid Interface System," IEEE Trans. Ind. Electron., vol. 64, pp 785-795, January 2017.

[11] R. Suman, "Finite Set Model Predictive Current Control Of a Grid Converter Equipped with an LCL Filter," M.S. thesis, Aalto University, Finland, 2017.

[12] N. Panten, N. Hoffmann, and F. W. Fuchs, "Finite Control Set Model Predictive Current Control for Grid-Connected Voltage-Source Converters With LCL Filters: A Study Based on Different State Feedbacks," IEEE Trans. Power Electron., vol. 31, pp 5189-5200, July 2016.

[13] L. Harnefors, A. G. Yepes, A. Vidal, and J. Doval-Gandoy, "Passivity-Based Controller Design of Grid-Connected VSCs for Prevention of Electrical Resonance Instability," IEEE Trans. Ind. Electron., vol. 62, pp 702-710, July 2014.

[14] X. Wang, F. Blaabjerg, and P. C. Loh, "Passivity-Based Stability Analysis and Damping Injection for Multiparalleled VSCs with LCL Filters," IEEE Trans. Power Electron., vol. 32, pp 8922-8935, November 2017.

[15] A. Reznik, M. G. Simoes, A. Al-Durra, and S. M. Muyeen, "LCL Filter Design and Performance Analysis for Grid-Interconnected Systems", IEEE Trans. Ind. Appl., vol. 50, pp 1225-1232, April 2014.

[16] C. Vecchio, "Sliding Mode Control: theoretical developments and applications to uncertain mechanical systems," Ph.D. Dissertation, Universidade de Pavia, Italy, 2008.

[17] F. Zhu, M. Xia, and P. J. Antsaklis, "Passivity Analysis and Passivation of Feedback Systems Using Passivity Indices," American Control Conference, USA, June 2014.

[18] H. S. Ramirez, and R. Ortega, "Passivity-Based Controllers for the Stabilization of DC-to-DC Power Converters," Conference on Decision & Control, USA, December 1995.

Design of a Low-Cost Phasor Measurement Unit (PMU) for Three-Phase Distribution Power Systems according IEEE C37.118.1

Alex Guamán, Marcelo Pozo[1], Isaac Pozo[1], Mario Pacas[2]
[1]Automation and Industrial Control Department
Escuela Politécnica Nacional-Quito, Ecuador
yamanta@hotmail.com,
{marcelo.pozo, isaac.pozo}@epn.edu.ec
[2]University of Siegen, Germany
pacas@uni-siegen.de

Ana Cabrera[3], Nataly Pozo[4]
[3]Facultad de Ingeniería en Ciencias Aplicadas
Universidad Técnica del Norte-Ibarra, Ecuador
akcabrera@utn.edu.ec
[4]Universidad San Franciso de Quito, Ecuador
npozov@usfq.edu.ec

Abstract— **Phasor Measurement Units (PMU) are widely used in the real-time monitoring of electrical power systems. This work presents the design and construction of a low-cost prototype of a PMU for three-phase distribution Power Systems according to the IEEE standard C37.118.1. This system allows to estimate phasors, frequency, rate of change of frequency and symmetric components with the time label corresponding to each value estimated in a three-phase network of a low power system. All these parameters are synchronized using the signals from a Global Positioning System-GPS as a time reference. The system uses as base a development platform controlled by a Digital Signal Processor (DSP) and the estimation of the synchrophasors is done using the Discrete Fourier Transform (DFT). Firstly, a phasor estimation method is presented, after the implemented hardware and software is described. Finally, the results obtained from the measurements are presented in order to verify the correct operation of the equipment. Additionally, operation and communication tests are performed to validate the prototype according to the IEEE C37.118.1 standard.**

Keywords— Digital Signal Processor, Discrete Fourier Transform, Phasor Measurement Unit, PMU, Phasor Estimation.

I. INTRODUCTION

Wide area monitoring networks (WAMS) are used to monitor, control and protect an electrical system. This system is composed of an interconnected network of devices. In modern power systems. the Synchrophasorial Measurement Units (PMU) are appropriate for obtaining the phasors of a point of the system at a reference time [2].

A PMU is a data acquisition device, which estimates synchrophasors, frequency and rate of change of frequency (ROCOF) with the time label, which shows the moment in which they are estimated [3]. The synchrophasor is a complex number, which shows the amplitude, and phase of a sinusoidal signal, where the phase angle is determined with respect to an absolute time reference, in this case the Coordinated Universal Time (UTC) [4]. The UTC allows to have a single reference for all the signals, which that are measured in a global area. In this way the analysis of phasors obtained in different points of the network of a very large area becomes easies, since they are related to the same instants.

In a PMU, the synchronization and time information are obtained from the global positioning system (GPS). By using a Phase-Locked Loop (PLL) a pulse signal is obtained that synchronizes the Analog to Digital Converters (ADC). Samples are sent to the Digital Signal Processor (DSP) where the phasor is estimated, and the time label is generated. This information is transmitted to a receiving device for a

subsequent analysis [5]. The schematic diagram of the basic structure of a PMU is shown in the Fig. 1 [7].

PMUs are commercial measuring instruments of high precision, which are usually located in the transmission of a power system. Currently, due to the introduction of smart grids, which require real-time monitoring of different electrical variables, the interest in PMUs for domestic use and for distribution and sub-distribution networks has significantly increased. For this reason, this work explains the development of a prototype and the performance tests of a low-cost PMU for low voltage systems [6].

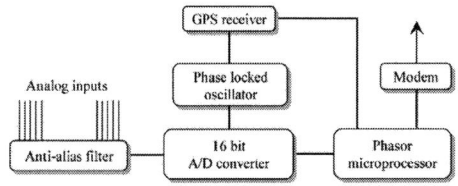

Fig. 1 Schematic diagram of the basic structure of a PMU [7]

II. PHASOR ESTIMATION

A. Phasor Estimation at Nominal Frequency

Given a sinusoidal function x(t) defined by:

$$x(t) = \sqrt{2} \cdot X_m (2\pi f_0 t + \varphi) \tag{1}$$

with X_m the RMS value of the function, f_0 its nominal frequency, and φ its phase angle.

The corresponding complex phasor is defined as:

$$\underline{X} = X_m[cos\varphi + j\,sin\varphi] = X_m\,e^{j\varphi} \tag{2}.$$

Given a sampling frequency Nf_0 frequency, the sampling intervals correspond to an angle $\theta = 2\pi/N$ and for the samples of the signal in a window of N samples follows:

$$x_n = \sqrt{2} \cdot X_n = \sqrt{2} \cdot X_m\,cos\,(n\theta + \varphi), \tag{2}$$

where: n = 0, 1, 2, …, N-1

By using the discrete Fourier transformation to perform the estimation of the complex phasor, a window of N samples of the input signal is defined. With each subsequent sample, the window is shifted by θ and the calculation process is repeated. Each phasor estimated in the process will have the same magnitude, but an angle of θ is offset from each other as shown in Fig. 2. To calculate the phasor 1 represented by $\underline{X}^{N-1}$ and the phasor 2 represented by $\underline{X}^N$, the following equations

978-1-7281-4181-7/19 $31.00 © 2019 IEEE

for extracting the fundamental of the wave by using a DFT together with a window evaluation are used: [8]

$$\underline{X}^{N-1} = \sum_{n=0}^{N-1} X_n[\cos(n\theta) - j\sin(n\theta)] = \frac{1}{N}\sum_{n=0}^{N-1} X_n e^{-jn\theta} \qquad (4)$$

$$\underline{X}^{N} = \sum_{n=1}^{N} X_n[\cos(n\theta) - j\sin(n\theta)] = \frac{1}{N}\sum_{n=1}^{N} X_n e^{-jn\theta} \qquad (5)$$

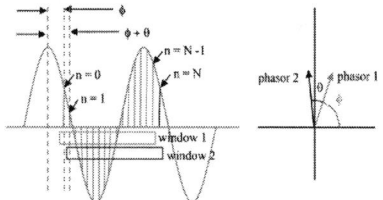

Fig. 2 Phasor estimation using nonrecursive DFT algorithm [8]

B. Phasor Estimation at Non-Nominal Frequency

The frequency of the electrical network fluctuates constantly. To perform the phasor estimation the DFT algorithm is used and an adjustment to the nominal frequency is needed. In Fig 3, a signal with a frequency lower than the nominal is illustrated.

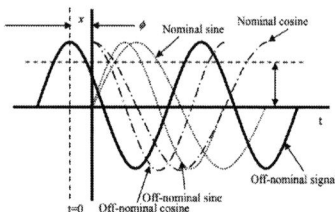

Fig. 3 Phasor estimation using DFT at off-nominal frequency input [8]

C. Estimation of the Symmetrical Components in Three-Phase System

A symmetrical and balanced three-phase system is one in which each phasor has an offset of 120 degrees and where the amplitude and frequency are the same. However, this is not always fulfilled due to the connection of unbalanced loads or system failures.

An unbalanced three-phase system can be decomposed into the sum of three symmetric groups of balanced phasors: positive-, negative- and zero-sequence [9].

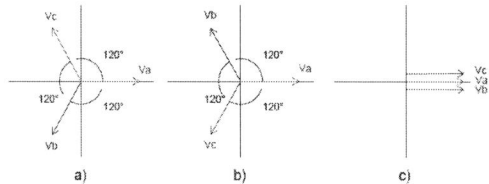

Fig. 3 Sequence of phases: a) positive sequence, b) negative sequence, c) zero sequence [9]

Given a three-phase system represented by: [9]

$$\begin{cases} \underline{X}_a = Re\left(\underline{X}_a e^{j\omega t}\right) \\ \underline{X}_b = Re\left(\underline{X}_b e^{j\left(\omega t - \frac{2\pi}{3}\right)}\right) \\ \underline{X}_c = Re\left(\underline{X}_a e^{j\left(\omega t + \frac{2\pi}{3}\right)}\right) \end{cases} \qquad (6).$$

Solving for a system with symmetric components, in which the phasors of each sequence are: [9]

$$\underline{X}_0 = \frac{1}{3}(\underline{X}_a + \underline{X}_b + \underline{X}_c) \qquad (7)$$

$$\underline{X}_+ = \frac{1}{3}(\underline{X}_a + \underline{X}_b e^{j\frac{2\pi}{3}} + \underline{X}_c e^{-j\frac{2\pi}{3}}) \qquad (8)$$

$$\underline{X}_- = \frac{1}{3}(\underline{X}_a + \underline{X}_b e^{-j\frac{2\pi}{3}} + \underline{X}_c e^{j\frac{2\pi}{3}}) \qquad (9)$$

In normal operation, the value of the positive sequence is larger than the other components, but in the presence of a fault, it can be smaller. The margin of variation between the negative and zero sequence with respect to the positive one allows determining the level of imbalance of the system.

III. Design and Implementation of a Low-Cost PMU for a Three-Phase Low Power System

Figure 5 shows the scheme of connection and interaction between the systems of the developed PMU. Red and orange represent power lines and analog circuits respectively. In green, the input signals are represented and in blue, the digital data buses are represented.

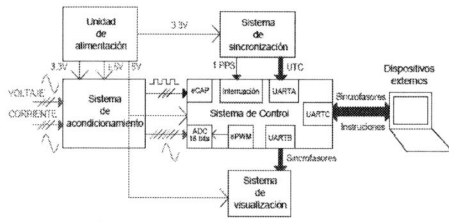

Fig. 5 scheme of connection and interaction between the systems of the developed PMU

In Fig. 6 shows the realized lab-prototype.

Fig. 6 PMU implemented

A. Hardware Devolpment

For the synchronization system, a DIGILENT GPS module is used, which has an UART communication and the possibility of connecting an outdoor antenna. This system allows obtaining a fixed time reference for applying a time stamp to each phasor [10].

978-1-7281-4181-7/19 $31.00 © 2019 IEEE

The variables, which to be measured are voltage, current and frequency of the electric network. The scheme of the developed circuit for the signal sensing can be seen in Fig. 7, which consists of three stages. The first reduces the input voltage to one that the microcontroller can handle, the second corresponds to a differential amplifier circuit with unity gain and the third a low pass filter.

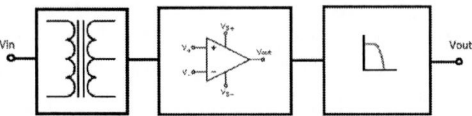

Fig. 7 Voltage sensing circuit

For the measurement of the current, a hall effect sensor T604004 from Vaccuumschmelze is used, which has galvanic isolation and a voltage output proportional to current input. The input current is supported from 0 to 15A [12].

For frequency measurement a zero-crossing detector, which sends pulses to the microcontroller every half the period of the signal was used.

The visualization of the data and the obtained results was implemented on a development card STM32F746Discovery of STMicroelectronics, which includes an LCD touch screen which communicates the visual part through a UART port and the tactile functions with I2C communication.

The program for the visualization on the STM32F746 card after configuring configures the peripherals enters an endless loop, which allows it to react to the entry of commands of the user. The user hast the option of displaying frequency, RCOF, UTC time, synchrophasors or symmetrical components.

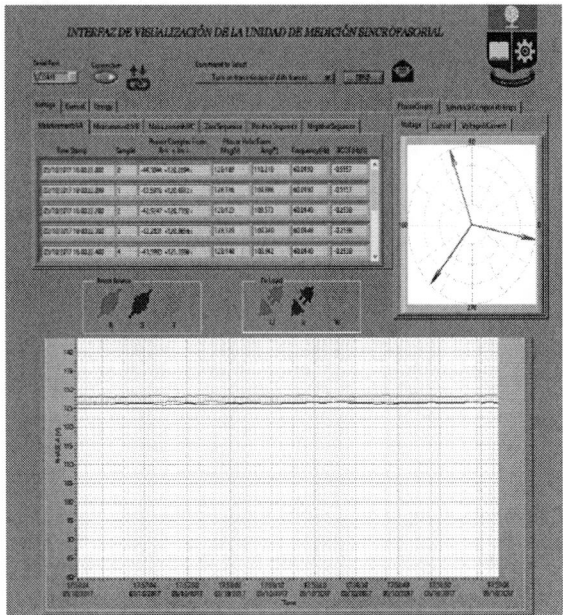

Fig. 98 HMI implemented in LabVIEW

B. Software Development

For the HMI, the LabVIEW platform was used, which shows the data obtained from the microcontroller in a table and the synchrophasors in a graph. It allows to send commands to enable or disable the data messages or set the transfer rate of measurements. In Fig. 98 the implemented interface is shown.

For the control system, the card F28379D of the C2000 series of Texas Instruments is used. This card features two cores, which can be separately programmed, thus the tasks are divided and can be performed in an efficient way. The first core is dedicated for signal sampling and estimation as shown in Fig. 98, while the second core is responsible for communications with external devices. [11]

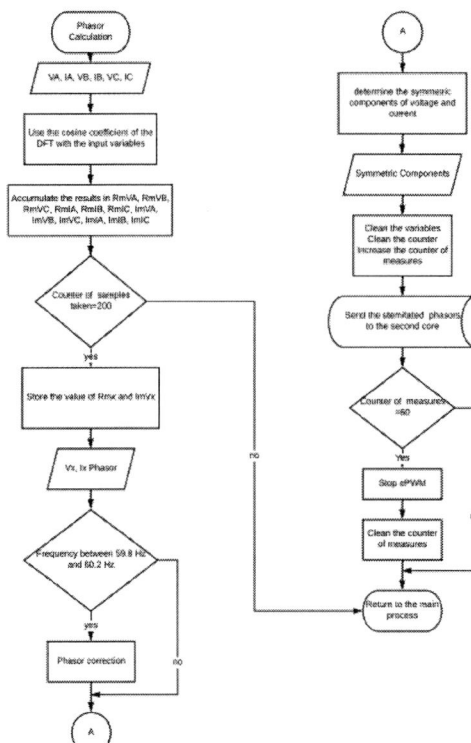

Fig. 98 Subroutine flow diagram for phasor calculation

IV. TESTS AND RESULTS

A. Tests for Synchrophasors referring to the IEEE C37 Standard

The TVE (Total Vector Error) is a characteristic value defined by the IEEE C37.118.1 standard and allows the evaluation evaluate of the difference between a theoretical phasor and the estimated by the PMU at the same of time and under known conditions. The TVE is defined as follows: [14]

$$TVE(\%) = \sqrt{\frac{(\overline{X_r} - X_r)^2 + (\overline{X_i} - X_i)^2}{(X_r - X_r)^2}} \qquad (10)$$

where X_r and X_i are the real and imaginary component respectively of the theoretical synchrophasor,and are compared with respective components of the synchrophasor estimated by the PMU. The conditions of the standard are presented in the following table: [14]

TABLE I.   EVALUATION CONDITIONS IN ACCORDANCE WITH IEEE C37.118.1 [16]

Evaluation parameters	Reference conditions	Range	TVE (%)
Frequency	60Hz	±2Hz	1,0
Voltage magnitude	120V	80% - 120% of ref.	1,0
Current magnitude	2A	10% - 200% of ref.	1,0
Phase Angle	0°	±180°	1,0
Harmonic Distortion	< 0,2%	1% to 50th	1,0

For the magnitude variation test, a constant frequency of 60Hz is maintained, the phase angle is 0 ° and an increase in amplitude between 80% and 120% of the nominal value for the voltage signal is simulated. The current signal is changed in the range 10% to 200%. The average result for the voltage magnitude was a TVE of 0.206% and for the current magnitude it was 0.551%

For the phase variation test, a constant voltage of 120 V, a constant current of 2 A and a frequency of 60 Hz must be maintained. The angle varies between ± 180 ° and an average TVE of 0.78% was obtained for the voltage and 0.34% for the current.

Like the previous tests, for the frequency variation test the magnitudes which are not going to be evaluated are kept constant and the frequency from 58Hz to 62Hz is changed. For this case, an average TVE of 0.511% for the voltage and 0.586% for the current was obtained.

For the test of harmonic distortion variation, a THD less than 0.2% must be obtained while all other parameters remain constant. The average TVE obtained for the voltage was 0.105% and for the current it was 0.371%.

B. Conventional Tests

In order to validate the values which are measured by the PMU, these are compared with a Fluke 87V multimeter with an accuracy of ± 0.5%, a Fluke 800i current test tip of accuracy ± 3% and the Tektronix TDS2022C oscilloscope. [13]

The frequency tests were performed with a function generator in sinusoidal mode, the average error result between the real voltage and the measured voltage was 1.213% and the average error between real frequency and measured frequency was 0.069%.

The voltage variation test was performed with an autotransformer at the entrance of the PMU and without load. The results are the following:

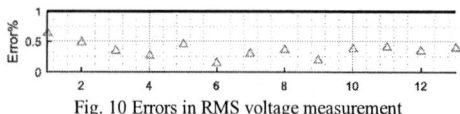

Fig. 10 Errors in RMS voltage measurement

For the current test, a load of resistive type with star connection with neutral is used to be able to vary the current in each phase. The results are shown in Fig. 11:

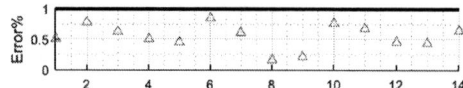

Fig. 11 Errors in current measurements with resistive load in phase A

For a resistive load it is expected that the voltage and current phasors are in phase. This result is shown in Fig 12.

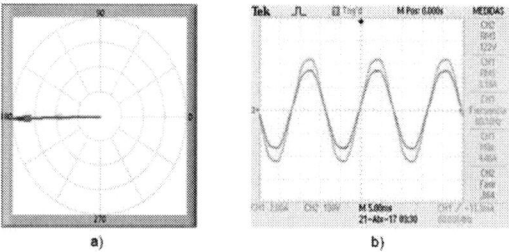

Fig. 12 Resistive load test: a) Phasorial diagram b) Waveforms

A capacitor bank in series with a rheostat was used for the capacitive test. The result of the RMS current measurements taken in phase A can be seen in Fig. 16.

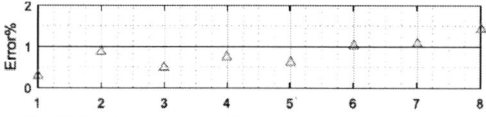

Fig. 16 Errors in current measurement with capacitive load

In Fig. 17 is shown that the current phasor (black) is delayed the voltage phasor (green).

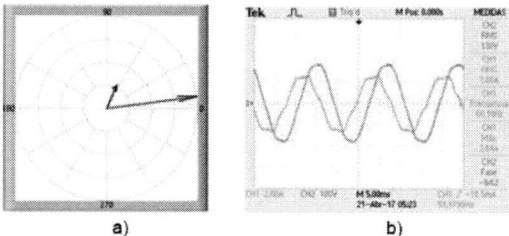

Fig. 17 Test with capacitive load: a) Phasorial diagram, b) Waveforms

For an inductive load, tests of voltage, current and power factor are carried out. Fig. 13 shows the results of the measurements taken of RMS current as a function of its error.

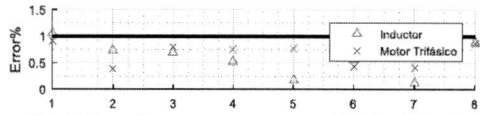

Fig. 13 Errors in current measurement with inductive load

Fig. 14 shows the lag due to an inductive load. In blue color the voltage phasor is represented and in black the current.

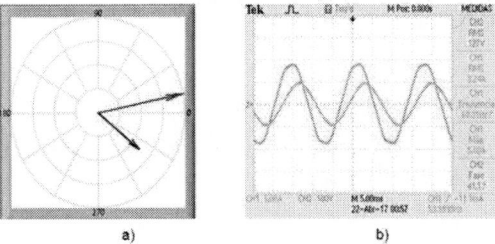

Fig. 14 Inductive load test: a) Phasorial diagram, b) Waveforms

In other case, a test was carried out with a three-phase motor and, as can be seen in Fig. 15, the voltages and currents are delayed with respect to their corresponding voltage phasor.

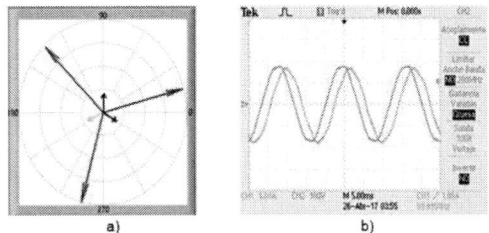

Fig. 15 Test with three-phase motor: a) Phasorial diagram, b) Waveforms

C. Symmetrical Components Estimation Test

The PMU must be able to estimate the symmetric components. For this test, the voltage signals were analyzed to obtain the components of zero, positive and negative sequence represented by the red, blue and green color respectively. Fig. 18 shows a slightly unbalanced positive sequence system. In Fig. 19 the diagram of a negative sequence is shown when exchanging the connection between phases B and C. For Fig. 20. Only a single phase is connected. In Fig. 21. An unbalanced network is shown where only 2 phases are connected.

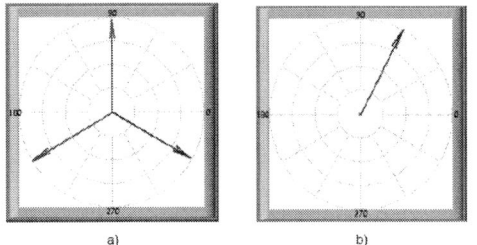

Fig. 18 Positive sequence phasor diagrams: a) Phasors, b) Symmetric components

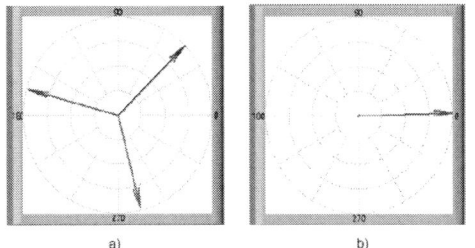

Fig. 19 Negative sequence phasor diagrams: a) Phasors, b) Symmetric components

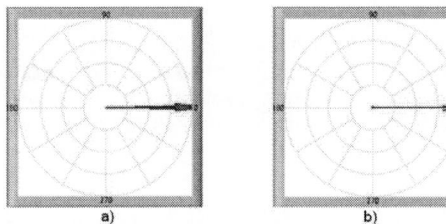

Fig. 20 Phasorial diagrams of zero sequence: a) Phasors, b) Symmetric components

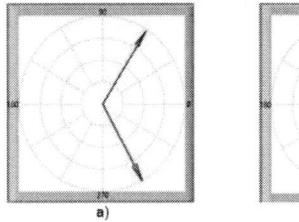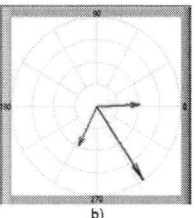

Fig. 21 Phasorial diagrams of an unbalanced system: a) Phasors, b) Symmetric components

V. CONCLUSIONS

A low power synchrophasorial meter was designed and implemented under the IEEE C37.118.1 standard, fulfilling the main objective of this project.

The algorithm of phasor estimation is based on the non-recursive DFT.

The DFT used in this equipment is adjusted to obtain the nominal frequency phasor. In the case when the signal has a different frequency, erroneous measurements are reported which represents a disadvantage that requires the application of the correction algorithm. This new algorithm represents more processing time.

The equipment can operate on balanced and unbalanced three-phase low voltage systems since, in addition to acquiring the phasors which describe the system, is possible to estimate the symmetric components. This feature is especially useful for analyzing unbalanced three-phase systems.

This PMU depends on the GPS reception system to operate and time synchronization. The measurements are carried out only while the synchronism pulse with the GPS is present. Is necessary to guarantee the correct reception of the GPS signal by placing the equipment in an area whith a minimum of interference.

The processing system must be fast because, being an embedded system, could oversees sensing the signals and processing the samples, executing the algorithms necessary to estimate the phasors. Additionally, managing the communication protocol in real time requires high availability of the device. As a result, the dual-core processor of this system is suitable because it allows handling communications and phasor measurement separately. In this way there are no conflicts in the execution of the processes.

Analyzing the results of the traditional tests, the developed equipment has an accuracy of 1% and does not incorporate high levels of distortion. The results for the distorted signals are not taken into account because according to the standard, these devices must operate in a system with low levels of harmonic content.

The TVE errors found by the tests in which a single parameter is modified according to the evaluation criteria of the IEEE C37.118.1 standard is below the established accuracy limit. Likewise, the software that allows verifying that the communication protocol is in accordance with the IEEE C37.118.2 standard. Therefore, it can be discerned that the equipment works in accordance with the EEE C37.118 standard.

REFERENCES

[1] A. R. Guamán, "Diseño e implementación de un medidor sincrofasorial de baja potencia bajo la norma IEEE C37.118.1," tesis de pregrado, DACI, Escuela Politécnica Nacional, Quito, Ecuador, 2018.

[2] "The Global Positioning System," GPS, 2017.

[3] The Data Conversion Handbook, 1st ed. Oxford, Analog Devices, Burlington, USA, 2005.

[4] M. Faúndez, "Módems," en Sistemas de Comunicaciones. Barcelona, España: Marcombo, 2001, ch. 8, pp. 208-242.

[5] P. Correia, Guía Práctica del GPS. Barcelona: Marcombo, 2002.

[6] "The Global Positioning System," GPS, 2017.

[7] The Data Conversion Handbook, 1st ed. Oxford, Analog Devices, Burlington, USA, 2005.

[8] M. Faúndez, "Módems," en Sistemas de Comunicaciones. Barcelona, España: Marcombo, 2001, ch. 8, pp. 208-242.

[9] P. Correia, Guía Práctica del GPS. Barcelona: Marcombo, 2002.

[10] "Fluke 80 Series V Digital Multimeters: The Industrial Standard," FLUKE.

[11] B. Hernández, "Diseño e implementación de un medidor fasorial síncrono normalizado," Tesis de maestría, Instituto Politécnico Nacional, México D.F.,2009.

[12] A. G. Phadke and J. S. Thorp, Synchronized Phasor Measurement and Their Aplications. New York: Springer, 2008.

[13] J.L. Kirtley, "Introduction To Symmetrical Components," Dept. Electrical Engineering and Computer Science, MIT, Supplementary Notes 4 2001.

[14] Linx, "Anntenna Factor by Linx," ANT-GPS-SH datasheet, Jan. 2014.

[15] "LAUNCHXL-F28379D Overview," Texas Instruments, User's Guide.

[16] Vacuumschmelze,"15 A Current Sensor for 5V- Supply Voltage," T60404-N4646-X662 datasheet, Aug. 2014

[17] "Fluke 80 Series V Digital Multimeters: The Industrial Standard," FLUKE.

[18] IEEE Standard for Synchrophasor Measurements for Power Systems, IEEE Standard C37.119.1, 2011

Controller Coefficients Tuning for a Single-Phase Photovoltaic System in Synchronous Reference Frame Through Genetic Algorithm

Marcos V. A. Vedovatte
Programa de Pós-graduação em
Engenharia Elétrica
Universidade Federal de Mato Grosso
do Sul
Campo Grande, Brazil
marcosvedovatte@gmail.com

Moacyr A. G. de Brito
Departamento de Engenharia Elétrica
Universidade Federal de Mato Grosso
do Sul
Campo Grande, Brazil
moa.brito@gmail.com

Luigi G. Junior
Departamento de Engenharia Elétrica
Universidade Federal de Mato Grosso
do Sul
Campo Grande, Brazil
lgalotto@gmail.com

Abstract— **This paper presents the design and simulation of a single-phase photovoltaic inverter connected to the low voltage power distribution grid. The system is controlled in the synchronous reference frame – dq axis mainly for active power injection. A genetic algorithm is adopted to find the best coefficients for the current controllers. Simulation results are presented using MatLab/Simulink® platform achieving superior performance to the conventional control strategy in the abc frame using Bode plot concepts.**

Keywords—controllers, dq system, active power, genetic algorithm.

I. INTRODUCTION

The usage of new energy sources, mainly renewable ones, for distributed generation has been increasingly common, standing out in this scenario the solar photovoltaic energy. However, the energy from the solar photovoltaic source has reduced voltage and current levels with continuous waveforms – DC type, while the distribution grid operate on alternating current – AC type. Thus, electronic power converters are required to adjust the energy for proper usage, especially the use of DC-DC and DC-AC converters.

The inverters are circuits that convert continuous energy to alternating current, and the single-phase full bridge inverter is widely used because of its inherent advantages and characteristics [1,2], such as ease of implementation and robustness.

For inverters to work properly, control systems are mandatory and different control methodologies can be applied. An unusual design methodology is the adoption of the synchronous reference frame for single phase systems (for three-phase this strategy is well-established). This strategy becomes attractive for research and development since it eliminates steady-state error, even for alternating systems, using controllers that are used for systems with continuous references. In [3,4,5] the synchronous reference technique has been applied for the control of local loads and harmonic currents mitigation with interesting results. Nevertheless, it is not used yet for grid-connection.

In that sense, this paper has the purpose to design a control system for a single-phase photovoltaic inverter in synchronous reference frame for the injection of active power into the electric power distribution grid. It is done controlling both id and iq currents in order to achieve superior performance when compared to the conventional methodology in abc frame with the common Bode plot

concepts. Additionally, in order to find the best controller coefficients a genetic algorithm is developed.

II. METODOLOGY

The VSI full bridge inverter has been adopted as the inverter topology, operating with unipolar modulation and connected to the grid through an inductive filter (L_{grid}). Fig. 1 shows the aforementioned approach. The energy source is provided by a set of photovoltaic solar panels operating at the MPP (Maximum power point) due to a voltage constant MPPT algorithm inserted into a common boost converter to provide the DC bus value (V_{BUS}) of 220V [6].

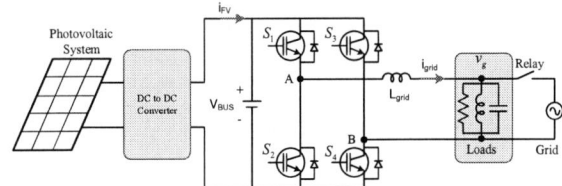

Fig. 1. Single-phase full bridge inverter - equivalent circuit in connection to the grid.

The paper focus in on the inverter stage and its control functions. So, table I shows the main inverter parameters. Equation (1) presents the obtainment of the modulation index m, while equation (2) presents the grid peak current I_{grid}.

TABLE I.   MAIN INVERTER PARAMETERS

Input Average Voltage V_{BUS} (V)	Grid Peak Voltage V_{peak} (V)	Grid Injected Power P_{AC} (W)	Switching Frequency fs (kHz)	Current Ripple (%)
220	180	400	30	10

$$m = \frac{V_{peak}}{V_{BUS}} \qquad (1)$$

$$I_{grid} = \frac{2.P_{AC}}{V_{peak}} \qquad (2)$$

In this specific conditions, the modulation index m is equal to 0,82 and the peak current to be injected into the grid I_{grid} is equal to 4,45 A, which at 127 V_{rms} provides 400 W of active power. The inductance filter (connection filter) is obtained as (5) based on the instantaneous voltage applied to the inductance when the mains voltage reaches its maximum instantaneous value (V_{peak}).

$$V_{grid\_inductance} = L_{grid} \cdot \frac{di_{grid}}{dt} \quad (3)$$

$$(V_{DC} - V_{peak}) = L_{grid} \cdot \frac{\Delta i_{grid}}{\Delta t} \quad (4)$$

$$L_{grid} = \frac{(V_{DC} - V_{peak}) \cdot \Delta t}{\Delta i_{grid}} \quad (5)$$

Considering Δt the time in which the switches remain in conduction (S_1 and S_4 connected) and that the period is the inverse of the switching frequency, therefore:

$$L_{grid} = \frac{(V_{DC} - V_{peak}) \cdot m \cdot T}{\Delta i_{grid}} \quad (6)$$

$$L_{grid} = \frac{(V_{DC} - V_{peak}) \cdot m}{\Delta i_{grid} \cdot f_c} \quad (7)$$

According to (7) and Table I data, a L_{grid} inductance of 2.45 mH is achieved.

Fig. 2 shows the system control block in dq synchronous reference frame. Equation (8) represents the system input equations. At this point, it is necessary to create a 90-degree lag of the real current, called i_β. Observing this diagram, when controlling the variable Ref_Id it is possible to change the value of the active power injected; Ref_Iq is set to zero so that there is no reactive injection.

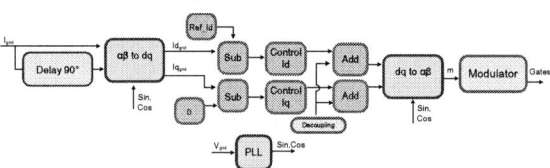

Fig. 2. Block diagram for the control system in synchronous reference frame - dq.

$$\begin{bmatrix} i_\alpha \\ i_\beta \end{bmatrix} = \begin{bmatrix} i_{grid}(\omega.t) \\ i_{grid}(\omega.t - \frac{\pi}{2}) \end{bmatrix} \quad (8)$$

In [5] some techniques for generation of i_β current are suggested. After analyzing the presented methods inserted into the control system of Fig. 2 the usage of a first-order all pass filter showed the best performance, i.e., faster and stable response. Equation (9) represents this transfer function, where ω_0 equals 377 rad/s.

$$\frac{i_\beta}{i_\alpha} = \frac{-s + \omega_0}{s + \omega_0} \quad (9)$$

The current control transfer functions for id and iq are obtained by (10), after conversion to the dq system, proper decoupling and after applying small signal analysis.

$$\frac{i_{d,q}(s)}{m(s)} = \frac{V_{DC}}{sL_{grid} + r_{L_{grid}}} \quad (10)$$

III. RESULTS AND DISCUSSION

A. Controllers Design by means of Bode Diagram

The inverter system was simulated in the Matlab/Simulink® software by adopting PI controllers with different gain crossover frequencies (ranging from 1.2 kHz to 7.5 kHz). The design was based on the Bode diagrams of module and phase. In Figs 3-a and 3-b the Bode diagrams of the compensated system are presented, adopting a crossover frequency of 3.3 kHz. Figs 4, 5 and 6 show, respectively, the currents i_α and i_β using the all pass filter, the direct axis current and the quadrature axis current of the analyzed system, pointing out that the PI controller had the zero allocated at -200, thus having the constants kp = 0.2311 and ki = 46.22. Lastly, the grid current (I_{grid}) is shown in Fig. 7, in perfect sinusoidal form and in phase with the reference. It has a Total Harmonic Distortion (THD) of 2.63%, much lower than the 5% required by the standards [7].

Considering the presented results, it was observed that, for the crossover frequency of 3.3 kHz, the system do not present overshoot at the initialization, has extremely reduced THD and has the lowest settling time among the analyzed controllers (varying the frequencies) for the initialization. At this point, no step reference change was applied during operation. Table II presents the summary of the obtained results for 6 different frequencies, drawing attention to the harmonic distortion of the mains current, always lower than 2.7%.

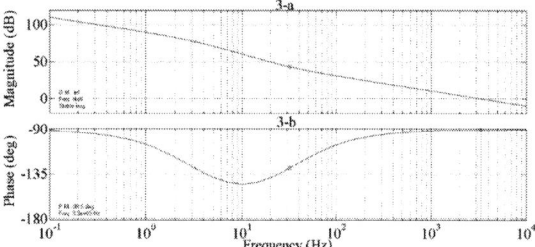

Fig. 3. Bode diagram for current control in dq frame, where a) Module and b) Phase.

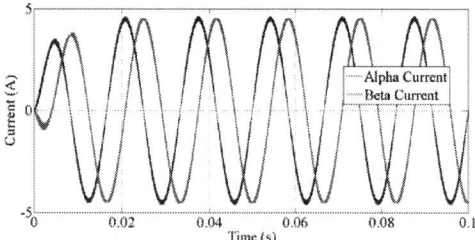

Fig. 4. Waveform corresponding to current i_α and i_β.

978-1-7281-4181-7/19 $31.00 © 2019 IEEE

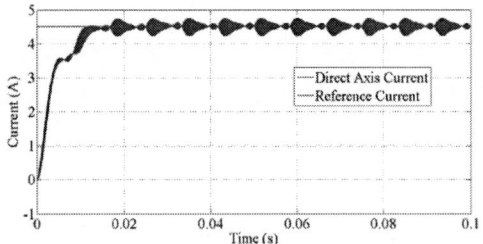

Fig. 5. Waveform corresponding to the direct axis current and its reference.

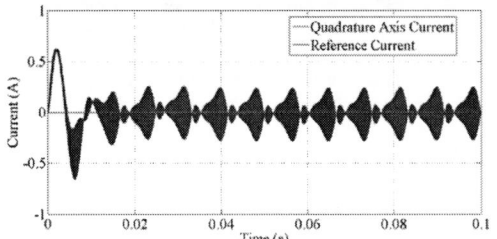

Fig. 6. Waveform corresponding to the quadrature axis current and its reference.

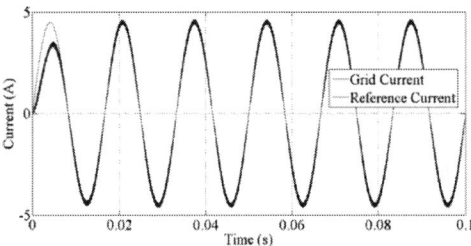

Fig. 7. Current injected into the grid along with the reference.

TABLE II.   SUMMARY OF THE VALUES OBTAINED FOR DIFFERENT CROSSOVER FREQUENCIES.

Crossover Frequency (kHz)	Coefficient kp	Coefficient ki	Harmonic Distortion - THD	Settling Time (ms)
7.50	0.5260	105.20	2.63 %	43
6.00	0.4210	84.20	2.63 %	43
4.65	0.3260	65.20	2.63 %	36
3.30	0.2311	46.22	2.63 %	27
1.95	0.1370	27.40	2.64 %	35
1.20	0.0840	16.80	2.65 %	52

B. Controllers Design by means of Genetic Algorithm

Genetic algorithms consist of a family regarding computational techniques based on genetics and species evolution theory. These techniques apply a parallel and structured but random search strategy, aiming at a potential solution of a problem, in a structure similar to that of a chromosome, and thus solving several problems in several areas of study, always considering the number of the population, generations, crossing over and genetic mutation rates [8,9,10].

Each stage of the genetic algorithm is called "genetic operator", which aims to carry out transformations in a population, making each and every new generation better, and contributing to the evolution of populations with each new generation [10]. Genetic operators have, briefly, the following classification:

1) Initialization of the population: It is a population of N individuals randomly generated. Each of the individuals represents a possible solution to the problem.

2) Calculation of aptitude: It is determined by means of a cost function, which depends on the project to be developed.

3) Selection: Based on the current generation of analysis, this phase consists of selecting the most suitable individuals. Each individual has a selection of probability, proportional to their abilities.

4) Reproduction: The individuals then selected are crossed as follows: they are shuffled randomly, thus creating a new list, called a list of partners. The chromosomes of each individual pair is partitioned at a point, called the cut point, randomly drawn, or another criteria.

5) Mutation: It is done by changing a value of a gene from a randomly drawn individual with a certain probability, called mutation probability. This means that a number of certain individuals in the new population may have their genes changed randomly. It is used to ensure a larger scanning space and thus prevent the genetic algorithm from converging to local values too soon [9].

C. Simulation Results with the Genetic Algorithm

For testing the genetic algorithm, it was considered 10 generations with 60 individuals each. Each individual is represented as a 2-point vector, containing kp and ki gains of the PI controller.

A decimal codification was adopted and the first individual has the coefficients of 1.2 kHz and the final individual has the coefficients of 7.5 kHz (60th individual) of crossing over frequency. Equation (11) presents the crossing over operator. Variable a may change between 0 and 1. x_m^1 and x_m^2 are the parents selected, while x_{m+1}^1 the offspring.

Equation (12) presents the cost function used to determine the aptitude of each individual. The THD, the average square error and the RMS value of modulation index are taken into account to optimize the coefficients selection. In order to maintain the most suitable individual for the next generation Elitism was adopted into the selection step [11]. A current reference step change is applied so it helps to optimize the values obtained by the cost function once the current reference is also input for the Simulink® model.

The cost function convergence is shown in Fig. 8, while Fig. 9 shows a detail as the system converges.

$$x_{m+1}^1 = (a) \cdot x_m^1 + (1 - a) \cdot x_m^2 \qquad (11)$$

$$cost = 5.THD(end) + mse(ID\_Error) + mse(IQ\_Error) + 10.RMS(Modulation) \qquad (12)$$

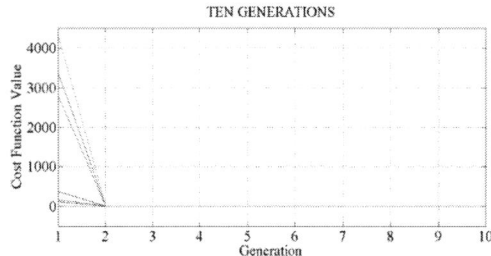

Fig. 8. Convergence of the cost function across generations.

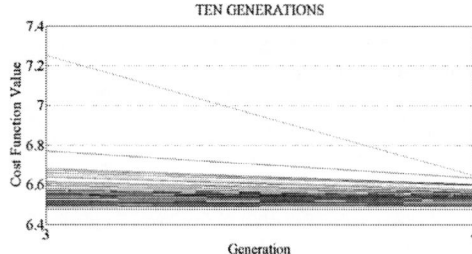

Fig. 9. Detail of the cost function behavior throughout generations.

Figs. 10, 11, 12, 13 and 14 present, respectively, the graph of kp for each individual for the last generation, the graph of ki for each individual also for the last generation, the optimal kp behavior over generations, the optimal ki behavior over generations, and the behavior of elite values (optimal cost function) over generations for a given run.

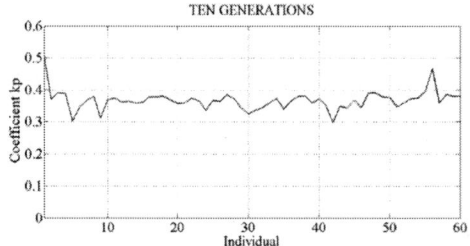

Fig. 10. Behavior of the curve of kp coefficients of the last generation.

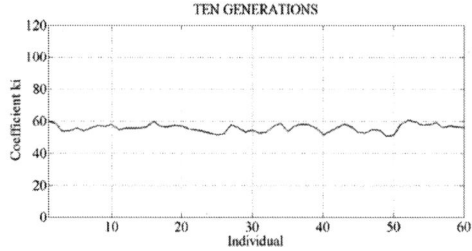

Fig. 11. Behavior of the curve of ki coefficients of the last generation.

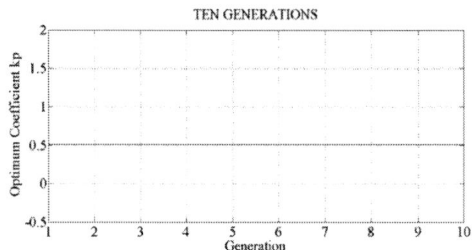

Fig. 12. Behavior of the optimal kp coefficients over the generations.

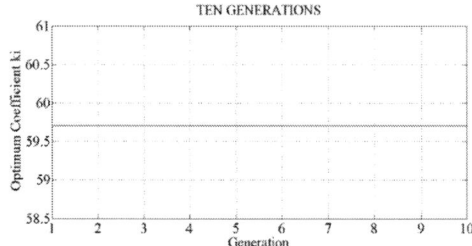

Fig. 13. Behavior of the optimal ki coefficients over the generations.

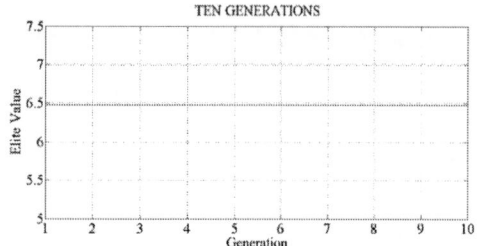

Fig. 14. Behavior of the elite value of the cost function over generations.

Once the genetic algorithm proposes a candidate solution for the best coefficients or a candidate in the vicinity of the best solution it has been performed a simple statistic analysis of the coefficients for ten different runs. The optimal kp and ki coefficients for each run are presented in Table III.

TABLE III.   RESULTS FIND FOR 10 GENETIC ALGORITHMS RUN

Run	Coefficient kp	Coefficient ki
1	0.4958	52.1328
2	0.5071	59.7072
3	0.5257	48.2180
4	0.5071	59.7072
5	0.5257	48.2180
6	0.4475	51.6019
7	0.4958	52.1328
8	0.5131	63.2207
9	0.4808	62.6152
10	0.5071	59.7072

Equations (13) and (14) present, respectively, the formulas for calculating the mean and standard deviation of the coefficients [12].

$$mean = \frac{\sum_{i=1}^{n} x_i}{n} \tag{13}$$

$$std = \sqrt{\frac{\sum_{i=1}^{n} (x_i - \bar{x})^2}{n-1}} \tag{14}$$

Applying (13) and (14) to the data in Table III, the mean values of kp and ki was, respectively, 0,5006 and 55,7261. The standard deviation of each was, respectively, 0,0236 and 5,8381. Considering the aforementioned kp and ki average values, novel simulations were performed as one can verify at Figs. 15 to 18. The same current step reference changes (at 1.25 cycle and at 3.25 cylce/ half-power to full power and vice-versa) are applied at the maximum instantaneous values to test the obtained controllers.

The behavior of iα and iβ variables are presented in Fig. 15, the behavior of id and iq currents are presented in Figs. 16 and 17, respectively. Fig. 20 presents the injected grid current.

978-1-7281-4181-7/19 $31.00 © 2019 IEEE   516

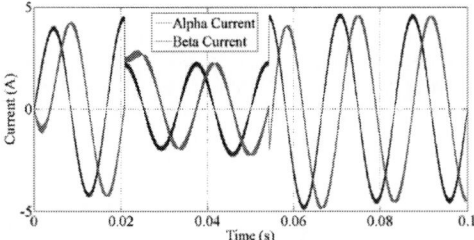

Fig. 15. Behavior of the currents iα and iβ.

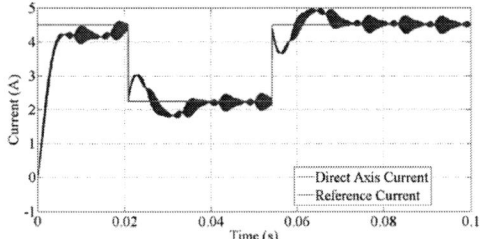

Fig. 16. Behavior of the direct axis current during step changes.

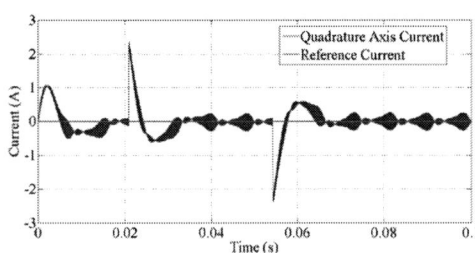

Fig. 17. Behavior of the quadrature current axis during step changes.

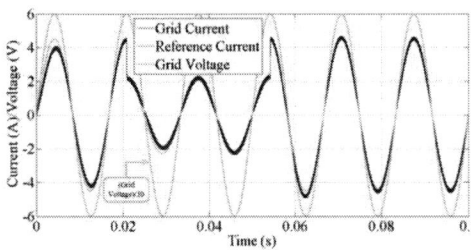

Fig. 18. Behavior of the grid current, along with the reference and grid voltage with step changes.

Analyzing the previous results, it is noticed that, despite the application of steps in the current reference, the optimized controller searched almost exactly the reference signal. One can verify the quality of the injected grid current at Fig. 19 with no overshoots considering the initialization. In Figs. 21, 21 and 22, without the application of the steps in the current reference, it is possible to observe the behavior of the variables in steady-state after the optimization through the genetic algorithm.

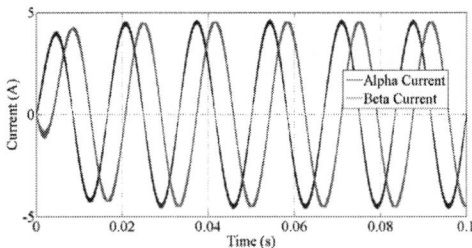

Fig. 19. Behavior of the currents iα and iβ.

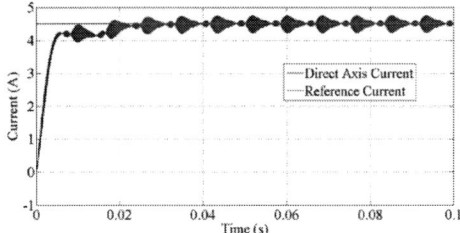

Fig. 20. Behavior of the direct axis current.

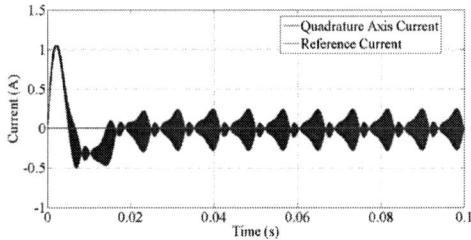

Fig. 21. Behavior of the quadrature axis current.

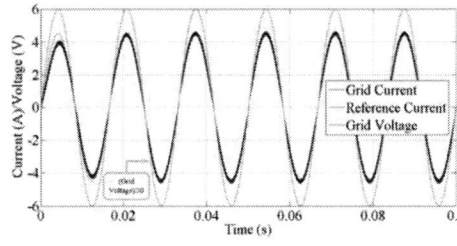

Fig. 22. Behavior of the grid current next to its reference and grid voltage.

In Fig. 23 it is possible to observe the active power injection and in Fig. 24 the reactive power injection in the grid, showing the quality of the proposed technique/controllers tuning procedure.

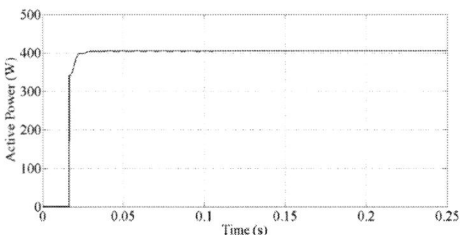

Fig. 23. Active power behavior.

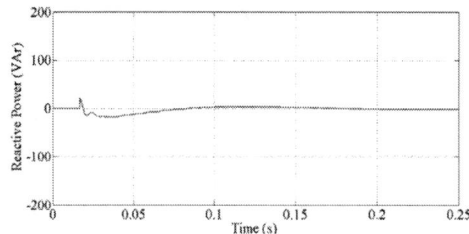

Fig. 24. Reactive power behavior.

Observing the initial tuning procedure through Bode diagrams and posterior optimization through the genetic algorithm, it is noticed that such presented proposal increased the overall system effectiveness. Observing Figs. 20 and 22, there are showed no overshoot and fast settling time, approximately equal to 30 ms and a THD of 2.6%.

IV. CONCLUSIONS

A suggestion was presented for the coefficients tuning of a PI controller for the current control in the synchronous reference frame for a single-phase photovoltaic system. In comparison with the tuning through Bode diagrams only, the posterior tuning through the genetic algorithm presented better results regarding the performance of the system, showing an alternative of excellent performance, analyzing that the injected current error in relation to its reference has been minimized as well as less settling time and lower THD.

REFERENCES

[1] A. Ahmed, Power Electronics. São Paulo: Pearson Prentice Hall, 2000.

[2] D. W. Hart, Power Electronics: Analysis and Circuits Project. Porto Alegre: AMGH, 2012.

[3] D. Baimel, J. Belikov, J. M. Guerrero, and Y. Levron, "Dynamic modeling of networks, microgrids, and renewable sources in the dq0 reference frame: a survey," IEEE Access, vol. 5, pp. 21323–21335, October 2017.

[4] I. R. Guizelini, Development of a Distributed Generation System Connected to the Single Phase Electric Network Using L and LCL Filters. Cornélio Procópio: UTFPR, 2017. In Portuguese.

[5] M. Monfared, S. Golestan, and J. M. Guerrero, "Analysis, Design, and experimental verification of a synchronous reference frame voltage control for single-phase inverters," IEEE Transactions On Industrial Eletronics, vol. 61, pp. 258–269, January 2014.

[6] M. A. G. Brito, L. G. Junior, L. P. Sampaio, G. A. Melo, and C. A. Canesin, "Evaluation of the main MPPT techniques for photovoltaic applications," IEEE Transactions On Industrial Eletronics, vol. 60, pp. 1156–1167, March 2013.

[7] ABNT – Brazilian Association of Technical Standards, NBR 16149: Photovoltaic Systems – Characteristics of the Connection Interface with the Distribution Electrical Network. Rio de Janeiro: 2013, pp. 12.

[8] A. C. P. Carvalho, Genetic Algorithms – An introduction. Available in: <http://conteudo.icmc.usp.br/pessoas/ andre/research/ genetic/>. Access in: January 10th, 2019.

[9] S. C. Hsieh, G. Y. Chen, J.W. Perng, "PI controller design based on genetic algorithm and parameter space", International Conference on Fuzzy Theory and Its Applications, pp. 1-6, 2013.

[10] T. D. O. Rosa, and H. S. Luz, "Practical concepts of genetic algorithms: theory and practice," Informatic Meeting, pp. 27–37, 2009.

[11] M. Obtiko, Genetic Algoritms, 1998. Available in: <http://www.obitko.com/tutorials/genetic-algorithms/portuguese /index.php>. Access in: February 8th, 2019.

[12] D. C. Montgomery, and G. C. Runger, Applied Statistics and Probability for Engineers. 5th ed., Rio de Janeiro: LTC, 2012.

Robust PID Controllers Optimized by PSO Algorithm for Power Converters

Lucas C. Borin, Everson Mattos, Caio R. D. Osorio, Gustavo G. Koch, Vinicius F. Montagner

Universidade Federal de Santa Maria

Santa Maria - RS, Brasil - 97105-900

lukascielo@gmail.com

Abstract—This paper provides an automated design to provide robust PIDs with fixed control gains, suitable to be applied in power converters whose parameters belong to real intervals. Differently from conventional PIDs which use only a nominal model to obtain the fixed control gains and, *a posteriori*, verify robustness, the proposed approach ensures, *a priori* (i.e. during the design stage), robust performance for a set of plant parameters. To illustratethe proposed procedure, two conventional PID controllers are given, to achieve phase margin and crossover frequency for a nominal model of a buck converter. An objective function based on frequency domain specifications is proposed. A particle swarm optimization algorithm is then used to find PIDs, in a large search space that include stable and unstable controllers, allowing to optimize this function for all cases of combinations of plant parameters. A case study for the buck converter illustrates the improvements of performance with the proposed method when compared to the conventional PID controllers. Additionally, the design is used in a more challenging application, for a buck-boost converter suitable for small satellites application, becoming a simple alternative for benchmarks for robust control of power converters.

Index Terms—DC-DC converter, Particle swarm optimization, Proportional-integral-derivative control, Robust control.

I. INTRODUCTION

Proportional-integral-derivative (PID) controllers are recognizedly important in industry applications. They can ensure, with a simple control structure, for several plants of practical interest, with suitable transient, steady state responses and good margins of stability [1]–[6]. PIDs can be designed from the well-known Ziegler-Nichols methods to the more advanced techniques as fuzzy and adaptive controllers [7]–[13]. In particular, these advanced control techniques allow to improve the performance, at the price of a more complex control implementation, when compared to simple fixed gains PIDs.

In the literature, there are several works that use PID controllers in power converters and some guidelines for the PID design can be found in [14], based on frequency domain features. The simpler choice for PID, is the fixed gain controller. In this direction, for output voltage control of buck

This study was financed in part by the Coordenação de Aperfeiçoamento de Pessoal de Nível Superior - Brasil (CAPES/PROEX) - Finance code 001. The authors would also like to thank the INCT-GD and the finance agencies (CNPq 465640/2014-1, CNPq Projeto 309536/2018-9, CAPES 23038.000776/2017-54 and FAPERGS 17/2551-0000517-1).

and boost converters, one has, for instance, the works in [15]–[19]. Other paradigm is the PID gain variable controllers. In this direction, one can cite, for instance, for output voltage control of buck and boost converters, the works [10], [20], [21], that use adaptive or fuzzy method for PID control. Note that adaptive or fuzzy controllers are best indicated in the scenario of unpredictable changes in system dynamics. However, there are situations where controllers with fixed gains are appropriate, such as in constant but unknown system dynamics [2].

In the context of PID fixed control gains, finding suitable gains is a challenge in the controller design, and the difficulty increases with the need to meet multiple control objectives. Moreover, structured parameter uncertainty leads to a set of plants to describe the control system and to the need of design of PIDs that can guarantee limits of performance for this whole set. In this scenario, metaheuristics such as genetic algorithms (GA) and particle swarm optimization (PSO) can be useful, as indicated in [22]. It is important to mention that GA and PSO algorithms have already been used in the design of PIDs in [23]–[25]. For instance, in [23] a PID control based on a multiobjective GA is proposed for a linear brushless DC motor, taking into account uncertainties, with an objective function based on the rise time, overshoot and steady-state error. In [24], a PSO is used to determine the gains of an optimal PID for an automatic voltage regulator system, and a comparison with a GA method is presented in this case. In [25], a PSO is used to find optimum values of the gains of a PID controller in order to reduce the current to the power-assisted steering in electric vehicles, having the objective of minimizing a mean square error function. As remarked in [22], the majority of the publication with metaheuristics in power electronics is on power quality and circuitry optimization. There is a lack of publications based on metaheuristics for control tuning in power converters. Recently, a few works are published using PSO to tuning a PID controller in [26]–[29], but not addressing the design of robust controllers. This makes the further investigation in this direction an important issue.

This paper provides as main contribution a procedure, based on PSO, to obtain fixed gains of robust PID controllers applied for voltage regulation of DC-DC converter with parameters not precisely known, but lying on uncertain intervals. The models of the plants used here contain interval uncertain parameters, leading to a set of linear models for each of the situations of extreme values of these uncertain parameters. As a first case study, two conventional PID designs were performed, for a buck converter, relying only on a nominal plant to meet phase margin and crossover frequency specifications, as commonly done in power electronics. One of these designs uses the *pidtune* function and the other uses the *sisotool*, both from MATLAB. The proposed algorithm is based on an automatic tuning of the PID gains, guided by the minimization of an objective function that takes into account phase margin and crossover frequency, for a set of 4 plants. In addition, an extension for a more complex converter is presented, specifically a buck-boost converter used in small satellites, to show the suitability of the proposed robust control design for DC-DC converters. The fixed control gains designed here completely offline, with good convergence in the performed case studies, allowing better performances than those obtained by conventionally designed PID controllers.

II. PROBLEM DESCRIPTION

The aim of this section is to illustrate two conventional PID designs and to describe the robust control problem to be addressed here. For sake of example, consider the output voltage control of a buck converter [14], where L is the filter inductor, C is the filter capacitor, R_L is the resistive load, V_g is the input voltage and V_o is the output voltage of the converter, to be regulated.

The buck converter has the transfer function

$$G_p(s) = \frac{R_L V_g}{s^2(CLR_L) + sL + R_L} \quad (1)$$

considering the duty cycle as input variable and the capacitor voltage as the output variable [14].

Differently from similar results in the literature, here some parameters of this converter are assumed, in the control design stage, as uncertain. Specifically, consider that the value of the load resistor R_L and the input voltage V_g are assumed as not precisely known, but belonging to real intervals for which only the upper and lower bounds are known, as given in Table I.

Table I
BUCK PARAMETERS.

Parameters	Values
Input voltage (V_g)	30 V $\pm 10\%$
Output voltage (on load) (V_o)	15 V
Switching frequency (f_s)	30000 Hz
Filter capacitor (C)	100 μF
Filter inductor (L)	100 μH
Resistive load (R_L)	3 Ω $\pm 50\%$

The control of the output voltage of the buck converter used here has a unit feedback, with the controller G_c being a fixed gain PID controller given by

$$G_c(s) = \left[\frac{K_d s^2 + K_p s + K_i}{s} \right] \times \left[\frac{pb}{pb + s} \right] \quad (2)$$

K_p, K_i and K_d are, respectively, the proportional, integral and derivative gains. pb is an additional pole placed at a high frequency to reduce interactions with the lead-lag action, and also to ensure a causal transfer function for the PID.

A. Conventional designs

As usually done for power converters, the gains of the PID can be computed in order to ensure 60 degree phase margin and a crossover frequency one decade below the converter switching frequency [14]. Two solutions for this control design problem are given in the sequence, one based on the function *pidtune* and the other based on iterative design in *sisotool*, both relying on MATLAB.

Choosing the case of maximum values of V_g and R_L to define the nominal plant for conventional control design, due to the fact that this is the plant with smaller damping factor, and the specifications of phase margin and crossover frequency given above, the MATLAB function *pidtune* provides the PID

$$G_c(s) = \frac{\left(1.89 \times 10^{-6} s^2 + 0.0226s + 64.9\right) 7.129 \times 10^4}{s(s + 7.129 \times 10^4)} \quad (3)$$

It is worth to mention that the *pidtune* returned an ideal PID, and then an additional pole was included one decade above the fastest zero of the controller to get the result in (3).

Now, as most commonly done by control engineers, using an interative design in *sisotool*, from MATLAB, considering the same nominal plant and the same design specifications used in *pidtune* example, one can get, by trial and error, the PID controller

$$G_c(s) = \frac{\left(4.16 \times 10^{-6} s^2 + 0.0315s + 56.29\right) 6.283 \times 10^4}{s(s + 6.283 \times 10^4)} \quad (4)$$

Figure 1 shows the step response of the closed-loop system with both conventionally designed controllers. One can see that the responses are quite similar, with a slight superior transient response from the *sisotool* based design.

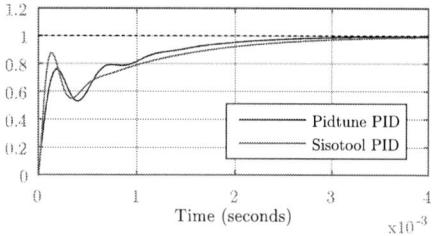

Figure 1. Converter with the MATLAB sisotool PID controller compared with the converter with the pidtune controller: Closed-loop system unit step response.

But the most important point to be remarked here is that these conventional designs need one nominal model of the plant to initiate the control design and, only *a posteriori*, the robustness and performance to intervalar parametric uncertainties are addressed. This can lead to a time consuming control design, based on trial and error sessions of a control engineer, which represents cost in industry.

This motivates the investigation in this paper, that aims to provide an automated tuning procedure for robust PID controllers, as described in the sequence.

B. Control problem definition

Find, by means of an automated tuning performed off-line, fixed gains K_p, K_I and K_d, and also the additional pole pb (i.e. four control parameters) for the PID in (2) such that, for all the extreme values of the plant model, the phase margin and the crossover frequency approach values prescribed by the control designer.

It is worth to mention that the controller solving the above design problem is a simple fixed gain PID, thus avoiding more complex implementations of PIDs, based on strategies that lead to time-varying control gains. Such controller, obtained in an off-line procedure can be useful as a benchmark for comparisons with other controllers.

III. PROPOSED DESIGN SOLUTION

In order to provide a solution for the above design problem, consider the vector of control parameters given by

$$K = [\; K_p \;\; K_I \;\; K_d \;\; pb \;] \tag{5}$$

To measure stability and the performance of the control system with these gains when affected by uncertain parameters, the following function are used here

$$f_j(K) = \delta_j(K)\,[\,|\,(PM^\star - PM_j(K))\,|\,\gamma_1 \\ + \,|\,(\omega_c^\star - \omega_{cj}(K))\,|\,\gamma_2\,]\,,\; j=1,\dots,p \tag{6}$$

where $PM^\star$ and $\omega_c^\star$ are reference values for desired phase margin and crossover frequency, respectively.

This function provide, for each one of the p possible conditions of extreme plant uncertain parameters, a weighted sum of the deviations of the phase margin and of the crossover frequency to their desired values.

The weights γ_1 and γ_2 are chosen, for instance, to compatibilize the size of the parcels in functions (6) or to set a higher importance for one of the perfomance indices (phase margin and crossover frequency) over the other. For each j, $\delta_j(K)$ is a scalar that plays the role of a penalty factor, being set to a high value when the control gain K under test produces instability, or being set to 1 when no unstable closed-loop pole is detected for any of the parameter situations $j=1,\dots,p$.

Notice that, for each control gain K under test, the values of $PM_j(K)$ and $w_{cj}(K)$ can be easily obtained, for instance, from function *margin*, of MATLAB, and the closed-loop poles can also be obtained from the function *pole*, of MATLAB, thus allowing to decide for the high or unit value for $\delta_j(K)$.

Defining the objective function

$$J(K) = \max\{f_j(K)\}\,,\, j=1,\dots,p \tag{7}$$

as the worst case value of the functions in (6), one has that the PID control gain vector that minimizes J in a given search space $\mathcal{K}$ (i.e. a given subset of $\mathcal{R}^4$), is optimal in the sense of ensuring the smallest weighted sum of deviations from the desired values of phase margin and crossover frequency in this space of controllers. In other words

$$K^\star = \arg\min_{K \in \mathcal{K}} J(K) \tag{8}$$

Notice that $K^\star$ that solves (8) will also provide a solution for the practical control problem in Section II.B.

A mathematical solution for the optimization control problem (8) is difficult to be obtained, due to the difficulty in expressing the gradient of the objective function and due to possible discontinuities of the objective function in a given broad search space. Particle swarm optimization has proven to be efficient to find minimum vaues of nonlinear objective functions, in search spaces with possible discontinuities and with possible multiple local minima [30].

A. Particle swarm optimization

PSO is a bio-inspired algorithm based on intelligent swarms, where the collective behavior of locally-interacting non-sophisticated agents creates global functional patterns, proposed by J. Kennedy and R. C. Eberhart in [30].

For the search of control gains using the PSO, the PID (9) can be seen as a particle i in the search space, that is, each K is associated with the position

$$s_i = [\; K_{pi} \;\; K_{Ii} \;\; K_{di} \;\; pb_i \;] \tag{9}$$

The particle swarm has a size of N particles, with N chosen sufficiently large to cover the search space.

The particles will evolve in the search space, that is, their positions will be updated from one epoch k to the next epoch $k+1$, until reaching the limit of epochs M for evolution.

Thus, at the epoch k, each particle has a position $s_i(k)$ and its own velocity $v_i(k)$, moving through the search space governed by the equations

$$s_i(k+1) \;=\; s_i(k) \;+\; v_i(k+1) \tag{10}$$

$$v_i(k+1) = \omega\,v_i(k) + \phi_1(rand_1(P_{i.best} - s_i(k))) \\ + \phi_2(rand_2(G_{best} - s_i(k))) \tag{11}$$

Each particle memorizes the position that got its best fitness, called $P_{i.best}$, and the swarm is also influenced by the particle that obtained the best fitness position among all particles, called G_{best}. This fitness is defined here by the objective function (7). ϕ_1 and ϕ_2 are the cognitive and social coefficients, respectively, ω the inertia and $rand_1$ and $rand_2$ random values between $[0,1]$. It is also noted that the position and velocity of each particle are represented by vectors [31].

To guide the choice of the PSO parameters, the number of particles and of epochs, respectively, N and M, and the coefficients ϕ_1 and ϕ_2 are chosen in order to ensure convergence of the fitness function with low computational effort.

Finally, in practical terms, it is worth to mention that one solution of the control problem in Section II can be achieved by the control designer choosing the worst case plant as the one more difficult to be stabilized (e.g. the one with smaller damping ratio or the one with smaller bandwidth). Then, a PID can be designed with a conventional technique, as the *pidtune*, exemplified in Section II. The stability and the performance of this controller with the other plant situations can be, *a posteriori*, checked. This strategy can lead to acceptable performance but there is no guarantee that better performances could not be found in a search space including this conventionally designed PID. If a search space is stablished around this viable controller, exhaustive grids would become rapidly prohibitive in terms of computational time. For instance, a simple cube in the search space, with side 1, discretized in 1000 points each side, would demand the test of 10^6 controllers. The proposed solution with PSO allows to highly improve the results obtained with conventional designs, as will be illustrated in the next section.

IV. PSO APPLIED TO THE BUCK CONVERTER CASE STUDY

In Section II, conventional PIDs were designed for a buck converter, using only one nominal model of the plant in the design. Now consider the same set of parameters given in Table I, for the design of a robust PID with the help of the PSO described in the previous section.

The design specifications remain the same: $PM^\star = 60°$ and $\omega_c^\star = 3000$ Hz, that is, one decade below the converter switching frequency. The weights in the terms of the objective function (6), which will guide the PSO, are $\gamma_1 = 1$ and $\gamma_2 = 10^{-2}$, and $\delta_j(K)$ is 10^3 in case of instability or 1 in the case of stability.

The search space chosen here is a hyper-rectangle around the parameters of the PID obtained from the *pidtune* function, given in (3), since this controller can be easily reproduced. The search space is described as $2.26 \times 10^{-5} \le K_p \le 22.6$, $6.49 \le K_I \le 6.49 \times 10^2$, $1.89 \times 10^{-9} \le K_d \le 1.89 \times 10^{-3}$ and, to keep the additional pole in high frequency range, $0.5 \times 7.1 \times 10^4 \le pb \le 1.5 \times 7.1 \times 10^4$. Even though parts of the search space represent PID parameters which will produce instability, the PSO will penalize these individuals with the high value of $\delta_j(K)$ in (6).

The algorithm starts with the following parameters: swarm size $N = 100$, maximum number of iterations (epochs) $M = 50$ and social and coefficient ϕ_1 and ϕ_2 with values of 0.5 each one. A random population of particles is initialized and then the PSO algorithm updates their position and velocity, based on (10) and (11), and evolves until the maximum number of epochs is reached or the stall criterion of 30 epochs without significant reduction in the objective function. The final value of the objective function obtained in one of the executions of the PSO is $J(K) = 11.06$, with the evolution of the fitness illustrated in Figure 2 (a).

The best particle found by the PSO, that is, the best PID control parameter vector with respect to the minimization of the objective function (7), is given by

$$G_c(s) = \frac{(5.31 \times 10^{-6}s^2 + 0.0571s + 292.8)\,8.266 \times 10^4}{s(s + 8.266 \times 10^4)} \tag{12}$$

For a performance comparison, the MATLAB step responses shown for the conventional controllers in Section II are reproduced now together with the response of the robust PID tuned by the PSO, in Figure 2 (b), where one can see the clear superiority of the performance with the controller obtained with the proposed method.

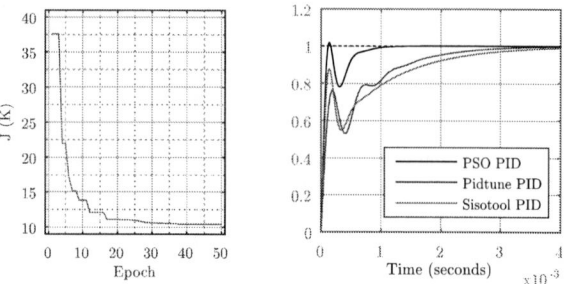

Figure 2. (a) Best fitness value in each epoch. (b) Comparison of the closed-loop system unit step response with the conventionally designed PIDs (3) and (4) and with the PSO (12).

To have a more realistic analysis of the PID controllers, comparisons of the other two conventionally designed PIDs obtained in Section II with the robust PID from PSO are carried out, based on the software PSIM with ideal components, including the effect of the PWM signal as input of the filter and load stage. Figure 3 (a) presents the startup transient response with the plant with maximum values of R_L and V_g, which is the situation used in the design of the conventional PID controllers. One can confirm the superiority of performance with the controller obtained with PID designed with the help of PSO.

For the result in Figure 3 (b), one has the system operating with each one of the PIDs, in steady state, in the mean parameter situation, and then the input voltage V_g is reduced in 10%. Again, one can see the superior transient recover from the PID tuned by the PSO. The result in Figure 3 (c) has the same initial conditions of the previous test, and a sudden reduction of 50% in R_L is applied (i.e. the load power consumption is suddenly increased). Again, the superior performance with the PSO based PID controller is confirmed.

To confirm robustness against parameter uncertainty, Figure 4 (a) shows the startup transient response of the closed-loop system with the PID tuned by the proposed PSO, for all the extreme load situation conditions. The line in blue represents the case of the plant with mean parameter values, and the lines in black represents the converter with the upper and lower bounds of R_L and V_g. It is noticed that all responses show good transient performance, due to the guaranteed phase margin and crossover frequency.

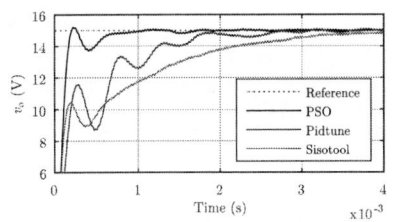 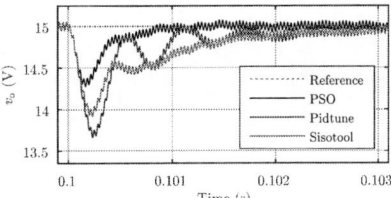 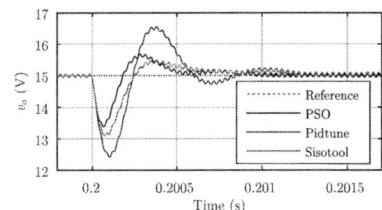

Figure 3. PSIM simulation of the closed-loop system for the converter with upper bounds of R_L and V_g: (a) System start-up. (b) 10% reduction in input voltage V_g in $t = 0.1\ s$. (c) 50% reduction in load resistance R_L in $t = 0.2\ s$.

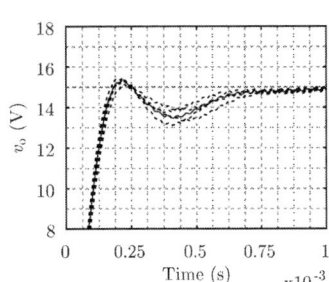 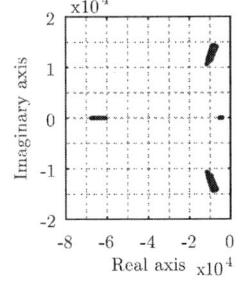

Figure 4. (a) Closed-loop start-up transient response system with nominal transfer function (in blue) and the upper and lower bounds of R_L and V_g (in black). (b) Closed-loop pole location.

Finally, Figure 4 (b) corroborates the stability of the system with the PID tuned with the help of PSO, for the domain of uncertain paramters, showing that all closed-loop poles remain on the lefthand side of the complex plane.

V. EXTENSION: BUCK-BOOST FOR SMALL SATELLITES

To apply the proposed control design to a more complex case, consider the converter in Figure 5, used in small satellites, which performs distributed bus power processing [32]. This converter operates in two different ways, which vary depending on the evolution of the satellite orbit. In the eclipse period when there is no solar energy in the photovoltaic panels, the converter regulates the output bus for the loads. In the sunlight period, when the photovoltaic panels are receiving solar radiation, the converter operates as a battery charger. For this example it was used only the eclipse converter plant, given by equation (13), which is represented by the averaged small signal model [33]–[35]

$$G_{v_o d}(s) = \frac{a_2 s^2 + a_1 s + a_o}{b_3 s^3 + b_2 s^2 + b_1 s + b_o} \quad (13)$$

being
$a_0 = R_b - R_o - D^2 R_o + 2DR_o$; $a_1 = L - C_b R_b R_o - C_b D^2 R_b R_o - C_i D^2 R_b R_o + 2C_b DR_b R_o + C_i DR_b R_o$;
$a_2 = C_b L R_b + C_i L R_b$; $b_o = R_b + R_o + D^2 R_o - 2DR_o$;
$b_1 = L + C_b R_b R_o + C_b D^2 R_b R_o + C_i D^2 R_b R_o - 2C_b DR_b R_o$;
$b_2 = C_b L R_b + C_i L R_b + C_i L R_o$; $b_3 = C_b C_i L R_b R_o$

For sake of comparison, a PID controller was designed with *pidtune* to achieve a phase margin of $60°$ and a crossover

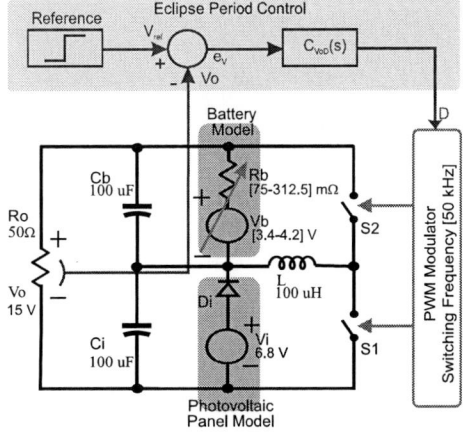

Figure 5. Bidirectional buck-boost converter with stacked input and output for small satellite application.

frequency of 400 Hz, and these requirements were also used to guide the design with the PSO, with the same parameters employed in Section IV (swarm size and number of iterations), leading to the controller

$$C_{v_o d}(s) = \frac{(4.95 \times 10^{-6} s^2 + 0.0045 s + 38.67) 4.755 \times 10^4}{s(s + 4.755 \times 10^4)}$$
$$(14)$$

Figure 6 shows the step responses for this controller and for the PID designed with *pidtune*, under the same design requirements. One can see the superiority of the PID obtained from the proposed procedure, providing the faster responses in Figure 6 (two upper curves), also for a more challenging application.

VI. CONCLUSION

This paper provided an alternative for automated design of robust (fixed gains) PIDs with application for power converters with uncertain parameters. The proposed solution relies on a PSO that minimizes an weighted sum of the deviations of desired values for phase margin and crossover frequency, allowing results superior than conventionally designed fixed gains PIDs. The main advantage of the proposed design procedure is to alleviate the control engineer from the time consuming task of obtaining the control gains by trial and error sessions. This task is delegated to an automated design

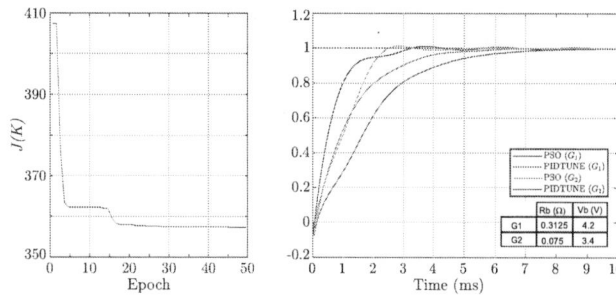

Figure 6. Step responses for the PID obtained from *pidtune* function and for the proposed PID in (14).

using PSO, which produce, in an offline procedure, the gains of robust PIDs, suitable for control of power converters.

REFERENCES

[1] S. Bhattacharyya, "Robust control under parametric uncertainty: An overview and recent results," *Annual Reviews in Control*, vol. 44, pp. 45–77, 2017.

[2] K. J. Åström, T. Hägglund, C. C. Hang, and W. K. Ho, "Automatic tuning and adaptation for PID controllers – a survey," *Control Engineering Practice*, vol. 1, no. 4, pp. 699–714, 1993.

[3] H. Hjalmarsson, M. Gevers, S. Gunnarsson, and O. Lequin, "Iterative feedback tuning: theory and applications," *IEEE control systems magazine*, vol. 18, no. 4, pp. 26–41, 1998.

[4] K. J. Astrom, T. Hagglund, and K. J. Astrom, *Advanced PID control*. ISA-The Instrumentation, Systems, and Automation Society Research Triangle ..., 2006, vol. 461.

[5] M.-T. Ho and C.-Y. Lin, "PID controller design for robust performance," *IEEE Transactions on Automatic Control*, vol. 48, no. 8, pp. 1404–1409, 2003.

[6] G. Feng, E. Meyer, and Y.-F. Liu, "A new digital control algorithm to achieve optimal dynamic performance in DC-to-DC converters," *IEEE Transactions on Power Electronics*, vol. 22, no. 4, pp. 1489–1498, 2007.

[7] J. G. Ziegler and N. B. Nichols, "Optimum settings for automatic controllers," *trans. ASME*, vol. 64, no. 11, 1942.

[8] Z.-Y. Zhao, M. Tomizuka, and S. Isaka, "Fuzzy gain scheduling of PID controllers," *IEEE transactions on systems, man, and cybernetics*, vol. 23, no. 5, pp. 1392–1398, 1993.

[9] R. K. Mudi and N. R. Pal, "A robust self-tuning scheme for PI- and PD-type fuzzy controllers," *IEEE Transactions on fuzzy systems*, vol. 7, no. 1, pp. 2–16, 1999.

[10] L. Guo, J. Y. Hung, and R. M. Nelms, "Evaluation of DSP-based PID and fuzzy controllers for DC–DC converters," *IEEE transactions on industrial electronics*, vol. 56, no. 6, pp. 2237–2248, 2009.

[11] M. Algreer, M. Armstrong, and D. Giaouris, "Adaptive PD+I control of a switch mode DC-DC power converter using a recursive fir predictor," 2010.

[12] H.-X. Li, L. Zhang, K.-Y. Cai, and G. Chen, "An improved robust fuzzy-PID controller with optimal fuzzy reasoning," *IEEE Transactions on Systems, Man, and Cybernetics, Part B (Cybernetics)*, vol. 35, no. 6, pp. 1283–1294, 2005.

[13] W. K. Ho, C. C. Hang, and J. Zhou, "Self-tuning PID control of a plant with under-damped response with specifications on gain and phase margins," *IEEE Transactions on Control Systems Technology*, vol. 5, no. 4, pp. 446–452, 1997.

[14] R. W. Erickson and D. Maksimovic, *Fundamentals of power electronics*. Springer Science and Business Media, 2007.

[15] L. Guo, J. Y. Hung, and R. Nelms, "PID controller modifications to improve steady-state performance of digital controllers for buck and boost converters," in *APEC. Seventeenth Annual IEEE Applied Power Electronics Conference and Exposition (Cat. No. 02CH37335)*, vol. 1. IEEE, 2002, pp. 381–388.

[16] V. Yousefzadeh, A. Babazadeh, B. Ramachandran, E. Alarcón, L. Pao, and D. Maksimovic, "Proximate time-optimal digital control for synchronous buck DC–DC converters," *IEEE Transactions on Power Electronics*, vol. 23, no. 4, pp. 2018–2026, 2008.

[17] S. Kapat and P. T. Krein, "Formulation of PID control for DC–DC converters based on capacitor current: A geometric context," *IEEE Transactions on Power Electronics*, vol. 27, no. 3, pp. 1424–1432, 2011.

[18] H.-H. Park and G.-H. Cho, "A DC–DC converter for a fully integrated PID compensator with a single capacitor," *IEEE Transactions on Circuits and Systems II: Express Briefs*, vol. 61, no. 8, pp. 629–633, 2014.

[19] E. W. Zurita-Bustamante, J. Linares-Flores, E. Guzman-Ramirez, and H. Sira-Ramirez, "A comparison between the GPI and PID controllers for the stabilization of a DC–DC "buck" converter: A field programmable gate array implementation," *IEEE Transactions on Industrial Electronics*, vol. 58, no. 11, pp. 5251–5262, 2011.

[20] C. Chang, Y. Yuan, T. Jiang, and Z. Zhou, "Field programmable gate array implementation of a single-input fuzzy proportional–integral–derivative controller for DC–DC buck converters," *IET Power Electronics*, vol. 9, no. 6, pp. 1259–1266, 2016.

[21] U. A. Shaikh, M. K. AlGhamdi, and H. A. AlZaher, "Novel product ANFIS-PID hybrid controller for buck converters," *The Journal of Engineering*, vol. 2018, no. 8, pp. 730–734, 2018.

[22] S. E. De León-Aldaco, H. Calleja, and J. A. Alquicira, "Metaheuristic optimization methods applied to power converters: A review," *IEEE Transactions on Power Electronics*, vol. 30, no. 12, pp. 6791–6803, 2015.

[23] C.-L. Lin, H.-Y. Jan, and N.-C. Shieh, "GA-based multiobjective PID control for a linear brushless DC motor," *IEEE/ASME transactions on mechatronics*, vol. 8, no. 1, pp. 56–65, 2003.

[24] Z.-L. Gaing, "A particle swarm optimization approach for optimum design of PID controller in AVR system," *IEEE transactions on energy conversion*, vol. 19, no. 2, pp. 384–391, 2004.

[25] R. A. Hanifah, S. F. Toha, S. Ahmad, and M. K. Hassan, "Swarm-intelligence tuned current reduction for power-assisted steering control in electric vehicles," *IEEE Transactions on Industrial Electronics*, vol. 65, no. 9, pp. 7202–7210, 2017.

[26] J. Darvill, A. Tisan, and M. Cirstea, "A novel PSIM and matlab co-simulation approach to particle swarm optimization tuning of PID controllers," in *2014 International Conference on Optimization of Electrical and Electronic Equipment (OPTIM)*. IEEE, 2014, pp. 784–789.

[27] E. Sahin, M. S. Ayas, and I. H. Altas, "A PSO optimized fractional-order PID controller for a PV system with DC–DC boost converter," in *2014 16th International Power Electronics and Motion Control Conference and Exposition*. IEEE, 2014, pp. 477–481.

[28] P. Farhang, A. M. Drimus, and S. Mátéfi-Tempfli, "New technique for voltage tracking control of a boost converter based on the PSO algorithm and LTspice," in *2015 56th International Scientific Conference on Power and Electrical Engineering of Riga Technical University (RTUCON)*. IEEE, 2015, pp. 1–6.

[29] P. Verma, N. Patel, N.-K. C. Nair, and A. Sikander, "Design of PID controller using cuckoo search algorithm for buck-boost converter of LED driver circuit," in *2016 IEEE 2nd Annual Southern Power Electronics Conference (SPEC)*. IEEE, 2016, pp. 1–4.

[30] R. Eberhart and J. Kennedy, "A new optimizer using particle swarm theory," in *In Proceedings of the Sixth International Symposium on Micro Machine and Human Science*. IEEE, 1995, pp. 39–43.

[31] M. Veerachary and A. R. Saxena, "Optimized power stage design of low source current ripple fourth-order boost dc–dc converter: A pso approach," *IEEE Transactions on Industrial Electronics*, vol. 62, no. 3, pp. 1491–1502, 2015.

[32] E. Mattos, A. M. Andrade, G. V. Hollweg, M. L. d. S. Martins, and J. R. Pinheiro, "Analysis and design of a stacked power subsystem on a picosatellite," *IEEE Aerospace and Electronic Systems Magazine*, vol. 33, no. 10, pp. 4–13, 2018.

[33] D. DePasquale and J. Bradford, "Nano/microsatellite market assessment," *Public Release, Revision A, SpaceWorks*, 2013.

[34] R. Burt, "Distributed electrical power system in cubesat applications," 2011.

[35] T. J. Doering, "Development of a reusable cubesat satellite bus architecture for the kysat-1 spacecraft," 2009.

Impacts of the Multi-Variables Modulation on Transformer and Soft-Switching of a DAB

1st Jeferson Fraytag
Power Electronics Institute (INEP)
Federal University of Santa Catarina
Florianópolis-SC, Brazil
jeferson.fraytag@gmail.com

2nd André Luís Kirsten
Power Electronics Institute (INEP)
Federal University of Santa Catarina
Florianópolis-SC, Brazil
kirsten.andre@ufsc.br

3rd Marcelo Lobo Heldwein
Power Electronics Institute (INEP)
Federal University of Santa Catarina
Florianópolis-SC, Brazil
heldwein@inep.ufsc.br

Abstract—Due to the current importance of Dual-Active-Bridge (DAB) converters, this work proposes an analysis framework to model the DAB static behavior when it is subjected to a multi-variables modulation. In this type of modulation, in addition to the angular-phase-shift between the DAB H-bridges, the primary and secondary-side voltage duty-cycles can also be varied. The current analysis proposes tools to define the most suitable operation point for a DAB converter, analyzing the impact of choosing this operating point on the transformer and switches. In this context, magnetic core volume, copper losses and soft-switching are considered. Experimental results for a 800 W prototype are presented to validate the proposed analyzes.

Index Terms—DAB Converter, Multi-Variables Modulation, Operation Point Impact, Power Analysis

I. INTRODUCTION

The most applications that feature bidirectional active power flow and galvanic isolation employ a two-stage interface composed of a grid connected inverter and an isolated dc-dc converter, from which the Dual-Active-Bridge converter (DAB) [1] is of great interest and use. This is because the DAB enables high power transfer levels, high conversion efficiency, galvanic isolation, and bi-directional power flow. It also features a good losses distribution among its components, and is currently proposed to be used in the galvanic isolation stage of solid state transformers (SSTs) for future power transmission and distribution systems [2], [3]. In addition to these applications, the DAB is also used in photovoltaic systems [4], [5], ultra capacitors [6], as battery interface [7], [8], electric automotive traction [9], [10], among others.

The DAB converter structure (see Fig. 1) is typically operated based on the phase-shift modulation [11], [12], where the actuation variable is the phase-shift angle ϕ between the pulses of the primary and secondary switches of the converter. Moreover, the turn on time at each switch is constant close to one half of the switching period and all the converter arms operate with a complementary feature. However, the present work considers the DAB converter operating not only with the angular phase-shift, but also inserting the voltages duty-cycles as additional control variables, which will determine the global converter behavior. This modulation, called here as "multi-variables", adds D_1 and D_2 into the analyzes. These are the duty-cycles of the primary voltage v_P and the secondary voltage v_S of the transformer, respectively. In

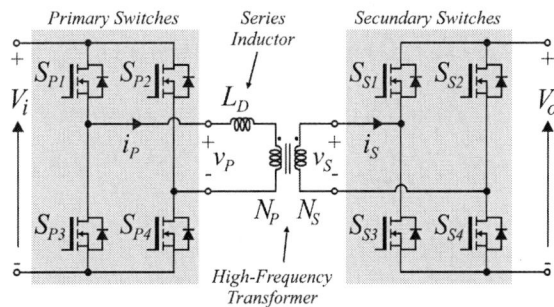

Fig. 1. Conventional DAB dc-dc converter structure with eight switches, divided into two H-bridges.

[13], for example, a similar modulation strategy is applied for the Three-Level Dual-Active-Bridge, where the authors present an expansion of the converter power area. In [14], the authors use this modulation proposal to obtain the conduction losses reduction with straightforward application into battery interface systems.

Although there are works related to the subject, the contribution of this work consists on presenting analytical tools to determine the different operating points possibilities when employing the multi-variables modulation. The analysis leads to the creation of an operational map, where the impact of the choice of operating point on the transformer core volume is observed, as well the conduction losses and the soft-switching feature of the switches. Novel equations for the relations between the transferred active power and the total transformer apparent power are derived. In this way, it is possible to develop analytical relationships that describe the converter behavior, which can be advantageously used in the expansion and optimization of its operation possibilities. Thus, from these relations, the converter designer will have tools that can assist in choosing the most appropriate operating point for a given application.

II. MULTI-VARIABLES MODULATION

The voltage pulse width applied to the transformer primary side (v_P), as seen in Fig. 2, is determined from the operation of the S_{P1}, S_{P2}, S_{P3} and S_{P4} switches. The same analysis is valid for the secondary side voltage (v_S), determined from the S_{S1}, S_{S2}, S_{S3} and S_{S4} switches. Thus, duty-cycles can

978-1-7281-4181-7/19 $31.00 © 2019 IEEE

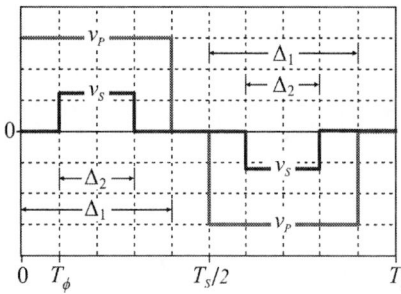

Fig. 2. Transformer primary voltage v_P and secondary voltage v_S for a specific point of operation.

be defined as $D_1 = \Delta_1/T_S$ and $D_2 = \Delta_2/T_S$, where T_S is the switching period, Δ_1 is the positive pulse voltage time in the primary side and Δ_2 on the secondary side of the transformer. From these considerations, the DAB converter operation and the multi-variables modulation influences can be analyzed.

A. Converter Operation Situations

The variation of the duty-cycles D_1 and D_2, as well as of the angular phase-shift ϕ, provide changes in the converter operation. In this context, different combinations of voltage and current waveforms appear and must be analyzed. Six different situations are considered, which are sufficient to describe the global behavior of the multi-variable topology, namely: Situation A for ($T_{D2} < T_{D1}$), Situation B for ($T_{D2} \geq T_{D1}$), Situation C for ($T_6 > 0$)$\wedge$($T_6 < T_{D1}$), Situation D for ($T_\phi < T_S/2$)$\wedge$($T_6 > T_{D1}$), Situation E for ($T_6 < T_{D1}$)$\wedge$($T_\phi > T_{D1}$) and, finally, Situation F valid if ($T_{D2} < T_S/2$)$\wedge$($T_\phi > T_{D1}$). Fig. 3 shows the voltage and current response for the six listed situations, where the time instants are defined as: $T_{D1} = D_1 T_S$, $T_{D2} = T_\phi + D_2 T_S$, $T_3 = T_{D1} + (T_S/2)$, $T_4 = T_\phi + (T_S/2)$, $T_5 = T_4 + D_2 T_S$, $T_6 = T_{D2} - (T_S/2)$ and $T_\phi = \phi(T_S/2\pi)$.

The combination values of ϕ, D_1 and D_2 determines in which of the six situations (A, B, C, D, E or F) the DAB converter operates. The variation of ϕ, D_1 and D_2 results in significant changes in the converter behavior. In this way, it is necessary to develop relationships that describe the power processing levels for each of the operating situations.

B. Converter Power Processing

The operation conditions set by the control variables ϕ, D_1 and D_2 define the energy transfer rate between the input and output ports of the topology. Thus, the power analysis is of fundamental importance for each of the six possible operating situations, resulting in different converter responses. Assuming ideal, i.e., lossless components and constant input and output DC voltages, the average transferred power for each operational situation P_j, with $j = \{A, B, C, D, E, F\}$, can be described as a function of the phase-shift angle ϕ and the duty-cycles D_1 and D_2 according to

$$P_A(\phi, D_1, D_2) = k_0 \left(\phi D_2 + \pi D_2{}^2 - \pi D_1 D_2 \right) \quad (1)$$

$$P_B(\phi, D_1, D_2) = k_0 \left(\phi D_1 + \pi(D_1 D_2 - D_1{}^2) - \frac{\phi^2}{4\pi} \right) \quad (2)$$

$$P_C(\phi, D_1, D_2) = k_0 \left(\phi D_1 - \pi D_1{}^2 - \phi D_2 - \frac{\phi^2}{2\pi} + k_1 \right) \quad (3)$$

$$P_D(\phi, D_1, D_2) = k_0 \left[D_1(\pi - \phi) + \pi(D_1{}^2 - D_1 D_2) \right] \quad (4)$$

$$P_E(\phi, D_1, D_2) = k_0 \left(-\phi D_2 - \frac{\phi^2}{4\pi} + k_1 \right) \quad (5)$$

$$P_F(\phi, D_1, D_2) = k_0 \left(\pi D_1 D_2 \right), \quad (6)$$

where the coefficients k_0 and k_1 are

$$k_0 = \frac{d V_i^2}{\pi f_S L_D} \quad (7)$$

$$k_1 = \pi D_2 (D_1 - D_2 + 1) - \frac{\pi}{4} + \frac{\phi}{2}, \quad (8)$$

and d corresponds to the converter voltage gain, which depends on the number of turns of the primary N_P and secondary N_S of the transformer [15], as defined by

$$d = \frac{V_o}{V_i} \left(\frac{N_P}{N_S} \right). \quad (9)$$

The mathematical interpolation among P_j expressions are depicted in Fig. 4, where the normalized active power $\bar{P}_o$ is the maximum transferable power point ($D_1 = D_2 = 1/2$ and $\phi = 90°$), given by

$$\bar{P}_o(d) = \frac{V_i^2 d}{8 L_D f_S}. \quad (10)$$

As can be seen in Fig. 4 representation, the reduction of the duty-cycles generally decreases the transferred active power. Moreover, the dependency of the phase-shift changes the points where the maximum power peaks occur. Considering an analysis with a fixed D_2 value and where D_1 is gradually reduced, it is observed that the active power also decreases, besides the tendency to shift the maximum power point towards lower ϕ angles. At an operation point with $D_2 = 0.4$ and $D_1 = 0.3$, for instance, the maximum power angle approaches $\phi \approx 75°$, while for $D_2 = 0.4$ and $D_2 = 0.2$ the maximum power angle reduces to $\phi \approx 52°$. Thus, the three variables can be used to control the active power.

Assuming null magnetizing current and lossless components, the total apparent power for the multi-variable system is defined in this work as being the sum of the apparent powers of both transformer windings. Thus,

$$S_T = v_P^{\text{rms}} i_P^{\text{rms}} + v_S^{\text{rms}} i_S^{\text{rms}}, \quad (11)$$

which can be rewritten as

$$S_{Tj}(\phi, D_1, D_2) = V_i \left(\sqrt{2D_1} + d\sqrt{2D_2} \right) \sqrt{\frac{1}{T_S} \int_0^{T_S} i_{Pj}{}^2 \, dt}, \quad (12)$$

where x^{rms} denotes the "Root-Mean-Square" value of a signal x. In addition to the dependency of v_P on the duty-cycle D_1, it is also a function of the input voltage V_i, the influence of the variation of v_S is taken into account through d and D_2.

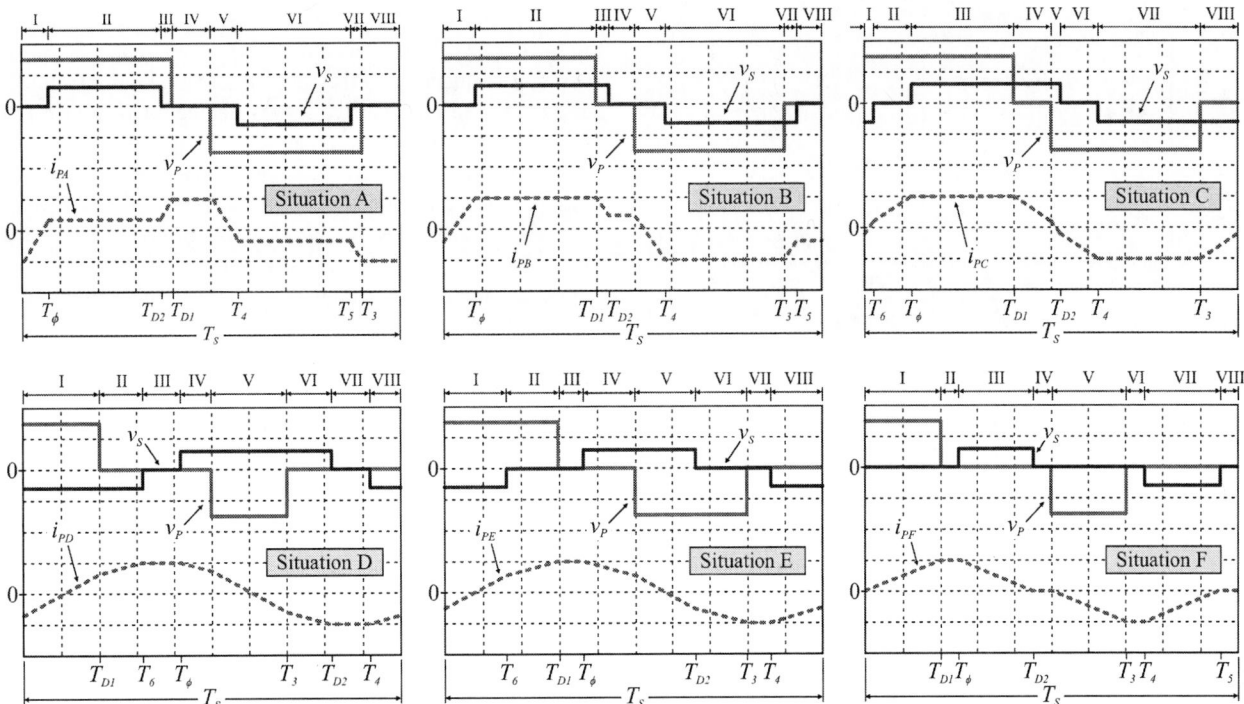

Fig. 3. Exemplification of the primary v_P and secondary v_S voltages and the primary current i_P: Situation A ($D_1 = 0.4$, $D_2 = 0.3$, $\phi = 30°$ and $d = 1$), Situation B ($D_1 = 0.4$, $D_2 = 0.35$, $\phi = 30°$ and $d = 1$), Situation C ($D_1 = 0.4$, $D_2 = 0.4$, $\phi = 45°$ and $d = 1$), Situation D ($D_1 = 0.2$, $D_2 = 0.4$, $\phi = 135°$ and $d = 1$), Situation E ($D_1 = 0.3$, $D_2 = 0.3$, $\phi = 130°$ and $d = 1$) and Situation F ($D_1 = 0.2$, $D_2 = 0.2$, $\phi = 90°$ and $d = 1$).

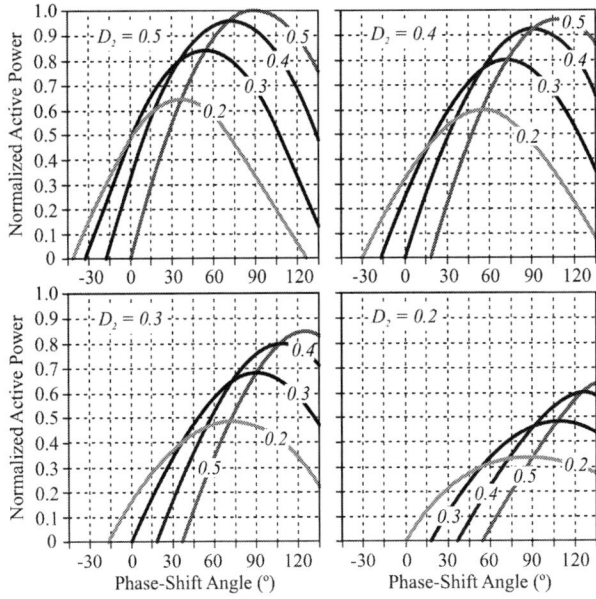

Fig. 4. Normalized direct active power response considering ϕ (-45 to 135°), D_1 (0.2 to 0.5), D_2 (0.2 to 0.5) and $d = 1$.

Thus, the total apparent power behavior for the multi-variables modulation is shown in Fig. 5, where the normalized apparent power $\bar{S}_T$ is defined with

$$\bar{S}_T(d) = \frac{V_i^2 (1 + d)}{L_D f_S} \sqrt{\frac{d^2 + 1}{48}} . \tag{13}$$

Based on the analysis of Fig. 5, the apparent power increases as the phase-angle is incremented. At this point, it is clear that there is more than one operating point providing the same active power with different apparent power levels. In this context, different combinations of ϕ, D_1 and D_2 are computed for all converter power possibilities, where the least apparent power point can be obtained. Considering, for example, an active power of 800 W, the converter can operate at $\phi = 13° \wedge D_1 = 0.4 \wedge D_2 = 0.4$, or at $\phi = 35° \wedge D_1 = 0.4 \wedge D_2 = 0.3$, among others points, as shown in Fig. 6. However, the total apparent power presents a large variation, depending on the chosen operating point, and the impacts of this will be seen in the following sections.

C. Operation Point Impact on Transformer

The presence of physical intrinsic effects in high-frequency transformers must be considered to analyze a converter efficiency. Skin and proximity effects change the current distribution in the wire, resulting in a resistance value increase with frequency [16], [17]. Thus, it is important to choose an operating point with reduced apparent power, which is associated with smaller rms current levels, in addition to typically lower magnetic volume and copper losses.

Considering the multi-variables modulation features, as well the set of parameters exemplified in the previous subsection, the relative behavior of the core volume and cooper losses can be established for variations of ϕ, D_1 and D_2, as shown in Fig. 7 for the 800 W active power. In these relations, the

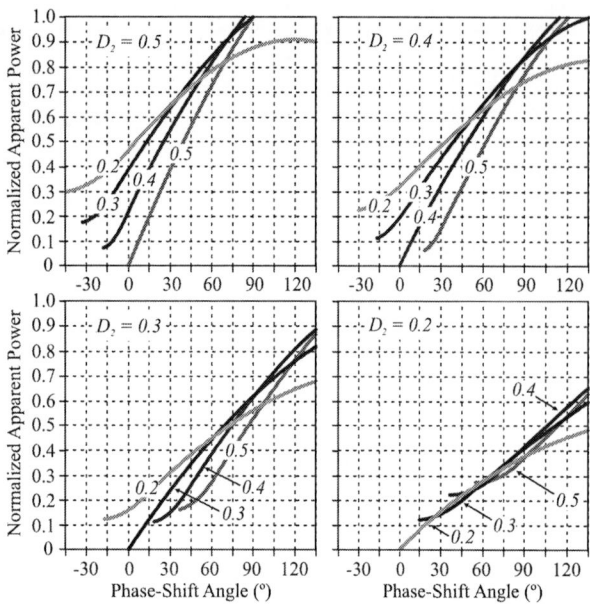

Fig. 5. Normalized total apparent power response considering ϕ (-45 to 135°), D_1 (0.2 to 0.5), D_2 (0.2 to 0.5) and $d = 1$.

design of the transformer is based on the analytical methods presented in [18] and [19].

Therefore, it is possible to verify that core volume and copper losses tend to decrease for operating points with lower $|\phi|$ angles. In general, the operating points with $D_1 = D_2$ and lower phase-shift angles tend to result in lower apparent power and, thus, core volume and copper losses can be achieved. Furthermore, in general, the best results was observed for an operation region between 15° and 30°.

D. Operation Point Impact on Switches

The operating point choice also influence the converter switches operation, where the conduction losses and the switching losses should be considered. The conduction losses are directly related to rms current value and, consequently, to the apparent power levels. Thus, this type of loss has the

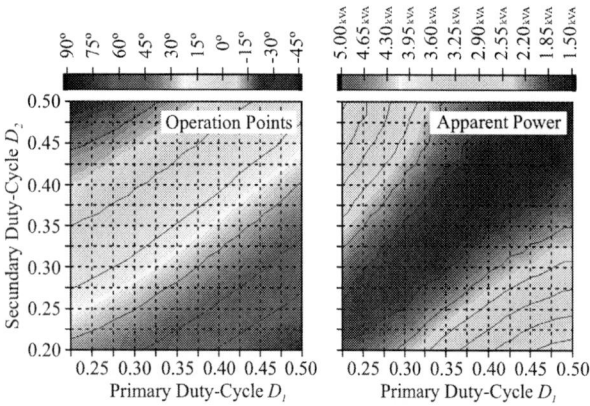

Fig. 6. Surfaces that demonstrate the different operating points for a same active power and the respective apparent power behavior. Example for $P_j = 800$ W, $V_i = 800$ V, $V_o = 400$ V, $f_S = 100$ kHz, $L_D = 220\,\mu$H and $d = 1$.

same behavior of the previously presented transformer copper losses. However, the switching losses are dependent on the operation frequency and the soft- or hard-switching feature of the switches. In this way, the soft-switching boundaries for the different operation points possibilities should be established.

The ZVS (Zero-Voltage-Switching) operation occurs in the DAB power semiconductors when there is a current through the antiparallel diodes of the switches prior to the turn-on signal. Thus, the switch voltage will be near zero during the switching turn-on transitions. Considering the variations of ϕ, D_1 and D_2 for different operating points, Fig. 8 shows the ZVS operation boundaries for the DAB converter to, both, primary and secondary switches. In these relations, the methodology presented in [15] is used for the multi-

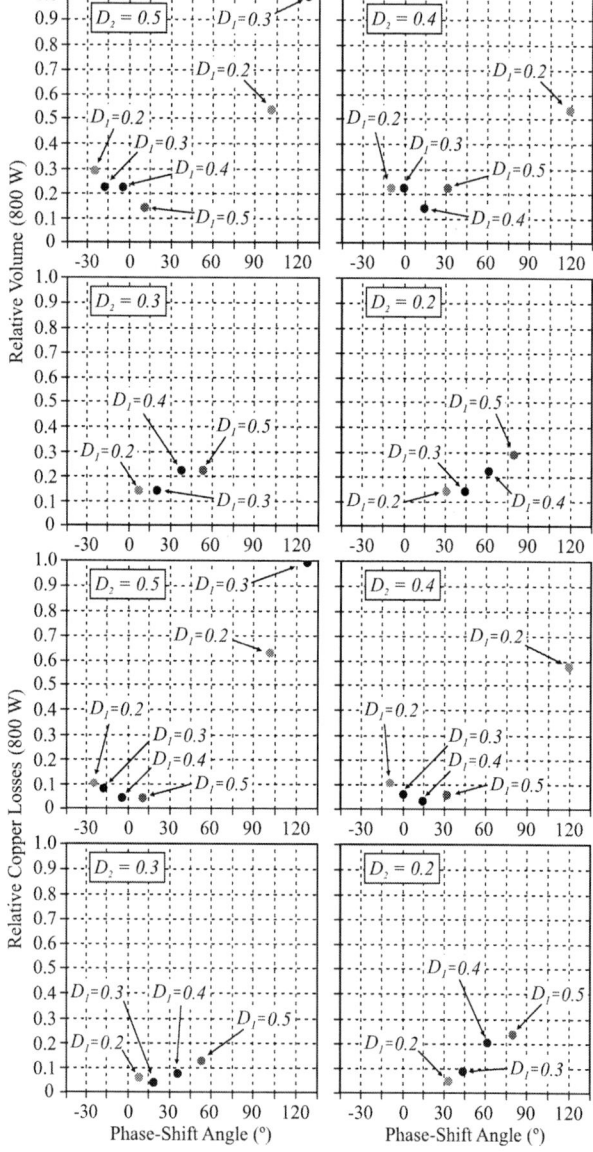

Fig. 7. Transformer core volume and copper losses for different operating points. Example for $P_j = 800$ W, $V_i = 800$ V, $V_o = 400$ V, $f_S = 100$ kHz, $L_D = 220\,\mu$H and $d = 1$.

978-1-7281-4181-7/19 $31.00 © 2019 IEEE

variables modulation. According to Fig. 8, there are operation points without the soft-switching feature for a same active power level. For the normalized active power of 0.3, e.g., this power level is obtained at the operating points with $\phi = -17° \wedge D_1 = 0.2 \wedge D_2 = 0.5$ or with $\phi = 93° \wedge D_1 = 0.2 \wedge D_2 = 0.5$. However, at the first point the converter will be out of a primary soft-switching region, which results in a switching losses increase. In this way, one can choose the most suitable operating point for the converter.

III. EXPERIMENTAL VALIDATION

A DAB converter prototype was developed in order to validate the multi-variable theoretical analyzes and the operation

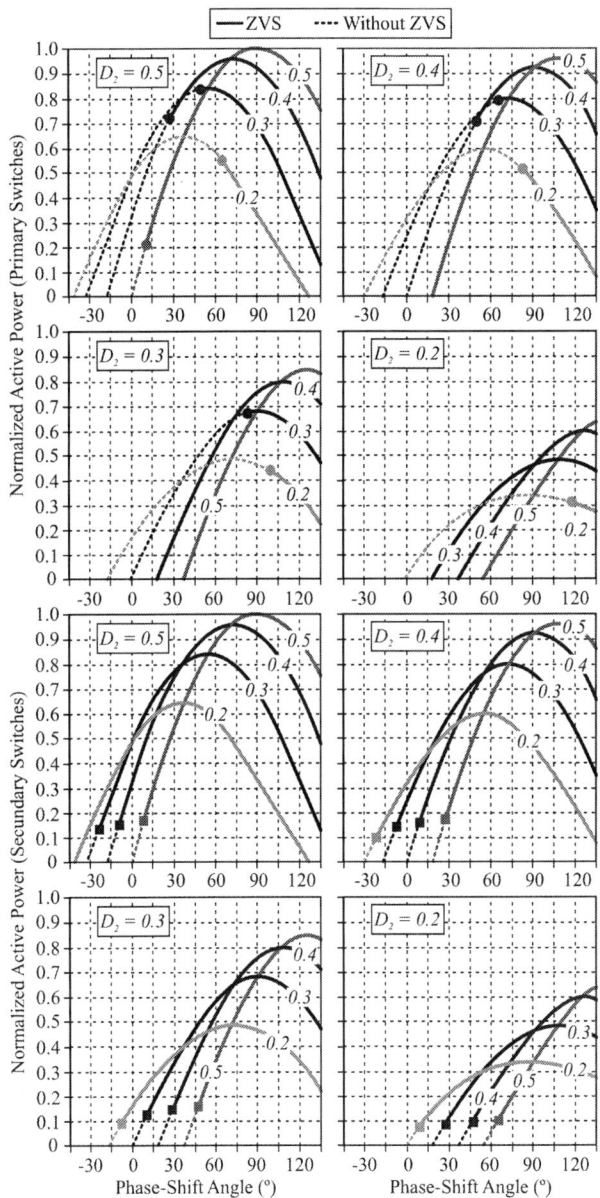

Fig. 8. ZVS regions of the primary and secundary switches for variations of ϕ, D_1 and D_2. Exemple for SiC-MOSFET SCT2450KE (1200 V-10 A) on primary and SiC-MOSFET UJC06505K (650 V-23 A) on secundary side.

point features, where the previous specifications were used. It should be noted that the dynamic analysis is out of the scope of this work.

The impact of the operating point on the transformer is shown in Fig. 9, where v_P waveforms is the primary-side voltage, v_S is the secondary-side voltage and i_P is the primary current for: (a) $\phi = 13° \wedge D_1 = 0.4 \wedge D_2 = 0.4$; and, (b) $\phi = 35° \wedge D_1 = 0.4 \wedge D_2 = 0.3$. At these operating points, the converter active power is approximately 800 W. However, there is a reduction difference of approximately 28% between both total apparent powers. The lowest value was obtained for operation with $D_1 = D_2$, according to the theoretical behavior previously presented.

Fig. 10 shows the impact of the operating point on the converter switches. In this representation, the ZVS feature on the primary side can be analyzed when the S_{P1} switch turns on, which characterizes the v_P voltage positive transition (crosshatched regions). At this time, if the current of the primary i_P is negative, the soft-switching on the primary is obtained. In the first point (a) $\phi = 34° \wedge D_1 = 0.4 \wedge D_2 = 0.3$; the converter provides a soft-switching operation for the primary and secondary switches. However, for the second point (b) where $\phi = 11° \wedge D_1 = 0.2 \wedge D_2 = 0.3$; it is possible to observe the occurrence of ZVS loss in the primary switches. The experimental soft-switching behavior is in accordance with the previously theoretical analyzes regions.

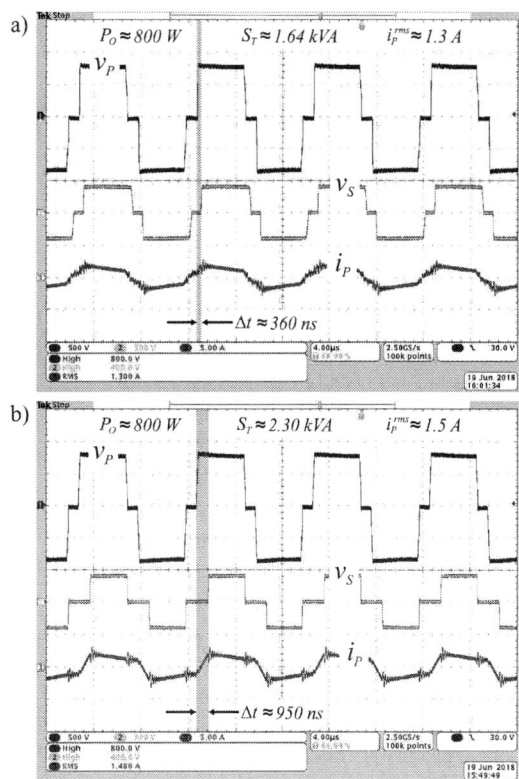

Fig. 9. Apparent power levels for different operating points in multi-variable modulation. Primary voltage v_P (Ch1 - 500 V/div.), secondary voltage v_S (Ch2 - 500 V/div.) and primary current i_P (Ch3 - 5 A/div.). Time: 4 μs/div.

978-1-7281-4181-7/19 $31.00 © 2019 IEEE

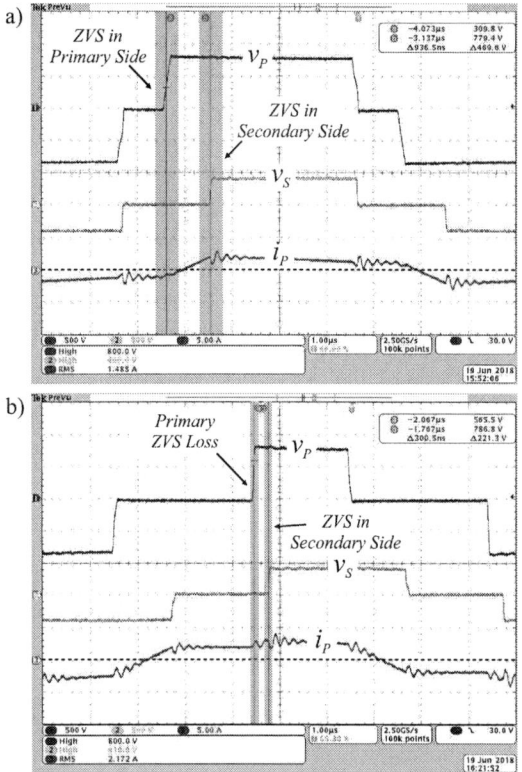

Fig. 10. ZVS features for different operating points in multi-variable modulation. Primary voltage v_P (Ch1 - 500 V/div.), secondary voltage v_S (Ch2 - 500 V/div.) and primary current i_P (Ch3 - 5 A/div.). Time: 1 μs/div.

IV. CONCLUSION

This work presented a contribution to the DAB converter analysis applying the multi-variable modulation. In this type of modulation, in addition to the phase-shift angle, the voltage duty-cycles associated with the primary and the secondary of the high-frequency transformer are also modified to enable different operation possibilities. These extra control variables provide peculiarities in the converter operation and result in six different operating situations to be analyzed. The transferrable active and apparent power levels were theoretically analyzed for each situation. It was observed that, according to the variation of the duty-cycles and the phase-shift angle, it is possible to minimize the apparent power for different active power ranges of the system. In general, the lowest apparent power is obtained with $D_1 \approx D_2$.

Considering the possibility of apparent power decrease, the impact on transformer volume and copper losses was analyzed. In this relation, there is a tendency of the core volume to increase for higher $|\phi|$ angles. The same conclusion is valid for the copper losses. In addition, the choice of operating point for switches was verified, establishing the regions where the loss of soft-switching occurs.

A DAB prototype was designed and tested for 800 W, with different operation points possibilities. It was observed that the obtained active power points were consistent with the theoretical predictions and with the apparent power analyzed behavior.

In general, the operation region between approximately 15° and 30° with $D_1 = D_2$, presented the lowest magnetic core volumes and the lowest copper losses levels, however, in some cases there are soft-switching loss. Thus, from the tools presented this work, the converter designer can choose the most appropriate operating point for each application.

REFERENCES

[1] M. N. Kheraluwala, R. W. Gascoigne, D. M. Divan and E. D. Baumann, "Performance Characterization of a High-Power Dual Active Bridge DC-to-DC Converter", IEEE Transactions on Industry Applications, 1992, 28, (6), pp. 1294–1301.

[2] U. Khalid, M. M. Khan, Z. Xiang and Y. Jianyang, "Bidirectional Modular Dual Active Bridge (DAB) Converter Using Multi-Limb-Core Transformer with Symmetrical LC Series Resonant Tank Based on Cascaded Converters in Solid State Transformer (SST)", China International Electrical and Energy Conference (CIEEC), 2017, pp. 627–632.

[3] C. Shuyu, V. B. Sriram, H. D. Tafti, K. V. R. Kishore, Y. H. Li and A. Tripathi, "Modular DAB DC-DC Converter Low Voltage Side DC Link Capacitor Two-Stage Charging-Up Control for Solid State Transformer Application", 2017 Asian Conference on Energy, Power and Transportation Electrification (ACEPT), 2017, pp. 1–7.

[4] R. Mirzahosseini and F. Tahami, "A Lifetime Improved Single Phase Grid Connected Photovoltaic Inverter", 2012 3rd Power Electronics and Drive Systems Technology (PEDSTC), 2012, pp. 234–238.

[5] M. A. Moonem and H. Krishnaswami, "Analysis and Control of Multi-Level Dual Active Bridge DC-DC Converter", 2012 IEEE Energy Conversion Congress and Exposition (ECCE), 2012, pp. 1556–1561.

[6] H. Zhou and A. M. Khambadkone, "Hybrid Modulation for Dual-Active-Bridge Bidirectional Converter with Extended Power Range for Ultracapacitor Application", IEEE Transactions on Industry Applications, 2009, 45, (4), pp. 1434–1442.

[7] F. Krismer, J. Biela and J. W. Kolar, "A Comparative Evaluation of Isolated Bidirectional DC/DC Converters with Wide Input and Output Voltage Range", 40th IAS Annual Meeting Conference Record of the 2005 Industry Applications Conference, 2005, pp. 599–606 Vol. 1.

[8] T. Ngo, J. Won and K. Nam, "A Single-Phase Bidirectional Dual Active Half-Bridge Converter", 2012 27th Annual IEEE Applied Power Electronics Conference and Exposition (APEC), 2012, pp. 1127–1133.

[9] M. Steiner and H. Reinold, "Medium Frequency Topology in Railway Applications", 2007 European Conference on Power Electronics and Applications, 2007, pp. 1–10.

[10] F. Krismer and J. W. Kolar, "Accurate Power Loss Model Derivation of a High-Current DAB for an Automotive Application", IEEE Transactions on Industrial Electronics, 2010, 57, (3), pp. 881–891.

[11] S. Inoue and H. Akagi, "A Bidirectional Isolated DC/DC Converter as a Core Circuit of the Next-Generation Medium-Voltage Power Conversion System", 37th IEEE Power Electronics Specialists Conf., 2006, pp. 1–7.

[12] R. T. Naayagi, A. J. Forsyth and R. Shuttleworth, "Performance Analysis of Extended Phase-Shift Control of DAB DC-DC Converter for Aerospace Energy Storage System", 2015 IEEE 11th International Conference on Power Electronics and Drive Systems, 2015, pp. 514–517.

[13] P. Liu, C. Chen, S. Duan and W. Zhu, "Dual Phase Shifted Modulation Strategy for the Three Level DAB DC-DC Converter", IEEE Transactions on Industrial Electronics, 2017, 64, (10), pp. 7819–7830.

[14] F. Krismer and J. W. Kolar, "Closed Form Solution for Minimum Conduction Loss Modulation of DAB Converters", IEEE Transactions on Power Electronics, 2012, 27, (1), pp. 174–188.

[15] A. L. Kirsten, F. G. Carloto, T. H. de Oliveira, J. G. P. Roncalio and M. A. Dalla Costa, "Phase-Shift Design Methodology for the DAB Converter", Power Electronics Magazine, 2014, 19, pp. 231–240.

[16] P. Meyer and Y. Perriard, "Skin and Proximity Effects for Coreless Transformers", 2011 International Conference on Electrical Machines and Systems, 2011, pp. 1–5.

[17] A. Roßkopf, E. Bär and C. Joffe, "Influence of Inner Skin and Proximity Effects on Conduction in Litz Wires', IEEE Transactions on Power Electronics, 2014, 29, pp. 5454–5461.

[18] W. T. McLyman, "Transformer and Inductor Handbook". 3rd ed. New York: Marcel Dekker, 1976.

[19] M. K. Kazimierczuk, "High-Frequency Magnetic Components". 2nd ed. Wiley: ISBN 978-1-118-71779-0, 2014.

978-1-7281-4181-7/19 $31.00 © 2019 IEEE

Developement of a Multilevel DVR with Battery Control and Harmonic Compensation

Bruno P.B. Guimarães
Itajubá Federal University
Itajuba, Brazil
brunopbp4@gmail.com

Wilson Cesar Santana
Gnarus Institute
Itajuba, Brazil
wilson_santana@ieee.org

Paulo F. Ribeiro
Itajubá Federal University
Itajuba, Brazil
pfribeiro@ieee.org

Robson Bauwelz Gonzatti
Itajubá Federal University
Itajuba, Brazil
bauwelz@gmail.com

Guilherme G. Pinheiro
Itajubá Federal University
Itajuba, Brazil
guilhermegpinheiro@gmail.com

Rondineli R. Pereira
Itajubá Federal University
Itajuba, Brazil
rondineli@unifei.edu.br

Fernando Nunes Belchior
Goias Federal University
Goiania, Brazil
fnbelchior@hotmail.com

Carlos Henrique da Silva
Ouro Preto Federal University
Joao Molevade, Brazil
carloschedas@gmail.com

Luiz Eduardo Borges da Silva
Itajubá Federal University
Itajuba, Brazil
leborgess@gmail.com

Abstract—**This article presents a dynamic voltage restorer (DVR) composed by a multi-level H-Bridge cascading topology, also operating as a series active filter and as a battery charger. The integration of these functions is discussed in detail, as well as the concepts involving the topology used, modulation technique and voltage sag compensation techniques. The voltage sag compensation and harmonic voltage controls were based on PR controllers while the battery banks charging control was based on an incremental technique. The analysis of the response of the DVR controls was done through the simulation of a single-phase system using Simulink software.**

Keywords—*DVR, Multilevel Converter, CHB, Active Filter, In-Phase Compensation*

I. INTRODUCTION

The growing energy demand in recent years has been accompanied by concern about power quality. This interest is due to the considerable increase of electronic devices in industrial processes, since they have greater sensitivity to power quality problems [1] - [3]. Among the many power quality problems, the most common are related to voltage disturbances (voltage sag, swell, flicker and harmonics), and it is agreed that voltage sags are the costliest to these electronic devices [4]. This disturbance is defined in standard IEEE 1159 [5] as the reduction of 0.1 to 0.9 pu of the RMS value of nominal voltage with duration of 0.5 cycles to 1 minute. Some factors responsible for the occurrence of these phenomena are: faults in the system, switching of heavy loads and large machine starts. Numerous power electronics-based devices are being developed and studied to improve network performance, as well as mitigate or reduce the damage caused by power quality problems. Among these devices, the DVR (Dynamic Voltage Restorer) has presented more economical and technically advanced resources for the compensation of voltage sags in distribution systems [1], [6], [7]. Although they are not as popular as the UPS devices because their development stage, the trend is that they become more popular due to the development of the power electronics industry and the cheapness of their devices [8]. The DVR consists of a VSI (Voltage Source Inverter) converter connected in series with the source upstream, so that upon detecting a voltage sag in the system, it injects a series voltage with a certain amplitude and a certain phase angle so that the load connected downstream do not notice the voltage variation. The basic scheme of a DVR is presented in Fig .1. Besides the power electronics converter, a LC filter is employed to suppress the switching noise and a coupling transformer is used to ensure galvanic insulation and to raise the voltage at the system level.

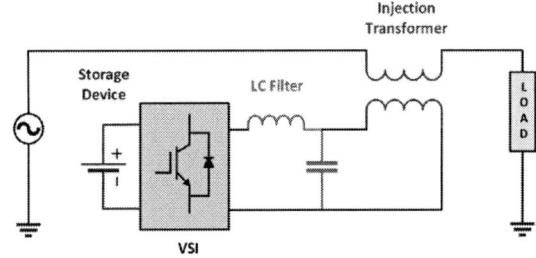

Fig. 1. Basic scheme of DVR configuration.

Considering that the DVR's utilization factor is low, since it operates only in moments of voltage sag, it is possible to increase its functionality with secondary functions during its standby period. Some of these functions are: voltage harmonic filtering, battery bank charging and inductance compensation of the line where it is inserted. Considering these possibilities of functionality, this work aims to simulate the active series filter function and the battery charge function, on a DVR. The simulation will consider a single-phase multilevel cascaded-H bridge DVR using the In-Phase Compensation technique [9], [10].

II. DVR COMPENSATION TECHNIQUES

There are basically three strategies of voltage compensation [9]. The first one is the In-Phase Compensation technique, which consists of injecting a voltage in phase with the source's voltage, not considering the pre-sag conditions. This technique is not able to compensate phase-jumps, since it acts only on the magnitude of the voltage noticed by the load. However, it does not require higher voltage rating of DC link as in the pre-sag compensation technique explained below, since the injected voltage is minimum [10], [11].

The second one consists of the pre-sag compensation technique, which injects the pre-sag and during-sag voltage difference, thus compensating both, module and voltage angle. This technique is able to handle phase-jumps, making

978-1-7281-4181-7/19 $31.00 © 2019 IEEE

it attractive for operation with loads sensitive to phase variations such as those controlled by thyristors. However, due to angular compensation, the equipment must be designed to handle higher voltage levels compared to In-Phase compensation [10], [12]. This implies the need for higher rated energy storage device and voltage injection transformer. Finally, the third technique is Energy Saving Compensation, which consists of avoiding the active power exchange between the converter and the system during compensation [11], [13] - [15]. The principle of operation is based on injecting and absorbing maximum reactive power to mitigate sag voltage, thus decreasing the amount of active power required in the DC link. Despite the advantage of not requiring active power, this technique has two disadvantages. One is due to the fact that, even if there is no phase-jump during the voltage sag, by avoiding the active power exchange, the DVR can cause a phase jump being harmful to sensitive loads. The other is due to the fact that the voltage imposed by the equipment can be higher than the last two techniques [10]. The fasorial diagrams of each sag compensation technique are shown in Fig.2. The phasor *Vpre* is the grid pre-sag voltage, in which is noticed by the load before de occurrence of sag. During the voltage sag, *Vpre* changes into *Vsag*, thus the DVR injects the voltage *Vinv* to compensate the dip. Lastly, during the voltage sag, the load notices the voltage *Vload*, that is the composition of *Vsag* and *Vinj*.

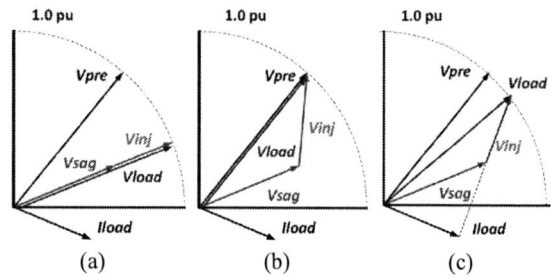

Fig.2. Phasor representation of sag compensation techniques [9][12]. (a) In-Phase Compensation b) Pre-Sag Compensation c) Energy Saving Compensation

III. MULTILEVEL TOPOLOGY

For the implementation of the DVR, the multi-level topology CHB (cascaded-H bridge) was chosen. This topology consists in n Bridge modules connected in series, reaching an output voltage of *2n + 1* levels. In this work a CHB with 3 modules was used, as shown in Fig. 3, totalizing in an output voltage of 7 levels. Due to the higher number of levels, the converter can synthesize the output voltage with smaller steps, resulting in lower THD, *dv/dt* and consequently lower EMI [14], [16], [17]. With the reduction of THD it is possible to use smaller LC filters, reducing losses and system cost [16]. The series configuration of the modules implies a lower stress on the power switches, increasing the system's capacity [18].

Another advantage of this topology is that due to its modular feature, it is relatively easy to expand to higher levels. For this purpose, is necessary just add new modules in series and maintaining the structure of the topology. Some cons like the huge number of switching devices and isolated DC sources are presents in the CHB topology [20].

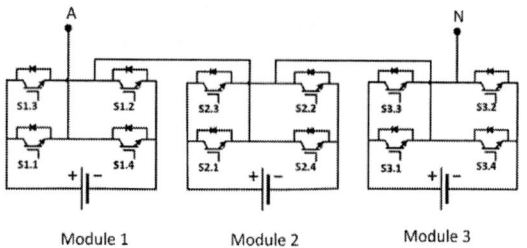

Fig.3. Topology CHB 7 levels

IV. PHASE-SHIFTED PWM

The adopted modulation technique is based on unipolar pwm modulation, so each module is controlled by the comparison between pairs of 180° displaced triangular carriers and one modulator. The number of carrier pairs to be used must be equal to the number of modules and the carrier displacement (ϕ_{cr}) is defined (1) [19], where m is the number of levels reached by the converter.

$$\phi_{cr} = \frac{360°}{(m-1)} \tag{1}$$

Fig.4 exemplifies the operation of the Phase-Shifted modulation considering the seven-level CHB converter shown in Fig.3 In this, the modulating signal and triangular carriers are shown, as well as the firing signals (GS1.1, GS1.3, GS2.1, GS2.3, GS3.1 and GS3.3) of the upper IGBTs of each module, the (VH1, VH2 and VH3) and the output voltage of the inverter (Vout). The firing signals of the lower IGBTs (GS1.2, GS1.4, GS2.2, GS2.4, GS3.2 and GS3.4) have not been shown since they operate inversely to their adjacent pairs. Since the Vout signal is composed by the signals VH1, VH2 and VH3, it is observed that an increase in the resulting switching frequency occurs at the output of the converter due to the switching pattern and the number of modules used. This feature allows the multilevel converter to operate at higher frequencies, but with less switching losses. In addition, this characteristic causes the displacement of spectrum harmonic to higher orders, being possible the reduction of the filter responsible for the switching filtering [20].

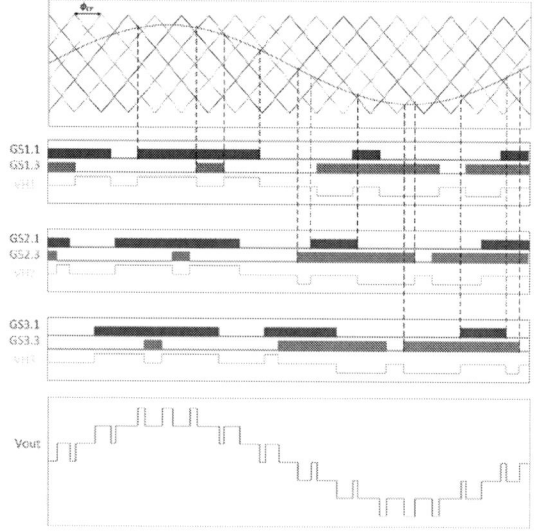

Fig.4. Phase-Shifted for seven-level CHB inverter [20]

V. VOLTAGE SAG AND VOLTAGE HARMONIC COMPENSATION CONTROL STRATEGY

The compensation of voltage sag and mitigation of voltage harmonics is done by the proportional resonant controller (PR). This controller is obtained by a mathematical transformation applied on a synchronous PI controller [21] - [23], transforming the DC controller into an AC controller. The transformation is performed by the expression (2) presented in [23].

$$C_{AC}(S) = \frac{C_{DC}(S+j\omega_0) + C_{DC}(S-j\omega_0)}{2} \qquad (2)$$

Applying (2) to the PI transfer function given in (3)

$$C_{DC}(S) = k_p + \frac{k_i}{S} \qquad (3)$$

the transfer function of the P + Resonant controller presented in (4) is obtained.

$$C_{AC}(S) = k_p + \frac{k_r S}{S^2 + \omega_0^2} \qquad (4)$$

In which k_p is the proportional gain, k_r the resonant gain and ω_0 is the resonant frequency. Due to the high gain in resonant frequency ω_0, the controller achieves zero steady-state error for a sinusoidal reference. Due to the ability to track a sinusoidal reference only through the signal error, it is not necessary to use PLL algorithms, thus reducing computational costs [21].

In order to compensate voltage sag, the controller was tuned to act on the fundamental frequency, while for the harmonic mitigation the tuning was done for harmonics of order 3, 5 and 7. In this way the transfer function (4) is now represented by (5).

$$C_{AC}(S) = k_p + \sum_{h=1,3,5,7} \frac{k_r S}{S^2 + (h.\omega_0)^2} \qquad (5)$$

The set-point of voltage sag compensation will be a sine wave with the phase angle (θ) of the voltage measured at the load and with amplitude equal to the peak of the nominal voltage of the system (V_p). In harmonic compensation, as no harmonic is desired in the load voltage, the set-point will be equal to 0. The control feedback is the voltage measured at the load (V_{load}). Fig. 5 shows the block diagram of the sag compensation and voltage harmonics mitigation control.

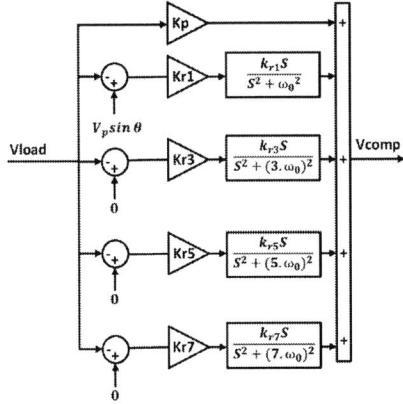

Fig.5. Voltage sag and harmonic mitigation control

VI. BATTERY CHARGING CONTROL STRATEGY

The lead acid battery presents 3 stages on the battery charge process as shown in Fig.6 [24], [25].

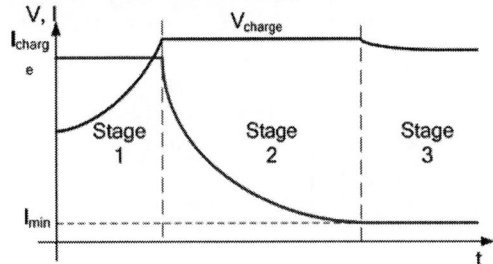

Fig.6. Voltage-current battery charging characteristic [24]

The first stage, known as the bulk phase, maintains a constant charge current until the battery reaches a certain voltage limit. The second stage is known as the absorption phase, where the battery voltage is kept constant while the charging current drops to a minimum value. Finally, the third stage known as floating phase, the battery is kept fully charged and feeded by a small current that maintains the voltage level [25]. In order to deal with the nonlinearity of each stage, an incremental battery charging strategy is adopted. This technique is similar to the one presented in [26] and has a good response in all battery charging stages.

To perform the battery charging, the converter must drain active power from the system, imposing a voltage portion in phase with the grid current. The amplitude of this voltage is defined according to the fluctuation voltage of the battery (V_{float}) and the reference current (I_{ref}) adopted for charging. These two quantities act as operational limits and can not be exceeded. In this way, if the current in the link DC (I_{dc}) is greater than (I_{ref}) and the voltage of the batteries (V_{dc}) is greater than (V_{float}) then the injected voltage (V_{ch}) is increased and drains more active power, otherwise the decrement is done. The variation of injected voltage (Δ_{Vch}) is done by the increment factor of index modulation (Δ_{ma}). The Fig.7 presents battery charging strategy flowchart.

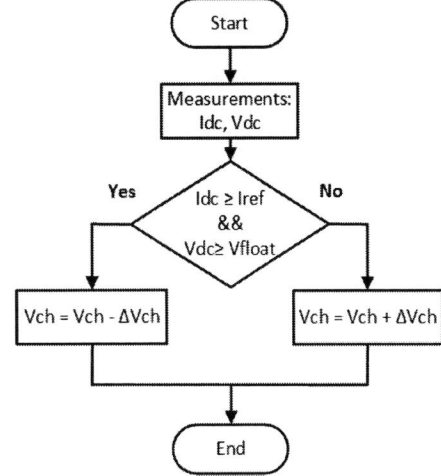

Fig.7. battery charging strategy flowchart

The complete control loop of the DVR is shown in Fig.8. In this, the control regarding voltage sag compensation and harmonic mitigation, as well as battery control are integrated. The output of the battery changing control ($V_{ch} \sin \theta$) is subtracted from the reference signal ($V_p \sin \alpha$) so that the PR controller does not compensate it. After this, it is added to the output signal of the PR controller (V_{comp}), where the resulting signal is normalized by gain k.

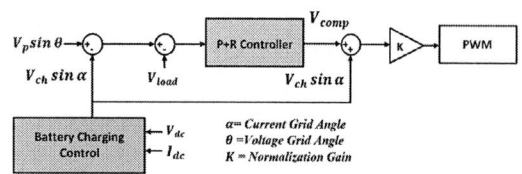

Fig.8. Complete Control Loop

VII. SIMULATED SYSTEM AND PARAMETERS

Simulation was performed using simulink software and controls were implemented through the S-builder block. In it was considered a low-power single-phase system, since this work will be used as the basis for future implementation in a experimental prototype. The system in which the DVR is inserted consists of an RL load powered by a controlled voltage source, which is responsible for emulating the voltage sag and to generate the harmonics of interest. The DVR is supplied by a Lead-Acid battery banks and is connected in series with the load by a coupling transformer, responsible for galvanic isolation and to raise the voltage to the system level. In addition, the equipment has an LC filter, to perform switching filtering. The parameters of the system components, DVR and control gains are presented in Table I, while Fig.9 shows the simulation screen.

TABLE I. SIMULATION PARAMETERS

System Parameters	
Nominal system voltage (RMS)	220 V
Load resistance	20 Ω
Load indutance	80 mH
DVR Parameters	
Battery Voltage	36 V
DC Link Capacitance	470µF
Switching Frequency	10 kHz
Transformer Ratio	1:4
Battery bank capacity	6.5 Ah
Proportional Resonant Controller Gains	
k_p	0.1
k_{r1}	850
k_{r3}	500
k_5	500
k_7	500
Battery Charge Control	
Δ_{ma}	0.0005
I_{ref}	600 mA
V_{float}	39 V

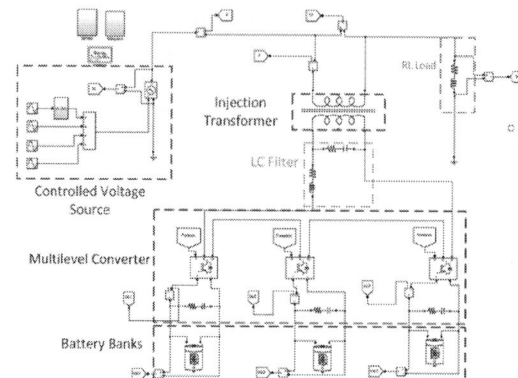

Fig.9. Simulated System

VIII. RESULTS AND DISCUSSION

Fig.10 presents an analysis of the DVR compensating voltage sag and harmonics. Harmonic voltage sources of order 3, 5 and 7 have been added in the proportion of 5%, 4% and 3.5% of the fundamental voltage, respectively. The inverter starts to operate at the instant 0.05s - initially compensating only the harmonic voltages. From the instant 0.15s until 0.35s a 80% voltage sag is applied to the grid. The top plot of Fig. 10 presents the grid voltage, the middle plot presents the load voltage and the bottom plot presents the voltage injected by the converter. It can be noticed that, during the whole simulation, the DVR maintains the load voltage free from the grid harmonics. Also, during the voltage sag, the DVR maintains the load voltage at near 100% of its nominal value. In addition, it can be noticed that the PR control presents a fast response, of less than half cycle of fundamental voltage. However, at the end of the sag, an overshoot of less than a quarter of cycle can be noticed.

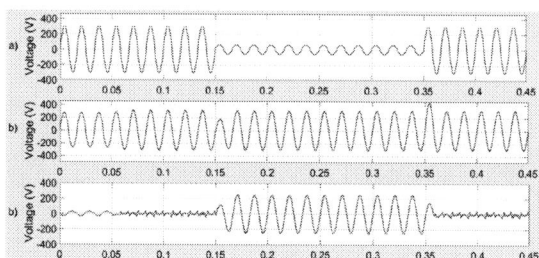

Fig.10. Voltage sag and Harmonic compensation. a) Voltage Grid b) Load Voltage c) Voltage Injected

Fig. 11 presents the harmonic spectrum analysis of both grid voltage (top plot) and load voltage (bottom plot). It can be noticed the THD at the load has been reduced from 7.30% to 2.46% and the harmonic components at the 3rd, 5th and 7th have become negligible.

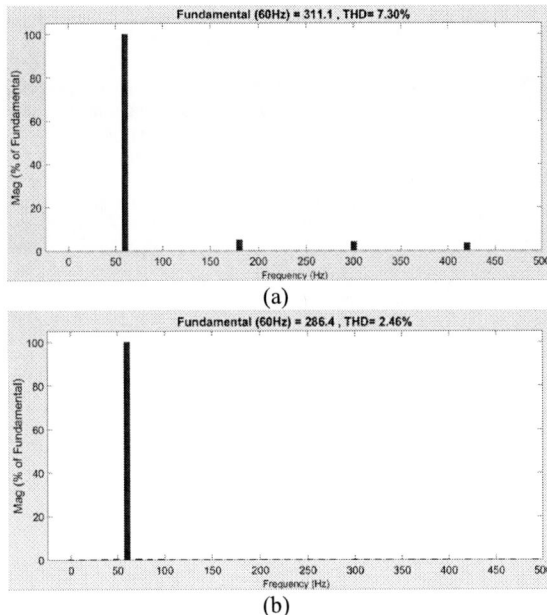

(a)

(b)

Fig.11. Spectrum Harmonic Analysis. a) voltage grid spectrum harmonic b) load voltage spectrum harmonic analysis

The same operating condition was adopted in the analysis of battery bank charging control, that is, the same harmonic ratio in the system voltage and the same critical sag of 80%. However, in this analysis, the simulation period is longer due to the battery banks slower response. In this test the voltage sag occurs in 0.6s and lasts until the moment 1.0s. Battery banks with a 99% state of charge were considered, since they will always remain approximately fully charged, being that the voltage sags have a short duration and the batteries will not be discharged considerably. Moreover, compensation of voltage harmonics also does not require considerable power from the banks. The Fig.12 shows the system voltage, the load voltage and the active power at the output of the converter. As can be analyzed, as long as voltage sag does not occur, active power is drained from the system in order to charge the battery banks. On the other hand, when the voltage sag occurs, active power is dispatched by the DVR.

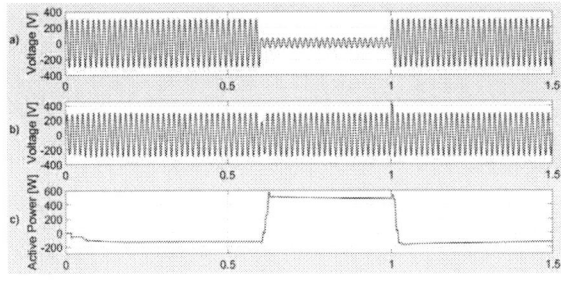

Fig.12. Analysis of power exchanged between converter and system. a) Voltage Grid b) Load Voltage c) active power at the output of the converter

Considering the negative current as being absorbed by the batteries and the positive one being supplied by them , Fig.13 depicts the fluctuations by showing the current and DC voltage behaviors. As occurs with active power, at times where there is no voltage sag, a DC current is injected into the battery banks at the maximum rate of 600mA. However, during the period of voltage sag, a current is drained from the

battery banks. The effectiveness of the operation of the battery charging control, is perceived when considering the battery banks voltage. It is observed that when the sag occurs the voltage on the banks decays. As soon as the sag is extinct, the voltage in the batteries starts to increase again.

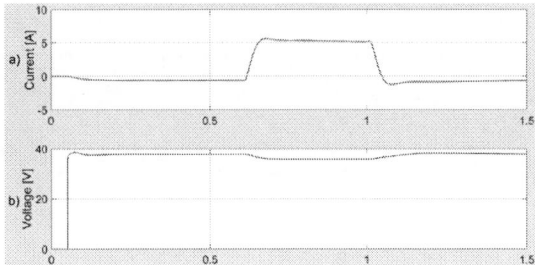

Fig.13. Average current and voltage in battery banks per module. a) Average DC current b) Average DC Voltage

IX. CONCLUSION

This paper presented the simulation of a multilevel DVR, also operating as an active series filter and charging the battery banks that supply it. The topology adopted in the converter consisted of the Cascaded H-bridge (CHB) with three modules, resulting in a 7-level configuration. The technique used to compensate voltage sags was In-Phase compensation. The voltage sag compensation and voltage harmonic mitigation controls were based on the Proportional Resonant Controller and were integrated with the battery bank charge control, which is based on an incremental technique.

The simulation was carried out using Simulink software, with the objective of analyzing the responses of the voltage sag compensation and voltage harmonic mitigation controls, as well as the battery banks charging control. For this, simulations were performed applying a voltage sag of 80% in a System containing voltage harmonics of order 3, 5 and 7. From these conditions, the response of the controls were evaluated.

It was observed a good response of the controls based on PR controller for harmonic mitigation and voltages sags compensation, since these reached a zero error in stead state and presented fast transient response. The battery banks control presented a good performance, since they respected the defined operation limits of voltage and current, effecting the control of battery banks voltage on efficient and stable way.

REFERENCES

[1] D. Vilathgamuwa, H. M. Wijekoon, and S. S. Choi, "A Novel Technique to Compensate Voltage Sags in Multiline Distribution System—The Interline Dynamic Voltage Restorer", IEEE Transactions on Industrial Electronics, vol. 53, no 5, p. 1603–1611, out. 2006.

[2] H. Awad, J. Svensson, and M. Bollen, "Mitigation of unbalanced voltage dips using static series compensator", in IECON'03. 29th Annual Conference of the IEEE Industrial Electronics Society (IEEE Cat. No.03CH37468), 2003, vol. 3, p. 2672-2677 Vol.3.

[3] F. B. Ajaei, S. Afsharnia, A. Kahrobaeian, and S. Farhangi, "A Fast and Effective Control Scheme for the Dynamic Voltage Restorer", IEEE Transactions on Power Delivery, vol. 26, no 4, p. 2398–2406, out. 2011.

[4] C. J. Melhorn, T. D. Davis, and G. E. Beam, "Voltage sags: their impact on the utility and industrial customers", IEEE Transactions on Industry Applications, vol. 34, no 3, p. 549–558, maio 1998.

[5] "IEEE Recommended Practice for Monitoring Electric Power Quality", IEEE Std 1159-2009 (Revision of IEEE Std 1159-1995), p. c1-81, jun. 2009.

[6] P. Kanjiya, B. Singh, A. Chandra, and K. Al-Haddad, "'SRF Theory Revisited' to Control Self-Supported Dynamic Voltage Restorer (DVR) for Unbalanced and Nonlinear Loads", IEEE Transactions on Industry Applications, vol. 49, no 5, p. 2330–2340, set. 2013.

[7] Y. W. Li, D. M. Vilathgamuwa, F. Blaabjerg, and P. C. Loh, "A Robust Control Scheme for Medium-Voltage-Level DVR Implementation", IEEE Transactions on Industrial Electronics, vol. 54, no 4, p. 2249–2261, ago. 2007.

[8] M.V.Kasuni Perera, "Control of a Dynamic Voltage Restorer to compensate single phase voltage sags", KTH Electrical Engineering, 2007.

[9] S. S. Choi, J. D. Li, and D. M. Vilathgamuwa, "A generalized voltage compensation strategy for mitigating the impacts of voltage sags/swells", IEEE Transactions on Power Delivery, vol. 20, no 3, p. 2289–2297, jul. 2005.

[10] A. K. Sadigh and K. M. Smedley, "Review of voltage compensation methods in dynamic voltage restorer (DVR)", in 2012 IEEE Power and Energy Society General Meeting, 2012, p. 1–8.

[11] C. N. M. Ho and H. S. H. Chung, "Fast Dynamic Control Scheme for Capacitor-Supported Dynamic Voltage Restorers: Design Issues, Implementation and Analysis", in 2007 IEEE Power Electronics Specialists Conference, 2007, p. 3066–3072.

[12] A. M. Rauf and V. Khadkikar, "An Enhanced Voltage Sag Compensation Scheme for Dynamic Voltage Restorer", IEEE Transactions on Industrial Electronics, vol. 62, no 5, p. 2683–2692, maio 2015.

[13] J. G. Nielsen, F. Blaabjerg, and N. Mohan, "Control strategies for dynamic voltage restorer compensating voltage sags with phase jump", in APEC 2001. Sixteenth Annual IEEE Applied Power Electronics Conference and Exposition (Cat. No.01CH37181), 2001, vol. 2, p. 1267–1273 vol.2.

[14] h Al-Hadidi, A. Gole, and D. Jacobson, "A novel configuration for a Cascade Inverter Based Dynamic Voltage restorer with reduced energy storage requirements", in 2008 IEEE Power and Energy Society General Meeting - Conversion and Delivery of Electrical Energy in the 21st Century, 2008, p. 1–1.

[15] H. Al-Hadidi, A. Gole, and D. Jacobson, "Minimum power operation of cascade inverter based dynamic voltage restorer", in 2008 IEEE Power and Energy Society General Meeting - Conversion and Delivery of Electrical Energy in the 21st Century, 2008, p. 1–1.

[16] S. Galeshi and H. Iman-Eini, "Dynamic voltage restorer employing multilevel cascaded H-bridge inverter", IET Power Electronics, vol. 9, no 11, p. 2196–2204, 2016.

[17] B. Wang and M. Illindala, "Operation and control of a dynamic voltage restorer using transformer coupled H-bridge converters", IEEE Transactions on Power Electronics, vol. 21, no 4, p. 1053–1061, jul. 2006.

[18] S. Wang, G. Tang, K. Yu, and J. Zheng, "Modeling and Control of a Novel Transformer-less Dynamic Voltage Restorer Based on H-Bridge Cascaded Multilevel Inverter", in 2006 International Conference on Power System Technology, 2006, p. 1–9.

[19] D. Sreenivasarao, P. Agarwal, and B. Das, "Performance evaluation of carrier rotation strategy in level-shifted pulse-width modulation technique", IET Power Electronics, vol. 7, no 3, p. 667–680, mar. 2014.

[20] B. Wu, "Cascaded H-Bridge Multilevel Inverters", in High-Power Converters and ac Drives, John Wiley & Sons, Inc., 2006, p. 119–142.

[21] R. B. Gonzatti, S. C. Ferreira, C. H. da Silva, L. E. B. da Silva, G. Lambert-Torres, and L. G. F. Silva, "PLL-less control for hybrid active impedance", in 2013 Twenty-Eighth Annual IEEE Applied Power Electronics Conference and Exposition (APEC), 2013, p. 2178–2185.

[22] Y. Sato, T. Ishizuka, K. Nezu, and T. Kataoka, "A new control strategy for voltage-type PWM rectifiers to realize zero steady-state control error in input current", IEEE Transactions on Industry Applications, vol. 34, no 3, p. 480–486, maio 1998.

[23] D. N. Zmood, D. G. Holmes, and G. Bode, "Frequency domain analysis of three phase linear current regulators", in Conference Record of the 1999 IEEE Industry Applications Conference. Thirty-Forth IAS Annual Meeting (Cat. No.99CH36370), 1999, vol. 2, p. 818–825 vol.2.

[24] I. Serban and C. Marinescu, "A look at the role and main topologies of battery energy storage systems for integration in autonomous microgrids", in 2010 12th International Conference on Optimization of Electrical and Electronic Equipment, 2010, p. 1186–1191.

[25] R. B. Gonzatti, S. C. Ferreira, C. H. da Silva, R. R. Pereira, L. E. B. da Silva. G. Lambert-Torres and R. M. R. Pereira, "Implementation of a grid-forming converter based on modified synchronous reference frame", in IECON 2014 - 40th Annual Conference of the IEEE Industrial Electronics Society, 2014, p. 2116–2121.

[26] W. C. Sant'Ana, R. B. Gonzatti, G. Lambert-Torres, E. L. Bonaldi, B. S. Torres, P. A. de Oliveira, R. R. Pereira, L. E. Borges-da-Silva, D. Mollica and J. Santana Filho, "Development and 24 Hour Behavior Analysis of a Peak-Shaving Equipment with Battery Storage", Energies, vol. 12.

978-1-7281-4181-7/19 $31.00 © 2019 IEEE

Modeling of Electrochemical Batteries Behavior and Lifetime Degradation for PV Applications

Rafael César Nolasco[1]

[1]Graduate Program in Electrical Engineering
Universidade Federal de Minas Gerais
Av. Antônio Carlos 6627, 31270-901
Belo Horizonte, MG, Brazil
rafaelcesarn@gmail.com

Victor Flores Mendes[1,2]

[2]Department of Electrical EngineeringUniversidade Federal de
Minas Gerais
Av. Antônio Carlos 6627, 31270-901
Belo Horizonte, MG, Brazil
victormendes@cpdee.ufmg.br

Abstract—**The interest in studying electrochemical batteries in many applications is rising in the last years, due to increasing applications of storage systems in which batteries have a crucial role. For example, one can cite electric vehicles and in the future power systems, where storage can be used for load leveling, power variation damping and power quality improvement. Although batteries are a component known for a long time, modelling them is a very complex task. There is not a general model easily applicable in every situation. To address this issue, this paper presents the approaches commonly used in battery modelling and develops one model to represent the behavior of battery voltage and capacity during a discharge and charge cycle. Furthermore, another model is used to predict the lifetime and the degradation. These models are developed in order to simulate small scale battery systems for Demand Side Management applications with PV generation.**

Keywords— **Battery modelling, energy storage systems, lifetime, renewable energy sources**

I. INTRODUCTION

Electrochemical batteries are an electrical component known for a long time. Back in 1859, before the existence of the first electric power systems, the French physicist Gaston Planté created the first lead-acid batteries that are still present nowadays. Nevertheless, there is still big challenges for developments in this field, and recently it is drawing more attention due to the various applications in which batteries will play a crucial role in the future electrical grids: electric vehicles, use as power sources in power systems for load leveling, power variation damping or transmission, and power quality improvement. Another aspect related it is the concerns about air quality and greenhouse gas emissions [1], once that energy storage systems are important for the grid integration of intermittent sources of power.

Batteries are non-linear components and their behavior are dependent on many internal and external variables, such as the discharge current, ambient temperature, state of charge (SOC), nominal capacity and age. Some of these variables must be estimated by linking them to other ones most of the times. That makes the task to precisely model batteries very complex and subjected to many errors. Li and Ke [2] affirm that the conventional mathematical models work only for specific applications and provide results with errors in the order of 5 to 20%. However, it is important to develop and validate models in order to make simulations and predictions in order to avoid expensive and prolonged laboratory tests.

Therefore, there is a wide variety of battery models suited for different purposes, as battery control systems, economic analysis, lifetime prediction and to simulate the dynamics. They also differentiate in their parameterization, complexity and limitations. For example, there are models developed for specific uses, like estimation of lifetime in electric vehicles [8]

and small-scale photovoltaic systems [7], or for predictive control of battery behavior [17]. In each application, some assumptions related to the battery usage are made to simplify the task of modelling the battery behavior. The models can also vary on their approach, like circuit-oriented models [3-4;19] and mathematical oriented models [5-6].

In this work, it is developed a battery model to integrate a system composed by a small-scale photovoltaic generator and battery connected in a consumer installation integrated to the electric grid. These systems are classified as a Demand Side Management (DSM) strategy, in which the consumer can gain an economic return increasing self-consumption of his own generation and certain independence from the electric grid for his own energy supply. Several benefits of these systems are being pointed by some studies [18-21], like grid relieving, enable more penetration of intermittent renewable sources, improve power quality and reduce power outages. To widespread these systems, there are challenges to overcome like the lack of metering, information and communication infrastructure, low competitiveness compared to other solutions and inappropriate market incentives [19]. Even so, these investments are becoming profitable with the fallen costs of storage, even without strong policy support [21].

In order to achieve the proposed model, an implementation in MATLAB is developed after a proper choice. This paper shows in Section II a summary of different battery models, which are divided in two categories: in first place, the models to represent the behavior of important variables during a battery cycle and then models to represent the degradation over time. In Section III these two models are developed. In Section IV the results are presented and discussed. Section V presents the conclusions.

II. BATTERY MODELLING

There are several different models to represent electrochemical batteries that differ from their conception to their application [3-10]. The existence of all these different models are explained by the fact that these components are very complex, with many different variables that affect the behavior and predicted outcome. Usually, the most important variables are the charge and discharge currents, battery capacity, state of charge, depth of discharges, temperature and age.

Therefore, it is important to know the model limitations to properly apply them. Different models may give very unequal results for the same application, as shown in the paper of Dufo-López [10], in which some lifetime prediction models are compared. While the Schiffer model [9] results in a lifetime of 5.8 years, other models predict from 14.9 to 19.8 years for the same application.

The battery modelling can be divided in two categories: modelling of battery behavior during a charge and discharge cycle, and the modelling of its lifetime. The lifetime of a battery is strongly dependent on its cycle conditions, but it can be evaluated separately because each cycle represents just a small fraction in the degradation processes.

Given the application, a proper model for the cycle behavior is important to understand the battery limitations and the integration with power converters and other components. The lifetime model will be used for better system sizing and component choice, regarding the system performance and economic return.

A. Modelling battery cycle behavior

An electrochemical battery is a device that store the electrical power into chemical energy and release the energy by chemical reaction when it is needed [2]. It is a non-linear power source because the voltage between the poles and the charge available depend on other factors like SOC, depth of discharge (DOD), temperature and the output current. The SOC and voltage are the most important variables to monitor during a discharge cycle. To simulate the dynamics during a cycle, the models can be divided in 3 classes: electrochemical, mathematical and circuit-oriented models.

The electrochemical models represent in a detailed way the components used to build the battery and its internal chemical reactions. In addition to basic parameters often present in other models, like temperature and discharge current, internal variables are also represented, like.g, acid concentration and active material resistance [9]. In general, most rechargeable battery systems can use the same types of electrochemical engineering models to represent the dynamics, but proper system parameters to account for different chemistries are necessary. However, for certain systems, like redox-flow batteries, which uses a pumped exterior electrolyte, the dynamics are very different, requiring a different approach for proper electrochemical modelling [12].

Although the electrochemical modeling is very important to understand the behavior of batteries at their core, they are usually very complex to implement and their validation depends on many laboratory tests, which are usually expensive and impractical. In order to cope with these complexities, simplified mathematical models were developed, which represent the fundamental relationships between the concerned variables based on experimental data. The mathematical battery models are developed based primarily on the Shepherd relation [6], which was improved in subsequent works by adding or modifying its terms to relax the assumptions behind the original proposed model [2]. The Shepherd relation is a single equation to find the voltage at the battery terminals in function of its capacity, extracted charge during the cycle and the output current.

Another notorious equation in battery modelling is the Peukert Equation [5], determined in 1987 to represent a battery cell capacity in function the discharge current. Using the Shepherd and Peukert equations it is possible to estimate with few parameters and variables the voltage dynamics and SOC during a discharge or charge cycle.

Another approach for battery modelling is to represent it by the means of electrical equivalent circuits using a combination of power sources, resistors and capacitors [2].

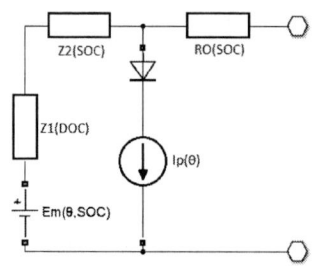

Fig. 1. Circuit model proposed by Ceraolo [3-4]

This approach helps the electrical engineers to use their know-how to analyze the battery behavior, simulate different conditions, connection with other electrical circuits and systems.I is possible to use and compute useful parameters such as short-circuit currents and power outputs. The basic electric circuit to represent a battery is composed by a voltage source and a resistor in series. Although it captures some important features of batteries, it misses most of batteries important characteristics. Ceraolo [3-4] proposes the model shown in Fig. 1 that has a current source in parallel with the main voltage source to represent parasitic currents inside the battery. There are also more impedances composed by parallel RC circuits to represent the dynamic behavior of battery when exposed to changes in its output current.

Usually, mathematical and circuit-oriented models are based on empirical data and are less detailed to provide better computational performance. However, these models use parameters that lack physical meaning. While they are easier to develop and use, they are not accurate out of certain conditions. If the response to different conditions, like ambient temperature and ageing effects are not included, it can make the model imprecise [21].

B. Modelling battery lifetime

Lifetime modelling consists in representing the ageing mechanisms of batteries that modify their behavior. Generally, the major effect observed over the time is the capacity reduction. The end of life (EOL) of a battery is usually defined when it reaches 80% of its rated capacity [9].

Each battery has its own ageing mechanisms. For example, in a valve regulated lead-acid (VRLA) battery the effects that typically determine its longevity are the positive grid corrosion, positive plate material breakdown, sulfation of negative electrode, water loss and electrolyte stratification. The corrosion effects are usually associated with standby and float charge applications while cycling tends to affect the positive material breakdown and negative electrode sulfation [13]. Therefore, it is important to know the battery usage to properly model the important ageing mechanisms

Similar to the battery behavior, the lifetime prediction models can be separated in 3 different approaches: physico-chemical, weighted Ah and event-oriented models.

In physico-chemical models, the chemical and physical processes that occur within the battery are modeled in a detailed way. They provide detailed information on local conditions such as temperature, potential, current, SOC and electrolyte concentration, which are result of operation conditions [10]. They are usually more detailed and complex, but simplifications can be made taken the application into account. For example, in the Schiffer work [9], which aims to estimate the lifetime of batteries in autonomous and renewable energy systems, the grid corrosion and the degradation of active material are considered, while other effects are simplified or even dismissed.

The weighted Ah models link the EOL to some parameters that can be determined such as Ah throughput, number of cycles or time elapsed since manufacturing [11]. These models assume that, under standard conditions, a battery can achieve a determined Ah throughput until it reaches the EOL. Deviations from these standard conditions, like the operation under different temperatures or currents, results in a decrease or increase in the battery lifetime [10]. These models are based on theoretical or experimental relations observed between these variables and the lifetime.

Event-oriented models are based on an approach very common in other areas of engineering, like the Wöhler or S-N curve used to study the fatigue of a material. These models add incremental losses in the studied component after the occurrence of some clearly defined events [10]. For battery lifetime estimation, each discharge or charge cycle is an event that causes a calculated damage to the battery, reducing its capacity for the next cycle.

III. MODEL IMPLEMENTATION

As stated in Section I, in this paper it will be developed a model for a small-scale battery-photovoltaic system to work as a demand side management system, allowing the consumer to store his generation during determined periods and use it as a supply for economic purposes. The developed model is based in the following premises:

- *Parameterization based on manufacturer data:* experimental tests to determine the model parameters are not always available, so it needs to rely on commonly available data;

- *Applicable to different battery chemistries:* in order to compare the performance of different technologies, it must be adaptable to different types. Shepherd [6] shows that most batteries have a similar behavior that can be represented by the same equation with different parameters;

- *Operation in regular cycles:* for a DMS application, the cycles usually happen in a very predictable way, since it usually depends on electricity rates that are previously known or have very low variance within time. This simplifies the task of evaluating the characteristics of the cycles and assumptions on the most important ageing mechanisms can be made.

For these reasons, the following development of the model is based on mathematical model based on the Shepherd relation to represent the behavior during a cycle, and an event-oriented model for the lifetime model, based primarily on the work of Narayan et al. [7]. In this model, each charge and discharge cycle are simulated separately and then the damage caused by that cycle is calculated and incorporated to the model.

In this work it is implemented a model of the battery Moura Clean 12MC220 based on its datasheet [14] and manual [15]. Its main characteristics are shown in Table 1.

A. Mathematical behavior modelling

In this paper the behavior of the battery is based on the simplified Shepherd relation shown in (1).

TABLE I. MOURA CLEAN 12MC220 MAIN CHARACTERISTICS

Technology	Valve regulated lead-acid (VRLA)	
	10h cycle	195Ah
Capacity at 25°C	20h cycle	220 Ah
	100h cycle	244 Ah
Rated voltage	12V	

$$E = E_S - K(Q/(Q - it))i - Ni + Ae^{(-Bit)} \qquad (1)$$

In (1) E is the voltage between the poles of the battery in volts, E_S is a constant voltage, K is the polarization coefficient in $\Omega.cm^2$, Q is the available active material in Ah/cm^2, i is the current density in A/cm^2, t is the elapsed time in the present charge or discharge cycle, N_d is the internal resistance in $\Omega.cm^2$ during a discharge, A and B are empirical constants to model an initial voltage drop in V and Ah^{-1}, respectively. The variables K, Q, i and N can also be expressed in Ω, Ah, A and Ω, respectively.

As stated before, the parameterization relies on manufacturer's data. To find Q, it is important to take into account that the charge that can be extracted from the battery depends on its current. The manufacturer gives a curve of the capacity in function of current. It was taken and modeled with two exponential functions as shown in (2).

$$Q(I_d) = p_1 e^{ip_2} + p_3 e^{ip_4} \qquad (2)$$

i is the output current and p_1, p_2, p_3 and p_4 are constants that makes the best fit of (3) on the manufacturer's provided data. Narayan [7] suggests a 4th order polynomial approximation, but the function presented in (2) presented a better fit.

N is found by dividing the difference in the initial voltage by the difference in the current in two different charge or discharge curves. To find the other parameters it is necessary to choose 3 points in one given curve: E_0 at the initial moment, E_{exp}, where the initial exponential voltage drop ends and E_{nom}, where the voltage starts to change abruptly. These points are shown in Fig. 2. Then the relations from (3) to (6) are applied in sequence.

$$A = E_{full} - E_{exp} \qquad (3)$$

$$B = 3/(i \times t_{exp}) \qquad (4)$$

$$K = (E_{full} - E_{exp} - A)(Q - i \times t_{nom})/(i \times t_{nom}) \qquad (5)$$

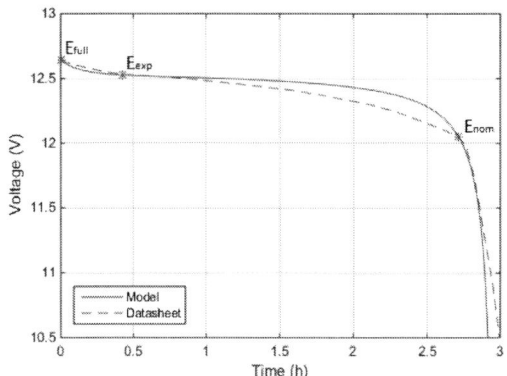

Fig. 2. Discharge curve of Moura Clean 12MC220 for a 5h cycle, and the points chosen for the parameterization

$$E_S = E_{full} + K + Ni - A \tag{6}$$

Equation (4) is obtained by subtracting E_{empty} from E_{nom} knowing the other parameters. This parameterization method is found in [22].

It is important to note that the choice of the curve for parameterization is very important. They must be representative of the discharge rates being studied. As stated in [2], when the extracted battery model is used under different conditions, large errors between prediction and measurements can occur.

In addition to these parameters, a maximum voltage (V_{max}) when the battery is fully charged was fed into the model in order to limit voltage in this condition. This information was taken directly from the datasheet.

B. Event-oriented lifetime modelling

The lifetime modelling is based on the work of Narayan [7] called Dynamic Capacity Fading Model. This model was simplified based on the assumption that the charge and discharge cycles are well known for the DSM application. In this application the rates and times when the battery is charged or discharged are well defined, with constant current and duration.

In the model, the degradation of the battery is represented by loss of capacity after the end of each cycle. Each cycle is defined by the time when the power output returns to or crosses the zero. After each cycle, it is evaluated DOD according to (7).

$$DOD_i = \int idt\ /Q_i \tag{7}$$

DOD_i is the depth of discharge in the ith cycle and Q_i is the capacity in the ith cycle calculate using (3) in function of i and accounting the degradation effects.

After each cycle, the damage to the battery on that cycle (D_i) is calculated based on manufacturer's curve of number of cycles in function of the DOD and temperature. The curve representing the effect of DOD is parameterized according to (8), similar to (2).

$$N(DOD) = p_5 e^{DOD p_6} + p_7 e^{DOD p_8} \tag{8}$$

N is the total number of cycles achieved by the battery and p_5 to p_8 are the constants that make the better fit of the curve.

To account for the temperature effects on lifetime, the curve of lifetime in function of temperature is fitted using a 4th order polynomial approximation, shown in (9), where p_9 to p_{12} are the constants that make the better fit of the curve.

$$N(T) = p_9 T^4 + p_{10} T^3 + p_{11} T^2 + p_{12} T + p_{13} \tag{9}$$

N(T) is parameterized considering that the lifetime achieved is equal to 100% when T is equal to the nominal battery temperature. Therefore, the total expected number of cycles N can be calculated as N(DOD) multiplied by N(T).

After each cycle, the damage in battery capacity is calculated as:

$$D = \sum D_i = \sum \frac{1}{N(DOD_i, T)} \tag{10}$$

Equation (10) states that the total damage in battery (D) is equal to the sum of the damage incurred each previous cycle (D_i). It is calculated the inverse of the number of cycles that can be achieved on the temperature and final DOD on that cycle. When D=1 the battery reaches its EOL.

In the next cycle, the battery capacity must be updated considering the damage incurred, as shown in (11).

$$Q_i = Q(I_d)(1 - 0,2D) \tag{11}$$

The constant 0,2 means that Q reaches 80% of its nominal capacity at the EOL, when D=1. The term (1-0,2D) is the state of health (SOH) of the battery.

IV. RESULTS

Using the methodology described in the previous section, the parameters to model the studied battery (Moura Clean 12MC220) are shown in Table 2.

Some results of the model were taken to show the general behavior of the model. In Fig. 3 it is shown the voltage seen in the battery terminals for discharge cycles of 3 and 5 hours, in comparison with the manufacturer's data. It can be noted that both curves have a good fitting. As expected, it is better adapted for the 5 hours cycle, in which the parameterization was made. In Fig. 4 the result for the voltage seen between its poles for a complete recharge cycle is shown. It can be noted that the fitting is not very good as it is for the discharge cycle.

If a better fitting is needed the model can be improved by the addition of another terms as it is demonstrated in [22]. Nevertheless, the presented model is accurate enough for the application that will be studied, in which the observed errors

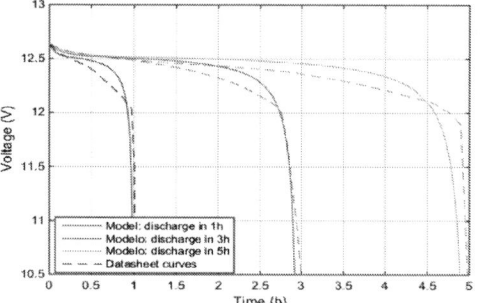

Fig. 3. Comparison between discharge curves in datasheet and the model

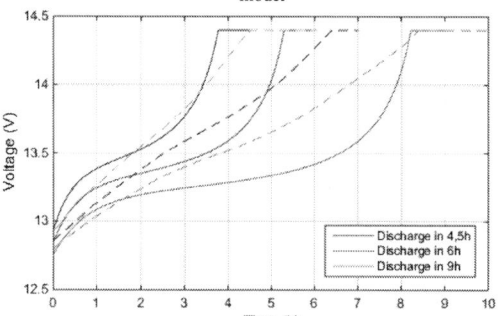

Fig. 4. Comparison between charge curves in datasheet and the model

will not have a great effect in the expected results as the main goal is to observe the expected lifetime and the general behavior of battery voltage in different conditions. A more complex model would require more elaborate parameterization methods and would have a higher computational cost.

As shown in the previous Section, this model is entirely based on manufacturer's data commonly available for the user, dismissing a deep knowledge about the internal operation of the battery. Although it is very practical, it has some problems. In first place, for currents very different from the one in it was parameterized it tends to have larger errors. For DODs closer to 100%, the errors also became larger. It restricts the usage to currents closer to the one in which it was parameterized and discharges lower than 80% (what is recommended in the operation lead-acid batteries). These problems can also be solved by adding terms to (1) or more elaborate parameterization methods.

Another drawback of the model is that the coulombic efficiency is considered equal to 100%. This means that there are no losses in the conversion of the current in electrical charge and vice-versa. A term accounting for the efficiency can be added in (7), relying on typical data or manufacturer's data. For example, in lead-acid batteries the round-trip efficiency is around 81% [21].

The model can also be improved by adding a thermal model to the battery, to simulate the temperature dynamics, as done by Ceraolo [3-4]. For this, a battery thermal resistance and capacitance would be needed, which is not always available.

TABLE II.   MOURA CLEAN 12MC220 PARAMETERIZATION

PARAMETERS FOR DISCHARGE						
A	0,1098	B	0,1368	E$_0$	12,577
K	0,0478	N	0		
PARAMETERS FOR CHARGE						
A	-0,4440	B	0,0584	E$_0$	12,9570
K	-0,0051	N	0,0061	V$_{max}$	14.4
$Q(I_d)$	$84.1e^{-0.04273I_d} + 169.5e^{-0.003021I_d}$				
$N(DOD)$	$12090e^{-0.1161DOD} + 1325e^{-0.1943DOD}$				
$N(T)$	$\begin{array}{l}(-9.63 \times 10^{-6}T^4 + 0.0013T^3 \\ \quad - 0.07934T^2 \\ \quad + 2.144T \\ \quad + 80.26) \\ /100\end{array}$				

Regarding the lifetime model implemented, Fig. 5 shows the expected lifetime of the battery in function for different discharge currents and temperatures. Fig. 6 shows the number of cycles when applied many different charge and discharge currents in different cycle times. In each simulations the current and time remain constant during all battery lifetime.

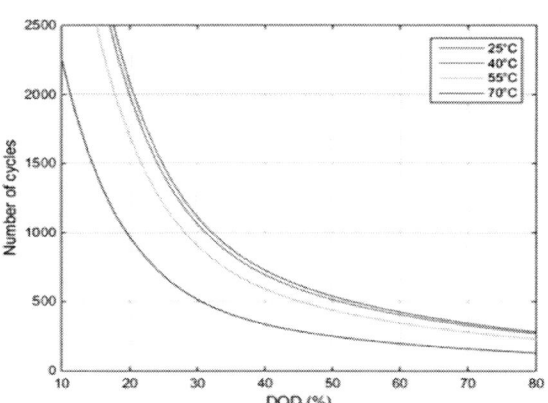

Fig. 5. Number of cycles for the Moura Clean 12MC220 battery in function of DOD and temperature

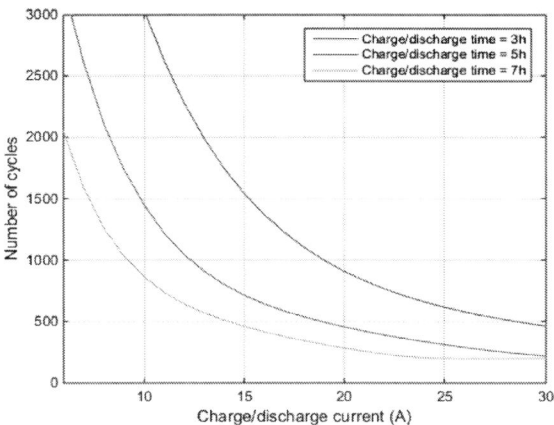

Fig. 6. Number of cycles obtained with different currents and charge/discharge times

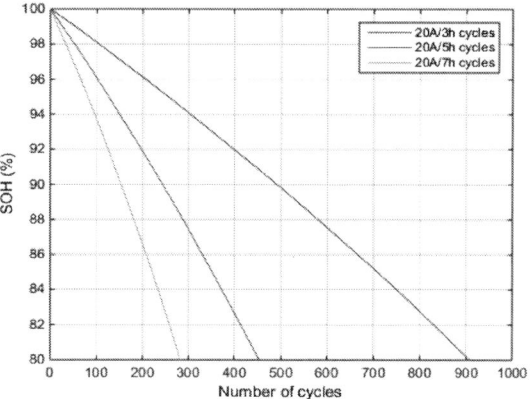

Fig. 7. Evolution of SOH obtained with different currents and charge/discharge times

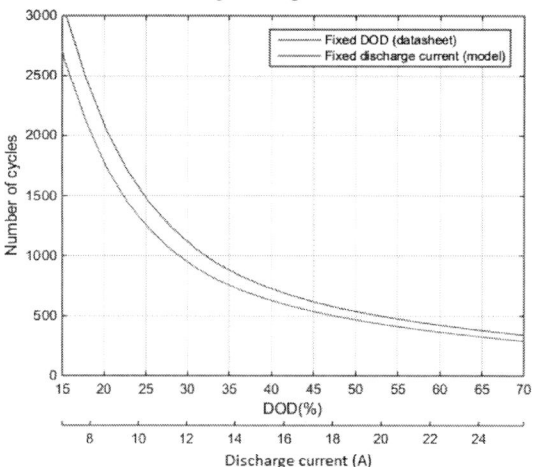

Fig. 8. Comparison between discharge curves in datasheet and the model

The effects on modifying the applied current or the cycle time can be seen in Fig. 6.

Fig. 7 shows the SOH along time when applied 20A in every cycle but with different discharge times. It can be noted that the battery loss of capacity tends to slight accelerate with time. That happens because applying the same current with decreasing capacities causes the obtained DOD to rise. With greater DOD the damage incurred in each cycle is higher.

In Fig. 8 is presented a comparison between the predicted lifetime on datasheet and the model. The implemented lifetime model has a good agreement with the manufacturer's data, but it tends to forecast lower lifetimes. This is explained by the fact that the manufacturer provides the information for a fixed DOD, and as stated before, the DOD obtained in each cycle increases along time.

V. CONCLUSION

In this paper, a model to represent the dynamic behavior of battery during a cycle was implemented alongside a model to determine its lifetime. These models were chosen among many others existent in literature because they are well suited for the concerned application, i.e., a Demand Side Management system.

The models were developed in separated ways even though they represent the same component. It can be noted that the simulation of the battery dynamics in a cycle can be decoupled from the evaluation of the degradation, considering that the individual cycles contribute to a very small part of battery degradation.

It is important to have in mind that the model presented in this paper has its limitations because it was build based on assumptions about the battery usage, in order to make the model more practical. As shown in this paper, battery modelling is a complex task in which these simplifications are a common approach. For a complete understanding of battery behavior in other conditions or applications another approaches, or more accurate modelling of battery dynamics and ageing mechanisms might be necessary.

The developed model as presented is validated with the provided data and ready to be used in the DSM application. The voltage behavior model will be useful to evaluate the link with the proper power converter, and the lifetime model to evaluate the expected lifetime according to the system configuration and costs related.

The next step is the integration of the battery model with the other system components like the power converters, photovoltaic generator and load profiles, which have their own models and maybe will require a refinement on the work presented in this paper.

VI. ACKNOWLEDGMENT

This work has the financial support of project P&D 0722 ANEEL/CEMIG and the Brazilian agencies CAPES, CNPQ and FAPEMIG.

REFERENCES

[1] V. A. Boicea, "Energy storage technologies: the past and the present," Proceedings of the IEEE. Volume: 102, Issue: 11. November 2014.

[2] S. Li and B. Ke, "Study of battery modeling using mathematical and circuit oriented approaches," IEEE Power and Energy Society General Meeting, 2011.

[3] M. Ceraolo, "New dynamical models of lead-acid batteries," IEEE Transactions on Power Systems. Volume: 15, Issue: 4, November 2000

[4] M. Ceraolo and S. Barsali, "Dynamical Models of Lead-Acid Batteries: Implementation Issues," IEEE Transactions on Energy Conversion, Vol.17(1), pp.16-23, March 2002.

[5] N. K. Medora, "Dynamic Battery Modelling of Lead-Acid Batteries Using Manufacturers' Data," Applied Energy, Vol.228, p.1629. October 2018.

[6] C. M. Shepherd, "Design of Primary and Secondary Cells – II. An Equation Describing Battery Discharge," Journal of Elechtrochemical Society, p 657-664, July 1965.

[7] N. Narayan et al, "Estimating battery lifetimes in Solar Home System design using a practical modelling methodology," Applied Energy, 15 October 2018, Vol.228, pp.1629-1639.

[8] V. Marano, S. Onori, Y. Guezennec, G. Rizzoni and N. Madella, "Lithium-ion batteries life estimation for plug-in hybrid electric vehicles," 2009 IEEE Vehicle Power and Propulsion Conference, September 2009.

[9] J. Schiffer, D.U. Sauer, H. Bindner, T. Cronin, P. Lundsager and R. Kaiser, "Model prediction for ranking lead-acid batteries according to expected lifetime in renewable energy systems and autonomous power-supply systems," Journal of Power Sources, Vol.168(1), pp.66-78, 25 May 2007.

[10] R. Dufo-López, J.M. Lujano-Rojas and J.L. Bernal-Agustín, "Comparison of different lead-acid battery lifetime prediction models for use in simulation of stand-alone photovoltaic systems," Applied Energy, Vol.115, pp.242-253, 15 February 2014.

[11] D. U. Sauer and H. Wenzl, "Comparison of different approaches for lifetime prediction of electrochemical systems - using lead-acid batteries as example," Journal of Power Sources, Vol.176(2), pp.534-546, 2008

[12] M. T. Lawder, "Battery energy storage system (BEES) and battery management system (BMS) for grid-scale applications," Proceedings of the IEEE, 2014, Vol.102(6).

[13] B. B. McKeon and J. Furukawa, "Advanced Lead–Acid Batteries and the Development of Grid-Scale Energy Storage Systems, " Proceedings of the IEEE. Volume: 102, Issue: 6. June 2014.

[14] Moura, " Catálogo Técnico Bateria Moura Clean," https://www.neosolar.com.br/media/pdf/manuais/moura_baterias_esta cionarioas_clean_pt.pdf. Accessed in June 2 2019.

[15] Moura, "Manual Técnico Batera Estacionária Clean Nano V1.3," https://moura-portal.s3.amazonaws.com/uploads/2017/07/MANUAL_CLEAN_NA NO_V13_06_DEZ_16.pdf. December 2016. Accessed in June 2 2019

[16] D. Newbery, "Shifting demand and supply over time and space to manage intermittent generation: The economics of electrical storage". Energy Policy, Vol.113, pp.711-720, February 2018.

[17] E. Raszmann, K. Baker, Y. Shi and D. Christensen, "Modelling stationary lithium-ion batteries for optimization and predictive control," IEE Power and Energy Conference, Champaign-Illinois, February 23-24 2017.

[18] J. Moshövel et al, "Analysis of the maximal possible grid relief from PV-peak-power impacts by using storage systems for increased self-consumption," Applied Energy, 1 January 2015, Vol.137, pp.567-575.

[19] G. Strbac, "Demand side management: benefits and challenges," Energy Policy, Dec, 2008, Vol.36(12), p.4419(8).

[20] K. Yang and A. Walid, "Outage-Storage Tradeoff in Frequency Regulation for Smart Grid With Renewables," IEEE Transactions on Smart Grid, March 2013, Vol.4(1), pp.245-252.

[21] J. Hoppmann, J. Volland; T. S. Schmidt, V. H. Hoffmann. "The economic viability of battery storage for residential solar photovoltaic systems – A review and a simulation model." Renewable and Sustainable Energy Reviews, November 2014, Vol.39, pp.1101-1118.

[22] O. Tremblay, L. Dessaint and A. Dekkiche.. "A Generic Battery Model for the Dynamic Simulation of Hybrid Electric Vehicles," 2007 IEEE Vehicle Power and Propulsion Conference, 9-12 September 2007, Arlington,

Comparison Between Two Modulation Strategies for the 3L-DC-SSI

Matheos C. Wermuth, Amanda C. Maia, Richard G.
Cornelius Guilherme S. da Silva.

GEPSEl – Electronic Systems Research Group

UNIPAMPA – Federal University of Pampa
Alegrete, Brazil
conetto197@gmail.com, amanda_maia15@hotmail.com
richardgc98@gmail.com, guilhermesds@gmail.com

Abstract— This paper presents the comparison of two modulation techniques, a carrier based PDPWM with third harmonic injection, and a double-sided modulation, for the Three Level Diode-Clamped Split-Source Inverter (3L-DC-SSI) and shows each benefits and drawbacks. The 3L-DC-SSI is applied to the electrical vehicle (EV) context since it has the advantage of operating with two isolated DC sources, being able to potentially ally the photovoltaic energy generation with an energy storing system commonly used in EVs. For this purpose, both modulation approaches are tested via simulation and the results are compared.

Keywords— 3L-DC-SSI, Electrical Vehicles, Modulation.

I. INTRODUCTION

Currently, electrical vehicles (EV) are becoming more common in the day-to-day lives of the modern society growing rapidly in number in big cities, especially in the public transportation context. In general, EVs are built with three main components, an electrical machine, a source of power (e.g., batteries banks or photovoltaic arrays) and a converting and control system which delivers power and controls the source of it and the machines involved in the process. EVs are a relatively clean mean of transportation, since they lack a combustion motor, the emission of greenhouse gasses is inexistent and the noise pollution is greatly decreased because electrical machines are considerably quieter than combustions ones [1].

Photovoltaic systems (PV) are becoming very attractive to vehicular use, due to its technological evolution and sustainability. However, PV systems have a slow instantaneous response to a power demand, so it is advantageous to combine PV systems to energy storage systems (ESS) which can withstand the power demands of the application [2].

The electrical machine normally needs a device to deliver the power from the ESS and PV systems to it. These devices are called power electronic inverters, which convert an dc input to an ac output. There are several topologies capable to perform this task, like the full-bridge converter, neutral-point clamped, Z-Source and many others. However, the topology presented by [3] known as the Three Level Diode-Clamped Split-Source inverter (3L-DC-SSI) is chosen due to its three operation levels, lower current stresses of the employed switches, lower passive component-count compared to several

topologies and the ability to be fed from two different and isolated dc sources [3]. The major drawback of this inverter is the high passive elements requirements since it generates low frequency components in the input currents and the dc bus voltage which the passive components have to endure [3].

In this paper two modulations strategies are compared when applied to the aforementioned converter, showing the positives and negatives sides of each one. The first strategy is discussed and presented in [4], this method is commonly referred to as a "Geometric Modulation" and consists in using a matrix to transform the phase voltage from the ABC coordinates to αβ coordinates, together with a variable called common mode voltage (V_o) generating a quasi-sinusoidal wave to modulate each leg of the inverter by comparing them with the two carriers.

The second modulation strategy is proposed by [5], this method is commonly called Double-Sided Modulation and proposes an operation without the low frequency ripple caused by the output current in the DC bus voltage by decomposing the modulation signal in two distinct waves and then comparing them with two carriers. This modulation also has the benefit of being able to of apply a modulation offset in each of the modulation signals. In both modulation strategies the Phase Disposition Pulse Width Modulation (PDPWM) method was used.

II. THE THREE LEVEL DIODE-CLAMPED SPLIT-SOURCE INVERTER

The 3L-DC-SSI, shown in Figure 1, was proposed in [3] and is a combination between a Split Source Inverter (SSI) proposed in [6] and a common Neutral Point Clamped Inverter (NPC). This combination uses two isolated dc sources, two inductors, two capacitors, six input diodes and twelve IGBT switches that will accomplish the boosting of the input voltage and the three-level modulating capability. This topology has some advantages when compared to other topologies such as lower current stress of the employed switches, lower passive component count and the ability to be fed from two different isolated dc sources (that can be unequal). However, it also has some drawbacks, such as significant low frequency components in the inverter voltage and high passive elements requirements.

978-1-7281-4181-7/19 $31.00 © 2019 IEEE   543

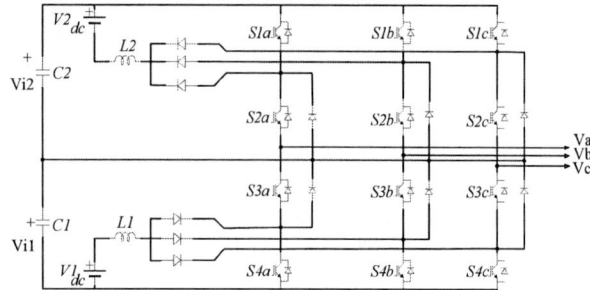

Fig. 1. Topology of the 3L-DC-SSI.

Both isolated dc sources are responsible for charging both capacitors in the dc bus connect between the inverter bridge, where $V2_{dc}$ charges the $C2$ capacitor through the $L2$ inductor and $V1_{dc}$ charges the capacitor $C1$ through the inductor L1. The operating stages of the 3L-DC-SSI are shown in Fig. 2 using the phase A as an example, is worth noting that the three phases (A, B and C) operate in the same manner.

The upper inductor $L2$ is charged when S1a or S1b or S1c is ON, as shown in Fig. 2 I and is discharged when S1a, S1b and S1c are simultaneously OFF as shown in Fig. 2 II and Figure 2 III. Meanwhile, the lower inductor $L1$ uses S4a, S4b or S4c in the ON state to charge, shown in Fig. 2 II and discharges when S4a, S4b and S4c are simultaneously OFF, shown in Fig. 2 I and Fig. 2 III.

III. MODULATION TECHNIQUES

The Phase Disposition Pulse Width Modulation (PDPWM) is used alongside both modulation techniques listed below. This method consists of comparing a modulation signal (e.g. a sine wave, cosine wave or any other control signal) with two repeating waves, called carriers, which have the same phase and amplitude being only dislocated in the cartesian plane. In PDPWM, the switching frequency is determined by the frequency of the carriers and the modulation signal determines the frequency of the output waves.

A. A carrier based PDPWM with third harmonic injection

This method is presented and discussed in [4] and was adapted to fit the 3L-DC-SSI behavior. Firstly, three sine waves are defined, corresponding to each leg and phase of the inverter, they were defined as shown in (1)

$$Va = M\sin(2\pi ft)$$
$$Vb = M\sin(2\pi ft - \frac{2\pi}{3})$$
$$Vc = M\sin(2\pi ft + \frac{2\pi}{3})$$
(1)

Where Va, Vb and Vc are the phase voltages, M is the modulation index and f is the system frequency. It is usual to use the αβ coordinates in three phase systems, to transform ABC coordinates to αβ a transformation matrix such as (2) can be used.

$$\begin{bmatrix} v_\alpha \\ v_\beta \end{bmatrix} = \frac{2}{3}\begin{bmatrix} 1 & -\frac{1}{2} & -\frac{1}{2} \\ 0 & \frac{\sqrt{3}}{2} & -\frac{\sqrt{3}}{2} \end{bmatrix}\begin{bmatrix} Va \\ Vb \\ Vc \end{bmatrix}$$
(2)

Where v_α and v_β are the phase voltage transformed to the αβ plane.

The modulation signals for each leg of the inverter are related to the phase voltage in the αβ plane by the following equations:

$$Van = v_\alpha + \frac{v_0}{2}$$
$$Vbn = -\frac{v_\alpha}{2} + \frac{\sqrt{3}v_\beta}{2} + \frac{v_0}{2}$$
$$Vcn = -\frac{v_\alpha}{2} - \frac{\sqrt{3}v_\beta}{2} + \frac{v_0}{2}$$
(3)

Where v_0 is called common mode voltage or zero-sequence voltage and can be used to maximize the utilization of the DC bus.

By determining that the Van, Vbn and Vcn will have a upper limit of 1 and a lower limit of -1 and then substituting these limits in (3), the following variables are defined:

$$Ra = -2v_\alpha$$
$$Rb = v_\alpha - \sqrt{3}v_\beta$$
$$Rc = v_\alpha + \sqrt{3}v_\beta$$
(4)

Using Ra, Rb and Rc, v_0 can be defined as:

$$v_0 = \frac{(\max(Ra, Rb, Rc) + \min(Ra, Rb, Rc))}{2}$$
(5)

Applying all the considerations above a modulation signal can be generated and is exemplified in Figure 3.

978-1-7281-4181-7/19 $31.00 © 2019 IEEE

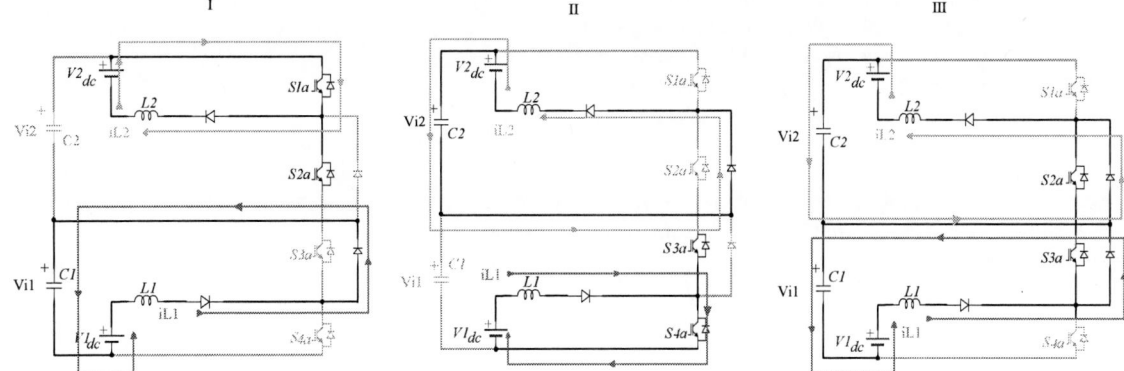

Fig. 2. Operating states of the 3L-DC-SSI.

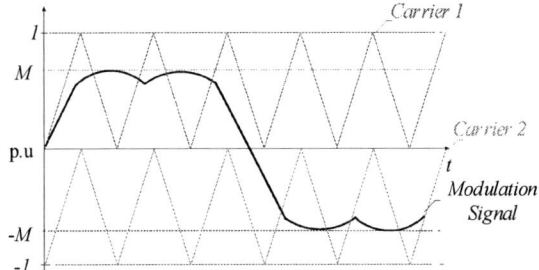

Fig. 3. Example of the "geometric modulation".

By comparing the modulation signal with both carriers, the switching logic is generated. The Carrier 1 shown in Figure 3 controls the switching of S1i and S3i (i = a,b,c) which have a complementary switching pattern and Carrier 2 controls the switching of S2i and S4i which also have the same pattern. If the amplitude of the modulation signal is greater than or equal to the carrier's the S1i and S2i switches are closed (ON) and the others will remain opened (OFF).

B. Double-Sided PDPWM

This modulation technique was presented by [5] and is designed to be used with neutral-point-clamped converters, such as the 3L-DC-SSI. Usually one modulation signal is used to control each phase of the converter, this technique proposes the use of two modulation signals for each phase to preserve the voltage balance in the dc bus and consequently reduce the low frequency ripples caused by the output currents.

To generate both modulation signals, some conditions must be followed:

$$
\begin{aligned}
Va' &= va - v_0' \\
Vb' &= vb - v_0' \\
Vc' &= vc - v_0'
\end{aligned}
\tag{6}
$$

Where va, vb and vc are the phase voltages for each leg of the converter and $v_{0'}$ is:

$$
v_0' = \frac{(\max(va, vb, vc) + \min(va, vb, vc))}{2}
\tag{7}
$$

The two modulation signals that will be generated must always obey the following condition:

$$
\begin{aligned}
Va' &= Vau + Val \\
Vb' &= Vbu + Vbl \\
Vc' &= Vcu + Vcl
\end{aligned}
\tag{8}
$$

Where $Viu \geq 0$ and $Vil \leq 0$, with i = (a,b,c). As described in [5], to maintain the voltage balance in the DC bus the averaged neutral point (NP) current must be zero. The chosen method to maintain the voltage balance is forcing the variables Viu and Vil to be zero for the maximum time possible, thus forcing some of the IGBT to not switch since none of the modulation signals crosses the carriers.

Considering the restrictions implied by (6) and (8), the modulation signals will be:

$$
\begin{aligned}
Viu &= \frac{vi - \min(va, vb, vc)}{2} \\
Vil &= \frac{vi - \max(va, vb, vc)}{2}
\end{aligned}
\tag{9}
$$

Where i = (a,b,c), and va, vb, and vc are the same signals as shown in (1).

Applying all considerations, the modulation signals generated for this technique are shown in Figure 4.

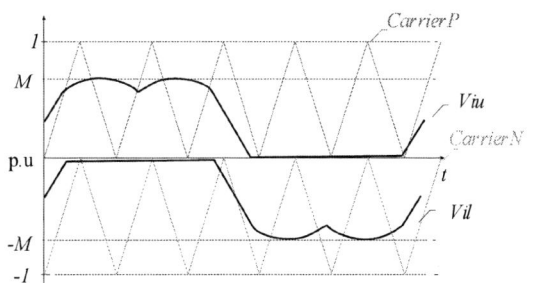

Fig. 4 Example of the Double Sided modulation.

One of the benefits of using this kind of modulation in the 3L-DC-SSI is the capability of applying a modulation offset *Mo* in each of the modulation signals. This offset refers to the boosting stage of the inverter, by increasing or decreasing this offset the voltage in the output can be increased or decreased almost linearly. However, the major drawback of using this offset is that it distorts considerably the output voltage.

The example of the Double-Sided modulation using the *Mo* offset is shown in Figure 5, where *M'* is the sum of the modulation index *M* and the offset *Mo*.

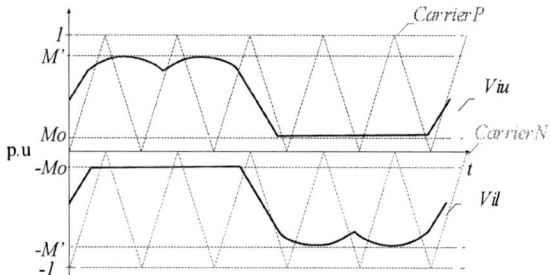

Fig. 5. Example of Double-Sided modulation with *Mo* offset.

The modulation offset *Mo* is applied in the modulation signals shown in equation (9) and can range from 0 to 1 - *M* since the amplitude of the modulation signals cannot be greater than the amplitude of the carriers. The new modulation signals, with the *Mo* offset are:

$$Viu' = \frac{vi - \min(va, vb, vc)}{2} + v_0$$

$$Vil' = \frac{vi - \max(va, vb, vc)}{2} - v_0 \quad (10)$$

$i = (a,b,c)$.

This modulation also has a unique switching logic applied to the inverter, which is the same for the signals with or without the aforementioned offset, firstly each modulation signal is compared to its corresponding carrier.

If $Viu > CarrierP$, then $Xi_+ = 1$; else, $Xi_+ = 0$.

If $Vil < CarrierN$, then $Xi_- = 1$; else, $Xi_- = 0$. (11)

$i = $ (a, b, c). Then, the switching signal for each phase is generated.

$$Xit = Xi_+ - Xi_- \quad (12)$$

Xit can have three values: 1, 0 or −1, which correspond to the possible output voltages for each phase. This control variable is applied to the converter as follows:

If $Xit = 1$, $S1i = S2i = 1$

If $Xit = 0$, $S2i = S3i = 1$ (13)

If $Xit = -1$, $S3i = S4i = 1$

If the conditions listed are not met, then *S1i, S2i, S3i* and *S4i* are zero. These values correspond to the state of the switches, being 1 equivalent to the ON state and 0 equivalent to the OFF state.

IV. SIMULATION RESULTS

The discussed modulation techniques were verified and compared by simulation. The values of the capacitors are $C1 = C2 = 1$ mF, for the inductors $L1 = L2 = 1$ mH, for the DC voltage sources $V1_{dc} = V2_{dc} = 100$ V. The inverter operates over a RL wye-connected load, with $L = 0.5$ mH and $R = 5$ Ω. The frequency of the carriers is 6 kHz in all results, the modulation index is $M = 0.65$ and when used, the modulation offset is $Mo = 0.15$.

A. Carrier based PDPWM with third harmonic injection modulation results

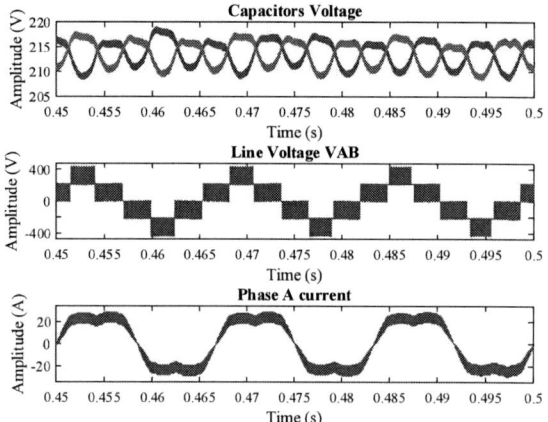

Fig. 6. Capacitor voltage, line voltage and phase current for the third harmonic injection modulation.

978-1-7281-4181-7/19 $31.00 © 2019 IEEE   546

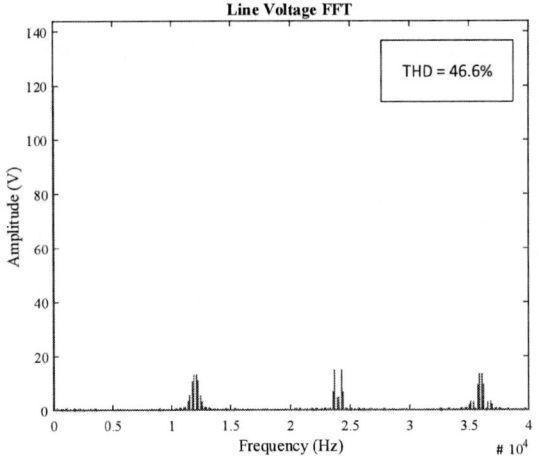

Fig. 7. Line voltage FFT for the third harmonic injection modulation.

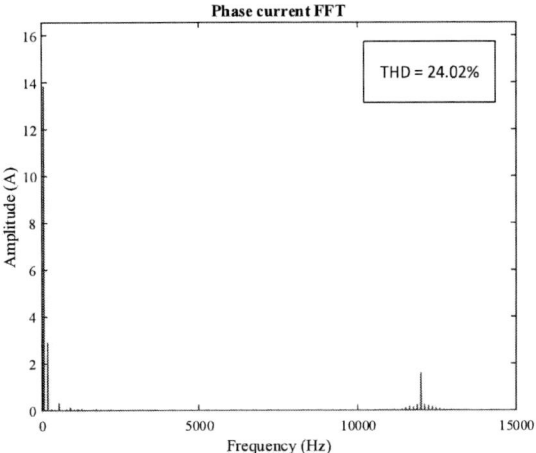

Fig. 8. Phase current FFT for the third harmonic injection modulation..

B. Double-Sided PDPWM results.

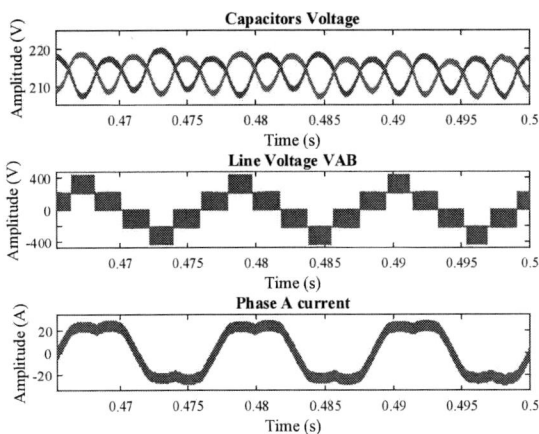

Fig. 9. Capacitor voltage, line voltage and phase current for the double-sided modulation.

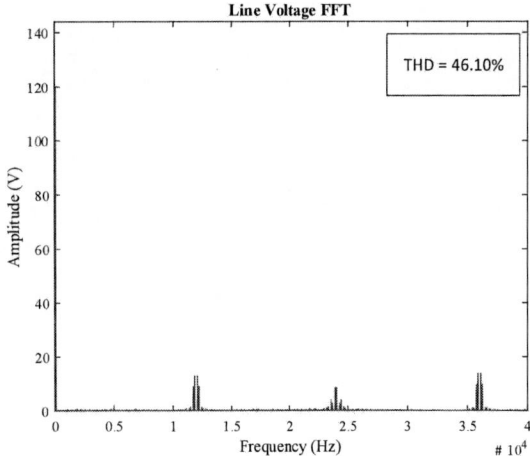

Fig. 10. Line voltage FFT for the double-sided modulation.

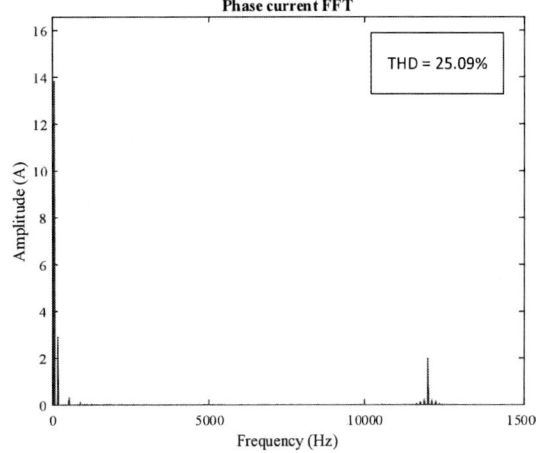

Fig.11. Phase current FFT for the double-sided modulation.

C. Double-Sided PDPWM with offset results.

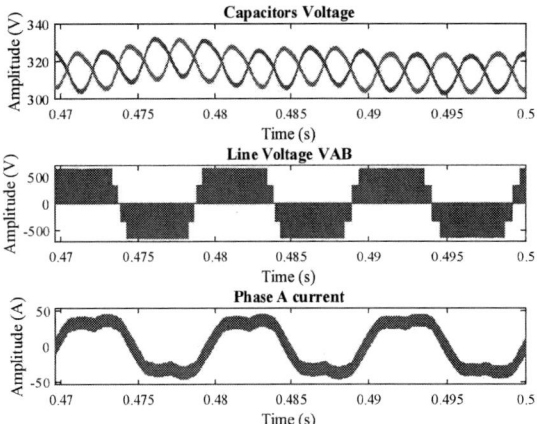

Fig. 12. Capacitor voltage, line voltage and phase current for the double-sided modulation with offset.

978-1-7281-4181-7/19 $31.00 © 2019 IEEE

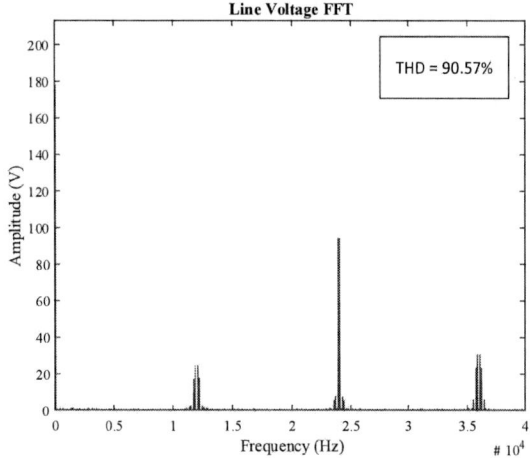

Fig. 13. Line voltage FFT for the double-sided modulation with offset.

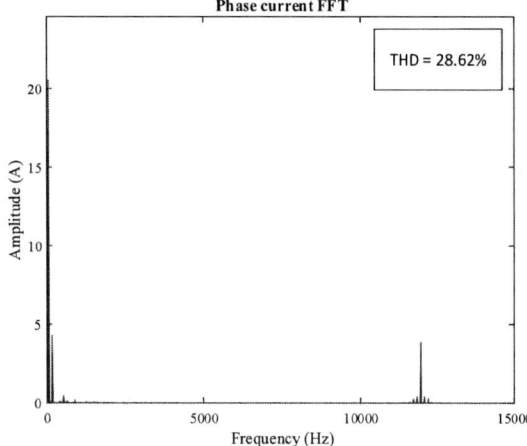

Fig. 14. Phase current FFT for the double-sided modulation with offset.

Firstly, is noticeable that the phase current is almost unchanged between the change of modulation techniques, as shown by their THD values, being the double-sided modulation with offset the one wich had a greater THD, this behavior is due to the load used in the simulation. The RL load, used to simulate an electrical machine, has a filter like effect in the phase current, altering the harmonic distortion by a small amount, as shown in Fig. 8, Fig. 11 and Fig. 14.

Secondly, when using the double-sided modulation, the voltage ripple in the capacitors is bigger, as noticed when comparing the Fig. 6 with Fig. 9 and Fig. 12, and shown in Table 1 where the upper capacitor was used to compare the peak ripple values in 180 Hz.

TABLE I.
PEAK VALUE OF THE UPPER CAPACITOR VOLTAGE RIPPLE FOR EACH MODULATION TECHNIQUE

Modulation Technique	Peak voltage ripple value in 180 Hz
Third harmonic injection	3.36 V
Double-Sided	4.86 V
Double-Sided with offset	9.99 V

Where, the increase in the peak voltage ripple value in the Double-Sided modulation with offset is due to the increase in the dc bus voltage, showing that the distortion in the dc voltage and line output voltage increases with the *Mo* offset.

The double-sided technique has the benefit of being able to use a modulation offset (*Mo),* which is very useful for the 3L-DC-SSI since the *Mo* is directly related to the boost operation of the converter, by increasing it, the voltage in the dc bus is increased. However, this modulation has a drawback of increasing the harmonic distortion in the output voltage and in the dc bus voltage.

V. CONCLUSION

When comparing both modulation techniques discussed in this paper it is noticeable both have benefits and disadvantages. The major benefit of the carrier based PDPWM with third harmonic injection modulation is the low harmonic distortion in both voltage and current outputs and the drawback is that the boost operation of the proposed inverter cannot be controlled via modulation. On the contrary, when using the double-sided modulation, the output current and voltage are distorted but the boost stage of the 3L-DC-SSI can be controlled via a simple modulation offset.

Applying these modulations in the aforementioned converter is important to understand it's behavior and work to improve the performance in the proposed conditions, since it can ally two isolated DC input sources (e.g., batteries banks or photovoltaic arrays) with a low voltage stress in the components making it desirable for electrical vehicles applications.

REFERENCES

[1]   J. de Santiago *et al.,* "Electrical Motor Drivelines in Commercial All-Electric Vehicles: A Review," in *IEEE Transactions on Vehicular Technology*, vol. 61, no. 2, pp. 475-484, Feb. 2012.

[2]   F. Khoucha, A. Benrabah, O. Herizi, A. Kheloui and M. E. H. Benbouzid, "An improved MPPT interleaved boost converter for solar electric vehicle application," *4th International Conference on Power Engineering, Energy and Electrical Drives*, pp. 1076-1081, Istanbul, 2013.

[3]   A. Abdelhakim and P. Mattavelli, "Analysis of the three-level diode-clamped split-source inverter," *IECON 2016 - 42nd Annual Conference of the IEEE Industrial Electronics Society*, pp. 3259-3264Florence, 2016.

[4]   F. B. Grigoletto, H. Pinheiro "Método de Modulação PWM para Equilíbrio das Tensões dos Cpacitores do Barramento CC em Conversores Multiníveis com Diodos de Grampeamento", *Eletrônica de Potência (Florianópolis)*, vol. 14, pp 63-74, 2009.

[5]   J. Pou *et al.*, "Fast-Processing Modulation Strategy for the Neutral-Point-Clamped Converter With Total Elimination of Low-Frequency Voltage Oscillations in the Neutral Point," in *IEEE Transactions on Industrial Electronics*, vol. 54, no. 4, pp. 2288-2294, Aug. 2007.

[6]   A. Abdelhakim, P. Mattavelli and G. Spiazzi, "Three-Phase Split-Source Inverter (SSI): Analysis and Modulation," in *IEEE Transactions on Power Electronics*, vol. 31, no. 11, pp. 7451-7461, Nov. 2016.

Robust Control of Switched Reluctance Generator In Connection With a Grid-Tied Inverter

Caio R. D. Osório, Filipe P. Scalcon, Rodrigo P. Vieira, Vinícius F. Montagner and Hilton A. Gründling

Power Electronics and Control Research Group - GEPOC

Federal University of Santa Maria - UFSM

Av. Roraima, 1000, Camobi, Santa Maria - RS

Email: caio.osorio@gmail.com, filipescalcon1@gmail.com

Abstract—This paper presents a robust control of a switched reluctance generator in connection with a L-filtered grid-tied inverter for wind power generation. A sliding mode controller is used in order to regulate the DC link voltage. The design procedure as well as the stability analysis are presented. A robust state-feedback controller is designed based on LMIs, aiming to synthesize a grid-injected current, ensuring suitable performance and stability for the entire range of uncertain grid inductance. Simulation result are shown in order to demonstrate the DC link voltage regulation as well as the grid injected currents.

Index Terms—Switched reluctance generator, grid-tied inverter, variable speed operation, robust control, wind power.

I. INTRODUCTION

The growing demand for energy associated with the shortage of fossil fuels brings out the need for diversification of the energy matrix, requiring the use of renewable energy sources, such as wind energy. Traditionally, doubly-fed induction generators and permanent magnet synchronous generators are used for this application [1]. These machines, however, present high weight and cost, respectively.

Another type of machine has gained attention on the field of wind power: the switched reluctance machine (SRM). SRMs are characterized by a low cost, simple and robust structure, the absence of permanent magnets or windings on the rotor and an inherent fault-tolerant structure. The switched reluctance generator (SRG) is also capable of operating in a wide range of speed, not requiring the use of a gearbox, which reduces the weight, cost and overall complexity of the system. Those features make the switched reluctance machine a viable option for several applications that demand variable speed and a harsh environment operation [2]–[5].

In the context of integration of renewable energy sources, grid-tied inverters (GTIs) are important elements, being responsible for ensuring high power quality injection into the grid, with grid-injected currents complying with stringent limits for harmonic distortion, such as the IEEE 1547 Standard [6], [7]. To attenuate harmonics from the pulse width modulation (PWM), these inverters require low-pass filters,

This study was financed in part by the Coordenação de Aperfeiçoamento de Pessoal de Nível Superior - Brasil (CAPES/PROEX) - Finance Code 001. The authors would also like to thank the INCT-GD and the finance agencies (CNPq 465640/2014-1, CNPq Projeto 424997/2016-9, CNPq 309536/2018-9, CAPES 23038.000776/2017-54 e FAPERGS 17/2551-0000517-1).

being the L filter a widely used alternative [8]. Moreover, these inverters must ensure suitable performance and stability even when operating against distorted grid voltages and parametric uncertainties, such as the uncertain grid impedance. In this scenario, robust current controllers are important features, being the state feedback current control designed by means of linear matrix inequalities (LMIs) an efficient tool, that has been successfully employed in the robust current control of GTIs [9]–[11].

Some studies aim to address the grid connection of a SRG [5], [12]–[14]. In [5], a control strategy for a grid connected SRG driven by a variable speed wind turbine is presented. A closed loop control strategy is proposed in order to ensure operation at the point of maximum aerodynamic efficiency. A PI controller is used to control the DC link voltage and a separate current controller is responsible for the grid connected converter. Experimental results are presented to demonstrate the system behavior subject to speed variations.

A control scheme for a SRG operating in a wide range of speed is presented in [14]. Two distinct direct power control algorithms are proposed, one for low and another for high speed operation. Optimal firing angles as well as a sliding mode controller are introduced in order to improve overall system performance.

In this paper, a robust control strategy is presented to a SRG connected to a grid-tied inverter. The generation system is comprised of a self-excited SRG operating with a hysteresis current regulator and a sliding mode controller, responsible for the DC link voltage control. An L-filtered GTI in used as an interface with the grid, and a robust current control based on LMIs is applied to ensure stable operation and suitable performance under uncertain grid parameters. Simulation results are presented to evaluate the system response and robustness.

II. SWITCHED RELUCTANCE GENERATOR

The switched reluctance machine works with DC pulsed current, demanding a static converter for proper operation. Its main characteristic is the double salient structure, responsible for many of the machine nonlinearities. The operation as a motor or generator is defined by the firing angles (θ_{on} and θ_{off}). Generator modeling as well as a sliding mode voltage controller will be detailed in this section.

A. Generator Modelling

A switched reluctance machine will operate as a generator if its phases are excited when the phase inductance is decreasing. Different converter topologies can be used to excite the SRG, however, the most commonly used is the asymmetric half-brigde converter (AHB), presented in Figure 1. The main reason for that is the fact that it allows the machine to be driven both as generator and motor [15], with minimal changes to the converter. The converter used in this paper is a self-excited configuration, where a battery is used to provide initial energy to the capacitor. After disconnecting the battery, the capacitor provides energy for further excitation to the phases and limits the DC link voltage ripple [16].

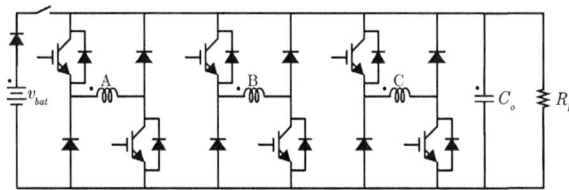

Figure 1. Self-excited asymmetric half-bridge converter.

For energy conversion to take place, firstly an external source of movement has to be connected to the SRG, such as a wind turbine. Once in motion, the generator phases are excited against their natural tendency, to seek rotor-stator alignment, by closing the AHB switches, as seen in Figure 2 (a). Once the switches are closed, the electromagnetic torque leads to a back electromotive force (EMF) that converts mechanical energy from the prime mover into electrical energy, defining the process of generation [2], and allowing the current to flow through the diodes, as seen in Figure 2 (b).

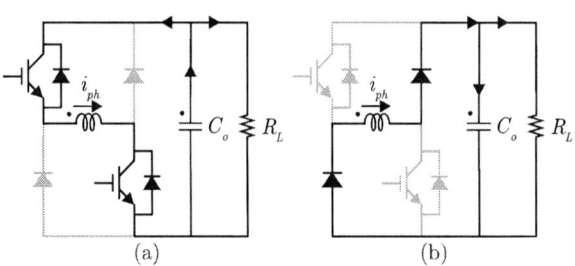

Figure 2. AHB converter operation. (a) Excitation. (b) Generation.

B. DC Link Voltage Control

The output voltage of a SRG is heavily dependant on the load connected to the machine. This happens because the SRG presents a current source characteristic, which differs from most generators, that have a voltage source behavior [17]. Therefore, an increase in the generator load leads to an increase in the output current and a decrease in the output voltage.

In order to regulate the DC Link voltage, regardless of the rotor speed and load condition, the magnetization level of the generator must be varied. The two most common methods of driving the SRG are the hysteresis and single-pulse mode. This paper will focus on the hysteresis mode, given it is used for medium and low speeds [2]. The control structure used is shown in Figure 3.

Considering hysteresis operation, the generator firing angles (θ_{on} and θ_{off}) are kept constant. The magnetizing current is controlled around a reference value during the excitation stage. Such reference is provided by another controller, which determines it based on the DC link voltage tracking error. Once a certain phase reaches θ_{on}, the converter switches for that phase are closed, and the current rises. When the current exceeds the maximum value of the hysteresis band, the switches are blocked. The current begins to decrease until it reaches the minimum value of the hysteresis. At this point, the switches are turned on again. This process is repeated until the end of the excitation, at θ_{off}.

The "SM Controller" block, in Figure 3, represents the sliding mode voltage controller, responsible for determining the reference current i_{ph}^* in order to minimize the output voltage tracking error e_v. The block "sig" is responsible for the switching logic, that based on θ_{on}, θ_{off} and θ, will determine which phase will be excited. The hysteresis regulator then generates the gate signals for each phase switches.

C. Design of the sliding mode controller

The design of the SM controller is based on the generation process described in Fig. 2 [18]. From the circuit presented, it is possible to apply Kirchhoff's Voltage and Current Laws in order to obtain the expressions bellow.

$$\frac{di_{ph}}{dt} = -\frac{1}{L_{ph}}V_{cc} - \frac{r_{ph}}{L_{ph}}i_{ph} - \frac{1}{L_{ph}}i_{ph}\omega_r\frac{dL_{ph}}{d\theta} \quad (1)$$

$$\frac{dV_{cc}}{dt} = -\frac{1}{R_L C_o}V_{cc} + \frac{1}{C_o}i_{ph} \quad (2)$$

where V_{cc} is the DC link voltage, i_{ph} is the current in a phase of the SRG, ω_r is the rotor speed, L_{ph} is the phase inductance, r_{ph} is the phase resistance, C_o is the DC link capacitor and R_L is the load equivalent resistance.

Defining x_1 as the voltage tracking error

$$x_1 = V_{ref} - V_{cc}, \quad (3)$$

and assuming that the voltage reference is kept constant, x_2 can be defined as the derivative of this error,

$$x_2 = \frac{dx_1}{dt} = -\frac{dV_{cc}}{dt} = \frac{1}{R_L C_o}V_{cc} - \frac{1}{C_o}i_{ph}. \quad (4)$$

The state-space model is given by

$$\begin{bmatrix} \frac{dx_1}{dt} \\ \frac{dx_2}{dt} \end{bmatrix} = \begin{bmatrix} 0 & 1 \\ 0 & -1/R_L C_o \end{bmatrix} \begin{bmatrix} x_1 \\ x_2 \end{bmatrix} + \begin{bmatrix} 0 \\ -1/C_o \end{bmatrix} u \quad (5)$$

where

$$u = \frac{di_{ph}}{dt}. \quad (6)$$

978-1-7281-4181-7/19 $31.00 © 2019 IEEE

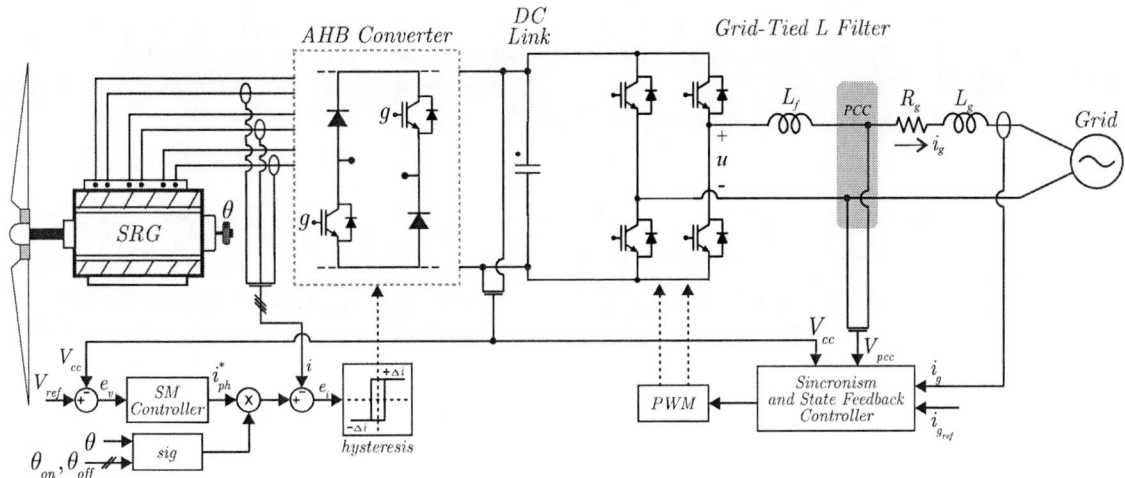

Figure 3. Proposed structure of a wind generation system based on a switched reluctance generator in connection with a grid-tied inverter.

The sliding surface is defined as

$$\sigma = kx_1 + x_2, \qquad (7)$$

where $k > 0$. The control law is defined as a combination of the states

$$u \triangleq \alpha x_1 + \beta x_2 \qquad (8)$$

Based on (6), the reference phase current can be obtained from the control law as

$$i_{ph}^* = \int u \, dt. \qquad (9)$$

Solving the integral and adding the switching function, the reference current can be written as

$$i_{ph}^* = \alpha \int x_1 dt + \beta x_1 + \gamma \, sign(\sigma). \qquad (10)$$

D. Stability analysis

Considering the Lyapunov candidate function

$$V = \frac{1}{2}\sigma^2. \qquad (11)$$

A sufficient condition for the existence of a sliding mode is achieved when the system guarantees the following condition

$$\dot{V} = \sigma\dot{\sigma} < 0 \qquad (12)$$

where

$$\dot{\sigma} = k\dot{x}_1 + \dot{x}_2 = kx_2 + \dot{x}_2 \qquad (13)$$

Solving (12) results in (14).

$$\dot{V} = \frac{1}{C_o} \left[\begin{array}{c} -\alpha k x_1^2 + x_2^2 \left(kC_o - \frac{1}{R} - \beta \right) - \gamma|\sigma| + \\ + \left(-\alpha + k^2 C_o - k\frac{1}{R} - k\beta \right) x_1 x_2 \end{array} \right] \qquad (14)$$

Therefore, it is possible to design the controller gains, α and β, in order to to ensure the stability of the Lyapunov function

(14), which results,

$$\begin{array}{l} \alpha > 0 \\ \beta > kC_o - \frac{1}{R} \\ \gamma > 0 \end{array} \qquad (15)$$

The reference current is implemented by equation (10) with the following gains: $\alpha = 10$, $\beta = 0.3$ and $\gamma = 1$. The switched function $sign(\sigma)$ was implemented using a *sigmoid* function and the surface gain is $k = 10$.

III. GRID-TIED INVERTER

Considering the L-filtered grid-tied inverter presented in Figure 3, this section presents its modelling and the robust control of the grid-injected currents. The power circuit parameters are given by $R_g = 0.5\Omega$, $L_f = 2$ mH, and L_g is an uncertain parameter lying in the interval from 1 mH to 5 mH.

A. State-Space Modelling

Note that the inductances of the filter (L_f) and of the grid (L_g) are associated in series, so they can be rewritten as

$$L = L_f + L_g \qquad (16)$$

Due to the uncertainty in L_g, L is also an uncertain parameter, lying in the interval given by

$$L \in [L_{\min}, L_{\max}]. \qquad (17)$$

Applying the Kirchhoff's Voltage Law, the continuous-time average model of the plant is given by

$$\frac{di_g}{dt} = -\frac{R}{L}i_g + \frac{1}{L}u - \frac{1}{L}v_g \qquad (18)$$

where u is the control input, v_g is the disturbance input, and i_g is the current injected into the grid (controlled output).

For the application of digital control, consider the discretization of the plant with a sufficiently small sampling period T_s. Then, insert an additional state θ to represent the delay in the application of the digital control signal and a

978-1-7281-4181-7/19 $31.00 © 2019 IEEE

resonant controller with internal states ξ, to ensure tracking of sinusoidal references and rejection of disturbances. Thus, an augmented model can be written as

$$i_g(k+1) = a(L)\, i_g(k) + b_u(L)\, \theta(k) - b_g(L)\, v_g(k)$$
$$\theta(k+1) = u(k)$$
$$\boldsymbol{\xi}(k+1) = \boldsymbol{R}\boldsymbol{\xi}(k) + \boldsymbol{T}i_{ref}(k) - \boldsymbol{T}i_g(k)$$

$$(19)$$

where

$$a(L) = 1 - T_s\frac{R}{L}, \quad b_u(L) = T_s\frac{1}{L}, \quad b_g(L) = T_s\frac{1}{L} \quad (20)$$

and

$$\boldsymbol{\xi} = \left[\begin{array}{c} \xi_1 \\ \xi_2 \end{array}\right], \ \boldsymbol{R} = \left[\begin{array}{cc} R_1 & 0 \\ 0 & R_2 \end{array}\right], \ \boldsymbol{T} = \left[\begin{array}{c} \tau_1 \\ \tau_2 \end{array}\right] \quad (21)$$

Systems affected by parametric uncertainties can be described by polytopic models. In this way (19) can be rewritten as

$$\left[\begin{array}{c} i_g(k+1) \\ \theta(k+1) \\ \boldsymbol{\xi}(k+1) \end{array}\right] = \left[\begin{array}{ccc} a(\alpha) & b_u(\alpha) & 0 \\ 0 & 0 & 0 \\ -\boldsymbol{T} & 0 & \boldsymbol{R} \end{array}\right]\left[\begin{array}{c} i_g(k) \\ \theta(k) \\ \boldsymbol{\xi}(k) \end{array}\right] +$$
$$\left[\begin{array}{c} 0 \\ 1 \\ 0 \end{array}\right]u(k) + \left[\begin{array}{c} -b_g(\alpha) \\ 0 \\ 0 \end{array}\right]v_g(k) + \left[\begin{array}{c} 0 \\ 0 \\ \boldsymbol{T} \end{array}\right]i_{ref}(k)$$

$$(22)$$

To obtain the parameters in model (22), the parameters in (20) were calculated for each extreme value (vertex) of L, and then convexically combined with a 2-vertex polytopic representation [19]

$$(a, b_u, b_d)(\alpha(k)) = \sum_{j=1}^{2} \alpha_j(k)(a_j, b_{uj}, b_{dj}) \quad (23)$$

$$\alpha_1 + \alpha_2 = 1, \quad \alpha_j(k) \geq 0, \ j = 1, 2. \quad (24)$$

In a compact form, (22) can be rewritten as

$$\rho(k+1) = \boldsymbol{G}(\boldsymbol{\alpha})\rho(k) + \boldsymbol{H_u}u(k) + \boldsymbol{H_g}(\boldsymbol{\alpha})v_g(k)$$
$$+ \boldsymbol{H_r}i_{ref}(k)$$

$$(25)$$

B. Robust Current Control

In order to provide a grid-injected current control robust against uncertain grid parameters, here it is employed a robust pole-location controller designed in terms of LMIs (for more details, please refer to Section III of [11], and to [10]).

In the discrete-time domain, considering a scalar pole location radius inside the unit circle, if there exist symmetric positive definite matrices $\boldsymbol{Q}$ and $\boldsymbol{J}$ that satisfy the LMIs, the state-feedback gain is given by

$$\boldsymbol{K} = \boldsymbol{J}\boldsymbol{Q}^{-1} \quad (26)$$

This gain ensures closed-loop robust stability for uncertainties and even arbitrarily fast variations on L_g, and limited transient responses. Moreover, considering the inclusion of one resonant controller on 60 Hz, based on the internal model principle, this gain ensures tracking of sinusoidal current

references and rejection of grid voltage disturbances on the frequencies for which they are designed.

For the parameter presented in the beginning of this section, and considering a pole location radios of 0.93, (26) leads to

$$\boldsymbol{K} = \left[\begin{array}{cccc} -25.1839 & -0.4714 & 112.2643 & -107.6470 \end{array}\right]. \quad (27)$$

IV. RESULTS

In order to test the proposed control scheme, simulation was performed in software Matlab/Simulink. The SRG analyzed in this paper has the following characteristics: 3 phases, 2 kW, 1500 rpm rated speed, 12 stator poles and 8 rotor poles (12x8 configuration). The AHB was constructed considering discrete components, being $C_0 = 2250 \ \mu F$.

The SRG's model used in this paper was obtained from experimental magnetization data, in the form of ITBL and TTBL look-up tables [20]–[22]. With the magnetic flux value and relative phase position it is possible to determine the current value using ITBL. Similarly, with the current value and relative phase position it is possible to determine the electromagnetic torque value. Figure 4 presents the block diagram of the SRG simulation model.

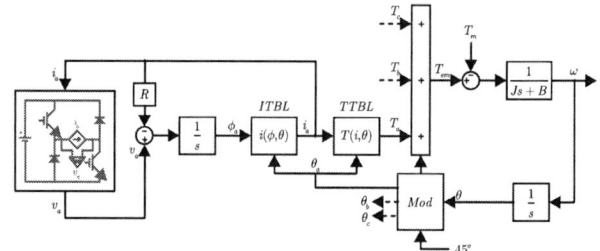

Figure 4. Block diagram of the SRG simulation model.

For the results presented below, the DC link is initially charged by the SRG connected to a resistive load. Once the DC link voltage reaches the reference value, the resistive load is disconnected and the grid connection is performed. The reference voltage is kept constant at 200V for all results.

Figure 5 shows the tracking capability of the system, with a change in current amplitude in $t = 0.8s$. It is possible to verify a good steady-state and transient response. The DC link voltage is also controlled successfully, with a maximum ripple of 4.5%. The SRG currents increase in order to maintain the DC link voltage, presenting a varying amplitude due to the oscillating power being injected into the grid.

Figure 6 presents the system response to parametric variation, from a weak grid ($L_g = 5mH$) to a stiff grid condition ($L_g = 1mH$). One can notice an increased current ripple after the grid variation, however, good performance for both situations is observed. Once more, the sliding mode controller is capable of controlling the DC link voltage, with a ripple smaller than 10V. Figure 7 shows the system capability of variable speed operation. At $t = 0.8s$ the rotor speed is reduced from 100 rad/s to 75 rad/s. Good transient response

978-1-7281-4181-7/19 $31.00 © 2019 IEEE

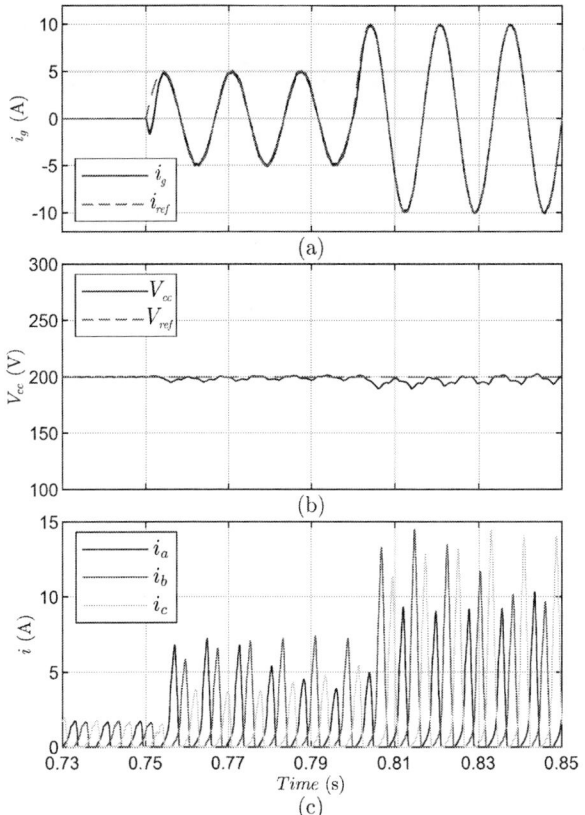

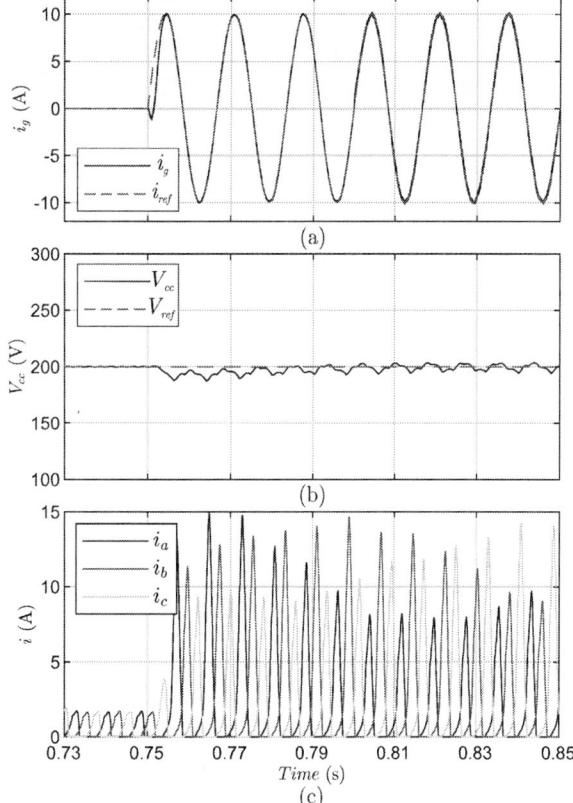

Figure 5. Simulation results for current reference variation ($i_{ref} = 5A$ to $i_{ref} = 10A$). (a) Grid current. (b) DC link voltage. (c) SRG phase currents.

Figure 6. Simulation results for grid inductance variation ($l_g = 5mH$ to $l_g = 1mH$). (a) Grid current. (b) DC link voltage. (c) SRG phase currents.

is observed, accompanied by a decrease in the frequency of the generator currents.

To verify the low harmonic distortion of the injected current in the grid, obtained with the proposed controller, Figure 8 presents the harmonic analysis of the current waveform shown in Figure 6 (a). A comparison with the limits of the IEEE 1547 standard [7] is carried out, from which it is clear that the grid-injected current is in accordance with the determined limits, for both weak and stiff grid conditions.

V. CONCLUSION

In this paper, a robust control of a switched reluctance generator in connection with a grid-tied inverter for wind power application was presented. A sliding mode technique was used to regulate the DC link voltage at the output of the switched reluctance generator. Considering two states, the output voltage tracking error and its derivative, a state-space description of the generator was obtained and a sliding mode control law was derived. A robust state-feedback control designed based on LMIs is employed to synthesize the grid-injected current, ensuring stability and suitable performance for the entire range of uncertain grid parameter. Simulation results validate the proposed technique, proving the system effectiveness to both regulate the DC link voltage and inject

power into the grid. Results with load disturbance, parametric variations and speed variations are presented.

REFERENCES

[1] S. Muller, M. Deicke, and R. W. De Doncker, "Doubly fed induction generator systems for wind turbines," *IEEE Industry Applications Magazine*, vol. 8, no. 3, pp. 26–33, May 2002.

[2] D. A. Torrey, "Switched reluctance generators and their control," *IEEE Transactions on Industrial Electronics*, vol. 49, no. 1, pp. 3–14, Feb 2002.

[3] T. Andre Santos Barros and E. Ruppert Filho, "Direct power control for switched reluctance. generator in wind energy," *IEEE Latin America Transactions*, vol. 13, no. 1, pp. 123–128, Jan 2015.

[4] E. Rahmanian, H. Akbari, and G. H. Sheisi, "Maximum power point tracking in grid connected wind plant by using intelligent controller and switched reluctance generator," *IEEE Transactions on Sustainable Energy*, vol. 8, no. 3, pp. 1313–1320, July 2017.

[5] R. Cardenas, R. Pena, M. Perez, G. Asher, J. Clare, and P. Wheeler, "Control system for grid generation of a switched reluctance generator driven by a variable speed wind turbine," in *30th Annual Conference of IEEE Industrial Electronics Society, 2004. IECON 2004*, vol. 2, Nov 2004, pp. 1879–1884 Vol. 2.

[6] F. Blaabjerg, R. Teodorescu, M. Liserre, and A. Timbus, "Overview of control and grid synchronization for distributed power generation systems," *IEEE Transactions on Industrial Electronics*, vol. 53, no. 5, pp. 1398 –1409, October 2006.

[7] "Ieee standard for interconnection and interoperability of distributed energy resources with associated electric power systems interfaces," *IEEE Std 1547-2018 (Revision of IEEE Std 1547-2003)*, pp. 1–138, April 2018.

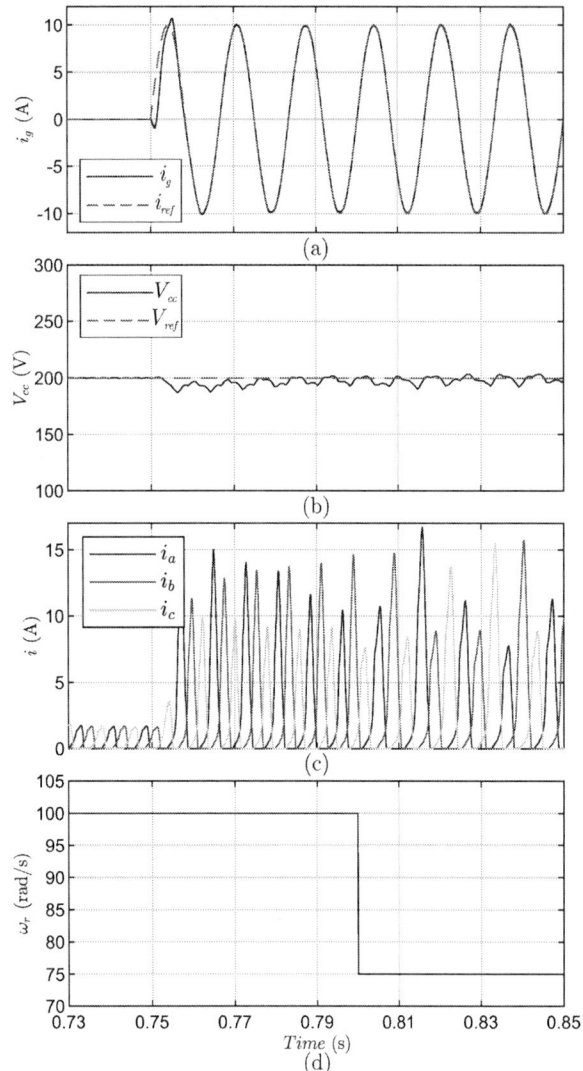

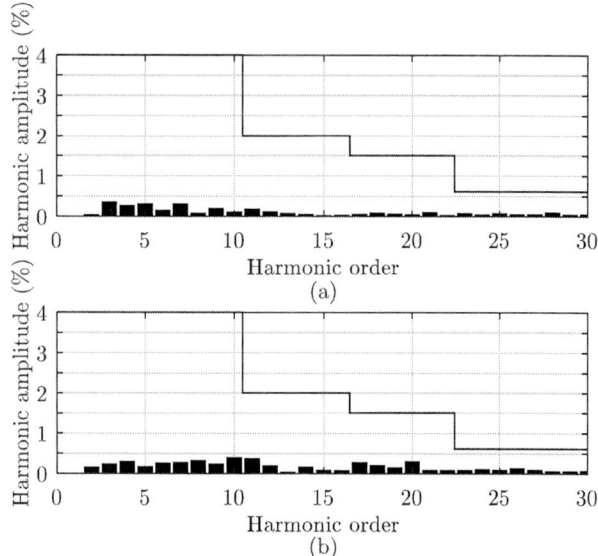

Figure 8. Harmonic content of the grid current, respecting the limits from IEEE Standard 1547. (a) THD= 1.24% for L_g = 5mH. (b) THD= 2.67% for L_g = 1mH.

Figure 7. Simulation results for rotor speed variation (ω_r = 100rad/s to ω_r = 75rad/s). (a) Grid current. (b) DC link voltage. (c) SRG phase currents. (d) SRG rotor speed.

[8] R. Teodorescu, M. Liserre, and P. Rodríguez, *Grid Converters for Photovoltaic and Wind Power Systems*, ser. Wiley - IEEE. John Wiley & Sons, 2011.

[9] C. Olalla, R. Leyva, A. El Aroudi, and I. Queinnec, "Robust LQR control for PWM converters: An LMI approach," *IEEE Transactions on Industrial Electronics*, vol. 56, no. 7, pp. 2548–2558, July 2009.

[10] L. A. Maccari, Jr., J. R. Massing, L. Schuch, C. Rech, H. Pinheiro, R. C. L. F. Oliveira, and V. F. Montagner, "LMI-based control for grid-connected converters with LCL filters under uncertain parameters," *IEEE Transactions on Power Electronics*, vol. 29, no. 7, pp. 3776–3785, July 2014.

[11] C. R. D. Osorio, G. G. Koch, L. C. Borin, I. Cleveston, and V. F. Montagner, "A robust quasi-deadbeat controller and relaxations applied to grid-connected inverters," *Brazilian Power Electronics Journal*, vol. 23, no. 3, pp. 320–329, 2018.

[12] G. P. Viajante, M. A. A. Freitas, J. A. Santos, V. R. Bernardeli, M. E. Oliveira, C. X. Rocha, D. A. Andrade, and L. C. Gomes, "Switched reluctance generator in connection with the three-phase power grid," in *2015 IEEE 15th International Conference on Environment and Electrical Engineering (EEEIC)*, June 2015, pp. 1527–1532.

[13] G. P. Viajante, E. N. Chaves, C. A. Queiroz, M. A. A. Freitas, L. C. Miranda, D. P. A. Silva, S. B. Silva, L. C. Gomes, and R. T. Fidelis, "A grid connection scheme of a switched reluctance generator using p+resonant controller," in *2017 IEEE International Conference on Environment and Electrical Engineering and 2017 IEEE Industrial and Commercial Power Systems Europe (EEEIC / I CPS Europe)*, June 2017, pp. 1–6.

[14] T. A. d. S. Barros, P. J. d. S. Neto, P. S. N. Filho, A. B. Moreira, and E. R. Filho, "An approach for switched reluctance generator in a wind generation system with a wide range of operation speed," *IEEE Transactions on Power Electronics*, vol. 32, no. 11, pp. 8277–8292, Nov 2017.

[15] A. Arifin, A. B. Ibrahim, and S. C. Mukhopadhyay, "State of the Art of Switched Reluctance Generator," *Energy and Power Engineering (EPE)*, vol. 04, no. 06, pp. 447–458, 2012.

[16] V. R. Bernardeli, D. A. Andrade, A. W. F. V. Silveira, L. C. Gomes, G. P. Viajante, and L. G. Cabral, "Self-excited switched reluctance generator," in *XI Brazilian Power Electronics Conference*, Sep. 2011, pp. 55–60.

[17] E. S. L. Oliveira, M. L. Aguiar, and I. N. Silva, "Strategy to Control the Terminal Voltage of a SRG Based on the Excitation Voltage," *IEEE Latin America Transaction*, vol. 13, no. 4, pp. 975–981, 2015.

[18] Y. Z. Liu, Z. Zhou, J. L. Song, B. J. Fan, and C. Wang, "Based on sliding mode variable structure of studying control for status switching of switched reluctance starter/generator," in *Chinese Automation Congress (CAC), 2015*, Nov 2015, pp. 934–939.

[19] S. Boyd, L. El Ghaoui, E. Feron, and V. Balakrishnan, *Linear Matrix Inequalities in System and Control Theory*. Philadelphia, PA: SIAM Studies in Applied Mathematics, 1994.

[20] T. A. Dos Santos Barros, P. J. Dos Santos Neto, M. V. De Paula, A. B. Moreira, P. S. Nascimento Filho, and E. Ruppert Filho, "Automatic characterization system of switched reluctance machines and nonlinear modeling by interpolation using smoothing splines," *IEEE Access*, vol. 6, pp. 26 011–26 021, 2018.

[21] C. R. D. Osorio, R. P. Vieira, and H. A. Grundling, "Sliding mode technique applied to output voltage control of the switched reluctance generator," in *IECON 2016 - 42nd Annual Conference of the IEEE Industrial Electronics Society*, Oct 2016, pp. 2935–2940.

[22] F. P. Scalcon, R. P. Vieira, and H. A. Gründling, "Sliding mode speed control applied to the switched reluctance motor," in *IECON 2018 - 44th Annual Conference of the IEEE Industrial Electronics Society*, Oct 2018, pp. 695–700.

978-1-7281-4181-7/19 $31.00 © 2019 IEEE

Analysis of Performance and Opportunity for Improvements in the Microgrid of Ilha Grande

Leonilson dos Santos Veras
Institute of Electrical Energy
Federal University of Maranhão
São Luís, Brazil
leonilsonsv@gmail.com

Hércules A. Oliveira
Institute of Electrical Energy
Federal University of Maranhão
São Luís, Brazil
hercules.oli@hotmail.com

José G. de Matos
Institute of Electrical Energy
Federal University of Maranhão
São Luís, Brazil
dematosjosegomes@gmail.com

Osvaldo Ronald Saavedra
Institute of Electrical Energy
Federal University of Maranhão
São Luís, Brazil
o.saavedra@ieee.org

Luiz A. de Sousa Ribeiro
Institute of Electrical Energy
Federal University of Maranhão
São Luís, Brazil
l.a.desouzaribeiro@ieee.org

Lucas de Paula Assunção Pinheiro
Energy Company of Maranhão - Cemar
Equatorial Energia SA
São Luís, Brazil
lucas.pinheiro@cemar-ma.com.br

Abstract— **Microgrids has emerged as an attractive solution to ensure electric power supply for remote locations, such as islands isolated from the continents. Sometimes, microgrids that have already been deployed need to be refurbished, either to keep up with the growing power demand or to replace end-of-life equipment. This paper reports the enhancement process that was carried out in a real microgrid system deployed in the northeast of Brazil. A comparative analysis of the microgrid operation performance is developed. The benefits of the refurbishment, both in terms of efficiency and power quality are also evaluated.**

Keywords—microgrid, isolated systems, PV micro generation.

I. INTRODUCTION

In 2015, only 0.3% of Brazilian population, which corresponds to more than six hundred thousand people, did not have access to electricity. In the state of Maranhão this index had reached 0.4% at this same year [1]. Furthermore, there are still some places where energy is provided precariously by diesel generators for about three to four hours a day [2]. The reasons for this deficit stem from geographical, economic, technical and often severe environmental constraints. A frequent situation is the communities located in islands, whose access is limited to small boats. In such cases it becomes impracticable to expand the conventional distribution system to serve these specific communities.

In this context, microgrids gain space as an important factor for the economic and social development of isolated communities. Supported by mature power electronics, they foment and strength the use of the renewable energy sources of each place, making the distribution power systems sustainable.

Microgrids are defined as electric systems containing loads and generation sources, that can be operated in a controlled and coordinated way, either connected to a main power network or in islanded operation mode [3].

In the case of an isolated operation, it is essential that the available energy resources be used close to consumption, a concept called distributed generation. The *Agência Nacional de Energia Elétrica* (ANEEL) defines distributed generation as power generation plants of any power capacity, with installations connected directly to the electric distribution system or through consumers facilities [4]. The idea of distributed generation basically consists of the existence of

generation sources connected to a main network, whether this main network is a microgrid or the national electric system.

This work presents an analysis of the performance of the microgrid of Ilha Grande, MA, where the technical improvements introducing during its operation, equipment performance, as well as indicators of energy quality delivered to consumers are pointed out. The microgrid is capable of providing electricity 24 hours a day, although its main source is photovoltaic.

The paper is organized as follows: in section II the microgrid of Ilha Grande is described; in section III the evolution of the microgrid is detailed, where the improvements added to the system are highlighted; in section IV the microgrid performance throughout its operation is presented and discussed; finally, the conclusions of this work are presented in section V.

II. CARACTERIZATION OF ILHA GRANDE MICROGRID

The Ilha Grande microgrid is located in the city of Humberto de Campos, in the state of Maranhão, Northeastern region of Brazil. A satellite view of Ilha Grande is depicted in Fig. 1. The island has 50 consumer units and approximately 200 inhabitants, mostly fishermen. In the community there is a school and a church, besides several trades. The average daily consumption is around 115 kWh. The domestic load is due to lighting and household appliances such as refrigerators, TV's, stereos, etc. A photograph of the entrance to Ilha Grande is shown in Fig. 2.

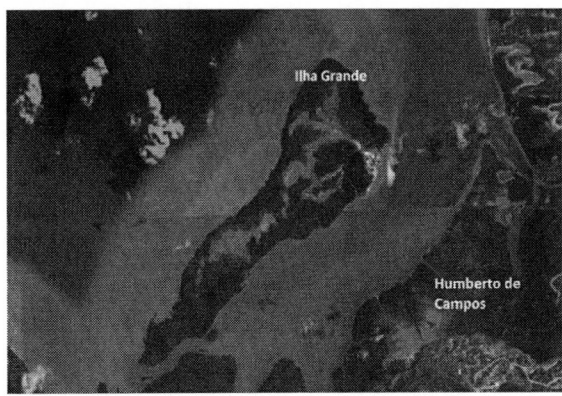

Fig. 1. Location of Ilha Grande (Google Maps).

978-1-7281-4181-7/19 $31.00 © 2019 IEEE

Fig. 2. Entrance to Ilha Grande.

Fig. 3 illustrates the original topology of the Ilha Grande microgrid. The system was installed in 2012 by Companhia Energética do Maranhão (CEMAR).

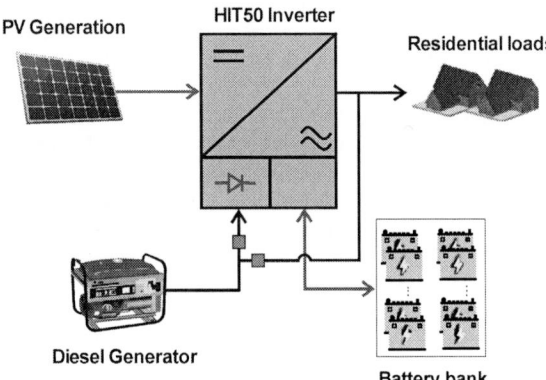

Fig. 3. Original topology of the Ilha Grande microgrid.

In the original topology of the system, a single Zigor HIT50 hybrid inverter [5] performed the functions of rectifier and inverter, functions necessary for charging the battery banks and powering the residential loads.

The photovoltaic panels were grouped into seven strings of 4410 Wp each. The diesel generator operated only in emergency (*stand-by*), that is, with the exclusive function of servicing the load in case the photovoltaic system runs out of power and low state of charge (SOC) of the battery banks, or maintenance or failure of the hybrid inverter.

The microgrid operated for less than a year, and its operation was finished in 2013. In 2014, the Institute of Electrical Energy (IEE) of the Federal University of Maranhão (UFMA) submitted to CEMAR a microgrid project that followed the standard installed in Lençóis Island, in operation since 2008 [6] [7] [8].

In 2015 the new system, revitalized and with new technological adaptations, came into operation assuming the configuration shown in Fig. 4. In this topology, the battery banks, photovoltaic generation and the diesel generator (via rectifier) are interconnected through a 120 Vdc bus. Connected to this same bus two inverters with efficiencies greater than 90% attend the load. The rectification process is also redundant, with two rectifiers of efficiency of 85% at full load.

The battery banks acts in order to compensate for variations in photovoltaic generation and to provide energy in the nighttime. The diesel generator acts as a dispatchable source to provide electricity in case the battery bank SOC

reaches a predefined low level, and for equalization charges to the battery banks Details of the original and current configuration are presented in the next section.

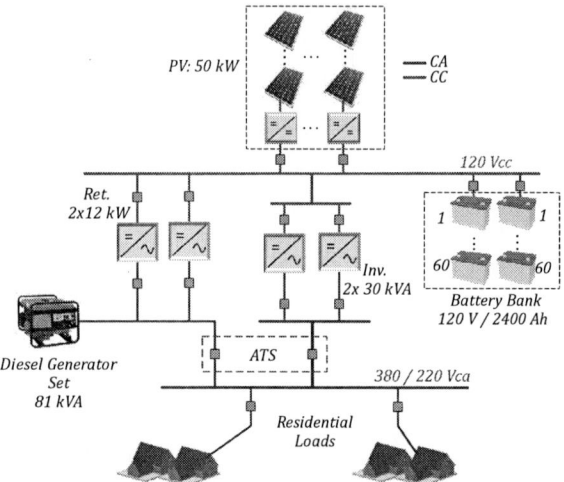

Fig. 4. Current topology of the Ilha Grande microgrid

III. EVOLUTION OF THE MICROGRID

The following is a review and comparison between the original and current topologies of the Ilha Grande microgrid, as well as their respective components. The improvements that were implemented were aimed at improving the power quality supplied to consumers and their increasing demand, as well as preserving the useful life of the installed equipment.

A. Solar Photovoltaic Generation

As previously mentioned, the first configuration of the Ilha Grande microgrid operated until 2013. The solar photovoltaic generator had a nominal power of approximately 30.87 kWp, with a total of 126 panels. In August 2018 further 60 panels of 330 Wp each were added, reaching the current power of 50.67 kWp. The details of the PV generators and the growth of installed power are presented in Table I.

TABLE I. GROWTH OF THE PHOTOVOLTAIC ARRANGEMENT.

Photovoltaic Arrangement		
Year	*2013*	*2019*
Number of panels	*126*	*186*
Cell types	*Polycrystalline Silicon*	*Polycrystalline Silicon*
Power of panels (Wp)	*245*	*245/330*
Efficiency of the panels	*14.8%*	*14.8% / 16.97%*
Dimensions per panel (mm x mm x mm)	*1665 x 991 x 50*	*1665 x 991 x 50/ 1960 x 992 x 35*
Power of the arrangement (kW)	*30.87*	*50.67*
% of capacity increase	*-*	*64%*

B. Storage system

The first installed bank was composed of OPzS batteries with nominal voltage of 2 V and 300 Ah, structured in two strings of 172 batteries. Therefore, the rated voltage of the bank was 344 DC with a capacity of 206.4 kWh, considering the depth of discharge (DOD) of 80% and manufacturer's life expectancy of 1500 cycles at 25° C (approximately 4 years).

The current bank is made up of higher capacity OPzS batteries [9], arranged in two strings of 60 elements in series. The two strings are connected in parallel forming a 120 DC bus. The rated power of the bank is 300 kWh. The DOD adopted is 40%, which at 25° C provides a life expectancy of about 3200 cycles (approximately 8 years).

The increase in the capacity of the storage system until the beginning of 2019 is shown in Table II. A significant increase in the capacity of the bank (more than 45%) is observed, with a consequent decrease in the number of batteries.

TABLE II.   CHANGES IN BATTERY BANK

Year	2014	2019
Number of strings	2	2
Batteries per string	172	60
Type	OPzS	OPzS
Rated voltage per cell (Vdc)	2	2
Rated voltage of the bank (Vdc)	344	120
Capacity (Ah) per element	300	1250
Capacity of the bank (KWh)	206.4	300
Depth of discharge (DOD)	80%	40%
% of capacity increase	-	45.35%

C. Converters

As mentioned earlier, the original configuration of the microgrid proposed in 2012 was based on the operation of a HIT50 hybrid inverter that accumulated rectifier and inverter functions in a single device. In the new topology of 2015 such functions were disaggregated, and two three-phase inverters were employed to supply AC voltage to the load. The 120 Vdc input voltage is modulated by a PWM space vector, which generates a line output RMS voltage of approximately 80 Vac. This voltage is then increased to a line RMS voltage of 380 Vac by a delta-wye transformer.

In addition, an N+1 level of security was added. The goal was to develop an inverter that, in addition to being highly reliable, also had the capacity to expand in order to keep up the load growth without the need to change the equipment already installed. In order to increase the reliability and robustness of the system, it was adopted the parallel operation of the inverters without communication between the units [2]. Similar configuration was designed for rectifiers.

The characteristics of the inverter used at the time of construction of the plant in 2010 and of the converters used to revitalize the system in 2015 are presented in Table III.

Parallel operation of the inverters allows, in case of loss of one of them, the other is capable of withstanding the entire load of the system, which would not occur in the original topology, since the loss of hybrid inverter would compromise all the system. In this case, the generator set could be directly connected to the load via a bypass switch, until the inverter has been fixed. However, this meant spending on diesel oil and greenhouse gas emissions to the environment.

Because inverters have sufficient capacity to withstand all community demands and system losses, they operate alternately, with alternation occurring in eight-day periods under command of a programmable logic controller (PLC). If demand exceeds 12 kW, both operate simultaneously until the load is less than 10 kW. With the loss of both inverters and rectifiers, the diesel generator could still meet the demand until the system operation is restored.

TABLE III.   CONVERTERS.

Zigor HIT50 Hybrid Inverter (2012)	
Max. continuous output power (KW)	50
Nominal output voltage (V)	380
Ac power frequency (Hz)	60
Max. efficiency	>96%
Quantity	1
Total harmonic distortion output voltage	<3%
Inverter (2015)	
Max. continuous output power (KW)	30
Nominal AC voltage (V)	380
Ac power frequency (Hz)	60
Max. efficiency	90%
Quantity	2
Total harmonic distortion output voltage	< 2%
Three-phase Rectifier (2015)	
Max. continuous output power (KW)	12
Nominal DC voltage (Vdc)	120
Max. efficiency	85%
Quantity	2

D. Diesel generator

By 2013, the Ilha Grande microgrid had a 30 kVA diesel generator as backup source. In 2015 this generator was replaced by a higher power, 48 kVA. In 2018, it was replaced by one of 73 kVA. This was intended for the generator to operate at less than 70% of its Prime Power capacity [10]. In that situation, it is expected that the useful life of generator would be extended. Some of the main specifications of the diesel generator sets are presented in Table IV. All reported values are given considering the "power prime" operation regime, including the values of fuel consumption and autonomy.

TABLE IV.   DIESEL GENERATOR.

Year	2013	2015	2019
Rated AC power (kVA)	30	48	73
Generator set (kW)	24	38	59
Fuel consumption (l/h)	7	11.5	17
Fuel tank (l)	160	200	200
Power Factor	0.8	0.8	0.8
Autonomy (h)	22	17	11
% of capacity increase	-	60%	52%

IV. MICROGRID PERFORMANCE

To evaluate the performance of the microgrid, data collected between 2016 and early 2019 were analyzed. During this period, it was possible to observe, among other things, the 26% increase in the community demanded load, as well as the participation of solar photovoltaic generation in the service to this load. At this point, it is important to observe that the gains in the microgrid performance are strongly dependent on employed technologies in solar panels, batteries and converters. Besides, the strategies adopted in the use of the diesel generator aim to preserve the useful-life of the batteries.

A. System expansion

The daily profile and solar photovoltaic generation from real data collected in September 2016 are presented in Fig. 5. The maximum load demanded in this period was 7.08 kW. The points of greatest demand are between 18 and 23 hours. Comparing this graph with others constructed from data collected in different months of the same year we can conclude that the generation and the demand curve follow a profile that does not undergo significant changes.

The generation curves and the load profile measured over a period of twenty-four hours in April 2019 are shown in Fig. 6. The load profile demanded is similar to that of Fig. 5, with the load peaks located between 18 and 23 hours. However, the average consumption increased 48.11% in relation to 2016, with the maximum demanded power of 8.94 kW.

It is also possible to observe the growth of the energy produced by the current photovoltaic generator in relation to 2016. In percentage terms, renewable generation registered a growth of 64.14%, which is compatible with the increase in the number of panels, as shown previously in Table I. It should be noted that the curves were constructed from measured data at periods with different climatic characteristics. While the curves of Fig. 5 were constructed from measured data in a period of high solar irradiation, the curves of Fig. 6 were constructed during a period of intense rainfall and low solar irradiation. Nevertheless, the results are relevant, as indicative of the microgrid operation.

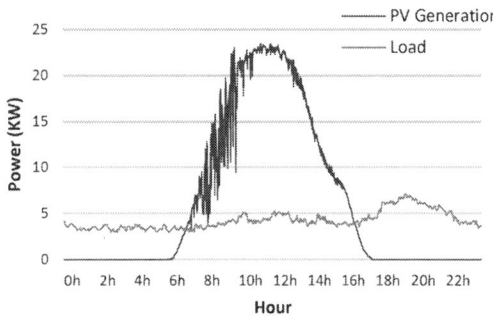

Fig. 5. Real operation of the Ilha Grande microgrid in September 2016.

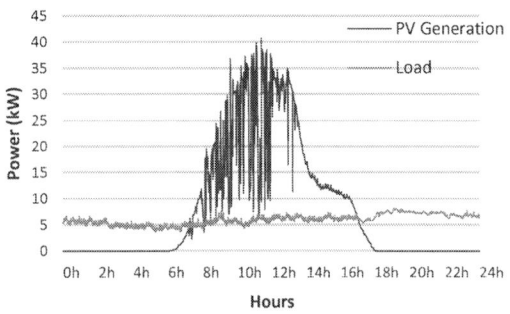

Fig. 6. Real opreation of Ilha Grande microgrid in January 2019 (own data)

B. Power quality

According to Brazilian legislation, the distribution system and the generation facilities connected to it must, under normal operating conditions and in steady state, operate within the frequency limits between 59.9 Hz and 60.1 Hz [11]. Frequency monitoring over a 24 hours period in June 2016 and April 2019 are shown in Fig. 7 and Fig. 8, respectively. As can be seen, at various times in Fig. 7 the frequency is greater than 60.1 Hz. In Fig. 8 the frequency does not exceed 60.106 Hz. In both figures, lower frequency values are located at peak times and reflect the action of the droop controller.

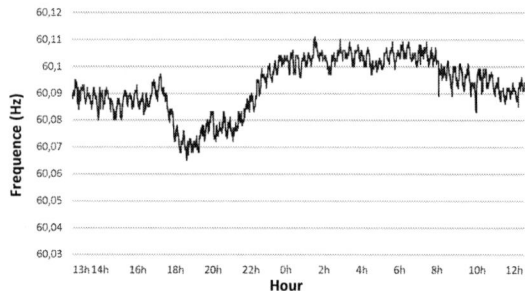

Fig. 7. Frequency deviation in June 2016.

Despite the fact that there are occasional frequency deviations, in 2019 the Frequency Performance in Steady State has been 97%, close to the 99% stipulated by ANEEL for the basic transmission system [12].

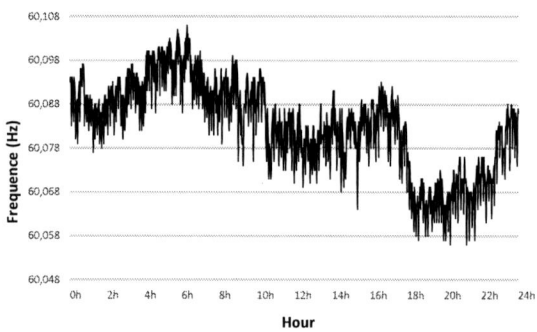

Fig. 8. Frequency deviation in April 2019.

The economic development of the community leads to frequent increases in load, resulting from the insertion of more equipment in the outlets, such as fluorescent lamps refrigerators and other non-linear loads. This phenomenon is responsible for inserting harmonics in the distribution network, deteriorating the voltage waveform and causing loss of quality.

The time evolution of current THD's (THD A L1, THD A L2 e THD A L3) and voltage THD (THD V) for the three phases of the system measured over a period of twenty-four hours in the years 2016 and 2019 are shown in Fig. 9 and Fig. 10, respectively. In both figures it is possible to observe the increase in the voltage and current THD values occurring mainly in the period between 18 and 23 hours, a period where there is a significant increase of loads in the system. The maximum values of current THD reached in 2019 do not exceed 45.79%, while those measured in 2016 are higher than 60%.

978-1-7281-4181-7/19 $31.00 © 2019 IEEE

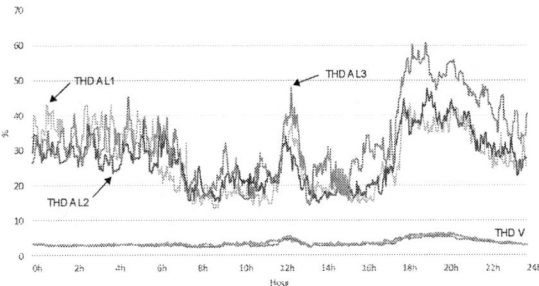

Fig. 9. Time evolution of voltage and current THD's in 2017.

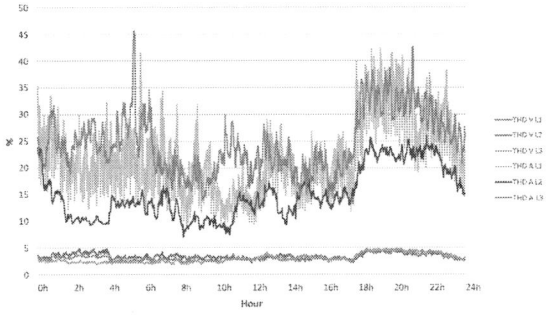

Fig. 10. Time evolution of voltage and current THD's in 2019.

Also, in 2019, there was a decrease in the voltage THD values compared to those measured in 2016. While in the latter the maximum value was 6.45% in 2019 this value was only 4.93%, well below that specified by EN50160 [13], which establishes a limit of 8% for harmonic distortion of voltage.

C. Operation

The average monthly consumption of diesel in 2017 and 2018 is presented in Fig. 11. It can be observed that there was a decrease in average fossil fuel consumption in the year 2018 compared to 2017.

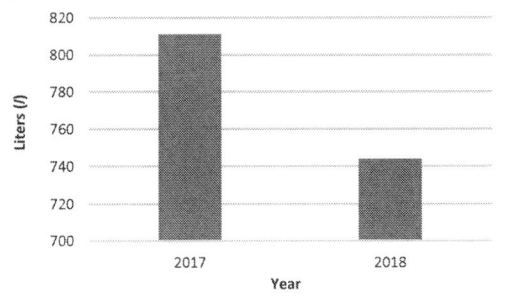

Fig. 11. Average monthly diesel consumption in the Ilha Grande micogrid in 2017 and 2018.

This fact can be explained by the increase in the number of photovoltaic panels which occurred in August 2018. We can observe in Table V that until July 2018, the system already had a saving of 206 liters of diesel compared to the same period in 2017. After August 2018 (month in which the new panels were added) fuel consumption has been reduced by more than 600 liters.

System efficiency gains are emphasized in Fig. 12 where in the month of September 2018 all the energy consumed was

of renewable origin, what did not occur in the same period of the previous year.

Despites the best performance presented by the system in 2018, the energy produced by the diesel generator is still large and corresponds to more than 25% of all energy produced during the year by the microgrid.

TABLE V. CONSUMPTION OF DIESEL.

Year	2017		2018	
Period	Jan to Jul	Aug to Dec	Jan to Jul	Aug to Dec
Consumption (l)	6826	2908	6620	2307
kWh Diesel/ kWh total	35%		25.76%	

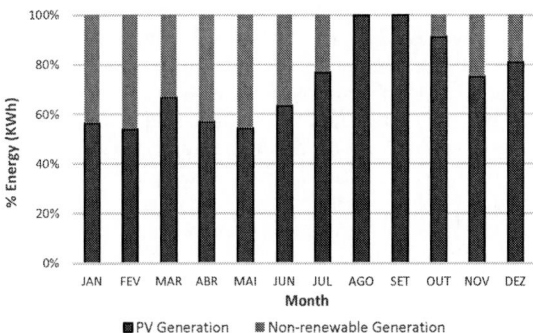

Fig. 12. Percentage of renewable and non-renewable generation in 2018.

D. Reliability

The system reliability degree is analyzed by observing, among other things, the number of interruptions in power supply and the durations of such interruptions. Table VI presents the DEC (equivalent duration of interruption per consumer unit) and FEC (equivalent frequency of interruption per consumer unit) measurements for the years 2017 and 2018. In 2017 supply interruptions were observed in the months of February and September. In 2018, there was interruption only in the month of May. It is possible to verify that the system presented a reduction in the number of interruptions, although the maintenance of the system occurs in a mostly corrective way. Despite this, the durations of the interruptions are high given the difficulties of access to the island by the maintenance teams.

TABLE VI. ANNUAL CONTINUITY INDICES.

Year	2017	2018
DEC (h)	42	48
FEC	2	1

E. The battery bank

Battery performance is severely affected by the intermittent cycles of solar generation, which makes it difficult to meet the operating restrictions specified by the manufacturer.

In [14] it was verified, from data from September 2016, that the aging factors of greater relevance in the Ilha Grande microgrid were the time between complete charges and the amount of Ah extracted. It is possible to verify this since, in the Ilha Grande microgrid, the power generated by photovoltaic panels is not always sufficient to maintain a minimum state of charge of the battery bank, so it is necessary to adopt strategies to keep the battery bank fully charged and to avoid the sulphation process. For this, the diesel generator is requested to equalize the battery bank, which occurs every ten days.

V. CONCLUSIONS

The current topology of the Ilha Grande microgrid has shown several advantages over its original configuration. The increase of the generation capacity and the adoption of better energy storage system technology associated with the high efficiency and flexibility resulting from parallel operation of the converters has enabled significant gains in performance and reliability.

We verified that there are still significant fuel costs for diesel generator operation, whose objective in the current topology is to avoid the reduction of the battery bank lifecycle due to the intermittent cycles of photovoltaic solar generation. However, experience has shown that the benefits of this procedure are still minimal, that is, banks are frequently replaced and diesel consumption is still large.

Even with all the improvements presented, the Ilha Grande microgrid still inspires care and investment in continuous improvements, especially with regard to diesel consumption and greater penetration of renewable energy in the microgrid power balance. Other aspects, such as those related to the improvement of reliability and maintenance planning, are subject to economic analysis, given the high costs involved in the implementation of adjustments in the maintenance strategies adopted.

REFERENCES

[1] IBGE - Instituto Brasileiro de Geografia e Estatística, "Objetivos de Desenvolvimento Sustentável," 15 abr. 2019. [Online]. Available: https://indicadoresos.ibge.gov.br/objetivo7/indicador711.

[2] L. A. S. Ribeiro, O. R. Saavedra, J. G. Matos, S. L. Lima and G. Bonan, "Making Isolated Renewable Energy System More Reliable," *Renewable Energy*, vol. 45, pp. 212-231, 2012.

[3] C. Marnay, S. Chatzivasileiadis and G. Joos, "Microgrid Evolution Roadmap. Engineering, Economics, and Experience," *International Symposium on Smart Electric Distribution Systems and Technologies (EDST15)*, pp. 1-6, 2015.

[4] ANEEL - Agência Nacional de Energia Elétrica, "Procedimento de distribuição de energia elétrica no sistema elétrico nacional – Prodist módulo 1 – Introdução," [Online]. Available: www.aneel.gov.br. [Accessed 16 Abr. 2019].

[5] Zigor, "Zigor HITC Hybrid Inverter," 2019. [Online]. Available: https://taspacenergy.co.nz/product/grid-hybrid-inverter-section. [Acesso em 24 may 2019].

[6] D. Q. Oliveira, S. L. Lima, O. R. Saavedra, L. A. S. Ribeiro and N. J. Camelo, "Confiabilidade de sistemas isolados baseados em Energias renováveis: requisitos, soluções e resultados," *Simpósio Brasileiro de Sistemas Elétricos*, 2010.

[7] O. R. Saavedra, S. H. Oliveira and J. G. Matos, "Energia solar e eólica como vetor de desenvolvimento de comunidades isoladas no Maranhão: o Projeto da Ilha dos Lençóis," *Revista INOVAÇÃO*, vol. 02, pp. 32-34, 05 Jul 2005.

[8] L. A. S. Ribeiro, O. R. Saavedra, S. L. Lima and J. G. de Matos, "Isolated micro-grids with renewable hybrid Generation: The case of Lençóis island," *IEEE Trans. Sustain. Energy*, vol. 2, pp. 1-11, 2010.

[9] Fulguris Tubular, "Manual técnico de baterias estacionárias ventiladas: Sistema fotovoltaico. Rev. C.," São Paulo, 2014.

[10] Cummins, "Engenharia de Aplicações - Manual de aplicações para grupos geradores arrefecidos a água," Power Generation, Guarulhos, 2016.

[11] ANEEL - Agência Nacional de Energia Elétrica, "Procedimento de distribuição de energia elétrica no sistema elétrico nacional – Prodist módulo 8 – Qualidade da Energia Elétrica," [Online]. Available: www.aneel.gov.br. [Accessed 11 Abr. 2019].

[12] ONS - Operador Nacional do Sistema Elétrico, "Procedimentos de Rede – Módulo 2 – Gerenciamento dos indicadores de qualidade da energia elétrica da Rede Básica," [Online]. Available: http://www.ons.org.br/paginas/sobre-o-ons/procedimentos-de-rede/vi¬gentes. [Accessed 16 Abr. 2019].

[13] European Committee for Electrotechnical Standardiz, "EN 50160. Voltage characteristics of electricity supplied by public electricity networks," 2010.

[14] P. B. L. Neto, "Contribuições Para a Operação Energética e Econômica de Microrredes Isoladas com Fontes Renováveis Diversificadas," Tese de Doutorado, São Luís, 2017.

[15] G. Bonan, A. M. Saccol, L. A. S. Ribeiro, O. R. Saavedra and J. G. Matos, "Parallel Connected Inverters Applied in Renewable Energy Systems," *Congresso Brasileiro de Eletrônica de Potência*, 2009.

[16] P. B. L. Neto, O. R. Saavedra and L. A. S. Ribeiro, "A Dual-Battery Storage Bank Configuration for Isolated Microgrids Based on Renewable Sources," *IEEE Transactions on Sustainable Energy*, vol. 9, pp. 1618-1626, Oct. 2018.

[17] F. de Bosio, A. C. Luna, L. A. d. S. Ribeiro, M. Graells, O. R. Saavedra and J. M. Guerrero, "Analysis and improvement of the energy management of an isolated microgrid in Lencois island based on a linear optimization approach," *2016 IEEE Energy Conversion Congress and Exposition (ECCE)*, pp. 1-7, 2016.

[18] L. A. S. Ribeiro, O. R. Saavedra, J. G. Matos, S. L. Lima, G. Bonan and A. S. Martins, "Design, control and operation of a hybrid electrical generation system," *Eletrônica de Potência*, vol. 15, pp. 313-322, Spt./Nov. 2010.

[19] R. Foster, R. Orozco and A. Rubio, "Lessons learned form Xcalak Village Hybrid System: A seven years retrospective," *International Solar Energy Society*, 1999.

[20] L. A. S. Ribeiro, O. R. Saavedra, J. G. Matos, G. Bonan and A. S. Martins, "Small renewable hybrid systems for stand alone applications," *IEEE Power Electronics and Machines in Wind Applications*, pp. 1-7, 2009.

[21] H. A. Oliveira, "Rede híbrida de distribuição de energia em CC e CA como uma solução alternativa para microrredes isoladas," São Luís, BR, 2017.

[22] H. A. Oliveira, J. G. de Matos, L. A. de Souza Ribeiros, A. S. Martins and G. C. Flores, "Operation of three-phase rectifiers with Diesel Generator Sets of similar power: Practical solution based on passive filters," *2017 IEEE 8th International Symposium on Power Electronics for Distributed Generation Systems (PEDG)*, pp. 1-5, 2017.

[23] M. S. Pereira, P. B. L. Neto, O. R. Saavedra and S. L. Lima, "Modelagem Via Redes Neurais da Capacidade de Sistemas de Armazenamento com Baterias em Smart Grids Isoladas com Fontes Renováveis," *12th Latin-American Congress on Electricity Generation and Transmission*, 2017.

Steady-State Characterization of the Three-Phase Isolated DC-DC Bidirectional Converter with LLC Resonant Tank

Kristian Pessoa dos Santos
Electrotechinique Departament
Federal Institute of Education, Science and Technology of Piauí
Parnaíba, Brazil
kristianpessoa@ifpi.edu.br

Hermínio Miguel Oliveira Filho
University for International Integration of the Afro-Brazilian
Lusophony - UNILAB
Redenção, Brazil
herminio@unilab.edu.br

Paulo Peixoto Praça
Power and Control Processing Group (GPEC)
Federal University of Ceará
Fortaleza, Brazil
paulopp@dee.ufc.br

Demercil de Souza Oliveira Júnior
Power and Control Processing Group (GPEC)
Federal University of Ceará
Fortaleza, Brazil
demercil@dee.ufc.br

Abstract— **This paper presents a theoretical and computational analysis of a novel three-phase DC-DC isolated bidirectional LLC resonant converter using frequency modulation to ensure the rated output voltage. The converter presents eighteen switches, three inductors and three high frequency single-phase transformers connected in open delta-wye configuration to isolation. The switching frequency preferably operates above the resonant frequency to ensure a ZVS soft switching operation. For the mathematical analysis of the converter, a single-phase model was used considering only the fundamental components and considering the main components as ideal. This converter is recommended for Energy Storage Systems (ESS) applications. Simulation results regarding a design example are realized to proof all theoretical analysis developed.**

Keywords—LLC converter, frequency modulation, three-phase DC-DC converter, bidirectional.

I. INTRODUCTION

With the increasing advances in power electronics technologies and microprocessor systems, the smart grids have been target of interest in research the development of technologies for photovoltaic panels, wind turbines, electric vehicles and electric power conversion systems to provide power flow or are energy consumers for EES as shows in Figure 1[1]. Actually, several topologies of DC-DC converters have been developed for integration of generation systems, storage or even consumers into a micro-grid. In addition, control strategies and modulation techniques are one of the main objects of researches to provide DC-DC converters with high efficiency and desirable characteristics such as bidirectionality and low harmonic content for the most diverse applications in the industry [2].

The resonant converters were developed in several possibilities with the main objective of reducing losses in semiconductors switches that operate in hard switching mode. In this way, it has been possible to achieve soft switching for many topologies and thus the power of the converters can gradually increase and, consequently, to increase also the frequency operation and the power density.

One of the most important resonant converters is the Series Resonant Converter (SRC), which has been extensively analyzed in the literature [1]-[5]. The maximum static gain is unitary and in low load situations using frequency modulation,

This work was support by National Council for Scientific and Technological Development (CNPq) and Cearense Foundation for Scientific and Technological Development Support (FUNCAP).

the SRC does not have voltage regulation capability since a large frequency range is required for voltage control, so it is not suitable for applications where there are large variations of the input voltage and load [6]-[7].

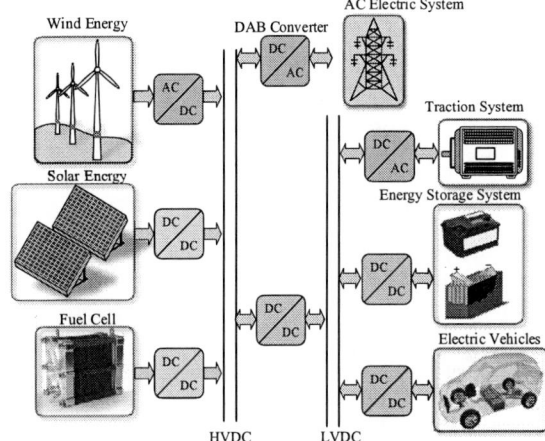

Fig. 1. Applications of bidirectional DC-DC converters with galvanic isolation in micro-grid [1].

To mitigate the disadvantages of SRC topology, the LLC converter, which has a 3th order resonant tank circuit, is more suitable for situations with high load variation. In addition, in a practical implementation, the parallel inductor is usually the high frequency transformer magnetization inductance and, as in SRC converter, the transformer leakage inductance can be used as all or part of the series inductor of the resonant circuit [7]-[9]. This work presents a novel propose of a resonant converter showed in Fig. 2 which is an improvement of the version of the topology introduced in [10]. The output voltage can be regulated by frequency modulation and power flow direction can be achieved with phase shift angle suitable [10]-[11].

II. ANALYSIS OF THE CONVERTER USING FUNDAMENTAL COMPONENTS

The propose converter uses a three-phase inverter with three single phase full-bridge circuits in parallel totalizing 12 semiconductor switches providing an increase in the power capability of the converter since the current stresses in the switches will be divided, in this way, this structure going to

978-1-7281-4181-7/19 $31.00 © 2019 IEEE

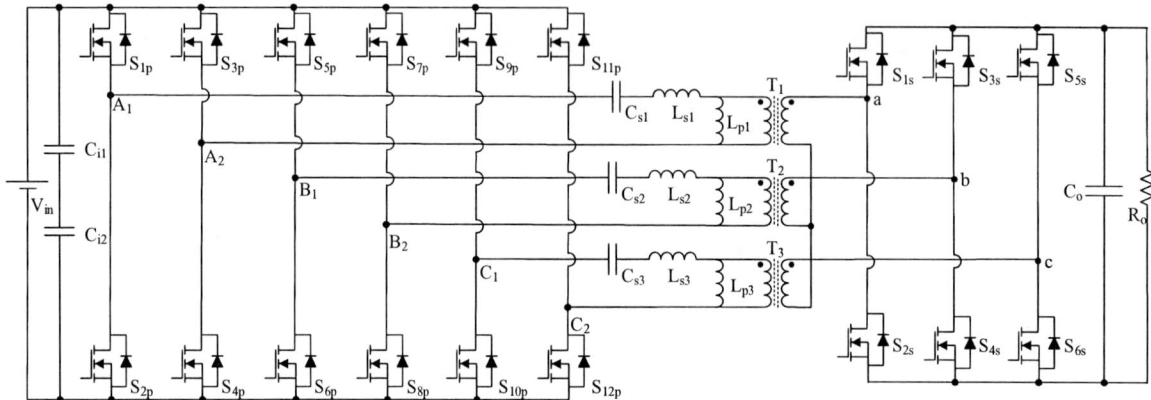

Fig. 2.   Three-Phase DC-DC Bidirectional LLC Resonant Converter.

ensure a better behavior than a bridge three-phase with six semiconductor switches. Besides, it decreases turn ratio due to a double output voltage by transformer open delta-wye connection. The duty cycle of converter is fixed at 50% and the switching frequency should stay above the resonant frequency to ensure ZVS over the entire operation range. The series capacitor is responsible to block DC current component which prevents the saturation of transformer [12]-[13].

In this work, a complete analysis, design example and simulation results are presented using Fundamental Harmonic Analysis (FHA) for the steady-state analysis, which only considers the fundamental component and neglects the dc element and higher order harmonic elements. All components are considered as ideal. This method is widely used in resonant converters to simplify the mathematical analysis and it ensures as low margin of error [12]-[14].

In Fig.2, the switches of each arm of the converter are phase shifted of 180°. The three full-bridges of primary circuit are phase shifted of 120° between them and the phase shift angle ϕ set the power flows direction. In other words, in the direct flow, the switches S_{1p} and S_{1s} have phase shift ϕ, switches S_{5p} and S_{3s} have phase shift $(120°+\phi)$ and switches S_{9p} and S_{5s} have phase shift $(240°+\phi)$.

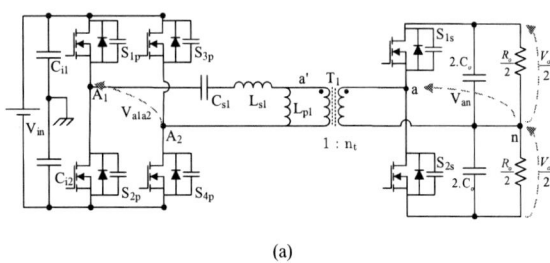

(a)

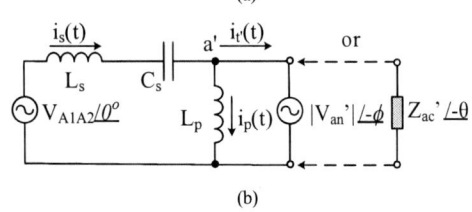

(b)

Fig. 3.   Equivalent circuit of the converter.

The equivalent circuit of phase A is depicted in Fig. 3, and phases B and C have the same equivalent circuit. In Fig. 3 (a)

it can be seen that the primary source of the converter $V_{A1A2}(t)$ is a square voltage source, since the switches operate with duty cycle equals 50% and the transformation ratio is $(1:n_t)$. In Fig. 3 (b) it can be seen that the secondary of the converter can be represented by a source with a phase-shift angle equal to $-\phi$ or by a complex impedance [7]. Moreover, due to use of the FHA, the primary side and secondary reflect by primary side sources will be represented by sinusoids, respectively, as shown by (1) and (2). The DC output voltage V_o reflected to the primary side and converter DC static gain M considering a duty cicle of 50% and which open delta-wye configuration has double the value of gain compared to a star connection in the primary circuit, are represented, respectively, by (3) and (4) [11].

$$V_{A1A2}(t) = \frac{4V_i}{\pi}\sin(\omega t) \tag{1}$$

$$V_{an'}(t) = \frac{4V_o'}{\pi}\sin(\omega t - \phi) \tag{2}$$

$$V_o' = \frac{V_o}{2n_t} \tag{3}$$

$$M = \frac{V_o'}{V_i} = \frac{V_o}{2n_t V_i} \tag{4}$$

Using the principle of superposition it is possible to find the currents $i_s(t)$, $i_p(t)$ e $i_t(t)$ shown in Fig.3(b). The angle α shown by (6) represents the phase angle in relation to (1) and X_s is the series impedance of L_s and C_s.

$$i_s(t) = \frac{4V_i\sqrt{M^2 - 2M\cos(\phi)+1}}{\pi X_s}\sin(\omega t - \alpha) \tag{5}$$

$$\alpha = tg^{-1}\left[\frac{1 - M\cos(\phi)}{M\sin(\phi)}\right] \tag{6}$$

978-1-7281-4181-7/19 $31.00 © 2019 IEEE

The current $i_p(t)$ through the parallel inductive reactance X_p consisting of the magnetizing inductance of the high frequency transformer is presented in (7).

$$i_p(t) = \frac{4V_o^{'}}{\pi X_p}\sin(\omega t - \phi - \frac{\pi}{2}) \qquad (7)$$

Once the value of series and parallel currents is determined by (5) and (7), respectively, the subtraction between them can determine the transformer current reflected to primary side presented in (8). The amplitude and phase angle in relation to (1) are, respectively, shown in (9) and (10).

$$i_t^{'}(t) = \left|I_t\right|\sin(\omega t - \beta) \qquad (8)$$

$$\left|I_t\right| = \frac{4V_i\sqrt{M^2 + \left|\dfrac{X_p}{X_s + X_p}\right|^2 - 2M\left|\dfrac{X_p}{X_s + X_p}\right|\cos(\phi)}}{\pi\left(\dfrac{X_s.X_p}{X_s + X_p}\right)} \qquad (9)$$

$$\beta = tg^{-1}\left[\frac{X_p - M\cos(\phi)(X_s + X_p)}{M\sin(\phi)(X_s + X_p)}\right] \qquad (10)$$

In order to express the output power of the single phase equivalent model shown in (11), the output voltage reflected to the primary side (3) is multiplied by the average value of (8) in a half-period. The active power of three-phase LLC converter proposed is shown in (12) and in the Fig.4 is represented graphically. It can be seen that the power of the converter varies according to the DC static gain, frequency and phase-shift angle, so to control the output power these quantities must be the variables of interest.

$$P_{o\_model} = \frac{8V_iV_o^{'}}{\pi^2 X_s}\sin(\phi) \qquad (11)$$

$$P_o = \frac{24MV_i^2}{\pi^2 X_s}\sin(\phi) \qquad (12)$$

For a more general analysis, all equations will be normalized to the base values.

$$V_{base} = V_i \qquad (13)$$

$$P_{base} = \frac{8V_{base}^2}{\pi X_s} \qquad (14)$$

$$Z_{base} = \sqrt{\frac{L_s}{C_s}} \qquad (15)$$

In addition, the power factor can be determined by the ratio between active power and apparent power of the converter and can be graphically represented by Fig.5. This last one can be obtained considering the single-phase model of the converter through the multiplication of the rms value of input voltage (1) and series current of resonant circuit (5).

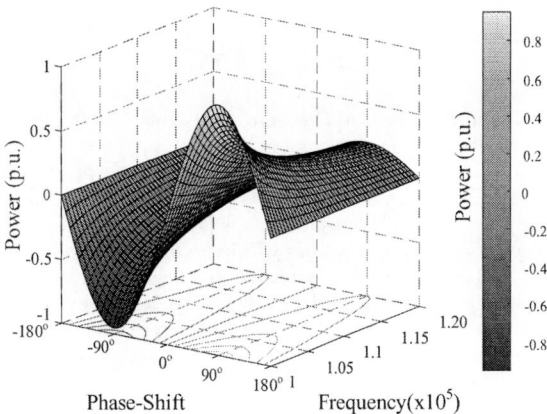

Fig. 4. Active power in pu of the converter versus phase shift angle and switching frequency.

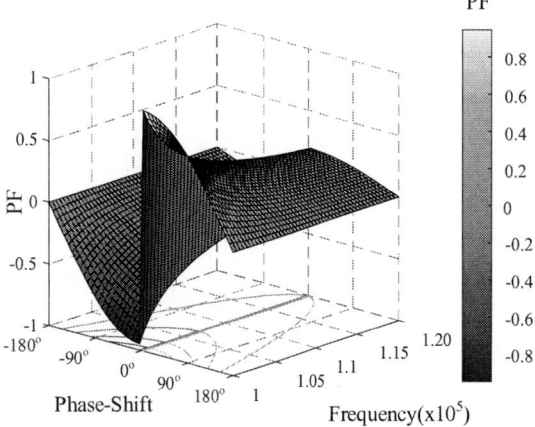

Fig. 5. Power factor of the converter versus phase shift angle and switching frequency.

The equivalent impedance presents in the Fig.3(b), which represents the secondary circuit reflected by primary side can be determined by ratio of *rms* value of (2) and (8) according (16). In particular case, where $\phi = \beta$, the secondary behaves as a resistive load in the same way as a diode bridge.

$$Z_{ac}^{'} = \left|\frac{V_{an'\_rms}}{I_{t\_rms}^{'}}\right|\angle\phi - \beta \qquad (16)$$

In the resonant converters, the normalized frequency $F = f_s / f_o$ is defined by ratio of the switching frequency to the resonant frequency. In addition, the quality factor Q and the ratio between the parallel and series inductances of the resonant tank circuit represented by λ are showed in (17) and (18), respectively, where $\omega_o = 2\pi f_o$ [7].

$$Q = \frac{\sqrt{\dfrac{L_s}{C_s}}}{Z_{ac}^{'}} = \frac{\omega_o.L_s}{Z_{ac}^{'}} \qquad (17)$$

$$\lambda = \frac{L_p}{L_s} \tag{18}$$

According to (16)-(18), the DC static gain M of the converter can be determined by (19). Fig. 6 shows the behavior of the voltage gain of the converter where in the region 01, the switching frequency is less than the resonant frequency and the converter has a capacitive characteristic. The semiconductor switches will be switched off with ZCS, but will be switched on with hard switching. In region 02, the switching frequency is intermediate between the open-circuit resonance frequency (lower) and the short-circuit resonance frequency (larger), however, in this region there is ZVS switching and depends exclusively on the load. In region 03, the switching frequency of the converter is always higher than the resonance frequency and presents essentially inductive behavior and the semiconductor switches operate with ZVS independent of the load conditions.

$$M = \left| \frac{V_o^{'}}{V_i} \right| = \left| \frac{\dfrac{Z_{ac}^{'} \cdot j\omega L_p}{Z_{ac}^{'} + j\omega L_p}}{j\left(\omega L_s - \dfrac{1}{\omega C_s}\right) + \left(\dfrac{Z_{ac}^{'} \cdot j\omega L_p}{Z_{ac}^{'} + j\omega L_p}\right)} \right| \tag{19}$$

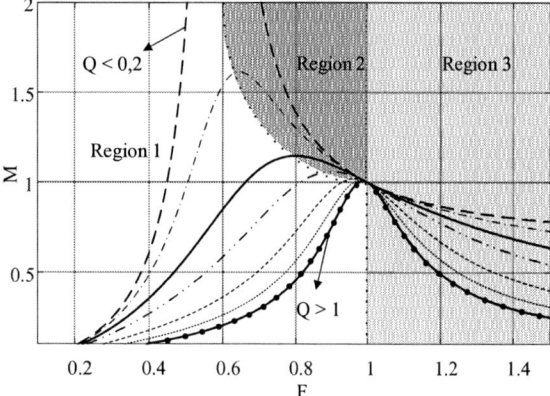

Fig. 6. Voltage gain M of the converter versus normalized frequency F.

The equivalent input impedance of the single-phase model shown in Fig. 3(b) can be determined by (20). The behavior of the module and phase of the input impedance with respect to the frequency is shown in Fig.7 considering values in *pu*.

$$Z_{in} = j\left(\omega L_s - \frac{1}{\omega C_s}\right) + \left(\frac{Z_{ac}^{'} \cdot j\omega L_p}{Z_{ac}^{'} + j\omega L_p}\right) \tag{20}$$

In Fig. 7, when Q> 1 and $f_s > f_o$, the converter behaves as an inductive load and therefore the current in the series capacitor C_s will be delayed in relation to the fundamental voltage $V_{A1A2}(t)$ which is required for switching ZVS. In this way, the *mosfets* antiparallel diodes turn off with low *di/dt* and do not generate reverse current peaks. Instead, when Q>1 and $f_s < f_o$, the converter behaves as a capacitive load and therefore the current will be advanced in relation to the voltage that is fundamental for ZCS switching. In this condition, the antiparallel diodes of the *mosfets* turn off with large *di/dt*, thus generating a high reverse current peak.

In situations where Q is large, the output voltage of converter will be more sensitive to small variations in the switching frequency. The increase of λ decreases the value of the static gain peak, which compromises the voltage regulation capacity for some load conditions and decreases the ZVS range below the resonance frequency.

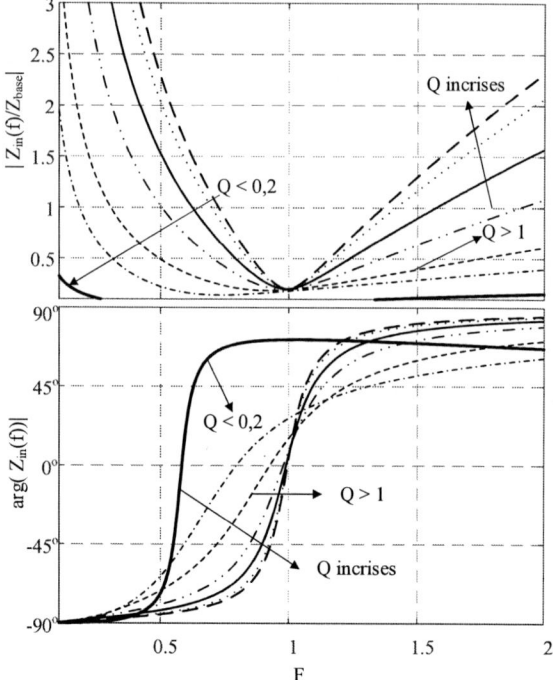

Fig. 7. Gain and phase of input impedance normalized versus normalized frequency F.

III. VALIDATION OF THE PROPOSED ANALYSIS

In Table I are presented the design specifications and assumed parameters of the converter which are used for the calculation of the components.

TABLE I. DESIGN SPECIFICATIONS OF THE EXAMPLE CONVERTER

Parameter		Value
V_i	Input voltage	96V
V_o	Output voltage	380V
f_s	Switching frequency	96 - 140 KHz
f_o	Resonant frequency	95.24 KHz
ϕ	Phase shift angle	-45° to 35°
P_o	Output power	3 kW

With the gradual increase of Z_{base} represented by (15), results in a larger switching frequency range which, consequently, can cause problems to develop a suitable project design. The ratio transformer of high frequency transformer is presented by (21) [10]-[11]. The main parameters of the converter are presented in Table II.

$$n_t = \frac{V_o}{4V_i(1-D)} = 1,979 \tag{21}$$

TABLE II. PARAMETERS OF CONVERTER

	Parameter	Value
D	Duty cicle	50%
M	Voltage gain	1
Q	Quality factor	3
λ	Indutance ratio	2
F	Normalized frequency	1.05
L_s	Series Inductor	61.6μH
C_s	Series Capacitor	45.33ηF

IV. SIMULATION RESULTS

To validate the proposed converter, simulation tests were performed with ideal components using the PSIM software. The analysis of the simulations will verify the operation of the steady state converter and the switching conditions. Fig. 8 shows the three-phase primary and secondary line voltages. The voltage V_{A1A2}(t) is the input voltage of the resonant tank circuit and it is displacement from the secondary voltage V_{an}(t) by phase-shift angle ϕ=30,8°. In theoretical analysis, these waveforms are analyzed considering only the fundamental component as presented in (1) and (2).

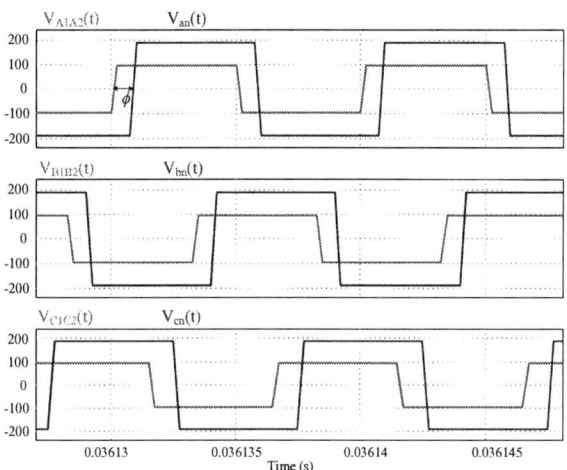

Fig. 8. Three-phase primary and secondary line voltages.

Fig. 9 shows the series inductors three-phase currents of the resonant tank circuit which are lagged at 120°from each other. The currents are balanced and have a peak and *rms* value, respectively, of 17.61A and 12.39A. The peak and rms values calculated by (5), correspond to 18.04A and 12.75A, respectively. Thus, the percentage error between the calculated and simulated values was 2.4% and 2.9%. Fig. 10 shows the series capacitors three-phase voltages of the resonant tank circuit which are lagged at 120° from each other. The voltages are balanced and they have a peak and *rms* value, respectively, of 621V and 439V. The peak and rms values correspond to 596V and 421V, respectively. Thus, the percentage error between the calculated and simulated values was 4% and 4.1%.

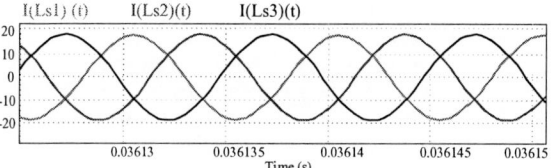

Fig. 9. Series inductors currents of resonant tank resonant.

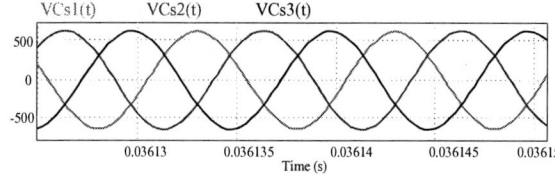

Fig. 10. Series capacitors voltages of resonant tank resonant.

Fig. 11 shows the active power and maximum apparent power of the converter. It can be seen that the active power of the converter was approximately 3 KW whereas the apparent power of the transformer has 3.23 KVA presenting a power factor of transformer equal to 0.913. The output voltage found was 380V and output current was 7.89A which are the values of the project.

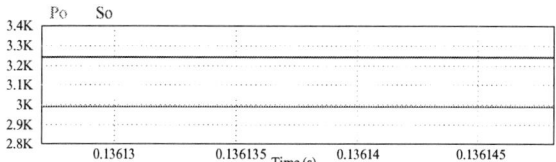

Fig. 11. Active power and apparent power of the converter in forward direction.

Fig.12 presents the line voltage of the primary circuit V_{A1A2} (t) is delayed in relation to the line voltage of the secondary circuit V_{an}(t) which characterizes the behavior opposite to that shown in Fig. 8. In addition, Fig.13 presents the active power and apparent power in the reverse direction of the power flow with the same frequency and phase-shift angle equal to -30.6° it is possible to observe that the same results verified in Fig.11 and calculated in design.

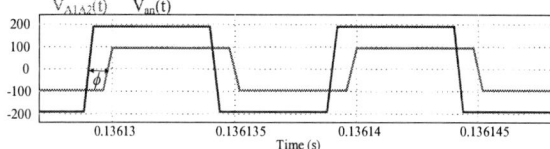

Fig. 12. Primary and secondary line voltage of primary circuit.

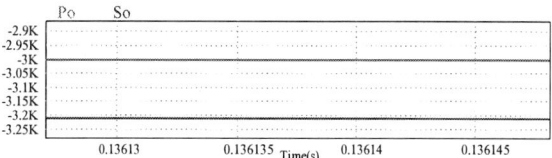

Fig. 13. Active power and apparent power of the converter in backward direction.

978-1-7281-4181-7/19 $31.00 © 2019 IEEE

Fig. 14 presents the voltage and current waveforms for upper and lower switches of the primary side in the forward direction of the power flow. When the voltage $VS1_p(t)$ in the upper switch $S1_p$ goes to zero, $S1_p$ is switched on and at the same instant the diode intrinsic to $S1_p$ is turned on and, thus, $S1_p$ operates with ZVS and hard switching when it is switched off. Similarly, the lower primary switch $S2_p$ has the same behavior but is displaced $180°$ in relation to $S1_p$.

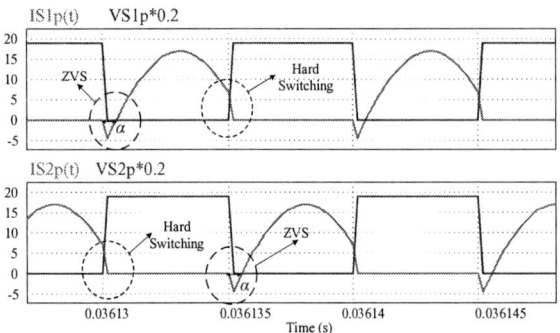

Fig. 14. Voltage and current waveforms of switches S1p and S2p of the primary side.

The same way, Fig. 15 presents the voltage and current waveforms for upper and lower semiconductors switches of the secundary side in the forward direction of the power flow. When the voltage $VS1_s(t)$ in the upper switch $S1_s$ goes to zero, $S1_s$ is switched on with ZCS. When the diode intrinsic current goes to zero, $S1_s$ is turned on and operates with ZVS and with hard switching when it is switched off. Similarly, the lower primary semiconductor switch $S2_s$ has the same behavior but is displaced $180°$ in relation to $S1_s$.

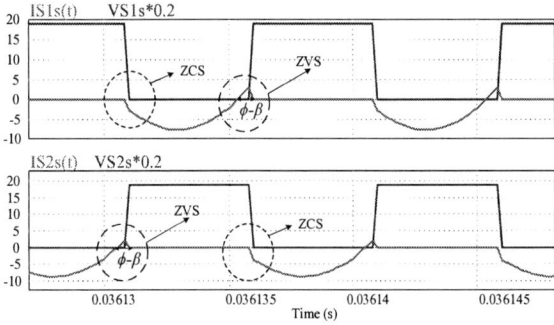

Fig. 15. Voltage and current waveforms of switches S1s and S2s of the secondary side.

V. CONCLUSION

This work has presented an analysis of a three-phase DC-DC bidirectional LLC resonant converter using FHA, and his behavior with frequency variable. The major contribution of this work is to present a novel propose of the topology proposed by [10] in a bidirectional resonant version using pulse frequency modulation. The resonant converter provides balanced sinusoidal currents in the high frequency transformer and it can operate with frequencies in the order of 100 KHz. Thus, the propose topology provides an improvement to the

reduction of the weight and volume of the magnetic and, consequently, to increase power density. In addition, the topology can operate with ZVS switching over a large load range. This topology can be used in several applications of industry like ESS for microgrids and vehicles electrics. A broader analysis is being developed in closed loop as well as experimental implementation and will be presented in the future.

ACKNOWLEDGMENT

The authors thank the technical suport by Power and Control Processing Group (GPEC) and Federal Institute of Piauí (IFPI) and also the financial support by the CNPq and FUNCAP.

REFERENCES

[1] M. Yaqoob, K.H. Loo, and Y. M. Lai, "Fully Soft-Switched Dual-Active-Bridge Series-Resonant Converter With Switched-Impedance-Based Power Control," IEEE Transactions on Power Electronics, vol. 33, nº 11, pp. 9267-9281, Nov. 2018.

[2] D.W. Hart, Power Electronics, 1th ed MC Graw Hill-Bookman, New York, 2011.

[3] X. Li and A.K.S. Bhat, "Analysis and Design of High-Frequency Isolated Dual-Bridge Series Resonant DC/DC Converter," IEEE Transactions on Power Electronics, vol. 25, nº 4, pp. 850-862, Apr. 2010.

[4] Y. Zhang, M. Lu and X. Li, "A Two-stage Control Scheme for a Dual-Bridge Series Resonant Converter," 9th International Conference on Power Electronics (ECCE) Asia - June 1 - 5, 2015, Seoul, Korea.

[5] R. Mirzahosseini and F. Tahami, "A Phase-Shift Three-Phase Bidirectional Series Resonant DC/DC Converter," 37th Annual Conference of the IEEE Industrial Electronics Society-(IECON) November 7 - 10, 2011, Melbourne, Australia.

[6] M.T.Outeiro, G.Buja, and D. Czarkowski, "Resonant Power Converters: An Overview with Multiple Elements in the Resonant Tank Network," IEEE Industrial Electronics Magazine, vol. 10, nº 2, pp. 21-45, Jun. 2016.

[7] M.K.Kazimierczuk, and D. CZARKOWSKI, Resonant Power Converters, 2th Ed. Ed.Wiley, New Jersey, 2011

[8] D.D. Nguyen, D.T. Nguyen, G. Fujita and T. Funabashi, "Dual-active-bridge series resonant converter: A new control strategy using phase-shifting combined frequency modulation," IEEE Energy Conversion Congres and Exposition (ECCE), September 20-24, 2015, Montreal, Canada.

[9] M.S. Almardy and A.K.S. Bhat, "Three-Phase (Lc)(L)-Type Series-Resonant Converter With Capacitive Output Filter," IEEE Transactions on Power Electronics, vol. 26, nº 4, pp. 1172-1183, Apr. 2011.

[10] C. Liu, A. Johnson, and J.Lai, "A Novel Three-Phase High-Power Soft-Switched DC/DC Converter for Low-Voltage Fuel Cell Applications," IEEE Transactions on Industry Applications, vol. 41, nº 6, pp. 1691-1697, Nov. 2005.

[11] H.M. Oliveira Filho, D.S. Oliveira Junior, and P. P. Praça, "Steady-State Analysis of a ZVS Bidirectional Isolated Three-Phase DC–DC Converter Using Dual Phase-Shift Control With Variable Duty Cycle," IEEE Transactions on Power Electronics, vol. 31, nº 03, pp. 1663-1872, Mar. 2016.

[12] F. Liu, Y. Chen, and X. Chen, "Comprehensive Analysis of Three-Phase Three-Level LC-Type Resonant DC/DC Converter With Variable Frequency Control—Series Resonant Converter" IEEE Transactions on Power Electronics, vol. 32, nº 07, pp. 5122-5131, Jul. 2017.

[13] R.W.De Donker, D.M.Divan and M.H..Kheraluwala, "A three-phase soft-switched high-power-density DC/DC converter for high-power applications", , IEEE Transactions on Industry Applications, vol. 27, n°. 1, pp. 63-73, Jan-Feb 1991.

978-1-7281-4181-7/19 $31.00 © 2019 IEEE

Gas Microturbines for Distributed Generation System

Walquíria do N. Silva, Janaína G. de Oliveira, Bruno H. Dias, Leonardo W. de Oliveira

Graduate Program in Electrical Engineering

Federal University of Juiz de Fora, Juiz de Fora - MG, Brazil 36036-900

walquiria.silva@engenharia.ufjf.br, janaina.oliveira@ufjf.edu.br bruno.dias@ufjf.edu.br, leonardo.willer@ufjf.edu.br

Abstract—This paper presents the operation and modelling of gas microturbines as Distributed Generation (DG) systems. The studied system consists of a gas microturbine linked to Permanent Magnet Synchronous Generator (PMSG), which is connected to a back-to-back converter. The purpose of this study is to analyze the dynamic behaviour of gas microturbines in a simple and operative model so that the results can be used to investigate the applicability of energy generation. The development of the system will be validated through simulation, using the Simulink-MATLAB® software, whose obtained results will verify the system's operation and suggest possible applications.

Index Terms—microturbine, distributed generation, control power, grid-connected mode

I. INTRODUCTION

Because of the current energy situation, more efficient power systems with lower cost and less environmental impact are proposed. In this context, renewable sources, such as solar and wind energy, systems of cogeneration and trigeneration coming from thermal energy from gas microturbines, internal combustion engines, fuel cells and small hydroelectric power plants systems have become important research topics. These systems aim at providing a sustainable supply of electricity, improving the reliability, stability and quality of distribution and generation systems [1].

Among these technologies, gas microturbines use has been increasing not only because it is a dispatchable generation source but also because it produces heat and cold. Thus, presenting an increase in the efficiency of those systems by cogeneration. They are used in applications where high quality in the power supply is required, is also used for the remote generation, in cases of suppression of voltage peaks. Also, it can be used in grid-connected or isolated system models, in medium and low voltage profiles [2].

Considering the increase in the use of gas microturbine, it is necessary to analyze some characteristics when it is connected to the electrical grid, such as transient stability of the system, harmonics and energy quality. Thus, it is essential to study this application regarding operability and impact on energy systems, to allude to the modelling of the microturbine and the technical aspects for connection to the distribution grid [3]. The dynamic model of gas microturbine has already been discussed in previous studies, in which they address the dynamics of the microturbine and control methodology for systems connected to the grid or isolated [4].

The most widespread turbine model has been developed by Rowen [5]. This model presents a simplified proposal for the analysis of the machine response when connected to large power systems. This prototype can be adapted to different applications, and it is possible to adapt the parameters of this dynamic model to represent small turbines, such as gas microturbines, by making some adaptations [6].

The microturbine is the denomination given to a small turbine with the potential to generate electricity within the 25 to 500 kW [7]. The microturbines have two different configurations, single-shaft or split-shaft. The single-shaft operates at high rotational speeds, generally between 90.000 and 120.000 rpm. The compressor, turbine and generator are connected on the same shaft [8].

Different authors have disseminated the study and implementation of the dynamic model of single-shaft gas microturbines for DG application in grid-connected and isolated systems. Thus, it is discussed in [9], [6], [7], [10] and [11] while a detailed study of the methodology is proposed in [5], which consists to apply speed, acceleration, temperature and fuel system controls of to generate energy in a turbine.Through this application, the authors suited the parameters of the dynamic model studied as they are feasible for different microturbines proposed in their respective works. As for the power electronic interface of the single-shaft microturbine connected to the grid or in an isolated system, there are different configurations available and this is a primary part for proper operation of the system [12].

The main power electronics topologies are the connection through three-phase diode rectifier and a three-phase PWM inverter to control the active and reactive powers to be injected into the grid as presented by [13], [14], [15]. Also, the other configuration used is presented by [12], [16], [17], [18], [6], is a back-to-back converter, where the three-phase controlled rectifier on the machine side controls the speed of the synchronous generator and the grid side inverter regulates the DC bus voltage and currents injected into the grid.

In this work, a single-shaft microturbine system connected to the three-phase grid is proposed. The dynamic modelling presented in [5] was adapted to the C30 microturbine model, for distributed generation systems applications. Considering the electronic interface, a back-to-back converter was adopted, with the control system given by the AC/DC converter controlled by the power to be injected into the grid. So there is only active power transfer. The DC/AC Converter operates the voltage control on the bus and consequently the currents to be injected into the grid. The simulation is realized to analyze the performance of the model when it is connected to the distribution grid.

This article is divided as follows: the second section describes the system under study and the applied methodology for the grid-connected microturbine components, presenting the mathematical model of operation and design of system controllers. Following, the third section, Simulation and Results, presents the main achievements obtained from the applied methodology. Finally, Conclusions are drawn in the last section.

II. METHODOLOGY

This section will describe the proposed system. Four essential parts are considered to model the grid-connected microturbine: (i) analysis of the dynamic behaviour of single-shaft gas microturbines, (ii) dynamics of the Permanent Magnet Synchronous Generator (PSMG), (iii) control methodology applied to back-to-back converter and grid-connected low pass filter design.

The proposed topology for the described application is presented in Fig.1.

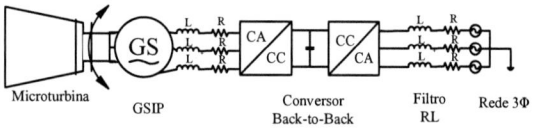

Fig. 1. Microturbine generation system.

Fig. 1 shows the implemented system. The power grid is a three-phase 220 V, 60 Hz power source. To obtain the pleaded characteristics for the system implemented concomitantly intending to give only the active power transfer to the grid, the parameters and the operation functionality of the constituent elements for correct and desired operation of the system. To perform the simulation of the proposed system was used software Simulink-MATLAB® and the use of the SIMPOWER library was made.

A. Microturbine

The microturbine considered in this study is the model C30 from Capstone, that generates a nominal power of about 30 kW. To perform realize the simulation in software, used the parameters presented in Table I, were used in the simulations, representing this microturbine according to the prototype proposed by [5].

The parameters presented in the Table I refer to the variables presented in Figure 2, where they mention the controller's gains and the time constants associated with each control system that make up the microturbine mathematical modelling where they are detailed in the works [5] and [6].

TABLE I
PARAMETERS CAPSTONE C30 USING IN ROWEN MODEL [6]

Parameters	Value	Parameters	Value
K	25		
T_1	0.40 s	T_{CD}	0.0341 s
T_2	0.05 s	T_{TD}	0.04 s
K_1	1.4455	K_4	0.85
K_2	0.50	K_5	0.15
K_3	0.3082	T_4	2.50 s
T_V	0.04 s	T_5	3.30
T_C	0.6675 s	T_R	275°C
$V_{CE_{MAX}}$	1.5	K_{T1}	454
$V_{CE_{MIN}}$	-0.10	K_{T2}	165
T_{CR}	0.005 s		

As shown in Fig.2 the model the microturbine consists of five control blocks: speed, acceleration, temperature, fuel system and turbine dynamics. The speed control will act on the load conjunctures, which may be isochron or droop type. The temperature control has the function of limiting the output power and the acceleration control has the function of avoiding that there is no speeding in the rotor decreasing mechanical stresses during start-up. These controllers act on the minimum selector block, whose output will be the smallest value among the three inputs that maintains the fuel flow at the appropriate level for the operation of the compressor-chamber-turbine set that represents the dynamics of the microturbine [11].

The compressor-chamber-turbine set presents the characteristics of the torque and the exhaust temperature, representing the power part of the system being responsible for the transformation of the chemical energy of the fuel into the thermal energy of the gases and thus in mechanical power over the shaft, defined as turbine torque in which triggers the electric generator [17].

The mathematical modelling of microturbine and the control system was analyzed using the model developed by Rowen [5], which presents a simplified proposal for machine response analysis when connected to large power systems. Thus for its implementation, only the operating parameters of the microturbine were adjusted according to Table I for model C30.

B. Permanent Magnet Synchronous Generator (PMSG)

The microturbine generates electric power through a high-speed generator, driven directly by the turbine shaft rotor. For microturbine configuration adopted in this work, a 2 pole permanent magnet synchronous generator of permanent magnets was used, with nominal output power of 30 kW and the line-to-line voltage of 480 V. The resistance of the stator windings is $R_s = 0.25 \ \Omega$ and inductance of direct and quadrature axis is $L_d = L_q = 0.6875$ mH.

The analysis of the PMSG is performed employing the dq0 axis theory, being possible to represent the electrical and mechanical parts of the generator by a space states model of second order. The equations expressed below make mention of the dynamics of the rotor and the electromechanical referential [12]. The electrical and mechanical equations are expressed as follows:

Electrical Equations:

978-1-7281-4181-7/19 $31.00 © 2019 IEEE

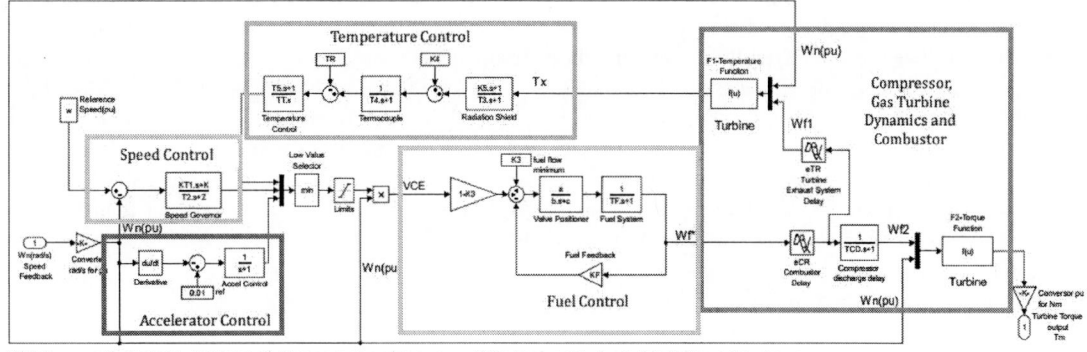

Fig. 2. Block diagram of microturbine system with control

$$\frac{\mathrm{d}}{\mathrm{dt}}i_d = \frac{1}{L_d}v_d - \frac{R}{L_d}i_d + \frac{L_q}{L_d}p\omega_r i_q \quad (1)$$

$$\frac{\mathrm{d}}{\mathrm{dt}}i_q = \frac{1}{L_q}v_q - \frac{R}{L_q}i_q + \frac{L_d}{L_q}p\omega_r i_d - \frac{\lambda p\omega_r}{L_q} \quad (2)$$

$$T_e = \frac{3}{2} \cdot p \cdot (\lambda i_q + (L_d - L_q)i_d i_q) \quad (3)$$

Mechanical equations:

$$T_e = p \cdot T_m \quad (4)$$

$$\frac{\mathrm{d}}{\mathrm{dt}}\omega_r = \frac{1}{J} \cdot (T_e - F\omega_r - T_m) \quad (5)$$

$$\frac{\mathrm{d}\theta}{\mathrm{dt}} = \omega_r \quad (6)$$

where: L_d and L_q inductance of the direct (d) and quadrature (q) axis, R is the resistance of the stator windings, i_d and i_q the currents of the (d) and (q) axis, v_d and v_q the voltages of the (d) and (q) axis, ω_r angular velocity of the rotor, λ is the flux induced by the permanent magnets in the stator windings, T_e electromagnetic torque, p is the number of pole pairs, J combined inertia of rotor and load, F combined viscous friction of rotor and load, θ rotor angular position and T_m the mechanical torque of the shaft.

The electrical frequency of this generator is 1600 Hz. Knowing that the grid operates at 60 Hz, it is necessary to use an electronic interface for grid connection. For this, a back-to-back power converter was used.

C. Power Electronics

Microturbines operate at a very high speed, and consequently generate electricity in alternating current with high frequency, requiring a reduction of the same for power generation applications. There are mechanisms to reduce the frequency so that it operates synchronised with the grid. Generally, for these applications, it makes use of AC/DC/AC power electronic converters.

The electronic interface that makes the connection of the microturbine with the grid is made by a back-to-back converter. This is a static converter that allows a bidirectional flux, being its topology consisting of two converters: an AC-DC converter operating as a rectifier and a DC-AC converter operating as the inverter, interconnected through a DC bus. This arrangement allows no communication between the control of the converters, as the DC bus gives a distinct control for each one, allowing the connection between the two AC systems with different operating characteristics.

The DC bus provides decoupling between converters, supplying the system power variation in the transitional regime. The voltage at this link must be higher than the peak voltages on the AC sides so that the bidirectional flux of active and reactive power is possible [19].

1) Machine Side Converter Control: Given that the PMSG speed is controlled by the microturbine, the machine output power is controlled by the rectifier (or Machine Side Converter Control) through current control, so that all transferred power is active. The control was made by controlling the dq0 synchronous coordinates (to implement in the software simulation, we used the following: 90 degrees behind phase A-axis- modified Park). To synchronize the voltage and frequency parameters for the Park transform, the electric angle converted from the mechanical angle from the rotor rotation of the PSMG was used as seen in Fig.3.

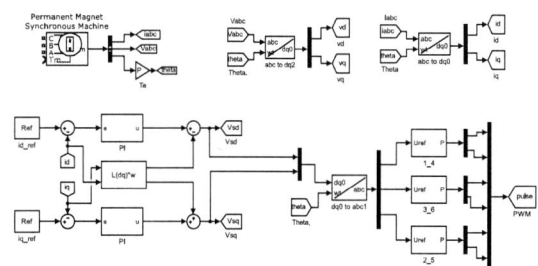

Fig. 3. Machine Side Converter Controller

For current control design, the dynamic behaviour of the system was observed and according to its stationary characteristics, and a PI controller was employed. It ensures that the system operates without a permanent regime error. For the tuning of the proposed controller used the methodology proposed in [20], which suggests that the parameters of K_p and K_i are obtained through the relationship of the values of resistance, inductance and the time constant of the compensated system.

2) Converter Control for Grid-Connected Mode: The control structure on the grid side is given through two meshes of the in-series control. The internal mesh carries out current control and the external mesh is the DC bus voltage control and aims at keeping the constant voltage at the DC link. The control topology for the grid-connected converter is shown in Fig. 4.

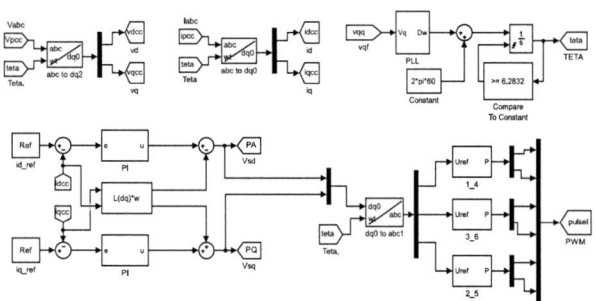

Fig. 4. Grid Side Converter Controller

The internal mesh control must be faster than the external mesh. The function that represents the dynamic behaviour of the voltage on the DC bus has a non-linear characteristic, so for the linear control project the system was analyzed in and a certain point of operation in a permanent regime and it was found that this, also has a stationary characteristic, so that the system operates expectedly made use of a PI type controller. To tune, the regulator adopted the methodology proposed in [21], which suggests that the parameters of K_p and K_i are obtained by comparing the closed-loop transfer function of the DC bus voltage squared to the canonical function of this comparison the gains of the controller are inferred.

For current control, the mode of regulation is analogous to the rectifier as they operate similarly which differs is the mode of synchronization of the currents. The sync structure adopted in this system was Phase-Locked Loop (PLL) which provides the phase angle of the voltage necessary for the transformations of Park acting following the characteristics of the grid.

The projects of the controllers proposed in this section were designed to obtain a satisfactory response concerning the transient response and the null stationary error assuring a desirable behaviour of the system.

D. Filter Design

According to [19], the connection of the three-phase converter to the mains can be given through an inductive filter.

The inductor is connected in series with the converter circuit and aims to attenuate the harmonics and the high-frequency components caused by the switching of semiconductors.

The criterion for the sizing of the inductor is to limit the peak of the current ripple. For this calculation despised the internal resistances that compose it and considered that the reference voltage of the converter is equal to the supply voltage [22]. The inductance of the filter can then be calculated by Equation 7:

$$L_f = \frac{V_{fase}}{2\sqrt{6}f_{sw}I_{r_{pico}}} \quad (7)$$

where V_{fase} is the phase voltage of the grid, f_{sw} is the switching frequency and I_{rp} is the desired peak current.

n practice it is not common to use the pure inductor for the coupling of the converter this presents an incorporated resistance. Therefore, it is customary to use resistance in series with the inductor. This must-have of approximately 10% of the value of the reactance referenced to the grid frequency [23].

III. SIMULATION AND RESULTS

In this section, the main results obtained from the simulation of the projected system are presented. Fig. 5 presents the main diagram of the system implemented in Simulink-MATLAB® to analyze the microturbine system operation performance in a grid-connected mode.

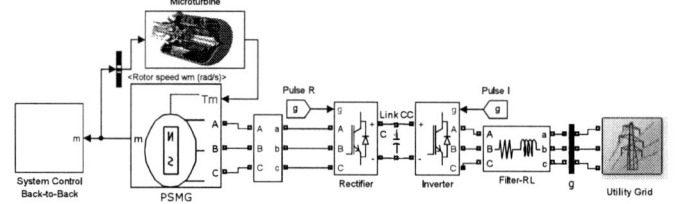

Fig. 5. Diagram of the grid-connected microturbine

Considering the analysis addressed in the methodology, the data used in the simulation are presented in Table II.

TABLE II
SIMULATION PARAMETERS

Parameter	Value
Switching frequency (f_{sw})	20 kHz
Voltage DC bus (V_{CC})	760 V
DC capacitance (C)	9200 μF
Inductance (L)	96 μH
Resistance (R)	3.6 mΩ
Sample Time (Ts)	1 μs

For the design considered that the parameters $\tau = 5$ ms, $\omega_n = 1.26 \cdot 10^3$ rad/s e $\zeta = 0.8$. The obtained gains are presented in Table III.

The variables inherent to the electrical generation from the simulation describing the microturbine system connected to the grid and the operation in a distributed generation were analyzed to corroborate the efficacy of the architected

978-1-7281-4181-7/19 $31.00 © 2019 IEEE

TABLE III
CONTROL SYSTEM PARAMETERS

Parameters	Value
$K_{p,r}$	1.3750 V/A
$K_{i,r}$	500 Vs/A
$K_{p,i}$	0.1901 V/A
$K_{i,i}$	7.2 Vs/A
$K_{p,dc}$	0.6866 A/V
$K_{i,dc}$	$1.0785 \cdot 10^4 As/V$

system. The parameters investigated in the simulation are those correlated to the electrical generation described below:

A. Voltage on the DC bus

The voltage behaviour of the DC bus is illustrated in Fig.6

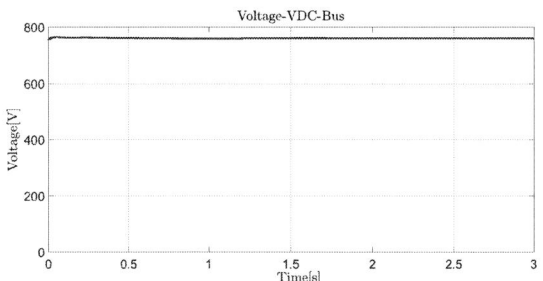

Fig. 6. Voltage at the DC-Bus

Fig. 6 represents the voltage on the back-to-back converter bus. It can be observed that the waveform obtained is a constant voltage at 760 V as drafted. The maximum voltage ripple is 1%.

The variations in the bus voltage occur only when there is an imbalance of the powers on the AC sides of the converter, in the transient.

B. Electrical voltage and current in the terminals of the grid

The three-phase voltage and current delivered to the grid are depicted in Fig.7 and Fig.8 respectively.

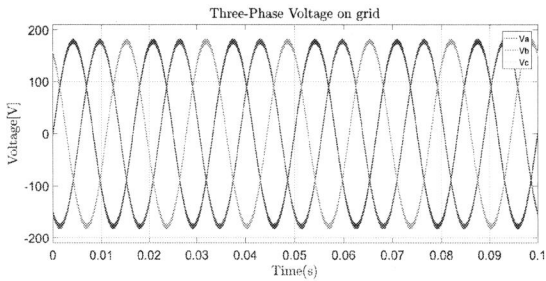

Fig. 7. Three-Phase Voltage on grid

As can be seen in Fig.7 and Fig. 8 the waveforms of the voltage and the current at the common coupling point (CCP) are balanced lagged at 120° and without distortions. The initial system transient takes about 0.2 s

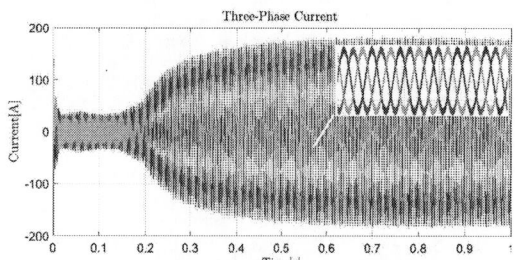

Fig. 8. Three-Phase Current on grid

C. Active power delivered to the grid

The current and the voltage delivered to the distributed generation system are illustrated in Fig. 9.

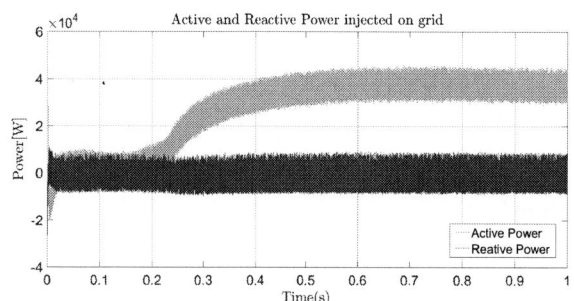

Fig. 9. Active and Reactive Power injected on the grid

Fig. 9 shows the active and reactive power injected into the electrical grid. It can be is noted that the reactive power is equal to zero as projected by the system control proving its efficacy. The active power injected into the grid is approximately 32 KW.

D. Compared Speed

To assure the correct operation of the microturbine, it is necessary to generate sufficient mechanical power for the mechanical torque to trigger the PSMG rotor and operate in normal and appropriate conditions. It should be checked whether the feedback rate comes from the generator output and to ensure that it is generated within the values of the nominal speed of the microturbine.

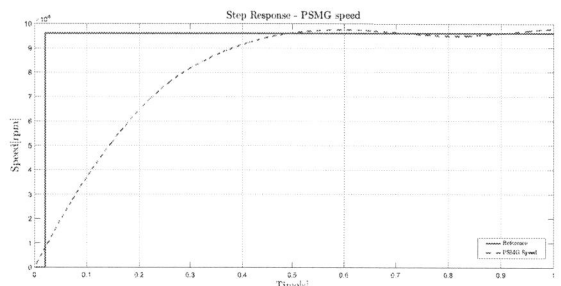

Fig. 10. Compared speed

Fig. 10 shows the dynamic response of PSMG speed to a microturbine speed step (9600 rpm). Thus, through the figure, it is possible to analyze the dynamic behaviour of PSMG speed.

IV. Discussion and Conclusion

The results show that the system projected presents the expected behaviour. The voltage on the DC bus remained constant at 760 V and within the projected value, that is, it has the highest value than the AC side voltages, allowing the bidirectional flow of active and reactive power in the system.

Regarding the behaviour of the current and the voltage delivered to the network, they reached the values expected by the designed control system. Through, reaching the adequate generation of active and reactive power.

Considering the microturbine, it was found that the model adopted has provided a satisfactory result for power generation. Also, this model can be adapted to different types of microturbine as presented in the literature. From this analysis, it was found that the difficulty in adapting the model to other types of microturbine is to correlate the parameters, that represent the dynamics of the prototype proposed by the author, to another model. Besides being necessary to improve the systems of Already pre-established controls.

Through the analyzes obtained in the previous section, it was possible to verify the dynamic behaviour of PSMG velocity about a step type input. Through this, it was found that the speed control of PSMG and microturbine operate in line. If this does not happen, it is necessary to improve the microturbine dynamic speed control system and/or even to add to the system a machine speed control as proposed by some authors.

As a future work intends to improve the control system Microturbine C30, add loads to the system and still make modifications to the control and structure of the system so that it operates in the form off-grid (isolated), check how the system of the Microturbine in front of an unbalanced system.

Acknowledgment

The authors would like to acknowledge CAPES and PPEE-UFJF for their support on this work. This study was financed in part by the Coordenação de Aperfeiçoamento de Pessoal de Nível Superior- Brasil (CAPES) - Finance Code 001

References

[1] D. Çelik and M. E. Meral, "Current control based power management strategy for distributed power generation system," *Control Engineering Practice*, vol. 82, pp. 72–85, 2019.

[2] P. Lilienthal, "Natural gas improves economic feasibility of hybrid distributed energy systems," *Natural Gas & Electricity*, vol. 35, no. 6, pp. 1–7, 2019.

[3] K. Srikanth, K. Naresh, L. N. Rao, and V. Ramesh, "Matlab/simulink based dynamic modeling of microturbine generator for grid and islanding modes of operation," *International Journal of Power Systems*, vol. 1, pp. 1–6, 2016.

[4] T. Sun, J. Lu, Z. Li, D. L. Lubkeman, and N. Lu, "Modeling combined heat and power systems for microgrid applications," *IEEE Transactions on Smart Grid*, vol. 9, no. 5, pp. 4172–4180, 2017.

[5] W. I. Rowen, "Simplified mathematical representations of heavy-duty gas turbines," *Journal of Engineering For Power*, vol. 105, no. 4, pp. 865–882, 1983.

[6] J. G. Rauber, "Avaliação de modelos de microturbina a gás single shaft para estudos de microgeração distribuída com cogeração térmica," dissertação de mestrado engenharia elétrica e computação, Universidade Estadual do Oeste do Paraná, 2016.

[7] H. Wei, Z. Jianhua, W. Ziping, and N. Ming, "Dynamic modelling and simulation of a micro-turbine generation system in the microgrid," in *2008 IEEE International Conference on Sustainable Energy Technologies*, pp. 345–350, IEEE, 2008.

[8] M. A. R. Maldonado, "Modelagem e simulação do sistema de controle de uma micro-turbina a gás. 2005. 149 f.," dissertação de mestrado engenharia elétrica e computação, Universidade Federal de Itajubá, Itajubá, MG, Brasil, 2005.

[9] W. I. Rowen, "Simplified mathematical representations of single shaft gas turbines in mechanical drive service," in *ASME 1992 International Gas Turbine and Aeroengine Congress and Exposition*, pp. 1–7, American Society of Mechanical Engineers, 1992.

[10] S. R. Guda, C. Wang, and M. Nehrir, "A simulink-based microturbine model for distributed generation studies," in *Proceedings of the 37th Annual North American Power Symposium, 2005.*, pp. 269–274, IEEE, 2005.

[11] D. M. Costa, "Aspectos técnicos e operacionais do uso de microturbinas conectadas aos sistemas elétricos para geração distribuída de energia," dissertação de mestrado em engenharia elétrica, Universidade Federal de Santa Maria, 2010.

[12] D. Gaonkar, R. Patel, and G. Pillai, "Dynamic model of microturbine generation system for grid connected/islanding operation," in *2006 IEEE International Conference on Industrial Technology*, pp. 305–310, IEEE, 2006.

[13] F. Ding, K. A. Loparo, and C. Wang, "Modeling and simulation of grid-connected hybrid ac/dc microgrid," in *2012 IEEE Power and Energy Society General Meeting*, pp. 1–8, IEEE, 2012.

[14] S. K. Nayak and D. N. Gaonkar, "Modeling and performance analysis of microturbine generation system in grid connected/islanding operation," *International Journal of Renewable Energy Research (IJRER)*, vol. 2, no. 4, pp. 750–757, 2012.

[15] F. Mohamed, "Microgrid modelling and simulation," *Helsinki University of Technology, Finland*, 2006.

[16] A. Kumar, K. Sandhu, S. Jain, and P. S. Kumar, "Modeling and control of micro-turbine based distributed generation system," *International Journal of Circuits, Systems and Signal Processing*, vol. 3, no. 2, pp. 65–72, 2009.

[17] S. Shakur and D. S. K. Jain, "Micro-turbine generation using simulink," *International Journal of Electrical Engineering*, vol. 5, no. 1, pp. 95–110, 2012.

[18] D. N. Gaonkar, "Performance of microturbine generation system in grid connected and islanding modes of operation," in *Distributed Generation*, pp. 185–208, InTech, 2010.

[19] K. S. Crispim, "Estudo do controle do conversor back-to-back para conversão de frequências.," 2018.

[20] D. Yazdani, A. Bakhshai, and P. K. Jain, "Grid synchronization techniques for converter interfaced distributed generation systems," in *2009 IEEE Energy Conversion Congress and Exposition*, pp. 2007–2014, IEEE, 2009.

[21] P. M. Almeida, A. A. Ferreira, H. A. Braga, and P. G. Barbosa, "Projeto dos controladores de um conversor vsc usado para conectar um sistema de geração fotovoltaico a rede elétrica," in *Anais do Congresso Brasileiro de Automática*, pp. 3960–3965, 2012.

[22] L. T. F. Soares, *Contribuiçao ao Controle de um Conversor Reversıvel Aplicado a um Aerogerador Sıncrono a Imas Permanentes*. Tese de doutorado, Universidade Federal de Minas Gerais, Belo Horizonte, MG, Brasil, 2012.

[23] R. Cutri, "Compensação de desequilíbrios de carga empregando conversor estático operando com modulação em largura de pulso.," dissertação de mestrado, Universidade de São Paulo, São Paulo, SP, Brasil, 2004.

Inherent Redundancy of SDBC-MMCC based STATCOM in the Overmodulation Region

D.C. Mendonça[a], A.F. Cupertino[b,c], H.A. Pereira[d], S.I. Seleme[c] and R. Teodorescu[e].

[a]Graduate Program in Electrical Engineering, Federal Center for Technological Education of Minas Gerais,
Belo Horizonte, MG, Brazil. dayane.mendonca@ufv.br
[b]Department of Materials Engineering, Federal Center for Technological Education of Minas Gerais
[c]Graduate Program in Electrical Engineering, Federal University of Minas Gerais,
Belo Horizonte, MG, Brazil. afcupertino@ieee.org, seleme@cpdee.ufmg.br
[d] Department of Electrical Engineering, Federal University of Viçosa, Viçosa, MG, Brazil. heverton.pereira@ufv.br
[e]Department of Energy Technology, Aalborg University, Aalborg, Denmark. ret@et.aau.dk

Abstract—The single delta bridge cell modular multilevel cascaded converter (SDBC-MMCC) has emerged as an interesting topology for Static Synchronous Compensator (STATCOM) applications. Since this converter is composed of tens/hundreds of components, reliability issues arise and redundant cells must be employed to fulfill the reliability requirements, increasing the initial cost of the STATCOM. In fact, there is a potential for fault tolerant operation when the overmodulation region is considered. This work aims to explore the inherent redundancy of SDBC-MMCC in the overmodulation region. Initially, the boundary conditions between linear and overmodulation region are analytically derived. Then, the inherent fault tolerance is evaluated based on 17 MVA/13.8 kV SDBC-MMCC based STATCOM case study with 24 cells per cluster. The results indicate that the converter is able to exchange rated power with the grid after 1 failure, with harmonic distortion within the limits suggested by the standards.

Index Terms—Modular Multilevel Cascaded Converter, STATCOM, Overmodulation Region

I. INTRODUCTION

The Modular Multilevel Cascaded Converter (MMCC) has emerged as an interesting topology for high power and medium/high voltage applications, as in High-Voltage Direct Current (HVDC) and Static Synchronous Compensator (STATCOM) [1], [2]. Compared to other topologies of multilevel converters or two-level converters, some features stand out as: modularity and scalability, high efficiency, superior harmonic performance and absence of dc-link capacitors [3], [4]. Furthermore, the SDBC-MMCC topology is widely used in STATCOM applications [5], [6]. Manufacturers like ABB, Siemens and GE already market the STATCOMs based on this converter topology [7]–[9].

In addition, the MMCC is composed of numerous components that can fail due to different reasons [10], [11]. In order to maintain the converter operation if one or more cells fail, additional cells called redundant cells are integrated in the MMCC [12]. Several papers in the literature propose redundancy strategies. Reference [13] presents a classification of redundancy strategies in hot and cold reserve. Reference

[14] proposes the standard redundancy operation where the capacitor voltage and the inner stress are reduced. Finally, an operational improvement of MMCC with redundancy cells, which can offer reduction of voltage harmonics and switching losses, is proposed in [15].

However, there is a cost associated with inserting redundant cells. It is possible to design the converter to withstand a voltage increase in the cells, but this implies a design with reduced utilization factor of semiconductors [16]. Thus, the derating strategy can be used during failures. The limitation of the derating strategy is that the converter is not able to exchange rated power with the grid.

On the other hand, there is a potential in the overmodulation region which was not explored in the literature yet. The converter operating in the nonlinear region is able to synthesize 10% more voltage ($\frac{2\sqrt{3}}{\pi} \approx 1.1$). Therefore, if this region is fully employed, a 10% redundancy factor can be obtained [17].

Reference [18] discusses the use of MMCC overmodulation techniques under voltage constraints through PWM modulation. Reference [19] shows a strategy of overmodulation based on a three-dimensional representation of the control voltages of the converter in the vector space. Although these references present studies of the overmodulation region, the boundaries between the linear and nonlinear regions of the converter were not studied.

Therefore, this work provides the following contributions:

- Analytical expressions for the limit between linear and non-linear region of the converter;
- Evaluation of the operation in the overmodulation region as a fault tolerance strategy in order to provide a cost effective design of SDBC-MMCC based STATCOM.

All the results are validated for a 17 MVA SDBC-MMCC based STATCOM case study. The outline of this paper is presented as follows. Section II presents the SDBC-MMCC based STATCOM and the control strategy used. Section III shows the limit between linear and non-linear region of the converter and discusses the sensitivity analysis. Section IV shows the simulation results. Finally, the conclusions are stated in section V.

II. SDBC-MMCC BASED STATCOM

The schematic of SDBC-MMCC based STATCOM is presented in Fig. 1. This topology presents three clusters. The inductance L_{cl} is responsible to reduce the high-order harmonics in the circulating current [1], [20]. C is the cell capacitance and L_f is the transformer inductance. A permanent bypass switch S_T in parallel with each cell is utilized in case of failures [21] and a bleeder resistor R_b is used to discharge the cell capacitor.

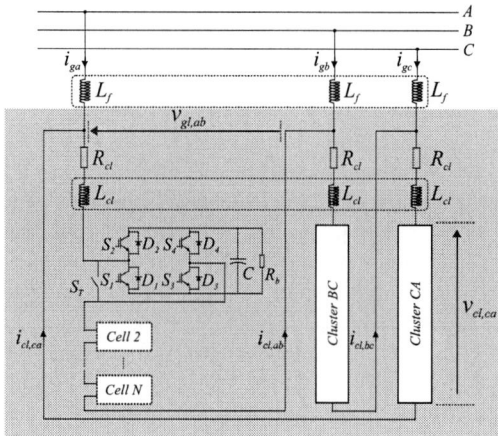

Fig. 1. Schematic of the SDBC-MMCC based STATCOM.

The output and the circulating currents can be calculated by the cluster currents, as follows:

$$\begin{cases} i_{ga} = i_{cl,ab} - i_{cl,ca}, \\ i_{gb} = i_{cl,bc} - i_{cl,ab}, \\ i_{gc} = i_{cl,ca} - i_{cl,bc}, \end{cases} \quad (1)$$

$$i_z = \frac{1}{3}(i_{cl,ab} + i_{cl,bc} + i_{cl,ca}). \quad (2)$$

The control strategy adopted is presented in [22] and is shown in Fig. 2. The circulating current (zero sequence current) is used to ensure capacitor voltage balancing. The circulating current reference is calculated by the clusters voltage balancing control [5], [22]. Moreover, the output current control is performed through the components in stationary reference frame controlling negative and positive sequence of the grid current [6].

The cluster and the average voltages are calculated by:

$$v_{cl} = \frac{1}{N} \sum_{j=1}^{N} v_{cell,j}, \quad (3)$$

$$v_{avg} = \frac{1}{N_T} \sum_{j=1}^{N_T} v_{cell,j}, \quad (4)$$

where N_T is the total of operating cells.

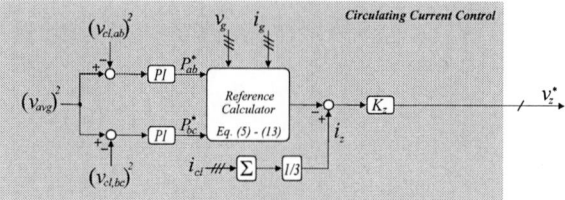

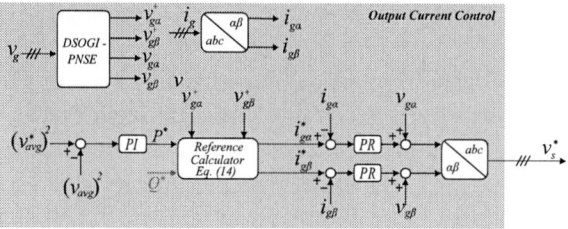

Fig. 2. Control strategy of the SDBC-MMCC based STATCOM.

The reference calculator in the circulating current control computes the power disturbances (P_{ab}^* and P_{bc}^*) in the clusters and calculates the circulating current reference [22], as follow:

$$i_z = \widehat{I}^z cos(\omega t + \phi^z), \quad (5)$$

where

$$I^z = \frac{P_{ab}^* - A}{X_1 cos(\phi^z) + X_2 sen(\phi^z)} = \frac{P_{bc}^* - B}{X_3 cos(\phi^z) + X_4 sen(\phi^z)}, \quad (6)$$

$$tan(\phi^z) = \frac{(P_{bc}^* - B)X_1 - (P_{ab}^* - A)X_3}{(P_{ab}^* - A)X_4 - (P_{bc}^* - B)X_2}, \quad (7)$$

and

$$A = \frac{\widehat{V}^+ \widehat{I}^-}{2} cos(\delta^+ - \phi^- + \frac{\pi}{3}) + \frac{\widehat{V}^- \widehat{I}^+}{2} cos(\delta^- - \phi^+ - \frac{\pi}{3}), \quad (8)$$

$$B = \frac{\widehat{V}^+ \widehat{I}^-}{2} cos(\delta^+ - \phi^- - \pi) + \frac{\widehat{V}^- \widehat{I}^+}{2} cos(\delta^- - \phi^+ + \pi), \quad (9)$$

$$X_1 = \frac{\sqrt{3}}{2} \widehat{V}^+ cos(\delta^+ + \frac{\pi}{6}) + \frac{\sqrt{3}}{2} \widehat{V}^- cos(\delta^- - \frac{\pi}{6}), \quad (10)$$

$$X_2 = \frac{\sqrt{3}}{2} \widehat{V}^+ sin(\delta^+ + \frac{\pi}{6}) + \frac{\sqrt{3}}{2} \widehat{V}^- sin(\delta^- - \frac{\pi}{6}), \quad (11)$$

$$X_3 = \frac{\sqrt{3}}{2} \widehat{V}^+ cos(\delta^+ - \frac{\pi}{2}) + \frac{\sqrt{3}}{2} \widehat{V}^- cos(\delta^- + \frac{\pi}{2}), \quad (12)$$

$$X_4 = \frac{\sqrt{3}}{2} \widehat{V}^+ sin(\delta^+ - \frac{\pi}{2}) + \frac{\sqrt{3}}{2} \widehat{V}^- sin(\delta^- + \frac{\pi}{2}). \quad (13)$$

Furthermore, the reference calculator in the output current control computes the active power that flows to the converter and calculates the grid current references in stationary reference frame. This principle is based on the instantaneous power theory developed by [23], as presented in (14).

$$\begin{bmatrix} i_{g\alpha}^* \\ i_{g\beta}^* \end{bmatrix} = \frac{1}{(v_{g\alpha}^+)^2 + (v_{g\beta}^+)^2} \begin{bmatrix} v_{g\alpha}^+ & v_{g\beta}^+ \\ v_{g\beta}^+ & -v_{g\alpha}^+ \end{bmatrix} \begin{bmatrix} P^* \\ Q^* \end{bmatrix} + \begin{bmatrix} i_{g\alpha}^- \\ i_{g\beta}^- \end{bmatrix}.$$
$$(14)$$

The reference voltages v_z^* and v_s^* are used for the modulation strategy. The Nearest Level Control (NLC) with cell tolerance band is considered [24]. The output voltage is synthesized based on the insertion index of each cluster, given by [6], [25]:

$$n = \frac{v_z^* + v_s^*}{v_c^\Sigma}, \tag{15}$$

where v_c^Σ represents the sum of the capacitor voltages of the cluster.

III. BOUNDARY BETWEEN LINEAR AND NON-LINEAR REGION

A. Minimum dc component of the sum of capacitor voltages

This section aims to calculate the boundary between linear and non-linear region of the converter. For the SDBC, the internal voltage reference v_z^* is neglected and (15) is summarized by:

$$n = \frac{v_s^*}{v_c^\Sigma}, \tag{16}$$

where v_s^* is the line voltage that the converter must synthesize, given by:

$$v_s^* = \sqrt{3}\widehat{V}cos(\omega t + \delta + \theta_v + \frac{\pi}{6}), \tag{17}$$

where θ_v assumes the values $(0, \frac{2\pi}{3}, \frac{-2\pi}{3})$ for phases A, B and C, respectively, δ is the angle of v_s, ω is the line frequency and $\widehat{V}$ is the line-neutral voltage amplitude.

The sum of capacitor voltages can be calculated by [6]:

$$v_c^\Sigma = v_d + \frac{N\widehat{V}\widehat{I}}{4\omega C v_d}sin\left(2\omega t + 2\theta_v + \delta + \phi + \frac{\pi}{3}\right), \tag{18}$$

where v_d is the dc component of v_c^Σ, C is the cell capacitance and ϕ is the angle of cluster current. The cluster current is given by:

$$i_{cl} = \frac{\widehat{I}}{\sqrt{3}}cos\left(\omega t + \theta_v + \phi + \frac{\pi}{6}\right) + i_z, \tag{19}$$

where i_z is neglected for the SDBC topology under balanced conditions [6].

The output voltage required for grid connected applications, when the converters injects current and assuming a negligible value of inductor resistance, is given by:

$$\widehat{V} = \sqrt{\left[\widehat{V}_g\left(1 + \Delta V_g\right) + x_{eq}\widehat{I}sin(\phi)\right]^2 + \left[x_{eq}\widehat{I}cos(\phi)\right]^2}, \tag{20}$$

where ΔV_g represents the percentual variation in the grid voltage and x_{eq} represents the output reactance of the converter, given by:

$$x_{eq} = \frac{x_{cl}}{3} + x_f, \tag{21}$$

where x_{cl} represents the cluster reactance and x_f represents the transformer reactance.

Finally, the maximum insertion index from (16) happens when $n = 1$ and $\omega t = -\delta - \frac{\pi}{6} - \theta_v$. Due to symmetry, only cluster AB is analyzed. Thus, $\theta_v = 0$. Then, from the equations (16), (17) and (18), the following relationship is obtained:

$$1 = \frac{\sqrt{3}\widehat{V}}{v_d + \frac{N\widehat{V}\widehat{I}}{4\omega C v_d}sin\left(\phi - \delta\right)}. \tag{22}$$

Solving (22) for v_d, the following second order polynomial is obtained:

$$\underbrace{4\omega C}_{a}v_d^2 \underbrace{-4\sqrt{3}\widehat{V}\omega C}_{b}v_d + \underbrace{N\widehat{V}\widehat{I}sin\left(\phi - \delta\right)}_{c} = 0. \tag{23}$$

The minimum v_d required is given by:

$$v_d = max\left(\frac{-b + \sqrt{b^2 - 4ac}}{2a}, \frac{-b - \sqrt{b^2 - 4ac}}{2a}\right). \tag{24}$$

Fig. 3 shows the minimum dc component of v_c^Σ, considering the parameters of Tab. I. As observed, the critical situation happens in the capacitive operation, because it requires 27.9% more dc component voltage than in the inductive region.

TABLE I
PARAMETERS OF THE MODULAR MULTILEVEL CONVERTER.

Parameter	Value
Grid voltage (v_g)	13.8 kV
Effective dc voltage (v_d)	22.8 kV
Rated power (S_n)	17 MVA
Transformer inductance (L_f)	4.75 mH
Transformer X/R ratio	18
Cluster inductance (L_{cl})	5 mH
Cluster inductor X/R ratio	17
Cell capacitance (C)	5 mF
Nominal cell voltage (v_{cell}^*)	950 V
Line frequency (f_n)	60 Hz
Sampling frequency (f_s)	6.48 kHz
Number of cells (N)	24 per cluster

The developed analytical model is validated through simulations in PLECS environment, as shown in Tab. II. Four operation conditions were analyzed. The largest relative error found between the analytical model and PLECS simulation is 4.5%.

Fig. 4 presents the inserted voltage and sum of capacitor voltages for 1 pu operation condition presented in Tab. II. Fig. 4(a) and Fig. 4(b) present the analytical model and PLECS simulation for capacitive operation. Fig. 4(c) and Fig.

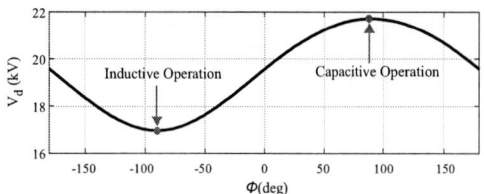

Fig. 3. Minimum dc component of the sum of the capacitor voltages.

TABLE II
MINIMUM v_d FOR OPERATION IN LINEAR REGION OF THE MODULATOR.

Operation Condition	Analytical	Simulation	Relative Error
$\widehat{I} = 1\text{pu}; \phi = \frac{\pi}{2}$	21.71 kV	22.25 kV	2.5%
$\widehat{I} = 1\text{pu}; \phi = -\frac{\pi}{2}$	16.98 kV	17.04 kV	0.38%
$\widehat{I} = 0.5\text{pu}; \phi = \frac{\pi}{2}$	20.66 kV	21.58 kV	4.5%
$\widehat{I} = 0.5\text{pu}; \phi = -\frac{\pi}{2}$	18.29 kV	17.99 kV	1.6%

4(d) present the analytical model and PLECS simulation for inductive operation.

As observed, the inserted voltage is limited by the sum of capacitor voltages in inductive and capacitive operation. The small differences observed are justified by 2 factors:

1) The capacitors in the simulation are not perfect balanced;
2) The internal voltage v_z, which was neglected in the analytical model.

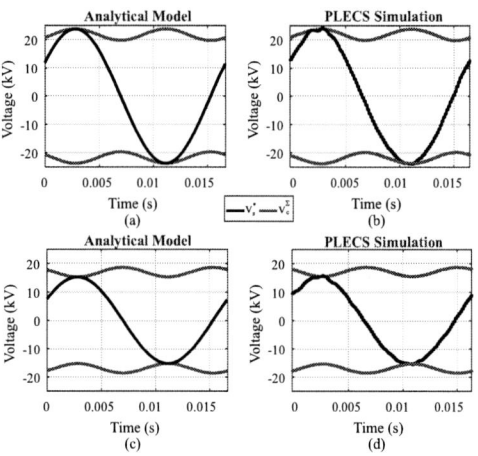

Fig. 4. Inserted voltage and sum of capacitor voltages in the limit of the modulator linear region for v_d presented in Tab. II, considering 1 pu of operation condition: (a) Analytical result for $\phi = \pi/2$; (b) Simulation result for $\phi = \pi/2$; (c) Analytical result for $\phi = -\pi/2$; (d) Simulation result for $\phi = -\pi/2$.

B. Sensitivity analysis

This section analyzes the effect of the variation of some parameters of the converter on the minimum dc component of the sum voltage capacitors, such as output current, cell capacitance and output reactance. These parameters are an important step for a cost-effective design.

The effect of the output current amplitude is presented in Fig. 5(a). As observed, when the current is reduced, the dc component voltage decreases in the capacitive region and increases in the inductive region.

The effect of the cell capacitance is presented in Fig. 5(b). The dc component voltage decreases in the inductive region and increases in the capacitive region when a capacitance is increased.

Finally, the effect of the output reactance is presented in Fig. 5(c). As observed, when the reactance is reduced, the dc component voltage increases in the inductive region and decreases in the capacitive region.

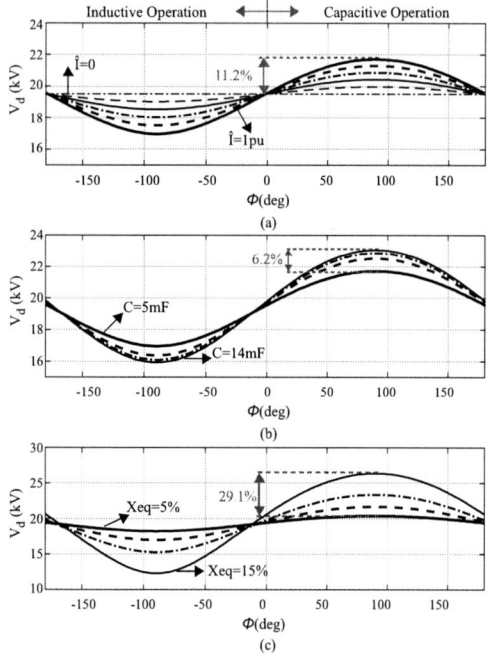

Fig. 5. Minimum v_d for variation of: (a) output current, (b) cell capacitance and (c) output reactance.

The SDBC-MMCC based STATCOM must be capable of operating at rated power for the two operating conditions. For the parameters adopted, $v_d \approx 21.71$ kV at nominal capacitive reactive power. In order to handle the parametric variations and the additional voltage component for circulating current control, an additional voltage margin must be taken into account. In this paper, a 5% margin is adopted. Therefore, $v_d \approx 22.8$ kV. Under such conditions, for $N = 24$, the reference voltage of the cell is $v_{cell}^* \approx 950$ V and power switches of 1.7 kV may be employed.

IV. SIMULATION RESULTS

The parameters of the designed SDBC-MMCC based STATCOM are presented in Tab. I. The simulations were developed in the PLECS environment in order to validate the inherent redundancy. Since the designed converter presents

978-1-7281-4181-7/19 $31.00 © 2019 IEEE

$N = 24$ cells, the operation in overmodulation region can guarantee the operation until 2 failures.

In order to analyze the operation of the converter in the overmodulation region, the compensation of 1 pu of reactive power in the capacitive region is simulated. At $t = 0.5$ second, the first failure happens. At $t = 1$ second, the second failure happens. This papers considers symmetric failures, which means that when one failure is identified in cluster AB, one cell of cluster BC and other of cluster CA are also bypassed.

Fig. 6 presents the instantaneous active and reactive power delivered to the grid. As observed, the converter can still deliver rated reactive power to the grid. After the second failure the oscillation in the instantaneous power increases due to the increase in the low order output current harmonics (typical in the overmodulation regions).

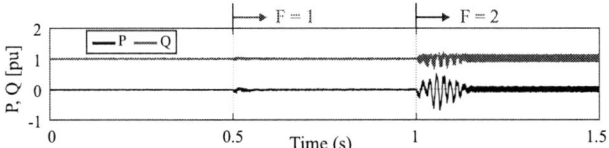

Fig. 6. Effect of the failures in instantaneous active and reactive power.

Fig. 7 illustrates the circulating current of the converter. As observed, after the second failure, the ripple increases significantly due to the increase of the low harmonic components synthesized by the converter.

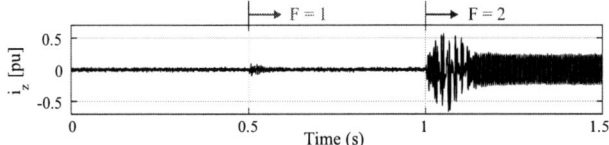

Fig. 7. Effect of the failures in circulating current.

The cell capacitor voltages are presented in Fig. 8. For the three phases, the effect of failures is shown. It is observed that the voltage of the capacitors remained within the tolerance band. Due to the bleeder resistor connected in each cell, it is possible to see the voltage of the faulty cells decreasing.

The output current is illustrated in Fig. 9. Fig. 9(a) shows in detail the current before the fault. Fig. 9(b) shows in detail the current after the first fault. Fig. 9(c) shows in detail the current after the second fault. As observed, after the first failure, the distortion in the output current does not increase significantlly. Since the converter is modulated with a stair-case strategy, the increase in the voltage and in the current distortion is low in the start of overmodulation region, as observed in Fig. 10(b). After the second failure, the converter is completely operating in the overmodulation region and the output current is clearly distorted, as observed in Fig. 10(c).

Finally, Fig. 11 presents the effect of failures in THD of the output current. As observed, after the second failure, the THD is greater than the 5% value recommended by IEEE 519-2014 [26].

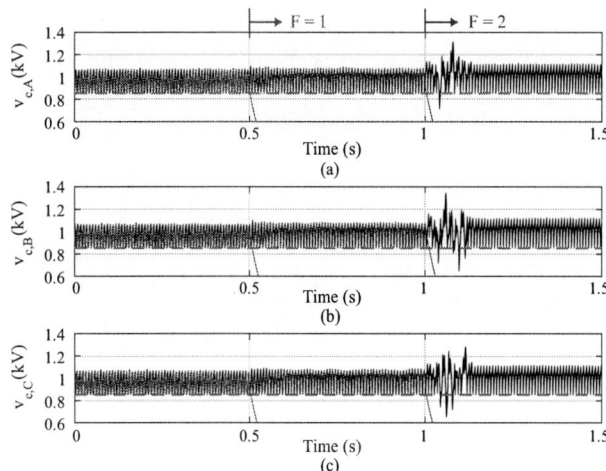

Fig. 8. Effect of the failures in cell voltage capacitors: (a) phase A, (b) phase B and (c) phase C.

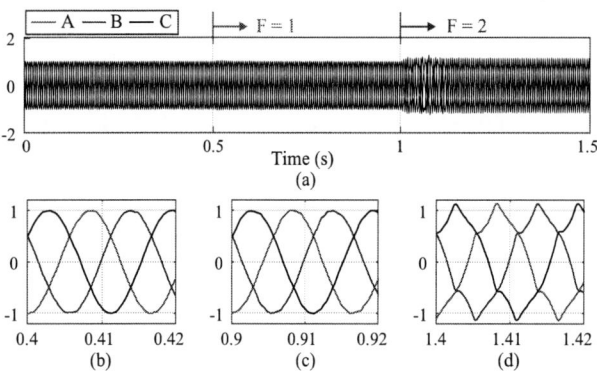

Fig. 9. Effect of the failures in: (a) output current, (b) zoomed view for F = 0, (c) zoomed view for F = 1 and (d) zoomed view for F = 2.

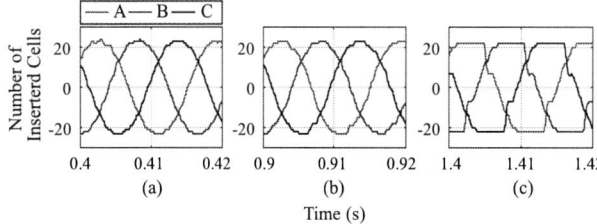

Fig. 10. Effect of the failures in the number of inserted cells: (a) zoomed view for F = 0, (b) zoomed view for F = 1 and (c) zoomed view for F = 2.

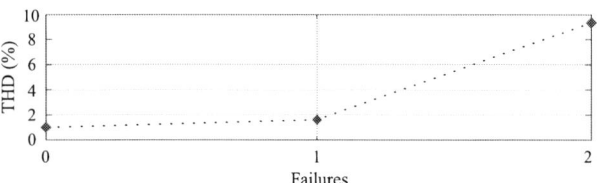

Fig. 11. Effect of the failures in THD of the output current.

V. CONCLUSION

This paper proposed to analyze the inherent redundancy of a SDBC-MMCC based STATCOM in the overmodulation region. Analytical expressions for the boundary conditions between linear and overmodulation region were derived. The minimum value for the dc component of the sum of capacitor voltages was analytically determined and validated through simulations. The sensitivity of the boundary conditions with respect to the converter parameters was also evaluated. As observed, there is a strong dependence of the limits of the linear region with respect to the inductance and the cell capacitance, since the capacitor voltage ripple waveform plays an important role in the maximum reachable output voltage in linear region.

Using the expressions derived, the required number of cells of a 17 MVA/13.8 kV STATCOM was determined. This case study is used to demonstrate the inherent fault tolerance of the SDBC-MMCC. The results indicate that the converter can operate after 2 cell failures. However, the output current THD and circulating current ripple can be an issue when the second cell failure happens. Therefore, considering the THD constraints, the obtained redundancy factor was $\frac{1}{24} \approx$ 4.2% without any additional cost. The authors believe that the redundancy can be increased if the converter with higher number of cells are taken into account.

ACKNOWLEDGMENT

This study was supported in part by the Coordenação de Aperfeiçoamento de Pessoal de Nível Superior - Brazil (CAPES) - Finance Code 001, in part by CNPq and part by FAPEMIG.

REFERENCES

[1] H. Akagi, "Classification, terminology, and application of the modular multilevel cascade converter (mmcc)," in *The 2010 International Power Electronics Conference - ECCE ASIA -*, June 2010, pp. 508–515.

[2] L. Zhang, Y. Zou, J. Yu, J. Qin, V. Vittal, G. G. Karady, D. Shi, and Z. Wang, "Modeling, control, and protection of modular multilevel converter-based multi-terminal hvdc systems: A review," *CSEE Journal of Power and Energy Systems*, vol. 3, no. 4, pp. 340–352, Dec 2017.

[3] S. Debnath, J. Qin, B. Bahrani, M. Saeedifard, and P. Barbosa, "Operation, control, and applications of the modular multilevel converter: A review," *IEEE Trans. Power Electron.*, vol. 30, no. 1, pp. 37–53, Jan 2015.

[4] K. Ilves, S. Norrga, L. Harnefors, and H. Nee, "On energy storage requirements in modular multilevel converters," *IEEE Trans. Power Electron.*, vol. 29, no. 1, pp. 77–88, Jan 2014.

[5] A. F. Cupertino, J. V. M. Farias, H. A. Pereira, S. I. Seleme, and R. Teodorescu, "Comparison of dscc and sdbc modular multilevel converters for statcom application during negative sequence compensation," *IEEE Trans. Ind. Electron.*, vol. 66, no. 3, pp. 2302–2312, March 2019.

[6] M. Hagiwara, R. Maeda, and H. Akagi, "Negative-sequence reactive-power control by a pwm statcom based on a modular multilevel cascade converter (mmcc-sdbc)," *IEEE Trans. Ind. Appl.*, vol. 48, no. 2, pp. 720–729, March 2012.

[7] ABB, *Static compensator (STATCOM)*. [Online]. Available: https://new.abb.com/facts/statcom

[8] Siemens, *Optimal dynamic grid stabilization*. [Online]. Available: https://new.siemens.com/global/en/products/energy/high-voltage/facts/portfolio/svcplus.html

[9] GE, *Static Synchronous Compensator (STATCOM) Solutions*. [Online]. Available: https://www.gegridsolutions.com/products/brochures/powerd_vtf/STATCOM_GEA31986_HR.pdf

[10] L. Hui, Z. Meimei, Y. Ran, L. Wei, D. Jili, L. Haiyang, and Q. Haitao, "Reliability modelling and analysis on mmc for vsc-hvdc by considering the press-pack igbt and capacitors failure," *The Journal of Engineering*, vol. 2019, no. 16, pp. 2219–2223, 2019.

[11] N. Ahmed, L. Ängquist, A. Antonopoulos, L. Harnefors, S. Norrga, and H. Nee, "Performance of the modular multilevel converter with redundant submodules," in *IECON 2015*, Nov 2015, pp. 003 922–003 927.

[12] G. S. Konstantinou, M. Ciobotaru, and V. G. Agelidis, "Effect of redundant sub-module utilization on modular multilevel converters," in *2012 IEEE International Conference on Industrial Technology*, March 2012, pp. 815–820.

[13] B. Li, Y. Zhang, R. Yang, R. Xu, D. Xu, and W. Wang, "Seamless transition control for modular multilevel converters when inserting a cold-reserve redundant submodule," *IEEE Trans. Power Electron.*, vol. 30, no. 8, pp. 4052–4057, Aug 2015.

[14] G. Liu, Z. Xu, Y. Xue, and G. Tang, "Optimized control strategy based on dynamic redundancy for the modular multilevel converter," *IEEE Trans. Power Electron.*, vol. 30, no. 1, pp. 339–348, Jan 2015.

[15] D. Kim, J. Kim, B. Han, and Y. Yoon, "Operational improvement of modular multilevel converter with redundancy sub-modules by new nlc scheme," in *2015 IEEE Power Energy Society General Meeting*, July 2015, pp. 1–5.

[16] J. V. M. Farias, A. F. Cupertino, H. A. Pereira, S. I. S. Junior, and R. Teodorescu, "On the redundancy strategies of modular multilevel converters," *IEEE Transactions on Power Delivery*, vol. 33, no. 2, pp. 851–860, April 2018.

[17] Y. Kwon, S. Kim, and S. Sul, "Six-step operation of pmsm with instantaneous current control," *IEEE Trans. Ind. Appl.*, vol. 50, no. 4, pp. 2614–2625, July 2014.

[18] M. López, F. Briz, A. Zapico, D. Diaz-Reigosa, and J. M. Guerrero, "Operation of modular multilevel converters under voltage constraints," in *2015 IEEE ECCE*, Sep. 2015, pp. 3550–3556.

[19] P. Briff, F. Moreno, and J. Chivite-Zabalza, "Extended controllability of hvdc converters in the vector space," in *2017 IEEE Manchester PowerTech*, June 2017, pp. 1–1.

[20] L. Harnefors, A. Antonopoulos, S. Norrga, L. Angquist, and H. Nee, "Dynamic analysis of modular multilevel converters," *IEEE Trans. Ind. Electron.*, vol. 60, no. 7, pp. 2526–2537, July 2013.

[21] B. Gemmell, J. Dorn, D. Retzmann, and D. Soerangr, "Prospects of multilevel vsc technologies for power transmission," in *IEEE/PES Transmission and Distribution Conference and Exposition*, April 2008, pp. 1–16.

[22] E. Behrouzian and M. Bongiorno, "Investigation of negative-sequence injection capability of cascaded h-bridge converters in star and delta configuration," *IEEE Trans. Power Electron.*, vol. 32, no. 2, pp. 1675–1683, Feb 2017.

[23] H. Akagi, E. H. Watanabe, and M. Aredes, *The Instantaneous Power Theory*. IEEE, 2007.

[24] A. Hassanpoor, L. Ängquist, S. Norrga, K. Ilves, and H. Nee, "Tolerance band modulation methods for modular multilevel converters," *IEEE Trans. Power Electron.*, vol. 30, no. 1, pp. 311–326, Jan 2015.

[25] L. Angquist, A. Antonopoulos, D. Siemaszko, K. Ilves, M. Vasiladiotis, and H. Nee, "Open-loop control of modular multilevel converters using estimation of stored energy," *IEEE Trans. Ind. Appl.*, vol. 47, no. 6, pp. 2516–2524, Nov 2011.

[26] "Ieee recommended practice and requirements for harmonic control in electric power systems," *IEEE Std 519-2014 (Revision of IEEE Std 519-1992)*, pp. 1–29, June 2014.

Active Cooling and Thermal Simulation Applied to an Extra-High Current COB LED

Dênis de Castro Pereira
Federal University of Juiz de Fora
Juiz de Fora, Brazil
denis.castro@engenharia.ufjf.br

Rúbio Campos Marques
Federal University of Juiz de Fora
Juiz de Fora, Brazil
rubio.campos@engenharia.ufjf.br

Pedro Santos Almeida
Federal University of Juiz de Fora
Juiz de Fora, Brazil
pedro.almeida@engenharia.ufjf.br

Guilherme Marcio Soares
Federal University of Juiz de Fora
Juiz de Fora, Brazil
guilherme.marcio@engenharia.ufjf.br

Fernando Lessa Tofoli
Federal University of São João del-Rei
São João del-Rei, Brazil
fernandolessa@ufsj.edu.br

Pedro Laguardia Tavares
Federal University of Juiz de Fora
Juiz de Fora, Brazil
pedro.laguardia@engenharia.ufjf.br

Henrique Antônio Carvalho Braga
Federal University of Juiz de Fora
Juiz de Fora, Brazil
henrique.braga@ufjf.edu.br

Abstract – **This work lies on the active cooling and thermal evaluation applied to an extra-high current (up to 12 A) chip-on-board (COB) light-emitting diode (LED). Simulations are performed employing computational fluid dynamics (CFD) so that the main thermal parameters can be obtained as close as possible to real results. Aiming at a thorough analysis of heat dissipation at very high current levels in floodlighting applications, a finite element analysis is carried out in order to establish a comparative approach concerning the convected air flow through a volume-reduced fin-heatsink. The simulated results show that the lighting system can be significantly improved by employing such active cooling technique for the device's power rated condition (i.e., 350 W). For this operating point, the manufacturer's heatsink physical volume could be decreased by 87% when applying the active cooling system with convected air flux, while also ensuring a safe thermal operation under maximum junction temperature of 120 °C.**

Keywords – **COB LEDs active cooling, Computational fluid dynamics, Extra-high current COB LEDs, Finite element analysis, Heatsink volume minimization.**

I. INTRODUCTION

Solid-state lighting (SSL) devices and applications are still arisen with great interest for both academia and industry due to distinctive characteristics of efficiency, high luminous efficacy, long lifespan, high reliability, and environmental friendliness. The Light-emitting diode (LED) technology is currently applied in several applications, such as indoor lights, backlights for displays devices, automotive and outdoor floodlighting for wider areas [1]. Typically, such large areas include but are not limited to sport stadiums, airport runways, and international borders, which require expressive high luminous flux levels. In last decade, most lighting applications have been based on series association of discrete LED chips, resulting in a high-voltage low-current module [2]. More recently, the chip-on-board (COB) technology for high currents and low voltage levels has also been applied due to intrinsic thermal advantages respected to several chip topologies, which brings new challenges to the lighting field [3].

A compact matrix structure of LEDs arranged on a single-substrate is called COB. In general, about 30% of the electric power is used for light emitting in a COB LED, while the remaining portion is converted into heat [4]. Moreover, for luminaires under increased input power levels, the device's junction temperature tends to rise to high values, which results in poor performance of energy efficiency, light intensity, and system lifespan. Under constant operation, the COB LEDs junction temperature is typically well above the ambient condition, *i.e.*, 25 °C, which can reduce the light output by about 10% or even below if compared to the nominal ratings described in their respective datasheets [5]. These values are found to be even worse for products with inadequate thermal design, so that a reliable thermal system is also mandatory in COB LED applications. Thus, to properly cooling the COB LED matrix structure, a well-designed heatsink system should be employed, while ensuring the correct operation for maximum junction temperature.

Cooling methods for COB LEDs are divided into active and passive. Passive methods are generally preferred for LED systems due to their high reliability from not requiring any additional moving parts [6]. These methods dissipate the heat generated by the COB structure under natural convection conditions. On the other hand, active cooling techniques apply some forced convection, as an air flowing cooler that can drive away the heat from a fin-heatsink. In extreme situations, where both volume and lighting efficiency are important design parameters, an efficient active cooling technique should be employed to ensure optimized heat dissipation from the COB LED/heatsink to the luminaire external environment.

In this work, an extra-high current (EHC) COB LED from Flip Chip Opto (FCOpto) is employed in the cooling system analysis [7]. Apollo 600 EHC COB LED model [8] has been chosen to be analyzed considering an active cooling thermal technique, while also establishing a comparison with its passive heat dissipation counterpart. Information regarding EHC COB LEDs as provided in datasheets is normally incomplete, in such a way a thorough static photoelectrothermal (PET) analysis [9] has been carried out for Apollo 600 model in [10].

The variations in thermal-related parameters when operating under very-high current and high junction temperature conditions are critical in the COB technology, so that this aspect is crucial for achieving a reliable thermal model. The design considerations for active cooling in EHC COB LEDs are significantly different and more complex when compared to traditional low-current discrete LEDs. In such cases, care must be taken to deeply understand the thermal metrics and heatsink performance.

This work aims to propose an active cooling system to improve the thermal dissipation in Apollo 600, whereas its original bulky heatsink can be proportionally reduced as a consequence. Within this context, Section II introduces the EHC COB LED device, while its main characteristics are highlighted. The finite element analysis (FEA) and computational fluid dynamics (CFD) are described in Section III, which can be used in active cooling by air flowing simulations. Section IV presents simulation results for the studied active cooling method applied to the EHC COB LED, where the main aspects are discussed. Besides, a volume-reduced heatsink system is estimated, while the feasibility of the employed active technique is shown. Finally, the main conclusions and contributions of this paper are summarized in Section V.

II. THE EXTRA-HIGH CURRENT COB LED

Aiming at volume and size minimization, modern high-power COB devices are designed featuring low-voltage and extra-high current levels [7], [8]. The analysis carried out in this work considers the EHC COB LED model Apollo 600, which is shown in Fig. 1 (a) mounted on a passive cooling fin-heatsink for power levels up to 600 W. One can see that, considering the manufacturer's heatsink, the main diameter is 34.1 cm, with height of 15 cm, and weight of 6.36 kg. Even considering this original passive cooling method, these quantities represent a significant reduction when compared to traditional associations of discrete elements to achieve high voltage levels [2]. Furthermore, Fig. 1 (b) shows the main dimensions of the EHC COB device, while Table I lists the main characteristics from datasheet regarding Apollo 600 [8].

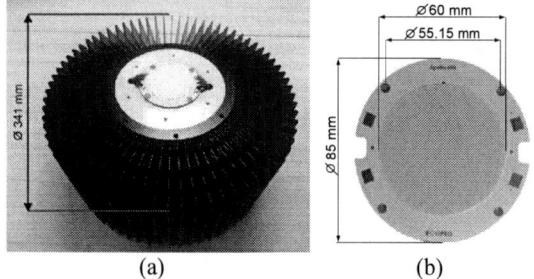

(a)   (b)

Fig. 1. Apollo 600 real picture mounted on the manufacturer's (FCOpto) passive fin-heatsink (a) and main dimensions (b).

TABLE I. APOLLO 600 MAIN PARAMETERS FROM DATASHEET [8]

Parameter (Symbol)	Value
Maximum output power (P_o)	608.4 W
Maximum dc forward current (I_F)	12 A
Nominal output power ($P_{o(nom)}$)	350 W
Threshold voltage (V_t)	40.5 V
Dynamic resistance (r_d)	0.95 Ω
Maximum junction temperature (T_j)	140 °C
Maximum luminous flux	60840 lm

As previously stated, the COB technology consists of a miniaturized LED chip matrix mounted on a substrate or circuit board. This design provides higher power density and uniform light output [3]. Considering extra-high current levels through the COB, the matrix composed by LED elements is mounted on a single substrate, whereas an improved thermal management is extremely relevant. Considering the EHC COB LEDs from FCOpto, high-power devices are conceived through a 3-pad patented technology, which is presented in Fig. 2 [8].

The mounting technique in Fig. 2 allows very-low junction-to-case thermal resistances (i.e., 0.008 °C/W in Apollo 600 model) and improved heat dissipation through a pillar-based direct connection. Thus, this technology provides extra-high current and luminous flux operation, bringing new challenges to the lighting field. Instead of the regular flip-chip LED, the 3-pad flip-chip presents an additional thermal-pad (T-pad), with better heat dissipation from the metal core printed circuit board.

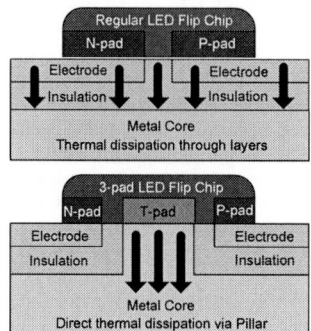

Fig. 2. 3-pad chip integration technology [8].

Although such improved mounting technique is applied in FCOpto devices, one can see that the major volume portion from the luminaire in Fig. 1 (a) is due to the bulky fin-aluminum heatsink specified for power levels up to 600 W. Therefore, the study carried out in this work aims to reduce the weight, volume, and main dimensions of this bulky heatsink by applying an active cooling method as shown in the schematic in Fig. 3. Thus, the analysis is performed so that a comparative evaluation can be finally considered for both passive and active techniques.

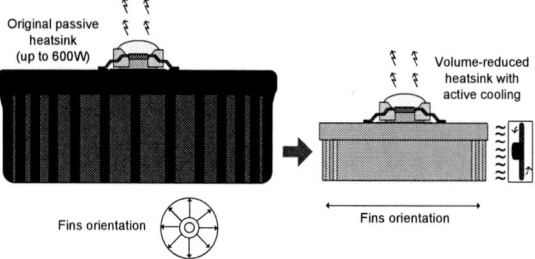

Fig. 3. Transition from passive to active heat dissipation systems aiming at luminaire volume minimization.

The circuit in Fig. 4 can be considered to evaluate the thermal domain in an EHC COB LED system. Initially, Q_{th} is the dissipated thermal power, which represents part of the total power delivered to the device. The device power is not entirely converted into luminous radiation, as P_d is the total power dissipation, and the coefficient k_h is the portion that is converted into heat. The ambient temperature is T_a, which is represented as a constant temperature source. Furthermore, T_c and T_{hs} are the case

978-1-7281-4181-7/19 $31.00 © 2019 IEEE

and heatsink temperatures, respectively. The thermal resistances in Fig. 4 are respected to the junction-to-case (R_{jc}), thermal paste (R_{tp}) and heatsink (R_{hs}), which are all considered to obtain the junction temperature (T_j) in (1). The junction temperature is an important parameter that will drastically change when applying such active cooling technique, as it must be kept under safe conditions based on datasheet maximum constraints (i.e., 140ºC for Apollo 600 model). For improved performance, is generally desired that T_j does not present values very close to this maximum, while it is reasonable to choose lower and safer T_j conditions accordingly to the active heat dissipation.

$$T_j = T_a + Q_{th}R_{hs} + Q_{th}R_{tp} + R_{jc}Q_{th} \qquad (1)$$

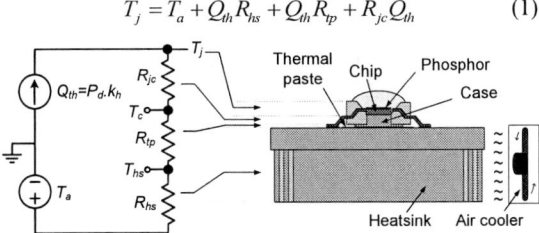

Fig. 4. Thermal circuit considered for an EHC COB LED lighting system with active cooling.

III. FINITE ELEMENT ANALYSIS FOR HEAT TRANSFER IN LED SYSTEMS

The Finite Element Modeling (FEM) is a numerical procedure that can be employed in order to obtain a particular solution for several engineering applications. This technique consists on dividing a continuous domain into several portions, which are properly discretized by interconnected elements that maintain the same properties of the original environment. These new elements are then described by differential equations and solved by mathematical models so that the quantified results can be obtained [11]. This numerical analysis is suitable to static and dynamic systems, from linear or nonlinear conditions, while allowing proper evaluation for mechanical stresses, heat transfer, fluid flux, and also other aspects [12].

Particularly for heat transfer analysis, FEM simulations are also applied by Computational Fluid Dynamics (CFD), which solves the conducted air flowing equations among a contour region. Within this case, the interconnection regarding all the discretized elements corresponds to the system loop. This loop must be programmed including the main materials and structural properties that define the system reactions to temperature variations [13]. Some particular regions with their respective heat density levels are then assigned to the thermal loop. Thus, some regions will receive high heat levels, with a high number of internal nodes, while some regions will receive low heat levels, with fewer internal nodes [14].

The heat density levels are defined from the air flow design and respective orientation, which will perform best cooling performance in specific areas for a general 3D model. Therefore, the accuracy in CFD results is directly related to the chosen loop and its intrinsic properties. Normally, a loop refinement can be used to improve the system results, although it also increases the total time simulation and computational effort [14].

The high intrinsic complexity in heat transfer analysis lies on the different dynamic thermal phenomena which are applied in the system. The heat transfer is performed in three different ways: conduction, i.e., the heat transferred through a solid mass via direct contact; convection, transferred through the movement of fluids; and radiation, transferred through electromagnetic field. Within an LED system, the major portions are due conduction and convection, whose heat transfer aspects depend on the material geometry and the surface in which it occurs [4]. Besides, it also depends on the physical fluid properties and respective temperature variation.

In order to ensure thermal results as close as possible to real applications, FEM simulations with CFD are highly recommended in heat transfer for LED systems. This technique simplifies the analysis from an initial complex system, while the related results are very close to reality. Thus, the CFD modeling is employed to quantify the heat transfer in an active cooling system for the EHC COB LED Apollo 600. Additionally, the junction temperature is supposed to be kept below 140 ºC in the simulations, thus ensuring proper and safe operation to the luminaire. The computational fluid dynamics has been performed using software SolidWorks®, which presents proper tools for both design and thermal simulations.

IV. RESULTS OF COOLING METHODS AND COMPARATIVE EVALUATION

The electrothermal profile of the lighting system is highly relevant in order to design a proper heatsink for an EHC COB LED, thus ensuring maximum luminous flux performance, minimized volume, and long lifespan at extra-high current levels [9]. The active cooling is generally employed to ensure that the maximum luminous flux and the minimal heatsink thermal resistance will be obtained with consequent improved design in terms of the physical volume. CFD simulations are then performed in this section to quantify the maximum junction temperature and also to define the heatsink physical volume difference when compared to the passive cooling system.

The real physically-reduced heatsink in Fig. 5 has been used for drawing its respective 3D-model on the designing software to be analyzed in heat transfer CFD simulations. This aluminum fin-heatsink model has been applied in an initial passive condition, and also for the active technique aiming at a comparative evaluation. One can see that its dimensions in Fig. 5 are significantly lower than the outer diameter from the manufacturer's heatsink in Fig. 1 (a).

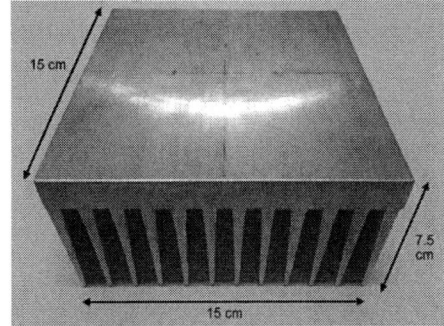

Fig. 5. Picture of the heatsink used as baseline for the 3D-model implemented in software for CFD analysis.

978-1-7281-4181-7/19 $31.00 © 2019 IEEE

The main parameters used in the thermal simulations are summarized in Table II. Firstly, the employed fan type in the active system is an axial model (9232-12H), which is placed horizontally to the finned-heatsink for improved air flowing and heat dissipation performance. This choice is basically based on heatsink fins orientation, which ensures the optimized way to dissipate the heat to the ambient [4], [5]. The cooler angular speed is chosen aiming at the maximum rpm for this particular system, while ensuring that the EHC COB LED junction temperature will be kept under 140 °C in this case. Table II also shows the external diameter for the cooler with its respective hub diameter, which are both important to be modeled in simulation software for a better mechanical approach close to real applications.

Other key parameters presented in Table II, which have been defined as the operating point in simulations, are the EHC COB LED active power and the maximum allowed junction temperature in the device. These values have been chosen considering trade-offs for maximum allowed active power and improved thermal performance considering the physically-reduced heatsink in Fig. 5. From Table II, a 350-W application with 125 °C maximum T_j has then been chosen to ensure reliable operation to the luminaire at nominal power under safe junction temperature conditions. Following, both passive and active techniques are thoroughly analyzed to establish a consequent comparative thermal study.

TABLE II. MAIN PARAMETERS EMPLOYED IN CFD THERMAL SIMULATIONS

Parameter	Model or Value
Fan Type	Axial 9232-12H
Reference air density	1.225 kg/m³
Rotor angular speed	418.87 rpm
Cooler outer diameter	9.2 cm
Hub diameter	5.08 cm
Direction of rotation	Counter-clockwise
EHC COB LED active power	350 W
Maximum allowed junction temperature in EHC COB LED	125 °C

A. Passive heat dissipation method

As previously stated, an initial passive cooling analysis has been performed aiming at a comparative evaluation. The heat dissipation by natural convection has then been considered for the EHC COB LED thermal system. The following results have been obtained from SolidWorks® while considering CFD for this particular case. Firstly, Fig. 6 presents the designed luminaire, which considers the cooling system inactive for passive dissipation. The heat generation rate in Fig. 6 corresponds to the amount of heat that is transferred per unit of time in the material (in dyn×cm/s). Its respective value is calculated by the own software environment considering the initial power conditions for the chosen operating point (350 W).

For the designed operating point (350 W), the thermal simulation has been performed in Solid Works®. Thus, Fig. 7 shows the thermal view from the software after CFD simulation in passive condition, from where one can see that the maximum junction temperature obtained in this case is well above the device's allowed T_j value. The results show that, for this volume-reduced fin-aluminum heatsink under passive convection for heat dissipation, the junction temperature is found to be 227.97 °C.

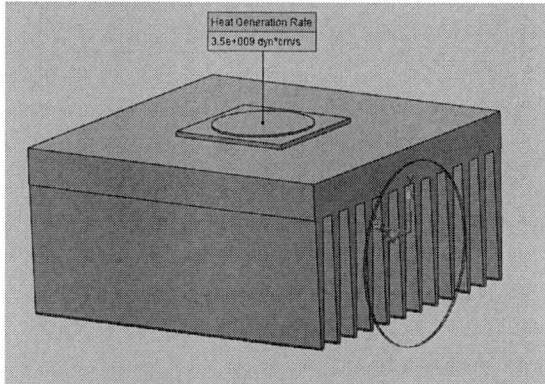

Fig. 6. EHC COB LED heat dissipation system with inactive cooler (CFD software environment).

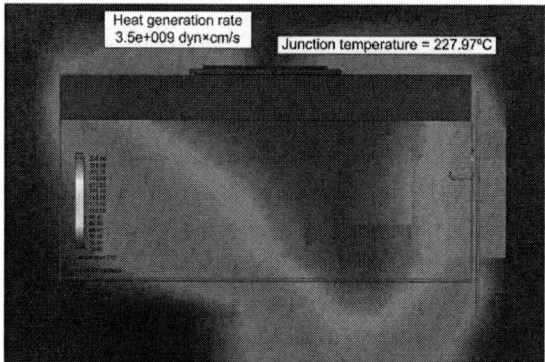

Fig. 7. Thermal view of the computational fluid dynamics simulation for the passive heat dissipation system.

In order to perform an in-depth analysis regarding the ambient air flow orientation for passive convection, Fig. 8 presents the respective conditions in this case. From Fig. 8, it can be seen a vertical thermal characteristic in the external environment, whereas the natural convection forces the heat dissipation mainly over and under the heatsink surface. Therefore, this heat dissipation method does not allow an improved performance in this high power application at the designed operating point, while an active solution shall be properly applied and evaluated.

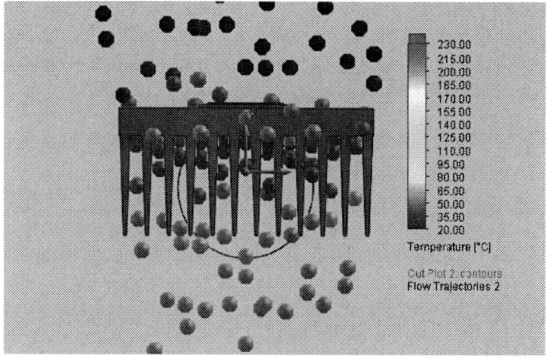

Fig. 8. Ambient air flow in natural convection from the EHC COB LED heat dissipation passive system (inactive cooler).

B. Active heat dissipation method

In order to improve the luminaire thermal performance, the system has been simulated considering an active heat dissipation system, which is employed to reduce the EHC COB LED junction temperature to satisfactory values for safe operation. In this particular case, the luminaire is composed by the EHC COB LED, the fin-aluminum heatsink from Fig. 5, and by the previously described axial fan, which is conceived to drive away the most part of heat generated by the lighting source. Fig. 9 presents the designed luminaire in software environment, while also showing the heat generation rate (3.5×10^9 dyn×cm/s) and the employed external fan (axial model 9232-12H). Considering such active heat dissipation technique, power consumption issues are expected to impact the luminaire efficiency, so care must be taken in this aspect in practical applications.

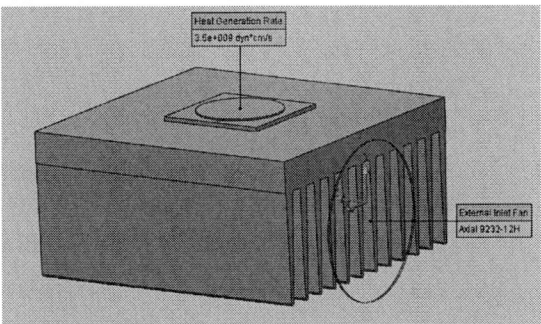

Fig. 9. EHC COB LED heat dissipation system with active cooling system (SolidWorks® software environment).

The CFD thermal simulation has also been performed in this case while considering the active cooling system operating in steady-state condition. In this case, it is important to highlight that the employed external inlet fan is always operated at constant rated speed. Fig. 10 shows the thermal results for this active heat dissipation technique. One can see that, for an active power of 350 W, the EHC COB LED junction temperature is properly kept under 125°C in steady-state. Precisely, the maximum junction temperature obtained in this case is 120.14°C, which ensures a safe condition for the device operation.

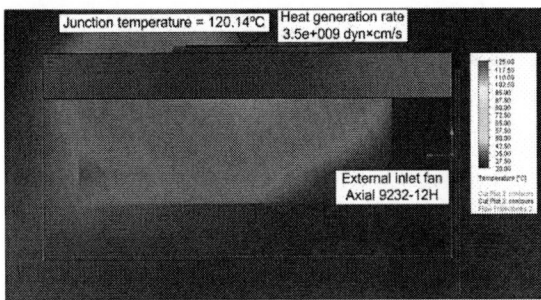

Fig. 10. Thermal view after CFD simulation for the active heat dissipation system by air flowing convection (active cooler).

The respective speed and direction characteristics for the convected air flux are presented in Fig. 11. From the simulated results in this active cooling case, one can see the improved thermal maintenance, where the generated heat can be driven away by the forced air through the own fins orientation from the heatsink. The forced air flowing denotes horizontal heat dissipation in this case, from the heatsink fins to the external ambient. Still in Fig. 11, the wind speed has also been quantified. The maximum value obtained from the active cooling flow simulation is 654.7 cm/s. The wind speed becomes lower as far it is from the mechanical cooler, i.e., the air flowing source.

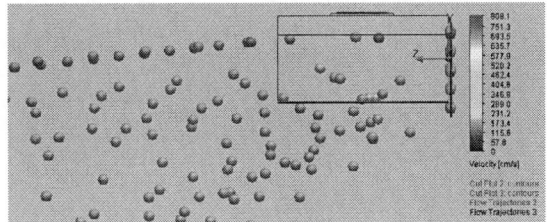

Fig. 11. Forced air flowing in the EHC COB LED active system (wind speed scale in cm/s).

C. Comparative analysis for the applied techniques

Aiming at a detailed evaluation between the described cooling techniques, the passive and active heat dissipation methods have been compared concerning their respective thermal quantities acquired from the previous CFD simulations. For this purpose, Table III presents the main simulated results, from where it can be seen the feasibility of the active cooling technique in this kind of high-power lighting application. From the previously results, it has been stated that the passive alternative with reduced heatsink dimensions is not sufficient to fulfill the safe junction temperature requirements in this case. Under an operating point of 350 W, the junction temperature could be decreased from 227.97°C to 120.14°C only by including the active system based on the forced air flow though the heatsink fins.

To also include the original manufacturer's heatsink previously presented in Fig. 1 (a) in the comparative analysis, its main parameters are also described in Table IV. This heatsink corresponds to a bulky element, which is used as a comparative background for the reduced heatsink employed in simulations. To solve such issue, the active cooling method along with the volume-reduced heatsink in Fig. 5 could perform optimized results. The thermal resistance for the active-cooled heatsink could be increased from 0.0785 °C/W (manufacturer's passive heatsink from Fig. 1 (a) [15]) to 0.622 °C/W (active-cooled 350-W heatsink from Fig. 5 [16]). Therefore, this value will correspond to a physically improved luminaire, thus leading to minimal overall lighting system volume.

From the summarized results in Table IV, one can see that the heatsink effective volume could be significantly decreased when applying such active cooling technique. The volume estimation performed in Table IV includes into the heatsink geometries the volume portion respected to the gaps between the several fins, thus turning their geometry into a basic 3-D solid for effective volume purposes. These simplifications can be done in order to consider the volume portion which is effectively occupied by the luminaire on a given application. The effective volume calculation has then been performed from a basic cylinder ($\pi \times radius^2 \times height$) and from a rectangular block (length×width×height), respectively to each heatsink basic structure. The obtained reduction percentage is 87.68%, which corresponds to an effective volume decreased from 13.7×10^3 cm^3 (original passive heatsink) to 1.68×10^3 cm^3 (active cooled heatsink). Furthermore, when the air cooler volume (i.e., ≈ 99.71 cm^3) is also taking into this account, the difference in percentage is minimal, reaching a reduction of 87% in this case. These results are quite expressive when implementing optimized heatsinks for EHC COB LEDs, which improves the luminaire physical adaptation to extra-high flux environments.

TABLE III. COMPARATIVE CFD ANALYSIS REGARDING COOLING TECHNIQUES APPLIED TO THE EHC COB LED

Parameter	Passive Method	Active Method
Maximum junction temperature @350 W	228.22 °C	120.14 °C
Heat generation rate @350 W	3.5×10^9 dyn×cm/s	3.5×10^9 dyn×cm/s
Air flowing orientation	Vertically by natural convection	Horizontally through fins
Maximum wind speed	-	654.7 cm/s

TABLE IV. COMPARATIVE ANALYSIS CONSIDERING PHYSICAL PARAMETERS OF THE EMPLOYED HEATSINKS

Parameter	Manufacturer's heatsink [15]	Active cooled heatsink [16]
Main dimensions [mm]	341 (ext. diameter)	150×150 (length×width)
Height [mm]	150	75
Weight [kg]	6.36	2.12
Heatsink effective volume [cm^3]	13.7×10^3	1.68×10^3
Amount of fins	72	12
Fins orientation	Radial	Straight-line
Thermal resistance from datasheet reference [°C/W]	0.0785	0.62
Cooler effective volume [cm^3]	-	99.71
Heatsink cost [US$]	200	35
Air cooler cost [US$]	-	5
Total cost [US$]	200	40

V. CONCLUSION

This paper has analyzed a novel technology of extra-high current COB devices, while also providing a finite element analysis concerning the heat transfer in such components. The EHC COB LED Apollo 600 has been thoroughly studied considering both passive and active cooling techniques, which have been applied to minimize the physical volume of the heatsink. Great research potential can be addressed to EHC COB LEDs, as they present even more complex thermal issues if compared with their low-current discrete counterparts.

CFD simulations have been performed to quantify the heat generated in such EHC application and to properly drive it out using active cooling techniques. Firstly, the passive cooling method has been applied to a heatsink with reduced volume and dimensions, while the results have shown that this technique is not suitable for the chosen operating point due to high values in the device's junction temperature. Otherwise, when applying the active cooling by convected air flowing, the EHC COB LED junction temperature could be kept under safe operating conditions, i.e., 125°C. Therefore, the device's heatsink can be significantly reduced when compared to the

manufacturer's original passive bulky element. When using the described active cooling technique, an effective volume minimization about 87% results for the heatsink system, which shows the feasibility of the proposed heat dissipation structure.

ACKNOWLEDGMENT

The authors would like to acknowledge CAPES, CNPq, FAPEMIG, and INERGE for the financial support; and FCOpto Company for the product support.

REFERENCES

[1] M. A. Reyes, J. J. Sammarco, S. Gallagher, and J. R. Srednicki. "Comparative evaluation of light-emitting diode lamps with an emphasis on visual performance in mesopic lighting conditions", *IEEE Transactions on Industry Applications*, vol. 50, no. 1, pp. 127-133, Jan 2014.

[2] Osram, "Floodlight 20 Maxi LED Module Generation 2". Osram Lighting, 2018. Available in: https://goo.gl/wGmf33. Accessed in 01/30/2019.

[3] N. Kafadarova N. Vakrilov, A. Andonova. "Study of high power COB LED modules with respect to topology of chips", In *Electronics Technology (ISSE), 2015, 38th International Spring Seminar*, vol. 2, pp. 108-113, 2015.

[4] H. H. Cheng, D. S. Huang, and M. T. Lin. "Heat dissipation design and analysis of high power LED array using the finite element method". *Microelectronics Reliability*, vol. 52, no. 5, pp. 905-911, 2012.

[5] B. Ahn, C. Jang, S. Leigh, S. Yoo, and H Jeong. "Effect of LED lighting on the cooling and heatsink loads in office buildings" *Applied Energy*, vol. 113, pp. 1484-1489, 2014.

[6] Q. Shen, D. Sun, Y. Xu, T. Jin, and X. Zhao. "Orientation effects on natural convection heat dissipation of rectangular fin heatsinks mounted on LEDs". *Int. Journal of Heat and Mass Transfer*, vol. 75, pp. 462-469, 2014.

[7] Flip Chip Opto, "FCOpto Starlite LED product catalogue", 2016. Available in: https://goo.gl/mxWZtx. Accessed in 02/03/2019.

[8] Flip Chip Opto, "Apollo 600 datasheet", 2016. Available in: https://goo.gl/CxkdGi. Accessed in 02/03/2019.

[9] R. Hui. Photo-electro-thermal Theory for LED Systems: Basic Theory and Applications, Cambridge University Press, 2017.

[10] D. C. Pereira, P. L. Tavares, P. S. Almeida, G. M. Soares, F. L. Tofoli, H. A. Braga. "Improved Photoelectrothermal Model with Thermal Parameters Variation Applied to an Extra-High Current COB LED." *Revista Eletrônica de Potência*, vol. 24, n° 2, pp. 147-156, 2019.

[11] R. Lewis, P. Nithiarasu, K. Seetharamu. "Fundamentals of Finite Element Method for Heat and Fluid Flow". Chinchester: John Wiley & Sons, 2004.

[12] S. Moaveni. "Finite Element Analysis". New Jersey: Prentice Hall, 1999.

[13] V. Nguyen. "Finite Element, Dimensional Model for Thermal Distribution". 2010.

[14] M. Ha. "Thermal Analysis of High Power LED Arrays". Georgia Institute of Technology, p. 150, 2009.

[15] Ursa Lighting/Starlite LED, "600-W Cold Forged Heatsink", 2016. Available in: https://goo.gl/LQ9RpN. Accessed in 06/28/2019.

[16] Brazelli Thermal Heatsinks, "Aluminum Cold Forged Heatsink", 2019. Available in: https://bit.ly/2NQqq9C. Accessed in 06/28/2019.

VOLTAGE REGULATION OF A REMOTE BUS OF A DISTRIBUTION NETWORK BY STATIC SYNCHRONOUS COMPENSATOR

Samuel N. Duarte, Bruno C. Souza, Pedro M. Almeida, Pedro G. Barbosa

Federal University of Juiz de Fora, Graduate Program in Electrical Engineering, Juiz de Fora–MG, Brazil

Email: samuel.neves@engenharia.ufjf.br, bruno.cortes@engenharia.ufjf.br,
pedro.machado@ufjf.edu.br, pedro.gomes@ufjf.edu.br

Abstract - **This paper presents a voltage control strategy for a static synchronous compensator (STATCOM). The control algorithm proposed in this paper allows the STATCOM to regulate the voltage at a remote bus of a power system. The mathematical modelings and control-loops developed for the static compensator are validated through results of digital simulations. The results presented show that the STATCOM can regulate the voltage at a remote bus in real-time. Moreover, it is shown also that, if the STATCOM regulates the voltage of the remote bus at the nominal value of the system, the voltages of the upstream buses of the power system may exceed its limits.**

Keywords – **STATCOM, Voltage Compensation, Remote bus.**

I. INTRODUCTION

The concept of smart grid has been changing the control and operation of the electric networks. Several researchers have published studies on the impacts on the electrical networks due to the increase in the number of battery chargers for electric vehicles [1] and distributed generation [2], such as the photovoltaic systems, connected to these networks. The high number of these equipment connected to the grid can result in bad operational conditions in terms of imbalance level, voltage security and power losses [3]. Therefore, the electricity companies have been constantly seeking for solutions to maintain the reliability of the distribution systems.

Several power electronic converters have recently been connected to the electric networks to improve the power quality delivered to customers [4]. The Static Synchronous Compensator (STATCOM) have emerged as a great solution for voltage regulation, improving the overall stability margin of the power systems by compensating reactive power at its AC terminals [5]. Since the STATCOM is shunt-connected to the grid it can be easily disconnected in case of a failure.

Although the static compensator is often designed to control the voltage at the bus it is connected, electric utilities and consumers may find difficult to install the STATCOM at the same bus where the voltage regulation is required. For example, it may be preferable to connect the STATCOM elsewhere due to easy-maintenance conditions. Furthermore, an industrial consumer may prefer to connect the STATCOM inside its substation area rather than at the end of the feeder where a priority load requires voltage regulation. Therefore, in these cases, the static compensator should be designed to control the voltage at a remote bus.

Nevertheless, the use of the STATCOM to control the voltage at a remote bus is not much discussed in the literature when compared to its common operation, in which the STATCOM controls the voltage at the Point of Common Coupling (PCC). In fact, there are a large number of works in the literature regarding the voltage control of a remote bus in terms of power flow modeling and steady-state analysis [6]. A typical application of remote voltage control is the use of on load tap changing (OLTC) transformers that executes this task through a Line Drop Compensator (LDC) [7].

In [8] numerical results are presented for the IEEE 118-bus and IEEE 300-bus systems, where a STATCOM is modeled in steady-state in order to perform several functions in the power system, including the voltage control at a remote bus. However, no modelings are presented for the time-domain.

In [9] the authors proposes the use of a STATCOM to mitigate voltage fluctuation in a weak power system. The STATCOM is modeled both in steady-state and transiently. The results shown that, although the STATCOM is commonly connected to the bus whose voltage must be regulated, in some cases the connection of the static compensator at a remote bus may be a good alternative from the cost-effectiveness point of view. Nevertheless, the mathematical modelings to develop the voltage control-loop of the STATCOM are not presented.

Thus, this paper presents a voltage control strategy for a STATCOM connected to a distribution network. The static compensator is used to regulate, in real-time, the voltage of a remote bus, while the static compensator is connected to the point of common coupling. The methodology of voltage control proposed in this paper is developed in such a way that it can be easily incorporated to the traditional voltage control-loops found in the literature. Moreover, the methodology presented can also be applied to any electric network topology. Therefore, an equivalent model for the electric network is presented in order to allow the implementation of the remote voltage control proposed in this paper. Results of digital simulations of a STATCOM connected to a distribution network are used to validate the proposed algorithm.

II. THE STATIC SYNCHRONOUS COMPENSATOR

Fig. 1 shows the topology of the STATCOM connected to an equivalent distribution network through a first-order passive filter. The equivalent electric network is modeled by three sinusoidal voltage-sources and by the equivalent resistance and inductance R_s and L_s, respectively. On the other hand, the parameters R_f and L_f represent the resistance and inductance

978-1-7281-4181-7/19 $31.00 © 2019 IEEE

of the interface filter, respectively. The equivalent capacitor of the STATCOM is modeled by the parameter C_{eq}. In a first moment, the STATCOM will be used to synthesize currents at its AC terminals in order to control the voltages at PCC. Thus, in the following two subsections the current and PCC voltage control-loops of the STATCOM will be developed based on mathematical models for the static compensator. Then, a modification in the PCC voltage control-loop will be presented in order to allow the voltage control at a remote bus.

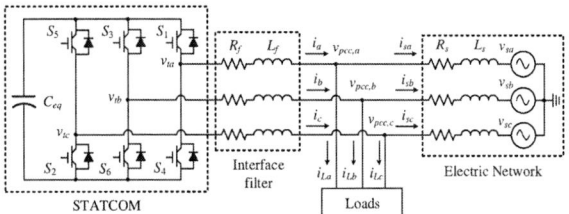

Fig. 1: Topology of the STATCOM connected to an equivalent distribution network.

A. The Current Control-Loop

The mathematical modeling of the circuit of Fig. 1, in the $\alpha\beta$ coordinates, allows to write the following dynamic system for the currents synthesized by the STATCOM:

$$\begin{cases} L_f \dfrac{di_\alpha}{dt} = -R_{eq}i_\alpha + v_{t\alpha} - v_{pcc,\alpha} \\[2mm] L_f \dfrac{di_\beta}{dt} = -R_{eq}i_\beta + v_{t\beta} - v_{pcc,\beta} \end{cases}, \qquad (1)$$

where i_α and i_β are the terminal currents of the STATCOM and $i_0 = 0$, $v_{pcc,\alpha}$ and $v_{pcc,\beta}$ are the PCC voltages, $v_{t\alpha} = m_\alpha(\frac{V_{dc}}{2})$ and $v_{t\beta} = m_\beta(\frac{V_{dc}}{2})$ are the terminal voltages of the STATCOM, V_{dc} is the DC bus voltage of the STATCOM and m_α and m_β are the modulation indexes. The equivalent resistance $R_{eq} = (R_{igbt} + R_f)$ is the series-association of the IGBTs and the interface filter resistances.

Based on (1), it is possible to build the block diagram for the α-axis current control-loop of the STATCOM, as shown in Fig. 2, where the superscript (*) is used to identify the reference signal. The block $K_i(s)$ represents a modified Proportional-Resonant (PR) controller [10] used to ensure that the STATCOM current track its reference signal. The transfer function of this controller is given by:

$$K_i(s) = k_{p,i} + k_{r,i} \frac{s\,\omega_{b,i}}{s^2 + s\,\omega_{b,i} + \omega_{r,i}^2}, \qquad (2)$$

where $k_{p,i}$ and $k_{r,i}$ are the proportional and resonant gains, respectively, $\omega_{r,i}$ is the resonant frequency and $\omega_{b,i}$ is the passband.

A control-loop similar to that shown in Fig. 2 can be built to control the current I_β. The reference currents I_α^* and I_β^* will be generated by an outer AC voltage control-loop and an outer DC voltage control-loop in order to control the PCC voltages and the DC bus voltage, respectively, as will be shown in the next sections.

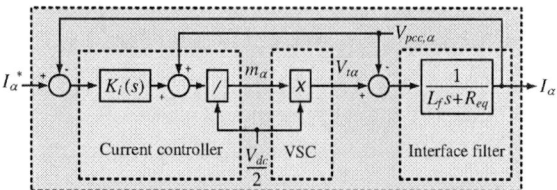

Fig. 2: Block diagram for the α-axis current control-loop of the STATCOM.

B. Principle of the PCC Voltage Compensation

Although the STATCOM can be used to compensate for negative and zero-sequence voltages in unbalanced distribution networks [11], in this paper it is desired the STATCOM compensates for only the positive-sequence voltages.

Fig. 3 shows the positive-sequence single-line equivalent circuit for the system of Fig. 1. In this figure the STATCOM is represented by a current source and the subindex $_{(1)}$ indicates the positive-sequence.

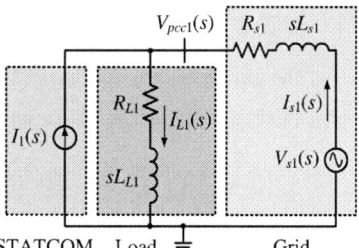

Fig. 3: Positive-sequence equivalent circuit.

Applying the superposition theorem in the circuit of Fig. 3, the contributions of the STATCOM and grid for the positive-sequence voltage at PCC (V_{pcc1}) can be calculated, as follows:

$$V_{pcc1}(s) = \frac{Z_{s1}Z_{L1}}{Z_{s1} + Z_{L1}}I_1(s) + \frac{Z_{L1}}{Z_{s1} + Z_{L1}}V_{s1}(s), \qquad (3)$$

where $Z_{s1}(s) = (R_{s1} + sL_{s1})$ and $Z_{L1}(s) = (R_{L1} + sL_{L1})$ are the impedances of the network and load, respectively, I_1 is the positive-sequence current of the STATCOM and V_{s1} is the positive-sequence voltage of the electric network.

It can be noted in (3) that both grid voltage (V_{s1}) and STATCOM current (I_1) can control the voltage V_{pcc1}. Thus, substituting s by $j\omega$ and considering $R_{s1} \approx 0$ and $R_{L1} \approx 0$ in (3), since the X/R ratio in distribution systems vary from 3 to 20 [12], the contribution of the STATCOM for the voltage V_{pcc1} is given by:

$$\dot{V}_{pcc1} = j\omega L_{eq}\dot{I}_1, \qquad (4)$$

where $L_{eq} = L_{s1}L_{L1}/(L_{s1} + L_{L1})$, $\dot{V}_{pcc1}$ is the PCC voltage phasor and $\dot{I}_1$ is the STATCOM current phasor.

Although (4) is derived from a steady-state mathematical modeling, the magnitude $\hat{V}_{pcc1}$ of the voltage $\dot{V}_{pcc1}$ and the magnitude $\hat{I}_1$ of the current $\dot{I}_1$ can also be calculated in the time domain through the concept of collective voltage/current [13]. Thus, as the objective of the STATCOM is to control

978-1-7281-4181-7/19 $31.00 © 2019 IEEE

the amplitude of the positive-sequence voltage at PCC through the amplitude of its terminal current, the following transfer function can be written by calculating the module of (4):

$$\frac{\hat{V}_{pcc1}(s)}{\hat{I}_1(s)} = \omega L_{eq}. \tag{5}$$

The knowledge of the transfer function ωL_{eq} makes it possible to build the block diagram of Fig. 4 for the voltage control-loop used to regulate the positive-sequence voltage at PCC. The block $K_v(s)$ represents the integral controller of the positive-sequence voltage control-loop, whose transfer function is given by $K_v(s) = k_v/s$. The error between the voltage $\hat{V}_{pcc1}$ and its reference ($\hat{V}^*_{pcc1}$) feeds the controller $K_v(s)$, whose output is the current $\hat{I}_1$.

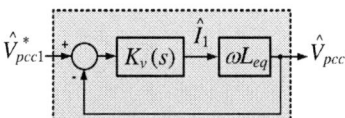

Fig. 4: Block diagram for the voltage control-loop of the STATCOM used to regulate the PCC voltage.

The calculation of the references for the current control-loop of Fig. 2 starts with the positive-sequence voltage detector of Fig. 5. Firstly, the voltages $v_{pcc,\alpha}$ and $v_{pcc,\beta}$, in the $\alpha\beta$ coordinates, are calculated by applying the Clarke transformation in the measured voltages, as shown bellow:

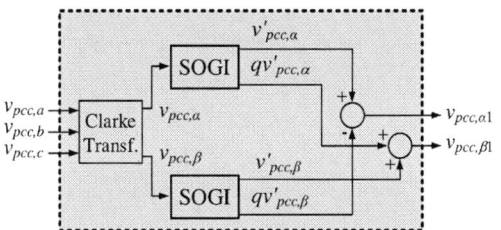

Fig. 5: Block diagram for the positive-sequence voltage detector.

$$\begin{bmatrix} v_{pcc,\alpha} \\ v_{pcc,\beta} \end{bmatrix} = \frac{2}{3} \begin{bmatrix} 1 & -\frac{1}{2} & -\frac{1}{2} \\ 0 & \frac{\sqrt{3}}{2} & -\frac{\sqrt{3}}{2} \end{bmatrix} \begin{bmatrix} v_{pcc,a} \\ v_{pcc,b} \\ v_{pcc,c} \end{bmatrix}. \tag{6}$$

The voltages $v_{pcc,\alpha}$ and $v_{pcc,\beta}$ feed two second-order generalized integrators (SOGI), as shown in Fig. 5. The output signals ($v'_{pcc,\alpha}$, $qv'_{pcc,\alpha}$, $v'_{pcc,\beta}$ and $qv'_{pcc,\beta}$) are combined to calculate the positive-sequence voltages $v_{pcc,\alpha1}$ and $v_{pcc,\beta1}$ in the $\alpha\beta$ coordinates. Fig. 6 depicts the internal structure of the SOGI circuit.

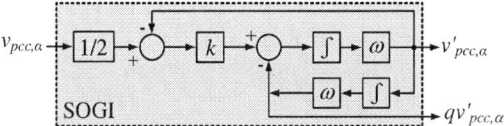

Fig. 6: Block diagram for the SOGI circuit.

The voltage $v_{pcc,\beta1}$, calculated using the positive-sequence

detector of Fig. 5, feed the block diagram of Fig. 7 used to calculate the reference currents for the current control-loop of Fig. 2. The SOGI circuit generates the quadrature voltages $v'_{pcc,\beta1}$ and $qv'_{pcc,\beta1}$ used to calculate the peak voltage $\hat{V}_{pcc1}$ [13]. Thus, knowing the STATCOM operates with 90-degrees lagged currents, the 90-degrees shifted voltage $v'_{pcc,\beta1}$ is used to generate the reference current i^*_α while the 180-degrees shifted voltage $qv'_{pcc,\beta1}$ is used to generate the reference current i^*_β.

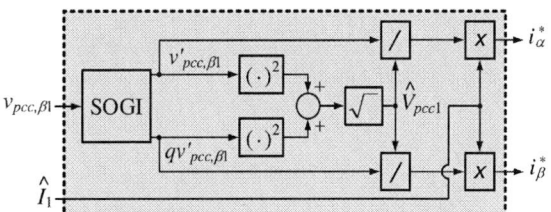

Fig. 7: Block diagram for the calculation of the reference currents of the STATCOM.

The voltage control-loop of Fig. 4 allows to control the voltage at the PCC, however, if it is necessary to control the voltage at a remote bus of a power system, a modification in the previous voltage control-loop must be done, as will be shown in the next section.

C. The Voltage Control of a Remote Bus

Fig. 8 shows the load connected to a remote bus, through an equivalent feeder, represented by the resistance R_g and the inductance L_g, that will be connected at PCC instead of the load of Fig. 1. Thus, now the function of the STATCOM is to compensate for the positive-sequence voltages at the remote bus ($v_{L,a1}$, $v_{L,b1}$ and $v_{L,c1}$).

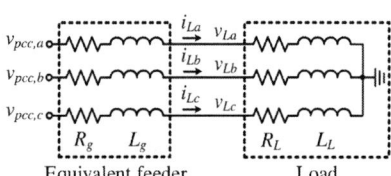

Fig. 8: Load connected to a remote bus through an equivalent feeder.

Based on the circuits of Fig. 1 and Fig. 8, the following transfer function can be written for the positive-sequence voltages at the remote bus and at the PCC:

$$U(s) = \frac{V_{L1}(s)}{V_{pcc1}(s)} = \frac{Z_{L1}(s)}{Z_{g1}(s) + Z_{L1}(s)}, \tag{7}$$

where $Z_{g1}(s) = (R_{g1} + sL_{g1})$ and $Z_{L1}(s) = (R_{L1} + sL_{L1})$ are the positive-sequence impedances of the equivalent feeder and load, respectively.

Substituting s by $\jmath\omega$ in (7), the module of $U(s)$, that relates the amplitudes of the positive-sequence voltages at the PCC and at the remote bus, is given by the following constant:

$$G = c\sqrt{A^2 + B^2}, \tag{8}$$

where $A = (\alpha_{L1}\alpha_{Lg1} + \omega^2)$, $B = \omega(\alpha_{Lg1} - \alpha_{L1})$, $c = (L_{L1}/L_{Lg1})/(\alpha_{Lg1}^2 + \omega^2)$, $\alpha_{L1} = (R_{L1}/L_{L1})$, $\alpha_{Lg1} = (R_{L1} + R_{g1})/(L_{L1} + L_{g1})$ and ω is the fundamental angular frequency of the electric network.

The knowledge of the constant G allows to build the voltage control-loop of Fig. 9 used to ensure that the voltages at the remote bus track its references. This control-loop is based on the voltage control-loop of Fig. 4. The error between the voltage measured at the remote bus and its reference signal is sent to the block $1/G$ whose output feeds the integral controller $K_v(s)$. From this point forward the control-loop is similar to that shown in Fig. 4. However, in the last part of the plant (equivalent electric network), it can be noted that the constant G relates the peak values of the voltages at the PCC and at the remote bus.

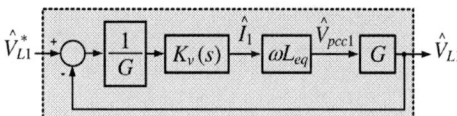

Fig. 9: Block diagram for the voltage control-loop of the STATCOM used to regulate the voltages at the remote bus.

The feeder's impedance $Z_g(s) = (R_g + sL_g)$ of Fig. 8 can represent an equivalent impedance of a generic network. The value of Z_g can be previously calculated in steady-state studies i) by injecting a current at the remote bus while open-circuiting the other buses; ii) through the inversion of the nodal admittance matrix to obtain the transfer impedance between the PCC and the remote bus; or iii) by using a proper reduction method that is able to achieve the circuit topology of Fig. 8 [14, 15].

D. The DC voltage control

The power of the DC capacitor of the STATCOM of Fig. 1 is given by:

$$P_c = \frac{1}{2}C_{eq}\frac{dV_{dc}^2}{dt}. \tag{9}$$

Applying the Laplace transform and performing a small-signal analysis in (9), the following transfer function can be written:

$$\frac{\tilde{V}_{dc}(s)}{\tilde{P}_c(s)} = \frac{1}{sC_{eq}\bar{V}_{dc}}, \tag{10}$$

where the symbols $(\tilde{\ })$ and $(\bar{\ })$ indicate the small-signal and steady-state variables, respectively.

Thus, it is possible to build the control-loop shown in Fig. 10 to control the DC voltage of the STATCOM, where the block $K_{vdc}(s) = (k_{p,vdc} + k_{i,vdc}/s)$ is a PI controller.

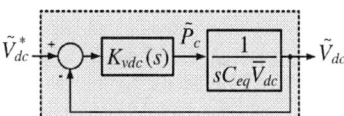

Fig. 10: Block diagram for the DC voltage control-loop of the STATCOM.

The output of the block K_{vdc} is the power $\tilde{p}_c$ used to generate the reference currents to control the DC voltage V_{dc}, as follows:

$$\begin{bmatrix} i_{\alpha,vdc}^* \\ i_{\beta,vdc}^* \end{bmatrix} = \frac{1}{v_{pcc,\alpha1}^2 + v_{pcc,\beta1}^2} \begin{bmatrix} v_{pcc,\alpha1} & v_{pcc,\beta1} \\ v_{pcc,\beta1} & -v_{pcc,\alpha1} \end{bmatrix} \begin{bmatrix} \tilde{p}_c \\ 0 \end{bmatrix}. \tag{11}$$

The currents $i_{\alpha,vdc}^*$ and $i_{\beta,vdc}^*$ are then added to the reference currents calculated in the block diagram of Fig. 7 before be sent to the current controller of Fig. 2.

III. RESULTS AND ANALYSIS

The circuits of Fig. 1 and Fig. 8, and all controllers of the STATCOM were modeled and simulated in an electromagnetic transient program in order to demonstrate and validate the voltage controllers proposed for the STATCOM. Initially, two distinct cases were simulated: (*i*) voltage compensation at the PCC and (*ii*) voltage compensation at the remote bus. The IGBTs of the STATCOM are switched with a sinusoidal pulse-width modulation strategy. Tables I, II, III and IV shows the parameters of the electric network, the STATCOM and its controllers and the load, respectively.

TABLE I

Grid parameters

Parameter	Value
RMS line voltage (V_s)	4.16 kV
Fundamental frequency (f_s)	60 Hz
Equivalent network inductance (L_s)	2.5 mH
Equivalent network resistance (R_s)	95 mΩ
Equivalent feeder inductance (L_g)	400 μH
Equivalent feeder resistance (R_g)	18 mΩ

TABLE II

STATCOM parameters

Parameter	Value
Equivalent capacitance (C_{eq})	3 mF
DC voltage (V_{dc})	10 kV
Inductance of the passive filter (L_f)	12 mH
Resistance of the passive filter (R_f)	45 mΩ
Resistance of the IGBTs (R_{igbt})	5 mΩ
Switching frequency (f_{sw})	9 kHz

TABLE III

Parameters of the STATCOM controllers

Controller	Parameter	Value
K_i	$k_{p,i}$	40 V/A
	$k_{r,i}$	50 V/A
	$\omega_{r,i}$	377 rad/s
	$\omega_{b,i}$	377 rad/s
K_v	k_v	35 A/(Vs)
K_{vdc}	$k_{p,vdc}$	1.8 A/V
	$k_{i,vdc}$	27 mA/(Vs)

TABLE IV

Load equivalent impedance

Phase	Value
"a"	$9.76 + j5.51\ \Omega$
"b"	$7.89 + j5.75\ \Omega$
"c"	$10.83 + j6.21\ \Omega$

A. Case 1

In this case, the STATCOM compensates for the positive-sequence voltages at the PCC by using the voltage control-loop of Fig. 4. Fig. 11 (a) and Fig. 11 (b) show the peak value of the system nominal voltage ($\sqrt{2/3} \times 4.16$ kV) used as reference for the voltage $\hat{V}_{pcc1}$ and the peak values of the positive-sequence voltages at the PCC ($\hat{V}_{pcc1}$) and at the remote bus ($\hat{V}_{L1}$), respectively. Fig. 11 (b) also shows the peak value of the positive-sequence at the remote bus ($\hat{V}'_{L1}$) calculated through the constant G and the voltage $\hat{V}_{pcc1}$ (Equations 7 and 8). The voltages $\hat{V}'_{L1}$ and $\hat{V}_{L1}$ were plotted together in order to validate the use of the constant G. It is possible to note that there are no discrepancies between the voltages $\hat{V}_{L1}$ and $\hat{V}'_{L1}$. In t = 0.35 s the STATCOM starts synthesizing currents in order to compensate for the positive-sequence voltages at the PCC. It can be noted in Fig. 11 (a) that the PCC voltages are increased until the nominal value, due to the STATCOM operation. Furthermore, it can be noted in Fig. 11 (b) that the peak value of the positive-sequence voltage at the remote bus ($\hat{V}_{L1}$) is also increased, although it is not directly controlled by the static compensator. Thus, if the voltage at the remote bus should be regulated to the nominal value, the control algorithm presented in Section II.C can be used to control the STATCOM, as will be shown in the next case.

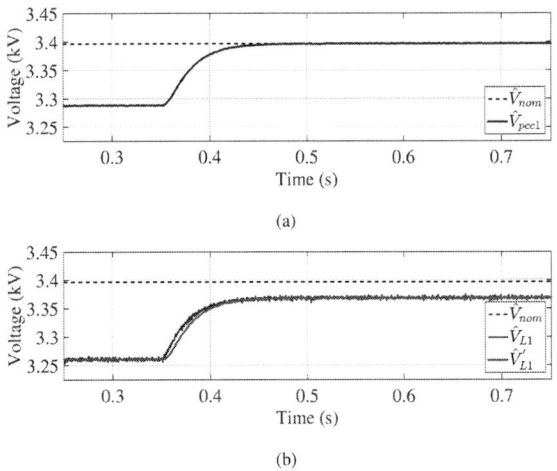

Fig. 11: Peak value of the positive-sequence voltage in Case 1: (a) PCC and (b) remote bus.

B. Case 2

In this second case the STATCOM compensates for the positive-sequence voltage at the remote bus, where the load is connected, by using the voltage control-loop of Fig. 9. Fig. 12 (a) and Fig. 12 (b) show the same waveforms of Fig. 11. In t = 0.35 s the STATCOM starts synthesizing currents in order to compensate for the positive-sequence voltages at the remote bus. It can be noted in Fig. 12 (b) that, after t = 0.35 s, the voltages at the remote bus are increased until the nominal value, as expected. However, it can also be noted in Fig. 12 (a) that the voltage at the PCC exceeds the nominal voltage of the system, since the PCC is more close

to the STATCOM then the remote bus. Thus, although the static compensator can control the voltage at a remote bus, it is important to verify if the voltages of the upstream buses are exceeding the limits established for the supply voltages.

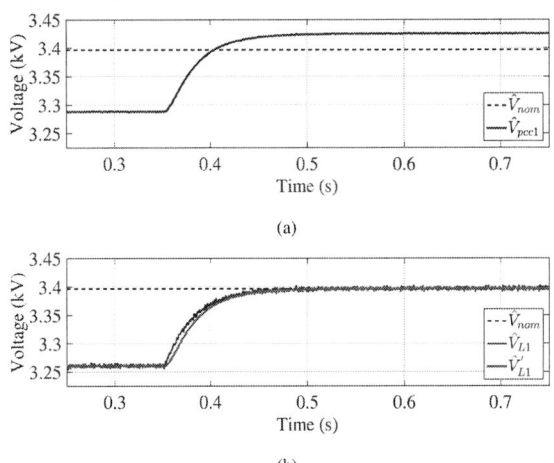

Fig. 12: Peak value of the positive-sequence voltage in Case 2: (a) PCC and (b) remote bus.

C. Variation of parameters

The load, equivalent feeder and grid parameters may vary during the STATCOM operation. However, it can be noted in Tables I and IV that the load have higher impedance values than the grid and equivalent feeder. Thus, in this last case it is investigated how these parameter variations can impact on the performance of the voltage control-loop proposed for the STATCOM. The load resistances and inductances were varied by ± 30 %. The STATCOM compensates for the positive-sequence voltage at the remote bus, as in the second case. Fig. 13 (a) and Fig. 13 (b) show the phase "a" current of the STATCOM and the peak value of the voltage at the remote bus, respectively, for three values of the load impedance ($0.7 \times Z_{L1}$, Z_{L1} and $1.3 \times Z_{L1}$). It can be noted in Fig. 13 (b) that, if the load impedance is lower than the nominal value, the voltage at the remote bus, before the STATCOM operation, is also lower due to the higher voltage drops in the grid impedances. Thus, it can be noted in Fig. 13 (a) that the current synthesized by the STATCOM is higher than the case in which the value of the load impedance is nominal. If the load impedance is higher than the nominal value the opposite occurs, as shown in Fig. 13 (a) and Fig. 13 (b). Moreover, it can be noted that, for all values of the load impedance, the peak value of the voltage at the remote bus tracks its reference.

IV. CONCLUSIONS

This paper presented a voltage control strategy for a static synchronous compensator used to regulate the voltage at a remote bus of a power system. However, firstly it was presented a voltage control-loop in order to compensate for the voltages at the point of common coupling. Then, a constant derived from a transfer function, that relates the voltages at the point of common coupling and at the remote bus, was inserted

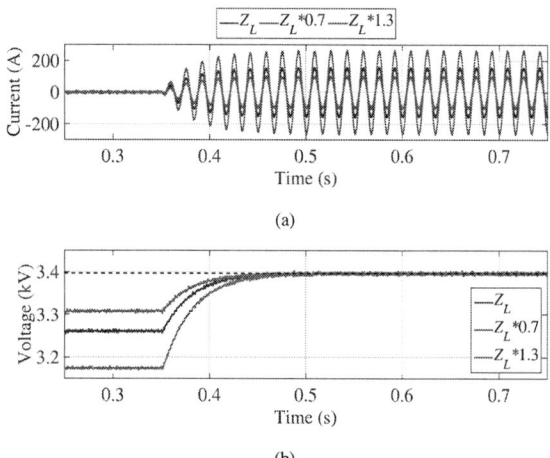

(a)

(b)

Fig. 13: Case 3 waveforms: (a) phase "*a*" current of the STAT-COM and (b) peak value of the positive-sequence voltage at the remote bus.

into the voltage controller in order to regulate the voltages at the remote bus. Results of digital simulations were also presented to demonstrate and validate the proposed control strategy. The results shown the effectiveness of the proposed voltage control-loop for the STATCOM. However, stability analysis for the voltage control-loop and/or a load-sensitive control approach, where the load model is adjusted according to the results of measurements, can be performed to improve the proposed methodology. Moreover, it was shown that, when the static compensator compensates the voltage of a remote bus at the nominal value, the upstream buses of the power system may exceed the limits established for the supply voltages.

ACKNOWLEDGEMENT

This study was financed in part by the Coordenação de Aperfeiçoamento de Pessoal de Nível Superior - Brasil (CAPES) - Finance Code 001, the National Council for Scientific and Technological Development (CNPq), the State Funding Agency of Minas Gerais (FAPEMIG) and the National Institute for Electric Energy (INERGE).

REFERENCES

[1] M. Yilmaz and P. T. Krein, "Review of battery charger topologies, charging power levels, and infrastructure for plug-in electric and hybrid vehicles," *IEEE transactions on Power Electronics*, vol. 28, no. 5, pp. 2151–2169, 2012.

[2] P. M. Almeida, K. M. Monteiro, P. G. Barbosa, J. L. Duarte, and P. F. Ribeiro, "Improvement of PV grid-tied inverters operation under asymmetrical fault conditions," *Solar Energy*, vol. 133, pp. 363–371, 2016.

[3] L. R. de Araujo, D. R. R. Penido, J. L. R. Pereira, and S. Carneiro, "Voltage security assessment on unbalanced multiphase distribution systems," *IEEE Transactions on Power Systems*, vol. 30, no. 6, pp. 3201–3208, 2015.

[4] H. Liao, S. Abdelrahman, and J. V. Milanović, "Zonal mitigation of power quality using FACTs devices for pro-

vision of differentiated quality of electricity supply in networks with renewable generation," *IEEE Transactions on Power Delivery*, vol. 32, no. 4, pp. 1975–1985, 2016.

[5] J. V. M. Farias, A. F. Cupertino, V. N. Ferreira, S. I. Seleme, H. A. Pereira, and R. Teodorescu, "Design and lifetime analysis of a DSCC-MMC STATCOM," in *2017 Brazilian Power Electronics Conference (COBEP)*. IEEE, 2017, pp. 1–6.

[6] K. M. Muttaqi, A. D. Le, M. Negnevitsky, and G. Ledwich, "A coordinated voltage control approach for coordination of OLTC, voltage regulator, and DG to regulate voltage in a distribution feeder," *IEEE Transactions on Industry Applications*, vol. 51, no. 2, pp. 1239–1248, 2015.

[7] C. R. Sarimuthu, V. K. Ramachandaramurthy, K. Agileswari, and H. Mokhlis, "A review on voltage control methods using on-load tap changer transformers for networks with renewable energy sources," *Renewable and Sustainable Energy Reviews*, vol. 62, pp. 1154–1161, 2016.

[8] X.-P. Zhang, E. Handschin, and M. Yao, "Multi-control functional static synchronous compensator (STATCOM) in power system steady-state operations," *Electric power systems research*, vol. 72, no. 3, pp. 269–278, 2004.

[9] C. Han, A. Q. Huang, M. E. Baran, S. Bhattacharya, W. Litzenberger, L. Anderson, A. L. Johnson, and A.-A. Edris, "STATCOM impact study on the integration of a large wind farm into a weak loop power system," *IEEE Transactions on Energy conversion*, vol. 23, no. 1, pp. 226–233, 2008.

[10] A. S. Ribeiro, A. d. O. Almeida, P. G. Barbosa, and P. M. de Almeida, "Analysis and design of proportional-resonant controllers based on pole placement approach," in *2018 Simposio Brasileiro de Sistemas Eletricos (SBSE)*. IEEE, 2018, pp. 1–6.

[11] S. N. Duarte, F. T. Ghetti, P. M. de Almeida, and P. G. Barbosa, "Zero-sequence voltage compensation of a distribution network through a four-wire modular multilevel static synchronous compensator," *International Journal of Electrical Power & Energy Systems*, vol. 109, pp. 57–72, 2019.

[12] A. A. Sallam and O. P. Malik, *Electric distribution systems*. Wiley-IEEE Press, 2018.

[13] H. Akagi, E. H. Watanabe, and M. Aredes, *Instantaneous power theory and applications to power conditioning*. John Wiley & Sons, 2017, vol. 62.

[14] F. Shen, P. Ju, M. Shahidehpour, Z. Li, and X. Pan, "Generalized discrete-time equivalent model for dynamic simulation of regional power area," *IEEE Transactions on Power Systems*, vol. 33, no. 6, pp. 6452–6465, 2018.

[15] Z. K. Pecenak, V. R. Disfani, M. J. Reno, and J. Kleissl, "Inversion reduction method for real and complex distribution feeder models," *IEEE Transactions on Power Systems*, vol. 34, no. 2, pp. 1161–1170, 2019.

978-1-7281-4181-7/19 $31.00 © 2019 IEEE

Single-Stage Single-phase AC/DC Converter with High Frequency Isolation Feasible to Microgeneration

Samanta Gadelha Barbosa
Electrical Engineering
Department
Federal University of Ceará
(UFC)
Fortaleza, Brazil
samantagadelha@dee.ufc.br

Bruno Ricardo de Almeida
Electrical Engineering
Department
University of Fortaleza
(UNIFOR)
Fortaleza, Brazil
almeida@unifor.br

Debora Pereira Damasceno
Electrical Engineering
Department
Federal University of Ceará
(UFC)
Fortaleza, Brazil
deborapd@alu.ufc.br

Demercil de S. Oliveira Jr.
Electrical Engineering
Department
Federal University of Ceará
(UFC)
Fortaleza,Brazil
demercil@dee.ufc.br

Abstract— **This paper proposes a multi-port, single-phase ac-dc converter with high frequency isolation and bidirectional power flow capability, which is suitable for distributed generation, interconnecting various sources and loads, and controlling the power flow between them in an integrated conversion stage. Three-state switching cells and interleaved converters are employed in order to achieve high power processing capability. In order to verify the feasibility of this topology, it is presented the principle of operation, theoretical analysis, control strategy and simulation waveforms. Some experimental results are presented for a 1 kW laboratory prototype under construction.**

Keywords — ac/dc converter, DAB, microgeneration, single phase, single stage.

I. INTRODUCTION

Nowadays, Renewable Energy Sources (RES) have an increasing influence on global electric power sector development. Besides environmental impacts reduction, RES generation has become more financially attractive [1]. The lower prices bring advantages to generation on residential level. According to [2], the increase prediction for residential generation on global scale in 10 years is from 94.9 MW (2016) to 3773.3 MW (2015). Fig.1 shows the growing of global electricity generation capacity between years 2007 and 2017, in which the largest capacity increase was observed in 2017, led by solar PV energy that accounted for nearly 55% of newly installed renewable power capacity [3]. Although the increase of distributed generation is promising, it comes along with a series of problems for energy systems, for example, power fluctuation due to intermittent characteristic of sources and softwares or hardwares not yet fully optimized [4]. Due to it, investment in research and new technologies are necessary in order to increase power quality for this kind of sources.

Different topologies of power electronic converters have been proposed for distributed generation applications. Those

topologies tend to present bidirectional power flow and usually connect ac grid with generation systems and energy storage, being capable of controlling the energy flow between those sources [5-8]. The ac-dc converters are increasingly developed in order to provide better power quality in terms of power factor correction, total harmonic distortion reduction and regulated dc voltage, with unidirectional and bidirectional modes of power flow [9-10]. The dual-active-bridge (DAB) converter is considered a good option when the power flow is bidirectional. Originally proposed in [11], it has characteristics such as voltage step up and down capability, bidirectional power flow and high-frequency isolation.

Within this context, this paper proposes a single-phase converter with an integrated stage as shown in Fig. 2. The proposed topology uses two three-state switching cells in its bidirectional version [12], operating in interleaving mode, associated with the DAB converter concept. It is suitable for distributed generation, interconnecting various sources and loads, and controlling the power flow between them in an integrated conversion stage.

II. PROPOSED POWER CONVERTER

The proposed topology is illustrated in Fig. 2 and was based on topologies mentioned in [12] and [13]. It can also be considered as a biphasic version of converter presented in [14] and consists of two interleaved dual-active-bridge (DAB) converters. Both sides have two interleaved bidirectional full-bridge topologies with three-state switching cells (3SSC), reducing the voltage and current stresses in the semiconductors and increasing the resulting frequency over the input and output filters [15]. Therefore, the higher number of semiconductors in the proposed structure is compensated

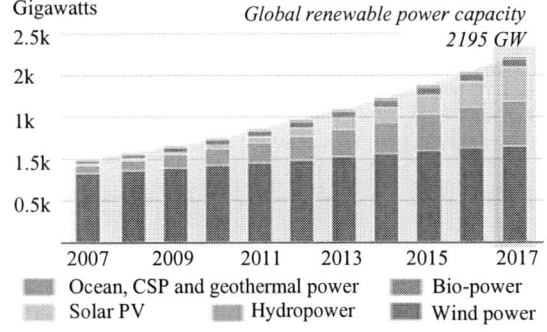

Fig. 1. Global renewable power capacity [3].

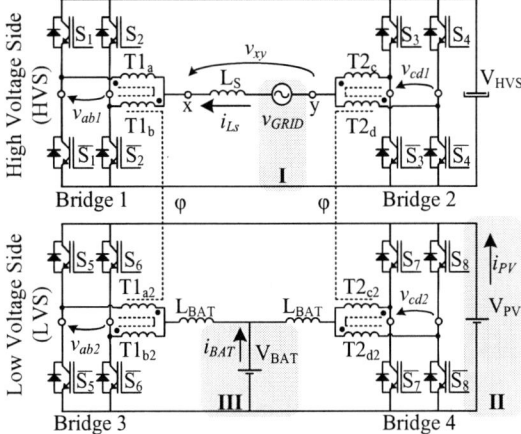

Fig. 2. Proposed converter.

by the reduction of the efforts on them and on the weight and volume of the magnetic ones due to the high frequency and the coupling through 3SSC's.

To simplify the distinction between isolated parts, the ac-dc side connected directly to the grid (bridges 1 and 2) is called High Voltage Side (HVS) and the dc-dc converter with two dc ports (bridges 3 and 4) is called Low Voltage Side (LVS), where it is possible to connect both battery bank and PV modules. The main features of the proposed topology are: Power Factor Correction (PFC); High-frequency galvanic isolation; Bidirectional power flow capability; and reduction of filter requirements due to a five-level voltage waveform resulting from the use of two three-state switching cells operating in interleaving mode.

A. Operating Principle

The HVS side has the same configuration as in [16], with a sinusoidal modulation that results in 14 different switching combinations. The same combinations are observed on LVS side switches, although it receives a continuous modulation. The switching combinations are presented in Table 1 with the respective resulting voltages.

TABLE I.   VOLTAGES OF THE FULL-BRIDGE CONFIGURATION

Switch State				Voltages		
$S_{1,5}$	$S_{2,6}$	$S_{3,7}$	$S_{4,8}$	Vab_x	Vcd_x	Vxy
1	1	0	0	0	0	Vdc_x
1	0	0	0	Vdc_x	0	$Vdc_x/2$
1	1	1	0	0	Vdc_x	$Vdc_x/2$
0	1	0	0	$-Vdc_x$	0	$Vdc_x/2$
1	1	0	1	0	$-Vdc_x$	$Vdc_x/2$
1	0	1	0	Vdc_x	Vdc_x	0
0	1	1	0	$-Vdc_x$	Vdc_x	0
0	1	0	1	$-Vdc_x$	$-Vdc_x$	0
1	0	0	1	Vdc_x	$-Vdc_x$	0
1	0	1	1	Vdc_x	0	$-Vdc_x/2$
0	0	1	0	0	Vdc_x	$-Vdc_x/2$
0	1	1	1	$-Vdc_x$	0	$-Vdc_x/2$
0	0	0	1	0	$-Vdc_x$	$-Vdc_x/2$
0	0	1	1	0	0	$-Vdc_x$

a. Vab_x and Vcd_x: index refers a both sides [x=1 (HVS) or x=2 (LVS)]

b. Vdc_x: index refers a both sides [x=1 (V_{HVS}) or x=2 (V_{PV})]

Hence, during one grid cycle the duty cycle of the 16 switches varies between 0 and 100% resulting in 3 voltage levels on the power transfer inductors between bridge legs (v_{abx} and v_{cdx}) and 5 voltage levels between bridges of HVS side (v_{xy}). Fig. 3 illustrates the resulting multilevel voltage (v_{xy}) using a sinusoidal modulation signal, where D_{S1} is the duty cycle on leg "a".

B. Modulation tecnique

As previously mentioned, the modulation strategy was based on the SPWM (Sinusoidal Pulse Width Modulation) technique. Bridge 1 and 2 receive the modulation signals ($m_{HVS(0°)}$) and ($m_{HVS(180°)}$), respectively, with phases shifted by 180°. The switches in a same bridge operate with triangular carriers shifted by 180°. The carriers in a given bridge are phase-shifted by 90° regarding the carriers in the other bridge. The leg "a" in bridge 1 is taken as a reference at 0°, and the legs "b", "c" and "d" are shifted by 180°, 90° and 270°, respectively. The modulation signals and their carriers are illustrated in Fig. 4.

In the LVS side a continuous modulation signal is used and the carriers signals are phase-shifted like in the HVS side, i.e.,

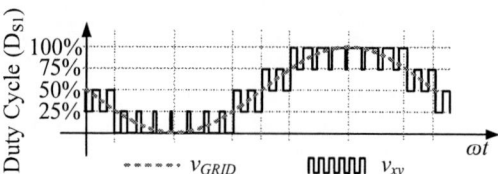

Fig. 3. Grid and five levels voltages on HVS side.

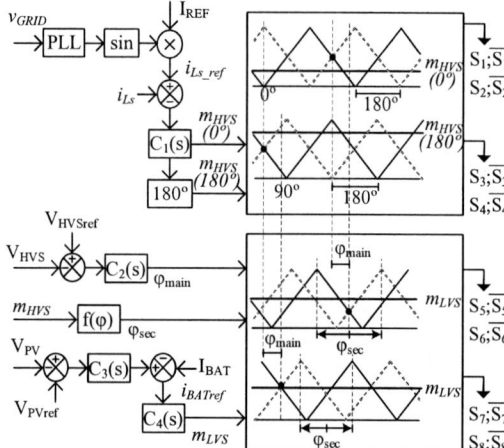

Fig. 4. Control strategy.

"a", "b", "c" and "d" are situated in 0°, 180°, 90° and 270° respectively. In order to obtain voltage pulses with the same width in both primary and secondary transformer sides, the modulation signals of HVS are used to vary the phase-shift between carriers signals on LVS. Due that, the reactive content in the transformer is reduced, similar to that presented in [17].

The power flux direction is defined by the phase-shift angle between voltage pulses on transformer windings (φ), according to a conventional DAB converter [11]. Topic D presents a power flow analysis more direct for the proposed topology.

C. Control Strategy

Fig. 4 shows the simplified control diagram for the proposed topology, where it is sought to validate the control of voltage and power flow between ports. The battery charging strategy and active power injected to, or extracted from the grid are supposed to be managed by the supervisory systems and is not object of this paper.

The control strategy on the HVS side consists in a current compensator (C1) which controls the ac current (i_{Ls}) in order to guarantee high power factor (PF) and low total harmonic distortion (THD). The current reference to be injected or requested from the grid comes from an external command, like an external supervisory system. The transfer function expression is based on a classical boost converter in the average current mode control and it is described in (1).

$$\frac{i_{Ls}(s)}{d(s)} = \frac{V_{HVS}}{s \cdot L_S} \quad (1)$$

The V_{HVS} voltage is controlled by the compensator (C2), that provides the phase-shift value to regulate the HVS bus on 400V. That response is used to shift the carriers phase directing the power flow between sides based on the phase-shift technique, where a positive value implies on the power flow direction from HVS to LVS, while a negative value causes power to flow from LVS to HVS. The transfer function used is described in (2), which is derived from the Gyrator theory applied to dual active bridge converters as in [18]. R_{HVS} is the resistive load value necessary to obtain the nominal power on V_{HVS} bus.

$$\frac{\hat{v}_{HVS}(s)}{\hat{\varphi}(s)} = \frac{V_{HVS}}{2 \cdot \pi \cdot fs \cdot (L_{disp})} \cdot \varphi \cdot \left(1 - \frac{|\varphi|}{\pi}\right) \cdot \frac{R_{HVS}}{R_{HVS} \cdot C_{HVS} \cdot s + 1} \quad (2)$$

To the LVS side, a voltage compensator (C3) is used for PV bus control and its response is the current reference necessary on the battery bus to maintain the V_{PV} on 96V. It works similar to a classical buck converter and the transfer function used is described by (3). The control of the current through the battery port uses a current compensator (C4) and the adopted transfer function is similar to (1) whose parameters were referred to the LVS side. To reduce the transformer reactive content a second phase-shift on the LVS carriers is used varying as a function of the m_{HVS} signal, similar to [17].

$$\frac{\hat{v}_{BAT}(s)}{\hat{i}_{BAT}(s)} = \frac{R_{SEC} \cdot V_{PV}}{1 + s \cdot R_{SEC} \cdot C_{SEC}} \quad (3)$$

As the ideal duty-cycle value on the LVS switches is 50%, but the PV and battery voltages require some variation, m_{LVS} is limited between 40% and 60%.

D. Power Flow

The power flow transfer between isolated parts is based on the Dual Active Bridge converter concept (DAB) [11]. The coupling between bridges 1 and 3, and 2 and 4 can be considered as two DAB converter operating separately with same operation modes, which allows to consider just one converter during analysis. Then, Fig. 5 illustrates the DAB resulted from bridges 1 and 3 which will be taken as reference.

The behaviour of the power flow on the proposed topology is similar to that presented in [14], which equivalent circuit is shown in Fig. 6a, where α is the relation between V_{PV} and V_{HVS} voltages and reflect the parameters to the HVS side. The pulse width of voltage v_{ab1} on HVS (Δd) as function of ac voltage angle (ωt) and the phase-shift ($\Delta \varphi$) are given by (4) and (5), respectively, and m_a is the modulation index that

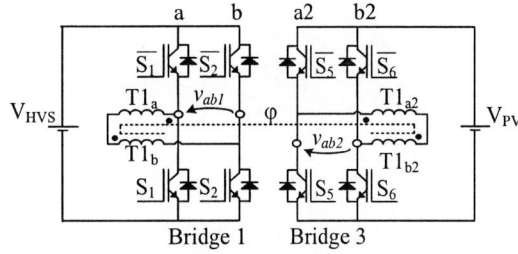

Fig. 5. Equivalent DAB converter.

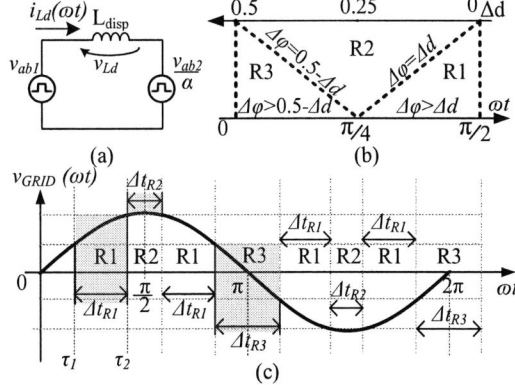

Fig. 6. DAB operation: a) equivalente circuit; b) regions limit; c) regions duration.

relates the ac voltage peak (V_{gPK}) with the dc voltage on HVS (V_{HVS}) calculated by (6).

$$\Delta d(\omega t) = \frac{1}{2} \cdot \left(1 - m_a \cdot |\sin(\omega t)|\right) \quad (4)$$

$$\Delta \varphi = \frac{\varphi}{2\pi} \quad (5)$$

$$m_a = \frac{|V_{gPK}|}{V_{HVS}} \quad (6)$$

Assuming φ value between 0 and 90º, the topology operation can be resumed to three operational regions. The boundaries between regions are defined by $\Delta \varphi$ and Δd, as illustrated in Fig. 6b. The time intervals where each operational region occurs are presented in Fig. 6c, and they are

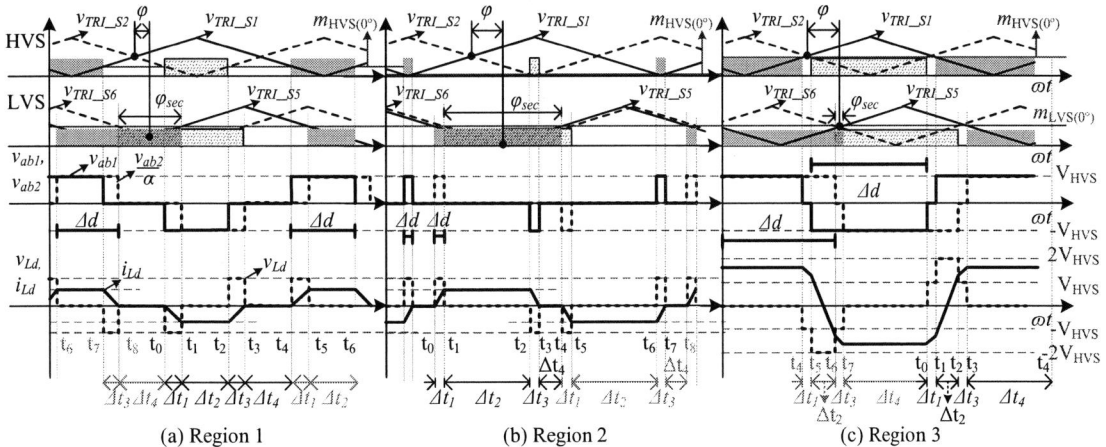

(a) Region 1   (b) Region 2   (c) Region 3

Fig. 7. Waveforms related to the operation regions.

defined by instants $\tau 1$ e $\tau 2$, calculated by (7) and (8), and depend on the terms m_a and $\Delta\varphi$.

$$\tau 1 = \frac{1}{\omega} \cdot \sin^{-1}\left(\frac{2\Delta\varphi}{m_a}\right) \quad (7)$$

$$\tau 2 = \frac{1}{\omega} \cdot \sin^{-1}\left(\frac{1-2\Delta\varphi}{m_a}\right) \quad (8)$$

Fig. 7 illustrates the converter operation in the 3 distinct operational regions with their respective modulation signals as well as the resulting voltages and current in the DAB stage. In this paper regions 1, 2 and 3 correspond respectively to regions 5, 6 and 8 of [14]. The expressions of current and duration for each region are summarized below from (9) to (20) and Ts is the sampling period.

$$i_{Ld}(t)_{R1} = \begin{cases} -\dfrac{v_{Ld}}{L_{disp}} \cdot (t_0 - t_1), & if\ t_0 < t < t_1 \\[2mm] \dfrac{v_{Ld}}{L_{disp}} \cdot \Delta\varphi \cdot Ts, & if\ t_1 < t < t_2 \\[2mm] -\dfrac{v_{Ld}}{L_{disp}} \cdot \Delta\varphi \cdot Ts + \dfrac{v_{Ld}}{L_{disp}} \cdot (t - t_0), & if\ t_2 < t < t_3 \\[2mm] 0, & if\ t_3 < t < t_4 \end{cases} \quad (9)$$

$$\Delta t_1 = \Delta t_3 = t_1 - t_0 = \Delta\varphi \cdot Ts \quad (10)$$

$$\Delta t_2 = t_2 - t_1 = \left[\Delta d(t) - \Delta\varphi\right] \cdot Ts \quad (11)$$

$$\Delta t_4 = t_4 - t_3 = \left[0.5 - \left(\Delta d(t) + \Delta\varphi\right)\right] \cdot Ts \quad (12)$$

$$i_{Ld}(t)_{R2} = \begin{cases} -\dfrac{v_{Ld}}{L_{disp}} \cdot \Delta d(t) \cdot Ts + \dfrac{v_{Ld}}{L_{disp}} \cdot (t - t_0), & if\ t_0 < t < t_1 \\[2mm] 0, & if\ t_1 < t < t_2 \\[2mm] \dfrac{v_{Ld}}{L_{disp}} \cdot (t - t_2), & if\ t_2 < t < t_3 \\[2mm] \dfrac{v_{Ld}}{L_{disp}} \cdot \Delta d(t) \cdot Ts, & if\ t_3 < t < t_4 \end{cases} \quad (13)$$

$$\Delta t_1 = \Delta t_3 = t_1 - t_0 = \Delta d(t) \cdot Ts \quad (14)$$

$$\Delta t_2 = t_2 - t_1 = \left[\Delta d(t) - 2 \cdot \Delta d\varphi\right] \cdot Ts \quad (15)$$

$$\Delta t_4 = t_4 - t_3 = \left[0.5 - \left(\Delta d(t) + \Delta\varphi\right)\right] \cdot Ts \quad (16)$$

$$i_{Ld}(t)_{R3} = \begin{cases} -\dfrac{v_{Ld}}{L_{disp}} \cdot \Delta\varphi \cdot Ts + \dfrac{v_{Ld}}{L_{disp}} \cdot (t - t_0), & if\ t_0 < t < t_1 \\[2mm] \dfrac{v_{Ld}}{L_{disp}} \cdot (\Delta\varphi + \Delta d(t) - 0.5) \cdot Ts + \dfrac{2v_{Ld}}{L_{disp}} \cdot (t - t_1), \\ \quad if\ t_1 < t < t_2 \\[2mm] \dfrac{v_{Ld}}{L_{disp}} \cdot (\Delta\varphi + \Delta d(t) - 0.5) \cdot Ts + \dfrac{v_{Ld}}{L_{disp}} \cdot (t - t_2), \\ \quad if\ t_2 < t < t_3 \\[2mm] \dfrac{v_{Ld}}{L_{disp}} \cdot \Delta\varphi \cdot Ts, & if\ t_3 < t < t_4 \end{cases} \quad (17)$$

$$\Delta t_1 = \Delta t_3 = t_1 - t_0 = \left(0.5 - \Delta d(t)\right) \cdot Ts \quad (18)$$

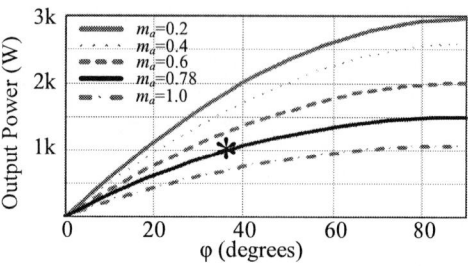

Fig. 8. Transferred power concerning to HVS side.

$$\Delta t_2 = t_2 - t_1 = \left[\left(\Delta d(t) + \Delta\varphi\right) - 0.5\right] \cdot Ts \quad (19)$$

$$\Delta t_4 = t_4 - t_3 = \left[\Delta d(t) - \Delta d\varphi\right] \cdot Ts \quad (20)$$

Given the specifications adopted in Table 2, and using the expressions (9) to (20), the transferred power considering different modulation indexes and phase-shifts are presented in Fig. 8. The nominal operating point chosen to the converter is marked on graph. It is possible to note that the high modulation index from the converter limits the power transfer between sides.

III. SIMULATION RESULTS

Simulation results were obtained using PSIM® software. Table 2 presents the specifications and parameters used for validation of the proposed topology. The main simulated waveforms that characterize its operation are shown.

TABLE II. SPECIFICATIONS AND PARAMETERS

Prototype Specifications and parameters	
RMS voltage (v_{GRID})	220 V/ 60 Hz
DC voltage on HVS (v_{HVS})	400 V
Voltage on port II (v_{PV})	96 V
Voltage on port III (v_{BAT})	48 V
Rated power (Po)	1000 W
Switching frequency (f_s)	50 kHz
HVS inductor (Ls)	0.5 mH
LVS inductor (L_{BAT})	2 x 0.25 mH
Power transfer inductance (L_{mag})	15 µH
Capacitive filter on HVS (C_{HVS})	3.76 mF
Capacitive filter on port II (C_{PV})	1.88 mF
Capacitive filter on port III (C_{BAT})	0.47 mF

Fig. 9 shows the steady state response of the converter when the HVS side works as a rectifier with a resistive load on V_{HVS} bus and a dc source of 48V at port III. Fig. 9a presents the waveforms of input and output current, grid voltage and dc bus voltage on HVS side. The resulting THD was 3.01% and PF equal to 0.9995. Fig. 9b shows the multilevel voltage (v_{xy}) with five levels and v_{GRID}. Fig. 9c presents both dc bus voltages regulated on LVS side. In all dc buses the voltage ripple was less than 1%.

Fig. 10 shows case 1, with HVS operating as an inverter with a positive step applied (50% to 75% in 0.6s), considering a constant power generation by port II. Fig. 10a shows the fast i_{Ls} response without significant distortion. As can be noticed in Fig. 10b, initially generated PV energy is injected into the grid and battery simultaneously, and in 0.6s the grid demand increases reducing the power injected into the battery. Fig. 10c

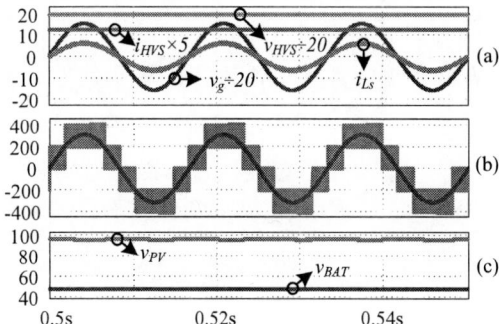

Fig. 9. Steady state response for HVS side as a rectifier.

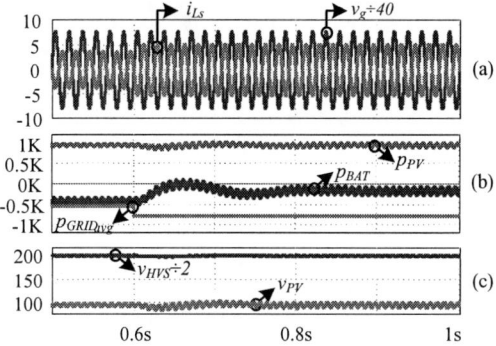

Fig. 10. Case 1: response on HVS side for a step in i_{Ls}.

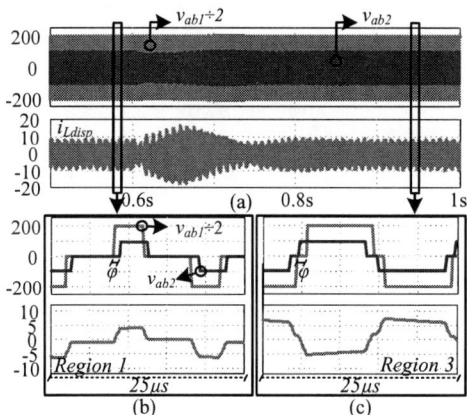

Fig. 11. Case 1: voltages and leakage current on transformer.

shows dc buses where it is verified an overshoot smaller than 5% and rapid stabilization, around 150 ms.

Still for case 1, Fig. 11a. presents the voltages and leakage current trough transformer T1. Fig. 11b and 11c show the DAB waveforms considering two distinct phase-shift angles and two distinct regions (regions 1 and 3).

Fig. 12 presents case 2: a reversion of power flow direction (50% to -50% of nominal power). Fig. 12a shows fast response of i_{Ls} without significant distortion. Fig. 12b presents the power flow reversion on grid and battery. The high overshoot of p_{BAT} is due to the aggressive step what usually doesn't occur in real situations, but proves the converter stability. Fig. 12c shows the v_{HVS} and v_{PV} buses stabilization with low overshoot and in short time, around 210 ms. Fig. 13a

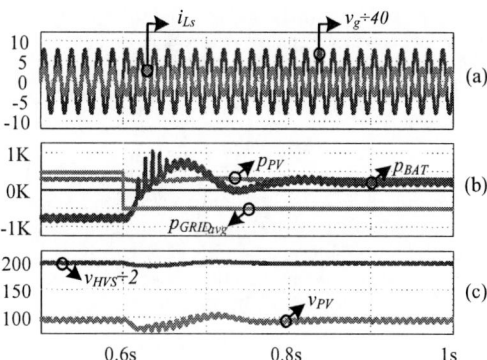

Fig. 12. Case 2: Reversion of power flow direction response.

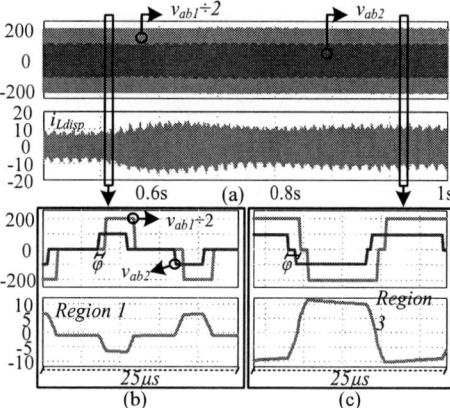

Fig. 13. Case 2: voltages and leakage current on transformer.

presents the voltages and leakage current on the transformer T1 to case 2, and Fig. 13b and 13c show the polarity of phase-shift angles to each power direction.

IV. EXPERIMENTAL RESULTS

Fig. 14 presents the prototype under construction, with two three-phase SiC transistor modules (Cree - CCS0220M12CM2 1200V 20A) and two isolated gate drivers (Cree - CGD15FB45P), that have been used in the assembly of the full-bridges, and one digital signal processor (DSP model TMS320F28377D by Texas Instruments) for system control. The experimental results presented refer to the HVS side, which validate the interleaved bidirectional full-bridge topology with three-state switching cells (3SSC).

Fig. 15 shows the nominal input and output steady state waveforms. A low frequency ripple and regulated values, as specified, can be noted. The ac current presents low distortion and is in phase with ac grid voltage. It was obtained a PF of

Fig. 14. Prototype.

0.986, THD of 5.31% and efficiency of 96.15%. Fig. 16 shows the five levels of v_{xy} in phase with v_{GRID}. Fig. 17 presents the magnetizing currents of the 3SSC's with the input current those are controlled to have a null average values.

V. CONCLUSION

This work proposes a single-phase ac-dc three-port converter feasible to distributed generation systems. The structure, modulation technique and the power flow characteristic were presented in detail. The control strategy is properly validated by simulation, where results were able to demonstrate the claimed advantages.

The prototype is under construction and some preliminary experimental results for interleaved bidirectional full-bridge topology with 3SSC are presented. It was obtained a FP of 0.986, THD of 5.31% and efficiency of 96.15%.

The use of a higher number of semiconductors allows an improved current stress and loss distribution, presenting advantages for operation with higher powers. The increased frequency on the filters and single-stage operation allow it to work with high power density.

ACKNOWLEDGMENT

The authors acknowledge the members of GPEC research group, and also research funding agencies CAPES and CNPq for the overall support.

REFERENCES

[1] I. A. Firsova, D. G. Vasbieva, A. V. Litvinov, O. E. Chernova and I. V. Telezhko, "Trends in the Development of the Global Energy Market," in International Journal of Energy Economics and Policy, vol. 9, no. 3, pp. 59-65, Mac. 2019.

[2] F. Hafiz, A. R. de Queiroz and I. Husain, "Solar Generation, Storage, and Electric Vehicles in Power Grids: Challenges and Solutions with Coordinated Control at the Residential Level," in IEEE Electrification Magazine, vol. 6, no. 4, pp. 83-90, Dec. 2018.

[3] Renewables 2018 Global Status Report. (2018), Available from: http://www.ren21.net/gsr-2018/chapters/chapter_01/chapter_01/. [Last accessed on 2019 Jun 20].

[4] S. V. Ratnera and R. M. Nizhegorodtseva , "Analysis of the World Experience of Smart Grid Deployment: Economic Effectiveness Issues" ISSN 0040-6015, Thermal Engineering, 2018, vol. 65, no. 6, pp. 387–399.

[5] B. Nordman and K. Christensen, "DC local power distribution: Technology, deployment, and pathways to success," IEEE Electrification Magaz., vol. 4, no. 2, pp. 29–36, Jun. 2016.

[6] Z. Weichao, L. Haifeng, B. Zhou, L. Wei, G. Ran, "Review of DC technology in future smart distribution grid," in Proc. IEEE PES Innov. Smart Grid Technol., Tianjin, China, 2012, pp. 1–4.

[7] S. Grillo, V. Musolino, L. Piegari, E. Tironi, and C. Tornelli, "DC islands in AC smart grids," IEEE Trans. Power Electron., vol. 29, no. 1, pp. 89–98, Jan. 2014

[8] K. M. Muttaqi, M. R. Islam and D. Sutanto, "Future Power Distribution Grids: Integration of Renewable Energy, Energy Storage, Electric Vehicles, Superconductor, and Magnetic Bus," in IEEE Transactions on Applied Superconductivity, vol. 29, no. 2, pp. 1-5, March 2019, Art no. 3800305.

[9] H. Mennicken, "Stromrichtersystem mit Wechselspannungszwischenkreis und seine Anwendung in der Traktionstechnik," Ph.D. thesis, RWTH Aachen, Germany, 1978.

[10] S. Ostlund, "Reduction of transformer rated power and line current harmonics in a primary switched converter system for traction applications," in Proc. 1993 5th Eur. Conf. Power Electron. Appl., Brighton, U.K., 1993, vol. 7, pp. 112–119.

[11] R. W. De Doncker, D. M. Divan, and M. H. Kheraluwala, "A threephase soft-switched high-power-density dc/dc converter for high-power applications," IEEE Trans. Ind. Appl., vol. 27, no. 1, pp. 63–73, Jan./Feb. 1991.

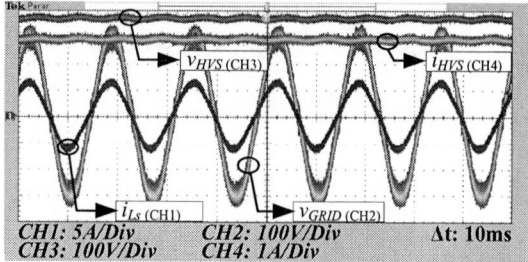

Fig. 15. Experimental steady state response to rectifier mode.

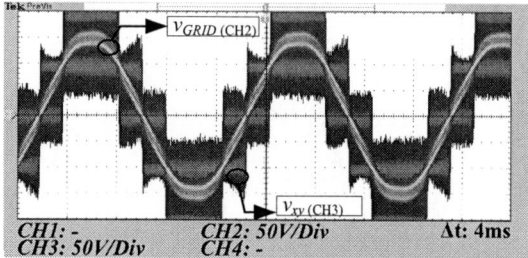

Fig. 16. Grid voltage and Five levels voltage on HVS.

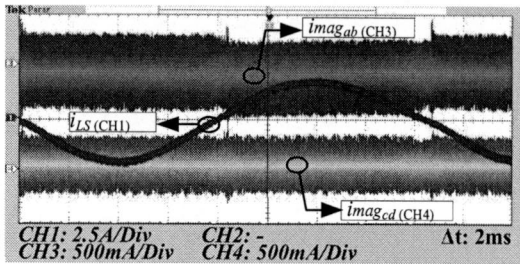

Fig. 17. Magnetizing control.

[12] D. S. Oliveira, M. I. V. Batista, L. H. S. C. Barreto and P. P. Praça, "A bidirectional single stage AC-DC converter with high frequency isolation feasible to DC distributed power systems", in 10th IEEE/IAS International Conference on Industry Applications, Fortaleza-CE-Brazil, pp. 1-7, 2012.

[13] O. Cipriano da Silva Filho and D. de Souza Oliveira, "Proposal of a new family of high frequency isolated single-phase AC-AC converters", in 12th IEEE International Conference on Industry Applications (INDUSCON), Curitiba-PR-Brazil, pp. 1-8, 2016. (verificar se troco pelo da revista).

[14] B. R. de Almeida, J. W. M. de Araújo, P. P. Praça and D. de S. Oliveira, "A Single-Stage Three-Phase Bidirectional AC/DC Converter With High-Frequency Isolation and PFC," in IEEE Transactions on Power Electronics, vol. 33, no. 10, pp. 8298-8307, Oct. 2018.

[15] G. V. T. Bascope and I. Barbi, "Generation of a family of non-isolated DC-DC PWM converters using new three-state switching cells", in 31st Annual Power Electronics Specialists Conference. Galway, vol.2, pp. 858-863, 2000.

[16] S. G. Barbosa, B. R. de Ameida, J. O. de Pacheco, D. d. S. Oliveira and P. P. Praça, "Multi-Port Single-Phase Converter Applied to Residential Microgeneration," 2018 13th IEEE International Conference on Industry Applications (INDUSCON), São Paulo, Brazil, 2018, pp. 1087-1093.

[17] A. U. Barbosa, B. R. de Almeida, D. d. S. Oliveira, P. P. Praça and L. H. S. C. Barreto, "Multi-port bidirectional three-phase AC-DC converter with high frequency isolation" in Applied Power Electronics Conference and Exposition (APEC), San Antonio-TX-USA, pp. 1386-1391, 2018.

[18] W. M. d. Santos and D. C. Martins, "Dual Active Bridge converter as gyrator", in Third International Conference on Sustainable Energy Technologies (ICSET), Kathmandu, pp. 169-176, 2012.

SmartBattery: An Active-Battery Solution for Energy Storage System

Lucas S. Araujo, Nicolas T. D. Fernandes, Danilo I. Brandao, Braz J. Cardoso Filho

Graduate Program in Electrical Engineering
Federal University of Minas Gerais (UFMG)
Av. Antônio Carlos 6627, 31270-901, Belo Horizonte, MG - Brazil
savoilucas@gmail.com, n.fernandes@ieee.org, dibrandao@ufmg.br, braz.cardoso@ieee.org

Abstract—The use of battery energy storage systems has gained wider attention in many sectors such as: power system, automotive, aircraft, oil and gas, among many others during the recent years. However, the association of batteries in a bank without the proper control may lead to problems to the batteries such as voltage imbalance, uneven degradation and incompletely use of the whole potential of the battery bank. This paper introduces a new concept of battery charger topology, named herein as *SmartBattery*, which is composed of a two-stage dc/dc converter integrated upon each battery pack. This solution allows each battery unit to be charged/discharged with different current profiles, even when they are series-connected. Then, this paper proposes a coordinated control strategy to balance the battery units in different stages of degradation and/or states of charge, without the need of bypassing any storage unit. Simulation results are performed to verify the equalization among batteries in a bank with different ages or states of charge during charging and discharging processes.

Index Terms—Battery chargers, battery management systems, coordinated control, energy storage.

I. Introduction

Energy storage system (ESS) has gained prominence with the steady growth of renewable energy sources, mainly due to their intermittent power generation, power flow contingencies and lack of inertia [1], [2]. Another area of application is electric mobility, which aims at reducing greenhouse gas emission, increasing vehicles efficiency, among other targets. The use of battery banks in these systems are appealing due to price and relatively low maintenance compared to others ESS solutions. [3]–[5].

The battery bank chargers are usually classified into two sorts: central and modular [6]. Due to the large number of battery cells in a bank, the central charger configuration is usually chosen. The problem related to this approach is that the central charger sees the bank as a set of battery cells considering them as a large equivalent battery, and then lacking the flexibility to handle each battery unit individually to apply different charge/discharge profile. Such absence of flexibility leads to problems like: overvoltage, voltage imbalance and undervoltage [7], which in turn accelerates the degradation of the batteries. In order to mitigate the voltage difference among the cells,

or units, strategies to balance the state of charge (SoC) are extensively studied. The goal of these studies is to mitigate voltage variations over the bank and therefore ensuring that each battery is properly charged.

Modular charger considers the difference among the battery units, and then coordinated control can be applied to sustain optimal charge control for every battery unit. In this manner, there is a trade-off between charge efficiency and cost of the system, in which many works have focused on to maximize the charge efficiency and reduce the overall financial costs [8].

The authors of [9] have presented a topology that encompasses a modular charger and an equalization circuitry. Such proposal simplifies the algorithm strategy; however, it includes extra costs. In the following subsection a detailed revision of charger strategies are shown.

A. Charge/Discharge Strategies

The authors of [10] propose a battery modular charger with a dc/dc converter to balance batteries voltage. The converter controls its output voltage that may be greater or lesser than the terminal voltage of the battery pack. For the discharge operation, the voltage regulation of the load, or the dc-link, is performed adjusting cooperatively the duty cycles of each dc/dc converter series-connected across the dc-link . Thus, the dc-link voltage is equal to the sum of the voltages across the output capacitors of dc/dc converters embedded upon each battery. On the other side, the average output current of every dc/dc converter is the same (series-connected converters). Typically, in this configuration, a central controller sends the duty cycle for each converter according to its algorithm [11].

In [12], the goal of the ESS control strategy for the discharge procedure is not only balance the SoC among the modules, but concomitantly provide voltage regulation at the dc-link. This is achieved defining different duty cycle values to each dc/dc converter series-connected, so the unit with higher voltage level runs with maximum duty cycle value, while the unit with lower voltage level operates with the lowest duty cycle. The duty cycles of the units vary around an average value defined initially, so the dc-link voltage slightly changes.

978-1-7281-4181-7/19 $31.00 © 2019 IEEE

The authors of [13] have also applied a centralized strategy in which the SoCs of the batteries are evaluated by their terminal voltage, and a mean value of voltage is calculated. The terminal voltage of each dc/dc converter is controlled by tracking the mean voltage value of the whole battery bank, and the charge balancing is performed by increasing or decreasing the current according to its local terminal voltage in comparison to the average voltage value. For stability constraints, one of the modules needs to be current-controlled, whereas the other ones can be voltage-controlled units. In this way, the current-controlled module defines the circulating current, while the other ones adjust their terminal voltages in order to keep the dc-link voltage constant. In [14], a similar approach is used, but each battery is connected to a dedicated inverter.

In [15], a decentralized strategy to control the dc-link voltage during the discharge procedure using modular cascaded converter (MCC) in battery bank is investigated. The current references for the converters are adjusted by a droop equation [16], for the sake of sharing the available power proportionally to the SoC value of each battery.

The main challenge in the solutions previously presented is that each battery unit is considered equally throughout the entirely process. Differences among batteries, such as degradation and uneven manufacture mismatches are neglected by the control charger. In this sense, those approaches are succeeded when the batteries are brand-new, and the discrepancy between the battery parameters are minimum. As the bank ages, the discrepancy becomes significant and to force maintaining the battery units SoC equalized will invariably accelerate the aging process of the most degraded units.

Thus, this paper proposes a charge/discharge control system applied to a battery bank based on SmartBattery (SB) concept in which every battery unit is endowed with a dc/dc converter. The central controller polls the SB status and then broadcasts a power command in order to *i*) meet each storage unit with their available capacity, and *ii*) guarantee their maximum capacity operation even considering batteries with uneven stages of degradation or different states of charge.

II. SMARTBATTERY ENERGY STORAGE SYSTEM

The concept of *SmartBattery* stands for a battery coupled with a power electronic converter, which can be assigned as an "electronic battery". The arrangement of SBs in series and/or in parallel sets up the *SmartBattery energy storage system (SBESS)*, as shown in Fig. 1. In Fig. 1 there are three SBs series-connected forming a SBESS. The SBESS is grid-connected through a power conditioning system (PCS), which interfaces the battery bank with the electrical network.

This concept fulfils the need of active storage elements, which allows more flexibility in term of operation, compared to existing solutions. In addition to dealing with the effects of equalization, SBESS also allows to enhance

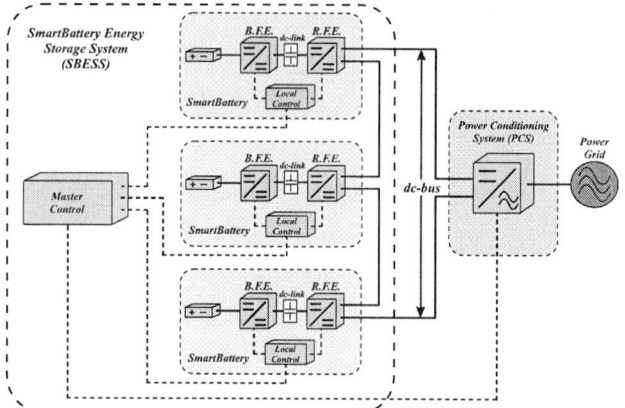

Fig. 1. SmartBattery Energy Storage System.

the charge procedure efficiency. It is achieved because each battery has its own SB, and then it can be charged considering a specific profile according to its stage of degradation and state of charge.

The SB function is to decouple the dynamics of the battery units with the dynamics of the entire storage system, or bank. In this way, the storage system would behave almost as an ideal source. Thereunto, it is necessary to introduce a local energy conditioning system between the battery and the PCS, so that it is possible to independently control the voltage and current that the PCS senses, and the voltage and current that charge and discharge each battery itself.

Thus, the SB hardware consists of a dual-stage dc/dc power converter, i.e., *Battery Front-End* (BFE) converter and *Rectifier Front-End* (RFE) converter. Between these two converters there is a capacitor which plays the role of decoupling storage element. Both the BFE and RFE are controlled by the *Local Control System* that perform measurements, signal conditioning and calculations.

Furthermore, when dealing with several SBs, it is necessary a *Master Control system* (MC) that manages and controls the entire storage system, as a global system, also communicating with the PCS. The MC can be associated with one of the SB or may be a independent unit.

A. Battery Front-End Converter Control

The BFE converter directly controls the battery current according to the best charge profile for a specific battery technology, and according to its SoC or state of health (SoH). Therefore, a current reference (i_B^*) is defined, and the battery current is regulated by the BFE converter.

Fig. 2 presents a compact version of the SBESS with the dc-dc converter representing both BFE and RFE, and showing the convention for positive voltage and current. The current reference i_B^* is generated by (1), where P_B^* is the power reference calculated in (2), and v_B is the measured battery voltage. The subscript "*j*" stands for a SB unit.

978-1-7281-4181-7/19 $31.00 © 2019 IEEE

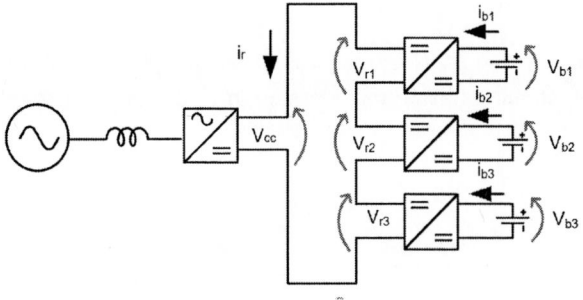

Fig. 2. *SmartBattery* modular converter concept

$$i_{Bj}^* = \frac{P_{Bj}^*}{v_{Bj}} \qquad (1)$$

$$P_{Bj}^* = P_{Bmaxj} \cdot \alpha \times \left[1 + \left(\overline{SoC} - SoCj\right) \cdot k\right] \qquad (2)$$

P_B^* is calculated by (2) based on the coefficient "α", on the SBs average SoC value ($\overline{SoC}$) - both broadcasted by the MC - and on its own SoC ($SoCj$). So, the MC steers the sharing of power proportionally to the SB rated power ($P_{Bmaxj} \cdot \alpha$), while an extra power portion is handled to balance the $SoCj$ among the jth-SB units $[P_{Bmaxj} \cdot \alpha \cdot \left(\overline{SoC} - SoCj\right) \cdot k]$. The gain "$k$" in (2) allows the SB to control the speed of SoC equalization and, if the SBESS is charging, it assumes positive values, otherwise it is negative. Note that (2) is composed by two terms, the first term is related to the proportional power sharing and the second term is related to the SoC balancing. Thereby, proportional power sharing among the SBs and batteries charge/discharge balance are performed.

The rated power (P_{Bmax}) is calculated differently under charging (3) or discharging (4) procedures, where Ic_{maxj} is the battery maximum charge current and Id_{maxj} is the battery maximum discharge current.

$$P_{Bmaxj_{Charging}} = V_{Bj} \cdot Ic_{maxj} \qquad (3)$$

$$P_{Bmaxj_{Discharging}} = V_{Bj} \cdot Id_{maxj} \qquad (4)$$

B. Rectifier Front-End Converter Control

Once the SB power reference is defined (2), it is necessary that the BFE and RFE converters process the same amount of power. Then the RFE converter controls the output voltage (v_R) according to the value of the series current circulating through the units (i_R), imposed by the PCS. Therefore, the voltage v_R is a consequence of maintaining the power balance between BFE and RFE, as shown in (5).

$$v_{Rj}^* = \frac{P_{Bj}^*}{i_R} \qquad (5)$$

C. Power-Conditioning System Control

The PCS is a static power converter that acts as a battery charger or discharger. This work considers a PCS as a current-controlled source. Hence, when the PCS is charging the SBESS, it maintains a controlled charge current throughout the charging process until the battery bank reaches its SoC of fully charged.

To discharge the SBESS, the PCS is an inverter with output current control loop. It means that an output power reference (P_{PCS}^*) is set, and the inverter must synthesize a controlled current flowing through its output filter inductance to track the power reference. For this, the inverter draws power from the dc bus and injects power into the ac grid. Therefore, the SBESS should provide the same amount of power for the dc bus that the inverter injects to the grid, characterizing the power balance between the SB units and PCS.

III. Charge/Discharge Balancing Control System

The charge/discharge balancing control is performed at the MC based on measured quantities of each SB unit. Both the charging and discharging procedures are considered using the same algorithm, since the PCS is considered as a controlled-current source. Let us consider the discharging process for the description herein presented. The quantities P_{Bj}, P_{Bmax_j} and SoC_j are transmitted by each SB_j to the MC.

The first step of the algorithm is to determine any system losses, mainly regarding to ohmic elements. The MC calculates the system losses (P_L) by (6), where P_{Bj} is the measured battery output power transmitted by each SB, and P_{PCS} is the PCS output power measured locally by the MC unit.

$$P_L = \sum_{j=1}^{N} P_{Bj} - P_{PCS} \qquad (6)$$

Next, it is defined the power reference that the PCS should process (P_{PCS}^*), and thereupon the coefficient "α" is calculated, where "N" is the total number of SB units:

$$\alpha = \frac{P_L + P_{PCS}^*}{\sum\limits_{j=1}^{N} P_{Bmax_j}}, \quad -1 \leq \alpha \leq 1 \qquad (7)$$

Finally, the average SoC is calculated as in (8):

$$\overline{SoC} = \frac{\sum\limits_{j=1}^{N} SoC_j}{N} \qquad (8)$$

The coefficient α and the average $\overline{SoC}$ are broadcasted to every SB units, in which their local control use those to define the power reference, P_{Bj}^*, as in (2). This configures different discharging current profiles for each SB even when series-connected.

A limitation of this strategy is that the charge/discharge balancing system considers the batteries as linear generator/load. Thus the charging mechanism works in the ohmic region of the charge profile. In Fig. 3, one can note the three regions of a battery during a charge profile. The first and third regions are non-linear, and thus inefficient, which can lead to degradation of the battery. Then, the operation of the storage system is highly desirable to be remained in the second region, which ranges from 20% to 90% of the SoC.

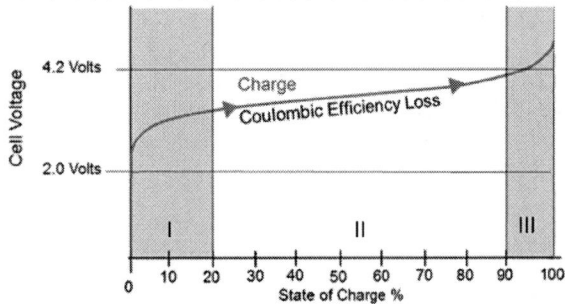

Fig. 3. Behaviour of a Lithium Ion battery during a charge process.

IV. SIMULATION RESULTS

For evaluation of the proposed balancing strategy and charger topology, a SBESS composed of three SBs based on lithium-ion batteries is simulated at *Matlab/Simulink* environment. The *Simulink* battery model based on [17] is used. The batteries have the same nominal voltage, rated capacity and maximum power, but they are aged differently, i.e., different equivalent full cycles for each battery are considered. The values of the battery parameters are shown in the Table I.

TABLE I
BATTERIES PARAMETERS

Parameter	Battery 1	Battery 2	Battery 3
Nominal voltage	7.2 V	7.2 V	7.2 V
Rated capacity	5.4 Ah	5.4 Ah	5.4 Ah
Max. power	70 W	70 W	70 W
Initial SoC	30%	30%	30%
Number of Cycles	2500	1000	0

The aging effect is set through the number of cycles as shown in Table I. This parameter affects the battery model varying its rated capacity and terminal resistance. Further information can be found in the *Matlab/Simulink* documentation [18]. The simulations performed do not consider high frequency effects, i.e., the power converters are modeled as ideal voltage (RFE. converter) and current (BFE converter and PCS) sources.

Two different situations are investigated in simulations: *i)* SoC balancing of batteries with different ages in charging mode; *ii)* operation at discharging mode with SoC

balancing even under perturbation, e.g., power reference variation. Since the point of interest is the battery dynamics, which is relatively slow compared to the converter dynamics, all the converters are modeled as ideal sources, without loss of generality.

Fig. 4 shows the SBESS charging processing a total

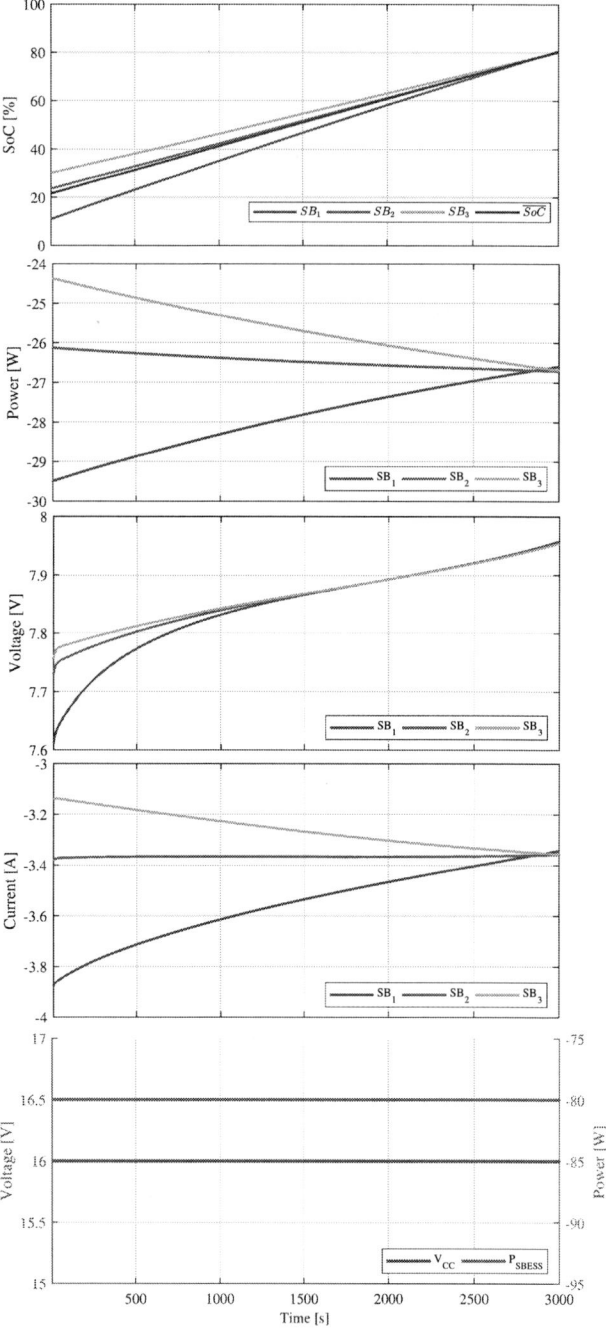

Fig. 4. Simulation results in charging mode: a) SoCs of the SBs; b) power of the SBs; c) voltage v_{Bj} of the SBs; d) current i_{Bj} of the SBs; and e) voltage and power of the SBESS.

power of 80 W. One can note the different stages of degradation among the batteries, which reflects in different values of SoC. The SB_1 is connected to an older and more degraded battery, so it holds the lowest SoC; as a consequence of the strategy, it absorbs a greatest amount of power. On the other hand, the SB_3 absorbs the least amount of power among the SB units. The value of the SoCs difference among the SBs is decreased throughout the charging process, when they reach a SoC of 80% the balancing is completed. The speed of the balancing process is dependent of the "k" factor of (2) which is considered $k = 0.01$.

The discharge operation is shown in Fig. 5. The same lithium-ion batteries are considered, however in this situation they are at the same age but considering different initial values of SoC. Similarly to the previous case study, the SB with the highest SoC value provides a greatest amount of power. The SBESS provides 100 W, and the SBs are almost with the same SoC at instant of 1500s. Thereafter, a power step is set to 200 W, and the results verify that the SoC balance still remains. One can note that when the SBs are with the same SoC value, they provide the same amount of power. In this case, it is considered $k = 0.1$, which evidences a faster SoC balance than the previous case study.

In both situations, it is achieved a well controlled voltage at the dc side of the inverter (V_{CC}), and also an accurate power flow control of the SBESS (P_{SBESS}).

The focus of these simulations is the bank operation, in which the proposed strategy proves to maintain the bank in a balanced condition even during charging/discharging procedures, and under transitions. Although, the authors are aware of keeping a battery bank balanced with batteries in different stages of degradation may not be the best condition, and may accelerate the aging effect on the worst battery units in the bank. But for operational purpose, it is quite worth.

V. CONCLUSIONS

This paper presented a new concept of battery charger topology, named *SmartBattery*. The proposal of using two-stage converter connected at each battery allows more flexibility of operation, so each battery can charge or discharge with different current values even if they are series-connected. A benefit of this proposed concept is associate batteries with different stages of degradation, capacities, or even batteries with different technologies – in this latter the SoC balancing would lose meaning.

As an example of the *SmartBattery* application, it was presented a SoC balancing strategy for a battery bank of the same technology. The results verified correct balance of the batteries SoCs during charge/discharge operation, and the batteries contribute proportionally to their current state, which ensures the equalized operation of the SB units, even with external disturbances. The strategy limitation is the operation in the ohmic region of the battery

and with the same battery maximum power. Other control strategies can still be developed to overcome these issues. Nonetheless, this method allows the gradual replacement of the batteries of a bank, instead of the necessity of all bank replacement as usual.

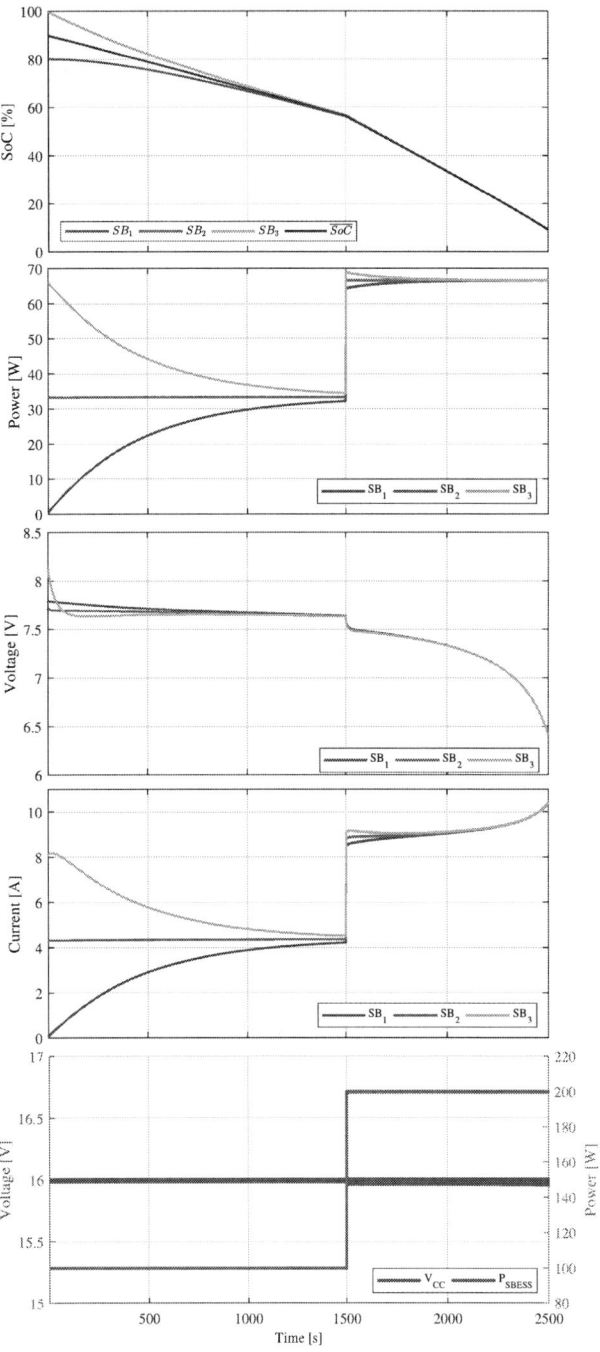

Fig. 5. Simulation results in discharging mode: a) SoCs of the SBs; b) power of the SBs; c) voltage v_{Bj} of the SBs; d) current i_{Bj} of the SBs; and e) voltage and power of the SBESS.

ACKNOWLEDGMENT

This work has been developed under the Research and Technological Development Program of the Electric Energy Sector regulated by ANEEL under the title "*Technical and commercial arrangements for the insertion of energy storage systems in the Brazilian electric sector*", project id ANEEL PD-00553-0046/2016, with Petrobras as the project proponent. The authors also thank the financial support from CAPES, FAPEMIG (grant APQ-02518-16) and CNPq (grant 420850/2016-3).

REFERENCES

[1] M. Braun, J. Brombach, C. Hachmann, D. Lafferte, A. Klingmann, W. Heckmann, F. Welck, D. Lohmeier, and H. Becker, "The Future of Power System Restoration: Using Distributed Energy Resources as a Force to Get Back Online," *IEEE Power and Energy Magazine*, vol. 16, no. 6, pp. 30–41, nov 2018. [Online]. Available: https://ieeexplore.ieee.org/document/8495076/

[2] C. Vartanian, R. Bauer, L. Casey, C. Loutan, D. Narang, and V. Patel, "Ensuring system reliability: Distributed energy resources and bulk power system considerations," *IEEE Power and Energy Magazine*, vol. 16, no. 6, pp. 52–63, Nov 2018.

[3] D. Akinyele and R. Rayudu, "Review of energy storage technologies for sustainable power networks," *Sustainable Energy Technologies and Assessments*, vol. 8, pp. 74–91, 2014.

[4] X. Luo, J. Wang, M. Dooner, and J. Clarke, "Overview of current development in electrical energy storage technologies and the application potential in power system operation," *Applied Energy*, vol. 137, pp. 511–536, 2015.

[5] J. Cho, S. Jeong, and Y. Kim, "Commercial and research battery technologies for electrical energy storage applications," *Progress in Energy and Combustion Science*, vol. 48, pp. 84–101, 2015.

[6] M. Bragard, N. Soltau, S. Thomas, and R. W. De Doncker, "The balance of renewable sources and user demands in grids: Power electronics for modular battery energy storage systems," *IEEE Transactions on Power Electronics*, vol. 25, no. 12, pp. 3049–3056, 2010.

[7] M. R. Palacín and A. De Guibert, "Batteries: Why do batteries fail?" *Science*, vol. 351, no. 6273, p. 1253292, feb 2016. [Online]. Available: http://www.ncbi.nlm.nih.gov/pubmed/26912708

[8] M. A. Hannan, M. M. Hoque, A. Hussain, Y. Yusof, and P. J. Ker, "State-of-the-Art and Energy Management System of Lithium-Ion Batteries in Electric Vehicle Applications: Issues and Recommendations," *IEEE Access*, vol. 6, pp. 19362–19378, 2018. [Online]. Available: https://ieeexplore.ieee.org/document/8320763/

[9] M. M. Hoque, M. A. Hannan, and A. Mohamed, "Optimal algorithms for the charge equalisation controller of series connected lithium-ion battery cells in electric vehicle applications," *IET Electrical Systems in Transportation*, vol. 7, no. 4, pp. 267–277, dec 2017. [Online]. Available: https://digital-library.theiet.org/content/journals/10.1049/iet-est.2016.0077

[10] C.-H. Hou, C.-T. Yen, T.-H. Wu, and C.-S. Moo, "A battery power bank of serial battery power modules with buck-boost converters," in *Power Electronics and Drive Systems (PEDS), 2013 IEEE 10th International Conference on*. IEEE, 2013, pp. 211–216.

[11] L.-R. Yu, Y.-C. Hsieh, W.-C. Liu, and C.-S. Moo, "Balanced discharging for serial battery power modules with boost converters," in *System Science and Engineering (ICSSE), 2013 International Conference on*. IEEE, 2013, pp. 449–453.

[12] C.-S. Moo, T.-H. Wu, C.-H. Hou, and Y.-C. Hsieh, "Balanced discharging of power bank with buck-boost battery power modules," in *Power Electronics Conference (IPEC-Hiroshima 2014-ECCE-ASIA), 2014 International*. IEEE, 2014, pp. 1796–1800.

[13] Y. Li and Y. Han, "A module-integrated distributed battery energy storage and management system," *IEEE transactions on power electronics*, vol. 31, no. 12, pp. 8260–8270, 2016.

[14] S. Lee, S. Baek, and C. Won, "Serial multi-module ups system control method considering battery module balancing," in *2015 IEEE International Telecommunications Energy Conference (INTELEC)*, Oct 2015, pp. 1–6.

[15] S. M. Chowdhury, M. E. Haque, A. Elrayyah, Y. Sozer, and J. De Abreu-Garcia, "An integrated control strategy for state of charge balancing with output voltage control of a series connected battery management system," in *2018 IEEE Energy Conversion Congress and Exposition (ECCE)*. IEEE, 2018, pp. 6668–6673.

[16] S. M. Chowdhury, M. Badawy, Y. Sozer, and J. A. D. A. Garcia, "A novel battery management system using a duality of the adaptive droop control theory," in *2017 IEEE Energy Conversion Congress and Exposition (ECCE)*. IEEE, 2017, pp. 5164–5169.

[17] N. Omar, M. A. Monem, Y. Firouz, J. Salminen, J. Smekens, O. Hegazy, H. Gaulous, G. Mulder, P. Van den Bossche, T. Coosemans *et al.*, "Lithium iron phosphate based battery–assessment of the aging parameters and development of cycle life model," *Applied Energy*, vol. 113, pp. 1575–1585, 2014.

[18] "Battery: Generic Battery Model Description," Available: https://www.mathworks.com/help/physmod/sps/powersys/ref/battery.html, last accessed 17th September 2019.

Feedforward Compensation of the ESS Low-Frequency Current Ripple in the Three-Ports ANPC Converter

Silvio Antonio Teston
Department of Physical Facilities
Federal University of Fronteira Sul
Chapecó-SC, Brazil
silvioteston@gmail.com

Kaio Vinicius Vilerá
Power Electronics and Control Research Group
Federal University of Santa Maria
Santa Maria-RS, Brazil
kaiovilera@gmail.com

Marcello Mezaroba
Electric Power Processing Group
University of Santa Catarina State
Joinville-SC, Brazil
marcello.mezaroba@gmail.com

Cassiano Rech
Power Electronics and Control Research Group
Federal University of Santa Maria
Santa Maria-RS, Brazil
rech.cassiano@gmail.com

Abstract—This paper proposes a feedforward control to mitigate the low-frequency current ripple in the secondary dc port current of the three-ports active neutral-point-clamped converter (ANPC-3P). This low-frequency current ripple is caused by the low-frequency voltage ripple at the inverter dc buses. Current ripple is an important design parameter for energy storage systems (ESS) and must be kept low to extend the ESS life cycle. Low-frequency current ripple is difficult to filter and thereby it should be avoided. In this paper, ESS current control system is composed of a proportional-integral (PI) controller associated with a proposed feedforward action. This control scheme is capable of generating a dc current without low-frequency ripple. The ac output current (grid current) is controlled by a resonant controller. The theoretical propositions are validated by simulation results.

Keywords—Three-Ports Active Neutral-Point-Clamped, Energy Storage Systems, Feedforward Compensation.

I. INTRODUCTION

Energy storage systems (ESS) have been widely used in uninterruptible power supplies (UPS) and off-grid renewable energy generation systems. Distributed generation (DG) can also benefit from the use of ESS because they add important functionalities such as load leveling, energy arbitrage, primary frequency regulation, and end-user peak shaving [1]. The connection of the ESS has many influences in the design, cost and operational aspects of the power electronics converters. ESS integration into the inverter topology allows the use of the same inverter power semiconductors, gate drivers, and passive components to simultaneously process the ESS power, thereby

This study was financed in part by the Coordenação de Aperfeiçoamento de Pessoal de Nível Superior - Brasil (CAPES/PROEX) - Finance Code 001. The authors thank INCT-GD, CNPq (processes 465640/2014-1, 306317/2015-0 and 427987/2018-0), CAPES (process 23038.000776/2017-54) and FAPERGS (17/2551-0000517-1) for the financial support.

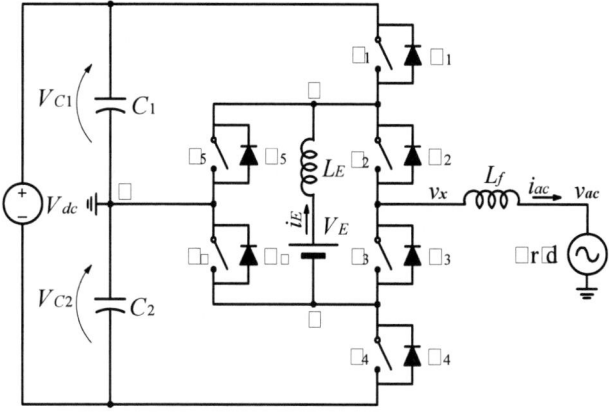

Fig. 1. ANPC-3P topology in a grid-tied application.

reducing part counts. Recently, several papers dealt with ESS integration in the inverter topology [2]–[9]. When the ESSs are connected directly to the dc bus, a complex modulation scheme is needed to generate sinusoidal currents from an unbalanced dc bus [5]–[8]. On the other hand, the Z-source-inverter (ZSI) presents relatively simple implementation, but it can integrate only one ESS [2]–[4].

Another topology capable of ESS integration is the three-ports active neutral-point-clamped (ANPC-3P) converter, proposed in [9]. This topology has the main dc port common to all legs, and, for each inverter leg, it has a secondary dc port and an ac port. Each secondary dc port of the ANPC-3P converter can be connected to an independent ESS. A single-phase ANPC-3P inverter in a grid-tied application is shown in Fig. 1. When the secondary dc port voltage (v_{AB}) is non-zero, its value can be imposed by the positive dc bus

978-1-7281-4181-7/19 $31.00 © 2019 IEEE

voltage (C_1) or by the negative dc bus (C_2) voltage. In most applications, the pulsed power on the dc bus causes these voltages to oscillate with the grid frequency. If not handled properly, this oscillatory behavior is transmitted to the ESS current i_E. The low-frequency current ripple is difficult to filter and can lead to a higher filter volume or a lower ESS life cycle.

Considering the issues regarding the ESS current in the ANPC-3P topology, this paper proposes a feedforward control action to mitigate the ESS low-frequency current ripple. The proposed control action uses the dc buses voltages and other internal parameters to calculate the duty cycle of the secondary dc port to cancel the low-frequency current ripple.

This paper is organized as follows. Section II reviews some aspects of the topology operation and introduces the secondary dc port static model. Section III presents the main aspects of the modulation scheme. Section IV deals with the control system and Section V includes some simulation results.

II. TOPOLOGY ANALYSIS

In the ANPC-3P topology, the ESS is integrated with the inverter leg. The integration is possible due to redundant switching states, which are presented in Table I. When a zero level is generated in v_x the voltage across nodes A and B (v_{AB}) can be zero (state 0UL) or $V_{dc}/2$ (states 0U1 and 0L1). By choosing the appropriate time intervals for levels zero and $V_{dc}/2$ on v_{AB} one can control the electrical parameters of a circuit connected to nodes A and B. Since the ESS has predominantly a voltage source characteristic, inductor L_E is added to limit the current ripple. Higher-order filters, such as LCL, could be considered to reduce overall filter volume. For the switching states 0U2 and 0L2, the voltage level in v_{AB} depends on the i_{ac} and i_E values. Therefore, these switching states are not used in the ANPC-3P topology [9].

TABLE I
SWITCHING STATES OF THE THREE-LEVEL ANPC-3P VSI.

State	Switches						v_x	v_{AB}
	S_1	S_2	S_3	S_4	S_5	S_6		
P	1	1	0	0	0	1	$V_{dc}/2$	$V_{C1} = V_{dc}/2$
0U4	0	1	1	0	1	0	0	0
0U3	0	1	0	0	1	1	0	0
0U2	0	1	0	0	1	0	0	−
0U1	0	1	0	1	1	0	0	$V_{C2} = V_{dc}/2$
0UL	0	1	1	0	1	1	0	0
0L1	1	0	1	0	0	1	0	$V_{C1} = V_{dc}/2$
0L2	0	0	1	0	0	1	0	−
0L3	0	0	1	0	1	1	0	0
0L4	0	1	1	0	0	1	0	0
N	0	0	1	1	1	0	$-V_{dc}/2$	$V_{C2} = V_{dc}/2$

Duty cycle d_z defines the duration of the zero level on v_{AB} and consequently the duration of the positive voltage level across L_E. From the steady-state analysis of the circuit shown in Fig. 1 and the waveforms presented in Fig. 2, the duty cycle d_z can be determined through volt-second balance in the inductor, so that:

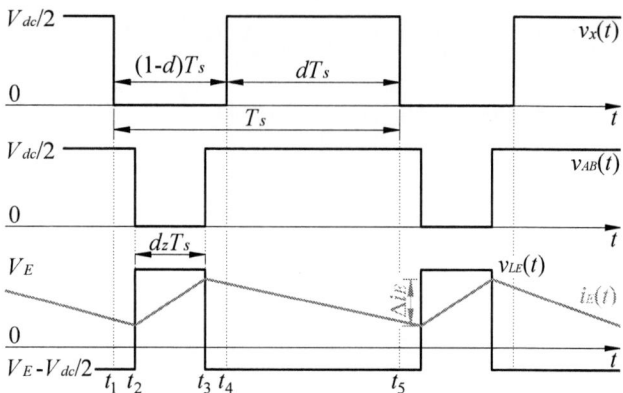

Fig. 2. Waveforms detailing the modulation of the secondary dc port.

$$d_z = 1 - \frac{V_E}{V_{dc}/2} \qquad (1)$$

Within certain limits, for each value of V_E and V_{dc}, duty cycle d_z can be adjusted to maintain a constant current i_E. Considering that the modulation of the dc port occurs while it is applied zero voltage level at the output ac port, d_z is upper limited to $1 - d_{max}$. Another constraint is related to the voltage across L_E, which must be sometimes positive and sometimes negative. Therefore, it is possible to state the range for the voltage V_E:

$$V_p < V_E < V_{dc}/2 \qquad (2)$$

where V_p is the peak value of the average voltage synthesized by the ac port when d reaches its maximum value (d_{max}).

A. Modulation

The modulation scheme presented in this paper is based on [9]. Three blocks are used to generate the gate signals for the six switches, as shown in Fig 3. The first block receives two carrier signals (v_{tri+} and v_{tri-}) in phase opposition disposition (POD) and signals $v_{m,ac}$ and $v_{m,dc}$, which are the modulating signals of the ac port and secondary dc port, respectively. This block is responsible for comparing the modulating signals with carriers and for generating the signals for the second block. The second block selects the appropriate switching state following the voltage levels requested by the first block and by the signal b_{dc}, which is generated by the dc bus control system to balance the power exchanged by the ESS with both dc buses. The third block generates the gate signals for the six switches, including the appropriate dead time.

When the ESS current is non-zero, it can help with the regulation of the dc bus voltages. This can be achieved by the proper selection of states 0U1 and 0L1. These states are redundant in terms of the output voltages, since they generate a zero level on v_x and a $V_{dc}/2$ level on v_{AB}. However, these states present opposed actions on the dc bus capacitors. Table II presents the logic to select 0U1 and 0L1 considering

978-1-7281-4181-7/19 $31.00 © 2019 IEEE

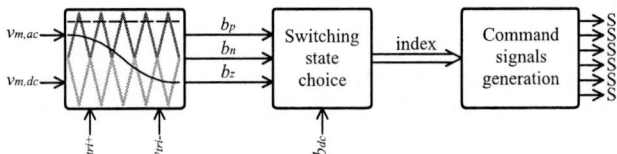

Fig. 3. Modulation block diagram.

the C_1 voltage and i_E polarity. For example, if $i_E > 0$ and $V_{C1} > V_{dc}/2$ state 0U1 is selected. This state causes current i_E to flow into C_2 through its positive pole. Consequently, C_2 is charged and the dc bus voltages tend to balance. On the other hand, if $i_E < 0$ and $V_{C1} > V_{dc}/2$ state 0L1 should be used. This state causes current i_E to flow through C_1 exiting from its positive pole. In this way, C_1 is discharged, trying to equalize the dc bus voltages.

TABLE II
LOGIC FOR SELECTING THE STATES 0U1 AND 0L1.

I_E	V_{C1}	State
> 0	$> V_{dc}/2$	0U1
> 0	$< V_{dc}/2$	0L1
< 0	$> V_{dc}/2$	0L1
< 0	$< V_{dc}/2$	0U1

As can be seen from Fig. 2, states 0U1 and 0L1 can be imposed on two time intervals of the switching cycle: $t_1 - t_2$ and $t_3 - t_4$. Since 0U1 and 0L1 have an opposite impact over the dc bus voltages, if both states were used, one for each time interval, their effect would be canceled out. However, when the same switching state is selected for both time intervals there is an effect on a dc bus voltage. In this paper, it is considered that b_{dc} is only evaluated at the beginning of the switching interval and, thus, only one of the states 0U1 and 0L1 is used.

III. CONTROL SYSTEM

This section presents the control system of the single-phase ANPC-3P inverter under analysis. This system can be divided into two subsystems, one responsible for the secondary dc port current control and the other responsible for the ac current control. The control system of the secondary dc port is composed of a PI controller to regulate the ESS current. A feedforward action is proposed and it is summed to the PI action to eliminate any low-frequency oscillatory component from i_E. On the inverter side, any well-established control strategy can be used. The controlled variable on the ac output is application dependent. For grid-tie inverters, the control system regulates the output currents. For UPS applications, the controlled variable is the output voltage. This paper considers a PV grid-tied application and a resonant controller (RES) is used to regulate the ac current. A block diagram of the complete control system is shown in Fig. 4, where each control block is explained in the following subsections.

Table III presents the main parameters used in the design.

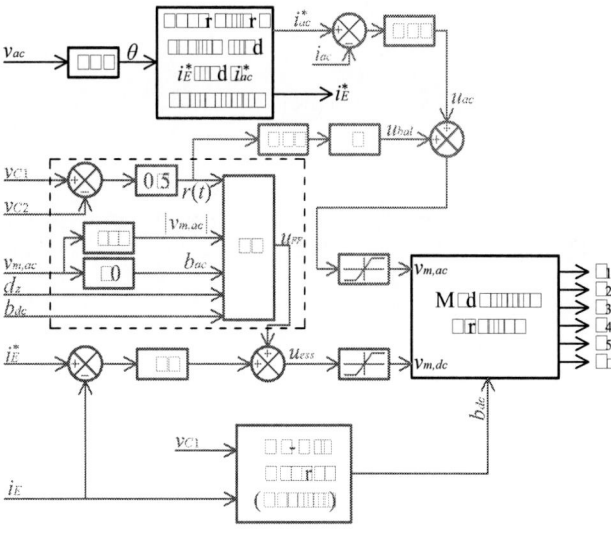

Fig. 4. Control system block diagram. The control subsystems of the secondary dc and ac ports are represented in blue and red colors, respectively.

TABLE III
MAIN PARAMETERS.

Parameter	Value
Output ac power (P_o)	1 kW
ESS power (P_E)	1 kW
Dc bus voltage (V_{dc})	550 V
Dc bus capacitance ($C_1 = C_2$)	500 μF
ESS voltage (V_E)	19 x 12 V VRLA batteries
ESS internal resistance (R_E)	0.5 Ω
Ac voltage (V_{ac})	127 V rms
Ac filter inductor (L_f)	6 mH ($R_s = 0.3\ \Omega$)
ESS filter (L_E)	10 mH ($R_{LE} = 0.5\ \Omega$)
Ac frequency (f_{ac})	60 Hz
Carrier frequency (f_c)	10.26 kHz
Amplitude modulation index (m_a)	0.653

A. Proposed ESS feedforward controller

Current i_E is defined by the two voltage sources (V_E and v_{AB}) connected across L_E terminals. In this sense, any low-frequency ripple superimposed in v_{AB} will result in low-frequency current components in i_E.

The feedforward (FF) action is designed to cancel any low-frequency component present in the voltage v_{AB}. Therefore, it is first necessary to obtain a mathematical model for the average value of v_{AB}. To derive the model, the voltages applied to v_{AB} during a switching period must be known. During $t_1 - t_2$, $t_3 - t_4$, and $t_4 - t_5$ voltage v_{AB} is non-zero and its value depends on the switching states. In intervals $t_1 - t_2$ and $t_3 - t_4$, the voltages are defined by the bit b_{dc} and are equal. The voltage v_{AB} during $t_4 - t_5$ depends on the polarity of the ac modulating signal. If $v_{m,ac} > 0$ the P state is used, otherwise N is used. In this way, it is useful to define a digital signal b_{ac} to indicate the polarity of $v_{m,ac}$. If $v_{m,ac} > 0$, $b_{ac} = 1$ and zero otherwise. With all these considerations, the average value of v_{AB} in a switching period can be calculated by:

$$\langle v_{AB}(t)\rangle_{T_s} = [V_{dc}/2 + r(t)]\,[b_{ac}d + b_{dc}(1 - d - d_z)] +$$
$$[V_{dc}/2 - r(t)]\,[(1 - b_{ac})d + (1 - b_{dc})(1 - d - d_z)] \quad (3)$$

where $d = |v_{m,ac}|$, and $r(t)$ is a ripple signal present in the dc bus capacitor voltages. By rearranging (3) it is possible to write:

$$\langle v_{AB}(t)\rangle_{T_s} = V_{dc}/2(1 - d_z) +$$
$$r(t)\,(d_z + 2b_{dc} - 2b_{dc}d - 2b_{dc}d_z + 2b_{ac}d - 1) \quad (4)$$

For steady-state operation, the average voltage across inductor L_E must be zero. The first term of (4) is equal V_E for d_z given by (1). When the dc buses present some voltage ripple ($r(t)$), the second term of (4) will impose a non-zero average voltage across L_E during a switching period. This is the cause of low-frequency current ripple in the ESS current. The purpose of the proposed FF controller is to inject a signal into d_z that is opposite to that represented by the second term of (4). This can be done by adding a FF action to the modulating signal $v_{m,dc}$ before applying it to the modulation system. Therefore, the FF action can be defined as:

$$u_{FF} = \frac{-r(t)}{V_{dc}/2}\,(d_z + 2b_{dc} - 2b_{dc}d - 2b_{dc}d_z + 2b_{ac}d - 1)$$
$$(5)$$

Signal $r(t)$ can be extracted from the dc bus capacitors voltages using:

$$r(t) = \frac{v_{C1}(t) - v_{C2}(t)}{2} \quad (6)$$

If the PV inverter varies the dc bus for maximum power point tracking (MPPT) purposes, the voltage $V_{dc}/2$ in (5) must be adjusted in real-time using the measurement of the dc bus capacitor voltages.

B. ESS PI controller

The feedforward action is designed to compensate the low-frequency ripple of current i_E. Another controller is added to achieve zero steady-state error for step inputs and to adjust the transient response of the closed-loop system.

The control-to-output transfer function of the secondary dc port is given by:

$$\frac{I_E(s)}{V_{m,dc}(s)} = -\frac{V_{dc}/(2L_E)}{s + R_s/L_E} \quad (7)$$

where resistance R_s is the sum of all series resistances, such as the inductor resistance (R_{LE}) and the ESS internal resistance (R_E), etc.

A PI controller can be used to achieve the design constraints. To calculate the gains of the controller, some specifications should be considered. It is desirable a fast response, in the order of a few milliseconds, with minimum overshoot for the

current i_E. The zero of the controller was placed to cancel the plant pole, i.e., $T_i = L_E/R_s$, where T_i is the integral time constant. Thus, the closed-loop system can be approximated by a first order system, and the proportional gain can be defined as:

$$k_p = -\frac{2L_E}{T_p V_{dc}} \quad (8)$$

where T_p is the time constant of the resulting closed-loop first order system. Considering the parameters shown in Table III and $T_p = 1$ ms, the resulting controller gains are: $k_p = -0.036$ and $T_i = 0.01$ s.

C. Ac Current Control

For the ac current control, a resonant controller is used. The supervisory block of Fig. 4 is responsible for generating the AC current reference. Grid voltage angle is determined by the phase-locked loop (PLL) block. The control-to-output transfer function is given by:

$$\frac{I_{ac}(s)}{D(s)} = \frac{V_{dc}/(2L_f)}{s} \quad (9)$$

The resonant controller is designed by adding a pair of complex conjugate poles at frequency f_{ac} with a damping factor of 0.001. To adjust the phase margin, another pair of complex conjugated zeros is added one decade below the crossover frequency with a damping factor of 0.7. Lastly, the gain is adjusted to result in a 1 kHz crossover frequency. In this frequency, the phase margin is 81.9 degrees.

D. Dc Bus Balancing Controller

It is fundamental to keep the dc bus voltages balanced. When i_E is non-zero it can be used to help to balance the dc bus voltages, as already explained in the modulation section. An on-off controller is used to generate the b_{dc} signal according to Table II. However, when i_E is zero the inverter should present some mechanism to balance the dc bus voltages. For the three-phase inverter, redundant switching states are available for synthesizing the same line-to-line voltage, and since they have an opposite effect on the dc bus voltages, they are usually chosen to balance the neutral point voltage [10], [11]. For the single-phase full-bridge NPC inverters, redundant switching states can also be used. However, single-phase half-bridge NPC inverters do not present such redundant switching states and a different approach should be considered. If the dc bus is generated by a front-end converter (MPPT), it can be designed to be responsible for balancing the dc bus voltages. Another common technique is to inject a dc value to the ac modulating signal [12], [13]. Obviously, this technique results in an undesired signal at the output voltage or current. Considering the injection of a dc balancing signal (u_{bal}) into the ac modulating signal and that i_{ac} is an ideal sinusoidal current source, u_{bal} affects the dc bus voltages according to the following approximated model:

$$\frac{V_{C_1}(s)}{U_{bal}(s)} = \frac{I_{ac,pk}\cos(\phi)}{\pi C}\frac{1}{s} \quad (10)$$

978-1-7281-4181-7/19 $31.00 © 2019 IEEE

where $C = C_1 = C_2$, $\cos(\phi)$ is the displacement power factor (DPF), and $I_{ac,pk}$ is the peak amplitude of the inverter output current.

To proper design the balancing controller it is needed to specify the transient response for a given $I_{ac,pk}$ and DPF. In this paper is used 10% inverter output power and unitary DPF to design the balancing controller. A simple proportional controller is used and the gain of the controller is adjusted to result in a closed-loop system with a time constant of $5\tau = 100$ ms. The resultant gain is 0.07 and it corresponds to a crossover frequency of 7.9 Hz.

Signal $r(t)$ can present a dc value and harmonic components, mainly fundamental and third harmonic. The balancing controller should act only on the dc value of $r(t)$ so that it does not affect the fundamental current and also does not introduce harmonics into the current synthesized by the inverter. In this paper two bandstop filters (BSF) are used to remove these unwanted signals.

IV. SIMULATION RESULTS

In this section, some simulation results are presented to prove the effectiveness of the proposed control system using the parameters given in Table III. The simulation is done using PSIM software.

Firstly, simulation results of the dc port control are discussed. The ac output load is nominal and $\phi = 0$ degrees. In Fig. 5, the FF action is initially off and the ESS current reference is zero. At 0.05 s the FF action is turned ON and the ESS low-frequency current ripple is immediately suppressed. Next, at 0.1 s a current step of 5 A is imposed in the ESS current reference. Current i_E follows the reference and it still does not introduce any appreciable low-frequency ripple. At 0.15 s the FF action is turned OFF and the low-frequency ripple is evident.

Another point to be highlighted is the non-influence of the feedforward action on the grid current THDi, as can be seen on Fig. 5(d). Modulation of the secondary dc port occurs during the zero levels synthesized at the ac port and therefore does not affect the synthesized voltage at this port. The only coupling between these ports is the dc bus voltage. The power of the secondary dc port is almost dc and therefore it does not insert harmonics on dc bus voltages. With these considerations, the FF action has no influence over the grid current THDi.

The simulation result shown in Fig. 6 demonstrates the effectiveness of the FF control action under unbalanced dc bus conditions. In this case, the ESS current reference is kept zero. The simulation starts with an ac load of 10% of the full inverter power and $\phi = 0$ degrees. At 0.05 s a perturbation of 10 V is inserted in each dc bus pole. For 35 ms the dc bus voltages are unbalanced, but this perturbation is compensated by the FF action and i_E is not affected. Next, in 0.2 s the ac output suddenly operates under full power. The ac load step also disturbs the dc bus and the FF action prevents this disturbance from affecting the ESS current. As can be seen, both perturbations on the dc bus are rejected by the FF

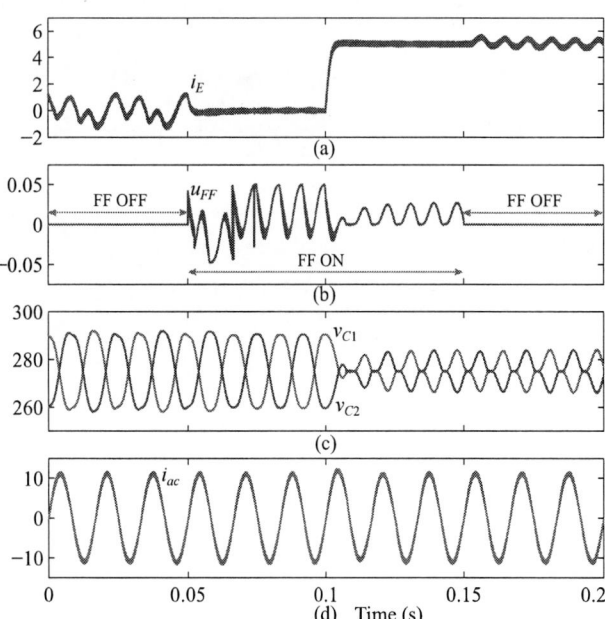

Fig. 5. Simulation results of the secondary dc port control showing: (a) ESS current (i_E), (b) FF control action (u_{FF}), (c) dc bus voltages, and (d) grid current (i_{ac}). FF action is turned ON at 0.05 s and turned OFF at 0.15 s. Voltages are given in volts and the currents in amperes.

controller. This simulation also demonstrates the operation of the ac port control system.

In Fig. 7, the dc bus balancing when i_E is non-zero is shown. In the beginning, the ac port load is at 10% of the full load power. At 0.05 s, the ESS starts recharging with a current of -5 A. At 0.1 s a perturbation of 10 V is inserted in each dc bus pole. Differently from the case shown in Fig. 6, in which the ac port balancing controller took approximately 35 ms to balance the dc bus voltages, in this case, the on-off balancing controller took less than 5 ms by using the current i_E. Obviously, when i_E is smaller the time will be longer. When current steps are imposed on the secondary dc port there is a disturbance on the dc bus and the balancing controller of the ac port causes distortions on i_{ac} as can be seen at times 0.05 s, 0.1 s, and 0.2 s. When i_E is non-zero the controlling action u_{bal} can be turned OFF to avoid these undesired distortions.

V. CONCLUSION

In this paper a complete control system for a single-phase ANPC-3P inverter is presented. A feedforward control action is proposed in this paper to compensate for the voltage ripple of the dc bus capacitors and to produce an ESS current composed only by a dc value and switching frequency harmonics. The feedforward action effectiveness is verified through simulation results. The single-phase case presents an extreme condition to the feedforward controller due to higher voltage ripple on the dc bus. The feedforward control action was also validated in severe conditions, with unbalanced dc bus voltages. In all simulated cases the ESS current presented only

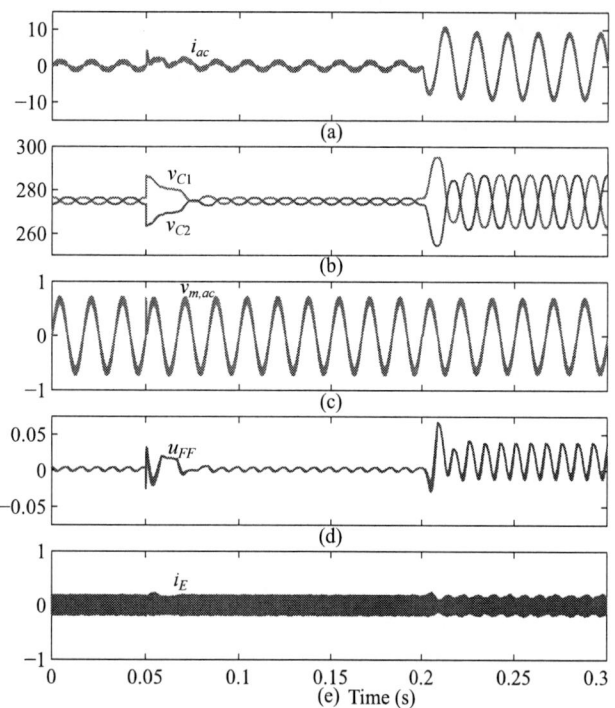

Fig. 6. Simulation results for a perturbation on the dc bus and an ac active power step. (a) grid current (i_{ac}), (b) dc bus voltages, (c) ac port modulating signal ($v_{m,ac}$), (d) FF control action (u_{FF}), and (e) ESS current (i_E). Voltages are given in volts and the currents in amperes.

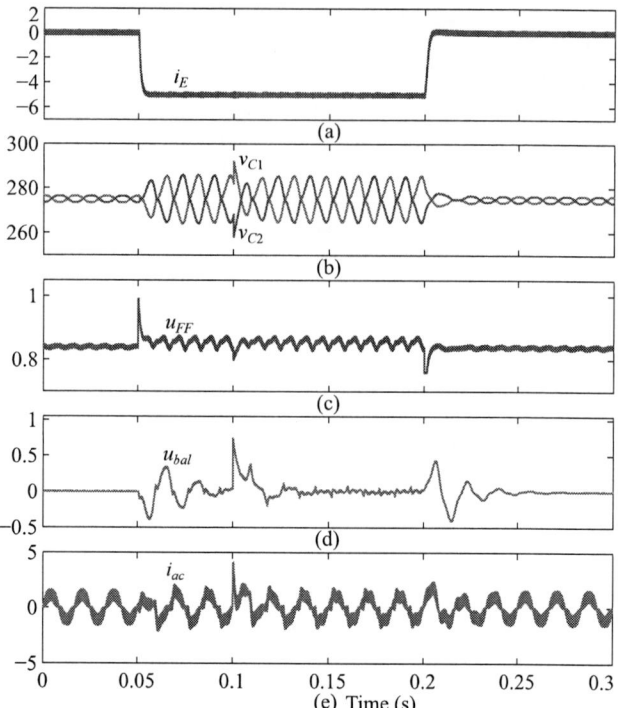

Fig. 7. Simulation results showing dc bus balancing with $i_E \neq 0$: (a) ESS current (i_E), (b) dc bus voltages, (c) secondary dc port modulating signal ($v_{m,dc}$), (d) ac port balancing action (u_{bal}), and (e) grid current (i_{ac}). Voltages are given in volts and the currents in amperes.

a dc value and switching harmonics, which are relatively easy to filter. The proposed ESS control system can also be used in single-phase full-bridge or three-phase inverters composed of ANPC-3P legs. In these cases, the inverter dc bus voltage balancing can be done by using space vector modulation with a mechanism for selecting the appropriate redundant switching states.

REFERENCES

[1] S. Vazquez, S. M. Lukic, E. Galvan, L. G. Franquelo, and J. M. Carrasco, "Energy storage systems for transport and grid applications," *IEEE Transactions on Industrial Electronics*, vol. 57, no. 12, pp. 3881–3895, dec 2010.

[2] J. G. Cintron-Rivera, Y. Li, S. Jiang, and F. Z. Peng, "Quasi-z-source inverter with energy storage for photovoltaic power generation systems," in *2011 Twenty-Sixth Annual IEEE Applied Power Electronics Conference and Exposition (APEC)*, March 2011, pp. 401–406.

[3] B. Ge, H. Abu-Rub, F. Z. Peng, Q. Lei, A. T. de Almeida, F. J. T. E. Ferreira, D. Sun, and Y. Liu, "An energy-stored quasi-z-source inverter for application to photovoltaic power system," *IEEE Transactions on Industrial Electronics*, vol. 60, no. 10, pp. 4468–4481, Oct 2013.

[4] S. Hu, Z. Liang, D. Fan, and X. He, "Hybrid ultracapacitor-battery energy storage system based on quasi-z-source topology and enhanced frequency dividing coordinated control for ev," *IEEE Transactions on Power Electronics*, vol. 31, no. 11, pp. 7598–7610, Nov 2016.

[5] S. D. G. Jayasinghe, D. M. Vilathgamuwa, and U. K. Madawala, "Diode-clamped three-level inverter-based battery/supercapacitor direct integration scheme for renewable energy systems," *IEEE Transactions on Power Electronics*, vol. 26, no. 12, pp. 3720–3729, Dec 2011.

[6] I. Vechiu, A. Etxeberria, H. Camblong, and J. M. Vinassa, "Three-level neutral point clamped inverter interface for flow battery/supercapacitor energy storage system used for microgrids," in *2011 2nd IEEE PES*

International Conference and Exhibition on Innovative Smart Grid Technologies, Dec 2011, pp. 1–6.

[7] H. R. Teymour, D. Sutanto, K. M. Muttaqi, and P. Ciufo, "Solar pv and battery storage integration using a new configuration of a three-level npc inverter with advanced control strategy," *IEEE Transactions on Energy Conversion*, vol. 29, no. 2, pp. 354–365, June 2014.

[8] Q. Tabart, I. Vechiu, A. Etxeberria, and S. Bacha, "Hybrid energy storage system microgrids integration for power quality improvement using four-leg three-level npc inverter and second-order sliding mode control," *IEEE Transactions on Industrial Electronics*, vol. 65, no. 1, pp. 424–435, Jan 2018.

[9] S. A. Teston, M. Mezaroba, and C. Rech, "ANPC inverter with integrated secondary bidirectional DC port for ESS connection," *IEEE Transactions on Industry Applications*, pp. 1–1, 2019.

[10] N. Celanovic and D. Boroyevich, "A comprehensive study of neutral-point voltage balancing problem in three-level neutral-point-clamped voltage source PWM inverters," *IEEE Transactions on Power Electronics*, vol. 15, no. 2, pp. 242–249, mar 2000.

[11] C. Wang and Y. Li, "Analysis and calculation of zero-sequence voltage considering neutral-point potential balancing in three-level NPC converters," *IEEE Transactions on Industrial Electronics*, vol. 57, no. 7, pp. 2262–2271, jul 2010.

[12] C. Newton and M. Sumner, "Neutral point control for multi-level inverters: theory, design and operational limitations," in *IAS '97. Conference Record of the 1997 IEEE Industry Applications Conference Thirty-Second IAS Annual Meeting.* IEEE, 1997.

[13] S. Cobreces, J. Bordonau, J. Salaet, E. Bueno, and F. Rodriguez, "Exact linearization nonlinear neutral-point voltage control for single-phase three-level NPC converters," *IEEE Transactions on Power Electronics*, vol. 24, no. 10, pp. 2357–2362, oct 2009.

SEPIC DC/DC converter control by observed-state feedback

Silas M. Sousa
dept. Electrical Engineering
Federal Center of Technological Education of Minas Gerais
(CEFET-MG), Belo Horizonte 30510-000, Brazil
silas-arcos@hotmail.com

Valter J. S. Leite
Dept. of Mechatronics Engineering
Federal Center of Technological Education of Minas Gerais
(CEFET-MG), Divinópolis 35503-822, Brazil
valter@ieee.com

Samir W. Fernandes
Dept. of Electrical Engineering
Federal University of Santa Catarina (UFSC)
Florianópolis 88040-900 , Brazil
samir_wf@hotmail.com

Isabel R. H. Oliveira
Dept. of Electrical Engineering
Federal University of São Joã del-Rei (UFSJ)
São João del-Rei 36309-034, Brazil
isabelrho@hotmail.com

Abstract—**DC-DC converters are used in photovoltaic inverters to step-up the solar plant voltage to the DC bus specification value. The single-ended primary inductor converter (SEPIC) is a DC-DC converter, and its modeling is performed in this work using the averaged switch and the small signals approach. On the SEPIC, there are four energy storage elements which hinder its control design. In this paper, the state feedback technique is applied to design the SEPIC controller without order reduction. An advantage of this approach is that no model order reduction is required, but the knowledge of inductors currents and capacitors voltages are necessary, which may be not feasible in practical applications. Therefore, a state observer provides estimated values of the states used in the feedback control. Output feedback with integral action is applied to reduce the steady-state errors. Tests were carried out to track the output signal reference by applying step and ramp variations and also load disturbances. The simulation results highlight that through the application of the observed-state feedback controller was possible to obtain the desired dynamics within the established control goals.**

Index Terms—**SEPIC converter, photovoltaic systems, small signal, averaged switch, state feedback, state observer.**

I. INTRODUCTION

As the technological development and worldwide consumption of electricity continue to grow, it is important to use alternative energy sources to supply the demand alongside with the traditional sources [1]. Regarding available alternative sources, photovoltaic energy presents promising characteristics [2]. Basically, the structure of a photovoltaic system is composed by a solar panel and a power converter. This system may be connected to the electrical grid, also known as grid-on system, or isolated, named grid-off, for example, to feed loads including battery power bank.

The structure of a photovoltaic system is composed of a solar panel and power converter. This system can be connected to the electrical grid (grid-on) or be isolated (grid-off), for example, to feed loads including battery power bank.

Photovoltaic inverters use different topologies of converters. For instance, in the Boost topology there are two energy

storage elements. Moreover, this converter is usually applied in cases where the photovoltaic string voltage level is lower than the voltage level required by the DC bus [3]. In another converter, the SEPIC, there are four energy storage elements, and the following advantages may be noticed: step-down as well as step-up converter, and the possibility of galvanically isolating the input from output [4].

The main contribution of this paper is applying state feedback to control the SEPIC converter by integral action and observed states. The use of observed states reduces the need of voltage and current measurements, which directly impact the implemantation costs turning the control strategy cheaper.

This paper is organized as follows. Section II describes the SEPIC converter modeling. Section III and IV present the controller design and results, respectively. Finally, Section V summarizes this paper conclusions.

II. SEPIC MODELING

The SEPIC DC-DC converter diagram, supplied by a solar plant and feeding a resistive load, is illustrated in Fig. 1.

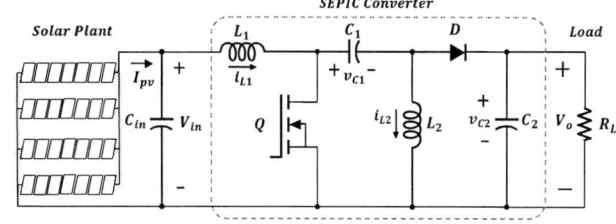

Fig. 1. SEPIC DC-DC converter circuit connected to the solar plant, feeding a resistive load. Adapted from [5].

The SEPIC operation principle in continuous conduction mode consists of two stages. On the first stage, Fig. 2 (a), the Metal Oxide Semiconductor Field Effect Transistor (MOS-FET) switch is triggered and represented as a connection. As the diode is blocked, it is represented by an open circuit.

978-1-7281-4181-7/19 $31.00 © 2019 IEEE

The L_1 inductor stores the source energy, thus, the inductor currents increase and the C_2 capacitor feeds the load (R_L). On the second stage, Fig. 2 (b), the switch opens and the diode conducts. This causes the inductors to transfer the stored energy to the C_2 capacitor and consequently to the load [6].

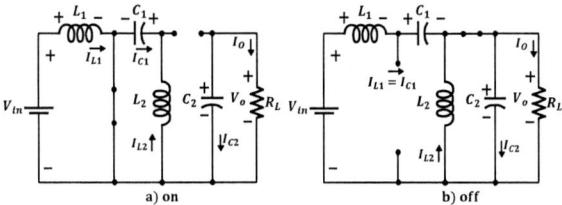

Fig. 2. Operation stages of SEPIC converter, with transistor: a) switched on and b) switched off. Adapted from [5].

According to the operational modes shown in Fig. 2, the dynamics of the electrical converter changes. With the switch in mode on, we have [7]:

$$
\begin{bmatrix} \dot{i}_{L_1} \\ \dot{i}_{L_2} \\ \dot{v}_{C_1} \\ \dot{v}_{C_2} \end{bmatrix} = \begin{bmatrix} 0 & 0 & 0 & 0 \\ 0 & 0 & \frac{1}{L_2} & 0 \\ 0 & -\frac{1}{C_1} & 0 & 0 \\ 0 & 0 & 0 & -\frac{1}{R_L C_2} \end{bmatrix} \begin{bmatrix} i_{L_1} \\ i_{L_2} \\ v_{C_1} \\ v_{C_2} \end{bmatrix} + \begin{bmatrix} \frac{1}{L_1} \\ 0 \\ 0 \\ 0 \end{bmatrix} V_{in},
\tag{1}
$$

while in the off mode, we get:

$$
\begin{bmatrix} \dot{i}_{L_1} \\ \dot{i}_{L_2} \\ \dot{v}_{C_1} \\ \dot{v}_{C_2} \end{bmatrix} = \begin{bmatrix} 0 & 0 & -\frac{1}{L_1} & -\frac{1}{L_1} \\ 0 & 0 & 0 & -\frac{1}{L_2} \\ \frac{1}{C_1} & 0 & 0 & 0 \\ \frac{1}{C_2} & \frac{1}{C_2} & 0 & -\frac{1}{R_L C_2} \end{bmatrix} \begin{bmatrix} i_{L_1} \\ i_{L_2} \\ v_{C_1} \\ v_{C_2} \end{bmatrix} + \begin{bmatrix} \frac{1}{L_1} \\ 0 \\ 0 \\ 0 \end{bmatrix} V_{in},
\tag{2}
$$

In these equations, i_{L_1} and i_{L_2} represent the inductors currents and v_{C_1} and v_{C_2} the capacitor voltages. The average behavior of the converter, considering the matrices: $A = A_{on}d + A_{off}(1-d)$ and $B = B_{on}d + B_{off}(1-d)$ is given by (3), where d is the duty cycle, A_{on} and B_{on} are switched on matrices and A_{off} and B_{off} switched off [6].

$$
\begin{bmatrix} \dot{i}_{L_1} \\ \dot{i}_{L_2} \\ \dot{v}_{C_1} \\ \dot{v}_{C_2} \end{bmatrix} = \begin{bmatrix} 0 & 0 & -\frac{(1-d)}{L_1} & -\frac{(1-d)}{L_1} \\ 0 & 0 & \frac{d}{L_2} & -\frac{(1-d)}{L_2} \\ \frac{(1-d)}{C_1} & -\frac{d}{C_1} & 0 & 0 \\ \frac{(1-d)}{C_2} & \frac{(1-d)}{C_2} & 0 & -\frac{1}{R_L C_2} \end{bmatrix} \begin{bmatrix} i_{L_1} \\ i_{L_2} \\ v_{C_1} \\ v_{C_2} \end{bmatrix} + \begin{bmatrix} \frac{1}{L_1} \\ 0 \\ 0 \\ 0 \end{bmatrix} V_{in}.
\tag{3}
$$

Equation (3) is nonlinear and can be linearized by inserting a small signal perturbation, so the variables are composed of a mean value, denoted by uppercase, plus a small signal perturbation, denoted by superscript "$\sim$" [6], [7] as follows.

$$
\begin{bmatrix} \dot{\tilde{i}}_{L1} \\ \dot{\tilde{i}}_{L2} \\ \dot{\tilde{v}}_{C1} \\ \dot{\tilde{v}}_{C2} \end{bmatrix} = \begin{bmatrix} 0 & 0 & -\frac{(1-D)}{L_1} & -\frac{(1-D)}{L_1} \\ 0 & 0 & \frac{D}{L_2} & -\frac{(1-D)}{L_2} \\ \frac{(1-D)}{C_1} & -\frac{D}{C_1} & 0 & 0 \\ \frac{(1-D)}{C_2} & \frac{(1-D)}{C_2} & 0 & -\frac{1}{RC_2} \end{bmatrix} \begin{bmatrix} \tilde{i}_{L1} \\ \tilde{i}_{L2} \\ \tilde{v}_{C1} \\ \tilde{v}_{C2} \end{bmatrix} +
$$
$$
\begin{bmatrix} \frac{V_{in}}{L_1(1-D)} \\ \frac{V_{in}}{L_2(1-D)} \\ \frac{-I_o}{C_1(1-D)} \\ \frac{-I_o}{C_2(1-D)} \end{bmatrix} \tilde{d}, \qquad y = \begin{bmatrix} 0 & 0 & 0 & 1 \end{bmatrix} \begin{bmatrix} \tilde{i}_{L1} \\ \tilde{i}_{L2} \\ \tilde{v}_{C1} \\ \tilde{v}_{C2} \end{bmatrix}.
\tag{4}
$$

The SEPIC elements design takes into account the Table I.

TABLE I
PARAMETERS FOR SEPIC ELEMENTS CALCULATION

Description	Variable	Value	Unity
Input voltage	V_{in}	185	V
Output voltage	V_o	390	V
Duty cycle	D	0.678	–
Input current	I_{L1}	7.92	A
Inductor L_2 / output currents	I_{L2}/I_o	3.76	A
Sample frequency	f_s	12	kHz
Input current ripple	ΔI_{L1}	3	$\%I_{L1}$
Capacitor C_1 voltage ripple	ΔV_{C1}	3	$\%V_{in}$
Capacitor C_2 voltage ripple	ΔV_{C2}	3	$\%V_o$

With these values it is possible to calculate L_1, L_2, C_1 and C_2 values, according to the following [4]:

$$
L_1 = \frac{V_{in}D}{\Delta I_{L_1} f_s}, \quad L_2 = \frac{L_1}{2}, \quad C_1 = \frac{I_{L2}D}{\Delta V_{C1} f_s}, \quad C_2 = \frac{I_o D}{\Delta V_{C2} f_s},
\tag{5}
$$

to obtain: $L_1 = 43.5\ mH$, $L_1 = 21.7\ mH$, $C_1 = 37.92\ \mu F$ and $C_2 = 17.99\ \mu F$. Substituting these values in (4) results:

$$
\begin{bmatrix} \dot{\tilde{i}}_{L_1} \\ \dot{\tilde{i}}_{L_2} \\ \dot{\tilde{v}}_{C_1} \\ \dot{\tilde{v}}_{C_2} \end{bmatrix} = \begin{bmatrix} 0 & 0 & -7.32 & -7.32 \\ 0 & 0 & 30.82 & -14.64 \\ 8390 & 17670 & 0 & 0 \\ 17690 & 17690 & 0 & -531 \end{bmatrix} \begin{bmatrix} \tilde{i}_{L_1} \\ \tilde{i}_{L_2} \\ \tilde{v}_{C_1} \\ \tilde{v}_{C_2} \end{bmatrix} +
$$
$$
\begin{bmatrix} 13060 \\ 26120 \\ -305100 \times 10^5 \\ -643100 \end{bmatrix} \tilde{d}, \quad y = \begin{bmatrix} 0 & 0 & 0 & 1 \end{bmatrix} \begin{bmatrix} \tilde{i}_{L_1} \\ \tilde{i}_{L_2} \\ \tilde{v}_{C_1} \\ \tilde{v}_{C_2} \end{bmatrix}.
\tag{6}
$$

III. CONTROLLER DESIGN

The design of the SEPIC consists on using the state feedback technique with the allocation of desired eigenvalues. In addition, a state observer and an integral gain output feedback are addressed as well. Regarding the implementation, the first step is verifying the controllability and observability of the system state matrices. Such procedures are needed to design the state feedback and observer. A system is indeed controllable and fully observable if the rank (ρ) of its controllability ($\mathcal{C}$) and the observability ($\mathcal{O}$) matrices is n (row and column dimensions of square matrix $A_{n \times n}$) [8]. A system with feedback and state observer can be described as a function of the states (x) and the estimation error (e), as follows.

$$
\begin{bmatrix} \dot{x} \\ \dot{e} \end{bmatrix} = \begin{bmatrix} A - BK & BK \\ 0 & A - LC \end{bmatrix} \begin{bmatrix} x \\ e \end{bmatrix} + \begin{bmatrix} C \\ 0 \end{bmatrix} r,
$$
$$
y = \begin{bmatrix} B & 0 \end{bmatrix} \begin{bmatrix} x \\ e \end{bmatrix},
\tag{7}
$$

where the matrix A of is a triangular block, so the eigenvalues of the state feedback controller are given by $A - BK$ and the observer by $A - LC$. This shows that the design of an observer (also called a state estimator) does not affect closed-loop poles of state feedback and can be done separately [8]. This fact is known as *principle of separation*.

978-1-7281-4181-7/19 $31.00 © 2019 IEEE

A. State feedback controller design

The state feedback control can handle the internal states of a system, assisting in the output control. An advantage of this type of controller is that through control's law $u(t) = r(t) + \mathbf{k}\mathbf{x}(t)$ it can arbitrarily allocate the desired closed-loop eigenvalues [9]. The main goals in a controller design are to ensuring stability, reducing the effects of parameter changes, and suppressing the effects of load disturbances and measurement noises. The equation of the system with closed-loop states feedback is presented:

$$\dot{\mathbf{x}}(t) = (\mathbf{A} - \mathbf{B}\mathbf{k})\mathbf{x}(t) + \mathbf{B}\,r(t), \\ y(t) = \mathbf{C}\mathbf{x}(t). \tag{8}$$

The corresponding diagram of (8) is illustrated in Fig. 3. In this diagram the system (plant) is inside the dashed line rectangle. The variable $\mathbf{k}$ is the feedback gain vector, $r(t)$ is the control reference and $u(t)$ the control signal.

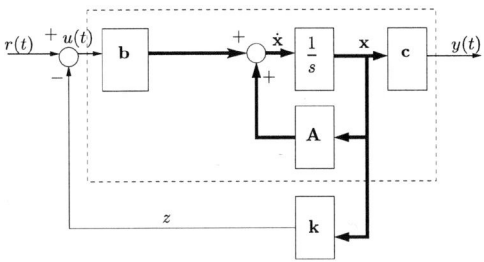

Fig. 3. State feedback diagram. Source: [8].

In order to track the input reference $r(t)$, on output y, with no steady-state error, an integral gain output feedback is required. The diagram of this system can be seen in Fig. 4.

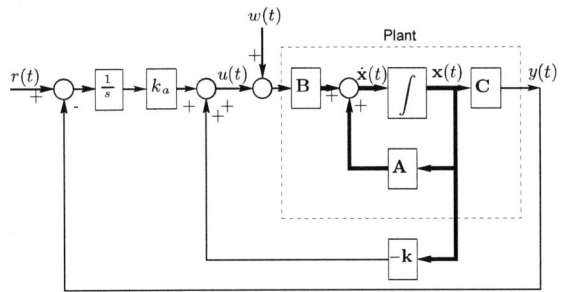

Fig. 4. System diagram with state feedback and output feedback with integral action. Source: [8].

B. State feedback and integral gains design

The state feedback and integral gains, k and k_a, respectively, can be projected in the same procedure, creating the augmented array $\mathbf{K} = [\text{-}\mathbf{k} \;\; k_a]$ [8]. Firstly, it is necessary to define the desired dynamic system, choosing the closed-loop eigenvalues. Such choice is based on the system performance requirements and may be guided by the region D illustrated in Fig. 5. Values within this region allow: a minimum response speed (σ), minimum damping (θ) and a maximum speed actuation of the control signal (r).

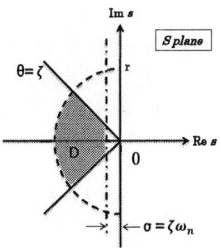

Fig. 5. D region with desirable characteristics for choice of eigenvalues placement in the s plane. Adapted from [8].

Based on the DC-DC converter dynamic response the following characteristics are specified:

- Accommodation time: $t_s = 0.3$ seconds;
- Rise time (t_r): smallest possible;
- Maximum overshoot: $M_p = 5$ %.

The feedback gain will be designed to obtain the closed-loop dynamic system as a typical second-order behavior. All eigenvalues must have negative real part so that the system is stable in closed-loop. The dominant poles of the system, for a damped oscillatory dynamics, are determined as following:

$$p_{1,2} = -\zeta\,\omega_n \pm j(\omega_n\sqrt{1 - \zeta^2}). \tag{9}$$

The accommodation time criteria (t_s) is used to calculate the time constant $\tau = \zeta\,\omega_n$, where ζ is the damping coefficient and ω_n is the system natural frequency. Equation (10) shows this calculation which will be applied to design the real part of the two desired dominant eigenvalues.

$$t_s = \frac{4}{\zeta\,\omega_n}, \qquad \tau = \zeta\,\omega_n = \frac{4}{t_s} = \frac{4}{0.3} = 13.33. \tag{10}$$

The maximum overshoot criteria is applied to calculate the imaginary part of the poles (9). Considering $M_p = 0.043$ (4.3 %), the damping coefficient ζ value is determined according to (11).

$$\zeta = \frac{-\ln(M_p)}{\sqrt{\pi^2 + \ln^2(M_p)}} = \frac{-\ln(0.043)}{\sqrt{\pi^2 + \ln^2(0.043)}} = 0.707 \tag{11}$$

Based on the damping coefficient (ζ) and time constant (τ) values, the natural frequency (ω_n) is obtained by (12).

$$\omega_n = \frac{\tau}{\zeta} = \frac{13.33}{0.707} = 18.85 \tag{12}$$

Thus it is possible to calculate the closed-loop poles imaginary part value (ω_d), according to (13).

$$\omega_d = \omega_n\sqrt{1 - \zeta^2} = 18.85\sqrt{1 - 0.707^2} = 13.33 \tag{13}$$

The closed-loop dominant eingenvalues are given by (14).

$$p_{1,2} = -\zeta\,\omega_n \pm j\omega_d = -13.33 \pm j13.33 \tag{14}$$

As the SEPIC is a fourth order system, four poles must compose the desired dynamic polynomial. Thus the other two non-dominant poles must be real and are chosen one decade below: $p_3 = -133.33$ and $p_4 = -166.67$. To complete the augmented matrix one more pole must be chosen for the dynamic integral gain, it was chosen: 173.33. Thus, it is possible to assemble the F matrix of desired eigenvalues:

$$F = \text{diag}\left\{ \begin{bmatrix} -13.33 & -13.33 \\ 13.33 & -13.33 \end{bmatrix}, \right.$$
$$\left. -133.33, -166.67, -173.33 \right\} \quad (15)$$

The next step is to define a value of the vector $\bar{k}$, that means the vector k without similarity transforming. The vector $\bar{k}$ can be chosen arbitrarily, provided that under one condition: the matrix formed by $[F \ \bar{k}]$ must be observable. The following vector was chosen: $\bar{k} = [1\ 1\ 1\ 1\ 1]$. With the presented values the observability matrix $\mathcal{O}(F, \bar{k})$ has full rank.

It is necessary to calculate the augmented matrices A and B by (16). Since the presented system has four states, the augmented matrix A_a must be 5×5 and B_a 5×1.

$$A_a = \begin{bmatrix} A & 0 \\ -C & 0 \end{bmatrix}, \quad B_a = \begin{bmatrix} B \\ 0 \end{bmatrix}. \quad (16)$$

Through the *Lyapunov* equation must find the function T that satisfies the equality $AT + TB + C = 0$ and that T is nonsingular. For this purpose, the Matlab software (*lyap* function) was used to calculate the T matrix. The T matrix represents the basis that makes the vector $\bar{k}$ be the vector K (solution of the present design). Then to find the vector K for similarity transformation is performed, as calculated:

$$K = \bar{k}F \quad (17)$$

Since the first n elements of the vector K are the feedback states gains (k) and the last element is the integral gain of the output feedback (k_a). The following gains were found:

$$k = \begin{bmatrix} -0.0031 & 0.0136 & 0.0011 & -0.0005 \end{bmatrix} \quad (18)$$

$$k_a = 5.0067 \times 10^{-6} \quad (19)$$

C. State observer design

State observers are needed when it is not possible measuring some type of state variable, for example, when the state is a mathematical variable without a physical sensing. Moreover, state observers may be applied even when a state can be measured, as sensors represent cost and values can be estimated. Since state feedback requires that all states must be available (measured or estimated), state observer technique is often used. Fig. 6 illustrates a block diagram of a state observer connected to a plant.

The observer gain design (L) is similar to the state feedback gain design, being a dual of this design procedure.

The first step in finding the L gain is to choose the desired F matrix. To choose the eigenvalues of L it is important to analyze the observer's role in state feedback. If the error of the state observer is large (bad estimation) the feedback will not be able to trace the reference value. This difference between the

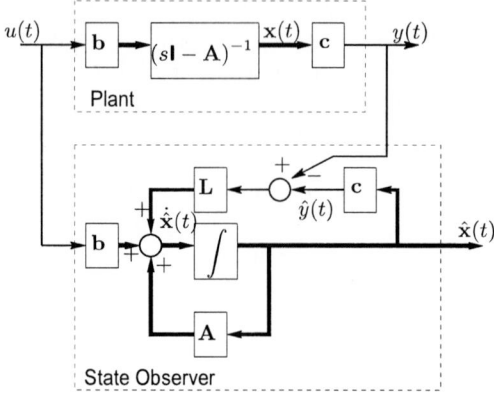

Fig. 6. Block diagram of the state estimator connected to a system. Source: [8].

estimated output and the actual output can occur, for example, when the system is started. For this, it is a good design practice to choose the eigenvalues observer 5 to 10 times faster than the dominants of the closed loop. The faster the eigenvalues, the worse they are when measuring noises.

$$F_{obs} = \text{diag}\left\{ \begin{bmatrix} -133.33 & -133.33 \\ 133.33 & -133.33 \end{bmatrix}, \right.$$
$$\left. -1333.33, -1343.33 \right\} \quad (20)$$

The next step is to determine a column vector L_{obs} so that the pair (F_{obs}, L_{obs}) is controllable. It has been found that the rank of the controllability matrix of the pair (F_{obs}, L_{obs}) is 4 (equal to n). Thus it can keep the values of the vector L_{obs}, and that T_{obs} is non-singular.

Using matrices A and C of SEPIC converter it will be necessary to find the Lyapunov matrix (T_{obs}) that meets the equality: $T_{obs}A - F_{obs}T_{obs} = L_{obs}C$.

IV. RESULTS

In order to test the control proposed in this paper, computational simulations were carried out through MATLAB/Simulink®. Firstly, the control was tested with the modeling system, being represented by a state space block. After that, the control system was applied to the SEPIC switched circuit, which is a batter representation of a real circuit.

A. Observer and state feedback tests using the SEPIC converter model

The first test was performed through the application of a step reference, while keeping the initial conditions of all states equal to zero. The applied voltage reference started with 0 V value and at the instant of 1 second it is changed to 5 V. The result of this test is presented in Fig. 7.

As it can be seen on the Fig. 7, there is an accommodation time, t_s, of about 0.4 second in the output signal. This result is 33 % greater than the projected value (0.3 second). The maximum overshoot presented is 4.06 %, meeting the

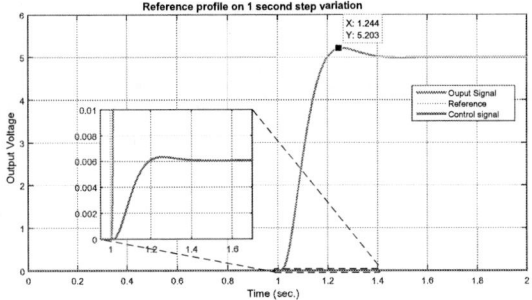

Fig. 7. Test with the SEPIC converter model for a variation of the step reference.

maximum overshoot requirement of less than 5 %. In the center of the Fig. 7 there is a zoomed detail to show the effect of the control signal. The behavior of this signal variation causes the output to track the reference.

In the second test a disturbance is added to the state variable x_1. For that, a value of 0.1 is added to x_1 from the instant of 10 seconds. Fig. 8 illustrates the behavior of the output signal and control due to the applied disturbance.

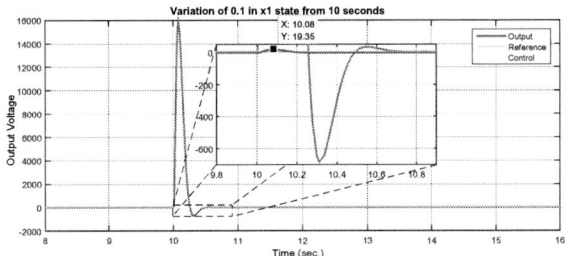

Fig. 8. Test with the SEPIC model for a disturbance in state x_1.

Fig. 8 shows that the perturbation generated a high-value output signal (close to $1,6 \times 10^5 \ V$). The control signal is shown in a zoom detail in the center of Fig. 8. The control signal presented the value of 19.35 and in an actual application would generate saturation. This would cause the stabilization time greater than 0.7 seconds (approximate value obtained in this simulation). For this test scenario it is shown in Fig. 9 the states measured, from model, and states estimated. Note that the estimated states are very close to the measured states, presenting estimation error close to zero.

In the third test, a noise signal was inserted into the state variable x_1. A random signal with amplitude ranging from -0.1 to 0.1 was added to the signal x_1, between 10 to 15 seconds.

By adding the noise, Fig. 10, the output signal varies with high voltages, reaching values of $2.4 \times 10^4 \ V$. The control signal goes beyond the value of 20. It is worth remembering that the control signal will act on the duty cycle of the SEPIC, so this signal saturates when passing the value of 1. A slower observer may improve such behavior, for example by choosing them twice slow as the dominant eigenvalues.

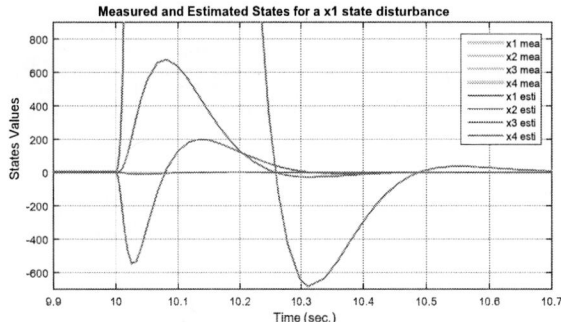

Fig. 9. Signals corresponding to the measured (model) and estimated (state observer) of the test applying pertubation in the state x_1.

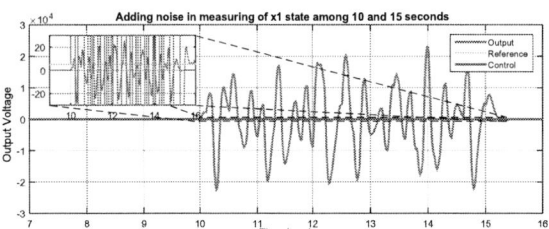

Fig. 10. Test with the SEPIC converter model by adding a random noise to the signal of the x_1 state variable.

B. Test with observer and state feedback applied to the SEPIC converter circuit

After testing the control in the system model, tests are performed with the switched circuit of the SEPIC. The complete circuit of the tests is shown in Fig. 11. The control signal $u(t)$ represents the duty cycle d which triggers the transistor when passing through the Pulse Width Modulation (PWM) block.

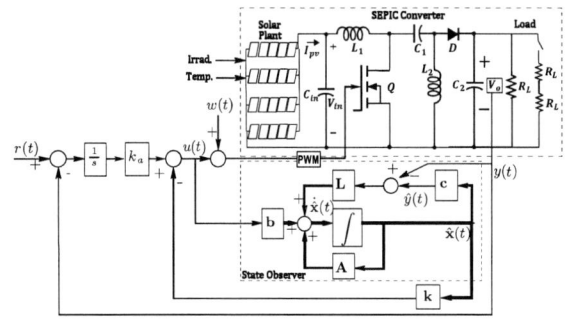

Fig. 11. Complete circuit of the test performed, containing SEPIC circuit, state observer, state feedback and output feedback with integral action.

The first test performed with the SEPIC circuit is varying the reference in the step form. The reference voltage is maintained at $390 \ V$ from 0 to 2 seconds. At 2 seconds the reference is changed to $400 \ V$, as shown in Fig. 12.

Oscillations in the output voltage (V_o) can be observed in Fig. 12, it is due to the transfering of voltages and currents between the converter elements. Such oscillations can be

978-1-7281-4181-7/19 $31.00 © 2019 IEEE

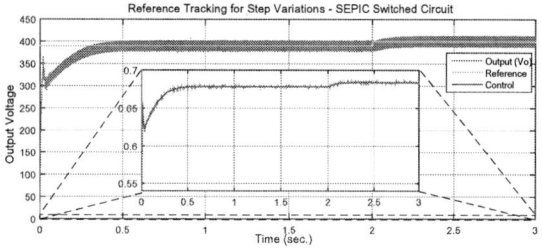

Fig. 12. Test with the SEPIC switched circuit, with reference in $390V$ from 0 to 2 seconds and step variation to $400V$ in the instant of 2 seconds.

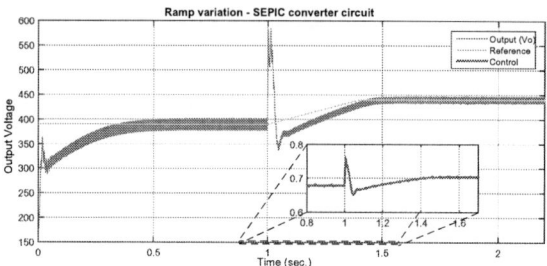

Fig. 14. Applying a ramp reference variation on the switched circuit of the SEPIC converter.

attenuated with a larger design value for the output capacitor V_{C2}. The effect of the tracking can be seen in Fig. 12. The accommodation time is about 0.4 second. A zoomed detail is presented in the center of the figure illustrating the control signal changing the value of the duty cycle, so that the system can track the reference value.

In order to verify the dynamic behavior of the designed controller, a load disturbance is performed. This test simulates the entry of a larger load into the system and it is checked if the control system is able to reject such disturbance. During the test the output voltage reference is maintained at $390\ V$. During the 0 to 1 seconds the system load is on its nominal value ($50\ \Omega$) and at the instant of 1 second it assumes the value of $100\ \Omega$.

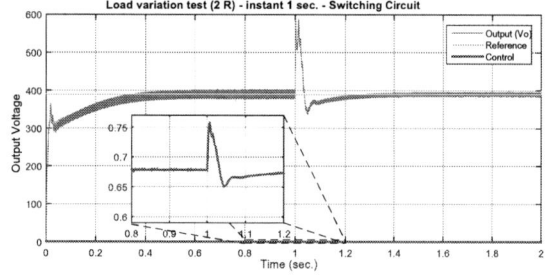

Fig. 13. Applying a load variation, where the initial load R changes to the value of $2R$, at 1 second for the SEPIC inverter.

Fig. 13 shows the behavior of the SEPIC to the load disturbance. At the moment that the load is increased, the value of the output voltage comes close to $600\ V$. That is due to the control signal that increases for the load variation.

The last test is performed for ramp reference variation. For that, the reference signal starts with a value of $390\ V$ and at the instant of 1 second the ramp variation starts to grow at $125\ V/\text{second}$. It starts in 1 second with a value of $390\ V$ e grows to the value of $440\ V$ at the instant of 1.4 seconds. Fig. 14 presented steady-state error for the ramp variation.

V. CONCLUSIONS

This work presented the modeling and design of a state feedback controller with state estimator for a SEPIC. The computational simulation using the model showed that the settling time for a step input exceeded the design requirement value.

Despite that, the system was able to reject the disturbance and noise, but in practical cases the control signal would saturate.

The ramp reference test showed that there is a steady-state error of approximately $0.1\ V$ for this type of input in the system. However, for a step-change in the reference signal, the system did not present a steady-state error. The tests showed proximity between the simulations using the system model in state-space equations and the switched circuit of the SEPIC.

The state feedback technique allowed the system to achieve, with reasonable proximity, the dynamics requirements designed in the control system. The use of the state observer dispensed the need to measure all state variables. With the integral action of the output feedback, the system did not present steady-state error for piecewise constant input, but rather for a ramp input.

Finally, the control strategy investigated in this paper has demonstrated the viability in controlling SEPIC, with the benefit of reducing cost by avoiding the measurement of all state variables. Moreover, the approach by a state-feedback control yields a more flexible set of specifications for the closed-loop dynamics, which is not the case with classical PI controllers.

REFERENCES

[1] G. J. Miranda, "Be prepared! [power industry deregulation]," *IEEE Industry Applications Magazine*, vol. 9, no. 2, pp. 12–20, Mar 2003.

[2] C. S. Chiu, "T-S fuzzy maximum power point tracking control of solar power generation systems," *IEEE Transactions on Energy Conversion*, vol. 25, no. 4, pp. 1123–1132, Dec 2010.

[3] D. C. Martins and I. Barbi, *Basics DC-DC converters non isolated*, 2nd ed., 2006.

[4] S. Durgadevi and M. G. Umamaheswari, "Analysis and design of single phase power factor correction with dcdc sepic converter for fast dynamic response using genetic algorithm optimised pi controller," *IET Circuits, Devices Systems*, vol. 12, no. 2, pp. 164–174, 2018.

[5] S. L. M. Confessor and E. R. L. Villarreal, *Comparative analysis of MPPT controllers in a system Photovoltaic: Comparison between the traditional method of perturbation and observation (P&O) and the method through Fuzzy Logic*. New Academic Editions., 2016.

[6] J. R. Britto, "Analysis, design and implementation of dc-dc converters with wide conversion range applied in solid state lighting." Doctoral thesis, Universidade Federal de Uberlândia, 2009.

[7] R. D. Middlebrook and S. Cuk, "A general unified approach to modelling switching-converter power stages," in *1976 IEEE Power Electronics Specialists Conference*, June 1976, pp. 18–34.

[8] C.-T. Chen, *Linear System Theory and Design*, 3rd ed. New York, NY, USA: Oxford University Press, Inc., 1998.

[9] J. Kautsky, N. K. Nichols, and P. V. Dooren, "Robust pole assignment in linear state feedback," *International Journal of Control*, vol. 41, no. 5, pp. 1129–1155, 1985.

Third Harmonic Injection Method for Reliability Improvement of Single-Phase PV Inverters

R.C. de Barros[a], R.P. Silva[b], D.B. da Silveira[b], W.C. Boaventura[a], A.F. Cupertino[c] and H.A. Pereira[b].

[a] Graduate Program in Electrical Engineering - Universidade Federal de Minas Gerais - Av. Antônio Carlos 6627, 31270-901, Belo Horizonte, MG, Brazil.
rodrigocdebarros@gmail.com, wventura@cpdee.ufmg.br
[b] Department of Electrical Engineering, Federal University of Viçosa, Viçosa, MG, Brazil.
rafdpaula@gmail.com, diogo.silveira@ufv.br, heverton.pereira@ufv.br
[c] Department of Materials Engineering, Federal Center for Technological Education of Minas Gerais.
afcupertino@ieee.org

Abstract—In the last decades, it is observed an increasing of Photovoltaic (PV) systems connected into the grid. In this context, the photovoltaic inverter reliability has became a target of study for many researchers and professionals in the area. In addition, the dc-link capacitor has been reported as one of the most critical components in the PV inverter. Regarding to the single-phase PV inverter, the voltage and current of the dc-link capacitor present a second harmonic component which directly affect its reliability. In order to solve this problem, a third harmonic current injection by the PV inverter is proposed in this work. Analysis of power losses, thermal stress and reliability are performed in the dc-link capacitor. In addition, the effects on the semiconductor devices are also considered. The results show the improvement in the dc-link capacitor reliability due to the proposed methodology. Also, it is shown that the third harmonic current insertion does not affect the PV inverter efficiency.

Index Terms—Photovoltaic System, Single-Phase PV Inverter, dc-link Capacitor, Third Harmonic Current Injection, Reliability.

I. INTRODUCTION

The Distributed Generation (DG) presented great growth in Brazil in recent years [1], being the installed capacity 273 MW in December 2017 and reaching 690 MW one year later [2]. The solar photovoltaic (PV) energy source is responsible for 84.4% of the DG in Brazil. This fact is explained based on the evolution of PV panel technology, which has improved its efficiency and made PV system prices lower [3]. In addition, developments related to the energy conversion efficiency, control and conditioning of electric power [4], [5].

The PV inverter is usually a critical component in PV systems, in terms of reliability, when compared to the others parts [6]. An industrial experience based survey indicates that more than 60% of the PV inverter failures from 10 to 20 years. This reference also identified the semiconductors devices and aluminum electrolytic capacitors as the main components responsible to the failures [7].

Electrolytic capacitors are used in power applications due the necessity of energy storage and filtering process. In a PV inverter, the dc-link capacitor presents voltage fluctuation,

which can be caused by the variation in the power demanded by the load. This voltage fluctuation directly affected the capacitor power losses and its reliability [8]. For this reason, the capacitors can have relative critical lifetime in some cases, reducing the reliability of the entire PV system [9].

In a single-phase PV inverter there is a second harmonic in the voltage and current in the dc-link capacitor, which is a typical behavior of this topology [10]. As discussed in [11], the capacitor power losses, and consequently its reliability, are directly related to the harmonic components in the capacitor current. Thus, the second harmonic current can reduce the dc-link reliability.

In order to solve this problem, this paper proposes the insertion of third harmonic component in the PV inverter current to reduce the second harmonic component in the dc-link capacitor current. This technique presents an interesting implementation, since the single-phase PV inverter usually has the strategy of control proportional resonant for the first and third harmonic currents [12], [13]. Different percentage values of the third harmonic are employed, considering the IEEE standards [14], and the effect on dc-link capacitors and semiconductor devices based on thermal stress are evaluated.

Thus, the main contributions of this paper are:

- Strategy implementation to reduce the second harmonic component in the voltage and current dc-link capacitor;
- Evaluation of the power semiconductors and capacitors thermal stresses when the proposed strategy is implemented;
- Evaluation of the potential of lifetime extension for the proposed technique;

This paper is divided in seven sections. Section II presents the single-phase inverter and proposed a third harmonic current injection method. Section III describes the effect on the dc-link capacitor power losses. In addition, section IV presents the reliability methodology for the capacitor and semiconductor devices. Section V presents the case study. Thus, section VI discusses the results obtained. The conclusions are stated in section VII.

II. SINGLE-PHASE PV INVERTER

The structure and control strategy of a grid-connected single-phase PV inverter are shown in Fig. 1. This system employs a conversion stage which interfaces the PV array and the electrical grid. The single-phase PV inverter model is a full bridge topology based on Insulated Gate Bipolar Transistors (IGBTs) with anti-parallel diodes. In addition, a LCL filter is used to attenuate the harmonic components generated by the inverter switching.

The control structure of the single-phase PV inverter has an outer loop which regulates the dc-link voltage. In addition, moving-average filter is used to obtain the dc-link voltage [15]. Besides, the control structure is based on conventional proportional-integral (PI) controller and it performs the maximum power point tracking (MPPT). Thus, the amplitude of the fundamental current that must be injected into the grid is provided. This signal is synchronized with the grid voltage [16] and subsequently added to the third harmonic current component injected by the inverter. Thus, the reference current of the inverter is generated.

The traditional solution for the current control is a resonant controller tuned in the fundamental frequency. Nevertheless, in the conventional operation of the single-phase PV inverter, the interaction of the second harmonic in the dc-link and the sinusoidal reference voltage generates a third harmonic component in the output current [17]. This component can be compensated by means of a third harmonic resonant controller. Therefore, a proportional multi-resonant (PMR) controller is employed in the inner loop. The PMR used is implemented to track the fundamental and suppress the third harmonic in the grid current. The same resonant controller is used to track the third harmonic component for dc-link voltage ripple reduction.

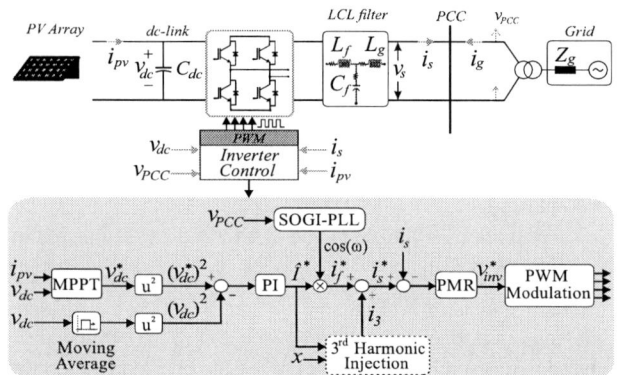

Fig. 1. Structure and control strategy of a grid-connected single-stage single-phase PV inverter system.

III. THE EFFECT OF THE THIRD HARMONIC CURRENT INJECTION ON THE DC-LINK CAPACITOR POWER LOSSES

Neglecting the voltage drop in the LCL filter and the switching harmonics, the output voltage $v_s(t)$ and the current injected into the grid $i_s(t)$ of a traditional PV inverter are denoted as:

$$v_s(t) = \widehat{V}\cos(\omega t), \tag{1}$$

$$i_s(t) = \widehat{I}\cos(\omega t + \phi), \tag{2}$$

where $\widehat{V}$ and $\widehat{I}$ correspond to the amplitudes of the grid voltage and the injected grid current, respectively; ω is the angular frequency of the grid and ϕ is the displacement angle. Therefore, the instantaneous power supplied by the inverter is calculated as follows:

$$p_s(t) = v_s(t)i_s(t) = \overbrace{\frac{\widehat{V}\widehat{I}}{2}\cos(\phi)}^{p_s'} + \overbrace{\frac{\widehat{V}\widehat{I}}{2}\cos(2\omega t + \phi)}^{p_s''(t)}, \tag{3}$$

where p_s' corresponds to the average power supplied by the inverter and $p_s''(t)$ represents a pulsation component at the double-line frequency present in $p_s(t)$ [18]. The power $p_s''(t)$ indicates that there is a second harmonic component in the dc-link voltage and current, which is one of the main contributors to the temperature increasing and reduction of the dc-link capacitor lifetime [8], [19].

In order to reduce the second harmonic component present in the dc-link voltage and current, a third harmonic current insertion method is proposed in this paper. In this case, the inverter output voltage remains as in (1) and the new inverter output current injected into the grid is given by:

$$i_s(t) = \widehat{I}\cos(\omega t + \phi) + \widehat{I}_3\cos(3\omega t + \alpha), \tag{4}$$

where $\widehat{I}_3$ corresponds to the magnitude of the third harmonic current component injected and α is the harmonic phase angle. In this context, the instantaneous power supplied by the inverter can be expressed as:

$$p_s(t) = \frac{\widehat{V}\widehat{I}}{2}\cos(\phi) + \frac{\widehat{V}\widehat{I}}{2}\cos(2\omega t + \phi) + \frac{\widehat{V}\widehat{I}_3}{2}\cos(2\omega t + \alpha)$$
$$+ \frac{\widehat{V}\widehat{I}_3}{2}\cos(4\omega t + \alpha). \tag{5}$$

As observed, there is a fourth harmonic component in the dc-link current. In addition, it is possible to reduce the total second harmonic component of $p_s(t)$ by choosing a specifics values to the current amplitude $\widehat{I}_3$. Thus, considering $\alpha = \phi$ and $\widehat{I}_3 = -x\widehat{I}$, $p_s(t)$ is given by:

$$p_s(t) = \overbrace{\frac{\widehat{V}\widehat{I}}{2}\cos(\phi)}^{p_s'} +$$

$$\overbrace{\frac{\widehat{V}\widehat{I}}{2}(1-x)\cos(2\omega t + \phi) + \frac{\widehat{V}\widehat{I}}{2}(-x)\cos(4\omega t + \phi)}^{p_s''(t)},$$

$$(6)$$

where x corresponds the percentage of the fundamental current amplitude which will be injected by the third harmonic current component.

Since the dc-link must absorb the power oscillation, $i_c(t) = p_s''(t)/v_{dc}(t)$. Considering that the voltage ripple is small and that the control follows the reference dc-link voltage, the capacitor current can be approximated by:

$$i_c(t) \approx \frac{\widehat{V}\widehat{I}}{2v_{dc}^*}(1-x)\cos(2\omega t + \phi) + \frac{\widehat{V}\widehat{I}}{2v_{dc}^*}(-x)\cos(4\omega t + \phi). \tag{7}$$

The power losses in a dc-link capacitor are determined by its Equivalent Series Resistance (ESR) and the RMS value of the current $i_c(t)$ using the following equation [11], [20]:

$$P_{c,losses} = I_{c(RMS)}^2 ESR. \tag{8}$$

The ESR parameter is affected by the temperature and frequency variation [11]. However, for sake of simplicity, the ESR value is considered constant with frequency equal 100 Hz. This value is the closest of the frequency of 120 Hz corresponding the highest harmonic amplitude of the dc-link capacitor current. Therefore, through equations (7) and (8) and after some mathematical manipulations, the dc-link capacitor power losses with the third harmonic current injection is calculated as follows:

$$P_{c,losses} \approx ESR \frac{\widehat{V}^2 \widehat{I}^2}{8v_{dc}^{*2}}(2x^2 - 2x + 1). \tag{9}$$

Equation (9) consists of a second order polynomial, which suggests that there are maximum and minimum power losses depending of the x value, as shown in Fig. 2. However, due to the limit of 5% for the total harmonic distortion (THD) recommended by IEEE, the minimum power losses, which occurs when $x = 50\%$, is not an achievable result [14]. Thus, based on the THD calculation, the parameter x is limited to 5%. As observed in Fig. 2, in the case where $x = 5\%$ the capacitor power losses is reduced in 9.5%, comparing to the power losses without the third harmonic current injection.

In this section, the effects of high frequency harmonics were disregarded. In addition, although it is known that the effects of the fourth harmonic component on the dc-link voltage and current ripple are lower compared to the effects caused by the second harmonic component, this part of the power losses was overestimated [21]. In this way, the analysis developed are a rough estimation of the dc-link power losses.

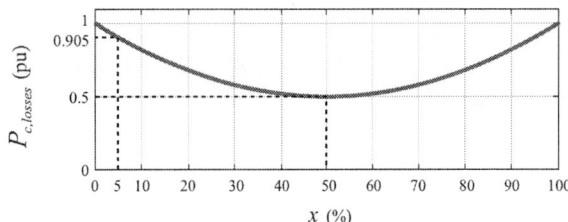

Fig. 2. Relationship between the dc-link capacitor power losses and the parameter x.

IV. RELIABILITY OF CAPACITOR AND SEMICONDUCTOR DEVICES

The electrical components operation conditions directly affects their reliability [22], [23]. Considering the dc-link capacitor, there are some methods in the literature which are able to estimate the time to failure of the component [24], [25]. Since the capacitor power losses are computed, it is possible to obtain the capacitor hot-spot temperature T_h by the following equation [24]:

$$T_h = T_a + R_c P_{c,losses}, \tag{10}$$

where T_a and R_c correspond to the ambient temperature and equivalent thermal resistance from hot-spot to ambient. In reference [26], the value for R_c, considering a electrolytic capacitor, is 6.0 K/W [27] .

A widely used model to estimate the time to failure of the electrolytic capacitors is given by [25]:

$$L = L_0 \left(\frac{V_c}{V_0}\right)^{-n} \times 2^{\frac{T_0 - T_h}{10}}, \tag{11}$$

where L and L_0 are the lifetime under operating and testing conditions, respectively. In addition, V_c is the operating voltage and V_0 is the voltage at test condition. T_h is the capacitor hot-spot temperature, calculate by (10), and T_0 is the temperature under test condition. As the lifetime of aluminum electrolytic capacitors quite depends on the voltage stress level, the value of the voltage stress exponent n used in this work is equal to 3. This value allows to obtain a linear equation, which it is more appropriate to describe the impact of voltage stress [25].

Regarding the semiconductor power device, the number of cycles until failure N_f can be estimated using the Bayerer model [28]. This model estimates the power devices N_f based on the junction temperature T_j, junction temperature variation ΔT_j and the temperature rising time t_{on}, as showed in (12). The parameters meaning and their limit considerations proposed by the Bayerer model are discussed with more details in [28].

$$N_f = A\left(\Delta T_j\right)^{\beta 1} \exp\left(\frac{\beta_2}{T_j + 273}\right) t_{on}^{\beta 3} I^{\beta 4} V^{\beta 5} D^{\beta 6}. \tag{12}$$

V. CASE STUDY

In this work, a 5 kW grid-connected single-phase PV inverter is employed. The main characteristics of the PV system utilized are presented in Table I.

TABLE I
PV SYSTEM PARAMETERS.

Parameter	Value
Nominal Power	5 kW
Grid Voltage	220 V
dc-link Voltage	390 V
Switching Frequency	12 kHz
Sampling Frequency	12 kHz
L_f Filter Inductor	1 mH
L_g Filter Inductor	1 mH
C_f Filter Capacitor	3.8 μF
PCC Voltage	220 V
Grid Frequency	60 Hz

The dc-link corresponds to a capacitor bank of 1680 μF which consists of three 560 μF (450 V) capacitors connected in parallel. Electrolytic capacitor used in this work are manufactured by TDK, part number B43512, and the IGBT used is manufactured by Infineon, part number IKW30N65H5.

In order to verify the effects of the third harmonic current injection on the capacitor and semiconductor power losses, the tests are divided in two cases. In the first case, denominated base case, the PV inverter operates without the third harmonic injection. In the second case, the PV inverter injects a third harmonic current component into the grid, considering the x value which respects the limit established for the output current THD. All simulations are implemented in PLECS environment.

In order to analyze the dc-link voltage reduction and the effect on the THD of the PV inverter output current, two different values of active power injected by the PV system are considered: 2.5 kW and 5 kW. In addition, considering the nominal PV inverter operation, the dc-link capacitor and semiconductor devices power losses are simulated.

Besides, measurements of the dc-link capacitor temperature and the junction temperature of the IGBTs, considering the nominal power of 5 kW and the ambient temperature of 40 °C, are performed. Thus, the methodology presented in the previous sections is applied to estimate the parameters N_f of the IGBT and L of the dc-link capacitor.

VI. RESULTS

In the first part of the section results, the value of the parameter x is chosen. Thus, Fig. 3 presents the relationship between x and the grid current THD. The analysis is made considering the PV inverter injecting the nominal active power to the grid. As observed, for $x = 5\%$, the THD presents value above the limit shown in the section III, which is 5%. It is important to note the difference between the theoretical estimated value ($x = 5\%$) and the simulation results. This fact is explained based on the other source of harmonic components, such as, the switching frequency. Thus, it is

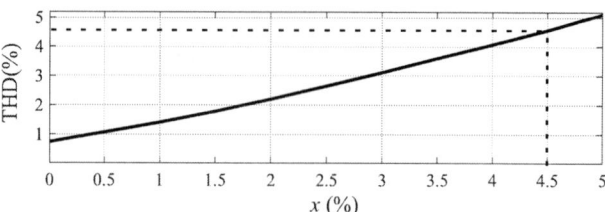

Fig. 3. Relationship between the variation of parameter x and the grid current THD.

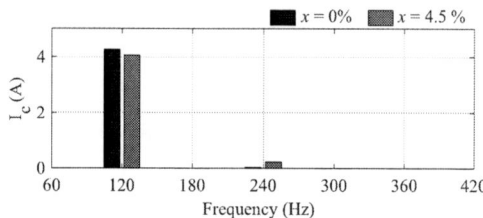

Fig. 4. Current capacitor spectrum considering the PV inverter nominal operation with $x = 0\%$ and $x = 4.5\%$.

necessary to choose the x value which presents THD lower than 5%. Thus, the x value used in this work is equal to 4.5% (THD = 4.56%).

The effect on the capacitor current for $x = 0\%$ and $x = 4.5\%$ third harmonic injection are analyzed. Fig. 4 presents the spectrum of the capacitor current I_c considering the base case ($x = 0$ %) and $x = 4.5\%$. As noticed, with the third harmonic injection, there is a current reduction of 0.22 A for frequency equal to 120 Hz. On the other hand, there is a current increase of 0.21 A for frequency equal to 240 Hz.

Fig. 5 analyzes the effects on the inverter current THD and the dc-link voltage due to the third harmonic current injection. The active power injected by the PV system into the grid is varied from 2.5 kW to 5.0 kW, as presented in Fig. 5 (a). In this context, the Fig. 5 (b) shows the inverter current for $x = 0\%$ and 4.5%. The inverter current THD for the base case, considering the active power equal to 2.5 kW and 5.0 kW, are 0.11% and 0.09%, respectively. Also, for $x = 4.5\%$, the inventer current THD are equal to 4.57% and 4.56% for active power equal to 2.5 kW and 5.0 kW, respectively.

Fig. 5 (c) presents the dc-link voltage. As observed, considering the third harmonic injection, there is a reduction in the voltage fluctuation compared to the base case. For the active power equal to 2.5 kW and 5.0 kW, the voltage reduction are equal to 4.92% and 4.75%, respectively.

In this context, the Fig. 6 (a) and (b) presents, respectively, the power losses dissipated in the semiconductors devices $P_{s,losses}$ and in the dc-link capacitor. As noticed, with the third harmonic injection, the power losses of the IGBT and diode increase with the increasing of the parameter x. Comparing the base case and for $x = 4.5\%$, the IGBT and diode power losses increase in 0.7% and 3.7%, respectively. On the other hand, the dc-link capacitor power losses decrease 9.1% when comparing the base case and $x = 4.5\%$.

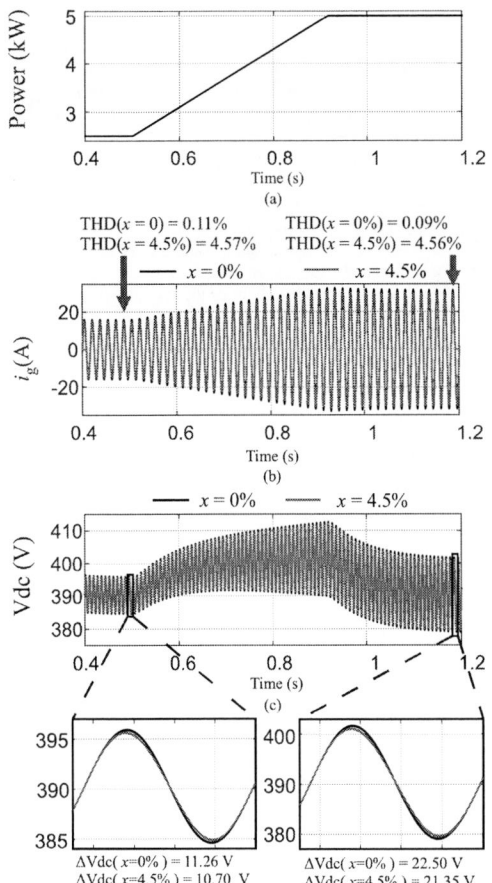

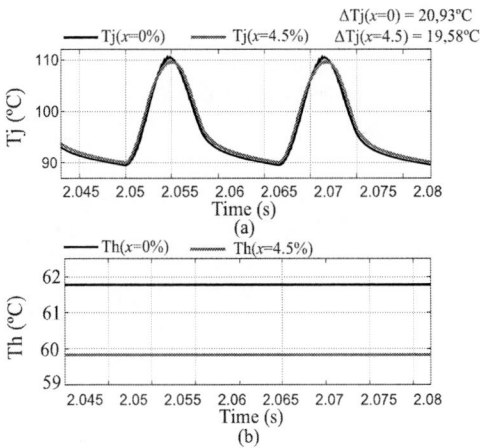

Fig. 7. (a) IGBT junction temperature; (b) Capacitor hot-spot temperature.

Table II presents the main parameters in the PV inverter affected by the third harmonic component current injection. In this table, 5 kW of active power injected into the grid was considered. As shown in the previous results, there is a reduction in the dc-link voltage fluctuation. This fact is explained based on the reduction of the capacitor second harmonic current component, as presented in Fig. 4. In this context, the power losses in the dc-link capacitor is reduced. Thus, the time to failure of the capacitor increases in 14.53%, from 52.50×10^5 to 60.13×10^5, compared to the base case.

TABLE II
COMPARATIVE ANALYSIS BETWEEN THE MAIN PARAMETERS OF THE PV INVERTER CONSIDERING THE THIRD HARMONIC INJECTION.

Parameters	$x = 0$ %	$x = 4.5\%$
$P_{c,losses}$ (W)	10.91	9.936
$P_{IGBT,losses}$ (W)	65.76	66.23
$P_{diode,losses}$ (W)	18.56	19.24
ΔV_{dc} (V)	22.50	21.35
THD (%)	0.09	4.56
ΔT_j	20.93	19.58
T_j	101.02	100.10
η (%)	98.10	98.09
N_f	1.84×10^9	2.52×10^9
L	52.50×10^5	60.13×10^5
Th	$61.82\,^{\circ}$C	$59.87\,^{\circ}$C

In addition, there is an increasing of 0.7% in the power losses dissipated in the IGBT. However, there is a reduction in 7.78% in the junction temperature fluctuation, which is an important parameter to estimate the number of cycle to failure of the semiconductor. In this context, with the third harmonic injection, the IGBT N_f goes from 1.84×10^9 to 2.52×10^9, which corresponds to an increase of 36.95%.

Based on the power losses of the dc-link capacitor and the semiconductor devices, it is possible to estimate the efficiency $\eta(\%)$ of the PV inverter. As noticed, comparing to the base case, the efficiency does not change with the third harmonic injection. This fact is explained based on the power losses variation dissipated in the electrical devices, while the values

Fig. 5. (a) Variation of the active power injected into the grid. (b) Grid current waveform considering $x = 0\%$ and $x = 4.5\%$. (c) Voltage waveform of the dc-link considering $x = 0\%$ and $x = 4.5\%$.

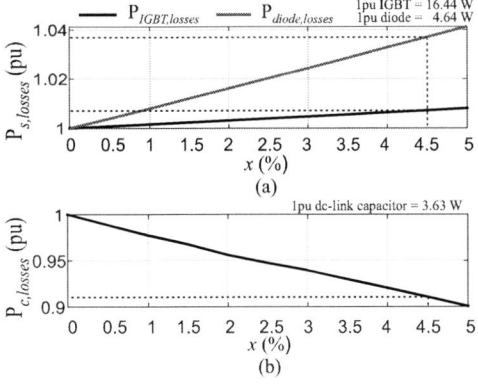

Fig. 6. Power Losses: (a) semiconductor devices and (b) dc-link capacitor.

Fig. 7(a) and (b) present, respectively, the temperature of the IGBT and dc-link capacitor hot-spot. As noticed, with the third harmonic injection, there is a reduction of 7.78% in the IGBT fluctuation junction temperature. In addition, there is a reduction in $1.95\,^{\circ}$C in the capacitor hot-spot temperature.

of $P_{IGBT,losses}$ and $P_{diode,losses}$ increase, the $P_{c,losses}$ is reduced.

VII. CONCLUSION

This work proposed to reduce the dc-link capacitor stress inserting the third harmonic current component. Analytical results from the proposed methodology were performed.

In terms of conversion efficiency, it is possible to observe that the third harmonic injection with $x = 4.5\%$ did not impact the performance of the PV inverter compared to the base case.

In addition, the simulation results shown that the power losses and the temperature of the dc-link capacitor are reduced with the third harmonic insertion. As a result, the lifetime of this component tends to increase. On the other hand, the third harmonic component increase the IGBT power losses. At the same time, the fluctuation junction temperature ΔT_j of the IGBT presents a decreasing. Between these two parameters, ΔT_j has greater influence on the semiconductor lifetime. Therefore, the number of cycles until failure N_f of the IGBTs increases. These results mean that the reliability of the PV inverter is improved in relation to the base case.

ACKNOWLEDGMENT

This work has been supported by the Brazilian agencies CAPES, FAPEMIG and CNPq. In addition, this work received financial support from CEMIG-D through the ANEEL (Brazilian Regulatory Agency) P&D program.

REFERENCES

[1] "Geração distribuída – regulamentação atual e processo de revisão," ANEEL - Agência Nacional de Energia Elétrica, 2019. [Online]. Available: http://www.aneel.gov.br/documents/655804/14752877/Geração+Distribuída+-+regulamentação+atual+e+processo+de+revisão.pdf/3def5a2e-baef -bb59-2ce1-4f69a9cb2d88

[2] "Informações gerenciais," ANEEL - Agência Nacional de Energia Elétrica, 2018. [Online]. Available: http://www.aneel.gov.br/documents/656877/14854008/Boletim+de+Informações+Gerenciais+-+4°+trimestre+de+20188/36e91555-141a-637d -97b1-9f6946cc61b3?version=1.2

[3] A. Sangwongwanich, Y. Yang, D. Sera, F. Blaabjerg, and D. Zhou, "On the impacts of pv array sizing on the inverter reliability and lifetime," *IEEE Trans. Ind. Appl.*, vol. 54, no. 4, pp. 3656–3667, July 2018.

[4] M. Molina, "Energy storage and power electronics technologies: A strong combination to empower the transformation to the smart grid," *Proc. IEEE*, vol. PP, pp. 1–29, 09 2017.

[5] R. STRZELECKI and G. Zinoviev, *Overview of Power Electronics Converters and Controls*, 08 2008, vol. 34, pp. 55–105.

[6] J. Flicker, R. Kaplar, M. Marinella, and J. Granata, "Pv inverter performance and reliability: What is the role of the bus capacitor?" in *2012 IEEE 38th Photovoltaic Specialists Conference*, June 2012, pp. 1–3.

[7] J. Falck, C. Felgemacher, A. Rojko, M. Liserre, and P. Zacharias, "Reliability of power electronic systems: An industry perspective," *IEEE Ind. Electron.. Mag.*, vol. 12, no. 2, pp. 24–35, June 2018.

[8] K. Zhao, P. Ciufo, and S. Perera, "Lifetime analysis of aluminum electrolytic capacitor subject to voltage fluctuations," in *Proceedings of 14th International Conference on Harmonics and Quality of Power - ICHQP 2010*, Sep. 2010, pp. 1–5.

[9] P. Spanik, M. Frivaldsky, and A. Kanovsky, "Life time of the electrolytic capacitors in power applications," in *2014 ELEKTRO*, May 2014, pp. 233–238.

[10] T. Wang and S. Lu, "A comprehensive analysis of dc-link current for single phase h-bridge inverter under harmonic output currents," in *2017 IEEE Energy Conversion Congress and Exposition*, Oct 2017, pp. 652–658.

[11] J. M. Lenz, J. R. Pinheiro, and H. C. Sartori, "Dc-link electrolyte capacitor lifetime analysis for a pv boost converter," in *2017 IEEE 8th International Symposium on Power Electronics for Distributed Generation Systems (PEDG)*, April 2017, pp. 1–6.

[12] R. Teodorescu, F. Blaabjerg, M. Liserre, and P. C. Loh, "Proportional-resonant controllers and filters for grid-connected voltage-source converters," *IEE Proceedings - Electric Power Applications*, vol. 153, no. 5, pp. 750–762, Sep. 2006.

[13] M. Ciobotaru, R. Teodorescu, and F. Blaabjerg, "Control of single-stage single-phase pv inverter," in *2005 European Conference on Power Electronics and Applications*, Sep. 2005, pp. 10 pp.–P.10.

[14] "Ieee recommended practice and requirements for harmonic control in electric power systems," *IEEE Std 519-2014 (Revision of IEEE Std 519-1992)*, pp. 1–29, June 2014.

[15] F. Sasongko, K. Sekiguchi, K. Oguma, M. Hagiwara, and H. Akagi, "Theory and experiment on an optimal carrier frequency of a modular multilevel cascade converter with phase-shifted pwm," *IEEE Trans. Power Electron.*, vol. 31, no. 5, pp. 3456–3471, May 2016.

[16] A. Timbus, M. Liserre, R. Teodorescu, and F. Blaabjerg, "Synchronization methods for three phase distributed power generation systems - an overview and evaluation," in *2005 IEEE 36th Power Electronics Specialists Conf.e*, June 2005, pp. 2474–2481.

[17] B. Li, S. Huang, and X. Chen, "Performance improvement for two-stage single-phase grid-connected converters using a fast dc bus control scheme and a novel synchronous frame current controller," *Energies*, vol. 10, p. 389, 03 2017.

[18] K. Mozaffari, M. Amirabadi, and Y. Deshpande, "A single-phase inverter/rectifier topology with suppressed double-frequency ripple," *IEEE Transactions on Power Electronics*, vol. 33, no. 11, pp. 9282–9295, Nov 2018.

[19] B. Wang, U. States, and Y. Song, "Single phase vsi with reduced-size dc-link capacitor," in *4th International Conf. on Power Engineering, Energy and Electrical Drives*, May 2013, pp. 1427–1430.

[20] G. I. Orfanoudakis, S. M. Sharkh, and M. A. Yuratich, "Analysis of dc-link capacitor losses in three-level neutral point clamped and cascaded h-bridge voltage source inverters," in *2010 IEEE International Symposium on Industrial Electronics*, July 2010, pp. 664–669.

[21] Y. Hu, X. Zhang, W. Mao, T. Zhao, F. Wang, and Z. Dai, "An optimized third harmonic injection method for reducing dc-link voltage fluctuation and alleviating power imbalance of three-phase cascaded h-bridge photovoltaic inverter," *IEEE Transactions on Industrial Electronics*, pp. 1–1, 2019.

[22] P. D. Reigosa, H. Wang, Y. Yang, and F. Blaabjerg, "Prediction of bond wire fatigue of igbts in a pv inverter under a long-term operation," *IEEE Trans. Power Electron.*, vol. 31, no. 10, pp. 7171–7182, Oct 2016.

[23] K. Ma, H. Wang, and F. Blaabjerg, "New approaches to reliability assessment: Using physics-of-failure for prediction and design in power electronics systems," *IEEE Power Electronics Magazine*, vol. 3, no. 4, pp. 28–41, Dec 2016.

[24] H. Wang, P. Davari, H. Wang, D. Kumar, F. Zare, and F. Blaabjerg, "Lifetime benchmarking of two dc-link passive filtering configurations in adjustable speed drives," in *2018 IEEE Applied Power Electronics Conf. and Exposition (APEC)*, March 2018, pp. 228–233.

[25] H. Wang and F. Blaabjerg, "Reliability of capacitors for dc-link applications in power electronic converters—an overview," *IEEE Trans. Ind. Appl.*, vol. 50, no. 5, Sep. 2014.

[26] A. Cupertino, L. Xavier, E. M. Brito, V. Mendes, and H. Pereira, "Benchmarking of power control strategies for pv systems under unbalanced conditions," *Elec. Power*, p. 335–345, 2019.

[27] H. Wang, P. Davari, H. Wang, D. Kumar, F. Zare, and F. Blaabjerg, "Lifetime estimation of dc-link capacitors in adjustable speed drives under grid voltage unbalances," *IEEE Trans. Power Electron.*, vol. 34, no. 5, pp. 4064–4078, May 2019.

[28] R. Bayerer, T. Herrmann, T. Licht, J. Lutz, and M. Feller, "Model for power cycling lifetime of igbt modules - various factors influencing lifetime," in *5th International Conf. on Integrated Power Electronics Systems*, March 2008.

978-1-7281-4181-7/19 $31.00 © 2019 IEEE

Didactic System For Control Of Electrical Machines In Education And Research Laboratories

Allan V. S. Andrade, Richard M. Stephan

Universidade Federal do Rio de Janeiro, Rio de Janeiro RJ, Brazil

e-mail: allanvinicius@poli.ufrj.br, richard@coe.ufrj.br

Abstract—This paper provides the development of a low-cost hardware solution capable of performing the control and drive of machines for the university's electrical machinery laboratory for the purpose of teaching and research students of undergraduate, master's and doctorate and be widely used in the academic environment.

Keywords—control system, electrical machines, hardware, DSP.

I. Introduction

The advancement of power electronics and the creation of microprocessors and DSPs, provided the development of digital control of electric machines. Such control allows the user to execute complex algorithms, reliability, performance, high precision and processing speed. Microprocessors have become the fundamental devices throughout this digital control process. With this, several hardware solutions were created, from several manufacturers, with very high costs, mainly taking into account the fact that they are foreign solutions and the need to import such equipment to the national soil, making costs even more expensive.

The demand for systems to control and activate electric machines mainly in the field of research has grown considerably. In [1] and [2] can be viewed some practical examples of the use of DSP for this kind of control. Environments, such as university laboratories, do not have the high financial resources required to purchase such solutions. The need versus the difficulty of acquisition motivated the creation of a low cost device capable of performing the same or at least the main functions of the commercial equivalents follow.

II. Hardware Structure

In figure 1 is possible to visualize the basic structure of how developed hardware will act connected to the system, in the case the electrical machines to be controlled.

The control hardware receives the electrical machine information obtained through the sensors, such as voltage, current, speed, torque and rotor position. With this information the real-time data processing is done and a control signal is sent to the actuator, which in the case of the design is an inverter electrically connected to the machine.

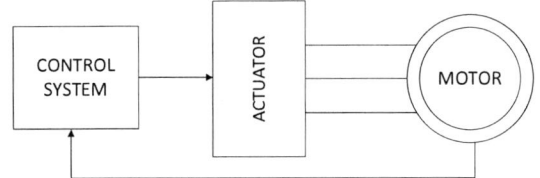

Fig. 1. Basic structure of the control system.

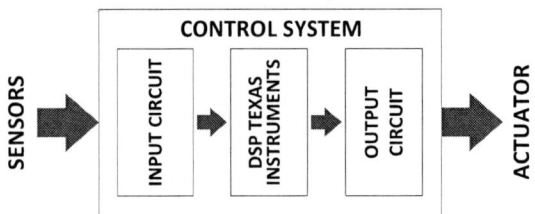

Fig. 2. Proposed control system.

The system consists basically of 3 components, the control part, which will deal with all logic; the actuator which basically is the power part supplying the power to the electric machine, and acts through the control system signal; and the electric machine.

In figure 2 is shown how the control system in question is constituted, and will be detailed in the following sections.

The control system is divided into 3 parts, which are: the input circuit; the DSP; and the output circuit. The data from the sensors first reach the input circuit, which is responsible for conditioning the signals in order to keep them within the acceptable ranges of the microcontroller input ports, and to protect against possible voltage surges. In the DSP, through the data obtained by the sensors and conditioned by the input circuit, all the logic of the control programmed by the software will be made, the data processing and then sending the control signal to the output circuit. The output circuit is responsible for conditioning the control signals from the microcontroller to an acceptable signal pattern by the actuator.

III. Actuator

The actuator used for the design is the WEG CFW 08 Plus power inverter [3] rated current 16 A and voltage 200-240 V. The purpose of this equipment is to receive the signal from the control system, and through that signal, provide power to power the machine. That is, in this type of arrangement the inverter does not perform any type of control under the machine.

For the inverter to operate in order to receive external control, it must be configured for a function called "remote mode". Through this function it is possible to define an analog signal input present in the equipment to receive a voltage level varying in scale from 0 to 10 V proportional to the frequency of the machine's power supply. The voltage level 0 V causes the machine to rotate at the minimum speed pre-set in the drive configuration parameters, see [3]. Already the voltage level of 10 V corresponds to the inverter causing the machine to rotate in its nominal capacity of rotation.

978-1-7281-4181-7/19 $31.00 © 2019 IEEE

Fig. 3. Control system connection diagram on the actuator.

IV. DSP

The microcontroller used is the Texas Instruments F28335 [4], it can be visualized in figure 3.

As can be seen in figure 3, the experimental kit TMS320F28335 was used, in addition to the F28335 microcontroller, the XDS100 emulator, which is a USB JTAG interface, allows the microprocessor to be connected to the computer through the USB port to send and receive information from the DSP through the computer, thus enabling programming and debugging of the microprocessor.

According to [4], the microcontroller F28335 is a 32-bit floating-point DSP. Its system works in the frequency of 150 MHz. It has 18 output PWM channels, 16 12-bit ADC channels, 2 32-bit channels for quadrature encoder, CAN, UART, McBSP, SPI and I2C serial ports, among other characteristics.

Most of the functions present in the F28335 will be used, since the idea is to make a device as generic as possible, being able to meet the needs of the most varied types of control developed in future studies. For this, the PWM outputs that will drive the inverter will be used; the ADC channels to capture the voltage and current signals, the quadrature encoder channel for the connection of the same, making it possible to measure the speed and position of the rotor axis of the electric machine; the eCAP port for measuring the frequency of a digital channel.

V. INPUT CIRCUIT

For the correct functioning of the control system, regardless of the type, it is necessary to obtain the values of voltage, current, speed, torque and rotor position, thus allowing closed-loop control of the system. The information must be updated in real time and sent to the microcontroller. Devices capable of making such measurements in real time are respectively voltage transducer, current transducer, torque sensor and encoder.

For each type of sensor used, there will be an associated circuit for the conditioning of its signal, keeping it within the standards accepted by the DSP, thus allowing its connection to the hardware, as shown in figure 4.

The Texas Instruments F28335 microcontroller has the following features on its input interface:

• Analog inputs: input voltage from 0 to 3 V;

• Digital inputs: input voltage from 0 to 3.3 V, with TTL logic (Transistor-Transistor Logic), for voltage level below 0.4 V set to "low", and level above 2.4 V defined as "high";

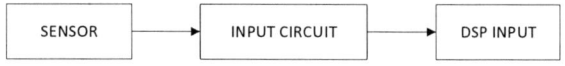

Fig. 4. Diagram connecting the sensors to the input circuit and DSP.

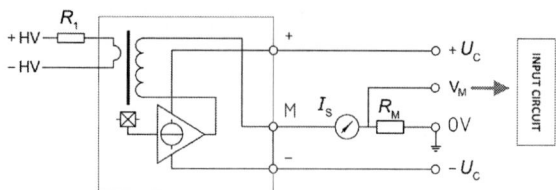

Fig. 5. Voltage transducer circuit.

A. Voltage transducer interface

The voltage transducer used was the LEM LV 20-P [5], which has a measuring capacity of 500 V RMS. It is a Hall Effect current transducer capable of measuring the voltage level through resistors R_1 and R_M connected in series to the primary and secondary transducer circuit, as shown in figure 5.

The values of R_1 and R_M can be obtained from (1), taking into account a maximum measuring limit of 500 V, nominal capacity of the transducer.

$$R_1 = \frac{V_{nominal\ measurement}}{I_{nominal\ in\ primary}} = \frac{500}{10\ x\ 10^{-3}} = 50\ k\Omega \qquad (1)$$

The minimum power that the resistor R_1 must support can be obtained through (2).

$$P_{R1\ min} \geq \frac{V_{nominal\ measurement}^2}{R_1} = \frac{500^2}{50\ K} = 5W \qquad (2)$$

For R_1, a power resistor with a minimum of 5W should be used. Taking into account the transformation ratio K_N and the resistance R_P of the transducer primary [5], the current in the secondary will be:

$$I_{PN} = \frac{V_{nominal\ measurement}}{R_1 + R_P} = \frac{500}{50K + 250} = 9.95\ mA \qquad (3)$$

$$I_{SN} = K_N\ x\ I_{PN} = 2.5\ x\ 9.95\ mA = 24.88\ mA \qquad (4)$$

And the value of R_M is obtained through (5).

$$R_M = \frac{V_{interface\ input}}{\sqrt{2}\ x\ I_{SN}} = \frac{1.5}{\sqrt{2}\ x\ 24.88\ x\ 10^{-3}} = 42.6\ \Omega \qquad (5)$$

$$R_M \rightarrow 39\ \Omega\ (selected\ value)$$

The interface circuit that will allow the connection of the voltage transducer to the DSP can be visualized in figure 6. It is compound of 2 cascade connected operational amplifiers. The component chosen for the operational amplifier is Texas Instruments TL082 because it has 2 amp-ops on a single circuit element. The TL082 also features such as low power consumption, low output signal distortion, high slew rate and high common mode rejection ratio.

Following the signal direction from the input to the circuit output, the first amp-op is connected in the voltage buffer configuration, isolating the input signal from the sensor of the rest of the circuit, with a high impedance at the input of the circuit, and low impedance at the output. The second amp-op is connected in the non-inverting summing amplifier configuration and allows the addition of a 1.5 V DC signal to the offset, without reversing the signal at the output.

978-1-7281-4181-7/19 $31.00 © 2019 IEEE

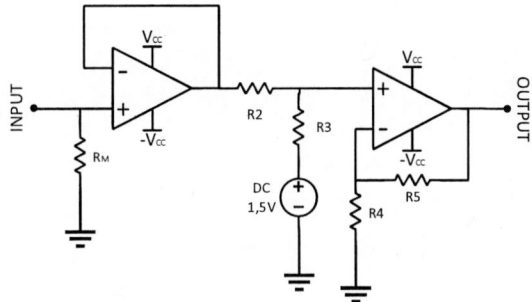

Fig. 6. Voltage transducer interface circuit.

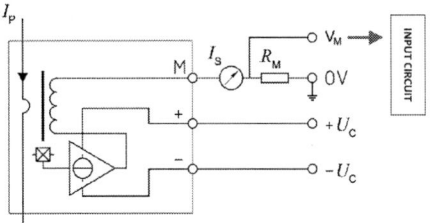

Fig. 7. Current transducer circuit.

Typical values were chosen for the resistors R_2, R_3, R_4, R_5:

$$R_2 = R_3 = R_4 = R_5 = 100\ k\Omega$$

For a three-phase voltage capture will require three circuits and three ADC channels.

B. Current transducer interface

In figure 7 can be viewed the current measurement transducer circuit of LEM LA-55P [6] used, which has a measuring capacity of 50 A RMS.

The same interface circuit of voltage transducer will be used with a different value of R_M. This value can be obtained on (6), taking into account the nominal measurement capacity of 70 A from [6], that gives a maximum value of I_{SN}= 70 mA.

$$R_M = \frac{V_{interface\ input}}{I_{SN}} = \frac{1,5}{70\ x\ 10^{-3}} = 21.4\ \Omega \qquad (6)$$

$$R_M \rightarrow 20\ \Omega\ (selected\ value)$$

For other resistors, were adopted the same values from voltage transducer interface circuit.

$$R_2 = R_3 = R_4 = R_5 = 100\ k\Omega$$

Again, for a three-phase current capture will require three circuits and three ADC channels.

C. Torque Interface Sensor

The torque measurement is given by a sensor physically coupled to the axis of the electric machine [7]. This sensor consists of 4 strain gauges connected in Wheatstone bridge, and its circuit can be visualized in figure 8.

According to the torque applied by the load on the axis of rotation, the strain gauges deform causing variations in the values of the R_1, R_2, R_3 and R_4 resistors of the Wheatstone

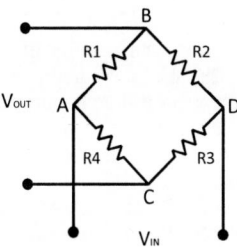

Fig. 8. Torque sensor circuit.

bridge, which are originally the same, for zero torque. By applying a V_{IN} voltage at terminals A and D of the bridge it is possible to obtain a voltage V_{OUT} at terminals B and C, which for each value of torque will have a respective associated voltage value, and for the absence of torque, zero torque, this value of V_{OUT} voltage will be zero.

The torque sensor is then connected to an interface with display that will condition the voltage signal V_{OUT} and translate it into its respective torque value. More information on the operation of this system can be found in [7]. In this interface it is also possible to extract, through an analogue output, voltage values proportional to the torque values obtained. Such analog output has a 10 V voltage signal excursion, which will be connected to the input circuit of the control hardware, as shown in figure 9.

The circuit for the torque sensor is shown in figure 10. In this circuit a connected op amp is used in the voltage buffer configuration, isolating the input signal from the DSP. For this circuit, the TL081 component of Texas Instruments was chosen, which has the same characteristics as TL082, but with 1 amp-op per component.

The 10 V signal reduction of the analog sensor output to 3 V of the DSP ADC input is done through a voltage divider at the input of the circuit, and the values of the selected resistors were:

$$R_1 = 22\ k\Omega$$

$$R_2 = 10\ k\Omega$$

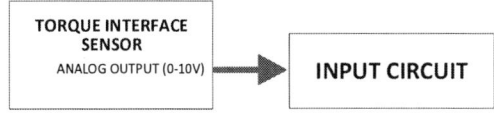

Fig. 9. Detail of torque sensor connection with control system.

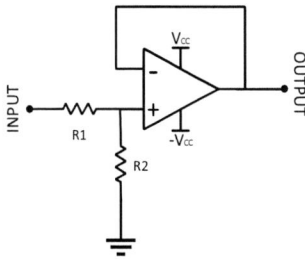

Fig. 10. Torque sensor interface circuit.

978-1-7281-4181-7/19 $31.00 © 2019 IEEE

D. Encoder Interface

The speed, position and direction of rotation of the electric machine axis can be measured via the encoder. An incremental 3-channel quadrature encoder was used [8]. The connection of the encoder to the DSP is made through the same circuit developed for the torque sensor, with a voltage divider in the input, followed by an op amp connected in the voltage buffer configuration, as shown in figure 10. To reduce the voltage level of 5 V of the encoder signal output to 3.3 V of the digital input of the DSP, the following resistors were used in the voltage divider:

$$R_1 = 69 \ k\Omega$$

$$R_2 = 100 \ k\Omega$$

Since the encoder has 3 output channels, 3 circuits and 3 DSP digital inputs are required.

VI. OUTPUT CIRCUIT

The output circuit will condition the digital control signals from the PWM output of the microcontroller to the analog input of the actuator, switching with a frequency of 5 kHz. Such conditioning will take place in 2 stages, one analog filter responsible for converting the digital signals to analog and another for the amplification of the signals in an acceptable range through the analog input port of the actuator, as shown in figure 11.

A. Analog filter

The purpose of the filter is to transform a switched signal into a DC signal for later connection to the analog input of the actuator. For this type of transformation a low-pass filter will be used in order to cut the high frequencies of the switched signal, thus obtaining only the DC level that will vary according to the duty cycle. The duty cycle will then be the variable to be controlled in the program.

The value of the DC component in the output $V_{DC \ out}$ will be the amplitude of the PWM signal at the filter input, times the duty cycle, as shown in (7).

$$V_{DC \ out} = Amplitude \ PWM \times duty \ cycle \quad (7)$$

To filter the high frequencies, the following parameters were taken into account in the filter design:

- Maximum 0.005 dB (A_{max}) attenuation at 100 Hz (f_p);

- Minimum of 70 dB (A_{min}) attenuation at 1 kHz (f_s);

The filter chosen to be used is "Butterworth" because it has a maximally flat signal bandwidth with no ripple. The desired response of the filter can be viewed on figure 12.

The first parameter to be calculated for the filter project is its n order. The following equations for attenuation in dB are obtained from [9] and [10]. The variable ϵ is the maximum variation in passband transmission.

Fig. 11. Diagram connecting the DSP to the actuator input circuit.

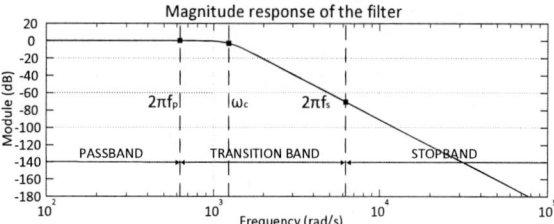

Fig. 12. Magnitude response of the designed filter

$$A_{max} = 20 \log \sqrt{1 + \epsilon^2} \therefore \epsilon = \sqrt{10^{A_{max}/10} - 1} \quad (8)$$

$$A_{min} = 10 \log[1 + \epsilon^2 (\omega_s/\omega_p)^{2n}] \quad (9)$$

From (8) and (9) the n order of the filter will be 5 and the cut-off frequency will be given by (10).

$$\omega_c = 2\pi f_p (1/\epsilon)^{1/n} \quad (10)$$

$$\omega_c = 1.2361 \ x \ 10^3 \ rad/s$$

To develop the filter, it's necessary to determine its transfer function. For 5-order filter, is set 5 poles of the function, and will be given according below:

$$p_1 = \omega_c(-\cos 72° + j \sin 72°)$$

$$p_2 = \omega_c(-\cos 36° + j \sin 36°)$$

$$p_3 = \omega_c(-\cos 0° + j \sin 0°)$$

$$p_4 = \omega_c(-\cos 36° - j \sin 36°)$$

$$p_5 = \omega_c(-\cos 72° - j \sin 72°)$$

From [9], the 5-order filter transfer function will be given by (11).

$$T(s) = \frac{\omega_c{}^5}{(s+\omega_c)(s^2+0.618\omega_c s+\omega_c{}^2)(s^2+1.618\omega_c s+\omega_c{}^2)} \quad (11)$$

The design was developed from [10] with 2 filters of second order in cascade with the topology of Sallen-Key, according to figure 13, and a filter of 1st order with the topology RC, according to figure 14, in cascade with the first ones.

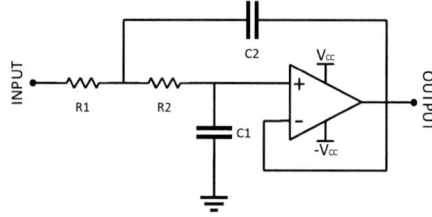

Fig. 13. Sallen-Key topology for 2nd order filters.

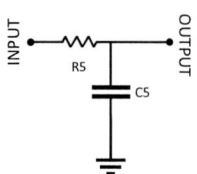

Fig. 14. RC topology for 1st order filters.

For the calculation of the parameters of the filters, it is necessary to calculate the quality factor of each filter through (12), for filter 1 (Q_1), and (13), for filter 2 (Q_2).

$$Q_1 = \frac{\sqrt{(-p_1)(-p_5)}}{(-p_1)+(-p_5)} = 1.6180 \tag{12}$$

$$Q_2 = \frac{\sqrt{(-p_2)(-p_3)}}{(-p_2)+(-p_3)} = 0.5257 \tag{13}$$

The components of the filter are obtained through the equations (14) and (15).

$$Q_n = \frac{1}{\omega_c(C_a(R_a+R_b))} \tag{14}$$

$$\omega_c = \frac{1}{\sqrt{C_aC_bR_aR_b}} \tag{15}$$

Then, for the first filter of 2nd order ($n = a \rightarrow 1 \text{ and } b \rightarrow 2$), choosing and setting the value of the resistors, the value of the capacitors C_1 and C_2 will be:

$$R_1 = 15\ k\Omega$$
$$R_2 = 39\ k\Omega$$
$$C_1 = 10\ nF$$
$$C_2 = 100\ nF$$

For the second filter of 2nd order order ($n \rightarrow 2, a \rightarrow 3 \text{ and } b \rightarrow 4$), using the same equations (14) and (15), but change the choosing and setting the value of the resistors, the value of the capacitors C_3 and C_4 will be:

$$R_3 = 5k6\ \Omega$$
$$R_4 = 10\ k\Omega$$
$$C_3 = 0,1\ \mu F$$
$$C_4 = 100\ nF$$

The first order filter can be obtained by choosing and setting the resistor value, and calculating the capacitor C_5 value in (16).

$$C_5 = \frac{1}{R_5 * \omega_c} \tag{16}$$
$$R_5 = 820\Omega$$
$$C_5 = 1\ \mu F$$

B. Amplifier

The amplifier adjusts the signal to the correct output operating range. The analog input of the actuator operates in the range of 0 - 10 V, since the DSP output signal operates in the range of 0 - 3.3 V, a gain of 3 in the amplifier circuit is required.

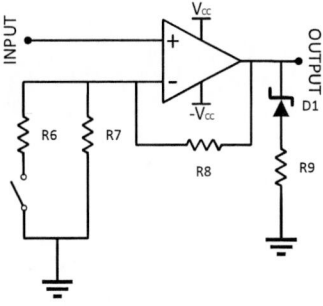

Fig. 15. Filter amplifier circuit developed.

The circuit was developed with an operational amplifier mounted in the non-inverting amplifier configuration. A 10 V Zener diode (D1) was also added, connected to the output to protect the actuator input from possible voltage rises above the permitted range. The circuit can be seen in figure 15.

The gain of the non-inverting amplifier configuration is given by (17).

$$Gain = 1 + \frac{R_8}{R_{eq}} \tag{17}$$

The equivalent resistance $R_{eq} = R_6//R_7$ when the jumper is closed, and $R_{eq} = R_7$ for the open jumper. This jumper setting allows you to vary the gain for different input types. The resistor R_9 was placed to polarize the Zener diode.

The values of the chosen resistors were:

$$R_6 = R_7 = R_8 = R_9 = 10\ k\Omega$$

For the configuration with the jumper closed, the gain of the amplifier circuit is 3, typical for the output of the 3.3 V DSP. For the configuration with the jumper open, the gain is 2, typical for the DSP outputs of 5 V.

VII. ESCALAR CONTROL

The control developed for the hardware test is the closed loop velocity scalar control [11] according to figure 16.

For this type of control, the speed of the motor axis n_{rotor} is compared with a reference velocity n_{ref} which, when passing through the PI controller and a limiter, generates the electric slip velocity $\omega_{2\ max}$. Slip velocity is added with the rotor speed feedback producing the electric angular frequency ω_1, which when divided by 2π results in the electric frequency f_1.

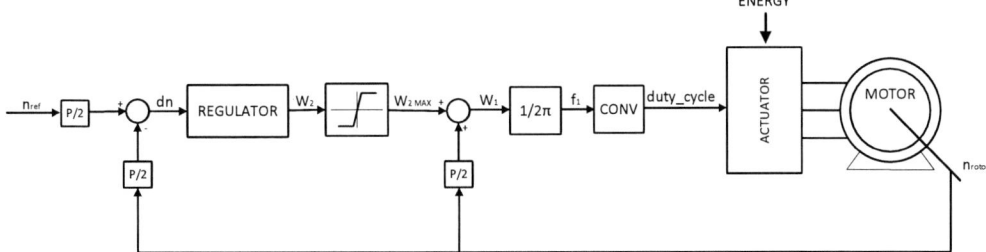

Fig. 16. Control diagram developed for system testing.

The latter passes through a converter, according to figure 17, to generate a duty cycle control signal at the output of the DSP.

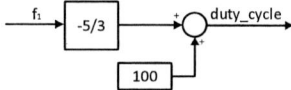

Fig. 17. Duty cycle frequency converter.

Using a machine with a moment of inertia $J_{eq} = 0.02\ kg.m^2$ and considering a torque constant $K_{torque} = 10.4\ \frac{N.m}{rad/s}$. The open-loop transfer function will be given by (18).

$$TF_{open-loop}(s) = \frac{520Ks + \frac{520K}{T_i}}{s^2} \qquad (18)$$

Adopting $T_i = 0.2\ s$ the closed loop transfer function will be given by (19).

$$TF_{closed\ loop}(s) = \frac{520Ks + 2600K}{s^2 + 520Ks + 2600K} \qquad (19)$$

For a percentage of overshoot $M_p = 25\%$ chosen for the project and the damping coefficient $\xi = 0.45$, the phase margin will be 45°. The poles of the closed loop of the system will be given by:

$$p_1 = -5 + j5$$
$$p_2 = -5 - j5$$

Through the obtained poles it is possible to obtain the denominator of the transfer function TF in (20).

$$FT_{den}(s) = s^2 + 10s + 50 \qquad (20)$$

Comparing the denominator of (19) with (20), the K gain of PI will be $K = 0.02$.

VIII. EXPERIMENTAL RESULTS

The experimental results response using the control hardware developed with the actuator on a WEG 6HP - 4 pole induction machine with a scalar control project can be observed in figure 18 and figure 19, which respectively show the response of a speed step and a torque perturbation.

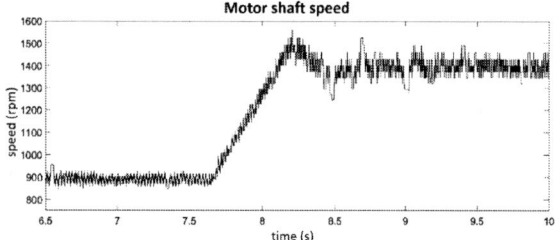

Fig. 18. Rotor speed for scalar control developed.

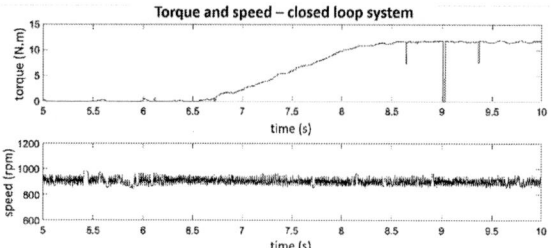

Fig. 19. Relation of the rotor speed with the applied torque for the scalar control.

As shown in figure 18, the scalar control simulation starts with the engine at a speed of 900 rpm, and at time t = 7.6 seconds a step is given in its reference to 1400 rpm. The engine stabilizes at the new speed after approximately 1 second, according to the design.

When applying a torque with a maximum value of approximately 11 N.m on the rotor shaft, it maintains with the initial speed of 900 rpm independent of the load applied on its axis.

IX. CONCLUSION

The system was proposed with the use of simple and low cost electronic components, reducing the final production value of the board. And with this hardware it was possible to develop the scalar control, as can be observed, the control acted in a satisfactory manner, reaching the desired reference speed, and subsequently maintained at a constant speed for a variation in the system load torque.

ACKNOWLEDGMENT

This study was financed in part by the Coordenação de Aperfeiçoamento de Pessoal de Nível Superior - Brasil (CAPES) - Finance Code 001.

REFERENCES

[1] TRZYNADLOWSKI, A. M., KAZMIERKOWSKI, M. P., GRABOWSKI, P. Z., BECH, M. M. "Three examples of DSP applications in advanced induction motor drives". In Proc. Amer. Control Conf., vol. 3, pp. 2139–2140, 1999.

[2] FRENCH, C. D., ACARNLEY, P. P. "Simulink real time controller implementation in a DSP based motor drive system". IEE Colloquium on DSP Chips in Real Time Measurement and Control , pp. 3/1-3/5, Leicester, UK, 1998.

[3] WEG. "Frequency Inverter Manual, Series: CFW-08, Version 0899.5242/09".

[4] TEXAS INSTRUMENTS. "TMS320F2833x, TMS320F2823x Digital Signal Controllers (DSCs)". Version SPRS439O June 2007, review April 2019.

[5] LEM. "Voltage Transducer LV 25-P datasheet".

[6] LEM. "Current Transducer LA 55-P datasheet".

[7] LEBOW. "Operating and Service Manual - Strain Gauge Indicator 7550 Series".

[8] S&E INSTRUMENTOS. "Manual de Instalação e Operação". Version 04, 2017.

[9] A. S. SEDRA AND K. C. SMITH. *Microelectronic Circuits*. Sixth Edition, New York, NY: Oxford University Press, 2013.

[10] B. RAZAVI. *Fundamentals of Microelectronics*. Second Edition, Wiley, 2008.

[11] R. STEPHAN. *Acionamento, Comando e Controle de Máquinas Elétricas*. Universidade Federal do Rio de Janeiro, Brasil, 2008.

On the Influence of Area Variations of the Photovoltaic Surface in Solar Cell Antennas

Eduardo Vicente Valdés Cambero
Engineering, Modeling and Applied
Social Sciences Center (CECS)
Universidade Federal do ABC
São Paulo, Brazil
eduardo.valdes@ufabc.edu.br

Humberto Pereira da Paz
Engineering, Modeling and Applied
Social Sciences Center (CECS)
Universidade Federal do ABC
São Paulo, Brazil
humberto.paz@ufabc.edu.br

Vinícius Santana da Silva
Engineering, Modeling and Applied
Social Sciences Center (CECS)
Universidade Federal do ABC
São Paulo, Brazil
vinicius.santana@ufabc.edu.br

Humberto Xavier de Araújo
Department of PPGMCS
Federal University of Tocantins
Palmas, Tocantins, Brazil
hxaraujo@uft.edu.br

Ivan Roberto Santana Casella
Engineering, Modeling and Applied
Social Sciences Center (CECS)
Universidade Federal do ABC
São Paulo, Brazil
ivan.casella@ufabc.edu.br

Carlos Eduardo Capovilla
Engineering, Modeling and Applied
Social Sciences Center (CECS)
Universidade Federal do ABC
São Paulo, Brazil
carlos.capovilla@ufabc.edu.br

Abstract—Some low power applications where antennas and solar cells are integrated into the same system have limitations and requirements imposed by the radiation characteristics that, usually, are fulfilled by varying solar cell physical parameters. The negative influences of those variations are not profoundly discussed in the literature that studies the extremely low-power systems, where radiofrequency and photovoltaic devices share the same structure. In this context, this work models a photovoltaic cell used as an antenna and analyzes the consequences of the solar cell area variation, imposed by the fabrication process of the radiating element. The parameters of the one-diode equivalent circuit model are extracted and used to simulated the I-V curves for different irradiance levels, validating the simulations with experimental measurements performed in the laboratory. Also, the solar cell fill factor, energy conversion efficiency, and current density are analyzed.

Index Terms—Solar cell modeling, solar cell and antenna integration, energy harvesting, low power devices.

I. Introduction

Renewable and nonpolluting energy sources can be seen as an essential factor to accelerate the development and sustainability of the industry and societies [1]. Nowadays, truly autonomous systems are the focus of attention in scientific research in Internet-of-Things solutions, wireless sensors networks (WSN) and medical treatment systems, among others [2], [3]. In literature, there are some integration examples of solar cells (SCs) and antennas into the same structure, for applications of extremely low-powers levels. SCs can be part of the antenna, either as a radiant element [4] or as the ground plane [5], achieving a high level of integration. For instance, a Vivaldi antenna was designed using two amorphous silicon SCs as radiant elements in [6]. In [3] and [7], an antenna array and a 3D antenna were designed, respectively, using

The authors would like to thank CAPES and CNPq (309848/2018-0).

SCs structures as the ground plane. Many of these integration solutions proposed are based on proofs-of-concept prototypes that involve degradation of the SC performance.

The operation frequency of the antennas is highly related to the dimensions of its radiating elements. Most of the time, the limitations imposed by the radiation requirements of the antennas and their figures-of-merits, including return losses, bandwidth, and resonance frequency, need the adjustment of the physical dimensions of the SCs. In this way, due to the area variations, the SCs may lose their optimum designed operating point, and the information provided by the manufacturer's data-sheets may no longer be valid in the applied context. Hence, it is necessary to extract the I-V curves of the SC with new characteristics, to assess the new SC performance. However, in literature can be found SCs antennas integration proposals like [4], [6], [8], [9], where the PV structure is adjusted to satisfy the radiation requirements and the SC resulting performance is not deeply analyzed. Therefore, a particular concern must be taken about suitability and feasibility of the integration process by comparing the resultant performance variations with the original SC.

In this context, here it is presented the design of a coplanar patch antenna using a SC as the radiating element. A commercial SC (CSC) was modified to achieve the radiation, and the I-V curves of the new SC were obtained in the laboratory under different irradiance levels. Then, three figureof-merits of SCs performance are calculated for each SC, evidencing the deterioration of the efficiency. The influence of the area variation imposed by the antenna fabrication process on the new SC performance was assessed by analyzing the parameters of the One-diode SC circuit model. This model can help to analyze and to predict the behavior of the electric components associated with the SC structure when there is no

978-1-7281-4181-7/19 $31.00 © 2019 IEEE

data available. Currently, several approaches can be found in the literature for SC modeling [10]–[12] however, an accurate methodology always implies the experimental measurement of the I-V curves and the extraction of circuit model parameters [13]. In that way, the I-V curves obtained in the laboratory were used as the starting point for the One-diode model parameters extraction. The approach used to obtain the circuit model parameters was validated by comparing the simulated I-V curves with those measured experimentally.

Besides this Introduction, Section II describes the area variation and SC performance deterioration. Section III presents the algorithm to extract the SC circuit model parameters. The SCA circuit model is described in Section IV. Simulated and measured I-V curves are compared, validating the extracted parameters. Also, the efficiency deterioration is explained through the differences between circuit model components. Finally, Section V exposes the conclusions of this work.

II. SC AREA REDUCTION

The CSC used to create the SCA patch radiator is the Black 21 NS6QL, commercialized by Neo Solar Power. This CSC has an area of 156.75x156.75 mm (see Fig. 1) and its conversion efficiency is 20.8% under Standard Test Conditions (STC). Its I-V curves are in [14] for STC, which establish the measurements are made with Air Mass coefficient AM=1.5, irradiance of 100 mW/cm^2 (1 sun), temperature of 300 K and wind speed of 0 m/s. Attending to the CSC I-V curves in [14], the short-circuit current (I_{sc}) and the open-circuit voltage (V_{oc}) are 9.65 A and 0.66 V, respectively. The wafer thickness is 200 μm with four 1 mm wide silver bus bars, separated by 39 mm from each other, and 1.5 mm of distance between the front grid fingers. An alkaline surface with dark blue silicon nitride anti-reflective coating is deposited over aluminum local back contact with 2.1 mm silver/aluminum pads. This CSC was used as a prove-of-concept in the integration process due to its easy access for our laboratory.

The SCA is a small rectangular piece, trimmed off from the CSC and its positive terminal (bottom plate) is used as a radiating element in a coplanar microstrip antenna (see Fig. 2). The SC used in the designed antenna (SCA) is a 32.5x22 mm

Fig. 1: CSC commercialized by Neo Solar Power.

Fig. 2: Picture of the built SCA.

piece, and all its measurements were realized with the SCA being already fixed on a substrate plate. The CSC is used as a reference for optimum operation performance, for analyzing the degradation on the SCA, which have the same material but different area.

The antenna design was carried out following the classical coplanar patch antenna transmission-line model [15]. The resonance frequency of the SCA is 2.45 GHz, that can be used for example in Wireless Local Area Networks (WLAN), and its bandwidth is approximately 300 MHz, taking as a reference the frequency band where more than 90% of the power is transmitted to the load. These magnitudes can be seen in the return losses curve shown in Fig. 3.

In order to compare each SC performance, both I-V curves need to be analyzed. Therefore, after designing the SCA, the I-V curves of the resulting SC shown in Fig. 4, were obtained experimentally using a Newport 96000 solar simulator. The irradiance level was controlled with the Optical Power/Energy Meter Newport 842-PE. The SCA was illuminated, and its I-

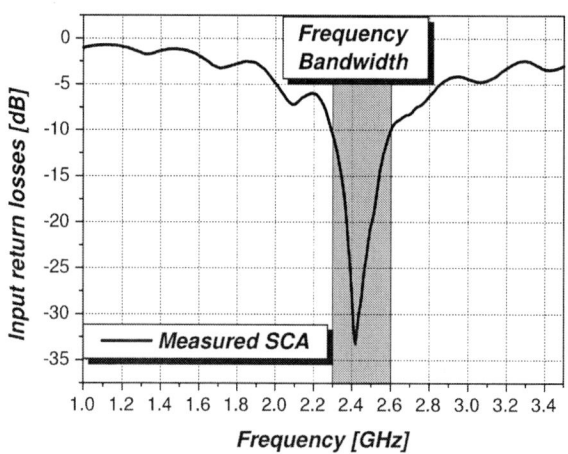

Fig. 3: Measured return losses of the SCA. The approximated frequency bandwidth for the operation frequency is highlighted.

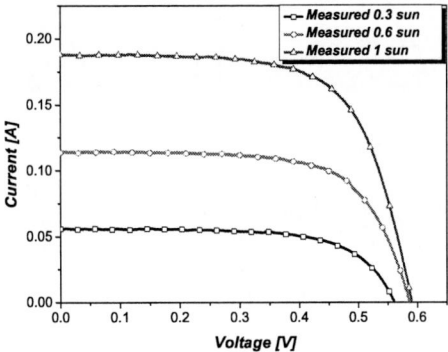

Fig. 4: Measured I-V curves of the SC-CPA for three different irradiance levels.

TABLE I: FF, η_{ec} and J results of both SC for STC.

Solar Cells	Operation conditions for 1 sun*					Analyzed Parameters		
	Area	I_{sc}	V_{oc}	MPP		η_{ec}	FF	J
	$[cm^2]$	[A]	[V]	I_{mp} [A]	V_{mp} [V]	[%]	[%]	$[mA/cm^2]$
CSC	245.70	9.65	0.66	9.16	0.55	20.5	79.1	37.3
SCA	7.15	0.188	0.59	0.167	0.45	10.5	67.8	23.4

*1 sun is equivalent to 100 mW/cm^2

V curves were recorded with the Precision Source/Measure Unit B2902A from Keysight. Since the approach is for indoor environments, the temperature effect over photo-generated current I_{ph} is not taken into consideration, and it can be considered as a constant for the settled ambient temperature (300 K) [11]. On the other hand, the I-V curves of the CSC were taken from the datasheet [14].

Three figure-of-merits of the CSC and the SCA were analyzed. The fill factor (FF), the energy conversion efficiency (η_{ec}) and the current density (J $[A/m^2]$) were calculated for 100 mW/cm^2 of irradiance. FF is defined in (1) as the ratio between maximum power delivered by a SC and the multiplication of the I_{sc} and V_{oc} [16]. The maximum power point (MPP) of a SC is defined by the maximum power current (I_{mp}) and the maximum power voltage (V_{mp}). Satisfactory FF values are expected to be between 60% and 85% [17]. This parameter is directly influenced by the series and shunt resistances (R_s and R_{sh}, respectively) helping to quantify the quality of a SC. By reducing the area, the optimized configuration of grid contacts, number of fingers and bus bars, and their spacing is completely changed. Then, new R_s and R_{sh} values will show up acting directly on the SC performance.

$$FF = \frac{I_{mp}V_{mp}}{I_{sc}V_{oc}} \tag{1}$$

The η_{ec} is a well-known parameter to evaluate energy sources performance and, in the case of a SC, it is defined by the ratio between the maximum power delivered and the total radiation power received by the PV surface [16]:

$$\eta_{ec} = \frac{I_{mp}V_{mp}}{GA_{pvs}} \tag{2}$$

where: G is the irradiance at operation condition, and A_{pvs} is the area of a SC exposed to the sunlight. The J is implicit in (2), being another magnitude affected by the area reduction.

Aiming to analyze how an area modification can decrease the quality of a SC performance, the CSC and the SCA are compared according to the FF, η_{ec} and the J. Taking as a reference the STC, the values of V_{oc}, I_{sc}, V_{mp}, I_{mp} for both

SCs and their areas are used to calculate each comparison parameter. The results can be found in Table I, where it can be seen that efficiency and FF decrease 10% and 11.3% respectively, with a drop of 37.2% of the initial J. These reductions occur due to the increase of the bulk, contact and conductor resistances, being all of them, components of the series resistance which must be kept as small as possible. The following Section presents the process of extracting the parameters that characterize the circuit model of the SCs. In this way, the causes of the SC performance degradation due to the active area reduction are demonstrated.

III. SCA PARAMETERS EXTRACTION

The circuitry analysis of SCs requires to know the behavior of the available output power under different operation conditions and, an accurate electrical model is an important element to be developed. Electrical modeling methods of SCs (one-diode or two-diode models) are often described in the literature involving analytical or iterative solutions, including the relatively recent development of neural networks for this propose [10]–[12]. These methods are based on the solution of a partially linear equivalent circuit, like presented in Fig. 5, which uses one-diode configuration to represent the I-V curves of a SC. For the circuit presented in Fig. 5, I_d is the diode current, I_s is the output current, R_{sh} is the shunt resistance, R_s is the series resistance, and V_o the output voltage.

The method used here to evaluate the equivalent circuit model is based on the Gauss-Seidel algorithm, described in [10], which is an iterative approach based on the V_{oc}, I_{sc}, and the MPP values of the SC, that can be obtained through manufacturer datasheet or experimental results for a given level of irradiance. The goal of the method is to obtain the magnitudes of R_s, R_{sh}, ideality factor (A), saturation current I_{sat} and I_{ph} associated to a specific G condition, making

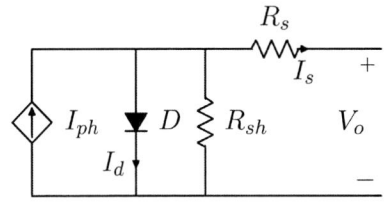

Fig. 5: Equivalent one-diode SC circuit.

978-1-7281-4181-7/19 $31.00 © 2019 IEEE

possible the one-diode SC circuit model determination.

Using the One-diode model, (3) and (4) were expanded for the three circuit operation point conditions above mentioned, to obtain a system of linear equations and determining the solution for I_s and V_o.

$$I_d = I_{sat}\left[e^{\frac{V_o}{AV_tN_s}} - 1\right] \tag{3}$$

$$I_s = I_{ph} - I_{sat}\left[e^{\frac{V_o+I_sR_s}{AV_tN_s}} - 1\right] - \frac{V_o + I_sR_s}{R_{sh}} \tag{4}$$

To derive the magnitudes of R_s, R_{sh}, A, I_{sat} and I_{ph}, were used (5) to (9), where V_t is the thermal voltage and N_s is the number of SCs. Furthermore, the manipulations $AV_t = X$ and $R_s + R_{sh} = R_t$ are done to clarify the equations comprehension. The N_s is equal to 1 due to the quantity of SC elements interconnected for the analysis proposed in this work.

$$R_s = \frac{V_{oc} - V_{mp} + X\ln\left[\frac{XI_{mp}R_t - XV_{mp}}{(I_{sc}R_t - V_{oc})(V_{mp} - I_{mp}R_s)}\right]}{I_{mp}} \tag{5}$$

$$R_{sh} = \frac{XR_t + R_s[I_{sc}R_t - V_{oc}]e^{\frac{I_{sc}R_s - V_{oc}}{X}}}{X + [I_{sc}(R_{sh} + R_s) - V_{oc}]e^{\frac{I_{sc}R_s - V_{oc}}{X}}} \tag{6}$$

$$A = \frac{1}{V_t}\frac{V_{mp} + I_{mp}R_s - V_{oc}}{\ln\left[\frac{I_{sc}R_t - I_{mp}R_t - V_{mp}}{I_{sc}R_t - V_{oc}}\right]} \tag{7}$$

$$I_{sat} = \frac{I_{sc}R_{sh} + I_{sc}R_s - V_{oc}}{R_{sh}e^{\frac{V_{oc}}{X}}} \tag{8}$$

$$I_{ph} = I_{sat}\left[e^{\frac{V_{oc}}{X}} - 1\right] + \frac{V_{oc}}{R_{sh}} \tag{9}$$

One requirement of this method is the necessity of an initial guess value for A, R_s and R_{sh}, so the following considerations are taken in order to find the solution. An ideal SC would be characterized by R_s as a short circuit and R_{sh} as a open circuit, it can be consider $R_s \ll R_{sh}$. This literal assumption may be demonstrated valid for high-efficiency SCs, but for low-efficiency, SCs the relation $R_s \ll R_{sh}$ may not be valid anymore and the algorithm can diverge. The solution proposed for this problem is to develop an approximation of (5), considering $R_{sh} \approx \infty$, and another for (6), setting $R_s \approx 0$. In this way, (10) and (11) are obtained and used as the initial guesses for the algorithm, and the initial value for A can also be set using (7).

$$R_s = \frac{V_{oc} - V_{mp} + V_t\ln\left[\frac{V_tI_{mp}}{I_{sc}V_{mp}}\right]}{I_{mp}} \tag{10}$$

$$Rsh = \frac{V_t}{I_{sc}e^{\frac{-V_{oc}}{V_t}}} \tag{11}$$

In this case, the flowchart of the parameters extraction algorithm is presented in Fig. 6 where Y represents the parameter value after each iteration, W is the weight used

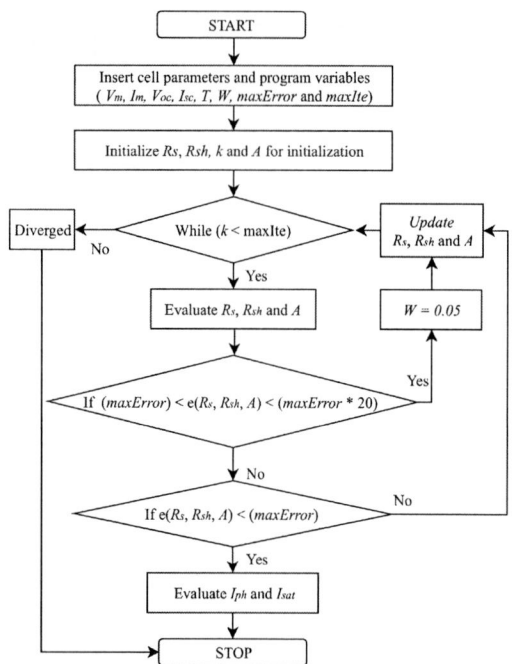

Fig. 6: Modified initial Gauss-Seidel algorithm.

in the algorithm, $maxError$ is the maximum acceptable error, k the iterations counter, $maxIt$ the maximum number of iterations and e is the error of a parameter. At each cycle, R_s, R_{sh} and A are evaluated through (12) for each variable, then e is estimated individually using the previous and the new evaluated values. The A value of the diode is limited by $1 < A < 2$. Furthermore, it is important to evidence the W adjustment during the process, which increases the convergence rate by decreasing the error estimation of the parameters. An initial guess for W is generally adopted from 0.1 to 0.3 [18], and in this case for the W is adjusted to be 0.05.

$$Y_{(k+1)} = [1 - W]Y_{(k-1)} - WY_{(k)} \tag{12}$$

Despite the satisfactory convergence of the algorithm, this method does not necessarily set the best initial guess, so the minimum number of iterations may not always be obtained, although the computational cost is kept low.

Obtaining the operation points used by the algorithm, three G values (100, 60 and 30 mW/cm^2) were generated with the solar simulator. After the experimental measurements under the three irradiance levels, three groups of V_{oc}, I_{sc}, and MPPs values are available to run into the algorithm. Then, by running the implemented code, it is possible to obtain three sets of values (one for each irradiance level) to fulfill the equivalent model proposed in this analysis and to generate the correspondent I-V curves. The resulting values of the circuit parameters given by the algorithm are grouped in the Table II. As can be seen, every initial condition is strictly related to the irradiance level, generating different values of

978-1-7281-4181-7/19 $31.00 © 2019 IEEE

TABLE II: SCA extracted parameters using the modified Gauss-Seidel algorithm.

Irradiance values [mW/cm^2]		Operation conditions				Extracted Parameters				
		I_{sc} [mA]	V_{oc} [V]	MPP		R_s [Ω]	R_{sh} [Ω]	A	I_{sat} [nA]	I_{ph} [mA]
				I_{mp} [mA]	V_{mp} [V]					
100	CSC	9650	0.66	9162	0.55	0.0029	3758210	1.09	0.56	9650
	SCA	188	0.59	167	0.45	0.22	122	1.72	283	188.6
60		114	0.59	100	0.46	0.16	128	1.8	328	114
30		56	0.56	49.6	0.43	0.59	349	1.66	106.6	56.1

each parameter for each irradiance level available. Once the conditions created by the incident energy density are defined, the algorithm converges to the same result.

IV. SCA MODELING

The one-diode equivalent circuit model is implemented in ADS software (Fig. 7) using the parameters extracted by the approach detailed in the previous section. Variable resistance is used as a load to generate the sweep voltage to bias the SC of the SCA. The A and the I_{sat} of the ADS diode model were modified, setting the values obtained by the parameter extraction algorithm.

Fig. 8 shows the measured and simulated I-V curves for 1, 0.6, and 0.3 sun, which are presenting a high similarity. The principal differences are located in the "knee region" around the MPP, where the R_s profoundly influences the I-V curve. The initial guessed R_s value used in [10] was chosen depending on the technology of the SC under test. Here is not possible to use that assumption due to the CSC characteristics were changed.

Showing also a good agreement between empirical and simulation processes, the Fig. 9 exhibits the measured and modeled I-V and power curves for 100 mW/cm^2. Having extracted the SC parameters of the SCA and modeled in a good concordance with the experimental results, it was used the same procedure to obtain the circuit model associated with the CSC. The circuit model CSC parameters are also shown

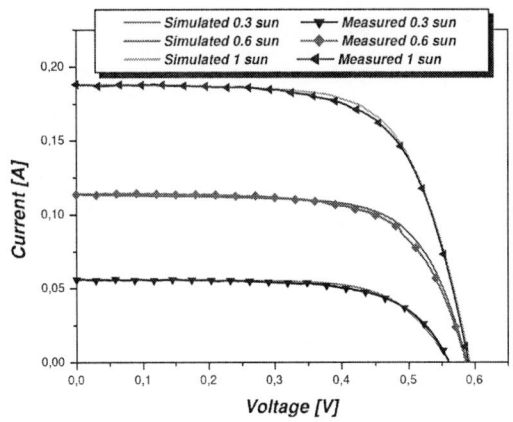

Fig. 8: Simulated and measured I-V curves of the SCA for three different irradiance levels.

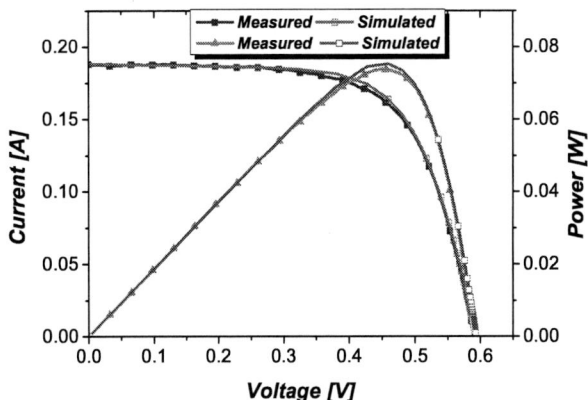

Fig. 9: Simulated and measured I-V and power curves of the SCA for STC.

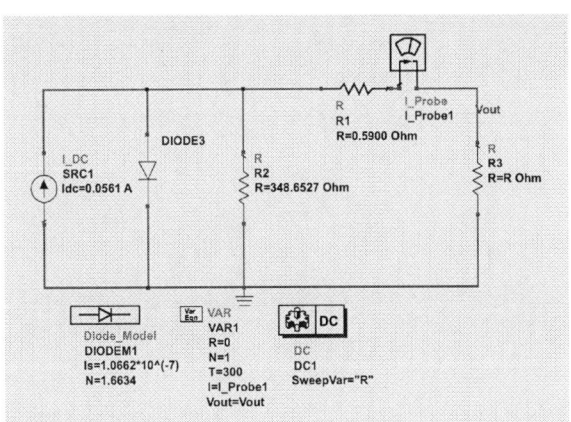

Fig. 7: One-diode SCA circuit model implemented in ADS.

in Table II. Fig. 10 shows the simulated I-V curve for STC of the CSC, which are also in concordance with the curve from the data-sheet [14]. After comparing the simulated and measured curves and observing the satisfactory concordance, the extraction method was validated. Therefore, it was possible to compare the values of parasitic resistances from both SCs and to demonstrate the decreasing causes in SC performance. In this way, the R_s and R_{sh} values of both cells were compared.

The series resistance value of the SCA (see Table II) is 100 times bigger than the R_s of the CSC, which is not desirable. On the other hand, R_{sh} is 1000 times smaller explaining the degradation on the FF, η_{ec} and the J. To improve the SC efficiency and FF it is necessary to keep R_s as small as possible, and R_{sh} must be kept as a high impedance.

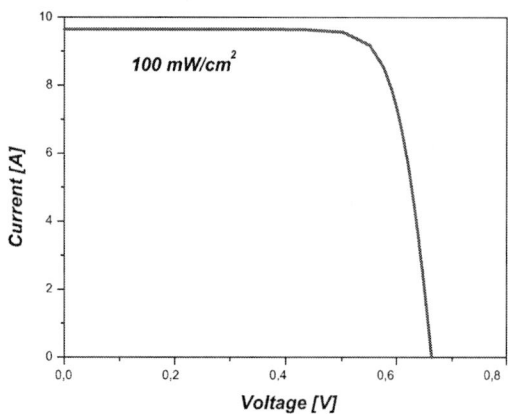

Fig. 10: Simulated I-V curves of the CSC.

V. CONCLUSION

A piece of SC trimmed of a CSC can be used as a radiating element in antennas for application with extremely low-power requirements, where the cost-benefits relation between efficiency and area are in agreement. As a result, a 238.55 cm^2 reduction in the SC area brought a 10% reduction in both efficiency and fill factor, as well as a current density dropping of 37.2%. Variations on the area of an optimized CSC bring as a consequence a considerable decrease of η_{ec} and FF that depending on the operation conditions could turn unfeasible the use of the SC component. The SCA equivalent circuit parameters were extracted, and the I-V curves were simulated, obtaining a high similarity between simulation and experimental measurements. The leading cause for this results is the degradation of the equivalent R_s and R_{sh} of the SC.

ACKNOWLEDGMENT

Also, the authors appreciate the supervision of the Prof. André S. Polo of the Universidade Federal do ABC on the I-V curves measurements process.

REFERENCES

[1] I. R. Casella, C. Capovilla, A. Sguarezi Filho, R. Jacomini, J. Azcue-Puma, and E. Ruppert, "An anfis power control for wind energy generation in smart grid scenario using wireless coded ofdm-16-qam," *Journal of Control, Automation and Electrical Systems*, vol. 25, no. 1, pp. 22–31, 2014.

[2] K. Niotaki, A. Collado, A. Georgiadis, S. Kim, and M. M. Tentzeris, "Solar/electromagnetic energy harvesting and wireless power transmission," *Proceedings of the IEEE*, vol. 102, no. 11, pp. 1712–1722, Nov 2014.

[3] Y. Tawk, J. Costantine, F. Ayoub, and C. G. Christodoulou, "A communicating antenna array with a dual-energy harvesting functionality [wireless corner]," *IEEE Antennas and Propagation Magazine*, vol. 60, no. 2, pp. 132–144, April 2018.

[4] J. Bito, R. Bahr, J. G. Hester, S. A. Nauroze, A. Georgiadis, and M. M. Tentzeris, "A novel solar and electromagnetic energy harvesting system with a 3-d printed package for energy efficient internet-of-things wireless sensors," *IEEE Transactions on Microwave Theory and Techniques*, vol. 65, no. 5, pp. 1831–1842, 2017.

[5] T. Yekan and R. Baktur, "Conformal integrated solar panel antennas: Two effective integration methods of antennas with solar cells." *IEEE Antennas and Propagation Magazine*, vol. 59, no. 2, pp. 69–78, Apr 2017.

[6] O. O'Conchubhair, K. Yang, P. McEvoy, and M. J. Ammann, "Amorphous silicon solar vivaldi antenna," *IEEE Antennas and Wireless Propagation Letters*, vol. 15, pp. 893–896, 2016.

[7] A. Rashidian, L. Shafai, and C. Shafai, "Miniaturized transparent metallodielectric resonator antennas integrated with amorphous silicon solar cells," *IEEE Transactions on Antennas and Propagation*, vol. 65, no. 5, pp. 2265–2275, May 2017.

[8] M. Elsdon, O. Yurduseven, and X. Dai, "Wideband metamaterial solar cell antenna for 5 GHz Wi-Fi communication," *Progress In Electromagnetics Research C*, vol. 71, pp. 123 – 131, 2017.

[9] F. Nashad, S. Foti, D. Smith, M. Elsdon, and O. Yurduseven, "Development of transparent patch antenna element integrated with solar cells for ku-band satellite applications," in *2016 Loughborough Antennas Propagation Conference (LAPC)*, Nov 2016, pp. 1–5.

[10] K. Et-torabi, I. Nassar-eddine, A. Obbadi, Y. Errami, R. Rmaily, S. Sahnoun, A. E. fajri, and M. Agunaou, "Parameters estimation of the single and double diode photovoltaic models using a gauss-seidel algorithm and analytical method: A comparative study," *Energy Conversion and Management*, vol. 148, pp. 1041 – 1054, 2017.

[11] A. F. Jaimes and F. R. de Sousa, "Simple modeling of photovoltaic solar cells for indoor harvesting applications," *Solar Energy*, vol. 157, pp. 792 – 802, 2017.

[12] M. Hadjab, s. Berrah, and A. Hamza, "Neural network for modeling solar panel," *INTERNATIONAL JOURNAL OF ENERGY*, vol. 6, pp. 9–16, Feb. 2012.

[13] M. G. Villalva, J. R. Gazoli, and E. R. Filho, "Comprehensive approach to modeling and simulation of photovoltaic arrays," *IEEE Transactions on Power Electronics*, vol. 24, no. 5, pp. 1198–1208, May 2009.

[14] N. S. Power, "Ns6ql black 21," https://www.enfsolar.com/pv/cell-datasheet/2039.

[15] C. A. Balanis, *Antenna theory: analysis and design*. Wiley-Interscience, 2005.

[16] P. Coppa, A. Vian, C. Vieira Tahan, E. Robba, M. Gouvea, M. Fernandes Gemignani, and A. Moretti, *A energia solar: tecnologia e regulação*. Oficio das Palavras, 2014.

[17] B. Ray and M. A. Alam, "Achieving fill factor above 80% in organic solar cells by charged interface," *IEEE Journal of Photovoltaics*, vol. 3, no. 1, pp. 310–317, Jan 2013.

[18] K. Mehrotra, C. Mohan, and S. Preface, "Elements of artificial neural nets," Jan 1997.

New semiconductor technologies for power electronics

Alisson Mengatto
Itaipu Binacional
Foz do Iguaçu, Paraná, Brazil
alisson_mengatto@hotmail.com

José Adriano Damacena Diesel
Electrical Engineering Department
Santa Catarina State University
Joinville, Santa Catarina, Brazil
jose.diesel@gmail.com

Pedro Henrique Thiesen de França
Electrical Engineering Department
Santa Catarina State University
Joinville, Santa Catarina, Brazil
pedro-franca@live.com

Joselito Anastácio Heerdt
Electrical Engineering Department
Santa Catarina State University
Joinville, Santa Catarina, Brazil
joselito.heerdt@udesc.br

Abstract—With the silicon power devices limits being reached, there is a need for new semiconductor technologies with higher efficiency, power density, voltage, current and switching speed. The materials that suit these requirements are called wide band semiconductors, featuring silicon carbide, gallium nitride, gallium arsenide and diamond. This paper presents the main characteristics of work of the gallium nitride transistor, with a brief comparison between the technologies mentioned above. Also, an experimental setup based on a double pulse tester is used to evaluate and compare the conduction and switching losses in different conditions.

Keywords—gallium nitride, silicon carbide, energy wide bandgap, switching.

I. INTRODUCTION

Gallium nitride (GaN), silicon carbide (SiC), gallium arsenide (GaAs) and diamond semiconductors are called wide bandgap semiconductors, also known as High Electron Mobility Transistor (HEMT), with the first two being possible candidates as silicon transistor substitutes. These devices have lower concentrations of intrinsic carriers, higher thermic conductivity, electrons saturation velocity and breakdown electric field than silicon [1][2]. Those properties are shown in Table 1.

Table 1: Wide bandgap semiconductors main characteristics

Properties	Si	GaAs	SiC 4H	GaN	Diamond
Bandgap, Eg (eV)	1.12	1.42	3.26	3.44	5.45
Electron saturation velocity, Vsat (x10^7 cm/s)	1.0	1.2	2.0	2.5	2.7
Electron mobility,μ_n (x10^3 cm^2/V.s)	1.35	8.5	1.9	2.00	3.80
Thermal conductivity, λ (W/cm. K)	1.5	0.46	4.9	1.3	22
Breakdown electric field, EC (MV/cm)	0.3	0.4	2.0	3.8	10
Relative dielectric constant, ε_r (x10^1)	1.18	1.31	1.00	0.95	0.55

Also, comparisons have been made between GaN devices and other types of technologies, showing superior performance [3][4]. A comparison between the three main semiconductor technologies is showed and Fig. 1. GaN transistors have inferior performance only in thermal conductivity, although have higher electronic mobility, electrons saturation velocity, bandgap energy, breakdown electric field and lower dielectric constant, being extremely promising within power electronics applications.

The electric properties shown in Table 1 can be defined as:

Breakdown electric field (E$_C$): Semiconductor devices with higher breakdown electric field value can block higher voltages, according with equation (1), that relates the breakdown electric field and block voltage for a diode.

$$V_B \approx \frac{\varepsilon . E_c^2}{2.q.N_D} \tag{1}$$

Where:

ε : dielectric constant
V$_B$: breakdown voltage (V)
E$_C$: breakdown electric voltage (MV/cm)
q: electron charge (1.6x10^{-19} C)
N$_D$: semiconductor material doping density(cm^{-3})

The blocking voltage can also be reduced increasing the doping density (ND), therefore decreasing the thickness of the drift region (WD), as a result, it decreases conduction resistance, in accordance with equations (1), (2) and (3). This drift region reduction reduces the parasitic capacitance, allowing that the semiconductor operates with higher switching speed.

$$W_D \approx \frac{2.V_B}{E_C} \tag{2}$$

Electron mobility (μn): semiconductor devices with higher electron mobility have less conduction resistance, in accordance with equation (3).

$$R_{on\_sp} = \frac{W_D}{q.\mu_n.N_D} \tag{3}$$

Where:

978-1-7281-4181-7/19 $31.00 © 2019 IEEE

R_{on_sp}: specific on-resistance ($\Omega.cm^2$)

W_D: drift region width

q: electron charge ($1.6x10^{-19}$ C)

μ_n: electron mobility ($cm^2/V.s$)

N_D: semiconductor material doping density (cm^{-3})

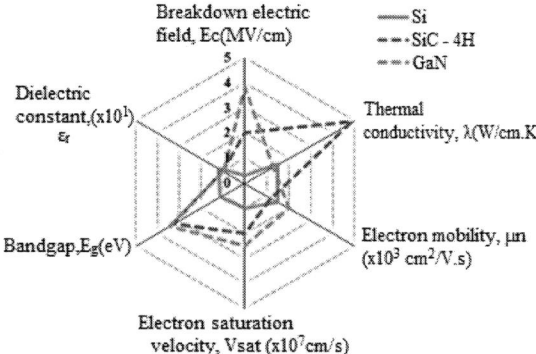

Figure 1: Comparison between semiconductor technologies for power electronics applications

Thermal conductivity (λ): The higher thermal conductivity is, the better will be the heat conduction of the semiconductor device to the environment, being able to work with higher temperatures, reducing the converter overall size by reducing the size of the heat sinks. However, a lower thermal conductivity doesn't forbid its operations in higher temperatures. A higher energy bandgap is what makes higher temperature operations possible, because it makes less likely that an electron passes through this bandgap with the increasing of the temperature, becoming more stable [5][6].

Dielectric constant (relative permeability – ε_r): with lower relative permeability, it's possible to build semiconductor devices with lower junction capacitances, in accordance to equation (4). With these lower capacitances, the device is able to switch faster.

$$C = \varepsilon_r \frac{\varepsilon_o.A}{d} \qquad (4)$$

Where:

C: capacitance (F)

ε_o: vacuum permittivity ($8.85x10^{-12}$ F/m)
a: area (m^2)

d: distance between plates (m)

ε_r: relative dielectric constant

Electrons saturation velocity (V_{sat}): The higher the saturation velocity, the faster the switching. A high electrons saturation velocity allows a faster removal of the depletion region charges, decreasing reverse recovery current and time, thus decreasing the switching losses and voltage oscillations across the switches, besides reducing the interaction with parasitic elements of the circuit, thus reducing electromagnetic interferences.

Energy bandgap (E_g): Devices with higher energy bandgap have less current leakage in high temperatures. The disadvantage is that diodes with high energy bandgap have higher voltage drop, in accordance equation (5).

$$V_d \approx \frac{E_g}{q} - \frac{k_B T}{q}.\ln\left(\frac{N_c.N_v}{N_a.N_d}\right) \qquad (5)$$

Where:

V_d: diode voltage drop (V)

E_g: energy bandgap (eV)

q: electron charge ($1.6x10^{-19}$ C)

k_B: Boltzmann constant ($1.38x10^{-23}$ J/K)

T: temperature (Kelvin)

N_c: effective electrons concentration (cm^{-3})

N_v: effective holes concentration (cm^{-3})

N_a: acceptor impurities concentration (cm^{-3})

N_d: donor impurities concentration (cm^{-3})

Using equation (1), the data of Table 1, and if the same doping density is being used for the semiconductors, it's possible to compare the blocking voltage of the normalized semiconductors in relation to silicon. The GaN semiconductor has a blocking voltage capability 129 times higher than silicon, and 3.43 times higher than SiC. Fig. 2 compares blocking voltages between technologies, while Fig. 3 shows a comparison between specific on-resistance and blocking voltage of different technologies [7].

One of the most important figures of merit for power semiconductors is the relationship between blocking voltage capability and specific on-resistance. A good power semiconductor must have high blocking voltage and low on-resistance [7]. By algebraic manipulation of equations (1), (2) and (3), the equation (6) is reached, relating both blocking voltage and specific on-resistance.

$$R_{on\_sp} = \frac{4.V_B^2}{\varepsilon.E_C^3.\mu_n} \qquad (6)$$

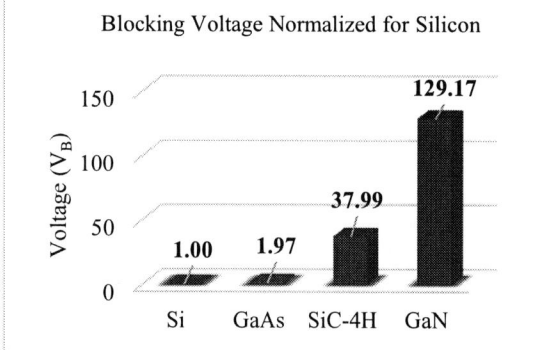

Figure 2: Blocking voltage normalized for silicon

978-1-7281-4181-7/19 $31.00 © 2019 IEEE

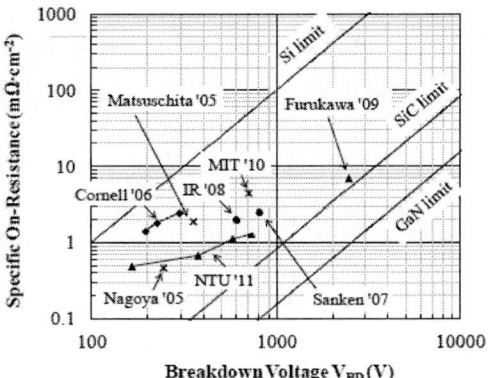

Figure 3: Comparison between specific on-resistance and blocking voltage [5].

II. GAN HEMT TRANSPHORM – TPH3205WS

This device is the first 600V GaN transistor using TO-247 encapsulation. It was designed for power converters that have continuous current on its inductors, such as inverters, totem-pole bridgeless rectifiers and power factor correction circuits [8]. This GaN transistor is formed by two transistors in a cascode configuration, with one being a low voltage normally-off silicon MOSFET and the other being a high voltage normally-on GaN HEMT transistor, as shown in Fig. 4. Its electrical characteristics are presented in Table 2 [9].

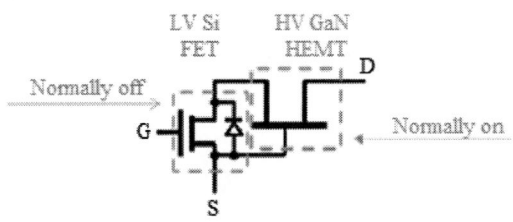

Figure 4: Thansphorm GaN transistor cascode configuration

Table 2: GaN Transphorm TPH3205WS electrical characteristics [9]

Symbol	Parameter	Value
$I_{D\,25°C}$	Continuous drain current @25°C	36 A
V_{DSS}	Blocking voltage (drain-source)	600 V
$R_{DS(ON)}$	Conduction resistance (Tj=25 °C)	52 mΩ
Vgs (th)	Conduction threshold voltage	2.1 V
Q_G	Total gate charge	28 nC
t_r	Rising time	7.5 ns
t_f	Falling time	4.5 ns
V_{SD}	Reverse voltage (Vgs =0 V, Is =12 A, Tj=25°C)	1.6 V
	(Vgs =0 V, Is =24 A, Tj=25°C)	2.2 V
t_{rr}	Reverse recovery time	30 ns
Q_{rr}	Reverse recovery charge	136 nC

The gate terminal of the GaN transistor is formed by the low voltage MOSFET gate (responsible for the conduction entrance and blocking of the GaN switch, which is a normally-on semiconductor), the source terminal is formed by the MOSFET source and by the GaN HEMT gate terminal, and finally the drain terminal being formed by the GaN HEMT drain. Internally, the GaN HEMT source is connected to the silicon MOSFET drain, although this connection isn't available.

The body diode is a p-n junction intrinsic to the MOSFET, that can conduct in the reverse way when the switch is blocked, acting as a freewheeling diode. Specially on high voltage MOSFETs, the anti-parallel diode stores a big quantity of minority charge in direct conduction mode, generating reverse recovery losses with high reverse recovery currents and long reverse recovery times. The Thansphorm GaN switch is composed by a device of majority charge and a low voltage MOSFET of minority charge, in a way that only a small quantity of minority charge is stored in this diode when it is in direct conduction, thus generating fewer losses and shorter reverse recovery times.

Fig. 5 shows a comparison between CoolMOS and GaN reverse recovery, showing that GaN transistors is superior in both reverse recovery time and current [10].

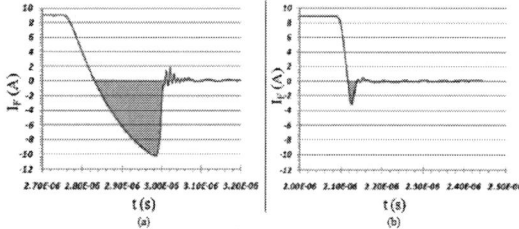

Figure 5: Reverse recovery comparison between CoolMOS and GaN [10]

III. PTH3205WS MODES OF OPERATION

The current conduction modes are shown in Fig. 6.

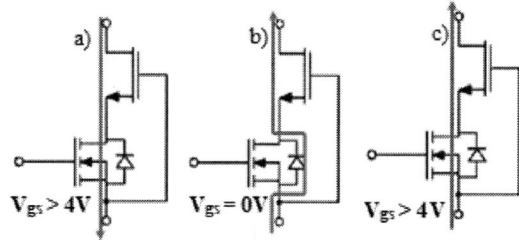

Figure 6: GaN HEMT cascode modes of operation

DIRECT CONDUCTION ($V_{GS} > V_T$ and $V_D, I_D > 0$)

It occurs when the MOSFET is turned on, with the voltage drop being determined by equation (7). In this mode of operation, the current flows through the drain-source terminals of the GaN HEMT and Silicon MOSFET.

$$V_{DS} = I_D(R_{DS(on),Si} + R_{DS(on),GaN}) \tag{7}$$

REVERSE CONDUCTION ($V_{DS} < 0$ and $V_{GS} = 0$)

It occurs when null voltage is applied, turning the Silicon MOSFET off, although it has freewheeling current flowing through the MOSFET's diode and the normally on GaN HEMT, thus applying a negative voltage at the GaN switch, composed by the voltage drops at the silicon diode and the

978-1-7281-4181-7/19 $31.00 © 2019 IEEE

GaN HEMT drain-source resistance, determined by equation (8).

$$V_{DS} = -(V_{D-Si} + I_F R_{DS(on)-GaN})$$ (8)

REVERSE CONDUCTION ($V_{DS} < 0$ and $V_{GS} > V_T$)

This mode of operation can be used to decrease reverse recovery losses, because reverse recovery losses at the MOSFET's diode is higher than reverse conduction losses at the MOSFET's drain-source resistance. In this case, the voltage drop across the device is determined by equation (9).

$$V_{DS} = -I_F (R_{DS(on)-Si} + R_{DS(on)-GaN})$$ (9)

In [11], it was obtained a voltage drop of approximately 0.8 V using the drain-source channel in reverse conduction instead of using the freewheeling diode.

IV. PCB LAYOUT AND MEASUREMENT PROBES FOR GaN TRANSISTORS

Although GaN switches offer significant advantages compared with silicon technologies, especially in relation with commutation speed, to operate in high commutation speed, great care must be taken during PCB layout and component placement in order to reduce trace inductances [12]. Parasitic elements form high frequency resonant circuits that can be excited by current or voltage transients and must be minimized by the layout designer.

The following considerations should be taken when design a circuit board for GaN switches:

- Minimize the inductances, leaving the components close one to another.
- Using ground plane.
- Detach the power circuit and ground circuit.
- Welding the gate driver integrated circuit ground pin close to the switch's source pin.
- Placing the gate driver pin close to the switch's gate.
- Using a decoupling capacitor at the gate driver feeding circuit.
- Avoid using wires at the current loop.
- Using a minimal number of probes for measurement.

V. PROPOSED DRIVER

To drive the GaN transistor, it was used a driver circuit proposed in [13], with a gate resistor of 300 Ohms associated in series with a 2200 Ω at 100MHz ferrite, that behaves like a resistor in high-frequencies, being used to minimize gate noises and avoid undesirable switching.

For the tests, it was used the circuit in Fig.7 with a freewheeling SiC diode, model C4D2012D from CREE (fast diode with low reverse recovery charge), being replaced afterwards for another GaN transistor to test its internal diode characteristics.

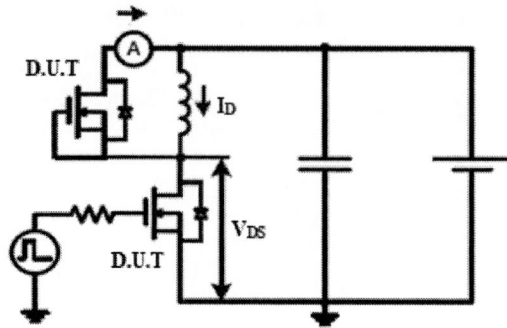

Figure 7: GaN transistor test circuit

Switching losses analysis were carried out by commanding the circuit with a pulse train in steps of 5A, recording the characteristics of turn-on, turn-off and at same current level. These tests were made for a junction temperature 100° C, with a switching frequency of 66 KHz.

The voltage measurements where made using a probe with a very short ground pin, to mitigate overvoltage caused by parasitic inductances. Fig. 8 shows wave forms of the switch's drain-source voltage and current for rated conditions (feeding voltage of 400V and a step current of 5A until it reaches 25A).

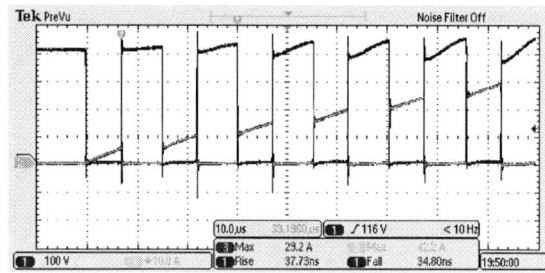

Figure 8: TPH3205 switching waveforms

The enlarged image of the switching entrance with a 25A current can be seen at Fig. 9 and blocking for the same level at Fig. 10.

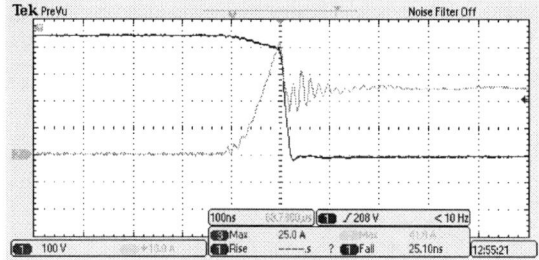

Figure 9: Conduction entrance for a 25A current

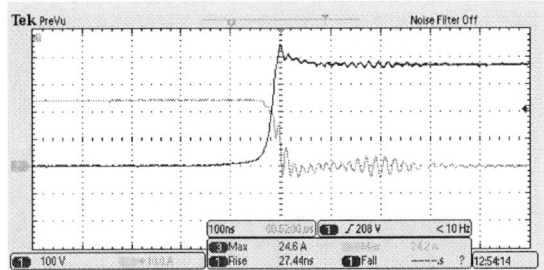

Figure 10: Conduction blocking for a 25A current

The switch conduction resistance measured was 55 and 89.65 mΩ for 25° and 100° C, respectively.

The GaN intrinsic diode reverse recovery current in shown at Fig. 11, in a different manner from other measurements used a shunt resistor with a gain of 0.79, so for the correct aquisition of the current, its measurement must be multiplied by it.

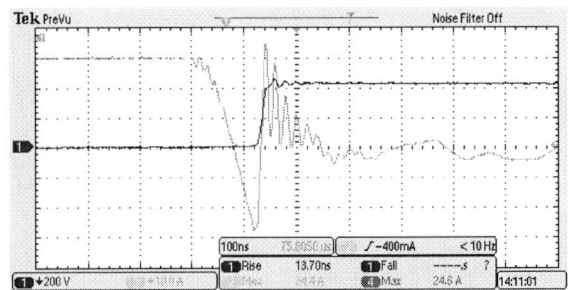

Figure 11: GaN intrinsic diode reverse recovery

Fig. 12 shows switch's intrinsic diode reverse recovery energy for different current levels. The calculated reverse recovery energy is the area between voltage and current when the current is negative, as shown in Fig. 11.

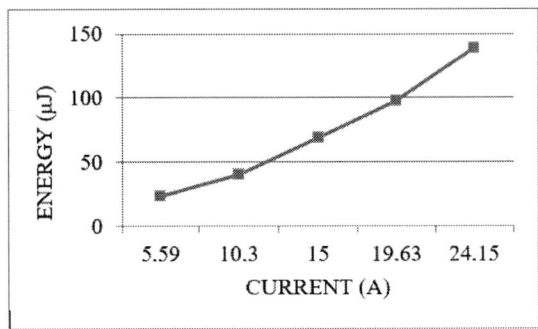

Figure 12: GaN intrinsic diode reverse recovery energy

For the conducted tests the voltage overshoot remained below 20%, as shown in Fig. 13, showing a good safety margin for a 600V transistor, tolerating a voltage overshoot of 800V for times lower than 1 microsecond for a switching frequency of 100 kHz.

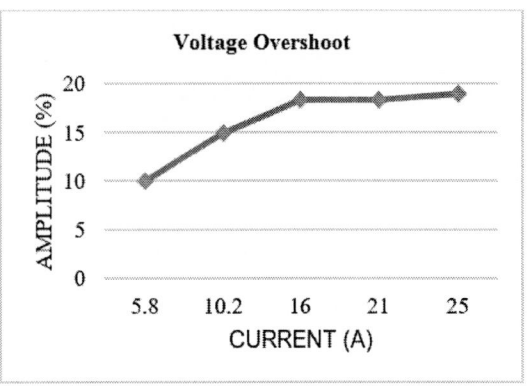

Figure 13: GaN switch voltage overshoot

The switching energy for different current levels can be seen at Fig. 14, becoming evident its low sensibility for temperature changes. The switch's losses concentrated mainly at the conduction entrance, being the blocking energy low, due to intrinsic characteristics of the cascode structure [14].

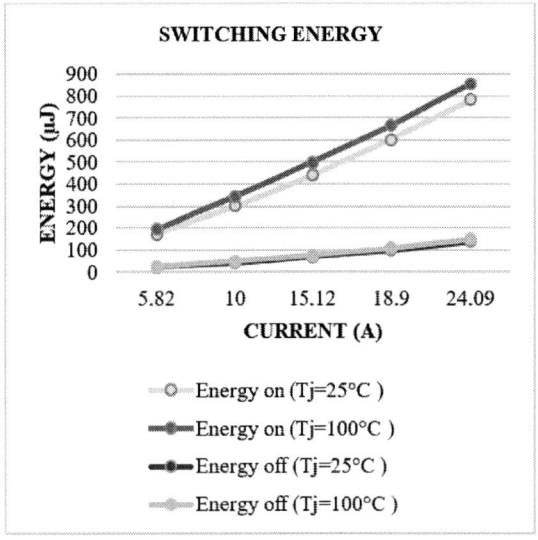

Figure 14: Switching energy

For comparison sake, Fig. 15 shows energy levels of a 600V Transphorm GaN (TPH3006PS) switch similar to the one used [13], showing the variation in relation to the gate resistor, in a way that the result presented are in the same order of physical quantity, giving lots of margin for improvement of the layout techniques and measurement, acknowledging that it wasn't possible to operate the converter with a lower value gate resistor.

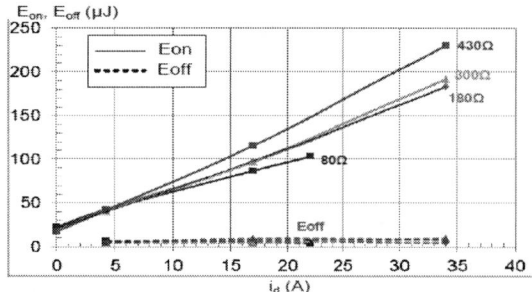

Figure 15: Switching energy for different gate resistor values [15]

VI. CONCLUSION

This paper presented a study about wide bandgap devices, focusing on the transistor GaN TPH3205WS from Transphorm for power electronics applications, in a way to better understand the characteristics and behavior of this transistor for future applications.

Energy wide-bandgap devices are superior in practically all physical properties in relation with silicon, being GaN and SiC characteristics similar between themselves, with the possibility of combining both technologies for even better semiconductor devices. Research in relation to SiC devices are in a more advanced stage than GaN, although GaN switches have better characteristics than SiC ones, predicting big advances for both technologies.

This technology has a broad possibility of development and already shows big advances in relation to SiC and silicon, however it brings big challenges for layout design, instrumentation and signal conditioning circuits because of noises caused by large current and voltage derivatives.

ACKNOWLEDGMENT

The authors are grateful to Transphorm for the donation of the devices necessary to the tests, to FITEJ and FAPESC, UDESC an NPEE for the space and equipment availability for the research, and to the group of collaborators of NPEE.

REFERENCES

[1] T.P Chow, "Progress in high voltage SiC and GaN power switching devices" in *Materials Science Forum*, 2014, pp.1077-1082.

[2] T.P Chow, "Wide bandgap semiconducor power devices for energy efficient system," in *IEEE 3rd Workshop on Wide Bandgap Power Devices and Applications (WiPDA)*, Blacksburg, VA, USA, 2015, pp.402-405.

[3] K. Li, P. Evans, M. Johnson, "SiC and GaN power transistors switching energy evaluation in hard and soft switching conditions". *IEEE 4th Workshop on Wide Bandgap Power Devices and Applications (WiPDA)*, Fayetteville, AR, USA, 2016, pp.123-128.

[4] L. C. M. Carrasco, A. J. Forsyth, "Energy analysis and performance evaluation of GaN cascode switches in an inverter leg configuration", *IEEE Applied Power Electronics Conference and Exposition (APEC)*, Charlotte, NC, USA, 2015, pp.2424-2431.

[5] K. Li, "Wide Bandgap (SiC/GaN) Power Devices Characterization and Modeling: application to HF Power Converters", Ph.D. dissertation, École Doctorale Sciences Pour L'Ingénieur, Université de Lille, Lille, France, 2014.

[6] C. Raynaud, D. Tournier, H. Morel, D. Planson, "Comparison of high voltage and high temperature performances of wide bandgap semiconductors for vertical power devices," *Diamond and related Materials*, vol. 19, pp.1-6, Jan. 2010.

[7] D. Visalli, "Optimization of GaN-on-Si HEMTs for high voltage application," Ph.D. dissertation, KU Leuven, Department Netuurkunde en Sterrenkunde, Leuven, Belgium, 2011.

[8] J. Honea, L. Zhou, Z. Wang, D. Kebort, J. Cortez, Y. Wu, "Packaging GaN in a TO-247," *Bodo's Power Systems*, pp.72-75, May 2015.

[9] Transphorm Inc., "600V Cascode GaN FET in TO-247", TPH3006PS datasheet, July 2018.

[10] Transphorm Inc., Application note AN-0002.

[11] Transphorm Inc., Application note AN-0004.

[12] Transphorm Inc., Application note AN-0003.

[13] Transphorm Inc., Application note AN-0008.

[14] A. Mengatto. "Análise de perdas em diferentes tipos de dispositivos semicondutores de potência e aplicação em conversor estático". Undergraduate dissertation, Dept. Elect. Eng. Santa Catarina State University, Joinville, SC, Brazil, 2015.

[15] Z. Wang, J. Honea, Y. Wu, "Design and Implementation of a High-efficiency Three-level Inverter Using GaN HEMTs". *Proceedings of PCIM Europe 2015; International Exhibition and Conference for Power Electronics, Intelligent Motion, Renewable Energy and Energy Management*, Nuremberg, Germany, 2015, pp.486-492.

Small Scale Compressed Air Energy Storage (SS-CAES) Strategies Overview

Luiz Fernando Martins Pastuch
Power Electronics Institute - INEP
Federal University of Santa Catarina - UFSC
Florianópolis, Brazil
luizfernandopastuch@gmail.com

Roberto Francisco Coelho
Power Electronics Institute - INEP
Federal University of Santa Catarina - UFSC
Florianópolis, Brazil
roberto@inep.ufsc.br

Telles Brunelli Lazzarin
Power Electronics Institute - INEP
Federal University of Santa Catarina - UFSC
Florianópolis, Brazil
telles@inep.ufsc.br

Marcos Antonio Salvador
Power Electronics Institute - INEP
Federal University of Santa Catarina - UFSC
Florianópolis, Brazil
marcos.salvador@inep.ufsc.br

Abstract—**Compressed air energy storage (CAES) is a technology to store electrical energy employed for decades, mainly through large scale systems. Today, small scale compressed air energy storage (SS-CAES) are also recently applied as an alternative to replace batteries in autonomous systems and as storage for intermittent renewable sources, promoting load leveling. These systems require compact and efficient power stages, with remarkable presence of power electronics. In this context, this article offers a comprehensive overview of SS-CAES systems, presenting the operating principles of various types of configurations, as well as information regarding energy density, efficiency, cost, limitations and challenges to be overcome in order to become an competitive alternative.**

Index Terms—**Compressed air energy storage (CAES), hybrid systems, compressed air**

I. INTRODUCTION

The current technology development and the concern on the enviroment have contributed to a noticeble increase in the energy generation by renewable sources. At the same time, energy storage have been approached by researches that propose a way to soften the oscillations of generation tipically linked to those sources and also related to the degradation of energy quality [1], [3]. Moreover, the problems associated to energy demand peaks and to electrical stability of power systems may be minimized by applying energy storage systems together with the power plants, in support to the transmission system, at various points in the distribution network and also on the consumer side [2].

The literature describes different ways of performing energy storage, with emphasis on the use of battery banks of various technologies and/or supercapacitors, as well as the use of fuel cells, flywheels, pumping techniques and storage of water, compressed air, among others [1].

Compressed air energy storage (CAES) allow energy storage in the air compression process, to posterior use during its expansion, being here considered as small ($\leq$ 100 kWh) and large (> 100 kWh) scale.

Over the years, different strategies related to CAES systems have been aproached aiming an increased efficiency and minimizing impacts to the environment, being proposed systems with reuse of the heat generated during the air compression process to reduce fuel burn during expansion [4]. Small energy storage hybrid systems based on pneumatic technology combined with supercapacitors, or hidropneumatic technologies associated with supercapacitors are also being proposed. Both require strategies to track the maximum power or maximum efficience point, lifting the efficiency of the CAES. [5].

Small scale compressed air energy storage (SS-CAES) are quoted as substitutes to some applications that usually use battery banks. SS-CAES systems have advantages from an ecological point of view and lifespan, when compared to commercial batteries, however, also present challenges to overcome, related to a smaller energy density and smaller yield. Such factores estimulate the search to increase the efficiency in the air expansion process [6].

The evolution of SS-CAES may place it as an option between the high density energy storages. Thus, it could be applied in already consolidated areas of power electronics, such as uninterruptible power supply (UPS); or emerging areas, such as isolated hybrid systems, active networks and distributed generation systems with energy storage, to solve problems as generation intermittence, load leveling, peak shaving, and others. Currently, all these applications use batteries. Furthermore, solutions are still sought in other forms of storage that are ecologically less aggressive and have a longer lifespan.

In this context, the present article offers a review regarding energy storage systems in the form of compressed air, with the purpose at pointing out the different approaches and highlighting the efforts made in search of new strategies for compressed air. Furthermore, the article demonstrates the potential and challenges of using power electronics applied to power processing in this type of storage system.

978-1-7281-4181-7/19 $31.00 © 2019 IEEE

II. COMPRESSED AIR ENERGY STORAGE

Atmospheric air consists of a mixture of gases, in the approximate proportion of 78% of Nitrogen, 21% of Oxygen and 1% of other elements. Being a gas, it has the property of compression and can be stored in reservoirs. In this condition, there is an increase in the number of air molecules per volume unit, thus, an increase in the internal pressure of the reservoir. The air compression requires the use of an external source of energy, part of which energy is stored and returned to the system during the expansion process [7].

Power electronics are often used as a link between the storage system and the power grid, making it possible to apply energy storage together with the concept of distributed generation [2]. Such possibilities have stimulated the development of SS-CAES systems with strategies for maximum efficiency point tracking (MEPT) [5] or for maximum power point tracking (MPPT) [6], since both power and efficiency of an SS-CAES vary depending on factors such as pressure, temperature and flow.

In addition, the application possibilities led to the development of hybrid solutions with their respective control strategies. Among the hybrid topologies, which will be demonstrated later in this review, are: CASCES (Compressed Air and Supercapacitors Energy Storage) and BOP (Battery with Oil-Hydraulics and Pneumatics) type A and type B. BOP type A systems have sealed gas compression/expansion cycles, while BOP type B systems are established with approximately isothermal atmospheric air compression and expansion cycles [5].

SS-CAES and hybrid SS-CAES solutions are structures based on static converters, but, as will be shown, studies addressing converter structures, modeling, control strategies and integration of these systems are still restricted. Therefore, taking as reference the storage systems based on the use of batteries, it is necessary to detail the characteristics of the load (CAES) to advance the development of this area.

A. SS-CAES System

Small scale compressed air energy storage (SS-CAES) systems have been studied as an alternative to battery replacement in autonomous systems, UPS, and distributed generation applications, being used in conjunction with renewable sources [4], [6].

Figure 1 shows a small scale system where air is typically compressed by a volumetric compression unit (compressor) and stored under pressure in open-air tanks. When released into the atmosphere, the stored air is used to move a pneumatic motor or a microturbine which, then, drives a dc generator. The energy generated is injected into the power grid or applied to a remote load using static converters, responsible for the processing of the generated electric energy.

SS-CAES systems do not burn fuel as in large systems, causing less negative impacts on the environment compared to electrochemical batteries, which generate toxic waste and have less longevity. However, the energy density and efficiency

of the SS-CAES systems are low, implying greater volume to supply the same amount of energy of the batteries [4]- [6].

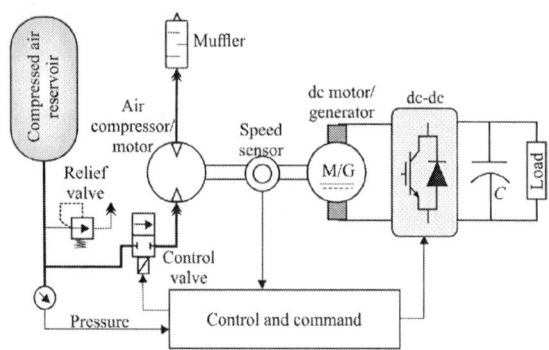

Fig. 1. SS-CAES system powering remote load [5].

As can be seen in Figure 1, a control system regulates the air reservoir discharge valve and provides parameters to the static converter, based on the measurement of quantities such as system pressure and generator shaft speed.

Generally, energy processing and storage systems search methods to diminish losses and raise yields in order to achieve maximum efficiency. In this regard, the studies targeted to SS-CAES systems have addressed strategies for the charging and discharging processes of the compressed air tank, as well as its application associated with photovoltaic arrays and wind generators. In [8], an air pump based on an air / liquid piston is proposed. This small scale compressor has a low compression ratio and low power, so its advantage is that it can be deployed together with photovoltaic systems in residential applications which output power is typically in the order of 160 W.

In [6], a control strategy for the discharge of compressed air from a SS-CAES system was developed, simulated and implemented by tracking the maximum power point (MPPT) of a small pneumatic vane motor. As can be seen in Figure 2, the system drives a permanent magnet dc generator which feeds a resistive load with a Buck converter.

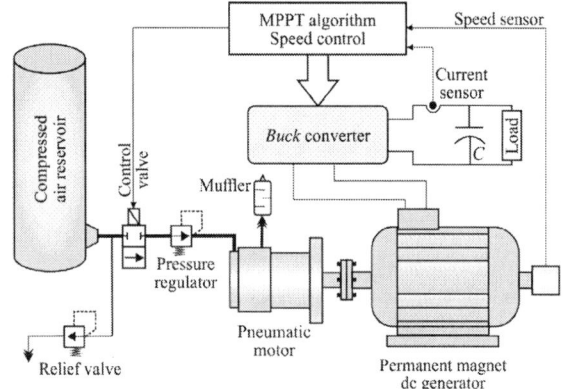

Fig. 2. Process of air discharge with MPPT in SS-CAES [6].

This system was analyzed using a small signal model and employing the perturb and observe method with small speed steps to seek convergence. The proposed control system does not need to monitor the compressed air pressure and flow, it only analyzes the generator shaft speed and the Buck converter output current to track the maximum power line shown in Figure 3 [6].

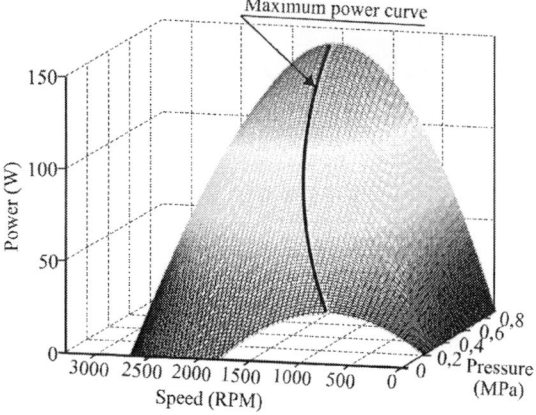

Fig. 3. Pneumatic motor maximum power line [6].

From the equations of torque, airflow and motor speed, it is possible to trace the surface correlated to the magnitudes of power, pressure and speed of the pneumatic motor, and establish the maximum power line. According to Figure 3, the maximum power line depends on the supply pressure and the motor speed [6].

In [5], the application of a pneumatic vane motor of 100 W for pneumatic/mechanical conversion was investigated, and a strategy of maximum efficiency point tracking (MEPT) was developed according to Figure 4.

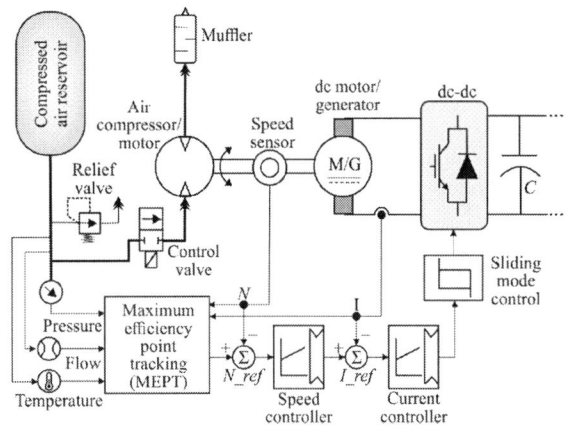

Fig. 4. SS-CAES system with MEPT strategy [5].

The pressure and load variations are respectively related to the air flow and the speed of the pneumatic motor, and directly affect its performance. The principle of the MEPT strategy of Figure 4 is to optimize the energy conversion based on the measurement of various quantities (pressure, flow, velocity, current, etc.) in order to determine the ideal speed according

to the maximum efficiency line shown in Figure 5 and thus use it in the speed control module [5].

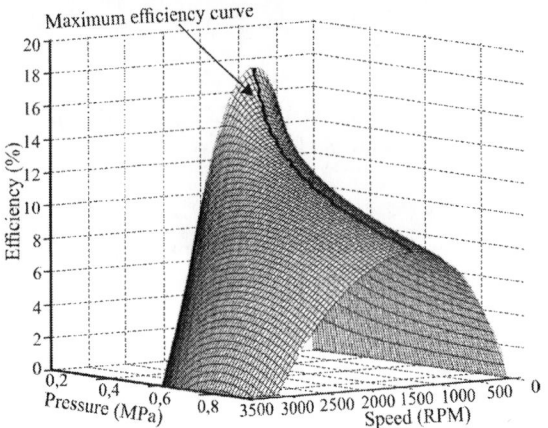

Fig. 5. Pneumatic motor maximum efficiency line [5].

Figure 5 shows that the pneumatic motor has a low efficiency, less than 20%. The reduced efficiency of a small scale system with a pneumatic motor is associated to heat losses related to the high area/volume ratio of such machines, given the irreversibility of a non-adiabatic process. For similar reasons and because of critical internal tolerances to reduce leakage losses, turbines are not used for small scale power generation.

B. CASCES System

The CASCES system (Compressed Air Supercapacitors Energy Storage) is shown in Figure 6.

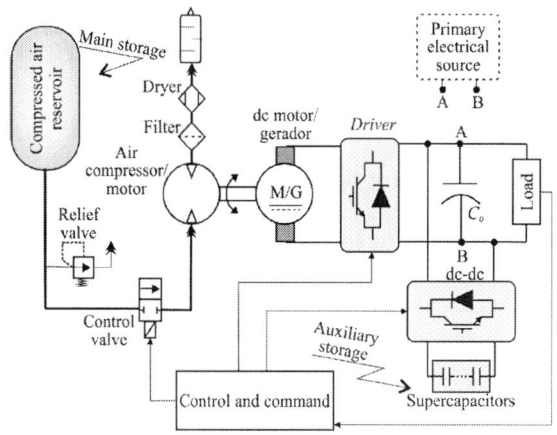

Fig. 6. CASCES hybrid storage system [5].

Such system consists of the combination of a high energy capacity storage (compressed air) and auxiliary devices with high energy density power reserve (supercapacitor) [5]. The components of the pneumatic/electric assembly are reversible, so the energy stored in this system comes from an external primary source, which may be the power grid or a photovoltaic array, for example.

978-1-7281-4181-7/19 $31.00 © 2019 IEEE

The combination of supercapacitors with the compressed air system allows better quality of energy delivered to the load, smoothing the imminent fluctuations of the output voltage due to possible load variations.

As can be seen in Figure 7, the supercapacitor is used to supply the load at times when the pneumatic motor is at rest and is also used as an assistant in the maintenance of the delivered power at times of peak load [5].

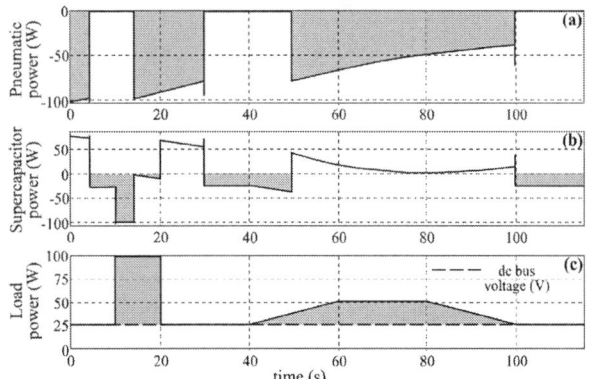

Fig. 7. Power curves of the CASCES system [5].

Figure 7(a) shows the power curve from the compressed air, converted and delivered to the system. The shaded regions of this figure correspond to the times when the pneumatic/electric conversion assembly is supplying the load and charging the supercapacitors. In this working mode, the supercapacitor bank voltage is monitored and the pneumatic/electric power conversion group is actuated when such voltage reaches the set minimum value, remaining on until the maximum voltage value is reached or until the resource of compressed air is exhausted. In Figure 7(b) the shaded regions correspond to the periods in which the supercapacitors supply energy to the load.

In Figure 7(c) there are two regions of power variation, and at the beginning of these variations the capacitor transfers all stored energy to the load and, subsequently, the pneumatic/electric assembly is activated to ensure that the supercapacitors recharge process occurs.

As suggested in Figure 7, the support of the supercapacitors to the pneumatic storage system can be quite interesting, especially in the case of variations in load power. However, the small pneumatic/electric conversion system continues with the same pneumatic motor efficiency constraints. To increase the overall efficiency of the SS-CAES system, hydropneumatic combinations are suggested below to replace the purely pneumatic part [5].

C. BOP-A System

The BOP-A (Battery with Oil-Hydraulics and Pneumatics type A) is a form of energy storage that combines the use of hydraulic oil and pneumatic devices. The gas compression and expansion processes occur in a closed cycle, that is, with sealed gas without air intake and exhaust from the environment [5], [9].

The design of this system is mainly due to the high efficiency of the engines and oil pumps. These hydraulic machines operate at high pressure levels, of the order of 10 to 35 MPa, and may exhibit efficiency above 90% at these pressure levels [5], [9].

The energy storage of the BOP-A system utilizes pressure reservoirs industrially known as hydraulic accumulators. These accumulators have two compartments, one for the gas and another for the liquid, separated by a membrane or by a free piston. When pumping oil into the accumulator, the pressure in the oil compartment begins to rise and thus the displacement of the membrane / piston increases, reducing the volume of the gas compartment, compressing it. In this type of application an inert gas such as nitrogen is usually used, since the combination of oxygen and oil under pressure can cause an explosive mixture. Figure 8 shows the accumulators and other basic components of the BOP-A storage system, including a supercapacitor bank used as an auxiliary reservoir with the same purpose described in the CASCES systems.

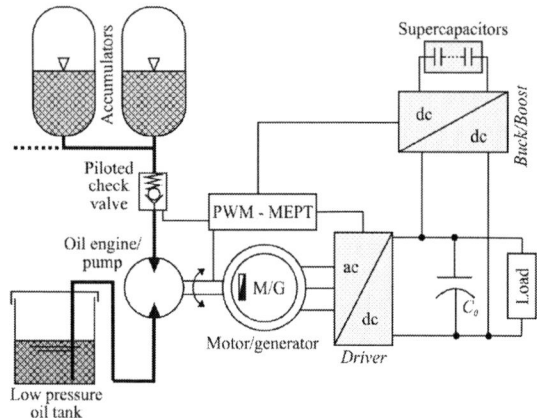

Fig. 8. BOP-A hybrid system with supercapacitors [5].

The components of the BOP-A system are bidirectional, so when the rotating machine (motor / generator) operates as a motor, it drives the hydraulic machine, which acts as a pump. The hydraulic pump is responsible for transferring the oil from a low pressure tank into the accumulators, compressing the pre-charged gas into the accumulator. At the end of this step, the control valve closes and the gas is compressed, storing energy. When the stored energy is requested, the control valve is opened and the compressed gas pushes the oil out of the accumulators. This oil flows into the low pressure tank, moving the hydraulic machine and making it run like a motor, which drives the generator, converting mechanical into electric energy.

In this system, power electronics are present in the converters used to condition the energy to the needs of the load and to drive the motor during the storage step, as well as in the processes of charging and discharging the supercapacitor. Therefore, the system uses bidirectional converter solutions in the motor/generator (dc-ac converters) and supercapacitor (dc-dc converter).

The development of suitable topologies for the power levels of the complete system and extraction of the maximum power are challenges in this area, considering the strategies of control of each converter. Initial solutions may be based on those already used in systems that employes batteries, however, the dynamics, modeling and existence of a point of maximum power in the operation of the system requires specific solutions for the case.

D. BOP-B System

The BOP-B (Battery with Oil-Hydraulics and Pneumatics type B) system is also a form of energy storage that combines the use of hydraulic oil and pneumatic devices. In this type, compression and expansion processes occur in open cycle, as compressed air is admitted from the environment during the compression stage and released to the atmosphere in the expansion [5], [9].

The atmospheric air compression and expansion processes performed in this system are based on a technology known as liquid piston [10]. This is a direct hydraulic/pneumatic transformation arrangement, which uses in the compression process a column of liquid to compress the gas present in a given fixed volume chamber. The liquid fluid, typically water or oil, is driven by a hydraulic pump, being that the air inlet and outlet of the chamber are controlled by pneumatic valves [10], [11].

In terms of power electronics, this solution differs from the previous one by the dynamic response of the storage system and maximum power point tracking. Thus, the choice of converters and control system needs to consider these new features.

In [11], [12] is possible to observe the size of the field of application of these technologies next to renewable sources, where an open cycle hydropneumatic energy storage system, similar to the described BOP systems, is proposed for application in wind turbines, using the architecture of the compression and expansion chambers of liquid piston, as per Figure 9.

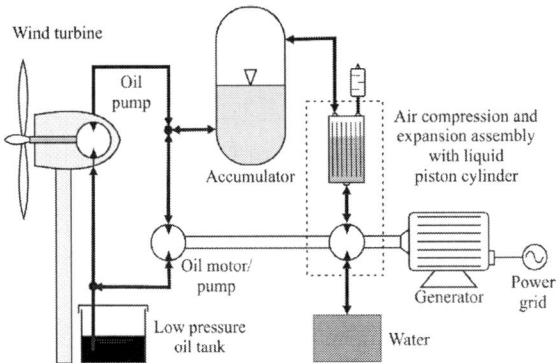

Fig. 9. Storage system for wind turbine [11].

Table I [9] presents an estimated cost comparison between BOP storage systems and batteries in an application with photovoltaic modules that power a dwelling. The case study considered an average power of 4 kW in a period of 7 h during the day, that is, 28 kWh. For the discharge of the system, a period of 5 h (between 18 h and 23 h) with an average power of 4 kW, that is, a supply of 20 kWh was considered at night. The economic framework also considered a need for autonomy of 3 days, therefore, the storage systems compared are sized to supply 60 kWh.

TABLE I
ESTIMATED COST COMPARATIVE FOR 60 KWH [9]

60 kWh		PbO$_2$ Battery	BOP-A	BOP-B
Technology	Storage	€18,000	€60,000	€4,500
	Conversion	/	€15,000	€18,000
	Total	€18,000	€75,000	€22,500
Operation and maintenance		(30%) €3,500	(20%) €15,000	(20%) €4,500
Total cost		€23,500	€90,000	€27,000
Life cycle		3,500 or 210,000 kWh	15,000 or 900,000 kWh	15,000 or 900,000 kWh
Cost of energy/kWh		€0.11/kWh	€0.1/kWh	€0.03/kWh

Table I shows that the highest storage cost is from the BOP-A system. This is mainly due to its low energy density, in the order of 2.5 Wh/kg. In general, the estimated costs of BOP-type applications are higher than lead-acid batteries, but since a much longer lifespan has been estimated for BOP systems, the cost of energy (kWh) is less than from the use of batteries [9].

The studies presented in [13] also suggest a favorable opinion on small scale air storage systems combined with the use of photovoltaic modules, with a payback in a period of 5 to 7 years for residential and hotel applications, pointing out that about 40% of the cost of the system is related to the value of the reservoirs.

III. FINAL CONSIDERATIONS

Based on the bibliography addressed and the content presented, it is possible to understand the applicability of compressed air storage systems, mainly in conjunction with renewable energy sources. In this way, more research in this field should be stimulated, such as the study of the hybrid set of Figure 10 [5], based on the BOP-B system with open cycle hydropneumatic architecture and supercapacitors, connected to the power grid and powered by photovoltaic modules.

One of the main advantages foreseen for the system of Figure 10 is to take advantage of the better performance of the hydraulic motor with a higher energy density when compared to the prototype of the BOP-A system, developed at the Federal Polytechnic School of Lausanne (EPFL, Switzerland) [5]. It should be noted that the BOP-A system requires a tank to store oil at low pressure, and its volume is directly related to the energy storage tanks, since about 50% of the volume of the hydraulic accumulators are occupied by oil in the step of compression.

The arrangement of Figure 10 needs a mechanical interface for threading oil and air fluids, of which operating cycle is identical to the BOP-B system, previously described. The

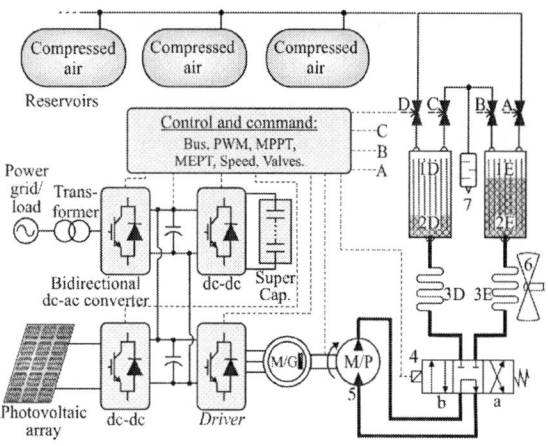

Fig. 10. Open cycle air storage system with hydraulic interface and superca-pacitors, adapted from [5].

electromechanical transformations are bidirectional, that is, performed by a rotating machine that can operate as a motor and generator. This bidirectionality is also required of the converter applied to the motor/generator, the dc-dc converter dedicated to the supercapacitor bank and the dc-ac converter arranged between the dc bus and the power grid. In the inverter mode, the converter enables the supply of electricity from the photovoltaic array and recover it from the storage system (supercapacitors and hydraulic interface pneumatic) to the grid. In rectifier mode, the converter can be used to recharge the storage system, at times of low demand or lack of photovoltaic generation (night periods).

The definition of the converter structures will depend on the power to be processed. The control strategy should optimize the use of stored power (MPPT), consider the dynamics of the system and control all modes of operation (power flow direction), such as: photovoltaic generation, charging and discharging of supercapacitor, charging and discharging of the compressed air system and injection of power in the electric network. In addition, studies of the overall efficiency of the system are required, considering restrictions such as operation cost, efficiency, uninterrupted operation, etc.

IV. CONCLUSIONS

Studies of small scale CAES systems show that pneumatic microturbines typically operate at low pressure levels and have low efficiency, which directly affects the efficiency of purely pneumatic structures. As an alternative to overcome this deficiency, mixed systems (oil/air) are approached using higher efficiency hydraulic motors/pumps and higher pressure levels. This combination requires an interface between fluids, oil and air, which makes the system more complex than purely pneumatic.

Power electronics are present in the various possibilities of application of small scale systems, acting as a link between the storage system and the power grid, in the presence of isolated loads and renewable sources. Bidirectional converters,

modeling, and multilevel control strategies will be required in the use of small scale CAES. The improvement of these systems requires and stimulates the development of the various elements that compose it, as a way of optimizing each one of them and increasing the overall efficiency.

ACKNOWLEDGEMENT

The authors would like to thank the CNPq for the financial support (process No. 422276/2016-2).

REFERENCES

[1] S. Vazquez, S. M. Lukic, E. Galvan, L. G. Franquelo, J. M. Carrasco, "Energy Storage Systems for Transport and Grid Applications", IEEE Transactions on Industrial Electronics, vol. 57, n° 12, pp. 3881-3895, December 2010.

[2] A. Mohd, E. Ortjohann, A. Schmelter, N. Hamsic, D. Morton, "Challenges in integrating distributed energy storage systems into future smart grid", in Proc. of IEEE International Symposium on Industrial Electronics, pp. 1627-1632, 2008.

[3] C. Xiaoguang, Z. Chenghui, K. Li, J. Yefei, "Dynamic modeling and efficiency analysis of the scroll expander generator system for compressed air energy storage", in Proc. of ICEMS - International Conference on Electrical Machines and Systems, pp. 1-5, 2011.

[4] A. Rogers, A. Henderson, X. Wang, M. Negnevitsky, "Compressed air energy storage: Thermodynamic and economic review", in Proc. of IEEE PES - Power and Energy Society General Meeting Conference and Exposition, pp. 1-5, 2014.

[5] S. Lemofouet, A. Rufer, "A hybrid energy storage system based on compressed air and supercapacitors with maximum efficiency point tracking (MEPT)", IEEE Transactions on Industrial Electronics, vol. 53, n° 4, pp.1105-1115, June 2006.

[6] V. Kokaew, S. M. Sharkh, M. Torbati, "Maximum Power Point Tracking of a Small Scale Compressed Air Energy Storage System", IEEE Transactions on Industrial Electronics, vol. 63, n° 2, pp. 985-994, September 2015.

[7] F. Fahy, Air: The Excellent Canopy, Woodhead Publishing, 1st edition, Cambridge , 2010.

[8] D. Villela, V. V. Kasinathan, S. De Valle, M. Alvarez, G. Frantziskonis, P. Deymier, K. Muralidharan, "Compressed-air energy storage systems for standalone off-grid photovoltaic modules", in Proc. of PVSC - Photovoltaic Specialists Conference, pp. 962-967, 2010.

[9] S. Lemofouet, A. Rufer, "Hybrid Energy Storage System based on Compressed Air and Super Capacitors with Maximum Efficiency Point Tracking", in Proc. of IEEE International Power Electronics and Applications Conference, pp.10, 2005.

[10] J. D. Van de Ven, P. Y. Li, "Liquid Piston Gas Compression," Applied Energy, vol. 86, n° 10, pp. 2183-2191, October 2009.

[11] M. Saadat, P. Y. Li, T. W. Simon, "Optimal Trajectories for a Liquid Piston Compressor/Expander in a Compressed Air Energy Storage System with Consideration of Heat Transfer and Friction", in Proc. of ACC -American Control Conference, pp. 1800-1805, 2012.

[12] M. Saadat, P. Y. Li, "Modeling and Control of an Open Accumulator Compressed Air Energy Storage (CAES) System for Wind Turbines", Applied Energy, vol. 137, pp. 603-616, Janeiro 2015.

[13] A. Tallini, A. Vallati, L. Cedola, "Applications of micro-CAES systems: energy and economic analysis" Energy Procedia, vol. 82, pp.797-804, December 2015.

Synchronous Reference Frame PLL Frequency Estimation under Voltage Variations

Eliabe Duarte Queiroz
School of Electrical and Computer Engineering
University of Campinas
Campinas, Brazil
eliabe.duarte.queiroz@gmail.com

José Antenor Pomilio
School of Electrical and Computer Engineering
University of Campinas
Campinas, Brazil
antenor@fee.unicamp.br

Abstract—In this paper, MIMO space state models that represent the voltage influence in the SRF-PLL over its frequency estimations is obtained. Throughout the model development, two estimation alternatives were explored. The model accuracy was verified by simulations of the linear models and then compared with the nonlinear models. The SRF-PLL linear model represents well the behavior under single and three phase voltage sags.

Index Terms—Phase locked loop (PLL), synchronous reference frame (SRF), Grid-connected converters, state-space model, linear systems.

I. INTRODUCTION

The presence of AC/DC and DC/AC power electronic converters is increasing continuously in the electric grid, as they are used to interface loads and distributed energy resources (DER) to the alternating current grid [1]-[4].

The control synchronization to the grid voltage is provided by a phase locked loop (PLL) scheme. Due to its large application to converter control and simplicity, the PLL considered in this paper is a synchronous reference frame PLL (SRF-PLL) [5]-[8], whose scheme is depicted in Fig. 1.

As the frequency, in a PLL, is estimated directly by observing the voltage signals, the PLL output is affected by voltage variations [9, 10]. Therefore, an accurate dynamic model is necessary to study the stability of a system composed by one or more converters, each one with its PLL, connected to a common electric network.

This paper aims to analyze two different PLL frequency estimations using the dynamic description in a synchronous reference frame. The focus is given on obtaining the PLL dynamic models that describe the influence of voltage perturbations over the frequency estimation.

Simulations using nonlinear and linearized models allow the comparison of the different strategies and models.

II. MODEL OF VOLTAGE INFLUENCES ON THE SRF-PLL ESTIMATIONS.

It is challenging to perceive little deviations on a sinusoidal voltage that varies over hundreds of volts, thus a Park transformation is desirable to easily analyze what happens in the

This work was funded by the National Council for Scientific and Technological Development (CNPq) , grant #302257/2015-2, and São Paulo Research Foundation (FAPESP), grant #2016/08645-9.

PLL under disturbances in the terminal voltage. Considering a system of one voltage source (VS) connected to a current source converter (CSC), synchronizing a Park transformation to the VS will lead to the existence of two synchronous rotating frames.

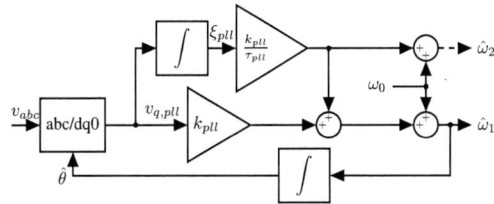

Fig. 1. SRF-PLL diagram.

The first (d_v and q_v) referenced on the voltage angular position of the VS; its voltage frequency is $\omega_v(t)$, a function of time.

The second frame (d_{pll} and q_{pll}) is the converter synchronous frame. The rotation frequency of this frame is a SRF-PLL estimation ($\hat{\omega}_1(t)$) that is also a function of time. These frames are depicted in Fig.. 2.

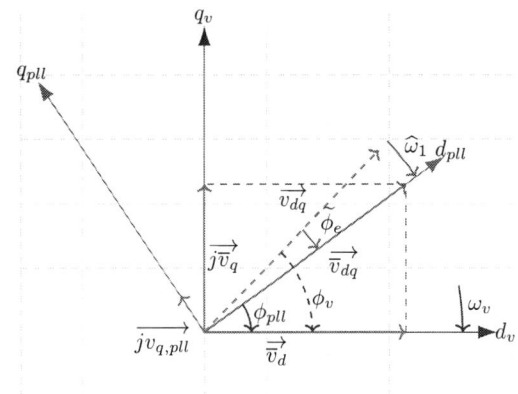

Fig. 2. Spatial vectors of the terminal voltages, the VS and the SRF-PLL synchronous reference frames.

The variables will be represented, generically by their value, x, composed by the steady value, $\overline{x}$ and the deviation, $\widetilde{x}$, as follows:

$$x = \overline{x} + \widetilde{x}$$

The SRF-PLL tracks the terminal voltage vector, $\vec{v}_{dq}$, represented in the VS ($d_v q_v$) frame. Considering the possibility of a feeder, the terminal voltage could have a phase difference to the VS reference frame.

Thus, the SRF-PLL angle estimation could also differ by a phase ϕ_{pll} to the VS reference frame. In steady state, ϕ_{pll} is equal to ϕ_v, the converter terminal voltage phase. These phases are depicted in Fig.. 2.

Considering the VS angle as the reference, the terminal voltage vector is given by:

$$\overrightarrow{v_{dq}} = v_d + j v_q = |\overrightarrow{v_{dq}}| e^{j\phi_v} \qquad (1)$$

Where ϕ_v is a phase difference between the direct axis d_v and the voltage vector:

$$\phi_v = \tan^{-1}\left(\frac{v_q}{v_d}\right) \qquad (2)$$

The variables v_d and v_q are the direct and quadrature voltage vector components in the VS reference frame.

During transients, the voltage vector will be the sum of the steady state and the deviation, denoted by $\overrightarrow{\overline{v}_{dq} + \widetilde{v}_{dq}}$ in Fig 2. The SRF-PLL reaction has a limited band and it does not follow exactly the voltage position. As a consequence, a phase error $\widetilde{\phi}_e$ between the terminal voltage vector phase, ϕ_v, and the converter frame phase, ϕ_{pll} will appear.

The voltage deviations can arise even in situations which the VS frequency and magnitude are constants by changes in the load, in the circuit topology or in the operating point of converters.

Considering that the direct and quadrature voltages have a module, V_{mod}:

$$V_{mod} = \sqrt{v_d^2 + v_q^2} \qquad (3)$$

The non-linear equation that describes the integral term of the SRF-PLL is given by:

$$\xi'_{pll} = v_{q,pll} = V_{mod}\sin(\widetilde{\phi}_e) \qquad (4)$$

The notation x' indicates the time derivative of x, as follows:

$$\xi'_{pll} = \frac{d\,\xi_{p}ll}{d\,t}$$

The terminal voltage phase, ϕ_v, in Fig. 2, is a component of the terminal voltage angle, θ_v, that is:

$$\theta_v = \int \omega_v dt + \phi_v \qquad (5)$$

Therefore, a deviation in ϕ_v affects θ_v and it has an influence on the frequency estimation [11].

The estimated angle used to create the SRF-PLL reference frame is:

$$\hat{\theta} = \int \hat{\omega}_1 dt \qquad (6)$$

The expression for the error between the terminal voltage and the PLL angle estimation is:

$$\widetilde{\phi}_e = \int (\omega_v - \hat{\omega}_1)dt + \phi_v \qquad (7)$$

The estimated frequency by the proportional-integral terms of the SRF-PLL is:

$$\hat{\omega}_1 = k_{pll} V_{mod}\sin(\widetilde{\phi}_e) + \frac{k_{pll}}{\tau_{pll}}\xi_{pll} + \omega_0 \qquad (8)$$

The expression for, ϕ_{pll}, the integral of the difference between $\hat{\omega}_1$ and ω_v, considering (8) is:

$$\phi'_{pll} = k_{pll} V_{mod}\sin(\phi_v - \phi_{pll}) + \frac{k_{pll}}{\tau_{pll}}\xi_{pll} + \omega_0 - \omega_v \qquad (9)$$

Linearizing (4) and (9) at the operating points of $\widetilde{\phi}_e$ and ξ_{pll}:

$$\overline{\xi}_{pll} = \frac{\tau_{pll}}{k_{pll}}(\overline{\omega}_L - \omega_0)$$

$$\overline{\phi}_e = 0$$

In steady state $\overline{\omega}_1$ equals $\overline{\omega}_L$, and the variations of the SRF-PLL states are given by:

$$\widetilde{\xi}'_{pll} = \overline{V}_{mod}\widetilde{\phi}_v - \overline{V}_{mod}\widetilde{\phi}_{pll} \qquad (10)$$

$$\widetilde{\phi}'_{pll} = k_{pll}\overline{V}_{mod}\widetilde{\phi}_v - k_{pll}\overline{V}_{mod}\widetilde{\phi}_{pll} + \frac{k_{pll}}{\tau_{pll}}\widetilde{\xi}_{pll} - \widetilde{\omega}_v \qquad (11)$$

For a system, as in Fig. 3, with a VS, a converter and a generic load, the phase difference ϕ_v, depends on the SRF-PLL voltage measurements, and, it is described by:

$$\widetilde{\phi}_v = t_d\widetilde{v}_d + t_q\widetilde{v}_q \qquad (12)$$

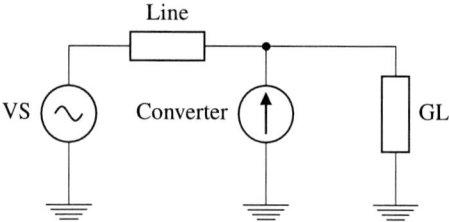

Fig. 3. Circuit with a VS conected to a converter and a generic load (GL).

Where t_d and t_q are the partial derivatives of the phase difference:

$$t_d = \frac{\partial \tan^{-1}\left(\frac{v_q}{v_d}\right)}{\partial v_d}\Bigg|_{\substack{v_d=\overline{v}_d \\ v_q=\overline{v}_q}} = -\frac{\overline{v}_q}{\sqrt{\overline{v}_d^2 + \overline{v}_q^2}}$$

$$t_q = \frac{\partial \tan^{-1}\left(\frac{v_q}{v_d}\right)}{\partial v_q}\Bigg|_{\substack{v_d=\overline{v}_d \\ v_q=\overline{v}_q}} = \frac{\overline{v}_d}{\sqrt{\overline{v}_d^2 + \overline{v}_q^2}}$$

It is possible to obtain a system that describes the estimation of the frequency by a SRF-PLL considering the terminal voltage influence replacing (12) in (4) and (9). Thus, the state-space model is given by:

$$\begin{cases} \widetilde{\mathbf{x}}'_{pll} = A_{pll}\widetilde{\mathbf{x}}_{pll} + B_{pll,v}\widetilde{\mathbf{u}}_{pll,v} + B_{pll,\omega}\widetilde{\mathbf{u}}_{pll,\omega} \\ \mathbf{y}_{pll} = C_{pll,1}\widetilde{\mathbf{x}}_{pll} + D_{pll,v1}\widetilde{\mathbf{u}}_{pll,v} \end{cases} \tag{13}$$

Where:

$$\widetilde{\mathbf{x}}_{pll} = \begin{bmatrix} \widetilde{\xi}_{pll} & \widetilde{\phi}_{pll} \end{bmatrix}^T$$

$$\widetilde{\mathbf{u}}_{pll,v} = \begin{bmatrix} \widetilde{v}_d & \widetilde{v}_q \end{bmatrix}^T$$

$$\widetilde{\mathbf{u}}_{pll,\omega} = \begin{bmatrix} \widetilde{\omega}_v \end{bmatrix}$$

$$\widetilde{\mathbf{y}}_{pll} = \begin{bmatrix} \widetilde{\hat{\omega}}_1 \end{bmatrix}$$

$$A_{pll} = \begin{bmatrix} 0 & -\overline{V}_{mod} \\ k_{pll}/\tau_{pll} & -k_{pll}\overline{V}_{mod} \end{bmatrix}$$

$$B_{pll,v} = \begin{bmatrix} \overline{V}_{mod}t_d & \overline{V}_{mod}t_q \\ k_{pll}\overline{V}_{mod}t_d & k_{pll}\overline{V}_{mod}t_q \end{bmatrix}$$

$$B_{pll,\omega} = \begin{bmatrix} 0 & -1 \end{bmatrix}^T$$

$$C_{pll,1} = \begin{bmatrix} \frac{k_{pll}}{\tau_{pll}} & -k_{pll}\overline{V}_{mod} \end{bmatrix}$$

$$D_{pll,v1} = \begin{bmatrix} k_{pll}\overline{V}_{mod}t_d & k_{pll}\overline{V}_{mod}t_q \end{bmatrix}$$

The estimated frequency ($\hat{\omega}_1$) is integrated to obtain the voltage estimated angle used to perform the Park transform. However, another frequency estimation output ($\hat{\omega}_2$), also shown in Fig. 1, can be obtained by using the output values of the quadrature voltage integrator:

$$\hat{\omega}_2 = \frac{k_{pll}}{\tau_{pll}}\widetilde{\xi}_{pll} \tag{14}$$

That will result in the matrices $C_{pll,2}$ and $D_{pll,v2}$,

$$C_{pll,2} = \begin{bmatrix} \frac{k_{pll}}{\tau_{pll}} & 0 \end{bmatrix}$$

$$D_{pll,v2} = \begin{bmatrix} 0 & 0 \end{bmatrix}$$

The matrix $D_{pll,2}$ with its terms equal to zero, means that there is no feed-through path in the PLL. Therefore, it limits the pass-band of this system for any stimuli that come from the converter terminal voltages. The system describing $\hat{\omega}_2$ is:

$$\begin{cases} \widetilde{\mathbf{x}}'_{pll} = A_{pll}\widetilde{\mathbf{x}}_{pll} + B_{pll,v}\widetilde{\mathbf{u}}_{pll,v} + B_{pll,\omega}\widetilde{\mathbf{u}}_{pll,\omega} \\ \mathbf{y}_{pll} = C_{pll,2}\widetilde{\mathbf{x}}_{pll} + D_{pll,v2}\widetilde{\mathbf{u}}_{pll,v} \end{cases} \tag{15}$$

Here, the SRF-PLL design prioritizes a PLL system that is the most damped and fastest given a integrator time constant (τ_{pll}). It is achieved in a situation of critical damping, where the two eigenvalues of A_{pll} are real and equal. The proportional constant to achieve that is:

$$k_{pll} = \frac{4}{\tau_{pll}\overline{V}_{mod}} \tag{16}$$

III. MODEL SIMULATIONS

The convergence between non linear and linear models are analyzed through simulations. A non linear SRF-PLL model was executed on Simulink, whose simulation schematic is depicted in Fig. 4. The linear model was reproduced through state-space models, (13) and (15), by the function lsim on Matlab.

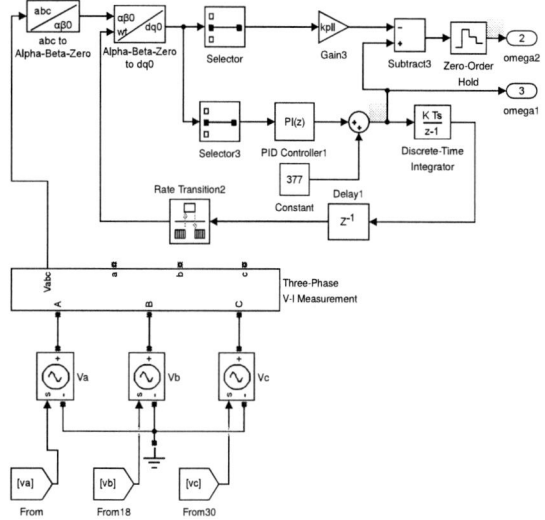

Fig. 4. Simulink schematic of the SRF-PLL non linear simulation.

Table I shows the parameters used in the simulations.

TABLE I
VALUES OF THE SRF-PLL PARAMETERS

Parameter	Value
τ_{pll}	0.0200
k_{pll}	1.1111
v_d	155.88
v_q	90
ω_0	376.99

The system response to steps, in both models, is used to compare the accuracy between them. Although occurrences of voltage and frequency steps on a real grid are unlikely, their simulation is desirable to verify the model response for a large frequency content.

In the following figures, the curves whose legend contain a subscript NLM, as in $\omega_{1,NLM}$, are variables obtained from non-linear models. The legends with a subscript SSM, as in $\omega_{1,SSM}$, are the ones obtained by the simulation of a state-space model.

To show the possibility of a phase difference from the VS to the converter terminal voltage, the initial voltage phase ϕ_v is 30°, as in the values of v_d and v_q shown in Fig. 5.

The frequency step of 1 Hz (6.28 rad/s) occurs at 0.8 s and the voltage steps in v_d and v_q of, respectively, -18 V and 25.7 V, at at 0.84 s and 0.88 s. The instants of the frequency and voltage steps are denoted, respectively, by I, II and II in Fig. 5, 6, 7 and 8. After these steps, ϕ_v becomes 40°.

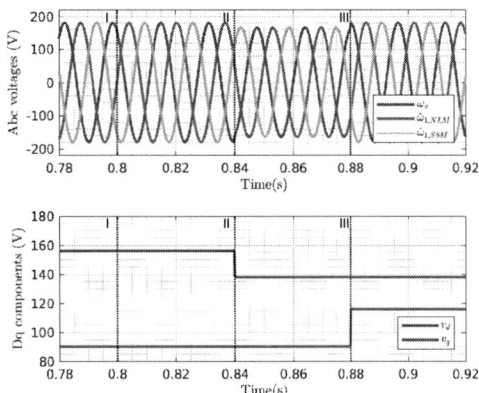

Fig. 5. Top: phase voltages in the PLL. Bottom: Dq voltage components in the PLL.

The frequency estimation $\hat{\omega}_1$ is shown in Fig. 6. The PLL response follows the frequency step with an overshoot.

At 0.84 s, the response due to the change in the direct voltage is a sharp peak. Another peak occurs at 0.88 s as a response to the quadrature voltage step. These peaks represent a high frequency content and occur due to a direct path from the voltages to the SRF-PLL output. This path is described by the matrix $D_{pll,v1}$ different from zero.

The errors between the Simulink simulation and the linear SRF-PLL model of the $\hat{\omega}_1$ estimation are shown in the second graphic of Fig. 6. The error is represented as a ratio of the difference between $\hat{\omega}_{1,NLM}$ and $\hat{\omega}_{1,SSM}$ to the deviation range in the non linear estimation. The model errors are much lower than the systems response showing that the linear model can fairly represent the SRF-PLL.

The $\hat{\omega}_2$ estimation, shown in Fig. 7, gives the response to the frequency step with the same settling time, when the signal error to the reference accommodates within 2% of the step value. However, this estimation has no overshoot, and there are no sharp peaks as in Fig. 6, that means less high frequency components on the estimation. The main difference in this system is the matrix $D_{pll,v2}$, equal to zero. Therefore, all components of the signal are subjected to an integration and the high frequency content is attenuated.

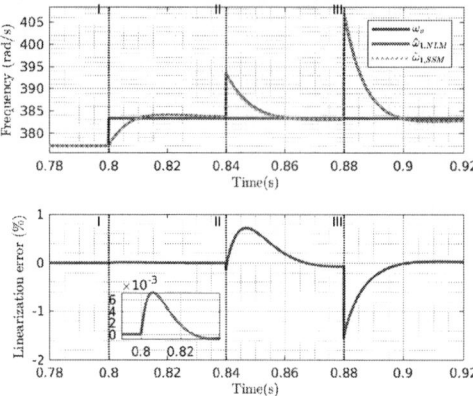

Fig. 6. Top: $\hat{\omega}_1$ frequency estimation. Bottom: Error between $\omega_{1,SSM}$ and the $\omega_{1,NLM}$ normalized to the range of $\omega_{1,NLM}$.

The difference between the linear and nonlinear models in the $\hat{\omega}_2$ estimation is shown in the bottom graph in Fig. 7. The outcome is similar, with a much lower error than the amplitude of the verified signal.

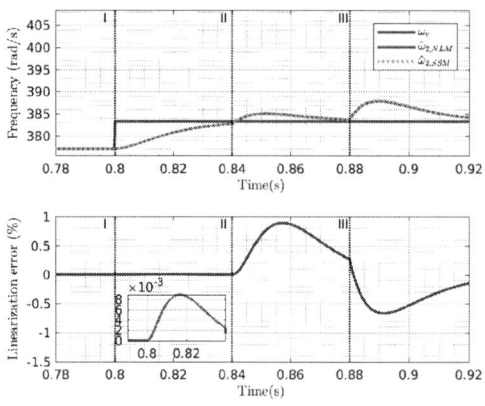

Fig. 7. Top: $\hat{\omega}_2$ frequency estimation. Bottom: Error between $\omega_{2,SSM}$ and the $\omega_{2,NLM}$ normalized to the range of $\omega_{2,NLM}$.

The model also represents the SRF-PLL reference frame phase difference ϕ_{pll}, the second state of $\tilde{\mathbf{x}}_{pll}$ in (13), that is shown in Fig. 8. In this figure, the v_{dq} vector voltage phase ϕ_v suffers deviations, due to changes in the grid frequency, v_d and v_q. As the SRF-PLL tries to follow the v_{dq} position, its phase difference $\phi_{pll,NLM}$, of the non linear model, suffers a deviation and after some time equals the phase ϕ_v. The same behavior happens in the SSM phase $\phi_{pll,SSM}$.

As controls of power electronics converters are sometimes performed in the SRF-PLL reference frame, changes in its phase difference affects the control outputs and they influences the converter stability [9]. The linear model fairly represents this phase difference with errors bellow 3%.

978-1-7281-4181-7/19 $31.00 © 2019 IEEE

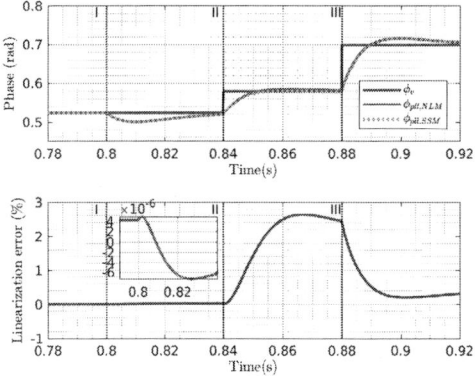

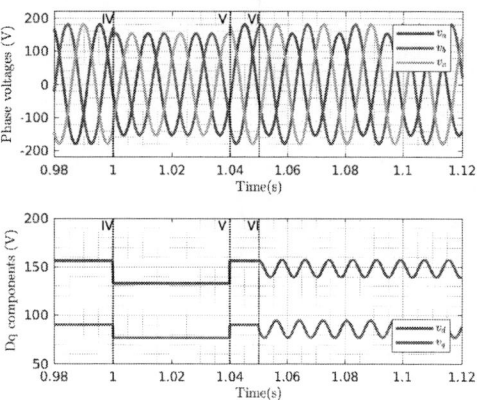

Fig. 8. Top: Phase difference between the VS and PLL reference frames. Bottom: Error between $\phi_{pll,SSM}$ and the $\phi_{pll,NLM}$ normalized to the range of $\phi_{pll,NLM}$.

Fig. 9. Voltage sags during frequency estimation.

Therefore, both models could be acceptable due to low errors. The $\hat{\omega}_2$ estimation presents less high frequency content as there are no sudden variations in its signal. This choice can be useful to generate the reference to power converter operation.

SRF-PLL state-space models, as described here, could be used to create more complex linear models to represent the overall converter dynamics even in a scenario of variable frequency.

A. Operation under voltage sags

Occurrence of voltage sags is common in the electric grid. This can happen due to a connection of a large load or by starting electrical machines. Therefore, simulations were done with the SRF-PLL model working under these condition.

Between the instants 1 s and 1.04 s (IV and V in Fig. 9 and 10) the system presents a three phase voltage sag of 15 %. After 1.05 s (VI in Fig. 9 and 10) the system presents a single phase voltage sag of the same magnitude. The input voltages are shown in Fig. 9.

In this figure, during the three phase voltage sag, the direct and quadrature voltages are reduced by the same proportion. During the single phase voltage sag, only v_a is reduced. The outcome of this is the appearance, in the direct and quadrature voltages, of a component in the double of the sinusoidal voltages frequency.

In Fig. 10, the SRF-PLL frequency estimation during the voltage sags is shown. The three phase sag is not harmful to the frequency estimations and both have a similar behavior to the nominal operation of the SRF-PLL.

However, one phase sag is very different due to the direct and quadrature voltages to present a content in the double nominal frequency. As the $\hat{\omega}_1$ has a direct path to the frequency components of the terminal voltage vector these frequencies are very significant in the frequency estimation. It can even be amplified by the k_{pll} gain.

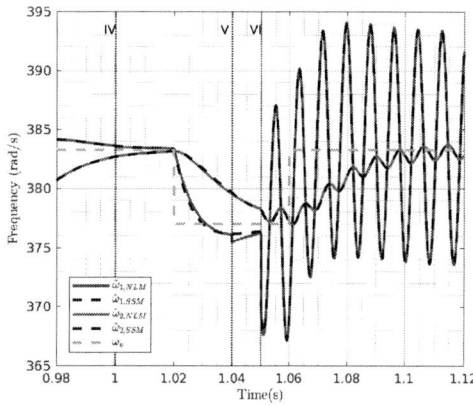

Fig. 10. Frequency estimations and linearized models under voltage sags

The $\hat{\omega}_2$ estimation attenuates the frequency content, however it is still visible in Fig. 10. There are possible solutions to the problem of tree phase voltage unbalancing, however the focus of this paper is to create a SRF-PLL model that reproduces the voltage influence in the estimated frequency [12] [13].

The SRF-PLL linear model works nicely even with unbalanced voltages as it can be represented in the direct and quadrature voltages. The linearized estimations can be seen in the black dashed lines that follow the lines of the non linear SRF-PLL estimations in Fig. 10.

IV. CONCLUSION

Disregarding the voltage influences over the SRF-PLL frequency estimation excludes some important dynamics of the SRF-PLL output, as the disturbances due to voltage variation present in Fig. 6 and 7 would not be represented. Therefore, a more complete model is needed.

The SRF-PLL models exploited in this paper fairly represents the PLL reference frame phase ϕ_{pll}, as well as the frequency estimations $\widehat{\omega}_1$ and $\widehat{\omega}_2$.

The two frequency estimation alternatives have similar settling time, although the frequency content on the $\widehat{\omega}_1$ estimation is larger than the one for $\widehat{\omega}_2$, what could indicate that a system could present different behaviors depending of which estimation was chosen.

A better behavior of the frequency $\widehat{\omega}_2$ is also expected in situations of single phase voltage sag. The linear models of both frequency estimation work well under voltage sag conditions as the unbalances can be represented in the Park transform.

ACKNOWLEDGMENT

Thanks to João Inácio Ota for his reviews and fruitful conversations about the topic. Thanks to Paloma Elias da Silva Pellegrini for reviewing English usage.

REFERENCES

[1] D. E. Olivares, A. Mehrizi-Sani, A. H. Etemadi, C. A. Cañizares, R. Iravani, M. Kazerani, A. H. Hajimiragha, O. Gomis-Bellmunt, M. Saeedifard, R. Palma-Behnke et al., "Trends in microgrid control," IEEE Transactions on smart grid, vol. 5, no. 4, pp. 1905–1919, 2014.

[2] E. A. A. Coelho, P. C. Cortizo, and P. F. D. Garcia, "Small-signal stability for parallel-connected inverters in stand-alone ac supply systems," IEEE Transactions on Industry Applications, vol. 38, no. 2, pp. 533–542, 2002.

[3] S. Khongkhachat and S. Khomfoi, "Droop control strategy of ac microgrid in islanding mode," in 2015 18th International Conference on Electrical Machines and Systems (ICEMS), Oct 2015, pp. 2093–2098.

[4] "Ieee standard for interconnection and interoperability of distributed energy resources with associated electric power systems interfaces," IEEE Std 1547-2018 (Revision of IEEE Std 1547-2003), pp. 1–138, April 2018.

[5] J. Rocabert, A. Luna, F. Blaabjerg, and P. Rodríguez, "Control of power converters in ac microgrids," IEEE Transactions on Power Electronics, vol. 27, no. 11, pp. 4734–4749, Nov 2012.

[6] H. Wu and X. Wang, "Transient stability impact of the phase-locked loop on grid-connected voltage source converters," in 2018 International Power Electronics Conference (IPEC-Niigata 2018-ECCE Asia). IEEE, 2018, pp. 2673–2680.

[7] Q.-C. Zhong and T. Hornik, Control of power inverters in renewable energy and smart grid integration. John Wiley & Sons, 2012, vol. 97.

[8] S. Gao and M. Barnes, "Phase-locked loops for grid-tied inverters: Comparison and testing," 2016.

[9] B. Wen, D. Boroyevich, P. Mattavelli, Z. Shen, and R. Burgos, "Influence of phase-locked loop on input admittance of three-phase voltage-source converters," in 2013 Twenty-Eighth Annual IEEE Applied Power Electronics Conference and Exposition (APEC), March 2013, pp. 897–904.

[10] S. Golestan, J. M. Guerrero, J. C. Vasquez, A. M. Abusorrah, and Y. Al-Turki, "A study on three-phase flls," IEEE Transactions on Power Electronics, vol. 34, no. 1, pp. 213–224, Jan 2019.

[11] R. M. Santos Filho, P. F. Seixas, P. C. Cortizo, L. A. B. Torres, and A. F. Souza, "Comparison of three single-phase pll algorithms for ups applications," IEEE Transactions on Industrial Electronics, vol. 55, no. 8, pp. 2923–2932, Aug 2008.

[12] L. Shi and M. L. Crow, "A novel pll system based on adaptive resonant filter," in 2008 40th North American Power Symposium, Sep. 2008, pp. 1–8.

[13] P. Rodríguez, J. Pou, J. Bergas, J. I. Candela, R. P. Burgos, and D. Boroyevich, "Decoupled double synchronous reference frame pll for power converters control," IEEE Transactions on Power Electronics, vol. 22, no. 2, pp. 584–592, 2007.

High Step-Up DC-DC Converter with Input Current Sharing Based on the Forward Converter

Víctor Ferreira Gruner
Departamente of Electrical and
Electronics Engineering
Federal University of Santa Catarina
Florianópolis, Brazil
victor.gruner@inep.ufsc.br

Lucas Fiamoncini
Departamente of Electrical and
Electronics Engineering
Federal University of Santa Catarina
Florianópolis, Brazil
lucas.fiamoncini00@gmail.com

Lenon Schmitz
Departamente of Electrical and
Electronics Engineering
Federal University of Santa Catarina
Florianópolis, Brazil
lenonsch@inep.ufsc.br

Denizar Cruz Martins
Departamente of Electrical and
Electronics Engineering
Federal University of Santa Catarina
Florianópolis, Brazil
denizar@inep.ufsc.br

Roberto Francisco Coelho
Departamente of Electrical and
Electronics Engineering
Federal University of Santa Catarina
Florianópolis, Brazil
roberto@inep.ufsc.br

Abstract—**This paper presents a DC-DC converter capable of dealing with low voltage and high input current sources. The proposed topology is based on Forward converters with inputs connected in parallel and outputs connected in series. This configuration allows a natural sharing of the input current and provides high static gain to the converter. The phase-shift modulation is applied to reduce size and volume of the single output filter presented at the topology. The study carried out in this work includes the converter operation stages, equation, theoretical waveforms and modelling. In addition, experimental results performed from a 1-kW prototype are presented and discussed in order to validate the proposed converter operation in closed loop.**

Keywords—Forward converter, High step-up, IPOS, Phase-shift, Single output filter.

I. INTRODUCTION

In recent decades, the improvement of power electronics devices, the availability of microprocessors with greater computational capacity, the development of new communication protocols, and the increase on the demand for electric energy have boosted the dissemination of renewable sources in the electric matrix. In this scenario, the electric power generation is modernizing by moving away from the conventional centralized power plants toward to the distributed generation units, where photovoltaic (PV) and wind (WT) have become suitable alternatives [1], [2].

One of the main drawbacks presented by these sources is the intermittent power levels caused by oscillations of weather conditions [3], [4]. As a possible solution for this problem, some authors have proposed the use of battery banks as a way of storing the primary energy resources (PV and WT) during intervals in which there is excessive generation, for later use. In many works [3], [5], the employment of fuel cells as a way to increase the autonomy of these systems is also cited as a proper possibility. Nevertheless, the usage of batteries and fuel cells are featured by the need of energy processing systems that promote high static gain, due to the low voltage level these sources are able to provide. Moreover, since the current supplied by batteries and fuel cells are usually high, solutions that allow sharing it between several stages are preferable to reduce the conduction losses [6].

In the literature there are several techniques that could be applied to lift the gain of dc-dc converters, with emphasis on the use of switched capacitors/inductors, voltage multipliers,

magnetic coupling, multilevel cells, cascading converters or converters with series-connected output [7], [8]. Clearly, such techniques do not guarantee the current sharing, usually achieved by applying interleaving techniques or converters with parallel-connected inputs [9].

In this context, [10] proposed a topology composed by Forward converters with inputs connected in parallel and outputs in series, as per Figure 1, simultaneously allowing high static gain, input current sharing, and modularity, as required for the application addressed in this paper. Despite the positive mentioned features, the topology proposed by [10] employs an LC filter at the output of each associated Forward converter. In addition, because of the pulsed input current, such topology also requires the use of an input filter, since fuel cells and batteries have their lifespan extended when they supply currents with low ripple.

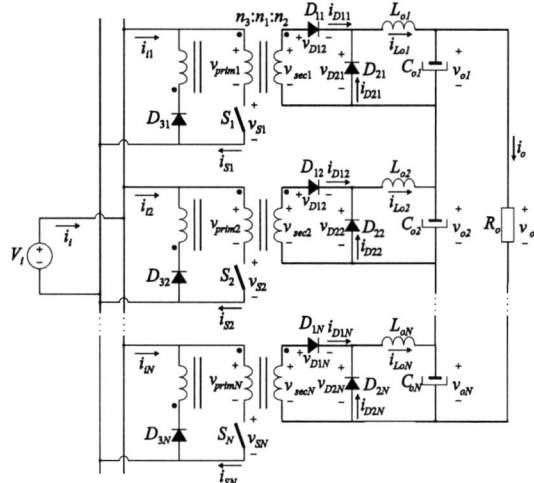

Fig. 1. Switched model of the converter proposed by [10].

Obviously, the high number of passive components present in the input and output filters of the topology proposed by [10] may be understood as a disadvantage because it implies considerable volume and low power density. Therefore, to overcome this disadvantage, this paper proposes modifications to be applied in that topology without compromising its original advantages: high static gain and input current sharing. Moreover, the paper presents the proposed converter operating principle, theoretical waveforms, equation, modeling and experimental results.

978-1-7281-4181-7/19 $31.00 © 2019 IEEE

II. PROPOSED DC-DC CONVERTER

The switched model of the proposed converter is illustrated in Fig. 2. The topology consists of Forward converters with the inputs parallel-connected to the source V_i. This connection allows sharing the input current I_{in} among the N associated Forward converters, that is:

$$I_{FWD} = \frac{I_{in}}{N} \tag{1}$$

where I_{FWD} is the input current of each Forward converter.

In addition, since the output terminals of the Forward converters are series-connected, the total output voltage V_o is given by:

$$V_o = NV_{FWD} \tag{2}$$

where V_{FWD} is the output voltage of each Forward converter.

Furthermore, the static gain G of the proposed topology can be determined accordingly to:

$$G = nND, \tag{3}$$

where n is the ratio between the number of turns of the secondary and primary windings of the transformer ($n=n_2/n_1$) of each Forward converter and D is its operating duty cycle.

In order to simplify the output filter of the converter proposed by [10], it is assumed that the N Forward converters are identical on from other. Under this consideration, the currents $I_{Lo1}, I_{Lo2}, I_{LoN}$ can be considered as being equal and thus the inductors $L_{o1}, L_{o2},$ and L_{oN} may be associated in series, resulting in the single output inductor illustrated in Figure 2, which value is:

$$L_o = L_{o1} + L_{o2} + ... + L_{oN} = NL_{oN} \tag{4}$$

Similarly, the current across the capacitors $i_{Co1}, i_{Co2}, i_{CoN}$ can also be considered equal, in a way such capacitors may be connected in series, resulting in:

$$C_o = \left(\frac{1}{C_{o1}} + \frac{1}{C_{o2}} + ... + \frac{1}{C_{oN}} \right)^{-1} = \frac{C_{oN}}{N} \tag{5}$$

In accordance with (4), the single output inductor of the proposed converter needs an inductance N times higher than the inductors required in the proposal of [10]. In other words, if on one hand we can associate N inductors in a single component, on another hand it will present higher size and volume. Nevertheless, this drawback may be overcome just by operating the Forward converters with phase-shift modulation. In this case, the apparent frequency of the current ripple f_{ripple} across L_o is given by:

$$f_{ripple} = Nf_s, \tag{6}$$

where f_s is the switching frequency of each Forward converter. Therefore, the inductance of the single output inductor is reduced by a factor N, which allows resizing it accordingly with:

$$L_o = L_{oN} \tag{7}$$

Additionally, the employment of phase-shifted modulation also allows reducing the input filter (if existing) since the apparent frequency related to the ripple of the input current I_{in} is also ruled by (6). Therefore, the input filter inductance and capacitance with phase-shifted modulation become N times lower than their respective values obtained from in-phase modulation.

Under phase-shift modulation and in continuous conduction mode (CCM), the proposed converter may operate in two states: with or without overlap of the gate signals. Since the overlap does not change the static gain determined by (3), the change from one state to another does not deteriorate the converter operation and permits the duty cycle to be varied as in a traditional Forward converter. Moreover, the maximum operating duty cycle D_{max} of each Forward converter must satisfy the time interval required to allow the transformer demagnetization:

$$D_{max} = \frac{1}{1 + n_3 / n_1}. \tag{8}$$

When the number of turns n_3 of the tertiary winding is equal to the number of turns n_1 of the primary winding, the maximum duty cycle is adjusted to 50%. Here, the demagnetization stage has been omitted from the analysis since it does not interfere in the transfer of energy to the load. Basically, in the operation of one of the Forward converters, while the switch S_N is turned on, the polarity of the voltage applied to its tertiary winding is inverted with respect to the primary winding, so that the diode D_{3N} remains blocked. After S_N is turned off, the polarity of the voltage across the tertiary winding is inverted and the diode D_{3N} is turned on, ensuring the continuity of the current in the magnetizing inductor by returning it to the power supply V_i and ending the demagnetization process of the transformer. This process will be identical for all of the N associated Forward converters.

In this paper, two states will be addressed: without overlap of gate signal and with overlap of two gate signals. For the first case, the duty cycle of the converter should be lower than $1/N$, while for the second case the duty cycle should be greater than $1/N$ and lower than $2/N$.

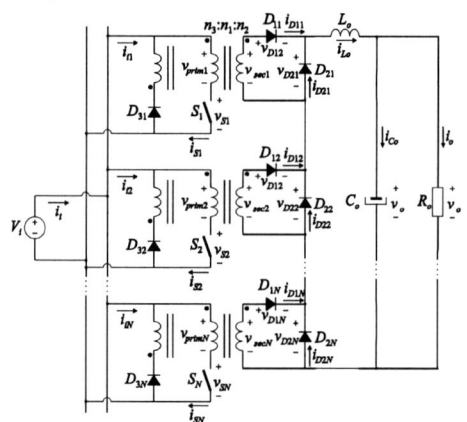

Fig. 2. Switched model of the proposed DC-DC converter.

A. Operation without overlap of gates signals (D<1/N)

The equivalent circuits and the main waveforms related to the operating stages of the proposed converter when using phase-shifted modulation without overlap of the gate signals are shown in Figure 3 and Figure 4, respectively.

- *First stage ($t_0 < t < t_1$):* starts when the switch S_1 is turned on, as Figure 3 (a). The polarities of the voltages applied to the primary and secondary windings of T_1 allow the energy to be transferred from the input source V_i to the load R_o through the diodes D_{11} and $D_{22}...D_{2N}$.

- *Second stage* ($t_1 < t < T_s/N$): starts when S_1 is turned off and the diode D_{11} is blocked, as per Figure 3 (b). At this time, the diode D_{21} assumes the inductor current along with the diodes $D_{22}...D_{2N}$, maintaining it in free wheel and in linear decreasing, while the load R_o is feed by the energy previously stored in the output capacitor and inductor.

- *Third stage* ($T_s/N < t < t_2$): starts when the switch S_2 is turned on. Similarly to the first stage, the polarities of the voltage applied to the windings of the transformer T_2 enable the energy to be transmitted from the source V_i to the load R_o; however, in this case, through the diodes D_{21}, D_{12} and D_{2N}, as per Figure 3 (c).

- *Fourth stage* ($t_2 < t < 2T_s/N$): starts when the switch S_2 is turned off. This stage, shown in Figure 3 (d), is identical to the second one, being featured as a free wheel stage, in which the inductor current decreases linearly through the diodes $D_{21}...D_{2N}$.

- $(2N-1)^{th}$ *stage* $[(N-1)T_s/N < t < t_N]$: starts when the switch S_N is turned on, in a way that the Nth Forward converter transmits energy to the load through the diodes D_{2N}, D_{22} and D_{21}, similarly to the first and third stages. This stage is depicted in Figure 3 (e).

- $2N^{th}$ *stage* ($t_N < t < T_s$): starts when the switch S_N is turned off. In this stage, illustrated in Figure 3 (f), the inductor current is in free wheel through the diodes D_{21}, D_{12} and D_{2N}. When this stage ends, the cycle is restarted from the first stage.

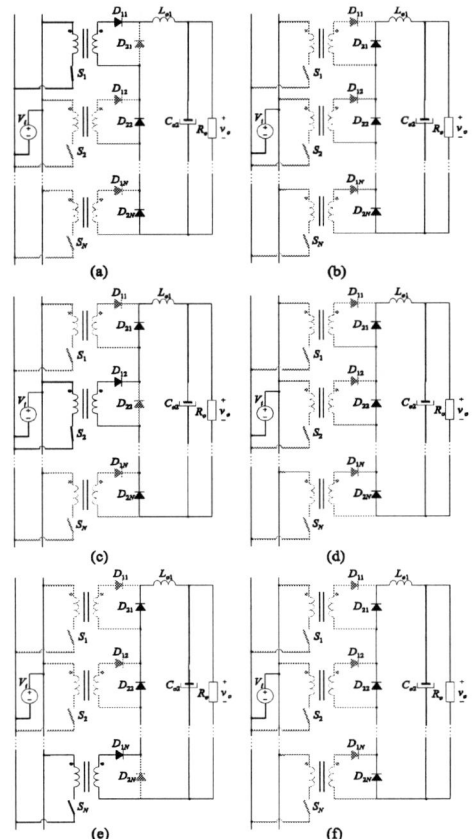

Fig. 3. Proposed converter operating stages under phase-shift modulation without overlap of gate signals.

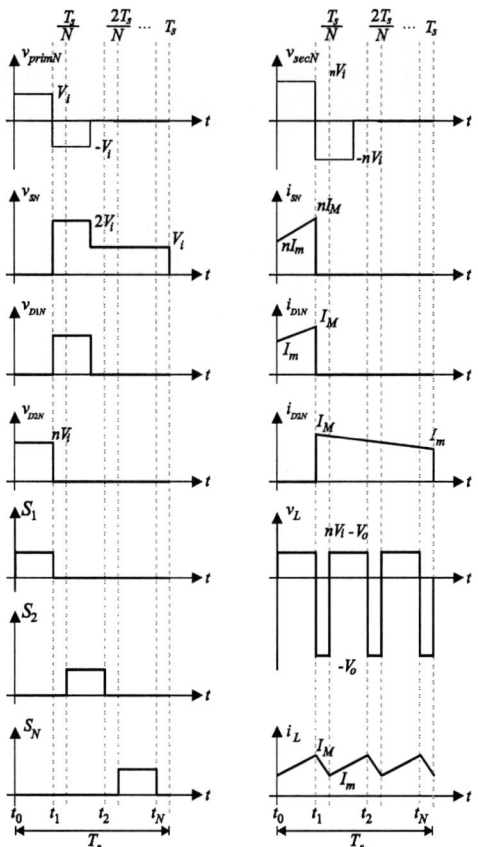

Fig. 4. Main waveforms without overlap of the gate signals.

B. Operation with overlap of two gate signals ($1/N<D<2/N$)

The equivalent circuits and the main waveforms related to the operating stages of the proposed converter when using phase-shifted modulation with overlap of two gate signals are shown in Figure 5 and Figure 6, respectively. In this case, the converter operating stages are described as follow:

- First stage ($t_0 < t < t_1$): starts when the switches S_1 and S_N are turned on. The polarities of the voltages applied to the primary and secondary windings of the transformers T_1 and T_N allow the energy to be transferred from the input source V_i to the load R_o through the diodes D_{11}, D_{1N} and D_{22}. During this stage, illustrated in Figure 5 (a), the inductor current increases linearly.

- Second stage ($t_1 < t < T_s/N$): starts when S_N is turned off. Immediately the current flowing through S_N is nulled and, as consequence, the D_{1N} diode is turned off. At this time, D_{2N} assumes the inductor current, which decreases linearly. The diodes D_{11}, D_{22} and D_{2N} transfer the energy from the source V_i to the load R_o, as per Figure 5 (b).

- Third stage ($T_s/N < t < t_2$): starts when the switch S_2 is turned on along whit S_1. Similarly to the first stage, the polarity of the voltage applied to the windings of the transformers T_1 and T_2 allows the energy to be transferred from the source V_i to the load R_o though the diodes D_{11}, D_{12} and D_{2N}, accordingly with Figure 5 (c).

978-1-7281-4181-7/19 $31.00 © 2019 IEEE

- **Fourth stage** ($t_2 < t < 2T_s/N$): starts when the switch S_1 is turned off, while the switch S_1 remains turned on. This stage, depicted in Figure 5 (d), is similar to the second one, nevertheless, the energy provided by the source V_i is transferred to the load R_o through D_{12}, D_{21} and D_{2N}.

- **$(2N\text{-}1)^{\text{th}}$ stage** $[(N-1)T_s/N < t < t_N]$: starts when S_N is turned on while S_2 is already turned on. Similarly to the first and third operating stages, the polarities of the voltages applied to the windings of the transformers T_2 and T_N allow the energy to be transferred from the source V_i to the load R_o through the diodes D_{12}, D_{1N} and D_{21}. This stage is represented by Figure 5 (e).

- **$2N^{\text{th}}$ stage** ($t_N < t < T_s$): starts when switch S_2 is turned off while the switch S_N remains turned on. This stage is similar to the second and forth ones, however, in this case the energy is transferred from the source V_i to the load through the diodes D_{1N}, D_{21} and D_{22}, as per Figure 5 (f). At the end of this stage the entire process is restarted.

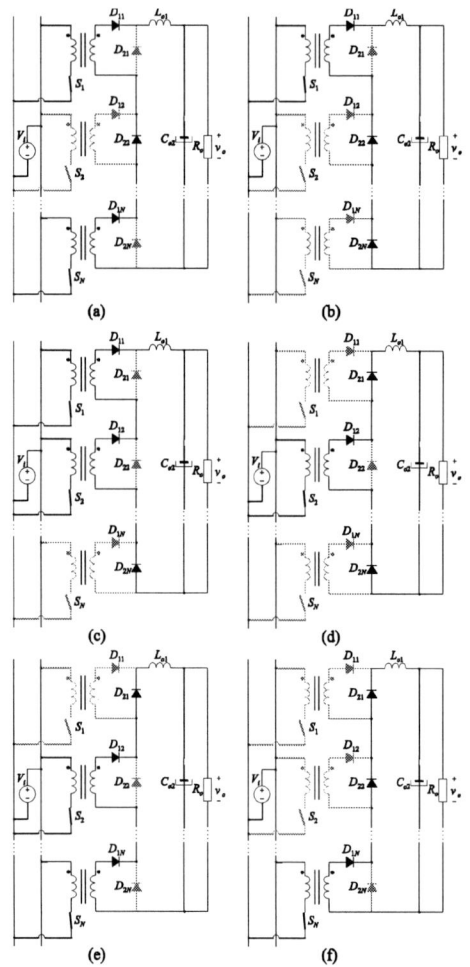

Fig. 5. Proposed converter operating stages under phase-shift modulation with overlap of two gate signals.

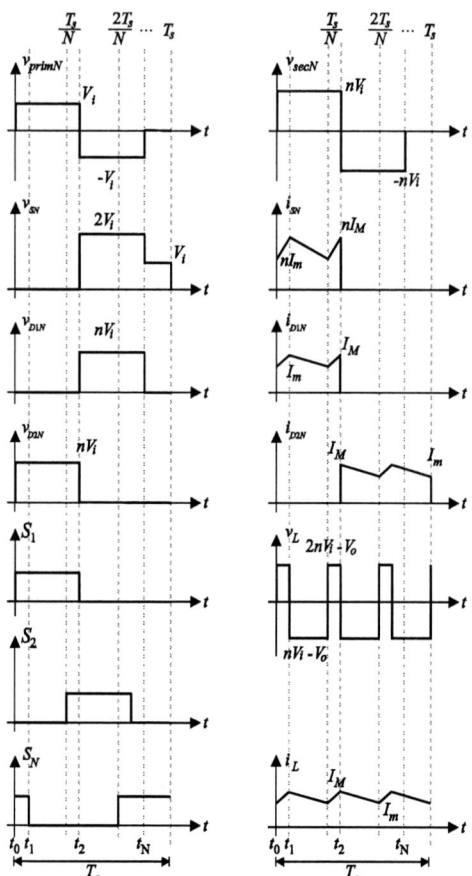

Fig. 6. Main waveforms with overlap of the two gate signals.

III. Control-Oriented Modeling

In order to define the transfer functions required to the design of the controllers, it is necessary to represent the converter by its average small-signal model. Thus, we propose the reduction of the proposed converter to a Buck converter with equivalent dominant dynamic behavior.

Since the inputs of the Forward converters are paralleled-connected, the secondary windings of the transformers $T_1 \ldots T_N$ are subjected to the same voltage. This assumption allows eliminating the transformers. In addition, in the classical Buck converter, the output voltage can be defined as the average value of the voltage applied to the diode. Nevertheless, as the diodes $D_1 \ldots D_N$ may be assumed as being in series, the output voltage is defined by N times the average value of the voltage across one of these diodes. As a result, the proposed converter can be reduced to an equivalent Buck converter with an input voltage of nNV_i. This last simplification eliminates the apparent high frequency (Nf_s) over the filters components, but the dominant low-frequency dynamic is not modified. After defining the equivalent Buck converter, it is possible to determine the transfer functions on using the average small-signal model [16], which results in:

$$G_i(s) = \frac{\hat{i}_{Lo}(s)}{\hat{d}(s)} = \frac{nNV_i(sR_oC_o+1)}{s^2R_oC_oL_o+sL_o+R_o} \qquad (9)$$

$$G_v(s) = \frac{\hat{v}_o(s)}{\hat{i}_{Lo}(s)} = \frac{R_o}{sR_oC_o+1} \quad . \qquad (10)$$

IV. DESIGN METHODOLOGY

Since the proposed topology is composed of Forward converters, the designing methodology adopted to size the power stage devices may follow the criteria already existing in the literature, thus, the switches, the diodes, and the transformers of each Forward converter were designed accordingly with [16].

Cleary, as the output filter was modified, a novel methodology to derive its components (L_o, C_o) needs to be derived. The filter inductance is calculated to ensure that the maximum specified current ripple is never exceeded, thus, it must be defined for the worst operating case. The general equation to determine the current ripple as a function of the number of Forward converters (N) and overlaps (n_{ov}) is plotted in Figure 7, being mathematically described by:

$$\Delta i_{Lo} = \frac{nV_i}{L_o f_s}\left[-ND^2 + D(2n_{ov}+1) - \frac{n_{ov}^2}{N} - \frac{n_{ov}}{N}\right]. \quad (11)$$

The maximum current ripple is determined by equaling the derivative of (8) with respect to D to zero, resulting in:

$$\Delta i_{Lomax} = \frac{1}{4N}. \quad (12)$$

Therefore, the minimum inductance is obtained by replacing (12) in (11):

$$L_{o_{min}} = \frac{nV_i}{4N\Delta i_{Lo}f_s}. \quad (13)$$

It can be observed from Figure 7 that there are some operating points where the output inductor current ripple becomes null. It occurs because in these points the sum of the voltages applied to the secondary windings of the transformers is equal to V_o:

$$V_o = (1+n_{ov})nV_i = nNDV_i. \quad (14)$$

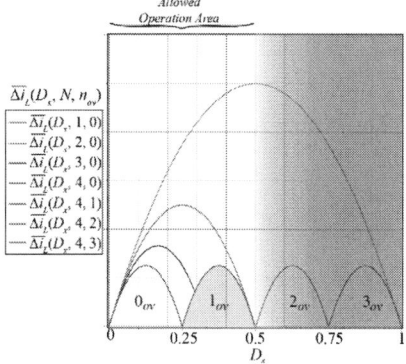

Fig. 7. Output inductor current ripple as a function of the duty cycle, for several combinations of nov and N.

As a result, the duty cycle values in which the converter operates with null output current ripple are:

$$D_{null\_ripple} = \frac{1+n_{ov}}{N} = \frac{1}{N}, \frac{2}{N}, ..., \frac{N}{N}. \quad (15)$$

As in closed-loop the duty cycle is defined by the control loop, it may vary along the converter operation, thus there is not possible to remove the output inductor even when the duty cycle at the rated power is in accordance with (15). Additionally, the output filter capacitance is determined by

analyzing the energy stored in the capacitor as a function of the maximum voltage ripple. In this approach, we assume that the alternate component of the current across the inductor entirely flows through the capacitor, thus:

$$C_{o_{min}} = \frac{\Delta q_{Co}}{\Delta v_{Co}} = \frac{nV_i}{8Nf_s^2 L_o \Delta v_{Co}}. \quad (16)$$

V. EXPERIMENTAL RESULTS

The specifications employed to design the prototype of Figure 8 are listed in Table 1. Due to the apparent high frequency (400 kHz) of the current in the filter, the capacitive effect of the transformers and MOSFETs become relevant, implying resonances that become visible in the waveforms.

TABLE I. POWER STAGE DESIGNING SPECIFICATIONS.

Symbol	Quantity	Values
N	Associated Forwards	4
V_o	Output voltage	400 V
P_o	Output power	1 kW
n_1 / n_3	Conversion ratio	1
n_1 / n_2	Conversion ratio	8,33
V_i	Input voltage	30 V
Δi_{Lo}	Maximum ripple of i_{Lo}	20%
Δv_{Co}	Maximum ripple of v_{Co}	1%
f_s	Switching frequency.	100 kHz
D	Operating duty cycle	0.4
D_{max}	Maximum allowed duty cycle	0.45

Fig. 8. Prototype of the proposed converter.

Figure 9 shows the input and output voltages and currents of the converter operating at the rated power. The stresses across the semiconductor located on the low voltage side, including the voltage and current applied to the switches and demagnetizing diode, are shown in Figure 10 (a). Moreover, Figure 10 (b) shows the voltage and current across the diodes D_1 and D_2, placed at the high voltage side.

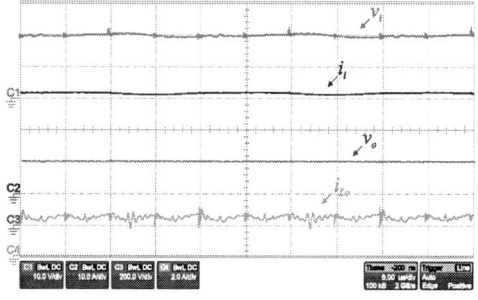

Fig. 9. Input and output voltages and currents at full load.

978-1-7281-4181-7/19 $31.00 © 2019 IEEE

In order to verify the proposed control-oriented modeling, a closed-loop control was design accordingly with [17]. Figure 11 (a) shows the input and output voltages and currents dynamics during a negative load step of 50%, while Figure 11 (b) shows the input and output voltages and currents behavior during a positive load step of 100%.

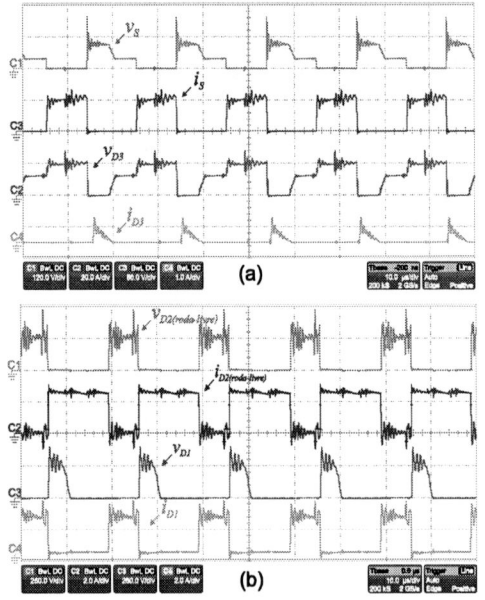

Fig. 10. Voltages and currents on: (a) switches and demagnetizing diode; (b) diodes D1 and D2.

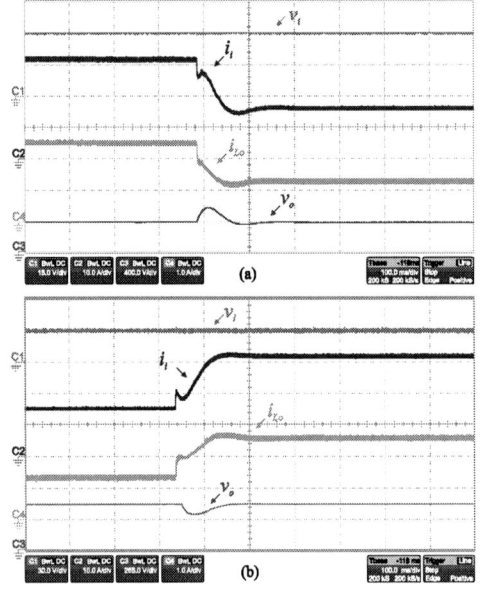

Fig. 11. Dynamic response under: (a) positive load step of 50%; (b) negative load step of 100%.

VI. CONCLUSION

This paper has presented a step-up isolated dc-dc converter with a promising application in the power processing of low voltage and high current sources. The proposed converter is based on the use of Forward converters with inputs connected in parallel and outputs in series (IPOS association).

The main advantages of this structure are the high static gain and the input current sharing, which allows the reduction of conduction loses in the power switches when compared to the conventional Forward converter. Another advantage is the use of phase-shift modulation, which results in an apparent frequency of the input and output filters higher than the switching frequency, allowing the reducing of the input and output filters components and increasing the converter power density, as well.

ACKNOWLEDGEMENT

The authors would like to thank the CNPq for the financial support (process No. 422276/2016-2)

REFERENCES

[1] A. Rohani, K. Mazlumi, H. Kord, "Modeling of a hybrid power system for economic analysis and environmental impact in HOMER", in *Proc. of ICEE*, pp. 819–823, May 2010.

[2] D. C. Martins, R. Demonti, "Photovoltaic energy processing for utility connected system", in *Proc. of IECON*, vol. 2, pp. 1292–1296, 2001.

[3] K. Agbossou, S. Kelouwani, A. Anouar, M. Kolhe, "Energy management of hydrogen-based stand-alone renewable energy system by using boost and buck converters", in *Proc. of IAS*, vol. 4, pp. 2786–2793, Oct. 2004.

[4] K. Agbossou, M. L. Doumbia, A. Anouar, "Optimal hydrogen production in a stand-alone renewable energy system", in *Proc. of IAS*, vol. 4, pp. 2932–2936, Oct. 2005.

[5] R. F. Coelho, L. Schimtz, D. C. Martins, "Grid-connected PV-wind-fuel cell hybrid system employing a supercapacitor bank as storage device to supply a critical DC load", in *Proc. of IINTEC*, pp. 1–10, Oct. 2011.

[6] J. Larmine, A. Dicks, Fuel cell systems explained, Wiley, 2000.

[7] M. Forouzesh, Y. P. Siwakoti, S. A. Gorji, F. Blaabjerg, B. Lehman, "Step-Up DC-DC Converters: A Comprehensive Review of Voltage Boosting Techniques, Topologies, and Applications", *IEEE Transactions on Power Electronics*, vol. PP, no. 99, pp. 1–1, 2017.

[8] B. Huang, I. Sadli, J. P. Martin, B. Davat, "Design of a High Power, High Step-Up Non-isolated DCDC Converter for Fuel Cell applications", in *Proc. of VPPC*, pp. 1–6, Sep. 2006.

[9] M. T. Zhang, M. M. Jovanovic, F. C. Y. Lee, "Analysis and evaluation of interleaving techniques in forward converters", *IEEE Transactions on Power Electronics*, vol. 13, pp. 690–698, Jul. 1998.

[10] L. Schmitz, R. F. Coelho, D. C. Martins, "High step-up DC-DC converter with input current sharing for fuel cell applications", in *Proc. of PEDG*, pp. 1–7, Jun. 2015.

[11] Y. Lian, G. Adam, D. Holliday, S. Finney, "Modular input-parallel output-series DC/DC converter control with fault detection and redundancy", *IET Generation, Transmission & Distribution*, vol. 10, no. 6, pp. 1361– 1369, 2016.

[12] S. Saravanan, N. R. Babu, "Design and Development of Single Switch High Step-Up DC-DC Converter", *IEEE Journal of Emerging and Selected Topics in Power Electronics*, no. 99, pp. 1–1, 2017.

[13] C. L. Shen, Y. C. Lee, J. C. Su, C. T. Tsai, "A high step-up DC/DC converter for PV panel application", in *Proc. of ISEEE*, vol. 2, pp. 1236–1240, Apr. 2014.

[14] S. J. Chen, S. P. Yang, C. M. Huang, C. K. Lin, "Interleaved high step-up DC-DC converter with parallel-input series-output configuration and voltage multiplier module", in *Proc. of ICIT*, pp. 119–124, Mar. 2017.

[15] J. Shi, L. Zhou, X. He, "Common-Duty-Ratio Control of Input-Parallel Output-Parallel (IPOP) Connected DC-DC Converter Modules With Automatic Sharing of Currents", *IEEE Transactions on Power Electronics*, vol. 27, no. 7, pp. 3277–3291, Jul. 2012.

[16] R. Erickson, D. Maksimovic, *Fundamentals of Power Electronics*, Norwell, MA: Kluwer, 2001.

[17] L. Dixon, Average Current Mode Control of Switching Power Supplies, Unitrode Power Supply Design Seminar Manual, pp. 1-12, 1990.

SRF-PLL Influence on the Stability of a Current Source Converter in Droop Mode

Eliabe Duarte Queiroz
School of Electrical and Computer Engineering
University of Campinas
Campinas, Brazil
eliabe.duarte.queiroz@gmail.com

José Antenor Pomilio
School of Electrical and Computer Engineering
University of Campinas
Campinas, Brazil
antenor@fee.unicamp.br

Abstract—A SRF-PLL model considering voltage influences is used to develop a state-space model of a current source converter operating on droop mode. The linear and non-linear models are simulated with two different frequency estimations. Due to the frequency estimation, an instability arises in the non-linear simulation and it is verified in the linear model. Comparisons are made between the linear and non-linear models.

Index Terms—Grid-connected converters, microgrids, phase locked loop (PLL), synchronous reference frame (SRF), power system stability, stability analysis, state-space model, system analysis and design.

I. Introduction

Power electronic converters are mostly used to interface loads and distributed power sources to the alternating current grid. Due to their fast response, higher frequency dynamics are introduced, and some caution in the controller design is necessary in order to avoid unstable interactions [1]. Instability can arise from the converter control feedback due to interaction with the feeder impedance [2][4]. The phase locked loop (PLL) dynamics foment it by increasing the converter quadrature admittance [3].

To prevent instabilities and improve power quality, converters can operate with added functions, as for example, frequency ride-trough capabilities. The IEEE Std 1547-2018 suggests over and under frequency ride-through capabilities as a frequency droop [5].

The droop mode is also used to provide power sharing among power sources [6]-[8]. Furthermore, a voltage droop of reactive power could be inserted to provide better voltage regulation [9][10].

The voltages in the point of common coupling (PCC) are influenced by the dynamics of distributed generation, loads, and feeder. This effect is more prominent in weak grids whose lines have significant impedance. As the voltage dynamics affect the PLL frequency estimation used in the active droop, it should be taken into account to analyze the system stability [11].

A synchronous reference frame PLL (SRF-PLL) model that considers voltage deviations is developed in [11] and it is used as a block to construct the state-space model (SSM) of a current source converter (CSC) connected to a feeder [12].

The converter SSM can be derived by applying the Park transform and subsequent linearizations. Finally, the main objective is to study the frequency estimation influence on the stability of CSCs operating in droop mode.

II. SRF-PLL MODEL CONSIDERING VOLTAGE INFLUENCE

Two SRF-PLL strategies are considered, as shown in Fig. 1. The first one, $\hat{\omega}_1$, is obtained by the SRF-PLL PI output. The second one, $\hat{\omega}_2$ considers only the integrator output.

From Fig. 1, a fraction of the $\hat{\omega}_1$ estimation is proportional to the $v_{q,pll}$, which is a transformation of the PCC voltages. Thus, it is clear that grid voltage disturbances directly affect the $\hat{\omega}_1$ estimation, while the impact on $\hat{\omega}_2$ is limited by the integrator. Note that even using the $\hat{\omega}_2$ output, the feedback continues using the $\hat{\omega}_1$ estimation.

Here, a variable x is composed by the steady value, $\overline{x}$, and the deviation, $\tilde{x}$ as follows:

$$x = \overline{x} + \tilde{x}$$

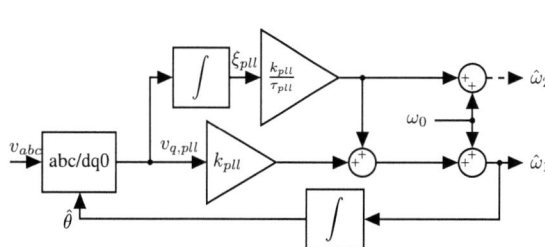

Fig. 1. SRF-PLL diagram.

The SSM, considering voltage influences to the $\hat{\omega}_1$ frequency estimation, is given by:

$$\begin{cases} \tilde{\mathbf{x}}'_{pll} = A_{pll}\tilde{\mathbf{x}}_{pll} + B_{pll,v}\tilde{\mathbf{u}}_v + B_{pll,\omega}\tilde{\mathbf{u}}_\omega \\ \mathbf{y}_{pll} = C_{pll,1}\tilde{\mathbf{x}}_{pll} + D_{pll,v1}\tilde{\mathbf{u}}_v \end{cases} \quad (1)$$

As for the matrices, they are given by:
$$A_{pll} = \begin{bmatrix} 0 & -\overline{V}_m \\ k_{pll}/\tau_{pll} & -k_{pll}\overline{V}_m \end{bmatrix}$$
$$B_{pll,v} = \begin{bmatrix} \overline{V}_m t_d & \overline{V}_m t_q \\ k_{pll}\overline{V}_m t_d & \overline{V}_m k_{pll} t_q \end{bmatrix}$$

This work was funded by the National Council for Scientific and Technological Development (CNPq), grant #302257/2015-2, and São Paulo Research Foundation (FAPESP), grant #2016/08645-9.

978-1-7281-4181-7/19 $31.00 © 2019 IEEE

$$B_{pll,\omega} = \begin{bmatrix} 0 & -1 \end{bmatrix}^T$$

$$C_{pll,1} = \begin{bmatrix} \frac{k_{pll}}{\tau_{pll}} & -k_{pll}\overline{V}_m \end{bmatrix}$$

$$\widetilde{\mathbf{x}}_{pll} = \begin{bmatrix} \widetilde{\xi}_{pll} & \widetilde{\phi}_{pll} \end{bmatrix}^T$$

$$\widetilde{\mathbf{u}}_v = \begin{bmatrix} \widetilde{v}_d & \widetilde{v}_q \end{bmatrix}^T$$

$$\widetilde{\mathbf{u}}_\omega = \begin{bmatrix} \widetilde{\omega}_v \end{bmatrix}$$

$$\widetilde{\mathbf{y}}_{pll} = \begin{bmatrix} \widetilde{\phi}_{pll} & \widetilde{\omega}_1 \end{bmatrix}^T$$

$$D_{pll,v1} = \begin{bmatrix} k_{pll}\overline{V}_m t_d & k_{pll}\overline{V}_m t_q \end{bmatrix}$$

Where: $t_d = -\dfrac{\overline{v}_q}{\sqrt{\overline{v}_d^2 + \overline{v}_q^2}}$

$t_q = \dfrac{\overline{v}_d}{\sqrt{\overline{v}_d^2 + \overline{v}_q^2}}$

$V_m = \sqrt{v_d^2 + v_q^2}$ - Voltage vector module,

ξ_{pll} - Integral of $v_{q,pll}$,

$v_{q,pll}$ - Quadrature voltage in the converter reference frame,

ϕ_{pll} - d axis phase of the converter reference frame,

ω_v - Voltage source frequency,

v_d and v_q - Respectively, direct and quadrature converter terminal voltages,

k_{pll} - SRF-PLL proportional constant,

τ_{pll} - SRF-PLL time constant,

$\widehat{\omega}_1$ - SRF-PLL frequency estimation with the sum of proportional and integral portions.

The frequency estimation $\widehat{\omega}_2$ is given by:

$$\begin{cases} \widetilde{\mathbf{x}}'_{pll} = A_{pll}\widetilde{\mathbf{x}}_{pll} + B_{pll,v}\widetilde{\mathbf{u}}_v + B_{pll,\omega}\widetilde{\mathbf{u}}_\omega \\ \widetilde{\mathbf{y}}_{pll} = C_{pll,2}\widetilde{\mathbf{x}}_{pll} + D_{pll,v2}\widetilde{\mathbf{u}}_v \end{cases} \quad (2)$$

$$C_{pll,2} = \begin{bmatrix} 0 & 1 \\ \frac{k_{pll}}{\tau_{pll}} & 0 \end{bmatrix}$$

$$D_{pll,v2} = \begin{bmatrix} 0 & 0 \\ 0 & 0 \end{bmatrix}^T$$

$$\widetilde{\mathbf{y}}_{pll} = \begin{bmatrix} \widetilde{\phi}_{pll} & \widetilde{\omega}_2 \end{bmatrix}^T$$

The $D_{pll,v2}$ matrix is equal to zero, which limits the pass band of the system. On the other hand, the $D_{pll,v1}$ matrix creates a path to forward the $\widetilde{\mathbf{u}}_v$ input to the output.

These models were developed and validated in [11], and are used to construct the converter model in the next sections. The estimated frequency and the respective SSM matrices will be referred as $\widehat{\omega}_x$, $C_{pll,vx}$, and $D_{pll,vx}$ for both possible estimations.

III. CURRENT CONTROL OF A CSC WITH INDUCTIVE OUTPUT FILTER

Let us consider a CSC with an inductive output filter, whose control is performed in direct and quadrature axes [13]. Its controller commands direct and quadrature voltages, $v_{i,d}$ and $v_{i,q}$, that are applied on the converter filter with inductance L_{cnv}. The direct and quadrature inductor currents, i_d and i_q, are:

$$i'_d = \frac{v_{i,d} - v_d}{L_{cnv}} + \omega_v i_q \quad (3)$$

$$i'_q = \frac{v_{i,q} - v_q}{L_{cnv}} - \omega_v i_d \quad (4)$$

The voltages $v_{i,d}$ and $v_{i,q}$ are composed by a proportional and an integral function of the direct and quadrature current errors, respectively, ϵ_d and ϵ_q to the references $i_{d,ref}$ and $i_{q,ref}$. The converter switching is disregarded. Also, an additional feed-forward compensation is added to these voltages to decouple the currents in the direct and quadrature axes.

In a general frame, the direct and quadrature inverter voltages suffer rotations due to ϕ_{pll} variations [3]. Transforming the controller voltages from the SRF-PLL frame to the general frame, as there are little rotations due to deviations in ϕ_{pll}, lead us to:

$$\begin{bmatrix} v_{i,d} \\ v_{i,q} \end{bmatrix} = \begin{bmatrix} k_{pi}(\epsilon_d + \frac{\xi_d}{\tau_{pi}}) \\ k_{pi}(\epsilon_q + \frac{\xi_q}{\tau_{pi}}) \end{bmatrix} \begin{bmatrix} \cos(\widetilde{\phi}_{pll}) & -\sin(\widetilde{\phi}_{pll}) \\ \sin(\widetilde{\phi}_{pll}) & \cos(\widetilde{\phi}_{pll}) \end{bmatrix}$$
$$+ \begin{bmatrix} \overline{\omega}_v i_q L_{cnv} \\ -\overline{\omega}_v i_d L_{cnv} \end{bmatrix} \quad (5)$$

Where $\xi_d = \int \epsilon_d dt$ and $\xi_q = \int \epsilon_q dt$ are the integrals of errors, k_{pi} and τ_{pi} are proportional and time constants of the current controller.

Thus, considering that the values of $\overline{\epsilon}_d$ and $\overline{\epsilon}_d$ are null, and $\widetilde{\phi}_{pll}$ is small, after linearization of (5), $v_{i,d}$ and $v_{i,q}$ deviations are:

$$\widetilde{v}_{i,d} = k_{pi}\left(\widetilde{\epsilon}_d + \frac{\widetilde{\xi}_d}{\tau_{pi}} - \frac{\overline{\xi}_q}{\tau_{pi}}\widetilde{\phi}_{pll}\right) + \overline{\omega}_v\widetilde{i}_q L_{cnv} \quad (6)$$

$$\widetilde{v}_{i,q} = k_{pi}\left(\widetilde{\epsilon}_q + \frac{\widetilde{\xi}_q}{\tau_{pi}} + \frac{\overline{\xi}_q}{\tau_{pi}}\widetilde{\phi}_{pll}\right) - \overline{\omega}_v\widetilde{i}_d L_{cnv} \quad (7)$$

A. The space state model of the current control

For a microgrid with inductive feeder, the droop function makes the active power reference, P_{ref}, to increase inversely to the difference between the estimated frequency and its nominal value, ω_0 [7]:

$$P_{ref} = P_0 - k_p(\widehat{\omega}_x - \omega_0) \quad (8)$$

Where k_d is the active droop gain and P_0 is the output active power reference at nominal frequency. Its deviation is:

$$\widetilde{P}_{ref} = -k_p\widetilde{\omega} \quad (9)$$

The injection of reactive power aims to attain terminal voltage values close to the nominal value, $V_{m,0}$. Thus, the reactive power reference Q_{ref} increases inversely to the difference between the grid voltage amplitude $V_{m,lp}$ to its nominal value:

$$Q_{ref} = Q_0 - k_q(V_{m,lp} - V_{m,0}) \quad (10)$$

Where k_q is the reactive droop gain and Q_0 is the output reactive power reference at nominal voltage. The grid voltage amplitude is a Cartesian sum of the filtered direct and quadrature voltages, $v_{d,m}$ and $v_{q,m}$:

$$V_{m,lp} = \sqrt{v_{d,m}^2 + v_{q,m}^2} \quad (11)$$

The reactive power deviation, $\widetilde{Q}_{ref}$, is given by:

$$\widetilde{Q}_{ref} = -k_q n_d \widetilde{v}_{d,m} - k_q n_q \widetilde{v}_{q,m} \quad (12)$$

Where n_d and n_q are the derivatives of $V_{m,lp}$, respectively to $v_{d,m}$ and $v_{q,m}$:

$$n_d = \frac{\overline{v}_d}{\sqrt{\overline{v}_d^2 + \overline{v}_q^2}}$$

$$n_q = \frac{\overline{v}_q}{\sqrt{\overline{v}_d^2 + \overline{v}_q^2}}$$

The active and reactive power reference operating points, $\overline{P}_{ref}$ and $\overline{Q}_{ref}$, are given by, respectively, (8) and (10), evaluated for the operating points of $\widehat{\omega}_x$ and $V_{m,lp}$.

Considering the active and reactive power references, P_{ref} and Q_{ref}, the converter output current components are:

$$i_{d,ref} = \frac{P_{ref} v_{d,m} + Q_{ref} v_{q,m}}{v_d^2 + v_q^2} \tag{13}$$

$$i_{q,ref} = \frac{P_{ref} v_q - Q_{ref} v_d}{v_d^2 + v_q^2} \tag{14}$$

Applying (13) and (14) in (6) and (7), subsequently applying these latter two in (3) and (4), for further linearization, the deviations around the operating points become:

$$\widetilde{i}_d' = -\frac{k_{pi} r_d k_p}{L_{cnv}} \widetilde{\omega}_x + \frac{k_{pi} d_{vd}}{L_{cnv}} \widetilde{v}_{d,m} - \frac{1}{L_{cnv}} \widetilde{v}_d + \frac{k_{pi} d_{vq}}{L_{cnv}} \widetilde{v}_{q,m}$$
$$- \frac{k_{pi}}{L_{cnv}} \widetilde{i}_d + \frac{k_{pi}}{\tau_{pi} L_{cnv}} \xi_d + \overline{i}_q \widetilde{\omega}_v - \frac{\overline{v}_{i,q}}{L_{cnv}} \widetilde{\phi}_{pll} \tag{15}$$

$$\widetilde{i}_q' = -\frac{k_{pi} r_q k_p}{L_{cnv}} \widetilde{\omega}_x + \frac{k_{pi} q_{vd}}{L_{cnv}} \widetilde{v}_{d,m} + \frac{k_{pi} q_{vq}}{L_{cnv}} \widetilde{v}_{q,m} - \frac{1}{L_{cnv}} \widetilde{v}_q$$
$$- \frac{k_{pi}}{L_{cnv}} \widetilde{i}_q + \frac{k_{pi}}{\tau_{pi} L_{cnv}} \xi_q - \overline{i}_q \widetilde{\omega}_v + \frac{\overline{v}_{i,d}}{L_{cnv}} \widetilde{\phi}_{pll} \tag{16}$$

Where:

$$r_d = \frac{\overline{v}_d}{\overline{v}_d^2 + \overline{v}_q^2}$$
$$r_q = \frac{\overline{v}_q}{\overline{v}_d^2 + \overline{v}_q^2}$$
$$s_{dd} = \frac{\overline{v}_q^2 - \overline{v}_d^2}{(\overline{v}_d^2 + \overline{v}_q^2)^2}$$
$$s_{dq} = \frac{-2\overline{v}_d \overline{v}_q}{(\overline{v}_d^2 + \overline{v}_q^2)^2}$$
$$s_{qq} = \frac{\overline{v}_d^2 - \overline{v}_q^2}{(\overline{v}_d^2 + \overline{v}_q^2)^2}$$
$$s_{qd} = \frac{-2\overline{v}_d \overline{v}_q}{(\overline{v}_d^2 + \overline{v}_q^2)^2}$$
$$d_{vd} = \overline{P}_{ref} s_{dd} + \overline{Q}_{ref} s_{qd} - r_q k_q n_d$$
$$d_{vq} = \overline{P}_{ref} s_{dq} + \overline{Q}_{ref} s_{qq} - r_q k_q n_q$$
$$q_{vd} = \overline{P}_{ref} s_{qd} - \overline{Q}_{ref} s_{dd} + r_d k_q n_d$$
$$q_{vq} = \overline{P}_{ref} s_{qq} - \overline{Q}_{ref} s_{dq} + r_d k_q n_q$$
$$\overline{v}_{i,d} = \overline{v}_d - \overline{\omega}_v L_{cnv} \overline{i}_q$$
$$\overline{v}_{i,q} = \overline{v}_q + \overline{\omega}_v L_{cnv} \overline{i}_d$$

In (15) and (16) there are two frequency values, the VS frequency (ω_v), that directly impacts on the behavior of the output current of the converter, and the estimated frequency ($\widehat{\omega}_x$), that yields the active power reference. Accordingly, there are two input vectors of frequency, $\widetilde{\mathbf{u}}_{ct,pll}$ for the estimated frequency and $\widetilde{\mathbf{u}}_{ct,\omega}$ for the actual voltage frequency. The same occurs to the voltages, a voltage input vector, $\widetilde{\mathbf{u}}_v$, influences directly the CSC currents and a voltage measurement input vector, $\widetilde{\mathbf{u}}_{vm}$, is used to generate the current control references.

The current controller can be described by:

$$\begin{cases} \widetilde{\mathbf{x}}_{ct}' = A_{ct} \widetilde{\mathbf{x}}_{ct} + B_{ct,v} \widetilde{\mathbf{u}}_v + B_{ct,vm} \widetilde{\mathbf{u}}_{vm} \\ \qquad\quad + B_{ct,pll} \widetilde{\mathbf{u}}_{ct,pll} + B_{ct,\omega} \widetilde{\mathbf{u}}_{ct,\omega} \\ \widetilde{\mathbf{y}}_{ct} = C_{ct} \widetilde{\mathbf{x}}_{ct} \end{cases} \tag{17}$$

Whose matrices are:
$$\widetilde{\mathbf{x}}_{ct} = \begin{bmatrix} \xi_d & \xi_q & \widetilde{i}_d & \widetilde{i}_q \end{bmatrix}^T$$
$$\widetilde{\mathbf{u}}_{ct,pll} = \widetilde{\mathbf{y}}_{pll}$$
$$\widetilde{\mathbf{y}}_{ct} = \begin{bmatrix} \widetilde{i}_d & \widetilde{i}_q \end{bmatrix}^T$$
$$\widetilde{\mathbf{u}}_{vm} = \widetilde{\mathbf{y}}_{lp,v}$$
$$A_{ct} = \begin{bmatrix} 0 & 0 & -1 & 0 \\ 0 & 0 & 0 & -1 \\ \frac{k_{pi}}{\tau_{pi} L_{cnv}} & 0 & -\frac{k_{pi}}{L_{cnv}} & 0 \\ 0 & \frac{k_{pi}}{\tau_{pi} L_{cnv}} & 0 & -\frac{k_{pi}}{L_{cnv}} \end{bmatrix}$$
$$B_{ct,vm} = \begin{bmatrix} 0 & 0 & d_{vd} & d_{vq} \\ 0 & 0 & q_{vd} & q_{vq} \\ 0 & 0 & \frac{k_{pi} d_{vd}}{L_{cnv}} & \frac{k_{pi} d_{vq}}{L_{cnv}} \\ 0 & 0 & \frac{k_{pi} q_{vd}}{L_{cnv}} & \frac{k_{pi} q_{vq}}{L_{cnv}} \end{bmatrix}$$
$$B_{ct,v} = \begin{bmatrix} 0 & 0 \\ 0 & 0 \\ -\frac{1}{L_{cnv}} & 0 \\ 0 & -\frac{1}{L_{cnv}} \end{bmatrix}$$
$$B_{ct,pll} = \begin{bmatrix} 0 & -r_d k_p \\ 0 & -r_q k_p \\ -\frac{\overline{v}_{i,q}}{L_{cnv}} & -\frac{k_{pi} r_d k_p}{L_{cnv}} \\ \frac{\overline{v}_{i,d}}{L_{cnv}} & -\frac{k_{pi} r_q k_p}{L_{cnv}} \end{bmatrix}$$
$$B_{ct,\omega} = \begin{bmatrix} 0 & 0 & \overline{i}_q & -\overline{i}_d \end{bmatrix}^T$$
$$C_{ct} = \begin{bmatrix} 0 & 0 & 1 & 0 \\ 0 & 0 & 0 & 1 \end{bmatrix}$$

$\mathbf{y}_{lp,v}$ - Voltage measurement through a low pass.

B. The converter state-space model

Filters in the direct and quadrature voltages measurements are used to avoid unstable behavior by attenuating higher frequencies components [14][15]. The SSM of such low pass filters is:

$$\begin{cases} \widetilde{\mathbf{x}}_{lp,v}' = A_{lp,v} \widetilde{\mathbf{x}}_{lp,v} + B_{lp,v} \widetilde{\mathbf{u}}_v \\ \widetilde{\mathbf{y}}_{lp,v} = C_{lp,v} \widetilde{\mathbf{x}}_{lp,v} \end{cases} \tag{18}$$

Where:
$$A_{lp,v} = \begin{bmatrix} -\omega_{lp,v} & 0 & 0 & 0 \\ \omega_{lp,v} & -\omega_{lp,v} & 0 & 0 \\ 0 & 0 & -\omega_{lp,v} & 0 \\ 0 & 0 & \omega_{lp,v} & -\omega_{lp,v} \end{bmatrix}$$
$$B_{lp,v} = \begin{bmatrix} \omega_{lp,v} & 0 \\ 0 & 0 \\ 0 & \omega_{lp,v} \\ 0 & 0 \end{bmatrix}$$
$$C_{lp,v} = \begin{bmatrix} 0 & 0 & 0 & 0 \\ 0 & 0 & 0 & 0 \\ 0 & 1 & 0 & 0 \\ 0 & 0 & 0 & 1 \end{bmatrix}$$

Thus, the complete converter is described by a state-space model, whose matrices are a combination from the ones in (1),(2), (17) and (18):

$$\begin{cases} \widetilde{\mathbf{x}}_{cnv}' = A_{cnv} \widetilde{\mathbf{x}}_{cnv} + B_{cnv,v} \widetilde{\mathbf{u}}_v + B_{cnv,\omega} \widetilde{\mathbf{u}}_\omega \\ \widetilde{\mathbf{y}}_{cnv} = C_{cnv} \widetilde{\mathbf{x}}_{cnv} \end{cases} \tag{19}$$

$$A_{cnv} = \begin{bmatrix} A_{lp,v} & 0 & 0 \\ 0 & A_{pll} & 0 \\ B_{ct,vm} C_{lp,v} & B_{ct,pll} C_{pll,x} & A_{ct} \end{bmatrix}$$
$$B_{cnv,v} = \begin{bmatrix} B_{lp,v}^T & B_{pll,v}^T & (B_{ct,pll} D_{pll,vx} + B_{ct,v})^T \end{bmatrix}^T$$
$$B_{cnv,\omega} = \begin{bmatrix} 0 & B_{pll,\omega}^T & B_{ct,\omega}^T \end{bmatrix}^T$$
$$C_{cnv} = \begin{bmatrix} 0 & 0 & C_{ct} \end{bmatrix}$$

IV. Model validation and analysis

Simulations were performed to evaluate the representation of the non-linear system by the linearized model. The non-linear model (NLM) was simulated in Simulink, and the linear model by the lsim function in Matlab.

A. Solution of non-linear equations

As the linearization works around an operating point, and some SSM parameters are functions of them. Therefore, the steady state needs to be calculated.

The set of non-linear equations can be simplified considering that, in steady state, the active and reactive power supplied by the CSC are equal to their references. This equality is true in a stable system in which the controller tracks the reference with a zero error.

Once in a real application, the inverter switching is harmful to the measurements. Aiming to attenuate it, a capacitor in the PCC is considered, and, moreover, it provides the model a terminal voltage state-space variable that simplifies it. Therefore, the three-phase electrical system is composed by an inductive feeder (L_F and R_F), a PCC capacitor (C_M) and the converter with an inductive filter (L_{cnv}). An unifilar diagram of the circuit is shown in Fig. 2.

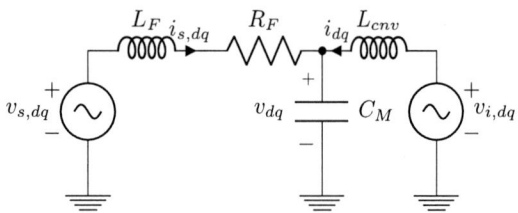

Fig. 2. Converter with an inductive filter and a measurement capacitor connected through an inductive feeder to an ideal source.

Considering the parameters of a circuit and the steady state of the input voltage in Table I, the steady values of v_d, v_q, i_{sd} and i_{sq}, are respectively 182.82 V, 3.9655 V, -11.083 A and 6.6323 A. Combing these values with those of the current controller, τ_{pi} and k_{pi}, respectively, 0.01 and 0.8, the converter matrices can be calculated.

TABLE I
VALUES OF PARAMETERS USED TO CALCULATE THE NON-LINEAR EQUATION SOLUTION.

Parameter	Value	Unity
R_F	30 m	Ω
L_F	1 m	H
C_M	100 μ	F
L_1	5 m	H
P^*	2000	W
Q^*	0	W
$v_{d,s}$	179	V
$v_{q,s}$	0	V
k_p	1591.5	
k_q	833.3333	
ω_L	375.74	rad/s
$\omega_{L,0}$	377	rad/s

B. The feeder and converter models

The SSM of a CSC connected by a feeder to a VS as in Fig. 2 is:

$$\widetilde{\mathbf{x}}'_{grid} = A_{grid}\widetilde{\mathbf{x}}_{grid} + B_{grid,v}\widetilde{\mathbf{u}}_v + B_{grid,\omega l}\widetilde{\mathbf{u}}_\omega \quad (20)$$

Where:
$$A_{grid} = \begin{bmatrix} A_{LF} & -B_{LF}C_{CM} & 0 \\ B_{CM}C_{LF} & A_{CM} & B_{CM}C_{cnv} \\ 0 & B_{cnv,v}C_{CM} & A_{cnv} \end{bmatrix}$$
$$B_{grid,v} = \begin{bmatrix} B_{LF,v}^T & 0 & 0 \end{bmatrix}^T$$
$$B_{grid,\omega l} = \begin{bmatrix} B_{LF,\omega l}^T & B_{CM,\omega L}^T & B_{cnv,\omega l}^T \end{bmatrix}^T$$
$$C_{grid} = \begin{bmatrix} 0 & 0 & C_{cnv} \end{bmatrix}$$
$$x_{grid} = \begin{bmatrix} x_{LF}^T & x_{CM}^T & x_{cnv}^T \end{bmatrix}^T$$

These matrices are composed by the converter matrices and, moreover, by the other elements in the circuit:

- The inductor matrices:

$$A_{LF} = \begin{bmatrix} \frac{R_F}{L_L} & \overline{\omega}_L \\ -\overline{\omega}_L & \frac{R_F}{L_F} \end{bmatrix} \qquad B_{LF,\omega l} = \begin{bmatrix} \overline{i}_{sq} \\ -\overline{i}_{sd} \end{bmatrix}$$
$$B_{LF,v} = \begin{bmatrix} \frac{1}{L_F} & 0 \\ 0 & \frac{1}{L_F} \end{bmatrix} \qquad C_{LF} = \begin{bmatrix} 1 & 0 \\ 0 & 1 \end{bmatrix}$$
$$\widetilde{\mathbf{x}}_{LF}^T = \begin{bmatrix} \widetilde{i}_{sd} & \widetilde{i}_{sq} \end{bmatrix}$$

- The capacitor matrices:

$$A_{CM} = \begin{bmatrix} 0 & \overline{\omega}_L \\ -\overline{\omega}_L & 0 \end{bmatrix} \qquad B_{CM,\omega l} = \begin{bmatrix} \overline{v}_q \\ -\overline{v}_d \end{bmatrix}$$
$$B_{CM,v} = \begin{bmatrix} \frac{1}{C_M} & 0 \\ 0 & \frac{1}{C_M} \end{bmatrix} \qquad C_{CM} = \begin{bmatrix} 1 & 0 \\ 0 & 1 \end{bmatrix}$$
$$\widetilde{\mathbf{x}}_{CM}^T = \begin{bmatrix} \widetilde{v}_d & \widetilde{v}_q \end{bmatrix}$$

C. NLM and SSM comparison

The linear model is compared with a simulation of the non-linear model using an averaged representation of the inverters. The non-linear simulation output is considered around the operating point, excluding an initial transitory. Both frequency estimations, $\widehat{\omega}_1$ and $\widehat{\omega}_2$, are employed in the active droop function to compare the simulation outputs and the stability of the system.

The systems, both in non-linear and linear simulations, are disturbed by steps around the operating points, as shown in Fig. 3. The inputs $\widetilde{v}_{s,d}$, $\widetilde{v}_{s,q}$ and $\widetilde{\omega}_L$ received steps of $3V$, $3V$ and $-0.314\ rad/s$ respectively at, $1s$, $1.1s$ and $1.2s$ (instants I, II, and III, respectively, in Fig. 3, 4, 5 and 6). The use of steps are mainly for model validation, as this kind of disturbances are not usual in real grids.

The direct and quadrature currents, in Fig. 4, show the effects of the source voltage and frequency deviations over the CSC. The quadrature current increases when there is a step in the direct voltage, at 1 s, as the reactive droop makes an effort to keep the grid voltage close to its nominal value. After a quadrature voltage step, at 1.1 s, that represents a phase jump in the VS, there is a growth in the estimated frequency, shown in Fig. 6. Therefore, the active droop function lower the reference to the direct current causing it to decrease. As expected, due the active droop operation, after a step down in frequency, at 1.2, the converter direct current grows, increasing the output active power.

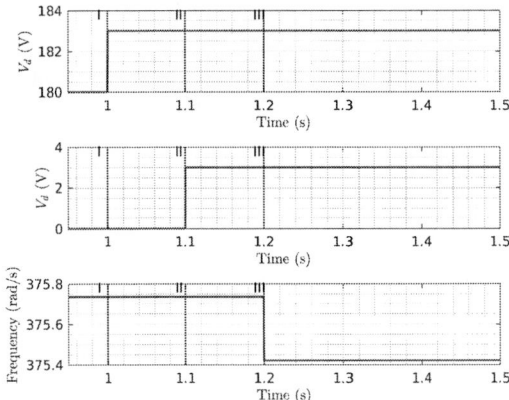

Fig. 3. Inputs to the non-linear and linearized simulations.

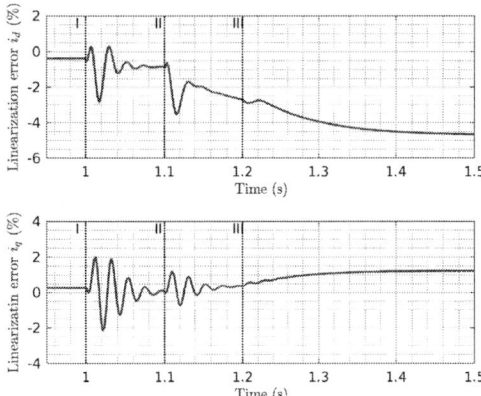

Fig. 5. Direct (top) and quadrature (bottom) current errors between the SSM to the NLM in percentage of the respective range of NLM current deviation.

The SSM accurately follows the non-linear operation, with errors lower than 5%. This is shown more clearly in Fig. 5 where the errors can be seen in percentage of the range current variation of the NLM.

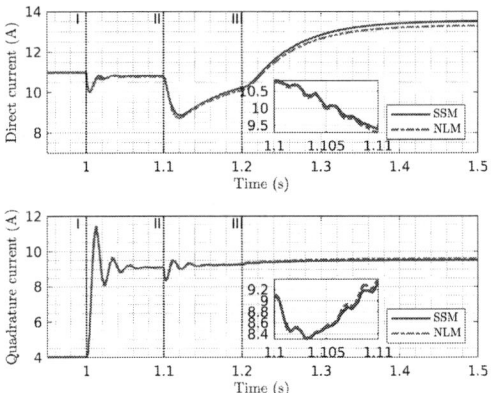

Fig. 4. Direct (top) and quadrature (bottom) currents of the droop mode converter in: the explored SSM, and in the NLM in Simulink.

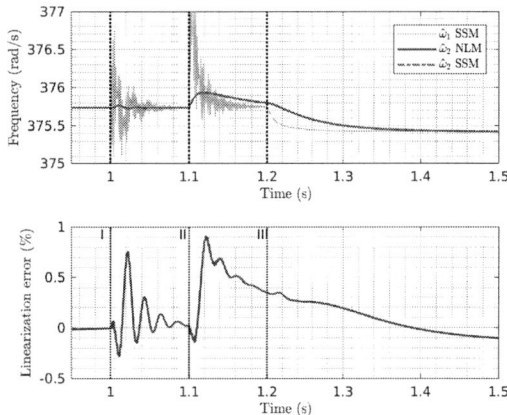

Fig. 6. SRF-PLL $\widehat{\omega}_1$ and $\widehat{\omega}_2$ frequency estimation (top) and the error between the SSM and the NLM $\widehat{\omega}_2$ estimation error (bottom) in percentage of the range of the NLM estimation.

D. Comparison of the $\widehat{\omega}_1$ and $\widehat{\omega}_2$ estimations

The frequency estimations are shown in Fig. 6 and they are represented in three lines of different colors: gray, blue and orange that are respectively, the $\widehat{\omega}_1$ in the SSM, $\widehat{\omega}_2$ in the NLM and $\widehat{\omega}_2$ in the SSM.

The $\widehat{\omega}_1$ estimation shows oscillations that come directly from the terminal voltages. While the frequency estimation $\widehat{\omega}_2$ comes from an integration, and presents less high frequency components. Thus, the droop mode could be more stable preferring the $\widehat{\omega}_2$ estimation over the $\widehat{\omega}_1$.

The consequences of the frequency estimation can be seen in Fig. 7 in which a linearized system, using the $\widehat{\omega}_1$ estimation

to the active droop, shows two eigenvalues in the right half-plane, therefore unstable. When using $\widehat{\omega}_2$, all the eigenvalues are in the left half-plane and the system is, accordingly, stable.

In the non-linear simulation, at 1.75 seconds, the PLL frequency estimation, that initially uses $\widehat{\omega}_2$, is switched to $\widehat{\omega}_1$. The outcome, as shown in Fig. 8 is an unstable behavior.

As only two eigenvalues in the right half-plane are expected, it is possible to estimate their position using a simple approximation. Through the values of the current variations from i_d in Fig. 8, the approximated frequency of the instability is 2785.1 rad/s and the real exponential part is roughly approximated by 656.2. The last value represents the instability growth, shown in the red dashed line ($\bar{i}_d \pm e^{656.2t}$) in Fig. 8.

This unstable behavior represents, in a SSM, eigenvalues in $656.2 \pm 2785.1j$. These values are close to the right plane eigenvalues of A_{grid} with the $\widehat{\omega}_1$ estimation, $654.9 \pm 2845.1j$.

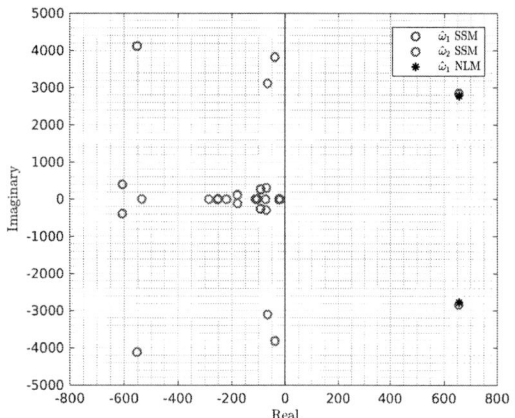

Fig. 7. Eigenvalues in the complex plane to frequency estimations $\widehat{\omega}_1$ and $\widehat{\omega}_2$ in the state-space models, $\widehat{\omega}_1$ SSM and $\widehat{\omega}_2$ SSM respectively, and the unstable modes in the Simulink simulation, $\widehat{\omega}_1$ NLM.

The difference between the response of linear and non-linear models could be due to non modeled behaviors, such as SRF-PLL and control discretization.

This instability is strongly related to the impedance of the feeder. The same CSC with the $\widehat{\omega}_1$ estimation would operate stably if the impedance was low enough. However, the $\widehat{\omega}_2$ estimation in this situation would still have a larger stability margin.

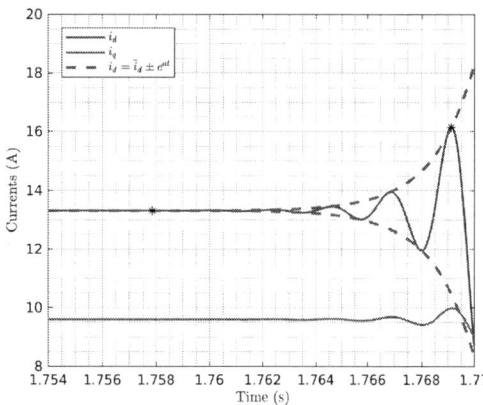

Fig. 8. Direct and quadrature current instabilities with the $\widehat{\omega}_1$ estimation.

Although there are inherent errors, the linear model fairly reproduces the main behavior of the system, and it could be used to better understand the stability of droop mode converters and their associations.

V. Conclusion

The linear model of the converter components satisfactorily represents the system behavior around the operating point, and it could be used to have a better guess of the small signal stability of droop systems.

If the voltage influences are accounted, for both linear and non-linear models where the frequency is estimated through $\widehat{\omega}_1$, instabilities arise more easily than in systems with a $\widehat{\omega}_2$ estimation. Thus, it is better to use the $\widehat{\omega}_2$ frequency estimation to the active droop function.

The system stable response is fairly reproduced by the SRF-PLL presented model, that shows better accuracy than a model not considering the voltage influences.

Acknowledgment

Thanks to João Inácio Ota for his reviews and fruitful conversations about the topic. Thanks to Paloma Elias da Silva Pellegrini for reviewing English usage.

References

[1] A. Ulbig, T. S. Borsche, and G. Andersson, "Impact of low rotational inertia on power system stability and operation," *IFAC Proceedings Volumes*, vol. 47, no. 3, pp. 7290–7297, 2014.

[2] X. Wang, L. Harnefors, and F. Blaabjerg, "Unified impedance model of grid-connected voltage-source converters," *IEEE Transactions on Power Electronics*, vol. 33, no. 2, pp. 1775–1787, 2017.

[3] B. Wen, D. Boroyevich, P. Mattavelli, Z. Shen, and R. Burgos, "Influence of phase-locked loop on input admittance of three-phase voltage-source converters," in *2013 Twenty-Eighth Annual IEEE Applied Power Electronics Conference and Exposition (APEC)*, March 2013, pp. 897–904.

[4] C. Li, "Unstable operation of photovoltaic inverter from field experiences," *IEEE Transactions on Power Delivery*, vol. 33, no. 2, pp. 1013–1015, April 2018.

[5] "Ieee standard for interconnection and interoperability of distributed energy resources with associated electric power systems interfaces," *IEEE Std 1547-2018 (Revision of IEEE Std 1547-2003)*, pp. 1–138, April 2018.

[6] D. E. Olivares, A. Mehrizi-Sani, A. H. Etemadi, C. A. Cañizares, R. Iravani, M. Kazerani, A. H. Hajimiragha, O. Gomis-Bellmunt, M. Saeedifard, R. Palma-Behnke *et al.*, "Trends in microgrid control," *IEEE Transactions on smart grid*, vol. 5, no. 4, pp. 1905–1919, 2014.

[7] E. A. A. Coelho, P. C. Cortizo, and P. F. D. Garcia, "Small-signal stability for parallel-connected inverters in stand-alone ac supply systems," *IEEE Transactions on Industry Applications*, vol. 38, no. 2, pp. 533–542, 2002.

[8] S. Khongkhachat and S. Khomfoi, "Droop control strategy of ac microgrid in islanding mode," in *2015 18th International Conference on Electrical Machines and Systems (ICEMS)*, Oct 2015, pp. 2093–2098.

[9] J. Rocabert, A. Luna, F. Blaabjerg, and P. Rodriguez, "Control of power converters in ac microgrids," *IEEE transactions on power electronics*, vol. 27, no. 11, pp. 4734–4749, 2012.

[10] Z. Liu, J. Liu, D. Boroyevich, R. Burgos, and T. Liu, "Small-signal terminal-characteristics modeling of three-phase droop-controlled inverters," in *2016 IEEE Energy Conversion Congress and Exposition (ECCE)*, Sep. 2016, pp. 1–7.

[11] E. D. Queiroz and J. A. Pomilio, "Voltage influences over srfpll frequency estimation," Accepted for publication on SPEC/COBEP 2019.

[12] N. Pogaku, M. Prodanovic, and T. C. Green, "Modeling, analysis and testing of autonomous operation of an inverter-based microgrid," *IEEE Transactions on power electronics*, vol. 22, no. 2, pp. 613–625, 2007.

[13] J. Rocabert, A. Luna, F. Blaabjerg, and P. Rodríguez, "Control of power converters in ac microgrids," *IEEE Transactions on Power Electronics*, vol. 27, no. 11, pp. 4734–4749, Nov 2012.

[14] M. Cespedes and J. Sun, "Impedance modeling and analysis of grid-connected voltage-source converters," *IEEE Transactions on Power Electronics*, vol. 29, no. 3, pp. 1254–1261, March 2014.

[15] S. Sumsurooah, M. Odavic, S. Bozhko, and D. Boroyevic, "Toward robust stability of aircraft electrical power systems: Using a ?-based structural singular value to analyze and ensure network stability," *IEEE Electrification Magazine*, vol. 5, no. 4, pp. 62–71, Dec 2017.

978-1-7281-4181-7/19 $31.00 © 2019 IEEE

A Low Cost Bi-directional Wireless Power Transfer System

Lei Wang
College of Electrical and Information
Engineering
Hunan University
Changsha, China
jordanwanglei@hnu.edu.cn

Udaya K. Madawala
Department of Electrical and Computer
Engineering
University of Auckland
Auckland, New Zealand
u.madawala@auckland.ac.nz

Man-Chung Wong
State Key Laboratory of Internet of
Tings for Smart City
University of Macau
Macau, China
mcwong@umac.mo

Abstract— **More and more consumer applications are powered using wireless power transfer (WPT) technology. This paper proposes a low cost WPT system for consumer applications that require bi-directional power flow. The proposed system uses two switches to regulate the wireless power flow in both directions. A mathematical model is presented, describing the different modes of operation of the WPT system. To validate the applicability of the proposed concept, both simulated and measured results are presented.**

I. INTRODUCTION

In recent years, wireless power transfer (WPT) technologies has gained substantial attention within the power electronics community. According to [1], the global market for wireless charging is expected to reach $37.2 billion by 2022, with an exponential growth in consumer electronics. Applications of WPT range [2]-[5] from medium and high power systems such as home appliances, material handling systems, street lighting systems, and electric vehicles (EVs) charging system to low-power systems such as biomedical implant, mobile device and measurement devices, etc.

The research trend has been to implement inexpensive, simple, efficient and compact converter systems for low power applications. The Class E resonant converter with a single switch can be considered as one of the classical converters [6], for low power and low cost applications. As reported in the literature, many modified topologies have been derived based on this converter, and these include Class DE [7], Class E^{-1} [8], Class EF2 [9], Class Φ [10], Class E with shunt filter [11], etc.

To extent the above class E-family converters into WPT systems, the class E-family converters have predominantly been developed for WPT applications [12]-[14]. However, these E-family converters [12]-[14] focus to provide wireless power flow only in one direction. The demand for WPT applications is endless and growing, and many consumer applications require wireless power transfer with bi-directional power flow. Therefore, bidirectional WPT (BD-WPT) systems have started to appear in the market place for low power applications. For instance, the mobile phone makers such as Huawei [15], Sony [16] and Samsung [16] have launched the latest models with the capability of charging one phone wirelessly from another. Apple Inc. [17] also has patented techniques related to BD-WPT, enabling device to device charging, including the iWatch, iPhone, iPad, laptop, etc. Another example is the class E BD-WPT system that has been proposed in [18]. In this system, the direction and amount of power flow was controlled by phase

shift control. In [19] and [20], the dual active single-ended converter, using a single switch, has been proposed for BD wireless V2G systems. The above BD-WPT systems have their own advantages and disadvantages, but the industry thrust is still to implement compact, low cost, and high efficiency BD-WPT systems for market dominance.

This paper proposes a new BD-WPT system for low power consumer applications. The proposed converter uses parallel compensation to reduce its VAR consumption, and is directly fed by a voltage source on each side, eliminating the need for extra an inductor and a diode. In comparison to existing converters [6]-[14], [18]-[20], the proposed converter uses less components. Therefore, the proposed converter can fit the converter developing trend of inexpensive, simple implement, efficient and compact. The paper describes the topology and operating principle of the proposed converter, and presents a comprehensive mathematical model to investigate its power flow. To demonstrate the applicability of the proposed converter topology, a 60W prototype is built and experimental results are presented in comparison to simulations.

II. THE PROPOSED BD-WPT SYSTEM

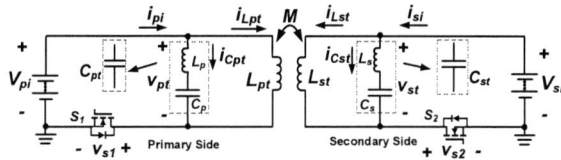

Fig. 1 The proposed BD-WPT system

The proposed BD-WPT system shown in Fig. 1, needs only 2 switches to regulate wireless power flow in both directions. The primary and secondary side coils, represented by self- inductances L_{pt} and L_{st}, are separated by an air-gap, but magnetically coupled through mutual inductance M. Self-inductance of the primary coil is parallel compensated by a series connected inductor (L_p) and capacitor (C_p) circuit, which is represented by an equivalent capacitor C_{pt}. The series inductor L_{pt} is used to limit the inrush current at start-up, and the circuit formed by L_{pt} and C_{pt} is excited by a voltage, controlled through switch S_1, to resonate at switching frequency f_s. Voltage and current of the primary side switch, inductor and capacitor are represented by v_{s1}, i_{pi}, i_{Lpt} and i_{Cpt}. Parameters of the secondary side of the system are represented in a similar manner by v_{s2}, i_{si}, i_{Lst} and i_{Cpt}, which is being identical to the primary side. Power transfer between the two voltage sources, represented by V_{pi} and V_{si}, takes

978-1-7281-4181-7/19 $31.00 © 2019 IEEE

place through magnetic coupling over the air-gap and can be regulated using the two switches S_1 and S_2.

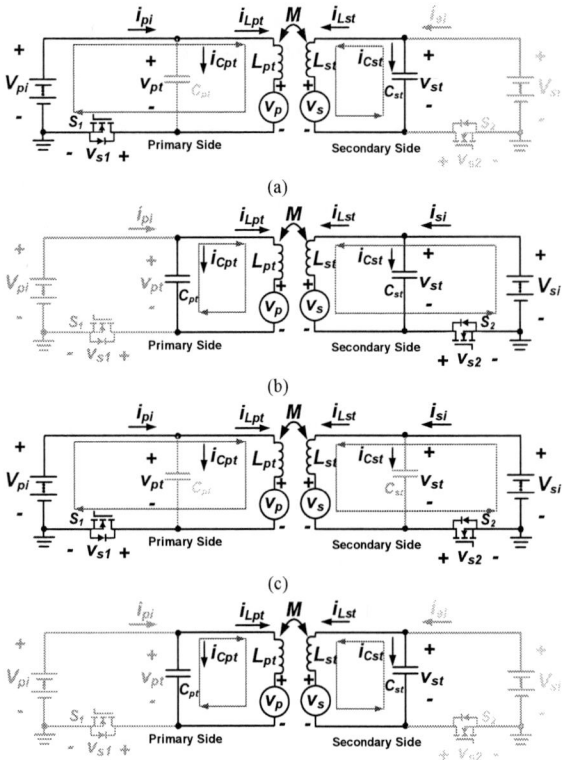

(a)

(b)

(c)

(d)

Fig. 2 Proposed operation modes of BD-WPT system (a) mode I, (b) mode II, (c) mode III, and (d) mode IV

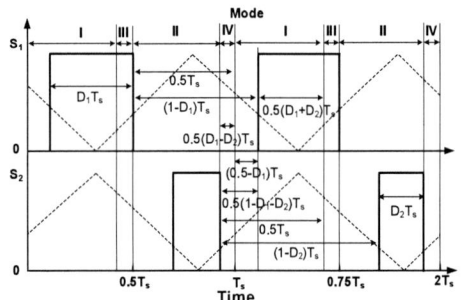

Fig. 3 The basic operation period of the different modes

The operation of the converter, shown in Fig. 1, can be categorized into 4 different modes, based on whether or not the two sources supply/sink currents i_{pi} and i_{si}. Fig. 2 shows these modes. The synchronized switching logics of both switches, for arbitrary duty cycles D_1 and D_2, are illustrated in Fig. 3 over two switching periods. For symmetry and to facilitate analyses, the trigger pulses of S_1 and S_2 are centered around the beginning and middle of the switching period, respectively.

Accordingly, the time durations of each mode can be represented as:

$$T_I = T_{II} = 0.5\left(1-|D_1-D_2|\right)\cdot T_s \tag{1}$$

$$T_{III} = T_{IV} = 0.5|D_1 - D_2|\cdot T_s \tag{2}$$

where T_s is the switching period and T_I, T_{II}, T_{III} and T_{IV} are the time duration of Mode I, II, III and IV, respectively.

III. MODELLING

The voltages induced on the primary and secondary coils due to mutual coupling are represented by $v_p(t)$ and $v_s(t)$, respectively. The primary side of the circuit in Fig. 2 can be mathematically represented in time domain as:

$$v_{pt}(t)=L_{pt}\frac{di_{Lpt}}{dt}+v_p(t)=L_{pt}\frac{di_{Lpt}}{dt}+M\frac{di_{Lst}}{dt} \tag{3}$$

$$v_{st}(t)=L_{st}\frac{di_{Lst}}{dt}+v_s(t)=L_{st}\frac{di_{Lst}}{dt}+M\frac{di_{Lpt}}{dt} \tag{4}$$

$$i_{Cpt}(t)=C_{pt}\frac{dv_{pt}(t)}{dt} \tag{5}$$

$$i_{Cst}(t)=C_{st}\frac{dv_{st}(t)}{dt} \tag{6}$$

$$i_{pi}(t)=i_{Lpt}(t)+i_{Cpt}(t) \tag{7}$$

$$i_{si}(t)=i_{Lst}(t)+i_{Cst}(t) \tag{8}$$

The secondary side of the system can also be represented in a similar manner.

Assuming no losses, the absolute value of the average power over a cycle on both sides of the converter are equal, which can be expressed by,

$$\frac{V_{pi}}{T_s}\int_{t=0}^{t=T_s}i_{pi}(t)dt=\frac{V_{si}}{T_s}\int_{t=0}^{t=T_s}i_{si}(t)dt \tag{9}$$

Accordingly, if only the average power P_p on the primary side is considered, it can be calculated from (9) for any given duty cycles and modes of operation.

$$P_p=\frac{V_{pi}}{T_s}\cdot\left[\int_{t=0}^{t_1=T_I}i_{pi}(t)dt+\int_{t_1=T_I}^{t_2=T_I+T_{III}}i_{pi}(t)dt+\int_{t_2=T_I+T_{III}}^{t_3=T_I+T_{III}+T_{II}}i_{pi}(t)dt+\int_{t_3=T_I+T_{III}+T_{II}}^{t_4=T_I+T_{III}+T_{II}+T_{IV}}i_{pi}(t)dt\right] \tag{10}$$

Using Fig. 2 and (1)-(8), $i_{pi}(t)$ in different modes is given by:

$$i_{pi}(t)=\begin{cases}\begin{aligned}&\frac{V_{pi}t}{L_{pt}}-\frac{M}{L_{pt}}A_1\sin(\omega_0\cdot t+\varphi_1)\\&+\frac{M}{L_{pt}}A_1\sin(\varphi_1)+i_{pi}(t=0)\end{aligned} & t\in\left[0,\ T_I=0.5\left(1-|D_1-D_2|\right)\cdot T_s\right]\\[2mm]\frac{V_{si}L_{st}-MV_{pi}}{L_{pt}L_{st}-M^2}\cdot(t-t_1)+i_{pi}(t=t_1) & t\in\left[T_I,\ T_I+T_{III}=\frac{T_s}{2}\right]\\[2mm]0 & t\in\left[T_I+T_{III},\ T_I+T_{III}+T_{II}=\left(1-0.5|D_1-D_2|\right)\cdot T_s\right]\\[2mm]0 & t\in\left[T_I+T_{III}+T_{II},\ T_I+T_{III}+T_{II}+T_{IV}=T_s\right]\end{cases} \tag{11}$$

As described in the previous section, primary or secondary side power (or voltage) of the converter can be regulated by controlling one of the two duty cycles while keeping the other constant. For example, the amount of power delivered to the secondary side is controlled by D_2 by keeping D_1 at 0.5.

For stable operation and to preserve resonance with no transient conditions at switching, D_1 is limited to 0.5. Now using (10) and (11), primary side power P_p can be expressed as:

$$P_p = \frac{V_{pi}}{T_s} \cdot \left[\int_{t_0=0}^{t_1=0.5(1-|D_1-D_2|)\cdot T_s} i_{pi}(t)dt + \int_{t_1=0.5(1-|D_1-D_2|)\cdot T_s}^{t_2=T_s/2} i_{pi}(t)dt + 0 \right]$$

$$= \frac{V_{pi}}{T_s} \cdot \begin{bmatrix} \frac{MA_1}{L_{pt}\omega_0}cos(\omega_0 \cdot t_1 + \varphi_1) - \frac{MA_1}{L_{pt}\omega_0}cos(\varphi_1) + \frac{V_{pi}t_1^2}{2L_{pt}} \\ + \left(\frac{M}{L_{pt}}A_1 sin(\varphi_1) + i_{pi}(t=0) \right) \cdot t_1 \\ + \frac{V_{si}L_{st} - MV_{pi}}{L_{pt}L_{st} - M^2} \cdot \frac{(0.5T_s - t_1)^2}{2} + i_{pi}(t=t_1) \cdot (0.5T_s - t_1) \end{bmatrix} \quad (12)$$

Now using boundary conditions and solving for the unknown constants, P_p can be expressed as:

$$P_p = \frac{V_{pi}}{T_s} \cdot \begin{bmatrix} \frac{MA_1}{L_{pt}\omega_0}cos(\omega_0 \cdot t_1 + \varphi_1) - \frac{MA_1}{L_{pt}\omega_0}cos(\varphi_1) + \frac{V_{si}L_{st} - MV_{pi}}{L_{pt}L_{st} - M^2} \cdot \frac{(0.5T_s - t_1)^2}{2} \\ + \left[\frac{V_{pi}t_1}{L_{pt}} - \frac{M}{L_{pt}}A_1 sin(\omega_0 \cdot t_1 + \varphi_1) - \frac{V_{pi}D_2T_s}{2L_{pt}} + \frac{M}{L_{pt}}A_1 sin\left(\omega_0 \cdot \frac{D_2T_s}{2} + \varphi_1 \right) \right] \cdot 0.5T_s \\ - \frac{V_{pi}t_1^2}{2L_{pt}} + \frac{M \cdot t_1}{L_{pt}}A_1 sin(\omega_0 \cdot t_1 + \varphi_1) \end{bmatrix}$$

$$(13)$$

where $t_1 = T_s/4 + D_2T_s/2$. The values of ω_0, φ_1 and A_1 are given by (14)-(16) as

$$\omega_0 = \sqrt{\frac{L_{pt}}{L_{pt}L_{st}C_{st} - C_{st}M^2}} \quad (14)$$

$$A_1 = \frac{V_{pi}T_s}{4(L_{pt} - M)} \quad (15)$$

$$\varphi_1 = \pi - \frac{\omega_0 \cdot D_2T_s}{2} \quad (16)$$

P_p can be calculated from (13) and the positive sign indicates that power is transferred from the primary to the secondary.

Based on (3) and parameters in Table I, the P_p in terms of D_2 can be plotted as Fig. 4.

Fig. 4. illustrates how the output power P_p can regulate through the duty cycle D_2 of the secondary side. Initially the converter is operated with equal duty cycles, thus the power transfer is near zero. The small difference in duty cycles and power are to account for system losses.

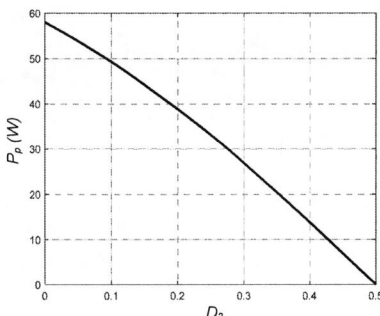

Fig. 4 P_p with the varying of D_2 (D_1=0.5)

IV. SIMULATION AND EXPERIMENTAL RESULTS

As described in the previous section, primary or secondary side power (or voltage) of the converter can be regulated by controlling one of the two duty cycles while keeping the other constant.

In order to verify both the proposed concept and developed theory, the converter was modeled and its behavior was investigated through simulations and experiments under different conditions, using the parameters given in Table I.

Table I parameters of the proposed converter.

Parameter	Value
Nominal power	60W
V_{pi}, (or V_{si}), r_{in}, (or r_{out}),	30V and/or 20V, 1x10⁻³Ω;
L_{pt}, (or L_{st}), r_{pt}, (or r_{st}),	17.8μH, 5.0mΩ;
C_{pt}, (or C_{st}), r_{Cpt}, (or r_{Cst}),	0.023μF, 5.0mΩ;
M, k	1.4x10⁻⁵, 0.78
f_s	360kHz

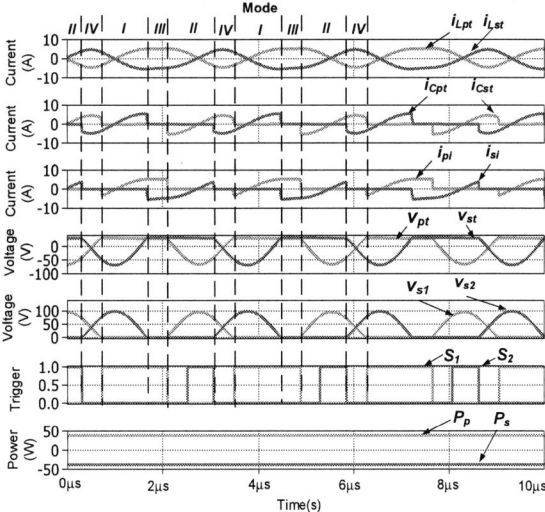

Fig. 5 Simulated steady state waveforms of the proposed converter when P_p=40W

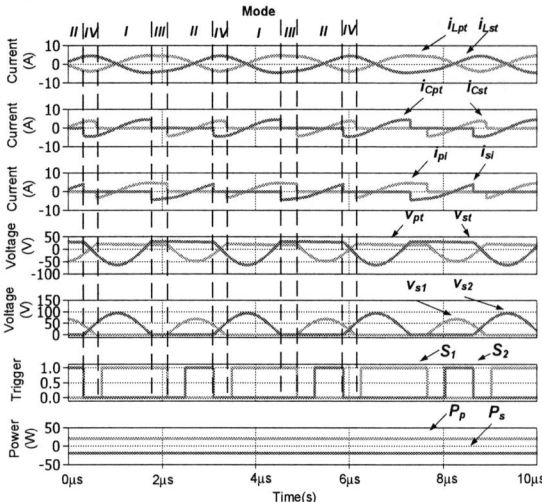

Fig. 6 Simulated steady state waveforms of each component in proposed converter with P_p=20W, V_{pi}=30V and V_{si}=20V

Fig. 5 shows the simulated steady-state behavior of the converter when it was operated with $D_1>D_2$. As predicted, the converter operates in all four modes in this situation and the average power transfer over a cycle is not zero. The primary side average power of about 40 W and it is positive while that on the secondary is negative, indicating that the power transfer is from the primary to secondary side.

The waveforms in Fig. 6 demonstrate how the converter can also be operated with different input and output voltages, V_{pi}=30 and V_{si}=20V, while regulating the amount power that is being transferred to around 20 W.

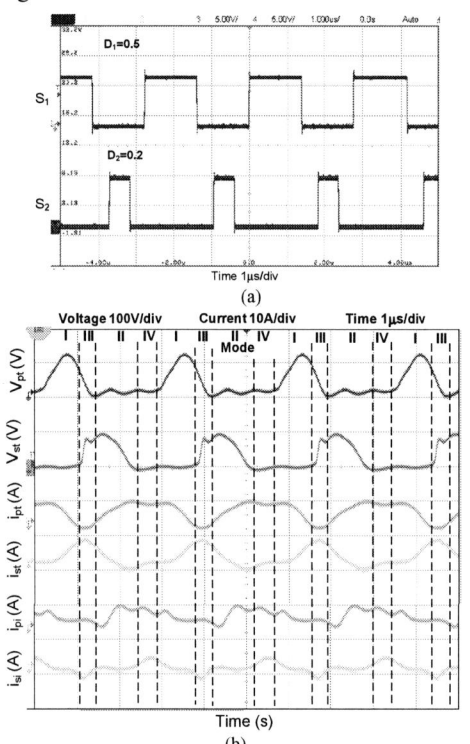

Fig. 7 Experimental waveforms of the Trigger signals and components: (a) D_1= 0.5 and D_2=0.2 (b) v_{pt}, v_{st}, i_{pt}, i_{st}, i_{pi} and i_{si}

The experimental waveforms that correspond to operation with $D_1>D_2$ are shown in Fig. 7. It can be observed from Fig. 7 that when the converter was operated with $D_1>D_2$, the power flow (35.2W) is from the primary to the secondary side. These waveforms convincingly indicate that proposed analysis and operation of the converter in different modes are accurate.

V.   CONCLUSIONS

A low-cost BD-WPT converter topology has been proposed for consumer applications that require bi-directional power flow. The proposed system uses two switches to regulate the wireless power flow in both directions. A mathematical model is presented, describing the different modes of operation of the WPT system. The both simulated and measured experimental results are presented to validate the applicability of the proposed concept.

VI.   ACKNOWLAGEMENT

Project supported by Qualcomm (USA) co-operation agreement, #34835.001 in part by the Science and Technology Development Fund, Macao SAR (FDCT) (FDCT 0026/2019/A1) and in part by the Research Committee of the University of Macau (MYRG2017-00038-FST and MYRG2018-00056-FST)

REFERENCES

[1] "Wireless charging market to reach $37.2 billion, globally by 2022," Allied Market Research. 2018. [Online]. Available: https://www.alliedmarket research.com/wireless-charging-market. Accessed on: Apr. 3, 2018.

[2] A. Ahmad, M. S. Alam and R. Chabaan, "A Comprehensive Review of Wireless Charging Technologies for Electric Vehicles," *IEEE Trans. Transport. Electrific.*, vol. 4, no. 1, pp. 38-63, March 2018.

[3] http://www.powercastco.com/

[4] M. Huang, Y. Lu and R. P. Martins, "A Reconfigurable Bidirectional Wireless Power Transceiver for Battery-to-Battery Wireless Charging," IEEE Trans. Power Electron., vol. 34, no. 8, pp. 7745-7753, Aug. 2019.

[5] R. W. Porto, V. J. Brusamarello, I. Müller, F. L. Cabrera Riaño and F. Rangel De Sousa, "Wireless power transfer for contactless instrumentation and measurement," *IEEE Trans. Instrum. Meas,* vol. 20, no. 4, pp. 49-54, August 2017.

[6] S. Park and J. R. Davila, "Duty Cycle and Frequency Modulations in Class-E DC-DC Converters for a Wide Range of Input and Output Voltages," *IEEE Trans. Power. Electron.*, vol. 33, no. 12, pp. 10524-10538, Dec. 2018.

[7] T. Inaba and H. Koizumi, "Class-DE$_M$ Tuned Power Amplifier," *IEEE Trans. Power Electron.*, vol. 34, no. 1, pp. 403-415, Jan. 2019.

[8] A. Sheikhi, M. Hayati and A. Grebennikov, "High-Efficiency Class E^{-1} and Class-F/E Power Amplifiers at Any Duty Ratio," *IEEE Trans. Ind. Electron.*, vol. 63, no. 2, pp. 840-848, Feb. 2016.

[9] S. Aldhaher, D. C. Yates and P. D. Mitcheson, "Load-Independent Class E/EF Inverters and Rectifiers for MHz-Switching Applications" *IEEE Trans. Power Electron.*, vol. 33, no. 10, pp. 8270-8287, Oct. 2018.

[10] J. M. Rivas, O. Leitermann, Y. Han and D. J. Perreault, "A Very High Frequency DC–DC Converter Based on a Class Φ Resonant Inverter," *IEEE Trans. Power Electron.,*vol. 26, no. 10, pp. 2980-2992, Oct. 2011.

[11] A. Grebennikov, "High-efficiency Class-E power amplifier with shunt capacitance and shunt filter," *IEEE Trans. Circuits Syst. I: Reg. Papers*, vol. 63, no. 1, pp. 12-22, Jan. 2016.

[12] S. Liu, M. Liu, S. Yang, C. Ma and X. Zhu, "A Novel Design Methodology for High-Efficiency Current-Mode and Voltage-Mode Class-E Power Amplifiers in Wireless Power Transfer systems," *IEEE Trans. Power Electron.*, vol. 32, no. 6, pp. 4514-4523, June 2017.

978-1-7281-4181-7/19 $31.00 © 2019 IEEE

[13] S. Liu, M. Liu, S. Han, X. Zhu and C. Ma, "Tunable ClassE^2DC – DC Converter with High Efficiency and Stable Output Power for 6.78-MHz Wireless Power Transfer," *IEEE Trans. Power. Electron.*, vol. 33, no. 8, pp. 6877-6886, Aug. 2018

[14] M. Liu, M. Fu and C. Ma, "Low-Harmonic-Contents and High-Efficiency Class E Full-Wave Current-Driven Rectifier for Megahertz Wireless Power Transfer Systems," *IEEE Trans. Power Electron.*, vol. 32, no. 2, pp. 1198-1209, Feb. 2017.

[15] https: //www.theverge.com/circuitbreaker/2018/10/16/17967012/ huawei-mate-20-pro-reverse-wireless-charging-smartphone

[16] https://www.wired.co.uk/article/sony-wireless-phone-charging

[16] https://techcrunch.com/2019/02/20/the-samsung-galaxy-s10-can-wirelessly-charge-other-phones/

[17] D. R Kasar, C. S Graham, E S. Jol, Inductive charging between electronic devices: U.S. Patent Application 15/925,410[P]. 2018-7-26.

[18] K. Li and S. Tan, "A Class E2 Inverter-Rectifier-Based Bidirectional Wireless Power Transfer System," *2018 IEEE 4th Southern Power Electronics Conference (SPEC)*, Singapore, Singapore, 2018, pp. 1-6.

[19] N. Mukaiyama, H. Omeri, N. Kimura, T. Morizane, M. Tsuno and M. Nakaoka, "A novel type of bidirectional IPT with a dual-active seamless controlled single-ended converter for wireless V2H," *2017 19th International Conference on Electrical Drives and Power Electronics (EDPE)*, Dubrovnik, 2017, pp. 53-58.

[20] T. Iwanaga, H. Omori, T. Takahashi, M. Tsuno, T. Morizane and N. Kimura, "A Novel Type of Wireless V2H with Single Switch Dual-Active Seamless Converter in a Smart House," *2018 IEEE 18th International Power Electronics and Motion Control Conference (PEMC)*, Budapest, 2018, pp. 46-51.

Wind Turbine Emulator with DC Motor

Tadeu F. dos Santos*, Igor V. Chacon*, Géssica C. de A. Souza*,
Guilherme A. P. de C. A. Pessoa*, Felipe O. S. Gama[†], Rodrigo de A. Teixeira* and Andrés O. Salazar[†]

*Department of Electrical Engineering (DEE)
Federal University of Rio Grande do Norte (UFRN)
[†]Department of Computer Engineering and Automation (DCA)
Federal University of Rio Grande do Norte (UFRN)
tadeu.felix@gmail.com,igorchacon@yahoo.com.br,gcas_@hotmail.com.br,guilhermepillon@hotmail.com,
felipe.gama@dca.ufrn.br,rodrigoandradeteixeira@gmail.com,andres@dca.ufrn.br

Abstract—The proposition of this work is to develop an emulation of the rotation characteristics of a wind turbine for later analysis of static and dynamic responses of the system to. Initially, for this purpose, the emulator system was simulated in a computational environment. Next, the practical implementation was executed. In the simulated system, the actual wind turbine turbine is replaced by a DC motor operating within controlled speed and torque. A DC/DC converter was used during the simulation step to power the machine and reproduce system characteristics. During the initial studies, the horizontal turbine model with three blades was considered. The results obtained through simulations and experimental tests demonstrate the feasibility of using emulation in the proposed system.

Index Terms—Wind turbine emulator, control system, DC machine.

I. Introduction

The increase in energy demand and environmental restrictions around the world has considerably increased the use of renewable sources of electricity generation. Among the renewable generation sources, wind energy has been showing great development and competitiveness in recent years [1]. Large wind turbines are complex and usually operate connected to the power grid using different control types, smaller wind turbines are simpler and can be used for standalone applications and connected to the local grid.

In this context, there are in the literature several works that suggest different topologies of electrical energy generation through the wind energy. In general, these topologies use mechanical structures to transform the speed involved in energy generation [2]. A new approach to couple the wind speed was proposed in [3]. This new approach multiplies the wind speed without using a gearbox or large converters. This new structure was called the electromagnetic frequency regulator (EFR). This work presents a structure to simulate wind behavior on wind turbines and it shall use this system in validation of the ERF.

The static and dynamic behavior of a wind turbine can be emulated in the laboratory for project optimization and analysis in a controlled environment. In [4] a system capable of emulating the production curve of a wind turbine was implemented using a DC/DC converter to control a DC machine. The DC machine emulating the wind resource is mechanical coupled to an asynchronous machine. In this case, the voltage

and current characteristics showed an electric stable behavior compatible with the load requirements.

In [5] a DC machine driving a synchronous generator is used for the emulation of a small wind generator. There, a digital controller of type PI is designed to impose to the motor a desired rotational speed according to a setpoint signal obtained directly from PC. To consider the inertia of the turbine rotor an inertia disk was incorporated into the emulated system. In [6] the static and dynamic characteristics of a wind emulator system are presented. For this purpose, a DC/DC converter was used to control a DC machine which in turn emulates the turbine behavior of the wind generator.

This work proposes a didactic emulator for the rotational characteristics of a small wind turbine, the system emulates the turbine rotation characteristics using a controlled converter that drives a DC machine magnetically coupled to a synchronous machine. Initially, the emulator system was implemented computationally for later practical implementation.

The remaining of this paper is organized as follows. Section II presents the wind emulation theory and the electromagnetic frequency regulator. The Section III presents the model of the wind emulation system proposed in this paper. The results and discussions are presented in Sections IV and V and the conclusion is drawn in Section VI.

II. Emulation System

In this section will be presented theories of ERF and of wind emulation on wind turbines.

A. Ectromagnetic Frequency Regulator

The EFR is equipment that receives variable speed and delivers constant speed. This equipment was developed from a machine that resembles an induction machine, and that allows the system to work with variable speed, coming from the winds, maintaining constant the speed of exit that, moving a synchronous generator. Contrary to other topologies, this machine uses a frequency inverter, which injects currents into the armature, which is primarily responsible for speed control. In this topology, the stator rotates in conjunction with the wind turbine. Already the rotor is responsible for generating the necessary speed in the generator to generate electricity with adequate nominal frequency for connecting to the distribution network. It is important to note, that the frequency inverter is powered by the DC created with the battery bank, feeds the asynchronous rotor armature and creates a rotating field from currents that vary at a frequency such as to determine the speed of this field. The frequency inverter aims to compensate for variations in wind speed, where it is operated through a rotor speed controller. Thus, the speed of the rotor will always be the sum of the speed of rotation of the turbine with the speed of the rotating field of the stator, less motor slip.

The main objective of this topology is to eliminate the gearbox that the other topologies use since the gears are the ones that have a greater need for exchange and maintenance in the system of generation of energetics through the wind.

Figure 1 illustrates the wind energy generation system with the EFR, in which the wind turbine identified as (1), the EFR (2), the synchronous generator (3) and the frequency inverter (4).

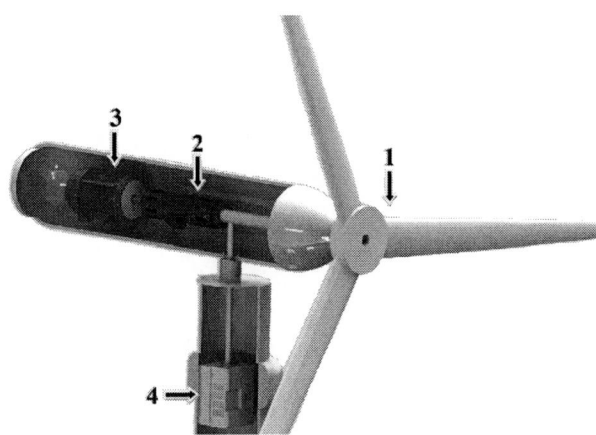

Figure 1. Wind energy generation system with the EFR.

B. Characteristics Wind Turbine

The wind turbine is the most important component of the wind power conversion system. The mathematical foundations of its modeling are widely diffused in the literature [6]. Based on these models the mechanical power at the output of a three-blade horizontal axis wind turbine can be written as:

$$P_m = 0.5\rho\pi R^2 C(\lambda, \beta)v^3. \tag{1}$$

Where πR^2 represents the area swept by the turbine rotor, the plot $C_p(\lambda, \beta)$ represents the area swept by the turbine rotor v characterizes wind speed. For small and medium wind power conversion systems the angle of inclination of blades β is kept constant and the power coefficient is a function only of speed rate $C_p(\lambda)$ Equation (2) which can be represented as a function of angular speed of turbine ω has given in (rad/s) and the wind speed v in (m/s) Equation (3).

$$C_p(\lambda) = 0.00044\lambda^4 - 0.012\lambda^3 + 0.097\lambda^2 - 0.2\lambda + 0.11 \tag{2}$$

$$\lambda = \frac{\omega R}{v} \tag{3}$$

The average torque developed by the turbine can be written as:

$$T = \frac{P_m}{\omega} \tag{4}$$

and therefore,

$$T = \frac{0.5\rho\phi R^2 C_p(\lambda)v^3}{\omega} \tag{5}$$

$$\omega = \frac{0.5\rho\phi R^2 C_p(\lambda)v^3}{T} \tag{6}$$

The speed described in Equation (6) plays the reference role of the rotor at specific wind speed, the torque pulses due to the shear wind and tower shadow influences can be represented by the Equation (7), with T_m and T_{mill} mills respectively representing the aerodynamic and average torques of a turbine.

$$T_m = T_{mill}(1 + 0.2\sin\omega_{mill}(t) + 0.4\sin 3\omega_{mill}(t)) \tag{7}$$

1) Proposed System: Proposed System: In the proposed system the wind turbine is replaced by a DC motor. This motor responsible for the generation of speed and torque for the synchronous generator is driven by a controlled DC-DC power converter. The motor is coupled via an electromagnetic frequency regulator to the synchronous generator, which in turn is connected to the load's Figure (2).

The projected power converter is shown in the model of Figure (2) consists of two IGBT switches connected to a voltage source composed of a capacitive link. The voltage is supplied from the drive to the DC motor through a half-bridge configuration. The output voltage of the drive has been connected to the DC motor armature terminals through one of IGBT (*Insulated Gate Bipolar Transistor*) arms driven by PWM (*Pulse Width Modulation*) pulses sent by the drive.

The electromagnetic frequency regulator is an optimizer system for wind generators, which replaces the gearbox used to interconnect the blade shaft to the axis of an electric generator [3] and [2]. The optimizer consists of a set of electromagnetic components, which in full operation ensure rotation in the the axis of a constant generator and within the limits considered acceptable.

Figure 3 illustrates the experimental bench, where you can see the three main modules of the system: the DC motor emulating the wind turbine, the EFR, and the synchronous generator. It is also important to note that all modules are mechanically connected.

978-1-7281-4181-7/19 $31.00 © 2019 IEEE

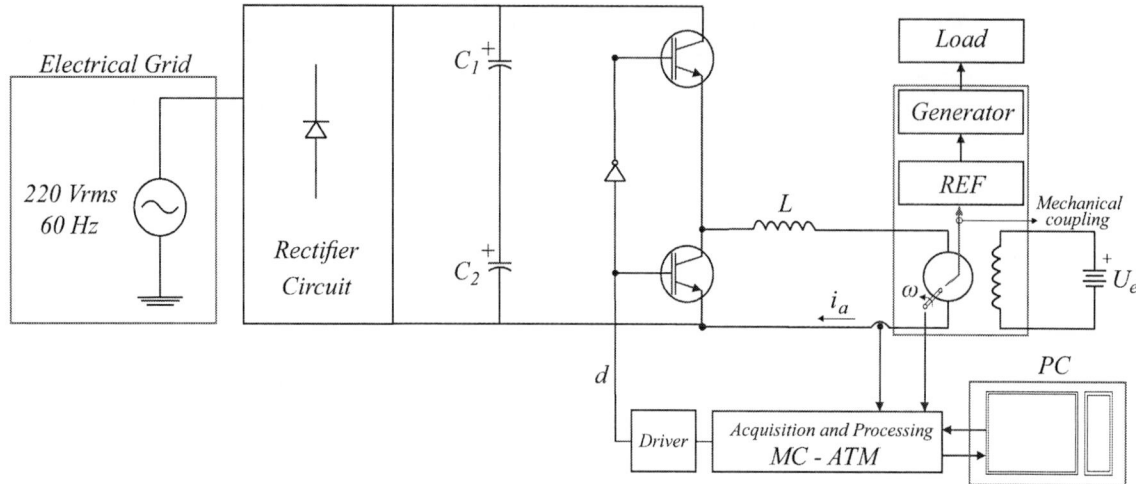

Figure 2. Schematic of the proposed system.

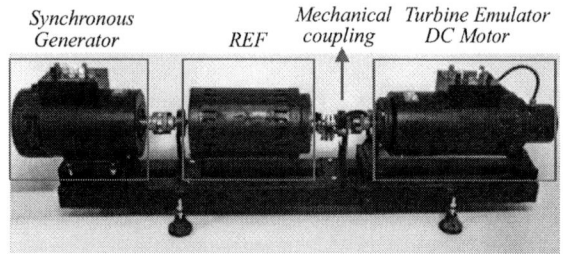

Figure 3. Workbench developed for experimental testing.

III. MACHINE MODEL

The DC motor equivalent circuit used to emulate the system turbine is shown in Figure (4). The armature circuit, including the compensating and switching poles of machine, maybe represented as the series association of a $f.e.m$ a resistor R_a and an inductance L_a. The field circuit, is equivalent to a coil with inductance L_e and resistance R_e [7]. Considering the permanently fixed flux, the electromechanical behavior of machine is completely described by the Equations (8), (9) e (10). Speed is dependent on the behavior of all involved mechanics, the Equation (11) supposes a moment of inertia J and a torque T_r corresponding to the effects of loading.

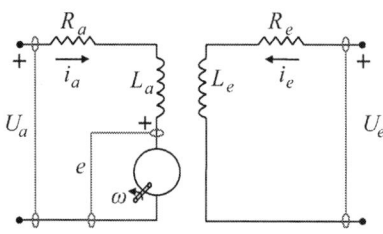

Figure 4. Equivalent machine circuit

$$u_a = e + R_a i_a + L_a \frac{di_a}{dt} \tag{8}$$

$$u_e = R_e i_e + L_e \frac{di_e}{dt} \tag{9}$$

$$\Phi = ki_e \tag{10}$$

$$T = J\frac{d\omega}{dt} + T_r \tag{11}$$

In steady-state the Equations described above give rise to the Equations (12) and (14). In the equation (12) the $f.e.m$ $k_a\Phi\omega$ becomes much larger than the parcel $R_a I_a$ as the speed becomes significant, therefore, $f.e.m$ is proportional to the spin speed ω and to the excitation flux Φ Equation (13). With constant flux the speed of machine is almost exclusively dependent on the tension applied to its armature which allows its speed regulation. The machine torque T from the electromechanical conversion is proportional to the flux, the armature current, and the proportionality constant k_a which depends on the constructive parameters of machine. Considering the constant flux the torque can be regulated by the action of armature current i_a Equation (14).

$$U_a = k_a\Phi\omega + R_a i_a \tag{12}$$

$$e = k_a\Phi\omega \tag{13}$$

$$T = k_a\Phi i_a \tag{14}$$

The DC motor parameters used in this work are shown in Table I below.

978-1-7281-4181-7/19 $31.00 © 2019 IEEE   670

Table I
MACHINE PARAMETERS

Parameters	Values
Nominal Armature Voltage	180 V
Nominal Armature Current	8 A
Nominal Active Power	0,932 KW
Nominal Speed	1800 RPM
Field Resistance	0,58 Ω
Armature Resistance	3,33 Ω
Field Inductance	1,99 mH
Armature Inductance	547 mH
Moment of Inertia	0,0540845 kgm^2
Coefficient Friction	0,0034445
Electromechanical Constant	0,1211

A. Control and Microprocessing System

In addition to the model based on the characteristics of the turbine, two PI type controllers were designed in the computational development environment for speed and current control of the DC machine. Initially, a model provides the reference signal as a function of the radius of rotor R, the density of air ρ, the wind speed υ and the control feedback signals. The calculated speed is used as the reference value for the speed controller, in turn, the speed controller generates the reference signal for the current control, the output signal of that controller is compared with a triangular carrier and the PWM, pulses are generated and sent to the converter IGBTs command. The control scheme used is made up of a classical structure that is widely used in other types of drives with an external speed controller Figure (6) and an internal current regulator Figure (5).

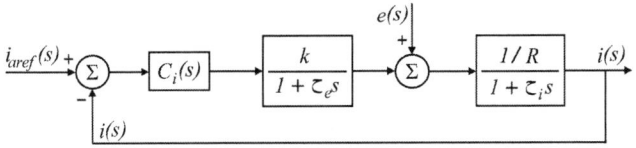

Figure 5. Current control

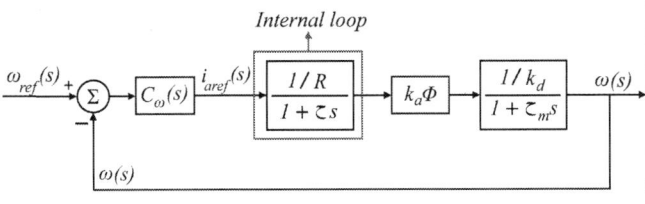

Figure 6. Speed control

The sizing of controllers was done in two steps, initially, the internal current controller was designed disregarding the disturbing $f.e.m$ due to its slow dynamics. The sizing of the speed controller considers the pair of conjugated complex poles of $i(s)/i_{ref}$ by a first-order approximation highlighted in Figure (6), so that the controller was scaled based on the Equation (12).

$$C_i(s) = K_e \frac{1 + \tau_e s}{s} \quad (15)$$

$$C_\omega(s) = K_\omega \frac{1 + \tau_v s}{s} \quad (16)$$

The embedded system was developed with microcontrollers of type Atmega328p used in the Arduino Uno. The Atmega 328p is a high performance low power 8-Bit microcontroller with 32 Kbytes InSystem programmable flash memory. The Arduino UNO is composed by the Atmega328 (model 328p is available too) chip and a recording circuit. The system offers support for C language, but other languages are possible to use with external softwares [8]. Figure (7) presents the layout of circuit employed in the project.

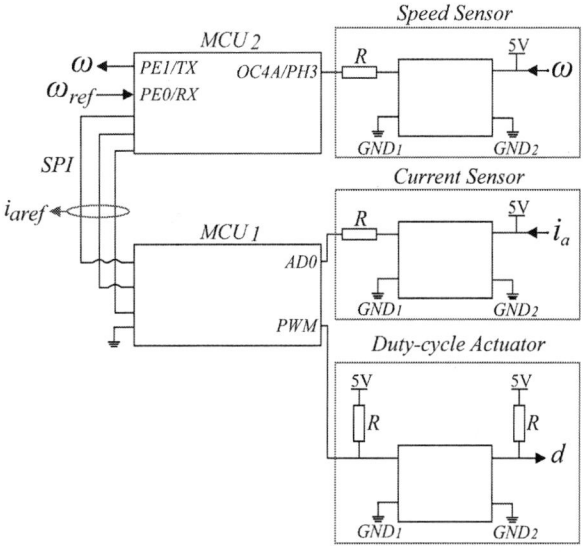

Figure 7. Schematic illustration of hardware module

The hardware module, illustrated in Figure 7 was based on two Atmega 328 microcontrollers, one of which (MCU-1) was responsible for execution of PI algorithm, measure armature current and generate signal PWM, while the function of other (MCU-2) was to measure the speed v from the DC motor, execution of PI algorithm and sending the reference armature current for (MCU-1).

The monitoring used an Interface Man Machine (IMM) with a Personal Computer connected with the Arduino UNO sending and receiving data. This IMM makes possible to control the parameters of wind, such as density and speed. These parameters are going to be used for control feedback.

In Figure 8, we present the experimental bench to validate the proposed wind speed emulation system on wind turbines. It is important to emphasize that the proposed system will use microcontrollers with low cost, enabling the implementation of the proposed system in other systems.

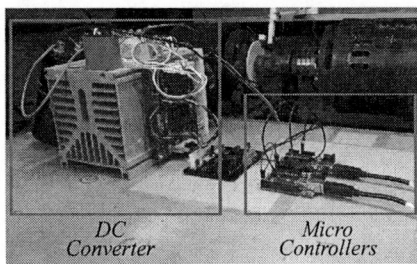

Figure 8. Workbench developed for experimental testing with the system control.

IV. SIMULATION RESULTS

This section presents some preliminary results of the control system designed to validate the proposal.

A. Speed Control

In Figure 9 the speed response of the system is shown when subjected to an acceleration ramp for a time interval of 2 seconds until it reaches the rated speed of the motor used. In evidence the instant of time that the engine reaches its rated speed.

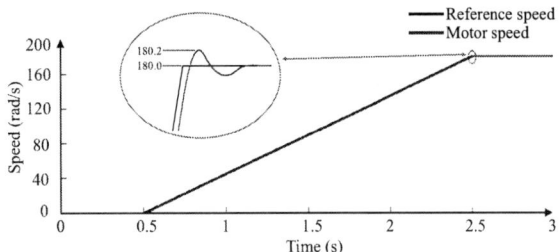

Figure 9. Speed controller

B. Current Control

Figure 10 shows the variation in the armature current of the machine during the acceleration imposed by the speed ramp. Figure 11 shows the current variation in the motor armature when it is subjected to a load of the order of 1N.m at steady-state and at rated speed.

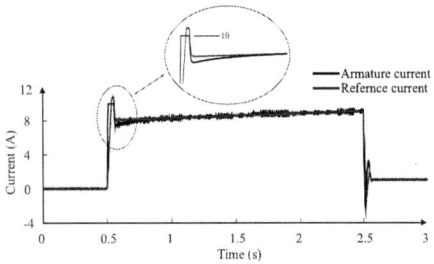

Figure 10. Armature current controller

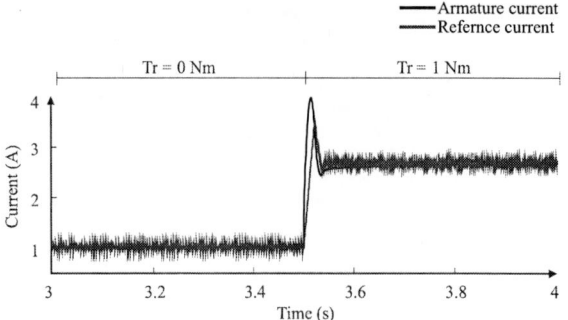

Figure 11. Armature current controller

V. EXPERIMENTAL RESULTS OBTAINED

This section presents the results obtained through practical experiments of the proposed control system. Experimental tests and open-loop computational simulations were performed by applying 15 volts to the armature and 200 mA of field current, in order to validate the motor model so that the cascade controllers can be tuned.

Figure 12 illustrates the results obtained by computer simulation and experimental test, in which it should be noted that the engine model obtained is similar to the actual dc engine. Thus, control tuning techniques were applied to obtain the controllers gains.

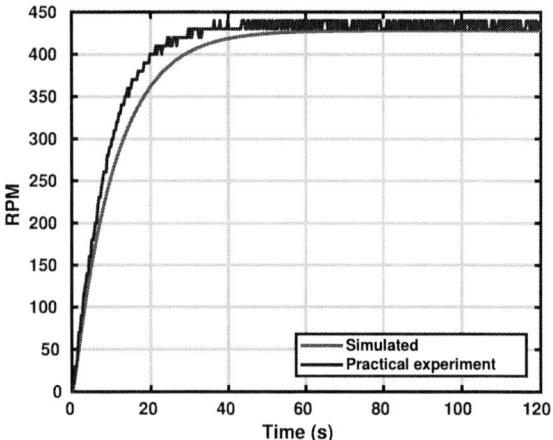

Figure 12. System open loop response to a step input 15 Volts.

After validating the engine model obtained and tuning the controllers, experimental tests of the proposed control system were performed. In order to verify the performance of the control system was verified the system response for several setpoints, which is illustrated in Figure 13. It can be observed in this figure that the proposed control system is controllable for low setpoints of rotations, which is justified since the rotations of the wind turbine blades are low.

978-1-7281-4181-7/19 $31.00 © 2019 IEEE

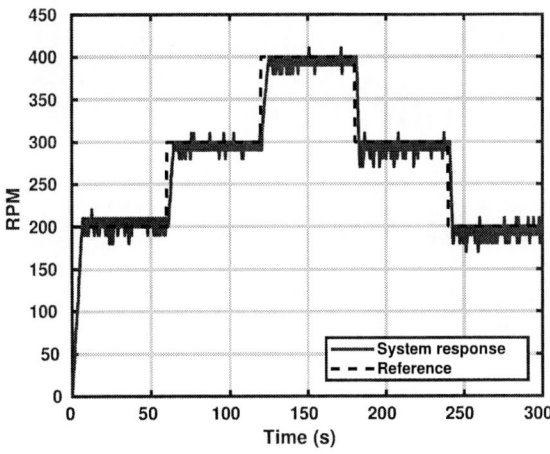

Figure 13. Proposed system response for multiple setpoint.

VI. CONCLUSION

This paper presents a control system for the implementation of an emulator of the rotation characteristics of a wind turbine using a DC motor. The turbine emulation algorithm is under development, however, the efficiency of the speed and current controllers used in the control system can be verified using simulation tools. The presented simulation and the algorithm under development served as basis for the application of the methods in a practical experiment. The practical results obtained validate the proposed system for the speed control of a dc motor coupled to the REF. Finally, this paper also has the contribution of proposing and implementing a dc motor speed control system using simple microntrollers instead of using DSP.

ACKNOWLEDGMENT

This work was suported by Comissão de Aperfeioamento de Pessoal do Nível Superior (CAPES) and Conselho Nacional de Desenvolvimento Científico e Tecnológico (CNPq).

REFERENCES

[1] DE PESQUISA ENERGÉTICA EMPRESA. Balanço energético nacional– 2016, 2016, 2018.

[2] Glauco George Cipriano Maniçoba. Análise e modelagem de um regulador eletromagnético de velocidade para sistemas de conversão de energia eólica. 2018.

[3] Paulo Vitor Silva. Regulador eletromagnético de frequência aplicado no controle de velocidade de geradores eólicos. 2015.

[4] N. Pozo F. Arvalo, P. Estrada and M. Pozo. Wind generation emulator using a dc machine. *2017 IEEE Second Ecuador Technical Chapters Meeting (ETCM)*, 1:1–6, 2017.

[5] M. T. Iqbal Md. Arifujjaman and J. E. Quaicoe. Emulation of a small wind turbine system. *Istanbul University Journal of Electrical Electronics Engineering*, 8(1):569–579, 2018.

[6] A.H.M.Yatim M.A.Bhayo, M.J.A.Aziz and N.R.N.Idris. Analyzing the static and dynamic characteristics of wind energy conversion system using wind turbine simulator. *IEEE*, 8:123–127, 2017.

[7] Jo ao C. P. Palma. *Accionamentos Electromecânicos de Velocidade Variável*, volume 1. Fund Caloust Gulbenkian, Lisboa, 1 edition, 1999.

[8] Felipe OS Gama, José KE da C Martins, Tiago F de Miranda, Willy M de F Tomé, Sérgio N Silva, and Marcelo AC Fernandes. Control of airflow in ventilation systems using embedded systems on microcontrollers. *Microsystem Technologies*, pages 1–10, 2019.

Generalized Mathematical Model for an N-cell Interleaved Boost Converter

1st Luana K. Melgaço Pereira
Department of Electronic Engineering
Federal University of Minas Gerais
Belo Horizonte, MG, Brasil
luanakrugerk@gmail.com

2nd Seleme I. Seleme Jr.
Department of Electronic Engineering
Federal University of Minas Gerais
Belo Horizonte, MG, Brasil
seleme@cpdee.ufmg.br

3rd João Lucas da Silva
Department of Electronic Engineering
Federal University of Itajubá
Itabira, MG, Brasil
joaolucassilva@gmail.com

Abstract—**This paper presents a generalized model for an N-cell interleaved boost converter (iBC) by using state-space averaging technique. The steady-state and small-signal models of the converter have been developed including the discontinuity shown in the duty-cycle. The validation of the obtained model and a frequency domain analysis are also developed, as well as the ripple characteristic in converter input current is presented for a variation of the number of cells in parallel.**

Index Terms—**Generalized model, Interleaved Boost Converter, small-signal modeling, state-space averaging technique.**

I. INTRODUCTION

In several applications, considering a wide range of power and switching frequencies, a series or parallel association of switching cells in power electronics is adopted. The structure of interleaved converters (parallel association of DC-DC converters) was originally proposed in [1] as a way of overcoming the technological limitations of components. These topologies reach higher voltage and/or current operational values, as well as increase the apparent switching frequency and, consequently, the reduction in filtering passive components, since they do not have to support high currents, like those that would circulate in converters without parallelism.

The interleaved power conversion is the interconnection between multiple switching cells with the exact frequency and duty-cycle, but different switching instants [2]. The interleaved converters adopt the Phase Shifted Pulse Width Modulation, PS-PWM, to reduce or cancel the ripple by phase shifting the pulses by $2\pi/N$, where N is the number of switching cells. Due to the interleaving of the control signals several harmonic components of the output voltage are canceled, thus the THD of the output voltage is smaller compared to that obtained in parallel converters without interleaving.

Investigations about the advantages of the interleaving technique in the literature mostly involve the use of an Interleaved Boost Converter (iBC) topology. Previous studies, such as in [3] and [4], had proposed the modeling of this converter for a limited number of cells, by using the state-space averaging technique. In [2] and [5] a frequency response analysis and development of control strategies is presented. As the number of converter cells increases, the analysis and investigation of the operational characteristics in the steady and transient states of the converter becomes more difficult. Although in [6] a

generalized modeling for the iBC using the state-space model is proposed, only one fixed duty-cycle interval is considered. In order to obtain the average model of the converter, the discontinuity presented for the duty-cycle value must be considered since there are two different average models for duty-cycles lower and higher than $\frac{1}{2}$ [3].

Thus, the goal of this paper is to determine the steady-state (DC) and dynamic (AC) models for an interleaved Boost converter operating in CCM, using the state-space averaging technique, for any number of cells, including discontinuity presented in the duty-cycle, without having to obtain the state matrix equations of the switched model. The validation of the obtained model and a frequency domain analysis are also presented, as well as the characterization of the input current ripple is made for a variation of the number of cells in parallel.

II. PRINCIPLE OF OPERATION

The topology for the Interleaved Boost Converter is presented in Fig. 1, which illustrates N parallel step-up converters. The iBC operating principle is similar to an ordinary boost converter. For high power applications the interconnection of parallel converters is recommended [7], in which the interleaved mode brings more benefits to the operation, as previously said.

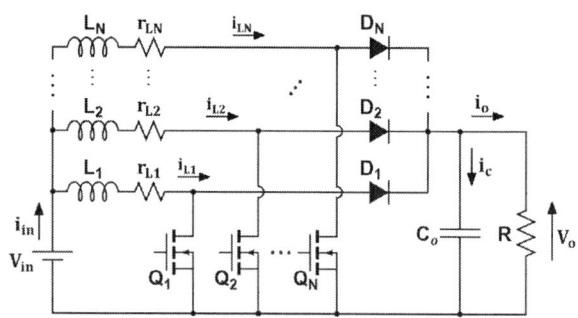

Fig. 1: Topology of an N-cell interleaved boost converter.

The principle of operation of this converter is based on the imposition of a charge rate and transfer of energy through a magnetic circuit. The understanding of its performance begins with the analysis of each cycle of operation. During the period

978-1-7281-4181-7/19 $31.00 © 2019 IEEE

in which the switch is conducting, the input inductor is directly connected to the DC-source increasing the energy stored in its magnetic circuit, just as the current in the inductor increases linearly with time, considering the high frequency switching. The current in the diode is zero, since it is reverse-biased. When the switch is turned off, the current in the inductor can not change immediately, so the diode is forward-biased and provides a path for the current, which keeps flowing in the same direction, charging the DC-link capacitor increasing the output voltage. This current will flow until the switch is turned back on, increasing the inductor current (continuous conduction mode, CCM) or until it reaches zero, discharging completely the energy stored in the magnetic circuit (discontinuous conduction mode, DCM).

As in CCM, a better use of the energy components happens with lower conduction losses and lower input ripple [2], in this study a converter operating in CCM is considered.

III. iBC MATHEMATICAL MODEL

The mathematical model of iBC is derived using the state-space averaging technique. The representation of the system in the state-space averaging method, characterized by the state equations of charge and discharge of energy of the inductor, results from the combination of the stages as a function of the duty-cycle dT_s, which determines the time that inductor stores energy, and $(1-d)T_s$, which characterizes the complementary stage. Thus, it is necessary to first obtain the converter state equations at each step of operation, i.e., the switched model is required, which describes its operation based on the states of the switches and is obtained using the state space representation as described in (1).

$$\begin{cases} \dot{x} = A_i x + B_i u \\ y = C_i x + E_i u \end{cases} \tag{1}$$

where the vectors x and u are given by $x = \begin{bmatrix} i_{L1} & i_{L2} & \cdots & i_{LN} & v_0 \end{bmatrix}'$, $u = \begin{bmatrix} v_{in} \end{bmatrix}'$, and the superscript $(')$ indicates transposed vector. The index i in the state and input matrices indicate the mode of operation that such matrices represent. The modes of operation are defined by the duty-cycle and therefore by the period in which the switches are in the conduction state and in the non-conduction state.

A. Discontinuity moment presented in the development of the mathematical model

In order to obtain the modes of operation, the discontinuity shown in the interleaved converters must be considered for the chosen duty-cycle value. For example, considering an iBC with two cells in parallel, the voltages in the switches are shown in Fig. 2, where $Q_j = 0$ when the switch is opened and $Q_j = 1$ when the switch is closed, where Q_j are the active function of the switches and $j = 1, 2$. The operating modes are set according to the states of the switches. For a duty-cycle lower than $\frac{1}{2}$, modes 2 and 4 are the same and relate to the instant the switches, Q_1 and Q_2, are in the non-conduction period. For a duty-cycle higher than $\frac{1}{2}$, modes 2 and 4 are also equal, but now consider the instant that the switches are

in the conduction period. Thus, the states matrices A_2 and A_4 which represent such modes of operation are different depending on the duty-cycle range chosen, i.e., there are two different models for duty-cycle lower and higher than $\frac{1}{2}$ [3].

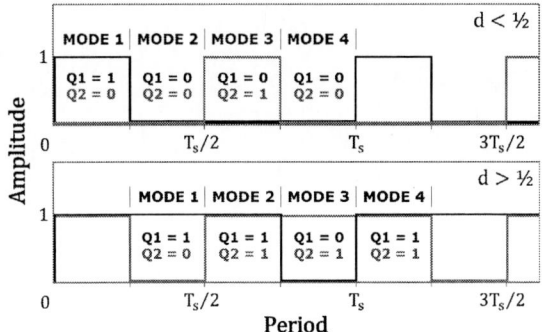

Fig. 2: Switches voltage for $d < \frac{1}{2}$ and $d > \frac{1}{2}$

To obtain the average model, the state matrix used according to the switches states is analyzed. Therefore, a set of linear equations for the equivalent circuits in each mode is developed by Kirchhoff's laws.

B. Non-linear Average Model

Considering that all the switching cells carry equal average current and are operated at the same duty-cycle and based on the waveforms of the inductors currents of a N-cell interleaved converter shown in Fig. 3, (2) and (3) is derived.

$$d_{2k-1} = d \tag{2}$$

$$d_{2k} = \frac{1}{N} - d \tag{3}$$

Where $k = 1, 2, ..., N$ and d is the duty-cycle in the switching cycle considered.

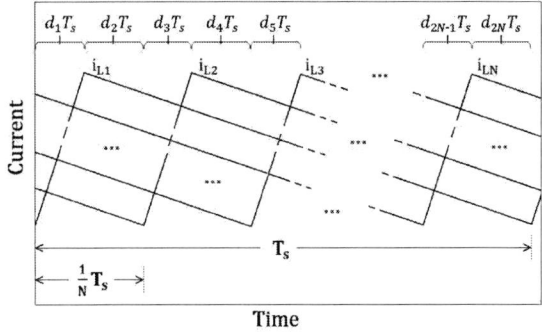

Fig. 3: Inductors Currents of an N-cell interleaved Boost converter. Adapted from [6].

The average state space model in a specific cycle can be written as shown in (4), in which the state and input matrices, A and B, respectively, represent the weighted sum of these matrices in each mode of operation of the converter, given by A_i and B_i, where $i = 1, 2, \cdots, 2N$.

978-1-7281-4181-7/19 $31.00 © 2019 IEEE

$$\dot{x} = Ax + Bu \qquad (4)$$

When determining (4), based on the current waveforms of Fig. 3 one obtains:

$$A = \sum_{i=1}^{2N} d_i A_i \qquad (5)$$

$$B = \sum_{i=1}^{2N} d_i B_i \qquad (6)$$

The complement of the duty-cycle, d', is defined as $\left(\frac{1}{N} - d\right)$. Using (2) and (3) on (5) and (6), state matrices of the average state space model can be written as (7) and (8) as shown in [6].

$$A = d\underbrace{\sum_{k=1}^{N} A_{2k-1}}_{\alpha_1} + d'\underbrace{\sum_{k=1}^{N} A_{2k}}_{\alpha_2} \qquad (7)$$

$$B = d\underbrace{\sum_{k=1}^{N} B_{2k-1}}_{\beta_1} + d'\underbrace{\sum_{k=1}^{N} B_{2k}}_{\beta_2} \qquad (8)$$

The generalized input matrix of the N-cells interleaved Boost converter is derived in (9). In order to simplify the equations of the model, since they present a discontinuity for different duty-cycles, as already mentioned, the variable d_x is introduced. This variable assumes the value d' for duty-cycles lower than $\frac{1}{2}$ and, for duty-cycles higher than $\frac{1}{2}$, it assumes the value d. In the equation (9), the identity matrix of size N is denoted by I_N.

$$A = \begin{bmatrix} \left(-\frac{r_L}{L}\right) I_N & \cdots & -\frac{d_x}{L} \\ \vdots & \ddots & \vdots \\ \frac{d_x}{C} & \cdots & \frac{1}{RC} \end{bmatrix}_{(N+1)\times(N+1)} \qquad (9)$$

$$d_x = \begin{cases} d', & \text{for } d < 0,5 \\ d, & \text{for } d > 0,5 \end{cases} \qquad (10)$$

Thus, a way of obtaining the average matrices for any number of cells or duty-cycle interval is presented, without the need to obtain the matrices in each mode of operation of the switched model. As established in [8], it is essential that in the power stage of the interleaved boost converter, the inductors and diodes should be identical to all cells. Therefore, $L = L_1 = L_2 = \cdots = L_N$. The average output matrix is the same as shown in the modes of operation of the switched model and given by $B = \begin{bmatrix} 1/L & \cdots & 1/L & 0 \end{bmatrix}'_{(N+1)\times 1}$.

C. Linearized Small-Signal Model

In steady state, the average current variations on the inductor and voltage on the capacitor are zero, i.e., $x = X$ and $\dot{X} = 0$. Resulting in the system of equations in (11).

$$\begin{cases} \frac{dX}{dt} = AX + BU = 0 \\ Y = CX \end{cases} \qquad (11)$$

where X and Y are the steady-state state and output vectors. Isolating the X and replacing it in Y, resulting in the static gain transfer function [9] of the converter in (12).

$$\begin{cases} X = -A^{-1}BU \\ Y = \left(-CA^{-1}B\right)U \end{cases} \qquad (12)$$

The matrices A and B are calculated in (7) and (8), respectively. The static gain transfer functions are given by the system of equation in (13).

$$X = \begin{bmatrix} I_{L1} \\ I_{L2} \\ \vdots \\ I_{LN} \\ V_0 \end{bmatrix} = \frac{V_{in}}{r_L + NRd_x^2} \begin{bmatrix} 1 \\ 1 \\ \vdots \\ 1 \\ NRd_x \end{bmatrix} \qquad (13)$$

The analysis done above concerns the behavior of the system for quiescent operating point, through static gain transfer functions related to steady state values (DC). In order to determine the dynamic response of the system, small signal analysis is used, resulting in the frequency domain transfer function through the Laplace Transform [9]. The foundations of this analysis is to study perturbations about the quiescent operating point.

To investigate the small-signals behaviour, perturbations in the input voltage, in the duty-cycle and in the states are introduced, as shown by the set of equations in (14), where the steady-state or DC term is represented in capital letters, the $\sim$ (tilde) quantity represents the AC term or small-signal perturbation.

$$\begin{aligned} x &= X + \tilde{x} \\ y &= Y + \tilde{y} \\ d &= D + \tilde{d} \\ u &= U + \tilde{u} \end{aligned} \qquad (14)$$

The AC model obtained is described as follows, where α_1, α_2, β_1 and β_2 are defined in equations (7) and (8) in the previous section.

$$\dot{X} + \dot{\tilde{x}}(t) = \left\{ \left[D + \tilde{d}(t)\right] \alpha_1 + \left[D' - \tilde{d}(t)\right] \alpha_2 \right\} [X + \tilde{x}(t)] + \left\{ \left[D + \tilde{d}(t)\right] \beta_1 + \left[D' - \tilde{d}(t)\right] \beta_2 \right\} [U + \tilde{u}(t)] \qquad (15)$$

In equation (16) the existence of terms of first and second order is verified. The AC variations are small in magnitude compared to the DC quiescent values, therefore the product of the small-signal terms will be neglected.

$$\dot{\tilde{x}}(t) = \underbrace{A\tilde{x}(t) + B\tilde{u}(t) + [(\alpha_1 - \alpha_2)X + (\beta_1 - \beta_2)U]\tilde{d}(t)}_{1^{st} \text{ order AC terms}} +$$

$$\underbrace{[(\alpha_1 - \alpha_2)\tilde{x}(t) + (\beta_1 - \beta_2)\tilde{u}(t)]\tilde{d}(t)}_{2^{nd} \text{ order AC terms (non-linear)}} \quad (16)$$

The linearization by constructing a small-signal model is shown in (17), which is used to obtain the small signal transfer functions.

$$\dot{\tilde{x}}(t) = A\tilde{x}(t) + B\tilde{u}(t) + B_d\tilde{d}(t) \quad (17)$$

where

$$B_d = \sum_{k=1}^{N}(A_{2k-1} - A_{2k})X + \sum_{k=1}^{N}(B_{2k-1} - B_{2k})U \quad (18)$$

Applying the Laplace Transform in (17), results in the system of equations in the frequency domain in (19).

$$\tilde{x}(s) = \begin{bmatrix}(sI - A)^{-1}B & (sI - A)^{-1}B_d\end{bmatrix}\begin{bmatrix}\tilde{v}_{in}(s)\\ \tilde{d}(s)\end{bmatrix} \quad (19)$$

Thus, in equation (19) we can obtain the transfer functions that describe the behavior of the converter in face of variations in the input voltage and variations in the duty-cycle. The transfer functions are presented in equations (20) - (22) for an N-cell converter, considering duty-cycle lower and higher than $\frac{1}{2}$ (for $r_L = 0$).

$$G_{v_o v_{in}} = \frac{\tilde{v}_0}{\tilde{v}_{in}}(s) = \frac{1}{d_x} \cdot \frac{\frac{Nd_x^2}{LC}}{s^2 + s\frac{1}{RC} + \frac{Nd_x^2}{LC}} \quad (20)$$

$$G_{v_o d} = \frac{\tilde{v}_0}{\tilde{d}}(s) = \frac{V_{in}}{d_x^2} \cdot \frac{\frac{Nd_x^2}{LC}}{s^2 + s\frac{1}{RC} + \frac{Nd_x^2}{LC}} \cdot \left(1 - s\frac{L}{NRd_x^2}\right) \quad (21)$$

$$G_{i_L d} = \frac{\tilde{i}_L}{\tilde{d}}(s) = \frac{V_{in}}{d_x} \cdot \frac{\frac{Nd_x^2}{LC}}{s^2 + s\frac{1}{RC} + \frac{Nd_x^2}{LC}} \cdot \left(\frac{2}{NRd_x^2} + s\frac{C}{Nd_x^2}\right) \quad (22)$$

IV. VALIDATION OF CONVERTER PLANT

The mathematical modeling of the system can be validated by comparing the frequency response of the electrical circuit with the response obtained with found plants. To that purpose, we use a feature of the electric simulator PSIM which allows to extract the transfer function from a disturbance. This feature, known as AC Sweep, performs an AC frequency scan and returns the plant bode diagram.

The PSIM schematic of the two-cell iBC circuit is shown in Fig. 4 and the converter specifications are shown in Table I. To validate the input-to-output transfer function $G_{v_o v_{in}}(s)$, for example, a disturbance (a sinusoidal source) was inserted in series with the input voltage, as also shown in Fig. 4. The

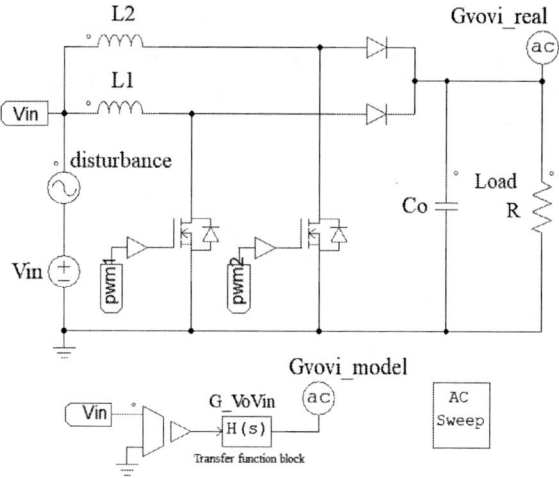

Fig. 4: iBC circuit used to validate the input-to-output transfer function.

TABLE I: iBC Parameters

PARAMETER	VALUES
Input voltage (V_{in})	160 V
Output voltage (V_0)	400 V
Output power (P_0)	400 W
Switching frequency (f_s)	20 kHz
Inductance (L_1, L_2)	4 mH
Capacitance (C_0)	1 μF

value used for the amplitude of the disturbance is 1% of the input voltage.

To monitor the output voltage, the AC sweep probe was placed at the output location and named $GVoVi\_real$. To compare the plants found in section III with the model that PSIM generates for the circuit, the *s-domain Transfer Function* block was added, in which the transfer function $G_{v_o v_{in}}(s)$ coefficients are inserted. This block receives as input signal V_{in} which passes through the frequency domain transfer function and returns the $GVoVi\_model$ response.

The frequency responses generated by the power circuit and the *s-domain Transfer Function* block are shown in Fig. 5, validating the model obtained in section III.

V. DYNAMICS ANALYSIS OF THE MODEL

Table II shows the damping ratio and natural frequency values for a conventional boost converter (one cell) and a N-cell interleaved boost converter.

TABLE II: Damping ratio and Natural frequency for a conventional boost converter and a N-cell parallel iBC

	Damping Ratio	Natural Frequency
Boost Converter	$\zeta_1 = \sqrt{LC}/2d_x RC$	$\omega_{n1} = d_x/\sqrt{LC}$
N-cell iBC	$\zeta_N = \sqrt{LC}/2d_x\sqrt{N}RC$	$\omega_n = d_x\sqrt{N}/\sqrt{LC}$

978-1-7281-4181-7/19 $31.00 © 2019 IEEE

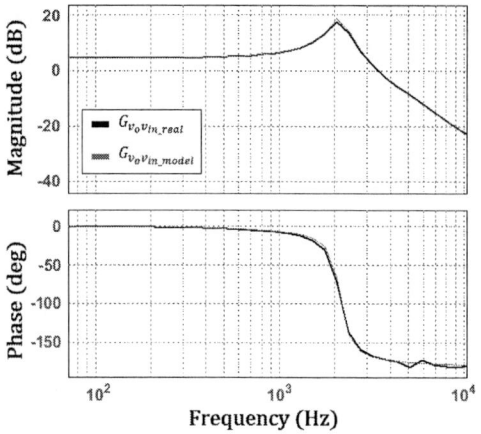

Fig. 5: Comparison between the diagrams generated by the s-domain Transfer Function block and generated by the power circuit.

Compared to the classical converter, the natural frequency for the N-cell converter is higher by $\sqrt{N}$, while the damping ratio is lower by $\sqrt{N}$, for both duty-cycle ranges. Therefore, the dynamic response of the interleaved converter is faster but more oscillatory. Using MATLAB simulation software, the step response of $G_{v_o v_{in}}$ is obtained for a one, two and three cell iBC, as seen in Fig. 6, which shows the dynamic response characteristic of the converter with increasing number of cells.

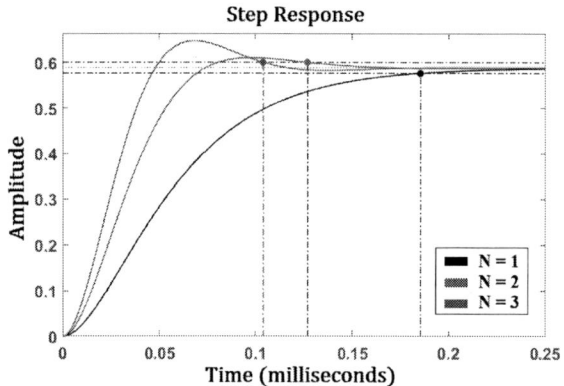

Fig. 6: Step responses of $G_{v_o v_{in}}(s)$.

A. Compensated system dynamics

Fig. 7 shows the open-loop Bode diagrams of the compensated system for a one, two and three cell iBC, also obtained via MATLAB simulation. The bandwidth increases as the number of cells increases.

In the step response shown in Fig. 8, it is verified that increasing the number of cells makes the dynamic response of the converter more oscillatory, as previously analyzed by decreasing the damping ratio. Although there is more oscillation, the overshoot is lower compared to the one cell converter. If compared to other responses, the arrangement with three cells may prove to be the safest and, depending on the system, this

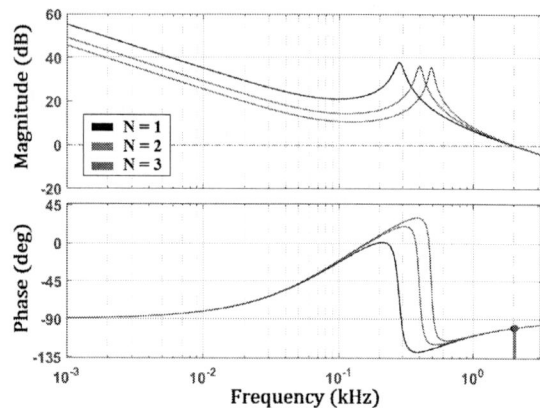

Fig. 7: Bode diagrams in open-loop system compensated.

becomes of great relevance (in practice). The step response of the proportional–integral compensated system approaches unity in the steady-state, therefore, the compensated system responds with zero steady-state error.

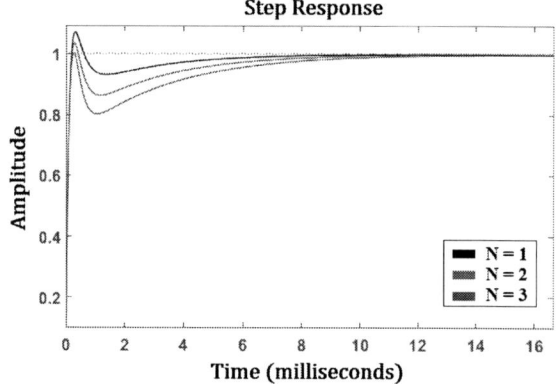

Fig. 8: Compensated closed-loop step response.

B. Input current ripple characteristic

As an example, a two-cell iBC is analyzed in this section. The simulated result of the interleaved boost converter is given by Fig. 9, in which the waveforms of the input current and the currents in the inductors are shown. Two switches are provided the gate signal which is out of phase by $\pi\ rad$. The effect of the interleaved operation is checked graphically: the decrease of the input current ripple amplitude.

In addition to ripple reduction, phase shifting the pulses causes the apparent switching frequency in the output is increased N times compared to each cell's frequency, as seen in Fig. 9. For the case of N-cell iBC, the frequency is increased N times. In this way, the equation can be presented for the input current ripple of the interleaved step-up converter:

$$\Delta I_L = \frac{\alpha \cdot (1 - \alpha)}{L f_s N} V_0 \tag{23}$$

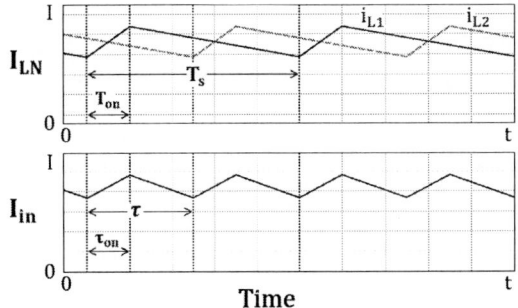

Fig. 9: Two-cell iBC currents.

where α is the apparent duty-cycle. Although the real duty-cycle d is the relation between T_{on} and T_s, the apparent duty-cycle is related to the apparent switching frequency noticed in the input current. The apparent duty-cycle is then obtained by the relation between the apparent switch-on (τ_{on}) and the apparent switching period (τ).

Since the energy stored in an inductor is given by $E = \frac{1}{2}I_L{}^2L$, for an interleaved converter the energy in each inductor is obtained by equation (24).

$$E_L = \frac{\left(\frac{I_L{}^2}{N}\right)^2 L}{2} = \frac{I_L{}^2 L}{2N^2} \quad (24)$$

As a consequence, the energy stored in the filter impedance can be reduced by a factor of N^2 for the same output ripple, due to the higher output frequency [10]. With smaller filter's passive components, the system overall dynamics is also improved and a better performance is achieved.

The current ripple characteristic is shown in Fig. 10. This decreases inversely as the number of cells increases, but the duty-cycle also interferes with their value. For example, considering the single-cell converter, the maximum ripple occurs at a duty-cycle of 50%. The ripple is cancelled when the duty-cycle is a multiple of $1/N$.

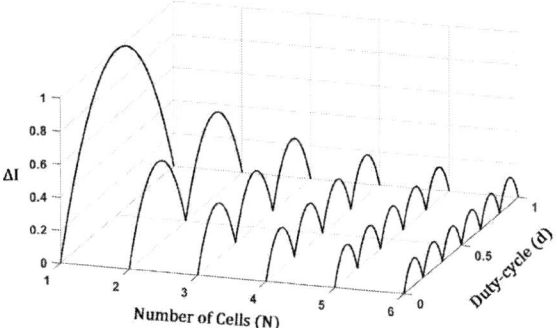

Fig. 10: Current ripple variation as the number of cells and duty-cycle change. Adapted from [2].

VI. CONCLUSIONS

A generalized modeling of the N-cell interleaved Boost converter, using the State-Space Averaging Technique, is pre-sented in this paper. The DC and AC models of the converter are derived including the discontinuity shown in the duty-cycle. With the mathematical model obtained for any number of switching cells, the converter can be designed to meet the performance specifications for any duty-cycle range, absent the need to obtain the operating modes for each of the ranges. Through the N-cell model, the advantages presented for the interleaved topology can be demonstrated as smaller filter's passive components, since the energy stored in the inductor is inversely proportional to the squared number of cells in parallel, and the consistent reduction current ripple, since the switching frequency is increased N times. The influence of the number of cells on the dynamic performance of the converter can also be verified through mathematical modeling, via analysis of the damping ratio and the natural frequency, being also verified by the step response obtained in simulation. The validation of the proposed mathematical model was made via the comparison of the power circuits and the converter plants frequency response.

ACKNOWLEDGMENT

The authors are grateful to UFMG and UNIFEI for their support of this project. And to CNPq for financial support conferred. This work has been supported by the Brazilian agency CAPES and project CAPES-COFECUB no. 876/2017.

REFERENCES

[1] D. R. Garth, W.J. Muldoon, G. C. Benson, and . N. Costague, "Multi-Phase, 2 Kilowatt, High Voltage, Regulated Power Supply," IEEE Power Electronics Specialists Conference, 1971.

[2] J. L. da Silva, "Design and Control of a Multicell Interleaved Converter for a Hybrid Photovoltaic-Wind Generation System," Ph.D. dissertation, University of Toulouse, 2017.

[3] P. A. de C. e Castro, M. H. da Silva Alves, J. H. D. G. Pinto, B. de J. Cardoso Filho, and S. I. Seleme, "Desenvolvimento de um Modelo Matemático Unificado para o Conversor Boost Entrelaçado," International Conference on Industry Applications - INDUSCON, vol. 1, pp. 1-6, São Paulo, 2018.

[4] M. R. Assunção, "Estudo e implementação de um conversor CC-CC Boost Entrelaçado em Regime de condução contínua," M.S. thesis, Federal University of Minas Gerais, 2014.

[5] J. L. da Silva, Reis, G., S. I. Seleme, and T. Meynard, "Control Design and Frequency Analysis of an Output Filter in Parallel Interleaved Converters," IEEE International Conference on Power and Energy (PECon), Melaka, Malaysia, Nov. 2016.

[6] N. Jantharamin and L. Zhang, "Analysis of multiphase interleaved converter by using state-space averaging technique," 6th International Conference on Electrical Engineering/Electronics, Computer, Telecommunications and Information Technology, vol. 1, pp. 288–291, May 2009.

[7] O. Hegazy, J. V. Mierlo, and P. Lataire, "Analysis, modeling, and implementation of a multidevice interleaved dc/dc converter for fuel cell hybrid electric vehicles," IEEE Transactions on Power Electronics, vol. 27, pp. 4445-4458 , Nov. 2012.

[8] R. Crews, "An-1820 lm5032 Interleaved Boost Converter," Texas AN-1820, May 2013.

[9] L. S. Garcia, "Controle Inversor dual de único estágio aplicado ao gerenciamento de energia através de um módulo fotovoltaico e uma célula a combustível," Ph.D. dissertation, Federal University of Uberlândia, 2014.

[10] J. L. da Silva, G. L. Dos Reis, R. M. Silva, S.I. Seleme, T. A. Meynard, and A. M. Llor, "Design, Modeling and Identification of the Generation Side Converter in an 11.7 kW Wind/Photovoltaic Hybrid Renewable Generation System," IEEE 8th International Symposium on Power Electronics for Distributed Generation Systems (PEDG), Florianopolis, Brazil, April 2017.

978-1-7281-4181-7/19 $31.00 © 2019 IEEE

AC-DC Converter with High-Frequency Isolation Operating Under ZVS

Bruno Alves Sousa da Silva
Electrical Engineering Department
Federal University of Ceará (UFC)
Fortaleza, Brazil
b.alves@dee.ufc

Dalton de Araújo Honório
Electrical Engineering Department
Federal University of Ceará (UFC)
Fortaleza, Brazil
dalton@dee.ufc.br

Demercil de Souza Olivera Júnior
Electrical Engineering Department
Federal University of Ceará (UFC)
Fortaleza, Brazil
demercil@dee.ufc.br

Bruno Ricardo de Almeida
Electrical Engineering Department
University of Fortaleza (UNIFOR)
Fortaleza, Brazil
almeida@unifor.br

Samanta Gadelha Barbosa
Electrical Engineering Department
Federal University of Ceará (UFC)
Fortaleza, Brazil
samantagadelha@dee.ufc.br

Luiz Henrique Silva Colado Barreto
Electrical Engineering Department
Federal University of Ceará (UFC)
Fortaleza, Brazil
lbarreto@dee.ufc.br

Caio Kerson O. Veras
Electrical Engineering Department
Federal University of Ceará (UFC)
Fortaleza, Brazil
caiokerson@dee.ufc.br

Abstract—**This paper presents the study of the single-stage rectifier with high-frequency isolation, focusing on its soft switching analysis in the primary side. The converter uses the concepts of interleaving by coupled-windings and the dual active bridge (DAB) structure, so that it is possible to obtain a power flow control through the delay angle applied on the secondary side with a good distribution of the semiconductor efforts. The soft switching investigation was carried out by the qualitative and quantitative analysis of the converter. The validation of such analysis is elaborated by spreadsheets and simulation results considering a 1 kW load. Experimental partial results is presented to validate the proposed converter.**

Keywords—*interleaving by coupled-winding, power factor correction, single-phase rectifier, delay angle*

I. INTRODUCTION

The power distribution is commonly realized with ac in the electric system. Therefore, if a dc load is used an AC-DC stage is necessary for the power adaptation in applications operating in dc. Moreover, the growth in the distributed generation systems, more particularly the distributed energy resources [1] and the smart-grid concept [2], created some wanted characteristics for the AC-DC conversion stage.

For instance, different generation plants should be connected in the same distribution system providing power for the loads, as well as, the controllability of the flow of such power among the generation system, energy store system (EES) and distribution system. In order to address that, bidirectional operation is necessary, narrowing the number of power converter able to provide such abilities, especially when isolation is required to increase reliability.

The AC-DC conversion with high-frequency isolation is well-known on industry, especially in power supplies for telecommunications [3] and [4], where high voltages are used due the large level of power. Therefore, power converter with high power density become more relevant.

Fig. 1 presents the power converter that is used in the dc link connections. In this kind of connection, other generation sources, like wind farms and PV-systems, are connected as well. The power flows from the generation plant to the load.

There are many power converter solutions [5] that create a point of connection (PC) in the dc-link, while provide [6] good losses distribution in the semiconductors, resulting in a high efficiency structure with low cost. To accomplish that, it is possible to employ some soft switching techniques, such as: zero voltage switching (ZVS) [7] and zero current switching (ZCS) [8].

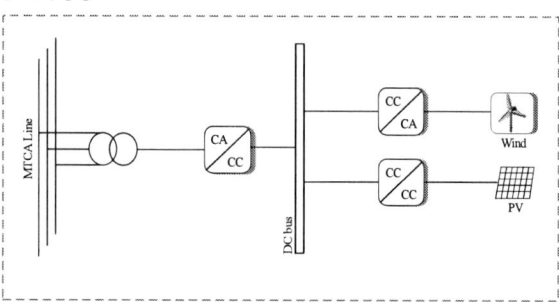

Fig. 1. Dc-link connection application.

Thus, the hard switching can be avoided and a high switching frequency can be used, reducing weight and volume of the magnetics.

In [9] a ZVS transformer-isolated step up/down DC link circuit for PWM unipolar single-phase inverter was proposed. In [10] a new interleaved AC-DC single-stage converter with proposed ZVS operation of its switches, and simple operation with conventional and standard control methods. In [11] a high efficiency two-stage AC/DC converter applied in on-board charger for electric vehicles is presented. This converter is a high efficiency ZVS half bridge power factor correction (PFC) topology. However, it can also be expensive as it requires two separate converters with two separate controllers to be implemented. Using the advantage of interleaving in [12] it is proposed a ZVS-PWM interleaved transformer- isolated boost dc/dc converter with a simple ZVS-PWM auxilary circuit during turn on and turn off the switches.

In this context, this work proposes an adaptation of the AC-DC single-stage converter proposed in [13], looking for

978-1-7281-4181-7/19 $31.00 © 2019 IEEE

to obtain the converter operation under ZVS in the switches at the primary side of the converter.

II. PROPOSED TOPOLOGY

The considered power converter consist in the adaptation of the power converter presented in [13], shown in Fig. 2. The converter is based on the DAB converter [14] and applies the concept of interleaving technique by coupled-windings [15].

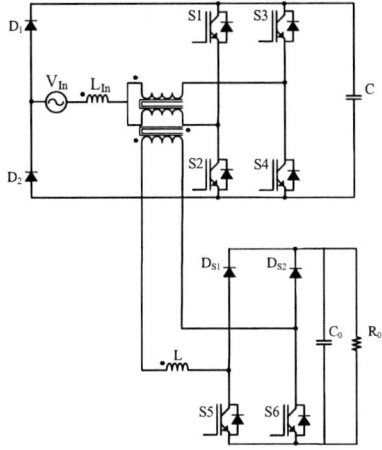

Fig. 2. Single-phase single-stage AC-DC isolated rectifier.

This technique provides good distribution of the current stress in the semiconductor, which combined with soft-switching operation helps to improve the losses distribution of the semiconductor in the primary side of the converter, while the secondary side aim to minimize the reactive power flowing by the transformer.

A. Operation Principals

The converter operates as a rectifier with power factor correction (PFC) and the dc output regulated.

The topology presents an h-bridge interleaved by coupled-windings in the primary side, while use a bridgeless converter in the secondary. The legs in the h-bridge operate with an 180° of phase-shifting between each other, with the switches pair in each leg working complementary, avoiding short-circuits.

The circuit analysis can be done considering only one half-cycle of the grid. Considering the positive half-cycle, the voltage across the primary side is null, since the switches $S2$ and $S4$ are on, as can be seen in Fig. 3a. Similar behavior is

presented in Fig. 3b, what is noted when the switches $S1$ and $S3$ are, also, on. Fig. 3c shows the converter configuration when the switches $S2$ and $S3$ are on, then the resulting voltage across the primary side is $-V_{ab}$. Finally, when switches $S1$ and $S4$ are on the voltage is equals to V_{ab}. Thus, it is possible to noted that the voltage in the primary side has a square waveform with three-levels ($-V_{ab}$, 0 and V_{ab}).

B. Control Strategy

The adopted control strategy in presented in Fig. 4. As mentioned before, two triangular carriers are used to generate the 180° phase-shift between the legs. The gate signals for h-bridge are obtained by the comparison of the carriers with the modulating signal from the control loops.

Thus, the PFC and the output voltage regulation are obtained, by the duty-cycle variations of the switches in h-bridge and the bridgeless structure, respectively. The current loop control was tuned with a crossing frequency about 12.5 kHz and phase margin of 60°, using the k-factor tuning method [16].

The gate signal of the secondary side is obtained by using a modulating signal delayed by angle φ according to a resulting in the delay between the voltage across the h-bridge in the primary side and the bridgeless structure in the secondary one. This allows the power-flow control loop to determinate the correct power amount requested by the load with the minimum reactive power demanded.

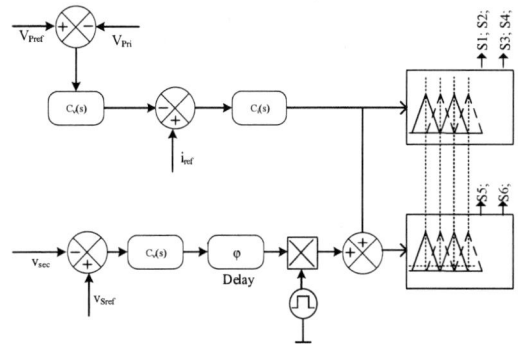

Fig. 4. Adopted control strategy.

This technique also allows the converter to operate under ZVS during turning on switches in the primary side. However, the switches in the secondary side operate in hard commutation.

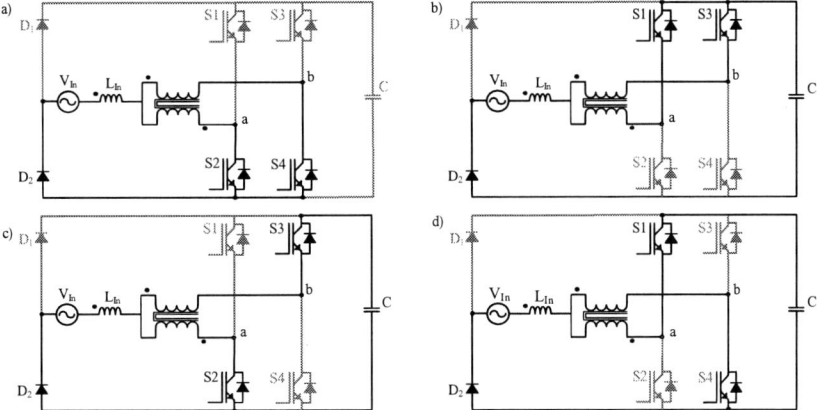

Fig. 3. Operational steps of the proposed topology.

C. Quantitative Analysis

In [17] is presented the basic dc-dc cell operating with 16 regions according to the φ and the duty-cycle of the semiconductors. Since the AC-DC converter performs PFC, the duty-cycle varies in a sinusoidal form and more than one operational region is used. However, the modulation technique combined with the angle φ can force the converter to operate only in one chosen operational condition, as presented in Fig. 5, and it will be described as follow.

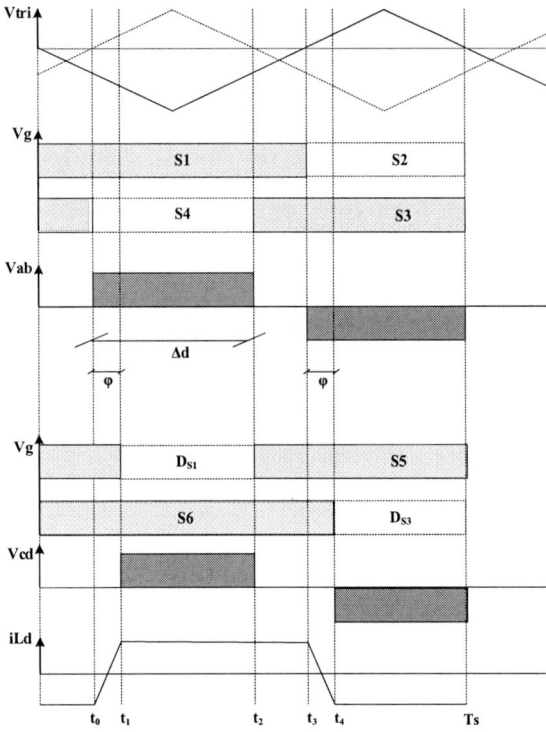

Fig. 5. Main theoretical waveform during a switching period, considering the operational steps.

1) Chosen operational region (Region 1): The region can be defined by eigth equations from the eyes of the grid period. The relationship of the current through the inductance in the secondary side of the transformer, i_L, is described by (1), considering the positive half-cycle.

$$
i_L = \begin{cases}
-\dfrac{V_{DC(ma)}}{2L}(\varphi - 2t), & \text{if } 0 < t > \varphi \\[2mm]
\dfrac{V_{DC(ma)}}{2L}(\varphi), & \text{if } \varphi < t < \Delta d \\[2mm]
\dfrac{V_{DC(ma)}}{2L}(T_s - 2t + \varphi), & \text{if } \Delta d < t < \Delta d + \varphi \\[2mm]
-\dfrac{V_{DC(ma)}}{2L}(\varphi), & \text{if } \Delta d + \varphi < t < T_s
\end{cases}
\tag{1}
$$

where $V_{DC(ma)}$ is the regulated voltage of the dc-link, L is the inductance in the secondary side of the transformer, φ is the phase-shift between the voltage waveforms of the primary side and secondary one and T_s is the switching period.

The voltage in the primary side and the one in the secondary have the same polarity and their form varies according to the duty-cycle, Δd, which can have its behavior represented by (2).

$$
\Delta d = \begin{cases}
(2ma\sin(\dfrac{2\pi t}{T_{grid}})\dfrac{T_s}{2}, & \text{if } 0 < t > \dfrac{1}{2ma} \\[2mm]
2 - (2ma\sin(\dfrac{2\pi t}{T_{grid}})\dfrac{T_s}{2}, & \text{if } \dfrac{1}{2ma} < t < \dfrac{T_s}{4}
\end{cases}
\tag{2}
$$

where ma is the modulation index and T_{grid} is the period of the grid.

2) Output power analysis: The power which flows from the primary to the secondary sides can be derived considering the instantaneous contribution of the described region 1, considering the system behavior symmetry. Thus, the instantaneuos power in the Region 1 can be obtained by the integration of (1) in each operational interval. The transfered power is described by (3) and considers the aforementioned integration for the whole grid period.

$$
P_{(\varphi,ma)} = \frac{V_{DC(ma)}^2 \varphi \left(-2T_s \sin(\dfrac{1}{2ma}) + \pi T_s - \pi\varphi \right)}{\pi L T_s} + \cdots
$$
$$
\cdots - \frac{V_{DC(ma)}^2 \varphi \left(-2T_s ma + 2T_s ma \sqrt{4 - \dfrac{1}{ma^2}} \right)}{\pi L T_s},
\tag{3}
$$

where $P_{(\varphi,ma)}$ is the transferred power. Fig. 6 shows the output power characteristic in terms of φ applied in different modulation indexes. It is possible to note for different power the higher transferred power amount happens when the modulation index decays and the φ increases. Therefore, the maximum transferred power point is achieved around $\varphi=50°$.

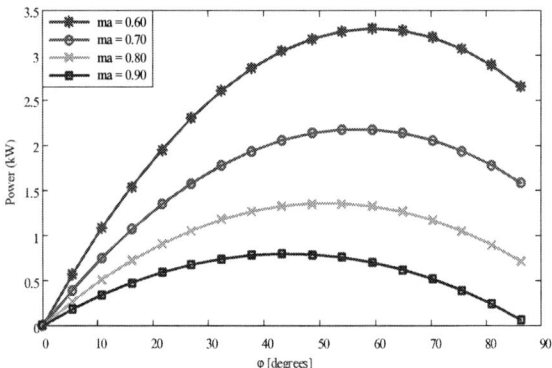

Fig. 6. Output power waveform considering different modulation indexes.

3) Soft-Switching Analysis: The power converter currents in the primary side are simillar, as well as, the ones in the secondary, once again considering the converter operation symmetry. Then, the currents through the switches can be described by (4) and (5) for the primary and secondary sides, respectivelly.

$$
is1(\varphi, ma, \tau) = iLs(\tau) - \frac{Iin(\varphi, ma)\sin(\tau)}{2},
\tag{4}
$$

$$
is5(\varphi, ma, \tau) = -iLs(\tau).n,
\tag{5}
$$

978-1-7281-4181-7/19 $31.00 © 2019 IEEE

where, $is1(\varphi,ma,\tau)$ the current through the switch $S1$, $is5(\varphi,ma,\tau)$ the current through the switch $S5$, $iLs(\tau)$ is the current through the secondary side referred to the primary one, and n is the turns ration of the transformer. From the equations in subsection *1)* the instantaneous current through switch $S1$, $is1$, is the instantaneous turning off current, considering the positive half-cycle, is given by (4). Moreover, during the negative half-cycle, $is1$ is now the one through switch $S4$, when $S4$ operates during the positive half-cycle. Fig. 7 presents the instantaneous behavior of the current through $S1$ in term of the maximum value of the input current and considering different values of phase-shift angles, as well as, modulation index. Analogously, the same behavior happens between switches $S2$ and $S3$.

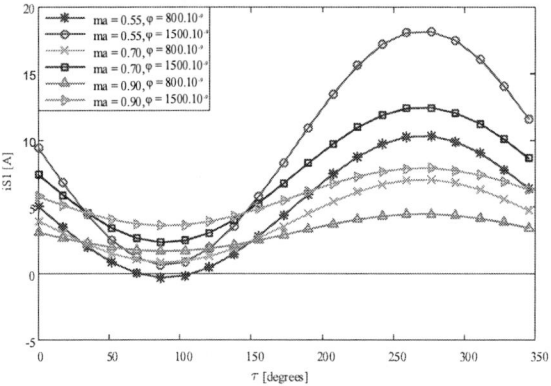

Fig. 7. Normalized turning-on current in switch $S1$.

In order to perform ZVS for all the switches in the primary side during the grid period, (4) must be greater than zero. Therefore, there is a critical condition for modulation index around 0.55. Thus, the power converter can operate with different values for ma and φ, allowing a broad operational range for the control loop without losing the ZVS characteristics.

Fig. 8 shows the instantaneous current through the switch $S5$, considering different modulation indexes and phase-shift angles. In (5) is described the instantaneous current of $S5$ during its turning on in the positive half-cycle. The same analysis is extrapolated for the instantaneous current in $S6$.

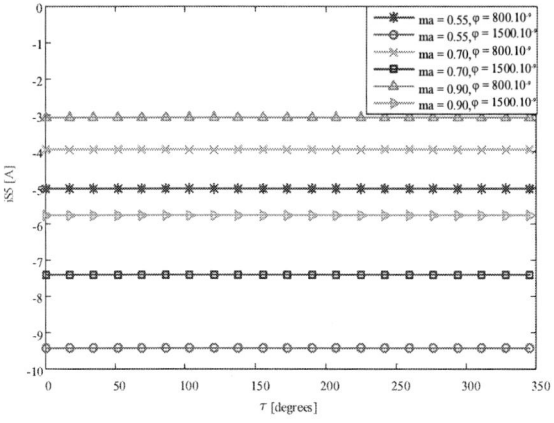

Fig. 8. Normalized turning-on current in switch $S5$.

It is possible to note, that the switches in the secondary side are switched in hard conditions, during the whole period of the grid.

III. Simulation Results

This section presents the simulation results achieved from a simulation model built in the PSIM (PowerSim®) to validate the analysis presented so far. The adopted parameters of the simulation model are presented in TABLE I.

TABLE I: Design Specifications and Parameters.

Rms grid voltage	220 V
Grid frequency	60 Hz
DC-link voltage	400 V
Rated power	1 kW
Switching frequency	50 kHz
Secondary capacitor	500 μF
Primary capacitor	500 μF
Secondary inductor	45 μH
Primary inductor	1 mH
Transformer turns ratio	1
Modulation index	0,777

The voltage across the dc-link in the secondary side is shown in Fig.9, where it is possible to see the good regulation of such voltage.

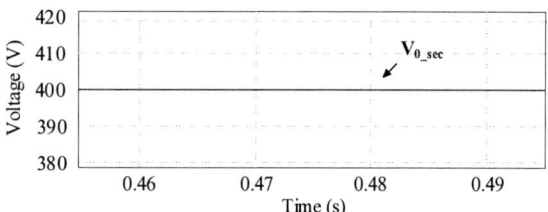

Fig. 9. Dc-link output voltage.

Fig 10 presents the behavior for both, the input voltage (*Vgrid*) and current (*i(Lin)*), validating the good performance of the PFC control loop, as the Power Factor (P.F.) = 0,998 and Total Harmonic Distortion of current (THDi) < 5%.

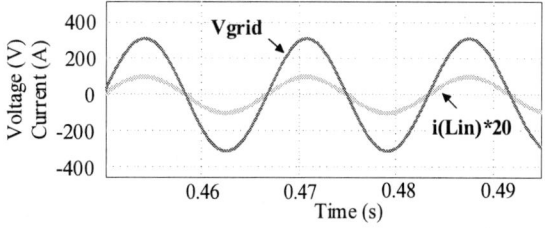

Fig. 10. Input current and grid voltage.

As mentioned before, the adopted modulation technique uses a delay between the voltages across the primary side and secondary one. Fig. 11 shows such difference, validating the functionality of the power flow control loop.

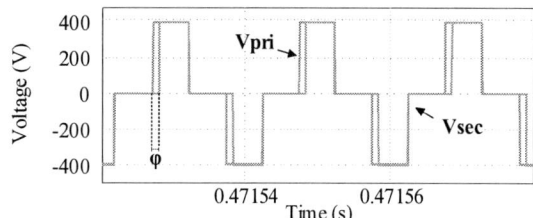

Fig. 11. Voltages across the primary side, *Vpri* and secondary one *Vsec*.

As presented in the last section, switches *S1* and *S2* operates under ZVS, then Fig. 12a and Fig. 12b show these behaviors for them, respectively. The ZVS conditions for switches *S3* and *S4* are presented by Fig. 12c and Fig. 12d, respectively.

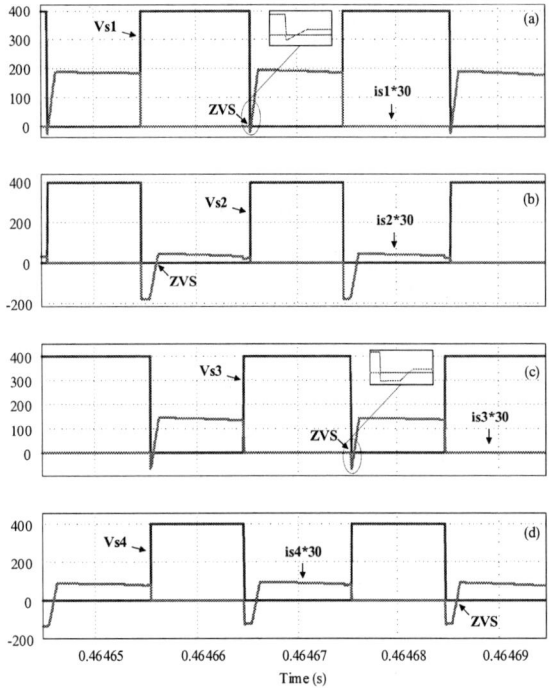

Fig. 12. Soft Switching ZVS (a) switch *S1*, (b) switch *S2*, (c) switch *S3* and (d) switch *S4*.

Although the switches in the primary side turning-off are in hard conditions, the soft-switching behavior in their turning-on results in a better switching losses conditions. For the secondary side, it is adopted the bridgeless structure with aims to contribute to the reduction of the losses in the converter. This solutions causes a slight increment of the conduction losses, however after the losses analysis, it is noted that the total loss is reduced.

IV. EXPERIMENTAL RESULTS

This section presents the partial experimental results in order to validate the proposed topology. Design specifications and parameters adopted are the same as those used in the simulation results presented in table I.

Fig. 13. Experimental prototype.

Fig. 13 shows the experimental prototype that consist in two three-phase SiC power module from CREE, model: CCS0220M12CM2 and a digital signal processor from Texas Instruments, model TMS320F28377D, where the modulation and control techniques are embedded.

Fig. 14 shows at nominal load condition grid voltage, input current and primary bus voltage. Current is in phase with grid voltage, presenting high power factor (PF) and low Total Harmonic Distortion current (THDi). DC bus regulated at design dc voltage 400V.

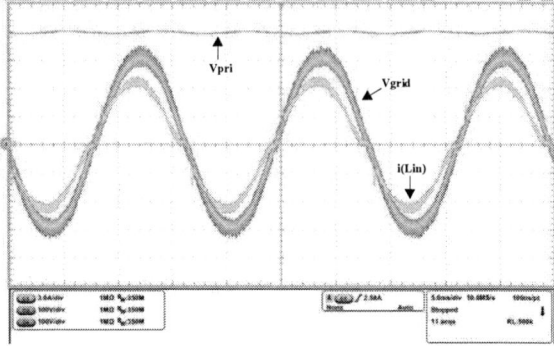

Fig. 14. Grid voltage, input current and primary bus voltage.

In the dynamic test to validate the designed voltage regulation control, the step load response test is perfomed from 70 to 100% and 100% to 70% of the load. It was possible to verify the good functioning closed loop current and voltage, the regulated bus voltage circuit, as shown in Fig 15 and Fig. 16.

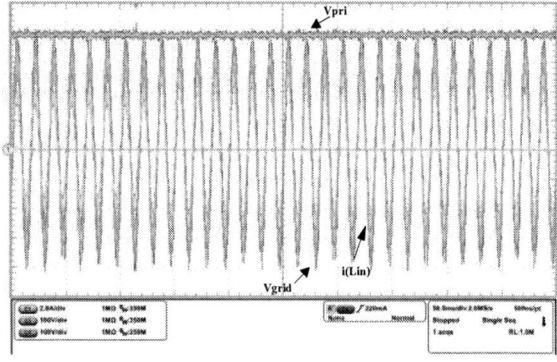

Fig. 15. Grid voltage, input current and primary bus voltage (step of 70% - 100%).

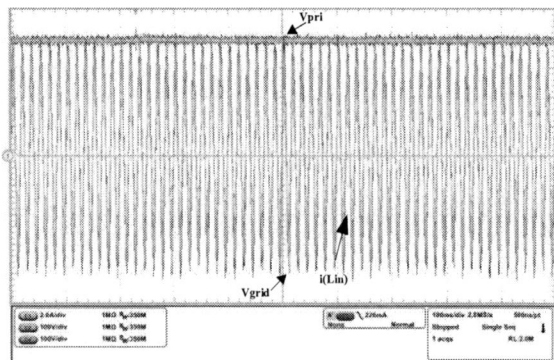

Fig. 16. Grid voltage, input current and primary bus voltage (step of 100% - 70%).

V. CONCLUSION AND FUTURE WORKS

This work presented the study of a soft-switching operation, achieving ZVS conditions for all switches in the primary side of the converter. The simulation model results, considering 1 kW, demonstrate a P.F. equal to 0.998 and good THDi. The power flow was carried out by the application of a modulation technique, which uses delay angle, achieving ZVS operation and low level of reactive power on the transformer. Moreover, it is possible to operate the converter in different values of ma and φ without lose the ZVS characteristic, with a critical point about ma=0.55. The good performance of the current control loop was noted by the result of the output voltage about 400 V in the dc-link.

The partial experimental result proved the good operation of the control meshes of the primary side of the converter. Drain current from the grid was in phase with input voltage. In the dynamic test, the primary bus voltage remained regulated after the load input and output step

ACKNOWLEDGMENT

The authors would like to thank the Federal University of Ceará (UFC), the Energy and Control Processing Group (GPEC). The Brazilian National Council for Scientific and Technological Development (CNPq) and Cearense Foundation for Scientific and Technological Development Support (FUNCAP) supported this work.

REFERENCES

[1] Normative resolution nº 482/2012 of 17 April 2012. 2012 B. Available em www.aneel.gov.br. Acess 20 de June de 2019.

[2] Vehbi C. Gungor, Dilan Sahin, Taskin Kocak, Salih Ergut, Concettina Buccella, Carlos Cecati and Gerhard P. Hancke. Smart grid technologies: Communication technologies and standards. IEEE transactions on Industrial informatics, v. 7, n. 4, p. 529-539, 2011.

[3] D. E. Burke, "Reflections on INTELEC 2014", Bodo's Power Systems Magazine (Electronics in Motion and Conversion), pp. 22, Novembro 2014.

[4] A. Pratt, P. Kumar, T. V. Aldridge, "Evaluation of 400V DC distribution in Telco and Data centers to Improve Energy Efficiency", in Proc. of INTELEC, pp. 32–39, 2007.

[5] D. S. Oliveira Jr, M. I. V. Batista, L. H. S. C. Barreto, P. P. Praca, "A bidirectional single stage AC-DC converter with high frequency isolation feasible to DC distributed power systems", in Proc. of INDUSCON, pp. 1-7, 2012.

[6] Maria D. Bellar, Tzong-Shiann Wu, Aristide Tchamdjou, Javad Mahdavi and M. Ehsani. A review of soft-switched DC-AC converters. IEEE Transactions on Industry Applications, v. 34, n. 4, p. 847-860, 1998.

[7] H. M. de Oliveira Filho, D. S. Oliveira Jr, C. E. de A e Silva, F. L. Tofoli, "ZVS bidirectional isolated three-phase DC-DC converter with dual phase-shift and variable duty cycle", in Proc. INDUSCON, pp. 1-8, 2012.

[8] Hang-Seok Choi and Bo Hyung Cho, "Novel zero-current-switching (ZCS) PWM switch cell minimizing additional conduction loss," in IEEE Transactions on Industrial Electronics, vol. 49, no. 1, pp. 165-172, 2002.

[9] WANG, Chien-Ming; CHEN, Guan-Yu. A ZVS-PWM Single-phase inverter using a ZVS transformer-isolated step-up/down dc link. In: 2017 IEEE 3rd International Future Energy Electronics Conference and ECCE Asia (IFEEC 2017-ECCE Asia). IEEE, 2017. p. 687-691.

[10] ABOSNINA, Adel Ali; KHODABAKHSH, Javad; MOSCHOPOULOS, Gerry. A Single-Stage ZVS AC-DC Boost Converter with Interleaving. In: 2018 IEEE Energy Conversion Congress and Exposition (ECCE). IEEE, 2018. p. 6790-6795.

[11] Siliang Zhang, Guixing Lan, Zezheng Dong, Xinke Wu. A High Efficiency two-stage ZVS AC/DC converter with all SiC MOSFET. In: 2017 IEEE 3rd International Future Energy Electronics Conference and ECCE Asia (IFEEC 2017-ECCE Asia). IEEE, 2017. p. 163-169.

[12] WANG, Chien-Ming; LIN, Chang-Hua; HSU, Shih-Yung. A ZVS-PWM interleaved transformer-isolated boost DC/DC converter with a simple zvs-pwm auxiliary circuit. In: 2012 IEEE Third International Conference on Sustainable Energy Technologies (ICSET). IEEE, 2012. p. 299-304.

[13] Bruno R. de Ameida, José W. M. de Araújo, Paulo P. Praça, Demercil S. Oliveira Jr. A Single-Stage Three-Phase Bidirectional AC/DC Converter With High-Frequency Isolation and PFC. IEEE Transactions on Power Electronics, v. 33, n. 10, p. 8298-8307, 2017.

[14] W. M. dos Santos, D. C. Martins, "Introdução ao conversor DAB monofásico", Eletrônica de Potência – SOBRAEP, vol. 19, nº 1, pp. 36-46, Fevereiro 2014.

[15] G. V. T. Bascope, I. Barbi, "Generation of a family of non-isolated DC-DC PWM converters using new three-state switching cells", in Proc. of PESC, pp. 858-863, vol. 2, 2000.

[16] H. D. Venable, "The k-factor: A New Mathematical Tool for Stability Analysis and Synthesis", in Proc. of POWERCON, pp. I2-1 - I2-17, 1983.

[17] L. C. S. Mazza, Luan, D. S. Oliveira Jr, F. L. M. Antunes, D. B. S. Alves, P. C. M. Campelo, F. J. L. Freire, "A Soft Switching Bidirectional DC-DC Converter with High Frequency Isolation Feasible to Photovoltaic System Applications", in Proc. of PCIM, pp.1-8, 2015.

Enhanced Power Management System for Droop Control in a Grid Connected DC Microgrid

Pedro José dos Santos Neto
Electrical and computer Eng. School
University of Campinas
Campinas, Brazil
pedrojsn@DSCE.fee.unicamp.br

Joao Pedro C. Silveira
Electrical and Computer Eng. School
University of Campinas
Campinas, Brazil
jpcarvalho@outlook.com.br

Tárcio André dos S. Barros
Faculty of Mechanical Engineering
University of Campinas
Campinas, Brazil
tarcioandre@fem.unicamp.br

Ernesto Ruppert Filho
Electrical and Computer Eng. School
University of Campinas
Campinas, Brazil
ruppert@fee.unicamp.br

Juan Carlos Vasquez
Department of Energy Technology
Aalborg University
Aalborg, Denmark
juq@et.aau.dk

Josep M. Guerrero
Department of energy technology
Aalborg University
Aalborg, Denmark
joz@et.aau.dk

Abstract—This paper presents an enhanced power management system for a DC microgrid operating in the grid connected mode using droop control. The proposed power management technique acts over the energy storage system (ESS) regulating the power flow according to the state of charge of the battery bank. A voltage source converter (VSC) is applied to interface the DC microgrid and the AC utility grid. Since droop control is applied, no communication between the VSC and the energy storage unity is required. Customer loads and distributed generation are considered as part of the DC microgrid. The droop characteristic of the energy management system is modified to force the ESS to operate autonomously, providing the energy required by the load, while the AC grid only supplies or absorbs power during deficit or surplus of energy. A comparison between the proposed method and a conventional power management for droop control is performed. Simulation results indicate that the proposed system leads to a better control over the DC microgrid power flow, in a smooth DC bus voltage variation and in a better ESS state of charge regulation.

Index Terms—Power management system, DC microgrid, droop control, energy storage system, state of charge.

I. INTRODUCTION

DC microgrids have gained research attention in the last decade due to the development of power electronics converters and control systems capable of overcoming initial difficulties in handling DC power [1], [2]. Compared to an AC microgrid, a DC microgrid presents higher efficiency since many AC-DC conversion steps are not required. In addition, no reactive power circulates in the cables, which reduces losses and cost [1], [3]. The operation in island mode is the main focus of research because many controllability problems arises when there is no support of the utility grid [4–6]. For an urban system, however, the grid connected mode is the most relevant one since the utility grid is presented in the vast majority of the time. An operation in island mode is seldom expected, although this mode is fundamental to achieve truly autonomous operation in a smart distribution system.

FAPESP proc. 2016/08645-9, 2017/21087-8, 2018/22076-2.

Droop control and its variants are by far the most employed technique used to manage DC microgrids. Its advantages include simplicity of implementation and the absence of communication link between the power electronic converters [7–9]. Adaptive droop is usually employed to achieve a state of charge (SOC) dependent behavior of the energy storage system. In the methods presented by [10–16], the droop coefficient varies as a function of the ESS SOC, which allows to adaptively control the output voltage. The cited methods are mainly applied to regulate power sharing among multiple energy storage systems. However, varying the droop characteristic with the state of charge does not reflect in a full control over the ESS output power and, thus, the DC microgrid power flow is compromised. Moreover, the cited methods usually are applied to isolated mode and grid connected operation is only proper considered by [15].

Droop control is especially useful for island operation, considering the distributed generation (DG) and the energy storage systems (ESS). For an urban DC microgrid in the grid connected mode, however, a conventional droop control forces the ESS to share the power flow with the AC utility grid. Nonetheless, if autonomous operation is willing to be achieved, the microgrid has to rely on the ESS and on the DGs to provide the power required by customer loads. Hence, the AC utility grid must act only absorbing surplus or injecting power when the other sources are not able to maintain the power flow.

Considering the discussed situation, this paper proposes an enhanced power management system to guarantee autonomous operation of a DC microgrid in the grid connected mode. The proposed power management technique acts directly over the ESS power in a DC microgrid with a VSC and the ESS operating in droop control. The objective is to achieve stable power flow considering the ESS state of charge limitations. In addition, the ESS droop curve is modified to force the battery bank to provide the required power to the system, while the

VSC only acts during transients and during mismatches of power inside the DC microgrid.

This paper is structured as follows. Section II presents the studied DC microgrid, including the considered elements of an urban distribution system. Section III presents the conventional and modified droop characteristics considered in this work. Section IV presents the proposed enhanced power management system and its main considerations. Simulation results showing the comparison between the conventional adaptive droop and the proposed technique are discussed in section V. Finally, the conclusions are summarized in section VI.

II. STUDIED DC MICROGRID

The studied DC microgrid is presented in Fig. 1. The system is formed by a voltage source converter (VSC) interfacing the AC utility grid and the DC bus, by a battery bank as ESS, by a distributed generator and by the customer loads. Four converters are employed in a cluster to create a minimum smart distribution system as a microgrid (or nanogrid). Those elements were chosen considering a future experimental setup under development.

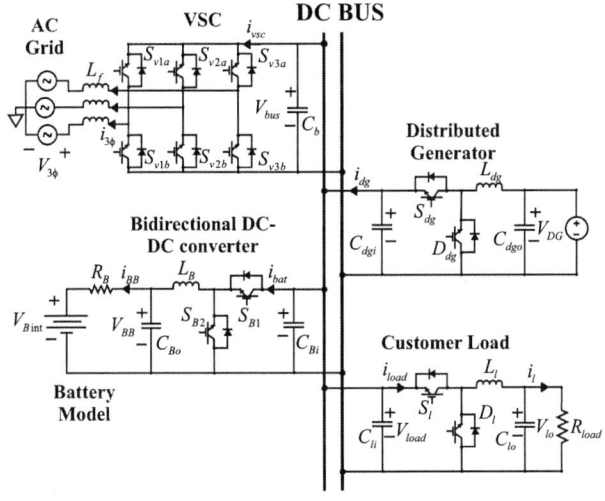

Fig. 1. Overview of the studied DC microgrid. The system is formed by a voltage source converter, a distributed generator, an energy storage system and a constant power load.

Initially, the DC bus is formed by the VSC operating as an active rectifier. Considering an AC utility grid of 220 V, 60 Hz, a DC 400 V level is chosen. After the initial grid forming operation, the DC bus is regulated by using droop control in a coordinated operation with the battery bank. For further experimental tests, the overall microgrid power capability is 5 kW.

A voltage source with a current controlled converter is used to model an ideal distributed generator operating as a grid feeding. The reason for that is for considering a wind or solar system, the DG should operates tracking the maximum power point, which can be modeled as a current source. In case the produced power surplus the load requirements, the ESS

should absorb the extra DG power. However, if the ESS is fully charged, then the extra power should go through the AC utility grid. The DG rated power is 2 kW.

A battery bank of 240 V is used as an energy storage system. A bidirectional DC-DC converter interfaces the ESS with the DC bus to allow charging and discharging operation. The rated power is also considered 2 kW in a way that the ESS is able to provide the total required power even when no energy is supplied by the DG. Coordinated operation between the VSC and the ESS is used. The VSC should act only during transients, or when a power mismatch occurs inside the DC microgrid. The ESS operates by a modified droop control, in which the required power absorbed or provide by the battery bank is a function of its state of charge (SOC) in an enhanced power management strategy.

The final element of the studied DC microgrid is the customer loads. In this paper, the DC loads are implemented by a DC-DC converter and a 2 kW resistive load. Load voltage is tightly regulated in 120 V. This is done to model the behavior of a constant power load (CPL), in which the load presents a negative impedance characteristic. Since CPLs are dominant in electronic loads, the effect of the negative impedance is one of the main concerns to the overall system stability [17].

III. DROOP CONTROL

As shown in Fig. 2(a), in a conventional droop control, the bus voltage linearly decreases with the output current and, consequently, with the power of the converter under droop operation. This behavior mimics the operation of a synchronous generator.

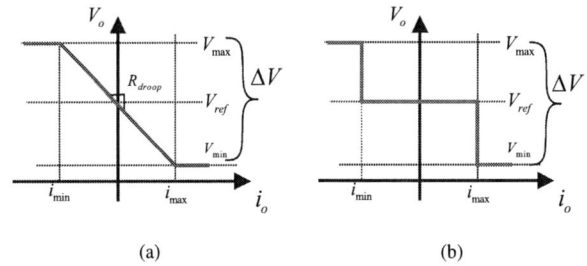

(a)   (b)

Fig. 2. Voltage droop characteristic as a function of the output current. a) VSC droop curve, b) ESS droop curve.

For a DC system, droop operation can be described as:

$$V_o = V_{ref} - R_{droop} i_o \qquad (1)$$

where V_o is the output voltage, V_{ref} is the desired reference voltage, R_{droop} is the droop coefficient, also called as virtual resistance, and i_o is the output current.

The virtual resistance can be calculated by:

$$R_{droop} = \left(\frac{V_{ref}}{2P_{\max}} \right) \Delta V \qquad (2)$$

where ΔV is the allowable voltage variation and P_{max} is the maximum power of the converter under droop control.

978-1-7281-4181-7/19 $31.00 © 2019 IEEE

The power management proposed in this paper considers that the ESS should provide the required power, while the grid supplies or absorbs only the unbalanced power. To do this, the ESS droop curve should behave as presented in Fig. 2(b). The ESS output voltage is kept at the reference level during charging and discharging mode, while the VSC controls the DC bus voltage according to the droop characteristics depicted in Fig. 2(a). If the current limits are achieved, then the maximum (charging) or minimum (discharging) allowed ESS output voltage limits are forced.

Droop control results in steady state error of the DC bus voltage. This characteristic is particularly problematic for data centers and telecommunication systems in which tight voltage regulation is required. Nonetheless, for urban DC distribution systems, a small bus voltage variation may be allowed, since other converters are employed to guarantee voltage and power regulation at the other point of connection. Hence, for small size microgrids as the one present in Fig. 1, in which only a couple of converters are considered, droop control is a viable option. This control is simple and reliable and no communications link is required.

IV. PROPOSED POWER MANAGEMENT SYSTEM

The control and operation of the VSC and ESS are presented in the next subsections. It is important to highlight that the DG is considered to be operating tracking the maximum power point, which is represented by controlling the current of a boost converter. The load voltage is regulated by controlling the output voltage of a buck converter. These controllers are usual and are not detailed in this paper.

A. VSC control

The control of the voltage source converter is presented in Fig. 3. A synchronous reference frame phase locked loop (SRF-PLL) is used synchronize the controller with the grid angle θ_s, thus the dq transformation can be obtained, as shown in the group A of Fig. 3.

For the VSC working as the grid forming converter, the dynamics of the DC bus voltage (V_{bus}) is given by:

$$\frac{dV_{bus}^2}{dt} = \frac{2}{C_b}\left\{P_{ext} - P_{loss} - \left[P_s + \left(\frac{2L_fP_s}{3V_s^2}\right)\frac{dP_s}{dt}\right]\right\} \quad (3)$$

where V_{bus} is the DC bus voltage, P_{ext} is an external active power acting as disturbance, P_{loss} is the VSC total losses, P_s is the power control input, C_b is the DC bus capacitance and V_s is the phase voltage peak.

Following the dynamics presented in (3), a PI controller is used to obtain the active VSC power reference (P_{ref}) when the VSC is working as the grid forming converter, as shown in group B of Fig. 3. The outer voltage loop is designed to have a slow dynamic response when compared with the inner current loop control.

After the DC bus voltage is established in 400 V, the voltage control is commuted to droop control, following the droop curve presented in Fig. 2(a). In this case, the VSC power reference generator is obtained trough:

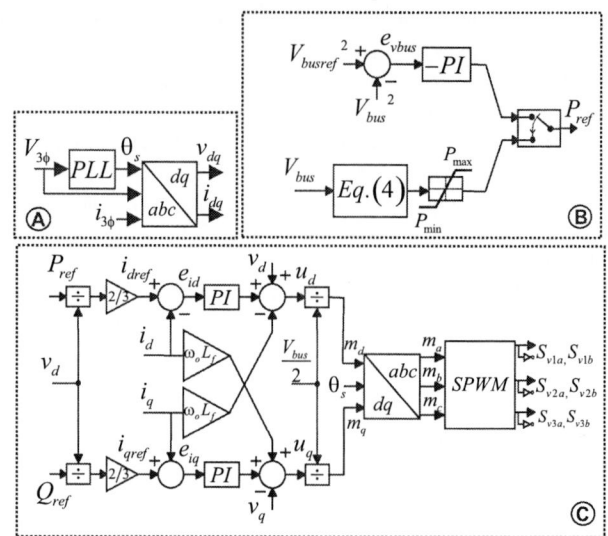

Fig. 3. Schematic of the VSC control. Group A shows the SRF-PLL. Group B shows the DC bus voltage control and droop control to generate the power reference. Group C shows the current controllers in decoupled dq frame.

$$P_{ref} = \left(\frac{2P_{vsc\,max}}{\Delta V_{bus}}\right) \times (V_{bus} - V_{busref}) \quad (4)$$

where V_{busref} is the DC bus voltage reference, $P_{vsc\,max}$ is the maximum power allowed for the VSC and ΔV_{bus} is the allowable voltage variation at the DC bus.

Considering a droop operation, the VSC is not any more responsible to stabilize the DC bus voltage. Instead, the voltage level will be achieved in agreement with the ESS droop characteristics. The objective of the voltage regulation mode and droop control is to obtain the active power reference for the VSC. The active and reactive power references are processed by the inner current loop control, as expressed in (5), which has a fast dynamic response.

$$\begin{cases} p_s(t) = \frac{3}{2}\left[v_{sd}i_d(t) + v_{sq}i_q(t)\right] \\ q_s(t) = \frac{3}{2}\left[-v_{sd}i_q(t) + v_{sq}i_d(t)\right] \end{cases} \quad (5)$$

where $p_s(t)$ and $q_s(t)$ are the active and reactive power, $i_d(t)$ and $i_q(t)$ are the dq currents and v_{sd} and v_{sq} are the terminal dq voltages.

The dq frame allows a known decoupling between active and reactive power. PI controllers are used to obtain zero steady state error for the current loop. In this paper, the reactive power is set to zero to work with high power factor. The dq modulation indexes (m_d, m_q) are obtained from the current controllers and are converted back to abc coordinates (m_a, m_b, m_c). Sinusoidal pulse width modulation (SPWM) is applied to activate the IGBTs. Switching frequency f_{sw} is chosen to achieve low harmonic distortion and fast response. VSC parameters are presented in Table I.

TABLE I
VSC PARAMETERS

Parameter	Value	Parameter	Value
P_s (rated)	5 kW	ω_0	377 rad/s
V_s	311 V	V_{busref}	400 V
L_f	10 mH	R	0.75 Ω
C_b	4 mF	f_{sw}	15 kHz
R_{droop}	4 Ω	ΔV_{bus}	40 V

B. ESS control

The implemented ESS is formed by the battery bank model and a bidirectional DC-DC converter. The internal voltage of the battery bank (V_{Bint}) and the internal resistance (R_B) are modeled to achieve the system requirements. Equation (6) describes the behavior of the battery bank.

$$V_{BB} = V_{Bint} - R_B i_{BB} \qquad (6)$$

where V_{BB} and V_{Bint} are the output and internal battery voltage, respectively, R_B is the internal impedance and i_{BB} is the battery electrical current.

The ESS state of charge (SOC) can be computed by using the Coulomb counting (ampere-hour balance) as given in (7).

$$SOC = SOC_0 + \frac{1}{C_{nom}} \int i_{BB} dt \qquad (7)$$

where SOC_0 is the initial state of charge and C_{nom} is the battery nominal capacity, usually informed by the manufacturer.

The main role of ESS converter is to regulate battery bank voltage and current. This is achieved by two control loops , as presented in Fig.4.

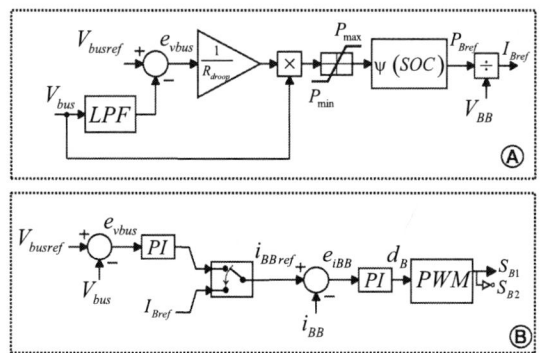

Fig. 4. Schematic of the ESS control. Group A shows the droop control to generate the current reference. Group B shows the voltage and current controllers.

In Fig. 4, from group A, one has the implemented ESS droop control derived by [15], [18]. The bus voltage V_{bus} is feedback trough a low pass filter (LPF) to attenuate transient behavior. The droop characteristics acts as a proportional controller with the gain as the inverse of droop coefficient, which established by the curve presented in Fig. 2(b). V_{bus} is multiplied by the output of the droop controller and, thus,

the ESS power reference is obtained. The closed loop transfer function for the DC bus voltage considering droop operation is given by [18]:

$$G_{clB}(s) = \frac{\left(\frac{1}{R_{droop}C_{Bi}}\right)(s + \omega_L)}{s^2 + \omega_L s + \frac{\omega_L}{R_{droop}C_{Bi}}} \qquad (8)$$

where R_{droop} is the battery ESS droop coefficient, C_{Bi} is the bidirectional DC-DC converter capacitance, and $\omega_L = 5$ Hz is the cutoff frequency of the low pass filter.

In droop control, the ESS role is to provide the required microgrid active power. A SOC-based function, $\psi(soc)$, is introduced to improve the life cycle of the battery bank. This introduced management function returns the ESS power reference as a function of its state of charge. $\psi(soc)$ is differently defined for ESS discharging and charging mode as presented in Figs. 5(a) and 5(b), respectively.

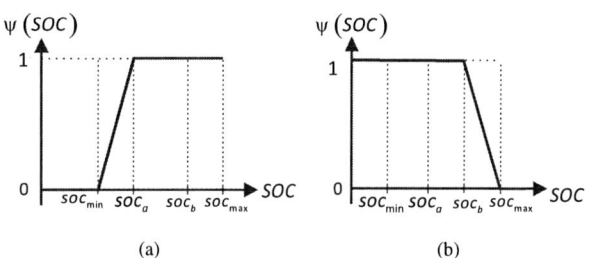

Fig. 5. Power management function based on the ESS state of charge. (a) ESS in discharging mode, (b) ESS in charge mode.

During the charging mode, the ESS must not provide any power if the minimum SOC limit (soc_{min}) is reached. Between soc_{min} and a specified point (soc_a), the reference power increases linearly. After soc_a, the ESS is allowed to provide the maximum power required by the microgrid. During the discharging mode, the ESS absorbs the maximum surplus power provided by the microgrid until a specified point (soc_b) is reached. Thus, the power absorbed by the ESS is linearly reduced until the maximum SOC limit is reached (soc_{max}). This action is important to avoid overcharging the battery bank.

Finally, the ESS current reference for the inner control loop (i_{Bref}) is obtained dividing the ESS power reference (P_{Bref}) by the battery bank voltage (V_{BB}). The droop control yields:

$$i_{Bref} = \left(\frac{\psi(soc) V_{bus}}{V_{BB} R_{droop}}\right) \left[V_{busref} - V_{bus}\left(\frac{\omega_L}{s + \omega_L}\right)\right] \qquad (9)$$

During the starting process, the ESS operates controlling the DC bus voltage by using an outer voltage loop control, as can be seen in Group B from Fig.4. In addition, this voltage control allows the DC microgrid to operate in island mode if the microgrid disconnects from the AC utility grid. A commutation between droop and voltage loop control is represented in Fig.4, Group B.

The dynamic equations of the control loops are given in (10). The first transfer function returns the relationship between current and duty cycle for the inner loop, while the second one returns the relationship between output voltage and current in the outer voltage loop. PI controllers are applied to achieve zero steady state error and PWM is employed to control switches S_{B2} and S_{B1}.

$$\begin{cases} \frac{\hat{i}_{LB}}{d_B} = \left(\frac{V_{BB}}{L_B}\right) \times \frac{1}{s} \\ \frac{\hat{v}_{BB}}{\hat{i}_{LB}} = \frac{R_{eqB}}{R_{eqB}C_B s + 1} \end{cases} \tag{10}$$

where $\hat{i}_{LB}$ and d_B are the dynamic current and duty cycle, respectively, V_{BB} is the controlled battery voltage, L_B is the converter inductance, $\hat{v}_{BB}$ is the dynamic output voltage, R_{eqB} is the equivalent output resistance and C_B is the converter capacitance.

It is important to highlight that, by adjusting properly the ESS droop coefficient according to the behavior presented in Fig. 2, the VSC does not provide power to the ESS and vice-versa. Thus, the proposed power management system only acts during the mismatch between the load requirements and the DG power. ESS parameters are given in Table II.

TABLE II
ESS PARAMETERS

Parameter	Value	Parameter	Value
L_B	1 mH	C_B	2 mF
f_{sw}	30 kHz	V_{BB}	220-260 V
V_{Bint}	240 V	R_B	4 Ω

V. SIMULATION RESULTS

Simulation results were obtained using MATLAB-SIMULINK software. The DC microgrid presented in Fig. 1 was implemented considering a sampling time of 30 kHz to discretize the simulation. VSC controller implementation is presented in Fig. 3 and ESS controller implementation is presented in Fig. 4. The performance of the system considering a predefined load consumption is presented in Fig. 6. The ESS nominal capacity is highly reduced to be possible to simulate its behavior. In a real system, it would take hours to achieve the ESS SOC limits.

The predefined load power is presented in Fig. 6(a). From 0 to 3 s, the initialization process of the microgrid is performed. During this period, the VSC operates as a grid forming converter to build the DC bus, while the other converter capacitors are being charged. At 4 s, after the DC link is established and all the converters are ready to operate, the load is inserted. It reaches the rated consumed power of 2 kW at 6 s. The DG starts to inject power into the microgrid at 10 s, as shown in Fig.6(b). The DG power increases in ramp until it reaches the rated generated power at 12 s. This profile of load and DG power were used to investigate different possibilities of power flow into a DC microgrid and verify the behavior of the VSC and ESS unities.

The performance of the proposed power management technique is verified in Fig. 6(c) and Fig. 6(d). Comparing with a

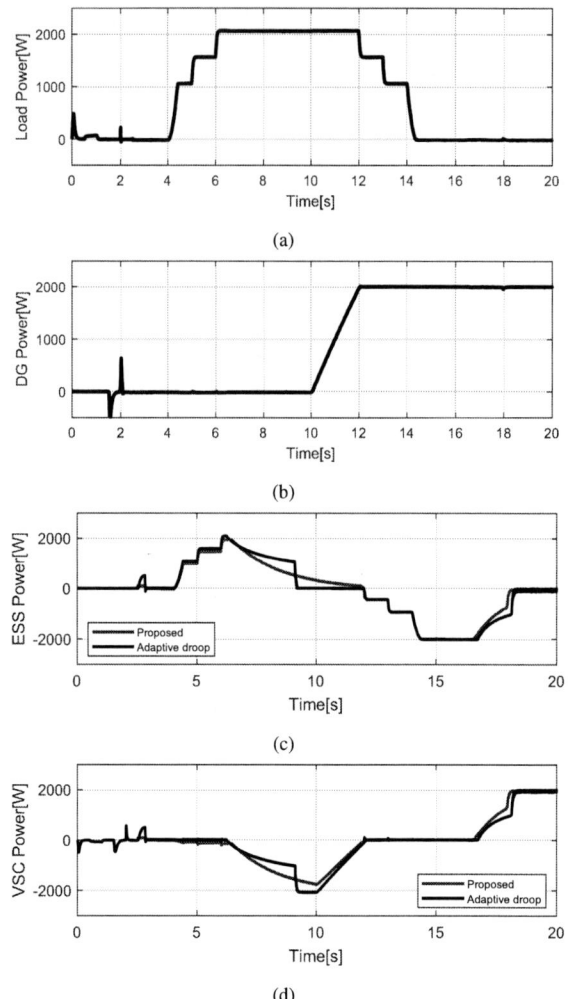

Fig. 6. Power results obtained for the grid connected DC microgrid and comparison results for a conventional power management method (black lines) and the proposed method (blue lines). (a) Applied load power for both methods (b) Applied DG power for both methods (c) Comparison between the obtained ESS powers, (d) Comparison between the obtained VSC powers.

conventional linear adaptive droop method, the proposed one returns a better control over the ESS power flow, as can be seen in Fig.6(c). The ESS supply the power required for the load until the battery bank reaches the SOC limit of 75% around 7 s. When this limit is reached, the ESS starts to reduce the supplied power in both methods and the VSC aids injecting power from the grid, as shown in Fig. 6(d). It is clear that the proposed technique allows a smooth power transition until the minimum ESS SOC of 50 % is achieved and the ESS stops supplying power into the DC microgrid.

After 10 s the DG starts to inject power into the microgrid. Since the ESS still discharged, the grid and the VSC work together to supply the load. At 14 s the load is totally removed and the DG provides its rated power. During this time, the ESS absorbs the full power until 16 s, when the SOC charging limit of 85 % is reached. Considering this SOC limit for both methods, the proposed method clearly performed better smoothing the ESS power behavior. Finally, at 19 s, the maximum SOC limit of 95 % is achieved and then the battery bank no longer absorbs power. The surplus power is totally injected into the grid.

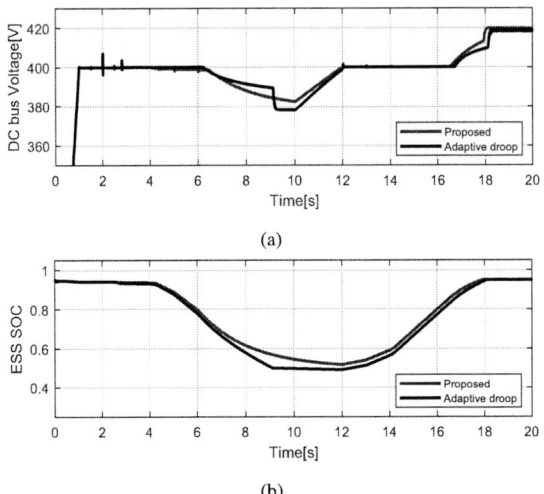

Fig. 7. Performance comparison of the proposed(blue line) a conventional(black line) power management system (a) DC Bus voltage (b) ESS State of Charge (SOC).

It is also convenient to evaluate the performance of the proposed system by inspecting the DC bus voltage presented in Fig. 7(a) and the ESS SOC behavior depicted in Fig. 7(a). From Fig. 7(a), it is possible to see that when the SOC limits are achieved, the DC bus voltage varies smoothly with the proposed method when compared to the adaptive droop. This achievement is highly important in a microgrid since abrupt voltage variations may lead to the system collapse. The proposed technique also performs better in regulating the battery bank state of charge, as shown in Fig.7(a). A better control over the state of charge leads to better utilization of the battery bank and improvements in its life cycle.

VI. Conclusion

This paper has presented an enhanced power management system for a droop controlled DC microgrid operating in the grid connected mode. The objective of the proposed technique is to achieve stable power flow considering the ESS state of charge limitations. The droop curve is also modified to force the ESS to provide the required power to the system, while the VSC only acts during transients and during mismatches of power inside the DC microgrid. Simulation results indicate that the proposed technique returns a better control over the microgrid power flow when compared with the conventional adaptive droop. A smooth DC bus voltage variation is achieved, which results in stable operation of the system. The ESS state of charge is well-regulated, which results in a better utilization of the battery bank. Therefore, for grid-connected mode, the proposed method leads to promising results to be confirmed experimentally.

References

[1] Dinesh Kumar, Firuz Zare, and Arindam Ghosh. DC Microgrid Technology: System Architectures, AC Grid Interfaces, Grounding Schemes, Power Quality, Communication Networks, Applications and Standardizations Aspects. *IEEE Access*, 5:1–1, 2017.

[2] Tomislav Dragicevic, Xiaonan Lu, Juan Vasquez, and Josep Guerrero. DC MicrogridsPart I: A Review of Control Strategies and Stabilization Techniques. *IEEE Transactions on Power Electronics*, 8993(c):1–1, 2015.

[3] Jackson John Justo, Francis Mwasilu, Ju Lee, and Jin Woo Jung. ACmicrogrids versus DC-microgrids with distributed energy resources: A review. *Renewable and Sustainable Energy Reviews*, 24:387–405, 2013.

[4] Chris S. Edrington, Tuyen V. Vu, Sanaz Paran, Fernand Diaz-Franco, and Touria El-Mezyani. An Alternative Distributed Control Architecture for Improvement in the Transient Response of DC Microgrids. *IEEE Transactions on Industrial Electronics*, 64(1):574–584, 2016.

[5] Alessio Iovine, Sabah Benamane Siad, Gilney Damm, Elena De Santis, and Maria Domenica Di Benedetto. Nonlinear Control of a DC MicroGrid for the Integration of Photovoltaic Panels. *IEEE Transactions on Automation Science and Engineering*, 14(2):524–535, 2017.

[6] Annette Werth, Alexis André, Daisuke Kawamoto, Tadashi Morita, Shigeru Tajima, Mario Tokoro, Daiki Yanagidaira, and Kenji Tanaka. Peer-to-Peer Control System for DC Microgrids. *IEEE Transactions on Smart Grid*, 9(4):3667–3675, 2018.

[7] Fatih Cingoz, Ali Elrayyah, and Yilmaz Sozer. Optimized Settings of Droop Parameters Using Stochastic Load Modeling for Effective DC Microgrids Operation. *IEEE Transactions on Industry Applications*, 53(2):1358–1371, 2017.

[8] Tuyen V. Vu, Dallas Perkins, Fernand Diaz, David Gonsoulin, Chris S. Edrington, and Touria El-Mezyani. Robust adaptive droop control for DC microgrids. *Electric Power Systems Research*, 146(2):95–106, may 2017.

[9] Xialin Li, Li Guo, Shaohui Zhang, Chengshan Wang, Yun Wei Li, Anwei Chen, and Yibin Feng. Observer-Based DC Voltage Droop and Current Feed-Forward Control of a DC Microgrid. *IEEE Transactions on Smart Grid*, 9(5):5207–5216, 2018.

[10] Tomislav Dragicevic, Josep M Guerrero, Juan C Vasquez, and Davor Skrlec. Supervisory Control of an Adaptive-Droop Regulated DC Microgrid With Battery Management Capability. *IEEE Transactions on Power Electronics*, 29(2):695–706, feb 2014.

[11] Xiaonan Lu, Kai Sun, Josep M Guerrero, Juan C Vasquez, and Lipei Huang. Double-Quadrant State-of-Charge Based Droop Control Method for Distributed Energy Storage Systems in Autonomous DC Microgrids. *IEEE Transactions on Smart Grid*, 6(1):147 – 157, 2015.

[12] Dan Wu, Fen Tang, Tomislav Dragicevic, Josep M. Guerrero, and Juan C. Vasquez. Coordinated control based on bus-signaling and virtual inertia for Islanded DC Microgrids. *IEEE Transactions on Smart Grid*, 6(6):2627–2638, 2015.

[13] Jianfang Xiao, Leonardy Setyawan, Peng Wang, and Chi Jin. PowerCapacity-Based Bus-Voltage Region Partition and Online Droop Coefficient Tuning for Real-Time Operation of DC Microgrids. *IEEE Transactions on Energy Conversion*, 30(4):1338–1347, 2015.

[14] Lexuan Meng, Qobad Shafiee, Giancarlo Ferrari Trecate, Houshang Karimi, Deepak Fulwani, Xiaonan Lu, and Josep M. Guerrero. Review on Control of DC Microgrids. *IEEE Journal of Emerging and Selected Topics in Power Electronics*, 5(3):1–1, 2017.

[15] Rodrigo A.F. Ferreira, Pedro G. Barbosa, Henrique A.C. Braga, and Andre A. Ferreira. Analysis of non-linear adaptive voltage droop control method applied to a grid connected DC microgrid. *2013 Brazilian Power Electronics Conference, COBEP 2013 - Proceedings*, pages 1067–1074, 2013.

[16] Thiago Ribeiro Oliveira, Waner Wodson Aparecido Gonçalves Silva, and Pedro Francisco Donoso-Garcia. Distributed secondary level control for energy storage management in DC microgrids. *IEEE Transactions on Smart Grid*, 8(6):2597–2607, 2017.

[17] Luis Herrera, Wei Zhang, and Jin Wang. Stability Analysis and Controller Design of DC Microgrids with Constant Power Loads. *IEEE Transactions on Smart Grid*, 8(2):881–888, 2017.

[18] Per Karlsson and Jorgen Svensson. DC bus voltage control for a distributed power system. *IEEE Transactions on Power Electronics*, 18(6):1405–1412, nov 2003.

Multi-Port System for Storage and Management of Regenerative Braking Energy in Diesel-Electric Locomotives

Caio G. da S. Moraes, Sergio L. B. Junior, Pedro P. Cavilha, Antonio L. S. Pacheco,
Marcelo L. Heldwein and Gierri Waltrich

Power Electronics Institute (INEP) – Departament of Electronics and Electrical Engineering
Federal University of Santa Catarina (UFSC), Florianópolis, SC, Brasil
caio.guimoraes@gmail.com, sergiobrock03@gmail.com, ppcavilha@gmail.com, schalata1966@gmail.com,
heldwein@inep.ufsc.br, gierri@gmail.com

Abstract—**In this paper, a multi-port system is proposed to recover the braking energy in a diesel-electric locomotive, using it to recharge a battery-supercapacitor based Energy Storage System (ESS) and to supply the locomotive auxiliary loads. To manage the system power flow, a single DC bus voltage controller generates current references for each converter. Also, an Energy Management System (EMS) decides the mode of operation, maintains the state of charge (SoC) of the battery and supercapacitor within a predefined range, and extends the battery lifetime by reducing its current stress. Numerical simulations are presented to validate the system concept.**

Index Terms—**Diesel-electric locomotive, multi-port system, energy storage, energy management.**

I. INTRODUCTION

The transportation sector efficiency is one of the fundamental pillars of current economics [1]. Although road transportation has been growing in the last decades, the railway transportation – due to its high capacity, reliability and safety features – is still the best land transportation option for large volumes of cargo over long distances [1], [2]. Furthermore, the possibility to obtain high autonomy electric locomotives, through electrified railway systems, makes this technology a sustainable alternative compared to other freight transportation modes, such as aircraft, trucks, and ships [3], [4].

However, in a country of continent-sized dimensions, as in the case of Brazil, where the transportation is usually carried out in isolated areas without electricity, the railway electrification becomes very expensive to build and maintain [4]. Because of that, diesel-electric locomotives are still widely used. As a matter of fact, the world has more than one million kilometers of railway track length served by diesel traction today [5].

Despite being widely used, diesel-electric locomotives generally do not have a system capable of regenerating or storing braking energy, which is dissipated in resistor banks. This waste corresponds to the main energy loss and contributes to further reduce the system efficiency, which is already poor due to the Internal Combustion Engine (ICE) [6], [11]. To improve the efficiency of diesel-electric locomotives, hybridization with

Energy Storage Systems (ESS) has been investigated [6]–[11]. Among the wide range of energy storage devices, the combination of lithium-ion batteries with supercapacitors is the most promising energy storage technology for diesel-driven rail vehicles, as reported in [11].

Having this in mind, this paper aims to propose a multi-port system capable of recovery the braking energy and uses it to charge batteries and supercapacitors, providing power for locomotive auxiliary loads and the traction system. The hybridization solution could be performed without any modification in the locomotive structure, besides reducing the fuel consumption and, consequently, the emission of polluting gases. To validate the system concept, numerical simulations are presented. Finally, conclusions and a future outlook for the research on this subject are included.

II. SYSTEM OVERVIEW

As regards the electrical power transmission within diesel-electric locomotives, there are different architectures. Basically, they are characterized by the type of current used by the main generator and traction motors, which may either be direct current (DC), alternating current (AC), or a combination of them [5]. In general, the generator is AC and the traction motors are DC [7], [9], [10]. The advantages of this traction system are the ease of control the speed and torque, and the ease of switching a motor to the generator mode for dynamic braking operations [5].

The principle of an AC-DC diesel-electric locomotive is summarized in Fig. 1, which also includes the proposed multi-port system. As can be seen, a generator system, composed of a diesel engine and two synchronous machines (Auxiliary and Main generators), supplies the traction system and electric auxiliaries.

Due to reversibility of DC electric machines, it is common to convert the motor into a braking generator dissipating the kinetic energy as heat in dedicated resistors [5]. Fig. 1 illustrates this dynamic braking system through the branch formed by the switch S_{DB} and the resistor R_{DB}.

978-1-7281-4181-7/19 $31.00 © 2019 IEEE

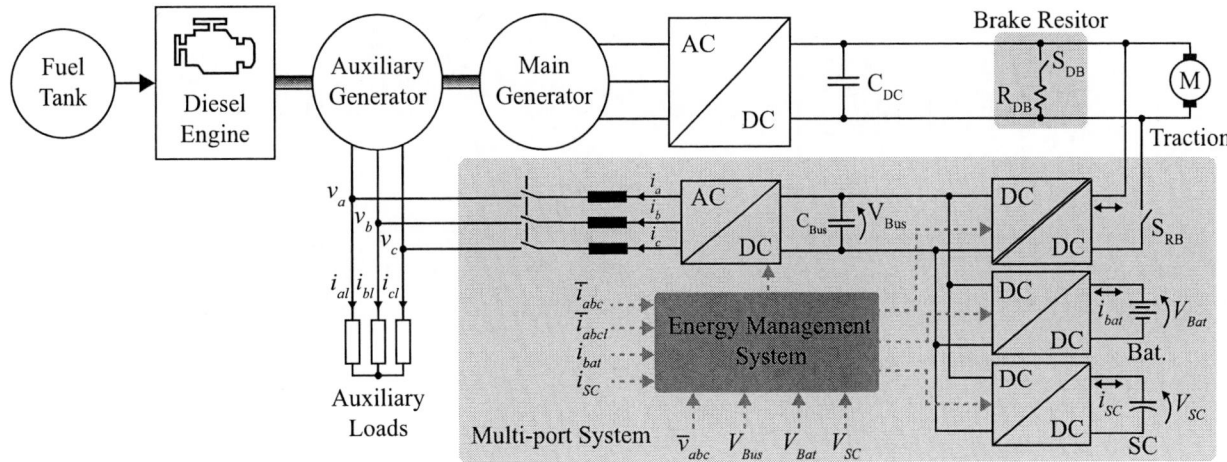

Fig. 1. Simplified AC-DC diesel-electric locomotive diagram with the proposed multi-port system.

In order to regenerate the braking energy, without the need to change the locomotive structure, it is proposed the implementation of a multi-port system, as shown in Fig. 1. When regenerative braking is used, the switch S_{RB} is turned on, allowing an isolated bidirectional DC-DC converter to recover energy from the traction motors and send it to a common DC bus. Then, a set of multiple static converters controls the power flow between all system ports. The DC-AC converter transfers real power from DC bus to the locomotive auxiliary loads, while the other bidirectional DC-DC converters charge/discharge the battery and supercapacitor units. Details of converter topologies are described in the following subsections.

It is important to note that the braking resistor cannot be removed from the locomotive for the sake of security. Braking power must be assured in all conditions, even if the energy storage system is fully charged and cannot absorb any more energy [6].

The bidirectional characteristic of this system allows it to both recover the braking energy and help the locomotive traction when the main generator could not supply sufficient power during extreme operation conditions (acceleration under high load during climbing, for example). Nevertheless, this paper focuses only on the application of the energy storage system to supply auxiliary loads, which means that neither the isolated dc-dc converter used in regeneration, nor the traction system will be addressed in this paper.

A. DC-DC converter topology

As mentioned before, the DC-DC converter is employed to interface the batteries and supercapacitors with the DC bus. For this purpose, three-phase DAB converter is a commonly used topology [13]. However, due to the large number of switches and gate driver circuits, it was decided to use a multistate switching cells (MSSC) based converter, as depicted in Fig. 2. This technique allows high overall efficiency and high power operation, besides increase of the switching frequency leading to reduction of the passive components employing the multi-interphase transformer (MIPT) [12].

B. DC-AC converter topology

To assist in auxiliary load supply, a conventional three-phase DC-AC converter with space vector modulation (SVM) and inductive output filter was selected.

III. CONTROL STRATEGY AND ENERGY MANAGEMENT SYSTEM

Various studies have been reported in the literature for energy management in different applications employing a battery-supercapacitor based hybrid ESS (HESS) [14]–[17]. In this work, the power balance in the system is achieved by maintaining constant DC bus voltage. This technique was first introduced in [16] and then also implemented and discussed in [17]. However, the focus of these works is on renewable grid integrated microgrid systems. Therefore, in this paper, some adaptations are made considering the operation and dynamics of a diesel-electric locomotive.

The simplified block diagram of the proposed multi-port system control strategy is shown in Fig. 3. A voltage feedback of the DC bus generates a total current reference (i_{tot}) that must be supplied or drawn by the system. This current reference is given by

$$i_{tot}(t) = K_p v_e(t) + K_i \int v_e(t)dt\,, \qquad (1)$$

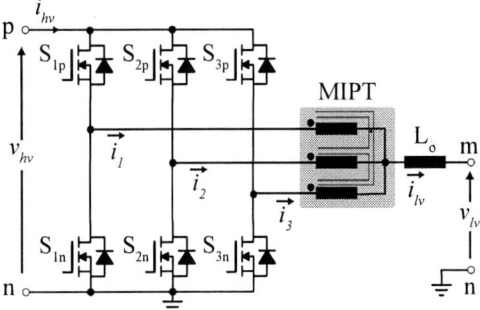

Fig. 2. Bidirectional DC-DC converter employing the multistate switching cell concept.

978-1-7281-4181-7/19 $31.00 © 2019 IEEE

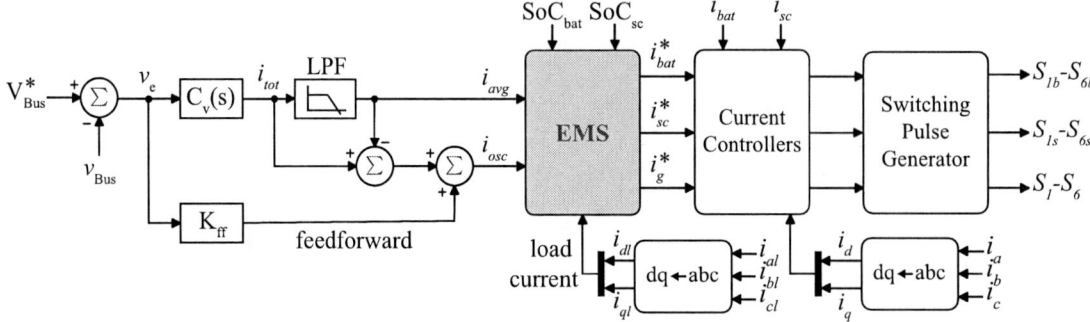

Fig. 3. Block diagram of the proposed multi-port system control strategy. Although not presented in this figure, all abc to dq transformations use the reference angle (θ) obtained from the phase-locked loop (PLL) routine. The PLL, in turn, computes the reference angle from line-to-neutral voltages (v_a, v_b and v_c), measured over the common connection point (c.f. Fig.1).

where $v_e(t) = V_{bus}^* - v_{bus}(t)$ is the error signal and K_p and K_i are proportional-integral (PI) controller gains.

To increase battery lifespan, the supercapacitor should respond to fast variations of current, while the battery should be charged or discharged more slowly [16]. Hence, it is necessary to extract the average and the transient components from the total current reference. The average component (i_{avg}) is obtained by using a first-order low-pass filter as

$$i_{avg}(t) = \frac{1}{s + \tau_c} i_{tot}(t). \tag{2}$$

where $f_c = 1/(2\pi\tau_c)$ is the cutoff frequency of the filter, which is selected to be $5\ Hz$.

It is observed from (2) that i_{avg} is negative during regenerative braking operation, because v_{bus} goes higher than V_{bus}^* due to the energy injected into the DC bus. Similarly, when there is no energy regenerated to the bus, i_{avg} is positive because v_{bus} goes below V_{bus}^*. Therefore, it is possible to determine the system operation mode based on the sign of i_{avg}.

The transient component (i_{osc}), on the other hand, is obtained according to

$$i_{osc}(t) = \left(1 - \frac{1}{1 + s\tau_c}\right) i_{tot}(t) + K_{ff}v_e(t). \tag{3}$$

A feedforward of the voltage error signal, represented by gain K_{ff}, is added to i_{osc} such that any power variations are directly transferred to the supercapacitor, improving the system dynamic and maintaining the DC bus voltage more stable.

Another important issue in applications with HESS, besides meeting the power balance in every operation mode, is maintaining the battery and supercapacitor state of charge (SoC) within limits. In this case, the Energy Management System (EMS) plays a fundamental role in reliable and continuous operation of the whole system [17].

A. Energy Management System

Keeping in view of the aforementioned issues, the proposed EMS decides the operating mode according to i_{avg} and has the following objectives: 1) to achieve power balance in every operation mode; 2) maintain battery and supercapacitor SoCs within their higher (H) and lower (L) limits; 3) reduce

battery current stress by allocating transient peak powers to supercapacitor and average powers to battery; and 4) supply as much as possible the auxiliary load in order to reduce fuel consumption. Thus, two operation mode are identified: regenerative braking mode ($i_{avg} < 0$) and load supply mode ($i_{avg} \geq 0$).

During regenerative braking mode (RBM) the control system prioritizes the charging of the battery and the supercapacitor units. As an example, when both SoCs are lower than H (first row of Table I), the average current component is used to recharge the battery. For this purpose, the function sat(i_{avg}) saturates the battery current reference (i_{bat}^*) obeying its charge rate. Then, the remaining excess power is injected into the locomotive utility grid, by making the grid current reference equal to $i_g^* = i_{avg} - \text{sat}(i_{avg})$. In addition, the supercapacitor is recharged with a current equivalent to I_{SC} and receives the transient power since its current reference is given by $i_{sc}^* = i_{osc} + I_{SC}$. Other operating conditions are shown in Table I. Furthermore, it is important to mention that in all

TABLE I
OPERATING CONDITIONS IN REGENERATIVE BRAKING MODE.

State of Charge	Regenerative Braking Mode
$SoC_B < H$ AND $SoC_{SC} < H$	$i_{bat}^* = \text{sat}(i_{avg})$ $i_{sc}^* = i_{osc} + I_{SC}$ $i_g^* = i_{avg} - \text{sat}(i_{avg})$
$SoC_B < H$ AND $SoC_{SC} > H$	$i_{bat}^* = \text{sat}(i_{avg})$, $i_{sc}^* = i_{osc}$ $i_g^* = i_{avg} - \text{sat}(i_{avg})$
$SoC_B > H$ AND $SoC_{SC} < H$	$i_{bat}^* = 0$, $i_g^* = i_{avg}$ $i_{sc}^* = i_{osc} + I_{SC}$
$SoC_B > H$ AND $SoC_{SC} > H$	$i_{bat}^* = 0$, $i_{sc}^* = i_{osc}$, $i_g^* = i_{avg}$

TABLE II
OPERATING CONDITIONS IN LOAD SUPPLY MODE.

State of Charge	Load Supply Mode
$SoC_B > L$ AND $SoC_{SC} > L$	$i_{bat}^* = \text{sat}(i_{avg})$, $i_{sc}^* = i_{osc}$ $i_g^* = i_{load}$
$SoC_B > L$ AND $SoC_{SC} < L$	$i_{bat}^* = \text{sat}(i_{avg})$, $i_g^* = i_{load}$ $i_{sc}^* = i_{osc} + I_{SC}$
$SoC_B < L$ AND $SoC_{SC} = X$	$i_{bat}^* = 0$, $i_{sc}^* = i_{osc}$, $i_g^* = 0$

case the dynamic braking is activated when the system is not able to process the excess regenerated power.

In load supply mode (LSM), as the name suggests, the energy stored in the ESS is used to supply the locomotive auxiliary loads. Thereby, the grid current reference is equal to the load current demand (i_{load}) until the battery reaches its lower SoC limit. At this point, the load is only supplied by the locomotive auxiliary generator, regardless of the supercapacitor SoC value, as shown in the last row of Table II. However, while the battery SoC is greater than L, the supercapacitor provides the load transient power until its SoC goes below the lower limit. In this case, the battery is used to recharge it with a current equivalent to I_{SC}. All these operating conditions are summarized in Table II.

B. Battery and supercapacitor current control

Due to nonidealities in the circuit components, the MSSC-based converters can present circulating currents trough the phases connected in parallel, which leads to the average value of the magnetic flux in the MIPT-cores not equal to zero [12]. In order to achieve a proper current sharing it is adopted a balancing technique by means of the Lunze's transformation.

Briefly, the Lunze's similarity transformation, for a N dimension coupled system, transforms the original circuit variables to $(N - 1)$ differential mode variables and one common mode variable [18], [19].

Fig. 4 shows the current balancing technique for the converter of Fig. 2. As indicated, the currents through the transformer windings are transformed to differential and common mode currents by means of the Lunze's matrix $\mathbf{T_L}$, then PI controllers are used to equalize winding currents. According to [12] the controller designed to common mode variable must be faster and its reference is proportional to output current i_{lv}, i.e. $i_{cm}^* = I_{lv}/N$. In contrast, controllers to the differential mode variables present slow dynamics and its reference is equal to zero (i.e. $i_{dm}^* = 0$) to avoid circulating current through the windings.

C. Grid current control

To control the current injected into the locomotive utility grid the abc reference frame is transformed into dq synchronous reference frame, applying Park's transformation. As a result, the control variables become DC quantities so that the filtering and controlling can be easily achieved [20].

A simplified schematic of the dq control structure is represented in Fig. 5. This strategy is normally associated with proportional-integral controllers since they have satisfactory behavior when regulating DC variables. For improving the performance of PI controllers, cross-coupling terms and voltage feedforward are used [20], [21], as highlighted in Fig. 5. A conventional phase-locked loop system based on the synchronous reference frame (SRF-PLL) [22], [23] is adopted for grid synchronization, generating the reference angle to the Park's transformation. Finally, a dq to $\alpha\beta$ transformation is employed to interface the dq control output with the SVM block. It is worth mentioning that the PLL strategy obtains its

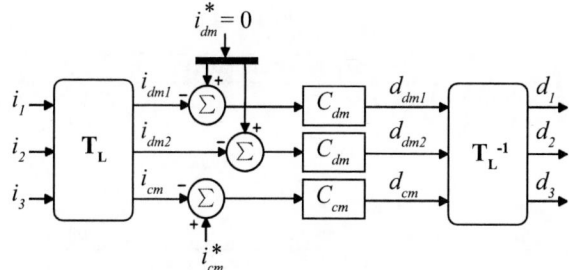

Fig. 4. Bidirectional DC-DC converter current control strategy. The common-mode current reference (i_{cm}^*) is equal to i_{bat}^* or i_{sc}^*, according to the respective converter.

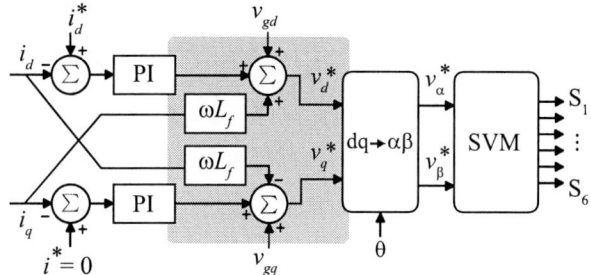

Fig. 5. Three-phase inverter current control strategy. The angle θ is obtained from SRF-PLL and the direct current reference (i_d^*) is obtained from the EMS, corresponding to i_g^*.

synchronization information from line-to-neutral voltages (v_a, v_b and v_c).

IV. SIMULATION RESULTS

In order to validate the proposed multi-port configuration, as well as the energy management scheme, a numerical simulation of a small scale system was carried out using PSIM software. The system parameters considered in this study are presented in Table III. For that, each converter was designed to be able to process 20 kW in all operation modes. The regenerative braking energy was emulated by a current source connected to the DC bus, making it possible to verify the steady-state and dynamic performances under various operating conditions. A branch consisting of a switch and a resistor was also introduced in the DC bus to emulate the dynamic braking process. Moreover, due to the system complexity, the average converter models were adopted to reduce the computational processing and the simulation time.

A. Regenerative Braking Mode (RBM)

Fig. 6 shows the system performance in RBM. In this test the locomotive auxiliary load was emulated by a fixed 19.3 kW resistive load. To verify the EMS logic functionality, an emulated regenerative current pattern was applied to the DC bus, as can be seen from the DC power (p_{DC}) in Fig. 6 (d). During the interval t_0 to t_1, both battery and supercapacitor SoCs are lower than H (c.f. first row of Table I). Then, the supercapacitor is charged with a current equal to I_{SC} and also suppress the sudden change in the DC bus voltage, smoothing the battery current variation, as shown in Fig. 6(a). Since the battery current is not greater than its maximum charge value, there is no excess power to be sent to the locomotive auxiliary

TABLE III
SYSTEM PARAMETERS.

Battery Pack Parameters	
Nominal Voltage (V_{Bat})	256 V
Ah Capacity	260 Ah
Supercapacitor Pack Parameters	
Nominal Voltage (V_{SC})	250 V
Equivalent Capacitance (C_{SC})	31 F
Absolute Maximum Current (I_{max})	1900 A
DC-DC Converter Parameters	
Switching Frequency (f_{sw})	15 kHz
MIPT self-inductance (L_s)	1.58 mH
MIPT muctual-inductance (L_M)	790 μH
Output Inductance (L_o)	63 μH
Three-phase Inverter Parameters	
Switching Frequency (f_{sw})	15 kHz
Filter Inductance (L_f)	1.92 mH
Locomotive Grid and DC Bus Parameters	
Grid Line-to-Line RMS Voltage (V_g)	380 V
Grid Frequency (f_g)	60 Hz
DC Bus Voltage (V_{Bus})	600 V
DC Bus Capacitance (C_{Bus})	2200 μF

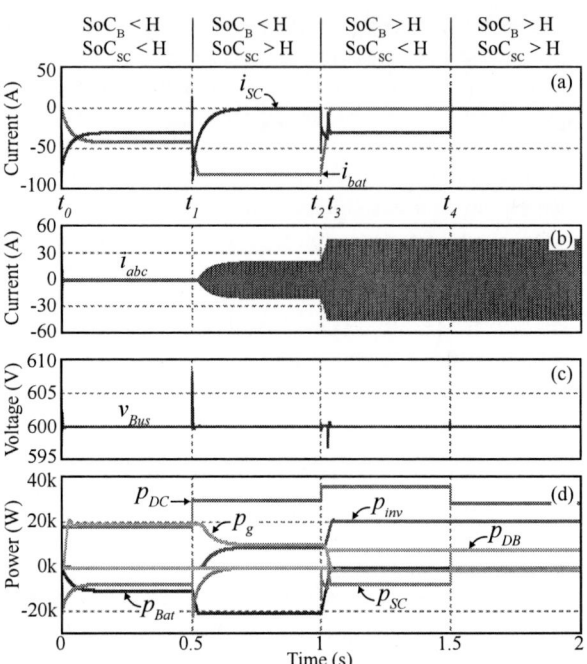

Fig. 6. System performance in RBM. (a) battery and supercapacitor currents; (b) inverter output currents; (c) DC bus voltage; (d) powers. Negative power means that the energy storage device are being charged.

loads. This can be confirmed in Fig. 6(d), which shows that the load is supplied only by the auxiliary generator (p_g), while the inverter output power (p_{inv}) is zero.

At $t = t_1$, the supercapacitor SoC becomes intentionally higher than H (c.f. second row of Table I), forcing its current reference to receive only the transient component, in agreement with Fig. 6(a). As a consequence, the battery is charged with its maximum rate and the excess power is used to supply a portion of the load, according to Fig. 6(b). In Fig. 6(d) it is possible to confirm that the auxiliary load is now supplied by both the generator and the inverter ($p_g \approx p_{inv}$).

From t_2 to t_4, the battery SoC $> H$ and the supercapacitor SoC $< H$ (made intentionally to test the control strategy, according to the third row of the Table I). In this case, the EMS forces the battery current reference to increase linearly until it reaches zero and the supercapacitor current reference to become equivalent to I_{SC}, recharging it as a result. When the battery current reaches zero, at $t = t_3$, the excess power extrapolates the load demand, which ends up activating the dynamic braking resistor, as indicated by the power p_{DB} in Fig. 6(d). In the same figure, it is noticed that the auxiliary load is supplied only by the inverter.

Finally, at $t = t_4$, both battery and supercapacitor SoCs are greater than H (c.f. fourth row of Table I). Therefore, their current references are zero, as shown in Fig. 6(a), and all the regenerated power is shared between the load and the braking resistor, according to Fig. 6(d).

B. Load Supply Mode (LSM)

The dynamic performance of the system in LSM is depicted in Fig. 7. During this operation mode there is no energy being regenerated to the DC bus, so that the power p_{DC} is kept null all the time, as shown in Fig. 7(d).

The system is started at $t = t_0$ with both battery and supercapacitor SoCs greater than L (c.f. first row of Table II).

Thus, the battery supplies a 9.65 kW auxiliary load through the three-phase inverter, while the supercapacitor provides peak current until the battery reaches steady state. A sudden change in load, from 9.65 kW to 19.3 kW, occurs at $t = t_1$ and, again, the supercapacitor supplies the transient power, smoothing the battery current transition, as can be seen in Fig. 7(a).

At $t = t_2$, the supercapacitor SoC becomes intentionally lower than L (c.f. second row of Table II). As a result, the

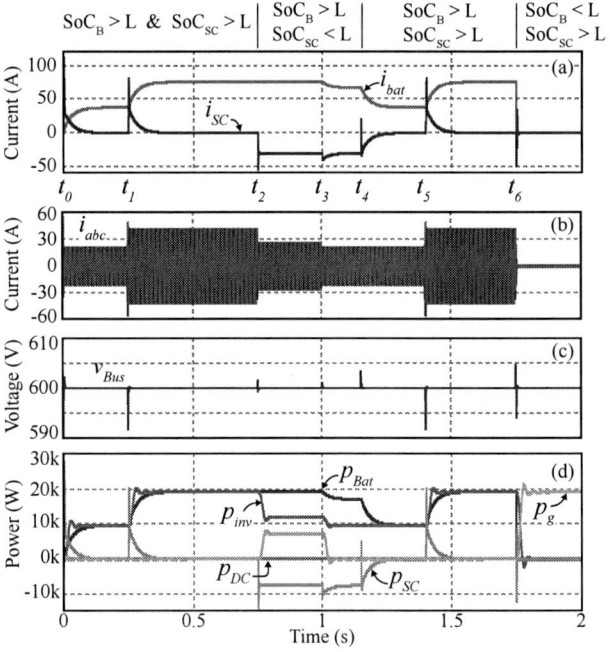

Fig. 7. System performance in LSM. (a) battery and supercapacitor currents; (b) inverter output currents; (c) DC bus voltage; (d) powers.

battery is used to recharge it with a current equal to I_{SC}. Since the battery can not supply the entire load and recharge the supercapacitor at the same time, a portion of the load is supplied by the locomotive auxiliary generator, as shown in Fig. 7(d).

A reduction in load, from 19.3 kW to 9.65 kW, occurs at $t = t_3$ in such a way that the battery then supplies all the power to the system. This condition lasts until $t = t_4$, when the supercapacitor SoC becomes higher than L, forcing its current reference to receive only the transient component.

Another sudden increase in load occurs at $t = t_5$, but in this case, the battery can provide all power to the load, since the supercapacitor is not charging. In the end, the battery SoC is made intentionally lower than L (c.f. third row of Table II) at $t = t_6$, which means that the ESS can not supply the auxiliary load. As a consequence, the multi-port system is disconnected from the locomotive utility grid and the auxiliary generator starts supplying the entire load, as can be confirmed in Fig. 7(d).

V. CONCLUSION

A multi-port system that takes advantage of a battery-supercapacitor based HESS was proposed in this work to recover the braking energy in a diesel-electric locomotive. The system presents itself as a simple and low-invasive solution, since it could be embedded in the railway vehicle without the need to change its structure. Simulation results showed that the energy management scheme employed to control the power flow is able to fast control the DC bus voltage, to achieve power balance in both operation modes, maintain the battery and supercapacitor SoCs within their limits, and to reduce the battery current stress. Future work will cover the overall performance of the system, considering the bidirectional converter used in regeneration process and its capacity to assist the locomotive traction. Moreover, it is proposed to validate the system concept in a practical prototype that is already being built.

ACKNOWLEDGMENT

This study was financed in part by the Coordenação de Aperfeiçoamento de Pessoal de Nível Superior – Brasil (CAPES) – Finance Code 001 and by CNPq. Furthermore, the authors would like to thank VALE SA for the resources and technical discussions throughout the project.

REFERENCES

[1] I. Dostál and V. Adamec, "Transport and its Role in the Society," in Transactions on Transport Sciences, vol. 4, no. 2, pp. 43-56, 2011.

[2] I. Dincer et al., "Clean Rail Transportation Options," Green Energy and Technology, Springer International Publishing, Switzerland, 2016.

[3] International Railway Association (UIC) and CER, "Rail Transport and Environment," Paris, 2015. ISBN 978-2-7461-2400-4.

[4] S. Frey, "Railway Electrification Systems & Engineering," first ed., White Word Publications, 2012. ISBN 978-81-323-4395-0.

[5] M. Spiryagin et al., "Design and Simulation of Rail Vehicles," Taylor & Francis Group, LLC, 2014. ISBN 978-11-380-7370-8.

[6] B. Destraz, P. Barrade and A. Rufer, "Power Assistance for Diesel - Electric Locomotives with Supercapacitive Energy Storage," in IEEE 35th Annual Power Electronics Specialists Conference, 2004.

[7] C. R. Akli, X. Roboam, B. Sareni and A. Jeunesse, "Energy management and sizing of a hybrid locomotive," in European Conference on Power Electronics and Applications, 2007.

[8] A. Jaafar, C. R. Akli, B. Sareni, X. Roboam and A. Jeunesse, "Sizing and Energy Management of a Hybrid Locomotive Based on Flywheel and Accumulators," in IEEE Transactions on Vehicular Technology, vol. 58, no. 8, pp. 3947-3958, October 2009.

[9] C. Mayet, J. Pouget, A. Bouscayrol and W. Lhomme, "Influence of an Energy Storage System on the Energy Consumption of a Diesel-Electric Locomotive," in IEEE Transactions on Vehicular Technology, vol. 63, no. 3, pp. 1032-1040, March 2014.

[10] T. Letrouvé, W. Lhomme, J. Pouget and A. Bouscayrol, "Different Hybridization Rate of a Diesel-electric Locomotive," in IEEE Vehicle Power and Propulsion Conference (VPPC), 2014.

[11] M. Meinert, P. Prenleloup, S. Schmid and R. Palacin, "Energy Storage Technologies and Hybrid Architectures for Specific Diesel Driven Rail Duty Cycles: Design and System Integration Aspects," in Applied Energy, vol. 157, pp. 619-629. Elsevier, 2015.

[12] F. J. B. Brito Jr., M. L. Heldwein and R. P. T. Bascopé, "Active Current Balancing Technique Employing the Lunze's Transformation for Converters Based on Multistate Switching Cells," in IEEE Brazilian Power Electronics Conference and 1st Southern Power Electronics Conference (COBEP/SPEC), 2015.

[13] N. H. Baars, H. Huisman, J. L. Duarte and J. Verschoor, "A 80 kW isolated DC-DC converter for railway applications," 16th European Conference on Power Electronics and Applications, Lappeenranta, 2014.

[14] J. Cao and A. Emadi, "A New Battery/UltraCapacitor Hybrid Energy Storage System for Electric, Hybrid, and Plug-In Hybrid Electric Vehicles," in IEEE Transactions on Power Electronics, vol. 27, no. 1, pp. 122-132, Jan. 2012.

[15] A. Lahyani, P. Venet, A. Guermazi and A. Troudi, "Battery/Supercapacitors Combination in Uninterruptible Power Supply (UPS)," in IEEE Transactions on Power Electronics, vol. 28, no. 4, pp. 1509-1522, April 2013.

[16] D. Paire, M. G. Simoes, J. Lagorse and A. Miraoui, "A Real-Time Sharing Reference Voltage for Hybrid Generation Power System," IEEE Industry Applications Society Annual Meeting, Houston, TX, 2010.

[17] S. Kotra and M. K. Mishra, "A Supervisory Power Management System for a Hybrid Microgrid With HESS," in IEEE Transactions on Industrial Electronics, vol. 64, no. 5, pp. 3640-3649, May 2017.

[18] J. Lunze, "Feedback control of large scale systems," Prentice-Hall international series in systems and control engineering, 1992.

[19] A. Garg, D. J. Perreault, G. C. Verghese, "Feedback control of paralleled symmetric systems, with applications to nonlinear dynamics of paralleled power converters," in Proc. 1999 IEEE Int. Symp. Circuits Syst., vol. 5, pp. 192-197 vol.5, 1999.

[20] F. Blaabjerg, R. Teodorescu, M. Liserre and A. V. Timbus, "Overview of Control and Grid Synchronization for Distributed Power Generation Systems," in IEEE Transactions on Industrial Electronics, vol. 53, no. 5, pp. 1398-1409, Oct. 2006.

[21] R. Teodorescu, M. Liserre and P. Rodríguez, "Grid Converters for Photovoltaic and Wind Power Systems," 2011 John Wiley & Sons, Ltd.

[22] V. Kaura and V. Blasko, "Operation of a phase locked loop system under distorted utility conditions," in IEEE Transactions on Industry Applications, vol. 33, no. 1, pp. 58-63, Jan.-Feb. 1997.

[23] S. Golestan, J. M. Guerrero and J. C. Vasquez, "Three-Phase PLLs: A Review of Recent Advances," in IEEE Transactions on Power Electronics, vol. 32, no. 3, pp. 1894-1907, March 2017.

Sensorless Control of Nonsinusoidal Back-EMF PMSM Based on State Observer

Thiago Lazzari[1], Filipe Scalcon[1], Cesar Volpato[1], Thieli Gabbi[1], Márcio Stefanelo[2] and Rodrigo P. Vieira[1]

Power Electronics and Control Research Group - GEPOC
[1]Federal University of Santa Maria - UFSM, Santa Maria - RS, Brazil
[2]Federal University of Pampa - UNIPAMPA, Bagé - RS, Brazil
e-mail: thiago.lazzari@hotmail.com

Abstract—**This paper presents a sensorless vector control method based on current and back-EMF state observers for a surface-mounted permanent magnet synchronous machine (PMSM) with nonsinusoidal back electromtive force (EMF). The mathematical modeling of the PMSM is presented in the stationary reference frame. The estimated rotor position is presented in terms of flux linkage in the stationary reference frame. Then, current and back-EMF observers are used, in order to obtain both actual and observed $\alpha\beta$ currents and $\alpha\beta$ EMF. The performance of the proposed approach for speed and current control is demonstrated by simulation results.**

Keywords – **Sensorless Control, Vector Control, State Observer, Permanent Magnet Synchronous Machine, Nonsinusoidal Back-EMF.**

I. Introduction

Currently, environmental concerns have placed electric vehicles (EV) as the focus of several research projects. Among the parts that consist of an EV, its propulsion system is composed of electric motors [1]. These motors can be positioned in different ways in the EV, on the rear axle, on the front axle, on the rear and front axle, or individualy, with one in-wheel motor for each wheel [2]. Among all these models, the in-wheel motors can apply almost all the power generated directly to the wheels, because there are no mechanical losses in transmission systems.

The electric motors must have a high power capacity in application where they are directly mounted in the EV wheels. A good candidate for this application is the permanent magnet synchronous motor (PMSM), which has about 1/3 of the mass of an induction motor for the same delivered power [3].

PMSM can be classified in two types, sinewave back-EMF motors and nonsinusoidal back-EMF motors. For the former, the classical control techniques applied are vector control techniques that require accurately knowledge of the rotor position [4]–[7]. For the latter, there are different techniques that can use hall effect sensors [8]–[10] or vector control techniques [11]–[13]. The use of vector control techniques presents better results in terms of efficiency and torque ripple reduction, justifying its application in PMSM with nonsinusoidal back-EMF [13]. However, since the in-wheel motors are single-shaft outer rotors, it means that only one side of the motor has an available shaft, making the use of precision position

sensors, which need to be mounted, making not feasible. Thus, sensorless techniques become attractive as they provide the estimated angular rotor position without the need for mechanical sensors.

In [14] the authors apply a sensorless technique based on adaptive observers for the control of a PMSM. The angular rotor position is obtained by means of the estimated back-EMF in the stationary reference frame. In [13], the authors apply a vector control technique using the back-EMF in the stationary reference frame as the basis for a new transformation matrix for the synchronous reference frame. The back-EMF are obtained by means of low order state observers. In [12], the authors apply a sensorless vector control technique in a brushless DC motor, where the estimated angular rotor position is corrected through a PLL technique and the back-EMF are estimated by Kalman filters.

Since the back-EMF of PMSM with nonsinusoidal back-EMF exhibit an irregular behavior, it is not possible to apply the same technique used in [14] to obtain the angular position of the rotor. The reason for this is that when applying the *atan* function in the back-EMF in a stationary reference frame, the rotor position presents undesirable ondulations, making the use of the classic Park Transform not viable.

This paper presents a sensorless vector control technique based on state observers applied to PMSM with nonsinusoidal back-EMF. The position obtained is not based on the back-EMF in the stationary reference frame, but in on the flux linkages. The rotor position obtained by this approach presents a behavior similar to the value of the actual rotor position, making possible the application of the Park Transformation in order to implement the vector control technique.

II. Mathematical Model of PMSM with Nonsinusoidal Back-EMF

The equations that describe the dynamic behavior of the surface-mounted PMSM are well consolidated in literature [15]–[17]. Thus, considering the equivalent electric circuit shown in Figure 1, it is possible to write the PMSM phase voltages as:

$$v_{abc} = R_s i_{abc} + L_s \frac{di_{abc}}{dt} + e_{abc}, \tag{1}$$

978-1-7281-4181-7/19 $31.00 © 2019 IEEE

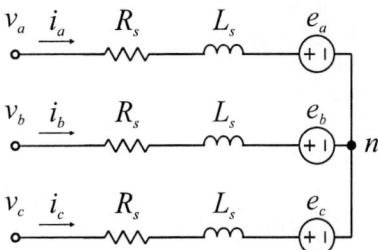

Figure 1. Equivalent circuit of the PMSM.

where $\boldsymbol{v}_{abc} = \begin{bmatrix} v_a & v_b & v_c \end{bmatrix}^T$, $\boldsymbol{i}_{abc} = \begin{bmatrix} i_a & i_b & i_c \end{bmatrix}^T$ and $\boldsymbol{e}_{abc} = \begin{bmatrix} e_a & e_b & e_c \end{bmatrix}^T$ are the phase voltages, currents and back-EMF, respectively. $\boldsymbol{R}_s$ e $\boldsymbol{L}_s$ are matrices of stator resistance (R_s) and stator inductance (L_s). Due to the symmetry of the Y-connected windings, $\boldsymbol{R}_s$ and $\boldsymbol{L}_s$ are represented respectively by:

$$\boldsymbol{R}_s = R_s \boldsymbol{I}_{3X3}, \tag{2}$$

$$\boldsymbol{L}_s = L_s \boldsymbol{I}_{3X3}, \tag{3}$$

where $\boldsymbol{I}_{3X3}$ is an identity matrix.

The back-EMF can be obtained through,

$$\boldsymbol{e}_{abc} = K_e \boldsymbol{f}_{abc}(\theta_r)\omega_r, \tag{4}$$

where K_e is the back-EMF constant and $\boldsymbol{f}_{abc}(\theta_r) = \begin{bmatrix} f_a & f_b(\theta_r + \frac{2\pi}{3}) & f_c(\theta_r - \frac{2\pi}{3}) \end{bmatrix}^T$ are normalized functions that represent the waveform of the back-EMF.

The rotor speed dynamic behavior is expressed by,

$$J\frac{d\omega_r}{dt} = T_e - B\omega_r - T_L \tag{5}$$

$$\frac{d\theta_r}{dt} = \omega_r \tag{6}$$

where J is the moment of inertia, T_e is the electromagnetic torque, T_L is the load torque, B is the coefficient of viscous friction and ω_r is the angular speed of the rotor. The mechanical angular position of the rotor, θ_r, is given by:

$$\theta_r = \frac{2}{P}\theta_e, \tag{7}$$

where P is the number of poles and θ_e is the electric angular position of the rotor.

Considering that cogging and reluctance torque can be neglected for modeling purposes, the electromagnetic torque can be determined by:

$$T_e = \frac{1}{\omega_r}(\boldsymbol{e}_{abc}^T \boldsymbol{i}_{abc}). \tag{8}$$

A. Stationary Reference Frame

The dynamic model of the PMSM with nonsinusoidal back-EMF can also be represented in the stationary reference frame

with the use of the Clarke Transform. The power invariant transformation matrix is given by:

$$T_{\alpha\beta0} = \sqrt{\frac{2}{3}} \begin{bmatrix} 1 & -\frac{1}{2} & -\frac{1}{2} \\ 0 & \frac{\sqrt{3}}{2} & -\frac{\sqrt{3}}{2} \\ \frac{1}{\sqrt{2}} & \frac{1}{\sqrt{2}} & \frac{1}{\sqrt{2}} \end{bmatrix}. \tag{9}$$

The voltage equations in the stationary reference frame, as well as the electromagnetic torque, are obtained by applying the transform into equations (1) and (8), as follows:

$$\boldsymbol{v}_{\alpha\beta} = R_s \boldsymbol{i}_{\alpha\beta} + L_s \frac{d\boldsymbol{i}_{\alpha\beta}}{dt} + \boldsymbol{e}_{\alpha\beta}, \tag{10}$$

$$T_e = \frac{1}{\omega_r}(\boldsymbol{e}_{\alpha\beta}^T \boldsymbol{i}_{\alpha\beta}), \tag{11}$$

where $\boldsymbol{v}_{\alpha\beta}$, $\boldsymbol{i}_{\alpha\beta}$ and $\boldsymbol{e}_{\alpha\beta}$ are the voltages, currents and back-EMF in the stationary reference frame, respectively.

The stator currents in the stationary reference frame can be represented by:

$$\frac{d\boldsymbol{i}_{\alpha\beta}}{dt} = \frac{1}{L_s}\boldsymbol{v}_{\alpha\beta} - \frac{R_s}{L_s}\boldsymbol{i}_{\alpha\beta} - \frac{1}{L_s}\boldsymbol{e}_{\alpha\beta}. \tag{12}$$

III. Sensorless Control Strategy

In the classic vector control structure for PMSM speed control, the system is usually composed of three control loops: two current control loops and a speed control loop. The dynamics of the speed loop is much slower, when compared to the current ones, resulting in a decoupled system. The current loops are controlled in the synchronous reference, that is, direct and quadrature axes. The classic transform used for this purpose is the Park's Transform, that converts sinusoidal variables into continuous variables, simplifying the system's control. The Park Transform is given by:

$$\boldsymbol{T}_{dq} = \begin{bmatrix} cos(\theta_e) & sin(\theta_e) \\ -sin(\theta_e) & cos(\theta_e) \end{bmatrix}. \tag{13}$$

As observed in (13), the Park's Transform is dependent on the electric angular position of the rotor θ_e. This variable can be obtained by using a position sensor coupled to the mechanical axis of the electric motor. However, in applications where it is not possible to use position sensors, the angular position must be estimated from other system variables.

According to equation (6) the angular position of the rotor can be obtained by the integral of ω_r. However, in sensorless techniques where the rotor speed is obtained by means of digital filters, the angular position of the estimated rotor will present a lag when compared to the actual angular position of the rotor, due to the natural characteristic of a filter to insert added phase to the system. Thus, when the estimated position is used in the Park Transform, the system becomes unstable.

For PMSMs with sinusoidal back-EMF, the angular position of the rotor θ_r can be obtained by means of $\boldsymbol{e}_{\alpha\beta}$. By applying the $atan$ function on these variables, it is possible to obtain the value of θ_r. On the other hand, if a PMSM with any nonsinusoidal back-EMF is considered, the $atan$ function results in a inaccurate value of θ_r. Figure 2(a) illustrates $\boldsymbol{e}_{\alpha\beta}$ of a PMSM with nonsinusoidal back-EMF, whereas in Figure 2(b) a comparison between the actual position of the rotor (red) and the estimated position (blue), which presents undesirable undulations. Due to this irregularity, it becomes impracticable to apply the Park Transform.

To overcome this problem, it is possible to integrate the variables $\boldsymbol{e}_{\alpha\beta}$, thus obtaining the flux linkages ($\boldsymbol{\lambda}_{\alpha\beta}$) of the motor. This way, the flux linkages will exhibit a more sinusoidal behavior when compared to $\boldsymbol{e}_{\alpha\beta}$, allowing the use of the *atan* function in a suitable way. A satisfactory behavior of the estimated angular position of the rotor θ_r (blue) can be visualized in Figure 2(d) when compared to the actual rotor angular position (red). The flux linkages $\boldsymbol{\lambda}_{\alpha\beta}$ are given in Figure 2(c).

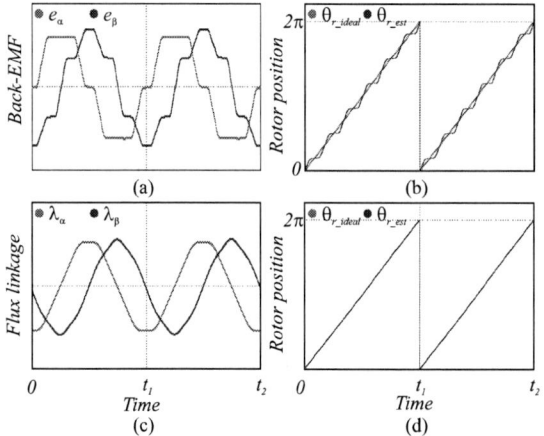

(a) (b)

(c) (d)

Figure 2. Rotor position based on back-EMF and flux linkage in stationary reference frame.

Although the estimated angular position of the rotor has performed well after the integration of $\boldsymbol{e}_{\alpha\beta}$, these variables are internal variables of the electric motor, that are difficult to measure. Thus, a solution to this problem is the use of state observers for both current and the back-EMF in the stationary reference frame. The estimated variables are both used to obtain the rotor angular position and rotor speed.

A. Current Observer

The current observer can be determined by:

$$\frac{d\hat{\boldsymbol{i}}_{\alpha\beta}}{dt} = \frac{1}{L_s}\boldsymbol{v}_{\alpha\beta} - \frac{R_s}{L_s}\hat{\boldsymbol{i}}_{\alpha\beta} + \hat{\boldsymbol{d}}_{\alpha\beta}, \tag{14}$$

where

$$\hat{\boldsymbol{d}}_{\alpha\beta} = -\frac{1}{L_s}\boldsymbol{v}_{\alpha\beta} + \frac{R_s}{L_s}\hat{\boldsymbol{i}}_{\alpha\beta} - h_1\tilde{\boldsymbol{i}}_{\alpha\beta}, \tag{15}$$

where $\boldsymbol{d}_{\alpha\beta}^T = -\frac{1}{L_s}\boldsymbol{e}_{\alpha\beta}^T$ is the feedback gain of the observer. The symbols $\hat{}$ and $\tilde{}$ represent the estimated variable and the error between the estimated value and the actual value,

respectively. In order to design the value of h_1, one must take into account the dynamic error of the currents, hence:

$$\frac{d\tilde{\boldsymbol{i}}_{\alpha\beta}}{dt} = -h_1\tilde{\boldsymbol{i}}_{\alpha\beta} - \frac{d\boldsymbol{i}_{\alpha\beta}}{dt}. \tag{16}$$

Thus, h_1 can be selected so that the error dynamic is faster than that the stator currents derivatives. If the gain is not large enough, the estimator error will impact the closed loop performance of the system, affecting the performance of the back-EMF observer. For the back-EMF observer is also necessary to take into account the variable error. Given the actual back-EMF value can not be measured, an equivalent back-EMF observer will be used. Assuming that the current observer converges to the actual value of the currents, the equivalent back-EMF observer can be defined as:

$$\boldsymbol{e}_{\alpha\beta}^* = -L_s\hat{\boldsymbol{d}}_{\alpha\beta}, \tag{17}$$

where the symbol * represents the equivalent value. Thus, due to the first-order behavior of equation (16), the estimated currents as well as the equivalent back-EMF can be obtained satisfactorily based on a suitable choice of h_1 .

B. Back-EMF Observer

For the development of the back-EMF observer, it is assumed that the mechanical dynamics are slower than the electrical dynamics, allowing the back-EMF model in the stationary frame to be written as:

$$\frac{de_\alpha}{dt} = -\omega_e e_\beta, \tag{18}$$

$$\frac{de_\beta}{dt} = \omega_e e_\alpha, \tag{19}$$

where ω_e is the electrical speed of the rotor. Thus, considering that the equivalent back-EMF observer has a satisfactory result, the observer for the back-EMF is given by:

$$\frac{d\hat{e}_\alpha}{dt} = -\hat{\omega}_e e_\beta^* - h_2\tilde{e}_\alpha, \tag{20}$$

$$\frac{d\hat{e}_\beta}{dt} = \hat{\omega}_e e_\alpha^* - h_2\tilde{e}_\beta, \tag{21}$$

where h_2 is the gain of the observer. Gain h_2 is determined from the error equations in the closed-loop system. Once again, it is assumed that the mechanical variables are much slower than the electrical dynamics of the motor, resulting in $\frac{d\omega_e}{dt} = 0$. Therefore, gain h_2 is determined by:

$$\frac{d\tilde{e}_\alpha}{dt} = -h_2\tilde{e}_\alpha - e_\beta^*\tilde{\omega}_e, \tag{22}$$

$$\frac{d\tilde{e}_\beta}{dt} = -h_2\tilde{e}_\beta + e_\alpha^*\tilde{\omega}_e. \tag{23}$$

The choice of the gain h_2 is a direct procedure, as it is done for h_1. With the appropriate value of this gain, the convergence of the back-EMF observer occurs.

978-1-7281-4181-7/19 $31.00 © 2019 IEEE

C. SVF for Speed Estimation

In order to obtain the rotor speed, the relationship between the back-EMF and its derivatives will be considered. Similar approaches can be found in [13], [18]. Differently from these papers, a State Variable Filter (SVF) will be considered where the filtered variable and the back-EMF derivatives in the steady reference frame are obtained [19]. The SVF transfer function can be represented by:

$$G_{SVF} = \frac{\omega_c^2}{s + \omega_c^2}, \tag{24}$$

where ω_c is the angular cutoff frequency of the filter. In order to avoid attenuation of the filtered value, ω_c must be 2 to 10 times the value of the input signal frequency. This way, equation (24) can be rewritten in terms of state space equations and subsequently discretized, according to equations (25) and (26), respectively. Euler's method for discretization was adopted.

$$\frac{d\boldsymbol{X}_{SVF}}{dt} = \boldsymbol{A}\boldsymbol{X}_{SVF} + \boldsymbol{B}\boldsymbol{u}_{in}, \tag{25}$$

$$\boldsymbol{X}_{SVF(k+1)} = (\boldsymbol{I}_{2X2} + \boldsymbol{A}t_s)\boldsymbol{X}_{SVF(k)} + \boldsymbol{B}t_s\boldsymbol{u}_{in(k)}, \tag{26}$$

where

$$\boldsymbol{A} = \begin{bmatrix} 0 & 1 \\ -\omega_c^2 & -2\omega_c \end{bmatrix}, \tag{27}$$

$$\boldsymbol{B} = \begin{bmatrix} 0 \\ \omega_c^2 \end{bmatrix}, \tag{28}$$

$\boldsymbol{X}_{SVF}$ is the state vector containing the filtered input and the derivatives of the input signals themselves. $\boldsymbol{u}_{in}$ is the input signal of the filter, that is, $\boldsymbol{u}_{in} = \begin{bmatrix} \hat{e}_\alpha & \hat{e}_\beta \end{bmatrix}^T$. t_s is the simulation period and k represents the discrete domain.

Since angular speed can be represented by equation (6), it also can be represented in terms of back-EMF in stationary reference frame, as:

$$\omega_r = \frac{d}{dt}atan(\frac{-e_\alpha}{e_\beta}) = \frac{e_\alpha \frac{de_\beta}{dt} - e_\beta \frac{de_\alpha}{dt}}{(e_\alpha)^2 + (e_\beta)^2}. \tag{29}$$

Then, with the filtered back-EMF ($\hat{e}_{\alpha\beta}^f$) and their derivatives ($\frac{d\hat{e}_{\alpha\beta}}{dt}$) from the SVF, it is possible to write the estimated speed ($\hat{\omega}_r$) as:

$$\hat{\omega}_r = \frac{\hat{e}_\alpha^f \frac{d\hat{e}_\beta}{dt} - \hat{e}_\beta^f \frac{d\hat{e}_\alpha}{dt}}{(\hat{e}_\alpha^f)^2 + (\hat{e}_\beta^f)^2}. \tag{30}$$

IV. SIMULATION RESULTS

In order to verify the performance of the proposed sensorless technique, a PMSM with trapezoidal back-EMF (BLDC) was used for simulation. The motor parameters are summarized in the table I. As sensorless techniques present difficulties for low speed operation, I-f open-loop control based on [20]

was used for startup. When the speed reaches an adequate value where back-EMF can be estimated, the sensorless technique is activated. This way, the following analisys will be performed in steady state, in order to verify if the proposed technique can properly estimate the angular position of the rotor and follow the speed reference. Figure 3 illustrates a block diagram of the proposed sensorless technique. The BLDC motor is driven by a voltage source inverter (VSI), using geometric modulation. A switching frequency of 10 kHz is used. The control of the BLDC motor is performed by the vector control technique, with two current loops (dq-axis) and a speed loop. A PI controller is used for each of the instances. The d-axis current reference is kept at $i_d^* = 0$, while the q-axis current reference (i_q^*) is generated from the external speed loop.

Table I
BLDC MOTOR PARAMETERS

Parameter	Symbol	Value
Stator Inductance	L_s	2.76 mH
Stator Resistance	R_s	11.9 Ω
Back-EMF Constant	k_e	0.1542 $\frac{V.s}{rad}$
Number of Poles	P	4
Moment of Inertia	J	7x10^{-6} $kg.m^2$
Friction Coefficient	B	0.001167 Nms

For the design of the PI controllers, it is considered that the current loop must be much faster than the speed loop for decoupling between the mechanical and electrical loops. In addition, the design of the controllers are based on canceling the pole of the plants. For current loops, the cutoff frequency is one tenth of the switching frequency. As for the speed loop, a cutoff frequency twice the speed of the mechanical dynamics was used.

In order to verify the performance of the sensorless technique throughout simulations, Figure 4 illustrates the behavior of $\hat{i}_\alpha$ and $\hat{e}_\alpha$ estimators with different gains for h_1 and h_2, respectively.

Considering the behavior of $\hat{i}_\alpha$ and $\hat{e}_\alpha$, gains h_1 and h_2 were set in 2.000 for the following simulations, as project choice. Figures 5 and 6 show the behavior of the current and back-EMF estimators in the stationary reference frame.

As observed in the obtained results, both the current and the back-EMF observer presented satisfactory results. Thus, the flux linkages obtained from the integral of the estimated back-EMF can be represented in Figure 7. The estimated angular position of the rotor is calculated by means of the *atan* function, and its result is compared to the actual angular position of the rotor in Figure 8.

The angular position of the estimated rotor showed very similar behavior with the actual position of the rotor. Thus, $\hat{\theta}_r$ can be substituted in the Park Transform to obtain the variables in the synchronous reference frame. In order to verify if the speed control follows the desired reference, four speed steps were applied. Initially, the reference speed was set to 100

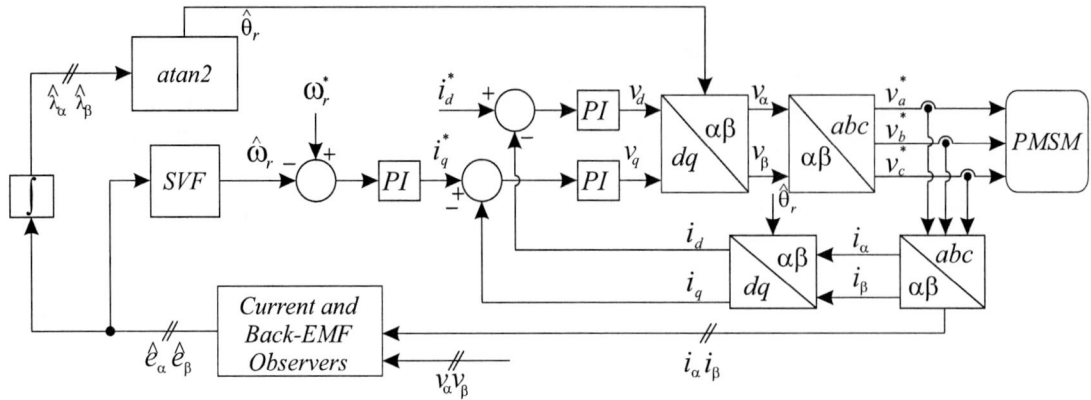

Figure 3. Block diagram of the control scheme.

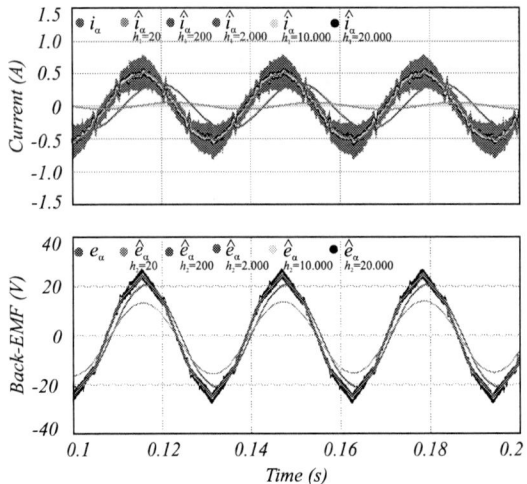

Figure 4. $\hat{i}_\alpha$ and $\hat{e}_\alpha$ estimators with different gains for h_1 and h_2.

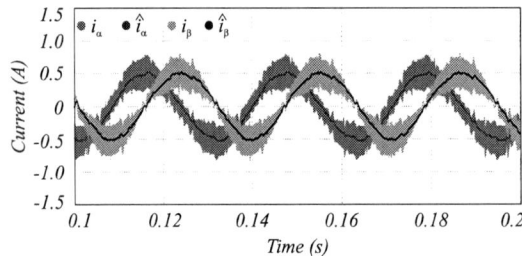

Figure 5. Estimated currents in stationary reference frame.

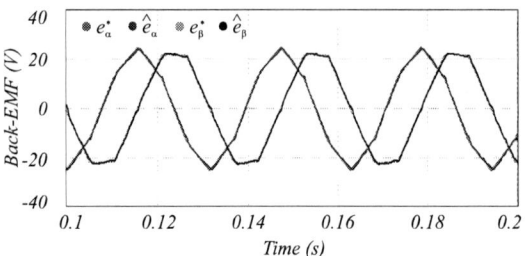

Figure 6. Estimated back-emf in stationary reference frame.

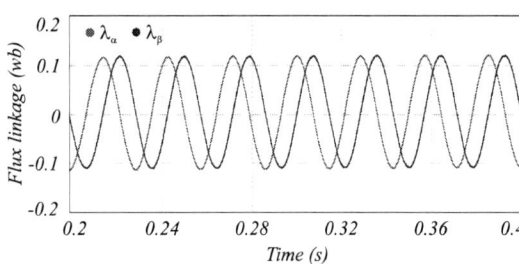

Figure 7. Flux linkage in stationary reference frame.

in Figure 10.

Based on the results of Figure 9, the estimated speed performs well when compared to the actual speed. In addition,

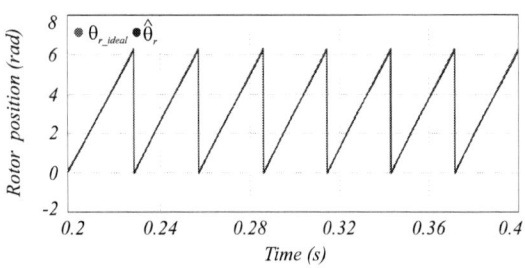

Figure 8. Estimated rotor position.

rad/s, at 0.15 s the first step was applied, raising the reference speed to 110 rad/s. At 0.25 s the speed reference becomes 120 rad/s, at 0.35 s the reference speed drops to 90 rad/s and at 0.45 s the reference speed returns to 100 rad/s. The behavior of the reference speed ($\omega_{r_{ref}}$), the actual speed (ω_r) and the estimated speed ($\hat{\omega}_r$) are shown in Figure 9. The behavior of the currents i_d and i_q for the respective speed steps is shown

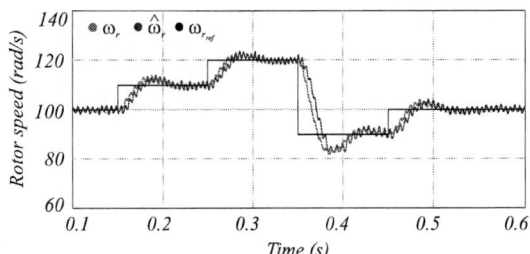

Figure 9. Estimated rotor speed.

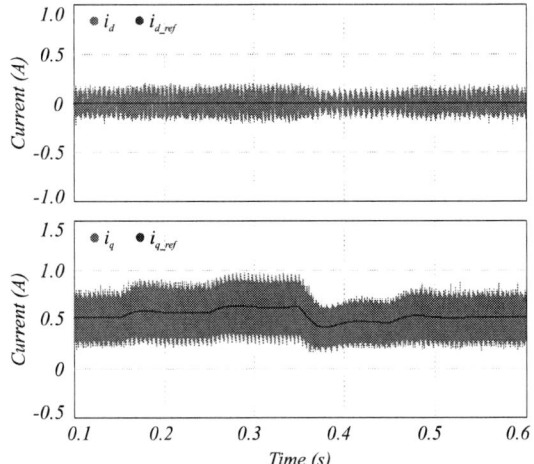

Figure 10. Currents at d and q axis.

the sensorless technique proved to be able to follow the speed references exhibiting a good convergence to the desired value.

V. CONCLUSION

This paper has presented a sensorless control method for permanent magnet motors with nonsinusoidal back-EMF. The sensorless technique consists of a current and a back-EMF observer in the stationary reference frame. The speed is obtained by means of a second order SVF. In addition, the flux linkages were used to obtain the angular position of the rotor, eliminating undesirable variations, which make the use of the Park's Transform not feasible. The design of the gains associated with the observers does not present complexity, given they present a first order behavior. The results showed that the proposed technique has a good feasibility of application, since it allows the use of the classic vector control structure for the control of permanent magnet synchronous motors with nonsinusoidal back-EMF.

ACKNOWLEDGEMENTS

This study was financed in part by the Coordenação de Aperfeiçoamento de Pessoal de Nível Superior - Brasil (CAPES/PROEX) - Finance Code 001, Conselho Nacional de Desenvolvimento Científico e Tecnológico (CNPq - Projeto, 422026/2016-6), the Fundação de Amparo à Pesquisa do Estado do RS (FAPERGS), the Programa de Pós-Graduação em Engenharia Elétrica da Universidade Federal de Santa Maria (PPGEE - UFSM), INCT-GD and the financing agencies (CNPq process 465640/2014-1, CAPES process No. 23038.000776/2017-54 and FAPERGS 17/2551-0000517-1).

REFERENCES

[1] C. C. Chan and Y. S. Wong, "Electric vehicles charge forward," *IEEE Power and Energy Magazine*, vol. 2, no. 6, pp. 24–33, Nov 2004.

[2] K. T. Chau, *Electric Vehicle Machines and Drives: Design, Analysis and Application.* Singapore: IEEE Press, 2015.

[3] J. F. Gieras and M. Wing, *Permanent Magnet Motor Technology: design and applications.* New York: CRC Press, 2002.

[4] M. Fatu, R. Teodorescu, I. Boldea, G. Andreescu, and F. Blaabjerg, "I-f starting method with smooth transition to emf based motion-sensorless vector control of pm synchronous motor/generator," in *2008 IEEE Power Electronics Specialists Conference*, June 2008, pp. 1481–1487.

[5] T. Bernardes, V. F. Montagner, H. A. Gründling, and H. Pinheiro, "Discrete-time sliding mode observer for sensorless vector control of permanent magnet synchronous machine," *IEEE Transactions on Industrial Electronics*, vol. 61, no. 4, pp. 1679–1691, April 2014.

[6] S. Dwivedi and B. Singh, "Vector control vs direct torque control comparative evaluation for pmsm drive," in *2010 Joint International Conference on Power Electronics, Drives and Energy Systems 2010 Power India*, Dec 2010, pp. 1–8.

[7] P. Kshirsagar, R. P. Burgos, J. Jang, A. Lidozzi, F. Wang, D. Boroyevich, and S. Sul, "Implementation and sensorless vector-control design and tuning strategy for smpm machines in fan-type applications," *IEEE Transactions on Industry Applications*, vol. 48, no. 6, pp. 2402–2413, Nov 2012.

[8] M. Lee and K. Kong, "Fourier-series-based phase delay compensation of brushless dc motor systems," *IEEE Transactions on Power Electronics*, vol. 33, no. 1, pp. 525–534, Jan 2018.

[9] C. Xia, Y. Wang, and T. Shi, "Implementation of finite-state model predictive control for commutation torque ripple minimization of permanent-magnet brushless dc motor," *IEEE Transactions on Industrial Electronics*, vol. 60, no. 3, pp. 896–905, March 2013.

[10] C. L. Baratieri and H. Pinheiro, "A novel starting method for sensorless brushless dc motors with current limitation," in *2012 XXth International Conference on Electrical Machines*, Sep. 2012, pp. 816–822.

[11] S. B. Ozturk and H. A. Toliyat, "Direct torque and indirect flux control of brushless dc motor," *IEEE/ASME Transactions on Mechatronics*, vol. 16, no. 2, pp. 351–360, April 2011.

[12] T. E. P. de Almeida, G. T. de Paula, A. G. de Castro, W. C. A. Pereira, and J. R. B. d. A. Monteiro, "Sensorless vector control for bldc machine," in *2017 Brazilian Power Electronics Conference (COBEP)*, Nov 2017, pp. 1–6.

[13] C. L. Baratieri and H. Pinheiro, "Sensorless vector control for pm brushless motors with nonsinusoidal back-emf," in *2014 International Conference on Electrical Machines (ICEM)*, Sep. 2014, pp. 915–921.

[14] C. J. V. Filho, F. P. Scalcon, T. S. Gabbi, and R. P. Vieira, "Adaptive observer for sensorless permanent magnet synchronous machines with online pole placement," in *2017 Brazilian Power Electronics Conference (COBEP)*, Nov 2017, pp. 1–6.

[15] P. Krause, O. Wasynczuk, and S. Sudhoff, *Analysis of Electric Machinery and Drive Systems*, 2nd ed. United States of America: Wiley-IEEE Press, 2002.

[16] C. L. Xia, *Permanent Magnet Brushless Dc Motor Drives and Controls*, 1st ed. China: John Wiley and Sons Singapore, 2012.

[17] R. Krishnan, *Permanent Magnet Synchronous and Brushless DC Motor Drives.* Virginia, U.S.A.: CRC Press, 2010.

[18] C. Lascu, I. Boldea, and F. Blaabjerg, "Comparative study of adaptive and inherently sensorless observers for variable-speed induction-motor drives," *IEEE Transactions on Industrial Electronics*, vol. 53, no. 1, pp. 57–65, Feb 2006.

[19] C. C. Gastaldini, R. P. Vieira, R. Z. Azzolin, and H. A. Gründling, "An adaptive feedback linearization control for induction motor," in *The XIX International Conference on Electrical Machines - ICEM 2010*, Sep. 2010, pp. 1–6.

[20] C. L. Baratieri and H. Pinheiro, "An i-f starting method for smooth and fast transition to sensorless control of bldc motors," in *2013 Brazilian Power Electronics Conference*, Oct 2013, pp. 836–843.

Power Demand Prediction Based on Mixed Driving Cycle Applied to Electric Vehicle Hybrid Energy Storage System

Lucas F. R. Lago, Silvana T. Faceroli, Rodrigo A. F. Ferreira, Márcio C. B. P. Rodrigues
Group of Power Electronics and Applications
Federal Institute of Education, Science and Technology of Southeast of Minas Gerais
Juiz de Fora, MG, Brazil
e-mail: lucas.lago@engenharia.ufjf.br, marcio.carmo@ifsudestemg.edu.br

Abstract—The use of multiple energy sources as power supply of an electric vehicle allows to improve its performance by increasing its autonomy and extending life cycle of on-board battery pack, which is the most expensive element of this type of automobile. In this work, it is proposed the use of computational intelligence techniques in the management of a hybrid energy storage system based on battery and supercapacitor, both embedded in an electric vehicle. For this purpose, a methodology for prediction and separation of the fractions of power demand, using an artificial neural network (ANN) based on the Non-Linear Autoregressive Model with Exogenous Inputs (NARX), is presented. A mixed driving cycle (MDC), composed by the combination of characteristics of different standard driving cycles, is used for NARX ANN training. This proposed MDC ANN training allows a better performance of power demand prediction, when compared to a single driving cycle ANN training approach, since it imposes a more diversified speed profile. From the simulations carried out and the adjustments of the network parameters, a very small error was found in relation to the predicted signals. Based on the obtained results, it is possible to conclude that the proposed method is effective and promising for the calculation of power demand in electric vehicles.

Index Terms – hybrid energy storage system, battery, supercapacitor, power predicition, NARX, electric vehicles

I. INTRODUCTION

Significant portion of greenhouse gases (GHG) emissions related to human activity is directly associated to the transportation sector [1–3]. Particularly, road transportation (which includes automobiles, trucks, buses, utility vehichles and motorcycles, driven by fossil fuels) can produce GHG emissions even greater than the whole industry sector [4]. In this context, the possibility of replacement of internal-combustion-motor-based vehicle fleet by electric vehicles comes up as being an important factor for improvement of environmental issues, since it allows reduction of atmospheric and sound pollutants in urban areas [3], [5].

Electric vehicles (EV) are vehicles in which partial or entire propulsion power are electrically provided [6]. Most of the automobile manufacturers have been producing electric propulsion models nowadays [7]. Several countries have a number of government incentives aiming at the diffusion of electric vehicle technology in the automotive market, such as

bonus to electric car buyers, tax discounts and adoption of restrictions on the use of conventional vehicles [3], [8].

However, the high acquisition cost, as well as lower autonomy than its fossil-fueled counterparts, whose are mainly related to EV battery pack, reduces the consumer interest on the electric vehicle technology. Several efforts and investments have been engaged in the research of battery technology in order to obtain enhanced batteries for EVs in recent years [9], [10]. Despite the advances in battery technology, the available batteries do not entirely meet the energy demands for EV power consumption [11].

Hybrid energy storage systems (HESS), combining batteries and supercapacitors (SC), have been proposed in order to improve EV performance and extend on-board battery pack lifetime [11–18]. Supercapacitors, also known as ultracapacitors, are electrochemical double-layer capacitors, which presents high power density and high degree of recyclability [19], [20].

Several power management strategies for battery/supercapacitor EV energy storage systems are found in literature. Generally, control and artificial intelligence techniques are combined in order to execute the HESS management tasks. Linear control strategies are used in [12]. The combination of a Fuzzy-logic-based supervisory system combined with linear control is the approach adopted in [21]. An adaptive Fuzzy control technique is proposed in [15]. Robust and non-linear control techniques are combined in [16], while the use of artificial neural networks (ANN) for HESS power management is approach used in [18].

In this context, a prediction methodology for power demand associated to EV driving is proposed in this paper, which aims the power management of battery/supercapacitor HESS. The proposed technique uses an ANN based on the Nonlinear AutoRegressive with eXternal inputs (NARX) [22]. Regarding the adopted power management strategy, which is described in next session, the contributions of this paper can be highlighted as the prediction and separation of instantaneous power associated to each HESS components, which are obtained directly from the predicted EV traction power. These information will be used as references for the control systems of the EV HESS power electronic converters. Also, the use of a mixed driving

978-1-7281-4181-7/19 $31.00 © 2019 IEEE

cycle (MDC), composed by the combination of characteristics of different standard driving cycles, in the NARX network training is another important contribution of this work. This proposed MDC ANN training allows a better performance of power demand prediction, when compared to a single driving cycle ANN training approach [23], since it imposes a more diversified speed profile, with different acceleration characteristics to the EV. System overview and preliminary considerations are presented in Section II. The proposed mixed driving cycle is described in Section III. Section V brings the obtained results and discussion, which is followed by this paper conclusion and references.

II. SYSTEM OVERVIEW AND PRELIMINARY CONSIDERATIONS

This section presents the methodology used for this work development. HESS configuration and control definition, EV traction power calculation, as well as main characteristics of the NARX ANN will be presented in this section.

A. HESS Configuration and Control

Configuration of the HESS impacts significantly on its performance. This work considers the configuration depicted in Fig. 1. It is a fully active configuration [11], [14], where two bidirectional DC/DC converters are responsible to couple battery pack and supercapacitor array to the DC bus, which feeds th EV traction system (referred as "load" in Fig. 1). Any control algorithm for power management can be implemented in this setup [11], which is widely employed in literature [16], [18], [21], [24], [25]. In order to handle with the different voltage levels of battery pack, supercapacitor array and DC bus, the bidirectional boost converter is an usual choice for this kind of application [25]. Entire development here presented is based on an ideal system, neglecting power losses. It means that traction power is shared between battery pack and supercapacitor array as defined by (1). Also, regenerative breaking is not covered in this paper.

$$P_t = P_{bat} + P_{sc}, \quad (1)$$

where P_t, P_{bat} and P_{sc} denote EV traction, battery pack and supercapacitor array instantaneous power, respectively.

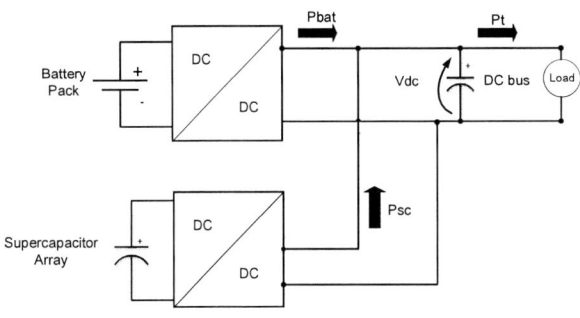

Figure 1: HESS fully active configuration.

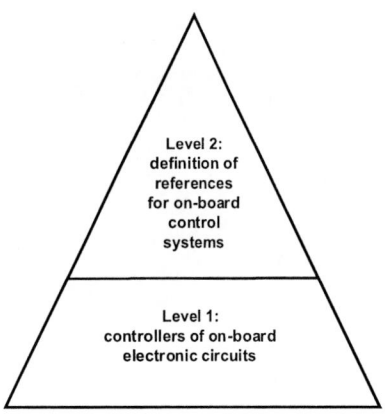

Figure 2: Battery/supercapacitor HESS power management hierarchical control structure.

This work considers that power management for the described HESS is performed by means of a two-level hierarchical control structure [26–29], as depicted in Fig. 2. Level 1 controllers perform current and voltage control of HESS DC/DC power converters. The ARN developed in this work is located at Level 2 control, which is responsible to define power demand associated to each HESS element. Definition of current and voltage references for Level 1 control loops is then calculated from power demand predicted by the ARN from Level 2. It is important to highlight that this paper focuses only on Level 2 ARN development and analysis.

B. NARX Artificial Neural Networks

Prediciton of EV traction power demand is performed in Level 2 hierarchical control structure by means of a NARX ANN. This paper presents the development and analysis of a time series prediction model for this HESS power management. Main goal of proposed approach is to split traction power demand estimation into two portions: *cruising power demand*, which is supplied from battery pack; and *peak power demand*, that must be provided by supercapacitor array. This power sharing strategy intends to smooth power (and current) demand variations for the battery pack, since frequent changes during battery discharging is harmful to its electrochemical processes [11]. Thus, EV traction power prediction structure provides P_{bat} output based on a average of power demand variation. Consequently, in compliance to system power balance described by (1), predicted P_{sc} is obtained as the difference between P_t and P_{bat} in order to supply peak traction power demand.

Considering a vehicular application of the proposed prediction model, due to safety issues, it is indispensable that references for the HESS control loops match demand traction power by the EV driver. Since the time response of an unalerted driver ranges around 2.5 seconds [30], [31], it has been defined a 0.5 second ahead prediction goal for the NARX ANN prediciton model (i.e. one-fifth of driver time response).

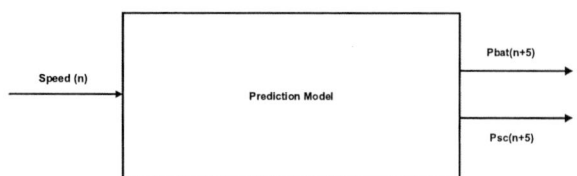

Figure 3: Prediction model.

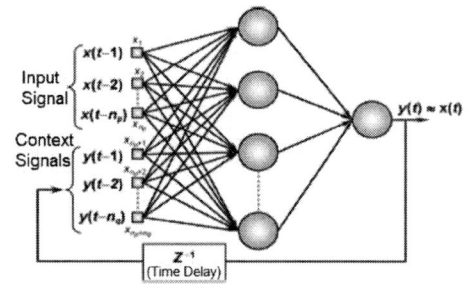

Figure 4: NARX ANN architecture (modified from [22]).

Sample time of Level 2 hierarchical control structure has been adopted as 0.1 second, which implies to a prediction model goal of 5 samples ahead, as depicted in Fig. 3, where *speed* is the current sample EV speed, and $P_{bat}(n+5)$ and $P_{sc}(n+5)$ respectively denote predicted power demands for battery pack and supercapacitor array, 0.5 second ahead.

Time series prediction structure here proposed uses a Perpectron ANN with multiple *layers* based on the Nonlinear AutoRegressive with eXternal inputs (NARX) [22] model. Feedback of ANN outputs, as shown in Fig. 4, is employed in this model in order to improve its nonlinear mapping ability. NARX ANN development, training and test processes have been implemented by using MATLAB software toolboxes. Activation function for hidden layer has been set as Tansig function, while Purelin function has been chosen for output layer. Input data of 84900 samples has been divided as follows: 70% for ANN training, 15% for test, and 15% for validation. LevenbergMarquardt algorithm has been chosen as machine learning strategy. NARX ANN performance has been evaluated for different number of hidden layer neuron, as discussed in Section IV.

C. EV Traction Power Calculation

Input data for the proposed NARX ANN prediction structure have been obtained by means of estimation of EV traction power regarding specific driving conditions. This power demand can be calculated by analyzing the forces that act in the vehicle surface and tires [23], [32]. From force diagram of Fig.5 it is possible to calculate the traction power, $P_t(t)$, demanded by a vehicle driven at speed $v(t)$ as described in (2). Speed profile $v(t)$ can be obtained from standard or real-world driving cycles for vehicle testing. In this paper, the use of a mixed driving cycle, composed by the combination of standard driving cycles, for traction power estimation is proposed. Once traction power at any time instant is known, it makes possible evaluate power demanded by each component of the HESS.

$$P_t(t) = [m.\frac{dv(t)}{dt} + m.g.sen(\alpha) + m.g.f_r.cos(\alpha) + \frac{1}{2}.\rho.A_f.C_D.(v(t) + v_w)^2].v(t), \quad (2)$$

where:

- P_t= power traction (W)
- m= vehicle mass (kg)
- v= speed $(\frac{m}{s})$
- g= gravity force $(\frac{m}{s^2})$
- α= slope angle (rad)
- f_r= rolling resistance coefficient
- ρ= air density $(\frac{kg}{m^3})$
- A_f= vehicle's frontal area (m^2)
- C_D= aerodynamics resistance coefficient
- V_w= wind speed $(\frac{m}{s})$

III. PROPOSED MIXED DRIVINF CYCLE

The proposed mixed driving cycle is shown on Fig. 6. It is composed by three full standard driving cycles: FTP-75 (Federal Test Procedure), WLTP (World Wide Harmonized Light-Duty Vehicles Test Procedure) class 2 and the JP10-15. The first one is a North-American urban driving cycle [33], also adopted in Brazil for automobile emissions tests, and is placed from 0 to 1875 s. WLTP class 2 driving cycle, which is used for vehicle emission certification approvals conducted in the European Union and other countries [33], also represent an urban-based speed profile and is placed in time interval (1875 s, 3352 s]. Japanese JP10-15 urban driving cycle for emission certification and fuel economy determination of light-duty vehicles completes the MDC composition. The whole proposed driving cycle is 4244 seconds long. With the proposed combination, a more diversified set of data can be offered for the ANN training and better results can be achieved in comparison to the ANN training based on a single standard driving cycle.

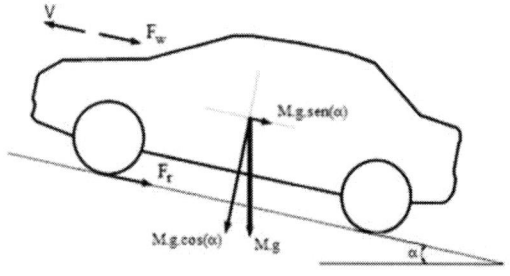

Figure 5: Forces acting over the vehicle [32].

978-1-7281-4181-7/19 $31.00 © 2019 IEEE

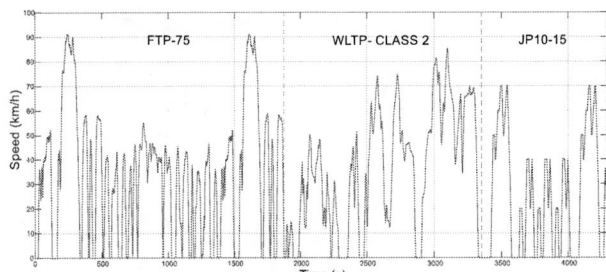

Figure 6: Proposed mixed driving cycle.

IV. RESULTS AND DISCUSSION

All of results and discussion here presented have obtained by using parameters described in Table I, which are based on EV model Tesla S P90D. Traction power calculation has been performed applying these parameters values on (2).

Table I: Parameters for traction power estimation.

Parameter	Value
m	2239 kg
g	9.81 m/s^2
α	0 rad
f_r	0.015
ρ	1.25 kg/m^3
A_f	2.39 m^2
V_ω	0 m/s

A. Number of Hidden Layer Neurons

Using the proposed mixed driving cycle it was possible to train the NARX ANN. First of all, the best number of neurons in the hidden layer has been evaluated by the analysis of mean squared error (MSE) for traction power prediction. As can be seen in Fig. 7, the use of 10 hidden layer neurons provided the minimum error.

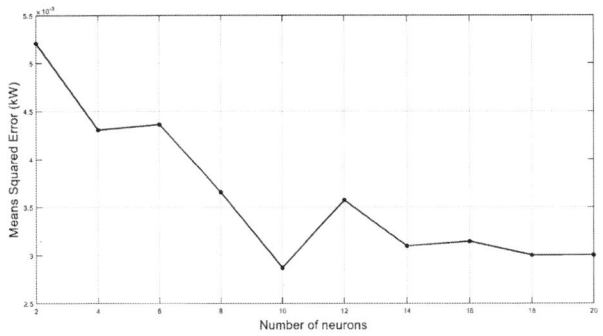

Figure 7: Traction power prediction MSE vs. number of hidden layer neurons.

B. Performance for the Best Instance

Once defined the number of hidden layer neurons, training of NARX ANN has been performed. Fig. 8 shows validation performance achieved in this case, which indicates high prediction accuracy.

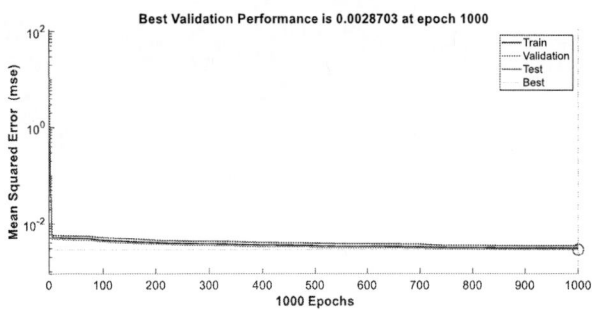

Figure 8: Performance validation (10 hidden layer neurons).

C. Power Demand Prediction

Fig. 9 show predicted HESS power demanded for the considered EV when it is submitted to proposed mixed driving cycle speed profile. As desired, it is possible to observe that battery pack power demand peaks, shown in Fig. 9(a), are considerably smaller than those imposed to supercapacitor array (Fig. 9(b)). For the latter, power peaks reach up to 12 kW, while the first one presents power peaks limited to 6 kW. Moreover, obtained results show that battery pack instantaneous power variation can be also substantially reduced, as depicted in Fig. 10, which details traction power demand and HESS power sharing during EV acceleration. This result evidences the ability of designed NARX ANN on assign the traction power variations to supercapacitor array, which leads to a smooth power demand profile imposed to battery pack.

In order to evaluate the prediction accuracy of designed NARX ANN, the comparison between predicted and calculated (by using (2)) traction power for different speed input conditions has been made. Fig. 11 shows a detailed view of a

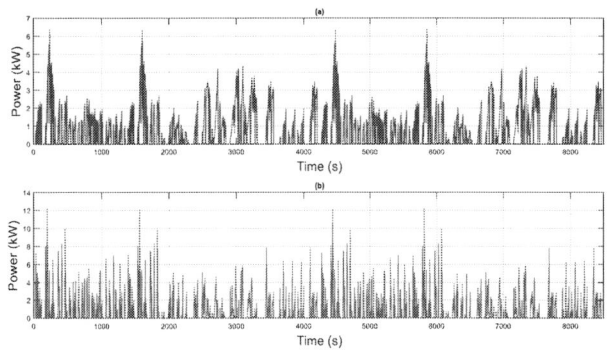

Figure 9: Predicted power demand associated to: (a) battery pack (b) supercapacitor array.

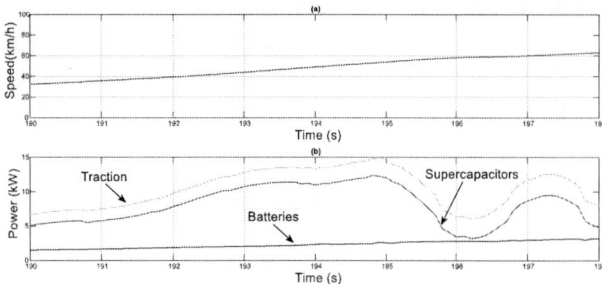

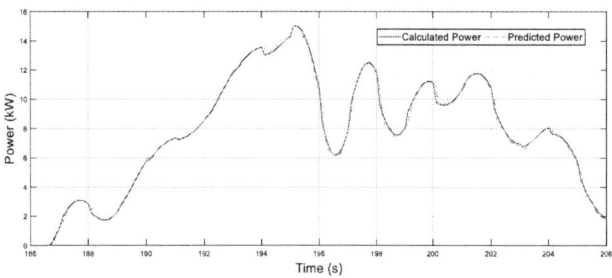

Figure 10: Detailed view of power demand prediction and HESS power sharing: (a) speed (portion of proposed MDC during vehicle acceleration); (b) Traction power (green) and demanded power from supercapacitor array (red) and battery pack (blue).

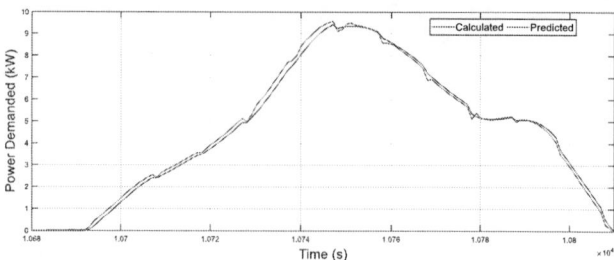

Figure 13: Predicted and calculated traction power (ANN input data: WLTP Class 3).

achieved.

V. CONCLUSION

A traction power prediction and electric vehicle HESS power sharing methodology has been presented in this paper. The proposed technique is based on a NARX ANN, which has been trained using a mixed driving cycle, composed by three full standard driving cycles. This proposed MDC ANN training allows a better performance of traction power demand prediction, when compared to a single driving cycle ANN training approach, as evidenced by obtained results.

It is possible to conclude, based on achieved results, that the proposed method was effective and promising for the calculation of traction power demand in electric vehicles.

Future work will include the evaluation of proposed technique using real-time hadware-in-the-loop simulation and implementation of on-line NARX ANN training for enhanced traction power prediction accuracy.

ACKNOWLEDGEMENT

Authors would like to thank CNPq and IF Sudeste MG for the financial support to this work.

Figure 11: Predicted and calculated traction power (ANN input data: proposed MDC).

portion of predicted and calculated traction power curves when the proposed MDC is considered as the NARX ANN input. Since this input data is the same used for ANN training, high prediction accuracy has been achieved. It is important to repeat a similar analysis for the case where an input data different from the one used for training is considered for ANN input. Two standard driving cycles have been arbitrarily chosen to perform this evaluation. Fig. 12 shows the comparison between predicted and calculated traction power using NEDC (New European Driving Cycle) as NARX ANN input data, while Fig. 13 illustrates the case for a WLTP Class 3 speed profile. In both cases, a quite good prediction accuracy have been also

REFERENCES

[1] Zeinab Rezvani, Johan Jansson, and Jan Bodin. Advances in consumer electric vehicle adoption research: A review and research agenda. *Transportation research part D: transport and environment*, 34:122–136, 2015.

[2] S. E. Lucena. A survey on electric and hybrid electric vehicle technology. In S. SOYLER, editor, *Electric Vehicles - The Benefits and Barriers*, pages 1–18. In Tech, 2011.

[3] Sami Kara, Wen Li, and Nikkita Sadjiva. Life cycle cost analysis of electrical vehicles in australia. *Procedia CIRP*, 61:767–772, 2017.

[4] CETESB. Primeiro inventário de emissões antrópicas de gases de efeito estufa diretos e indiretos. Technical report, Companhia Ambiental do Estado de São Paulo - CETESB, 2011.

[5] S. R. Reis and E. A. Silva. Motores eltricos flex a etanol: uma nova era no setor automotivo mundial. *Revista de Cihncias Exatas e Tecnologia*, 12(12):45–48, 2017.

[6] M. Ehsani, Y. Gao, and A. Emadi. *Modern Electric Hybrid Electric and Fuel Cell Veihcles*. CRC Press, 2010.

[7] D. Welch and T. Ebhardt. Montadoras criam "guerra do carro eltrico" carssima por causa da tesla, 2017.

[8] Y. Vasconcelos. A asceno dos eltricos. *Revista Pesquisa FAPESP*, (258):19–27, Agosto 2017.

[9] Dominic Bresser, Kei Hosoi, David Howell, Hong Li, Herbert Zeisel, Khalil Amine, and Stefano Passerini. Perspectives of automotive battery r&d in china, germany, japan, and the usa. *Journal of Power Sources*, 382:176–178, 2018.

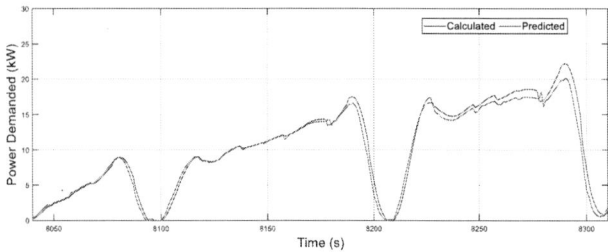

Figure 12: Predicted and calculated traction power (ANN input data: NEDC).

[10] Wenhua Zuo, Ruizhi Li, Cheng Zhou, Yuanyuan Li, Jianlong Xia, and Jinping Liu. Battery-supercapacitor hybrid devices: recent progress and future prospects. *Advanced Science*, 4(7):1600539, 2017.

[11] Lia Kouchachvili, Wahiba Yaïci, and Evgueniy Entchev. Hybrid battery/supercapacitor energy storage system for the electric vehicles. *Journal of Power Sources*, 374:237–248, 2018.

[12] Z. Amjadi and S. S. Williamson. Prototype design and controller implementation for a battery-ultracapacitor hybrid electric vehicle energy storage system. *IEEE Transactions on Smart Grid*, 3(1):332–340, March 2012.

[13] A. A. Ferreira, J. A. Pomilio, G. Spiazzi, and L. de Araujo Silva. Energy management fuzzy logic supervisory for electric vehicle power supplies system. *IEEE Transactions on Power Electronics*, 23(1):107–115, Jan 2008.

[14] G. Udhaya Sankar, C. Ganesa Moorthy, and G. RajKumar. Smart storage systems for electric vehicles a review. *Smart Science*, 7(1):1–15, 2019.

[15] Victor Herrera, Aitor Milo, Haizea Gaztañaga, Ion Etxeberria-Otadui, Igor Villarreal, and Haritza Camblong. Adaptive energy management strategy and optimal sizing applied on a battery-supercapacitor based tramway. *Applied Energy*, 169:831–845, 2016.

[16] Ziyou Song, Jun Hou, Heath Hofmann, Jianqiu Li, and Minggao Ouyang. Sliding-mode and lyapunov function-based control for battery/supercapacitor hybrid energy storage system used in electric vehicles. *Energy*, 122:601–612, 2017.

[17] Pablo García, Juan P Torreglosa, Luis M Fernández, and Francisco Jurado. Control strategies for high-power electric vehicles powered by hydrogen fuel cell, battery and supercapacitor. *Expert Systems with Applications*, 40(12):4791–4804, 2013.

[18] Hristiyan Kanchev, Nikolay Hinov, Bogdan Gilev, and Bruno Francois. Modelling and control by neural network of electric vehicle traction system. *Elektronika ir Elektrotechnika*, 24(3):23–28, 2018.

[19] Lei Zhang, Xiaosong Hu, Zhenpo Wang, Fengchun Sun, and David G Dorrell. A review of supercapacitor modeling, estimation, and applications: A control/management perspective. *Renewable and Sustainable Energy Reviews*, 81:1868–1878, 2018.

[20] A. A. Ferreira, J. Antenor Pomilio, EP Silva, and Diego Vaz Pontes Cambra. Metodologia para dimensionar mltiplas fontes de suprimento de energia de veculos eltricos. In *ABVE-VE 2007*. 2007.

[21] A. A. Ferreira, J. A. Pomilio, G. Spiazzi, and L. de Araujo Silva. Energy management fuzzy logic supervisory for electric vehicle power supplies system. *IEEE Transactions on Power Electronics*, 23(1):107–115, Jan 2008.

[22] Apolinar Reynoso-Hernández, José E Rayas-Sánchez, Lina M Aguilar-Lobo, José R Loo-Yau, Susana Ortega-Cisneros, and Pablo Moreno. Application of the narx neural network as a digital predistortion technique for linearizing microwave power amplifiers. 2015.

[23] L. F. R. Lago. *Gestão do sistema híbrido de energia de um veículo elétrico utilizando redes neurais artificiais*. Monography (undergraduation), Instituto Federal de Educação, Ciência e Tecnologia do Sudeste de Minas Gerais, 2019.

[24] Pablo García, Juan P Torreglosa, Luis M Fernández, and Francisco Jurado. Control strategies for high-power electric vehicles powered by hydrogen fuel cell, battery and supercapacitor. *Expert Systems with Applications*, 40(12):4791–4804, 2013.

[25] M. C. B. P. Rodrigues. *Integração de Filtro Ativo de Potência Monofásico e Bifásico ao Sistema de Propulsão de um Veículo Elétrico*. Doctoral thesis, Universidade Federal de Juiz de Fora, 2014.

[26] Elkhatib Kamal, Lounis Adouane, Rustem Abdrakhmanov, and Nadir Ouddah. Hierarchical and adaptive neuro-fuzzy control for intelligent energy management in hybrid electric vehicles. *IFAC-PapersOnLine*, 50(1):3014–3021, 2017.

[27] CN Papadimitriou, EI Zountouridou, and ND Hatziargyriou. Review of hierarchical control in dc microgrids. *Electric Power Systems Research*, 122:159–167, 2015.

[28] Allal M Bouzid, Josep M Guerrero, Ahmed Cheriti, Mohamed Bouhamida, Pierre Sicard, and Mustapha Benghanem. A survey on control of electric power distributed generation systems for microgrid applications. *Renewable and Sustainable Energy Reviews*, 44:751–766, 2015.

[29] D. C. Silva Junior. Modelagem e controle de funes auxiliares em inversores inteligentes para suporte a microrredes ca - simulao em tempo real com controle hardware in the loop. Master's thesis, Universidade Federal de Juiz de Fora, Juiz de Fora, 2017.

[30] George T Taoka. Brake reaction times of unalerted drivers. *ITE journal*, 59(3):19–21, 1989.

[31] Heikki Summala. Brake reaction times and driver behavior analysis. *Transportation Human Factors*, 2(3):217–226, 2000.

[32] T. Tanaka, T. Sekiya, H. Tanaka, M. Okamoto, and E. Hiraki. Smart charger for electric vehicles with power-quality compensator on single-phase three-wire distribution feeders. *IEEE Transactions on Industry Applications*, 49(6):2628–2635, 2013.

[33] A. Rosca. Light duty vehicle test cycle generation based on real-world data. Master thesis, Instituto Superior Técnico, 2013.

Symmetrical Hybrid Multilevel VSI and CSI Inverters Derived from Dc-Dc Converters

Domingo Ruiz-Caballero, Luis Colque Miranda, Carlos Paredes, Javier Riedemann, Werner Jara Montecinos[*], Marcelo Lobo Heldwein, and Samir Ahmad Mussa[**]

[*]Pontifical Catholic University of Valparaíso
School of Electrical Engineering – EIE
Power Electronic Laboratory – LEP
Av. Brasil 2147, P.O. BOX 4059,
Valparaíso, CHILE.
Phone: +56-32-2273695
domingo.ruiz@ucv.cl

[**]Federal University of Santa Catarina
Department of Electrical Engineering/Power
Electronics Institute – INEP
P. O. BOX 5119 – 88040-970
Florianópolis – SC – BRAZIL
Phone: +55-48-3331.9204/Fax: +55-48-3234.5422
samir@inep.ufsc.br

Abstract – **This work presents various multilevel converters obtained from basic dc-dc converters for high-voltage and high-power applications. Several cells are proposed, which have a single switching pattern realized via a simple modulator circuit that compares triangular carriers with a sine modulator. One of the main features of the proposed circuits is that some transistors work at low frequency and withstanding high voltages or high currents, while others are switched at high frequency and low voltage or low current levels. The topologies can be classified as hybrid. In addition, taking account that having the slow bridge as an output circuit means that a reduced number of semiconductor components leads to a relatively high number of levels.**

Keywords: **DC-AC converters, Hybrid Inverters, Symmetrical Multilevel Inverters.**

I. INTRODUCTION

It can be observed from the various specialized papers that there is a trend to have more modular multilevel circuits [4-6]. A second observable trend in multilevel converters is to have a minimum of constituent switches for a given number of levels in specific applications [7] [11].

From the specialized literature it is observed that, until recently, various topologies and studies have been introduced [1] - [14]. Of them, the most popular are three multilevel voltage converters based on voltage source converter synthesis, which are: NPC, flying capacitor and cascade H-bridge inverters with separate dc sources and within this latter classification the asymmetric hybrid inverters [4]. Fig. 1(a) shows a seven-level asymmetric hybrid multilevel converter. This is based on the binary configuration of voltage sources, which means that this circuit can synthesize (2^{F+1} -1) voltage levels at the load side, where F is the number of dc sources. The upper H-bridge is composed of high voltage blocking switches operating at low operating frequency (GTO, for example), as for the lower H-bridge it is switched at high frequency. However, the switches can be rated for low voltage (IGBT as an example). Using an appropriate modulation strategy (Fig. 1 (b)) it is possible to synthesize a seven-level waveform: -3E, -2E, -E, 0, E, 2E and 3E. The modulated output voltage for this circuit is

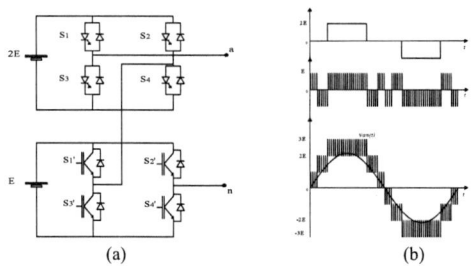

Fig. 1. (a) Voltage source inverter hybrid asymmetric multilevel.
(b) Modulation strategy applied to this inverter.

seen in Fig. 1(b). This modulation strategy defines the concept of hybrid modulation, which is based on single pulse modulation in conjunction with sinusoidal pulse width modulation (SPWM). Under this modulation strategy, the slow frequency switches are modulated to operate at the fundamental frequency, while the fast switches are modulated at high frequency. An improvement in the quality of the output waveform results. The spectral response of the output voltage regarding its high frequency behavior depends on the fast switches, while the high power and voltage generation low frequency components depend on the slow switches [7].

This work employs the hybrid modulation and topology principles to propose novel structures that are based on basic modules switching at high frequency and an inverter cell operating at low frequency. The characteristics of the basic modules are inherited by the full converter structure accordingly. Thus, current source and voltage source inverters can be created that promote the current and/or voltage sharing among modules.

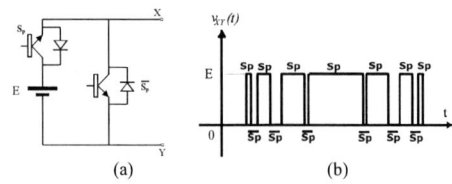

Fig. 2. (a) Buck Cell. (b) Voltage through x-y.

978-1-7281-4181-7/19 $31.00 © 2019 IEEE

II. VOLTAGE SOURCE CELLS TOWARD THE LOAD

This section discusses the generation of hybrid converters based on voltage source cells.

A. Buck Cell

Fig. 2(a) shows the basic cell of this inverter showing in Fig. 2(b) the theoretical waveform of the voltage between X and Y. This is clearly a voltage source cell type at the output. Several popular topologies such as H-Bridge [6] and MMC's [5] use this kind of cell.

If an H-bridge modulated at low frequency is connected between X and Y points (Fig. 3 (a) and (b)), it is possible to obtain a properly modulated three-level ac signal that can be applied to an inductive load.

The modulation of these cells is done in the same way between them, i.e. to drive a number of C cells, so as to obtain N levels in the output uses the comparison of P triangular carriers, phase-shifted by a suitable angle between them (θ), with a modulating signal that for fast switches is a rectified sinusoid signal and for the slow ones the same non-rectified sinusoid, see Fig.3(c).

If these cells are connected in series and with a slight change in the angle of the carriers and latter inverted by the slow H-bridge (S_{L1} and S_{L2}), multilevel voltages are obtained in the output, where the number of levels is directly linked to the number of connected cells, as shown in Fig. 4 (a) and (b). Five levels are obtained with two cascaded cells, with four there are nine levels with six thirteen levels and so on, i.e., if C is the number of cells connected in cascade and N the number of levels, then

$$N = 2 \cdot C + 1 \tag{1}$$

Note that due to the use of the slow H-bridge, a total converter with a reduced number of semiconductors is obtained, when compared with the cascaded H-Bridge or MMC topologies for the same number of levels of the output voltage. In the case of the MMC, there would be twice as many switches, since there would be a positive branch of cascaded cells and a negative one in order to obtain both half-cycles of the output voltage [3], it can be established that this form of connection there is no circulation current.

B. Boost-Buck or Cuk Cell

Fig. 5 shows the basic Boost-Buck or Cuk cell. With this cell it is possible to obtain two-level voltages between X and Y by pulse width modulation in the same way to that obtained with the Buck cell. With the difference that now these voltages can be greater or smaller than E, because the gain of the Cuk converter allows to raise or reduce the voltage at the X and Y points [13]. Note that the voltage reflected at these points is that over the storage capacitor (C_a).

Adding the slow H-Bridge between the X and Y points, the structure can synthesize this waveform with a continuous mean value greater than zero in an alternating signal of three levels with zero mean value (see figure 5

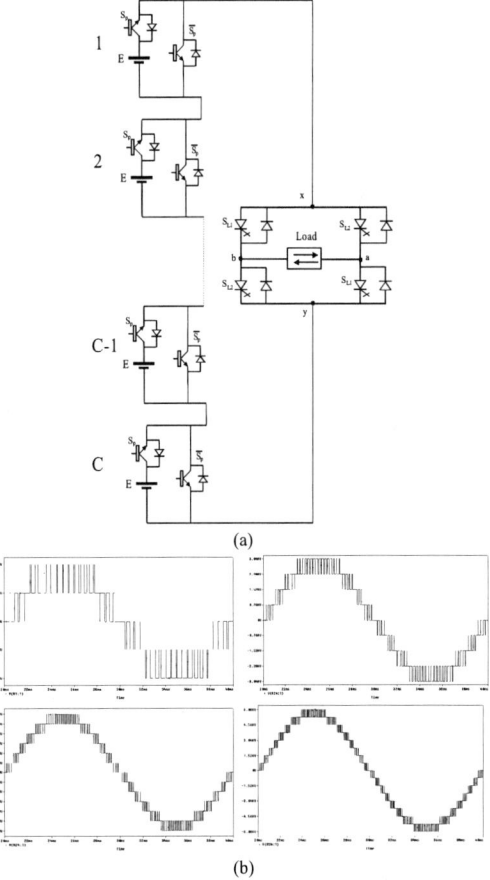

(a)

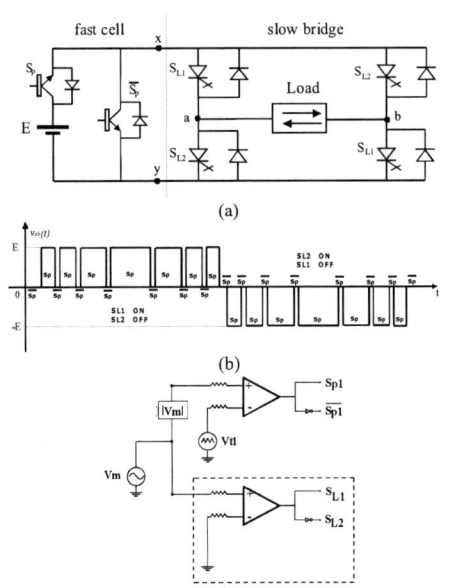

Fig. 3. (a) Inverter of three levels based in the Buck converter. (b) Load voltage (c) Basic modulator Circuit.

Fig. 4 (a) Single-phase Buck- cell Symmetrical Hybrid multilevel Inverter with own PWM circuit, (b) Output (V_{ab}) voltage waveform for two, four , six and eigth cascaded cell respectively.

978-1-7281-4181-7/19 $31.00 © 2019 IEEE

(b)).

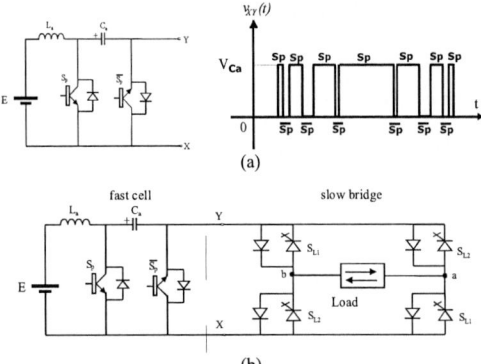

(a)

(b)

Fig. 5 (a) Boost-Buck cell and Voltage X and Y (b) Three level Boost-Buck circuit.

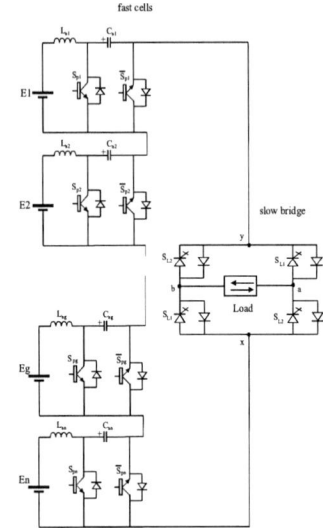

Fig. 6 Single-phase multi-level inverter Boost-Buck or Cuk.

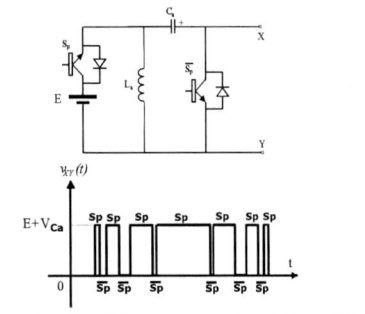

Fig. 7 Zeta basic cell and its voltage waveform between X and Y.

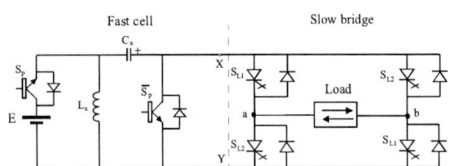

Fig. 8. (a) Single-phase Symmetrical Hybrid multilevel Inverter proposed based in three-level cell.

To obtain a multilevel voltage it is sufficient to connect several of these basic cells in series and connected them to the load through a single H-bridge inverter with slow modulation as shown in Fig. 6. The bridge has to able to bear the voltage stress of the full dc-link.

C. Buck-Boost/Buck cell (or Zeta cell)

The Buck-boost/Buck or Zeta based converter can be generated in the same way to what is done with the Buck and the Cuk. The basic cell is shown in Fig. 7. Continuing with what has been done to the two previous circuits to obtain the alternate signal, the slow H-bridge is accordingly connected as seen in Fig. 8 [12].

Connecting these basic cells in cascade and interconnecting them with a bridge-H operated in low frequency has a converter of 'N' levels with output voltage higher or lower than the source 'E'.

From these structures, it is observed that if the load is current source type it is not necessary to connect an inductor in the output circuit, which is necessary if the load is a voltage source type.

D. Voltage Source Topologies Modulator

The modulation strategy for the switches is based on the Phase-Shifted Disposition (PSD) technique. The modulator circuit applied to all the previous cells, which synthesize the command signals to all the switches of the fast cell is shown in Fig. 9. The switches of the cells are switched at high frequency and the H-bridge at low frequency, i.e., at the operating frequency of the load. Thus, for operation as a multilevel voltage, it is sufficient to connect C basic cells, properly modulated and keep the H-bridge operated at low frequency.

The carriers (Vtn) are triangular signals shifted between them by an angle θ, given by

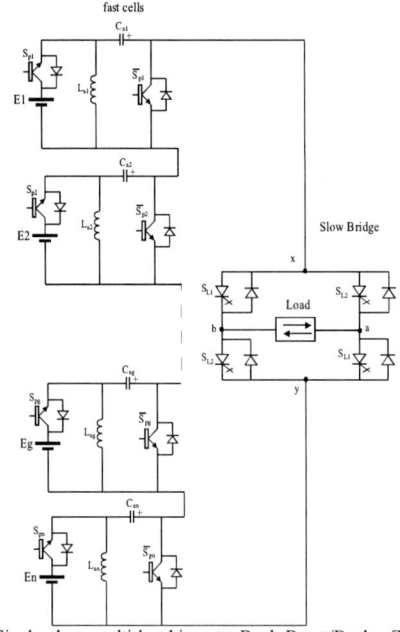

Fig. 9. Single-phase multi-level inverter Buck-Boost/Buck o Zeta.

978-1-7281-4181-7/19 $31.00 © 2019 IEEE   712

$$\theta = \frac{360^o}{P}. \qquad (2)$$

This is due to the applied unipolar modulation, where $P = (N-1)/2$ represents the number of carriers needed to obtain N levels. The modulating signal is the desired voltage at the output. In addition, the output high frequency (F_{S1}) of the inverter with this modulation is related to the switching frequency of the fast cell switches (Fsw) by

$$F_{s1} = P \cdot F_{SW}, \qquad (3)$$

This can be observed in Fig. 4.

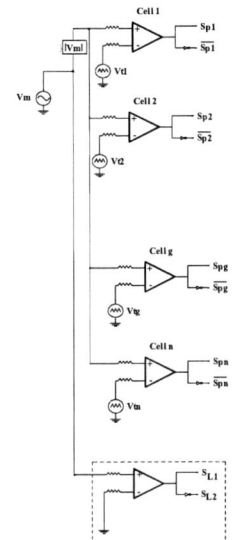

Fig. 10 Generic Modulator Circuit.

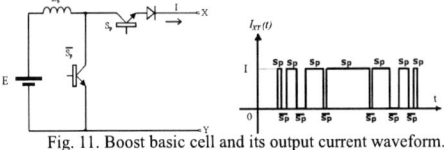

Fig. 11. Boost basic cell and its output current waveform.

III. CURRENT SOURCE CELLS (CSI) TOWARDS THE LOAD

A. Boost Cell

Fig. 11 shows the basic Boost cell as the dual circuit of the Buck cell [14]. Square currents with adjustable pulse width (SPWM) are obtained due to the current source nature of the cell. Thus, the load must be a voltage source type or present a capacitive behavior.

By connecting an H-bridge modulated at low frequency between the X and Y points, it is possible to obtain a modulated alternating current with three levels through the load or output filter. The complete circuit and its modulator are presented in Fig. 12.

The modulation strategy is similar to the voltage source inverters. The only difference is that the signal that operates the main Buck switch now enables the auxiliary transistor and vice versa.

The parallel connection of the cells can be done to enable a current multilevel operation as shown in Fig. 13,

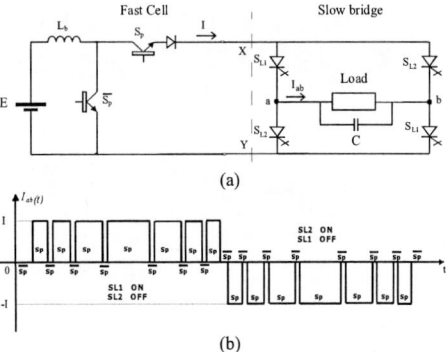

Fig. 12. (a) Single-phase three-level CSI Symmetrical Hybrid multilevel Inverter proposed based on the Boost cell; (b) Load current waveform (I_{ab}).

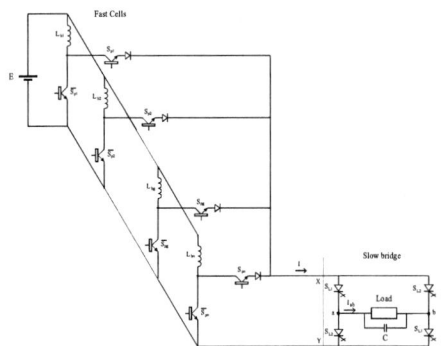

Fig. 13. Single-phase N-level CSI Symmetrical Hybrid multilevel Inverter proposed based in Boost converters.

where each cell is activated by comparing a unipolar modulation signal with a carrier signal with appropriate phase. The carrier phase is also governed by (1) as for the voltage source inverter cascade-type circuits.

B. Buck-boost cell

Fig. 14(a) shows the basic Buck-boost cell and from this basic cell it is possible to obtain modulated two-level currents through the X and Y points connected load in the same way as they were obtained for the Boost cell.

The inversion of the current is obtained with a zero average value and three levels by connecting the slow H-Bridge between the X and Y points. In order to obtain the multilevel currents in the load, as in the Boost case, the cells are connected in parallel obtaining a current in X and Y of N levels. This is shown in Fig. 16. The modulation is identical to the previously applied to the Boost converter.

C. Boost / Buck-boost (or SEPIC) cell

Fig. 16 shows the basic SEPIC cell, just as the Boost cell and Buck-boost behaves as a current source to the load, which must, therefore, be capacitive or a voltage source.

The way to modulate the output current is identical to the Boost and Buck-boost converter. When connecting the H-bridge, a zero-value signal is obtained, as seen in Fig. 17.

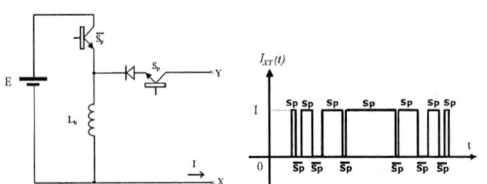

Fig. 14   Buck-boost basic cell and its output current waveform.

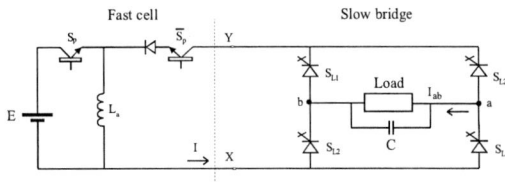

Fig. 15   (a) Single-phase CSI Symmetrical Hybrid multilevel Inverter proposed based in Buck-boost three-level cell

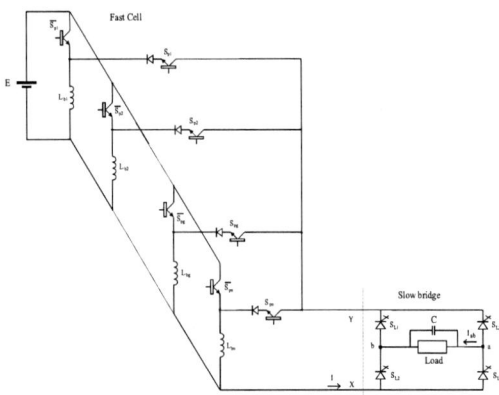

Fig. 16. Single-phase N-level CSI Symmetrical Hybrid multilevel Inverter proposed based in Buck-Boost converters.

In order to operate as a current multilevel inverter, the parallel connections of the cells must be made and these must be correctly modulated. Fig. 18 shows this, where a multilevel converter of *N* current levels is obtained. The modulator circuit is the same as it applies for all circuits.

D. Modulator Circuit

The modulator circuit applied to all the previous cells, which synthesizes the command signals for the transistors of the fast cell is the same as in Fig. 10. The transistors of the cells are switched at high frequency and on the H-bridge at low frequency (see Fig. 10).

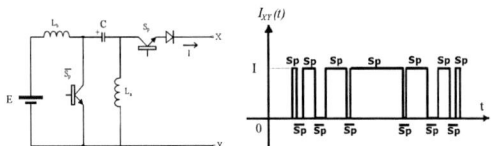

Fig. 17. Basic SEPIC cell and its output current waveform.

IV. Simulation Results

In order to verify the operation of the different single-phase cells, all circuits were simulated for a disposition of seven cells (for the VSI′s connected in series and for the CSI′s connected in parallel). For the voltage source inverters, an inductor L=10 mH was inserted in the load (R=10 Ω). The simulations were performed for all cases with the following design specifications: Fs=850 Hz and a modulation index of 0.9454545.

The results are shown in the following: for the Buck converter according to Fig. 20, the Boost-Buck (Cuk) in Fig. 21, and the Buck-Boost/Buck (Zeta) in Fig. 22, where you can observe the phase voltage generated by the inverter (V_{ab}) and the current in the load. The voltage of the Buck modular multilevel inverter is the one that better reproduces the modulating signal since, both, the Cuk multilevel inverter and the Zeta reflect on the load, the storage capacitor voltage, instead of the voltage at the input source, as for the Buck inverter.

However, both, Cuk- and Zeta-based converters generate higher values than the Buck. However, in the Cuk converter it is possible to observe that the output voltage is the reflection of the accumulation capacitor voltage and in the case of Zeta it will be the sum of the input voltage and that of the accumulation capacitor. Thus, these will have a greater degree of distortion, which is a product of the ripple in the capacitor voltage. Figures 23 to 25 show the load voltage and current (times 6) for the CSI inverters at steady state with seven modules in parallel and a modulation index of 0.9454545. For the current source inverters, a capacitor $C = 8$ uF was inserted in

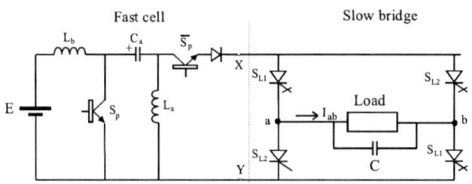

Fig. 18   (a) Single-phase CSI Symmetrical Hybrid multilevel Inverter proposed based in SEPIC three-level cell.

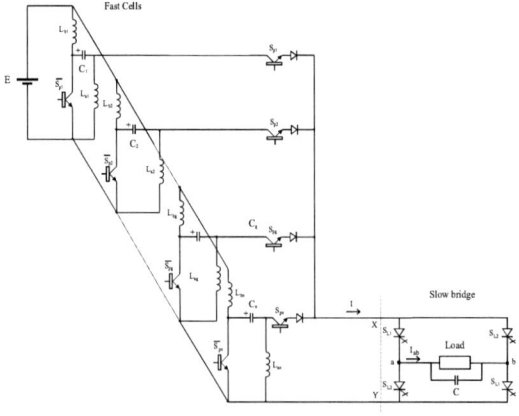

Fig. 19. Single-phase N-level CSI Symmetrical Hybrid multilevel Inverter proposed based in SEPIC converters.

978-1-7281-4181-7/19 $31.00 © 2019 IEEE

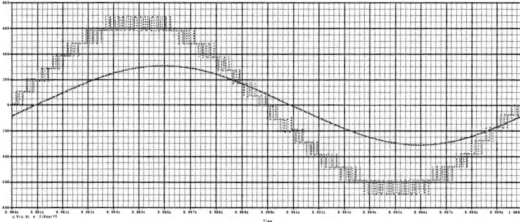

Fig. 20 – Waveforms generated by simulation of a single- phase multilevel Buck inverter, mf=17, mi=0.9454545.

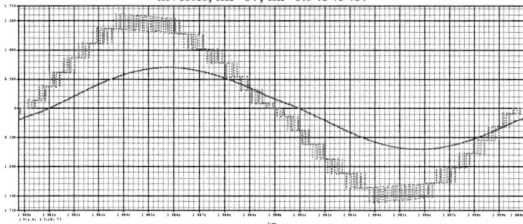

Fig. 21. Waveforms generated by simulation of a single- phase multilevel Cuk inverter, mf=17, mi=0.9454545.

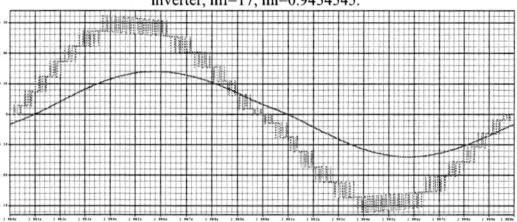

Fig. 22. Waveforms generated by simulation of a single- phase multilevel Zeta inverter, mf=17, mi=0.9454545.

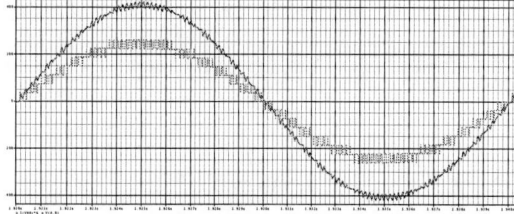

Fig. 23. Waveforms generated by simulation of a single- phase multilevel Boost inverter, mf=17, mi=0.9454545.

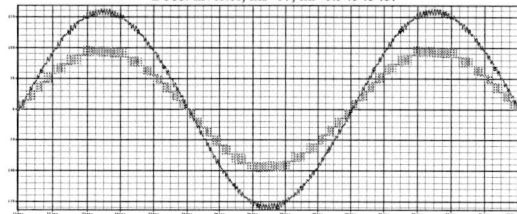

Fig. 24. Waveforms generated by simulation of a single- phase multilevel Buck-Boost inverter, mf=17, mi=0.9454545.

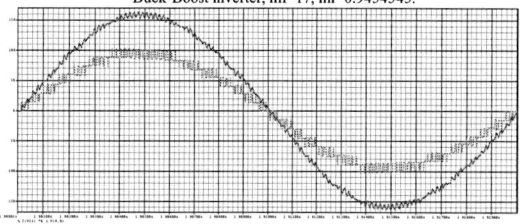

Fig. 25. Waveforms generated by simulation of a single- phase multilevel SEPIC inverter, mf=17, mi=0.9454545.

parallel with the load (R=10 Ω), where the phase PWM current generated by the inverter and the load voltage are observed. Results are shown for the Boost inverter in Fig. 23, the Buck-Boost in Fig. 24, and the Boost/Buck-Boost in Fig. 25. The output current of the Boost multilevel inverter is nearest to the modulating signal, since the SEPIC multilevel inverter reflects on the accumulation inductor and the accumulation capacitor voltage, instead of the input source voltage.

V. CONCLUSIONS

Two families of multilevel inverters were introduced in this work. These were generated from basic dc-dc converters and include voltage source and current source inverters with multilevel output waveforms. The main characteristics of these circuits are that they are symmetric hybrids, and, with the exception of the Buck and Boost converter, are able to, both, reduce and increased the output voltage with respect to the input voltage level depending on the desired modulation index. The proposed circuits operate from a single extended modulator to obtain an N-level circuit. The hybrid nature of the circuits cause that different voltage or current ratings occur in the power semiconductor devices. On the other hand, if compared to the other modular multi-level converters they present a strong reduction in the number of constituent switches for a given number of output levels.

REFERENCES

[1] A. Nabae, I. Takahashi and H. Akagi. "A New Neutral-Point-Clamped PWM Inverter". IEEE Transactions on Industry Applications, vol. IA-17, No. 5, pp 518-523, 1981.

[2] Richard Baker, US patent US5568371, "Bridge Converter Circuit", published 05/26/1981.

[3] D. A. Ruiz-Caballero, R. M. Ramos-Astudillo, S.A. Mussa, and M. L. Heldwein, "Symmetrical Hybrid Multilevel DC-AC Converters with Reduced Number of Insulated DC Supplies," *IEEE Trans. on Ind.Electron.*, vol. 57, pp. 2307-2314, July. 2010.

[4] Madhav D. Manjrekar, Peter Steimer and Thomas A. Lipo, "Hybrid Multilevel Power Conversion System: a competitive solution for high power applications". IEEE Transaction on Industry Applications, Vol. 36, No3, pp.834-841 May/june 2000.

[5] R. Marquardt and A. Lesnicar "New Concept for High Voltage – Modular Multilevel Converter", IEEE PESC 2004, Aachen.

[6] Jih-Sheng Lai and Fang Zheng Peng. "Multilevel Converters – A New Breed of Power Converters". IEEE Transactions on Industry Applications, vol. 32, N° 3, May/june 1996.

[7] Hector Vergara, René Sanhueza, Miguel López, Marcelo L. Heldwein, Samir Mussa and Domingo Ruiz-Caballero, "Symmetrical Hybrid Multilevel DC-AC Converters in Cascade ", ECCE' 2010 - IEEE Congreso de Exposición y Conversión de Energía, Atlanta Georgia - EEUU, Septiembre de 2010.

[8] Fang Zheng Peng, "A Generalized Multilevel Inverter Topology with Self Voltage Balancing". IEEE Transactions on Industry Applications, vol. 37, N° 2, march/april 2001.

[9] Thierry A. Meynard, Henri Foch, Francois Forest, Christophe Turpin, Frédéric Richardeau, Laurent Delmas, Guillaume Gateau, and Elie Lefeuvre. "Multicell Converters: Derived Topologies". IEEE Transactions on Industrial Electronics, vol. 49, N° 5, October 2002.

[10] Daniel Korbes , Samir Mussa and Domingo Ruiz-Caballero, "Modified Hybrid Symmetrical Multilevel Inverter", APEC' 2012 - Conferencia de Electrónica de potencia Aplicada, Long Beach - EEUU, Febrero de 2012.

[11] D.Ruiz- Caballero, René Sanhueza, Sebastian Arancibia, Miguel López, Samir Mussa and Marcelo Lobo Heldwein, "Symmetrical Hybrid Multilevel Inverter with 'N´Cells connected in Parallel employing Multi-state Switching", COBEP' 2011 - Congreso Brasileño de Electrónica de Potencia, Natal-RN - Brasil, Septiembre de 2011.

[12] Luis Colque, R. Sanhueza, M. López, Samir Mussa and Domingo Ruiz-Caballero, "High-Gain Symmetrical Hybrid Multilevel DC-AC Converters - Single-Phase circuits ", COBEP' 2013 - Congreso Brasileño de Electrónica de Potencia, Gramado-RS - Brasil, 27-31 de Octubre de 2013..

[13] Franco Chiarella, R. Sanhueza, M. López, S. Fingerhuth, Samir A. Mussa and Domingo Ruiz-Caballero, "Symmetrical Hybrid Multilevel DC-AC Step- Up/Down Converters - Single-Phase circuits", COBEP' 2013 - Congreso Brasileño de Electrónica de Potencia, Gramado-RS - Brasil, 27 - 31 de Octubre de 2013.

[14] Carlos Paredes, Tiago Kommers, Marcelo L. Heldwein, Samir Mussa and Domingo Ruiz-Caballero, "A Current Symmetrical Hybrid Multilevel DC-AC Converter", EPE' 2011 - Conferencia Europea de Electrónica de potencia, Birmingham- UK, 30 de Agosto al 1 de Septiembre de 2011.

Structural and Performance Comparison Between Harmonic Selective Repetitive Controllers for Shunt Active Power Filter

R. C. Neto, F. A. S. Neves, E. V. Stangler, F. Bradaschia
Power Electronics and Drives Research Group
Universidade Federal de Pernambuco
Recife/PE, Brazil
rafael.cavalcantineto@ufpe.br, fneves@ufpe.br,
eduardo.vasconcelosstangler@ufpe.br, fabricio.bradaschia@ufpe.br

H. E. P. de Souza
Department of Industry
Instituto Federal de Educação,
Ciência e Tecnologia de Pernambuco
Pesqueira/PE, Brazil
helberelias@pesqueira.ifpe.edu.br

Abstract—The repetitive controller (RC) represents a very effective solution for the current control of shunt active power filters. Using this structure as basis, several authors have proposed harmonic selective schemes that are more advantageous than the conventional RC when the control system references only contain a well-known family of harmonic components, such as $6k \pm 1$ or $4k \pm 1$ (all odd). In this sense, this paper presents a structural and performance comparison between real and complex harmonic selective RC-based solutions. Experimental results were used to validated the comparative study.

Index Terms—repetitive controller, digital control, harmonic compensation, active power filter.

I. INTRODUCTION

With the advancement of microelectronics, semiconductor devices have become increasingly present in power control equipment for electrical installations and machines. The operation of these electronic equipments causes disturbances in the electrical system in which they are connected, which happens due to the fact that these equipments are seen by the system as non-linear loads. In order to mitigate the negative effects caused by such devices, active power filters (APF) have emerged as a promising solution to power systems designers.

Originally, active power filters were divided into two classes of performance: shunt APFs and series APFs. The first usually operates as a controlled current source, injecting a current into the system that attenuates the current disturbances in the network and compensates for reactive power [1]. In this sense, the proper choice and design of the current controller in shunt APFs determine the performance characteristics of the entire system.

Several authors have proposed current control schemes for shunt APFs, some of the most relevant proposals are: the proportional-integral controller in synchronous reference frame [2]; multiple reference integrators [3]; and structures based on resonant controllers in parallel [4]. In order to understand the similarities and differences between these and

other control strategies applied to shunt APFs, Limongi et al. [5] published a comparative survey that pointed out the repetitive controller also as a viable and competitive solution for this application.

The repetitive controllers (RCs) are characterized by having a periodic signal generator in its structure, which allows them to compensate periodic signals with harmonics of any order. However, with a in-depth analysis of this control strategy, several authors proposed harmonic selective solutions (such as [6]–[10]) for applications where it is not desired to compensate all the harmonic components. Due to this fact, a comparative survey must be done in order to clarify the difference between those proposals.

In this context, the present paper aims to compare the structural and performance characteristics of the main harmonic selective RCs proposed in the literature. Therefore, it is organized as follows: Section II presents the fundamentals of repetitive controllers; in Section III, the evaluated harmonic selective RCs are introduced and their structural similarities and differences are also approached; in Section IV, experimental results are used to compare the performance of these harmonic selective RCs; finally, conclusions are presented in Section V.

II. FUNDAMENTALS OF REPETITIVE CONTROL

In 1981, Inoue et al. [11] proposed a high accuracy control scheme, also known as repetitive controller, whose main characteristic was the application of high gain for a selected frequency (chosen by the control system designer) and for all its harmonics (including 0 Hz). The magnitude plot of the frequency response of this control strategy is shown in Fig. 1. As a matter of fact, this characteristic occurs due to the RC's periodic signal generator, which, according to the internal model principle [12], also allows the system to asymptotically track periodic references.

Years after Inoue's proposal, Hara et al. [13] developed a input-output stability analysis of the RC. As result, a second direct path with gain a was added in parallel with the RC's

---

The authors would like to thank *Conselho Nacional de Desenvolvimento Científico e Tecnológico* - CNPq and *Fundação de Amparo à Ciência e Tecnologia do Estado de Pernambuco* - FACEPE, for the financial support.

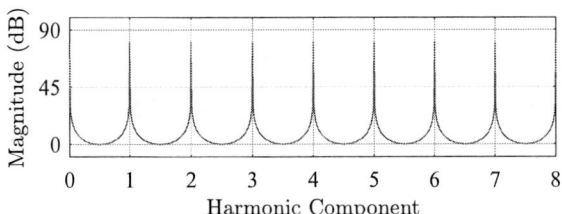

Fig. 1: Magnitude plot of the repetitive controller.

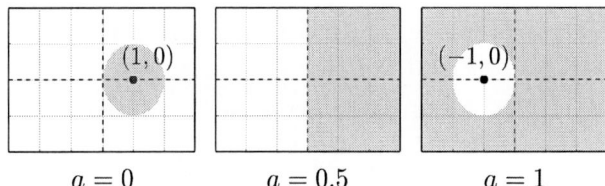

Fig. 3: Stability domains (shaded area) as proposed in [13].

periodic signal generator (Fig. 2), which allowed to select the RC implementation configuration as will be shown below. From this point, the authors will refer to Hara's proposal as conventional RC (CRC), whose transfer function is

$$C_1(z) = K_{rc}\left(a + \frac{z^{-N}}{1 - z^{-N}}\right), \tag{1}$$

in which N is the number of samples per cycle of the fundamental period and K_{rc} represents the RC gain.

According to Hara et al. [13], when considering a RC system formed by unit negative feedback system with only the CRC and the plant ($G(z)$) in its direct path, two conditions must be satisfied in order to obtain input-output stability:

1) $G_m(z)[1 + a \ G_m(z)]$ must be a proper stable rational transfer function ($G_m(z) = K_{rc} \ G(z)$); and
2) The Nyquist diagram of $G_m(z)$ must be contained in a area of the complex plane called by stability domain of the RC system [13].

Due to this fact, by changing the parameter a (a real constant gain) the control system designer can select the RC system's stability domain suitable for each desired application. As mathematically proven in [13], the stability domain can be enlarged if the parameter a is increased (Fig. 3).

Although the parameter a can be used to increase the stability domain, the complex plane origin (point $(0, 0)$) is not included for any selected a. This aspect restricts conventional RC systems to plants of zero relative degree [13], which is minority in electrical systems. In order to solve this problem, a low-pass filter with finite-impulse response (FIR) $Q(z)$ (or a constant attenuation) can be used in cascade with the RC's delay element (z^{-N}), resulting in a CRC that has greater applicability [14] whose transfer function is

$$C_1'(z) = K_{rc}\left(a + \frac{Q(z) \ z^{-N}}{1 - Q(z) \ z^{-N}}\right). \tag{2}$$

When using a causal and symmetric FIR filter, all poles allocated by the CRC experience a linear frequency shift. This happens because the FIR filter causes a phase angle displacement on its output signal, which must be correctly

compensated for the proper operation of the RC system [15]. In this sense, when the transfer function of the symmetric FIR filter is

$$Q(z) = \begin{aligned} &b_0 + b_1 z^{-1} + \cdots + b_{\frac{L}{2}} z^{-\frac{L}{2}} + \cdots \\ &+ b_1 z^{-(L-1)} + b_0 z^{-L} \end{aligned}, \tag{3}$$

with L being an even number, the effect described above can be mitigated by changing the number of samples delayed N to $(N - \frac{L}{2})$ in the RC, as demonstrated in [10].

III. HARMONIC SELECTIVE REPETITIVE CONTROLLERS

In order to improve the performance of RC systems, several authors proposed RC-based control schemes that change the family of harmonic components in which the CRC applies high gain. In fact, by varying the amount of samples delayed by the RC's periodic signal generator, without changing the frequency of the reference signal, the CRC starts to act in a new family of harmonic components. Examples of this feature can be seen in [16], in which RC-based schemes that applies high gain only in odd harmonic components are proposed.

Several harmonic selective RC schemes (also known as $nk \pm m$ RC and $nk + m$ RC) have been proposed in the literature. All these schemes present structural advantages (e. g., less number of memory elements required) and performance advantages (e. g., shorter response time) when compared to CRCs. On the other hand, they are more restrictive with respect to the input signal, i. e., a $nk \pm m$ RC system can only control references that contain exclusively harmonic components of the family $H = \{nk \pm m | k \in \mathbb{N}\}$. For example, a $6k \pm 1$ RC system can only control references with harmonic content in the components $H = \{6k \pm 1 | k \in \mathbb{N}\} = \{1, 5, 7, 11, 13, \cdots\}$.

Since it is complex to evaluate the influence of the RC's parameters in harmonic selective RC schemes, these controllers can be decomposed into associations of elementary structures, whose frequency responses are well known. An example of RC-based elementary structure, named primitive repetitive cell (PRC), is proposed in [8] (Fig. 4).

The PRC shown in Fig. 4 allows the application of high gain in harmonic components of the family $H_s = \{nk + m | k \in \mathbb{Z}\}$. Note that, since a PRC is a complex control structure, it is commonly used to control space-vector references, in which positive-sequence harmonic components are represented as positive spectrum while negative-sequence harmonic components are shown in the negative spectrum [10]. For example, a $6k + 1$ RC system can only control references with harmonic content in the components $H_s = \{6k + 1 | k \in \mathbb{Z}\} = \{\cdots, -11, -5, +1, +7, +13, \cdots\}$. In addition, $nk \pm m$ RCs

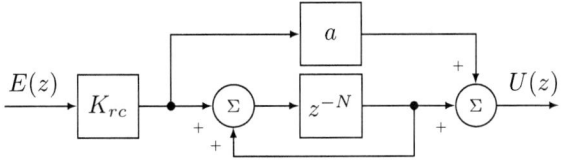

Fig. 2: Block diagram of the discrete-time conventional RC.

978-1-7281-4181-7/19 $31.00 © 2019 IEEE

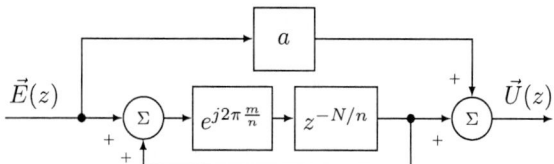

Fig. 4: Block diagram of the PRC proposed in [8].

can be decomposed into a $nk + m$ RC in parallel with a $nk - m$ RC.

From Fig. 4, it is important to highlight the following three parameters of the PRC:

1) **The generic delay** $z^{-N/n}$: which enables compensation of the family $H_s = \{nk|k \in \mathbb{Z}\}$ instead of all harmonic components (CRC);
2) **The complex gain** $e^{j2\pi\frac{m}{n}}$ **cascaded with** $z^{-N/n}$: when it is used together with the generic delay, the parameter m is chosen so that it allows the PRC to control the family $H_s = \{nk + m|k \in \mathbb{Z}\}$ (with $n > m \geq 0$); and
3) **A second direct path with gain** a: as done in [13], when selecting a constant real value for $0 \leq a \leq 1$, it becomes possible select the PRC's stability domain.

Then, its transfer function is given by

$$PRC_{nk+m}(a, z) = a + \frac{e^{j2\pi\frac{m}{n}} z^{-N/n}}{1 - e^{j2\pi\frac{m}{n}} z^{-N/n}}. \quad (4)$$

The concept of stability domains proposed in [13] for CRCs can be extended to the PRCs, being therefore useful in the discussion below.

Back to the harmonic selective RC schemes, the main strategies proposed in the literature are briefly discussed in the following subsections.

A. $nk \pm m$ RC proposed by Lu and Zhou [6]

As can be seen in Fig. 5, Lu and Zhou's proposal uses a total number of memory cells of $2N/n$ times the number of axes (NoA), i. e., $NoA = 1$ for single-phase applications, $NoA = 2$ for three-phase applications using $\alpha\beta$ reference frame and $NoA = 3$ for three-phase applications using abc reference frame. In addition, it presents a maximum time delay of $2N/n$ samples [6].

With respect to this controller's stability domain, its transfer function can be decomposed in two PRCs as follows:

$$C_2(z) = K_{rc} \frac{1 - z^{-2N/n}}{1 - 2\cos(2\pi\frac{m}{n}) z^{-N/n} + z^{-2N/n}}, \quad (5)$$

$$C_2(z) = K_{rc} [PRC_{nk+m}(a = 0.5, z) + PRC_{nk-m}(a = 0.5, z)]. \quad (6)$$

Thus, since $a < 1$, this strategy is not based in the largest possible stability domain.

B. $nk \pm m$ RC proposed by Lu et al. [7]

The harmonic selective RC scheme proposed in [7] is shown in Fig. 6. In order to make a fair comparison between the control strategies approached in this paper, this controller's

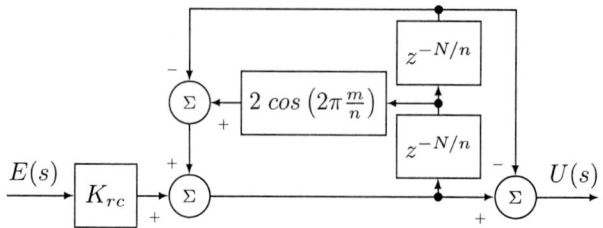

Fig. 5: $nk \pm m$ RC proposed by Lu and Zhou [6].

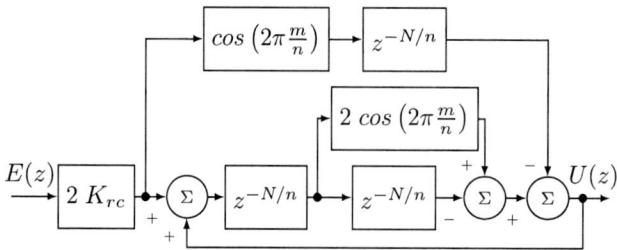

Fig. 6: $nk \pm m$ RC proposed by Lu et al. [7].

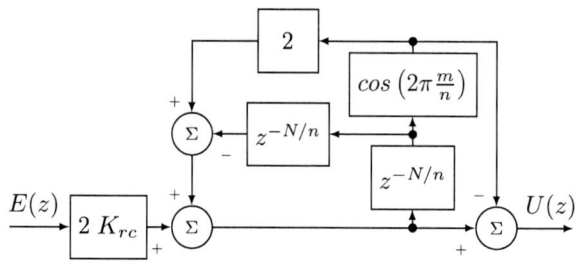

Fig. 7: $nk \pm m$ RC proposed by Neto et al. [8].

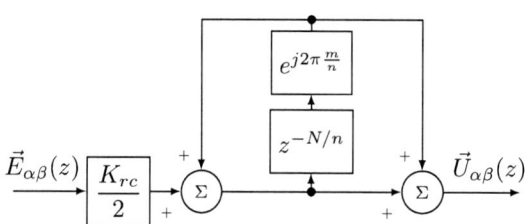

Fig. 8: $nk + m$ RC proposed by Luo et al. [9].

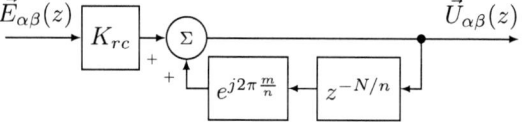

Fig. 9: $nk + m$ RC proposed by Zimann et al. [10].

gain was changed from K_{rc} to $2 K_{rc}$. Therefore, its discrete-time transfer function is expressed as follows:

$$C_3(z) = 2 K_{rc} \frac{\cos(2\pi\frac{m}{n}) z^{-N/n} - z^{-2N/n}}{1 - 2\cos(2\pi\frac{m}{n}) z^{-N/n} + z^{-2N/n}}, \quad (7)$$

which also can be rewritten in terms of PRCs as

$$C_3(z) = K_{rc} [PRC_{nk+m}(a = 0, z) + PRC_{nk-m}(a = 0, z)]. \quad (8)$$

As can be seen in (8), this controller is based in PRCs with $a = 0$, which results in very restrictive stability conditions.

978-1-7281-4181-7/19 $31.00 © 2019 IEEE

Although it has the same characteristic equation of the solution presented in Subsection III-A, its structure is distinct from the previous one. In fact, this controller causes a different zeros allocation and requires $3N/n$ memory cells per number of axes (NoA) implemented. It is important to highlight that Lu et al. [7] indicate this controller as a plug-in RC solution that should be used together with a phase compensation filter, whose main function is to mitigate the effect of the unit time computational delay from digital implementation.

C. $nk \pm m$ RC proposed by Neto et al. [8]

While evaluating the decomposition of harmonic selective RC schemes into PRCs in parallel, Neto et al. [8] proposed a $nk \pm m$ RC whose main characteristic is the use of PRCs with larger stability domains as elementary cell. In fact, as can be seen in Fig. 7, its discrete-time transfer function can be decomposed in two PRCs with $a = 1$:

$$C_4(z) = 2\,K_{rc}\,\frac{1 - cos(2\pi\frac{m}{n})\,z^{-N/n}}{1 - 2\,cos(2\pi\frac{m}{n})\,z^{-N/n} + z^{-2N/n}}, \quad (9)$$

$$C_4(z) = K_{rc}\,[PRC_{nk+m}(a = 1, z) + \\ PRC_{nk-m}(a = 1, z)]. \quad (10)$$

As consequence, Neto et al.'s proposal results in a better trade-off between stability and performance [8].

Regarding the total number of memory cells, this control strategy requires $2N/n$ memory cells times the number of axes (NoA) that must be controlled.

D. $nk + m$ RC proposed by Luo et al. [9]

In 2016, the first complex RC-based controller was proposed by Luo et al. [9]. While real solutions apply high gain in the harmonic components of the family $H = \{nk \pm m | k \in \mathbb{N}\}$, being able to control real periodic signals with harmonic content in this family, such as the currents of a three-phase diode rectifier, the complex solutions present an asymmetric frequency response with respect to the axis of ordinates, applying high gain the components $H_s = \{nk + m | k \in \mathbb{Z}\}$.

Another relevant feature of complex RC-based schemes is that, since the input and output signals of a complex controller should be complex signals, space-vectors can be used as reference signals (as can be seen in Figs. 8 and 9). Therefore, as discussed for PRCs, high gain negative frequencies represent the controlled negative-sequence components while high gain positive frequencies indicate the controlled positive-sequence components. In this sense, a real control system should be used to control the phase currents of a three-phase diode rectifier (whose harmonic spectrum is shown in Fig. 10), while a complex control system can be used to control the space-vector signal obtained from these currents (whose harmonic spectrum is shown in Fig. 11).

When compared to the real solutions described before, the complex $nk + m$ RC proposed by Luo et al. [9] presents a smaller total amount of memory cells required for three-phase implementation (N/n elements for the real part plus N/n

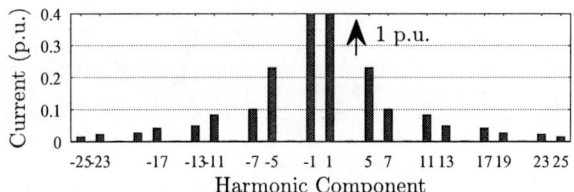

Fig. 10: Harmonic spectrum of the a phase current of a three-phase diode rectifier.

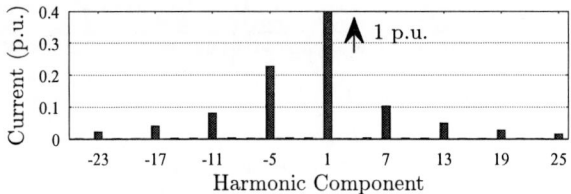

Fig. 11: Harmonic spectrum of the space-vector obtained from the three-phase rectifier input currents.

elements for the imaginary part of the complex signal). Its discrete-time transfer function is described as

$$\vec{C}_5(z) = \frac{K_{rc}}{2}\,\frac{1 + e^{j2\pi\frac{m}{n}}\,z^{-N/n}}{1 - e^{j2\pi\frac{m}{n}}\,z^{-N/n}}, \quad (11)$$

$$\vec{C}_5(z) = K_{rc}\,PRC_{nk+m}(a = 0.5, z). \quad (12)$$

Thus, similarly to Lu and Zhou's proposal [6], this strategy is based on PRC with $a = 0.5$.

Since it is a complex controller, its implementation is done through a real multiple-input multiple-output (MIMO) system in $\alpha\beta$ reference frame (α being the real component of the space-vector and β being the imaginary one), presenting a coupling between the stationary axes [9] (implementation block diagram can be observed in the original paper).

During the design procedure, the real MIMO control scheme can be simplified as a complex single-input single-output (SISO) controller (Fig. 8). Therefore, conventional design methodologies are applicable to this control strategy, with the advantage of using a single structure to control all three-phase quantities rather than an individual controller for each phase.

E. $nk + m$ RC proposed by Zimann et al. [10]

In 2019, Zimann et al. [10] proposed a complex $nk + m$ RC that is equivalent to a PRC with $a = 1$, i. e.,

$$\vec{C}_6(z) = K_{rc}\,\frac{1}{1 - e^{j2\pi\frac{m}{n}}\,z^{-N/n}}, \quad (13)$$

$$\vec{C}_6(z) = K_{rc}\,PRC_{nk+m}(a = 1, z), \quad (14)$$

whose block diagram is shown in Fig. 9.

This control strategy has a better trade-off between stability and performance than the solution presented in Subsection III-D. All characteristics about complex controllers presented in the previous subsection are also valid for this controller.

TABLE I: Summary of the structural similarities and differences between the control schemes presented in Section III.

Control Scheme	Structural Comparison				
	Maximum time delay in samples	Total number of memory cells	Target application	Coupling between stationary axes	Improved stability domain
Conventional RC [13]	N	$N \times NoA^*$	1ϕ and 3ϕ	✗	Configurable
(Lu and Zhou, 2011) [6]	$2N/n$	$2N/n \times NoA^*$	1ϕ and 3ϕ	✗	✗
(Lu et al., 2014) [7]	$2N/n$	$3N/n \times NoA^*$	1ϕ and 3ϕ	✗	✗
(Neto et al., 2018) [8]	$2N/n$	$2N/n \times NoA^*$	1ϕ and 3ϕ	✗	✓
(Luo et al., 2016) [9]	N/n	$2N/n$	3ϕ	✓	✗
(Zimann et al., 2019) [10]	N/n	$2N/n$	3ϕ	✓	✓

*$NoA = 3$ for implementation using abc reference frame; $NoA = 2$ for implementation using $\alpha\beta$ reference frame.

F. Structural Comparison

The main structural characteristics of the CRC and the control schemes presented in Section III are shown in Table I. From this table, it is possible to realize that among all evaluated real RC-based schemes, those proposed by Lu et al. [6] and Neto et al. [8] require the smaller number of memory cells, the last one being more advantageous due to its improved stability domain (which results in a better trade-off between stability and performance). In addition, even though complex RC-based strategies are only applicable to three-phase control systems, they have shorter maximum delay in samples and lower total number of memory cells than real RC-based solutions. Zimann et al.'s proposal [10] also stands out for having improved stability domain.

It must be highlighted that, in order control plants of non-zero relative degree, FIR filters $Q(z)$ should be used in cascade with each generic delay element ($z^{-N/n}$) for all approached harmonic selective RC schemes. However, the number of delayed samples should be changed to $\left(\frac{N}{n} - \frac{L}{2}\right)$ to avoid the frequency shift of the controller's poles, as done for the CRC.

IV. EXPERIMENTAL RESULTS AND COMPARISON

A three-phase shunt APF is used to experimentally evaluate the control structures described in Section III. The diagram of the system used for the performance comparison is shown in Fig. 12. For control purposes, the APF output filter is modeled as a three-phase plant, being represented by the following transfer function:

$$G(s) = \frac{\vec{I}_f(s)}{\vec{D}(s)} = \frac{V_{dc}/L_f}{s + R_f/L_f}. \tag{15}$$

In this equation, $\vec{D}(s)$ is the space-vector control action and $\vec{I}_f(s)$ is the space-vector obtained from the APF output currents. The prototype is shown in Fig. 13 and its parameters are indicated in Table II.

Since the harmonic components of the load currents are in the family $H = \{6k \pm 1 \mid k \in \mathbb{N}\}$ (Fig. 10), they lead to the grid currents shown in Fig. 14. Therefore, the parameters used for implementing the real harmonic selective RC schemes are $m = 1$ and $n = 6$. On the other hand, since the harmonic components of the load currents space-vector are in the family

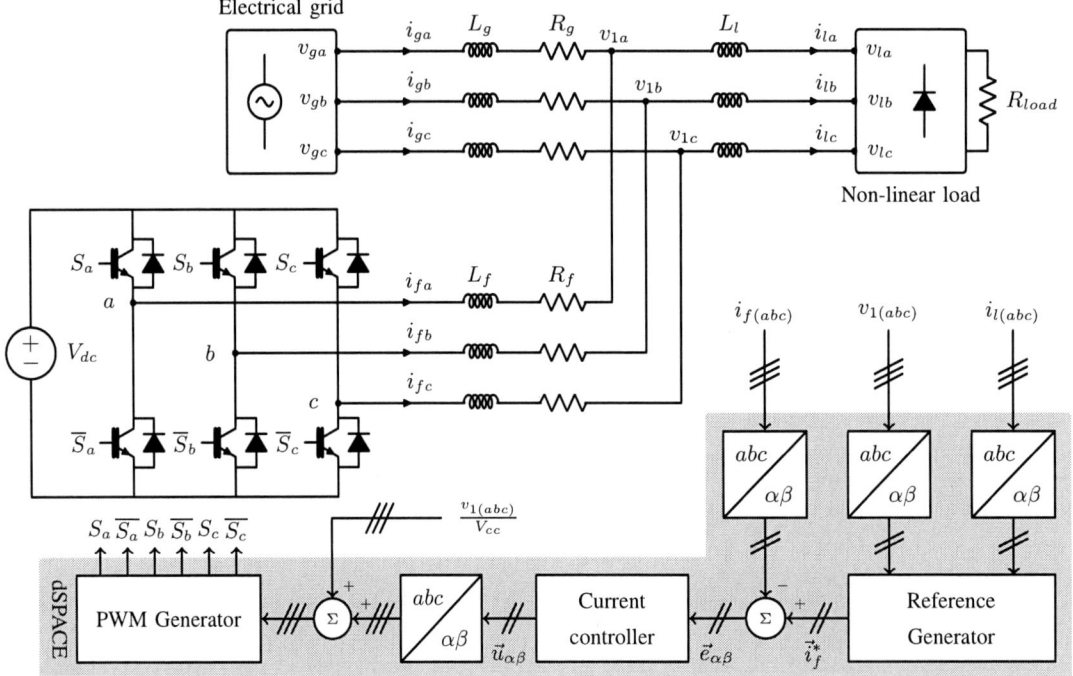

Fig. 12: Diagram of the system used for performance comparison of the control schemes presented in Section III.

TABLE II: Parameters of the experimental setup and the evaluated controllers.

Experimental Setup										RC-Based Controllers	
$V_{g(line)}$	L_g	R_g	L_l	L_f	R_f	R_{load}	V_{dc}	f_s *	f_g	N/n	$L/2$
380 V_{rms}	186.17 μH	31.7 mΩ	1.483 mH	2.563 mH	307.5 mΩ	48.4 Ω	600 V	17.28 kHz	60 Hz	48	3

*Sampling and switching frequency: $f_s/f_g = N$ must be integer and f_s must be multiple of 32 because of the PLL used for reference generation [20].

TABLE III: Summary of the performance comparison between the RC-based schemes presented in Section III.

Performance Comparison						
Control Scheme	RC gain	Proportional gain	0 dB gain crossover frequency	Sensitivity index [18]	VTHD of grid currents in steady-state operation [19]	Settling Time (5%)
$C_1(a=1,z)$ [13]	0.060	———	1.7 kHz	0.32	1.63 %	32.1 ms
$C_2(z)$ [6]	0.060		—————————— UNSTABLE ——————————			
$C_2(z)$ [6] $+P$	0.013	0.025	1.9 kHz	0.32	4.30 %	13.9 ms
$C_3(z)$ [7]	0.060		—————————— UNSTABLE ——————————			
$C_3(z)$ [7] $+P$	0.016	0.034	1.9 kHz	0.32	4.98 %	10.2 ms
$C_4(z)$ [8]	0.060		—————————— UNSTABLE ——————————			
$C_4(z)$ [8]	0.039	———	1.9 kHz	0.32	2.88 %	7.8 ms
$\vec{C}_5(z)$ [9]	0.060		—————————— UNSTABLE ——————————			
$\vec{C}_5(z)$ [9] $+P$	0.030	0.025	1.9 kHz	0.32	3.63 %	5.5 ms
$\vec{C}_6(z)$ [10]	0.060	———	1.9 kHz	0.32	2.65 %	4.2 ms

$H_s = \{6k+1 \mid k \in \mathbb{Z}\}$ (Fig. 11), $m = 1$ and $n = 6$ are also used for the complex schemes. In addition, due to the chosen sampling and grid frequencies (f_s and f_g in Table II), the number of samples per cycle of the fundamental period is $N = f_s/f_g = 288$. For the experiment, the grid voltages were approximately symmetrical and had low harmonic distortion.

Since $G_m(s) = G(s) \cdot K_{rc}$ has non-zero relative degree, FIR filters must be used to enlarge the stability domain of the evaluated controllers to make the RC system stable. Also, in order to provide a fair comparison, all evaluated controllers were designed using the method proposed in [17], so that they have similar 0 db gain crossover frequencies and sensitivity index [18] (Table III). From this method, the RCs gain were obtained (Table III) and the lead compensator $H_l(z)$, which was used in cascaded with the harmonic selective RC schemes to attenuate the computational delay effect [10], was defined.

Therefore, all evaluated strategies used FIR filters, of order $L = 6$ and cutoff frequency $f_c = 1.8$ kHz, whose transfer function is:

$$Q(z) = 0.0127 + 0.07715z^{-1} + 0.2415z^{-2} + 0.3372z^{-3} + 0.2415z^{-4} + 0.07715z^{-5} + 0.0127z^{-6}, \quad (16)$$

and lead compensator

$$H_l(z) = \frac{0.6526z - 0.4301}{z - 0.08271}. \quad (17)$$

With respect to the target device used to implement the controllers, all results were obtained using a dSPACE platform (as indicated in Fig. 13), model DS1005, featuring a processor running at 1 GHz.

The controllers presented in Section III are compared below through two indicators: the vector total harmonic distortion (VTHD) [19], which can be used to evaluate the system performance in steady-state operation; and the settlement time, which can be used to evaluate the transient response. It is important to note that the VTHD of the grid currents before APF operation is 27.12 %, as indicated in Fig. 14.

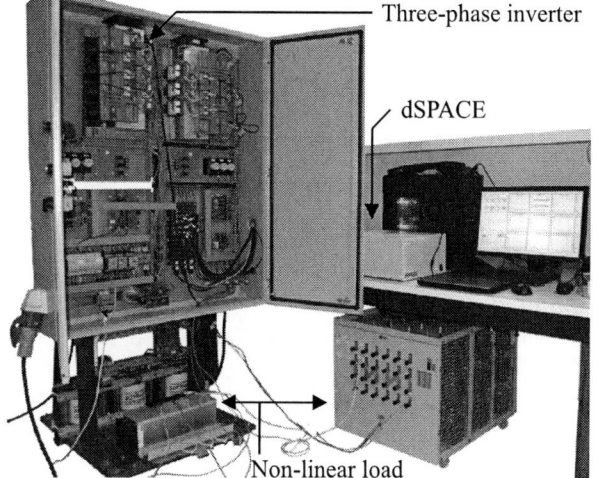

Fig. 13: Experimental set-up.

A. Performance Comparison

The tuning of the RC gains was firstly done for the CRC ($C_1(a = 1, z)$ [13]). As shown in Table III, the selection of $K_{rc} = 0.060$ results in reasonable stability and performance characteristics. When applying this same gain to the evaluated harmonic selective RC-schemes, only $C_5(z)$ [9] and $\vec{C}_6(z)$ [10] are stable. However, even though it is stable, the solution

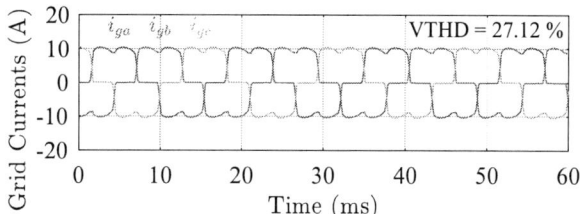

Fig. 14: Grid currents before APF operation.

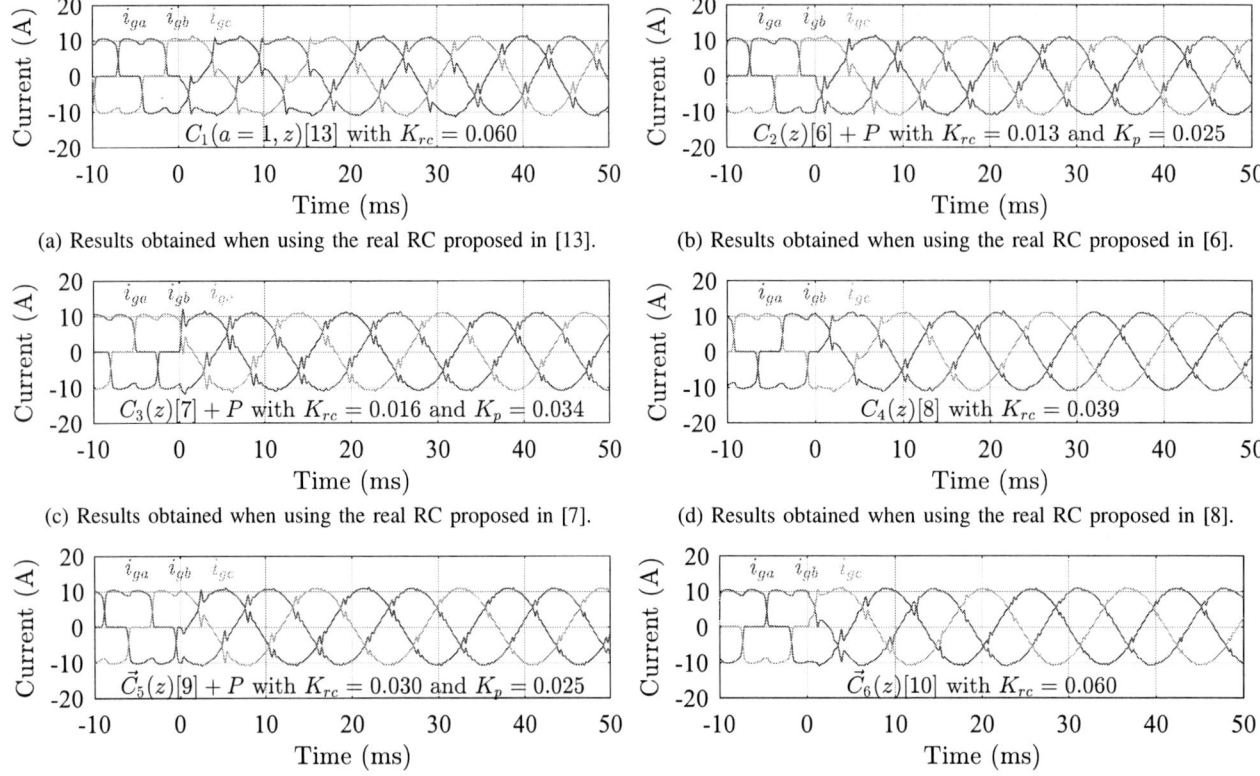

(a) Results obtained when using the real RC proposed in [13].

(b) Results obtained when using the real RC proposed in [6].

(c) Results obtained when using the real RC proposed in [7].

(d) Results obtained when using the real RC proposed in [8].

(e) Results obtained when using the complex RC proposed in [9].

(f) Results obtained when using the complex RC proposed in [10].

Fig. 15: Grid currents during control startup. Experimental results obtained using a dSPACE plataform.

proposed in [9] did not allow the control of high-frequency harmonic components when using $K_{rc} = 0.060$.

Then, the RC gain of all unstable solutions was reduced until they had sensitivity index and 0 dB gain crossover frequency comparable to $C_1(a = 1, z)$ [13] and $\vec{C}_6(z)$ [10]. By doing so, only $C_4(z)$ [8] managed to reach such conditions. A proportional action (P) was added to the other controllers so that they presented similar stability and bandwidth characteristics.

All results are summarized in Table III while the grid currents during control startup are shown in Fig. 15. It is observed that the complex $nk + m$ RC based on a PRC with $a = 1$ ($\vec{C}_6(z)$ [10]) results in better transient and steady-state characteristics than the real $nk \pm m$ RC based on PRCs with $a = 1$ ($C_4(z)$ [8]), both with same sensitivity index and 0 db gain crossover frequency. This effect is also observed for the control schemes based on PRCs with $a = 0.5$ ($C_2(z)$ and $\vec{C}_5(z)$). This happens due to the fact that complex RC-based solutions allow the use of higher RC gains. On the other hand, it should be noted that since the reference currents may have harmonics that are not controlled by the selected harmonic selective RC scheme, which may happen in unbalanced systems, the CRC presented the lowest VTHD because it controls all the harmonic components, however having the worst response time. All VTHD data presented in Table III were computed few seconds after the control startup.

Finally, it can be seen that Zimman et al.'s proposal [10]

provides the best transient and steady-state characteristics among all evaluated harmonic selective RC schemes, while Neto et al.'s proposal [8] stands out among the real ones. This happens because both are based on PRCs with $a = 1$.

V. CONCLUSION

In this paper, several complex and real harmonic selective RC-based schemes are compared in terms of structural and performance characteristics. After theoretical and experimental evaluation, it can be observed that complex RC-based controllers present better steady-state, transient and memory requirement aspects than the real ones based on the same PRCs, however, they can not be used in single-phase applications.

REFERENCES

[1] M. Aredes, "Active power line conditioners," Ph.D. dissertation, Technische Universitat Berlin, 1996.

[2] S. Bhattacharya, T. M. Frank, D. M. Divan, and B. Banerjee, "Parallel active filter system implementation and design issues for utility interface of adjustable speed drive systems," in *31st IEEE IAS Annual Meeting*, vol. 2, Oct. 1996, pp. 1032–1039.

[3] M. Bojrup, P. Karlsson, M. Alakla, and L. Gertmar, "Multiple rotating integrator controller for active filters," in *EPE Conf. Proc.* EPE Association, 1999.

[4] X. Yuan, W. Merk, H. Stemmler, and J. Allmeling, "Stationary-frame generalized integrators for current control of active power filters with zero steady-state error for current harmonics of concern under unbalanced and distorted operating conditions," *IEEE Trans. Ind. App.*, vol. 38, no. 2, pp. 523–532, Mar. 2002.

978-1-7281-4181-7/19 $31.00 © 2019 IEEE

[5] L. R. Limongi, R. Bojoi, G. Griva, and A. Tenconi, "Digital current-control schemes," *IEEE Ind. Electron. Magazine*, vol. 3, no. 1, pp. 20–31, Mar. 2009.

[6] W. Lu and K. Zhou, "A novel repetitive controller for nk $\pm$ m order harmonics compensation," in *Proc. 30th Chin. Control Conf.*, Jul. 2011, pp. 2480–2484.

[7] W. Lu, K. Zhou, D. Wang, and M. Cheng, "A generic digital nk $\pm$ m-order harmonic repetitive control scheme for pwm converters," *IEEE Trans. Ind. Electron.*, vol. 61, no. 3, pp. 1516–1527, Mar. 2014.

[8] R. C. Neto, H. E. P. de Souza, C. Rech, and F. A. S. Neves, "A nk $\pm$ m-order harmonic repetitive control scheme with improved stability characteristics," in *27th IEEE ISIE*, Jun. 2018, pp. 465–470.

[9] Z. Luo, M. Su, J. Yang, Y. Sun, X. Hou, and J. M. Guerrero, "A repetitive control scheme aimed at compensating the 6k + 1 harmonics for a three-phase hybrid active filter," *Energies*, vol. 9, no. 10, p. 787, Set. 2016.

[10] F. J. Zimann, R. C. Neto, F. A. S. Neves, H. E. P. de Souza, A. L. Batschauer, and C. Rech, "A complex repetitive controller based on the generalized delayed signal cancelation method," *IEEE Trans. Ind. Electron.*, vol. 66, no. 4, pp. 2857–2867, Apr. 2019.

[11] T. Inoue, M. Nakano, T. Kubo, S. Matsumoto, and H. Baba, "High accuracy control of a proton synchrotron magnet power supply," *IFAC Proceedings Volumes*, vol. 14, no. 2, pp. 3137 – 3142, Aug. 1981.

[12] B. Francis and W. Wonham, "The internal model principle for linear multivariable regulators," *Appl. Math. Optim.*, vol. 2, no. 2, pp. 170–194, Jun. 1975.

[13] S. Hara, Y. Yamamoto, T. Omata, and M. Nakano, "Repetitive control system: a new type servo system for periodic exogenous signals," *IEEE Trans. Automatic Control*, vol. 33, no. 7, pp. 659–668, Jul. 1988.

[14] T. Inoue, "Practical repetitive control system design," in *29th IEEE CDC*, vol. 3, Dez. 1990, pp. 1673–1678.

[15] A. W. Osburn and M. A. Franchek, "Designing robust repetitive controllers," *Journal of Dynamic Systems, Measurement, and Control*, vol. 126, no. 4, pp. 865–872, Mar. 2005.

[16] G. Escobar, P. R. Martinez, J. Leyva-Ramos, and P. Mattavelli, "A negative feedback repetitive control scheme for harmonic compensation," *IEEE Trans. Ind. Electron.*, vol. 53, no. 4, pp. 1383–1386, Jun. 2006.

[17] R. C. Neto, F. A. S. Neves, H. E. P. de Souza, F. J. Zimann, and A. L. Batschauer, "Design of repetitive controllers through sensitivity function," in *27th IEEE ISIE*, Jun. 2018, pp. 495–501.

[18] A. G. Yepes, F. D. Freijedo, . Lopez, and J. Doval-Gandoy, "Analysis and design of resonant current controllers for voltage-source converters by means of nyquist diagrams and sensitivity function," *IEEE Trans. Ind. Electron.*, vol. 58, no. 11, pp. 5231–5250, Nov. 2011.

[19] F. A. S. Neves, H. E. P. de Souza, M. C. Cavalcanti, F. Bradaschia, and E. J. Bueno, "Digital filters for fast harmonic sequence component separation of unbalanced and distorted three-phase signals," *IEEE Trans. Ind. Electron.*, vol. 59, no. 10, pp. 3847–3859, 2012.

[20] F. A. S. Neves, M. C. Cavalcanti, H. E. P. de Souza, F. Bradaschia, E. J. Bueno, and M. Rizo, "A generalized delayed signal cancellation method for detecting fundamental-frequency positive-sequence three-phase signals," *IEEE Trans. Power Del.*, vol. 25, no. 3, pp. 1816–1825, Jul. 2010.

AUTHOR INDEX

Abrantes-Ferreira, Armando J. G...........................1010
Afonso, Leonardo Carlos.............................445
Aguiar, Maria Júlia Rosa 826
Albuquerque, Lorena 838
Albuquerque, Lorena Lorraine Oliveira...............1040
Aller, José Manuel 1289
Almeida, Andrei De O. 25
Almeida, Antonio D. D...........................868
Almeida, Felipe A. F. 1469
Almeida, Pedro M. 19, 585, 1360
Almeida, Pedro S...................730, 1052, 1118
Almeida, Pedro Santos...................370, 579
Alonso, Augusto Matheus Dos Santos 269, 445
Altuna, José A. Torrico 952
Alves, Fábio 880
Alves, Marcos Henrique Da Silva 160
Alves, Max Dannyel De Carvalho.....................1447
Amaral, Fernando Venâncio.......................1004
Amorim, W. C. S. 790
Andersen, Romero 838
Andersen, Romero Leandro.........................1040
Andrade, Allan V. S. 621
Andrade, António M. S. S........................856
Andrade, Lucas Mendonça 1082
Andrea, Cristiano Quevedo.......................998
Antunes, Fernando.............................1124
Antunes, Fernando L. M. 1118, 1331
Araújo, Leandro R.......................19, 1360
Araújo, Lucas 435
Araujo, Lucas S............................597
Araujo, Renato G. 1307
Archetti, João A. G. 742
Aredes, Mauricio............................184
Aredes, Maurício............................880
Arenas, Luís De Oro 1136, 1211, 1487
Attuati, Gabriel 1064
Ayala, M. 1441
Azcue, José L.........................347, 1166
Azeredo, Lucas F. S..........................898
Azevedo, Gustavo M. S..........................1390
Bacheti, Gabriel G. 772
Balbino, Anderson José 886
Barbi, Ivo................................917
Barbosa, Paula Stael S........................1028
Barbosa, Pedro G. 19, 25, 263, 585, 910, 1360
Barbosa, Pedro Gomes 13, 423
Barbosa, Samanta Gadelha......................591, 680
Barbosa, Victor E. S.........................472, 993
Barcelos, R. P..............................969
Barcelos, Silvangela L........................400
Barcelos, Silvangela L. S. L.....................1034
Barreto, Luiz Henrique Silva Colado.................680
Barros, Luciano Sales 748
Barros, Tárcio André Dos S. 686

Bartsch, Arthur G. 923, 1325, 1384
Bartsch, Camila R. S. 923
Batista, Edson Antonio 998
Batista, Oureste E.......................898, 1378
Batista, Oureste Elias 1
Batschauer, A. L. 220
Batschauer, Alessandro L........................724
Batschauer, Alessandro Luiz......................208
Bekoski, Kleber Chan 245
Belchior, Fernando Nunes 531
Bellinaso, Lucas Vizzotto....................406, 940
Bento, Aluísio Alves De Melo 1396, 1458
Bernardeli, Victor Regis 305
Bertoldi, B. 417
Bessa, Isaías V..............................172
Bessa, Iury V. 172, 1076
Bettiol, Arlan Luiz 917
Bisogno, Fábio E............................233
Blaabjerg, Frede 1
Boaventura, W. C. 615
Bonaldi, Erik Leandro 85
Bonaldo, Jakson Paulo 269
Bonifacio, Joao.............................55
Borchardt, Mauricio 376
Borges, Victor Luiz Flor 917
Borges-Da-Silva, Luiz Eduardo 85
Borin, Lucas C. 202, 519
Bradaschia, F. 220, 716
Braga, Alexandre Viana..........................49
Braga, Henrique Antônio Carvalho...............370, 579
Braga, Mateus Freitas 423
Branco, Carlos Gustavo C. 1307
Brandao, Danilo I. 364, 466, 597
Brandao, Danilo Iglesias 445
Brasil, Thiago Americano Do 934
Bravo, P.................................1402
Bressan, Marcos Vinicius 724
Brinker, Tobias.............................975
Brito, Moacyr..............................1195
Brito, Moacyr A. G...........................501
Bueno, Alexandre Galvão 311
Bueno, Mateus De F. 376
Busarello, Tiago Davi Curi 125
Buso, Simone..............................143
Buticchi, Giampaolo..........................257
Cabrera, Ana 507
Cacau, Ronny Glauber De Almeida....................886
Cagnini, Paulo R............................993
Caicedo, Jorge 880
Caldeira, Carolina A. 820, 868
Caldognetto, Tommaso 143
Calil, Henrique A. M. 92
Caliman, João Olímpio 844
Callegari, João M. S. 478

AUTHOR INDEX

Cambero, Eduardo Vicente Valdés 627, 1349
Campos, Rafael De Farias...299
Campos, Rafael Espino...1130
Capovilla, Carlos E. 952
Capovilla, Carlos Eduardo................................627, 1349
Carati, Emerson G. 472, 993, 1148
Carati, Emerson Giovani.......................................1142, 1420
Cárdenas, Roberto 61
Cardoso, Rafael 472, 993, 1142, 1148, 1223, 1420
Cardoso, Ronnan De B. 1295
Carralero, Leandro L. O. 778, 964
Carvalho, Edivan Laercio 1142, 1420
Carvalho, Henrique T. M. 288
Carvalho, Itaiara F. 1372
Casella, Ivan Roberto Santana.......................627, 1349
Cavalca, Mariana S. M. 1325
Cavalcante, Levy R. 958
Cavalcanti, Dayse M. 1325
Cavalcanti, Marcelo C. 1390
Cavilha, Pedro P. 692
Chacon, Igor V. 668
Chub, Andrii 1178, 1493
Ciarnoscki, Pábulo F. 1172
Coelho, Ernane A. A. 98, 288, 1253
Coelho, Roberto F. 251
Coelho, Roberto Francisco 451, 639, 651
Colque, Juan C. 347
Conceição, Claudio Alvares 1004
Conde, Eliomar R.D..73
Conde, Manoel 411
Cordero, Raymundo 1195, 1453
Cornelius, Richard G...461, 543
Corrêa, Douglas Rosa 281
Corrêa, Henrique Pires 1414
Correa, Maurício B. R. 1106, 1372
Corrêa, Rafael Di Lorenzo 49
Côrtes, Hyghor Miranda 1366
Cortizo, P. C. 178
Costa, Fabiano F. 778, 964
Costa, Jean Patric 1142
Costa, Louelson A. 1106
Costa, Paulo Júnior Silva 1301
Costabeber, Alessandro 257
Cupertino, A. F..573, 615, 790
Cupertino, Allan F. 466, 478, 850
Da Câmara, Raphael A..904
Da Costa, André E. L...1295
Da Costa, Hugo Álisson Alves..................................1265
Da Costa, Jean P. 472, 1148
Da Costa, Jean Patric 993, 1420
Da Fonseca, Brayan Sobral 946
Da Luz, Caio Meira Amaral..239
Da Paz, Gilielson F. 820, 1100, 1475
Da Paz, Humberto Pereira 627, 1349

Da Rocha, Antonia F......................................1118, 1331
Da Rocha, Lorrana F. 1058
Da Rosa, Murilo Brunel 1217
Da Silva, Alexandre G. F. 1100, 1475
Da Silva, Bruno Alves Sousa 680
Da Silva, Carlos Henrique 531
Da Silva, Edison R. C. 1295
Da Silva, Edison Roberto C. 1022, 1154
Da Silva, Eduardo Luiz Santos....................................382
Da Silva, Gabriel M. 1172
Da Silva, Guilherme S. 461, 543
Da Silva, João Lucas 674
Da Silva, Luiz Eduardo Borges 531
Da Silva, Mauricio M. 856
Da Silva, Newton 125, 1343
Da Silva, Rogerio Luiz 917
Da Silva, Sérgio Augusto Oliveira.................484, 1301
Da Silva, Simplicio A. 1475
Da Silva, Vinícius Santana........................627, 1349
Da Silva, Werbet Luiz Almeida.................................1283
Da Silveira, D. B. 615
Da Siveira, Augusto W. Fleury Veloso 981
Dacol, Rodrigo Patrício..341
Dai, Hang 67
Dalmolin, Thiago Brezolin..406
Damasceno, Debora Pereira 591
Dantas, Marcos...838
De Aguiar, Manoel L...108
De Aguiar, Manuel Luis..892
De Albuquerque, Vinicius M. 730, 1052
De Almeida, Antonio D. D. 820
De Almeida, Bruno Ricardo 591, 680
De Almeida, Pedro M.......................25, 263, 910, 1058
De Almeida, Rogério Gaspar.......................................987
De Almeida, Thales E. P. 149
De Alvarenga, Bernardo Pinheiro.............................1070
De Andrade, Darizon Alves 305
De Andrade, Guilherme Giglio 934
De Andrade, Jessika Melo 451
De Andrade, Khristian M. 149
De Araújo, Humberto Xavier 627, 1349
De Araújo, Rafael Magalhães Nóbrega 1265
De Barros, R. C. 615
De Bessa, Isaías Valente..190
De Bessa, Iury Valente..190
De Brito, Moacyr A. G. 513, 832
De Camargo, Robinson Figueiredo.............................1064
De Campos, Luis F. F. 923
De Carvalho, Gabriel Ubirajara 1223
De Carvalho, Marenice M. 1076
De Castro, Allan G. 108
De Castro, Allan Gregori 335, 892
De Castro, Marcelo 910
De Figueiredo, Hugo F. M. 814, 874

AUTHOR INDEX

De França, Pedro Henrique Thiesen 633
De Freitas, Isaac S. 1100, 1475
De Freitas, Isaac Soares 1040
De Freitas, Luiz C. G. 288, 1253
De Freitas, Marcos Antonio Arantes 305
De Freitas, Tiara Rodrigues Smarssaro 844
De Lacerda, Macklyster Lãnucy Scherre Stofel 455
De Lima, Gustavo B. 288
De Lucena, Lucas Fabrício M. 1022
De Matos, José G. 555, 1229
De Matos, José Gomes 275, 400
De Medeiros, Luiz Otávio Campos 49
De Morais, Aniel Silva 79, 281
De Morais, Fernanda Carnielutti 929
De Novaes, Yales R. 376
De Novaes, Yales Rômulo 214, 784
De Oliveira, André L. P. 478
De Oliveira, Azauri A. 484
De Oliveira, Carlos M. R. 108, 335
De Oliveira, Carlos Matheus R. 892
De Oliveira, Cássio Alves 981
De Oliveira, Fellipe André Lucena 1040
De Oliveira, Janaína G. 567, 742, 910, 1028
De Oliveira, José 299, 923, 1384
De Oliveira, Júlio Cesar Lopes 495
De Oliveira, Leonardo W. 567, 742
De Oliveira, Pedro Andrade...................................85
De Oliveira, Raphael A. Bispo 1366
De Oliveira, Rodrigues 1046
De Paula, Geyverson T. 149, 335
De Sá, Marcos Victor Dantas 1040
De Seixas, Falcondes José Mendes......1136, 1211, 1487
De Sousa, Gean J. M. 802
De Sousa, R. O. 790
De Souza, H. E. P. 716
De Souza, Igor D. N..263, 910
De Souza, Marina Hassen.......................................1112
De Souza, Ramon Henriques.................................1313
De Toledo, Henrique 1166
De Vasconcelos, Carlos Henrique Silva.................1271
Delamare, Guillaume 1408
Denardin, Gustavo W. 814, 874
Denardin, Gustavo Weber.................................245, 1223
Dias, Bruno H. 567
Dias, Mateus P. 766
Dias, Robson F. Da S. 1010
Dias, Robson F. S. 1034
Dicler, Felipe Novaes Francis 184
Diesel, José Adriano Damacena 633
Dill, Gustavo Kaefer 490
Dohler, Jessica S. 1028, 1058
Dos Reis, Mauro Sandro...934
Dos Santos, Ângelo Marcílio M..............................862
Dos Santos, Cássia C. C. 1148

Dos Santos, Kristian Pessoa............................561, 1447
Dos Santos, Rafael...1435
Dos Santos, Rodrigo L. 1118, 1331
Dos Santos, Stefan T. C. A. 108
Dos Santos, Tadeu F..668
Dos Santos, Walbermark Marques........455, 844, 1426
Dragicevic, T. 1441
Drexhage, Paul..1277
Duarte, Jorge L. 43
Duarte, Samuel N. 19, 585, 1360
Duarte, Samuel Neves 423
Eckstein, Rafael H. 1431
El Kattel, Menaouar Berrehil....................1217, 1354
Enomoto, Kelly 411
Escobar, Gerardo 376
Espindula, Carla J..898
Estrabis, Thyago 1453
Estrabis, Thyago Vasconcelos 998
Ewerling, Marcos Vinícius Mosconi 358
Fabricio, Edgard L. L. 868
Faceroli, Silvana T..704
Faistel, Tiago M. K. 856
Fajardo, Marco..1289
Fardin, Jussara Farias 1337
Farias, João. V. M. 850
Fauth, Leon 975
Feltrin, Valesca Bettim 406
Feretti, Paulo Henrique 166
Fernandes, Andressa Da S. 1331
Fernandes, Darlan A. 1295
Fernandes, Darlan Alexandria 987
Fernandes, Filipe 1384
Fernandes, Marcelo C. 263
Fernandes, Marcelo De C. 730
Fernandes, Nicolas T. D. 597
Fernandes, Samir W. 609
Fernández-Ramírez, Luis M...................................904
Ferreira, Andre 1028
Ferreira, Andre A..1058
Ferreira, André Augusto 826
Ferreira, Julio Cesar 934
Ferreira, Kaique 1426
Ferreira, Myrlena R. M. 1229
Ferreira, Rodrigo A. F. 704, 1241
Ferst, Matheus K..814, 874
Fiamoncini, Lucas 651
Filho, A. J. Sguarezi 73
Filho, Alfeu J. Sguarezi 778, 1183
Filho, Antonio Venâncio De M. Lacerda 1295
Filho, Braz De J. Cardoso............................160, 1004
Filho, Braz J. Cardoso 597
Filho, Ernesto Ruppert 686
Filho, Hermínio Miguel Oliveira 561
Filho, João Inácio Da Silva 1366

AUTHOR INDEX

Filho, José A. Olimpio 1435
Filho, José De Arimatéia Olímpio.........................269
Filho, Joselino Santana....................................85
Filho, Marcos José De Moraes................................981
Flores, Mendes...1046
Fogli, Gabriel A.............................263, 364, 910
Fonseca, Jean M. L..1307
Font, Carlos Henrique Illa 358
Fontes, Guillaume 1408
Foster, João G. L..131
Foster, Joao Gabriel Luppi 85
França, Bruno 880
Fraytag, Jeferson 525
Freitas, Alisson 1124
Freitas, Isaac 838
Freitas, Isaac S. 868
Freitas, Luiz C. G..98
Friebe, Jens...975
Furlan, André..1408
Furtado, Pablo C. De S. 13
Gabbi, Thieli..698
Gajo, Clovis N. L...1277
Galea, Michael...257
Galotto, Luigi...832
Galottojunior, Luigi..1195
Gama, Felipe O. S...668
García, C..1402
García, Raymundo Cordero....................................998
García-Triviño, Pablo 904
Gehrke, Camila S..820
Gil, Eric...1435
Gobbato, Cassio 245
Godoy, Ruben Barros 832
Gomes, Hugo M. C..778
Gomes, Lucas Do N...1010
Gomes, Luciano Coutinho.............................305, 981
Gomes, Marcos Paulo Brito 1112, 1313
Gonçalves, Flávio A. S......................................1435
Gonçalves, Flávio Alessandro Serrão..........1343, 1469
Gonzatti, Robson B. 131
Gonzatti, Robson Bauwelz............................85, 531
Grabovski, E. F. C......................................417, 1016
Grassi, Márcio Afonso Soleira 998
Gregor, R..1441
Griffiths, Alison..227
Grigoletto, Felipe Bovolini 929
Gründling, Hilton A..549
Gruner, Víctor Ferreira 651
Guamán, Alex..507
Guazzelli, Paulo R. U..............................108, 335
Guazzelli, Paulo Roberto U. 892
Guerra, Felipe 1469
Guerreiro, Joel F..766
Guerreiro, Joel Filipe..........................736, 754

Guerrero, Josep M. 686
Guillardi, Hildo 766
Guimarães, Alexandre Magnus F. 1265
Guimarães, Bruno P. B......................................531
Guo, Yanjie 104
Ha, Hengxu 227
Harikumaran, Jayakrishnan 257
Hartmann, Lucas V. 820
Hayashi, Augusto.. 1195, 1453
Hayashi, Pedro 411
He, Zhengyou 441
Heerdt, Joselito Anastácio 341, 633
Heldwein, M. L. 417, 1016
Heldwein, Marcelo L...................692, 802, 1408
Heldwein, Marcelo Lobo...............525, 710, 1259
Henning, Luiz Fernando 495
Herrera, Felipe 61
Hoch, Henrique J. 856
Honório, Dalton De Araújo 680
Huang, Shili 1160
Husev, Oleksandr 1481
Idalgo, Gabriel F. 952
Ikhide, Monday 227
Imai, Fernando 214
Inácio, Cleber Onofre 1004
Jacobina, Cursino Brandão 1022
Jacoboski, Marcos José 1259
Jacomini, Rogério Vani....................................1183
Jahns, Thomas M. 67
Junior, Ademir Da S. T...................................... 910
Júnior, Ademir S. T..263
Junior, Dalmo C. Silva 1058
Júnior, Demercil De Souza Oliveira 561, 680
Junior, Emerson Abreu Bastos 239
Junior, Florindo A. C. 1076
Junior, Florindo A. C. Ayres 172, 190
Junior, Guilherme Penha Da Silva 748
Júnior, Hildo Guillardi 736, 754
Junior, Ildenor Davi S.....................................862
Junior, Josemar Alves Dos Santos 981
Junior, Lourenço Matakas 1189
Junior, Luigi G. 513
Junior, Nery De Oliveira....................................49
Júnior, Paulo R. M. 850
Junior, Sergio L. B. 692
Junior, Sérgio Luiz Sambugari 125
Kawahara, Yoshihiro 137
Kennel, Ralph 55
Khera, Fatma A. 196
Kirsten, André Luís 382, 525
Klumpner, Christian 196
Koch, Gustavo G. 202, 519
Koleff, Lucas............................... 411, 429, 435
Komatsu, Wilson 411, 429, 435, 1189

AUTHOR INDEX

Korkh, Oleksandr 1178
Kouro, Samir....................................1178
Kraemer, Rodrigo A. S.............................472
Kumar, Dinesh 1160
Kutt, Lauri....................................1481
Lafay, Jean M. S. 993
La-Gatta, Filipe A. 1241
Lago, Lucas F. R. 704
Lai, Jih-Sheng 1493
Lambert, Gustavo 784
Lambert-Torres, Germano 85, 131
Lange, A. B. 417
Lange, André De Bastiani 1259
Laurindo, Bruno 880
Lazzari, Thiago 698
Lazzarin, T. B..................................969
Lazzarin, Telles B. 388, 394, 796, 1431
Lazzarin, Telles Brunelli 358, 451, 639, 886
Leal, Daniel Franco..............................1313
Lee, Woongkul 67
Leguizamón, Daniel M. Barrera 294
Leite, Valter J. S...............................609
Lemes, Lucas José...............................305
Lessa, Tofoli..................................1046
Li, Fang......................................353
Li, Shufan 353
Lima, Antonio Marcus Nogueira 1463
Lima, Francisco Kleber A. 1307
Lima, Gustavo B. 98
Lima, Mateus L. 742
Limongi, Leonardo R..............................1390
Lin, Siqi.....................................975
Lino, Fernando.............................73, 1183
Liu, Qing.....................................143
Liu, Yeran....................................441
Liu, Zhimeng 104
Lopes, Juliano De Pelegrini 245
López, Diego A. Bautista 294
López, Fabián R. Jiménez 294
Lucas, Kevin E. 31, 172, 190, 1076
Luiz, Alex-Sander Amável 1112, 1206, 1313
Lumertz, Mateus M...............................335
Lumertz, Mateus Moro 892
Macedo, Elienai O...............................1034
Macedo, Tatiana Saviato 455
Machado, Sebastián De J. Manrique....................484
Madawala, Udaya K. 441, 663
Mai, Leonardo S. 796
Mai, Ruikun...................................441
Maia, Amanda C. 461, 543
Maqueda, E. 1441
Marafão, Fernando P..............................1435
Marafão, Fernando Pinhabel 269, 445
Marcilio, Wagner 155

Marin, Caroline 1378
Marques, Luciana 429
Marques, Rúbio Campos 579
Marques, Vanessa Da Costa 987
Marra, Enes Goncalves...........................1070
Martins, Débora De Souza 1337
Martins, Denizar C. 251
Martins, Denizar Cruz 455, 651
Martins, Eduardo 155
Martins, Mário L. Da S............................856
Matakas, Lourenço.................411, 429, 435, 1195
Matias, Rafael Rocha 1319, 1447
Mattavelli, Paolo 766
Mattos, Everson 519
Mayer, Robson...............................1217, 1354
Medeiros, Lucas Taylan P. 862, 958
Medeiros, Renan L. P..................31, 172, 190, 1076
Medina, Augusto César Rueda........................1426
Melo, Fernando C. 288, 1253
Melo, Victor F. M. B. 1100, 1475
Mendes, Mariana A...........................898, 1378
Mendes, Mariana Altoé...............................1
Mendes, Victor F. 364, 466, 478
Mendes, Victor Flores 537
Mendonça, D. C. 573, 790
Mendonça, Gabriel A..............................850
Mendonça, Lucas S. 233
Meneghetti, Luiz H...............................993
Meneghetti, Luiz Henrique......................1142, 1420
Mengatto, Alisson 633
Mertens, Thomas M. 37
Mesquita, Daniel De Bastos 1130
Meynard, Thierry...............................1408
Mezaroba, M. 220
Mezaroba, Marcello 208, 603
Michels, Leandro............................406, 940
Miranda, Luis Colque 710
Mollica, Denis 85, 131
Montagner, Vinícius F. 202, 519, 549
Montecinos, Werner Jara 710
Monteiro, Amanda Thayla S 1319
Monteiro, Amanda Thayla Silva 1447
Monteiro, Felipe Alexandre 1453
Monteiro, José R. B. A. 108, 335, 484
Monteiro, Jose Roberto B. A. 892
Monteiro, Marina V. C. 1028
Moraes, Caio G. Da S. 388, 394, 692
Moraes, Cassiano F. 1148
Moraes, Cassiano Ferro 245, 1223
Morais, Douglas Carvalho...............1136, 1211, 1487
Morais, L. M. F. 178, 790
Moreira, Adson B. 958
Moreira, Adson Bezerra 862
Moreira, Ana Carolina 495

AUTHOR INDEX

Moreira, Hugo Soeiro 1130
Moreira, Marcos 155
Morentin, Alvaro 1408
Mota, Denisia De V. 958
Mussa, S. A. 1016
Mussa, Samir A. 796
Mussa, Samir Ahmad 710
Musse, Bernardo F. 742
Nadal, Zeno L. I. 993
Nadal, Zeno Luiz Iensen 1142, 1420
Naidon, Thiago C. 233
Narusue, Yoshiaki 137
Nascimento, Felipe C. Do 1331
Nascimento, Saulo O. 730
Neira, S. 1402
Neres, Fábio De P. 1118
Neto, Antônio O. Costa 98
Neto, Augusto Nery De Lima 317
Neto, João Amin Moor 934
Neto, João T. De Carvalho 1265
Neto, Jose Antonio Dos Santos 1319
Neto, Pedro Jose Dos Santos 686
Neto, R. C. 716
Neves, F. A. S. 220, 716
Neves, Francisco De Assis Dos Santos 208
Neves, Marcello 880
Neves, Vitor G. 772
Nicolai, Ulrich 1235
Nicolini, André Miguel 1366
Nied, Ademir 299, 923, 1325, 1384
Nolasco, Rafael César 537
Norambuena, Margarita 929
Nunes, Evandro Ailson De Freitas 1283
Ogoulola, Christel Enock Ghislain 49
Oliveira, Alexandre Cunha 1463
Oliveira, Demercil De S. 591, 904
Oliveira, Erik C. 1390
Oliveira, Fellipe 838
Oliveira, Hércules A. 555
Oliveira, Hércules Araújo 275
Oliveira, Isabel R. H. 609
Oliveira, Janaina G. 730, 1052, 1058
Oliveira, Matheus H. M. Zanchetta 1004
Oliveira, Sérgio Vidal Garcia 1217, 1354
Oliveira, Tatiane Martins 79, 239
Oliveira, Yago F. 1241
Omine, Leandro T. 501
Onofre, João 1195, 1453
Orige, Mateus C. 376
Oshiro, Marcos Roberto 832
Osório, Caio R. D. 202, 519, 549
Ota, João I. Y. 766
Ota, João Inácio Yutaka 736
Pacas, Mario 507

Pacheco, Antonio L. S. 692
Pacher, J. 1441
Pagano, Daniel J. 31, 802, 1172
Pagano, Daniel Juan 382
Paredes, Carlos 710
Paredes, Helmo Kelis Morales 269
Paredes, Marina G. S. P. 1201
Parreiras, Thiago Morais 160
Pastuch, Luiz Fernando Martins 639
Pelizari, Ademir 952
Pellini, Eduardo 411, 429, 435
Pereda, J. 1402
Pereira, Dênis De Castro 370, 579
Pereira, H. A. 573, 615, 790
Pereira, Heverton A. 466, 478, 850
Pereira, Luana K. Melgaço 674
Pereira, Rondineli R. 131, 531
Pereira, Rondineli Rodrigues 85
Pereira, Thiago A. 251
Pessoa, Guilherme A. P. De C. A. 668
Pessoa, Guilherme Afonso Pillon De C. A. 1283
Piccoli, Daniel Cesar 495
Pimenta, Marcio Annibal 1189
Pimentel, Sergio Pires 1070, 1481
Pineda, C. 1402
Pinheiro, Guilherme G. 531
Pinheiro, Humberto 929
Pinheiro, Lilian V. 742
Pinheiro, Lucas De Paula Assunção 555
Pinheiro, Ricardo Ferreira 1283
Pinheiro, Vinícius Marcos 981
Piontkewicz, Rodrigo Jose 495
Pomilio, José A. 766
Pomilio, José Antenor 311, 317, 645,
657, 736, 754, 1201, 1247
Pompermayer, Daniel C. 1378
Possamai, Carlos Eduardo 917
Possamai, Maicon Douglas 1354
Pozo, Isaac 507
Pozo, Marcelo 507
Pozo, Nataly 507
Praça, Paulo 1124
Praça, Paulo P. 904
Praça, Paulo Peixoto 561
Prado, Ricardo Alves Do 1235
Puma, J. L. Azcue 73
Qiu, Hao 137
Queiroz, Eliabe Duarte 645, 657
Queiroz, Fernando 1124
Queiroz, Luann G. O. 1378
Quizhpi, Flavio 1289
Rabelo, Everton Bernard Figueiredo 370
Rael, Victor 429
Ramos, Gabriel Vilkn 1112, 1313
Ramos, Marcos Leonardo 49

AUTHOR INDEX

Raposo, Rafael Fernandes 233
Rech, Cassiano 603, 724, 940, 1094
Regina, Bruno De Almeida 826
Rego, Rosana C. B. 120
Reis, Victor Camargo 1271
Rendón, Manuel A. 730, 1052
Resende, Ênio C. 288
Rezek, Angelo José Junqueira 49
Ribeiro, Brunno Monteiro Guimarães 1206
Ribeiro, Enio Roberto 166
Ribeiro, Luiz A. De S. 275, 400, 555, 1229
Ribeiro, Luiz Felipe Corrêa De Sá Santos 184
Ribeiro, Paulo F. 531
Ricciotti, Antonio Carlos Duarte 1366
Ricciotti, Viviane B. Da S. Duarte 1366
Riedemann, Javier 710
Rios, Lara Ana Rodarte 79, 239
Rios, Nicolas Parma 808
Rivera, M. 1441
Rivera, Marco 8, 61
Riveros, José A. 8, 61
Rocha, Filipe V. 1154
Rocha, Nady 868, 987, 1022, 1154
Rocha-Osorio, C. M. 73
Rodrigues, André Augusto 1088
Rodrigues, Andressa De Melo 760
Rodrigues, Gabriel Sales Lins 1463
Rodrigues, Gleice M. S. 868
Rodrigues, Gleice Mylena Da S. 1154
Rodrigues, Gleice Mylena Da Silva 987
Rodrigues, José Carlos Grilo 49
Rodrigues, Márcio C. B. P. 704, 1052, 1241
Rodrigues, Marcus Vinícius Maia 1343
Rodríguez, J. 1402
Rodriguez, Jose 929, 1481
Roe, Maurice G. L. 43
Roig, Mateo D. G. 388, 394
Rojas, Jeimy C. Sanabria 294
Rolim, Luís G. B. 1010
Rolim, Luis Guilherme Barbosa 184
Roman, Kosenko 1178
Romero, C. 1441
Rospirski, Alexsandra 796
Ruiz, Flávia P. 958
Ruiz-Caballero, Domingo 710
Ruppert, Ernesto 347
Sá, Edilson M. 1118, 1331
Saavedra, Osvaldo Ronald 275, 555
Sacco, Francesco 411
Saccol, Gabriel Avila 940
Sakô, Elson Yoiti 1130
Sakurai, Takayasu 137
Salazar, Andrés O. 668
Salazar, Andres Ortiz 1265, 1283

Salvador, Marcos Antonio 639
Salvadori, Fabiano 820
Sampaio, Leonardo Poltronieri 1301
Sant'Ana, Wilson C. 85, 131
Santana, Luan S. 778
Santana, R. A. S. 178
Santana, Wilson Cesar 531
Santiago, Raphael Perci 1396, 1458
Santisteban, José Andrés 946
Santos, Hugo E. 149
Santos, João 155
Sarlioglu, Bulent 67
Sarrias-Mena, Raúl 904
Scalcon, Filipe 698
Scalcon, Filipe P. 549
Schardong, Charles 940
Scherer, Lucas Giuliani 1064
Schlickmann, Henrique R. 820
Schmidt, Gustavo B. K. 1420
Schmidt, Luiz H. T. 802
Schmitz, Lenon 651
Schowantz, G. S. 969
Schuetz, Dimas Alã 929
Schulter, Wolfgang 952
Seixas, P. F. 178
Seleme, S. I. 573
Seleme, Seleme I. 674, 772
Silva, Andre Felicio De Sousa 1070
Silva, Danilo P. E 1337
Silva, Gabriel S. Barbara Da S. E 964
Silva, Jailson Leite 1319, 1447
Silva, João Lucas De Souza 1130
Silva, Laylla Fernandes 760
Silva, Leonardo P. S. 862, 958
Silva, Luciano De Souza Da Costa E 760,
    1136, 1211, 1487
Silva, R. P. 615
Silva, Rafael M. 364, 772
Silva, Ranoyca Nayana Alencar
    Leão E 1447
Silva, Renata C. 772
Silva, Sidelmo Magalhães 1004
Silva, Vinicius Zimmermann 49
Silva, Walquíria Do N. 567
Silveira, Joao Pedro C. 686
Simões, Marcelo Godoy 269
Simonetti, Domingos 844
Simonetti, Domingos S. L. 329
Singh, Mukhtiar 1499
Soares, Ana L. 98
Soares, Débora M. 92
Soares, Emerson L. 1022
Soares, Guilherme Marcio 370, 423, 579, 808
Soares, Marcus Vieira 784
Solís-Chaves, J. S. 73

AUTHOR INDEX

Sousa, Clodualdo V. 364, 772
Sousa, Silas M. 609
Sousa, Thiago William Pires 490
Souza, Bruno C. 19, 585, 1360
Souza, Géssica C. De A. 668
Souza, Lucas Carvalho 760, 1136, 1211, 1487
Souza, Marcus E. T. 1253
Spiazzi, Giorgio 114, 143
Stangler, E. V. 220, 716
Stangler, Eduardo Vasconcelos 208
Stefanelo, Márcio 698
Stein, Carlos M. 993
Stein, Carlos M. O. 472, 1148
Stein, Carlos Marcelo De Oliveira 1420
Stein, Carlos Marcelo Oliveira 1142
Stepenko, Serhii 1481
Stephan, Richard M. 37, 92, 621
Stopa, Marcelo M. 850
Stopa, Marcelo Martins 1112, 1206, 1313
Suárez, José H. 778
Tahim, André P. N. 964
Takamiya, Makoto 137
Tao, Chengxuan 104, 353
Tavares, Pedro Laguardia 579
Tedeschi, Elisabetta 445
Teixeira, Estêvão Coelho 808
Teixeira, Rodrigo De A. 668
Teixeira, Vanessa S. C. 958
Teixeira, Vanessa Siqueira De C. 862
Tennakoon, Sarath 227
Teodorescu, R. 573
Teston, Silvio Antonio 603, 1094
Tian, Haonan 323
Tibola, Gabriel 43
Tofoli, Fernando Lessa 79, 166, 239, 281, 370, 579, 1082, 1088
Toledo, S. 1441
Tonini, Luiz G. R. 898, 1378
Toniolo, Francesco 143
Torres, Bruno Silva 85
Torres, Renato A. 67
Torres, Vitor C. S. 1052
Tripathi, Anshuman 323
Vaca, David A. 1076
Vaisambhayana, Sriram 323
Valentim, Gustavo 429
Vargas, Murillo C. 898, 1378
Vargas, Murillo Cobe 1
Vargas, Rodrigo Z. 347
Vasquez, Juan Carlos 686
Vaz, Jerson Rogério Pinheiro 275
Vedovatte, Marcos V. A. 513
Veras, Caio Kerson O. 680
Veras, Leonilson Dos Santos 555
Viajante, Ghunter Paulo 305

Vicente, Eduardo Moreira 239, 1082, 1088
Vicente, Paula Dos Santos 1082, 1088
Vieira, Flávio Henrique Teles 1414
Vieira, Rodrigo P. 549, 698
Vilela, Wellington M. 149
Vilerá, Kaio Vinicius 603, 1094
Vilkn, P. H. J. 178
Villalva, Marcelo Gradella 1130
Vinnikov, Dmitri 1178, 1481, 1493
Viola, Julio 1289
Vitorino, Montiê A. 1106
Volpato, Cesar 698
Waltrich, Gierri 341, 692, 1431
Wang, Haoran 1160
Wang, Huai 1160
Wang, Lei 663
Wang, Lifang 104, 353
Wang, Liye 353
Wang, Qi 8
Watanabe, Edson H. 1034
Wermuth, Matheos C. 461, 543
Wheeler, P. 8, 61, 257, 1441
Wheeler, Pat W. 196
Wintrich, Arendt 1277
Wong, Man-Chung 663
Xavier, Lucas S. 466, 478
Yang, Yongheng 1
Yin, Chengliang 104
Yuan, Xibo 329
Zacarias, Joice D. S. 466
Zambon, Mário 435
Zhang, Ya 43
Zhu, Guorong 1160
Zimann, F. J. 220
Zimann, Felipe Joel 208
Zucuni, Jordan 929

2019 IEEE 15th Brazilian Power Electronics Conference and 5th IEEE Southern Power Electronics Conference (COBEP/SPEC 2019)

Santos, Brazil
1-4 December 2019

Pages 724-1504

IEEE Catalog Number: CFP1977F-POD
ISBN: 978-1-7281-4181-7

**Copyright © 2019 by the Institute of Electrical and Electronics Engineers, Inc.
All Rights Reserved**

Copyright and Reprint Permissions: Abstracting is permitted with credit to the source. Libraries are permitted to photocopy beyond the limit of U.S. copyright law for private use of patrons those articles in this volume that carry a code at the bottom of the first page, provided the per-copy fee indicated in the code is paid through Copyright Clearance Center, 222 Rosewood Drive, Danvers, MA 01923.

For other copying, reprint or republication permission, write to IEEE Copyrights Manager, IEEE Service Center, 445 Hoes Lane, Piscataway, NJ 08854. All rights reserved.

*** This is a print representation of what appears in the IEEE Digital Library. Some format issues inherent in the e-media version may also appear in this print version.

IEEE Catalog Number: CFP1977F-POD
ISBN (Print-On-Demand): 978-1-7281-4181-7
ISBN (Online):   978-1-7281-4180-0
ISSN: 2165-0454

Additional Copies of This Publication Are Available From:

Curran Associates, Inc
57 Morehouse Lane
Red Hook, NY 12571 USA
Phone: (845) 758-0400
Fax: (845) 758-2633
E-mail: curran@proceedings.com
Web: www.proceedings.com

TABLE OF CONTENTS

SIMPLIFIED SINGLE-PHASE PV GENERATOR MODEL FOR DISTRIBUTION FEEDERS WITH HIGH PENETRATION OF POWER ELECTRONICS-BASED SYSTEMS 1

Mariana Altoé Mendes ; Murillo Cobe Vargas ; Oureste Elias Batista ; Yongheng Yang ; Frede Blaabjerg

MODULATED MODEL PREDICTIVE CURRENT CONTROL FOR PMSM OPERATING WITH THREE-LEVEL NPC INVERTER 8

Qi Wang ; Marco Rivera ; Jose A. Riveros ; Patrick Wheeler

MODEL PREDICTIVE CONTROLLER FOR TWO-PHASE THREE-WIRE GRID-CONNECTED CONVERTERS 13

Pablo C. De S. Furtado ; Pedro Gomes Barbosa

A NEW ZERO-SEQUENCE VOLTAGE COMPENSATION ALGORITHM FOR A DSTATCOM BASED ON CONSUMER UNBALANCE 19

Bruno C. Souza ; Samuel N. Duarte ; Pedro M. Almeida ; Pedro G. Barbosa ; Leandro R. Araújo

DESIGN OF RESONANT CONTROLLERS FOR COMPENSATION OF THIRD HARMONIC RIPPLE IN THE DC CAPACITORS VOLTAGES OF NPC CONVERTERS 25

Andrei De O. Almeida ; Pedro M. De Almeida ; Pedro G. Barbosa

SINGLE PHASE-SHIFT CONTROL OF DAB CONVERTER USING ROBUST PARAMETRIC APPROACH 31

Kevin E. Lucas ; Daniel J. Pagano ; Renan L. P. Medeiros

OPERATION BOUNDARIES OF A SINGLE PHASE THYRISTOR DRIVEN DC-MOTOR 37

Thomas M. Mertens ; Richard M. Stephan

INTEGRATED LOCAL CONTROL OF ACTIVE POWER AND VOLTAGE SUPPORT FOR THREE-PHASE THREE-WIRE CONVERTERS 43

Ya Zhang ; Gabriel Tibola ; Maurice G. L. Roe ; Jorge L. Duarte

IMPLEMENTATION OF A DIDACTIC PLATFORM FOR A GENERIC LOAD TORQUE EMULATOR USING INDUCTION MACHINES AND PWM INVERTERS 49

Luiz Otávio Campos De Medeiros ; José Carlos Grilo Rodrigues ; Angelo José Junqueira Rezek ; Nery De Oliveira Junior ; Rafael Di Lorenzo Corrêa ; Alexandre Viana Braga ; Christel Enock Ghislain Ogoulola ; Vinicius Zimmermann Silva ; Marcos Leonardo Ramos

ENERGY EFFICIENT CONTROL OF SYNCHRONOUS MACHINES IN DEEP FIELD-WEAKENING OPERATION INCLUDING SATURATION EFFECTS 55

Joao Bonifacio ; Ralph Kennel

PREDICTIVE VOLTAGE CONTROL OPERATING AT FIXED SWITCHING FREQUENCY OF A NEUTRAL-POINT CLAMPED CONVERTER 61

Felipe Herrera ; Roberto Cárdenas ; Marco Rivera ; José A. Riveros ; Patrick Wheeler

DEVELOPMENT OF CURRENT-SOURCE-INVERTER-BASED INTEGRATED MOTOR DRIVES USING WIDE-BANDGAP POWER SWITCHES 67

Renato A. Torres ; Hang Dai ; Woongkul Lee ; Thomas M. Jahns ; Bulent Sarlioglu

POWER CONTROL OF A DOUBLY FED INDUCTION WIND GENERATOR EMPLOYING A TAKAGI-SUGENO FUZZY LOGIC CONTROLLER 73

C. M. Rocha-Osorio ; J. S. Solís-Chaves ; Eliomar R. Conde D. ; J. L. Azcue Puma ; Fernando Lino ; A. J. Sguarezi Filho

NONISOLATED DC-DC QUADRATIC CUK CONVERTER FOR WIDE CONVERSION RANGE APPLICATIONS 79

Tatiane Martins Oliveira ; Lara Ana Rodarte Rios ; Fernando Lessa Tofoli ; Aniel Silva De Morais

IMPLEMENTATION OF AUTOMATIC BATTERY CHARGING TEMPERATURE COMPENSATION ON A PEAK-SHAVING ENERGY STORAGE EQUIPMENT 85

Wilson Cesar Sant'Ana ; Robson Bauwelz Gonzatti ; Germano Lambert-Torres ; Erik Leandro Bonaldi ; Pedro Andrade De Oliveira ; Bruno Silva Torres ; Joao Gabriel Luppi Foster ; Rondineli Rodrigues Pereira ; Luiz Eduardo Borges-Da-Silva ; Denis Mollica ; Jos

CASCADE CONTROL VS FULL-STATE FEEDBACK 92

Débora M. Soares ; Henrique A. M. Calil ; Richard M. Stephan

DESIGN AND PERFORMANCE ANALYSIS OF ISOLATED CUK CONVERTER EMPLOYED IN MULTIPLE PULSE RECTIFIER SYSTEMS 98

Ana L. Soares ; Antônio O. Costa Neto ; Gustavo B. Lima ; Luiz C. G. Freitas ; Ernane A. A. Coelho

A RESEARCH ON CONSTANT VOLTAGE OUTPUT CHARACTERISTICS OF WIRELESS POWER TRANSFER SYSTEM WITH A DC-DC CONVERTER 104

Zhimeng Liu ; Lifang Wang ; Chengliang Yin ; Yanjie Guo ; Chengxuan Tao

FINITE CONTROL SET MODEL BASED PREDICTIVE CONTROL OF GRID-TIED SIX-SWITCH CONVERTER APPLIED TO INDUCTION GENERATOR 108

Paulo R. U. Guazzelli ; Allan G. De Castro ; Stefan T. C. A. Dos Santos ; Carlos M. R. De Oliveira ; José R. B. A. Monteiro ; Manoel L. De Aguiar

APPLYING COUPLED INDUCTORS TO THE CLAMPED-RESONANT INTERLEAVED BOOST CONVERTER 114

Giorgio Spiazzi

LPV MODELING OF BOOST CONVERTER AND GAIN SCHEDULING MPC CONTROL 120

Rosana C. B. Rego

ZERO-CROSSING DETECTION FREQUENCY ESTIMATOR METHOD COMBINED WITH A KALMAN FILTER FOR NON-IDEAL POWER GRID 125

Tiago Davi Curi Busarello ; Sérgio Luiz Sambugari Junior ; Newton Da Silva

A REVIEW OF FCS-MPC IN MULTILEVEL CONVERTERS APPLIED TO ACTIVE POWER FILTERS 131

João G. L. Foster ; Rondineli R. Pereira ; Robson B. Gonzatti ; Wilson C. Sant'Ana ; Denis Mollica ; Germano Lambert-Torres

DISTANCE DETECTION SYSTEM FOR DIGITAL TRANSMITTER COIL ACHIEVING DISTANCE-VARIATION-TOLERANT WIRELESS POWER TRANSFER 137

Hao Qiu ; Yoshiaki Narusue ; Yoshihiro Kawahara ; Takayasu Sakurai ; Makoto Takamiya

DIGITAL CURRENT CONTROL FOR A BIDIRECTIONAL INTERLEAVED BOOST CONVERTER WITH COUPLED INDUCTORS 143

Francesco Toniolo ; Qing Liu ; Tommaso Caldognetto ; Simone Buso ; Giorgio Spiazzi

PEMSYN: A FREE SOFTWARE TO ASSIST THE DESIGN AND PERFORMANCE ASSESSMENT OF PERMANENT MAGNETS SYNCHRONOUS MACHINES 149

Khristian M. De Andrade ; Hugo E. Santos ; Wellington M. Vilela ; Thales E. P. De Almeida ; Geyverson T. De Paula

DEVELOPMENT OF LINEAR GENERATOR PROTOTYPE AS PART OF A POINT ABSORBER WAVE ENERGY CONVERTER 155

Eduardo Martins ; Wagner Marcilio ; Marcos Moreira ; João Santos

THE TRUE UNITY POWER FACTOR CONVERTER APPLIED TO PHOTOVOLTAIC APPLICATIONS 160

Marcos Henrique Da Silva Alves ; Thiago Morais Parreiras ; Braz De Jesus Cardoso Filho

HIGH-VOLTAGE STEP-UP DC-DC CONVERTER EMPLOYING THE FOUR STATE SWITCHING CELL AND VOLTAGE MULTIPLIER CELLS 166

Paulo Henrique Feretti ; Enio Roberto Ribeiro ; Fernando Lessa Tofoli

STABILIZATION OF DC MICROGRIDS WITH POINT-OF-LOAD CONVERTERS AS CONSTANT POWER LOADS 172

Isaías V. Bessa ; Renan L. P. Medeiros ; Iury V. Bessa ; Florindo A. C. Ayres Junior ; Kevin E. Lucas

INDUCTOR DESIGN METHODOLOGY FOR POWER ELECTRONICS APPLICATIONS 178

P. H. J. Vilkn ; L. M. F. Morais ; R. A. S. Santana ; P. C. Cortizo ; P. F. Seixas

REAL-TIME IMPLEMENTATION OF A DC CONVERTER USING MODIFIED NODAL ANALYSIS, SPARSITY HANDLING AND PARALLELISM ON A DSP PLATFORM 184

Luiz Felipe Corrêa De Sá Santos Ribeiro ; Felipe Novaes Francis Dicler ; Luis Guilherme Barbosa Rolim ; Mauricio Aredes

INVESTIGATION OF CONTROL STRATEGIES TO MITIGATE THE OSCILLATION EFFECTS CAUSED BY INTERCONNECTED BUCK CONVERTERS 190

Isaías Valente De Bessa ; Renan L. P. Medeiros ; Iury Valente De Bessa ; Florindo A. C. Ayres Junior ; Kevin E. Lucas

INTEGRATING A SINGLE Z-SOURCE NETWORK WITH A MODULAR MULTILEVEL CONVERTER FOR VOLTAGE BOOSTING 196

Fatma A. Khera ; Christian Klumpner ; Pat W Wheeler

OPTIMIZATION OF ROBUST PI CONTROLLERS FOR GRID-TIED INVERTERS 202

Caio R. D. Osório ; Lucas C. Borin ; Gustavo G. Koch ; Vinícius F. Montagner

INVESTIGATION OF VOLTAGE REGULATION WITH ACTIVE AND REACTIVE POWER WITH DISTRIBUTED LOADS ON A RADIAL DISTRIBUTION FEEDER 208

Felipe Joel Zimann ; Alessandro Luiz Batschauer ; Marcello Mezaroba ; Eduardo Vasconcelos Stangler ; Francisco De Assis Dos Santos Neves

DESIGN AND ASSEMBLY OF A BIPOLAR MARX GENERATOR BASED ON FULL-BRIDGE TOPOLOGY APPLIED TO ELECTROPORATION 214

Fernando Imai ; Yales Rômulo De Novaes

IMPLEMENTATION OF A IUPQC CONTROL SCHEME FOR ENSURING AN IMPROVED COMPENSATION PERFORMANCE 220

E. V. Stangler ; F. A. S. Neves ; F. Bradaschia ; M. Mezaroba ; F. J. Zimann ; A. L. Batschauer

ASPECTS OF TRAVELLING WAVE BASED PROTECTION PHILOSOPHY FOR CONSIDERATION IN DC GRIDS OF THE FUTURE................227
Monday Ikhide ; Sarath Tennakoon ; Alison Griffiths ; Hengxu Ha

AN UNIT-LESS MATHEMATICAL MODEL FOR ANALYSIS AND DESIGN OF CLASS-E RESONANT CONVERTERS................233
Lucas S. Mendonça ; Thiago C. Naidon ; Rafael Fernandes Raposo ; Fábio E. Bisogno

A CURVE TRACER FOR PHOTOVOLTAIC MODULES BASED ON THE CAPACITIVE LOAD METHOD................239
Emerson Abreu Bastos Junior ; Caio Meira Amaral Da Luz ; Tatiane Martins Oliveira ; Lara Ana Rodarte Rios ; Eduardo Moreira Vicente ; Fernando Lessa Tofoli

LED DRIVER WITH REDUCED REDUNDANT POWER PROCESSING AND DIMMING FOR STREET LIGHTING APPLICATIONS................245
Kleber Chan Bekoski ; Cassio Gobbato ; Cassiano Ferro Moraes ; Gustavo Weber Denardin ; Juliano De Pelegrini Lopes

ACTIVE-CAPACITOR FOR POWER DECOUPLING IN SINGLE-PHASE GRID-CONNECTED CONVERTERS................251
Thiago A. Pereira ; Denizar C. Martins ; Roberto F. Coelho

RELIABILITY ANALYSIS OF AIRCRAFT STARTER GENERATOR DRIVE CONVERTER................257
Jayakrishnan Harikumaran ; Giampaolo Buticchi ; Michael Galea ; Alessandro Costabeber ; Pat Wheeler

MULTIVARIABLE CONTROL OF A GRID FORMING SYSTEM BASED ON BACK-TO-BACK TOPOLOGY................263
Igor D. N. De Souza ; Gabriel A. Fogli ; Marcelo C. Fernandes ; Ademir S. T. Júnior ; Pedro G. Barbosa ; Pedro M. De Almeida

3-PHASE MULTI-FUNCTIONAL GRID-TIED INVERTER FOR COMPENSATION OF OSCILLATING INSTANTANEOUS POWER................269
José De Arimatéia Olímpio Filho ; Helmo Kelis Morales Paredes ; Augusto Matheus Dos Santos Alonso ; Jakson Paulo Bonaldo ; Fernando Pinhabel Marafão ; Marcelo Godoy Simões

COMPARATIVE STUDY OF DIFFERENT CORRECTION METHODS TO ANALYZE WIND TURBINE PERFORMANCE................275
Hércules Araújo Oliveira ; Luiz Antonio De Souza Ribeiro ; Jerson Rogério Pinheiro Vaz ; Osvaldo Ronald Saavedra ; José Gomes De Matos

NONISOLATED QUADRATIC SEPIC CONVERTER WITHOUT ELECTROLYTIC CAPACITORS FOR LED DRIVER APPLICATIONS................281
Douglas Rosa Corrêa ; Aniel Silva De Morais ; Fernando Lessa Tofoli

A PERFORMANCE ANALYSIS OF ACTIVE ANTI-ISLANDING METHODS BASED ON FREQUENCY DRIFT................288
Ênio C. Resende ; Henrique T. M. Carvalho ; Fernando C. Melo ; Ernane A. A. Coelho ; Gustavo B. De Lima ; Luiz C. G. De Freitas

SIMULATION OF THE MODEL, DESIGN AND CONTROL OF A CURRENT SOURCE INVERTER WITH UNIPOLAR SPWM MODULATION................294
Jeimy C. Sanabria Rojas ; Daniel M. Barrera Leguizamón ; Diego A. Bautista López ; Fabián R. Jiménez López

ANALYSIS AND COMPARISON OF THE DYNAMIC RESPONSE OF DIRECT AND INDIRECT ROTOR FLUX CONTROL APPLIED TO AN ASYMMETRICAL TWO-PHASE INDUCTION MOTOR................299
Rafael De Farias Campos ; José De Oliveira ; Ademir Nied

DYNAMIC ANALYSIS OF SELF-EXCITED SRG OPERATING IN OPEN LOOP................305
Lucas José Lemes ; Victor Regis Bernardeli ; Luciano Coutinho Gomes ; Darizon Alves De Andrade ; Ghunter Paulo Viajante ; Marcos Antonio Arantes De Freitas

THREE-PHASE, FOUR-WIRE PWM RECTIFIER APPLIED TO VARIABLE FREQUENCY AC SYSTEMS IN AIRPLANE ELECTRIC GRID UNDER FAULT CONDITIONS................311
Alexandre Galvão Bueno ; José Antenor Pomilio

MODEL, SIMULATION AND ANALYSIS OF BLDCM FOR A DIFFERENTIAL CONTROLLED ELECTRIC-POWERED WHEELCHAIR................317
Augusto Nery De Lima Neto ; José Antenor Pomilio

ANALYSIS AND OPTIMAL DESIGN OF MAGNETIC COMPONENTS IN DUAL-ACTIVE-BRIDGE CONVERTER FOR 1 MVA SOLID-STATE TRANSFORMER................323
Haonan Tian ; Sriram Vaisambhayana ; Anshuman Tripathi

CENTER-TAPPED π-TYPE SINGLE-PHASE CELL................329
Domingos S. L. Simonetti ; Xibo Yuan

NOVEL MTPA APPROACH FOR IPMSM WITH NON-SINUSOIDAL BACK-EMF................335
Allan Gregori De Castro ; Paulo R. U. Guazzelli ; Mateus M. Lumertz ; Carlos M. R. De Oliveira ; Geyverson T. De Paula ; José R. B. A. Monteiro

NON-ISOLATED HIGH CURRENT BATTERY CHARGER WITH PFC SEMI-BRIDGELESS RECTIFIER 341

Rodrigo Patrício Dacol ; Joselito Anastácio Heerdt ; Gierri Waltrich

COMPARATIVE ANALYSIS BASED ON THE SWITCHING FREQUENCY OF MODULATION TECHNIQUES FOR MMC APPLICATIONS 347

Juan C. Colque ; Ernesto Ruppert ; Rodrigo Z. Vargas ; José L. Azcue

DESIGNING OF THE TRANSMITTING COILS AND COMPENSATION NETWORK OF A SEGMENTED DWPT SYSTEM 353

Shufan Li ; Lifang Wang ; Chengxuan Tao ; Fang Li ; Liye Wang

PROPOSAL OF AN ISOLATED TWO-SWITCH DC-DC SEPIC CONVERTER 358

Marcos Vinícius Mosconi Ewerling ; Telles Brunelli Lazzarin ; Carlos Henrique Illa Font

COMPARISON AMONG TWO-PHASE THREE-WIRE AC OFF-GRID POWER SYSTEMS 364

Rafael M. Silva ; Danilo I. Brandao ; Gabriel A. Fogli ; Victor F. Mendes ; Clodualdo V. Sousa

EFFICIENCY ANALYSIS FOR INTERLEAVED BUCK CONVERTERS EMPLOYED AS EXTRA-HIGH CURRENT COB LED DRIVERS 370

Dênis De Castro Pereira ; Everton Bernard Figueiredo Rabelo ; Pedro Santos Almeida ; Guilherme Marcio Soares ; Fernando Lessa Tofoli ; Henrique Antônio Carvalho Braga

MODELING OF A THREE-LEVEL QUADRATIC BOOST CONVERTER 376

Mauricio Borchardt ; Mateus C. Orige ; Mateus De F. Bueno ; Gerardo Escobar ; Yales R. De Novaes

DISCRETE SPS CONTROL OF A DAB CONVERTER USING PARTIAL FEEDBACK LINEARIZATION 382

Eduardo Luiz Santos Da Silva ; André Luís Kirsten ; Daniel Juan Pagano

STEADY-STATE ANALYSIS OF A SINGLE-PHASE MODIFIED BRIDGELESS BOOST RECTIFIER IN DCM 388

Mateo D. Roig G. ; Caio G. Da S. Moraes ; Telles B. Lazzarin

A TWO-STAGE BATTERY CHARGER WITH ACTIVE POWER DECOUPLING CELL FOR SMALL ELECTRIC VEHICLES 394

Caio G. Da S. Moraes ; Mateo D. Roig G. ; Telles B. Lazzarin

MODELLING AND ANALYSIS OF THE ISOLATED MICROGRID INSTALLED AT THE LENÇÓIS ISLAND USING PSCAD/EMTDC 400

Silvangela L. Barcelos ; José Gomes De Matos ; Luiz Antonio De Souza Ribeiro

METHODOLOGY FOR EXPERIMENTAL DETERMINATION OF EQUIVALENT GRID IMPEDANCE BY USING EXTERNAL COMMANDS OF PV INVERTERS 406

Valesca Bettim Feltrin ; Thiago Brezolin Dalmolin ; Lucas Vizzotto Bellinaso ; Leandro Michels

DEVELOPMENT OF A FPGA-BASED CONTROL SYSTEM FOR MODULAR MULTILEVEL CONVERTER APPLICATIONS 411

Lucas Koleff ; Manoel Conde ; Pedro Hayashi ; Francesco Sacco ; Kelly Enomoto ; Eduardo Pellini ; Wilson Komatsu ; Lourenço Matakas

VARIABLE-STEP DFT ALGORITHM TO SPEED UP OPTIMIZATION ROUTINES APPLIED TO A THREE-PHASE INTERLEAVED VIENNA RECTIFIER 417

B. Bertoldi ; E. F. C. Grabovski ; A. B. Lange ; M. L. Heldwein

DESIGN METHOD TO REDUCE THE DC LINK VOLTAGE OF A THREE-WIRE THREE-PHASE HYBRID ACTIVE POWER FILTER 423

Mateus Freitas Braga ; Samuel Neves Duarte ; Guilherme Márcio Soares ; Pedro Gomes Barbosa

DEVELOPMENT OF A MODULAR OPEN SOURCE POWER ELECTRONICS DIDACTIC PLATFORM 429

Lucas Koleff ; Gustavo Valentim ; Victor Rael ; Luciana Marques ; Wilson Komatsu ; Eduardo Pellini ; Lourenço Matakas

FLEXIBLE DIDACTIC PLATFORM FOR THYRISTOR-BASED CIRCUITS 435

Lucas Koleff ; Lucas Araújo ; Mário Zambon ; Wilson Komatsu ; Eduardo Pellini ; Lourenco Matakas

CONTROL OF WIRELESS POWER TRANSFER SYSTEMS UNDER LARGE COIL MISALIGNMENTS 441

Yeran Liu ; Udaya K. Madawala ; Ruikun Mai ; Zhengyou He

CONSIDERATIONS ON COMMUNICATION INFRASTRUCTURES FOR COOPERATIVE OPERATION OF SMART INVERTERS 445

Augusto Matheus Dos Santos Alonso ; Leonardo Carlos Afonso ; Danilo Iglesias Brandao ; Elisabetta Tedeschi ; Fernando Pinhabel Marafão

250 W SINGLE STAGE STEP-UP INVERTER CONNECTED TO THE GRID 451

Jessika Melo De Andrade ; Roberto Francisco Coelho ; Telles Brunelli Lazzarin

EXPERIMENTAL ANALYSIS FOR LOW POWER SERIES-SERIES COMPENSATED INDUCTIVE POWER TRANSFER SYSTEM 455

Macklyster Lânucy Scherre Stofel De Lacerda ; Tatiana Saviato Macedo ; Denizar Cruz Martins ; Walbermark Marques Dos Santos

METHOD TO TRACE THE PHOTOVOLTAIC CHARACTERISTIC CURVE WITH REVERSE VOLTAGE FOR SHADING CONDITIONS 461

Richard G. Cornelius ; Amanda C. Maia ; Matheos C. Wermuth ; Guilherme S. Da Silva

HARMONIC COMPENSATION STRATEGIES APPLIED TO MULTIFUNCTIONAL PHOTOVOLTAIC INVERTERS 466

Lucas S. Xavier ; Joice D. S. Zacarias ; Allan F. Cupertino ; Heverton A. Pereira ; Danilo I. Brandao ; Victor F. Mendes

EVENT MANAGER AND CONTROL STRUCTURE FOR HIGH PERFORMANCE THREE-PHASE GRID-TIED INVERTERS 472

Victor E. S. Barbosa ; Rodrigo A. S. Kraemer ; Emerson G. Carati ; Jean P. Da Costa ; Rafael Cardoso ; Carlos M. O. Stein

PARTIAL HARMONIC CURRENT COMPENSATION APPLIED TO MULTIPLE PHOTOVOLTAIC INVERTERS IN A RADIAL DISTRIBUTION LINE 478

André L. P. De Oliveira ; Lucas S. Xavier ; João M. S. Callegari ; Allan F. Cupertino ; Victor F. Mendes ; Heverton A. Pereira

ENHANCED SPACE-STATE REDUCED-ORDER MICROGRID MODEL IN COMMON DQ-REFERENCE FRAME 484

Sebastián De J. Manrique Machado ; Sérgio Augusto Oliveira Da Silva ; José R. B. A. Monteiro ; Azauri A. De Oliveira

OPTIMAL VOLTAGE COORDINATED CONTROL FOR GRID-CONNECTED PHOTOVOLTAIC SYSTEMS 490

Thiago William Pires Sousa ; Gustavo Kaefer Dill

COMPARATIVE STUDY OF RC SNUBBER CONFIGURATIONS IN SWITCHING CIRCUITS 495

Ana Carolina Moreira ; Daniel Cesar Piccoli ; Júlio Cesar Lopes De Oliveira ; Luiz Fernando Henning ; Rodrigo Jose Piontkewicz

COMPARATIVE STUDY OF CONTROL SYSTEMS FOR A PHOTOVOLTAIC INVERTER WITH LCL FILTER 501

Leandro T. Omine ; Moacyr A. G. Brito

DESIGN OF A LOW-COST PHASOR MEASUREMENT UNIT (PMU) FOR THREE-PHASE DISTRIBUTION POWER SYSTEMS ACCORDING IEEE C37.118.1 507

Alex Guamán ; Marcelo Pozo ; Isaac Pozo ; Mario Pacas ; Ana Cabrera ; Nataly Pozo

CONTROLLER COEFFICIENTS TUNING FOR A SINGLE-PHASE PHOTOVOLTAIC SYSTEM IN SYNCHRONOUS REFERENCE FRAME THROUGH GENETIC ALGORITHM 513

Marcos V. A. Vedovatte ; Moacyr A. G. De Brito ; Luigi G. Junior

ROBUST PID CONTROLLERS OPTIMIZED BY PSO ALGORITHM FOR POWER CONVERTERS 519

Lucas C. Borin ; Everson Mattos ; Caio R. D. Osorio ; Gustavo G. Koch ; Vinicius F. Montagner

IMPACTS OF THE MULTI-VARIABLES MODULATION ON TRANSFORMER AND SOFT-SWITCHING OF A DAB 525

Jeferson Fraytag ; André Luís Kirsten ; Marcelo Lobo Heldwein

DEVELOPEMENT OF A MULTILEVEL DVR WITH BATTERY CONTROL AND HARMONIC COMPENSATION 531

Bruno P. B. Guimarães ; Wilson Cesar Santana ; Paulo F. Ribeiro ; Robson Bauwelz Gonzatti ; Guilherme G. Pinheiro ; Rondineli R. Pereira ; Fernando Nunes Belchior ; Carlos Henrique Da Silva ; Luiz Eduardo Borges Da Silva

MODELING OF ELECTROCHEMICAL BATTERIES BEHAVIOR AND LIFETIME DEGRADATION FOR PV APPLICATIONS 537

Rafael César Nolasco ; Victor Flores Mendes

COMPARISON BETWEEN TWO MODULATION STRATEGIES FOR THE 3L-DC-SSI 543

Matheos C. Wermuth ; Amanda C. Maia ; Richard G. Cornelius ; Guilherme S. Da Silva

ROBUST CONTROL OF SWITCHED RELUCTANCE GENERATOR IN CONNECTION WITH A GRID-TIED INVERTER 549

Caio R. D. Osório ; Filipe P. Scalcon ; Rodrigo P. Vieira ; Vinícius F. Montagner ; Hilton A. Gründling

ANALYSIS OF PERFORMANCE AND OPPORTUNITY FOR IMPROVEMENTS IN THE MICROGRID OF ILHA GRANDE 555

Leonilson Dos Santos Veras ; Hércules A. Oliveira ; José G. De Matos ; Osvaldo Ronald Saavedra ; Luiz A. De Sousa Ribeiro ; Lucas De Paula Assunção Pinheiro

STEADY-STATE CHARACTERIZATION OF THE THREE-PHASE ISOLATED DC-DC BIDIRECTIONAL CONVERTER WITH LLC RESONANT TANK 561

Kristian Pessoa Dos Santos ; Hermínio Miguel Oliveira Filho ; Paulo Peixoto Praça ; Demercil De Souza Oliveira Júnior

GAS MICROTURBINES FOR DISTRIBUTED GENERATION SYSTEM 567

Walquíria Do N. Silva ; Janaína G. De Oliveira ; Bruno H. Dias ; Leonardo W. De Oliveira

INHERENT REDUNDANCY OF SDBC-MMCC BASED STATCOM IN THE OVERMODULATION REGION.................573
D. C. Mendonça ; A. F. Cupertino ; H. A. Pereira ; S. I. Seleme ; R. Teodorescu

ACTIVE COOLING AND THERMAL SIMULATION APPLIED TO AN EXTRA-HIGH CURRENT COB LED.................579
Dênis De Castro Pereira ; Rúbio Campos Marques ; Pedro Santos Almeida ; Guilherme Marcio Soares ; Fernando Lessa Tofoli ; Pedro Laguardia Tavares ; Henrique Antônio Carvalho Braga

VOLTAGE REGULATION OF A REMOTE BUS OF A DISTRIBUTION NETWORK BY STATIC SYNCHRONOUS COMPENSATOR.................585
Samuel N. Duarte ; Bruno C. Souza ; Pedro M. Almeida ; Pedro G. Barbosa

SINGLE-STAGE SINGLE-PHASE AC/DC CONVERTER WITH HIGH FREQUENCY ISOLATION FEASIBLE TO MICROGENERATION.................591
Samanta Gadelha Barbosa ; Bruno Ricardo De Almeida ; Debora Pereira Damasceno ; Demercil De S. Oliveira

SMARTBATTERY: AN ACTIVE-BATTERY SOLUTION FOR ENERGY STORAGE SYSTEM.................597
Lucas S. Araujo ; Nicolas T. D. Fernandes ; Danilo I. Brandao ; Braz J. Cardoso Filho

FEEDFORWARD COMPENSATION OF THE ESS LOW-FREQUENCY CURRENT RIPPLE IN THE THREE-PORTS ANPC CONVERTER.................603
Silvio Antonio Teston ; Kaio Vinicius Vilerá ; Marcello Mezaroba ; Cassiano Rech

SEPIC DC/DC CONVERTER CONTROL BY OBSERVED-STATE FEEDBACK.................609
Silas M. Sousa ; Valter J. S. Leite ; Samir W. Fernandes ; Isabel R. H. Oliveira

THIRD HARMONIC INJECTION METHOD FOR RELIABILITY IMPROVEMENT OF SINGLE-PHASE PV INVERTERS.................615
R. C. De Barros ; R. P. Silva ; D. B. Da Silveira ; W. C. Boaventura ; A. F. Cupertino ; H. A. Pereira

DIDACTIC SYSTEM FOR CONTROL OF ELECTRICAL MACHINES IN EDUCATION AND RESEARCH LABORATORIES.................621
Allan V. S. Andrade ; Richard M. Stephan

ON THE INFLUENCE OF AREA VARIATIONS OF THE PHOTOVOLTAIC ;SURFACE IN SOLAR CELL ANTENNAS.................627
Eduardo Vicente Valdés Cambero ; Humberto Pereira Da Paz ; Vinícius Santana Da Silva ; Humberto Xavier De Araújo ; Ivan Roberto Santana Casella ; Carlos Eduardo Capovilla

NEW SEMICONDUCTOR TECHNOLOGIES FOR POWER ELECTRONICS.................633
Alisson Mengatto ; José Adriano Damacena Diesel ; Pedro Henrique Thiesen De França ; Joselito Anastácio Heerdt

SMALL SCALE COMPRESSED AIR ENERGY STORAGE (SS-CAES) STRATEGIES OVERVIEW.................639
Luiz Fernando Martins Pastuch ; Roberto Francisco Coelho ; Telles Brunelli Lazzarin ; Marcos Antonio Salvador

SYNCHRONOUS REFERENCE FRAME PLL FREQUENCY ESTIMATION UNDER VOLTAGE VARIATIONS.................645
Eliabe Duarte Queiroz ; José Antenor Pomilio

HIGH STEP-UP DC-DC CONVERTER WITH INPUT CURRENT SHARING BASED ON THE FORWARD CONVERTER.................651
Víctor Ferreira Gruner ; Lucas Fiamoncini ; Lenon Schmitz ; Denizar Cruz Martins ; Roberto Francisco Coelho

SRF-PLL INFLUENCE ON THE STABILITY OF A CURRENT SOURCE CONVERTER IN DROOP MODE.................657
Eliabe Duarte Queiroz ; José Antenor Pomilio

A LOW COST BI-DIRECTIONAL WIRELESS POWER TRANSFER SYSTEM.................663
Lei Wang ; Udaya K. Madawala ; Man-Chung Wong

WIND TURBINE EMULATOR WITH DC MOTOR.................668
Tadeu F. Dos Santos ; Igor V. Chacon ; Géssica C. De A. Souza ; Guilherme A. P. De C. A. Pessoa ; Felipe O. S. Gama ; Rodrigo De A. Teixeira ; Andrés O. Salazar

GENERALIZED MATHEMATICAL MODEL FOR AN N-CELL INTERLEAVED BOOST CONVERTER.................674
Luana K. Melgaço Pereira ; Seleme I. Seleme ; João Lucas Da Silva

AC-DC CONVERTER WITH HIGH-FREQUENCY ISOLATION OPERATING UNDER ZVS.................680
Bruno Alves Sousa Da Silva ; Dalton De Araújo Honório ; Demercil De Souza Oliveira Júnior ; Bruno Ricardo De Almeida ; Samanta Gadelha Barbosa ; Luiz Henrique Silva Colado Barreto ; Caio Kerson O. Veras

ENHANCED POWER MANAGEMENT SYSTEM FOR DROOP CONTROL IN A GRID CONNECTED DC MICROGRID.................686
Pedro Jose Dos Santos Neto ; Joao Pedro C. Silveira ; Tárcio André Dos S. Barros ; Ernesto Ruppert Filho ; Juan Carlos Vasquez ; Josep M. Guerrero

MULTI-PORT SYSTEM FOR STORAGE AND MANAGEMENT OF REGENERATIVE BRAINING ENERGY IN DIESEL-ELECTRIC LOCOMOTIVES ... 692

Caio G. Da S. Moraes ; Sergio L. B. Junior ; Pedro P. Cavilha ; Antonio L. S. Pacheco ; Marcelo L. Heldwein ; Gierri Waltrich

SENSORLESS CONTROL OF NONSINUSOIDAL BACK-EMF PMSM BASED ON STATE OBSERVER ... 698

Thiago Lazzari ; Filipe Scalcon ; Cesar Volpato ; Thieli Gabbi ; Márcio Stefanelo ; Rodrigo P. Vieira

POWER DEMAND PREDICTION BASED ON MIXED DRIVING CYCLE APPLIED TO ELECTRIC VEHICLE HYBRID ENERGY STORAGE SYSTEM ... 704

Lucas F. R. Lago ; Silvana T. Faceroli ; Rodrigo A. F. Ferreira ; Marcio C. B. P. Rodrigues

SYMMETRICAL HYBRID MULTILEVEL VSI AND CSI INVERTERS DERIVED FROM DC-DC CONVERTERS ... 710

Domingo Ruiz-Caballero ; Luis Colque Miranda ; Carlos Paredes ; Javier Riedemann ; Werner Jara Montecinos ; Marcelo Lobo Heldwein ; Samir Ahmad Mussa

STRUCTURAL AND PERFORMANCE COMPARISON BETWEEN HARMONIC SELECTIVE REPETITIVE CONTROLLERS FOR SHUNT ACTIVE POWER FILTER ... 716

R. C. Neto ; F. A. S. Neves ; E. V. Stangler ; F. Bradaschia ; H. E. P. De Souza

IMPACT OF CAPACITOR DESIGN METHODOLOGY ON FC INVERTERS ... 724

Marcos Vinicius Bressan ; Cassiano Rech ; Alessandro L. Batschauer

MODELING AND CONTROL OF A BACK-TO-BACK SYSTEM FOR TURBOELECTRIC PROPULSION ... 730

Saulo O. Nascimento ; Vinicius M. De Albuquerque ; Marcelo De C. Fernandes ; Manuel A. Rendón ; Janaína G. Oliveira ; Pedro S. Almeida

AN ENHANCED THÉVENIN EQUIVALENT CIRCUIT OF A RESONANT-CONTROLLER-BASED UTILITY-INTERFACE ... 736

Joel Filipe Guerreiro ; Hildo Guillardi Júnior ; João Inácio Yutaka Ota ; José Antenor Pomilio

REAL TIME SIMULATION IN A DISTRIBUTION SYSTEM INCLUDING PV INVERTER AND VOLTAGE REGULATOR: VOLTAGE IMPACT ANALYSIS ... 742

João A. G. Archetti ; Lilian V. Pinheiro ; Mateus L. Lima ; Bernardo F. Musse ; Janaína. G. De Oliveira ; Leonardo. W. De Oliveira

USING SYNCHRONVERTER IN DISTRIBUTED GENERATION FOR FREQUENCY AND VOLTAGE GRID SUPPORT ... 748

Guilherme Penha Da Silva Junior ; Luciano Sales Barros

DESIGN PROCEDURES AND PROTOTYPING OF A FULL-BRIDGE HIGH FREQUENCY POWER INVERTER ... 754

Joel Filipe Guerreiro ; Hildo Guillardi Júnior ; José Antenor Pomilio

SVC OPERATING AS AN UNBALANCE COMPENSATOR WITH CONTROL SYSTEM BASED ON THE STEINMETZ METHOD AND THE INSTANTANEOUS POWER THEORY ... 760

Andressa De Melo Rodrigues ; Laylla Fernandes Silva ; Luciano De Souza Da Costa E Silva ; Lucas Carvalho Souza

DC CURRENT REDISTRIBUTOR FOR ELECTRIC AIRCRAFT SYSTEM ... 766

Mateus P. Dias ; Hildo Guillardi ; Joel F. Guerreiro ; João I. Y. Ota ; José A. Pomilio ; Paolo Mattavelli

TWO-STAGE STAND ALONE PHOTOVOLTAIC SYSTEM FOR WATER PUMPING SYSTEM ... 772

Renata C. Silva ; Gabriel G. Bacheti ; Rafael M. Silva ; Vitor G. Neves ; Cloduado V. Sousa ; Seleme I. Seleme

AN IMPROVED IMPEDANCE ESTIMATION METHOD BASED ON POWER VARIATIONS IN GRID-CONNECTED INVERTERS ... 778

José H. Suárez ; Hugo M. C. Gomes ; Luan S. Santana ; Alfeu J. Sguarezi Filho ; Leandro L. O. Carralero ; Fabiano F. Costa

HYBRID SWITCHED CAPACITOR DC-DC CONVERTER BASED ON MMC ... 784

Marcus Vieira Soares ; Gustavo Lambert ; Yales Rômulo De Novaes

THERMAL STRESS EVALUATION OF A MULTIFUNCTIONAL MODULAR MULTILEVEL CONVERTER – STATCOM OPERATING AS ACTIVE FILTER ... 790

R. O. De Sousa ; W. C. S. Amorim ; D. C. Mendonça ; A. F. Cupertino ; L. M. F. Morais ; H. A. Pereira

TOTEM-POLE BRIDGELESS PFC CONVERTER IN DCM WITH SYNCHRONOUS RECTIFICATION ... 796

Leonardo S. Mai ; Alexsandra Rospirski ; Samir A. Mussa ; Telles B. Lazzarin

PASSIVE CAPACITOR VOLTAGE BALANCING IN MODULAR MULTILEVEL CONVERTER DURING ITS PRECHARGE: ANALYSIS AND DESIGN ... 802

Luiz H. T. Schmidt ; Gean J. M. De Sousa ; Marcelo L. Heldwein ; Daniel J. Pagano

DESIGN OF AN INTEGRATED CIRCUIT FOR LED DRIVING IN VISIBLE LIGHT COMMUNICATION APPLICATIONS ... 808

Nicolas Parma Rios ; Guilherme Márcio Soares ; Estêvão Coelho Teixeira

CONNECTION TIME IN MODBUS/TLS FOR SECURE COMMUNICATIONS ON PHOTOVOLTAIC SYSTEMS ... 814

Matheus K. Ferst ; Hugo F. M. De Figueiredo ; Gustavo W. Denardin

MODELING AND SIMULATION OF THE BATTERY ENERGY STORAGE SYSTEM FOR ANALYSIS IMPACT IN THE ELECTRICAL GRID. ... 820

Carolina A. Caldeira ; Henrique R. Schlickmann ; Antonio D. D. De Almeida ; Lucas V. Hartmann ; Camila S. Gehrke ; Fabiano Salvadori ; Gilielson F. Da Paz

COMPREHENSIVE AND DIDACTIC DC SERVOMOTOR CONTROL PLATFORM ... 826

Bruno De Almeida Regina ; Maria Júlia Rosa Aguiar ; André Augusto Ferreira

PERFORMANCE ANALYSIS OF ACTIVE ANTI-ISLANDING TECHNIQUES FOR PHOTOVOLTAIC APPLICATION ... 832

Marcos Roberto Oshiro ; Ruben Barros Godoy ; Moacyr A. G. De Brito ; Luigi Galotto

A HYBRID BIDIRECTIONAL PUSH-PULL DC-DC CONVERTER WITH A LADDER SWITCHED-CAPACITOR CELL ... 838

Marcos Dantas ; Fellipe Oliveira ; Lorena Albuquerque ; Isaac Freitas ; Romero Andersen

HIGH POWER FACTOR THREE-PHASE THREE-SWITCH STEP-DOWN CONVERTER ... 844

João Olímpio Caliman ; Walbermark Marques Dos Santos ; Tiara Rodrigues Smarssaro De Freitas ; Domingos Simonetti

SELECTION OF THE NUMBER OF LEVELS OF A MODULAR MULTILEVEL CONVERTER FOR AN ELECTRIC DRIVE ... 850

Paulo R. M. Júnior ; João. V. M. Farias ; Allan F. Cupertino ; Gabriel A. Mendonça ; Marcelo M. Stopa ; Heverton A. Pereira

HIGH VOLTAGE GAIN DC-DC CONVERTER BASED ON A SIMPLE CONFIGURATION OF SWITCHED CAPACITOR AND COUPLED INDUCTOR ... 856

Henrique J. Hoch ; Tiago M. K. Faistel ; Mauricio M. Da Silva ; António M. S. S. Andrade ; Mário L. Da S. Martins

WIND POWER SYSTEM CONNECTED TO THE GRID FROM SQUIRREL CAGE INDUCTION GENERATOR (SCIG) ... 862

Ângelo Marcílio M. Dos Santos ; Lucas Taylan P. Medeiros ; Leonardo P. S. Silva ; Ildenor Davi S. Junior ; Vanessa Siqueira De C. Teixeira ; Adson Bezerra Moreira

SINGLE-PHASE TO THREE-PHASE AC-DC-AC CONVERTER BASED ON CASCADED TRANSFORMERS RECTIFIER AND OPEN-END WINDING INDUCTION MOTOR ... 868

Antonio D. D. Almeida ; Nady Rocha ; Edgard L. L. Fabricio ; Carolina A. Caldeira ; Gleice M. S. Rodrigues ; Isaac S. Freitas

AN OVERVIEW ABOUT DETECTION OF CYBER-ATTACKS ON POWER SCADA SYSTEMS ... 874

Hugo F. M. De Figueiredo ; Matheus K. Ferst ; Gustavo W. Denardin

COMPARATIVE STUDY BETWEEN VIRTUAL SYNCHRONOUS MACHINE AND VIRTUAL IMPEDANCE TECHNIQUES FOR TWO PARALLELED INVERTERS SHARING A LOAD ... 880

Bruno Laurindo ; Fábio Alves ; Marcello Neves ; Jorge Caicedo ; Bruno França ; Maurício Aredes

ANALYSIS OF PARTIAL-POWER PROCESSING CONVERTERS FOR SMALL WIND TURBINES SYSTEMS ... 886

Anderson José Balbino ; Ronny Glauber De Almeida Cacau ; Telles Brunelli Lazzarin

PROPORTIONAL WAVELET SLIDING MODE CONTROLLER FOR TORQUE RIPPLE REDUCTION IN BLDC MOTOR ... 892

Mateus Moro Lumertz ; Carlos Matheus R. De Oliveira ; Allan Gregori De Castro ; Paulo Roberto U. Guazzelli ; Manuel Luis De Aguiar ; Jose Roberto B. A. Monteiro

VOLTAGE REGULATOR BEHAVIOR ON POWER DISTRIBUTION GRIDS WITH HIGH INTEGRATION OF PVDG ... 898

Lucas F. S. Azeredo ; Luiz G. R. Tonini ; Mariana A. Mendes ; Murillo C. Vargas ; Oureste E. Batista ; Carla J. Espindula

AN APPLICATION OF THE MULTI-PORT BIDIRECTIONAL THREE-PHASE AC-DC CONVERTER IN ELECTRIC VEHICLE CHARGING STATION MICROGRID ... 904

Raphael A. Da Câmara ; Luis M. Fernández-Ramírez ; Paulo P. Praça ; Demercil De S. Oliveira ; Pablo García-Triviño ; Raúl Sarrias-Mena

UNIFIED ROBUST CONTROL DESIGN FOR BTB-VSC SUBJECT TO UNCERTAINTIES IN GRID EQUIVALENT CIRCUIT ... 910

Marcelo De Castro ; Igor D. N. De Souza ; Gabriel A. Fogli ; Ademir Da S. T. Junior ; Pedro M. De Almeida ; Janaína G. De Oliveira ; Pedro G. Barbosa

A ISOP AC-AC HYBRID SWITCHED-CAPACITOR SRC FOR SOLID STATE TRANSFORMER APPLICATIONS ... 917

Victor Luiz Flor Borges ; Rogerio Luiz Da Silva ; Carlos Eduardo Possamai ; Arlan Luiz Bettiol ; Ivo Barbi

LOW COMPUTATIONAL COST TECHNIQUE FOR SPMSM SENSORLESS DRIVE USING ACTIVE FLUX CONCEPT...923
Camila R. S. Bartsch ; Luis F. F. De Campos ; Arthur G. Bartsch ; José De Oliveira ; Ademir Nied

A MODEL PREDICTIVE CONTROL APPLIED TO SINGLE-PHASE PACKED-U-CELLS CONVERTER...929
Jordan Zucuni ; Dimas Alã Schuetz ; Felipe Bovolini Grigoletto ; Fernanda Carnielutti De Morais ; Margarita Norambuena ; José Rodriguez ; Humberto Pinheiro

DESIGN OF ROBUST STRUCTURED CONTROL STRATEGY FOR SINGLE-PHASE DYNAMIC VOLTAGE RESTORER..934
João Amin Moor Neto ; Guilherme Giglio De Andrade ; Thiago Americano Do Brasil ; Mauro Sandro Dos Reis ; Julio Cesar Ferreira

REFERENCE GRID IMPEDANCE FOR TESTS OF GRID-CONNECTED POWER CONVERTERS FOR DISTRIBUTED ENERGY RESOURCES: THE BRAZILIAN CASE.....................940
Gabriel Avila Saccol ; Charles Schardong ; Leandro Michels ; Lucas Vizzotto Bellinaso ; Cassiano Rech

AN ELECTRONIC DRIVE FOR A SWITCHED RELUCTANCE MOTOR USING A DSC.................946
Brayan Sobral Da Fonseca ; José Andrés Santisteban

DC MOTOR MODEL FOR WINDOWS PINCH PROTECTION APPLICATIONS.........................952
Gabriel F. Idalgo ; José A. Torrico Altuna ; Carlos E. Capovilla ; Ademir Pelizari ; Wolfgang Schulter

POWER CONTROL AND HARMONIC CURRENT MITIGATION FROM A WIND POWER SYSTEM WITH PMSG...958
Leonardo P. S. Silva ; Denisia De V. Mota ; Flávia P. Ruiz ; Levy R. Cavalcante ; Lucas Taylan P. Medeiros ; Vanessa S. C. Teixeira ; Adson B. Moreira

PV EMULATOR BASED ON A FOUR-SWITCH BUCK-BOOST DC-DC CONVERTER.................964
Leandro L. O. Carralero ; Gabriel S. Barbara Da S. E Silva ; Fabiano F. Costa ; André P. N. Tahim

SINGLE-STAGE BRIDGELESS AC-DC PFC FLYBACK INTERLEAVED...................................969
G. S. Schowantz ; R. P. Barcelos ; T. B. Lazzarin

PREMAGNETIZED INDUCTORS IN SINGLE PHASE DC-AC AND AC-DC CONVERTERS...........975
Jens Friebe ; Siqi Lin ; Leon Fauth ; Tobias Brinker

THREE-PHASE INDUCTION MOTORS EFFICIENCY ANALYSIS USING A PROGRAMMABLE POWER SUPPLY...981
Cássio Alves De Oliveira ; Josemar Alves Dos Santos Junior ; Marcos José De Moraes Filho ; Vinícius Marcos Pinheiro ; Augusto W. Fleury Veloso Da Siveira ; Luciano Coutinho Gomes

REACTIVE POWER CONTROL OF DISTRIBUTED PHOTOVOLTAIC GENERATION SYSTEM IN LOW VOLTAGE ELECTRICAL GRIDS...987
Vanessa Da Costa Marques ; Rogério Gaspar De Almeida ; Nady Rocha ; Darlan Alexandria Fernandes ; Gleice Mylena Da Silva Rodrigues

MICROINVERTER WITH REDUCED NUMBER OF SEMICONDUCTOR SWITCHES..................993
Paulo R. Cagnini ; Luiz H. Meneghetti ; Victor E. S. Barbosa ; Emerson G. Carati ; Carlos M. Stein ; Zeno L. I. Nadal ; Jean M. S. Lafay ; Jean Patric Da Costa ; Rafael Cardoso

APPLICATION OF MODEL PREDICTIVE CONTROL IN A RESOLVER-TO-DIGITAL CONVERTER...998
Thyago Vasconcelos Estrabis ; Raymundo Cordero García ; Edson Antonio Batista ; Cristiano Quevedo Andrea ; Márcio Afonso Soleira Grassi

ON THE APPLICATION OF A POWER ELECTRONICS-BASED ARC-FLASH SUPPRESSOR.......1004
Fernando Venâncio Amaral ; Matheus H. M. Zanchetta Oliveira ; Claudio Alvares Conceição ; Sidelmo Magalhães Silva ; Cleber Onofre Inácio ; Braz De J. Cardoso Filho

DERIVING STABILITY CONDITION FOR ONE-CYCLE CONTROL WITH TRIANGULAR CARRIER BY POINCARÉ MAPS..1010
Armando J. G. Abrantes-Ferreira ; Lucas Do N. Gomes ; Robson F. Da S. Dias ; Luís G. B. Rolim

THREE-PHASE ADAPTIVE FREQUENCY ESTIMATOR WITH A DELAYED SIGNAL CANCELLATION PRE-FILTER UNDER HEAVILY DISTORTED GRID CONDITIONS.................1016
E. F. C. Grabovski ; M. L. Heldwein ; S. A. Mussa

ENHANCED PHASE-SHIFTED CARRIER PWM APPLIED TO 3-PHASE MULTILEVEL COUPLED INDUCTORS INVERTERS..1022
Emerson L. Soares ; Lucas Fabrício M. De Lucena ; Nady Rocha ; Cursino Brandão Jacobina ; Edison Roberto C. Da Silva

DIMENSIONING AND DEVELOPEMENT OF AN AC MICROGRID IN THE UFJF CAMPUS.......1028
Paula Stael S. Barbosa ; Marina V. C. Monteiro ; Jessica S. Dohler ; Andre Ferreira ; Janaina G. De Oliveira

UNBALANCED VOLTAGE MITIGATION USING D2VC WITH PROPORTIONAL RESONANT CONTROLLER IN Aß-FRAME..1034
Elienai O. Macedo ; Robson F. S. Dias ; Silvangela L. S. L. Barcelos ; Edson H. Watanabe

DYNAMIC MODELING AND CONTROL OF A THREE-PORT ZVS-PWM THREE-PHASE PUSH PULL DC-DC CONVERTER....................1040

Lorena Lorraine Oliveira Albuquerque ; Marcos Victor Dantas De Sá ; Fellipe André Lucena De Oliveira ; Isaac Soares De Freitas ; Romero Leandro Andersen

THERMAL MODELING OF CONVERTERS FOR WIND CONVERSION SYSTEMS EMPLOYING DFIG TECHNOLOGY....................1046

Rodrigues De Oliveira ; Tofoli Lessa ; Mendes Flores

INTERLEAVED BIDIRECTIONAL DC-DC CONVERTER FOR APPLICATION IN HYBRID PROPULSION SYSTEM: MODELING AND CONTROL....................1052

Vitor C. S. Torres ; Vinicius M. De Albuquerque ; Manuel A. Rendón ; Pedro S. Almeida ; Janaina G. Oliveira ; Márcio C. B. P. Rodrigues

ANALYSIS AND OPERATION OF A PV-BATTERY SYSTEM USING A MULTI-FUNCTIONAL CONVERTER....................1058

Jessica S. Dohler ; Lorrana F. Da Rocha ; Dalmo C. Silva Junior ; Pedro M. De Almeida ; Andre A. Ferreira ; Janaina G. Oliveira

STATOR CURRENT CONTROLLER FOR HARMONIC AND UNBALANCE COMPENSATION APPLIED TO SEIG BASED SYSTEMS....................1064

Gabriel Attuati ; Robinson Figueiredo De Camargo ; Lucas Giuliani Scherer

ENERGY-BALANCE BASED VOLTAGE REGULATION METHOD FOR MULTIPLE DC-LINKS IN ASYMMETRICAL CASCADED MULTILEVEL INVERTERS....................1070

Andre Felicio De Sousa Silva ; Sergio Pires Pimentel ; Enes Goncalves Marra ; Bernardo Pinheiro De Alvarenga

COMPARISON OF THE PLL CONTROL TECHNIQUES APPLIED IN PHOTOVOLTAIC SYSTEM....................1076

Marenice M. De Carvalho ; Renan L. P. Medeiros ; Iury V. Bessa ; Florindo A. C. Junior ; Kevin E. Lucas ; David A. Vaca

A CRITICAL ANALYSIS OF PSO AND ITS VARIATIONS APPLIED TO MPPT FOR PV SYSTEMS UNDER PARTIAL SHADING CONDITION....................1082

Lucas Mendonça Andrade ; Paula Dos Santos Vicente ; Fernando Lessa Tofoli ; Eduardo Moreira Vicente

EVALUATION OF TECHNIQUES TO REDUCE THE EFFECTS OF PARTIAL SHADING ON PHOTOVOLTAIC ARRAYS....................1088

André Augusto Rodrigues ; Paula Dos Santos Vicente ; Fernando Lessa Tofoli ; Eduardo Moreira Vicente

ANALYSIS OF NEUTRAL-POINT VOLTAGE BALANCING IN THREE-PORTS ACTIVE NEUTRAL-POINT-CLAMPED CONVERTER....................1094

Kaio Vinicius Vilerá ; Cassiano Rech ; Silvio Antonio Teston

DIRECT TORQUE CONTROL SCHEME FOR A NINE-PHASE INDUCTION MOTOR WITH REDUCED CURRENT HARMONIC....................1100

Gilielson F. Da Paz ; Isaac S. De Freitas ; Victor F. M. B. Melo ; Alexandre G. F. Da Silva

SINGLE-PHASE AC-DC-AC FIVE-LEVEL X-TYPE CURRENT SOURCE CONVERTER....................1106

Louelson A. Costa ; Montiê A. Vitorino ; Maurício B. R. Corrêa

APPLICATION OF A SHE-PWM MODULATION FOR A LOW SWITCHING FREQUENCY MOTOR DRIVE WITH HARMONIC INVESTIGATION USING THE DTFT....................1112

Marcos Paulo Brito Gomes ; Marina Hassen De Souza ; Gabriel Vilkn Ramos ; Alex-Sander Amável Luiz ; Marcelo Martins Stopa

A RESONANT-SWITCHED-CAPACITOR STEP-DOWN DC–DC CONVERTER IN CCM OPERATION AS AN LED DRIVER....................1118

Fábio De P. Neres ; Antonia F. Da Rocha ; Rodrigo L. Dos Santos ; Pedro S. Almeida ; Fernando L. M. Antunes ; Edilson M. Sá

HIGH-GAIN BIDIRECTIONAL DC-DC CONVERTER FOR BATTERY CHARGING IN DC NANOGRID OF RESIDENTIAL PROSSUMER....................1124

Fernando Queiroz ; Paulo Praça ; Alisson Freitas ; Fernando Antunes

CONCEPTS AND CASE STUDY OF MISMATCH LOSSES IN PHOTOVOLTAIC MODULES....................1130

Elson Yoiti Sakô ; João Lucas De Souza Silva ; Daniel De Bastos Mesquita ; Rafael Espino Campos ; Hugo Soeiro Moreira ; Marcelo Gradella Villalva

NOVEL BUCK-BOOST PFC CONVERTER WITH THREE-STATE SWITCHING CELL....................1136

Douglas Carvalho Morais ; Falcondes José Mendes De Seixas ; Luís De Oro Arenas ; Lucas Carvalho Souza ; Luciano De Souza Da Costa E Silva

CONTROL STRATEGY FOR MULTIFUNCTIONAL PV CONVERTER....................1142

Luiz Henrique Meneghetti ; Edivan Laercio Carvalho ; Emerson Giovani Carati ; Jean Patric Costa ; Carlos Marcelo Oliveira Stein ; Zeno Luiz Iensen Nadal ; Rafael Cardoso

PHOTOVOLTAIC BOOST CONVERTER CONTROL OPERATING IN THE MPPT AND LPPT MODES....................1148

Cássia C. C. Dos Santos ; Cassiano F. Moraes ; Jean P. Da Costa ; Carlos M. O. Stein ; Emerson G. Carati ; Rafael Cardoso

PREDICTIVE CONTROL FOR A HALF-CONTROLLED BOOST RECTIFIER 1154
Gleice Mylena Da S. Rodrigues ; Nady Rocha ; Edison Roberto C. Da Silva ; Filipe V. Rocha

LIFETIME INVESTIGATION OF DC-LINK CAPACITORS IN MULTIPLE SLIM DRIVES SYSTEM 1160
Shili Huang ; Haoran Wang ; Dinesh Kumar ; Guorong Zhu ; Huai Wang

DIRECT POWER CONTROL WITH SPACE VECTOR MODULATION APPLIED FOR THE BRUSHLESS DC MOTOR 1166
Henrique De Toledo ; José L. Azcue

MODELING AND CONTROL OF A FORWARD DC-DC CONVERTER FOR BATTERY VOLTAGE BALANCING 1172
Pábulo F. Ciarnoscki ; Daniel J. Pagano ; Gabriel M. Da Silva

ENERGY YIELD ASSESSMENT METHODOLOGY FOR PHOTOVOLTAIC MICROINVERTERS 1178
Andrii Chub ; Kosenko Roman ; Oleksandr Korkh ; Dmitri Vinnikov ; Samir Kouro

DIRECT POWER CONTROL STRATEGY TO ENHANCE THE DYNAMIC BEHAVIOR OF DFIG DURING VOLTAGE DIP 1183
Fernando Lino ; Rogério Vani Jacomini ; Alfeu J. Sguarezi Filho

ELECTRICAL SIMULATION OF TRACTION SUBWAY SYSTEM FOR ENERGY RECOVERY AND ENERGY SAVING STUDIES 1189
Marcio Annibal Pimenta ; Wilson Komatsu ; Lourenço Matakas Junior

MODELING AND SIMULATION OF A STIRLING-ENGINE-BASED GENERATOR CONNECTED TO THE GRID 1195
Augusto Hayashi ; Moacyr Brito ; João Onofre ; Lourenço Matakas ; Raymundo Cordero ; Luigi Galottojunior

COMPARATIVE STRATEGIES OF CONTROL FOR REGENERATIVE BRAKING IN ELECTRIC VEHICLES 1201
Marina G. S. P. Paredes ; José Antenor Pomilio

THEORETICAL SOLUTION OF THE OUTPUT VOLTAGE HARMONIC SPECTRA OF DUAL-INVERTER FED OPEN-END WINDING LOADS WITH DEAD TIME EFFECT 1206
Brunno Monteiro Guimarães Ribeiro ; Marcelo Martins Stopa ; Alex-Sander Amável Luiz

BRIDGELESS BUCK-BOOST PFC CONVERTER WITH THREE-STATE SWITCHING CELL 1211
Douglas Carvalho Morais ; Falcondes José Mendes De Seixas ; Luís De Oro Arenas ; Lucas Carvalho Souza ; Luciano De Souza Da Costa E Silva

EXPERIMENTAL RESULTS OF A BIDIRECTIONAL COUPLED INDUCTOR DC-DC CONVERTER 1217
Murilo Brunel Da Rosa ; Menaouar Berrehil El Kattel ; Robson Mayer ; Sérgio Vidal Garcia Oliveira

DESIGN AND TEST OF A SRF-PLL BASED ALGORITHM FOR POSITIVE-SEQUENCE SYNCHROPHASOR MEASUREMENTS 1223
Gabriel Ubirajara De Carvalho ; Gustavo Weber Denardin ; Rafael Cardoso ; Cassiano Ferro Moraes

ECONOMIC ANALYSIS OF A PEAK SHAVING SYSTEM WITH DIESEL GENERATOR 1229
Myrlena R. M. Ferreira ; Luiz A. De S. Ribeiro ; José G. De Matos

THYRISTOR TRIGGERING, STATIC AND DYNAMIC CHARACTERISTICS 1235
Ricardo Alves Do Prado ; Ulrich Nicolai

DEVELOPMENT OF A FPGA-BASED REAL-TIME SIMULATION SYSTEM 1241
Yago F. Oliveira ; Filipe A. La-Gatta ; Rodrigo A. F. Ferreira ; Márcio C. B. P. Rodrigues

POWER ELECTRONICS LAB: CONVERGING KNOWLEDGE AND TECHNOLOGIES 1247
José Antenor Pomilio

INTEGRATION OF SOLAR PHOTOVOLTAIC (PV) SYSTEMS WITH CCM INVERTERS INTO VCM DROOP-CONTROLLED ISLANDED AC MICROGRIDS 1253
Marcus E. T. Souza ; Fernando C. Melo ; Ernane A. A. Coelho ; Luiz C. G. De Freitas

CLOSED-FORM SOLUTIONS FOR CORE AND WINDING LOSSES CALCULATION IN SINGLE-PHASE BOOST PFC RECTIFIERS 1259
Marcos José Jacoboski ; André De Bastiani Lange ; Marcelo Lobo Heldwein

DEVELOPMENT OF A HYBRID PV-THERMOELECTRIC SYSTEM 1265
Rafael Magalhães Nóbrega De Araújo ; Hugo Álisson Alves Da Costa ; João T. De Carvalho Neto ; Alexandre Magnus F. Guimarães ; Andrés Ortiz Salazar

EXPERIMENTAL WORKBENCH: A TOOLS TO HELP TO TEACHING THE TECHNIQUES OF DRIVES OF ELECTRIC MACHINES 1271
Victor Camargo Reis ; Carlos Henrique Silva De Vasconcelos

PREDICTING THE LIFE TIME OF POWER SEMICONDUCTOR MODULES 1277
Clovis N. L. Gajo ; Arendt Wintrich ; Paul Drexhage

WORKBENCH FOR MONITORING AND OPERATION OF ELECTRIC MACHINES 1283
Guilherme Afonso Pillon De C. A. Pessoa ; Evandro Ailson De Freitas Nunes ; Werbet Luiz Almeida Da Silva ;
Ricardo Ferreira Pinheiro ; Andres Ortiz Salazar

**DC-LINK VOLTAGES BALANCE METHOD FOR SINGLE-PHASE NPC INVERTER
OPERATING WITH REACTIVE POWER COMPENSATION**.. 1289
Marco Fajardo ; Julio Viola ; José Manuel Aller ; Flavio Quizhpi

**NEW FIVE-LEVEL FLYING CAPACITOR INVERTER FED BY A BOOST-FLYBACK DC-DC
VOLTAGE SOURCE**... 1295
Antonio Venâncio De M. Lacerda Filho ; André E. L. Da Costa ; Edison R. C. Da Silva ; Ronnan De B. Cardoso ;
Darlan A. Fernandes

**INTEGRATED ZETA INVERTER APPLIED IN A SINGLE-PHASE GRID-CONNECTED
PHOTOVOLTAIC SYSTEM** 1301
Leonardo Poltronieri Sampaio ; Sérgio Augusto Oliveira Da Silva ; Paulo Júnior Silva Costa

**A NOVEL PHASE-LOCKED LOOP WITH POSITIVE AND NEGATIVE SEQUENCE
DETECTION CAPABILITY**.. 1307
Jean M. L. Fonseca ; Francisco Kleber A. Lima ; Carlos Gustavo C. Branco ; Renato G. Araujo

**PERFORMANCE ANALYSIS OF ALTERNATIVES SWITCHING COMMANDS FOR THREE-
LEVEL BOOST RECTIFIER WITH HYSTERESIS CURRENT CONTROL.**.. 1313
Gabriel Vilkn Ramos ; Alex-Sander Amável Luiz ; Marcelo Martins Stopa ; Marcos Paulo Brito Gomes ; Ramon
Henriques De Souza ; Daniel Franco Leal

HIGH POWER FACTOR RECTIFIER USING ONE CYCLE CONTROL STRATEGY 1319
Amanda Thayla S Monteiro ; Rafael Rocha Matias ; Jailson Leite Silva ; Jose Antonio Dos Santos Neto

**A COMPARISON AMONG DIFFERENT FINITE CONTROL SET APPROACHES AND CONVEX
CONTROL SET MODEL-BASED PREDICTIVE CONTROL APPLIED IN A THREE-PHASE
INVERTER WITH RL LOAD** 1325
Arthur G. Bartsch ; Dayse M. Cavalcanti ; Mariana S. M. Cavalca ; Ademir Nied

**DESIGN AND ANALYSIS OF OUTPUT FILTER WITH LONG LIFETIME E-CAP FOR AC-DC
LED DRIVER** 1331
Andressa Da S. Fernandes ; Felipe C. Do Nascimento ; Antonia F. Da Rocha ; Rodrigo L. Dos Santos ; Fernando
L. M. Antunes ; Edilson M. Sá

HYBRID MODEL OF ELECTRIC VEHICLE 1337
Débora De Souza Martins ; Danilo P. E Silva ; Jussara Farias Fardin

**STATIC TRANSFER SWITCH APPLIED TO SINGLE-PHASE UNINTERRUPTIBLE POWER
SUPPLY**... 1343
Marcus Vinícius Maia Rodrigues ; Newton Da Silva ; Flávio Alessandro Serrão Gonçalves

**ANALYSIS OF RECTIFIERS FOR RF ENERGY HARVESTING AIMING LOW POWER
SENSING APPLICATIONS**.. 1349
Humberto Pereira Da Paz ; Vinícius Santana Da Silva ; Eduardo Vicente Valdés Cambero ; Humberto Xavier De
Araújo ; Ivan Roberto Santana Casella ; Carlos Eduardo Capovilla

**A SIMPLIFIED ANALYSIS OF BUCK-TYPE INTERLEAVED DC-DC CONVERTER FOR
BATTERY CHARGERS APPLICATION**... 1354
Menaouar Berrehil El Kattel ; Robson Mayer ; Maicon Douglas Possamai ; Sérgio Vidal Garcia Oliveira

**PERFORMANCE OF A MULTI-GROUNDED DISTRIBUTION NETWORK WITH A FOUR-
WIRE THREE-PHASE POWER CONDITIONER**.. 1360
Samuel N. Duarte ; Bruno C. Souza ; Pedro M. Almeida ; Leandro R. Araújo ; Pedro G. Barbosa

**A NEW STRATEGY OF MODULATION BASED ON SPACE VECTOR MODULATION AND
ANNOTATED PARACONSISTENT LOGIC FOR A THREE-PHASE CONVERTER**....................... 1366
Antonio Carlos Duarte Ricciotti ; João Inácio Da Silva Filho ; Raphael A. Bispo De Oliveira ; Viviane B. Da S.
Duarte Ricciotti ; Hyghor Miranda Côrtes ; André Miguel Nicolini

**TECHNIQUES OF SOLAR IRRADIANCE ESTIMATION FROM DATASHEET INFORMATION
OF PHOTOVOLTAIC PANELS** 1372
Itaiara F. Carvalho ; Maurício B. R. Correa

**EXTRA REACTIVE POWER ANALYSIS ON A DISTRIBUTION GRID WITH HIGH
INTEGRATION OF PV GENERATION**.. 1378
Daniel C. Pompermayer ; Caroline Marin ; Mariana A. Mendes ; Luann G. O. Queiroz ; Luiz G. R. Tonini ;
Murillo C. Vargas ; Oureste E. Batista

ALTERNATIVE FCS-MPC CONCEPTS FOR CASCADE FREE MOTOR SPEED CONTROL 1384
Filipe Fernandes ; José De Oliveira ; Ademir Nied ; Arthur G. Bartsch

**COMPARATIVE STUDY OF THE POWER SHARING TECHNIQUES FOR MICROGRIDS IN
AUTONOMOUS MODE**.. 1390
Gustavo M. S. Azevedo ; Erik C. Oliveira ; Marcelo C. Cavalcanti ; Leonardo R. Limongi

HYBRID ONE-CYCLE CONTROL STRATEGY FOR BUCK+BOOST PFC BATTERY CHARGER 1396
Aluísio Alves De Melo Bento ; Raphael Perci Santiago

TRAPEZOIDAL CURRENT MODE FOR BIDIRECTIONAL HIGH STEP RATIO MODULAR MULTILEVEL DC-DC CONVERTER 1402

C. Pineda ; J. Pereda ; S. Neira ; P. Bravo ; J. Rodríguez ; C. García

HOMOTHETIC METHOD TO COMPUTE WINDING LOSSES IN THE DESIGN OF POWER INDUCTORS 1408

André Furlan ; Alvaro Morentin ; Guillaume Fontes ; Guillaume Delamare ; Marcelo L. Heldwein ; Thierry Meynard

EVALUATION OF POWER QUALITY IMPACTS DUE TO PHOTOVOLTAIC PENETRATION IN DISTRIBUTION GRIDS VIA TIME-DOMAIN SIMULATION 1414

Henrique Pires Corrêa ; Flávio Henrique Teles Vieira

CONTROL STRATEGY AND POWER MANAGEMENT FOR MULTIFUNCTIONAL INVERTERS WITH BESS AND REACTIVE POWER COMPENSATION 1420

Luiz Henrique Meneghetti ; Edivan Laercio Carvalho ; Gustavo B. K. Schmidt ; Emerson Giovani Carati ; Jean Patric Da Costa ; Carlos Marcelo De Oliveira Stein ; Zeno Luiz Iensen Nadal ; Rafael Cardoso

SIZING OF SUPERCAPACITOR AND BESS FOR PEAK SHAVING APPLICATIONS 1426

Kaique Ferreira ; Walbermark Marques Dos Santos ; Augusto César Rueda Medina

TWO-STAGE SEPIC-BUCK TOPOLOGY FOR NEIGHBORHOOD ELECTRIC VEHICLE CHARGER 1431

Rafael H. Eckstein ; Telles B. Lazzarin ; Gierri Waltrich

MODELING BATTERY ENERGY STORAGE SYSTEM OPERATING IN DC MICROGRID WITH DAB CONVERTER 1435

Rafael Dos Santos ; Flávio A. S. Gonçalves ; José A. Olimpio Filho ; Fernando P. Marafão ; Eric Gil

MULTI-MODULAR SCALABLE DC-AC POWER CONVERTER FOR CURRENT INJECTION TO THE GRID BASED ON PREDICTIVE VOLTAGE CONTROL 1441

S. Toledo ; M. Rivera ; E. Maqueda ; M. Ayala ; J. Pacher ; C. Romero ; R. Gregor ; T. Dragicevic ; P. Wheeler

FSC-MPC CURRENT CONTROL OF A 5-LEVEL HALF-BRIDGE/ANPC HYBRID THREE-PHASE INVERTER 1447

Jailson Leite Silva ; Rafael Rocha Matias ; Max Dannyel De Carvalho Alves ; Ranoyca Nayana Alencar Leão E Silva ; Amanda Thayla Silva Monteiro ; Kristian Pessoa Dos Santos

MODELING AND SIMULATION OF A STIRLING ENGINE IN SCILAB 1453

Raymundo Cordero ; Thyago Estrabis ; João Onofre ; Felipe Alexandre Monteiro ; Augusto Hayashi

A LOW COST PHOTOVOLTAIC PANEL EMULATOR 1458

Aluísio Alves De Melo Bento ; Raphael Perci Santiago

COMBINING MODEL-BASED AND EXTREMUM SEEKING CONTROL FOR FAST TRACKING THE MAXIMUM POWER POINT OF LUNDELL ALTERNATOR 1463

Gabriel Sales Lins Rodrigues ; Alexandre Cunha Oliveira ; Antonio Marcus Nogueira Lima

Z-SOURCE INVERTER FOR PHOTOVOLTAIC MICROGENERATION 1469

Felipe A. F. Almeida ; Felipe Guerra ; Flávio Alessandro Serrão Gonçalves

IFOC FOR A NINE-PHASE INDUCTION MOTOR DRIVE WITH CURRENT HARMONIC INJECTION 1475

Alexandre G. F. Da Silva ; Isaac S. De Freitas ; Simplicio A. Da Silva ; Victor F. M. B. Melo ; Gilielson F. Da Paz

A COMPARISON OF A DISCRETE-TIME PI AND AN INDIRECT MPC CURRENT CONTROLLERS FOR A SINGLE-PHASE GRID-CONNECTED INVERTER OPERATING WITH DISTORTED GRID AND SIGNIFICANT COMPUTATION FEEDBACK DELAY 1481

Sergio Pires Pimentel ; Oleksandr Husev ; Dmitri Vinnikov ; Serhii Stepenko ; Lauri Kutt ; Jose Rodriguez

A NON-ISOLATED DC-DC BOOST CONVERTER WITH THREE-STATE SWITCHING CELL 1487

Lucas Carvalho Souza ; Falcondes José Mendes De Seixas ; Luís De Oro Arenas ; Douglas Carvalho Morais ; Luciano De Souza Da Costa E Silva

INPUT VOLTAGE RANGE EXTENSION METHODS IN THE SERIES-RESONANT DC-DC CONVERTERS 1493

Andrii Chub ; Dmitri Vinnikov ; Jih-Sheng Lai

EXTENDED KALMAN FILTER BASED SPEED ESTIMATION FOR THE CONTROL OF PMSG 1499

Mukhtiar Singh

Author Index

Impact of Capacitor Design Methodology on FC Inverters

1st Marcos Vinicius Bressan
Department of Electrical Engineering
Santa Catarina State University (UDESC)
Joinville, Brazil
bressan.marcosvinicius@gmail.com

2nd Cassiano Rech
Department of Electrical Energy Processing
Federal University of Santa Maria (UFSM)
Santa Maria, Brazil
rech.cassiano@gmail.com

3rd Alessandro L. Batschauer
Department of Electrical Engineering
Santa Catarina State University (UDESC)
Joinville, Brazil
alessandro.batschauer@udesc.br

Abstract—This work evaluates two capacitors design methodologies for Flying Capacitors Inverters. The first design methodology is developed through analysis of one switching period of the flying capacitor inverter. The second one analyses a complete output period of the inverter using the double Fourier series. Bolt methodologies use of the Phase-Shift Pulse Width Modulation. Experimental and numeric results verify the impacting of the two methodologies design in a five-level flying capacitor inverter.

Index Terms—Capacitors Design, Flying Capacitor Converter, Multilevel Inverters.

I. Introduction

In the applications with dc-ac energy conversion, the multilevel inverters has been getting attention due to its better characteristic when comparing with two-level voltage-source inverter (VSI), such as: high quality of the output waveform, electromagnetic compatibility and semicondutor's lower blocking voltage [1]–[3]. However, the multilevel inverters are more complexity than VSI, spending more time and efforts to develop [4], [5]. So, the design of multilevel inverters must be based on a theoretical background, for optimize the development time and to provide good support for the hardware designer.

One of the main topologies of the multilevel converters is the Neutral Point Clamped (NPC) converter. The simplicity design allows its widely used by the industry, particularly in motor drives and power systems applications [6], [7].

Another multilevel converter employed in motor drive and power conditioners applications is the Cascaded H-Bridge Converter (CHB). The CHB structure is widely explored and well known, and it is made by H-bridge converters in series connection. The new challenges for the CHB converters are related to the control techniques, as the power flow control [8]–[10].

The Modular Multilevel Converter (MMC) presents an ideal structure for high-voltage applications, its modular structure allows to use of the low-voltage semiconductors in high-voltage applications [11]–[15]. The MMC has been the main focus of the a lot researches in the last years and its structure is consolidated (based on half-bridge or full-bridge converters). The new aim of the researchers is in the development of new applications and control strategies [16], [17].

The Flying Capacitor Converter (FC) was introduced in [18] and, recently, many papers have explored the new techniques to control the flying capacitor voltages [19]–[21]. However, a few papers have been explored the design the flying capacitors, especially when the FC is employed as an inverter. In [22] is proposed a design methodology of flying capacitor for FC converter in dc-dc energy conversion, and some works [23]–[27] have used this methodology to design the flying capacitors in dc-ac energy conversion applications. However, to design the flying capacitors in dc-ac energy conversion other aspects must be considered, such as: the output current phase angle and the amplitude modulation index. These two aspects are considered in [28], where the design of the flying capacitors are performed using the double Fourier series in the analysis.

In this paper design methodologies proposed for FC inverters are explained and compared, shown the impacting in design of flying capacitors for two different applications of the FC, as voltage source inverter and reactive power conditioner.

Following the sections of this paper the FC are analyzed in Section II. Section III shows the design methodologies and its the impacting in the size of the flying capacitors. Section IV includes some experimental results for a single-phase 5L-FC-FC.

II. Flying Capacitor Converter

The 5L-FC uses the eight switches and three flying capacitors, as shown in Fig. 1 [18].

978-1-7281-4181-7/19 $31.00 © 2019 IEEE

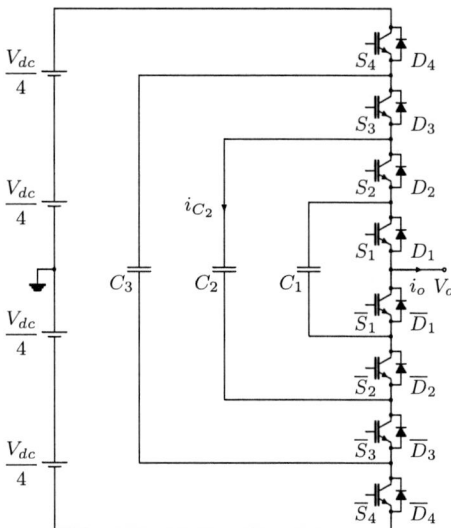

Fig. 1. 5L-FC Converter.

For operation of the 5L-FC, the flying capacitors voltages must be stable and in nominal rating, as defined by

$$V_{C_k} = \frac{k\,V_{dc}}{4}, \tag{1}$$

where $k \in [1,\ldots,3]$ and V_{dc} is the dc source voltage. However, each switching state has a different impact in the charging and discharging of the flying capacitors. This away, the flying capacitor current i_{C_k} is defined by the output current i_o and by the switching functions of the nearest switches, S_k and S_{k+1}, given by

$$i_{C_k}(t) = [S_{k+1}(t) - S_k(t)]\,i_o(t). \tag{2}$$

The stability condition for operation oh the FC inverter is obtained if the average current through the flying capacitors is equal to zero. This condition is achieved if the switches $S_{k+1}(t)$ and $S_k(t)$ have the same duty cycle in a switching period, if the output current can be considered constant in a switching period (the output's fundamental period is larger than the switching period). The PS-PWM modulation achieves the naturally stability for the flying capacitor voltages and it is commonly used in the FC converters [29], [30]. This reason takes the PS-PWM to be the focus of this work and all the analyses are for the FC converter operating with this modulation.

III. Design of the Flying Capacitors

There are different types and sizes of the capacitors available in the market by different manufacturers and the hardware designer has a difficult choice to define which power capacitor should be employed. The impact of this choice is directly linked with capacitor's lifetime, the safe and proper operation of the FC inverter. So, the design methodology used must be in agreement with the application of the FC converter, for dc-dc or dc-ac

energy conversion. Different of the dc-dc energy conversion applications, in dc-ac energy conversion, two important variables must be considered: the output current phase angle and the index modulation value. It can be proved by analyzing the design methodologies proposed in [22] and [28].

In [22] the design methodology is evaluated for one switching period, analyzing the ripple voltage and the current through in the flying capacitors. Thus, few variables are considered in the design methodology, such as: the number of output voltage levels n, the voltage ripple of the flying capacitors ΔV_C, the carrier frequency f_c and the maximum output current I_p. The capacitance of the flying capacitors is given by

$$C_k = \frac{I_p}{(n-1)\,\Delta V_{C_k}\,f_c}. \tag{3}$$

The simplistic design methodology proposed by [22] is a good alternative for FC converter in dc-dc energy conversion applications. Once the methodology only considers in one switching period in its analyses.

Therefore, in dc-ac energy conversion applications, it is important to analyze the impacts of the switching operation for complete output period as have been done in [28]. In [28], the design methodology considers the impact of the switching operation through of the harmonic representation using of the double Fourier series analysis. This analysis is presented in [29] where the switching functions are defined as

$$S_k(t) = \frac{C_{00}}{2} + C_{01}\cos\left(\omega_o\,t + \theta_o\right)$$
$$+ \sum_{g=1}^{\infty}\sum_{h=-\infty}^{\infty} C_{gh}\cos\left[\begin{array}{c} g\left(\omega_c\,t + \theta_{c,k}\right) \\ + h\left(\omega_o\,t + \theta_o\right) \end{array}\right], \tag{4}$$

whereas the ω_c, ω_o, θ_c and θ_o, are the carrier and reference frequencies, and phase angles, respectively. As shown in [31] the baseband, carrier, and side-band harmonic magnitudes for the PS-PWM are given by:

$$C_{00} = 1, \tag{5}$$

$$C_{01} = \frac{M_a}{2}, \tag{6}$$

$$C_{gh} = \frac{2}{g\,\pi}\sin\left[(g+h)\,\frac{\pi}{2}\right] J_h\left(g\,\frac{\pi}{2}\,M_a\right), \tag{7}$$

for $g = 1, 2, \ldots, \infty$, $h = -\infty, \ldots, -1, 0, 1, \ldots, \infty$, where $J_h(x)$ is a Bessel function of the first kind with argument and order h and M_a is the amplitude index modulation defined by

$$M_a = \frac{V_{o_{pp}}}{V_{dc}}, \tag{8}$$

where $V_{o_{pp}}$ is the output peak-to-peak voltage.

978-1-7281-4181-7/19 $31.00 © 2019 IEEE

Replacing (4) into (2), the current through the flying capacitors is given by

$$i_{C_k}(t) = I_p \sin(\omega_o t - \Phi)$$
$$\times \sum_{g=1}^{\infty} \sum_{h=-\infty}^{\infty} C_{gh} \left\{ \cos \begin{bmatrix} g(\omega_c t + \theta_{c_{k+1}}) \\ + h(\omega_o t + \theta_o) \end{bmatrix} \right. $$
$$\left. - \cos \begin{bmatrix} g(\omega_c t + \theta_{c_k}) \\ + h(\omega_o t + \theta_o) \end{bmatrix} \right\}. \tag{9}$$

Therefore, because of the complex mathematical operations some graphics are generated to design the flying capacitors. Using the graphics to determine phase angle factor correction F_A and frequency factor correction F_C, the capacitance are given by

$$C_k = \frac{I_p F_A F_C}{\Delta V_{C_k} - 2 I_p R_{ESR}}, \tag{10}$$

where R_{ESR} is the capacitor series resistance. The factors F_A and F_C are obtained in Fig. 2 for a 5L-FC converter. For FC converters with different numbers of the output levels, others graphics must be generated, as presented in [28].

In spite of the design of the flying capacitors is more complex from [28], important variables are included. So, disagreement in the results of these two design methodologies presented are expected, especially in conditions when the FC converter is delivering active or reactive power. In the Table I, it can be verified the capacitance value for 5L-FC operating as dc-ac converter for these two points of operation.

For the Cases 1 and 2, the FC converter operates with low value of the output current phase angle, in other words, it just delivers active power to a load or into the grid, the reactive power energy processed is negligible, as in applications for grid-connected photovoltaic systems [27], [32]. In the cases 3 and 4, the design of the capacitor is done for the case where the FC converter just delivers reactive power energy to a load or into the grid (high value of the output current phase angle), e.g., operating as reactive power conditioners (STATCOM) [33], [34].

As shown in the Table I, both design methodologies present different capacitance values for each case, it means that the flying capacitor can be undersized or oversized depending of the converter application. To verify the differences of the previous analyses, experimental results are obtained using a 5L-FC-FC converter.

IV. Experimental Results

The experimental results have been obtained with 5L-FC-FC converter, as shown in Fig. 3. The structure of the 5L-FC-FC converter is basically the H-bridge connection of the two 3L-FC converter. This structure in H-Bridge connection uses the same PS-PWM as in the 5L-FC, it

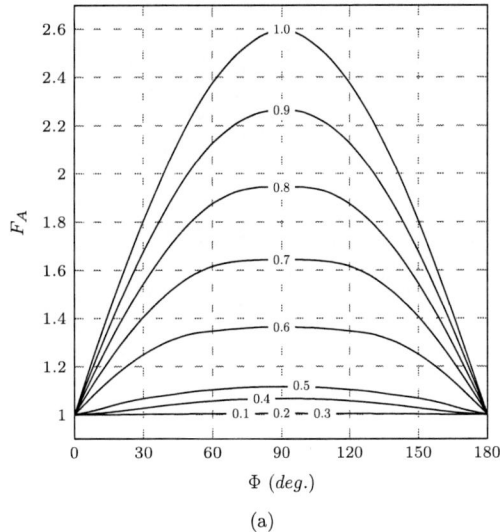

(a)

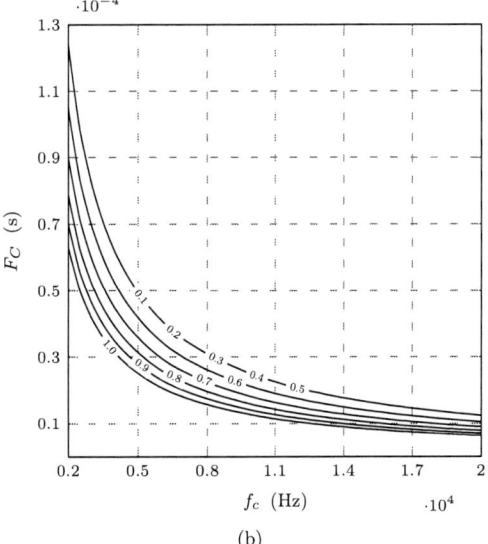

(b)

Fig. 2. (a) Phase angle factor correction F_A and (b) carrier frequency factor correction F_C for the 5L-FC.

means the graphics obtained for the 5L-FC can be used for the 5L-FC-FC. Table II shows the prototype parameters.

Due to the dead-time and delays in the switching commands and to improve the response at transient loads it is used an active control of the flying capacitor voltages. Thus, imbalances in the flying capacitor voltages are avoided and it ensures stability for the operation of the converter. The active control is done with little variations in modulation signals to keep the flying capacitors voltages stables. In steady-state operation conditions, these little variations are negligible, and it does not interfere in the experimental results obtained. The active control used is similar that present [35], and as the design of the active control is not the goal of this paper, it will not be introduced.

TABLE I
Comparing of design methodologies for cases operation of the 5L-FC.

Design Parameters	Case 1	Case 2	Case 3	Case 4
dc voltage source	$V_{dc} = 400\,V$	$V_{dc} = 400\,V$	$V_{dc} = 400\,V$	$V_{dc} = 400\,V$
Flying cap. volt. ripple	$\Delta V_{C_k} = 10\,V$	$\Delta V_{C_k} = 10\,V$	$\Delta V_{C_k} = 10\,V$	$\Delta V_{C_k} = 10\,V$
Output peak current	$I_p = 6.428\,A$	$I_p = 6.428\,A$	$I_p = 6.428\,A$	$I_p = 6.428\,A$
Output frequency	$f_o = 50\,Hz$	$f_o = 50\,Hz$	$f_o = 50\,Hz$	$f_o = 50\,Hz$
Carrier frequency	$f_c = 2\,kHz$	$f_c = 2\,kHz$	$f_c = 2\,kHz$	$f_c = 2\,kHz$
Modulation index	$M_a = 0.7$	$M_a = 0.9$	$M_a = 0.7$	$M_a = 0.9$
Phase angle	$\Phi = 0° - 5°$	$\Phi = 0° - 5°$	$\Phi = 75° - 90°$	$\Phi = 75° - 90°$
Flying cap. ESR	$R_{ESR} = 120\,\Omega$	$R_{ESR} = 120\,m\Omega$	$R_{ESR} = 120\,m\Omega$	$R_{ESR} = 120\,m\Omega$
Proposed in [22]	$C_k = 80.35\,\mu F$	$C_k = 80.35\,\mu F$	$C_k = 80.35\,\mu F$	$C_k = 80.35\,\mu F$
Proposed in [28]	$C_k = 66.9\,\mu F - 72.2\,\mu F$	$C_k = 52.44\,\mu F - 59.26\,\mu F$	$C_k = 109.7\,\mu F - 109.7\,\mu F$	$C_k = 117.47\,\mu F - 119.1\,\mu F$

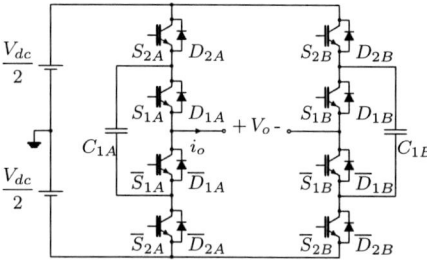

Fig. 3. H-bridge connection of 3L-FC converters (5L-FC-FC).

TABLE II
Parameters of the 5L-FC-FC prototype.

Parameters	Value
dc voltage source	$V_{dc} = 400\,V$
Flying capacitor voltages	$V_{C_k} = 200\,V$
Flying capacitors rating	$C_k = 120\,\mu F$
Flying cap. ESR	$R_{ESR} = 120\,m\Omega$
Carrier frequency	$f_c = 2\,kHz$
Output frequency	$f_o = 50\,Hz$

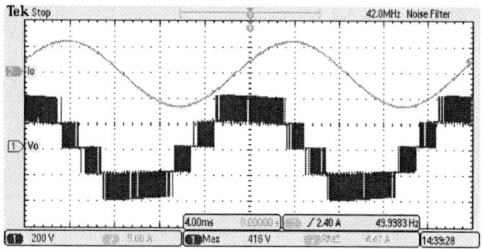

(a) Output voltage $V_o(t)$ and output current $i_o(t)$.

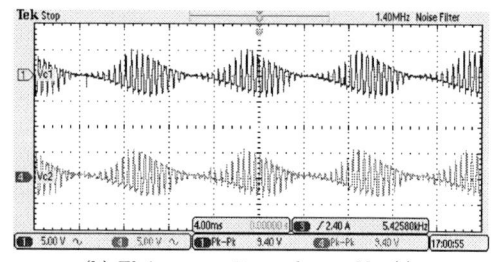

(b) Flying capacitor voltages $V_{C_k}(t)$.

Fig. 4. Experimental results of the 5L-FC-FC.

Fig. 4 shows the experimental results of the output current and voltage of the 5L-FC-FC in steady-state conditions. The voltage waveform presents five-level steps due to the operating of the converter with $M_a = 0.9$, and the output current presents low ripple due to the high inductance of the RL load, with phase angle of the $\Phi = 67°$. In steady-state conditions, the voltage in the flying capacitors maintain stable, with average value of $200\,V$ and the ripple value of $9.4\,V$. The experimental results show appropriate operation of the 5L-FC-FC, keeping the capacitor voltages within the nominal ratings.

These operating conditions are nearby of the "Case 4", in Table I, and once the capacitance value of the prototype capacitors ($C_k = 120\,\mu F$) are closer of the capacitance value of the methodology proposed in [28] ($C_k = 117.47\,\mu F$) than design methodology proposed in [22] ($C_k = 80.35\,\mu F$), showing divergence of the two design methodologies.

Others experimental results for different operating points of the inverter are evaluated and shown in

Fig. 5. The experimental results have been obtained from measuring the output current and the flying capacitor voltages, and through analysis by computer software, they are presented in Fig. 5.

The little differences between the experimental results and the theoretical graphics can be explained due to measuring disturbances of the voltage ripple in the flying capacitors, which turn the measured noise into a source of imprecision.

V. Conclusion

This paper verified the impact of the design methodologies of the flying capacitors when the FC converter is used in dc-ac conversion. Two design methodologies were evaluated for different cases of operation, when the FC converter delivers active or reactive energy to a load or into the grid. All the cases, the design methodologies presented different results for the capacitance value.

978-1-7281-4181-7/19 $31.00 © 2019 IEEE

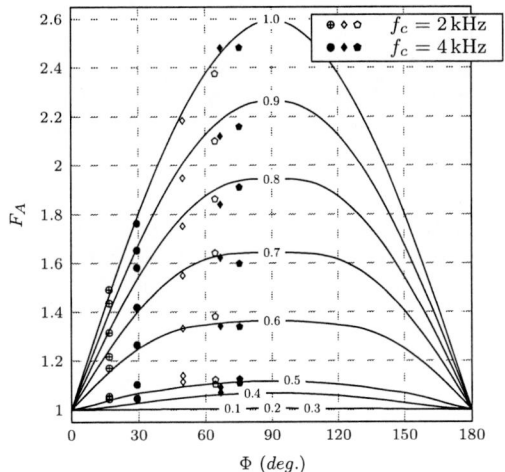

Fig. 5. Experimental results of the phase angle factor correction as a function of the output phase angle and amplitude modulation index using different loads.

The experimental results obtained with the 5L-FC-FC for different operation points, verified that the design methodology proposed in [28] presented the better results for the design of flying capacitors, once this methodologies consider in its analyses two important variables, as the index modulation value and the output current phase angle.

The methodology proposed in [28] has proven to be a good alternative for design the flying capacitor when the FC converter is used in dc-ac energy conversion. Thus, ensuring the flying capacitor do not be undersized or oversized.

Acknowledgment

The authors would like to thank, the FAPESC, FAPERGS, CNPq, INCT-GD, Capes, Procad, FITEJ and UDESC for the financial support.

References

[1] J. Rodriguez, J.-S. Lai, and F. Z. Peng, "Multilevel inverters: a survey of topologies, controls, and applications," IEEE Transactions on Industrial Electronics, vol. 49, no. 4, pp. 724–738, Aug. 2002.

[2] L. G. Franquelo, J. Rodriguez, J. I. Leon, S. Kouro, R. Portillo, and M. A. M. Prats, "The age of multilevel converters arrives," IEEE Industrial Electronics Magazine, vol. 2, no. 2, pp. 28–39, Jun. 2008.

[3] S. Kouro, M. Malinowski, K. Gopakumar, J. Pou, L. G. Franquelo, B. Wu, J. Rodriguez, M. A. Perez, and J. I. Leon, "Recent advances and industrial applications of multilevel converters," IEEE Transactions on Industrial Electronics, vol. 57, no. 8, pp. 2553–2580, Aug. 2010.

[4] K. K. Gupta, A. Ranjan, P. Bhatnagar, L. K. Sahu, and S. Jain, "Multilevel inverter topologies with reduced device count: A review," IEEE Transactions on Power Electronics, vol. 31, pp. 135–151, Jan. 2016.

[5] M. Norambuena, S. Kouro, S. Dieckerhoff, and J. Rodriguez, "Reduced multilevel converter: A novel multilevel converter with a reduced number of active switches," IEEE Transactions on Industrial Electronics, vol. 65, no. 5, pp. 3636–3645, May 2018.

[6] S. Madhusoodhanan, K. Mainali, A. Tripathi, D. Patel, A. Kadavelugu, S. Bhattacharya, and K. Hatua, "Harmonic analysis and controller design of 15 kv sic igbt-based medium-voltage grid-connected three-phase three-level npc converter," IEEE Transactions on Power Electronics, vol. 32, no. 5, pp. 3355–3369, May 2017.

[7] C. Bharatiraja, R. Selvaraj, T. R. Chelliah, J. L. Munda, M. Tariq, and A. I. Maswood, "Design and implementation of fourth arm for elimination of bearing current in npc-mli-fed induction motor drive," IEEE Transactions on Industry Applications, vol. 54, no. 1, pp. 745–754, Jan. 2018.

[8] A. Marzoughi, R. Burgos, D. Boroyevich, and Y. Xue, "Design and comparison of cascaded h-bridge, modular multilevel converter and 5-l active neutral point clamped topologies for motor drive application," IEEE Transactions on Industry Applications, pp. 1–1, 2017.

[9] H. Geng, S. Li, C. Zhang, G. Yang, L. Dong, and B. Nahid-Mobarakeh, "Hybrid communication topology and protocol for distributed-controlled cascaded h-bridge multilevel statcom," IEEE Transactions on Industry Applications, vol. 53, no. 1, pp. 576–584, Jan. 2017.

[10] E. Behrouzian and M. Bongiorno, "Investigation of negative-sequence injection capability of cascaded h-bridge converters in star and delta configuration," IEEE Transactions on Power Electronics, vol. 32, no. 2, pp. 1675–1683, Feb. 2017.

[11] H. Akagi, "Classification, terminology, and application of the modular multilevel cascade converter (mmcc)," IEEE.

[12] E. Solas, G. Abad, J. Barrena, S. Aurtenetxea, A. Carcar, and L. Zajac, "Modular multilevel converter with different submodule concepts - part ii: Experimental validation and comparison for hvdc application," IEEE.

[13] M. Glinka and R. Marquardt, "A new ac/ac multilevel converter family," IEEE Transactions on Industrial Electronics, vol. 52, no. 3, pp. 662–669, June. 2005.

[14] L. M. Cunico, Y. R. D. Novaes, A. Nied, and S. V. G. Oliveira, "Inner current control method for modular multilevel converter applied in motor drive," in 10th IEEE/IAS International Conference on Ind. Application, 2012, pp. 1–8.

[15] T. Nakanishi and J. I. Itoh, "High power density design for a modular multilevel converter with an h-bridge cell based on a volume evaluation of each component," IEEE Transactions on Power Electronics, vol. 33, no. 3, pp. 1967–1984, Mar. 2018.

[16] A. Nami, J. Liang, F. Dijkhuizen, and G. D. Demetriades, "Modular multilevel converters for hvdc applications: Review on converter cells and functionalities," IEEE Transactions on Power Electronics, vol. 30, no. 1, pp. 18–36, Jan. 2015.

[17] J. Wang, J. Liang, C. Wang, and X. Dong, "Circulating current suppression for mmc-hvdc under unbalanced grid conditions," IEEE Transactions on Industry Applications, vol. 53, no. 4, pp. 3250–3259, Jul. 2017.

[18] T. A. Meynard and H. Foch, "Multi-level conversion: high voltage choppers and voltage-source inverters," in Power Electronics Specialists Conference, 1992, pp. 397–403.

[19] A. M. Y. M. Ghias, J. Pou, M. Ciobotaru, and V. G. Agelidis, "Voltage-balancing method using phase-shifted pwm for the flying capacitor multilevel converter," IEEE Transactions on Power Electronics, vol. 29, no. 9, pp. 4521–4531, Sept. 2014.

[20] A. K. Sadigh, V. Dargahi, and K. Corzine, "Calculation of conduction power losses in double flying capacitor multicell converter," in IEEE Application Power Electronics Conference and Exposition, 2015, pp. 2351–2357.

[21] A. M. Y. M. Ghias, J. Pou, G. J. Capella, P. Acuna, and V. G. Agelidis, "On improving phase-shifted pwm for flying capacitor multilevel converters," IEEE Transactions on Power Electronics, vol. 31, no. 8, pp. 5384–5388, Aug. 2016.

[22] F. Hamma, T. Meynard, F. Tourkhani, and P. Viarouge, "Characteristics and design of multilevel choppers," in IEEE Power Electronics Specialists Conference, 1995, pp. 1208–1214.

[23] S. S. Fazel, S. Bernet, D. Krug, and K. Jalili, "Design and comparison of 4-kv neutral-point-clamped, flying-capacitor, and series-connected h-bridge multilevel converters," IEEE Transactions on Industry Applications, vol. 43, no. 4, pp. 1032–1040, Jul. 2007.

978-1-7281-4181-7/19 $31.00 © 2019 IEEE

[24] M. R. Islam, Y. Guo, J. G. Zhu, and D. Dorrell, "Design and comparison of 11 kv multilevel voltage source converters for local grid based renewable energy systems," in 37th Annual Conference of the IEEE Industry Electronics Society, 2011, pp. 3596–3601.

[25] Z. Lim, A. I. Maswood, and G. H. P. Ooi, "Modular-cell inverter employing reduced flying capacitors with hybrid phase-shifted carrier phase-disposition pwm," IEEE Transactions on Industrial Electronics, vol. 62, no. 7, pp. 4086–4095, Jul. 2015.

[26] C. B. Barth, T. Foulkes, W. H. Chung, T. Modeer, P. Assem, P. Assem, Y. Lei, and R. C. N. Pilawa-Podgurski, "Design and control of a gan-based, 13-level, flying capacitor multilevel inverter," in IEEE 17th Workshop on Control and Modeling for Power Electronics, 2016, pp. 1–6.

[27] Y. Lei, C. Barth, S. Qin, W. C. Liu, I. Moon, A. Stillwell, D. Chou, T. Foulkes, Z. Ye, Z. Liao, and R. C. N. Pilawa-Podgurski, "A 2-kw single-phase seven-level flying capacitor multilevel inverter with an active energy buffer," IEEE Transactions on Power Electronics, vol. 32, no. 11, pp. 8570–8581, Nov. 2017.

[28] M. V. Bressan, C. Rech, and A. L. Batschauer, "Design of flying capacitors for n-level fc and n-level smc," International Journal of Electrical Power & Energy Systems, vol. 113, pp. 220 – 228, 2019.

[29] B. P. McGrath and D. G. Holmes, "Analytical modelling of voltage balance dynamics for a flying capacitor multilevel converter," IEEE Transactions on Power Electronics, vol. 23, no. 2, pp. 543–550, Mar. 2008.

[30] T. A. Meynard, M. Fadel, and N. Aouda, "Modeling of multi-level converters," IEEE Transactions on Industrial Electronics, vol. 44, no. 3, pp. 356–364, Jun. 1997.

[31] D. Holmes and T. Lipo, Pulse Width Modulation For Power Converters. Piscataway: Wiley-IEEE Press, 2003.

[32] Y. P. Siwakoti and F. Blaabjerg, "A novel flying capacitor transformerless inverter for single-phase grid connected solar photovoltaic system," in 2016 IEEE 7th International Symposium on Power Electronics for Distributed Generation Systems (PEDG), June 2016, pp. 1–6.

[33] C. Hochgraf, R. Lasseter, D. Divan, and T. A. Lipo, "Comparison of multilevel inverters for static var compensation," in Proceedings of 1994 IEEE Industry Applications Society Annual Meeting, Oct 1994, pp. 921–928 vol.2.

[34] C. J. Nwobu, I. B. Efika, O. J. K. Oghorada, and L. Zhang, "A modular multilevel flying capacitor converter-based statcom for reactive power control in distribution systems," in 2015 17th European Conference on Power Electronics and Applications (EPE'15 ECCE-Europe), Sep. 2015, pp. 1–9.

[35] M. Khazraei, H. Sepahvand, K. A. Corzine, and M. Ferdowsi, "Active capacitor voltage balancing in single-phase flying-capacitor multilevel power converters," IEEE Transactions on Industrial Electronics, vol. 59, no. 2, pp. 769–778, Feb. 2012.

978-1-7281-4181-7/19 $31.00 © 2019 IEEE

Modeling and Control of a Back-to-Back system for turboelectric propulsion

Saulo O. Nascimento, Vinicius M. de Albuquerque, Marcelo de C. Fernandes,
Manuel A. Rendón, Janaína G. Oliveira, Pedro S. Almeida

Federal University of Juiz de Fora (UFJF)
Modern Lightining Research Group(NIMO)
Juiz de Fora,MG, Brazil
oliveira.saulo@engenharia.ufjf.br,vinicius.albuquerque@engenharia.ufjf.br, decasm3@rpi.edu
manuel.rendon@ufjf.edu.br,janaina.oliveira@ufjf.edu.br,pedro.almeida@ufjf.br

Abstract—**This paper studies the aspects of turbo-electric system control regarding the power conversion topology. Both Modeling and control of a back-to-back system connected to synchronous machines in both ends are considered. An instantaneous power technique in the $\alpha\beta$-frame is used to control the active rectifier connected to the permanent magnet synchronous generator (PMSG) while the inverter associated in the motor speed control uses synchronous (dq) frame robust control technique. The system considers a synchronous motor of $42\,kW$ connected to a $220\,V$ PMSG by a $670\,V$ DC bus.Mathematical modeling and simulation are used to provide insights on the whole system.**

I. Introduction

Turboelectric propulsion uses electrical generators to convert mechanical energy from a turboshaft turbine into electrical energy and electric motors to produce mechanical work on an output shaft [1]. The first turboelectric drive was presented in the USA in the year 1917 [2]. The question of weight in the turboelectric system was always problematic making its use very restricted in the beginning. However, the use of turboelectric has increased with new technologies for warships, rail locomotives (using gas turbines) and the advantage of the turboelectric system is providing electricity to warships and trains and communication equipment.

In the present work, a back-to-back converter structure will be used because of the necessity of the connection of a system in alternating current that works on a specific frequency (given by a PMSG), and another AC system working on another frequency (for PMSM speed regulation, a variable frequency is required), connected by a DC bus. Each subsystem will be represented by a bidirectional 3-phase voltage source converter. Each of the subsystems will have a specific function. The first subsystem consists of a 60 Hz PMSG connected to a PWM rectifier and the second subsystem consists of a PWM inverter connected to a permanent magnet synchronous motor (PMSM), a motor that became very popular nowadays because of the compact size, high efficiency and lower weight specially important in

aircraft environment [3].In this article, The PWM Inverter will be primarily responsible for the control of direct axis current to ensure maximum torque on the shaft and the mechanical angular speed of the PMSM. On the Inverter-motor side, state feedback technique and feedback gain calculations by robust control (through LMI resolution) will be used.The PWM Rectifier will be responsible for maintaining high power factor and performing bus voltage control at the PMSG-rectifier side. For the PMSG side, the p-q control will be used, where the active and reactive powers are controlled directly through instantaneous power theory, a reliable control technique capable of ensuring balanced sinusoidal currents are drained from the machine. With these two forms of control, the desired functionability will be achieved.

The paper is organized as it follows: Section II deals with in higher detailed with the system in question. In section II, the mathematical modeling of both subsystems are presented. In section III the controllers' design is detailed for both sides, allowing the simulations results in section IV which show the system behavior under disturbances. Finally, conclusions will be drawn in Section V.

II. Methodology

A. System under study

The following complete topology of a back-to-back converter connected to synchronous machines on both sides is shown in Fig. 1.

B. Mathematical Model of the PMSM Subsystem

This subsystem is composed of an inverter connected PMSM.The control of inverter switching that powers the PMSM is responsible for speed control as well as current control to achieve maximum torque.The state space model of the PMSM is described by the differential equations of the system, considering the representation of the PMSM in the direct and quadrature axes. The system dynamics is described by Eq.(1) [4]:

978-1-7281-4181-7/19 $31.00 © 2019 IEEE

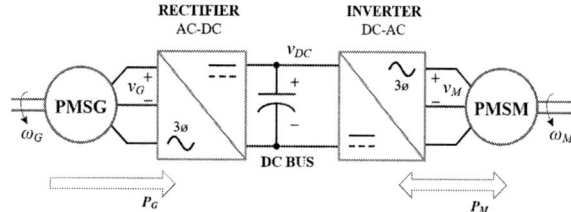

Fig. 1. The complete topology of the system under study - Back-to-Back system

$$
\begin{cases}
\dfrac{di_d}{dt} = \dfrac{v_d}{L_d} - \dfrac{R_M i_d}{L_d} + \dfrac{L_q P \omega_m i_q}{L_d} \\[2mm]
\dfrac{di_q}{dt} = \dfrac{v_q}{L_q} - \dfrac{R_M i_q}{L_q} - \dfrac{P \omega_m L_d i_d}{L_q} - \dfrac{P \omega_m \phi_{pm}}{L_q}
\end{cases}
\tag{1}
$$

The variables i_d and i_q are the direct and quadrature currents, respectively; R_M represents the PMSM internal resistance, L_q and L_d are the quadrature and direct axis inductances; ϕ_{pm} represents the permanent magnet flux linkage; ω_m is the rotor angular speed and P are pairs of poles. Once the mechanical angular speed is a desired control variable, its derivative must be described as a state variable. In order to find the third equation, electromagnetic torque needs to be found by Eq.(2):

$$
T_{em} = \frac{3}{2} P(\phi_{pm} i_q + (L_d - L_q) i_d i_q)
\tag{2}
$$

As can be seen, if the saliency of the machine is ignored such that the direct and quadrature inductance are almost the same, the direct axis current presents no effect on the torque magnitude, meaning that the torque can be controlled only by the component iq. The machine studied in this paper, however, does presents slightly values of L_d and L_q and this effects will be considered on the modeling and simulation. Yet, specifically for torque maximization the simplification will be considered once the inductance values (see table I) are very similar. Therefore, maximum torque equation can be simplified to Eq.(3).

$$
T_{em} = \frac{3}{2} P(\phi_{pm} i_q)
\tag{3}
$$

Thus, the PMSM rotation is described through Eq.(4).

$$
J_M \frac{d\omega_m}{dt} = T_{em} - B_M \omega_m - T_m
\tag{4}
$$

Where B_M is the viscous friction coefficient and J_M is the moment of inertia. Thus, the third equation of the mathematical model of the PMSM with inverter is found as Eq.(5).

$$
\frac{d\omega_m}{dt} = \frac{3 P \phi_{pm} i_q}{2 J_M} - \frac{T_m}{J_M} - \frac{B_m \omega_m}{J_M}
\tag{5}
$$

The state-space equations of the system are described in its matricial form by Eq.(6):

$$
\frac{d}{dt}
\begin{bmatrix} i_d \\ i_q \\ \omega_m \end{bmatrix}
=
\begin{bmatrix}
-\frac{R_M}{L_d} & P\omega_m \frac{L_q}{L_d} & 0 \\
-P\omega_m \frac{L_d}{L_q} & \frac{-R_M}{L_q} & -P\frac{\phi_{pm}}{L_q} \\
0 & \frac{3P\phi_{pm}}{J_M} & \frac{-B_M}{J_M}
\end{bmatrix}
\begin{bmatrix} i_d \\ i_q \\ \omega_m \end{bmatrix}
+
$$

$$
+
\begin{bmatrix}
\frac{1}{L_d} & 0 \\
0 & \frac{1}{L_q} \\
0 & 0
\end{bmatrix}
\begin{bmatrix} v_d \\ v_q \end{bmatrix}
+
\begin{bmatrix} 0 \\ 0 \\ \frac{-1}{J_M} \end{bmatrix}
\begin{bmatrix} T_m \end{bmatrix}
\tag{6}
$$

The product of state variables by control variables in the Eqs. 1 returns a non-linear set, making linearization necessary. A procedure for system decoupling will be adopted according to [5]. An alternative way to linearize the matrix is shown in [6]. Firstly, it can be considered the relation provided by Eq.(7).

$$
\begin{bmatrix} v_d \\ v_q \end{bmatrix}
= K_{inv}
\begin{bmatrix} m_d \\ m_q \end{bmatrix}
\tag{7}
$$

To perform the system decoupling of the system described, new variables will be created as higlited in (8).

$$
\begin{cases}
u_{d0} = -\dfrac{P\omega_m \phi_q}{K_{inv}} \\[2mm]
u_{q0} = \dfrac{P\omega_m \phi_d}{K_{inv}}
\end{cases}
\tag{8}
$$

Where ϕ_d and ϕ_q are the flux linkage in the direct and quadrature axes, respectively, given by Eq.(9).

$$
\begin{cases}
\phi_d = L_d i_d + \phi_{pm} \\
\phi_q = L_q i_q
\end{cases}
\tag{9}
$$

The variable K_{inv} represents the gain of the inverter, given by $K_{inv} = \dfrac{V_{DC}}{2}$. In this way, by inserting the new variables of Eq.(8) into Eq. (1) leads to Eq. (10).

$$
\begin{cases}
K_{inv} v_{dd} = K_{inv}(m_d - u_{d0}) = \frac{R_M i_d}{L_d} + \frac{d\phi_d}{dt} \\[2mm]
K_{inv} v_{qq} = K_{inv}(m_q - u_{d0}) = \frac{R_M i_q}{L_q} + \frac{d\phi_q}{dt}
\end{cases}
\tag{10}
$$

By executing the linearization of the system through the change of variable described by equation 10, we find the new decoupled system by Eq.(11).

$$
\frac{d}{dt}
\begin{bmatrix} i_d \\ i_q \\ \omega_m \end{bmatrix}
=
\begin{bmatrix}
-\frac{R_M}{L_d} & 0 & 0 \\
0 & \frac{-R_M}{L_q} & 0 \\
0 & \frac{3\phi_{pm}P}{2J_M} & \frac{-B_M}{J_M}
\end{bmatrix}
\begin{bmatrix} i_d \\ i_q \\ \omega_m \end{bmatrix}
+
$$

$$
+
\begin{bmatrix}
\frac{V_{DC}}{2L_d} & 0 \\
0 & \frac{V_{DC}}{2L_q} \\
0 & 0
\end{bmatrix}
\begin{bmatrix} v_{dd} \\ v_{qq} \end{bmatrix}
+
\begin{bmatrix} 0 \\ 0 \\ \frac{-1}{J_M} \end{bmatrix}
\begin{bmatrix} T_m \end{bmatrix}
\tag{11}
$$

978-1-7281-4181-7/19 $31.00 © 2019 IEEE

C. Mathematical model of PMSG subsystem

The modeling for the subsystem represented by the rectifier connected to a 60 Hz PMSG (Modeled as a 3-phase source) is described using the differential equations of the system. The rectifier is represented by a three-phase voltage source converter (VSC) system. As pq-theory will be applied, the subsystem can be represented in the coordinate frame $\alpha\beta$ expressing terminal voltages in terms of its modulation indices through Eq.(12).

$$
\begin{cases}
L_G \dfrac{di_{\alpha G}}{dt} = -R_G i_{\alpha G} - \dfrac{V_{DC}}{2} m_{\alpha G} + v_{\alpha,pac} \\[2mm]
L_G \dfrac{di_{\beta G}}{dt} = -R_G i_{\beta G} - \dfrac{V_{DC}}{2} m_{\beta G} + v_{\beta,pac}
\end{cases}
\tag{12}
$$

Where $i_{\alpha G}$, $i_{\beta G}$ are the PMSG currents in the $\alpha\beta$ frame, respectively. R_G is the PMSG line resistance; $v_{\alpha,pac}$ and $v_{\beta,pac}$ are the voltages at the $\alpha\beta - frame$ that can not be controlled [4]. However, the equations of the currents in $\alpha\beta$-frame are not sufficient to describe the whole subsystem. In this way, it is necessary to pay attention to the power balance:

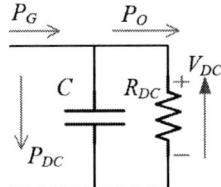

Fig. 2. PMSG power balance

Writing the equation:

$$
P_G = P_{DC} + P_o
\tag{13}
$$

Where P_{DC} represents a power consumed by the capacitor, given by Eq. (14)

$$
P_{DC} = C_{DC}(V_{DC})^2
\tag{14}
$$

And P_o is the power consumed by the load and can be given by Eq.(15).

$$
P_o = \frac{V_{DC}^2}{R_{DC}}
\tag{15}
$$

In this way, we can rewrite the Eq.(13) as Eq.(16).

$$
P_G = p = V_{DC}\left(C_{DC}\frac{dV_{DC}}{dt} + \frac{V_{DC}}{R_{DC}}\right)
\tag{16}
$$

III. Control Design

A. PMSM control design

There are several control strategies for certain PMSM applications. Some strategies are summarized as follows [7]:

- Direct shaft current control at zero for maximum torque;
- Flux weakening control;
- Optimum torque control per amper.

The PMSM control strategy will be to control the direct axis current at zero, maintaining the 90° torque angle, which represents the angle between the rotor flux and the stator magnetic field that provides the maximum torque [7]. To ensure that the steady-state error is null for references of continuous values, the control loop must contain two added integrators (one for each controlled variable), invariably creating new state variables x_{id} and x_{ω_m}. Thus, the added states are described by Eq.(17).

$$
\frac{d}{dt}\begin{bmatrix} x_{id} \\ x_{\omega_m} \end{bmatrix} = \begin{bmatrix} -1 & 0 & 0 \\ 0 & 0 & -1 \end{bmatrix}\begin{bmatrix} i_d \\ i_q \\ \omega_m \end{bmatrix}
\tag{17}
$$

Equation (18) represents the augmented system [12].

$$
\frac{d}{dt}\begin{bmatrix} i_d \\ i_q \\ \omega_m \\ x_{id} \\ x_{\omega_m} \end{bmatrix} = \begin{bmatrix} -\frac{R_M}{L_d} & 0 & 0 & 0 & 0 \\ 0 & \frac{-R_M}{L_q} & 0 & 0 & 0 \\ 0 & \frac{3\phi_{pm}P}{J_M} & \frac{-B_M}{J_M} & 0 & 0 \\ -1 & 0 & 0 & 0 & 0 \\ 0 & 0 & -1 & 0 & 0 \end{bmatrix}\begin{bmatrix} i_d \\ i_q \\ \omega_m \\ x_{id} \\ x_{\omega_m} \end{bmatrix} +
$$

$$
+ \begin{bmatrix} \frac{V_{DC}}{2L_d} & 0 \\ 0 & \frac{V_{DC}}{2L_q} \\ 0 & 0 \\ 0 & 0 \\ 0 & 0 \end{bmatrix}\begin{bmatrix} v_{dd} \\ v_{qq} \end{bmatrix} + \begin{bmatrix} 0 \\ 0 \\ \frac{-1}{J_M} \\ 0 \\ 0 \end{bmatrix}\begin{bmatrix} T_m \end{bmatrix}
\tag{18}
$$

In this way, it is possible to calculate the feedback gain using MATLAB® Robust control toolbox(by solving linear matrix inequalities, or linear matrix inequality - LMI), which allows you to create an S region where the controller will be stable and also obey the desired performance parameters for every operating point [11],The result is highlighted in Eq.(19).

$$
i_d\,loop \qquad\qquad\quad \omega_m\,loop
$$
$$
\begin{cases} k_{11}i_d = & 1.8\cdot10^{-3} \\ k_{12}i_d = & 0 \\ k_{13}i_d = & 0 \\ k_{14}i_d = & -0.26 \\ k_{15}i_d = & 0 \end{cases}
\begin{cases} k_{21}\omega_m = & 0 \\ k_{22}\omega_m = & 2.2\cdot10^{-3} \\ k_{23}\omega_m = & 3.6\cdot10^{-2} \\ k_{24}\omega_m = & 0 \\ k_{25}\omega_m = & -3.15 \end{cases}
\tag{19}
$$

B. PMSG control Design

The control mode used in this work will be the current control mode, where initially the AC side current is controlled and the active and reactive powers are controlled by the phase angle and the line current amplitude of the VSC. An advantage of this current control mode is, the VSC is protected against overload conditions. Thus, by returning to the set of equations given in Eq.(12), we can obtain the transfer function of the plant through Eq.(20).

$$G_{i\alpha\beta}(s) = \frac{V_{\alpha G} - \dfrac{m_{\alpha G} V_{DC}}{2}}{R_G + sL_G} \qquad (20)$$

By developing the above equation for small signals (noises), knowing that one can neglect the product between two small-signal elements (resulting in a value close to zero) we obtain the transfer function $G_{i\beta}(s)$ through Eq.(21).

$$G_{i\alpha\beta}(s) = \frac{\tilde{i}_{\alpha\beta G}}{\tilde{m}_{\alpha\beta G}} = -\frac{\dfrac{V_{DC}}{2}}{R_G + sL_G} \qquad (21)$$

Returning to Eq.(16), by executing the Laplace transform for analysis of controller design in the frequency domain, Eq.(22) can be found representing the instantaneous active power required to keep the capacitor charged. Isolating V_{DC}^2:

$$V_{DC}^2 = p \frac{R_{DC}}{1 + sR_{DC}C_{DC}} \qquad (22)$$

In this way, the transfer function $G_v(s)$ is characterized by Eq.(23).

$$G_v(s) = \frac{V_{DC}^2}{p} = \frac{R_{DC}}{1 + sR_{DC}C_{DC}} \qquad (23)$$

With the well-defined transfer functions $G_{i\alpha\beta}(s)$ and $G_v(s)$, the respective compensators can be designed. For $G_{i\alpha\beta}(s)$, a compensator $C_{i\alpha\beta}(s)$ represented by a PR controller, and a PI controller for $Gv(s)$. In the control system, there is an external loop that calculates reference currents $i_{\alpha G}$ and $i_{\beta G}$ given the values of active and reactive power specified, according to power theory presented in [8]. The internal loop is represented by a current control loop. A complete block diagram of the control system is displayed in Fig 3.

Treating $CLTF_i$ as the slower closed-loop transfer function of the internal current loop (by reducing the block diagram), a block diagram of the outer control system for the rectifier can be represented as shown in Fig. 4, where $T_i(s) = C_{i\alpha\beta}(s).G_{i\alpha\beta}(s)$ and $G_v(s)$ is the control plant. The compensator $C_{i\alpha\beta}$ is resonant controller, capable of following sinusoidal references. Equation (24) shows its transfer function.

$$C_{i\alpha\beta}(s) = \frac{K_p s^2 + 2K_i s + \omega_0^2 K_p}{s^2 + \omega_0^2} \qquad (24)$$

Considering that the internal current loop has a slower dynamics of at least 1/10 of the switching frequency, the natural frequency of the current loop is given by Eq.(25).

$$\omega_{CL} = 0.1 \cdot 2\pi \cdot f_s \qquad (25)$$

Given the desired switching frequency equal to 20 kHz, the natural frequency of the loop is given by $\omega_{CL} = 12566.370\,rad/s$. The calculation of K_p proceeds [9] according to Eq.(26).

$$K_p(s) = \frac{\omega_{CL} L_G}{2V_{DC}} \qquad (26)$$

With $\omega_{CL} = 12566.370 rad/s$, the inductance of the PMSG as $L_G = 4mH$ and considering a voltage bus of $V_{DC} = 670V$, the gain is found to be $K_p(s) = 3.75 \cdot 10^{-2}$. Based on the calculation of $K_p(s)$ in Eq.(26), the integer gain K_i is found through Eq.(27), where t_r is represented by the controller response time, evaluated in a range of $10 - 90\%$.

$$K_i(s) = \frac{2.2K_p(s)}{t_r} \qquad (27)$$

Considering $t_r = 1ms$, the value of gain $K_i(s)$ is given by $K_i(s) = 82.52$.

The voltage controller, represented by $C_v(s)$, has an PI transfer function given by Eq.(28).

$$C_v(s) = K \frac{1 + sT}{sT} \qquad (28)$$

Where K is the gain k of PI controller and T is the time constant. Using the Sisotool toolbox from MATLAB®, the chosen values for the PI that yields desired conditions are $K = 0.0049$ and $T = 2.6 \cdot 10^{-4}\,s$.

IV. SIMULATION RESULTS

A. PMSM and VSC Inverter Simulation Results

For the PMSM, the PMSM parameters were included in the PSIM® software model to simulate the voltage source converter (VSC Inverter) controlling the motor speed and current, using the previously mentioned control strategy that attempts to cancel the d-axis current. Consider the Table I with the simulation parameters, including data of the motor used, the EMRAX 228 [10].

The simulation was performed such that a mechanical load of $T_m = 50\,Nm$ is connected to the motor in stationary state at $t = 0\,s$ but with an acceleration of $1000\,rpm/s$ such that the desired operational speed in steady state is reached at $t = 0.2\,s$. At $t = 0.3\,s$ an event is simulated such that a disturbance causes the load to be reduced to $T_m = 30\,Nm$. Nominal conditions are returned at $t = 0.6\,s$. The simulation response for PMSM can be seen in Fig. 5.

As can be seen, the controlled system responds to ramp references of velocity, allowing a soft start. During this time, once acceleration is required, the electromagnetic torque T_{em} is different and higher than the load torque

978-1-7281-4181-7/19 $31.00 © 2019 IEEE

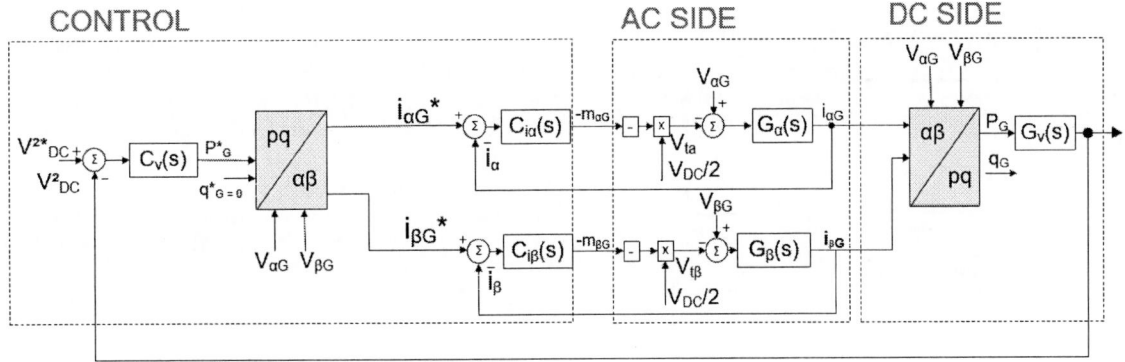

Fig. 3. Control block diagram

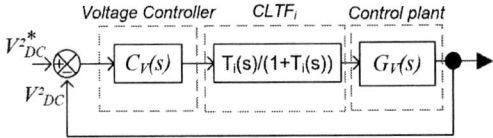

Fig. 4. Reduced control block diagram

TABLE I
SIMULATION PARAMETERS FOR PMSM

Description	Symbol	Value
Switching Frequency	f_s	$20\,kHz$
DC bus Voltage	V_{DC}	$670\,V$
Stator resistance	R_M	$18m\,\Omega$
d-axis Inductance	L_d	$0.183\,mH$
q-axis Inductance	L_q	$0.177\,mH$
Pairs of Poles	P	10
Moment of Inertia	J_M	$421\,kgcm^2$
Magnetic Flux - Axial	ϕ_{pm}	$0,0542\,V.s$
Viscous friction coefficient	B_M	$0.005\,Nms$
Nominal Speed	ω_m	$2000\,rpm$
Max Load Torque	T_m	$50\,Nm$

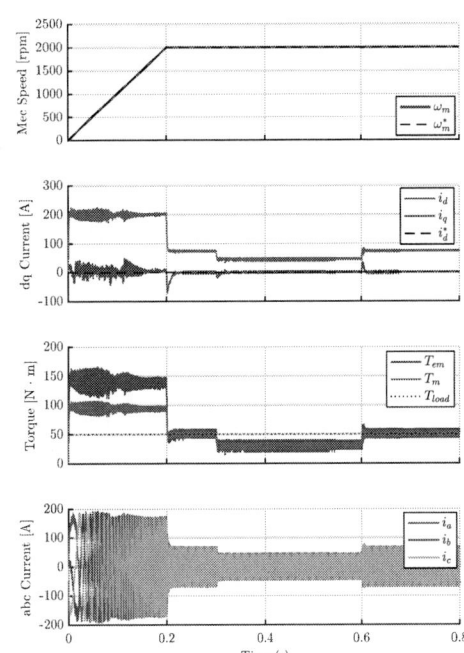

Fig. 5. Simulation response for the PMSM mechanical speed control.

T_m. Once steady state is reached, the T_m and T_{em} are the same, with value equal of the load in steady state. Reducing the load reduces the value of both torques once the machine remains at constant peed and on acceleration is required. The system returns to nominal conditions and no significant overshoot is noticed.

B. PMSG and Rectifier Simulation Results

The simulation of the system was done in MATLAB®
to calculate the gains of the proportional and resonant controllers. Finally, the gains obtained were inserted in the second program developed in the PSIM® in order to validate the developed project. Table II highlights the simulation parameters of the PMSG-Rectifier system.

The simulation results can be seen in Fig. 6. The system is initiated at $t = 0\,s$ with the capacitor charged and the rectifier delivering nominal power to the load. At time $t = 0.25\,s$ a disturbance was added reducing the power in half, with the system returning no nominal values at $t = 0.6\,s$.

As can be seen, the capacitor voltage , albeit suffering from significant spikes due to the violent load transition, remains controlled at nominal value. Using the information of the voltage variation, the controller is fully capable to calculate the required active power delivered in order to assure nominal voltage and the active power adapts to the

978-1-7281-4181-7/19 $31.00 © 2019 IEEE

TABLE II
SIMULATION PARAMETERS FOR PMSG-RECTIFIER

Description	Symbol	Value
Switching frequency	f_s	$20\,kHz$
Capacitance DC	C_{DC}	$250\,\mu F$
DC bus Voltage	V_{DC}	$670\,V$
Nominal Resistive Load	R_{DC}	$24\,\Omega$
Fundamental Frequency	f_f	$60\,Hz$
PMSG Resistance	R_G	$0.1\,\Omega$
PMSG Inductance	L_G	$4\,mH$

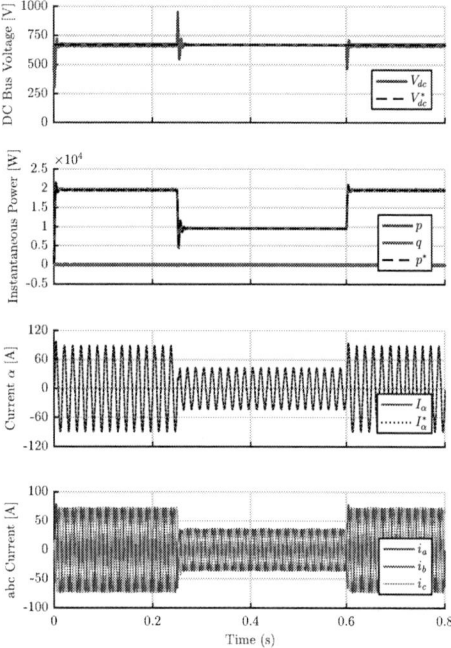

Fig. 6. Simulation response for the PMSG bus voltage control.

The control system adopted was shown to be efficient in calculating required power delivered in order to keep voltage regulated, yet the fast response adds the drawback of high voltage overshoots. More sensitive systems should benefit from a slower-response compensator.

On the other hand, the motor side of the back-to-back was design in order to control the motor speed while maintaining null direct current for maximum torque. Such control was achieved using the robust control technique which ensures stability at multiple operating points, through LMIs resolution [11].A $42\,kW$ motor was simulated with a speed of up to $2000\,rpm$ and even at fast load disturbances no significant overshoot was detected.

Futurely, this project is intented to pursue experimental validation starting with the motor control using a DC source directly feeding the DC bus and inverter, with the control operating using the DSP contained in Texas Instruments' C2000 launchpad, a user-friendly but powerful microprocessor. The final version of this work will present advancements towards that goal, with the merging of the currently separated simulations into a single system containing the back-to-back system as a whole as well as practical studies regarding the control-to-power interface.

VI. ACKNOWLEDGMENT

Authors thank the funding agencies CAPES, FAPEMIG (grants APQ-01378/16 and APQ-03593/17), CNPq (grant 432307/2016-8), UFJF and Embraer S. A.

REFERENCES

[1] K.Davies, "Review of Turboelectric Distributed Propulsion Tecnologies for N+3 Aircraft Electrical Systems",2013.
[2] Choi.B. Benjamin,"Propulsion Powertrain Simulator:Future Turboelectric distributed-propulsion",2014.
[3] M.J. Mojibian, "Modeling and Control of an Anti Rotational back to back dual PMSMs for electrical propulsion systems.",2013.
[4] Yazdani and Iravani,"Voltage-sourced converters in power systems", New Jersey: Hoboken, 2010.
[5] L. M. Grzesiak and T. Tarczewski, "PMSM servo-drive control system,pp.367–382,2012.
[6] Suntio, T., Messo, T. and Puukko, J.Power Electronic Converters: Dynamics and Control in Conventional and Renewable Energy Applications, John Wiley & Sons,2018.
[7] R. Krishnan,"Electric Motor Drives:Modeling,Analysis and Control",2001.
[8] H. Akagi, "Instantaneous Power Theory and applications to power conditioning", New Jersey: Hoboken, 2007.
[9] Simone Buso,"Digital Control in Power Electronics",In Lectures on power electronics, volume 1, pages 1–158. Morgan & Claypool Publishers, 2006.
[10] EMRAX Innovative E-Motors,"EMRAX 228 Technical Data Table", User's Manual for Advanced Axial Flux Synchronous Motors and PMSGs,August 2018.
[11] K.-Z. Liu and Y. Yao, "Robust Control: Theory and Applications. Wiley,2016".
[12] Fernandes, M., "Desenvolvimento de Controladores Multivariáveis LQR e Robusto para Integração de um Microrrede à Rede Elétrica",CBA,2018.

new load requirement and calculated reference. This can be seen affecting the calculated current in the $\alpha - \beta$ frame as well as in the actual currents demanded from the PMSG. During the whole process, the reactive power remains at zero, indicating that no reactive power is being added to the PMSG.

V. CONCLUSION

In the article under study, the modeling and control of the different frequency AC subsystems that make up a turboelectric system was performed, where a 60 Hz PMSG was connected to a PMSM through a back-to-back converter. The model of both sides of the back-to-back converter was described as well as the required means of closed-loop control. The PMSG side operated with an active rectifier controlled through instantaneous power technique such that a unit power factor was maintained while the bus voltage was controlled even during load disturbance.

An Enhanced Thévenin Equivalent Circuit of a Resonant-Controller-Based Utility-Interface

Joel Filipe Guerreiro, Hildo Guillardi Júnior, João Inácio Yutaka Ota and José Antenor Pomilio, *Senior, IEEE*

University of Campinas - UNICAMP

School of Electrical and Computer Engineering

Campinas-SP, Brazil

{joel.engeletrica; hildogjr}@gmail.com; {yutaka; antenor}@fee.unicamp.br

Abstract—This work presents an alternative method to simulate an Utility Interface (UI) consisting of a Power Electronics Converter (PEC). The proposed method is based on an equivalent resistor-inductor-capacitor (RLC) passive network circuit that resembles the output impedance of the UI, taking into account the internal resonant control characteristics of the UI. Although a controllable voltage source may emulate an UI converter, introducing an RLC network leads to a more accurate approximation of the complete UI simulated system. Comparisons are made using the frequency response of the complete system and the RLC-network circuit, and simulations are carried out on PSIM software. Results show a satisfactory dynamic and steady-state performance with a notable reduction of the elapsed simulation time. Therefore, the RLC-network-based UI could be applicable to fasten and to simplify simulations when multiples PECs are being analyzed within a complex environment, such as a microgrid.

Index Terms—Circuit simulation, Control systems, Equivalent Circuits, Power Supply Units.

I. INTRODUCTION

Microgrids have been recognized as a key concept element for the automation and further development of distribution power systems and Smart Grids [1], [2]. A microgrid consists of local generation, such as renewable energy sources, loads and energy storage systems, and it can operate autonomously when disconnected from its main generation branch or distribution power system. In this context, local generation and distributed energy storage systems within the microgrid may dispatch cooperatively to balance load supply and demand, as well to managing system stability and power quality [3].

The interconnection of power electronics converters (PECs) within a microgrid can be analyzed and shaped through the input impedance [4], [5]. Once the analysis is done, usually simulations are carried out to evaluate the microgrid operation connected to a distribution system or in islanded mode [6]. Although a complete system simulation is advantageous due to its accuracy, a power electronics converter (PEC) usually has a controller with many internal loops and synthesizes switched

This work was financially supported by project grants #302257/2015-2 and #401216/2016-0 from the National Council for Scientific and Technological Development (CNPq), project Grant #2016/08645-9, #2017/05565-7, and #2018/13993-1 from São Paulo Research Foundation (FAPESP). This study was financed in part by the *Coordenação de Aperfeiçoamento de Pessoal de Nível Superior - Brasil* (CAPES) - Finance Code 001.

voltage waveforms by the means of pulse-width modulation methods. As a result, computer simulation of multiples PECs can be expensive in time as the complexity of the controllers increases. Nevertheless, designers can count on alternative methods to perform faster simulations, such as Hardware in the Loop (HIL) real-time simulation [7]. However, costs of HIL real-time simulation may be expensive, once it requires purchase of additional hardware [6].

Alternatively, a simple method to obtain a faster simulation is to replace PECs by controllable voltage and current sources according to their main functions, such as grid forming or injection of current. Such method is reasonable because power electronics engineers are usually interested in evaluating the design and performance of only one or two PECs within a microgrid. Nonetheless, simply replacing PECs by controllable voltage or current sources may bring inaccuracies to the simulation, as the controller of the PEC affects the synthesis of voltage or current significantly. Thus, a certain equivalence is highly desirable even if a simulation of a microgrid is to be simplified for the sake of simulation resources.

Reference [8] presents a method to model a microgrid system seen by inverters operating as voltage sources. Namely, a non-dynamic model and a dynamic model based on Thévenin equivalent circuits are proposed, therefore Distributed Energy Resources and inverters can be simplified in simulation. However, the methods do not achieve high accuracy because a reduced order system model is used.

A method to replace grid-forming PECs is discussed in this work. An approach is presented to replace a network of resonant controllers by an equivalent circuit consisting of a controllable voltage source and an impedance network of resistors, inductors, and capacitors, hereinafter referred as "RLC-network circuit". Differently from a standard Thévenin controllable voltage source, the impedance network emulates accurately the dynamic behavior of a complete grid-forming PEC model. Moreover, once the RLC-network circuit is made exclusively by passive components, a reduced number of calculations is performed for every simulation step. As a result, the elapsed simulation time is significantly reduced.

The work is organized as follows: Sec. II describes the grid-forming PEC; Sec. III presents the modeling of the equivalent RLC-network circuit; Sec. IV shows the

978-1-7281-4181-7/19 $31.00 © 2019 IEEE

simulated results of the complete grid-forming PEC and the RLC-network circuit; finally, Sec. V concludes the work.

II. DESCRIPTION OF THE UTILITY INTERFACE PEC

A. Study Case: Simplifying a Grid-Forming Utility-Interface

A microgrid must have a grid-forming element, hereinafter denominated Utility Interface (UI), which consists of a PEC and an non-intermittent energy source. Besides the UI, a microgrid contains loads and multiple PECs, which have the role of interfacing Renewable Energy Sources (RES) and Energy Storage Systems (ESS) mainly. When a microgrid operates in islanded mode, the UI is a fundamental element since it generates the voltage reference for the loads and the remaining PECs.

To evaluate the performance of an specific PEC connected to a microgrid, e.g. a Power Electronics Converter connected to a photovoltaic array [9], proper simulations must be carried out, which must consider critical situations and operation points of the microgrid, such when the microgrid operates in islanded mode.

Simulations must take into account the dynamic response of the elements of the microgrid, not only the steady-state conditions. As a consequence, internal controllers of PECs, which consist of time-consuming complex algorithms based on feedback and integration loops, must be included in simulations. The replacement of grid-forming PECs and current-injection PECs by equivalent Thévenin-model voltage sources and equivalent Norton-model current sources, respectively, may reduce the simulation burden [8]. However, accuracy may be lost when evaluating the stability and the operation points of the UI, because such models do not include the dynamics of the internal controllers of the PECs.

An enhanced Thévenin equivalent circuit model is then proposed by this work. Even though it consists of passive components, the proposed circuit model is able to emulate the dynamics of the internal controllers. To evaluate this method, a study case is proposed by analyzing the system consisting of a grid-forming UI, which has internal impedance-based resonant controllers [9], and a load connected through a distribution cable. Hereinafter, the focus of this work is how to obtain the enhanced equivalent circuit model for the UI, and to evaluate its performance.

B. Impedance-based control of the Utility Interface

Fig. 1 shows a system consisting of an UI supplying power to a non-linear load. The UI consists of a single-phase half-bridge inverter with a split dc bus. An LC filter is connected to the ac terminal. Note that the distribution cable has a non-negligible impedance represented by Z_{eq}^{cable}. The non-linear load consists of a single-phase diode rectifier, with an inductive filter L, a capacitor C, and a resistance load R.

Fig. 2 shows the overall control scheme of the UI. The ac voltage reference v_{UI}^* is set to form the grid voltage of the microgrid. The control system has two control loops, a current control loop and a voltage control loop. The current control loop allows the impedance-based control of the UI

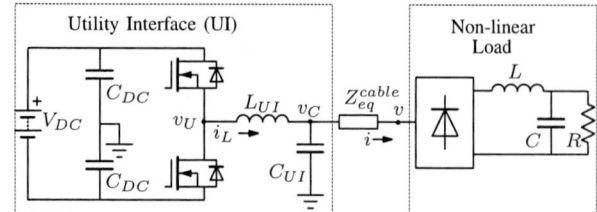

Fig. 1. Single-phase system consisting of an Utility Interface, a cable and a non-linear Load.

by regulating the current through the inductor L_{UI}. The gain k_{dc} is the static gain of the UI, and k_{pi} imposes a virtual resistance to the UI [10]. The voltage control loop regulates the ac voltage synthesized by the UI by the means of a resonant control block [11], [12]. The voltage on the capacitor is acquired to regulate the ac voltage. The transfer function $F_r(s)$ corresponds to the resonant controllers, and it is given by

$$F_r(s) = \sum_{n=1,3,5,7,9} \frac{2sk_n\Delta\omega}{s^2 + 2s\Delta\omega + n^2\omega^2}, \tag{1}$$

where n represents the nth-order harmonic, k_n is the resonant gain, ω is the fundamental frequency, and $\Delta\omega$ is the bandwidth around the central frequency of each resonant controller. Note that a feedforward compensation is added to v_{UI}^*. The proportional integral controller may be equally used if it is placed in a synchronous dq-frame, as derived in [13].

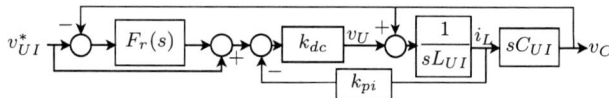

Fig. 2. Control block diagram of the UI.

It is expected that the UI can filter the harmonics produced by the non-linear load up to the 9th harmonic. Further details on the UI design are presented in [9].

III. MODELING OF IMPEDANCE-BASED GRID

A. Obtaining Thévenin impedance Z_g

The proposed method consists in modeling the UI as a Thévenin equivalent circuit, which an alternative Thévenin impedance is obtained from the output impedance of the UI.

Fig. 3 shows the equivalent circuit representation of the system in Fig. 1. The UI is represented by a voltage source, v_{UI}^*, an impedance Z_C corresponding to the output capacitance of the UI, and an impedance Z_{UI} representing the output impedance of the UI.

The Thévenin voltage source v_{UI}^* corresponds to the voltage reference of the UI, as seen in Fig. 2. The impedance Z_{UI} is obtained by inspection of Fig. 2:

$$Z_{UI}(s) = \frac{v_C(s)}{i_L(s)} = \frac{sL_{UI} + k_{pi}k_{dc}}{1 + k_{dc}F_r(s)}. \tag{2}$$

978-1-7281-4181-7/19 $31.00 © 2019 IEEE   737

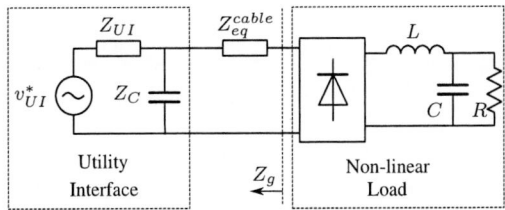

Fig. 3. Equivalent circuit diagram of the microgrid.

Considering negligible the effect of the output impedance of the non-linear load, an equivalent Thévenin impedance Z_g can be obtained from Fig. 3:

$$Z_g(s) = \frac{Z_{UI}(s) Z_C(s)}{Z_{UI}(s) + Z_C(s)} + Z_{eq}^{cable}(s), \quad (3)$$

which includes the effect of the cable impedance.

B. The RLC-Network Circuit of Z_g

Fig. 4 shows an equivalent circuit diagram of the microgrid which includes the "RLC-network" circuit block representing the Thévenin impedance $Z_g(s)$. The RLC-network circuit consists of three main branches: the UI input impedance, the input capacitor of the UI, and the equivalent impedance of the cable. The input impedance Z_{UI} in Fig. 3 can be converted into a multiple RLC-branch circuit through inference analysis of the transfer function in (2). Next, the branches of Z_{UI} are detailed.

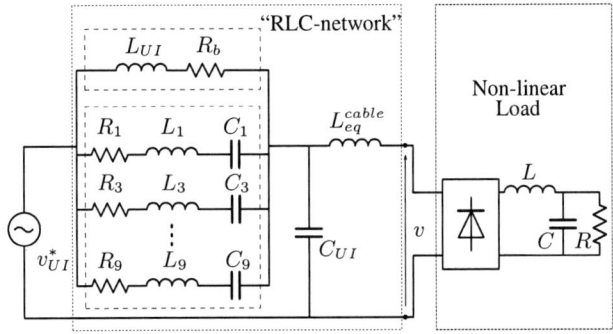

Fig. 4. Equivalent circuit diagram of the microgrid including the RLC-network circuit.

1) Resistor-Inductor Branch of Z_{UI}: Initially, a gain around 0 dB is calculated for the frequency response of Z_g. Since this gain is considered to be constant over the frequency range, a resistor branch is chosen to emulate it. The resistance R_b can be calculated by

$$R_b = ||Z_g(s)|_{s=0}||, \quad (4)$$

where $Z_g(s)$ is given by (3). The inductance L_{UI} is connected in series to R_b.

2) RLC-based Branch of Z_{UI}: Resonant controllers are used for regulating the 1st, 3rd, 5th, 7th, and 9th harmonics of the ac voltage of the UI. Each n-th-harmonic-order resonant controller is modeled as a resistor-inductor-capacitor (RLC)

branch, and it is connected in parallel to the resistor-inductor branch. Considering a n-th resonant controller as an individual second-order band-reject filter with a local gain of $||Z_g(s)|_{s=jn\omega}||$, the individual RLC-branch equivalent impedance $Z_n(s)$ can be modeled by

$$Z_n(s) = \frac{s^2 L_n C_n + s R_n C_n + 1}{s C_n}$$
$$= ||Z_g(s)|_{s=jn\omega}|| \frac{s^2 + s 2\Delta\omega + n^2\omega^2}{s 2\Delta\omega}, \quad (5)$$

where the elements of the n-th RLC branch are calculated by

$$R_n = ||Z_g(s)|_{s=jn\omega}||, \quad (6)$$

$$L_n = \frac{R_n}{2\Delta\omega}, \quad (7)$$

$$C_n = \frac{1}{L_n n^2 \omega^2}. \quad (8)$$

Note that the RLC network resembles the dynamics of the UI in a specific frequency operation point, which is the frequency of the UI controller stationary frame.

IV. COMPARISON OF THE COMPLETE SIMULATION AND THE RLC-NETWORK CIRCUIT

A. Comparing Frequency Response of $Z_g(s)$ and the RLC-Network Model

Table I describes the circuit parameters of Fig. 1 and the gains in Fig. 2 and (1), respectively.

TABLE I
PARAMETERS OF FIG. 1, FIG. 2, AND (1).

Parameter	Value
V_{DC}	400 V
C_{DC}	746 µF
L_{UI}	633 µH
C_{UI}	27 µF
L_{eq}^{cable}	18.5 µH
Switching frequency	12 kHz
k_{dc}	200 V
k_n	$100 \dfrac{4}{127\sqrt{2} n\pi}$ V V^{-1} ($n = 1, 3, 5, 7, 9$)
k_{pi}	$\dfrac{1}{25}$ V A^{-1}
$\Delta\omega$	π rad s^{-1} ($\Delta f = 0.5$ Hz)
ω	120π rad s^{-1} ($f = 60$ Hz)

Fig. 5 shows the frequency response of $Z_g(s)$, which is obtained from (3), and of the RLC-network model for the values presented in Table I. Resonance valleys appear in magnitude plot of the frequency response at the 1st-, 3rd-, 5th-, 7th-, and 9th-harmonic of the 60-Hz fundamental frequency. A valley is also visible around 7 kHz, which corresponds to the resonance frequency of Z_C and Z_{eq}^{cable}. Despite some minimal variations in the magnitude and phase frequency response, the impedance of the RLC-network circuit represents very well the UI behavior for the frequency range considered.

978-1-7281-4181-7/19 $31.00 © 2019 IEEE   738

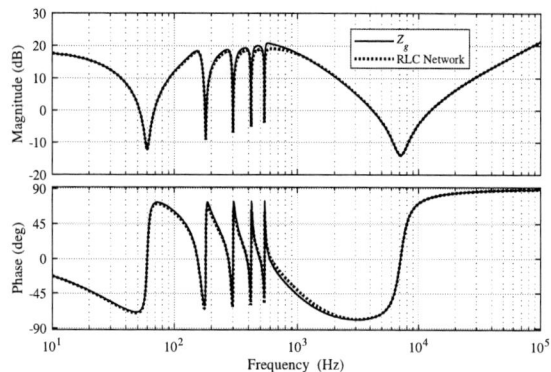

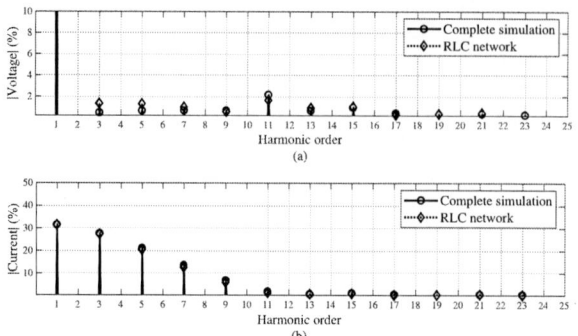

Fig. 5. Frequency response of Z_g and of the equivalent circuit containing the RLC network.

Fig. 7. Harmonic content of (a) v and (b) i of Fig. 6. Fundamental frequency is 60 Hz.

B. Simulated Results of the Microgrid and the RLC-Network Model

To verify the dynamics of the proposed RLC-network circuit, simulations for the complete system of Fig. 1 and for the RLC-network-based circuit of Fig. 4 are performed. The UI voltage v_{UI}^* is set to 127 V rms and 60 Hz.

1) Case 1: Figs. 6 and 7 show simulated results when the complete system and the equivalent RLC-network circuit are compared accordingly to parameters in Table I, and the non-linear load contains no inductor, $R = 42.5\,\Omega$, and $C = 450\,\mu\text{F}$. Fig. 6 shows voltage v and current i waveforms of the UI, and Fig. 7 shows the harmonic content of v and i.

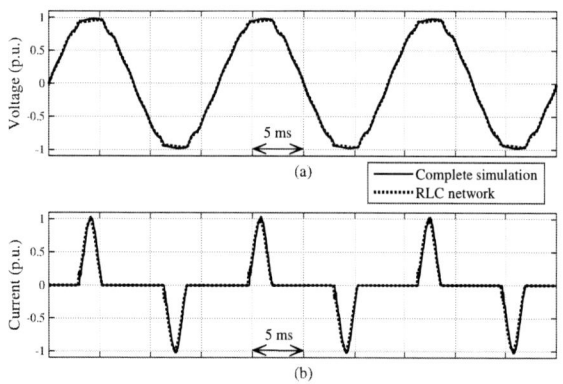

Fig. 6. Simulated waveforms of steady-state response for Case 1. (a) Voltage v for a base rms value of 127 V. (b) Current i for a base rms value of 18 A.

Current i has high harmonic content, as shown in Fig. 6(b) and Fig. 7(b), and distortion on the voltage appears due to the characteristics of the non-linear load, as seen Fig. 6(a) and Fig. 7(a). Besides forming the grid, the UI operates to regulate the voltage, as explained in Section II-B. Therefore, the 3rd-, 5th-, 7th-, and 9th-harmonic components are reduced, as seen in Fig. 7(a). Note that, in Fig. 7(a), voltage harmonics

higher than 11th-order are not regulated by the UI, therefore v presents small distortion, as seen in Fig. 6(a).

When comparing the complete system and the equivalent RLC-network, small differences are visible for the lower-order voltage harmonics mainly, as shown in Fig. 7(a), and currents harmonics are very similar in Fig. 7(b).

2) Case 2: Figs. 8 and 9 show simulated results when the complete system and the equivalent RLC-network circuit are compared accordingly to parameters in Table I, and the non-linear load contains $R = 4.25\,\Omega$, $L = 1\,\text{H}$, and $C = 450\,\mu\text{F}$. Exaggerated inductance is chosen to synthesize a quasi-square current waveform.

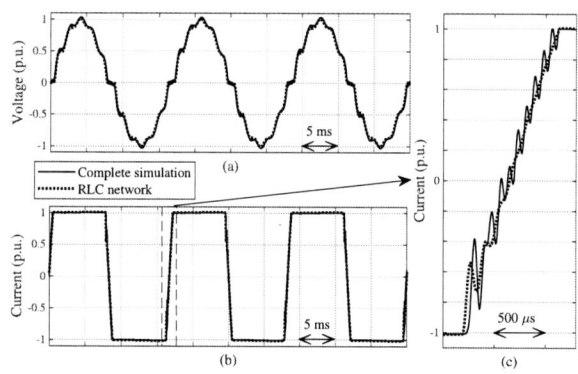

Fig. 8. Simulated waveforms of steady-state response for Case 2. (a) Voltage v for a base rms value of 127 V. (b) Current i for a base rms value of 18 A. (c) Zoomed detail of current i for a base rms value of 18 A.

For Case 2, i has a higher harmonic content than Case 1, by comparing Fig. 9(a) and Fig. 9(b). As a result, the effects of the voltage regulation of the UI are more perceptible, as seen in Fig. 8(a) and Fig. 9(a).

Fig. 8(c) zooms a step change in i. Note that the average step response of the RLC network is very similar to the complete simulation, despite the noticeable, but small, difference due to the switching effect of the complete simulation.

For Case 2, as also noticed in Case 1, currents harmonics in Fig. 7(b) are very similar when comparing the complete

978-1-7281-4181-7/19 $31.00 © 2019 IEEE

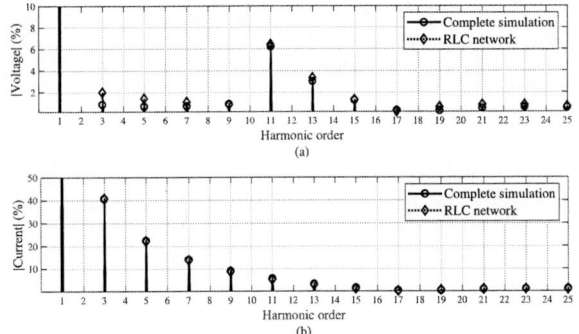

Fig. 9. Harmonic content of (a) v and (b) i of Fig. 8. Fundamental frequency is 60 Hz.

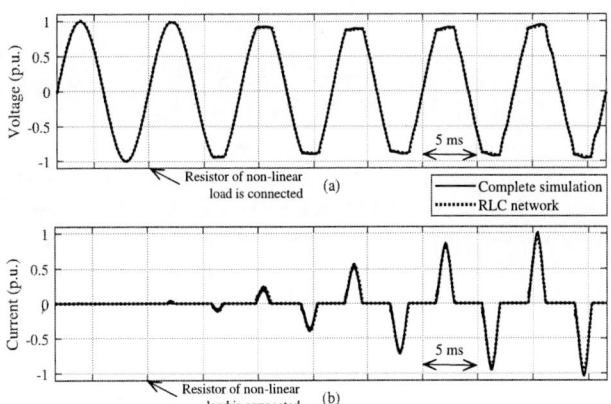

Fig. 10. Simulated waveforms of transient response for non-linear load of Case 1. (a) Voltage v for a base rms value of 127 V. (b) Current i for a base rms value of 18 A.

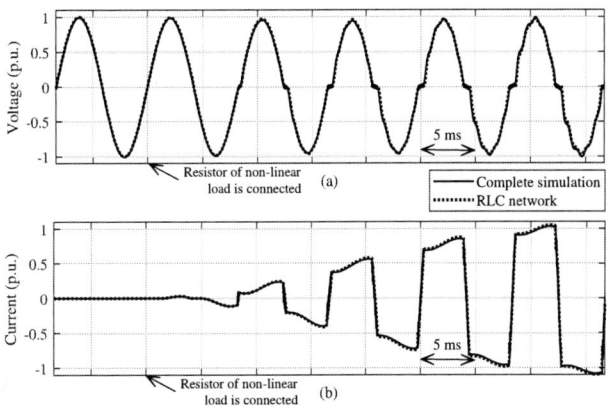

Fig. 11. Simulated waveforms of transient response for non-linear load of Case 2 with $L = 100$ mH. (a) Voltage v for a base rms value of 127 V. (b) Current i for a base rms value of 18 A.

system and the equivalent RLC network. The small differences for the lower-order voltage harmonics are noticeable as well, as shown in Fig. 9(a).

Although the comparison of the complete system and the equivalent RLC-network is made through simulations, non-linear aspects of the complete simulation may be still not contemplated in the proposed RLC network. Nevertheless, the simulated results suggest that the RLC-network approach is much better than a simple voltage source with an inductive impedance for an UI equivalent model, once that the main control features of the UI are preserved in the RLC network.

3) Transient Response: Figs. 10 and 11 show simulated results comparing the complete system and the equivalent RLC-network when a transient change is perfomed by inserting the resistor R of the non-linear load after one cycle of the UI voltage is completed. Fig. 10 shows the transient response for the non-linear load described in Case 1. Voltage and current reach the steady state in five cycles approximately. Fig. 11 shows the transient response for the non-linear load described in Case 2. Note that the inductance is changed to $L = 100$ mH to reduce the transient time to five cycles approximately. The inductance originally in Case 2 is very large, which yields a very slow transient response. From Figs. 10 and 11, note that current and voltage of the equivalent RLC network are very accurate to the ones of the complete system for both transient responses.

4) Limitations of Equivalent RLC Network: The equivalent RLC network presented in this paper emulates well and accurately the UI for both the steady state and the transient response. Nevertheless, the RLC network consists of linear passive components fundamentally, which implies that a proper emulation is expected within the linear operation of the UI. Once the UI operates in non-linear region (i.e., current demanded from UI is higher than nominal value), the RLC might not emulate the UI properly.

C. Comparison of Elapsed Simulation Time

Table II shows the average elapsed time to simulate the complete simulation and the equivalent RLC network for Case 2 of Section IV-B. For both cases, total simulation time is set to 2 s for a time step of 1 μs. The circuit simulation software PSIM, version 11.0.3 64-bits, is used in a Intel Core™ i5-7400 3.00-GHz 8-GB-RAM desktop computer. Note that the elapsed simulation time of the RLC network is much lower than the complete simulation.

TABLE II
COMPARISON OF ELAPSED SIMULATION TIME IN PSIM

	Average elapsed simulation time
Complete simulation	104 s
RLC-network simulation	8 s

V. CONCLUSION

The work has presented an method to obtain an enhanced Thévenin equivalent passive "RLC-network circuit" of an UI when considering simulations in a complex environment, such

978-1-7281-4181-7/19 $31.00 © 2019 IEEE

as a microgrid. The proposed method considers the controller of the UI to obtain an Thévenin equivalent output impedance. Performance of the RLC-network is compared by the means of frequency response of the equivalent output impedance and simulations for two cases of non-linear load. Simulated results have shown small differences between the complete simulation and the equivalent RLC network, mainly in the voltage lower-order harmonics. Nevertheless, such difference can be dismissed considering that the main control features of the UI are preserved and the elapsed simulation time is expressively reduced . This method is to be used by power electronics engineers, or even power engineers, when accurate simulations of PECs within their nominal ratings are demanded, but the grid-forming UI is not the converter under test in a complex simulation such as of a microgrid.

ACKNOWLEDGMENT

This work was financially supported by project grants #302257/2015-2 and #401216/2016-0 from the National Council for Scientific and Technological Development (CNPq), project Grant #2016/08645-9, #2017/05565-7, and #2018/13993-1 from São Paulo Research Foundation (FAPESP).

This study was financed in part by the *Coordenação de Aperfeiçoamento de Pessoal de Nível Superior - Brasil* (CAPES) - Finance Code 001.

REFERENCES

[1] R. H. Lasseter, "Microgrids," in *2002 IEEE Power Engineering Society Winter Meeting. Conference Proceedings (Cat. No.02CH37309)*, vol. 1, Jan 2002, pp. 305–308 vol.1.

[2] N. Hatziargyriou, H. Asano, R. Iravani, and C. Marnay, "Microgrids," *IEEE Power and Energy Magazine*, vol. 5, no. 4, pp. 78–94, July 2007.

[3] M. Soshinskaya, W. H. Crijns-Graus, J. M. Guerrero, and J. C. Vasquez, "Microgrids: Experiences, barriers and success factors," *Renewable and Sustainable Energy Reviews*, vol. 40, no. C, pp. 659–672, 2014.

[4] J. Sun, "Impedance-based stability criterion for grid-connected inverters," *IEEE Transactions on Power Electronics*, vol. 26, no. 11, pp. 3075–3078, Nov 2011.

[5] J. He, Y. W. Li, D. Bosnjak, and B. Harris, "Investigation and active damping of multiple resonances in a parallel-inverter-based microgrid," *IEEE Transactions on Power Electronics*, vol. 28, no. 1, pp. 234–246, Jan 2013.

[6] R. O. Salcedo, J. K. Nowocin, C. L. Smith, R. P. Rekha, E. R. Corbett, E. G.and Limpaecher, and J. M. LaPenta, "Development of a real-time hardware-in-the-loop power systems simulation platform to evaluate commercial microgrid controllers," MIT Lincoln Laboratory Lexington United States, Lexington, Massachusetts, Tech. Rep. AD1033882, Feb. 2016.

[7] O. Lucia, I. Urriza, L. A. Barragan, D. Navarro, O. Jimenez, and J. M. Burdio, "Real-time fpga-based hardware-in-the-loop simulation test bench applied to multiple-output power converters," *IEEE Transactions on Industry Applications*, vol. 47, no. 2, pp. 853–860, March 2011.

[8] M. Naderi, Y. Khayat, Q. Shafiee, and H. Bevrani, "Modeling of voltage source converters in microgrids using equivalent thevenin circuit," in *2018 9th Annual Power Electronics, Drives Systems and Technologies Conference (PEDSTC)*, Feb 2018, pp. 510–515.

[9] J. F. Guerreiro, H. Guillardi, Jr., and J. A. Pomílio, "An approach to the design of stable distributed energy resources," in *2018 IEEE 19th Workshop on Control and Modeling for Power Electronics (COMPEL)*, June 2018, pp. 1–8.

[10] Q.-C. Zhong and T. Hornik, *Control of Power Inverters in Renewable Energy and Smart Grid Integration*. Wiley-IEEE Press, 2012, ch. 7: "Control of Inverter Output Impedance", pp. 149–163.

[11] R. Teodorescu, F. Blaabjerg, M. Liserre, and P. C. Loh, "Proportional-resonant controllers and filters for grid-connected voltage-source converters," *IEE Proceedings - Electric Power Applications*, vol. 153, no. 5, pp. 750–762, September 2006.

[12] J. P. Bonaldo, H. K. Morales Paredes, and J. A. Pomilio, "Control of single-phase power converters connected to low-voltage distorted power systems with variable compensation objectives," *IEEE Transactions on Power Electronics*, vol. 31, no. 3, pp. 2039–2052, March 2016.

[13] C. Zou, B. Liu, S. Duan, and R. Li, "Stationary frame equivalent model of proportional-integral controller in dq synchronous frame," *IEEE Transactions on Power Electronics*, vol. 29, no. 9, pp. 4461–4465, Sep. 2014.

Real Time Simulation in a Distribution System Including PV Inverter and Voltage Regulator: Voltage Impact Analysis

João A. G. Archetti[1], Lilian V. Pinheiro[1], Mateus L. Lima[1], Bernardo F. Musse[1], Janaína. G. de Oliveira[1], Leonardo. W. de Oliveira[1]

[1]Universidade Federal de Juiz de Fora, Juiz de Fora – MG, Brazil

e-mail: joaoarchetti@gmail.com, lilian.venturi@engenharia.ufjf.br, mateus.lopes@engenharia.ufjf.br

Abstract—**This paper presents an analysis of the voltage profile in a distribution system with high photovoltaic (PV) penetration rate. The system used was the IEEE 13 Node Test Feeder with some modifications. The system was modelled in RSCAD software and simulated using Real-Time Digital Power System Simulator (RTDS). PV inverters were modelled with their respective control loops in the software mentioned above, performing the simulations aimed to analyze the voltage variation of the load center, considering the impact of a Voltage Regulator (VR). The VR control was embedded in Digital Signal Processing and Control Engineering (dSPACE), which communicates with RTDS performing a Hardware In the Loop (HIL) simulation. The PV inverter control was modelled and implemented with details. Results show that the effectiveness of the VR control depends on the length of the distribution lines. In this work, conclusions present that for medium length lines, standard VR control cannot limit the voltage profile between the allowed values.**

Keywords— **PV inverters, PV penetration, voltage variation, RTDS, Hardware-in-the-loop.**

I. INTRODUCTION

Currently, there is an increase in demand for new energy sources, especially among those that are consolidated as clean and renewable energy [1]. The restructuring of the electric sector was accompanied by the insertion of alternative sources of energy, aimed at supplying this input with environmental sustainability and operational efficiency [2] - [3].

The installation of photovoltaic panels is being done mostly in Brazil by residential consumers in the low voltage (LV) network. This new form of energy generation is called Distributed Generation (DG) [4] and has been boosted by several advantages, among them the proximity of electric power generation to consumer units, thus reducing losses and congestion in transmission [5] - [6].

The DG has a growth projection in Brazil [3], and brings with it new challenges to be analyzed, such as harmonic distortions, protection problems, changes in the power factor of the system, voltage variations in the electric grid busses and overcurrent in stretches [7], [8]. Overvoltages occur due to reverse flows of energy through the distribution grid at times of high penetration of DGs, when this energy exceeds the local demand of the system [7], [9] and can cause the burning and reduction of the useful life of equipment [10], making it necessary to study it at the distribution level.

In [7] it is presented in more detail the problems that can be caused in the distribution systems when high levels of PV penetration are inserted, assigning the issue to the unidirectional characteristic of the distribution system, which is due to the high ratio R / X / line impedance. Already, [8], [9],

[10], and [11] show that one of the most critical problems to be addressed is overvoltage due to the stochastic nature of solar radiation, PV is not dispatchable, with an increase in distribution line voltage at times of high generation.

Photovoltaic panels generate energy in direct current; that is, there is a need to use power electronics to condition this energy in alternating current for use by end consumers [12]. According to Brazil's regulatory standards [13], it is necessary that the studies and planning of the electrical systems take into account the DGs, as well as their connection to the distribution/transmission grid, thus, the dynamics of the devices used to carry out this process, are relevant at the research level [12].

There are several topologies of electronic devices to connect PV power to the grid, being this connection made through the use of converters [14]. Among the topologies, two stand out, the topology with two stages, where there is a DC/DC converter to stabilize the DC bus voltage from the photovoltaic panels and the single stage topology, which uses only a DC/AC converter for the conditioning of the generated energy, the difference between these topologies is the number of components used and the control strategy addressed in each one [14].

The present article aims to analyze the impact on the voltage profile due to the high penetration of renewables in distribution systems. For this, the IEEE 13 bus system was used with some modifications. By the review proposed in the introduction, it is noticed that several studies can be found analyzing these characteristics, however, the authors did not find articles that present a methodology of simulation in real time using Hardware In the Loop (HIL) simulation that allows evaluating the impact the size of the distribution line and the operating limit of the traditional voltage regulators, located in distribution transformers. Also, this article brings the modelling and control of single-stage PV inverters, developed in RSCAD software and simulated in real time with the aid of RTDS and dSPACE simulators, performing a hardware-in-the-loop (HIL) simulation. It used real data of load and radiation for different levels of penetration and distances of the feeders. Another significant contribution of the work is to verify the real need for new control strategies to maintain the voltage in the system bus at levels established by regulatory standards [13].

This work is divided into five sections, section II presents the description of the whole system and equipment used, section III discusses the methodology adopted to carry out the work, deducing the modeling and control of the PV inverters, the strategy of the follower of the Maximum Power Point Tracking (MPPT) used and the synchronization circuit with the

978-1-7281-4181-7/19 $31.00 © 2019 IEEE

grid, section IV presents the results found in the simulations and addresses some discussions to the work and section V comments conclusively the results obtained, in addition to showing some future work perspectives.

II. System description and Methodology

The system implemented in RSCAD and used to obtain the results presented in this article can be visualized in Fig. 1. The complete system consists of the IEEE 13 bus system, with some modifications: load curves with time-varying behavior were implemented, changes in the feeder size of the low voltage bus. Also, PV penetration levels entered from 0 to 40% of the total system demand in the LV 634 bus. Each of the components involved in the system presented above will be described in greater detail in the following subsections.

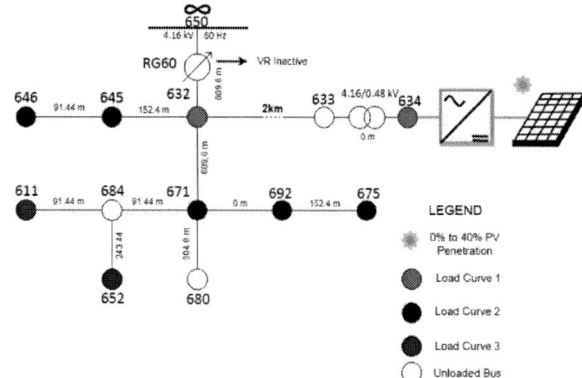

Fig.1. IEEE 13 Bus System Adapted and Used in Simulation.

A. IEEE system 13 node test feeder

The IEEE-13 bus system consists of a small distribution grid with short feeders and unbalanced phases, its demand is approximately 4 MVA, operating at medium voltage (MV) of 4.16 kV nominal and a bus (634) of low voltage (LV) at 0.48 kV, besides being composed of three-phase, biphasic and single-phase buses [15]. Adaptations were made in the original system, these adaptations can be found in more detail in [16], with the difference that due to the IEEE 13 bus system have short feeders, was proposed in present work a line increase in the LV bus feeder (632-633), from 152 m to 2 km, which is acceptable for an energy distribution analysis, due to the chosen feeder size [17]. The load legend shown in Fig. 1 is associated with the housing load models by color, detailed in Fig. 2. Also, Fig. 2 shows the solar radiation curve used in work in normalized dimensions.

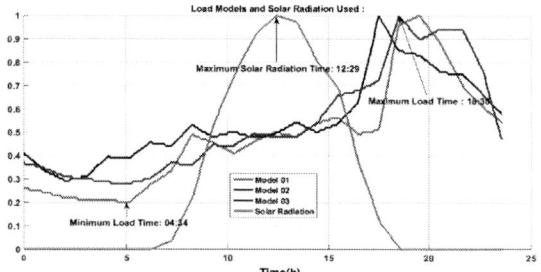

Fig. 2. Load Models and Solar Radiation Curves Used.

The IEEE 13 bus system loads are connected in star and delta [12]. Based on the original load values, the present article associates its behavior according to the load curves of Fig. 2, that is, the initial load values are proportional over time to the models of the residential load curves, which are real and have been assigned by an energy concessionaire (Energisa) [12] - [18].

The curve of solar radiation presented by Fig. 2, and used in this work, refers to a real solar radiation curve obtained by the Solar Photovoltaic Laboratory (LabSolar) of the Federal University of Juiz de Fora.

B. Photovoltaic Array (PV)

A photovoltaic array is a series-mounted and/or parallel structure of photovoltaic panels, which in turn are made up of photovoltaic cells [19]. For the present article, the mathematical model of photovoltaic arrangements proposed in [19] has been used and implemented in a block of language C, in RSCAD.

The Kyocera KC200GT photovoltaic panel model was used, with a nominal power of 200 Wp. PV penetrations range from 0 to 40% of the total system demand, and the characteristic values of the panel, along with the number of modules connected in series (N-M-S) and parallel (N-M-P) are described in Table I.

TABLE I
PANEL CHARACTERISTICS AND NUMBER OF MODULES FOR EACH PV PENETRATION LEVEL

Parameters	Penetration level PV (%)		
	0 %	20 %	40%
C-C-C	8,21 A	698,2 A	1396,4 A
T-C-A	32,9 V	1546,3 V	1546,3 V
T-MPPT	26,3 V	1236,1 V	1236,1 V
C-MPPT	7,61 A	647,2 A	1294,4 A
N-M-S	1	47	47
N-M-P	1	85	170

The abbreviations described in Table I are: Short Circuit Current (C-C-C); Open Circuit Voltage (T-C-A); Maximum power point voltage (T-MPPT); Maximum power point current (C-MPPT). To understand the parameters of Table I, the necessary voltage value in the DC link (V_{DC}), so that the inverter can maintain the output voltage at the Point of Common Coupling (PCC) [20], is described in (1). The voltage V_{DC}, is calculated by the MPPT algorithm and extracted from the photovoltaic panels.

$$V_{DC} = \sqrt{3}V_{pcc} + (0,3 \ to \ 0,5 \ )\sqrt{3}V_{pcc} \qquad (1)$$

Where V_{pcc} is the voltage in the PCC. In the system under study, the PCC is the 634 bus, of LV that operates in 0.48 kV. Selecting arbitrarily the scaling factor (0.5) for the constant of Equation 1, $V_{DC} = 1247 \ V$. Thus, according to the parameter of Table I (T-MPPT), responsible for calculating the N-M-S, DC link voltage returns to 47,44 modules in series of photovoltaic panels, thus, the integer value of 47 modules was selected in series, which results in a new voltage $V_{DC} = 1236,1 \ V$, according to Table I. From this analysis, it is possible to calculate the N-M-P, because its relation is associated with the power to be injected into the PCC and with the C-MPPT, if you want to inject into the PCC 1 MW, there is a need for a current 809 A, this already set the value of V_{DC} and the unit power factor. In Table I, the C-MPPT = 7.61 A, which results in the need for 107 modules in parallel.

C. PV Inverter

DC/AC converters, also known in the literature as inverters, are semiconductor key bridges whose main objective is to transform direct current into alternating current through the switching of electronic devices.

In this work, a two-phase, three-phase VSC (Voltage source converter) connected to the LV bus (634) was used. A PWM (Pulse Width Modulation) frequency of 2.8 kHz was stipulated due to the limitations found in the converter used in the RSCAD software for real-time simulations. The inverter modeling and controller parameters will be discussed in the next section.

D. Voltage Regulator (VR)

The Voltage Regulator (VR) is an equipment installed in distribution grid to keep the value of the output voltages in its bus in preset and supposedly constant values. This is done by controlling its shape taps automatic [21].

The VR with line-fall compensation used in this work may assume different configurations, the most common of which is a step voltage regulator with 32 taps, where an autotransformer with a tap-change mechanism is controlled by a relay [22]. The adjustment capacity of this type of regulator is around 10% of the nominal bus voltage [16].

For the present article, an VR was used, compensating line drop model, which acts on the load center of the system (671). Its operation is triggered according to the comparison of the voltage level in the load center, and the voltage stipulated as nominal (1 pu), if the voltage at the load center is higher than nominal, a command is sent to decrease the autotransformer, otherwise, the tap value will be increased. The VR was modeled in the Simulink / MATLAB® software, and accessed by dSPACE, through a library present in Simulink.

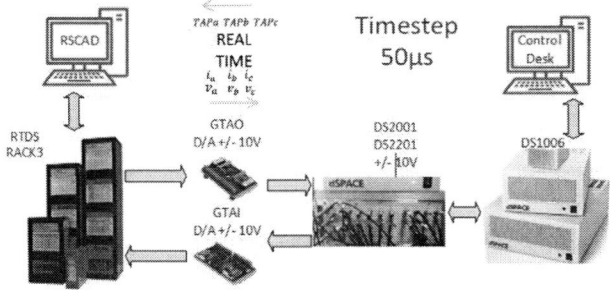

Fig. 3. Schematic of HIL Communication.

E. RTDS, dSPACE and HIL Simulation

The RTDS is a world standard equipment for simulation of real-time energy systems [23]. The dSPACE is an equipment

capable of processing and implementing algorithms, for the most part, of control systems [24]. Fig. 3 shows more clearly the HIL process.

As shown in Fig. 3 and explained in the article, the IEEE 13 bus system model was modeled in RSCAD and simulated in RTDS along with the VR and PV inverter. The RT control has been developed in Simulink and is being accessed by dSPACE. Thus, voltage and current values of the load center (671) are sent to the dSPACE by the RTDS analog output cards (GTAO). After receiving the RTDS information the logic implemented in dSPACE treats the data and returns the control action with the value of the VR tap to the load center parameters, the information of the tap values are received by the analog input cards (GTAI) of the RTDS. The HIL simulation consists of the exchange of information between hardware and software at a rate of 50µs, which allows to emulate more reality in simulation systems and to provide a greater proximity of the experimental tests, once it validates the hardware control in HIL characteristics, makes possible experimental tests with the same control strategies and equipment.

III. METHODOLOGY OF CONTROL

For the PV inverter to inject power into the grid correctly, it was necessary to model and calculate the synchronous circuit controllers with the grid and the current and voltage loops, so that Table I was satisfied and the dynamics system was fast enough to be simulated in real time

The MPPT algorithm for incremental conductance was implemented in RSCAD [20]. MPPT's function is to track the output voltage and current values of the photovoltaic arrangements to maximize the potential and cause the components to be allocated. The MPPT output voltage will serve as the reference voltage for the PV inverter voltage loop, as will be better shown later.

A. SRF-PLL Synchronous Circuit

The operation of power electronics inverters connected to the grid must be synchronous so that at the moment of connection of the PV inverter the voltage, frequency and phase angle amplitude must be synchronized with that of the grid in the PCC, for which the synchronism circuit (SRF-PLL), present in [25], was used.

B. Current Control Loop

In the operation of the PV inverter connected to the grid, there is a need to control the active and reactive power flowing through the VSC converter. The strategy used in this work is called control in current mode and can be found in [20], [25]. The deductions use the Park transform, which passes the dynamics of the AC side of the PV inverter, to dq coordinates. Knowing that the switching frequency (fc) chosen is 2.8 kHz, the time constant τ_i required to obtain the current loop controllers was calculated, considering the time constant τ_i ten times slower than the frequency of inverter PV [20], we have that $\tau_i = 3.57$ ms.

It is known from the deductions of [25] that after the decoupling of the loops of the direct and quadrature axis, we obtain the same dynamics for both, the same PI controller can be used to control the current loop. In (2) and (3) show how to get the parameters of the current loop controllers.

The dynamics of the current loop is responsible for generating the modulations that trigger the semiconductor switches of the PV inverter, by PWM.

$$K_{P,i} = \frac{L_f}{\tau_i} \qquad (2)$$

$$K_{I,i} = \frac{R_f}{\tau_i} \qquad (3)$$

Where L_f is the AC side inductor of the inverter PV, and R_f is the line resistance plus the intrinsic resistance of the semiconductor switches [25]. The values of the parameters of the controllers, together with the values of the components used, are shown in Table II. With the current loop controllers obtained, the reference currents I_{dref} and I_{qref} need to be generated by an external voltage loop, to inject power from the PV panels.

C. Voltage Loop Control

To obtain the reference current for the current loop, and assuming that the PV energy injection has a unitary power factor, i.e., $I_{qref} = 0$. It was proposed in work the use of a slower voltage control loop, where its dynamics should be at least ten times slower than the current loop. Fig. 4 shows the proposed system.

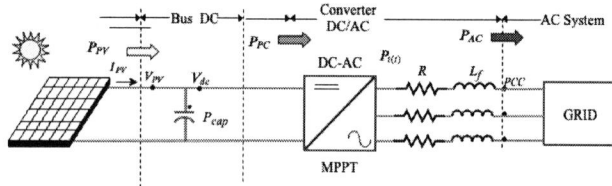

Fig. 4. System Used For connecting the PV Inverter to the grid.

From Fig. 4 and considering the law of conservation of energy:

$$P_{Cap} = P_{PV} - P_{cc} \qquad (4)$$

Where P_{Cap} is the capacitor power of the DC link, P_{PV} is the PV power, and P_{CC} is the DC link power. Considering the system without losses:

$$P_{cc} = P_t(t) \qquad (5)$$

Where Pt(t) is the instantaneous power on the AC side of the inverter. Through the Park transform, and applying differential equations in (4):

$$C_f \frac{dV_c}{dt} V_c = P_{PV} - \frac{3}{2}(V_d I_d + V_q I_q) \qquad (6)$$

V_d and I_d are the voltage and current of the direct axis of the AC side, together with V_q and I_q in quadrature, as already mentioned, the quadrature current is zero. Applying the Laplace transform to (7) and manipulating it, we have:

$$V_c^2(s) = \frac{2P_{PV}(s)}{C_f s} - \frac{3V_d(s)I_d(s)}{C_f s} \qquad (7)$$

Knowing that V_c is the voltage of the capacitor C_f, delivered by the MPPT algorithm, the purpose of the system is to make the voltage control loop to be scanned, and that the voltage produced by the PV panel is the same as the algorithm. For this, Fig. 5 shows the system control loop.

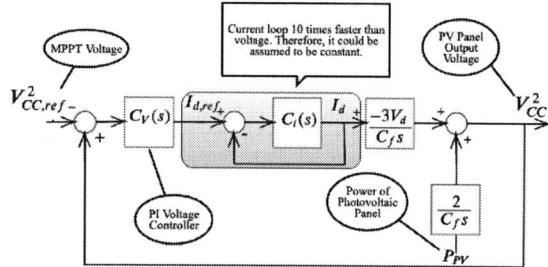

Fig. 5. Block Diagram of the Voltage Loop Control

By manipulating the block diagram, the transfer function in closed loop is:

$$FT = \frac{(3V_d K_{p,v} + 2)s + 3K_{i,v}V_d}{C_f s^2 + 3V_d K_{p,v}s + 3K_{i,v}V_d} \qquad (8)$$

To facilitate the calculations, and to arrive at the canonical formula expressed in (10):

$$K = \frac{(3V_d K_{p,v} + 2)}{C_f} \qquad (9)$$

$$K\left(\frac{s + \dfrac{3K_{i,v}V_d}{(3V_d K_{p,v} + 2)}}{s^2 + \dfrac{3V_d K_{p,v}}{C_f}s + \dfrac{3K_{i,v}V_d}{C_f}}\right) = \frac{s + a}{s^2 + 2\zeta\omega_n s + \omega_n^2} \qquad (10)$$

Where K is a proportional gain, and does not interfere with the dynamics of the system,

$$K_{p,v} = \frac{2\zeta\omega_n C_f}{3V_d} \qquad (11)$$

$$K_{i,v} = \frac{\omega_n^2 C_f}{3V_d} \qquad (12)$$

As mentioned above, the voltage loop is ten times slower than the current loop, hence $\omega_n = 176$ rad / s and $\zeta = 1.41$ chosen according to [20].

D. Structure of Simulations

As shown in [16], the critical bus of the system are the places where the PV penetration is coupled, i.e, the PV inverter of the present work is coupled to the bus 634, and this makes it the critical bus. The analyses carried out were aimed at comparing the voltage variation in bus 634 when the size of its feeder was increased, aiming to analyze if only the VR can keep the system voltages within the regulatory norms [13].

The total demand of the system passes a fluctuation, and instantly its value is changed seen from the substation, this new curve is denominated Duck curve. With this curve, it is possible to analyze the critical times of the system and to draw a simple conclusion of the moment in which the reverse flow of energy in the distribution can occur [26] - [27].

IV. RESULTS AND DISCUSSIONS

Table II shows the values adopted in the design and the controllers of the current and voltage loops.

TABLE II
PROJECT PARAMETERS AND CONTROLLER VALUES

Project Parameters		Controller Values	
C_f	1500 μF	$K_{P,i}$	0,28
R_f	2,07 mΩ	$K_{I,i}$	0,5796
L_f	1 mH	$K_{p,v}$	0,000063
f_c	2.8 kHz	$K_{i,v}$	0,039

Through the data mentioned in Table II, it is possible to analyze the behavior of the PV insertion in the IEEE-13 bus system, Fig. 6 shows the PV penetration curve inserted in the grid by the inverter to 20 and 40% (Yellow and Black, respectively), in addition to the total system demand curve (Blue) and the duck curves resulting from power penetrations in the system (Green to 20% and red to 40% PV penetration).

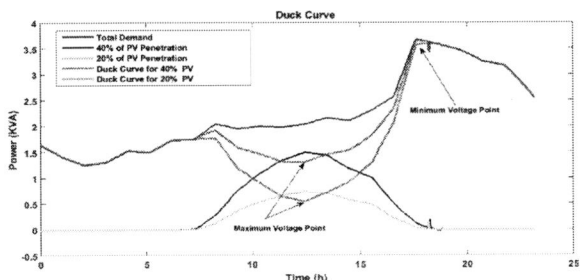

Fig. 6. Duck Curve for insertion of 20 and 40% PV Penetration.

As shown in Fig. 6, when PV penetration is inserted into the grid, the maximum and minimum voltage times of the system are highlighted by the curves of the resulting duck, since the voltage tends to sink at maximum load and to raise at minimum load [18]. The time of maximum voltage occurs very close to the time of maximum solar radiation because it is precisely the time, where more power is injected into the grid.

The black and yellow curves of Fig.6 show the PV power injection in the grid, at a level of 40% and 20% of the total system demand, respectively. To plot the duck curve, one must subtract the PV penetration curve from the whole demand curve

of the system. The resulting curve shows the new behaviour of the system in front of the DG. The higher the penetration level (black curve Fig. 6), the lower the local demand for energy at that time (red curve Fig. 6) [26] - [27].

Another important analysis is the reverse flow that happens in the system with 40% penetration, because the power injected into the system (in black), becomes greater than the new instantaneous demand of the system (in red), which results in overvoltage in the system period of penetration PV.

Fig. 7 shows the values of the voltage variation when the percentage of DG is changed and compares the impact on the grid with and without the voltage regulator for the size of the original line of the IEEE 13 bus, the phase C of the LV bus was used to show the results, due to being less charged and suffering a greater impact [15] - [16].

According to Fig. 7, the impacts observed without the VR for the low voltage bus, extrapolate greatly the voltage values stipulated by the norm that regulates the quality of energy in Brazil [13], which can cause losses and damages to the equipment. In contrast, VR was able to maintain the system's critical bus voltage within acceptable levels, even at high levels of penetration.

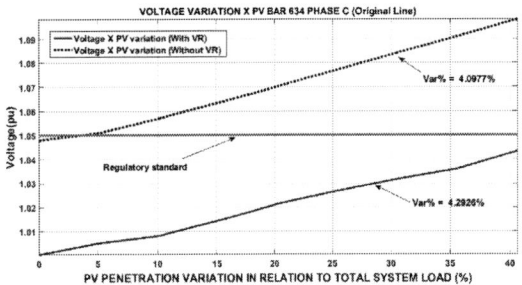

Fig. 7. Variation of Voltage in Bus 634, Phase C, for Original Line, and Bus Impact with and without RT, for a 0 to 40% PV Penetration Level.

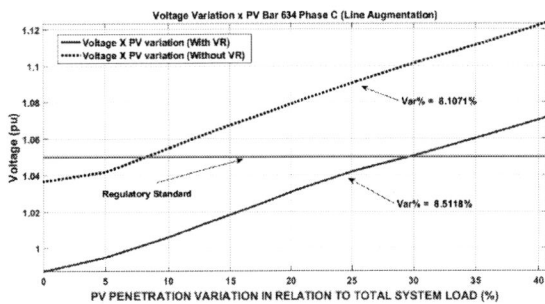

Fig. 8. Variation of Voltage in Bus 634, Phase C, with a Line of LV Increased to 2 km, and Impact in Bus with and without VR, for a Level of 0 to 40% of PV Penetration.

Considering the increase in the distance of the LV feeder from the IEEE 13 bus system, in order to analyze the impact of the penetration of the system in grids farthest from the load center, the results presented in Fig. 8 were obtained.

Analyzing the new system, with a perspective of having a bus away from the load center and that bus is the critical bus of the system, it is perceived in its results that the impacts suffered in its environment are severe, where for the analysis without the use of the voltage regulator, 40% of photovoltaic penetration results in a voltage of more than 1.12 pu, and the use of VR

978-1-7281-4181-7/19 $31.00 © 2019 IEEE

does not become more effective from the 30% PV penetration, which reinforces the analysis made of Fig. 6.

V. CONCLUSIONS

The paper presented simulation analyzes of the IEEE distribution system, with penetration of renewable energy based on photovoltaic generation, including analyzes obtained from real-time HIL simulation. Showing the voltage behavior analysis for a bus farthest from the load center of the system, and proving that with increasing feeder size the voltage at the PV coupling bus exceeds the levels set by the standard, whether with or without the VR.

The present article still brings as a contribution the control of the PV inverter in a particular way, using a method that the voltage loop is slower than the current loop, precisely to ensure that the reference current stabilizes before the new control action of the voltage loop. This facilitates the deductions and works perfectly, as can be seen in Fig. 6, in the power injection curves (in black and yellow).

As proposed future work, it is necessary to study and implement new control strategies to maintain the voltage level within acceptable levels for any feeder size and PV penetration level, such as the use of a synchronous machine (VSM), as a Volt/Var control strategy, or adaptive control, where the voltage level can be adjusted according to each system variable.

ACKNOWLEDGEMENT

The authors thank FAPEMIG, CNPq, CAPES, Finep, Inerge and the Graduate Program in Electrical Engineering (PPEE) for supporting this project.

REFERENCES

[1] L. W. de Oliveira and T. C. J. Maria, *Planning of Renewable Generation in Distribution System Considering Daily Operating Periods.* IEEE Latin America Transactions, V.15, nº. 5, p .901-907, 2017.

[2] J. C. de M. V. Junior, *Detecção de Ilhamento de Geradores Distribuidos: Uma Revisão Bibliografica Sobre o Tema.* Salvador, UNIFACS, Revista Eletrônica de Energia, V.1, nº.1, p. 3-14, 2011.

[3] World Energy Scenarios 2017 – LATIN AMERICA & THE CARIBEAN Energy Scenarios, 2017.

[4] M. E. Baran, et al, *Accommodating high PV penetration on distribution fedders.* IEEE Transactions on smart grids, V. 3, nº. 2, p. 1039 – 1046, 2012.

[5] S. P. dos Santos and R. Rther, *The potencial of building-integregrated (BIPV) and building-applied photovoltaics (BAPV) in single-family, urban residences at low latitudes in Brazil.* Ebergy and Buildings, V.50, p. 290-297, 2012.

[6] W.P. B. Filho and A. C. S. de Azevedo, *Geração Distribuida : Vantagens e Desvantagens,* II Simpósio de estudos e pesquisas em ciências ambientais na Amazônia, 2013.

[7] L. Mukwekwe, et al, *A review of the impacts and mitigation strategies of high PV penetration in low voltage networks.* IEEE PES powerAfrica, p. 274 -279, 2017.

[8] Y. Hou, *impact on voltage rise of photovoltaic generation in Swedish urban areas with high PV population.* 2014.

[9] R. Tonkoski, L. Lopes and T. H. M. El-Fouly, *Impact of high PV penetration on voltage profiles in residential neighborhoods.* IEEE Transactions on Sustainable Energy, V. 3, nº. 3, p. 518 – 527, 2012.

[10] A. Anzalchi, et al, *Power quality and voltage profile analyses of high penetration grid-tied photovoltaics: A case study.* Industry Applications Society Annual Meeting, IEEE, 2017.

[11] R. Tonkoski, L. Lopes and T. H. M. El-Fouly, *Coordinated active power curtailment of grid connected PV inverters for overvoltage prevention.* IEEE Trans. Sustain. Energy, V. 2, nº. 2, p. 139 – 147,2011.

[12] B. F. Musse, et al, *Controller-hardware-in-the-loop simulation of a distribution system with PV penetration using RTDS and dSPACE,* Brazilian Power eletronics conference (COBEP), 2017.

[13] Procedimentos de Distribuição de Energia Elétrica no Sistema Elétrico Nacional – PRODIST. ANEEL, Módulo 8, p. 29-30, 2010.

[14] M. M. Casaro and D. C. Martins, *processamento eletrônico da energia solar fotovoltaica em sistemas conectados à rede elétrica.* Revista Controle & Automação, v. 21, nº. 2, 2010.

[15] IEEE 13 Node Test Feeder, IEEE PES Distribution System Analysis Subcommitee's Distribution Test Feeder Working Group, 2000.

[16] J. A. G. Archetti, et al, *Simulations and Analysis of a Distribution System With Penetration PV Using RTDS,* Simpósio Brasileiro de Sistemas Elétricos, 2018.

[17] C. M. Luiz, *Avaliação dos Impactos da Geração Distribuída para a Proteção dos Sistemas Elétricos,* Universidade Federal de Minas Gerais-UFMG, Programa de Pós-Graduação em Engenharia Elétrica-PPGEE,2012.

[18] A. A. Francisquini, *Estimação de curvas de carga em pontos de consume e em transformadores de distribuição,* Master Thesis, Universidade Estadual Paulista – UNESP, 2006.

[19] M. M. Casaro and D. C. Martins, *Modelo de Arranjo fotovoltaico destinado a Analíses em Eletrônica de Potência via Simulação.* Eletrônica de Potência, V. 13. nº. 3, p. 141-146, 2008.

[20] P. M. de Almeida, *Modelagem e Controle de Conversores Estáticos Fonte de Tensão Utilizados em Sistemas de Geração Fotovoltaicos Conectados à Rede Elétrica de Distribuição.* Universidade Federal de Juiz de Fora – UFJF, Programa de Pós-Graduação em Engenharia Elétrica-PPEE,2011.

[21] D. H. Spatti, et al, *Regulação Automática de Tensão em Transformadores de Subestação de DistribuiçãoUsando Implementação fuzzy,* sba: Controle & Automação Sociedade Brasileira de Automatica, V. 22, nº. 2, p. 169-183, 2011.

[22] W. H. Kersting, *Distribution System Modeling and Analysis.* CRC Press, 3. Ed. 2001.

[23] M. D. O. Faruque, et al, *Real-time Simulation Technologies for Power System Design, Tensting and Analysis.* IEEE Power and Energy Technology Systems Journal, V. 2, nº. 2, p. 63-73, 2015.

[24] Digital Signal Processing and Control Engineering. 2017. Disponível em: < https://www.dspace.com/>.

[25] A. Yazdani and R. Iravani, *Voltage-Sourced Converters in Power System: Modeling, Control and Applications,* Wiley IEEE Prees, 2010.

[26] H. or R. Howlader, et al, *Duck Curve Problem Soolving Strategies With Thermal Unit Commitment by Introducing Pumped Storage Hydroelectricity & Renewable Energy.* IEEE, 2018.

[27] D. Lew and N. Miller, *Reaching new solar heights: integrating high penetration PV into the power system.* IET Renewable Power Generation, V. 11, p. 20 – 26, 2017.

Using Synchronverter in Distributed Generation for Frequency and Voltage Grid Support

Guilherme Penha da Silva Junior
Federal University of Rio Grande do Norte (UFRN)
Natal, Brazil
gpsilvajr@gmail.com

Luciano Sales Barros
Federal University of Rio Grande do Norte (UFRN)
Natal, Brazil
lsalesbarros@dee.ufrn.br

Abstract—The high integration of Distributed Generation (DG) into the conventional electric system brings many challenges to the operating sector, in view of the growing concern associated with the reliability of the system and the quality of the energy generated, due to main sources of DG do not have control over the production of active power, and besides being dependent on the of MPPT (Maximum Power Point Tracking) algorithm. On the other hand, DG are normally coupled to the power grid through fast-response power converters, which do not possess any inertia, affecting stability of electric systems. This paper presents a control method for the grid-side voltage source inverter (VSI) of DG units based on virtual synchronous generator (VSG). The technique to be used is the Synchronverter, enabling control over the active and reactive powers generated, by acting in grid support as ancillary services. Simulation results suggest the effectivity of the Synchronverter.

Index Terms—Distributed Generation, voltage source inverter, Virtual Synchronous Generator, Synchronverter.

I. INTRODUCTION

Nowadays, the harmful effects of environment pollution caused by the fossil fuel combustion in thermal, diesel, and gas-based power generating plants is something apparent to everyone [1]. Due to the growing concern in energy crisis and environmental issues, the penetration of renewable energy sources (RES) (e. g. photovoltaics and wind turbines) in power systems is necessary and inevitable [2].

In 2017, RES accounted for 34% of global electricity generation; it is believed that they will have reached 54% by 2040 [3]. The diversity of the RES and its strong dependence on geological location and meteorological situation propitiate conditions for electricity to be generated more and more by small distributed generation (DG) units [4].

However, with a larger number of DG units with higher capacities, the overall dynamics of power systems are significantly affected [5], [6]. In particular, power system inertia provided by the rotating masses of synchronous generators continues to decrease [7]. The reason is that RES are normally coupled to the power grid through fast-response power converters, which do not possess any inertia [8].

Another problem is the fact that the main sources of DG do not have control over the production of active power, besides being dependent on the MPPT (Maximum Power Point Tracking) algorithm which seeks to extract the maximum power for a certain speed produced by wind or solar irradiation. These parameters cannot be controlled and accurately predicted, and they are of stochastic nature. The lack of energy control dispatched by GD is one of the greatest challenges for microgrid (MG) operation, due to the fact that sudden variations in load cause an imbalance between generation and load, affecting the frequency stability of grid.

In [9], the transient stability of the power system with increased DG penetration is studied. Since DG penetration causes permanent replacement of conventional generators, therefore, the existing system inertia is reduced. Simulation results have shown that increased DG penetration results in rotor speed deviation, rotor speed oscillation duration, and deviation of electrical frequency increases as overall system inertia decreases with DG penetration.

During the last decade, rapid development in wind turbine technologies has made wind energy the least expensive among renewable energy sources, therefore, the integration of wind power generation into the grid has increased significantly [10] [11]. In [12] it is demonstrated that the system frequency stability is affected if conventional synchronous generators are replaced by the increasing penetration of wind system.

Another problem is that most of these VSIs are not sized to provide any reactive power at full output. The inability to provide reactive power in contingency situations is a major factor causing system voltage instability.

The key problem here is how to control the VSI (Voltage Source Inverter) in distributed power generators. There are two options: the first is to redesign the whole power system and to change the way it is operated (e.g., establish fast communication lines between generators and possibly central control) and the second is to find a way by which these VSIs can be integrated into the existing system and behave the same way as large synchronous generators (SG) do [13].

For economic reasons, the second option is more feasible, which gives a greater autonomy to DG. Thus, the latter option can solve this problem by using the concept of virtual synchronous machine (VSM). Such is a control algorithm to make an VSI operate as a conventional synchronous machine, but in order to have that it is necessary a storage system, battery banks [14]. The basic idea of the VSM bases itself on reproducing the static and dynamic properties of a real synchronous machine on a power electronic interface between a DG unit and the grid [15], inheriting the advantages of a synchronous machine in consideration of power system stability, such as adjustable active and reactive power and acting in the ancillary service of the system. The basic idea is illustrated in Fig. 1.

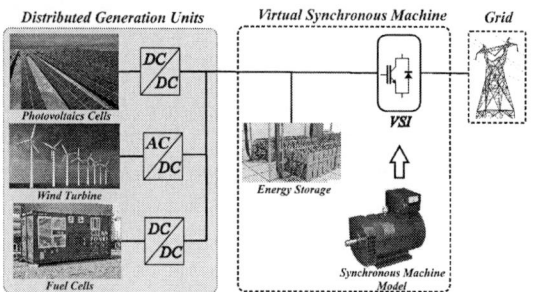

Fig. 1: Basic idea of the machine virtual synchronization machine.

There are several study groups that have developed the VSM concept: the VSYNC project (the project formed by several European companies and universities) [17], [18]; the ISE Laboratory in Osaka University in Japan [19], [20]; the Institute of Electrical Power Eng. (IEPE) in Germany [21], [4]; and the Synchronverter technique developed by [15].

This process aims to emulate the operation of the VSI to that of a conventional synchronous machine; then, the Synchronverter technology is a promising solution to this problem. Thus, it has made possible to operate as generator/motor with controlled active and reactive powers. The powers too can be automatically shared using the well-known frequency and voltage-drooping mechanisms, acting in maintaining grid stability.

This paper aims to demonstrate the behavior of Synchronverter acting as ancillary service, being organized as it follows: Section II presents a model of Synchronverter used for simulations. Section III presents the control systems regarding the Synchronverter. Results are presented in Section IV, and then in the conclusions.

II. SYNCHRONVERTER

This section presents the idea of operating a VSI in order to make a synchronous generator of performance. This is done by using the mathematical model of the synchronous generator for the development of a control algorithm implemented for the VSI.

A. Electric Model

The mathematical model of the synchronous machine can be found in several sources, as [22], [23], [24] and [25]. In order to obtain a Synchronverter operating as a synchronous generator, its control must incorporate its characteristics. In this work, the model defined in [13] was the one used.

The synchronous generator has three windings in the armature (stator) and one winding in the rotor. The schematic sketch of the windings of the three-phase cylindrical rotor synchronous generator/motor is shown in Fig. 2.

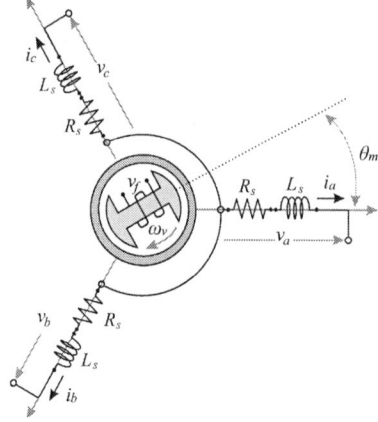

Fig. 2: Schematic synchronous generator (being $p = 1$).

The flux linkages of the armature phases a, b and c and the field winding f are expressed in terms of inductances and currents, such as:

$$\Phi_a = L_{aa}i_a + L_{ab}i_b + L_{ac}i_c + L_{af}i_f \quad (1)$$
$$\Phi_b = L_{ba}i_a + L_{bb}i_b + L_{bc}i_c + L_{bf}i_f \quad (2)$$
$$\Phi_c = L_{ca}i_a + L_{cb}i_b + L_{cc}i_c + L_{cf}i_f \quad (3)$$
$$\Phi_f = L_{af}i_a + L_{bf}i_b + L_{cf}i_c + L_{ff}i_f. \quad (4)$$

The items above subscripted with equal letters (L_{xx}) indicate self-inductances and the ones with different letters (L_{xy}) indicate mutual inductance between two windings. Being i_a, i_b and i_c the phase currents of the stator and i_f the excitation current.

The mutual inductances between stator and rotor vary instantaneously with θ_m, which is the virtual mechanical angle between the axis of the field winding and the phase a, as it is shown in Fig. 2. To obtain the mutual inductance, there is the need to calculate the electric angle between the magnetic axis of the field winding and the phase a, which is given by $\theta_v = p\theta_m$. Where θ_v is the virtual electric angle and p is the number of pole pairs, thus:

$$L_{af} = M_f \cos(\theta_v) \quad (5)$$
$$L_{bf} = M_f \cos(\theta_v - \frac{2\pi}{3}) \quad (6)$$
$$L_{cf} = M_f \cos(\theta_v - \frac{4\pi}{3}). \quad (7)$$

For a synchronous machine as a cylindrical rotor, the self-inductances of the stator and rotor windings do not depend on θ_v. Then, the self-inductances are constant. Therefore:

$$L_{aa} = L_{bb} = L_{cc} = L_{aac} + L_{aad} \quad (8)$$
$$L_{ff} = L_{ffc} + L_{ffd}. \quad (9)$$

978-1-7281-4181-7/19 $31.00 © 2019 IEEE

Where L_{aac} and L_{ffc} are the components of the self-inductances due to the main fluxes of the stator windings and field. So, L_{aad} and L_{ffd} are the additional components due to the leakage flux [26].

The mutual inductances of the armature windings are related directly to the main flux. As the armature windings displaced by 120° electric and as $\cos(120°) = -1/2$, then:

$$L_{ab} = L_{ba} = L_{ac} = L_{ca} = L_{bc} = L_{cb} = -\frac{1}{2}L_{aac}.$$
(10)

Replacing (8) and (10) in the expression of the flux linkages of phase a for (1), then:

$$\Phi_a = (L_{aac} + L_{aad})i_a - \frac{1}{2}L_{aac}(i_b + i_c) + L_{af}i_f.$$
(11)

As the armature currents are balanced, then:

$$i_a + i_b + i_c = 0$$
(12)
$$i_b + i_c = -i_a.$$
(13)

Then replacing in (11), thus:

$$\Phi_a = (\frac{3}{2}L_{aac} + L_{aad})i_a + L_{af}i_f.$$
(14)

By categorizing $\frac{3}{2}L_{aac} + L_{aad}$ as a synchronous inductance (L_s), it is concluded that:

$$\Phi_a = L_s i_a + L_{af}i_f.$$
(15)

Doing the same for all other phases of the armature winding and setting:

$$\vec{\Phi}_{abc} = \begin{bmatrix} \Phi_a \\ \Phi_b \\ \Phi_c \end{bmatrix}, \widetilde{cos}(\theta_v) = \begin{bmatrix} \cos(\theta_v) \\ \cos(\theta_v - \frac{2\pi}{3}) \\ \cos(\theta_v - \frac{4\pi}{3}) \end{bmatrix},$$

$$\vec{i}_{abc} = \begin{bmatrix} i_a \\ i_b \\ i_c \end{bmatrix}, \widetilde{sen}(\theta_v) = \begin{bmatrix} sen(\theta_v) \\ sen(\theta_v - \frac{2\pi}{3}) \\ sen(\theta_v - \frac{4\pi}{3}) \end{bmatrix}.$$

With this, the flux linkages can be rewritten as:

$$\vec{\Phi}_{abc} = L_s\vec{i}_{abc} + M_f i_f \widetilde{cos}(\theta_v)$$
(16)

$$\Phi_f = L_f i_f + M_f \left\langle \vec{i}_{abc}, \widetilde{cos}(\theta_v) \right\rangle.$$
(17)

Assuming that the stator winding resistance is R_s, then the voltages at the stator terminals of the synchronous generator, shown in Fig. 3, are given by $\vec{v}_{abc} = \begin{bmatrix} v_a & v_b & v_c \end{bmatrix}^T$ and they can be obtained by summing the voltage drops in the resistor R_s and the induced voltage. In this context, the induced voltages can be calculated by Faraday's law using (16), thus:

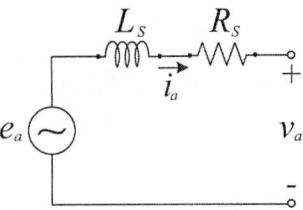

Fig. 3: Reference generator.

$$\vec{v}_{abc} = -R_s\vec{i}_{abc} - \frac{d\vec{\Phi}_{abc}}{dt} = -R_s\vec{i}_{abc} - L_s\frac{d\vec{i}_{abc}}{dt} + \vec{e}_{abc}$$
(18)

where $\vec{e}_{abc} = \begin{bmatrix} e_a & e_b & e_c \end{bmatrix}^T$ represents the back electromotive force (EFM). The vector $\vec{e}_{abc}$ is given by:

$$\vec{e}_{abc} = M_f\frac{d[i_f\widetilde{cos}(\theta_v)]}{dt}$$

$$= M_f i_f \omega_v \widetilde{sen}(\theta_v) + M_f\frac{di_f}{dt}\widetilde{cos}(\theta_v).$$
(19)

Considering a constant excitation current ($\frac{di_f}{dt} = 0$) and $\Phi_v = M_f i_f$ as virtual air gap flux of the Synchronverter. Thus:

$$\vec{e}_{abc} = \Phi_v \omega_v \widetilde{sen}(\theta_v).$$
(20)

Similarly, as it occurred in (18), the voltages at the terminals of the field winding are calculated as it follows:

$$v_f = -R_f i_f - \frac{d\Phi_f}{dt}$$
(21)

where R_f is the resistance of the rotor windings.

B. Mechanical Model

The mechanical model of the virtual synchronous machine is associated with the swing equation of the conventional synchronous machine, thus:

$$J\frac{d\omega_v}{dt} = T_m - T_e - D_p\omega_v.$$
(22)

Where J is the moment of inertia of all the rotating parts with the rotor, T_m is the mechanical torque, T_e is the electromagnetic torque, and D_p is the damping factor.

III. SYNCHRONVERTER CONTROL SYSTEM

In this section, the details about how to implement a VSI as a Synchronverter will be described, as proposed in [13]. The structure of the Synchronverter can be divided into two parts: a power circuit and a control circuit.

In the power circuit there is the VSI, which consists of 6 (six) IGBTs (Insulated Gate Bipolar Transistor), being two per each one of the three legs. The switches of a leg cannot be triggered simultaneously. It also consists of a three-phase LC filter (design is proposed in [28]), which is used to attenuate

978-1-7281-4181-7/19 $31.00 © 2019 IEEE

switching ripple, where the resistance R_s and the inductance L_s represent the impedance of the armature windings of the virtual synchronous generator and L_g and R_g are parameters of the grid. The power circuit is represented in Fig. 4.

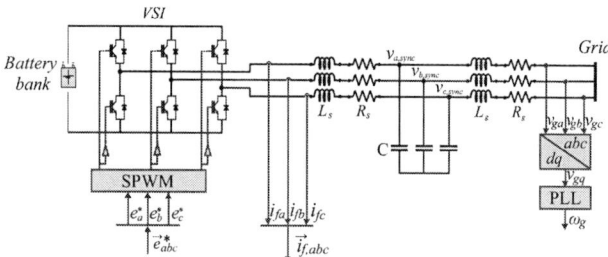

Fig. 4: Power circuit of the Synchronverter.

The control circuit can be implemented via software. Thus, the state variables of the Synchronverter are $\vec{i}_{abc}$ on the VSI terminals, alongside the virtual angle θ_v and the virtual angular speed ω_v. The control input signals are the virtual magnetic flux Φ_v and the mechanical torque T_m. In order to operate the Synchronverter stably, it is needed a controller that generates the T_m and Φ_v signals, so that system stability is maintained and the expected values of amplitude and frequency of the voltage $\vec{e}^*_{abc}$ are obtained. Then, this reference voltage is processed, and signals are then sent to switch the VSI switches via PWM (Pulse Width Modulation), achieving the desired values of the voltages e_a, e_b and e_c. Fig. 5 shows the block diagram for implementation of the Synchronverter via software.

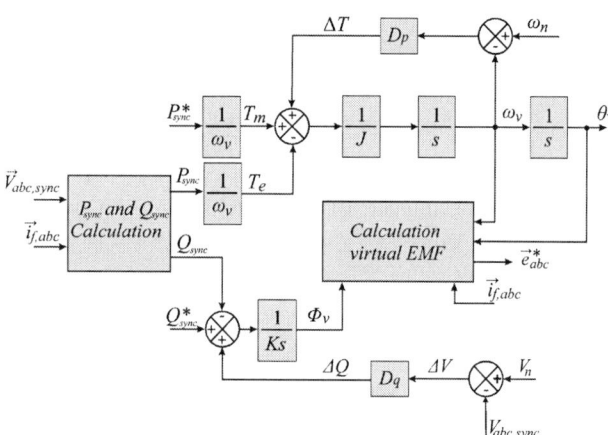

Fig. 5: Block diagram of the control of the Synchronverter.

To active power-frequency droop control, the parameters to be tuned are coefficients D_p and the virtual inertia J for the active power loop, where D_p does not only represent the virtual friction of the machine, but also the active power frequency droop coefficient of the controller. T_m can be directly calculated from the active power setpoint P^*_{sync} simply by dividing it by virtual frequency ω_v.

The reactive power-voltage droop control have the coefficients D_q and the factor K of the reactive power loop.

Its activity reacts to a voltage deviation ΔV from its nominal/reference value with a change regarding the reactive power setpoint ΔQ according to the droop coefficient D_q. Just as the instantaneous reactive power measured (Q_{sync}) at the output of the converter is then compared to its setpoint (Q^*_{sync}), it is added to the signal coming from the voltage droop (ΔQ). The resulting quantity is processed through an integrator with gain $1/K$ producing the virtual mutual flux Φ_v.

IV. SIMULATION RESULTS

The proposed control strategy was tested with simulations carried out in Matlab/Simulink. The sampling frequency of the controller was 100 kHz and the switching frequency of the VSI was 10 kHz. The Synchronverter was connected to a three-phase 690 V/50 Hz grid, and the parameters used in the simulations are given in Tab. I.

TABLE I: Parameters Used in Simulation

Parameters	Values	Parameters	Values
D_p	80	L_g	0.045 H
J	0.0075	R_g	0.15 mΩ
D_q	500	L_s	0.04 mH
K	100000	R_s	0.03 mΩ

In this simulation with VSI connected to the grid, the method required the PLL (Phase Locked Loop) to perform synchronization, as it is proposed in [13].

Initially, the active (P^*_{sync}) and reactive (Q^*_{sync}) powers have null references. The PLL application ensures that the phase of the voltage generated by the VSI (θ_v) is in phase with the grid voltage (θ_g). On the other hand, the difference between the amplitudes of the rated voltage and the voltages generated by the Synchronverter produce a virtual excitation flux (Φ_v), seeking to achieve the balance between them.

A. Variation in Grid Frequency

In the first scenario, initially the grid frequency is 50 Hz; in the range from 1 to 2 seconds there is a frequency variation to 51 Hz, and in from 3 to 4 seconds frequency variation to 49Hz, according to Fig. 6a. In this case, there is no variation in grid voltage, then, the voltage loop will produce the virtual flux (Φ_v) until the balance between the voltages, in Fig. 6b, it is possible to observe the gradual increase of the voltage generated by the Synchronverter.

At $t = 0.5s$ of simulation, the set-point of P^*_{sync} was changed to 600 kW, at $t = 1s$ it has been observed that, with increase frequency automatically attenuates the active power generation delivered by the Synchronverter to the grid. On the other hand, when the grid frequency decreases ($t = 3s$), the generated active power is increased automatically, as noted in Fig. 6c; ultimately, at $t = 4.5s$ the set-point of P^*_{sync} was changed to 400 kW.

When it reached $t = 0.5s$, Q^*_{sync} has been changed to -200 kvar, as it is shown in Fig. 6d. It is important to note

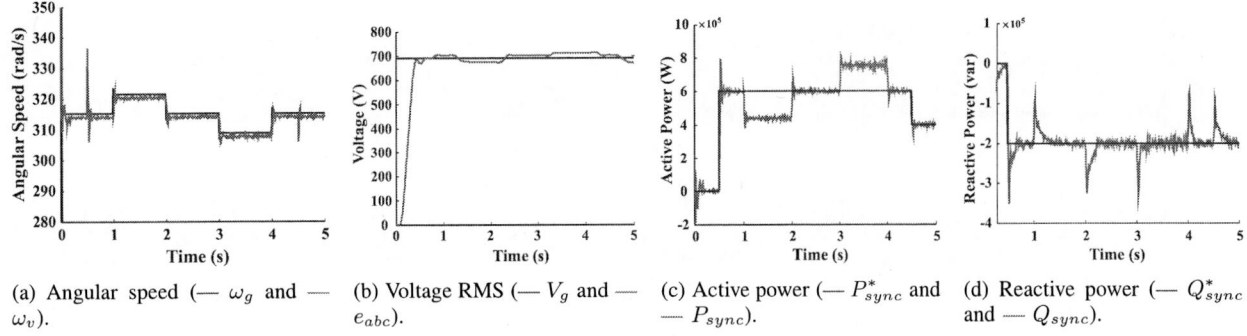

(a) Angular speed (— ω_g and — ω_v).

(b) Voltage RMS (— V_g and — e_{abc}).

(c) Active power (— P^*_{sync} and — P_{sync}).

(d) Reactive power (— Q^*_{sync} and — Q_{sync}).

Fig. 6: Simulated results for grid frequency variations.

that there is perfect decoupling between the active and reactive powers.

It is also possible to operate Synchronverter in ancillary service operating as a motor, or in battery bank charging mode, as shown in Fig. 7. At $t = 0.5s$ of simulation, the set-point of P^*_{sync} was changed to -600 kW and at $t = 4.5s$ for -400 kW; thus, the frequency variation of the grid was similar to the generator mode. It can be observed that the variation of the consumed active power is the opposite compared to the generator mode.

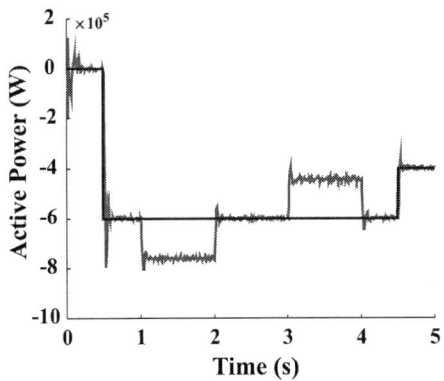

Fig. 7: Active power in battery charging mode (— P^*_{sync} and .... P_{sync}).

B. Variation in Grid Voltage

In the second scenario, the initial main voltage initially is 690 V, in the range from 1 to 2 seconds there is a voltage variation to 710 V, and in the range from 3 to 4 seconds to 660 V, according to Fig. 8b. In this case, there is no variation in grid frequency, as noted in Fig. 8a.

The acting of reactive power-voltage droop response with different characteristics for over and under- voltage conditions. At $t = 0.5s$ of simulation, the set-point of Q^*_{sync} was changed to 100 kvar. When at $t = 1s$ the grid voltage increases, the reactive power delivered by Synchronverter decreases automatically. On the other hand, at $t = 3s$ the grid voltage decreases the reactive power delivered by the Synchronverter

increases, as observed in Fig. 8d; ultimately, at $t = 4.5s$ the set-point of Q^*_{sync} was changed to 200 kvar. This is a similar behavior to that of the voltage control of the conventional synchronous generator.

The active power reference (P^*_{sync}) was changed at $t = 0.5s$, as there was no change in the grid frequency, the active power remained at its reference value.

V. CONCLUSIONS

This paper presents the idea of operating a VSI as a conventional synchronous generator, through virtual synchronous machine method, using the Synchronverter technique to improve the power quality and stability of the grid. The results obtained demonstrate that the technique which was implemented in the control of the VSI of the DGs can be used to keep the stability of power systems because it has been shown that it is possible to control the active and reactive powers of DG. The Synchronverter is controlled to generate the required amount of both active and reactive power, which includes the effects of the grid frequency and grid voltage by droop control mechanisms. The ancillary services are increasingly important as the requirements for operational flexibility grow, aiming for energy-efficient, electrical-safe and reliable operation.

REFERENCES

[1] T. V. Kumar, V. Thomas, S. Kumaravel and S. Ashok, *Performance of virtual synchronous machine in autonomous mode of operation*, 2018 5th International Conference on Renewable Energy: Generation and Applications (ICREGA), Al Ain, 2018, pp. 310-314.

[2] F.S. Rahman, T. Kerdphol, M. Watanabe, Y. Mitani *Active Power Allocation of Virtual Synchronous Generator Using Particle Swarm Optimization Approach*, Energy and Power Engineering, 2017, vol. 4, pp.

[3] IEA, *World Energy Outlook 2018*, International Energy Agency. Paris, 2018.

[4] Y Chen, R Hesse and H Beck, *Comparison of Methods for Implementing Virtual Synchronous Machine on Inverters*, International Conference on Renewable Energies and Power Quality, 2012. vol. 1, pp. 414-424.

[5] M. Reza, P. H. Schavemaker, J. G. Slootweg, W. L. Kling and L. van der Sluis, *Impacts of Distributed Generation Penetration Levels on Power Systems Transient Stability*, IEEE Power Engineering Society General Meeting, 2004., Denver, CO, 2004, pp. 2150-2155 Vol.2.

[6] R. Majumder, *Some Aspects of Stability in Microgrids*, IEEE Transactions on Power Systems, vol. 28, no. 3, pp. 3243-3252, Aug. 2013.

[7] J. Fang, H. Li, Y. Tang and F. Blaabjerg, *Distributed Power System Virtual Inertia Implemented by Grid-Connected Power Converter*, in IEEE Transactions on Power Electronics, vol. 33, no. 10, pp. 8488-8499, Oct. 2018.

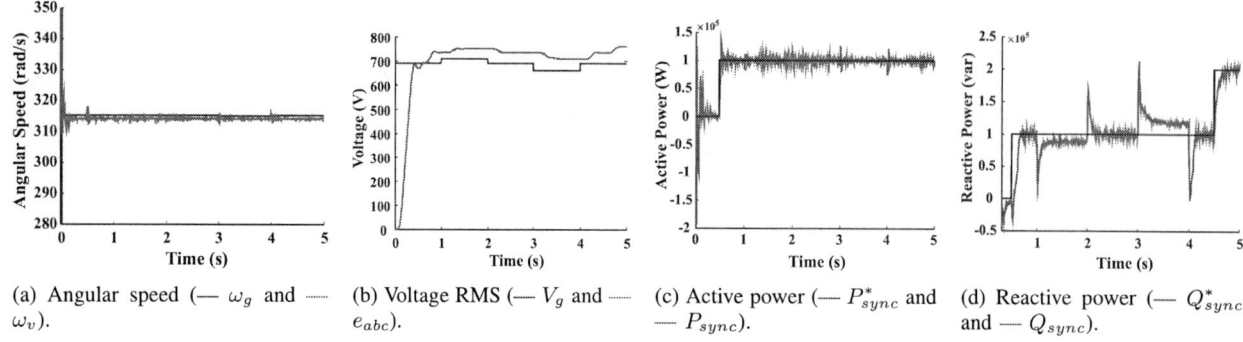

(a) Angular speed (— ω_g and ···· ω_v).

(b) Voltage RMS (— V_g and ···· e_{abc}).

(c) Active power (— P_{sync}^* and — P_{sync}).

(d) Reactive power (— Q_{sync}^* and — Q_{sync}).

Fig. 8: Simulated results for grid voltage variations.

[8] F. Blaabjerg, R. Teodorescu, M. Liserre and A. V. Timbus, *Overview of Control and Grid Synchronization for Distributed Power Generation Systems*, in IEEE Transactions on Industrial Electronics, vol. 53, no. 5, pp. 1398-1409, Oct. 2006.

[9] U. Datta, A. Kalam and J. Shi, *Power system transient stability with aggregated and dispersed penetration of hybrid distributed generation*, 2018 Chinese Control And Decision Conference (CCDC), Shenyang, 2018, pp.4217-4222.

[10] REN21, *Renewables 2018 Global Status Report*, Renewable Energy Policy Network for the 21st Century. Paris, 2018.

[11] H. Pingping, D. Ming and L. Binbin, *Study on Transient Stability of Grid-Connected Large Scale Wind Power System*, The 2nd International Symposium on Power Electronics for Distributed Generation Systems, Hefei, 2010, pp. 621-625.

[12] L. Meegahapola and D. Flynn, *Impact on transient and frequency stability for a power system at very high wind penetration*, IEEE PES General Meeting, Providence, RI, 2010, pp. 1-8.

[13] Q. Zhong and G. Weiss, *Synchronverters: Inverters That Mimic Synchronous Generators*, IEEE Transactions on Industrial Electronics, vol. 58, no. 4, pp. 1259-1267, April 2011.

[14] H. Bevrani, T. Ise and Y. Miura, *Virtual synchronous: A survey and new perspectives*, International Journal of Electrical Power e Energy Systems, vol. 54, pp. 244-254, January 2014.

[15] Q. C. Zhong, *Virtual Synchronous Machines: A Unified Interface for Smart Grid Integration*, IEEE Power Electronics Magazine, vol. 3, n. 4, pp. 18-27, December, 2016.

[16] L. Chen, Y. Wang, L. Yang, Y. Si, T. Chen and S. Mei, *Consensus control strategy with state predictor for virtual synchronous generators in isolated microgrid*, IEEE International Conference on Power System Technology (POWERCON), Wollongong, NSW, 2016.

[17] T. V. Van and K. Visscher and J. Diaz and V. Karapanos and A. Woyte and M. Albu and J. Bozelie and T. Loix and D. Federenciuc, *Virtual synchronous generator: An element of future grids*, IEEE PES Innovative Smart Grid Technologies Conference Europe (ISGT Europe), pp. 1-7, October 2010.

[18] V. Karapanos and Z. Yuan and S. Haan and K. Visscher, *A Control Algorithm for the Coordination of Multiple Virtual Synchronous Generator Units*, June, 2014.

[19] K. Sakimoto and Y. Miura and T. Ise, *Stabilization of a power system with a distributed generator by a Virtual Synchronous Generator function*, 8th International Conference on Power Electronics - ECCE Asia, pp. 1498-1505, May, 2011.

[20] J. Liu and Y. Miura and T. Ise, *Dynamic Characteristics and Stability Comparisons between Virtual Synchronous Generator and Droop Control in Inverter-Based Distributed Generators*, International Power Electronics Conference, pp. 1536-1543, May, 2014.

[21] H. Beck and R. Hesse, *Virtual synchronous machine*, 9th International Conference on Electrical Power Quality and Utilisation, pp. 1-6, October, 2007.

[22] P. Kundur, *Power System Stability and Control*, McGraw-Hill, 1994.

[23] J. Machowski, *Power System Dynamics: Stability and Control*, WileySons, 2008.

[24] K. Padiyar, *Power System Dynamics: Stability and Control*, WileySons, 2008.

[25] P. M. Anderson, *Power System Control and Stability*, WileySons, 2003.

[26] A. E. Fitzgerald and C. Kingsley, *Electric Machinery*, Mc Graw Hill, 6th ed, 2003.

[27] Q. C. Zhong and G. Weiss, *Static Synchronous Generators for Distributed Generation and Renewable Energy*, IEEE/PES Power Systems Conference and Exposition, pp. 1-6, March, 2009.

[28] M. Liserre and F. Blaabjerg and S. Hansen, *Design and control of an LCL-filter-based three-phase active rectifier*, IEEE Transactions on Industry Applications, pp. 1281-1291, Sep, 2005.

Design Procedures and Prototyping of a Full-Bridge High Frequency Power Inverter

Joel Filipe Guerreiro, Hildo Guillardi Júnior and José Antenor Pomilio, *Senior, IEEE*
University of Campinas
School of Electrical and Computer Engineering
Campinas-SP, Brazil
{joel.engeletrica; hildogjr}@gmail.com; antenor@fee.unicamp.br

Abstract—Nowadays, power electronics inverters are everywhere, from customer electronics to industry applications. In the heart of these converters there are discrete semiconductor switches. Most applications make use of Insulated Gate Bipolar Transistors (IGBTs), which are a first choice due to their switching capabilities (from several thousands of hertz to some tens of kilohertz's) and power (from some tens of watts to megawatts). However, some applications may require faster switching frequencies, such as aeronautical and automotive electrical systems. In this scenario, FET-Based devices are a suitable choice due to their extremely fast switching characteristics. Yet, high speed switching creates problematic effects such as voltage and current oscillations which are disparately addressed in literature. Thus, this work incorporates various design advice to elaborate a methodology for the design of a 5 kVA - 100 kHz FET-Based full-bridge inverter. Recommendations are given for components selection, gate driver realization and layout of the power tracks. Simulations and experimental results are shown to validate the proposed methodology.

Index Terms—MOSFETs, Power Converter, Inverter, High Switching Frequency.

I. Introduction

World-wide, research and development efforts are focused on upgrading infrastructure in renewable energies, industry, customer electronics and so on. Among the technologies that supports this growth, power electronics represents a major enabler since several loads are turning to be powered by some sort of electronic circuitry [1]. The power inverter is the heart of power electronics applications such as power supplies, motor drives, renewable energy interfaces and so on. The volume of these systems are mostly due to the dc link capacitor and the output filter, which is generally composed of an association of inductors and capacitors. The size of these passive components is inversely proportional to the switching frequency, as shown in [2], [3]. Therefore, higher switching frequency results in low footprint and cost. Another way to reduce footprint is to use higher power density components such as Wide Band-Gap (WBG)

This study was financed in party by the *Coordenação de Aperfeiçoamento de Pessoal de Nível Superior - Brasil* (CAPES) - Finance Code 001, and project grants #302257/2015-2 and #401216/2016-0 from the National Council for Scientific and Technological Development (CNPq) and project Grant #2016/08645-9 and #2017/05565-7 from São Paulo Research Foundation (FAPESP).

semiconductor technologies [4]. These devices present lower losses and higher conductivity what leads to smaller heatsink and more dense layouts. Unfortunately, this advantage is accompanied by undesirable and severe switching oscillations and electromagnetic emissions [5].

The aeronautical industry is one attractive application niche for high frequency power inverters. Regarding the fact that the modern Aircraft Electrical Power System (AEPS) frequency may sweep between 360 and 800 Hz, an enhanced filtering technique could be required [6]. For instance, active filters, when applied to AEPSs cannot operate at common switching frequencies (*e.g.* 12 kHz). In this system, harmonics can reach very high frequencies (*e.g.* 20 kHz for a fifteenth order harmonic at 400 Hz nominal frequency). In this scenario, an active filter requires at least 200 kHz of switching frequency.

Furthermore, switching frequencies as high as hundreds of kilohertz impose severe limitations to the transition speed of the semiconductor. For instance, transition rates of tens of nano seconds must be achieved to allow short dead times and low losses. However, fast transitions combined with parasitic inductive and capacitive effects are the source of oscillations created by intrinsic resonant circuits [7], [8]. The literature addresses several simulations and experimental examples of half-bridge converters as in [8]–[10]. However, few documents exploit full-bridge hardware implementation. In this topology, one leg may interact with the other. For instance, the oscillations generated by one leg are reflected to the other leg though due to current oscillations.

The major motivation of this work resides in the limited availability of commercial high frequency inverters for grid connected applications, such as renewable energy and AEPS's solutions. Only one inverter solution was found for purchase after an extensive research in the Brazilian market. However, laboratory experiments have shown that the gate driver imposes high dead-times and propagation delays, consequently, there is an impact on the performance of applications that require low duty cycles and sampling rates.

This work is organized as follows: Section II presents recommendations to the design of a FET-based full-bridge inverter and driver; Section III presents a detailed simulation; Section IV shows the experimental results of the inverter switching at 100kHz; finally, Section V concludes the work.

978-1-7281-4181-7/19 $31.00 © 2019 IEEE

II. Component Selection, Gate Driver and Layout

When designing a high frequency converter, great attention must be given to optimize the layout, minimize parasitic effects, increase switching speed and so on. The right choice of passive and active components, as well as component quality may drastically affect switching behavior. The authors have extensively researched typologies, techniques and methods to design high frequency power converters. This section summarizes the most important breakthroughs.

A. Discrete semiconductors

FETs characteristics are extremely variable depending on the material, voltage and current levels. There are some application notes [11] which provides guidelines to select a power MOSFET. Some characteristics were found to be crucial, as follows:

1) Material: Considering the frequency of operation, and power rating, the semiconductor material could be either Silicon, Silicon Carbide or Gallium Nitride accordingly to [12]. For a first prototype, the N-channel silicon MOSFET will be chosen due to its reduced price and high availability. Other devices will be used in a future work;

2) Voltage: World-wide nominal distribution voltage is in the range of $100\,\text{V}$ to $240\,\text{V}$, these levels also include AEPSs [13]. If the ac system voltage is $240\,\text{V}$, the peak voltage is $340\,\text{V}$. For a 0.7 modulation index, the dc bus voltage must be about $480\,\text{V}$. Nevertheless, it is known that MOSFETs are prone to over voltages due to switching oscillations [7]. Assuming that in the worst case scenario, a $30\,\%$ overvoltage is possible, the MOSFET shall withstand at least $630\,\text{V}$;

3) Current: For instance, a $5\,\text{kVA}$ application is considered, which drains a load current of about $21\,\text{A}$ at $240\,\text{V}$. As a matter of approximation, the MOSFET must withstand around $30\,\text{A}$ peak current at the temperature of $100\,^\circ\text{C}$. Such current level is large for a single switch. Fortunately, it is relatively simple to parallelize MOSFETs thanks to its positive thermal coefficient [14]. Therefore two parallel MOSFETs are considered, each one to handle half of the rated current. It is important to recall that the maximum MOSFET current must respect the package safe operation area, provided in the datasheet;

4) Performance: Firstly, the MOSFET should present low on resistance (for low losses) and a relatively high gate threshold voltage (in the vicinity of $5\,\text{V}$) to prevent spurious turn on. Secondly, low gate charge and output capacitance account for good dynamics such as fast on and off times and low parasitic oscillations [15]. Additionally, low intrinsic gate resistance is good for a better gate driving. Finally, the source drain diode should have low reverse recovery time and charge to avoid excessive negative current during dead time [16];

5) Package: When using the above mentioned characteristics to select MOSFETs, there are few package options. Some common options are the TO-247 and the TO-220, from these, the TO-247 is chosen due to its higher power, isolation and mechanical robustness.

Table I lists 3 devices and some of the important parameters regarded to facilitate comparison.

TABLE I. Comparison of Power MOSFETs

MOSFET:	STW28N65M2	IPW60R160P6	SiHG22N60AE
Max. Voltage	$650\,\text{V}$	$650\,\text{V}$	$650\,\text{V}$
Max. Current	$13\,\text{A}$ @ $100\,^\circ\text{C}$	$15\,\text{A}$ @ $100\,^\circ\text{C}$	$12\,\text{A}$ @ $100\,^\circ\text{C}$
$R_{ds(On)}$	$0.18\,\Omega$	$0.16\,\Omega$	$0.15\,\Omega$
$V_{gs(th)}$	$3\,\text{V}$	$4\,\text{V}$	$3\,\text{V}$
R_g	$4.9\,\Omega$	$1.6\,\Omega$	$0.6\,\Omega$
C_{gs}/C_{ds}	25.8	23.37	21.26
Q_{rr}	$8.2\,\mu\text{C}$	$5.3\,\mu\text{C}$	$4.9\,\mu\text{C}$

From Table I, $R_{ds(On)}$ is the maximum on resistance, $V_{gs(th)}$ is the gate threshold voltage, R_g is the intrinsic gate resistance, C_{gs}/C_{ds} is the figure of merit of gate source capacitance to drain source capacitance and diode Q_{rr} is the reverse recovery charge. All of the three MOSFETs are suitable for the application. The model STW28N65M2 has a higher intrinsic gate resistance, however its lower cost is attractive, therefore it is selected for this implementation.

B. Decoupling Capacitors

Passive components, such as capacitors, have internal parasitic effects that could degrade converter efficiency and stability. Therefore, capacitors must be chosen with caution. For this work, high quality capacitors are needed due to the relatively high frequency.

1) DC Bus capacitor: The electrolytic capacitors are to decouple the high currents drained from the switching stage. The capacitance range around tens to hundreds of micro Faraday depending on the application. Some attention should be given to parameters, such as, maximum dc voltage, internal resistance and maximum current. High voltage electrolytic capacitors could be expensive, one solution is to place these capacitors in series with a resistor network to allow proportional voltage division. In this project VISHAY® MAL2159 capacitors are selected to be placed in two-by-two series-parallel in each bridge;

2) DC Bus polyester capacitor: It is of great importance to place a smaller and faster capacitor close to the MOSFETs leg. A good capacitor may minimize the inductance path of the high frequency currents components, what leads to reduced oscillations while switching [17]. This capacitor material is generally polyester or ceramic, and it is fundamental to select models with very low equivalent series inductance and resistance as well as high capacitance density and thermal robustness. The EPCOS™ B32674 capacitors seemed to be suitable for the purpose of dc link decoupling and should be placed close to each leg.

978-1-7281-4181-7/19 $31.00 © 2019 IEEE

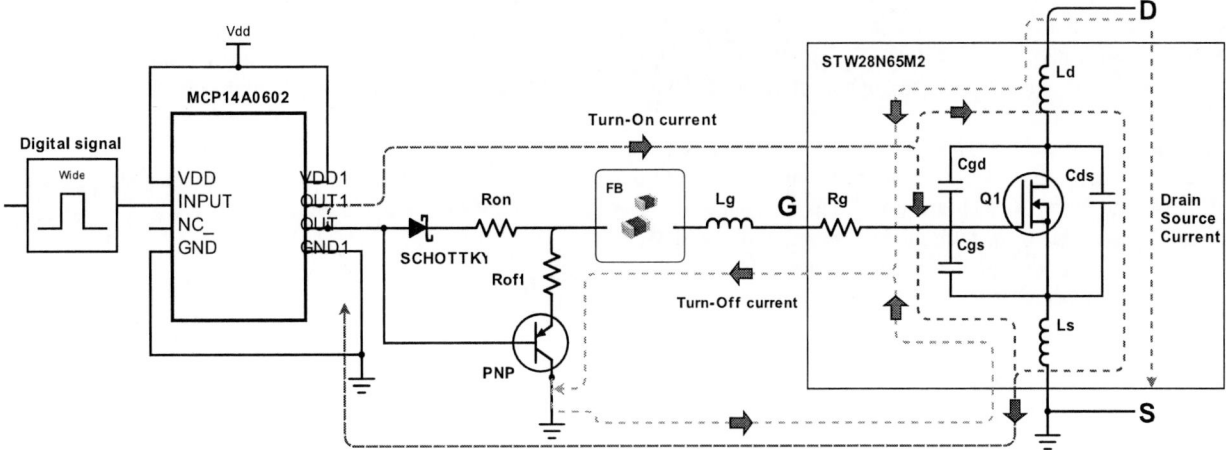

Fig. 1. Gate driver circuitry and current paths.

C. Gate Driver topology

The gate driver is responsible to trigger the power MOSFET on and off as follows:

1) Turn On: During turn on, the gate capacitance of the MOSFET (C_{gs}) is charged, this procedure can be described in two most important stages. Firstly, the gate source voltage (v_{gs}) starts to climb. When the voltage reaches the MOSFET threshold level ($V_{gs(th)}$), the drain current starts to increase until the Miller plateau $V_{gs,Miller}$ is reached. Secondly, as the gate reaches this plateau, the drain to source voltage (v_{ds}) starts to drop while the gate voltage remains unchanged. During this period, the gate drain capacitance C_{gd} discharges and this capacitance may raise two to three digits in value (e.g. from 2 pF to 1 µF as for the STW28N65M2 device, the opposite happens to the gate capacitance (C_{gs});

2) Turn Off: The turn off transition happens basically following backwards the turn on. Firstly v_{gs} drops while a reverse current flows from the gate, this current is supplied mainly by C_{gs} and C_{gd}. When the device reaches $V_{gs,Miller}$, the V_{ds} starts to increase. As soon as the drain voltage reaches its maximum value, the gate voltage drops, alongside the current reduces drastically as $V_{gs(th)}$ is approached. The values of C_{gd} and (C_{gs}) also return to the original.

Certainly, the most harmful effect on switching performance is due to large stray inductance. Some suppliers offer special designs of packages to overcome this problem [18]. This inductance is formed by the bond wire internally integrated into the MOSFET package and the gate driver common ground. Consequently, during the Miller plateau stage, when the current quickly rises or falls, a severe oscillation happens in the source inductance. This effect is hard to control and drastically harmful for the driving, as the gate voltage becomes unstable and ringing effects may appear [19].

In this work, the gate circuit must be able to handle the above mentioned transitions in a very fast manner since the switching frequency of 100 kHz should be achieved. There are numerous types of gate driver circuits, from which, the most popular would be the ground referenced and the booststrap gate drivers [20]. Moreover, the right choice of gate driver will depend on the application and printed circuit board layout. Despite its lower price, the bootstrap-based driver is not attractive for high performance operation. This circuit requires large tracking distance between the driver and the MOSFET, what is disadvantageous due to high inductive paths.

Following recommendations presented in [21], [22] and [23], an isolated low-side gate driver is selected for both the top and bottom switches for the inverter, as presented in Fig. 1. The right side of this figure shows a simplified equivalent circuit of a MOSFET and some parasitic components. Those are the parasitic inductances: L_d drain and track inductance; L_s the source and track inductance; L_g the gate track inductance and R_g the gate intrinsic resistance. Considering that two MOSFETs are placed in parallel, one group of transistors, schottky diodes, ferrite beads (FB) and on/off resistors must be used for each one [24], while only one driver is used to both circuits. The PNP transistor acts as a turn off enhancement circuit, shortening the path of the current for the off transition. This device must be placed as close as possible to the gate and source of the MOSFET. The ferrite bead act as a low pass filter blocking high oscillation frequencies that may appear in the gate. The on-off transitions can cause oscillations in the range of tens to hundred of megahertz due to resonant circuits. The on-off resistors must be placed to tune the resonant circuits in the gate driver. Selecting a good gate resistor value is fundamental for minimizing losses. In one hand, low values result in gate voltage overshoot and fast turn-on speed. In the other hands, excessively high resistors may mitigate oscillations while extending switching times and increasing losses and dead time requirement, see design equation in [25].

Lastly, the driver will be supplied by a 15 V source to guarantee the minimum on resistance of the MOSFET. Yet, the driver chip must have fast fall and rise times and high source current. A budget and yet high performance device is the MCP14A0602, which has schmitt triggered inputs to prevent high frequency driving as well as an enable pin for protection. The gate driver is equipped with a protection and dead time circuit, which are not discussed in this work. Essentially, the dead time circuitry imposes 500 ns delay between the transitions of the top and bottom FETs, while the protection disables the driver in overcurrent, overvoltage or overtemperture events.

D. Power and Driver Boards

The power and driver boards were designed using the open source Kicad® software, see the 3D image in Fig. 2. Note that this layout is a double full-bridge converter or a four leg converter. The MOSFETs are placed on the bottom with the tab down. The driver is in a separate board and it is placed vertically very close to the MOSFETs tabs. The PWM pulses are send to the board using RJ45 connectors. There is a polyester capacitor (in orange) for each half-bridge. A bank of electrolytic capacitors is placed in the center. The power inputs and outputs are made with M6 screws.

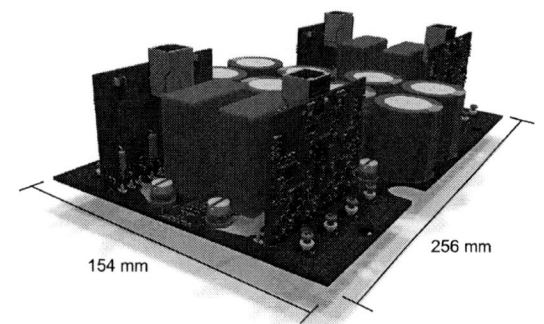

Fig. 2. Power board and gate drivers.

III. FULL-BRIDGE CONVERTER SIMULATIONS

Most application notes and MOSFET datasheets addresses half-bridge implementations. Nevertheless, in full-bridge mode, one leg disturbs the other. This faulty behavior is caused by fast transitions and parasitic components.

A. Full-Bridge Study Case

In this mode, the unipolar modulation will be applied, as shown in Fig. 3 (a). Note that the control signals "D" and "-D" are opposed and have sinusoidal form. The pulses $Q_{1,\cdots,4}$ are the PWM gate signals. When this modulation is applied to a full-bridge converter it doubles the switching frequency at the load, as shown in the waveform of V_L. For simplicity, one positive current switching cycle will be analysed, as highlighted in Fig. 3 (a). Note that within this cycle, there are four stages of the current, as shown in Fig. 3 (b) and summarized below:

1) Q_1 and Q_4 are handling the current flowing through the load;
2) Q_4 is turned off and Q_3 is on. During dead time the current is forced to flow through the anti-parallel diode of Q_3 and remain there until the next stage;
3) Q_4 is on again and the current return to raise;
4) Q_1 is off and the current flows though the anti-parallel diode of Q_2 until the end of the cycle.

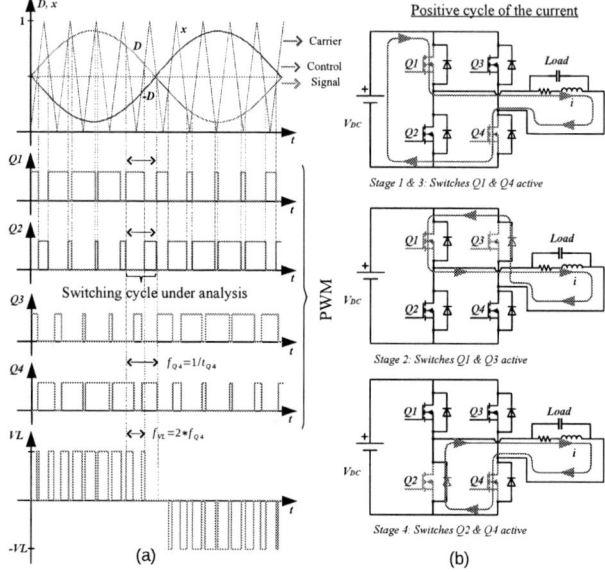

Fig. 3. Gate signals of the four MOSFETs.

B. Simulation Results

The aforementioned case study was implemented in PSIM® 11.0.3. This software supports two level of components simulations. Level one represents the principal characteristic of a component (e.g. the resistance of a resistor). Level two components also consider parasitic parameters (e.g. the parasitic inductance and capacitance of the resistor). Level two simulations are yet more precise if a small simulation step is used, thus 1 ns simulation step was used. The parasitic components (capacitance and inductance) were obtained from the datasheets. Remark that the output capacitance (C_{oss}), and the reverse transfer capacitance (C_{rss}) of the MOSFET are variable, consequently, troublesome to simulate. Hence, the dynamic parameters provided in the datasheet should used instead. The MOSFET's stray (L_S) and drain (L_D) inductance values were considered to be 9 nH and 6 nH respectively [7], while the gate inductance was estimated to be 17 nH. The load is a 150 μH ferrite-based inductor in series with a 50 Ω resistor. The load generally has a very small parasitic capacitance. However, as the transition times are in the order of nano seconds, even small values can be relevant. In this case 100 pF is considered. This capacitance creates a short current path during high frequency transitions, therefore, the voltage in one half-bridge may be affected by the other as shown in Fig. 4.

978-1-7281-4181-7/19 $31.00 © 2019 IEEE

Fig. 4 shows the waveforms of the simulation. Sub-figures (a) and (d) show V_{gs} of each MOSFET synchronized to V_{ds} shown in (b) and (e) and the current waveforms in (c) and (f). Note that "$i_{D_{Q_2}}$" and "$i_{D_{Q_3}}$" are intrinsic diode currents, as discussed in Fig. 3 (b). It can be observed that the most spurious oscillations happens when the current flowing though the diode of one MOSFET, e.g. Q_2, is redirected to the other, e.g. Q_1, when the last is switched on. Essentially, the MOSFETs internal capacitances are voltage dependent, therefore variable accordingly to the polarization. These capacitances create a resonant circuit with the parasitic inductance of the package and tracks. When the current changes drastically, the intrinsic resonances are excited and the oscillations start, leading to gate driving issues. Hence, the ultimate solution for this issue is to minimize inductance paths in the layout as well as to select better MOSFETs with lower output capacitance. Turn-off snubbers are a possible, yet troublesome solution for switching oscillations, however these devices are not addressed in this work.

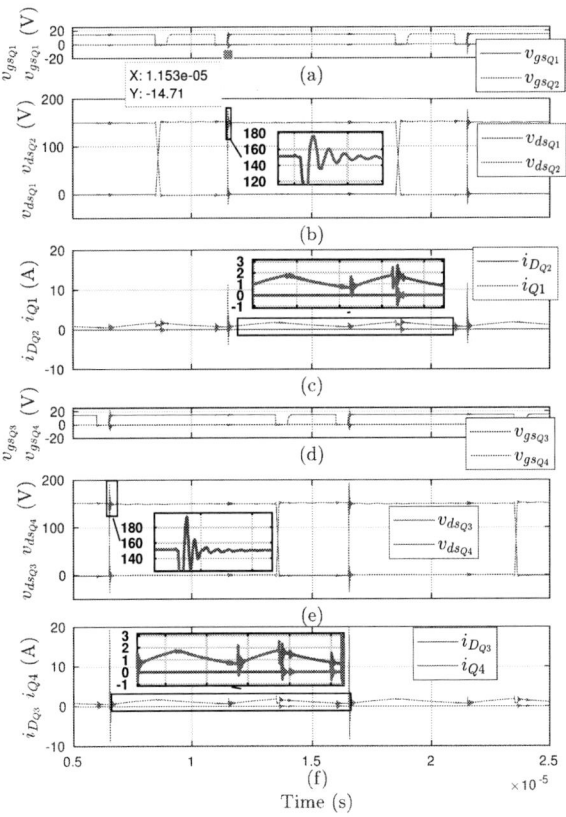

Fig. 4. Simulation results.

IV. PROTOTYPE AND EXPERIMENTATION

For validating the simulation, a prototype of the full-bridge converter was built. Due to the compact layout, the current will be measured only in the load. Yet, the MOSFET voltages are easy to acquire, and for all measurements the oscilloscope probes are placed as recommended in [21].

A. Half-Bridge Implementation

As a starting point, the prototype inverter was tested in half-bridge, being the load connected to the middle point of the electrolytic capacitor bank. For the following experiments, the resistor selected for the on/off transitions were $2.2\,\Omega$ and $0\,\Omega$, respectively. Fig. 5 shows the transitions of the MOSFET Q_2 for a $440\,V$ dc link and $100\,kHz$ switching frequency. Note that the gate voltage has an overshoot, this is due to the ferrite bead in series to the driver on/off resistance and gate source capacitance. Remark that the voltage at the MOSFET, as well as the current, do not present unexpected behavior. Remark that the correct oscilloscope scale is written in the legend for all oscilloscope plots in this work.

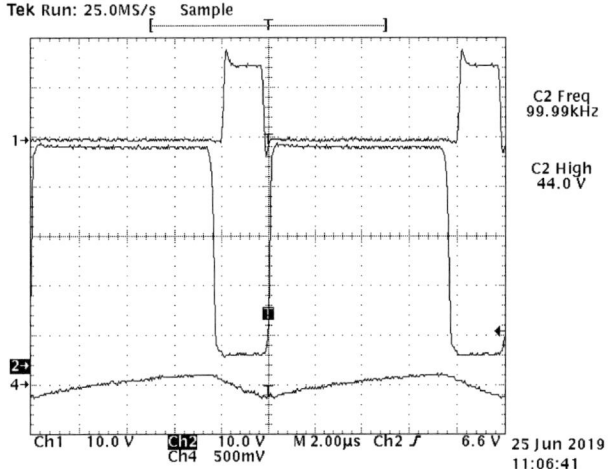

Fig. 5. Half-bridge waveforms. V_{gs} in Q_2 (Ch1 - 10 V/div), V_{ds} in Q_2 (Ch2 - 100 V/div) and I_l (Ch4 - 5 A/div).

B. Full-Bridge Implementation

Following the same structure presented in the simulations, Fig. 6 shows the results for $150\,V$ at the dc link and switching frequency of $100\,kHz$. All measurements are made in the Q_2 MOSFET during the positive cycle of the current. When Q_1 is switched off, the current decreases, going through the diode of Q_2. When Q_1 turn on again, the voltage of Q_2 raises rapidly creating an oscillation in the load current and in the gate voltage. Note that if the gate voltage of the MOSFET reaches the $V_{gs(th)}$, there is a risk of shoot through in the leg.

Fig. 7 shows a situation that triggered the driver protection of Q_1 due to high current. Note that V_{ds} increases for a very short time. This effect started to happen around $200\,V$ at the dc link and it was provoked by the transition of the other leg. Experiments concluded that the current was excessively high, reaching values above $50\,A$, which is the protection threshold. As illustrated in Fig. 3 and Fig. 4, the transition form stage 2 to stage 3 creates a very high current oscillation. The magnitude of this oscillation is proportional to the dc link voltage due to the charge of the parasitic capacitors. A priori, leg oscillations can be reduced by the usage of faster decoupling capacitors and MOSFETs with lower output capacitance.

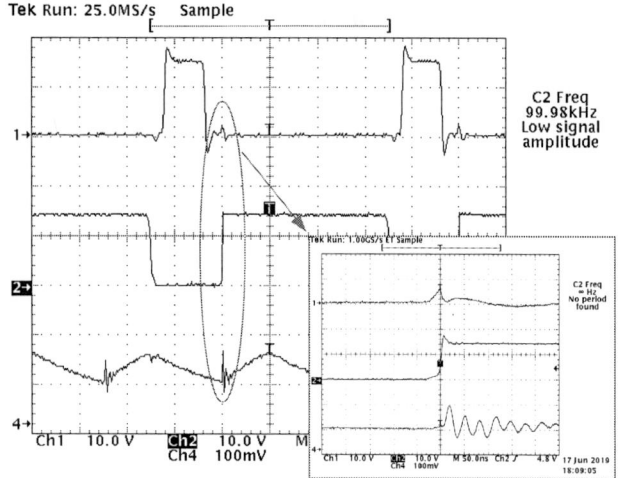

Fig. 6. Full-bridge waveforms. V_{gs} in Q_2 (Ch1 - 10 V/div), V_{ds} in Q_2 (Ch2 - 100 V/div) and I_L (Ch4 - 1 A/div).

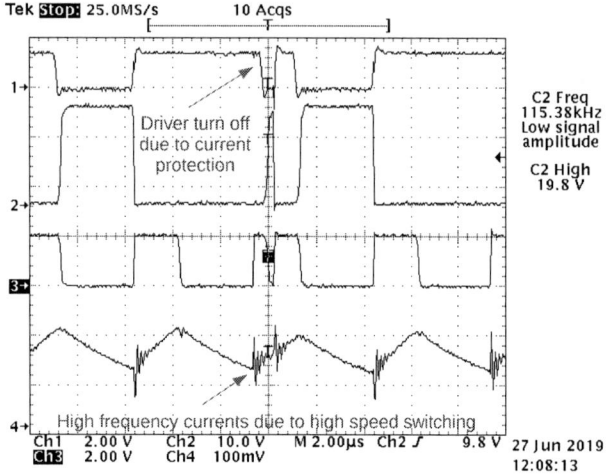

Fig. 7. Driver protection due to high current in the leg. V_{gs} in Q_1 (Ch1 - 20 V/div), V_{ds} in Q_1 (Ch2 - 100 V/div), V_L (Ch3 - 200 V/div) and I_L (Ch4 - 1 A/div).

V. CONCLUSIONS

This work presented the design and implementation of a high frequency full-bridge inverter. The selection of high quality components is crucial for a performing design, therefore, a number of figures of merit are given for the selection of MOSFETs, capacitors and gate driver topology. Neglecting design rules leads to the appearance of parasitic components, which are the first source of shoot through and driving issues. Thus, some guidance are given for a careful printed circuit board layout. Simulations further showed up some oscillations which were validated by experiment. The operation in half-bridge exhibited robustness, validating the design criteria. Nevertheless, in full-bridge mode, the same inverter presented a problematic behavior related to overcurrent during transitions. Prototyping a high frequency

inverter from scratch is a hard task, every progress leads to a new challenge. A second inverter version is under production and is expected to overcome the herein presented issues.

REFERENCES

[1] F. Blaabjerg and D. M. Ionel, "Renewable energy devices and systems – state-of-the-art technology, research and development, challenges and future trends," *Taylor and Francis*, pp. 1319 – 1328, 07 2015.

[2] Y. Tang, P. C. Loh, P. Wang, F. H. Choo, F. Gao, and F. Blaabjerg, "Generalized design of high performance shunt active power filter with output lcl filter," *IEEE Transactions on Industrial Electronics*, vol. 59, no. 3, pp. 1443–1452, March 2012.

[3] J. F. Guerreiro, H. G. Júnior, and J. A. Pomílio, "An approach to the design of stable distributed energy resources," in *2018 IEEE 19th Workshop on Control and Modeling for Power Electronics (COMPEL)*, June 2018, pp. 1–8.

[4] D.-P. SADIK, "On reliability of sic power devices in power electronics," Ph.D. dissertation, KTH School of Electrical Engineering, Stockholm, Sweden, 2017.

[5] J. Wang, R. T. Li, and H. S. Chung, "An investigation into the effects of the gate drive resistance on the losses of the MOSFET–snubber–diode configuration," vol. 27, no. 5, pp. 2657–2672, 05 2012.

[6] J. F. Guerreiro, J. A. Pomilio, and T. Davi Curi Busarello, "Design and implementation of a multilevel active power filter for more electric aircraft variable frequency systems," in *2013 Brazilian Power Electronics Conference*, Oct 2013, pp. 1001–1007.

[7] T. Liu, R. Ning, T. T. Y. Wong, and Z. J. Shen, "Modeling and analysis of sic mosfet switching oscillations," *IEEE Journal of Emerging and Selected Topics in Power Electronics*, vol. 4, no. 3, pp. 747–756, Sep. 2016.

[8] L. Zhang, S. Guo, X. Li, Y. Lei, W. Yu, and A. Q. Huang, "Integrated sic mosfet module with ultra low parasitic inductance for noise free ultra high speed switching," in *2015 IEEE 3rd Workshop on Wide Bandgap Power Devices and Applications (WiPDA)*, Nov 2015, pp. 224–229.

[9] S. Jahdi, O. Alatise, J. Ortiz-Gonzalez, P. Gammon, L. Ran, and P. Mawby, "Investigation of parasitic turn-on in silicon igbt and silicon carbide mosfet devices: A technology evaluation," in *2015 17th European Conference on Power Electronics and Applications (EPE'15 ECCE-Europe)*, Sep. 2015, pp. 1–8.

[10] Z. Zhang, J. Dix, F. F. Wang, B. J. Blalock, D. Costinett, and L. M. Tolbert, "Intelligent gate drive for fast switching and crosstalk suppression of sic devices," *IEEE Transactions on Power Electronics*, vol. 32, no. 12, pp. 9319–9332, Dec 2017.

[11] *Power MOSFET Selecting - MOSFETs and Consideration for Circuit Design*, Toshiba, 7 2018.

[12] A. Bhalla, "White paper: Practical considerations when comparing sic and gan in power applications," United Silicon Carbide, Inc, Monmouth Junction, NJ, Tech. Rep.

[13] V. Madonna, P. Giangrande, and M. Galea, "Electrical power generation in aircraft: Review, challenges, and opportunities," *IEEE Transactions on Transportation Electrification*, vol. 4, no. 3, pp. 646–659, Sep. 2018.

[14] *Thermal Stability of MOSFETs*, ON Semiconductors, 1 2014, rev. 1.

[15] *Parasitic Oscillation and Ringing of Power MOSFETs*, Toshiba Electronic Devices & Storage Corporation, 7 2018.

[16] *Hard Commutation of Power MOSFET OptiMOS TM FD 200V/250V*, Infineon Technologies AG, 3 2014, rev. 1.

[17] *Ceramic Capacitor Technology*, EPCOS TDK.

[18] *CoolMOS TM C7 650V Switch in a Kelvin Source Configuration*, Infineon Technologies, 5 2013, rev. 1.0.

[19] *Parasitic Oscillation and Ringing of Power MOSFETs*, Toshiba, 7 2018, rev. 1.0.

[20] *Fundamentals of MOSFET and IGBT Gate Driver Circuits*, Texas Instruments Incorporated, 10 2018, rev. 2.

[21] *Driving and Layout Design for Fast Switching Super-Junction MOSFETs*, ON Semiconductor, 11 2014, rev. 1.0.1.

[22] *PCB layout guidelines for MOSFET gate driver*, Infineon Technologies, 1 2018, rev. 1.0.

[23] *Limits and hints how to turn off IGBTs with unipolar supply*, Semikron, 6 2015, rev. 1.0.

[24] H. J., S. F., and V. E., *CoolMOS^{TM} C7: Mastering the Art of Quickness: A Technology Description and Design Guide*, Infineon.

[25] *External Gate Resistor Design Guide for Gate Drivers*, Texas Instruments, 5 2018, rev. 1.0.

978-1-7281-4181-7/19 $31.00 © 2019 IEEE

SVC OPERATING AS AN UNBALANCE COMPENSATOR WITH CONTROL SYSTEM BASED ON THE STEINMETZ METHOD AND THE INSTANTANEOUS POWER THEORY.

Andressa de Melo Rodrigues
Federal Institute of Education, Science and Technology of Goiás
Jataí, Brazil
ee.amelor@gmail.com

Laylla Fernandes Silva
Federal Institute of Education, Science and Technology of Goiás
Jataí, Brazil
layllaescoteira@hotmail.com

Luciano de Souza da Costa e Silva
Federal Institute of Education, Science and Technology of Goiás
Jataí, Brazil
lucianocosta_@hotmail.com

Lucas Carvalho Souza
Federal Institute of Education, Science and Technology of Goiás
Jataí, Brazil
lucas.souza@ifg.edu.br

Abstract— This paper proposes a digital control system, based on the Steinmetz method and instantaneous power theory, applied to the static var compensator to compensate the unbalance between the phases and the power factor of loads. The proposal incorporates an additional digital filtering system for elimination of harmonic components and for the extraction of positive-sequence voltage. The experimental results demonstrated good precision in the compensation of the unbalance and in the correction of the power factor. The experimental tests indicate reduction of the unbalance factor of 44.9% to 3.3% and correction of the fundamental power factor to the unit.

Keywords—unbalance, fundamental power factor, SVC, Instantaneous Power Theory, Steinmetz Method.

I. INTRODUCTION

The expansion of the three-phase electric distribution system, due to the increase in energy consumption, has caused numerous obstacles due to the large installation of single-phase or two-phase loads to the system. It is incumbent upon the electric utilities to regulate the services that are offered, as well as establish the voltage levels of electric power and set their limits of variations. In this sense, the main cause of the voltage unbalance at a point of common coupling (PCC) of an electric power distribution system is the load unbalance. The voltage drops across the feeders is related to the short-circuit currents of the lines and the load currents, which, when unbalance, also unbalance the three-phase voltages in the PCC.

The unbalance causes failures and complications in the electric power distribution system and equipment installed in this, mainly motors, causing overheating and consequently losses by joule effect, decreasing the useful life of equipment and conductors, poor operation of protection and control systems, and even saturation of transformers [1]. The most significant unbalance is produced by power single-phase industrial loads around of hundreds of kilowatts or megawatts. Welding equipment, light rail vehicles and arc furnaces are asymmetric loads that demand high power and considerably increase the number of negative-sequence components in the electrical grid, as well as cause other disturbances such as harmonic distortions, fluctuations and voltage sags [2].

In order to mitigate such perturbations, researchers present a variety of topologies capable of compensating for unbalance.

In particular, the shunt compensator has current source behavior [3] and enable the insertion of capacitive reactive and/or inductive reactive current in the system. Although they are commonly used as voltage regulators, they can operate as unbalanced loads compensators in distribution systems [4]. Static Var Compensator (SVC) has been widely used as an unbalanced load compensator for high power applications, due to robustness and rapid response to dynamic reactive power compensation.

The work proposed in [5] presents a computational simulation study on the dynamic performance of SVC shunt compensator as unbalance compensator. The authors describe that the positive and negative sequence components of the voltage are extracted by combining the signal obtained in the last cycle with the same delayed signal by a quarter of a cycle. This orthogonal signal generation process is used to obtain the symmetrical components by the classic Fortescue theorem. Such process, for not being implemented instantly, can deteriorate the compensation dynamics.

The compensator proposed in [6] develops a control system based on the Steinmetz method and the instantaneous power theory. However, as the authors perform the calculations of the positive and negative sequence powers in d-q coordinates, the requirement of three Park transformations for the practical digital implementation would increase the complexity of the system and would require large capacity of the hardware used. Also, if there are distortions or unbalance in the voltage, the computation of the implemented compensation susceptances may accumulate errors, since the calculation of the positive and negative sequence powers by the instantaneous power theory considers symmetrical, balanced, and sinusoidal supply voltages.

The work developed in [7] also presents the same problem of complexity evaluated in [6]. This fact is evident when evaluating the implementation of the prototype, which uses two TMS320F2812 Digital Signal Processor (DSP) with 150MHz clock frequency, in addition to a Field Programmable Gate Array (FPGA). In the development of the prototype, a DSP is dedicated exclusively to obtaining PLL angles and calculations of trigonometric functions, both necessary in the implementation of the Park transform.

The control system proposed in [8] also ensures unbalance compensation with excellent performance, but from a much less complex algorithm than the proposals [5-7]. The digital implementation of the compensation system is simplified because the calculations are performed based on the Clarke transform with the addition of Low Pass Filter (LPF), and do not require the application of PLL and Park transformations. Although simpler, this control system also does not consider the occurrence of distortions and unbalance in the supply voltages, which may cause errors in the determination of compensation susceptances. Furthermore, an additional feedback loop based on the Fuzzy PID controller is applied to reduce the compensation error, which implies an increase in the number of sensors.

Therefore, this work has as main objective to implement an automatic unbalance and power factor compensator with digital control system based on the Steinmetz method and the instantaneous power theory. The power stage will consist of an SVC, and the proposed control system will be embedded on the Texas Instruments TMS320F28069M microcontroller. The control system proposal differs from [8] by the additional implementation of digital filtering systems for elimination of harmonic components and for the extraction of positive-sequence voltage in order to ensure excellent performance in compensating for unbalanced load, even in distorted environments with unbalanced voltages. The proposal also uses a single direct loop, which implies the use of less current sensors.

II. SVC AS AN UNBALANCE COMPENSATOR

Fig. 1 shows the arrangement of three-phase SVC connected in delta at PCC, in parallel with a three-phase load fed by the electrical power distribution system. The control system is presented in "black box", as it will be detailed in section III.

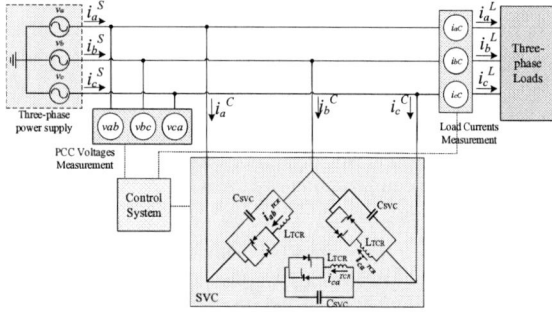

Fig. 1. Schematic of the SVC connected in parallel with a three-phase load

The SVC can operate as a thyristor delay angle - controlled current source due to its behavior of variable susceptance and to the voltage measurement at PCC [3]. Thus, in order for the SVC to act as an unbalance compensator, it is necessary for each phase to operate individually. Therefore, the thyristors of each phase are fired with different delay angles in order to operate with the reference susceptances calculated by the algorithm of the control system. In this case, the compensating currents generate are added to the unbalanced load currents, result in positive sequence balanced line currents [5].

A. Steinmetz Equivalent Circuit

Steinmetz's equivalent circuit proposes that any three-phase unbalanced load system, when connected in parallel with an adjustable reactive power compensator, can become a balanced three-phase load, without this process altering the active power flow between the system and the load [6].

Consider the line-to-line voltages of the electrical power system balanced, represented in (1), where v_{ab}, v_{bc} e v_{ca} represent the instantaneous functions of these voltages, and V is the RMS value of the line-to-neutral voltages.

$$\begin{cases} v_{ab} = \sqrt{2}\sqrt{3}V\cos(\omega t + 30°) \\ v_{bc} = \sqrt{2}\sqrt{3}V\cos(\omega t - 90°) \\ v_{ca} = \sqrt{2}\sqrt{3}V\cos(\omega t + 150°) \end{cases} \quad (1)$$

Since the load is connected to three wires, as shown in Fig. 1, there is no zero-sequence component. Therefore, the load currents can be described in (2) as a function of their positive and negative sequence symmetric components.

$$\begin{cases} i_a^L = \sqrt{2}I_+^L\cos(\omega t + \phi_+) + \sqrt{2}I_-^L\cos(\omega t + \phi_-) \\ i_b^L = \sqrt{2}I_+^L\cos(\omega t + \phi_+ - 120°) + \sqrt{2}I_-^L\cos(\omega t + \phi_- + 120°) \\ i_c^L = \sqrt{2}I_+^L\cos(\omega t + \phi_+ + 120°) + \sqrt{2}I_-^L\cos(\omega t + \phi_- - 120°) \end{cases} \quad (2)$$

Where i_a^L, i_b^L e i_c^L are the instantaneous line currents in the load, I_+^L is the load positive-sequence RMS current and I_-^L is the load negative-sequence RMS current. The "+" and "-" indices will be used to refer as the positive and negative sequence components, respectively, in the course of this paper.

Since SVC consists only of reactive elements, the instantaneous compensation currents of each phase of the compensator depend only on the equivalent susceptances of the respective phases and the voltage imposed on them. Where B_{ab}^C, B_{bc}^C e B_{ca}^C are compensator equivalent susceptance per phase.

$$\begin{cases} i_{ab}^C = v_{ab}(\omega t + 90°)B_{ab}^C = \sqrt{2}\sqrt{3}V\cos(\omega t + 120°)B_{ab}^C \\ i_{bc}^C = v_{bc}(\omega t + 90°)B_{bc}^C = \sqrt{2}\sqrt{3}V\cos(\omega t)B_{bc}^C \\ i_{ca}^C = v_{ca}(\omega t + 90°)B_{ca}^C = \sqrt{2}\sqrt{3}V\cos(\omega t - 120°)B_{ca}^C \end{cases} \quad (3)$$

The Steinmetz method is summarized in the obtention of the compensator phase susceptances, necessary for the mitigation of the negative-sequence component of the load current and for the correction of the fundamental power factor [8].

Applying the Fortescue theorem in (2) and (3) and relating the positive and negative sequence components of the load with the expected positive and negative sequence components in the compensated line, it obtains in (4) the compensator susceptance.

$$\begin{cases} B_{ab}^C = \dfrac{1}{V_L^2}\left(-\dfrac{1}{3}Q_+^L + \dfrac{1}{\sqrt{3}}P_-^L + \dfrac{1}{3}Q_-^L\right) \\ B_{bc}^C = \dfrac{1}{V_L^2}\left(-\dfrac{1}{3}Q_+^L - \dfrac{2}{3}Q_-^L\right) \\ B_{ca}^C = \dfrac{1}{V_L^2}\left(-\dfrac{1}{3}Q_+^L - \dfrac{1}{\sqrt{3}}P_-^L + \dfrac{1}{3}Q_-^L\right) \end{cases} \quad (4)$$

Where:

$$Q_+^L = -\sqrt{3}V_L I_+ \operatorname{sen}\phi_+ \quad (5)$$

$$P_-^L = \sqrt{3}V_L I_- \cos\phi_- \quad (6)$$

$$Q_-^L = -\sqrt{3}V_L I_-^L \operatorname{sen}\phi_- \quad (7)$$

By compensating (5) the power factor correction is guaranteed and by canceling (6) and (7), the unbalance is eliminated.

978-1-7281-4181-7/19 $31.00 © 2019 IEEE

B. SVC Reactive Element Specification

The fixed capacitors and reactors of each phase of the SVC must be dimensioned according to the reactive power they need to process. Although it does not contribute to the resulting three-phase reactive power, the powers due to the negative-sequence components impact on the specification of the SVC phase components [6].

When the firing angle is minimum, $\alpha = 90^{0}$, the thyristors operate turned-on for the entire grid cycle, resulting in the maximum admittances and inductive reactive power. This relation is described in (8) as a function of the maximum unbalance factor capable of being compensated for inductive negative-sequence currents in the load, UF_L, due to the maximum apparent power in positive-sequence demanded by the load, S^L_+, and as a function of the maximum inductive reactive power in positive-sequence processed by the compensator, $Q^C_{+L(max)}$, responsible for the compensation of possible capacitive reactive power demands.

$$\alpha = 90^{0} \rightarrow Q^C_{L(max)} = Q^C_L - Q^C_C = \frac{Q^C_{+L(max)} + 2UF_L.S^L_+}{3} \quad (8)$$

Where:

- $Q^C_{L(max)}$ represents the maximum inductive reactive power processed by SVC;
- Q^C_L represents the reactive power processed by the SVC fixed inductor;
- Q^C_C represents the reactive power processed by the SVC fixed capacitor;
- $Q^C_{+L(max)}$ represents the maximum positive-sequence inductive reactive power processed by SVC;

When the thyristors operate with $\alpha = 180^{0}$, as a turned-off switch, the current in the TCR is zero for the entire grid cycle. Therefore, the SVC equivalent susceptance is maximum capacitive and equal to the fixed capacitor's susceptance. Analogously to (8), the maximum inductive reactive power processed in each phase of the SVC is presented in (9).

$$\alpha = 180^{0} \rightarrow Q^C_{C(max)} = -Q^C_C = \frac{-Q^L_+ - 2.UF_C.S^L_+}{3} \quad (9)$$

Where:

- $Q^C_{C(max)}$ represents the maximum capacitive reactive power processed by SVC;
- Q^L_+ represents the maximum positive-sequence inductive reactive power of the load to be compensated for correction of fp_1;

Taking (8) and (9) and considering the three-phase operation of the compensator, the capacitance C and the fixed inductance L per phase of the SVC are defined in (10) and (11).

$$C = \frac{Q^L_+ + 2.UF_C.S^L_+}{3.\omega.V^2_L} \quad (10)$$

$$L = \frac{3V^2_L}{\omega\left(Q^C_{+L(max)} + Q^C_C\right) + 2\omega UF_L S^L_+} \quad (11)$$

III. DIGITAL CONTROL BASED ON INSTANTANEOUS POWER THEORY

The microcontroller used in the control system of this project is from Texas Instruments, model PICCOLO TMS320C28069M. The block diagram of the proposed control system, responsible for the reduction of the unbalance factor and correction of the fundamental power factor, is presented in Fig. 2.

The method of Steinmetz, theoretical basis for implementation of the control system, has its foundation built on the hypothesis that the system does not present unbalanced voltage, neither harmonic distortions in the voltages and currents. In this way, digital notch filters are implemented to extract the fundamental components of voltage and current in the input stage of the control system. In the case of the voltages, for the applications in systems with asymmetries in the feeding, the extraction of the positive-sequence three-phase instantaneous voltages in the PCC is still implemented.

The methodology used for the extraction of the positive-sequence voltage is described in [9] and [10]. These are digital filters operating as phase shifters (H_{PS}), as a transfer function in domain z described in (12), where α represents the displacement angle and ω_d the discrete frequency.

$$H_{PS}(\alpha, z^{-1}) = \frac{\sin(\omega_d + \alpha) - \sin(\alpha).z^{-1}}{\sin(\omega_d)} \quad (12)$$

Applying the transformation in symmetric components through the Fortescue Theorem, the generalized positive-sequence instantaneous component Y^+_{ab} is obtained as a function of the unbalanced signals X_{ab}, X_{bc} e X_{ca}.

$$Y^+_{ab}(z) = \frac{1}{3}\left(X_{ab}(z) + H_{PS}(120^{0}, z^{-1})X_{bc}(z) + H_{PS}(-120^{0}, z^{-1})X_{ca}(z)\right) \quad (13)$$

Therefore, the algorithm responsible for the extraction of the positive-sequence instantaneous voltage v_{ab} at PCC is based on the difference obtained from (12) and (13). The other voltages representing the three-phase system are obtained considering the respective relation of symmetry in positive-sequence.

The next step of the control system refers to the digital processing of the powers defined in (5), (6) and (7) from the instantaneous power theory. The p-q theory has been diffused in the analysis and designs in power systems because it is a very dynamic, flexible method that allows the compensation of unbalanced and distorted systems.

Accordingly, through instantaneous line-to-line voltage and line currents, it is possible to calculate in (14) and (15) the active and reactive three-phase instantaneous power according to [11].

$$p_{3\phi}(t) = v_{ab}.i_a - v_{bc}.i_c \quad (14)$$

$$q_{3\phi}(t) = \frac{1}{\sqrt{3}}\left(v_{ab}.i_c + v_{bc}.i_a + v_{ca}.i_b\right) \quad (15)$$

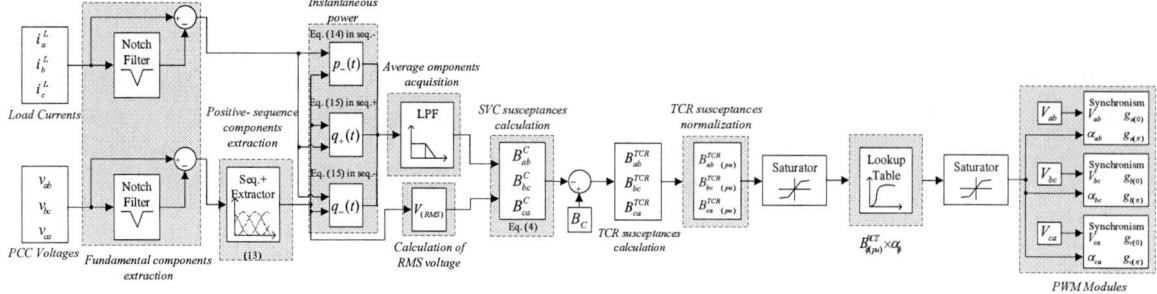

Fig. 2. Block diagram representation of the control loop.

Both instantaneous powers can be divided into a part of average value, a consequence of the product of voltage and current of the same sequence and harmonic order, and another oscillating portion resulting from the product between different voltages and sequence currents and harmonic orders [11].

The results of the procedures adopted for the determination of the active and reactive powers due to the positive and negative sequence components of the load currents are based on [8]. Taking (1) and (2), respectively line-to-line voltages and unbalanced load currents, and substituting them into (14), it defines in (16) the positive-sequence instantaneous active power, $p_+(t)$.

$$p_+(t) = \sqrt{3}V_L I_+^L \cos(\phi_+) + \sqrt{3}V_L I_-^L \cos(2\omega t + \phi_-) = \bar{p}_+ + \tilde{p}_- \quad (16)$$

From (16), the average active power in the load due to the load positive-sequence current, $\bar{p}_+$, stands out, respectively, in (17) and (18), and the oscillating active power in the load due to the load negative-sequence current, $\tilde{p}_-$.

$$\bar{p}_+ = \sqrt{3}.V_L.I_+^L.\cos(\phi_+) \quad (17)$$

$$\tilde{p}_- = \sqrt{3}.V_L.I_-^L.\cos(2\omega t + \phi_-) \quad (18)$$

Similarly, by applying the instantaneous power theory, but by changing the phase sequence of the load currents, it has in (19) and (20) the average active power due to the load negative-sequence current, $\bar{p}_-$, and the oscillating active power due to the load positive-sequence current, $\tilde{p}_+$.

$$\bar{p}_- = \sqrt{3}.V_L.I_-^L.\cos(\phi_-) = P_-^L \quad (19)$$

$$\tilde{p}_+ = \sqrt{3}.V_L.I_+^L.\cos(2\omega t + \phi_+) \quad (20)$$

By repeating the process performed to find the active powers of (17) to (20), but now, substituting (1) and (2) into (15), there are, respectively, from (21) to (24), the average reactive power in the positive-sequence, $\bar{q}_+$, the oscillatory reactive power in the negative-sequence, $\tilde{q}_-$, the average reactive power in the negative-sequence, $\bar{q}_-$, and the oscillatory reactive power in the positive-sequence, $\tilde{q}_+$.

$$\bar{q}_+ = -\sqrt{3}.V_L.I_+^L.\sin(\phi_+) = Q_+^L \quad (21)$$

$$\tilde{q}_- = \sqrt{3}.V_L.I_-^L.\sin(2\omega t + \phi_-) \quad (22)$$

$$\bar{q}_- = -\sqrt{3}.V_L.I_-^L.\sin(\phi_-) = Q_-^L \quad (23)$$

$$\tilde{q}_+ = \sqrt{3}.V_L.I_+^L.\sin(2\omega t + \phi_+) \quad (24)$$

The power values used in the Steinmetz method for calculations of equivalent susceptances per phase of the compensator, described in (4), are numerically identical to the values of the average component of instantaneous powers defined in (19), (21) and (23). Therefore, by associating digital Low-Pass Filters (LPF) with the simple processes of the instantaneous power's calculation, defined in (14) and (15),

can obtain the average powers required for the calculation of compensation susceptances. The RMS of line-to-line voltage, also necessary for the calculation of susceptances, is obtained from a simple average quadratic process applied to the positive-sequence instantaneous voltage, resulting from the process defined in (13).

From the calculated reference susceptances, responsible to cancel the unbalance and to correct the fundamental power factor, it is possible to obtain, through lookup table, the values of the firings angles, α_{ab}, α_{bc} and α_{ca}, which will be loaded into the sync modules for the generation of pulses. These modules are implemented by synchronized PWM channels, which are part of the microcontroller's own peripheral resources. The generated trigger pulses are received by the gate driver board, which isolates the signals and guarantees the current gain necessary for the safe firing of the thyristors.

IV. EXPERIMENTAL RESULTS

The load used in the tests to represent the unbalance and the power demand consists of a resistors bank and three three-phase induction motors operating no load, both connected in delta. As well as a bank of resistors connected between phases A and C. Specifications of the physical elements of the SVC are described in Table I.

TABLE I.   ELEMENTS OF SVC

Element	Parameter	Value
Semikron SKKT 26/16E	Forward voltage	1.8 V
	Forward resistance	12 mΩ
Capacitor	Capacitance	53.85 μF
	Equivalent Series Resistance (ERS)	0.215 Ω
Inductor	Inductance	109.60 mH
	Inductor series resistance	5.86 Ω

The loads rated characteristics for the experimental tests are shown in Table II.

TABLE II.   LOADS CHARACTERISTICS

Symbol	Description	Value
V_L	Supply effective line voltage	220 V
f	Fundamental frequency	60 Hz
S_+^L	Load three-phase apparent power in positive-sequence	1680 VA
Q_+^L	Load three-phase reactive power in positive-sequence	1200 VAr
UF_C^{max}	Maximum capacitive unbalance factor	54.2%
UF_L^{max}	Maximum inductive unbalance factor	13.5%
$Q_{+C(max)}^{SVC}$	SVC maximum capacitive reactive power in positive-sequence	1200VAr
$Q_{+L(max)}^{SVC}$	SVC maximum inductive reactive power in positive-sequence	0

978-1-7281-4181-7/19 $31.00 © 2019 IEEE

Table III shows the characteristics required to implement the embedded digital control to the SVC and the respective filters mentioned above.

TABLE III.   SPECIFICATIONS OF THE DIGITAL CONTROL SYSTEM

System	Symbol	Description	Value
A/D	f_s	Sampling frequency	15.36k Hz
	T_s	Sampling period	65.10 µs
	$G_{A/D}$	A/D Converter Gain	4095/3.3
Notch Filter	Q_{NTC1}	Quality factor	5
	Q_{NTC2}	Quality factor	5×10^4
	n	Filter order	2°
	k	Gain at notch frequency	1
	f_c	Notch frequency	60 Hz
	ω_c	Notch angular frequency	376.99 rad/s
Positive-Sequence Components Extractor	ω_c	Angular cut-off frequency	376.99 rad/s
LPF	Q_{LPF}	Quality factor	$1/\sqrt{2}$
	n	Filter order	2°
	k	Bandwidth gain	1
	f_c	Cut-off frequency	6 Hz
	ω_c	Angular cut-off frequency	37.7 rad/s

In Fig. 3 it is possible to visualize the experimental setup of SVC operating as unbalance compensator.

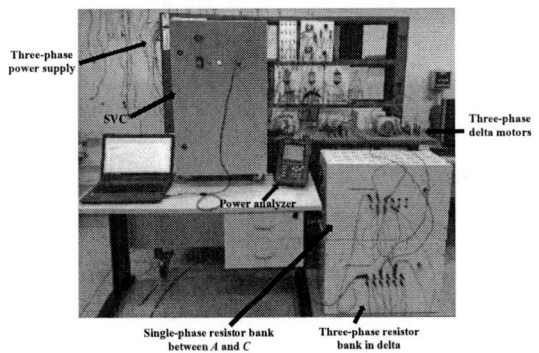

Fig. 3.   Experimental setup.

The prototype control system is shown in detail in Fig. 4, composed of analog voltage and current processing boards and the PICCOLO TMS32C28069M microcontroller responsible for digital signal processing.

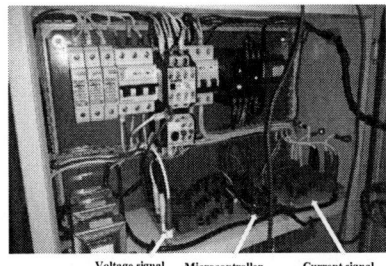

Fig. 4.   Vision of analog and digital signal processing.

Fig. 5(a) shows the line currents of the set of unbalanced loads. Fig. 5(b) shows the phasors of the fundamental components of the unbalanced line currents and the respective *UF* (%). It is verified that the *UF* (%), in the image referred to as *Aunb*, is significant and reaches a value of 44.9%. The *UF* shown on the Power Pad® 8335 power analyzer screen follows the IEC/TR 61000-3-14 standard [13], which is based on the transformation into symmetrical components, and relates the negative and positive sequence components.

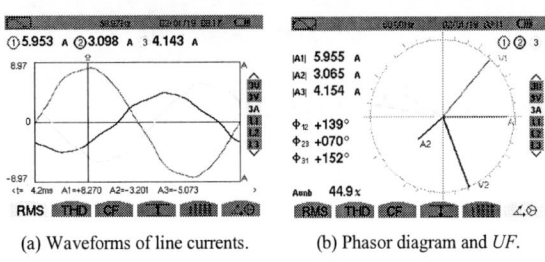

(a) Waveforms of line currents.   (b) Phasor diagram and *UF*.

Fig. 5.   Unbalanced load currents.

It is also observed, through the waveforms of Fig. 5(a), that the currents present harmonic content. This disturbance is due to the fact that the feed voltages at PCC are not purely sinusoidal, thus, they lead to non-sinusoidal currents, despite the linearity of the loads used.

The result in the compensated line is illustrated in Fig. 6. The *UF* (%) decreased from 44.9% to 3.3%, meeting project expectations. The currents are practically balanced with RMS value around 3.3A and phase displacement of approximately 120°. The waveforms of the line currents of the system can be analyzed in Fig. 6a). Due to the non-linear characteristic of the SVC, such currents present harmonic distortions.

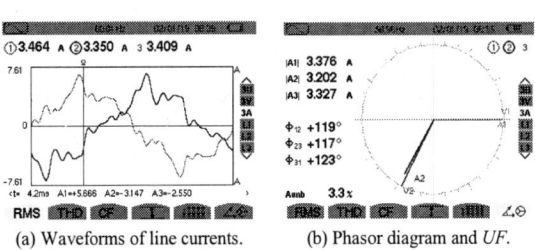

(a) Waveforms of line currents.   (b) Phasor diagram and *UF*.

Fig. 6.   Line unbalance compensation.

There was no concern with mitigation of harmonic components, since the work objective is to evaluate the system responsible for the unbalance compensation. However, several works present techniques to mitigate the distortion caused by the operation of the SVC, such as connections via multi-pulse transformers. In [7] and [12], the authors present the addition of passive filters to solve this disturb.

Fig. 7(a) shows the line-to-neutral voltage waveform, v_c, and the AC phase TCR current, i_{ca}^{TCR}. Since it is an inductive reactive element, the phase current is lag in relation to the line-to-neutral voltage. Fig. 6(b) shows the three-phase currents in the phases of the TCR operating asymmetrically. Such a feature ensures the automatic injection of the negative-sequence current required to compensate for unbalance.

978-1-7281-4181-7/19 $31.00 © 2019 IEEE

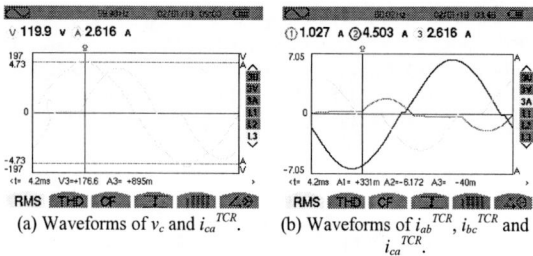

(a) Waveforms of v_c and i_{ca}^{TCR}.　　(b) Waveforms of i_{ab}^{TCR}, i_{bc}^{TCR} and i_{ca}^{TCR}.

Fig. 7.　Experimentally waveforms in the TCR.

Fig. 8(a) shows the fundamental power factor in the load.

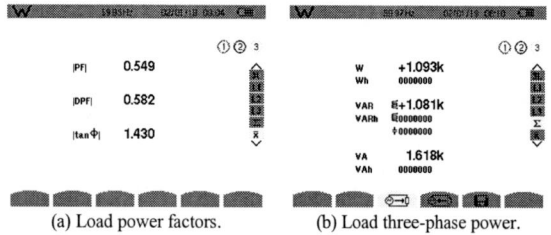

(a) Load power factors.　　(b) Load three-phase power.

Fig. 8.　Factors and power consumption in the load.

The low value of the power factor is due to the no load operation of the motors. It is possible to see how much this load is demanding inductive reactive power of the system in Fig. 8(b).

Fig. 9(a) shows fp_1, defined as |DPF| by the analyzer, and the global power factor, fp, indicated |PF| in the equipment, both measured on the side of the compensated line.

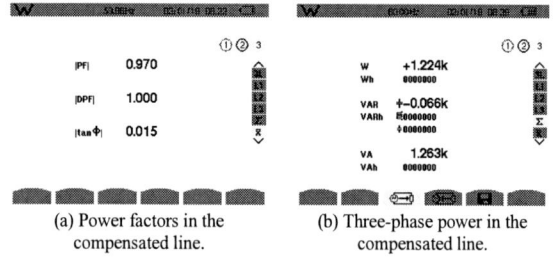

(a) Power factors in the compensated line.　　(b) Three-phase power in the compensated line.

Fig. 9.　Factors and power consumption in the compensated line.

Comparing the global power factor with the fundamental power factor, there is a small difference between the two. This is due to the harmonic distortion generated in the SVC operation. Also, note that the reactive power does not fail to zero, even with unitary fundamental power factor. This occurs because the energy analyzer considers in the calculation of the reactive power what the IEEE-1459 standard [14] defines as nonactive power. Therefore, the distortion powers are computed in the calculation of the reactive power, hence the difference.

V. CONCLUSIONS

In this work, a solution was proposed and implemented to compensate for unbalanced load current, as well as fundamental power factor correction, using a static VAR compensator, automatically and dynamically controlled by the embedded digital system. In general, the closed-loop control system presented good results, guaranteeing accuracy in the compensation of disturbances, even in nonideal feeding conditions. The expected behavior in the theoretical analyzes was validated by the experimental results, taking the unbalance factor to a minimum value and a fundamental power factor to the unit.

ACKNOWLEDGMENT

The present work was carried out with the physical and personal support of the Federal Institute of Education, Science and Technology.

REFERENCES

[1] R. Cutri, "Compensation of unbalanced load employing static converter operating with pulse width modulation," Department of Energy Engineering and Electrical Automation – Polytechnic School of the University of Sao Paulo (EPUSP), Sao Paulo, 2004.

[2] A. Pana, "Active load balancing in a three-phase network by reactive power compensation," in Power Quality Monitoring, Analysis and Enhancement. INTECH Open Access Publisher, 2011.

[3] N. G. Hingorani and L. Gyugyi, "Understanding facts: concepts and technology of flexible AC transmission systems," IEEE Series on Power Engineering Society; Mohamed E. El-Hawary, Series Editor, 2000.

[4] E. H. Watanabe, P. G. Barbosa, K. C. Almeida, and G. N. Taranto, "Tecnologia FACTS-tutorial," SBA Controle & Automação, 1th ed., vol. 9, 1998.

[5] A. Alsulami, M. Bongiorno, K. Srivastava, and M. Reza, "Balancing asymmetrical load using a static var compensator," IEEE PES Innovative Smart Grid Technologies Europe (ISGT Europe), Istanbul, Turkey, October 2014, pp.1-6.

[6] D. Wang, C. Yang, X. Zhang, J. Wang, and G. Li, "Research on application of TCR+FC typed SVC in power quality integrated management for power traction system," International Conference on Sustainable Power Generation and Supply (SUPERGEN 2012), Hangzhou, China, September 2012, pp. 1-5.

[7] F. Ji, M. M. Khan, and C. Chen, "Static var compensator based on rolling synchronous symmetrical component method for unbalance three-phase system," 2005 IEEE International Conference on Industrial Technology, Hong Kong, China, December 2005, pp. 621-626.

[8] J. Wang, C. Fu, and Y. Zhang, "SVC control system based on instantaneous reactive power theory and fuzzy PID," IEEE Transactions On Industrial Electronics, vol. 55, n. 4, pp. 1658-1665, April 2008.

[9] L. O. Arenas, G. A. Melo, and C. A. Canesin, "FPGA-based power meter implementation for three-phase three-wire and four-wire power systems, according to IEEE 1459-2010 standard," 2017 Brazilian Power Electronics Conference (COBEP), Juiz de Fora, Brazil, November 2017, pp. 1-6.

[10] P. D. Poljak, M.D. Kušljević, and J. J. Tomić, "Power components estimation according to IEEE standard 1459–2010 under wide-range frequency deviations," IEEE Transactions On Instrumentation And Measurement, vol. 61, n. 3, pp. 636-644, March 2012.

[11] H. Akagi, E. H. Watanabe, and M. Aredes, "Instantaneous power theory and applications to power conditioning," IEEE PRESS, Piscattaway, 2007.

[12] P. Vigneau, J. Destombes, R. Grünbaum, T. Gustafsson, J. Hasler, and A. Persson, "SVC for load balancing and maintaining of power quality in an island grid feeding a nickel smelter," IECON 2006 - 32nd Annual Conference on IEEE Industrial Electronics, Paris, France, November 2006, pp. 1981-1986.

[13] IEC, "Electromagnetic compatibility (EMC) - Part 3-13: Limits - Assessment of emission limits for the connection of unbalanced installations to MV, HV and EHV power systems," Technical Report n. TR 61000-3-13, 2008.

[14] IEEE, "IEEE standard definitions for the measurement of electric power quantities under sinusoidal, nonsinusoidal, balanced, or unbalanced conditions," Resolution n. 1459, New York, 2010.

978-1-7281-4181-7/19 $31.00 © 2019 IEEE

DC Current Redistributor for Electric Aircraft System

Mateus P. Dias, Hildo Guillardi Jr., Joel F. Guerreiro,
João I. Y. Ota, José A. Pomilio
School of Electrical and Computer Engineering
University of Campinas
Campinas, Brazil
{mateuspinheirodias, hildogjr, joel.engeletrica}@gmail.com;
{yutaka, antenor}@fee.unicamp.br

Paolo Mattavelli
Dept. of Management and Engineering
University of Padova
Vicenza, Italy
paolo.mattavelli@unipd.it

Abstract—The More Electric Aircraft (MEA) concept focuses on the replacement of pneumatic and hydraulic actuators by electromechanical actuators and increases the electric power demand in the aircraft. This article aims to the study and the implementation of a Power Electronic Conditioner (PEC) to mitigate the unwanted effects found in the dc aircraft internal feeders. The PEC must deal with the current unbalance between the positive and negative poles of the dc bus to comply with the standards. A mathematical dynamic model and control of PEC were simulated in the software PSIM®.

Index Terms—More Electric Aircraft; DC Current Redistributor; Power Electronic Conditioner.

I. Introduction

The electrical power consumption in commercial aircrafts has increased during the years, mainly due to the replacement of hydraulic and pneumatic actuators by electromechanical devices. This replacement has been called as the More Electric Aircraft (MEA) [1], [2], and it contributes for decreasing aircraft weight and the fuel consumption. The MEA purpose is to minimize losses in the system and to be able to comply with the ac and dc grids standards. Furthermore, this advancement has motivated researches in power electronics for more reliable, less maintenance effort and more efficient equipment [3]–[6], which include power distribution, power generation, engine gearbox etc. Power electronics devices have an important contribution for the development of a new Power Distribution Systems (PDS), including protection devices like circuits breaker and fault circuits breaker [7], [8].

Fig. 1 shows a modern commercial aircraft based on the MEA concept, which has an electrical grid consisting of many subsystems including different ac and dc voltage grids [1].

The first generation of ac grids on airplanes works at a fixed frequency (400 Hz 115/200 V) [9], including, for example, the EMB190 and 195 (Embraer). The second generation, with variable frequency (230/400 V 360-800 Hz) includes Boeing

This work was supported by the São Paulo Research Foundation (FAPESP) under grants 2017/05565-7, 2017/11623-0, 2018/13993-1, 2014/11720-7 and 2018/21436-5, and National Council for Scientific and Technological Development (CNPq) under grants 302257/2015-2 and 401216/2016-0). This study was financed in part by the *Coordenação de Aperfeiçoamento de Pessoal de Nível Superior- Brasil* (CAPES) - Finance Code 001.

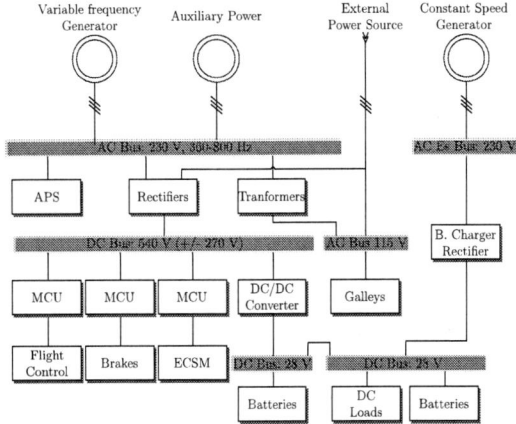

Fig. 1. A typical architecture of a modern MEA [1].

787 [1]. Both ac and dc grids must comply with Power Quality (PQ) standards. The MIL-STD-704F [10] standard establishes the voltage limits in dc and ac buses and the spectral distortion. On the other hand, the standard RCTA-DO106F [11] describes the procedures for testing equipment or loads that will be installed in the airplane.

With the implementation of MEA concept, the number of electric and electronic devices in aircraft has grown significantly, and the dynamic interactions among them have become more complex. As a consequence, complying with the power quality standards has become a challenge in MEA systems. Especially in the dc bus, current unbalance due to unbalanced loads, may increase common-mode current on neutral wire. The increase of common-mode current may cause serious problems on the MEA system mainly to the ac generator since the neutral point of the symmetrical dc bus is tied to the ac three-phase generator neutral. A dc redistributor is an electronic device that shall be able to mitigate unbalanced currents, this minimizing common-mode current.

This article presents the study and implementation of a dc redistributor for a bipolar dc bus. The dc redistributor is based on a dc-dc converter, which is modeled and simulated for unbalanced load conditions of the dc bus grid.

978-1-7281-4181-7/19 $31.00 © 2019 IEEE

II. BIPOLAR DC BUS

The usual topology to supply energy through a bipolar dc bus is a Transformer Rectifier Unit (TRU) with multipulse diode rectifiers due to the harmonics restrictions, mainly the 3th, 5th and 7th harmonic components. For the 540 V dc bus, with +/- 270 V symmetric voltages, the 12-pulse diode rectifier appears as a simple solution once it has no-actives elements, simple structure and low harmonic content. However, it does not allow the dc voltage regulation [12]. The PWM-rectifier is an alternative due to their high efficiency [13]–[15], high power factor, low harmonic content [16], output dc voltage regulator, since it is an controlled rectifier. For the purpose of this paper, the rectifier topology is not important. The relevant aspects are the availability of the symmetrical dc bus and the unbalanced dc loads on the rectifier.

Consider a 12-pulse rectifier, built by the series connection of two 6-pulse diode rectifier. The transformer that feeds the rectifier has two secondary windings, one star connected and the other a delta connected. Unbalanced loads will result different currents at each bridge, and the effective cancellation of the 5th and 7th harmonies will not happen at the transformer primary side.

For the PWM rectifier, the issue is the zero sequence current component that, in case of unbalanced dc load will flow through the neutral wire, which is tied to the neutral point of the ac generators, both at the ground reference. Common mode current affects the magnetizing flux of the generator, increasing losses and reducing the quality of the generated voltages.

Fig. 2 shows a simplified model of the dc bus, including ideal dc sources, positive and negative; the feeder resistances (positive, negative, and neutral), and the positive and negative loads.

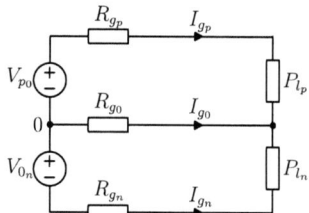

Fig. 2. Simplified Circuit of dc bus [16].

This article investigates a PEC for the dc bus aimed to compensate unbalanced loads. Fig. 3 shows a Current Redistributor Converter (CR), the structure for balancing the bipolar currents, i_{g_p} and i_{g_n} [13]. The CR is also known as current equalizer.

A. DC-DC converters as Current Redistributor

The main function of the Current Redistributor (CR) is to guarantee that the neutral current, i_{g_0}, is null. Fig. 4 shows the dc-dc converters which can operate as a CR.

The CR can do other functions, such as:

1) Voltage dynamic regulation;

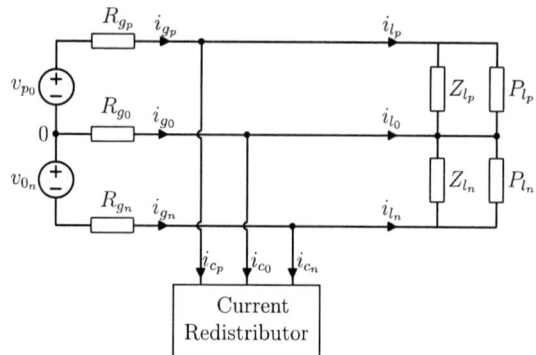

Fig. 3. Bipolar dc bus architecture including a CR and lumped loads.

2) Active damping;
3) Absorbing regenerate energy ;
4) Interface with dc energy storage (battery bank and super-capacitors) and being able to operate with loss of one dc line [17];

Fig. 4a shows a bidirectional buck converter, which is the simplest topology able to redistribute the current in the dc bus. It contains one inductor and two transistors, however the currents in the positive and negative feeders are not filtered and none of the other functions can be implemented [16].

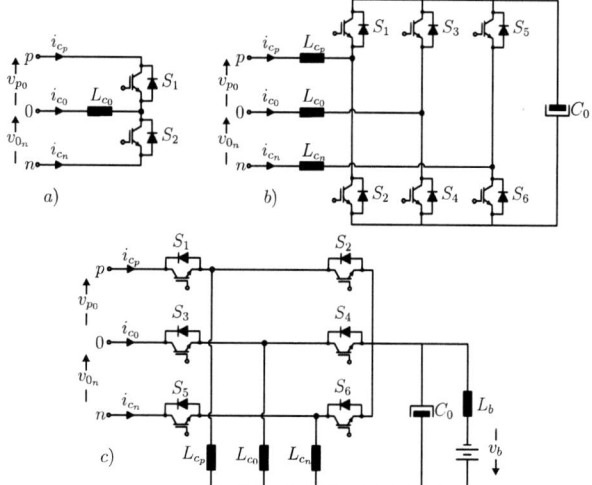

Fig. 4. DC-DC converters that can operated as a CR.

Fig. 4b shows a bidirectional boost converter, which is able to perform not only the current balance but also the functions 1, 2, 3 in the dc bipolar network. This topology enables active power injection if the inner dc bus voltage is higher than the voltage of the dc bus. This converter is the chosen dc redistributor topology in this study.

Fig. 4c shows a bidirectional buck-boost converter, which is able to perform the current balance and the function 4 in the dc bipolar network. However, the input currents in the upper and lower dc-side are not filtered, introducing high frequency disturbance in the dc bus.

978-1-7281-4181-7/19 $31.00 © 2019 IEEE

III. System description

The presence of multiple power converters in the grid, with the respective control structures, requires the study of the stability of the complete grid. Many of these converters behave as constant power loads (CPL), which means that they will absorb a higher current when the input voltage decreases and vice-versa. The result is a dynamic characteristic expressed by a negative resistance, affecting the system stability [17].

An example of CPL is a dc-dc voltage regulator feeding resistive load. The input power of the dc-dc converter is constant due to the output voltage regulation. Other example is a dc-ac converter feeding a motor, under a speed control. In steady state, a controller regulates the speed at a given operating point. Thus, the torque is constant and the system behaves as a CPL.

Fig. 5 shows a block diagram of a power converter that feeds an ideal CPL. The load is modeled as a controlled current source. The system stability can be analysed using the Nyquist Criterium, based on the ratio between the CPL input impedance (Z_i) and the converter output impedance (Z_o) [18].

Fig. 6 shows that the Current Redistributor Converter, associated with the dc loads can work as a CPL. In this sense, the CR dynamic model must be developed.

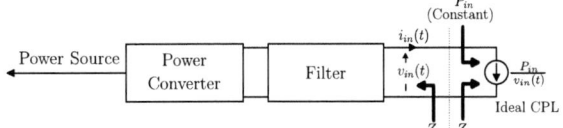

Fig. 5. Ideal CPL [18].

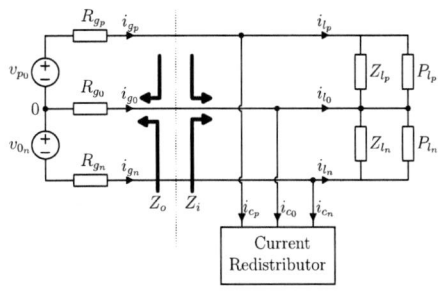

Fig. 6. Bipolar dc Bus architecture including a CR.

IV. Current and Voltage Models

This section shows the development of the CR model proposed in [16]. The converter uses pulse width modulation (PWM), with a switching frequency f_s. The duty cycles are based on the ratio of dc bus voltage,V_{pn}, and dc inner voltage, V_o, that can be obtained by (1), D_δ.

With dc bus symmetrical voltages, $V_{p0} = V_{0n} = V_{pn}/2$, the duty cycle of the switch connected to the central line is chosen $D_o = 1/2$. The upper, D_p, and lower, D_n, line duty cycle switches are represented by (2) and (3). Equation (4) shows their relation.

$$D_\delta = \frac{V_{pn}}{2V_0} \tag{1}$$

$$D_p = \frac{1}{2} + D_\delta \tag{2}$$

$$D_n = \frac{1}{2} - D_\delta \tag{3}$$

$$D_p - D_0 = D_0 - D_n = D_\delta = \frac{V_{pn}}{2V_0} \tag{4}$$

A. Current model

Fig. 7 shows the circuit of the dc redistributor. The transistors S_1 to S_6 can be represented as a single switch with two positions. In the following equations, the index x represents either p, 0 or n. Equations (5) and (6) show the commutation function $s_x(t)$ and the voltage between the switches and the reference T. When $s_x = 1$ (S_x is in position 1) the voltage v_{xT} assumes v_o, when $s_x = 0$ (S_x is in position 0) the voltage v_{xT} assumes 0.

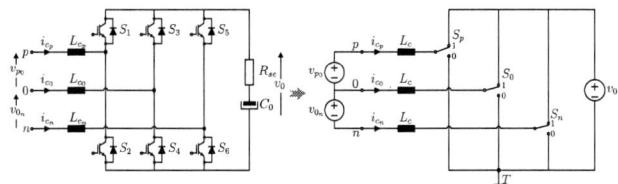

Fig. 7. The chosen dc redistribuitor topology.

$$s_x(t) = \begin{cases} 1, & S_x \text{ is on} \\ 0, & S_x \text{ is off} \end{cases} \tag{5}$$

$$v_{xT} = s_x v_o \tag{6}$$

Fig. 8 shows the equilibrium-point obtained applying the quasi-instantaneous average definition, (7), in the model variables and linearizing. Fig. 9 shows the steady state operation point of the converter.

Equation (8) shows the quasi-instantaneous value of the commutation function, known as duty cycle. It is composed by the D_x, steady state value, and the perturbation $\hat{x}$, as shown in (9).

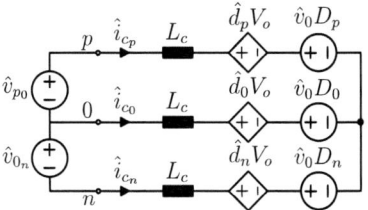

Fig. 8. Equilibrium point model

$$\bar{x} = \langle x \rangle_{T_s} = \int_{\tau - T_s}^{\tau} x(t) dt \tag{7}$$

$$d_x = \langle s_x(t) \rangle_{T_s}, \qquad 0 \le d_x \le 1 \qquad (8)$$

$$\begin{cases} d_0 = D_0 + \hat{d}_0 \\ d_p = D_p + \hat{d}_p \\ d_n = D_n + \hat{d}_n \end{cases} \qquad (9)$$

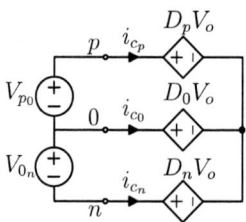

Fig. 9. Steady state operation point model.

The voltages $\hat{v}_{p_0}$ and $\hat{v}_{0_n}$ are describe in (10) and (11).

$$\hat{v}_{p_0} = V_o(\hat{d}_p - \hat{d}_0) + \hat{v}_0(D_p - D_0) \qquad (10)$$

$$\hat{v}_{0_n} = V_o(\hat{d}_0 - \hat{d}_n) + \hat{v}_0(D_0 - D_n) \qquad (11)$$

B. Current decoupling method

In the dc circuit, one of the duty cycles can be chosen arbitrarily, as mentioned before. To obtain a decoupled system, the superposition method can be applied.

Let $\hat{d}_p \ne 0$, $\hat{d}_n = 0$ and $\bar{i}_{c_n} = 0$. Fig. 10 shows the equivalent circuit. To satisfy $\bar{i}_{c_n} = 0$, the central leg duty cycle must follow (12).

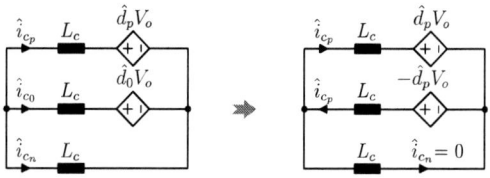

Fig. 10. Upper leg decoupling circuit.

$$\hat{d}_0 = -\hat{d}_p \qquad (12)$$

Let $\hat{d}_n \ne 0$, $\hat{d}_p = 0$ and $\bar{i}_{c_p} = 0$. Fig. 11 shows the equivalent circuit. To satisfy $\bar{i}_{c_p} = 0$, the central leg duty cycle must follow (13).

$$\hat{d}_0 = -\hat{d}_n \qquad (13)$$

The perturbation is the result of the superposition. Therefore, (14) shows the pattern of perturbation.

$$\hat{d}_0 = -\hat{d}_p - \hat{d}_n \qquad (14)$$

Rewriting (14), (15) shows the pattern of perturbation.

$$\hat{d}_0 = (D_p + D_n + D_0) - d_p - d_n \qquad (15)$$

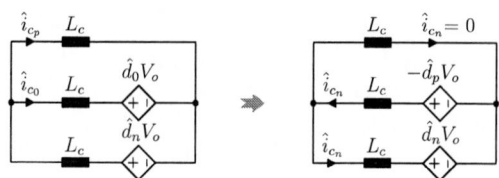

Fig. 11. Lower leg decoupling circuit.

Replacing $\hat{d}_0$ in (10) and (11), eq. (16) and (17) describe the voltages $\hat{v}_{p_0}$ and $\hat{v}_{0_n}$. Fig. 12 shows the decoupled circuit.

$$\hat{v}_{p_0} = V_o(2\hat{d}_p + \hat{d}_n) + \hat{v}_0(D_p - D_0) \qquad (16)$$

$$\hat{v}_{0_n} = V_o(-\hat{d}_p - 2\hat{d}_n) + \hat{v}_0(D_0 - D_n) \qquad (17)$$

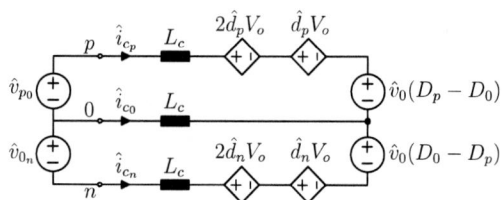

Fig. 12. Decoupled circuit.

Considering the bus voltage as a perturbation, eq. (18) and (19) show the transfer function of the upper leg current $\hat{i}_{c_p}$ and lower leg current $\hat{i}_{c_n}$ [16].

$$G_{i_{c_p}}(s) = \frac{\hat{i}_{c_p}}{\hat{d}_p} = -\frac{V_o}{sL} \qquad (18)$$

$$G_{i_{c_n}}(s) = \frac{\hat{i}_{c_n}}{\hat{d}_n} = -\frac{V_o}{sL} \qquad (19)$$

C. Voltage model

Fig. 13 shows the voltage model of the dc redistribuitor, which is evaluated following the same procedure.

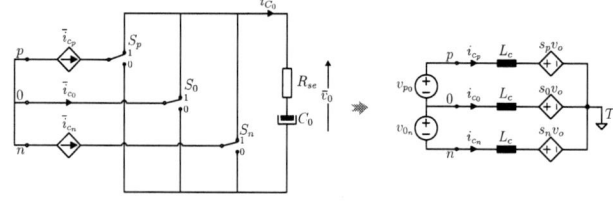

Fig. 13. Equivalent voltage model.

Owning to the linear relationship of the dc redistributor currents, eq. (20) describes $\bar{i}_{c_0}$.

$$\bar{i}_{c_0} = \bar{i}_{c_p} - \bar{i}_{c_n} \qquad (20)$$

Eq. (21) evaluates $\bar{i}_{C_0}$:

978-1-7281-4181-7/19 $31.00 © 2019 IEEE

$$\bar{i}_{C_0} = \bar{i}_{c_p} D_p + (-\bar{i}_{c_p} - \bar{i}_{c_n})D_0 + \bar{i}_{c_n} D_n$$
$$= D_\delta(\bar{i}_{c_p} - \bar{i}_{c_p}) \tag{21}$$

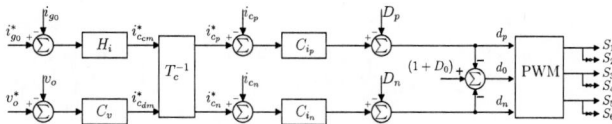

Fig. 15. Block Diagram of propose control.

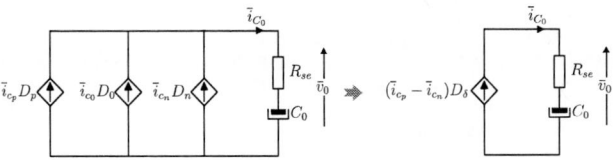

Fig. 14. Simplified circuit.

Fig. 14 shows the simplified circuit. The voltage v_o is evaluated from the circuit on the right-hand side. Expanding (22), considering only the perturbation component, (23) shows the voltage $\hat{v}_o$.

$$\bar{v}_o = \bar{i}_{C_0} Z_c \tag{22}$$

$$\hat{v}_o(s) = \frac{V_{p_n}}{2V_o} \frac{(sC_o R_{se} + 1)}{sC_o}(\hat{i}_{c_p} - \hat{i}_{c_n}) \tag{23}$$

Eq. (24) and (25) describe the dc differential current and the dc common-mode current, respectively.

$$\hat{i}_{c_{dm}} = \frac{\hat{i}_{c_p} - \hat{i}_{c_n}}{2} \tag{24}$$

$$\hat{i}_{c_{cm}} = -\frac{\hat{i}_{c_p} + \hat{i}_{c_n}}{2} \tag{25}$$

Eq. (26) and (27) describe the dc positive current and the negative current, respectively.

$$\hat{i}_{c_p} = \hat{i}_{c_{dm}} - \hat{i}_{c_{cm}} \tag{26}$$

$$\hat{i}_{c_n} = -\hat{i}_{c_{dm}} - \hat{i}_{c_{cm}} \tag{27}$$

Applying (24) in (23), eq. (28) shows the transfer function of the voltage model.

$$G_v(s) = \frac{\hat{v}_o}{\hat{i}_{c_{dm}}} = \frac{V_{p_n}}{2V_o} \frac{(sC_0 R_{se} + 1)}{2sC_o}$$
$$= \frac{D_\delta(sC_0 R_{se} + 1)}{2sC_o} \tag{28}$$

V. CR CONTROL SYSTEM

Fig. 15 shows the propose control of the dc current redistributor. The references are the neutral grid current $i_{g_0}^*$, that must be zero, and the dc bus redistributor voltage v_0^*. T_c^{-1} represents the inverse transformation of common and differential modes into the positive and negative current references, which are represented in (26) and (27). Next, the decoupling method described in (15) takes place and the duty cycles for the PWM switching are obtained.

The blocks C_v, C_{i_p}, C_{i_n} correspond to the voltage, upper and lower current controllers, respectively. These controllers are proportional integrator (PI) obtained from the SISOTOOL MATLAB®.

VI. SIMULATION RESULTS

Table I shows the parameters used for this simulation in PSIM®. Fig. 16 shows the circuit diagram for the simulation of the aircraft system containing the dc bus, the dc current redistribuitor and the unbalanced loads, including CPL. The CPL is emulated by an ideal controlled current source.

TABLE I
SIMULATION PARAMETERS FOR DC REDISTRIBUITOR.

Parameters	Value
dc bus voltages ($V_{p0} = V_{0_n}$)	270 V
dc redistributor bus voltage(V_0)	800 V
Positive feeder Load (R_{lp1})	20 Ω
Negative feeder Load (R_{ln1})	20 Ω
Additional negative feeder Load (R_{ln2})	10 Ω
Power rating of Ideal CPL (w_q)	2000 W
Switching frequency (f_s)	10 kHz
dc redistributor Capacitor (C_o)	1 mF
Redistributor Inductor (L_c)	2 mH
Current cut-off frequency (f_c)	998 Hz

Fig. 17 shows the simulation results for different loads configurations. The top waveforms represent the dc bus grid currents, i_{g_0}, i_{g_p} and i_{g_n}. The bottom waveforms represent the dc CR inner voltage V_0 and its reference, V_{0ref}.

Before $t = 1.5$ s, the dc bus feeds a 20 Ω balance load and the dc redistributor operates. At $t = 1.5$ s, a 2000 W CPL load is added to the positive branch. The dc redistribuitor adjusts the current flowing in the dc bus grid, therefore i_{g_0} remains zero and i_{g_p} and i_{g_n} are balanced, i.e. $i_{g_p} = -i_{g_n}$. At $t = 2.5$ s, an extra 40 Ω parallel load is added to the negative branch. The dc redistribuitor keeps the balance between i_{g_p} and i_{g_n}. At both $t = 1.5$ s and $t = 2.5$ s, the voltage controller keeps v_0 to its reference.

At $t = 3.5$ s, the inner voltage reference goes to 850 V. Even with this step in voltage reference, the dc CR is able to regulate the inner CR voltage at its new reference.

VII. CONCLUSION

This work presents the modeling and simulation of a power electronics converter, working as a current equalizer, applied to the symmetrical dc bus of the electric network boarded in an aircraft. The equalization of the currents between the positive and negative buses is necessary to guarantee correct operation of the rectifier, in order to comply with aeronautical power quality standards. The bidirectional boost converter is chosen due to its capacity to perform the current balance

978-1-7281-4181-7/19 $31.00 © 2019 IEEE

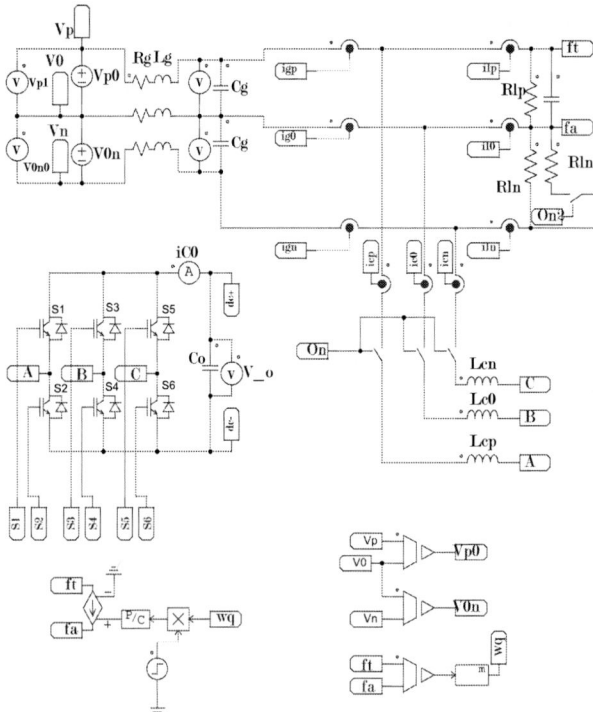

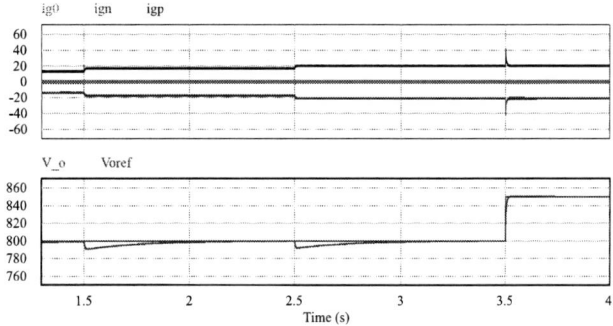

Fig. 16. Diagram of the power circuit in PSIM.

Fig. 17. Simulated results from different loads.Top: dc bus grid currents, i_{g_0}, i_{g_p} and i_{g_n}; Bottom: dc CR inner voltage V_0 and its reference, V_{0ref}.

between the positive and negative buses, active damping, regeneration of energy and dc energy storage interaction. From the dynamic model of the bidirectional boost converter, the current controllers as well as the internal dc voltage regulator can be designed. Simulation results show that the converter is capable of balancing the currents and maintaining its dc voltage regulated in situations of load variation, whether they are resistive or have a constant power (CPL) behavior. In conclusion, the topology chosen is a potential and suitable candidate for the current redistributor applied in More Electric Aircraft environment. Future experimental results shall be carried out and confirm the effectiveness of the proposed current redistributor.

REFERENCES

[1] X. Roboam, B. Sareni, and A. D. Andrade, "More electricity in the air: Toward optimized electrical networks embedded in more-electrical aircraft," IEEE Industrial Electronics Magazine, vol. 6, no. 4, pp. 6–17, Dec 2012.

[2] J. F. Guerreiro, J. A. Pomilio, and T. D. C. Busarello, "Design and implementation of a multilevel active power filter for more electric aircraft variable frequency systems," in 2013 Brazilian Power Electronics Conference, Oct 2013, pp. 1001–1007.

[3] M. Guacci, D. Bortis, I. F. Kovačević-Badstübner, U. Grossner, and J. W. Kolar, "Analysis and design of a 1200 v all-sic planar interconnection power module for next generation more electrical aircraft power electronic building blocks," CPSS Transactions on Power Electronics and Applications, vol. 2, no. 4, pp. 320–330, December 2017.

[4] P. Zanchetta, M. Degano, J. Liu, and P. Mattavelli, "Iterative learning control with variable sampling frequency for current control of grid-connected converters in aircraft power systems," IEEE Transactions on Industry Applications, vol. 49, no. 4, pp. 1548–1555, July 2013.

[5] M. Odavic, M. Sumner, and P. Zanchetta, "Control of a multi-level active shunt power filter for more electric aircraft," in 2009 13th European Conference on Power Electronics and Applications, Sept 2009, pp. 1–10.

[6] J. F. Guerreiro, J. A. Pomilio, and T. D. C. Busarello, "Design of a multilevel active power filter for more electrical airplane variable frequency systems," in 2013 IEEE IEEE Aerospace Conference, 2013, pp. 1–12.

[7] D. Izquierdo, A. Barrado, C. Raga, M. Sanz, and A. Lazaro, "Protection devices for aircraft electrical power distribution systems: State of the art," IEEE Transactions on Aerospace and Electronic Systems, vol. 47, no. 3, pp. 1538–1550, July 2011.

[8] A. V. Dias, J. A. Pomilio, and S. Finco, "A current limiting switch for applications in space power systems," in 2017 IEEE Southern Power Electronics Conference (SPEC), Dec 2017, pp. 1–6.

[9] R. Abdel-fadil, A. Eid, and M. Abdel-Salam, "Electrical distribution power systems of modern civil aircrafts," in 2nd International Conference on Energy Systems and Technologies – ICEST2013, Cairo, Egypt, 02 Feb, 2013, pp. 201–210.

[10] D. of Defense interface Standard, "Mil-std-704f aircraft electric power characteristics," 12 march 2004.

[11] RTCA, "Environmental condtions and test procedures for airborne equipment," in Radio Tech. Comm. Aeronaut. DO-160, Dec,2007.

[12] L. A. Vitoi, J. A. Pomilio, and D. I. Brandão, "Analysis of 12-pulse diode rectifier operating in aircraft systems with constant frequency," in 2017 Brazilian Power Electronics Conference (COBEP), Nov 2017, pp. 1–6.

[13] J. Moia, J. Lago, A. J. Perin, and M. L. Heldwein, "Comparison of three-phase pwm rectifiers to interface ac grids and bipolar dc active distribution networks," in 2012 3rd IEEE International Symposium on Power Electronics for Distributed Generation Systems (PEDG), June 2012, pp. 221–228.

[14] Guanghai Gong, M. L. Heldwein, U. Drofenik, J. Miniböck, K. Mino, and J. W. Kolar, "Comparative evaluation of three-phase high-power-factor ac-dc converter concepts for application in future more electric aircraft," IEEE Transactions on Industrial Electronics, vol. 52, no. 3, pp. 727–737, June 2005.

[15] M. Hartmann, J. Miniboeck, H. Ertl, and J. W. Kolar, "A three-phase delta switch rectifier for use in modern aircraft," IEEE Transactions on Industrial Electronics, vol. 59, no. 9, pp. 3635–3647, Sep. 2012.

[16] J. Lago, "Current redistributor for bipolar dc distribution networks," mathesis, Santa Catarina Federal university, 2011.

[17] S. Sumsurooah, M. Odavic, S. Bozhko, and D. Boroyevic, "Toward robust stability of aircraft electrical power systems: Using a -based structural singular value to analyze and ensure network stability," IEEE Electrification Magazine, vol. 5, no. 4, pp. 62–71, Dec 2017.

[18] X. Liu, A. Forsyth, H. Piquet, S. Girinon, X. Roboam, N. Roux, A. Griffo, J. Wang, S. Bozhko, P. Wheeler, and o. , "Power quality and stability issues in more-electric aircraft electrical power systems," MOET Forum, 01 Sept, 2009.

978-1-7281-4181-7/19 $31.00 © 2019 IEEE

Two-stage stand alone photovoltaic system for water pumping system

Renata C. Silva
Graduate Program in Electrical Engineering
Universidade Federal de Minas Gerais
Belo Horizonte, Brazil
renatacristina@ymail.com

Gabriel G. Bacheti
Dept of Electrical Engineering
Universidade Federal de Itajubá
Itabira, Brazil
gabrielbacheti@outlook.com

Rafael M. Silva
Graduate Program in Electrical Engineering
Universidade Federal de Minas Gerais
Belo Horizonte, Brazil
rafaellmario@ufmg.br

Vitor G. Neves
Dept of Electrical Engineering
Universidade Federal de Itajubá
Itabira, Brazil
vitorgomes22neves@yahoo.com.br

Clodualdo V. Sousa
Dept of Electrical Engineering
Universidade Federal de Itajubá
Itabira, Brazil
clodualdosousa@unifei.edu.br

Seleme I. Seleme Jr.
Dept of Electronic Engineering
Universidade Federal de Minas Gerais
Belo Horizonte, Brazil
seleme@cpdee.ufmg.br

Abstract—**This paper presents a two-stage photovoltaic system based on a three-phase induction machine for water pumping purposes. The system is designed to work in isolated mode as it is intended to use in remote areas. A control algorithm is proposed for a DC-link voltage regulation and boost current control in a cascaded loop. The induction machine has a speed control based on V/F scalar method. Experimental results demonstrate the behavior of the systems for different levels of irradiance, temperature and load.**

Index Terms—**Photovoltaic system, three-phase inverter, pumping system, boost control**

I. INTRODUCTION

The photovoltaic (PV) power systems has a wide range of applications, one of each is the use of solar energy to power water pumping systems. As a motivation for the use of this system, we can look to the small farmers at Brazil. Although responsible for 70% of the food consumed in the country, this type of agriculture generally has little access to the electric power distribution [1]. They need to pump water for irrigation, to livestock and for domestic use. These farmers need a system that can meet their needs of energy and it is low cost. A system that is powered by renewable energy sources (solar and wind) seems suitable for this application. This system has an advantage of being reliable, with low maintenance and long lifespan.

Usually, a PV system consists of solar panels, energy storage and the motor. The storage component generally is a battery bank. However, these elements present a high cost and short lifetime, making it unfeasible its application in real systems. As energy storage substitute, one can use a water storage tank, which has a lower cost to implement and simpler maintenance.

There are several studies, based on simulation or experimental, about photovoltaic pumping system for use in irrigation or domestic use. The simplest system consists of panels connected directly to a DC motor. Since it has no control, the volume of water pumped is proportional to the irradiance [2]. An improved technique where is used an single-stage PV system has also been analyzed in [3]. The power from the PV system is conditioned by the inverter and transferred to the motor. The functions of maximum power point tracker (MPPT) and speed control are performed by the inverter [4]. Aditionally, in a two-stage topology, the first stage extracts the maximum power from the panels and the second stage control the power flow to the load. [5]. A single stage has an advantage of only using the inverter and eliminating the losses of the DC-DC converter. However, the second stage can offer a more stable voltage for the inverter.In [9], to improve the intermittent nature of a PV system is added a converter that connect the system to the grid (when it is available).

The pump is driven by an electric motor. The AC motor most utilized is the induction machine. These types of motor are low cost, widely used, and reliable. The machines are either controlled using scalar control [6] or by closed-loop vector control, eventually with speed estimation [7] [8].As for the scalar control, that has been proposed an improved speed control that provides zero steady-state error [10]. For DC motor, one uses the permanent magnet brushless DC machine, which has as characteristic high efficiency and high power density [9].

The main objective of this article is to present a photovoltaic water pumping system based on an induction motor. In the present work, we use V/F to control the speed of the motor. The system has a two-stage topology where the inverter side is decouple from the PV panels, and two independent control structures can be implemented.

II. MODELING OF THE PV PUMPING SYSTEM

The proposed system is shown in Fig. 1. The system is composed of solar panels that convert solar irradiance into DC power. The PV array is made up of PV modules connected in series and parallel to achieve a suitable voltage and power.

978-1-7281-4181-7/19 $31.00 © 2019 IEEE

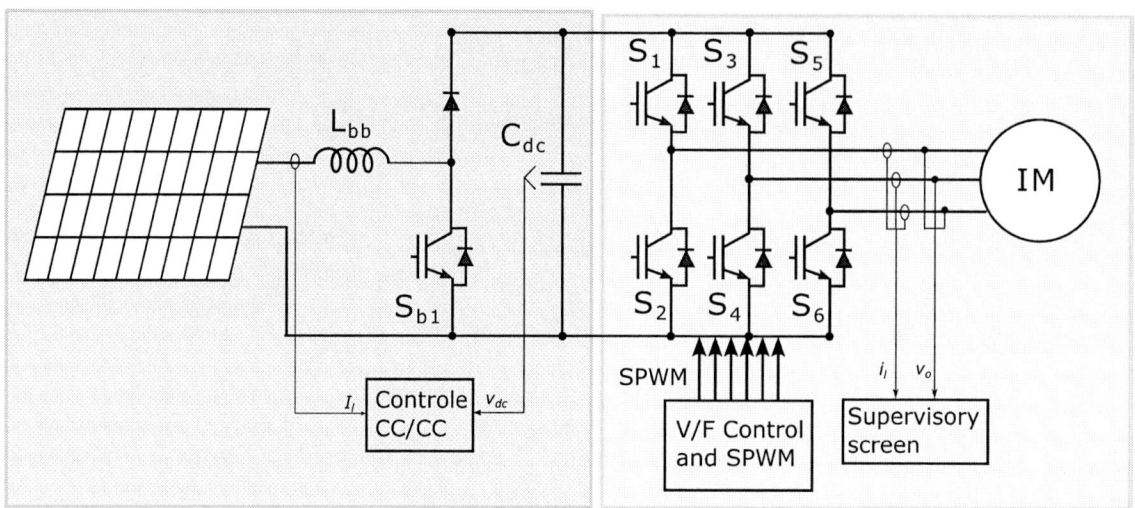

Fig. 1: Proposed system structure

However, the PV array voltage is not high enough to connect directly to the inverter. The panels are then connected to the boost converter to obtain higher stable voltage level for the DC-link.

In the second stage of the system, the VSI inverter (voltage source inverter) controls the induction motor speed by scalar control. This technique allows to control the motor speed by altering the frequency and amplitude of the supply voltage if the ratio amplitude/frequency is kept constant. The ac machine pumps water to a water tank. The pump speed is controlled so much or less water is pumped. No energy storage or MPPT is needed. The energy source is the PV panel that suplly only the power being demanded by the motor.

A. Photovoltaic Cell

The I-V characteristic of the PV array is shown in Fig. 2. By controlling either the current or the voltage, the power from the panel can be controlled. However, the I-V curve also depends on external influences as irradiation level and temperature. Each condition of irradiation and temperature influences the light-generated current according to the following equation [12]:

$$I_pv = (I_{pv,n} + K_I\Delta_t)\frac{G}{G_n} \qquad (1)$$

where $I_{pv,n}$ is the current at STC, K_i is the short circuit current/temperature coefficient, Δ_t is difference from the nominal temperature $(T - Tn)$ [in Kelvin], and G, G_n is the actual and nominal irradiance, respectively. As the system is exposed to the environment conditions, the system should be tested for various temperature and irradiance.

III. CONTROL OF THE INVERTER

The scalar control is a speed control of induction motors implemented in open-loop. The objective is to maintain the ratio of voltage to frequency constant therefore keeping the

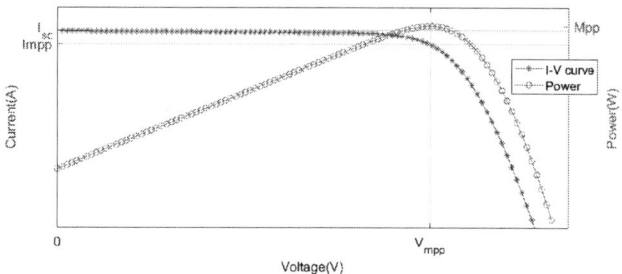

Fig. 2: I-V and P-V curves of the photovoltaic cell

air-gap flux at its nominal value, and torque is independent of the speed value [13]. The synchronous speed is given by:

$$N_s = \frac{120f}{p} \qquad (2)$$

where N_s is the speed in RPM and p is the number of poles. Thus, it can be seen that the change in frequency is proportional to the change in speed. Additionally, the amplitude of the output voltage has the following relationship:

$$V = V_0 + K_0(f_s) * f_s \qquad (3)$$

where V_o is the minimum voltage to cover the voltage at the stator resistance at low speeds and the term $K_o(f_S) * f_s$ is directly proportional to the flux wave [13]. Furthermore, if the stator resistance is eliminated, the equation 3 becomes a linear relationship between voltage and frequency.

$$K_0 = \frac{V}{f_s} \qquad (4)$$

If this ratio constant is kept constant, the air-gap flux is kept constant and the speed can be changed without affecting the torque. For this reason, this method is simple to be implemented and is widely used to control speed of the induction motor.

978-1-7281-4181-7/19 $31.00 © 2019 IEEE

The range of frequencies and voltage that can be generated to maintain the ratio voltage/frequency is defined by the PWM and the DC-link voltage, being that the last one should be sufficiently high to provide the nominal voltage to the motor.

Fig. 3 presents the block diagram of the open loop scalar control implemented in an FPGA. The reference speed is the input to the loop. Next, the frequency and voltage are calculated from nominal values of the motor. Then, the FPGA generates three digital sinusoidal refernce signals with the required amplitude and frequency. A digital PWM is also implemented [14]. and comparing the two signals generates the logic state for each switch.

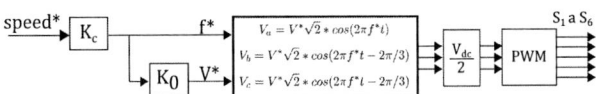

Fig. 3: diagram of V/F control

IV. BOOST CONVERTER MODEL

The DC control manages the energy transfer between the PV panel and the DC-link. In this work, the control structure implemented is composed of an outer voltage control to control the DC-link voltage and an inner current control loop responsible for controlling the current at the inductor and that power demanded from the panels. By controlling the current, we set which point at the curve the PV panel works, and the power generated is going to match that required by the load. The control is implemented with PI controllers and the topology is shown in Fig. 4. To design a controller,

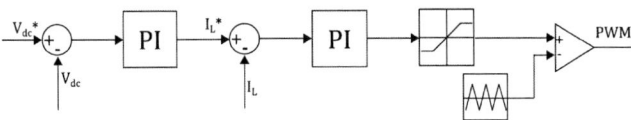

Fig. 4: Boost control

first step is to elaborate an adequate model describing the system dynamics. For the converter operating in continuous conduction mode, the transfer function obtained by state-space averaged model is [16]:

$$\frac{d}{dt}\begin{bmatrix} \tilde{i}_L \\ \tilde{v}_c \end{bmatrix} = \begin{bmatrix} -\frac{r_L}{L} & \frac{-(1-D)}{L} \\ \frac{(1-D)}{C_{dc}} & -\frac{1}{C_{dc}R_o} \end{bmatrix} \begin{bmatrix} \tilde{i}_L \\ \tilde{v}_c \end{bmatrix}$$
$$+ \begin{bmatrix} \frac{1}{L} & -\frac{R_oV_i+(1-D)}{L*\alpha} \\ 0 & -\frac{V_i}{C_{dc}*\alpha} \end{bmatrix} \begin{bmatrix} \tilde{v}_i \\ \tilde{d} \end{bmatrix} \quad (5)$$

$$y = \begin{bmatrix} 0 & 1 \end{bmatrix} \begin{bmatrix} \tilde{v}_i \\ \tilde{d} \end{bmatrix} \quad (6)$$

where

$$\alpha = (R_oD^2 - 2DR_o + R_o + r_l)$$

where R_o is the load resistance, r_l is the equivalent series resistance of the inductor L, v_i is the input voltage, v_c is the output voltage at the capacitor, and D the duty cycle. The

quantities $\tilde{i}_l$, $\tilde{v}_c$, $\tilde{v}_i$, $\tilde{d}$ represent the small ac variations about the equilibrium solution given by:

$$x = -A^{-1}BU \quad (7)$$

$$y = (-CA^{-1}B + E)U \quad (8)$$

Based on this state space model, the current to output and voltage to output transfer function can be found to be:

$$\frac{i_l(s)}{\tilde{d}(s)} = \frac{V_i}{(1-D)} \frac{sR_oC+2}{s^2R_oLC+sL+R_o(1-D)^2} \quad (9)$$

$$\frac{v_c(s)}{\tilde{d}(s)} = \frac{V_i}{(1-D)} \frac{-Ls+R_o(1-D)^2}{s^2R_oLC+sL+R_o(1-D)^2} \quad (10)$$

And the voltage to current transfer function can be found from (10) and (9) as:

$$\frac{v_c(s)}{\tilde{I_l}(s)} = \frac{1}{(1-D)} \frac{-Ls+R_o(1-D)^2}{sR_oC+2} \quad (11)$$

PI controllers tuning were based on the frequency response procedure [16]. The gains are chosen to provide a reasonable overshoot and transient response.

V. EXPERIMENTAL RESULTS

The proposed system is assessed by means of experimental results accomplished using a inverter developed according to the structure from Fig.1. The experimental setup is shown in Fig. 5 and its parameters are given in Table I. The induction motor pumps water from the bottom water tank to the upper water tank. The DC and AC voltages and currents are measured by hall sensors. Control and supervisory is programmed in a National Instruments sbRIO 9606 FPGA with Labview software. The conversion system is constituted by two Semikron Stacks, each with three half-bridges made with IGBTs, capacitors forming the DC link, and one arm with one IGBT to discharge the capacitor though an external resistor. Three inductors was installed to use with the three arms of the converter, however only one is used for the boost. The PV array has been emulated by a programmable DC source. The parameters for a real PV array are entered at the DC source that can produce current and voltage according to the PV curves. In order to evaluate the control performance, the system was tested for changes in the set point of the DC-link control, and changes in temperature, irradiance and load. The experimental curves are shown after 20s, because that is the time the system needs to be in initialized. All set points are changed in ramps. Fig. 6 shows the PV curves used for the tests.

A. Control response to change in reference of the DC-link voltage.

The regulation at the DC-link is tested by changing the set point of the control loop. The PV is configured to be at the curve for 300 W/m^2 and temperature at 35^oC. The motor is kept at 1800 RPM. As shown in Fig. 7, the set point is changed multiples times. In all occasions the control is stable. The current and power are approximately constant.

Fig. 5: Experimental setup: Pumping system, controller (top) and inverter plus converter DC/DC (bottom)

TABLE I: System Parameters

Inverter Parameters	
Parameter	Value
Nominal Power	10kVA
AC nominal voltage	220V
fundamental frequency f	60Hz
DC voltage V_{dc}	450V
Capacitor C_{dc}	$3060\mu F$
Inductance L	$3.2mH$
Inductor's resistance r_l	0.015Ω
Switching frequency f_{sh}	$5000Hz$
Machine Parameters	
Parameter	Value
Power	1.5kW
Nominal line voltage	220
Nominal current	5.55
Nominal frequency	60
Poles	4
Nominal speed (RPM)	3460

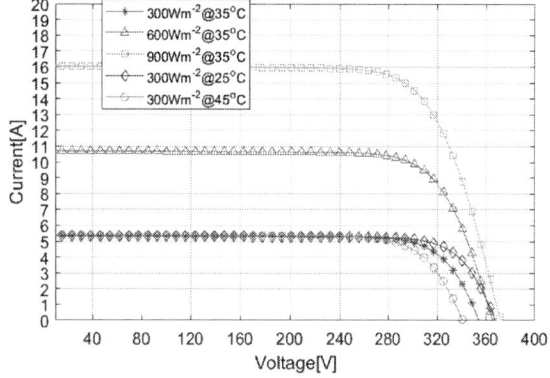

Fig. 6: PV curves used for tests

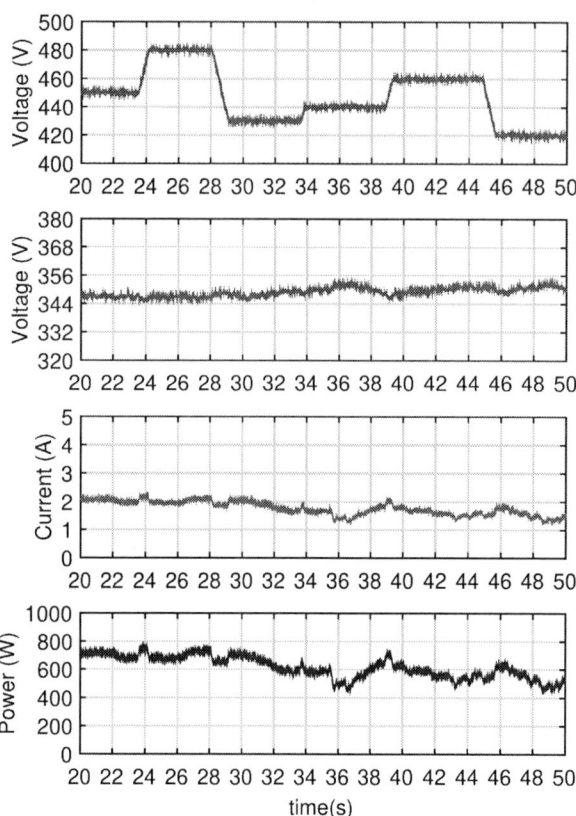

Fig. 7: Response of DC-link to a change in set point. Voltage at DC-link, voltage at the PV array, current at the inductor and power from the PV.

B. System response to a change in the irradiance

The results at Fig. 8 show how the system responds when there is a change in irradiance. It can be seen that when the change occurs at $t \approx 30\ s$, the voltage of the PV panel changes and at $t = 42\ s$ the current changes. Because the PV curve changes, the point of operation must change so that the PV array can produced that same power. The power curve at Fig. 8 has some fluctuations but in the end the system was capable to regulate the power to the initial value.

C. System response to a change in the temperature

When the temperature at a PV array changes so does the power produced. Thus, Fig. 9, presents the system response for a change in temperature for a PV curve of $300\ Wm^{-2}$. At the time the temperature is changed, the system responds by changing the voltage to adapt to the new condition. Because an mppt algorithm is not implemented, the power provided by the PV is not controlled. It can be seen in the Fig. 9 that power drops but it stay between 660 W and 46 0W, which is any case sufficient to feed the load.

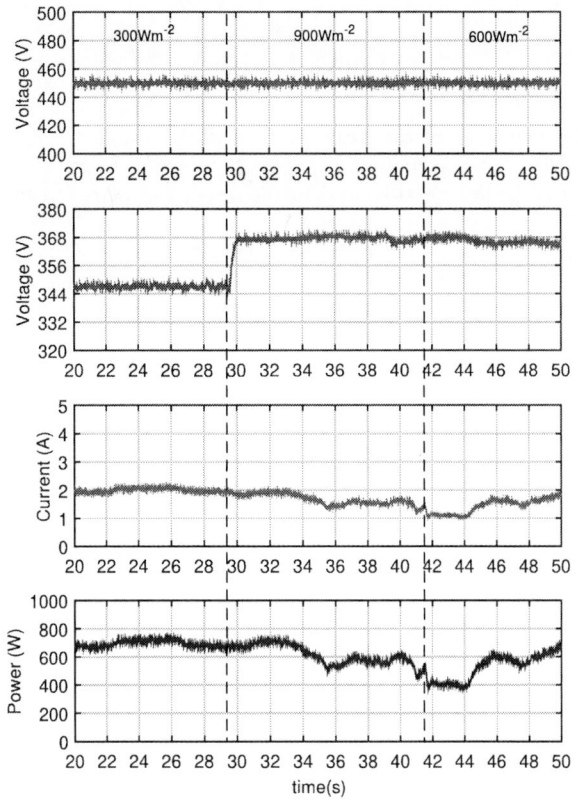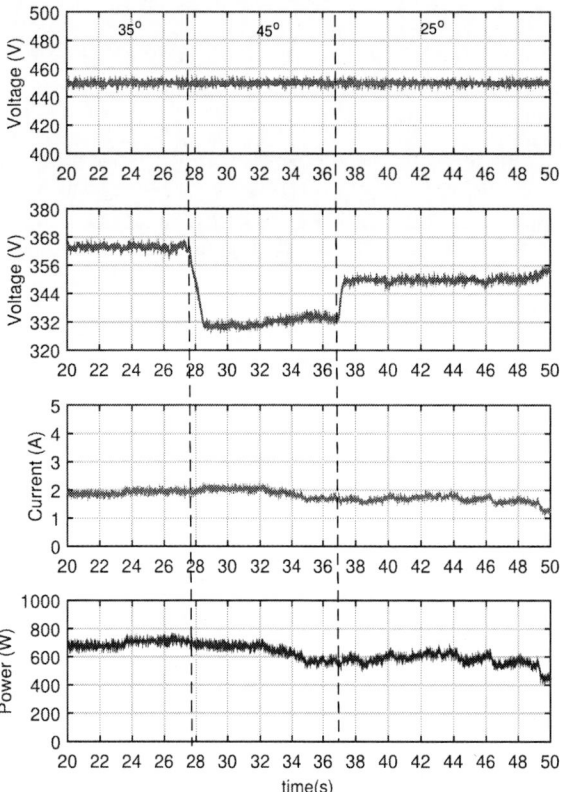

Fig. 8: System response to irradiance change. Voltage at DC-link, voltage at the PV array, current at the inductor and power from the PV.

Fig. 9: System response to temperature change. Voltage at DC-link, voltage at the PV array, current at the inductor and power from the PV.

D. System response to load change

The results presented in Fig. 10 for the case of load change, show that the power generated by the PV is proportional to the load. As the speed of the motor increases, the torque increases because more water is pumped. The system operates at $300\ Wm^{-2}$ and $35°$. At the beginning, the motor's speed is 1200 RPM and the PV system generates approximately 140 W. Next, the speed reference is changed to 2400 RPM and the PV generates 728 W and, lastly, for a speed of 1800 RPM the PV system generates 377 W. The PV panel is able to supply power to the motor because it is operating below the maximum power point. The maximum power point for $300\ Wm^{-2}$ and $35°C$ is 1.49kW for V=296.3V and I=5.03A. In table II, the voltage, frequency and speed measured at the motor are shown for various speed. It can be observed that the V/F algorithm works, and the motor operates at the desirable speed, slip considered, and frequency. However, the measured voltage has an error because of the harmonics present in the signal. Fig. 11 show the speed and frequency measurements at the induction motor.

TABLE II: Load Test

Speed Reference	Variable	Reference	Measured
1200RPM	Voltage	73.33 V	63.22 V
	Frequency	20 Hz	20.1 Hz
	Speed	1200 RPM	1194 RPM
2400RPM	Voltage	146.67 V	128.30 V
	Frequency	40 Hz	40.0 Hz
	Speed	2400 RPM	2322 RPM
3000RPM	Voltage	183.33 V	163,03 V
	Frequency	50 Hz	50.30 Hz
	Speed	3000 RPM	2926 RPM

VI. CONCLUSION

Control strategies for a two-phase PV inverter for water pumping system are presented in this paper. The proposed control approach is based in a cascaded control loop of the DC-link voltage and inductor current. The system is implemented without MPPT, and the PV array generates only the energy demanded by the load. A scalar control for the induction motor is presented. The experimental results demonstrate that the performance of the controllers is satisfactory under steady state. The system has been assessed for different operating conditions which replicates situations that could happen in a real system.

ACKNOWLEDGMENT

The authors thanks all the members of the Conversion and Control of Electric Energy - CCEE Research Group at UNIFEI, Campus Itabira.

This work has been supported by the Brazilian agency CAPES.

REFERENCES

[1] "Agricultura familiar produz 70% dos alimentos consumidos por brasileiro," Governo do Brasil. [Online]. Available: http://www.brasil.gov.br/economia-e-emprego/2015/07/agricultura-familiar-produz-70-dos-alimentos-consumidos-por-brasileiro. [Accessed: 03-Jun-2019].

[2] T. A. Vicentin, O. J. Seraphim, R. J. Halmeman, J. F. Presenço, and A. J. de Oliveira Júnior, "Análise no acionamento de motoombas atrav's de sistemas fotovoltaicos," EnergAgric, vol. 31, no. 1, p. 72, Apr. 2016.

[3] C. Ramulu, P. Sanjeevikumar, R. Karampuri, S. Jain, A. H. Ertas, and V. Fedak, "A solar PV water pumping solution using a three-level cascaded inverter connected induction motor drive," Engineering Science and Technology, an International Journal, vol. 19, no. 4, pp. 1731–1741, Dec. 2016.

[4] E. Souza de Santana, J. Jesus Fiais Cerqueira, and T. Silva Flanklin, "Fuzzy and PI controllers in pumping water system using photovoltaic electric generation," IEEE Latin Am. Trans., vol. 12, no. 6, pp. 1049–1054, Sep. 2014.

[5] B. Singh and S. Murshid, "A Grid-Interactive Permanent-Magnet Synchronous Motor-Driven Solar Water-Pumping System," IEEE Transactions on Industry Applications, vol. 54, no. 5, pp. 5549–5561, Sep. 2018.

[6] Dongliu Jiang et al., "Design of photovoltaic water-pump control system based on TMS320F2812," in 2011 International Conference on Materials for Renewable Energy & Environment, Shanghai, China, 2011, pp. 147–150.

[7] A. Chikh and A. Chandra, "Optimization and control of a photovoltaic powered water pumping system," in 2009 IEEE Electrical Power & Energy Conference (EPEC), Montreal, QC, Canada, 2009, pp. 1–6.

[8] S. G. Malla, C. N. Bhende, and S. Mishra, "Photovoltaic based water pumping system," in 2011 International Conference on Energy, Automation and Signal, Bhubaneswar, India, 2011, pp. 1–4.

[9] R. Kumar and B. Singh, "Grid interactive solar PV based water pumping using BLDC motor drive," in 2016 IEEE 7th Power India International Conference (PIICON), 2016, pp. 1–6.

[10] A. Munoz-Garcia, T. A. Lipo, and D. W. Novotny, "A new induction motor V/f control method capable of high-performance regulation at low speeds," IEEE Trans. on Ind. Applicat., vol. 34, no. 4, pp. 813–821, Aug. 1998.

[11] D. Sera, R. Teodorescu, and P. Rodriguez, "PV panel model based on datasheet values," in 2007 IEEE International Symposium on Industrial Electronics, 2007, pp. 2392–2396.

[12] M. G. Villalva, J. R. Gazoli, and E. R. Filho, "Comprehensive Approach to Modeling and Simulation of Photovoltaic Arrays," IEEE Transactions on Power Electronics, vol. 24, no. 5, pp. 1198–1208, May 2009.

[13] I. Boldea and S. A. Nasar, The induction machines design handbook, 2nd ed. Boca Raton, FL: CRC Press/Taylor & Francis, 2010.

[14] S. Buso and P. Mattavelli, "Digital Control in Power Electronics," in Digital Control in Power Electronics, 2006.

[15] R. F. Bastos, "Sistema de gerenciamento para carga e descarga de baterias (chumbo-ácido) e para busca do ponto de máxima potência gerada em painéis fotovoltaicos empregados em sistemas de geração distribuída," Mestrado em Sistemas Dinâmicos, Universidade de São Paulo, São Carlos, 2013.

[16] R. W. Erickson and D. Maksimovic, Fundamentals of Power Electronics, 2ed ed. Springer, 2001.

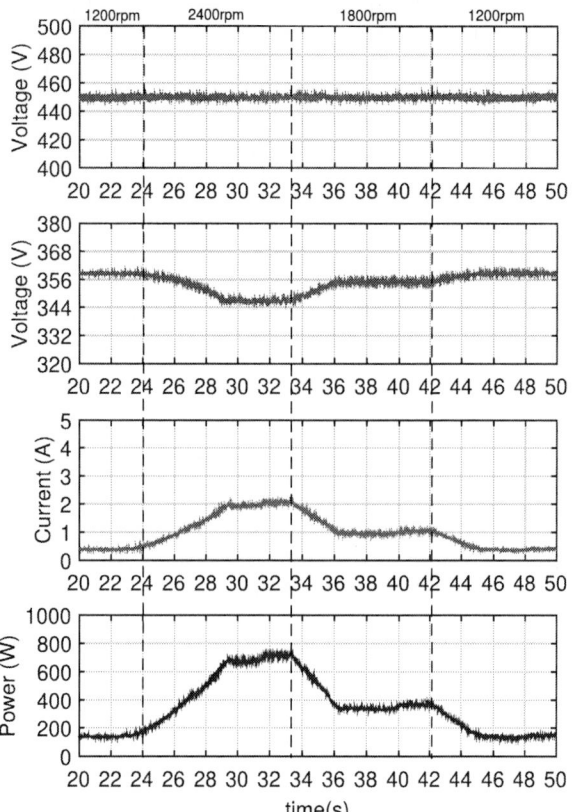

Fig. 10: System response to load change. Voltage at DC-link, voltage at the PV array, current at the inductor and power from the PV.

 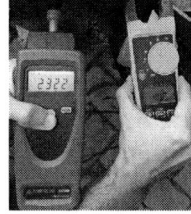 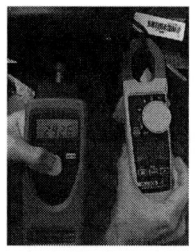

(a) 1200 RPM   (b) 2400 RPM   (c) 3000 RPM

Fig. 11: Measure of speed and frequency for three different speeds

AN IMPROVED IMPEDANCE ESTIMATION METHOD BASED ON POWER VARIATIONS IN GRID-CONNECTED INVERTERS

José H. Suárez
Department of Electrical Engineering
Federal University of Bahia
Salvador, Brazil
jsuarez@ufba.br

Hugo M. C. Gomes
Department of Electrical Engineering
Federal University of Bahia
Salvador, Brazil
hugo.cotrim@ufba.br

Luan S. Santana
Pos-graduate program in Mechatronics
Federal University of Bahia
Salvador, Brazil
santana.luan@ufba.br

Alfeu J.Sguarezi Filho
Centre Eng. Model. Appl. Social Scien.
Federal University of ABC
Santo André, Brazil
alfeu.sguarezi@ufbc.edu.br

Leandro L. O. Carralero
Department of Electrical Engineering
Federal University of Bahia
Salvador, Brazil
leandro.oro@ufba.br

Fabiano F. Costa
Department of Electrical Engineering
Federal University of Bahia
Salvador, Brazil
fabiano.costa@ufba.br

Abstract—This paper proposes an improved version of a grid-impedance estimation method based on imposed variations of active and reactive power injected by inverters into the grid. The variations cause changes in the synchronous voltage v_d and current i_q at the point of common coupling (PCC) between the inverter and the grid. These variations are used to calculate the grid impedance. The proposed improvement relies on a positive sequence estimator enhanced by a Fourier filter which provides harmonic rejection capability to the method. A complete evaluation of the proposed method is performed through simulations in Matlab/Simulink environment. The results show that the method notably refines the impedance computation in moderate weak grids.

Index Terms—Impedance Estimation, Distributed Generation, Positive Sequence Estimation, Half Cycle Fourier Algorithm.

I. INTRODUCTION

In the last decade, a significant number of Distributed Generation Systems (DGS) have been integrated into the distribution power grid. The suitable integration of these generators into the grid must meet technical requirements of quality and safety. Towards this purpose, several electrical variables are controlled at the PCC, one of them being the grid impedance which is paramount to the inference of the DGS [1]. Moreover, it can be used for detecting the DSG islanding condition [2], [3], or it can be applied to augmenting the control performance against voltage harmonic distortions [4].

Grid impedance estimation methods are classified as non-invasive or invasive. The first approach is based on measurements of voltages and currents that can naturally vary whenever the grid experiments any change. However, in most cases, variations in such quantities may not occur in the range and rate necessary to correctly estimate the impedance [5]. An invasive approach is to disturb the grid intentionally, and then perform signal acquisition and processing. Although in

this approach, perturbations are introduced into the grid, the invasive methods have predetermined characteristics regarding repeatability and intensity, which allow a higher accuracy for the grid impedance estimation. Thus, there are many invasive techniques found in the literature [6], [7], [8], [9].

A method for grid impedance estimation based on the variation of the active and reactive power imposed by a controlled inverter is presented in [10]. The proposal in [10] allows minimizing the impact caused by invasive methods on power quality. Using this method, it is possible to separately analyze the voltage and current in the synchronous axes dq. Then, the values of resistance (R) and inductance (L) of the grid can be determined through the intercalated variations of the reference currents i_d^* and i_q^* injected by the inverter. The method performs well under non-distorted and balanced grid conditions,

This paper ameliorates the proposition discussed in [10]. It is robust against harmonic distortions and unbalances grid voltages. To accomplish these features, one incorporates to the method proposed by in [10] a positive sequence estimator on the grid voltages and currents measured at the PCC. These actions mitigate the interferences and flatten the synchronous dq values, enhancing the accuracy of the grid impedance estimation. It is important to point out that the approach for grid-impedance estimation adopted in this paper has an important advantage over the ones that involves some kind of non-characteristic signals injected into the grid because it does not affect the grid power quality.

II. IMPEDANCE ESTIMATION METHOD

Consider a three-phase distributed generation system connected to the grid, as represented in the diagram shown in Figure1. In this figure, v_a, v_b, v_c and i_a, i_b, i_c represent the

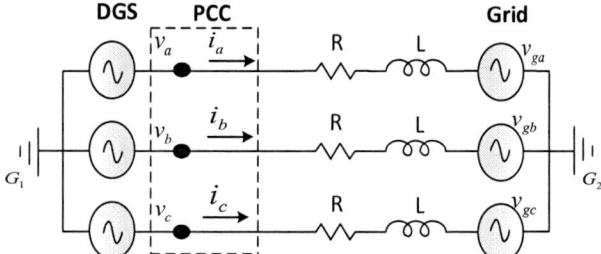

Fig. 1. Simplified circuit of a three-phase distributed generation system connected to the grid.

voltages and currents per phase at the PCC and v_{ga}, v_{gb}, v_{gc}, the grid phase voltages. The grid impedance is represented by a resistance R and an inductance L.

Based on Figure 1, one can write

$$v_a = Ri_a + L\frac{di_a}{dt} + v_{ag}, \qquad (1)$$

$$v_b = Ri_b + L\frac{di_b}{dt} + v_{bg}, \qquad (2)$$

$$v_c = Ri_c + L\frac{di_c}{dt} + v_{cg}. \qquad (3)$$

The above equations can be rewritten in matrix form as:

$$\begin{bmatrix} v_a \\ v_b \\ v_c \end{bmatrix} = \begin{bmatrix} R & 0 & 0 \\ 0 & R & 0 \\ 0 & 0 & R \end{bmatrix} \begin{bmatrix} i_a \\ i_b \\ i_c \end{bmatrix} + \begin{bmatrix} L & 0 & 0 \\ 0 & L & 0 \\ 0 & 0 & L \end{bmatrix}$$
$$\frac{d}{dt} \begin{bmatrix} i_a \\ i_b \\ i_c \end{bmatrix} + \begin{bmatrix} v_{ga} \\ v_{gb} \\ v_{gc} \end{bmatrix}. \qquad (4)$$

Through the Park's transformation, it is possible to related the dq variables to phase abc variables as:

$$\begin{bmatrix} v_d \\ v_q \\ v_0 \end{bmatrix} = T \begin{bmatrix} v_a \\ v_b \\ v_c \end{bmatrix}, \qquad (5)$$

where the matrix T is defined by:

$$T = \frac{2}{3} \begin{bmatrix} cos(\theta_g) & cos(\theta_g - \frac{2\pi}{3}) & cos(\theta_g + \frac{2\pi}{3}) \\ -sin(\theta_g - \frac{2\pi}{3}) & -sin(\theta_g - \frac{2\pi}{3}) & -sin(\theta_g + \frac{2\pi}{3}) \\ \frac{1}{2} & \frac{1}{2} & \frac{1}{2} \end{bmatrix}, \qquad (6)$$

and θ_g is the phase for the voltage v_a, provided by a PLL algorithm. Hence, applying (5) in (4), one obtains the following equations:

$$v_d = Ri_d - L\omega_0 i_q + L\frac{di_d}{dt} + v_{dg}, \qquad (7)$$

$$v_q = Ri_q + L\omega_0 i_d - L\frac{di_q}{dt} + v_{qg}, \qquad (8)$$

where v_d, v_q and i_d, i_q are the voltages and currents at the PCC in the synchronous dq axes and v_{dg}, v_{qg}, the grid voltages, also in the synchronous axes. The grid fundamental frequency is

represented by ω_0. As already mentioned, the grid impedance estimation method proposed here is based on the variations of the active and reactive power injected by the inverter. These variations causes, in turn, variations in dq voltages and currents at the PCC. The active power is related to the reference current i_d^* and the reactive power is associated with the reference current i_q^*. These currents are imposed by the inverter control system.

In the proposed method, the variations in currents i_d^* and i_q^* are measured considering two specific instants t_1 and t_2. Also, the derivatives in (7) and (8) can be disregarded since at steady-state the synchronous dq currents and voltages are constants. Hence, it is possible to rewrite (7) and (8) for t_1 and t_2. For $t = t_1$:

$$v_{d_1} = Ri_{d_1} - L\omega_0 i_{q_1} + v_{dg_1}, \qquad (9)$$

$$v_{q_1} = Ri_{q_1} + L\omega_0 i_{d_1} + v_{qg_1}, \qquad (10)$$

and for $t = t_2$:

$$v_{d_2} = Ri_{d_2} - L\omega_0 i_{q_2} + v_{dg_2}, \qquad (11)$$

$$v_{q_2} = Ri_{q_2} + L\omega_0 i_{d_2} + v_{qg_2}. \qquad (12)$$

Subtracting (9) from (11) and (10) from (12), one obtains:

$$\Delta v_d = R\Delta i_d - L\omega_0 \Delta i_q, \qquad (13)$$

$$\Delta v_q = R\Delta i_q + L\omega_0 \Delta i_d, \qquad (14)$$

where

$$\Delta v_d = v_{d_2} - v_{d_1} \quad ; \quad \Delta v_q = v_{q_2} - v_{q_1}, \qquad (15)$$

$$\Delta i_d = i_{d_2} - i_{d_1} \quad ; \quad \Delta i_q = i_{q_2} - i_{q_1}, \qquad (16)$$

and the variations for the grid voltage are assumed to be null.

It is demonstrated in [10] that, due to the phase-locked-loop (PLL), the variation in Δv_q leads to an error in the estimation of the grid impedance. Therefore, here, only the Δv_d is considered in grid impedance estimation. To determine the resistance R, one imposes variation only in i_d, keeping the i_q constant. Hence, from (13), one obtains:

$$R = \frac{\Delta v_d}{\Delta i_d}. \qquad (17)$$

To estimate the inductance L, one varies i_q^*, maintaining i_d^* constant. Hence:

$$L = \frac{-\Delta v_d}{\omega_0 \Delta i_q}. \qquad (18)$$

978-1-7281-4181-7/19 $31.00 © 2019 IEEE

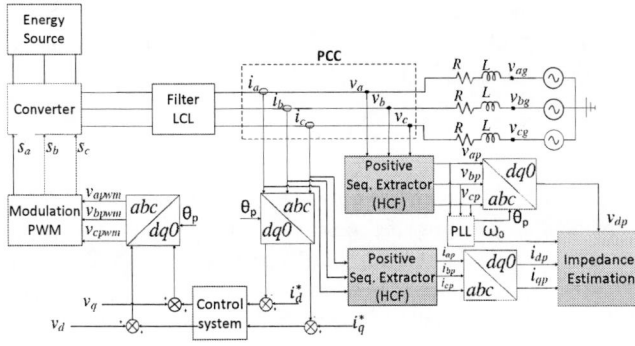

Fig. 2. Three-phase distributed generation system.

TABLE I
SYSTEM PARAMETERS.

Parameters	Value
(1)	(2)
Grid Line Voltage (rms)	220 V
Grid Line Voltage (rms)	220 V
Grid Inductance (L)	16 mH
Grid Resistance (R)	2 Ω
Link DC(inverter)	500 V
Grid Frequency	60 Hz
Inductance Filter L_1	20 mH
Inductance Filter L_2	1 uH
Capacitor Filter	4 uF
Resistance Filter	3.5 Ω
Switching Frequency	18 kHz

III. SCOPE OF APPLICATION

To evaluate the proposed method, a basic platform of a three-phase distributed generation system is simulated in Matlab/Simulink environment. The electrical diagram of the simulated DGS is depicted in Figure 2. In this figure v_a, v_b and v_c are the phase voltages and i_a, i_b and i_c are the line currents at the PCC, which are transformed to dq variables. The currents i_{ap}, i_{bp} and i_{cp} as well as v_{ap}, v_{bp} and v_{cp} are currents and voltages filtered by the positive sequence extractor. The characteristics of the simulated system are summarized in Table I.

The simulations are carried out with the system shown in Figure 2 under harmonic distortion (5^a and 11^a orders) and unbalanced conditions. The total harmonic distortion (THD) value reaches 7.13%. This value is within the limits set by IEEE 519-2014. A voltage unbalance corresponding to the maximum amount allowed in three-phase systems of 2% according IEC 61000-2-12. The proposed method is suitable for weak power grids [11]. Here the short circuit ratio (SCR) value is 2.73.

IV. PROPOSED METHOD

In the proposed method, a positive sequence estimator is applied to three-phase voltages, and another positive-sequence

estimator is applied to three-phase currents. This strategy of applying these structures greatly reduces the distortions presented in the voltages and currents. Therefore, it is possible for the algorithm to correctly identify the variations in the currents Δi_d, Δi_q and in the voltage Δv_d. It is essential that the range of these variations, particularly V_d, is high enough to ensure that the current and voltage sensors are able to detect it clearly.

For the grid resistance, the estimation starts at an instant of time t_1. The values for v_d and i_d immediately before this instants are already stored in v_{d1} and i_{d1}. Also at instant t_1, the reference for the current is set in i_{d2}^*. Then the implemented algorithm keeps accessing whether there is a variation in v_d greater than a given tolerance. When, at instant t_2, this variation is detected, the algorithm stores this instant and the v_{d2} value. The current i_{d2} should follow the reference imposed by the inverter at instant t_1. The values for Δi_d and Δv_d are taken and the grid resistance can be computed using (17). For the estimation of the grid inductance, the same approach is taken. The difference is that the variation is imposed in i_q.

In the following, it is discussed the positive sequence estimator applied to improve the grid impedance estimator, earlier discussed.

A. Positive Sequence Estimator

The estimation sequence estimator used in this work is based on voltages v_α and v_β, which are computed from the instantaneous values of the phase voltages v_a, v_b e v_c. The three-phase voltages v_a, v_b e v_c can be represented by a synchronous vector $\vec{v}_s$ in the α and β plane, whose angular velocity is ω. Thus, ω and $\vec{v}_s$ are positive constants and $\vec{v}_s$ rotates counterclockwise. In case of an unbalance in the electric grid, the vector $\vec{v}_s$ can be defined by a sum of two vectors: one of positive sequence, $\vec{v}_p$, rotating counterclockwise with velocity ω, and another of negative sequence, $\vec{v}_n$ that rotates clockwise, with angular velocity $-\omega$. Each of these vectors has constant magnitudes V_p e V_n. The angular positions are given by $\theta_p = \omega t + \phi_p$, for the positive sequence, and $\theta_n = -\omega t + \phi_n$, for the negative sequence, as can be seen in Figure 3. Examining this figure, it is possible to write the voltages v_α and v_β as [12]:

$$v_\alpha = X_1^c cos\omega t + X_1^s sen\omega t, \tag{19}$$

$$v_\beta = Y_1^c cos\omega t + Y_1^s sen\omega t, \tag{20}$$

where

$$\begin{cases} X_1^c = V_p cos\phi_p + V_n cos\phi_n = V_{p\alpha0} + V_{n\alpha0}, \\ X_1^s = -V_p sin\phi_p + V_n sin\phi_n = -V_{p\beta0} + V_{n\beta0}, \\ Y_1^c = V_p sin\phi_p - V_n sin\phi_n = V_{p\beta0} - V_{n\beta0}, \\ Y_1^s = V_p cos\phi_p - V_n cos\phi_n = V_{p\alpha0} - V_{n\alpha0}, \end{cases} \tag{21}$$

and $V_{p\alpha0} = V_p cos\phi_p$, $V_{p\beta0} = V_p sen\phi_p$, $V_{n\alpha0} = V_n cos\phi_n$ and $V_{n\beta0} = V_n sen\phi_n$. Hence, once the v_α and v_β are given (through the Clarke transformation on the phase voltages), it is straightforward to determine X_1^c, X_1^s, Y_1^c and Y_1^s by means of a linear estimation. These variables, in turn, can be used

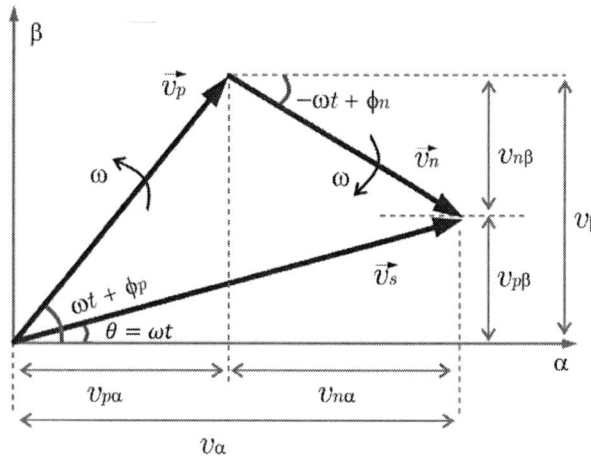

Fig. 3. Decomposition of the synchronous vector in the $\alpha\beta$ plane.

to compute $V_{p\alpha0}$, $V_{p\beta0}$, $V_{n\alpha0}$ and $V_{n\beta0}$ which are employed to estimate the positive and negative sequences. Moreover, as v_α and v_β are sinusoid signals that can be contaminated by harmonics, in this work, the half-cycle Fourier (HCF) algorithm is applied to filter out the odd-order harmonics from v_α and v_β [13], [14]. Therefore, one can determine X_1^c, X_1^s, related to the fundamental sinusoid of v_α, and Y_1^c, Y_1^s, associated with the fundamental of v_β. Rearranging (21), with X_1^c, X_1^s and Y_1^c, Y_1^s, it is possible to obtain the following relationships:

$$V_{p\alpha0} = V_p \cos\phi_p = \frac{1}{2}(X_1^c + Y_1^s), \quad (22)$$

$$V_{p\beta0} = V_p \operatorname{sen}\phi_p = \frac{1}{2}(Y_1^c - X_1^s), \quad (23)$$

$$V_{n\alpha0} = V_p \cos\phi_n = \frac{1}{2}(X_1^c - Y_1^s), \quad (24)$$

$$V_{n\beta0} = V_n \operatorname{sen}\phi_n = -\frac{1}{2}(Y_1^c + X_1^s). \quad (25)$$

Considering (25), the magnitudes of positive and negative sequences are provided by:

$$V_p = \sqrt{V_{p\alpha0}^2 + V_{p\beta0}^2}, \quad (26)$$

$$V_n = \sqrt{V_{n\alpha0}^2 + v_{n\beta0}^2}, \quad (27)$$

and the initial phases by:

$$\phi_p = \arctan\left(\frac{V_{p\beta0}}{V_{p\alpha0}}\right), \quad (28)$$

$$\phi_n = \arctan\left(\frac{V_{n\beta0}}{V_{n\alpha0}}\right). \quad (29)$$

The $\alpha\beta$ voltages associated with the positive sequence are given by:

$$v_{p\alpha} = V_{p\alpha0} \cos(\omega t + \phi_p), \quad (30)$$

$$v_{p\beta} = V_{p\alpha0} \cos(\omega t + \phi_p - 90^o). \quad (31)$$

Finally, the voltages $v_{p\alpha}$ and $v_{p\beta}$ can be used to construct the positive sequence for the phase voltages, with the inverse of Park's transformation.

V. RESULTS

The results presented in this section highlight the improvements provided by the positive sequence extractor to the standard method of grid impedance estimation based on the imposed variations of the active and reactive powers. The proposed method is tested in conditions where the grid presents harmonics and voltage unbalance in simulations performed with Matlab/Simulink. For this purpose, two scenarios were considered. In the first the impedance estimation method is evaluated under weak grid conditions with harmonics of 5ª and 11ª orders. The grid unbalance is 2%, the maximum permitted value [15]. In the second scenario, the method, enhanced by sequence estimator and the Fourier technique, is applied to the same power grid.

A. First scenario- Impedance estimation without the sequence estimation

Figures 4, 5 and 6 illustrate the standard method inefficacy for situations where the grid is affected by harmonics and unbalance conditions in estimating L_{est} and R_{est}. In these figures, the blue curves identifie the instants and intensities of variations in currents i_{dq} and voltage v_d. The only variation which is properly detected is in the current i_d. Figures 7 show that variation for current i_q and voltage v_d are properly detected and estimated. In Figures 4 and 6, that the algorithm continues to detect successive variations in i_q and v_d. This leads to several estimates of values for the grid inductance and resistance. Figure 7 shows the estimative for the inductance which are clealy affected by the multiple false detections of variations in voltages and currents. The true value of 16 mH is never reached.

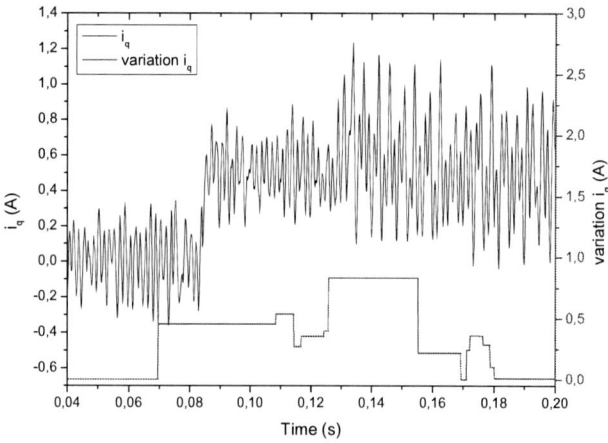

Fig. 4. Estimating the variation of i_q without filters.

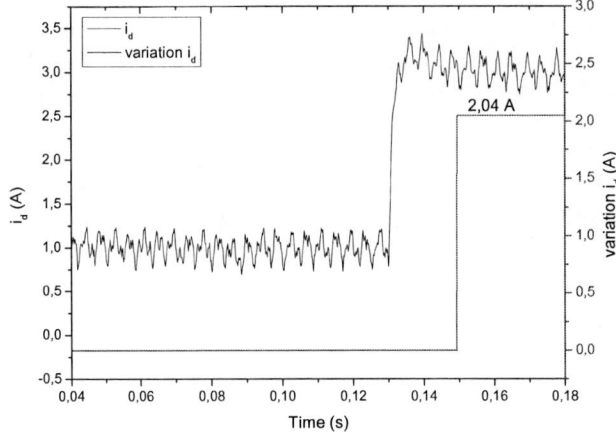

Fig. 5. Estimating the variation of i_d without filters.

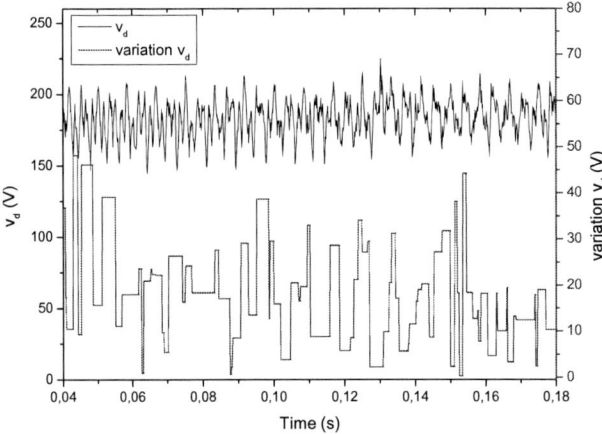

Fig. 6. Variation of v_d in PCC.

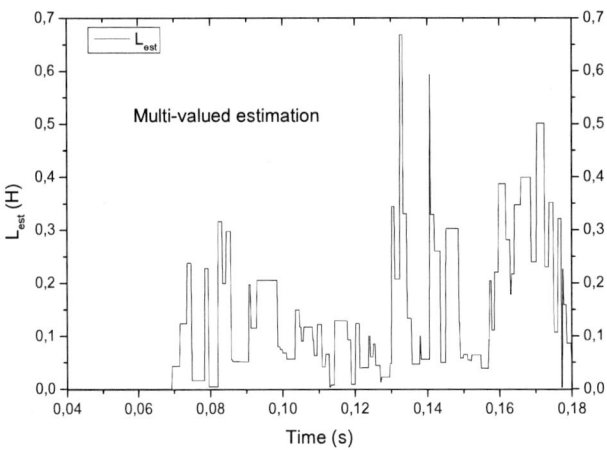

Fig. 7. Estimating of L_{est}.

B. Second scenario- Impedance estimation with the sequence estimation in the voltage and current at the PCC

To eliminate the interferences presented in the voltage v_d and the currents i_q and i_d, in this section, one incorporate the

Forier technique to the standard method. Once the signals v_d, i_q and i_d have been filtered, the variations in these signals can be determined with greater accuracy. Figures 8 and 9 illustrate how the method computes the grid inductance L_{est}. In 8, the grid-phase voltages are processed by the sequence estimator, enhance by the Fourie technique and the voltage v_d is displayed by the black curve, while its variation is depicted in blue. The results related to the current i_q are shown in 9. The variations, in these figures, are properly filtered and the estimation for the inductance is shown in 10. The same procedure is carried out, envolving the variations in the current i_d and, again, the voltage v_d. The estimation for the grid resistance R_{est} is depicted in 11.

The percentage errors for the estimate of L_{est} and R_{est} are 0 % and 0.5%, respectively. These results are shown in Figures 10 and 11.

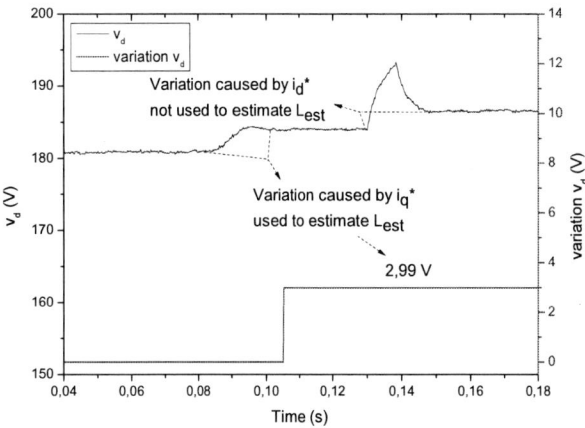

Fig. 8. Variation of v_d caused by i_q in the PCC.

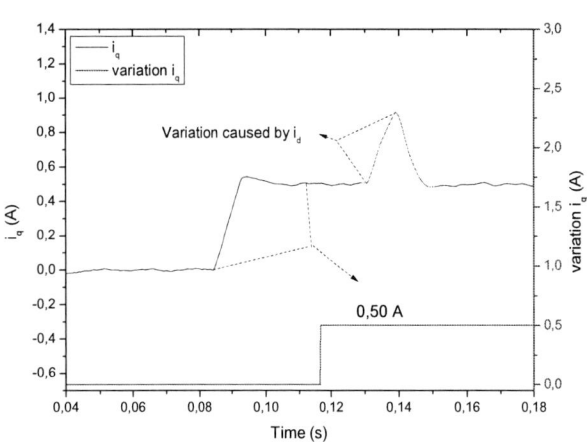

Fig. 9. Estimating the variation of i_q with filters.

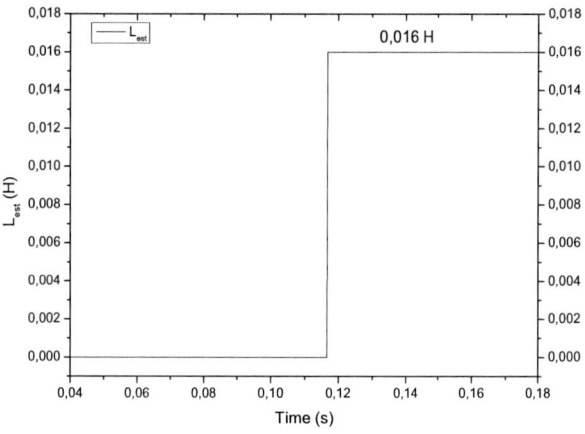

Fig. 10. Estimating of L_{est}.

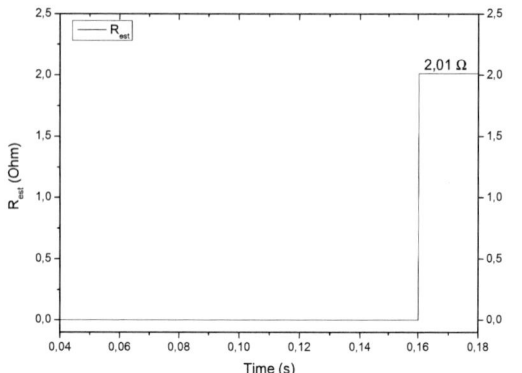

Fig. 11. Estimating of R_{est}.

C. Third scenario - Impedance estimation for different grid SCR values.

The proposed method was evaluated for different grid SRC values. If $SCR > 2$ the grid is strong, for $2 \leq SCR \leq 3$, the grid is weak, and when $SCR < 2$, the grid is very weak [16]. Thus, the estimation for L and R was obtained for networks with strong, weak and very weak characteristics. The results obtained are shown in the Table II.

TABLE II
L AND R ESTIMATION FOR DIFFERENT SRC VALUES.

SCR	Inductance			Resistance		
	L (mH)	L_{est} (mH)	Error (%)	R (Ohm)	R_{est} (Ohm)	Error (%)
4.43	10	10.03	0.3	1	1.006	0.6
2.73	16	16	0.0	2	2.010	0.5
1.88	22	22.13	0.5	4	4.028	0.7

VI. CONCLUSIONS

This work has proposed an improved version of the grid impedance estimation method based on dq synchronous current variations injected by a inverter into the power grid.

The improvement consistes in using a positive sequence estimator to filter the voltages and currents measured from the power grid. The method is robust against harmonic distortion and unbalanced grid conditions. The results showed that the method works properly for strong, weak and very weak grids improving the accuracy. The results showed that the proposed sequence estimator improves the accuracy of the grid impedance estimation.

REFERENCES

[1] A. Ghanem, et al., "Grid impedance estimation for islanding detection and adaptive control of converters", *IET Power Electronics*, vol. 10, no. 11, pp. 1279-1288, 2017.

[2] K. Jia, et al., "An Islanding Detection Method for Multi-DG Systems Based on High-Frequency Impedance Estimation", *IEEE Trans. Sustain. Energy*, vol. 8, no. 1, pp. 74-83, 2017.

[3] IEEE:2000, "IEEE Recommended Practice for Utility Interface of Photovoltaic (PV) Systems", *IEEE Standard 929-2000*.

[4] A. Micallef, et al., "Mitigation of Harmonics in Grid-Connected and Islanded Microgrids Via Virtual Admittances and Impedances", *IEEE Transactions on Smart Grid*, vol. 8, no. 2, pp. 651-661, 2017.

[5] K. O. H. Pedersen, A. H. Nielsen and N. K. Poulsen, "Short-circuit impedance measurement", *IEEE Proceedings - Generation, Transmission and Distribution*, vol. 150, no. 2, pp. 169-174, 2003.

[6] J. P. Rhode, A. W. Kelley and M. E. Baran, "Complete characterization of utilization-voltage power system impedance using wideband measurement", *IEEE Trans. Ind. Appl.*, vol. 33, no. 6, pp. 1472-1479, 1997.

[7] A. V. Timbus, et al., "Online grid measurement and ENS detection for PV inverter running on highly inductive grid", *IEEE Power Electron Lett.*, vol. 2, no. 3, pp. 77-82, 2004.

[8] ASIMINOAEI, Lucian et al. "A digital controlled PV-inverter with grid impedance estimation for ENS detection," IEEE Transactions on Power Electronics, v. 20, n. 6, p. 1480-1490, 2005.

[9] Asiminoaei, Lucian and Teodorescu, Remus and Blaabjerg, Frede and Borup, Uffe, "A digital controlled PV-inverter with grid impedance estimation for ENS detection", IEEE Transactions on Power Electronics, vol. 20, no. 6, pp. 1480–1490, 2005.

[10] J. H. Cho, et al., "A novel P-Q variations method using a decoupled injection of reference currents for a precise estimation of grid impedance", *IEEE Energy Conversion Congress and Exposition (ECCE)*, pp. 5059-5064, 2014.

[11] Abeyasekera, Tusitha and Johnson, C Mark and Atkinson, David J and Armstrong, Matthew, "Suppression of line voltage related distortion in current controlled grid connected inverters", The Journal of Engineering, no. 13, pp. 2015–2020, 2017.

[12] D. A. Fernandes, S. R. Naidu and C. A. E. Coura Jr., "Instantaneous sequence-component resolution of 3-phase variables and its application to dynamic voltage restoration", *IEEE Transactions on Instrumentation and Measurement*, vol. 58 pp. 2580-2587, 2009.

[13] L. H. S. Silva, et al., "A sequence components estimation technique applied for distributed generation", *IEEE Energy Conversion Congressand Exposition (ECCE)*, pp. 11901195, 2014.

[14] K. M. Silva ; F. A. O. Nascimento, "Modified DFT-Based Phasor Estimation Algorithms for Numerical Relaying Applications", *IEEE Transactions on Power Delivery*, vol.33, pp.3, 2018.

[15] IEC 61000-2-12, "Part 2: Environment, Section 12: Compatibility Levels for Low Frequency Conducted Disturbances and Signaling in Public Medium Voltage Power Supply Systems", *IEC 61000-2-12*, 2004.

[16] IEEE Std 1204-1997, "IEEE Guide for Planning DC Links Terminating at AC Locations Having Low Short-Circuit Capacities", IEEE Std 1204-1997, pp. 1–216, 1997.

Hybrid Switched Capacitor DC-DC Converter Based On MMC

Marcus Vieira Soares
Dept. of Electrical Engineering
Santa Catarina State University
Joinville, Brazil
marcus.sov@gmail.com

Gustavo Lambert
Dept. of Electrical Engineering
Santa Catarina State University
Joinville, Brazil
gustavolambert@outlook.com

Yales Rômulo de Novaes
Dept. of Electrical Engineering
Santa Catarina State University
Joinville, Brazil
yales.novaes@udesc.br

Abstract—**Different types of dc-dc converters have been subject of research to fulfill the need of interconnecting renewable generation sources, existing or new grids, electronic loads and battery based systems. To deal with high voltage levels and high power, Modular Multilevel Converter based structures have been proposed. On the other hand, for lower power ratings, Switched Capacitor and Hybrid Switched Capacitor converters have emerged due to their simple structures and self voltage balancing characteristics. In this scenario, a non-isolated Hybrid Switched Capacitor dc-dc converter based on Modular Multilevel Converter is proposed and its unidirectional structure is examined through circuit analysis. In the proposed converter, the voltage static gain is independent of the number of submodules, allowing a good regulation range of the output voltage for high voltage applications. Moreover, the Switched Capacitor behavior provides automatic clamp of the arms voltages, avoiding the need of ac loops to perform the arms voltage balance. A converter design methodology based on current spikes limitation is also proposed. All the theoretical analysis and the design methodology are validated through comparisons with simulated waveforms.**

Index Terms—**Switched Capacitor, Modular Multilevel Converter, MMC, Hybrid Switched Capacitor, Non-Isolated DC-DC Converters.**

I. INTRODUCTION

The growth of concentrated and distributed energy generation via renewable sources, the increasing use of electronic loads and batteries in general society, going from different types of industries to data-centers and domestic applications, and the continuous development of semiconductors able to operate with higher voltages and currents has led to the research of dc-dc converters capable to adapt the different voltage levels used on these scenarios.

In order to deal with medium/high voltage and high power processing, Modular Multilevel Converter (MMC) based dc-dc topologies have been investigated in literature. Isolated converters, such as the Front-to-Front Modular Multilevel Converter [1], feature high modularity, galvanic isolation and high voltage gain provided by the medium frequency transformer (MFT). However, the MFT has to process the total output power and its design is still a challenge, mainly when high dv/dt is applied on the windings [2].

This work was developed with the suport of the *Programa Nacional de Cooperação Acadêmica da Coordenação de Aperfeiçoamento de Pessoal de Nível Superior* – CAPES/Brazil. The authors would like to thank FAPESC and UDESC for their financial support.

Also based on MMC, the dc auto-transformer converters proposed in [3] and [4] allows the reduction of the power processed by the transformer creating a direct dc power path between the dc input and the dc output of the converter. The lower the difference between the input and output voltages, the lower is the transformer processed power. Despite the presence of the transformer in the structure, creating the ac power path to perform the dc voltage links balancing, there is no galvanic isolation between the dc input and output voltages.

A variety of transformerless high power dc-dc converters based on MMC has been presented in [5]–[11], where the ac power path needed to balance the dc voltage links is composed by capacitors, submodules (SM) or inductors. These structures tend to be less bulky than the dc-dc auto-transformer converters. However, the gain provided by the transformer turns ratio is lost which increases the losses when there is a large difference between the input and output dc voltages.

Switched Capacitor (SC) and Hybrid Switched Capacitor (HSC) converters have been subject of studies to be used in different applications where the power rating is higher than the usual for these types of converters. The scalability, the capacitors voltage clamp and the self-voltage balancing are the main features of these structures. However, the output voltage control limitations and the need to attenuate the current spikes are important drawbacks of SC and HSC converters [12]–[15].

A Hybrid Switched Capacitor dc-dc converter based on MMC is presented in this paper. The unidirectional version of the converter is used to describe its topological stages and to extract the main static characteristics of the structure. A preliminary design methodology based on the limitation of the components current spikes is also presented and validated by simulation.

II. PROPOSED DC-DC CONVERTER

The proposed structure presented in this work is a non-isolated and bidirectional dc-dc converter based on hybrid switched-capacitor converter. It can be seen as a three-level Flying Capacitor dc-dc converter with the switches replaced by power SMs. The SMs of the upper arms ($A1$ and $A2$) can assume different configurations, such as Half-Bridge and Full-Bridge, where each configuration has its particularities

978-1-7281-4181-7/19 $31.00 © 2019 IEEE

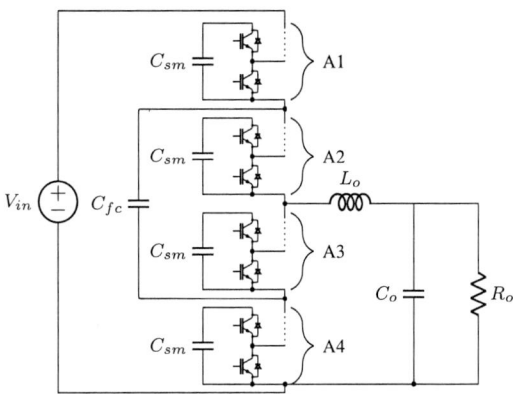

Fig. 1. DC-DC converter

regarding losses and short-circuit protection. Fig. 1 presents a representative diagram of the proposed converter.

Each arm of the converter is composed of N series connected SM and behaves as a switch. It means that, out of the topological stage transitions, all the arm SMs must have their capacitors either inserted or bypassed, emulating an open switch and a closed switch, respectively. Therefore, the arms can be seen as switches and they must respect the following logic to avoid short-circuits and voltage inconsistencies: $A1 = \overline{A4}$ and $A2 = \overline{A3}$.

Similar to the three-level Flying Capacitor dc converter, the proposed structure is composed of a flying capacitor (C_{fc}) and an LC output filter (L_o and C_o). In order to keep the capacitor with an average voltage equal to half the input voltage ($V_{in}/2$), the flying capacitor must be charged and discharged equally. This proper operation allows the generation of a voltage with twice the switching frequency ($2f_s$) on the LC filter input, providing size reduction of L_o and C_o. However, a control loop must be used to regulate the C_{fc} average voltage on the correct level, which increases the system complexity.

Assuming that the C_{fc} average voltage is regulated on $V_{in}/2$ and the arms to behave as complementary switches ($A1 = \overline{A4}$ and $A2 = \overline{A3}$), it can be shown that the total voltage of each arm is clamped either by the set of input voltage source V_{in} and C_{fc} or just by C_{fc}. This clamping has a switched-capacitor characteristic, since the input voltage source, the flying capacitor and a set of SM capacitors are periodically connected in parallel. This operation technique keeps each arm total voltage automatically balanced, including the external arms $A1$ and $A4$ voltages, since the C_{fc} voltage is controlled. It avoids the need of an ac power path to achieve this balancing. However, the capacitors, the on-resistances of the semiconductors and the switching frequency must be carefully chosen to ensure operation with low losses and low current spikes.

As other dc based Modular Multilevel converters, the proposed structure is able to operate with a high input voltage by the series connection of several low voltage SMs. However, natural differences on the capacitance values and on the switching times of the same arm SM's can cause the SMs capacitor voltage to deviate from each other. Therefore, a SM voltage balance method must be applied to ensure the proper operation of the converter.

The SM voltage balance method chosen for this structure is the Quasi-Two-Level modulation [16], where a stepped transition between the topological stages is made by using the SMs individually. The steps must respect an order so that the capacitors of the upper arms SMs with the lowest voltage can stay in the output current-source path for a longer time, charging more than the other capacitors of the same arm. On the other hand, the capacitors of the lower arms SMs with the highest voltage must stay in the output current-source path for a longer time, discharging more than the other capacitors of the same arm. With this procedure, the capacitors which compose an arm achieve the voltage balance among them. Moreover, the stepped transitions also reduce the dv/dt over the inductor L_o when compared to the standard Two-Level modulation, since the voltage applied on the input of the LC filter inherits the stepped transition waveform. This characteristic is important to ensure a good lifetime for the inductor insulation material and to avoid parasitic currents on the converter, mainly when high dc voltage levels are applied on the converter input.

A. Topological stages

Some considerations are made in order to simplify the analysis of the converter topological stages:

- A unidirectional type of the converter is used;
- All the SMs are equal and configured as a Half-Bridge circuit (bidirectional on the upper arms and unidirectional on the lower arms);
- The MOSFET switches have an on-resistance $R_{DSon} = R$;
- All the diodes operate with a forward conduction resistance $R_{fon} = R$ and zero forward voltage drop;
- Each arm is composed by N series-connected SMs, which are represented by an equivalent SM per arm;
- The capacitance of the flying capacitor C_{fc} is at least ten times higher than the total arm capacitance $C_{eqx} = C_{sm}/N$, where C_{sm} is the capacitance of each SM and $x = \{1, 2, 3, 4\}$ is the arm number. This simplification is assumed to be suitable because the converter is directed to applications with high input voltage, which requests a relevant number of series SMs per arm. Therefore, the total arm capacitance tends to be much lower than C_{fc};
- The topological stages transition time (t_{tr}) is much lower than half the switching period ($T_s/2$), making the transition effects negligible;
- The converter is operated only with positive inductor current;
- Only the converter operation for $V_o/V_{in} \leq 0.5$ is approached.

The topological stages are described in Table I, where $D = 2t_{on}/T_s$ is the duty cycle ($0 \leq D \leq 1$), t_{on} is the du-

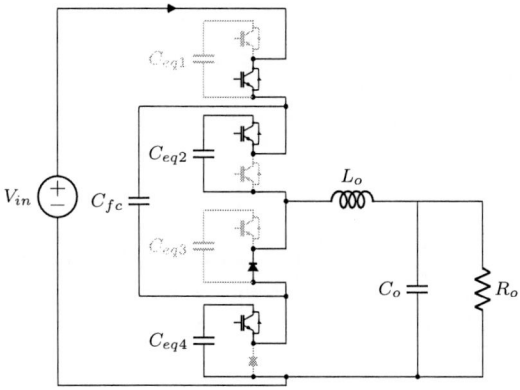

Fig. 2. Topological state 1

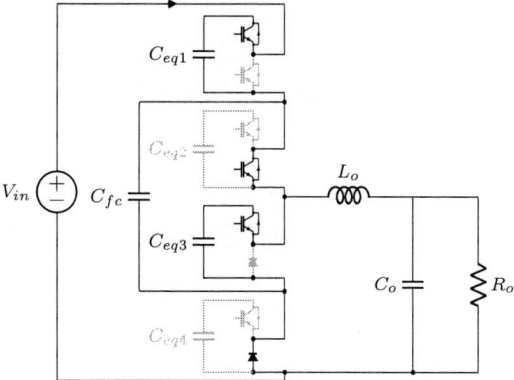

Fig. 4. Topological state 3

component that compose the converter. This analysis results in important average equations used to understand and to design the converter.

The state space representation of each topological stage is created from the Kirchoff's Voltage and Current Law (KVL and KCL, respectively). The input is V_{in} and the state variables are v_{Cfc}, v_{Ceq1}, v_{Ceq2}, v_{Ceq3}, v_{Ceq4}, i_{Lo} and v_{Co}, resulting in the following state space equation, where $i = \{1, 2, 3, 4\}$ represents the topological stage

$$\dot{\boldsymbol{x}} = \boldsymbol{A_i} \cdot \boldsymbol{x} + \boldsymbol{B_i} \cdot \boldsymbol{u} \qquad (2)$$

In steady state, the average value of each state variable is constant. Therefore, it is possible to define $\dot{\boldsymbol{x}} = 0$. Moreover, the matrices that represent the state space for average values are written by multiplying the topological stages matrices by their respective duration time and dividing by the switching period, which results in

$$\boldsymbol{A} = \boldsymbol{A_1} \cdot \frac{D}{2} + \boldsymbol{A_2} \cdot \frac{(1-D)}{2} + \boldsymbol{A_3} \cdot \frac{D}{2} + \boldsymbol{A_4} \cdot \frac{(1-D)}{2} \quad (3)$$

$$\boldsymbol{B} = \boldsymbol{B_1} \cdot \frac{D}{2} + \boldsymbol{B_2} \cdot \frac{(1-D)}{2} + \boldsymbol{B_3} \cdot \frac{D}{2} + \boldsymbol{B_4} \cdot \frac{(1-D)}{2} \quad (4)$$

Using (3) and (4) in (2), the state space representation of the converter for average values becomes

$$0 = \boldsymbol{A} \cdot \boldsymbol{X} + \boldsymbol{B} \cdot V_{in} \qquad (5)$$

Isolating $\boldsymbol{X}$ in (5), the average values of the state variables are obtained by

$$\boldsymbol{X} = -\boldsymbol{A}^{-1} \cdot \boldsymbol{B} \cdot V_{in} \qquad (6)$$

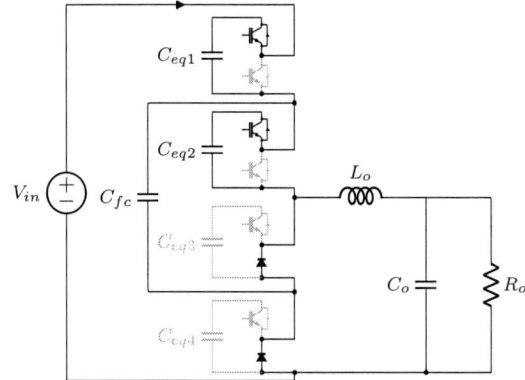

Fig. 3. Topological state 2

ration time of the topological stages 1 and 3 and the time intervals are defined by

$$\begin{cases} \Delta t_1 = 0 \le t \le DT_s/2 \\ \Delta t_2 = DT_s/2 \le t \le T_s/2 \\ \Delta t_3 = T_s/2 \le t \le (1+D)T_s/2 \\ \Delta t_4 = (1+D)T_s/2 \le t \le T_s \end{cases} \qquad (1)$$

TABLE I
TOPOLOGICAL STAGES

	Stage 1	Stage 2	Stage 3	Stage 4
Interval	Δt_1	Δt_2	Δt_3	Δt_4
C_{eq1}	Bypassed	Inserted	Inserted	Inserted
C_{eq2}	Inserted	Inserted	Bypassed	Inserted
C_{eq3}	Bypassed	Bypassed	Inserted	Bypassed
C_{eq4}	Inserted	Bypassed	Bypassed	Bypassed
C_{fc}	Charge	-	Discharge	-
L_o	Charge	Discharge	Charge	Discharge

B. Average Value Static Analysis

In this section, a static analysis via average value is demonstrated in order to describe the voltages and currents of each

978-1-7281-4181-7/19 $31.00 © 2019 IEEE

$$
\mathbf{X} = \begin{bmatrix} V_{Cfc} \\ V_{Ceq1} \\ V_{Ceq2} \\ V_{Ceq3} \\ V_{Ceq4} \\ I_{Lo} \\ V_{Co} \end{bmatrix} = \begin{bmatrix} \dfrac{V_{in}}{2} \\ \dfrac{V_{in}}{2}\left(1 + \dfrac{DNR}{2NR+R_o}\right) \\ \dfrac{V_{in}}{2}\left(1 + \dfrac{DNR}{2NR+R_o}\right) \\ \dfrac{V_{in}}{2}\left(1 - \dfrac{DNR}{2NR+R_o}\right) \\ \dfrac{V_{in}}{2}\left(1 - \dfrac{DNR}{2NR+R_o}\right) \\ \dfrac{V_{in}D}{2(2NR+R_o)} \\ \dfrac{V_{in}DR_o}{2(2NR+R_o)} \end{bmatrix} \tag{7}
$$

where R_o is the load equivalent resistance.

Since V_{Co} is the converter output voltage, the converter static gain (M_{VDC}) is defined by

$$
M_{VDC} = \frac{V_{Co}}{V_{in}} = \frac{DR_o}{2(2NR+R_o)} \tag{8}
$$

Manipulating the equation for I_{Lo} obtained in (7), the following of equation is achieved

$$
\frac{DV_{in}}{2} = I_{Lo}(2NR + R_o) \tag{9}
$$

which represents the output characteristic of the converter. It can be seen that it is similar to the output characteristic of Buck-type converters, with the difference that the number of SMs per arm influences on the equivalent output impedance ($2NR$). It is also noticeable that the equivalent output impedance is independent of the duty cycle, unlike the hybrid switched capacitor converter described in [15].

The static average state model described is valid for the cases where the current peaks caused by the switched capacitor characteristic of the converter are low. It is important to operate the converter in this condition because the switching losses, the EMI and the RMS current of the semiconductors and capacitors tend to be reduced.

C. Flying Capacitor Peak Current Estimation

To ensure that the converter components operate with low current peaks, an analysis of the capacitor C_{fc} initial current during the topological stage 3 is done (the topological stage 1 can also be used), since this is the stage where the worst current peak case occurs. Using KCL and KVL, the C_{fc} peak current during stage 3 can be described by

$$
i_{Cfc\_pk}^{st3} = \frac{-\Delta V_{Cfc} + V_{Ceq30}^{st3} - V_{Ceq10}^{st3}}{2NR} \tag{10}
$$

where ΔV_{Cfc}, V_{Ceq10}^{st3} and V_{Ceq30}^{st3} are the C_{fc} voltage ripple, the C_{eq1} stage 3 initial voltage and the C_{eq3} stage 3 initial voltage.

Since $C_{fc} \geq 10C_{eqx}$ and the time constant $2\pi\sqrt{L_oC_o}$ tends to be high regarding half the switching period, the C_{fc} voltage can be approached as a trapezoidal waveform with period T_s, whereas the L_o current has a triangular waveform with period

$T_s/2$. Therefore, the C_{fc} voltage ripple and the L_o current ripple are defined by

$$
\Delta v_{Cfc} = \frac{I_{Lo}DT_s}{2C_{fc}} \tag{11}
$$

$$
\Delta i_L = \frac{V_o(1-D)T_s}{2L_o} \tag{12}
$$

Considering the v_{Cfc} and i_{Lo} trapezoidal and triangular waveforms, respectively, the equations that describe the behavior of v_{Ceq1} and v_{Ceq3}, presented in (2), can be developed to achieve the equations for V_{Ceq10}^{st3} and V_{Ceq30}^{st3}.

Replacing the equations of V_{Ceq10}^{st3} and V_{Ceq30}^{st3} in (10), the resulting equation can be parameterized by I_{Lo}, providing the information of how high the C_{fc} peak current is as regards the output current in function of f_s, D, $2NRC_{fc}$ and p, where $p = \Delta i_{Lo}/I_{Lo}$.

At the beginning of the stage 3, the voltage difference between C_{fc} and C_{eq1} and between C_{fc} and C_{eq3} tends to increase when $D \to 1$ and p is high, also increasing $i_{Cfc\_pk}^{st3}$. However, analyzing (12), it is noticeable that $p \to 0$ when $D = 1$ and, since p stops to influence on $i_{Cfc\_pk}^{st3}$, this is not the worst peak current case. Therefore, considering a worst case where $D = 0.99$ and $p = 1$, two-dimensional graphs of $i_{Cfc\_pk}^{st3}/I_{Lo}$ can be plotted, showing the highest C_{fc} peak current cases for different values of $2NRC_{fc}$ and f_s. The graphs are presented in Fig. 5 and it can be seen that the normalized peak current is inversely proportional to $2NRC_{fc}$ and f_s.

D. Design Methodology

The design methodology used to perform the simulation of the converter is presented in this section. From Fig. 5 it can be noticed that for all the switching periods, in the worst case, $i_{Cfc\_pk}^{st3} \approx 1.25I_{Lo}$ when $2NRC_{fc} \approx 4T_s$, which is an acceptable peak current value for C_{fc}. Therefore, C_{fc} can be defined by

$$
C_{fc} = \frac{2T_s}{NR} \tag{13}
$$

Respecting the restriction $2NRC_{fc} \approx 4T_s$, the proposed converter operates with low peak values for the semiconductors and capacitors currents. Moreover, these peak values decrease slowly during the topological stages intervals. Therefore, this converter operation can be seen as equivalent to the Partial Charge (PC) or No Charge (NC) modes of a conventional SC converter, defined in [17].

As explained in Section II-A, the capacitance C_{sm} can be defined by

$$
C_{sm} = \frac{NC_{fc}}{10} \tag{14}
$$

Knowing the specification of V_o and Δi_{Lo}, the design of L_o and D can be done using the equations (12) and (8), respectively.

Considering that all the ac component of i_{Lo} circulates through the output capacitor and that this current has a

978-1-7281-4181-7/19 $31.00 © 2019 IEEE

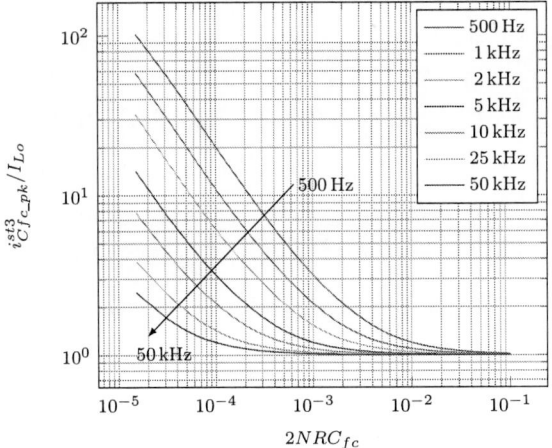

Fig. 5. Abacus relating $i_{Cfc\_pk}^{st3}/I_{Lo}$, $2NRC_{fc}$ and f_s for $D = 0.99$ and $p = 1$

triangular waveform, the capacitance C_o can be obtained by calculating the area of the positive part of i_{Co}, which represents the total amount of charge received by C_o, and dividing it by Δv_o, that is the desired output voltage ripple. The resultant equation is given by

$$C_o = \frac{V_o(1-D)T_s^2}{32\Delta v_o L_o} \quad (15)$$

III. SIMULATION RESULTS

In order to verify the theoretical analysis and the design methodology, a simulation of the converter is performed using the software PSIM®. The specification of the converter is shown in Table II as well as the values of the parameters calculated using the equations specified in Section II-D. The converter is operated in open-loop and with equal SM capacitances.

TABLE II
CONVERTER SPECIFICATION

Parameter	Value
Output power (P_o)	1666 W
Input voltage (V_{in})	1 kV
Output voltage (V_o)	400 V
Number of SMs per arm (N)	2
Semiconductors on-resistance (R)	100 mΩ
Switching frequency (f_s)	10 kHz
Duty cycle (D)	0.803
Output voltage ripple (ΔV_o)	1% of V_o
Inductor current ripple (Δi_{Lo})	25% of I_{Lo}
Flying capacitor capacitance (C_{fc})	1 mF
SM capacitors capacitance (C_{sm})	200 μF
Output capacitor capacitance (C_o)	1.63 μF
Inductor inductance (L_o)	3.78 mH

The results of the simulation are shown in Fig. 6 and Fig. 7. The average output voltage achieves 399.83 V with 4.06 V of voltage ripple, whereas the average inductor current achieves 4.16 A with 1.05 A of current ripple. Compared to the design specification, the voltage ripple differs -0.15% and the current ripple differs 0.48%. This error is attributed to the ac components with frequency $2f_s$ present in both output voltage and inductor current which modifies the average value.

The average voltage of the capacitors in the arms $A1$ and $A2$ are stabilized in 250.42 V each, whereas the average voltage of capacitors in the arms $A3$ and $A4$ are in 294.57 V. It matches with the values obtained from (7) ($V_{Ceq1} = V_{Ceq2} = 500.83$ V and $V_{Ceq3} = V_{Ceq4} = 499.15$ V), since the simulated arms average voltage are $V_{Ceq1} = V_{Ceq2} = 500.85$ V and $V_{Ceq3} = V_{Ceq4} = 499.17$ V. It can be noticed that arms voltage are automatically clamped by the flying capacitor, avoiding the need of an ac power path to balance the voltage of the upper and lower arms.

Analyzing the flying capacitor voltage it is noticeable that it has a trapezoidal waveform as expected, since $C_{fc} = C_{sm}/10N$, with an average value of $V_{in}/2$. This average value must be controlled in a real converter to ensure that it does not deviate from the correct value during the operation. As regards the C_{fc} current, it can be seen that its peak is low due to the chosen $2NRC_{fc}$. The peak value is 4.51 A, which is about 12.5% the output current. Since $D < 0.99$ and $p < 1$, it was expected that the peak value was lower than 25% of I_{Lo} because the graphs from Fig. 5 consider the worst peak current case.

As a matter of comparison, the converter was simulated with $2NRC_{fc} = 4T_s/5$ by making $C_{fc} = 200$ μF. Consequently, $C_{sm} = 40$ μF. Fig. 8 shows v_o and i_{Cfc} for this case. The average output voltage is 399.79 V and C_{fc} current peak value is 6.37 A, which is 52.9% higher than I_{Lo}. Since the output current flows mostly through C_{fc}, its capacitance reduction results in a neglegible increase of the semiconductors RMS current. Consequently, the average output voltage remains pratically unchanged. On the other hand, the higher current peak values will increase the switching losses and the EMI, jeopardizing the efficiency and the operation of the converter.

IV. CONCLUSION

The Hybrid Switched Capacitor DC-DC Converter based on MMC has been presented in this paper. Observing the theoretical analysis developed, validated by simulation, the proposed converter has a Buck-type output characteristic even with the switched capacitor operation at the LC filter input. This feature is interesting by the point-of-view that the equivalent output resistor is independent of the duty cycle, allowing the converter to be designed with some freedom regarding the voltage static gain. The $2f_s$ frequency of the output current ac component also helps to decrease the output inductance, which tends to be high due to the Buck output characteristic. Moreover, the switched capacitor behavior provides self arm voltage balancing, avoiding the need of inductor at the converter input.

The capacitance of the flying capacitor needs to be carefully designed to avoid high peak and RMS currents on the

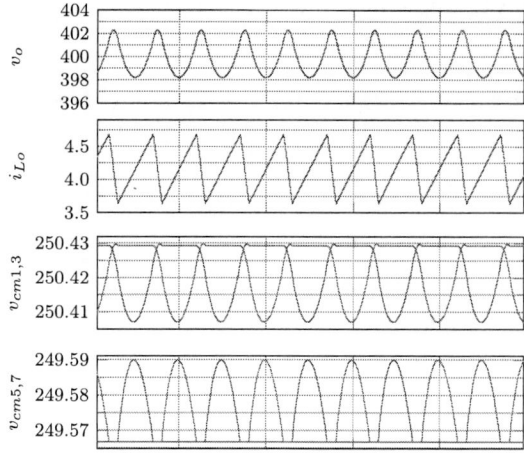

Fig. 6. Output voltage (v_o), inductor current (i_{Lo}), $A1$ and $A2$ SM capacitors voltages ($v_{cm1,3}$), $A3$ and $A4$ SM capacitors voltages ($v_{cm5,7}$)

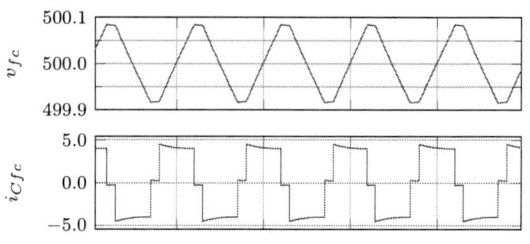

Fig. 7. Flying capacitor voltage (v_{Cfc}) and current (i_{Cfc}).

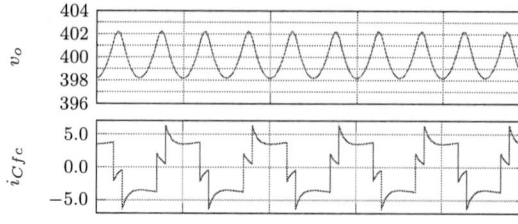

Time (100 µs/div)

Fig. 8. Output voltage (v_o) and flying capacitor current (i_{Cfc}) with $2NRC_{fc} = 4T_s/5$.

semiconductors and capacitors. It can cause this capacitance to reach high values depending on the switching frequency, the number of SMs and the semiconductors on-resistance.

Experimental results were not presented in this paper because the converter is still under investigation. The research is being conducted towards a prototype construction of the bidirectional converter version to validate experimentally the theoretical analysis. However, the simulation software used has shown to be a reliable tool to understand and design hybrid switched capacitor converters.

For limited space reasons, further details regarding the topological stages, the losses analysis and the issues related to capacitors and semiconductors technologies will be addressed in future papers.

REFERENCES

[1] S. Kenzelmann, A. Rufer, D. Dujic, F. Canales, and Y. R. de Novaes, "Isolated dc/dc structure based on modular multilevel converter," *IEEE Transactions on Power Electronics*, vol. 30, no. 1, pp. 89–98, Jan 2015.

[2] S. Milovanović and D. Dujić, "High-power dc–dc converter utilising scott transformer connection," *IET Electric Power Applications*, April 2019. [Online]. Available: https://digital-library.theiet.org/content/journals/10.1049/iet-epa.2018.5746

[3] A. Schön and M.-M. Bakran, "A new hvdc-dc converter for the efficient connection of hvdc networks," in *PCIM Europe 2013 : International Exhibition & Conference for Power Electronics, Intelligent Motion, Renewable Energy and Energy Management*. Nürnberg: VDE Verlag GmbH, 2013, pp. 525–532.

[4] I. A. Gowaid, G. P. Adam, B. W. Williams, A. M. Massoud, and S. Ahmed, "A dc autotransformer design for medium and high voltage dc transmission systems," in *2018 IEEE 12th International Conference on Compatibility, Power Electronics and Power Engineering (CPE-POWERENG 2018)*, April 2018, pp. 1–6.

[5] S. Du, B. Wu, K. Tian, D. Xu, and N. R. Zargari, "A novel medium-voltage modular multilevel dc-dc converter," *IEEE Transactions on Industrial Electronics*, vol. 63, no. 12, pp. 7939–7949, Dec 2016.

[6] S. Du and B. Wu, "A transformerless bipolar modular multilevel dc-dc converter with wide voltage ratios," *IEEE Transactions on Power Electronics*, vol. 32, no. 11, pp. 8312–8321, Nov 2017.

[7] S. Du, B. Wu, and N. R. Zargari, "A transformerless high-voltage dc-dc converter for dc grid interconnection," *IEEE Transactions on Power Delivery*, vol. 33, no. 1, pp. 282–290, Feb 2018.

[8] S. H. Kung and G. J. Kish, "A modular multilevel hvdc buck-boost converter derived from its switched-mode counterpart," *IEEE Transactions on Power Delivery*, vol. 33, no. 1, pp. 82–92, Feb 2018.

[9] G. J. Kish, M. Ranjram, and P. W. Lehn, "A modular multilevel dc/dc converter with fault blocking capability for hvdc interconnects," *IEEE Transactions on Power Electronics*, vol. 30, no. 1, pp. 148–162, Jan 2015.

[10] H. You and X. Cai, "A three-level modular dc/dc converter applied in high voltage dc grid," *IEEE Access*, vol. 6, pp. 25 448–25 462, 2018.

[11] ——, "Stepped 2-level operation of non-isolated modular dc/dc converter applied in high voltage dc grid," *IEEE Journal of Emerging and Selected Topics in Power Electronics*, vol. PP, no. 99, pp. 1–1, 2017.

[12] R. L. da Silva, T. B. Lazzarin, and I. Barbi, "Reduced switch count step-up/step-down switched-capacitor three-phase ac–ac converter," *IEEE Transactions on Industrial Electronics*, vol. 65, no. 11, pp. 8422–8432, Nov 2018.

[13] T. B. Lazzarin, R. L. Andersen, and I. Barbi, "A switched-capacitor three-phase ac–ac converter," *IEEE Transactions on Industrial Electronics*, vol. 62, no. 2, pp. 735–745, Feb 2015.

[14] R. L. Andersen, T. B. Lazzarin, and I. Barbi, "A 1-kw step-up/step-down switched-capacitor ac–ac converter," *IEEE Transactions on Power Electronics*, vol. 28, no. 7, pp. 3329–3340, July 2013.

[15] M. D. Vecchia, M. A. Salvador, and T. B. Lazzarin, "Hybrid nonisolated dc–dc converters derived from a passive switched-capacitor cell," *IEEE Transactions on Power Electronics*, vol. 33, no. 4, pp. 3157–3168, April 2018.

[16] I. A. Gowaid, G. P. Adam, A. M. Massoud, S. Ahmed, D. Holliday, and B. W. Williams, "Quasi two-level operation of modular multilevel converter for use in a high-power dc transformer with dc fault isolation capability," *IEEE Transactions on Power Electronics*, vol. 30, no. 1, pp. 108–123, Jan 2015.

[17] S. Ben-Yaakov and M. Evzelman, "Generic average modeling and simulation of the static and dynamic behavior of switched capacitor converters," in *2012 Twenty-Seventh Annual IEEE Applied Power Electronics Conference and Exposition (APEC)*, Feb 2012, pp. 2568–2575.

978-1-7281-4181-7/19 $31.00 © 2019 IEEE

Thermal Stress Evaluation of a Multifunctional Modular Multilevel Converter - STATCOM Operating as Active Filter

R.O. de Sousa[a], W.C.S. Amorim[b], D.C. Mendonça[b], A.F. Cupertino[a,c], L.M.F. Morais[d] and H.A. Pereira[e].

[a] Graduate Program in Electrical Engineering, Federal University of Minas Gerais,
Belo Horizonte, MG, Brazil. (renata.sousa@ufv.br, afcupertino@ieee.org)
[b] Graduate Program in Electrical Engineering, Federal Center for Technological Education of Minas Gerais,
Belo Horizonte, MG, Brazil. (william.caires@ufv.br, dayane.mendonca@ufv.br)
[c] Department of Materials Engineering, Federal Center for Technological Education of Minas Gerais
[d] Department of Electronic Engineering, Federal University of Minas Gerais,
Belo Horizonte, MG, Brazil. (lenin@cpdee.ufmg.br)
[e] Department of Electrical Engineering, Federal University of Viçosa, Viçosa, MG, Brazil. (heverton.pereira@ufv.br)

Abstract—**The Modular Multilevel Converter (MMC) has proved to be an attractive technology in medium and high voltage Static Synchronous Compensator (STATCOM) applications. Nevertheless, once reactive power compensation profile is not constant over time, there are periods that a MMC-STATCOM works below its nominal capacity. In these periods, the converter can be used to perform ancillary services, such as Active Power Filtering. This paper analyzes the thermal stresses in the capacitors and semiconductor devices (i.e. IGBTs and diodes). The analyses take into account reactive compensation and harmonic current compensation of 5th and 7th harmonic components of a MMC based STATCOM. In addition, the limits of harmonic compensation for 5th and 7th harmonic orders are analytically derived. The results demonstrated lower thermal stresses during Active Power Filtering than the traditional STATCOM operation. Moreover, the 5th harmonic compensation is the most stressed one among the cases studied in this work.**

Index Terms—**Modular Multilevel Converter, STATCOM, Multifunctional, Active Power Filter, Harmonic Compensation.**

I. INTRODUCTION

The power quality of distribution and transmission systems is a major concern which has to be examined with caution in order to achieve a reliable power system. Nevertheless, these systems have become very complex structures with a huge variety of loads [1]. Moreover, most of these loads are non-resistive or fluctuating, which can generate reactive current (non-unitary power factor), grid voltage variations (flickers) and power quality deterioration. These phenomena result in inefficient usage of power and they are the major source of abrupt equipment failures [2]. In addition, once these loads are connected together in the system, the effect of any issue caused by them is very large. For this reason, these phenomena have to be controlled [1], [2].

In order to minimize the voltage variations problems, reactive power compensation is indicated. In this context,

the Static Synchronous Compensators (STATCOMs) based on voltage source converters have been highlighted [3]. Among the STATCOM topologies developed over the decades, the Modular Multilevel Converter (MMC) has proved to be an attractive technology in medium and high voltage applications [4], [5]. Indeed, some advantages of the MMC are high-efficiency, superior harmonic performance, low-voltage semiconductor technologies and low filtering requirement [6].

Nevertheless, reactive power compensation may vary over time. Therefore, there are moments in which the MMC-STATCOM works below its nominal capacity. In these periods, the converter could perform ancillary services. Among the ancillary tasks that perform, the harmonic current compensation is highlighted [7]. Indeed, the features of the MMC make it promising for Active Power Filter (APF) applications [8].

Several works study the MMC-APF application [8]–[12], in which only [7] explores its multifunctional capacity of reactive compensation and harmonic mitigation. Nevertheless, neither of these works evaluated the thermal stress in the MMC components caused by the harmonic compensation. Therefore, this paper analyzes the thermal stress in the capacitors and semiconductor devices (i.e. IGBTs and diodes) during STATCOM and APF operations. The analyses are made during reactive and harmonic current compensation of 5th and 7th harmonic orders. For this purpose, the limits of harmonic current are computed numerically.

This paper is outlined as follows. Section II introduces the MMC control algorithm and the methodology. Section III presents the case study and the parameters of the simulated model. Section IV presents the simulation results. Finally, the conclusions of this work are stated in Section V.

978-1-7281-4181-7/19 $31.00 © 2019 IEEE

II. METHODOLOGY

A. MMC Topology and Control Strategy

Fig. 1 illustrates the MMC-STATCOM topology employed. As observed, each phase of the converter is connected in the grid by the coupling of the upper and lower arm, joined by the arm inductors (L_{arm}). L_{arm} is responsible to limit the currents during possible faults and reduce the harmonics in the circulating current [13], [14]. In addition, each arm is composed by N submodules (SMs), in which each SM contains a capacitor (C) and four semiconductor devices (S_1, S_2, D_1, D_2,) in the half-bridge configuration. Moreover, there is a switch S_T in parallel with each SM, which bypasses the SM in case of failures.

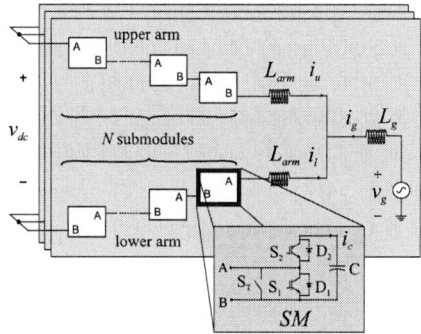

Fig. 1. Schematic of the Modular Multilevel Converter.

The system configuration is composed by the MMC-STATCOM in shunt with a non-linear load, as illustrated in the Fig. 2.

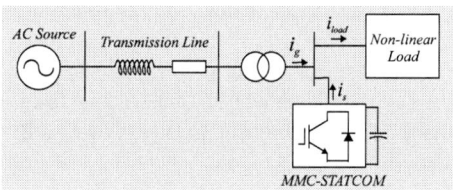

Fig. 2. System configuration.

The MMC-STATCOM control overview is presented in Fig. 3. This control strategy is composed by the grid current control and the circulating current control [15]. In addition, the modulation strategy Nearest Level Control with Cell Tolerance Band algorithm (NLC-CTB) is employed [3]. Moreover, the arm current reference for the CTB is estimated following the methodology proposed by [16].

The outer loop controls the average voltage square (v_{avg}^2) of all converter SMs. The average voltage is defined by:

$$v_{avg} = \frac{1}{6N} \sum_{i=1}^{6N} v_{SM,i},\qquad(1)$$

where $v_{SM,i}$ is the ith SM voltage.

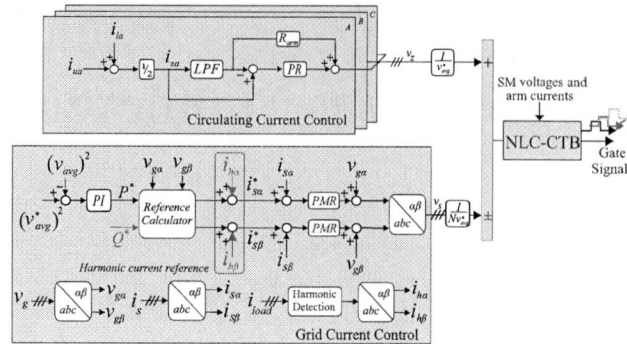

Fig. 3. MMC control with harmonic compensation structure.

The average voltage reference v_{avg}^* and the SM reference voltage v_{SM}^* are expressed by:

$$v_{avg}^* = v_{SM}^* = \frac{v_{dc}}{N},\qquad(2)$$

where v_{dc} is the dc-link voltage .

The reference calculator in Fig. 3 is used to compute the grid current in stationary ($\alpha\beta$) reference frame. Using the instantaneous power theory [17], it is possible to express the grid current by:

$$\begin{bmatrix} i_{g\alpha}^* \\ i_{g\beta}^* \end{bmatrix} = \frac{1}{v_{g\alpha}^2 + v_{g\beta}^2} \begin{bmatrix} v_{g\alpha} & v_{g\beta} \\ v_{g\beta} & -v_{g\alpha} \end{bmatrix} \begin{bmatrix} P^* \\ Q^* \end{bmatrix} + \begin{bmatrix} i_{h\alpha} \\ i_{h\beta} \end{bmatrix}.\qquad(3)$$

In order to perform harmonic current mitigation, it is added a harmonic current ($i_{h\alpha}$ and $i_{h\beta}$) in stationary reference frame in the grid current reference. These harmonic components are detected from the load current (i_{load}). Proportional Multiresonant controllers (PMR) are used in the grid current control strategy.

The circulating current control, presented in Fig. 3 inserts a damping in the converter dynamic response through a proportional controller. Moreover, a resonant controller tuned in the second harmonic is used to compensate the second order harmonic in the circulating current which is typical in MMC. The circulating current is computed per phase as follows:

$$i_{z,n} = \frac{i_{u,n} + i_{l,n}}{2}.\qquad(4)$$

where $i_{u,n}$ and $i_{l,n}$ ($n = a, b, c$) are the upper and lower arm current, respectively. In addition, for the voltage reference (v_s), the injection of 1/6 pu of third harmonic voltage is used in order to increase the linear operational area of the modulation curve [18].

B. MMC Harmonic Current Mitigation: Operation Limits

The harmonic compensation capacity of the MMC is limited by the maximum current and voltage which can be synthesized by the converter. This section aims to investigate the operational limits of the converter which can be evaluated based on the simplified average model. In this simplification,

978-1-7281-4181-7/19 $31.00 © 2019 IEEE

the MMC is represented by the equivalent arm inductor per phase ($L_{arm}/2$) and the voltage synthesized by the converter (v_s). The grid model is represented by the transformer inductance (L_g) and the grid voltage source (v_g).

Considering that the converter injects a single harmonic current into the grid, i_s in phase A is giving by:

$$i_{s,a} = \widehat{I}_h cos(h\omega t + \phi), \tag{5}$$

where $\widehat{I}_h$ is the current harmonic amplitude, h is the harmonic order and ϕ is the harmonic phase angle.

In order to inject harmonic current components (5) into the grid, the converter voltage is approximately given by:

$$v_{s,a} = v_g + h\widehat{I}_h\omega(\frac{L_{arm}}{2} + L_g)cos(h\omega t + \phi + \frac{\pi}{2}), \tag{6}$$

v_g is the grid voltage, given by:

$$v_g = \widehat{V}_g cos(\omega t). \tag{7}$$

From the presented equation, an iterative algorithm that searches for the maximum harmonic current that can be synthesized by the MMC is implemented [19]. The input variables are the MMC nominal power, the dc-link voltage, the harmonic orders and the equivalent output impedance. This algorithm is based on the voltage limit that the converter can synthesize taking into account the linear region of the modulator, given by:

$$\widehat{V}_{s,max} = \frac{v_{dc}}{\sqrt{3}}. \tag{8}$$

The 5th and 7th harmonic current are compensated in this work, without the fundamental component. In this analysis, the parameters of Tab. I are considered.

For the analysis of each (5th or 7th) and both (5th and 7th) harmonic components, it is presented in Fig. 4 the maximum output current for a range of harmonic phase angle between 0 and 2π. In this case, it is varied at the same time the grid current and the harmonic phase angle. In the case of 5th and 7th the same amplitude and phase angle was considered for both during the variation. As observed, with the harmonic order increasing, there is a reduction in the harmonic amplitude which can be compensated.

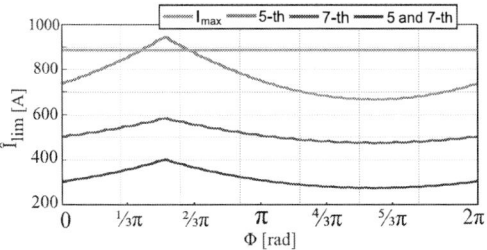

Fig. 4. Limitation imposed by the maximum voltage synthesized by the MMC in the injected harmonic current.

C. Semiconductor Devices and Capacitors Thermal Analysis

The thermal stress of the SM components (i.e. semiconductor devices and capacitors) are evaluated separately. In order to perform these analyses, the SM components are approximated to commercial values. Regarding the semiconductor devices, an ABB IGBT module, part number 5SND 0500N 330300, of 3.3kV-500A is employed. In addition, the SM capacitance is based on Electronicon Film Capacitors, part number E50.S34-115NT0, of 1.06 mF and 1.8 kV-120 A, per module 7 capacitors in parallel are employed.

For the semiconductor devices, the power losses are obtained from lookup tables based on the datasheet curves. In addition, the stress factors are related to the junction and case temperature. The hybrid thermal model is employed to estimate these temperatures [20], as shown in Fig. 5 (a). In addition, a heatsink per cell is considered and the impedances are calculated following the methodology proposed by [21]. Thus, it is considered the heatsink-to-water resistance of 0.0069 K/W and the water cooling-to-ambient thermal resistance of 0.0956 K/W.

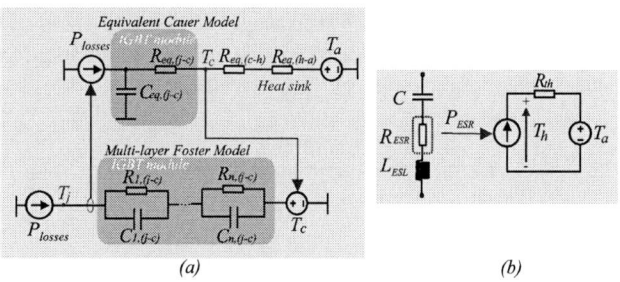

Fig. 5. Thermal model: (a) Hybrid thermal model based on Foster and Cauer models; (b) Capacitor thermal model.

For the SM capacitor, the hotspot temperature (T_h) is calculated by the power dissipation in R_{ESR} presented by Fig. 5 (b). Thus, the capacitor power losses can be obtained by:

$$P_{ESR} = R_{ESR}(f_i)I_{C,rms}^2, \tag{9}$$

where f_i is the i-*th* frequency. In order to simplify the methodology, only the fundamental frequency is considered in this work and the frequency dependence of ESR is neglected. In addition, the thermal resistance R_{th} of the capacitor is given in the datasheet as 0.9 K/W.

III. CASE STUDY

The main circuit parameters of the designed MMC-STATCOM are presented in Tab. I. These parameters were obtained following the methodology proposed in [15]. The base values are 15 MVA and 13.8 kV. In addition, the simulation results of this work were performed in PLECS environment.

Four case studies were evaluated: one case of STATCOM operation in rated inductive mode and three cases of APF

TABLE I
MODULAR MULTILEVEL CONVERTER PARAMETERS.

Parameter	Value
dc-link voltage (v_{dc})	28 kV
Rated power (S_n)	15 MVA
Grid voltage (v_g)	13.8 kV
Grid frequency (f_g)	60 Hz
Sampling frequency (f_{samp})	9.72 kHz
Transformer inductance (L_g)	1.35 mH (0.04 pu)
Transformer X/R ratio	18
Arm inductance (L_{arm})	5.1 mH (0.15 pu)
Arm resistance (R_{arm})	0.065Ω (0.005 pu)
Capacitance (C)	7.42 mF
Number of SMs per arm (N)	18
Nominal SM voltage ($v_{SM,n}^*$)	1.56 kV

operation with 5th and 7th harmonic current compensation. The selection criterion of current compensation respect the maximum current presented in Fig. 4 and the Total Harmonic Distortion (THD), lower than 5% (recommended by [22]). The values of amplitude and harmonic phase angle are presented in Tab. II for each case (the base current is the nominal grid current).

For all case studies, a resistive load of 10 MVA connected in parallel with the non-linear load is considered. In addition, the ambient temperature of 25°C is considered.

TABLE II
OPERATION CONDITION OF THE CASE STUDIES.

Case	I_5 (pu)	φ_5 (deg)	I_7 (pu)	φ_7 (deg)	Q (pu)
1	0	0	0	0	1.0
2	0.8500	70.25	0	0	0
3	0	0	0.6565	91.33	0
4	0.4511	91.04	0.4511	91.04	0

IV. SIMULATION RESULTS

Figs. 6-9 present the dynamic responses of the MMC-STATCOM for all cases. For the cases with APF operation, the non-linear load is connected to the system in 0.5 s and the converter starts to compensate the harmonic components in 0.8 s.

Fig. 6 (a)-(d) show the MMC output current behavior for all case studies. Furthermore, Fig. 6 (e)-(h) present the zoomed views of each case after the compensation period start. As observed, the converter injects current with frequency according to the operation mode, i.e. grid frequency in STATCOM operation and the compensated harmonic frequency in APF operation.

Fig. 7 (a)-(d) show the grid current dynamic response for all case studies. In addition, Fig. 7 (e)-(h) present the detail of each case after the compensation period start. As observed, the grid current distortion reduces after the compensation start. Tab. III shows the current THD which is lower than 5% in all cases.

In terms of circulating current, the dynamic responses are presented in Fig. 8. As observed, the maximum variation of

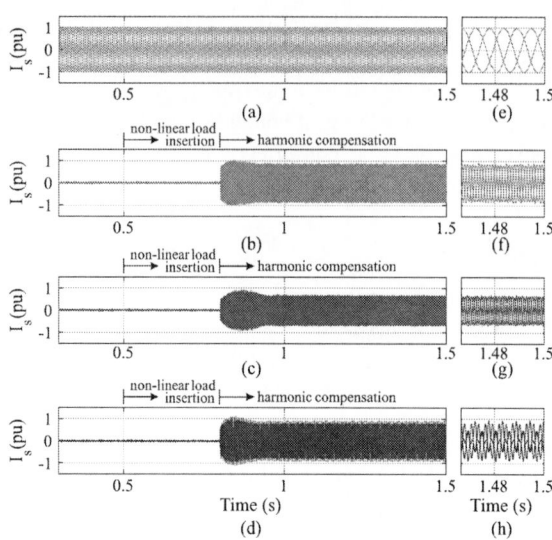

Fig. 6. MMC output current: (a) STATCOM operation in rated inductive mode (Case 1); (b) APF operation in 5th harmonic compensation (Case 2); (c) APF operation in 7th harmonic compensation (Case 3); (d) APF operation in 5th and 7th harmonic compensation (Case 4); (e) detail of Case 1; (f) detail of Case 2; (g) detail of Case 3; (h) detail of Case 4.

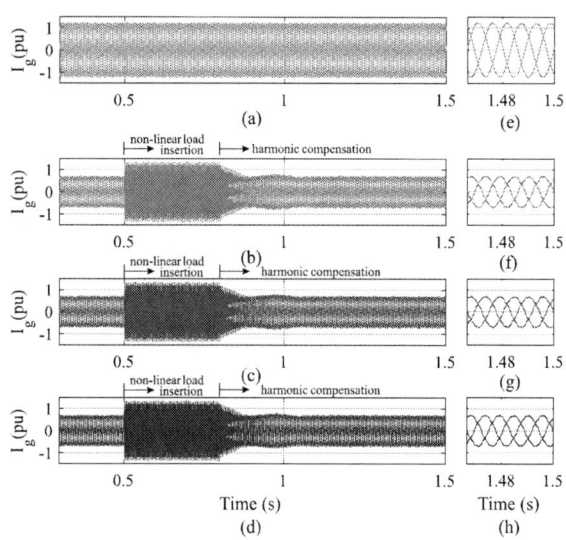

Fig. 7. Grid current: (a) STATCOM operation in rated inductive mode (Case 1); (b) APF operation in 5th harmonic compensation (Case 2); (c) APF operation in 7th harmonic compensation (Case 3); (d) APF operation in 5th and 7th harmonic compensation (Case 4); (e) detail of Case 1; (f) detail of Case 2; (g) detail of Case 3; (h) detail of Case 4.

TABLE III
GRID CURRENT TOTAL HARMONIC DISTORTION.

Case	1	2	3	4
THD (%)	1.17	4.21	4.21	4.17

this current (0.0405 pu) is 11% higher than that considering the APF compensation (0.0365 pu).

Fig. 9 presents the SM capacitor voltages. As observed in

978-1-7281-4181-7/19 $31.00 © 2019 IEEE

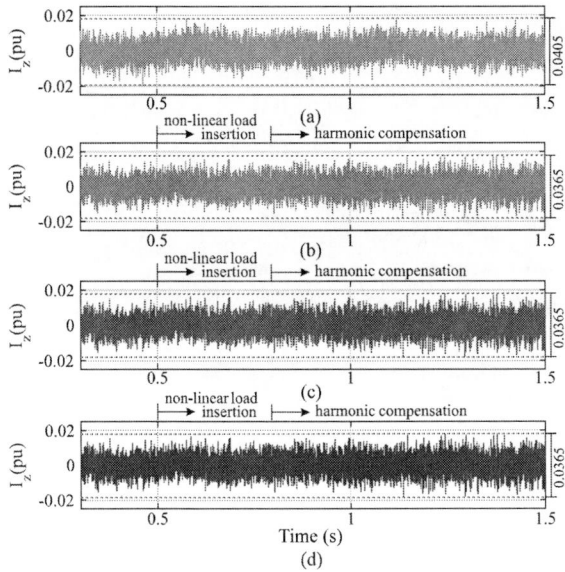

Fig. 8. Circulating current: (a) STATCOM operation in rated inductive mode (Case 1); (b) APF operation in 5th harmonic compensation (Case 2); (c) APF operation in 7th harmonic compensation (Case 3); (d) APF operation in 5th and 7th harmonic compensation (Case 4).

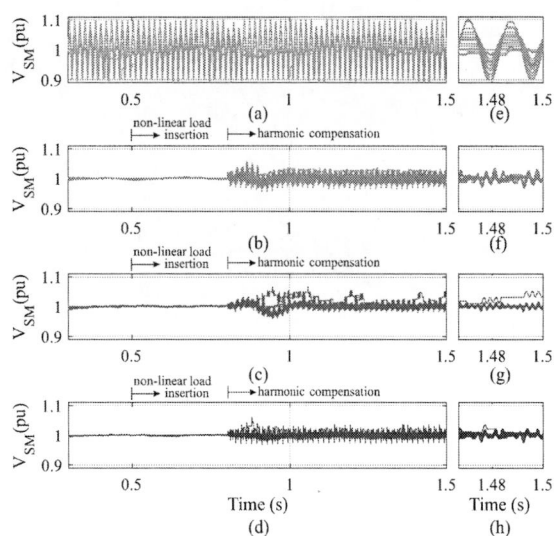

Fig. 9. SM Capacitor Voltages: (a) STATCOM operation in rated inductive mode (Case 1); (b) APF operation in 5th harmonic compensation (Case 2); (c) APF operation in 7th harmonic compensation (Case 3); (d) APF operation in 5th and 7th harmonic compensation (Case 4); (e) detail of Case 1; (f) detail of Case 2; (g) detail of Case 3; (h) detail of Case 4.

the details of Fig. 9 (e)-(h), the voltage ripple is lower in the cases of APF operation. These phenomena is explained by the fact that capacitor impedance is inversely proportional to the frequency. In addition, the highest amplitude in the spectrum of the SM capacitor voltages corresponds to the harmonic frequency compensated by the converter.

As observed, the STATCOM operation results in a higher voltage ripple than the converter operating in APF mode. This conclusion indicates that the energy storage requirements of the converter in order to guarantee a certain voltage ripple must be defined for STATCOM operation.

Regarding the thermal stress analyses, Fig. 10 presents the junction temperature for all the semiconductor devices (i.e. IGBTs and diodes). As observed, the STATCOM operation presents higher thermal stress in all semiconductor devices when compared to the cases with APF operation. Additionally, the thermal cycling increase in APF operation due to the harmonic content also present in the temperature. Among the cases of harmonic compensation, the Case 2 is the most stressed, followed by Case 3.

Fig. 11 presents the hotspot temperature of the capacitors. This result demonstrates the same pattern of the junction temperatures, in which the Case 1 is the most stressed, followed by Case 2, Case 3 and Case 4, respectively.

V. Conclusions

This work analyzed the thermal stresses of a MMC-STATCOM operating in the harmonic current compensation mode. The maximum harmonic current which can be compensated was numerically computed.

The results showed THD values lower than 5 % for all case studies and lower circulating current ripple and SM capacitor

voltage ripple during APF operation. Nevertheless, the cases of APF operation presented other harmonic component caused by the harmonic compensation. Regarding the thermal stress, the case of STATCOM operation presented the higher temperatures for all cell components (i.e. semiconductor devices and capacitors). Among the cases of APF operation, the case of 5th harmonic compensation was the most stressed.

Therefore, the results shown the MMC-STATCOM potential of multifunctional operation. Indeed, once this converter can operate as APF with lower thermal stress, this ancillary service can contribute with the electrical system power quality when rated reactive power compensation is not required.

Acknowledgment

This study has been supported by the Brazilian agency CAPES, the Project CAPES-COFECUB 876/2017, CNPq and FAPEMIG.

References

[1] F. O. Igbinovia, G. Fandi, J. Švec, Z. Müller, and J. Tlusty, "Comparative review of reactive power compensation technologies," in *16th Inter. Scientific Confer. on Electric. Power Engineering*, May 2015, pp. 2–7.

[2] S. Khan, B. Singh, and P. Makhija, "A review on power quality problems and its improvement techniques," in *Innovations in Power and Advanced Computing Technologies*, April 2017, pp. 1–7.

[3] K. Sharifabadi, L. Harnefors, H. Nee, S. Norrga, and R. Teodorescu, *Design, Control and Application of Modular Multilevel Converters for HVDC Transmission Systems*. John Wiley & Sons, Incorporated, 2016.

[4] M. Pereira, D. Retzmann, J. Lottes, M. Wiesinger, and G. Wong, "Svc plus: An mmc statcom for network and grid access applications," in *IEEE Trondheim PowerTech*, June 2011, pp. 1–5.

[5] A. Lesnicar and R. Marquardt, "An innovative modular multilevel converter topology suitable for a wide power range," in *IEEE Bologna Power Tech Confer. Proc.*, vol. 3, June 2003, pp. 6 pp. Vol.3–.

978-1-7281-4181-7/19 $31.00 © 2019 IEEE

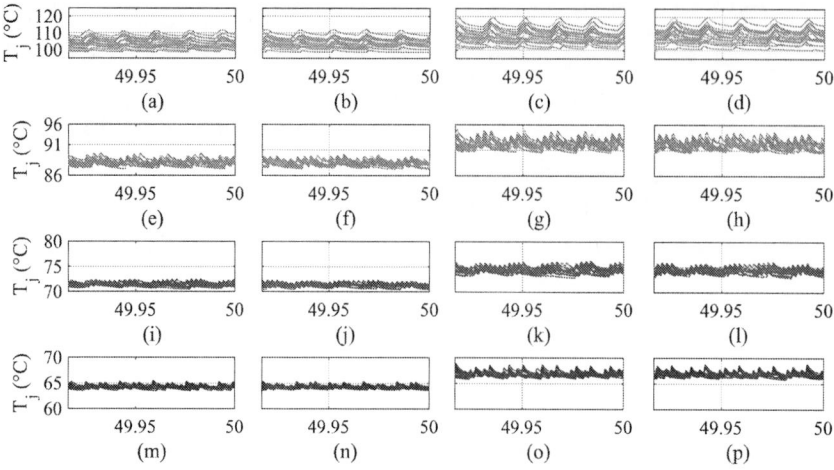

Fig. 10. Junction temperature: (a) S1 in STATCOM operation (Case 1); (b) S2 in Case 1; (c) D1 in Case 1; (d) D2 in Case 1; (e) S1 in APF operation with 5th harmonic compensation (Case 2); (f) S2 in Case 2; (g) D1 in Case 2; (h) D2 in Case 2; (i) S1 in APF operation with 7th harmonic compensation (Case 3); (j) S2 in Case 3; (k) D1 in Case 3; (l) D2 in Case 3; (m) APF operation with 5th and 7th harmonic compensation (Case 4); (n) S2 in Case 4; (o) D1 in Case 4; (p) D2 in Case 4.

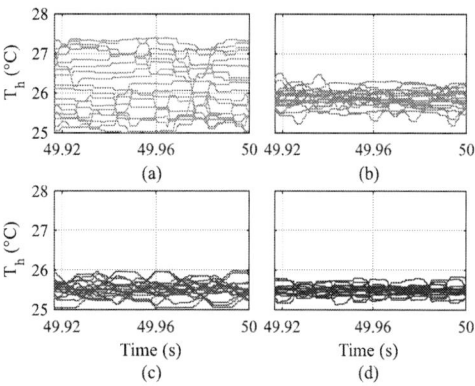

Fig. 11. Hotspot temperature: (a) STATCOM operation in rated inductive mode (Case 1); (b) APF operation in 5th harmonic compensation (Case 2); (c) APF operation in 7th harmonic compensation (Case 3); (d) APF operation in 5th and 7th harmonic compensation (Case 4).

[6] A. Dekka, B. Wu, R. L. Fuentes, M. Perez, and N. R. Zargari, "Evolution of topologies, modeling, control schemes, and applications of modular multilevel converters," *IEEE Journal of Emerg. and Sel. Topics in Power Electron.*, vol. 5, no. 4, pp. 1631–1656, Dec 2017.

[7] E. Kontos, G. Tsolaridis, R. Teodorescu, and P. Bauer, "High order voltage and current harmonic mitigation using the modular multilevel converter statcom," *IEEE Access*, vol. 5, pp. 16 684–16 692, 2017.

[8] J. Wu, X. Xu, Y. Liu, and D. Xu, "Compound control strategy of active power filter based on modular multilevel converter," in *Proc. of the 11th World Congress on Intell. Control and Autom.*, June 2014, pp. 4771–4777.

[9] M. S. Hamad, K. H. Ahmed, and A. I. Madi, "Current harmonics mitigation using a modular multilevel converter-based shunt active power filter," in *IEEE Inter. Confer. on Renewable Energy Research and Appl.*, Nov 2016, pp. 755–759.

[10] F. T. Ghetti, A. A. Ferreira, H. A. C. Braga, and P. G. Barbosa, "A study of shunt active power filter based on modular multilevel converter (mmc)," in *10th IEEE/IAS Inter. Confer. on Ind. Appl.*, Nov 2012, pp. 1–6.

[11] Z. Shu, M. Liu, L. Zhao, S. Song, Q. Zhou, and X. He, "Predictive harmonic control and its optimal digital implementation for mmc-based active power filter," *IEEE Trans. on Ind. Electron.*, vol. 63, no. 8, pp.

5244–5254, Aug 2016.

[12] A. I. Madi, M. S. Hamad, R. A. Hamdy, and I. F. El-Arabawy, "Mmc-based hpf migitating the mediumvoltage motor harmonic currents," in *Intl Conf on Advanced Control Circuits Syst.and Intl Conf on New Paradigms in Electr. Inf. Technol.*, Nov 2017, pp. 337–343.

[13] H. Akagi, "Classification, terminology, and application of the modular multilevel cascade converter (mmcc)," in *The 2010 International Power Electron. Confer.*, June 2010, pp. 508–515.

[14] L. Harnefors, A. Antonopoulos, S. Norrga, L. Angquist, and H. Nee, "Dynamic analysis of modular multilevel converters," *IEEE Trans. Ind. Electron.*, vol. 60, no. 7, pp. 2526–2537, July 2013.

[15] A. F. Cupertino, J. V. M. Farias, H. A. Pereira, S. I. Seleme, and R. Teodorescu, "Dscc-mmc statcom main circuit parameters design considering positive and negative sequence compensation," *Journal of Control, Automation and Electrical Systems*, vol. 29, no. 1, pp. 62–74, Feb 2018.

[16] E. Behrouzian and M. Bongiorno, "Dc-link voltage modulation for individual capacitor voltage balancing in cascaded h-bridge statcom at zero current mode," in *20th European Conf. on Power Electr. and Appl.*, Sep. 2018, pp. P.1–P.10.

[17] H. Akagi, E. H. Watanabe, and M. Aredes, *The Instantaneous Power Theory*. IEEE, 2007.

[18] K. Ilves, S. Norrga, L. Harnefors, and H. P. Nee, "On energy storage requirements in modular multilevel converters," *IEEE Trans. Power Electronics*, vol. 29, no. 1, pp. 77–88, Jan 2014.

[19] V. M. R. de Jesus, L. S. Xavier, A. F. Cupertino, V. F. Mendes, and H. A. Pereira, "Operating limits of three-phase multifunctional photovoltaic converters applied for harmonic current compensation," in *IEEE 8th Inter. Symposium on Power Electr. for Distributed Generation Syst.*, April 2017, pp. 1–8.

[20] Q. Tu and Z. Xu, "Power losses evaluation for modular multilevel converter with junction temperature feedback," in *IEEE Power and Energy Society General Meeting*, July 2011, pp. 1–7.

[21] P. R. Júnior, A. F. Cupertino, G. A. Mendonça, and H. A. Pereira, "On lifetime evaluation of medium-voltage drives based on modular multilevel converter," *IET Electric Power Appl.*, March 2019.

[22] "IEEE recommended practice and requirements for harmonic control in electric power syst." *IEEE Std 519-2014*, pp. 1–29, June 2014.

978-1-7281-4181-7/19 $31.00 © 2019 IEEE

Totem-Pole Bridgeless PFC Converter in DCM with Synchronous Rectification

Leonardo S. Mai, Alexsandra Rospirski, Samir A. Mussa, Telles B. Lazzarin.
Federal University of Santa Catarina
Power Electronics Institute (INEP)
Florianópolis, SC - Brazil
E-mail: leonardo.mai@inep.ufsc.br

Abstract—**This paper presents the mathematical modeling and simulation used for Totem-Pole Bridgeless topology design, operating in a discontinuous conduction mode with the addition of Synchronous Rectification and applied to power factor correction. The operational steps of this topology, and thus, the modeling of the equations necessary for its project are presented. Also, focus will be given to the zero-crossing switching problems, intrinsic to this topology, and mitigation techniques will be addressed. Lastly, the simulation results will be presented using the software PSIM.**

Keywords—Power Factor Correction, Totem-Pole Bridgeless, Discontinuous Conduction Mode, Synchronous Rectification, Zero-Crossing.

I. INTRODUCTION

In recent years, due to the intense technological advance, there has been a increase in the quantity of switched power supplies loads connected to the grid. Televisions, computers and microwave ovens are just a few examples of the abundance of electronics present in homes, commercial establishments and industries. However, this type of equipment usually presents low Power Factor (PF) and high levels of harmonic distortion, phenomena resulting from the interruption of the input current in the periods of commutation.

[1] and [2] highlight that in linear systems, the load drains a pure sine-wave current from the source, so the FP can be determined simply by the phase difference between the voltage and the current.

However, in power electronics systems, the non-linear behavior of the switching devices means that only angular phase difference analysis is not sufficient. Because of the interruption of the current, caused by switching, current harmonic spectrum analyzes is required [3].

Based on this information, this paper will address the use of the topology Totem-Pole Bridgeless as a power factor correction (PFC) converter. This topology is advantageous since it has few components in series with the current path, improving the efficiency [4]. However, during the moments when the current in the inductor is zero, this component stops to impose a current direction, allowing the energy stored in the output capacitor to be discharged at the input source, causing large current spikes on the semiconductors [5]. This effect occurs in zero crossing of the input voltage and, in converters

operating with synchronous rectification, at each switching period when the current in the inductor is extinguished.

This article will discuss ways to mitigate these problems, addressing the switching logic required for proper switching operation. It will also be demonstrated the equation of the converter, determining its transfer function and equations to calculate currents in the semiconductors. Lastly, simulation results will be presented in PSIM software, validating the design equations, transfer function and circuit operation.

II. POWER FACTOR CORRECTION

The techniques used for power factor correction are divided into two large groups, passive topologies and active topologies. Passive topologies are composed of capacitors and inductors, used to correct the power factor at high power levels, such as capacitor banks in transmission lines. On the other hand, the active topologies stand out over the passive ones by the ability to correct both the power factor and the harmonic distortion.[3]

Among the techniques of the active topologies used, DC-DC converters are predominant, as the Totem-Pole Bridgeless topology, the focus of this paper. The most commonly used topology for power factor correction is the Boost converter, since it has the inductor located at the input of the circuit. This characteristic will cause the current drained by the circuit to be continuous. In addition, the Boost converter's voltage-boosting nature ensures that it will operate correctly for a wide input voltage range. Starting from the boost converter it is possible to change the position of the inductor to before the bridge rectifier and replace two of the diodes with switches, this being the Boost Bridgeless converter.

A. Totem-Pole Bridgeless

The Boost Bridgeless topology has the drawback that the rectification diodes exhibit high frequency switching. With a simple change of position of the circuit's constituent components, it is possible to solve the presented problem. By placing the switches on the same arm and the diodes in the adjacent arm, the rectifying diodes switch on the same frequency as the input source, thus forming the Totem-Pole Bridgeless topology shown in Figure 1 (a) and the same topology with the addition of Synchronous Rectification is shown in (b).

Although the number of constituent components is the same, compared to the previous topology, the resulting efficiency

978-1-7281-4181-7/19 $31.00 © 2019 IEEE

will be improved by the reduction of switching losses on the diodes. In Table 1 the constituent components of the previously presented topologies are shown for comparison.

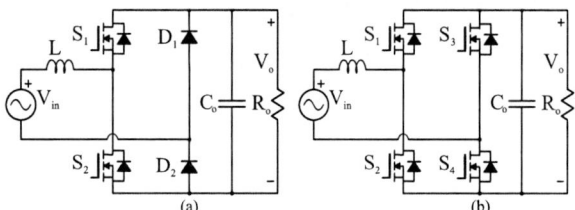

Fig. 1. Totem-Pole Bridgeless Topology for Power Factor Correction

This topology, however, presents special characteristics in its switching logic, besides presenting switching problems due to the positioning of the switches. These characteristics will be approached in this article, as well as techniques to mitigate the presented problems.

Table I. NUMBER OF COMPONENTS USED IN EACH TOPOLOGY.

	Slow Diode	Fast Diode	Switch
Boost	4	1	1
Boost Bridgeless	0	2	2
Totem-Pole Bridgeless	2	0	2

III. MODELING OF THE TOTEM-POLE BRIDGELESS IN DCM

A Totem-Pole Bridgeless PFC operating in discontinuous conduction mode has as characteristic the complete discharge of the energy stored in the inductor during the conduction of the switch S_1 in the positive half-cycle of the input voltage. Analogously, for the negative half-cycle of the voltage, the discharging of the inductor occurs by conducting the switch S_2.

The main waveforms of the operational values in the inductor, during all stages of operation, are shown in Figure 2. Hereafter, in Figure 3, all the steps of operation of the Totem-Pole Bridgeless PFC in DCM are shown.

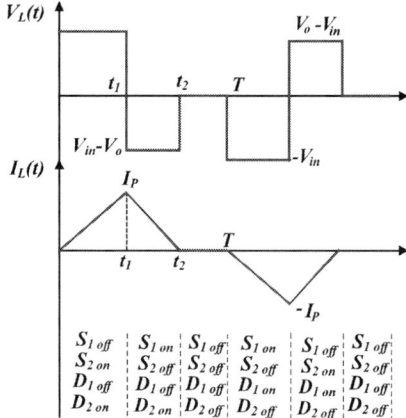

Fig. 2. Voltage and current waveforms in the inductor during the operation steps

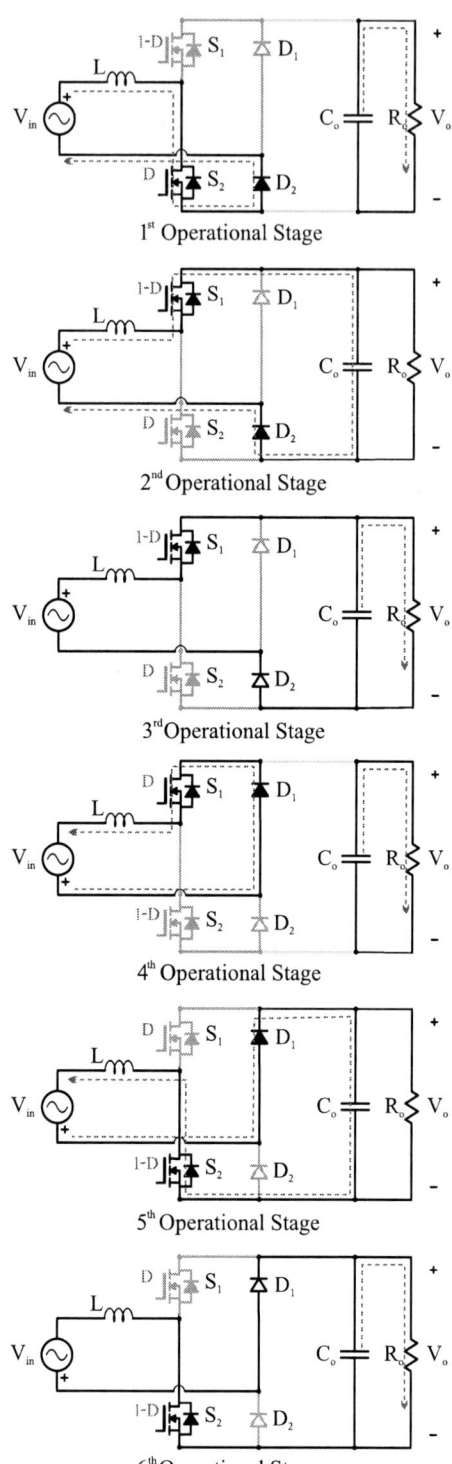

Fig. 3. Stages of Operation for the PFC Totem-Pole Bridgeless in DCM

978-1-7281-4181-7/19 $31.00 © 2019 IEEE

A. Static Gain

During the inductor discharge stage, the current in the diode is equal to the current in the inductor. Also, the average current in the diode is equal to the current in the load. Therefore, the equations representing the current in the diode and load are represented by:

$$i_D(t) = Iin_{pk}\frac{V_o - V_p sin(\omega t)}{L}t \qquad (1)$$

$$I_D(t) = I_o(t) = \frac{Iin_{pk}\Delta t_2}{2T} \qquad (2)$$

As Iin_{pk} and $\Delta t2$ are defined in (3) and (4) respectively, by performing algebraic manipulations one can define the Totem-Pole Bridgeless gain as (5).

$$Iin_{pk} = \frac{V_{in}}{L}\Delta t_1 = \frac{V_p sin(\omega t)}{L}\Delta t_1 \qquad (3)$$

$$\Delta t_2 = \frac{V_p sin(\omega t)}{V_o - V_p sin(\omega t)}\Delta t_1 \qquad (4)$$

$$\frac{V_o}{V_p sin(\omega t)} = 1 + \frac{V_p sin(\omega t)D^2}{2LfI_o} \qquad (5)$$

B. Simplified Transfer Function

As previously mentioned, the equivalent circuit of a Totem-Pole Bridgeless PFC can be represented by a Boost converter. Thus, to find the desired transfer function, the circuit model of a Boost converter will be used. In a simplified modeling, it is considered that the current in the output diode can be replaced by a current source defined by I_D. This result in the transfer function shown in (6).

$$G_{V_o d}(s) = \frac{\Delta V_o}{\Delta d} = \left(\frac{DV_p Y(\alpha)}{\pi L f_s}\right)\left(\frac{R_o}{C_o R_o s + 1}\right) \qquad (6)$$

Where α, $Y(\alpha)$ and D are defined as:

$$\alpha = \frac{V_p}{V_o} \qquad (7)$$

$$Y(\alpha) = -2 - \frac{\pi}{\alpha} + \frac{2}{\alpha\sqrt{1-\alpha^2}}\left(\frac{\pi}{2} + atan\frac{\alpha}{\sqrt{1-\alpha^2}}\right) \qquad (8)$$

$$D = \sqrt{\frac{2\pi f_s LI_o}{V_o \alpha Y(\alpha)}} \qquad (9)$$

IV. Components Characterization

To perform the characterization of the inductor and capacitor, the same method commonly used for the Boost converter can be used in this converter, due to its similar behavior, as seen in the previous section.

A. Diodes Current Rating

Because the input voltage of the PFC is a sinusoidal source, it is necessary to integrate in two time intervals to perform stress calculations of the circuit components. The first interval will refer to the switching frequency and the second to the frequency of the grid.

The diodes of the circuit, although operating in different stages of operation, present the same value for average and RMS current, due to the symmetry of operation for both half-cycles of the grid.

For the calculation of the average and RMS currents in the diode it is necessary to define the time in which it operates during the period of commutation and its current in this interval. Thus, during the charging of the inductor the diode current is defined by (10) and the diode conducts during the period from 0 to D. At the inductor discharge period, the current in the diode is established by (11), this leads to the total discharging of the inductor at time t_2, represented by (12).

$$I'_D = \frac{V_{inp}sin(\omega t)}{L} \qquad (10)$$

$$I_D" = I_p - \frac{V_o - V_{inp}sin(\omega t)}{L} \qquad (11)$$

$$T_2 = \frac{\alpha sin(\omega t)}{1 - \alpha sin(\omega t)}DT_s \qquad (12)$$

Thus, the average current in the diode during the switching period is:

$$I_{Davg}(\omega t) = \frac{1}{T_s}\left(\int_0^{DT_s} I'_D d(t) + \int_{DT_s}^{T_2} I_D" d(t)\right) \qquad (13)$$

For the mains period it is determined by:

$$I_{Davg} = \frac{1}{2\pi}\int_0^\pi I_{Davg}(\omega t)d(\omega t) \qquad (14)$$

For the calculation of the RMS current during switching period yields:

$$I_{D_{RMS}}(\omega t) = \sqrt{\frac{1}{T_s}\left(\int_0^{DT_s}(I'_D)^2 d(t) + \int_0^{T_2}(I_D")^2 d(t)\right)} \qquad (15)$$

And for the mains period:

$$I_{D_{RMS}} = \sqrt{\frac{1}{2\pi}\int_0^\pi (I_{D_{RMS}}(\omega t))^2 d(\omega t)} \qquad (16)$$

978-1-7281-4181-7/19 $31.00 © 2019 IEEE

B. Switches Current Rating

The transistors presents a complementary switching between them, but an equivalent waveform, thus resulting in the same average and efficient current between both. Due to this equivalent behavior, the operating periods of the S_2 switch will be analyzed for the calculations. Thus, the equations that represent its waveform in the first and second operating periods are presented below.

$$I_S' = \frac{V_{inp}sin(\omega t)}{L} \tag{17}$$

$$I_S" = -I_p + \frac{V_o - V_{inp}sin(\omega t)}{L} \tag{18}$$

Substituting the equations in the integral for the switching period leads to:

$$I_{S1avg}(\omega t) = \frac{1}{T_s} \int_0^{DT_s} I_S' d(t) \tag{19}$$

$$I_{S2avg}(\omega t) = \frac{1}{T_s} \int_{DT_s}^{T_2} I_S" d(t) \tag{20}$$

In the same way, during the mains period yields:

$$I_{Savg} = \frac{1}{2\pi} \left(\int_0^\pi I_{S1avg}(\omega t)d(t) + \int_\pi^{2\pi} I_{S2avg}(\omega t)d(t) \right) \tag{21}$$

The resulting equation for the average current in the switches will return a value close to zero due to the symmetry of the waveform in the positive and negative half-cycles. This value will therefore be considered null.

The equations for the RMS current in the switches for the switching period are:

$$I_{S1_{RMS}}(\omega t) = \sqrt{\frac{1}{T_s} \int_0^{DT_s} (I_S')^2 d(t)} \tag{22}$$

$$I_{S2_{RMS}}(\omega t) = \sqrt{\frac{1}{T_s} \int_0^{T_2} (I_S')^2 d(t)} \tag{23}$$

And for the mains period:

$$I_{S_{RMS}} = \sqrt{\frac{1}{2\pi} \left(\int_0^\pi (I_{S1_{RMS}}(\omega t))^2 d(t) + \int_0^\pi (I_{S2_{RMS}}(\omega t))^2 d(t) \right)} \tag{24}$$

Analyzing the realized calculations, it is observed that the equations of the RMS currents, in the diodes and in the switches, present the same result. This occurs because in the equivalent circuit of the topology, such components, in given half-cycle, are in series, therefore, the current that circulates through both is the same.

C. Inductor Current Rating

The current in the inductor presents a sinusoidal behavior, following the envelope of the input voltage, so the average current will be null. The RMS current in the inductor is defined by the square root of the sum of the squares of the RMS current in the diodes and in the switches, resulting in:

$$I_{L_{RMS}} = \sqrt{(I_{D_{RMS}})^2 + I_{S_{RMS}})^2} \tag{25}$$

D. Voltage Ratings on Components

Both the semiconductors present during their respective blocking periods the maximum voltage on their terminals equal to the output voltage V_o.

V. TOPOLOGY CHARACTERISTICS

A. Synchronous Rectification

The losses in the diode are the main responsible for reducing the efficiency of the converter, being part of the conduction losses since they vary according to the current demanded by the load. These losses are due to the existence of a potential difference between the terminals of the diode, usually $0.7V$ for silicon diodes and these can reach $1V$ in $600V$ rectification diodes, being this characteristic intrinsic to the device .

In order to reduce this set of losses, it is possible to replace the rectification diodes by MOSFET transistors that will be triggered according to the half-cycle of the electrical grid. This practice improves the circuit efficiency, however, it does not completely eliminate the losses, since the MOSFET presents certain series resistance when conducting, in addition to introducing switching losses.

The introduction of the concept of synchronous rectification in the present circuit together with the discontinuous conduction mode causes a new problem that must be addressed. At the moment the current in the inductor is zero, it will no longer force the direction of the current in the circuit. Because the synchronous rectifier MOSFET is still triggered, it maintains a path for the current to flow by discharging the output capacitor at the input source. Such an occurrence has multiple negative effects on the circuit, as well as causing current spikes, it can destabilize the control loop.

To correct this problem, the inclusion of a current sensing circuit to monitor the current flowing to the load is required. This sensor will have the function of detecting the moment when the current flowing from the inductor to the load approaches zero, commanding the opening of the synchronous rectifying MOSFET. In this way, the internal diode of the MOSFET will assume the conduction of the current and when it becomes negative the diode will go through the natural switching process.

This adopted solution, however, will make it necessary to switch the MOSFET in all the switching periods, thus increasing the switching losses. Even with the addition of these switching losses, synchronous rectification technique remains more efficient when compared to use of conventional diodes.

978-1-7281-4181-7/19 $31.00 © 2019 IEEE

B. Switching Issues

A disadvantage of the Totem-Pole Bridgeless topology is the occurrence of current spikes during the zero crossing of the mains voltage. These current spikes are generated by the slow recovery of the internal diodes and excessive output capacitance (C_o) of the switches S_3 and S_4, and by the sudden change of duty cycle. These current peaks, in addition to being potentially harmful to the hardware due to their amplitude, will also increase the total harmonic distortion (THDi), consequently, reducing the power factor [8].

In order to circumvent this problem, a logic must be implemented which identifies the input zero voltage crossing and momentarily pauses the switching. In this way, the circuit must rely on a reliable zero-crossing detection system, since an erroneous switch can cause damage to the components.

In addition to pausing the switching of the transistors, a soft-start initialization sequence must be adopted in the mains half-cycle change. Due to the low input voltage near the zero cross point, the duty cycle will be maximum. This topology dictates that the transistors switch function every half-cycle, so the transistor that acts as the main switch will operate as a synchronous rectifier and vice versa. In this way, the duty cycle applied to each switch will vary sharply if no soft-start techniques are applied.

C. Ganfet Use

As already described, the Totem-Pole Bridgeless topology presents as a disadvantage the occurrence of current peaks during the switching process. Although the occurrence of this problem in the rectification arm (S_3 and S_4) can be remedied by soft-start techniques, a more improved solution is required in the switching arm (S_1 and S_2). Since these current peaks are caused by the reverse recovery charge of the internal diode of the MOSFET transistor, it is proposed in the literature the use of Gallium Nitride transistors, GaNFET, as a solution, since it does not have an internal diode.[5] [9]

However, GaN transistors are capable of performing the same function as MOSFETs with the presence of diodes, but in a different way. Assuming that the voltage between gate and source is zero, there is an absence of electrons below the gate region. At the moment the voltage in the drain contact is reduced, a positive voltage point is created in the gate of the device, thereby attracting electrons to this region. When there are enough electrons, a conductive channel between drain and source will form. The advantage of this mechanism is that the reverse recovery time will be zero. Although the absence of reverse recovery time represents less losses to the device, the GaN transistors must charge and discharge the gate every cycle, even when there is a reverse current [10].

Circuits using GaN transistors, then, must have optimized deadtime to minimize such losses. Usual deadtime values for these devices are less than $10ns$, which translates into lower losses by driving in the third quadrant [10].

D. Switching Logic

Although the Totem-Pole Bridgeless topology equivalent circuit is the same as a conventional boost converter, additional care must be taken in regard to the switching logic of the switches. First of all, because of the function switch of the main transistors (S_1 and S_2), one must detect the mains cycle and apply the correct command signal, D or $1 - D$, to each transistor. Mains cycle detection is also used to determine which synchronous rectifying MOSFET (S_3 or S_4) must be triggered.

Next, the issue of switching during zero crossing of the input voltage must be addressed. As the input voltage is already being sampled for the detection of the mains cycle, it is only necessary to add a logic of comparison to this signal, in order to determine the period in which the voltage will be within the range that is desired to pause the switching .

Finally, a current sensor is used to monitor the current flowing to the load. This current signal is then used in a comparison logic that identifies the zero approximation of the current and interrupts the switching of the synchronous rectifiers. Upon interruption of the switching, the MOSFETs will remain open and the rest of the current stored in the inductor will be conducted by the internal diodes.

VI. PSIM SIMULATION

For circuit simulation the software PSIM was used, Table 2 shows the values used for the components.

Table II.   VALUES USED IN SIMULATION

Input Peak Voltage (V_{inp})	156	V
Output Power (P_o)	500	W
Output Voltage (V_o)	400	V
Switching Frequency (f_s)	50k	Hz
Input Inductor (L)	100 μ	H
Output Capacitor (C_o)	1 000 μ	F
Input Filter Inductor	16.2 μ	H
Input Filter Capacitor	14.8 μ	F

The results obtained in simulation are presented below. The Figure 4 shows the response to the load step of 50% → 100% → 50% in $85V$, $110V$, $220V$ e $265V$ respectively

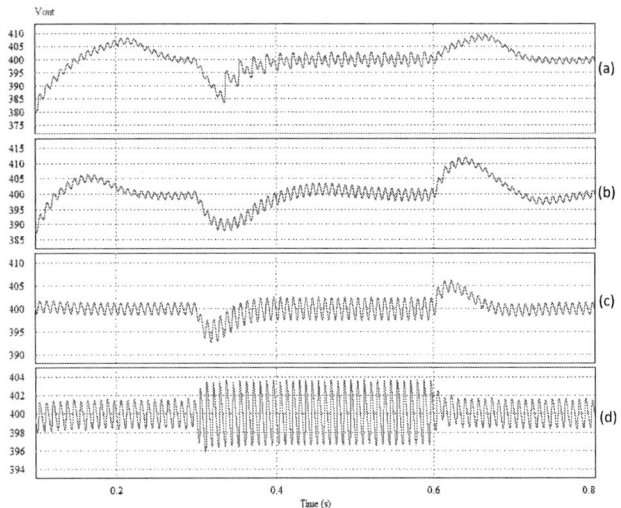

Fig. 4.   Load step transient for $V_{in} = 85V$(a), $V_{in} = 110V$(b), $V_{in} = 220V$(c) and $V_{in} = 265V$(d)

The figure 5 shows the current and voltage at the input. Where (a) shows the current spikes on zero-crossing when no correction techniques are used and (b) with the use of these corrections. The resulting power factor was 0.986 and the total harmonic distortion of the current resulted in 6.8% for the converter operating at rated load.

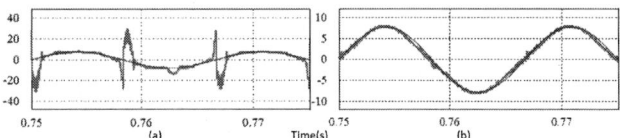

Fig. 5.   Voltage and current at the input of the simulated circuit

Figure 6 illustrates the currents in the rectifier MOSFETs, GaNFETs and in the inductor, respectively.

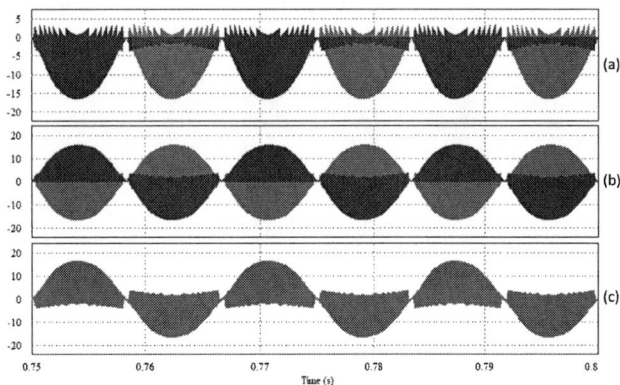

Fig. 6.   Currents in the MOSFETs(a), GanFETs(b) and Inductor(c)

The Figure 7 presents the comparison of the resulting current harmonics, using both diode and MOSFET as rectification, with the maximum values established by the norm IEC 61000-3-2.

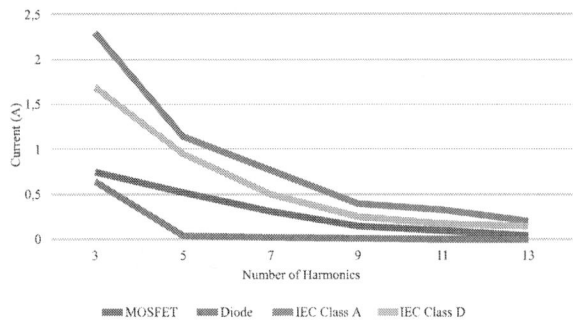

Fig. 7.   Current harmonics in comparison with IEC 610000-3-2 limits

VII. CONCLUSION

This paper describes the mathematical analysis in addition to the simulation of a power factor correction device using the Totem-Pole Bridgeless topology in discontinuous conduction mode. Afterwards, the characteristics of the topology are presented detailing the adaptations necessary for the correct operation with the use of the synchronous rectification technique. ·

It is concluded that this topology, in discontinuous conduction mode, presents some complexity for its correct operation, however, it is possible to carry out its implementation and the resulting performance is satisfactory. With the advantage of reducing the required inductance, when compared to the same topology in critical or continuous mode of operation.

Regarding efficiency, the Totem-Pole Bridgeless topology will exhibit lower losses when compared to other PFC topologies, due to the reduced number of components in the current path, use of synchronous rectification and the use of Gallium Nitride transistors in the switching arm.

Acknowledgment

The authors would like to thank the Brazilian National Counsel of Technological and Scientific Development (CNPq) for the financial support.

REFERENCES

[1] BATARSEH, I; WEI, H. *Power eletronics handbook.* 3. ed. 2011: Butterworth-Heinemman. pg. 523 – 548.

[2] GHANDI, B; EZHILMARAN, M. *Achieving high input power factor for DCM boost converters by controlling variable duty cicle.* ICCPEIC 2013. pg. 18 – 20. 2013.

[3] POMÍLIO, J.A. *Pré-Reguladores de Fator de Potência.*, 2007. [Online] Available: http://www.dsce.fee.unicamp.br/ antenor/pfp.html.

[4] SU, B.; ZHANG, J.; LU, Z. *Totem-Pole Boost Bridgeless PFC Rectifier With Simple Zero-Current Detection and Full-Range ZVS Operating at the Boundary of DCM/CCM* IEEE Transactions on Power Electronics, vol. 26, no. 2, pp. 427-435, Feb. 2011.

[5] ZHOU, L.; WU, Y.; HONEA, J.; WANG, Z. *High-efficiency True Bridgeless Totem Pole PFC based on GaN HEMT: Design Challenges and Cost-effective Solution.* 2015. PCIM Europe 2015, Nuremberg, Germany

[6] RASHID, M. H. *Power Electronics Handbook: Devices, circuits, and applications*, 3rd ed. Pensacola, Florida, USA. Elsevier, 2011.

[7] ZHOU, Bo., *Totem Pole Bridgeless PFC with Ultra Fast IGBT.* Thesis. Virginia Polytechnic Institute and State University. 2014.

[8] SUN, B. *Control challenges in a totem-pole PFC.* In: Analog Applications Journal 2Q. 2017. [Online] Available: http://www.ti.com/lit/an/slyt718/slyt718.pdf.

[9] GAN SYSTEMS. *High Efficiency CCM Bridgeless Totem Pole PFC Design using GaN E-HEMT.* 2016. [Online] Available: https://www.mouser.com/pdfDocs/gs665btp-refrev170411.pdf.

[10] EPC. *Fundamentals of Gallium Nitride Power Transistors.* 2011. [Online] Available: http://epc-co.com/epc/Portals/0/epc/documents/product-training/Appnote-GaNfundamentals.pdf

Passive Capacitor Voltage Balancing in Modular Multilevel Converter During its Precharge: Analysis and Design

Luiz H. T. Schmidt[1], Gean J. M. de Sousa[2], Marcelo L. Heldwein[2], Daniel J. Pagano[1]

Federal University of Santa Catarina (UFSC) — 88040-900, Florianópolis, Brazil

[1]Department of Automation and Systems, [2]Power Electronics Institute

Email: luiz.schmidt@posgrad.ufsc.br, gean@inep.ufsc.br, marcelo.heldwein@ufsc.br, daniel.pagano@ufsc.br

Abstract—The modular multilevel converter (MMC) is one of the main converter topologies used in High Voltage Direct Current and is also used in other applications including medium voltage drives and renewable energy sources integration. Some advantages of the MMC are that it can be used without bulky transformers and output filters and its modular design based on the connection of submodules (SMs), which makes it easy to manufacture. It is advantageous that each SM local signal electronics circuits are self powered by means of a local auxiliary power supply fed from the SM dc-link. However, during the precharge stage this configuration might not lead to balanced nor stable SM voltages. The balancing of the MMC's capacitor voltages during its precharge stage is crucial to its correct operation. A passive balancing strategy that consists in adding a balancing resistance in parallel to each one of the SMs is analyzed in this work in light of its non linear dynamics. The stability of the desired operation equilibrium point is analytically proven for an arbitrary number of submodules. Guidelines for the design of the resistance and characterization of the system nonlinear dynamics are presented from simulation results. Experiments confirm the analytical and simulation results.

Index Terms—Modular Multilevel Converter, Precharge, Capacitor Voltage Balancing, Passive, Stability, Nonlinear Dynamics

I. INTRODUCTION

The modular multilevel converter (MMC) and its variants have become the standard solution in VSC-HVDC systems [1]. This is mainly due to an ideally balanced distribution of voltage and current stresses among its power semiconductors. This allows operation in high voltages without series connected devices and all the related challenges. Instead, series connections of identical submodules—half-bridge and full-bridge types are the most common—compose each of the six branches, or arms, of the topology. The resulting symmetry makes the MMC (see Fig. 1(a)) a highly modular and scalable converter, easing the design of new stations with different specifications. Medium voltage applications can also benefit from the MMC features, which has encouraged manufactures to develop MMC based machine drives [2], and researchers to propose new MMC topologies and applications in MV dc-dc

This study was founded in part by the Personnel Improvement Coordination of Superior Level - Brazil (CAPES) - Finance Code 001.

conversion [3], wind power [4], energy storage systems [5], static compensation, among others [6].

Fig. 1(a) also shows the parts that constitute a SM. Besides the main components, the switching cell (S_1 and S_2) and the energy storage element $C_{j,k}$, a real converter submodule, to properly operate, needs many other components: sensors, controllers, communication related devices, gate drivers and an auxiliary power supply (APS) to feed all these circuits. Each APS can be powered from the own submodule capacitor or externally through a high voltage isolated power supply system, each approach having advantages and disadvantages. The latter allows the submodule controllers to perform test routines before the main power is connected, enabling early detection of failures in the control and communication systems. The always-on controller also simplifies the challenges of hot swapping submodules [7]. The APS cost and complexity, however, are high, since its insulation has to withstand medium/high working voltages. An APS designed to be powered from its own submodule has to insulate only the much lower capacitor voltage. On the other hand, the controllers can only perform test routines after the main power connection, decreasing chances of early detecting failures that could potentially lead to damage.

Independently of the type of APS, the connection of the converter to the grid can only be performed (by K_{ac} in Fig. 1(a)) after precharging all capacitors to minimum voltage levels that prevents inrush currents. Researches have proposed different precharge schemes, which can be classified into two main categories: using a low voltage source to charge submodules one by one, which requires an externally fed APS [8], [9]; or precharging all submodules simultaneously by series current limiting resistors from the ac or dc side [10], [11]. The strategy analyzed in this work, shown in Fig. 1(a), employs a dc side precharge scheme. Contactor K_1 is used to start the procedure, while K_2 bypasses the current limiting resistor R_{dc} once the sum of the capacitor voltages has reached an appropriate value.

Considering the case where the APS is fed internally, the precharge procedure can be divided into three stages, namely:

1) *Uncontrolled and inactive*—This stage starts with K_1 closing. The capacitor voltages rise rapidly up to ap-

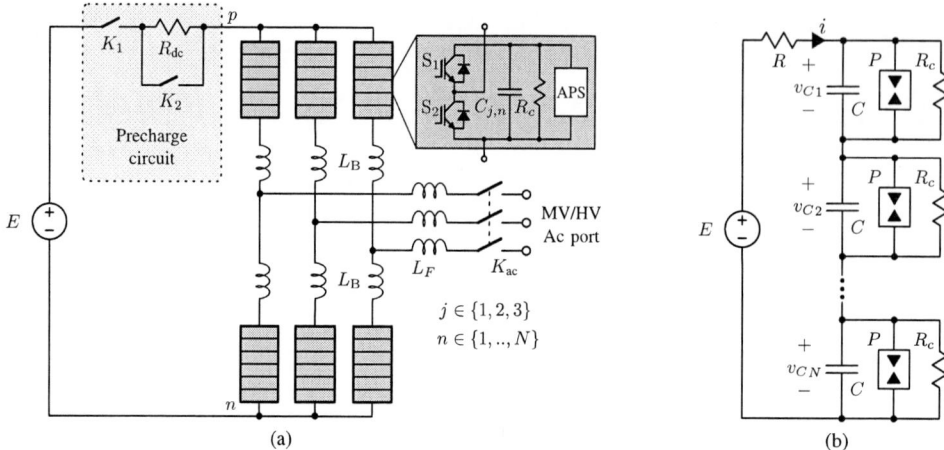

Fig. 1: (a) MMC converter and its precharge circuit. (b) Equivalent phase circuit during the *Uncontrolled and active* precharge stage.

proximately $v_{Cj,k} \approx E/N$. The actual value depends mainly on the capacitances considering actual tolerance values, the smaller ones ending with a higher voltage. The APSs start turning on after $v_{Cj,k}$ has reached a minimum value, which is design dependent. It is also common that APSs have an inertia to start operation even after the full voltage is present due the dynamics of the startup circuit, which can change the duration of this stage. The bypass contactor K_2 can be closed after detecting that v_{pn} is close to the maximum value or after a suitable time.

2) *Uncontrolled and active*—This stage starts when all APSs are on. Modeling each APS as a constant power drain, the equivalent circuit of one phase becomes as shown in Fig. 1 (b), which is known to be unstable [12]. A way to counteract the instability is to add a resistance R_c in parallel with each capacitor [13]. The voltage drops across arm inductors are small and have been neglected.

3) *Controlled*—In this stage the control system commands the switches S_1 and S_2 of each submodule according to an appropriate control strategy that targets bringing all capacitor voltages to their nominal values. At this stage, R_c can be disconnected from the circuit.

The aim of this work is to present an in-depth analysis of the stability of the second stage and to provide a guideline for choosing R_c. The presented analysis is based on the system non linear dynamics and has not been previously demonstrated. The condition for stability for a single phase MMC is derived for the general case with any number of submodules. However, to gain more insight about the system dynamics through graphical analysis, the case $N = 2$ is explored in more detail. Lastly, experimental results are presented for $N = 10$.

II. MMC Pre-charging Equivalent Circuit Model

The main goals are to understand the capacitor voltages balancing and also finding out which value of balancing resistance R_c makes the system operation point stable. The model states are the SMs voltages v_{Ci}. E, R, C and P are considered model parameters, and the discharging resistance R_c is here treated as a design parameter of the circuit. The resulting model is

$$\frac{dv_{Ci}}{dt} = \frac{1}{C}\left(\frac{E - \sum_{i=1}^{N} v_{Ci}}{R} - \frac{v_{Ci}}{R_c} - \frac{P}{v_{Ci}} \right). \tag{1}$$

III. The Stability Condition

The critical value of R_c that is needed to maintain the operation equilibrium point stable is derived in this section. This is made first for the two submodules case in order to reduce complexity and then for an arbitrary number of submodules. The desired operation point is $v_{C1} = v_{C2} = \cdots = v_{CN}$. By applying standard nonlinear analysis the Jacobian matrix and its eigenvalues are computed for the system linearized at the desired operation equilibrium point and the stability condition arises from that.

For the two-dimensional case, the model equations are,

$$\frac{dv_{C1}}{dt} = \frac{1}{C}\left(\frac{E - (v_{C1} + v_{C2})}{R} - \frac{P}{v_{C1}} - \frac{v_{C1}}{Rc} \right), \tag{2}$$

$$\frac{dv_{C2}}{dt} = \frac{1}{C}\left(\frac{E - (v_{C1} + v_{C2})}{R} - \frac{P}{v_{C2}} - \frac{v_{C2}}{Rc} \right), \tag{3}$$

and the resulting equilibrium point is found for $\boldsymbol{v_{Ce}} = (v_{Ce1}, v_{Ce2})$ by solving $\frac{d\boldsymbol{v_C}}{dt} = \left[\frac{dv_{C1}}{dt} \frac{dv_{C2}}{dt} \right]^T = \left[\begin{array}{cc} 0 & 0 \end{array} \right]^T$. The line equation $v_{Ce1} = v_{Ce2} = v_{Ce}$ satisfies the equilibrium point equation. Thus, at least one of the equilibrium is located somewhere over this line. This point corresponds to the desired operation point, where the voltages are balanced. The Jacobian from (2) and (3) at this equilibrium point is

$$A = \frac{1}{C}\begin{bmatrix} \dfrac{P}{v_{Ce}^2} - \dfrac{1}{R} - \dfrac{1}{R_c} & -\dfrac{1}{R} \\[2ex] -\dfrac{1}{R} & \dfrac{P}{v_{Ce}^2} - \dfrac{1}{R} - \dfrac{1}{R_c} \end{bmatrix} \tag{4}$$

978-1-7281-4181-7/19 $31.00 © 2019 IEEE   803

and standard equilibria stability conditions for planar systems are $Det(\boldsymbol{A}) > 0$ and $Tr(\boldsymbol{A}) < 0$, which can be expressed as

$$\frac{v_{Ce}^2}{R_c} > P \tag{5}$$

$$\frac{v_{Ce}^2}{R} + \frac{v_{Ce}^2}{R_c} > P. \tag{6}$$

As (5) is more restrictive than (6) for the maximum value of R_c, Equation (5) is a sufficient condition needed to delimit this value.

This stability analysis can be generalized for the case with an arbitrary number N of SMs. For the model generalized equations in (1) the Jacobian matrix $\boldsymbol{A}$ as

$$\boldsymbol{A} = \begin{bmatrix} a_{11} & a_{12} & \cdots & a_{1N} \\ a_{21} & a_{22} & \cdots & a_{2N} \\ \vdots & \vdots & \ddots & \vdots \\ a_{N1} & a_{N2} & \cdots & a_{NN} \end{bmatrix}, \tag{7}$$

calculated at the operating equilibrium point where $v_{Ce1} = v_{Ce2} = \cdots = v_{CeN}$, whose whose elements are defined in

$$a_{ij} = \begin{cases} \dfrac{1}{C_i}\left(-\dfrac{1}{R} + \dfrac{P_i}{v_{C_i}^2} - \dfrac{1}{R_i}\right) & , \text{ if } i = j \\[2ex] -\dfrac{1}{RC_j} & , \text{ if } i \neq j \end{cases} \tag{8}$$

with $j \in \{1, 2, \cdots, N\}$. Its eigenvalues λ_i at the operation equilibrium point, are

$$\lambda_1 = \lambda_2 = \cdots = \lambda_{N-1} = \frac{1}{C}\left(\frac{P}{v_{Ce}^2} - \frac{1}{R_c}\right) \tag{9}$$

$$\lambda_N = \frac{1}{C}\left(-\frac{N}{R} - \frac{P}{v_{Ce}} - \frac{1}{R_c}\right) \tag{10}$$

from which it is possible to establish the following stability conditions:

$$\frac{v_{Ce}^2}{R_c} \geq P \tag{11}$$

$$\frac{v_{Ce}^2}{R_c} \geq P - \frac{Nv_{Ce}^2}{R} \tag{12}$$

As we are interested in the maximum value of R_c, the condition (11) is more restrictive than (12). Thus, the power consumed by the equilibrium resistances R_c must be at least of the same magnitude of the power consumed in the constant power load. Voltage v_{Ce} can be calculated from

$$E - Ri_R - Nv_{Ce} = 0 \tag{13}$$

$$Nv_{Ce}i_R = (1 + \gamma)PN, \tag{14}$$

which are the Kirchoff Voltage Law applied to the circuit of Fig.1(b) and the power consumed in the SMs with γ being the ratio between the power consumed in R_c and P, both assuming that the system operates at the equilibrium point. The solutions are:

$$v_{Ce} = \frac{E \pm \sqrt{E^2 - 4R(1+\gamma)PN}}{2}. \tag{15}$$

N [ ]	E [V]	R [Ω]	C [mF]	P [W]	R_c [KΩ]
2	320	20	2.82	10	2.2889

TABLE I: Circuit parameters for simulation of the stable operation equilibrium point case. R_c is dimensioned so that $\gamma = 1.1$ when the system reaches the equilibrium point.

Assuming the solution from (15) that is closer to E/N leads to an R_c value as in

$$R_c = \frac{v_{Ce}^2}{\gamma P}. \tag{16}$$

It is indispensable to notice that the stability of the equilibrium point does not implies that the system trajectories will reach it, that is, the capacitor voltages will converge to the desired point. That will depend if the initial condition, given by the first stage of the precharge process is inside the attraction domain of this equilibrium point. Therefore, it is advisable to design R_c to consume a power greater than P by some safety margin (this phenomenon is illustrated in the following section). The voltage divergences on the first stage of the precharge process are given mainly due to uncertainties or tolerances on the circuit parameters values. These parameter variations are not taken in account in this work. In future works we plan to study the necessary conditions that guarantee that the system's voltages will converge under these assumptions.

IV. SIMULATION RESULTS

Simulation results for two cases are discussed here assuming a system with two SMs in order to use the phase plane tool to better understand and characterize its dynamics. The cases are: one with a value of R_c that accounts for a stable operating equilibrium point and another with a value that makes the operating equilibrium point unstable. The stability of the system equilibrium points and time responses related to some of the phase plane trajectories are also presented.

A. Case I: Stable Operation Equilibrium Point ($\gamma > 1$)

This simulation uses the parameters given in Table I. These were obtained from an existing lab prototype and the R_c value was specified so that $\gamma = 110\%$ of the power consumed in the APS, assuring the stability condition (11).

The simulation result is seen in Fig. 2. Fig. 2(a) shows the phase plane, where four equilibrium points are present, one of which is the desired operation point, which is stable. The time responses associated with trajectories (b) and (c) from the phase plane Fig. 2(b) and (c) are plotted, showing one case where the capacitor voltages converge and one where they diverge to/from the operation equilibrium point. Since the system is locally stable, the convergence of the trajectories to the equilibrium point depends on whether the initial conditions are inside the attraction domain of the operation equilibrium point. That means for some strong unbalance in the capacitor voltages the system can not converge to the desired operation equilibrium point.

978-1-7281-4181-7/19 $31.00 © 2019 IEEE   804

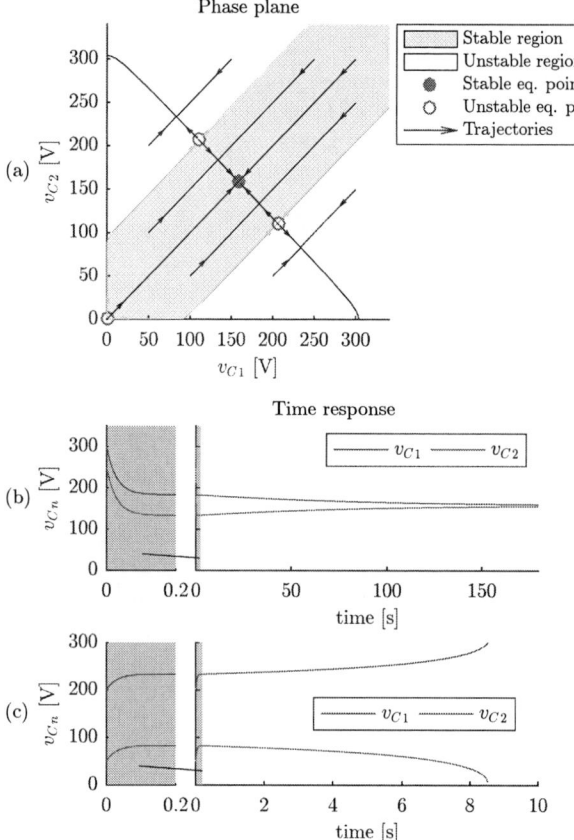

Fig. 2: Phase portrait (a) and time responses (b) and (c) for Case I ($\gamma = 1.1$). In (b) it is illustrated is the time response with initial condition $v_{C0} = (300, 250)$ V. In (c) it is shown the time response with initial condition $v_{C0} = (200, 50)$ V. The gray area in (b) and (c) corresponds to the simulation first time instants.

The stability of the equilibrium points as well as some import information about the system dynamics can be understood looking at the dat on Table II. The equilibrium point $(v_{Ce1}, v_{Ce2}) = (159, 159)$ V is the desired operation point of the system. With both real and negative eigenvalues, it is classified as a nodal sink (stable). Notice that its associated eigenvector v_1 direction is perpendicular to the line $v_{C1} + v_{C2} = E$ and that v_2 is parallel to the same line. Also, that the magnitude of $\lambda_1 = -35.5$ (eigenvalue associated with v_1) has a magnitude order a thousand times bigger than $\lambda_2 = -0.0141$ (eigenvalue associated with v_2). That means that in the neighborhood of this equilibrium point the system trajectories in the phase plane tend to approximate the line $v_{C1} + v_{C2} = E$ about a thousand times faster than they reach $v_{C1} = v_{C2}$. In other words, the sum of the capacitors voltages tends to add up to the voltage source voltage a thousand times faster than the capacitor voltages balance between themselves.

The second and third equilibrium points, respectively

$(v_{Ce1}, v_{Ce2}) = (111, 206)$ V and $(v_{Ce1}, v_{Ce2}) = (206, 111)$ V are classified as saddle points (unstable), since one of its eigenvalues is real and positive, and another is real and negative. The stable manifold that reaches these points are marked in green in Fig. 2 and delimits the attraction domain of the stable operation equilibrium point. This stable manifold is tangent to one of the eigenvectors associated to each equilibrium point. Notice that as in the previous equilibrium point, the magnitude of the eigenvalues associated to a given point differ by a magnitude of thousand between themselves. Also, these points associated eigenvectors that point perpendicularly to the $v_{C1} + v_{C2} = E$ line are associated with the bigger eigenvalue.

The last calculated equilibrium point is $(v_{Ce1}, v_{Ce2}) = (0.627, 0.627)$ V. Its eigenvalues are both real with positive real part so it is classified as an unstable source. The dynamics around this equilibrium point are not valid for the real case because when the capacitor voltages are too low, the APS corresponding to the constant power loads are turned off.

B. Case II: Unstable Operating Equilibrium Point

The circuit is now simulated with the same parameters from the previous case except for the R_c value, which is now dimensioned to a corresponding $\gamma = 0.9$ at the operation equilibrium point and, thus, does not verify the stability condition (5). The circuit parameters are listed in Table III.

Simulation results are seen in Fig. 3. In Fig. 3(a) there are only two equilibrium points, compared to four in the previous one. Also, these two are unstable. The operation point is a saddle point (its eigenvalues are displayed in Table II) and only the trajectories with initial conditions where $v_{C01} = v_{C02}$ will reach this equilibrium point. This is very unlike, if not impossible in practical terms, so the system tends not to reach the desired operation equilibrium point. One example of this situation can be seen in Fig. 3(b), whereas the most likely and realistic situation is shown in Fig. 3(c), where the capacitor voltages tend to diverge. The second equilibrium is a nodal source (unstable) as in the previous case.

The problem of this system operating equilibrium point stability regarding the variation of R_c can be interpreted by looking to the constant power load behavior. As the power consumed is constant, the lower the voltage in a given capacitor is, the more current is drained out of it and vice-versa. That leads to unbalance. The balancing resistance has an opposite effect, i.e., the less the voltage there is in a given capacitor, the less current it drains out of this capacitor. Also, the more voltage in a given capacitor, the less it drains out, contributing that way to the voltage balance. Thus, the smaller R_c is, the more current it drains and the bigger the stable region gets. In all cases, if the unbalance is too severe, that is, the states are out of the stable region, the trajectory does not converge to the desired equilibrium point.

C. Attraction Domain of the Operation Equilibrium Point

The limits of the attraction domain of the stable operating point for several values of γ are computed here based on the parameters from Table IV. The according results are presented

Case	$v_{Ce}[V]$	λ_1	λ_2	v_1	v_2
Stable	(159, 159)	−35.5	−0.0141	$[0.707\ 0.707]^T$	$[-0.707\ 0.707]^T$
	(111, 206)	0.0310	−35.41	$[0.709\ -0.705]^T$	$[0.705\ 0.709]^T$
	(206, 111)	−35.4	0.03101	$[-0.709\ -0.705]^T$	$[0.705\ -0.709]^T$
	(0.627, 0.627)	8970	9000	$[-0.707\ -0.707]^T$	$[-0.707\ 0.707]^T$
Unstable	(159, 159)	−35.4	0.0141	$[0.707\ 0.707]^T$	$[-0.707\ 0.707]^T$
	(0.627, 0.627)	9000	8970	$[0.707\ -0.707]^T$	$[0.707\ 0.707]^T$

TABLE II: Equilibrium points coordinates and associated eigenvalues and eigenvectors from linearized system at these points, for both stable and unstable operation point simulated cases.

$N\,[\,]$	$E\,[V]$	$R\,[\Omega]$	$C\,[mF]$	$P\,[W]$	$R_c\,[K\Omega]$
2	320	20	2.82	10	2.8021

TABLE III: Circuit parameters for simulation of the unstable operation equilibrium point case. R_c is dimensioned so that $\gamma = 0.9$ when the system reaches the equilibrium point.

$N\,[\,]$	$E\,[V]$	$R\,[\Omega]$	$C\,[mF]$	$P\,[W]$
2	320	20	2.82	10

TABLE IV: Circuit parameters for simulation of the attraction domain variation in as a funtion of γ

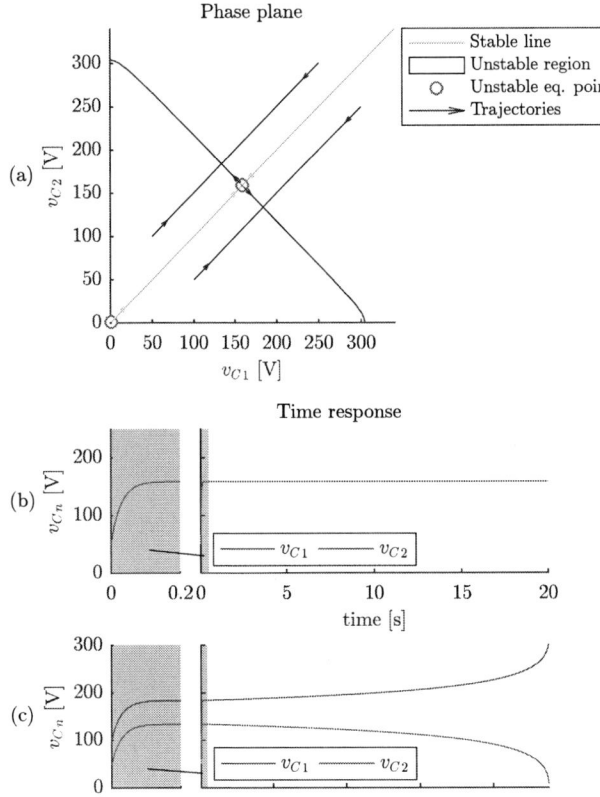

Fig. 3: Phase portrait (a) and time responses (b) and (c) for Case I ($\gamma = 0.9$). In (b) the time response with initial condition $v_{C0} = (50, 50)$ V is plotted. The time response with initial condition $v_{C0} = (100, 50)$ V is in (c). The gray area in (b) and (c) corresponds to the simulation first time instants.

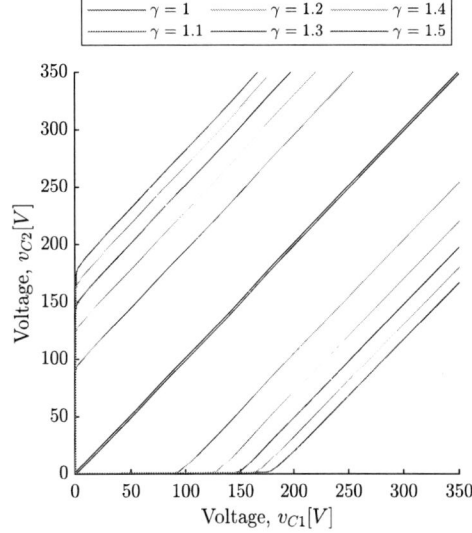

Fig. 4: Boundaries of the attraction Domain of the stable operation equilibrium points for several values of γ

V. EXPERIMENTAL RESULTS

A MMC prototype with 10 submodules per phase has been used to test the passive balancing method analyzed in this work. The precharge circuit is connected to only one of the phases, while the other two remained unconnected. Each APS draws, on average, $P \approx 10.74$ W from the precharge circuit. Table V shows other relevant parameters.

Fig. 5 shows the capacitor voltages obtained by the voltage measurement system embedded in each submodule during the precharge procedures. Before the APS of submodule n turns

$N\,[\,]$	$E\,[V]$	$R\,[\Omega]$	$C\,[mF]$	$P\,[W]$
10	740	$100 \pm 5\%$	$2.82 \pm 20\%$	10.74

TABLE V: Circuit parameters for the experimental case. R_c was tested in open circuit ($R_c \rightarrow \infty$) and $R_c = 375\ \Omega$.

in Fig. 4. It is seen that influences of γ value in the limits of the attraction domain tend to diminish as γ increases.

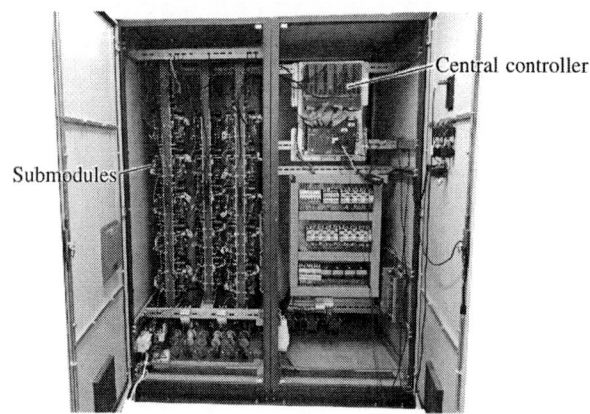

Fig. 6: Photograph of the MMC prototype used in this work.

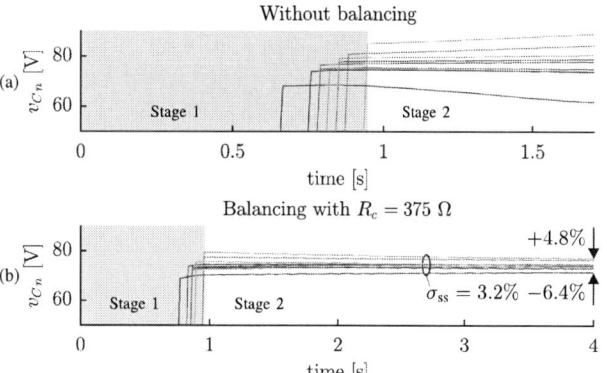

Fig. 5: Experimental results for the (a) precharge without balancing and (b) with passive balancing ($R_c = 375\ \Omega$).

on and its control and communication systems initialize, v_{Cn} is unknown and its value is assumed to be zero in Fig. 5. The capacitance dispersion results in a high spread of the voltages during stage 1, which, together with the APS own parameter spread, results in very different startup times. After all submodules have been initialized, stage 2 begins. For the case where no balancing resistor is used (see Fig. 5(a)), the voltages diverge until one of them reaches a threshold value that trips a minimum or maximum voltage protection.

For the case with passive balancing, $\gamma = 1.2$ is chosen to ensure a sufficiently wide domain of attraction for the stable operating point. This value is approximately achieved with $R_c = 375\ \Omega$. Fig. 5(b) shows that the capacitor voltages converge almost monotonically for this configuration once the system has entered stage 2. The final values spread in the range $[65.3, 73.2]$ V ($[-6.4\%, +4.8\%]$ in respect to average), corresponding to calculated power dissipations in the range $[11.37, 14.3]$ W. Standard deviation in steady state is $\sigma_{ss} = 3.2\%$. This unbalance is likely to be linked with the dispersion in the values of P and R_c and further uncertainties in the measurements, which are obtained from the own submodule

acquisition system based on a micro-controller. Fig. 6 shows a picture of the prototype used in this experiment.

VI. CONCLUSION

A strategy for passive voltage balancing of the MMC capacitors during precharge stage was analyzed is this work. The strategy consists in adding a balancing resistance in parallel with the capacitor of each SM. From this analysis, analytical stability conditions for the system operating point depending on the balancing resistance value were established.

A characterization of the nonlinear dynamics of the system for the case with two SM was obtained also through simulation results. From this study, a guideline to design of the balancing resistance R_c was proposed. Experimental results obtained in a MMC prototype allow to confirm the analytical and simulation findings.

REFERENCES

[1] H. J. Knaak, "Modular multilevel converters and HVDC/FACTS: A success story," in *Proc. 14th European Conf. Power Electronics and Applications*, Aug. 2011, pp. 1–6.

[2] "Sinamics perfect harmony GH150," 2017. [Online]. Available: www.siemens.com/sinamics-perfect-harmony-gh150

[3] G. P. Adam, I. A. Gowaid, S. J. Finney, D. Holliday, and B. W. Williams, "Review of dc–dc converters for multi-terminal HVDC transmission networks," *IET Power Electronics*, vol. 9, no. 2, pp. 281–296, 2016.

[4] J. Lyu, X. Cai, and M. Molinas, "Optimal design of controller parameters for improving the stability of MMC-HVDC for wind farm integration," *IEEE Journal of Emerging and Selected Topics in Power Electronics*, vol. 6, no. 1, pp. 40–53, Mar. 2018.

[5] A. Lachichi, "Modular multilevel converters with integrated batteries energy storage," in *2014 International Conference on Renewable Energy Research and Application (ICRERA)*. Institute of Electrical and Electronics Engineers (IEEE), oct 2014.

[6] T. Geyer, G. Darivianakis, and W. van der Merwe, "Model predictive control of a STATCOM based on a modular multilevel converter in delta configuration," in *Proc. 17th European Conf. Power Electronics and Applications (EPE'15 ECCE-Europe)*, Sep. 2015, pp. 1–10.

[7] D. Cottet, F. Agostini, T. Gradinger, R. Velthuis, B. Wunsch, D. Baumann, W. Gerig, A. Rüetschi, D. Dzung, H. Vefling, A. E. Vallestad, D. Orfanus, R. Indergaard, T. Wien, and W. van der Merwe, "Integration technologies for a medium voltage modular multi-level converter with hot swap capability," in *Proc. IEEE Energy Conversion Congress and Exposition (ECCE)*, Sep. 2015, pp. 4502–4509.

[8] A. Lesnicar and R. Marquardt, "An innovative modular multilevel converter topology suitable for a wide power range," in *Proc. IEEE Bologna Power Tech*, vol. 3, Jun. 2003, pp. 6 pp. Vol.3–.

[9] K. Tian, B. Wu, S. Du, D. Xu, Z. Cheng, and N. R. Zargari, "A simple and cost-effective precharge method for modular multilevel converter by using a low-voltage DC source," *IEEE Transactions on Power Electronics*, pp. 1–1, 2015.

[10] L. Zhang, J. Qin, X. Wu, S. Debnath, and M. Saeedifard, "A generalized precharging strategy for soft startup process of the modular multilevel converter-based HVDC systems," *IEEE Transactions on Industry Applications*, vol. 53, no. 6, pp. 5645–5657, nov 2017.

[11] B. Li, D. Xu, Y. Zhang, R. Yang, G. Wang, W. Wang, and D. Xu, "Closed-loop precharge control of modular multilevel converters during start-up processes," *IEEE Transactions on Power Electronics*, vol. 30, no. 2, pp. 524–531, feb 2015.

[12] L. Luo, Y. Zhang, L. Jia, and N. Yang, "A novel method based on self-power supply control for balancing capacitor static voltage in mmc," *IEEE Trans. Power Electron.*, vol. 33, no. 2, pp. 1038–1049, Feb. 2018.

[13] J. Li, B. Zhao, Q. Song, Y. Huang, and W. Liu, "Minimum voltage tracking balance control based on switched resistor for modular cascaded converter in MVDC distribution grid," *IEEE Transactions on Industrial Electronics*, vol. 63, no. 9, pp. 5437–5441, sep 2016.

978-1-7281-4181-7/19 $31.00 © 2019 IEEE

Design of an Integrated Circuit for LED Driving in Visible Light Communication Applications

Nicolas Parma Rios
NIMO – Modern Lighting Group
Federal University of Juiz de Fora
Juiz de Fora, Brazil
nicolas.parma@engenharia.ufjf.br

Guilherme Márcio Soares
NIMO – Modern Lighting Group
Federal University of Juiz de Fora
Juiz de Fora, Brazil
guilherme.marcio@ufjf.edu.br

Estêvão Coelho Teixeira
NIMO – Modern Lighting Group
Federal University of Juiz de Fora
Juiz de Fora, Brazil
estevao.teixeira@ufjf.edu.br

Abstract—**This paper presents the design of an integrated circuit devised to control an LED driver with visible light communication (VLC) capability. The chip was designed for a standard 0.5-μm CMOS technology and it is applied to control a buck converter. The adopted communication strategy is the variable pulse position modulation (VPPM), which allows the LEDs dimming and the data transmission at rates up to 266.6 kbit/s. The building blocks of the proposed integrated circuit are described, and SPICE simulation results are shown.**

Keywords—**LED lighting, Visible light communication, Integrated Circuit, CMOS, Variable pulse position modulation, Buck converter.**

I. INTRODUCTION

The solid-state lighting uses power light-emitting diodes (LEDs) instead of conventional lighting devices as incandescent, halogen and fluorescent bulbs [1]. Among the several advantages presented by LEDs over other artificial lighting sources, it has been listed their longer lifetime, greater luminous efficiency, the simplicity for implementing the dimming capability (adjustment of brightness) and the absence of heavy metals in their composition, which makes them more suitable to the concept of sustainable buildings and cities.

The LED lighting is an area in continuous development, including research lines like the development of driver circuits with longer lifetime [2] or high power factor [3], more efficient electronic switches [4], and a more recent field that study the data transmission through visible light communication (VLC) [5].

The VLC is defined by the use of lighting devices to transmit data in restricted enviroments [6]. Since the human eye is uncapable to notice the modulation applied to the LEDs to transmit data, the devices can perform both functions of lighting and transmitting data, which must be received by an appropriate type of sensor [7].

The VLC technology is presented as an alternative to radiofrequency communication, which suffers from bandwidth signal and jamming constraints. The fact that the VLC technology was devised to be used in restricted enviroments (since receivers must be in the same enviroment as lighting devices) associated with the non-necessity of a license to operate makes this technology suitable for many applications [8].

The IEEE Standard 802.15.7 had its first edition published in 2011, establishing different protocols for the VLC communication [9]. One of the allowed modulation strategy is the variable pulse position modulation (VPPM), which allows to transmit data at rates up to 266.6 kbits/s (with a 400 kHz optical data rate) and dimming the LEDs simultaneously.

One of the main mechanisms to implement the VPPM modulation in LED drivers is derived from PWM dimming strategies and it is illustrated in Fig 1. This approach is based on the series PWM dimming [10], to which a switch is added in series with the LED string, allowing the possibility to block the current circulation through the devices. In this case the current source is implemented by controlling the current output of a dc-dc buck converter. If the switch is controlled by a VPPM-modulated high frequency logic signal, the luminous flux of the LED will be controlled, allowing the data transmission according to the aforementioned strategy.

Fig. 2 shows the switch S command signals waveforms for three diferent dimming levels, applying the VPPM technique. It is possible to observe that using this strategy, the difference between the transmission of a '0' and '1' is given by the pulse position, which can be at the beginning or at the end of the bit transmission period, T_{VLC}. On the other hand, the brightness will be controlled by the VLC signal duty cycle, $d_{VLC} = T_{on}/T_{VLC}$.

This paper presents the design of an integrated circuit (IC) devised in order to drive a LED string with VLC capability. The chip implements the VLC modulator and the circuit to control the power converter. Besides that, the chip is intended to be applied on a buck converter, operating in discontinuous conduction mode (DCM).

The paper is structured as follows: In Section II, the overall system is described, which can explain the interaction between the chip and the employed power converter. In Section III, the main blocks of the chip are described in detail. Secion IV presents briefly the design parameters of the buck converter. In Section V, some SPICE simulation results are shown, attesting the funcionality of the chip. The main conclusions of the work are presented in Section VI.

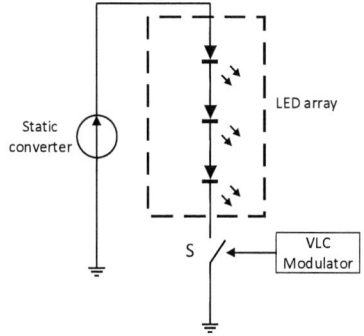

Fig. 1. Simplified circuit of an array of LEDs for VLC communication.

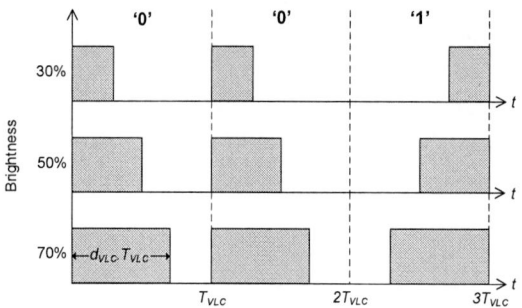

Fig. 2. Command signals using the VPPM for different brightness levels.

II. OVERALL SYSTEM DESCRIPTION

The system comprising both the proposed IC and the buck converter is illustrated in Fig. 3. The dashed line surrounds the main blocks of the chip. It is noticeable, at the converter output, the LED string (represented by an ideal diode in series with the equivalent threshold voltage, V_t, and the dynamic resistance, r_D). The string is in series with the switch S_2, driven by the VPPM signal. A current sensing resistor, R_{CS}, converts the current i_{LED} into a voltage signal, v_{CS}, used by the control circuit, which must be able to maintain the specified maximum value of the current flowing through the LEDs.

The input data are transmitted to the chip by a host computer, or a microcontroller, via serial peripheral interface (SPI). The VPPM modulator generates, from this input package, the VLC signal that drives the switch S_2. This VPPM signal is also used to drive a PWM modulator, that generates the signal v_{MOD} (only internal to the chip), whose mean value is given by

$$V_{MOD} = V_{REF} \cdot d_{VLC} \qquad , \qquad (1)$$

where d_{VLC} is the duty cycle of the VPPM signal. The signal v_{MOD} is applied to the positive input of an operational transconductance amplifier (OTA). On the other hand, the voltage signal v_{CS} is applied to the negative input of the OTA, which is used to build, with the capacitor C_p (external to the chip), an OTA-C integrator. If G_m is the transconductance of the OTA, then the integrator gain, K_i, is given by

$$K_i = \frac{G_m}{C_p} \qquad (2)$$

Therefore, the signal v_{INT} is

$$v_{INT} = \frac{G_m}{C_p} \cdot \int \left(v_{MOD} - v_{CS} \right) dt \qquad (3)$$

Equation (3) shows that a purely integral (I) controller is implemented on the chip. This controller is able to avoid steady-state errors, maintaining the peak of the LEDs current at a constant level. It is important to highlight that by keeping the LED current peak constant, the LED chromatic coordinates also tend to remain in the same level [3].

Fig. 4 can illustrate the relationship between signals v_{MOD} and i_{LED}, for steady-state operation. It can be noted that the mean value of i_{LED} ($= I_{MAX}.d_{VLC}$) will be proportional to V_{MOD} ($= V_{REF}.d_{VLC}$). Hence, external reference V_{REF} is proportional to the peak value of the LED current, I_{MAX}.

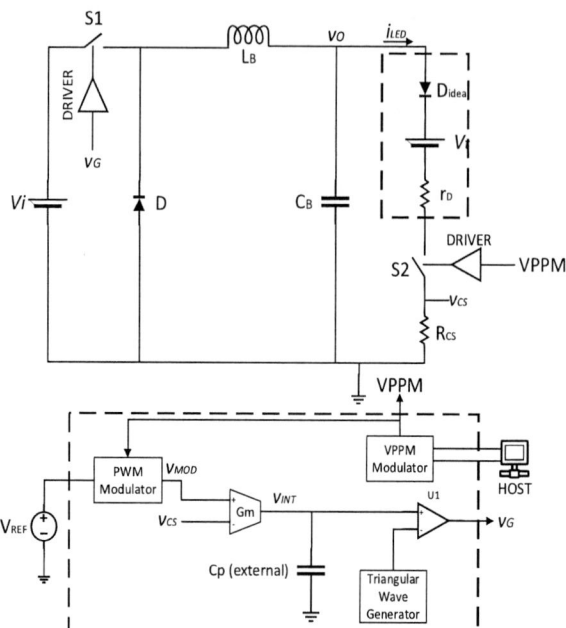

Fig. 3. Diagram illustrating the proposed IC and the buck converter. The chip blocks are surrounded by dashed lines.

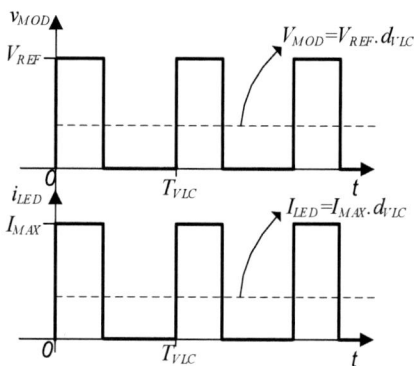

Fig. 4. Relationship between signals v_{MOD} and i_{LED}.

The signal v_{INT} is compared with a triangular waveform, generated by the chip with frequency f_s, which can be defined by means of an external capacitor, as will be shown in Section III. The output of comparator is v_G, a PWM signal that drives the main converter switch, S_1.

III. INTEGRATED CIRCUIT DESCRIPTION

The integrated circuit was designed for the ON Semiconductor C5N process (0.5-µm, 5-V, N-Well, triple metal, two polysilicon layers, standard CMOS technology). By the exposed on the Section II, it is a mixed-signal circuit, designed for a 5-V single supply, and composed by four main blocks: (i) VPPM modulator; (ii) PWM modulator; (iii) operational transconductance amplifier; (iv) triangular waveform generator.

978-1-7281-4181-7/19 $31.00 © 2019 IEEE

A. VPPM Modulator

The VPPM modulator is a digital-only circuit whose block diagram, shown in Fig. 5, is based on [11]. Each block on the figure was designed at transistor level, by using MOSFETs logic gates based on a free library of standard cells, available in [12].

The circuit is composed by one decade counter, one 4-bit comparator, one 4-bit subtractor, one 4-bit 2x1 multiplexer (MUX), two single-bit 2x1 MUX and one D flip-flop.

The modulator operates from an external 4-MHz clock signal, which is applied to the decade counter, whose 4-bit output ($Comp[3..0]$) is compared to the output of the 4-bit 2x1 MUX, $MUX1\_Out[3..0]$. The selector of this MUX is the input serial signal D_{TX}, and its two inputs are given by the desired dimming value, $N_{TR}[3..0]$, or by ('1010'$_2$ − $N_{TR}[3..0]$).

The word $N_{TR}[3..0]$ can range from '0000'$_2$ to '1010'$_2$. However, for actually applying the VPPM, $N_{TR}[3..0]$ must range from '0001'$_2$ to '1001'$_2$ (corresponding to nine discrete dimming levels), since $N_{TR}[3..0]$ = '0000'$_2$ or '1010'$_2$ would maintain the modulator output at a permament '0' or '1', respectively.

The comparator output signal, $Comp\_Out$, is used as a selection signal for a 2x1 MUX, which has as inputs, the bit to be transmitted, D_{TX}, or its complement. In addition, the output of this MUX ($MUX2\_Out$) passes through another 2x1 MUX, which can force a Reset level, if desired (the Reset level is given by the logic value of the external signal EX). In sequence, the signal passes through a D flip-flop, which is employed to eliminate glitches in the output signal, $VPPM\_Out$. An idealized waveform diagram illustrating the operation of the VPPM Modulator, for $N_{TR}[3..0]$ = '0011'$_2$ (30% dimming level) is shown in Fig. 6.

In order to provide test options to the user, four sequence generators were created to be used as the D_{TX} transmission signal: a pseudo-random sequence generator, a bit 1 generator followed by a 0 bit sequence, a generator of bit 0 followed by a sequence of bits 1 and a generator of alternated bits.

B. PWM Modulator

The control circuit needs to limit the maximum value of the current through the LEDs and not its mean value. Therefore, it is necessary to convert the maximum reference value into a mean value according to the dimming level. A modulator driven by the VPPM generated signal is able to make the conversion. Two analog switches, driven by the VPPM signal, as shown in Fig. 7, compose the PWM modulator.

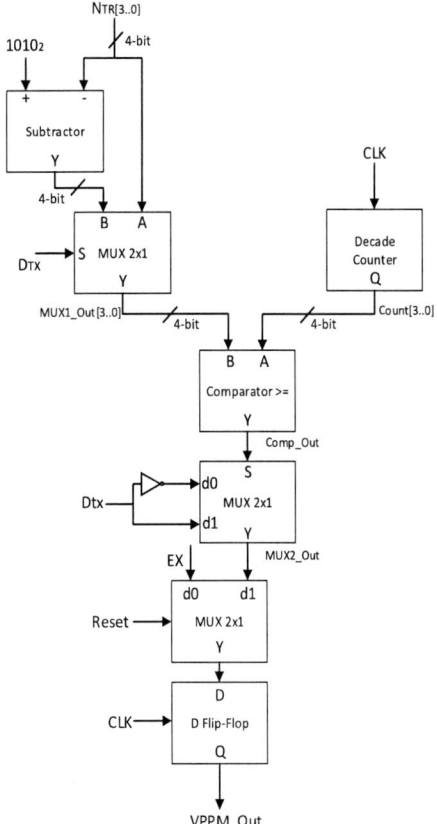

Fig. 5. VPPM Modulator block diagram.

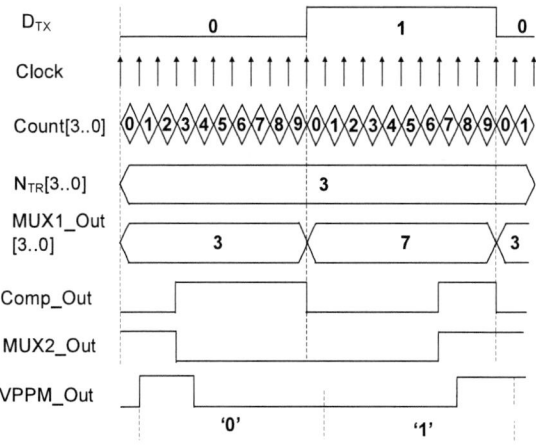

Fig. 6. Idealized waveforms illustrating the operation of the VPPM modulator, for $N_{TR}[3..0]$ = '0011'$_2$ (30% dimming level).

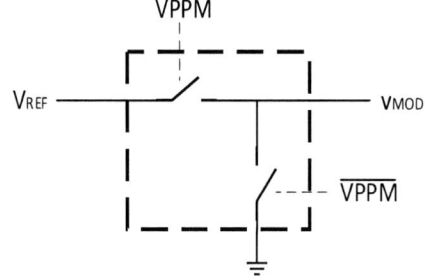

Fig. 7. Structure of the PWM modulator.

C. Transconductance Amplifier

The Operational Transconductance Amplifier (OTA) is designed for a 5-V, single supply operation. The circuit schematic is illustrated in Fig. 8.

It is composed by a PMOS differential pair (M_1 and M_2) with a source degenerating scheme that provides good linearity over a wide range of the input voltages, as exposed in [13]. This needs a resistor of about 50 kΩ, which could be implemented on-chip by using a high-resistance polysilicon layer, available in the C5N process.

The high compliance current mirror (composed by M_3, M_5, M_7 and M_9) and the cascode mirror (built with M_{11}, M_{12}, M_{13} and M_{14}) gives to this OTA a high output impedance (of about 15 MΩ), a desirable feature for implementing the integrator. A transconductance G_m of 95 µA/V was achieved on this project.

D. Triangular Waveform Generator

The integrated circuit design proposes, besides the data modulation for the VLC transmission, the control of the power converter exemplified in Fig. 1. To do this, it will be necessary to implement a triangular wave generator circuit that will be used to generate a PWM signal that drives S_1, in Fig. 3. The simplified circuit used to implement the triangular wave is shown in Fig. 9.

During a period, the external capacitor C_{ext} is charged by the current source I_{CAP} (implemented with PMOS current mirrors), and the rate of the voltage variation in the capacitor is equal to I_{CAP}/C_{ext}. When the voltage reaches 4 V, one of the comparators applies a Set pulse to the SR latch, which enables the *Down* signal. This connects the programmable current source between the node V_{triang} (node which the triangular wave is obtained) and the ground, with current equal to $2I_{CAP}$. Then, the capacitor starts to charge with a current given by $I_{CAP} - 2I_{CAP} = -I_{CAP}$, and the voltage variation rate becomes negative, but with the same module of the previous rate. Therefore, the signal at the positive terminal of the capacitor is the desired triangular wave. When it reaches a value lower than 1 V, the source of value $2I_{CAP}$ is turned off, and the capacitor is recharged with a positive current I_{CAP}. The values of C_{ext} and I_{CAP} are appropriately adjusted to have a triangular signal with approximate frequency of 20 kHz, oscillating between 1 and 4 V.

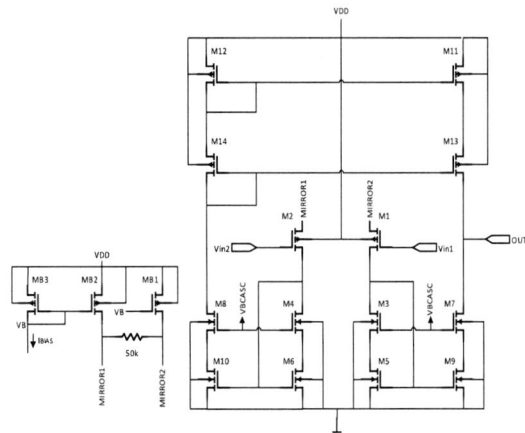

Fig. 8. Schematic of the operational transconductance amplifier.

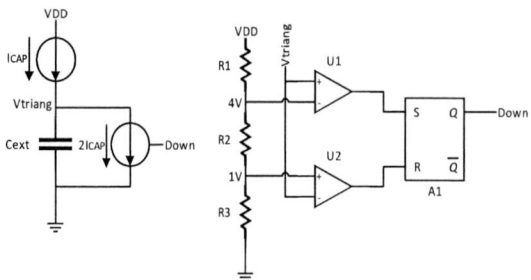

Fig. 9. Triangular wave generator circuit.

IV. DESIGN OF THE POWER CONVERTER

As already mentioned, the buck converter was designed to operate in DCM in order to reduce the size of the inductor. By using the design parameters presented in Table I, the inductor L_b and the capacitor C_b were designed using the methodology presented in [14] and considering that the switch S_2 is permanently on (maximum power), which yields $L_b = 700$ µH and $C_b = 100$ µF. The threshold voltage V_t and dynamic resistance r_D of the LED string were obtained from [15]. It is important to highlight that the design of the buck converter is well known and is not the main scope of this paper, therefore, some design details were suppressed.

TABLE I. DESIGN PARAMETERS

Parameters	Value
Input Voltage, V_I	48 V
Output Voltage, V_o	15.75 V
Nominal current, I_o	350 mA
Output Power, P_o	5.5 W
LED string threshold voltage, V_t	14.5 V
LED string dynamic resistance, r_D	2.57 Ω
Shunt sensing resistor, Rcs	1 Ω
Switching frequency, f_S	20 kHz
Buck converter nominal duty cycle, D_b	0.32
Buck converter conduction parameter, K [14].	0.62
Maximum Output current ripple, ΔI_o	15 mA (< 5%)

V. SIMULATION RESULTS

This Section shows the transient simulation results for two blocks of the chip, as well as a simulation of the system comprising the integrated controller and the buck converter. The LTSpice XVII platform was used for all the simulations, owing to its ability to support the BSIM3 MOSFET models for the C5N CMOS process.

A. VPPM Modulator

The results of the VPPM modulator simulation for $d_{VLC} = 30\%$ are shown in Fig. 10. These results can be compared with the idealized waveforms depicted in Fig. 6. The figure shows an input clock of frequency 4 MHz and an example of sequence with '0-1-0-0' as input data (D_{TX}). The comparator output, the *MUX2_Out* signal and the *VPPM_Out* signal (to be applied at S_2 in Fig. 3) are also shown. The transmission period, T_{VLC}, is 2.5 µs, corresponding to an optical data rate of 400 kHz.

978-1-7281-4181-7/19 $31.00 © 2019 IEEE

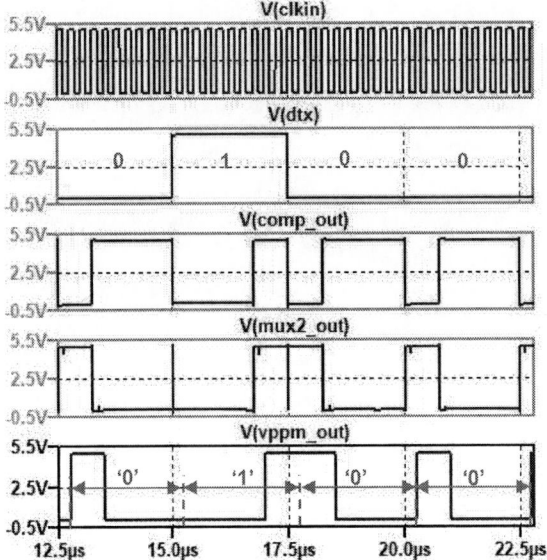

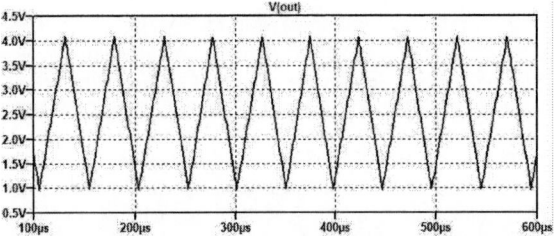

Fig. 11. Transient simulation of the triangular waveform generator output voltage. The achieved frequency was 20.39 kHz.

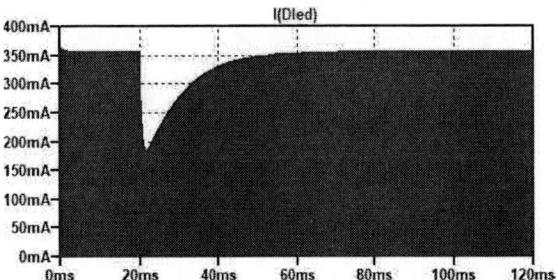

Fig. 12. LED current for the buck converter controlled by the chip, initially with d_{VLC} = 30%. At time t = 20 ms, d_{VLC} is changed to 70%. Reference current I_{REF} = 350 mA.

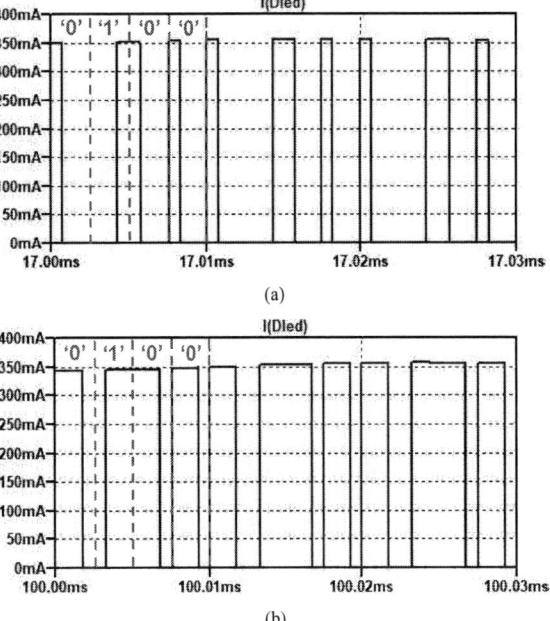

Fig. 10. Simulation results for the VPPM modulator. From top to bottom: input clock (CLK_{IN}), input data sequence (D_{TX}), comparator output ($Comp\_Out$), MUX2 output ($MUX2\_Out$) and output signal ($VPPM\_Out$).

It can be noted that, as expected, the signal *VPPM_Out* is delayed from *MUX2_Out* by one clock cycle, due to the D flip flop (shown in the block diagram of Fig. 5). This does not affect the transmission of the symbols, while, by the other hand, eliminates the glitches (present at *MUX2_Out*) from the output signal, what is a desirable feature.

B. Triangular waveform generator

The transient simulation of the triangular waveform generator shown in Fig. 9 is depicted in Fig. 11, where I_{CAP} = 60 μA and C_{ext} = 470 pF. The achieved frequency was of 20.39 kHz.

C. Simulation of the buck converter with the control circuit

In order to simulate the buck converter with the integrated control circuit, some blocks of the chip were replaced by idealized or parametrized elements. This strategy allowed for the reduction of the simulation time, which would be excessively long if the chip was entirely simulated at transistor level. The simplification was done for the VPPM modulator, which was replaced by a simplified logic using pulse voltage sources to make the example '0-1-0-0' bit sequence. In addition, the comparator was replaced by a parametrized operational amplifier, and the triangular waveform generator replaced by a 20-kHz triangular voltage source. The PWM modulator and the OTA were simulated at transistor level.

Fig. 12 shows the LED current for the circuit operating initially with d_{VLC} = 30% (the initial transient response was skipped by setting initial conditions for voltages at the output capacitor and integrator capacitor). d_{VLC} is then changed to 70% at t = 20 ms. The details for the current on the LED for two intervals of time are plotted in Fig. 13. Considering the obtained OTA transconductance, G_m = 95 μA/V, the chosen capacitance C_p for these simulations was 470 nF, yielding an integral gain $K_i \approx 200$ s^{-1} according to (2). It is worth mentioning that since the design of the control circuit is out of the scope of this work, the integral gain was defined empirically aiming an overdamped response of the closed-loop system, which prevents overshoots in the output current.

Fig. 13. Details on the LED current for the buck converter controlled by the chip, with (a) d_{VLC} = 30%; and (b) d_{VLC} = 70%.

VI. CONCLUSIONS

The visible light communication (VLC) uses lighting devices to transmit data in restricted environments, and is a research area with many applications, especially in cases where radiofrequency communications gets affected by some constraints. One of the modulation strategies is the variable pulse position modulation (VPPM), which can establish data transmission rates up to 266.6 kbps at same time as it allows the dimerization of the LEDs.

This paper presented the design of an integrated circuit devised to apply the VLC to a LED string, using the VPPM. The chip comprises both the VPPM modulator, based on digital standard cells, and an analog control circuit to be applied on a buck converter that feeds the LED arrangement. The control circuit must ensure that a set maximum value of LED current is maintained for different dimerization levels. The main block of the controller is an operational transconductance amplifier (OTA), which implements an OTA-C integrator. Also, a PWM modulator and a triangular waveform generator were described in the text.

Therefore, the proposed IC is a mixed-signal circuit, suitable for fabrication on a 0.5-μm, 5-V, standard CMOS process. The transistor level design of the VPPM modulator, as well as more details on the design of the PWM modulator and the OTA, can be further described in future publications.

The buck converter is intended to operate in discontinuous conduction mode, and its main design parameters were presented. The SPICE simulation results shows the operation of the VPPM modulator and the triangular waveform generator, and corroborate the ability of the chip on applying the VPPM to a LED string and controlling its maximum current, using the adopted converter topology.

ACKNOWLEDGMENT

The authors thank to the PROPP/UFJF for the partial support to the project by means of scientific initiation scholarships granted in the XXX BIC/UFJF – 2017/2018 and XXXI BIC/UFJF – 2018/2019 Programs.

REFERENCES

[1] F. J. Nogueira, l. A. Vitoi, l. H. Gouveia et al., "Street lighting LED luminaires replacing high pressure sodium lamps: Study of case". In: 2014 11th International Conference on Industry Applications (INDUSCON 2014), Juiz de Fora, Brazil, 2014.

[2] G. M. Soares, P. S. Almeida, J. M. Alonso, H. A. C. Braga, "Capacitance minimization in integrated off-line LED drivers using an active-ripple-compensation technique". IEEE Transactions on Power Electronics, vol. 32, pp. 3022-3033, 2017.

[3] P. S. Almeida, G. M. Soares, H. A. C. Braga, "A novel single-switch high power factor LED driver topology with high-frequency PWM dimming capability". Eletrônica de Potência (in portuguese), vol. 18, pp. 855-863, 2013.

[4] D. C. Pereira, W. J. Paula, H. A. C. Braga, "SPICE simulation and evaluation of a GaN-based synchronous full-bridge resonant converter". In: 2016 12th International Conference on Industry Applications (INDUSCON 2016), Curitiba, Brazil, 2016.

[5] M. L. G. Salmento et al., "Application of a flyback converter and variable pulse position modulation for visible light communication". In: XIV Congresso Brasileiro de Eletrônica de Potência (COBEP), Juiz de Fora, Brazil, 2017.

[6] K. Modepalli, L. Parsa, "Dual-purpose offline LED driver for illumination and visible light communication". IEEE Transactions on Industry Applications, vol. 51, n. 1, pp.406-419, 2015.

[7] J. J. Souza, S. L. Stevan, M. A. C. Pompermaier et al., "Project of a communication system by visible light comunication (VLC) based on LED lighting". Iberoamerican Journal of Applied Computing, vol. 3, n. 3, pp. 20-29, 2013.

[8] F. Che, L. Wu, B. Hussain et al., "A fully integrated IEEE 802.15.7 visible light communication transmitter with on-chip 8-W 85% efficiency boost LED driver". Journal of Lightwave Technology, vol. 34, n. 10, pp. 2419-2430, 2016.

[9] Institute of Electrical and Electronics Engineers (IEEE), *IEEE Standard for Local and Metropolitan Area Networks–Part 5.7: Short-Range Wireless Optical Communication Using Visible Light*, IEEE Std. 802.15.7-2011, pp. 1–309, 2011.

[10] D. Gacio et al., "PWM series dimming for slow-dynamics HPF LED drivers: The high-frequency approach". IEEE Transactions on Industrial Electronics, vol. 59, n. 4, pp. 1717-1727, 2011.

[11] J. Jeong, S. Lim, I. Jang, M. Kim, "Novel architecture for efficient implementation of dimmable VPPM in VLC lightings". ETRI Journal, vol. 36, n. 6, pp. 905-912, December 2014. Available in: <http://dx.doi.org/10.4218/etrij.14.0114.0396>. Accessed in: 04/09/2018.

[12] Static Free Soft, "CMOS Standard Cells". Available in: <https://staticfreesoft.com/productsLibraries.html>. Accessed in: 12/09/2018.

[13] P. Chu (editor), *Advances in Solid State Technologies*. InTechOpen, 2010. Available in: < https://www.intechopen.com/books/advances-in-solid-state-circuit-technologies>.

[14] R. W. Erickson, D. Maksimovic, *Fundamentals of Power Electronics*. Springer Science & Business Media, 2007.

[15] P. S. Almeida, A. L. C. Mello, V. M. Alguquerque, G. M. Soares, D. P. Pinto, H. A. C. Braga, "Improved state-space averaged representation of LED drivers considering the dynamic model of the load". In: XII Congresso Brasileiro de Eletrônica de Potência (COBEP), Gramado, Brazil, 2013.

Connection Time in Modbus/TLS for Secure Communications on Photovoltaic Systems

Matheus K. Ferst, Hugo F. M. de Figueiredo and Gustavo W. Denardin

Postgraduate Program in Electrical Engineering (PPGEE)
Federal University of Technology - Paraná (UTFPR)
Pato Branco, Brazil
Email: matheus.ferst@gmail.com, hugoffigueiredo1@gmail.com,
gustavo@utfpr.edu.br

Abstract—**The growth of Renewable Distributed Generations (RDG) presents both, a solution to common problems of traditional energy sources - such as the environmental impact and losses in transmission and distribution - and a challenge to appropriate control a large number of Distributed Energy Resources (DERs). The feasibility of such control depends on adequate communication protocols that, among other features, need to address the security requirements of DER applications. Modbus/TLS was presented in previous work as a possible protocol for direct DER communications and analyzed with respect to bulk data transfer. This paper extends such analysis, evaluating the connection time for multiple key exchange algorithms configurations, discussing possible impacts and proposing alternatives for environments where the observed connection times are prohibitive.**

Index Terms—**Modbus, TLS, SunSpec, Photovoltaic, Microgrid**

I. INTRODUCTION

The ever-growing demand for energy [1], the losses related to transmission and distribution [2], and the environmental impact of traditional energy sources [3] have made the Renewable Distributed Generations (RDGs) a suitable option for the power grid expansion [4]. Among these options, photovoltaic (PV) systems are an attractive solution for small scale installations [5].

However, additional problems arise with the increase of RDGs connected to the grid, such as voltage rise, harmonic current emissions, and false islanding [6]. The overcome of such obstacles involve the control of a large number of Distributed Energy Resources (DERs), which additionally provides an opportunity to increase the power quality and reliability of the grid [7].

The SunSpec Alliance was founded in 2009, aiming to accelerate the growth of Distributed Energy Resources (DERs) in the PV field [8]. With the premise that the energy industry would follow the same development of internet technology, the power generation would be distributed across the grid, to provide efficiency, resilience, and security to the system operation. Development of de facto standards, the use of open

This work was funded by the Research and Development project PD 2866-0468/2017, granted by the Brazilian Electricity Regulatory Agency (ANEEL) and Companhia Paranaense de Energia (COPEL).

source licensing, cooperative development, and coordination with official standardization entities would provide interoperability, scalability and sustainable growth [9].

SunSpec had published 14 specification and 52 information models until 2016, covering most aspects of DERs, including device models and field bus protocols [9]. Furthermore, their workgroups have developed reference implementations and other documents, such as best practices guides.

In particular, [10] discusses the security requirements of DER systems - such as availability, safety, confidentially of client sensible data, non-repudiation, and accountability – and their placement in the SunSpec Reference Architecture, which is shown in a simplified form in Figure 1. Additionally, application protocols involved in DER operation are discussed,

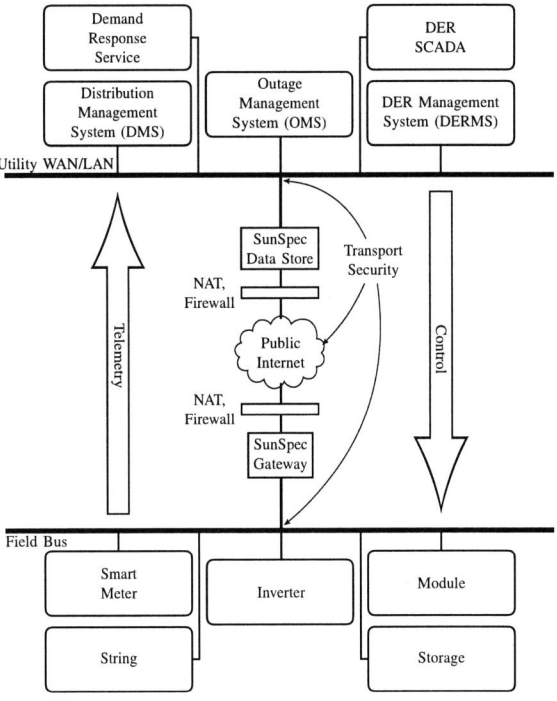

Fig. 1. Simplified representation of the SunSpec Reference Architecture.

evincing the Modbus Protocol as the "weak link" in the control architecture.

The Modbus protocol was developed by Modicon in 1979 and has become a *de facto* standard for serial communications in industry [11]. It was initially created for asynchronous serial transmissions and posteriorly extended to packet-switched networks, such as TCP/IP.

However, this enhances also granted a new avenue for attacks and opened a whole new level of vulnerabilities to the protocol. The original Modbus serial transmission has intrinsic isolation between the enterprise level and the control level of an industry network [12] and therefore ensuring the physical security of the perimeter where the system operates should be a sufficient measure to mitigate the protocol flaws [10]. The TCP version, on the other hand, enables remote exploitation and, therefore, additional measures need to be taken to ensure operational safety.

As a solution, [10] suggests protecting the transport layer of Modbus with the Transport Layer Security (TLS) protocol, creating a Modbus/TLS protocol. A previous work [13] explores the Modbus/TLS alternative and analyses the impact of TLS on bulk data transfer taking average latency and goodput as evaluation metrics. More recently, a Modbus/TCP Security Protocol specification was issued by Modbus Organization, standardizing the minimal TLS version and required cipher suites and TLS extensions [14].

TLS is a protocol that provides communications security over the Internet [15]. It was created by Netscape Communications in 1993 aiming mainly the security of the World Wide Web (WWW) [16]. However, with its evolution, the protocol has started to effectively describe a framework for the development and deployment of cryptographic protocols [17].

This paper assess the connection time for the same implementation. Although the previous analysis should be enough to characterize the protocol operation in normal circumstances, the connection time should evince possible new problems while operating in unstable networks, which are prone to frequent reconnections.

II. SunSpec Modbus

Initially designed for asynchronous serial lines (RS-232 and RS-485) [18], the Modbus protocol was developed by Modicon in 1979 as an open, royalty-free protocol [19]. The company, founded in 1968, had been part of the Gould Electronics since 1977 and was sold to the German company AEG in 1989. In 1994 AEG and the French Groupe Schneider combined to create the AEG Schneider Automation, which became just Groupe Schneider in 1996 and was finally renamed to Schneider Electric in 1999 [20].

In the same year, the TCP/IP version of the protocol was developed, seeking the benefits of Ethernet standard such as scalability, data transmission at rates of 10/100 Mbps and the simple integration with other systems [19]. In 2004 the protocol was transferred to the Modbus Organization,

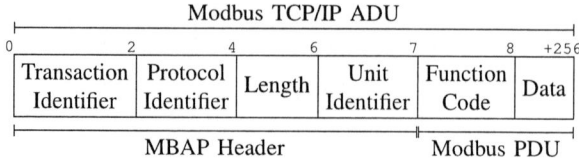

Fig. 2. Modbus TCP Frame

an independent, member-based, nonprofit, organization that manages the protocol evolution [21].

The Modbus/TCP frame is shown in Figure 2. The frame starts with a Transaction Identifier used for transaction pairing. The response frame has the same transaction number of its request. The Protocol Identifier is used to multiplex Modbus/TCP and Modbus Plus, the Schneider's token passing proprietary protocol. This field is always zero for Modbus/TCP [22].

The Length field delimits the end of the frame, since TCP is a stream-based protocol and, thus, does not preserve message boundaries. The Unit Identifier is used to enable communications with multiple slaves through devices that use a single IP address, such as gateways, bridges, and proxies [22].

Finally, the Function Code and Data fields compose the de Modbus Protocol Data Unit (PDU), that is shared between the serials and TCP versions of the protocol and implements the Modbus Application Protocol [11].

The application layer of the protocol is defined by the Modbus Application Protocol. It specifies a request-reply protocol, where slave devices offer services identified by function codes in the range of 1-127. The Data Model of Modbus defines four tables with different content sizes and access types:

- The Discrete Inputs table has bit length entries that are read-only;
- The Coils table, also known as Bit table, has bit-sized elements with read-write access;
- The Input Register table has 16-bits read-only words and entries;
- The Register table, also referenced as Holding Register table, has 16-bits elements with read-write access.

Each table has up to 65 535 entries that are accessed and manipulated with specifics function codes [11].

The Modbus protocol does not define the meaning of each element of a certain table and, thus, the Master Terminal Unit (MTU) needs previous knowledge of the information mapping to interact with the Remote Terminal Unit (RTU). This mapping is sometimes referenced as the Modbus Profile of the device [23].

The SunSpec Alliance defines Information Models to compose the Modbus Profile of PV applications devices [23]. Their models define not only the data mapping but also new data types structured over 16-bits registers, such as strings, scale factors, and float point numbers [24].

978-1-7281-4181-7/19 $31.00 © 2019 IEEE

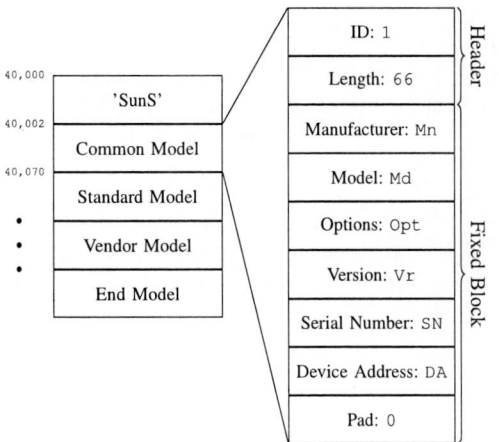

Fig. 3. SunSpec Information Models mapped in Modbus Holding Registers

A SunSpec Device Definition is a collection of two or more Information Models. Each Model starts with a uniquely, well-known identifier and its length, followed by one or more Blocks. A Block is composed of a collection of Points, that are values encoded according to the SunSpec data types. Fixed Blocks contain points that are unique in the model while repeating Blocks could have multiple instances of its point collections [24].

The Model set of a device is mapped in the Holding Register map as shown in Figure 3. The mapping starts preferably at 40 000, and alternatively at 50 000 or 0. The constants 0x5375 and 0x6e53 (ASCII values for the string 'SunS') appears at the beginning of the map to identify a SunSpec compatible device. They are followed by the Common Model and then by any other standard model. In the sequence, vendor-specific models are placed. The final of the mapping is denoted by the End Model, that consist of a zero-length model with identification field as 0xFFFF [24].

In a typical application, the SunSpec Gateway shown in Figure 1 or any other kind of SunSpec Modbus MTU would be connected to the field bus or local area network (LAN) of an asset with multiple SunSpec Modbus RTUs. For serial buses, the MTU would broadcast the Modbus command "Report Server ID" (0x11) to discover the RTU addresses, but for Modbus/TCP the MTU needs either previous knowledge of the present slaves or other protocol to discover them.

Once a new RTU is discovered, the MTU will scan the possible base addresses of the model mapping, searching for the 'SunS' string. If found, the following registers are read, expecting the Common Model header. The length of the model is added to the address of the last read register to found the next model. The process is repeated until the End Model is detected. In the end, the master will have the Modbus Profile of the scanned slave. With this information, the MTU could start to issue commands and polling this device for information.

If the network where such operation takes place present instabilities, the MTU may lose the connection with the RTU

after a number of undelivered or damaged packets. At this point, the slave cannot provide any service, and the master cannot control or observe the evolution of the slave states. The operation will be compromised if the RTU cannot act autonomously, and partial unavailability would happen until the MTU reconnects. After the reconnection, the slave device usually is rescanned, which increases the downtime generated by the instability.

If the source of such instability is unknown, intrinsic to the environment or otherwise unaddressable, reconnections could become frequent, and the time need for the connection establishment becomes a critical factor to enable the protocol deployment in such environment.

III. TRANSPORT LAYER SECURITY

The initially called Security Socket Layer (SSL) protocol was created by Netscape Communication in 1993 to address the security of online transactions in the World Wide Web (WWW) with a trade-off between ease of use and end-to-end security. The Internet Engineering Task Force (IETF) created the TLS Work Group to standardize the protocol in 1996. The final specification was released in 1999, with few differences from the last version of SSL and renaming the protocol to Transport Layer Security (TLS) [16].

TLS can be characterized as a stateful, connection-oriented, client-server protocol that aims to provide communications security over the Internet [15]. Moreover, the protocol evolution has granted flexibility that turned it in a framework for cryptographic protocols deployment [17].

Its operation can be placed between the Application and Transport layers of the Internet Model [16] or in the sixth layer of the OSI Model [17]. Additionally, the protocol is subdivided in two layer and five subprotocols, as shown in Figure 4.

A. TLS Handshake

A TLS connection starts with the TLS Handshake, a process where the session parameters are negotiated between the communicating pairs [17]. These parameters are comprised of the Session Identifier, the Peer Certificate, the compression

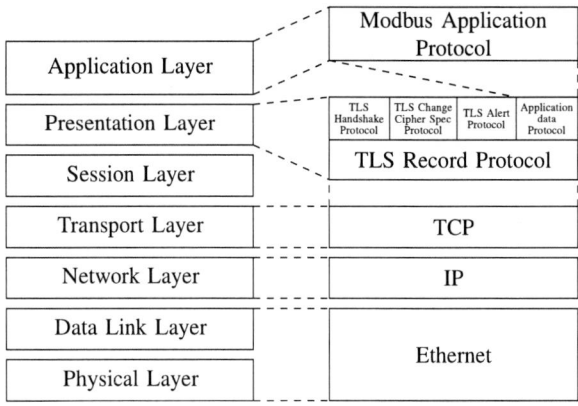

Fig. 4. Modbus/TLS in the OSI Model.

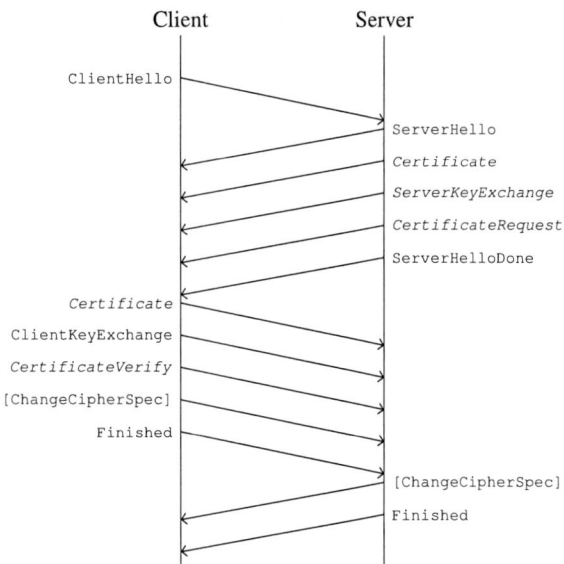

Fig. 5. TLS Handshake Process

method, the Cipher Spec, the Master Secret, and if the session is resumable or not [15].

The underlying TLS Record Protocol uses such information to properly select the algorithms and keys to compress, authenticate, and encrypt the higher layers data. At the beginning of the connection, the identity function is used for these operations to enable TLS Record operation during the initial handshake [16], [17].

An overview of the Handshake process is shown in Figure 5. Emphasized names represent optional messages and messages between brackets belongs to the TLS Change Cipher Spec Protocol. The whole processes take six to thirteen messages, depending on the client and server configurations [17].

The `ClientHello` message is first issued by the client to initiate the handshake [16]. Its content informs the server of the client capabilities and preferences, such as the best protocol version supported and preference-ordered lists of supported cipher suites, compression methods, and extensions. Also, a 32-bytes random value is provided, which is the random data contribution of the client to the handshake [15]. Finally, if the client wishes to resume a previous session, this message also carries the corresponding session identifier [17].

The server then responds with a `ServerHello` message, composed by the its choice among the capabilities offered by the client [16]. If the selected cipher suite does use authentication and the authentication method relies on certificates, the server proceed with the handshake sending a `Certificate` message [17], that contains its X.509 certificate chain encoded in ASN.1 DER [15].

If the key exchange algorithm in use needs more information than the provided by the certificates, the server sends a `ServerKeyExchange` message with the re-

quired additional data [16]. The server can then send a `CertificateRequest` if the client authentication is desired [15]. The `ServerHelloDone` message then signals that the server has sent all intended messages.

If the server requested client authentication, the client continues the handshake process by issuing its `Certificate` message. Then again, according to the key exchange method in use, a correspondent `ClientKeyExchange` may be sent to complete the required client information [16].

If the client sent a `Certificate` message, it must send a `CertificateVerify` message to prove the possession of the corresponding private key. The client sends a `ChangeCipherSpec` message indicating the possession of enough information to switch to an encrypted communication [15]. This message is, in fact, not part of the TLS Handshake Protocol, but the only message of the TLS Change Cipher Spec Protocol [16].

The last message sent by the client is of `Finished` type. It is the first encrypted message exchanged between the communicating peers with the new agreed session parameters and contains a hash of all exchanged messages until this point [17].

The server then responds with a `ChangeCipherSpec` and its `Finished` message. PRF determinism allows both sides of the communication to derive the expected output of the other side `Finished` message. Since this message is encrypted and authenticated with the negotiated parameters, its content can be used to check the success of the handshake process [16], [17].

IV. DEVELOPMENT

Proceeding from the same setup described in the previous paper [13], an STM32F749 microcontroller running FreeRTOS with the FreeModbus implementation of the Modbus protocol was used as the server and a Raspberry Pi 3B running Raspbian GNU/Linux with the libmodbus library was the client. Figure 6 shows the network which the tests were run.

The mbedTLS library was used as the TLS implementation. The PI-MBUS-300 [25] based test suite was adapted to provide the connection timing information, with calls to `clock_gettime` before and after the call to `modbus_connect`.

Twelve certificates sets were generated to the tests: RSA certificates with 1024, 2048, and 4096-bits length keys [26] and Elliptic-Curve Cryptography (ECC) certificates with the brainpool [27], [28] curves P256r1, P384r1, and P512r1 and

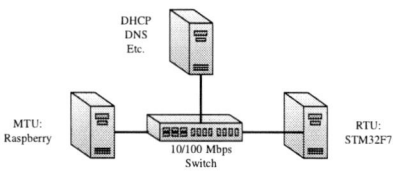

Fig. 6. Network environment employed for the tests

978-1-7281-4181-7/19 $31.00 © 2019 IEEE

secp [29], [30] curves 192k1, 224k1 224r1, 256k1, 384r1, and 521r1.

V. RESULTS

In all, 79 cipher suites were tested. The results are divided according to the type of security offered: Ephemeral key exchange algorithms provide forward secrecy, but require the generation of temporary keys at each connection [17], while non-ephemeral alternatives are faster, but if the certificate's private key is discovered, all previous session keys are compromised [16]. The connection times of non-ephemeral and ephemeral algorithms with each certificate set are shown in Figures 7 and 8, respectively.

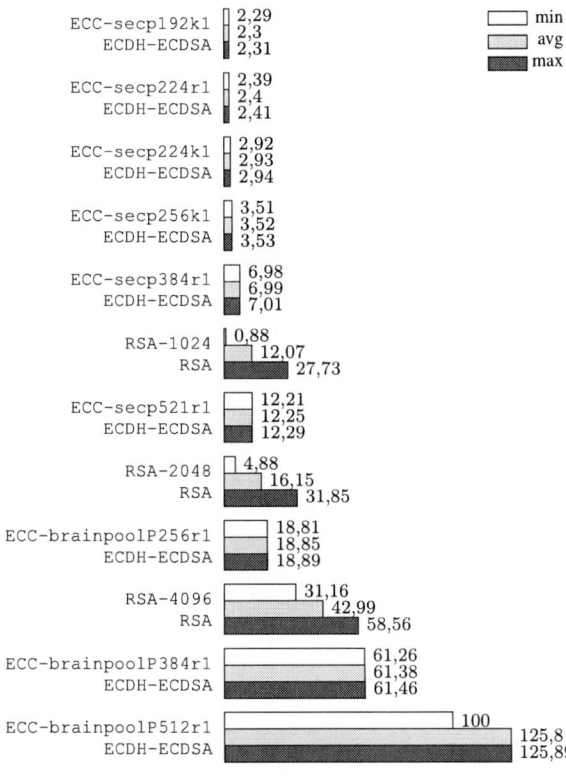

Fig. 7. Observed connection time (s) for non-ephemeral key exchange algorithms

As expected, elliptic key cryptography outperforms RSA with the secp curve. The brainpool curve, however, uses a pseudorandom method to select the primes for the base field, preventing the use of some fast modular arithmetic methods in favor of better security [27].

The key generation in ephemeral alternatives triples the connection times for secp192k1 and secp224r1. The secp224k1 and secp256k1 presents a 2.5 times increase, while the secp384r1 and secp521r1 shows 1.7 and 1.4 increase, respectively. For brainpool, the additional overhead is less significant

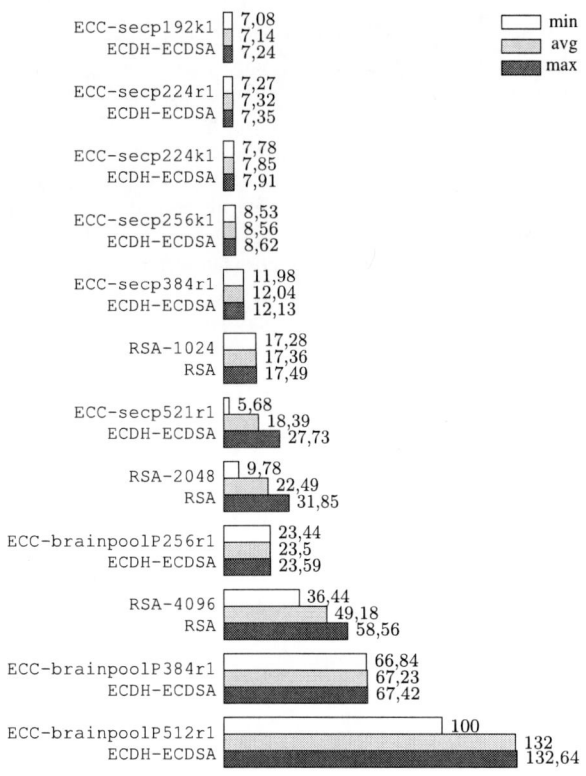

Fig. 8. Observed connection time (s) for ephemeral key exchange algorithms

compared with the original connection time, but yet present an increase of 4.6, 5.8, and 6.2 seconds for the P256r1, P384r1, and P512r1 curves. Finally, the RSA ephemeral options present 1.5, 1.4 and 1.1 times slower connection time.

As discussed in [13], channel encryption is not always required. In this kind of application and others that do not prioritize privacy but still require authenticity, forward secrecy may be neglected for better connection times.

In cases where these reconnection delays still unacceptable, the session resumption feature of the handshake protocol presents an alternative to reduce de connection time. This option, although, require the server the store thenegotiated session parameters. In Modbus/TLS applications the embedded system implements the server and, therefore, memory constraints may inviabilize the approach.

[31] presents a TLS extension that enables session resumption without a server-side state, possibly circumventing these restrictions. Another option is the use of Pre-Shared Keys (PSK) cipher suites to reduce the number of asymmetric primitives operations [32] and, therefore, reduce the connection time. This alternative also avoids the complexity of deployment of a Public Key Infrastructure (PKI), but add problems related to share and revoke symmetric keys between the communication pair.

VI. CONCLUSION

The expansion of grid-connected RDGs to address problems, such the environmental impact of traditional energy sources and the losses related to transmission and distribution, requires the proper control of a huge number of devices to avoid rising of other problems such as voltage rise and power quality degradation. To enable such control, proper standard and communication protocols should be developed and deployed.

These protocols need to consider the security requirements of DER applications. Addressing Modbus, the "weak link" of the control architecture, [13] analyzed the impact of a TLS based Modbus implementation on bulk data transfer. While this should be enough to characterize Modbus/TLS in normal operation, abnormal situations such as an unstable network may cause multiple reconnections, thereby reducing the link goodput.

In such cases, lower connection times becomes relevant to enable the use of the proposed protocol. This paper takes the same network environment to assess this characteristic in multiple key exchange algorithms configurations. The obtained results show connection time varying from 2.3 s for the fastest secp curve to more than two minutes for the brainpoolP512r1 curve in non-ephemeral algorithms and from 7.1 to 132 s for the ephemeral versions.

If these times are unacceptable for certain environments, other solutions are presented and discussed, such as the TLS session resumption and the use of PSK. Future works should apply a deeper analysis for these alternatives to determine their connection time and other significant characteristics.

ACKNOWLEDGMENT

This work was funded by the Research and Development project PD 2866-0468/2017, granted by the Brazilian Electricity Regulatory Agency (ANEEL) and Companhia Paranaense de Energia (COPEL). The authors also would like to thank to FINEP, SETI, CNPq, Fundação Araucária, CAPES and UTFPR for additional funding.

REFERENCES

[1] B. Bezerra, S. Mocarquer, L. Barroso, and H. Rudnick, "Expansion pressure: Energy challenges in brazil and chile," *IEEE Power and Energy Magazine*, vol. 10, no. 3, pp. 48–58, May 2012.

[2] B. M. Eid, N. A. Rahim, J. Selvaraj, and A. H. El Khateb, "Control methods and objectives for electronically coupled distributed energy resources in microgrids: A review," *IEEE Systems Journal*, vol. 10, no. 2, pp. 446–458, June 2016.

[3] F. Blaabjerg, Y. Yang, D. Yang, and X. Wang, "Distributed power-generation systems and protection," *Proceedings of the IEEE*, vol. 105, no. 7, pp. 1311–1331, July 2017.

[4] J. Feng, B. Zeng, D. Zhao, G. Wu, Z. Liu, and J. Zhang, "Evaluating demand response impacts on capacity credit of renewable distributed generation in smart distribution systems," *IEEE Access*, vol. 6, pp. 14 307–14 317, 2018.

[5] P. IEA-PVPS, "PVPS report snapshot of global pv 1992-2013," *Report IEA-PVPS T1-24*, 2014.

[6] E. Demirok, P. C. González, K. H. B. Frederiksen, D. Sera, P. Rodriguez, and R. Teodorescu, "Local reactive power control methods for overvoltage prevention of distributed solar inverters in low-voltage grids," *IEEE Journal of Photovoltaics*, vol. 1, no. 2, pp. 174–182, Oct 2011.

[7] N. Hatziargyriou, H. Asano, R. Iravani, and C. Marnay, "Microgrids," *IEEE Power and Energy Magazine*, vol. 5, no. 4, pp. 78–94, July 2007.

[8] S. Alliance, "About the sunspec alliance," SunSpec Alliance, 2017, acesso em: 05 jun. 2019. [Online]. Available: https://sunspec.org/sunspec-about/

[9] ——, "Sunspec alliance: Information standards for distributed energy: Corporate backgrounder," *SunSpec Alliance*, 2016.

[10] J. Blair, J. Nunneley, R. Kaisler, B. Fox, F. Nagy, B. Randle, L. Linse, and T. Tansy, "Security recommendations," Best Practices, SunSpec DER Cybersecurity Workgroup, Best Practices, June 2013.

[11] I. Modbus, "Modbus application protocol specification v1.1b," *North Grafton, Massachusetts (www.modbus.org/specs.php)*, 2006.

[12] I. N. Fovino, A. Carcano, M. Masera, and A. Trombetaa, "Design and implementation of a secure modbus protocol," *Critical Infrastructure Protection*, vol. 3, pp. 83–96, 2009.

[13] M. K. Ferst, H. F. M. de Figueiredo, G. Denardin, and J. Lopes, "Implementation of secure communication with modbus and transport layer security protocols," in *2018 13th IEEE International Conference on Industry Applications (INDUSCON)*, Nov 2018, pp. 155–162.

[14] Modbus Organization, "Modbus/TCP security protocol specification," Modbus Organization, Tech. Rep., jul 2018.

[15] T. Dierks and E. Rescorla, "The transport layer security (tls) protocol version 1.2," Internet Requests for Comments, RFC Editor, RFC 5246, August 2008, http://www.rfc-editor.org/rfc/rfc5246.txt. [Online]. Available: http://www.rfc-editor.org/rfc/rfc5246.txt

[16] R. Oppliger, *SSL and TLS: Theory and Practice*. Artech House, 2016.

[17] I. Ristic, *Bulletproof SSL and TLS: Understanding and Deploying SSL/TLS and PKI to Secure Servers and Web Applications*. Feisty Duck, 2014.

[18] Modbus Org, "Modbus over serial line specification & implementation guide v1.02."

[19] Modbus FAQ, "About the protocol," *FAQ. Modbus Organization inc.*, 2017.

[20] Schneider Eletric, "Modicon is now schneider electric," *Our Brand History*, 2019.

[21] Modbus FAQ, "About the modbus organization," *FAQ. Modbus Organization inc.*, 2017.

[22] Schneider Automation, "Modbus messaging on tcp/ip implementation guide v1.0b," *MODBUS Organization, last accessed June*, p. 46, 2006.

[23] SMA, "Sunspec modbus interface for sunny boy / sunny tripower," SMA Solar Technology AG., Tech. Rep., 2015.

[24] S. Alliance, "Sunspec informatio model specifications," *SunSpec Alliance*, 2015.

[25] I. Modicon, "Modicon modbus protocol reference guide," PI–MBUS–300 Rev. J, Tech. Rep., 1996.

[26] K. Moriarty, B. Kaliski, J. Jonsson, and A. Rusch, "PKCS #1: RSA Cryptography Specifications Version 2.2," RFC 8017, Nov. 2016. [Online]. Available: https://rfc-editor.org/rfc/rfc8017.txt

[27] M. Lochter and J. Merkle, "Elliptic curve cryptography (ecc) brainpool standard curves and curve generation," Internet Requests for Comments, RFC Editor, RFC 5639, March 2010, http://www.rfc-editor.org/rfc/rfc5639.txt. [Online]. Available: http://www.rfc-editor.org/rfc/rfc5639.txt

[28] Y. Nir, S. Josefsson, and M. Pegourie-Gonnard, "Elliptic curve cryptography (ecc) cipher suites for transport layer security (tls) versions 1.2 and earlier," Internet Requests for Comments, RFC Editor, RFC 8422, August 2018, http://www.rfc-editor.org/rfc/rfc8422.txt. [Online]. Available: http://www.rfc-editor.org/rfc/rfc8422.txt

[29] Certicom Research, "Sec 2: Recommended elliptic curve domain parameters," Standards for Efficient Cryptography 2 (SEC2), Standard, Jan. 2010. [Online]. Available: http://www.secg.org/sec2-v2.pdf

[30] B. Moeller, N. Bolyard, V. Gupta, S. Blake-Wilson, and C. Hawk, "Elliptic Curve Cryptography (ECC) Cipher Suites for Transport Layer Security (TLS)," RFC 4492, May 2006. [Online]. Available: https://rfc-editor.org/rfc/rfc4492.txt

[31] P. Eronen, H. Tschofenig, H. Zhou, and J. A. Salowey, "Transport Layer Security (TLS) Session Resumption without Server-Side State," RFC 5077, Jan. 2008. [Online]. Available: https://rfc-editor.org/rfc/rfc5077.txt

[32] H. Tschofenig and P. Eronen, "Pre-Shared Key Ciphersuites for Transport Layer Security (TLS)," RFC 4279, Dec. 2005. [Online]. Available: https://rfc-editor.org/rfc/rfc4279.txt

Modeling and Simulation of the Battery Energy Storage System for Analysis Impact in the Electrical Grid.

Carolina A. Caldeira, Henrique R. Schlickmann, Antonio D. D. de Almeida, Lucas V. Hartmann,
Camila S. Gehrke, Fabiano Salvadori and Gilielson F. da Paz

Federal University of Paraíba (UFPB)

João Pessoa - PB - Brazil

e-mails: (carolina.caldeira)(henrique.raldi)(antonio.almeida)(lucas.hartmann)(camila)(salvadori.fabiano)(gilielson)@cear.ufpb.br

Abstract—**With increasing use of intermittent renewable energy sources, energy storage is needed to maintain the balance between demand and supply. The renewable energy sources, e.g. solar and wind energy sources, are characterized by their intermittent generation, causing fluctuations in power generation, and, similarly, demand may vary. There may be fluctuations in power generation, and, similarly, demand may vary. Then, for these new sources become completely reliable as primary energy sources, energy storage is a crucial factor. This work uses real-time simulation to analyze the impact of battery-based energy storage systems on electrical systems. The simulator used is the OPAL-RT/5707™ real-time simulator, from OPAL-RT Technologies company. The simulated system consists of a three-phase inverter connected to a BESS (battery energy storage system) and to the electrical grid with variable loads. The obtained results from real-time simulations prove its effectiveness for this type of analysis and open the possibility to perform PHIL (power hardware in the loop) simulations.**

Index Terms—**real-time simulation, battery energy storage systems, impact analysis**

I. INTRODUCTION

The models, whether physical, analog or mathematical, are constructed with the purpose of analyzing the system behavior according to external interferences. A system model or process can be defined as any representation of the relation between the quantities involved in it. The usual forms of representation include mathematical expressions, block diagrams, templates, tables, diagrams and graphs. As system quantities, both the constants and variables are considered, including, for example, time, position, velocity, temperature, number of items in a set, voltage, current and electrical resistance.

The most important application of a model is to make inferences about the system behavior, and based on them inform the strategic decision-making process. The large majority of physical processes have a high complexity relationship between their magnitudes, often making inference difficult enough for decision making to be no longer relevant. Thus, in order to facilitate the inference process, most models are a simplification of the complete system, and represent only a subset of phenomena, parameters, or parameter values of interest.

One of the ways to understand the complex systems behavior is to study their model response during disturbances or parametric variations. Computational simulation is one way to produce these responses, which can be studied by observing instantaneous values on time domain, time domain RMS values, or frequency response components.

The simulations can be divided into conventional or real-time simulations. In the conventional simulation, the model complexity is adjusted to obtain the desired level of the process representativity, the simulation time period is not considered relevant and it is a consequence only of the ratio between the model complexity and the available computational capacity. In real-time simulations, the course of time in the simulated process must be identical to the one in the outside world, allowing, with the appropriate equipment, the connection between the physical system and the simulated one. This makes it possible to use as physical processes every easily accessible parts and complex models, while hard-to-access parts and simple models can be maintained in simulation. On the other hand, in the real-time simulation, the processing capacity imposes limitations on the simulated model complexity, in which the first one must overcome the second, and only with recent advances this has become popular.

Several studies in the literature discuss about real-time simulation. In [1] the authors propose a real-time simulation method for photovoltaic generation systems under real climatic conditions, using a real-time digital simulator (RTDS™). The current and voltage curves of a real photovoltaic panel are tested, and it is created in the simulator a hypothetical grid of the tested photovoltaic panel. The actual weather conditions, insulation and the photovoltaic panel temperature are interconnected through the simulator's analog input ports for real-time simulation. [2] presents the dynamic modeling and simulation of a wind turbine connected to the grid using PSCAD/EMTDC™ (Power Systems Computer Aided Design/Electromagnetic Transients including DC). The system composed of the wind turbine, the generator and the converter are modeled for dynamic analysis. The control strategy is designed to capture the maximum energy from the wind

978-1-7281-4181-7/19 $31.00 © 2019 IEEE

speed variation, maintaining the reactive power generation at a predetermined level for a constant power factor or a certain level of voltage regulation. In [3] the authors propose two power flow control algorithms for a voltage source converter connected to the grid. The operation of both algorithms was verified inside the structure of a HIL (hardware in the loop) simulator (Typhoon^{TM}), with ultra low latency, which makes the converter behavior analysis accurate, safe and easily implementation for any electrical grid conditions.

In [4], the authors analysis different battery energy storage technologies the which would assist 50% wind power penetration in Denmark (by 2025). They concludes..."the battery storage technology will play a major role in the reliable and economic operation of smart electric grids with significant amounts of renewable power". Furthermore the development of battery technology, it is mandatory to develop analysis tools to examine the technical and economic feasibility of integrating battery (BESS) in electrical grid.

The authors in [5] proposed a mixed integer linear programming model for optimal battery energy storage system operation in electrical grids. The proposed model considers various parts of the battery energy storage system including battery pack, inverter, and transformer. The proposed model is applied to the IEEE 33-bus test case and the results prove the accuracy and efficiency of the proposed model.

In this paper was performed the modeling and real-time simulation to analyze the impact of battery-based energy storage systems in the electrical systems. The simulator used is the OPAL-RT/5707^{TM} real-time simulator, from OPAL-RT Technologies company. The use of real-time simulation allows to perform tests of difficult execution in a real physical system. Operating conditions that could damage a physical system can be accomplished using a real-time simulator.

II. BATTERY ENERGY STORAGE SYSTEM

A battery energy storage system is a system that stores energy through the battery technologies for further use. There is a wide range of battery technologies that can be used, among which can be mentioned:

1) VRLA (valve regulated lead-acid batteries)
2) Lithium-ion batteries
3) Vanadium flow batteries
4) Sodium nickel batteries
5) Liquid metal batteries

A model that provides data on the battery state of charge (SoC), in addition to its current and voltage, allows a charge/discharge process with low risk of overload, improved management of the power supplied by the main source, as well as extending the battery service life and increase the system's efficiency powered by it ([6]).

This work uses the model described in [7], which is based on the open circuit voltage (E) representation of each cell (Eq. 1). The lead-acid battery was chosen to be implemented in the system due to its high energy efficiency (around 85 to 90 %), low self-discharge rate, low maintenance, long service life

(1200 to 1800 cycles or 5 to 15 years) and low implantation cost ([8]).

$$E = E_0 - K \left(\frac{Q}{Q - it} \right) (i^* + it) + Aexp(-Bit) \quad (1)$$

where E_0 is the battery constant voltage (V), K is the polarization constant (V/Ah), Q is the the battery maximum capacity (A/h), A is the voltage amplitude at the exponential zone (V), i is the battery current (A), it is the extracted capacity (Ah), i^* represents the low frequency current (A) and B is the exponential capacity (Ah^{-1}).

For this model the following characteristics were adopted:

- The parameters were considered the same for the battery charge and discharge cycles;
- The internal resistance was considered constant for both cycles;
- The temperature effects were not considered in the model;
- Both the battery memory and self-discharge effects were not considered.

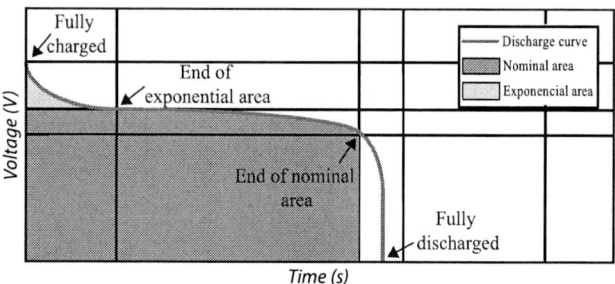

Fig. 1. Typical battery discharge curve.

A typical discharge curve of a battery can be seen in Fig. 1. This curve is composed by three regions: the first represents the exponential voltage drop existing when the battery is fully charged, the second region represents the charge that can be obtained until the battery voltage is lower than its nominal voltage, and the third represents the total battery discharging.

III. SYSTEM MODEL

The proposed system is composed of a conventional and balanced three-phase source, which is connected to a non-linear load and a three-phase bidirectional inverter, forming the common coupling point (CCP). The inverter is connected to a DC link, which in turn is connected to the BESS's terminals, as can be seen in Fig.2.

The three-phase inverter acts as an APF (active power filter), representing an efficient solution for harmonic reduction and power factor control in the electrical grid. The technique used in the active filtering makes it possible to measure the harmonic currents of one or more phases of the input source and generates opposite-phase currents to those measurements. In this way, the harmonics normally originated by non-linear loads are canceled out by the reactive compensation. In Brazil, the Agência Nacional de Energia Elétrica (ANEEL) establishes

978-1-7281-4181-7/19 $31.00 © 2019 IEEE

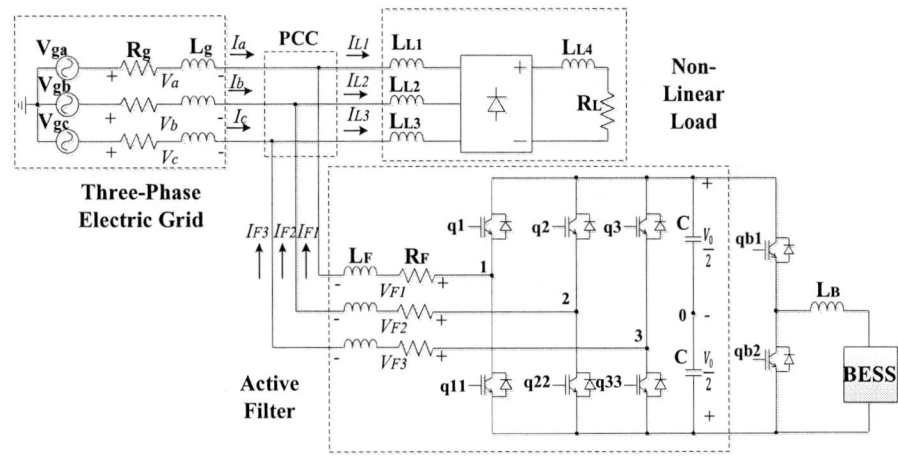

Fig. 2. Simulated system.

in its normative resolution that the electricity utilities power factor must be greater than or equal to 0.92, which also implies the adjustment the total harmonic distortion (THD) rate of the currents, according to the power system's size. In this study, the international standard states that the THD must be less than 5 %.

According to the currents direction and the voltage analysis in one loop, as shown in Fig. 2, it was possible to obtain the system model, as described in (2), (3), (4) and (5).

$$-V_{ga} + V_a - V_{F1} + V_{10} - V_{20} + V_{F2} - V_b + V_{gb} = 0 \quad (2)$$

$$V_j = i_j R_g + L_g \frac{di_j}{dt} \quad (3)$$

where: j = a or b.

$$V_{Fk} = i_{Fk} R_F + L_F \frac{di_{Fk}}{dt} \quad (4)$$

where: k = 1 or 2.

$$V_{k0} = (2q_k - 1) \frac{V_0}{2} \quad (5)$$

IV. CONTROL STRATEGY

The control strategy used for this work consisted in the use of a PI controller along with both the Clarke's ($\alpha\beta$) and Park's Transform in its algorithm, in order to minimize the inherent complexity to the AC-DC inverters mathematical modeling, according to (6), (7) and (8). This control was carried out according to the method proposed in [9], which is characterized by measuring only the currents and voltages of the input source.

$$\begin{bmatrix} V_\alpha \\ V_\beta \end{bmatrix} = \sqrt{\frac{2}{3}} \begin{bmatrix} 1 & -\frac{1}{2} & -\frac{1}{2} \\ 0 & \frac{\sqrt{3}}{2} & -\frac{\sqrt{3}}{2} \end{bmatrix} \begin{bmatrix} V_{ga} \\ V_{gb} \\ V_{gc} \end{bmatrix} \quad (6)$$

$$\theta = arctg \left(\frac{V_\beta}{V_\alpha} \right) \quad (7)$$

$$\begin{bmatrix} I_d \\ I_q \end{bmatrix} = \sqrt{\frac{2}{3}} \begin{bmatrix} cos(\theta) & -sen(\theta) \\ cos(\theta - \frac{2\pi}{3}) & -sen(\theta - \frac{2\pi}{3}) \\ cos(\theta - \frac{4\pi}{3}) & -sen(\theta - \frac{4\pi}{3}) \end{bmatrix}^T \begin{bmatrix} I_a \\ I_b \\ I_c \end{bmatrix} \quad (8)$$

In this system, the quadrature axis reference current (I_{qref}) was defined as null, since the quadrature current must be equal to zero so that the converter has a power factor around 1. On the other hand, the direct axis reference current (I_d) was generated from the voltage controller, that is, it comes from the difference between the voltage measured at the bus terminals (V_0) and the reference voltage (600 V), based on the rms voltage from the main power supply (V_g). After conducting the PI control, the dq voltages were obtained and then were transformed from the dq plane to the $\alpha\beta$ coordinates, and finally, into the ABC coordinates. Finally, it was applied the sine-triangle PWM technique, with a 12 kHz switching frequency. The control strategy can be visualized in the block diagram of Fig. 3. The BESS control is implemented using a PI controller, whose input is the error between the measured current that circulates through the BESS (I_{bat}) and the reference current (I_{bref}). The controller output is the BESS voltage (V_{bat}), responsible to drive the inverter switches that control the voltage, the SoC and the current in the accumulator.

V. SYSTEM'S DESCRIPTION

Using a high capacity simulator to perform the calculations in a short time step, is denominated the real-time simulation, where the simulator performs calculations faster than the physical passage of time [10]. Therefore, it is possible to configure a time step so that the simulator can generate the real-time results, which allows the use of digital and analog inputs and outputs, as well as the communication between physical data (motors, photovoltaic plants, battery bank, power grid or other emulator modules) and mathematically modeled data.

In this work the real-time simulator OPAL-RT/5707 is used. The simulator, with capacity for up to 32 processing units, allows real-time simulations in a short time step that, along with

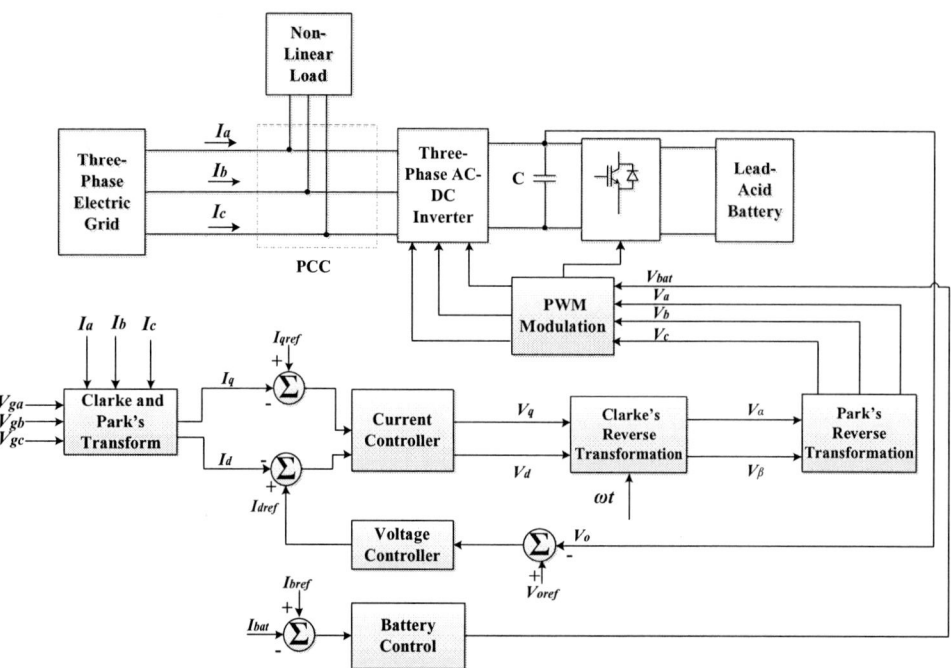

Fig. 3. Complete scheme of the proposed control.

the appropriate system modeling, guarantees the reliability of the obtained results. Simulink/MATLAB™ software was used to model the system.

The simulator has 2 OP5330 modules (64 analog inputs' and outputs' channels) and 2 modules OP5353 (64 digital inputs' and outputs' channels). Each module is divided into 4 groups with 16 channels each. The analog outputs have a voltage range of ± 16 V and the digital inputs allow signals with a peak value of ± 20 V. Both outputs have 16-bit resolution.

I/O modules and real-time results allows the hardware in-the-loop (HIL) simulation. In this case, the model can be separated into physical parts and modeled parts, referred as subsystems, where part of the model will be emulated by the simulator. For example, the simulator can emulate an operational electrical grid, connected through the simulator's inputs and outputs, with devices such as photovoltaic panels and energy storage systems. This simulation allows in-depth studies and approximates the idealization of the system under study to its prototype, decreasing the risks to the real power grid in operation.

In this work, real-time simulations were divided into two steps:

- Step 1 - The network, non-linear load, inverter and BESS models (Fig. 2) were loaded into the real-time simulator, characterizing a SIL (*software in-the-loop*) simulation.
- Step 2 - The BESS embedded model was replaced by a 1 kW analog bipolar power supply, operable in 4 quadrants, forming a HIL configuration (KEPCO) to emulate the BESS. The source has a 14-bit resolution for both output's current and voltage, with input voltage ranging

from 176-264 V, maximum input current between 9.5-6.4 A, switching frequency 70 kHz and nominal frequency between 47-63 Hz. It can be configured in series, parallel and in both combinations (identical units - 2x2 or 3x2).

In both steps it is possible to change simulation parameters and verify in real time the impact of the BESS insertion in the electrical grid.

VI. REAL-TIME SIMULATION RESULTS

A. Software in-the-loop

To validate the system's operation, the configuration uses the simulator to emulate the entire system and enables the user to change parameters and verify in real time the BESS's behavior in different operation stages.

At first, the SIL simulation consisted of the system's response without the active filter and the BESS. Fig. 4(a) illustrates the input current waveform in phase A to $R_L = 50 \ \Omega$. Fig. 4(b) illustrates its respective frequency analysis, through the FFT function. With the harmonic components analysis, it was calculated the THD, which resulted in 26,67 %.

Afterward, the real-time simulation was performed according to the following steps:

- In 1.05 s, the BESS is connected an the load changes from 100 Ω to 50 Ω.
- The BESS assists the input source injecting 30 A.
- In order to analyze the PI control that drives the battery, when SoC < 86 %, the load value return to 100 Ω and the system starts charging the battery with 15 A. This SoC range was chosen due to hardware limitations while simulating at Simulink™.

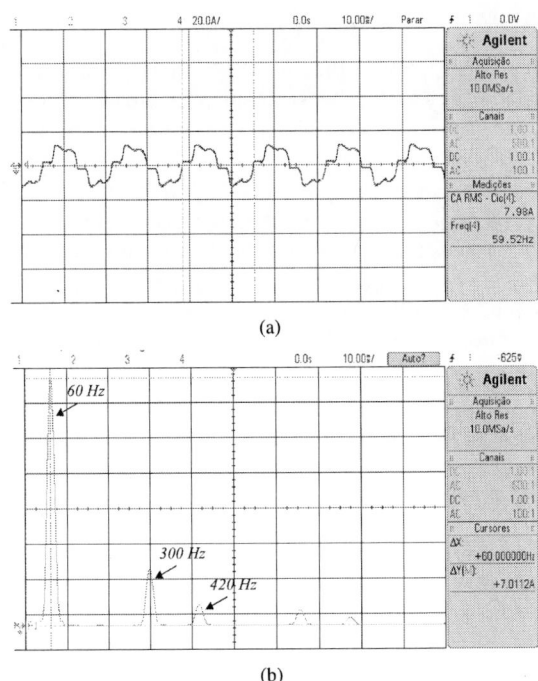

(a)

(b)

Fig. 4. Input/Load current with $R_L = 50$ Ω. (a) Wave shape. (b) FFT.

- When SoC > 89 %, the load variation occurs again ($R_L = 50$ Ω) and the cycle restarts.

The DC link voltage is shown in Fig. 5 while varying the load with the BESS insertion. It is possible to verify the proper operation of the DC link voltage control to adjust the voltage at 600 V, in the face of the 50 % load change.

The Fig. 6 shows the input's voltage and current during the active power injection by the battery in the system, thus reaching a power factor of 99.95 %. The phase shift around 180° indicates the provision of ancillary services by injecting active power into the electrical grid.

The input current waveform in phase A while the battery is injecting active power into the system and the FFT signal harmonic analysis is shown in Fig. 7(a). In Fig. 7(b) the magnitude and frequency spectrum of the most significant harmonics are presented, during the period in which the load variation occurs concurrently with the active power injection into the system by the battery.

B. Hardware in-the-loop

For HIL simulation, the real-time simulator controls the programmable source through its analog outputs, enabling KEPCO™, provided by Korea Electric Power Corporation, to emulate the BESS in real time, according to the scheme of Fig. 8. The battery model had a 50 V nominal voltage and the dc-ac converter was represented by an equivalent resistor of 17.8 Ω. The current flowing through the resistors was measured and then sent to the OPAL-RT, which controlled the battery's charging and discharging. Finally, the battery output was applied to the KEPCOs source analog outputs to

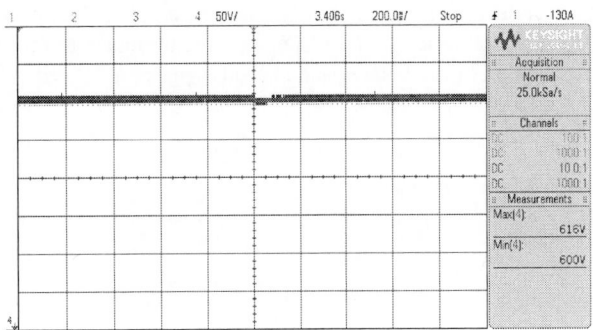

Fig. 5. DC link voltage with load variation and with BESS discharge ($R_L = 50$ Ω).

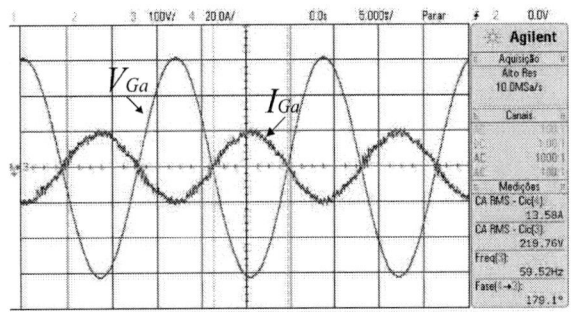

Fig. 6. Input's voltage and current during battery's discharging.

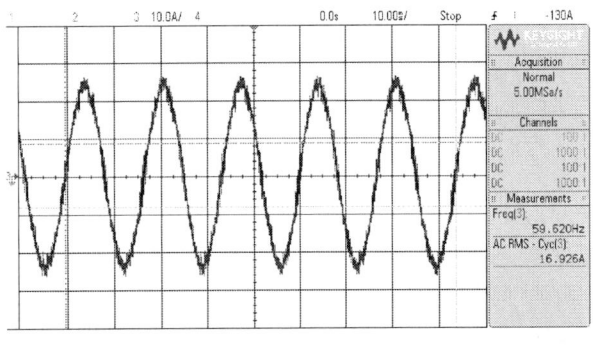

(a)

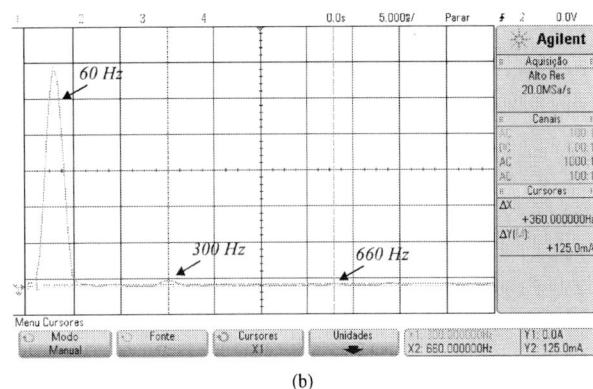

(b)

Fig. 7. Input current with battery discharging. (a) Wave shape. (b) FFT.

control its voltage and current. Moreover, a 220 μF capacitor has been connected to the KEPCO source terminals to reduce oscillations in the voltage and current signals.

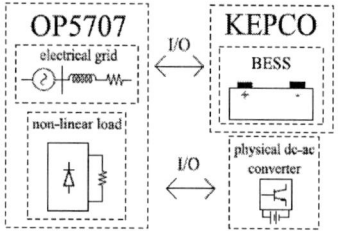

Fig. 8. Schematic diagram of the simulated system with the KEPCO source emulating the SAEB - HIL structure.

Fig. 9 illustrates the result of the current and voltage control performed on the KEPCO source, starting with battery charging and, after approximately 20 s, its discharge was started. The source was adequately controlled, since the control followed the imposed references of approximately 3 A and 50 V. Fig. 10, in turn, illustrates the experimental platform used.

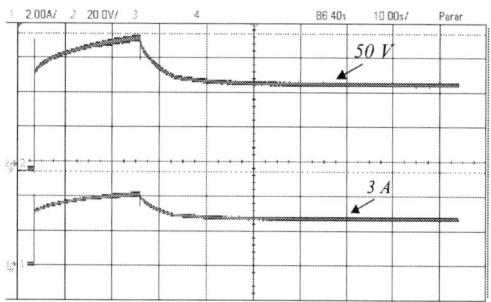

Fig. 9. Voltage and current control results from KEPCO source using HIL simulation.

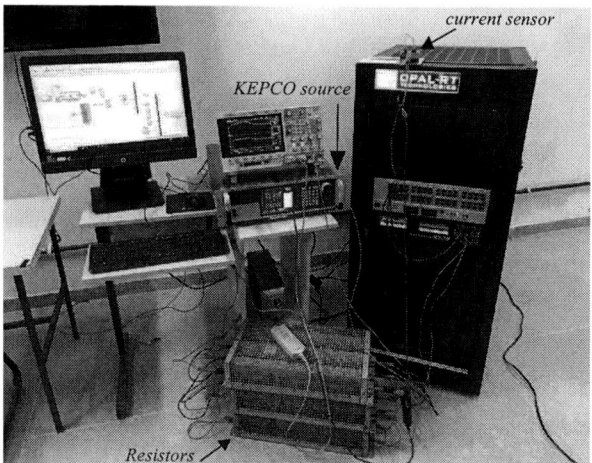

Fig. 10. Experimental platform used in the HIL simulation.

VII. CONCLUSION

In this paper was presented the modeling and real-time simulation of a system consisting of a three-phase bidirectional inverter (active filter) connected to a BESS, which in turn provided ancillary services, reducing the consumer spending on electricity and providing active power during peak hour load variations, besides maintaining the power factor and THD levels according to the current regulations. The control strategy allows the BESS optimal performance, with SOC between 86 and 90 %, thus ensuring a longer service life and the control of the drained current from the power grid. The real-time simulation allowed to evaluate the impacts of the BESS insertion in the electrical grid for diverse operational conditions, approaching the system study of its real operation. Both the results obtained with the entire system embedded in the simulator (step 1) and the results obtained in the HIL configuration, using the KEPCO source (step 2), showed that the results in the real-time simulations prove the analysis effectiveness and it opens up the possibility to perform PHIL simulations in the sequence of this work.

ACKNOWLEDGEMENT

The authors thank, to Coordenação de Aperfeiçoamento de Pessoal de Nível Superior - Brasil (CAPES) - Finance Code 001, CNPq, project #312646/2017-8, and, the companies ALSOL and CEMIG-D (R&D ANEEL sectorial fund).

REFERENCES

[1] M. Park and I.-K. Yu, "A novel real-time simulation technique of photovoltaic generation systems using rtds," *IEEE Transactions on Energy Conversion*, vol. 19, no. 1, pp. 164–169, March 2004.

[2] S. Kim and E. Kim, "Pscad/emtdc-based modeling and analysis of a gearless variable speed wind turbine," *IEEE Transactions on Energy Conversion*, vol. 22, no. 2, pp. 421–430, June 2007.

[3] Z. R. Ivanović, E. M. Adžić, M. S. Vekić, S. U. Grabić, N. L. Čelanović, and V. A. Katić, "Hil evaluation of power flow control strategies for energy storage connected to smart grid under unbalanced conditions," *IEEE Transactions on Power Electronics*, vol. 27, no. 11, pp. 4699–4710, Nov 2012.

[4] K. Divya and J. Østergaard, "Battery energy storage technology for power systems—an overview," *Electric Power Systems Research*, vol. 79, no. 4, pp. 511 – 520, 2009. [Online]. Available: http://www.sciencedirect.com/science/article/pii/S0378779608002642

[5] H. Mehrjerdi and R. Hemmati, "Modeling and optimal scheduling of battery energy storage systems in electric power distribution networks," *Journal of Cleaner Production*, vol. 234, pp. 810 – 821, 2019. [Online]. Available: http://www.sciencedirect.com/science/article/pii/S0959652619321572

[6] L. W. Yao, J. A. Aziz, P. Y. Kong, and N. R. N. Idris, "Modeling of lithium-ion battery using matlab/simulink," *IECON 2013 - 39th Annual Conference of the IEEE Industrial Electronics Society*, pp. 1729–1734, 2013.

[7] C. M. Shepherd, "Design of primary and secondary cells," *Journal of The Electrochemical Society*, vol. 112, pp. 657–664, 1965.

[8] X. Hu, C. Zou, C. Zhang, and Y. Li, "Technological developments in batteries: a survey of principal roles, types, and management needs," *IEEE Power and Energy Magazine*, vol. 15, pp. 20–31, 2017.

[9] D. Casadei, G. Grandi, U. Reggiani, and C. Rossi, "Control methods for active power filters with minimum measurement requirements," *Fourteenth Annual Applied Power Electronics Conference and Exposition*, vol. 2, pp. 1153–1158, 1999.

[10] J. Belanger, P. Venne, and J. N. Paquin, "The what, where, and why of real-time simulation," *Planet RT*, p. 37–49, 2010.

978-1-7281-4181-7/19 $31.00 © 2019 IEEE

Comprehensive and Didactic DC Servomotor Control Platform

Bruno de Almeida Regina
Energy Systems
Federal University of Juiz de Fora
Juiz de Fora, Brazil
almeida.bruno@engenharia.ufjf.br

Maria Júlia Rosa Aguiar
Energy Systems
Federal University of Juiz de Fora
Juiz de Fora, Brazil
maria.aguiar@engenharia.ufjf.br

André Augusto Ferreira
Electronic Systems
Federal University of Juiz de Fora
Juiz de Fora, Brazil
andre.ferreira@engenharia.ufjf.br

Abstract—This paper proposes a modular and graphical user interface programmable control plant that follows the new academic guidelines for engineering courses in Brazil. It is a project that allows interdisciplinarity and dynamism in Control Systems laboratory classes, and is assembled with affordable and user friendly materials. The test bench is composed of a DC servomotor, a L298n Dual H-bridge driver and an Arduino board, whose angle position and velocity can be programmed graphically and real-time controlled with Simulink software. Cascade Control was successfully implemented in the model, and the proposed plant was considered an useful resource to be used alongside commercial ones in the teaching of control.

Index Terms—control, parameter estimation, DC servomotor, PBL, didactic plant.

I. Introduction

In April 2019, the Brazilian Ministry of Education approved new academic guidelines for engineering courses that were proposed in a technical advice from the National Council of Education (*Conselho Nacional de Educação* - CNE). The guidelines were created with the aim of guiding the design and the planning of undergraduate courses with the insertion of technological innovations and methodologies, greater integration between companies and universities, and the enhancement of interdisciplinarity throughout the classes. Extension projects, scientific initiation and development of prototypes should be encouraged, as well as the composition of laboratory activities and their different forms of implementation [1].

Among the new forms of methodologies suggested by the CNE are Project-Based Learning (PBL), that aims to develop competencies in collaborative learning and interdisciplinarity. The main point is to give more meaning, dynamism and autonomy to the learning process in Engineering through the student's engagement in practical activities [1]. Because of its characteristics, PBL can join technical learning and more transversal competences, such as creativity, leadership, collaborative work and critical thinking, preparing the students to truly meet the demand of employers [2].

In engineering courses, Control Theory is usually taught through analyzes with control loops in ideal models and, although they have satisfactory results, sometimes may distance themselves from the reality seen in the field. The professional

life of the students begins in the universities through the learning in classrooms and classes of laboratories. In this way, the use of didactic plants is necessary, as it complements the practical training of students with models closer to the real ones that are used in industries. They generate more confidence for the future engineers, since they already had contact with the experiment, being able to use in the job market the notions obtained through it [3].

There are universities which use commercial didactic control plants, such as the one showed in [4], used for level control of a water tank. These plants are usually expensive and have pre-established laboratory scripts and control structures that does not allow to expand to the other control strategies of the students and teachers. Other non-commercial didactic plants were proposed, such as the ones in [5], [6], whose experiments relied on unusual equipment (like a quadrotor test bench) and whose focus was only in the Control knowledge of the students. In [7], [8] low-cost didactic control plants of a DC servomotor were proposed, but the approached contents were also limited to those one might expect from a Control Theory class. In [8], the projects were significantly simplified for the students when circuit boards were developed to connect sensors to the Arduino. In [9], software and hardware are built into an electronic workbench capable of field-oriented control of a three-phase induction motor.

This paper aims to compile and review different contents that can be taught in a hands-on and intuitive control plant of a DC servomotor using Arduino and board programmed with Simulink, which are tools to aid in the teaching of control, mainly in laboratory classes. This control plant can be assembled with materials that are quite easy to obtain, and with it experiments can be conducted for estimation of the parameters of the model. Those parameters could be used to regulate position and velocity of the DC servomotor through Cascade Control, for instance. The proposed didactic plant links practical knowledge in Power Electronics and Control Theory, as well as experience in software and toolboxes which may be essential to graduate resourceful engineers, as was proposed by CNE's Engineering Undergraduate Courses Guidelines.

This paper is structured as follows: Section II presents the dynamic model and the theoretical equations of the DC servomotor. Section III is focused on the components used in the assembled test bench. Section IV deals with the estimation of the parameters used in the model. Section V details the communication between Simulink and Arduino board.. Section VI presents information on the Cascade Control implemented in this work. Section VII shows the results obtained with the control. Finally, Section X presents the conclusions of this work.

II. DYNAMIC MODEL

The general model of a DC servomotor is as depicted by Fig. 1, where the applied voltage $v_t(t)$ regulates the mechanical torque τ_m on the motor's shaft, and consequently its angular velocity $\omega(t)$ and position angle $\theta_m(t)$. Equations (1) to (4) represent the mathematical relations of the model.

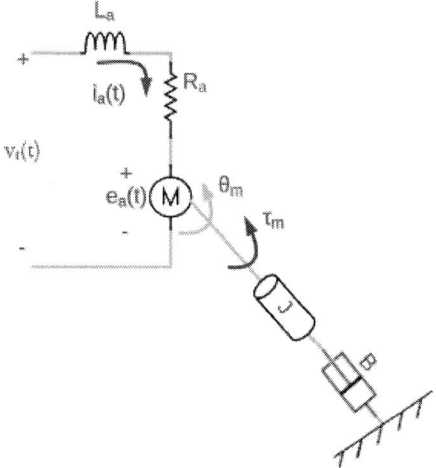

Fig. 1. DC servomotor model.

$$v_t(t) = R_a.i_a(t) + L_a.\frac{di_a(t)}{dt} + e_a(t) \quad (1)$$

$$e_a(t) = k_e.\frac{d\theta_m(t)}{dt} = k_e.\omega(t) \quad (2)$$

$$\tau_m(t) = k_t.i_a(t) \quad (3)$$

$$J.\frac{d\omega(t)}{dt} + B.\omega(t) = \tau_m(t) \quad (4)$$

where R_a is the armature resistance, L_a is the armature inductance, $i_a(t)$ is the electric current in the circuit and $e_a(t)$ is the voltage that the motor is subjected to. k_e and k_t are the electric and torque constants of the DC servomotor, respectively. J and B correspond to the equivalent inertia and damping of the system.

III. EXPERIMENTAL MODEL

A. Test Bench Components

The main components used in the experiment were: Arduino Mega 2560 R3; 12V DC servomotor with quadrature encoder; Dual H Bridge L298n; a Breadboard; 2 pull-up resistors ($3k\Omega$); 9V external source; a Protractor; an angle indicator. The connection between the elements can be seen in the circuit diagram in Fig. 2 and in the assembled test bench seen in Fig. 3.

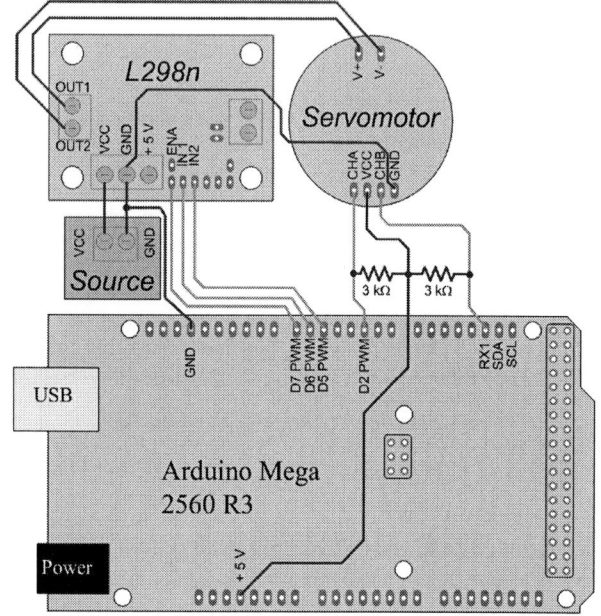

Fig. 2. Circuit diagram of elements used in the control plant.

Fig. 3. Experimental test bench used as the didactic control plant.

The assembly of the electrical components is an important part of the proposed didactic control plant. It enables the students to have close contact with Power Electronics Building Blocks, learning relevant concepts (such as the role of the pull-up resistor and how incremental encoder works) in the process. The students are encouraged to use multimeters to

978-1-7281-4181-7/19 $31.00 © 2019 IEEE

verify electrical connections and to better understand how the modification of certain electric parameters (e.g. source voltage) leads to changes in the DC servomotor behavior.

B. Commercial Motor Driver Shield

In order to activate the DC servomotor, the commercial L298n motor driver shield is used. It consists of 8 transistors without power and signal isolation, which can rotate the motor in both reverse and forward directions [10]. For it to operate, it receives information on three channels: one with the assigned value of PWM (Pulse Width Modulation) which controls speed and the other two (IN1 and IN2) containing data with its direction. This last parameter is controlled by sending a HIGH or LOW signal to the drive for each channel. For example, a HIGH to IN1 and a LOW to IN2 will cause it to turn in one direction, and a LOW and HIGH will cause it to turn in the other direction. This logic must be adapted when programming in Arduino board in Simulink environment.

IV. PARAMETER ESTIMATION

Didactic control plants usually have known parameters (inertia, damping coefficients, armature resistance, etc.), however, in practical situations, this may not always true. A didactic plant in which the students need to find these parameters based on the dynamic response of a motor and then project its control and verify its viability experimentally can help consolidate knowledge on the area. With the help of Simulink Design Optimization toolbox, it is possible to perform parameter estimation of the model using measured data by the real system.

A set of experiments are conducted, with voltage being applied as input in the experimental model in order to get its output data (rotation speed).

Fig. 4 shows the block diagram made in Simulink with the motor model. The representation of both electrical and mechanical aspects of the model is done by its transfer functions. In this model, the parameters to be optimized are J, B, R_a, L_a, k_t and k_e. The first step to obtain those parameters is to run the simulation with arbitrary values of variables in order to get an initial response of the model. In the toolbox Parameter Estimation, it is possible to specify the constraints of the parameters, import input and output experimental data and start the simulation of estimation process.

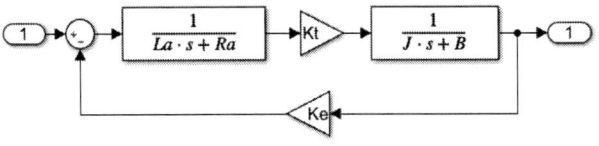

Fig. 4. DC servomotor block diagram

The input used in the experimental tests (Fig. 5) are applied to the Simulink model, whose parameters are optimized until the experimental outputs curve fits the model response with minimum error (Fig.6). It can be seen that when receiving low voltage, the motor actuates in a nonlinear region. Because of

TABLE I
ESTIMATED PARAMETERS OF THE DIDACTIC PLANT

Parameters	Values
J $(Kg.m^2)$	0.023
B (N.m.s/rad)	0.19
R_a (Ω)	0.12
L_a (H)	0.010
k_t (N.m/A)	4.47
k_e (V.s/rad)	0.00038

that, the measured rotation gets smaller than the predicted by the parameter estimation. The values of the variables can be followed during the simulation, as shown in Fig. 7 and the final result is obtained when the solution converges to an optimal value. The best parameters found in the estimation can be seen in Table I.

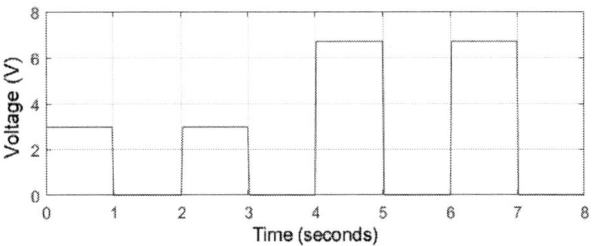

Fig. 5. Input applied in the experimental model in order to estimate its parameters.

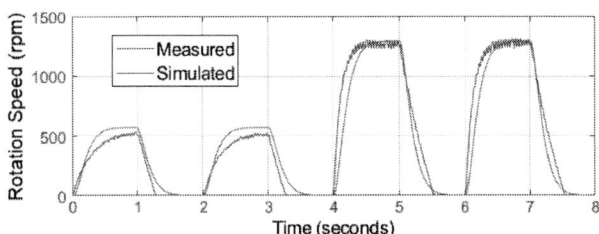

Fig. 6. Measured Experimental data and Simulated Data for the output (above) according to input signal (below).

V. SOFTWARE-HARDWARE COMMUNICATION

Simulink is used to design the control of the DC servomotor, to configure hardware and to program the Arduino board. Simulink serves its purpose mainly as an user interface to the real-time application, as it makes the display of the model variables intuitive and simplifies the modification of parameters. The blocks used in the model are compiled in C language into the target hardware, Arduino, which is responsible for all of the processing, and that is only possible through the use of Simulink in its External Mode [13].

The connection between software and hardware is done through serial communication and all the information to do

978-1-7281-4181-7/19 $31.00 © 2019 IEEE

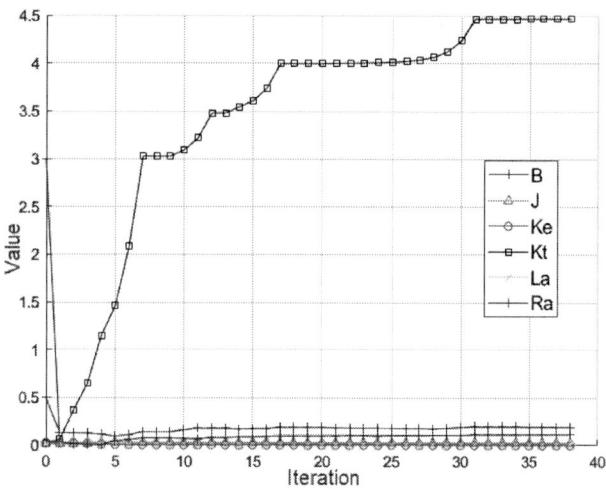

Fig. 7. Parameter Estimation results

so is withheld on the blocks of the diagram. A free toolbox called "Simulink Support Package for Arduino Hardware" is used to modify pin values and, therefore, to send data from Simulink to Arduino. To acquire data from the encoder, a code in C language was implemented in a special block (S-Function Builder) that allows to program algorithms that run directly on the target hardware. With that information it was possible to calculate the speed of the servomotor through the counting of encoder pulses, without having to send the data back to the computer. If this block is not used, there is loss of information through the USB cable due to the allowed sample time rate (0.005 s).

VI. CASCADE CONTROL

In order to demonstrate the use of Simulink to program Arduino board and to control a DC servomotor, this section shows a cascade control example. However, it is simple to change blocks in order to build and to test different (classical or modern) digital control strategies. The cascade control has an architecture that reduces effects of disturbances in the control variables, besides improving the performance of the dynamic control process [11], [12]. The internal loop controller regulates the speed, and works at a faster rate than the the external loop, which controls the position angle. The external loop output is the setpoint for the secondary controller.

A proportional controller was chosen to regulate the angle position, and a Proportional-Integral one was used to adjust the speed of the DC servomotor. The gains were obtained through the Successive Loop Closure technique [14] with its values being: $K_p = 50.27$ for the external loop and $K_p = 1.12$ and $K_i = 162.6$ for the internal loop. The control diagram on Simulink can be seen in Fig. 8.

The S-Function Builder block ("Encoder - Plant Data") acquires information of the experiment, which is used to compute the error of the system. The integral control action can grow to a magnitude that no longer applies to the problem, and

thereafter an Anti-windup block with saturation was used to minimize this effect. The values that come from the controller can be negative or positive, representing the rotation direction. A subsystem of this diagram is represented separately in Fig. 9, and is used to set the orientation of the DC servomotor according to the logic seen in Section III-B. The absolute magnitude of the PWM parameter is then taken, because the L298n driver does not accept negative values as input. Because the DC servomotor tends to operate in a nonlinear region when low voltage is applied to it, as seen in Section IV, the minimum value of the PWM was limited in the simulation.

VII. RESULTS

Two tests were made in the didactic control plant: position and speed control of the DC servomotor. The controller gains can be better adjusted if the obtained responses are not satisfactory to the project. By changing manually the desired setpoint in the Simulink diagram to values of 90 and 180 degrees, it is possible to obtain the dynamic response of the model and verify it in the test bench with the use of the protractor and the angle indicator. Results of the control position can be seen in Fig. 10, where after an overshoot the system stabilizes itself according to the adopted setpoint.

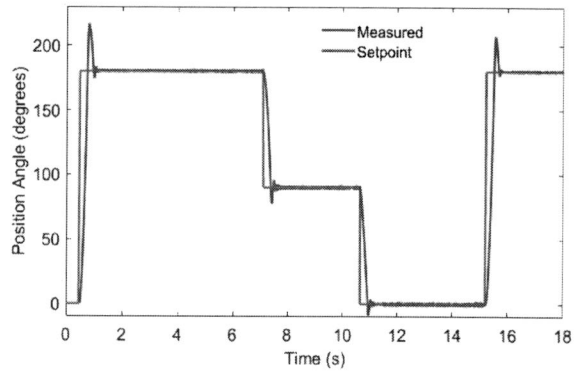

Fig. 10. Position Control

The speed control does not follow exactly the desired reference. As can be seen in Fig. 11, for small values of the motor speed it operates in its nonlinear region (as shown in Fig. 6). Furthermore, high velocities can cause loss of encoder pulse data, due to the allowed sample time of the simulation (Section V). This phenomenon leads to noise when reading information of the sensor.

Other phenomena that limit the speed control of the test bench are step functions or extremely high setpoints as the input of the model. These events lead the hardware current to values which can generate error in the serial communication and may cause Simulink to crash unexpectedly.

978-1-7281-4181-7/19 $31.00 © 2019 IEEE

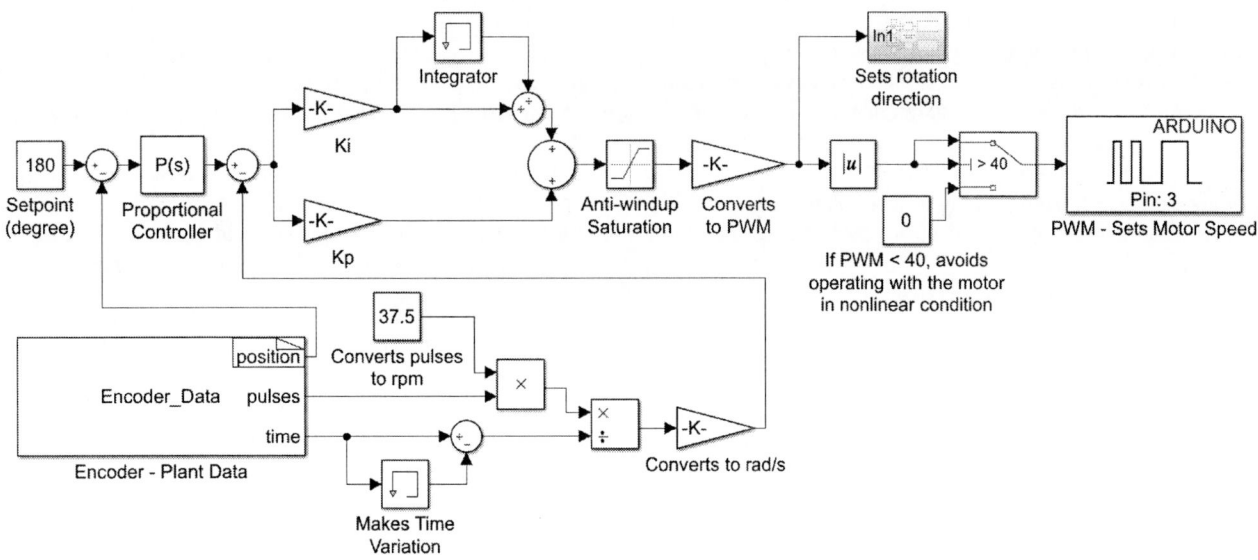

Fig. 8. Control scheme implemented in Matlab Simulink®.

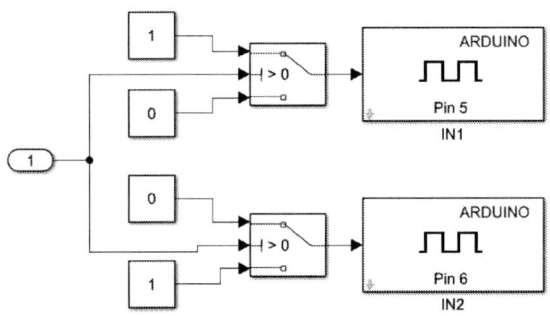

Fig. 9. Subsystem of Simulink® model which sets rotation direction for the DC servomotor.

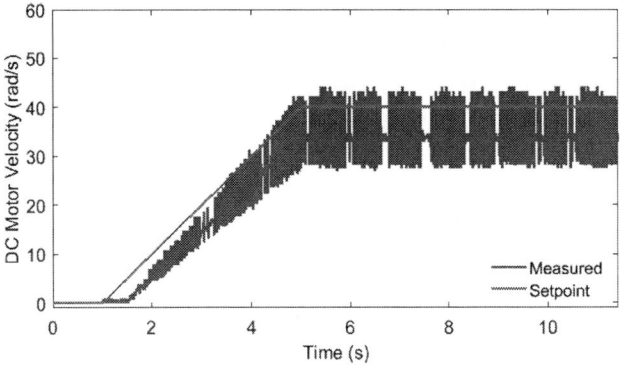

Fig. 11. Speed Control

VIII. SUGGESTED SCRIPT

The project related to this work shows how the system can be implemented in the laboratory classes of a control discipline, and suggested steps to follow are outlined below:

- Step 1 - Model and calculate system equations: analyze the system and express the equations in their Laplace form;
- Step 2 - Develop a physical model: perform physical system assembly for DC motor control with all necessary components;
- Step 3 - System performance analysis: verify open-loop system behavior and collect data for parameter identification;
- Step 4 - Motor Parameter Identification: use data measured by the actual system in the Motor parameter identification performed in the Simulink Design Optimization toolbox;
- Step 5 - Determining Controller Gains: design Cascade controller gains for speed and position control ;
- Step 6 - Design system block diagram in MATLAB Simulink: create the block diagram graphical representation with all of the parts of the system and simulate it until design controller project achieves desired performance;
- Step 7 - Program motor encoder reading: implement a code in C language in a special block (S-Function Builder) in Simulink to acquire data from the encoder;
- Step 8 - Run Experiments: perform experiments during the evaluations, changing the control setpoints and the control variables;
- Step 9 - Results analysis.

IX. COMPARISON BETWEEN DIDACTIC AND COMMERCIAL PLANTS

The control plants used in laboratory classes, whether commercial or made by students and teachers, have their advantages and disadvantages and differ from one another.

A platform such as the one proposed in this paper, besides introducing a new teaching methodology, brings innumerable innovations such as: practical work; interdisciplinarity; design and construction of prototypes; reuse of electronic components; low cost of design; research and experimentation. Furthermore, the didactic plants are open, that is, they will not work as a black box. Commercial plants and platforms are great teaching tools that usually have the functionality needed for good student learning. Many of these plants have industrial components making their applications more interesting due to their proximity to the market. The disadvantage of this type of platform is that students are not aware of the entire process of making the product, having only contact with it already assembled and following the manuals that accompany them. In addition, commercialized plants generally are quite expensive.

Because of their differences, the two types of platforms can be used together in the classroom, complementing each other. Commercial plants are a good alternative for introducing classes for students to begin to understand concepts and their application. By the time students already have more complete theoretical load and greater knowledge, a didactic platform can be used for the students to engage in comprehensive projects, allowing a better understanding of the process and concepts.

X. CONCLUSION AND FUTURE WORKS

The proposed didactic control plant is a relevant tool for teaching Control Systems to undergraduate students in an interdisciplinary manner. Programming, Power Electronics, Theory of Control and topics of optimization can be approached interactively.

Arduino and Simulink are easy tools to introduce students to programming and microcontroller hardware. Both of them have extensive accessibility, with a lot of documentation available on the internet, which makes them relevant choices for teaching control.

Commercial didactic control plants usually used in laboratory classes sometimes have scripts that do not follow labour market requirements, because the students do not always comprehend the control process as a whole. Federal University of Juiz de Fora, for example, uses a commercial control plant with known model parameters, pre-made laboratory scripts and a companion software that does a considerable part of the job for the students. The price of the experiments can also be a relevant aspect to be taken into account. A single unity of the commercial control plant of the laboratories found in Federal University of Juiz de Fora can cost up to US$11,000.00. Considering all the hardware and software (Matlab Student Edition, Simulink, Simulink Design Optimization, Simulink Coder) used in the didactic control plant proposed in this work, it would cost approximately US$100.00.

As can be seen in Section VII, this DC servomotor test bench has its limitations. Speed control results are not the ones expected, due to physical restrictions and problems with the serial communication. The proposed Didactic DC Servomotor Platform Control can be used in association with a commercial one for effective learning, according to the CNE's National Engineering Undergraduate Course Guidelines. While the commercial one can approach Control Theory topics which the other cannot, the proposed one has the advantages seen in Project-Based Learning. The latter allows more interdisciplinarity, is more thought-provoking and requires logical reasoning which is similar to the ones needed in industry applications.

In future works, an Arduino Shield electric current sensor can be applied to the test bench, so another loop can be added in the Cascade Control, which should make the responses smoother. A filter could also be applied to eliminate noise in the curves of the speed control.

XI. ACKNOWLEDGMENT

The authors would like to thank the Graduate Program of Electrical Engineering (*Programa de Pós-Graduação em Engenharia Elétrica* - PPEE) of Federal University of Juiz de Fora and CNPq (*Conselho Nacional de Desenvolvimento Científico e Tecnológico*). They would also like to thank Rodolfo Lacerda Valle for the valorous contributions during this project.

REFERENCES

[1] Conselho Nacional de Educação. Brazilian National Engineering Undergraduate Course Guidelines. Technical Advice 1/2019, approved in January 23 2019. (CNE/ CEB), Brasília, 2019.

[2] S. de Castro Lobato, D. A. Carvalho, A. A. Ferreira and V. F. Montagner, "Didactic prototype to model and to design linear control applied to a RLC plant," 2017 Brazilian Power Electronics Conference (COBEP), Juiz de Fora, 2017, pp. 1-6.

[3] L. Mescolin de Oliveira, D. Pinheiro Teixeira, A.Rocha de Oliveira, M.J. do Carmo, L.O. de Araújo Junior, "Utilização de uma Planta didática SMAR para Complementação do Ensino de Engenharia de Controle e Automação", XL Congresso Brasileiro de Educação em Engenharia (COBENGE), Belém - PA , 2012, pp. 1-11.

[4] D. R. G. da Silva et al. "Controle de nvel por aquisio de imagem numa planta didtica/Control of level by image acquisition in a didactic plant". Brazilian Applied Science Review, v. 2, n. 1, p. 252-261, 2018.

[5] W. L. Torres et al. "Mathematical Modeling and PID Controller Parameter Tuning in a Didactic Thermal Plant". IEEE Latin America Transactions, v. 15, n. 7, p. 1250-1256, 2017.

[6] S. Khan et al. "Teaching tool for a control systems laboratory using a Quadrotor as a plant in Matlab". IEEE Transactions on Education, v. 60, n. 4, p. 249-256, 2017.

[7] D. S. Castanho, F. C. Corrêa, "Retroactive control teaching and learning module from a didactic DC servomotor". Journal— MESA, v. 9, n. 4, p. 499-506, 2018.

[8] S. Dale, C. R. Costea, S. Zsolt. "Functionality Extension for a DC Motor Speed Control Didactic System with PID Controller". The Scientific Bulletin of Electrical Engineering Faculty, v. 18, n. 2, p. 59-62, 2018.

[9] Rolim, L.G.; Suemitsu, W.; Stephan, R. M.; M. B. Medeiros "An Experimental Setup for the Study of Field Oriented control of AC Machines ". Congresso Brasileiro de Automtica, 2000.

[10] O. Ulkir, I. Ertugrul, N. Akkus. "Embedded System Based Real Time Position Control Of A DC servomotor Using Matlab Motor Using Matlab". International Journal of Research in Engineering & Advanced Technology. 2016.

[11] P. Somkane, V. Kongratana, S. Gulpanich, V. Tipsuwanporn and N. Wongvanich, "A study of flow-level cascade control with WirelessHARTTM transmitter using LabVIEW", 2017 17th International Conference on Control, Automation and Systems (ICCAS), Jeju, 2017, pp. 856-861.

[12] R.Bhavina,N.Jamliya,K.Vashishtha,"Cascade Control of DC servomotor with Advance Controller",International Journal of Industrial Electronics and Electrical Engineering,Vol 1, Issue-1, 2013, pp. 18-20.

[13] MATHWORKS. External Mode. MathWorks, 2019. Disponivel em: https://www.mathworks.com/help/supportpkg/rtlsdrradio/ug/run-model-in-external-mode.html. Acesso em: 27 Junho 2019..

[14] R. W. Beard and T. W. Mclain, Small unmanned aircraft: Theory and practice. Princeton university press, 2012.

Performance Analysis of Active Anti-islanding Techniques for Photovoltaic Application

Marcos Roberto Oshiro
FAENG
Federal University of Mato
Grosso do Sul
Mato Grosso do Sul, Brazil
oshiromarcosroberto@gmail.com

Ruben Barros Godoy
FAENG
Federal University of Mato
Grosso do Sul
Mato Grosso do Sul, Brazil
ruben.godoy@ufms.br

Moacyr A. G. de Brito
FAENG
Federal University of Mato
Grosso do Sul
Mato Grosso do Sul, Brazil
moacyr.brito@ufms.br

Luigi Galotto Junior
FAENG
Federal University of Mato
Grosso do Sul
Mato Grosso do Sul, Brazil
luigi.galotto@ufms.br

Abstract— In this paper some anti-islanding detection methods are evaluated through MatLab/Simulink® models. The methods are embedded into the control system of a voltage source inverter that is used to inject power to the grid from a set of photovoltaic modules. Due to low efficiency of the passive methods in detection of islanding and to the high cost of implementing the remote methods, only the active methods will be evaluated. In that sense, this paper presents comparative analysis regarding Active Frequency Drift (AFD), Slip-Mode Frequency Shift (SMS) and Sandia Frequency Shift (SFS) methods. Commonly, the active methods are implemented combining the passive sub/overvoltage and sub/overfrequency protection methods. The simulations presented comply with the voltage and frequency limits established by IEEE Std 929-2000 and harmonic distortion established by IEEE Std 519-2014.

Keywords—Islanding Detection Method, Active Frequency Drift, Slip-Mode Frequency Shift, Sandia Frequency Drift, Total Harmonic Distortion.

I. INTRODUCTION

According to the IEEE Std 929-2000, the islanding phenomena is defined as a condition in which a part of the electrical grid that contains loads and distributed resources remains energized while isolated from the rest of the electrical system [1].

The ideal islanding concept is depicted in Fig. 1. The DC-AC converter is responsible to interface the photovoltaic (PV) system to the grid, providing power to it. A local RLC load is connected at the Point of Common Coupling (PCC) and may receive power from both PV and grid [2].

The islanding phenomena occurs when the distribution grid automatic recloser is opened, but the PV source continues supplying power to the grid section between the recloser and the PCC. In this scenario, the distributed generation will operate in an independent condition without grid voltage and frequency references, leaving the local load subject to changes in the PCC voltage outside the limits allowed by the IEEE Std 929-2000 [1], [2].

In order to detect the phenomenon of islanding, the Islanding Detection Methods (IDM) resident at the inverter are commonly applied, once they have lower implementation costs and high effectiveness. Such methods are classified into active and passive ones. The passive methods seek to detect an abnormality in the PCC voltage and do not introduce disturbances in the system, resulting in large non-detection zones (NDZ) [3], [4]. On the other hand, the active methods cause perturbations in the system in order to detect the islanding even when there is balance between the photovoltaic generation and the consumption (worst case condition), resulting in a small NDZ [5], [6].

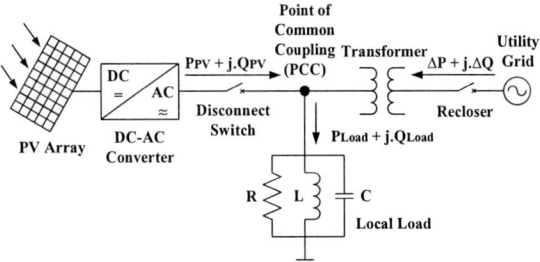

Fig. 1. Photovoltaic system and local parallel RLC load for anti-islanding simulations.

In such context, three active anti-islanding detection methods are evaluated namely, Active Frequency Drift (AFD), Slip-Mode Frequency Shift (SMS) and Sandia Frequency Shift (SFS). These are easy of implementing and cause changes in the frequency of the voltage in the PCC [7]. Comparisons regarding the amount of injecting disturbances, computational efforts and effectiveness are also presented to analyze the advantages and disadvantages of each active method.

II. ACTIVE ANTI-ISLANDING DETECTION METHODS

The active methods have the purpose of injecting a small disturbance signal into the PCC and observe the grid voltage response. The disturbance influence over the grid voltage will become substantial when the distribution grid is no longer connected [8].

A. Active Frequency Drift (AFD)

In this method the current waveform is slightly distorted (Fig. 2). The implementation is done seeking to increase the frequency of the voltage at the PCC by adding a frequency deviation (δf) and thus reaching values higher than the resonance frequency of the local load (f_0) [6], [9] - [11].

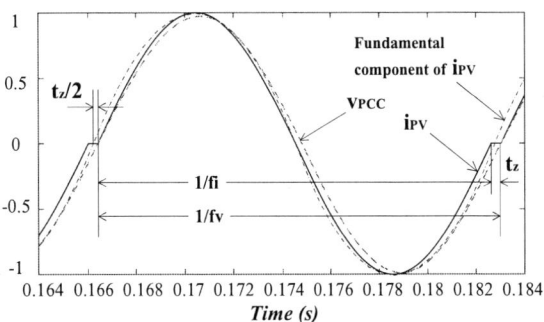

Fig. 2. AFD method – voltage waveform v_{PCC} and current waveform i_{inv}.

The AFD current reference is shown in (1); I_{invMAX} is the peak current and f_{PCC} is the voltage frequency at the PCC [6].

$$i_{ref} = I_{invMAX}.\sin[2.\pi.(f_{PCC} + \delta f).t] \qquad (1)$$

The phase angle between the inverter current (i_{inv}) and the PCC voltage (v_{PCC}) for the AFD method (θ_{AFD}) is presented in (2); t_z represents the zero current time [6].

$$\theta_{AFD} = \pi.f_{PCC}.t_Z = \frac{\pi.\delta f}{f_{PCC} + \delta f} \qquad (2)$$

B. Slip-Mode Frequency Shift (SMS)

In this method, the current phase angle is controlled to be a function of the frequency deviation of the voltage at the PCC of the last cycle in relation to the nominal grid frequency (f_{grid}) [6], [9], [12], [13].

The reference inverter current that implements the SMS method is calculated by (3) [6].

$$i_{ref} = I_{invMAX}.\sin(2.\pi.f_{PCC}.t + \theta_{SMS}) \qquad (3)$$

The SMS phase shift (θ_{SMS}), obtained through (4), is adjusted to be a sinusoidal deviation function of frequency from the base f_{grid} frequency; θ_m represents the maximum phase shift in degrees (°) and f_m the frequency at witch θ_m occurs [6].

$$\theta_{SMS} = \frac{2.\pi}{360}.\theta_m.\sin\left(\frac{\pi}{2}.\frac{f_{PCC} - f_{grid}}{f_m - f_{grid}}\right) \qquad (4)$$

In Fig. 3 it is possible to identify f_{PCC} in steady state through the intersection between the load angle (θ_{load}) and the θ_{SMS} curve, resulting in an operating point that may be stable or unstable. The stable points (A for $Q_f = 5$ and B and C for $Q_f = 1.5$) indicates the steady state frequency of the islanded system, and shall be outside the frequency limits allowed by the IEEE Std 929-2000 standard in order to detect islanding [14].

C. Sandia Frequency Shift (SFS)

This method is an extension of the AFD method. It presents two zero current segments (Fig. 4) and uses the f_{PCC} as positive feedback to increase the chopping fraction (cf), that is defined as the ratio of t_Z and the PCC voltage period (T_v) given by $cf = 2.(t_Z/T_v)$. The SFS chopping fraction can be defined as a linear function between f_{PCC} and f_{grid} (5). In (5) cf_0 is the initial chopping fraction and k_{SFS} is the accelerator gain [6], [9] - [11].

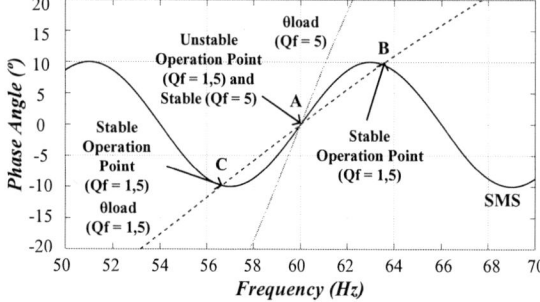

Fig. 3. Phase curves of SMS and the RLC loads for different quality factors.

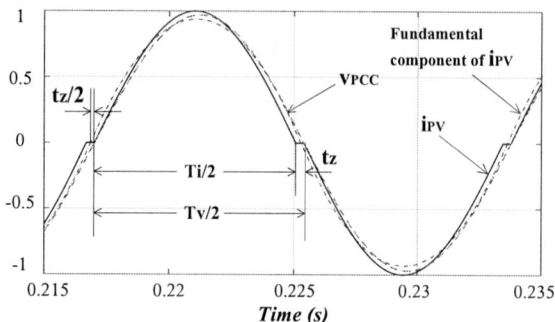

Fig. 4. SFS method - voltage waveform v_{PCC} and current waveform i_{inv}.

$$cf = cf_0 + k_{SFS}.(f_{PCC} - f_{grid}) \qquad (5)$$

The reference inverter current that implements the SFS method is calculated by (6) [6].

$$i_{ref} = I_{invMAX}.\sin\left(2.\pi.\frac{f_{PCC}}{1-cf}.t\right) \qquad (6)$$

The phase angle between i_{inv} and v_{PCC} (θ_{SFS}) is obtained through (7) [6].

$$\theta_{SFS} = \pi.f_{PCC}.t_Z = \frac{\pi.cf}{2} \qquad (7)$$

III. SIMULATION MODELLING

In the simulation carried out in MatLab/Simulink® platform (Fig. 5), the photovoltaic arrangement was simplified by a DC voltage source. The system is connected at the PCC by means of a current controlled single-phase voltage source inverter. The local load is a parallel RLC load connected also at the PCC.

The control circuit contain four fundamental blocks: Calculus of frequency and RMS values, Under Voltage Protection and Over Voltage Protection (UVP/OVP) and Under Frequency Protection and Over Frequency Protection (UFP/OFP), Islanding Detection Methods (IDMs) and PI Current Control with Modulator. The main simulation parameters are presented in Table I.

TABLE I

MAIN SIMULATION PARAMETERS

Sampling interval	$T_S = 10^{-6}$ s
Filter inductance	$L_{line} = 3\ mH$
Filter resistance	$R_{line} = 0.01\ \Omega$
DC bus voltage	$V_{CC} = 220\ V$
Grid voltage	$V_{grid} = 127\ Vrms$
Grid frequency	$f_{grid} = 60\ Hz$
Switching frequency	$f_C = 3\ kHz$
Active power	$P_{load} = 800\ W$
Load quality factor	$Q_f = 2.5$
Load resistance	$R_{load} = 20.16\ \Omega$
Load inductance	$L_{load} = 21.4\ mH$
Load capacitance	$C_{load} = 328.92\ \mu F$

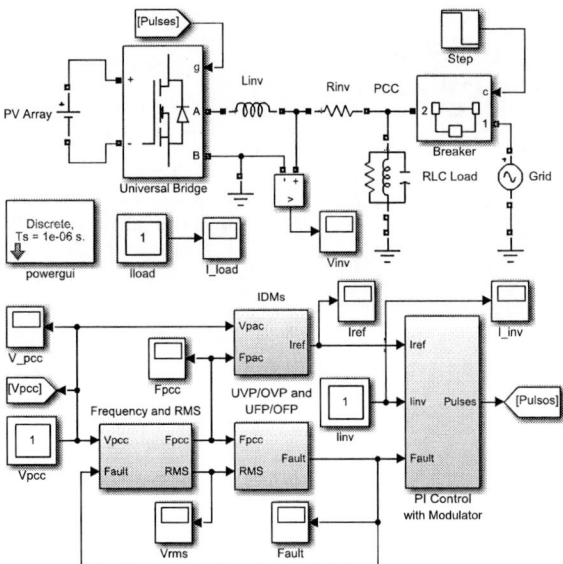

Fig. 5. Model under analysis at Matlab/Simulink®

A. Frequency and RMS block

In the Frequency and RMS block it is performed the calculus of frequency and RMS value of the voltage at the PCC.

As presented by [14], a frequency measurement method was used based on the quantity of samples in a complete cycle, containing the zero crossings of the mains voltage. The frequency is obtained by (8); N is the number of samples by complete cycle and T_S the sampling interval.

$$f = \frac{1}{N.T_S} \qquad (8)$$

Depending on the frequency value, or its modification during operation, the N value probably will not be an integer number. So this calculus must obey (9) and depends on the values of the two samples around the first zero crossing ($|y_{-1}|$ and $|y_0|$), the values of the last two samples around the last zero crossing ($|y_k|$ and $|y_{k+1}|$) and of the whole number of samples between these zero crossings (k).

$$N = K_1 + K_2 + K_3 = k + \frac{|y_0|}{|y_0|+|y_{-1}|} + \frac{|y_k|}{|y_k|+|y_{k+1}|} \qquad (9)$$

Fig. 6 shows the frequency measurement block. A Switch block is added so that the initial frequency is equal to the grid nominal frequency.

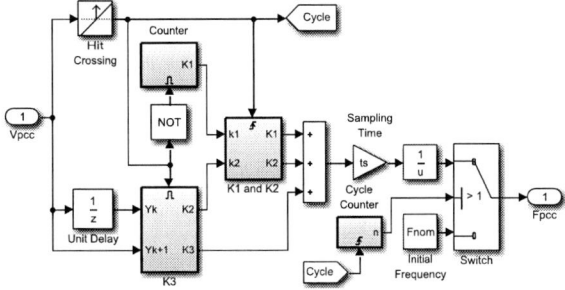

Fig. 6. Frequency block model at Matlab/Simulink®

B. UVP/OVP and UFP/OFP Block

In this block (Fig. 7), the RMS value and the voltage frequency at the PCC are sent to the subsystems Voltage Thresholds e Frequency Thresholds. If the RMS or frequency values exceeds the limits established by IEEE Std 929-2000 standard the Cycles Counter is enabled to count the abnormal voltage or frequency cycles. This bock is reseted if the voltage or frequency returns to their normal values or the islanding is detected.

C. Islanding Detection Methods (IDMs) Blocks

In this block the inverter reference current is calculated with the application of active AFD, SMS or SFS methods. Fig. 8 shows one possibility to implement the active anti-islanding AFD method.

Likewise, Fig. 9 and Fig. 10 present one implementation possibility for the SMS and SFS algorithms, respectively.

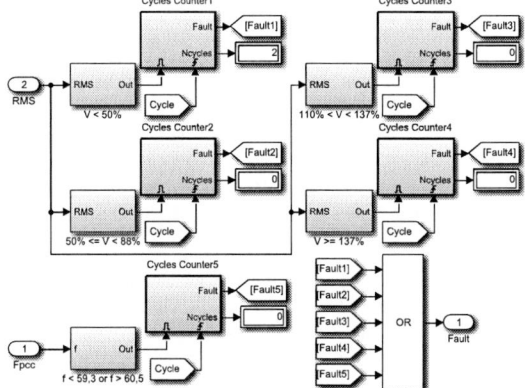

Fig. 7. UVP/OVP and UFP/OFP block models at Matlab/Simulink®

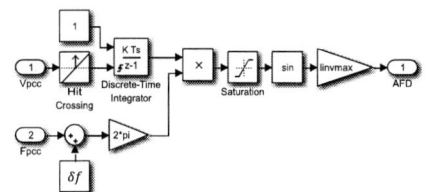

Fig. 8. Active AFD method at Matlab/Simulink®

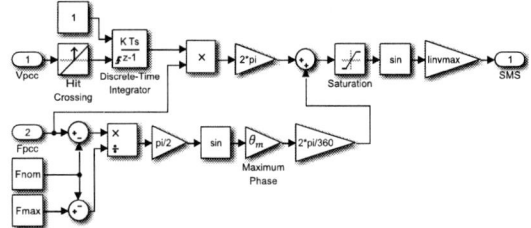

Fig. 9. Active SMS method at Matlab/Simulink®

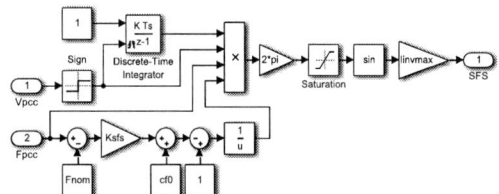

Fig. 10. Active SFS method at Matlab/Simulink®

IV. SIMULATION RESULTS

To compare the performance of AFD, SMS and SFS active methods, it was necessary to set the same disturbance in the inverter current reference for each method by matching the values calculated in (1), (3) e (6) making fair comparisons.

Considering f_{PCC} equals to the over frequency limit of 60.5 Hz, it was obtained $cf_0 = 0.03$ and $k_{SFS} = 0.01$ for the SFS method. In order to match (1) and (6) it is possible to find $\delta f = 2.1943$ for the AFD method. Equaling (3) and (6) and considering $f_m - f_{grid} = 3\ Hz$ it is found $\theta_m = 50.45°$ for the active SMS method.

In the case of these simulations, the inverter was disconnected as soon as the frequency reach the limits of UFP/OFP thresholds presented by the standards.

A. Active Frequency Drift (AFD)

Fig. 11 shows the simulation response of the active AFD method (with Table I parameters) for v_{PCC} voltage, i_{inv} current, failure signal (multiplied by 150) and f_{PCC} frequency.

The disconnection of the distribution grid occurs in 66.7 ms, i.e., at the end of the fourth cycle of the voltage at the PCC. For $\delta f = 2.1943$, the photovoltaic inverter stops supplying the local load in 83.1 ms, that is, islanding detection occurs after 16.4 ms when the frequency f_{PCC} reaches the over frequency limit of 60.5 Hz considering IEEE Std 929-2000 standard.

Fig. 12 presents the odd harmonics of i_{inv} current for the fourth cycle. The obtained THD (Total Harmonic Distortion) is about 3.54%, lower than the 5% allowed by the IEEE Std 519-2014 for $I_{SC}/I_L < 20$, and the highest distortion value equal to 5.08% is obtained for the second harmonic due to the asymmetry in the waveform by the presence of the time t_Z (Fig. 13) [15]. However, the IEEE Std 519-2014 also recommends that the sum of even harmonics must be lower than 25% of the sum of odd harmonics.

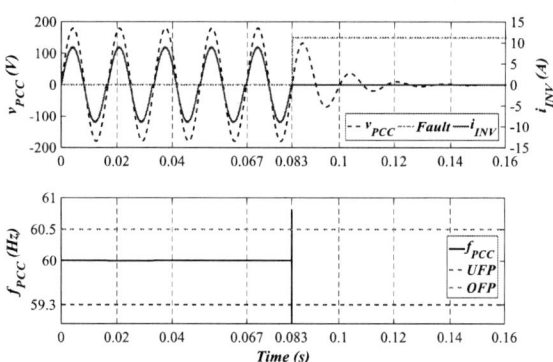

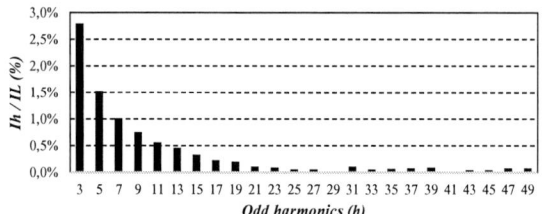

Fig. 11. AFD simulation results for $\delta f = 2.1943$ (fault signal is multiplied by 150).

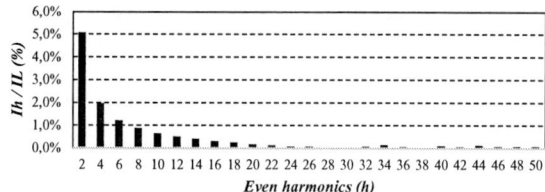

Fig. 13. Even harmonics of i_{inv} current for the AFD method.

B. Slip-Mode Frequency Shift (SMS)

Fig. 14 shows the response of the active SMS simulation method for v_{PCC} voltage, i_{inv} current, failure signal (multiplied by 150) and f_{PCC} frequency.

The disconnection of the distribution grid occurs in 66.7 ms, i.e., at the end of the fourth cycle of the voltage at the PCC. For $\theta_m = 50.45°$, the photovoltaic inverter stops supplying the local load in 117.2 ms, that is, the islanding detection occurs after 50.5 ms when the frequency f_{PCC} reaches the under frequency limit of 59.3 Hz for the IEEE Std 929-2000 standard.

Fig. 15 presents the odd harmonics of i_{inv} current for the fourth cycle. The obtained THD (Total Harmonic Distortion) is about 0.2%, far lower than the 5% allowed by the IEEE Std 519-2014 for $I_{SC}/I_L < 20$ [15]. Similar to the AFD, it is worth mentioning that the highest distortion value is obtained for the second harmonic, however, a value of 0.01%, lower than the active AFD method (Fig. 16).

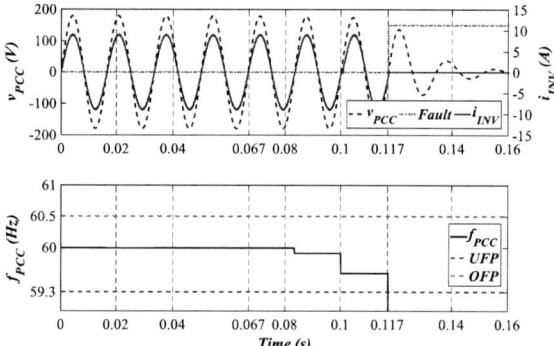

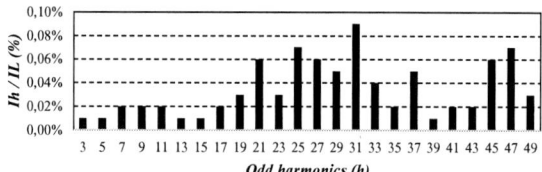

Fig. 14. SMS simulation results for $\theta_m = 50,45°$ e $f_m - f_{grid} = 3\ Hz$ (fault signal is multiplied by 150).

Fig. 15. Odd harmonics of i_{inv} current for the SMS method.

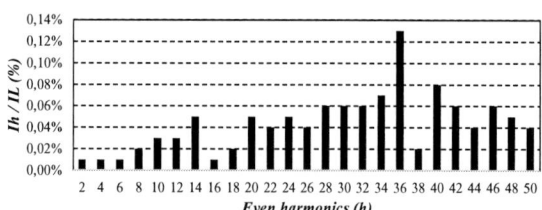

Fig. 16. Even harmonics of i_{inv} current for the SMS method.

C. Sandia Frequency Shift (SFS)

Fig. 17 shows the simulation response of the active SFS method for v_{PCC} voltage, i_{inv} current, failure signal (multiplied by 150) and f_{PCC} frequency.

The disconnection of the distribution grid occurs in 66.7 ms, i.e., at the end of the fourth cycle of the voltage at the PCC. For $cf_0 = 0.03$ and $k_{SFS} = 0.01$, the photovoltaic inverter stops supplying the local load in 116.3 ms, that is, the islanding detection occurs after 49.6 ms when the frequency f_{PCC} reaches the over frequency limit of 60.5 Hz for the IEEE Std 929-2000 standard.

Fig. 18 presents the odd harmonics of i_{inv} current for the fourth cycle. The obtained THD (Total Harmonic Distortion) is about 3.21%, lower than the 5% allowed by the IEEE Std 519-2014 for $I_{SC}/I_L < 20$ [15]. However, the highest distortion value of 2.38% is obtained for the third harmonic and a distortion of 0.01% for the second harmonic, being much lower than the active AFD method due to the presence of the two time segments t_Z (Fig. 19).

Fig. 20 compares the three methods with respect to the last cycle before the grid disconnection. For the 3rd to 9th harmonics, a THD of 3.43% is obtained for the active AFD method, 0.03% for the active SMS method and 2.96% for the active SFS method, that is, the highest THD is obtained for the active AFD method and the lowest for the active SMS method. For the other harmonic intervals, the highest THD are for the active SFS method and the lowest for the active SMS method. Thus, the highest distortions for the active AFD and SFS methods occur for the 3rd to 9th harmonics (Fig. 12 and Fig. 18) and in the active SMS method the highest distortions are distributed in the higher harmonics (Fig. 15).

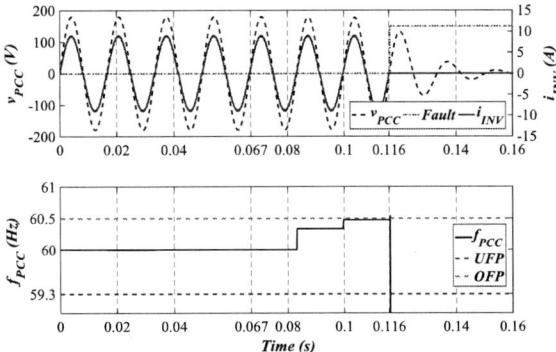

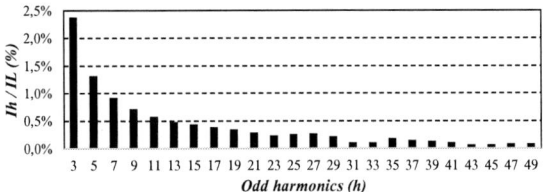

Fig. 17. SFS simulation results for $cf_0 = 0.03$ and $k_{SFS} = 0.01$ (fault signal is multiplied by 150).

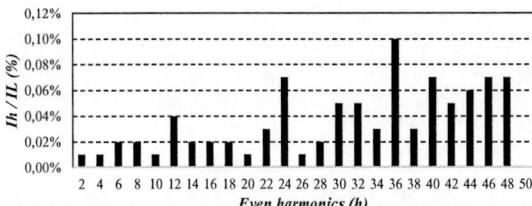

Fig. 19. Even harmonics of i_{inv} current for the SFS method.

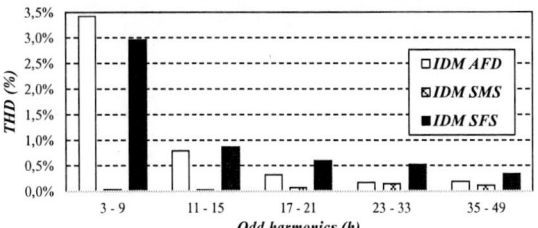

Fig. 20. Comparison of odd harmonics for i_{inv} current.

Table II presents comparisons over AFD, SMS e SFS methods for the mentioned characteristics [9], [16], [17]:

1) Detection time: According to the results shown in Fig. 11, Fig. 14 and Fig. 17, one may conclude that the AFD method is the fastest regarding detection time, while SMS and SFS methods have very close detection times. However, in [18] another option is presented for the SMS method using a non-sinusoidal function, resulting in a smaller NDZ and being more reliable for large load variations.

2) Injected current THD: According to the results shown in Fig. 12, Fig. 15 and Fig. 18, one can verify that the AFD method present the greatest THD. In AFD and SFS methods, the biggest distortions are concentrated in the lower harmonics. In SMS case, the highest distortions are distributed in the higher harmonics, contributing to the reduction of the THD. However, in [19]-[21] other options of current waveforms for the SFS are proposed trying to reduce the THD.

3) Number of mathematical operations: As shown in Fig. 8, Fig. 9 and Fig. 10, at AFD method there are executed 6 mathematical operations, being 2 of addition/subtraction, 3 of multiplication/division and 1 trigonometric. Considering the SMS method, 13 mathematical operations are applied, being 4 of addition/subtraction, 7 multiplication/division and 2 trigonometric. For the SFS method, there are executed 12 operations, 4 of addition/subtraction, 7 of multiplication/division and 1 trigonometric. In this way, it is possible to conclude that the SMS and SFS methods are those that require a greater amount of operations, which implies in longer simulation time and consequently higher computational cost.

TABLE II

MAIN CHARACTERISTICS OF THE EVALUATED METHODS

	AFD	SMS	SFS
Detection Frequency	60.5 Hz	59.3 Hz	60.5 Hz
Detection Time	16.4 ms	50.5 ms	49.6 ms
THD of injected current	3.54%	0.20%	3.21%
Mathematical operations	6	13	12
Simulation time	846.3 ms	847.1 ms	848.5 ms

4) Simulation time: To obtain the simulation time for the AFD, SMS and SFS methods for a complete cycle the tic-toc function of Matlab® was adopted. The simulations were redone 1000 times in order to obtain the minimum simulation time spent by the algorithms. Considering that the algorithms of the analyzed methods are easy to implement, there is a small difference among the simulation times obtained. However, the highest values are obtained for the simulation of the SMS and SFS methods, which can be explained due to the greater amount of mathematical operations required.

CONCLUSION

This paper presented the modeling and simulation of a single-phase photovoltaic inverter connected to the distribution grid in low voltage with the objective of analyzing and comparing the active AFD, SMS and SFS anti-islanding detection methods.

Regarding the AFD method, although it has been the fastest algorithm with the shortest time for islanding detection and shorter simulation time, it has the disadvantage of having the highest THD value, and it is observed that the greatest current distortion occurs in the second harmonic.

Regarding the SMS method, despite being the one with the lowest THD value because it presents no additional distortions in the current waveform and consequently a distribution of distortion in the higher harmonics, it has the disadvantage of requiring the greatest amount of mathematical operations and time to detect islanding.

With respect to the SFS method, it has the advantage of presenting a distortion value for the second harmonic much lower than the AFD method and the disadvantage of presenting a high THD and greater simulation time in the studied simulation case.

Thus, in order to do the analyzed active methods more efficient regarding the islanding detection, studies have been developed mainly to reduce the NDZ and decrease the THD of the injected grid current.

For future work, the suggestion is the experimental development of the algorithms for the anti-islanding methods analyzed, aiming to implement them in a low cost microcontroller and obtain the results in order to validate the analysis via simulation. Also, seek to improve the anti-islanding techniques evaluated by modifying the inverter reference current waveform distortion to decrease the non-detection zone (NDZ) and decrease the total harmonic distortion (THD) of the inverter output current.

REFERENCES

[1] IEEE Recommended Pratice for Utility Interface of Photovoltaic (PV) Systems, IEEE Std 929-2000.

[2] M. R. Islam, W. Xu and F. Rahman, Advances in Solar Photovoltaic Power Plants, Berlin: Springer, 2016.

[3] P. Mahat, Zhe Chen and B. Bak-Jensen, "Review of Islanding Detection Methods for Distributed Generation," 2008 Third International Conference on Electric Utility Deregulation and Restructuring and Power Technologies, Nanjing, 2008, pp. 2743-2748.

[4] Zhihong Ye, A. Kolwalkar, Yu Zhang, Pengwei Du and Reigh Walling, "Evaluation of Anti-Islanding Schemes Based on Nondetection Zone Concept," in IEEE Transactions on Power Electronics, vol. 19, no. 5, pp. 1171-1176, Sept. 2004.

[5] M. E. Ropp, M. Begovic, A. Rohatgi, G. A. Kern, R. H. Bonn and S. Gonzalez, "Determining the Relative Effectiveness of Islanding Detection Methods Using Phase Criteria and Nondetection Zones," in IEEE Transactions on Energy Conversion, vol. 15, no. 3, pp. 290-296, Sept. 2000.

[6] L. A. C. Lopes and H. Sun, "Performance assessment of active frequency drifting islanding detection methods", IEEE Transactions on Energy Conversion, vol. 21, no. 1, pp. 171-180, March 2006.

[7] W. Bower and M. Roop, "Evaluation of islanding detection methods for photovoltaic utility-interactive power systems," Task V Report IEA-PVPS T5-09, March 2002.

[8] R. Teodorescu, M. Liserre and P. Rodríguez, Grid Converters for Photovoltaic and Wind Power Systems, United Kingdom: John Wiley & Sons, 2011.

[9] M. A. G. de Brito, M. G. Alves, L. P. Sampaio and C. A. Canesin, "Anti-island Strategies Applied At Distributed Generation Systems," Power Electronics Review, vol. 23, no. 2, pp. 226-234, June 2018.

[10] M. E. Roop, M. Begovic and A. Rohatgi, "Analysis and performance assessment of the active frequency drift method of islanding prevention", IEEE Transactions on Energy Conversion, vol. 14, no. 3, pp. 810-816, September 1999.

[11] Y. Jung, J. Choi, B. Yu, J. So, G. Yu and J. Choi, "A Novel Active Frequency Drift Method of Islanding Prevention for the grid-connected Photovoltaic Inverter," 2005 IEEE 36th Power Electronics Specialists Conference, Recife, 2005, pp. 1915-1921.

[12] F. Liu, Y. Kang, Y. Zhang, S. Duan and X. Lin, "Improved SMS islanding detection methods for grid-connected converters", IET Renewable Power Generation, vol. 4, iss. 1, pp. 36-42, 2010.

[13] Bahador, M. Pahlevani, S. Makhdoomi Kaviri and P. Jain, "Advanced Slip Mode Frequency Shift Islanding Detection Method for Single Phase Grid Connected PV Inverters," 2016 IEEE Applied Power Electronics Conference and Exposition (APEC), Long Beach, CA, 2016, pp. 378-385

[14] S. Huili, "Performance Assessment of Islanding Detection Methods Using the Concept of Non-Detection Zones," M. A. Sc. Thesis, Dept. Elect. Comput. Eng. Montreal, QC: Concordia University, February 2005.

[15] IEEE Recommended Pratice and Requirements for Harmonic Control in Electric Power Systems, IEEE Std 519-2014.

[16] R. S. Kunte and W. Gao, "Comparison and Review of Islanding Detection Techniques for Distributed Energy Resources," 2008 40th North American Power Symposium, Calgary, AB, 2008, pp. 1-8.

[17] M. El-Moubarak, M. Hassan and A. Faza, "Performance of Three Islanding Detection Methods for Grid-tied Multi-Inverters," 2015 IEEE 15th International Conference on Environment and Electrical Engineering (EEEIC), Rome, 2015, pp. 1999-2004.

[18] B. Mohammadpour, M. Pahlevaninezhad, S. M. Kaviri and P. Jain, "A New Slip Mode Frequency Shift Islanding Detection Method for Single Phase Grid Connected Inverters," 2016 IEEE 7th International Symposium on Power Electronics for Distributed Generation Systems (PEDG), Vancouver, BC, 2016, pp. 1-7.

[19] A. Yafaoui, B. Wu and S. Kouro, "Improved Active Frequency Drift Anti-islanding Detection Method for Grid Connected Photovoltaic Systems," in IEEE Transactions on Power Electronics, vol. 27, no. 5, pp. 2367-2375, May 2012.

[20] W. Chen, G. Wang, X. Zhu and B. Zhao, "An Improved Active Frequency Drift Islanding Detection Method with Lower Total Harmonic Distortion," 2013 IEEE Energy Conversion Congress and Exposition, Denver, CO, 2013, pp. 5248-5252.

[21] H. H. Zeineldin and S. Kennedy, "Sandia Frequency-Shift Parameter Selection to Eliminate Nondetection Zones," in IEEE Transactions on Power Delivery, vol. 24, no. 1, pp. 486-487, Jan. 2009.

978-1-7281-4181-7/19 $31.00 © 2019 IEEE

A Hybrid Bidirectional Push-Pull DC-DC Converter with a Ladder Switched-Capacitor Cell

Marcos Dantas, Fellipe Oliveira, Lorena Albuquerque, Isaac Freitas and Romero Andersen
Department of Electrical Engineering
Federal University of Paraiba
João Pessoa, Paraíba
Email: marcos.sa@cear.ufpb.br, fellipe.oliveira@cear.ufpb.br, lorena.albuquerque@cear.ufpb.br, isaacfreitas@cear.ufpb.br, romero@cear.ufpb.br

Abstract – **Dc-dc power conversion has been an important topic research in the last years, due to increased use of distributed generation, electric vehicles, energy storage and uninterruptible power supplies. In these applications, it is common to have situations where there's no specific load or power supply side, that is, both sides can behave like a power supply or a load, which highlights the importance of studying bidirectional converters. This paper proposes a new hybrid push-pull dc-dc converter with a switched capacitor ladder cell and will be presented the converter analysis, deduction of mathematical equations of static gain and design, control strategy and simulation results. The advantages of this topology are bidirectionality, soft switching, high gain, natural voltage clamping, reduced switch voltage and high-frequency isolation.**

Keywords – *Bidirectional converter, push-pull converter, ladder switched-capacitor cell.*

I. INTRODUCTION

The global energy sector is undergoing major changes as conventional energy sources are becoming scarce and expensive to extract and the rising of greenhouse gases are accelerating global warming. These two aspects have led the world to adopt renewable resources as sources of energy and to transform this energy in places close to the points of use and because of this the microgrids have been seen as a viable option to provide electricity to areas where the electric network does not reach [1].

Since most renewable sources are high current low voltage applications it is preferable to use high frequency isolated DC-DC converters. In addition, current fed converters are desired in high conversion gain applications using low turn ratio transformers compared to voltage fed converters [2].

Conventional topologies that use boost converters must operate at high duty cycle to achieve high voltage gain which leads to increased stress on the switches resulting in high conduction and reverse diode recovery losses [3].

Another problem in isolated current-fed converters is the overshoot voltage at the turn-off of switches caused by mismatch between the currents of an input inductor and a leakage inductance of a transformer [4], [5].

This paper proposes a new bidirectional isolated high gain hybrid converter with high frequency isolation composed of a push-pull on the low-voltage side, which ensures the natural clamping of the switches by eliminating the overshoot voltage, and a ladder capacitor-switched cell on the side of high voltage capable of dividing up to 2 times the bus voltage and thus reducing the stress on the switches.

This study was financed in part by the Coordenação de Aperfeiçoamento de Pessoal de Nível Superior - Brasil (CAPES) - Finance Code 001.

II. PROPOSED TOPOLOGY

As shown in Fig. 1, the bidirectional and isolated topology consists of a push-pull converter located on the low-voltage side and a ladder switched-capacitor cell on the high-voltage side. In addition, the circuit contains eight bidirectional switches represented as S_1-S_8, one main inductor represented as L_1, four very small inductors represented as L_{LK1}, L_{LK2}, L_{H1} and L_{H2} and five capacitors represented as C_P and C_1-C_4.

In spite of the number of switches present in the converter, only two control signals (V_{G1} and V_{G2}), 180° phase-shifted, and their respective complementary signals are required.

Figure 2 shows how command signals are obtained and their formats. V_{G1} controls the switch S_4 and V_{G2} the switch S_3, V_{G1c} controls the switches S_2, S_6 and S_8 while V_{G2c} controls switches S_1, S_5 and S_7.

In this topology, the inductors L_{LK}, L_{H1} and L_{H2} guarantee that the waveforms of the currents in the capacitors will be linear, but the L_{LK} leakage inductances generate loss of duty cycle, like a traditional ZVS-PWM push-pull.

Another important feature of this topology is the bi-directionality without the need to change the command signals.

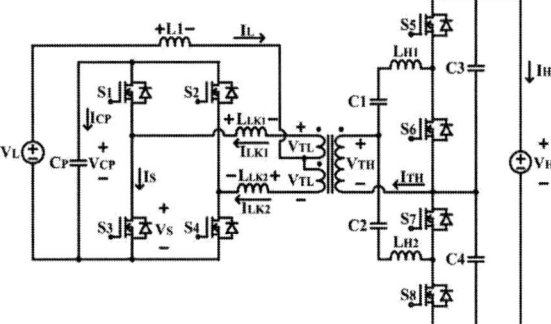

Fig. 1. Proposed converter.

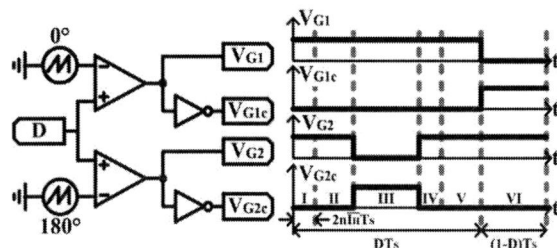

Fig. 2. Generation of command signals.

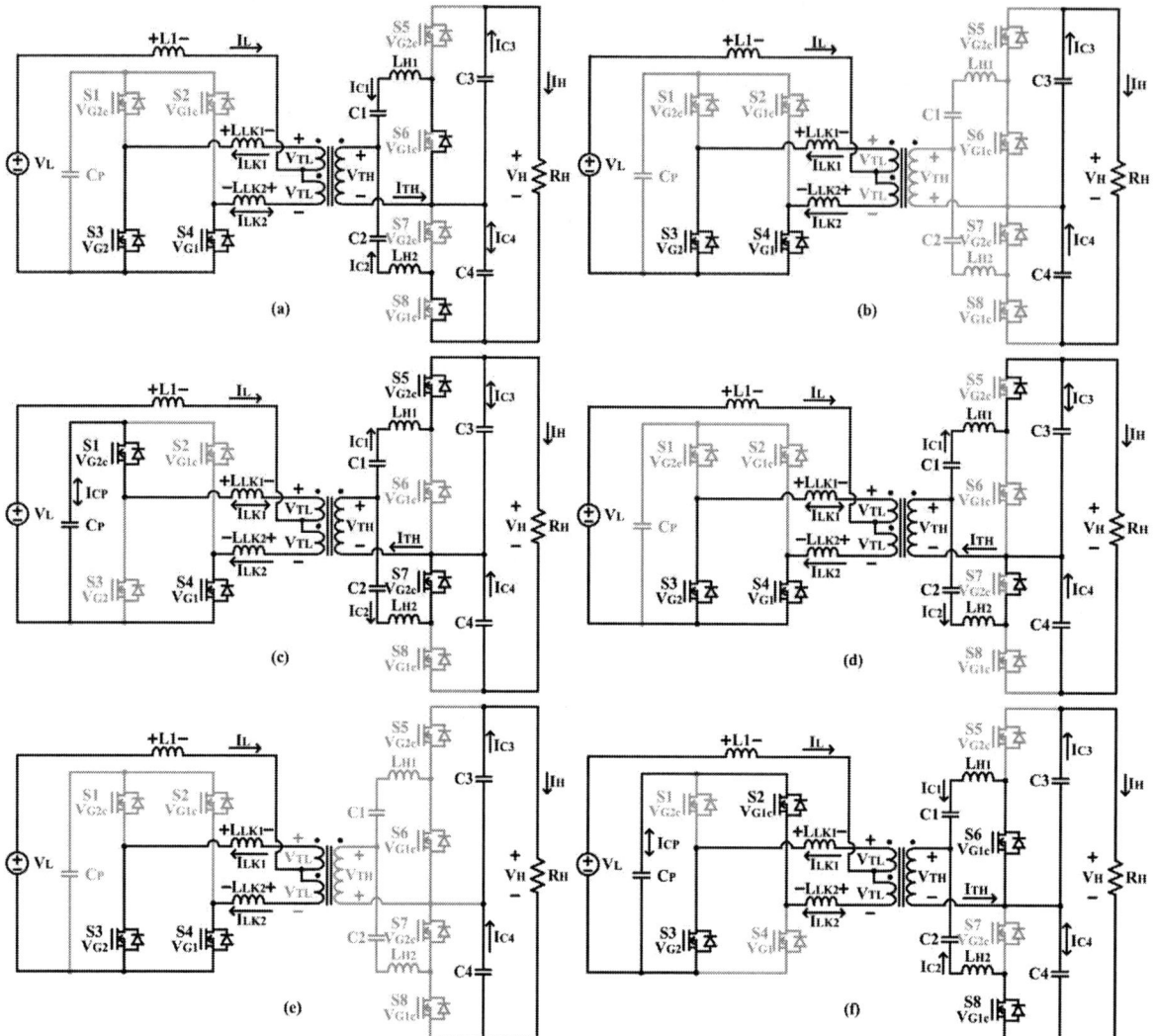

Fig. 3. Six stages of operation: (a) stage I, (b) stage II, (c) stage III, (d) stage IV, (e) stage V, (f) stage VI.

If the source and load sides are reversed the power flow becomes from the high side to the low voltage side and all the stages of operation are the same, but the directions of the currents are all inverted. If there are voltage sources on both sides of the transformer, then the bidirectionality is ensured by simply changing the value of the duty cycle of the control signals.

III. PRINCIPLE OF OPERATION

For simplicity, the six stages of operation, shown in Fig. 3, will be explained for the circuit operating as a step-up converter with a unity turns ratio.

In stages I, II, IV and V the main inductor L_1 is being magnetized and in stages III and VI the inductor is demagnetized because it is supplying energy for the load.

If unity turns ratio is considered, the high side winding current of the transformer I_{TH} is the difference between I_{LK2} and I_{LK1} for all stages.

A. Stage I

The first stage of operation has the equivalent circuit shown in Fig. 3(a). The current of inductance L_{LK1} decreases

linearly and the current of inductance L_{LK2} increases linearly and the high side transformer current circulates through the intrinsic diodes of the switches S_6 and S_8.

B. Stage II

For the second stage, the L_{LK} inductance currents converge to $I_L/2$ which is constant and their difference is zero, so capacitors C_3 and C_4 provide energy for the load, as shown in Fig. 3(b).

C. Stage III

In the third stage, switches S_1 and S_4 are conducting on the low voltage side, the equivalent circuit is shown in Fig. 3(c) and the L_1 current decreases linearly.

During the entire third stage, switches S_5 and S_7 conduct the high side winding current of the transformer.

D. Stage IV

The fourth stage is similar to the first, as can be seen in the Fig. 3(d), the difference occurs in the currents of the leakage inductors since the current of L_{LK1} is now increasing

978-1-7281-4181-7/19 $31.00 © 2019 IEEE

and the current of L_{LK2} decreases while S5 and S7 conduct I_{TH} current.

E. Stage V

The equivalent circuit of the fifth stage is shown in Fig. 3(e), where it is possible to see that the same of stage II occurs.

F. Stage VI

The circuit of the sixth stage is shown in Fig. 3(f), which is similar to stage III with opposite current directions.

IV.   THEORETICAL WAVEFORMS

The main theoretical waveforms of the topology are presented in this section in Figs. 4 and 5. The first shows the voltages and currents of the switches and the inductor located on the low voltage side, and the second contains the waveforms of switches in the high voltage side, as well the voltages and currents in the transformer.

To obtain these graphs the voltages in the inductors L_{H1} and L_{H2} were neglected and thus the voltage at the terminals of the transformer on the high side is considered to be a quarter of the voltage of the high voltage bus ($V_H/4$) during the time interval of stages I and IV.

In addition, V_X is the voltage between the center tap of the transformer winding, on the low side, and the negative terminal of the power supply.

V.   STATIC CHARACTERISTICS OF THE CONVERTER

To calculate the static gain of the converter, the low-voltage side seen in Fig. 1, which is configured as a push-pull converter, is analyzed.

Since the voltages in the inductors and the transformer have a zero average value, the average voltage of the switch S_2 is given by (1), and the voltages V_{CP} and V_L are constant, as seen in Fig. 4, so the average values of the voltages are related according to (2).

$$V_{S2} = D \cdot V_{CP} \tag{1}$$

$$V_{CP} = V_{S2} + V_L \tag{2}$$

Substituting (1) in (2) and rearranging the terms it is possible to find (3).

$$V_{CP} = V_L / (1 - D) \tag{3}$$

In order to calculate the static gain, the volt-second balance of V_{L1} shown in Fig. 4 is applied, so we can find the peak value of the voltage V_X through (4).

$$V_L \cdot (D - 0.5) \cdot T_S = (V_{Xp} - V_L) \cdot (1 - D) \cdot T_S \tag{4}$$

Rearranging the terms of (4), V_{Xp} is calculated by (5).

$$V_{Xp} = \frac{V_L}{2 \cdot (1 - D)} = \frac{V_{CP}}{2} \tag{5}$$

Now, by analyzing the circuit related to the third stage of operation shown in Fig. 3(c), equation (6) is extracted and the low and high side voltages at the transformer terminals

are given by (7) and (8), respectively, where n is the turns ratio of the transformers.

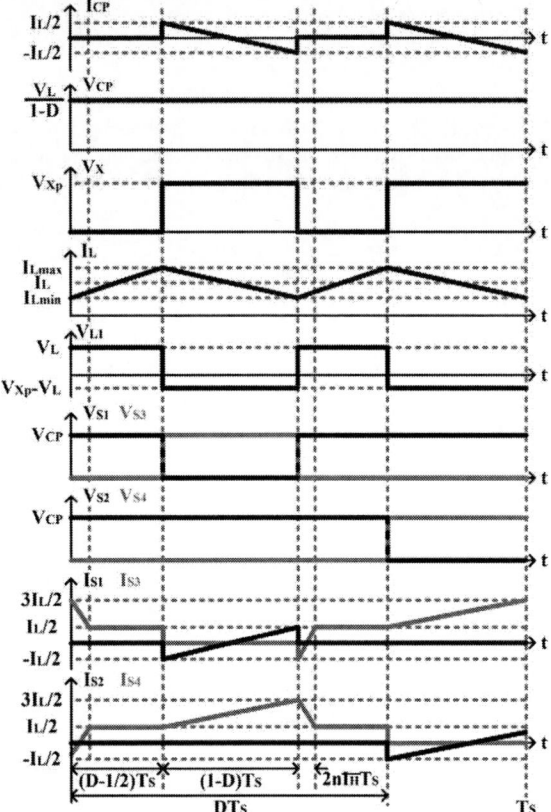

Fig. 4. Voltage and current waveforms for the low voltage side.

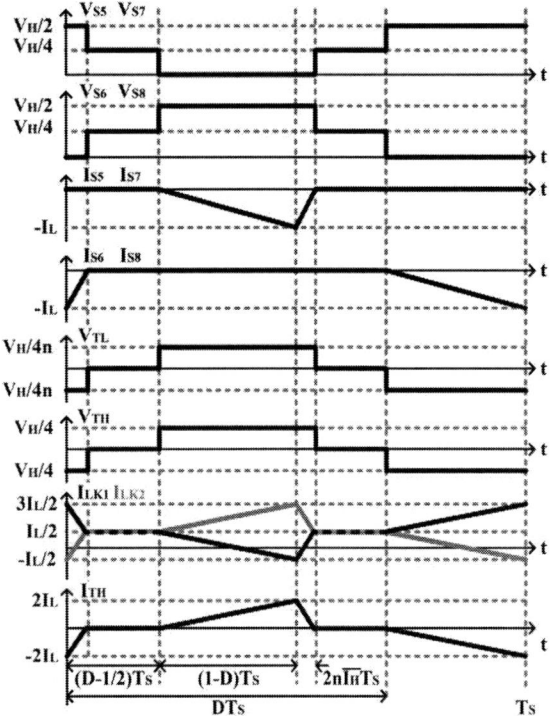

Fig. 5. Voltage and current waveforms for the high voltage side.

$$V_{CP} = 2 \cdot V_{LK} + 2 \cdot V_{TL} \qquad (6)$$

$$V_{TL} = V_{TH}/n \qquad (7)$$

$$V_{TH} = V_H/4 \qquad (8)$$

Substituting (7) and (8) into (6) and using (3) it is possible to obtain (9) which provides the voltage in the leakage inductances during the third stage of operation.

$$V_{LK} = \frac{V_L}{2 \cdot (1-D)} - \frac{V_H}{4n} \qquad (9)$$

In steady-state, the charge balance of capacitor C_P has to be met. During stages III and VI its current shown in Fig. 4 is equal to the leakage inductors currents shown in Fig. 5. Thus, the average value of I_{CP} can be calculated by (10).

$$I_{CP} = \frac{2}{T_S} \cdot \int_0^{(1-D)T_S} \left(\frac{I_L}{2} - \frac{V_{LK} \cdot t}{L_{LK}} \right) dt = 0 \qquad (10)$$

Solving the integral, using (9) and knowing that I_L can be calculated by (11), it is possible to obtain (12).

$$I_L = \frac{P_H}{V_L} = \frac{V_H \cdot I_H}{V_L} \qquad (11)$$

$$\frac{V_H}{V_L} = \frac{V_L}{2 \cdot L_{LK} \cdot f_S \cdot I_H} - \frac{V_H}{4n \cdot L_{LK} \cdot f_S \cdot I_H} \qquad (12)$$

Normalizing the current of the high voltage bus through (13) and considering q as the relation between V_H and V_L, then the static gain can be calculated by (14). This equation has a term representing the loss of duty cycle related to stages I and IV, which is given by (15).

$$\bar{I}_H = \frac{2 \cdot L_{LK} \cdot f_S \cdot I_H}{V_L} \qquad (13)$$

$$q = \frac{V_H}{V_L} = \frac{2n}{(1-D) + 2n\bar{I}_H} \qquad (14)$$

$$D_{loss} = 2n\bar{I}_H \qquad (15)$$

VI. INDUCTANCE AND CAPACITANCES CALCULATIONS

A. Inductor L_1

The inductor voltage can be calculated multiplying its inductance by the ratio of its current ripple (ΔI_L) to the time interval ($(D-0.5)T_S$) in which it occurs, so the inductance value is obtained by (16), which proves the condition that the duty cycle should be greater than 50%, otherwise the inductance value would be negative, which is not possible.

$$L_1 = \frac{V_L \cdot (D - 0,5)}{f_S \cdot \Delta I_L} \qquad (16)$$

B. Capacitors C_1 and C_2

The voltage across capacitors C_1 and C_2 reach their maximum and minimum values approximately in the time interval of $(1-D)T_S$, which corresponds to the third and sixth operating stages.

In the third stage, the current in capacitor C_2 is half of the difference between currents I_{LK2} and I_{LK1}, so the DC value of both currents cancel each other and the capacitor C_2 current will have the same inclination as I_{LK2}. Thus, the voltage ripple across capacitor C_2 can be calculated as (17). Capacitor C1 has the same voltage ripple.

$$\Delta V_{C1} = \Delta V_{C2} = \left| \frac{1}{C_{1,2}} \int_0^{(1-D)T_S} \left(\frac{V_{LK}}{L_{LK}} t \right) dt \right| \qquad (17)$$

Solving the integral, applying (9) and rearranging the terms it is possible to find the capacitance value of C1 and C2 through (18).

$$C_{1,2} = \frac{(1-D)^2}{2 \cdot \Delta V_{C1,2} \cdot L_{LK} \cdot f_S^2} \cdot \left[\frac{V_L}{2(1-D)} - \frac{V_H}{4n} \right] \qquad (18)$$

C. Capacitors C_3 and C_4

Capacitors C_3 and C_4 have the same voltage ripple. For capacitor C_4 the voltage ripple occurs at the stages II to V.

The current flowing through C_4 is equal to the high voltage side bus current, but the direction is inverted.

Thus, the voltage ripple in capacitors C_3 and C_4 can be calculated by (19). Then, to calculate the capacitances C_3 and C_4, simply solve the integral to obtain (20).

$$\Delta V_{C3} = \Delta V_{C4} = \left| \frac{1}{C_4} \cdot \int_{2n\bar{I}_H T_S}^{DT_S} (-I_H) dt \right| \qquad (19)$$

$$C_{3,4} = \frac{I_H}{f_S \cdot \Delta V_{3,4}} \cdot (D - 2n\bar{I}_H) \qquad (20)$$

D. Push-Pull Clamping Capacitor (C_P)

The maximum voltage ripple of this capacitor occurs during the half of the third stage of operation. Then, the voltage ripple value can be obtained through (21) and, solving the integral, we can calculate the value of the push-pull capacitance with (22).

$$\Delta V_{CP} = \left| \frac{1}{C_P} \int_0^{\frac{(1-D)T_S}{2}} \left(\frac{I_L}{2} - \frac{V_{LK}}{L_{LK}} t \right) dt \right| \qquad (21)$$

$$C_P = \frac{(1-D)}{4 \cdot f_S^2 \cdot \Delta V_{CP}} \cdot \left[I_L \cdot f_S - \frac{V_L}{4 \cdot L_{LK}} - \frac{V_H \cdot (1-D)}{8n \cdot L_{LK}} \right] \qquad (22)$$

VII. CURRENT CONTROL STRATEGY

A simple inductor current control loop is used to demonstrate the bidirectionality of the converter as shown in Fig. 6.

In order to describe the dynamic behavior of the converter and correctly adjust the PI controller parameters a transfer function was obtained as follows.

978-1-7281-4181-7/19 $31.00 © 2019 IEEE

The main inductor voltage can be expressed as (23) for stages I and II and as (24) for stage III as shown in Fig. 4.

$$v_{L1} = v_L \tag{23}$$

$$v_{L1} = v_L - \frac{v_{CP}}{2} \tag{24}$$

Now, using the same modeling method of [6], knowing that the low-side bus voltage and the push-pull capacitor voltage are constant ($\hat{v}_L(t) = \hat{v}_{CP}(t) = 0$) and applying the Laplace transform in (25) the transfer function of the converter current plant is given by (26).

$$L_1 \frac{d\hat{i}_L(t)}{dt} = \frac{\hat{v}_L(t)}{2} - \frac{\left(\hat{v}_{CP}(t)D' - \hat{d}(t)V_{CP}\right)}{2} \tag{25}$$

$$\frac{\hat{i}_L(s)}{\hat{d}(s)} = \frac{V_{CP}}{s \cdot 2 \cdot L_1} \tag{26}$$

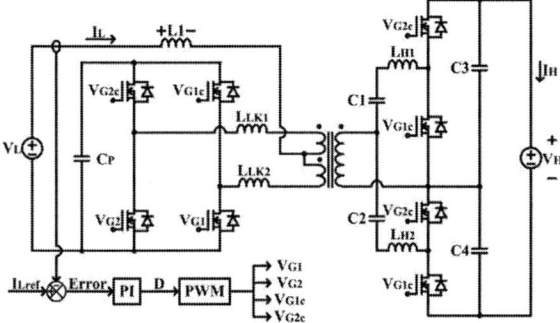

Fig. 6. Block diagram of current control loop.

VIII. SIMULATION RESULTS

Table 1 shows all the specifications used for the simulation of the proposed converter and that allow the calculation of the other parameters presented in table 2.

With all parameters obtained, a PI controller was used to control the low voltage side bus current and thus generate the waveforms that prove the correct operation of the converter.

A. Bidirectionality

The first characteristic analyzed is the converter bidirectionality that is perceived through the current in the inductor L_1, as shown in Fig. 7.

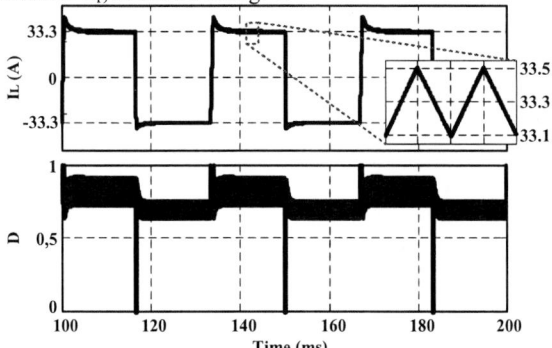

Fig. 7. Simulation waveforms of I_{L1} and duty cycle D.

TABLE I. MAIN SPECIFICATIONS

Specification	Value
Switching frequency (f_s)	30 kHz
Low side bus voltage (V_L)	60 V
High side bus voltage (V_H)	400 V
Leakage inductance (L_{LK})	2 uH
High side inductance (L_H)	2 uH
High side capacitors voltage ripple (ΔV_c)	1 V
Push-Pull capacitor voltage ripple (ΔV_{cp})	0.25 V
Main inductor current ripple (ΔI_L)	0.4 A
Rated Power (P)	2000 W
Turns ratio (n)	1

TABLE II. CALCULATED CONVERTER PARAMETERS

Parameter	Value
Low side current bus (I_L)	33.3 A
High side current bus (I_H)	5 A
Normalized high side current bus	0.01
Power source inductance (L_1)	1.1 mH
Duty cicle (D)	0.72
Capacitors C_1 and C_2	155.5 uF
Capacitors C_3 and C_4	116.7 uF
Push pull capacitor (C_P)	155.5 uF
Load resistance (R_H)	80 Ω

The current value is alternated between 33.3 A and -33.3 A by the controller and its ripple is 0.4 A as desired. So the converter is able to work bidirectionally because its current can be both positive and negative and with only a small change in the value of the switches command signal duty cycle.

B. Capacitors

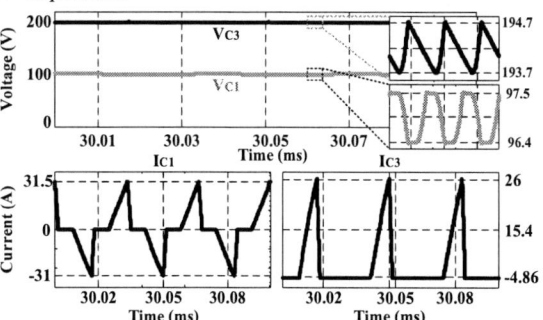

Fig. 8. Capacitors C1-C3 voltages and currents.

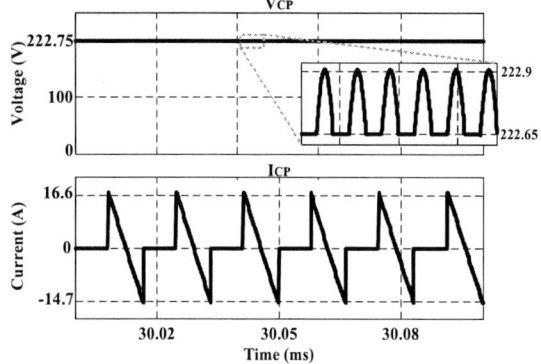

Fig. 9. Push-pull capacitor voltage and current.

978-1-7281-4181-7/19 $31.00 © 2019 IEEE   842

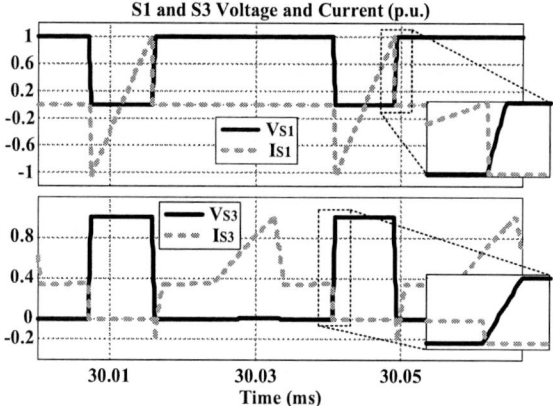

Fig. 10. Soft switching detailed in S1 and S3.

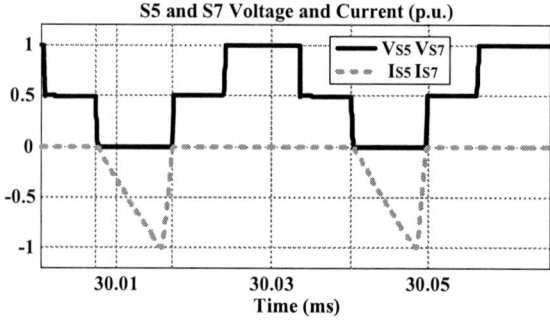

Fig. 11. Voltages and currents of switches S5 and S7.

The voltages and currents of the capacitors C_1 and C_3 are presented in Fig. 8 and it is noticed that the voltage ripples are close to 1 V and the currents obey the format of the theoretical waveforms. For the push-pull capacitor, the voltage ripple was 0.25 V and its current behaved as desired, as shown in Fig. 9.

In addition, it is possible to see that in all capacitors there is no presence of high current peaks due to the small inductors added to the switched capacitor cell and the leakage inductances.

C. Soft Switching

By choosing appropriate dead-time values for the switches control signals, soft switching can be achieved, as shown in Fig. 10. For switch S1 its maximum voltage and current are 222.75 V and 16.6 A and for S3 are 222.75 V and 49 A.

For the high voltage side switches which are the ones suffering from the highest voltage stresses the maximum voltage on the proposed converter switches S_5 and S_7 is half the bus voltage and its value is 200V.

In addition, it is shown in Fig. 11 that the high voltage side switches also have soft switching and the peak current value passing through these switches is 32.2 A.

The voltages and currents of C_2, C_4, S_2, S_4, S_6 and S_8 have the same formats presented in C_1, C_3, S_1, S_3, S_5 and S_7, respectively, however they are 180 ° phase shifted.

IX.   CONCLUSIONS

In this paper was presented the connection of a push-pull converter with a ladder capacitor-switched cell, forming a single hybrid converter capable of solving the voltage overshoot problems on the low-voltage switches present in current-fed converters, as well as reducing the stress on the high voltage side switches, which is a common problem for high gain converters.

In addition, the proposed converter demonstrated good performance and simplicity to obtain the switches control signals, and the insertion of the small inductances for conditioning the currents in the capacitors produced the desired result by eliminating the current peaks which is a characteristic often present in capacitor-switched cell.

REFERENCES

[1] Savitha K.P and P. Kanakasabapathy, "Multi-port DC-DC converter for DC microgrid applications," *2016 IEEE 6th International Conference on Power Systems (ICPS)*, New Delhi, 2016, pp. 1-6.

[2] S. Bansal, "Current-Fed Bidirectional Isolated DC/DC Converters for Hybrid Energy System," *2018 International Conference on Power Energy, Environment and Intelligent Control (PEEIC)*, Greater Noida, India, 2018, pp. 409-414.

[3] K. R. Kothapalli, M. R. Ramteke, H. M. Suryawanshi and N. K. Reddi, "Dual-input Single Inductor Current -Fed Isolated Soft-switched High Gain DC to DC Converter for Hybrid Energy Systems," *2018 IEEE International Conference on Power Electronics, Drives and Energy Systems (PEDES)*, Chennai, India, 2018, pp. 1-5.

[4] A. Chub, R. Kosenko, A. Blinov, V. Ivakhno, V. Zamaruiev and B. Styslo, "Full soft-switching bidirectional current-fed DC-DC converter," *2015 56th International Scientific Conference on Power and Electrical Engineering of Riga Technical University (RTUCON)*, Riga, 2015, pp. 1-6.

[5] U. R. Prasanna, A. K. Rathore and S. K. Mazumder, "Novel Zero-Current-Switching Current-Fed Half-Bridge Isolated DC/DC Converter for Fuel-Cell-Based Applications," in *IEEE Transactions on Industry Applications*, vol. 49, no. 4, pp. 1658-1668, July-Aug. 2013.

[6] M. V. D. de Sá and R. L. Andersen, "Dynamic modeling and design of a Cúk converter applied to energy storage systems," *2015 IEEE 13th Brazilian Power Electronics Conference and 1st Southern Power Electronics Conference (COBEP/SPEC)*, Fortaleza, 2015, pp. 1-6.

High Power Factor Three-Phase Three-Switch Step-Down Converter

João Olímpio Caliman
Dep. of Electrical Engineering
Federal University of Espírito Santo- UFES
Vitória, Brazil
calimanbs@hotmail.com

Walbermark Marques dos Santos
Dept. of Electrical Engineering
Federal University of Espirito Santo –UFES
Vitória, Brazil
walbermark.santos@ufes.br

Tiara Rodrigues Smarssaro de Freitas,
Dept. of Electrical Engineering
Federal University of Espirito Santo - UFES
Vitória, Brazil
tiara.freitas@ufes.br

Domingos Simonetti
Dept. of Electrical Engineering
Federal University of Espirito Santo - UFES
Vitória, Brazil
d.simonetti@ele.ufes.br

Abstract— A high-power-factor three-phase rectifier is proposed in this paper. It operates as a step-down converter, and employs three switches driven by the same signal. A complementary-drive switch is series-connected to the output, avoiding a short-circuit loop whenever main switches are on. Operation analysis is presented, and design equations are derived. Simulation and experimental results are also included, to validate the analysis.

Keywords—Converter, rectifier, power factor, preregulator, three-phase.

I. INTRODUCTION

With the technological development, a significant variety of residential, commercial and industrial electric / electronic devices have appeared, either using alternated current (AC) or direct current (DC) and with linear characteristics or not. According to [1], it was possible by three factors: The emergence of high frequency semiconductors operating as switch, the advent of new advanced control techniques and the introduction of the digital signal processors.

Some loads bring the necessity of using rectifiers and / or inverters that match their current and voltage waveforms to the supply requirements. However, to electrical systems, the set formed by converters and loads usually deteriorates the waveform of grid's voltages due to its power factor or by injecting harmonics.

On this, power factor preregulator (PFP) converters were idealized to improve power factor and reduce harmonics injected to the grid. Controlled converters, especially basic PWM ones (boost, buck, buck-boost, Cuk, sepic, zeta) can operates as single-phase rectifiers presenting high power factor, since well designed and adequate control loop [2-5].

The operation in continuous conduction mode requires the use of two control loops, the input current and the output voltage. But for operation in discontinuous conduction mode (DCM), only the output voltage loop is need because the input current naturally follows the input voltage. For three-phase systems, some converters also can operate as PFP under DCM [6-12]. A three-phase topology is presented in literature based on Single-Ended Primary-Inductor Converter (SEPIC) making use of three single-phase SEPIC converters [13].

This study was financed in part by the Coordenação de Aperfeiçoamento de Pessoal de Nível Superior - Brasil (CAPES) - Finance Code 001.

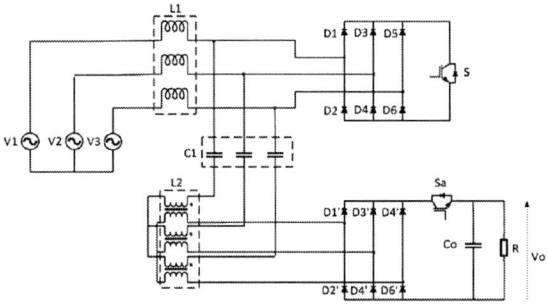

Fig. 1. Sepic-type converter presented by [14].

The three-phase PFP topology presented by [14] and showed in Fig. 1, uses the SEPIC converter operation principle. It operates in DCM and employs two diode bridges and one or two controlled switches in addition to inductors and capacitors. The switch Sa is not required if, from the primary side, the PFP operates boosting the voltage.

A new approach for the above converter operating as a step-up one was presented in [15], where a similar converter is derived using three switches, Fig. 2, which are activated by the same drive signal. This improvement decreases the number of semiconductors on the current path while switches are activated, decreasing conduction losses. However, the converter as presented only works in an application with output voltage greater than peak of line input voltage. In the topology, trying to operate with an output DC voltage lower than the peak of input line voltage, makes appear a short-circuit through the capacitor after the output diode bridge whenever the main switches are activated. This happens because the output diodes are directly polarized.

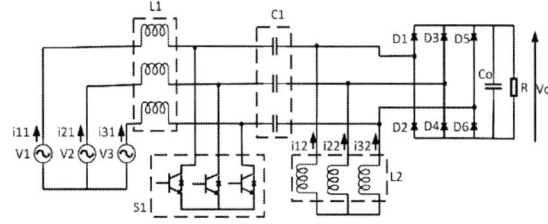

Fig. 2. Three-phase SEPIC-type topology proposed by [15].

This paper presents a single-drive three-phase step down converter derived of the one proposed in [15]. The short-circuit loop is broken inserting a series switch in the full diode bridge's output, operating in a complementary manner to the main ones. The next section performs the analysis of the proposed topology and its operation stages. Section III highlights main design equations, followed by some simulation results in section IV and experimental results in section V. The conclusions are available in section VI.

II. ANALYSIS OF PROPOSED TOPOLOGY

Figure 3 shows the proposed topology that is analyzed in this paper.

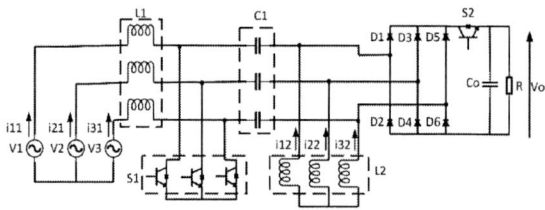

Fig. 3.   Topology proposed with S2 switch.

The S1 switches are turned on and off at the same time and the S_2 switch operates in a complementary manner to them. The analysis is carried out for $60° < \omega t < 90°$ and can be extended to any other interval. It's considered sinusoidal and symmetrical input voltages, Fig. 4. The equations that describe the input voltages are:

$$V_1 = V_{pk} * \sin(\omega t)$$
$$V_2 = V_{pk} * \sin(\omega t - 120°) \qquad (1)$$
$$V_3 = V_{pk} * \sin(\omega t + 120°)$$

Where, V_{pk} represents the peak voltage to each phase, so its rms value is V_{pk} divided by $\sqrt{2}$.

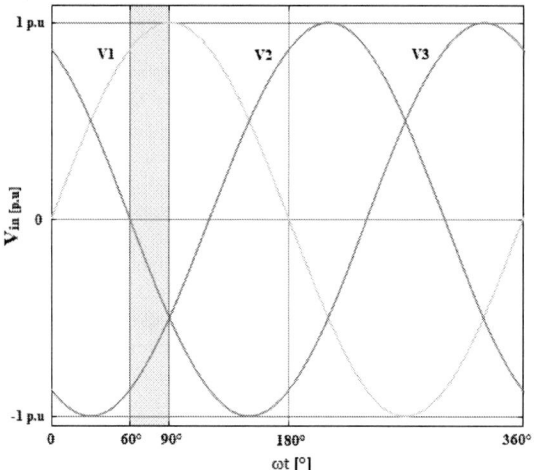

Fig. 4. Input voltages and interval under analysis.

The input voltages will be considered constant during the commutation period, and the capacitors ripple voltages are disregarded.

The converter operates in DCM, and its operation can be divided in three stages. The first can be considered as energy accumulation, the second consists in energy transfer to the load and the third one is the free-wheeling. As the input voltage is different from one phase to other, the second and third stages are not equal among the phases.

A. First Stage

Figure 5 illustrates the equivalent circuit. The switches S_1 are turned on and there is no current passing through the load due to the S_2 being off. The L_1 currents: i_{11} (phase A), i_{21} (phase B), and i_{31} (phase C) grows from the free-wheeling values in module, as can be verified looking to Figure 6. This stage duration is given by the product between the duty cycle (d) and the period of operation (T_s), given by (2):

$$T_{on} = d * T_s \qquad (2)$$

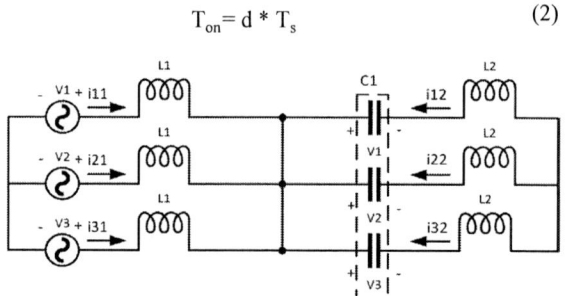

Fig. 5. Turn on equivalent circuit.

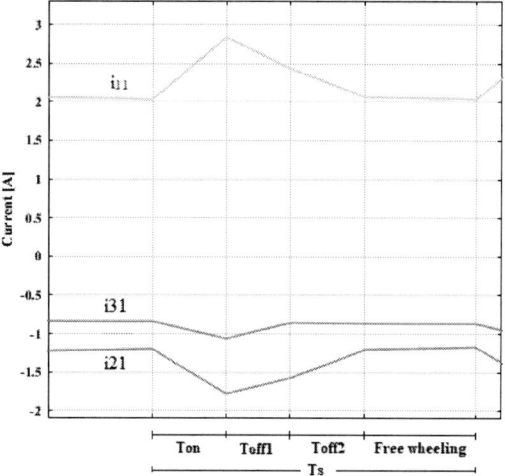

Fig. 6. Input currents behavior during a switching period.

B. Second Stage

This stage starts when the main switches are open and the complementary switch is turned on. One important fact to be mentioned is that the currents pass through the load in this stage and it can be divided into two intervals, "toff1" and "toff2". The L_1 and L_2 inductors currents decrease in module.

Looking to the Fig. 6, that represents the currents behavior during a switching period inside the interval highlighted in Fig. 4, it is remarkable the negative slope of phase 1 current that pass through L1 (i_{11}) and the positive slope of phase 2 (i_{21}) and 3 (i_{31}) currents passing through the others L_1 inductors.

In the Fig. 6 is possible to see the Toff1 interval that occurs until that the lowest current, in module, reaches the free-wheeling value. The Toff1 equivalent circuit is illustrated in Fig. 7.

978-1-7281-4181-7/19 $31.00 © 2019 IEEE   845

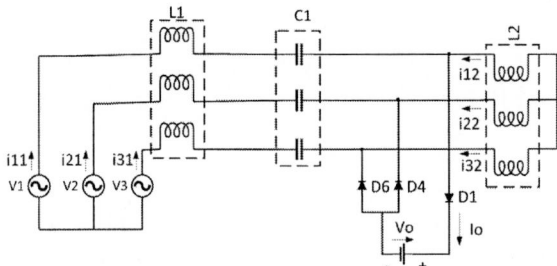

Fig. 7. Toff1 equivalent circuit.

Applying Kirchhoff laws to the mesh formed by phases 2 and 3 and them to the mesh formed by phases 1 and 3 is possible to find (3) doing derivative approximations considering high frequency operation.

$$T_{off1} = \frac{3 * d * T_s * \sin(\omega t - 60°)}{M} \quad (3)$$

Where:

- "d" is the duty cycle.
- "T_s" is the switching period.
- $M = \frac{V_o}{V_{pk}}$, the ratio output voltage / peak phase input voltage.

Equation 3 is almost equal to that presented by [14], but with small changes considering a different topology and a different interval of analysis.

The duration of T_{off2} period is given by (4) and its equivalent circuit is illustrated in Fig. 8. During T_{off2}, to the period of analysis, i_{31} is already in free-wheeling value, so, it's possible obtain (4) just analyzing the mesh formed by phases 1 and 2.

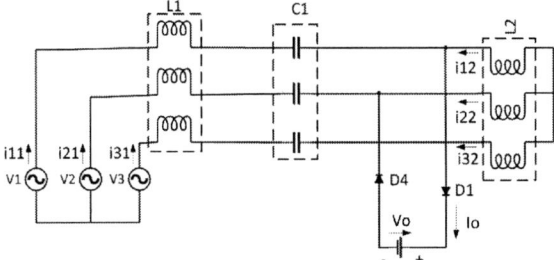

Fig. 8. T_{off2} equivalent circuit.

$$T_{off2} = \frac{2 * \sqrt{3} * d * T_s * \cos(\omega t)}{M} \quad (4)$$

The end of T_{off2} match with second stage end, when all the currents reach the free-wheeling values. It can be seen also in Figure 6.

C. Third stage

In the SEPIC third stage, Fig. 9, all the currents are in their free-wheeling values and keep it that way until the main switches are turned back on. Again, there is no current passing through the load, but it is because the inverted polarization to which the diodes are subjected. Is possible to know the interval of free-wheeling stage (T_{fw}) beyond (5).

$$T_{fw} = T_{on} - T_{off1} - T_{off2} \quad (5)$$

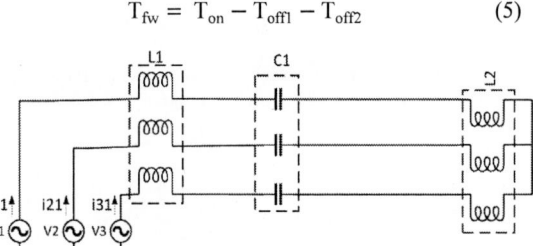

Fig. 9. Third Stage equivalent circuit.

D. Step Down Operation

To make clear the need of the switch S2 in the proposed topology (Fig. 11), the simulation result of the V_{S2} voltage in a few switching periods (top), along with S_1 gate drive (bottom) is shown in Fig. 10. It is noticeable that while switches S_1 are on, the switch S_2 is over a direct voltage between drain and source. In the case S_2 is not present, a short-circuit occurs. All the simulation results in this paper were obtained with Simulink tool in the MATLAB software.

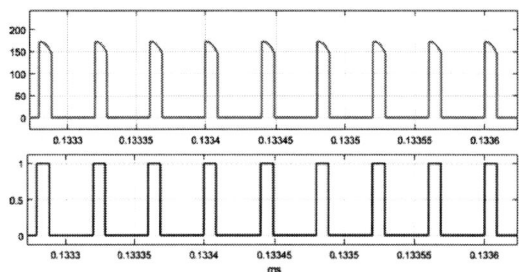

Fig. 10 - V_{S2} voltage (red) and S_1 V_{GS} drive (blue).

Figure 11 represent the equivalent circuit in this situation in the new topology. The switch S_2 voltage is given by:

$$V_{s2} = V_1 - V_o - V_2 \quad (6)$$

So,

$$V_{s2} = V_{pk} * \sqrt{3} - V_o \quad (7)$$

If V_{S2} is positive, the diodes D_2 and D_3 are directly biased, showing that switch S_2 is required when we have $V_o < (V_{pk} * \sqrt{3})$.

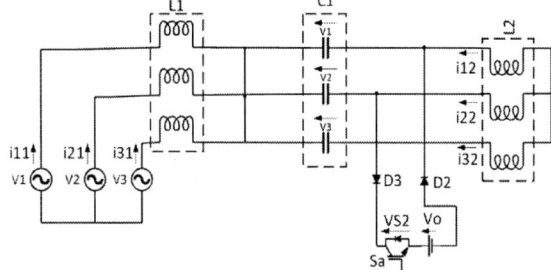

Fig. 11. Turn on circuit analysis.

III. CONVERTER DESIGN AND SIMULATION RESULTS

There are a few points to take care when dimensioning this converter. First, it operates in DCM, and this depends on the duty cycle. To guarantee DCM the duty cycle must be:

978-1-7281-4181-7/19 $31.00 © 2019 IEEE

$$d < \frac{M}{M+\sqrt{3}} \qquad (8)$$

The second point to observe is the converter input current ripple, that is related to the converter power (P), the duty cycle (d), switching period (T_s) and input inductor (L_1). The current ripple is given by (9).

$$\text{ripple}_{L1}(\%) = \frac{3 * d * T_s * V_{pk}^2}{2 * P * L_1} * 100 \qquad (9)$$

The average output current is given by (10). Since the average output current is constant, the power injected to the load will also be constant. This makes easy to dimension the inductors to be used, (11):

$$I_o = \frac{3 * d^2 * T_s * V_{pk}^2}{4 * V * L_{eq}} \qquad (10)$$

$$L_{eq} = \frac{3 * d^2 * T_s * V_{pk}^2}{4 * P} \qquad (11)$$

The equivalente Inductance (L_{eq}) is:

$$L_{eq} = \frac{L_1 * L_2}{L_1 + L_2} \qquad (12)$$

A converter was designed with the following characteristics:

- P= 500 W.
- V_{pk}= 180 V.
- Vo= 150 V.
- $\text{ripple}_{L1}(\%)$= 8%.
- Switching frequency= 25Khz.

The converter was projected to be fed by a symmetrical three-phase source, as described by (1). So, using the previous presented design equations, the values showed in the Table 1 are found.

The chosen value of capacitors C_1 was 4.7 µF such that its voltage ripple was minimal. The load is resistive with 45 ohms to absorb the power of 500W from the 150V DC output. The duty-cycle is d=0.22.

TABLE I.   CONVERTER COMPONENTS VALUES

L_1 inductors	L_2 inductors	C_o capacitor	C_1 capacitor	R_o resistance
12 mH	95 uH	100 uF	4.7 uF	45 ohms

Figure 12 shows input voltage and current for phase A. Fig. 13 shows the output voltage Vo and S2 current, whereas in Fig. 14 voltage and current on switches S2 and S1 are shown.

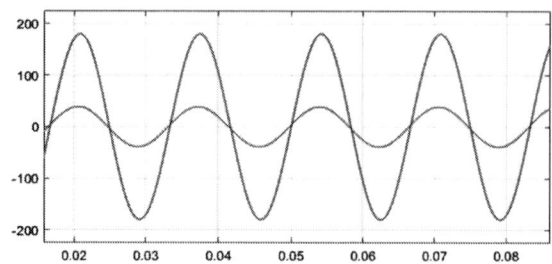

Fig. 12 – Phase A voltage (red, V) and current (blue, Ax20).

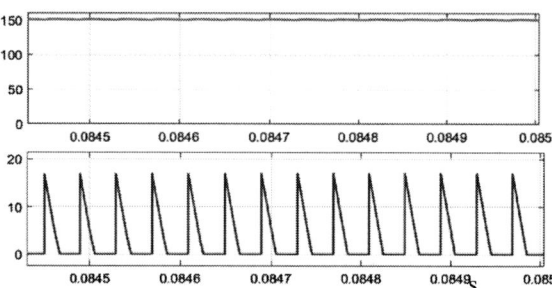

Fig. 13 – Output voltage (top, V) and S_2 current (bottom, A).

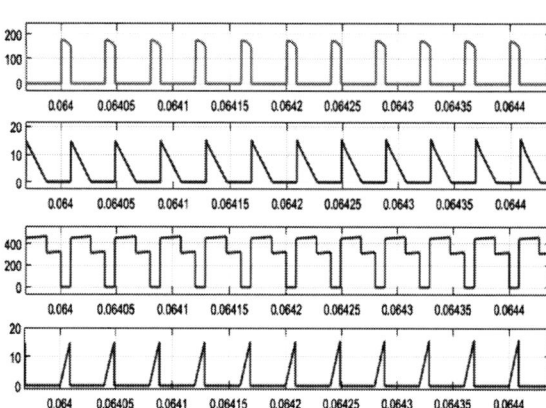

Fig. 14 – From top to bottom: S_2 voltage (V); S_2 current (A); Phase A S_1 voltage (V) and its current (A).

IV. EXPERIMENTAL RESULTS

The following experimental analysis was performed at the Power Electronics and Drives Laboratory of the Federal University of Espírito Santo, Vitória, Brazil.

Due to a practical problem in the collector-emitter voltages of the switches, it was only possible to power the converter by a 50Vrms three-phase sine wave, but the problem is still being studied to be solved.

The Fig. 15 illustrates the input current (yellow) and voltage (blue). It can be seen that the input current has a similar waveform to the input voltage. It also can be seen the converter output voltage (purple), 50V DC.

978-1-7281-4181-7/19 $31.00 © 2019 IEEE   847

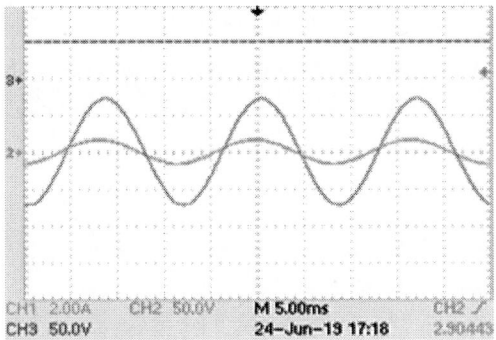

Fig. 15. Experimental result of: input voltage (blue); input current (orange) and output voltage (purple).

The L_2 currents are shown in Fig. 16 and expanded in Fig.17.

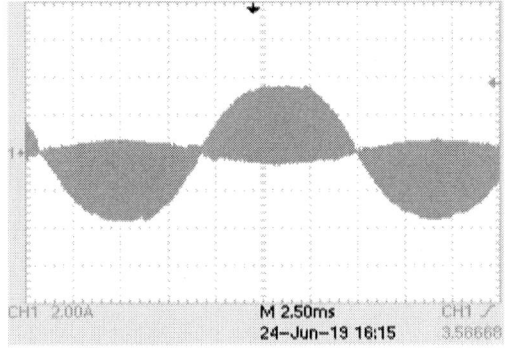

Fig. 16. Experimental result of: L_2 current.

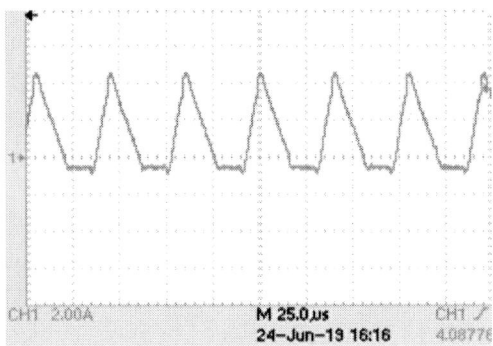

Fig. 17. Experimental L_2 current in a few switching periods.

Figure 18 compares the input voltage to respective C_1 capacitor voltage, showing that the C_1 voltage follows the input voltage.

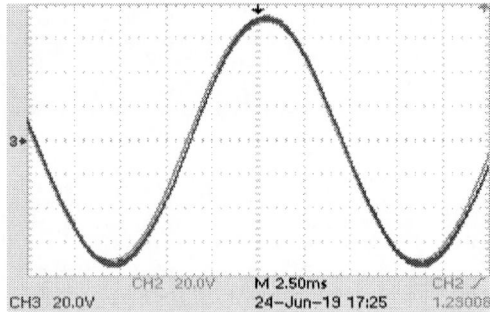

Fig. 18. Comparing input phase voltage (blue) and capacitor C_1 voltage (purple).

V. CONCLUSION

In this paper it is proposed a new three-phase high power factor step-down rectifier. Its operation is similar to SEPIC operation. To work properly it requires a single drive to main switches and a complementary signal to an auxiliary switch. The converter operates in DCM with constant switching frequency, and keeping the output voltage a constant duty cycle is need for each output power level. Equations to design the converter are easy to use. Input currents naturally follow input voltage, and with a ripple defined in the design stage. Simulation and experimental results are presented to corroborate the approach. The topology is a good alternative to achieve a three-phase high power factor rectifier if an output voltage lower than the peak of input voltage is needed.

REFERENCES

[1] Zhu, Z.Q; Jiabing, Hu. "Electrical machines and power-electronic systems for high-power wind energy generation applications Part II - power electronics and control systems" - Special Issue Paper. COMPEL International Journal for Computation and Mathematics in Electrical and Electronic Engineering, v. 32 n. 1, p. 34-71, 2013.

[2] K. H. Liu and Y. L. Lin, "Current waveform distortion in power factor correction circuits employing discontinuous mode boost converters," in Proc. IEEE Power Electron. Spec. Conf. (PESC), 1989, pp. 825–829.

[3] R. Erickson, M. Madigan, and S. Singer, "Design of a simple highpower-factor rectifier based on the flyback converter," in Proc. IEEE APEC, 1990, pp. 792–801.

[4] Simonetti, D. S. L., Sebastian, J., & Uceda, J. (1997). The discontinuous conduction mode Sepic and Cuk power factor preregulators: analysis and design. IEEE Transactions on Industrial Electronics, 44(5), 630-637.

[5] García, O., Cobos, J. A., Prieto, R., Alou, P., & Uceda, J. (2003). Single phase power factor correction: A survey. IEEE Transactions on Power Electronics, 18(3), 749-755.

[6] Borges, Altamir Ronsani; BARBI, Ivo. "Study of a single stage buck-boost three-phase rectifier with high power fator operating in discontinuous conduction mode (DCM)" in Proc. COBEP, 2009, p. 870-877.

[7] Ismail, Esam H.; Erickson, Robert W.. "Single-switch 3 phase PWM low harmonic rectifiers". IEEE Trans. Power Electron., vol. 11, n.2, p. 338-346. Mar. 1996.

[8] Jang, Yungtaek; Erickson, Robert W.."New single-switch three-phase high power fator rectifiers using multi-resonant zero current switching". IEEE Trans. Power Electron., vol. 13, n.1, p. 194-201, Jan.1998.

[9] Prasad, A. S.; Ziogas, Phoivos D.; Manias, Stefanos. "An active power fator correction technique for three-phase diode rectifiers". IEEE Trans. Power Electron., vol. 6, n.1, p. 83-92, Jan. 1991.

[10] Singh, Bhim; Sing, Brij N.; Chandra, Ambrish.; Al-HaddadL, Kamal; Pandey, Ashish; Kothary, Dwarka P. "A review of three-phase improved power quality AC-DC converters". IEEE Trans. Ind. Appl., vol. 51, n. 3, p. 641-660, Mai/Jun 2004.

[11] Yao, K., Meng, Q., Bo, Y., & Hu, W. (2015). Three-phase single-switch DCM boost PFC converter with optimum utilization control of switching cycles. IEEE Transactions on Industrial Electronics, 63(1), 60-70.

[12] Gangavarapu, S., & Rathore, A. K. (2018). Three-Phase Buck–Boost Derived PFC Converter for More Electric Aircraft. IEEE Transactions on Power Electronics, 34(7), 6264-6275.

[13] Tibola, Gabriel; Barbi, Ivo. (2013). Isolated Three-Phase High Power Factor Rectifier Based on the SEPIC Converter Operating in Discontinuous Conduction Mode. IEEE Transactions on Power Electronics, vol. 28, n. 11, p. 4962-4969.

[14] Simonetti, Domingos S. L.; Sebastián J.; Uceda J. "A novel three-phase ac-dc power factor preregulator". In: 26th annual IEEE power electronics specialists conference. Atlanta 1995. Anais do IEEE pesc 1995, p.979-984.

[15] de Freitas T.R.S., Antunes H.M.A., Vieira J.L.F., Ferreira R.T., Simonetti DSL. A DCM three-phase SEPIC converter for low-power PMSG. In: Proceedings of 10th IEEE/IAS International Conference on Industry Applications (INDUS-CON). (Fortaleza, Brazil); November 2012. p. 1–5.

Selection of the Number of Levels of a Modular Multilevel Converter for an Electric Drive

Paulo R. M. Júnior[a], João. V. M. Farias[a], Allan F. Cupertino[b,c], Gabriel A. Mendonça[c],
Marcelo M. Stopa[a] and Heverton A. Pereira[d]

[a]Department of Electrical Engineering, Federal Center for Technological Education of Minas Gerais,
Belo Horizonte, MG, Brazil. paulomatiaspq@gmail.com, joaofariasgv.jvmf@gmail.com, marcelo@cefetmg.br
[b]Department of Materials Engineering, Federal Center for Technological Education of Minas Gerais,
Belo Horizonte, MG, Brazil. afcupertino@ieee.org
[c]Graduate Program in Electrical Engineering, Federal University of Minas Gerais,
Belo Horizonte, MG, Brazil. gforti@gmail.com
[d]Department of Electrical Engineering, Federal University of Viçosa,
Viçosa, MG, Brazil. heverton.pereira@ufv.br

Abstract—**The modular multilevel converter (MMC) is an inherently fault-tolerant topology and an interesting solution for medium-voltage (MV) electrical drives, especially when quadratic loads are employed. In order to select the best cost-effective solution, this paper presents a design procedure and comparison of MMC based MV electrical drives. The designs with the best IGBTs utilization factor for blocking voltages of 3.3 kV, 4.5 kV and 6.5 kV were chosen for comparison. The comparison is based on cost, performance, complexity and energy losses metrics. Manufacturer data were used in the power losses estimation procedure. These metrics are normalized to an index, introduced in order to select the best cost-effective MMC design. The methodology and comparison are exemplified through a 13.8 kV - 16 MW three-phase induction motor which drives an exhaust fan based on real industrial mission profile. The simulation results show that the 29-level MMC has the best performance and lowest drive energy losses. Despite of its higher complexity at a moderate cost, it was evaluated as the best cost-effective solution based on the application mission profile.**

Index Terms—**Modular multilevel converter, electric drive, design, cost-effective solution.**

I. INTRODUCTION

In the recent years, there is a growing concern with energy efficiency in industrial electrical installations. The use of electric drives with variable speed has become an interesting option, since this technology allows to reduce the total energy consumption [1]. For MV electrical drives, the MMCs has become a promising family of converters [2]. Its main features include low dv/dt, high efficiency, modularity and low harmonic distortion in the output variables [3], [4].

About MMC topologies, the Double-Star Half Bridge (DSHB) is widely used in electric drives which load torque is a quadratic function of the motor speed [5]. This type of load is easily found in the industrial field, accounting for approximately 70 % of the market for MV electrical drives [6]. However, this converter topology presents some limitations due to the high voltage oscillations of the submodule (SM) capacitors when the motor operates at low speed and high

torque [4]. In order to improve the MMC-DSHB dynamics at low speeds, it is necessary to use some techniques to mitigate the voltage ripple of the SM capacitors [4], [7]. Companies like Siemens and Benshaw solved this problem and already have this converter topology in the market for pumps, blowers and compressors [5].

An important issue in the MMC design is the blocking voltage of commercially available power IGBTs in the market, ranging from 0.6 to 6.5 kV [8]. Some papers evaluate the optimal converter design for a given application, and different voltage and power rating. Reference [9] proposes the design of a cascade converter for renewable energy integration, taking into account system performance, control complexity and semiconductor cost. The results show that the 19 level topology is the optimal choice for an 11-kV power conversion system. In [10], the MMC and the three-level neutral-point-clamped (3L-NPC) converter are compared and benchmarked for battery energy storage system (BESS). The comparison is based on power modules numbers and ratings, the filter elements, system efficiency, harmonic content and investment cost. Reference [11] presents the design procedure and comparison of power converters used in MV drives for different voltage and power levels. Nevertheless, these references do not present one clear systematic methodology for optimal design in MMC based electric drive system.

The optimal selection of the blocking voltage based on the application mission profile (motor speed and ambient temperature data) and the number of levels are very important for the best performance/cost ratio of the MV electric drive system. Therefore, this work proposes a methodology to select the number of levels for MMC based electric drives. The following contributions are provided:

- Development of a systematic methodology for optimal design on MMC for an electric drive system;
- Selection of the most suitable design based on the mission profile application, considering the system complexity,

performance, cost and energy losses.

The methodology is exemplified through a 13.8 kV - 16 MW three-phase induction motor which drives an exhaust fan. The mission profile data is obtained from the steel industry in southeastern Brazil. This paper is outlined as follows. Section II introduces the MMC based electric drive system and presents the control strategies. The parameter design of the system is presented in Section III. Section IV presents the case study and the parameters of the MMC based electric drive system. Furthermore, the obtained simulation results are discussed in Section V. Finally, Section VI draws the conclusions of this paper.

II. MMC Based Electric Drive

The circuit of the DSHB-MMC based electric drive system is shown in Fig. 1. As observed, this topology contains a cascade association of N half-bridge SMs per arm, with two IGBTs, S_1 and S_2, two diodes, D_1 and D_2, and a SM capacitor, C. R_b refers to the bleeder resistor. The thyristor based switch, S_T, is responsible for bypassing the SM if failure occurs [12]. The arm inductors, L_{arm}, reduce the harmonic distortion in the arm currents [13]. Coupled inductors are considered in this paper due to their reduced volume and weight. These inductors present minimal inductance for the fundamental output current, being theoretically null. Thus, the motor speed dynamics is kept unaffected [4].

Regarding the control scheme, the strategy is based on [4]. The control structure can be divided into: averaging voltage and circulating current control, individual balancing control and a rotor field oriented control (RFOC), as presented in Fig. 2. The averaging voltage and circulating current control is presented ind Fig. 2 (a). The external loop controls the average voltage of all SMs per phase and calculates the circulating current required by the inverter phase. The inner loop is responsible for controlling the circulating current in order to mitigate the second harmonic and introduces damping in the converter dynamics. This control is based on proportional resonant (PR) controller, to suppress the second harmonic component, which is typical ind DSHB topology [14]. In order to reduce the capacitor voltage ripple under low speeds, a common-mode injection (CMI) is employed. This strategy

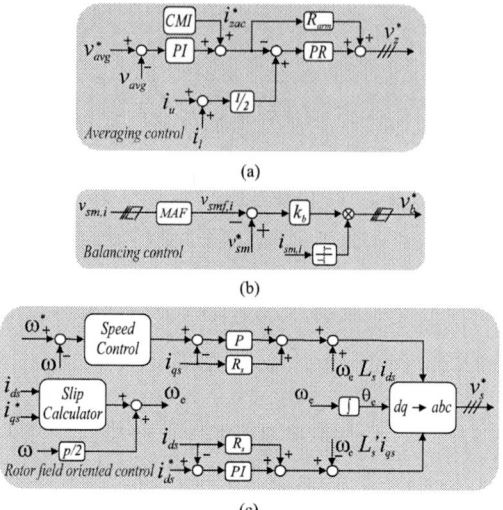

Fig. 2. Control scheme for MMC based electric drive system: (a) Average and circulating current control; (b) Individual balancing control; (c) Rotor field oriented control. *CMI-Common-mode injection. MAF-Moving average filter.*

consists of inserting an alternating circulating current and a common-mode voltage in the reference signals [4].

Furthermore, the individual balancing control, Fig. 2 (b), is used to guarantee the voltage balance of SM capacitors. A moving averaging filter (MAF) is used to mitigate the capacitors voltage ripple and to improve the individual balance performance. Finally, the traditional RFOC, Fig. 2 (c) is responsible for controlling the motor speed, and it is based on [15]. The control signals are summed up, normalized and compared by the voltage modulator, in which the phase-shifted pulse width modulation (PS-PWM) with third harmonic injection is considered in this paper [16]. The normalized reference signals per phase are given by:

$$v_u^* = v_b^* + \frac{v_z^*}{v_{sm,u}^*} - \frac{v_s^*}{v_{sm,u}^* N} + \frac{v_{com}^*}{v_{sm,u}^* N} + \frac{1}{2}, \quad (1)$$

$$v_l^* = v_b^* + \frac{v_z^*}{v_{sm,l}^*} + \frac{v_s^*}{v_{sm,l}^* N} + \frac{v_{com}^*}{v_{sm,l}^* N} + \frac{1}{2}, \quad (2)$$

where v_b^* is the reference of the balancing control, v_z^* is the voltage generated by the control of the circulating current, v_s^* is the reference voltage of the RFOC, $v_{sm,u}^*$ and $v_{sm,l}^*$ are the reference voltages of the SMs of the upper and lower arm, respectively. The v_{com}^* refers to the common mode voltage.

III. MMC Design

The first action to design a MMC based electric drive application is determining the pole-to-pole voltage. The minimum pole-to-pole voltage can be computed by:

$$V_{dc} = \sqrt{2} v_s k_{res}, \quad (3)$$

where v_s is the rated rms line-to-line stator voltage and k_{res} is the percentage of reserve voltage. For typical industrial MV

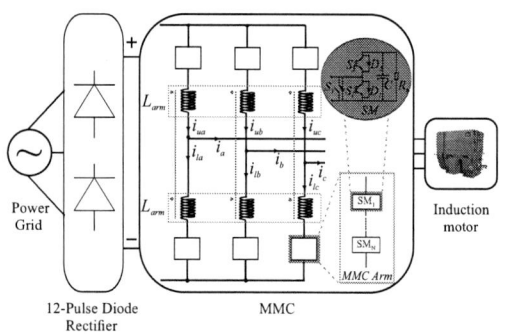

Fig. 1. MMC based electric drive system.

electrical drives, the k_{res} is usually 1.2 [17]. The percentage of reserve voltage is used to guarantee proper operation of the MMC under transient conditions and also to compesate the voltage drop on arm and stray inductances [11].

The best cost/effective design for number of SM is primarily dictated by the blocking voltage capability of the employed IGBTs. In this paper, IGBTs with blocking voltage, V_{bv}, in the range of 3.3 to 6.5 kV are considered. The device utilization factor, f_u can be calculated from:

$$f_u = \frac{V_{sm}^*}{V_{bv,100fit}}, \qquad (4)$$

where $V_{bv,100fit}$ is the IGBT blocking voltage for a device reliability of 100 failures in time (FIT) due to cosmic radiation [9]. The reference SM voltage V_{sm}^* can be computed as:

$$V_{sm}^* = \frac{V_{dc}}{N}. \qquad (5)$$

A higher f_u is key for an improved cost/effective design, since the semiconductor cost is a significant figure in MV converter applications [18]. Tab. I summarizes the reference SM voltage and utilization factor for some SM numbers. Considering the availability of the power semiconductors devices, only a few SM number of 7, 11 and 14 may give higher f_u. Thus, such values were considered for further analysis. A low f_u means the use of unnecessarily high-cost semiconductors. In general, to use the active switching devices cost-effectively, a design must have a f_u of 0.9 or above [18].

The SM capacitance can be calculated based on the MMC energy storage requirements. According to [19], the minimum SM capacitance is given by:

$$C = \frac{NS_nW_{conv}}{3V_{dc}^2}, \qquad (6)$$

where S_n is the MMC apparent power and W_{conv} is the required energy storage per MVA. A typical values of W_{conv} is approximately 60 kJ/MVA for MMC based drive [5].

As mentioned in the previous section, this paper considers the PS-PWM. In this modulation technique, N triangular carriers are used per arm, displaced by $360°/N$, featuring an effective output frequency of [20]:

$$f_{ef} = 2Nf_{sw}, \qquad (7)$$

where f_{sw} is the carrier frequency [20]. In order to maintain constant stator current THD throughout the evaluated topologies, f_{ef} and L_{arm} are kept constant for all designs. The MMC ratings for each MMC design is shown in Tab. II.

IV. CASE STUDY

The electrical drive used in the tests carried out in this work consists of a 16 MW - 13.8 kV induction motor which drives an exhaust fan. The motor parameters are reported in Tab. III.

Simulations are performed in PLECS/MATLAB aiming to compare the designs in terms of performance, efficiency, complexity and cost of each design. The performance analysis is done through the MMC output voltage THD, the efficiency derived from the daily energy loss, and the drive complexity from the total number of arithmetic and logic operations (ALO) performed by the processor. The ABB IGBTs part number 5SNA 0800N330100 of 3.3 kV - 800 A, 5SNA 0800J450300 of 4.5 kV - 800 A and 5SNA 0800J450300 of 6.5 kV - 750 A are selected for this application. The conduction, switching losses and thermal impedance are extracted from look-up table based on the datasheets. The heatsink-to-ambient thermal resistance is calculated in order to have the same junction temperature, T_j, for the most stressed device in all the three designs, within a safety limit (the junction and case temperature below 115 °C and 100 °C, respectively). The mission profile based on the motor speed and ambient

TABLE II
MMC RATINGS FOR SELECTED DESIGNS.

Parameters	MMC Design		
	I	**II**	**III**
N	7	11	14
Number of levels	15	23	29
V_{dc}	24 kV	24 kV	24 kV
V_{bv}	6.5 kV	4.5 kV	3.3 kV
V_{sm}^*	3.43 kV	2.18 kV	1.71 kV
S_n	20 MVA	20 MVA	20 MVA
C_{sm}	5 mF	7.86 mF	10 mF
L_{arm}	7.7 mH	7.7 mH	7.7 mH
R_{arm}	0.1451 Ω	0.1451 Ω	0.1451Ω
f_{sw}	945 Hz	602 Hz	473 Hz
f_{ef}	13.23 kHz	13.23 kHz	13.23 kHz

TABLE I
UTILIZATION FACTOR FOR DIFFERENT SM NUMBER.

N	v_{sm}^* (kV)	V_{bv} (kV)	$V_{d,100fit}$ (kV)	f_u
7	3.43	6.5	3.6	0.95
8	3	6.5	3.6	0.83
9	2.67	6.5	3.6	0.74
10	2.4	6.5	3.6	0.67
11	2.18	4.5	2.25	0.97
12	2	4.5	2.25	0.89
13	1.85	4.5	2.25	0.82
14	1.71	3.3	1.8	0.95
15	1.6	3.3	1.8	0.89
16	1.5	3.3	1.8	0.82

TABLE III
PARAMETERS OF THE INDUCTION MOTOR.

Parameter	Value
Rated active power (P)	16 MW
Rated rms line-to-line stator voltage (v_s)	13.8 kV
Rated stator current (i_s)	801 A
Rated frequency (f)	60 Hz
Rated rotational speed (w_m)	1795 rpm
Rated power factor	0.9
Rated efficiency (η)	97.4 %
Number of poles (p)	4

temperature measurements were collected from a steel industry in southeastern Brazil and are presented in Fig. 3.

The investment cost are evaluated according to the methodology presented in [21]. The cost of SMs capacitors are 150 €/kJ. Furthermore, the dominant initial investment cost, the power electronics cost (semiconductors, wiring, cabinets, control), are estimated in 3.5 €/kVA of installed switching power, P_{sw}, which is given by [10]:

$$P_{sw} = 3.5 N_{semi} V_{bv} I_c, \qquad (8)$$

where N_{semi} is the number of semiconductors devices and I_c is the device collector current. In order to compare the MMC designs with different and often descriptive or nonnumerical indicators, a normalized index, k_x, is calculated by:

$$k_x = \frac{x - x_{min}}{x_{max} - x_{min}}, \qquad (9)$$

where x is the scored value (performance, efficiency, complexity and cost), and x_{min} and x_{max} are the minimum and maximum value of the indicator, respectively. To compare the MMC designs, the total normalized index is calculated by:

$$k_t = k_{performance} + k_{efficiency} + k_{complexity} + k_{cost}. \qquad (10)$$

In this paper, it was considered that all normalized index have the impact factor 1. However, instead of a straight sum of all the k's indices, a weighted sum can be applied to reflect different design emphasis/objectives.

V. RESULTS

A. Dynamic Performance

Initially, the start-up of the MMC for all designs are compared in terms of the dynamic behavior considering a ramp speed profile and a load torque derived as a quadratic function of the motor speed. The results for the design I are presented in Fig. 4. As observed in Fig. 4 (a), the speed control does not present overshoot or oscillations. The line-to-line stator voltages are illustrated in Fig 4 (b). The oscillations observed

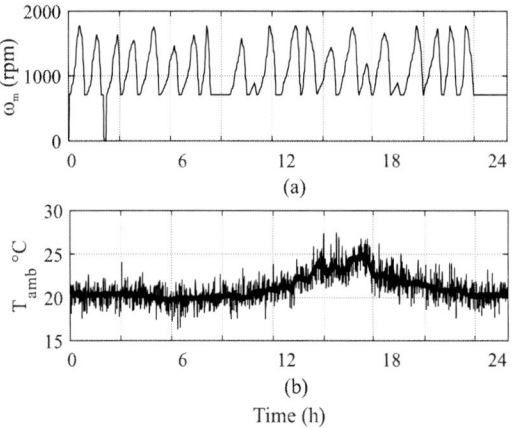

Fig. 3. Mission profiles: (a) Motor speed; (b) Ambient temperature.

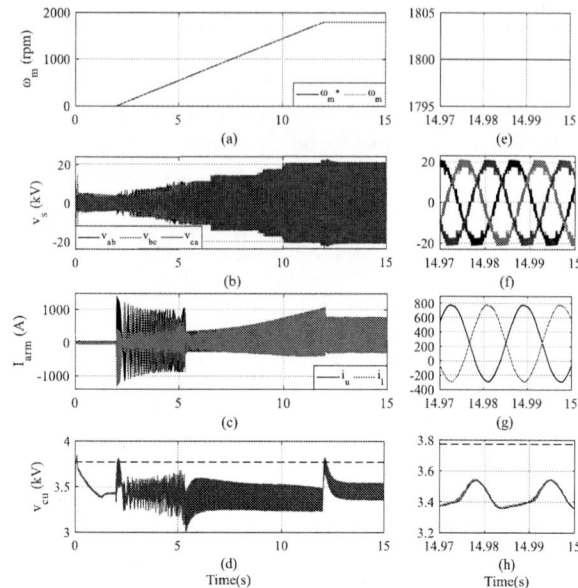

Fig. 4. Start-up of the MMC based electric drive system for design I: (a) Speed control; (b) Stator voltages; (c) Arm currents; (d) Capacitors voltages of the upper arm; (e) Detail of speed control; (f) Detail of stator voltages; (g) Detail of arm currents; (h) Detail of capacitors voltages of the upper arm.

up to $t = 5.3$ s are due the CMI, which is used to mitigate the voltage capacitor ripples at low speed. In steady-state, the stator voltage THD is 9.77 %.

Moreover, Fig. 4 (c) illustrates the insertion of the CMI in the arm currents. This component is injected until $t = 5.3$ s, when the frequency applied to the motor is lower than 20 Hz. The SMs capacitor voltages of the upper arm for the phase a are shown in Fig. 4 (d). As observed, the capacitor voltages in steady-state have a ripple of 3 %, that is, within the dashed band in the figure. Usually a safe margin of 10 % of capacitors voltage ripple is adopted. During the transient, the instantaneous value reaches a maximum of 10.5 %.

The results for the designs II and III are shown in Fig. 5 and Fig. 6, respectively. As observed in Fig. 5 (a) and 6 (a), speed control is attained without oscillations at ramp-up and steady-state for these designs. Fig. 5 (b) and Fig. 6 (b) indicates that with the increase of SMs, the voltage waveform presents more levels, and a subsequently lower THD of 7.29 % and 4.48 % for the design II and III, respectively. The arm currents are represented in Fig. 5 (c) and Fig. 6 (c), from which no significant differences were observed as compared with design I. Furthermore, for the capacitor voltages in Fig. 5 (d) and Fig. 6 (d), the steady-state and transient ripple is also 3 % and 10.5 %, respectively, due the same energy storage energy requirement adopted in the design.

The total semiconductor power losses for the full speed range is illustrated in Fig. 7. As observed, the design that contains the IGBT with the lowest blocking voltage has the lowest power losses. At the rated speed, the design III has 18 % less power losses than to the design II and 42 % less power losses than to the design I. This is because devices

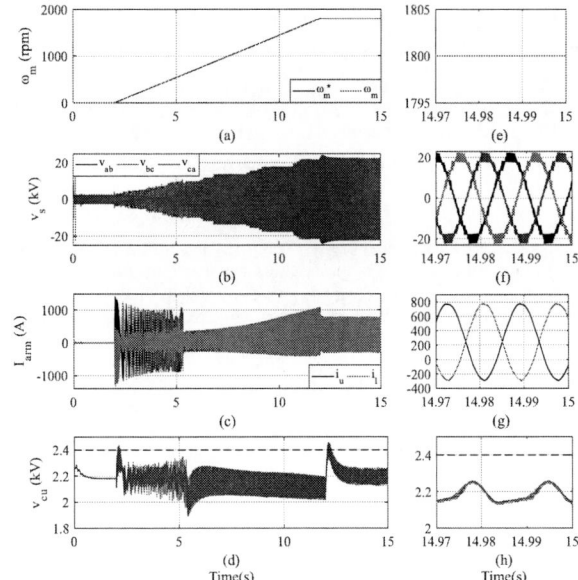

Fig. 5. Start-up of the MMC based electric drive system for design II: (a) Speed control; (b) Stator voltages; (c) Arm currents; (d) Capacitors voltages of the upper arm; (e) Detail of speed control; (f) Detail of stator voltages; (g) Detail of arm currents; (h) Detail of capacitors voltages of the upper arm.

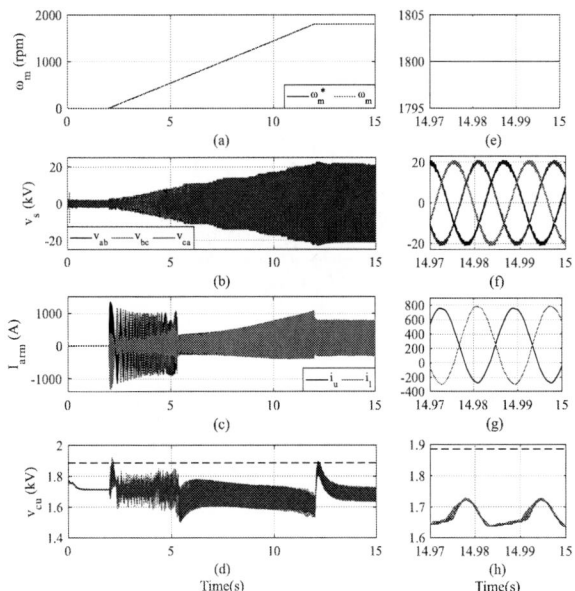

Fig. 6. Start-up of the MMC based electric drive system for design III: (a) Speed control; (b) Stator voltages; (c) Arm currents; (d) Capacitors voltages of the upper arm; (e) Detail of speed control; (f) Detail of stator voltages; (g) Detail of arm currents; (h) Detail of capacitors voltages of the upper arm.

with a higher voltage blocking capacity tend to have more conduction losses due to the larger width of the chip to limit the electric field in the device at the same level.

Finally, considering the mission profile data, the junction temperature is illustrated in Fig. 8. As observed, the junction temperature variations follow closely the mission profile, Fig.

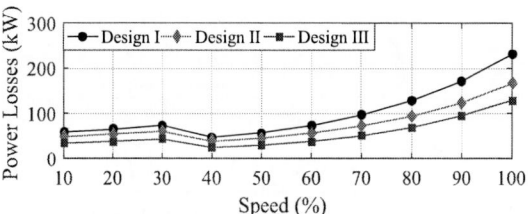

Fig. 7. Semiconductors power losses in the MMC based electric drive system as function of the motor speed (percentage of the nominal value).

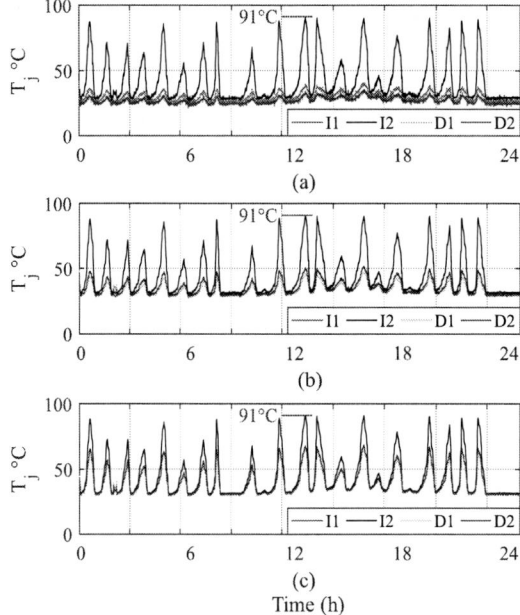

Fig. 8. Junction temperature: (a) Design I; (b) Design II; (c) Design III.

3 (a). The switch $S2$ presents highest junction temperature compared the others semiconductors. The maximum junction temperature is 91 °C for all designs, as shown in Fig. 8 (a), (b) and (c). This indicates a proper and consistent heatsink design, since the maximum temperature is kept below the limit and has the same value for different designs.

B. Benchmarking

In order to select the best cost-effective design, the decision making is based on system efficiency, performance, cost and complexity. Tab. IV shows the overall comparison of each MMC design. The design III presents the highest number of IGBTs, sensors (voltage and current), and ALO, since it increases according to the number of SM. According to the economic evaluation, the design II has the highest cost (semiconductors and capacitors). About the heatsink thermal resistance, design I has the smallest value, because when it is desired to maintain the same junction temperature in all three designs, the one with the highest power losses should have the small heatsink thermal resistance. As desired, the stator current THD remained constant for all three designs. The

TABLE IV
OVERALL COMPARISON OF EACH DESIGN.

Parameters	MMC Design		
	I	II	III
N	7	11	14
Number of IGBTs	84	132	168
f_u	0.95	0.97	0.95
Number of sensors	48	72	90
Cost	1,618,380 €	1,848,330 €	1,737,450 €
R_{h-a}	1 K/kW	8 K/kW	28.5 K/kW
THD i_s	0.553 %	0.547 %	0.543 %
THD v_s	9.77 %	7.29 %	4.48 %
ALO	1388	1904	2291
Daily energy losses	1968 kWh	1540 kWh	1092 kWh

stator voltage THD has decreased according to the increase of SM, since it is proportional to the number of levels of the MMC output voltage. Regarding the daily energy losses, the design III has the highest value, since it presents the highest power losses, as observed in Fig. 7.

According to (9) and based on Tab. IV, the overall performance graphs are plotted and shown in Fig. 9. In order to choose the design most cost-effective, it will be chosen which presents the lowest index k_t, since a metric with index 0 is the best option. Thus, the design III presents the lowest total index value, because of its improved high output power quality, efficiency and moderate cost. This compensated the higher complexity related to component number and control complexity and which increases with the number of levels. Therefore, the design III, with 29-level output voltage, is the best cost-effective choice for the application used in this work.

VI. CONCLUSIONS

This paper proposes the systematic methodology for optimal number of levels on MMC based electric drive system. Furthermore, the designs with the best utilization factor of IGBTs with the blocking voltage of 3.3 kV, 4.5 kV and 6.5 kV were chosen for comparison. The comparison is based on cost, performance, energy losses, complexity metrics. The best cost-effective design is chosen based on the total normalized index, with each metric has the same impact factor.

A 13.8 kV - 16 MW induction motor which drives an exhaust fan based on real mission profile is considered. The simulation results indicate that the design III, with 29-level output voltage, has the best performance, lowest energy losses,

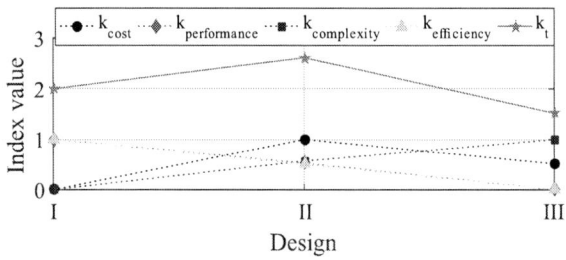

Fig. 9. Selection of the MMC based electric drive design.

moderate cost and a high complexity is the best cost-effective solution based on the application mission profile.

ACKNOWLEDGMENT

This study was financed by CAPES - Finance Code 001, CNPq, FAPEMIG and CEFET-MG.

REFERENCES

[1] S. Kouro, J. Rodriguez, B. Wu, S. Bernet, and M. Perez, "Powering the future of industry: High-power adjustable speed drive topologies," *IEEE Ind. App. Mag.*, vol. 18, no. 4, pp. 26–39, July 2012.

[2] Y. S. Kumar and G. Poddar, "Control of medium-voltage ac motor drive for wide speed range using modular multilevel converter," *IEEE Trans. on Ind. Electron.*, vol. 64, no. 4, pp. 2742–2749, April 2017.

[3] A. Antonopoulos, L. Ängquist, S. Norrga, K. Ilves, L. Harnefors, and H. Nee, "Modular multilevel converter ac motor drives with constant torque from zero to nominal speed," *IEEE Trans. on Ind. Appl.*, vol. 50, no. 3, pp. 1982–1993, May 2014.

[4] M. Hagiwara, I. Hasegawa, and H. Akagi, "Start-up and low-speed operation of an electric motor driven by a modular multilevel cascade inverter," *IEEE Trans. on Ind. Appl.*, vol. 49, pp. 1556–1565, July 2013.

[5] H. Akagi, "Multilevel converters: Fundamental circuits and systems," *Proc. of the IEEE*, vol. 105, no. 11, pp. 2048–2065, Nov 2017.

[6] B. Wu and M. Narimani, *Introduction.* IEEE, 2017. [Online]. Available: https://ieeexplore.ieee.org/document/7827497

[7] Y. S. Kumar and G. Poddar, "Medium-voltage vector control induction motor drive at zero frequency using modular multilevel converter," *IEEE Trans. on Ind. Electron.*, vol. 65, no. 1, pp. 125–132, Jan 2018.

[8] J. E. Huber and J. W. Kolar, "Optimum number of cascaded cells for high-power medium-voltage ac–dc converters," *IEEE J. of Emerging and Select. Topics in Power Electron.*, vol. 5, pp. 213–232, March 2017.

[9] M. R. Islam, Y. Guo, and J. Zhu, "A high-frequency link multilevel cascaded medium-voltage converter for direct grid integration of renewable energy systems," *IEEE Trans. on Power Electron.*, vol. 29, no. 8, pp. 4167–4182, Aug 2014.

[10] H. A. B. Siddique, A. R. Lakshminarasimhan, C. I. Odeh, and R. W. De Doncker, "Comparison of modular multilevel and neutral-point-clamped converters for medium-voltage grid-connected applications," in *Int. Conf. on Renewable Energy Research and Appl.*, Nov 2016, pp. 297–304.

[11] A. Marzoughi, R. Burgos, D. Boroyevich, and Y. Xue, "Design and comparison of cascaded h-bridge, modular multilevel converter, and 5-l active neutral point clamped topologies for motor drive applications," *IEEE Trans. on Ind. Appl.*, vol. 54, no. 2, pp. 1404–1413, March 2018.

[12] B. Gemmell, J. Dorn, D. Retzmann, and D. Soerangr, "Prospects of multilevel vsc technologies for power transmission," in *IEEE/PES Transmission and Distrib. Conf. and Expo.*, April 2008, pp. 1–16.

[13] L. Harnefors, A. Antonopoulos, S. Norrga, L. Angquist, and H. Nee, "Dynamic analysis of modular multilevel converters," *IEEE Trans. on Ind. Electron.*, vol. 60, no. 7, pp. 2526–2537, July 2013.

[14] J. V. M. Farias, A. F. Cupertino, H. A. Pereira, S. I. S. Junior, and R. Teodorescu, "On the redundancy strategies of modular multilevel converters," *IEEE Trans. on Power Del.*, vol. 33, April 2018.

[15] D. W. Novotny and T. Lipo, *Vector Control and Dynamics of AC Drives.* Clarendon Press, 1996.

[16] M. Hagiwara and H. Akagi, "Control and experiment of pulsewidth-modulated modular multilevel converters," *IEEE Trans. on Power Electron.*, vol. 24, no. 7, pp. 1737–1746, July 2009.

[17] *Voltage ratings of high power semiconductors*, ABB Switzerland Ltd Semiconductors, 8 2013, application note 5SYA 2051.

[18] J. Z. Md. Rabiul Islam, Youguang Guo, *Power Converters for Medium Voltage Networks.* Springer-Verlag Berlin Heidelberg, 2014.

[19] K. Ilves, S. Norrga, L. Harnefors, and H. Nee, "On energy storage requirements in modular multilevel converters," *IEEE Trans. on Power Electron.*, vol. 29, no. 1, pp. 77–88, Jan 2014.

[20] A. Marzoughi, R. Burgos, and D. Boroyevich, "Investigating impact of emerging medium-voltage sic mosfets on medium-voltage high-power industrial motor drives," *IEEE J. of Emerging and Selec. Topics in Power Electron.*, vol. 7, no. 2, pp. 1371–1387, June 2019.

[21] S. P. Engel, M. Stieneker, N. Soltau, S. Rabiee, H. Stagge, and R. W. De Doncker, "Comparison of the modular multilevel dc converter and the dual-active bridge converter for power conversion in hvdc and mvdc grids," *IEEE Trans. on Power Electron.*, vol. 30, Jan 2015.

978-1-7281-4181-7/19 $31.00 © 2019 IEEE

High Voltage Gain DC-DC Converter based on a Simple Configuration of Switched Capacitor and Coupled Inductor

Henrique J. Hoch
Federal University of Santa Maria
Cachoeira do Sul, Brazil
henrique.j.h99@gmail.com

Tiago M. K. Faistel
Federal University of Santa Maria
Santa Maria, Brazil
tiagofaistel@yahoo.com.br

Mário L. da S. Martins
Federal University of Santa Maria
Santa Maria, Brazil
mariolsm@gmail.com.br

Mauricio M. da Silva
Universidad Tecnologica
Durazno, Uruguay
mauricio.mendes@utec.edu.uy

António M. S. S. AndradeFederal
University of Santa Maria
Cachoeira do Sul, Brazil
antoniom.spencer@gmail.com

Abstract— Due to the low voltage of the solar panels source, DC-DC high voltage gain converters are required for photovoltaic generation systems. So a new high voltage gain dc–dc converter is proposed in this paper. The proposed converter is generated by association of standard boost converter with switched-capacitor technique and a coupled inductor. This proposed converter achieves high voltage gain with low duty cycle and low turns ratio of the coupled inductor. In addition, it may be pointed out that these features are achieved with a low number of components, current and voltage stress. To evaluate the performance of the proposed converter, a 200-W prototype was simulated.

Keywords—boost converter, DC-DC converter, high step-up.

I. INTRODUCTION

In recent years, interest in renewable energy systems, such as solar power systems, has increased among researchers and industry as an alternative to fossil fuel power generation systems [1-2]. To provide electric power with PV to the grid tie (220 V), initially it is necessary to raise the voltage supplied by the FV (< 50 V) to a higher voltage bus (400 V). So that it can be connected to an inverter [3]. On the other hand, the PV can be connected in series, thus reaching a higher voltage. However, the voltage of PVs generators when exposed to the shading effect, there is a decrease in power supply of the photovoltaic array [4-5].

The module integrated converter (MIC) consists of a high voltage gain DC-DC converter, which increases the voltage of the PV to the DC bus. The second stage, DC-AC inverter, must ensure the voltage regulation of the DC and track the maximum power point (MPPT) of the photovoltaic panel [6-7]. One of the advantages of this type of system is the independent operation of each module, that is, if any panel is exposed to the shading effect, this does not performance of the other PV. On the other hand, one of the great challenges is to achieve high efficiency at the stage of the high voltage gain DC-DC converter [8-9].

As it has been studied in the literature, the boost converter does not achieve a high efficiency when the voltage gain is high. This is because as the duty cycle increases, the losses in the intrinsic resistors of the converter increase, which compromises the performance of the converter [10-11]. To overcome this challenge, some converter alternatives have been proposed. One of the simple techniques is the coupled inductor (Fig. 1(a)) associated with

the boost converter. Among these combinations, it can highlight the boost converter with coupled inductor in cascaded [12], parallel [13] and stacked [14]. In addition, it can also be highlighted the boost with coupled inductor and: voltage multiplier [15]; built-in transformer [16]; double-tapped inductor boost converter [17], among others. To ensure good performance, these converters aim to reduce the duty cycle and transformation ratio of the coupled inductor. Since, if the turns ratio of the coupled inductor is high, the leakage inductance will have high losses in the switch and the diodes.

In other way, another simple technique that can be associated to the boost converter is the capacitor switched (Fig. 1(b)), according to [18]. In recent years, many converters based on the boost converter and switched capacitor was proposed, such as: super-lift [19]; voltage-lift [20]; modified Dickson charge pump [21]; extendable switched capacitor [22]. These sets of converters have the disadvantages that present high peak currents in the switches and diodes, which compromise the efficiency of these converters.

In this sense, the present paper aims to propose a new high voltage gain DC-DC based on boost converter witch coupled inductor and switched capacitor. It has the following advantages: simple and low number of components; simplicity in operating mode; low voltage and current stress in the components; high voltage gain; high efficiency. The contents of this paper are presented as follows: In section is made the synthesis and comparison of the proposed converters and the original topologies. Section III evaluates the operating mode and design methodology of the proposed converter chosen in section II. Finally in section IV the main conclusions of the work.

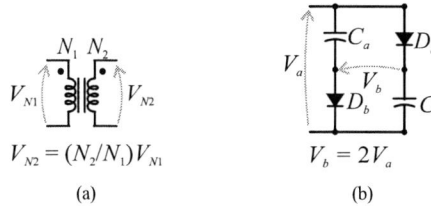

Fig. 1. Voltage gain techniques. (a) coupled inductor. (b) switched capacitor.

978-1-7281-4181-7/19 $31.00 © 2019 IEEE

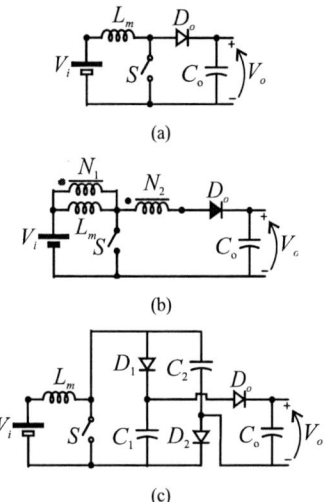

Fig. 2. DC-DC converter. (a) Standard boost converter. (b) boost with coupled inductor. (c) boost converter with switched capacitors.

II. HIGH VOLTAGE GAIN TECHNIQUES APPLIED TO STANDARD BOOST CONVERTER

To obtain a better understanding of the proposed high step-up DC-DC converters, this section presents the general characteristics of the two simple techniques employed to standard boost converter to achieve a high voltage gain.

A. Topology Derivation

Initially, the standard boost DC-DC converter is presented in Fig. 2(a). This converter is defined as single PWM cell and it will be used in the following analyses. To increase de voltage gain of the standard boost converter, the coupled inductor can be associated in this converter, as showed in Fig. 2(b). To achieve high voltage gain, the turns ratio N of the coupled inductor has to increase. Still, loss in the copper wire and leakage inductance (L_k) increase, consequently, deteriorates the efficiency of the converter. In other hand, the switched capacitor (Fig. 1(b)) has been successfully used in standard boost converter, vide Fig. 2(c). The output voltage of this converter is the sum of the voltage of capacitors V_{C1} and V_{C2}, where $V_{C1} = V_{C2} = 1/(1-D)$. To achieve a high voltage gain, the duty cycle of the switch must be high. This causes losses in the MOSFET to be high,

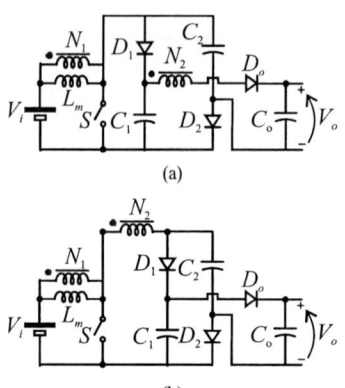

Fig. 3. Proposed High step-up DC-DC converter. (a) Type I. (b) Type II.

which compromises the efficiency of the converter.

Thus, to overcome these problems and to guarantee the simplicity of these converters, in Fig. 3 is proposed two high voltage gain DC-DC converters. As can be seen, the proposed converters are similar to the original ones. Only the position of the secondary of the coupled inductor changes. This features causes the different voltage gain, current and voltage stresses. Thus, in the next section these characteristics will be evaluated.

B. Comparative Evaluation

In order to evaluate the advantages and disadvantages of the proposed converters, some comparative analysis is done in this section.

The voltage gain of each converter is showed in Table I. Fig. 4(a) shows the voltage gain vs duty cycle of each converters using a turns ratio of the coupled inductor $N = 3$. As can be seen the proposed converters have the greeter's voltage gain's. From this, it is evident that the proposed topologies have an advantage in this aspect.

In relation to the switch, in Table I the voltage stress of the switch of each converter is showed. As can be seen, the voltage stress of the switch is the same for all converters in Table I. So to evaluate the voltage stress on the switch, Fig. 4(b) was generated. From the Fig. 4(b) it is evident that for the same voltage gain, in the entire range of the duty cycle, the proposed converter Type II presents lower voltage stress. This allows a MOSFET with a low $R_{DS(on)}$ to be used, which

Table I – Comparative Evaluations

Converters	Gain $M = \dfrac{V_o}{V_i}$	Switch Voltage Stress $M = \dfrac{V_s}{V_i}$	Diodes D_1, D_2 Voltage Stress $\dfrac{V_{D1}}{V_i} = \dfrac{V_{D2}}{V_i}$	Output Diode D_0 Voltage Stress $\dfrac{V_{Do}}{V_i}$
Boost	$\dfrac{1}{1-D}$	$\dfrac{1}{1-D}$	---	$\dfrac{1}{1-D}$
Boost with Coupled Inductor	$\dfrac{ND+1}{1-D}$	$\dfrac{1}{1-D}$	---	$\dfrac{2ND+1-N}{1-D}$
Boost with Switched Capacitor	$\dfrac{2}{1-D}$	$\dfrac{1}{1-D}$	$\dfrac{1}{1-D}$	$\dfrac{1}{1-D}$
Proposed converter Type I	$\dfrac{ND+2}{1-D}$	$\dfrac{1}{1-D}$	$\dfrac{1}{1-D}$	$\dfrac{N+1}{1-D}$
Proposed converter Type II	$\dfrac{N+ND+2}{1-D}$	$\dfrac{1}{1-D}$	$\dfrac{N+1}{1-D}$	$\dfrac{ND+1}{1-D}$

978-1-7281-4181-7/19 $31.00 © 2019 IEEE

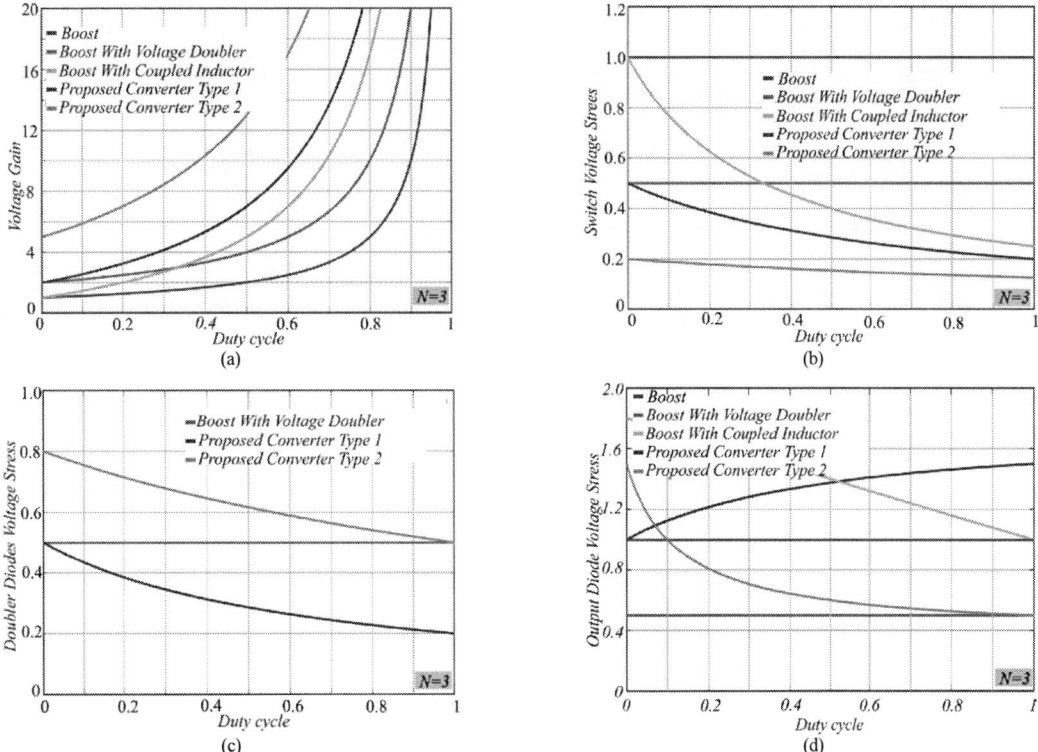

Fig. 4. Comparative evaluation. (a) Static Voltage gain. (b) Normalized switch voltage stress. (c) Normalized diode (D_1 and D_2) voltage stress. (d) Normalized output diode voltage stress.

preserves the efficiency of the converter. On the other hand, in relation to diodes D_1 and D_2, only three converters have these diodes as can be seen in Table I. Fig. 4 (c) was generated by evaluating the voltage stress in these diodes. In relation to these diodes, the proposed Type I converter has lower voltage stresses. So the diodes used in the proposed converter Type I will have lower forward voltage (v_f). Thus, the losses in these diodes for the proposed converter Type I will be smaller. Finally, the voltage stress in the output diode of each converter is evaluated, as can be seen in Fig. 4 (d). As can be seen, the boost converter with switched capacitor presents smaller voltage stresses on the output diode (D_o) throughout the duty cycle range. Also emphasizing that the proposed converter Type II presents the second best result. From the evaluations made, it was concluded that the proposed converter Type II presents the greatest characteristics. Thus, in the next section the evaluations of the operation of this converter, design methodology and experimental evaluations are made.

III. PROPOSED BOOST WITH COUPLED INDUCTOR AND SWITCHED CAPACITORS CONVERTER

Initially the proposed converter Type II with leakage inductance (L_k) is presented in Fig. 5(a). This will allow the theoretical analyzes to be closer to the experimental results.

A. Principle of Operation

In one complete switching period, the proposed converter Type II has two operation modes, as shown is Fig. 5(b) and (c). In order to perform the steady-state analysis of the proposed converter Type II in continuous conduction mode (CCM), the following assumptions are made:

- All power devices (S, D_1, D_2 and D_o) are ideal, i.e., lossless;

- All capacitors (C_1, C_2 and C_o) are large enough to assume their voltage are constant;

- the relation between the primary and secondary of the coupled inductor is given by $N = N_2/N_1$;

- The components that are in gray are OFF.

Fig. 6 depicted the key waveforms of the proposed converter Type II in one switching period. The principle of operation of the proposed converter is given as follows:

Stage 1 *[$t_0 - t_1$, Fig. 5(b)]:* This stage begins when switch S is turned ON. The inductor (L_m) of the coupled inductor is magnetizing with voltage V_i. So, the current i_{Lm} increase linearly, as given in (1). In relation to the leakage inductance (L_k), its voltage is approximately zero ($V_{Lk} \cong 0$), so the current (i_{Lk}) can be assumed as (2). Consequently, the current through the switch S is given by the sum of the inductors current (L_m and L_k), as assumed by (3). In this stage the diodes (D_1 and D_2) are OFF, so their currents (i_{D1} and i_{D2}) are given by (4). The current through the diode D_o is equal to the current of leakage inductance of the coupled inductor, defined by (5).

$$i_{Lm} = \frac{V_i}{L_m}t + I_{Lm}(t_o) \qquad (1)$$

$$i_{Lk} = \frac{I_i - i_{Lm}}{N} = \frac{I_i}{N} - \frac{V_i}{NL}t - \frac{I_{Lm}(t_o)}{N} \qquad (2)$$

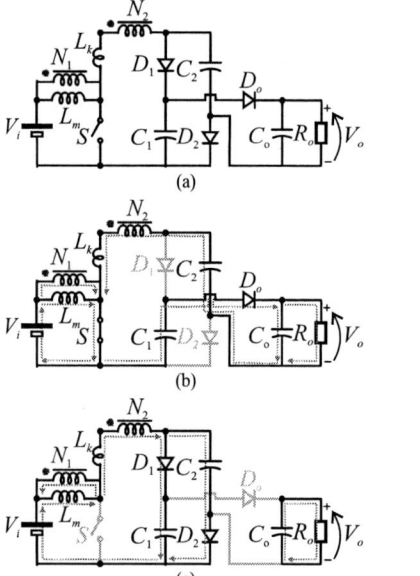

$$i_{D1} = i_{D2} = \frac{I_i}{2} \tag{9}$$

$$i_{Do} = 0 \tag{10}$$

B. Voltage Gain Derivation

In the steady state, the time integral of the inductor (L_m) voltage of the coupled-inductor over one time period must be zero, that is:

$$\int_0^{Ts} v_{Lm}\,dt = 0 \therefore \int_{t_o}^{t_1} v_{Lm}\,dt + \int_{t_1}^{Ts} v_{Lm}\,dt = 0 \tag{11}$$

where

$$\int_{t_o}^{t_1} v_{Lm}\,dt = V_i D T_s \tag{12}$$

$$\int_{t_1}^{Ts} v_{Lm}\,dt = \left(\frac{V_i - V_{C1}}{N+1}\right)(1-D)T_s \tag{13}$$

Substituting (12) and (13) in (11) the voltage V_{C1} and V_{C2} can be found as,

$$V_{C1} = V_{C2} = \frac{ND+1}{1-D}V_i \tag{14}$$

As can be seen, in the first stage the capacitor C_o is charged, as defined by:

$$V_o = V_{C1} + V_{C2} + V_{N2} \tag{15}$$

where $V_{N2} = NV_i$.

So, reorganizing (15) yields in,

$$V_o = \frac{ND+1}{1-D}V_i + \frac{ND+1}{1-D}V_i + NV_i \tag{16}$$

Finally the static voltage gain of the proposed converter Type II is given by:

$$M = \frac{V_o}{V_i} = \frac{N+ND+2}{1-D} \tag{17}$$

C. Design Methodology

The specifications of the proposed converter are given in Table II. From this, the determination of the components can be done.

Duty Cycle (D) and Turns Ratio of Coupled Inductor (N): As can be seen in Table II, the voltage gain is 13.33. From (17), the value of D and N can be found. In order to reduce conduction losses of the switch and losses of the coupled inductor, the value $D = 0.51$ and $N = 3$ was established.

Magnetizing Inductance: From the theoretical waveforms and the operation stage, the inductance (L_m) value can be found by (18).

$$L_m = \frac{DV_i}{f_s \Delta I_{Lm}} = 230\ \mu F \tag{18}$$

Switch and Diodes: The voltage stress value in the switch and diodes can be obtained from (19), (20) and (21), respectively.

$$V_S = \frac{1}{1-D}V_i = 64\ V \tag{19}$$

Fig. 5. Stage of operation. (a) Proposed converter Type II with L_k. (b) Stage 1. (c) Stage 2.

$$i_S = i_{Lm} + (N+1)i_{Lk} = \frac{N+1}{N}I_i - \left(\frac{V_i}{L_m}t + I_{Lm}(t_o)\right) \tag{3}$$

$$i_{D1} = i_{D2} = 0 \tag{4}$$

$$i_{Do} = i_{Lk} = \frac{I_i}{N} + \frac{V_i}{NL_m}t + \frac{I_{Lm}(t_o)}{N} \tag{5}$$

Stage 2 [$t_1 - T_s$, Fig. 5(c)]: At time $t = t_1$, switch S is turned OFF. The inductor L_m of the coupled inductor is demagnetized, $V_{Lm} = (V_i - V_{C1})/(N+1)$. The current i_{Lm} is given by (6). In relation to the leakage inductance (L_k), its current is equal to the input current, as defined in (7). Switch S is OFF, so its current is zero, as can be seen in (8). The current through the diodes D_1 and D_2 is given by (9). Finally, the current of the output diode is given by (10).

$$i_{Lm} = \frac{V_i - V_{C1}}{L_m(N+1)}t + I_{Lm}(t_1) \tag{6}$$

$$i_{Lk} = I_i \tag{7}$$

$$i_S = 0 \tag{8}$$

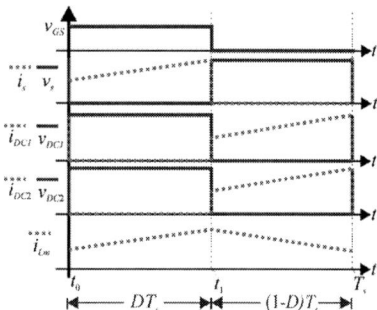

Fig. 6. Key waveforms.

978-1-7281-4181-7/19 $31.00 © 2019 IEEE   859

Table II – Specification of the Proposed Converter Type II

Symbol	Name	Value
P_i	Input Power	200 W
M	Voltage Gain	13.33
V_i	Input Voltage	30 V
V_o	Output Voltage	400 V
N	Turns Ratio of the Coupled-Inductor	3
D	Duty Cycle	0.51
f_s	Switching Frequency	50 kHz
S	Switch	IRFP4668PbF (200 V, 130 A)
D_1 and D_2	Diodes	C3D04060A (600 V, 6 A)
D_o	Diode	MBR20200CT (200 V/ 20 A)
L_m	Magnetizing Inductor	230 μH
L_k	Leakage Inductance	30 μH
ΔIL_m	Current ripple	20 %
C_1 and C_2	Capacitors	1.2 μF (Film Capacitor)
ΔIV_C	Voltage ripple	5 %
C_o	Capacitors	620 nF (Film Capacitor)
ΔIV_{Co}	Voltage ripple	2 %

$$V_{D1} = V_{D1} = \frac{N+1}{1-D}V_i \simeq 245\,V \tag{20}$$

$$V_{Do} = \frac{ND+1}{1-D}V_i \simeq 155\,V \tag{21}$$

Capacitors: The capacitor values can be obtained as follows:

$$C_1 = C_2 = \frac{(1-D)i_{Lm}}{2(N+1)f_s\Delta V_C} \simeq 1.05\,\mu F \tag{22}$$

$$C_o \simeq \frac{(1-D)I_o}{f_s\Delta V_o} \simeq 613\,nF \tag{23}$$

IV. SIMULATION RESULTS OF PROPOSED CONVERTER

To verify the performance of the proposed converter in the PV system, a prototype circuit is simulated in the PSIM® and Matlab® according to the specifications given in Table II.

Fig. 7 depicted the key simulation results. Fig. 7(a) shows the control signal of the switch (D), the input voltage ($V_i = 30$ V) and output voltage of the proposed converter ($V_o = 400$ V). As can be seen, the voltage gain $M = 13.33$ is achieved. Fig. 7(b) the control signal of the switch (D), the input voltage ($V_i = 30$ V), capacitor C_1 voltage ($V_{C1} = 155$ V) and capacitor C_2 voltage ($V_{C2} = 155$ V). In relation of voltage stress, in Fig. 7(c) is depicted the control signal of the switch (D), switch S voltage ($V_s = 64$ V), diode D_1 voltage ($V_{D1} = 245$ V) and the diode D_2 voltage ($V_{D2} = 245$ V). Fig. 7(d) shows the current waveforms of the primary winding of the coupled inductor ($i_{Lm} + i_{N1}$) and secondary of the proposed converter.

Finally, Fig. 8 shows the behavior of theoretical voltage gain versus simulation voltage gain of the proposed converter for different duty cycle values. For a duty cycle range between 0.22 to 0.65, the simulation voltage gain presents similar values to the theoretical voltage gain. So, it can be concluded that the proposed converter is operating as expected.

V. CONCLUSION

This paper proposed a high step-up dc–dc converter. The aim is to use simple techniques, switched capacitor and

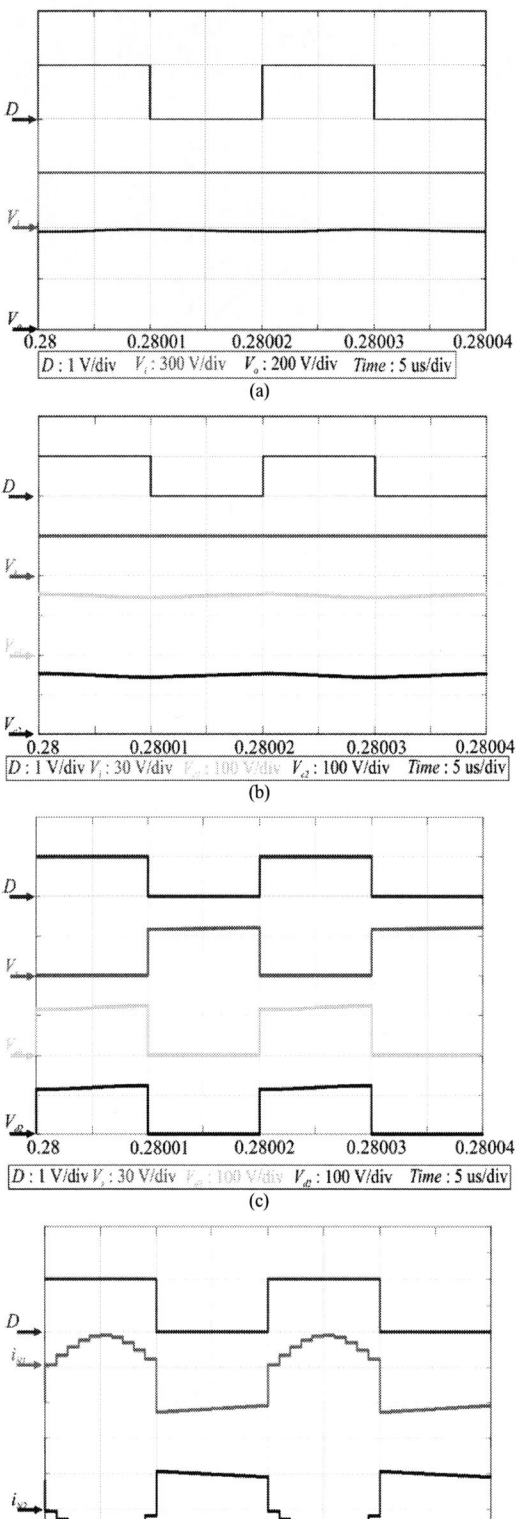

Fig. 7 Simulation results. (a) Voltage gain. (b) Voltage of Capacitors. (c) Semiconductors Voltage Stress. (d) Current waveforms of coupled inductor

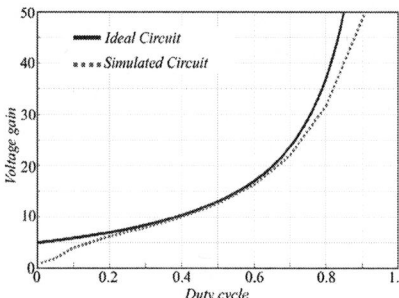

Fig. 8 Voltage gain comparison

coupled inductor, with low number of components to achieve high voltage gain. The maim features of the proposed converter are high voltage, low number of components, low complexity, single switch and low current and voltage stresses. To evaluate the proposed converter was simulated in the PSIM®. The simulation results prove and validate the converter operation and features.

ACKNOWLEDGMENT

This study was financed by "Conselho Nacional de Desenvolvimento Científico (CNPq)" Finance code 425155/2018-8.

REFERENCES

[1] M. Forouzesh, Y. Shen, K. Yari, Y. P. Siwakoti and F. Blaabjerg, "High-Efficiency High Step-Up DC–DC Converter With Dual Coupled Inductors for Grid-Connected Photovoltaic Systems," *IEEE Trans. Power Electron.*, vol. 33, no. 7, pp. 5967-5982, July 2018.

[2] K. Tseng, C. Huang and C. Cheng, "A High Step-Up Converter With Voltage-Multiplier Modules for Sustainable Energy Applications," *IEEE Journal of Emerg. and Selected Topics in Power Electron.*, vol. 3, no. 4, pp. 1100-1108, Dec. 2015.

[3] A. M. S. S. Andrade, L. Schuch, and M. L. da Silva Martins, "High stepup PV module integrated converter for PV energy harvest in FREEDM systems," *IEEE Trans. Ind. Appl.*, vol. 53, no. 2, pp. 1138–1148, Mar./Apr. 2017.

[4] A. M. S. S. Andrade, R. C. Beltrame, L. Schuch and M. L. d. S. Martins, "PV module-integrated single-switch DC/DC converter for PV energy harvest with battery charge capability," 2014 11th IEEE/IAS International Conference on Industry Applications, Juiz de Fora, 2014, pp. 1-8.

[5] D. Pera, J. A. Silva, S. Costa and J. M. Serra, "Investigating the impact of solar cells partial shading on photovoltaic modules by thermography," *IEEE 44th Photovoltaic Spec.t Conf. (PVSC)*, Washington, DC, 2017, pp. 1979-1983.

[6] Y. Shen, A. Chub, H. Wang, D. Vinnikov, E. Liivik and F. Blaabjerg, "Wear-Out Failure Analysis of an Impedance-Source PV Microinverter Based on System-Level Electrothermal Modeling," *IEEE Trans. Ind. Electron.*, vol. 66, no. 5, pp. 3914-3927, May 2019.

[7] D. Vinnikov, A. Chub, E. Liivik, R. Kosenko and O. Korkh, "Solar Optiverter—A Novel Hybrid Approach to the Photovoltaic Module Level Power Electronics," *IEEE Trans. Ind. Electron.*, vol. 66, no. 5, pp. 3869-3880, May 2019.

[8] F. Edwin, W. Xiao and V. Khadkikar, "Topology review of single phase grid-connected module integrated converters for PV applications," *38th Annual Conf. IEEE Ind. Electron. Society*, Montreal, QC, 2012, pp. 821-827.

[9] M. Forouzesh, Y. P. Siwakoti, S. A. Gorji, F. Blaabjerg and B. Lehman, "Step-Up DC–DC Converters: A Comprehensive Review of Voltage-Boosting Techniques, Topologies, and Applications," *IEEE Trans. Power Electron.*, vol. 32, no. 12, pp. 9143-9178, Dec. 2017.

[10] A. M. S. S. Andrade, L. Schuch and M. L. da Silva Martins, "Analysis and Design of High-Efficiency Hybrid High Step-Up DC–DC Converter for Distributed PV Generation Systems," *IEEE Trans. Ind. Electron.*, vol. 66, no. 5, pp. 3860-3868, May 2019.

[11] A. M. S. S. Andrade, H. L. Hey, L. Schuch and M. L. da Silva Martins, "Comparative Evaluation of Single Switch High-Voltage Step-Up Topologies Based on Boost and Zeta PWM Cells," *IEEE Trans. Ind. Electron.*, vol. 65, no. 3, pp. 2322-2334, March 2018.

[12] V. Fernao Pires, A. Cordeiro, D. Foito and J. Fernando Silva, "High Step-up DC-DC Converter for Fuel Cell Vehicles Based on Merged Quadratic Boost Ćuk," IEEE Trans. Vehicular Techn., early acess.

[13] A. M. S. S. Andrade and M. L. d. S. Martins, "Quadratic-Boost With Stacked Zeta Converter for High Voltage Gain Applications," *IEEE Jour. Emerg. and Selected Topics in Power Electron.*, vol. 5, no. 4, pp. 1787-1796, Dec. 2017.

[14] Q. Zhao, F. Tao and F. C. Lee, "A front-end DC/DC converter for network server applications," *IEEE 32nd Annual Power Electron. Specialists Conf*, vol. 3, Vancouver, BC, 2001, pp. 1535-1539.

[15] A. M. S. S. Andrade, E. Mattos, L. Schuch, H. L. Hey and M. L. da Silva Martins, "Synthesis and Comparative Analysis of Very High Step-Up DC–DC Converters Adopting Coupled-Inductor and Voltage Multiplier Cells," *IEEE Trans. on Power Electronics*, vol. 33, no. 7, pp. 5880-5897, July 2018.

[16] T. Nouri, N. Vosoughi, S. H. Hosseini, E. Babaei and M. Sabahi, "An Interleaved High Step-Up Converter With Coupled Inductor and Built-In Transformer Voltage Multiplier Cell Techniques," *IEEE Trans. Ind. Electron.*, vol. 66, no. 3, pp. 1894-1905, March 2019.

[17] H. Liu, H. Hu, H. Wu, Y. Xing and I. Batarseh, "Overview of High-Step-Up Coupled-Inductor Boost Converters," *IEEE Jour. of Emerg. and Selec. Top. Power Electron.*, vol. 4, no. 2, pp. 689-704, June 2016.

[18] K. Li, Y. Hu and A. Ioinovici, "Generation of the Large DC Gain Step-Up Nonisolated Converters in Conjunction With Renewable Energy Sources Starting From a Proposed Geometric Structure," *IEEE Trans. Power Electron.*, vol. 32, no. 7, pp. 5323-5340, July 2017.

[19] F. L. Luo and H. Ye, "Super-lift boost converters," *IET Power Electron.*, vol. 7, no. 7, pp. 1655-1664, July 2014.

[20] F. Mohammadzadeh Shahir, E. Babaei and M. Farsadi, "Analysis and design of voltage-lift technique-based non-isolated boost dc–dc converter," *IET Power Electron.*, vol. 11, no. 6, pp. 1083-1091, 2018.

[21] B. P. Baddipadiga and M. Ferdowsi, "A high-voltage-gain dc-dc converter based on modified dickson charge pump voltage multiplier," *IEEE Trans. Power Electron.*, vol. 32, no. 10, pp. 7707-7715, Oct. 2017.

[22] A. Amir, H. S. Che, A. Amir, A. E.Khateb, N. A. Rahim, "Transformerless high gain boost and buck-boost DC-DC converters based on extendable switched capacitor (SC) cell for stand-alone photovoltaic system," *Solar Energy*, vol 171, no 9, pp 212-222, Sep. 2018.

Wind power system connected to the grid from Squirrel Cage Induction Generator (SCIG)

1st Ângelo Marcílio M. dos Santos
PPGEEC/UFC, Campus Sobral
Federal University of Ceará
Sobral-CE, Brazil
angelomarcilio@alu.ufc.br

2nd Lucas Taylan P. Medeiros
Electrical Engineering, Campus Sobral
Federal University of Ceará
Sobral-CE, Brazil
lucastaylanp@gmail.com

3rd Leonardo P. S. Silva
Electrical Engineering, Campus Sobral
Federal University of Ceará
Sobral-CE, Brazil
leonardo85pires@gmail.com

4th Ildenor Davi S. Júnior
Electrical Engineering, Campus Sobral
Federal University of Ceará
Sobral-CE, Brazil
juniordavid1208@gmail.com

5th Vanessa Siqueira de C. Teixeira
PPGEEC/UFC, Campus Sobral
Federal University of Ceará
Sobral-CE, Brazil
vanessasct@gmail.com

6th Adson Bezerra Moreira
PPGEEC/UFC, Campus Sobral
Federal University of Ceará
Sobral-CE, Brazil
adsonbmoreira@gmail.com

Abstract—The main contribution of this paper is to present a methodology proposal for the development of active and reactive power control of a wind power generation system connected to the grid using the three-phase induction generator with a squirrel cage rotor (SCIG). The SCIG stator terminals are connected to electric grid before the inductive filter using an AC/DC/AC power converter topology, called back-to-back, while the rotor terminals are short circuited, so that the SCIG's active and reactive power control techniques are presented. The control of grid side converter (GSC) is presented, responsible for the power control and also for keeping the DC bus voltage constant. The control of induction generator side converter (IGSC) is presented also, in which vector control is used in rotor variables, so that the SCIG is controlled through the torque references and magnetizing current. The studied system was mathematically modeled and simulated using Matlab/Simulink software.

Index Terms—electric power generation, wind Energy, SCIG, electronic back-to-back converter, controllers design

I. INTRODUCTION

With the progressive decline of fossil fuels reserves and ever-increasing demand of energy, the production of electric energy from renewable resources, such as solar, wind and tidal streams oceans, has now been accepted as a potentially promising solution to the energy problem [8]. Of these, wind power is one of an effective mitigation measures, the development of wind power generation has increasingly attracted attention in various countries [9].

There are several types of electric generators used in wind power system. This way, the choice of wich type of electric generator depends on the application of a distributed machine, wind farms, electric transmission, machine power and cost [11].

This work was supported by FUNCAP grant code BP3-0139-00022.01.00/18.

In this paper a squirrel cage induction generator (SCIG) is used because are extensively employed for power generation from wind energy due to simple and rugged rotor construction, low cost, almost nil maintenance and generator operation without the need of DC supply [7].

Besides that, wind turbines based on (SCIG) with back-to-back voltage source converter are becoming increasingly popular. Compared with the wind turbines using fixed-speed induction generators, the SCIG-based wind turbines offers not only the advantages of variable speed operation and four-quadrant active and reactive power capabilities, but also separation between grid and generator by conversion of AC-DC-AC [10].

The back-to-back converter is composed of two voltage source converters connected together via a DC bus capacitor. One converter stays between the SCIG and the DC bus capacitor called induction generator side converter (IGSC), which has the function to produce the machine flux for SCIG and to optimum the energy capture from the wind. The second converter stays between the DC bus capacitor and the grid called grid side converter (GSC), which has the function to regulate the DC bus voltage [5].

In this paper, the employed methodology obtains results based on mathematical modeling of the system, shown in Fig. 1. The studies will be conducted through computer simulation with mathematical models of studied system for validation of control strategies, and the SCIG connected to power grid. For the simulations we will use computational tool *SimPowerSystems*© of *Matlab/Simulink*©.

The paper is structured as follows: Section II describes the modeling of the GSC control system. The section III describes the modeling of the IGSC control system. The section IV describes the analysis and results. Finally, in the section V

978-1-7281-4181-7/19 $31.00 © 2019 IEEE

exposes the conclusions.

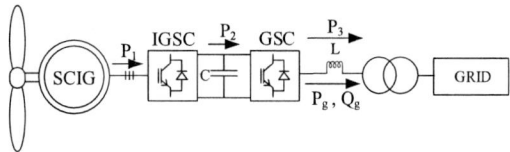

Fig. 1. Wind system studied.

II. GRID SIDE CONVERTER - GSC

The grid side converter, a three-phase DC-to-AC electronic converter, controls the DC bus voltage and the current injected into the grid [2].

GSC control is performed by the block diagram of Fig. 2, in which the current loop is shown in Fig. 2 (a) and the voltage loop in Fig. 2 (b) [3].

The PLL (phase locked loop) is responsible for maintaining the synchronism between the voltages of the grid and those produced by the inverter, generating an angle θ in phase with the grid voltage, where V_A, V_B and V_C are grid line voltages.

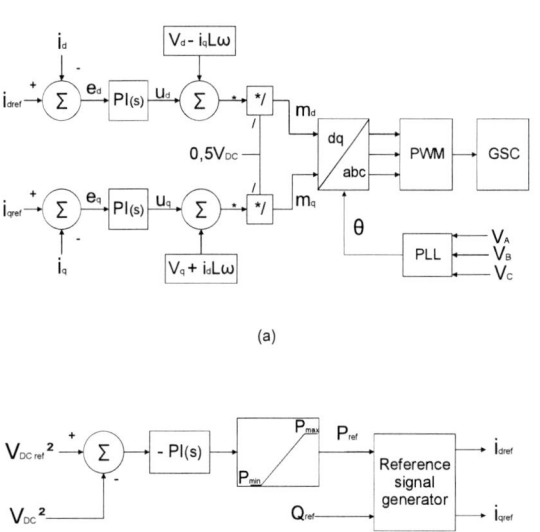

(a)

(b)

Fig. 2. Grid side converter control scheme (current loop (a) and voltage loop (b)).

The current control loop I_d (direct axis current) shows I_{dref} (direct axis reference current) as a reference from the DC bus voltage control. In the current control loop I_q (quadrature axis current)), assumed I_{qref} (quadrature axis reference current) = 0, making the converter operate with unit power factor. The signal reference generator produces current references, (I_{dref} and I_{qref}), from (1) and (2), where V_{sd} and V_{sq} are the phase voltages of the grid in dq coordinates.

$$P_{sref} = \frac{3}{2}\left[V_{sd}\, i_{dref} + V_{sq}\, i_{qref}\right] \tag{1}$$

$$Q_{sref} = \frac{3}{2}\left[-V_{sd}\, i_{qref} + V_{sq}\, i_{dref}\right] \tag{2}$$

Since $V_q = 0$, (1) and (2) can be simplified as (3) and (4).

$$i_{dref} = \frac{2}{3V_{sd}}\, P_{sref} \tag{3}$$

$$i_{qref} = -\frac{2}{3V_{sd}}\, Q_{sref} \tag{4}$$

A. GSC DC bus voltage and current control

The design of the controllers is based on the frequency response, so it must have a gain margin (GM) greater than 6 dB and a phase margin (PM) between 30º and 60º [1], [4].

The block representation of the GSC direct and quadrature axis current controllers is shown in Fig. 3, in which the block $PI(s)$ is a PI controller, $G_p(s)$ characterizes the dynamics of the GSC PWM and $G_c(s)$ is the GSC plant with L filter. $G_i(s)$ is the product of blocks $G_p(s)$ and $G_c(s)$ cascading.

The transfer functions $G_p(s)$ e $G_c(s)$ are given, respectively, by (5) and (6), in which T_s is the sampling time, L is the inductance of the filter that connects to the grid to the converter, and R is the electrical resistance present in the inductor [1].

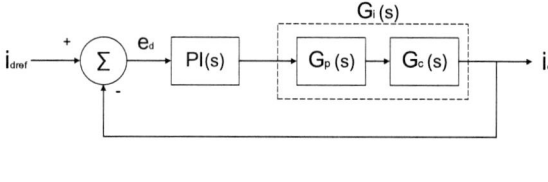

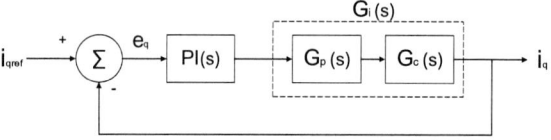

Fig. 3. Simplified block diagram of GSC current controller with L filter at dq coordinates.

$$G_p(s) = \frac{1 - s\left(\frac{T_s}{4}\right)}{1 + s\left(\frac{T_s}{4}\right)} \tag{5}$$

$$G_c(s) = \frac{1}{Ls + R} \tag{6}$$

The transfer function of the PI controller is given in (7).

$$PI(s) = k_p\left(1 + \frac{1}{T_i\, s}\right) \tag{7}$$

For $s = j\omega$, it has (8) and (9).

$$PI(j\omega) = k_p\left(1 + \frac{1}{T_i\, j\omega}\right) \tag{8}$$

978-1-7281-4181-7/19 $31.00 © 2019 IEEE   863

$$\angle PI(j\omega) = -\arctan \frac{1}{T_i \, \omega} \tag{9}$$

The desired phase margin, PM_d, is calculated from (10), where ω_c is the gain crossing frequency.

$$PM_d = \pi + \angle G_i(j\omega_c) + \angle PI(j\omega_c) \tag{10}$$

When applying (9) in (10) and when isolating T_i, it is determined (11), the first condition of the controller project.

$$T_i = \frac{1}{\omega_c \cdot \tan\left(\pi + \angle G_i(j\omega_c) - PM_d\right)} \tag{11}$$

According to [1], the magnitude of the open-loop transfer function of a controlled system is the unit at the crossing frequency, obtaining (12). Substituting (8) into (12) and isolating k_p, the second controller project condition is given by (13).

$$|PI(j\omega_c)| \cdot |G_i(j\omega_c)| = 1 \tag{12}$$

$$k_p = \frac{1}{|G_i(j\omega_c)| \cdot \left|1 - \frac{j}{\omega_c T_i}\right|} \tag{13}$$

From (11) and (13), to $\omega_c = 1000$ rad/s and $PM = 60°$, $k_p = 4.9747$ and $T_i = 0.0014$ were obtained for the current controller of the GSC.

The dynamics of the DC bus voltage controller of the GSC is represented by Fig. 4, where block $PI(s)$ is a PI controller, $G_{if}(s)$ is the closed loop of GSC electric current control and $G_{vcc}(s)$ characterizes the DC voltage dynamics of the DC bus.

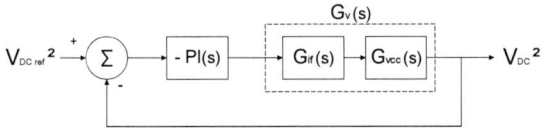

Fig. 4. Simplified block diagram of DC bus voltage controller.

The transfer function $G_{vcc}(s)$ is given by (14), where C is the equivalent capacitance of the AC-DC converter and τ is described in (15), with P_{exto} being the active power.

$$G_{vcc}(s) = -\left(\frac{2}{C}\right)\frac{\tau s + 1}{s} \tag{14}$$

$$\tau = \frac{2L \, P_{exto}}{3V_{sd}^2} \tag{15}$$

Applying the project methodology of the PI controllers and adopting $\omega_c = 202 \ rad/s$ and $PM = 60°$, $k_p = 0.3143$ and $T_i = 0.0143$ were obtained for the GSC voltage controller.

III. INDUCTION GENERATOR SIDE CONVERTER - IGSC

This section describes the main control loop of induction generator side converter. The outputs of the systems control loop are voltage reference used to command the PWM (pulse wide modulation).

The inner of the systems loop controls current by using the model equations of SCIG represented in terms of d-axis, q-axis rotor flux reference frame. The currents are controlled using a standard indirect vector control [6].

A. Squirrel Cage Induction Generator (SCIG)

The equations that govern the dynamics of the AC generator are represented as (16) to (23), where the dispersion factors of the stator (σ_e) and the rotor (σ_r) are described in (24) and (25) [3].

$$\frac{d\lambda_{ed}}{dt} = V_{ed} - R_e \, i_{ed} \tag{16}$$

$$\frac{d\lambda_{eq}}{dt} = V_{eq} - R_e \, i_{eq} \tag{17}$$

$$\frac{d\lambda_{rd}}{dt} = V_{rd} - R_r \, i_{rd} \tag{18}$$

$$\frac{d\lambda_{rq}}{dt} = V_{rq} - R_r \, i_{rq} \tag{19}$$

$$\lambda_{ed} = L_m \left[(1+\sigma_e) \, i_{ed} + e^{j\theta_r} \, i_{rd}\right] \tag{20}$$

$$\lambda_{eq} = L_m \left[(1+\sigma_e) \, i_{eq} + e^{j\theta_r} \, i_{rq}\right] \tag{21}$$

$$\lambda_{rd} = L_m \left[(1+\sigma_r) \, i_{rd} + e^{-j\theta_r} \, i_{ed}\right] \tag{22}$$

$$\lambda_{rq} = L_m \left[(1+\sigma_r)i_{rq} + e^{-j\theta_r} \, i_{eq}\right] \tag{23}$$

$$\sigma_e = \frac{L_{le}}{L_m} - 1 \tag{24}$$

$$\sigma_r = \frac{L_{lr}}{L_m} - 1 \tag{25}$$

V_{ed} and V_{eq} are the stator voltages, V_{rd} and V_{rq} are the voltages in the rotor, i_{ed} and i_{eq} are the stator currents, i_{rd} and i_{rq} are the currents in the rotor, λ_{ed} and λ_{eq} are the stator concatenated fluxes, λ_{rd} and λ_{rq} are the fluxes concatenated in the rotor, in dq coordinates respectively, R_e and R_r are the stator and rotor resistances, L_m is the magnetization inductance, and θ_r is the rotation angle of the rotor.

In SCIG the rotor terminals are shorted and the rotor current is not measured. Then, $V_{rdq} = 0$ and i_{rdq} are not measured. Thus, the rotor currents in dq coordinates are given by (26) and (27), in which $\hat{i}_{mr}$ is the magnitude of the magnetizing current [3].

$$i_{rd} = \frac{\hat{i}_{mr} - i_{ed}}{1 + \sigma_r} e^{-j\theta_r} \tag{26}$$

$$i_{rq} = \frac{-i_{eq}}{1 + \sigma_r} \, e^{-j\theta_r} \tag{27}$$

The electric torque (T_e) and the rotor time constant (τ_r) are defined in (28) and (29).

$$T_e = \frac{3}{2} \frac{L_m}{1 + \sigma_r} \, \hat{i}_{mr} \, i_{eq} \tag{28}$$

$$\tau_r = \frac{L_m \, (1 + \sigma_r)}{R_r} \tag{29}$$

The rotating magnetic field velocity and rotor speed are determined in (30) and (31), in which ρ is the angle of the rotating axis of stator voltages and currents.

$$\omega_m = \frac{d\rho}{dt} \tag{30}$$

$$\omega_r = \frac{d\theta_r}{dt} \tag{31}$$

B. Flux observer

In SCIG the flux observer is required to obtain the magnetizing current, the rotating field angle and the speed of the rotating field.

The flux observer is performed by the block diagram of Fig. 5, and is based on (32) and (33), where $\hat{i}_{mr}$ is set to a constant value and $\hat{i}_{mr} = i_{ed}$, which is required for a linear torque control by i_{eq} [3].

$$\tau_r \frac{d}{dt} \left[\hat{i}_{mr} \right] = -\hat{i}_{mr} + i_{ed} \tag{32}$$

$$\omega_m = \frac{i_{eq}}{\tau_r \, \hat{i}_{mr}} + \omega_r \tag{33}$$

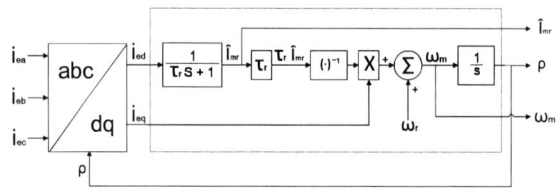

Fig. 5. Flux observer block diagram.

C. Vector control of SCIG in rotor field coordinates

The vector control of the SCIG as a function of the coordinates of the rotor field is shown in the block diagram of Fig. 6. This control is implemented from (28) and (32) to obtain i_{edref} and i_{eqref} of the generator.

The reference magnetizing current and the reference electric torque are obtained by (34) and (35), in which V_{en} is the nominal line rms voltage of the stator and ω_{m0} is the nominal angular velocity of the generator ($\omega_{m0} = 2\pi f_{m0}$) and P_{ref} is the desired reference power.

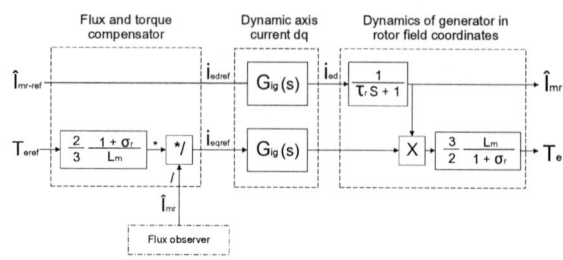

Fig. 6. Block diagram of the vector control of the SCIG as a function of the coordinates of the rotor field.

$$\hat{i}_{mr-ref} = \sqrt{\frac{2}{3}} \frac{V_{en}}{(1 + \sigma_e) \, L_m \, \omega_{m0}} \tag{34}$$

$$T_{eref} = \frac{P_{ref}}{\omega_r} \tag{35}$$

D. IGSC current control

The block diagram representation of the dq axis current controllers of the IGSC is presented in Fig. 7, while Fig. 8 shows the simplified diagram.

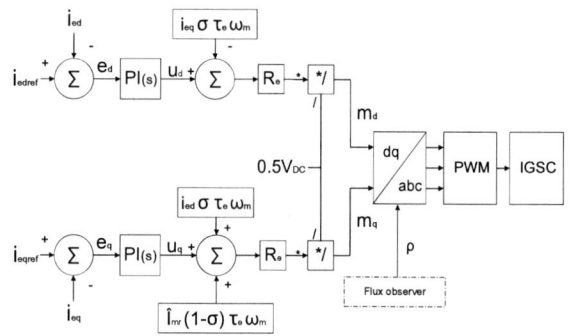

Fig. 7. IGSC Current Control System.

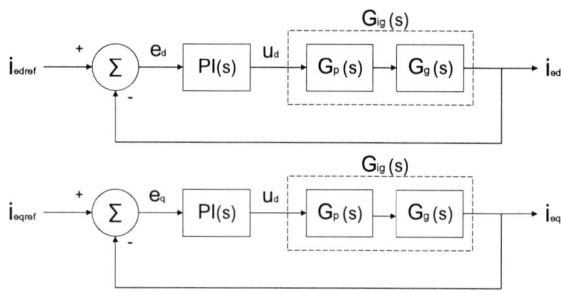

Fig. 8. Simplified block diagrams of IGSC electrical current controllers at dq coordinates.

The block $PI(s)$ is a PI controller, $G_p(s)$ represents the PWM dynamics of the DC-AC electronic converter, and $G_g(s)$ is the IGSC plant. $G_{ig}(s)$ is the product of blocks $G_p(s)$ and $G_g(s)$ in cascade. $G_g(s)$ is defined in (36).

978-1-7281-4181-7/19 $31.00 © 2019 IEEE

$$G_g(s) = \frac{1}{\sigma \tau_e s + 1} \qquad (36)$$

Applying the design methodology of the PI controllers and adopting $\omega_c = 500 \; rad/s$ and $PM = 60°$, it was obtained $k_p = 6.3986$ and $T_i = 0.0028$ for the IGSC current controllers.

IV. Results

The wind power generation system shown in Fig. 1 was implemented in software $Matlab/Simulink^{©}$. The generation system parameters used in the simulation are shown in Table I, so that the IGSC parameters were obtained in [5]. The switching frequency of the converters is 10 kHz.

The simulation of the wind generation system was carried out with variable speed, in which the generator operates with a speed of 150 rad/s and 200 rad/s, according to Fig. 9.

TABLE I
SIMULATION PARAMETERS

	Parameters	Values
SCIG	P_n, V_n, F_n, P	$15kW$, $460V$, $60Hz$, 4
SCIG	$R_e, R_r, L_e = L_r$	$276.1m\Omega$, $164.5m\Omega$, $78.3mH$
SCIG	L_m, L_{le}, L_{lr},	$76.14mH$, $2.191mH$, $2.191mH$
Converter	C_{DC}, V_{DC}	$3500\mu F$, $800V$
Grid	V_{LL}, F, R_g, L_g	$380V$, $60Hz$, 0.8Ω, $6mH$

The GSC control consists of the components: DC bus voltage control and current control.

The Figures 10, 11 and 12 show the results obtained for DC bus voltage control and current control.

The Fig. 10 shows the reference voltage of the DC bus and the measured voltage on the DC bus. It was verified that the DC bus voltage control remains stable and regulated at 800 V even with the speed variation of the electric generator.

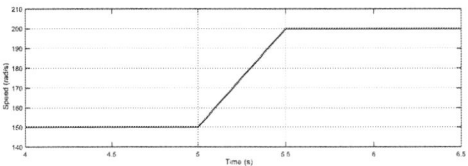

Fig. 9. SCIG rotor speed ramp from 150 rad/s to 200 rad/s.

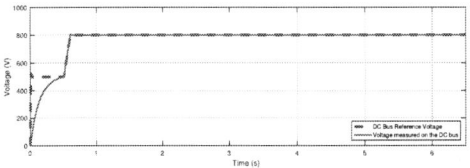

Fig. 10. Reference voltage of the DC bus and measured voltage of the DC bus.

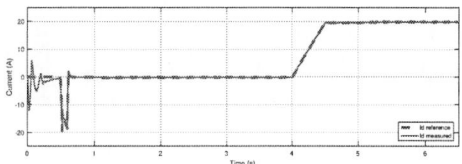

Fig. 11. Reference direct axis current and measured direct axis current.

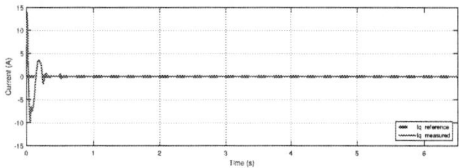

Fig. 12. Reference quadrature axis current and measured quadrature axis current.

The Figures 11 and 12 show the responses of the GSC currents control, i_d and i_q, following their references, i_{dref} and i_{qref}, showing the good functioning of the control.

In Fig. 10, it observes i_d presents negative values from 0.5s to 0.6s, since in this time interval the DC bus voltage reaches 800V, and current from the grid to the converter that requires power. After the DC bus voltage stabilizes at 0.6s, i_d remains constant at zero up to 4.0s. From 4.0s, i_d presents an increasing positive value when the generator starts supplying active power to the grid because i_d is directly proportionally to the active power delivered to the grid. The current i_q is maintained at 0A, keeping the reactive power sent to the grid at zero, being the reactive power controlled by i_q.

Figures 13 and 14 illustrate the measured magnetizing current of the generator and its reference, and measured torque of the generator and its reference. As can be seen, the magnetization current and the torque follow its references, which proves the suitable functioning of the control. It is also observed that the torque changes according to the variation of the generator speed.

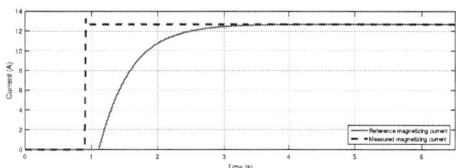

Fig. 13. Reference magnetizing current and magnetising current measured from generator.

Figures 15 and 16 show the responses of IGSC current control loops, i_{ed} and i_{eq}, following their references, i_{edref} and i_{eqref}.

It is verified that i_{eq} is controlled by torque. When torque varies i_{eq} changes. It is also observed that i_{ed} follows its reference from the IGSC current control.

The Figures 17, 18 and 19 illustrate a comparative of the system powers during the speed variation of the generator

978-1-7281-4181-7/19 $31.00 © 2019 IEEE

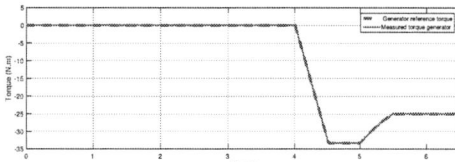

Fig. 14. Torque reference and measured torque of the generator.

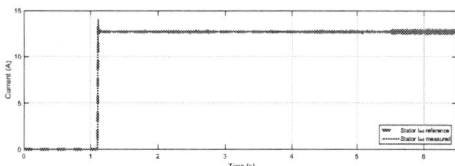

Fig. 15. Direct stator reference axis current and stator direct axis current measured.

shown in Fig. 9, where P1 is the power delivered by the generator to the IGSC, P2 is the power received by the GSC, and P3 is the power delivered to the power grid.

From Figures 17, 18 and 19 it is possible to observe the difference of the power values during the operation of the system. The power values are decreasing during the course of the system, which is due to the switching losses of the converters.

In addition, it is noted that the powers increase with increasing speed, and the Figures 17, 18 and 19 show that the higher the speed of operation of the generator, the higher the power delivered to the grid.

V. Conclusions

This research investigated the behavior of a wind power system with SCIG. The controllers of the wind power system were calculated by the presented methodology and it was verified a suitable behavior of the control, which was observed

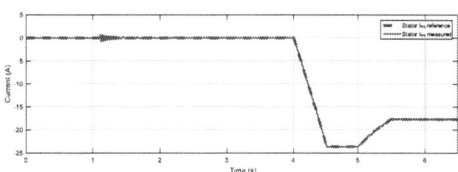

Fig. 16. Stator reference quadrature axis current and stator measured quadrature axis current.

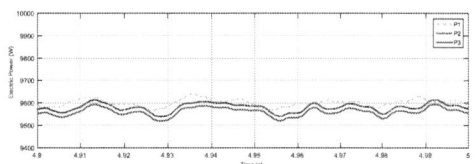

Fig. 17. Comparison of powers measured throughout the system between 4.90s and 5.0s.

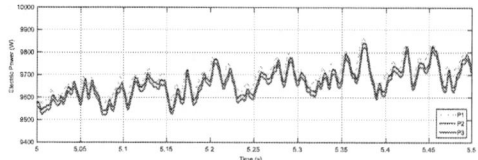

Fig. 18. Comparison of powers measured throughout the system between 5.0s and 5.5s.

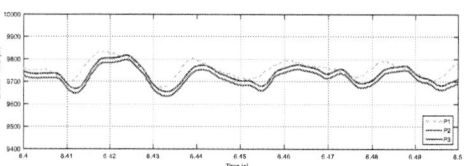

Fig. 19. Comparison of powers measured throughout the system between 6.40s and 6.50s.

with the results obtained from the DC bus voltage control, GSC current control and IGSC current control. From the results it was also observed the active power delivered to the grid, noting that the increase of the generator speed caused the increase of the power generated and delivered to the grid.

References

[1] Adson Bezerra Moreira, Tárcio A. S. Barros, Vanessa S. C. Teixeira, Paulo S. Nascimento Filho, and Ernesto Ruppert Filho, "Controle de Potências para Geração Eólica e Filtragem de Corrente Harmônica com Gerador de Indução Duplamente Alimentado," An. do VI Simpósio Bras. Sist. Elétricos, 2016.

[2] T. A. Lipo, "Vector control and dynamics of AC drives," [S.l.]: Oxford university press, 1996.

[3] A. Yazdani and R. Iravani, "Voltage-sourced converters in power systems: modeling, control, and applications," [S.l.]: John Wiley Sons, 2010.

[4] K. Ogata, "Engenharia de Controle Moderno," 5ªed., [S.l.]: Pearson, 2011.

[5] M. HEYDARI, A. Y. VARJANI and M. MOHAMADIAN, "A Novel Variable-Speed Wind Energy System Using Induction Generator and Six-Switch AC/AC Converter," 2012 3rd Power Electronics and Drive Systems Technology (PEDSTC), 2012.

[6] W. Suebkinorn and B. Neammanee, "An implementation of field oriented controlled SCIG for variable speed wind turbine," Proc. 2011 6th IEEE Conf. Ind. Electron. Appl. ICIEA 2011, pp. 39–44, 2011.

[7] S. Mahajan, S. K. Subramaniam, K. Natarajan, A. G. Nanjappa Gounder, and D. V. Borru, "Analysis and control of induction generator supplying stand-alone AC loads employing a Matrix Converter," Eng. Sci. Technol. an Int. J., vol. 20, no. 2, pp. 649–661, 2017.

[8] H. Merabet Boulouiha, A. Allali, M. Laouer, A. Tahri, M. Denaï, and A. Draou, "Direct torque control of multilevel SVPWM inverter in variable speed SCIG-based wind energy conversion system," Renew. Energy, vol. 80, pp. 140–152, 2015.

[9] Y. Chen, Y. Yang, L. Wang, Z. Jia, and W. Wu, "Grid-connected and control of MPPT for wind power generation systems based on the SCIG," CAR 2010 - 2010 2nd Int. Asia Conf. Informatics Control. Autom. Robot., vol. 3, no. 5, pp. 51–54, 2010.

[10] M. Benchagra, M. Maaroufi, and M. Ouassaid, "Study and analysis on the control of SCIG and its responses to grid voltage unbalance," Int. Conf. Multimed. Comput. Syst. -Proceedings, no. 4, pp. 1–5, 2011.

[11] M. A. H. Navas, J. L. A. Puma, and A. J. S. Filho, "Direct torque control for squirrel cage induction generator based on wind energy conversion system with battery energy storage system," 2015 IEEE Work. Power Electron. Power Qual. Appl. PEPQA 2015 - Proc., no. July 2018, 2015.

978-1-7281-4181-7/19 $31.00 © 2019 IEEE

Single-phase to three-phase ac-dc-ac converter based on cascaded transformers rectifier and open-end winding induction motor

Antonio D. D. Almeida[1], Nady Rocha[1], Member, IEEE,
Edgard L. L. Fabricio[2], Member, IEEE,
Carolina A. Caldeira[1], Member, IEEE,
Gleice M. S. Rodrigues[1], Member, IEEE and Isaac S. Freitas[1], Member, IEEE
[1]Federal University of Paraíba (UFPB) - João Pessoa - PB - Brazil
[2]Federal Institute of Paraíba (IFPB) - João Pessoa - PB - Brazil
e-mails: [antonio.almeida, nadyrocha, carolina.caldeira, gleice.rodrigues, isaacfreitas]@cear.ufpb.br,
edgard.fabricio@ifpb.edu.br

Abstract—A single-phase to three-phase ac-dc-ac converter is proposed and analyzed in this paper for application in electric rail systems and in rural distribution grid. This system is composed of two cascaded transformers multilevel converter and two conventional two-level three-phase inverter feeding an open-end winding induction motor. Model and a control strategy of the system are developed. Compared to the conventional single-phase to three-phase converter, the proposed topology has lower voltage and power ratings in power switches. In addition, the control system ensures balanced dc-link voltages and sinusoidal grid current with high power factor. Simulated and experimental results are also presented to demonstrate the feasibility of the proposed circuit.

Index Terms—multilevel, single-phase to three-phase, OEWIM

I. INTRODUCTION

Most electrical grid in countryside is not 3-phase, but 1-phase, which makes the use of more efficient three-phase motors difficult. Three-phase applications present better dynamic and they are able to achieve high power levels. Single-phase to three-phase converters appear as a possibility to overcome the lack of three-phase grid. Furthermore, electrical transportation brings energy-efficient appeal. Most modern railway system are based on ac single-phase high voltage, then power converters are also required to supply electrical three-phase motors [1] and [2]. Many countries around the world, mainly in Europe and Asia, have been world have been investing in electrical railway system to transport people and cargo [3] and [4]. This topic has been also the object of study for several years in applications such as three-phase motor drive [5]–[7] uninterrupted power supply (UPS) [8], distributed generation system [9], and power quality conditioner [10] and [11].

In high power/medium voltage drives, the multilevel converters have been widely applied as a viable solution to overcome current and voltage limitations of power switches. The multilevel converters are able to generate output voltage waveforms with large number of levels, using multiple power

sources and switching devices with lower voltage ratings. In addition, the harmonic distortion of waveforms and dv/dt stress on the switches are reduced. Furthermore, the electromagnetic interference (EMI) and switching losses are also reduced [12]. Overall, multilevel converters seek a better quality of energy from grid to load side. Some important issues in this concern are: common mode voltage (CMV), leakage current, passive components size. In this context, topologies originated from series connections of basic modules are good options for high voltage applications. Among them, the most popular topology is the cascaded h-bridge with cascaded transformers [13] and [14] and without [15]–[17]. Besides that, open-end winding (OEW) concept appears as an option to achieve multilevel features through series-connection of three-phase converter [18].

The conventional topology of single-phase to three-phase active power filter is presented in Fig. 1. Based on this topology, this paper proposes an ac-dc-ac topology that brings together some multilevel features found in the literature, to provide three-phase grid from a single-phase one. The rectifier side of this topology is composed of two isolation transformers series-connected at primary side supplying two h-bridge converters through a single-phase grid. The inverter side is composed of two three-phase VSI converters series-connected feeding an OEW Induction Motor (OEWIM), as shown in Fig. 2. Series-connected converters are required for high-voltage application by distributing voltage stress among cells and achieving higher quality of waveforms. In this way, Even with higher number of power switches compared with conventional topology, the proposed topology appears as a suitable solution for many applications. Overall, this topology brings multilevel features at rectifier and inverter sides with galvanic isolation avoiding circulating current. Besides that, it ensures power factor close to one, dc-link control, and controlled output voltages. This paper develops the system model of proposed topology, presents a control algorithm to balance dc-link voltages and to ensure sinusoidal current at grid side, and simulation and experimental results are presented to illustrate the correct operation of proposed topology.

II. SYSTEM MODEL

The proposed topology, shown in Fig. 2, is composed of two cascaded full bridge rectifiers (Converters A and B) isolated by two transformers (T_a and T_b) and two three-phase inverters feeding an OEWIM.

The transformers ensure electrical isolation between the grid and the rectifiers. They are required in high-voltage railway systems allowing connections of many motor drive systems at secondary. The input and magnetizing inductance of each transformer can cause input current mismatch related to the mains voltage. Thus to perform the correct voltage and current synchronization, a phase correction control is required as performed in this paper.

The system model equations of input side can be found at the primary side of the transformers. The equations are shown below:

$$e_g = r_g i_g + l'_g \frac{di_g}{dt} + v_g \tag{1}$$

$$l'_g = l_g + 2l_t \tag{2}$$

$$i_{gk} = \eta i_g \tag{3}$$

$$v_g = \eta(v_{ga} + v_{gb}) \tag{4}$$

$$v_{gk} = v_{gk10} - v_{gk20} \tag{5}$$

where $k = a, b$, e_g is the grid voltage, i_g is the grid current, i_{gk} are the input currents of the rectifiers, η is the transformation ratio of transformers, v_{gk} are the voltages generated by rectifiers, v_{gk10} and v_{gk20} are the rectifier pole voltages, r_g and l_g represent the resistance and inductance of the input filter, l_t and l'_g represent the leakage inductance of each transformer viewed from the grid side, considering non-ideal transformers, and the equivalent input inductance, respectively.

The converters P and N are feeding an OEWIM. Considering the voltages at the induction machine as v_{sj}, where $j = 1, 2, 3$, the converter voltages can be written as:

$$v_{srj} = v_{pj0_a} - v_{nj0_b} \tag{6}$$

$$v_{srj} = v_{sj} - v_{0_a0_b} \tag{7}$$

$$v_{0_a0_b} = \frac{1}{3}\sum_{j=1}^{3}(v_{spj} - v_{snj}) \tag{8}$$

where $v_{0_a0_b}$ is the voltage between the midpoints of the dc links, v_{pj0_a} and v_{nj0_b} are the pole voltages of converters P and N, respectively. The voltage $v_{0_a0_b}$ is given by (8).

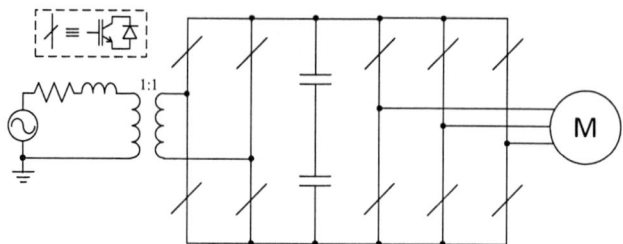

Fig. 1. Conventional single-phase to three-phase converter.

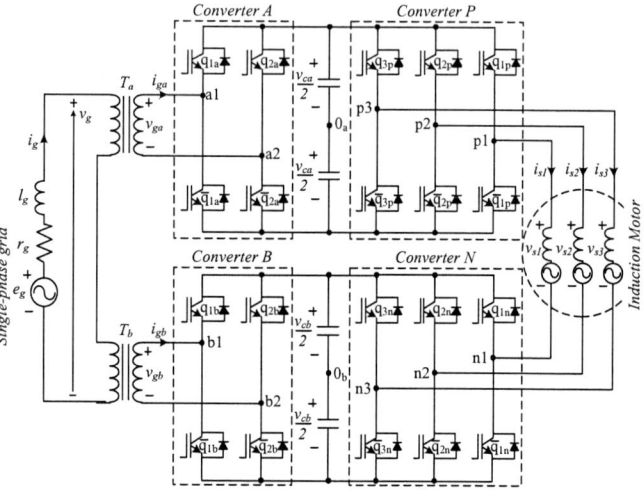

Fig. 2. Proposed single-phase to three-phase converter with two dc-link capacitors.

III. CONTROL STRATEGIES

The Fig. 3 presents the control block diagram for the proposed configuration. The two dc-link voltages are controlled initially through the average dc-link capacitor voltages $[v_c = (v_{ca} + v_{cb})/2]$. v_c is adjusted to its reference value v_c^* by using the conventional Proportional-Integral type controller (PI). Such controller provides the amplitude of the grid current reference i_g^* (primary of the transformer). In order to obtain balanced dc-link voltages, the difference voltage between the dc-links is ensured null through another PI controller, that provides the variable ΔV. Then, the desired reference voltages v_{ga}^* and v_{gb}^* in the converters A and B, respectively, can be defined from ΔV and v_g^*, where $v_{ga}^* = (1 - \Delta V)v_g^*$ and

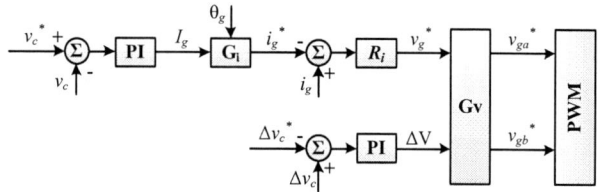

Fig. 3. System block control diagram for grid side proposed configuration.

978-1-7281-4181-7/19 $31.00 © 2019 IEEE

$v_{gb}^* = (1 + \Delta V)v_g^*$.

The high power factor, at the grid side, is achieved by the grid current control R_i. The instantaneous reference current i_g^* is synchronized (via Phase Lock Loop scheme - PLL) with the grid voltage e_g. The current control is achieved by resonant control as shown in [19], and the voltage v_g^* is the output of current controller.

IV. MODULATION STRATEGY

The modulation strategy applied in the proposed topology is based on duty-cycle functions. This modulation is about the choice of vectors that will be applied, this technique is interesting for multilevel applications since it replaces the need of level shifted carriers. Initially considering the converters A and B, the duty-cycles are calculated from voltages from reference voltages v_{gk}^* (with k = a, b) which are provided by controllers. In this paper, the modulation of the converters A and B is treated individually. Then, as each single-phase converter can produce three voltage levels (v_C, 0 and $-v_C$), the duty-cycle of each switch (d_{k1} and d_{k2}) can be determined for two regions of the reference voltages v_{gk}^*, as shown in Fig. 4. Each region is delimited by the two nearest voltage levels, one of higher value V_{high} and another of lower value represented by V_{low}. For converter A, when $v_{ga}^* > 0$, the voltages levels V_{high} is produced by state $[q_{a1}, q_{a2}] = 10$ and V_{low} is produced by state $[q_{a1}, q_{a2}] = 00$ or $[q_{a1}, q_{a2}] = 11$ (redundant states). On the other hand, when $v_{ga}^* < 0$, the voltages levels V_{high} is produced by state $[q_{a1}, q_{a2}] = 00$ or $[q_{a1}, q_{a2}] = 11$ (redundant state) and V_{low} is produced by $[q_{a1}, q_{a2}] = 01$. The duty-cicle d_{k1} and d_{k2} is the average value of v_{gk}^* between V_{high} and V_{low} normalized between 0 and 1. The assignment of values for the duty-cycle follows the following logic:

1) If the state of the switch remained 1, then $d_k = 1$;
2) If the state of the switch remained 0, then $d_k = 0$;
3) If the state of the switch changes from 0 to 1, then

$$d_k = \frac{v_{gk}^* - V_{high}}{V_{high} - V_{low}} \quad (9)$$

4) If the state of the switch changes from 1 to 0, then

$$d_k = \frac{V_{high} - v_{gk}^*}{V_{high} - V_{low}} \quad (10)$$

As an example, in region 1 ($v_{ga} > 0$), there is only a state of $V_{high} = V_2$, i.e., $[q_{a1}, q_{a2}] = 10$ and choosing the state $[q_{a1}, q_{a2}] = 00$ for $V_{low} = V_0$. At the transition from V_2 to V_0, q_{a1} changes from 1 to 0. Thus, the third condition is used and duty-cycle d_{a1} is calculated by (9). Since q_{a2} is maintained at 0, satisfying the first condition, the duty-cycle is $d_{a2} = 0$. Gating signals can be obtained by a comparison of the duty-cycles with a high frequency triangular carrier signal. In this paper, gating signals are obtained comparing duty-cycle with double high frequency triangular carrier signals, i.e., double-carrier-based PWM. In the case of double-carrier-based PWM, the phase shift of the triangular carrier signals, between converters A and B, is $180°$.

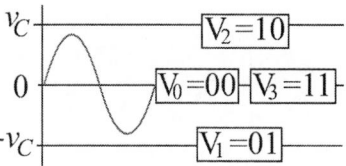

Fig. 4. One-dimension modulation spaces.

There are different solutions to modulation of open-end winding converter with isolated dc-link voltages, such as phased-disposition PWM (PD-PWM), level-shifted PWM (LS-PWM), space-vector PWM (SV-PWM), etc. The modulation strategy applied in converters P and N (open-end winding) is also based on duty-cycle functions. As each phase is treated individually, then for generation the voltage v_{srj}, there are four possible states $[q_{p1}, q_{n1}] =00$, 01, 10 and 11, as well as for the converters A and B. In this way, the same methodology to generate the duty-cycle of converters A and B can be applied.

V. SIMULATIONS RESULTS

Simulation of the proposed single-phase to three-phase OEWIM was carried out for a 2 kW 220V/60Hz three-phase induction motor. A volt/hertz control was used with a frequency of 25 Hz. The other parameters of the system and induction motor are, respectively, presented in the Tables I and II. The simulation results are shown in Fig. 3. The grid voltage as well as the supplied current are shown in Fig. 5(a), where the grid current are in phase with the grid voltage with a PF (Power Factor) of 99.81 %. In Fig. 5(b) the five-level input voltage is a result of the use of cascaded transformer multilevel rectifier. In Fig. 5(c), is shown the dc-link voltages (v_{ca} and v_{cb}). Notice that the balancing between the dc-link voltages was effective, following their reference values. Figs. 5(d) and 5(e) show voltages and currents of the load, respectively. Notice that the load voltages have nine levels.

Fig. 6 presents the total harmonic distortion of grid current for different input inductance (l_g') values. Overall, the proposed topology requires a inductance four times lower than conventional topology for a THD of 2%. The results were obtained in open-loop with RL load to obtain a more objective comparison between the proposed and the classic topology.

VI. EXPERIMENTAL RESULTS

The experimental results were obtained using an experimental platform as depicted in Fig. 7. The main parameters are provided in Table III. Figs. 8 and 9 show the experimental results of proposed topology using a 0.75kW induction motor. Fig. 8(a), shows the waveforms of the voltage and current of the single-phase grid. Notice that the power factor is close to unity. Fig. 8(b) ilustrates the five-level converter input voltage. The dc-link voltages (v_{ca} and v_{cb}) are controlled according to voltage reference of 105 V, as shown in Fig. 8(c). These voltages present a natural oscillation (second order harmonic), which is common in single-phase installations. The motor

978-1-7281-4181-7/19 $31.00 © 2019 IEEE

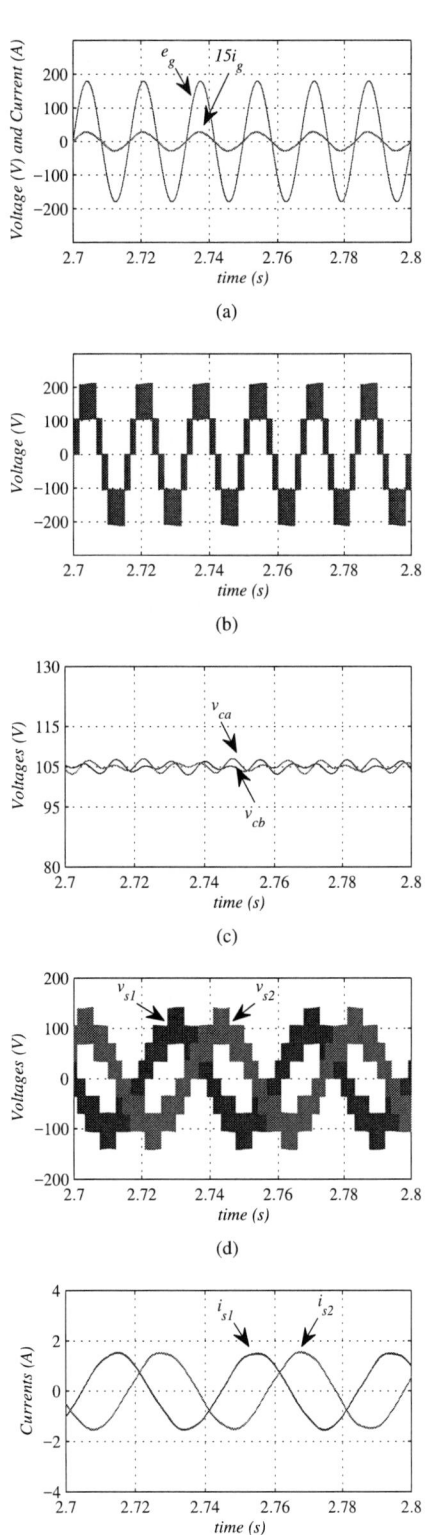

(a)

(b)

(c)

(d)

(e)

Fig. 5. Simulation results. (a) Voltage e_g and current i_g of the grid. (b) Input voltage v_g. (c) Voltage v_{ca} and v_{cb}. (d) Machine voltages v_{s1} and v_{s2}. (e) Machine currents i_{s1} and i_{s2}.

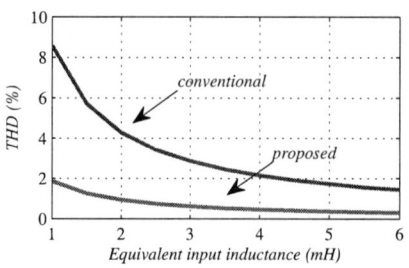

Fig. 6. THD simulation results of input current.

voltages v_{s1} and v_{s2} as in Fig. 9(a) where, the nine distinct voltage levels are clearly defined. In Fig. 9(b) shows the output currents i_{s1} and i_{s2} obtained from OEWIM.

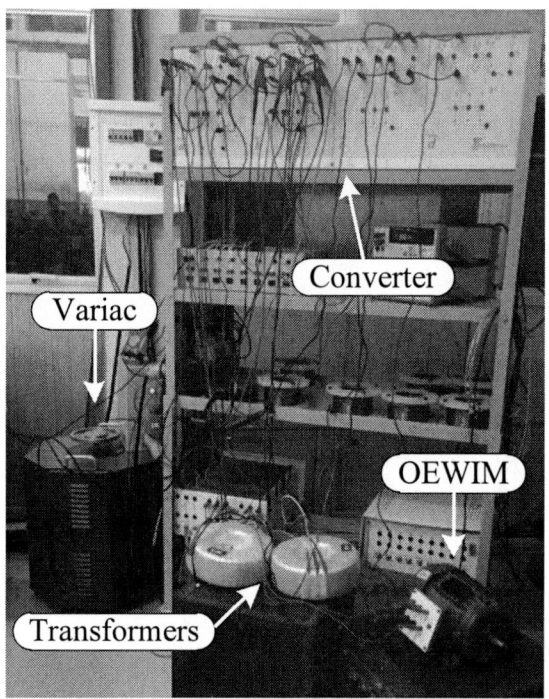

Fig. 7. Experimental platform.

TABLE I
PARAMETERS EMPLOYED ON THE SIMULATION RESULTS.

Parameter	Value	Parameter	Value
e_g	127 V (RMS)	Transformers	1:1
l_g	10 mH	r_g	0.1 Ω
v_{ca}, v_{cb}	105 V	dc-link	2200 μF
Switching freq.	10 kHz	Grid freq.	60 Hz
Motor freq.	25 Hz		

978-1-7281-4181-7/19 $31.00 © 2019 IEEE

TABLE II
MOTOR PARAMETERS EMPLOYED IN THE SIMULATION RESULTS.

Parameter	Value	Parameter	Value
R_s	3 Ω	L_s	14.9 mH
R_r	2.99 Ω	L_r	14.9 mH
L_m	599.2 mH	p (poles)	2
Load	1 N.m	Moment of Inertia	0.005

TABLE III
SETUP USED FOR EXPERIMENTAL RESULTS.

Parameter	Value	Parameter	Value
DSP	TMS320F28379D	Driver	SKHI-23
Power switches	SKM50GB123D	Fundamental freq.	60 Hz
e_g	127 $V_{(RMS)}$	Transformers	1:1
l_g	6.7 mH	r (load)	50 Ω
v_{ca}, v_{cb}	105 V	dc-link	2200 μF
Switching freq.	10 kHz	Induction Motor	0.75 kW
Motor freq.	25 Hz		

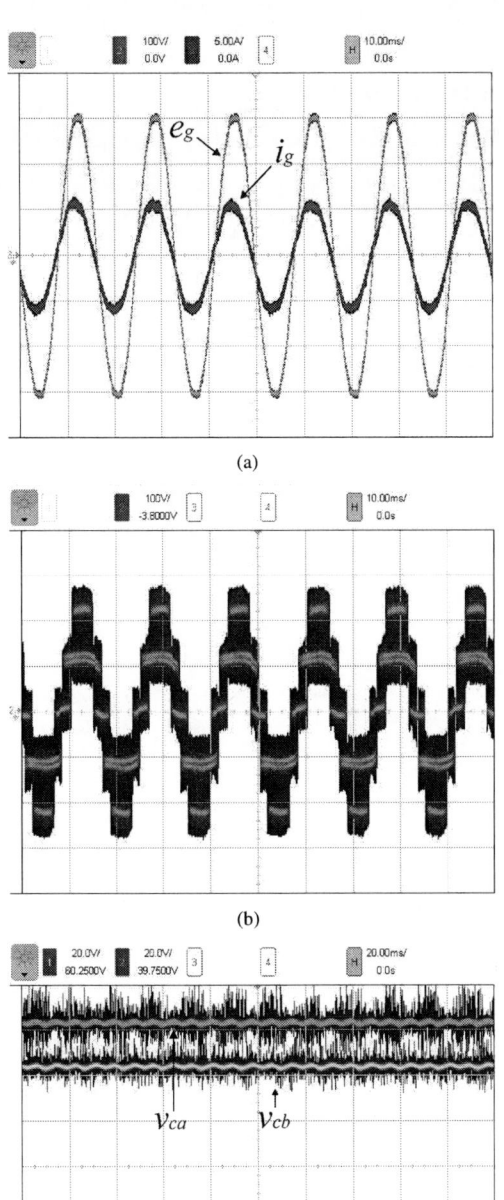

Fig. 8. Experimental results. (a) Voltage e_g and current i_g of the grid. (b) Input Voltage v_g. (c) Voltages v_{ca} and v_{cb}.

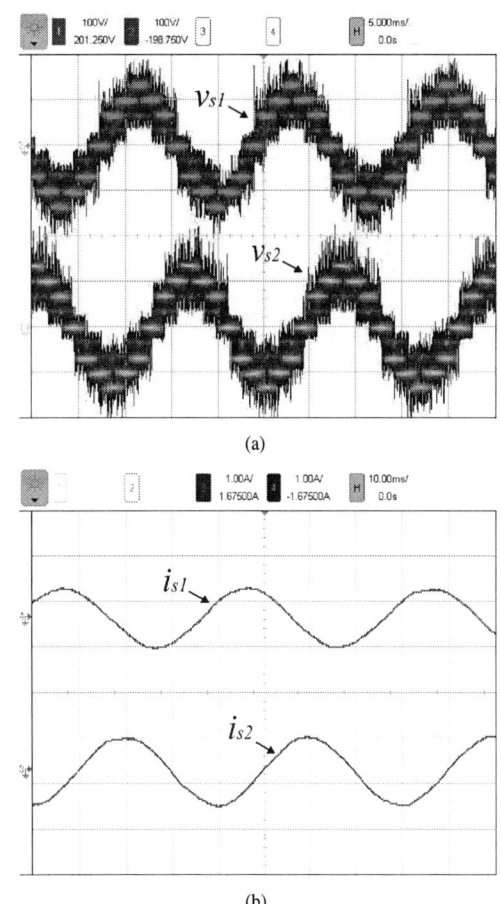

Fig. 9. Experimental results. (a) Machine voltage v_{s1} and v_{s2}. (b) Machine currents i_{s1} and i_{s2}.

VII. CONCLUSIONS

A single-phase to three-phase ac-dc-ac converter was proposed in this paper. The converter consists of rectifiers isolated by cascade transformers by raising the safety of the grid in case of failure of the converters. This topology ensures five-level input voltage as shown in the results. The dc-links, voltages were controlled with the balancing technique proposed in the paper. The input current was controlled, keeping in phase with the grid voltage, ensuring power factor close to unity. A feasible PWM strategy was applied ensuring, lower total harmonic distortion. The experimental and simulation results compared with conventional topology validate of the proposed converter.

ACKNOWLEDGEMENT

The authors would like to thank in part the Coordenação de Aperfeiçoamento de Pessoal de Nível Superior - Brasil (CAPES) and the Conselho Nacional de Desenvolvimento Científico e Tecnológico CNPq for the financial support.

REFERENCES

[1] E. Pilo de la Fuente, S. K. Mazumder, and I. G. Franco, "Railway electrical smart grids: An introduction to next-generation railway power systems and their operation." *IEEE Electrification Magazine*, vol. 2, no. 3, pp. 49–55, Sep. 2014.

[2] P. Drabek, Z. Peroutka, M. Pittermann, and M. Cedl, "New configuration of traction converter with medium-frequency transformer using matrix converters," *IEEE Transactions on Industrial Electronics*, vol. 58, no. 11, pp. 5041–5048, Nov 2011.

[3] R. J. Hill, "Electric railway traction. part 3. traction power supplies," *Power Engineering Journal*, vol. 8, no. 6, pp. 275–286, Dec 1994.

[4] H. Glickenstein, "Israel railways orders new electric locomotives [transportation systems]," *IEEE Vehicular Technology Magazine*, vol. 11, no. 1, pp. 21–24, March 2016.

[5] A. K. Adapa and V. John, "Active-phase converter for operation of three-phase induction motors on single-phase grid," *IEEE Transactions on Industry Applications*, vol. 53, no. 6, pp. 5668–5675, Nov 2017.

[6] V. Verma and A. Kumar, "Cascaded multilevel active rectifier fed three-phase smart pump load on single-phase rural feeder," *IEEE Transactions on Power Electronics*, vol. 32, no. 7, pp. 5398–5410, July 2017.

[7] N. B. de Freitas, C. B. Jacobina, A. C. N. Maia, and A. C. Oliveira, "Six-leg single-phase to three-phase converter," *IEEE Transactions on Industry Applications*, vol. 53, no. 6, pp. 5527–5538, Nov 2017.

[8] R. Q. Machado, S. Buso, and J. A. Pomilio, "A line-interactive single-phase to three-phase converter system," *IEEE Transactions on Power Electronics*, vol. 21, no. 6, pp. 1628–1636, Nov 2006.

[9] E. C. D. Santos, C. B. Jacobina, N. Rocha, J. A. A. Dias, and M. B. R. Correa, "Single-phase to three-phase four-leg converter applied to distributed generation system," *IET Power Electronics*, vol. 3, no. 6, pp. 892–903, Nov 2010.

[10] E. C. dos Santos, C. B. Jacobina, J. A. A. Dias, and N. Rocha, "Single-phase to three-phase universal active power filter," *IEEE Transactions on Power Delivery*, vol. 26, no. 3, pp. 1361–1371, July 2011.

[11] S. A. O. da Silva and F. A. Negrão, "Single-phase to three-phase unified power quality conditioner applied in single-wire earth return electric power distribution grids," *IEEE Transactions on Power Electronics*, vol. 33, no. 5, pp. 3950–3960, May 2018.

[12] S. Sau, S. Karmakar, and B. G. Fernandes, "Modular transformer-based regenerative-cascaded multicell converter for drives with multilevel voltage operation at both input and output sides," *IEEE Transactions on Industrial Electronics*, vol. 65, no. 7, pp. 5313–5323, July 2018.

[13] Y. Suresh and A. K. Panda, "Research on a cascaded multilevel inverter by employing three-phase transformers," *IET Power Electronics*, vol. 5, no. 5, pp. 561–570, May 2012.

[14] M. R. Banaei, H. Khounjahan, and E. Salary, "Single-source cascaded transformers multilevel inverter with reduced number of switches," *IET Power Electronics*, vol. 5, no. 9, pp. 1748–1753, November 2012.

[15] F. V. Amaral, T. M. Parreiras, G. C. Lobato, A. A. P. Machado, I. A. Pires, and B. de Jesus Cardoso Filho, "Operation of a grid-tied cascaded multilevel converter based on a forward solid-state transformer under unbalanced pv power generation," *IEEE Transactions on Industry Applications*, vol. 54, no. 5, pp. 5493–5503, Sep. 2018.

[16] J. Shi, W. Gou, H. Yuan, T. Zhao, and A. Q. Huang, "Research on voltage and power balance control for cascaded modular solid-state transformer," *IEEE Transactions on Power Electronics*, vol. 26, no. 4, pp. 1154–1166, April 2011.

[17] M. Najjar, A. Moeini, M. K. Bakhshizadeh, F. Blaabjerg, and S. Farhangi, "Optimal selective harmonic mitigation technique on variable dc link cascaded h-bridge converter to meet power quality standards," *IEEE Journal of Emerging and Selected Topics in Power Electronics*, vol. 4, no. 3, pp. 1107–1116, Sep. 2016.

[18] E. G. Shivakumar, K. Gopakumar, S. K. Sinha, A. Pittet, and V. T. Ranganathan, "Space vector pwm control of dual inverter fed open-end winding induction motor drive," in *APEC 2001. Sixteenth Annual IEEE Applied Power Electronics Conference and Exposition (Cat. No.01CH37181)*, vol. 1, March 2001, pp. 399–405 vol.1.

[19] C. B. Jacobina, M. B. de R. Correa, T. M. Oliveira, A. M. N. Lima, and E. R. C. da Silva, "Current control of unbalanced electrical systems," *IEEE Trans. Ind. Electron.*, vol. 48, no. 3, pp. 517–525, June 2001.

978-1-7281-4181-7/19 $31.00 © 2019 IEEE

An Overview About Detection of Cyber-Attacks on Power SCADA Systems

Hugo F. M. de Figueiredo, Matheus K. Ferst, and Gustavo W. Denardin

Postgraduate Program in Electrical Engineering (PPGEE)
Federal University of Technology - Paraná (UTFPR)
Pato Branco, Brazil
hugoffigueiredo1@gmail.com, matheus.ferst@gmail.com, gustavo@utfpr.edu.br

Abstract—The power SCADA systems have been undergoing several upgrades over the years. At first, these systems were based only on isolated processes, without the need to connect in a global network, such as the Internet. In recent years, the access to such an open network has been critical for the SCADA systems, due to the need for data exchange across large areas, between geographically remote industrial plants, or even because of the demand for remote applications. Despite the SCADA systems have advantages by being now connected to a network such as the Internet, several problems can arise due to that. For example, such a connection makes it possible for an intruder to compromise the whole system. SCADA systems are essential in the new concept of smart grids and are widely used in the monitoring and control of Distributed Energy Resources (DER) based on photovoltaics. In that context, this paper presents an overview of the main attacks found and documented in the literature, as well as possible techniques to detect such attacks.

Index Terms—SCADA Systems, Smart Grids, Security, Intrusion Detection Systems

I. INTRODUCTION

The environmental concerns have incentivized the search for Distributed Generations (DG), which has made this one of the most interesting research topics in recent years [1]. Photovoltaic generation is one of these DGs made by renewable resources that has received a lot of attention in the literature. Additionally, the photovoltaic energy is already widespread in the market, with several trading solutions being used in residential, commercial, and industrial sectors [2].

The photovoltaic sector, however, is now focused on improving and expanding the quality of the power grid [3]. In this context, one of the main topics of interest refers to Supervisory Control and Data Acquisition (SCADA). SCADA systems are used for command and supervise gas industry plants, water treatment stations, power generation systems, among many other processes [4].

Such a technology solution is considered crucial for power systems [5]. These systems can provide the monitoring and supervision of devices and variables associated with the power system, as well as implementing decision-making based on the reading and writing of the grid data.

Fig. 1 shows one generic SCADA structure developed for DGs. The control center is responsible for system operation

This work was funded by the Research and Development project PD 2866-0468/2017, granted by the Brazilian Electricity Regulatory Agency (ANEEL) and Companhia Paranaense de Energia (COPEL).

and monitoring. Usually, this level is comprised of an HMI (Human-Machine Interface) and work stations, which are the devices capable of sending and receiving commands from the remote sites.

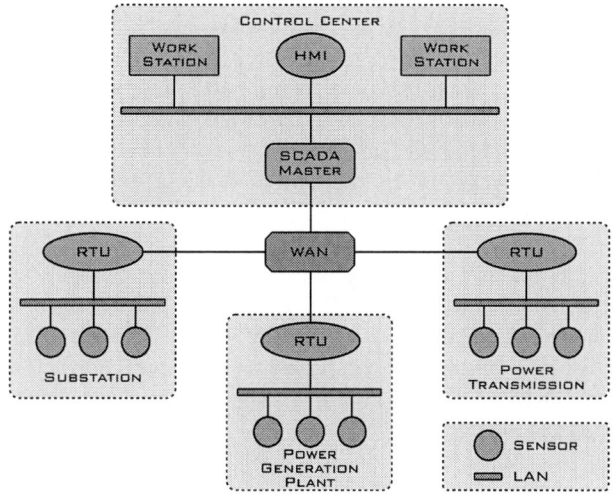

Fig. 1. Generic Power SCADA System.

The Remote Terminal Unit (RTU) shown in Fig. 1 provides the information on substations, power generation plants and power transmission lines. RTU refers to a microprocessor-controlled electronic device that can connect components to a SCADA system using telemetry and able to transfer data to a master system.

The photovoltaic Distributed Energy Resources (DER) industry has made considerable progress over the years, enabling remote access to the inverters and other devices that make a solar generation system, which allows agents physically disconnected from several DER plants to monitor and control them. Thereby, the connection of the system to the Internet becomes important and can provide many advantages, for example, the economy about the network infrastructure.

In most cases, SCADA systems have access to a corporate network realized by existing technologies, such as the Virtual Private Networks (VPN) [6]. However, the growth of connected photovoltaic generation systems on the Internet has exposed the SCADA system to several vulnerabilities [7].

The manipulation of exclusive system data by cyber-attacks has become a common and serious event [8]. For example, if an invader gets total control of a utility network, all private corporate information is readily available for such an external agent, which makes it necessary the avoid unsecured channels on the corporate network [6]. Also, it can be possible to interfere with the provided service, making it unstable or unavailable.

Invasions in SCADA systems composed of unsafe channels have reached considerable levels in recent years [4]. The disturbances that these attackers can generate reflects directly in the system reliability, in the process carried out by the software and the accuracy of the system [9]. Therefore, to avoid such security glitches, it is recommended to implement security models on corporate networks to evaluate whether there has been a change by a malicious external agent.

Although several new exploits are trying to take advantage of a vulnerability, an attack has been considered a milestone for the security of information in SCADA systems, the discovery of the Stuxnet worm in 2010 [10]. Stuxnet targets programmable logic controllers (PLCs), which allow the automation of electromechanical processes, such as those used to control machinery and industrial processes. This worm has become known by the invasion of the control system of a uranium enrichment centrifuge in Iran, causing substantial damage to Iran's nuclear program [11].

The Stuxnet had the ability to reprogram control logics and hide all the changes made [12]. This malicious agent exploited the vulnerability of a controller designed by SIEMENS [13]. Also, this worm made use of modern techniques compared to known forms of attack and was considered the most complex malware observed so far [13].

This worm was designed to work on vulnerabilities that were unknown at that time [12]. Furthermore, its purpose was physically deteriorating the target architecture. This attacker was not exclusively destinated to SCADA systems [13]. However, after the Iran attack in which the affected system had an incorporated SCADA architecture, it became clear that such systems are susceptible to invasions. Thus, the search for vulnerabilities in SCADA systems was stimulated.

Following the Stuxnet attack in 2010, cyber-attacks targeting SCADA systems have grown significantly. In recent years, another relevant attack has occurred in Ukraine, becoming known as BlackEnergy [14]. This form of attack refers to a web-based DDoS bot that was used by Russian hackers [15]. Also, in the Ukraine episode, the attackers have exploited relevant information of a power system to hijack the SCADA DMS (Distribution Management System) performing false actions to cause a power outage [14].

Due to the growth of grid-connected DGs in the conventional power system, several adversities were generated, such as false islanding detection, voltage elevation, the overload of network devices, among others [16]. These problems can be increased by cyber-attacks, due to the connection of these systems to the Internet.

A system connected to the Internet, in addition to having vulnerabilities to malicious software known as malware, spyware, and computer viruses are also susceptible to other possible attacks [17]. These attacks can have many different forms, such as the Man-In-The-Middle (MITM), Denial of Service (DoS), Replay, False Data Injection, among several other attacks. The first step to counter-attack the malicious intruders is based on the implementation of mechanisms that can identify the occurrence of inadequate activities on the network.

An option to defend a SCADA system of these attacks is the use of an Intrusion Detection System (IDS). Such a system is intended to verify and observe security-related occurrences of the process to recognize possible malicious acts that could compromise the system [18], [19]. Intrusion detection is currently indispensable in SCADA structures [20]. However, to handle the problem of attacks on a smart grid it is necessary the use of attack mitigation systems.

Nowadays, the mitigation of attacks is another interesting topic in the literature. The main techniques available to mitigate attacks are based on the protection of the lower levels of a network, such as TLS (Transport Layer Security) [21], [22], and IPsec (IP Security Protocol) [23], [24]. The IDS is insufficient to completely solve the security problems in a SCADA system but is the last defense line to countermeasure the attacks, and thus reduce the damage of attacks in the structure.

To improve the current methods of detection, it becomes necessary the expand of the study on this subject. Therefore, this paper addresses the most critical attacks to power SCADA systems and some of their respective detection methods based on Intrusion Detection Systems.

II. FORMS OF CYBER-ATTACKS

This section approaches non-exhaustively some of the main forms of the cyber-attack present in current systems. These attacks work by making alterations of the communication network. These modifications that able to cause severe physical damages on the system [25], [26]. These models operate on the software making the change of data and logic and also work directly on the hardware by attacking the MTU (Master Terminal Unit) and RTU drives.

A. Man-In-The-Middle

The man-in-the-middle attack works the control of information exchanged between two platforms [27]. It provides to the invader an ability to mask the communication on the network. Thereby, this brings to the communication system a false impression that the extremities are communicating normally, making the system unstable.

This camouflage allows the attacker to detect the data sent and perform the transmission of false data [28]. Furthermore, it grants the exclusion of packages without informing the network of any sign of divergence in the information. When the attack is accomplished in photovoltaic DER systems, the malicious user acts in the modification of the information on

978-1-7281-4181-7/19 $31.00 © 2019 IEEE

distribution and control of energy by inserting modified data into substations, meters, inverters, among other devices [29]. Consequently, the system is forced to operate with untrusted references.

Man-in-the-middle attacks can cause considerable damages on an electric grid [28]. However, among all the possible adversities generated by invasions to these systems, it is interesting to highlight some of them, such as the intermittencies generated in the grid, the loss of the system response rate, malfunction in devices and even a system blackout [30].

Fig. 2 illustrates how an attack carried out by the MITM model works. This attack acts by modifying information transmitted between the two victims. Thus, after the intruder completes his objective, the attack compromises the channel on which the data was carried, and all system information needs to be passed by the middleman first, which in this case is the attacker.

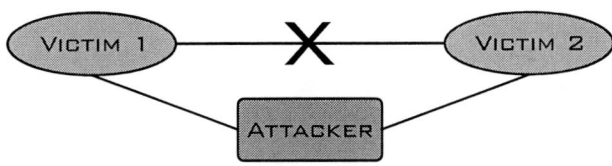

Fig. 2. Man-In-The-Middle Attack.

In the MITM attack, the victims try to establish secure communication by sending public keys to each other. However, when an attacker is present on the network, this connection is not completed. Then, in this case, the invader retains the information and returns other data composed of their public keys to the users.

After the data manipulated by the attacker is sent, the victim encrypts the message by the attacker's key and sends it to another victim. But the invader intercepts the data and decrypts it using a private key. Then, the malicious agent sends to the target victim false information composed by the public key of the user [27]. Thus, this makes the system work perfectly without the victims recognizing that they are being invaded and that the information exchanged between them has been changed.

B. Replay Attack

Replay attack refers to the invasion in which the attacker can read and modify the current system data [31]. Furthermore, the attacker can manipulate the system data to retransmit several times an old data generated by an affected device, providing a misleading behavior in the system [32].

The basic purpose of the invader is to make the modified data appears correct. Thus, during a replay attack, the reliability of the data being transmitted/received in the system can not be guaranteed. The main method to detect such an attack is to identify it at the time the attack is being carried out. If the identification can be realized, then the attack counter measurements will be possible [33].

C. False Data Injection Attack

False data injection attacks in SCADA systems are being considered as a constantly evolving scenario [34]. This attack has been developed to strike Energy Management Systems (EMS), mainly modern smart grids [35]. A smart EMS is composed of state estimations. The purpose of state estimation is ensuring that the power grid is operating in the intended conditions [36].

FDI attacks have proved to be efficient in modifying the state estimation of these systems [37]. FDIA works in the modification of information that is generated by the SCADA system, to produce negative effects on the structure [38].

If the control system can not identify malicious changes in the state estimation, mistaken actions will be realized by the system operator, compromising the system security. However, even if the invasion can be detected, part of the system information is lost during the attack, which makes the state estimator generate wrong values, and thus cause any physical or economic effects in the grid [39].

After the invasion has been completed and the entire system is reached, it is impossible to prevent the invader from working in the majority of the process. A false state estimation can misleading all control and operation functions of an EMS [39]. Because of this, the false data injection attack is considered one of the most severe threats that can occur in a smart grid [40].

D. Denial of Service and Distributed Denial of Service

DoS and DDoS attacks on DER systems are based on causing network congestion. These invasions aim to transform the current system data into obsolete or even inaccessible [41], [42]. These attacks can cause considerable failures in the systems, such as suppression of energy supply for example [43].

Fig. 3 demonstrates how DoS and DDoS attacks work in a SCADA structure. These attacks can occur in an RTU or MTU module in the network. However, the remote terminal unit is the most vulnerable component of the SCADA system, because one system can have several RTUs, which can be positioned in different locations around the plant being monitored and controlled [25]. Let us suppose that a DoS or DDoS attack have occurred to one or more RTUs. Therefore, the attacker can congest the network, and the system can not identify this problem, compromising the communication. Due to the disrupt in the communication, the MTU level is also compromised, accomplishing the purpose of the attack.

The main differences between DoS and DDoS attacks are in the accomplishment method of these invasions. The denial of service attack is occasioned by only one invader. However, the invader is destined to send multiple packets, thus causing the overload of the destination device and making it impossible to use [44].

On the other hand, in distributed denial of service mode, the attack is carried out by multiple sources synchronously, causing amplified damage and making the protection even more difficult to be accomplished [42]. As nowadays smart

978-1-7281-4181-7/19 $31.00 © 2019 IEEE

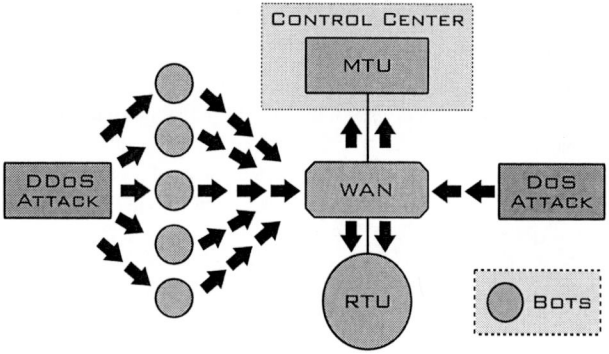

Fig. 3. DoS and DDoS Attack.

grids are connected to the internet, any SCADA system is vulnerable to DDoS attacks [45], [46].

The DDoS attack aims to impair the decision-making capacity that occurs in the communication network. The attack is designed to diminish the network performance, making the modes of operation selected by allowed clients in the structure can not be accepted [47].

III. DETECTION ATTACK METHODS

Some of these detection techniques are designed to identify a specific attack, while others implement the detection of several forms of attacks simultaneously. In this section, we present an overview of the most-known methods and some innovative techniques for detecting MITM, FDI, DoS and Replay attacks in power SCADA systems.

A. False Data Injection Attack

In order to detect false data injection attacks, several techniques have been developed. As the Cumulative SUM (CUSUM) detector [48], Support Vector Machine (SVM) algorithm [49], State Forecasting Detection (SFD) format [50] and SFD with clustering algorithms [40], and deep learning techniques [51].

B. DoS and DDoS Attack

DoS detection approaches investigated in this paper uses several techniques, some of them will be presented next. The common methods to identify denial of service attacks in the literature are based on the clusterings algorithms [4], neural networks [52], and the trust-based methods, such as the MDR (Monitoring, Detecting and Rehabilitation) [8].

C. Hybrid Forms of Attacks

The vulnerability of a SCADA system in the most times is not only associated with one specific form of attack. Because an unsafe system is susceptible to several invaders, including all existing forms of attack at the same time. Therefore, this has incentivized the development of works to detect distinct attacks simultaneously, some of these works will be presented below.

The main hybrid methods to identify attacks in the literature are based in SVM, such as the One-Class Support Vector

Machine (OCSVM), called IT-OCSVM [53], which detect man-in-the-middle and denial of service attacks. As well as the technique of detection and isolation of intrusions in the SCADA systems based in SVM [26], which can detect several different attacks, such as replay, false data injection, among other attacks.

IV. DISCUSSION OF METHODS

A. CUSUM Algorithm

The CUSUM algorithm shows valid performance to high rate attacks, but their efficiency reduces in low rate attacks [54]. Another adversity in the CUSUM detector is that the method induces the control center to read simultaneously all raw data from the meter, which can cause a communication overhead in long-distance networks [48], therefore, it is necessary to establish some cares during the development of technique. Another significant disadvantage of this method refers to a long time to detect the invader in wide-area systems [40].

This detection method, however, can demonstrate a robust performance against several invaders [55], not only in an FDI attack. Another advantage associated with this approach is the elevated accuracy in the detection of false data in ideal and non-ideal conditions [40].

B. Machine Learnings

The problem of all techniques based on supervised, semisupervised, and unsupervised machine learning is to recognize the instances of the system [26]. Thus, it is necessary to perform a test to recognize the instances before the method application.

The system efficiency is directly related to the instances of the base, then, when there are many bases in the instances, the system's effectiveness becomes better. Also is required the addition of several attack examples in the base during the training step, to acquire considerable malicious event detection.

Systems submitted to an effective training can extrapolate the classified events, thus being possible to adapt and identify attacks not compounded in the base. However, a base of the instances comprised of many references can submit the system to recognize only know events, which makes the system inefficient to extrapolate the information obtained during the training period, this process is called overfitting. Overfitting can generate an undesirable behavior in the system because even if a normal event occurs outside of the base, it will be considered the malicious act.

C. Trust-based System

The trust-based system is able to perform secure routing and detect malicious nodes in the system [56]. However, this technique needs the recognizing of network topology, which is a disadvantage.

This approach has a limitation, which arises during the divergences of situations, such as in moments where are many malicious nodes in the system, which makes system

978-1-7281-4181-7/19 $31.00 © 2019 IEEE

performance not be reliable [56]. Another problem to this technique happens relevant to highlight, such as the possibility of a system with this method can provide support to malicious acts, according to the trust in the references, and even when they are malicious events.

D. State Forecasting Detection

The SFD format has an interesting performance over normal and abnormal situations that can occur in a power system [57]. However, one adversity that can occur in this method is related to prediction, because despite that systems can reach a significant precision and accuracy, some times the accomplish of prediction technique can require a long detection time [50], [57].

V. Conclusion

The power SCADA systems connected to the Internet are vulnerable to several attacks and the implement of models that can amplify their security is essential. The Intrusion Detection System is one of the mechanisms that can increase security, as well as is the last step to countermeasure the cyber-attacks. However, the attack detections need to be based on several attack types, because power SCADA systems are vulnerable to many threats, and detect only one of the possible attacks is not sufficient to guarantee their security. Also, only IDS can not solve completely the problem of security of these structures, so for this, the inclusion of a mitigation system would be required.

Cyber-attacks can cause several damages in a power SCADA system, such as the intermittencies in the grid, electricity theft, the loss of the system response rate, suppression of energy supply, malfunction in devices, system blackout, among others. This paper shows a non-exhaustive overview of cyber-attacks in a power SCADA system. Firstly, an introduction to this problem is presented. Secondly, the theoretical basis of some of the main cyber-attacks methods that can occur in this type of system is introduced. Then, the paper discusses the predominant techniques to detect these attacks. This work motivates possible research approached in the middle. These orientations aim to detection methods that can identify FDI, MITM, DoS and replay attacks, as well as the hybrid methods.

Acknowledgment

This work was funded by the Research and Development project PD 2866-0468/2017, granted by the Brazilian Electricity Regulatory Agency (ANEEL) and Companhia Paranaense de Energia (COPEL). The authors also would like to thank to FINEP, SETI, CNPq, Fundação Araucária, CAPES and UTFPR for additional funding.

References

[1] D. Q. Hung, N. Mithulananthan, and R. Bansal, "Analytical strategies for renewable distributed generation integration considering energy loss minimization," *Applied Energy*, vol. 105, pp. 75–85, 2013.

[2] O. M. Toledo, D. Oliveira Filho, and A. S. A. C. Diniz, "Distributed photovoltaic generation and energy storage systems: A review," *Renewable and Sustainable Energy Reviews*, vol. 14, no. 1, pp. 506–511, 2010.

[3] N. Hatziargyriou, H. Asano, R. Iravani, and C. Marnay, "Microgrids," *IEEE Power and Energy Magazine*, vol. 5, no. 4, pp. 78–94, 2007.

[4] A. Almalawi, A. Fahad, Z. Tari, A. Alamri, R. Alghamdi, and A. Y. Zomaya, "An Efficient Data-Driven Clustering Technique to Detect Attacks in SCADA Systems," *IEEE Transactions on Information Forensics and Security*, vol. 11, no. 5, pp. 893–906, 2016.

[5] Y. Zhang, L. Wang, Y. Xiang, and C.-W. Ten, "Power system reliability evaluation with scada cybersecurity considerations," *IEEE Transactions on Smart Grid*, vol. 6, no. 4, pp. 1707–1721, 2015.

[6] S. Sridhar and G. Manimaran, "Data Integrity Attacks and their Impacts on SCADA Control System," *IEEE PES General Meeting*, pp. 1–6, 2010.

[7] Y. Yang, K. McLaughlin, S. Sezer, T. Littler, E. G. Im, B. Pranggono, and H. F. Wang, "Multiattribute scada-specific intrusion detection system for power networks," *IEEE Transactions on Power Delivery*, vol. 29, no. 3, pp. 1092–1102, 2014.

[8] G. K. Chalamasetty and S. Member, "Secure SCADA Communication Network for Detecting and Preventing Cyber-Attacks on Power Systems," *2016 Clemson University Power Systems Conference (PSC)*, pp. 1–7, 2016.

[9] H. Hilal and A. Nangim, "Network Security Analysis SCADA System Automation on Industrial Process," *2017 International Conference on Broadband Communication, Wireless Sensors and Powering (BCWSP)*, 2017.

[10] R. Langner, "To kill a centrifuge: A technical analysis of what stuxnet's creators tried to achieve," *The Langner Group*, 2013.

[11] T. Chen and S. Abu-Nimeh, "Lessons from stuxnet," *Computer*, vol. 44, no. 4, pp. 91–93, 2011.

[12] N. Falliere, L. O. Murchu, and E. Chien, "W32. stuxnet dossier," *White paper, Symantec Corp., Security Response*, vol. 5, no. 6, p. 29, 2011.

[13] R. Langner, "Stuxnet: Dissecting a cyberwarfare weapon," *IEEE Security & Privacy*, vol. 9, no. 3, pp. 49–51, 2011.

[14] D. U. Case, "Analysis of the cyber attack on the ukrainian power grid," *Electricity Information Sharing and Analysis Center (E-ISAC)*, 2016.

[15] E. Alomari, S. Manickam, B. Gupta, S. Karuppayah, and R. Alfaris, "Botnet-based distributed denial of service (ddos) attacks on web servers: Classification and art," *International Journal of Computer Applications*, 2012.

[16] E. Demirok, P. C. González, K. H. B. Frederiksen, D. Sera, P. Rodriguez, and R. Teodorescu, "Local reactive power control methods for overvoltage prevention of distributed solar inverters in low-voltage grids," *IEEE Journal of Photovoltaics*, vol. 1, no. 2, pp. 174–182, 2011.

[17] L. A. Maglaras, J. Jiang, and T. J. Cruz, "Combining ensemble methods and social network metrics for improving accuracy of OCSVM on intrusion detection in SCADA systems," *Journal of Information Security and Applications*, vol. 30, pp. 15–26, 2016. [Online]. Available: http://dx.doi.org/10.1016/j.jisa.2016.04.002

[18] P. I. Radoglou-Grammatikis and P. G. Sarigiannidis, "Securing the smart grid: A comprehensive compilation of intrusion detection and prevention systems," *IEEE Access*, vol. 7, pp. 46 595–46 620, 2019.

[19] J. Zuniga-Mejia, R. Villalpando-Hernandez, C. Vargas-Rosales, and A. Spanias, "A linear systems perspective on intrusion detection for routing in reconfigurable wireless networks," *IEEE Access*, vol. 7, pp. 60 486–60 500, 2019.

[20] V. T. Alaparthy and S. D. Morgera, "A multi-level intrusion detection system for wireless sensor networks based on immune theory," *IEEE Access*, vol. 6, pp. 47 364–47 373, 2018.

[21] R. Oppliger, *SSL and TLS: Theory and Practice*. Artech House, 2009.

[22] T. Dierks and E. Rescorla, "The transport layer security (tls) protocol version 1.2," Tech. Rep., 2008.

[23] C. R. Davis, *IPSec: Securing VPNs*. McGraw-Hill Professional, 2001.

[24] N. Doraswamy and D. Harkins, *IPSec: the new security standard for the Internet, intranets, and virtual private networks*. Prentice Hall Professional, 2003.

[25] R. Kalluri and L. Mahendra, "Simulation and Impact Analysis of Denial-of-Service Attacks on Power SCADA," *2016 National Power Systems Conference (NPSC)*, no. 1, pp. 1–5, 2016.

[26] V. L. Do, L. Fillatre, U. Côte, S. Antipolis, I. Nikiforov, and U. D. Technologie, "Feature Article : Security of SCADA Systems Against Cyber-Physical Attacks," *IEEE Aerospace and Electronic Systems Magazine*, vol. 32, no. 10, pp. 28–45, 2017.

[27] M. Conti, S. Member, N. Dragoni, and V. Lesyk, "A Survey of Man In The Middle Attacks," *IEEE Communications Surveys & Tutorials*, vol. 18, no. 3, pp. 2027–2051, 2016.

[28] J. Kim and L. Tong, "On topology attack of a smart grid: Undetectable attacks and countermeasures," *IEEE Journal on Selected Areas in Communications*, vol. 31, no. 7, pp. 1294–1305, 2013.

[29] B. Kang, P. Maynard, K. Mclaughlin, S. Sezer, F. Andrén, C. Seitl, F. Kupzog, and T. Strasser, "Investigating Cyber-Physical Attacks against IEC 61850 Photovoltaic Inverter Installations," *2015 IEEE 20th Conference on Emerging Technologies & Factory Automation (ETFA)*, pp. 1–8, 2015.

[30] L. Langer, P. Smith, M. Hutle, and A. Schaeffer-filho, "Analysing Cyber-physical Attacks to a Smart Grid : A Voltage Control Use Case," *2016 Power Systems Computation Conference (PSCC)*, pp. 1–7, 2016.

[31] B. Chen, D. W. Ho, G. Hu, and L. Yu, "Secure fusion estimation for bandwidth constrained cyber-physical systems under replay attacks," *IEEE transactions on cybernetics*, vol. 48, no. 6, pp. 1862–1876, 2017.

[32] T. Irita and T. Namerikawa, "Detection of Replay Attack on Smart Grid with Code Signal and Bargaining Game," *2017 American Control Conference (ACC)*, pp. 2112–2117, 2017.

[33] Y. Mo, R. Chabukswar, and S. Member, "Detecting Integrity Attacks on SCADA Systems," *IEEE Transactions on Control Systems Technology*, vol. 22, no. 4, pp. 1396–1407, 2014.

[34] Y. Liu, P. Ning, and M. K. Reiter, "False data injection attacks against state estimation in electric power grids," *ACM Transactions on Information and System Security (TISSEC)*, vol. 14, no. 1, p. 13, 2011.

[35] Y. Wang, M. M. Amin, J. Fu, and H. B. Moussa, "A novel data analytical approach for false data injection cyber-physical attack mitigation in smart grids," *IEEE Access*, vol. 5, pp. 26 022–26 033, 2017.

[36] Q. Yang, J. Yang, W. Yu, and D. An, "On False Data-Injection Attacks against Power System State Estimation : Modeling and Countermeasures," *IEEE Transactions on Parallel and Distributed Systems*, vol. 25, no. 3, pp. 717–729, 2014.

[37] J. Zhao, G. Zhang, Z. Y. Dong, and K. P. Wong, "Forecasting-aided imperfect false data injection attacks against power system nonlinear state estimation," *IEEE Transactions on Smart Grid*, vol. 7, no. 1, pp. 6–8, 2015.

[38] G. Hug and J. A. Giampapa, "Vulnerability Assessment of AC State Estimation With Respect to False Data Injection Cyber-Attacks," *IEEE Transactions on Smart Grid*, vol. 3, no. 3, pp. 1362–1370, 2012.

[39] G. Liang, J. Zhao, F. Luo, S. R. Weller, and Z. Y. Dong, "A review of false data injection attacks against modern power systems," *IEEE Transactions on Smart Grid*, vol. 8, no. 4, pp. 1630–1638, 2016.

[40] R. Xu, R. Wang, Z. Guan, L. Wu, J. Wu, and X. Du, "Achieving efficient detection against false data injection attacks in smart grid," *IEEE Access*, vol. 5, pp. 13 787–13 798, 2017.

[41] R. C. Diovu and J. T. Agee, "Quantitative Analysis of Firewall Security under DDoS Attacks in Smart Grid AMI Networks," *2017 IEEE 3rd International Conference on Electro-Technology for National Development (NIGERCON)*, pp. 1–6, 2017.

[42] R. Diovu and J. Agee, "A cloud-based openflow firewall for mitigation against ddos attacks in smart grid ami networks," in *2017 IEEE PES PowerAfrica*. IEEE, 2017, pp. 28–33.

[43] R. K. Pandey and S. M. Ieee, "Cyber Security Threats - Smart Grid Infrastructure," *2016 National Power Systems Conference (NPSC)*, pp. 1–6, 2016.

[44] T. A. Rizzetti, P. Wessel, A. S. Rodrigues, B. M. da Silva, R. Milbradt, and L. N. Canha, "Cyber security and communications network on SCADA systems in the context of Smart Grids," *2015 50th International Universities Power Engineering Conference (UPEC)*, pp. 1–6, 2015.

[45] C. Foglietta, D. Masucci, C. Palazzo, R. Santini, S. Panzieri, L. Rosa, T. Cruz, and L. Lev, "From detecting cyber-attacks to mitigating risk within a hybrid environment," *IEEE Systems Journal*, no. 99, pp. 1–12, 2018.

[46] S. A. Yadav, S. R. Kumar, S. Sharma, and A. Singh, "A Review of Possibilities and Solutions of Cyber Attacks in Smart Grids," *2016 International Conference on Innovation and Challenges in Cyber Security (ICICCS-INBUSH)*, no. Iciccs, pp. 60–63, 2016.

[47] K. Wang, M. Du, S. Maharjan, and Y. Sun, "Strategic honeypot game model for distributed denial of service attacks in the smart grid," *IEEE Transactions on Smart Grid*, vol. 8, no. 5, pp. 2474–2482, 2017.

[48] S. Li, Y. Yılmaz, and X. Wang, "Quickest detection of false data injection attack in wide-area smart grids," *IEEE Transactions on Smart Grid*, vol. 6, no. 6, pp. 2725–2735, 2014.

[49] M. Esmalifalak, L. Liu, N. Nguyen, R. Zheng, and Z. Han, "Detecting stealthy false data injection using machine learning in smart grid," *IEEE Systems Journal*, vol. 11, no. 3, pp. 1644–1652, 2017.

[50] J. Zhao, G. Zhang, M. La Scala, Z. Y. Dong, C. Chen, and J. Wang, "Short-term state forecasting-aided method for detection of smart grid general false data injection attacks," *IEEE Transactions on Smart Grid*, vol. 8, no. 4, pp. 1580–1590, July 2017.

[51] Y. He, G. J. Mendis, and J. Wei, "Real-time detection of false data injection attacks in smart grid: A deep learning-based intelligent mechanism," *IEEE Transactions on Smart Grid*, vol. 8, no. 5, pp. 2505–2516, 2017.

[52] W. Gao, T. Morris, B. Reaves, and D. Richey, "On SCADA Control System Command and Response Injection and Intrusion Detection," *2010 eCrime Researchers Summit*, pp. 1–9, 2010.

[53] T. Cruz, L. Rosa, J. Proença, L. Maglaras, M. Aubigny, L. Lev, J. Jiang, and P. Simoes, "A cybersecurity detection framework for supervisory control and data acquisition systems," *IEEE Transactions on Industrial Informatics*, vol. 12, no. 6, pp. 2236–2246, 2016.

[54] V. A. Siris and F. Papagalou, "Application of anomaly detection algorithms for detecting syn flooding attacks," in *IEEE Global Telecommunications Conference, 2004. GLOBECOM'04.*, vol. 4. IEEE, 2004, pp. 2050–2054.

[55] M. Ashrafuzzaman, Y. Chakhchoukh, A. Jillepalli, P. Tosic, D. Conte de Leon, F. Sheldon, and B. K. Johnson, "Detecting stealthy false data injection attacks in power grids using deep learning," 2018, pp. 219–225.

[56] A. Alsumayt, J. Haggerty, and A. Lotfi, "Performance, analysis, and comparison of mrdr method to detect dos attacks in manet," in *2015 European Intelligence and Security Informatics Conference*, 2015, pp. 121–124.

[57] A. M. Leite da Silva, M. B. Do Coutto Filho, and J. F. de Queiroz, "State forecasting in electric power systems," *IEE Proceedings C - Generation, Transmission and Distribution*, vol. 130, no. 5, pp. 237–244, 1983.

Comparative Study Between Virtual Synchronous Machine and Virtual Impedance Techniques for Two Paralleled Inverters Sharing a Load

Bruno Laurindo
PPGEET
UFF
Niterói, Brazil
brunoml@poli.ufrj.br

Fábio Alves
PEE
COPPE/UFRJ
Rio de Janeiro, Brazil
fabioleitealves@lemt.ufrj.br

Marcello Neves
PEE
COPPE/UFRJ
Rio de Janeiro, Brazil
marcello@lemt.ufrj.br

Jorge Caicedo
PEE
COPPE/UFRJ
Rio de Janeiro, Brazil
jrgcaicedo@lemt.ufrj.br

Bruno França
PPGEET
UFF
Niterói, Brazil
bwfranca@id.uff.br

Maurício Aredes
PEE
COPPE/UFRJ
Rio de Janeiro, Brazil
aredes@lemt.ufrj.br

Abstract—The growing penetration of renewable sources of energy, storage systems and other technologies based on power electronics devices in Electric Power Systems (EPS) request changes in the way that those systems are controlled and operated. Power electronics equipment, which do not have natural load sharing behavior, operating together can be harmful for both themselves and the system connected to them. The objective of this paper is to study and compare two of the most recent load-sharing control techniques for power electronics converters. The controls known as Virtual Synchronous Machine and Synchronous-Reference-Frame Virtual Impedance will be analyzed and simulated in PSCAD-EMTDC environment, based on two existing 30 kVA converters in the Laboratory of Power Electronics and Medium Voltage Applications (LEMT-UFRJ). The obtained results confirm that both techniques are efficient and able to share different types of loads in microgrids.

Keywords— power electronics converters, microgrids, load sharing, virtual impedance, virtual synchronous machine

I. INTRODUCTION

Microgrids are electric subsystems that efficiently integrate various types of energy generators with local loads. A microgrid consisting of diverse type of generation units, such as photovoltaic panels, wind turbine emulator, battery bank storage system, is being built in LEMT. With a considerable number of converters operating together, a load sharing capability in these systems is needed, ensuring stability and reliability to the loads.

Therefore, the objective of the research group is to study two of the most reliable and flexible methods of load sharing: Synchronous-Reference-Frame Virtual Impedances (SRFVI) and Modified Virtual Synchronous Machine (MVSM). Their mathematical models will be discussed, as well as implementation in a PSCAD-EMTDC simulation environment, where two 30 kVA back-to-back converters will be in parallel feeding different types of loads.

Synchronous machines have been used for power generation since the growth of EPS around the world. As a consequence, both the generation and its control are widely documented in the literature and dominated in their applications. From this point, in [1],[2] and [3] a control for three-phase converters is proposed that aims to simulate the behavior of the synchronous machine, combining the positive points of the synchronous generator with those of power electronics converters.

The SRFVI method, presented in [4], [5] and [6], is a recent control technique when compared to the Virtual Synchronous Machine (VSM). Therefore, it is natural that a comparison between this new method and a conventional one should be done, to elucidate the vantages, disadvantages and features of each method. It must be clear that the comparison to be done is between the SRFVI and the MVSM. The contribution of this work is the dynamical comparison between aforementioned techniques, with a particular focus on load sharing capabilities.

Thus, the paper is organized as follows: after this introduction, in Section II is presented important concepts for the proper understanding of this document as fundamentals on load sharing, especially in distribution networks. Section III and IV present theoretical and mathematical details of the load-sharing algorithms that are studied: In section V the simulations of the control techniques, operating in parallel in two identical converters are verified. And finally, section VI shows the conclusions and possible future work.

II. NETWORK LOAD SHARING

Load sharing in electrical networks is deeply related to the impedance of transmission lines that interconnect the system. Fig. 1 denotes a generic system where V_a is the terminal voltage of the generator or converter, with load angle δ, connected to the network with voltage V_b, through the impedance $Z = R + jX$.

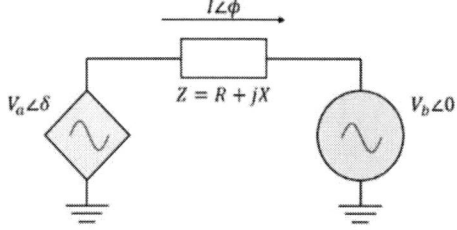

Fig. 1: Generic system - V_a is the controlled voltage and V_b is the voltage of the common connection point.

The transmitted P power and Q injected by the generator can be calculated by the following algebraic manipulations:

978-1-7281-4181-7/19 $31.00 © 2019 IEEE

$$S_a = \dot{V_a}\dot{I}^* = P + jQ \tag{1}$$

$$P_a = Re(S_a) = \frac{V_a}{R^2 + X^2}[R(V_a - V_b \cos\delta) + XV_b \sin\delta] \tag{2}$$

$$Q_a = Im(S_a) = \frac{V_a}{R^2 + X^2}[X(V_a - V_b \cos\delta) + RV_b \sin\delta] \tag{3}$$

Equations (2) and (3) are the generic, injected by source a active and reactive power expressions for the system exposed on Fig. 1. The next two items show the simplification for each operational situation.

A. Transmission

Transmission lines operate at high voltage (69 kV to 138 kV) or extra high voltage (138 kV or above) levels. To improve the efficiency of large transmission systems, where any loss is undesirable, better quality materials are used to construct their components (transformers, generators, transmission lines). Because these large equipment are made of better materials, they have a predominantly inductive characteristic ($X >> R$), propagating this aspect to the entire transmission system.

From the assumptions made, one can simplify equations (2) and (3) to:

$$P_a \approx \frac{V_a V_b \sin\delta}{X} \tag{4}$$

$$Q_a \approx \frac{V_a(V_a - V_b \cos\delta)}{X} \tag{5}$$

So, assuming small angle deviations around zero:

$$\delta \approx \frac{X P_a}{V_a V_b} \tag{6}$$

$$V_a - V_b \approx \frac{X Q_a}{V_a} \tag{7}$$

As a result, variations in the load angle δ are directly related to the flow P_a, while voltage variations are directly related to the injection of Q_a. Through these findings, the droop equations are written for mostly inductive lines:

$$f - f_0 = -k_p(P - P_0) \tag{8}$$

$$V - V_0 = -k_q(Q - Q_0) \tag{9}$$

The load sharing in high voltage generating units will occur according to their respective droop curves, where k_p and k_q are the slopes of the lines that govern the fitting. Fig. 2 shows graphically equations (8) and (9).

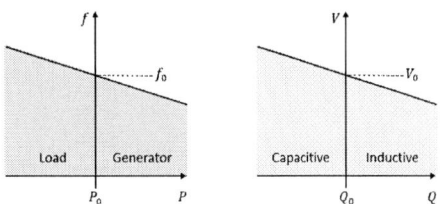

Fig. 2: *P-f* and *Q-V* droop curves.

B. Microgrids

In microgrids, other approaches must be assumed to the electric system behavior. In the case of installations that are mostly of low voltage, in addition to the physical and electrical proximity of the distributed resources, the system tends to be mostly resistive, *i.e.*, $R >> X$. The injections of P_a and Q_a can be simplified to:

$$P_a \approx \frac{V_a(V_a - V_b \cos\delta)}{R} \tag{10}$$

$$Q_a \approx -\frac{V_a V_b \sin\delta}{R} \tag{11}$$

Rewriting (10) and (11) after the trigonometric simplifications:

$$\delta \approx -\frac{R Q_a}{V_a V_b} \tag{12}$$

$$V_a - V_b \approx \frac{R P_a}{V_a} \tag{13}$$

Instead the transmission systems, the variations in the load angle δ cause a change in the Q_a injection, while the voltage difference between transmitter and receiver causes variations in the flow of P_a. Hence, the droop equations for low voltage grids are:

$$V - V_0 = -k_p(P - P_0) \tag{14}$$

$$f - f_0 = k_q(Q - Q_0) \tag{15}$$

In Fig. 3 it is possible to note that the droop curve *Q-f* is the only one that has a positive slope when compared to the previous curves.

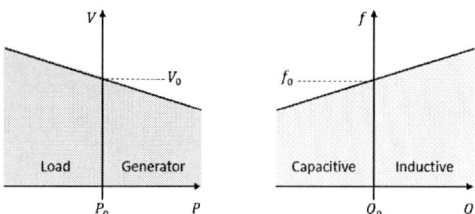

Fig. 3: Droop curves *P-V* and *Q-f* in the resistive case.

The droop control is responsible for the division of load between the various generating units of a microgrid, and generally is implemented as the primary control in a microgrid. Within this hierarchical level, there are three large families of converters based on their behaviors: grid-forming, grid-feeding and grid-supporting power converters [7]. Grid-forming are converters that behave as voltage source. In the case of microgrids, when operating in island mode, it is necessary that at least one generating unit is grid-forming to maintain the stable voltage and frequency levels of the system.

III. SYNCHRONOUS-REFERENCE-FRAME VIRTUAL IMPEDANCE

The main characteristic of SRFVI control scheme is that it is not necessary to calculate the instantaneous powers to apply the classical droop technique [4]. To perform a similar function, the load current of the converter is decomposed into the *dq* axes, where each of the axes will have virtual resistances, which are responsible for the load-sharing rate of the current components of each converter. Such resistive droop technique was inspired by similar techniques used for division of load in dc microgrids [9].

A. Principle of Control

Fig. 4 illustrates the control principle of SRFVI, which consists of a cascade controller and the virtual impedance loop. The inner loop controls the output current, while the outer loop controls the output voltage of the converter. The converter to be controlled is a three-phase inverter with LC filter at its output terminals.

The measured voltages in the capacitors and currents of the filter inductors are transformed to the reference $\alpha\beta$, and they are used as feedback to the controller. Thus, it is necessary to use Proportional-Resonant (PR) controllers to ensure zero steady state error. The load current is transformed to dq, and is multiplied by the resistive droop in synchronous reference, becoming the virtual voltage drop. This value will make the reference composition for the voltage controller, after being transformed to $\alpha\beta$ reference. The inner current controller consists of a Proportional gain.

The qPLL is responsible to provide the grid phase and frequency reference for both the dq transformations and the voltage reference generator (v_{ref}). The magnitude of the voltage reference is commonly obtained by a superior level of control. Due to the fact that those superior controls have a slower dynamic [8] and the focus of the study is only the SRFVI dynamic, the voltage magnitude is considered as a constant in this analyses.

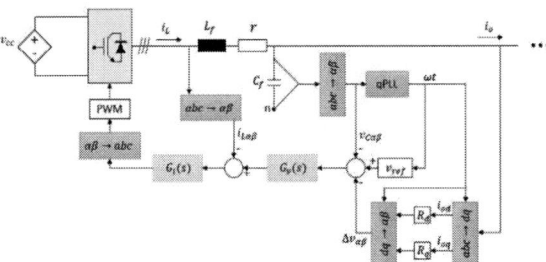

Fig. 4: Unifilar diagram of the Synchronous-Reference-Frame Virtual Impedance – Based on [3].

The control effort performed by the internal current loop is transformed into the abc system to be used as a reference for the PWM. The virtual impedances in the dq axes are responsible for the injection of active and reactive powers of the converter. In order to understand these relations, it is necessary to model the system properly, according to Fig. 6.

The virtual impedances implemented are resistors, and predominant in the total impedance of the line [4], so this control can be placed in the P-V droop category, according to equations (14) and (15). This feature will be explained below.

The relation between active power, in this case, referenced with the direct-axis, and voltage amplitude can be interpreted as follows: if the converter is feeding a resistive load, there will be I_{0d} current flowing through the system; this current will flow through R_{vird}, resulting in a virtual voltage drop on the direct-axis, which will reduce the voltage amplitude in the abc system. This ensures the P-V coupling. The relation between reactive power, in the quadrature-axis,

with the frequency can be explained analogously to the d-axis. The voltage drop on the q-axis will accelerate the frequency of the PLL, even if it tries to synchronize with the common AC bus. Thus, the voltage generated by the converter will advance in relation to the common bus, increasing the injection of the reactive power in the system, ensuring the coupling Q-f.

B. Virtual Impedance Design

The proper design of the virtual impedance is vital for system stability. The calculation presented as well as the equations exposed were based on [4]. The R_{vird} resistance, responsible for regulating voltage and active injected power, is designed from the desired maximum voltage variation. Assuming the case where the converter is feeding a purely resistive load, there will be only d-axis current (I_{0d}). In this scenario, the direct-axis voltage variation (ΔV_{maxd}) will be equal to the total variation (ΔV_{max}). The upper limit of virtual resistance (R_{maxd}) can be calculated by taking I_{0d} as the nominal current I_{nom} of the converter. Mathematically:

$$R_{maxd} = \frac{\Delta V_{max}}{I_{nom}} \qquad (16)$$

The R_{virq} resistance regulates the frequency variation and reactive injection in the network. Assuming the extreme case, when the converter is feeding a purely reactive load, the maximum allowable value of virtual resistance R_{maxq} is obtained. However, different from the R_{maxd} calculation, the equation that governs R_{maxq} is not a simple relation, as in (16), since the frequency of the system depends on the filter parameters, PR controllers and the inverter itself. Equation (17), proposed in [3], explains the relation of these quantities to obtain R_{maxq}.

$$R_{maxq} = \frac{1}{\zeta} V_{ref} \omega_{max}(\omega_{max}^2 - \omega_0^2) \cdot \left[k_{rv}\left(1 + C_f L_f \omega_{max}^2\right) \right.$$
$$\left. + C_f K_{pv}\left(K_{pi}K_{pwm} + r\right)(\omega_{max}^2 - \omega_0^2) \right] \quad (17)$$

Where,

$$\zeta = I_{nom}K_{pi}K_{pwm}[\omega_{max}^2\left(K_{rv}^2 + K_{pv}^2\omega_{max}^2\right) + k_{pv}^2\omega_0^2(\omega_0^2 - 2\omega_{max}^2)]$$

In this equation, V_{ref} is the amplitude of the reference voltage, ω_0 is the nominal angular frequency of the system, ω_{max} is the maximum angular allowed frequency, K_{pwm} is the gain of the PWM switching, K_{pv} is the gain of the proportional voltage controller, K_{rv} is the voltage resonant controller gain, K_{pi} the current proportional controller gain, C_f the filter capacitance, L_f the filter inductance. R_s is the intrinsic resistance of this inductor.

IV. VIRTUAL SYNCHRONOUS MACHINE

From the mathematical model of synchronous machines, widely known in the literature, it is possible to develop the VSM control proposed in [1]. The main behavior added to the converter's capability is the inherent droop mechanism of the synchronous machine (P-f and Q-V). Adverse effects such as losses and non-linarites due to magnetic hysteresis are neglected.

Fig. 5 shows the complete control of the VSM. The active power loop is responsible for emulating the mechanical equation of the generator. The control reference is given by P_{set}, which when divided by ω, becomes the virtual mechanical torque T_m. This value is subtracted from T_e, virtual electric torque, and T_d, virtual torque due to droop. T_d has a relevant role as it implements the relation between active power and frequency. Both ω and ωt are feedback to the generator's equations in order to keep the internal calculations updated.

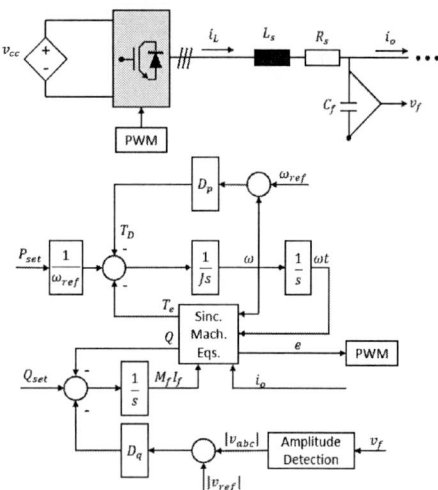

Fig. 5: Control diagram of the VSM and its power circuit - Based on [5].

The reactive power loop of the VSM is responsible for implementing the *P-V* droop mechanism. $M_f i_f$ will also be, besides other functions, the voltage amplitude generated by the VSM when multiplied by the virtual speed ω, emulating the generator speed voltage, thus ensuring the *Q-V* coupling.

A. Modified-Virtual Synchronous Machine

In this version of VSM the output is not the internal voltage e. The VSM loop responsible for the mechanical equation and droop *P-f* will generate the signals of ω and ωt, like the conventional VSM. However, instead of feedback into the block of equations of the VSM, they are used as input to a signal generator in $\alpha\beta$ reference. Similarly, the reactive power loop of the VSM, responsible for droop *Q-V*, generates the reference amplitude in $\alpha\beta$ coordinates. With the amplitude and frequency signals synthesized by the droop loops of the MVSM, the signal generator is used as a reference for the cascade controller, where the voltage controller is the outer loop and the inner loop is the current controller. The voltages of the filter capacitors are the controlled variables of the voltage loop PR, while the output currents of the inverter are controlled variables of the current loop P.

Fig. 6 shows the diagram of the Modified-Virtual Synchronous Machine (MVSM), where the *P-f* and *Q-V* loops are analogous to the classic VSM loops.

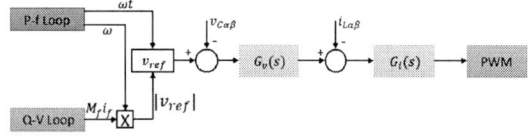

Fig. 6: Diagram of the Modified-Virtual Synchronous Machine.

In this way, the implementation of this control differs from what was proposed in [1]. The calculation of the reactive power generated internally in the VSM can be problematic [10], since it may occur differences in the reactive power at the converter terminals if L_s is high. To overcome this problem, in this work it is used the Instantaneous Power Theory to calculate the active and reactive powers at the converter terminals. In addition, all the measures are normalized, so the controller operates in the per unit (pu) system. As a consequence, the electrical torque is numerically equal to the active power. In steady state and with small frequency disturbances:

$$T_e(pu) = P_e(pu) \qquad (18)$$

V. SIMULATION AND RESULTS

The simulated system consists in two identical 30 kVA inverters connected in parallel, supplying a 1 pu load with power factor equals to 0.707 lagging. The electrical values of a single converter, as well as their controller gains, are shown in Table I. Thus, the following scenarios were simulated:

- Two inverters with identical MVSM;

- Two inverters with identical SRFVI.

TABLE I. CONVERTER'S ELECTRICAL VALUES AND CONTROLLER'S GAINS.

Line-to-Line Voltage (V_{rms})	220
DC Link Voltage (V)	406
Rated Current (A_{rms})	78.73
Rated Power (kVA)	30
Fundamental Frequency (Hz)	60
Switching Frequency (Hz)	5940
L_{conv} (mH)	0.6
R_{conv} ($m\Omega$)	185
L_{grid} (mH)	0.17
R_{grid} ($m\Omega$)	185
C_{filter} (μF)	30
$Z_{load}(\Omega)$	$1.025 + j1.025$
K_{pi} (pu)	0.004325
K_{pv} (pu)	0.5205
$K_{rv}(pu)$	58100

The simulated converter, modeled from the existing one, has an output LCL filter, not a LC, as supposed to the mathematical development of the MVSM and SRFVI techniques. Therefore, the grid inductance and grid resistance of the filter are considered part of the line impedance that interconnects the system.

A. Modified-Virtual Synchronous Machine Results

Table II shows the values of the MVSM. The reactive power droop is not implemented, because it was desirable that the amplitude of the load voltage does not change during the tests. Hence, v_{ref} is a constant equal to 1 pu.

TABLE II. MVSM PARAMETERS

Virtual Inertia Constant (s)	100
Virtual Droop Constant (pu)	10000

Figs. 7, 8, 9 and 10 show the results of voltage, current and active power of both converters when the circuit-breaker that connects the resistive load is closed, respectively.

In Fig. 7, the voltage signals of each phase are overlapping, as was expected in the case of converters with the same control. At 0.9 seconds, the load is connected and a small distortion appears, but the signal remains with a peak value of 1 pu.

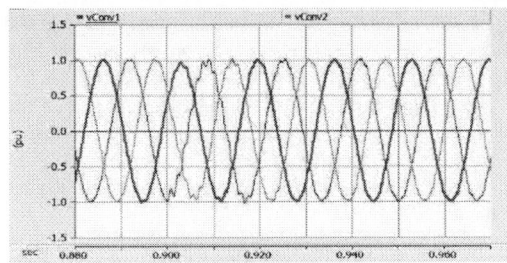

Fig.7: Three-phase output voltages of the inverters with MVSM control during the connection of the load.

In Fig. 8 the three-phase currents supplied by the inverters are plotted, with amplitude of 0.5 pu. The currents of phase *abc* are shown. In red, the current in phase *a* of the load, with amplitude of 1 pu, showing that the inverters are sharing the load equally.

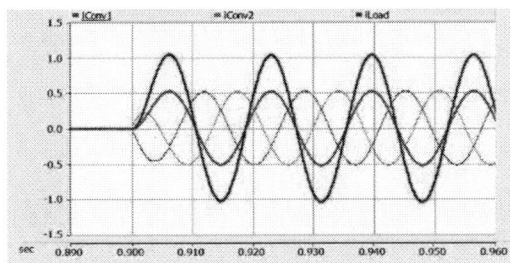

Fig.8: Three-phase converter currents with MVSM control, after connection of the load. In red, the current of phase a of the load fed.

Fig. 9 and 10 show the output active and reactive power of each converter, respectively, with MVSM. Both active and reactive power of the converters are overlapping, and their values are half the power of the load.

Fig. 11 exhibits both converters frequency during the whole simulated time. From 0.0 to 0.3 s, the MVSM is initialized. At 0.9 s, the load is connected. There are no major changes between the frequency before and after the load connection.

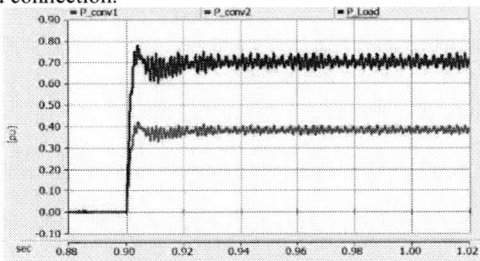

Fig.9: In brown, the active power flowing to the load. In green, the power supplied by the converter 2, overlapping with the power of the converter 1, both with MVSM control.

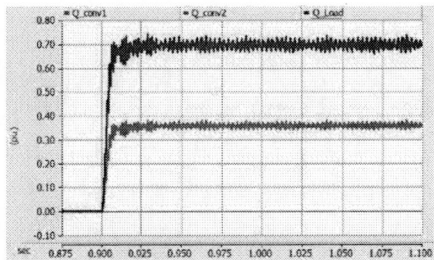

Fig. 50: In brown, the reactive power injected to the load. In green, the power supplied by the converter 2, overlapping with the power of the converter 1, both with VSM control.

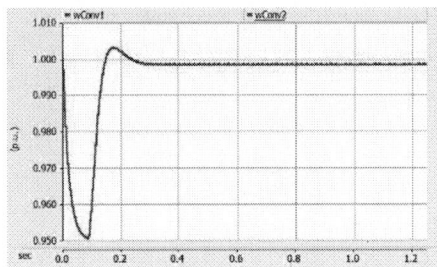

Fig. 11: Converter's frequency during the whole simulated time.

B. Synchronous-Reference-Frame Virtual Impedance Results

Table III shows the resistance values of the SRFVI, calculated from equations (16) and (17).

TABLE III. VISRF PARAMETERS

R_{vird} (Ω)	0.5584
R_{virq} (Ω)	0.5375

Figs. 12, 13, 14 and 15 show the results of voltage, current, active and reactive power of both converters when the circuit-breaker that connects the load is closed, respectively.

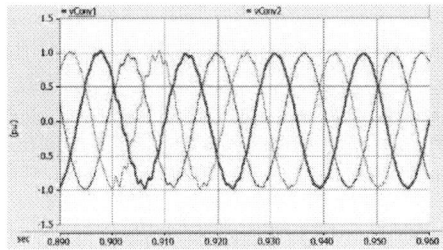

Fig. 62: Three-phase output voltages of the inverters with SRFVI control during the connection of the load.

In Fig. 13 it is exposed only the phase *a* of the load and both converters.

978-1-7281-4181-7/19 $31.00 © 2019 IEEE

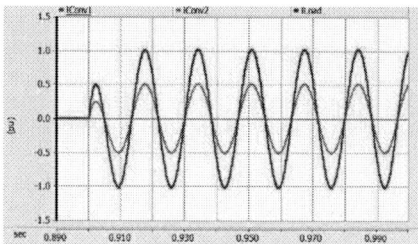

Fig. 73: Three-phase currents of the inverters with SRFVI control, after connection of the resistive load. In red, the current of phase a of the load fed.

Figs. 14 and 15 show the output active and reactive power of each converter.

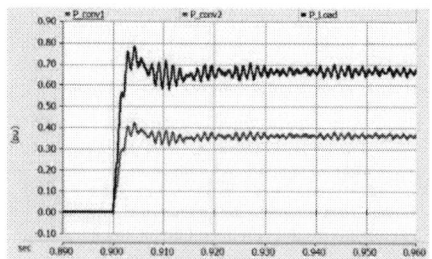

Fig.14: In brown, the active power flowing to the load. In blue, the power supplied by the converter 1, overlaped with the power of the converter 2, both with SRFVI control.

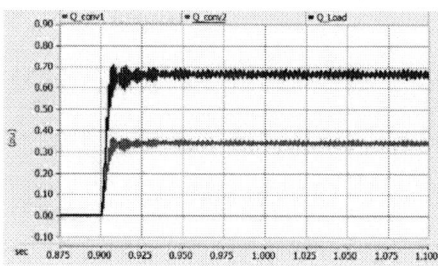

Fig.15: In brown, the reactive power flowing to the load. In blue, the power supplied by the converter 1, overlaped with the power of the converter 2, both with SRFVI control.

Fig. 16 shows both converters' frequency before and after the load connection. After the connection, both converters inject reactive to the load, forcing the frequency to rise, as expected.

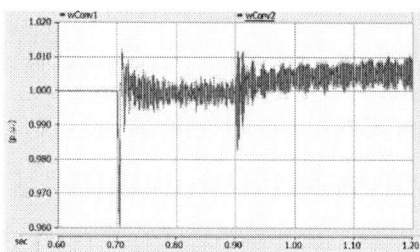

Fig. 16: Converter's frequency before and after the connection of the load, at 0.90 s.

VI. CONCLUSIONS

The results obtained in this study show that for LEMT microgrid environment, where the equipment is physically near to each other and low voltage, both the MVSM control method and the SRFVI control method were able to share loads between the generating units. MVSM achieved better results, but this result was expected, since MVSM is a more mature and documented technology among the authors.

The implementation of the SRFVI showed that it is necessary a better dimension of its parameters in order to reduce the noises present in the frequencies measured by their respective PLLs. However, even with this problem, the SRFVI showed to be able to share loads efficiently.

For the future, a deeper understanding of the SRFVI technology is needed, such as stability studies and optimization of the parameters. The experimental bench application will also be done in order to validate the results obtained in this paper.

ACKNOWLEDGMENT

This study was financed in part by the Coordenação de Aperfeiçoamento de Pessoal de Nível Superior – Brasil (CAPES) – Finance Code 001.

REFERENCES

[1] Q. Zhong and G. Weiss, "Synchronverters: Inverters That Mimic Synchronous Generators,"in IEEE Transactions on Industrial Electronics, vol. 58, no. 4, pp. 1259-1267, April 2011.

[2] Q. C. Zhong, "Virtual Synchronous Machines: A unified interface for grid integration", IEEE Power Electron. Mag., vol. 3, nº 4, p. 18–27, dez. 2016.

[3] Qing-Chang Zhong, Phi-Long Nguyen, Zhenyu Ma, e Wanxing Sheng, "Self-Synchronized Synchronverters: Inverters Without a Dedicated Synchronization Unit", IEEE Trans. Power Electron., vol. 29, nº 2, p. 617–630, fev. 2014.

[4] Y. Guan, J. M. Guerrero, X. Zhao, J. C. Vasquez and X. Guo, "A New Way of Controlling Parallel-Connected Inverters by Using SynchronousReference-Frame Virtual Impedance Loop - Part I: Control Principle,"in IEEE Transactions on Power Electronics, vol. 31, no. 6, pp. 4576-4593, June 2016.

[5] Y. Guan, J. M. Guerrero, X. Zhao, e J. C. Vasquez, "Comparison of a synchronous reference frame virtual impedance-based autonomous current sharing control with conventional droop control for parallel-connected inverters", in 2016 IEEE 8th International Power Electronics and Motion Control Conference (IPEMC-ECCE Asia), Hefei, China, 2016, p. 3419–3426.

[6] Y. Guan, W. Feng, J. Lu, J. M. Guerrero, e J. C. Vasquez, "A Novel Grid-Connected Harmonic Current Suppression Control for Autonomous Current Sharing Controller-Based AC Microgrids", in 2018 IEEE Energy Conversion Congress and Exposition (ECCE), Portland, OR, 2018, p. 5899–5904.

[7] J. Rocabert, A. Luna, F. Blaabjerg and P. Rodríguez, "Control of Power Converters in AC Microgrids,"in IEEE Transactions on Power Electronics, vol. 27, no. 11, pp. 4734-4749, Nov. 2012.

[8] J. M. Guerrero, J. C. Vasquez, J. Matas, L. G. de Vicuna and M. Castilla, "Hierarchical Control of Droop-Controlled AC and DC Microgrids—A General Approach Toward Standardization,"in IEEE Transactions on Industrial Electronics, vol. 58, no. 1, pp. 158-172, Jan. 2011.

[9] M. S. Neves, "Desenvolvimento do modelo de uma máquina cc virtual aplicada a microrredes em corrente contínua", Dissertação de M.Sc., Programa de Engenharia Elétrica, COPPEUFRJ, 06/2018.

[10] B. W. França, "Static synchronous generator with sliding droop control for distributed generation in microgrids", tese de D.Sc., Programa de Engenharia Elétrica, COPPE-UFRJ, 07/2016.

Analysis of Partial-Power Processing Converters for Small Wind Turbines Systems

Anderson José Balbino
Power Electronics Institute
Federal University of Santa Catarina
Florianopolis, Brazil
anderson.balbino@inep.ufsc.br

Ronny Glauber de Almeida Cacau
Power Electronics Institute
Federal University of Santa Catarina
Florianopolis, Brazil
rgacacau@inep.ufsc.br

Telles Brunelli Lazzarin
Power Electronics Institute
Federal University of Santa Catarina
Florianopolis, Brazil
telles@inep.ufsc.br

Abstract—In order to connect small wind turbines (SWT) systems in the single-phase grid, three-stages configurations (rectifier, dc-dc converter and inverter) have been suitable due to the power decoupling between the inverter dc-bus and the rectifier output, adding facility to realize the maximum power point tracking. However, the losses of the system are increased due to the additional dc-dc converter. To reduce this effect, partial-power converters (PPCs), in which only a part of the power generated by the SWT system is processed by the converter, can be used. In this method, the characteristics of topology (boost or buck), power and voltage levels, and the operating range can impact in the amount of active power handled by the converter. Since the analysis of PPC applied to wind systems remains to be investigated in the literature, this paper analyzes the Full-Bridge Phase Shift (FBPS) with Zero-Voltage Switching (ZVS) operating as PPC in SWT systems connected to the single-phase grid. To verify the theoretical analysis and evaluate the performance of the proposed system, simulation results are presented for a 1.5 kW SWT system, in which the FBPS PPC processed only 70% of the generated power.

Index Terms—DC-DC converter, Full-Bridge Phase Shift, Partial-Power Converters, Small Wind Turbines.

I. INTRODUCTION

Conventional power generation plants using fossil fuels is determined as unsuitable in long-term strategic plans. Consequently, many researches have been developed in order to improve renewable energy sources technologies [1]. Some, e.g. photovoltaics (PVs) and high-power wind turbines (WTs), are already reaching a mature point of growth. However, another field still needs more development and requires attention as the low-power small wind turbines (SWT) systems [2]. Some proposals with different characteristics has been suggesting to improve the conversion efficiency, minimize costs, and increase the realiability of SWT systems [3].

Three-stage power processing systems, designed by a rectifier, a dc-dc converter, and a inverter, can enhance the maximum power point tracking (MPPT) range in grid-connected SWT systems. This feature is available since the power decoupling between the inverter dc-bus voltage and rectifier [4]. Although, the additional energy processing stage results in a increase in the number of components, cost, size, and adds more losses to the system [5]. The ac-dc, dc-dc, and the dc-ac conversion stages processes all the power generated by

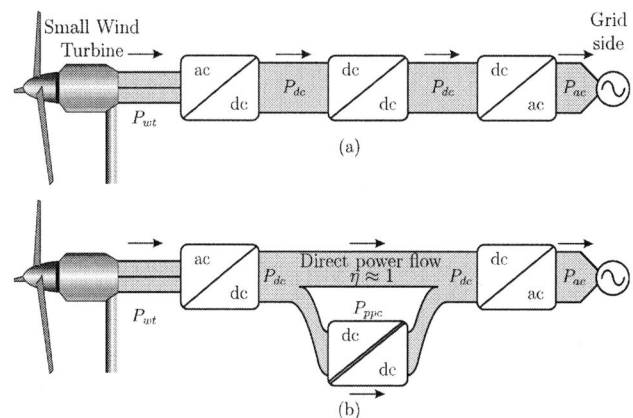

Fig. 1. Power flow of a three-stage SWT system employing a: (a) FPC and (b) PPC.

the SWT system, which is known as the full power converters (FPC), as shown in Fig. 1(a).

In this method, the application of partial-power converters (PPC) guarantee the increase of dc-dc stage efficiency. The idea of partial-power processing (PPP) is established on a fraction of the power being handled by the converter, whereas the remaining power flows through the source to the load without conversion, i.e., with unitary efficiency [6], [7]. The PPC power flow in a three-stage SWT system is depicted in Fig. 1(b). In this scenario, only the amount of power processed by the PPC can lead to losses, which will be naturally lower than with FPC. Consequently, higher efficiency and power density are obtained, as well as lower volume.

The PPP idea can be performed, basically, in two types: via series-connected PPC (S-PPC) or parallel-connected PPC (P-PPC) [8]. The S-PPC can realize voltage regulation for multistring inverter PV systems connected to a common dc bus [10]. Moreover, the P-PPC makes current regulation by providing a parallel current path, i.e., allowing singular MPPT in strings of PV modules under partial shading conditions [9].

Recently, many researches have been published applying S-PPC in PV systems, resulting in higher efficiency and reduced power rating compared to standard FPC topologies [11]–[15]. However, the analysis of PPC employed to SWT systems remains to be investigated in the literature. Thus, the purpose of this paper is to evaluate the use of S-PPC in SWT systems connected in the single-phase grid. In order to choose the best S-PPC topology, possible connection schematics, active power (AP) and efficiency, nonactive power (nAP), converter characteristic (step-up or step-down), and the operating range of SWT system will be analyzed. Simulation results are carried out with the objective to verify the theoretical analysis.

II. S-PPC APPLIED TO SWT SYSTEMS

In order to analyze the quantity of power handled by a dc-dc converter, it is important to explain the power processing concept and know how it is associated to power-loss [8]. The AP is the averaged instantaneous power processed to the output, and the nAP is a evaluation of the circulating power that is determined from the stored energy in inductors and capacitors of a switched converter [16]. Specifically, the nAP is associated with the current ripple in inductors and voltage ripple in capacitors.

The application of S-PPC configurations adds an parallel path to AP flow from source to the load, which can be associated in two ways [17]: with a series connection to the output (type I) or with its input (type II), as depicted in Fig. 2 (a) and Fig. 2 (b), respectively.

It should be highlighted that due to the series connection between input and output, the application of any isolated topology results in the loss of galvanic isolation [8]. Hence, structures based on this concept requires a topology with galvanic isolation, in order to avoid a potential short-circuit.

A. Acitve Power and Efficiency

The S-PPC output AP ($P_{C,out}$) is the average value of instantaneous power in a time interval at the output terminals. In the S-PPC shown in Fig. 2(a), if disregarded current and voltage ripples, the AP $P_{C,out}$ is given by:

$$P_{C,out} = V_C I_{out}, \quad (1)$$

and the output AP for the whole dc-dc stage (P_{out}) is defined by:

$$P_{out} = V_{out} I_{out}. \quad (2)$$

Therefore, the ratio between $P_{C,out}$ and P_{out} in type I configuration is described by:

$$\frac{P_{C,out}}{P_{out}} = \frac{V_C I_{out}}{V_{out} I_{out}} = \frac{V_{out} - V_{in}}{V_{out}} = 1 - \frac{V_{in}}{V_{out}} = 1 - k, \quad (3)$$

where k is the ratio between the input (V_{in}) and output (V_{out}) voltages.

On the other hand, the relationship between $P_{C,out}$ and P_{out} for the S-PPC type II, illustrated in Fig. 2(b), is given by:

$$\frac{P_{C,out}}{P_{out}} = \frac{V_{out} I_{C,out}}{V_{out} I_{out}} = \frac{I_{out} - I_{in}}{I_{out}} = 1 - \frac{I_{in}}{I_{out}} = 1 - \frac{1}{k}. \quad (4)$$

In order to illustrate this effect, Fig. 3 shows the AP handled by the S-PPC versus parameter k. The curves were calculated by means of (3) and (4) for S-PPC type I and type II, respectively.

According to Fig. 3, as lower is the difference between source and load voltages, smaller will be the AP processed by the S-PPC in comparison to the total output AP. When the voltage gain ($M = V_{out}/V_{in}$) is close to one, the S-PPC AP approaches to zero. In addition, it is possible to conclude that type I configuration is more suitable when boost characteristic (step-up) is desired, whereas the type II configuration is more suitable for buck characteristic (step-down). In addition, an important feature of type I configuration is that the AP in the S-PPC varies linearly with the input voltage variation, as can be seen in (3).

The S-PPC efficiency ($\eta_{regulator}$) is defined by the ratio between its AP at the output and input terminals, given by:

$$\eta_{regulator} = \frac{P_{C,out}}{P_{C,in}}. \quad (5)$$

The global efficiency (η_{global}) is the ratio between the AP at the load and source terminals for the entire dc-dc system. Once $P_{C,out}$ is smaller than P_{out}, and all losses are concentrated in the S-PPC, η_{global} is greater than $\eta_{regulator}$, and can be calculated as [11]:

$$\eta_{global} = \frac{P_{out}}{P_{in}} = 1 - \frac{P_{C,out}}{P_{out}} \left(1 - \eta_{regulator} \right). \quad (6)$$

Equation (6) led to the analysis that reducing the ratio $P_{C,out}/P_{C,in}$ is enough to increase the overall efficiency in relation to a FPC [7], [14]. However, the $\eta_{regulator}$ does not

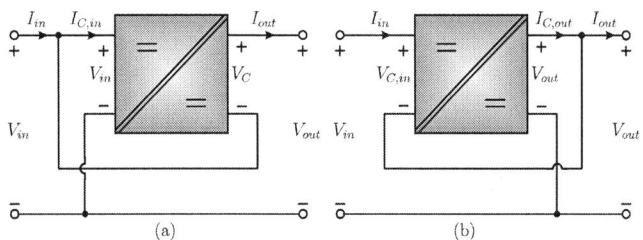

Fig. 2. S-PPC connections schemes [17]. (a) Type I. (b) Type II.

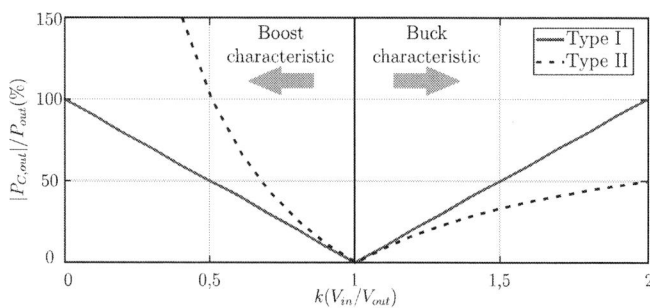

Fig. 3. Active power in S-PPC type I and type II plotted in function of input voltage variation.

depend only of the AP processed in the S-PPC, but it depends of the losses related upon both AP and nAP as well [8].

B. Nonactive Power

The semiconductor swithicing in the power converter results in a quantity of power flowing between inductors and capacitors, but it is does not mean that all power is transformed into AP. As the fundamental frequency is equal to zero, this power flow is named nAP (N), according to the definition of IEEE Standard 1459-2010 [18]. The nAP is the power flow that does not produces AP at any frequency, and its measured in reactive volt-amp (var).

Considering that the AP of a switched converter does not comprehends all the power being handled, the fraction of nAP processed in the converter can be an essential figure of merit for analyze. For instance, it can be checked between different converter topologies which one processes more or less power [19].

C. Comparison of Nonactive Power in Different Converters

In order to evaluate the performance of different S-PPC topologies in terms of nAP, [8] developed an analytic solution to analyze the nAP processed by a full-power boost converter in relation to a Flyback S-PPC, Forward S-PPC, and Full-Bridge Phase-Shift (FBPS) S-PPC. The important parameters are the nAP in inductors (N_L), capacitors (N_C), and the input of the converter (N_{in}), which are determined in terms of the input voltage. The output nAP (N_{out}) is equal to zero, once the methodology considers a resistive load in the output.

The expressions are depicted in Table I, where P_{in} is the input AP, d is the duty cycle, and n is the turns ratio. It is important to note that these solutions allow calculating nAP in terms of generic parameters between different topologies, regardless the size of filters.

The S-PPC total nAP is the sum of the nAP in inductors, capacitors, input, and output, as written in:

$$N_{total} = N_L + N_C + N_{in} + N_{out}. \qquad (7)$$

Fig. 4 shows the total nAP handled regarding the k parameter for all analyzed topologies. In these analytical results, all converters operate with the same input power (expressed in per unit system), and the nAP in a boost converter is included for comparison. It is important to highlight that the minimum n for each topology is defined by its maximum duty cycle, and the ratio between input and output voltages is increased from 70% to 100%.

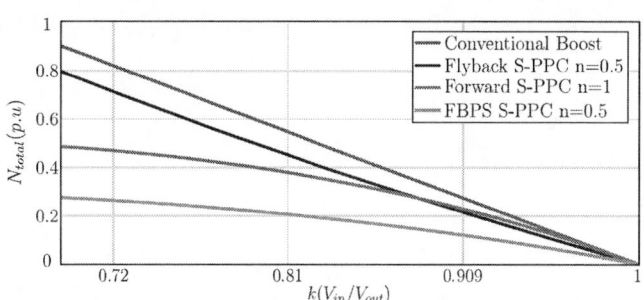

Fig. 4. Comparison of the total nAP in Flyback S-PPC, Forward, S-PPC, and FBPS S-PPC in relation to boost converter.

A S-PPC is considered a truly PPC when it processes less nAP than boost converter for the same operational point [8]. It is possible observe in Fig. 4 that the maximum difference of the Flyback S-PPC and Forward S-PPC relative to the boost converter is 12% and 46%, respectively. In relation to FBPS S-PPC, even with a larger active and passive number of components, compared to boost converter, the total nAP processed is, approximately, 70% smaller than the reference, making an improvement performance in comparison to a FPC. In addition, the turns ratio can be optimized for each operational point according to the voltage-gain range.

Thus, due to the advantages in terms of nAP of FBPS S-PPC in comparison to Flyback S-PPC and Forward S-PPC (Fig. 4), the Full-Bridge Phase-Shift is selected in this paper as the S-PPC topology.

III. OPERATING RANGE OF SWT SYSTEM

In this study, the wind power is extracted from a 1.5 kW SWT model Gerar 246, manufactured by Enersud, with horizontal axis, three blades and active stall, connected to an axial flux permanent-magnet synchronous generator.

The power delivered by the SWT system can be estimated in terms of its angular rotational speed (ω_r) for different wind speeds (v_ω). Therefore, the characteristic between mechanical power as a function of angular speed $P_{mec}(\omega_r)$ is created, as depicted in Fig. 5. For each value of v_ω, there is an optimum operational point, in which the maximum quantity of mechanical power is obtained [20]. The P_{max} curve, which connects these maximum power points, shows the range of ideal operation.

In relation to the project of FBPS S-PPC, the wind speed and the output voltage levels of the three-phase non-controlled must be considered. The S-PPC input voltage range specify the gain necessary to realize the MPPT, and therefore, the S-PPC AP rate, as demonstrated in (3).

Plotting the rectifier output power characteristic as a function of FBPS S-PPC input voltage, obtained through simulation, it is analyzed that the SWT presents a single maximum power point for each rotational speed, as illustrated in Fig. 6. Then, it is available to define the minimum and maximum parameters of voltage and power in the rectifier output, used to define the S-PPC operational range.

TABLE I
ANALYTIC EXPRESSIONS OF NAP [8].

	Boost	Flyback S-PPC	Forward S-PPC	FBPS S-PPC
N_L	$2P_{in}d$	$2P_{in}\dfrac{d}{nd-d+1}$	$2P_{in}\dfrac{nd(1-d)}{1+nd}$	$2P_{in}\dfrac{nd(1-d)}{1+nd}$
N_C	$2P_{in}d$	$2P_{in}\dfrac{nd^2}{nd-d+1}$	$\cong 0$	$\cong 0$
N_{in}	$\cong 0$	$2P_{in}\dfrac{nd}{\frac{nd}{1-d}+1}$	$2P_{in}\dfrac{nd(1-d)}{1+nd}$	$2P_{in}\dfrac{nd(1-d)}{1+nd}$

978-1-7281-4181-7/19 $31.00 © 2019 IEEE

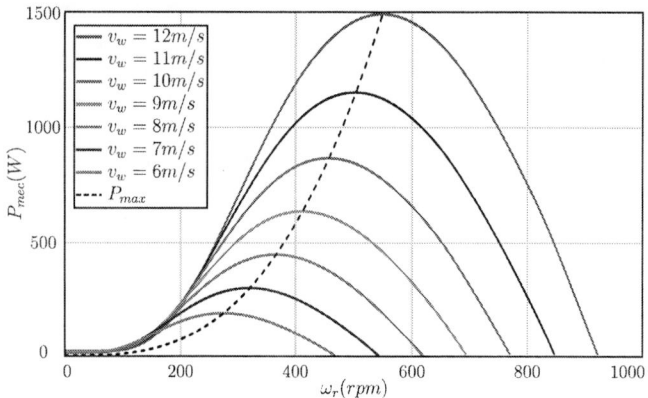

Fig. 5. Curves of mechanical power in function of angular rotational speed $P_{mec}(\omega_r)$ for wind speeds from 6 to 12 m/s.

IV. FULL-BRIDGE S-PPC APPLIED TO SWT SYSTEMS

The Full-Bridge Phase-Shift S-PPC topology is shown in Fig. 7. In [11], the FBPS has been applied as an S-PPC topology with great performance of efficiency and power density. This circuit includes an output filter inductor, as well as, a transformer that transfers energy directly from the primary to the secondary. The switches in the primary side are controlled by phase-shift modulation, and the diodes in the secondary side play as a rectifier [21]. Once the converter output voltage (V_C) is positive, the S-PPC operates as step-up in type I connection. Furthermore, the FBPS topology has the gain of allowing zero-voltage switching (ZVS) for a wide load range.

Considering that the dc bus voltage (V_{out}) needed by the inverter is imposed by the electrical grid (close to 220 V for a single-phase grid of 127 V_{rms}), and the rectifier output voltage varies according to Fig. 6, the FBPS S-PPC AP range can be determined as:

$$P_{C,max} = P_{in,max}\left(1 - \frac{V_{in,max}}{V_{out}}\right), \qquad (8)$$

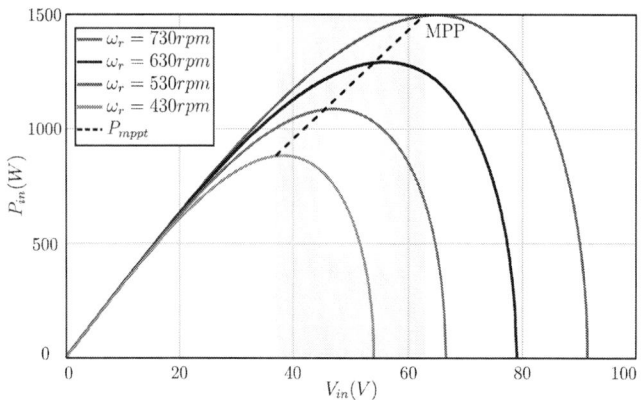

Fig. 6. Electrical power curves (P_{in}) as a function of the FBPS S-PPC input voltage (V_{in}) for angular speeds from 430 to 730 rpm.

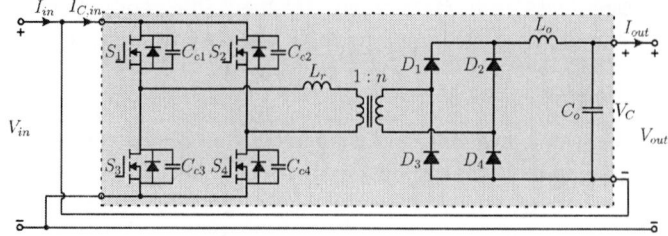

Fig. 7. FBPS dc-dc topology connected as S-PPC.

$$P_{C,min} = P_{in,min}\left(1 - \frac{V_{in,min}}{V_{out}}\right). \qquad (9)$$

Based on (8) and (9) and the input voltage variation shown in Fig. 6, it is defined in Fig. 8 the input power, AP handled by the FBPS S-PPC, and direct power flow for each operating point of the SWT system.

It is observed in Fig. 8 that, for nominal operation with rotational speed of 730 rpm ($P_{in} = 1.5\ kW$), the AP processed by the FBPS S-PPC (P_C) is, approximatelly, 1050 W, whereas the direct power flow (P_{dir}) is around 450 W. As a result, the PPP concept applied to SWT systems provides a reduction of 30% in the AP handled by the dc-dc converter at rated power operation, resulting in a 1 kW design for the FBPS S-PPC.

V. SIMULATION RESULTS

In order to corroborate the theoretical analysis, Fig. 9 depicts a general overview of the proposed system. It is observed the integration of rectifier with the FBPS S-PPC of Fig. 7, as well as, the inverter and the control strategy.

The design specifications of FBPS S-PPC are presented in Table II. The static and dynamic analysis of FBPS dc-dc converter, design equations and transfer functions were derived from [21]. In addition, the small-signal analysis were developed by modelling the effects introduced by the phase-shift modulation and the application of transformer leakage inductance to resonate with the junction capacitance of MOS-FETs to achieve ZVS.

With the objective to obtain the maximum power available from the SWT Gerar 246, the generator reference speed (ω_r^*) must be controlled by the MPPT algorithm. Thus, measurements of electrical power and rotational speed are necessary in order to determine the maximum power point (MPP) [22].

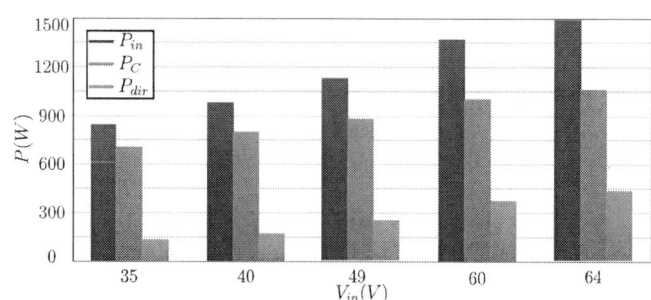

Fig. 8. Input power (P_{in}), AP processed (P_C), and direct power (P_{dir}) in FBPS S-PPC for different input voltages.

978-1-7281-4181-7/19 $31.00 © 2019 IEEE

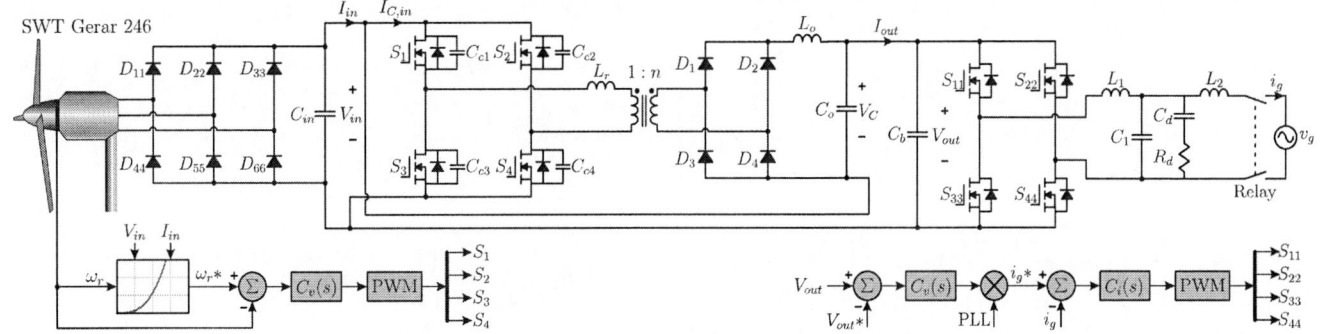

Fig. 9. General overview of the proposed system.

Taking the MPP parameters for each value of v_ω (Fig. 5), an optimum power curve can be traced, which is stored in a lookup table. Thus, according to the measured power, the MPP can be calculated.

In order to connect the dc bus to the grid, a inverter have to be included to the system. In this study, the full-bridge topology with three-level sinusoidal PWM modulation was selected. In order to attenuate high frequency componentes due to the switching, an LCL low-pass filter was employed. The inverter stage were designed and modeled according to [23].

The inverter stage in Fig. 9 is able to inject a sinusoidal current in the grid with low total harmonic distortion (THD), and in phase with the grid voltage. Also, the dc bus voltage can be regulated by controlling the amplitude of grid current.

Fig. 10 presents the simulation results of input voltage (V_{in}), FBPS S-PPC output voltage (V_C), dc bus voltage (V_{out}), AP processed (P_C), direct power (P_{dir}), and power in the FBPS S-PPC input (P_{in}). It should be noted the system initialization routine, which comprehends: PLL synchronism ($0\ s < t < 0.23\ s$); ramp reference to realize the dc bus charge up to 220 V ($0.23\ s < t < 0.36\ s$); zero crossing detector to close the relay connection ($0.36\ s < t < 0.4\ s$); enabling ramp reference of MPPT and inverter control ($0.4\ s < t < 0.5\ s$); systems stabilization ($0.5\ s < t < 0.8\ s$); nominal AP injection in the grid ($t > 0.8\ s$).

Operating at rated power and steady state, it is observed

TABLE II
DESIGN SPECIFICATION OF FBPS S-PPC.

Parameter	Specification
Input voltage range (V_{in})	$35 - 64\ V$
DC bus voltage (V_{out})	$220\ V$
Maximum output voltage (V_C)	$185\ V$
Maximum S-PPC AP ($P_{C,out}$)	$1\ kW$
Output AP (P_{out})	$1.5\ kW$
Switching frequency (f_s)	$40\ kHz$
Capacitor voltage ripple (Δ_{VC})	1%
Inductor current ripple (Δ_{IL})	5%
Inductor value (L_o)	$1.378\ mH$
Capacitor value (C_o)	$47\mu F$
Ressonant inductor (L_r)	$0.615\mu H$
Turns ratio (n)	8

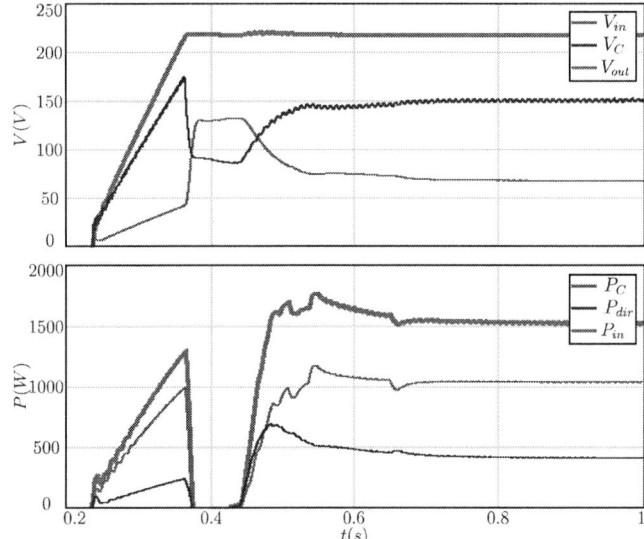

Fig. 10. Simulation results of input voltage (V_{in}), FBPS S-PPC output voltage (V_C), dc bus voltage (V_{out}), AP processed (P_C), direct power (P_{dir}), and power in the diode bridge output (P_{in}).

that the dc bus voltage V_{out} is fixed in 220 V, while the input voltage V_{in} is equal to 65 V (according to the MPP of Fig. 6). Moreover, FBPS S-PPC output voltage V_C is stabilized in 155 V, which is validating (3).

In relation to the AP handled by the FBPS S-PPC, it is confirmed through Fig. 10 that P_C is only 70% of the input power, whereas the direct power P_D is, approximately, equal to 450 W. The simulation results are in agreement with the theoretical values, corroborating the proposed analysis.

As depicted in Fig. 11, the rotational speed is around 700 rpm in nominal power, in accordance with the rectifier output characteristic in Fig. 6. In addition, from $t = 0,8\ s$, the inverter injects a sinusoidal current on the grid with a THD equal to 2.82%, and in phase with grid voltage.

VI. CONCLUSIONS

This paper analyzed the appliance of S-PPC to improve the efficiency of the dc-dc stage in SWT systems connected in the single-phase grid. The main advantages are related

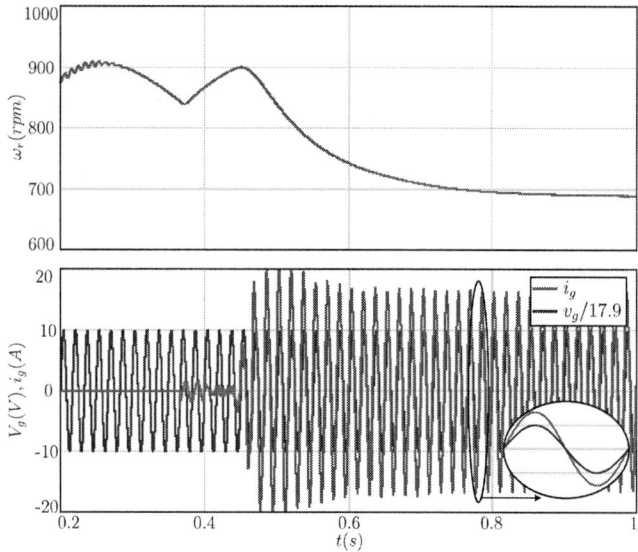

Fig. 11. Simulation results of rotational speed (ω_r), injected grid current (i_g), and voltage grid (v_g).

with the fraction of power processed by the converter, which reduces the total losses of the system. In order to choose the best S-PPC topology, possible connection schematics, AP and efficiency, nAP, converter characteristic, and the operating range of SWT system were discussed.

The analysis of power processing characteristics revealed that the FBPS S-PPC topology can minimize the nAP processing in comparion to a full-power converter. According to the voltage levels of the maximum power generated by the SWT system, the operating range of FBPS S-PPC were determined.

To corroborate order to validate the theoretical analysis, simulation results were presented herein for a 1.5 kW SWT system. The FBPS S-PPC processed around 70% of the generater power, since 30% flows from input to output with unitary efficiency. With a view to reduce even more the AP processed by the S-PPC, SWT with higher output voltages should be employed.

ACKNOWLEDGEMENT

The authors would like to thank the CNPq for the financial support (process No. 422276/2016-2).

REFERENCES

[1] F. Blaabjerg, Y. Yang, D. Yang, X. Wang. "Distributed Power-Generation Systems and Protection". Proceedings of the IEEE, vol. 105, no. 7, pp. 1311-1331, May 2017.

[2] M. Malinowski, A. Milczarek, R. Kot, Z. Goryca, J. T. Szuster. "Optimized Energy-Conversion Systems for Small Wind Turbines: Renewable energy sources in modern distributed power generation systems". IEEE Power Electronics Magazine, vol. 2, no. 3, pp. 16-30, Sep. 2015.

[3] F. Blaabjerg, M. Liserre, K. Ma. "Future on Power Electronics for Wind Turbine Systems". IEEE Journal of Emerging and Selected Topics in Power Electronics, vol. 1, no. 3, pp. 139-152, Aug. 2013.

[4] S. Kouro, J. I. Leon, D. Vinnikov, and L. G. Franquelo. "Grid-connected photovoltaic systems: An overview of recent research and emerging PV converter technology". IEEE Industrial Electronics Magazine, vol. 9, no. 1, pp. 47-61, Mar. 2015.

[5] J. W. Zapata, S. Kouro, G. Carrasco, H. Renaudineau, and T. A. Meynard. "Analysis of Partial Power DC–DC Converters for Two-Stage Photovoltaic Systems". IEEE Journal of Emerging and Selected Topics in Power Electronics, vol. 7, no. 1, pp. 591-603, March 2019.

[6] H. Chen, K. Sabi, H. Kim, T. Harada, R. Erickson, and D. Maksimovic, "A 98.7% efficient composite converter architecture with application-tailored efficiency characteristic," IEEE Transactions on Power Electronics, vol. 31, no. 1, pp. 101-110, Jan. 2016

[7] M. Kasper, D. Bortis, and J. W. Kolar. "Classification and comparative evaluation of PV panel-integrated DC–DC converter concepts". IEEE Trans. on Power Elec., vol. 29, no. 5, pp. 2511-2526, May 2014.

[8] J. R. R. Zientarski, M. L. da S. Martins, J. R. Pinheiro, and H. L. Hey. "Evaluation of Power Processing in Series-Connected Partial-Power Converters," IEEE Journal of Emerging and Selected Topics in Power Electronics, vol. 7, no. 1, pp. 343-351, March 2019.

[9] K. A. Kim, P. S. Shenoy, and P. T. Krein. "Converter rating analysis for photovoltaic differential power processing systems," IEEE Transactions on Power Electronics, vol. 30, no. 4, pp. 1987-1997, Apr. 2015.

[10] J. R. R. Zientarski, M. L. da S. Martins, J. R. Pinheiro, and H. L. Hey. "Series-Connected Partial-Power Converters Applied to PV Systems: A Design Approach Based on Step-Up/Down Voltage Regulation Range," IEEE Trans. on Power Elec., vol. 33, no. 9, pp. 7622-7633, Sept. 2018.

[11] B.-D. Min, J.-P. Lee, J.-H. Kim, T.-J. Kim, D.-W. Yoo, and E.-H. Song. "A new topology with high efficiency throughout all load range for photovoltaic PCS". IEEE Transactions on Industrial Electronics, vol. 56, no. 11, pp. 4427-4435, Nov. 2009.

[12] J. P. Lee, B. D. Min, T. J. Kim, D. W. Yoo, and J. Y. Yoo, "A novel topology for photovoltaic DC/DC full-bridge converter with flat efficiency under wide PV module voltage and load range," IEEE Transactions on Industrial Electronics, vol. 55, no. 7, pp. 2655-2663, Jul. 2008.

[13] M. S. Agamy, M. H.- Todorovic, A. Elasser, S. Chi, R. L. Steigerwald, J. A. Sabate, A. J. McCann, L. Zhang, and F. J. Mueller. "An efficient partial power processing DC/DC converter for distributed PV architectures". IEEE Transactions on Power Electronics, vol. 29, no. 2, pp. 674-686, Feb. 2014.

[14] H. Zhou, J. Zhao, and Y. Han. "PV balancers: Concept, architectures, and realization". IEEE Transactions on Power Electronics, vol. 30, no. 7, pp. 3479-3487, Jul. 2014.

[15] M. Chen, F. Gao, R. Li, and X. Li. "A dual-input central capacitor DC/DC converter for distributed photovoltaic architectures," IEEE Trans. on Industry Applications, vol. 53, no. 1, pp. 305-318, Jan. 2017.

[16] R. W. Erickson and D. Maksimović. "Fundamentals of Power Electronics," 2nd ed., Kluwer Academic Publishers, 2001.

[17] J. Zhao, K. Yeates, and Y. Han. "Analysis of high efficiency DC/DC converter processing partial input/output power". IEEE 14th Workshop on Control and Modeling for Power Electronics (COMPEL), pp. 1-8, Jun. 2013.

[18] IEEE Standard 1459-2010. "IEEE Standard Definitions for the Measurement of Electric Power Quantities Under Sinusoidal, Nonsinusoidal, Balanced, or Unbalanced Conditions", pp. 1–50, Mar. 2010.

[19] J. R. R. Zientarski, M. L. da S. Martins, J. R. Pinheiro, and H. L. Hey. "Understanding the partial power processing concept: A case-study of buck-boost dc/dc series regulator," 2015 IEEE 13th Brazilian Power Electronics Conference and 1st Southern Power Electronics Conference (COBEP/SPEC), pp. 1-6, Dec. 2015.

[20] H. Wang, C. Nayar, J. Su, and M. Ding. "Control and Interfacing of a Grid-Connected Small-Scale Wind Turbine Generator," IEEE Transactions on Energy Conversion, vol. 26, no. 2, pp. 428-434, April 2011.

[21] V. Vlatkovic, J. A. Sabaté, R. B. Ridley, F. C. Lee, and B. H. Cho. "Small-Signal Analysis of the Phase-Shifted PWM Converter," IEEE Transactions on Power Electronics, vol. 7, no. 1, pp. 128-135, Jan. 1992.

[22] E. Koutroulis, and K. Kalaitzakis. "Design of a maximum power tracking system for wind-energy-conversion applications," IEEE Transactions on Industrial Electronics, vol. 53, no. 2, pp. 486-494, April 2006.

[23] G. C. Knabben. "Microinversor fotovoltaico não isolado de dois estágios," M. S. Thesis, Federal University of Santa Catarina, Florianópolis, Brazil, 2017.

978-1-7281-4181-7/19 $31.00 © 2019 IEEE

Proportional Wavelet Sliding Mode Controller for Torque Ripple Reduction in BLDC Motor

Mateus Moro Lumertz
School of Engineering of São Carlos
University of São Paulo
São Carlos, Brazil
mateuslumertz@usp.br

Carlos Matheus R. de Oliveira
School of Engineering of São Carlos
University of São Paulo
São Carlos, Brazil
carlosmro@usp.br

Allan Gregori de Castro
School of Engineering of São Carlos
University of São Paulo
São Carlos, Brazil
allangregori@usp.br

Paulo Roberto U. Guazzelli
School of Engineering of São Carlos
University of São Paulo
São Carlos, Brazil
paulo.ubaldo@usp.br

Manuel Luis de Aguiar
School of Engineering of São Carlos
University of São Paulo
São Carlos, Brazil
aguiar@sc.usp.br

Jose Roberto B. A. Monteiro
School of Engineering of São Carlos
University of São Paulo
São Carlos, Brazil
jrm@sc.usp.br

Abstract—**Permanent magnetic synchronous machines with trapezoidal back-electromotive force and six-step operation, also known as brushless direct current motors, have high torque undulations because needed current waveform can't be provided. Proper vector control techniques used to reduce chattering, are high dependent on machine parameters and are complex to implement on real applications. Sliding Mode and Pseudo Sliding Mode are controllers with high dynamical performance and simple implementation, but increase torque undulations in machine control applications. On the other hand, wavelet multiresolution controllers have the feature of chattering reduction, but are complex to tune and analyze, due to the number of parallel detail and approximation coefficients used. This works proposes a simplified wavelet controller, replacing approximation coefficients with a proportional controller. This controller was used to design a proper sliding manifold for a sliding mode speed control, on a BLDC motor, without using machine parameters. A genetic algorithm was used to determine an optimum mother wavelet for a fixed decomposition level, leaving just one parameter to tune in future works. In results, an improvement to the proposed controller was suggested, for a controller with high dynamical and steady state performance, without torque undulations and easier to tune than wavelet multiresolution controllers.**

Index Terms—**brushless motor, sliding mode control, wavelet transform.**

I. Introduction

Brushless Direct Current (BLDC) motors are lower-cost with higher power/weight relationship machines than induction ones [1]. Also known as Permanent Magnetic Synchronous Machines (PMSM), their difference from conventional synchronous machines is the replacement of field windings by permanent magnets.

Magnets geometry and disposition results in different formats of machine's back-Electromotive Force (EMF). Sinusoidal back-EMF permanent magnetic machines are natural substitutes to synchronous machines, and are widely used in

This work is supported by São Paulo Research Foundation (FAPESP), grant #2006/04226-0.

eolic generation systems. However, wave formats with high third harmonic component, like trapezoidal back-EMFs, results in machines with higher power density [2], being smaller with lower cost.

However, in machines with conventional six-step operation and trapezoidal back-EMFs, expected current waveform to produce a constant electromagnetic torque cannot by provided, resulting in chattering, inefficient operation and wear of mechanical parts. Vector control techniques have been proposed to solve this problem, but they have implementation difficulties and are highly dependent on machine parameters [3][4][5].

Sliding Mode (SM) and Pseudo Sliding Mode are controllers with high dynamical performance and simple implementation. In machine control applications, limited switching frequency of real switches can cause chattering, which can be reduced by smooth switching functions. Major of efforts to solve the problem, and turn SM suitable to real applications, propose modifications on the sliding manifold [6][7].

Wavelet Transform is a data processing tool, most used for transients analysis and associated to neural networks for pattern recognition. Wavelet Multiresolution based controllers are reported in literature to have the feature of chattering reduction [8][9]. Several detail and approximation coefficients are used by this controllers, to create a suitable control action.

Detail coefficients output is most dependent on transients and high frequency terms, so they act more directly on chattering. However, they are generated by high-pass filters and as pure derivative controllers, don't eliminate steady state error. Approximation coefficients are introduced to improve steady state performance, since they are generated by low-pass filters.

Multiple decomposition levels of this coefficients as output of wavelet controllers, results in a system with a large number of parameters to tune, being complex and difficult to analyze and implement.

A simplified wavelet controller, with one level detail coefficient parallel to a proportional block is proposed in this work, to process a suitable sliding manifold for a sliding mode controller, for speed control in a BLDC motor. System with high dynamical performance, reduced chattering, with just one gain to tune is expected in results. Genetic algorithm was used to find an optimum mother wavelet and controller gain.

II. MACHINE MODEL

In this work, the term BLDC motor is used to describe a three phase system with a six-step converter, connected to a DC source, and used to control a surface-mount permanent magnet synchronous machine, with trapezoidal back-Electromotive Force.

Relations between phase voltages, currents and back-EMF are described by

$$
\begin{bmatrix} V_a - V_n \\ V_b - V_n \\ V_c - V_n \end{bmatrix} = \begin{bmatrix} e_a \\ e_b \\ e_c \end{bmatrix} + \begin{bmatrix} L & M & M \\ M & L & M \\ M & M & L \end{bmatrix} \frac{d}{dt} \begin{bmatrix} i_a \\ i_b \\ i_c \end{bmatrix} + \\ R \begin{bmatrix} i_a \\ i_b \\ i_c \end{bmatrix}
$$

(1)

where

V_a, V_b, V_c are phase voltages;
V_n is the motor neutral voltage;
e_a, e_b, e_c are trapezoidal back-EMFs;
L is the stator phase indutance;
M is the stator mutual indutance;
i_a, i_b, i_c are stator phase currents; and
R is stator phase resistance.

Machine's stator and rotor are relationed by Lenz's Law, described by

$$
\begin{bmatrix} e_a \\ e_b \\ e_c \end{bmatrix} = -N \frac{d}{dt} \begin{bmatrix} \phi_a \\ \phi_b \\ \phi_c \end{bmatrix} = -N \omega_e \frac{d}{d\theta} \begin{bmatrix} \phi_a \\ \phi_b \\ \phi_c \end{bmatrix}
$$

(2)

where ϕ_a, ϕ_b, ϕ_c are magnetic fluxes, generated by rotor magnets and linked by stator windings, N is the number of turns in the stator windings, θ is the rotor angle, and ω_e is the electrical speed.

Electromagnetic torque is given by

$$
T_e = n_{pp} \left(i_a \frac{d\phi_a}{d\theta} + i_b \frac{d\phi_b}{d\theta} + i_c \frac{d\phi_c}{d\theta} \right)
$$

(3)

where n_{pp} is machine number of pole pairs.

According to (2), $d\phi_a/d\theta$, $d\phi_b/d\theta$ and $d\phi_c/d\theta$ should have trapezoidal wave form. So, to produce constant electromagnetic torque according to (3), phase currents should have 120° square wave form. However, stator inductance doesn't allow

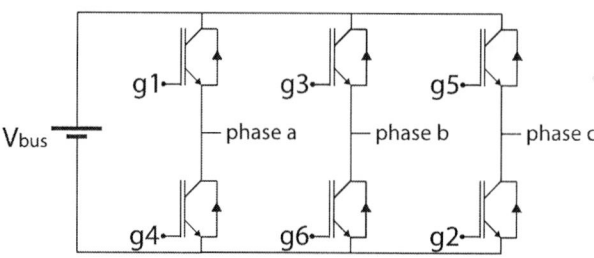

Fig. 1. Six-step converter

currents to vary instantly, so electromagnetic torque suffer undulations, generating chattering in the machine rotor.

Conventional six-step operation are defined by converter shown in Fig. 1, and the following commutation logic:

$0^o \leq \beta < 60^o$: g_1 and g_2 are activated;
$60^o \leq \beta < 120^o$: g_3 and g_2 are activated;
$120^o \leq \beta < 180^o$: g_3 and g_4 are activated;
$180^o \leq \beta < 240^o$: g_5 and g_4 are activated;
$240^o \leq \beta < 300^o$: g_5 and g_6 are activated;
$300^o \leq \beta < 360^o$: g_1 and g_6 are activated.

Where V_{bus} are the DC voltage source, and β is the machine electrical angle, defined by

$$
\beta = n_{pp} \theta
$$

(4)

Machine is controlled by Pulse Width Modulation (PWM) in g_1, g_3 and g_5, while g_4, g_6 and g_2 maintain full duty cycle.

III. SLIDING MODE CONTROLLER

Sliding Mode (SM) Controller can be implemented as the signal function of the input S, multiplied by a gain K, as expressed by

$$
y = K.sgn(S)
$$

(5)

where S is a function of input error, known as sliding manifold.

High dynamical performance and control frequency are main features of SM. However, applications in machine control with basic conceptions of SM results in torque ripple, increasing vibration, noise and losses.

Soft actuation functions can be used instead of signal function to reduce torque ripple, as

$$
y = K.tanh(S/\phi)
$$

(6)

where ϕ is a parameter used to adjust the saturation region.

Stability of SM controllers can be analyzed through it's phase plane. System is stable (Lyapunov stability) if it's motion in phase plane converge or orbit an equilibrium point.

IV. WAVELET TRANSFORM

Wavelet Transform is a signal processing tool capable of analysis in both time and frequency domains. Identifies and locate transients in time-varying systems, common applications are faults analysis and transients pattern recognition.

978-1-7281-4181-7/19 $31.00 © 2019 IEEE

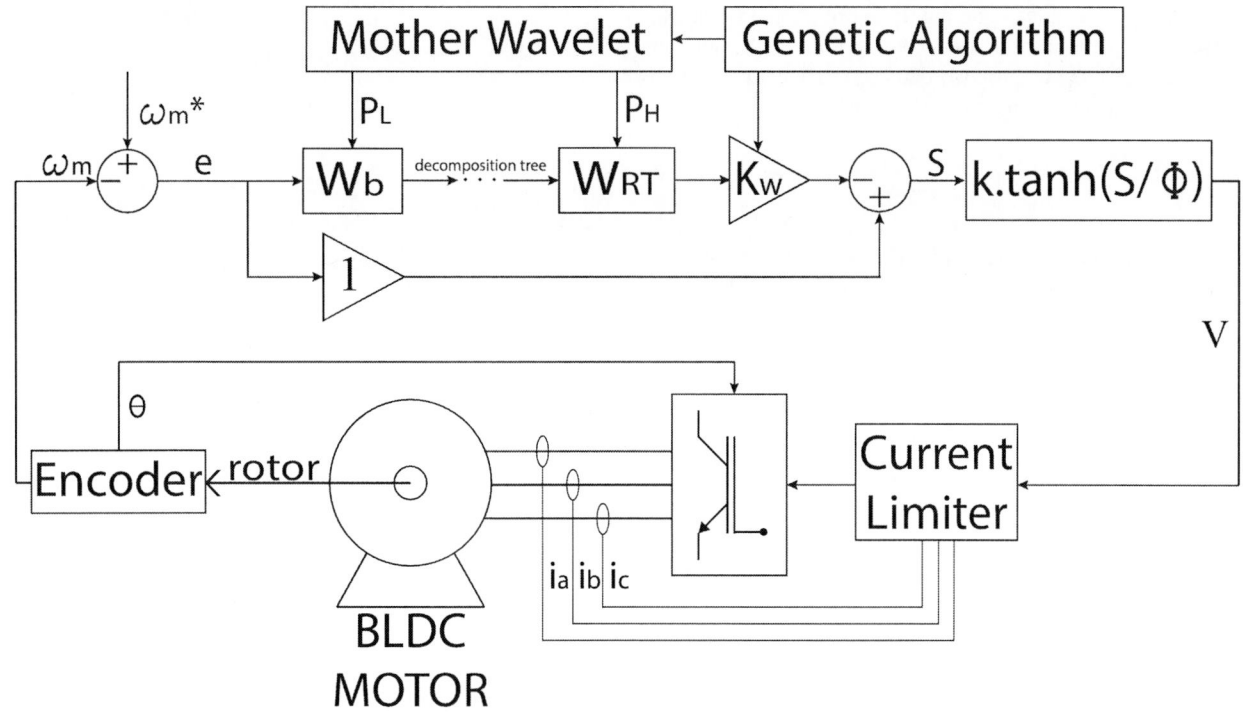

Fig. 2. Proportional Wavelet Sliding Mode Controller

In continuous time domain, WT is defined as the inner product of functions $s(t)$ and $\psi_{a,b}(t)$ in L^2 space, calculated by

$$< s, \psi_{a,b} > = \int_{-\infty}^{\infty} s(t) \bar{\psi}_{a,b}(t) dt \qquad (7)$$

where s is the input signal, a and b parameters of scale and translation, $\psi_{a,b}$ the mother wavelet and $\bar{\psi}_{a,b}$ it's complex conjugate.

Wavelets are functions in L^2 space with compact support which satisfy the following condition

$$C_\psi = \int_0^{\infty} \frac{|\hat{\psi}_{a,b}(\omega)|^2}{|\omega|} d\omega \qquad (8)$$

where $\hat{\psi}_{a,b}$ is the wavelet in frequency domain. Mother wavelet modify the output pattern of the transform, it's choice depend of desired matched signal.

V. MULTIRESOLUTION ANALYSIS

Discrete-time implementation of WT can be done through Multiresolution Analysis (MRA), where a filter bank with low-pass and high-pass filters corresponding to the mother wavelet is created. Low-pass filter response is the approximation coefficients, and high-pass filter response is the detail coefficients.

Wavelet MRA applies convolutions between input signal and low-pass and high-pass filter vectors, in each decomposition level. Filtered responses are the input signals for next

decomposition levels, designing a decomposition tree. This operation is calculated by

$$(f * g)(n) = \sum_{j=-\infty}^{\infty} f(j).g(n-j) \qquad (9)$$

where f and g are the input vectors. The discrete convolution is analog to multiplication of polynomials with coefficients equal to elements of f and g.

Polynomial representation for a given input signal $s = [s_n, s_{n-1}, ..., s_1, s_0]$, where s_i are the samples in a fixed sampling window and s_0 the newest element, are defined by

$$p(s, x) = s_n.x^n + s_{n-1}.x^{n-1} + ... + s_1.x + s_0 \qquad (10)$$

Convolution operation applied by MRA, in polynomial form, is calculated by

$$W(s * P)(x) = \sum_{i=0}^{n} s_i.x^i . \sum_{j=0}^{m} P_j.x^j \qquad (11)$$

where n are the length of the input signal s, m the length of the polynomial P with coefficients P_j defined by the filter bank. Polynomial P are fixed for each mother wavelet and different for low-pass (P_L) or high-pass (P_H) application.

Resultant polynomial converted to a signal vector, and down-sampled by half, is the response for one operation in the decomposition tree. However, elements in the edge of the vector don't convolve with the entire filter polynomial, and the

978-1-7281-4181-7/19 $31.00 © 2019 IEEE   894

element with lower degree will be expressed by $W_0 = s_0 P_0$, being simply the input value multiplied by the constant P_0.

Elements with degree lower than m are removed from response, in order to improve data processing in the vector border. So, the output polynomial is calculated by

$$W_b(s * P)(x) = \frac{1}{x^m}\left[\sum_{i=0}^{n} s_i x^i \sum_{j=0}^{m} P_j x^j - \sum_{k=0}^{m-1} s_k x^k P_{m-k-1}\right]$$
(12)

Real time application of MRA can be implemented by considering the coefficient with lower degree in (12) on final node of the decomposition tree as output, as expressed by

$$W_{RT}(s) = \sum_{i=0}^{m} s_{m-i}.P_i$$
(13)

VI. PROPORTIONAL WAVELET SLIDING MODE CONTROLLER

Wavelet and Sliding Mode cascading control can be interpreted as SM post processing the Wavelet controller, or the Wavelet Transform to design sliding manifold of SM controller.

However, MRA detail coefficients are generated by high-pass filters, and as pure derivative controllers, don't eliminate steady state error in control applications. Some works uses multiple parallel detail and approximation coefficients to produce a suitable controller. However, this approach results in a controller with many degrees of freedom, being complex, unpredictable and hard to tune.

This work proposes a simplified wavelet controller, with detail coefficients parallel with a proportional controller, as illustrated in Fig. 2, where ω_m is mechanical speed, ω_m^* the speed controller reference, e the input error and V the mean voltage of six-step converter, controlled by PWM.

In Proportional Wavelet Sliding Mode (PWSM) controller, the proportional gain is fixed, the relation between proportional and wavelet blocks can be adjusted by SM saturation coefficient ϕ.

Proportional controller is simpler and have less computational cost than wavelet approximation coefficients. With a single output in the wavelet block, it's easier to tune empirically or with optimization/adaptive algorithms.

VII. GENETIC ALGORITHM AND OPTIMIZATION

Several optimization algorithms have been proposed in the literature for controllers tuning and self-tuning [10],[11],[12].

Genetic Algorithm (GA) is inspired by theory of evolution, of biology. All possible solutions to the problem are considered as individuals in a population, variable parameters such as controller gains are obtained from the chromosomes of these individuals, which are part of their genetic codes.

Major steps of GA are illustrated in Fig. 3. An initial population with random chromosomes values is generated. Each individual is tested, where their performance is evaluated by a objective function f_o, which is the optimized function.

Based on it's performance, individuals who will be parents for next generation are selected, however, both good and bad

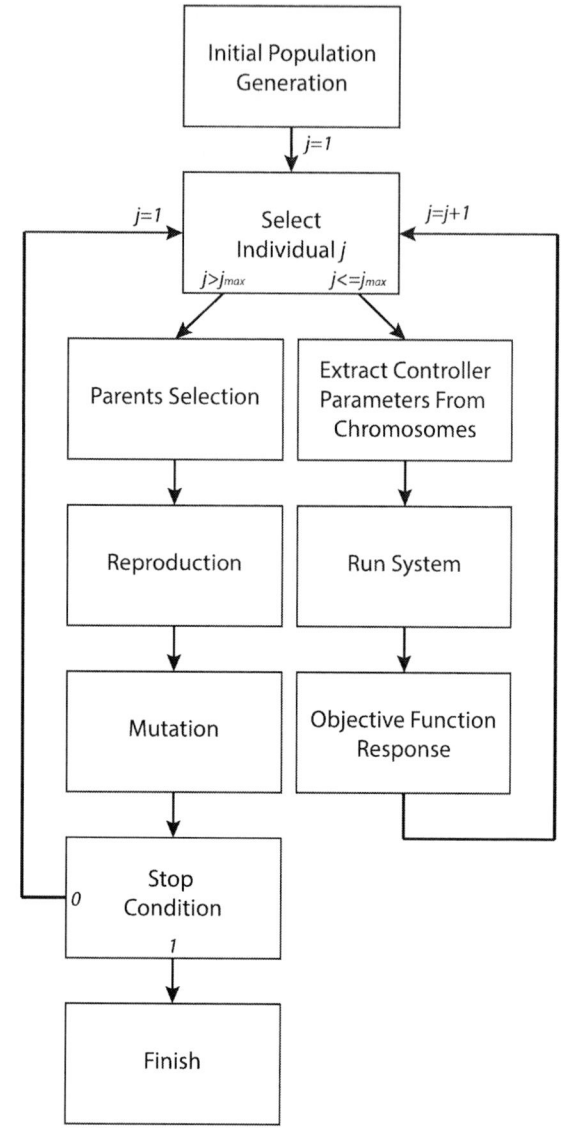

Fig. 3. Genetic Algorithm Implementation

individuals should be chosen to preserve the genetic variability of the population.

In the reproduction stage occurs the creation of children individuals, from the commutation of chromosomes of the parents. Each chromosome has a small probability to suffer a random modification, known as mutation.

Stopping conditions for GA can be generation limit, tolerance on the objective function error, or population convergence. On online adaptive applications, stop conditions are not used.

In this work, GA is used to determine an optimum mother wavelet and K_W gain in PWSM controller, for a fixed decomposition level. Objective function used was

$$f_o = k_1|e_s| + k_2\sqrt{e_{ds}}$$
(14)

978-1-7281-4181-7/19 $31.00 © 2019 IEEE

where

$$e_s = \sum_{j=1}^{L_W F_s} S^* - S(j) \tag{15}$$

and

$$e_{ds} = \sum_{j=2}^{L_W F_s} (S(j) - S(j-1))^2 \tag{16}$$

being L_W the length of analyzed window, e_s the component used to minimize steady state error, e_{ds} the component used to reduce chattering, k_1 and k_2 constants which define the desired operation point. In this work chattering reduction was prioritized, so $k2 > k1$.

VIII. RESULTS AND DISCUSSION

Simulated BLDC motor parameters was: $R = 2.4\Omega$, $L = 12.4mH$, $npp = 3$, nominal load of $2.6Nm$ and nominal speed of $100rad/s$.

Genetic algorithm optimization lasted 46 generations, keeping a constant population of 20 individuals. Algorithm stopped when it didn't find a better solution for 10 consecutive generations. Second decomposition level was fixed, and the optimum solution found was function Daubechies 2 as mother wavelet, with Daubechies 1 to 6 as candidate functions, and $K_W = 169$.

Comparison between SM controller defined in (6) and optimum PWSM controller is shown in Fig. 4. PWSM controller output presented higher frequency components, because input error presented low undulations, with some high frequency terms.

Both controllers presented steady state error, as reference speed was $80rad/s$. PWSM controller has the feature of torque and speed undulations reduction, for better performance on steady state an integrator can be added in parallel to the wavelet block, or in a Wavelet Super Twisting controller.

Frequency-domain comparison between both controllers is shown in Fig. 5. SM presented high 2nd, 3rd and 6th harmonics of the electrical speed in the electromagnetic torque, while PWSM presented reduction in all harmonic components, with some remain 3rd, 6th and 12th components.

Fig. 6 shows phase plane for PWSM in steady state operation.

Fig. 7 shows the machine start-up until $0.9s$, with a constant load of $2N.m$. Speed reference was changed on instant $0.33s$, from $80rad/s$ to $60rad/s$, and on instant $0.66s$ load was increased to $4N.m$. PWSM and SM controllers presented similar dynamic performance through this variations, therefore, Wavelet Transform for reduced torque ripple did not interfere in the SM controller qualities.

IX. CONCLUSION

This work introduced a simplified wavelet controller, replacing multiple detail and approximation coefficients on output by one detail coefficient parallel to a proportional controller. Presented topology is easier to tune, and was implemented as sliding manifold of a sliding mode controller.

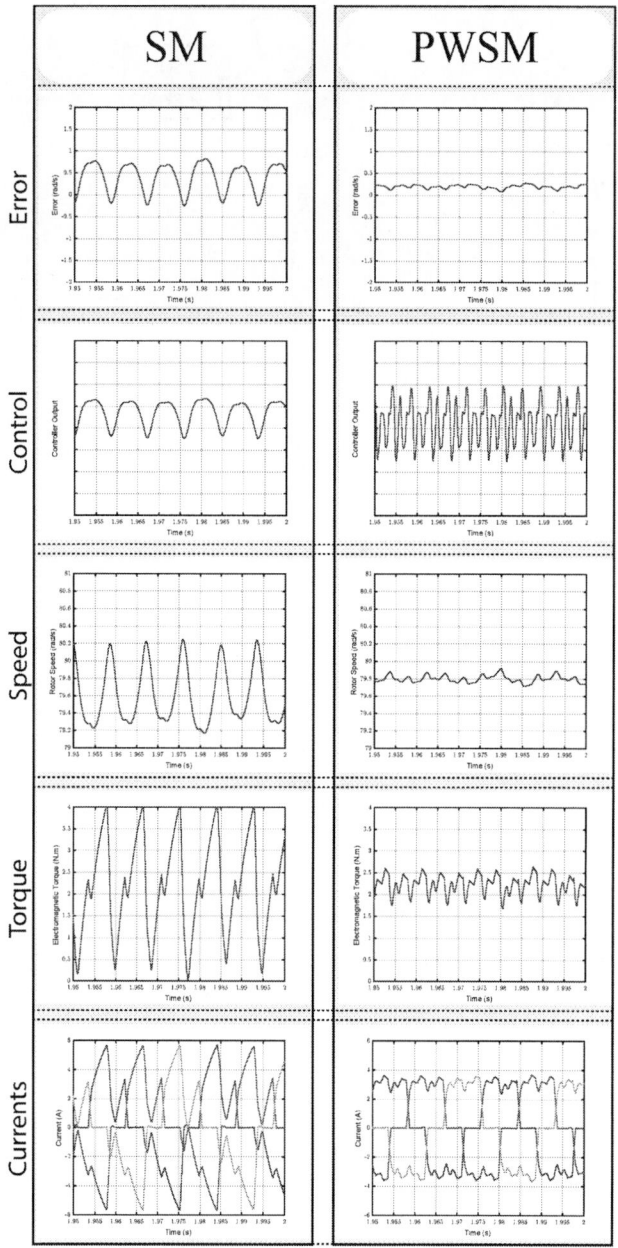

Fig. 4. Controllers Comparison

Genetic algorithm was used to find an optimum mother wavelet and controller gain for a fixed decomposition level, where function Daubechies 2 presented optimum performance. In future works this mother wavelet can be used with the proposed controller topology, remaining only one parameter to be tuned.

Proposed controller was used in speed control of a BLDC motor, in a control system completely independent of machine parameters. Comparison with conventional sliding mode controller, showed the feature of chattering reduction without

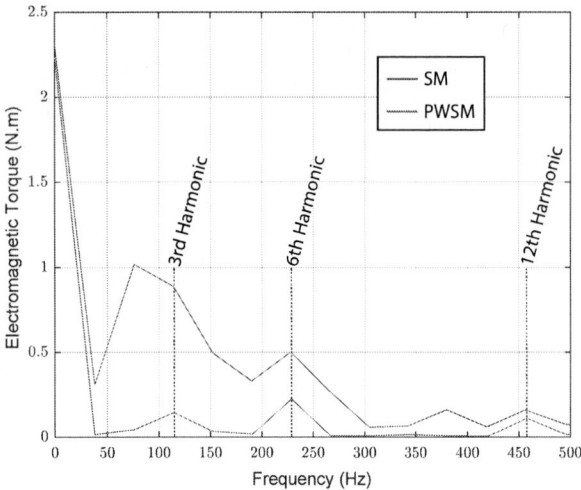

Fig. 5. Frequency-Domain Response of Electromagnetic Torque

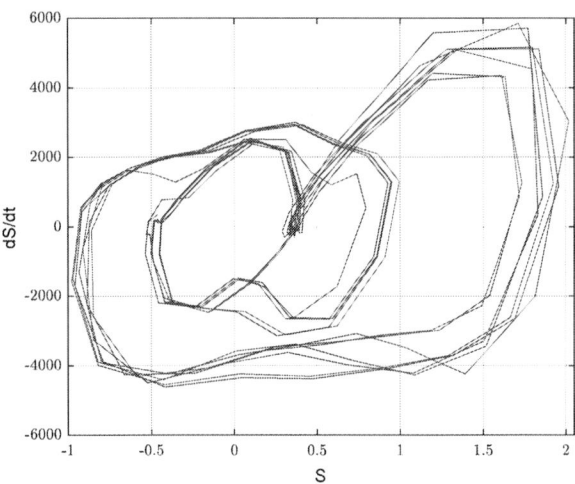

Fig. 6. PWSM Phase Plane

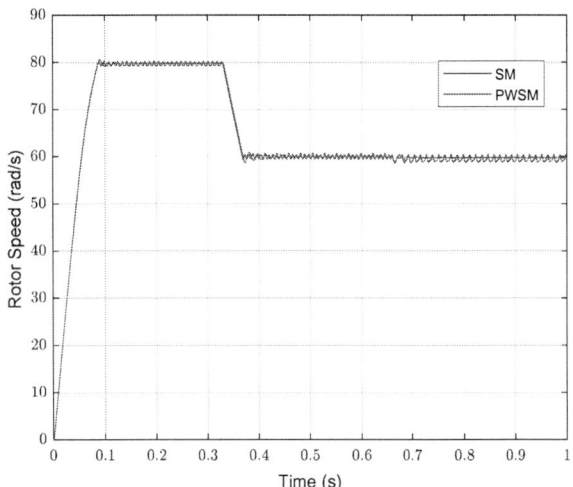

Fig. 7. Controllers Dynamics

interfering on dynamical performance.

Both compared controllers presented steady state error, considering mean value, in future works a Wavelet Integral Sliding Mode, or Wavelet Super Twisting controller can be used for dynamical and steady state performance, with reduced chattering in machine control applications.

REFERENCES

[1] M. J. Melfi, S. Evon and R. Mcelveen, "Induction versus permanent magnet motors," IEEE Industry Applications Magazine, vol. 15, pp. 28 - 35, 2009.

[2] K. Wang, Z. Q. Zhu and G. Ombachn, "Torque enhancement of surface-mounted permanent magnet machine using third-order harmonic," IEEE Transactions on Magnetics, vol. 50, 2013.

[3] J. R. B. A. Monteiro, A. G. Castro and T. E. Almeida, "Sliding mode vector control of non-sinusoidal permanent magnet synchronous machine," Brazilian Power Electronics Conference (COBEP). Juiz de Fora, Brazil: IEEE, 2017, pp. 1 - 7.

[4] J. R. B. A. Monteiro, A. A. Oliveira Jr., M. L. Aguiar and E. R. Sanagiotti, "Electromagnetic torque ripple and copper losses reduction in permanent magnet synchronous machines," European Transactions on Electrical Power, vol. 22, pp. 627 - 644, 2012.

[5] C. L. Baratieri and H. Pinheiro, "Hybrid orientation for sensorless vector control of nonsinusoidal back-EMF PMSM," 40th Annual Conference of the IEEE Industrial Electronics Society (IECON). Dallas, USA: IEEE, 2014, pp. 621 - 627.

[6] C. M. R. Oliveira, M. L. Aguiar, W. C. A. Pereira, A. G. Castro, T. E. P. Almeida and J. R. B. A. Monteiro, "Integral sliding mode controller with anti-windup method analysis in the vector control of induction motor," 12th IEEE International Conference on Industry Applications (INDUSCON), 2016, pp. 1 - 7.

[7] G. Bartolini, A. Ferrara and E. Usai, "Chattering avoidance by second-order sliding mode control," IEEE Transactions on Automatic Control, vol. 43, pp. 241 - 246, 1998.

[8] M. A. S. K. Khan and M. A. Rahman, "Implementation of a New Wavelet Controller for Interior Permanent-Magnet Motor Drives," Industry Applications Annual Meeting. New Orleans, USA: IEEE, 2007, pp. 1280 - 1287.

[9] L. W. Lee and I. H. Li, "Wavelet-based adaptive sliding-mode control with H∞ tracking performance for pneumatic servo system position tracking control," IET Control Theory & Applications, vol. 6, pp. 1699 - 1714, 2012.

[10] P. N. Menon and R. Anasraj, "Particle Swarm Optimized Sliding Mode Controller for an AC-DC Boost Converter," 3rd International Conference on Eco-friendly Computing and Communication Systems. Mangalore, India: IEEE, 2014.

[11] F. A. Patakor, N. Jantan, Z. Salleh and M. Sulaiman, "Auto-tuning sliding mode control for induction motor drives," 8th Computer Science and Electronic Engineering (CEEC). Colchester, UK: IEEE, 2016.

[12] A. B. Rad, W. L. Lo and K. M. Tsang, "Self-tuning PID controller using Newton-Raphson search method" IEEE Transactions on Industrial Electronics, vol. 44, pp. 717 - 725, 1997.

978-1-7281-4181-7/19 $31.00 © 2019 IEEE

Voltage Regulator Behavior on Power Distribution Grids with High Integration of PVDG

Lucas F. S. Azeredo
Postgraduate Program in Electrical Engineering, PPGEE
Federal University of Espírito Santo, UFES
Vitória, Brazil

Luiz G. R. Tonini
Postgraduate Program in Electrical Engineering, PPGEE
Federal University of Espírito Santo, UFES
Vitória, Brazil

Mariana A. Mendes
Postgraduate Program in Electrical Engineering, PPGEE
Federal University of Espírito Santo, UFES
Vitória, Brazil

Murillo C. Vargas
Postgraduate Program in Electrical Engineering, PPGEE
Federal University of Espírito Santo, UFES
Vitória, Brazil

Oureste E. Batista
Postgraduate Program in Electrical Engineering, PPGEE
Federal University of Espírito Santo, UFES
Vitória, Brazil

Carla J. Espindula
Postgraduate Program in Electrical Engineering, PPGEE
Federal University of Espírito Santo, UFES
Vitória, Brazil

Abstract—In consequence to the worldwide growth of distributed generation (DG), especially photovoltaic (PV), several studies discuss on the impacts on the voltage profile of the distribution feeders. However, they do not apply Brazilian real electricity regulation for DG and PV systems. This article proposes to study the voltage Profile of the IEEE 13-Node radial test feeder with high PV penetration level with the presence of the voltage regulator (OLTC) connected to the substation output. The results obtained for the case without voltage regulator and with the presence of the regulator will be compared under the regulation of electricity of Brazilians Regulatory Agency of Electric Energy (ANEEL) and Electrical System Operator (ONS). The results show that PV generators operating with fixed PF (PF = 1), and with the presence of the voltage regulator (OLTC) the voltage values and the electricity quality levels are directly affected.

Index Terms—photovoltaic generation; voltage profile; regulator.

I. INTRODUCTION

Since April 17, 2012, Brazilian consumers can generate their own electricity from renewable sources or qualified cogeneration and even supply the surplus to their local distribution network [1].

Given this, it is important to highlight the micro and distributed generation of electricity, which are innovations that can combine financial economy, social and environmental awareness and self-sustainability.

In addition, the incentives for distributed generation (DG) are justified by the potential benefits that this modality can provide to the electric system. This includes postponing investments in transmission and distribution system expansion, low environmental impact, reduced grid load, loss minimization and diversification of the energy matrix.

Among the various types of DG, photovoltaic (PV) system generation is the most popular. This is due to the Because it is a system without moving parts, which reduces maintenance costs and frequency, has no significant noise, the Sun is an abundant source and it is relatively easy to install panels [2], [3].

It is important to note that, although the growth of photovoltaic generation in Brazil is evident, there are countries that are already living another reality in distributed generation.

Considering that Germany is already experiencing a high concentration of generators connected to distribution grids, mainly photovoltaic.

Given this reality found in Germany, it is important to highlight that, according to the German legislation for DG, some strategies are made to control the voltage levels [4], [5], among which we can highlight:

- Reinforcement in distribution;
- Limitation of supply of active power at 70capacity installed;
- Voltage Buffer Control energy
- Supply of reactive power
- Automatic voltage limiting limited by a control dynamic power
- Automatic voltage limiting limited by a control dynamic active and reactive power
- Distribution transformers with tap change under load.

With the insertion of the DG in the electric power system, some network parameters such as voltage profile, nominal current, short-circuit current and losses can be modified [6]–[9]. This can occur because the system will have multiple generators injecting current, which can modify the flow of energy.

The active energy and the injection of reactive energy in the network can cause increase of voltage, and the consumption of reactive energy can cause voltage drops. In addition, it is known that electricity regulations define the maximum and minimum limits of service voltage to meet the levels of electricity quality [10].

Thus, knowing that DG can influence the behavior of the voltage, it is important to study the voltage variation through the feeder in this scenario and the conformity with the levels of electricity quality. Although several similar studies have been carried out on the impacts of DG on the voltage profile [10], this research aims at analyzing the voltage regulator impact on the voltage profile for DG in Brazil [1] and Network Procedures in Brazil (PRORED E) [11], prepared by the National Electric System Operator (ONS) and ANEEL, to configure PVG operation in a distribution feeder.

This study attempts to simulate a real high PV scenario penetration with the presence of voltage regulator (OLTC). The results of the stress profile are discussed. and evaluated according to ANEEL voltage parameters, which are classified into three levels: precarious, proper and critical.

II. IEEE 13-Node Test Feeder

The IEEE 13-Node Test Feeder circuit model is very small and used to test common features of distribution analysis software, operating at 4.16 kV. It is characterized by being short, relatively highly charged, a single voltage regulator in the substation, overhead and underground lines, bypass capacitors, an in-line transformer and unbalanced load.

This network is available in MATLAB. The network has twelve load flow bus blocks used to calculate an unbalanced load flow in a model that represents the IEEE 13 Node Test Feeder circuit, originally published in the Subcommittee's Report on Systems Analysis IEEE distribution. It is important to note that the model includes the regulating transformer between nodes 650 and 632 of the reference test model.

For the article in question, which seeks to analyze the impact of distributed generation, it was necessary to connect the DG to the nodes of the network. This connection is shown in figure 1.

In this one-line diagram, you can see all points connection. Note that figure 1 does not contain the voltage regulator. However, the insertion of the regulator is of paramount importance to analyze its impact in networks with high DG presence and important to compare the voltage profile with the presence of the regulator and the absence of the voltage regulator.

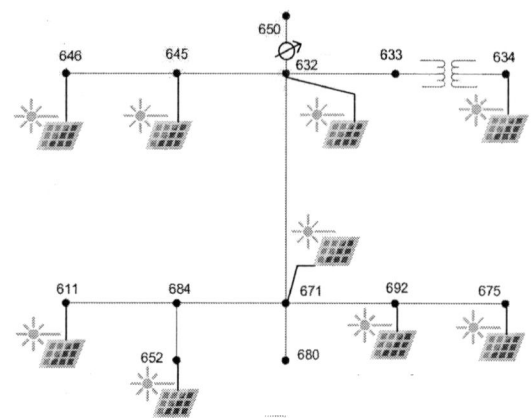

Fig. 1: IEEE 13-Node Test Feeder with DG

On the other hand, it is important to point out that the distributed generation tends to promote an improvement in the voltage profile, regardless of the presence of the voltage regulator.

III. THREE-PHASE OLTC REGULATING TRANSFORMER

The on-load tap changer(OLTC) transformer model provided in the Simulink library. The figure 2 shows one phase of the three-phase regulating transformer.

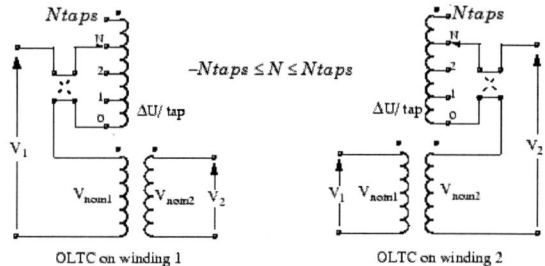

Fig. 2: One Phase of the Regulating Transformer

Each phase consists of the main windings 1 and 2 and a control winding with tapping. The voltage ratios in the two windings are expressed by equations 1 and 2 which are given below:

$$\frac{V_2}{V_1} = \frac{1}{1 + N * \Delta U} * \frac{V_{nom2}}{V_{nom1}} \tag{1}$$

$$\frac{V_2}{V_1} = (1 + N * \Delta U) * \frac{V_{nom2}}{V_{nom1}} \tag{2}$$

978-1-7281-4181-7/19 $31.00 © 2019 IEEE

where:

N = tap position

ΔU voltage per tap in pu of nominal voltage of winding 1 or winding 2.

Voltage regulation is performed by varying the transformer ratio V2/V1 by means of the OLTC. The OLTC can be connected either on winding 1 (left diagram) or on winding 2 (right diagram).

The OLTC can select any touch of position 0 (no voltage correction) to the maximum tap (Ntap position) that indicates the maximum voltage correction.

The OLTC is also equipped with a reversing switch which allows connecting the regulation winding either in additive or subtractive polarity. The factor multiplying Vnom2/Vnom1 is the voltage correction factor.

IV. SIMULATION AND METHODOLOGY

To perform the proposed simulation, it was necessary to insert the OLTC block in the 13 node model available in simulink, requiring an update in the application of the OLTC model. To meet the IEEE 13 network configurations, it was necessary to create a single-phase voltage regulator from the combination of the three-phase OLTC voltage regulator blocks available on simulink.

This methodology was necessary because the voltage regulator specified for the 13-node network is a single-phase regulator. The voltage regulator has its variable parameters parameterized according to the data provided for the transformer between nodes 650 and 632. The parameters were also readjusted to obtain values close to the IEEE report for the 13-node network.

On the other hand, it is important to note that ANEEL classifies the voltage limits as precarious, critical and appropriate. The classification according to ANEEL for a network with a nominal voltage of 4.16 kV is shown in table I [12].

TABLE I: ANEEL nominal voltage limits at pcc for 4,16 kV

Classification	Voltage Range for Vpcc = 4,16 kV
Proper	0.93 pu <V pcc <1.05 pu
Precarious	0.90 pu <V pcc <0.93 pu
Critical	Vpcc <0.90 pu or Vpcc>1.05 pu

In addition, the DG 634 is in a network in which the voltage is 480, because there is a downstream transformer 4.16 kV / 480 V between nodes 633 and 634 it is necessary to analyze the voltage classification for that value. The table for the voltage at 480 is shown in table II [12].

TABLE II: ANEEL nominal voltage limits at pcc for 480 V

Classification	Voltage Range for Vpcc =480 V
Proper	0.92 pu <V pcc <1.05 pu
Precarious	0.87 pu <V pcc <0.92 pu
Critical	Vpcc <0.87 pu or Vpcc>1.06 pu

On the other hand, the data provided by the IEEE for the 13-node test feeder are of great relevance for analyzing the network voltage profile for the case where there is no distributed generation. The data can be visualized in table III.

TABLE III: IEEE 13 Node Test Feeder

IEEE 13 Node Test Feeder			
With regulator			
Phase	A	B	C
650	1.021	1.042	1.017
632	1.018	1.040	1.015
633	0.994	1.022	0.996
634	0	1.033	1.016
645	0	1.031	1.013
646	0.990	**1.053**	0.978
671	0.990	**1.053**	0.978
680	0.988	0	0.976
684	0	0	0.976
611	0.983	0	0
652	0.990	**1.053**	0.977
692	0.984	**1.055**	0.976
675	0.935	1.008	0.940

According to table III, it is possible to verify that at nodes 671, 652 and 692 the voltage values are exceeding the limit of the range of appropriate voltage values.

This is an interesting factor because the IEEE 13-node network has a voltage regulator and this equipment influences the voltage profile of this distribution network.

For the cases studied, photovoltaic DGs were allocated in the 13 nodes of the network and only two cases of operation were considered. The case of maximum DG operation and the case where the DG is disabled [13].

Based on this, the simulation model was separated into three cases for analysis. The three cases are:

- Case 1: all active DGs;
- Case 2: DGs 646,645,632,634 active;
- Case 3: DGs 611, 684, 671,692,675 active.

Case 1 operates with all active DGs on all nodes. In case 2, the DGs higher and closer to the regulator are in operation and the others are inactive. In case 3, the lower and furthest DGs of the regulator are in operation and the others are inactive.The cases are presented through tables IV, V and VI, respectively.

TABLE IV: Simulation in Case 1

CASE 1						
	Without regulator			With regulator		
Phase	A	B	C	A	B	C
650	1.000	1.000	1.000	1.000	1.000	1.000
632	0.973	0.993	0.982	1.030	1.048	1.029
633	0.972	0.992	0.981	1.029	1.047	1.028
634	0.960	0.982	0.971	1.015	1.035	1.017
645	0	0.991	0.980	0	1.045	1.028
646	0	0.991	0.979	0	1.045	1.027
671	0.948	0.997	0.976	1.003	**1.052**	1.020
680	0.948	0.997	0.976	1.003	**1.052**	1.020
684	0.947	0	0.976	1.002	0	1.021
611	0	0	0.977	0	0	1.021
652	0.946	0	0	1.001	0	0
692	0.948	0.997	0.976	1.003	**1.052**	1.020
675	0.948	0.997	0.976	1.002	**1.053**	1.020

TABLE V: Simulation in Case 2

CASE 2

Phase	Without regulator			With regulator		
	A	B	C	A	B	C
650	1.000	1.000	1.000	1.000	1.000	1.000
632	0.966	0.999	0.954	1.023	**1.053**	1.000
633	0.964	0.998	0.953	1.021	**1.051**	0.999
634	0.953	0.988	0.944	1.008	1.039	0.989
645	0	0.997	0.952	0	1.049	0.999
646	0	0.997	0.951	0	1.049	0.997
671	0.936	1.009	**0.919**	0.991	**1.063**	0.963
680	0.936	1.009	**0.919**	0.991	**1.063**	0.963
684	0.934	0	**0.917**	0.989	0	0.962
611	0	0	**0.915**	0	0	0.960
652	**0.929**	0	0	0.984	0	0
692	0.936	1.009	**0.919**	0.991	**1.063**	0.963
675	0.930	1.011	**0.917**	0.985	**1.065**	0.962

TABLE VI: Simulation in Case 3

CASE 3

Phase	Without regulator			With regulator		
	A	B	C	A	B	C
650	1.000	1.000	1.000	1.000	1.000	1.000
632	0.969	0.982	0.984	1.026	1.037	1.031
633	0.966	0.981	0.981	1.023	1.035	1.028
634	0.943	0.963	0.963	0.999	1.016	1.009
645	0	0.974	0.982	0	1.027	1.029
646	0	0.972	0.980	0	1.026	1.027
671	0.944	0.984	0.978	1.000	1.039	1.021
680	0.944	0.984	0.978	1.000	1.039	1.021
684	0.943	0	0.978	0.999	0	1.022
611	0	0	0.979	0	0	1.022
652	0.943	0	0	0.998	0	0
692	0.944	0.984	0.978	1.000	1.039	1.021
675	0.944	0.985	0.978	0.999	1.041	1.022

V. SIMULATION RESULTS AND DISCUSSION

A. Case 1

In case 1, where all DGs are connected, it is possible to verify, from Table IV, that for the case without voltage regulator, the voltage levels are suitable according to table I and II. The figure 4 shows the voltage profile of phase A.

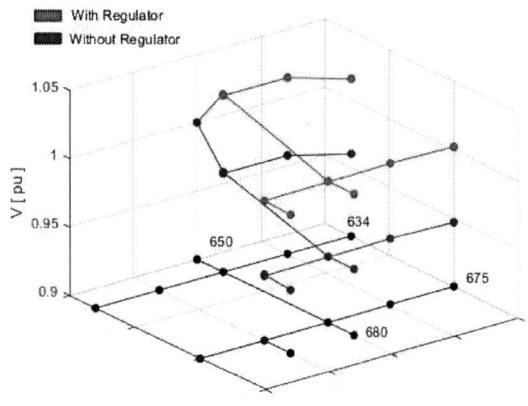

Fig. 3: Voltage profile of phase A in case 1

On the other hand, when the voltage regulator was inserted in the simulation, there was a voltage increase in phase B, and it has been found that the voltage levels in phase B are above range of values, therefore, included in the critical range according to ANEEL. Figure 5 shows the voltage profile in phase B.

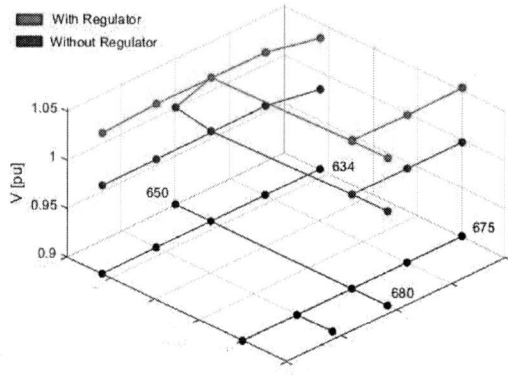

Fig. 4: Voltage profile of phase B in case 1

This fact occurs at nodes 671, 652, 692. It can be seen from table IV that these nodes are further away from the regulator and this factor is relevant in the high voltage value at these nodes.

In phase C, the voltage values increased. However, these values are within the appropriate voltage range. The voltage profile in phase C is shown in figure 6.

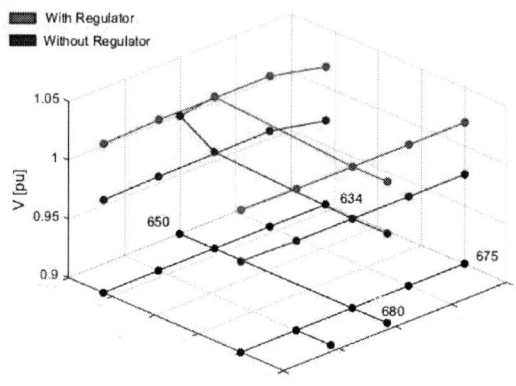

Fig. 5: Voltage profile of phase C in case 1

B. Case 2

In case 2, the simulation model was changed to the case where the upper DGs, which are closer to the regulator, are active. Firstly, for the case without voltage regulator, it is possible to verify, by table IV that phase A has a voltage value at the pre-stressed voltage interval in node 652. In addition, in phases C, there are several values which are within the precarious voltage range.

When the voltage regulator was inserted, it was the voltage phase A values have been improved and are within the range of appropriate voltage values.

The voltage profile in phase A can be seen in figure 7.

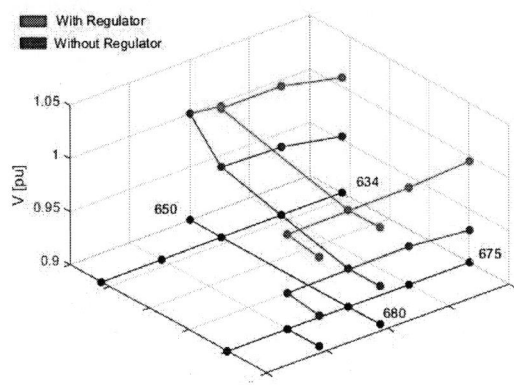

Fig. 6: Voltage profile of phase A in case 2

In phase B, the voltage values without voltage regulator increased in some nodes (632, 633, 671, 680, 692 and 675) and thus remained within the critical voltage range. Is important Note that, unlike Case 1, the voltage values were higher in the case where only the upper DGs are active and as lower DGs are inactive, their voltage values have influence on your voltage profile when inserting the regulator in the simulation. This can be viewed in the figure 8.

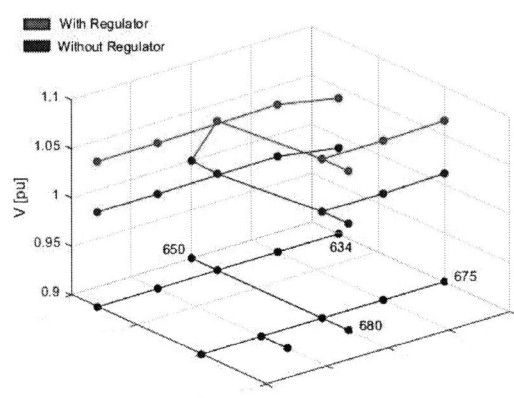

Fig. 7: Voltage profile of phase B in case 2

In Phase C, the voltage values that were in the precarious voltage range were increased and are in the appropriate voltage range. This fact shows the effectiveness of the voltage regulator to realize an improvement in the voltage profile of

this phase. Considering that in phase B the regulator raised the voltage beyond the permitted limits. The figure 9 shows the voltage profile of phase C in case 2.

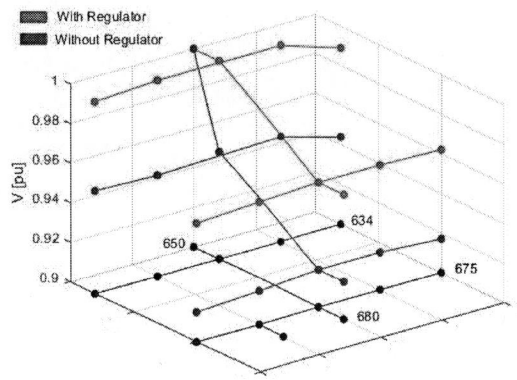

Fig. 8: Voltage profile of phase C in case 2

C. Case 3

In case 3, the simulation model was changed to the lower DGs, which are further away from the regulator, are active. For the case where there is no voltage regulator, it is possible to check in Table VI that all voltage values are within the appropriate voltage ranges.

When a voltage regulator is added in the simulation, it is possible to check an increase in voltage at all phase A. The figure 10 shows voltage profile in phase A for case 3.

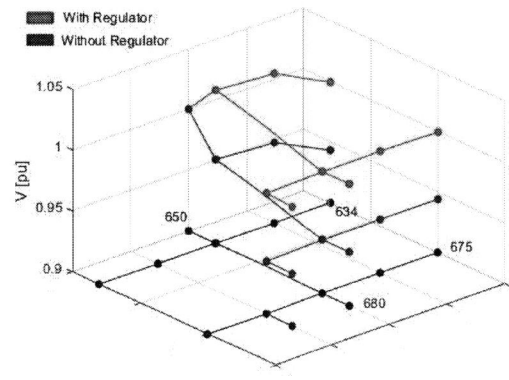

Fig. 9: Voltage profile of phase A in case 3

In addition, it is possible to verify that, unlike case 2, phase B obtained values within the range of appropriate voltage values for all nodes. This fact demonstrates that DG, when active, improves the voltage profile of the network. This can be seen in the figure 11.

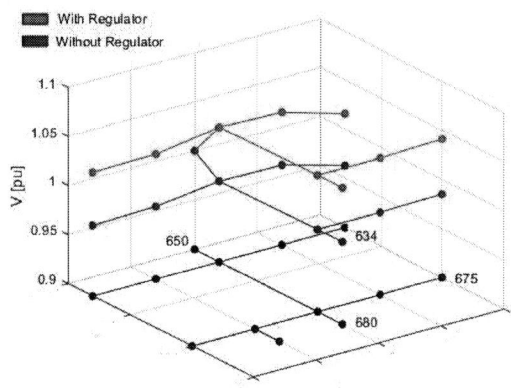

Fig. 10: Voltage profile of phase B in case 3

In phase C, the voltage values are increased and are within the appropriate voltage range. The figure 12 shows the voltage profile in phase C.

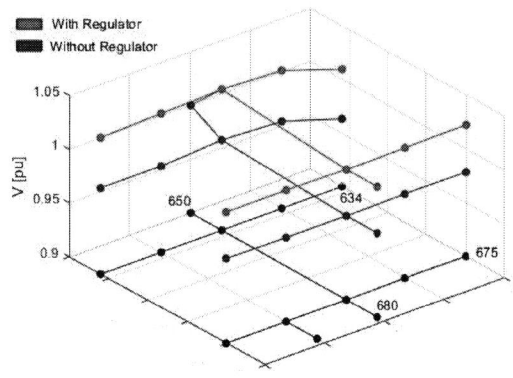

Fig. 11: Voltage profile of phase C in case 3

VI. CONCLUSION

This work presents an analysis on the impact of the voltage regulator (OLTC) in a network with high presence of distributed photovoltaic generation.

In this work, the behavior of the voltage profile in this voltage regulator network was analyzed, since the DGs already contribute to improve the voltage profile, it was extremely important to analyze the real case in which the DGs are connected in the network together with the regulators.

The results were of great importance and showed that distributed photovoltaic generation contributes to improve the voltage profile of the network. This can be seen in the analysis of case 3, where the lower DGs were activated and there was an improvement of the tension profile in comparison to case 2.

Another relevant factor for the high values in phase B is due to the fact that the regulator was parameterized to approximate the values that were supplied to the 13-node network and according to these data, Table III, it is possible to verify that the phase B has values close to the appropriate voltage limit. In addition, installing other voltage regulators may be an important factor to further improve the voltage profile of this network.

Finally, the regulators used in conjunction with distributed photovoltaic generation contribute to improve the voltage profile of the network and can provide efficient network operation.

ACKNOWLEDGEMENTS

This study was financed in part by the Conselho Nacional de Desenvolvimento Científico e Tecnológico - Brazil (CNPq).

REFERENCES

[1] "Unidades cosumidoras com geração distribuída [consumer units with distributed generation],," Brazilian Electricity Regulatory Agency (ANEEL),, 2018. [Online]. Available: http://www2.aneel.gov.br/area.cfm?idArea=757idPerfil=2

[2] S. Kouro, J. I. Leon, D. Vinnikov, and L. G. Franquelo, "Grid- connected photovoltaic systems: An overview of recent research and emerging pv converter technology,," *IEEE Industrial Electronics Magazine*, vol. 9, no. 1, pp. 47–61, 2015.

[3] T. Adefarati and R. Bansal, "Integration of renewable distributed generators into the distribution system: a review," *IET Renewable Power Generation,*, vol. 10, no. 7, pp. 873–884, 2016.

[4] T. Stetz and F. M. an M. Braun, "Improved low voltage grid-integration of photovoltaic systems in germany," *Power and Energy Society General Meeting*, 2013.

[5] J. Cappelle, J. Vanalme, S. Vispoel, T. V. Maerhem, B. Verhelst, C. Debruyne, and J. Desmet, "Introducing small storage capacity at residential pv installations to prevent overvoltages," *Smart Grid Communications (SmartGridComm) - IEEE International Conference*, 2011.

[6] R. Walling, R. Saint, R. Dugan, J. Burke, and L. Kojovic, "Summary of distributed resources impact on power delivery systems," *IEEE Transactions on Power Delivery*, vol. 10, no. 7, pp. 873–884, 2016.

[7] G. Pepermans, J. Driesen, D. Haeseldonckx, R. Belmans, and W. D'haeseleer, "Distributed generation: definition, benefits and issues," *Energy Policy*, vol. 33, no. 6, pp. 787–798, 2005.

[8] P. Barker and R. D. Mello, "Determining the impact of distributed generation on power systems. i. radial distribution systems," *in 2000 Power Engineering Society Summer Meeting*, vol. 3, pp. 1645–1656, 2000.

[9] K. Balamurugan, D. Srinivasan, and T. Reind, "Impact of distributed generation on power distribution systems," *Energy Procedia*, vol. 25, pp. 93–100, 2012.

[10] M. C.Vargas, M. A. Mendes, and O. E. Batista, "Impacts of high pv penetration on voltage profile of distribution feeders under brazilian electricity regulation," *13th IEEE International Conference on Industry Applications*, 2018.

[11] "Procedimentos de rede (prorede) - submódulo 3.6 - requisitos técnicos mínimos para a conexão às instalações de transmissão [network procedures (prorede) - submodule 3.6 - minimum technical requirements for connection to transmission facilities],," Brazilian National Electrical System Operator (ONS). [Online]. Available: http://ons.org.br/pt/paginas/sobre-o-ons/ procedimentos-de-rede/vigentes

[12] "Procedimentos de distribuição de energia elétrica no sistema elétrico nacional (prodist) - módulo 8: Qualidade da energia elétrica [procedures for power delivery in the national electric system (prodist) - module 8: Electric power quality]," Brazilian Electricity Regulatory Agency (ANEEL), 2018. [Online]. Available: http://www.aneel.gov.br/documents/656827/14866914

[13] M. A. Mendes, "Análise dos impactos da alta inserção de geração distribuída fotovoltaica na proteção de sobrecorrente temporizada," Master's thesis, Universidade Federal do Espírito Santo, Espírito Santo - Brasil, 2018.

978-1-7281-4181-7/19 $31.00 © 2019 IEEE

An Application of the Multi-Port Bidirectional Three-Phase AC-DC Converter in Electric Vehicle Charging Station Microgrid

Raphael A. da Câmara
Control and Energy Processing Group
Dept. of Electrical Engineering
Federal University of Ceara
Fortaleza, Brazil
raphael@dee.ufc.br

Luis M. Fernández-Ramírez
Department of Electrical Engineering
Higher Polytechnic School of Algeciras
University of Cádiz
Algeciras, Spain
luis.fernandez@uca.es

Paulo P. Praça
Control and Energy Processing Group
Dept. of Electrical Engineering
Federal University of Ceara
Fortaleza, Brazil
paulopp@dee.ufc.br

Demercil de S. Oliveira Jr.
Control and Energy Processing Group
Dept. of Electrical Engineering
Federal University of Ceara
Fortaleza, Brazil
demercil@dee.ufc.br

Pablo García-Triviño
Department of Electrical Engineering
Higher Polytechnic School of Algeciras
University of Cádiz
Algeciras, Spain
pablo.garcia@uca.es

Raúl Sarrias-Mena
Department of Engineering in
Automation, Electronics and Computer
Architecture & Networks
Higher Polytechnic School of Algeciras
University of Cádiz
Algeciras, Spain
raul.sarrias@uca.es

Abstract—**This paper presents an application of the multi-port bidirectional three-phase ac-dc converter as interface between a microgrid composed by several power sources and an electric vehicle charging station (EVCS). The main advantage of using this converter is that it can integrate multiple power sources and loads into a single power conversion stage and thus control the power flow between them reducing the number of power conversion stages and / or devices as well as weight and volume of the entire system and the control architecture does not require communication strucure as main current solutions in this field present. The microgrid of this study was composed of a photovoltaic system, a battery energy storage system, two 48 kW fast charging units for electricle vehicles, a connection to the local grid and the multi-port bidirectional converter with minor changes to be able in this type of application with a rated power at 100 kW. Simulation results obtained from a system model are presented and discussed in order to validate that under different power sources conditions the converter operates effectively confirming the feasibility of using this type of application in EVCS microgrid technology.**

Keywords— *ac-dc power conversion, charging station, electric vehicles, microgrid, single-stage topologies.*

I. Introduction

Fossil fuels such as oil, natural gas and coal are the main energy supplier of the world economy, especially in the transport sector [1, 2]. But it is aware of all that the environmental issues caused by these energy sources, such as climate changes and CO_2 and NO_x emissions, for example. For this reason, it makes humanity look for energy resources that are less polluting to the environment. For the transport sector, the promotion of plug-in hybrid electric vehicles and electric vehicles (EVs) are a feasible solution to deal this problem [3].

The electric vehicle charging stations (EVCSs) nowadays are not generally avaible and must be distributed in the cities, equivalent to petrol stations, to supply a fast charge to the batteries of the EVs, which implies a high-power demand. The local electrical grid must be capable of supporting this high-power demand and the impacts related to the connection of dc systems (harmonics, voltage outages and fluctuations) [4] - [6]. Also this high-power demand in the mains affects electricity market operations and planning.

Basically, the structure of these EVCSs present two types of topologies. One based on a common ac bus feeding ac-dc EV chargers and another one based on a common dc bus feeding dc-dc chargers. Both type of topologies are regulated [7, 8].

However, considering that the current structure of the local ac grid would take a long time and a considerable cost, and considering the interest in clean technologies, it leads to the further development of power distribution systems and smart grids that use renewable power sources [9, 10]. In addition, the increasing penetration of distributed generation (DG) systems at the consumption level provides a distribution of dc power with important advantages over its ac counterpart, since photovoltaic (PV) panels, battery banks as energy storage system (ESS), fuel cells and even wind turbines based on permanent magnet synchronous generators can be easily connected to a dc grid [11, 12]. In this context, this type of DG system or microgrid with dc power is very interesting for EVCSs dc bus applications.

On the other hand, if dc grids become more accessible, more common electronic loads can easily adapt. In this case, there are numerous options for the ac-dc conversion through the power electronics [13]. This fact can also lead to the reduction in the number of conversion stages that exist in the chargers of electric vehicles. In this case, ac-dc converters with bidirectional characteristics would be essential in this process [14] - [16].

The main current solutions in dc power EVCSs using microgrid have in their basic topology a common dc bus where the power DG sources, the main grid and the EVCS are connected [3], [17] - [21]. The interface between the dc bus and each microgrid component is through different and separated individual power converters, which are not isolated and / or with low frequency isolation, making the volume and weight of the microgrid system higher, several power conversion stages are necessary as well as centralized control architecture with requiring communication structure.

The proposal of this paper is presents an application of the multi-port birectional three-phase ac-dc converter [22], [23] with a rated power at 100 kW as interface between a microgrid based on [24] composed of the PV panels, a Li-ion battery as

978-1-7281-4181-7/19 $31.00 © 2019 IEEE

energy storage system (BESS), a connection to the mains, and two fast charging units for EVCS. For this microgrid application the multi-port converter will undergo minor changes to adapt to the EVCS requirements. The main advantage on this application is connect several power sources using a single power conversion stage and control the power flow between them reducing the number of power conversion stages and / or devices (electronics and passive components), volume and weight of the entire system and control architecture with no requiring communication structure. Simulation results showed that under different power sources conditions the converter operated effectively, which confirms the feasibility of using this application in microgrid of dc power EVCS technology.

II. Description of the Proposed EVCS Microgrid

The topology of the multi-port converter applied in this proposed EVCS microgrid is presented in Fig. 1 with its ports and external converters to connect the EVCSs highlighted.

The configuration of the proposed EVCS microgrid is shown in Fig. 2, where it is observed that each port is connected to a given power source/load. It consists of a connection to the local grid (Port I), two fast charging units with external dc-dc buck converters (Port II), a PV system (Port III), and a Li-ion BESS (Port IV). Comparing with the basic topology within a common dc bus, the proposed system has much less power conversion stages and a smaller number of components involved which are the main goals of this study.

At the mains (Port I) the converter controller is able to manage the power flow between the sources and the grid. Thus, depending of the power generated by PV system and / or the level of the BESS state-of-charge (SOC), the converter works as a rectifier delivering power from the grid or as an inverter receiving power to the grid with a high power factor and a low THD.

The two fast charging units have a rated power of 48 kW each and belong to Mode 4, dc level 2, according to the IEC 61851-1 with the follow characteristics: fast charging with an external charger in dc; voltage inferior to 500 V, and current inferior to 200 A [25]. The external dc-dc buck converter control the charging of the EV while the converter controller is supose to keep constat the voltage in the Port II. It is

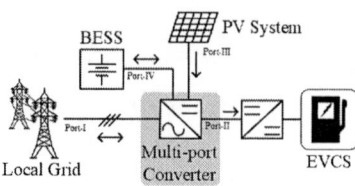

Fig. 2. Example of a typical microgrid with EVCS application.

observed that there is no galvanic isolation between mains and EVCS despite being recommended [26, 27]. However, initially it is intended to maintain this configuration and a galvanic isolation will be provided later in the next steps of the study.

The PV system must be able to deliver at least the rated power of EVCS without another power source involved. Thus, considering the selected PV system [28], the peak power is 119.7 kWp composed by an array of 7 modules connected in series (207.2 V) with 76 parallel branches totaling 532 modules. The converter controller is supose to keep constant the voltage in the Port III tracking the maximum power point (MPP) of the PV panels power generation.

Considering the possibility of fully charge an EVCS and taking into account the fast charging unit maximum continuous current (during the charging of the EV, the charge current is 120 A in the current contant mode), the selected BESS [29] has a rated capacity of 24.57 kWh composed by an array of 2 batteries connected in series (96 V) with 6 parallel branches totaling 12 batteries with a maximum continuous current of 150 A. The converter controller will charging and discharging the BESS according of its SOC.

III. Multi-port Converter Control Strategy

The schematic of the multi-port converter control strategy is shown in Fig. 3 [23]. The converter primary side control is based on the dq theory [30 - 33]. For power factor correction (PFC), the reference current i_{d_ref} is grounded. The compensators $C_1(s)$ and $C_2(s)$ have the same components values and are responsible for the current control where the reference current i_{q_ref} comes from Port II voltage loop control by compensator $C_3(s)$. The primary side modulating signals (m_a, m_b, and m_c) are obtained from inverse Park transform (dq—abc) and applied to modulator generating the primary side switches signals.

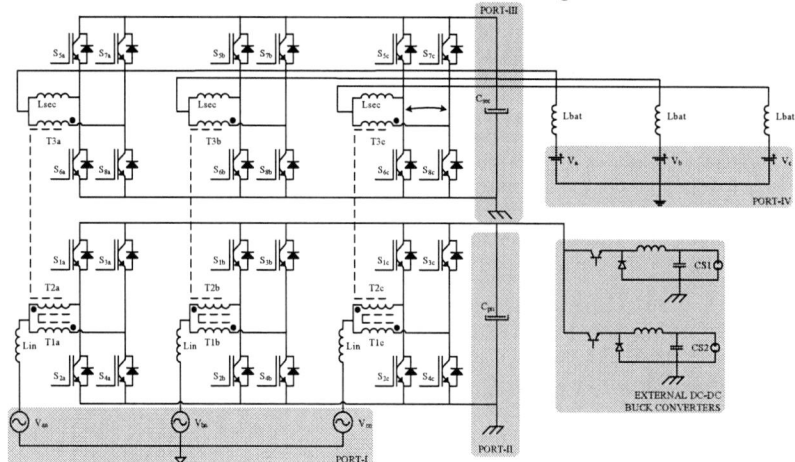

Fig. 1. Topology of the multi-port converter applied in this EVCS microgrid proposal.

978-1-7281-4181-7/19 $31.00 © 2019 IEEE

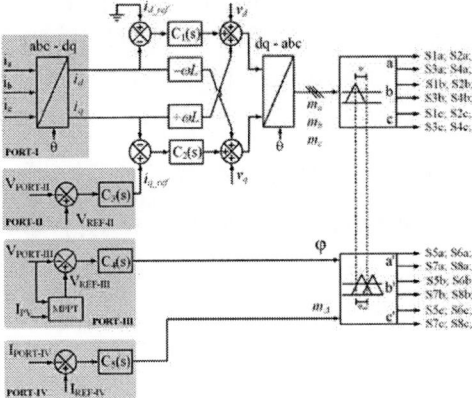

Fig. 3. Control strategy schematic.

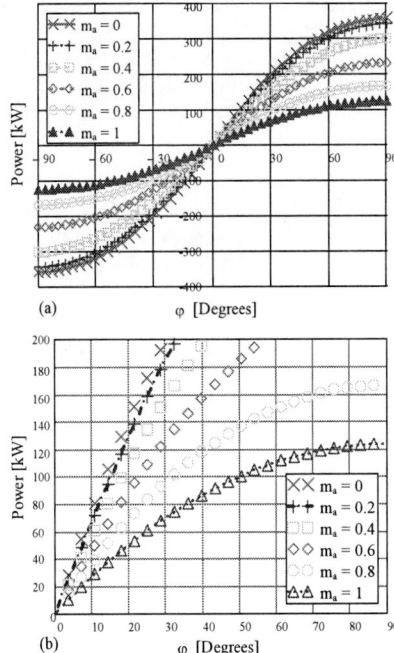

Fig. 4. Ouput characteristic curves.

The power flow on the multi-port converter, i.e., if the converter operates as a rectifier or as an inverter is achieved controlling the angle between primary and secondary windings voltages. This control is known as phase-shift technique, defined as angle φ and calculated by compensator $C_4(s)$, which controls the voltage across the Port III. If the angle is positive the converter operates as rectifier, otherwise operates as inverter automatically. The Port III voltage reference $V_{REF-III}$ comes from the PV system MPPT control.

The Port IV current is controlled by compensator $C_5(s)$, where depending of BESS SOC, the reference current I_{REF-IV} is positive to discharging and negative to charging the BESS. This current reference can be defined by an upper-level energy management system (EMS) that coordinates the operation of all the elements in the microgrid. This EMS can perform a battery SOC regulation by defining the adequate I_{REF-IV}. The resultant variable of $C_5(s)$ defines the value of the duty cycle in the switches on the secondary side. At last, the synchronism angle θ is obtained by the q-PLL circuit.

IV. MULTI-PORT CONVERTER DESIGN

The multi-port converter has 16 possible operating regions that depending on the phase-shift angle φ and the duty cycle [34]. Due to the sinusoidal variation of the duty cycle, the phase-shift angle can be either positive and negative.

Considering the existing symmetry and operation within the range ± 90°, the converter analysis was made in [22] and the output power (P_o) is given by the equation (1), where V_p and V_{ph} are defined in equations (2) and (3), respectively, where f_s is the switching frequency, and m_a is the modulation index for primary side of transformer.

$$P_o(\varphi) = 3 \cdot \left(V_p^2 + V_{ph}^2/2\right) \cdot [1/(4 \cdot \pi \cdot f_s \cdot L_{sec}) \cdot \sin(\varphi)] \quad (1)$$

$$V_p = \left(2 \cdot V_{port-I}/\pi\right) \cdot \{1 + sin[(\pi/2) \cdot (1 + m_a)]\} \quad (2)$$

$$V_{ph} = \left(2 \cdot V_{port-I}/\pi\right) \cdot \{1 - sin[(\pi/2) \cdot (1 + m_a)]\} \quad (3)$$

From equation (1) is possible to determine throughout the output characteristic curves for some modulation index (see Fig. 4(a)) the phase-shift angle for the converter rated power, and from that one, determine the power transfer inductor L_{sec} to achieve this rated power (see Fig. 4(b)). Thus, with a modulation index around 0.8 and a rated power of 100 kW, the phase-shift angle is equal to 30° and the chosen L_{sec} is equal to 280 nH to reach that power.

The EVCS microgrid main parameters are presented in Table I. For all multi-port converter controllers design were used the k-factor method, compensator type 2 [35], and a chosen phase margin equal to 60°.

For control the external dc-dc buck converter is used the well-know average voltage and current mode control [36, 37]. The multi-port converter controllers design are based on each port transfer function [23] and described as follows.

A. Port I

The Port I controller is a current loop control based on dq theory as mentioned before. The transfer function that relates dq currents with duty cycle is given by (4).

$$G_{i_{dq}}(s) = V_{PortII}/(s \cdot L_{in} + r_{Lin}) \quad (4)$$

TABLE I.   CONVERTER PARAMETERS

Parameters	Value
Fast charging units	2x48 kW
PV system	119.7 kWp
BESS system	24.57 kWh
Multi-Port Converter Output Power	100 kW
Port I rms voltage	380 Vac
Mains frequency	60 Hz
Port II voltage	700 Vdc
Port III voltage	200 Vdc
Port IV voltage	100 Vdc
Switching frequency	50 kHz
Primary inductor per phase	0.5 mH
Secondary inductor per phase	280 nH
Port IV inductor per phase	10 mH
Port II capacitor	1 mF
Port III capacitor	1 mF
Transformer turns ratio	0.2971
External buck output current	120 A

978-1-7281-4181-7/19 $31.00 © 2019 IEEE

Where L_{in} and r_{Lin} are the inductance and resistance values at this port respectively. Thus, designing a compensator type 2 for this controller, the Bode diagrams for the open loop and closed loop transfer functions are shown in Fig. 5 (a) where it's observed a cross frequency of 12.5 kHz, a fast current loop design.

B. Port II

The Port II controller is a voltage loop control where the simplified transfer function relates the Port II voltage with the Port II output current and it is given by (5), where C_{pri} is the value of the capacitor at Port II..

$$G_{v_{portII}}(s) = 1/(s \cdot C_{pri}) \qquad (5)$$

The open loop and closed loop transfer functions Bode diagrams are shown in Fig. 5 (b), where cross frequency was 36 Hz, too much slower than current loop.

C. Port III

For the Port III controller is considered the phase-shift effect. The plant transfer function that relates the Port III voltage and the phase-shift is given by (6), where C_{sec} is the value of the capacitor and Ro_s is the value of output rated load at Port III.

$$G_{v_{portIII}}(s) = (\varphi - |\varphi|^2/\pi) \cdot V_{portIII}/(2 \cdot \pi \cdot f_s \cdot L_{sec}) \cdot Ro_s/(s \cdot C_{sec} \cdot Ro_s + 1) \qquad (6)$$

The Bode diagrams for the open loop and closed loop transfer functions are shown in Fig. 6 (a). Due to the algorithm for MPPT, a cross frequency of 500 Hz was chosen.

D. Port IV

Finally, for the Port IV controller the current loop control is modelled as same as the Port I current loop control changing

only the Port IV voltage and the inductance and resistance values (L_{bat} and r_{Lbat} respectively). The transfer function is given by (7).

$$G_{i_{portIV}}(s) = V_{portIV}/(s \cdot L_{bat} + r_{Lbat}) \qquad (7)$$

The Bode diagrams for the open loop and closed loop transfer functions are shown in Fig. 6 (b). The chosen cross frequency also is the same of Port I current loop control, 12.5 kHz.

E. MPPT Controller

The PV system is connected in Port III where to achieve the maximum panels power generation, a MPPT algorithm is implemented. In this study the adopted method was the well-know disturb and observe algorithm [38].

V. SIMULATION RESULTS

To validate the application of the multi-port converter on the EVCS microgrid, simulation results were obtained under different power sources conditions emulating possible real situations using PSIM® software. The proposed situations are shown in Fig. 7 and described as follows:

a) Simulation mode I: The mains is responsible to charge the EVs and the BESS, since the PV panels do not provide any power. This is a common situation happening at night. This case study starts with only one EV connected and a second one EV is plugged-in (see Fig. 7 (a)).

b) Simulation mode II: No EV are connected in this case study, and the PV system and BESS are providing power to the mains (see Fig. 7 (b)). This possibility is the most often where the microgrid contributes to the mains demand.

The availability of energy storage devices can allow soft mode transitions if the energy flows are modified progressively instead of instantaneously. Moreover, the

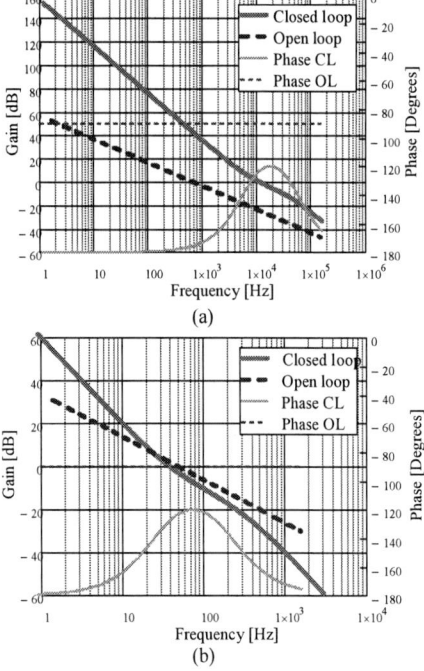

Fig. 5. The Bode diagrams for Port I and II controllers.

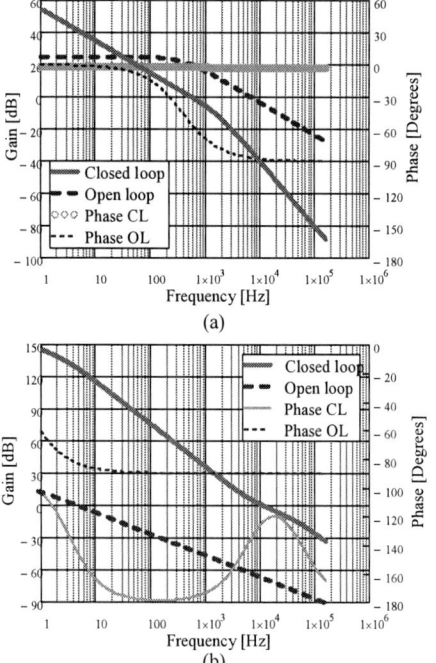

Fig. 6. Bode diagrams for Port III and IV controllers.

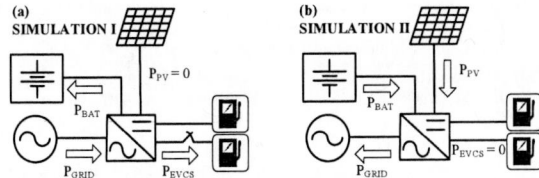

Fig. 7. Simulation modes of operation.

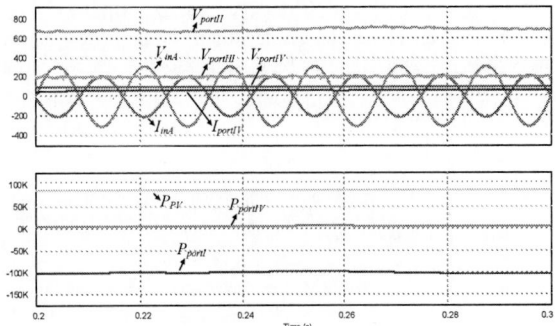

Fig. 9. Simulation results for mode II.

contribution of the PV panels also changes progressively as the sun radiation increases and decreases every day. Therefore, this parameter is not expected to change abruptly, easing a soft transition between different operation cases.

Several options of situations for run simulation would be possible here but these two contemplate the most relevant characteristics of the converter and the microgrid.

For the simulation mode I, the multi-port converter performance is presented in Fig. 8 where it can be seen the system operating as a rectifier with a low THD input current (about 1.394 %) and a high power factor (around 0.99996), in accordance with the requirements of the IEC 61000-3-2 standard [39]. It can also be seen a load step change at 150 ms and the subsequent response of the controller. As seen, the current loop control responds in 30.7 ms and the Port II voltage in 72.9 ms, thus proving that the current loop is faster than voltage loop according to the controllers designed. Futhermore, a negative current in the BESS indicates that this device is being charged in this mode. It can be noticed that despite the decrease in the Port II dc bus voltage during the load step, the output voltages of the external buck converters remain constant in the EV charging port, showing that a variation on the input voltage does not imply a change on its output voltage.

For the simulation mode II, the behavior of the multi-port converter is shown in Fig. 9. It can be noticed that the converter is operating as an inverter injecting power into the mains. Voltages at all ports comply with the expected behavior, and the positive current at Port IV indicates the discharge of the BESS, which provides about 6.5 kW. The PV system is supplying approximately 90 kW, which sums the system rated power (100 kW), as shown at the bottom of the Fig. 9, where the negative input power means that power is being delivered to the mains.

VI. CONCLUSIONS AND FUTURE WORK

This paper has presented an application of the multi-port bidirectional three-phase ac-dc converter in electric vehicle charging station microgrid. The main converter advantage in this type of application is that it can integrate multiple power sources and loads into a single power conversion stage reducing the number of power conversion stages as well as devices compared to main solutions in this field. Simulation results obtained from a microgrid composed of a PV system, BESS, local grid, EVCS and a 100 kW multi-port converter rated power under different power sources conditions validate the converter operation, where simulation waveforms have proven the effectiveness of the control scheme during transients confirming the feasibility of using this converter in EVCS microgrid applications.

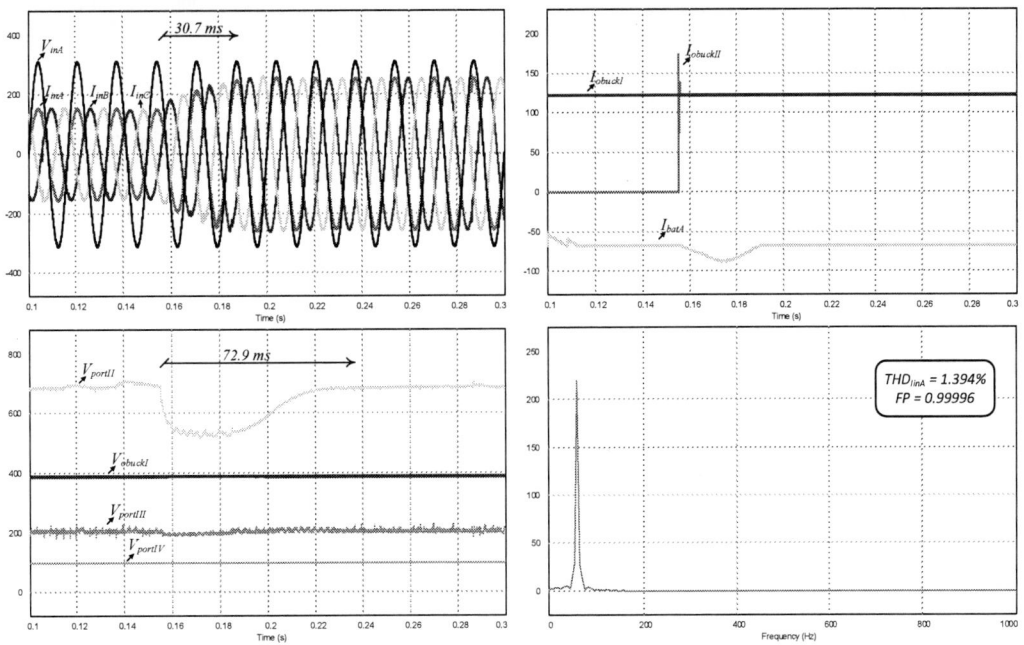

Fig. 8. Simulation results for simulation mode I.

A supervisory EMS for this EVCS microgrid is currently being developed based on a descentralized fuzzy logic control as well as galvanic isolation for external dc-dc buck converters and another improvements for futher results that will be presented in future papers.

REFERENCES

[1] U.S. Energy Information Administration (EIA), "Annual Energy Outlook 2019 with projections to 2050", 2019. [Online]. Disponible: www.eia.gov/aeo.

[2] International Energy Agency (IEA), "2018 Key World Energy Statistics", 2019. [Online]. Disponible: www.iea.org/statistics/kwes.

[3] P. Goli, W. Shireen, "PV powered smart charging station for PHEVs", Renew: Energy, vol. 66, pp. 280-287, June 2014.

[4] L. Kütt, E. Saarijärvi, H. Mõlder, J. Niitsoo, and M. Lehtonen, "A review of the harmonic and unbalance effects in electrical distribution networks due to EV charging," in Proc. EEEIC, Wroclaw, Poland, May 2013, pp. 556–561.

[5] K. Clement-Nyns, E. Haesen, and J. Driesen, "The impact of charging plug-in hybrid electric vehicles on a residential distribution grid," IEEE Trans. Power Syst., vol. 25, no. 1, pp. 371–380, February 2010.

[6] L. Pieltain Fernández, T. Gómez San Román, R. Cossent, C. Mateo Domingo, and P. Frías, "Assessment of the impact of plug-in electric vehicles on distribution networks," IEEE Trans. Power Syst., vol. 26, no. 1, pp. 206–213, February 2011.

[7] J. Ying, V. K. Ramachandaramurthy, K. Miao and N. Mithulananthan, "Bi-directional electric vehicle fast charging station with novel reactive power compensation for voltage regulation," Int. J. Electr. Power Energy Syst. , vol. 64, pp. 300-310, January 2015

[8] M. Yilmaz and P. T. Krein, "Review of charging power levels and infrastructure for plug-in electric and hybrid vehicles," Proc. IEEE Int. Electric Vehicle Conf., 2012, pp. 1-8.

[9] R. Lawrence and S. Middlekauff, "The New Guy on the Block", IEEE Industry Applications Magazine, vol. 11, nº 1, pp. 54-59, Jan./Feb. 2005.

[10] F. Li et al, "Smart Transmission Grid: Vision and Framework", IEEE Transactions on Smart Grid, vol. 1, nº 2, pp. 168-177, September 2010.

[11] E. Rodriguez-Diaz, F. Chen, J. C. Vasquez, J. M. Guerrero, R. Burgos, and D. Boroyevich, "Voltage-level selection of future two-level LVdc distribution grids: A compromise between grid compatibiliy, safety, and efficiency," IEEE Electrification Magaz., vol. 4, no. 2, pp. 20–28, June 2016.

[12] A. Jhunjhunwala, A. Lolla, and P. Kaur, "Solar-dc microgrid for Indian homes: A transforming power scenario," IEEE Electrification Mag., vol. 4, no. 2, pp. 10–19, June 2016.

[13] P. Wang, L. Goel, X. Liu, and F. H. Choo, "Harmonizing AC and DC: A hybrid AC/DC future grid solution," IEEE Power Energy Mag., vol. 11, no. 3, pp. 76–83, May/Jun. 2013.

[14] B. Nordman and K. Christensen, "DC local power distribution: Technology, deployment, and pathways to success," IEEE Electrification Mag., vol. 4, no. 2, pp. 29–36, June 2016.

[15] Z. Weichao, L. Haifeng, B. Zhou, L. Wei and G. Ran, "Review of DC technology in future smart distribution grid," in Proc. IEEE PES Innov. Smart Grid Technol., Tianjin, China, 2012, pp. 1–4.

[16] S. Grillo, V. Musolino, L. Piegari, E. Tironi, and C. Tornelli, "DC islands in AC smart grids," IEEE Trans. Power Electron., vol. 29, no. 1, pp. 89–98, January 2014.

[17] G. Preetham and W. Shireen, "Photovoltaic charging station for plug-in hybrid electric vehicles in a smart grid environment," in Proc. IEEE PES Innov. Smart Grid Technol. (ISGT), Jan. 2012, pp. 1–8.

[18] N. Liu, Q. Chen, X. Lu, J. Liu, and J. Zhang, "A charging strategy for PV-based battery switch stations considering service availability and self-consumption of PV energy," IEEE Trans. Ind. Electron., vol. 62, no. 8, pp. 4878–4889, August 2015.

[19] T. Dragičević, S. Sučić, J. C. Vasquez, and J. M. Guerrero, "Flywheelbased distributed bus signalling strategy for the public fast charging station," IEEE Trans. Smart Grid, vol. 5, no. 6, pp. 2825–2835, November 2014.

[20] Y. Liu, Y. Tang, J. Shi, X. Shi, J. Deng, and K. Gong, "Application of small-sized SMES in an EV charging station with DC bus and PV system," IEEE Trans. Appl. Supercond., vol. 25, no. 3, Jun. 2015, Art. no. 5700406.

[21] M. O. Badawy and Y. Sozer, "Power flow management of a grid tied PV-battery powered fast electric vehicle charging station," in Proc. IEEE ECCE, Montreal, QC, Canada, Sep. 2015, pp. 4959–4966.

[22] B. R. de Almeida, J. W. M. de Araújo, P. P. Praça and D. S. Oliveira Jr., "A Single-stage three-phase bidirectional AC/DC converter with high-frequency isolation and PFC", IEEE Trans. on Power Electronics, vol. 33, no. 10, pp. 8298-8307, October 2018.

[23] A. U. Barbosa, B. R. de Almeida, D. S. Oliveira Jr., P. P. Praça and L. H. S. C. Barreto, "Multi-port didirectional three-phase AC-DC converter with high frequency isolation," Proc. IEEE Applied Power Electron. Conf. and Expo., 2018, pp. 1386-1391.

[24] P. García-Triviño, J. P. Torreglosa, L. M. Fernández-Ramírez and F. Jurado, "Decentralized Fuzzy Logic Control of Microgrid for Electric Vehicle Charging Station," IEEE Journal of Emerging and Selected Topics in Power Electronics, vol. 6, no. 2, pp. 726-737, June 2018.

[25] International Electrotechnical Commision. IEC 61851-1 Electric vehicle conductive charging system – Part 1 General requirements, 2010.

[26] CHAdeMO Association. Technology Overview. [Online]. Disponible: https://www.chademo.com/technology/technology-overview/ (accessed June 12, 2019).

[27] K. Stengert, "On-board 22 kW fast charger "NLG6"," Proc. Int. Battery, Hybrid and Fuel Cell Electric Vehicle Symposium, 2013, pp. 1-11.

[28] Suntech oiwer, Suntech STP225-20/Wd datasheet. [Online]. Disponible:http://www.pannellisolari.bologna.it/files/Suntech_STP225-20_Wd_225Wp_EN[1].pdf (accessed June 12, 2019).

[29] Discover Energy Copr. 13-48-2000 datasheet. [Online]. Disponible: https://discoverbattery.com/products/lifepo4-batteries-aes/ (accessed June 12, 2019).

[30] E. H. Watanabe, R. M. Stephan and M. Aredes, "New concepts of instantaneous active and reactive powers in electrical systems with generic loads," Power Delivery, IEEE Transactions on, vol. 8, pp. 697-703, 1993.

[31] M. Aredes, H. Akagi, E. H. Watanabe, E. Vergara Salgado and L. F. Encarnacao, "Comparisons Between the p--q and p--q--r Theories in Three-Phase Four- Wire Systems," Power Electronics, IEEE Transactions on, vol. 24, pp. 924-933, 2009.

[32] Sasso, E. M., Sotelo, G., Ferreira, A., Watanabe, E. H., Aredes, M., Barbosa, P. G. "Investigação dos Modelos de Circuitos de Sincronismo Trifásicos Baseados na Teoria das Potências Real e Imaginária Instantâneas (p-PLL e q-PLL)", in Proceedings of the 14th Brazilian Automatic Control Conference, pp. 480-485 (in Portuguese), 2002.

[33] H. Guan-Chyun, J. C. Hung, "Phase-Locked Loop Techniques. A Survey", IEEE Transactions on Industrial Electronics, vol. 43, pp. 609–615, 1996.

[34] L. C. S. Mazza, et al., "A Soft Switching Bidirectional DC-DC Converter with High Frequency Isolation Feasible to Photovoltaic System Applications," Proc. PCIM Europe 2015 International Exhibition and Conf. for Power Electronics, Intelligent Motion, Renewable Energy and Energy Management, 2015. p.1-8.

[35] H. D. Venable, The k-factor: A New Mathematical Tool for Stability Analysis and Synthesis. Powercon, 1983.

[36] V. Vorperian, "Simplified analysis of PWM converters using model of PWM switch. Continuous conduction mode," IEEE Trans. on Aerospace and Electr. Syst., vol. 26, pp. 490-496, May 1990.

[37] A. J. Forsyth and S. V. Mollov, "Modelling and control of DC-DC converters," Power Engineering Journal, vol. 12, pp. 229-236, October 1998.

[38] A. M. Bhandare, P. J.Bandekar and S. S. Mane, "Wind energy maximum power extraction algorithms: A review," Proc. IEEE Intern. Conf. on Energy Efficient Technologies for Sustainability, 2013.

[39] IEC, "IEC 61000-3-2: Electromagnetic Compatibility (EMC) – Part 3: Limits – Section 2: Limits for Harmonic Current Emissions (Equipment input current < 16 A per phase)," vol. Emenda A14, International Electrotechnical Commission Ed., 2001.

978-1-7281-4181-7/19 $31.00 © 2019 IEEE

Unified Robust Control Design for BTB-VSC Subject to Uncertainties in Grid Equivalent Circuit

1st Marcelo de Castro
Dept. of Electrical, Computer and Systems Engineering
Rensselaer Polytechnic Institute
Troy, USA
Dept. of Electrical Engineering
Federal University of Juiz de Fora
Juiz de Fora, Brazil
decasm3@rpi.edu

2nd Igor D. N. de Souza
Dept. of Electrical Engineering
Federal University of Ouro Preto
João Monlevade, Brazil
Dept. of Electrical Engineering
Federal University of Juiz de Fora
Juiz de Fora, Brazil

3rd Gabriel A. Fogli
Dept. of Electronic Engineering
Federal University of Minas Gerais
Belo Horizonte, Brazil

4th Ademir da S. T. Junior, 5th Pedro M. de Almeida
6th Janaína G. de Oliveira, 7th Pedro G. Barbosa
Dept. of Electrical Engineering
Federal University of Juiz de Fora
Juiz de Fora, Brazil

Abstract—This paper presents the design of a unified model and control system for a voltage source converter (VSC)-based Back-To-Back (BTB). The controller based on robust pole placement is designed taking into account that the BTB converter deals with bidirectional power flow and uncertainties related to grid equivalent resistance and inductance at the Point of Common Coupling (PCC). Non-linear and linearized models for the BTB converter along with the design of a state feedback control are presented in a state-space formulation. A BTB-VSC system is assembled and tested in a digital environment. Results for simulations performed are presented showing that the controller can meet desired performance requirements for bidirectional power flow and for different grid equivalent impedance values.

Index Terms—Back-to-Back, Voltage Source Converter, Robust Control, Multi-variable Control.

I. Introduction

Power electronics devices have major importance in the current power system. Many solutions proposed to enhance the power system are based on controlled converters that are designed in order to make the grid more resilient and robust. Among these converters, the Back-To-Back (BTB) is the most common topology used in industry to perform low-power and low-voltage connection between two AC systems [1].

The usage of BTB is increasing due to some particular characteristics such as: (i) the provision of wide control in bidirectional power flow and (ii) independent control of active and reactive power allowing a flexible operation [1]. Besides, these two characteristics combined lead to the interesting property of rapid manipulation of power flows in order to provide the balance between two different systems [2].

The Back-to-Back based on Voltage Source Converter (BTB-VSC) consists in two VSCs sharing a common link on the DC side while performing the interconnection of two different AC systems. The BTB-VSC is frequently used in

Unified Power Quality Conditioners (UPQC) and in applications related to renewable energy generation such as wind generators connected to the grid [3], [4].

It is evident that the control approach is essential in order to take advantage of the important properties associated with the BTB-VSC. The conventional control strategy designed for BTB converters is based on two loops: the inner and the outer controllers [2]. The inner controller usually deals with the AC current injected by each VSC while the outer controller addresses the DC link capacitor voltage. In order to this design to work properly and the controllers to be designed independently, the inner loop may have a much faster time constant than the outer loop [5]. This conventional configuration of two control loops is shown in [6] to perform the control of a BTB converter dealing with bidirectional power flow.

Although the conventional control approach is well established, it is rather usual to see alternative control strategies being proposed in the literature. An illustration of alternative control strategy is the full state-feedback. In [2] a unified multi-variable control strategy is proposed to control a BTB converter with LCL output filters. The proposed state feedback strategy is found to deal very well with transient mismatches of power flow and to outperform the traditional approach. In [7] a robust control strategy is proposed to a grid-connected inverter. Results show that the controller ensure a stable operation and good performance for a set of uncertainties related to the grid equivalent inductance and resistance at the Point of Common Coupling (PCC). In the same context, the present paper aims to propose a multi-variable control, based on robust pole placement [8], [9], to a BTB-VSC dealing with uncertainties related to the grid equivalent impedance at the PCC.

The rest of this paper is organized in the following manner.

978-1-7281-4181-7/19 $31.00 © 2019 IEEE

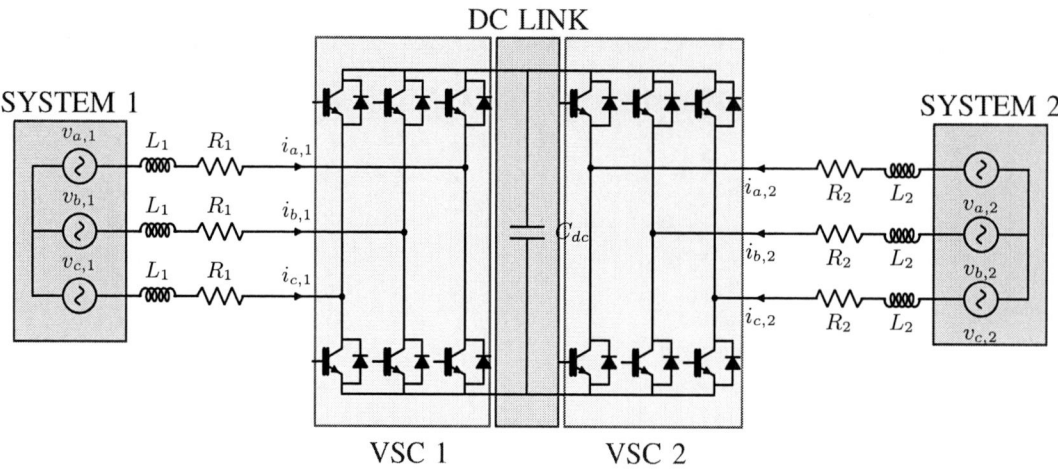

Figure 1. Diagram representing the interconnection of the BTB-VSC to Systems 1 and 2.

Section II describes the non-linear and the linear model of the BTB converter. Section III discusses the robust pole placement control strategy. Section IV presents the test system used in this paper and Section V gives the results from simulation. Finally, Section VI brings the concluding remarks for this paper.

II. MODELING OF THE BACK-TO-BACK

In this section, the dynamic model of a BTB converter based on VSC is presented. The diagram of a BTB-VSC used to connect two different AC systems, System 1 and System 2, is depicted in Figure 1. Each VSC is connected to its respective AC system through a RL branch, that is, a resistor and an inductor connected in series. Note that each RL branch have a different set of parameters R_1 and L_1 for VSC 1 or R_2 and L_2 for VSC 2. Furthermore, the voltage in the DC-link is maintained using a capacitor with capacitance C_{dc}.

A. Nonlinear Averaged Model

As it was already mentioned, each VSC is connected to its respective AC grid through a RL branch such as the one depicted in Figure 2. The VSC is expected to behave as controlled voltage source and it is represented by v_t. In this study, for System 1, $R = R_1$ and $L = L_1$ are the filter parameters and are known. On the other hand, for System 2, $R = R_2 = R_f + R_g$ which is the sum of the known filter parameter R_f with the grid equivalent resistance r_g at the point of common coupling. The latter is a variable subject to uncertainty. Similarly, the inductance $L = L_2 = L_f + L_g$ is also subject to the uncertainties coming from the term L_g.

The differential equation derived for the circuit depicted in Figure 2 is shown in (1). Note that this equation shows the behavior of each phase current to any variation of terminal or grid voltages v_t and v_g, respectively.

$$\frac{d}{dt}i = -\frac{R}{L}i + \frac{1}{L}v_t - \frac{1}{L}v_g, \qquad (1)$$

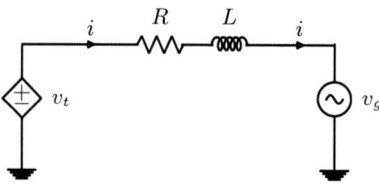

Figure 2. Diagram representing the connection between each VSC and its respective AC grid for each phase.

However, the terminal voltage v_t, which is the input of this system, can be written as a function of modulation index m and dc link capacitor voltage v_{dc} when an averaging operator is used. For a sinusoidal Pulse Width Modulation (PWM) switching strategy, this relation becomes $v_t = m\frac{v_{dc}}{2}$ the one shown in equation (2) [5].

$$v_t = m\frac{v_{dc}}{2}, \qquad (2)$$

Equation (1) only represents one phase instead of the whole three-phase system. In order to cover the modeling for a three-phase system, space phasor notation may be used to get to a synchronous frame representation [5], [10]. Adopting the synchronous coordinate system and observing equation (2), it is possible to write system of equations (3) for the BTB system as a whole [5].

$$\begin{cases} \dfrac{d}{dt}i_{d,1} = +\omega_1 i_{q,1} - \dfrac{R_1}{L_1}i_{d,1} + \dfrac{m_{d,1}}{2L_1}v_{dc} - \dfrac{1}{L_1}v_{g,d,1}, \\[2mm] \dfrac{d}{dt}i_{q,1} = -\omega_1 i_{d,1} - \dfrac{R_1}{L_1}i_{q,1} + \dfrac{m_{q,1}}{2L_1}v_{dc} - \dfrac{1}{L_1}v_{g,q,1}, \\[2mm] \dfrac{d}{dt}i_{d,2} = +\omega_2 i_{q,2} - \dfrac{R_2}{L_2}i_{d,2} + \dfrac{m_{d,2}}{2L_2}v_{dc} - \dfrac{1}{L_2}v_{g,d,2}, \\[2mm] \dfrac{d}{dt}i_{q,2} = -\omega_2 i_{d,2} - \dfrac{R_2}{L_2}i_{q,2} + \dfrac{m_{q,2}}{2L_2}v_{dc} - \dfrac{1}{L_2}v_{g,q,2}. \end{cases}$$
$$(3)$$

978-1-7281-4181-7/19 $31.00 © 2019 IEEE

Now, the only component that lacks the dynamic equation is the voltage v_{dc} of the DC link capacitor. Its dynamic is closely related to the power balance in the BTB. In order to make this analysis more simple, is convenient to represent the power flow in the BTB as it is depicted in Figure 3. Note that the adopted model has resistance r_p connected in parallel with capacitor of capacitance C_{dc}.

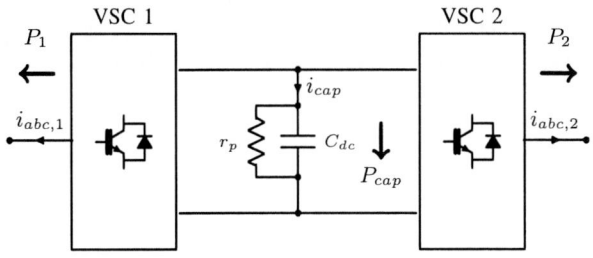

Figure 3. Diagram showing the power flow through the BTB-VSC.

Analyzing Figure 3 and assuming that both VSC have negligible power loss, it is possible to state that the power flow to System 1, P_1, summed with the power flow to System 2, P_2, and the power in the capacitor, P_{cap}, must be equal to 0, as it is written in equation (4).

$$0 = P_{cap} + P_1 + P_2. \tag{4}$$

The power flowing to the capacitor, P_{cap} can be written as a sum of power in the resistor in parallel with power of the capacitor itself, as shown in equation (5).

$$P_{cap} = \frac{v_{dc}^2}{r_p} + C_{dc}\left(\frac{d}{dt}v_{dc}\right)v_{dc}. \tag{5}$$

For this analysis, it is necessary to remember that resistance r_p has a typical high value, being responsible for just a small percentage of P_{cap} [5]. Therefore, it will be neglected in the following steps. In equation (4) output powers P_1 and P_2 can be written as function of the product of each respective terminal voltages with branch currents, all in synchronous reference. However, it is necessary to remember that each converter's terminal voltage is a function of the modulation index and the DC link voltage. Thus, it is possible to write equation (6).

$$\left(\frac{d}{dt}v_{dc}\right)v_{dc} = -\frac{3m_{d,1}v_{dc}}{4C_{dc}}i_{d,1} - \frac{3m_{q,1}v_{dc}}{4C_{dc}}i_{q,1}$$
$$- \frac{3m_{d,2}v_{dc}}{4C_{dc}}i_{d,2} - \frac{3m_{q,2}v_{dc}}{4C_{dc}}i_{q,2}. \tag{6}$$

Dividing the whole equation for v_{dc} equation (7) can be obtained.

$$\frac{d}{dt}v_{dc} = -\frac{3m_{d,1}}{4C_{dc}}i_{d,1} - \frac{3m_{q,1}}{4C_{dc}}i_{q,1} - \frac{3m_{d,2}}{4C_{dc}}i_{d,2} - \frac{3m_{q,2}}{4C_{dc}}i_{q,2}. \tag{7}$$

Hence, combining the set of equations (3) with equation (7) it is possible to assemble the nonlinear set of equations which describes the averaged model of a BTB-VSC. The nonlinear characteristic comes from the multiplication of state variables and control input variables [12].

B. Model Linearization

In last subsection the system model was shown to be nonlinear and, in order to use linear controllers, it is necessary to linearize the equations. The nonlinear system can have its dynamic behavior approximated to a linear system in a region around the equilibrium [11]. Therefore, for the linearization process the value of the variables in the operation point is necessary and any generic variable (state, input or disturbance) $z(t)$ can be written as the sum of its steady state component z with its small-signal variation $\tilde{z}(t)$.

$$z(t) = Z + \tilde{z}(t) \Rightarrow \tilde{z}(t) = z(t) - Z. \tag{8}$$

In the equilibrium, the derivative terms are equal to zero and, hence, the states, inputs and disturbances can be calculated in their steady state value. By using an appropriate PLL system, quadrature axis voltages $V_{g,q,1}$ and $V_{g,q,2}$ are equal to zero while the direct axis voltages $V_{g,d,1}$ and $V_{g,d,2}$ are equal to the peak value of the wave in steady state [13]. The voltage over the capacitor V_{dc} is kept constant in the operation of the BTB in a predetermined value. In addition, in this study, the reference value of reactive power is set to be zero on both sides. Hence, the quadrature voltages $I_{q,1}$ and $I_{q,2}$ are also equal to zero. Hence, by setting a power reference P_1^*, the steady state values of all other variables can be calculated as listed in (9) [12].

$$\begin{cases} I_{d,1} = \frac{2P_1}{3V_{g,d,1}}, \\ M_{d,1} = \frac{2V_{g,d,1}+2R_1I_{d,1}}{V_{dc}}, \\ M_{q,1} = \frac{2L_1I_{d,1}\omega_1}{V_{dc}}, \\ M_{d,2} = \frac{V_{g,d,2}+\sqrt{V_{g,d,2}^2+2V_{dc}R_2M_{d,1}I_{d,1}}}{V_{dc}}, \\ I_{d,2} = -\frac{M_{d,1}I_{d,1}}{M_{d,2}}, \\ M_{q,2} = \frac{2L_2I_{d,2}\omega_2}{V_{dc}}. \end{cases} \tag{9}$$

The resulting set of equations, written in a state-space representation, that shows the small-signal behavior for the whole system is shown in equation (10) below.

$$\frac{d}{dt}\mathbf{x_s} = \mathbf{A_{conv}}\mathbf{x_s} + \mathbf{B_{conv}}\mathbf{u} + \mathbf{B_G}\mathbf{w}, \tag{10}$$

where

$$\mathbf{x_s} = \begin{bmatrix} \tilde{i}_{d,1} & \tilde{i}_{q,1} & \tilde{i}_{d,2} & \tilde{i}_{q,2} & \tilde{v}_{dc} \end{bmatrix}^T,$$

$$
\mathbf{A_{conv}} = \begin{bmatrix}
-\frac{R_1}{L_1} & \omega_1 & 0 & 0 & \frac{M_{d,1}}{2L_1}, \\
-\omega_1 & -\frac{R_1}{L_1} & 0 & 0 & \frac{M_{q,1}}{2L_1} \\
0 & 0 & -\frac{R_2}{L_2} & \omega_2 & \frac{M_{d,2}}{2L_2} \\
0 & 0 & -\omega_2 & -\frac{R_2}{L_2} & \frac{M_{q,2}}{2L_2} \\
-\frac{3M_{d,1}}{4C} & -\frac{3M_{q,1}}{4C} & -\frac{3M_{d,2}}{4C} & -\frac{3M_{q,2}}{4C} & 0
\end{bmatrix},
$$

$$
\mathbf{B_{conv}} = \begin{bmatrix}
-\frac{V_{dc}}{2L_1} & 0 & 0 & 0 \\
0 & -\frac{V_{dc}}{2L_1} & 0 & 0 \\
0 & 0 & -\frac{V_{dc}}{2L_2} & 0 \\
0 & 0 & 0 & -\frac{V_{dc}}{2L_2} \\
-\frac{3I_{d,1}}{4C} & -\frac{3I_{q,1}}{4C} & -\frac{3I_{d,2}}{4C} & -\frac{3I_{q,2}}{4C}
\end{bmatrix},
$$

$$
\mathbf{u} = \begin{bmatrix} \tilde{m}_{d,1} & \tilde{m}_{q,1} & \tilde{m}_{d,2} & \tilde{m}_{q,2} \end{bmatrix}^T,
$$

$$
\mathbf{B_G} = \begin{bmatrix}
-\frac{1}{L_1} & 0 & 0 & 0 & 0 \\
0 & -\frac{1}{L_1} & 0 & 0 & 0 \\
0 & 0 & -\frac{1}{L_2} & 0 & 0 \\
0 & 0 & 0 & -\frac{1}{L_2} & 0
\end{bmatrix}^T, \quad (11)
$$

$$
\mathbf{w} = \begin{bmatrix} \tilde{v}_{g,d,1} & \tilde{v}_{g,q,1} & \tilde{v}_{g,d,2} & \tilde{v}_{g,q,2} \end{bmatrix}^T.
$$

III. MODELING THE CONTROLLER

The model was written using a synchronous reference frame and, therefore, the designed controllers must be able to follow a constant-value reference [5]. Consequently, an integral action is sufficient to meet a null steady state error condition [8]. In this study, the reactive power of both sides must be kept under the desired condition, which means that $i_{q,1}$ and $i_{q,2}$ must be controlled. In addition, the voltage over the DC link, v_{dc} must be controlled to be a constant value during the operation. In addition, it is necessary to met a reference value for power set to one side of the BTB-VSC. If the first side is chosen, the current $i_{d,1}$ must be controlled, which is the case in this study. If the state variable is chosen to be the output of the integral blocking acting on the error, it is possible to write equation (12) for the controller.

$$
\frac{d}{dt} \overbrace{\begin{bmatrix} x_1 \\ x_2 \\ x_3 \\ x_4 \end{bmatrix}}^{\mathbf{x_c}} = \overbrace{\begin{bmatrix} -1 & 0 & 0 & 0 & 0 \\ 0 & -1 & 0 & 0 & 0 \\ 0 & 0 & 0 & -1 & 0 \\ 0 & 0 & 0 & 0 & -1 \end{bmatrix}}^{\mathbf{A_{cont}}} \begin{bmatrix} \tilde{i}_{d,1} \\ \tilde{i}_{q,1} \\ \tilde{i}_{d,2} \\ \tilde{i}_{q,2} \\ \tilde{v}_{dc} \end{bmatrix}
$$
$$
+ \underbrace{\begin{bmatrix} 1 & 0 & 0 & 0 \\ 0 & 1 & 0 & 0 \\ 0 & 0 & 1 & 0 \\ 0 & 0 & 0 & 1 \end{bmatrix}}_{\mathbf{I_4}} \overbrace{\begin{bmatrix} i_{d,1}^* - I_{d,1} \\ i_{q,1}^* - I_{q,1} \\ i_{q,2}^* - I_{q,2} \\ v_{dc}^* - V_{dc} \end{bmatrix}}^{\mathbf{u_{ref}}}. \quad (12)
$$

where over-script (*) means the reference value and $\mathbf{I_4}$ is the identity matrix with dimension four by four. Combining equations (10) and (12), it is possible to write the augmented equation (13) below.

$$
\frac{d}{dt} \begin{bmatrix} \mathbf{x_s} \\ \mathbf{x_c} \end{bmatrix} = \overbrace{\begin{bmatrix} \mathbf{A_{conv}} & \mathbf{0}_{5 \times 4} \\ \mathbf{A_{cont}} & \mathbf{0}_{4 \times 4} \end{bmatrix}}^{\mathbf{A}} \overbrace{\begin{bmatrix} \mathbf{x_s} \\ \mathbf{x_c} \end{bmatrix}}^{\mathbf{x}} + \overbrace{\begin{bmatrix} \mathbf{B_{conv}} \\ \mathbf{0}_{4 \times 4} \end{bmatrix}}^{\mathbf{B}} \mathbf{u}
$$
$$
+ \underbrace{\begin{bmatrix} \mathbf{B_G} \\ \mathbf{0}_{4 \times 4} \end{bmatrix}}_{\mathbf{B_w}} \mathbf{w} + \underbrace{\begin{bmatrix} \mathbf{0}_{5 \times 4} \\ \mathbf{I_4} \end{bmatrix}}_{\mathbf{B_{ref}}} \mathbf{u_{ref}}. \quad (13)
$$

A. State Feedback

The control law of a state-feedback controller can be defined as equation (14) below.

$$
\mathbf{u} = \mathbf{Kx}. \quad (14)
$$

Thus, the linear system with this sort of control is described as equation 15.

$$
\dot{\mathbf{x}} = \mathbf{Ax} + \mathbf{Bu} = (\mathbf{A} + \mathbf{BK})\mathbf{x}. \quad (15)
$$

The linearized system may have its stability assessed through the analysis of matrices $\mathbf{A}$ and $\mathbf{B}$ which characterizes the system. The system is said to be asymptotically stable if exists a positive matrix $\mathbf{P}$, and matrix $\mathbf{W} = \mathbf{KP}$ that satisfies the LMI presented in equation (16) [8].

$$
\mathbf{AP} + \mathbf{PA}^T + \mathbf{BW} + \mathbf{W}^T\mathbf{B}^T < 0. \quad (16)
$$

B. Robust Pole Placement

This type of state feedback control is concerned with finding a gain matrix $\mathbf{K}$ which guarantees that all poles are located inside a region $\mathcal{D}$ in the complex plane while the LMI (16) is satisfied. This can be done by assembling a system of equations that must be solved simultaneously. A typical region $\mathcal{D}$ is represented in Figure 4. As one can see, the region is completely defined by three parameters: α, θ and ρ.

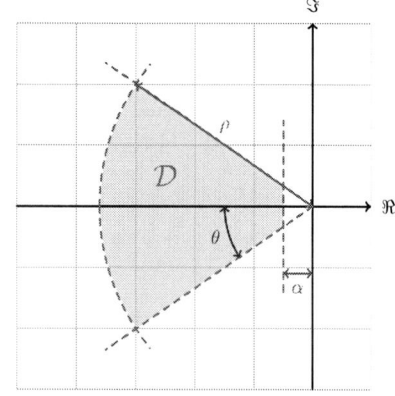

Figure 4. Convex $\mathcal{D}$ region and its parameters α, ρ e θ.

This work aims to design a controller for a BTB-VSC allowing it to deal with bidirectional power flow and with

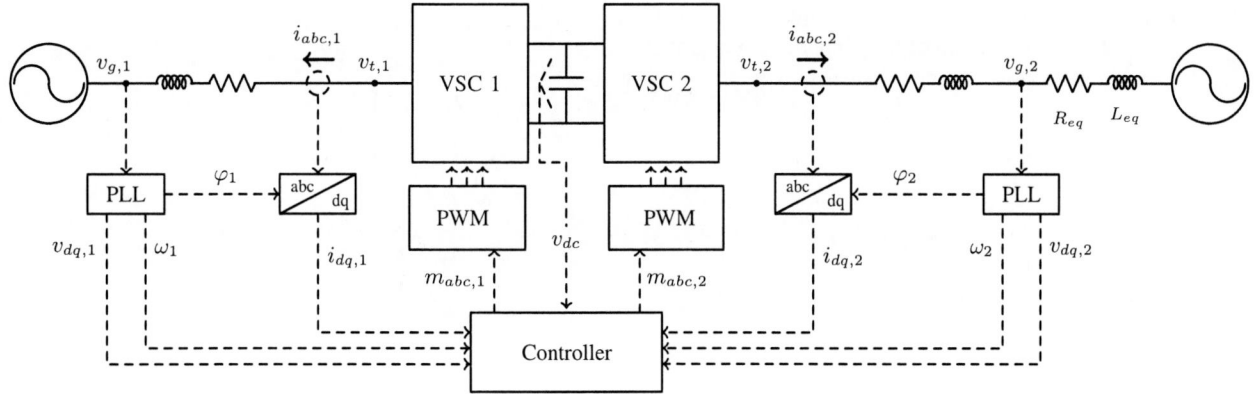

Figure 5. Diagram for the studied system.

uncertainties related to the grid equivalent resistance and inductance at the PCC. Let us say that the reference power may vary in in the range $[P_{min}\ P_{max}]$, the grid resistance may vary from $[R_{g,min}\ R_{g,max}]$ and the inductance from $[L_{g,min}\ L_{g,max}]$. It is possible, then, to combine them in eight cases, or equilibrium points. For each case, matrices depending on the values of those variables should be calculated. For all cases, the closed loop poles for the entire system must be located in region $\mathcal{D}$. Hence, one must assemble and solve a linear matrix inequality (LMI) system composed of equations (17), where matrices $\mathbf{A}$ and $\mathbf{B}$ are calculated for each case.

$$2\alpha\mathbf{P} + \mathbf{M} < 0,$$
$$\begin{bmatrix} -\rho\mathbf{P} & \mathbf{AP} + \mathbf{BW} \\ \mathbf{PA}^T + \mathbf{W}^T\mathbf{B}^T + & -\rho\mathbf{P} \end{bmatrix} < 0, \qquad (17)$$
$$\begin{bmatrix} \mathbf{M}\sin\theta & \mathbf{L}\cos\theta \\ \mathbf{N}\cos\theta & \mathbf{M}\sin\theta \end{bmatrix} < 0,$$

where

$$\mathbf{M} = \mathbf{AP} + \mathbf{PA}^T + \mathbf{BW} + \mathbf{W}^T\mathbf{B}^T,$$
$$\mathbf{L} = \mathbf{AP} - \mathbf{PA}^T + \mathbf{BW} - \mathbf{W}^T\mathbf{B}^T, \qquad (18)$$
$$\mathbf{N} = -\mathbf{AP} + \mathbf{PA}^T - \mathbf{BW} + \mathbf{W}^T\mathbf{B}^T.$$

IV. STUDIED SYSTEM AND CONTROLLER DESIGN

The studied system consists of a BTB-VSC interconnecting two AC system of same line-to-line voltage V_{rms} and nominal frequency f. In order to synchronize each VSC terminal voltage with the respective system voltage, it is necessary to use a phase-locked loop (PLL). In this study, a synchronous reference frame based PLL, with a Double Second Order Generalized Integrator (DSOGI) filter is used [13]. The system is depicted in Figure 5 while its parameters are shown in Table I. The parameters used for linearization and control design are also listed in Table I.

Table I
PARAMETERS USED IN SIMULATION.

System Parameters	
Description	Value
RMS Line Voltage (V_{rms})	220 V
Grid Frequency (f)	60 Hz
DC Link Capacitor (C_{dc})	2 mF
DC Link Voltage (V_{dc})	500 V
System 1 Inductance (L_1)	2 mH
System 1 Resistance (R_1)	75 $m\Omega$
System 2 Inductance (L_2)	[2.2 4] mH
System 2 Resistance (R_2)	[80 150] $m\Omega$
Switching Frequency (f_s)	20 kHz
Controller Parameters	
Description	Value
Maximum Power (P^{max})	30 kW
Minimum Power (P^{min})	-30 kW
Reference ($I_{q,1}^*$)	0 A
Reference ($I_{q,2}^*$)	0 A
Variable (α)	121.2 s^{-1}
Minimum Damping (θ)	50 $^\circ$
Maximum Radius (ρ)	12566,4 s^{-1}

A MATLAB toolbox for solving LMIs was used in order to obtain gain matrix $\mathbf{K}$, presented in equation (19). Matrix $\mathbf{K_s}$ represents the gain matrix that multiplies the states from the converter, presented in vector $\mathbf{x_s}$. Matrix $\mathbf{K_c}$, on the other hand, represents the gain matrix that multiplies the states from the controller, listed in $\mathbf{x_c}$.

$$\mathbf{K} = \begin{bmatrix} \mathbf{K_s} & \mathbf{K_c} \end{bmatrix}. \qquad (19)$$

where

$$\mathbf{K_s} = \begin{bmatrix} 0.0575 & 0.0017 & 0.0462 & -0.0014 & -0.0874 \\ -0.0028 & 0.0442 & 0.0017 & 0.0007 & -0.0007 \\ -0.0030 & -0.0001 & 0.0257 & 0.0042 & 0.0028 \\ -0.0089 & 0.0014 & -0.0081 & 0.0669 & 0.0116 \end{bmatrix}. \qquad (20)$$

$$\mathbf{K_c} = \begin{bmatrix} -1.1280 & 0.1700 & -0.0164 & 11.1748 \\ 0.2385 & -6.1297 & -0.1163 & 0.0027 \\ 4.3244 & 0.0155 & 0.0937 & -1.6905 \\ 0.8086 & -0.2023 & -11.0824 & -1.8159 \end{bmatrix}. \qquad (21)$$

Using the gain matrix $\mathbf{K}$, it is possible to calculate the closed loop poles of the system for all of the eight cases. The poles are presented as blue asterisks in 6, while the region $\mathcal{D}$ is presented with a dashed contour. It is important to note that all poles are, in fact, inside the determined $\mathcal{D}$ region.

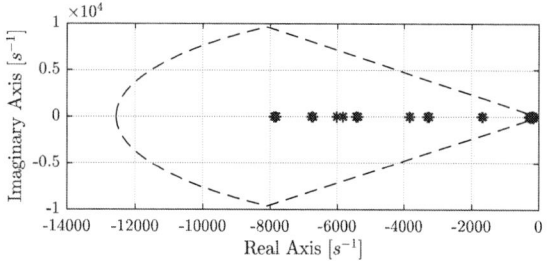

(a) Region $\mathcal{D}$ and poles.

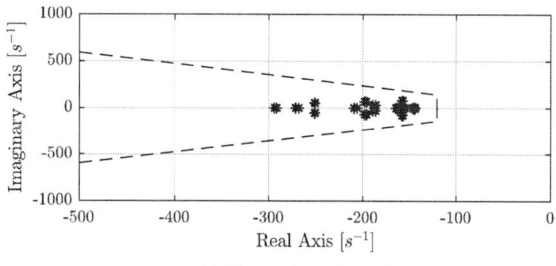

(b) Slower dynamics poles.

Figure 6. Closed loop poles for the studied system.

V. RESULTS

The system and the controllers described in the previous section are tested under two different test conditions that are further explained. The experiment is made in a digital simulation software called PSIM and the both simulations last for a total of 0.6 second. The largest TDD in the current was of 0.26%, below the limits recommended by [14].

In the first case, the grid resistance and inductance are maintained constant throughout the simulation and with value equal to $R_2 = 100m\Omega$ and $L_2 = 3.2mH$. The power reference changes from 0 to $10kW$ and then to $30kW$ in $t = 0.1\ s$ and $t = 0.2\ s$, respectively. The power reference changes again to $-30kW$ in a ramp starting from $t = 0.3$ and ending at $t = 0.4\ s$. In Figure 7 it is possible to observe the results for power on the terminal of VSC 1, current injected in System 2 and the voltage over the DC capacitor. Note that the power reference is met with a smooth behavior. Consequently, the current also changes, due to the reference change, without any overshoot. In addition, observe that the voltage has less than $10V$ variation for any power reference change.

In the second case, power reference changes from 0 to $30kW$ in $t = 0.1\ s$ and is kept constant in the rest of the simulation. At $t = 0.32\ s$, grid inductance value in the PCC changes abruptly from $L_2 = 2.2$ to $L_2 = 4$, while resistance is constant and equal to $R_2 = 100m\Omega$. In Figure 8, results for power at the terminal of VSC 1, currents injected in System 2 and the DC link voltage are presented. Note that the abrupt grid parameter change is translated in small changes in all quantities presented in Figure 8. The changes in the current,

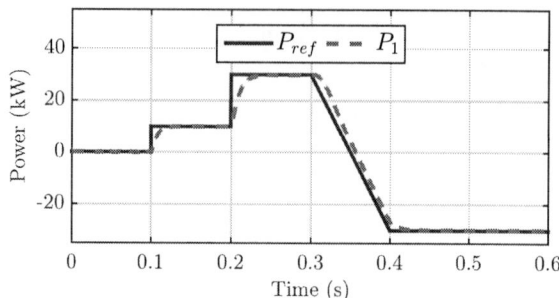

(a) Power behavior in terminal 1 of VSC.

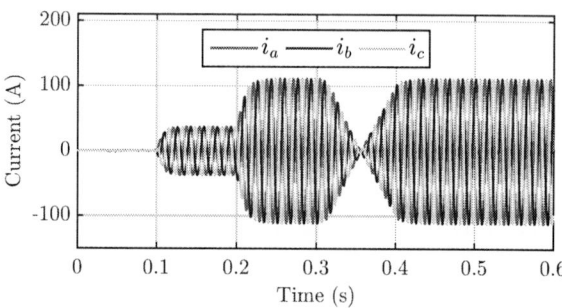

(b) Current injected in System 2.

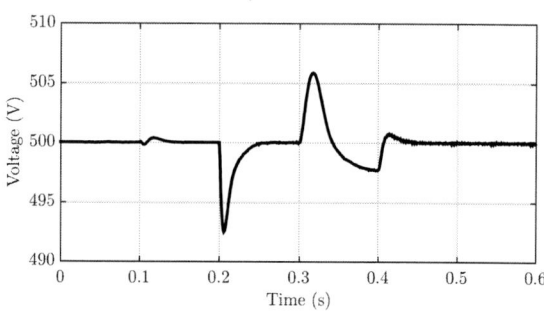

(c) Voltage over capacitor in DC link.

Figure 7. Behavior of variables for bidirectional power flow.

shown in (b), are almost negligible while the change in DC-link voltage is less than $5V$, showing that the controller deals very well with abrupt changes in grid impedance.

VI. CONCLUSION

In this paper, the state-space modeling of a BTB-VSC interconnecting two AC systems was developed. Furthermore, the modeling of control strategy based on robust pole placement for the BTB-VSC is provided in this paper. The control approach is focused on allowing bidirectional power flow over the converter, and on making the operation of the converter able to deal with uncertainties in grid equivalent inductance and resistance at the PCC.

The controller is designed taking the dynamic behavior of both AC sides and the DC-link capacitor at once. This fact is extremely important for its robust performance. The process to get to the LMI system is not straight-forward but

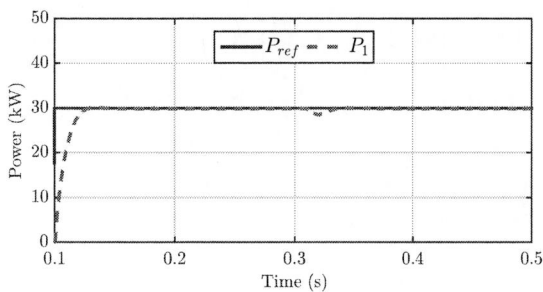

(a) Power behavior in terminal 1 of VSC.

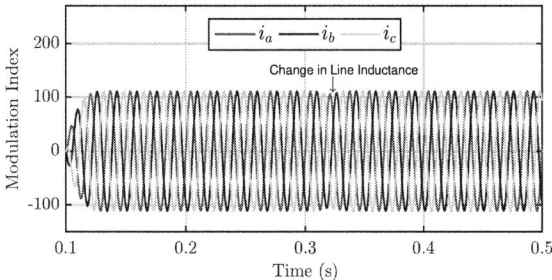

(b) Current injected in System 2.

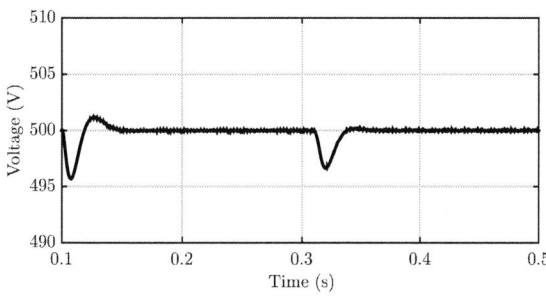

(c) Voltage over capacitor in DC link.

Figure 8. Behavior of variables for abrupt change in grid inductance.

the implementation of the controller is rather simple. This simplicity in implementing such control action should be highlighted as another advantage of the robust control system.

After the calculation of matrix $\mathbf{K}$ and its implementation, the suggested control strategy allowed the BTB-VSC to operate successfully under different grid conditions and power reference values without any issue. Results provided in this paper show that the designed controller based on the robust pole placement has a very good performance under bidirectional power flow and uncertainties related to grid equivalent impedance. Controlled variables presented a smooth dynamic with small overshoot for step changes in reference values and for an abrupt change in the grid equivalent inductance value. Therefore, considering all the advantages that were observed, the designed controller is shown to be a very promising alternative control strategy for BTB-VSC systems.

ACKNOWLEDGEMENTS

This study was financed in part by the Coordenação de Aperfeiçoamento de Pessoal de Nível Superior - Brasil (CAPES) - Finance Code 001 and FAPEMIG, project code APQ-01666-15. The authors would like to also acknowledge the financial support in part of CNPq, and express gratitude for the educational support of Federal Universisty of Juiz de Fora, Brazil, and the Department of Electrical, Computer and Systems Engineering, Rensselaer Polýtechnic Institute, USA.

REFERENCES

[1] Friedli, T., Kolar, J. W., Rodriguez, J. and Wheeler, P. W., *Comparative evaluation of three-phase AC–AC matrix converter and voltage DC-link back-to-back converter systems,* IEEE Transactions on industrial electronics, vol. 59,no. 12, pp. 4487–4510, 2011, IEEE.

[2] Rodríguez-Cabero, A., Sánchez, F. H. and Prodanovic, M., *A unified control of back-to-back converter,* 2016 IEEE Energy Conversion Congress and Exposition (ECCE), pp. 1–8, 2016, IEEE.

[3] Blaabjerg, F., Liserre, M. and Ma, K., *Power electronics converters for wind turbine systems,* IEEE Transactions on industry applications, vol. 48, no. 2, pp. 708–719, 2011, IEEE.

[4] Chen, Z., *An overview of power electronic converter technology for renewable energy systems,* Electrical Drives for Direct Drive Renewable Energy Systems, pp. 80–105, 2013, Elsevier.

[5] Yazdani, A. and Iravani, R., *Voltage-sourced converters in power systems: modeling, control, and applications.* 2010, John Wiley & Sons.

[6] Alcalá, J., Cardenas, V., Ramirez-Lopez, A. R. and Gudino-Lau, J., *Study of the bidirectional power flow in Back-to-Back converters by using linear and nonlinear control strategies,* 2011 IEEE Energy Conversion Congress and Exposition, pp. 806–813, 2011, IEEE.

[7] Osório, C. R. D., Koch, G. G., Borin, L. C., Cleveston, I. and Montagner, V. F., *Controlador Robusto Quasi-Deadbeat e Relaxações com Aaplicação em Inversores Conectados à Rede,* Eletrônica de Potência, vol. 23, no. 3, pp. 320–329, 2018, SOBRAEP.

[8] Liu, K.-Z. and Yao, Y.*Robust Control: Theory and Applications.* 2018, John Wiley & Sons.

[9] Duan, G. R. and Yu, H. H.,*LMIs in control systems: analysis, design and applications.* 2013, CRC press.

[10] de Almeida, P. M., Ferreira, A. A., Braga, H. A. C. and Barbosa, P. G. *Projeto dos controladores de um conversor VSC usado para conectar um sistema de geração fotovoltaico à rede elétrica.* Congresso Brasileiro de Automática, 2012, SBA.

[11] Khalil, H. K. and Grizzle, J. W., *Nonlinear systems,* vol. 3, 2002, Prentice Hall Upper Saddle River.

[12] Suntio, T., Messo, T. and Puukko, J. *Power Electronic Converters: Dynamics and Control in Conventional and Renewable Energy Applications.* 2018, John Wiley & Sons.

[13] Rodriguez, P., Teodorescu, R., Candela, I., Timbus, A. V., Liserre, M., Blaabjerg, F.*New positive-sequence voltage detector for grid synchronization of power converters under faulty grid conditions.* Power Electronics Specialists Conference, 37th PESC'06, p. 1-7, 2006, IEEE.

[14] The Institute of Electrical and Electronics Engineers, *IEEE Std 519-2014.* IEEE Recommended Practice and Requirements for Harmonic Control in Electric Power Systems, 2014, IEEE Power and Energy Society.

A ISOP AC-AC Hybrid Switched-Capacitor SRC for Solid State Transformer Applications

Victor Luiz Flor Borges, Rogerio Luiz da Silva Jr, Carlos Eduardo Possamai, Arlan Luiz Bettiol, Ivo Barbi

Brazilian Institute of Power Electronics and Renewable Energies - IBEPE

Florianopolis, Brazil

victor@ibepe.org, rogerio@ibepe.org, carlos@ibepe.org, arlan@averodomino.com.br, ivobarbi@ibepe.org

Abstract—This paper presents the one phase of a three-phase solid-state transformer (SST), which will be employed in a 13.8 kV / 220 V distribution system, realized as input-series output-parallel (ISOP) arrangement of multiple AC-|AC| modules and by a |AC|-AC unfolding stage. The modules are compounded by a front-end AC-|AC| active rectifier and by a |AC|-|AC| hybrid switched-capacitor series resonant converter (HSCSRC). The auto-balancing among the modules, the absence of large DC link capacitors and the simple modulation scheme are some advantages of the proposed SST. Experimental results are presented for a laboratory prototype of an ISOP arrangement of two modules with the unfolding stage for theoretical validation. The system rated power is 3.4 kVA with 2.3 kV input voltage and 220 V output voltage. The prototype holds a maximum efficiency of 96.8%.

Index Terms—Solid-State Transformer, Resonant Converter, ISOP, Isolated AC-AC Converter, Switched-Capacitor.

I. INTRODUCTION

The renewable energy growth in the last few decades introduced the concept of the smart grid, which requires resources for power flow management and transmission that are not available on the conventional distribution electrical grid. Therefore, the need for smart equipment and systems capable to operate on the smart grids, that will be responsible to process, monitor and control the distributed generation sources with high efficiency. In this scenario, the solid-state transformer (SST) appears as the main power processor. The SST can connect a different kind of energy sources, integrate the control and intelligence of the distribution grid, control the power flow between medium voltage (MV) and low voltage (LV) sides, providing galvanic isolation at medium or high frequencies, power factor correction and LV side voltage regulation.

The first SSTs prototypes designed by companies, such as: Siemens, Bombardier, Alstom and ABB, were for electric traction systems applications. The advantages of weight and volume reduction provided by SSTs are very attractive in this environment. The prototypes developed for applications in distribution systems (UNIFLEX [1], GE [2], ETHz [3], FREEDM [4] and EPRI [5]) are recent. It is a consensus for some authors that the application of SSTs in distribution systems is still a major challenge [6], [7].

This work is supported by Companhia Energética de Brasília (CEB) as part of R&D program ANEEL 001/2016.

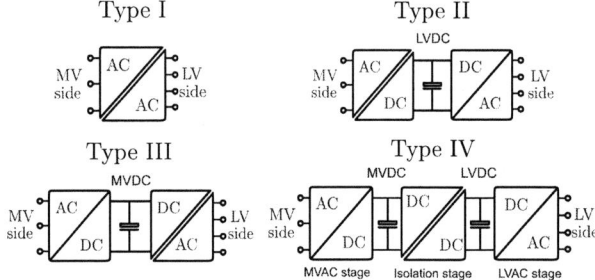

Fig. 1. Solid state transformer architectures classification [6].

The SST can be classified according to the existence or position of the DC link in its structure, it has been categorized in [6] as presented in Fig. 1. The Type I do not use the DC link, and, typically, matrix-type converters and/or cycloconverters are used. This type of structure has the disadvantage of using a large number of semiconductors, since bidirectional switches are required. An alternative is the solution presented in this paper, which employs a front-end active rectifier and a back-end unfolder, both operating at line-frequency, enabling the use of unidirectional switches in the high-frequency stage, which reduces the total switch count. The second type (Type II) is formed by two conversion stages and the low-voltage DC link (LVDC) is used. The first conversion stage performs the insulation and the input rectification, while the second inverts the DC link voltage. The DC link presence has some advantages, i.e., it allows the control of the output frequency and power factor correction. However, the DC link volume is relatively high regarding the total converter volume. Moreover, when using electrolytic capacitors, the converter's life can be compromised. Type III is similar to Type II, the main difference is the DC link position, wherein Type II uses it on the medium voltage side (MVDC). This position change allows the control of the input reactive power of the SST. Finally, Type IV uses both high and low voltage links, which means that three conversion stages are used. The employment of three stages allows various combinations of topologies, however, it considerably increases the costs and project complexity.

Since the maximum voltage levels of the high-frequency switches available on the market are lower than the requirements for MV SST applications and high-voltage (HV) wide

978-1-7281-4181-7/19 $31.00 © 2019 IEEE

bandgap switches (in the order of tens of kV) are undergoing experimentation [2], [8], there is a need to combine converters for semiconductors voltage stresses reduction. Therefore, the MV stages described to different types of SST architectures (Fig. 1) can be compounded by the association of several modules. The input-series output-parallel (ISOP) modules arrangement is the most common solution used to reduce voltage stress on semiconductors in power electronics converters [1], [7], [9].

The Fig. 2 shows the single-phase concept of a three-phase SST that will be employed in a distribution system (13.8 kVAC/ 60 kVA/ 60 Hz). The module principle of operation is presented in Section II. In Section III, the ISOP arrangement using the hybrid switched-capacitor series resonant converter (HSCSRC) topology is studied. Finally, in Section IV, experimental results from the highlighted circuit in Fig.2 are reported. The prototype has processed 3.4 kVA with input and output rms voltages equal to 2.3 kV and 220 V, respectively.

II. MODULE PRINCIPLE OF OPERATION

The modules of the ISOP system are compounded by a front-end AC-|AC| active rectifier operating at the line frequency, also known by folder, and by a |AC|-|AC| high-

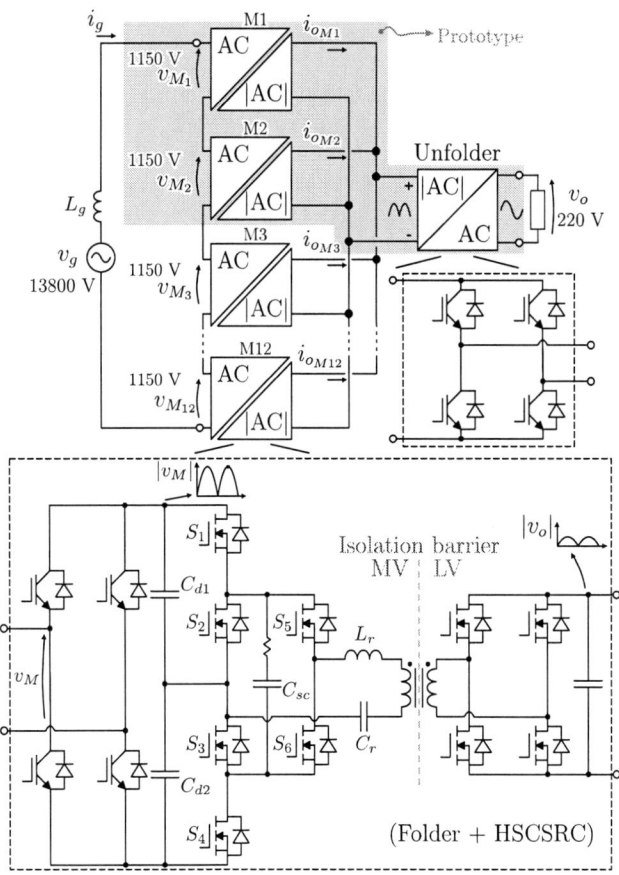

Fig. 2. ISOP arrangement of twelve modules (Folder + |AC|-|AC| HSCSRC) with the unfolding stage.

frequency stage HSCSRC converter. The HSCSRC's application in the ISOP configuration is the main contribution of the proposed SST architecture.

The series resonant converter (SRC) operates at the resonant frequency of the series elements L_r and C_r. Then, the switching frequency (f_s) is, ideally, equal to the resonant frequency (f_o), and, consequently, the series elements impedance is null, resulting in a unitary gain of the resonant stage. The ladder switched-capacitor cell integrated with the SRC stage does not modify the SRC operation mode, and his main function is for voltage stresses reduction across the MV and high-frequency switches (S_1 to S_6, see Fig. 2). The voltage stresses are reduced by half of the module input voltage (v_m), and the voltage balance among the capacitors C_{d1} and C_{d2} are guaranteed by the switched capacitor C_{sc} operation. Thus, the voltages over the switches S_1 to S_6 are auto-balanced as well. The capacitive input power factor is a special attribute of the proposed SST since the inductive load is predominant in the industrial plants and distribution systems. The detailed analysis of the HSCSRC converter, such as commutation study, was presented in [10].

A. Topological stages and static gain

The converter works in open loop, with both constant switching frequency and duty cycle (50%). The main waveforms at switching period, and at grid period, are illustrated in Fig. 3, such as the high-frequency topological stages with the LV elements referred to the MV side by the transformer turns ratio (1), in which N_s and N_p are the secondary and primary turns, respectively.

$$a = \frac{N_s}{N_p} \tag{1}$$

In the first topological stage (Fig. 3(a)), the odd switches are turned on (Fig. 3(c)), resulting in both positives v_{ab} voltage and i_{L_r} current. As defined that $f_s = f_o$, the i_{L_r} current is sinusoidal and it is in phase disposition with v_{ab} voltage, in the way that it starts and ends half of the switching period at zero (Fig. 3(d)), what characterizes the ZCS commutation because the switches currents are related with i_{L_r}.

The second topological stage (Fig. 3(b)) starts when the odd switches are turned off and the even are turned on (Fig. 3(c)). Thus, the i_{L_r} current and v_{ab} voltage are negative, and, again, the current reaches zero at the end of the switching period. The ZCS commutation of the switches S_1 and S'_{o2} can be seen in Fig. 3(f).

The voltages across the L_r inductor and the C_r capacitor are illustrated in Fig. 3(e), and they are in phase-quadrature, what implies in a null impedance of the L_r and C_r pair (the addition result of v_{L_r} and v_{C_r} voltages are zero), which means that the resonant stage static gain is unitary. Therefore, the static gain of the hybrid structure is, ideally, dependent on the transformer turns ratio (2).

$$G_{HSCSRC} = \frac{a}{2} \tag{2}$$

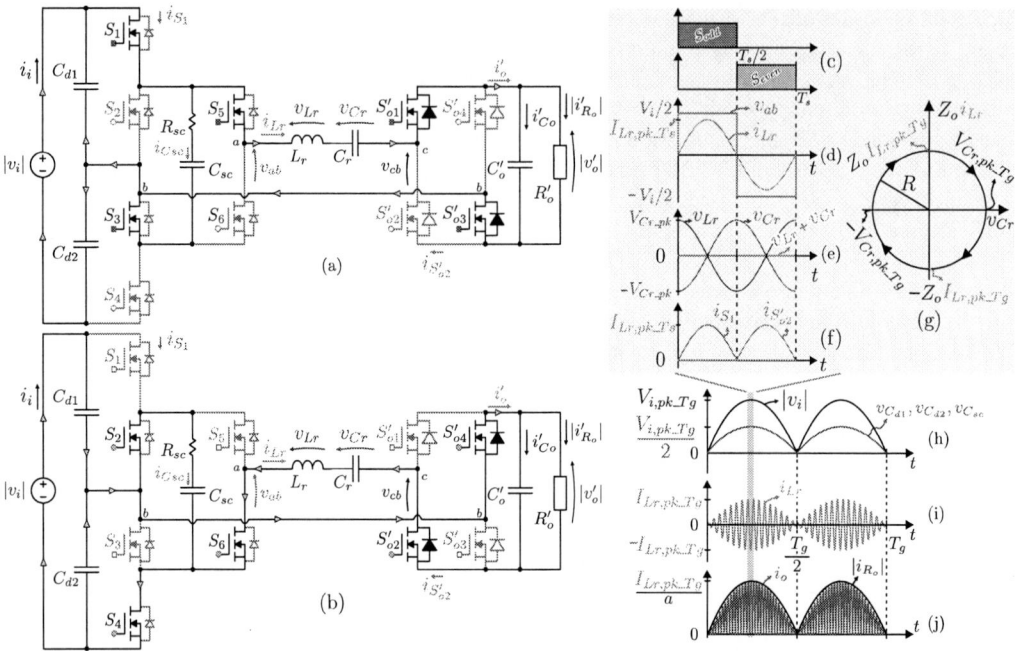

Fig. 3. |AC|-|AC| HSCSRC high frequency topological stages. (a) First and (b) second stages, (c) switches gate signals, (d) v_{ab} voltage and current through L_r at switching frequency, (e) voltages across L_r and C_r at switching frequency, (f) current through switches S_1 and S'_{o2} at switching frequency, (g) state plane at grid peak voltage, (h) input voltage and voltage across capacitors C_{d1}, C_{d2} and C_{sc} at grid frequency, (i) current through inductor L_r at grid frequency and (j) output and load currents at grid frequency.

The peak value of the i_{L_r} current (Fig. 3(i)) at the grid period (I_{Lr,pk_Tg}) can be determined from the calculus of the average value of the output current (i_o) of the LV high frequency rectifier, which envelope is the modulus of the load current (Fig. 3(j)). Thus, I_{Lr,pk_Tg} is calculated by

$$I_{Lr,pk\_Tg} = \frac{2\pi P_o}{V_{i,pk\_Tg}} \frac{f_o}{f_s}, \quad f_o = \frac{1}{2\pi\sqrt{L_r C_r}}. \quad (3)$$

From the state-plane analysis (Fig. 3(g)), and stating that Z_r is the characteristic impedance of L_r and C_r pair, the peak value, at the grid period, of the voltage across C_r is determined by

$$V_{Cr,pk\_Tg} = Z_r I_{Lr,pk\_Tg} \rightarrow V_{Cr,pk\_Tg} = \frac{P_o}{f_s C_r V_{i,pk\_Tg}}. \quad (4)$$

The equations (3) and (4) are fundamental for designing the converter, because the current stresses through the components are calculated as a function of (3) and the voltage stress of the C_r capacitor is determined by (4).

B. Transformer characteristics

The electrical isolation standards applicable to the distribution low-frequency transformer (LFT) should be extended to the SSTs, as already reported [11]. Therefore, the slotted bobbin assembled in an ETD (Economic Transformer Design) core was chosen for this purposes, and this kind of transformer assembly normally presents a large leakage inductance. Thus, the L_r inductor of the resonant tank can be the leakage

inductance of the high-frequency transformer itself, which minimizes the cost and volume, as well as increase the structure efficiency because a magnetic element is eliminated from the converter. This solution is well known in the literature by integrated tank resonant [12], usually applied in LLC converters. Then, the HSCSRC's static gain considering the resonant tank integrated to the transformer, being a_{int} the turns ratio of the integrated transformer, is determined by

$$a_{\mathrm{int}} = \frac{a}{\sqrt{1-\lambda}}, \quad \lambda = \frac{L_r}{L_m} \quad \rightarrow \quad G_{HSCSRC}(a,\lambda) = \frac{a_{\mathrm{int}}(a,\lambda)}{2} \quad (5)$$

in which L_r and L_m are leakage and magnetizing inductances, respectively, all primary side referred, and a is the ideal transformer turns ratio (1). Before, G_{HSCSRC} was defined by (2), and now, is a function of L_r and L_m, as can be seen on (5). The leakage inductance "boost" the transformer gain, because, in general, $L_r < L_m$ and $\lambda < 1$.

The magnetizing inductance is a project parameter to achieve ZVS commutation [10], and the ratio between L_r and L_m should be considered to get the desired output voltage by choosing the correct secondary turns. If no ZVS commutation is required, the magnetizing inductance can be chosen larger than L_r and, then, the integrated transformer gain (5), with $L_m >> L_r$, converges to the ideal relationship between primary and secondary turns (1). If $L_m >> L_r$, the LLC resonant current, with $f_s \cong f_o$, approaches even more the SRC sinusoidal current (as described in Fig. 3(d)), because the magnetizing current can be neglected.

978-1-7281-4181-7/19 $31.00 © 2019 IEEE

The SRC converter with a transformer is the LLC converter itself because the magnetizing inductance is an intrinsic parameter of a real transformer. Nevertheless, the switching frequency approaches the resonance frequency of the L_r and C_r series elements, and, in this case, the LLC converter has the same attributes of an SRC converter working at resonance.

III. ISOP ARRANGEMENT

The ideal ISOP auto-balance is reached when the input voltages of the modules are equal. However, parametric deviations among the modules may cause an imbalance in the input voltages, and, then, unbalance the power division of the modules. The steady-state input voltage imbalance is caused by deviation of the static gain of each module from the ideal one. In spite of that, the ISOP solution has a stable operation and the system does not diverge.

As presented in (5), the converter static gain (G_{HSCSRC}) is related with the ideal transformer turns ratio (1) and it is a function of the magnetizing (L_m) and leakage (L_r) inductances, both related by λ parameter. This is the main cause of static gain deviations of the proposed structure because manufacturing identical transformers is a hard mission.

The rectified output voltages of all modules are equal, so the input peak voltage of each module can be related by the static gain (6) for a generic number of modules (n).

$$V_{o,pk} = \frac{a_{\text{int}1}\, V_{i,pk\_m_1}}{2} = ... = \frac{a_{\text{int}\,n}\, V_{i,pk\_m_n}}{2} \quad (6)$$

The addition of the modules input voltages must be equal to the total input voltage of the ISOP system (7) ($V_{i,pk\_isop,n}$), which depends on the number of modules. Then, writing the relation of the input voltage of each module as function of the nth module voltage from (6) and applying it into (7), the nth module peak input voltage is determined (8).

$$V_{i,pk\_isop,n} = \sum_{j=1}^{n} V_{i,pk\_m_j} \quad (7)$$

$$\frac{V_{i,pk\_m_n}}{V_{i,pk\_isop,n}} = \frac{1}{a_{\text{int}\,n}\,(a_n, \lambda_n) \cdot \sum\limits_{j=1}^{n} \frac{1}{a_{\text{int}\,j}(a_j, \lambda_j)}} \quad (8)$$

For comparisons purposes, the voltage ratio (8) is normalized as function of the idealized voltage of the nth module (9).

$$V_{ratio,m_n\_norm} = \frac{V_{i,pk\_m_n}}{V_{i,pk\_isop,n}} - \frac{1}{n} \quad (9)$$

The results of the equation (9) for two and twelve modules with different λ values are shown in Fig. 4. It is important to note that it shows the module n voltage variation regarding the total input voltage with the λ and a parameters of the others ($n-1$) modules fixed. Hence, only the module n has the parameter λ changed, because it would be impossible to make conclusions changing the parameter λ for all modules.

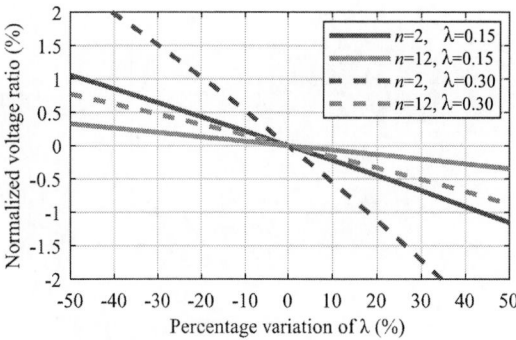

Fig. 4. Percentage voltage ratio between module n and total ISOP input voltage to different λ values of the module n.

Thus, from the analysis of Fig. 4, it can be seen that for an increase of the λ of all modules, the voltage of the nth module becomes more sensitive to variations of his λ parameter, and for a different number of modules the effect is the same. Moreover, when the number of modules increases, the effect of the λ deviations in the nth module voltage is minimized, and it is interesting for a medium voltage SST which employs the ISOP concept with a large number of modules.

IV. EXPERIMENTAL RESULTS

A reduced-scale prototype of the SST's single-phase was built employing two modules. As can be observed in Table I, there are slight differences between their transformers' parameters. As explain in [10], the input bus capacitors (C_{d1}, C_{d2} and C_{sc}), as also the output capacitor C_o, have low capacitance values, and modifies the resonance frequency, because of that the switching frequency is lower than the tank resonance frequency. The module's picture and the reduced-scale ISOP system under experimentation are shown in Fig. 5(a) and Fig. 5(c), respectively. Module 1 is identified by M1 and module 2 by M2. The Fig. 6 displays the waveforms of the ISOP arrangement at nominal voltages with 1.2 kVA output power.

TABLE I
MODULES' DESIGN SPECIFICATIONS.

Rated power (P_o)	1.667 kW
Input rms voltage (V_{i_rms})	1150 V
Output rms voltage (V_{o_rms})	220 V
Grid frequency (f_g)	60 Hz
Switching frequency (f_s)	75 kHz
Resonance frequency (f_o)	81 kHz
($f_s\tau$)	0.24
Input power factor (PF)	0.96
Transformer turn ratio (a)	0.3584
Module 1 Resonant inductor (L_{r1})	124.15 μH
Module 1 Magnetizing inductance (L_{m1})	919.87 μH
Module 2 Resonant inductor (L_{r2})	124.92 μH
Module 2 Magnetizing inductance (L_{m2})	932.99 μH

An evidence of the soft switching in both modules is detected from the fact that the $i_{Lr_{M1}}$ e $i_{Lr_{M2}}$ currents are, respectively, in phase with $v_{ab_{M1}}$ and $v_{ab_{M2}}$ voltages, as shown

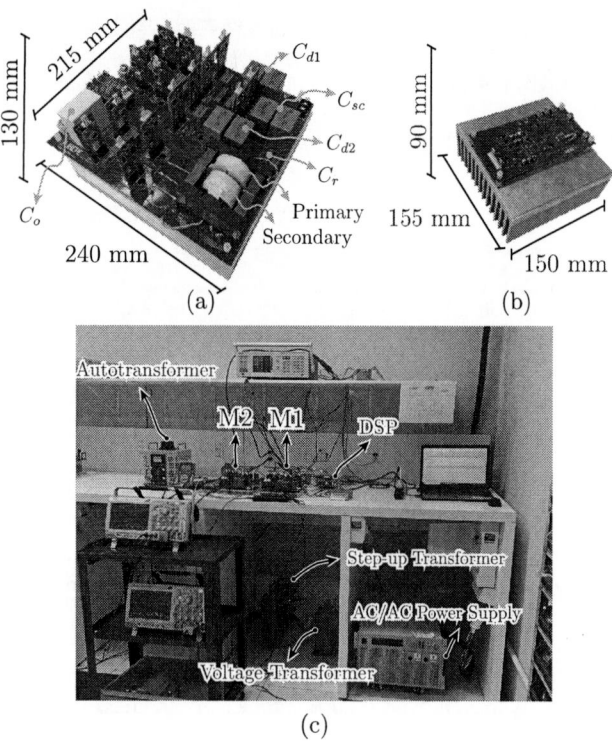

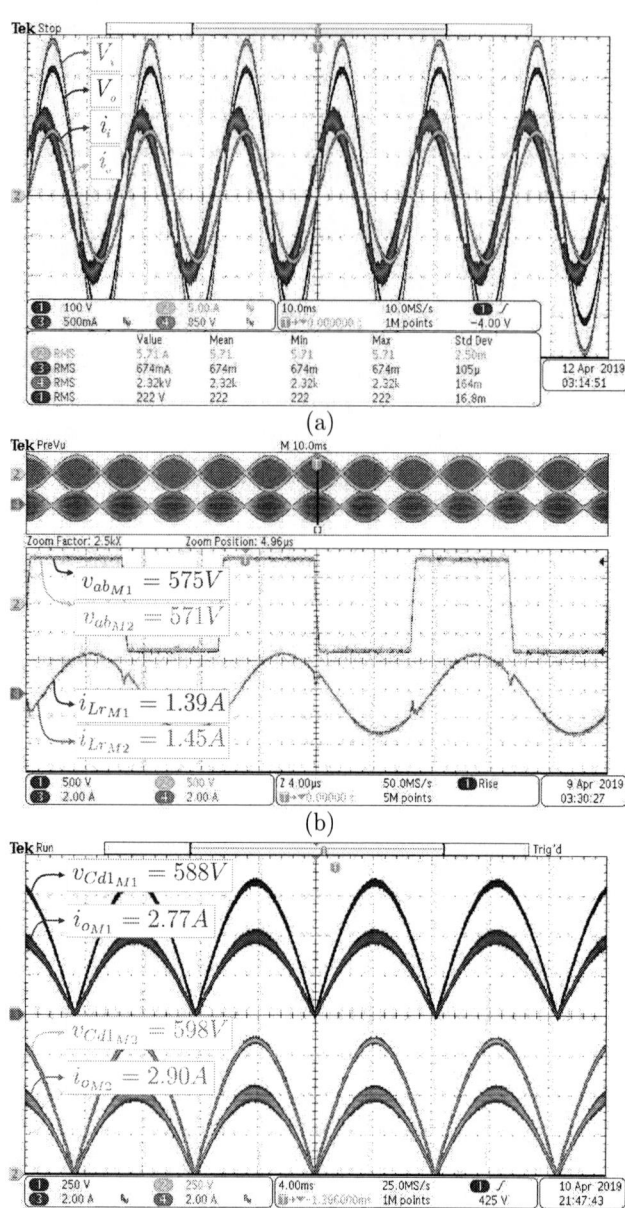

Fig. 5. Prototypes and setup test. (a) Module M1, (b) unfolder and (c) laboratory equipments and ISOP system under test.

Fig. 6. Experimental waveforms of the ISOP configuration of two modules. (a) Input and output voltages and input and load currents at grid period, (b) v_{ab} voltages and i_{L_r} currents of both modules at grid and switching frequencies, (c) voltages across capacitors $C_{d1_{M1}}$ and $C_{d1_{M2}}$ and output currents of both modules.

in Fig. 6(b). It is worth highlighting that there is power balancing between the modules, since, by visual inspection, the waveforms are superimposed.

The voltage balance among the modules can be seen across the voltage capacitor C_{d1} of the modules M1 and M2 in Fig. 6(c). In the same figure, the output currents ($i_{o_{M1}}$ and $i_{o_{M2}}$) of the modules are shown.

The modules' input voltages M1 and M2 were measured from the voltages across the capacitors C_{d1} and C_{d2} by using precision multimeters. Thus, the equation (8) was validated using the Keysight multimeters, model U1242B, which have error measurements of 0.15% for DC and 1% for AC. Two experiments were performed, the first one using the transformers with L_r e L_m shown in Table I and the second one with the air gap of the M1 transformer readjusted for increasing the magnetizing inductance. The new parameters are shown in Table II.

A comparison is made between the input voltages measurements and the theoretical values obtained using (8) in Table III. It is impossible to validate (8) using these results because the small difference between the measurements could be within the multimeter's error range. For this reason, a new transformer has been employed in M1. Table IV shows the modules' input voltage measurements when the transformers of Table II had been used. It can be observed that there is a 40.21 V difference between the modules voltages, where the module with the smaller voltage value has the largest λ. The

difference between the measured and theoretically estimated voltages is less than 0.1%, which validates the equation (8).

If the modules' input voltages are equal in the ISOP arrangement, then all modules will process the same power. This allows the theoretical conclusion that the efficiencies of one module and the ISOP arrangement of n modules are the same. The system experimental efficiency curves with one and two modules associated with the unfolder are presented in Fig. 7. The results were obtained from the Tektronix PA3000

978-1-7281-4181-7/19 $31.00 © 2019 IEEE

TABLE II

TRANSFORMERS CHARACTERISTICS TO THE SECOND EXPERIMENT.

Transformer turn ratio (a)	0.3584
Module 1 Resonant inductor (L_{r1})	126.85 μH
Module 1 Magnetizing inductance (L_{m1})	1.73 mH
Module 2 Resonant inductor (L_{r2})	124.92 μH
Module 2 Magnetizing inductance (L_{m2})	932.99 μH

TABLE III

THE MODULES' EXPERIMENTAL AND THEORETICAL INPUT VOLTAGE IN
THE FIRST EXPERIMENT.

	Experimental (rms)	Theoretical (rms)	Error (%)
Input voltage of the module 1	1147.93 V	1147.74 V	0.016
Input voltage of the module 2	1148.26 V	1148.45 V	0.016

TABLE IV

THE MODULES' EXPERIMENTAL AND THEORETICAL INPUT VOLTAGE IN
THE SECOND EXPERIMENT.

	Experimental (rms)	Theoretical (rms)	Error (%)
Input voltage of the module 1	1168.81 V	1168.12 V	0.059
Input voltage of the module 2	1128.6 V	1129.3 V	0.061

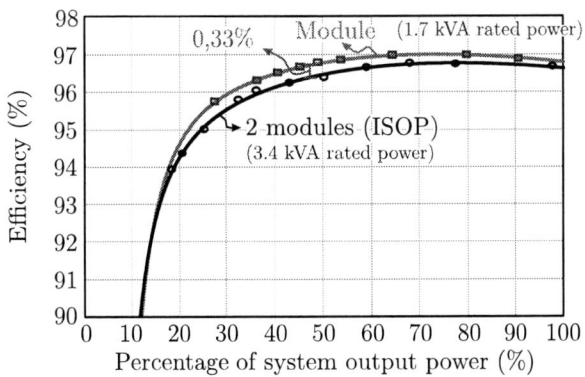

Fig. 7. Experimental results of efficiency of the unique module and the ISOP arrangement of two modules.

power analyzer, where the input voltage was measured using a voltage transformer (VT) with a measurement error less than or equal to 0.3%. The largest difference between the results was 0.33%, a value that can be attributed to small differences in input voltages, increased power processing by the unfolder, or even deviations in the VT measurements.

V. CONCLUSIONS

One phase of a three-phase SST has been proposed and analyzed, where the theoretical studies have been verified from a reduced scale prototype employing two modules with the unfolding stage. The system efficiency was higher than 95% for a wide power range, with a maximum of 96.8% and 96.6% at rated power. The parameters that affect the voltage balance of the proposed architecture were identified and the theoretical analysis was validated with the experimental measurements, with errors less than 1%.

The main advantages of the proposed SST are the simple modulation scheme, auto-balance of the input voltages and power of the modules, high efficiency and the absence of large DC link capacitors.

There are some challenges to overcome until the full-scale prototype to be implemented. The main challenge is related to the high voltage insulation between the transformer's medium and low voltage side. The projects of the auxiliary power supply, the high-frequency transformer, and the gate signals' transmission are being developed to guarantee safety electrical insulation.

REFERENCES

[1] S. Bifaretti, P. Zanchetta, A. Watson, L. Tarisciotti, and J. C. Clare, "Advanced Power Electronic Conversion and Control System for Universal and Flexible Power Management," *IEEE Transactions on Smart Grid*, vol. 2, no. 2, pp. 231–243, Jun. 2011.

[2] D. Grider, M. Das, A. Agarwal, J. Palmour, S. Leslie, J. Ostop, R. Raju, M. Schutten, and A. Hefner, "10 kV/120 A SiC DMOSFET half H-bridge power modules for 1 MVA solid state power substation," in *2011 IEEE Electric Ship Technologies Symposium*, Apr. 2011, pp. 131–134.

[3] J. E. Huber, J. Böhler, D. Rothmund, and J. W. Kolar, "Analysis and Cell-Level Experimental Verification of a 25 kW All-SiC Isolated Front End 6.6 kV/400 V AC-DC Solid-State Transformer," *CPSS Transactions on Power Electronics and Applications*, vol. 2, no. 2, pp. 140–148, Jun. 2017.

[4] L. Wang, Q. Zhu, W. Yu, and A. Q. Huang, "A Medium-Voltage Medium-Frequency Isolated DC-DC Converter Based on 15-kV SiC MOSFETs," *IEEE Journal of Emerging and Selected Topics in Power Electronics*, vol. 5, no. 1, pp. 100–109, Mar. 2017.

[5] J. S. Lai, W. H. Lai, S. R. Moon, L. Zhang, and A. Maitra, "A 15-kV class intelligent universal transformer for utility applications," in *2016 IEEE Applied Power Electronics Conference and Exposition (APEC)*, Mar. 2016, pp. 1974–1981.

[6] X. She, A. Q. Huang, and R. Burgos, "Review of Solid-State Transformer Technologies and Their Application in Power Distribution Systems," *IEEE Journal of Emerging and Selected Topics in Power Electronics*, vol. 1, no. 3, pp. 186–198, Sep. 2013.

[7] J. E. Huber and J. W. Kolar, "Volume/weight/cost comparison of a 1mva 10 kV/400 V solid-state against a conventional low-frequency distribution transformer," in *2014 IEEE Energy Conversion Congress and Exposition (ECCE)*, Sep. 2014, pp. 4545–4552.

[8] A. Q. Huang, Q. Zhu, L. Wang, and L. Zhang, "15 kV SiC MOSFET: An enabling technology for medium voltage solid state transformers," *CPSS Transactions on Power Electronics and Applications*, vol. 2, no. 2, pp. 118–130, 2017.

[9] L. F. Costa, G. Buticchi, and M. Liserre, "Quad-active-bridge as cross-link for medium voltage modular inverters," in *2015 IEEE Energy Conversion Congress and Exposition (ECCE)*, Sep. 2015, pp. 645–652.

[10] D. G. Bandeira, V. L. F. Borges, R. L. da Silva, and I. Barbi, "AC-AC Hybrid Switched-Capacitor Series Resonant Converter," in *2018 13th IEEE International Conference on Industry Applications (INDUSCON)*. São Paulo, Brazil: IEEE, Nov. 2018, pp. 1107–1114. [Online]. Available: https://ieeexplore.ieee.org/document/8627229/

[11] S. Zhao, Q. Li, F. C. Lee, and B. Li, "High-Frequency Transformer Design for Modular Power Conversion From Medium-Voltage AC to 400 VDC," *IEEE Transactions on Power Electronics*, vol. 33, no. 9, pp. 7545–7557, Sep. 2018. [Online]. Available: https://ieeexplore.ieee.org/document/8113593/

[12] S. De Simone, C. Adragna, and C. Spini, "Design guideline for magnetic integration in LLC resonant converters," in *2008 International Symposium on Power Electronics, Electrical Drives, Automation and Motion*. Ischia, Italy: IEEE, Jun. 2008, pp. 950–957. [Online]. Available: http://ieeexplore.ieee.org/document/4581225/

Low computational cost technique for SPMSM sensorless drive using active flux concept

Camila R. S. Bartsch
Electrical Engineering Department
Santa Catarina State University
Joinville, SC, Brazil
c.scalabrin@gmail.com

Luis F. F. de Campos
Electrical Engineering Department
South Santa Catarina University
Tubarão, SC, Brazil
luisf_ferreiraa@hotmail.com

Arthur G. Bartsch
Electrotechnology Department
Federal Institute of Santa Catarina
Jaraguá do Sul, SC, Brazil
arthur.bartsch@ifsc.edu.br

José de Oliveira
Electrical Engineering Department
Santa Catarina State University
Joinville, SC, Brazil
jose.oliveira@udesc.br

Ademir Nied
Electrical Engineering Department
Santa Catarina State University
Joinville, SC, Brazil
ademir.nied@udesc.br

Abstract—**In this paper, permanent magnet synchronous motor flux control is performed, involving the active flux concept. This flux control is associated with a scalar drive (V/f), being a native sensorless technique, without a special approach. Moreover, the applied strategy has low computational cost and can operate at very reduced speeds, tracking references and rejecting disturbances. The focused application is direct-drive washing machines. Experimental results are presented, including a very low-speed test. The main contribution of this paper is related to the control design.**

Index Terms—**sensorless, scalar control, PMSM, low computational cost**

I. INTRODUCTION

Modern washing machines often use a direct-drive system to move its load. In this type of drive system, the motor is directly coupled to the load, eliminating the gearbox. However, the electronic motor drive needs to allow motor operation in a relatively large speed range and high torque in starting [1]–[3].

Permanent magnet synchronous motors (PMSMs) fulfill these conditions [3]. Furthermore, such motors are more efficient, if compared with induction motors. The motor drive system needs to be reliable, i.e. fails should be minimized. Also, it needs to have low cost and to be capable of operating in a variable speed range. Therefore, one of the most important requirement is a low computational cost drive strategy, reducing the processor cost [1]–[3]. Also, a technique without speed sensors, i. e. sensorless or encoderless, since, besides avoiding the high cost, removing these components reduces elements subjected to fails [1]–[3].

Among the most commons motor drive techniques, for PMSMs, there are the field-oriented control (FOC) [2] and the direct torque control (DTC) [4], [5]. These strategies are developed to track the torque (or speed) reference, following the maximum torque per ampere (MTPA) condition [6]. This condition can be summarised into: eliminate unnecessary flux linkage of the motor core. Due to this, a flux or extra current

control is often applied. Both techniques could be used with or without speed sensors. In this second case, the computational complexity would increase, implying in an expensive processor. There are several other techniques, further DTC and FOC, as model-based predictive control [7], internal model control [8], fuzzy control [8] among others. However, all of them need a high-level processor, considering a sensorless approach [9].

There are three main sensorless drive strategies [10]–[12]. The first involves injecting high-frequency signals in the drive [10]. It is properly used in low-speed applications [11], [13]. The second idea is related to the back-electromotive force estimation, suitable for medium to high-speed applications [11]. Finally, the third main approach is based on robust estimators, as sliding mode observers, model reference adaptive systems or Kalman filters. These strategies can be applied in a large speed range, but they significantly increase the computational cost, due to the estimation [12].

Considering the exposed, the active flux estimator and control was proposed to operate with a large range of techniques [14]–[18]. The active flux concept is to control the flux (or the active flux) objecting to employ only the necessary flux in the motor. In other words, this approach allows the drive system naturally to obey the MTPA condition. An advantage of this theory is the minimum dependence of motor parameters (which are time-varying). Also, the general concept of the technique allows controlling both salient poles and non-salient poles PMSM cases similarly.

One interesting application of the active flux concept is associating it with the scalar drive (V/f or I-f) [15], [17]. The active flux control is used to stabilize the scalar drive, allowing it to operate in the MTPA condition [15], [16]. However, the scalar drive is naturally sensorless, since the synchronous motor does not have slip (the imposed electrical frequency is directly related to motor speed, considering the number of poles). Consequently, with the flux control, the scalar drive can impose a large speed range operation, without

978-1-7281-4181-7/19 $31.00 © 2019 IEEE

switching estimator models, increasing control complexity or even using the Park's transformation. This condition implies a low computational cost control with a great variety of control resources.

However, considering the application of scalar drive and flux control at the same time, one of the main existing difficulties lies in the control design. A clear methodology of control design (or, at least, a model) is absent in the main references of this technique [15]. Thus, the main objective of this paper, and the main advance in relation to [15], is to present control models for the flux control design. Also, an innovative control scheme is developed, considering there is a simplification in the proposed control loop if the scheme were compared with previous works [15], [16]. This innovation is a result of the application of the technique in a non-salient PMSM.

Therefore, this paper is organized as follows: Section II presents the PMSM modeling, in the stationary reference frame, shows active flux concept and presents the drive description. The proposed flux models, for improving the control design, are presented in Section III. Simulation and experimental results are presented in Section IV, respectively. The conclusion appears in Section V.

II. MOTOR MODELING, ACTIVE FLUX AND DRIVE SYSTEM

The PMSM dynamical behavior can be mathematically described using a concentrated parameters model. Therefore, considering the PMSM an electrical three-phase circuit, symmetric and balanced, with neglecting saturation, hysteresis and Eddy currents effects, in stationary reference frame, it can be expressed with [11]:

$$v_{\alpha s}(t) = R_s i_{\alpha s}(t) + \frac{\mathrm{d}}{\mathrm{d}t}\lambda_{\alpha s}(t) \tag{1}$$

$$v_{\beta s}(t) = R_s i_{\beta s}(t) + \frac{\mathrm{d}}{\mathrm{d}t}\lambda_{\beta s}(t) \tag{2}$$

where R_s is the circuit resistance, $v_{\alpha s}(t)$ and $v_{\beta s}(t)$ are the stationary voltages, $i_{\alpha s}(t)$ e $i_{\beta s}(t)$ are the stationary currents and $\lambda_{\alpha s}(t)$ and $\lambda_{\beta s}(t)$ are the stationary linkage fluxes. The fluxes components are expressed by

$$\lambda_{\alpha s}(t) = L_s i_{\alpha s}(t) + \varphi_{\alpha s}(t) \tag{3}$$

$$\lambda_{\beta s}(t) = L_s i_{\beta s}(t) + \varphi_{\beta s}(t) \tag{4}$$

where L_s is the inductance in stationary reference frame and $\varphi_{\alpha s}(t)$ e $\varphi_{\beta s}(t)$ are the stationary permanent magnet flux linkages, given by:

$$\varphi_{\alpha s}(t) = \varphi \cos(\theta_e(t)) \tag{5}$$

$$\varphi_{\beta s}(t) = \varphi \sin(\theta_e(t)) \tag{6}$$

where φ is the permanent magnet linkage flux per poles pair and $\theta_e(t)$ is the electrical angular position, referred to stator flux.

The mechanical dynamics is expressed with:

$$\tau_e(t) - \tau_L(t) = b_m \omega_m(t) + J_m \frac{\mathrm{d}\omega_m(t)}{\mathrm{d}t} \tag{7}$$

where $\tau_L(t)$ is the motor load torque, b_m is the friction, J_m is the inertia, $\omega_m(t)$ is the mechanical speed and $\tau_e(t)$ is the electrical torque, which for surface PMSM (SPMSM), is given with

$$\tau_e(t) = \frac{3}{2}n_p[i_{\beta s}(t)\varphi_{\alpha s}(t) - i_{\alpha s}(t)\varphi_{\beta s}(t)] \tag{8}$$

where n_p is the pair of poles.

The motor electrical position $\theta_e(t)$ is obtained by:

$$\theta_e(t) = n_p \int_0^t \omega_m(t')\,\mathrm{d}t'. \tag{9}$$

Active flux $(\overline{\lambda})$ theory proposes that only the useful flux linkage, for torque production, should stay in the motor core [14]. This is a generic theory, which could be applied in any alternate current (AC) machine. Particularly, for PMSMs, the active flux components can be calculated with:

$$\overline{\lambda}_{\alpha s}(t) = \lambda_{\alpha s}(t) - L_s i_{\alpha s}(t) \tag{10}$$

$$\overline{\lambda}_{\beta s}(t) = \lambda_{\beta s}(t) - L_s i_{\beta s}(t). \tag{11}$$

According to (8), only the permanent magnet linkage fluxes $\varphi_{\alpha s}(t)$ and $\varphi_{\beta s}(t)$ are useful for torque production in non-salient PMSM [14]. Therefore, for this specific case,

$$\overline{\lambda}_{\alpha s}(t) = \varphi_{\alpha s}(t) \tag{12}$$

$$\overline{\lambda}_{\beta s}(t) = \varphi_{\beta s}(t) \tag{13}$$

$$\overline{\lambda}(t) = \varphi = \sqrt{\varphi_{\alpha s}(t)^2 + \varphi_{\beta s}(t)^2}. \tag{14}$$

For other AC machines, it is possible to control the active flux directly [14]. For non-salient PMSM, it is necessary to establish the active flux (permanent magnet flux) as the flux linkage amplitude reference. Therefore, the non-useful flux can be eliminated, implying operation in MTPA condition [14].

From (1) and (2), stationary motor fluxes can be estimated. This way:

$$\lambda_{\alpha s}(t) = \int_0^t [v_{\alpha s}(t') - R_s i_{\alpha s}(t')]\,\mathrm{d}t' \tag{15}$$

$$\lambda_{\beta s}(t) = \int_0^t [v_{\beta s}(t') - R_s i_{\beta s}(t')]\,\mathrm{d}t'. \tag{16}$$

The integrals from (15) and (16) can be replaced by low-pass filters, with low cutoff frequencies (around 2.0 Hz), in the estimation process. However, a good filter gain adjustment is needed to avoid signal distortion. Also, using the filter, the control system will only present good operation for frequencies a decade greater than cutoff filter frequency. The direct integration also can be used, with a compensation term or without [14], [15].

With the active flux components value it is possible to estimate the electrical position, given by [15]

$$\theta_e[k] = \arctan\left(\frac{\overline{\lambda}_{\beta s}[k]}{\overline{\lambda}_{\alpha s}[k]}\right) \tag{17}$$

and the mechanical speed, given by

$$\omega_m[k] = \frac{1}{n_p}\frac{\overline{\lambda}_{\alpha s}[k-1]\overline{\lambda}_{\beta s}[k] - \overline{\lambda}_{\beta s}[k-1]\overline{\lambda}_{\alpha s}[k]}{t_s \overline{\lambda}^2} \tag{18}$$

978-1-7281-4181-7/19 $31.00 © 2019 IEEE   924

where k means the discrete time and t_s indicates the sampling time.

In this work, the estimations given by (17) and (18) are not used, which implies a simplification in the control scheme.

As said in the Introduction, the active flux concept can be utilized to design several sensorless drive, since it is possible directly estimate speed and position as seen in (17) and (18) [17], [18]. However, it is also possible to use this theory with an inherent sensorless control, the scalar drive. In this case, a flux control loop is employed to stabilize the drive system [15].

The scalar drive V/f functions applying a balanced three-phase sinusoidal voltage with an electrical frequency equivalent to the desired mechanical speed. If there is sufficient voltage, the motor will run in the desired mechanical speed since the motor is synchronous. If there is not a sufficient voltage, the system will become unstable. Due to this, in scalar drive systems often there is an excess of energy, which implies low efficiency.

However, the flux control loop can be employed to minimize the energy that source the motor, using only the need to maintain the motor operation. Therefore, the scalar drive uses an acceleration ramp with proportional voltage v_p and frequency, whose relation is given by [15]:

$$v_p(t) = \kappa_{\mathrm{vf}} \frac{\omega_e^*(t)}{2\pi} + v_0 \tag{19}$$

where κ_{vf} is the proportionality constant, ω_e^* the electrical angular speed reference (which, in steady state, will become the motor electrical speed) and $v_0 = R_s i_n$ is a initial voltage for load starting, being i_n the rated motor current.

The stabilization process of scalar drive is performed with the flux control. It is possible to manipulate (19), considering that flux control is responsible for a small variation in the voltage amplitude. This variation will imply the stabilization and the MTPA condition operation. Thus,

$$v_p^*(t) = v_p(t) + \Delta v \tag{20}$$

with Δv being calculated by the flux controller.

The voltage reference angle, which is used to generate the symmetric sinusoidal phase voltages, is calculated by:

$$\theta_e^*(t) = \int_{t_0}^{t} \omega_e^*(t') \, dt'. \tag{21}$$

III. FLUX CONTROL DESIGN METHODOLOGY

In this section, two models for the control design are presented. After, the control design method is presented. Finally, the control evaluation procedure is defined. As in works [15] there is no information about the control design, this section can be considered an advance in this area since the developed procedure avoids the empirical control adjust.

A. Phenomenological model

Assuming that:

$$v_s(t) = R_s i_s(t) + \frac{d}{dt} \lambda_s(t) \tag{22}$$

and that:

$$i_s(t) = \frac{1}{L_s}(\lambda_s(t) - \varphi_s(t)), \tag{23}$$

it is possible to substitute (23) in (22), resulting in:

$$v_s(t) = \frac{R_s}{L_s}(\lambda_s(t) - \varphi_s(t)) + \frac{d}{dt}\lambda_s(t). \tag{24}$$

Using Laplace's transformation in (24) results in:

$$v_s(p) = \frac{R_s}{L_s}(\lambda_s(p) - \varphi_s(p)) + p\lambda_s \tag{25}$$

where p is the Laplace operator.

Considering, in (25), that the term $-\frac{R_s}{L_s}\varphi_s(p)$ is a disturbance, (25) can be rewritten as a transfer function

$$\frac{\lambda_s(p)}{v_s(p)} = \frac{t_b}{1 + t_b p} \tag{26}$$

where t_b is the system time base ($t_b = L_s/R_s$).

Therefore, (26) is a useful model to design the flux control. The model may look simple but it is not present in the main references of this strategy application [15], [17].

B. Identification model

The numerical identification method consists in to obtain a model from experimental samples of a tested plant [19]. Therefore, it is possible to impose a voltage input and to analyze the flux linkage amplitude output.

Assuming the existence of a response flux vector Λ and a voltage input vector V, both with n_a samples, obtained with a t_s sampling time, it is possible to get a first order model. For this, given a matrix H

$$H = \begin{bmatrix} \Lambda[k-1] & V[k-1] \end{bmatrix}, \tag{27}$$

it is possible to apply the least square method, with discrete coefficients κ_{ak} and κ_{ak} are given with:

$$\begin{bmatrix} \kappa_{ak} \\ \kappa_{bk} \end{bmatrix} = (H^{\mathrm{T}}H)^{-1}H^{\mathrm{T}}\Lambda[k]. \tag{28}$$

Therefore, supposing a continuous model given by

$$\frac{\lambda_s(p)}{v_s(p)} = \frac{\kappa_b}{p + \omega_0} \tag{29}$$

the continuous parameters can be calculated approximately with

$$\omega_0 = \frac{1 - \kappa_{ak}}{t_s} \tag{30}$$

$$\kappa_b = \frac{\kappa_{bk}}{t_s}. \tag{31}$$

This approach is pretty interesting in an experimental platform since it allows including nonlinear effects associated with the drive system in the coefficient values.

Algorithm 1 Algorithm for controller

1: Read currents $i_{\alpha s}$ and $i_{\beta s}$
2: Calculate $\alpha\beta$ flux components using (15) and (16) and/or a low-pass filter.
3: Calculate the active flux with (14)
4: Calculate Δv with PI controller over active flux error
5: Integrate the reference angular position using (21)
6: Calculate $v_p(t)$ and $v_p{}^*(t)$ using (19) and (20)
7: Generate symmetrical co-sinusoidal functions with correct peak voltage and the electrical angular position and send to modulator
8: Wait for next sampling period

C. Control design technique and algorithm

For the control/estimator design, it is important to emphasize some points in this peculiar application:

- the flux control cannot interfere in the V/f drive, which means it cannot be so fast that the V/f could not provide energy for the motor. Due to this, it is interesting an open-loop procedure;
- the flux control cannot be so slow that implies the system instability with a load disturbance;
- the flux estimator needs to be fast enough to converge, giving the correct information for the flux control, but it cannot be so fast at the point to be noise susceptible;
- if the control does not start with the drive, the estimator may be slower than the condition when the control starts with the drive since it can estimate the flux when the control is not operating.

Considering these issues, a good approach for the control and estimator design is the frequency response technique. For the estimator design, a pure integrator (with some type of offset compensation) or a low-pass filter is enough, if the control is not activated at the start. The low-pass filter can have a low cutoff frequency since it ideally should be an integrator.

The controller frequency band is a function of the motor time base. However, a good control choice is to place the control zero about 5 to 10 times slower than the plant natural frequency (to allow the control some filter characteristics) and the control cutoff frequency about 2 to 5 times slower than the plant natural frequency, to avoid actuates over V/f scalar drive. Algorithm 1 presents an algorithm for the implemented control.

IV. RESULTS

A numerical simulation was develop to test the controller. In the developed simulation, the phenomenological model was used to design the control. Parameters of evaluated PMSM are presented in Table I.

A proportional-integrative (PI) controller, tuned with 5 Hz cut-off frequency was used. The sampling time adopted was 0,1 ms. The V/f constant $\kappa_{\mathrm{vf}} = 2.5$ V/Hz. The start was considered without load. A load torque of 3.2 Nm was inserted at 5 s. Figure 1 shows simulated speed and torque dynamics.

TABLE I
EVALUATED PMSM PARAMETERS

Parameter	Value	Parameter	Value
R_s	16.1 Ω	Rated power	800 W
L_s	3.2e-2 H	Rated speed	100 rpm
φ	2.33e-1 Wb	Rated current	2.3 A
J_m	5.2e-2 N m s^2	Rated voltage	220 V
b_m	9.8e-4 N m s	Pair of poles (n_p)	24

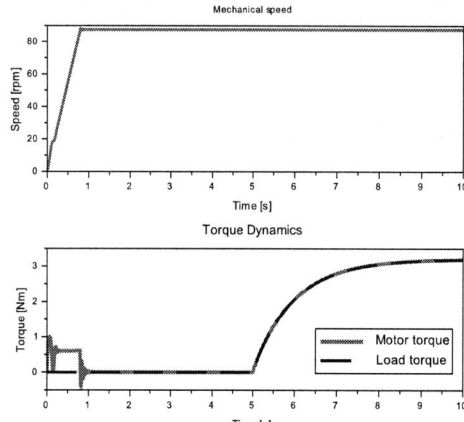

Fig. 1. Simulation of speed and torque

From Figure 1, it is possible to see that even with load insertion, the speed was maintained approximately constant, due to the proposed control methodology design, showing its validity. It occurs since the flux is reduced with the load insertion, then the controller increases the voltage to compensate for this. This last remark induces the speed loop is stabilized by the controller, in V/f drive: it has low current with no load condition and enough current to reject torque disturbance when it occurs.

The dSPACE 1103 was used to control implementation in the experimental setup. The STEVAL IHM 023v1 inverter was used. A 2 Hz low-pass filter was used instead of the integrator, for the flux estimation. The filter gain established a 5.6 Vs/rad reference for the flux. It was not compensated to reduce processing. A pulse-width modulation (PWM), operating at 16 kHz, was used to drive the inverter. The controller was tuned with the identified model. This model considered the gain change in the system, after the filter, the PWM modulation, and other practical issues. None extra dynamics were considered. Therefore, the identified model was:

$$\frac{\lambda_s(p)}{v_s(p)} = \frac{98.14}{p + 1013}. \tag{32}$$

Two experimental tests of the scalar drive with the flux stabilization loop are presented. In the first, only one ramp (followed by a constant) reference is imposed on the motor. Then, a load torque of 3.2 Nm is inserted to the drive, at 5 s. The flux controller is only activated after the ending of the acceleration ramp. The second is a low-speed test, where the system is submitted to eight low-speed degrees.

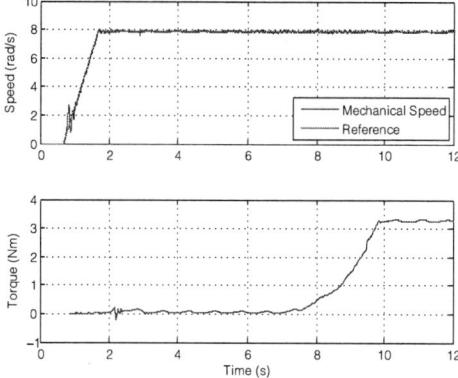

Fig. 2. Speed and torque response

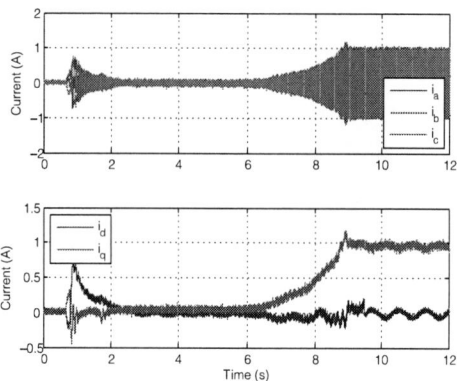

Fig. 3. Phase and dq current response

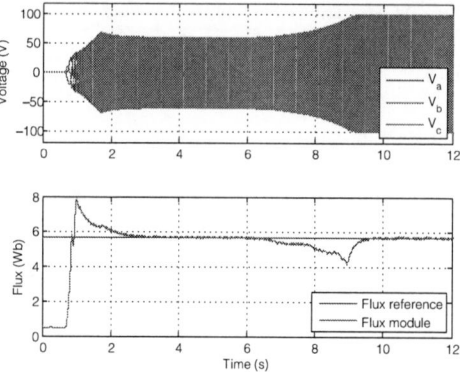

Fig. 4. Voltage and flux response

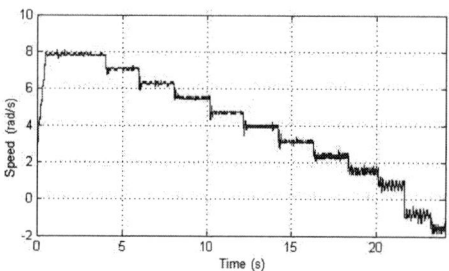

Fig. 5. Low speed test dynamics

Figures 2-4 present the dynamic speed and torque response, current, voltage and flux dynamics in the scalar drive with flux stabilizing loop. Figures 5-6 show the low-speed test and the current dynamics obtained in this second test.

In Figure 2, it is clear that the speed achieves correctly in the reference, which means that the scalar drive works suitable. Moreover, the disturbance practically unaffected the speed, since the flux control was capable to actuate in the system and rejects the load disturbance.

In Figure 3, it is possible to see that the flux controller yields to the current reduction while there is not load in the motor. This condition implies an improvement in motor efficiency since unnecessary flux is removed. The currents only increase when there is load disturbance, which allows the disturbance rejection. In this same Figure 3, the dq currents are shown (they were not used at any moment in the drive). These currents demonstrate that the MTPA condition of SPMSMs was achieved by the system (the direct axis current is almost zero in all moments after the flux stabilization).

In Figure 4, the voltage decreases after the flux control to be activated (with the acceleration ramp end). Concomitantly, the flux is converging for its reference. With the load torque input, there is a perturbation in the flux and its control actuates

in a way to return it to the reference. Consequently, there is an augment in the voltage level and the load disturbance rejection.

From Figure 5, it is perceived that the proposed drive system was capable to operate the motor in a very low-speed condition and also performs a reversion in this condition. The motor was maintained stable even subjected to cogging torque, inductance and resistance variation, friction variation among others. In Figure 6, it is possible to see that the current was maintained minimal until the reference of 6.0 rpm (2.4 electrical Hz). With lower references, the current filter of 2 Hz cutoff frequency was not more actuating as a filter, degrading the controller performance and, also, violating the MTPA condition. Furthermore, in this low-speed condition, given the accentuated nonlinearities, the own model used for control design is not valid. Besides it, the control was robust enough to operate the system.

After the low-speed test, the motor load curve was obtained, with voltage per speed, presented in Figure 7. In this figure, it is possible to note that the correct region for the control actuation (with speeds higher than 6.0 rpm) and the nonlinear region (speeds lower than 6.0 rpm). Also, there is a difference in the curves with and without load condition, since the voltage increases with the load disturbance to reject it.

Usually, sensorless techniques have difficulties with the low-speed operation and, even, with starting. The proposed technique was capable to deal with both conditions, with good dynamic behavior, accomplishing the MTPA condition. Also, the drive rejected suitably disturbances. It is noted that

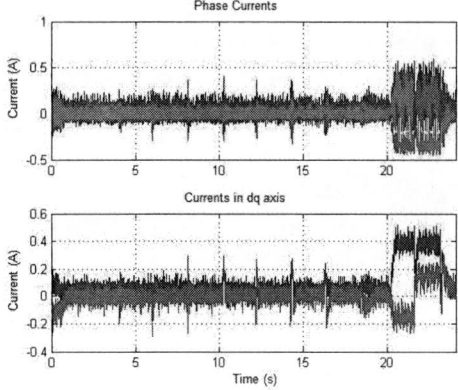

Fig. 6. Current dynamics in low speed test

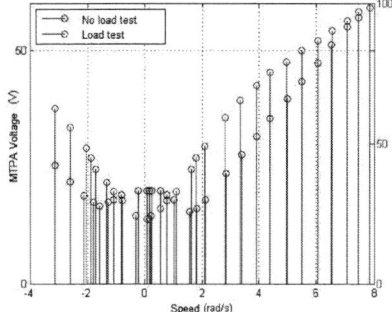

Fig. 7. Motor load curve

the proposed technique has only one flux PI, one filter and two sinusoidal functions, while FOC needs, at least, eight sinusoidal functions, three PIs and a lot of effort with control tuning.

V. CONCLUSION

In this work, an alternative scalar drive with flux stabilization loop was proposed, mainly improving the presented in [15], for the SPMSM specific case. The control design, with control models, also was a contribution of this work. The proposed technique has low processing and hardware cost. The strategy was very interesting for washing machine application since it attends some condition as MTPA, low-speed operation, load disturbance rejection, sensorless operation (without an increase of computational cost) and speed reference tracking. As a future step, a direct comparison with FOC is proposed.

ACKNOWLEDGMENT

The authors thank STM, CAPES, FAPESC, "Programa de Bolsas Universitárias de Santa Catarina UNIEDU" (FUMDES) and UDESC for financial support for this work realization.

REFERENCES

[1] R. Bojoi, B. He, F. Rosa, and F. Pegoraro, "Sensorless direct flux and torque control for direct drive washing machine applications," in *2011 IEEE Energy Conversion Congress and Exposition*, 2011, pp. 347–354.

[2] M. A. Bevilaqua, A. Nied, and J. de Oliveira, "Labview fpga foc implementation for synchronous permanent magnet motor speed control," in *2014 11th IEEE/IAS International Conference on Industry Applications*, 2014, pp. 1–8.

[3] S. Chi, Z. Zhang, and L. Xu, "Sliding-mode sensorless control of direct-drive pm synchronous motors for washing machine applications," *IEEE Transactions on Industry Applications*, vol. 45, no. 2, pp. 582–590, 2009.

[4] G. Buja and M. Kazmierkowski, "Direct torque control of pwm inverter-fed ac motors – a survey," *IEEE Transactions on Industrial Electronics*, vol. 51, no. 4, pp. 744–757, Abril 2004.

[5] F. Niu, B. Wang, A. Babel, K. Li, and E. Strangas, "Comparative evaluation of direct torque control strategies for permanent magnet synchronous machines," *IEEE Transactions on Power Electronics*, vol. 31, no. 2, pp. 1408–1424, Fevereiro 2016.

[6] M. Preindl and S. Bolognani, "Model predictive direct torque control with finite control set for pmsm drive systems, part 1: Maximum torque per ampere operation," *IEEE Transactions on Industrial Informatics*, vol. 9, no. 2, pp. 1912–1921, Novembro 2013.

[7] J. Rodríguez, M. P. Kazmierkowski, J. R. Espinoza, P. Zanchetta, H. Abu-Rub, H. A. Young, and C. A. Rojas, "State of the art of finite control set model predictive control in power electronics," *IEEE Transactions on Industrial Informatics*, vol. 9, no. 2, pp. 1003–1016, Maio 2013.

[8] M. Seilmeier, A. Boehm, I. Hahn, and B. Piepenbreier, "Identification of time-variant high frequency parameters for sensorless control of pmsm using an internal model principle based high frequency current control," in *2012 XXth International Conference on Electrical Machines*, 2012, pp. 987–993.

[9] L. Rovere, A. Formentini, A. Gaeta, P. Zanchetta, and M. Marchesoni, "Sensorless finite-control set model predictive control for ipmsm drives," *IEEE Transactions on Industrial Electronics*, vol. 63, no. 9, pp. 5921–5931, 2016.

[10] M. A. Ghazimoghadam and F. Tahami, "Flux estimation by asymmetric carrier injection for sensorless direct torque control of pmsm," in *2012 3rd Power Electronics and Drive Systems Technology (PEDSTC)*, 2012, pp. 44–50.

[11] O. Lehmann, N. K. Nguyen, J. Schuster, and J. Roth-Stielow, "Optimized sensorless dtc of pmsm for electric vehicles by using a switching command synchronized evaluation at standstill and low speed," in *2015 9th International Conference on Power Electronics and ECCE Asia (ICPE-ECCE Asia)*, 2015, pp. 1386–1393.

[12] Z. Guo, G. Xiang, M. Pu, and S. Dian, "Sensorless drive of direct-torque-controlled pmsms based on robust extended kalman filter," in *2016 35th Chinese Control Conference (CCC)*, 2016, pp. 4711–4716.

[13] P. Kshirsagar, R. P. Burgos, J. Jang, A. Lidozzi, F. Wang, D. Boroyevich, and S. K. Sul, "Implementation and sensorless vector-control design and tuning strategy for smpm machines in fan-type applications," *IEEE Transactions on Industry Applications*, vol. 48, no. 6, pp. 2402–2413, 2012.

[14] I. Boldea and S. C. Agarlita, "The active flux concept for motion-sensorless unified ac drives: A review," in *International Aegean Conference on Electrical Machines and Power Electronics and Electromotion, Joint Conference*, 2011, pp. 1–16.

[15] A. Moldovan, S. C. Agarlita, G. D. Andreescu, and I. Boldea, "Wide speed range v/f with stabilizing loops control of tooth-wound ipmsm drives," in *2012 13th International Conference on Optimization of Electrical and Electronic Equipment (OPTIM)*, 2012, pp. 424–431.

[16] G. D. Andreescu, C. E. Coman, A. Moldovan, and I. Boldea, "Stable v/f control system with unity power factor for pmsm drives," in *2012 13th International Conference on Optimization of Electrical and Electronic Equipment (OPTIM)*, 2012, pp. 432–438.

[17] I. Boldea, A. Moldovan, and L. Tutelea, "Scalar v/f and i-f control of ac motor drives: An overview," in *2015 Intl Aegean Conference on Electrical Machines Power Electronics (ACEMP), 2015 Intl Conference on Optimization of Electrical Electronic Equipment (OPTIM) 2015 Intl Symposium on Advanced Electromechanical Motion Systems (ELECTROMOTION)*, 2015, pp. 8–17.

[18] F. J. H. Kalluf, A. S. Isfănuţi, L. N. Tutelea, A. Moldovan-Popa, and I. Boldea, "1-kw 2000-4500 r/min ferrite pmsm drive: Comprehensive characterization and two sensorless control options," *IEEE Transactions on Industry Applications*, vol. 52, no. 5, pp. 3980–3989, 2016.

[19] K. J. Astrom and B. Wittenmark, *Computer-controlled systems theory and design*, 2nd ed. Prentice Hall International Editions, 1990.

A Model Predictive Control Applied To Single-Phase Packed-U-Cells Converter

Jordan Zucuni*, Dimas Alã Schuetz*, Felipe Bovolini Grigoletto[†],
Fernanda Carnielutti de Morais*, Margarita Norambuena[‡], José Rodriguez[§], Humberto Pinheiro*
*Federal University of Santa Maria - UFSM, Santa Maria, Brazil
{jzucuni, dimasschuetz96, fernanda.carnielutti and humberto.ctlab.ufsm.br}@gmail.com
[†]Federal University of Pampa - UNIPAMPA, Alegrete, Brazil, grigoletto@gmail.com
[‡]Universidad Tecnica Federico Santa Maria, Valparaíso, Chile, margarita.norambuena@gmail.com
[§]Universidad Andres Bello, Santiago, Chile, jose.rodriguez@unab.cl

Abstract—This paper proposes a Modulated Model Predictive Control technique for a Single Phase Packed-U-Cells Converter, resulting in fixed switching frequency while maintaining the basic features of the standard Model Predictive Control approach, such as inclusion of model and control nonlinearities, constraints and multiobjective optimization on a straightforward way. Specifically, the proposed algorithm arranges the converter voltage vectors in a switching sequence to be applied over a sampling period in order to control the output current, balance internal capacitors voltages and minimize the number of switches commutations. Simulation results are shown comparing the proposed technique with the standard one in terms of output current and voltage, regulation of the internal capacitor voltage, harmonic content and number of switch commutations.

Index Terms—Model Predictive Control, Packed-U-Cells Converter, Optimization.

I. INTRODUCTION

The Packed-U-Cells (PUC) is a recent topology of multilevel converters that has less components when compared to other multilevel converter topologies [1]. As a disadvantage, this topology presents asymmetric reverse voltage across the semiconductor switches. The PUC converter can generate seven or five voltage levels according to the DC capacitor voltages, but the configuration with seven levels requires a more complex control for the DC bus capacitor. As a result, for a five-level voltage, also named PUC5, is a promising topology that can be employed in applications such as drives and renewable energies [2].

Recently, some modulation strategies have been developed for the PUC5 converter. In [3] a Space Vector Modulation (SVM) is applied to a single-phase sensorless PUC5 converter. Although sensorless operating is desirable, this strategy results in larger voltage oscillations on the DC bus, requiring a large capacitance to limit the voltage ripple. In [4], a SVM strategy has been applied to a three-phase PUC5 in order to minimize the commutation losses and maintain the DC capacitor voltage balanced and regulated. This approach uses the three nearest converter voltage vectors, selected in order to minimize a cost function. Also, Pulse Width Modulation (PWM) is proposed for sensorless operating of PUC5 converter [5], [6]. As disadvantage, large DC capacitors are required to guarantee acceptable levels of the voltage ripple.

Another interesting alternative that presents good performance is the Model Predictive Control (MPC) [7]. It consists of predicting future values of the controlled outputs according to the control action, that is, in this case the converter voltage vectors, or switching states. This predictions are analysed by means of a cost function, which reflects the design objectives. As described in [7], a cost function is evaluated for each possible converter voltage vector, and the control action that presents the minimum cost is applied for a sampling period T_s. This is a very interesting approach that can handle nonlinearities and constraints, but brings a well-known disadvantage, which consists of a variable switching frequency.

Different MPC strategies for the PUC converter have already been presented in the literature. In [8] an MPC algorithm is proposed for a PUC converter with seven levels. This algorithm selects just one vector in each sampling time and the objective is to minimize the grid current error and maintain the capacitor voltage regulated. On the other hand, MPC is also applied to the PUC5 converter, as described in [9]. In this work, a dynamic voltage restorer, DVR, is proposed in order to be controlled by a low-complexity MPC algorithm. In addition, in [10] an MPC strategy is presented for a PUC5 converter. In this approach, a fixed switching frequency algorithm is used for minimizing only the current grid error. However, it can be observed that there is no feedback of the internal capacitor voltage.

In order to overcome the problem of variable switching frequency, one solution was proposed by [11], that brings an MPC technique based on SVM, called Modulated Model Predictive Control - M^{2}PC. This strategy uses a generic cost function, as in standard MPC, that can encompass many terms, nonlinearities and restrictions as desired, but differs from the fact that the switching vectors are arranged in sectors, resembling the SVM approach, from which the term "Modulated" derives. Thus, it preserves the main advantages of standard MPC while resulting in fixed switching frequency. As a result, in this paper this methodology is extended for a PUC5 converter, controlling the output current, balancing internal capacitor voltages as well as diminishing the number of commutations. Then, the results are compared to the standard MPC approach for the proposed converter.

978-1-7281-4181-7/19 $31.00 © 2019 IEEE

This paper is presented as follows: Section II presents the single-phase PUC5 topology and its switching states; Section III describes the proposed M²PC algorithm; in Section IV, extensive simulation results are shown in order to demonstrate the good performance of the algorithm; finally, Section V presents some discussions and the conclusions of this paper.

II. DESCRIPTION OF THE SINGLE-PHASE PUC5 TOPOLOGY

The single-phase PUC converter, represented in Fig. 1, is composed of three sets of semiconductors, where the switches of each set are complementary. Depending on the relation between the capacitor and source voltages, it can generate different numbers of output voltage levels. The term PUC5 refers to the converter in which the value of capacitor voltage is regulated to half the DC voltage source amplitude E, that is, $E/2$, generating five levels on the output voltage.

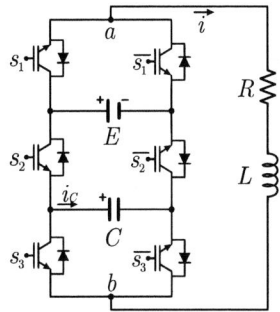

Fig. 1. Single-Phase PUC5 Converter.

The PUC5 converter has eight different switching combinations/states, called switching vectors, where some different switching vectors result in the same output voltage. These states are called redundancies. Table I shows all the switching states for PUC5 converter, where it is possible to see that some output levels can be synthesized by two different switching vectors, resulting in opposite currents flowing through the capacitor. These states can be used in order to regulate the capacitor voltage to the desired value.

TABLE I
SWITCHING STATES OF A SINGLE-PHASE PUC5 CONVERTER.

State	s_{1x}	s_{2x}	s_{3x}	i_{Cx}	Output Voltage (v_{ab})	v_x/E_x
1	0	1	1	0	$-E_x$	-1
2	0	1	0	$-i_x$	$v_{Cx} - E_x$	-1/2
3	0	0	1	i_x	$-v_{Cx}$	
4	0	0	0	0	0	0
5	1	1	1		0	
6	1	0	1	i_x	$E_x - v_{Cx}$	1/2
7	1	1	0	$-i_x$	v_{Cx}	
8	1	0	0		E_x	1

III. DESCRIPTION OF THE PROPOSED M²PC STRATEGY

First of all, three objectives for the proposed M²PC strategy, whose block diagram is presented in Fig. 2, can be highlighted: (i) control the output current in order to minimize the error between the actual and reference current; (ii) regulation of the DC bus capacitor voltage to the desired value, that is, $E/2$ and (iii) minimization of the number of switch commutations.

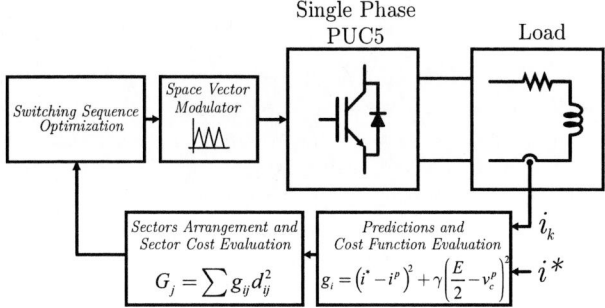

Fig. 2. Block Diagram of the proposed M²PC strategy.

In order to implement the M²PC approach, first, for each of the eight switching states presented in Table I, the cost function (1) is evaluated:

$$g = (i^* - i^p)^2 + \gamma \left(\frac{E}{2} - v_c^p \right)^2 \tag{1}$$

where i^* is the reference current and i^p and v_c^p are the current and the capacitor voltage predicted for the ith switching state, respectively, at $k+1$ sampling instant. In the standard MPC, the switching state that results in the minimum value for the cost function is applied to the converter. However, it is possible to have the same vector selected for more than one sampling period, which results in a variable switching frequency. To overcome this problem, M²PC arranges the switching vectors into sectors. In order to understand it, let us consider the space vector diagram for the single-phase PUC5 converter shown in Fig. 3.

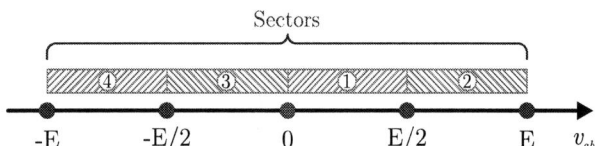

Fig. 3. Space vector diagram for the single-phase PUC5 converter.

The idea of M²PC technique consists of selecting the optimum sector of Fig. 3 instead of choosing just the vector which presents the minimum cost. In order to do that, after Eq. (1) is evaluated for each vector, a cost is assigned to the sectors via a manipulation of all individual vector costs that define the sector. Then, the sector with the minimum cost is selected and applied to the converter just like in the conventional SVM.

A key point here is how the manipulation is made to attribute a cost for each sector from the individual vector costs. This formulation is explained in [11] and will be briefly reviewed. First of all, it must be observed that as long as the exposed technique aims at resembling a conventional SVM, in each sampling period T_s, a sequence of vectors is applied, having each vector an associated duty cycle. The cost function (1) can be thought of as a squared quadratic norm of the error prediction, i.e, $g_i = e_i^2$, where

$$e_i = \left\| \begin{bmatrix} 1 & \gamma \end{bmatrix} \left(\begin{bmatrix} i \\ E/2 \end{bmatrix}^* - \begin{bmatrix} i \\ V_c^p \end{bmatrix} \right) \right\| \quad (2)$$

It can be observed that if T_s is considered sufficiently small, the average error introduced by the ith vector can be calculated by $d_i g_i$, $i \in \{1..8\}$, where d_i represents the duty cycle associated to vector of cost g_i.

Then, the RMS value of the duty cycle weighted errors associated to the vectors that constitute the sectors is minimized. As a result, the problem is reduced to minimizing the sector cost G_j in (3), where j represents the sector of jth index, formed by the vectors with indexes into the set called S_j, which defines each sector.

$$\min_{d_{ij}} \quad G_j = \sum_{i \in S_j} g_{ij} d_{ij}^2$$
$$s.t. \quad \sum_{i \in S_j} d_{ij} = 1 \quad 0 \le d_{ij} \le 1 \quad \forall i \in S_j \quad (3)$$

As long as this minimization is a convex problem and as it satisfies Karush-Kuhn-Tucker [12] conditions, we have:

$$d_{ij} = \frac{Q_j}{g_{ij}}, \qquad Q_j = \frac{1}{\sum_{i \in S_j} g_{ij}^{-1}} \quad (4)$$

As a result, an algorithm based on (3) and (4) is necessary to calculate the costs and duty cycles for each sector j, $j \in \{1..4\}$, selecting the sector with the minimum cost. Once the sector is selected, a switching sequence $\mathbf{S}_\omega$ needs to be defined. Here comes an important issue: the sequence is selected via a routine that choose the vectors in order to minimize the number of commutations. The minimization is done looking just to the next vector, i.e., the algorithm searches for the vector that implies on the lowest number of commutations in relation to the previous one. In this way, switching losses tend to be minimized.

IV. SIMULATION RESULTS

In this section, simulation results are shown comparing the proposed M^2PC algorithm to standard variable switching frequency MPC, utilizing the cost function (1) . The single-phase PUC5 converter is connected to an RL load, considering $E = 100$ V, $T_s = 100$ μs, R = 3 Ω, L = 1 mH, inner capacitor C = 1 mF, $\gamma = 2$ and fundamental frequency of 50 Hz.

Figs. 4 and 5 show, respectively, the output voltage and current with the corresponding spectra for the proposed M^2PC with switching sequence optimization. On other hand, Figs. 6 and 7 show the results for the same MPC technique but without switching sequence optimization.

It can be observed that for the case of switching sequence optimization - Fig. 5, there is a different output current shape from that encountered in the corresponding case without switching sequence optimization - Fig. 7. The spectra in the case of switching optimization is also more polluted. But, on

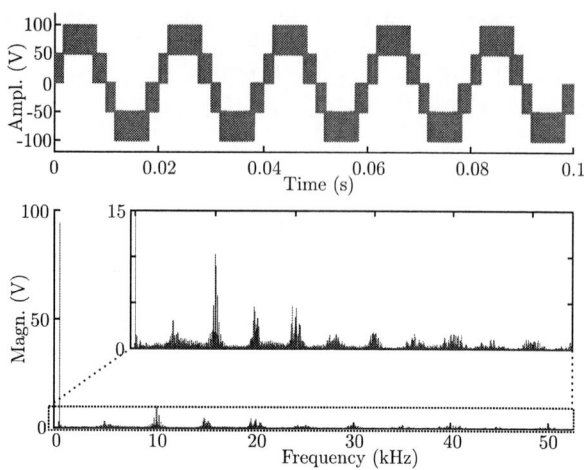

Fig. 4. Output voltage and their spectra for the proposed fixed frequency M^2PC algorithm - with switching sequence optimization.

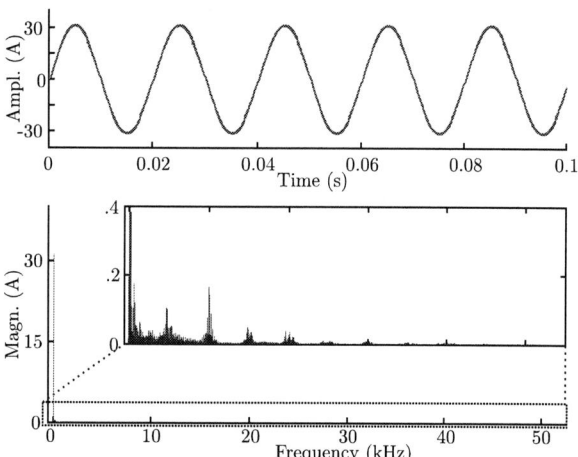

Fig. 5. Output current and their spectra for the proposed fixed frequency M^2PC algorithm - with switching sequence optimization.

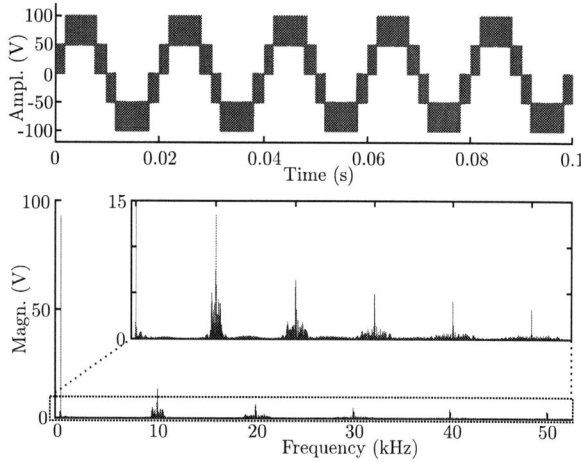

Fig. 6. Output voltage and their spectra for the proposed fixed frequency M^2PC algorithm - without switching sequence optimization.

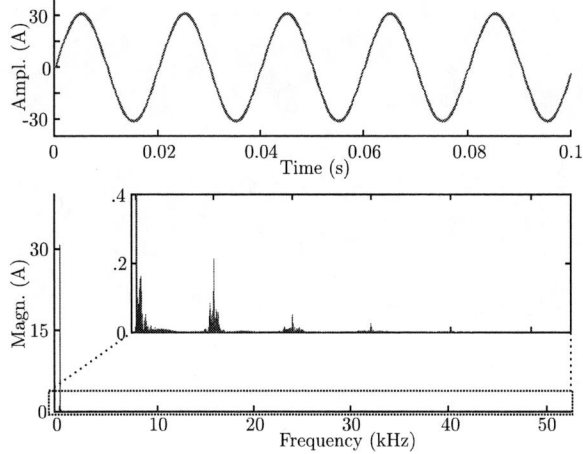

Fig. 7. Output current and their spectra for the proposed fixed frequency M²PC algorithm - without switching sequence optimization.

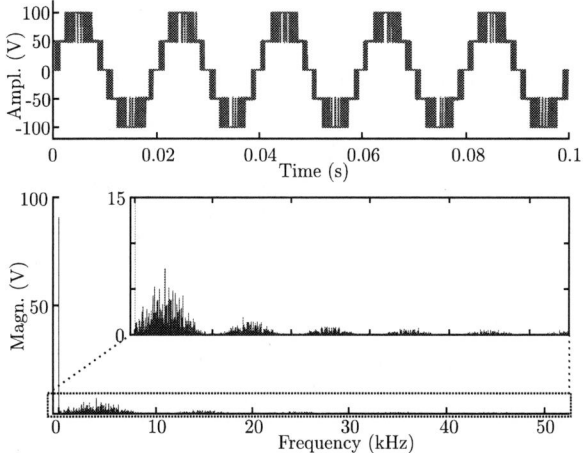

Fig. 8. Output voltage and their spectra for the one vector standard MPC - one vector per sampling period.

the other hand, the sum of commutations of all switches is 47% lower compared to the case without switching optimization. As a result, it presents the advantage of increasing the efficiency.

Comparing both M²PC cases with the standard MPC - Figs. 8 and 9, a major difference can be noted in the harmonic content. As expected, M²PC presents fixed switching frequency with concentrated harmonics at the sampling frequency and at its multiples. On the other hand, standard MPC presents spread harmonics across the entire frequency range, implying on a negative impact on the output filter design.

In addition, it can be observed from Figs. 5 and 9 that as the standard MPC applies just one vector per sampling period, it shows a higher distortion on the output current when compared to M²PC. However, the sampling period does not change, which is advantageous in terms of signal processing, although M²PC shows a higher computational burden.

In order to test the dynamic response of the controller,

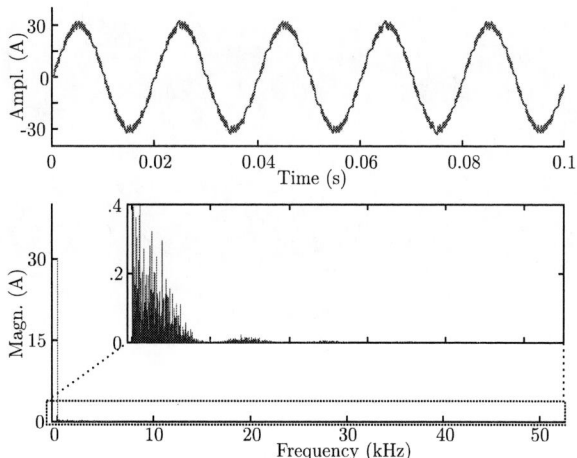

Fig. 9. Output current and their spectra for the one vector standard MPC - one vector per sampling period.

TABLE II
THD AND WTHD FOR OUTPUT VOLTAGE AND CURRENT FOR EACH MPC TECHNIQUE

TECHNIQUE	THD_v	$WTHD_v$	THD_i	$WTHD_i$
M²PC - Switching Sequence Optimization	32.49	0.62	2.57	0.56
M²PC - Without Switching Sequence Optimization	31.58	0.51	2.16	0.46
Standard MPC	31.18	0.81	5.33	0.58

Fig. 10 shows the behavior of the output voltage and current when there is reference current steps. It shows a fast transient response of the controller as well as good performance of the capacitor voltage regulation. Finally, Fig. 11 concentrates on the capacitor voltage regulation, starting from an unbalanced point, observing the behavior when in the presence of a DC bus voltage step. Once again, the regulation shows good performance.

In this paper, the Total Harmonic Distortion - THD and Weighted Total Harmonic Distortion - WTHD indexes are used for comparing the harmonic content results of the output voltages and currents. The THD index represent the ratio of the sum of the magnitude value of all harmonic components to the magnitude value of the fundamental component. In the same hand, the WTHD index takes into account the order of each harmonic component. These indexes are calculated as:

$$\text{THD}(\%) = \frac{100}{V_1}\sqrt{\sum_{h=2}^{\infty} V_h} \tag{5}$$

$$\text{WTHD}(\%) = \frac{100}{V_1}\sqrt{\sum_{h=2}^{\infty} \frac{V_h}{h}} \tag{6}$$

Table II ends this section presenting a summary of THD and WTHD for the output voltage (v) and current (i) for each case.

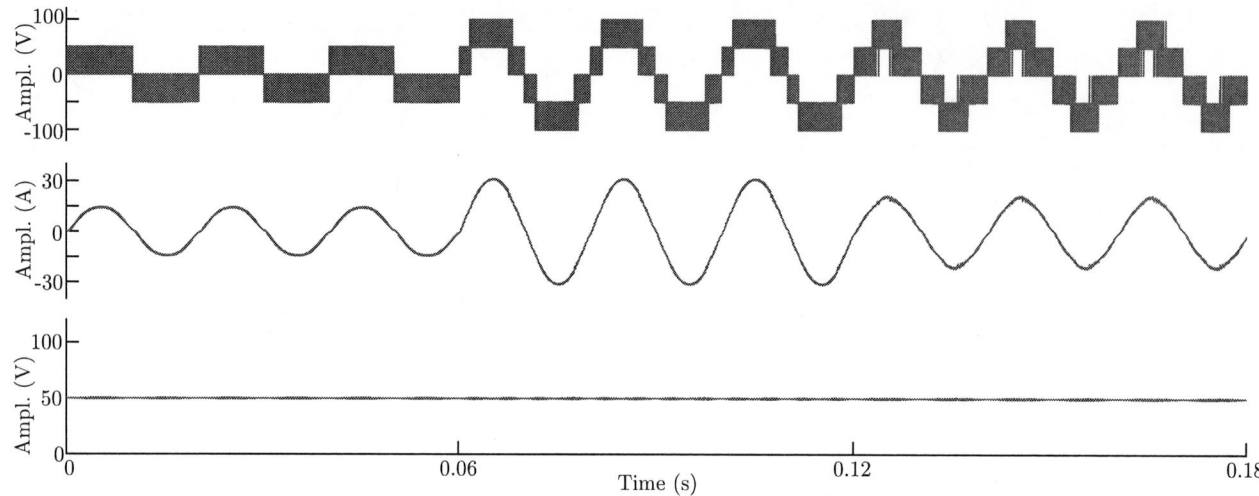

Fig. 10. Results during 50% increase and 33% decrease in reference current amplitude. From top to bottom: output phase voltage; output phase current; voltages of the DC bus capcitor.

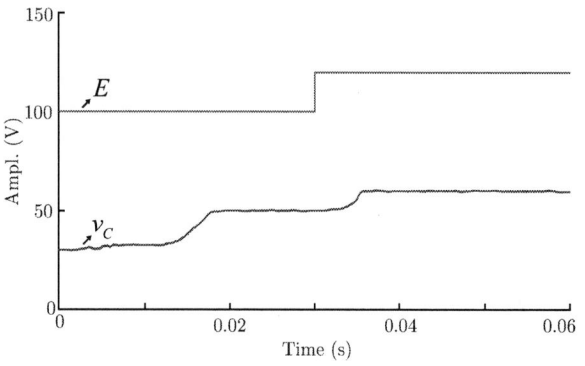

Fig. 11. Capacitor voltage started from unbalanced value and after an increase of 20% in the DC source (E).

V. CONCLUSION

This paper discussed the M²PC technique for PUC5 converter, implementing a switching sequence optimization in order to minimize switching losses. The proposed technique has the advantage of fixed switching frequency, resulting in a harmonic spectrum with a more defined shape, having a positive impact in the output filter design, when compared to the standard MPC. It also brings a simple solution to deal with the redundant switching vectors in order to handle internal capacitor voltage balancing, giving the designer the freedom of easily handling the cost function.

VI. AKNOWLEDGEMENTS

Jose Rodriguez acknowledges projects FB0008 and Fondecyt 1170167 from the Chilean Research Fund. This study was financed in part by the Coordenação de Aperfeiçoamento de Pessoal de Nível Superior Brasil (CAPES/PROEX) Finance Code 001.

REFERENCES

[1] Y. Ounejjar, K. Al-Haddad, and L. Gregoire, "Packed u cells multilevel converter topology: Theoretical study and experimental validation," *IEEE Trans. on Ind. Electron.*, vol. 58, pp. 1294–1306, April 2011.

[2] H. Vahedi and K. Al-Haddad, "Puc5 inverter - a promising topology for single-phase and three-phase applications," in *IECON 2016 - 42th Ann. Conf. of the IEEE Ind. Electron. Soc.*, Oct 2016, pp. 6522–6527.

[3] S. Arazm, H. Vahedi, and K. Al-Haddad, "Space vector modulation technique on single phase sensor-less puc5 inverter and voltage balancing at flying capacitor," in *IECON 2018 - 44th Annual Conference of the IEEE Industrial Electronics Society*, Oct 2018, pp. 4504–4509.

[4] F. B. Grigoletto, D. Schuetz, L. A. Junior, F. de M. Carnielutti, and H. Pinheiro, "Space vector modulation for packed-u-cell converters (puc)," in *IECON 2018 - 44th Ann. Conf. of the IEEE Ind. Electron. Soc.*, Oct 2018, pp. 4498–4503.

[5] H. Vahedi, P. Labb, and K. Al-Haddad, "Sensor-less five-level packed u-cell (puc5) inverter operating in stand-alone and grid-connected modes," *IEEE Ind. Electron. Inf.*, vol. 12, no. 1, pp. 361–370, Feb 2016.

[6] S. Arazm, H. Vahedi, and K. Al-Haddad, "Phase-shift modulation technique for 5-level packed u-cell (puc5) inverter," in *2018 IEEE 12th International Conference on Compatibility, Power Electronics and Power Engineering (CPE-POWERENG 2018)*, April 2018, pp. 1–6.

[7] J. Rodriguez, J. Pontt, C. A. Silva, P. Correa, P. Lezana, P. Cortes, and U. Ammann, "Predictive current control of a voltage source inverter," *IEEE Trans. on Ind. Electron.*, vol. 54, no. 1, pp. 495–503, Feb 2007.

[8] M. Trabelsi, S. Bayhan, K. A. Ghazi, H. Abu-Rub, and L. Ben-Brahim, "Finite-control-set model predictive control for grid-connected packed-u-cells multilevel inverter," *IEEE Trans. on Ind. Electron.*, vol. 63, no. 11, pp. 7286–7295, Nov 2016.

[9] M. Trabelsi, H. Vahedi, H. Komurcugil, H. Abu-Rub, and K. Al-Haddad, "Low complexity model predictive control of puc5 based dynamic voltage restorer," in *2018 IEEE 27th Int. Symp. on Ind. Electron. (ISIE)*, June 2018, pp. 240–245.

[10] F. Sebaaly, H. Vahedi, H. Y. Kanaan, and K. Al-Haddad, "Experimental design of fixed switching frequency model predictive control for sensorless five-level packed u-cell inverter," *IEEE Trans. on Ind. Electron.*, vol. 66, no. 5, pp. 3427–3434, May 2019.

[11] F. Donoso, A. Mora, R. Crdenas, A. Angulo, D. Sez, and M. Rivera, "Finite-set model-predictive control strategies for a 3l-npc inverter operating with fixed switching frequency," *IEEE Trans. on Ind. Electron.*, vol. 65, no. 5, pp. 3954–3965, May 2018.

[12] S. Boyd and L. Vandenberghe, *Convex Optimization*. Cambridge U.K.: Cambridge Univ. Press, 2004.

Design of Robust Structured Control Strategy for Single-Phase Dynamic Voltage Restorer

João Amin Moor Neto
DEELE
CEFET-RJ
Rio de Janeiro, Brazil
jamoor.neto@gmail.com

Guilherme Giglio de Andrade
DEELE
CEFET-RJ
Rio de Janeiro, Brazil
g_giglio@live.com

Thiago Americano do Brasil
DEELE/COPPE
CEFET-RJ/UFRJ
Rio de Janeiro, Brazil
thiago.abrasil@gmail.com

Mauro Sandro dos Reis
DEELE
CEFET-RJ
Rio de Janeiro, Brazil
mauro.s.reis@gmail.com

Julio Cesar Ferreira
DEELE
CEFET-RJ
Rio de Janeiro, Brazil
julio.carvalho@cefet-rj.br

Abstract— **In this paper a single-phase transformerless Dynamic Voltage Restorer (DVR) with uncertain filter parameters is presented. A robust control strategy is implemented through a non-smooth H∞ technique by weighting selections. Besides concerning a robust performance in presence of parametric uncertainty, it also considers voltage sag/swell and harmonics compensation. The performance of the DVR compensation was evaluated from the impact caused in the load voltage from the variation in the parameters of the passive converter filter. Extensive simulation results are presented to verify and validate the effectiveness of proposed control strategy, showing that the structured controller is suitable to diverse scenarios increasing the operation point of the system.**

Keywords — **Control Strategy, DVR, Harmonics, non-smooth H∞, Robustness, sag-swell.**

I. Introduction

The technological evolution in the field of power electronics allowed the development and application in different approaches of the Custom Power concept [1]. Custom power devices are used to enhance power quality in recent times. Custom power offers flexible solutions to many power quality problems.

The importance of power quality is associated with the growth in the use of sensitive loads. These devices are frequently susceptible to malfunctioning caused by voltage disturbances. Worsening in voltage quality may impact the performance of these equipment, increasing the overall risk of damage. Utility voltage variations that were considered tolerable until few years ago are now considered to drastically affect the performance of these sensitive devices [2]. Although modern automated processes and electro-electronic equipment generally require good quality power supplies, they also are a source of non-linearities in the distribution networks, contributing significantly to loss of quality of the supply voltage. One can imagine the problem in an industrial park with several consumers with modern manufacturing processes.

In this way, solutions to mitigate the consequences of voltage disturbances have attracted attention of the electrical industry in recent years. Voltage sag and swells are one of the most severe disturbances to sensitive loads. Due to the great occurrence in electrical systems, another important representation of the energy quality is the harmonic distortion. The advancement in power electronics technology has enabled favorable conditions to seek a high quality power supply for customers in general. The Dynamic Voltage Restorer (DVR) is one of the most effective Custom Power devices to mitigating voltage sag-swell and harmonics [3-6].

The voltage harmonics limits are recommended on IEEE standard 519 guidelines. Based on the IEEE standard 519, harmonic voltage distortion on power system 69 kV and below is limited to 5.0% total harmonic distortion (THD%) with each individual harmonic limited to 3.0% [7]. As presented in standard C84.1, the steady-state voltage tolerances for an electrical power system of 600V (low voltage) and below are ±5% of their rated value [8]. Voltage sags and swells are defined in different manners based on their individual characteristics. There are two main standards, IEC 61000-4-30 [9] and IEEE 1159-1995 [10], which define voltage sag, interruption and swell.

Different power circuit topologies alternatives of DVR can be used both for three-phase circuits and for single-phase circuits [11-13]. The control strategies for DVR are also analyzed [14-17]. For a successful compensation, the DVR control strategy must be able to detect voltage disturbances and control the converter to prevent against sags, swells and harmonics [4]. Thus, a suitable reference signal with proper synchronization with the supply voltage must be generated fast and accurately ensuring high immunity to voltage sag/swells or harmonic distortions in order to control the DVR.

In this paper, a robust PWM PI control for a single-phase a transformerless DVR is proposed, along with a simplified reference voltage synthesizing algorithm to compose a control strategy for harmonic and sag/swell compensation [18]. H∞ design technique for the PWM PI control is presented and the robustness of the proposed control strategy against variations in the system parameters is achieved through simulations using MATLAB Simulink.

II. General Description of Single-Phase DVR

A. Structure and Components

The DVR is connected in series with the grid point of common coupling and the sensitive load, synthesizing a reference voltage to ensure that the load voltage is close enough to the rated one. The general structure of a transformerless constant dc-link DVR is depicted in Figure 1. It consists of a passive filter, a H-bridge voltage source inverter (VSI), and a rechargeable dc power supply.

978-1-7281-4181-7/19 $31.00 © 2019 IEEE

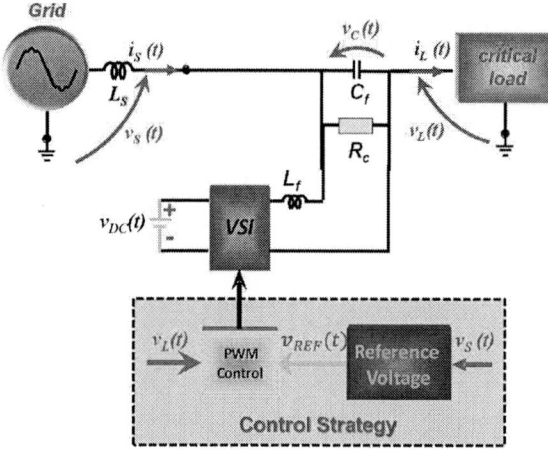

Fig. 1. Single-Phase DVR under study.

B. Modelling of transformerless DVR

According to [19], after obtain the average model being, the state-space small-signal model can be derived by linearizing it. The proposed configuration can be described through (1) to (4):

$$\dot{x} = Ax + Bu \quad (1)$$

$$y = Cx + Du \quad (2)$$

$$\frac{d}{dt}\begin{bmatrix} \bar{I}_f \\ \bar{V}_c \end{bmatrix} = \begin{bmatrix} 0 & \frac{-1}{L_f} \\ \frac{1}{C_f} & \frac{-1}{R_cC_f} \end{bmatrix} \begin{bmatrix} \bar{I}_f \\ \bar{V}_c \end{bmatrix} + \begin{bmatrix} \frac{V_{DC}}{L_f} \\ 0 \end{bmatrix} \bar{m} \quad (3)$$

$$y = \begin{bmatrix} 0 & 1 \end{bmatrix} \begin{bmatrix} \bar{I}_f \\ \bar{V}_c \end{bmatrix}, \quad (4)$$

where L_f represents the uncertain VSI filter inductor, C_f denotes the uncertain VSI filter capacitor and R_c the parallel discharge resistor as shown in Fig. 1. Furthermore, $\bar{I}_f$ is the average current from direct current source (V_{DC}), $\bar{V}_c$ is the average DVR output voltage and $\bar{m}$ is the PWM control signal modulation.

C. Control Strategy

In general, the process control of DVR includes three steps:

1. Detection of voltage disturbance occurrence in the system;

2. Comparison with the reference voltage signal;

3. Generation of gate pulses to the voltage source inverter (VSI) to generate the DVR output voltages which compensates the voltage sag/swells or harmonic distortions.

The two blocks that represent the DVR voltage control strategy are shown in Fig.1. The block diagram of the reference voltage algorithm that represents the sag/swell and harmonics detection techniques is shown in Fig. 2. The FCD algorithm is the main component of the reference voltage determination algorithm where its main function is to extract the fundamental component of grid voltage $v_{S1}(t)$ and the magnitude value of the fundamental component V_{S1}, which are used to determine the target signal of the output voltage [20]. The FCD block diagram is shown inf Fig.3. It presents

good trade off characteristics between immunity to harmonics distortions and fast dynamic response.

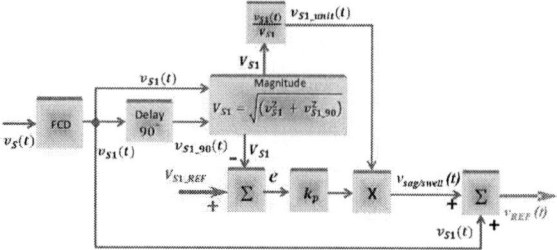

Fig. 2. Reference voltage determination block diagram.

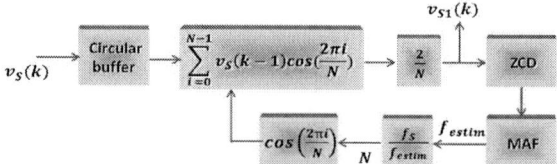

Fig. 3. Fundamental Component Detector block diagram.

The fundamental component detector (FCD) estimates the instantaneous fundamental component $v_{S1}(t)$ of the input signal $v_S(t)$ always tracked with the fundamental phase of this input. The magnitude value of the fundamental component estimated by the FCD is calculated by means of (5):

$$V_{S1} = \sqrt{v_{S1}^2 + v_{S1\_90}^2} \quad (5)$$

The unit magnitude signal synchronized with the fundamental component of the grid voltage $v_{s1_{unit}}(t)$ is obtained by (6). Subsequently, the sag/swell sinusoidal signal is calculated by (7).

$$v_{S1\_unit} = sin(\omega_1(t) + \emptyset_1) = \frac{v_{S1}(t)}{V_{S1}} \quad (6)$$

$$v_{sag/swell}(t) = e \cdot v_{S1\_unit}(t) \quad (7)$$

Finally, to compensate voltage harmonics it is necessary to add the fundamental component of grid voltage $v_{S1}(t)$ to the sag/swell sinusoidal signal $v_{sag/swell}(t)$:

$$v_{REF}(t) = v_{sag/swell}(t) + v_{S1}(t) \quad (8)$$

The PWM control block shown in Fig. 1 is responsible for controlling the state of the power inverter switches in order to produce the voltage determined by the reference voltage algorithm proposed. The SPWM technique was the switching method adopted, with a fixed switching frequency of 9 kHz. The block diagram of the PWM scheme adopted is represented in Fig. 4.

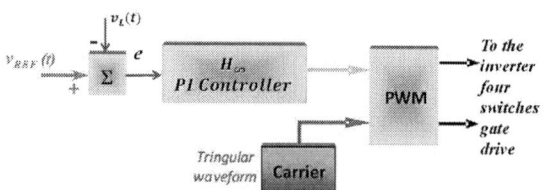

Fig. 4. PWM Voltage Control.

978-1-7281-4181-7/19 $31.00 © 2019 IEEE

In this way, this control method accounts for all the uncertainties in designing of the control system. This paper presents a robust H∞ control for compensating voltage disturbances. The simulation results shown prove that the designed controller can achieve faster responses and better external disturbance rejection capabilities when compared to conventional PI controllers.

III. ROBUST PWM CONTROLLER

The system considered in this paper has issues that makes classic modern control design method a troublesome approach to achieve all requirements. Due to the uncertainty and performance specification, control designers may choose complex control architecture to deal with these problems [19],[21]. In our approach to reach design requirements, addressing uncertainties and disturbances issues, a nonsmooth H_∞ design technique is used [22]. This technique was chosen, among others, due to its capability to synthesize a structured control (PID, lead-lag, etc.).

In our approach to design a controller (K), it was utilized a standard plant augmentation structure for mixed-sensitivity (S/KS/T) (MATLAB function *augw*). This plant augmentation (P(s)) can be represented as (9):

$$P(s) = \begin{bmatrix} Z_1 \\ Z_2 \\ Z_3 \\ e \end{bmatrix} = \begin{bmatrix} W_1 & -W_1 G \\ 0 & W_2 \\ 0 & W_3 G \\ I & -G \end{bmatrix} \begin{bmatrix} u_1 \\ u_2 \end{bmatrix} \quad (9)$$

where I is the identity matrix, Z's are the weighted output, e is the error, u_1 and u_2 are the exogenous input and the control output respectively. These weights (W) are made of penalizing signals, thus it is possible to insert design requirements and uncertainties into the nominal plant (G).

The uncertainty of the system is represented by W_3 as a multiplicative output uncertainty. Considering a good range of possible scenarios that our system can reach, we can establish our weight as follow (10):

$$\Delta(s) = \bar{\sigma}\left(\frac{G_\Delta - G}{G}\right) \quad (10)$$

where $\bar{\sigma}$ denotes the peak value of the transfer function maximum singular value norm. G_Δ is the plant with a uncertain value.

The weight W_3 is settled as a bound of the worst case of $\Delta(s)$, furthermore it was also included a bit of conservatism and assign it as 105% of its original value. Fig. 5 shows a range of $\Delta(s)$ in dotted black line, the estimated weight in solid red line and cross green line W_3. It is possible to note that in low frequencies our weight is conservative, however it is close to the uncertainty lump and consequently a good fit for worst case. The transfer function of W_3 employed in this paper is $(0.8781 s + 8.489)/ (s + 6124)$.

To represent system requirements a suitable transfer function must be designed for W_1. Based on [19], [23] definitions of a proper weight, it is possible to build a satisfactory W_1. Thus the weight W_1 applied here is $(4.419e4s+3.37e5)/(s^2+226.2s+1.421e5)$. For W_2 it was utilized a small value of 1e-6.

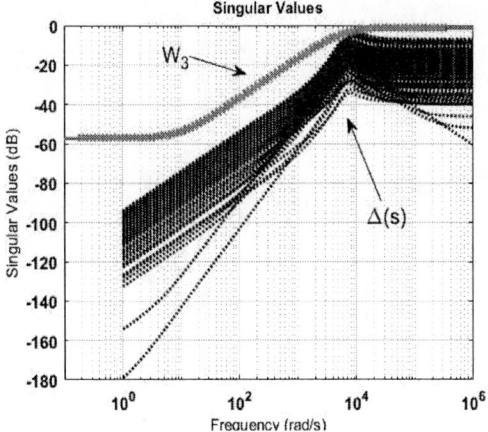

Fig. 5. Singular values of Δ(s) (black), estimated weight (red) and W₃ (green).

Finally, after defining the weights transfer functions, the structured control can be synthesized by the non smooth H_∞ design technique [22] (MATLAB function *hinfstruct*). In this work a PI structure control was assigned as the controller. The resulting controller (K) is (11):

$$K = 0.135 + \frac{3.31\times10^{-8}}{s} \quad (11)$$

IV. COMPUTER SIMULATIONS

A. System Modeling

The proposed DVR control strategy was implemented in PSCAD/EMTDC. The simulations are based on the test case shown in Figure 1 and the details of the system parameters are presented in Table I.

TABLE I.   SYSTEM PARAMETERS

Parameters	Component Values
Power Supply V_S, f_S, L_S	127 Vrms, 60 Hz, 50 µH
Linear load S, PF - ($R_L + jX_L$)	2100 VA and PF = 82.5%, (6 + j4.14) Ω
DC link voltage (V_{DC})	138 V
LC a filter ripple (L_f, C_f, R_C)	5mH; 4 µF; 50 Ω
Switching frequency f_{SW}	9 kHZ
Robust PWM PI controller ($K_{P\text{-}Rpwm}$, $K_{I\text{-}Rpwm}$)	0.135; 3.31x10⁻⁸
Proportional controller of the sag/swell compensation algorithm (K_P)	1.20

The system robustness was evaluated considering variations in some systems parameters:

- Power supply: Grid inductance Ls varies from 50 µH to 400 µH, a step increment of 8 times. In this particular case, the source impedance Ls varies while the DVR is operating;

- Passive Filter: Variations of ±20% for the capacitor Cf and inductor Lf.

The supply voltages present the following characteristics:

- Voltage harmonics: The total harmonic distortion (THD) is 18.21% (3rd = 10.91%, 5th= 13.91% and 7th= 4.36%);

- Voltage sag: The grid fundamental component voltage decrease to 52,2% during 100 ms (0.5s to 0.6 s);

- Voltage swell: The grid fundamental component voltage increase to 117.39% of its normal value during 100ms (0.5s to 0.6 s).

B. Simulation Results with Voltage sag and voltage harmonics

Case1: $C_f = 4$ µF, inductor $L_f = 5$ mH and Ls varies at 0.45s. The performance of the proposed control algorithm in this scenario is shown in Fig. 6.

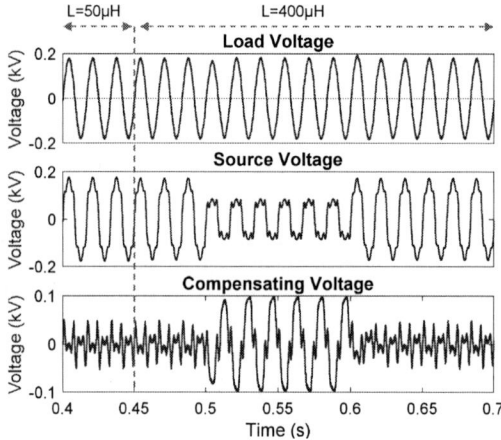

Fig. 6. Mitigation of voltage harmonics and voltage sag for case 1. (a) Load voltage, (b) Grid voltage, (c) Compensation voltage.

As can be verified in Fig. 6, the load voltage remains within specifications regarding harmonic distortion and fundamental component value due to the compensating voltage injected by the DVR. Thus, we can say that the tension in the load was not affected by the disturbances here considered.

Case 2: $C_f = 3.2$ µF, inductor $L_f = 5$ mH and Ls varies at 0.45s. The results of this simulation are shown in Fig. 6 and details of the worst case transition points at 0.45s are showing Fig. 7.

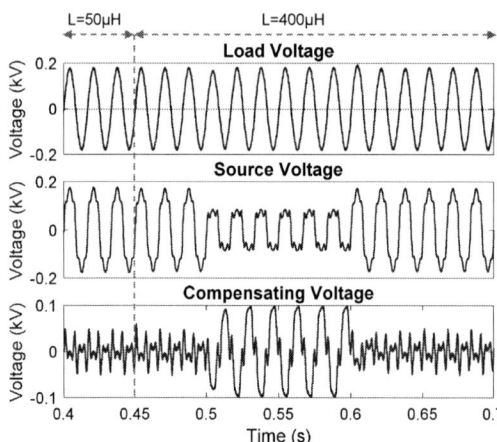

Fig. 7. Mitigation of voltage harmonics and voltage sag for case 2. (a) Load voltage, (b) Grid voltage, (c) Compensation voltage.

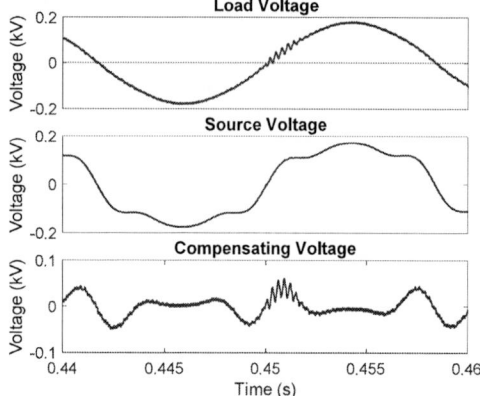

Fig. 8. Mitigation of voltage harmonics and voltage sag, case2, detailed. (a) Load voltage, (b) Grid voltage, (c) Compensation voltage.

In this case, it can be seen from Fig. 7 that the load voltage was weakly affected by the source voltage disturbances due to the compensating voltage injected by the DVR. In Fig. 8 one can see the impact of the induction variation representing the short circuit capacity of the power supply. Note that the performance of the DVR was not sensitive to this fact.

Case 3: $C_f = 4$ µF and, inductor $L_f = 4$mH and Ls varies at 0.45s.

The results of this simulation are shown in Fig. 9. The performance in compensating for voltage disturbances was similar to previous scenarios.

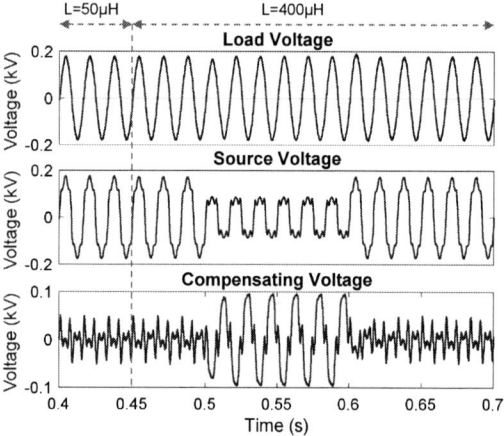

Fig. 9. Mitigation of voltage harmonics and voltage sag for case 3. (a) Load voltage, (b) Grid voltage, (c) Compensation voltage.

C. Simulation Results with Voltage swell and voltage harmonics

Case 4: $C_f = 3.2$ µF, inductor $L_f = 5$ mH and Ls varies at 0.45s. The performance of the proposed control algorithm in this scenario is shown in Fig. 10. **Case 5:** $C_f = 3.2$ µF, inductor $L_f = 4$ mH and Ls varies at 0.45s. The performance of the proposed control algorithm in this scenario is shown in Fig. 11. **Case 6**: $C_f = 4$ µF, inductor $L_f = 4$ mH and Ls varies at 0.45s. The performance of the proposed control algorithm in this scenario is shown in Fig. 12. **Case 7**: $C_f = 4.8$ µF, inductor $L_f = 6$ mH and Ls varies at 0.45s. The performance of the

978-1-7281-4181-7/19 $31.00 © 2019 IEEE

proposed control algorithm in this scenario is shown in Fig. 13.

Table II shows the harmonic distortions values after compensation of the voltage disturbances, before and during voltage sag/swell period. These values represent the mean values of all cases.

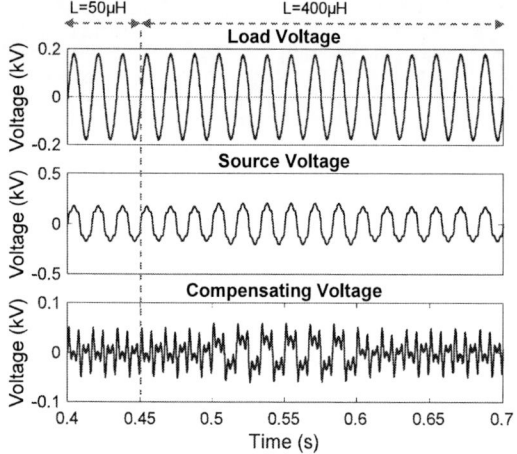

Fig. 10. Mitigation of voltage harmonics and voltage sag for case 4. (a) Load voltage, (b) Grid voltage, (c) Compensation voltage.

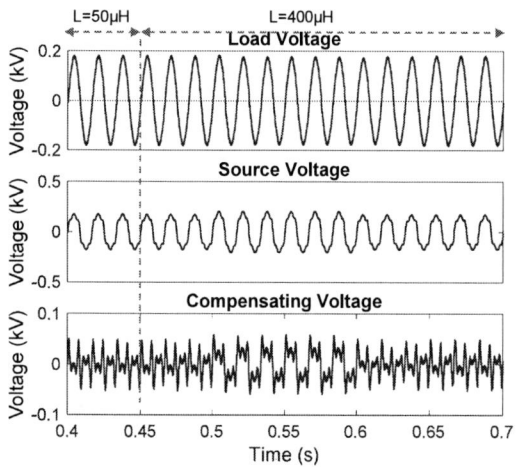

Fig. 11. Mitigation of voltage harmonics and voltage sag for case 5. (a) Load voltage, (b) Grid voltage, (c) Compensation voltage.

In case 4, the DVR was analyzed with a lower capacitance value. For case 5, a lower value was imposed on the passive filter inductance too. In case 6, there was a reduction in the passive filter inductance value. Based on the waveforms illustrated in Figs. 10, 11 and 12, it is evident that the DVR compensates for the harmonics in the source voltage as well as compensates for the voltage swell that occurred in t = 0.5s.

In case 7 an assessment was performed in a scenario where the values of the two passive filter components increased. Again, as can be seen from the load voltage waveform shown in Fig. 13 that the existing source voltage waveform disturbances shown in Fig. 13 have been greatly reduced to a desired level.

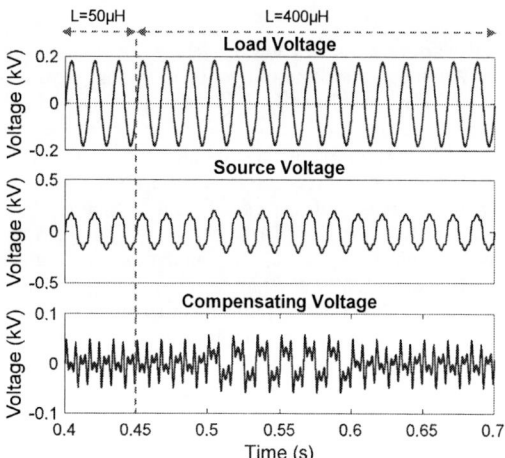

Fig. 12. Mitigation of voltage harmonics and voltage sag for case 6. (a) Load voltage, (b) Grid voltage, (c) Compensation voltage.

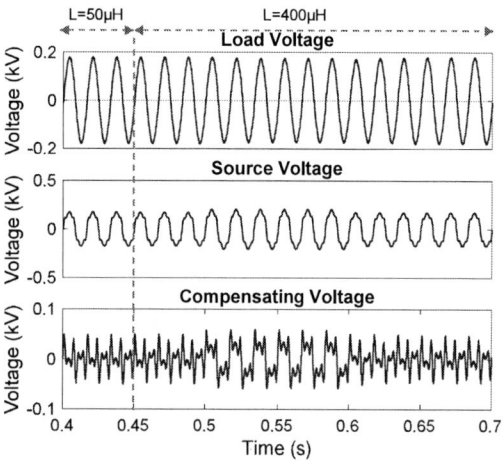

Fig. 13. Mitigation of voltage harmonics and voltage sag for case 7. (a) Load voltage, (b) Grid voltage, (c) Compensation voltage.

TABLE II. SYSTEM THD AND HARMONIC VALUES DURING THE PERIOD WHICH THE PCC VOLTAGE WERE DISTORTED.

Voltage harmonic distortions	PCC voltage	PCC voltage during Sag/swell period	Load voltage DVR connected	Load voltage DVR connected during sag/swell period
THD(%)	17.7	34.20	1.79	2.85
3rd(%)	10.85	21.19	1.18	1.63
5th(%)	13.05	25.44	1.25	2.19
7th(%)	4.34	8.49	0.50	0.81

Regarding the performance of the DVR in the sag and sweel compensation imposed in the simulations performed, we can summarize the values of the fundamental component of the load voltage, as follows: The lowest peak voltage value during sinking was 165 V in the first semicycle and 175 V in the second semicycle and later the peak voltage value was 179 V. Already at the end of the sinking in the first semicycle the peak voltage value reached 189 V.

978-1-7281-4181-7/19 $31.00 © 2019 IEEE

V. CONCLUSIONS

In this paper a robust control strategy was proposed for sag-swell compensation and mitigation of voltage harmonics distortions. The nonsmooth H_∞ technique, with proper weights, guarantees a robust controller, with classical structure (Proportional-Integral), simultaneously with PWM method was able to achieve a robust performance increasing the operating range of system. For the control system validation different values of passive filter (C_f and L_f) were applied in MATLAB/Simulink simulation environment. Considering the established premise of analyzing the impact of change in passive filtering element parameters, it was possible to verify that the simulation results demonstrated the effectiveness of suggested approach, which stability and performance were evaluated by contrasting with system's performance specification.

REFERENCES

[1] Acha, et al. Power Electronic Control in Electrical Systems, Newness Power Engineering Series, 2002.

[2] El Mofty, A.; Youssef, K., "Industrial power quality problems," Electricity Distribution, 2001. Part 1: Contributions. CIRED. 16th International Conference and Exhibition on (IEE Conf. Publ No. 482), vol.2, no., pp.5 2001.

[3] Dalmo C. Silva ; Josué L. Silva ; Janaína G. de Oliveira ; Marlon J. Carmo ; Matusalém M. Lanes, "Compensation of disturbances from the electrical network using the Dynamic Voltage Restorer (DVR)", 2017 Brazilian Power Electronics Conference (COBEP), pp. , nov. 2017.

[4] Samet Biricik, and Hasan Komurcugil, Optimized Sliding Mode Control to Maximize Existence Region for Single-Phase Dynamic Voltage Restorers, IEEE Transactions on Industrial Informatics, Vol. 12, No. 4, pp. 1486-1497, August 2016.

[5] Pal, R.; Gupta, S. State of the Art: Dynamic Voltage Restorer for Power Quality Improvement. Electrical & Computer Engineering: An International Journal (ECIJ) Volume 4, Number 2, June 2015

[6] Resmi, R.; Reshmi, V.; Jacob, J. Mitigation of Voltage Sag Swell and Harmonics by Dynamic Voltage Restorer using Matrix Converter. Int. J. Adv. Res. Electr. Electron. Instrum. Eng. 2013, 2, 297–304.

[7] IEEE Std 519-1992, "IEEE Recommended Practices and Requirements for Harmonic Control in Electric Power Systems", Institute of Electrical and Electronics Engineers, Inc. 1993.

[8] ANSI C84.1-2011 ELECTRIC POWER SYSTEMS AND EQUIPMENT - VOLTAGE RANGES (American National Standard for Electric Power Systems and Equipment – Voltage Ratings (60 Hertz).

[9] IEC 61000-4-30. Electromagnetic Compatibility (EMC) -Part 4-30: Testing and measurement techniques -Power quality measurement methods.

[10] IEEE Std. 1159-1995. IEEE recommended practice for monitoring electric power quality. Technical report, The Institute of Electrical and Electronics Engineers, Inc., 1995.

[11] John Godsk Nielsen, Frede Blaabjerg, "A Detailed Comparison of System Topologies for Dynamic Voltage Restorers," IEEE Transactions on Industry Applications, Vol. 41, No. 5, pp. 1272-1280, September/October 2005.

[12] A. Elserougi, A. M. Massoud, A. S. Abdel-Khalik, S. Ahmed, and A. A. Hossam-Eldin, "An interline dynamic voltage restoring and displacementfactorcontrollingdevice(IVDFC),"IEEETrans.PowerElec tron.,vol.29, no. 6, pp. 2737–2749, Jun. 2014.

[13] E. Babaei, M. F. Kangarlu, and M. Sabahi, "Dynamic voltage restorer based on multilevel inverter with adjustable dc-link voltage," IET Power Electron., vol. 7, no. 3, pp. 576–590, Mar. 2014.

[14] Y. W. Li, D. M. Vilathgamuwa, F. Blaabjerg, and P. C. Loh, "A robust control scheme for medium – voltage level DVR implementation," IEEE Trans. Ind. Electron., vol. 54, no. 4, pp. 2249–2261, Aug. 2007.

[15] F. B. Ajaei, S. Afsharnia, A. Kahrobaeian, and S. Farhangi, "A fast and effective control scheme for the dynamic voltage restorer," IEEE Trans. Power Del., vol. 26, no. 4, pp. 2398–2406, Oct. 2011.

[16] C.Kumar and M.K.Mishra,"Predictive voltage control of transformerless dynamic voltage restorer," IEEE Trans. Ind. Electron., vol. 62, no. 5, pp. 2693–2697, May 2015.

[17] S. Jothibasu and M. K. Mishra, "A control scheme for storageless DVR based on characterization of voltage sags," IEEE Trans. Power Del., vol. 29, no. 5, pp. 2261–2269, Oct. 2014.

[18] João A. Moor Neto, Vinicius de M. Brown, Júlio Cesar Ferreira, Mauro S. Reis, Thiago A. do Brasil, Fast and Robust Control Strategy for Single-Phase Dynamic Voltage Restore, International Journal of Applied Engineering Research ISSN 0973-4562 Vol. 13, Number 22 (2018) pp. 15860-15871.

[19] Alireza Javadi, Abdelhamid Hamadi, Lyne Woodward, and Kamal Al-Haddad, Experimental Investigation on a Hybrid Series Active Power Compensator to Improve Power Quality of Typical Households, IEEE Transactions on Industrial Electronics, VOL. 63, NO. 8, pp.4849-4858, August 2016.

[20] A. D. Sousa, J. A. Moor Neto, J. C. Ferreira, M. S. dos Reis, L. Peres, Fast and robust fundamental component estimation algorithm for a single phase system. 2017 Brazilian Power Electronics Conference (COBEP), Brazil., pp-1-6.

[21] Y. W. Li, D. M. Vilathgamuwa, F. Blaabjerg and P. C. Loh, "A Robust Control Scheme for Medium-Voltage-Level DVR Implementation," in IEEE Transactions on Industrial Electronics, vol. 54, no. 4, pp. 2249-2261, Aug. 2007.

[22] P. Apkarian and D. Noll, "Nonsmooth H ∞ Synthesis," in IEEE Transactions on Automatic Control, vol. 51, no. 1, pp. 71-86, Jan. 2006.

Reference Grid Impedance for Tests of Grid-connected Power Converters for Distributed Energy Resources: The Brazilian Case

Gabriel Avila Saccol, Charles Schardong, Leandro Michels, Lucas Vizzotto Bellinaso and Cassiano Rech

Power Electronics and Control Research Group - GEPOC
Federal University of Santa Maria - UFSM
Santa Maria, Brazil
saccol.gabriel@gmail.com, schardong.charles@gmail.com, michels@gepoc.ufsm.br,
lucas@gepoc.ufsm.br, rech.cassiano@gmail.com

Abstract—**The Brazilian standard ABNT NBR 16149 defines the tests and procedures for the connection of photovoltaic inverters to the low voltage electrical grid. According to the ABNT NBR 16149, the flicker tests must be performed based on the international standards IEC 61000-3-3, IEC 61000-3-11 and IEC 61000-3-5, wich define a reference impedance for the equipment connection to the low voltage grid. However, the Brazilian grid presents different voltage levels and transformer configurations, so that the reference impedance defined in the international standards are not always compatible with the Brazilian systems. In this sense, this paper proposes a new classification for the inverters, according to the voltage levels and characteristics of the grid. A new reference impedance is also proposed and simulation results are included to demonstrate the impact of this choice for the operation and tests of grid connected inverters.**

Index Terms—**Brazilian standard, photovoltaic grid-connected inverter, flicker, reference impedance.**

I. INTRODUCTION

The adoption of sustainable practices for electrical energy generation is a featured theme currently. The probable shortage of fossil fuel, coupled with the great pollution rates of these energy sources, makes the need of adoption of renewable energy sources increasingly clear. In this scenario, the integration between centralized and distributed energy systems is a very relevant issue for expansion of the Brazilian electric system [1].

In the last years, the advances in semiconductor technologies for the development of inverters have strongly contributed to making photovoltaic (PV) energy more competitive on the world stage. It is estimated that 181 GW associated with renewable energy sources were installed around the world from the year 2012 with 55% of the total capacity corresponding to solar energy [2]. China and the United States were the countries that most invested in photovoltaic energy systems [3].

In Brazil, about 60% of electric energy is produced by hydraulic supplies [4], so that the country pluvial regime significantly affects the energy production. As a way to diversify the national energy matrix, the use of grid-connected

photovoltaic systems is a promising alternative, due to the high solar irradiation rates in Brazil, especially in the northeast region. It is estimated that for the São Francisco Valley, the annual average solar irradiation is equal to 6 kWh / m^2day [5], higher than most European countries.

On the other hand, according to Brazilian Institute of Metrology, Quality and Technology (INMETRO), all photovoltaic inverters connected to the Brazilian low voltage grid must comply with ABNT NBR 16149 [6], [7]. Among the criteria adopted by this regulation, those of flicker are based on the international standards IEC 61000-3-3 [8], IEC 61000-3-11 [9] and IEC 61000-3-5 [10], which specify the test conditions and voltage variation limits for grid connected equipment. Therefore, the reference impedance for the inverter connection to the grid is also specified in these standards, based on the Technical Report IEC/TR 60725 [11], which performs an in-depth study of the impedances of the electrical distribution systems of several countries for the definition of a reference impedance that can be used in the tests.

However, voltage levels and configurations between transformers are not standardized in the Brazilian electrical system and present variations between states and municipalities. Thus, the adoption of reference impedance values as defined in international standards is not always the most appropriate for inverters in Brazil. In this sense, adjustments to the impedance values defined in the IEC 61000-3-3 and 61000-3-11 standards may be necessary to make the inverter tests compatible with the Brazilian grid.

Therefore, the objective of this paper is to propose a reference impedance wich is compatible with the Brazilian low voltage distribution grid, making the tests feasible in the national territory. So, a new classification of the inverters is also proposed, according to the voltage levels of the grid and the connection with the transformers.

This paper is organized as follows: Section II deals with the flicker criteria defined for photovoltaic systems connected to the Brazilian grid, and the standard impedance for conducting the tests is presented. In Section III it is suggested a new classification of the inverters, according to the operating

978-1-7281-4181-7/19 $31.00 © 2019 IEEE

voltage of the electric grid and the type of connection with the distribution system. Section IV presents the simulation results for the input voltages of grid connected inverters and a new impedance is proposed based on these results.

II. ABNT NBR 16149

The regulation of the use of photovoltaic systems has become a necessity in Brazil against the increasing use of this energy source, as a way of not causing impacts on the power quality provided by the grid. ABNT NBR 16149 deals with the interface and connection characteristics between photovoltaic systems and the Brazilian low voltage distribution grid. According to INMETRO Ordinance N° 357 [12], inverters up to 10 kW connected to the grid must comply with the limits imposed by NBR. Some topics covered by this standard are described in the following sections.

A. Network compatibility

According to NBR 16149, the quality of power injected to the grid by PV systems must meet certain specifications for flicker, distortion of the injected current, the amount of reactive power injected/consumed by the PV system, among others. In general, the harmonic distortion of the current can not exceed 5% of its fundamental component, while the power factor is defined according to the operating power of the converter. The flicker standards are defined by the NBR according to international standards IEC 61000-3-3 (for systems with current less than 16 A), IEC 61000-3-11 (for systems with current greater than 16 A and less than 75 A) and IEC 61000-3-5 (for systems with current greater than 75 A).

The standards IEC 61000-3-3 and IEC 61000-3-11 deal with international criteria that regulate voltage fluctuations and flicker produced in the electrical network by external equipment, wich operate with currents of up to 75 A. In other words, these regulations establish the procedures, circuits and evaluation methods that should be used in equipment voltage fluctuation tests. For the accomplishment of these tests that define the maximum values of the grid voltage fluctuation, any equipment must be connected to the electric grid according to the circuit of Fig. 1 where v_R, v_S and v_T represent the three-phase nominal voltage of the grid, Z_F is the phase conductor reference impedance and Z_N is the neutral conductor reference impedance, defined according to the IEC TR 60725, summarized in Table I. However, these standards have been developed to meet a power supply system where the phase-to-neutral voltage has a limit between 220 V and 250 V, with an operating frequency of 50 Hz, which may be inadequate in certain situations when applied to the Brazilian electricity distribution system.

B. Personal safety and photovoltaic system protection

Criteria for protection of the photovoltaic system and personal safety must be guaranteed to prevent damages caused by electrical surges and situations of risk and accidents. Thus, ABNT NBR 16149 establishes voltage limits for the operation of the photovoltaic system, according to Table II. If these limits

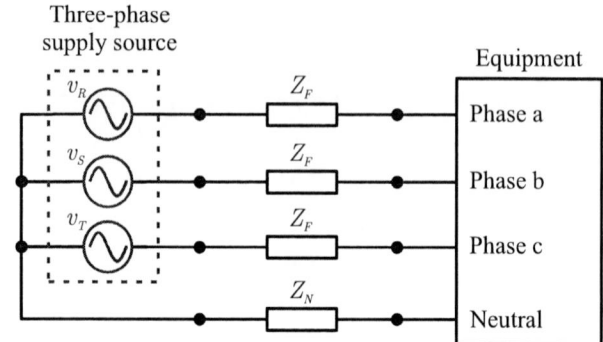

Fig. 1. Electric grid for the tests of the standards IEC 61000-3-3 and IEC 61000-3-11 [8].

TABLE I
REFERENCE IMPEDANCE VALUES [11].

Parameter	Conductors current capacity	
	Inferior to 100 A	Superior to 100 A
Z_F	$0.24 + j0.15\ \Omega$	$0.15 + j0.15\ \Omega$
Z_N	$0.16 + j0.10\ \Omega$	$0.10 + j0.10\ \Omega$

are not met, the photovoltaic system must cease the power supply, according to the times specified in Table II. Therefore, the reference impedance used for the voltage fluctuation and flicker tests shall ensure that the voltages at the inverter terminals comply with the personal safety and protection limits defined in the NBR. Otherwise, the photovoltaic system is automatically disconnected from the grid during the flicker test.

III. NETWORK CONFIGURATIONS IN BRAZIL

According to the National Electric Energy Agency (ANEEL), the consumer units connected to the distribution system in Brazil are classified according to the voltage level [13]:

- High Voltage Distribution Systems (SDAT - *Sistemas de Distribuição em Alta Tensão*) - 230 kV, 88kV to 138 kV and 69 kV;
- Medium Voltage Distribution Systems (SDMT - *Sistemas de Distribuição em Média Tensão*) - Above 1 kV to 44 kV;
- Low Voltage Distribution Systems (SDBT - *Sistemas de Distribuição em Baixa Tensão*) - Less or equal than 1 kV.

However, the voltage levels in SDBT are not standardized in Brazil and suffer variations between states and municipali-

TABLE II
RESPONSE TO ABNORMAL GRID VOLTAGE CONDITIONS [6].

Voltage in the common connection point (% to the nominal voltage)	Maximum shutdown time
V < 80 %	0.4 s
80 % < V < 110 %	Normal operation
V > 110 %	0.2 s

Configuration I:

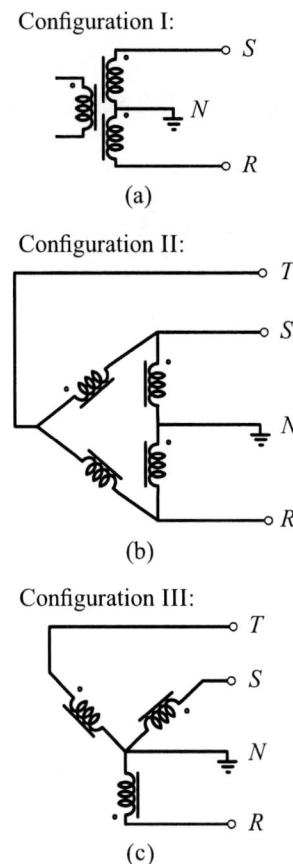

(a)

Configuration II:

(b)

Configuration III:

(c)

Fig. 2. Transformer connections in Brazil. (a) Split-phase transformer. (b) Three-phase delta-to-neutral transformer. (c) Three-phase wye transformer.

TABLE III
PROPOSED INVERTER CLASSIFICATIONS.

Group	Grid Connection	Configuration	Voltage (V)
A_{1a}	R-N, S-N	I, II	110, 115, 120, 127
A_{2a}	R-N, S-N, T-N	III	120, 127
A_{3a}	R-S	I, II	220, 230, 240, 254
A_{4a}	R-S, R-T, S-T	III	208, 220
A_{5a}	R-S-N	I, II	110, 115, 120, 127
A_{1b}	R-S-N, R-T-N, S-T-N	III	120, 127
A_{1c}	R-S-T	II	230
A_{2c}	R-S-T	III	208, 220
A_{3c}	R-S-T-N	III	120, 127
B_{1a}	R-N, S-N	I, II	220
B_{2a}	R-N, S-N, T-N	III	220
B_{3a}	R-S	I, II	440
B_{4a}	R-S, R-T, S-T	III	380
B_{5a}	R-S-N	I, II	220
B_{1b}	R-S-N, R-T-N, S-T-N	III	220
B_{1c}	R-S-T	II	380
B_{2c}	R-S-T	III	380
B_{3c}	R-S-T-N	III	220

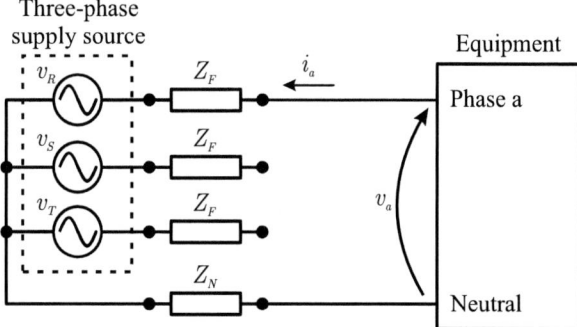

Fig. 3. Single-phase two-wire configuration (phase and neutral connection).

ties. In addition, there are different connection configurations between the transformers, according to Fig. 2, so that the inverters must be designed separately for each region to satisfy the requirements of where they will be used. Based on the different configurations shown in Fig. 2, this paper proposes a classification that encompasses similar groups of inverters, according to Table III. Thus, the inverters are divided into groups A and B, according to the voltage levels of the grid at which they will be connected. Possible connections between the network and the inverters are specifically detailed in the following sections.

A. Single-phase two-wire configuration (phase and neutral connection)

In the single-phase two-wire configuration, only one phase of the inverter is connected to the grid, and the neutral conductor is actively used to conduct current, according to Fig. 3. In this case, the current injected into the grid must be in phase with the incoming voltage v_a of the inverter, so that the circuit operates at a power factor close to the unit.

B. Single-phase two-wire configuration (phases connection)

Unlike the previous system, the single-phase configuration with two wires connected between the phases of the grid is not

connected to the neutral conductor, as shown in Fig. 4. Thus, the current i_a synthesized by the inverter must be in phase with the line voltage v_{ab}. This ensures that the power factor of the inverter is unitary. Consequently, a portion of reactive power is processed in the electrical network, since the current injected into the network is not in phase with the respective phase voltages.

C. Single-phase three-wire configuration

The single-phase three-wire configuration, shown in Fig. 5 can be used in split-phase systems, as shown in Fig. 2 (a) or between the terminals R and S of Fig. 2 (b), so that the voltages of each of the phases of the electric grid are delayed by 180°, in relation to the neutral conductor. Thus, the inverter provides power to the grid in a manner similar to a full bridge topology.

D. Two-phase configuration

Similarly to the single-phase configuration with two wire, the two-phase configuration is connected in two phases of

978-1-7281-4181-7/19 $31.00 © 2019 IEEE

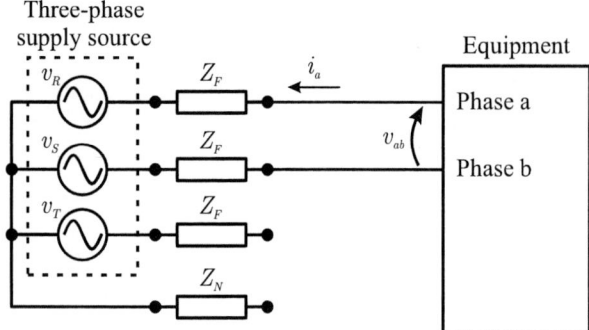

Fig. 4. Single-phase two-wire configuration (phases connection).

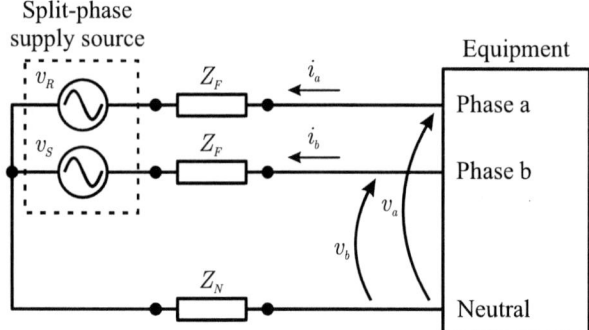

Fig. 5. Single-phase three-wire connection.

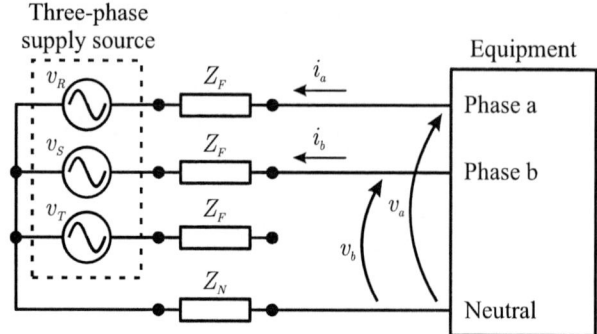

Fig. 6. Two-phase configuration.

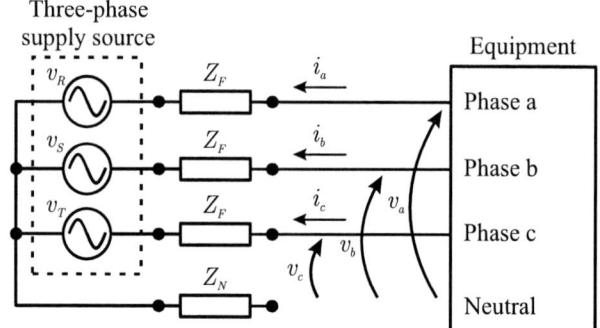

Fig. 7. Three-phase three-wire configuration.

the electric grid, according to Fig. 6. However, this structure has the grid neutral point connected to the inverter and is characterized by the current i_a and i_b injection in phase with the v_a and v_b voltages, respectively. Because it is an unbalanced system, a part of current flows through the neutral conductor in this system.

E. Three-phase three-wire configuration

The three-phase configuration with three wires is shown in Fig. 7. In this case, the injected currents i_a, i_b and i_c are in phase with the v_a, v_b and v_c voltages but without the neutral conductor availability, the grid synchronization is made through the line voltages, resulting in a perfectly balanced system.

F. Three-phase four-wire configuration

The three-phase configuration with four wires has one conductor for each phase and one neutral conductor connected to the inverter circuit, according to Fig. 8. This configuration is particularly interesting on photovoltaic applications, since it allows current leakage reduction from modules through the neutral conductor connection to the center point of the inverter bus, considering that the three-phase system is perfectly balanced.

IV. ANALYSIS OF GRID IMPEDANCE

Note that the input voltage of the PV inverter for the presented configurations depends on the grid voltages v_R,

v_S or v_T and their respective impedance value Z. Therefore, there will be divergences between the equipment input voltage for the different voltage levels of the Brazilian power supply system. For example, a vectorial analysis of the three-phase configuration presented in Fig. 8 [14], [15] can be performed to determine the input voltage value v_a, according to Fig. 9, considering that the equipment operates with a displacement factor $\cos(\theta)$. The variable φ represents the phase shift angle between the grid voltage and its respective equipment input voltage.

Thus, the equations that determine the rms value of the equipment input voltage and the phase shift angle φ are given by:

$$\begin{cases} v_a = v_R\cos(\varphi) + r_l i_a \cos(\theta) - x_l i_a \sin(\theta) \\ 0 = v_R\sin(\varphi) + r_l i_a \sin(\theta) + x_l i_a \cos(\theta) \end{cases} \quad (1)$$

where r_l and x_l are the resistance and the reactance of the grid impedance Z.

In this work, the connection tests between the inverter and the grid are carried out considering that the reference impedance proposed in the IEC/TR 60725 represent the real impedance of the Brazilian distribution system, for a 220 V rms voltage between the phase and neutral conductor, since no detailed analysis were found for the Brazilian electric grid. Thus, Fig. 10 shows the percentage variation of the input voltage ΔV of the three-phase inverter connected according to Fig. 8, considering that the grid operates with the 5%

978-1-7281-4181-7/19 $31.00 © 2019 IEEE

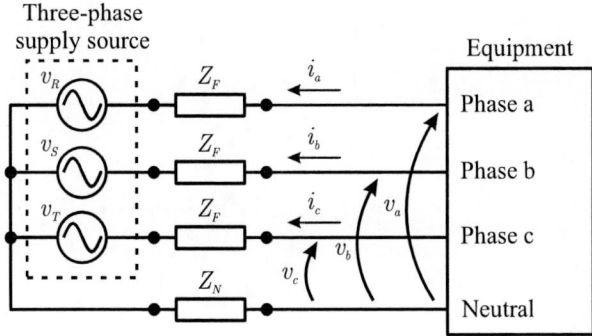

Fig. 8. Three-phase four-wire configuration.

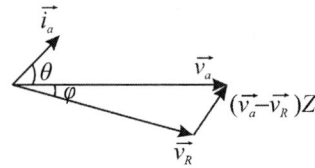

Fig. 9. Phasor diagram for the Phase a of the three-phase configuration presented in Fig. 8.

TABLE IV
PROPOSED VALUES FOR THE REFERENCE IMPEDANCE.

Power	Voltage	Reference impedance
Inferior to 25 kW	Inferior to 150 V	$Z_F = 0.12 + j0.15\ \Omega$ $Z_N = 0.08 + j0.10\ \Omega$
Superior to 25 kW	Inferior to 150 V	$Z_F = 0.075 + j0.15\ \Omega$ $Z_N = 0.05 + j0.10\ \Omega$
Inferior to 50 kW	Superior to 200 V	$Z_F = 0.24 + j0.15\ \Omega$ $Z_N = 0.16 + j0.10\ \Omega$
Superior to 50 kW	Superior to 200 V	$Z_F = 0.15 + j0.15\ \Omega$ $Z_N = 0.10 + j0.10\ \Omega$

maximum allowable voltage variation defined by PRODIST [13]. The green and blue lines represent the voltage variation by changing the inverter power factor to 0.9 (over-excited and under-excited operation), which is considered as the reactive power injection limit defined by ABNT NBR 16149, for inverters with a power greater than 6 kW. The red line symbolizes the minimum power factor that enables the equipment operation in accordance with the variations limits defined in Table II, for current levels of 100 A and 200 A, respectively.

According to ABNT NBR 16149, the inverter should control the reactive power levels injected to the grid up to the power factor limit of 0.9, for over-excited or under-excited operation, in order to reduce the inverter input voltage. However, through the analysis presented in Fig. 10, the inverter would stop the power supply to the grid as the current levels approach 100 A, for a reference impedance $Z_F = 0.24 + j0.15\ \Omega$ and $Z_N = 0.16 + j0.12\ \Omega$, since the reactive power consumption is not sufficient to ensure that the equipment does not exceed the limit of 10% of the grid nominal voltage. In this sense, for these current limits, the equipment remains disconnected from the grid under normal operating conditions. Note that the percentage value variation ΔV tends to increase for the group A inverters defined in Table III, due to the lower operating voltages of these equipment. For this case, the inverters would disconnect from the grid for a rms current value of 45.1 A and 125.7 A, for each of the reference impedance values. However, in Brazilian low voltage electrical distribution systems that operate with 127 V rms voltage, the power utilities use twice the diameter of conductors used in 220 V rms installations [16]. Thus, the resistance values of the cables are reduced by half for 127 V, compared to 220 V systems, reducing the voltage drop of the conductors. The inductive reactance

of the cables remains unchanged, as the cable arrangement is maintained. Therefore, it is possible to establish a new relationship between the voltage variation ΔV and the current injected into the grid, according to Fig. 11.

Thus, to provide an impedance that is compatible with the Brazilian reality and that can be easily used for the flicker tests, it is suggested to redefine the reference impedance values defined in IEC/TR 60725, maintaining the maximum compatibility with this standard. In addition, it is proposed that the values used for the reference impedance should be defined according to the distribution system power levels, due to the wide range of voltage configurations in the Brazilian low voltage grid. Thus, the new classification was performed based on the three-phase inverter groups defined in Table III, for the 75 A rms equipment current limit defined in IEC-61000-3-11. Therefore, the resistance of the reference impedance must be halved for the inverters of the group A of Table III, while the resistance for the group B inverters can be maintained in accordance with the values defined in the international standards, as presented in Table IV.

V. CONCLUSIONS

This paper presented a brief review of the standards and criteria adopted to perform flicker tests in Brazil. Thus, it was verified that the international standards on which the Brazilian standards are based do not contemplate the wide variety of grid configurations existing in Brazil. In this way, a new order of classification of the inverters was suggested, separated by the type of connection to the grid and its voltage levels.

Thus, it was found that in most cases, the use of reference impedances for flicker tests was not the most suitable, since the use of this impedance results in abnormal voltage conditions at the connection point of the inverter to the grid. In addition, considering that the reference impedance defined in IEC/TR 60725 represent the Brazilian electric grid, the inverters are not able to operate for certain current values injected to the grid and the inverter is automatically disconnected. Thus, in this paper a new value for the reference impedance for the inverter operation and for the flicker tests was proposed, so that it is compatible with the local grid, enabling the operation in the different configurations of the power distribution system.

ACKNOWLEDGMENT

This study was financed by Hi-Mix Eletronicos S/A and by the Coordenação de Aperfeiçoamento de Pessoal de Nível Su-

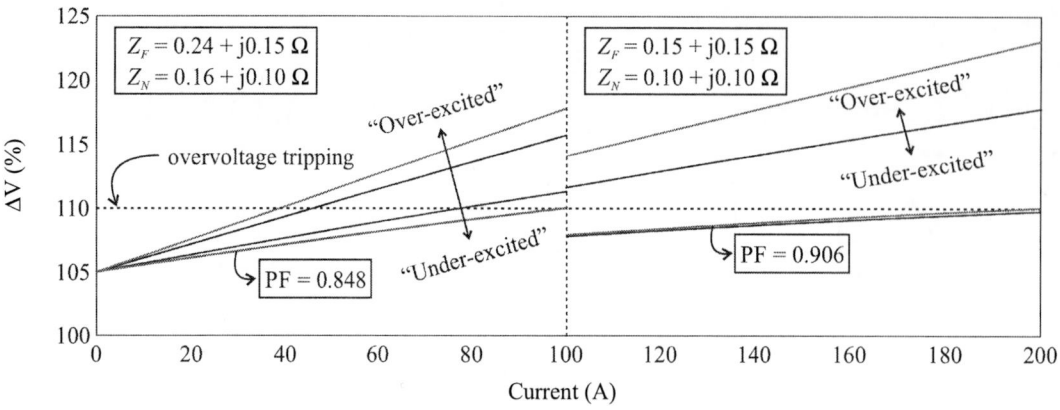

Fig. 10. Variation of the three-phase inverter of Fig. 8 input voltage ΔV in relation to the rated nominal voltage of 220 V of the grid.

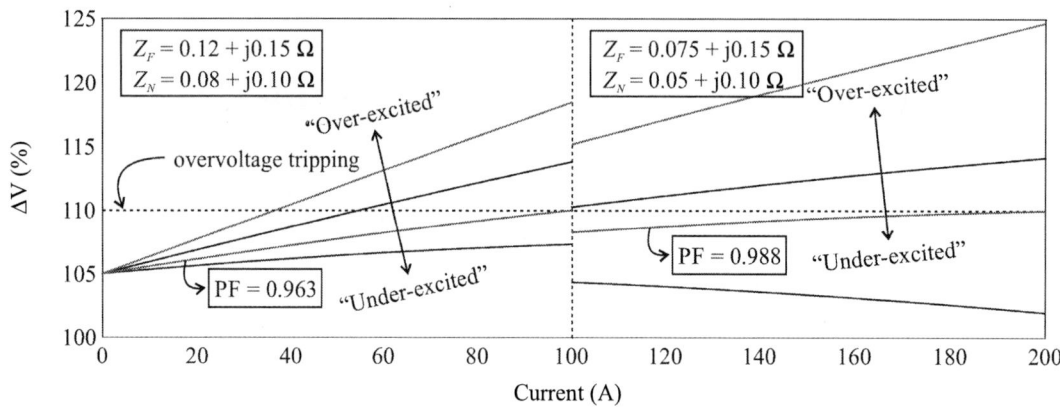

Fig. 11. Variation of the three-phase inverter of Fig. 8 input voltage ΔV in relation to the rated nominal voltage of 127 V of the grid.

perior - Brasil (CAPES/PROEX) - Finance Code 001. The authors also thank INCT-GD and CNPq (process 465640/2014-1 and 306317/2015-0), CAPES (process 23038.000776/2017-54) and FAPERGS (17/2551-0000517-1) for the financial support.

REFERENCES

[1] R. Ferreira, P. H. Corredor, H. Rudnick, X. Cifuentes and L. Barroso, "Electrical Expansion in South America: Centralized or Distributed Generation for Brazil and Colombia," in IEEE Power and Energy Magazine, vol. 17, no. 2, pp. 50-60, March-April 2019.

[2] REN21; (2019). Renewables 2019 global status report. REN21. [Online]. Available: https://www.ren21.net/status-of-renewables/global-status-report.

[3] EPIA. Global Market Outlook for Solar Power 2018-2022. Brussels. 2018.

[4] EPE. Anuário Estatístico de Energia Elétrica 2018: ano base 2017. Rio de Janeiro.

[5] ANEEL. Atlas de Energia Elétrica do Brasil 2002. Brasil.

[6] ABNT Standard Photovoltaic (PV) Systems - Characteristics of the Connection Interface with the Distribution Grid. ABNT NBR Std. 16149, 2012.

[7] H. H. Figueira, H. L. Hey, L. Schuch, C. Rech and L. Michels, "Brazilian grid-connected photovoltaic inverters standards: A comparison with IEC and IEEE," 2015 IEEE 24th International Symposium on Industrial Electronics (ISIE), Buzios, 2015, pp. 1104-1109.

[8] IEC Consolidated Version Electromagnetic Compatibility (EMC) - Part 3-3: Limits - Limitation of voltage changes, voltage fluctuations and flicker in public low-voltage supply systems, for equipment with rated

current ≤ 16 A per phase and not subject to conditional connection, IEC Std. 61000-3-3, 2017.

[9] IEC International Standard compatibility (EMC) - Part 3-11: Limits - Limitation of voltage changes, voltage fluctuations and flicker in public low-voltage supply systems - Equipment with rated current ≤ 75 A and subject to conditional connection, IEC Std. 61000-3-11, 2017.

[10] IEC Technical Specification Electromagnetic compatibility (EMC) - Part 3-5: Limits - Limitation of voltage fluctuations and flicker in low-voltage power supply systems for equipment with rated current greater than 75 A, 2009.

[11] IEC Technical Report Consideration of reference impedance and public supply network impedances for use in determining the disturbance characteristics of electrical equipment having a rated current ≤ 75 A per phase, 2012.

[12] INMETRO Ordinance N 357, 2014.

[13] ANEEL Distribution Procedures of Electric Energy in the National Electric System, 2016.

[14] R. Z. Scapini, C. Rech, T. B. Marchesan, L. Schuch, R. F. de Camargo and L. Michels, "Capability analysis of a D-STATCOM integrated to a single-phase to three-phase converter for rural grids," 2014 IEEE 23rd International Symposium on Industrial Electronics (ISIE), Istanbul, 2014, pp. 2560-2565.

[15] R. Z. Scapini, C. Rech, T. B. Marchesati, L. Schuch, R. F. de Camargo and L. Michels, "Distribution STATCOM integrated to a single-phase to three-phase converter," IECON 2014 - 40th Annual Conference of the IEEE Industrial Electronics Society, Dallas, TX, 2014, pp. 1423-1429.

[16] L. Queiroz, "Estimação e Análise das Perdas Técnicas na Distribuição de Energia Elétrica", Ph.D. Dissertation, UNICAMP, Campinas, BR, 2010.

An Electronic Drive for a Switched Reluctance Motor Using a DSC

Brayan Sobral da Fonseca
Graduate Program in Electrical Engineering
COPPE - Universidade Federal do Rio de Janeiro
Rio de Janeiro, Brazil
brayanuff@hotmail.com

José Andrés Santisteban
Graduate Program in Electrical and Telecommunications
Engineering - PPGEET
Universidade Federal Fluminense
Niterói, Brazil
josesantisteban@id.uff.br

Abstract—**With the technological advances using electronic circuits based on microprocessors, the drive and control of electrical machines in electromechanical systems have become increasingly efficient. In this paper, for an axial magnetic flux reluctance motor, for which it is not possible to supply its energy directly from the utility, the developing of an electronic drive and its control is described. Unlike other alternatives, a development module, named DSC (Digital Signal Controller), which is based on a modern microcontroller, allows the reduction of discrete components, like operational amplifiers, and wires. The workbench comprises a three-phase power converter that is equipped with high-speed semiconductor switches, current sensors, voltage sensors and an USB interface to allow communication with a computer. The effectiveness of this approach is verified through open loop and closed loop experimental tests, where the waveforms of the trigger pulses generated by the DSC, the motor currents and their speed were monitored.**

Keywords— Switched Reluctance Motor, Drives, Control, Digital Signal Controller

I. INTRODUCTION

With the evolution of electronics and computational advancements, drives and controls of electromechanical systems have been continuously improved. There are many industrial applications and research in this area like in electric vehicles [1]. Electronic drives may change the electrical frequency, voltages and currents of the motors through a power converter. Some of the forms to control it is through a device having at least one microprocessor, such as a computer, or a microcontroller or even a system based on a Digital Signal Processor (DSP) [2, 3]. In this way, the electrical to mechanical energy conversion is possible to be more efficient [4].

Some years ago, some experimental results respect to the control of an axial magnetic flux switched reluctance motor (AFSRM) were reported in [5]. At that time, its electronic control system was implemented using several discrete components like operational amplifiers, discrete TTL digital circuits, voltage regulators and a fixed point PIC microcontroller. Consequently, many wires were necessary to link all the components.

Nowadays, the use of high-speed microprocessors is a common practice to improve modern electronic drives and motor controllers, however, the price is proportional to the speed capacity. In order to reduce the customer investment, several companies offer compact devices comprising a microprocessor and appropriate peripherals such as digital and analog input/output pins, counters and comparators by instance. Thus, according to the sophistication one can found some devices named as microcontrollers, Digital Signal Controllers (DSC) and digital signal processors (DSP).

In this sense, the main motivation of this work was the developing of a compact, modern and economic drive system for the AFSRM. Thus, a DSC based development module was chosen as the price of a DSP based one, with similar peripherals, is three or more times higher.

The workbench comprises an angular position sensor, a power converter with six MOSFETs, forming three half-bridge structures, current sensors and some power supplies. The angular sensor generates pulses that are sent to the DSC. As well known, the stator windings are energized in a sequential way according to the rotor position, producing an electromagnetic torque that controls the speed of the rotor.

Finally, in order to validate our approach, experimental results in open loop and closed loop speed control are shown.

II. SWITCHED RELUCTANCE MOTOR

A. SRM (Switched Reluctance Motor)

The functioning principle of the switching reluctance motor (SRM) is not easy to formulate because of its nonlinear electromagnetic relationships. In addition, there are many design problems because they do not follow the classical techniques used for conventional DC and AC motors.

The SRM is an electrical machine with copper coils concentrated at the stator poles and without coils or magnets at the rotor poles. All the poles are axially salient [5, 6]. The prototype here used is of the 6/4 type, that is, they have six poles on the stator and four poles on the rotor. The air gap magnetic flux is axial. Both stator and rotor cores are made of SAE-1020 steel and the shaft is made of stainless steel. The relative magnetic permeability of the stainless steel is equal to the air.

B. Formulation

Taking into account all of the windings of the SRM and that two coils form a winding, the circuit analysis leads to a matrix differential equation shown in (1). Where $\mathbf{V}$ is the voltage vector applied to the windings, $\mathbf{R}$ is the resistance vector for each coil, $\mathbf{I}$ is the current vector, $\mathbf{L}$ is the inductance matrix, which is a function of the rotor position θ and ω is the angular speed of the rotor. It is straightforward to demonstrate that if the currents are imposed, the net electromagnetic torque is proportional to the variation of the inductance $\mathbf{L}$ with the change of θ [5, 7].

$$[\mathbf{V}] = 2\mathbf{R}[\mathbf{I}] + \partial/\partial t\,([\mathbf{L}(\theta)].[\mathbf{I}]\,) =$$

$$2\mathbf{R}[\mathbf{I}] + \omega.\partial/\partial\theta\,([\mathbf{L}(\theta)].[\mathbf{I}] + [\mathbf{L}(\theta)].\,\partial/\partial t\,[\mathbf{I}]). \qquad (1)$$

C. Power Converter

Each of the half bridge in the converter can be driven according to the chosen strategy. The most simple consists in the use of a hysteresis current controller in such a way that the semiconductor switches commutate when the measured winding current reach predefined limits. Alternatively, the pulse width modulation (PWM) technique to impose voltage to the windings can be used [2, 5, 7]. In this case, pulses with high constant frequency are generated so less acoustic noise is produced [8]. In both cases, the switches are closed or open according to the position of the rotor. In this work, to drive the converter using the DSC, both strategies were tested.

III. THE DIGITAL SIGNAL CONTROLLER

With the technological advances, the silicon industry has been able to pack, in a small piece of silicon, everything that is needed for an automatic control system be implemented by a microcomputer. Thus, there are families of microcontrollers with different internal registers, set of instructions, number of peripherals and internal memory space. All these products in the market can be easily found.

Nowadays, many of the equipment and products that are part of the houses, automobiles, etc. They use several kinds of microcontrollers. As an example, a modern car has about 80 microcontrollers to perform all the electronic functions like the ABS system, central locking and locks, electric windows, mirror adjustments, among other equipment or devices that require more and more controllers to enhance comfort and increase the energy efficiency [9].

The TMS320F28377S is a microcontroller that belongs to the C2000 family with CPU that combines analog and digital peripheral control in a single device with a single-core design. In general, the systems implemented with these devices are efficient and inexpensive. The core has a 200 MHz operating speed that provides a complete real-time control and high-performance signal processing solution in power systems such as industrial motors, inverters and many other applications [10].

IV. THE WORKBENCH

In addition to the aforementioned SRM, the three half bridges and the electronic position sensor, the experimental workbench is comprised of Hall Effect current sensors, power supplies and a development card (LAUNCHXL-F28379D) based on the TMS320F28377S, connected to a computer through an USB cable. In addition, a TEKTRONIX oscilloscope captures the waveforms generated by the system to be analyzed. A block diagram of the circuit interconnections between the SRM and the other components is shown in Fig. 1 and in Fig. 2 there is a photography of the workbench where the experiments were carried out. It includes the SRM, the DSC, and the auxiliary circuits inside boxes. In table I, some important parameters of the motor used in this work are found.

In operation, the DSC captures the signal from the position sensor through the (ECAP) input and then resolves an algorithm, which is updated in 100 μs, time that is programmed in the interrupt register (ISR). At the end of the algorithm, pulses are sent to the converter in order to generate the correspondent voltages or currents needed to hold the rotor in movement.

In Fig. 3, an example of the current waveform in one of the SRM windings being driven, in open loop, by pulses of 10 kHz, generated by the DSC during the interruptions, is shown. The train of PWM pulses generates a current with some noise due to the switching of the corresponding MOSFETs. As the pulse width decreases, because the duty cycle becomes less and less, reductions in the coil currents and in the machine speed and power are obtained. While current is circulating in a winding, it remains zero in the other ones.

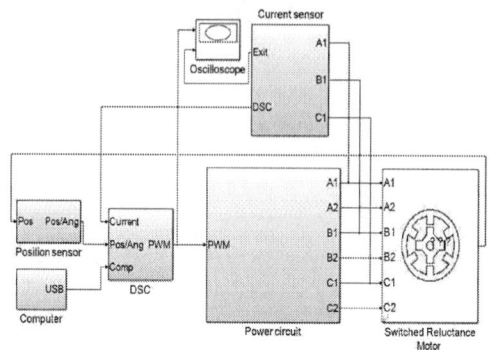

Fig. 1. Block diagram of the workbench.

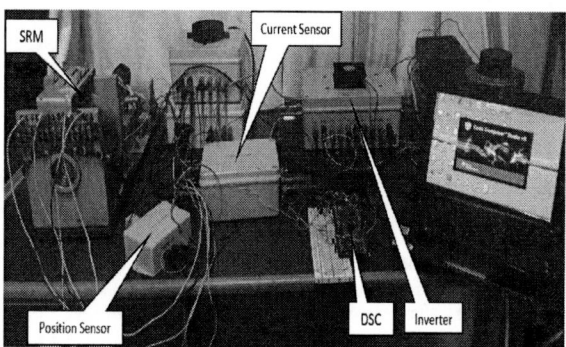

Fig. 2. Picture of the system.

Along the operation time, the DSC receives rectangular pulses from the rotor angular position sensor, which consists of a disc that has of 180 little holes and is fixed to the rotor shaft to generate pulses each two degrees of angular displacement. At the beginning of the motor starting, a winding is energized to take the rotor to its initial position and then the DSC estimates the correct sector to supply the appropriated winding to maintain the rotor in movement. Respect to the control algorithm, this is implemented in an interrupt routine using C language instructions.

TABLE I. IMPORTANT PARAMETERS OF THE AXIAL FLUX SRM

Parameter Value	
Outer diameter of rotor and stator	126 mm
Inner diameter of rotor and stator poles	63 mm
Shaft Diameter	40 mm
Stator and rotor poles arc	40^0
Stator poles area (axial cross-section)	1039 mm^2
Number of turns per stator poles	175
Turn wire	24 AWG
Coil resistance	2.3 Ω
Stator and rotor cores material	Steel SAE-1020
Number of phases	3
Number of poles	6/4

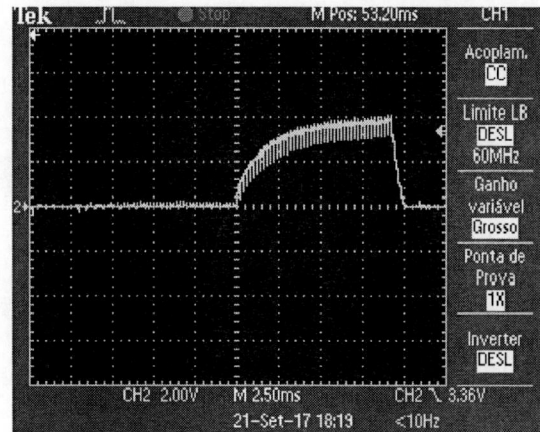

Fig. 3. Current waveform in one phase.

V. EXPERIMENTAL RESULTS

A. Open loop results

As a first test to verify the capacity of the DSC to drive a three-phase SRM, an open loop speed control was implemented. In this case, with the aid of the position sensor, a revolution (360°) is divided in four sectors of 45 pulses and then subdivided in 15 pulses by phase, which is equivalent to the mechanical angle of 30°. It is divided in four sectors because the rotor has four poles. The maximum speed is obtained when the correspondent MOSFETs remain in the "on" state during all the 30°. The other phases are activated in a sequence just to give the desired shaft direction, clockwise or counterclockwise.

To reduce the speed, the 30° pulse is split in several pulses with variable duty cycle. As the average voltage is reduced, the average current is also reduced. In Fig. 4, the voltage applied to a phase (in yellow), after a trigger pulse with 100% of duty cycle is applied to the converter, and its correspondent current (in blue) are shown. As the current sensor was calibrated to the ratio 1 V/ 1 A, the maximum current is 4 A. In Fig. 5, for a voltage supply of 30V, currents of two phases are shown. As noted, comparing with Fig. 4, the frequency of pulses is proportional to the DC link voltage.

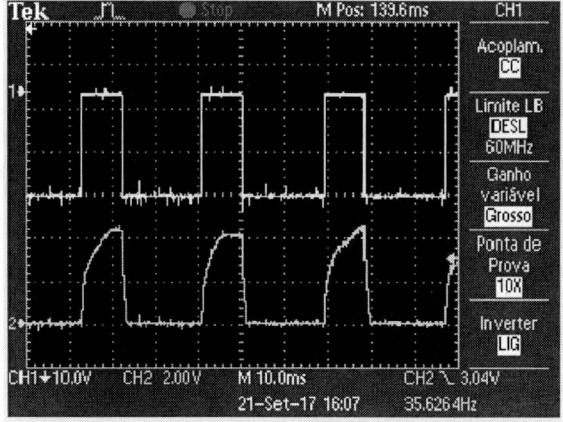

Fig. 4. Voltage and current waveforms for a phase using 100% of duty cycle.

Additional tests were also executed holding the DC link voltage but reducing the duty cycle. The level of the currents was also reduced. For this SRM, with 24V, the minimum duty cycle to maintain the shaft in movement was 80%.

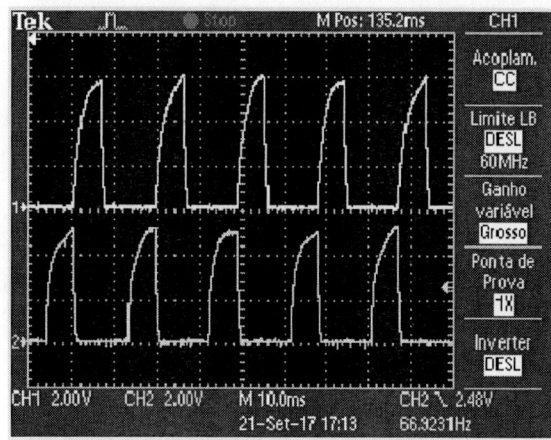

Fig. 5. Measurements of currents in two different phases.

For the last condition, Fig. 6 presents the waveforms for the voltage applied in the winding and its respective current. It is clear from the figure that several pulses compose the voltage. The frequency of these pulses is 10 kHz, which is in accordance with the interruption cycles of 0.1 ms. O the other hand, it should be noted the reduction of the correspondent current, limited to 2 A. The final effect is the reduction of the torque and consequently the reduction rotor speed, as can be verified by the time scales. In Fig. 4 is 10 ms/div while in Fig. 6 is 25 ms/div.

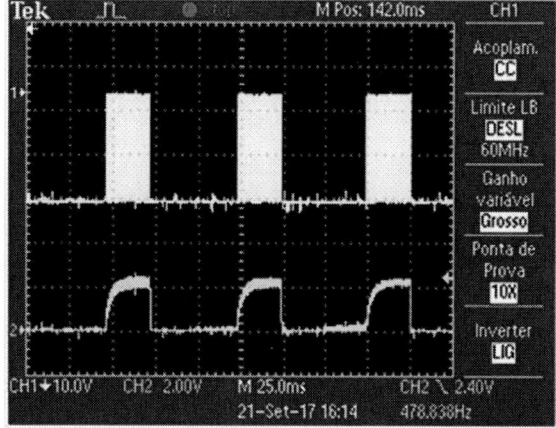

Fig. 6. Voltage and current waveforms for a phase using 80% of duty cycle.

Fig. 7 summarize the results obtained without shaft load. There were plotted the obtained speeds for the SRM for different duty cycles. According to this figure, a linear behavior can be observed, where the maximum speed was 720 RPM and the minimum 227 RPM. It was also verified that in an open loop control the rotor speed variation as a function of the pulse width (duty cycle) is limited to a narrow band. In our case, it varies from 100% to 80%. For lower values, the SRM operates unstable or does not rotate due to the currents gets too small. This is expected, as the low magnetic field generated by lower currents are not sufficient to overcome the frictional forces offered by the mechanical bearings that support the shaft.

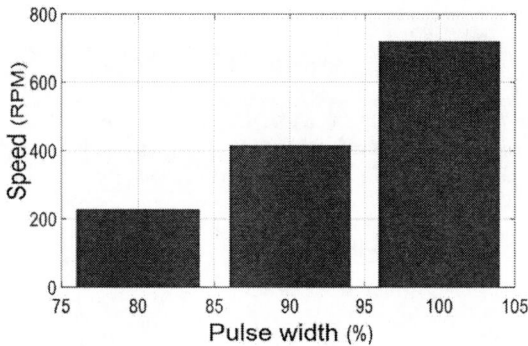

Fig. 7. Speed (RPM) as a function of the duty cycle (%).

B. Closed loop results

The second test for the DSC was executed to verify its capacity to operate in a closed loop condition. In this case, to control the SRM shaft speed. For this, Fig. 8 depicts the basic control strategy. First, from the reading of the position sensor the rotor speed is estimated. Using a previous established reference speed an error is calculated which is further processed by a digital PI controller. Its output is understood as the reference for the electrical current that should be imposed to the three windings, in a sequential manner that depends on the position sector and the direction of rotation.

To complete the control system, the DSC reads the signals of the current sensors that are compared with their reference current [11, 12]. In this case, a single hysteresis comparator algorithm is used. Thus, the appropriate semiconductor switch is triggered when the current is lower than the lower limit and is not triggered when it is higher than the higher limit [8].

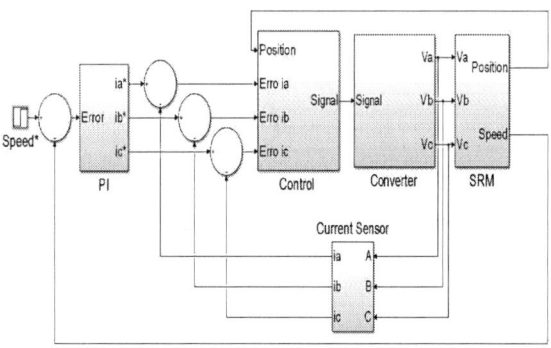

Fig. 8. Closed loop control strategy.

Fig. 9 shows the experimental results with no load at the shaft from the stop condition up to the reference speed of 330 RPM. The speed and current scales were, respectively, 330 RPM/V and 1.0 A/V. A variation of the frequency in the waveform of the current during starting is observed because a gradual increase of the speed of the rotor takes place and with this a decrease in the time between the switching of the respective sectors. One can observe that when the reference speed of 330 RPM is reached, at approximately 1.25 s, the current amplitude decreases, going from approximately 5 A to 3 A.

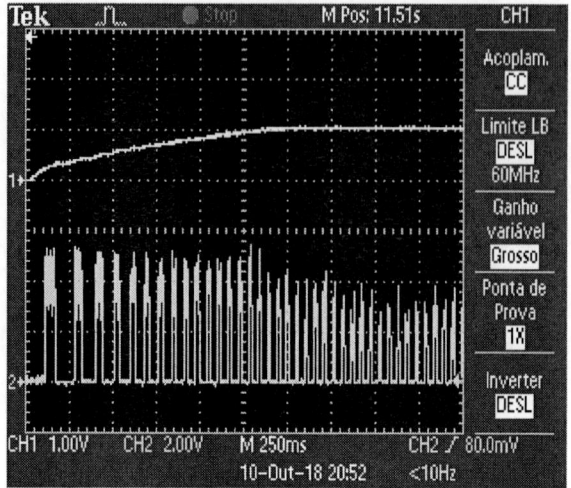

Fig. 9. Response of speed (yellow) and current (blue) during starting.

Fig. 10 shows the experimental results, with no load, for a speed variation from 330 RPM to 530 RPM. It is observed an increase in the current during the shaft acceleration. Fig. 11 shows this same variation but decreasing, i.e. the shaft speed ranging from 530 to 330 RPM. During this variation, the current in the phases is zero, because the speed error is negative and the reference current must be zero as a half bridge is used. A complete bridge inverter would reduce the braking time during this variation as negative voltage could be imposed.

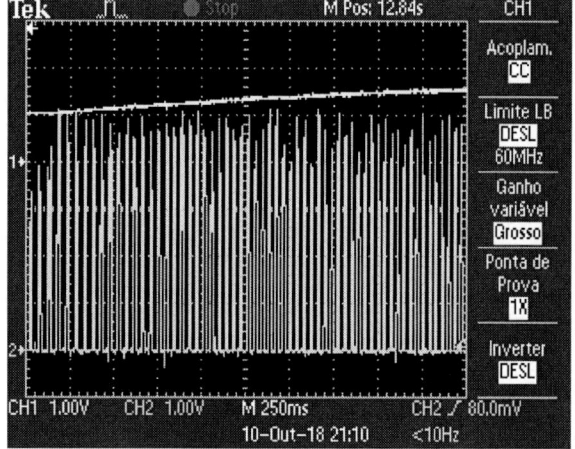

Fig. 10. Response of speed (yellow) and current (blue) during positive speed variation.

978-1-7281-4181-7/19 $31.00 © 2019 IEEE   949

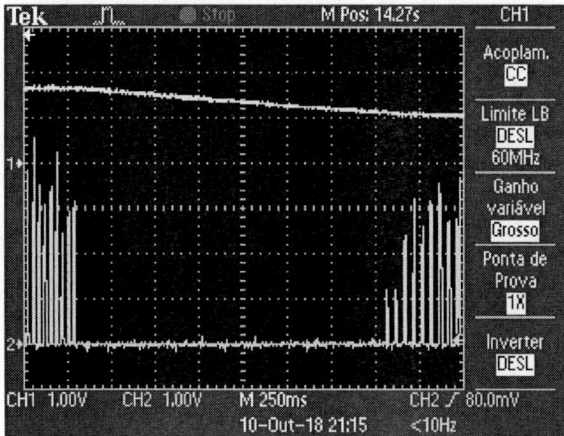

Fig. 11. Response of speed (yellow) and current (blue) during negative speed variation.

Fig. 12 shows the measured current (blue) in one of the stator phases following the reference current (yellow) determined by the PI controller. The figure shows the details of the currents when the rotor changes the speed at the instant 250 ms (center).

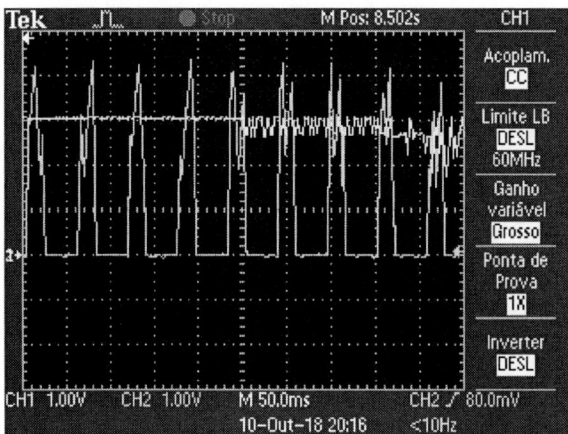

Fig. 12. Response of the real current (blue) and reference current (yellow).

Fig.13 and Fig. 14 show the experimental results of the stator phase currents at the moment of coupling and decoupling of a load on the shaft, represented by an intense disturbance caused by rubbing down the motor shaft. In both figures after the disturbances, the current changes perceptibly (center). In the case of coupling, in Fig. 13, the current is increased and, in the case of uncoupling, in Fig. 14, the current decreases. In both cases, the speed remained unchanged and equal to 330 RPM due to the effective closed loop of the SRM speed control despite of the shaft disturbances.

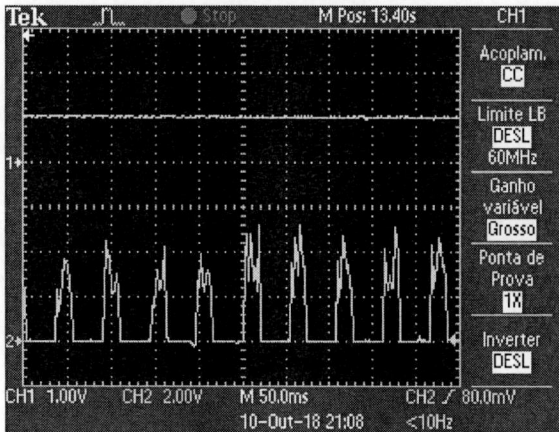

Fig. 13. Response of speed (yellow) and current (blue) at the time of load coupling.

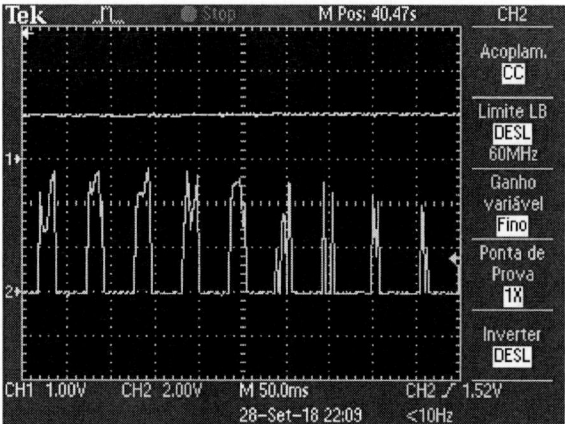

Fig. 14. Response of speed (yellow) and current (blue) at the time of load decoupling.

Finally, Fig.15 shows the experimental results of the measured and reference currents during the friction disturbance on the shaft. In this figure, after approximately 250 ms (center), the increase of the reference current by the controller and consequent increase of the stator phase current due to the load coupling on the shaft is noted.

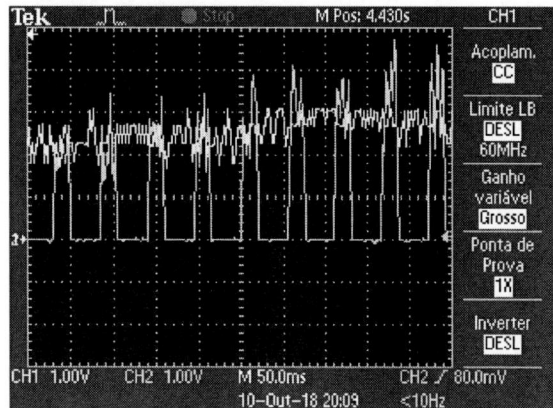

Fig. 15. Response of the measured (blue) and reference currents (yellow) at the load coupling moment.

VI. CONCLUSIONS

In this work, an electronic drive system for a SRM based in a development card with a Texas Instruments DSC was developed.

For our axial flux SRM, in an open loop control the rotor speed variation as a function of the pulse width is limited to variations from 100% to 80% of the duty cycle. For lower values, the SRM operates unstable or does not rotate. This is expected, as the low magnetic field generated by lower currents are not sufficient to overcome the frictional forces offered by the bearings that support the shaft.

In a closed loop, a speed control system was also developed and implemented with the DSC. From the speed error between the measured and the one desired by the user, a classical control PI algorithm determines the reference current amplitude that the phases must have. Some measured results of speed and electrical currents were obtained with and without load at the shaft. In all the cases, the results were the expected.

In this way, the modernization of a SRM drive using a development card based on a Digital Signal Controller has shown to be an effective solution bringing as advantage a reduction of components and volume without costly investment. In addition, due to its higher speed, it could be tested other sophisticated control algorithms which demand a higher number of computational instructions.

REFERENCES

[1] Z. Omaç, M. Polat, E. Öksüztepe, "Design, analysis, and control of in-wheel switched reluctance motor for electric vehicles", Electrical Engineering, Vol 100, Issue 2, pp 865–876, June 2018. https://doi.org/10.1007/s00202-017-0541-3.

[2] S. Paramasivam, R. Arumugan, "Hybrid fuzzy controller for speed control of switched reluctance motor drives", Energy Conversion and Management 46 1365–1378, 2005.

[3] X. Li et al., "Design of Switched Reluctance Motor Speed Control System Base on DSP", Applied Mechanics and Materials, Vols. 462-463, pp. 727-734, 2014.

[4] X. Gao, X. Wang , Z. Li1 and Y. Zhou1, "A Review of Torque Ripple Control Strategies of Switched Reluctance Motor", International Journal of Control and Automation, Vol. 8, No. 4, pp. 103-116, 2015, http://dx.doi.org/10.14257/ijca.2015.8.4.13.

[5] E. S. Sanches, E.S, J. A. Santisteban, "Comparative Study of Conventional, Fuzzy Logic and Neural PID Speed Controllers with Torque Ripple Minimization for an Axial Magnetic Flux Switched Reluctance Motor", En-gineering, 6, 655-669. DOI: 10.4236/eng.2014.611065.

[6] E. S. Sanches, J. A. Santisteban, "Mutual Inductances Effect on the Torque of an Axial Magnetic Flux Switched Reluctance Motor", IEEE Latin America Transaction, Vol. 13, No. 7, July 2011.

[7] E. S. Sanches and J. A. Santisteban, "Implementing a neural PID speed controller for a single stator axial flux switched reluctance motor aiming torque ripple minimization", Brazilian Power Electronics Conference, Gramado 2013, pp. 929-934. DOI: 10.1109/COBEP.2013.6785226.

[8] K. B. Bose, Modern Power Electronics and AC Drives. Prentice Hall. United States of America, 2001.

[9] F. Bormann, "Introduction to TMS320F28335", OpenStax-CNX module: m36699, Jan 2011.

[10] K. W. Schachter, "The TMS320F2837xD Architecture: Achieving a New Level of High Performance", Technical Brief SPRT720 Texas Instruments, February 2016.

[11] P. Jeevananthan, C. Sathish Kumar, R. K. Nithya, "Fuzzy Logic Based Field Oriented Control of Permanent Magnet Synchronous Motor (PMSM)", International Journal of Advanced Research in Electrical, Electronics and Instrumentation Engineering, vol. 1, Issue 5, November 2012.

[12] L. O. A. P. Henriques, W. I. Suemitsu, "Development and Experimental Tests of a Simple Neurofuzzy Learning Sensorless Approach for Switched Reluctance Motors", IEEE Transactions on Power Electronics, Vol. 26, No. 11, November 2011.

DC MOTOR MODEL FOR WINDOWS PINCH PROTECTION APPLICATIONS

Gabriel F. Idalgo
CECS - UFABC
Santo André, SP-Brazil
gabriel.idalgo
@ufabc.edu.br

José A. Torrico Altuna
CECS - UFABC
Santo André, SP-Brazil
jose.torrico
@ufabc.edu.br

Carlos E. Capovilla
CECS - UFABC
Santo André, SP-Brazil
carlos.capovilla
@ufabc.edu.br

Ademir Pelizari
CECS - UFABC
Santo André, SP-Brazil
ademir.pelizari
@ufabc.edu.br

Wolfgang Schulter
Hochschule Ravensburg-
Weingarten - Germany
wolfgang.schulter
@rwu.de

Abstract—**This paper implements a time variant dynamical DC motor model for windows pinch protection. The model includes the commutator effect in the armature current that is frequently used in anti-pinch protection of automotive power window lifters. The variations of armature resistance and inductance are also modeled in the time variant model. Experimental results obtained are used to validate the implemented model.**

Index Terms—**DC Motor, Power Window Lifter, Automotive Window, Power Electronics, Control System, Torque Estimation.**

I. INTRODUCTION

Automotive industry uses DC motor commutator effect in armature current for application in automotive power window lifter. The main approach processes armature current waveform to determine obstacles, torque and size of the automotive window. Brazilian market applications use experimental data to solve all the existing automotive windows lifters, spending many hours of engineering. Unfortunately, this approach is unable to verify all real situations. As a severe consequence, this method generates many false reversals and reversion forces without satisfying current Brazilian standards [1]. In this context, it is implemented the DC machine model based on a time variant dynamic model. The model based approach allows to add mathematically electrical motor parameters variations, noise and uncertainties. It is also possible to include the mechanical model equations to improve control and estimation algorithms. Normally, current researches in this area address mainly linear model of DC motor and acquired armature current waveform.

Practical anti-pinch protection system based on infinite H filter was proposed in [2]. Through a Hall Effect sensor coupled in the motor to obtain the angular velocity measurements, system calculates the angular velocity through a proposed algorithm with the measurement noise reduction logic. The algorithm anti-pinch detection makes use of torque and evaluates information. For it, the torque rate is increased to the system model and the torque rate estimator is derived

by applying the recursion of the H-filter at steady state to the model. The approach comes from the idea that the torque rate is less sensitive to the uncertainties of the engine parameters. The proposed scheme minimizes the exposure of the anti-crushing window control system and false reversal.

A practical torque estimator based on the Kalman filter was proposed for low-cost anti-pinch protection of power window control in [3]. To obtain the precise angular velocity of the Hall effect sensor measurements, the angular velocity calculation algorithm is proposed including the measurement noise reduction logic. The proposed anti-pinch detection algorithm makes use of the torque value variation, to do this, the torque rate is increased for the system model and the torque rate estimator is derived from the a posteriori estimation of the co-variance matrix of the Kalman filter at steady state to the model. The approach comes from the torque rate is less sensitive to the uncertainties of the engine parameters. Thus, the proposed scheme minimizes the anti-pinch window and controls the exposure of the false reversal.

A system aiming to effectively reduce the problem hidden in window electric automotive was proposed in [4]. The article shows an anti-pinch protection system without sensor based on a window control module, which uses the micro-controller PIC18F258 to sample and process the current of the electric window motor. The method uses the combination of the current amplitude and the value of the integral current and through the analysis of these values, determines if there are obstacles in the window, thus implementing the anti-pinch protection function.

The Time-Variant DC Motor Model [5] proposes a two values representation of armature resistance and inductance, based on the size and position of the brushes. It is also shown simulation results with the model. The present work uses this model and compares its results with test bench results of a commercial application.

978-1-7281-4181-7/19 $31.00 © 2019 IEEE

II. MATHEMATICAL MODEL OF THE DC MOTOR

A. Linear DC Motor Model

In most of the power windows lifter applications, the magnetic field is generated by permanent magnet whose geometric form generates uniform magnetic field through all the armature circuit. In Fig. 1 is shown the equivalent electrical circuit of DC motor armature, R_a is armature resistance, L_a is armature inductance, U_q is induced voltage and R_k is line resistance.

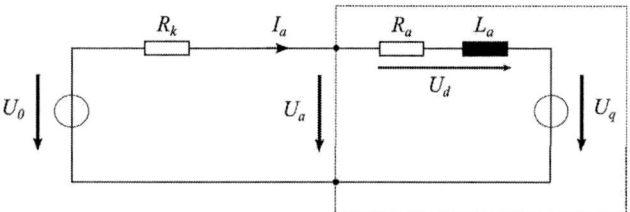

Fig. 1. DC motor armature equivalent circuit

(1) shows the dynamic of DC motor armature equivalent circuit. (3) as a function of current I_a and voltage U_d and (4) in frequency domain.

$$U_0 = R_k.I_a + R_a.I_a + L_a\frac{dI}{dt} + U_q \tag{1}$$

$$\text{With:} \quad U_d = U_a - U_q \tag{2}$$

$$I_a(t) + T_a(t).\frac{dl_a(t)}{dt} = U_d(t).R_a^{-1}(t) \tag{3}$$

$$H_a(s) = \frac{I_a(s)}{U_d(s)} = \frac{R_a^{-1}(s)}{1 + s\frac{L_a(s)}{R_a(s)}} = \frac{R_a^{-1}(s)}{1 + sT_a} \tag{4}$$

$$\text{With:} \quad T_a = \frac{L_a}{R_a} \tag{5}$$

As magnetic flux Φ is constant, the constants $C_1\Phi$ and $C_2\Phi$, of equation 6 and 7, can be determined by the motor design.

$$C_1\Phi = \frac{M_a}{I_a} \quad \text{Torque constant} \tag{6}$$

$$C_2\Phi = \frac{U_q}{n_m} \quad \text{Main motor constant} \tag{7}$$

The armature voltage is always proportional to mechanical speed without load. The static torque M_a is always proportional to the current armature current I_a. There is a simple dependency between the two motor constants $C_2\Phi = 2\pi.C_1\Phi$ while C_2 is a constant of design and Φ is the magnetic flux. The equations of armature circuit can be expressed in the frequency domain or in the time domain as the corresponding differential equation.

The mechanical equation is presented in (8), where J_a is the moment of inertia and make $M_d = M_a - M_L$ can be expressed in (9) in the frequency domain. [6], [7], [8].

$$M_a(t) - M_L(t) = 2\pi J_a.\frac{dn_m(t)}{dt} \tag{8}$$

$$H_b(s) = \frac{n_m(s)}{M_d(s)} = \frac{1}{2\pi J_a.s} \tag{9}$$

The torque difference can be expressed as the equivalent current difference, with I_l as current load per (10):

$$M_d(s) = M_a(s) - M_L(s) =$$

$$C_1\Phi.(I_a(s) - I_L(s))$$

$$\text{with} \quad I_L \text{ Load Current} \tag{10}$$

$$M_d(t) = M_a(t) - M_L(t) =$$

$$C_1\Phi.(I_a(t) - I_L(t)) = C_1\Phi.I_a(t) - M_L(t)$$

In Fig. 2 is show the block diagram of DC motor dynamic model. The state of the motor is determined by the armature current I_a and the rotation speed n_m. The dynamics is determined mainly by the transfer function armature $H_a(s)$ and the torque integration H_b (s). The state of the motor determines the rotor speed and the armature current can be expressed by (13) and (14).

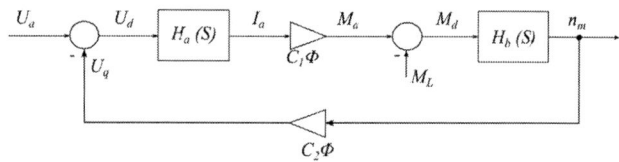

Fig. 2. Block diagram of the equivalent system with the two motor input quantities U_a and charge torque M_L.

The integration of the rotor speed produces the rotational angle φ_m of the rotor - (Fig. 3). In (11) we have the relation in the frequency domain and in (12) it appears in the time domain.

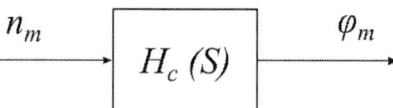

Fig. 3. Integration of rotational speed results into the rotation count φ_m (= rotational angle divided by 2π resp. per 360 °) of the armature (rotor). [6]

$$H_c(s) = \frac{\varphi_m(s)}{n_m(s)} \tag{11}$$

$$\frac{d}{dt}\varphi_m(t) = n_m(t) \tag{12}$$

978-1-7281-4181-7/19 $31.00 © 2019 IEEE

$$n_m(s) = \left[\frac{U_a(s)}{C_2\Phi} - \frac{M_L(s)}{C_1C_2\Phi^2}.R_a.(1 + sT_a) \right].H_s(s) \quad (13)$$

$$I_a(s) = \frac{U_a(s)}{R_a.(1 + sT_a)}.[1 - H_s(s)] + \frac{M_L(s)}{C_1\Phi}.H_s(s) \quad (14)$$

With the function of the system $H_s(s)$, which can be written in function of τ_s e T_a.

$$H_s(s) = \frac{1}{1 + s\tau_s + s^2 T_a \tau_s} \quad (15)$$

$$\tau_s = \frac{2\pi J_a R_a}{C_1 C_2 \Phi^2} = \frac{4\pi^2 J_a R_a}{(C_2\Phi)^2} \quad (16)$$

$$T_a = \frac{L_a}{R_a} \quad (17)$$

In most motors, the system time constant τ_s is much longer than the armature time constant T_a, such that $H_s(s)$ can be approximated by a first system order. [6]

$$H_s(s) \approx \frac{1}{1 + s\tau_s} \quad \text{with: } \tau_s \gg T_a \ , \ T_a \approx 0 \quad (18)$$

With simplification (18), the equations of the motor state (13) and (14)

$$C_2\Phi.n_m(s) = U_q(s) = U_a(s) - R_a.I_a(s)$$

$$C_2\Phi.n_m(t) = U_q(t) = U_a(t) - R_a.I_a(t) \quad (19)$$

With: $\qquad \tau_s \gg T_a \qquad , \qquad T_a \approx 0$

B. Time-Discrete Linear Motor Model

Using the numerical integration Euler's method, differential (3) and (4) can be sampled in discrete time t $\leftarrow$ n.T_s (= sampling with sample period T_s).

$$s.X(s) \rightarrow \frac{dx(t)}{dt} \approx \frac{x(t) - x(t - T)}{T} =$$

$$\frac{x[n] - x[n - 1]}{T} \quad (20)$$

The Euler's method applied to the transfer function of the armature (4).

$$I_a(t) + T_a \frac{d}{dt} I_a(t) = R_a^{-1}.U_d(s)$$

$$I_a[n] + \frac{I_a[n] - I_a[n - 1]}{T_s} = R_a^{-1}.U_d[n] \quad (21)$$

$$\rightarrow \quad I_a[n] = \frac{R_a^{-1}.U_d[n]\frac{T_a}{T_s}.I_a[n - 1]}{1 + \frac{T_a}{T_s}}$$

$$I_a[n] = \underbrace{\frac{T_s R_a^{-1}}{T_s + T_a}}_{-a_1}.U_d[n] + \underbrace{\frac{T_a}{T_s + T_a}}_{b_0}.I_a[n - 1],$$

$$\qquad\qquad (22)$$

$$U_d[n] = U_m[n] - U_q[n]$$

The motor current (22) is thus represented as a recursive system of 1^a. order with input $U_d[n]$ as input.

The Euler's method applied to the torque integration (4), [9].

$$2\pi J_a n_m(t) = M_a(s) - M_L(s)$$

$$2\pi J_a \frac{n_m[n] - n_m[n - 1]}{T_s} =$$

$$M_a[n] - M_L[n] = C_1\Phi.(I_a[n] - I_L[n]) \quad (23)$$

$$n_m[n] = \frac{C_1\Phi.(I_a[n] - I_L[n])}{\frac{2\pi J_a}{T_s}} + n_m[n - 1] =$$

$$T_s.\frac{C_2\Phi.(I_a[n] - I_L[n])}{4\pi^2 J_a} + n_m[n - 1]$$

$$n_m[n] = T_s.\frac{C_2\Phi}{4\pi^2 J_a}.I_d[n] + n_m[n - 1], \quad (24)$$

$$I_d[n] = I_a[n] - I_L[n]$$

Finally, equation (25) shows that the current mechanical speed depends on the previous mechanical speed and the current $I_d[n]$.

$$\frac{d}{dt}\varphi_m(t) \ \rightarrow \ \frac{\varphi_m[n] - \varphi_m[n - 1]}{T_s} = n_m[n]$$

$$\qquad\qquad (25)$$

$$\varphi_m[n] = T_s.n_m[n] + \varphi_m[n - 1]$$

C. Time-Variant DC Motor Model

The armature of mechanically commutated DC motor consists of a winding which is connected at exactly P contacts to a specific segment on a commutator ring (Fig. 4), With P is referred to as slot number [9], [10], [11]. The electrical equivalent circuit of the rotor is shown in (Fig. 5) [5].

With Rotor equivalent circuit of a DC motor with P=8 segments. Each string element Z_S represents a series connection of R_S, L_S and a voltage source U_S for the induced voltage due to the rotation string (coil) within the magnetic field Φ. The switches represent the coal brushes, successively short-circuiting pairs of string elements due to rotation. The brushes cover a segment partly, which is expressed by a relative

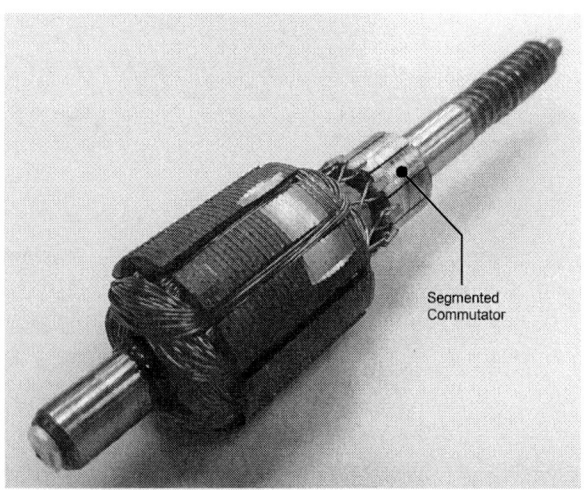

Fig. 4. Rotor of a DC motors with mechanical commutation with a segmented commutator ring and coal brushes, arranged in a typical 180° angle.

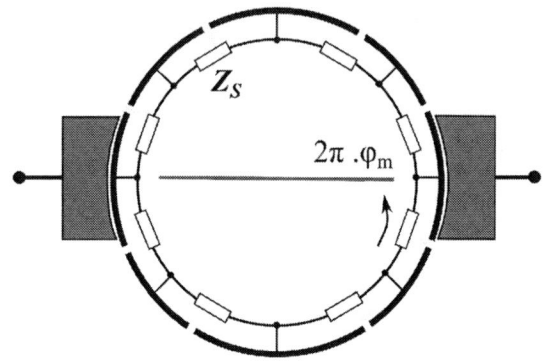

Fig. 5. Rotor equivalent circuit of a DC motor with P=8 segments.

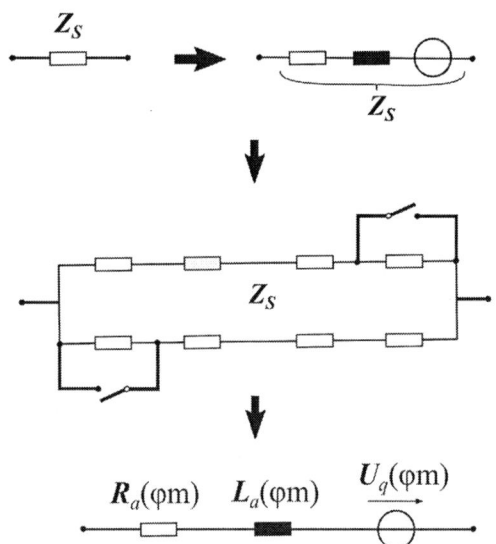

Fig. 6. Z_S represents a series connection of Rs, L_S.

coverage Cb = 0.7 to 1.0. Thus, the total rotor impedance $R_a(\varphi m)$, $L_a(\varphi m)$ is rotation angle dependent and also time dependent.

As a 1st order approximation, the rotor resistance $R_a(\varphi m)$ at the two Motor connectors can be expressed by means of two extreme values R_{amin} and R_{amax}.

Both elements $R_a(\varphi m)$ and $L_a(\varphi m)$ are depending on the rotation angle (rotation count). (26)-(32) describe the switching between 2 extreme values, respecting the brush coverage Cb. [12], [9], [10].

$$R_a(\varphi m) = \left\{ \begin{array}{ll} R_{amin} & \text{if P } .\varphi\text{m} < \text{Cb} \\ R_{amax} & \text{else} \end{array} \right\} \quad (26)$$

$$R_{amin} = \frac{\frac{P}{2}-1}{2}.R_s = \frac{P-2}{4}.R_s \quad (27)$$

$$R_{amax} = \frac{\frac{P}{2}}{2}.R_s = \frac{P}{4}.R_s$$

$$G_a = \frac{1}{\overline{R_a}} = \frac{Cb}{P.R_{amin}} + \frac{1-Cb}{P.R_{amax}} \quad (28)$$

$$\frac{4Cb}{R_s(P-2)} + \frac{4(1-Cb)}{R_s} = \frac{4}{R_s}\left[\frac{Cb}{P-2} + \frac{1-Cb}{P}\right] \quad (29)$$

$$R_s = \overline{R_a}.4.\left[\frac{Cb}{P-2} + \frac{1-Cb}{P}\right] \quad (30)$$

The same approach can be used for the inductance La(φ m) at the motor connectors:

$$L_a(\varphi m) = \left\{ \begin{array}{ll} L_{amin} & \text{if P}.\varphi\text{m} < \text{Cb} \\ L_{amax} & \text{else} \end{array} \right\} \quad (31)$$

$$L_{amin} = \frac{\frac{P}{2}-1}{2}.L_s = \frac{P-2}{4}.L_s$$

$$\quad (32)$$

$$L_{amax} = \frac{\frac{P}{2}}{2}.L_s = \frac{P}{4}.L_s$$

Equation (33) shows that the induced voltage $U_q(\varphi m)$ can be approximated by a DC term plus a relative AC term with a relative ripple amplitude Aq. The induced voltage depends on the rotation angle and some initial angle $\Phi q0$. Thus, the armature ripple current is mainly depend on (33) in high speeds (low currents) while it is mainly depend on (32) in low speeds (high currents)

$$Uq(t) = C_2\Phi.n_m(t).\left(1 + A_q.\sin\left(2\pi\varphi_m(t) + \phi_{q0}\right)\right) \quad (33)$$

This time variant DC motor model describes the dynamical system behavior of the rotational speed/count as a function of the input quantities as ripple current.

978-1-7281-4181-7/19 $31.00 © 2019 IEEE

III. SIMULATION SETUP

The equations for modeling the DC machine with ripple were inserted at Matlab/Simulink environment as shown in (Fig. 7).

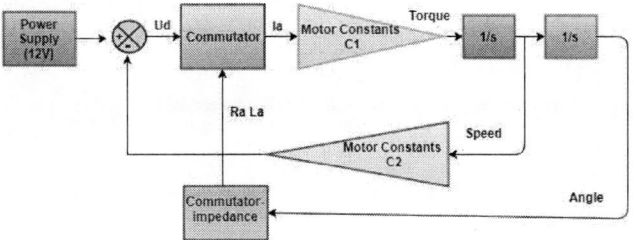

Fig. 7. Block diagram at simulation setup.

TABLE I
PARAMETERS OF THE DC MACHINE

Rotor Resistance	$R_a = 500m\ \Omega$
Rotor Inductance	$L_a = 1\ mH$
Mass Moment of Inertia	$0.002\ Kg.m^2$
Slot Number	$P = 8$

A voltage of 12v is applied to armature circuit of the model of the machine. A torque profile insert in the model, according to Fig. 8, at t = 0s and a mechanical torque is 3.5N/m apply, at t = 4s a mechanical torque is 25N/m , these levels of torque use to simulate profile a automotive window load.

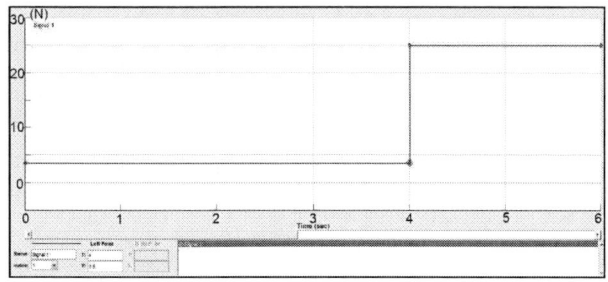

Fig. 8. Mechanical load profile applied in the simulation.

IV. BENCH TESTS

The environment used in the test is shown in Fig. 9, and consists of the following items:
1 - Automotive Windows (mechanical load);
2 - Oscilloscope;
3 - DC Power supply - 12 V - simulating vehicle battery;
4 - Control module - typical commercial solution;
5 - Bridge for current measurement 100mV / 1A;

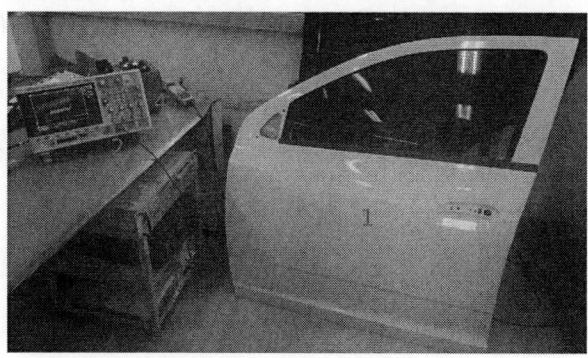

Fig. 9. Test environment for the collection of motor current profile in the automotive windows.

V. EXPERIMENTAL RESULTS

In this section will present the results generated by the model comparing them with the acquired data in the experimental tests. Fig. 10 and Fig. 11 is the starting of motor in open-loop without mechanical load. Small differences were observed in the curve generated by the model and the one obtained in the experimental tests, it is treated that the mechanical load in the model is zero, and in the open motor presents a load due to its internal gear, inherent in its construction.

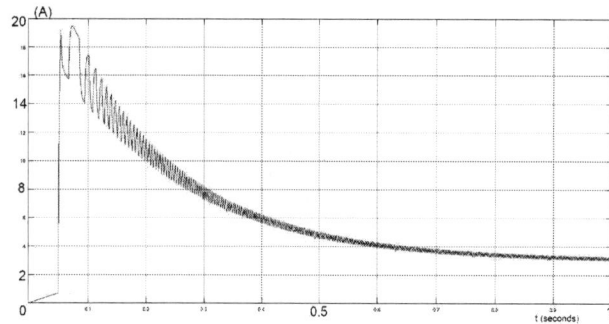

Fig. 10. DC motor model - Starting without mechanical load

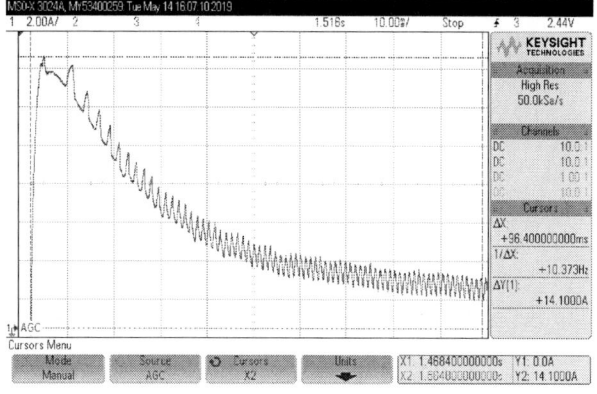

Fig. 11. Real Motor - Starting without mechanical load.

In Fig. 12 and Fig. 13, when comparing the simulation versus the experimental curve, also small differences were

978-1-7281-4181-7/19 $31.00 © 2019 IEEE

observed in the commutator effect ripple format, this is due to the fact that the actual load is not linear.

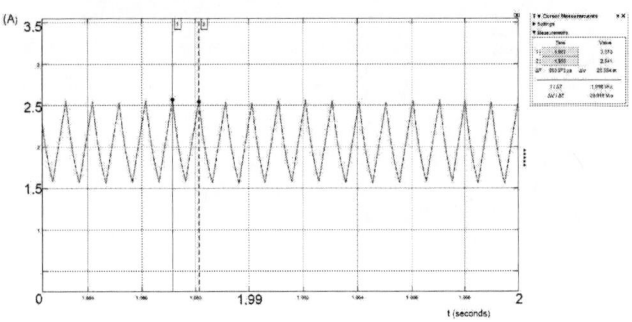

Fig. 12. DC motor model - commutator effect ripple.

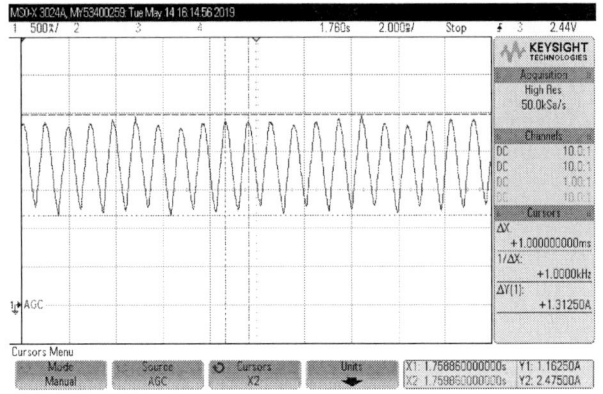

Fig. 13. Real motor - commutator effect ripple.

Fig. 15 deals with the current profile of the motor in a movement of the automotive window. This curve was acquired in the environment presented in the Fig. 9, and when it is compared with the curve generated by the model (Fig. 14) can be observed small differences in the profile of the curve due to mechanical load generated by the automotive window. In a future work it will be included the mechanical load model closer to that observed in the automotive window.

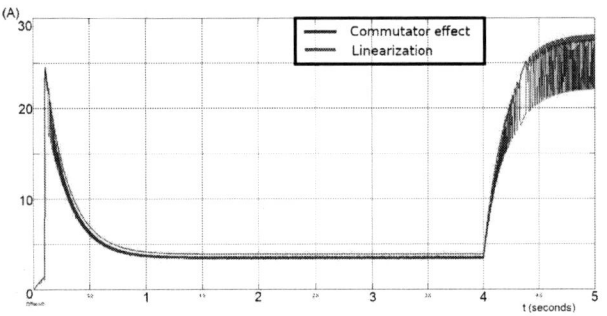

Fig. 14. Current profile of the DC motor with "ripple" generating by the simulation.

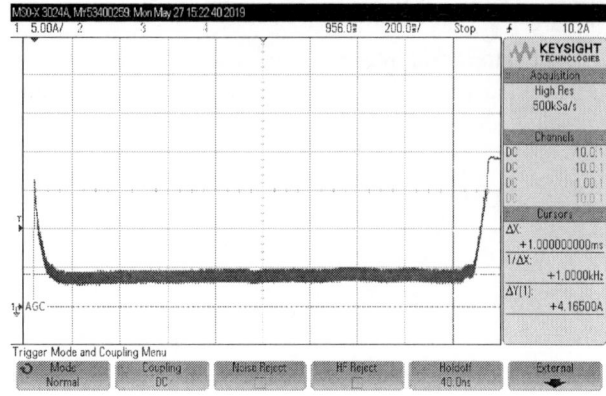

Fig. 15. Motor current profile during full windows movement.

VI. CONCLUSIONS

It was validated the time variant DC motor model in comparison with experimental bench tests. The experimental results were very close to the results of the simulation. This approach of DC motor model could be very useful in the automotive area, especially in power windows lifters. The model includes the effect of armature resistance and inductance variation not included in traditional models. It was shown that the armature ripple current depends not only on induced voltage but also on armature inductance variation. By adding the mechanical load model, it could result in a powerful tool to design new control and estimation algorithms.

REFERENCES

[1] *RESOLUÇÃO N⁰ 468,*, CONTRAN - Conselho Nacional de Trânsito Std. RESOLUÇÃO N⁰ 468, December,2013.

[2] H.-J. Lee, T.-S. Yoon, W.-S. Ra, and J.-B. Park, "Practical pinch detection algorithm for low-cost anti-pinch window control system," in *2005 IEEE International Conference on Industrial Technology*, Dec 2005, pp. 995–1000.

[3] H.-J. Lee, W.-S. Ra, T.-S. Yoon, and J.-B. Park, "Robust pinch estimation and detection algorithm for low-cost anti-pinch window control systems," in *31st Annual Conference of IEEE Industrial Electronics Society, 2005. IECON 2005.*, Nov 2005, pp. 6 pp.–.

[4] Y. Qiu, J. Yin, Y. Wang, and Q. Liu, "The research about sensor-less anti-pinch system for automotive electric window," in *2010 International Conference on Computational and Information Sciences*, Dec 2010, pp. 808–811.

[5] W. Schulter, "Aec-07 - pwm-dc-motors-emc," HOCHSCHULE RAVENSBURG-WEINGARTEN, Germany, Tech. Rep. AEC-07, Jan. 2016.

[6] A. Fitzgerald, A. Fitzgerald, C. Kingsley, and S. Umans, *Electric Machinery*, ser. Electrical Engineering Series. McGraw-Hill Companies,Incorporated, 2003.

[7] R. Fischer, *Elektrische Maschinen*. Gebundene Ausgabe, 1995.

[8] K. Kammeyer, *Digitale Signalverarbeitung*. Auflage, Teubner, 2002.

[9] S. Chapman, *Electric Machinery Fundamentals*, ser. Electric machinery fundamentals. McGraw-Hill Companies,Incorporated, 2005.

[10] S. M. Atkin, "Development of a ripple rejection controller for dc commutated motors," Utah State University, Tech. Rep. 23, 8 2012, master of Science (MS).

[11] D. Gerling, *Electrical Machines: Mathematical Fundamentals of Machine Topologies*. Springer, 2014.

[12] P. Krause, O. Wasynczuk, S. Sudhoff, and I. P. E. Society, *Analysis of electric machinery and drive systems*, ser. IEEE Press series on power engineering. IEEE Press, 2002.

978-1-7281-4181-7/19 $31.00 © 2019 IEEE

Power Control and Harmonic Current Mitigation from a Wind Power System with PMSG

Leonardo P. S. Silva, Denisia de V. Mota, Flávia P. Ruiz, Levy R. Cavalcante, Lucas Taylan P. Medeiros, Vanessa S. C. Teixeira, Adson B. Moreira

Electrical Engineering
Campus Sobral, Federal University of Ceará
Sobral, Brazil

leonardo85pires@gmail.com, denisia.vasconcelos07@gmail.com, flaviaperozaruiz@gmail.com, levyrodrigues8@gmail.com, lucastaylanp@gmail.com, vanessasct@gmail.com, adsonbmoreira@gmail.com

Abstract— This paper describes a wind power system working with control of active and reactive power as well as harmonic current compensation. Power control is made through stator flux oriented in machine-side converter. The active filtering is done by grid-side converter (GSC) causing to grid current waveform without distortions. The algorithm applied for harmonic mitigation has shown to be effective in improving of power quality. Simulation results confirm the effectiveness of the proposed research, since through the power control of permanent magnet synchronous generator (PMSG) and harmonic compensation on GSC was possible to deliver of power to electric grid, satisfactorily.

Keywords—Permanent magnet synchronous generator, grid-side converter, machine-side converter, power control and harmonic compensation.

I. INTRODUCTION

Non-linear loads need non sinusoidal currents to work. They have been connected to the electric grid and affected power quality, so harmonic currents appear on the grid and require more reactive power, then the power factor of the grid is reduced. Therefore, the copper losses increase, along with voltage sag and fluctuation [1], underutilizing the installed capacity, in which the stability of electrical system is damaged. Due to the kind of load that is connected, even when using wind farms connected to the grid the mentioned problems persist.

Wind turbines with permanent magnet synchronous generator (PMSG) have been used in wind farms when the generation system has a variable speed. It is becoming popular for its loss reduction, less maintenance requirements costs, higher power density, optimal energy efficiency and reliability [2]-[4].

Paper [5] presents a strategy using the machine-side converter (MSC) to control its speed, reaching maximum power point tracking (MPPT). The generator speed varies between 85 and 100 rad/s and the total harmonic distortion (THD) reduced from 28% to 2,3% in the best case using partial load active support. Authors in [3] studied a control strategy of PMSG based on the vector control (VC) theory to regulate grid side to achieve unity power factor and machine converter to extract maximum power when the system works at different wind speeds while utilizing MPPT controller. As it is shown in [6], a 14,03% THD reduction can be obtained using the double fed induction generator (DFIG) and active filtering through grid-side converter (GSC) in the point of common coupling (PCC) with the electrical grid.

To distribute a better power quality, an active and reactive power control is developed according to the technique of stator flux oriented to the wind power generation system using DFIG with active power filter [7]. An active power filter is employed in harmonic filtering using dq variables applied to synchronous reference frame (SRF) theory to provide the amount of harmonic current that non-linear loads connected to the PCC request from the grid, so that distortions do not appear. Both operations occur simultaneously.

In this research, a control strategy of a wind generation system with PMSG is proposed, in Fig. 1. The PMSG is connected to the electrical grid through a back-to-back converter with the function of filtering harmonic currents in the presence of non-linear load. Two case studies analyze the behavior of the system. Results for power control and harmonic mitigation are presented for generator speeds at 170 and 180 rad / s.

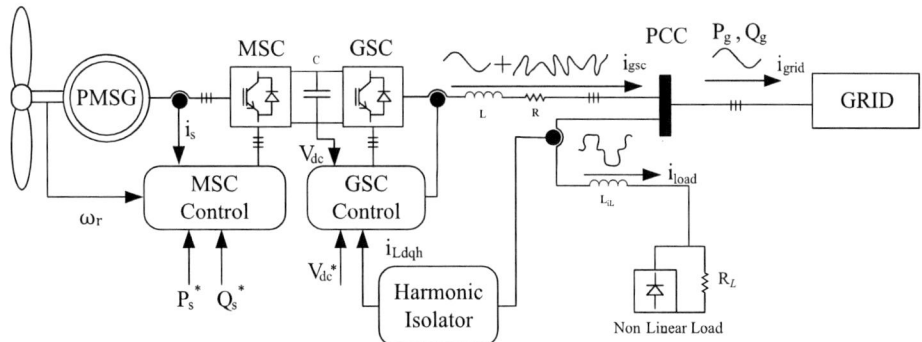

Fig. 1. Operation design for the proposed PMSG/APF.

II. POWER CONTROL OF THE PERMANENT MAGNET SYNCHRONOUS GENERATOR

The power control of the PMSG is controlled by the orientation of the stator flux in the reference of synchronous rotation directed along the axis d. With that strategy, the active and reactive powers of the stator are decoupled [8]. The dynamic model of a PMSG in the dq reference frame can be represented by equations (1), (2) and (3) [8].

$$V_{sd} = R_M i_{sd} + L_{sd} \frac{di_{sd}}{dt} - \omega_r L_{sq} i_{sq} \tag{1}$$

$$V_{sq} = R_M i_{sq} + L_{sq} \frac{di_{sq}}{dt} + \omega_r L_{sd} i_{sd} + \omega_r \lambda_m \tag{2}$$

where V_{sd} and V_{sq} are the stator voltage components, R_M is the machine resistance, L_{sd} and L_{sq} are the stator inductances components, in a round rotor machine ($L_{sd} = L_{sq}$), i_{sd} and i_{sq} are the stator current components, ω_r is generator speed and λ_m is magnetic flux.

The electromagnetic torque in the rotor can written as:

$$T_e = \left(\frac{3}{2}\right)\left(\frac{p}{2}\right)\left[(L_{sd} - L_{sq})\right]i_{sq}i_{sd} + \lambda_m i_{sq} \tag{3}$$

where T_e is the electromagnetic torque and p is the number of poles. For the control of the generator, two new control variables are introduced from (1) and (2):

$$u_d = V_{sd} + \omega_r L_{sq} i_{sq} \tag{4}$$

$$u_q = V_{sd} - \omega_r L_{sd} i_{sq} - \omega_r \lambda_m \tag{5}$$

For stator current components regulation, (1) and (2) are rewrite to represent two decoupled, first-order, single-input-single-output (SISO):

$$L_{sd} \frac{di_{sd}}{dt} + R_M i_{sd} = u_d \tag{6}$$

$$L_{sq} \frac{di_{sq}}{dt} + R_M i_{sq} = u_q \tag{7}$$

Equations (6) and (7) regulate i_{sd} and i_{sq} for their references i_{sd}^* and i_{sq}^* through the scheme shown in the Fig. 2.

The stator active and reactive power is defined by (8) and (9), respectively.

$$P_s = \frac{3}{2}(v_{sd} i_{sd} + v_{sq} i_{sq}) \tag{8}$$

$$Q_s = \frac{3}{2}(v_{sq} i_{sd} + v_{sd} i_{sq}) \tag{9}$$

The control of the generator is done by the quadrature axis and considering $i_{sd} = 0$, therefore, the stator active and reactive power can be rewritten:

$$P_s = \frac{3}{2} v_{sq} i_{sq} \tag{10}$$

$$Q_s = \frac{3}{2} v_{sd} i_{sq} \tag{11}$$

The PLL (Phase Locked Loop) block generat the grid angle through voltage measurement on PCC to according [7].

The scheme represented in the Fig. 3 show the control of GSC control, where the currents reference are given by (12) and (13).

$$P_{sref} = \frac{3}{2}[v_d i_d^* + v_q i_q^*] \tag{12}$$

$$Q_{sref} = \frac{3}{2}[-v_d i_q^* + v_q i_d^*] \tag{13}$$

where v_d and v_q are the grid voltage components and i_d and i_q are the grid current components.

Considering $v_q = 0$ (12) and (13) can be rewritten as (14) and (15).

$$i_d^* = \frac{2}{3 v_d} P_{sref} \tag{14}$$

$$i_q^* = \frac{2}{3 v_d} Q_{sref} \tag{15}$$

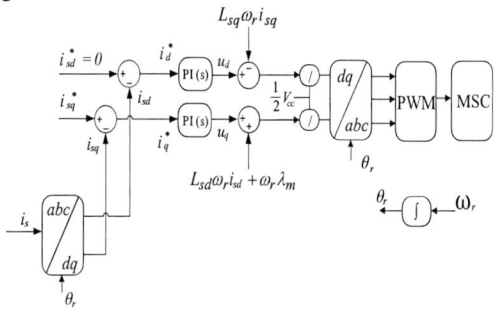

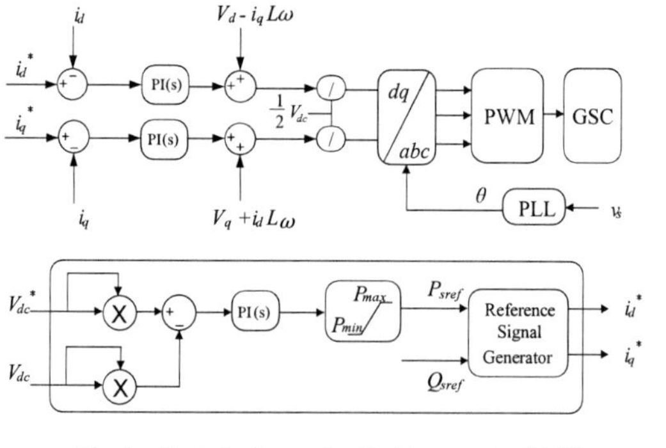

Fig. 2.   Control scheme of machine-side converter (MSC).

Fig. 3.   Control scheme of grid-side converter (GSC).

III. ACTIVE FILTER ON THE GRID-SIDE CONVERTER

The basic idea of harmonic compensation is to implement an active power filter (APF) and do its current control in a closed loop, so the back-to-back converter acts as an APF injecting an equal current in magnitude, but displaced by 180º, so that when added with the load current will result in a sinusoidal waveform in the grid.

In order to control the APF, the control structure of the grid-side converter is changed, i.e, the harmonic currents from the nonlinear load, i_{Lhd} and i_{Lhq}, are added with the reference currents of the control, produced by the reference signal generator in the current control scheme i_d and i_q. This change allows regulating the voltage on the DC bus, beyond to mitigate the harmonic components of the grid current, as shown in Fig. 4.

The calculation of reference currents for harmonic compensation, i_d^r and i_q^r, is determined by (16) and (17):

$$i_d^r = i_{Lhd} + i_d^* \tag{16}$$

$$i_q^r = i_{Lhq} + i_q^* \tag{17}$$

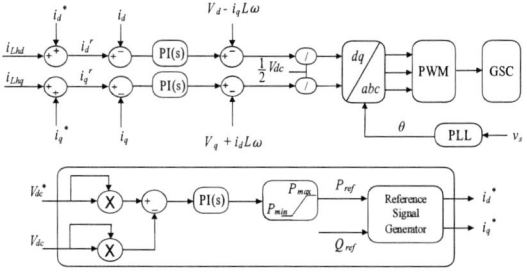

Fig. 4. Proposed control scheme of active filtering

For system filtering purposes, an harmonic extractor is implemented. This is done by transforming three-phase load currents in the direct quadrature synchronous reference frame (dq), as follows (18):

$$\begin{bmatrix} i_{Ld} \\ i_{Lq} \end{bmatrix} = \frac{2}{3} \begin{bmatrix} \cos(\theta) & \cos\left(\theta - \frac{2\pi}{3}\right) & \cos\left(\theta - \frac{2\pi}{3}\right) \\ -\sin(\theta) & -\sin\left(\theta - \frac{2\pi}{3}\right) & -\sin\left(\theta - \frac{2\pi}{3}\right) \end{bmatrix} \begin{bmatrix} i_{La} \\ i_{Lb} \\ i_{Lc} \end{bmatrix} \tag{18}$$

where i_{La}, i_{Lb} and i_{Lc} are currents measured at three-phase time-domain, i_{Ld} and i_{Lq} are load currents in the direct quadrature synchronous reference frame.

This transformation in (2) is done because in dq0 rotating reference frame, the fundamental component becomes a constant which can pass through a low pass filter to remove the high frequency components. Therefore, this transformation allows the currents i_{Ld} and i_{Lq} to pass through the low pass filter, as a consequence, only the fundamental remains in the waveform (Fig 5).

In harmonic mitigation the high frequency components are canceled, when added. For this it is necessary that an isolation should be made only of high frequency current components. It is possible by subtracting the current with all components, i_{Ld} and i_{Lhq}, of the filter's output current, i_{Lfd} and i_{Lfq}. The expression for this, is given in (19) and (20).

$$i_{Lhd} = i_{Ld} - i_{Lfd} \tag{19}$$

$$i_{Lhq} = i_{Lq} - i_{Lfq} \tag{20}$$

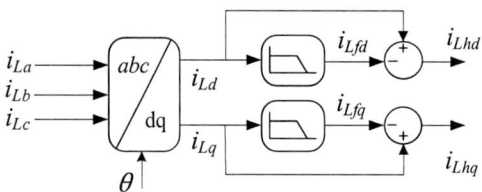

Fig. 5. Harmonic Identifier.

Then currents, i_{Lhd} and i_{Lhq}, are added with the current references generated by the reference signal generator. Consequently, when this addition is made, the harmonic components will be mitigated.

IV. SIMULATION RESULTS AND DISCUSSION

The proposed control strategy for PMSG/APF under speed variation is simulated using Matlab/Simulink®. The circuit shown in Fig. 1 consists of a wind power system with PMSG, a back-to-back converter, electric grid and a three-phase rectifier feeding a non-linear load. The switching frequency of converters is equal to 15 kHz. The other parameters used in the simulation are shown in Table I.

TABLE I. SIMULATION PARAMETERS

	PARAMETERS	VALUES
PMSG	P_n, V_{LL}, frequency, R_M, L_M, λ_m, p	12kW, 380V, 60Hz, 2.895Ω, 8.5mH, 0.175Wb, 4 poles
GSC	L, R, C	10mH, 5Ω, 9000µF
Load	R_L, L_{iL},	15Ω, 10mH
Grid	V_{LL}, frequency	380V, 60Hz

A. Case 1

In this case, a permanent magnet synchronous generator operating was analyzed in generator mode, supplying a power of 12 kW. A speed variation is included in the analysis to shown a system closer to the real.

The power produced by PMSG delivered to the electric grid (Fig. 6) increases in accordance with the machine-side converter control, where the reactive power maintains the unit power factor through the grid-side converter with $V_q = 0$ and $iq = 0$.

The dc-link voltage (V_{dc}) is set to 800 V according to its reference (V_{dc}^*) established in grid-side converter control, which is shown in Fig. 7.When the generator speed (ω_r) is increased to 180 rad/s, as presented in the Fig. 8, the dc-link

978-1-7281-4181-7/19 $31.00 © 2019 IEEE

voltage, such as active and reactive power remains stable without overshoot.

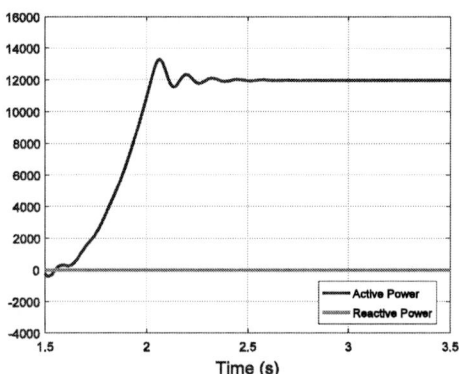

Fig. 6. Response to the active and reactive power delivered to the electric grid.

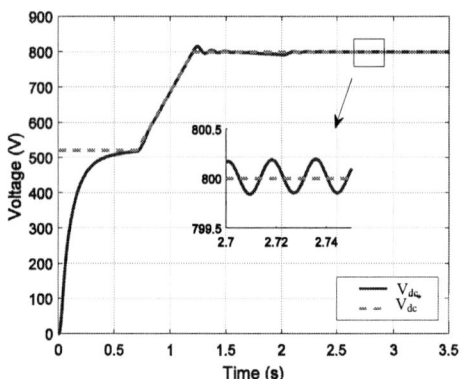

Fig. 7. Response to the dc-link voltage.

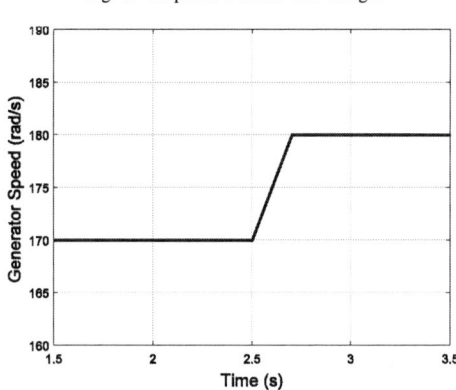

Fig. 8. Generator Speed.

Current waveforms in MSC are shown in Fig. 9 in dq-axis and phase A. Generator control is done by the quadrature axis, where i_d follows its reference (i_{sd}^*) and i_q also follows its reference (i_{sq}^*) causing the phase A current (i_{am}) enhancing in the same proportion.

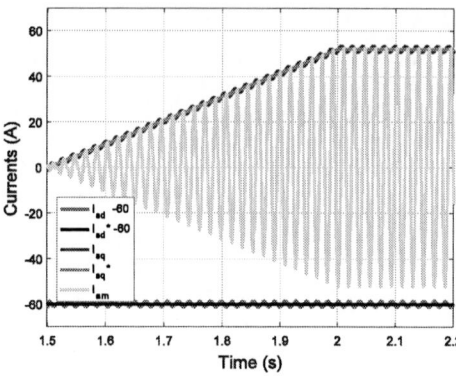

Fig. 9. Current waveforms at machine-side converter (MSC) in abc-axis (phase A) and dq-axis.

Current waveforms of the load, electric grid and GSC are obtained when the generator feeds a non-linear load for the speed generator of 170 rad/s without harmonic current mitigation (Fig. 10).

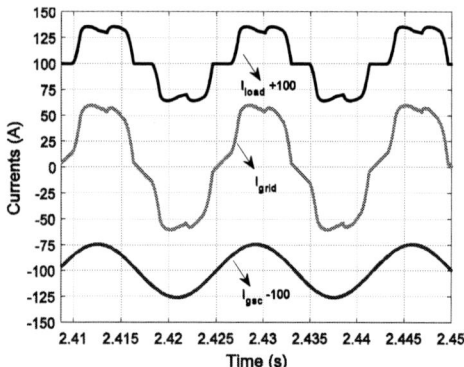

Fig. 10. Current waveforms of the load, electric grid and grid-side converter currents operating at the traditional speed of 170 rad/s without harmonic compensation.

The load current waveform is distorted and its harmonic spectrum is shown in Fig. 11. Thereby, total harmonic distortion for the non-linear load current is aproximately 24,68%.

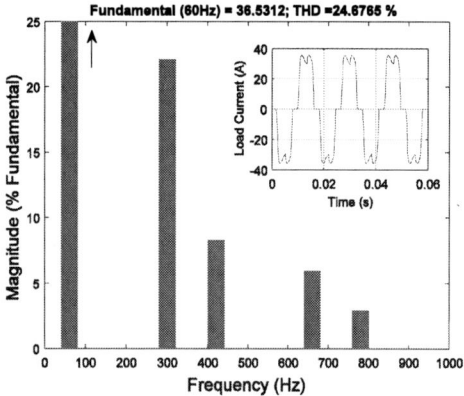

Fig. 11. Spectrum of the load current.

978-1-7281-4181-7/19 $31.00 © 2019 IEEE

For the harmonic spectrum of the grid current operating at 170 rad/s, which is shown in Fig. 12, presents a distortion of 14,60% and for 180 rad/s (Fig. 13) is 14,60%. Thus, these values are not according to the limits reccomended by the international standarts for generators connecting in low-voltage grid [9].

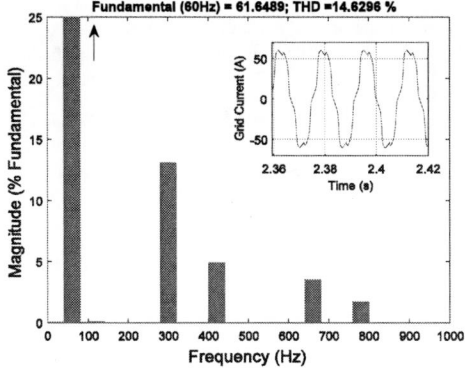

Fig. 12. Spectrum of the grid current without harmonic compensation at the speed of 170 rad/s.

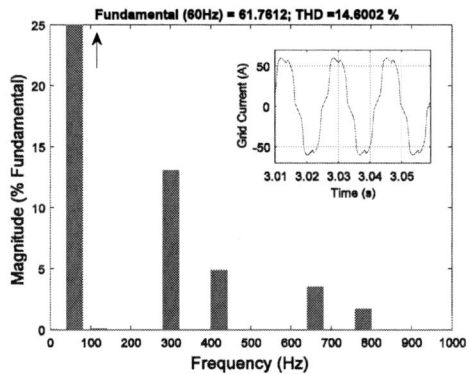

Fig. 13. Spectrum of the grid current without harmonic compensation at the speed of 180 rad/s.

Harmonic content of the grid current shows the main harmonics order that contributes to this distortion are odd harmonics (5, 7 and 11). For different speeds, THD$_i$ values almost does not show variation. For consumers connect at PCC, this deformation may cause a distorted voltage.

B. Case 2

For this case, harmonic compensation and power control of PMSG are done. The generator works in variable speed supplying a power of 12 kW. Results obtained in Figs. 6 until 9 are the same for both cases, once GSC control is equal.

Current waveform of the load, grid and for GSC for the wind system operating with harmonic mitigation at the traditional speed of 170 rad/s, feeding a non-linear load are shown in Fig. 14. Thus, grid current has a sinusoidal waveform when generator working in active filtering mode.

Harmonic spectrum of grid current with harmonic compensation at the speed generator of 170 rad/s, in Fig. 15,

presents a THD of 3,51%, and Fig. 16 shows the grid current THD at the speed generator of 180 rad/s of 3,50%.

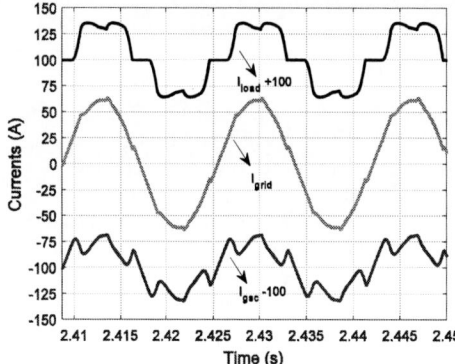

Fig. 14. Current waveforms of the load, grid and grid-side converter currents operating at the traditional speed of 170 rad/s with harmonic compensation.

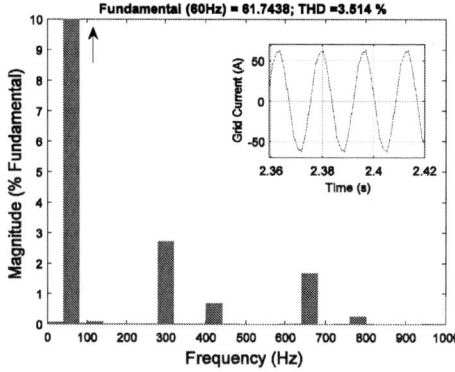

Fig. 15. Spectrum of the grid current with harmonic compensation at the speed of 170 rad/s.

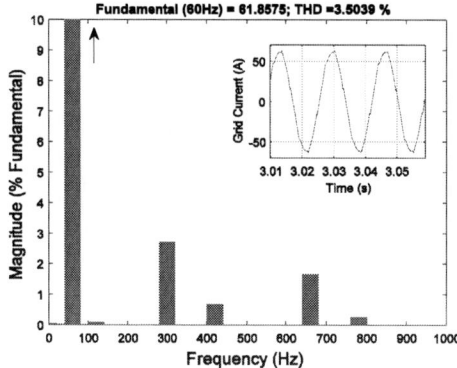

Fig. 16. Harmonic content of the grid current with harmonic compensation at the speed of 180 rad/s.

Comparing the case 1 with case 2 is possible to note the harmonic compesation of the grid current has attenuation, due its odd harmonic components were attenuated, as verified in Fig 12 and 15, which THD$_i$ was reduced from 14,63% to 3,51% , Fig. 13 and 16 from 14,60% to 3,50%, respectively.

V. Conclusion

The purpose of this paper is to investigate a wind power system with a PMSG under speed variation operating in traditional mode of power generation (case 1), and a system working with harmonic current mitigation of the grid and power control of generator (case 2). The power delivered to the grid and dc-link voltage are maintained the same in both case. Thereby, the technique based on stator flux oriented proved to be efficient for the power generation.

Comparing harmonic content of the grid before the harmonic compensation and after implementation of the active filtering algorithm, for 170 rad/s was decreased of 14,63% to 3,51% and 180 rad/s of 14,60% to 3,50%. Therefore, is possible to confirm that the THD has been reduced to acceptable levels in accordance with international standards. Thus, the harmonic compensation strategy incorporated into a wind generator with PMSG improves power quality.

Acknowledgements

This work was supported by FUNCAP grant code BP3-0139-00022.01.00/18.

References

[1] A. B. Moreira., T. A. S. Barros, V. S. C. Teixeira , P.S.F NASCIMENTO, E. Ruppert, "Metodologia e projeto de controle de potências para geração eólica e filtragem de corrente harmônica com gerador de indução duplamente alimentado". In: CBA2016 - XXI Congresso Brasileiro de Automática, 2016, Vitória-ES. CBA2016 - XXI Congresso Brasileiro de Automática, 2016.

[2] R. M. Pindoriya, A. Usman, B. S. Rajpurohit, K. N. Srivastava, "PMSG based wind energy generation system: Energy maximization and its contro", 2017 7th International Conference on Power Systems.

[3] Y. Errami, M. Ouassaid, M. Maaroufi, "Control of a PMSG based wind energy generation system for power maximization and grid fault conditions", *Energy Procedia* vol. 42, pp. 220 – 229, 2013.

[4] D. M. Miao, Y. Mollet, J. Gyselinck, and J. X. Shen, "Direct voltage field-oriented control for permanent-magnet synchronous generator systems with an active rectifier," unpublished, will be presented at the IEEE Int. Energy Conf., Leuven, Belgium, 2016.

[5] M. Singh, V. Khadkikar, and A. Chandra, "Grid synchronization with harmonics and reactive power compensation capability of a permanent magnet synchronous generator-based variable speed wind

energy conversion system," IET Power Electron., vol. 4, no. 1, pp. 122–130, Jan. 2011.

[6] A. B. Moreira, T. A. D. S. Barros, V. S. D. C. Teixeira, R. R. D. Souza, M. V. D. Paula and E. R. Filho, "Control of Powers for Wind Power Generation and Grid Current Harmonics Filtering From Doubly Fed Induction Generator: Comparison of Two Strategies," in *IEEE Access*, vol. 7, pp. 32703-32713, 2019.

[7] A. B. Moreira., T. A. S. Barros, V. S. C. Teixeira , P.S.F NASCIMENTO, E. Ruppert, "Controle de potências para geração eólica e filtragem de corrente harmônica com gerador de indução duplamente alimentado". In: V Simpósio Brasileiro de Sistemas Elétricos, 2016, Natal-RN. V Simpósio Brasileiro de Sistemas Elétricos, 2016.

[8] Amirnaser Yazdani, R.I., 2010. *Voltage-Source Converters in Power Systems - Modeling, Control, and Applications*, WILEY IEEE.

[9] IEEE, "IEEE Recommended Practices and Requirements for Harmonic Control in Electric Power System Project IEEE-519.," 1991.

[10] F. S. Dos Reis, S. Islam, K. Tan, J. V. Ale, F. D. Adegas, and R. Tonkoski ´ Jr., "Harmonic mitigation in wind turbine energy conversion systems," in Proc. IEEE PESC 2006, Jeju, Korea, Jun. 18–22, pp. 2748–2754.

[11] A. B. Moreira, T. A. S. Barros, V. S. C. Teixeira, E. Ruppert, "Compensação harmônica de corrente e controle de potência para geração eólica com gerador duplamente alimentado", In: INDUSCON 2014-11th IEEE/IAS International Conference on Industry Applications 2014, v. 14, Juiz de Fora, 2014.

[12] Z. Chen, J. M. Guerrero, and F. Blaabjerg, "A review of the state of the art of power electronics for wind turbines," IEEE Trans. Power Electron., vol. 24, no. 8, pp. 1859–1875, Aug. 2009.

[13] S. H. Qazi, et al., "Current Harmonics Mitigation from Grid Connected Variable Speed Wind Turbine due to Nonlinear Loads using Shunt Active Power Filter" ,l Jurnal Teknologi, vol. 4, pp. 45-53, 2017.

[14] J. Tsai and K. Tan. "H APF harmonic mitigation technique for PMSG wind energy conversion system" IEEE Trans on. Power Engineering, AUPEC 2007, pp. 1-6.

[15] Jinbo, Maro & Cardoso, Ghendy & Farret, F.A. & Gustavo Trapp, Jordan & Ribeiro dos Santos, Edson & Pereira Machado, Jawilson. (2015). "Sistema eólico de velocidade variável com pmsg conectado à rede elétrica". 487-496. 10.5151/mathpro-cnmai-0084.

[16] K. Tan, K. B. Ng, S. Sugiarto & H. H. Tumbelaka, "Shunt Active Power Filter Harmonic Mitigation Technique in a Grid-Connected Permanent Magnet Wind Generation System," presented Australian Universities Power Engineering Conference (AUPEC), Perth, Australia, 2006.Trans. Roy. Soc. London, vol. A247, pp. 529–551, April 1955. *(references)*

PV Emulator Based on a Four-Switch Buck-Boost DC-DC Converter

Leandro L. O. Carralero
Energy Efficiency Lab
Federal University of Bahia
Salvador, Brazil
leandro.oro@ufba.br

Gabriel S. Barbara da S. e Silva
Energy Efficiency Lab
Federal University of Bahia
Salvador, Brazil
gsantabarbara@ufba.br

Fabiano F. Costa
Energy Efficiency Lab
Federal University of Bahia
Salvador, Brazil
fabiano.costa@ufba.br

André P. N. Tahim
Energy Efficiency Lab
Federal University of Bahia
Salvador, Brazil
atahim@ufba.br

Abstract—**Currently, there is a growing need to develop and evaluate power converters that integrates photovoltaic (PV) systems to the grid. It is a complex task to develop such converters using PV panels exposed to environmental conditions that may vary abruptly. Thus, it becomes difficult to evaluate developing converters in all operating points during the sunshine hours. One solution is the use of panel emulators that provide controlled output independent of external conditions. This work proposes a low-cost PV emulator using a four-switch buck-boost (FSBB) dc-dc converter. As the voltage/current output can be controlled in a dc-dc converter, an algorithm emulates the I-V curve at the converter's terminals based on the estimated load connected to it. From this value, a linear approximation of the I-V curve is used in association with the Newton method to obtain the voltage reference. The voltage reference value dictates the FSBB operation mode (buck, boost, buck-boost) and PI controllers are used to track it. Finally, simulation results of the proposed PV emulator working in different operation modes are reported to validate the effectiveness of the approach.**

Index Terms—**PV emulator, FSBB, dc-dc converter, I-V curve.**

I. INTRODUCTION

IN recent years the use of PV systems has increased significantly throughout the world [1]. Investments have been applied in the research of power converters that integrate the PV systems and the grid to improve efficiency and reliability of such systems. Thus, there is a growing need to evaluate power converters and the stability of systems with distributed generation. The main problem arises during the development and performance evaluation of PV power converters due to the great variability of the weather conditions, since temperature and irradiation are not controlled and they may vary, sometimes abruptly. Therefore, it's necessary a means of evaluating the system without the influence of these variations at any time and in a controlled way. A useful tool for solving this problem consists of PV emulators, which emulate the electrical characteristics of a PV panel under different weather conditions in a laboratory environment.

Several approaches to design PV emulators are found in literature: low-power reference solar cells [2]; dc power supply with variable resistances [3]; and switch-mode dc-dc power converters [4]–[8]. Some of them are not flexible to emulate different types of modules or PV array configuration, while some, despite the flexibility are expensive and preclude the

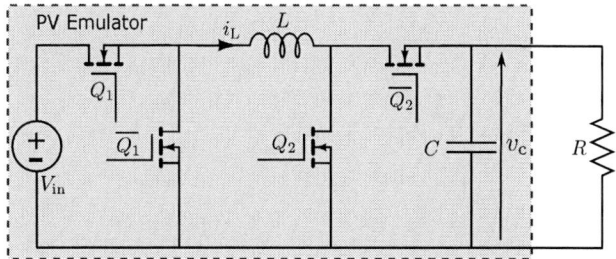

Fig. 1. PV emulator based on the four-switch buck-boost (FSBB) converter.

use in small projects. Therefore, a low power PV emulator is proposed to emulate the PV terminals for irradiation and temperature variations. This type of emulator is useful for small energy harvesting projects.

The main goal is to emulate a PV panel through a synchronous dc-dc converter, reproducing at the output terminals the electrical characteristics of this panel. For this, a circuit formed by a four-switch buck-boost (FSBB) converter that can behave like buck, boost or buck-boost is used, as depicted in Fig. 1. The topology has a pair of synchronous switches, where Q_1 and Q_2 are the active switches, and $\overline{Q_1}$ and $\overline{Q_2}$ are the synchronous rectifiers [9]. This topology allows the output voltage to operate in a wider range than other types of non-isolated converters. Furthermore, this configuration is suitable for applications where the operating point varies over a wide range of voltages and currents (load variation), such as the I-V characteristic curve of photovoltaic panels.

The remaining sections are organized as follows: Section II describes the mathematical model of the FSBB converter; the proposed PV emulator design and control are presented in Section III; Section IV is dedicated to the analysis of simulation results; and the final comments are in Section V.

II. MATHEMATICAL MODELING

A. FSBB Converter Model

To perform the modeling, the average model for dc-dc converters was used [10]. Thus, are analyzed the FSBB converter separately as shown in the Table I. Also, Orellana et al. [11] have obtained the state-space equations of the FSBB topology (Fig 1), using this method,

978-1-7281-4181-7/19 $31.00 © 2019 IEEE

TABLE I
CONVERTER STATES FOR EACH OPERATION MODES.

Operation Mode	Converter State	A Matrix	B Matrix
BUCK	ON	$\begin{bmatrix} 0 & -\frac{1}{L} \\ \frac{1}{C} & -\frac{1}{RC} \end{bmatrix}$	$\begin{bmatrix} \frac{1}{L} \\ 0 \end{bmatrix}$
	OFF	$\begin{bmatrix} 0 & -\frac{1}{L} \\ \frac{1}{C} & -\frac{1}{RC} \end{bmatrix}$	$\begin{bmatrix} 0 \\ 0 \end{bmatrix}$
BOOST	ON	$\begin{bmatrix} 0 & 0 \\ 0 & -\frac{1}{RC} \end{bmatrix}$	$\begin{bmatrix} \frac{1}{L} \\ 0 \end{bmatrix}$
	OFF	$\begin{bmatrix} 0 & -\frac{1}{L} \\ \frac{1}{C} & -\frac{1}{RC} \end{bmatrix}$	$\begin{bmatrix} \frac{1}{L} \\ 0 \end{bmatrix}$
BUCK-BOOST	ON	$\begin{bmatrix} 0 & 0 \\ 0 & -\frac{1}{RC} \end{bmatrix}$	$\begin{bmatrix} \frac{1}{L} \\ 0 \end{bmatrix}$
	OFF	$\begin{bmatrix} 0 & -\frac{1}{L} \\ \frac{1}{C} & -\frac{1}{RC} \end{bmatrix}$	$\begin{bmatrix} 0 \\ 0 \end{bmatrix}$

$$\begin{bmatrix} \dot{v_C} \\ \dot{i_L} \end{bmatrix} = \begin{bmatrix} -\frac{1}{RC} & 0 \\ 0 & 0 \end{bmatrix} \begin{bmatrix} v_C \\ i_L \end{bmatrix} + \begin{bmatrix} 0 \\ \frac{1}{L} \end{bmatrix} v_{in} D_1 +$$
$$+ \begin{bmatrix} 0 & \frac{1}{C} \\ -\frac{1}{L} & 0 \end{bmatrix} \begin{bmatrix} v_C \\ i_L \end{bmatrix} (1 - D_2), \tag{1}$$

where, v_{in} and v_c are the input and output voltages respectively. The current of inductor L is represented by i_L, C is the capacitor value and R is the resistive load.

The FSBB converter has two duty cycles to be controlled, D_1 for Q_1 and D_2 for Q_2, resulting in two degrees of freedom. The switches $\overline{Q_1}$ and $\overline{Q_2}$ are synchronous and complementary of Q_1 and Q_2, respectively. The equilibrium point of (1), output voltage and inductor current, is given by

$$(v_{C_{eq}}, i_{L_{eq}}) = \left(V_{in} \frac{D_1}{1 - D_2}, \quad \frac{V_{in}}{R} \frac{D_1}{(1 - D_2)^2} \right). \tag{2}$$

Note that if $D_2 = 0$ (Q_2 turned OFF and $\overline{Q_2}$ turned ON), the FSBB converter behaves like a buck converter, shown in Fig. 2(a), and if $D_1 = 1$ (Q_1 turned ON and $\overline{Q_1}$ turned OFF), this converter has a similar behavior to a boost converter, illustrated in Fig. 2(b). Another possibility is when D_1 and D_2 are equal ($D_1 = D_2$). In such condition, the converter operates as a buck-boost converter, as depicted in Fig. 1.

Applying Laplace Transform in each system (1) are obtained the transfer functions, for each operation mode, that relates the duty cycle [$D(s)$] to the output voltage [$V_C(s)$] in (3), (5) and (7), and to the inductor current [$I_L(s)$] in (4), (6) and (8). These transfer functions are used to design three controllers for the FSBB converter, one for each operation mode.

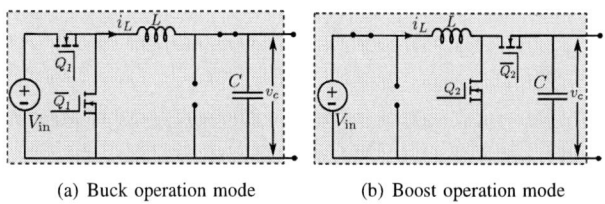

(a) Buck operation mode (b) Boost operation mode

Fig. 2. Operation modes of FSBB converter: (a) buck; (b) boost.

FSBB converter model in buck operation mode:

$$\frac{V_C(s)}{D(s)} = \frac{V_{in}}{LCs^2 + \frac{L}{R}s + 1} \tag{3}$$

$$\frac{I_L(s)}{D(s)} = \frac{\frac{V_{in}}{R}(1 + RCs)}{LCs^2 + \frac{L}{R}s + 1} \tag{4}$$

FSBB converter model in boost operation mode:

$$\frac{V_C(s)}{D(s)} = \frac{\frac{V_{in}}{(1 - D_2)^2}\left[1 - \frac{L}{R(1 - D_2)^2}s\right]}{\frac{LC}{(1 - D_2)^2}s^2 + \frac{L}{R(1 - D_2)^2}s + 1} \tag{5}$$

$$\frac{I_L(s)}{D(s)} = \frac{\frac{V_{in}}{R(1 - D_2)^3}[2 + RCs]}{\frac{LC}{(1 - D_2)^2}s^2 + \frac{L}{R(1 - D_2)^2}s + 1} \tag{6}$$

FSBB converter model in buck-boost operation mode (In this situation $D_1 = D_2 = D$):

$$\frac{V_C(s)}{D(s)} = \frac{\frac{V_{in}}{(1 - D)^2}\left[1 - \frac{LD}{R(1 - D)^2}s\right]}{\frac{LC}{(1 - D)^2}s^2 + \frac{L}{R(1 - D)^2}s + 1} \tag{7}$$

$$\frac{I_L(s)}{D(s)} = \frac{\frac{V_{in}}{R(1 - D)^3}[(1 + D) + RCs]}{\frac{LC}{(1 - D)^2}s^2 + \frac{L}{R(1 - D)^2}s + 1} \tag{8}$$

B. I-V and R-V Characteristic curves of the PV panel

The typical representation of the output characteristic of a PV panel is called current-voltage (I-V) curve, illustrated in Fig. 3. Ortiz-Rivera and Peng [12] present a review of approaches to obtain the I-V characteristic curve of photovoltaic panels. Furthermore, the authors propose an I-V equation model which is used in this work as the I-V reference curve that the emulator must reproduce

$$I(V) = \alpha I_{sc} \tau_i \left[1 - e^{\frac{V}{b(\alpha\gamma + 1 - \gamma)(V_{oc} + \tau_v)} - \frac{1}{\beta}} \right], \tag{9}$$

where, I_{sc} is a short-circuit current, V_{oc} is the open-circuit voltage, α represents the effective percentage of irradiation, γ is the shading factor, τ_i is the rate of change of the I_{sc} according to temperature ($A/°C$), τ_v is the rate of V_{oc} variation according to the temperature ($V/°C$) and β is a constant that can be calculated from the datasheet of the panel.

Furthermore, from the I-V curve it is possible to obtain two additional important curves: P-V curve (relates output power to the voltage); and the R-V curve (relates the output load with the voltage). The R-V curve can be obtained by measuring current and voltage at the converter output terminals and applying the Ohm's Law:

$$R(V) = \frac{V}{I(V)}. \tag{10}$$

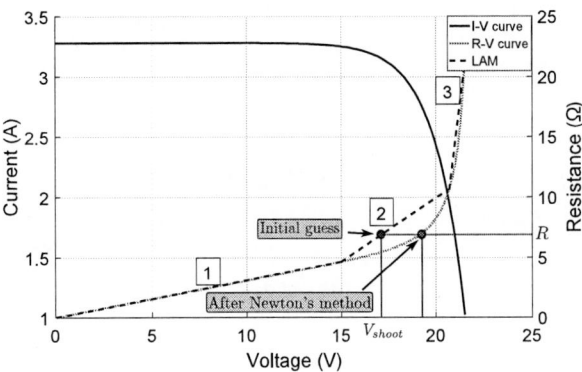

Fig. 3. I-V and R-V curves with linear approach method (LAM).

For each resistance value, there is a unique I-V pair for a fixed weather condition. Therefore, by estimating the output load value connected to the converter, it is possible to control the output voltage of the converter to impose the corresponding I-V pair on the output. Since I-V and R-V curves are transcendental equations and require an unbearable computational effort for embedded processors, the R-V curve is approximated by lines, named the linear approach method (LAM) [4], depicted in Fig. 3. These lines are the first step to obtain the voltage reference value used by the controllers to mimic a PV panel.

III. EMULATOR DESIGN

A. Voltage Reference Algorithm

The original I-V or R-V curves are convoluted to obtain in an embedded system. Thus a LAM is used to reduce computational complexity. First, using (10), 3 lines are calculated offline and embedded into the microcontroller, as depicted in Fig 3. The output voltage and current are measured at the converter terminals to estimate the load value (R). The corresponding line to this R value is used to obtain an initial

guess of the output voltage reference (V_{shoot}), (see Fig. 3). This initial guess, that is very close to the true root, is fed into the Newton's method to reduce error. Only three iterations are necessary to find a voltage value near the original curve to use as reference. Finally, knowing the voltage reference value that must be imposed at the output terminals, the operation mode is selected (*buck if $V_{ref} \leq V_{buck}$, buck-boost between V_{buck} and V_{boost} and boost if $V_{ref} \geq V_{boost}$*) and the corresponding control acts on the switches. This method is presented through the flowchart in Fig. 4.

B. FSBB Power Circuit Design

In the power circuit design, inductors and capacitors need to be sized to guarantee a continuous conduction mode (CCM) and the ripple (ΔV) of output voltage does not exceed 5% at the switching frequency (20 kHz). Using (11) and (12) [13], the minimum values for each operation mode of the converter, varying the duty cycle for each mode and using different resistance values of the R-V curves, are obtained: $L = 400 \ \mu H$ and $C = 1 \ mF$.

$$L_{min} \geq 2 \ max \left[\frac{(1 - D_1)R}{2f}, \quad \frac{D_2(1 - D_2)^2 R}{2f} \right] \quad (11)$$

$$C_{min} \geq 2 \ max \left[\frac{1 - D_1}{8L\Delta V f^2}, \quad \frac{D_2}{R\Delta V f} \right] \quad (12)$$

C. Control System Design

In this stage, a cascade control is designed to regulate output voltage using an internal current loop and an external voltage loop, as shown in Fig. 5. Three different controllers are tuned, one for each operation mode of the converter. It is worth saying that the response time of internal loop must be faster than the external loop in order for the cascade control to be successful. Using equations (3)–(8) it is possible to obtain the transfer function that relates the output voltage to the inductor current by dividing the two that relate both to the duty cycle.

The PI controllers, $C_1(s)$ and $C_2(s)$, are tunned using the frequency domain and root locus method. Controllers were designed for each operation mode using the same requirements: settling time - 1 ms using the 5 % criterion. The internal loop must be able to filter the system response at the switching frequency, so the crossover frequency used is 2 kHz. The values of the controller parameters K_p and T_i obtained are represented in Table II.

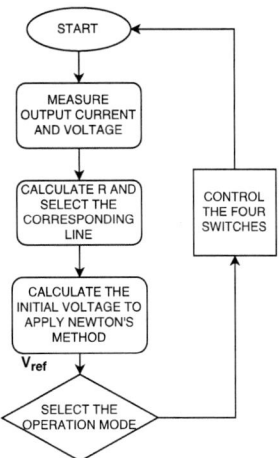

Fig. 4. Flowchart of the PV emulator control algorithm.

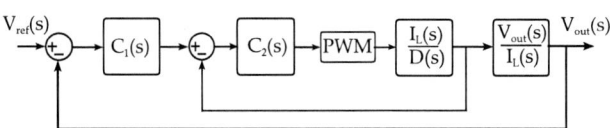

Fig. 5. Cascade control system to regulate the FSBB converter for each operation mode.

TABLE II
PI CONTROLLERS - 3 OPERATION MODES.

Operation Mode	Operation Point		External Loop		Internal Loop	
	V (V)	R (Ω)	K_p	T_i	K_p	T_i
BUCK	6	1.83	3.03	0.0017	0.412	0.0035
BUCK-BOOST	12	3.66	3.65	0.0023	0.2085	0.0069
BOOST	18	5.9	3.33	0.0028	0.2085	0.0112

TABLE III
ELECTRICAL PARAMETERS OF THE YL055P PANEL.

Yingli YL055P 17b	
Maximum power (Pmax)	55 W
Voltage at Pmax (Vmppt)	17.83 V
Current at Pmax (Imppt)	3.08 A
Open Circuit Voltage (Voc)	22.07 V
Short Circuit Current (Isc)	3.28 A
Isc Temperature Coefficient (T_{CI})	60 mA/ °C
Voc Temperature Coefficient (T_{CV})	-330 mV/ °C

IV. SIMULATION RESULTS

In this work, the solar module YL55P-17b, manufactured by Yingli Solar, is used as reference. Its electrical parameters at Standard Test Conditions (STC) are presented in Table III. Using (9), (10), data of Table III, Matlab R2016a and PSIM softwares, the I-V and R-V curves were obtained varying irradiation and temperature. Through the values obtained in these curves, the linear approximation method used to emulate I-V curves of YL55P panel.

To verify the proposed algorithm, two tests were simulated where, from a fixed temperature, the irradiation is varied, and another where the temperature varied using the maximum irradiation. In both cases the fixed output load is 10 Ω and the emulation time chosen was 0.1 s to simulate abrupt changes in the weather conditions. In order to develop the first test, the standard temperature of 25°C was used and the irradiation was varied according to the Table IV, comparing the analytical model with the values obtained in simulation of PSIM at the output terminals of the converter, shown in Fig. 6. In such case, it shows that the control is able to change the operation mode of the emulator in a stable way. Thus, in the second test, using the maximum irradiation, the temperature was varied according to the values of Table V, comparing the analytical model with the values obtained in the converter terminals from the simulation in PSIM, shown in Fig. 7. Acceptable values can be observed in these tables, since there is a satisfactory error between the value emulated with the theoretical value.

TABLE IV
COMPARISON BETWEEN THE DEVELOPED MODEL AND THE PV
EMULATOR OUTPUT, VARYING IRRADIANCE AT 25°C.

	Simulation Results			
G (W/m^2)	I-V curve model		PV Emulator	
	I (A)	V (V)	I (A)	V (V)
250	0,82	8,2	0,818	8,18
500	1,61	16,1	1,607	16,07
750	1,96	19,6	1,962	19,62

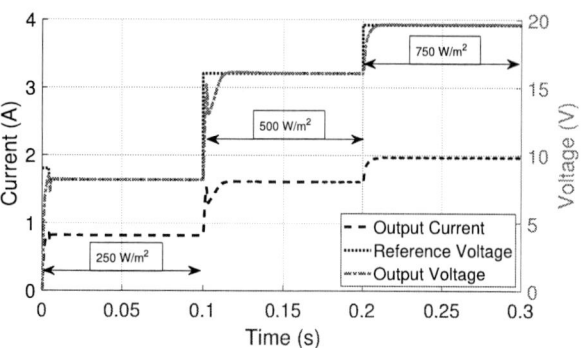

Fig. 6. PV emulator output varying irradiation at 25°C.

Therefore, it can be said that the PV emulator, for these case studies, operates correctly.

TABLE V
COMPARISON OF THE DEVELOPED MODEL WITH THE PV EMULATOR
OUTPUT, VARYING TEMPERATURE IN THE MAXIMUM IRRADIATION.

	Simulation Results			
T (°C)	I-V curve model		PV Emulator	
	I (A)	V (V)	I (A)	V (V)
10	2,145	21,45	2,147	21,47
50	1,905	19,05	1,907	19,07
75	1,75	17,5	1,748	17,48

Then, a last test was made to simulate several points on the same I-V curve. Since $V_{in} = 12V$, three operating points where the converter operates as buck, buck-boost and boost were selected (*buck if $V_{ref} \leq 9V$, buck-boost if V_{ref} between 9 V and 15 V and boost if $V_{ref} \geq 15V$*). Then, varying the output load forces the converter to work at different operation modes. The simulated load values are: 2 Ω for buck mode, 4 Ω for buck-boost mode and 12 Ω for boost mode, represented in Fig. 8. Simulation results of the proposed PV emulator working in each operation mode, illustrated in Fig. 9, are compared with the I-V curve selected, represented in Fig. 8. These results are found in Table VI, demonstrating

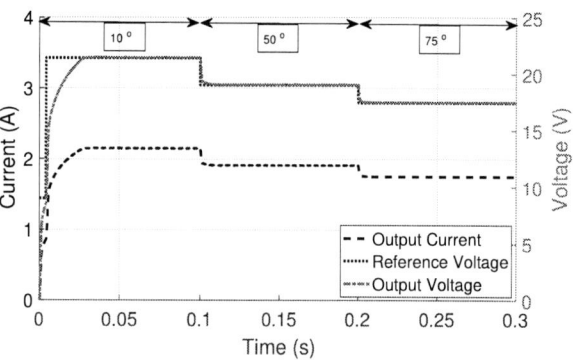

Fig. 7. PV emulator output varying temperature at 1000 W/m^2.

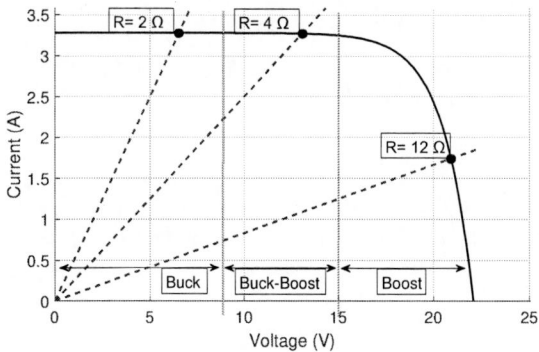

Fig. 8. I-V curve with the three different operation points.

TABLE VI
RESULTS OBTAINED FROM THE SELECTED I-V CURVE AND THE PROPOSED
PV EMULATOR FOR EACH OPERATION MODE.

| Operation Mode | $R(\Omega)$ | Simulation Results | | | |
| | | I-V curve model | | PV emulator | |
		I (A)	V (V)	I (A)	V (V)
BUCK	2	3.28	6.6	3.28	6.585
BUCK-BOOST	4	3.27	13.1	3.268	13.11
BOOST	12	1.75	20.9	1.748	20.91

the effectiveness of the proposed emulator. It's observed that each value at the output of the FSBB converter for different operation modes corresponds to selected values of the chosen I-V curve.

V. CONCLUSIONS

This paper presents the design of a PV emulator based on a four-switch buck-boost (FSBB) dc-dc converter. The emulator uses the linear approach and Newton's method to determine the reference voltage that corresponds to the I-V curve, with low complexity and error. This reference voltage with adequate accuracy is tracked by a cascade structure composed by two PI controllers. The proposed algorithm was verificated for abrupt changes in the weather condintions. This emulator has been simulated using different resistive loads forcing the FSBB converter to work in each operation modes (buck, buck-boost and boost). Simulation results demonstrate the effectiveness of the emulator in reaching the values of the I-V curve model under different load conditions.

ACKNOWLEDGEMENT

The authors would like to thank CAPES for the financial support and the colleagues of Energy Efficiency Lab (LABEFEA) for assistance provided.

REFERENCES

[1] P. G. V. Sampaio and M. O. A. Gonzalez, "Photovoltaic solar energy: Conceptual framework," *Renewable and Sustainable Energy Reviews*, vol. 74, pp. 590 – 601, 2017.

[2] S. Armstrong, C. K. Lee, and W. G. Hurley, "Investigation of the harmonic response of a photovoltaic system with a solar emulator," in *European Conference on Power Electronics and Applications*, Sept 2005, pp. 8 pp.–P.8.

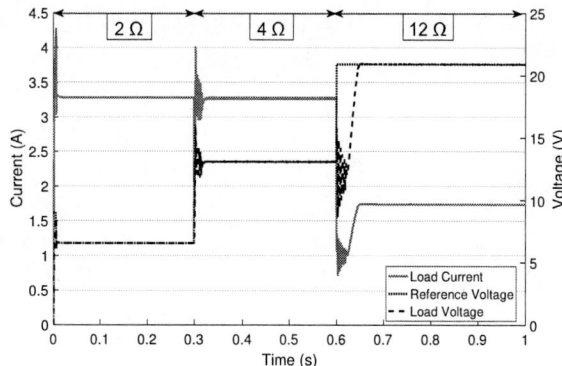

Fig. 9. Simulation results of the PV emulator under load variation.

[3] A. Mukerjee and N. Dasgupta, "Dc power supply used as photovoltaic simulator for testing mppt algorithms," *Renewable Energy*, vol. 32, no. 4, pp. 587 – 592, 2007.

[4] D. Lu and Q. Nguyen, "A photovoltaic panel emulator using a buck-boost dc/dc converter and a low cost micro-controller," *Solar Energy*, vol. 86, p. 1477–1484, May 2012.

[5] A. Sanaullah and H. A. Khan, "Design and implementation of a low cost solar panel emulator," in *IEEE 42nd Photovoltaic Specialist Conference (PVSC)*, June 2015, pp. 1–5.

[6] A. F. Cupertino, G. V. Santos, H. A. Pereira, S. R. Silva, and V. F. Mendes, "Modeling and control of a flexible photovoltaic array simulator," in *IEEE 24th International Symposium on Industrial Electronics (ISIE)*, June 2015, pp. 318–324.

[7] A. Cordeiro, D. Foito, and V. F. Pires, "A pv panel simulator based on a two quadrant dc/dc power converter with a sliding mode controller," in *International Conference on Renewable Energy Research and Applications (ICRERA)*, Nov 2015, pp. 928–932.

[8] U. K. Shinde, S. G. Kadwane, S. P. Gawande, and R. Keshri, "Solar pv emulator for realizing pv characteristics under rapidly varying environmental conditions," in *IEEE International Conference on Power Electronics, Drives and Energy Systems (PEDES)*, Dec 2016, pp. 1–5.

[9] X. Ren, X. Ruan, H. Qian, M. Li, and Q. Chen, "Three-mode dual-frequency two-edge modulation scheme for four-switch buck-boost converter," *IEEE Transactions on Power Electronics*, vol. 24, no. 2, pp. 499–509, Feb 2009.

[10] R. W. Erickson and D. Maksimovic, *Fundamentals of Power Electronics*, 2nd ed. Kluwer Academic Publishers, 2004.

[11] M. Orellana, S. Petibon, B. Estibals, and C. Alonso, "Four switch buck-boost converter for photovoltaic dc-dc power applications," in *IECON - 36th Annual Conference on IEEE Industrial Electronics Society*, Nov 2010, pp. 469–474.

[12] E. I. Ortiz-Rivera and F. Z. Peng, "Analytical model for a photovoltaic module using the electrical characteristics provided by the manufacturer data sheet," in *IEEE 36th Power Electronics Specialists Conference*, June 2005, pp. 2087–2091.

[13] S. Bacha, I. Munteanu, and A. I. Bratcu, *Power Electronic Converters Modeling and Control*. Springer-Verlag London, 2014.

Single-Stage Bridgeless AC-DC PFC Flyback Interleaved

Schowantz, G. S., Barcelos R. P., Lazzarin, T. B.

Department of Electrical Engineering, Federal University of Santa Catarina, Florianopolis - SC, Brazil
e-mail: guilherme.schowantz@inep.ufsc.br, renan.barcelos@inep.ufsc.br, telles@inep.ufsc.br

Abstract—This paper presents a full analysis on a single-phase PFC (Power Factor Correction) Flyback rectifier. The study approaches all design equations, including input filter and leakage inductance loss, for operation in DCM (Discontinuous Conduction Mode) and in CCM (Continuous Conduction Mode). Both operation modes are compared in the paper using a designed example. In addition, it is introduced a single-stage bridgeless Flyback rectifier and it is compared with other isolated and non-isolated similar topologies. Finally, in order to make the bridgeless version a modular solution, it is designed to operate in interleaved with 3 modules, tripling the rated power and reducing the input filter.

Index Terms—Flyback, Buck-boost, Bridgeless, Interleaved, PFC, DCM, CCM.

I. INTRODUCTION

Non-linear loads such as diode rectifiers, thyristor rectifiers, cycloconverters and other switching circuits, produce a high THD in the current. There are some methods that are commonly used to eliminate these harmonics, like using a set of passive filters or an active filter installed near the non-linear load. Another solution is to install a PFC converter after the rectifier, changing the original non-linear load into a fully controlled converter capable of emulating a resistor and delivering a good and stable DC voltage for the load. Most of single-phase solutions are based on Boost rectifier, which is a non-isolated and a step-up topology.

The Flyback converter is also used as a PFC rectifier, as seen in [1]–[7], and it has some advantages such as: it is a step-down and step-up topology; its output has the characteristics of a voltage source; it has a transformer so it can provide galvanic isolation between input and output, which eliminate problems with common mode; it can operate with large differences between input and output voltage, adjusting the turns ratio between primary and secondary of the transformer.

However, it brings some challenges, such as the pulsating input current, like the Buck converter, which requires an input filter; and another challenge is the leakage inductance of the transformer, which produces over-voltages on the switch if the switch is not protected by a passive of active clamper. Besides that, the Flyback topology has often been used in low power applications due to low cost, simplicity and wide operation range.

In the following chapters, the conventional AC-DC PFC Flyback operating in DCM and CCM will be explained and compared with each other. Then, it will be introduced the Bridgeless version, showing its differences and its design to

The authors would like to thank the CNPq for the financial support (process No. 422276/2016-2).

operate at up to 500W. Finally it is introduced the interleaved version of the Bridgeless Flyback to operate at up to 1500W.

II. AC-DC PFC FLYBACK IN DCM

The conventional AC-DC Flyback PFC converter is presented in Fig. 1. The input filter is required, both in DCM and in CCM, to avoid injecting in the main grid high frequency harmonics of the input current of the converter. When the filter is adequately designed, it does not change much the input voltage of the converter.

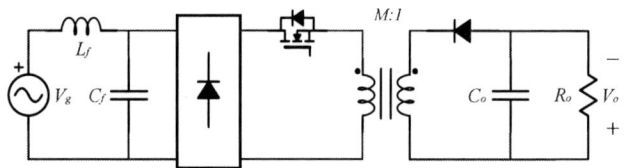

Fig. 1. Conventional AC-DC PFC Flyback.

A. Stages of Operation

1) Discontinuous Conduction Mode: The converter operating in DCM will have three stages. During first the first stage (Δt_1) the switch is turned on and the transformer's magnetizing inductance is charging. In the second stage (Δt_2), the inductor transfer the energy to the output. And the third stage (Δt_3) is the discontinuous stage.

The current in the inductor starts and ends in zero. Applying the volt-second law in the inductor results in:

$$V_g \cdot \Delta t_1 - V_o \cdot M \cdot \Delta t_2 + 0 \cdot \Delta t_3 = 0 \qquad (1)$$

Δt_1 can be defined as $\Delta t_1 = D \cdot T_s$, as long the converter operates with a fixed switching frequency, where T_s is the switching period and D is the Duty Cycle. Besides, V_g can be written as $V_g = V_{gp} \cdot sin(\omega_0 \cdot t)$, and a new variable α can be defined as where $\alpha = \dfrac{V_{gp}}{V_o \cdot M}$. Using these definitions, Δt_2 can be isolated in the last equation and rewritten as:

$$\Delta t_2 = \alpha \cdot D \cdot T_s \cdot sin(\omega_o \cdot t) \qquad (2)$$

The maximum value of Δt_2 from the last equation happens when $sin(\omega_o t) = 1$, and it is also known that it cannot be bigger than $(1-D) \cdot T_s$, or else it will operate in CCM rather than DCM. As result, the maximum value for D that ensures the converter to operate in DCM is:

$$D_{max} = \frac{V_o \cdot M}{V_o \cdot M + V_{pk}} = \frac{1}{1+\alpha} \qquad (3)$$

Finally, Δt_3 can be written as:

$$\Delta t_3 = T_s(1 - D(1 + \alpha \cdot sin(\omega_o \cdot t))) \qquad (4)$$

978-1-7281-4181-7/19 $31.00 © 2019 IEEE

2) Continuous Conducting Mode: When operating in CCM, the stage Δt_3 does not exist. The current in the inductor acquires a continuous shape, while the current in the switch and diode remains discontinuous. The stages Δt_1 and Δt_2 are now defined by $\Delta t_1 = D \cdot T_s$ and $\Delta t_2 = (1-D) \cdot T_s$. Applying the volt-second law in the inductor results in:

$$\frac{V_o \cdot M}{V_g} = \frac{D}{1-D} = Gain \tag{5}$$

$$D = \frac{V_o \cdot M}{V_o \cdot M + V_{gp} \cdot \sin(\omega_0 \cdot t)} = \frac{1}{1 + \alpha \cdot \sin(\omega_o \cdot t)} \tag{6}$$

Differently from the DCM, the Duty Cycle is now time-variant, and its value change from $D = 1$ when the grid voltage is zero, to $D = \frac{1}{1+\alpha}$ when the grid voltage is at its peak.

The current waveform in the inductor can be calculated as:

$$i_L = \frac{2 \cdot P_o}{V_{pk}} \left(\sin(\omega_o \cdot t) + \alpha \cdot \sin(\omega_o \cdot t)^2 \right) \tag{7}$$

B. Mutual Inductance and Output Capacitor

1) Discontinuous Conducting Mode: The maximum value of mutual inductance of the flyback transformer, to guarantee the operation in DCM during the whole period of the grid, can be calculated as:

$$L_{m_{max}} = \frac{V_{pk}^2 \cdot D_{max}^2}{4 \cdot P_o \cdot f_s} = \frac{V_{pk}^2}{4 \cdot P_o \cdot f_s \cdot (1+\alpha)^2} \tag{8}$$

It is advised to choose a value of inductance a little bit lower than the maximum, to avoid the converter entering in CCM due to small deviations in the circuit. After choosing a value for the mutual inductance, the value of Duty Cycle in steady state can be calculated as:

$$D = \sqrt{\frac{4 \cdot P_o \cdot f_s \cdot L_m}{V_{pk}^2}} \tag{9}$$

The output capacitor is designed to keep the 120 Hz fluctuation of V_o within the specified limits:

$$C_{o_{min}} = \frac{P_o}{2 \cdot \pi \cdot f_o \cdot V_o^2 \cdot \Delta V_{o\%}} \tag{10}$$

2) Continuous Conducting Mode: The output capacitor is designed in the same method as before, and its equation is the same. The mutual inductance, however, is now designed to maintain the current in CCM and to keep the ripple of the current within the specified limits:

$$L_{m_{min}} = \frac{V_{pk}^2}{2 \cdot P_o \cdot f_s \cdot (1+\alpha)^2 \cdot \Delta I_{L\%}} \tag{11}$$

C. Voltage and Current Stress on Components, PF and THD

Table I shows the equations for the stress in the main components of the PFC Flyback converter, also for the PF and THD.

TABLE I
GENERAL EQUATIONS FOR AN AC-DC FLYBACK CONVERTER.

	DCM	CCM
$V_{S_{max}}$ (V)	$V_{pk} + M \cdot V_o$	$V_{pk} + M \cdot V_o$
$V_{D_{max}}$ (V)	$\frac{V_{pk}}{M} + V_o$	$\frac{V_{pk}}{M} + V_o$
$I_{L_{max}}$ (A)	$\frac{V_{pk} \cdot D}{L_m \cdot f_s}$	$\frac{2 \cdot P_o \cdot (1+\alpha)}{V_{pk}}$
$I_{L_{med}}$ (A)	$\frac{V_{pk} \cdot D^2}{L_m \cdot f_s} \left(\frac{4 + \pi \cdot \alpha}{4 \cdot \pi} \right)$	$\frac{P_o \cdot (4 + \alpha \cdot \pi)}{\pi \cdot V_{pk}}$
$I_{S_{med}}$ (A)	$\frac{V_{pk} \cdot D^2}{\pi \cdot L_m \cdot f_s}$	$\frac{4 \cdot P_o}{\pi \cdot V_{pk}}$
$I_{D_{med}}$ (A)	$\frac{V_{pk}^2 \cdot D^2}{4 \cdot V_o \cdot L_m \cdot f_s} = \frac{P_o}{V_o}$	$\frac{P_o}{V_o}$
$I_{L_{rms}}$ (A)	$\frac{V_{pk} \cdot D}{L_m \cdot f_s} \cdot K_1$	$\frac{P_o}{V_{pk}} \cdot K_3$
$I_{S_{rms}}$ (A)	$\frac{V_{pk} \cdot D}{L_m \cdot f_s} \sqrt{\frac{D}{6}}$	$\frac{P_o}{V_{pk}} \sqrt{\frac{6 \cdot \pi + 16 \cdot \alpha}{3 \cdot \pi}}$
$I_{D_{rms}}$ (A)	$\frac{V_{pk} \cdot M \cdot D}{L_m \cdot f_s} \sqrt{\frac{4 \cdot D \cdot \alpha}{9 \cdot \pi}}$	$\frac{M \cdot P_o}{V_{pk}} \cdot K_4$
$I_{C_{rms}}$ (A)	$\frac{V_{pk} \cdot M \cdot D}{L_m \cdot f_s} \cdot K_2$	$\frac{P_o}{V_o} \sqrt{\frac{3 \cdot \pi \cdot \alpha + 32}{6 \cdot \pi \cdot \alpha}}$
PF	$\sqrt{\frac{3 \cdot D}{4}}$	$\sqrt{\frac{3 \cdot \pi}{3 \cdot \pi + 8 \cdot \alpha}}$
THD	$\sqrt{\frac{4 - 3 \cdot D}{3 \cdot D}}$	$\sqrt{\frac{8 \cdot \alpha}{3 \cdot \pi}}$

$$K_1 = \sqrt{\frac{D \cdot (3 \cdot \pi + 8 \cdot \alpha)}{18 \cdot \pi}} \ , \ K_2 = \sqrt{\frac{D \cdot \alpha \cdot (64 - 9 \cdot \pi \cdot D \cdot \alpha)}{144 \cdot \pi}} \ ,$$

$$K_3 = \sqrt{\frac{9 \cdot \pi \cdot \alpha^2 + 64 \cdot \alpha + 12 \cdot \pi}{6 \cdot \pi}} \ , \ K_4 = \sqrt{\frac{9 \cdot \pi \cdot \alpha^2 + 32 \cdot \alpha}{6 \cdot \pi}} \ .$$

D. Power Loss due to Leakage Inductance

All kind of transformers presents a leakage inductance, and in the Flyback converter this inductance is in series with the switch. For this reason, this converter have to use a clamper circuit to clamp the voltage across the switch below a limited value. For high efficiency converters it is used an active clamping, which absorbs the energy of the leakage inductance to use it later. There is, however, a simpler solution that is a passive clamper, which simply dissipates the energy stored in the leakage inductance and, for the sake of simplicity, will be the clamp presented here.

Two additional parameters are needed to calculate the power loss here: the value of leakage inductance (L_k) which, usually, is approximately 1% the value of self inductance (this information can be used as an initial step in the project); and the clamp voltage (V_Z) which necessarily needs to be higher than the value of $V_{pk} + M \cdot V_o$, and will affect the time needed to discharge the leakage inductance.

Fig. 2 shows two passive clamping circuits across the switch, one is an RCD clamp, and the other is a TVS diode (which acts similar to a zener diode). It is important to notice that, if the clamp circuit is across the transformer instead of the switch, the equations and losses will be different from those presented in this paper.

978-1-7281-4181-7/19 $31.00 © 2019 IEEE

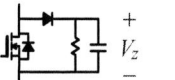

Fig. 2. Example of passive clamping circuits.

1) Discontinuous Conduction Mode:

$$P_{Lk} = \frac{2 \cdot P_o \cdot L_k}{\pi \cdot L_m} \int_0^\pi \frac{V_Z \cdot sin(\omega t)^2}{V_Z - M \cdot V_o - V_{pk} \cdot sin(\omega t)} \; d\omega t \quad (12)$$

The last equation can be reasonably approximated (the higher the value of V_Z, the better is the approximation) to:

$$P_{Lk} = \frac{P_o \cdot L_k}{L_m} \left(\frac{V_Z}{V_Z - M \cdot V_o - V_{pk}} \right) \quad (13)$$

2) Continuous Conduction Mode:

$$P_{Lk} = \frac{P_o \cdot L_k \cdot (2 + \Delta I_{L\%})^2}{4 \cdot \pi \cdot L_m \cdot \Delta I_{L\%} \cdot (\alpha + 1)^2} \int_0^\pi K_5(\omega t) \; d\omega t \quad (14)$$

where $K_5(\omega t) = \dfrac{V_Z \cdot sin(\omega t)^2 \cdot (\alpha \cdot sin(\omega t) + 1)^2}{V_Z - M \cdot V_o - V_{pk} \cdot sin(\omega t)}$.

As well as in DCM, the last equation can be reasonably approximated to:

$$P_{Lk} = \frac{P_o \cdot L_k \cdot V_Z \cdot (\Delta I_{L\%} + 2)^2 \cdot (9 \cdot \alpha^2 + 64 \cdot \frac{\alpha}{\pi} + 12)}{96 \cdot L_m \cdot \Delta I_{L\%} \cdot (\alpha + 1)^2 \cdot (V_Z - M \cdot V_o - V_{pk})} \quad (15)$$

E. Input Filter

1) Discontinuous Conduction Mode: The input filter used in this work is a low-pass LC filter. The initial requirement of this filter is to keep the voltage ripple at the input of the converter within the specified limit. Therefore, the minimal capacitance required can be calculated as:

$$C_{f\,min} = \frac{D^2 \cdot (2 - D)^2}{8 \cdot L_m \cdot f_s^2 \cdot \Delta V_{f\%}} \quad (16)$$

The inductor can be calculated as:

$$L_f = \frac{1}{C_f \cdot (\omega_f)^2} \quad (17)$$

Where ω_f is the resonance frequency of the LC filter. A good approach is to choose the resonance frequency 5 times lower than the switching frequency. In that case:

$$\omega_f = 2 \cdot \pi \cdot \frac{f_s}{5} \quad (18)$$

2) Continuous Conduction Mode: The approach is the same as before, but the equation for the minimal capacitance is now defined as:

$$C_{f\,min} = \frac{2 \cdot P_o}{f_s \cdot V_{pk}^2 \Delta V_{g\%}} \cdot \frac{\alpha}{(1 + \alpha)} \quad (19)$$

With the equations presented so far it is possible to make a comparative table between the design of an AC-DC PFC Flyback working in DCM and in CCM, for different values of turns ratio (M), which is done in Table II. It is important

to notice that, in the table, the values of D and L_m for DCM represent the maximum value, and for CCM they represent the minimum value, in the same way they were introduced in equations (3), (6), (8) and (11).

TABLE II
COMPARISONS FOR THE AC-DC PFC FLYBACK.

$P_o = 500W$, $V_o = 100V$, $V_{pk} = 100V$, $f_0 = 60Hz$, $f_s = 20kHz$, $\Delta V_{o\%} = 5\%$, and $\Delta I_{L\%} = 20\%$.					
	DCM	CCM	DCM	CCM	DCM	CCM
M	0.4	0.4	1	1	2.5	2.5
α	2.5	2.5	1	1	0.4	0.4
D	0.29	0.29	0.5	0.5	0.71	0.71
$L_m(\mu H)$	20.4	204	62.5	625	128	1276
$C_o(mF)$	2.65	2.65	2.65	2.65	2.65	2.65
$V_{S_{max}}(V)$	140	140	200	200	350	350
$V_{D_{max}}(V)$	350	350	200	200	140	140
$I_{S_{max}}(A)$	70	38.5	40	22	28	15.4
$I_{D_{max}}(A)$	28	15.4	40	22	70	38.5
$I_{L_{avg}}(A)$	18.9	18.9	11.4	11.4	8.37	8.37
$I_{S_{avg}}(A)$	6.37	6.37	6.37	6.37	6.37	6.37
$I_{D_{avg}}(A)$	5	5	5	5	5	5
$I_{L_{rms}}(A)$	27	22.3	15.7	13.1	11.2	9.48
$I_{S_{rms}}(A)$	15.3	12.5	11.5	9.61	9.66	8.18
$I_{D_{rms}}(A)$	8.9	7.38	10.6	8.94	14.1	12
$I_{C_{rms}}(A)$	7.36	5.43	9.39	7.41	13.2	10.9
$PF(\%)$	46.3	56.6	61.2	73.5	73.2	86.4
$THD(\%)$	191	146	129	92.1	93.1	58.3

$\Delta V_{f\%} = 20\%$, $\omega_f = 0.4 \cdot \pi \cdot f_s$, $L_k = 0.01 \cdot L_m$, and $V_Z = 1.5 \cdot (Vpk + M \cdot V_o)$.					
$C_f(\mu F)$	18.4	17.9	14.1	12.5	10.3	7.1
$L_f(\mu H)$	86.2	88.7	113	127	153	221
$V_Z(V)$	210	210	300	300	525	525
$P_{Lk}(W)$	12.8	32.5	13.3	35.4	13.9	38.9

F. Transfer Function

1) Discontinuous Conduction Mode: The control of output voltage in DCM is very simple and it is done by voltage loop. The transfer function of the output voltage for duty cycle is given by:

$$\frac{\delta v_o}{\delta d} = G_{vd}(s) = \left(\frac{V_o}{D} \right) \cdot \frac{1}{\left(\frac{C_o \cdot V_o^2}{2 \cdot P_o} \right) \cdot s + 1} \quad (20)$$

2) Continuous Conduction Mode: The control of output voltage in CCM is more complex and is done by two different control loops working simultaneously. The internal loop (fast response) controls the instantaneous value of current in the inductor. The external loop (slow response) controls the output voltage. The transfer functions are given by:

$$\frac{\delta i_L}{\delta d} = G_{id}(s) = \frac{2 \cdot V_{pk} + \pi \cdot V_o \cdot M}{\pi \cdot L_m} \cdot \left(\frac{1}{s} \right) \quad (21)$$

$$\frac{\delta v_o}{\delta i_L} = G_{vi}(s) = \frac{\alpha \cdot \pi \cdot R_o \cdot M}{2 \cdot \pi \cdot \alpha + 8} \cdot \left(\frac{1}{\left(\frac{C_o \cdot V_o^2}{2 \cdot P_o} \right) s + 1} \right) \quad (22)$$

978-1-7281-4181-7/19 $31.00 © 2019 IEEE

III. BRIDGELESS PFC FLYBACK

Fig. 3 shows a gray box which contains an AC-DC PFC converter. This section will bring variations to the conventional Flyback rectifier.

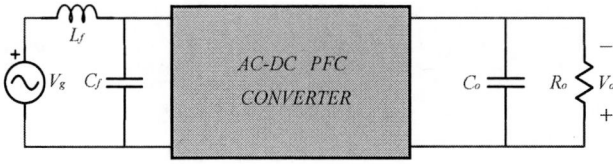

Fig. 3. Generic AC-DC PFC converter.

A. Topologies

Fig. 4 presents six different variations of the topology studied in this paper, and they all could be inside the gray box of Fig. 3.

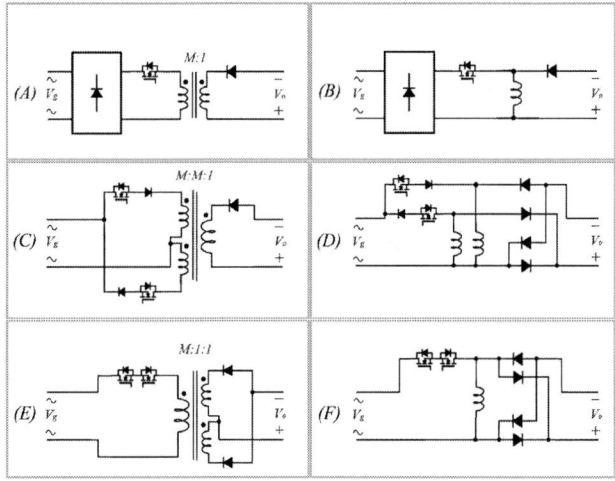

Fig. 4. Variations of the AC-DC Flyback converter.

From Fig. 4:

(A) is the conventional Flyback which was used in the whole section II;

(B) is the non-isolated version of the Flyback converter, best known as Buck-Boost converter. As long as the turns ratio of the transformer is unitary ($M = 1$) and the leakage inductance is neglected ($L_{lk} = 0$), the circuit analyses and equations are identical in both converters;

(C) is a bridgeless version of (A). The rectifier bridge + switch + primary winding is replaced by two sets of switch + diode + primary winding, with one set working during the positive half-wave of the grid voltage, and the other set working during the negative half-wave, while the secondary side of the transformer remains unchanged. The diode in series with the switch is necessary because the switch can not block reverse voltage, but it makes possible for the same gate signal to be applied in both switches, and the maximum voltage over this diode is $V_{pk} \cdot 2$;

(D) is the non-isolated version of (C);

(E) is another bridgeless version of (A). The rectifier bridge + switch is replaced by a bidirectional switch in the primary side of the transformer, and two sets of diode + secondary winding at the secondary side. Like in (C), one set will work during the positive half-wave of the grid voltage, and the other set will work during the negative half-wave. The maximum voltage over the diodes on the secondary side is $2 \cdot V_o$;

(F) is the non-isolated version of (E).

The analysis and equations presented in Section II are 100% valid for these converters, with exception to the current stress in components that work only during one half-wave of the grid voltage (in that case, the value of average current should be divided by 2, and the value of RMS current, divided by $\sqrt{2}$).

The main advantage of using a bridgeless circuit is having fewer semiconductors in the path of current, leading to lower conduction losses. Also, some components have lower current stresses. On the other hand, it increases the complexity of the circuit by raising the number of controlled devices, isolated drivers, clamper circuits and windings at the transformer.

Between the two bridgeless versions, (C) is more appropriate when $M < 1$ and (E) is more appropriate when $M > 1$, because in that way there will be more components to share the conduction loss in the side of the transformer with higher current, and fewer components in the side of the transformer with higher voltage.

It is worth to notice one risk using the converters (E) and (F): if at any moment, the input voltage (V_{Cf}) is higher than the output voltage ($V_o \cdot M$), the Switch cannot be turned on, otherwise the diode in the secondary side will be forward biased and enter conduction, resulting in a kind of "short-circuit" between the input and the output.

B. Design Specification

Fig. 5 shows the converter used in this project. It is based on the converter (C) from Fig. 4 and it will be designed to operate in DCM.

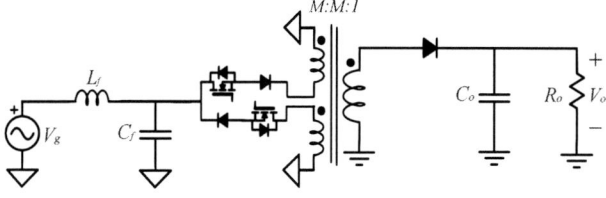

Fig. 5. Bridgeless AC-DC PFC Flyback

The specifications of the converter are to operate in DCM and: $P_o = 500W$, $V_o = 400V$, $V_{pk} = 170V$, $f_0 = 60Hz$, $f_s = 50kHz$ and $\Delta V_{o\%} = 5\%$.

C. Choosing the Parameters

For this project, the best value found was $M = 0.3$, and the following parameters were calculated or chosen: $R_o = 320\Omega$, $\alpha = 1.417$, $C_o = 180\mu F$, $L_m = 45\mu H$ and $D = 0.395$.

978-1-7281-4181-7/19 $31.00 © 2019 IEEE

This value of self inductance (L_m) is about 10% lower than $L_{m_{max}}$, given by Eq. 8, to guarantee the operation in DCM with small deviations. And, using this value of capacitance (C_o) in Eq. 10, the output voltage ripple is $\Delta V_{o\%} = 4.6\%$, or $\Delta V_o = 18.42V$, at twice the frequency of the grid.

To keep track of the desired output voltage, a voltage control was implemented.

For this first simulation, the input filter and leakage inductance are neglected. Fig. 6 shows the behavior of output voltage (V_o), in the first diagram, and of Duty Cycle (D), in the second diagram, when the reference voltage (green line) changes. The red lines correspond to the measurements on the commutated circuit, and the blue lines correspond to the measurements on the blocks diagram, which uses the transfer function of Eq. 20. It can be concluded that the transfer function is correct, and the voltage controller works.

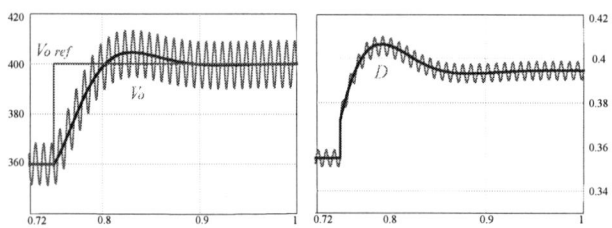

Fig. 6. Behavior of V_o and D with variation of voltage reference.

Using the equations presented in Section II, the chosen components of the input filter are $C_f = 6\mu F$ and $L_f = 60\mu H$. Using this capacitor in Eq. 16, the input voltage ripple is $\Delta V_{f\%} = 7.43\%$, or $\Delta V_f = 12.63V$, at the switching frequency.

For the leakage inductance, it was chosen the value $L_k = 0.01 \cdot L_m$, or $450nH$. The voltage of the clamping circuit was chosen to be $V_Z = 450V$, so a 500V switch could be used in this application. Using these values in Eq. 12, the power loss is $P_{Lk} = 12.37W$.

D. Full Simulation

In Section II it was presented the equations needed to calculate various parameters of the circuit, in the steady state. The converter was simulated in a simulation software, and the measurements were compared with the values given by the equations.

Table III shows the comparison between values calculated via equations and values measured via simulation. The first column shows the quantity that is being compared; the second column shows the values using the equations; the third column shows the measured values, when the input filter and leakage inductance are neglected; and the fourth column shows the measured values with the complete converter.

In Table III, the values of $I_{S_{rms}}$ are for one switch only. To find the rms value of current for the two switches, those values should be multiplied by $\sqrt{2}$. Furthermore, for the column with Measurement 2: the value of $V_{S_{max}}$ increases to the value of the voltage on the clamping circuit; the value of $V_{D_{max}}$

TABLE III
COMPARISON BETWEEN CALCULATED AND MEASURED VALUES.

	Calc.	Meas. 1	Meas. 2
D	0.3946	0.3944	0.3948
ΔV_o (V)	18.42	18.53	18.51
$V_{S_{max}}$ (V)	290	290.3	450
$V_{D_{max}}$ (V)	966.67	967.64	980.78
$I_{S_{max}}$ (A)	29.81	29.83	29.92
$I_{D_{max}}$ (A)	8.944	8.951	8.903
$I_{S_{rms}}$ (A)	5.406	5.410	5.468
$I_{D_{rms}}$ (A)	2.515	2.515	2.506
$I_{C_{rms}}$ (A)	2.183	2.183	2.174
FP (%)	54.40	54.41	99.68
THD (%)	154.24	154.18	3.64
ΔV_f (V)	12.63	0	13.19
P_{Lk} (W)	12.37	0	12.52

increases a bit as consequence of the voltage ripple from the capacitor C_f; the duty cycle and switch current increase a bit to compensate the loss from leakage inductance; and the PF and THD_i improves a lot as consequence of the input filter. Therefore, Table III indicates that the equations developed in Section II are very accurate.

IV. BRIDGELESS FLYBACK INTERLEAVED

A. Bridgeless PFC Flyback

There are several reasons to use an interleaved structure, like: to provide reliability and redundancy to critical loads (if one converter stops working there will be others to transfer power to the load); to use components with lower current ratings (more available); to decrease the total THD of the current, as will be shown in this section; etc.

In the previous section, a single-stage bridgeless AC-DC PFC Flyback was designed, in DCM, for a load of $500W$. In this section it will be used an interleaved version of that converter, with three modules, as shown in Fig. 7, for a load three times higher, or $1500W$.

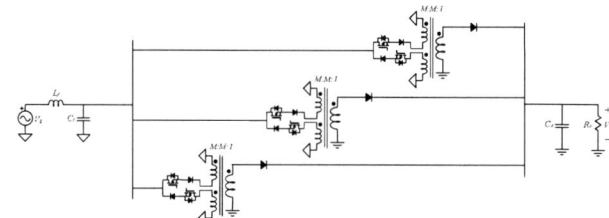

Fig. 7. Interleaved Bridgeless AC-DC PFC Flyback.

In DCM, the amount of power processed is automatically well distributed among each module, even with small differences between the components, so the same controller can be applied to all modules (even the same gate signal). On the other hand, this is not true, and each module must have its own current control to ensure that the module will not process more power and take over the others.

The circuit in Fig. 7 is working in DCM with $P_o = 1500W$ and each module is processing $500W$, so the same components

978-1-7281-4181-7/19 $31.00 © 2019 IEEE

and parameters used in last Section are also used in this circuit, except for C_o which needs to be 3 times larger, as it is directly proportional to the output power, as shows equation 10 .

One special advantage in using the interleaved structure in DCM, is that the input filter can be drastically reduced, using a feature of phase-shift of the gate signals between the modules. When using an interleaved structure with 3 modules, if the duty cycle is compared with one triangular carrier for each module, and these carriers are equally phase-shifted from each other, (120), the frequency of the resultant current ripple in the filter will be 3 times higher than the switching frequency, thus, improving a lot the performance of the filter. Table IV shows the effects of this feature, where Situation (1) refers to the single converter designed in Section III.

TABLE IV
FILTER PERFORMANCE.

Situation	PF	THD_i	ΔV_f	ΔI_g
(1)	99.67 %	3.62 %	7.72 %	9.87 %
(2)	99.89 %	3.68 %	23.33 %	9.98 %
(3)	99.95 %	1.04 %	2.24 %	0.30 %
(4)	99.97 %	1.44 %	6.78 %	2.81 %

(1) $6\mu F$, $60\mu H$, 1 module, 500W ;
(2) $6\mu F$, $60\mu H$, 3 modules, 1500W, same gate signals;
(3) $6\mu F$, $60\mu H$, 3 modules, 1500W, phase-shifted signals;
(4) $2\mu F$, $20\mu H$, 3 modules, 1500W, phase-shifted signals.

In Table IV is possible to see a huge difference in the filter performance from situations (2) and (3) when both converters are physically the same, the only thing that changed was the gate signals. In situation (4) the input filter reduces a lot and the performance is still better than in situation (1).

V. SIMULATION RESULTS

The converter in Fig. 7 was simulated using these conditions: $P_o = 1500W$; $V_o = 400V$; $V_{pk} = 170V$; $f_0 = 60Hz$; $f_s = 50kHz$; $R_o = 106.7\Omega$; $C_o = 540\mu F$; $L_m = 45\mu H$; $D = 0.395$; $C_f = 2\mu F$; $L_f = 20\mu H$; $L_k = 0.45\mu H$; $V_Z = 450V$; the same PI of Fig. 6 and 120 phase-shifted gate signals between the three modules. The result is seen in Fig. 8:

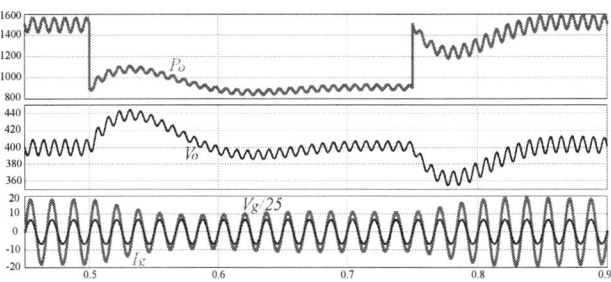

Fig. 8. Behavior of output power, output voltage, input current and input voltage.

In Fig. 8, the first curve shows the output power, which is at steady state in 1500W at the beginning, then there is a step of -600W and it stabilizes at 900W, then there is a step of +600W

and it stabilizes again at 1500W. The second curve shows the behavior of the output voltage with these disturbances. The third curve shows the grid current in red and the grid voltage, at scale, in blue.

In Fig. 9, the red curve shows the sum of the currents in all modules, the blue curve shows the current in the switch for one module, and the green curve shows the current in the diode of the same module.

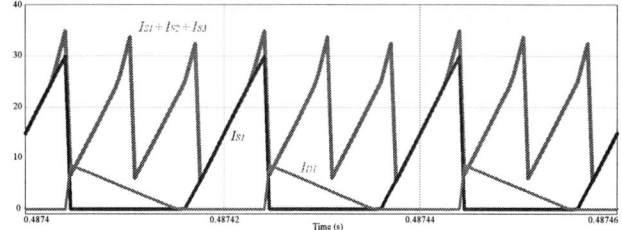

Fig. 9. Sum of the current in the three active switches, current in only one switch and current in only one diode.

VI. CONCLUSION

This work presented all equations needed to design an AC-DC PFC Flyback converter, including input filter, leakage inductance loss and control of output voltage, for the operation in DCM and in CCM. Later, both operation modes were compared using an example design.

The single-stage bridgeless topology was approched and compared to different versions, all of them that uses the same equations developed for the common Flyback, with a few exceptions that were well described.

It was then chosen one of the single-stage bridgeless topology, and decided to work from there only in DCM. After, the equations were compared using a simulation software, validating the equations.

Finally, an interleaved version of the bridgeless topology was introduced, with the objective to increase the output power without using semiconductors with higher current ratings. Exploring a feature available in this structure when working in DCM, it was possible to greatly reduce the input filter while still reducing the voltage ripple in the input capacitor and THD of the grid current.

REFERENCES

[1] R. Watson, G. C. Hua, F. C. Lee, "Characterization of an active clamp flyback topology for power factor correction applications"
[2] J. Baek, J. Shin, P. Jang, B. Cho, "A critical conduction mode bridgeless flyback converter"
[3] R. P. Bascope, B. B. Chaves, S. G. Barbosa, B. A. Da Silva, C. K. O. Veras, J. J. S. Souza, D. D. S. Oliveira, "Interleaved two-switch flyback converter with power factor correction for UPS Applications"
[4] S. W. Lee, H. L. Do, "Single-Stage Bridgeless ACDC PFC Converter Using a Lossless Passive Snubber and Valley Switching"
[5] K. T. Mok, Y. M. Lai, K. H. Loo, "A singlestage bridgeless power-factor-correction rectifier based on flyback topology"
[6] J. W. Shin, J. B. Baek, B. H. Cho, "Bridgeless isolated PFC rectifier using bidirectional switch and dual output windings"
[7] J. W. Shin, S. J. Choi, B. H. Cho, "High-efficiency bridgeless flyback rectifier with bidirectional switch and dual output windings"

Premagnetized Inductors in Single Phase dc-ac and ac-dc Converters

Jens Friebe
Institute for Drive Systems and Power Electronics
Leibniz University Hannover
Welfengarten 1, 30167 Hanover, Germany
Friebe@ial.uni-hannover.de

Siqi Lin
Institute for Drive Systems and Power Electronics
Leibniz University Hannover
Welfengarten 1, 30167 Hanover, Germany
Siqi.Lin@ial.uni-hannover.de

Leon Fauth
Institute for Drive Systems and Power Electronics
Leibniz University Hannover
Welfengarten 1, 30167 Hanover, Germany
Leon.Fauth@ial.uni-hannover.de

Tobias Brinker
Institute for Drive Systems and Power Electronics
Leibniz University Hannover
Welfengarten 1, 30167 Hanover, Germany
Tobias.Brinker@ial.uni-hannover.de

Abstract—**Premagnetization can be used to reduce the size, weight and losses of inductors. Due to the non-symmetric LI-curve of these inductors, they are typically only used in unidirectional dc converters. This paper introduces and discusses three topologies with premagnetized inductors for dc-ac conversion, including measurement results for one topology.**

Index Terms—**Premagnetization, Inductor, Magnetics Design, Inverter, Converter**

I. INTRODUCTION

The premagnetization of dc inductors is a known possibility for the reduction of the size of inductors in various applications [1], [2], [12], [13]. The typical and most suitable applications are low frequency (below some kHz) and medium frequency (up to around 100 kHz) dc inductors with a saturation limit in a conventional design. While also thermally limited inductor designs can be optimized due to reduced dc-bias core losses, this paper focuses on the saturation limited designs. Fig.1 shows an example of the size reduction of a dc inductor for a boost converter for photovoltaic applications.

Generally, premagnetization of dc inductors is based on an appropriate hard magnetic material in the airgaps of a (soft) magnetic circuit. Combinations of soft and hard magnetic materials can be e.g. ferrite and polymer bonded neodymium, electrical steel and samarium-cobalt magnets and many others [1]–[3], [12], [13]. Attention should be taken with regards to eddy currents in the hard magnetic materials if they are not polymer bonded [1], [10], [12]. The premagnetization leads to a shift of the LI-curve, making it asymmetrical. Therefore, the effective saturation flux limit can nearly be doubled for unipolar applications. Obviously, the drawback of the premagnetization is a limited usability of the inductance for both current directions, examples are shown in Fig.2.

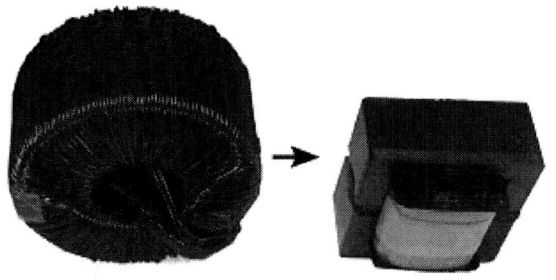

Fig. 1. Reduction of the size of a dc inductor for a boost converter in photovoltaic applications – shown for a reference design based on a ring core and an inductor based on a standard E70 core with premagnetization, revised and updated from [1]

This work was funded by the Ministry of Science and Culture of Lower Saxony and the Volkswagen Foundation

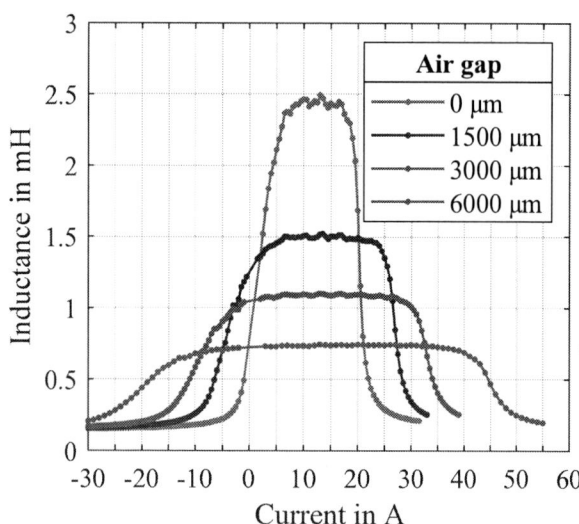

Fig. 2. LI-curves of a premagnetized inductor with E70-Ferrite core and the turn number N=70 and different additional air gaps for different inductance and saturation currents, revised and updated from [1]

978-1-7281-4181-7/19 $31.00 © 2019 IEEE

With the relationship between the saturation flux density B_{sat} of the soft magnetic material, the inductance L, the saturation current I_{sat}, the turn number N and the core cross sectional area A_C, the possible impact on the design can be seen in (1).

$$B_{sat} = \frac{L \cdot I_{sat}}{N \cdot A_C} \qquad (1)$$

Therefore, the premagnetization can be used to reduce the core cross sectional area A_C or the turn number N by the factor of two, double the inductance L or the saturation current I_{sat} or to realize any combination of these four possibilities for an optimization of the design.

The asymmetric LI-curve excludes the possibility of premagnetization in the optimization for inductors in typical voltage fed dc-ac and ac-dc converters. Nevertheless, at least three voltage feed dc-ac topologies allow for inductors with non-symmetric LI-curve, in particular also with a strong benefit for the utilization of the performance of newest wide-bandgap semiconductors. These topologies will be presented and discussed with the focus on the premagnetization of the inductors. One of the topologies will be assembled to verify its functionality in conjunction with an optimized inductor design based on the premagnetization of the magnetic core.

II. Topology Analysis

The three suggested topologies are shown in Fig. 3, Fig. 4 and Fig. 5. The first topology is known as a H4-Topology with unipolar driving scheme and a capacitive filter against the dc link and will be referred to as unipolar H4. The second topology was introduced in 2010 for photovoltaic systems and will be named split full bridge [3]. The third topology is known as a single phase Z-Source topology based on the general Z-Source approach for impedance-source power conversion and will be named single-phase Z-source [4], [5]. For the single-phase Z-source inverter only the inductors of the impedance-source network can be used for premagnetization while the inductor in series to the ac-load not. Therefore, the premagnetization is also applicable to the well-known Z-source inverter for three-phase conversion [5]. The semiconductors are either depicted as IGBT (slow switch) or FET (fast switch) for easy assignment of the typical corresponding frequency (low/high) needed for a cost and efficiency optimized operation.

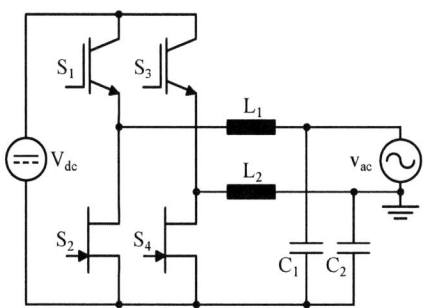

Fig. 3. First topology for dc-ac conversion with premagnetized inductors: unipolar H4

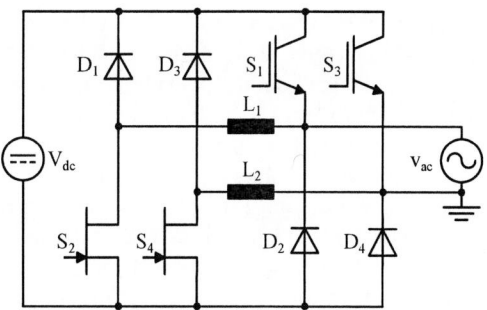

Fig. 4. Second topology for dc-ac conversion with premagnetized inductors: split full bridge

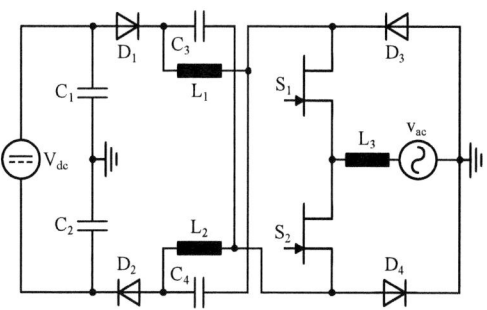

Fig. 5. Third topology for dc-ac conversion with premagnetized inductors: single phase Z-source

The operation of the unipolar H4 and the split full bridge corresponds to two buck-, respective boost-converters in parallel. At each time, just one of the half-bridges is switching at high frequency, while the other half-bridge switches based at the loads frequency and vice versa. This enables an arrangement with switches optimized for either low conduction losses (exemplary depicted as IGBT) or low switching losses (exemplary depicted as FET). It has to be taken care of suitable diode behavior of the low conduction loss switch as being part of the critical commutation cell. The operation of the single phase Z-source incorporates a capacitive voltage divider combined with a half bridge for buck-operation and ac-commutation and additionally a Z-Source configuration for a boosting of the input voltage to cover the peak of the ac voltage. Following this systematic of splitting up the functionality of the parts of a topology, one can see that it leads to the conclusion, that indirect current source inverter also have the intrinsic possibility of premagnetization, including unfolding bridge based topologies [6], [7].

The first two topologies come with the drawback, that only one of the inductors attenuates the current at a time. Therefore, they require double the inductance (in sum), compared to a standard unipolar switched topology and equal inductance, compared to a bipolar switched full bridge [3]. The difference of both topologies is the current path for loads frequency. While current at the loads frequency has to pass both inductors in the unipolar H4, it only passes one inductor at a time in the split full bridge topology while the current at switching frequency pass only one inductor at a time in both topologies. Therefore, the inductors can be designed with at least twice the specific core losses during switching operation of their corresponding half-bridge. As a direct consequence of this, the inductor design of both inductors will lead to a higher frequency of which the saturation flux will limit the design instead of the thermal limit [1], [8]. This makes them more advantageous for measures increasing the saturation flux limitation of the soft magnetic material of the inductors. Due to this, the unipolar H4 and the split full bridge will be investigated further with the focus on the premagnetization of the inductors. The advantages of the topologies are as follows:

- Both Topologies
  - Higher specific core losses realizable at same thermal conditions
  - Higher specific winding losses at switching frequency realizable at same thermal conditions
- Unipolar H4
  - Soft switching with triangular current mode possible
  - Usual full bridge design, simplifying the commutation cell design and choice of available semiconductor modules
  - Typical connection of the inductors for usual system layout of the full converter stage
- Split full bridge
  - No high side driver at high frequency capacitive node
  - Half rms current at loads frequency in each inductor, further increasing the possible specific core losses and the specific winding losses at switching frequency

III. SIMULATION RESULTS

Both topologies haven been build up with PLECS for a circuit simulation. Fig. 6 shows the current waveforms of the inductor and the output for unipolar H4 and Fig. 7 the current waveforms of the inductor and the output for split full bridge in dc-ac conversion, exemplary with 20 A effective current, 16 kHz switching Frequency, 250 μH inductance. Fig. 8 and Fig. 9 show the current waveforms of the inductor and the output for unipolar H4 and split full bridge respectively in ac-dc conversion. The simulation was done with a current control due to the difficulty of discontinuous conduction mode during the voltage zero crossings of the mains while working with variable input voltage and in boost mode of the respective driving scheme. This control leads to some noise on the currents though this has no effect on the general outcome regarding the current direction in each inductor. One ac period is shown in both figures. The required sum of the output filter capacitors is the same for both setups, due to better capacitor utilization of the split full bridge, the output current noise is less for this topology (not depicted in the figures due to simplicity. The placement of the capacitor is typically line-to-line for the split full bridge).

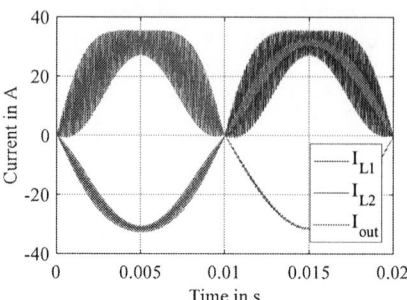

Fig. 6. Current waveforms for the unipolar H4 in dc-ac conversion

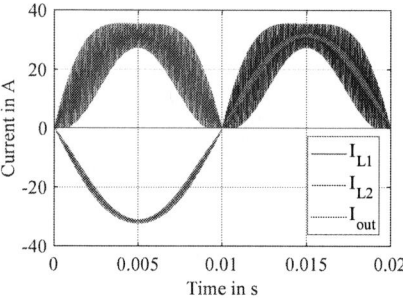

Fig. 7. Current Waveforms for the split full bridge topology in dc-ac conversion

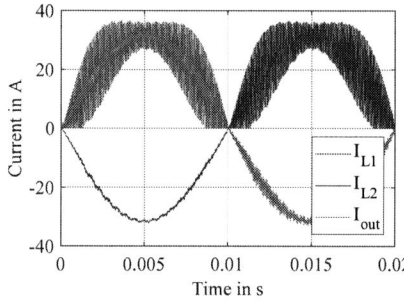

Fig. 8. Current waveforms for the unipolar H4 in ac-dc conversion

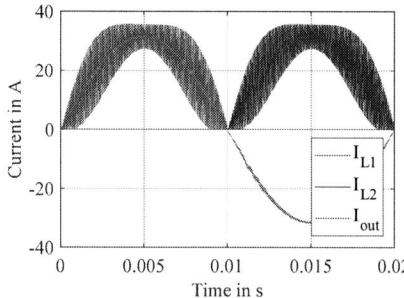

Fig. 9. Current Waveforms for the split full bridge topology in ac-dc conversion

Because of the focus on increasing the saturation flux limit, the value of the inductors is selected to realize a current at maximum power with the maximum of its envelope following a trapezoidal like form. Therefore, the peak current can be designed close to the saturation current over around 50% of the time in this operation. Any further reduction of the inductance would lead to significant increase of the peak current and also a high impact of triangular current mode like operation for the unipolar H4 topology or significant discontinuous current mode for the split full bridge topology.

Of particular importance is the current direction for both inductors in both topologies. In the unipolar H4, the current through each inductor is unidirectional for one half of the fundamental period. For the other half, the respective inductor is bypassed by its corresponding capacitor except for the low frequency component of the current. Therefore, an inductance is not required in this current direction. In fact, this has also a considerable impact on triangular current mode because of the reduced inductance for low and negative currents of a premagnetized inductor (also see Fig. 2) which has already been discussed for dc-dc topologies [9]. While typically this would be considered a negative effect due to the higher slew rate of the current it helps reaching the required negative current for soft switching for triangular current mode, reducing the required reactive power for soft switching.

Reactive power capability for both topologies is generally limited to applications where only a low capacitive dc-side with respect to earth is present [3]. This is because the change of the driving scheme at zero voltage respectively zero current crossings will lead to a change of the common mode voltage of the dc-side against earth. This is the case for applications, where additional galvanic isolation is used at the dc-link. Nevertheless, because both operation modes, dc-ac and ac-dc conversion, allow the use of premagnetized inductors, also reactive power capability is given.

IV. Measurement Setup

Doubling the effective saturation flux density based on premagnetization will allow for the four optimization possibilities shown in Fig. 10 and also mentioned in the introduction. Two examples are shown in Fig. 10. The reduction of the size can be done with a significant decrease of the core cross sectional area A_C to half. The other example is a hybrid optimization with regards to winding optimization with 25% less winding turns with still 50% more inductance. Due to comparability, the measurement setup has been built up with identical inductor and just twice the inductance due to different air gap length (respectively magnet thickness). As reference also an identical inductance, but with half saturation current for the non-premagnetized inductor has been built up. Fig. 11 shows the inductor in PQ35 shape with polymer bonded neodymium magnets (36 turns, magnet thickness of 2.5 mm, material: Neofer 41/100p, see [11]) inside of the airgaps of all three limbs. Fig. 12 shows the LI-curves of the inductor with airgap and with premagnetization, enabling twice the saturation current due to the premagnetization.

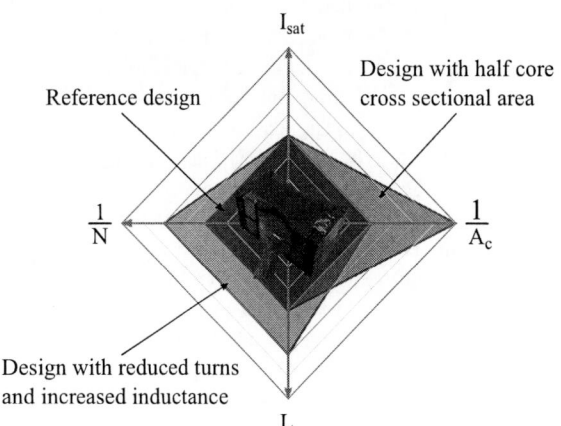

Fig. 10. Example of optimized designs under different optimization goals in comparison to a reference design. An example picture of a parallel premagnetized inductor in PQ32/20 shape is shown

Fig. 11. Premagnetized inductor with polymer bonded neodymium magnet in the airgaps

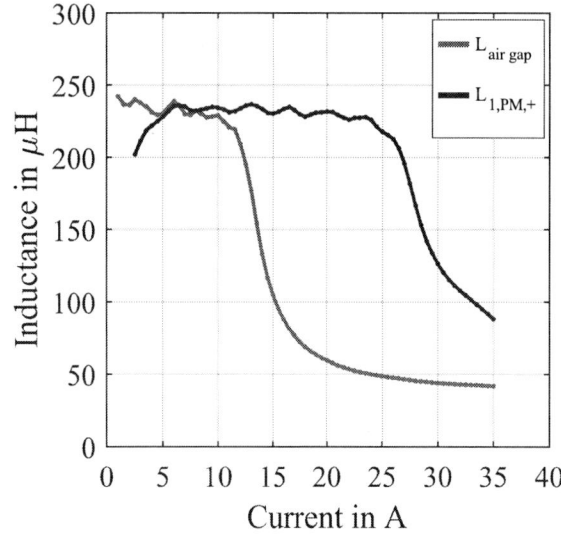

Fig. 12. LI-curve of the conventional reference inductor and the premagnetized inductor with same inductance but slightly more than twice the saturation current

It has to be mentioned that a premagnetization leading to saturation of the inductance for small currents up to 20% of the saturation current might be advantageous for inherent protection against lifetime and overtemperature issues [1]. The inverter stage is designed for switching frequencies in the range of 100 kHz to 400 kHz and a power rating of up to 600 W. The semiconductors are 650 V GaN-switches (GS66508B) with half-bridge drivers switching with 0 V and 6 V and about 100 ns dead time. One high side semiconductor is a fast 650 V SiC-MOSFET (C3M0065090J) for other research purposes [14] and can be neglected within the scope of this paper. The polymer-bonded rare earth NdFeB hard magnetic material Neofer 41/100p having a remanence of $B_r = 460$ mT is used for the premagnetization of the inductors based on premagnetization in all three limbs [11]. Fig. 13 shows the converter used for the measurements, including the inductors.

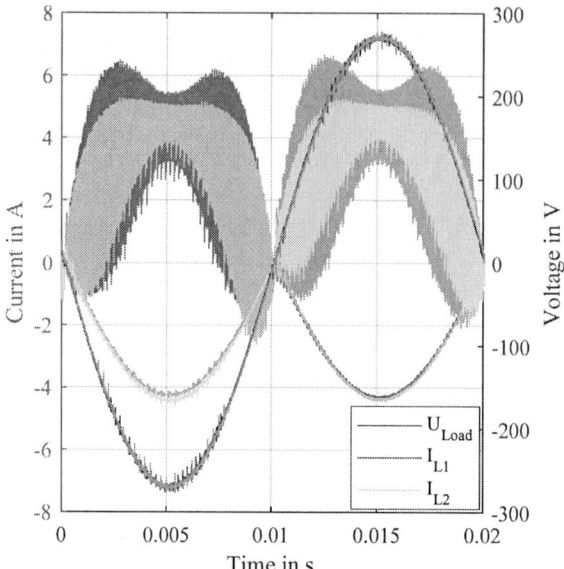

Fig. 14. Inductor current and output voltage waveforms for the unipolar H4 in dc-ac conversion, comparison of premagnetized (220 µH, lighter colors) and non premagnetized inductors (110 µH, darker colors), 100 kHz, 600 W

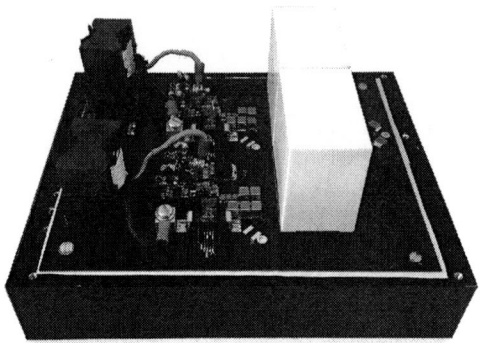

Fig. 13. Measurement Setup for the analysis of the unipolar H4, including dc-link and inductors

The setup was run at switching frequencies of 100 kHz, 200 kHz and 400 kHz and at output power from 150 W to 600 W. The upper power limit is based on the design of the converter, while the lower power limit comes from the available loads in laboratory. The half-bridges where operated in synchronous rectification in all tests to reduce reverse conduction losses, especially because of the use of GaN-switches. Fig. 14 and Fig. 15 show the inductor currents and the output voltage for the operation at 100 kHz and 400 kHz respectively. Due to high slew rates of approximately 50 V/ns the current and voltage measurements show some disturbances, especially when on-switching occurs. This is the case when the lower envelop of the inductor currents is positive. Two main results can be derived:

- As simulated, each inductor only needs to attenuate the current in one direction. Therefore this verifies the approach of the premagnetization in dc-ac operation.
- The reduced inductance value in negative current direction increases the current ripple. Because of the synchronous operation mode and the high ratio of switching frequency to mains frequency, this has only neglectable influence on the output voltage, as can be seen in both Figures.

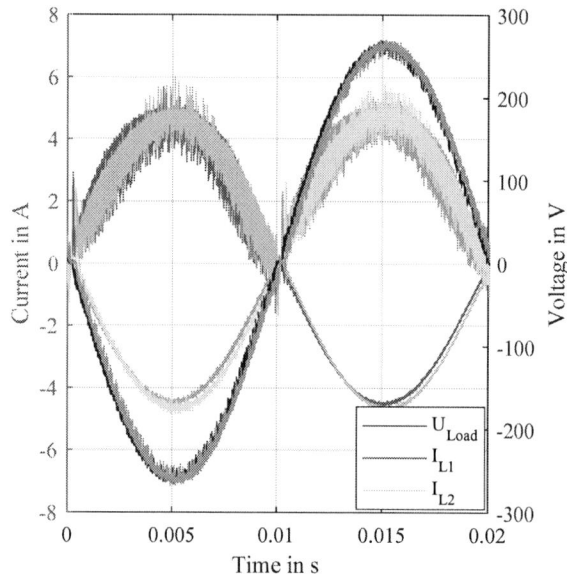

Fig. 15. Inductor current and output voltage waveforms for the unipolar H4 in dc-ac conversion, comparison of premagnetized (220 µH, lighter colors) and non premagnetized inductors (110 µH, darker colors), 400 kHz, 600 W

Efficiency measurements were done for the investigation of the influence of the premagnetization on the losses of the inverter. Fig. 16 shows the results of the measurements, including the before mentioned reference inductor with identical inductance but reduced saturation current still enabling the same full output power. It can be seen that the premagnetized inductor leads to around 0.2% increased efficiency for all switching frequencies measured, also verifying the benefit of the premagnetization.

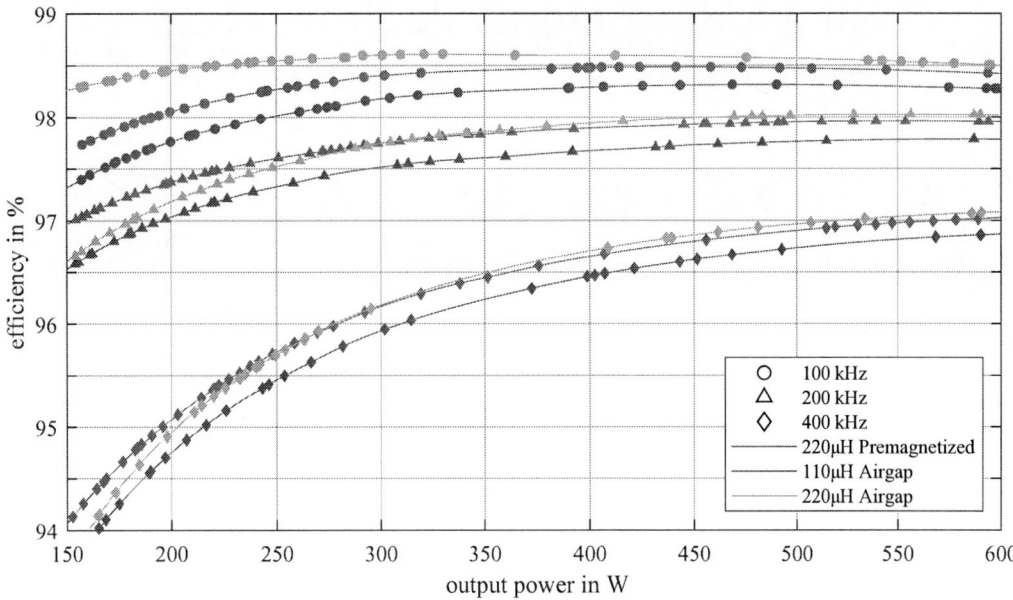

Fig. 16. Measured efficiency of the measurement setup with 300 V dc input and 200 V ac output voltage. The output was slightly reduced against standard 230 V ac mains due to the limited availability of dc voltage sources.

Nevertheless, the measurements with the reference inductor with the same inductance but half the saturation current (and therefore half the maximum usable range of the LI-curve) even show slightly higher efficiencies. The reason this reference is used here is to show the possible significant influence of any parameter of the inductor design. Especially in context of an overall converter optimization, also the impact of the inductor design on the switching performance has to be taken into account. Therefore, the approach of the premagnetization of the inductors in dc-ac or ac-dc converters could be validated in this paper, but its application should be analyzed in every application on its own with regards to the specific optimization goal.

V. Conclusion

In this paper, the possibilities of the premagnetization of inductors in dc-ac and ac-dc topologies are discussed. Three possible topologies are presented and one is selected and build up with a premagnetized inductor. Reactive power capability is taken care of in in the simulation while the measurement setup shows the impact and the utilization of wide-band-gap semiconductors. The impact on the efficiency of the utilization of premagnetized inductors is shown. The comparison is done against the same inductor, just with half the inductance. Future work should also indentify potentials in the different optimization of the inductors, e.g. different size, core shapes, winding material, turn number, etc.

References

[1] J. Friebe, "Permanentmagnetische Vormagnetisierung von Speicherdrosseln in Stromrichtern", Dissertation, University Kassel, Kassel, Germany, 2014.

[2] Y. Haijun, C. Wei and L. Zengyi, "Scheme and Design on High-frequency Power Inductor with Magnet Pre-bias", Proceedings of the CSEE, Vol.31, No.24, August 2011.

[3] S. V. Araujo, P. Zacharias, R. Mallwitz, "Highly Efficient Single-Phase Transformerless Inverters for Grid-Connected Photovoltaic Systems", IEEE Transactions on Industrial Electronics, Volume: 57, No. 9, September 2010

[4] P. Zacharias, L. M. Menezes, J. Friebe, "2 New Topologies for Transformerless Grid Connected PV-Systems with Minimum Switch Number", PCIM 2008, Nuremberg, Germany

[5] F. Z. Peng, "Z-Source Inverter", Conference Record of the 2002 IEEE Industry Applications Conference. 37th IAS Annual Meeting, 2002, Pittsburgh, USA

[6] B. Sahan, "Wechselrichtersysteme mit Stromzwischenkreis zur Netzanbindung von Photovoltaik-Generatoren", Dissertation, University Kassel, Kassel, Germany, 2010

[7] J. Friebe, O. Prior, "Method for Operating an Inverter with Reactive Power Capability Having a Polarity Reverser, and Inverter with Reactive Power Capability Having a Polarity Reverser", Patent, US9793812B2, 2017

[8] T. Kleeb, S. V. Araujo, P. Zacharias, "Size and performance optimization of filter inductors for highly efficient and compact power conversion circuits", 15th European Conference on Power Electronics and Applications (EPE), 2013

[9] J. Friebe, K. Rigbers, S. Lederer, "Bidirectional DC Converter Comprising a Core with Permanent Magnetization", Patent, US9755513B2, 2017

[10] S. Lin, J. Friebe, S. Langfermann, M. Owzqreck, "Premagnetization of High-Power Low-Frequency DC-Inductors in Power Electronic Applications", PCIM 2019, Nuremberg, Germany, 2019

[11] Magnetfabrick Bonn, "Datasheet: Neofer 41/100p", https://www.magnetfabrik.de/magnetfabrik_de/download_pdf.php?id=22, visited June 30th 2019

[12] R. Wrobel, N. McNeill, P. H. Mellor, , "Design of a high-temperature pre-biased line choke for power electronics applications", 2008 IEEE Power Electronics Specialists Conference, 2008

[13] T. Fujiwara, H. Matsumoto, "A new downsized large current choke coil with magnet bias method", The 25th International Telecommunications Energy Conference, INTELEC'03, 2003

[14] L. Fauth, T. Brinker, S. Lin, J. Friebe, "Asymmetric Half-Bridge Configurations in Power Electronics Converters", EPE'19 ECCE Europe, Genova, 2019

Three-Phase Induction Motors Efficiency Analysis Using a Programmable Power Supply

Cássio Alves de Oliveira
Laboratory of Electrical Drives (LAcE)
Federal University of Uberlândia (UFU)
Uberlândia-MG, Brazil
kass-07@hotmail.com

Josemar Alves dos Santos Junior
Nucleus of Research in Energy Systems (NUPSE)
Federal Institute of Goiás (IFG) – Campus Itumbiara
Itumbiara-GO, Brazil
josemarjr@gmail.com

Marcos José de Moraes Filho
Laboratory of Electrical Drives (LAcE)
Federal University of Uberlândia (UFU)
Uberlândia-MG, Brazil
marcos.jmf@hotmail.com

Vinícius Marcos Pinheiro
Laboratory of Electrical Drives (LAcE)
Federal University of Uberlândia (UFU)
Uberlândia-MG, Brazil
viniciusmarcospinheiro@hotmail.com

Augusto W. Fleury Veloso da Siveira
Laboratory of Electrical Drives (LAcE)
Federal University of Uberlândia (UFU)
Uberlândia-MG, Brazil
gutofleury@gmail.com

Luciano Coutinho Gomes
Laboratory of Electrical Drives (LAcE)
Federal University of Uberlândia (UFU)
Uberlândia-MG, Brazil
lcgomes@ufu.br

Abstract — **This paper presents a three-phase induction motors efficiency study based on technical standards. An experimental bench was developed using a high precision programmable power supply. Many experiments were performed applying different and predetermined loads. The machine's currents and voltages were measured to define the efficiency. Thus, an experimental setup was developed and the tests were performed with a high precision programmable power supply so that the acquired data is very reliable. The experimental setup was designed to perform tests at specifics operating points, allowing to apply variable loads.**

Keywords — *Three-phase induction motor, efficiency, technical standards, tests.*

I. INTRODUCTION

The electric motors are essential in industrial processes, since they move all kinds of machines and equipment. It is estimated that in the world there are more than 300 million motors, which consume about 7400 Terawatt-hours (TWh), equivalent to approximately 40% of the world's electric power production [1]. The three-phase induction motor is considered to be the main electromechanical converter, a true workhorse of the industry [2]. This type of machine is robust, cost effective, easy to maintain, high efficiency at rated operating point and adaptable to different load situations.

The majority of three-phase induction machines used in industrial processes have a cage rotor. This type of rotor has no commutator, brushes or slip rings, therefore inaccessible to the external environment. This configuration fosters the machine reliability and reduced sparking risks, enabling safe use in harsh environments such as areas containing potentially explosive atmospheres. The rotor works at high speed and withstands large mechanical and electrical overloads [3].

The performance of the three-phase induction motor is influenced by several factors such as undersized loads, incorrect shaft alignment, rewinding, harmonic distortion, among others [4]. Undersized loads cause a reduction in both, efficiency and power factor; incorrect shaft alignment, rewinding, and incorrect repairs take the motor to lower efficiency points [4]. Harmonic distortions can also influence directly the machine performance.

Rewound motors lose their original electrical and mechanical characteristics. These machines may have their efficiency compromised, have shortened service life and loss of original energy efficiency. Laboratory testing to verify electrical and mechanical behavior is recommended [5].

The growing demand for more economical processes, with rational and efficient use of electricity, has led to detailed studies to optimize industrial processes [6]. Industries have been striving to optimize production processes. Thus, the identification and correction of electrical and mechanical problems associated with induction machines are fundamental to increase systems reliability levels and increase operational capacity.

The use of efficient motors can increase profitability because lower efficiency leads to higher operating costs. Therefore, accurate and reliable machines testing is essential. To accurately measure motor efficiency, it is necessary to calculate losses and verify design parameters such as current, voltage, power etc. [7].

To contribute to the analysis of three-phase induction machines operation in situations close to those found in real applications, this work presents a platform that uses a programmable power supply to evaluate motor performance, allowing data acquisition and operation monitoring, aiming at the accomplishment of tests established in the norms of the Brazilian Association of Technical Standards (ABNT).

II. TECHNICAL STANDARDS

According to the international definition, a standard is a "document stated by consensus and approved by a recognized organization, which provides, for common use and repetitive, rules, guidelines or characteristics for activities or their results, aiming at obtaining an optimal degree of ordering in a given context". To this definition can be added the recommendation that "standards should be based on the consolidated results of science, technology and accumulated experience, aiming at optimizing benefits for the community".

978-1-7281-4181-7/19 $31.00 © 2019 IEEE

In other words, technical standards provide world-class specifications for products, services and systems for the purpose of ensuring quality, safety and efficiency. A standard is, in principle, voluntary use, but is almost always used because it represents the consensus on the state of the art of a given subject, obtained from experts of the interested parties.

The Brazilian Association of Technical Standards is responsible for the elaboration of the Brazilian Standards (ABNT NBR), prepared by its Brazilian Committees (ABNT/CB), Sectorial Standardization Bodies (ABNT/ONS) and Special Study Committees (ABNT/CEE).

The standard that prescribes applicable tests for the determination of performance characteristics of induction motors is NBR 17094, which is divided into four parts:

- Part 1: Three-phase induction motors – Requirements;

- Part 2: Single-phase induction motors – Requirements;

- Part 3: Three-phase induction motors – Test methods;

- Part 4: Single-phase induction motors – Testing methods.

Procedures and standards must be followed to ensure the proper functioning of the motors and the safety of the people involved in the operation, avoiding possible personal and / or material damage.

III. THE THREE-PASE INDUCTION MOTOR

A. Constructive aspects

The operation of electric motors is associated with the interaction of forces of electromagnetic origin between a fixed part, the stator of the machine, and a moving part, called of rotor.

The stator of the three-phase induction motor is structured using a laminated ferromagnetic material with grooves to store the winding. The blades are insulated to minimize the effects of the parasitic currents and the winding is three-phase, formed by three coils delimited, each other, from 120º.

The movable part of the induction motor can be of the wound rotor type or the squirrel cage type.

A wound rotor has a complete set of three-phase windings like to the stator windings. The terminations of the three phases of the windings are connected to sliding rings on the rotor shaft. The rotor windings are short-circuited by means of brushes that rest on the sliding rings. The currents circulating in the rotor can be accessed through the brushes and external resistors can be inserted into the circuit to modify the torque versus motor speed characteristic [8]. Wound rotor induction motors are only used in specific applications because they require more maintenance due to the abrasion associated with the brushes and the sliding rings, as well as being more expensive.

The squirrel cage rotor contains a series of conductive rods which are engaged within grooves in the surface of the rotor and short-circuited by conductive rings at both ends. It is the type of rotor most used in the manufacture of induction machines because it has remarkable advantages, such as simplicity and robustness in its construction.

B. Equivalent Circuit

The induction motor is called a single excitation machine, since the power is supplied only to the stator circuit. Thus, the equivalent circuit of the three-phase induction motor can be obtained by the transformers theory and introducing the necessary modifications to consider the fact that the rotor circuit operates at a frequency different from that of the stator as a function of the speed difference between the rotor and the magnetic field produced in the stator.

All parameters of the equivalent circuit are expressed per phase, regardless of the type of connection (star or delta), assuming that the winding is symmetric and the machine is fed with balanced three-phase voltages. The determination of the values of the equivalent circuit parameters can be done by a no-load test, a locked-rotor test and the measurements of the resistance of the stator windings. The single-phase equivalent circuit of the three-phase induction motor is shown in Fig. 1.

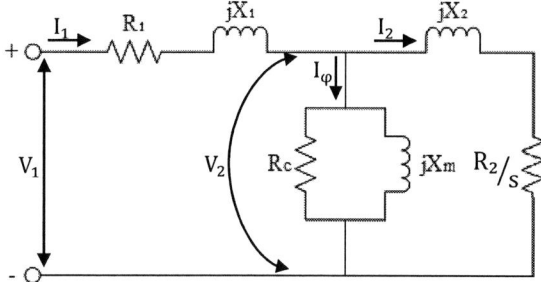

Fig. 1. Single-phase equivalent circuit of the three-phase induction motor.

The circuit parameters of Figure 1 are defined as follows:

- R_1: Stator resistance;

- X_1: Stator leakage reactance;

- R_c: Core-loss resistance;

- X_m: Magnetizing Reactance;

- R_2: Rotor resistance;

- X_2: Rotor leakage reactance;

- s: Slip.

C. Power flow and losses

The induction machine stator receives input power from the power grid, which will be converted to mechanical output power, which is the power supplied on the machine shaft. If all input power were transferred to the axis, the efficiency of the transformation would be 100%. However, in any transformation system, part of the energy is dissipated, which implies loss of power. The losses define the performance of the machine and significantly influence its operation.

The efficiency of a three-phase induction motor can be defined as the ratio of output power to input power. The output power is obtained by subtracting the input power from the internal losses. Consequently, if two of these three variables (output power, input power or losses) are known, the efficiency can be determined as follows:

$$\eta_\% = \frac{P_{input} - P_{losses}}{P_{input}} \cdot 100 \qquad (1)$$

$$\eta_\% = \frac{P_{output}}{P_{output} + P_{losses}} \cdot 100 \qquad (2)$$

In an induction motor, the losses are separated in fixed and variable, where the variable losses depending on the percentage of load on the machine shaft. The first losses obtained are losses in the stator windings. Still in the stator, a certain amount of power is lost in the hysteresis cycle and due to parasitic currents. The remaining power is transferred to the machine rotor through the air gap. After the transfer, part of the power is eliminated in the rotor conductors and the remainder is converted from the electric form to the mechanical form. Finally, mechanical and additional losses are subtracted and the remaining power is the output power of the motor. Figure 2 shows the power flow through the three-phase induction motor.

The losses in the core of the induction motor are partly from the stator circuit and partly from the rotor circuit. Since the induction motor normally operates at a speed close to the synchronous speed, the relative motion of the magnetic fields on the rotor surface is very slow and the losses in the rotor core are very small compared with the stator core losses [8].

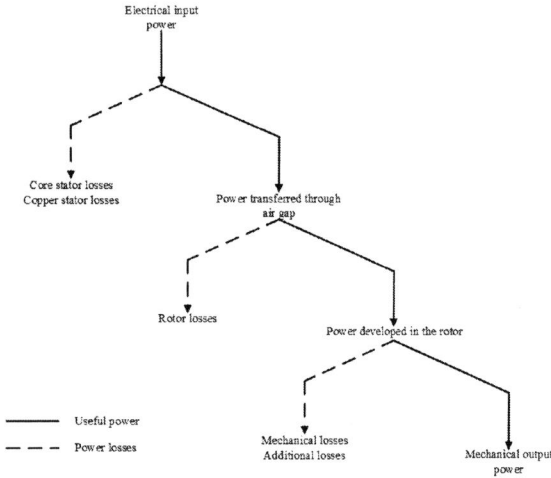

Fig. 2. Power flow diagram through the three-phase induction motor.

Nowadays, high-efficiency induction motors are produced, and various techniques are used to increase the efficiency of these motors compared to traditional motors. Placing more steel and copper in the stator construction, increasing the length of the stator and rotor cores and the caution to create an uniform air gap are some of the strategies that allow the reduction of losses and, consequently, increasing the efficiency.

IV. Methodology Used and Experimental Results

A. The test-bench

The experimental test-bench, shown in Fig. 3, was developed to make feasible the performance analysis of three-phase induction motors through several tests. In order to facilitate the handling of all the equipment necessary to carry out the work, the platform counts on the main devices of

protection against eventual failures that may occur during the execution of the tests.

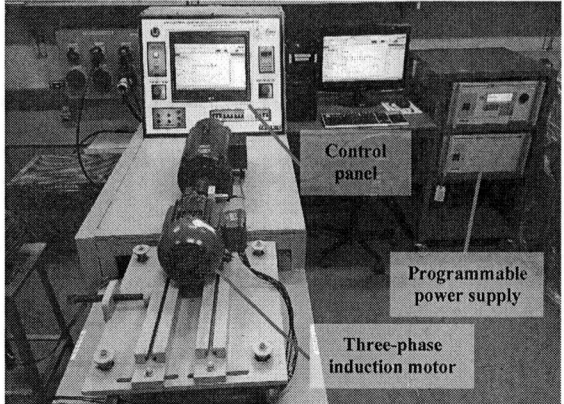

Fig. 3. The test-bench.

The platform has a programmable power supply, developed by AMETEK Programmable Power. It is a high efficiency equipment that provides accurate output with low distortion. The California Instruments CSW5550 model allows parallel connection with up to eight modules of the same model to increase output power. The device also has the remote interfaces GPIB, RS232 and USB.

The programming of the source can be done through the keypad of the front panel or through a specific software. In the latter case, the manufacturer provides two interface options for Windows operating system. Software programming offers the advantages of storing configurations for future use.

Data reading is shown on the front panel display of the programmable power supply and by software screen. In addition, the source also has options for storing the read data in text file with possibility of setting the time interval for updating this data, as the tests are performed. Figure 4 shows the programmable power supply software data reading screen.

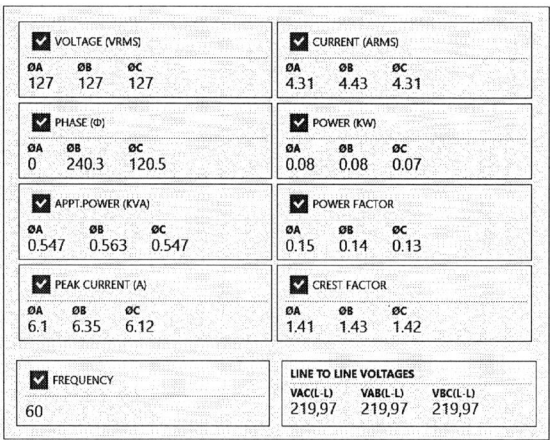

Fig. 4. Programmable power supply software data reading interface.

In the test-bench, the tested three-phase induction motor will be driven through the programmable power supply. The

978-1-7281-4181-7/19 $31.00 © 2019 IEEE

main characteristics of the three-phase induction motor under test are shown in Table I.

TABLE I. CHARACTERISTICS OF TRHEE-PHASE MACHINE UNDER TESTING

Three-phase induction motor - Cage rotor	
Manufacturer	WEG
Output power	2.2 (3.0) kW (hp)
Voltage (Δ/Y)	220 / 380 V
Current (Δ/Y)	8.39 / 4.86 A
Frequency	60 Hz
Speed	3450 rpm – 2 poles
Efficiency	81.9%
Power Factor	0.84
Class	N

In order to demonstrate the capabilities of the experimental test-bench, tests were performed to determine the efficiency of the three-phase induction motor, with it operating in conditions that may occur in industrial environments, and failure to monitor certain situations can cause severe damage to the systems.

B. Test for efficiency determination

According to NBR 17094-3:2018, the efficiency must be determined for nominal voltage and frequency, unless otherwise specified. The standard lists ten methods that can be used to perform the test, and for this work, the 5th method was chosen due to limitations in the experimental platform.

The 5th method consists of the measurement of the input power, and the output power is obtained by subtracting the total losses of the input power. The total losses represent the sum of the stator and rotor losses corrected to a specified temperature for the correction of resistance, core losses, mechanical losses and additional losses.

When using the 5th method, the additional loss is determined according to the Table II for nominal load. For another point different to the nominal load, it must be assumed that the additional loss is proportional to the rotor current squared:

$$P_{add} = P'_{add} \cdot \left(\frac{I_2}{I'_2}\right)^2 \qquad (3)$$

Where:

P_{add} – Value of the additional loss for a load point different to the nominal.

P'_{add} – Value of the additional loss corresponding to the value of the current I'_2.

I_2 – Rotor current appropriate to the point of loading for which additional loss is to be determined.

I'_2 – Rotor current value corresponding to the nominal load.

TABLE II. ESTABLISHED VALUES OF THE ADDITIONAL LOSS

Motor nominal power		Additional loss (percentage of nominal output power)
kW	hp	
0.75 – 90	1 – 125	1.8
91 – 375	126 – 500	1.5
376 – 1839	501 – 2499	1.2
1840 and over	2500 and over	0.9

The procedures for execution the efficiency test are as follows:

- Perform a no-load test;

- Perform a load test. In order to obtain the required data is necessary to couple the motor to a variable load system. For each of the load points, measure the input power, line current, applied voltage, speed, ambient temperature and stator winding resistance or temperature;

- Determine stator resistive losses (I^2R_s);

- Determine rotor resistive losses (I^2R_r);

- Determine core losses;

- Determine mechanical losses;

- Calculate the rotor current corresponding to each load point;

- Calculate the additional loss for each load point;

- Determine the efficiency for each load point.

No-load test, whose procedures and results are presented in [9], determines the core and mechanical losses, which are classified as fixed losses. Therefore, all other losses vary according to the load percentage.

The load test, also described in [9], should be performed for load points approximately equally spaced and in descending order as recommended by NBR 17094-3: 2018. The points were chosen in 140%, 125%, 100%, 75% and 50%. Table III shows the results of the load test.

TABLE III. VARIABLE LOAD TEST RESULTS

	Percentage load				
	140%	125%	100%	75%	50%
Input power (kW)	3.803	3.442	2.732	1.803	0.853
Line current (A)	11.356	10.530	8.738	6.613	4.908
Line voltage (V)	219.95	219.95	219.95	219.95	219.95
Speed (rpm)	3275	3340	3432	3508	3564
Ambient temperature (°C)	25	25	25	25	25
Stator resistance (Ω)	3.75	3.45	3.15	2.85	2.55

Stator resistive losses are calculated for each load point using the line current and stator resistance values shown in Table III.

To obtain rotor resistive losses is necessary to initially calculate the slip for each load point, using the velocity values of Table III in the equation:

$$s = \frac{n_s - n}{n_s} \qquad (4)$$

Where:

s – Slip.

n_s – Synchronous speed (rpm).

n – Mechanical speed (rpm).

Once the slip is calculated, the losses in the rotor are defined as follows:

$$P_{rotor} = \left(P_{input} - P_{stator} - P_{core}\right) \cdot s \qquad (5)$$

Where:

P_{rotor} – Rotor resistive loss ($I^2 R_r$).

P_{stator} – Stator resistive loss ($I^2 R_s$).

P_{core} – Core loss.

In order to calculate the additional loss, it is necessary first to know the value of the rotor current for each load point, which can be determined by equation:

$$I_2 = \sqrt{I^2 - I_0^2} \qquad (6)$$

Where:

I_2 – Rotor current.

I – Suitable stator line current for which additional loss is being determined.

I_0 – No-load stator current.

Using Table II is possible to find the value of the additional loss at the nominal load. For all other load points, simply apply (6) and (3).

The values of all internal losses of the motor tested were calculated for each loading point and are shown in Table IV.

TABLE IV.   INTERNAL LOSSES FOR EACH LOAD POINT

	Percentage load				
	140%	125%	100%	75%	50%
Stator resistive loss (W)	483.595	382.539	240.511	124.636	61.426
Core loss (W)	95.196	95.196	95.196	95.196	95.196
Rotor resistive loss (W)	291.074	214.086	111.827	40.459	6.964
Mechanical loss (W)	90	90	90	90	90
Additional loss (W)	84.654	70.749	44.186	19.09	3.975
Total (W)	1044.519	852.57	581.72	369.381	257.561

With the values of the losses available, one can use (1) to calculate the efficiency. The result is shown in Table V.

TABLE V.   DETERMINATION OF PERCENTAGE EFFICIENCY

	Percentage load				
	140%	125%	100%	75%	50%
Input power (W)	3803	3442	2732	1803	853
Total losses (W)	1044.519	852.57	581.72	369.381	257.561
Output power (W)	2758.481	2589.43	2150.28	1433.619	595.439
Efficiency ($\eta_\%$)	75.53%	75.23%	78.70%	79.51%	69.80%

Analyzing the results of Tables IV and V, it is possible to verify the influence of the stator and rotor losses and the additional losses in the efficiency value.

C. Efficiency evaluation with harmonic distortion

The characterization of the presence of harmonics can be done through individual or total treatment. The total harmonic distortion represents the joint action of all the harmonic frequencies present in the voltage and / or current signals, expressed by a quadratic composition of the individual distortions:

$$THD_V = \frac{\sqrt{\sum_{h=2}^{hmax} V_h^2}}{V_f} \cdot 100 \qquad (7)$$

Where:

THD_V – Total harmonic distortion of voltage.

V_h – Individual harmonic voltage of h order.

V_f – Voltage at fundamental frequency.

In three-phase induction motors, the presence of harmonic content in the supply voltage causes the copper and iron losses to rise. The increase of the losses in the iron is a consequence of higher levels of parasitic currents in the rotor and stator sheet and the additional heat generation in the iron of the machine [10]. Copper losses are also considerable because of variations in winding resistance and increase in total root mean square (rms) current value. The increase in losses is reflected in the decrease in the efficiency and the useful life of the machine.

The tests performed in the experimental test-bench with insertion of voltage harmonics had as main objective the experimental verification of the total losses in the three-phase induction motor. The programmable power supply has been set to enter a 10% of THD and the motor has been driven with nominal load. Table VI shows the effective values of the harmonics inserted in the voltage waveform and Table VII shows the measurements made during the test.

TABLE VI.   ROOT MEAN SQUARE (RMS) VALUE OF VOLTAGE HARMONICS

Harmonic order	Percentage	RMS value
Fundamental	100%	220 V
5	8%	17.6 V
7	5%	11 V
11	3%	6.6 V
13	1%	2.2 V
17	1%	2.2 V

TABLE VII.   MEASUREMENTS DURING THE TEST WITH HARMONIC DISTORTION OF VOLTAGE

	Phase A	Phase B	Phase C
Voltage (V)	220	220	220
Current (A)	8.004	8.211	8.318
Input power (kW)	0.804	0.815	0.844
Power factor	0.79	0.78	0.80
THD_V(%)	9.91	9.91	9.91
THD_I(%)	15.62	15.39	15.23

For this test, the efficiency was calculated according to [9] and [11] and the value found was 59.62%. Compared with the value for nominal load (Table I), it is possible to verify that there was a reduction of more than 25%, which proves that the

978-1-7281-4181-7/19 $31.00 © 2019 IEEE

presence of harmonics significantly increases the internal losses.

Figure 5a shows the voltage waveform and Fig. 5b shows the current waveform of each phase of the motor for this test. The images were obtained by a digital oscilloscope.

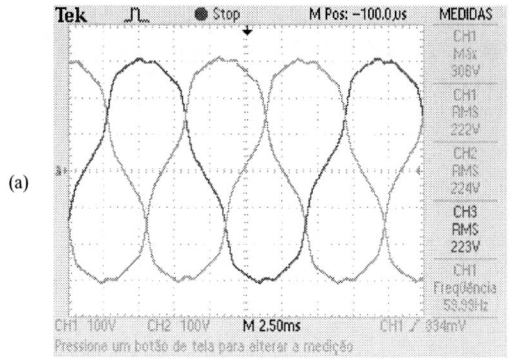

(a)

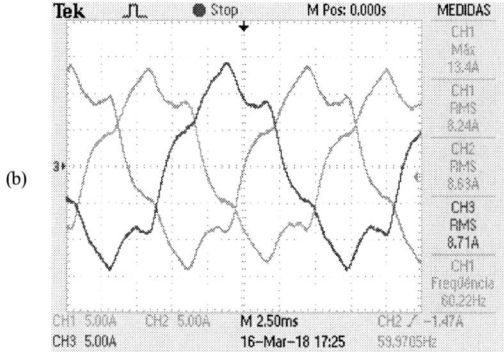

(b)

Fig. 5. Test with harmonic distortion: (a) Voltage; (b) Current.

V. CONCLUSION

Studying the efficiency of three-phase induction motors at various load points allows evaluate internal machine losses as well as to investigate some operating situations that occur in industrial environments, such as harmonic distortions.

Compliance with the recommendations contained in the technical standards of the electrical machinery segment guarantees quality, safety and efficacy for the various systems. The consensus on the use of standards ensures the proper functioning of the motors and the safety of people and equipment, seeking to minimize or avoid possible faults.

The developed test-bench can be used to test new or rewound motors and even those that have been used for some time in industrial plants. Ensuring efficiency significantly reduces machine losses and contributes to minimizing

operating costs. A key factor in cost effectiveness is the correct selection of the electric motor for a given application.

ACKNOWLEDGMENTS

The authors thanks to the Conselho Nacional de Desenvolvimento Científico e Tecnológico (CNPq) and the Fundação de Amparo à Pesquisa do Estado de Minas Gerais (FAPEMIG) for their financial support, the Coordenação de Aperfeiçoamento de Pessoal de Nível Superior (Capes) for the scholarship, and the Electrical Drive Laboratory of the Universidade Federal de Uberlândia for the permanent incentive to research.

REFERENCES

[1] Cartilha WEG – Gestão Eficiente da Energia Elétrica: Motores Elétricos, Inversores de Frequência e Geração Solar [Online]. Disponível em: http://ecatalog.weg.net/files/wegnet/WEG-cartilha-weg-uso-eficiente-da-energia-eletrica-50030292-catalogo-portugues-br.pdf.

[2] W. F. Godoy, I. N. Silva, A. Goedtel, R. H. C. Palácios and T. D. Lopes, Application of intelligent tools to detect and classify broken rotor bars in three-phase induction motors fed by an inverter, IET Electric Power Applications, vol. 10, no. 5, pp. 430-439, June 2016.

[3] R. Bulgarelli, Proteção Térmica de Motores de Indução Trifásicos Industriais, Dissertação de Mestrado, Escola Politécnica da Universidade de São Paulo, 2006.

[4] V. P. Silva, Análise Comparativa do Desempenho do Motor de Indução Trifásico de Alto Rendimento e Linha Padrão em Condições de Alimentação Ideal e não Ideal, Dissertação de Mestrado, Universidade Federal de Uberlândia, 2012.

[5] P. H. O. Rezende e D. Bispo, "Análise Econômica em Motores de Indução Trifásicos", Revista Horizonte Científico, vol. 9, no. 1, Maio 2015.

[6] T. Izhar, M. Ali and A. Nazir, Development of a Motor Test Bench to Measure Electrical/Mechanical Parameters. 2017 International Conference on Energy Conservation and Efficiency, pp. 22-23, November 2017.

[7] C. A. Oliveira, Estudo do Desempenho do Motor de Indução Trifásico Acionado a Velocidade Variável com Utilização de Técnicas Digitais, Trabalho de Conclusão de Curso, Universidade Federal de Uberlândia, 2015.

[8] S. J. Chapman, Fundamentos de Máquinas Elétricas, 5 ed., Porto Alegre, AMGH Editora Ltda., 2013.

[9] C. A. Oliveira, Plataforma para Ensaios de Motores de Indução Trifásicos e Simulação de Cargas Mecânicas: Acionamento, Operação e Monitoramento com Auxílio de Fonte Programável, Dissertação de Mestrado, Universidade Federal de Uberlândia, 2018.

[10] A moderna eficientização energética e seus possíveis efeitos sobre o desempenho operacional de equipamentos e instalações elétricas: Distorções harmônicas – Uma revisão de conceitos gerais [Online]. Disponível em: http://www.engeparc.com.br/cariboost_files/4-Harmonicas.pdf.

[11] C. A. Oliveira et al., Plataforma para Ensaios com Motores de Indução Trifásicos: Operação e Monitoramento com Auxílio de Fonte Programável, 2018 13th IEEE International Conference on Industry Applications (INDUSCON), 2018.

[12] Associação Brasileira de Normas Técnicas, Máquinas Elétricas Girantes – Parte 3: Motores de Indução Trifásicos – Métodos de Ensaio, 2018.

Reactive Power Control of Distributed Photovoltaic Generation System in Low Voltage Electrical Grids

Vanessa da Costa Marques
Departamento de Engenharia Elétrica
Universidade Federal da Paraíba
João Pessoa, Brazil
vanessa.marques@cear.ufpb.br

Rogério Gaspar de Almeida
Departamento de Engenharia Elétrica
Universidade Federal da Paraíba
João Pessoa, Brazil
rogerio@cear.ufpb.br

Nady Rocha
Departamento de Engenharia Elétrica
Universidade Federal da Paraíba
João Pessoa, Brazil
nadyrocha@cear.ufpb.br

Darlan Alexandria Fernandes
Departamento de Engenharia Elétrica
Universidade Federal da Paraíba
João Pessoa, Brazil
darlan@cear.ufpb.br

Gleice Mylena da Silva Rodrigues
Departamento de Engenharia Elétrica
Universidade Federal da Paraíba
João Pessoa, Brazil
gleice.rodrigues@cear.ufpb.br

Abstract—**This work analyzes the impact of high levels of penetration of the distributed photovoltaic generation on the low voltage electrical grids ($127V$ / $220V$ / $380V$) of the electric power distribution system. It also presents a proposal of control based on $\alpha\beta$-dq coordinates for distributed generation with solar panels, which aims to inject reactive power to adjust the voltage at the connection point of the generation system with the electric grid during imbalances between generation and consumption. The simulations were performed using the Matlab/Simulink® software.**

Index Terms—**Photovoltaic Panel, Overvoltage, Reactive Power Control, Distributed Generation**

I. INTRODUCTION

Among the known technologies in distributed generation, the photovoltaic source has presented accelerated growth compared to the other sources, such as biogas, wind and thermal [1]. The implementation of distributed grid-connected photovoltaic systems in Brazil has been motivated by ANEEL 687/2015 resolution that encourages the use of distributed generation [2]. The accelerated growth of photovoltaic panels connected to low voltage systems ($127V$ / $220V$ / $380V$) may cause problems related to the variation of the network voltage levels due to the high injected energy in the distribution networks. The energy produced is injected at the point of common connection - (PCC) between the panel, the load and electrical grid.

One of the main concerns about network-connected photovoltaic systems is voltage fluctuations, which can lead to voltage limit violations established by United Distribution Norm 013 from the power distributor company Energisa/PB [3] in Brazil, the norm provides that if the mains grid voltage measured in rms at the common connection point is 10% higher than the rated voltage, photovoltaic generation must be

The author Vanessa da Costa Marques thanks support from the Brazilian agency FAPESQ-PB grant number 005/19.

disconnected from the distribution system with a maximum time of 0.2s.

In [4], it was presented that, without voltage control, the distribution network had an increase in its voltage around noon. In this period, one observes a high irradiance and photovoltaic generator generates more energy as shown in the Fig. 1.

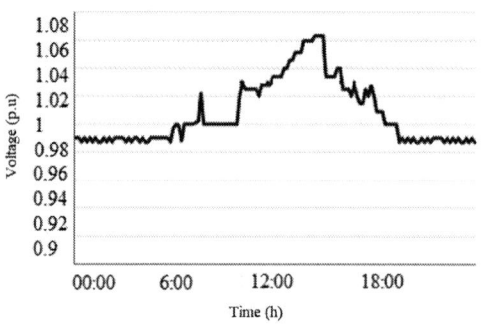

Fig. 1. Voltage profiles per days without voltage control [4].

According to [4], controlling the voltage of the low voltage distribution system is a critical issue in the presence of a lot of photovoltaic generation because the X/R ratio is low, so that the resistive effect can not be neglected. In order to reduce the voltage the production of photovoltaic energy must be reduced or the surplus energy must be used. Knowing the evolution projections of the solar distributed generation, it is of fundamental importance to study solutions to control and minimize the voltage variations of the electrical grid to take advantage of the maximum power that this source can generate without having to disconnect the inverters according United Distribution Norm 013.

This work aims to define a control strategy to act on the inverter of the photovoltaic system to minimize the overvoltage in the electrical network.

978-1-7281-4181-7/19 $31.00 © 2019 IEEE

II. DISTRIBUTED PHOTOVOLTAIC GENERATION

According to [5], distributed generation is defined as small power plants connected directly to the distribution network. There are many forms of connection for photovoltaic systems with the low voltage electrical network. In the case of this paper the main components of a photovoltaic system connected to the single phase grid are shown in Fig. 2. The photovoltaic panel is connected to the grid via a DC-DC converter and a DC-AC PWM inverter. The boost converter is controlled to track the maximum power point tracking (MPPT) of the photovoltaic generator. The inverter contains mandatory function that is the anti-islanding system that serve to guarantee the safety of operators and users of the electrical system. Then, the photovoltaic generation systems must be turned off when the power grid is turned off.

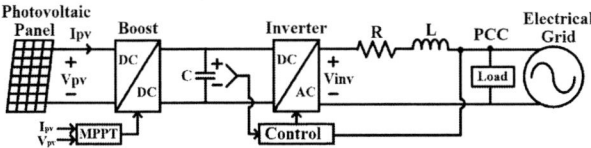

Fig. 2. Photovoltaic System – Single Phase Grid.

A. Impact of the photovoltaic generation distributed in electrical grid

The demand of consumers connected to the electrical grid is very dynamic and there are periods of the day in which the photovoltaic power curve is greater than the load demand curve. In [6] it was presented the load demand curve over a day and the photovoltaic power curve as shown in the Fig. 3. In situations such as this, overvoltage may occur. The voltage rise

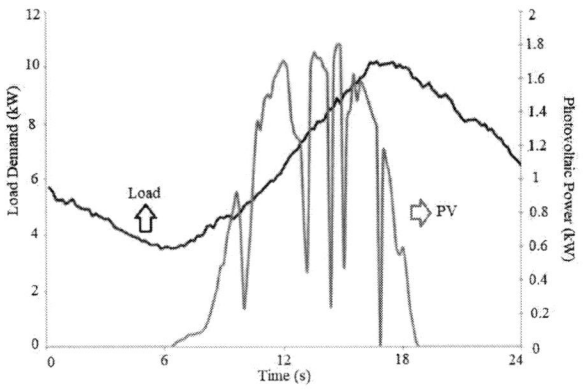

Fig. 3. Load Demand and Photovotaic Power in a over a day [6].

in the low voltage distribution system has become one of the most negative effects due to the large number of photovoltaic panels connected in the distribution system.

To solve the overvoltage problem this article suggests the use of control based in dq componentes. The control used in this work is based on [7] which allows the decoupling of the active and reactive power injection in the electric network through the decoupling of the current components of the inverter.

B. Control

1) α-β and dq Control Loops: The control based in dq components consists of the independent control of the active and reactive power produced by the photovoltaic panel. The independent control of the active and reactive power of the inverter occurs by decoupling the inverter current components at coordinates dq. Thus, it is not necessary to disconnect the PV system in case of overvoltage, as the control will contribute through the injection of reactive power for the reduction of the voltage of the electric network. Knowing that the demand of consumers connected to the grid is very dynamic through the d component of the inverter current, the control acts in a manner that the voltage in the DC bus do not change with the charge variation seen by the inverter.

For the definition of the reactive control strategy in the photovoltaic system, one may use of the equivalent circuit of Fig. 2 shown in the Fig. 4.

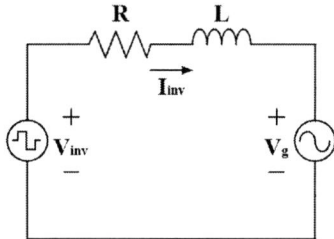

Fig. 4. Equivalent Circuit.

Based on the Fig. 4, the model of system is defined by (1):

$$V_{inv} = RI_{inv} + L\frac{dI_{inv}}{dt} + V_g, \tag{1}$$

which V_{inv} is the inverter voltage, I_{inv} is the inverter current, R and L are the resistance and inductance of the photovoltaic system and V_g is the grid voltage.

Equation (1) can be represented in coordinates α-β as shown in (2). The component β is 90° shifted of component α.

$$\begin{bmatrix} V_{\alpha inv} \\ V_{\beta inv} \end{bmatrix} = \begin{bmatrix} R & 0 \\ 0 & R \end{bmatrix}\begin{bmatrix} I_{\alpha inv} \\ I_{\beta inv} \end{bmatrix} + \begin{bmatrix} L & 0 \\ 0 & L \end{bmatrix}\frac{d}{dt}\begin{bmatrix} I_{\alpha inv} \\ I_{\beta inv} \end{bmatrix} + \begin{bmatrix} V_{\alpha g} \\ V_{\beta g} \end{bmatrix}, \tag{2}$$

where $V_{\alpha inv}$ and $V_{\beta inv}$ are the α-β components of the inverter voltage, $I_{\alpha inv}$ and $I_{\beta inv}$ are the α-β components of the inverter current, $V_{\alpha g}$ and $V_{\beta g}$ are the α-β components of the grid voltage.

Equation (2) is transformed to the axis of the coordinates dq through the relations shown in (3), which come from the Fig. 5.

$$\begin{bmatrix} d \\ q \end{bmatrix} = [T_{dq}]\begin{bmatrix} \alpha \\ \beta \end{bmatrix} \quad \text{and} \quad \begin{bmatrix} \alpha \\ \beta \end{bmatrix} = [T_{dq}^{-1}]\begin{bmatrix} d \\ q \end{bmatrix}, \tag{3}$$

where $[T_{dq}]$ and $[T_{dq}]^{-1}$ are represented respectively by (4) and (5).

$$[T_{dq}] = \begin{bmatrix} \sin\theta & -\cos\theta \\ \cos\theta & \sin\theta \end{bmatrix} \quad (4)$$

$$[T_{dq}]^{-1} = \begin{bmatrix} \sin\theta & \cos\theta \\ -\cos\theta & \sin\theta \end{bmatrix} \quad (5)$$

Designing the components $\alpha\beta$ in the dq axis through the (3), assuming that V_{inv} and V_g are coincident with the d axis and applying the lapace transform, (6) and (7) are obtained.

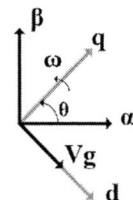

Fig. 5. Transformation axis.

$$V_d = (R + sL)I_d - \omega L I_q + V_g \quad (6)$$
$$V_q = (R + sL)I_q + \omega L I_d \quad (7)$$

Considering $(R + sL)I_d = V'_d$ and $(R + sL)I_q = V'_q$ are obtained the (8) and (9) that describe mathematically the control model used.

$$V_d = V'_d - \omega L I_q + V_g \quad (8)$$
$$V_q = V'_q + \omega L I_d \quad (9)$$

Where V_d represents the inverter voltage in the d-axis, V_q represents the inverter voltage in the q-axis, L and R are the resistance and inductance of the photovoltaic system, I_d represents the inverter current in the d-axis, I_q represents the inverter current in the q-axis, ω represents the frequency and V_g represents the grid voltage that is in phase and has the same angular frequency of the d-axis.

In the implementation of the α and β components, a second order generalized integrator (SOGI) for quadrature-signals generation was used [8]. The SOGI-QSG diagram is shown in the Fig. 6, where ω' and K set resonance frequency and damping factor of the SOGI respectively, with $\omega' = 2\pi 60$ and $K = \sqrt{2}$.

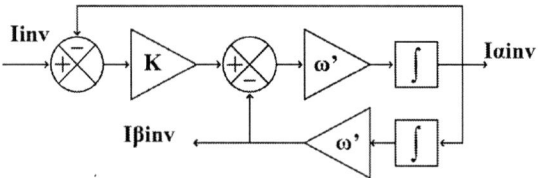

Fig. 6. SOGI-QSG Scheme.

The diagram of the control based in dq components is shown in Fig. 7. This system also has a Phase-Locked Loop (PLL) to capture the angle of the electrical grid and synchronize the PWM inverter with the power grid, whose structure is shown in [8] and proportional-integral controllers (PI).

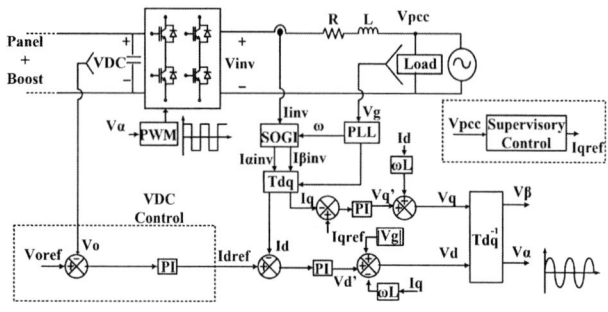

Fig. 7. Diagram of the control based in dq components

Currently in Brazil, there are no standards that deal with the reduction of overvoltage by applying photovoltaic systems. In case of overvoltage, the photovoltaic generation is disconnected from the distribution network [3]. Hence, it will be used the grid code of wind generators used by European countries as reference.

2) Supervisory Control: The increasing number of wind generators in European countries required the distribution companies to update the network procedures called grid codes, with the purpose of ensuring the optimal operation of the electrical grid with the appropriate voltage levels. In order to keep the aerogenerators connected in the electrical grid in situations in which the voltage at the PCC is less than $0.8pu$ and to make the wind turbines survive voltage sags, specific standards were created by the distribution companies and by manufacturers of generators such as the German company E.ON [9]–[11].

Similarly as it is proposed in this article for photovoltaic solar energy, the wind generators used of Europe must have the capacity to generate or absorb reactive energy, in order to control the voltage level in the PCC by increasing or decreasing the voltage in the electrical grid. The wind generators are designed to regulate the profile of the terminal voltage of the wind farm through the injection of reactive power. Usually, they inject approximately 90% of the inverter capacity [12], [13]. The Table I shows the control strategy for the injection of reactive current. Depending whether there is overvoltage or undervoltage in the electrical grid, the generator will inject reactive current of the inductive or capacitive type, respectively. The reference of reactive current for this article is based on the grid codes requirements for wind power integration in Europe.

III. RESULTS AND DISCUSSIONS

To evaluate the impacts of distributed generation with photovoltaic panels connected in only a single phase of the

TABLE I
SUPERVISORY CONTROL STRATEGY

i_{qref}	Conditions
$+i_{qref}$ A	$V_{PCC} > 1.05V_{nom}$
0 A	$0.95V_{nom} \leq V_{PCC} \leq 1.05V_{nom}$
$-i_{qref}$ A	$V_{PCC} < 0.95V_{nom}$

TABLE III
PARAMETERS AWG/CA 380/220 V

Resistance in 55°C	$0.975\ ohm/km$
Line Length	35 m
Frequency	60 Hz
X/R	100

TABLE IV
DISTRIBUTION OF LOADS PER PHASE

Phase	Loads
A	$2x7.5kVA + 5kVA + 6kVA + 8kVA$
B	$7.5kVA + 12.5kVA + 6kVA + 8kVA$
C	$7.5kVA + 12.5kVA + 6kVA + 8kVA$

distribution grid, it was created a scenario composed by the 13.8kV grid, a 13.8kV / 380V transformer of 150kVA that feeds single-phase loads connected to the three-phase grid system and distribution lines with three AWG / AC cables. The Fig. 8 represents the system simulated in Matlab/simulink, adapted from [14].

In the proposed project it was used a photovoltaic system consisting of a set of 25 panels, with 5 panels in series and 5 panels in parallel (see its parameters in Table II), which is able to inject approximately 38.6 A RMS, operating at a temperature of 35 °C. The gains of the current controllers used were $K_p = 48.36$ and $K_i = 10.14$. The controller gains used for the voltage control loop were $K_p = 0.414$ and $K_i = 3.313$.

TABLE II
PARAMETERS OF PHOTOVOLTAIC SYSTEM

Components	Capacity
Photovoltaic Array	8 kW
Inverter	9 kVA
Capacitor	2400 uF

To implement simulation scenario, we used the Matlab/Simulink® software. The parameters of the electrical grid (Bus 1), which is considered an infinite bus for the distribution network, are Vrms = 13.8 kV, ratio X/R = 1000 and 60 Hz. Table III shows the parameters of the distribution network formed by $(AWG\,/\,AC)$ cables $(380V\,/\,220V)$. In the distribution grid, each phase contains 34 kVA (see them in Table IV). All of them are equally distributed with power factor of 0.92.

From the system developed in Fig. 8, to evaluate the impact of the distributed photovoltaic distributed generation in the electrical grid of low voltage, three cases were defined:

- Case 1: Distribution network without photovoltaic distributed generation (PDG).
- Case 2: Distribution network with PDG distributed in Phase A, however with decreasing of charges and without injection of reactive current (photovoltaic system without power control).
- Case 3: Distribution network with PDG distributed in Phase A, with decreasing of charges and injection of reactive current (photovoltaic system with power control) only through the photovoltaic array of Bus 3. Only the photovoltaic system conected between Bus 2 and Bus 3 contains the control based in dq components.

For the studies presented in this paper, it is considered that the distribution network presents voltage drop and that the transformer has reached its voltage regulation limit, because

it is intended to demonstrate the ability of distributed photovoltaic systems to assist in the system voltage regulation.

For the case 1, all keys of the loads are closed $(L1, L2, ..., L5)$ and all keys of the photovoltaic panels are open $(S1, S2, ..., S5)$. The peak voltage of the nearest and and farthest bus from the transformer, number 2 and 6 respectively, are measured. The voltage of the three phases of the distribution network are shown in Fig. 9.

For the case 2, there is no control based in dq components and all the keys of the photovoltaic panels remain open, except $S1$. First one consider that all the keys of loads $(L1, L2, ..., L5)$ are closed totalizing 34 kVA in each phase. From the 0.33 seconds of simulation, load variation occurs. So, the switches $L3$, $L4$ and $L5$ are opened and lead to a decrease of 19 kVA in phase A. In the 0.5 second of the simulation, two loads are removed from phase A by opening the switches $L1$ and $L2$, totalizing a decrease of 15 kVA.

It is interesting to measure the peak voltages of phase A of the bus 2, which is the first one, and also of the bus 6 which is the furthest one from the transformer, considered a critical bus. The peak voltages of the phase A of the distribution network are shown in Fig. 10. We can conclude that, if the load decreases and the generation remains constant, without control, overvoltage may occur; in the farthest bus, (bus 6), it will attain its highest value.

The Fig. 11 shows the peak voltages of phases A, B and C of case 2. According to the simulation, phase A, which contains distributed generation presents a peak voltage of approximately 310 V, while the other phases engender around 303 V. Note that only the fact of the phase A possesses distributed generation makes it present a better voltage regulation when compared to the phases B and C and to case 1.

For case 3, initially the keys of loads are closed $(L1, L2, ..., L5)$, totalizing 34 kVA in each phase. From the 0.33 seconds of simulation, load variation occurs and the switches $L3, L4$ and $L5$ are open and then decrease 19 kVA of phase A. In the 0.5 second of the simulation, two loads are removed from phase A by opening the switches $L1$ and $L2$, totalizing a decrease of 15 kVA. In this case, the inverter of the photovoltaic panel connected to bus 3 presents control based in dq components, and only it can inject or absorb

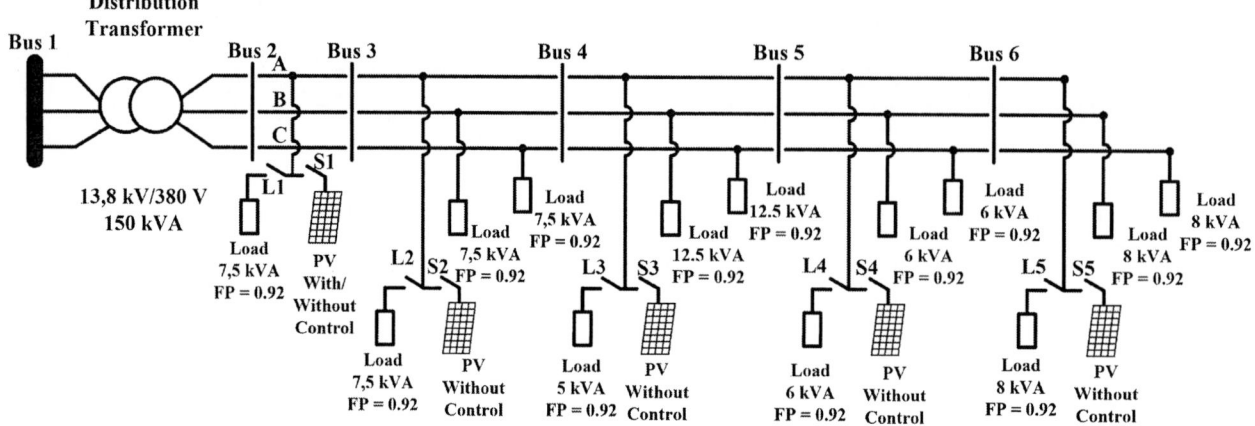

Fig. 8. Scenario of Simulation with Distributed Generation.

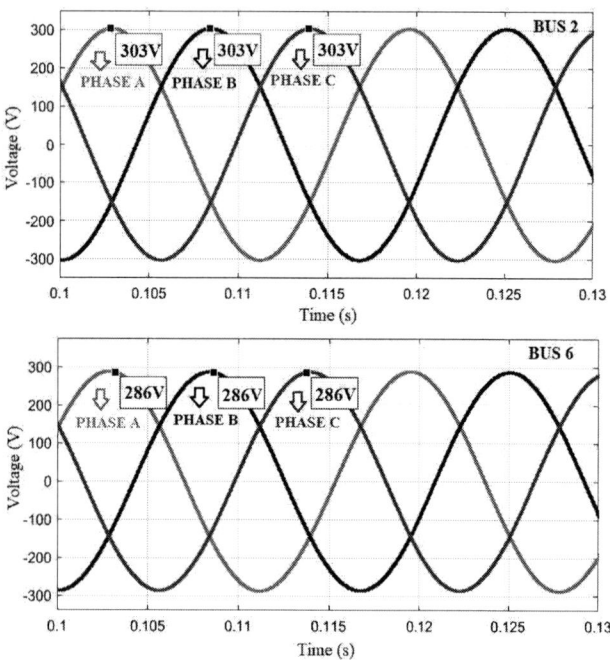

Fig. 9. Case 1: Voltages of the Bus 2 (TOP). Voltage of the Bus 6 (Buttom).

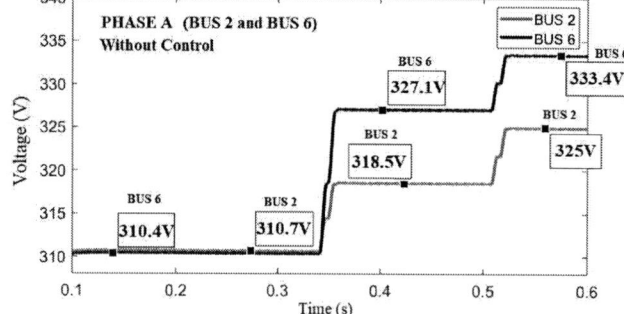

Fig. 10. Case 2: Peak Voltages of Phase A

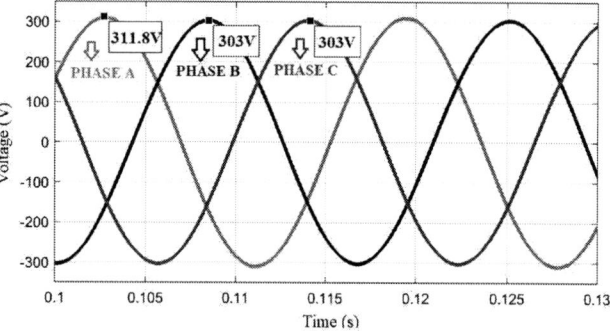

Fig. 11. Case 2: Voltages of the Bus 2

reactive. Considering the nominal power of the inverter and the characteristics of the study network (load and impedance), the inverter is designed to inject a fixed reactive current value of 30 A (peak), following the philosophies adopted for wind generator.

At 0.33 seconds of simulation, in which there is a decrease of loads, the inverter begin to inject inductive current and both the power grid and the inverter start to operate with a greater current. Fig.12 shows the inductive current, i_q, injected by the inverter from the moment the system detects the overvoltage. The peak voltages of the phase A of the Bus 2 and Bus 6 are shown in Fig. 13.

The comparison of the peak voltage values of case 2 and case 3 is made in Table V. Is verified that the control based in dq components is able to control the overvoltage in the distribution network and that in the case where there is inductive current injection the voltage is lower. Although the reduction of the grid voltage was only 3V when the photovoltaic systems present the control strategy proposed in this work, it was possible to contribute to the improvement

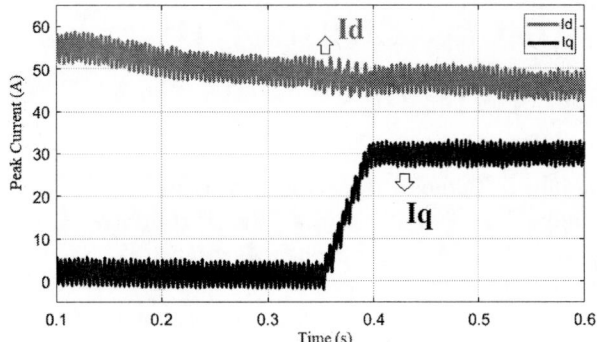

Fig. 12. Case 3: Inverter´s current: dq axes.

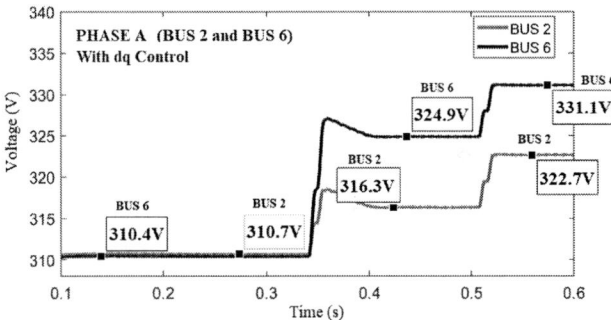

Fig. 13. Case 3 - Peak Voltages of Phase A

of the grid voltage level to meet the maximum voltage limit allowed by the utility. In this way, the ideal would be that all the photovoltaic that were connected to the distribution network had control based in dq components.

TABLE V
COMPARISON OF VOLTAGE VALUES OF CASE 2 AND CASE 3.

Load	BUS 2 (Case 2)	BUS 2 (Case 3)
34 kVA	310.7 V	310.7 V
15 kVA	318.5 V	316.3 V
0 kVA	325 V	322.7 V
Load	BUS 6 (Case 2)	BUS 6 (Case 3)
34 kVA	310.4 V	310.4 V
15 kVA	327.1 V	324.9 V
0 kVA	333.4 V	331.1 V

IV. CONCLUSIONS

It can be concluded that distributed photovoltaic generation compensates the voltage drop in the electrical grid, since phase A (which contains photovoltaic generation) remained closer to the voltage regulation required by the standards than those without distributed photovoltaic generation.

However, it was possible to verify that, in situations of low demand and high photovoltaic generation, overvoltage may occur, and to solve this problem without disconnecting the photovoltaic system from the power grid, the inverter of the

photovoltaic system must contain a control to inject inductive reactive current.

The use of the proposed control based in dq components contributed to the improvement of the grid voltage level to meet the maximum voltage limit allowed by the utility. It was seen that the farthest bus had higher voltage levels, however, it is worth mentioning that in this work only the panel connected between bus 2 and 3 had control. Thus, all panels should make use of the control so that buses are within the permitted voltage range.

REFERENCES

[1] Ministério de Minas e Energia, Empresa de Pesquisa Energética, "Plano Decenal de Expansão de Energia 2026," Brasília, 2017.

[2] Agência Nacional de Energia Elétrica, "Resolução Normativa nº 687, de 24 de Novembro de 2015," ANEEL, Brasília, 2015.

[3] Energisa Distribuidora de Energia, "NDU-013 - Critérios para a conexão de acessantes de geração distribuída ao sistema de distribuição da Energisa - Conexão em Baixa Tensão," December 2012.

[4] P. Chaudhary and M. Rizwan, "Voltage regulation mitigation techniques in distribution system with high PV penetration: A review," September 2017.

[5] P.S. Georgilakis and N.D. Hatziargyriou, "Optimal Distributed Generation Placement in Power Distribution Networks: Models, Methods, and Future Research," IEEE transactions on power systems, 2012.

[6] J. W. Smith, R. Dugan and W. Sunderman, "Distribution Modeling and Analysis of High Penetration PV," IEEE Power and Energy Society General Meeting, 2011.

[7] S. Samerchur, S. Premrudeepreechacharn, Y. Kumsuwun, and K. Higuchi, "Power Control of Single-Phase Voltage Source Inverter for Grid-Connected Photovoltaic Systems," IEEE, 2011.

[8] P. Rodriguez, R. Teodorescu, I. Candela, A. V. Timbus, M. Liserre, F. Blaabjerg, "New positive-sequence voltage detector for grid synchronization of power converters under faulty grid conditions," Proc. IEEE PESC, pp. 1-7, 2006-Jun.

[9] Pedro F. Marques and J. A. Peças Lopes, "Procedimentos de Rede para Aceitação de Produção Eólica e Especificação de Ride Through Defaul," Trabalho de Consultoria para o Operador Nacional do Sistema Eléctrico Brasileiro – ONS, Julho, 2004.

[10] Pedro F. Marques, J. A. Peças Lopes, Ângelo Mendonça and Rogério Almeida, "Avaliação do Comportamento Dinâmico da Rede Eléctrica Portuguesa num Cenário de Grande Integração de Produção Eólica," ENER'05, Conferência sobre Energias Renováveis e Ambientais em Portugal, 5-7 de Maio 2005, Figueira da Foz, Portugal.

[11] J.P. Sucena Paiva, J.M. Ferreira de Jesus, Rui Castro, Pedro Correia, João Ricardo, A. Reis Rodrigues, João Moreira and Bruno Nunes, "Transient Stability Study of the Portuguese Transmission Network with a High share of Wind Power," XI ERIAC CIGRÉ – Undécimo Encuentro Regional Iberoamericano de Cigré, Paraguay, May 2005.

[12] C. Sourkounis, and P. Tourou, "Grid code requirements for wind power integration in Europe," Conference Papers in Energy, Article ID 437674, vol. 2013.

[13] C. Chompoo-Inwai, C. Yingvivatanapong, K. Methaproyoon, and Wei-Jen Lee, "Reactive Compensation Techniques to Improve the Ride-Through Capability of Wind Turbine During Disturbance," IEEE Transactions on Industry Applications, vol. 41, no. 3, May/June, 2005.

[14] A. Safayet, P. Fajri, and I. Husain, "Reactive Power Management for Overvoltage Prevention at High PV Penetration in a Low-Voltage Distribution System," IEEE Transactions On Industry Applications, vol. 53, no. 6, Novembro, 2017.

Microinverter with reduced number of semiconductor switches

Paulo R. Cagnini[1], Luiz H. Meneghetti[1], Victor E. S. Barbosa[1], Emerson G. Carati[1],
Carlos M. Stein[1], Zeno L. I. Nadal[2], Jean M. S. Lafay[1], Jean Patric da Costa[1], Rafael Cardoso[1].
[1]Universidade Tecnológica Federal do Paraná (UTFPR), Pato Branco, Brazil
[2]Copel Distribuição S.A., Curitiba, Brazil
paulocagnini@alunos.utfpr.edu.br, luiz_lhm2@hotmail.com, victorbarbosa@alunos.utfpr.edu.br,
emerson@utfpr.edu.br, cmstein@utfpr.edu.br,zeno.nadal@copel.com,
jean.utfpr@gmail.com, jpcosta@utfpr.edu.br, rcardoso@utfpr.edu.br.

Abstract—**This paper proposes the use of a topology of a positive-input boost converter and symmetrical positive and negative outputs for the DC-DC stage of a microinverter connected to the electrical network with a small number of semiconductor switches. To validate and prove the operation of the system, simulations were performed in the PSIM software. All controllers and frequency-locked-loop (FLL) are exposed in the discrete form to facilitate implementation in a digital signal processor. For the reliability and adherence of the simulations with real cases, real data of solar irradiance and temperature of panels were experimentally collected and used.**

Index Terms—**Photovoltaic Systems, Micro Inverter, DC-DC Converter**

I. INTRODUCTION

Enterprises in distributed generation (DG) are growing year after year. Brazil is a country where this growth is recent and is only at the beginning. Brazil, together with other countries such as Saudi Arabia and Egypt have excellent climatic conditions for the photovoltaic sector, are the countries with the best growth prospects for the coming years [1], [2].

This growth is a result of more flexible regulations, entrepreneurship, and reduction of equipment and installation costs. The biggest costs involved in acquiring photovoltaic systems are in panels and inverters.

Panels have suffered constant reductions in production costs, but in the Brazilian case, import taxation remains high and the domestic industry still does not completely produce a solar panel, which makes price reduction slower. In the case of inverters, the domestic industry already produces them, and the semiconductor components are constantly being improved, with the continuous reduction of production costs and increase of efficiency, improving the use of the energy generated by the panels [3].

One of the high-cost components and responsible for most of the electrical losses in inverters is precisely the semiconductor switches used. This work proposes the topology of a micro-inverter with a reduced number of semiconductor switches using only one switch in the DC-DC step-up stage and two switches for DC-AC conversion [4], [5].

This paper is organized as follows: Section I describes the proposed microinverter and its stages of operation. Section II presents the control strategies used. Simulation results are presented in section III. Finally, section IV concludes the paper.

II. PROPOSED SYSTEM

In the case of photovoltaic systems, installations using micro-inverters have better energy utilization, compared to installations with centralized power processing. The present work suggests a topology different from those normally observed in the literature for application in photovoltaic microinverters [14], [15]. These converters are used directly connected to panels, making the extraction of power either individually or in small sets, increasing the efficiency of the system as a whole [13].

Fig. 1 shows a basic block diagram of a micro-inverter. Some micro-inverters have only the DC-DC converter and are used to track the maximum power point and inject energy into a DC bus, which is followed by a DC-AC inverter that concentrates the energy of several microsystems and is also responsible by the injection of energy in the utility grid [13], [14]. The proposed microinverter has both DC-DC and DC-AC stages.

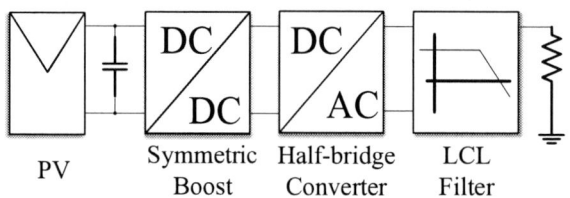

Fig. 1: Block diagram of a microinverter.

A. Micro Inverter

Fig. 2 shows the topology of the proposed circuit. The microinverter has a simple arrangement and a reduced number of components, especially semiconductor switches. This reduces implementation cost and complexity. Converters such as those proposed in [15] and [16] are much more complex.

The photovoltaic panel is connected to the micro-inverter by a capacitor C_{pv} which serves to decouple the dynamic behavior between the panel and the DC-DC converter [3], [14].

978-1-7281-4181-7/19 $31.00 © 2019 IEEE

The DC-DC stage uses a boost converter that has a positive input and symmetric positive and negative outputs [5]. This is an advantage that allows the use of a half-bridge inverter, reducing the need of semiconductor switches that would be required by full-bridge inverters, such as those described in [12] and [13]. The symmetrical output also allows the creation of a high voltage DC bus without the use of coupled inductors or transformers to increase the gain as those described in [7] and [8].

A half-bridge converter is used as inverter in addition to an LCL filter used to attenuate the harmonic content of the output voltage.

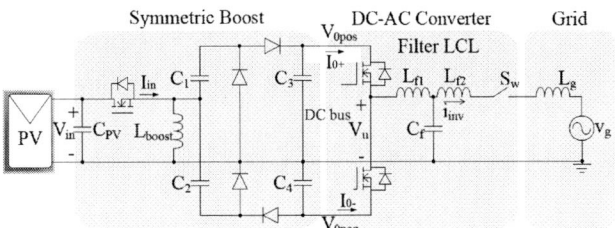

Fig. 2: Proposed microinverter topology.

III. SYMMETRIC BOOST CONVERTER

The symmetrical boost utilizes only one switch and one inductor, like the conventional boost. However, it uses four diodes and two capacitors. But these extra components have lower costs when compared to active semiconductors. In the literature, there are several topologies of step-up converters with two or more outputs, but in most cases, as in [10], the outputs have the same polarity, while the symmetrical boost has symmetric outputs [5].

According to [5], the static gain of the symmetrical boost is given by

$$V_{0_{pos}} = |V_{0_{neg}}| = \frac{D}{1-D}V_{in}. \tag{1}$$

The duty cycle of the symmetric boost is given by

$$D = 1 - \frac{V_{in}}{V_o}, \tag{2}$$

where, $V_o = V_{0_{pos}} = |V_{0_{neg}}|$, since it is a symmetric circuit.

The converter operates with constant output voltage V_o. Depending on the irradiance the control will act on the duty cycle D to track the maximum power point and to keep the output voltage constant for different values of V_{in}.

The equations for calculating the inductor and capacitors values are the same as those used in a conventional boost and are given by

$$L_{boost} = \frac{V_{in}D}{\Delta I_L f}, \tag{3}$$

$$L_{boost_{crit}} = \frac{V_{in}D}{2I_{in}f}, \tag{4}$$

and

$$C_3 = C_4 = \frac{I_o D}{\Delta V_C f}, \tag{5}$$

where D represents the symmetrical boost duty cycle, ΔI_L is current ripple on L_{boost} inductor, f is the switching frequency, I_{in} is current in L_{boost} inductor, I_o is the boost output current, ΔV_C is the voltage ripple on C_3 and C_4 capacitors and $L_{boost_{crit}}$ is the critical inductor that makes the converter to operate in discontinuous mode.

Capacitors C_3 and C_4 have the function of stabilizing the voltage for each of the outputs. These and other equations for calculating symmetric boost components can be found in [5].

IV. CONTROL STRATEGY

Fig. 3 illustrates the block diagram of the control system of the proposed converter. The inverter is responsible for DC bus control and the methodology for LCL filter design is found in [22]. The symmetric boost is used to step-up the voltage of the photovoltaic array and to ensure the operation at the maximum power point. This strategy is detailed in what follows.

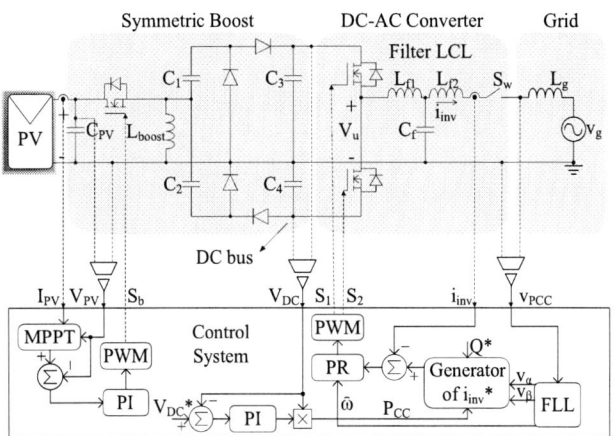

Fig. 3: Block diagram of the control system of the proposed converter.

A. MPPT

The MPPT (Maximum Power Point Tracking) algorithm uses the Disturb and Observe (P&O) method [23]. A proportional-integral controller (PI) is used to control the boost converter as depicted in figure 3.

B. Grid Synchronization

The grid synchronization is performed by a Frequency Locked Loop (FLL). The adopted FLL consists of a second order generalized integrator (SOGI), as described in [17]. The FLL also provides an estimation of the grid frequency $\hat{\omega}$.

The equations that describe the FLL are given by:

$$V_\alpha[n] = (K_1(V_{PCC}[n] - V_{PCC}[n-2]) + \\ -K_3 V_\alpha[n-1] - K_4 V_\alpha[n-2])/K_2, \tag{6}$$

$$V_\beta[n] = (K_5(V_{PCC}[n] + 2V_{PCC}[n-1] + \\ +V_{PCC}[n-2]) - K_3V_\beta[n-1] - K_4V_\beta[n-2])/K_2 \quad (7)$$

and

$$\hat{\omega} = \omega_s[n] + \omega_g, \quad (8)$$

where,

$$\omega_s[n] = \omega_s[n-1] + 0.5T_s\omega_{in}[n], +0.5T_s\omega_{in}[n-1], \quad (9)$$

$$\omega_{in}[n] = \frac{-\gamma K_e\hat{\omega}V_\beta[n]}{V_\alpha^2[n] + V_\beta^2[n]}(V_{PCC}[n] - V_\alpha[n]) \quad (10)$$

and

$$\begin{cases} K_1 = 2T_sK_e\hat{\omega}, \\ K_2 = T_s^2\hat{\omega}^2 + 2K_eT_s\hat{\omega} + 4, \\ K_3 = 2T_s^2\hat{\omega}^2 - 8, \\ K_4 = T_s^2\hat{\omega}^2 - 2K_eT_s\hat{\omega} + 4, \\ K_5 = T_s^2K_e\hat{\omega}^2. \end{cases} \quad (11)$$

In the aforementioned equations, ω_g is the nominal grid frequency and T_s is the sampling period.

C. Reference Current Generation

The reference current generation is based on [17] and [18]. Since it is a single-phase system, the i_β component of the output current of the inverter can be disregarded. Hence, the inverter reference current $i_{inv}^* = i_\alpha^*$ is given by

$$i_{inv}^*[n] = 2\frac{V_\alpha[n](P^* - P_{cc}) - V_\beta[n]Q^*}{V_\alpha^2[n] + V_\beta^2[n]}, \quad (12)$$

where, P_{cc} is the power variation due to voltage oscillations in the DC bus.

D. Output Current Control

The output current control system is implemented in the stationary reference frame $\alpha\beta$. A proportional-resonant controller (PR) was used. Since the PR has infinite gain at frequency ω, it is possible to zero the reference tracking error in a closed-loop system [17] and [18]. The transfer function of the PR controller is given by

$$G_{PR}(s) = K_p + \frac{2K_Is}{s^2 + \hat{\omega}^2} \quad (13)$$

where $\hat{\omega}$ is the estimated grid frequency provided by the FLL.

In the discrete form, using the Tustin method for discretization, the differential equations of the PR controller are obtained. That is,

$$V_P[n] = e_i[n]K_P, \quad (14)$$

$$V_R[n] = \frac{K_{ib}(e_i[n] - e_i[n-2]) + K_{ic}V_R[n-1]}{K_{ia}} - V_R[n-2], \quad (15)$$

and

$$V_{PR}[n] = V_P[n] + V_R[n], \quad (16)$$

where, $e_i[n] = i_{inv}^*[n] - i_{inv}[n]$, $V_P[n]$ represents the proportional term of the control law, V_R the resonant term of the control law and V_{PR} the control action. In addition,

$$\begin{cases} K_{ia} = 4 + T_s^2 + \hat{\omega}^2, \\ K_{ib} = 4K_IT_s, \\ K_{ic} = 8 - 2T_s^2\hat{\omega}^2. \end{cases} \quad (17)$$

E. DC Bus Voltage Control

The control of the DC bus voltage is carried out using the current i_{inv}. The voltage oscillations in the DC bus are related to the available power in the PV panels. For this, a PI controller is used, that is,

$$P_{cc} = (e_{DC} \cdot K_{PDC} + e_{DC} \cdot \frac{K_{IDC}}{s})V_{DC}. \quad (18)$$

In the discrete form, (18) can be implemented using

$$u_{PDC}[n] = e_{DC}[n]K_{PDC}, \quad (19)$$

$$u_{IDC}[n] = u_{IDC}[n-1] + \frac{K_{IDC}T_s(e_{DC}[n] + e_{DC}[n-1])}{2} \quad (20)$$

and

$$P_{cc} = (u_{PDC}[n] + u_{IDC}[n])V_{DC}, \quad (21)$$

where, $e_{DC}[n] = V_{DC}^*[n] - V_{DC}[n]$.

V. RESULTS

Table I presents the parameters and values of components used in the microinverter simulations.

TABLE I: Parameters of the proposed system.

Grid		
Voltage	V_g	$127\ V_{RMS}$
Frequency	f_g	$60\ Hz$
Equivalent inductance	L_g	$1\ mH$
DC-AC Converter		
Filter inductors	L_{f1} - L_{f2}	$570\ \mu H$ - $43\ \mu H$
Filter capacitor	C_f	$3.3\ \mu F$
DC bus		
Voltage	V_{DC}	$400\ V_{AVG}$
DC Bus Capacitor	$C_3 = C_4$	$1000\ \mu F$
Symmetric boost		
Inductor	L_{boost}	$800\ \mu H$
Capacitors	$C_1 = C_2$	$22\ \mu F$
PV		
Power (MPP)	P_{PV}	$680\ W$
Voltage (MPP)	V_{PV}	$75.2\ V$
Current (MPP)	I_{PV}	$9.05\ A$
Filter capacitor	C_{PV}	$100\ \mu F$
Switching frequency	f_{sw}	$40\ kHz$
Sampling frequency	f_s	$40\ kHz$

All simulations were performed using PSIM software with irradiance and temperature obtained from real data. The data was obtained based on 1500 measurements taken every 1 second during 25 minutes in a day with large variations in solar incidence.

To avoid time demanding simulations, the data was scaled down. This means that 25 minutes of information has been condensed in 5 seconds, which makes the variations even more

978-1-7281-4181-7/19 $31.00 © 2019 IEEE

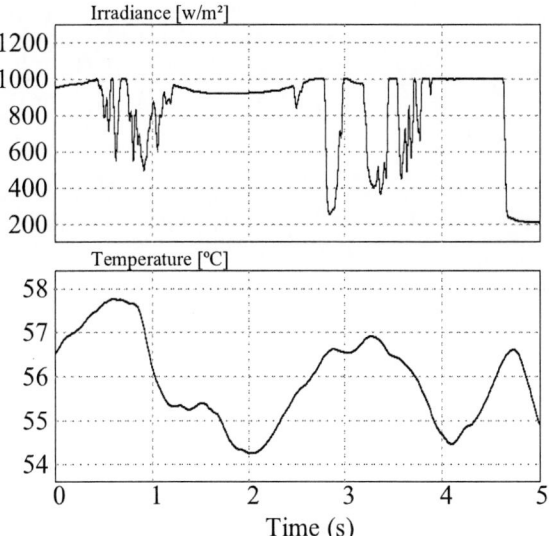

Fig. 4: Actual irradiance and temperature curves used in the simulations.

severe and exposes control to even greater efforts. The scaled data is depicted in Fig 4 .

Fig. 5 shows the output voltages of the symmetric boost converter. In the upper subpicture, the two symmetrical voltages can be seen while the lower subpicture depicts the differential voltage between the outputs. It shows the high gain voltage obtained with this topology.

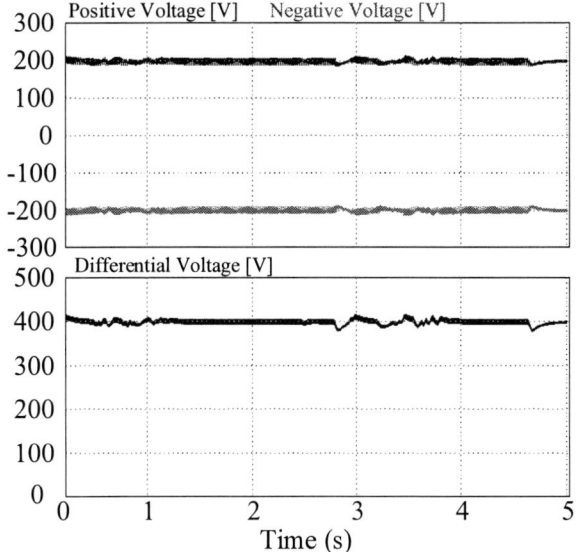

Fig. 5: Positive, negative and differential output voltages of the symmetric boost converter.

Due to the connection of the inverter to the grid, a periodic ripple at 120 Hz on each output of the DC-DC converter is observed. Due to the differential characteristic of the output

of the boost converter, this ripple is attenuated.

Fig. 6 shows the input power (power of the photovoltaic panels) and output power of the microinverter. It can be seen that the control system adequately tracks the maximum power point of the panels. Fig. 6 also depicts the current injected into the grid. It can be observed that the amplitude of the current follows the power profile of the PV generation. The detail of the current also shows that the output current is sinusoidal with low harmonic distortion. The THD of the current is 1.48 %, which is well below the limit defined in international standards (IEC 61000-3-2, 4 and IEEE Std 519TM – 2014).

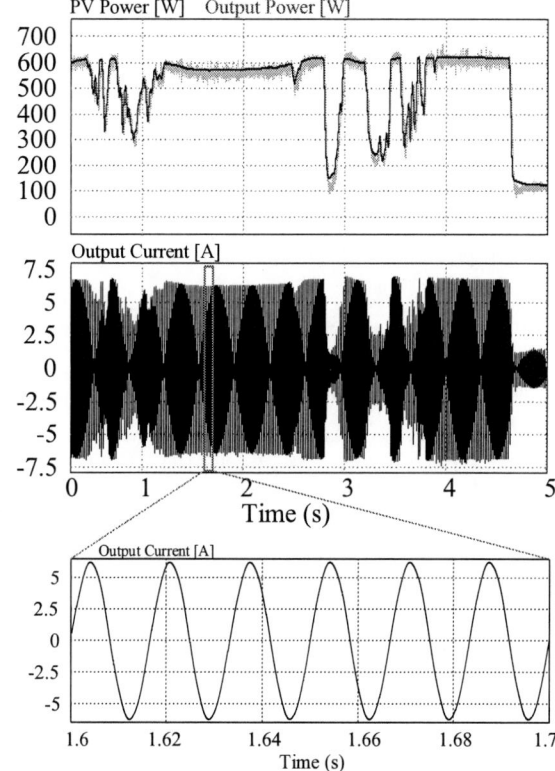

Fig. 6: Microinverter input and output power and current injected into the grid.

VI. CONCLUSION

This paper presented a microinverter topology and its control system applied to photovoltaic generation. Its main advantage is the reduced number of active switches used in its implementation. The DC-DC converter is based on a symmetrical boost that provides high voltage gain and symmetrical output. It allows the use of a half-bridge inverter as the output stage of the converter.

The simulations corroborated the proposal. The converter was capable to provide high DC-DC gain and to track the maximum power point. At the same time, the output inverter current presented low THD, complying with the standards.

In addition, the use of real data for irradiance and temperature allowed a more realistic analysis of the behavior of the converter. The irradiance and temperature curves chosen were curves with extreme conditions of variations, which allowed to expose the control system to great efforts that proved their effectiveness.

ACKNOWLEDGMENT

This work was funded by the Research and Development project PD 2866-0468/2017, granted by the Brazilian Electricity Regulatory Agency (ANEEL) and Companhia Paranaense de Energia (COPEL). The authors also would like to thank to FINEP, SETI, CNPq, Fundação Araucária, CAPES and UTFPR for additional funding.

REFERENCES

[1] TOLMASQUIM, Mauricio T. ROSA, Luiz P. SZKLO, Alexandre S. Tendências da Eficiência Elétrica no Brasil: Indicadores de Eficiência Energética. Rio de Janeiro: COPPE/UFRJ, 1998.

[2] SOUZA, Angelo R. R. Conexão de Geração Distribuída em Redes de Distribuição. Departamento de Engenharia Elétrica- UFP, Curitiba, 2009.

[3] D. W. Hart, Eletrônica de Potência: Análise e Projeto de Circuitos, 1 ed. AMGH, Porto Alegre, 2012.

[4] M. H Rashid. Eletrônica de potência: dispositivos, circuitos e aplicações. 4 ed., Pearson, São Paulo, 2015

[5] K. I. Hwu, Y. T. Yau and J. Shieh, "Dual-output boost converter with positive and negative output voltages under single positive voltage source fed," The 2010 International Power Electronics Conference - ECCE ASIA -, Sapporo, 2010, pp. 1034-1037.

[6] Y. Kanthaphayao and C. Boonmee, "Dual-output DC-DC power supply without transformer," 2004 IEEE Region 10 Conference TENCON 2004., Chiang Mai, 2004, pp. 41-44 Vol. 4.

[7] P. Klimczak and S. Munk-Nielsen, "A single switch dual output non-isolated boost converter," 2008 Twenty-Third Annual IEEE Applied Power Electronics Conference and Exposition, Austin, TX, 2008, pp. 43-47.

[8] K. I. Hwu, Y. T. Yau and Jenn-Jong Shieh, "Dual-output boost converter," 2011 IEEE Ninth International Conference on Power Electronics and Drive Systems, Singapore, 2011, pp. 940-943.

[9] K. I. Hwu, Y. T. Yau and J. Shieh, "Dual-output buck-boost converter with positive and negative output voltages under single positive voltage source fed," The 2010 International Power Electronics Conference - ECCE ASIA -, Sapporo, 2010, pp. 420-423.

[10] Siho Park, Honnyong Cha, H. Kim, J. Kim and Jintae Cho, "A novel dual output boost converter with output voltage balancing," 2016 IEEE 8th International Power Electronics and Motion Control Conference (IPEMC-ECCE Asia), Hefei, 2016, pp. 2198-2203.

[11] D. Dah-Chuan Lu, M. Wu and T. Cheng, "Using cross regulation in single-switch single-inductor dual-output CCM boost converter to simplify controller design," 2016 IEEE International Conference on Industrial Technology (ICIT), Taipei, 2016, pp. 390-395.

[12] C. Liao, W. Lin, Y. Chen and C. Chou, "A PV Micro-inverter With PV Current Decoupling Strategy," in IEEE Transactions on Power Electronics, vol. 32, no. 8, pp. 6544-6557, Aug. 2017.

[13] S. Strache, R. Wunderlich and S. Heinen, "A Comprehensive, Quantitative Comparison of Inverter Architectures for Various PV Systems, PV Cells, and Irradiance Profiles," in IEEE Transactions on Sustainable Energy, vol. 5, no. 3, pp. 813-822, July 2014.

[14] Mahinda Vilathgamuwa; Dulika Nayanasiri; Shantha Gamini, "Power Electronics for Photovoltaic Power Systems," in Power Electronics for Photovoltaic Power Systems , , Morgan & Claypool, 2015, pp

[15] Hossam A. Gabbar, "Microinverter Systems For Energy Conservation In Infrastructures," in Energy Conservation in Residential, Commercial, and Industrial Facilities , , IEEE, 2018, pp.

[16] M. Rajeev and V. Agarwal, "Analysis and Control of a Novel Transformer-Less Microinverter for PV-Grid Interface," in IEEE Journal of Photovoltaics, vol. 8, no. 4, pp. 1110-1118, July 2018.

[17] R. Teodorescu, M. Liserre, P. Rodrígues, "Grid Converters for Photovoltaic a and nd Wind Power Systems," John Wiley & Sons, New Jersey, 2011.

[18] R. A. Junior, E. G. Carati, J. P. da Costa, R. Cardoso, C. M. Stein, "Robust Design of Active Damping with Current Estimator for Single-Phase Grid-Tied Inverters", IEEE Transactions on Industry Applications, vol. 54, pp. 4672– 4681, 2018.

[19] R. A. Liston Junior, E. G. Carati, J. P. da Costa, R. Cardoso, C. M. Stein, "Single-Phase Grid-Tied Inverters: Guidelines for Smoother Connection and Suitable Control Structure", in 12th IEEE International Conference on Industry Application – INDUSCON 2016, Curitiba, 2016.

[20] F. Li, X. Zhang, H. Zhu, H. Li, C. Yu, "An LCLLC filter for grid-connected converter: Topology, parameter,andanalysis",IEEE

[21] Transactions on Power Electronics, vol. 30, no. 9, pp. 5067–5077, 2015, doi: 10.1109/TPEL.2014.2367135.

[22] F. Li, X. Zhang, H. Zhu, H. Li and C. Yu, "An LCL-LC Filter for Grid-Connected Converter: Topology, Parameter, and Analysis," in IEEE Transactions on Power Electronics, vol. 30, no. 9, pp. 5067-5077, Sept. 2015.

[23] A. N. A. Ali, M. H. Saied, M. Z. Mostafa and T. M. Abdel- Moneim, "A survey of maximum PPT techniques of PV systems," 2012 IEEE Energytech, Cleveland, OH, 2012, pp. 1-17.

978-1-7281-4181-7/19 $31.00 © 2019 IEEE

Application of Model Predictive Control in a Resolver-to-Digital Converter

Thyago Vasconcelos Estrabis
Electrical Engneering Department
Federal University of Mato Grosso do Sul
Campo Grande, Brazil
thyago.estrabis@gmail.com

Raymundo Cordero García
Electrical Engneering Department
Federal University of Mato Grosso do Sul
Campo Grande, Brazil
rcorderog@gmail.com

Edson Antonio Batista
Electrical Engneering Department
Federal University of Mato Grosso do Sul
Campo Grande, Brazil
edson.ufms@gmail.com

Cristiano Quevedo Andrea
Electrical Engneering Department
Federal University of Mato Grosso do Sul
Campo Grande, Brazil
quevedo_unesp@yahoo.com.br

Márcio Afonso Soleira Grassi
Electrical Engneering Department
Federal University of Mato Grosso do Sul
Campo Grande, Brazil
marciograssi14@gmail.com

Abstract—A resolver is an angular position sensor used in applications that demands robustness and reliability, such as electric vehicles. However, getting the angular position from resolver output signals is a difficult task, and many algorithms were proposed to achieve that task. This paper describes the application of model predictive control (MPC) to get the angular position from resolver signals. Synchronous demodulation is used to get the envelopes of the resolver outputs and get the estimation error. The structure of the conventional model predictive controller was modified to be used as an angle tracking observer (ATO). Simulations show the performance of the proposed approach. According to the bibliographic review of the authors, it is the first time that model predictive control is applied as an observer to get the angular position from resolver signals.

Keywords—Angle tracking observer, model predictive control, generalized predictive control, resolver.

I. INTRODUCTION

Applications based on electrical motors drives, such as electric/hybrid vehicles (EV/HEVs), aircrafts, CNCs and robotics work under difficult conditions [1], [2]. In order to guarantee reliability of the speed and position control, it is required a robust and accurate angular position sensor. Between the different position sensors available in the industry, resolver is one of the most used for harsh applications. This sensor can resist higher temperatures, shocks and vibrations than encoders [3].

A resolver generates two amplitude-modulated output voltages: a high-frequency voltage is modulated in amplitude by the sine and cosine of the mechanical angle. For that reason, getting the angular position from resolver signals is a difficult task. Observers called resolver-to-digital converters (RDC) are used to estimate the angular positions. Nowadays, many RDCs are software algorithms that are implemented in digital processors [4].

Many types of RDCs were proposed in literature. Most of them are closed-loop observers based on PID regulators [4], [5], second-order observers [6], [7], type III observers [8], software based PLL [9], and others [10]. These algorithms are tuned using techniques such as Bode diagram or Ackermann

formula. Indeed, the development of robust and precise RDCs is still an open question.

Nowadays, model predictive control (MPC) becomes an interest topic of research due to its fast response and robustness [11]–[14]. It was in the 60's in which the modern control theory began to diffuse by the need to control more complex plants [15]–[16]. Interest in MPC begins in the late 1970's with emergence of several papers, and thus consolidating MPC in the industry [17]. This interest existed for his behavior in the time domain and for its robustness [14], since the industrial engineers have a tendency to use robust control [14]. Clarke in [18]–[19], presents a generalized predictive control (GPC) model, one of MPC's robust methods [13]. Camacho in [13], quotes that the MPC controller can be interpreted as being a compensator based on stable state observer and that its performance and robustness is determined by the number of poles in the observer (which can be directly adjusted by the setting parameters) and the regulator poles.

This paper explores the use of MPC in the development of an angle tracking observer (ATO) which is the main part of a RDC. In first place, synchronous demodulation is applied to get the envelopes of the resolver outputs. These signals are proportional to the sine and cosine of the angular position. After that, these envelopes are used to get a signal which depends on the angle estimation error. The MPC receives this error and gives an estimative of the angular position. Usually, a MPC controller receives the reference to be tracked. However, in the case of a RDC, there is no reference to be tracked, just an estimative of the estimation error. In order to use the MPC in a RDC, the structure of the conventional MPC system was modified to estimate the angular position using the signal proportional to the estimation error. Simulations show the performance of the proposed approach.

According to the bibliographic review of the authors, it is the first time that model predictive control is applied as an observer to get the angular position from resolver signals. Thus, the final objective of this research is to introduce the MPC technique in the development of RDCs.

II. THEORETICAL FOUNDATIONS

A. Resolver

The Resolver has an excitation rotor winding coupled to the motor shaft and two output stator windings, according to Fig. 1 [8]. The excitation winding receives a high frequency sinusoidal excitation voltage $v_e(t)$ through a rotary transformer [8]. Two voltages, $v_s(t)$ and $v_c(t)$, are induced in the output windings [8]:

$$v_e(t) = a_e cos(2\pi f_e t) \quad (1)$$

$$v_s(t) = k_e a_e cos(2\pi f_e t) sin(\theta) \quad (2)$$

$$v_c(t) = k_e a_e cos(2\pi f_e t) cos(\theta) \quad (3)$$

where a_e is the excitation amplitude, f_e is the excitation frequency (1 to 10 kHz), t denotes time, k_e is the transformation ratio and θ is the angular position of the motor shaft. Equations (2) and (3) indicate that the resolver outputs are amplitude-modulated signals: the high-frequency voltage is modulated by the sine and cosine of the angular position.

B. Synchronous Demodulation

According to (2) and (3), the envelopes (modulating signals) of the resolver outputs are the sine and cosine of the angular position. Synchronous demodulation is a simple technique to get these envelopes [7]. In this technique, the resolver outputs are sampled at the peaks and/or the valleys of the resolver excitation, i.e., when $|cos(2\pi f_e t)| = 1$, as illustrated in Fig. 2. If the samples are acquired when the $v_e(t)$ is negative, then the samples are multiplied by −1. Thus, the demodulated signals d_s and d_c are:

$$d_s = k_e a_e sin(\theta) \quad (4)$$

$$d_c = k_e a_e cos(\theta) \quad (5)$$

Observe that, using both the peaks and valleys of $v_e(t)$, the sampling frequency of the resolver (f_s) signals are twice the excitation frequency, due to two samples are acquired in each period of the resolver excitation signal.

$$f_s = 2f_e \quad (6)$$

Synchronous demodulation requires the synchronization between the excitation signal and the data acquisition system. This requirement can be achieved because the excitation voltage can be generated by the same digital processor that acquires the resolver outputs.

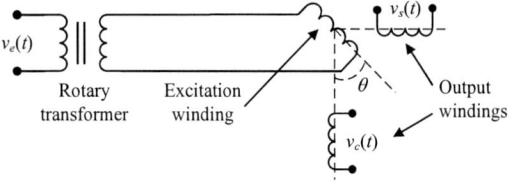

Fig. 1. Structure of the resolver.

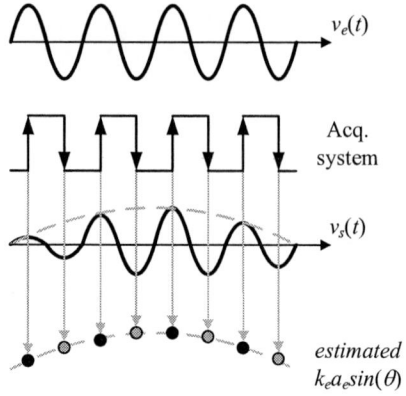

Fig. 2. Synchronous demodulation.

C. Model Predictive Control

Nowadays, MPC is used in different applications, such as control and estimation in power electronics devices (e.g. three-phase motors and converters). There exist many types of MPC approaches. The choice of a specific approach depends on the application [20].

The MPC general formulation of state-space in discrete time, assuming that the plant is a SISO, can be described by (7) and (8):

$$x_m(k+1) = A_m x_m(k) + B_m u(k) \quad (7)$$

$$y(k) = C_m x_m(k) \quad (8)$$

According from the principle of receding horizon control, the input will not affect the output $y(k)$ at the same time, so the equation can be defined by (9) and (10).

$$\Delta x_m(k+1) = A_m \Delta x_m(k) + B_m \Delta u(k) \quad (9)$$

$$y(k+1) - y(k) = C_m A_m \Delta x_m(k) + C_m B_m \Delta u(k) \quad (10)$$

where $\Delta x_m(k+1)$, $\Delta x_m(k)$, and $\Delta u(k)$ are describe in (11), (12) and (13).

$$\Delta x_m(k+1) = x_m(k+1) - x_m(k) \quad (11)$$

$$\Delta x_m(k) = x_m(k) - x_m(k-1) \quad (12)$$

$$\Delta u(k) = u(k) - u(k-1) \quad (13)$$

In this paper, it is considered the generalized MPC (GPC) model proposed by Wang [14]. This method has an integrator embedded, where $y(k)$ is the system output, $x_m(k)$ is a state variable vector and $u(k)$ is input..

The Fig. 3 is a diagram that Wang in [14] represented his augmented GPC.

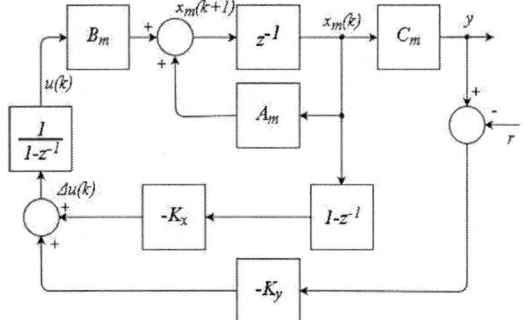

Fig. 3. Shows the diagram used in [14] to represent the augmented GPC.

The state-space equation (7), (8) of this Wang's model [14] can be described in matrix form (14) (15). Note that the matrix A, B and C are named augmented matrices and they have this format because of the embedded integrator.

The variable o_m is a zero vector which your dimension depends on the number of rows in the matrix A.

$$\overbrace{\begin{bmatrix} \Delta x_d(k+1) \\ y(k+1) \end{bmatrix}}^{x(k+1)} = \overbrace{\begin{bmatrix} A_m & o_m{}^T \\ C_m A_m & 1 \end{bmatrix}}^{A} \overbrace{\begin{bmatrix} \Delta x_m(k) \\ y(k) \end{bmatrix}}^{x(k)} + \underbrace{\begin{bmatrix} B_m \\ C_m B_m \end{bmatrix}}_{B} \Delta u(k) \quad (14)$$

$$y(k) = \underbrace{\begin{bmatrix} o_m & 1 \end{bmatrix}}_{C} \begin{bmatrix} \Delta x_m(k) \\ y(k) \end{bmatrix} \quad (15)$$

It is necessary to perform sequential calculations, presented in (16), (17), (18), (19).

$$x(k_i + 1|k_i) = Ax(k_i) + B\Delta u(k_i)$$
$$x(k_i + 2|k_i) = Ax(k_i + 1|k_i) + B\Delta u(k_i + 1)$$
$$= A^2 x(k_i) + AB\Delta u(k_i) + B\Delta u(k_i + 1)$$
$$\vdots$$
$$x(k_i + N_P|k_i) = A^{N_P} x(k_i) + A^{N_P-1} B\Delta u(k_i) +$$
$$A^{N_P-2} B\Delta u(k_i + 1) + \ldots + A^{N_P-N_c} B\Delta u(k_i + N_c - 1) \quad (16)$$

$$y(k_i + 1|k_i) = CAx(k_i) + CB\Delta u(k_i)$$
$$y(k_i + 2|k_i) = CA^2 x(k_i) + CAB\Delta u(k_i) + CB\Delta u(k_i + 1)$$
$$= CA^3 x(k_i) + CA^2 B\Delta u(k_i) + CAB\Delta u(k_i + 1) +$$
$$CB\Delta u(k_i + 2) \quad (17)$$
$$\vdots$$
$$y(k_i + N_P|k_i) = CA^{N_P} x(k_i) + CA^{N_P-1} B\Delta u(k_i) +$$
$$CA^{N_P-2} B\Delta u(k_i + 1) + \ldots + CA^{N_P-N_c} B\Delta u(k_i + N_c - 1)$$

By the equations (16), (17), it can observed that the predicted variable depend on the current state and the movement of future control. Let define the output vector Y and future control vector:

$$Y = [y(k_i + 1|k_i) \quad \ldots \quad y(k_i + N_P|k_i)]^T \quad (18)$$

$$\Delta U = [\Delta u(k_i) \quad \Delta u(k_i + 1) \quad \ldots \quad \Delta u(k_i + N_c - 1)]^T \quad (19)$$

The future control ΔU is present in the output Y, as the evaluated system is a single input single output, the dimensions of the output Y and the future control ΔU are

defined by N_P and N_c, and expressed in (20) with their compact matrices in (21) and (22).[14].

$$Y = Fx(k_i) + \Phi \Delta U \quad (20)$$

The variables N_P and N_c are prediction horizon and control horizon.

$$F = \begin{bmatrix} CA \\ CA^2 \\ CA^3 \\ \vdots \\ CA^{N_P} \end{bmatrix} \quad (21)$$

$$\Phi = \begin{bmatrix} CB & 0 & 0 & 0 \\ CAB & CB & 0 & 0 \\ & & \vdots & \\ CA^{N_P-1} & CA^{N_P-2}B & \ldots & CA^{N_P-N_c}B \end{bmatrix} \quad (22)$$

The MPC is based in a minimization of cost function [18][21], in other words, this parameters is designed to find the 'best' control parameter vector of future control ΔU [14].

The cost function (J) is defined as follows:

$$J = (R_s - Y)^T (R_s - Y) + \Delta U^T \overline{R} \Delta U \quad (23)$$

The cost function J reflects the control objective [14]. Assume that the data vector R_s^T contains the set-point information $R_s^T = \overbrace{[1 \; 1 \ldots 1]}^{N_P} r(k_i)$, where $r(k_i)$ is the set-point. The variable $\overline{R}$ is diagonal matrix that contains weight information $\overline{R} = r_w I_{N_c x N_c}$, where r_w is a tuning parameters for closed-loop performance.

To find the optimal ΔU, it's necessary to derive the cost function J assuming that $\partial J / \partial \Delta U = 0$, after that, the optimal ΔU is defined by (23).

The gain state feedback K_{mpc} is described in (24), where $K_{mpc} = [K_x \quad K_y]$ is a vector where K_x is a vector gain for feedback state and K_y is gain for feedback error. Besides, K_y is the last element of K_{mpc} vector.

$$K_{mpc} = \overbrace{[1 \; 0 \ldots 0]}^{N_c} (\Phi^T \Phi + \overline{R})^{-1} (\Phi^T F) \quad (24)$$

On the gain vector K_{mpc}, it's important to note that it directly influences the output Y of the MPC controller.

III. ANGLE ESTIMATION USING MPC

Fig. 4 shows the proposed RDC system to estimate the angular position from resolver output signals. It is composed by the synchronous demodulation system explained previously, and the angular tracking observer (ATO) based on MPC. In order to simplify the mathematical operations, let $k_e = 1$ and $a_e = 1$, i.e. we will work with normalized resolver signals. Thus:

$$d_s = sin(\theta) \quad ; \quad d_c = cos(\theta) \quad (25)$$

Based on (25), Fig. 1 and trigonometric properties, the signal g in Fig. 4 has the following value:

978-1-7281-4181-7/19 $31.00 © 2019 IEEE

$$g = d_s \cdot cos(\theta_e) - d_c \cdot sin(\theta_e)$$

$$= sin(\theta) \cdot cos(\theta_e) - cos(\theta) \cdot sin(\theta_e) \quad (26)$$

$$= sin(\theta - \theta_e)$$

If x is small, then $sin(x) \approx x$. Applying this property, and assuming that $\theta - \theta_e$ is small, then g is, in fact, the angle estimation error:

$$g = sin(\theta - \theta_e) \approx \theta - \theta_e \quad (27)$$

Thus, the angle tracking observer (ATO) can be linearized as indicated in Fig. 5. The gain K was added to improve the response of the MPC.

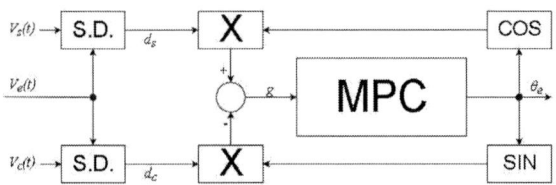

Fig. 4.   Proposed RDC algorithm, where the MPC acts as an ATO.

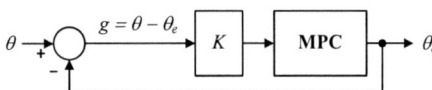

Fig. 5.   Linearized proposed ATO.

A. Discrete Plant for the GPC-based ATO

According to (7), (8) and Fig. 4, it is necessary to define a plant to define the GPC controller. In most papers, the ATO has, at least, one integrator [4]-[9]. For that reason, in this paper, two GPC-based ATOs were designed the plants $1/s$ and $1/s^2$, which are continuous SISO systems. In order to define the GPC controller, these plants must be discretized, due to GPC is defined for discrete systems [14].

The ZOH (Zero Order Hold) method is used to discretize these plants. These plants were discretized through MATLAB, considering a sampling time of $T = 50$ μs.

B. Modified GPC Model :

Observe in Fig. 5 that the information that the MPC block in the ATO receives is the estimation error, not the plant reference (as in Fig. 3). Hence, a modification of the MPC controller was done.

Initially, let $K = 1$. Observe that the output of the gain block K_y in Fig 3 is $K_y(y - r)$. Note that, in the linearized ATO in Fig. 5, the output of the MPC is the estimated angular position, $y = \theta_e$, while the reference is the angular position, $r = \theta$. Thus:

$$K_y(y - r) = K_y(\theta - \theta_e) = -K_y \cdot g \quad (32)$$

Based on (32), the proposed GPC model for the ATO uses the estimation error $g = \theta - \theta_e$. Fig. 6 shows the structure of the modified GPC-based ATO.

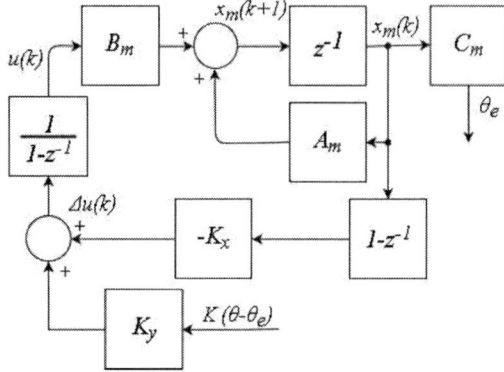

Fig. 6.   Adapted GPC used in the ATO.

IV. RESULTS

The proposed approach based on MPC to estimate an error position with synchronous demodulation was simulated in MATLAB/SIMULINK. Due to the excitation frequency of the resolver of 10 kHz, according to (6), the sampling time for the discrete systems is 20 kHz (a sampling time of 50 μs).

The Table I and Table II show, respectively, the error result of the first and the second order plant simulation with noise after speed goes into steady state.

TABLE I.   FIRST ORDER PLANT SIMULATION WITH NOISE

Parameters	Np = 2 Nc =1	Np = 10 Nc = 1	Np = 10 Nc = 5
std$_e$	0.0091	0.0034	0.0146
mean$_e$	$-7.9582 * 10^{-5}$	$1.5510 * 10^{-4}$	$-1.4974 * 10^{-4}$
median	$1.6876 * 10^{-4}$	$-6.0595 * 10^{-5}$	$-5.2535 * 10^{-4}$
min	-0.0256	-0.0077	-0.0328
max	0.0279	0.0104	0.0381

TABLE II.   SECOND ORDER PLANT SIMULATION WITH NOISE

Parameters	Np = 2 Nc =1	Np = 10 Nc = 1	Np = 10 Nc = 5
std$_e$	0.0146	0.0031	0.0146
mean$_e$	$-2.5487 * 10^{-6}$	$2.1468 * 10^{-4}$	$2.5487 * 10^{-6}$
median	$-4.723 * 10^{-4}$	$2.8976 * 10^{-5}$	$-4.7230 * 10^{-4}$
min	-0.0325	-0.0093	-0.0325
max	0.0389	0.0062	0.0389

The negative averages in the first and second order plant occur by the various negative peaks in the error. Note that standard deviation value, minimum and maximum values in Table I is near close to the values obtained in Table II.

The considered mechanical speed, for each plant, is shown in Fig. 7.

The Fig. 8 and Fig. 12 show the estimation errors, without noise, for the plants 1/s and 1/s², respectively, and for different values of Np and Nc. On the other hand, the Figs. 9 to 11 and Figs. 13 to 15 show the estimation errors with noise, for the plants 1/s and 1/s², respectively. For better visualization, the units of the axis of Figs. 8 to 15 were not shows. The x-axis unit is second, while the y-axis units is rad.

978-1-7281-4181-7/19 $31.00 © 2019 IEEE

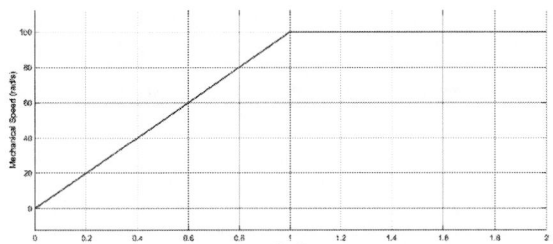

Fig. 7.   Mechanical speed used for the plants 1/s and 1/s².

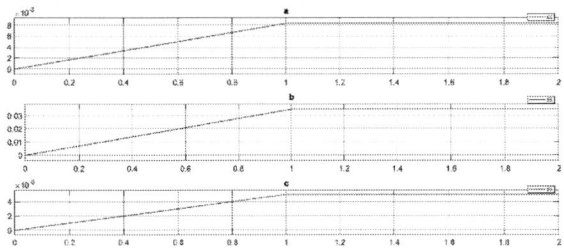

Fig. 8.   Simulation results for the plant 1/s and for different values of Np and Nc: (a) Np = 2 Nc = 1; (b) Np = 10 Nc = 1; (c) Np = 10 Nc = 5.

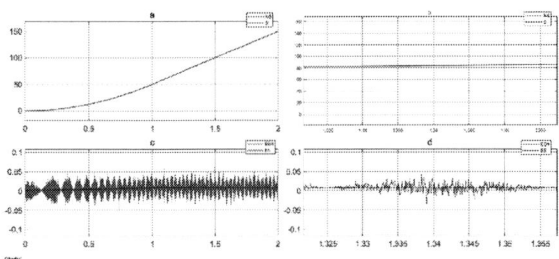

Fig. 9.   Simulation results, with noise, for the plant 1/s and for Np = 2 and Nc = 1: (a) Angle Estimation Error; (b) Angle Estimation Error Zoom; (c) Estimation error; (d) Estimation Error Zoom .

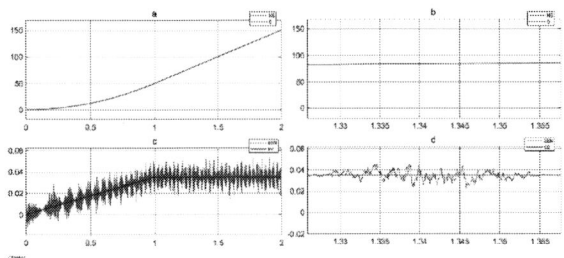

Fig. 10. Simulation results, with noise, for the plant 1/s and for Np = 10 and Nc = 1: (a) Angle Estimation Error; (b) Angle Estimation Error Zoom; (c) Estimation error; (d) Estimation Error Zoom .

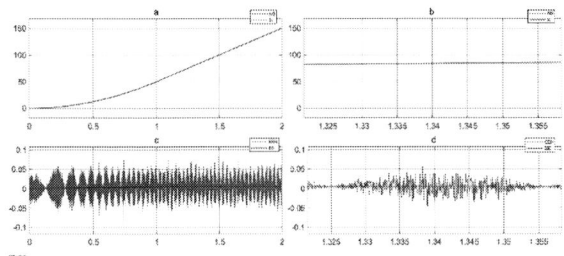

Fig. 11. Simulation results, with noise, for the plant 1/s and for Np = 10 and Nc = 5: (a) Angle Estimation Error; (b) Angle Estimation Error Zoom; (c) Estimation error; (d) Estimation Error Zoom.

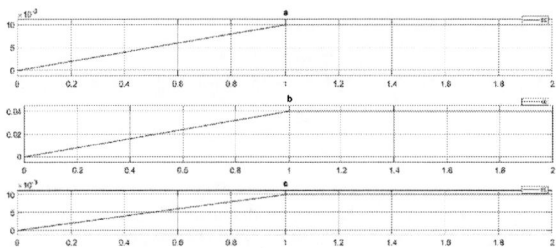

Fig. 12. Simulation results for the plant 1/s² and for different values of Np and Nc: (a) Np = 2 Nc = 1; (b) Np = 10 Nc = 1; (c) Np = 10 Nc = 5.

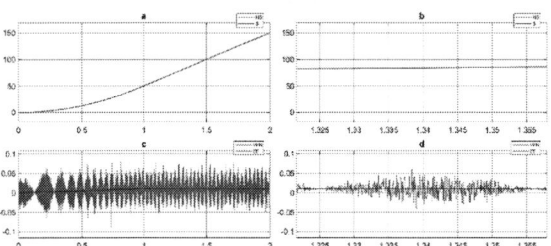

Fig. 13. Simulation results, with noise, for the plant 1/s² and for Np = 2 and Nc = 1: (a) Angle Estimation Error; (b) Angle Estimation Error Zoom; (c) Estimation error; (d) Estimation Error Zoom .

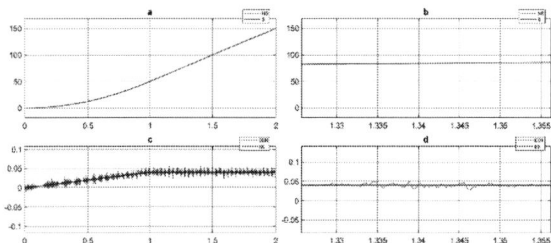

Fig. 14. Simulation results, with noise, for the plant 1/s² and for Np = 10 and Nc = 1: (a) Angle Estimation Error; (b) Angle Estimation Error Zoom; (c) Estimation error; (d) Estimation Error Zoom .

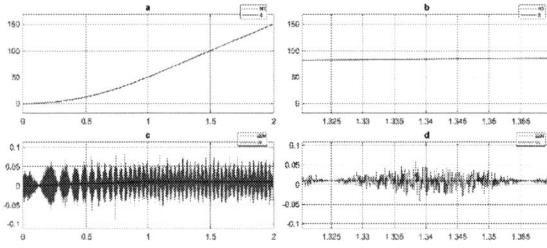

Fig. 15. Simulation results, with noise, for the plant 1/s² and for Np = 10 and Nc = 5: (a) Angle Estimation Error; (b) Angle Estimation Error Zoom; (c) Estimation error; (d) Estimation Error Zoom .

In the Figs. 10 and 13, it is possible to observe a smaller noise amplitude but the error is bigger than the other simulations of their respective plants. In the simulation whose the Nc is near Np, or the reason between Np and Nc less than 50 %, showed better result for the noisy plant. However, the best choice would be Np = 2 and Nc = 1, which showed lower peak values, steady state, than the other simulation.

V. Conclusion

In this paper, a modified GPC algorithm for the implementation of a RDC was proposed to estimate the angular position from the resolver signals. This is the first time MPC theory is applied to create an angle tracking observer (ATO). It was found that the steady state error is negligible, even during acceleration. As future work, find way to reduce noise interference, as for the model used [14], GPC does not become as robust as it should be for high frequency rates.

Acknowledgment

Authors want to thanks Graduation Program in Electrical Engineering of the Federal University of Mato Grosso do Sul, PPGEE-UFMS, for the support to this research.

References

[1] Noori, N. and Khaburi, D. A. Diagnosis and compensation of amplitude imbalance, imperfect quadrant and offset in resolver signals. 7th Power Electronics and Drive Systems Technologies Conference, 2016, pp. 76–81.

[2] Staebler, M. and Verma, A. TMS320F240 DSP Solution for Obtaining Resolver Angular Position and Speed, Texas Instruments Application Report SPRA605A, 2017.

[3] Jin, C.-S., Jang, I.-S., Bae, J.-N., Lee, J. and Kim, W.-H. "Proposal of Improved Winding Method for VR Resolver". IEEE Trans. on Magn., vol. 51, No. 3, pp. 1–4, 2015.

[4] Kaewjinda, W. and Konghirun, M. A DSP – Based Vector Control of PMSM Servo Drive Using Resolver Sensor. IEEE Region 10 Conference, 2006, pp. 1 –4.

[5] Carusso, M. di Tomasso, A. O., F. Genduso, F., Miceli, R. and Galluzzo, G. R., A DSPBased Resolver-to-Digital Converter for High Performance Electrical Drive Applications. IEEE Trans. Ind. Electron., vol. 63, No. 7, pp. 4042–4051, 2016.

[6] Qamar, N. A., Hatziadoniu, C. J. and Wang, H. Speed Error Mitigation for a DSP-Based Resolver-to-Digital Converter Using Autotuning Filters. IEEE Trans. Ind. Electron., vol. 62, No. 2, pp. 1134–1139, 2015.

[7] Idkhajine, L., Monmasson, E. , Naouar, M. W., Prata, A. and Boullaga, K. Fully Integrated FPGA-based Controller for Synchronous Motor Drive. IEEE Trans. Ind. Electron., Vol. 56, No. 10, pp. 4006–4017, 2009.

[8] Cordero, R., Pinto, J. O. P., Suemitsu, W. I. and Soares, J. O., Improved Demultiplexing Algorithm for Hardware Simplification of Sensored Vector Control Through Frequency-Domain Multiplexing. IEEE Trans. Ind. Electron., vol. 64, No. 8, pp. 6583–6548, 2017.

[9] Bergas-Jané, J., Ferrater-Simón, C., Gross, G.,Ramírez-Pisco, R., Galceran-Arellano, S. and Rull-Duran, J., High-Accuracy All-Digital Resolver-to-Digital Conversion. IEEE Trans. Ind. Electron., vol. 59, no. 1, pp. 326–333, 2012.

[10] Sarma, S., Agrawal, V. K. and S. Udupa, S. Software-based Resolver-To-Digital Conversion Using a DSP. IEEE Trans. Ind. Electron., Vol. 55, No. 1, pp. 371 –379, 2008.

[11] V. Smidl, S. Janous, Z. Peroutka and L. Adam, "Time-optimal current trajectory for predictive speed control of PMSM drive," in Proc. 2017 IEEE Int. Symp. Predictive Control of Electrical Drives and Power Electron., 2017, pp. 83–88.

[12] J. C. Moreno, J. M. Espi Huerta, R. G. Gil and S. A. Gonzalez, "A robust predictive current control for three-phase grid-connected inverters," IEEE Trans. Ind. Electron., vol. 56, no. 6, pp. 1993–2004, June 2009.

[13] E. F. Camacho and C. Bordons. Model Predictive Control. Springer, New York, 2004, pp. 13–124.

[14] Wang, L.,Model Predictive Control System Design and Implementation Using Matlab®. London: Springer, 2009.

[15] K. Ogata, "Modern Control Engineerign," 4th edition, Prentice Hall, vol. 2, pp2.

[16] Ruchika, Neha Raghu. Model Predictive Control: History and Development. International Journal of Engineering Trends and Technology (IJETT) – Volume 4 Issue 6- June 2013.

[17] S. J. Qin,, Thomas A. Badgwell 2003. A survey of industrial model predictive control technology. Control Engineering Practice 11, 2003, pp.733 – 764.

[18] D.W. Clarke, C. Mohtadi, and P.S. Tuffs. Generalized Predictive Control. Part I.The Basic Algorithm. Automatica, 23(2):137–148, 1987.

[19] D.W. Clarke, C. Mohtadi, and P.S. Tuffs. Generalized Predictive Control. Part II. Extensions and Interpretations. Automatica, 1987, 23(2):149–160.

[20] Rossiter, J. A., Model-based Predictive Control: a practical approach. Boca Raton: CRC, 2004.

[21] Grassi, M.A.S., Batista, E.A. Andrea, C.Q., Garcia, R.C., Do Nascimento, E.S.,. Controladores Preditivos para um sistema Turbina-Gerador: Testes em Cossimulação e FPGA-In-The-Loop, 2018.

On the Application of a Power Electronics-based Arc-Flash Suppressor

Fernando Venâncio Amaral
Electronics and Biomedical Department
CEFET-MG
Belo Horizonte, Brazil
famaral@cefetmg.br

Matheus H. M. Zanchetta Oliveira
Electrical Engineering Department
CEFET-MG
Belo Horizonte, Brazil
matheuszanchetta96@gmail.com

Claudio Alvares Conceição
Department of Electrical Maintenance
Petrobras
Betim, Brazil
claudioac@petrobras.com.br

Sidelmo Magalhães Silva
Electrical Engineering Department
UFMG
Belo Horizonte, Brazil
sidelmo@ufmg.br

Cleber Onofre Inácio
Research and Development Department
Petrobras
Rio de Janeiro, Brazil
cleberoi@petrobras.com.br

Braz de J. Cardoso Filho
Electrical Engineering Department
UFMG
Belo Horizonte, Brazil
braz.cardoso@ieee.org

Abstract—This paper presents an original investigation on the application of a thyristor-based bidirectional switch for arc-flash mitigation in industrial and commercial power systems. This study includes the modeling and digital simulation of the proposed solution under each one of the possible combinations of shunt arcing faults in these systems. The results are presented in order to allow an initial evaluation of the electrical stresses that the thyristors would be subjected to, as well as the effectiveness of the solution. The simulation of a real industrial, low-voltage (LV) motor control center (MCC) was performed in the ATP-EMTP™ software. The results show that besides the variability in the currents that each thyristor conducts depending on the specific fault condition, the value is clearly within the surge capability of commercially available devices. Moreover, it is shown that the proposed solution is effective in reducing the voltage across the arcing path, guaranteeing a safe environment for workers and preserving the MCC integrity.

Index Terms—Industry applications, Accident prevention, Electrical safety, Fault protection, Power semiconductor switches

I. INTRODUCTION

According to the *Occupational Safety and Health Administration* (OSHA), arc-flash (AF) events are the cause of approximately 80% of electrically-related accidents and fatalities among qualified electrical workers in industry in USA [1], [2]. Reference [3] states that 50% of the people in burn care units in the USA have AF injuries, and between one and two out of five of these individuals do not survive these lesion. According to *Electricite de France*, 21% of electrical injuries, including AF, tend to be permanent [4].

Between April 1984 and June 2007, 37 LV burn-related electrical fatalities and 298 LV burn-related injuries were revealed by a set of records made by OSHA [5]. For both

This work has been funded by brazilian agencies CAPES, CNPq, and FAPEMIG, and by Petroleo Brasileiro S.A. under research grant ANEEL PD-00553-0052/2017.

cases, the most of the registered events - 93.29% and 96.97%, respectively - corresponded to 480-V power systems.

A. Arcing Faults in Low-voltage Power Systems

An electrical power system is, in general, composed of a wide range of equipment, which make it reasonably complex and subjected to failures. Within this context, faults are a critical concern. It has been shown that 95% of all faults involve ground and that single line-to-ground fault currents can exceed three-phase fault current levels [6]. Moreover, single-phase faults can evolve into three-phase faults within 5 ms [7]. These types of faults are referred to as *shunt faults*. Other types of fault conditions that may be of interest include the so-called *series faults*, for example one or two lines open in case of connection failure, poor joint between the conductors of the same phase, loose cable lugs, etc. [8]. Bolted fault studies are the first to be taken into account when a power system is being analysed, despite the fact that arcing fault is the type that occurs the most [9]. In a bolted fault, the fault resistance is zero and there is no arcing, and no arc-flash hazard. In an arcing fault, however, the current path can be modeled as a high resistance with a non-linear time-varying component, which represents a fast, irregular change in arcing geometry due to convection, plasma jet, electromagnetic forces, among other effects as transient recovery voltage between conductors of a three-phase busbar [10].

B. Commercially Available Arc-flash Solutions

Arc-flash quenching can not rely on traditional protective devices like circuit-breakers (CB) or fuses, since the clearing times of such devices are not low enough to reduce the incident energy to an acceptable level. The fastest electromechanical CB nowadays operates no faster than three power-frequency cycles, for example [11]. Various manufacturers have different technologies or methods applicable for arc-flash suppression.

All of these technologies rely on a detection system which is able to initiate a trip within 1-2 ms [2]. Once triggered, a high-speed short-circuit device operates within a few milliseconds thereafter. These include pyrotechnical pressure elements, Thomson coils, micro gas cartridges, arcing chambers, or spring mechanisms assisted by an electromagnetic repulsion system [11], [12]. Unfortunately, these technologies are based on electromechanical devices, which are not fast enough to the complete mitigation of incident energy. Moreover, the replacement of at least part of their components are necessary after an operation, increasing the overall cost of the solution and leading to extra time required for maintenance and additional unavailability of the power system [12].

C. Proposed Solution

The application of electronic devices to operate as an arc elimination switch is an effective advance from electromechanical devices, since it is fast enough to manage the fault, practically zeroing incident energy value, besides being more promising in terms of withstanding as many operations as it can be verified during the lifetime of an industrial switchgear or MCC. In this sense, the main contribution of this paper is to introduce the application of power thyristors as the building block of an AC switch applicable to arc-flash suppression.

II. Electronic Arc-flash Suppressor Realization

A. Discrete-time Modeling and Simulation

The proposed AC switch is composed by two antiparallel-connected thyristors, as shown in figure 1, and the overall solution is composed by three AC switches, each one connected from one of the system phases to ground. Once an arcing fault is detected, the proper group of switches necessary to quench the arcing path are operated. Since the closing of the AC switches results in a bolted fault on the involved phases, a high surge current flows in the thyristors. The power system is generally designed to withstand this stress, so must be the AC switches.

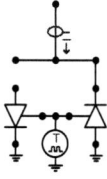

Fig. 1. Proposed electronic AC switch.

Figure 2 presents a simplified schematic of a typical LV industrial MCC busbar feeding three induction motors (groups "mot"). Possible arcing faults are represented by the groups "arc", and the arc-flash suppressor is represented by the groups called "sup".

An industrial 480-V system of an oil refinery was modeled. It consists of a MCC busbar which is fed by a 2 MVA, 13.8-0.48 kV Δ-Y transformer with a high resistance grounded secondary, which limits phase-to-ground bolted fault current to

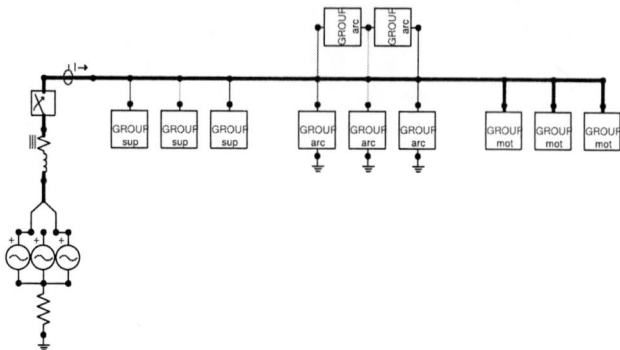

Fig. 2. Typical low-voltage power system, arc-flash and suppression system.

5 A. This busbar feeds three grid-connected induction motors through insulated cables. Table I presents some important parameters of the system.

TABLE I
MOTOR PARAMETERS

Motor	Nominal power	Short-circuit current*	Starting time
1	100 cv	30.1 kA	15 s
2	125 cv	30.1 kA	5 s
3	60 cv	14.2 kA	1.46 s

*Symmetrical, at the point of motor connection.

Figures 3 and 4 present arcing path implementation within ATPDraw [13] - the last one was developed using the *Transient Analysis of Control Systems* (TACS) tool to calculate arcing path conductance g according to Cassie's model in order to take into account its dynamic behavior, as shown in equation (1):

$$g = g_{min} + \frac{e_a \cdot i_a}{E_0^2} - \theta \cdot \frac{dg}{dt} \qquad (1)$$

where e_a and i_a are arcing path voltage and current, respectively. The parameter E_0 is a momentarily constant steady-state arc voltage, which strongly governs the level of arc voltage at high currents [14]. The constant g_{min} is a finite, small value of conductance between any two electrodes when the arc is absent. Finally, θ is the arcing path time constant, where

$$\theta = \theta_0 + \theta_1 \cdot e^{-\alpha \cdot |i_a|} \qquad (2)$$

and

$$i_a = g \cdot e_a \qquad (3)$$

The constants θ_0, θ_1 and α define the arcing voltage waveform [14]. Table II shows typical values for a 480-V MCC, considering a clearing distance of 25 mm between live parts and between live parts and ground. IEEE Std 1584-2018 - *Guide for Performing Arc Flash Calculations* adopts default bus gaps of 32 mm for LV switchgears and 25 mm for LV MCCs [15].

978-1-7281-4181-7/19 $31.00 © 2019 IEEE

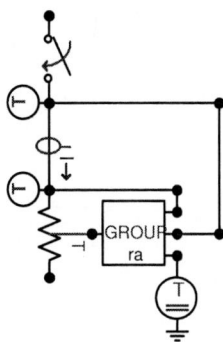

Fig. 3. Arcing fault implemented in ATPDraw.

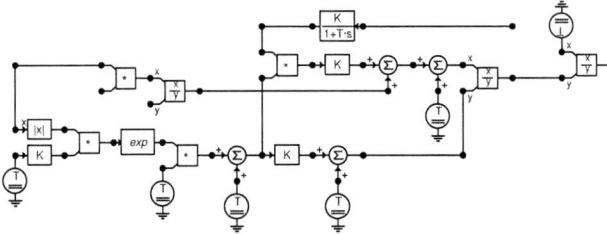

Fig. 4. Discrete-time calculation of arcing fault resistance in TACS.

B. Operation Strategy

The operation of the proposed arc-flash suppressor is ideally dependent on knowing which type of fault is taking place so the proper set of phase-to-ground AC suppression switches can be closed to mitigate the flash. However, arcing fault analysis is much harder to be accomplished than bolted fault. Each arcing fault is unique and cannot be repeated, since there are many variables involved. The probability of two identical arcing faults in the real world is a physical impossibility because of the random nature of the arcing in a plasma cloud [16]. Even if current transformers are properly installed to detect phase current increasing, one must keep in mind that LV arcing fault magnitude is always lower than that of a bolted fault. Arcs for ground fault can be sustained even at low values of ground fault currents of the order of 800 A, which may not be any more than the load current [17]. Finally, it should be highlighted the fact that if an arcing fault is not properly eliminated at its initial instants, it can evolve to a three-phase fault, which could increase further the harm to people and equipment. In this sense, the safest approach consists on closing the three AC switches disregard the type of fault. An arc-flash relay detects the arcing fault and immediately signals it to the control system of the electronic AC switch.

TABLE II
ARCING FAULT PARAMETERS

Parameter	g_{min}	E_0	θ_0	θ_1	α
Value	0.008 Ω^{-1}	200 V	50 μs	10 μs	0.0005

C. Interaction with the Existing Protective System

Since an arc-flash condition will result in a three-phase, bolted short-circuit on the power system, it is expected that the existing protective devices operate to clear this fault. Besides the current fed into the AC switches by the grid, motor contribution must be accounted for. The initial current magnitude depends upon the instant of the short-circuit. Once the fault starts, the *Magneto-Motive Force* (MMF) in the motor windings will decay depending on motor and load inertia, as well as the magnitude of the current back feed into the AC switches [17]. Since the complete opening of the best commercially available CB takes a time interval around three power-frequency cycles, this will be the duration of the surge current that the AC switches must be designed to withstand.

III. RESULTS

The model presented in figure 2 was simulated for each one of the shunt faults possibilities involving phase A: single-phase-to-ground, multiple-phase-to-ground, phase-to-phase, and three-phase. The main reasons for this choice are: (i) shunt faults are the most common in a power system, especially those involving the ground; and (ii) this set of simulations allow a sensitivity analysis comparing the electrical stresses that the SCRs of a phase AC switch are subjected to. The results are presented below. In addition to the current waveforms, the arc energy $e_a \cdot i_a \cdot \Delta t$ and the Joule integral $i_T^2 \cdot \Delta t$ of thyristors of phase A are calculated and plotted. At the time the fault starts, motors *1, 2* and *3* are running at rated speed and 50%, 21% and 42% load, respectively (the motors were preset to be running near rated speed at the beginning of the simulation). The fault starts at $t = 750$ ms. It is considered that the suppressor operates immediately after tripped (just 1 ms after arcing fault starting, i.e., $t = 751$ ms) and that CBs (both power system's and motor's) are tripped at the same time as the suppressor, but takes three 60-Hz cycles after fault starting to open (on zero crossing of phase currents at $t = 800$ ms).

A. Phase-to-ground Faults

Figures 5 to 8 show relevant results regarding an phase-to-ground arcing fault. Figure 5 demonstrates suppressor effectiveness, since arcing energy is kept below 0.015 cal. The threshold value of 1.2 cal/cm^2 is internationally recognized as the energy density above which a second-degree burn may occur. Figure 6 shows the current surge that the electronic suppressor is subjected to. Such a high amplitude, high DC level, were expected since a bolted three-phase short-circuit is produced by the suppressor. Figure 7 shows individual motor contribution along with grid contribution for the suppressor current. As it can be noticed, it vanishes in a couple of cycles after fault starting. It is important to highlight the fact that the worst case for the suppressor could be obtained with motor fully running or turned off, depending on the phase difference between the currents of the motors and and between the currents of the motors and the grid. Figure 8 shows motor deceleration during the suppressor operation.

978-1-7281-4181-7/19 $31.00 © 2019 IEEE

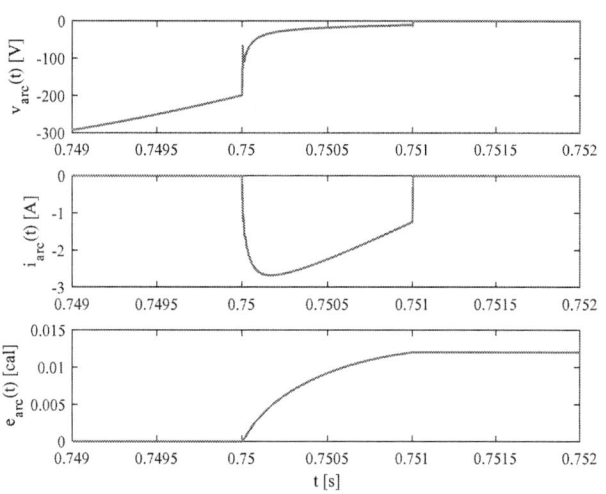

Fig. 5. Phase A to ground arcing fault - arcing voltage, current and energy.

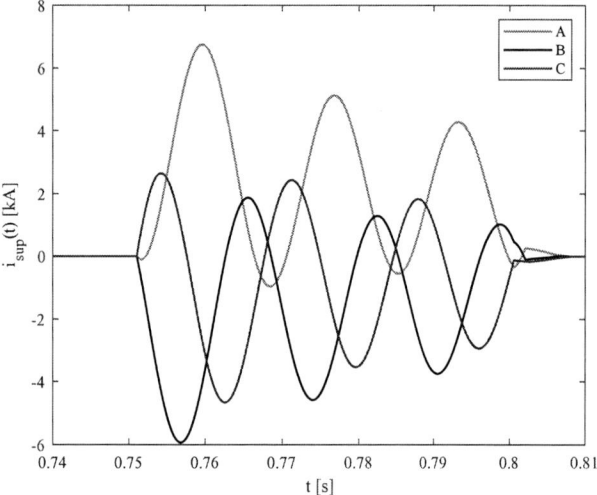

Fig. 6. Phase A to ground arcing fault - suppressor currents.

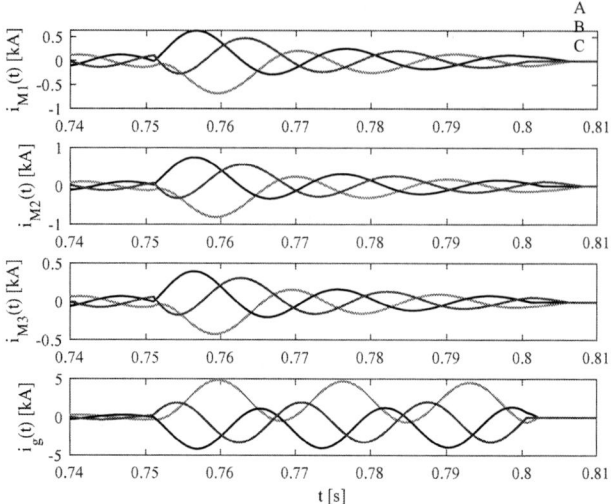

Fig. 7. Phase A to ground arcing fault - motors and grid contributions to short-circuit current.

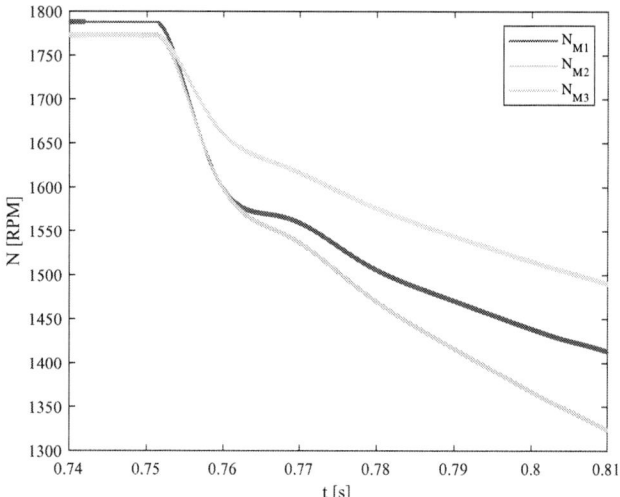

Fig. 8. Phase A to ground arcing fault - motor deceleration.

Figure 9 presents current and Joule's integral waveforms on each thyristor during the suppressor operation. The calculated values will be compared against those obtained for other arcing fault types, as it will be shown next.

Arcing voltages, currents and energy are presented in figures 10 and 11 for a multiphase-to-ground arcing fault. Resultant energy on involved phases (A and B) are kept around 0.1 cal, and both the currents and the Joule's integral at the thyristors are practically the same as the ones verified for an single-phase-to-ground fault.

Finally, figure 12 shows the results obtained for a three-phase-to-ground arcing fault. Notice that in this case, resultant energy at phase C, which is almost at its maximum voltage when the fault starts, is approximately 4 cal, while for phases A and B it is around 2 cal. The Joule's integral waveforms on the thyristors are identical to those of figure 9 and it will not be

shown for convenience. This result was already expected since the suppressor operates almost under the same conditions of the previous case, which is true regardless the type of fault as long as it is not of series type.

B. Phase-to-phase Faults

Figures 13 and 14 presents arcing currents, voltages and energy waveforms verified when an A-to-B and an three-phase (not involving ground) faults takes place, respectively. In the first case, the energy dissipated on the arcing path is around 0.15 cal, which is a very low value. In the second case, though, the energy dissipated on the phase-to-phase (B to C) arcing path, which voltage is around its maximum value when the fault starts, is significantly higher than in all the other simulated cases, reaching approximately 14 cal. Again,

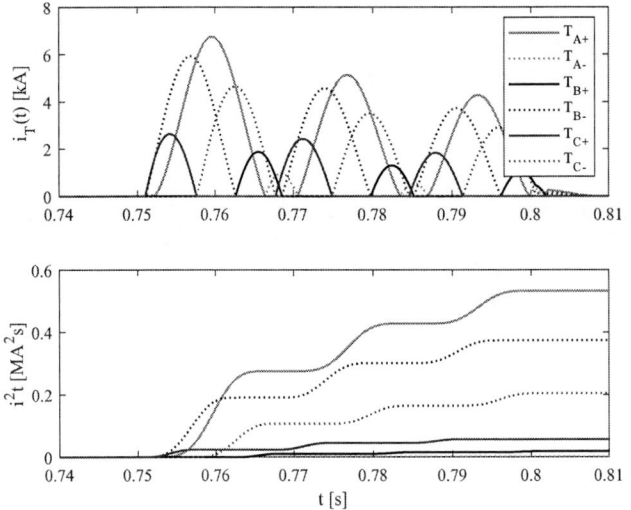

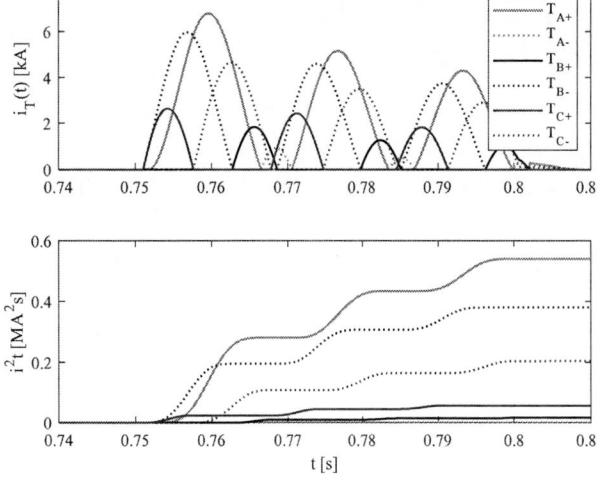

Fig. 9. Phase A to ground arcing fault - current and Joule's integral at suppressor thyristors.

Fig. 11. Phases A and B to ground arcing fault - current and Joule's integral at suppressor thyristors.

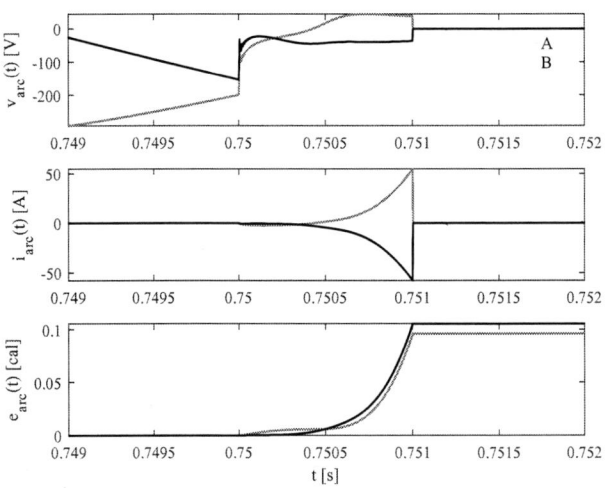

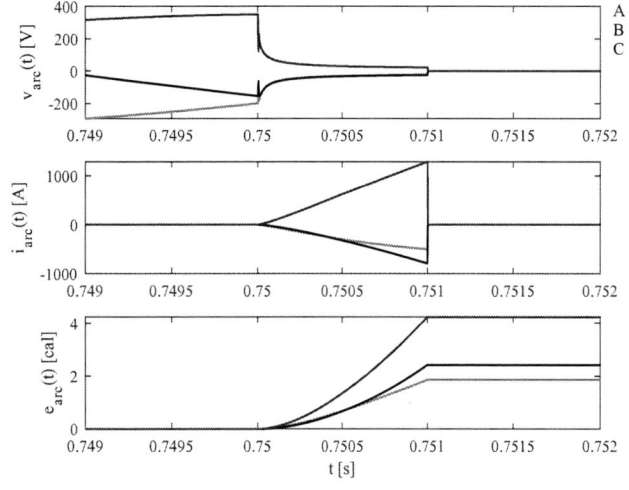

Fig. 10. Phases A and B to ground arcing fault - arcing voltage, current and energy.

Fig. 12. Phases A, B and C to ground arcing fault - arcing voltage, current and energy.

current and Joule's integral at the suppressor thyristors have the same waveforms as those shown before in figures 9 or 11.

C. Comparative Analysis

As it can be verified on the waveforms, DC component of short-circuit current will lead to a huge increase in the current that a thyristor must withstand. Some considerations on this respect have been made in reference [18]. Nevertheless, simultaneous, invariably operation of the electronic switches connected to the three power system phases has the advantage of resulting in no voltage increase on the non-faulty phases in non-effectively grounded systems, besides guaranteeing arcing fault elimination.

Arcing fault energy was verified to be the highest for three-phase instead of single-phase arcing faults. Moreover, the

phases that have the greatest instantaneous voltage result in the greatest energy as well. The calculated values quantify electrical energy, not incident energy. That last one demands that an electro-thermal energy conversion process is completed, which in turn results in an energy density in cal/cm^2. Considering that the operator is 5 cm away from the arcing path and that all the electrical energy is uniformly converted into heat at all directions (i.e., spherically), then a density equivalent to 0.2 cal/cm^2 results, which is greatly below the threshold of 1.2 cal/cm^2. Furthermore, since the majority of three-phase arcing faults start as a single-phase to ground, such an energy value will rarely be a reality provided that the suppressor is fast enough to avoid fault escalation, which is accomplished by the proposed electronic scheme.

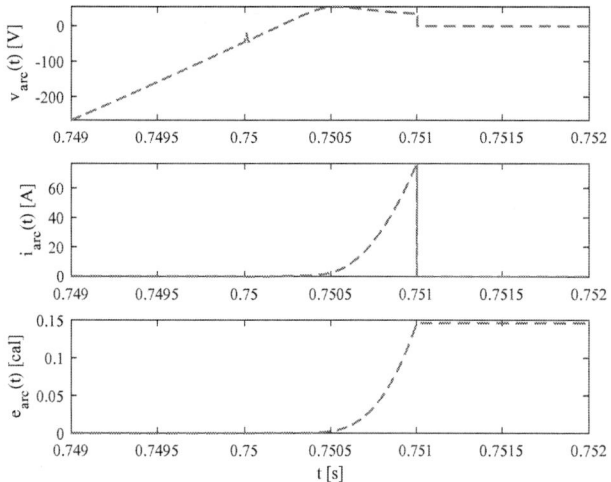

Fig. 13. Phase A to B arcing fault - arcing voltage, current and energy.

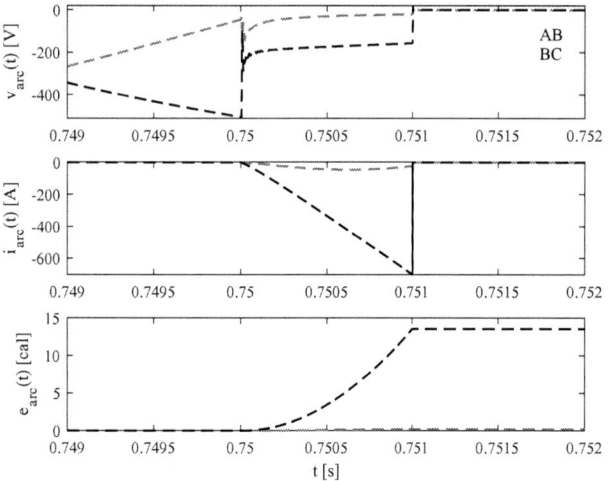

Fig. 14. Three-phase arcing fault - arcing voltage, current and energy.

IV. CONCLUSION

From the suppressor point of view, it does not matter which arcing fault is taking place or what is its conductance value, since a three-phase, bolted fault current will be produced after it is triggered. The currents on the AC switches are actually dependent just on the instant the suppressor starts operating and on the power system and motors operative conditions at this instant. This is a great consideration towards suppressor design, although it is not enough since series arcing faults must be considered as well. Notice that the thyristors that will be subjected to the highest peak current will depend on the specific electrical conditions, but these semiconductors must be selected to account for the worst possible scenarios.

It has been shown in this paper that the proposed electronic arc-flash suppressor is effective for shunt arcing fault elimination in industrial, low-voltage power systems. The presented results did not include series arcing faults, which will be evaluated in a future paper.

ACKNOWLEDGMENT

The authors would like to acknowledge the engineer Armando Guedes for his contribution on the information necessary to set the computational simulations. An acknowledgement is also due in advance to the engineers Gustavo Fontoura and Antonio Gamaliel for their valuable efforts to build the experimental setup that will serve as the basis for the continuity of this research.

REFERENCES

[1] Littelfuse, "How protection relays solve electrical problems," Application note, 2015.
[2] J. Seedorff, "Arc-Flash Protection - Key Considerations for Selecting an Arc-Flash Relay," Application note, 2015.
[3] G. Rocha, E. Zanirato, F. Ayello, and R. Taninaga, "Arc-flash protection for low- and medium-voltage panels," 2011 Record of Conference Papers Industry Applications Society 58th Annual IEEE Petroleum and Chemical Industry Conference (PCIC), Sept. 2011.
[4] Honeywell, "Arc Flash Statistics - The surprising costs of an arc flash," Application note, 2014.
[5] J. P. Nelson, J. D. Billman, and J. E. Bowen, "The effects of system grounding, bus insulation, and probability on arc flash hazard reduction - The missing links," IEEE Transactions on Industry Applications, vol. 50, n. 5, pp. 3141-3152, Sept. 2014.
[6] P. E. Dev Paul and P. B. R. Chavdarian, "Ground fault current protection of a 480Y/277V power system compliance with NEC design requirements," in Industry Applications Society 60th Annual Petroleum and Chemical Industry Conference, Sept. 2013.
[7] J. P. Nelson, J. D. Billman, J. E. Bowen, and D. A. Martindale, "The effects of system grounding, bus insulation, and probability on arc flash hazard reduction - Part 2: Testing," IEEE Transactions on Industry Applications, vol. 51, n. 3, pp. 2665-2675, May 2015.
[8] IEEE Std 399, "Recommended Practice for Industrial and Commercial Power Systems Analysis (Brown Book)," 1998.
[9] M. D'Mello, M. Noonan, H. Aulakh, and J. Mirabent, "Arc flash energy reduction - Case studies," IEEE Transactions on Industry Applications, vol. 49, n. 3, pp. 1198-1204, May 2013.
[10] S. A. Saleh, A. S. Aljankawey, R. Errouissi, and M. A. Rahman, "Experimental performance of the phase-based digital protection against arc flash faults," 2015 IEEE Industry Applications Society Annual Meeting, Oct. 2015.
[11] T. Smith, C. Burnette, and M. Valdes, "Does every millisecond really count - a comparison of protection based arc flash mitigation techniques," 2016 IEEE Pulp, Paper Forest Industries Conference (PPFIC), June 2016.
[12] K. Ahn, Y. Jeong, S. Lee, S. Park, and Y. Kim, "Development of arc eliminator for 7.2/12 kV switchgear," 2015 3rd International Conference on Electric Power Equipment – Switching Technology (ICEPE-ST), Oct. 2015.
[13] B. Bonatto, "EMTP Modelling of Control and Power Electronic Devices," Doctorate Thesis, The University of British Columbia, 2001.
[14] K. Tseng, Y. Wang, and D. Vilathgamuwa, "An experimentally verified hybrid Cassie-Mayr electric arc model for power electronics simulations," IEEE Transactions on Power Electronics, vol. 12, n. 3, pp. 429-436, May 1997.
[15] IEEE Std 1584, "Guide for Performing Arc-Flash Hazard Calculations," 2018.
[16] J. P. Nelson, J. D. Billman, and J. E. Bowen, "The Effects of System Grounding, Bus Insulation, and Probability on Arc Flash Hazard Reduction—The Missing Links," IEEE Transactions on Industry Applications, vol. 50, n. 5, pp. 3141-3152, Sept. 2014.
[17] J. Das, "Arc Flash Hazard Analysis and Mitigation," Wiley, 2012.
[18] F. V. Amaral, S. M. Silva, J. A. S. Brito, and B. J. C. Filho, "Analysis and characterization of an active bypass switch for series connected power conditioners," 2015 IEEE 13th Brazilian Power Electronics Conference and 1st Southern Power Electronics Conference (COBEP/SPEC), Fortaleza, 2015, pp. 1-5.

Deriving Stability Condition for One-Cycle Control with Triangular Carrier by Poincaré Maps

Armando J. G. Abrantes-Ferreira, Lucas do N. Gomes, Robson F. da S. Dias, Luís G. B. Rolim
Universidade Federal do Rio de Janeiro, Rio de Janeiro - RJ, Brasil
abrantes.ferreira@coe.ufrj.br, lucasng@coe.ufrj.br, dias@dee.ufrj.br, rolim@ufrj.br

Abstract—**This work analyzes the stability conditions for the OCC technique with triangular carrier based on Poincaré Maps, defining a minimum inductance value for duty cycle convergence of the control system. Simulations are presented to verify the analytically derived conditions. Comparison between sawtooth-carrier-based and triangular-carrier-based OCC strategies shows that besides the simplification of control circuit and avoiding the need of any averaging scheme for the sensed current, the use of triangular carrier brings improvements on system stability, with lower mandatory minimum inductance.**

Index Terms—**analog control, one-cycle control technique, stability, poincaré maps, boost converter**

I. INTRODUCTION

One-Cycle Control is a nonlinear control technique suitable for control of static converters in several applications, including Power Factor Corrected (PFC) Rectifiers [1], [2], Active Power Filters (APF) [3] and Grid Connected Inverters (GCI) [4]. The technique can be analog or digital based [5] and is distinguishably characterized by having no explicit current control loop and constant switching frequency. Furthermore, on one hand conventional OCC strategies for PFC Rectifiers and APF do not employ grid voltage sensor nor analog multipliers [6]. On the other hand, the classical approach of OCC for GCI demands the employment of grid voltage sensors [7]. In [8] was proposed an OCC-based strategy for light load operation and GCI without grid voltage sensors, where the sensed voltage signals were replaced by an approximation calculated based in quasi-steady state input-output relationship with the duty cycle reconstructed from filtering of the switching output signal. The main advantages of the OCC are simplicity, fast response, stability and robustness [9], [10].

In the modulation process, the carrier has linear (sawtooth or triangular) waveform when controlling the input current [11] even though in control of single-phase topologies operating in CCM the carrier can be nonlinear when controlling diode (quadratic waveform) or switch current (parabolic waveform) [12], [13]. The conventional OCC uses a sawtooth carrier which leads to distortions due to peak current control effect [14]. In [2] the sawtooth carrier was replaced by a triangular carrier, allowing the simplification of control circuitry and avoiding the need of any averaging scheme for the sensed current.

Another issue related to carrier waveform concerns about the minimum conditions for proper control operation which defines a mandatory inductance value for convergence of duty

cycle [14], [15], [16]. In [3] it was presented an expression for the minimum inductance value for triangular carrier based OCC strategies but it was not shown the derivation of that expression. So far, no study has been published concerning the stability analysis of the OCC technique using the triangular carrier.

In this context, this work analyzes the stability conditions for the OCC technique with triangular carrier based on Poincaré Maps. Simulations are presented to verify the analytically derived conditions.

Section II presents the General Principle of analog OCC applied to DC-DC operation of a Full-Bridge Boost Converter, but the results can be extended to AC-DC or DC-AC operation as well. Section III starts with a brief introduction of Poincaré Maps Approach and then applies it to derivation of conditions for stability of the technique.

II. ONE-CYCLE CONTROL TECHNIQUE

Considering the DC-DC Full-Bridge Boost Converter shown in Fig. 1, where v_s, i_s and L_s are the input voltage, current and inductor, respectively. On output side, C_{DC} is the DC-bus capacitor and V_o is the output voltage. R_s is the grid current sensor output gain, which for example for shunt resistive current sensor represents its output resistance.

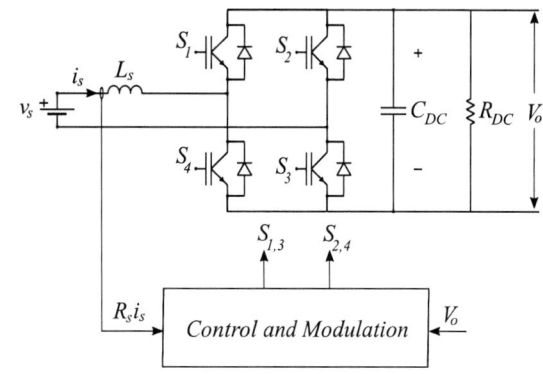

Fig. 1. Model of DC-DC Full-Bridge Boost Converter.

Switches $S_{1,...,4}$ are operated in bipolar modulation mode, where the converter has its switches' states defined by the switching function $s(t)$ referred to switches S_2 and S_4 expressed by (1).

$$s(t) = \begin{cases} 1 & 0 < t < T_{on} \\ 0 & T_{on} < t < T_s \end{cases}, \qquad (1)$$

where:

- T_s is the switching period;
- T_{on} is time that switches S_2 and S_4 remain closed.

The duty cycle $d = T_{on}/T_s$ is defined respecting the limits

$$0 \le d \le 1. \qquad (2)$$

The OCC-based strategies rest on cycle-by-cycle integration of control variables and their control laws are derived from the quasi-steady-state approach, which is built based on the assumption that input voltage being constant within a switching cycle.

The input-output voltage relationship in quasi-steady operation for the system from Fig. 1 is given by

$$V_o = \frac{v_s}{1 - 2d} \qquad (3)$$

where d is the duty cycle referred to the switches S_2 and S_4.

Assuming the converter emulates a resistor R_e such that

$$R_e = \frac{v_s}{\bar{i}_s}, \qquad (4)$$

where $\bar{i}_s$ is the averaged input current.

Manipulating and substituting (4) in (3) and accounting R_s, the OCC law is obtained

$$R_s \bar{i}_s = (1 - 2d)V_m = v_c, \qquad (5)$$

where

$$V_m = R_s V_o / R_e \qquad (6)$$

is the output of the DC-bus voltage regulator in steady-state operation which defines the amplitude of the carrier v_c.

The control law (5) considers the average of input current for one switching period. Therefore, in order to obtain the emulated resistance R_e averaging schemes are normally employed in conventional OCC strategies [1], [7].

The modulation process is illustrated in Fig. 2: the averaged sensed input current $R_s \bar{i}_s$ is compared to the carrier v_c to define the duty cycle d for the switches.

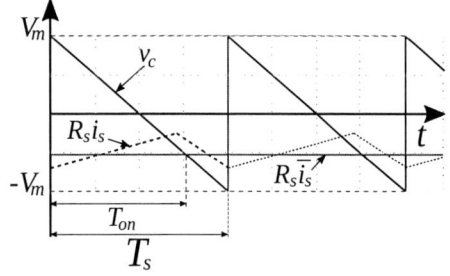

Fig. 2. Carrier of Control with Sawtooth-based Bipolar OCC.

The carrier waveform is obtained by replacing d by t/T_s in (5), which results in a negative slope linear carrier with peak value V_m described by

$$\begin{cases} v_c(t) = Vm(1 - 2t/T_s), & 0 \le t < T_s; \\ v_c(t + T_s) = v_c(t). \end{cases} \qquad (7)$$

The schematic for analog implementation of (5) is shown in Fig. 3-(a). The circuit consists of a PI controller, an integrator with reset, a low pass filter, a comparator (PWM block), a constant frequency clock and a flip-flop circuit for switch drive. A great advantage of this method is that it requires only DC bus voltage and input current sensors, making it very suitable for cost sensitive applications.

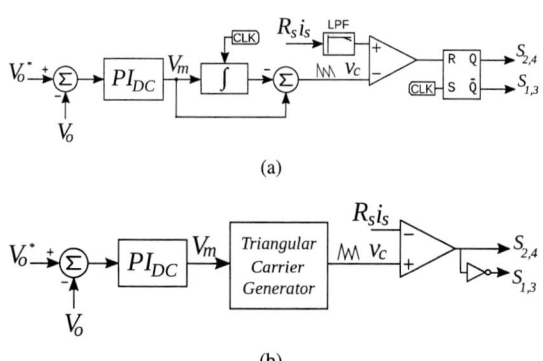

(a)

(b)

Fig. 3. Schematics of One-Cycle Control with analog implementation: (a) Sawtooth Carrier Based and (b) Triangular Carrier Based.

The PWM block can also be done with a triangular carrier [2] as depicted in Fig. 3-(b). The employment of triangular carrier allows to remove the flip-flop together with the clock generator and also avoids the need for averaging schemes for the sensed line current. Furthermore, as it is going to be seen in the next section, it enhances control system stability.

III. STABILITY OF ANALOG OCC STRATEGIES

The stability analysis of control methods can be carried out using the Poincaré Maps, which is an effective approach for analyzing the dynamics and stability of nonlinear systems [17], [18]. In [19] it is studied the stability for the convergence of duty cycle of converters with nonlinear control with sawtooth carrier considering the peak current case. It results in a mandatory inductance value for system stability, which states that operation at null input current is prohibitive

$$L > \frac{1}{2f_s} \frac{V_s}{I_{smin}} = L_{minSAW}. \qquad (8)$$

The stability analysis for the convergence of duty cycle for converters controlled by OCC with triangular carrier is not found in literature. This is going to be carried out in this section. First the concept of Poincaré Maps is introduced, together with the duty cycle convergence ratio λ. Then this approach is applied to derive the minimum inductance for duty cycle convergence of OCC with triangular carrier.

A. Poincaré Maps

A generic n-order continuous-time dynamic system is described by the differential equation

$$\frac{d}{dt}\mathbf{x} = f(t,\mathbf{x}), \tag{9}$$

where $\mathbf{x}$ is a n-dimensional state vector as a function of the independent variable time t.

The state $\mathbf{x}(T_s)$ at time $t = T_s$ is obtained by means of integration of function $f(\cdot)$ within a period T_s starting from an initial state $\mathbf{x}(0)$ at time $t = 0$, therefore

$$\mathbf{x}(T_s) = \int_0^{T_s} f(\tau,\mathbf{x})\,d\tau + \mathbf{x}(0). \tag{10}$$

For systems with periodic dynamics or chaotic behavior their state spaces are bounded in a definite region in phase plane. For these systems defining a cross section transversal to state flow where given an initial condition whithin that section, the orbit will return to the same section, so that the system dynamics can be described by samples of the states at the instants when the state motion crosses the section, giving rise to a lower-dimensional discrete dynamical system that describes the dynamics from crossing to crossing. This approach is called Poincaré Map.

The dynamic behavior of an unidimensional system can be described by the Poincaré Map as shown in Fig. 4 for a transient cycle, where the system starts from Poincaré Section $\sum$ at x_n in time instant $t = nt_s$ and returns to the section at point x_{n+1} at $t = (n+1)T_s$ by the state flow γ.

$\mathbf{P}(\cdot)$ is a mapping relationship from x_n to x_{n+1}, normally nonlinear with respect to x_n, such that [20]

$$x_{n+1} = \mathbf{P}(x_n), n = 0,1,2,\dots \tag{11}$$

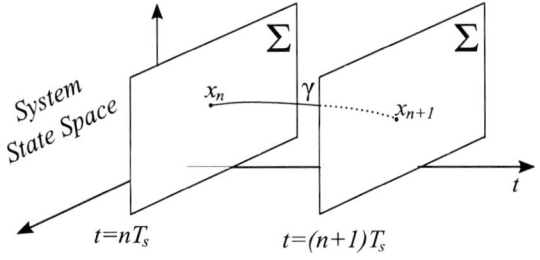

Fig. 4. Poincaré Map and Boundary Value Condition - Adapted from [20].

For steady-state operation the values x_n and x_{n+1} are equal, i.e.

$$x_{n+1} = \mathbf{P}(x_n) = x^*. \tag{12}$$

The point x^* is called fixed point, since its value is not modified by the mapping $\mathbf{P}(\cdot)$. If for any x_n the orbit converges to $x_{n+1} = x^*$ this point is called attracting fixed point, on the other hand, if the orbit diverges the fixed point is called non-attracting. An attracting fixed point of a mapping

relationship corresponds to a steady-state stable point of a continuous time system [21].

It is possible to analyze the stability of a system by means of a mapping relationship $\mathbf{P}(\cdot)$ [22]. For a n-order system the condition for stability of fixed points is that the eigenvalues of the Jacobian Matrix of $\mathbf{P}(\cdot)$ are placed inside the unitary cycle.

For a first order system

$$\lambda = \frac{d}{dx_n}\mathbf{P}(x_n)\Big|_{x_n=x^*}, \tag{13}$$

$$|\lambda| < 1. \tag{14}$$

For a convergent mapping, i.e., $|\lambda| < 1$, there exist three distinct cases for which examples of duty cycle convergence are presented in Fig. 5:

- $-1 < \lambda < 0$ the convergence to x^* occurs in a oscillatory manner;
- $\lambda = 0$ the convergence to x^* occurs in one period T_s;
- $0 < \lambda < 1$ the convergence to x^* occurs in monotonically increasing behavior.

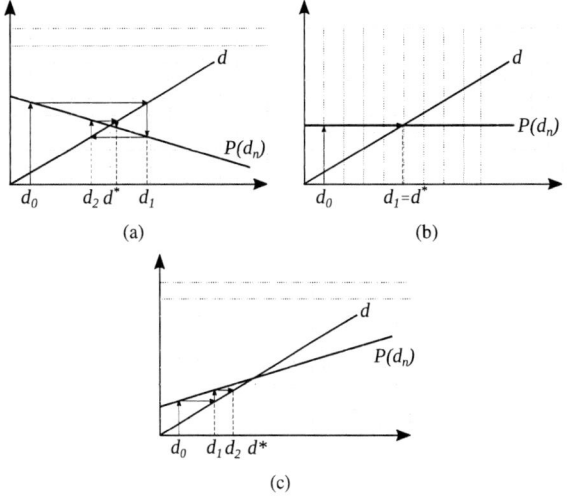

Fig. 5. Duty Cycle Convergence of Convergent Mapping Systems: (a) $-1 < \lambda < 0$, (b) $\lambda = 0$ and (c) $0 < \lambda < 1$.

B. One-Cycle Control with Triangular Carrier

Considering the DC-DC Full-Bridge Boost Converter shown in Fig. 1 controlled by a triangular-based OCC strategy, the evolution of the signals of sensed current $R_s i_s$, triangular carrier v_c and duty cycle d_n of switches $S_{2,4}$ during a transient are shown in Fig. 6.

The absolute values of inductor current rising slope m_1 and falling slope m_2 and the carrier rising/falling slope m_c are given by

$$\begin{cases} m_1 = \frac{|v_s+V_0|}{L}, \\ m_2 = \frac{|v_s-V_0|}{L}, \\ m_c = 4V_m f_s. \end{cases} \tag{15}$$

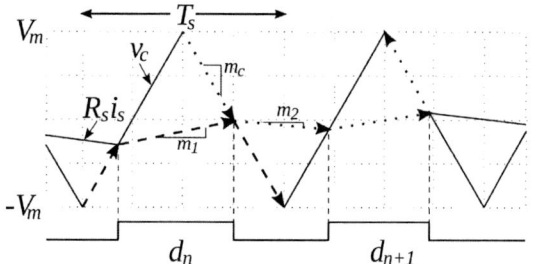

Fig. 6. Waveforms of Sensed Current and Triangular Carrier During Transient.

From Fig. 6, equaling the vertical distances of the segments from the dashed path, and also doing the same procedure with the dotted path, with simple algebra manipulation the relationship that maps the duty cycle d_{n+1} from d_n is derived

$$d_{n+1} = \frac{2m_2 m_c}{(m_1 + m_c)(m_2 + m_c)} + \frac{(m_1 - m_c)(m_2 - m_c)}{(m_1 + m_c)(m_2 + m_c)} d_n,$$
(16)

in the form

$$d_{n+1} = \mathbf{P}(d_n)$$
(17)

The equilibrium point is obtained from $\mathbf{P}(d^*) = d^*$

$$d^* = \frac{m_2}{m_1 + m_2} = \frac{v_s}{V_0}.$$

Defining λ the duty cycle convergence ratio from (13)

$$\lambda = \frac{(m_1 - m_c)(m_2 - m_c)}{(m_1 + m_c)(m_2 + m_c)}.$$
(18)

Considering the stability criterion $|\lambda| < 1$

$$-1 < \frac{(m_1 - m_c)(m_2 - m_c)}{(m_1 + m_c)(m_2 + m_c)} < 1.$$
(19)

Solving

$$\begin{cases} (m_c)^2 + m_1 m_2 > 0; \\ m_c(m_1 + m_2) > 0. \end{cases}$$
(20)

From (15), since the values of m_c, m_1, m_2 are positive, conditions (20) are theoretically always satisfied. Furthermore, according to (18) $\lambda \to 1$ when $m_c \gg m_1, m_2$ ($f_s \to \infty$ or $L \to \infty$) or $m_1, m_2 \gg m_c$ ($f_s \to 0$ or $L \to 0$), so that from the theoretical point of view the system is stable for any finite value of L and f_s. However, L and f_s should be chosen to keep λ away from unity proportionally to model uncertainty. Moreover, as it is going to be seen in this section, there will be minimum f_s and L for duty cycle convergence due to physical limitations of the switch and comparator devices.

First, considering the converter from Fig. 1 with $v_s = 100.0V$, $V_0 = 200.0V$ e $R_{DC} = 40.0\Omega$. The output of the DC bus controller is calculated from (6) assuming $R_s = 1$

$$V_m = 20.$$
(21)

The behavior of λ according to variations in L and f_s are shown in Fig. 7 for $f_s = 30$ kHz and Fig. 8 for L = 550.0 μH, respectively. It can be defined three different regions of convergence:

$$\begin{cases} I: & m_1 < m_c \text{ and } m_2 \leq m_c, \ 0 \leq \lambda < 1; \\ II: & m_1 > m_c \text{ and } m_2 \leq m_c, \ \lambda < 0; \\ III: & m_1 > m_c \text{ and } m_2 > m_c, \ 0 \leq \lambda < 1. \end{cases}$$
(22)

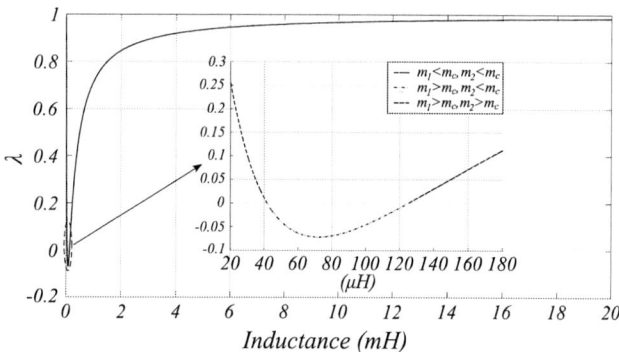

Fig. 7. Behavior of λ with Variation of Inductance L for OCC with Triangular Carrier.

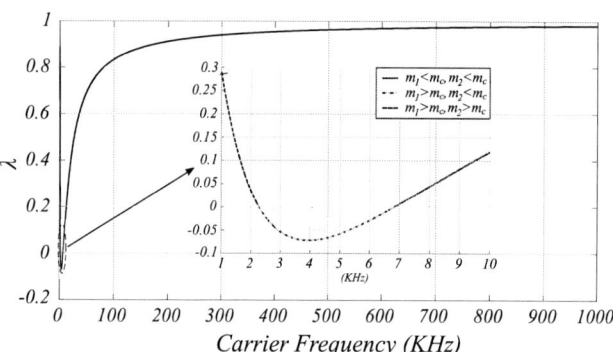

Fig. 8. Behavior λ with Variation of Switching Frequency f_s for OCC with Triangular Carrier.

Following, the system is analyzed in each region of convergence with the aid of simulation results of the system for load step response from null-load to full-load operation at $t = 0$, considering that the DC-bus voltage regulator responds instantaneously after the load transient.

a) Region I: In this case, the current rising slope m_1 is smaller ($\lambda > 0$) or equal ($\lambda = 0$) to the carrier slope m_c and current falling slope m_2 is always smaller than m_c. The convergence of the duty cycle d_n must occur in one switching cycle for $\lambda = 0$ or more switching periods for $\lambda > 0$.

The results for the converter operating with $\lambda = 0.5$ ($fs = 26.69$ kHz) are shown in Fig. 9. From Fig. 9-c it can be seen the exponential decaying behavior on d, as expected. Taking in account the 5% criteria the convergence of the duty cycle occurs after five periods. The current ripple is 25%.

978-1-7281-4181-7/19 $31.00 © 2019 IEEE

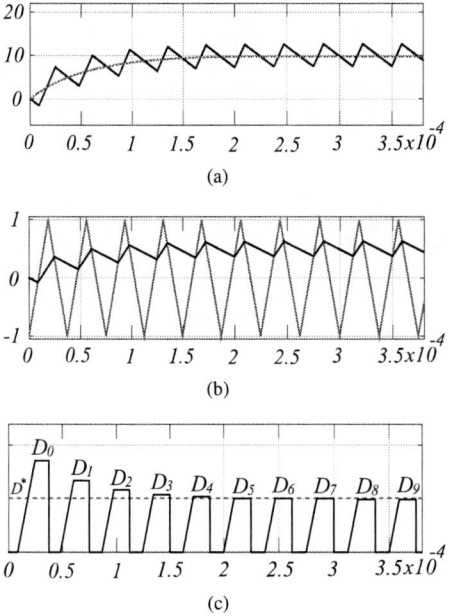

Fig. 9. Converter response for $\lambda = 0.5$: Input current (black) and reference current from equivalent RL circuit (grey). (b) Behavior of duty cycle over time. (c) Carrier (grey), current (black).

In Fig. 10 the results for the converter operating with $\lambda = 0$ ($f_s = 6.82$ kHz) are shown. The convergence occurs in one switching cycle. However, the current ripple becomes 100%, demanding increase in input passive filter, which would delay the convergence time. Hence, there is a trade-off between the response time and current ripple.

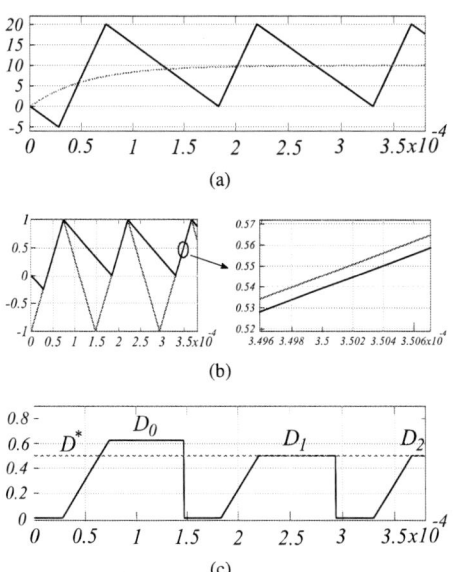

Fig. 10. Converter Response for $\lambda = 0$. (a) Input Current (Solid) and Reference Current from Equivalent RL Circuit (Dashed), (b) Carrier (Grey), Current (black), (c) Behavior of Duty Cycle Over Time.

b) *Region II:* In this region, $-1 < \lambda < 0$ with $m_1 > m_c$ and $m_2 < m_c$, i.e., during the interval of rising slope of the carrier signal its dynamics is slower than that of the sensed current signal, so that the converter will switch as many times as the devices - switches and comparator - allow, which is an undesired situation, since the switch losses will increase. In the falling slope, the carrier is still faster than the current. This situation is depicted in Fig. 11.

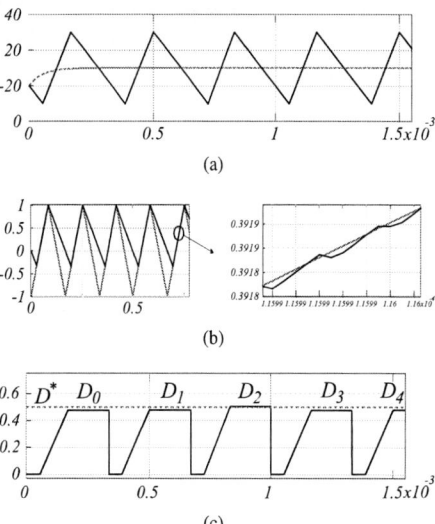

Fig. 11. Converter Response for $\lambda < 0$. (a) Input Current (Solid) and Reference Current from Equivalent RL Circuit (Dashed), (b) Carrier (Grey), Current (Black), (c) Behavior of Duty Cycle Over Time.

c) *Region III:* This region, with $\lambda > 0$ with $m_1 > m_c$ and $m_2 > m_c$, has the worst case scenario, where the sensed input current signal is faster than the carrier during the entire switching period, as illustrated in Fig. 12.

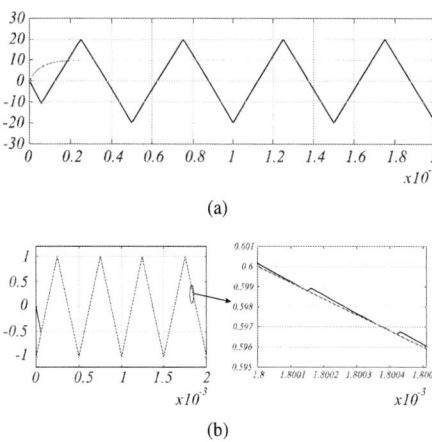

Fig. 12. Converter Response for $\lambda > 0$ in Region III. (a) Input Current (Solid) and Reference Current from Equivalent RL Circuit (Dashed), (b) Carrier (Grey), Current (Black).

1) *Mandatory Minimum Inductance:* Even though from theoretical analysis from (14) there is duty cycle convergence

978-1-7281-4181-7/19 $31.00 © 2019 IEEE

in Regions II and III, these regions should be avoided since the switching frequency would be limited only by the speed response of PWM circuitry. Hence only the Sector I is suitable for proper operation of the converter in the studied conditions.

Therefore, a stability condition for the OCC with triangular carrier can be obtained considering (15) and (22)

$$m_c \geq m_1, \tag{23}$$

$$f_s \geq \frac{m+1}{4L} R_e, \tag{24}$$

where m is the modulation ratio.

The mandatory minimum inductance is defined as

$$L > \frac{m+1}{4f_s} \frac{V_s}{I_{smin}} = L_{minTRI}. \tag{25}$$

From comparison of (8) and (25), and since $0 < m < 1$, it can be seen that $L_{minTRI} < L_{minSAW}$ so that for the DC-DC full-bridge converter controlled by OCC the use of triangular carrier improves the stability of the system, since it requires a lower minimum inductance value, compared to the strategy with sawtooth carrier.

IV. Conclusions

This work has discussed about some features related to the carrier waveform in One-Cycle Control strategies. Compared to sawtooth carrier based OCC strategies, triangular carrier based OCC strategies bring the advantages of simplified control circuitry, absence of averaging schemes for sensed input current and, as derived in this text, it requires a lower minimum inductance value for duty cycle convergence. The determination of this minimum inductance is the main contribution of this work and it was carried with the aid of Poincaré Maps approach, based on the relationships of current and carrier slopes. From the results and discussion, it can be concluded that the application of triangular carrier waveform in modulation process of OCC is a compelling alternative for analog implementation.

Acknowledgment

Authors acknowledge the Coordenação de Aperfeiçoamento de Pessoal de Nível Superior Instituto Nacional de Energia Elétrica (INERGE) and CYTED/MEIHAPER by the support. This study was financed in part by Coordenação de Aperfeiçoamento de Pessoal de Nível Superior - Brasil (CAPES) Finance Code 001,

References

[1] A. S. Lock, E. R. da Silva, M. E. Elbuluk, and D. A. Fernandes. A hybrid current control for a controlled rectifier. In *2010 IEEE Energy Conversion Congress and Exposition.*

[2] A. A. M. Bento, P. K. P. Vieira, and E. R. C. da Silva. Application of the one-cycle control technique to a three-phase three-level npc rectifier. *IEEE Transactions on Industry Applications*, 50(2):1177–1184, August 2013.

[3] A. A. M. Bento, L. F. C. Monteiro, and E. R. C. da Silva. Fast response one-cycle control strategy for three-phase shunt active power filter. In *2017 IEEE Southern Power Electronics Conference (SPEC).*

[4] C. Qiao and K. M. Smedley. Three-phase grid-connected inverters interface for alternative energy sources with unified constant-frequency integration control. In *Conference Record of the 2001 IEEE Industry Applications Conference. 36th IAS Annual Meeting (Cat. No.01CH37248).*

[5] A. J. G. Abrantes-Ferreira. Digital one-cycle control technique with grid voltage measurement applied to three-phase power factor corrected rectifiers and active power filters. M.Sc. dissertation, Federal University of Rio de Janeiro, Rio de Janeiro, Brazil, 2019.

[6] C. Qiao, T. Jin, and K. M. Smedley. Unified constant-frequency integration control of three-phase active-power-filter with vector operation. In *2001 IEEE 32nd Annual Power Electronics Specialists Conference.*

[7] C. Qiao and K. M. Smedley. Unified constant-frequency integration control of three-phase standard bridge boost rectifiers with power-factor correction. *IEEE Transactions on Industrial Electronics*, 50(1):100–107, January 2003.

[8] A. A. Ghodke and K. Chatterjee. One-cycle-controlled bidirectional three-phase unity power factor ac-dc converter without having voltage sensors. *IET Power Electronics*, 5(9):1944–1955, November 2012.

[9] Taotao Jin, Lihua Li, and Keyue Smedley. A universal vector controller for three-phase pfc, apf, statcom, and grid-connected inverter. In *Nineteenth Annual IEEE Applied Power Electronics Conference and Exposition, 2004. APEC'04.*, volume 1, pages 594–600. IEEE, 2004.

[10] K.M. Smedley and S. Cuk. One-cycle control of switching converters. *IEEE Transactions on Power Electronics*, 10(6):625–633, November 1995.

[11] J. Rajagopalan, F.C. Lee, and P. Nora. A general technique for derivation of average current mode control laws for single-phase power-factor-correction circuits without input voltage sensing. *IEEE Transactions on Power Electronics*, 141(4):663–672, July 1999.

[12] Dragan Maksimovic, Yungtaek Jang, and Robert Erickson. Nonlinear-carrier control for high power factor boost rectifiers. In *Proceedings of 1995 IEEE Applied Power Electronics Conference and Exposition-APEC'95*, volume 2, pages 635–641. IEEE, 1995.

[13] Z. Lai and K.M. Smedley. A family of continuous-conduction-mode power-factor-correction controllers based on the general pulse-width modulator. *IEEE Transactions on Power Electronics*, 13(3):501–510, May 1998.

[14] Chongming Qiao and Keyue Smedley. Three-phase bipolar mode active power filters. *IEEE Transactions on Industry Applications*, 38(1):149–158, 2002.

[15] C. Qiao, T. Jin, and K. M. SMedley. One-cycle control of three-phase active power filter with vector operation. *IEEE Transactions on Industrial Electronics*, 51(2):455–463, April 2004.

[16] A. A. M. Bento and L. F. C. Monteiro. Considerations on one cycle control technique. In *2017 Brazilian Power Electronics Conference (COBEP).*

[17] John Guckenheimer and Philip Holmes. *Nonlinear oscillations dynamical systems, and bifurcations of vector fields.* Springer-Verlag, New York, NY, 1 edition, 1983.

[18] S. Banerjee and G. C. Verghese. *Nonlinear Phenome in Power Electronics: Attractors, Bifurcations, Chaos, and Nonlinear Control.* John Wiley & Sons, Hoboken, NJ, 1 edition, 2001.

[19] K. M. Smedley. Tricks of the trade: Poincare stability analysis of switching converters with nonlinear control. *IEEE Power Electronics Society NEWSLETTER*, 14(1):5–6, January 2002.

[20] Toshiji Kato, Kaoru Inoue, and Yuki Takami. Stability analysis using poincaré map in the time-domain for grid-connected inverter. In *2017 IEEE 18th Workshop on Control and Modeling for Power Electronics (COMPEL)*, pages 1–7. IEEE, 2017.

[21] David C Hamill, Jonathan HB Deane, and David J Jefferies. Modeling of chaotic dc-dc converters by iterated nonlinear mappings. *IEEE transactions on Power Electronics*, 7(1):25–36, 1992.

[22] Shi-Ping Hsu, Art Brown, Loman Rensink, and RD Middlebrook. Modelling and analysis of switching dc-to-dc converters in constant-frequency current-programmed mode. In *1979 IEEE Power Electronics Specialists Conference*, pages 284–301. IEEE, 1979.

Three-Phase Adaptive Frequency Estimator with a Delayed Signal Cancellation Pre-Filter Under Heavily Distorted Grid Conditions

E. F. C. Grabovski, M. L. Heldwein, S. A. Mussa
Federal University of Santa Catarina (UFSC)
Department of Electronics and Electrical Engineering (EEL)
Power Electronics Institute (INEP)
88040-970 PO box:5119 Florianópolis, SC, Brazil
e-mail: edhuado.celli@grad.ufsc.br; heldwein@inep.ufsc.br; samir.ahmad.mussa@gmail.com

Abstract—A reliable and precise frequency detection algorithm is useful for different control strategies, especially on microgrid and aerospace applications. The widely-linear adaptive frequency estimator is based on the Augmented Complex Least Mean Square Filter and, thus inherits its fast dynamic response. However, it presents a high susceptibility to input noise, which might present an unexpected behaviour under distorted grid conditions. Therefore, this work proposes a Delayed Signal Cancellation pre-filter stage to eliminate most of the harmonics to achieve fast estimation response and high rejection of harmonic distortions.

Index Terms—Frequency estimation, signal processing, adaptive filtering.

I. INTRODUCTION

THE frequency of a power system is an important parameter which may regulate power injection, as well as detect faults and limit the operation of the power converter in extreme conditions. The insertion of such power converter units on the electric grid is extremely dependable on synchronization algorithms, which aim to ensure a safe operation of the interface between source and grid and respect grid codes.

Currently, the most employed methods in power electronics are typically associated with frequency and phase-locked loops [1] (FLLs and PLLs, respectively). Other approaches, such as using least mean square (LMS) adaptive filters, Kalman filters [2] and Fourier-transform based methods [3], [4] have been proposed in the literature and present faster dynamic responses if compared to traditional methods. However, such methods are optimally employed in balanced grid conditions or single-phase systems and may present a low numerical stability or extremely complex practical implementation. Methods based on linear adaptive filters are also usually susceptible to non-circular second order noise, which is present in unbalanced systems, and deterministic noise and disturbances, such as grid harmonics.

A solution to non-circular second order noise is presented in [5], [6], although the high implementation complexity might introduce numerical errors due to a complex implementation as addressed in [7]. Such methods are also heavily impacted by quantization errors. A solution to improve the robustness

of linear adaptive algorithms is through the use of a pre-filter stage, which sacrifices the algorithm's rate of convergence.

Another issue is that the trajectory of grid electrical quantities are assumed to be ellipsoidal, i.e., distortions that might provoke unexpected behaviour are neglected, in special during grid faults. A solution to this problem is through the use of analog or digital filters, which may insert an additional dynamic to the estimation, or through the use of a higher order estimation as described in [8]. A disadvantage is the implementation complexity, as well as the slower rate of convergence due to a higher number of local minima for the gradient descent algorithm.

A solution is the use of a delayed signal cancellation (DSC) structure [9]–[12] to pre-condition the electrical quantities for the algorithm, with its drawback lying in the estimation settling time and the implementation complexity of such filters. Other adaptive solutions have been proposed in [10], [13]–[15], although the observed performance is heavily impacted by the frequency estimation technique used in the adaptive filter stage, e.g., the slow dynamic of the line filter of a Synchronous Reference Frame Phase-Locked Loop (SRF-PLL) heavily impacts the final frequency estimation dynamic.

In this sense, these pre-filter structures might represent a solution to the inherent lack of robustness of first order adaptive filters. Therefore, this work proposes a FPGA implementation of a Widely-Linear Adaptive Frequency Estimator (WLE) based on an Augmented-Complex Least Mean Square (ACLMS) adaptive filter using a DSC pre-filter stage as to improve the robustness of the frequency estimation of linear adaptive filters. An inherent characteristic of the proposed algorithm is the high immunity to unbalances and harmonic distortions of the electrical quantities while preserving a relatively fast response of approximately one grid cycle with satisfactory precision.

II. DELAYED-SIGNAL CANCELLATION PRE-FILTER

The DSC filter consists in a network comprised of cascaded delays and gains, which results in a selective filter with the structure illustrated in Fig. 1. Another interpretation of the

978-1-7281-4181-7/19 $31.00 © 2019 IEEE

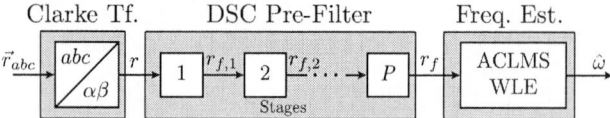

Fig. 1. Block diagram of the proposed frequency estimation technique employing a DSC Pre-Filter and a WLE based on the ACLMS filter.

cascaded delay network is through a series of rotations and time-delays of a complex-valued signal $r \in \mathbb{C}$ spanning over $\mathbb{R}^2$.

In three-phase power systems, a complex-valued signal can be used to express the behaviour of the projection of a given electrical quantity over the $\alpha\beta$ plane. This is obtained via the suppression of the zero-axis component in Clarke's Transformation, as in

$$r = \frac{2}{3} \begin{bmatrix} 1 & e^{-\iota\frac{2}{3}\pi} & e^{\iota\frac{2}{3}\pi} \end{bmatrix} \begin{bmatrix} r_a \\ r_b \\ r_c \end{bmatrix}, \qquad (1)$$

where the set $\{a, b, c\}$ are the phases of a given electrical quantity pertaining to $\mathbb{R}^3$ and with $\iota^2 \triangleq -1$.

The signal r can also be expressed through its Fourier series according to

$$r(n) = \sum_{k \in \mathfrak{K}} r_k(n) = \sum_{k \in \mathfrak{K}} R_k(n)e^{\iota\Omega n}, \qquad (2)$$

with T as a constant sampling time, $\Omega = 2\pi fT = 2\pi/N$ as the discrete angular frequency of r, N as the number of samples per grid period and $n \in \mathbb{N}$. Also, let $\mathfrak{K} \in \mathbb{Z}$ be the set of harmonics, with $\mathfrak{K} \cap \mathbb{Z}_+^*$ as the positive sequence harmonics and $\mathfrak{K} \cap \mathbb{Z}_-^*$ as the negative sequence harmonics, and let $R_k \in \mathbb{R}$ the amplitude of the kth harmonic component.

A delayed signal can then be written as

$$r(n - d) = \sum_{k \in \mathfrak{K}} R_k e^{-\iota k \Omega d} e^{\iota k \Omega n} \qquad (3)$$

with a generic delay $d \in \mathbb{N}$. The objective of the DSC filter is to find a linear combination between the signal and its delayed counterpart as to selectively cancel the harmonic $k_s \in \mathfrak{K}$, resulting in

$$r_f(n) = ar(n) + br(n - d) = \sum_{\substack{k \in \mathfrak{K} \\ k \neq k_s \pm \frac{N}{d}}} R_k e^{\iota k \Omega n}, \qquad (4)$$

with complex-valued constants a and b. The periodicity of $\exp\{-\iota k_s \Omega d\}$ makes the harmonic cancellation effective for $k_s \pm N/d$. Substituting (2) and (3) in (4) results in

$$a = \frac{1}{1 - e^{\iota(k_s - 1)\Omega_s d}}, \quad b = -ae^{\iota k_s \Omega_s d}, \qquad (5)$$

where Ω_s is the filter discrete angular frequency, which ideally equals the input signal fundamental frequency. In addition, constants a and b can be interpreted as complex rotations of the input signal and delayed input, respectively.

However, if the input signal has a small frequency deviation from the filter rejection, i.e., $\Omega_s = \eta\Omega$, the harmonic cancellation is hindered and the positive sequence fundamental component can be written as

$$r_{f,1}(n) = \left(\frac{1 - \cos\left((k_s - 1)\Omega d\right)}{1 - \cos\left((k_s - \eta)\Omega d\right)} \right) R_1(n) \exp\left\{\iota\Omega n + \varphi_{k_s}\right\}$$

$$\varphi_{k_s} = \tan^{-1}\left(\frac{\sin\left((k_s - \eta)\Omega d\right)}{1 - \cos\left((k_s - \eta)\Omega d\right)} \right)$$

$$- \tan^{-1}\left(\frac{\sin\left((k_s - 1)\Omega p\right)}{1 - \cos\left((k_s - 1)\Omega d\right)} \right). \qquad (6)$$

for a single DSC stage. Since the harmonic cancelling is not perfect, the trajectory of the filtered r_f is not necessarily circular. If the DSC filter is employed as a pre-processing stage, the frequency estimation algorithm must be able to reject noncircular second-order noise, i.e, eventual voltage unbalances, as to provide an accurate estimation.

Most of the harmonic content present in r can be rejected through the use of multiple DSC stages. However, the insertion of each DSC stage inserts an according dynamic in the system, which is directly tied to the number of samples used in the delay structure. Some work also propose the use of cascaded adaptive DSC structures as to make $\eta = 1$ and, thus, provide a better harmonic cancellation. The following section provides an analysis of a fast adaptive frequency estimator, which might also improve the dynamic behaviour of adaptive DSC structures.

III. WIDELY LINEAR FREQUENCY ESTIMATOR

The Widely Linear Frequency Estimator is base on a Widely Linear Estimator (WLE) adaptive filter. A WLE is able to reject second order circular noise, which translates to elliptical trajectories of the three-phase vector projection on the $\alpha\beta$ plane. The idea is to estimate an augmented system using an Augmented Complex LMS (ACLMS) filter, with its domain defined by $\{r, r^*\}$.

Assuming a generic signal r representing a certain electric quantity and containing only the fundamental harmonic (positive and negative sequence), i.e., $\mathfrak{K} = \{-1, 1\}$, the generic signal can be described by

$$r(n) = R_1(n)e^{\iota\Omega n} + R_{-1}(n)e^{-\iota\Omega n} \qquad (7)$$

in accordance with (2).

Since the input vector positive and negative sequence components are orthogonal, this technique provides a mean to estimate two distinct orthogonal systems, named here $\mathbf{h}$ and $\mathbf{g}$. The formulation and stability analysis of the ACLMS-based frequency estimator was presented in [6]. Hence, the ACLMS algorithm is given by

$$\hat{r}(n + 1) = \mathbf{h}^H(n)r(n) + \mathbf{g}^H(n)r^*(n)$$
$$e(n) = r(n + 1) - \hat{r}(n + 1)$$
$$\mathbf{h}(n + 1) = \mathbf{h}(n) + \mu e^*(n)r(n) \qquad (8)$$
$$\mathbf{g}(n + 1) = \mathbf{g}(n) + \mu e^*(n)r^*(n),$$

978-1-7281-4181-7/19 $31.00 © 2019 IEEE

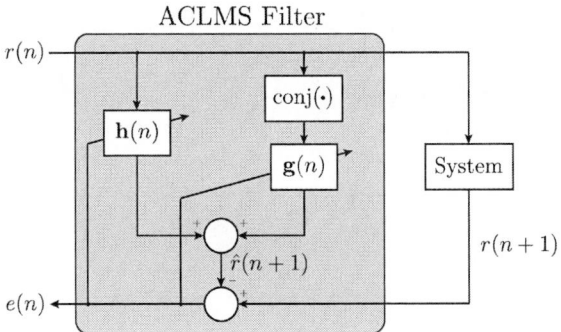

ACLMS Filter

Fig. 2. Augmented Complex Least Mean Square (ACLMS) filter block diagram, which is the basis of the Widely-Linear adaptive frequency estimator.

where $\{\cdot\}^H$ is the hermitian operator, r is the system input, $\hat{r}$ is an input estimate and μ is a learning rate. In this case, the hermitian operator translates into the conjugate operator $\{\cdot\}^*$. A block diagram of the algorithm is presented in Fig. 2.

By applying some algebraic manipulations as in [6], the estimated discrete frequency can be described by

$$\hat{\Omega} = \tan^{-1}\left(\frac{\sqrt{\Im\{\mathbf{h}(n)\}^2 - \|\mathbf{g}(n)\|^2}}{\Re\{\mathbf{h}(n)\}} \right). \quad (9)$$

From inspection of equations (8) and (9), the higher the sampling frequency, the smaller the imaginary part of $\mathbf{h}$. Hence, there are inherent problems with fixed-point representations, as a large number of bits is necessary to accurately represent the frequency, and the utilization of look-up tables to perform calculations is not recommended, as a large number of resources might be required. Quantization errors and the calculation of inverse trigonometric function, which is non-monotonic, limit the numerical stability of the algorithm. Details regarding these implementation aspects, such as quantization errors, fixed-point representations and a method to calculate the inverse trigonometric function are presented in the following section.

IV. IMPLEMENTATION

The proposed strategy was fully implemented in an Altera Cyclone V 5CSEMA4U23C6N FPGA. The DSC-filter sampling frequency was chosen as 160 kHz as to improve the filtering capabilities for higher order frequencies and reduce quantization noise in the subsequent structures. A 16-bit reference source was implemented using an Opal-RT OP5700 real-time simulator connected to the FPGA. The experimental results were obtained using the SignalTap II Logic Analyzer directly from the FPGA, as other methods would mask the precision and numerical stability, which is also present in most estimation algorithms.

The time delays chosen for the pre-filter are shown in Table I and were chosen according to [14] to filter the most common grid harmonics while minimizing the number of operations required, with a as a constant and $n \in \mathbb{Z}$. The time delays can be chosen to alter the algorithm dynamic. Hence, a

TABLE I
DELAYED SIGNAL CANCELLATION (DSC) TIME DELAYS

Pre-filter Stage	Delay	Filtered Harmonics	a	b
1	$T/2$	$2 \pm 2n$	0.5	$0.5e^{\iota\pi}$
2	$T/6$	$3 \pm 6n$	0.5	$0.5e^{\iota\frac{\pi}{2}}$
3	$T/6$	$5 \pm 6n$	0.5	$0.5e^{\iota\frac{\pi}{4}}$
4	$T/12$	$7 \pm 12n$	0.5	$0.5e^{\iota\frac{\pi}{8}}$
5	$T/24$	$13 \pm 24n$	0.5	$0.5e^{\iota\frac{\pi}{16}}$

faster response can be obtained while sacrificing the harmonic rejection of the pre-filter.

As the WLE struggles with resolution problems for high sampling frequencies [7], especially due to the small rotation angles, the output of the DSC-filter was downsampled to a 10 kHz frequency for the frequency estimation algorithm. A Moving Average FIlter (MAF) with a buffer width of 16 samples was implemented as an anti-aliasing filter while also improving the discretization noise rejection.

A block diagram of the proposed DSC implementation is shown in Fig. 3.a, which presents a buffer of length d_p for the p-esim DSC stage. A single multiplexed unit was implemented to perform the signal rotations as to reduce the resource consumption. A state machine was also implemented to control the buffer and multiplexer behaviour, which is shown in Fig. 3.b.

The implementation complexity of the ACLMS algorithm lies in performing the square root and inverse trigonometric functions with a high precision, as a minimal variation results in a large estimation disturbance. Therefore, a 32-iteration pipelined CORDIC-based algorithm with three stages per pipeline operating in hyperbolic mode was implemented to perform the square-root function with, aiming to improve the numerical stability and minimize the number of memory blocks used by the implementation, as a 32-bit look-up table (LUT) is not feasible due to a large memory consumption. Also, a recursive 32-iteration CORDIC operating in rotating mode was implemented to perform the atan2$\{\cdot, \cdot\}$ function with satisfactory numerical accuracy.

V. EXPERIMENTAL RESULTS

The experimental results were separated into steady-state and dynamic behaviour as to analyze both numerical precision and dynamic response of the proposed algorithm under different grid disturbances. All the signals were acquired with a 16-bit numerical accuracy, and the learning rate for all cases was fixed in 0.0625, and the DSC rated frequency was defined as 50 Hz. A trade-off between numerical precision and dynamic behaviour was observed, as a higher learning rate leads to a faster although less precise behaviour.

The numerical precision of the proposed algorithm can be evaluated through the steady-state response for different grid

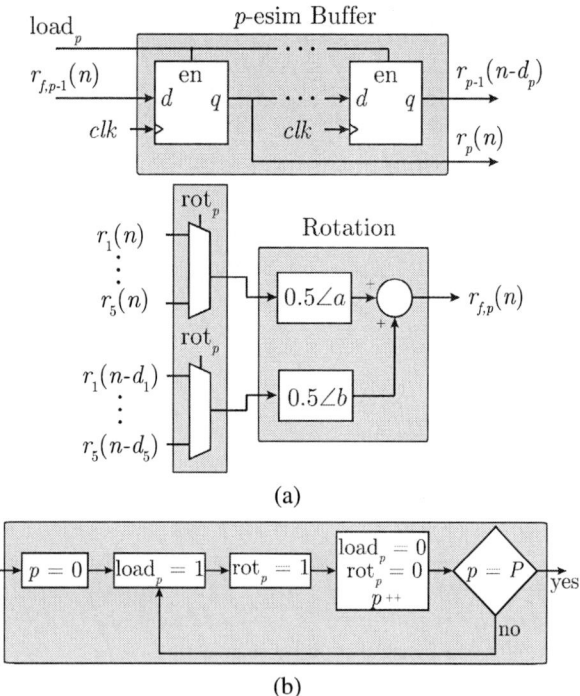

Fig. 3. (a) Block diagram of the DSC pre-filter implementation. (b) Finite-state machine diagram. The FSM states Load$_p$ and Rot$_p$ refer to delay and rotation of the p-esim pre-filter stage.

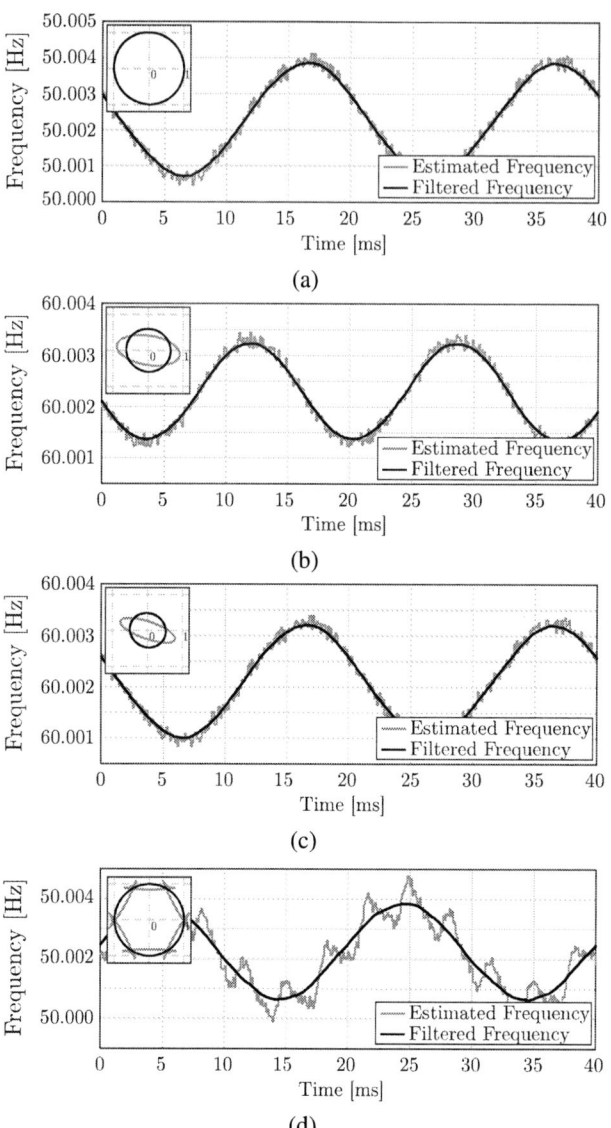

Fig. 4. Steady-state estimated frequency for four different conditions: (a) a circular trajectory signal operating at rated frequency (50 Hz). (b) an unbalanced signal with three-phase amplitude vector given by (1.0, 0.6, 0.4) p.u operating at non-rated frequency (60 Hz). (c) a short-circuit operation with three-phase amplitude vector given by (1.0, 0.6, 0.0) operating at non-rated frequency (60 Hz). (d) a distorted signal of amplitude vector (0.75, 0.75, 0.75) contaminated with 2% 2nd, 5% 3rd, 1% 4th, 30% 5th, 20% 7th, 10% 11th, 5% 13th and 2% 17th harmonics operating at rated frequency (50 Hz).

conditions, as shown in Fig. 4, while the dynamic behaviour can be assessed through Fig. 5.

A. Balanced, non-distorted grid condition

The estimation precision for a pure sinusoidal grid condition operating with the rated DSC frequency of 50 Hz is shown in Fig. 4.a, with a maximum estimation error of 0.008% and an estimation bias of 0.004%, mostly due to quantization errors. The dynamic of a frequency step from 50 Hz to 55 Hz is illustrated in Fig. 5.a, demonstrating a settling time of approximately 20 ms.

B. Unbalanced conditions at non-rated frequency

In this case, the three-phase vector amplitude was set to (1.0, 0.6, 0.4) p.u. with phase displacements of $-15°$ in phase b and $+12°$ in phase c, describing an ellipsoidal trajectory in $\alpha\beta$ coordinates. The filtered signal trajectory presented an eccentricity of 0.98634 due to the phase mismatch of the pre-filter, which demonstrates the need of a WLE approach. The estimation in this case demonstrated to be more precise, with a maximum estimation error of 0.0055%, as shown in Fig.4 b, as the lower amplitude of the three-phase vector has a similar effect of a slower having a smaller learning rate.

C. Short-circuit at non-rated Frequency

Similarly to the previous case, the three-phase vector amplitude was set to (1.0, 0.6, 0), with a maximum frequency estimation error of 0.0055%, as shown in Fig. 4.c. The

filtered signal also demonstrated an ellipsoidal trajectory with an eccentricity of 0.98948. This condition demonstrates the algorithm behaviour under a grid short-circuit. It is worth noting that the algorithm cannot operate with two short-circuited phases.

The dynamic behaviour of a frequency step from 50 Hz to 55 Hz is shown in Fig. 5.b, which in this case is slower than the balanced case due to the smaller signal amplitude.

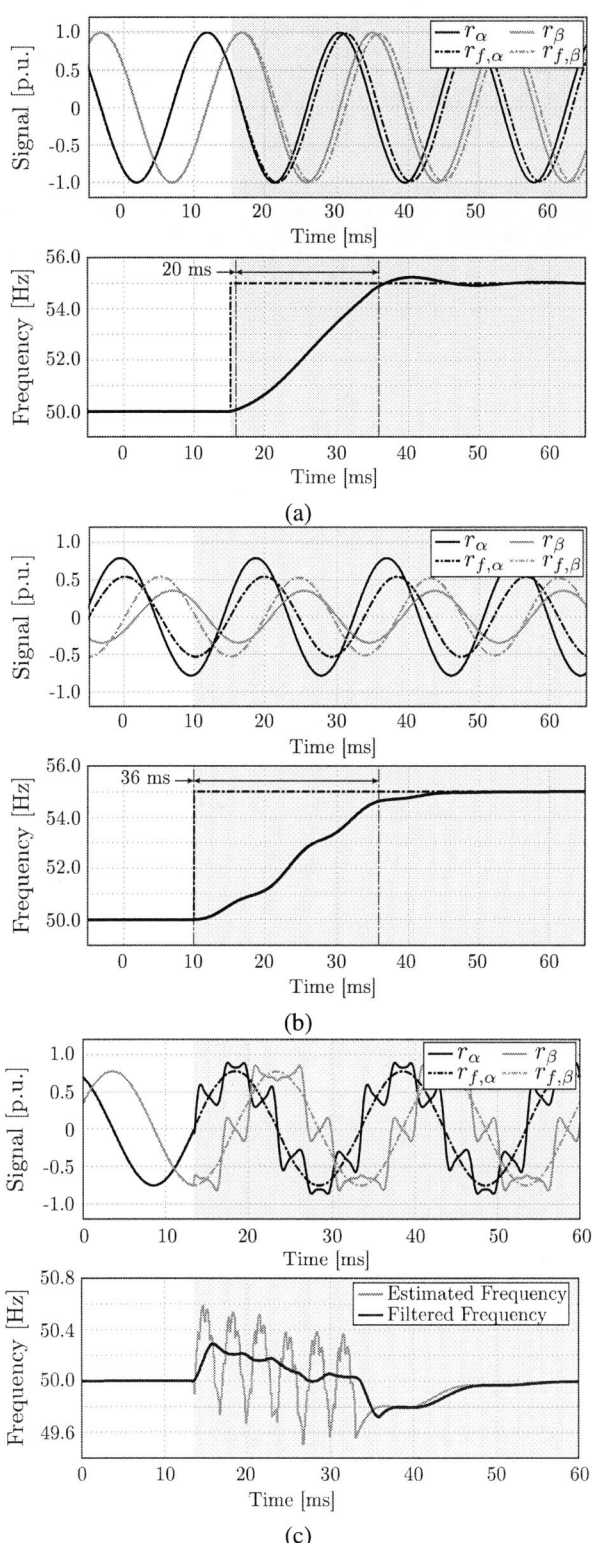

Fig. 5. Dynamic behaviour of the estimated frequency for three different conditions: (a) a circular trajectory signal operating at rated frequency (50 Hz) to 55 Hz. (b) a short-circuit operation with three-phase amplitude vector given by (1.0, 0.6, 0.0) at rated frequency (50 Hz) to 55 Hz. (c) from a circular trajectory signal of amplitude vector (0.75, 0.75, 0.75) to a distorted condition with 2% 2nd, 5% 3rd, 1% 4th, 30% 5th, 20% 7th, 10% 11th, 5% 13th and 2% 17th harmonics operating at rated frequency (50 Hz).

D. Distorted grid conditions

A (0.75, 0.75, 0.75) p.u. signal contaminated with the following grid harmonics: 2% 2nd, 5% 3rd, 1% 4th, 30% 5th, 20% 7th, 10% 11th, 5% 13th and 2% 17th, aiming to emulate a heavily distorted grid condition. The steady-state frequency estimation is shown in Fig. 4.d, with a maximum error of approximately 0.008%, and the dynamic of a change from non-distorted to distorted can be seen in Fig. 5.c, with a maximum averaged frequency deviation of approximately 0.3 Hz during the transient reponse.

VI. CONCLUSION

This work presents a solution to LMS based frequency estimators through the use of a Delayed Signal Cancellation pre-filter, as the WLE does not function properly under grid distortions. The adaptive algorithm presents a fast dynamic response, which is significantly hindered by the pre-filtered stage. However, the settling time of approximately one grid cycle was considered adequate for practical applications, although it can be improved by choosing different time delays for the pre-filter stage.

The maximum steady-state frequency variation of approximately 0.008% even under heavily distorted grid conditions demonstrate an adequate operation, especially for Phase Measurement Units due to the contraint of a high resolution A/D converter. A more thorough evaluation might be necessary for lower resolution measurements due to the quantization noise. However, the pre-filter and the anti-aliasing filter before the decimation processes might be sufficient to lower the Signal-to-Noise Ratio (SNR) of the WLE input signal.

REFERENCES

[1] S.-K. Chung, "A phase tracking system for three phase utility interface inverters," *IEEE Transactions on Power Electronics*, vol. 15, no. 3, pp. 431–438, May 2000.

[2] A. Routray, A. K. Pradhan, and K. P. Rao, "A novel kalman filter for frequency estimation of distorted signals in power systems," *IEEE Transactions on Instrumentation and Measurement*, vol. 51, no. 3, pp. 469–479, June 2002.

[3] J.-Z. Yang and C.-W. Liu, "A precise calculation of power system frequency and phasor," *IEEE Transactions on Power Delivery*, vol. 15, no. 2, pp. 494–499, April 2000.

[4] F. A. S. Neves, H. E. P. de Souza, F. Bradaschia, M. C. Cavalcanti, M. Rizo, and F. J. Rodriguez, "A space-vector discrete fourier transform for unbalanced and distorted three-phase signals," *IEEE Transactions on Industrial Electronics*, vol. 57, no. 8, pp. 2858–2867, Aug 2010.

[5] Y. Xia and D. P. Mandic, "Widely linear adaptive frequency estimation of unbalanced three-phase power systems," *IEEE Transactions on Instrumentation and Measurement*, vol. 61, no. 1, pp. 74–83, Jan 2012.

[6] ——, "A full mean square analysis of clms for second-order noncircular inputs," *IEEE Transactions on Signal Processing*, vol. 65, no. 21, pp. 5578–5590, Nov 2017.

[7] E. F. C. Grabovski and S. A. Mussa, "Three-phase frequency estimator in smart grid applications: Practical issues using fpga," in *2017 IEEE 26th International Symposium on Industrial Electronics (ISIE)*, June 2017, pp. 175–179.

[8] Y. Xia, L. Qiao, Q. Yang, W. Pei, and D. P. Mandic, "Widely linear adaptive frequency estimation for unbalanced three-phase power systems with multiple noisy measurements," in *2017 22nd International Conference on Digital Signal Processing (DSP)*, Aug 2017, pp. 1–5.

[9] Y. N. Batista, H. E. P. de Souza, F. A. S. Neves, R. F. D. Filho, and F. Bradaschia, "Variable-structure generalized delayed signal cancellation pll to improve convergence time," *IEEE Transactions on Industrial Electronics*, vol. 62, no. 11, pp. 7146–7150, Nov 2015.

[10] S. Golestan, J. M. Guerrero, and J. C. Vasquez, "Hybrid adaptive/nonadaptive delayed signal cancellation-based phase-locked loop," *IEEE Transactions on Industrial Electronics*, vol. 64, no. 1, pp. 470–479, Jan 2017.

[11] Y. F. Wang and Y. W. Li, "Three-phase cascaded delayed signal cancellation pll for fast selective harmonic detection," *IEEE Transactions on Industrial Electronics*, vol. 60, no. 4, pp. 1452–1463, April 2013.

[12] F. A. S. Neves, M. C. Cavalcanti, H. E. P. de Souza, F. Bradaschia, E. J. Bueno, and M. Rizo, "A generalized delayed signal cancellation method for detecting fundamental-frequency positive-sequence three-phase signals," *IEEE Transactions on Power Delivery*, vol. 25, no. 3, pp. 1816–1825, July 2010.

[13] S. Golestan, F. D. Freijedo, A. Vidal, A. G. Yepes, J. M. Guerrero, and J. Doval-Gandoy, "An efficient implementation of generalized delayed signal cancellation pll," *IEEE Transactions on Power Electronics*, vol. 31, no. 2, pp. 1085–1094, Feb 2016.

[14] P. S. B. Nascimento, H. E. P. de Souza, F. A. S. Neves, and L. R. Limongi, "Fpga implementation of the generalized delayed signal cancelation – phase locked loop method for detecting harmonic sequence components in three-phase signals," *IEEE Transactions on Industrial Electronics*, vol. 60, no. 2, pp. 645–658, Feb 2013.

[15] S. Golestan, J. M. Guerrero, J. C. Vasquez, A. M. Abusorrah, and Y. Al-Turki, "Research on variable-length transfer delay and delayed-signal-cancellation-based plls," *IEEE Transactions on Power Electronics*, vol. 33, no. 10, pp. 8388–8398, Oct 2018.

Enhanced Phase-Shifted Carrier PWM Applied to 3-Phase Multilevel Coupled Inductors Inverters

Emerson L. Soares
Dept. of Electrical Engineering
Federal University of Campina Grande
Campina Grande, Brazil
emerson.soares@ee.ufcg.edu.br

Lucas Fabrício M. de Lucena
Dept. of Electrical Engineering
Federal University of Paraíba
João Pessoa, Brazil
lucas.lucena@cear.ufpb.br

Nady Rocha
Dept. of Electrical Engineering
Federal University of Paraíba
João Pessoa, Brazil
nadyrocha@cear.ufpb.br

Cursino Brandão Jacobina
Dept. of Electrical Engineering
Federal University of Campina Grande
Campina Grande, Brazil
jacobina@dee.ufcg.edu.br

Edison Roberto C. da Silva
Dept. of Electrical Engineering
Federal University of Paraíba
João Pessoa, Brazil
ercdasilva@gmail.com

Abstract—The use of coupled inductor inverters (CII) is attracting due of its several advantages as: 3-level output voltage, reduced number of controlled switches and no dead-time requirement. However, while using the conventional phase-shifted pulse width modulation (PS-PWM) shows low performance, space vector techniques presented in the literature requires a complex algorithm implementation. This paper applies an enhanced phase-shifted PWM (EPS-PWM) in order to improve quality of CII phase and line-to-line voltages, also the phase currents. The CII operation principles are discussed, and its performance using the PS- and EPS-PWM are compared in therms of voltage and current harmonic distortions. Simulations and experimental results are presented in order to validate the presented modulation performance.

Index Terms—Multilevel converters, three-phase inverters, split-wound coupled inductor inverters (CII), enhanced phase-shifted carrier modulation (EPS-PWM).

I. INTRODUCTION

Over the past few years, several topologies of multilevel converters have been studied for many applications [1], [2]. They have been presented in order to improve technical features such as: output voltage quality, components quantity and size, and power losses reduction [2], [3]. Among them, the cascaded H-bridge (CHB), the flying capacitor (FC), and the neutral point-clamped (NPC) inverters are the remarkable multilevel topologies [2], [4], [5]. Despite the traditional multilevel inverters are mostly used in high-voltage and high-power applications, there are some studies for low-voltage and low-power applications, as microgrids and photovoltaic systems [6], [7]. In this scenario, multilevel topologies using coupled inductors have also been investigated [8]–[10].

The basic structure of a coupled inductor inverter (CII) is shown in Fig. 1. The use of CII is attracting due: its output voltage has the same number of voltage levels when comparing with the NPC inverter, with only half of the number of controlled switches and no need of dc-link capacitors voltage balance; also no need of dead-time requirement in order to avoid dc-link short-circuit; smaller output filters; and its circulating currents immunity [1]–[3],

This study was financed in part by the Coordenação de Aperfeiçoamento de Pessoal de Nível Superior - Brasil (CAPES) - Finance Code 001.

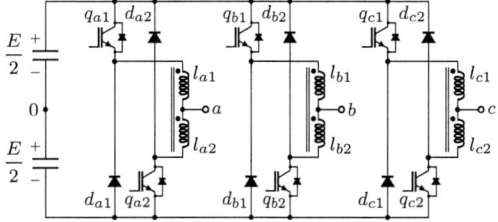

Fig. 1. Three-phase multilevel coupled inductor inverter.

[11], [12]. Furthermore, when using the CII, the interleaved pulse width modulation (PWM) switching of the upper and lower switches on each leg of the inverter is enabled, and are operated with complementary modulating signals [1], [3], [9].

Previous works demonstrate how the PWM modulation strategy greatly interferes with CII performance [1], [3], [13]. The interleaved space-vector PWM (SV-PWM) presented in [1], as well as the discontinuous space-vector PWM (DSV-PWM) presented in [3], reduces inductive losses and high-frequency current ripple by using high-effective inductance switching states. However, the presented methods have disadvantages as: lower quality output currents; and the need to find and select switching states using mapping techniques, which elevates algorithm complexity. Furthermore, there are carrier based methods, as the interleaved phase-shifted PWM (PS-PWM), which applies phase-shifted multi-carrier signals with the same magnitude and frequency, evenly phase shifted within a switching period [14]–[16]. PS-PWM shows low complexity implementation, however the quality of line voltage and phase currents are reduced [3], [14].

In this way, an interleaved phase-shifted PWM (EPS-PWM) strategy is present for multilevel CII in this paper. This very simple technique performs the line-to-line voltages switching exclusively between adjacents level. The EPS-PWM uses two sets of two dynamically allocated phase-shifted carriers, only depending on the pole voltage reference signals location. Therefore, the quality of the line-to-line and phase output voltages, in addition to the output current, is

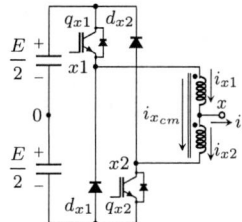

Fig. 2. Coupled inductor converter leg, for $x = a, b, c$.

improved at all levels of modulation when compared to the traditional PS-PWM.

The rest of this paper is organized as follows. First, basic concepts of the coupled inductor inverter is discussed in Section II. Then, the interleaved PS-PWM is reviewed Section III, and the proposed EPS-PWM is presented in Section IV. Simulation and experimental results that illustrate the improvements achieved in terms of total harmonic distortion (THD) and waveform quality in the line-to-line voltages are given in Section V and VI. Finally, conclusions from this paper are presented in Section VII.

II. COUPLED INDUCTOR INVERTER MODEL

Fig. 2 shows a CII leg in detail. There are in total six switching power devices, two for each leg (q_{x1} and q_{x2}); six diodes, also two for each leg (d_{x1} and d_{x2}); and three split-wound coupled inductors, one for each leg (l_x, constituted by l_{x1} and l_{x2}); while x is related to the proper leg (a, b or c). The switches states are defined as: open when $q_{x1} = 0$; and closed when $q_{x1} = 1$. The output current (i_x) and the common mode current ($i_{x_{cm}}$) are defined as:

$$i_x = i_{x1} - i_{x2} \qquad (1)$$

$$i_{x_{cm}} = \frac{i_{x1} + i_{x2}}{2} \qquad (2)$$

The common mode current described in (2) produces a dc flux that can be reduced through a different coupled inductor magnetic core design [17]. There are four continuous conduction modes, which occur when $i_{x1} > 0$ and $i_{x2} > 0$; and five discontinuous conduction modes, which occur when neither a diode nor a switch conducts, also when only one diode or switch conducts [18]. As operation in discontinuous conduction modes deteriorates the output voltage quality, only the conduction in continuous conduction mode is considered.

The voltage between the points $x1$ and $x2$ to the DC-link capacitor midpoint 0 are defined as v_{x10} and v_{x20}. They can be determined observing the operation states shown in Fig. 3. They can also be written as a function of the switches states q_{x1} and q_{x2}:

$$v_{x10} = \frac{E}{2}(2q_{x1} - 1) \qquad (3)$$

$$v_{x20} = \frac{E}{2}(2q_{x2} - 1) \qquad (4)$$

As l_{x1} and l_{x2} have the the same equivalent reluctance circuit, the output voltage v_{x0} can be defined as:

$$v_{x0} = \frac{1}{2}(v_{x10} + v_{x20}) \qquad (5)$$

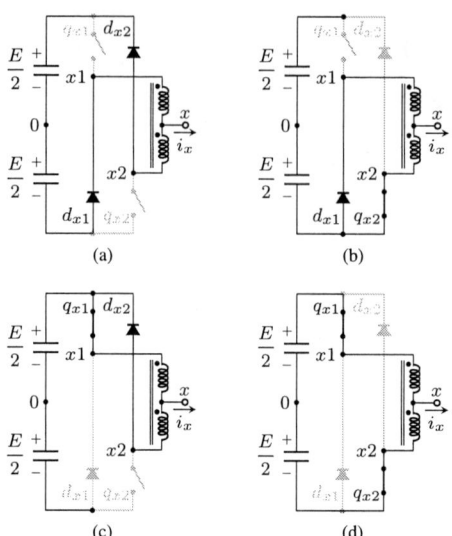

Fig. 3. CII operation in continuous conduction mode, considering the switches states (q_{x1}, q_{x2}): (a) (0,0); (b) (0,1); (c) (1,0); (d) (1,1).

TABLE I
CII SWITCHES STATES AND OUTPUT VOLTAGES.

	q_{x1}	q_{x2}	v_{x10}	v_{x20}	v_{x1}	v_{x2}	v_{x0}
(a)	0	0	$-E/2$	$E/2$	0	0	0
(b)	0	1	$-E/2$	$-E/2$	0	$E/2$	$-E/2$
(c)	1	0	$E/2$	$E/2$	$E/2$	0	$E/2$
(d)	1	1	$E/2$	$-E/2$	$E/2$	$E/2$	0

therefore:

$$v_{x0} = (v_{x1} - v_{x2}) \qquad (6)$$

From (3) to (6), v_{x1}, v_{x2} and v_{x0} can be written as a function of the switches states:

$$v_{x1} = \frac{E}{2}(q_{x1}) \qquad (7)$$

$$v_{x2} = \frac{E}{2}(q_{x2}) \qquad (8)$$

$$v_{x0} = \frac{E}{2}(q_{x1} - q_{x2}) \qquad (9)$$

Table I summarizes the output voltages in function of the switches states (a)-(d), as described in Fig. 3, when operating in continuous conduction mode. There are three output levels: 0, $E/2$ and $-E/2$. Fig. 4 shows the CII simplified leg model and the three-phase equivalent circuit model.

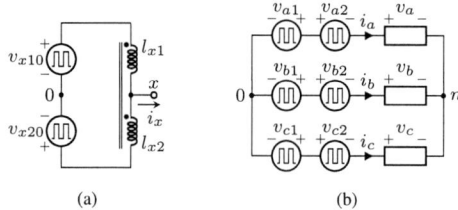

Fig. 4. CII simplified models. (a) Leg model; (b) Three-phase equivalent circuit model.

978-1-7281-4181-7/19 $31.00 © 2019 IEEE

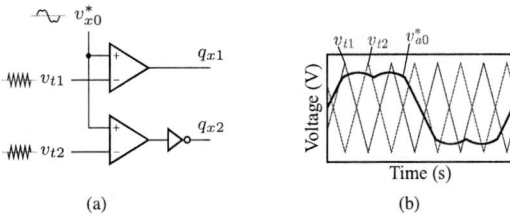

Fig. 5. PS-PWM scheme. (a) Simplified diagram. (b) Reference pole voltage and the set of carriers.

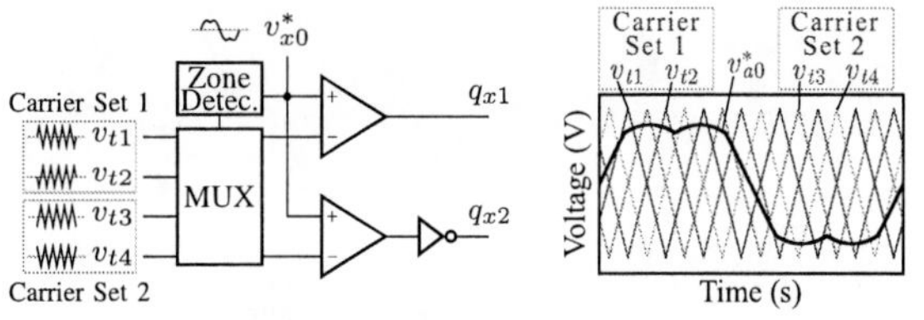

Fig. 6. EPS-PWM scheme. (a) Simplified diagram. (b) Reference pole voltage and the set of carriers.

III. PHASE-SHIFTED CARRIER PWM

The traditional phase-shifted PWM (PS-PWM) uses double high frequency triangular carrier (v_{t1} and v_{t2}) in order to obtain the switches gating signals (q_{x1} and q_{x2}). The carriers have the same frequency and magnitude, but are 180° phase shifted signals. Fig. 5 shows a generalized operating principle of the PS-PWM strategy. The reference pole voltages are compared with the carriers in order to generate the gating signals for the power switches. Still considering $x = \{a, b, c\}$, with x denoting each of the three phases of the system, the CII reference pole voltages can be defined as:

$$v_{x0}^* = v_x^* + v_\mu^* \tag{10}$$

where v_a^*, v_b^* and v_c^* are the reference phase voltage, that were adjusted in function of the desired modulation index (m). The variable v_μ^* is an injected zero-sequence signal for a three-phase inverter, in order to reduce the voltage distortion [19], [20]. The reference value v_μ^* can be calculated taking into account the high and low limits that the reference phase voltage can assume:

$$v_{\max}^* = \frac{E}{2} - \max(v_a^*, v_b^*, v_c^*) \tag{11}$$

$$v_{\min}^* = -\frac{E}{2} - \min(v_a^*, v_b^*, v_c^*) \tag{12}$$

$$v_\mu^* = \mu v_{\max}^* + (1 - \mu)v_{\min}^* \tag{13}$$

The apportioning factor μ can be adjusted ($0 \leq \mu \leq 1$) in order to reduce the total harmonic distortion (THD) of the converter [20].

PS-PWM strategy is a simple implementation method, although there are switching between adjacent levels, degrading the output line- and phase- voltage. This issue can be fixed using an appropriate dynamic selected set of carriers, that will be discussed in the next section.

IV. ENHANCED PHASE-SHIFTED CARRIER PWM

The enhanced phase-shifted PWM (EPS-PWM) was originality proposed in [14], taking into account conventional multiphase voltage source inverters with interleaved parallel-connected legs. There was presented a general method, valid for any number of parallel legs, also for any number of phases. In this paper, this method was used specifically considering a three-phase system and two connected parallel legs for each phase. This consideration is valid, as long as the phase leg representation (as seen in Fig. 2) can be visualized as a two parallel-connected legs phase.

As well as in the PS-PWM strategy, in the EPS-PWM method the gating signals of the two connected parallel legs (q_{x1} and q_{x2}) are obtained comparing the pole voltages, as defined in (10), with a double-carrier PWM. Although, EPS-PWM method uses two sets of carriers: Set 1 is made up of two phase-shifted carriers v_{t1} and v_{t2}, that have the interleaving angle equal to 0° and 180°, respectively; Set 2 is made up of other two phase-shifted carriers v_{t3} and v_{t4}, that have the interleaving angle equal to 90° and 270°, respectively. The selected set of carriers depends on the instantaneous normalized value (ranges from -1 to 1) of the reference pole voltages (v_{a0}^*, v_{b0}^* and v_{c0}^*).

Fig. 6 shows a generalized operating principle of the PS-PWM strategy, explaining how the comparisons are made up. The zone detector block defines which set of carriers should be chosen for each phase, taking into account the expression:

$$sel_x = \begin{cases} 1, & \text{if } v_x^* \geq 0 \\ 2, & \text{otherwise} \end{cases} \tag{14}$$

The variable sel_x stores the information about which carrier set must be used for each phase, and controls the multiplexers that dynamically route carrier Set 1 or Set 2. The EPS-PWM reference zones and the sel_x values for each zone and each phase are illustrated in Fig. 7.

V. SIMULATION RESULTS

In order to verify the effectiveness of the studied modulation, the three-phase CII, as illustrated in Fig. 1, was simulated. The simulated parameters are shown in Table II. In the scheme, a three-phase RL load was connected to the inverter output. The DC-link voltage was adjusted to 400 V by a constant voltage source.

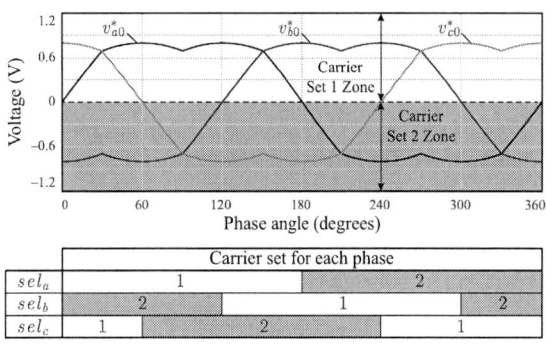

Fig. 7. EPS-PWM reference zones and carrier set selection.

978-1-7281-4181-7/19 $31.00 © 2019 IEEE

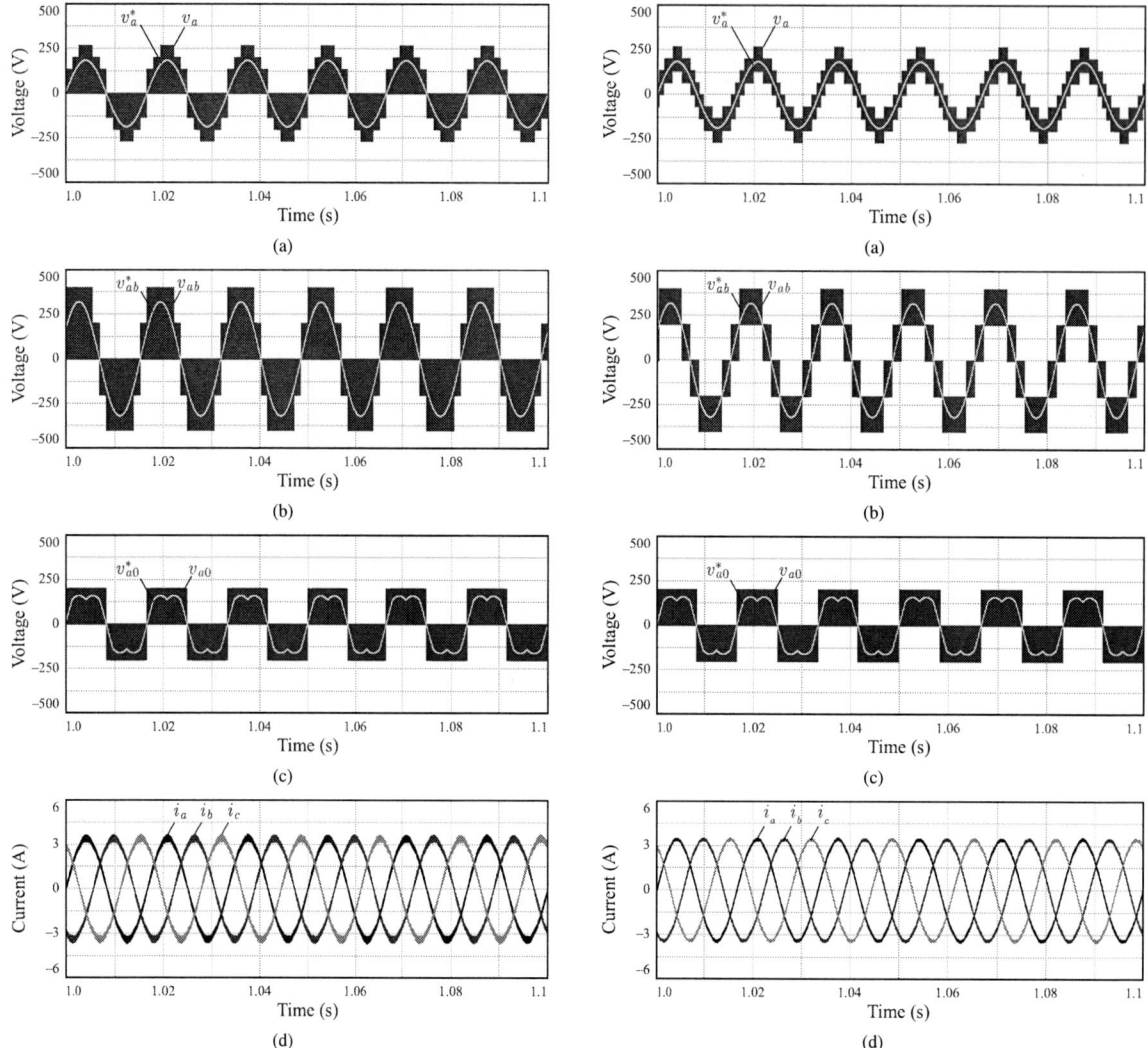

Fig. 8. Simulation results using conventional PS-PWM for $E = 400\,\text{V}$, $f_s = 10\,\text{kHz}$ and $m = 0.8$. (a) Phase voltage (v_a) and its reference (v_a^*). (b) Line-to-line voltage (v_{ab}) and its reference (v_{ab}^*). (c) Pole voltage (v_{a0}) and its reference (v_{a0}^*). (d) Phase currents (i_a, i_b and i_c).

Fig. 9. Simulation results using EPS-PWM for $E = 400\,\text{V}$, $f_s = 10\,\text{kHz}$ and $m = 0.8$. (a) Phase voltage (v_a) and its reference (v_a^*). (b) Line-to-line voltage (v_{ab}) and its reference (v_{ab}^*). (c) Pole voltage (v_{a0}) and its reference (v_{a0}^*). (d) Phase currents (i_a, i_b and i_c).

Both conventional PS-PWM and EPS-PWM approaches were implemented, as shown in Fig. 8 and Fig. 9, respectively. From top to the bottom: phase voltage (v_a) and reference phase voltage (v_a^*); line-to-line voltage (v_{ab}) and reference

line-to-line voltage (v_{ab}^*), pole voltage (v_{a0}) and reference pole voltage (v_{a0}^*); and the phase currents (i_a, i_b and i_c). Although the output pole voltage waveform has three levels in both modulation strategies (0, $E/2$ and $-E/2$), the line-to-line voltage only switches among adjacent levels when using the EPS-PWM. There are also significant improvements in phase voltage switching and in the phase currents ripple.

Fig 10 shows the frequency analysis results comparing both PS- and EPS-PWM strategies. Considering first the phase voltage, despite the carrier frequency was set to $10\,\text{kHz}$, the main switching frequency component can be seen in $20\,\text{kHz}$ for PS-PWM. Using EPS-PWM, there are a significant reduction at this point, and effective switching frequency becomes at $40\,\text{kHz}$. The same reduction in the $20\,\text{kHz}$ component can be seen in the line-to-line voltages and the phase currents when using EPS-PWM.

In order to analyse the multilevel voltage distortion, the

TABLE II
PARAMETERS USED IN THE SIMULATIONS AND EXPERIMENTAL RESULTS.

	Parameters	Values
R	Load resistor	$54\,\Omega$
L	Load inductor	$6\,\text{mH}$
l_{x1}	Coupled inductors impedance	$1.5\,\text{mH}$
l_m	Coupled inductance	99%
E	DC-link voltage	$400\,\text{V}$
m	Modulation index	0.8
f_{v_a}	Phase voltages frequency	$60\,\text{Hz}$
μ	Apportioning factor	0.5
f_s	Switching frequency	$10\,\text{kHz}$

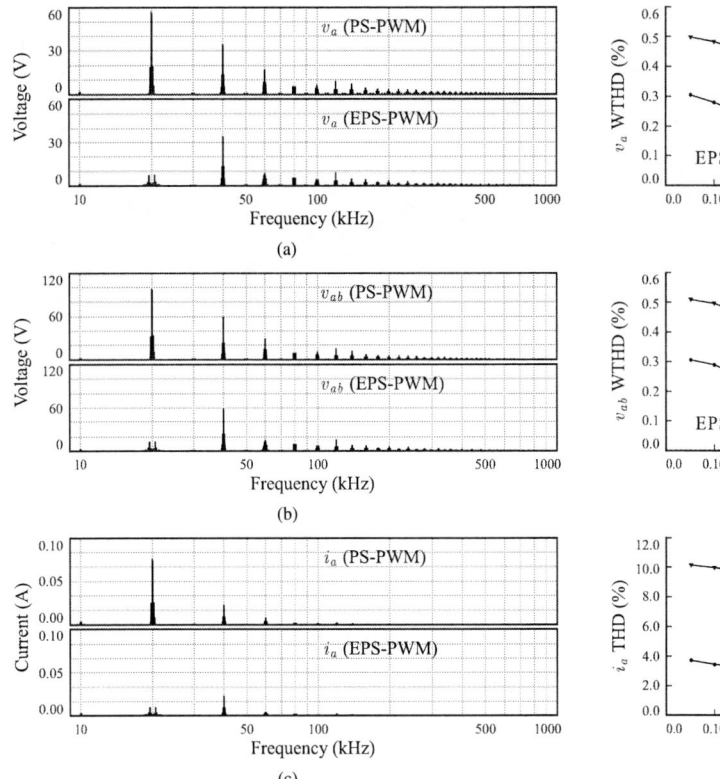

Fig. 10. Frequency analysis results comparing both PS- and EPS-PWM strategies. (a) Phase voltage (v_a). (b) Line-to-line voltage (v_{ab}). (c) Phase currents (i_a, i_b and i_c).

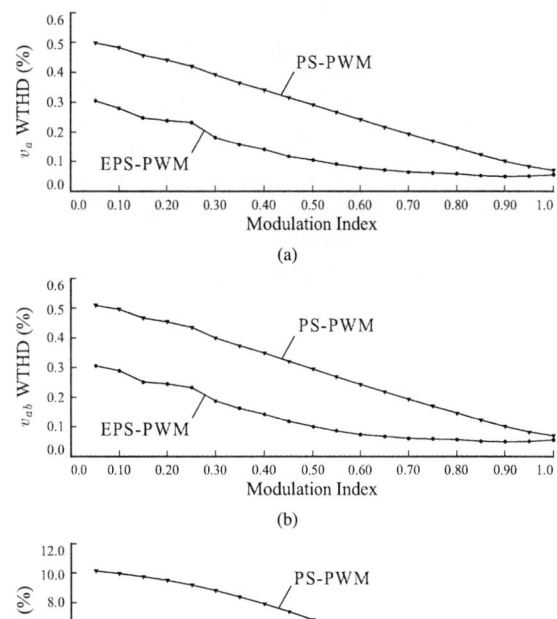

Fig. 11. Simulation results comparing both PS- and EPS-PWM strategies in therms of WTHD and THD, as a function of the modulation index. (a) Phase voltage (v_a) WTHD. (b) Line-to-line voltage (v_{ab}) WTHD. (c) Phase current (i_a) THD.

weighted total harmonic distortion (WTHD) is a better approach than THD [21]. The WTHD weights the voltage harmonics with its frequency [22], and was calculated considering the first 1000 harmonics. Fig 11 shows the WTHD for the phase and line-to-line voltage as a function of the modulation index, as well the THD for the phase currents. It can be noted that the EPS-PWM method shows better performance than PS-PWM in all cases, with best results considering the low modulation index region.

VI. Experimental Results

In order to validate the EPS-PWM results, the same system considered in simulations was implemented in laboratory, as shown in Table II. The experimental setup was based on a digital signal processor (DSP), model TMS320F28335, with a microcomputer equipped with appropriate plug-in boards and sensors. The converters use power switches model SKM50GB123D. The results were obtained by an Agilent oscilloscope, model DSO-X 3014A.

As well as the simulations, both conventional PS-PWM and EPS-PWM approaches were implemented, as shown in Fig. 12 and Fig. 13, respectively. From top to the bottom: phase voltage (v_a), line-to-line voltage (v_{ab}), pole voltage (v_{a0}); and the phase currents (i_a, i_b and i_c). All the experimental results are compatible with the simulations, with both approaches showing three levels in the pole voltage output (0, $E/2$ and $-E/2$). The main points are the switching only among adjacent levels when using the EPS-PWM, also significant

improvements in phase voltage switching and in the phase currents ripple.

VII. Conclusion

In this paper, the application of EPS-PWM in multilevel CII was analyzed. EPS-PWM method uses two carrier sets that are dynamically allocated, with selected set of carriers depends on the instantaneous normalized reference values. The application of this modulation was possible as the inverter was visualized as having two parallel-connected legs phase. Simulation and experimental results were presented, showing that the quality of the output voltages and currents were improved: line-to-line voltages switching exclusively among adjacent levels; lower WTHD in phase- and line-voltages; and improvements in the phase currents THD.

References

[1] B. Vafakhah, J. Salmon, and A. M. Knight, "Interleaved discontinuous space-vector PWM for a multilevel PWM VSI using a three-phase split-wound coupled inductor," *IEEE Transactions on Industry Applications*, vol. 46, no. 5, pp. 2015–2024, Sep. 2010.

[2] S. Salehahari, E. Babaei, and M. Sarhangzadeh, "A new structure of multilevel inverters based on coupled inductors to increase the output current," in *The 6th Power Electronics, Drive Systems Technologies Conference (PEDSTC2015)*, Feb 2015, pp. 19–24.

[3] B. Vafakhah, A. M. Knight, and J. Salmon, "Improved interleaved discontinuous carrier-based PWM strategy for 3-level coupled inductor inverters," in *2011 IEEE Energy Conversion Congress and Exposition*, Sep. 2011, pp. 2095–2101.

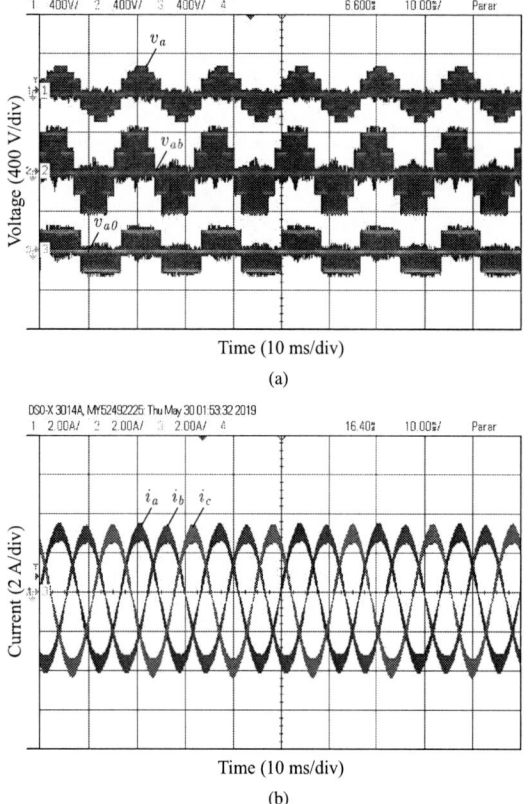

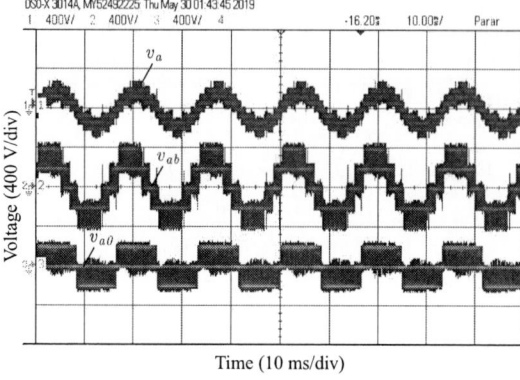

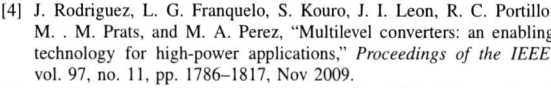

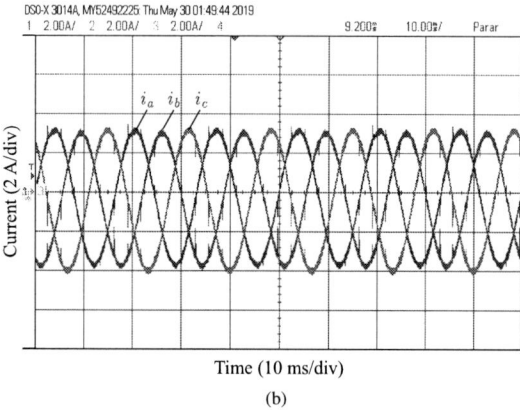

Fig. 12. Experimental results using conventional PS-PWM for $E = 400$ V, $f_s = 10$ kHz and $m = 0.8$. (a) Phase (v_a), line-to-line (v_{ab}) and pole (v_{a0}) voltages. (b) Phase currents (i_a, i_b and i_c).

Fig. 13. Experimental results using EPS-PWM for $E = 400$ V, $f_s = 10$ kHz and $m = 0.8$. (a) Phase (v_a), line-to-line (v_{ab}) and pole (v_{a0}) voltages. (b) Phase currents (i_a, i_b and i_c).

[4] J. Rodriguez, L. G. Franquelo, S. Kouro, J. I. Leon, R. C. Portillo, M. . M. Prats, and M. A. Perez, "Multilevel converters: an enabling technology for high-power applications," *Proceedings of the IEEE*, vol. 97, no. 11, pp. 1786–1817, Nov 2009.

[5] S. Salehahari and E. Babaei, "A new hybrid multilevel inverter based on coupled-inductor and cascaded h-bridge," in *2016 13th International Conference on Electrical Engineering/Electronics, Computer, Telecommunications and Information Technology (ECTI-CON)*, June 2016.

[6] J. Mei, B. Xiao, K. Shen, L. M. Tolbert, and J. Y. Zheng, "Modular multilevel inverter with new modulation method and its application to photovoltaic grid-connected generator," *IEEE Transactions on Power Electronics*, vol. 28, no. 11, pp. 5063–5073, Nov 2013.

[7] G. Buticchi, D. Barater, E. Lorenzani, C. Concari, and G. Franceschini, "A nine-level grid-connected converter topology for single-phase transformerless PV systems," *IEEE Transactions on Industrial Electronics*, vol. 61, no. 8, pp. 3951–3960, Aug 2014.

[8] S. Salehahari, R. Pashaei, E. Babaei, and C. Cecati, "Coupled-inductor based multilevel inverter," in *2017 14th International Conference on Electrical Engineering/Electronics, Computer, Telecommunications and Information Technology (ECTI-CON)*, June 2017, pp. 907–910.

[9] A. M. Knight, J. Ewanchuk, and J. C. Salmon, "Coupled three-phase inductors for interleaved inverter switching," *IEEE Transactions on Magnetics*, vol. 44, no. 11, pp. 4119–4122, Nov 2008.

[10] S. Salehahari, "A new coupled inductor multilevel inverter based on switched DC sources," in *2016 13th International Conference on Electrical Engineering/Electronics, Computer, Telecommunications and Information Technology (ECTI-CON)*, June 2016.

[11] J. Salmon, A. M. Knight, and J. Ewanchuk, "Single-phase multilevel PWM inverter topologies using coupled inductors," *IEEE Transactions on Power Electronics*, vol. 24, no. 5, pp. 1259–1266, May 2009.

[12] Y. Li, Y. Wang, and B. Q. Li, "Generalized theory of phase-shifted carrier PWM for cascaded h-bridge converters and modular multilevel converters," *IEEE Journal of Emerging and Selected Topics in Power Electronics*, vol. 4, no. 2, pp. 589–605, June 2016.

[13] B. Vafakhah, J. Salmon, and A. M. Knight, "A new space-vector PWM with optimal switching selection for multilevel coupled inductor inverters," *IEEE Transactions on Industrial Electronics*, vol. 57, no. 7, pp. 2354–2364, July 2010.

[14] G. J. Capella, J. Pou, S. Ceballos, G. Konstantinou, J. Zaragoza, and V. G. Agelidis, "Enhanced phase-shifted PWM carrier disposition for interleaved voltage-source inverters," *IEEE Transactions on Power Electronics*, vol. 30, no. 3, pp. 1121–1125, March 2015.

[15] J. Lee, K. Lee, and Y. Ko, "An improved phase-shifted PWM method for a three-phase cascaded h-bridge multi-level inverter," in *2017 IEEE Energy Conversion Congress and Exposition (ECCE)*, Oct 2017, pp. 2100–2105.

[16] E. C. dos Santos and S. Sajadian, "Fault-tolerant DC-AC converter with split-wound coupled inductors," in *2013 Brazilian Power Electronics Conference*, Oct 2013, pp. 30–35.

[17] J. Salmon, J. Ewanchuk, and A. Knight, "PWM inverters using split-wound coupled inductors," in *2008 IEEE Industry Applications Society Annual Meeting*, Oct 2008, pp. 1–8.

[18] C. Chapelsky, J. Salmon, and A. M. Knight, "High-quality single-phase power conversion by reconsidering the magnetic components in the output stage-building a better half-bridge," *IEEE Transactions on Industry Applications*, vol. 45, no. 6, pp. 2048–2055, Nov 2009.

[19] E. R. C. da, Silva, E. C. dos, Santos, and C. B. Jacobina, "Pulsewidth modulation strategies," *IEEE Industrial Electronics Magazine*, vol. 5, no. 2, pp. 37–45, June 2011.

[20] E. Santos and E. R. Silva, *Advanced power electronics converters: PWM converters processing AC voltages*. John Wiley & Sons, 2014, vol. 46.

[21] T. Lipo, "An improved weighted total harmonic distortion index for induction motor drives," in *Proc. Int. Conf. OPTIM*, vol. 2, 2000, pp. 311–322.

[22] D. G. Holmes and T. A. Lipo, *Pulse width modulation for power converters: principles and practice*. John Wiley & Sons, 2003, vol. 18.

978-1-7281-4181-7/19 $31.00 © 2019 IEEE

Dimensioning and Developement of an AC Microgrid in the UFJF Campus

1st Paula Stael S. Barbosa
School of Electrical Engineering
Federal University of Juiz de Fora
Juiz de Fora - MG, Brazil
paula.stael@engenharia.ufjf.br

2nd Marina V. C. Monteiro
School of Electrical Engineering
Federal University of Juiz de Fora
Juiz de Fora - MG, Brazil
marina.vidal@engenharia.ufjf.br

3rd Jessica S. Dohler
School of Electrical Engineering
Federal University of Juiz de Fora
Juiz de Fora - MG, Brazil
jessica.dohler@engenharia.ufjf.br

4th Andre Ferreira
School of Electrical Engineering
Federal University of Juiz de Fora
Juiz de Fora - MG, Brazil
andre.ferreira@ufjf.edu.br

5th Janaina G. de Oliveira
School of Electrical Engineering
Federal University of Juiz de Fora
Juiz de Fora - MG, Brazil
janaina.oliveira@ufjf.edu.br

Abstract—This paper aims at designing and developing an AC microgrid to be implemented at the Faculty of Engineering of the Federal University of Juiz de Fora (UFJF). Microgrids can be defined as an integrated system that involves distributed energy resources and various electric loads operating as an autonomous grid, whether parallel or islanded from the main power grid. One of the great advantages of working with microgrids is the possibility of testing devices and controls necessary for the development of smart grid, with lower risk and cost. In this context, the proposed system was modeled and simulated using PSIM software considering load, generation and storage existent at the Solar laboratory of the Faculty of Engineering, at UFJF. Focus was given to the design and simulation of power electronics devices. In this model the following cases were simulated: islanded system and grid-connected mode. Results show that, for dimensioned values, the microgrid can operate in both cases with stability in voltage and current levels, with a total harmonic distortion of 0.57% for voltage in islanded mode, indicating a good operation for the dimensioned system.

Index Terms—Dimensioning, Microgrid, Photovoltaic System, Simulation

I. INTRODUCTION

In the processes of generation and transmission of electricity, it has been increasingly required to improve energy efficiency, diversification of the energy matrix with low environmental impact, minimization of losses and reduction in grid loading. From this context, emerged the concept of Distributed Generation (DG) [1].

Distributed Generation can be characterized as a source of electrical energy connected to the distribution network or to the point of consumption, such as generators with low nominal capacity, with energy possibly from renewable sources [2]. In the scope of distributed generation, microgrids are presented as a group of distributed energy resources manageable, with defined limits, able to attend loads, operating connected or isolated from the main electric grid. The microgrids provide verification of devices and controls necessary for the development of smartgrids, with lower risk and cost.

For the design of a microgrid, energy sources are mostly used in small units such as microturbines, photovoltaic panels and fuel cells. These energy sources are connected to the microgrids through interfaces based on power electronics [3]. In addition to the generation devices, there are also devices for energy storage to meet the energy needs of consumers in cases of failure in the local generation system or in the distribution grid, which are all located and connected in a distributed way in the system electric medium voltage (distribution system) [4].

Some universities in Brazil have developed microgrids as research equipment to allow the analysis of small smartgrids behaviors in an economical and safe way. In 2017, the Federal University of Ceara (UFC) developed the design and implementation of the structure of a microgrid connected in low voltage (220 V/60 Hz). This structure was installed in the laboratory of Smartgrids, in the Department of Electrical Engineering [5].

The Federal University of Santa Catarina, in partnership with CERTI (Innovative Technologies Reference Center) and Engie Brasil Energia, has developed a project which integrates micro generators, energy storage, controllable loads and a hybrid microgrid (AC/DC). Control strategies were developed for systems integration and energy management [6].

The State University of Campinas (Unicamp) has a photovoltaic plant on the Barão Geraldo campus. The generation of solar energy is a subproject of the Sustainable Campus, a program in partnership with CPFL Energia and with the support of Aneel (Agência Nacional de Energia Elétrica) and the company BYD, which donated part of the panels. In addition to providing savings in electricity consumption, the solar panels integrate the "living laboratory", which allows the development of research in the area of solar energy generation, modeling of modules, energy simulation methodologies, evaluation of system performance, solarimetry, among others [7].

978-1-7281-4181-7/19 $31.00 © 2019 IEEE

In view of the above-mentioned microgrids, it is known that there is a tendency and a great interest of the scientific community for such research. In this way, the Federal University of Juiz de Fora (UFJF) is developing a microgrid, planned to operate autonomously (isolated or connected from the electric grid) with approved resources. This paper aims to present the design and simulation of this microgrid for islanded or non-islanded mode of operation, already considering existing equipment within the Solar laboratory of the Faculty of Engineering, at UFJF.

This paper is divided as it follows: Section II brings a description of the system under study, followed by Section III, Converters Topology and Modeling, where the mathematical model and controllers are suggested for the AC side of the microgrid. Finally, Results and Conclusions are brought in Sections IV and V.

II. SYSTEM UNDER STUDY

A. Simulated System

The system is composed of solar panels, unbalanced loads located at the Solar Photovoltaic Laboratory (LabSolar) of the UFJF with an estimated value of 16971 W, power electronics devices and an energy storage (system batteries), besides measurements and protection, which are not discussed whithin the scope of this paper. The microgrid will operate in two modes: islanded or grid-connected. In the islanded mode of operation, a grid-forming inverter is used to monitor the utility's grid status, establish the isolated grid in the event of a fault, and resynchronize the utility grid in case of grid failure correction and reestablishment of the network. The implementation of the microgrid will take place in the interconnection of these equipments, through protection and measurement units. The configuration of the proposed microgrid is shown in Fig. 1.

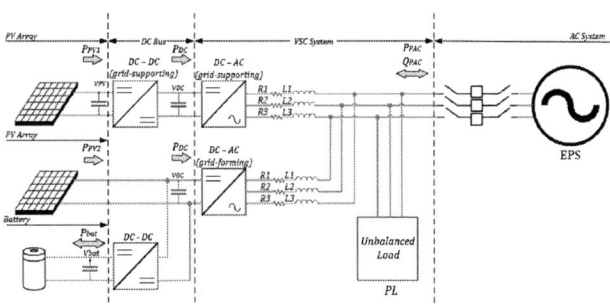

Fig. 1. Overview of the simulated system.

B. Photovoltaic System

Photovoltaic systems are formed by module connections or solar panels. These, in turn, are formed by the connection of cells in series and parallel in order to guarantee desired voltage and current levels [8]. The sizing of the system components was proposed based on the size of the photovoltaic system existing in the UFJF LabSolar.

UFJF has one of the largest photovoltaic arrangements in the country at a university of 30 kWp, half of which is connected

to the grid and dedicated to power generation for the UFJF, and the other 15 kWp are available for the development of scientific research at the university, as shown in Fig.2. The photovoltaic plant generates energy using 264 photovoltaic panels grouped in 11 independent arrays with a capacity of 404 V and 7.12 A per arrangement, under maximum power conditions. Each array has 24 panels which are connected in parallel, of 12 panels associated in series [9]. The panels are of the SX120U model of BP solar. The generation of electric energy by the LabSolar of the UFJF during the month of January 2019 is shown in Fig. 3.

Fig. 2. Solar Photovoltaic Laboratory of the UFJF.

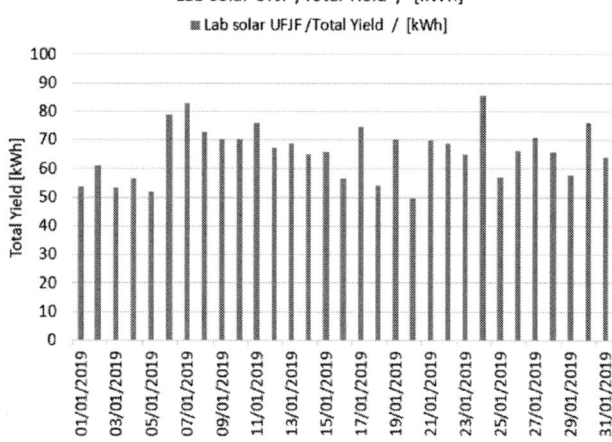

Fig. 3. Energy generated in the Solar Photovoltaic Laboratory of the UFJF during the month of January, 2019.

In order for the microgrid to have a good energy efficiency, it is necessary to extract the maximum power from the photovoltaic panels. The Maximum Power Point Tracking (MPPT) varies non-linearly according to the available irradiance and temperature [4]. Thus, it is necessary to implement a code that can track the MPP in real time. In the literature there are different MPPT techniques, in this paper the Perturb and Observe (P&O) technique will be used.

P&O is a simple technique, based on the voltage and current curves of the photovoltaic panel. The algorithm applies a small increase in voltage or current and observes the power variation. If there is an increase in panel power, the algorithm continues to apply perturbations in the same direction, but if there is a decrease in power, the disturbance will happen in the opposite direction in the next iteration of the algorithm. Thus, the algorithm works by varying the reference values close to the maximum power point [10].

C. Storage System

For the proper choice of the energy storage system used for microgrid applications, it becomes necessary to be aware of the power characteristics and the energy supply time of the respective storage system. Batteries are the most commonly used devices in energy storage systems applications. Among modern batteries, Nickel-Cadmium (NiCd) batteries and Lithium-Ion (Li-Ion) batteries [11] stand out.

In this paper, a bank of lithium-ion batteries (Li-Ion) will be used. Its advantages include greater energy density, greater conversion efficiency and absence of memory effect, being able to perform several cycles of loading and unloading per day. The choice of battery is due to these characteristics, together with the objective of the work, which demands a storage system with a duration of half an hour. The photovoltaic system must be able to supply the load connected to the electric grid, regardless of the variations of photovoltaic panel generation. Thus, a battery bank of 48 V Lithium with storage capacity of 194.5 Ah was estimated.

III. CONVERTERS TOPOLOGY AND MODELING

Power electronic converters can perform distinct functions in a microgrid. These are: grid-feeding, grid-forming and grid-supporting [12].

The grid-feeding converter acts as a current source in the active and reactive power mode and still manages to partially or totally supply the electric charges of the microgrid [12]. The grid-forming converter is used when the microgrid operates in the islanded mode, i.e. without the presence of the main grid. It has as main function to detect the islanding and to establish voltage and frequency in the terminals of the microgrid. In this way, guaranteeing stability of operation of the isolated system as a whole. And it should be emphasized that the grid-forming must operate with some energy storage system or primary source that is available at any time to ensure its operation as a source of controlled voltage. The grid-supporting converter has the role of providing grid support, such as voltage and frequency regulation, active filtering, power factor correction, among other functions to improve the quality of electric power [13].

The focus of this paper in on the application of Grid-forming and grid-feeding converters in AC microgrids.

A. Grid Forming Converter

The grid-forming converter will act by establishing at its output a reference voltage and frequency. The operation of the forming is given by an Uninterruptible Power Supply (UPS), which manages to supply an electric load independently of the presence of the main power grid, using a bank of stationary batteries.

As batteries store energy in a continuous way, so for applications in electric power systems, it is necessary to use static converters based on battery power electronics. An interleaved bidirectional DC-DC converter connected to the battery was used, as it could provide lower distortion and ripple on the output waveforms [14].

In this way, the DC bus voltage is maintained through the power flow of the battery and the bidirectional DC-DC converter used. This guarantees the uninterrupted supply of the loads when the microgrid is operating in an islanded operating mode, that is, in the grid-forming mode. The interleaved technique was chosen in this paper due to the fact that it has a lower input current ripple, it has an improved efficiency of the converter and faster transients in response to load variations.

The modeling of the bidirectional interleaved DC-DC converter, as well as the gains of the controllers that were calculated by the function *lqr* of the MATLAB software, are described in details in [14].

B. Two-stage Grid Feeding Converter

Grid-feeding converter is a two-stage converter and, in the case of the presente work, represents a commercial inverter, existent at the Solar Laboratory. It consists of (i) a Boost DC-DC converter, responsible for raising the maximum voltage of the PV array in order to extract the maximum power from the panels, the MPPT - and (ii) a three-phase voltage source converter DC/AC interface with the power grid responsible for injecting the active power available in the microgrid.

(i) Modeling of the DC-DC Converter: To control the photovoltaic panel at maximum power, the Boost converter is added to the panel terminals, operating in continuous drive mode. The maximum power tracking algorithm is inserted into the converter. MPPT provides the voltage and current reference to the controller, which in should regulate the desired voltage and current at the terminals of the device. As previously described, the MPPT algorithm used in this model was the perturbation and P&O observation [15].

The transfer function of the closed-loop Boost converter is:

$$G(s) = \frac{V_0 - V_D + R_s i L}{sL + D_0 R_s + R_L - V_0 - V_D + R_s I_L} \quad (1)$$

The values of k_p and k_i are:

$$\left| \frac{-K_p(s + 0.1)(V_0 + V_D - R_S I_L)}{s(s_L + (D_0 R_S + R_L))} \right|_s = 1 \quad (2)$$

$$k_i = 520 k_p \quad (3)$$

(ii) Inverter Modeling and Control: The Voltage Source Converter (VSC) is the most widespread topology in the world when it comes to supplying energy from an array of photovoltaic panels to the three-phase power grid [8]. In Fig. 4 the block diagram of the VSC connected to the AC grid and the PV panel is shown. The inverter is composed of six complete bridge switches. The voltage and current of the DC bus comes from the output of the Boost converter terminals.

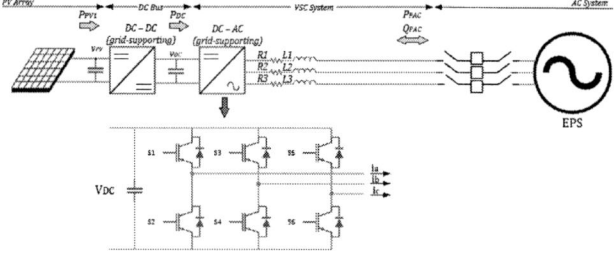

Fig. 4. Block diagram of the VSC connected to the grid and to the Photovoltaic Panel. Adapted from [16].

In this paper, the DC/AC converter control was implemented in the synchronous coordinate system dq. The dq system is obtained through the Park transformation, guaranteeing a steady behavior for current and voltage signals of the converter. The input are the three-phase currents of lines i_a, i_b and i_c which results in the components i_d and i_q at the output. In the same way for the voltage, at the input one has the three-phase line voltages V_a, V_b and V_c, resulting in the voltages V_d and V_q in the output. To make this transformation, it was necessary to use a phase angle ρ for the transformation abc in coordinates dq.

Thus, in order for the energy from the photovoltaic array to be properly injected into the grid, a synchronization system becomes indispensable. For this purpose one can use phase-angle detection circuits known as Phase-Locked Loop (PLL) [8].

The main purpose of this circuit is to have the quadrature axis voltage V_q, pac stabilize at 0. For simplicity of implementation, Synchronous Reference Frame - Phase Locked Loop (SRF-PLL) has been used. For the parameters of the SRF-PLL controller we have:

$$
\begin{cases}
k_p, PLL = \frac{(2*(2\pi\omega_n)*\zeta)}{V_{pac}} \\
\tau_{PLL} = \frac{k_{p,PLL}*V_{pac}}{(2\pi\omega_n)^2}
\end{cases}
\tag{4}
$$

where ω_n is the cutoff frequency and ζ is the smoothing factor.

To generate the current i_{dref} a voltage control is required. A voltage control of the DC bus is performed, and the voltage and current harmonics generated by the inverter switching have been neglected.

In (4) is the equation that represents the control. The equation to find (5) can be seen in [8].

$$
\frac{(\Delta Vcc^2)^*}{(\Delta Vcc)^2} = \frac{\frac{3k_{p,v}V_{do,pac}}{C_{eq}}s + \frac{3k_{i,v}V_{do,pac}}{C_{eq}}}{s^2 + \frac{3k_{p,v}V_{do,pac}}{C_{eq}}s + \frac{3k_{i,v}V_{do,pac}}{C_{eq}}}
\tag{5}
$$

where: ΔV_{CC}^{2*} is the reference value for the square of the DC voltage, ΔV_{CC}^2 is the the square of the DC voltage, $k_{p,v}$ and $k_{i,v}$ are the proportional and integral gains of the PI

respectively, $V_{do,pac}$ common coupling point voltage in the d axis and C_{eq} DC capacitor of the converter.

The canonical form of a transfer function is:

$$
H(s) = \frac{2\zeta\omega_n s + \omega_n^2}{s^2 + 2\zeta\omega_n s + \omega_n^2}
\tag{6}
$$

Comparing (5) with (6), the gains of the PI for the control of the voltage in the DC bus can be obtained:

$$
k_{p,v} = \frac{2\zeta\omega_n C_{eq}}{3V_{do,pac}}
\tag{7}
$$

$$
k_{i,v} = \frac{\omega_n^2 C_{eq}}{3V_{do,pac}}
\tag{8}
$$

where ω_n is the cutoff frequency and ζ is the smoothing factor [8].

To obtain the transfer function and for the design of the controller the decoupling of the currents is done and the values of the voltages at the common coupling point are also compensated.

As can be seen in (9), the variables of the d and q axis are decoupled, that is, i_d and i_q can be directly controlled by v_d and v_q, respectively. For the control, the voltages v_d and v_q are outputs of two compensators, thus the direct axis compensator processes: $e_d = i_d^* - i_d$, where a similar process is done for the quadrature axis [8].

The closed loop transfer function is:

$$
\begin{cases}
L\frac{di_d}{d_t} = -(R_{eq})i_d + v_d \\
L\frac{di_q}{d_t} = -(R_{eq})i_q + v_q
\end{cases}
\tag{9}
$$

where i_d^* is the current reference on the axis d of $\tau_i = L/k_{(p,i)}$.

Thus, the compensator gains can be obtained:

$$
k_{p,i} = \frac{L}{\tau_i}
\tag{10}
$$

and

$$
k_{i,i} = \frac{R_{eq}}{\tau_i}
\tag{11}
$$

The parameter τ_i (compensating system time constant) must be small enough to ensure a rapid plant response. In [8] it is advised that the time constant should have its value between $0.5\ ms$ and $5\ ms$.

IV. RESULTS

With the description of the system and the modeling performed, it is possible to simulate the analyzed scenarios. This section describes the simulation processes as well as the previously scaled project reference values of the project. For the validation of the proposed system, variations in the irradiation and consequently in the active power converted by the PV array are performed over time.

Assuming the system parameters presented in Table II and III and considering the control parameters: $\tau = 0.5\ ms$, $\omega_n =$

$2\pi15\ rad/s$ and $\zeta = 0.707$ to design the previously modeled controllers, we have the values of the gains shown in Table IV.

TABLE I
GRID FEEDING CONVERTER PARAMETERS

Grid Supplying Converter Parameters	Value
Useful phase voltage in the grid	127 V
Fundamental frequency of the grid	60 Hz
Panel Output Capacitor	47 μF
DC Link Capacitor	4.7 mF
VSC Output Filter Inductor	1.5 mH
VSC Output Resistance	0.3 Ω
Boost Converter Inductor	16.5 mH
Switching frequency of DC-DC converter	5 kHz
Switching frequency of the DC-AC converter	20 kHz

TABLE II
GRID FORMING CONVERTER PARAMETERS

Grid Forming Converter Parameters	Value
Useful phase voltage in the grid	127 V
Fundamental frequency of the grid	60 Hz
DC Link Capacitor	5500 μF
VSC Output Filter Inductor	2 mH
VSC Output Resistance	0.07 Ω
Interleaved Inductor	2.5 mH
Interleaved Resistance	1 mΩ
Interleaved Capacitor	5500 μF
Battery Resistance	0.00147 mΩ
Battery Capacitor	100 μF
Battery voltage	250 V

TABLE III
PARAMETERS OF PHOTOVOLTAIC PANEL BP SX 120

Parameters	Symbols	Value
Maximum power	P_{MP}	120 W
Current at maximum power point	I_{MP}	3.56 A
Maximum power point voltage	V_{MP}	33.7 V
Short Circuit Current	I_{SC}	3.87 A
Open Circuit Voltage	V_{OC}	42.1 V
I_{sc} Temperature Coefficient	α_T	0.065 A/°C
Temperature coefficient of V_{oc}	β_T	160 mV/°C
Normal Operating Temperature	NOCT	47.9 °C

TABLE IV
GRID FEEDING CONVERTER PARAMETERS OF THE CONTROLLERS

Parameters	Value
Proportional Gain of PLL	2.97 rad/V_s
Integral Gain of PLL	267 rad/V_{s2}
Proportional gain of current	3 V/A
Integral gain of current	600 V/A_s
Proportional gain of voltage	0.0012 A/V
Integral gain of voltage	0.077 A/V_s
Proportional gain (Boost)	4 A/V
Integral gain (Boost)	2080 A/V_s

The system was simulated in the PSIM version 9.1.1 in the two modes of operation proposed, feeding a linear load of 15 kW. A simulation step of 10^{-6} s and duration of 1 s was used. In Figure 5, at time 0 to 0.5 s, the system is connected to the main grid, operating only on the power feeding mode. On this mode, the DC bus voltage control is performed via the Boost converter connected to the PV array. The inverter controls active and reactive power on the AC side. Power support for the load is supplied by the grid when the renewable source is missing.

At the instant of 0.5 to 1 s the switches that connect the grid to the distributed generation are opened. The system starts operating in the islanded mode, i.e. the control mode of the grid forming is active. In this mode, the grid forming converter coupled with the renewable energy system manages to supply the load, the excess energy is absorbed by the storage system. It is possible to observe that during this transition an insignificant disturbance occurs in the current of the load, but it returns to normal afterwards.

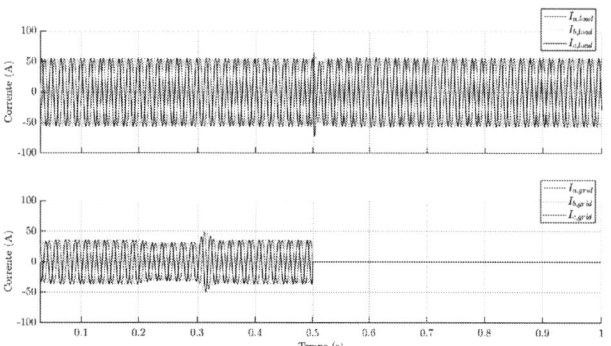

Fig. 5. System response in islanded and non-islanded mode.

In the interval of 0.2 s, a disturbance occurs in the irradiation of the distributed generator. The maximum power tracking algorithm regulates the desired DC voltage and current on the DC bus. The current response to irradiation disturbance is shown in Figure 6 and voltage response is shown in Figura 7. Both results are related to the grid feeding converter.

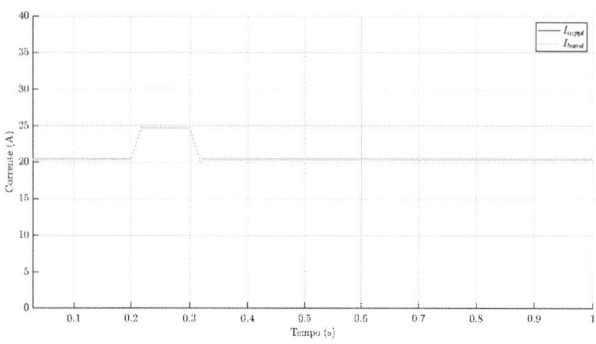

Fig. 6. Current response to irradiation disturbance on the grid feeding converter.

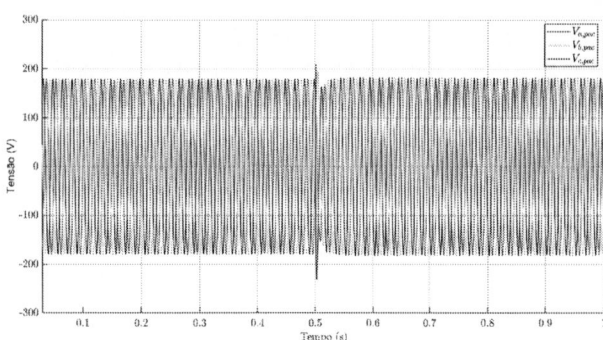

Fig. 7. Voltage response to irradiation disturbance on the grid feeding converter.

The voltages at the common coupling point in islanded mode have a total harmonic distortion of 0.57%. In this way, the good functioning of the model is verified in Figure 8.

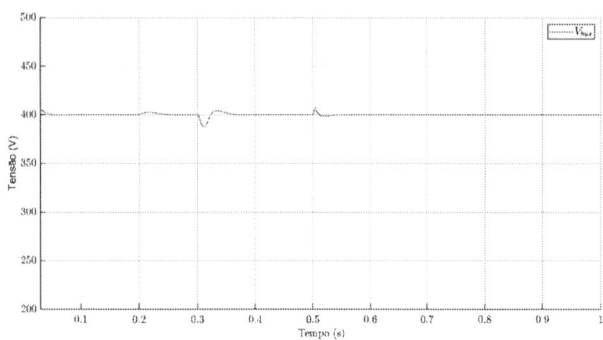

Fig. 8. System response at the common coupling point before and after the disconnection of the main grid.

V. Conclusions

In this paper a microgrid was designed for the campus of the Federal University of Juiz de Fora. The proposed system consists of solar panels, power electronic devices and a battery bank. The system feeds the loads of the Solar Photovoltaic Laboratory. The microgrid was simulated in two modes of operation, islanded to the main power grid and non-islanded mode.

When the system is connected to the main grid, the DC/AC inverter operates in the supply mode. In this case, the inverter controls the active and reactive power on the AC side and the Boost converter controls the voltage on the DC bus. In the island mode the grid-forming converter starts to act, it keeps the grid parameters and supplies the loads. The excess energy is absorbed by the storage system.

In order to test the model, a disturbance in the irradiation and consequently in the active power converted by the PV array was performed. In this case the system was able to follow the voltage reference on the DC bus.

In conclusion, the modeling of the converter plant satisfactorily represents the design of the hybrid microgrid of the Federal University of Juiz de Fora.

Acknowledgment

The authors are grateful to the Federal University of Juiz de Fora, Graduate Program in Electrical Engineering - PPEE / UFJF, and to the development agencies: FAPEMIG, CAPES, CNPq, INERGE and Finep for the support of this project.

References

[1] P. Chiradeja, "Benefit of distributed generation: A line loss reduction analysis," in *Proc. IEEE/PES Transmission Distribution Conf. Exposition: Asia and Pacific*, Aug. 2005, pp. 1–5.

[2] J. Driesen and R. Belmans, "Distributed generation: challenges and possible solutions," in *Proc. IEEE Power Engineering Society General Meeting*, Jun. 2006, pp. 8 pp.–.

[3] D. M. FALCÃO, "Smart grid e microredes: o futuro já é presente." in *VIII Simpósio de automação de sistemas elétricos – SIMPASE*, Rio de Janeiro, Brasil, Aug. 2009.

[4] W. Chiang, H. Jou, and Jinn-Chang Wu, "Maximum power point tracking method for the voltage-mode grid-connected inverter of photovoltaic generation system," in *Proc. IEEE Int. Conf. Sustainable Energy Technologies*, Nov. 2008, pp. 1–6.

[5] A. V. CARNEIRO, "Projeto, desenvolvimento e implementação de microrrede em campus universitário com tecnologia solar fotovoltaica e de armazenamento." Master's thesis, Programa de Pós-Graduação em Engenharia Elétrica do Centro Tecnológico da Universidade Federal do Ceará – UFC, Fortaleza, 2017.

[6] *UFSC (Santa Catarina). Laboratório de Microrredes Inteligentes (uGridLab)*, Std., disponivel em: http://ugridlab.paginas.ufsc.br/ Acesso em: 01jun.2019.

[7] *UNICAMP (Campinas- SP). Jornal da Unicamp – Edição Web.*, Std., disponível em: https://www.unicamp.br/unicamp/ju/noticias/ Acesso em: 01unh.2019.

[8] D. C. CASARO, M.M.; MARTINS, "Modelo de arranjo fotovoltaico destinado a análises em eletrônica de potência via simulação." E. de Potência, Ed., vol. 13, no. 3, 2008, pp. 141–146.

[9] M. RODRIGUES, "Microrrede híbrida cc/ca baseada em fontes de energia renovável aplicada a um edíficio sustentável." in *XIX CONGRESSO BRASILEIRO DE AUTOMÁTICA, CBA.* Campina Grande - PB: CBA, 2012, pp. 1 – 9.

[10] A. Y. J. R. S. ATALLAH, Ahmed M.; ABDELAZIZ, "Implementation of perturb and observe mppt of pv system with direct control method using buck and buck-boost converters. emerging trends in electrical." in *Electronics & Instrumentation Engineering: An International Journal (eeiej).*, Cairo, Egypt, 2014, pp. 1–14.

[11] M. G. FARRET, F. A.; SIMOES, "Integration of alternative sources of energy," 2006, john Wiley & Sons.

[12] G. M. S. Azevedo, M. C. Cavalcanti, F. A. S. Neves, L. R. Limongi, and F. Bradaschia, "A control of microgrid power converter with smooth transient response during the change of connection mode," in *Proc. Brazilian Power Electronics Conf*, Oct. 2013, pp. 1008–1015.

[13] R. B. Gonzatti, S. C. Ferreira, C. H. da Silva, R. R. Pereira, L. E. B. da Silva, G. Lambert-Torres, and R. M. R. Pereira, "Implementation of a grid-forming converter based on modified synchronous reference frame," in *Proc. IECON 2014 - 40th Annual Conf. of the IEEE Industrial Electronics Society*, Oct. 2014, pp. 2116–2121.

[14] J. S. DOHLER, "Analysis and operation of a hybrid system using a converter multifunctional." in *Power Electronics Conference (COBEP)*, Brazilian, 2019.

[15] A. CUPERTINO, "A grid-connected photo-voltaic system with a maximum power point tracker using passivity-based control applied in a boost converter," 10th IEEE/IAS International Conference, Ed. 10th IEEE/IAS International Conference: Industry Applications (INDUSCON), 2012, pp. 1–8.

[16] P. M. ALMEIDA, "Modelagem e controle de conversores estáticos fonte de tensão utilizados em sistemas de geração fotovoltaicos conectados à rede elétrica de distribuição," Master's thesis, Tese de Mestrado, UFJF, Juiz de Fora, MG, 2011.

Unbalanced Voltage Mitigation using D²VC with Proportional Resonant Controller in αβ-Frame

Elienai O. Macedo*, Robson F. S. Dias*, Silvangela L. S. L. Barcelos†, Edson H. Watanabe*

*Federal University of Rio de Janeiro - Electrical Engineering Program. - COPPE, Rio de Janeiro-RJ, Brazil
†Federal University of Maranhão - Electrical Engineering Dept., - UFMA São Luís-MA, Brazil
email: elienai@coe.ufrj.br, dias@dee.ufrj.br, silvangela.barcelos@ufma.br, watanabe@coe.ufrj.br

Abstract—**This paper proposes a control algorithm for unbalanced voltage compensation using the Dynamic Direct Voltage Controller (D²VC). An advantage of this device is the possibility of a online direct voltage control in distribution systems, rather than on-load tap changers or other indirect control schemes. The proposed algorithm is based on the instantaneous positive and negative sequence components in αβ-frame, dismissing the use of a Unified Three-Phase Processor (UTSP), consequently avoiding its complexity and excessive computational efforts, and the compensation reference signals are obtained using proportional plus resonant (PR) controllers. There is no need of using any synchronism algorithm, e.g. PLL, which means that the D²VC's performance is independent of the dynamics of this type of synchronization circuit. Simulation results carried out in time domain using PSCAD/EMTDC are presented to show the effectiveness of the proposed solution.**

Index Terms—**Direct Voltage Control, Unbalanced Voltage, Voltage Sourced Converter (VSC), Alpha-Beta Reference Frame, Sequence Components.**

I. INTRODUCTION

An Electrical Power System should provide electrical energy with reliability and quality. Among other characteristics, it can be understood as continuous feeding service with voltages free of harmonics, and magnitude and frequency in adequate levels. In Brazil those levels are indicated by the Electrical Energy National Agency (ANEEL) [1]. The lack of such characteristics indicates a poor electrical energy quality in the system, which causes not only technical problems but financial losses as well [2], [3]. The voltage deviations can be even more common in power distribution grids due to their low short circuit levels (weak grids) and the resistance (R) predominant over the inductive reactance (X) [4]. Moreover, in a utility grid it is very common to have unbalanced load connection, as a consequence, the three-phase grid faces voltage imbalances, both magnitude and phase angle [5].

Many research works have proposed solution for mitigation of the unbalanced voltage in power system [6]–[13]. Basically, the proposed solutions could be comprised in two categories: direct and indirect voltage control.

In order to maintain the voltage within a desirable range, the voltage compensation based in indirect control consists in the management of reactive power injection through the connection of a shunt capacitor or a reactor in a critical point of the grid, or by injection of reactive currents using static compensators, e.g., STATCOM or SVC [11]–[14]. The limitation of the indirect voltage control is that it depends on the inductive characteristic of the line, i.e., high X/R relation, which happens mostly in power transmission systems [5], [14], making the voltage amplitude highly coupled to reactive power. Nevertheless, this sensibility is lower in distribution systems, which has a low X/R relation, sometimes even lower than unity. It makes the indirect control ineffective and highly disadvantageous for utility grids. For those situations, direct control methods are more suitable than indirect control.

In direct control, the voltage is regulated by adding a voltage compensation directly in series to the sensitive load or to the bus that should have its voltage controlled. The voltage compensation could be generated by a voltage source converter (VSC) and connected in series to the grid through a transformer, such as the Dynamic Voltage Restorer (DVR) [7]–[9], [15]. In [9], [16] DVR's action have been studied regulating voltage individually, solving the unbalanced voltage problem, although by connecting three single-phase VSC, one in each phase. On the other hand, in [17] it is shown the connection of the DVR device to the grid with just one three-phase VSC, reducing the number of IGBTs by half, regulating voltage even under unbalanced situations. However, this DVR-based imbalance mitigation requires an energy storage system, such as batteries, which configures significant costs as drawback. The voltage regulation could also be achieved by means of an On-Load Tap Changer Transformer (OLTC), which changes the secondary tap maintaining voltages in normal levels, being able to mitigate unbalanced voltages too [18], [19]. Despite of this positive aspect, such equipment has its mechanical changing process as a huge setback, which has a slow response time associated to its operation (millisecond to seconds of interval), leaving the voltage levels outside the acceptable range for too long, not to mention its limited number of maneuvers.

Another power electronics device, the Dynamic Direct Voltage Controller D²VC, can act performing voltage regulation in utility grids. Its conception is inspired on a Unified Power Flow Controller (UPFC), which was originally proposed to enhance power flow [20], [21], but in recent works have been studied to compensate voltage fluctuations in balanced grids [22] and also in unbalanced ones [23], [24]. The D²VC generates three-phase compensation voltages through a VSC, galvanically isolated from the grid [22], or through three single-phase VSCs, compensating voltages individually [6]. Generally, a galvanic isolation from the grid would be re-

978-1-7281-4181-7/19 $31.00 © 2019 IEEE

quired, in order to guarantee safe connection and generate voltage through the windings of its transformer. Using D^2VC in utility grids, such need can be dismissed, in view of the low voltage levels.

In [6], it is proposed the direct voltage control using the Unified Three Phase Processor (UTSP) for calculation of the instantaneous positive and negative sequence voltage components [25]. Nevertheless, the effectiveness of the proposed solution for unbalanced voltage compensation and the control strategy is dependent on the UTSP, which is a very complex and cumbersome algorithm. In order to avoid these dependency and complexity, in this work is proposed a control algorithm for the D^2VC performed directly in the $\alpha\beta$-Frame using proportional plus resonant controllers. The positive and negative sequence voltage components are calculated instantaneously and the negative sequence is fully compensated, whereas the amplitude of the positive sequence is regulated at a desired level.

The rest of the paper is organized in four sections: Section II shows specifics about the D^2VC, Section III depicts the proposed control for voltage regulation and unbalanced voltage compensation. Section IV presents evidences of the proposed solution through simulation results. Section V shows the main reached conclusions of this study.

II. DYNAMIC DIRECT VOLTAGE CONTROLLER (D^2VC)

D^2VC details, shown in Fig. 1, illustrate two VSCs, connected in back-to-back, with a very similar structure of an UPFC, and may assume some other configurations, as individual shunt and series VSC's in each phase, or simply a three-phase VSC, as studied in this paper. But the purpose of voltage compensation remains the same, either individually or not.

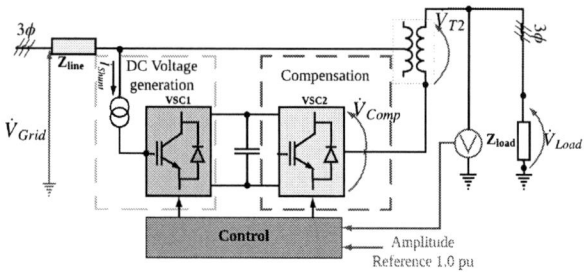

Fig. 1. D^2VC single line diagram

The VSC_1 is shunt connected to the grid and it operates regulating the dc voltage at a constant level so as the VSC_2 may synthesize the series compensation voltage adequately. A conventional hysteresis current control could be applied and is not in the scope of this work. More details could be find in [6].

The VSC_2 is connected to the grid at a low voltage level, in series with the secondary windings of a pre-existing utility transformer. Its worthy to mention that the DC voltage is

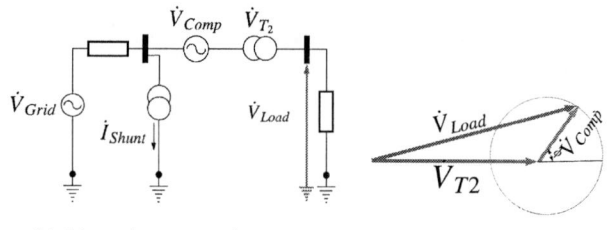

(a) Schematic compensation concept.   (b) Phasor representation.

Fig. 2. Compensation concept.

provided by the VSC_1, and therefore, a storage device is not necessary.

The concepts about the direct voltage compensation are to be seen in Fig. 2a. In order to regulate the voltage at the load bus, $\dot{V}_{Load}$, the D^2VC generates a compensation voltage, $\dot{V}_{comp}$, to be added to the transformer's secondary voltage, $\dot{V}_{T2}$, as shown in (1). D^2VC is able to compensate voltage amplitude and shift the voltage phase angle at the load terminal synthesizing the appropriate amplitude voltage compensation and phase angle displacement, see (2). Fig. 2b shows the phasor diagram of the resultant voltage compensation for one phase.

$$\dot{V}_{Load} = \dot{V}_{T2} + \dot{V}_{Comp} \tag{1}$$

$$\dot{V}_{Comp} = |V_{Comp}|e^{j\phi} \tag{2}$$

The compensation voltage can be different in each phase of a three phase system, so as it is possible to compensate unbalanced voltage at the load. In the following section it is proposed a new method of controlling VSC_2 to compensate voltage deviations on the grid even under unbalanced situations.

III. UNBALANCED VOLTAGE COMPENSATION

This section presents a control strategy for regulating the amplitude of the positive sequence voltage components and for imbalance mitigation due to negative sequence components appearance. The zero sequence component is not considered since it is assumed that the load and the D^2VC are three wire devices. Fig. 3 shows the block diagram of the proposed control scheme.

The three-phase grid voltages, v_a, v_b, v_c, are measured at the point whose voltage should be regulated, for instance, at the terminals of a sensitive load. The sequence components of the grid voltage are calculated in $\alpha\beta$ frame using the Clarke Transformation, (3), and applying (4) and (5). For generating a quadrature signal of the $\alpha\beta$ components and calculation of sequence components in time domain, a the Double Second-Order Generalized Integrator with a Frequency Locked Loop (DSOGI-FLL) is utilized. The DSOGI-FLL structure is comprised by two SOGI-FLL and the Positive Negative Sequence Calculation block (PNSC) [26]–[28], as shown in Fig. 3.

978-1-7281-4181-7/19 $31.00 © 2019 IEEE

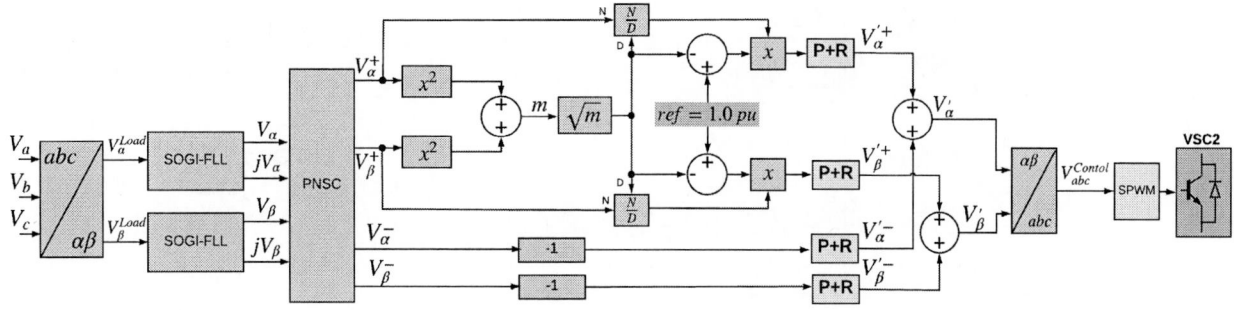

Fig. 3. Block diagram of the proposed control scheme.

$$\mathbf{v}_{\alpha\beta} = \frac{2}{3} \begin{bmatrix} 1 & -1/2 & -1/2 \\ 0 & \sqrt{3}/2 & -\sqrt{3}/2 \end{bmatrix} \begin{bmatrix} v_a \\ v_b \\ v_c \end{bmatrix} \quad (3)$$

$$\mathbf{v}_{\alpha\beta}^+ = \frac{1}{2} \begin{bmatrix} 1 & -j \\ j & 1 \end{bmatrix} \mathbf{v}_{\alpha\beta}. \quad (4)$$

$$\mathbf{v}_{\alpha\beta}^- = \frac{1}{2} \begin{bmatrix} 1 & j \\ -j & 1 \end{bmatrix} \mathbf{v}_{\alpha\beta}. \quad (5)$$

$\mathbf{v}_{\alpha\beta}^+$ and $\mathbf{v}_{\alpha\beta}^-$ are the positive and negative sequence components in $\alpha\beta$ frame on the time domain, and j is a 90°-lagging phase-shifting operator also applied on the time domain.

The comparison between $\mathbf{v}_{\alpha\beta}$ and the reference $\mathbf{v}*_{\alpha\beta}$ generates the error signal to the PR controller (Figure 4). Hence, it demands that the comparison must be made with a reference set of $\mathbf{v}*_{\alpha\beta}$, synchronized to the grid, leading to the necessity of a Phase Locked Loop (PLL) algorithm to guarantee the synchronization of such references. The PLL would add more computational calculations and associated delays, not to mention the settling time that it would apply to the reference for every disturbance occurred in the grid. The novel control scheme proposed in this work does not have these issues because it dismisses the use of PLL for these reference sinusoidal wave forms.

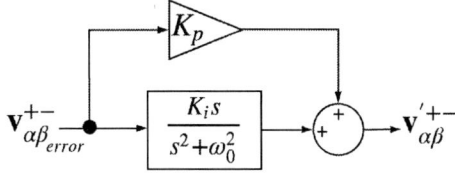

Fig. 4. Proportional plus resonant [29].

The amplitude error of $\mathbf{v}_{\alpha\beta}^+$ is calculated by the comparison with a scalar reference, which is set to 1.0 pu, for normal operation point. The amplitude of the positive sequence components $|\mathbf{v}_{\alpha\beta}^+|$ is estimated using (6).

$$|\mathbf{v}_{\alpha\beta}^+| = \sqrt{(v_\alpha^+)^2 + (v_\beta^+)^2} \quad (6)$$

The input of each PR controller for positive sequence amplitude regulation is calculated as the amplitude error multiplied by unitary sinusoidal in-quadrature signals synchronized to $\mathbf{v}_{\alpha\beta}^+$. The unitary sinusoidal signal is obtained by the normalization: positive sequence components by the amplitude estimated in (6), i.e., $\frac{\mathbf{v}_{\alpha\beta}^+}{|\mathbf{v}_{\alpha\beta}^+|}$. Thus, the sinusoidal inputs to each proportional resonant controller for positive component are given by (7) and (8), respectively.

$$v_{\alpha\ \text{error}}^+ = (1 - |\mathbf{v}_{\alpha\beta}^+|)\frac{v_\alpha^+}{|\mathbf{v}_{\alpha\beta}^+|} \quad (7)$$

$$v_{\beta\ \text{error}}^+ = (1 - |\mathbf{v}_{\alpha\beta}^+|)\frac{v_\beta^+}{|\mathbf{v}_{\alpha\beta}^+|} \quad (8)$$

A slightly different analysis could be done for unbalanced voltage compensation. Since the objective is to fully compensate the negative sequence, its negative content must be sourced to the PR controller input, i.e., the negative sequence component from the PNSC block times -1. This is equivalent to compare the amplitude of the negative sequence component to a scalar reference equal to zero. The input errors of the PR controllers for negative sequence compensation are given by (9) and (10).

$$v_{\alpha\ \text{error}}^- = -v_\alpha^-. \quad (9)$$

$$v_{\beta\ \text{error}}^- = -v_\beta^-. \quad (10)$$

The outputs of the PR controllers are the compensation voltages that should be synthesized by the converter, $\mathbf{v}'_{\alpha\beta}^+$ and $\mathbf{v}'_{\alpha\beta}^-$. For the SPWM firing, the voltages reference in abc-frame $\mathbf{v}'_{abc}$ are obtained applying the inverse Clarke Transformation.

The choice for a proportional plus resonant controller is that it can be tuned exactly at a desired resonant frequency, with an infinite gain, rejecting all other frequencies that the input may have and will not introduce any stationary error [29]. The resonant may be calculated dynamically through the FLL used in the pre-filtering DSOGI-FLL. In this work, it is used a fixed resonant frequency equal to the utility frequency, i.e., 60 Hz.

TABLE I
ELECTRICAL PARAMETERS FOR SIMULATIONS

Parameters		Value
Grid	Grid Voltage	13.8 kV
	Line Series Impedance	$28.032 + j20.5837\Omega$
	Tranformer power rating	100 MVA
	Transformer relation	13.8/0.69 kV
	Leakage reactance	0.00125 pu
LCL Filter	L1 inductor	3.152 mH
	Lf inductor	0.315 mH
	Cf capacitor	$30.0585\mu F$
	Rd resistence	6.2Ω
Balanced Load	Three-phase	3.2+j0.008
Unbalanced Load	Phase A	$7.3+j0.008\Omega$
	Phase B	$5.1+j0.008\Omega$
	Phase C	$3.2+j0.008\Omega$
Inverter	Switching Frequency	10 kHz

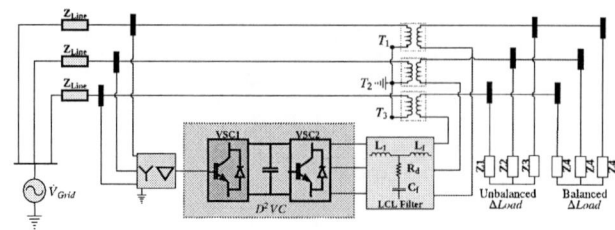

Fig. 5. Distribution grid test

IV. ASSESSMENT OF THE PROPOSED VOLTAGE COMPENSATION ALGORITHM

Fig. 5 shows an equivalent circuit of a utility grid that is adapted to evaluate the proposed algorithm. In order to assess the proposed algorithm, simulations on time domain using detailed models of the VSC are performed using PSCAD/EMTDC. Both compensation functionalities of the proposed algorithm are assessed, i.e., *i* amplitude voltage regulation of the positive sequence component, and *ii* unbalanced voltage compensation due to the negative sequence components. For the amplitude regulation, a balanced three-phase load is connected in order to cause a voltage sag. The imbalance is caused by the connection of a three-phase unbalanced load, and the objective is to evaluate the restoration of balanced voltage condition through the compensation of the negative sequence based on the proposed control scheme.

The D^2VC device, depicted in Fig. 5, is connected to the grid through a LCL passive filter. The series VSC_2 is switched based on an unipolar SPWM strategy, aiming further better design of filter and harmonics elimination. The electrical parameters are based on [6], and shown in (see Table IV), but for the imbalance assessment, the resistive load values are redefined to impose a higher unbalanced voltage condition before compensation, and the X/R relation is decreased under the unity, to assess the effectiveness in weak grids. The load connection is done in 0.2s, and the system experiences its influence during 0.4s, until the D^2VC compensation starts to act in 0.6s. The simulation results are shown from Fig. 6 to 12.

A. Positive Sequence Voltage Amplitude Regulation

This section presents the effectiveness evaluation of the proposed algorithm for voltage amplitude regulation of the positive sequence component. Figure 6 shows the voltage sag caused by the connection of the balanced load at 0.2s, and lasts for 0.4s, before the D^2VC control starts to compensate the voltage sag. The D^2VC synthesizes a three-phase voltage in series and in phase to the secondary voltage of the transformer so as keeping the amplitude at the desired level, i.e., 1.0 p.u. (Figure 6b). It can be notice that the proposed compensation

algorithm regulates the rms value of the load voltage properly. This result confirms its effectiveness for voltage balance regulation.

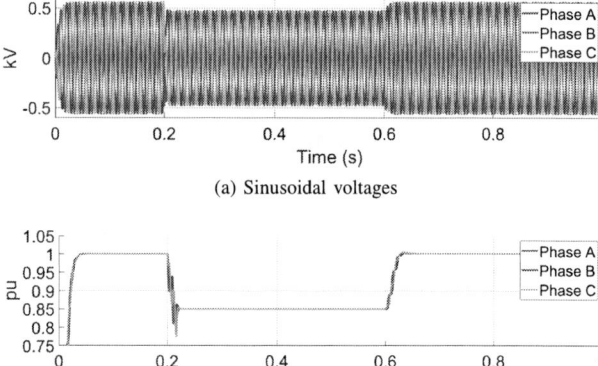

(a) Sinusoidal voltages

(b) rms values (p.u.)

Fig. 6. Voltage profile in balanced conditions

B. Unbalanced Voltage Compensation

This section presents the effectiveness evaluation of the proposed algorithm for unbalanced voltage compensation. In Figure 7a is shown the voltage profile at the load terminal when the unbalanced load is connected. Figure 7b shows accurately the rms value, in p.u., during the unbalanced load influence, it can be seen that all three phases are impacted in different levels. Phase C is the most affected one.

Figure 8 shows the negative sequence components in $\alpha\beta$-frame due to the unbalanced load connection. The load connection affects both the positive and the negative sequence components at the load terminal as shown in Figure 9.

The input error of the PR controllers $\mathbf{v}_{\alpha\beta}^+$ and $\mathbf{v}_{\alpha\beta}^-$ components, calculated by (7) – (10), are shown in Figure 10. It can be noticed that before the unbalanced load connection, both errors, once null, arise after imbalance condition, which means that the load affects both positive and the negative sequence components. For the positive sequence error, the control algorithm should compensate and regulate the amplitude, based on (7) and (8), considering a rms reference equal to 1.0 p.u. For negative sequence components, the control algorithm should fully compensate these components, based on (9) and (10). As it can be seen in Figure 10, after t = 0.6s, the D^2VC is turned

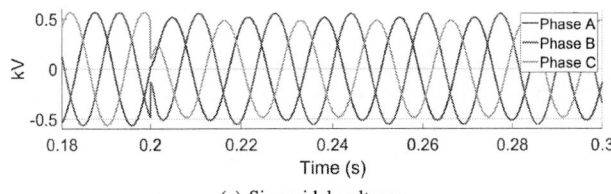

(a) Sinusoidal voltages

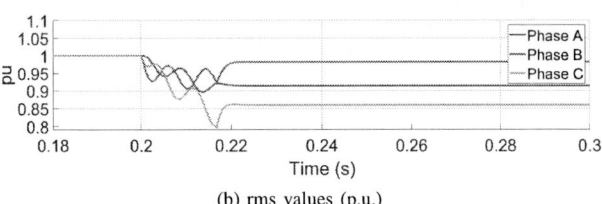

(b) rms values (p.u.)

Fig. 7. Voltage profile in unbalanced conditions

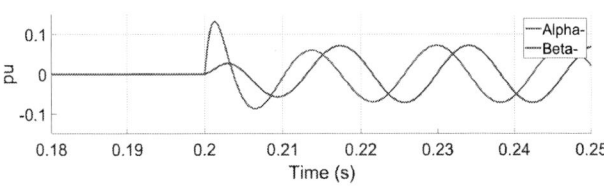

Fig. 8. Negative sequence components

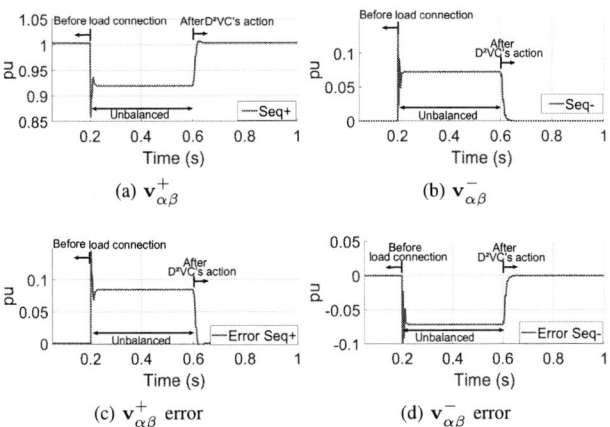

(a) $\mathbf{v}_{\alpha\beta}^{+}$

(b) $\mathbf{v}_{\alpha\beta}^{-}$

(c) $\mathbf{v}_{\alpha\beta}^{+}$ error

(d) $\mathbf{v}_{\alpha\beta}^{-}$ error

Fig. 9. Amplitudes before and after D²VCs action

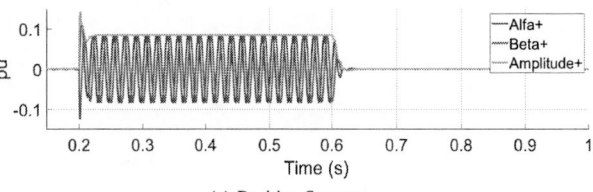

(a) Positive Sequence

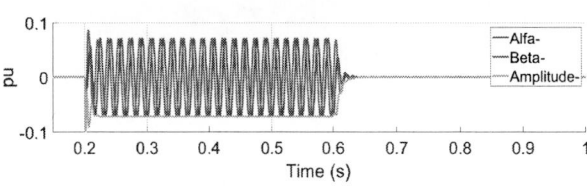

(b) Negative Sequence

Fig. 10. $\alpha\beta$ reference components

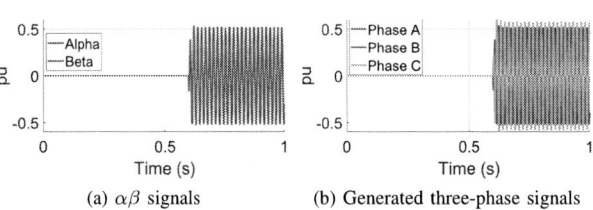

(a) $\alpha\beta$ signals

(b) Generated three-phase signals

Fig. 11. Reference signals for SPWM

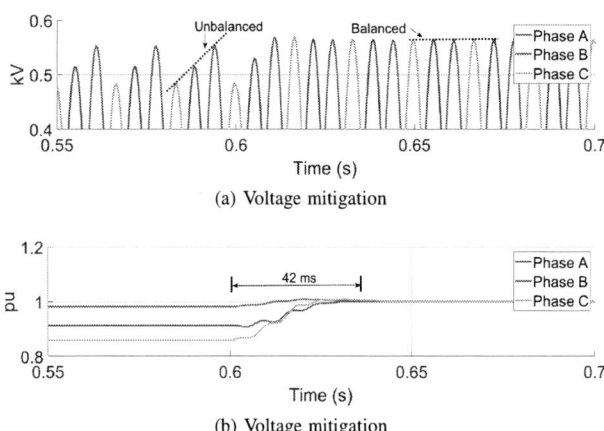

(a) Voltage mitigation

(b) Voltage mitigation

Fig. 12. D²VC mitigates unbalanced voltages

on and both errors go to zero after a short time. Figure 11 shows the SPWM reference signals. The PR controller gains (Figure 4) are K_p=1 and K_i=1000, based on [23], [29], and refined to get better dynamics.

The D²VC is able to mitigate the imbalance on the load bus, since the voltages are brought back to normal operation, as shown in Figure 12, within some milliseconds after the start of operation as detailed in Figure 12b. It is worthy to mention that the start up operation of the D²VC is not optimized and there is not any feed-forward signal to improve the transient performance in order to reduce the time response. Another point that should be highlighted is that the rms values are calculated using the FFT block in the PSCAD, so there may

be a delay due to the sliding window of the algorithm.

As a figure of merit for imbalance measurement in the system, [1] establishes the imbalance factor (IF), defined as the ration between the amplitudes of negative and positive sequence components. In accord to [1], IF should be below 3% in distribution systems, so as to be considered a balanced voltage system. In Figure 13 it is shown the imbalance factor, it can be seen that the load is highly submitted to a unbalanced condition, and the IF is almost 3 times higher than [1] prerogatives, until the D²VC starts operation. Hence, the D²VC accomplishes its task to mitigate unbalanced voltage and regulate the positive sequence amplitude at the load terminal, according to ANEEL regulations.

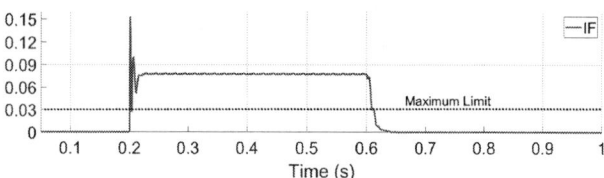

Fig. 13. Imbalanced factor

V. Conclusions

This paper has investigated the voltage profile in weak distribution systems and proposes a control algorithm for voltage amplitude regulation and for mitigation of unbalanced voltage due to negative sequence using the D^2VC. The main characteristic of the proposed algorithm is that it is based on a PR Controller in $\alpha\beta$frame without the need of any extra synchronization algorithm.

The simulation results have shown the effectiveness of D^2VC as device to regulate voltage in distribution systems based on the proposed algorithm. Further investigation will be carried out to improve the transient time response and to include protection against low voltage ride through and high current. Next step is the implementation of a prototype for experimental trials.

Acknowledgment

Authors acknowledge the National Council for Scientific and Technological Development and the Instituto Nacional de Energia Elétrica (INERGE), for funding in part this investigation.

References

[1] Agência Nacional de Energia Elétrica (ANEEL), "Procedimentos de Distribuição de Energia Elétrica no Sistema Elétrico Nacional – PRODIST Módulo 8 – Qualidade de Energia Elétrica Seção 8 . 0 – Introdução e Conceituação," p. 90, 2016.
[2] V. Jayalakshmi and N. O. Gunasekar, "Implementation of discrete PWM control scheme on Dynamic Voltage Restorer for the mitigation of voltage sag /swell," *2013 International Conference on Energy Efficient Technologies for Sustainability, ICEETS 2013*, pp. 1036–1040, 2013.
[3] H. S. Hatami H, Shahnia F, Pashaei A, "Operation of D-STATCOM for Voltage Control in Distribution Networks with a New Control Strategy," *2007 IEEE Lausanne POWERTECH, Proceedings (2007)*, pp. 1–4, 2007.
[4] H. Laaksonen, P. Saari, and R. Komulainen, "Voltage and Frequency Control of Inverter," *2005 International Conference on Future Power Systems*, no. 2, 2005.
[5] M. G. Kundur, Prabha and Balu, Neal J and Lauby, *Power System Stability And Control*. McGraw-hill New York, 1994.
[6] S. L. d. S. L. Barcelos, "UPFC para controle de tensão," Ph.D. dissertation, Universidade Federal do Rio de Janeiro, 2013.
[7] S. a. Mohammed, A. G. Cerrada, and B. Hasanin, "Dynamic Voltage Restorer (DVR) System for Compensation of Voltage Sags , State-of-the-Art Review," *International Journal Of Computational Engineering Research (ijceronline.com)*, vol. 3, no. 1, pp. 177–183, 2013.
[8] Z. Shuai, P. Yao, Z. J. Shen, C. Tu, F. Jiang, and Y. Cheng, "Design considerations of a fault current limiting dynamic voltage restorer (FCL-DVR)," *IEEE Transactions on Smart Grid*, vol. 6, no. 1, pp. 14–25, 2015.
[9] L. F. Meloni, Â. J. Rezek, and Ê. R. Ribeiro, "Small-signal modeling of a single-phase DVR for voltage sag mitigation," *Proceedings of International Conference on Harmonics and Quality of Power, ICHQP*, vol. 2016-Decem, pp. 55–59, 2016.
[10] K. Li, J. Liu, Z. Wang, and B. Wei, "Strategies and Operating Point Optimization," vol. 22, no. 1, pp. 413–422, 2007.
[11] K. Chamundeswari and Y. B. Raju, "Adaptive fuzzy control strategy of STATCOM for the regulation of voltage," *Proceedings - 7th International Conference on Communication Systems and Network Technologies, CSNT 2017*, pp. 309–313, 2018.
[12] N. Takahashi and Y. Hayashi, "Centralized voltage control method using plural D-STATCOM with controllable dead band in distribution system with renewable energy," *IEEE PES Innovative Smart Grid Technologies Conference Europe*, pp. 1–5, 2012.
[13] Y. Sugahara and T. Takeshita, "Suppression control of module capacitor voltage fluctuation for cascade STATCOM," *3rd International Conference on Renewable Energy Research and Applications, ICRERA 2014*, pp. 722–727, 2014.
[14] A. J. Monticelli, "Fluxo de Carga em Redes de Energia Elétrica," p. 164, 1983.
[15] A. V. Ital and S. A. Borakhade, "Compensation of voltage sags and swells by using Dynamic Voltage Restorer (DVR)," in *International Conference on Electrical, Electronics, and Optimization Techniques, ICEEOT 2016*. Institute of Electrical and Electronics Engineers Inc., nov 2016, pp. 1515–1519.
[16] H. Heydari, S. M. Oladali, E. Heydari, and A. Nazarzadeh, "Smart optimal control of DVR to compensate voltage sag of the critical loads in the distribution power system," *2014 Smart Grid Conference (SGC)*, pp. 1–6, 2015.
[17] P. Boonchiam and N. Mithulananthan, "Dynamic control strategy in medium voltage DVR for mitigating voltage sags/swells," *2006 International Conference on Power System Technology, POWERCON2006*, pp. 0–4, 2007.
[18] J. Hu, M. Marinelli, M. Coppo, A. Zecchino, and H. W. Bindner, "Coordinated voltage control of a decoupled three-phase on-load tap changer transformer and photovoltaic inverters for managing unbalanced networks," *Electric Power Systems Research*, vol. 131, pp. 264–274, 2016. [Online]. Available: http://dx.doi.org/10.1016/j.epsr.2015.10.025
[19] M. M. Rahman, A. Arefi, G. M. Shafiullah, and S. Hettiwatte, "A new approach to voltage management in unbalanced low voltage networks using demand response and OLTC considering consumer preference," *International Journal of Electrical Power and Energy Systems*, vol. 99, no. December 2017, pp. 11–27, 2018.
[20] A. E. l. Gyugyi, C. D. Schauder, S. L. Williams, T. R. Rietman, D. R. Torgerson, "The unified power flow controller: a new approach to power transmisson control," vol. 10, no. 2, pp. 1085–1097, 1995.
[21] M. Hingorani, Narain G and Gyugyi, Laszlo and El-Hawary, "Understanding FACTS: concepts and technology of flexible AC transmission systems." IEEE press New York, 2000.
[22] P. B. Panumat Sanpoung and B. Plangklang, "Analysis and control of UPFC for voltage compensation using ATP/EMTP," *Asian Journal on Energy and Environment*, vol. 10, no. 04, pp. 241–249, 2009.
[23] S. L. Lima, R. F. Dias, and E. H. Watanabe, "Direct voltage control in grids with intermittent sources using UPFC," *2013 Brazilian Power Electronics Conference, COBEP 2013 - Proceedings*, pp. 974–980, 2013.
[24] D. W. Lee, S. J. Ahn, and S. I. Moon, "A study on coordinated control of UPFC and voltage compensators using voltage sensitivity," *IEEE Power and Energy Society 2008 General Meeting: Conversion and Delivery of Electrical Energy in the 21st Century, PES*, pp. 1–6, 2008.
[25] H. Karimi, A. Yazdani, and R. Iravani, "Negative-Sequence Current Injection for Fast Islanding Detection of a," *IEEE Transactions on Power Electronics*, vol. 23, no. 1, pp. 298–307, 2008.
[26] P. Rodríguez, R. Teodorescu, I. Candela, A. V. Timbus, M. Liserre, and F. Blaabjerg, "New positive-sequence voltage detector for grid synchronization of power converters under faulty grid conditions," *PESC Record - IEEE Annual Power Electronics Specialists Conference*, 2006.
[27] C. Rocha, J. Camacho, E. Coelho, and W. Parreira, "Selective Three-phase Current Reference Generation Using Multi-resonant Method For Shunt Active Power Filter," *Eletrônica de Potência*, vol. 22, no. 1, pp. 19–30, 2017.
[28] P. Rodríguez, A. Luna, I. Candela, R. Teodorescu, and F. Blaabjerg, "Grid synchronization of power converters using multiple second order generalized integrators," *IECON Proceedings (Industrial Electronics Conference)*, pp. 755–760, 2008.
[29] R. Teodorescu, F. Blaabjerg, and M. Liserre, "Proportional-resonant controllers. A new breed of controllers suitable for grid-connected voltage-source converters," *OPTIM 2004, Brasov, Romania*, pp. 1–6, 2004.

Dynamic Modeling and Control of a Three-Port ZVS-PWM Three-Phase Push Pull DC-DC Converter

Lorena Lorraine Oliveira Albuquerque
Department of Electrical Engineering
Federal University of Paraíba
João Pessoa, PB, Brasil
lorena.albuquerque@cear.ufpb.br

Marcos Victor Dantas de Sá
Department of Electrical Engineering
Federal University of Paraíba
João Pessoa, PB, Brasil
marcos.sa@cear.ufpb.br

Fellipe André Lucena de Oliveira
Department of Electrical Engineering
Federal University of Paraíba
João Pessoa, PB, Brasil
fellipe.oliveira@cear.ufpb.br

Isaac Soares de Freitas
Department of Electrical Engineering
Federal University of Paraíba
João Pessoa, PB, Brasil
isaacfreitas@cear.ufpb.br

Romero Leandro Andersen
Department of Electrical Engineering
Federal University of Paraíba
João Pessoa, PB, Brasil
romero@cear.ufpb.br

Abstract— **This paper proposes the mathematic model and the control of the current and voltage loops of the ZVS-PWM three-phase current-fed push-pull DC-DC converter. The converter has three ports, two bidirectional ports and one unidirectional port and can be used in various applications such as electric vehicles, battery charging system, fuel cell, isolated mini grids and micro grids. The proposed control is able to invert power flow by imposing main inductor current value and direction as well as maintaining voltage at a given port under load variations. The control structure for each case is presented, as well as the deduction of the main mathematical expressions. In order to validate the performance of the proposed control strategy, simulation results for a 500 W converter with 50V in port 1 and 200 V in port 3 and a switching frequency of 40 kHz are presented here.**

Keywords—three-phase dc-dc converter, multi-port, bidirectionality, three-port converter.

I. INTRODUCTION

With the expansion of distributed systems and the development of smart grid technology in the last decades, bidirectional topologies have become more widely used and capable of controlling the flow of energy between various sources of energy and storage devices such as batteries [1-2].

The converter proposed in [3] is suitable for hybrid fuel cell / battery power systems, in addition, the stress voltage on the switches is half of the voltage on the high voltage side, so the converter can be used in high-voltage applications.

A novel three-port three-phase dc-dc converter is proposed in [4] to interface with low voltage fuel cell and ultracapacitor (UC), the major features are the achievement zero-voltage-switching (ZVS), phase-shift-modulation (PSM) control, draws and injects smooth current from the fuel cell and UC, reduce core number and current ripple with improved efficiency.

Bidirectional multi-ports converters can also be used in hybrid electric vehicles and uninterruptible power systems. In order to avoid wasting energy in the braking process it is possible to use the converter to reverse the energy flow back to the storage system, characterizing the regenerative braking process. In [5] it is proposed the use of multi-port bidirectional converters for charging stations of electric vehicles.

The ability to control the power flow direction is of extreme importance in a mini grid. In [6] a bidirectional energy converter with an inverter is modeled and designed, whose function in a mini grid consists of making an interface between a storage system and a distribution network, controlling the voltage and frequency of the mini grid and ensuring the integrity of the system.

It is possible to use multi-port converter in mini grids to eliminate the interconnection buses of the various converters, connecting all the loads, the generation units and the storage systems in a single equipment. One of the main advantages is the use of a smaller number of components, reducing costs and size.

In order to contribute to this area, this paper presents the development of a voltage and current loop control system of the ZVS-PWM three-phase current-fed push-pull dc-dc converter [7] and explores its possibilities as a three-port converter.

II. THE ZVS-PWM THREE-PHASE CURRENTE-FED PUSH-PULL DC-DC CONVERTER

The push-pull converter shown in Fig. 1 consists of the following main components: clamping capacitor C_g, six active switches (S_1', S_2', S_3', S_1, S_2, S_3), commutation inductances (L_d) to maintain the current during switching intervals, filter capacitor C_o, boost inductance L and three-phase high frequency transformer connected to diode rectifier bridge (D_1, D_2, D_3, D_4, D_5, D_6).

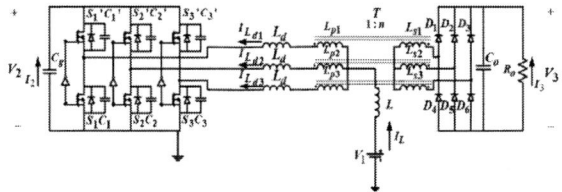

Fig. 1. The ZVS-PWM three-phase current-fed push-pull dc-dc converter.

978-1-7281-4181-7/19 $31.00 © 2019 IEEE

This topology presents three different operating regions depending on the duty cycle D. The gate signals applied to the switches of different legs are delayed 120° and the signals applied to the switches of the same leg are complementary. The mathematical expressions were deduced in [7].

The used converter has three ports, with ports 1 and 2 bidirectional and port 3 unidirectional. It is possible to replace the clamping capacitor C_g on port 2 with a source V_2 and a L_2C_2 filter. In section III is demonstrated the dynamic modeling of the equations of current and voltage plants. In section IV, the current and voltage control structure in port 1 will be presented. In section V, the voltage control structure in port 3 will be demonstrated, since the current loop is the same as in section IV. In section V and VI are presented the design example and simulation results.

III. CONVERTER MODELING

A. Current Plant Model G(s)

The model of the plant for the inductor current control is found analyzing the equivalent circuit of Fig. 2. In this circuit, v_x represent the average voltage between the primary neutral point of the transformer and the reference.

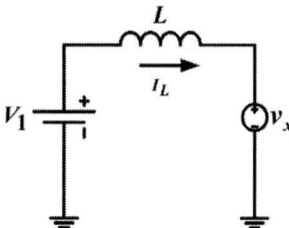

Fig. 2. Equivalent electric circuit.

From the equivalent circuit of Fig. 2, (1) is obtained. The behavior of the voltage waveform v_x [7] is shown in Fig. 3.

$$\frac{L \cdot di_L(t)}{dt} = V_1 - v_x \qquad (1)$$

Knowing that $\Delta t_3 = (1-d) \cdot T_s$, from Fig. 3, it is possible to obtain (2).

$$v_x = (1-d) \cdot V_2 \qquad (2)$$

Substituting (2) in (1), (3) is obtained.

$$\frac{L \cdot di_L(t)}{dt} = \left[V_1 - (1-d(t)) \cdot V_2 \right] \qquad (3)$$

Applying a small perturbation in the system and considering V_1 and V_2 to be constant, (4) is obtained.

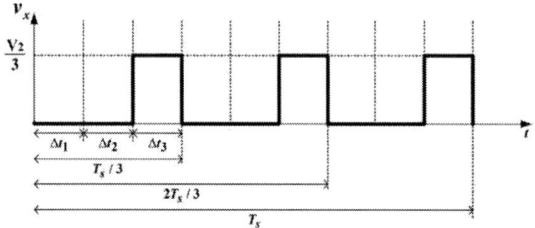

Fig. 3. Waveform of the voltage V_X.

$$\frac{L \cdot d\left[I_L + \hat{i}_L(t) \right]}{dt} = V_1 - \left[1 - (D + \hat{d}(t)) \right] \cdot V_2 \qquad (4)$$

Considering the average values of (1) and (2) and knowing that the average voltage at the inductor is zero, the DC terms of (4) cancel out, and the non-linear AC terms are neglected because they are very small, then (4) becomes (5).

$$\frac{L \cdot d\left(\hat{i}_L(t) \right)}{dt} = \hat{d}(t) \cdot V_2 \qquad (5)$$

Applying the Laplace transform in (5) it is possible to obtain the transfer function of the current plant for the ports 1 and 3 which is given by (6).

$$G(s) = \frac{\hat{i}_L(s)}{\hat{d}(s)} = \frac{V_2}{s \cdot L} \qquad (6)$$

The transfer function of (6) shows how current I_L is affected by a change in duty cycle and can be used to design a current controller.

B. Voltage plant model H(s)

The following procedure aims to obtain the transfer functions of the plant for the control of voltages V_1 and V_3. Using average values on port 1 the circuit can be represented as shown in Fig. 4(a), the deduction of the mathematical expression is given by the following equations.

$$i_1 = i_{C_1} + i_{R_1} \qquad (7)$$

$$i_1(t) = C_1 \cdot \frac{d(v_1(t))}{dt} + \frac{v_1(t)}{R_1} \qquad (8)$$

Applying the Laplace transform in (8) and rearranging the terms as follows, gives the model of the plant for the voltage control in (12).

$$I_1(s) = s \cdot C_1 \cdot V_1(s) + \frac{V_1(s)}{R_1} \qquad (9)$$

$$R_1 \cdot I_1(s) = s \cdot R_1 \cdot C_1 \cdot V_1(s) + V_1(s) \qquad (10)$$

$$R_1 \cdot I_1(s) = V_1(s) \cdot (s \cdot R_1 \cdot C_1 + 1) \qquad (11)$$

$$H_1(s) = \frac{V_1(s)}{I_1(s)} = \frac{R_1}{s \cdot R_1 \cdot C_1 + 1} \qquad (12)$$

The same procedure can be done for port 3, leading to the equivalent circuit of Fig. 4(b) and the transfer function (13).

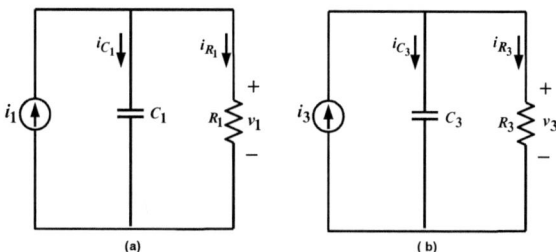

Fig. 4. (a) Equivalent circuit of average values of port 1. (b) Equivalent circuit of average values of port 3.

978-1-7281-4181-7/19 $31.00 © 2019 IEEE

$$H_3(s) = \frac{V_3(s)}{I_3(s)} = \frac{R_3}{s \cdot C_3 \cdot R_3 + 1} \qquad (13)$$

IV. VOLTAGE AND CURRENT CONTROL

A. Port 1

The proposed structure to control the voltage V_1 and the current at port 1 is shown in Fig. 5. The voltage error ε_v is generated by subtracting the measured value of voltage on port 1 ($V_{1\text{-meas}}$) from reference voltage value ($V_{1\text{-ref}}$) and is applied to the voltage compensator $C_V(s)$ generating the control signal of the voltage loop V_{CV}, which is used as reference for the current loop.

The current error ε_i is generated by subtracting from V_{CV} the measured value of current on port 1($I_{L\text{-meas}}$) and is applied in the current compensator $C_I(s)$ generating the control signal of the current loop, which is compared with a sawtooth signal with frequency f_s and amplitude V_s, generating the PWM pulses for switches activation.

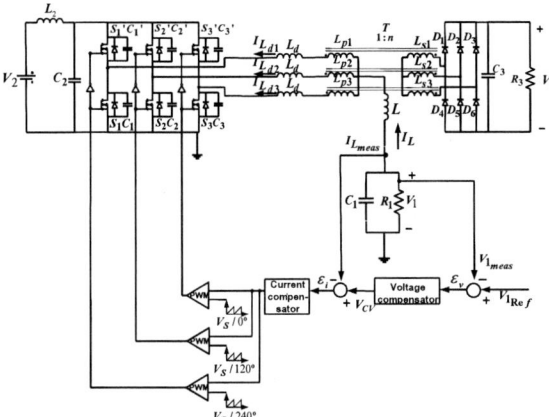

Fig. 5. Basic structure of current and voltage control at port 1.

The open-loop transfer functions of the current and voltage loops are given by (14) and (15) respectively. The gains associated with these functions are represented in Fig. 6, where K_{IS} and K_{VS} are respectively the gains of the current sensor and voltage sensor. In this design, since the current loop is faster than the voltage loop, the current loop can be seen as a gain K in the voltage transfer function (15).

$$OLTF_i = C_I(s) \cdot G_{PWM} \cdot G(s) \cdot K_{IS} \qquad (14)$$

$$OLTF_{v1} = C_V(s) \cdot K \cdot H_1(s) \cdot K_{VS} \qquad (15)$$

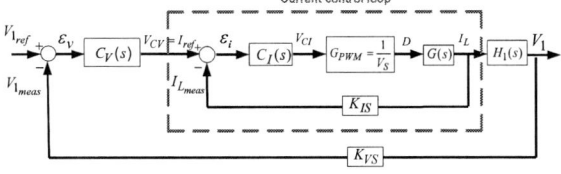

Fig. 6. Block diagram of the current and voltage control on port 1.

An application for port 1 would be to place a battery bank in its terminals, so it would be possible to charge the battery bank through port 2 or, if the batteries were already

charged and no power source in port 2, port 1 will supply power to the load located in port 3, as shown in fig.7.

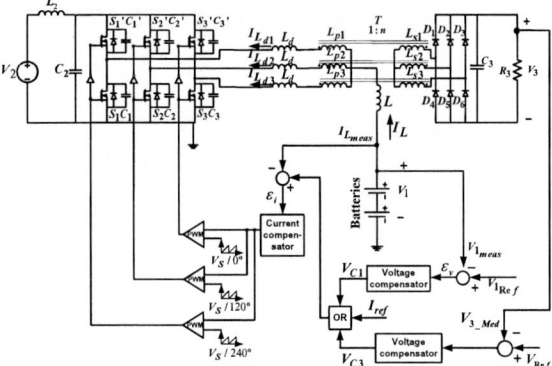

Fig. 7. Simulation circuit of the battery charge method in port 1.

B. Port 3

Using the same control method in port 1 for port 3, it is possible to control the current I_L (port 1) and the voltage V_3. The control scheme is shown in Fig 8.

The open loop transfer function of the voltage is given by (16), where the gains associated with this function are represented in Fig. 9.

In this case, a gain K_3 is necessary to relate the current I_L and the current I_3 in port 3 whose voltage is being controlled. In the previous section this gain was unitary, because the current I_L is equal to the current I_1 in port 1.

The current control loop is the same as that used in the previous section, including the current plant model G(s).

$$OLTF_{v3} = C_V(s) \cdot K \cdot K_3 \cdot H_3(s) \cdot K_{VS} \qquad (16)$$

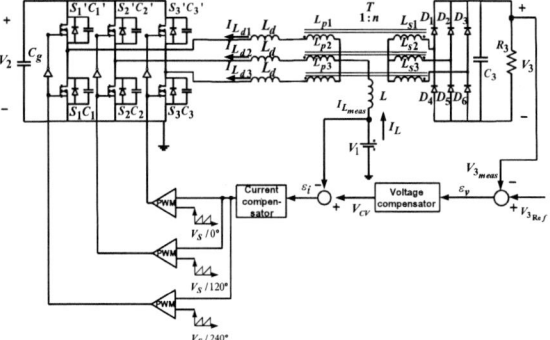

Fig. 8. Basic structure of current and voltage control at port 3.

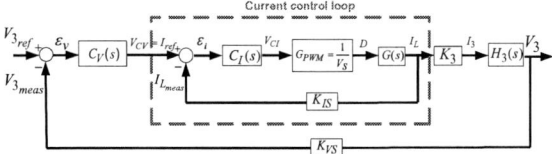

Fig. 9. Block diagram of the current and voltage control on port 3.

V. DESIGN EXAMPLE

Table I shows the values of the main parameters used in the simulation.

TABLE I. SPECIFICATIONS OF THE PROPOSED CONVERTER.

Parameter	Value
Voltage on port 1	$V_1 = 50$ V
Expected efficiency	$\eta = 0.93$
Voltage on port 3	$V_3 = 200$ V
Rated power on port 3	$P_3 = 500$ W
Switching frequency	$f_s = 40$ kHz
Relative voltage ripple at 120 Hz	$\Delta V_3 = 0.02$
Current ripple	$\Delta I_L = 0.08$
Capacitor on port 3	$C_3 = 1000$ µF
Capacitor on port 1	$C_1 = 1000$ µF
Current in boost inductance	$I_L = 10.753$ A
Resistor on port 3	$R_3 = 80$ Ω
Resistor on port 1	$R_1 = 4.65$ Ω
Voltage on port 2 (Clamping voltage)	$V_2 = 187.5$ V
Duty cycle	$D = 0.733$
Effective duty cycle reduction	$I_{no} = 5\%$
Turns ratio of the transformer	$n = 1.267$
Boost inductance on port 1	$L = 96.875$ µH

In Fig. 10, Fig. 11 and Fig. 12 the gain and phase Bode diagrams of the open loop transfer function of the current and voltage control on port 1 and voltage control on port 3 are represented according to the equations (14), (15) and (16) deduced previously. The gain crossover frequencies of the current and voltage loops on port 1 and 3 transfer functions are respectively 8 kHz, 100 Hz, 800 Hz and the values of all the phase margins demonstrate that the system is stable.

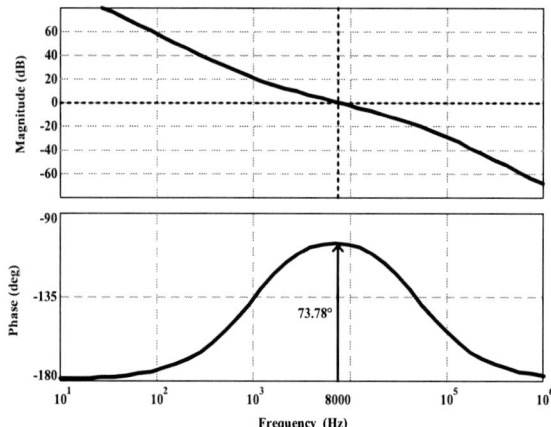

Fig. 10. Gain and phase Bode diagrams of the open loop transfer function of the current control on port 1.

VI. SIMULATION RESULTS

In this section are presented the simulation results for the controllers obtained in this paper, demonstrating that they are working as desired, and a process of charge and discharge of a battery bank applied to the terminals of port 1.

A. Current Control on Port 1

Initially a simulation was performed using only the current control. The inductor current I_L is 10.7A. A simulation was performed using a V_2 voltage source with an L_2C_2 filter in place of the clamping capacitor used initially,

with this configuration the same result as the previous was obtained. During the operation of the circuit a change was made in the reference current as can be observed in Fig. 13, besides that it is possible to reverse the current flow through the inductor, in this case, voltage source V_2 acts by injecting current and power at source V_1. This result indicates the bidirectionality of port 1.

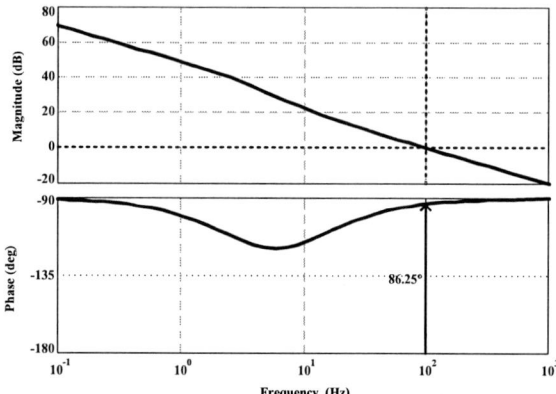

Fig. 11. Gain and phase Bode diagrams of the open loop transfer function of the voltage control on port 1.

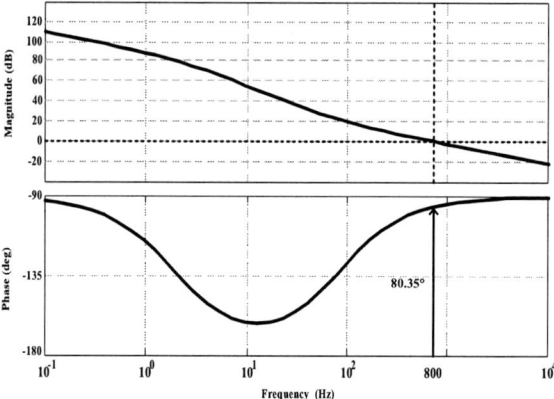

Fig. 12. Gain and phase Bode diagrams of the open loop transfer function of the voltage control on port 3.

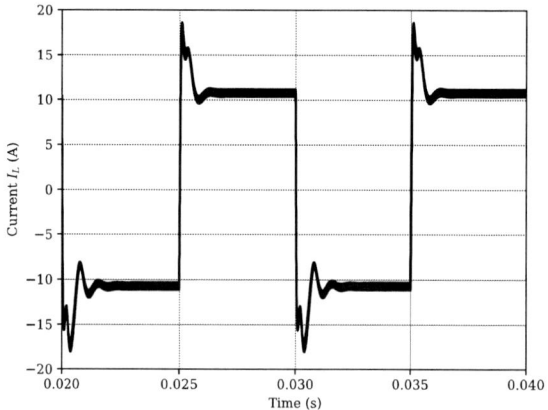

Fig. 13. Current I_L demonstrating bidirectionality on port 1 with V_2 as the main power source.

B. Current and Voltage Control on Port 1

The average values of currents I_1 and I_2, voltages V_1 and V_3 are 10.7A, 5.56A, 49.95 V and 199.6V, respectively. During the operation of the circuit, a load step was applied on port 1, to visualize the behavior of the structure when subjected to load transients.

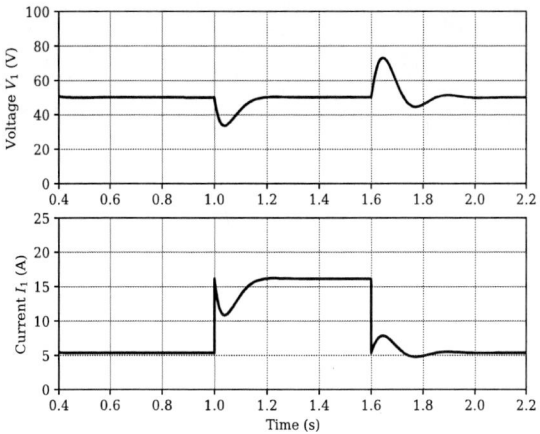

Fig. 14. Waveforms of the simulation with load step on port 1. (a) Voltage V₁. (b) Current I₁.

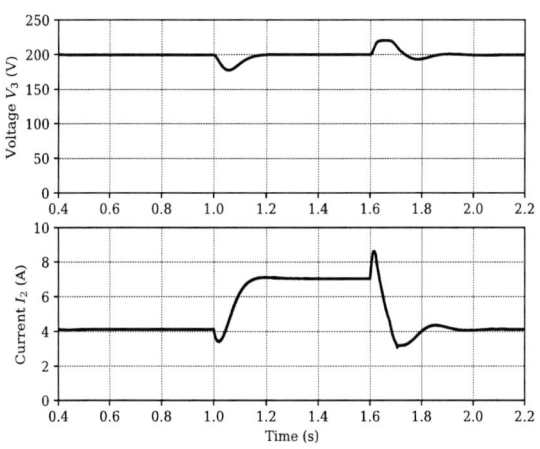

Fig. 15. Waveforms of the simulation with load step on port 1. (a) Voltage V₃. (b) Current I₂.

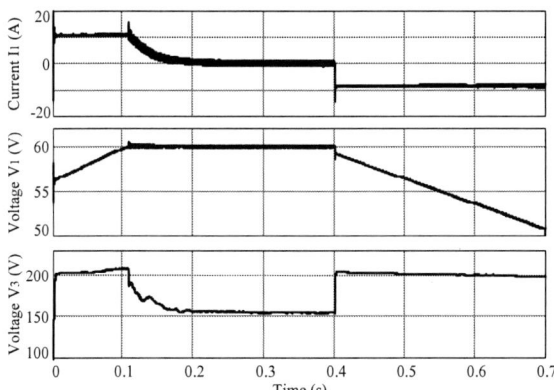

Fig. 16. Constant current constant voltage method for battery charge.

The voltage control maintain the value of V_1 even with the load variation, as shown in Fig. 14, the behavior of the current on port 1 is also represented. In Fig. 15 is shown the behavior of the current on port 2 when subjected to load transients. The proposed control can also maintain the voltage on port 3.

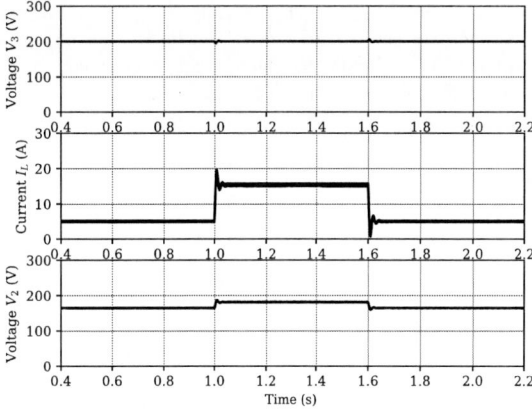

Fig. 17. Waveform of the simulation with load step on port 3. (a) Voltage V₃. (b) Current I_L. (c) Voltage V₂.

C. Battery charge method in port 1

As an example of the proposed controls applied to a system that includes a battery bank, it can be mentioned battery charging methods.

The method used in this paper is constant current and constant voltage, which consists of keeping the current in the battery bank with a constant value until its voltage reaches its equalization value that is higher than the normal voltage in port 1 (previously defined in Table I).

The equalization value was chosen as 60 V, so as long as the voltage at port 1 is less than that, the current control acts alone and the reference current value is fixed at 10.753 A, which is the nominal value specified for this port.

After the battery voltage achieves 60 V, the voltage control acts keeping the voltage at this value and generating the reference for the current control, so the current gradually decreases to zero. As current I_1 decreases, current I_2 also decreases, thus the voltage on port 3 is unregulated during the charge process and its voltage is less than its nominal value, as shown in Fig. 16.

Finally, when the battery is charged and there is no voltage source in port 2, the battery bank is commuted to supply power to the load in port 3, so the voltage control loop of port 3 generates the reference for inductor L current.

It is possible to see in Fig. 16 that when the battery is charging its current is positive and after 0.4 seconds the battery starts to supply energy and its current becomes negative and the voltage begins to decrease, because it is discharging.

D. Current and Voltage Control on Port 3

The current I_L behavior is the same as in the previous section. The average value of V_3 is 200V. During the operation of the circuit a load step was applied to visualize the behavior of the structure when subjected to load transients. As there is a voltage control, the voltage V_3 is regulated as shown in Fig. 17. In this figure, the behavior of the current I_L and voltage V_2 when subjected to load transients is also represented.

CONCLUSIONS

The article proposes the current and voltage control loops of a ZVS-PWM three-phase current-fed push-pull DC-DC converter with three ports that because of the bidirectionality characteristic can be used in applications such as electric vehicles and aircraft, fuel cell, battery charging system, isolated mini grids and micro grids.

By the waveforms it is possible to observe that the values obtained by simulation agree with the theoretically calculated values. Using the current loop is possible to impose the current on the inductor of port 1 and to reverse the power flow.

Using the voltage control loop, it is possible to control the value of the voltages of port 1 and 3 even though there is a load variation, which demonstrates that the converter works as expected. Using a battery and voltage and current control is possible to demonstrate a battery charge method on port 1 and when there is no power source in port 2, port 1 can supply power to the load located in port 3.

REFERENCES

[1] H. M. de Oliveira Filho, D. S. Oliveira Jr, C. E. de A e Silva, F. L. Tofoli, "ZVS bidirecional isolated three-phase DC DC converter with dual phase-shift and duty cycle", in Proc. INDUSCON, pp. 1-8, 2012.

[2] M. Stieneker and R. W. De Doncker, "Dual-active bridge dc-dc converter systems for medium-voltage DC distribution grids," *2015 IEEE 13th Brazilian Power Electronics Conference and 1st Southern Power Electronics Conference (COBEP/SPEC)*, Fortaleza, 2015, pp. 1-6.

[3] K. Jin, M. Yang, X. Ruan and M. Xu, "Three-Level Bidirectional Converter for Fuel-Cell/Battery Hybrid Power System," in *IEEE Transactions on Industrial Electronics*, vol. 57, no. 6, pp. 1976-1986, June 2010.

[4] D. Liu and H. Li, "A Three-Port Three-Phase DC-DC Converter for Hybrid Low Voltage Fuel Cell and Ultracapacitor," *IECON 2006 - 32nd Annual Conference on IEEE Industrial Electronics*, Paris, 2006, pp. 2558-2563.

[5] H. Tao, A. Kotsopoulos, J. L. Duarte, and M. A. M. Hendrix, "Family of multiport bidirectional DC-DC convertrs", Eletric Power Application, IEE Proceedings, v. 153, pp.451-458, 2006.

[6] F. S. F. de Silva, "Inversor bidirecional para controle de fluxo de potência em minirredes com geração distribuída", Universidade Federal do Maranhão, Centro de Ciências Exatas e Tecnologia, Maranhão, Abril 2014.

[7] R. L. Andersen and I. Barbi, "A ZVS-PWM Three-Phase Current-Fed Push–Pull DC–DC Converter," in *IEEE Transactions on Industrial Electronics*, vol. 60, no. 3, pp. 838-847, March 2013.

Thermal Modeling of Converters for Wind Conversion Systems Employing DFIG Technology

Rodrigues de Oliveira, Igor
Graduate Program in Electrical Engineering - Universidade Federal de Minas Gerais - Av. Antônio Carlos 6627, 31270-901, Belo Horizonte, MG, Brazil
igorrooli@gmail.com

Tofoli Lessa, Fernando
Department of Electrical Engineering Federal University of São João del- Rei São João del- Rei, Brazil
fernandolessa@ufsj.edu.br

Mendes Flores, Victor
Graduate Program in Electrical Engineering - Universidade Federal de Minas Gerais - Av. Antônio Carlos 6627, 31270-901, Belo Horizonte, MG, Brazil
victormendes@cpdee.ufmg.br

Abstract— With the increasing use of wind energy conversion systems, it is important to study different aspects of the different technologies employed in these systems. The key of the variable speed technologies is the power converter, thus the properly design and evaluation of its behavior is important. In this work, a thermal modeling of a wind energy conversion system (WECS) using doubly fed induction generator (DFIG) technology is developed, which evaluates the behavior of the junction temperature of the semiconductors devices that compose the power converters of the system. The system is modeled using the PLECS software and results of a 2.5 MW system are presented in different operation conditions to demonstrate the modeling.

Keywords—power converters, DFIG, wind energy, thermal analysis, control

I. INTRODUCTION

As one of the promising alternatives to solve the energy crisis worldwide, wind energy has been growing in the last years and they are object of several studies [1]. Wind energy conversion systems (WECS) are generally composed of the wind turbine, an electric generator, a static power converter and the corresponding control system.

The electric machine with the most widespread use in industrial scale is the three-phase induction machine, also known as asynchronous electric machine, whose application industry was introduced in 1880. Among the inductions machines used as wind generators, the DFIG (doubly-fed induction generator) is the most used as it allows the operation with variable speed turbines. It grants the advantage in active and reactive power capabilities using the converter rated for only a small fraction (generally 30%) of the rated power. However, DFIG are very sensitive to grid disturbances, especially to voltages dips. The abrupt drop of the grid voltage causes overvoltages and overcurrents in the rotor windings that could even destroy the converter if no protection elements are included [2, 3].

In this context, there are important studies in the literature on the behavior of DFIG during voltages sags, in order to minimize the harmful effects and to develop new strategies to improve the DFIG LVRT capability. Examples of these studies can be seen in [4-8].

As one the most vulnerable component in the system is the power electronic, efforts have been recently devoted to the reliable behavior of the power semiconductor due to the increased cost and time for repair after failures [9-12].

Notably the thermal profile is an important indicator of the lifetime of the power semiconductor and it has an influence on the reliable operation [13, 14]. The number of energy cycles and the average junction temperature are very relevant factors for the joint temperature fluctuation [15-18]. The thermal behavior of the power devices during grid fault is, for example, evaluated in [19].

The present work aims to perform a thermal modeling of the semiconductors devices present in the converters of a WECS based on DFIG technology, investigating the general behavior and thermal stresses that these devices are submitted during the working regime. Section II shows the structure of a WECS using DFIG, as well as an explanation of the parts of this system and also the control strategies of the power converters. In addition section III brings all the modeling of the system, both the study from the point of view of the control of the converters as well as the thermal study. All the steps in the design of the IGBTs switches are described to carry out the study that will be done through a simulation of a wind power conversion system of 2.5 MW using the program Plexim: Electrical Engineering Software (PLECS). Subsequently the results obtained are described in section IV. Finally, in section V presents the conclusions pertinent to the study carried out by this work.

II. STRUCTURE OF SYSTEM

The basic structure of a wind turbine based on a DFIG is show in Fig. 1.

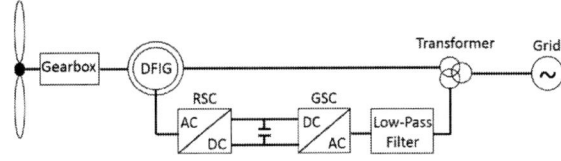

Fig. 1. Wind turbine with DFIG.

A. Turbine

The energy extracted from the wind depends on the constructive characteristics of the turbine, such as the radius, and also the wind speed, the rotational speed of the turbine and the pitch angle. The static power characteristic of the wind turbine (P) is described by the equation(1).

$$P = \frac{1}{2}\rho A V^3 C_p(\lambda, \beta) \qquad (1)$$

978-1-7281-4181-7/19 $31.00 © 2019 IEEE

where ρ is the density of air, V is the wind speed, A is the swept area by the turbine and $C_p(\lambda,\beta)$ is the so-called power coefficient. In [20] the $C_p(\lambda,\beta)$ is described with a non-polynomial function as can be seen in equation(2):

$$C_p(\lambda,\beta) = 0.22\left(\frac{116}{\lambda_i} - 0.4\beta - 5\right)e^{\frac{-12.5}{\lambda_i}} \quad (2)$$

β is the pitch angle of the turbine blade and the factor λ_i is indicated in equation (3):

$$\frac{1}{\lambda_i} = \frac{1}{\lambda + 0.88\beta} - \frac{0.0035}{\beta^3 + 1} \quad (3)$$

where λ is the tip speed ratio, described by equation(4):

$$\lambda = \frac{\omega_t R}{V} \quad (4)$$

being that ω_t is a rotational speed of turbine in (rad/s) and R the radius of the blades. A typical curve of the power coefficient in function of the speed ratio for different pitch angles is shown in the Fig. 2.

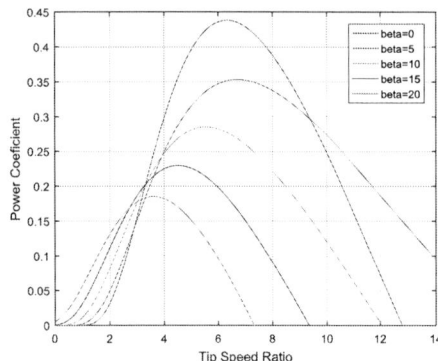

Fig. 2. Power coefficient for the wind turbine.

Associating equation (1) to (4) are the characteristic power curves are shown in Fig. 3, which show the power behavior of the wind turbine, when subject to different wind conditions.

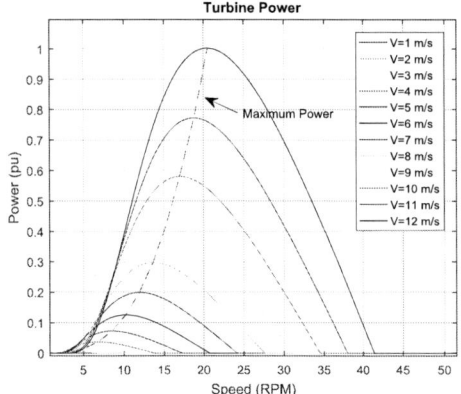

Fig. 3. Turbine power as a function of shaft speed for different winds.

The dashed line of graph of Fig. 3 marks the points where the power is maximum for each wind. These points must be reached by the generator control using the so-called Maximum Power Point Tracker (MPPT).

B. Doubly-fed induction generator (DFIG)

DFIG converts the mechanical energy generated by the turbine wind power transmission, which is transmitted to the grid by stator and rotor windings. The windings the rotor are connected to an AC/DC/AC converter and the stator windings are directly connected to the mains power. The classical induction motor model in a synchronous reference is presented [21]:

$$\vec{v}_s = R_s\vec{i}_s + \frac{d\vec{\psi}_s}{dt} + j\omega_s\vec{\psi}_s \quad (5)$$

$$\vec{v}_r = R_r\vec{i}_r + \frac{d\vec{\psi}_r}{dt} + j\omega_r\vec{\psi}_r \quad (6)$$

where the variables are in the synchronous reference frame and the parameters are referred to the stator. The subscripts s and r refer to the variables and parameters of the stator and the rotor, respectively, so that the meanings of these are described further:

v_s and v_r are voltages, i_s and i_r are currents, ω_s and ω_r are angular speeds, ψ_s and ψ_r are flux linkages, R_s e R_r are resistances.

The stator and rotor flux linkages, ψ_s and ψ_r, are given as:

$$\vec{\psi}_s = L_s\vec{i}_s + L_m\vec{i}_r = (L_{\sigma s} + L_m)\vec{i}_s + L_m\vec{i}_r \quad (7)$$

$$\vec{\psi}_r = L_r\vec{i}_r + L_m\vec{i}_s = (L_{\sigma r} + L_m)\vec{i}_r + L_m\vec{i}_s \quad (8)$$

where L_s, L_r are self-inductances, $L_{\sigma s}$, $L_{\sigma r}$ are leakage inductances and L_m is the magnetizing inductance.

The mechanical differential equation is written as:

$$J\frac{d\omega_m}{dt} + k_f\omega_m = T_e - T_{mec} \quad (9)$$

where J is the inertia constant, k_f is the friction coefficient and T_{mec} is the mechanical torque provided by a wind turbine. T_e is the machine electromagnetic torque calculated through:

$$T_e = \frac{3}{2}P\frac{L_m}{L_s}\text{Im}(\vec{\psi}_s\hat{\vec{i}}_r) \quad (10)$$

where the superscript "^" indicates the complex conjugate.

C. AC/DC/AC Converter

The AC/DC/AC converter is composed of two converters in the back-to-back configuration: the rotor side converter (RSC) and the grid side converter (GSC). Both use IGBTs to synthesize an AC voltage from a DC link. In essence, the RSC controls the active and reactive power generation of the DFIG and the GSC regulates the DC link voltage of the capacitor and the reactive power absorbed by the grid. The conventional control mechanism for these converters is based on the concept of axes decoupling vector control dq. Then, the control strategies used in the GSC e RSC are described hereafter.

The classical GSC control strategy uses internal loops controlling the grid currents, which are oriented in the angle of the grid voltage. This orientation permits a decoupled control of active and reactive power [22]. External to the direct axis current loop it is implemented the DC-link voltage control, because with this orientation the direct axis component (d) is responsible for the active power. The reactive power is controlled by a controller external to the quadrature (q) current control loop. A block diagram representation of the GSC control can be seen in Fig. 4.

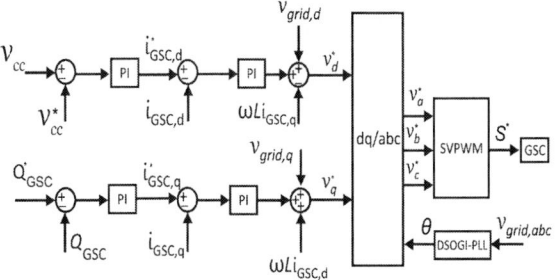

Fig. 4. Block diagram representation of the GSC control structure.

As illustrated in Fig. 4, the angle of the grid voltage vector ($\hat{\theta}_N$), which is used in the orientation of the converter currents, is estimated using the so-called phase-locked loop (PLL). In [23] it is possible to see the entire PLL structure, as well as its block diagram.

The GSC controls keeps a constant DC-link voltage, permitting the generator control the RSC which regulates the rotor currents.

The active and reactive powers are calculated as [24]:

$$P_s = \frac{3}{2} v_{s_d} i_{s_d} = -\frac{3}{2} \frac{L_m}{L_s} v_{s_d} i_{r_d} \qquad (11)$$

$$Q_s = -\frac{3}{2} v_{s_d} i_{s_q} = \frac{3}{2} \frac{v_{s_d}^2}{\omega_s L_s} + \frac{3}{2} \frac{L_m}{L_s} v_{s_d} i_{r_q} \qquad (12)$$

The active stator power depends on the direct axis rotor current and the reactive power depends on the quadrature axis rotor current. Therefore, decoupled rotor currents internal control loops with external active and reactive power controls loops are used in the RSC, as depicted in Fig. 5.

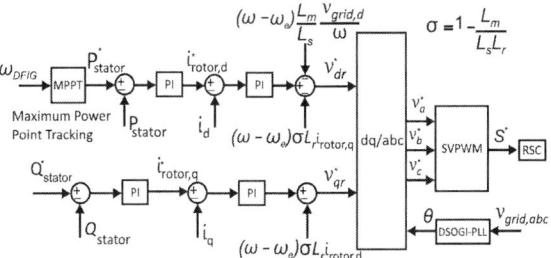

Fig. 5. Block diagram representation of the RSC control structure.

D. LCL Filter

In order to minimize the grid harmonics injected by GSC, it is common to use a LCL (inductor-capacitor-inductor) filter to connect the converter to the grid. A schematic diagram of the LCL filter can be seen in the Fig. 6.

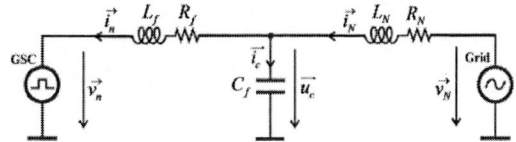

Fig. 6. LCL filter schematic circuit.

L_f and R_f represent the filter inductance and resistance at the converter side, respectively, and L_N and R_N are the filter inductance and resistance at the grid side which can also include the grid impedance.

Using the vector notation and considering a three-phase balanced system, the filter response, evaluated in the grid frequency, can be modeled as simple LR circuit in the synchronous reference frame, as follow:

$$\vec{v}_N^g = R_f \vec{i}_n^g + L_f \frac{d\vec{i}_n^g}{dt} + j\omega_N L_f \vec{i}_n^g + \vec{v}_n^g \qquad (13)$$

where in i_s the filter current, v_N is the grid voltage, v_n is the voltage imposed by the GSC converter and the grid voltage angular frequency is ω_N.

III. THERMAL MODELING

At a first moment the complete system described in section II was implemented using the computer program PLECS®. The TABLE I shows the specifications of the simulated system.

TABLE I. SPECIFICATIONS OF THE SIMULATED SYSTEM

Parameter	Value
Stator/GSC Voltage	690 V/380 V
Grid frequency	60 Hz
Output rated power	2.5 MW
DC-link voltage	700 V
Switching frequency	6 kHz

On the basis of these data, a module IGBT able of supporting 60% more of the DC-link voltage was chosen, then the module must withstand a voltage of 1120 V. With respect to current, it was determined that the module must be able to support the current referring to the rotor, because this is the largest current of the system, which at nominal conditions reaches approximately 2400 A RMS. This module is composed of an Insulated Gate Bipolar Transistor (IGBT) and a free-wheeling diode in antiparallel. So, the chosen module was the FF1500R12IE5 from the manufacturer Infineon [27] that supports a reverse voltage of 1200 V and a current of 1500 A. Since this current is less than the maximum current from the rotor, modules will be used in parallel in order to ensure the proper functioning of the modules by sending their overheating at IGBTs junctions. The heatsink that will be used for cooling the modules is the

P16/300 from the manufacturer SEMIKRON, which have an air cooling forced using the SKF16B-230-01 fan [25].

The data presented in the characteristics curves in [26] that relate to the losses by conduction and switching of the switches present in the module were loaded in the thermal library of the PLECS. The way these curves are loaded and worked inside the thermal library is illustrated in Fig. 7.

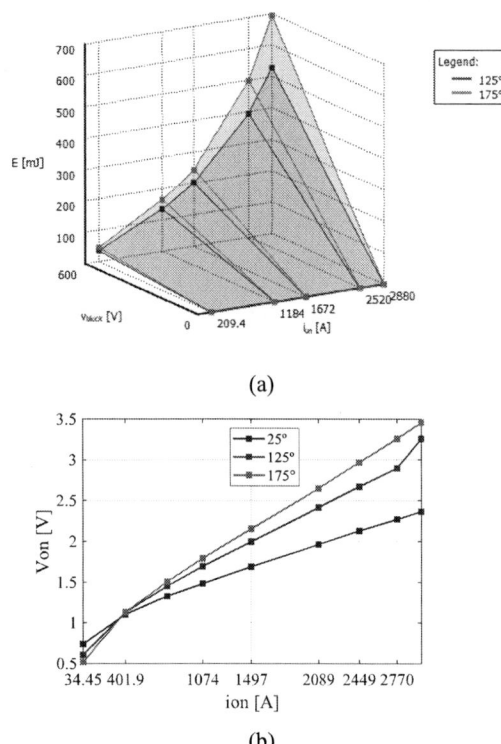

(a)

(b)

Fig. 7. Characteristics curves of the module loaded in the thermal library for: (a) Turn-on loss e (b) Conduction loss.

The thermal impedance inside the power module is usually tested by the power semiconductor manufacturer, and the value is provided in terms of multi-layer Foster structure, as listed in TABLE II.

TABLE II. JUNCTION TO CASE THERMAL IMPEDANCE OF POWER MODULE

		1st	2nd	3rd	4th
IGBT	R (K/kW)	0.527	8.61	8.74	1,63
	τ (s)	0.0012	0.0271	0.0739	0.967
Diode	R (K/kW)	2.72	13.4	16.5	2.35
	τ (s)	0.0012	0.0221	0.0782	1.53

The thermal impedance of the heat-sink is basically not included in the power semiconductor manufacturer datasheet. However, it can normally be provided from the cooling manufacturer. The heatsink thermal resistance is 0.05 K/W and the thermal capacitance was neglected, since the thermal loading of interest is the one of the power semiconductors. Therefore, it is possible to simulate the behavior of the system in seconds, analyzing he stresses which the power semiconductors are subjected to. The preliminary results obtained are presented in the next section.

IV. RESULTS

In order to investigate the behavior and stresses that the switches are subjected to in a WECS based on the DFIG technology, a simulation was developed in the PLECS, aiming the monitoring of the junction temperature of these semiconductor devices, based on the data presented in TABLE I and II.

Fig. 8 shows the results for GSC found when the system is operating in the nominal condition and the Fig. 9 illustrated the results for RSC at the same conditions.

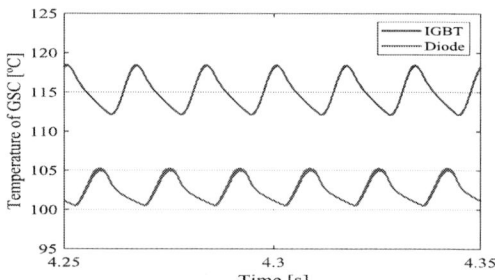

Fig. 8. Junction temperature device of GSC in the nominal condition of the DFIG: the green curve for IGBT and the red curve for diode.

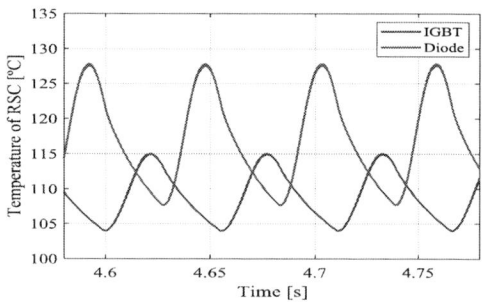

Fig. 9. Junction temperature for devices of RSC in the nominal condition of the DFIG: the green curve for IGBT and the red curve for diode.

It is noted the devices are able to maintain the temperature bellow the limit values that the manufactures determine, in this case about 150ºC, when the system operates in the nominal power condition of the DFIG, this is, 2.5 MW and speed 2340 RPM.

The Fig. 10 shows the curves of the active power of the stator of the DFIG, GSC and grid for the nominal operating conditions of the system.

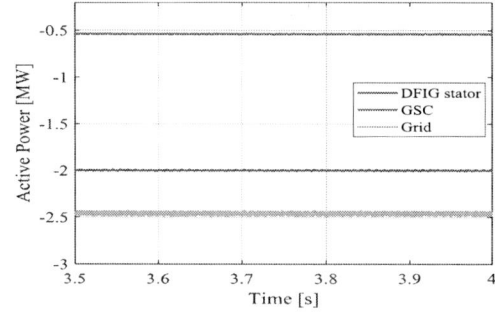

Fig. 10. Active powers: the green curve for stator of the DFIG, red curve for GSC and the blue curve for grid.

Looking at Fig. 10 it can be seen that the active power of the GSC is about 30% of the DFIG stator, highlighting the advantage in using such technology in the WECS, this is, the converter is dimensioned for 30% of the machine power. All powers are negative indicating the injection of energy into the grid.

The currents of GSC and RSC converters can be seen in Fig. 11. The GSC currents have higher harmonic content, since they are measured in the converter terminals. In the grid this high frequency harmonics are attenuated by the filter capacitor.

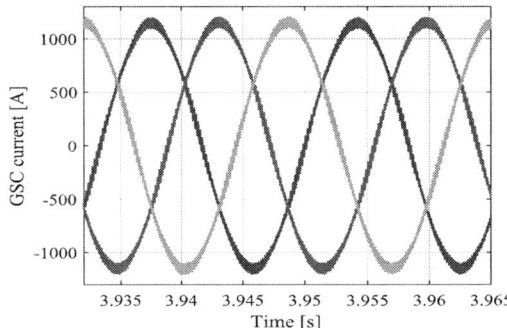

(a)

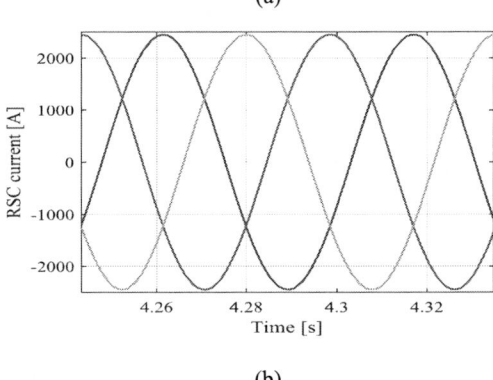

(b)

Fig. 11. Currents in converters: (a) GSC and (b) RSC.

The previous test was considered in the super synchronous speed (slip = -30%) varying the speed of the rotor shaft to 1260 RPM (slip = +30%), a sub synchronous operation, and setting the correspondent active power to 312 kW, Fig 12 presents the thermal behavior of the semiconductors devices.

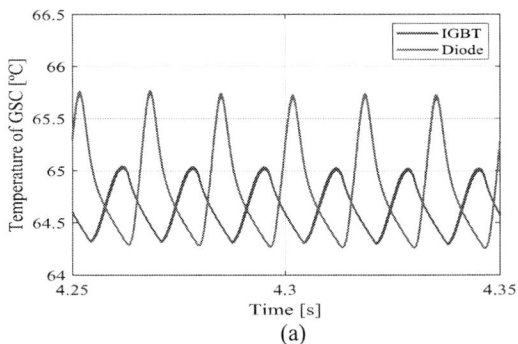

(a)

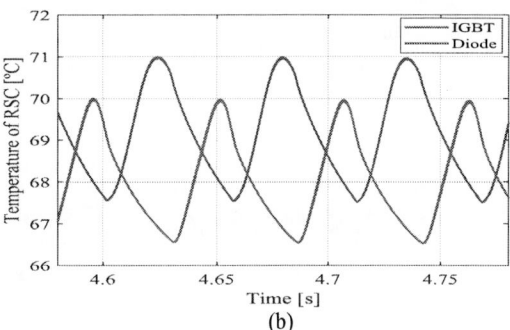

(b)

Fig. 12. Junction temperature in the converters for 1260 RPM: (a) GSC e (b) RSC: the green curve for IGBT and the red curve for diode.

One can be seen in Fig. 12 that the temperatures behavior is much lower to the previous case, since the power is lower. Nevertheless, since the system is operating in slip=+30%, 30% of the active power also flows through the converter. However, in the sub synchronous operation the rotor is consuming power whereas in the super synchronous it delivers power. Therefore, one can be seen the difference in the device stressed (diode or IGBT) in comparison with the super synchronous operation.

A performance test was also performed where a constant speed was set to 2100 RPM on the rotor shaft and after 2.5 s a step was given in the active power of the DFIG, leading to this initial value of 1 MW for 2 MW. The purpose of this test is to verify how the junction temperature of the switches behaves in response to variations in the active power of the system. The results are presented in Fig. 13.

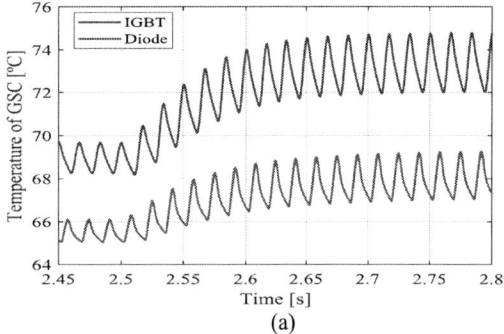

(a)

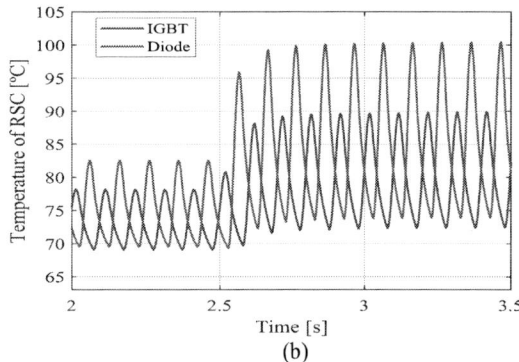

(b)

Fig. 13. Behavior of the junction temperature of the switches: (a) GSC e (b) RSC: the green curve for IGBT and the red curve for diode.

These results demonstrated that the semiconductor devices operate bellow the limit temperature that the power module supports, as well as the efficiency of the heatsink in cooling the module.

V. CONCLUSION

This work presented a thermal modeling of a WECS based on the DFIG technology in which the junction temperatures of the semicondutors devices present in the system converters were monitored in order to verify the thermal stresses that they are subjected to. The preliminary results show the operation of the system in different conditions, indicating that the chosen devices are appropriate for the simulated wind energy conversion system.

With this simulation model, the next steps in this work will be evaluate the DFIG thermal behavior during voltage sags and evaluate different ride through strategies from the point of view of the thermal stresses.

VI. ACKNOWLEDGMENT

This work has the financial support of project P&D 0722 ANEEL/CEMIG and the Brazilian agencies CAPES, CNPQ and FAPEMIG.

REFERENCES

[1] M. Cui, D. Ke, Y. Sun, D. Gan, J. Zhang, and B.-M. Hodge, "Wind power ramp event forecasting using a stochastic scenario generation method," *IEEE Transactions on sustainable energy,* vol. 6, pp. 422-433, 2015.

[2] D. Xiang, L. Ran, P. J. Tavner, and S. Yang, "Control of a doubly fed induction generator in a wind turbine during grid fault ride-through," *IEEE transactions on energy conversion.,* vol. 21, pp. 652-662, 2006.

[3] J. López, E. Gubía, E. Olea, J. Ruiz, and L. Marroyo, "Ride through of wind turbines with doubly fed induction generator under symmetrical voltage dips," *IEEE Transactions on Industrial Electronics,* vol. 56, pp. 4246-4254, 2009.

[4] A. Kasem, E. El-Saadany, H. El-Tamaly, and M. Wahab, "An improved fault ride-through strategy for doubly fed induction generator-based wind turbines," *IET Renewable Power Generation,* vol. 2, pp. 201-214, 2008.

[5] L. G. Meegahapola, T. Littler, and D. Flynn, "Decoupled-DFIG fault ride-through strategy for enhanced stability performance during grid faults," *IEEE Transactions on Sustainable Energy,* vol. 1, pp. 152-162, 2010.

[6] J. Morren and S. W. De Haan, "Ridethrough of wind turbines with doubly-fed induction generator during a voltage dip," *IEEE Transactions on energy conversion,* vol. 20, pp. 435-441, 2005.

[7] M. Rahimi and M. Parniani, "Efficient control scheme of wind turbines with doubly fed induction generators for low-voltage ride-through capability enhancement," *IET Renewable Power Generation,* vol. 4, pp. 242-252, 2010.

[8] J. Yang, J. E. Fletcher, and J. O'Reilly, "A series-dynamic-resistor-based converter protection scheme for doubly-fed induction generator during various fault conditions," *IEEE Transactions on Energy conversion,* vol. 25, pp. 422-432, 2010.

[9] F. Blaabjerg, Z. Chen, and S. B. Kjaer, "Power electronics as efficient interface in dispersed power generation systems," *IEEE transactions on power electronics,* vol. 19, pp. 1184-1194, 2004.

[10] F. Blaabjerg, K. Ma, and D. Zhou, "Power electronics and reliability in renewable energy systems," in *2012 IEEE International Symposium on Industrial Electronics,* 2012, pp. 19-30.

[11] S. Yang, A. Bryant, P. Mawby, D. Xiang, L. Ran, and P. Tavner, "An industry-based survey of reliability in power electronic converters," *IEEE transactions on Industry Applications,* vol. 47, pp. 1441-1451, 2011.

[12] H. Polinder, J. A. Ferreira, B. B. Jensen, A. B. Abrahamsen, K. Atallah, and R. A. McMahon, "Trends in wind turbine generator systems," *IEEE Journal of emerging and selected topics in power electronics,* vol. 1, pp. 174-185, 2013.

[13] D. Zhou and F. Blaabjerg, "Dynamic thermal analysis of DFIG rotor-side converter during balanced grid fault," in *2014 IEEE Energy Conversion Congress and Exposition (ECCE),* 2014, pp. 3097-3103.

[14] D. Zhou, F. Blaabjerg, M. Lau, and M. Tonnes, "Thermal behavior of doubly-fed induction generator wind turbine system during balanced grid fault," in *2014 IEEE Applied Power Electronics Conference and Exposition-APEC 2014,* 2014, pp. 3076-3083.

[15] A. Wintrich, U. Nicolai, W. Tursky, and T. Reimann, "Application manual power semiconductors. Semikron international GmbH," ed: ISLE Verlag, Illmenau, 2011.

[16] F. Richardeau and T. T. L. Pham, "Reliability calculation of multilevel converters: Theory and applications," *IEEE Transactions on Industrial Electronics,* vol. 60, pp. 4225-4233, 2012.

[17] H. Behjati and A. Davoudi, "Reliability analysis framework for structural redundancy in power semiconductors," *IEEE Transactions on Industrial Electronics,* vol. 60, pp. 4376-4386, 2012.

[18] J. Berner, "Load-cycling capability of HiPak IGBT modules," *ABB Application Note 5SYA 2043-02,* p. 201, 2012.

[19] D. Zhou and F. Blaabjerg, "Thermal analysis of two-level wind power converter under symmetrical grid fault," in *IECON 2013-39th Annual Conference of the IEEE Industrial Electronics Society,* 2013, pp. 1904-1909.

[20] V. AKHMATOV, "Analysis of dynamic behavior of electric power systems with large amount of wind power," *PhD Thesis, Orsted, DTU,* 2003.

[21] P. K. Kovács, "Transient phenomena in electrical machines," *394 pp,* 1983.

[22] R. Pena, J. Clare, and G. Asher, "Doubly fed induction generator using back-to-back PWM converters and its application to variable-speed wind-energy generation," *IEE Proceedings-Electric Power Applications,* vol. 143, pp. 231-241, 1996.

[23] P. Rodriguez, R. Teodorescu, I. Candela, A. V. Timbus, M. Liserre, and F. Blaabjerg, "New positive-sequence voltage detector for grid synchronization of power converters under faulty grid conditions," in *2006 37th IEEE Power Electronics Specialists Conference,* 2006, pp. 1-7.

[24] V. F. Mendes, "Ride-through fault capability improvement through novel control strategies applied for doubly-fed induction wind generators," Escola de Engenharia- EE, Universidade Federal de Minas Gerais- UFMG, 2013.

[25] SEMIKRON, "Datasheet of heatsink P16," S.I. ed.

[26] Infineon, "Datasheet of module FF1500R12IE5," ed: S.I.

978-1-7281-4181-7/19 $31.00 © 2019 IEEE

Interleaved Bidirectional DC-DC Converter for Application in Hybrid Propulsion System: Modeling and Control

Vitor C. S. Torres, Vinicius M. de Albuquerque,
Manuel A. Rendón, Pedro S. Almeida,
Janaina G. Oliveira
Group of Electromechanical Energy Conversion
Federal University of Juiz de Fora
Juiz de Fora, Brazil
vitor.torres@engenharia.ufjf.br

Márcio C. B. P. Rodrigues
Group of Power Electronics and Applications
Federal Institute of Southeast of Minas Gerais
Juiz de Fora, Brazil

Abstract—This paper presents the development of a state-space-based modeling approach for an interleaved bidirectional DC-DC converter, which is part of an aircraft hybrid propulsion system. The proposed model is presented in order to obtain relationships of voltage and current control loops of two operational modes of this interleaved DC-DC converter: buck mode, which controls voltage and current during on-board battery pack charging; and boost mode, responsible for aircraft hybrid propulsion system DC bus regulation. Simulation results are used to validate the developed model as well as to evaluate control system performance.

Index Terms – **hybrid aircraft, bidirectional interleaved DC-DC converter, boost mode, buck mode**

NOMENCLATURE

v_b Battery voltage.
L_1 Inductor.
C_B DC bus capacitance.
R_{DC} DC bus resistance.
i_1 Current.
V_{DC} DC bus voltage.
v_{b0} Internal battery voltage.
r_b Internal battery resistance.
d Duty cycle for buck mode.
D Duty cycle for boost mode.

I. INTRODUCTION

Environmental concerns from nowadays, related to climate change and other issues, increase the demand for a less polluting and more sustainable society. Common energy-efficient enterprises aimed at such purpose such as the emergence of viable electric cars[1] and the development of energy storage technologies, as well as renewable energy sources harnessing [2], in turn, require a massive presence of Power Electronics.

Several types of power electronic converters are found in literature. Among them, special focus can be given to DC-DC converters, since this kind of power electronic converter are widely used in many applications, such as: speed control of

DC motors [3], battery charging [4, 5], and power management in hybrid energy storagem systems [6]. DC-DC converters can be defined as electronic circuits capable to convert a DC input voltage to a different DC voltage level, always providing a regulated output [7]. Some of the most common types of DC-DC converters are the buck converter (step-down), the boost converter (step-up) and the buck-boost converter (step-down/step-up).

This paper is focused on the analysis of an interleaved bidirectional DC/DC converter, which can operate either in buck or boost mode. Several applications can be found in literature for this topology, such as battery charging/discharging control [8], public lighting systems [9] automotive applications [10], and renewable energy systems [11]. When compared to their conventional counterparts, interleaved DC-DC converters propitiate current and voltage outputs with reduced ripple. The interleaved DC-DC converter topology treated here is composed by three half-bridge arms, that can be seen as three independent DC-DC converters which work cooperatively.

More specifically, this paper deals with the modeling and control of a bidirectional interleaved DC-DC converter, which works as a part of a hybrid aircraft propulsion system, described in next section. In this context, this interleaved DC-DC converter is responsible to interface on-board battery pack and DC bus, as well as to control battery charging and discharging processes. This article is organized as follows: Section 2 presents system description and modeling. Obtained simulation results are discussed in Section 3, which is followed by paper conclusions and references.

II. INTERLEAVED BIDIRECTIONAL DC-DC CONVERTER MODELING AND CONTROL

A. System Description and Preliminary Considerations

An overview of the aircraft hybrid propulsion system considered in this work is shown in Fig. 1. An electric generator, denoted as "G", which is driven by an aircraft turbine and connected to the DC link through a controlled rectifier controls

978-1-7281-4181-7/19 $31.00 © 2019 IEEE

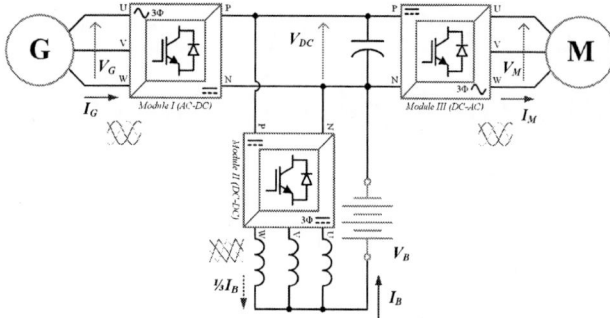

Figure 1: Overview of the aircraft hybrid propulsion system considered in this work.

the DC bus voltage. An inverter module drives the propulsion motor, "M". A battery pack is connected to DC bus by means of an interleaved bidirectional DC-DC converter, whose analysis and control are presented in this paper. As already mentioned, this DC-DC converter controls battery charging and discharging.

Due to its characteristic of current bidirectionality, the DC-DC converter under study presents two operation modes: the battery charging mode (buck mode)[12]; and the discharging mode (boost mode)[13]. Thus, two equivalent circuits has been considered for mathematical modeling of, one for each operation mode. System modeling has been developed in the state space [14], using the average current mode technique [15, 16].

B. Modeling of the boost operation mode

Fig. 2 shows the equivalent circuit for boost mode, which is associated to battery pack discharging. Since battery voltage dynamics is substantially slower than time constants of circuit currents and voltages, battery pack is assumed as constant DC source in this model. Each half-bridge of DC-DC converter has its switches driven complementary. Additionally, there is a phase lag of 120 degrees between each converter half-bridge.

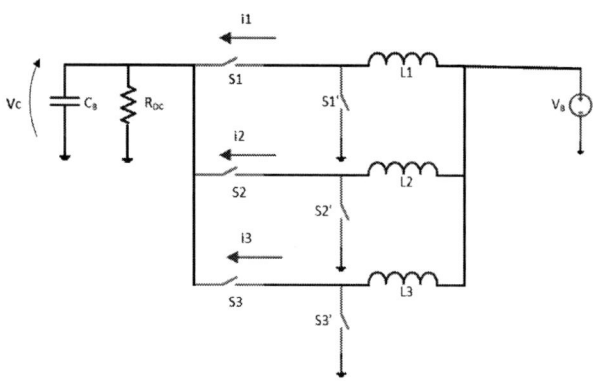

Figure 2: Boost mode equivalent circuit.

Circuit modeling has been developed by means of the analysis of the two possible states for switches from each half-bridge of converter in order to obtain the average model as presented in [17]. The same equivalent circuits have been obtained for each arm of converter. Thus, it is possible to perform the modeling of each arm of the considered interleaved DC-DC converter independently, as depicted in Fig. 3.

First step of the analysis of equivalent circuit of Fig. 3 considers S_1' ON and S_1 OFF, what leads to the equivalent circuit of Fig. 4. Equations (1) and (2) describe current and voltage relations of this equivalent circuit. State space equations are described in (3) and the state matrices for this case are highlighted in (4).

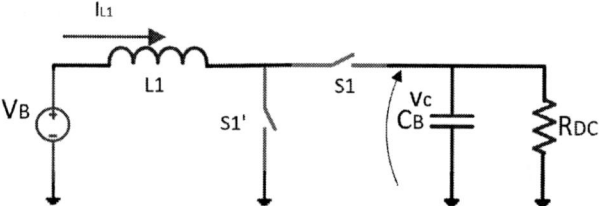

Figure 3: Equivalent circuit for one arm of DC-DC converter (boost mode)

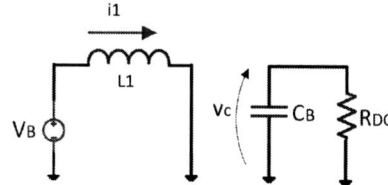

Figure 4: Equivalent circuit for one arm of DC-DC converter (boost mode): S_1' ON and S_1 OFF.

$$L_1 \cdot \frac{di_1}{dt} = v_B \tag{1}$$

$$C_B \cdot \frac{dv_c}{dt} = -\left(\frac{v_{DC}}{R_{DC} * C_B}\right) \tag{2}$$

$$\mathbf{X} = \mathbf{A_1} \cdot \mathbf{x} + \mathbf{B_1} \cdot \mathbf{u} \tag{3}$$

$$\mathbf{A_1} = \begin{bmatrix} 0 & 0 \\ 0 & \dfrac{-1}{C_B \cdot R_{DC}} \end{bmatrix} \qquad \mathbf{B_1} = \begin{bmatrix} \dfrac{1}{L} \\ 0 \end{bmatrix} \tag{4}$$

The equivalent circuit for S_1' off and S_1 is depicted in Fig. 5. Current and voltage for this case are described by (5) and (6). Also, state space equations are defined in (7) with state matrices highlighted in (8).

$$L_1 \cdot \frac{di_1}{dt} = v_B - v_{DC} \tag{5}$$

978-1-7281-4181-7/19 $31.00 © 2019 IEEE   1053

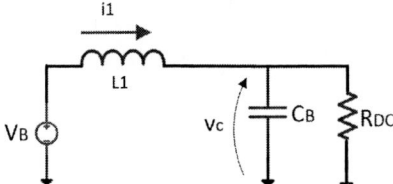

Figure 5: Equivalent circuit for one arm of DC-DC converter (boost mode): S_1' OFF and S_1 ON.

$$C_B \cdot \frac{dv_c}{dt} = i_1 - \left(\frac{v_{DC}}{R_{DC}}\right) \tag{6}$$

$$\mathbf{X} = \mathbf{A_2} \cdot \mathbf{x} + \mathbf{B_2} \cdot \mathbf{u} \tag{7}$$

$$\mathbf{A_2} = \begin{bmatrix} 0 & \dfrac{-1}{L} \\ \dfrac{1}{C_B} & \dfrac{-1}{C_B \cdot R_{DC}} \end{bmatrix} \quad \mathbf{B_2} = \begin{bmatrix} \dfrac{1}{L} \\ 0 \end{bmatrix} \tag{8}$$

With the matrices obtained for both cases, it is necessary to obtain the average model for the operation of the circuit, as described in (9).

$$\begin{cases} D \cdot \mathbf{X} = \mathbf{A_1} \cdot D \cdot \mathbf{x} + \mathbf{B_1} \cdot D \cdot \mathbf{u} \\ (1 - D) \cdot \mathbf{X} = \mathbf{A_2} \cdot (1 - D) \cdot \mathbf{x} + \mathbf{B_2} \cdot (1 - D) \cdot \mathbf{u} \end{cases} \tag{9}$$

Applying results of previous equations to (9) yields to state-space averaged model of boost operation mode.

$$\mathbf{A} = \begin{bmatrix} 0 & \dfrac{-(1-D)}{L} \\ \dfrac{(1-D)}{C_B} & \dfrac{-1}{C_B \cdot R_{DC}} \end{bmatrix} \quad \mathbf{B} = \begin{bmatrix} \dfrac{1}{L} \\ 0 \end{bmatrix} \tag{10}$$

A small-signal analysis of the system is becomes necessary due to non-linearity of circuit under study [18]. From this analysis, the small-signal $\mathbf{B_d}$ matrix can be obtained:

$$\mathbf{B_d} = \begin{bmatrix} \dfrac{V_{DC}}{L} \\ \dfrac{V_{DC}}{C_B \cdot R_{DC} \cdot (1-D)} \end{bmatrix} \tag{11}$$

Transfer functions can be obtained from state-space averaged model. Inductor current and impedance transfer functions are described in (12) and (13), respectively.

- Current Transfer function:

$$G_i(s) = \frac{V_{DC} \cdot (L \cdot s + R_{DC} \cdot D^2)}{D \cdot (C_B \cdot L \cdot R_{DC} \cdot s^2 + L \cdot s + R_{DC} \cdot D)} \tag{12}$$

- Impedance transfer function:

$$Z_0(s) = \frac{R_{DC}}{C_B \cdot R_{DC} \cdot s + 1} \tag{13}$$

Using obtained transfer functions, it is possible to design voltage and current controllers for the boost operation mode. Block diagram of Fig. 6 shows that control strategy is based on an internal current loop and an external voltage loop. For correct operation, it is necessary that the internal current loop has to be faster than the external voltage loop. Since block diagram of represents the entire interleaved DC-DC converter, its transfer functions for input current and output impedance are defined as (14) and (15).

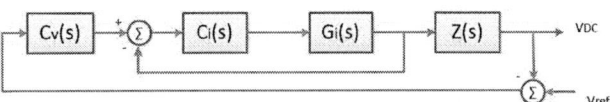

Figure 6: Block diagram for the boost operation mode

$$G_i(s) = \frac{3 * V_{DC} \cdot (L \cdot s + R_{DC} \cdot D^2)}{D \cdot (C_B \cdot L \cdot R_{DC} \cdot s^2 + L \cdot s + R_{DC} \cdot D)} \tag{14}$$

$$Z_0(s) = \frac{R_{DC}}{C_B \cdot R_{DC} \cdot s + 1} \tag{15}$$

Two proportional-integral (PI) controllers were designed for each of the loops in the block diagram using MAT-LAB®SISOTool. Internal current loop controller (C_i) and external voltage loop controller (C_v) are described by transfer functions (16) and (17), respectively.

$$C_i = \frac{2.5 \cdot (0.015 \cdot s + 1)}{s} \tag{16}$$

$$C_v = \frac{120 \cdot (0.00038 \cdot s + 1)}{s} \tag{17}$$

C. Modeling of the Buck operation mode

Consider the buck operation mode equivalent circuit of Fig. 7, where battery pack is modeled as a voltage source connected in series to a resistance. Current and voltage relations of this circuit are described in equations (18) to (21).

$$L_1 \cdot \frac{di_1}{dt} = d \cdot V_{DC} - v_B \tag{18}$$

$$L_2 \cdot \frac{di_2}{dt} = d \cdot V_{DC} - v_B \tag{19}$$

$$L_3 \cdot \frac{di_3}{dt} = d \cdot V_{DC} - v_B \tag{20}$$

$$C_B \cdot \frac{dV_B}{dt} = (i_1 + i_2 + i_3) - \left(\frac{v_B - b_{B0}}{r_B}\right) \tag{21}$$

From the above equations, state space matrices that describes the operation of the averaged converter model for buck operation mode can be described as

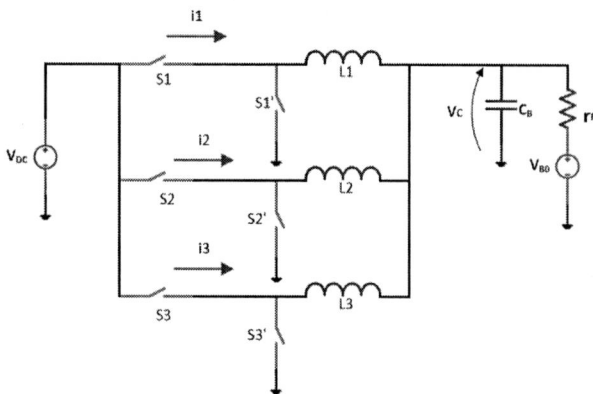

Figure 7: Buck mode equivalent circuit.

$$\mathbf{A} = \begin{bmatrix} 0 & 0 & 0 & \dfrac{1}{L_1} \\[6pt] 0 & 0 & 0 & \dfrac{1}{L_2} \\[6pt] 0 & 0 & 0 & \dfrac{1}{L_3} \\[6pt] \dfrac{1}{C_B} & \dfrac{1}{C_B} & \dfrac{1}{C_B} & -\dfrac{1}{r_B \cdot C_B} \end{bmatrix}. \quad (22)$$

The matrix $\mathbf{B_v}$ is the matrix obtained from the average modeling of the converter while the matrix $\mathbf{B_d}$ is obtained from the small-signal model (linear model) of the system. Both are described as

$$\mathbf{B_v} = \begin{bmatrix} \dfrac{d}{L_1} & 0 \\[6pt] \dfrac{d}{L_2} & 0 \\[6pt] \dfrac{d}{L_3} & 0 \\[6pt] 0 & \dfrac{1}{r_B \cdot C_B} \end{bmatrix}; \quad \mathbf{B_d} = \begin{bmatrix} \dfrac{v_{DC}}{L_1} & 0 \\[6pt] \dfrac{v_{DC}}{L_2} & 0 \\[6pt] \dfrac{v_{DC}}{L_3} & 0 \\[6pt] 0 & \dfrac{1}{r_B \cdot C_B} \end{bmatrix}. \quad (23)$$

Thus, since state-space matrices have been defined, it is possible to obtain the transfer functions for inductor current and converter impedance as displayed in (24) and (25) respectively.

- Transfer function for inductor current in each arm of converter:

$$G_{il}(s) = \frac{v_{DC} \cdot (1 + C_B \cdot r_B \cdot s)}{C_B \cdot L \cdot r_B \cdot s^2 + L \cdot s + 3 \cdot r_B} \quad (24)$$

- Transfer function for the converter impedance:

$$Z_0(s) = \frac{r_B}{1 + s \cdot r_B \cdot C_B} \quad (25)$$

Transfer functions for the block diagram of Fig. 8 are defined by equations (26) and (27).

-

$$G_i(s) = \frac{3 * v_{DC} \cdot (1 + C_B \cdot r_B \cdot s)}{C_B \cdot L \cdot r_B \cdot s^2 + L \cdot s + 3 \cdot r_B} \quad (26)$$

Figure 8: Block diagram for buck operation mode.

-

$$Z_0(s) = \frac{r_B}{1 + s \cdot r_B \cdot C_B} \quad (27)$$

Designed internal current loop controller (C_i) and external voltage loop controller (C_v) are described by transfer functions (28) and (29), respectively.

$$C_i = \frac{50.456 \cdot (0.00041 \cdot s + 1)}{s} \quad (28)$$

$$C_v = \frac{2000}{s} \quad (29)$$

III. SIMULATION RESULTS

In order to verify closed-loop system performance, voltage and current step response for boost and buck operation modes of interleaved DC-DC converter have been evaluated. Simulations here presented have been conducted using the parameters described in Table I.

Table I: Simulation parameters.

DC Bus		
DC Bus Voltage	V_{DC}	$700\,V$
DC Bus Capacitance	C_B	$250\,\mu F$
DC Bus Load (discharge)	R_{DC}	$17.5\,\Omega$
Battery		
Battery Voltage	v_{b0}	$250\,V$
Battery resistance	r_B	$0.1\,\Omega$
Converter		
Inductance	L	$1\,\mu H$
Switching Frequency	f_s	$16\,kHz$

Figs. 9 and 10 are related to the boost operation mode. Figure 9 displays the system response to a reference change of 25 % at the battery voltage. As can be seen, the system behaves properly and zero steady-state error has been achieved. Inductor currents behaves as expected and batery current presents the expected decrease in current variation. Control signals are highlited to show non-saturation.

Additionally, the system was tested during DC bus load variation, displayed in Fig. 10. The load resistance was decreased by half, and therefore output power was doubled. Therefore, the interleaved DC-DC converter, with properly designed controllers, is able to regulate DC bus of hybrid aircraft propulsion system according adjusted set-point.

A similar analysis procedure was adopted to evaluate the operation of interleaved DC-DC converter in buck mode. Firstly, a 25 % voltage increase was imposed to the reference. Fig. 11 shows the system response to the reference change with currents at the inductors and at the battery behaving as

978-1-7281-4181-7/19 $31.00 © 2019 IEEE

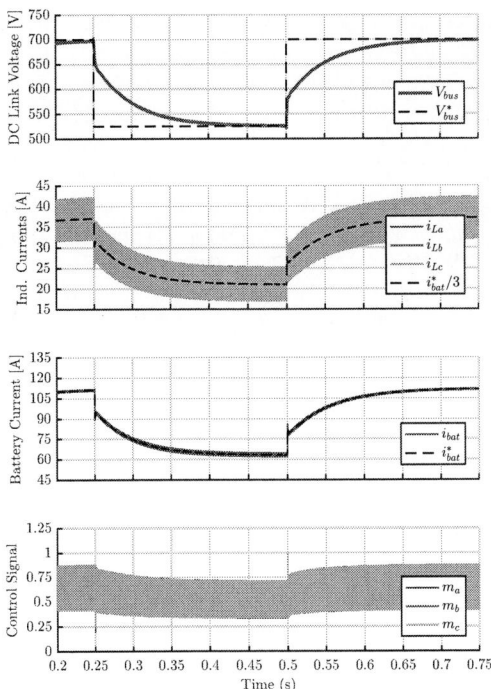

Figure 9: Boost system response to DC bus voltage reference variation.

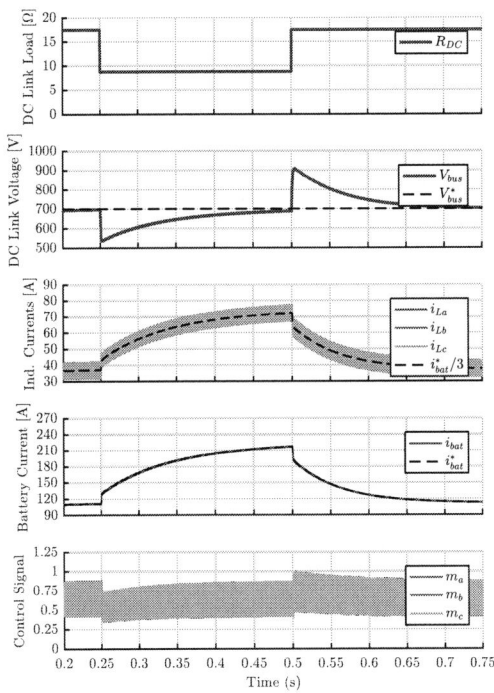

Figure 10: Boost system response to DC bus increase load.

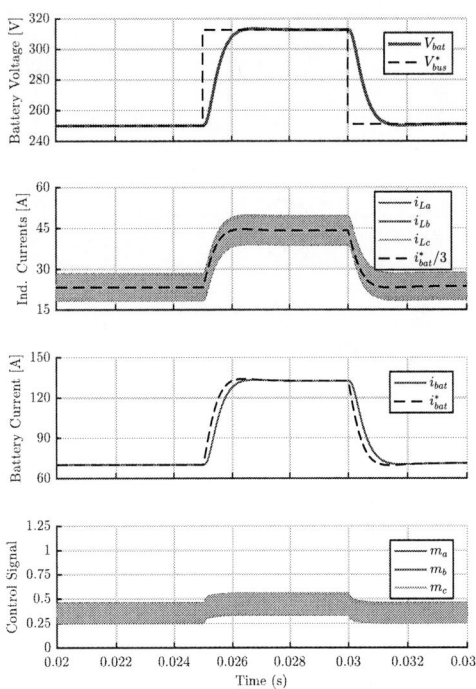

Figure 11: Buck system response to reference change in battery voltage.

simulation results and control is shown to be sufficient for disturbance rejection with a very fast response. In both cases, it is possible to verify the ability of interleaved DC-DC converter to perform voltage regulation at the battery pack even during disturbance.

IV. CONCLUSION

This paper presented a state-space approach for modeling of an interleaved bidirectional DC-DC converter. Control loops have been proposed in order to regulate DC bus and battery voltages and currents of an aircraft hybrid propulsion system. Obtained simulations results provide evidence of the validity of developed power electronic converter model. Also, these simulation results show the good performance of designed controllers for the application considered in this paper. Future work will include experimental implementation of described system.

V. ACKNOWLEDGMENT

Authors would like to thank FAPEMIG and EMBRAER (Project TEC-APQ-03593-17), as well as CAPES and CNPq, for the financial support to this work.

REFERENCES

[1] G. Balen, A. R. Reis, H. Pinheiro, and L. Schuch. Modeling and control of interleaved buck converter for electric vehicle fast chargers. In *2017 Brazilian Power Electronics Conference (COBEP)*, pages 1–6, Nov 2017.

expected. Also, signal control signals are displayed to prove non-saturation. Finally, a disturbance test was performed by simulating a 25 % voltage loss at the DC bus. Figure 12 depicts

978-1-7281-4181-7/19 $31.00 © 2019 IEEE

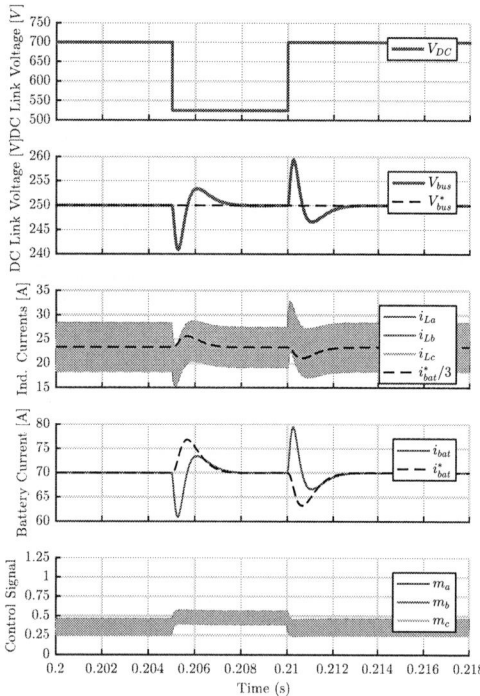

Figure 12: Buck system response to DC bus voltage disturbance.

[2] M. Rashed, J. Le Peuvedic, and S. Bozhko. Conceptual design of battery energy storage for aircraft hybrid propulsion system. In *2016 International Conference on Electrical Systems for Aircraft, Railway, Ship Propulsion and Road Vehicles International Transportation Electrification Conference (ESARS-ITEC)*, pages 1–6, Nov 2016.

[3] D. Potnuru and J. S. V. S. Kumar. Design of a front-end dc-dc converter for a permanent magnet dc motor using fuzzy gain scheduling. In *2017 IEEE International Conference on Power, Control, Signals and Instrumentation Engineering (ICPCSI)*, pages 1502–1505, Sep. 2017.

[4] R. Kushwaha and B. Singh. A unity power factor converter with isolation for electric vehicle battery charger. In *2018 IEEMA Engineer Infinite Conference (eTechNxT)*, pages 1–6, March 2018.

[5] Y. Wang, F. Qin, and Y. Kim. Bidirectional dc-dc converter design and implementation for lithium-ion battery application. In *2014 IEEE PES Asia-Pacific Power and Energy Engineering Conference (APPEEC)*, pages 1–5, Dec 2014.

[6] G. Udhaya Sankar, C. Ganesa Moorthy, and G. RajKumar. Smart storage systems for electric vehicles – a review. *Smart Science*, 7(1):1–15, 2019.

[7] Daniel W Hart. *Eletrônica de Potência: análise e projetos de circuitos*. 1st edition edition, 2012.

[8] M. A. Mughis, I. Sudiharto, I. Ferdiansyah, and D. S. Yanaratri. Design and implementation of partial m-type zero voltage resonant circuit interleaved bidirectional dc-dc converter (energy storage and load sharing). In *2018 International Electronics Symposium on Engineering Technology and Applications (IES-ETA)*, pages 123–128, Oct 2018.

[9] F. S. Alargt, A. S. Ashur, and A. H. Kharaz. A novel structure of a modular interleaved dc/dc converter for renewable street lighting systems. In *2018 53rd International Universities Power Engineering Conference (UPEC)*, pages 1–6, Sep. 2018.

[10] M. O. Younsi, M. Bendali, T. Azib, C. Larouci, C. Marchand, and G. Coquery. Current-sharing control technique of interleaved buck converter for automotive application. In *7th IET International Conference on Power Electronics, Machines and Drives (PEMD 2014)*, pages 1–6, April 2014.

[11] S. Nahar, M. R. Ahmed, and A. Afrin. Performance analysis of solar mppt (inc algorithm) based three phase interleaved boost converter using coupled inductor for brushless dc motor. In *2018 International Conference on Advancement in Electrical and Electronic Engineering (ICAEEE)*, pages 1–4, Nov 2018.

[12] Suryanarayana K and H. N. Nagaraja. Modeling and design of 700w digital average current mode controlled multiphase bidirectional dc-dc converter. In *2015 IEEE Asia Pacific Conference on Postgraduate Research in Microelectronics and Electronics (PrimeAsia)*, pages 142–147, Nov 2015.

[13] M. A. Shrud, A. Kharaz, A. S. Ashur, M. Shater, and I. Benyoussef. A study of modeling and simulation for interleaved buck converter. In *2010 1st Power Electronic Drive Systems Technologies Conference (PEDSTC)*, pages 28–35, Feb 2010.

[14] A. C. Schittler, D. Pappis, C. Rech, A. Campos, and M. A. Dalla Costa. Generalized state-space model for the interleaved buck converter. In *XI Brazilian Power Electronics Conference*, pages 451–457, Sep. 2011.

[15] R. Li, T. O'Brien, J. Lee, J. Beecroft, and K. Hwang. A comparative study of small-signal average models for average current mode control. In *Proceedings of The 7th International Power Electronics and Motion Control Conference*, volume 3, pages 1729–1736, June 2012.

[16] Y. Yan, F. C. Lee, and P. Mattavelli. Analysis and design of average current mode control using a describing-function-based equivalent circuit model. *IEEE Transactions on Power Electronics*, 28(10):4732–4741, Oct 2013.

[17] Amirnaser Yazdani and Reza Iravani. *Voltage-sourced converters in power systems*, volume 34. Wiley Online Library, 2010.

[18] W. Tang, F. C. Lee, and R. B. Ridley. Small-signal modeling of average current-mode control. In *[Proceedings] APEC '92 Seventh Annual Applied Power Electronics Conference and Exposition*, pages 747–755, Feb 1992.

ANALYSIS AND OPERATION OF A PV-BATTERY SYSTEM USING A MULTI-FUNCTIONAL CONVERTER

Jessica S. Dohler [1], Lorrana F. da Rocha[1], Dalmo C. Silva Junior[1], Pedro M. de Almeida[1],
Andre A. Ferreira[1], Janaina G. Oliveira[1]

[1]Graduate Program in Electrical Engineering, Federal University of Juiz de Fora, MG, Brazil.
e-mail: jessica.dohler@engenharia.ufjf.br, lorrana.faria@engenharia.ufjf.br, dalmo.cardoso@engenharia.ufjf.br,
janaina.oliveira@ufjf.edu.br, pedro.machado@engenharia.ufjf.br e andre.ferreira@engenharia.ufjf.br.

Abstract - **This work presents the development and results of a hybrid system (PV-Battery) that will be implemented in the UFJF solar laboratory, Campus Juiz de Fora. The system consists of a photovoltaic panel and an energy storage system based on a battery bank connected/isolated from the mains through a multi-functional converter. The main objective is to maintain the power balance within the system regardless of the consumption profile of the connected load. This goal is achieved by using a power converter functionality divided into three categories: grid conditioning, grid forming, and grid supplying, depending on the function to be performed. Simulation results will be shown, highlighting the multi-functional inverter performance, the control of the energy storage system and the islanding detection technique.**

Keywords – **Multi-functional Converter, Microgrid, Distributed Generation, Grid-connected and Islanded Operation Mode.**

I. INTRODUCTION

As a result of concerns about environmental issues, the possible shortage of fossil fuels, as well as the search for increased reliability in the energy generation system, has led to greater use of renewable sources to meet the new energy challenges evidenced today [1]. Among such sources of energy are hydro, solar, wind and biomass energy, which are sources of energy under increasing attention due to the expansion of distributed generation [2].

Distributed generation brings some benefits to the electrical system, among them: reduction of energy losses in the transmissions lines, increase of energy efficiency by the co-generation and use of waste, development of the national industry, taking advantage of regional vocations, the possibility of urban decentralization and diversification of the energy matrix [3]. With the insertion of the GDs in the electric power system, a new concept called microgrid emerges. According to [4], this system is serviced directly by a distributed generation unit, electrical loads, energy storage elements and control devices, forming a controllable power supply system.

The electronic power converters found in these microgrids usually have Voltage Sourced Converter (VSC) topology inverters, and can be classified into three groups, depending on the function to be performed in the system [5–7]. One of these

functions is the grid supplying that works by injecting active power into the power grid and is usually controlled as a current source in the active and reactive power mode (PQ) [8]. The second converter functionality is the grid forming, which operates as a voltage source in the voltage-frequency mode (V-f), supplying electric charges in the islanded mode [9]. Finally, the grid conditioning converter that works, providing auxiliary services to the microgrid, contributing to improving the quality of electric power [10].

In this context, the objective of this work is to describe the modeling and control of an electricity distribution network, being directly attended by a photovoltaic generation unit and an energy storage system, according to Figure 1. In order to test the studied system behavior, two methods of operations will be approached: (i) acting connected to the electric grid and (ii) in islanding mode, where the inverter plays the role of grid supplying and grid forming, respectively. The same converter might still be able to offer conditioning to the [11] power grid, but this functionality is not exploited in this article.

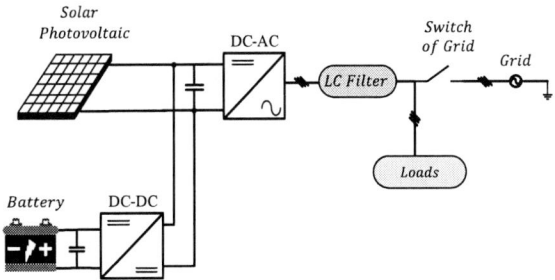

Fig. 1. Hybrid system.

Section 2 presents the description of the main components used in the system, as well as their functionalities. Section 3 discusses the control methods used. Section 4 presents the results of the simulation and analysis of the system. Finally, section 5 addresses the final comments of the paper.

II. SYSTEM DESCRIPTION

Proper choice of a converter that operates as a grid forming, grid supplying, and grid conditioning in one microgrid, enables the same converter to perform all of the necessary system functionalities. The transition of these modes should be rapid and undisturbed in order to avoid loss of quality energy [12]. The modeling of the two modes of operation of the multi-functional converter, as well as the modeling of the DC bus control performed by the storage system will be presented in the next section.

A. Grid Supplying Converter

The inverter used in the system is a two-level three-phase voltage source converter, shown in Figure 2. When connected to the electrical grid, with grid-switch closed, a voltage, and frequency at the Point of Common Coupling (PCC) are imposed by the grid and the injected current is controlled by the converter. As a result, the converter modeling when operated as a grid supply involves the equation of the three-phase currents.

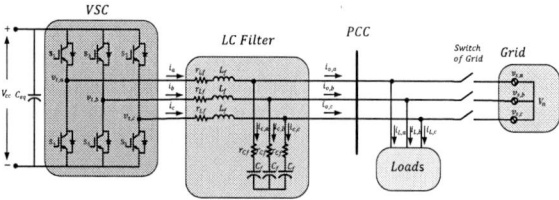

Fig. 2. Grid-connected converter.

Considering that the AC electrical grid is balanced, the current components of the three-phase VSC can be calculated for each phase. Thus, applying Kirchhoff's law of voltage in the circuit, as follows:

$$\frac{d}{dt}\begin{bmatrix} i_a \\ i_b \\ i_c \end{bmatrix} = \frac{1}{L_f}\left(\begin{bmatrix} v_{t,a} \\ v_{t,b} \\ v_{t,c} \end{bmatrix} - r_{Lf}\begin{bmatrix} i_a \\ i_b \\ i_c \end{bmatrix} - \begin{bmatrix} v_{s,a} \\ v_{s,b} \\ v_{s,c} \end{bmatrix} \right) \quad (1)$$

where i_x is the converter inductor current; $v_{t,x}$ is the terminal stresses thereof; $v_{s,x}$, is the voltage of the AC electrical grid. A passive filter is used between the grid and the output of the inverter, allowing attenuation of the switching ripple of the inverter. The current in the capacitor is given by:

$$\frac{d}{dt}\begin{bmatrix} v_{c,a} \\ v_{c,b} \\ v_{c,c} \end{bmatrix} = \frac{1}{C_f r_{Cf}}\left(\begin{bmatrix} v_{s,a} \\ v_{s,b} \\ v_{s,c} \end{bmatrix} - \begin{bmatrix} v_{c,a} \\ v_{c,b} \\ v_{c,c} \end{bmatrix} \right) \quad (2)$$

where $v_{c,x}$ is the instantaneous phase voltage at the VSC capacitor terminals. Applying now Kirchhoff's law of capacitor terminal currents gives the following equations:

$$\frac{d}{dt}\begin{bmatrix} i_{o,a} \\ v_{o,b} \\ v_{o,c} \end{bmatrix} = \begin{bmatrix} i_a \\ i_b \\ i_c \end{bmatrix} + \frac{1}{r_{Cf}}\left(\begin{bmatrix} v_{c,a} \\ v_{c,b} \\ v_{c,c} \end{bmatrix} - \begin{bmatrix} v_{s,a} \\ v_{s,b} \\ v_{s,c} \end{bmatrix} \right) \quad (3)$$

According to [13], the system in natural coordinates can be represented by the relation in synchronous coordinates, as follows:

$$\begin{cases} \frac{d(i_{Ld})}{dt} = \omega_s i_{Lq} + \frac{1}{L_f}\left(v_{t,d} - r_{Lf} i_{Ld} - v_{s,d} \right) \\ \frac{d(i_{Lq})}{dt} = -\omega_s i_{Ld} + \frac{1}{L_f}\left(v_{t,q} - r_{Lf} i_{Lq} - v_{s,q} \right) \\ \frac{d(v_{c,d})}{dt} = \omega_s v_{cq} + \frac{1}{C_f r_{Cf}}\left(v_{s,d} - v_{c,d} \right) \\ \frac{d(v_{c,q})}{dt} = -\omega_s v_{cd} + \frac{1}{C_f r_{Cf}}\left(v_{s,q} - v_{c,q} \right) \end{cases} \quad (4)$$

In the switching of the circuit, the sinusoidal modulation (Sinusoidal Pulse Width Modulation, SPWM) was considered, with:

$$V_{t,dq} = \frac{m_{dq} V_{CC}}{2} \quad (5)$$

The system (4) can be represented by the following generic state space system:

$$\begin{cases} \dot{\mathbf{x}}_n(t) = \mathbf{A}_n\mathbf{x}_n(t) + \mathbf{B}_n\mathbf{u}_n(t) + \mathbf{F}_n\mathbf{w}_n(t) \\ \mathbf{y}_n(t) = \mathbf{C}_n\mathbf{x}_n(t) + \mathbf{D}_n\mathbf{u}_n(t) + \mathbf{G}_n\mathbf{w}_n(t) \end{cases} \quad (6)$$

The matrix output vectores are:

$$\begin{aligned} \mathbf{x_1} &= \begin{bmatrix} i_d & i_q & v_{c,d} & v_{c,q} \end{bmatrix}^T \\ \mathbf{u_1} &= \begin{bmatrix} m_d & m_q \end{bmatrix}^T \\ \mathbf{w_1} &= \begin{bmatrix} v_{s,d} & v_{s,q} \end{bmatrix}^T \end{aligned} \quad (7)$$

$$\mathbf{A_1} = \begin{bmatrix} -\frac{r_{Lf}}{L_f} & \omega_s & 0 & 0 \\ -\omega_s & -\frac{r_{Lf}}{L_f} & 0 & 0 \\ 0 & 0 & -\frac{1}{C_f r_{Cf}} & \omega_s \\ 0 & 0 & -\omega_s & -\frac{1}{C_f r_{Cf}} \end{bmatrix}$$

$$\mathbf{B_1} = \begin{bmatrix} \frac{V_{CC}}{2L_f} & 0 \\ 0 & \frac{V_{CC}}{2L_f} \\ 0 & 0 \\ 0 & 0 \end{bmatrix} \quad \mathbf{F_1} = \begin{bmatrix} -\frac{1}{L_f} & 0 \\ 0 & -\frac{1}{L_f} \\ \frac{1}{C_f r_{Cf}} & 0 \\ 0 & \frac{1}{C_f r_{Cf}} \end{bmatrix} \quad (8)$$

$$\mathbf{C_1} = \begin{bmatrix} 1 & 0 & \frac{1}{r_{Cf}} & 0 \\ 0 & 1 & 0 & \frac{1}{r_{Cf}} \end{bmatrix} \quad \mathbf{D_1} = \begin{bmatrix} 0 & 0 \\ 0 & 0 \end{bmatrix}$$

$$\mathbf{G_1} = \begin{bmatrix} -\frac{1}{r_{Cf}} & 0 \\ 0 & -\frac{1}{r_{Cf}} \end{bmatrix}$$

The Phase-Locked Loop (PLL) algorithm is required for the operation and control of the grid supplying converter. Among some existing structures, Double Second Order Generalized Integrator (PLL-DSOGI) stands out for its good performance, it generates the angle and amplitude of the fundamental component of a positive and negative sequence of the grid voltage observed in the PCC with precision [13].

In addition, in the supplier mode, an islanding detection technique was used for the change of control to the forming mode. This technique proposed in [14], is based on the injection of negative sequence current in the system by the converter and, quantifying the corresponding negative sequence voltage in the PCC, to detect the isolation with the electrical grid.

B. Grid Forming Converter

In this context, it is possible to define a new operation mode, where in Figure 2 the grid keys are open and the system operates islanded. The grid former converter acts by imposing a reference voltage and frequency on its output. The principle of operation of the grid former comes from an Uninterruptible Power Supply (UPS), which can supply a given electrical load regardless of the presence of the power grid because it has a stationary bank of batteries. Applying the Kirchhoff voltage and current laws to the inverter output and the LC filter, the following equations are obtained:

978-1-7281-4181-7/19 $31.00 © 2019 IEEE

$$
\begin{cases}
\frac{d(i_{Ld})}{dt} = \omega_s i_{Lq} + \frac{1}{L_f}\left(m_d \frac{V_{CC}}{2} - r_{Lf} i_{Ld} - v_{o,d} + r_{Cf} i_{o,d} - r_{Cf} i_d\right) \\
\frac{d(i_{Lq})}{dt} = -\omega_s i_{Ld} + \frac{1}{L_f}\left(m_q \frac{V_{CC}}{2} - r_{Lf} i_{Lq} - v_{c,q} + r_{Cf} i_{o,q} - r_{Cf} i_q\right) \\
\frac{d(v_{c,d})}{dt} = \omega_s v_{c,q} + \frac{1}{C_f}\left(i_{Ld} - i_{o,d}\right) \\
\frac{d(v_{c,q})}{dt} = -\omega_s v_{c,d} + \frac{1}{C_f}\left(i_{Lq} - i_{o,q}\right)
\end{cases}
\tag{9}
$$

In this way, the state equations obtained for VSC can be represented by (4). The matrix output vectores are:

$$
\begin{aligned}
\mathbf{x_2} &= \begin{bmatrix} i_d & i_q & v_{c,d} & v_{c,q} \end{bmatrix}^T \\
\mathbf{u_2} &= \begin{bmatrix} m_d & m_q \end{bmatrix}^T \\
\mathbf{w_2} &= \begin{bmatrix} i_{o,d} & i_{o,q} \end{bmatrix}^T
\end{aligned}
\tag{10}
$$

$$
\mathbf{A_2} = \begin{bmatrix}
-r_{Lf} - \frac{r_{Cf}}{L_f} & \omega_s & -\frac{1}{L_f} & 0 \\
-\omega_s & -r_{Lf} - \frac{r_{Cf}}{L_f} & 0 & -\frac{1}{L_f} \\
\frac{1}{C_f} & 0 & 0 & \omega_s \\
0 & \frac{1}{C_f} & -\omega_s & 0
\end{bmatrix}
$$

$$
\mathbf{B_2} = \begin{bmatrix}
\frac{V_{cc}}{2L_f} & 0 \\
0 & \frac{V_{cc}}{2L_f} \\
0 & 0 \\
0 & 0
\end{bmatrix}
\quad
\mathbf{F_2} = \begin{bmatrix}
\frac{r_{Cf}}{L_f} & 0 \\
0 & \frac{r_{Cf}}{L_f} \\
-\frac{1}{C_f} & 0 \\
0 & -\frac{1}{C_f}
\end{bmatrix}
\tag{11}
$$

$$
\mathbf{C_2} = \begin{bmatrix} r_{Cf} & 0 & 1 & 0 \\ 0 & r_{Cf} & 0 & 1 \end{bmatrix}
\quad
\mathbf{D_2} = \begin{bmatrix} 0 & 0 \\ 0 & 0 \end{bmatrix}
$$

$$
\mathbf{G_2} = \begin{bmatrix} -r_{Cf} & 0 \\ 0 & -r_{Cf} \end{bmatrix}
$$

C. Energy Storage System

The energy storage system is of extreme importance in islanded mode operation due to the intermittent character of the renewable energy source. For power storage system applications, the power flow through the battery is controlled using a bidirectional converter, shown in Figure. 3. The battery power flow enables the DC bus voltage to be maintained at all times when the battery is charged. The bidirectionality of power provided by the DC/DC converter is what ensures uninterrupted supply to the loads in the network forming mode [15].

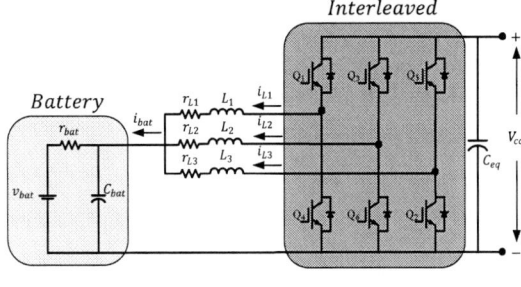

Fig. 3. Bidirectional interleaved converter DC/DC.

The bus voltage control grid is responsible for providing the magnitude of the reference current that will be used in the current loop. According to [16], the DC side dynamics of the converter can be completely equated by applying the principle of conserving energy at the converter DC capacitor terminals, shown in Figure 4:

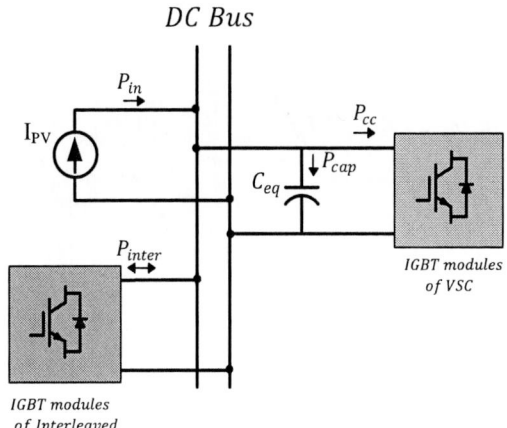

Fig. 4. Hybrid system DC bus.

$$
P_{in} = P_{cap} + P_{cc} \pm P_{inter}
\tag{12}
$$

where P_{in} is the power supplied by the photovoltaic panel, P_{cc} is the power supplied by the DC side of the inverter, P_{inter} is the power of the interleaved and P_{cap} is the power in the capacitor. The modeling of the six-arm interleaved converter is described in [17], the operating point being defined by:

$$
\begin{cases}
I_{L1} = I_{L2} = I_{L3} = \frac{I_{vsc}}{3} - \frac{I_{pv}}{3} \\
V_{cb} = V_{bat} - I_{pv} r_{bat} + I_{vsc} r_{bat} \\
D = \left(\frac{3V_{bat} - I_{pv} r_L + I_{vsc} r_L - 3 I_{pv} r_{bat} + 3 I_{vsc} r_{bat}}{3 V_{cc}}\right)
\end{cases}
\tag{13}
$$

In Figure 5 is depicted the complete controller structure of the system proposed.

III. CONTROLLERS DESIGN

All control meshes were performed with state feedback, which makes the plant input a sum of each of the system states, multiplied by an appropriate gain. The feedback gains were found using the Linear Quadratic Regulator (LQR). They are defined by values imposed for the matrices $\mathbf{Q}$ and $\mathbf{R}$, which determine the weights for the state variables and for the control variables, respectively, in the controller action. The parameters used in the simulation of this system are found in Table I:

A. Grid Supply Converter

For this system, the total of two states is verified, being they refer to the direct axis and quadrature currents in the inductors of the converter. By virtue of the chosen detection method, matrices and gains are obtained for the positive and negative sequence due to the different operating condition. With the matrices shown in (14), the controller gains were computed by the *lqr* function of *software* MATLAB©:

$$
\begin{aligned}
\mathbf{Q} &= \mathrm{diag}\left(\begin{bmatrix} 1e^0 & 1e^0 & 1e^{-1} & 1e^{-1} & 1e^4 & 1e^4 \end{bmatrix}\right) \\
\mathbf{R} &= \mathrm{diag}\left(\begin{bmatrix} 2e^3 & 2e^3 \end{bmatrix}\right)
\end{aligned}
\tag{14}
$$

Matrices and Equivalent Gains for Negative Sequence:

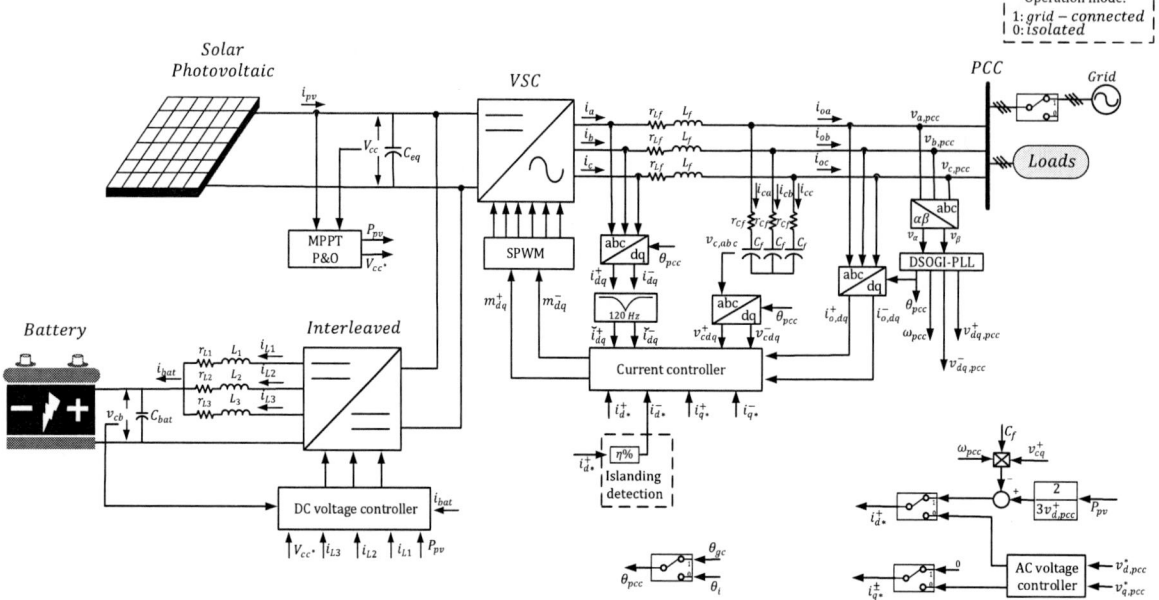

Fig. 5. General system.

TABLE I
System parameters

PV array model SX 120U of BP solar	
Parameter	Value
$I_{pv,mp}$	39,16A
$V_{pv,mp}$	404,4V
Inverter	
Parameter	Value
$V_{s\ rms}$	220 V
r_{Lf}	0.078 Ω
L_f	2 mH
C_f	85 μF
r_{Cf}	0.001 Ω
f_r	60 Hz
f_s	16 kHz
Interleaved	
Parameter	Value
L_{123}	2,5 mH
$r_{L,123}$	1 mΩ
C_{eq}	5500 μF
f_s	20kHz
Battery	
Parameter	Value
r_{bat}	0,0147 Ω
C_{bat}	100 μF
v_{bat}	250 V

$$\mathbf{Q} = \mathrm{diag}\left(\begin{bmatrix} 1e^0 & 1e^0 & 1e^{-1} & 1e^{-1} & 1e^4 & 1e^4 \end{bmatrix}\right)$$
$$\mathbf{R} = \mathrm{diag}\left(\begin{bmatrix} 2e^3 & 2e^3 \end{bmatrix}\right) \tag{15}$$

B. Grid Former Converter

The control of the inverter operating in the island mode is related to the currents in the inductors and the voltages of the grid. Thus, the equivalent matrices and gains for this operation are as follows:

$$\mathbf{Q} = \mathrm{diag}\left(\begin{bmatrix} 1e^{-2} & 1e^{-2} & 1e^{-2} & 1e^{-2} & 2.5e^2 & 2.5e^2 \end{bmatrix}\right)$$
$$\mathbf{R} = \mathrm{diag}\left(\begin{bmatrix} 1e^3 & 1e^3 \end{bmatrix}\right) \tag{16}$$

C. Energy Storage System

In this stage, the total of five states is verified, three being the currents in the inductors of the converter, the fourth related to voltage in the battery and the fifth the voltage in the DC bus. Therefore, the matrices and equivalent gains for this operation:

$$\mathbf{Q} = \mathrm{diag}\left(\begin{bmatrix} 1e^{-2} & 1e^{-2} & 1e^{-2} & 1e^{-2} & 1e^{-2} & 3e^4 \end{bmatrix}\right)$$
$$\mathbf{R} = \mathrm{diag}\left(\begin{bmatrix} 5e^2 \end{bmatrix}\right) \tag{17}$$

IV. SIMULATION RESULTS

From the modeling and design of the presented controllers, the system was simulated in the two modes of operation proposed, feeding linear loads. The system was simulated in the PSIM software with a time step of $1e^{-6}$. Already the controller's design was done in MATLAB©.The results for the scenarios are illustrated below, with the respective operating sequence defined as:

- t = 0 s: connection to microgrid, with initial conditions of solar irradiance 1000 $\frac{W}{m^2}$ and load of 15 kW;
- t = 0.25 s and 0.35 s: variation of simulated solar irradiance fo 1200 $\frac{W}{m^2}$ with load of 15 kW;
- t = 0.45 s and 0.65 s: variation of simulated load for 20 kW with solar irradiance 1000 $\frac{W}{m^2}$;
- t = 1 s: islanding;
- t = 1.35 s and 1.45 s: variation of simulated solar irradiance fo 1200 $\frac{W}{m^2}$ with load of 15 kW;
- t = 1.55 s and 1.75 s: variation of simulated load for 20 kW with solar irradiance 1000 $\frac{W}{m^2}$.
- t = 1.9 s: end of simualation.

A. Grid-connected mode

First, the circuit was simulated in the supply mode. In this mode, the interleaved power flow is used only for voltage control on the DC bus and the inverter controls the active and reactive power on the AC side. In Figure 6 the variation of simulated solar irradiance is during the periods $0.25\ s$ and $0.35\ s$. It is possible to notice that the increase of the current injection, caused by the change of the irradiation, causes an increase in the production of active power. Thus, with the converter operating in the grid supplying mode ensures that this excess active power is injected in the grid.

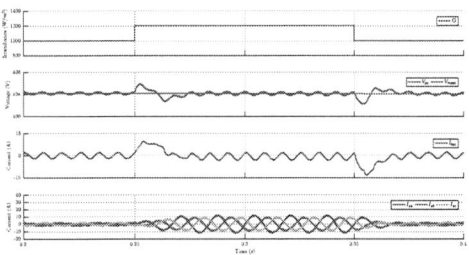

Fig. 6. System response to disturbance in irradiance of the solar panel in the grid supply mode.

As the simulated photovoltaic system is a single stage, the algorithm that tracks the maximum power of the panel directly provides the voltage reference for voltage control on the DC bus. The ripple present in the bus voltage is due to the negative sequence current injection in the AC side, used for islanding detection. This injection causes the emergence of an oscillating power, according to instantaneous power theory.

At the instant of time $0.45\ s$ Figure 7, the second load of $5\ kW$ is connected to the system, being noticeable by the variation of the amplitude in the currents of the network. In the supplying mode, the power support for the load, missing from the renewable source, is made by the grid. As a result, it was necessary for the grid to inject active power into the system to meet the load demand, equivalent to $20\ kW$.

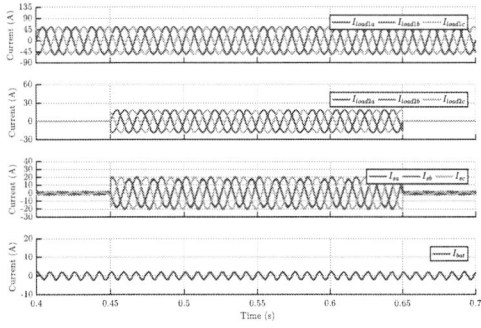

Fig. 7. System response to a load variation in the supply mode.

At the moment of $t = 1\ s$, the switches that connects the network to the distributed generation are opened. Figure 8

shows three-phase PCC voltage, current and frequency during grid-connected to isolated mode transitions. After sag detection the operation mode is switched from grid supplying to grid forming mode. The algorithm spent $11\ ms$ to detect using negative sequence injection method and to change the operation mode. In this mode, it performs voltage control and internally the control of the currents, without the injection of negative sequence.

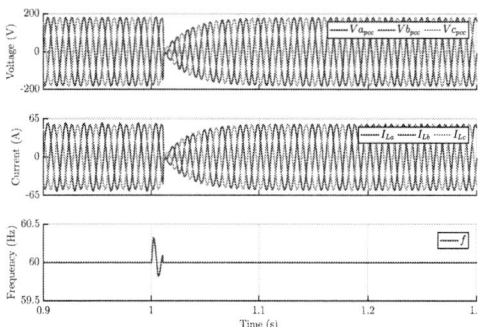

Fig. 8. Frequency response and direct axis negative sequence voltage at the time of the islanding.

B. Islolated mode

In this scenario, the multi-functional converter operates as a grid forming, Figure 9. In the interval of $1.35\ s$, a disturbance occurs in the irradiation of the distributed generator, where all the power of the load is supplied by the renewable energy and the excess of energy is absorbed by the storage system.

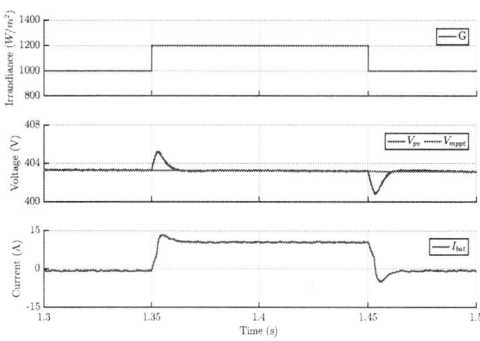

Fig. 9. System response to a disturbance in solar radiation.

Figure 10 illustrates the response of the system when a second charge, of $5\ kW$, is added to the circuit at the instant of time of $1.6\ s$. Note that the active power injected by the photovoltaic panel is not enough to power both loads, corresponding together at $20\ kW$, being necessary that the battery supply the remaining power.

V. CONCLUSION

In this work, the importance of a multi-functional converter in a microgrid was discussed. At first, a grid supplying con-

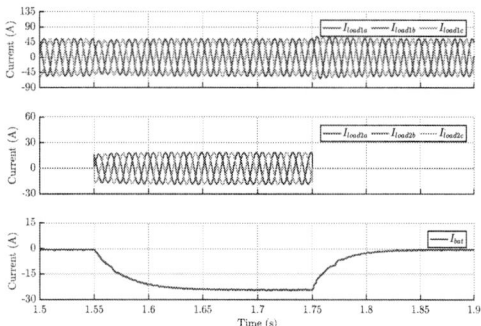

Fig. 10. System response to a load variation in the grid former.

verter and some simulation results of its operation as a current source were defined in the PQ model. The grid former converter was also defined, highlighting its importance to the microgrid in the islanding mode with simulation results supplying linear loads. The converters were modeled by small-signal in state space and the control technique was state feedback, whose gains were calculated by the LQR method. In both modes of operation of the converter, the control proved to be efficient to stabilize the output currents and the DC bus voltage for disturbance rejection based on the time domain analyzes.

ACKNOWLEDGEMENT

The authors would like to thank the Federal University of Juiz de Fora - UFJF, Graduate Program in Electrical Engineering - PPEE, FAPEMIG, CAPES, CNPq, INERGE and Finep for supporting this work.

REFERENCES

[1] Stanley R Bull. Renewable energy today and tomorrow. *Proceedings of the IEEE*, 89(8):1216–1226, 2001.

[2] José Carlos de Melo Vieira Jr. Detecção de ilhamento de geradores distribuídos: uma revisão bibliográfica sobre o tema. *Revista Eletrônica de Energia*, 1(1), 2011.

[3] Frederico AS MARQUES, Jesus A MORAN, Lísias ABREU, et al. Impactos da expansão da geração distribuída nos sistemas de distribuição de energia elétrica. *Procedings of the 5th Encontro de Energia no Meio Rural*, 2004.

[4] Xuesong Zhou, Tie Guo, and Youjie Ma. An overview on microgrid technology. In *2015 IEEE international conference on mechatronics and automation (ICMA)*, pages 76–81. IEEE, 2015.

[5] Chinmay Shah, Mehdi Abolhassani, and Heidar Malki. Fuzzy controlled vsc of battery storage system for seamless transition of microgrid between grid-tied and islanded mode. In *2017 International Joint Conference on Neural Networks (IJCNN)*, pages 3224–3227. IEEE, 2017.

[6] Ling Xu, Zhixin Miao, and Linglnig Fan. Control of a back-to-back vsc system from grid-connection to is-

landed mode in microgrids. In *IEEE 2011 EnergyTech*, pages 1–6. IEEE, 2011.

[7] Daming Zhang and John Fletcher. Operation of autonomous ac microgrid at constant frequency and with reactive power generation from grid-forming, grid-supporting and grid-feeding generators. In *TENCON 2018-2018 IEEE Region 10 Conference*, pages 1560–1565. IEEE, 2018.

[8] Sahar Zafar, M Awais Amin, Beenish Javaid, and Hassan Abdullah Khalid. On design of dc-link voltage controller and pq controller for grid connected vsc for microgrid application. In *2018 International Conference on Power Generation Systems and Renewable Energy Technologies (PGSRET)*, pages 1–6. IEEE, 2018.

[9] Amirnaser Yazdani and Reza Iravani. *Voltage-sourced converters in power systems: modeling, control, and applications*. John Wiley & Sons, 2010.

[10] Dalmo Cardoso da Silva Júnior et al. Modelagem e controle de funções auxiliares em inversores inteligentes para suporte a microrredes ca-simulação em tempo real com controle hardware in the loop. 2017.

[11] Jessica S Dohler, Pedro M de Almeida, Janaina G de Oliveira, et al. Droop control for power sharing and voltage and frequency regulation in parallel distributed generations on ac microgrid. In *2018 13th IEEE International Conference on Industry Applications (INDUSCON)*, pages 1–6. IEEE, 2018.

[12] Hélio Marcos André Antunes. Conversor multifuncional recongurável e tolerantes a falhas para microrredes de energia elétrica. 2018.

[13] Pedro Machado de Almeida et al. Modelagem e controle de conversores fonte de tensão utilizados em sistemas de geração fotovoltaicos conectados à rede elétrica de distribuição. 2011.

[14] Houshang Karimi, Amirnaser Yazdani, and Reza Iravani. Negative-sequence current injection for fast islanding detection of a distributed resource unit. *IEEE Transactions on power electronics*, 23(1):298–307, 2008.

[15] Lucas Paulis Mendonça. Introdução às microrredes e seus desafios. *Projeto para obtenção do grau de engenheiro eletricista. Escola Politécnica da Universidade Federal de Rio de Janeiro, Universidade Federal de Rio de Janeiro, Rio de Janeiro, RJ, Brasil*, 2011.

[16] Pedro M. de Almeida, Pedro G. Barbosa, André A. Ferreira, and Henrique A. C. Braga. Projeto dos controladores de um conversor vsc usado para conectar um sistema de geração fotovoltaico à rede elétrica. In *Congresso Brasileiro de AutomÃ¡tica*. SBA, Campina Grande, 2012.

[17] N Jantharamin and L Zhang. Analysis of multiphase interleaved converter by using state-space averaging technique. In *2009 6th International Conference on Electrical Engineering/Electronics, Computer, Telecommunications and Information Technology*, volume 1, pages 288–291. IEEE, 2009.

Stator Current Controller for Harmonic and Unbalance Compensation Applied to SEIG Based Systems

Gabriel Attuati, Robinson Figueiredo de Camargo, Lucas Giuliani Scherer

Power Electronics and Control Group - GEPOC
Federal University of Santa Maria - UFSM
Santa Maria - RS, Brazil
ga.attuati@gmail.com, robinson.camargo@gmail.com, lgscherer@gmail.com

Abstract—This papers proposes a proportional-resonant (PR) stator current controller applied on a self excited induction generator (SEIG) based system. It is considered a three-phase, three-wire isolated system, composed by an induction generator (IG) driven by a constant speed prime mover, and regulated by a three-phase three-leg distribution synchronous static compensator (DSTATCOM). The control system is implemented on the DSTATCOM, which directly controls the SEIG stator currents, allowing for harmonic and unbalanced current compensation without the need of measuring the load currents. Experimental results are presented to demonstrate the feasibility of the proposed system.

Index Terms—Load current sensorless, SEIG, DSTATCOM

I. INTRODUCTION

Self-excited induction generators are a feasible option in isolated microgrids with sources such as microhydro and biomass [1], [2]. Features such as robustness, high energy density, low cost of maintenance, brushless construction and self-protection against short circuits suggest its use. They present poor regulation of frequency and voltage, requiring additional systems to control these variables. Among the proposed topologies, the use of a DSTATCOM in parallel to the system stands out [3]–[5].

In SEIG applications, the DSTATCOM is used primarily for the AC voltage regulation of the isolated system, but it can also perform nonlinear and unbalanced load compensation [1]. It is controlled through the use of multi-loop systems, with external loops for the SEIG AC and DSTATCOM DC voltage regulation, and internal loops for the current control. Regarding current loop, the hysteresis controller is the most proposed method for this type of application [6]–[8]. Despite its advantages such as robustness, good reference tracking and good dynamic capacity, it presents problems such as the variable switching frequency, low quality and high current ripples, and also a greater difficulty for the filter design [9].

Controllers that work together with conventional modulation techniques for voltage source PWM converters, such as sinusoidal PWM or space vector are an alternative. Well defined switching pattern and harmonic spectrum, and optimum utilization of the DC bus are some advantages of their utilization [9]. In this context, the PR controller is a possibility, and it

was proposed for the DSTATCOM current control in SEIG applications in 3-wire systems [5], and 4-wire systems [10].

These approaches, however, were proposed for the control of the DSTATCOM filter current. In this configuration, it is necessary the measurement of the load currents, extraction of the disturbances, and the generation of references in opposing phase and in order to perform harmonic or unbalanced load current compensation. Meanwhile, load current sensorless methods have been proposed for grid-connected active filters [11], [12], where, instead of the filter current, the control variable is the grid source current. This could be extended to be applied on a DSTATCOM in a SEIG system, with the control variable becoming the generator source current.

In [13] it is presented a stator current control scheme for a SEIG based system using a generalized impedance controller, where load current sensors are not employed for the voltage regulation and load unbalance compensation. However, the SEIG rotor speed versus stator current characteristics must be previously obtained, and for the scheme to work, it is also needed a mechanical sensor for the measurement of the SEIG rotor speed. Further, the paper does not address the case when harmonic content is present.

Therefore, in this paper it is proposed a SEIG stator current control system applied to a DSTATCOM, with the objective of voltage regulation, harmonic and unbalanced load disturbance compensation. For this, it is proposed a PR SEIG stator current controller, allowing for the rejection of disturbances in the generator currents. Analysis, design and implementation of the control system is presented, as well as experimental results from a laboratory setup of the proposed topology.

II. SYSTEM CONFIGURATION

The proposed system's topology is shown in Fig. 1. It consists of a constant speed primary machine (PM), a SEIG, a bank of excitation capacitors and a three-phase DSTATCOM, which is connected to the generator terminals by an inductive filter. The capacitor bank is designed so that nominal voltage is obtained by the generator in a no load condition.

The DSTATCOM compensates the voltage variations produced by the load perturbations, generating capacitive or

978-1-7281-4181-7/19 $31.00 © 2019 IEEE

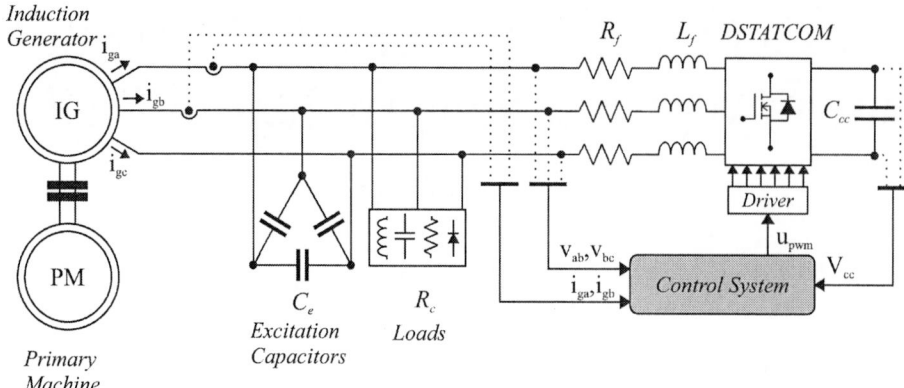

Fig. 1. Topology of the stand-alone system based on a self excited induction generator.

inductive reactive power to maintain the SEIG voltage at its nominal value. In addition, the DSTATCOM must also compensate for load current unbalanced or harmonic content, preventing these disturbances from being reflected in the currents and voltages of the SEIG, thus contributing to the energy quality of the isolated system.

A. System Modeling

The model of the system is obtained by considering the DSTATCOM as an ideal voltage source, with its phase voltages being denominated u_a, u_b and u_c. These are the voltages that must be synthesized by the modulation of the converter. The DSTATCOM output filter is modeled by the inductances L_f and resistances R_f associated with it, identical for the three phases. The excitation capacitances are denoted as C_{exc}, and the loads are modeled as a resistance R_c. In order to ease the modeling process, a Δ-Y transformation is performed on the excitation capacitor bank and on the loads, resulting in the C_e and R_c variables which are used in the model.

Applying Kirchoff's current law on the system of Fig. 1, the following dynamic equations can be obtained in the stationary $(\alpha\beta)$ reference frame:

$$\dot{v}_\alpha(t) = -\frac{1}{C_e R_c} v_\alpha(t) + \frac{1}{C_e} i_\alpha(t) - \frac{1}{C_e} i_{g\alpha}(t) \quad (1)$$

$$\dot{v}_\beta(t) = -\frac{1}{C_e R_c} v_\beta(t) + \frac{1}{C_e} i_\beta(t) - \frac{1}{C_e} i_{g\beta}(t) \quad (2)$$

$$\dot{i}_\alpha(t) = -\frac{1}{L_f} v_\alpha(t) - \frac{R_f}{L_f} i_\alpha(t) + \frac{1}{L_f} u_\alpha(t) \quad (3)$$

$$\dot{i}_\beta(t) = -\frac{1}{L_f} v_\beta(t) - \frac{R_f}{L_f} i_\beta(t) + \frac{1}{L_f} u_\beta(t). \quad (4)$$

From (1) to (4), the system can be represented by the following state-space model:

$$\dot{\mathbf{x}_{v\alpha\beta}}(t) = \mathbf{A}_{v\alpha\beta}\mathbf{x}_{v\alpha\beta}(t) + \mathbf{B}_{v\alpha\beta}\mathbf{u}_{\alpha\beta}(t) \quad (5)$$

$$\mathbf{x}_{v\alpha\beta}(t) = \begin{bmatrix} v_\alpha & v_\beta & i_\alpha & i_\beta \end{bmatrix}^T,$$
$$\mathbf{u}_{\alpha\beta}(t) = \begin{bmatrix} u_\alpha & u_\beta \end{bmatrix}^T, \quad (6)$$

$$\mathbf{A}_{v\alpha\beta} = \begin{bmatrix} \frac{-1}{C_e R_c} & 0 & \frac{1}{C_e} & 0 \\ 0 & \frac{-1}{C_e R_c} & 0 & \frac{1}{C_e} \\ \frac{-1}{L_f} & 0 & -\frac{R_f}{L_f} & 0 \\ 0 & \frac{-1}{L_f} & 0 & -\frac{R_f}{L_f} \end{bmatrix}, \quad (7)$$

$$\mathbf{B}_{v\alpha\beta} = \frac{1}{L_f} \begin{bmatrix} 0 & 0 & 1 & 0 \\ 0 & 0 & 0 & 1 \end{bmatrix}^T. \quad (8)$$

From (5), the transfer function with the voltage dynamics of the system can be obtained according to (9),

$$G_v(s) = \frac{v_\alpha(s)}{u_\alpha(s)} = \frac{v_{\beta(s)}}{u_{\beta(s)}} = \mathbf{C}_{v\alpha\beta}(s\mathbf{I} - \mathbf{A}_{v\alpha\beta})^{-1}\mathbf{B}_{v\alpha\beta} \quad (9)$$

where $\mathbf{C}_{\alpha\beta} = \begin{bmatrix} 1 & 0 & 0 & 0 \end{bmatrix}$ or $\mathbf{C}_{\alpha\beta} = \begin{bmatrix} 0 & 1 & 0 & 0 \end{bmatrix}$.

The SEIG is modeled using the induction machine stator-current and rotor-flux motion equations [14]:

$$\dot{i}_{g\alpha}(t) = -\gamma i_{g\alpha}(t) + \beta\eta\lambda_{r\alpha}(t) + \beta\omega_r\lambda_{r\beta}(t) + \frac{1}{\sigma L_s}v_\alpha(t) \quad (10)$$

$$\dot{i}_{g\beta}(t) = -\gamma i_{g\beta}(t) - \beta\omega_r\lambda_{r\alpha}(t) + \beta\eta\lambda_{r\beta}(t) + \frac{1}{\sigma L_s}v_\beta(t) \quad (11)$$

$$\dot{\lambda}_{r\alpha}(t) = \eta L_m i_{g\alpha}(t) - \eta\lambda_{r\alpha}(t) - \omega_r\lambda_{r\beta}(t) \quad (12)$$

$$\dot{\lambda}_{r\beta}(t) = \eta L_m i_{g\beta}(t) + \omega_r\lambda_{r\alpha}(t) - \eta\lambda_{r\beta}(t) \quad (13)$$

where

$$\sigma = 1 - \frac{L_m^2}{L_s L_r}, \beta = \frac{L_m}{\sigma L_s L_r}, \quad (14)$$

$$\eta = \frac{r_r}{L_r}, \lambda = \frac{r_s + \frac{L_m^2}{L_r^2}}{\sigma L_s}. \quad (15)$$

978-1-7281-4181-7/19 $31.00 © 2019 IEEE

The equations (10) to (13) can be expressed in the state-space form according to the following:

$$\mathbf{x_{g\alpha\beta}}(t) = \mathbf{A_{g\alpha\beta}}\mathbf{x_{g\alpha\beta}}(t) + \mathbf{B_{g\alpha\beta}}\mathbf{v_{\alpha\beta}}(t) \tag{16}$$

where

$$\mathbf{x_{g\alpha\beta}}(t) = \begin{bmatrix} i_{g\alpha} & i_{g\beta} & \lambda_{r\alpha} & \lambda_{r\beta} \end{bmatrix}^T,$$
$$\mathbf{v_{\alpha\beta}}(t) = \begin{bmatrix} v_\alpha & v_\beta \end{bmatrix}^T, \tag{17}$$

$$\mathbf{A_{g\alpha\beta}} = \begin{bmatrix} -\gamma & 0 & \beta\eta & \beta\omega_r \\ 0 & -\gamma & -\beta\omega_r & \beta\eta \\ \eta L_m & 0 & -\eta & -\omega_r \\ 0 & \eta L_m & \omega_r & -\eta \end{bmatrix}, \tag{18}$$

$$\mathbf{B_{g\alpha\beta}} = \frac{1}{\sigma L_s}\begin{bmatrix} 1 & 0 & 0 & 0 \\ 0 & 1 & 0 & 0 \end{bmatrix}^T. \tag{19}$$

From the model (16), the SEIG stator current dynamics transfer function is obtained according to (20),

$$G_{ig}(s) = \frac{i_{g\alpha}(s)}{v_\alpha(s)} = \frac{i_{g\beta}(s)}{v_\beta(s)} = \mathbf{C_{g\alpha\beta}}(s\mathbf{I} - \mathbf{A_{g\alpha\beta}})^{-1}\mathbf{B_{g\alpha\beta}} \tag{20}$$

where $\mathbf{C_{\alpha\beta}} = \begin{bmatrix} 1 & 0 & 0 & 0 \end{bmatrix}$ or $\mathbf{C_{\alpha\beta}} = \begin{bmatrix} 0 & 1 & 0 & 0 \end{bmatrix}$. The transfer functions (9) and (20) are used for the design of the internal current loop PR controllers.

The outer SEIG voltage control loop is performed on the synchronously rotating (dq) reference frame. As the SEIG i_{gq} current corresponds to a reactive power component, its control allows for the AC voltage regulation of the system. Therefore, in order to be obtained a transfer function appropriate for controller design, an $\alpha\beta$-dq transformation can be applied to Equations (1)-(2), resulting in the following state-space system:

$$\mathbf{x_{dq}}(t) = \mathbf{A_{dq}}\mathbf{x_{dq}}(t) + \mathbf{B_{dq}}\mathbf{w_{dq}}(t) \tag{21}$$

where

$$\mathbf{x_{dq}}(t) = \begin{bmatrix} v_d & v_q \end{bmatrix}^T,$$
$$\mathbf{w_{dq}}(t) = \begin{bmatrix} i_d & i_q & i_{g_d} & i_{g_q} \end{bmatrix}^T, \tag{22}$$

$$\mathbf{A_{dq}} = \begin{bmatrix} \dfrac{-1}{C_e R_c} & \omega \\ -\omega & \dfrac{-1}{C_e R_c} \end{bmatrix}, \tag{23}$$

$$\mathbf{B_{dq}} = \frac{1}{C_e}\begin{bmatrix} 1 & 0 & -1 & 0 \\ 0 & 1 & 0 & -1 \end{bmatrix}. \tag{24}$$

Thus, a transfer function with the relationship between the SEIG v_d voltage and i_{gq} current can be obtained through (25), where $\mathbf{C_{dq}} = \begin{bmatrix} 1 & 0 \end{bmatrix}$.

$$G_{vd}(s) = \frac{v_d(s)}{i_{gq}(s)} = \mathbf{C_{dq}}(s\mathbf{I} - \mathbf{A_{dq}})^{-1}\mathbf{B_{dq}} \tag{25}$$

Similarly, the SEIG i_{gd} current corresponds to an active power component, and its control allows for the voltage regulation of the DSTATCOM DC bus. The transfer function between the DSTATCOM V_{cc} voltage and the SEIG i_{gd} current can be obtained through the power balance of the system, being given by (26).

$$G_{vcc}(s) = \frac{v_{cc}^2(s)}{i_{gd}(s)} = \frac{-2v_d}{C_e s} \tag{26}$$

III. CONTROL STRUCTURE

In Fig. 2 it is shown the system proposed for the control of the DSTATCOM. Five system variables are sampled: the SEIG line voltages v_{ab} and v_{bc}, the SEIG stator currents i_{ga} and i_{gb}, and the DSTATCOM DC bus voltage V_{cc}. The current i_{gc} is obtained internally in the DSP by the composition of the other two measured phases.

The inner current control loops are implemented on the stationary $\alpha\beta$ reference frame, while the external voltage regulation loops are implemented on the synchronously rotating dq reference frame. The synchronization angle θ_s is obtained through a PLL (Phase-Locked Loop) based on the Kalman filter.

In order to control the voltage on the DC bus, the error between the reference V_{cc}^* and the sampled value V_{cc} is applied to a PI controller, which consequently generates the reference i_{gd}^*. Likewise, in order to control the voltages of the SEIG, a PI controller receives the error between the reference v_d^* and the sampled value v_d, resulting in the reference i_{gq}^*.

The internal SEIG stator current loops are comprised by proportional-resonant controllers, and receive the current references in the stationary frame, $i_{g\alpha}^*$ and $i_{g\beta}^*$, producing as the control action the reference voltages u_α and u_β. These are synthesized by the DSTATCOM through a carrier based geometric modulation approach.

IV. CONTROLLER DESIGN

A. Inner current loop

The inner current control loop structure is show in Fig. 3-(a), being consisted of two PR controllers and an active damping

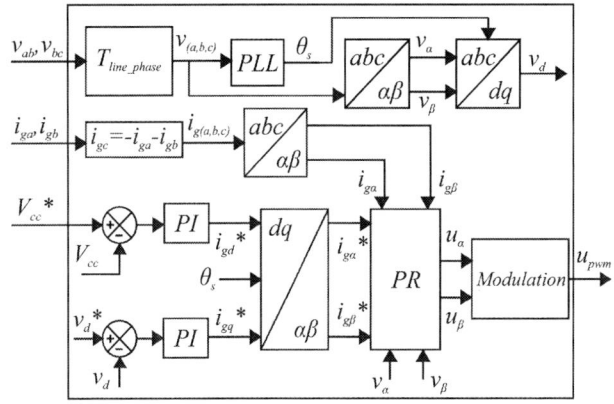

Fig. 2. Control system diagram.

loop. It is implemented both on the α and β axes, with the tracking of $i_{g\alpha}{}^*$ and $i_{g\beta}{}^*$ generating the u_α and u_β control actions, respectively.

Given the system's weak grid parameters and the non-minimum phase zero characteristic of the plant, it becomes difficult to directly design a single stator current controller with sufficient bandwidth in order to perform the harmonic compensation. Therefore, it is used here a cascaded approach [15], where the control problem is split in two, with an external current controller generating a reference for an internal voltage controller, which then produces the control actions. If the dynamics of the outer loop is designed to be slower than that of the inner loop, then the two loops can be designed separately, and the external loop can be designed under the assumption that the inner loop is in steady state [15].

The PR controller is chosen because of its reference tracking and disturbance rejection charecteristics [16]. For this work, it is used the following notation:

$$G_{PR(I,V)}(s) = k_p + \sum \frac{2k_{rh}\zeta\omega_n hs}{s^2 + 2\zeta\omega_n h + (\omega_n h)^2}, \quad (27)$$

where h is the order of the harmonic, k_p is the proportional gain, $0 < \zeta \leq 1$ is the damping factor, ω_n the resonant frequency and k_r the resonant gain. Besides the fundamental frequency, necessary for reference tracking and unbalance load disturbance compensation, the resonant controllers are designed for the compensation of the 5th and 7th harmonics.

For the design of the outer PR controller, the transfer function (20) is used as the model of the system. The resulting open loop frequency response of the compensated system is shown in Fig. 4, with a crossover frequency (CF) of 523 Hz and phase margin (PM) of 54.4°.

For the inner PR controller, the transfer function (9) is used. It presents a resonance peak that degrades control performance, hence needing attenuation. Therefore, it is proposed an active damping solution by applying the SEIG voltages v_α and v_β to a lead compensator. The filter is given by (28), and its output is afterwards subtracted to the output of the inner controller.

$$G_{ad}(s) = k_d \frac{s}{s + p_1} \quad (28)$$

The lead compensator must be designed so that it provides its maximum phase lead around the resonance frequency (f_{res}) of the system [17]. After including the active damping, the open loop transfer function used for the inner resonant controller design becomes:

$$G_{v\_ad}(s) = \frac{G_v(s)}{1 + G_{ad}(s)}. \quad (29)$$

Additionaly, to improve the phase-margin of the inner loop, it is proposed a phase compensator, according to (30). The pole and zero are designed so the required maximum phase-lead occurs at the desired crossover frequency.

$$G_{PD}(s) = \frac{1 + f_z}{1 + f_p} \quad (30)$$

The active damping must be designed first, and then with the transfer function (29) obtained, the PR controller parameters can be chosen. Finally, the phase compensator can be designed so adequate phase-margin is obtained. The resulting open loop frequency response of the compensated inner loop is shown in Fig. 5, with a CF of 750 Hz and PM of 33.2°.

The outer and inner resonant controllers, phase compensator and active damping parameters are shown in Table I. Afterwards, they are discretized using the *Tustin* method for digital implementaion, using a sample time of $T_s = 0.0001$ s.

B. Outer voltage loops

The outer loops structures are shown in Fig. 3-(b), for the DSTATCOM voltage regulation, and Fig. 3-(c), for the SEIG voltage control. As the external loops voltage dynamics are much slower than that of the currents, the inner loop is considered a unitary gain. From the models of equations (25) and (26), the controllers are designed so the bandwidths of both loops are set at 10 Hz. The obtained gains are expressed on Table I.

V. RESULTS

The proposed control system was implemented on a TMS320F28335 DSP and tested on a laboratory prototype

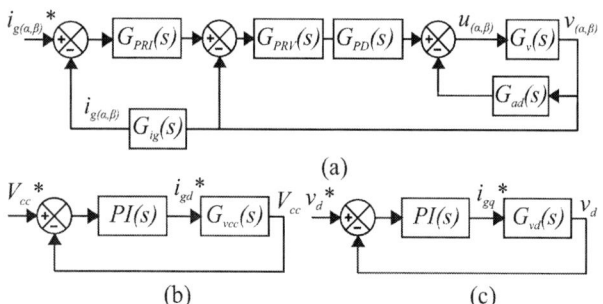

(a)

(b)　　　　　　　　　(c)

Fig. 3. Control loops, (a) SEIG current loop, (b) DSTATCOM DC bus voltage loop (c) SEIG AC voltage loop

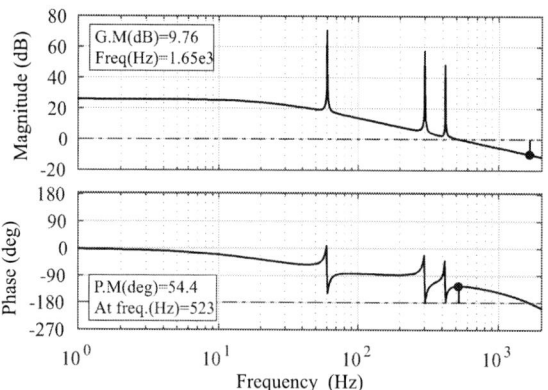

Fig. 4. Open loop bode plot of $G_{PRI}(s)$

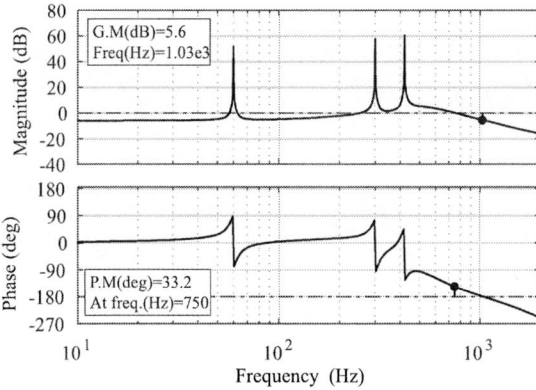

Fig. 5. Open loop bode plot of $G_{PRV}(s)$

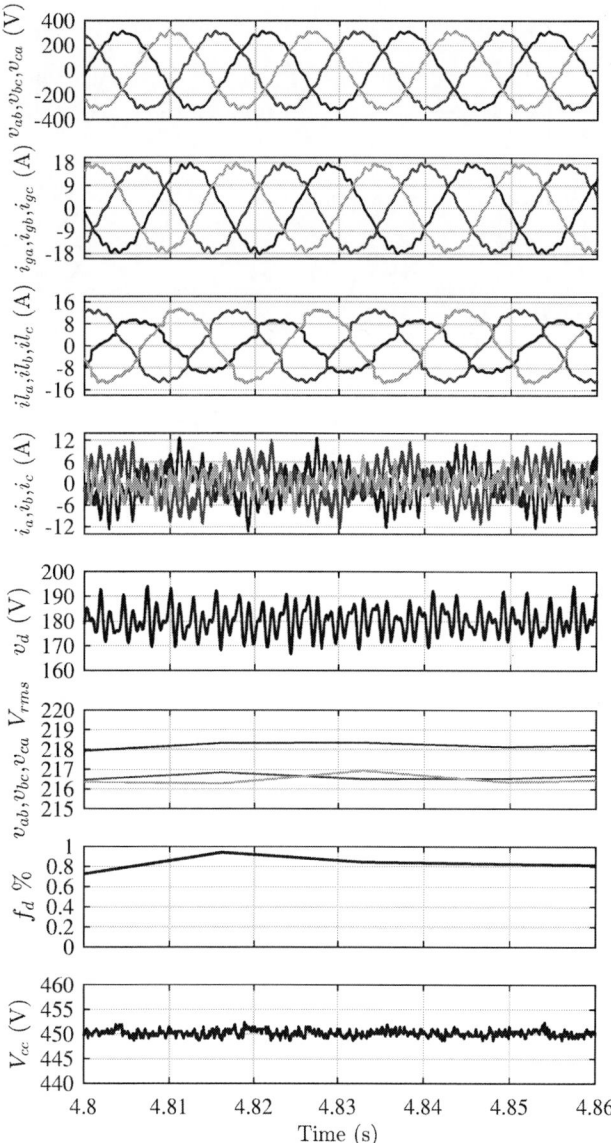

Fig. 6. System variables for the connection of a combination of three-phase linear, single-phase linear and three-phase nonlinear load.

of the system presented in Fig. 1. The experimental setup parameters are displayed in Table II.

Experimental results of the steady-state performance of the system feeding a combination of a 1.8 kW three-phase linear load, 0.6 kW single-phase linear load, and a 1.2 kW three-phase nonlinear load from a diode rectifier, are presented in Fig. 6. The waveforms show that the SEIG line voltages, (v_{ab}, v_{bc}, v_{ca}), and stator currents, (i_{ga}, i_{gb}, i_{gc}), are balanced, with a THD of 6.47% and 7.47%, respectively. The voltages are regulated at the nominal value of $v_d = 180$ V, with an unbalance factor f_d of 0.8 %. The DSTATCOM DC voltage

TABLE I
CONTROLLER PARAMETERS

Parameter	Value
Current PR controller	$\omega_n = 376.8$, $k_p=7$, $\zeta=0.0001$, $k_p=14$ k_{r1}=6,000, k_{r5}=6,000, k_{r7}=3,000
Voltage PR controller	$\omega_n = 376.8$, $k_p=7$, $\zeta=0.0001$, $k_p=7$ k_{r1}=5,000, k_{r5}=5,000, k_{r7}=5,000
Phase compensator	$f_z = 1,313$, $f_p = 18,290$
Active Damping	$k_d = 5$, $p_1 = 15,000$
SEIG voltage PI controller	$k_p = -0.02615$, $k_i = -10.98$
DSTATCOM voltage PI controller	$k_p = -0.0001895$, $k_i = -0.001289$

TABLE II
SYSTEM PARAMETERS

Parameter	Value
Power	P_n = 3.7 kW
Line Voltage	V_n = 220 V, v_d^*= 180 V
Frequency	f= 60 Hz
Excitation Capacitor	C_{exc}= 40 μF, C_{eq}= 120 μF
Load Resistance	R_c= 39 Ω, R_{eq}= 13 Ω
DSTATCOM output filter	L_f= 2.5 mH, R_f= 0.05 Ω
DC bus capacitance	C_{cc}= 4,700 μF
DSTATCOM switching frequency	f_{sw}= 10 kHz
SEIG parameters	R_s= 0.66 Ω, R_r= 0.15 Ω L_m= 55.7 mH, L_s=L_r=58.2 mH

(V_{cc}) is regulated at the nominal value of 450 V, presenting oscillations which are caused by the unbalance of the DSTATCOM currents, (i_a, i_b, i_c), resulting from the compensation of the unbalanced load currents (il_a, il_b, il_c).

The SEIG inner stator current control loop performance is shown in Fig. 7, for the outer current PR controller, and Fig. 8, for the inner voltage PR controller. It is noticed that the designed controllers track the references accordingly.

VI. CONCLUSIONS

The design of a proportional-resonant current controller applied on a DSTATCOM to the control of the stator currents of a SEIG was presented in this paper. Results from the connection of a combination of different loads demonstrated

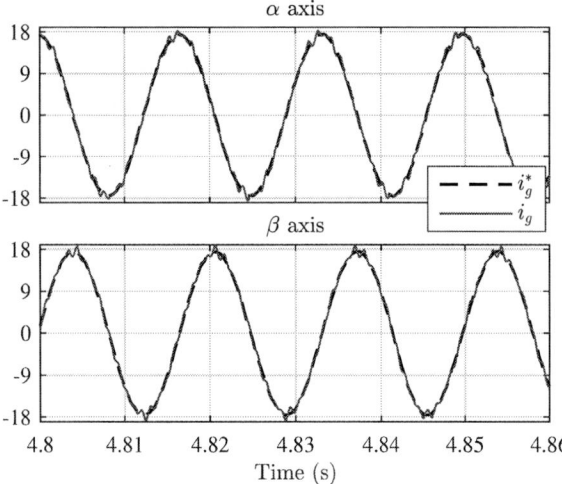

Fig. 7. SEIG current tracking.

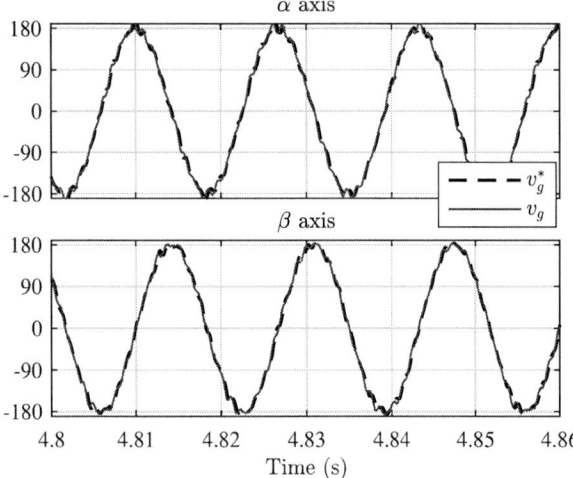

Fig. 8. SEIG voltage tracking.

the proper functioning of the proposed technique. The currents are tracked accordingly. The advantage of the proposed control system is eliminating the measurement of the load currents in order to perform unbalanced and harmonic load current compensation, reducing 2 sensors in the system. However, this is done at the expense of a more complex controller design in comparison to other works in the literature where the load current measurement is employed.

ACKNOWLEDGMENT

The authors thank INCT-GD, CNPq nᵒ 306490/2017-0, CAPES nᵒ 23038.000776/2017-54 and FAPERGS nᵒ 17/2551-0000517-1. This study was financed in part by the Coordenação de Aperfeiçoamento de Pessoal de Nível Superior - Brasil (CAPES/PROEX) - Finance Code 001.

REFERENCES

[1] G. K. Kasal and B. Singh, "Decoupled voltage and frequency controller for isolated asynchronous generators feeding three-phase four-wire loads," *IEEE Transactions on Power Delivery*, vol. 23, no. 2, pp. 966–973, April 2008.

[2] A. Braga, A. Rezek, V. Silva, A. Viana, E. Bortoni, W. Sanchez, and P. Ribeiro, "Isolated induction generator in a rural brazilian area: Field performance tests," *Renewable Energy*, vol. 83, pp. 1352 – 1361, 2015.

[3] B. Singh, S. S. Murthy, and S. Gupta, "Statcom-based voltage regulator for self-excited induction generator feeding nonlinear loads," *IEEE Transactions on Industrial Electronics*, vol. 53, no. 5, pp. 1437–1452, Oct 2006.

[4] R. R. Chilipi, B. Singh, S. S. Murthy, S. Madishetti, and G. Bhuvaneswari, "Design and implementation of dynamic electronic load controller for three-phase self-excited induction generator in remote small-hydro power generation," *IET Renewable Power Generation*, vol. 8, no. 3, pp. 269–280, April 2014.

[5] C. B. Tischer, J. R. Tibola, L. G. Scherer, and R. F. de Camargo, "Proportional-resonant control applied on voltage regulation of standalone seig for micro-hydro power generation," *IET Renewable Power Generation*, vol. 11, no. 5, pp. 593–602, 2017.

[6] U. K. Kalla, B. Singh, and S. S. Murthy, "Slide mode control of microgrid using small hydro driven single-phase seig integrated with solar pv array," *IET Renewable Power Generation*, vol. 11, no. 11, pp. 1464–1472, 2017.

[7] V. C. Sekhar, K. Kant, and B. Singh, "Dstatcom supported induction generator for improving power quality," *IET Renewable Power Generation*, vol. 10, no. 4, pp. 495–503, 2016.

[8] S. Kewat, B. Singh, and I. Hussain, "Power management in pv-battery-hydro based standalone microgrid," *IET Renewable Power Generation*, vol. 12, no. 4, pp. 391–398, 2018.

[9] M. P. Kazmierkowski and L. Malesani, "Current control techniques for three-phase voltage-source pwm converters: a survey," *IEEE Transactions on Industrial Electronics*, vol. 45, no. 5, pp. 691–703, Oct 1998.

[10] L. G. Scherer, C. B. Tischer, and R. F. de Camargo, "Voltage regulation of stand-alone micro-generation seig based system under nonlinear and unbalanced load," in *2015 IEEE 24th International Symposium on Industrial Electronics (ISIE)*, June 2015, pp. 428–433.

[11] Y. Tang, P. C. Loh, P. Wang, F. H. Choo, F. Gao, and F. Blaabjerg, "Generalized design of high performance shunt active power filter with output lcl filter," *IEEE Transactions on Industrial Electronics*, vol. 59, no. 3, pp. 1443–1452, March 2012.

[12] M. B. Ketzer and C. B. Jacobina, "Virtual flux sensorless control for shunt active power filters with quasi-resonant compensators," *IEEE Transactions on Power Electronics*, vol. 31, no. 7, pp. 4818–4830, July 2016.

[13] P. J. Chauhan and J. K. Chatterjee, "A novel speed adaptive stator current compensator for voltage and frequency control of standalone seig feeding three-phase four-wire system," *IEEE Transactions on Sustainable Energy*, vol. 10, no. 1, pp. 248–256, Jan 2019.

[14] P. C. Krause, O. Wasynczuk, and S. D. Sudhoff, *Analysis of Electric Machinery and Drive Systems*. Wiley-IEEE Press, 2002.

[15] Q. Zhong and T. Hornik, "Cascaded current–voltage control to improve the power quality for a grid-connected inverter with a local load," *IEEE Transactions on Industrial Electronics*, vol. 60, no. 4, pp. 1344–1355, April 2013.

[16] R. Teodorescu, M. Liserre, and P. Rodriguez, *Grid Converters for Photovoltaic and Wind Power Systems*. Wiley-IEEE Press, 2011.

[17] M. Liserre, A. Dell'Aquila, and F. Blaabjerg, "Stability improvements of an lcl-filter based three-phase active rectifier," in *2002 IEEE 33rd Annual IEEE Power Electronics Specialists Conference. Proceedings (Cat. No.02CH37289)*, vol. 3, June 2002, pp. 1195–1201 vol.3.

978-1-7281-4181-7/19 $31.00 © 2019 IEEE

Energy-Balance Based Voltage Regulation Method for Multiple DC-links in Asymmetrical Cascaded Multilevel Inverters

Andre Felicio de Sousa Silva, Sergio Pires Pimentel, Enes Goncalves Marra, and Bernardo Pinheiro de Alvarenga

Federal University of Goias (UFG) – Goiania, GO - Brazil

E-mail: and.felicio.sousa@gmail.com

Abstract—**This paper presents a dc-link voltage regulation method for the asymmetrical cascaded multilevel inverter (ACMI). Such method is based on energy-balance analysis and is here evaluated in the context of a single-phase grid-connected photovoltaic (PV) system. However, the proposed strategy can be easily extended to different kinds of applications related to distributed generation with multiple primary power sources. There are several works that deal with the dc-link voltage control for conventional PWM inverters, but such methods cannot be applied to multilevel topologies due to the presence of multiple and isolated dc-links in the ACMI topology. Regarding the proper operation of this topology, the dc-link voltages must be strictly held in a particular set of scalable values. Therefore, a method for regulating the dc-link voltages is required. Although the here proposed method could be applied to any N number of h-bridge inverter series-connected cells, this work analyses a 3 h-bridge ACMI case, which is able to provide an (up to) 19 levels output voltage. A computational model built in the MATLAB/SIMULINK® platform was considered in order to evaluate the dynamic performance of the proposed control method on continuous time-domain under two different scenarios: with proportional (P) controllers; and with a set of proportional and PI (P+PI) controllers. Simulation results show that the proposed method is suitable, offering appropriate transient and steady-state responses even during load and available power variations.**

Keywords—*Multilevel Inverter, Asymmetrical Cascaded, Distributed Generation, Power Electronics, Control Systems.*

I. INTRODUCTION

In recent years it could be noticed a significant increase in the presence of distributed generation (DG) plants composing the power systems around the world. Such category of power generation can utilize primary renewable power sources such as hydraulic, wind, biomass, solar energy etc., showing itself as a good alternative to the use of fossil fuel plants or large hydroelectric systems [1]. In this context, photovoltaic (PV) distributed generation systems are receiving significant attention, specially the grid-connected systems [2]. In this regard, multilevel inverters correspond to an attractive alternative to the use of conventional inverters in DG applications as they are capable to deal with a great amount of power, can provide high-quality output waveforms, with low switching frequency, without the need of complex filter topologies, since multilevel inverters are capable to provide staircase output voltage with very low total harmonic distortion (THD) [3-6].

In the scenario of multilevel inverters, the cascaded topologies can be highlighted due to its capacity to provide an output voltage with a greater number of levels, having a reduced number of switches, and submitting them to low voltage stress, which is an important feature concerning the durability and efficiency of the system [7]. Regarding such cascaded topologies, they can be of classified as:

symmetrical cascaded multilevel inverter (SCMI) or asymmetrical cascaded multilevel inverter (ACMI), which the main difference between them is based on the values of the isolated dc voltage sources. The cascaded h-bridge cells from SCMI is composed by the same dc voltage value, while the ACMI has different dc voltage values. Despite that, those dc voltage values are important for generating generate the expected number of levels on inverter output voltage. These two topologies of cascaded inverters can be operated by the use of distinct modulation strategies, being them based on PWM [8], space vector modulation (SVM) [9] or staircase waveform generation.

In this framework, this paper deals with the dc-link voltage regulation of an ACMI composed by 3 cells, being controlled by a staircase modulation strategy that involves a pure PWM applied only to the lower voltage h-bridge cell [10]. Such modulation strategy is capable to provide an output voltage of (up to) 19 levels when the dc-link voltages are scaled as (1:2:6).

As said before, the dc-link voltage regulation strategies applied to conventional PWM inverters are not suitable to be applied in the ACMI, due to the presence of multiple dc-links and the internal power distribution in the topology. In [11-12] energy-balance control methods are presented for conventional PWM inverters. Reference [13] presents an adaptive dc-link voltage control method for a two-stage PV system based on a full-bridge inverter, but such methods, as mentioned earlier, are not suitable for cascaded multilevel topologies. In [14] a dc-link voltage regulation method is used for the ACMI presenting asymmetrical energy distribution in its cells, a case not covered by the method here proposed and that deals with variations in the inverter modulation indices. In [15] the capacitor voltage control of an ACMI with 3 cells, for active power filter applications is addressed. Although in [15] the control method presented satisfactory capacitor voltage regulation results, the inverter output voltage can present significant harmonic distortions, for some operation conditions, since the control strategy is based on variations in the time interval that the inverter switches remain open or closed, which can lead to the output voltage to not have 19 levels or to such levels to not be equally spaced [15].

Against this background, the main aim from this work is to present a dc-link voltage regulation method for the ACMI topology based on energy-balance principles, without the need of a direct interference on the switching behavior of each h-bridge cell of the inverter, which can cause distortions on the output voltage. In the Section II of this paper, the DG system based on an ACMI is briefly described. In Section III the dc-link of the system is modeled, and two sets of linear controllers are then designed and analyzed. In order to assess the performance of the proposed voltage regulation strategy,

978-1-7281-4181-7/19 $31.00 © 2019 IEEE

simulation results are presented and discussed in Section IV. Finally, Section V presents some conclusions from this work.

II. SYSTEM DESCRIPTION

A. Power Stages

In this work, a case study concerning a single-phase grid-connected PV system with two power stages is presented. Fig. 1 shows the configuration of the connection between the two power conversion stages: the first stage regards the dc/dc power conversion and the second, the ACMI, concerns the dc/ac power conversion. The first power stage is intended to realize the conditioning of the power provided by the PV arrays and to continuously perform the maximum power point tracking (MPPT) for each PV module, individually. The dc/dc converters operation and the applied MPPT algorithm are not discussed in this paper, since its main objective concerns only the dc-link voltage regulation in the 3 h-bridge inverter cells.

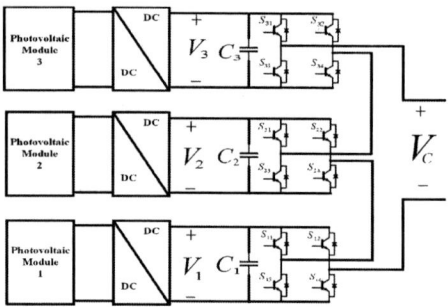

Fig. 1. Two-stage single-phase PV system based on the asymmetrical cascaded multilevel inverter.

The second power stage is responsible for the dc/ac conversion and is composed by the ACMI. As one can observe in the Fig. 2, the second stage performs the grid current injection control, and is also responsible to feed power to a non-linear local load. That said, the objective of the proposed voltage regulation method is to allow the PV system to feed power to a local load connected to the point of common coupling (PCC), while regulating the power factor (PF) and the current injection in the utility grid. Such accomplishment would be successfully possible only if the voltages on each of the capacitors C_1, C_2, and C_3 would be kept close enough to their reference values. It means that the voltages V_1, V_2, and V_3 must be scaled by (1:2:6), corresponding to $V_2 = 2V_1$ and $V_3 = 6V_1$ values.

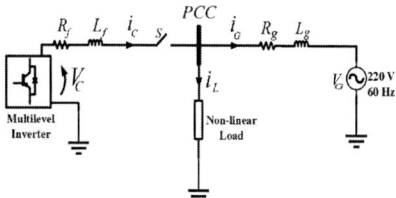

Fig. 2. AC interface between the grid-connected ACMI and the PCC.

B. Dc-links

The main contribution of the voltage regulation method presented in this work is to design individual linear controllers to act over the power reference generation that should be injected in the utility grid through the inverter. In this sense, it is possible to regulate the voltages V_1, V_2, and V_3 using combined linear controllers to act over the available power reference, P_G^*, which represents the total power generated by all the PV modules and that has to be delivered to the local-load or injected in the utility grid. In Fig. 3 it is possible to observe a generalized circuit representing the Nth inverter h-bridge cell.

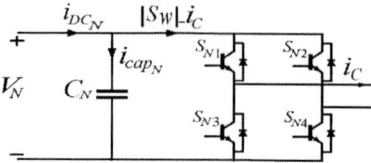

Fig. 3. Generalized dc-link circuit connection from a generic Nth h-bridge inverter cell.

As one can observe in Fig. 3, it is possible to control the cycles of charge and discharge of the dc-link capacitors by means of the current i_C, which can be indirectly done using controllers to act modifying the value of P_G^*, which is provided by the MPPT algorithms (P_{MPPT}) and the feedback of the load power consumption (P_L), leading the system to inject more or less power in the PCC according to the necessity of the system to increase or decrease the voltages V_1, V_2, and V_3 on the capacitors. In Fig. 4 one can see the cascaded control diagram that illustrates the proposed strategy. It is important to notice that the values of ΔP_1, ΔP_2, and ΔP_3 do not interfere significantly in the amount of power injected into the grid, since the parcel of energy used to charge (or discharge) the capacitors is small when compared to the amount of total generated power. The both design of the current and active power controllers are out of scope in this paper, the main objective is only to design controllers that are capable to provide values of ΔP_1, ΔP_2, and ΔP_3 suitable to achieve a satisfactory regulation of the dc-link voltages, and that ensures the DG system stable operation.

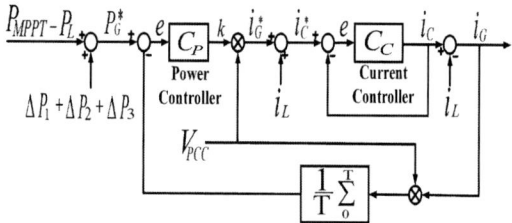

Fig. 4. Proposed control diagram of the system.

C. Modulation Strategy

The modulation strategy applied in the inverter operation plays an important role regarding the number of voltage levels that the ACMI is capable to provide, the behavior of the switching function S_W, showed in Fig. 4, and in the distribution of the power flowing through the h-bridge inverter cells. Since the cascaded converter has its dc-link voltages unequal, the internal power distribution in the inverter is also unequal. Against this background, in order to develop a voltage regulation method based on the energy-balance of the capacitors, it is important to fully characterize the power flow behavior in the ACMI.

The internal power distribution of a 27-level ACMI is addressed in [16], in which the proportions of the power provided by each h-bridge are described using Fourier series decomposition of the staircase output voltage given by such inverter. Applying the same methodology to the 19-level ACMI here presented, it is possible to conclude that the for the ACMI scaled in (1:2:6), a convenient arrangement for the PV modules connected to the inverter cells is to scale them in a way that the Module 3 is responsible to deliver approximately 83% of total the power, the Module 2 with 14% and the Module 1 with 3%. Although such percentages are always complementary to each other, they can vary depending on the applied modulation index. In Fig. 5 a voltage reference and the 19-level output voltage provided by the ACMI using the staircase modulation are shown.

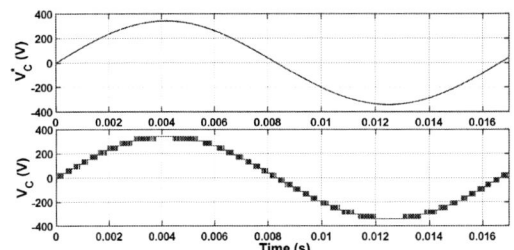

Fig. 5. Voltage reference and the ACMI 19-level output voltage, V_C, operating with the staircase modulation.

III. SYSTEM MODELING AND CONTROLLERS DESIGN

As mentioned in section II, linear controllers can be designed to accomplish the dc-link voltage regulation task. Such controllers can be arranged to act together by means of generating variations in the reference of the available power to be injected P_G^*, causing the dc-link capacitors to charge or discharge accordingly, thus establishing the required voltage regulation for the multiple ACMI dc-link voltages.

In order to design such controllers, the generalized dc-link circuit, depicted in Fig. 3, can be analyzed. From such circuit, it is possible to concluded that

$$V_N(t) = V_0(t) + \frac{1}{C_N}\int i_{cap_N}(t)dt \qquad (1)$$

where V_0 represents the initial value of capacitor voltage or the previous voltage during the last cycle of charge/discharge. It is important to notice that the integral term in (1) represents exactly the voltage variation intended to be imposed in order to regulate the V_N value. The relation presented in (1) can be modified to express $V_N(t)$ in terms of the available power of the respective PV module: P_{MPPT_N}, the inverter current i_C, and the switching function S_W, that relates i_C and the current that leaves the dc-link. Such modified expression can be seen in (2).

$$V_N(t) = V_0(t) + \frac{1}{C_N}\int \left[\frac{P_{MPPT_N}(t)}{V_N(t)} - |S_W|i_C(t) \right]dt \quad (2)$$

The relation (2) can be expressed in the frequency domain, as one can see in (3).

$$V_N(s) = V_0(s) + \frac{1}{C_N}\cdot\frac{1}{s}\cdot\frac{P_{MPPT_N}(s)}{V_N(s)} + $$
$$- \frac{1}{C_N}\cdot\frac{1}{s}\cdot|S_W|i_C(s) \qquad (3)$$

The terms in (3) can be rearranged in a way that one can get (4).

$$V_N(s) - V_0(s) = \Delta V_N = \frac{1}{C_N}\cdot\frac{1}{s}\cdot\frac{P_{MPPT_N}(s)}{V_N(s)} + $$
$$- \frac{1}{C_N}\cdot\frac{1}{s}\cdot|S_W|i_C(s) \qquad (4)$$

The analysis of (4) allows to concept a control diagram for the dc-link voltage of the Nth inverter h-bridge cell. Such diagram can be seen in Fig. 6. Hence, in order to regulate all the ACMI capacitor voltages individual controllers can be designed, based in the diagram depicted in Fig. 6, and disposed in a control configuration so as to combine all the ΔP_N individual components provided by the linear controllers, as indicated in the Fig. 4 by the sum of ΔP_1, ΔP_2, and ΔP_3.

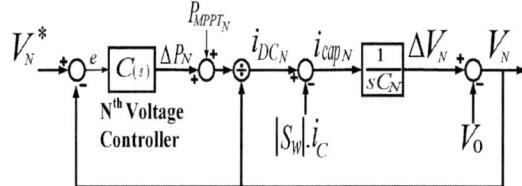

Fig. 6. DC-link voltage control diagram for the Nth ACMI h-bridge.

The proposed control diagram, showed in Fig. 6, allows the individual design of linear controllers for each h-bridge inverter cell in order to achieve their dc-link voltage regulation. Then, in order to evaluate the voltage regulation strategy, two sets of controllers can be designed: one composed only by proportional (P) controllers and another set composed by a mix of proportional and PI controllers (P+PI). During the design of such controllers it is important to consider that h-bridge cells with lower voltages tend to charge and discharge their capacitors faster than the others. In that sense, for the case study here presented, the controller responsible for the regulation of the voltage V_1 has to be faster than the one regulating V_2, that also has to be faster the controller for V_3, being this last the slowest.

By using the software MATLAB® and considering the control diagram depicted in Fig. 6, the system parameters values presented in Tab. I. It is important to notice, in Table I, that the rated power of each PV module was chosen approximately in accordance to the proportions of 83%, 14% and 3% of the system total power, for the PV modules 3, 2 and 1 respectively.

In this work, the procedure of controllers design It was based on the cut-off frequency of the others cascaded controllers, shown in Fig. 4, has to be taken into account. That said, considering that the active power controller has a cut-off frequency of 45 Hz, none of the voltage controllers may have a cut-off frequency larger than that. As mentioned earlier, the controller responsible for the regulation of the h-bridge with the lowest voltage needs to be the fastest. The controller with the intermediate voltage needs to be slower and so the controller for the highest voltage. From this perspective the set of P controllers can be designed, all having a phase margin of 85° and cut-off frequencies of: 25 Hz, 6 Hz and 0.7 Hz for the capacitors C_1, C_2 and C_3 respectively. The gains of this set of controllers are presented in Tab. II. It is important to state the cut-off frequencies of the voltage controllers were chosen by a trial-and-error method considering the control diagram shown in Fig. 6, which aimed to achieve controllers capable to operate in compliance with the PV system demands, and also presenting speeds that matches the characteristics mentioned earlier regarding each h-bridge.

A set of controllers having an arrangement of P and PI controllers can be designed in order to determinate if it is possible to improve the dynamic and steady state behavior of the dc-link voltages in the ACMI, comparing to the performance provided by a set of proportional controllers only. After realizing simulation tests and analysis, it was observed that PI controllers can only be used to regulate the lowest voltage.

Such characteristic takes place due to the difference among the inverter h-bridges regarding the speed dynamics and the amount of power each one is capable to provide. When PI controllers are used in the two other inverter cells, the integrating component of the controllers are not capable to nullify the error, implying in ever growing ΔP terms, causing the entire system to collapse. For that reason, only the lowest voltage inverter cell operates with a PI controller in the mixed set of controllers, and the P controllers were chosen as the same presented in Tab. II. The PI controller was designed also having a phase margin of 85° and a cut-off frequency chosen to be 30 Hz, which means that such controller is a little bit faster (in transient time-response) than the proportional ones but is yet under the 45 Hz. The mixed P + PI controllers set can be seen Tab. III. The here proposed trial-and-error control design strategy does not indicate analytical expressions for the controllers design. Such design methodology could be improved and is not the only one that can be used in order to project the controllers. The main goal of this paper is to propose controllers whose control action is based in the analysis of each h-bridge circuit dynamics and energy-balance principles.

IV. RESULTS AND DISCUSSION

The main objective of the voltage regulation method here proposed is to guarantee that the dc-link voltages of the

ACMI remain as close as possible to their reference values, indicated in Table I, and within the scale factor of (1:2:6).

In the next subsections, two sets of linear controllers are going to have its results analyzed: a set of P controllers and another set of mixed P+PI controllers. The goal of such comparison is to observe which set is going to provide better voltage regulation dynamic results and is more suitable for DG demands, such as load and available power variations. The case study here presented regards a model of the grid-connected PV system built in the MATLAB/SIMULINK® environment.

It is important to highlight that all the simulation results presented in this section were done closing the switch S, indicated in Fig. 2, after 1 cycle of 60 Hz and the irradiance is provided for the PV modules as a ramp time function. Such measures were taken in order to carry the simulations not overloading the system and to not cause instability while connected to the utility grid. Except for the simulations regarding variations in the irradiance, all results are shown considering standard test conditions (STC) for the PV system.

Since the method was developed based on energy-balance principles and is intended to be used for distributed generation applications, there is a minimum amount of power that the system needs to be provided by the primary sources for the voltage regulation to properly work. For the case study here presented, concerning a DG PV system, it was found that the proposed method can regulate the ACMI dc-link voltages for irradiance values varying from $G = 100 \ W/m^2$ to $G = 1000 \ W/m^2$.

The first result of interest is the dc-link voltages behavior when the system operates without any regulation on the capacitors voltages (without the presence of ΔP_1, ΔP_2 and ΔP_3, from the control diagram shown in Fig. 4). As one can see in the Fig. 7, the dc-link voltages V_1, V_2 and V_3, represented in per-unit values (according to their respective reference values shown in Tab. I), deviate significantly from its reference values, causing the system operation to a collapse condition as the voltages were not regulated to kept the relation (1:2:6). Such result also shows that the lower voltage V_1, as said before, is the more sensible, therefore demanding the fastest regulation action.

A. Proportional Voltage Controllers

The first set of controllers that will be evaluated is the

TABLE I.   SYSTEM PARAMETERS.

Description	Parameter	Value
Nominal Power	P_{MPP}	7.1 kW
PV 1 Power	P_{MPP_1}	0.210 kW
PV 2 Power	P_{MPP_2}	1 kW
PV 3 Power	P_{MPP_3}	5.9 kW
Peak Inverter Voltage	$V_{C_{peak}}$	$1.1 * 220\sqrt{2}$
Inverter Module Voltage	V_1	38 V
Inverter Module Voltage	V_2	76 V
Inverter Module Voltage	V_3	228 V
Inverter dc-Link Capacitors	$C_1 = C_2 = C_3$	6.8 mF
Inductor Filter	L_f	2.5 mH
Inductor Filter Resistance	R_f	50 mΩ
Inverter swithing frequency	f_{sw}	18 kHz
Load Nominal Power	P_L	1.6 kW
Grid RMS Voltage	V_G	220 V
Grid Fundamental Frequency	f_G	60 Hz
Grid Equivalent Resistance	R_G	0.1 Ω
Grid Equivalent Inductance	L_G	0.5 mH

TABLE II.   P CONTROLLERS FOR THE VOLTAGE REGULATION OF THE ACMI.

Gain	Value
P_{C1}	160
P_{C2}	20
P_{C3}	14

TABLE III.   P + PI CONTROLLERS FOR THE VOLTAGE REGULATION OF THE ACMI.

Gain	Value
P_{C1}	180
I_{C1}	1080
P_{C2}	20
P_{C3}	14

proportional (P), whose gains are presented in Tab. II. In Fig. 8 it is possible to observe that the P controllers are capable to regulate the dc-link voltages, maintaining the PV system stable (feeding power to the local load and injecting the surplus in the grid). As it was expected, the proportional controllers are not capable to eliminate the steady-state reference tracking error, hence the dc-link voltages tend only to stabilize its values on a stable operation point.

Although such set of P controllers were not able to eliminate the steady-state error, they were capable to regulate the voltages presenting average percentage errors of 5.17%, 0.9% and 2.9% for V_1, V_2 and V_3 respectively. It could be a simple solution that satisfies a variety of ACMI applications.

In the Fig. 9 one can see the dc-link voltage behavior when there are variations on the non-linear local load connected to the PCC, indicated in Fig. 2. As it is possible to notice from Fig. 9, although the set of P controllers were capable to maintain the system stability, returning to a stable point of operation after the applied load variations, the dynamic behavior of the dc-link voltage V_1 is not satisfactory, dropping to a value near to 0.6 p.u.

In Fig. 10 the dynamic behavior of the ACMI voltages are showed when the distributed generation system is submitted to variations in the available power. Such variations are here realized through changes in the incident irradiance (G) on the PV modules. As one can see, as the irradiance G changes, the DC-link voltages tends to stabilize in different operational points, which is an expected behavior for the designed proportional controllers.

B. P+PI Voltage Controllers

As mentioned earlier, a refinement in the proposed voltage regulation method can be done. As one can see in the Fig. 8, the voltage V_1 presents significant sensibility, implying in a tendency to deviates from its reference value, which can cause system instability. Also, Figs. 9 and 10 shows that the dynamic behavior of the DC-link voltages, for

the operation with the proportional controllers, need improvement. In that sense, a PI controller can be used to regulate the lowest ACMI voltage, V_1. Simulation tests were done using PI controllers for the control of voltages V_2 and V_3, but the results showed that the method proposed in this work cannot use PI controllers to nullify the reference tracking error of such ACMI cells.

In the Fig. 11 it is possible to notice the voltage regulation provided by the mixed P+PI controllers. It is possible to observe that such set of controllers is capable to regulate the dc-link voltages of the ACMI with significant lower average percentage errors: 0.63%, 1% and 2,8% for V_1, V_2 and V_3 respectively, implying in a negligible deviation from the required proportion of (1:2:6) when compared to the results of Fig. 8. Such result shows that this set of P+PI controllers is slightly better than that previous set with only P controllers.

In Fig. 12 it can be noticed the behavior of dc-link voltages considering the same load variations as those shown in Fig. 9. Such results indicate that, once compared to the one showed in Fig. 9, that the P+PI set of controllers provides better transient behavior and is better able to guarantee the overall stability of the system. Fig. 13 presents the ACMI voltages for variations on the available power, showing that the added PI controller to regulate the voltage V_1 improved the system operational conditions for DG applications, such as the case of study presented in this work.

Another important point to be observed, regards the controllers performance in terms of current Total Harmonic Distortion (THD) behavior for different values of available power in the PV modules of the system. In Tab. IV one can observe the THD values for different values of available power considering the two compared set of controllers. As it is was expected, the P+PI set of controllers provide lower THD values, since such controllers are capable to maintain smaller average percentage errors regarding their reference tracking capacity.

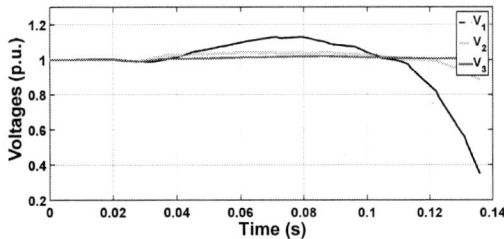

Fig. 7. Dc-link voltages of the ACMI operating without any voltage regulation (without controllers).

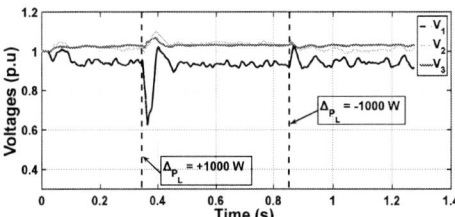

Fig. 9. Dc-link voltages behavior operating with the P set of controllers for load variations.

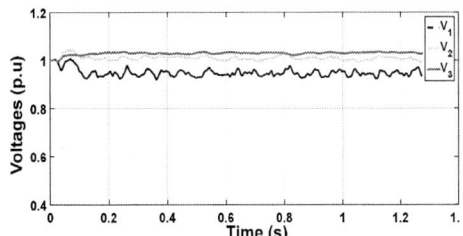

Fig. 8. Dc-link voltages of the ACMI operating with the P set of controllers.

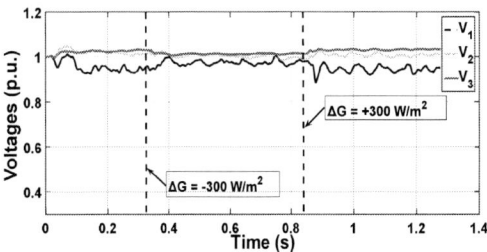

Fig. 10. Dc-link voltages behavior operating with the P set of controllers for available power variations.

978-1-7281-4181-7/19 $31.00 © 2019 IEEE   1074

V. CONCLUSIONS

This work presented a dc-link voltage regulation method for the ACMI concerning distributed generation applications. Although a grid-connected PV system was addressed, the control design methodology presented can be applied to a great variety of ACMI based DG systems. The proposed method utilizes the energy-balance of the ACMI dc-link capacitors in order to guide the design of linear controllers. Two sets of controllers were designed to assess the performance of the proposed strategy. Such evaluation determined that a set of P+PI controllers provided better dynamic results in the face of load and available power variations. Simulation results also shows that the presented voltage regulation strategy can be applied to ACMI topologies with any N number of h-bridge cells, which indicates the success of the presented dc-link modeling strategy and the relevancy of the proposed voltage regulation method. It is expected that further experimental results would confirm the results of the proposed control system and validate the utilized theoretical concepts.

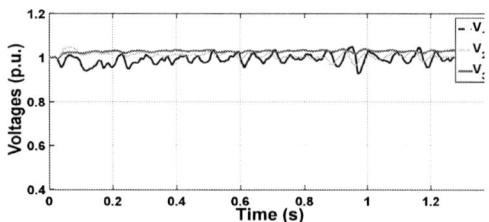

Fig. 11. Dc-link voltages of the ACMI operating with the P+PI set of controllers.

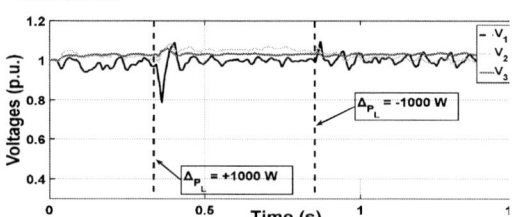

Fig. 12. Dc-link voltages behavior operating with the P+PI set of controllers for load variations.

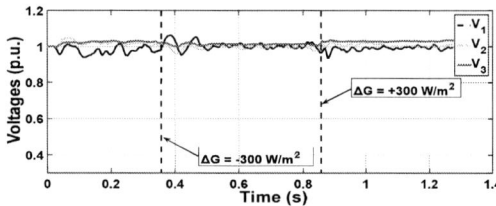

Fig. 13. Dc-link voltages behavior operating with the P+PI set of controllers for available power variations.

TABLE IV. THD VALUES OF GRID CURRENT VS. AVAILABLE POWER CONDITIONS CONSIDERING THE TWO SETS OF CONTROLLERS

Irradiance (W/m²)	THD of i_G	
	(P) Set	(P+PI) Set
1000	1.12%	1.1%
800	1.17%	1.1%
600	1.62%	1.32%
400	2.2%	1.7%

ACKNOWLEDGMENT

Authors would like to thank for the financial support provided by the Brazilian agency CAPES (*Coordenação de Aperfeiçoamento de Pessoal de Nível Superior*).

REFERENCES

[1] Keyhani, A., Marwali, M.N., and Dai, M.: "Integration of Green and Renewable Energy in Electric Power Systems" (Wiley, 2009).

[2] S. Kouro, J. I. Leon, D. Vinnikov and L. G. Franquelo, "Grid-Connected Photovoltaic Systems: An Overview of Recent Research and Emerging PV Converter Technology," in *IEEE Industrial Electronics Magazine*, vol. 9, no. 1, pp. 47-61, March 2015.

[3] B. Xiao, L. Hang, J. Mei, C. Riley, L. M. Tolbert and B. Ozpineci, "Modular Cascaded H-Bridge Multilevel PV Inverter With Distributed MPPT for Grid-Connected Applications," in *IEEE Transactions on Industry Applications*, vol. 51, no. 2, pp. 1722-1731, March-April 2015.

[4] H. Akagi, "Multilevel Converters: Fundamental Circuits and Systems," in *Proceedings of the IEEE*, vol. 105, no. 11, pp. 2048-2065, Nov. 2017.

[5] R. R. Karasani, V. B. Borghate, P. M. Meshram, H. M. Suryawanshi and S. Sabyasachi, "A Three-Phase Hybrid Cascaded Modular Multilevel Inverter for Renewable Energy Environment," in *IEEE Transactions on Power Electronics*, vol. 32, no. 2, pp. 1070-1087, Feb. 2017.

[6] S. Amamra, K. Meghriche, A. Cherifi and B. Francois, "Multilevel Inverter Topology for Renewable Energy Grid Integration," in *IEEE Transactions on Industrial Electronics*, vol. 64, no. 11, pp. 8855-8866, Nov. 2017.

[7] C. Rech, H. Pinheiro, H. A. Grundling, H. L. Hey and J. R. Pinheiro, "Analysis and comparison of hybrid multilevel voltage source inverters," *2002 IEEE 33rd Annual IEEE Power Electronics Specialists Conference*. Proceedings, Cairns, Qld., Australia, 2002, pp. 491-496 vol.2.

[8] J. A. Aziz and Z. Salam, "A PWM strategy for the modular structured multilevel inverter suitable for digital implementation," *VIII IEEE International Power Electronics Congress, 2002*. Technical Proceedings. CIEP 2002., Guadalajara, Mexico, 2002, pp. 160-164.

[9] A. R. Bakhshai, H. R. Saligheh Rad and G. Joos, "Space vector modulation based on classification method in three-phase multi-level voltage source inverters," *2001 IEEE Industry Applications Conference. 36th IAS Annual Meeting*, Chicago, IL, USA, 2001, pp. 597-602 vol.1.

[10] Manjrekar, M.D.; Steimer, P.K.; Lipo, T.A.; "Hybrid Topology for Multilevel Power Convertions," U. S. Patent 6 005 788, 1999.

[11] A. Kotsopoulos, J. L. Duarte and M. A. M. Hendrix, "Predictive DC voltage control of single-phase PV inverters with small DC link capacitance," *2003 IEEE International Symposium on Industrial Electronics*, Rio de Janeiro, Brazil, 2003, pp. 793-797 vol. 2.

[12] C. Meza, J. J. Negroni, D. Biel and F. Guinjoan, "Energy-Balance Modeling and Discrete Control for Single-Phase Grid-Connected PV Central Inverters," in *IEEE Transactions on Industrial Electronics*, vol. 55, no. 7, pp. 2734-2743, July 2008.

[13] G. Ding *et al.*, "Adaptive DC-Link Voltage Control of Two-Stage Photovoltaic Inverter During Low Voltage Ride-Through Operation," in *IEEE Transactions on Power Electronics*, vol. 31, no. 6, pp. 4182-4194, June 2016.

[14] H. Iman-Eini, S. Bacha and D. Frey, "Improved control algorithm for grid-connected cascaded H-bridge photovoltaic inverters under asymmetric operating conditions," in *IET Power Electronics*, vol. 11, no. 3, pp. 407-415, 20 3 2018.

[15] L. A. Silva, S. P. Pimentel and J. A. Pomilio, "Analysis and Proposal of Capacitor Voltage Control for an Asymmetric Cascaded Inverter," *2005 IEEE 36th Power Electronics Specialists Conference*, Recife, 2005, pp. 809-815.

[16] J. Dixon, J. Pereda, C. Castillo and S. Bosch, "Asymmetrical Multilevel Inverter for Traction Drives Using Only One DC Supply," in *IEEE Transactions on Vehicular Technology*, vol. 59, no. 8, pp. 3736-3743, Oct. 2010.

Comparison of the PLL Control techniques applied in Photovoltaic System

Marenice M. de Carvalho[1], Renan L. P. Medeiros[1], Iury V. Bessa[1],
Florindo A. C. Junior[1], Kevin E. Lucas[2] and David A. Vaca[3]

[1]Department of Electricity, Federal University of Amazonas (UFAM), Manaus, Brazil
[2]Department of Automation and Systems, Federal University of Santa Catarina (UFSC), Florianópolis, Brazil
[3]ESPOL Polytechnic University, Faculty of Electrical and Computer Engineering, Campus Gustavo Galindo, Guayaquil, Ecuador
Emails: { marenice, renanlandau, iurybessa, florindoayres} @ufam.edu.br, kevin.lucas@posgrad.ufsc.br, devaca@espol.edu.ec

Abstract—**The increasing demand for low-carbon power and concerns about the environment resulted in the proliferation of alternative sources of energy. Among these sources, the photovoltaic systems stand out. However, these systems produce energy in direct current and require power inverters to connect them to the electric grid. For this connection to be realized the generated signal must have the same parameters of amplitude, phase, and frequency of the electrical network. In this sense, Phase-Locked Loop (PLL) controllers are used to synchronize the output signal with the network signal. For this type of system there are several topologies, such as PLL Synchronous Reference Frame (SRF-PLL), Second-Order Generalized Integrator PLL (SOGI-PLL), Enhanced PLL (EPLL) and Quadrature PLL (QPLL). This work proposes to compare these four controller topologies applied to the inverter of the photovoltaic system connected to the electric grid.**

Index Terms—**PLL, SRF-PLL, SOGI-PLL, QPLL, EPLL, Comparation**

I. INTRODUCTION

The use of electric energy is growing a lot, and one way to meet the increase of this consumption is to find new ways to generate energy. With this, and considering the environmental issue, the use of renewable systems has been much approached, among them, we have the photovoltaic systems. However, the photovoltaic systems produce direct current, therefore, it is necessary to use a DC/AC converters in order to extend its use for grid-connected systems.

For establishing such connection between the photovoltaic system and the electrical grid, it is necessary to ensure that the properties of the converter output signal meet the grid standards, e.g., the phase, frequency and amplitude. There are some problems that can degrade the power quality of the electrical network. For instance, harmonic distortions, voltage sag, and frequency disturbances.

In [1] and [2], a comparison between different PLL topologies applied in converters is presented, and [3] compares the techniques to a single-phase network. In [4], the SOGI-type scheme is discussed, in two formats, of the frequency-adaptive and frequency-fixed type. In [5], the transfer function that relates the input and output of the SRF-PLL is presented. The QPLL are described in [6], showing the filtering capacity

of such system, whose structure is based on estimating in-phase and quadrature-phase amplitudes of the fundamental component of the input signal.

Different from the works presented [1], [2] and [3], in this paper, it is performed a comparison between the different topologies of the PLL controller, that are SRF-PLL, SOGI-PLL, EPLL, QPLL, but applied to the converter of a photovoltaic system connected to the grid. A brief approach will be made to the PLL Reference Souce Frame (RSF-PLL), Second Order Generation Integrarion PLL (SOGI-PLL), Enhanced Phase-Looped Loop (EPLL) and Quadrature Phase-Looped Loop (QPLL).

The rest of this paper is organized as follows: Section 2 addresses the theoretical aspects about the topologies of the PLL controllers; Section 3 presents the methodological procedures used in this work; Section 4 presents the experiments results and the comparative analysis; and Section 5 presents the main conclusions of the study.

II. BACKGROUND

A. PLL

The Phase Locked Loop is a control technique widely used to perform phase synchronization in systems connected to the network. This technique has already been used in communication systems and allowed the improvement of the performance and reliability of electronic systems [7]. With the development of digital technology, there has been a growth in the implementation and development of new types of controllers.

The basic principle of PLL operation is to tune the signal components according to a reference signal. It has three well-defined blocks, one to detect the phase of the input signal of the system, comparing it with the output signal of the VCO, a block with low pass filter to attenuate noise and high frequency, and another one that uses as parameter the signal detected in the first block to generate an output signal, this is used to feed the system. Fig. 1 shows the block diagram of the PLL.

In this technique variations made in the PD block give rise to the following PLL topologies: Synchronous Refer-

978-1-7281-4181-7/19 $31.00 © 2019 IEEE

ence Frame (SRF-PLL), Second-Order Generalized Integrator (SOGI-PLL), Enhanced PLL (EPLL) and Quadrature PLL (QPLL).

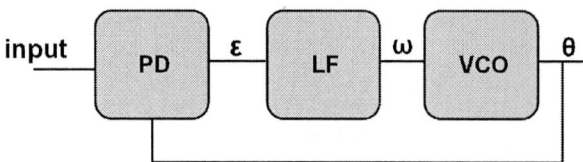

Fig. 1: Scheme of PLL.

B. SRF-PLL

Within the topologies of the PLL one of the most employed is the SRF-PLL [1], due to its robustness and simplicity [5]. Fig. 2 shows the construction of this topology. The input three-phase signal is transformed by means of the Park transform into two lagged signals of $\frac{\pi}{2}$ rad of each other, Vq and Vd. Where Vd contains the amplitude measure of the three-phase signals and Vq contains information of the error of phase [8]. In sequence this error signal passes through the PI and VCO filter blocks, returning the phase angle used for system feedback

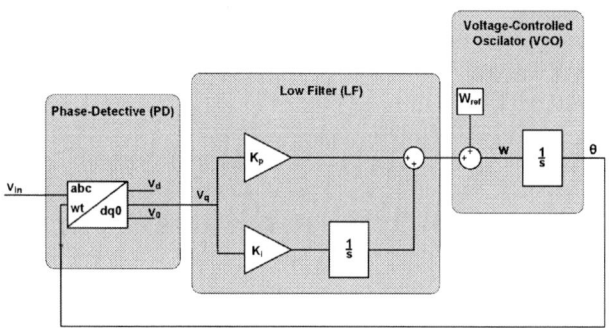

Fig. 2: Scheme of SRF-PLL.

C. SOGI-PLL

The topology of the Second-Order Generalized Integrator for three-phase input is shown in Fig. 3. It has the same filter construction and VCO having only a different phase detector. V_α and V_β are lagged in $\frac{\pi}{2}$ rad, these signals are obtained from a transfer function defined by

$$H(s) = \frac{\omega s}{s^2 + \omega^2} \quad (1)$$

where ω represents the resonance frequency of the SOGI [3].

The two components generated by the SOGI show v' and qv', the first with the same phase of the input signal, v, and the second one challenged by 90 degrees. Your transfer functions are

$$H_d(s) = \frac{v'}{v} = \frac{k\omega s}{s^2 + k\omega s + \omega^2} \quad (2)$$

$$H_q(s) = \frac{qv'}{v} = \frac{k\omega^2}{s^2 + k\omega s + \omega^2} \quad (3)$$

where k is the parameter that adjusts the filtering capacity of the system. A small k narrows the bandwidth of the filter, in contrast, it leaves the dynamic response slower [9]. The SOGI structure is a simple integrals scheme that already has a pair of already-filtered orthogonal signals.

For the three-phase input, SOGI is organized into a DUAL structure as seen in Fig. 3. Each SOGI block has it's v' and qv' element, the first SOGI block is fed with the alpha component, the second SOGI block is fed with the beta component, both obtained from the Clarke transform [1].

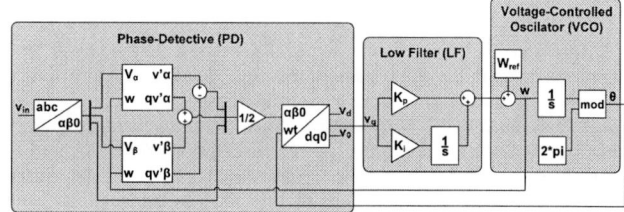

Fig. 3: Scheme of SOGI-PLL

D. EPLL

The Enhanced PLL structure is based on a nonlinear filter. This topology improves PD performance. The filter follows the fundamental frequency of the network. It allows greater flexibility compared to conventional PLL, informing amplitude and phase angle. And it is also able to estimate the frequency.

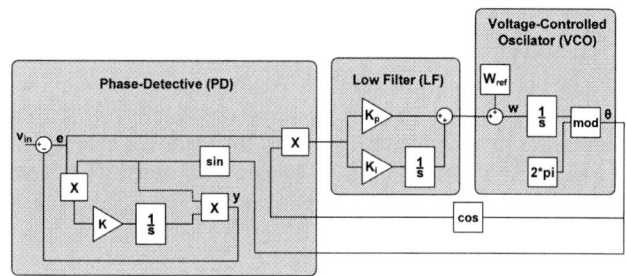

Fig. 4: Scheme of EPLL.

The set of equations presented in (4), provide the frequency, $\omega(t)$, phase angle, $\theta(t)$, amplitude, A(t), fundamental component, y(t), and error signal, e(t). The three-phase structure is obtained with [1].

$$
\begin{aligned}
\dot{\omega}(t) &= k_i \cdot e(t) \cdot cos(\theta(t)) \\
\theta(t) &= \frac{\dot{\omega}(t)}{k_i} \cdot k_p + \omega_r(t) \\
\dot{A}(t) &= k \cdot e(t) \cdot cos\left(\theta(t) - \frac{\pi}{2}\right) \\
y(t) &= A(t) \cdot cos\left(\theta(t) - \frac{\pi}{2}\right) \\
e(t) &= u(t) - y(t)
\end{aligned}
\quad (4)
$$

978-1-7281-4181-7/19 $31.00 © 2019 IEEE

When K, K_p, K_i are gains that define system behavior. K is the speed of convergence of amplitude, and K_i and K_p is the speed of convergence of frequency and phase.

E. QPLL

Quadrature PLL is a PLL method that in the PD block generates an input signal based on estimates for both the phase and quadrature amplitudes, as well as phase and frequency variations [6].

A nonlinear structure in the definition of the error signal. QPLL is controlled by the following internal parameters: μ_s, μ_c which controls in-phase and quadrature amplitude convergence and μ_f which controls frequency convergence [3]. Considering $\mu_s = \mu_c$, it is possible to simplify internal parameters as follows $2\mu_s = 2\mu_c = \mu_a$ and $\mu_f/\mu_s = \mu_f/\mu_c$ and the system takes the form shown in [2].

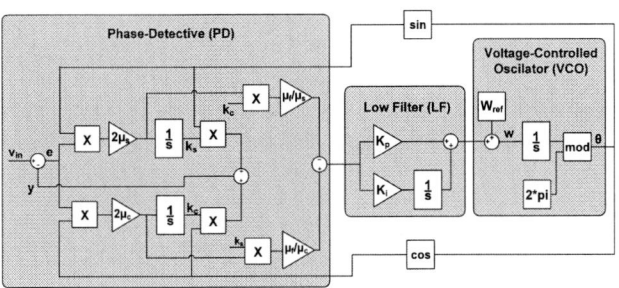

Fig. 5: Scheme of QPLL.

The set of equations presented in (5), provide the quadrature amplitudes $K_s(t)$ and $K_c(t)$, frequency variation, $\dot{\triangle}\omega(t)$, phase angle, $\dot{\theta}(t)$, fundamental component, y(t), and error signal, e(t). The three-phase structure is obtained with [1].

$$
\begin{aligned}
\dot{k}_s(t) &= 2\mu_s \cdot e(t) \cdot sin\theta(t) \\
\dot{k}_c(t) &= 2\mu_c \cdot e(t) \cdot cos\theta(t) \\
\dot{\triangle}\omega(t) &= 2\mu_f \cdot e(t) \cdot [k_s \cdot cos\theta(t) - k_c \cdot sin\theta(t)] \\
\dot{\theta}(t) &= \omega_0 + \triangle\omega(t) \\
y(t) &= k_s \cdot sin\theta(t) + k_c \cdot cos\theta(t) \\
e(t) &= u(t) - y(t)
\end{aligned}
\tag{5}
$$

III. METHODOLOGICAL PROCEDURES

A. Experimental Environment

The simulations are performed in the Matlab/Simulink environment with fixed-step size of 10^{-4} and solver ode4 (Rugekutta).

B. Description of experiments

Three types of tests are performed in the four PLL topologies presented. Test of frequency variation, voltage sag, and harmonic distortion rate.

1) Change of frequency: To verify the behavior of the systems when there is variation in frequency from 60Hz to 50Hz in the range of 0.2s to 0.4s.

2) Voltage sag: A voltage sag is performed during the time interval of 0.2s to 0.4s. This test simulates an increase of loads in parallel in the system, requiring more current to meet the demand, consequently generating a voltage sag.

3) Harmonic Distorcion: The harmonics currents are caused by equipment that has a non-linear charge. The odd harmonics are the most common to appear because of some physical characteristics. Harmonics cause power losses in the network and increase the level of current and voltage in the network. In this test is inserted signals with harmonics of the 5th and 7th order.

Table 1 shows the value of the parameters that were used in the four PLL topologies, which were obtained by the Ziegler-Nichols method.

TABLE I: Parameter used for PLL topologies

Configuration	K_p	K_i	K	μ_a	μ_f/μ_s
SRF-PLL	100	4228	-	-	-
SOGI-PLL	154.8	7871.2	1.2	-	-
EPLL	180	5202.3	100	-	-
QPLL	9.9	1440	-	100	1

IV. RESULTS

1) Test 1: Fig. 6 to 9 show the results of the frequency variation tests for the four topologies mentioned in this work. It can be observed that the SRF-PLL, SOGI-PLL and QPLL topologies showed good results for settling time, while EPLL was the slowest. SOGI-PLL showed some oscillation while EPLL and QPLL topologies showed more oscillations and SRF did not oscillate. The table 2 summarizes the results obtained with for overshoot and settling time values were analyzed for frequency response curves.

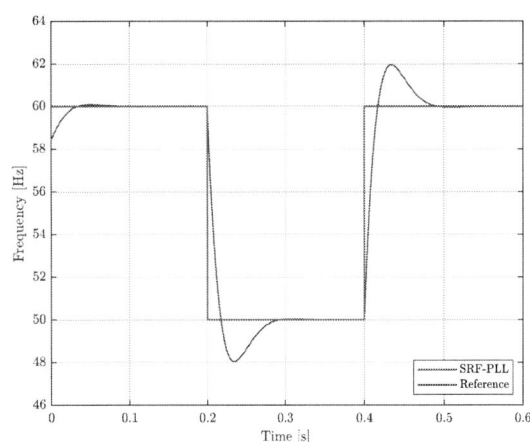

Fig. 6: Frequency variation of SRF-PLL.

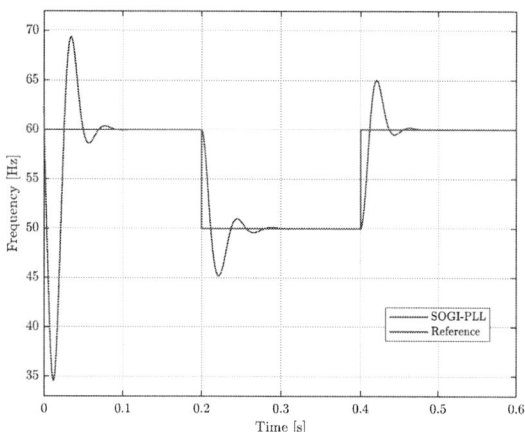

Fig. 7: Frequency variation of SOGI-PLL.

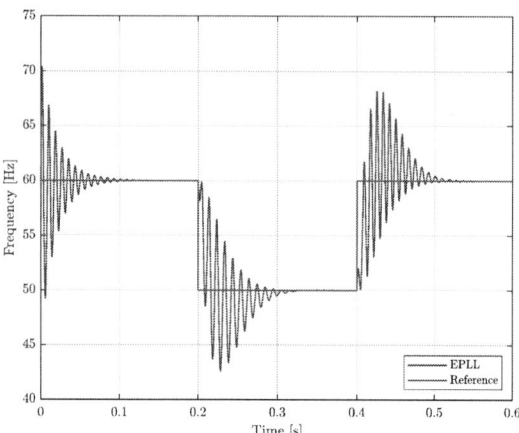

Fig. 8: Frequency variation of EPLL.

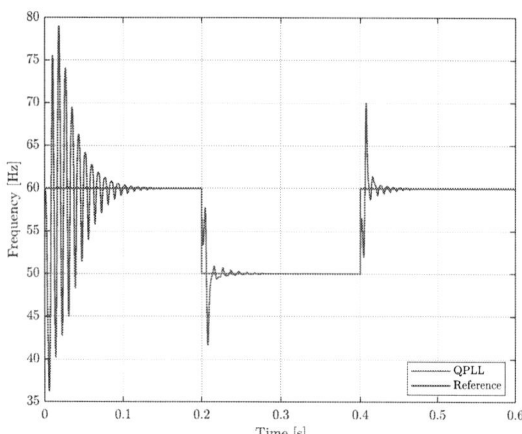

Fig. 9: Frequency variation of QPLL.

TABLE II: Comparative performance analysis

Topologies	Overshoot	Settling time
SRF-PLL	3.94 %	53 ms
SOGI-PLL	10.00 %	34 ms
EPLL	14.68 %	84 ms
QPLL	16.58 %	17 ms

2) Test 2: Fig. 10 to 13 presents the results for a 20% voltage sag in the range of 0.2s to 0.4s. The voltage drop suggests the insertion of loads in parallel in the system, and it can be observed that all models presented variation in phase A maintaining the voltage level during the interval of 0.2s to 0.4s. In this test, none of the topologies showed overshoot in the voltage sag period, and all accommodated quickly in a timeless than 100ms, and neither showed oscillations.

3) Test 3: Fig. 14 to 17 shows the ability of harmonic filtering models. In this test was inserted harmonics of 5th and 7th order at the input of the system. The graphs show the input with the presence of harmonics from 0.1s and then its output signal with the filtered harmonics.

The EPLL topologie presented higher filtering capacity, the QPLL still presented a slight disturbance in the output, the SOGI-PLL was able to reduce the presence of the harmonics, but still one can perceive a great distortion in the output and SRF-PLL presented the worst result with no filter capability. The summary of this result can be seen in table 3.

TABLE III: Comparative THD

Topologies	TDH input	TDH output
SRF-PLL	12.73 %	12.73 %
SOGI-PLL	12.73 %	6.36 %
EPLL	12.73 %	0.83 %
QPLL	12.73 %	3.81 %

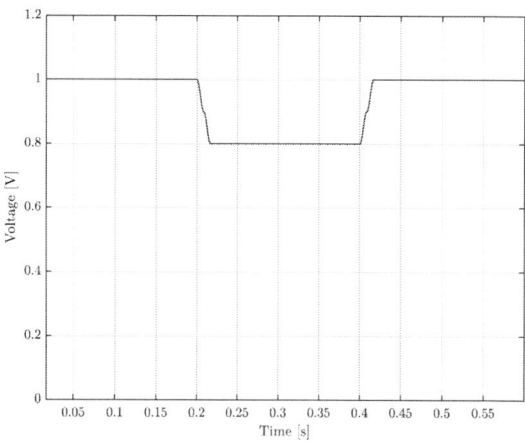

Fig. 10: Voltage sag of SRF-PLL.

978-1-7281-4181-7/19 $31.00 © 2019 IEEE

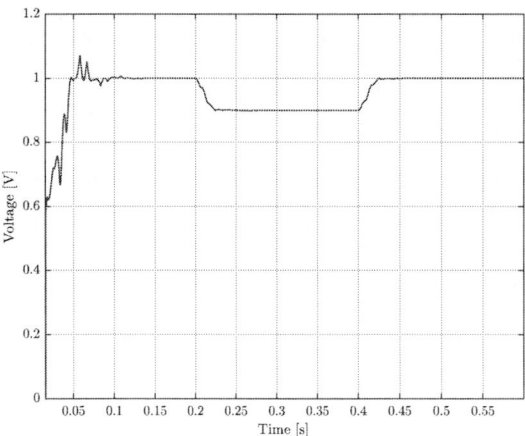

Fig. 11: Voltage sag of SOGI-PLL.

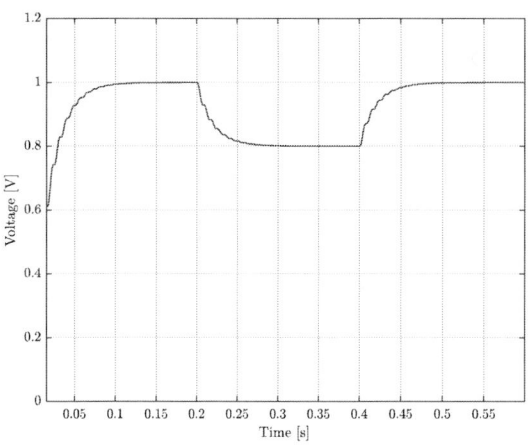

Fig. 12: Voltage sag of EPLL.

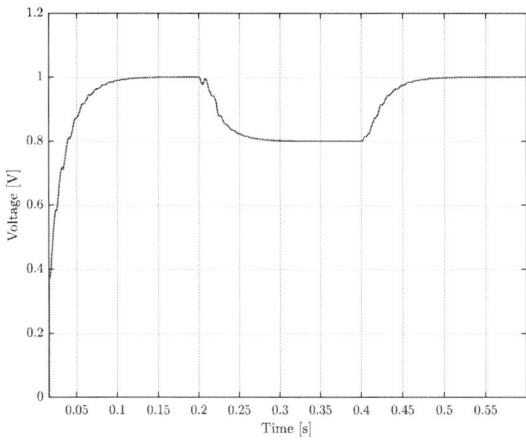

Fig. 13: Voltage sag of QPLL.

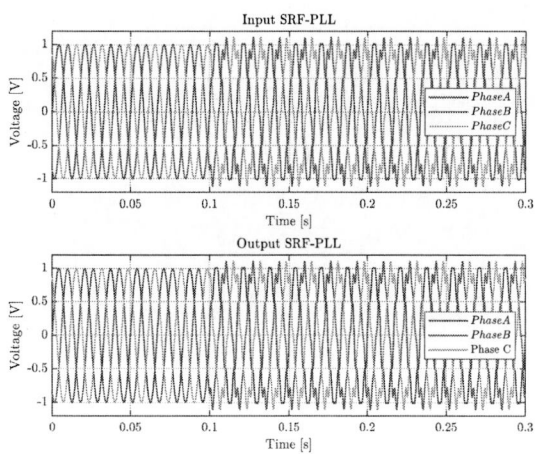

Fig. 14: Filtering capability of the SRF-PLL.

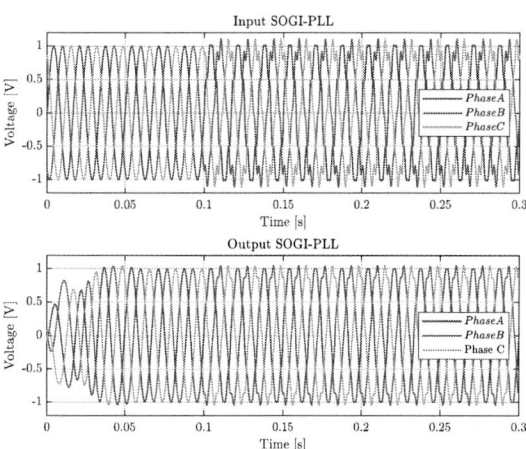

Fig. 15: Filtering capability of SOGI-PLL.

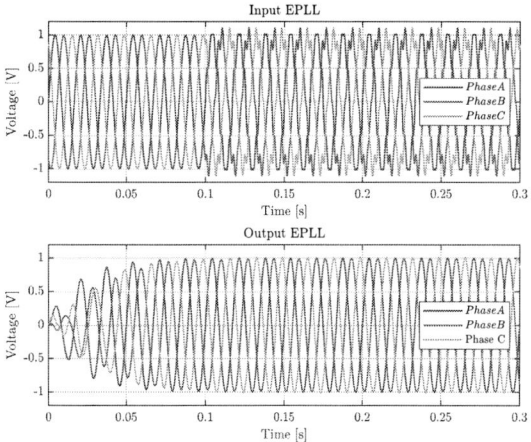

Fig. 16: Filtering capability of EPLL.

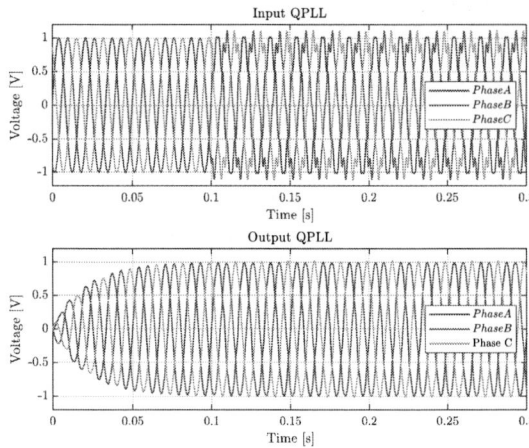

Fig. 17: Filtering capability of QPLL.

V. CONCLUSION

A comparison was proposed for four PLL topologies. They are Syncrhonous Reference Frame PLL, Second Order Generated PLL Integrator, Enhanced PLL and Quadrature PLL. In order to evaluate these different types of topologies, we performed tests of frequency variation, voltage sag, and also harmonic distortion.

In the first test, it can be observed that all topologies had a settling time less than 100ms, the best result for overshoot was SRF-PLL, EPLL and QPLL showed large oscillations that can be improved with a better tuning of their parameters. In the second test, all showed a good response to voltage drop. For the harmonic variation, the best response was the EPLL, followed by the QPLL, SOGI-PLL and finally the SRF-PLL, which was the only one that could not filter the inserted harmonics.

ACKNOWLEDGMENT

This work was supported in part by CNPq under grant 432341/2018-8, in part by CAPES, FAPEAM, and in part by the PROPG-CAPES/FAPEAM Scholarship Program.

REFERENCES

[1] A. Nicastri, and A. Nagliero, "Comparison and evaluation of the PLL techniques for the design of the grid-connected inverter systems," pp. 3865–3870, July 2010.

[2] A. Nagliero, R. Mastromauro, M. Liserre, "Monitoring and Synchronization Techniques for Single-Phase PV Systems," International Symposium on Power Electronics, Electrical Drives, Automation and Motion, 2010, pp.1404–1409.

[3] R. J. Ferreira, R. E. Araújo, and J.A. Peças Lopes, "A comparative analysis and implementation of various PLL techniques applied to single-phase grids," 3rd International Youth Conference on Energetics, 2011, pp. 1–8.

[4] F. Xiao, L. Dong, L. Li, and X. Liao, "A Frequency-Fixed SOGI Based PLL for Single-Phase Grid-Connected Converters," Transactions on Power Electronics, vol. 8993, 2016, pp. 1–6.

[5] S. Golestan, and J. M. Guerrero, "Conventional Synchronous Reference Frame Phase-Locked Loop Is An Adaptive Complex Filter," IEEE Transactions on Industrial Electronics, vol. 1, pp. 1–4, 2014.

[6] M. Karimi-Ghartemani, H. Karimi, and M. R. Iravani, "A Magnitude/Phase-Locked Loop System Based on Estimation of Frequency and In-Phase/Quadrature-Phase Amplitudes" IEEE Transactions on Industrial Electronics, .vol 51, no. 2, april 2004, pp 511–517.

[7] G. Hsieh, and J. C. Hung, "Phase-Locked Loop Techniques-A Survey" IEEE Transactions on Industrial Electronics, .vol 45, no. 6, december 1996, pp 609–615.

[8] S. Golestan, J. M. Guerrero and J. C. Vasquez, "Three-Phase PLLs: A Review of Recent Advances," IEEE Transactions on Power Electronics, vol. 32, pp. 9013–9030, 2017.

[9] M. Ciobotaru, R. Teodorescu and F. Blaabjerg, "A New Single-Phase PLL Structure Based on Second Order Generalized Integrator," 2006 37th IEEE Power Electronics Specialists Conference, pp. 1–6, 2006.

978-1-7281-4181-7/19 $31.00 © 2019 IEEE

A critical analysis of PSO and its variations applied to MPPT for PV Systems under Partial Shading Condition

Lucas Mendonça Andrade, Paula dos Santos Vicente, Fernando Lessa Tofoli and Eduardo Moreira Vicente
Department of Electrical Engineering
Federal University of São João del-Rei - UFSJ
São João del-Rei, Brazil
andrademendoncalucas@gmail.com, paulasantos@ufsj.edu.br, fernandolessa@ufsj.edu.br, eduardomoreira@ufsj.edu.br

Abstract—**The maximum power that a photovoltaic system is able to provide depends on the climate conditions in which it is submitted. In order to extract the maximum available energy from the system, independent of weather conditions, a Maximum Power Point Tracking (MPPT) system must be used. Several methods can be used to track the maximum power, but the efficiency of tracking depends on the chosen technique. Moreover, when the array is under a partial shading condition, new power peaks arise, which makes some techniques unfeasible. In this paper, an analysis of the evolutionary algorithm Particle Swarm Optimization (PSO) as an MPPT method under partial shading condition is evaluated. Simulations of this technique and its variations, such as DPSO, in different partial shading patterns are performed. As the other MPPT techniques, the PSO and its variations have inconveniences that decrease their efficiency under partial shading conditions. Among these, the wrong power measurement and blocking of particles in DPSO need to be considered because they result in high power losses. To avoid these conditions, computational adjustments are proposed and their efficiency is analyzed through simulations.**

Keywords—*PSO, Global Maximum Power Point Tracking (GMPPT), PV systems, Buck-Boost converter.*

I. INTRODUCTION

The operation of photovoltaic (PV) systems depends on climate conditions. The variation of these parameters, mainly the irradiance, which is the measure of solar power intensity, and the PV module temperature, directly influences the characteristic curve, which represents the relation between the power generated by the array and its voltage, the P-V curve.

A single peak of maximum power, the Maximum Power Point (MPP), at uniform irradiance levels, characterizes the P-V curve. To operate at this point, regardless of climate conditions, a Maximum Power Point Tracking (MPPT) system is necessary. It consists of a DC-DC converter coupled to the PV system, whose purpose is to vary the operation voltage until the MPP is reached.

In residential systems, generally due to the surrounding constructions or cloud movement, the modules can present different irradiance levels. This phenomenon is known as Partial Shading Condition (PSC). Under these conditions, the modules that are subjected to lower incident irradiance, consequently, produce less current, which implies in the behavior of these elements as load to the rest of the string [1]. This fact results in the overheating of these modules' cells, a phenomenon known as hot spots. To prevent the possible damages caused by the PSC, by-pass diodes are connected in parallel with the modules. Despite the protection, the P-V curve is modified, presenting several power peaks, in such a way that the MPP becomes to be known as the Global Maximum Power Point (GMPP).

The occurrence of more than one power peak during the PSC compromises the utilization of the conventional MPPT methods, such as Perturb and Observe (P&O) and Incremental Conductance. These methods are incapable to distinguish the different power peaks, which results commonly in the tracking of a local peak instead of the GMPP, resulting in energy losses [2].

Under those conditions, techniques that use artificial intelligence concepts, such as Fuzzy Logic and Artificial Neural Network [3]-[5] track the global maximum in most weather conditions, but with a high complexity degree and high computational effort. Consequently, methods that achieve similar performances, with lower complexity degree, have received prominence, such as evolutionary algorithms.

The evolutionary algorithms are based on mechanisms of nature and mainly on the survival of the fittest. Among these, the technique known as Particle Swarm Optimization (PSO) deserves attention, because it can track, in most conditions, the GMPP in a short time and with low steady-state oscillation. This method is also classified as a metaheuristic since it uses randomness and previous results to apply tracking in the search space with the aim of locating the best result. This method is based on the social behavior of bird flocks, fish shoals, among others, and was initially proposed in [6].

Despite the method efficiency, some errors can occur due to its random characteristic and equations. With the purpose of solving these errors or improve the method operation, some adjustments are proposed in the literature. In [7], [8] and [9], the PSO is used with the P&O, which makes it a hybrid technique. In [7], the PSO equation is modified and in [8] and [10], some variables are adjusted throughout iterations, in order to decrease the convergence time and to improve tracking.

Regardless of the modifications proposed by the new techniques, some errors persist, as the wrong power measurement, and new errors arise as the blocking of particles in the method proposed in [7]. In order to fix these errors, this paper proposes computational corrections for the PSO technique and its variations, with the aim to make the methods reach the GMPP in a larger number of cases, increasing the PV system efficiency under partial shading conditions.

II. PARTICLE SWARM OPTIMIZATION (PSO)

In the PSO applied to MPPT, a set of different duty cycles is used to track the GMPP. The value of each duty cycle is modified according to the best power values acquired individually or by the set. This change occurs until all the duty cycles values get close to the one that result in the maximum power.

In literature, some terms are used to describe this technique. Each duty cycle receives the name of particle and its value is represented by the particle's position. The

978-1-7281-4181-7/19 $31.00 © 2019 IEEE

variation of the position is known as velocity and the term fitness is used to represent the power value obtained by a particle in a certain position.

An important point to be defined before the implementation of the PSO is the number of particles of the set or swarm. A high number results in a slow but effective tracking, while a small one results in a quick but not always effective tracking. Therefore, the definition of the swarm size depends on the system and on the designer, but it is common the use of three particles in literature, such as in [7] and [10].

In Fig. 1, the particles' dynamic in the operation of the method is demonstrated. The P-V curve considered belongs to a certain PV system and three particles were used in the method. The indexes near to the particles demonstrate the number of the position. Hence, the particles' initial positions are represented by the index 1, for example. In the PSO, the initial positions are generated randomly. In general, the particles are restricted to a certain search space, defined mostly between the values of duty cycles equal to 0.1 and 0.9. This restriction decreases the convergence time because it restricts the space where the particles will search for the MPP.

After the definition of the initial positions, the converter operates with each duty cycle during a specific period, also known as sampling time. The power values are obtained after this time and are stored as the fitness of each position. Since it is the first iteration, each particle's power is its best result and, consequently, the respective duty cycle is the best individual position. All power values are compared and the highest is defined as the best value of the swarm and its duty cycle is defined as the best position. In the example of the Fig. 1, the particle P2 is the one that represents the swarm best value in the first position.

With the positions, as well as the best individual and global values defined, the changes in the particle's position must be performed to obtain better results. The step size, or the particle's velocity, is performed as (1). The position is modified as (2).

$$v_i^{k+1} = \omega v_i^k + c_1 r_1 \left(D_{besti} - D_i^k \right) + c_2 r_2 (GD_{best} - D_i^k) \quad (1)$$

$$D_i^{k+1} = D_i^k + v_i^{k+1} \quad (2)$$

In (1) and (2), i represents the particle's identification number, k represents the iteration number, v is the particle's velocity, D_i^k is the present particle's position, D_{besti} is the best individual duty cycle and GD_{best} is the swarm best position. Because it is the first iteration to define velocity, the last velocity value v_i^k is considered as zero. The parameter ω is known as the inertia coefficient or inertia weight, because its mathematical meaning refers to the physical principle. This factor determines the influence of the last step on the next one. Its value is not well defined in literature and depends on the system. The parameters c_1 and c_2 are the cognitive and social coefficients, respectively. The value of the cognitive coefficient indicates the influence of the best particle position on its velocity. Similarly, c_2 indicates the influence of the best swarm position on the velocity definition and it is defined according to the system. The parameters r_1 and r_2 are random values between zero and one generated every iteration.

For the particle that presented the best swarm's position in the first iteration, the present, individual and global best positions are equal. This fact, associated with the previous null velocity, implies the cancellation of the new step, so the particle does not influence the method until its power value is overcome by other particle.

With the velocity of each particle calculated and incremented to the last position, the converter will operate with the new duty cycles and new power outputs will be obtained. The new fitness of each particle is compared with its best value until then. If it is greater than the actual best value, the new power and position replace the best position values. Finally, the best power values of each particle are compared with the swarm best value and if any of them is greater than it, the new position and the new power output are stored as the swarm best values.

In Fig. 1, for example, the particle P2 presented the best power in the first iteration and thus its calculated velocity is zero. In contrast, particles P1 and P3 presented non-zero velocities and their positions changed. In the next iterations, the initial best particle continues to present a null velocity and the others, due to the mathematical structure of the equations, oriented to the position that presents the best power until then, the P2's position. However, in the P1's fourth position, the fitness obtained is greater than the obtained by the stuck particle and, therefore, P2 starts to contribute with the GMPP search.

As all particles tends to the same position, the proximity between them results in almost equal values of swarm and individual best position, which reduces the second and third terms of the velocity equation, causing (1) to behave as (3).

$$v_i^{k+1} = \omega v_i^k \quad (3)$$

As the adopted value of ω is, in general, less than one, the step tends to decrease and the result is that the particles tend to move toward the same position, the GMPP, in low velocities, as demonstrated in the last particles' positions represented in Fig. 1. In most implementations of PSO in MPPT, when the difference between the particles' positions achieves a predetermined value, it is considered that the method converged and it is terminated. That is, the converter starts to operate with the duty cycle that presented the best power and it is not changed until the power is modified above the defined limit, which characterizes the restart of the method. Although, as the particles tend to the same position with low velocities, the array voltage remains almost constant [11] as well as its power. Therefore, some authors suggest the extension of the method until the condition of reset occurs, without using only one duty cycle after the convergence.

As demonstrated, the PSO applied to MPPT does not need complex equations or high computation effort. Meanwhile, despite the great efficiency, the method presents tracking errors related to its operation logic. Among these errors, the structure of (1) and its random characteristic deserves attention.

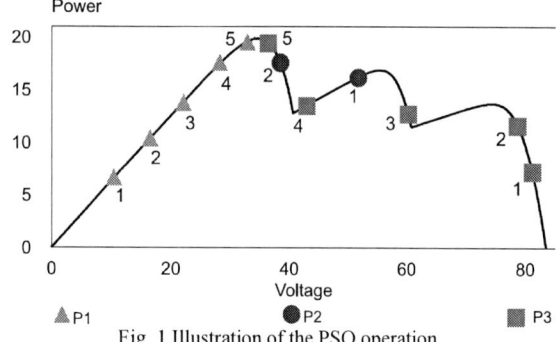

Fig. 1 Illustration of the PSO operation.

978-1-7281-4181-7/19 $31.00 © 2019 IEEE

In relation to (1), the definition of the values of ω, c_1 and c_2 must be the result of a study of the impact of these variables on the method and system, which makes the use of PSO more complex. For example, the velocity is directly proportional to ω and therefore the adopted value implies in the dynamics of the tracking. Other issue is the null velocity of the particle with the best position in the initialization, which delays the method.

The non-limitation of the velocity value may result in the tracking of a local peak or in a premature convergence. Both conditions occur when particles go through peaks without been influenced by their values, which is due to high steps values. If the particles are not influenced by the GMPP, the tracking of a local peak may occur. The premature convergence occurs when none of the peaks influences the particles. In this case, the particles tend to move toward the swarm best position, which simulates a convergence.

In relation to the random characteristic, if the value of r_1 or r_2 is close to zero, slower tracking or velocities, in the opposite direction of the optimization, may be evidenced. For the randomness in the particles' initial position, the tracking can be confined to a small space or the search can become slower if the positions are distant. To correct the initialization problem, some techniques can be used, as [12], in which the Lagrange Interpolation is used to determine the proximity of the GMPP to then initialize the particles close to it.

In order to improve the dynamics of the method or to prevent a wrong tracking, modifications in PSO are presented, as in [7], [8] and [10]. Although the proposed modifications are substantial, the original method is rarely improved and these variations present the same errors. Therefore, to evaluate the efficacy of these adjustments, a comparison between the PSO and its variations is necessary.

III. Deterministic Particle Swarm Optimization (DPSO)

The DPSO was proposed in [7] with the objective of correcting the errors caused by (1) and by the method's randomness. For this purpose, the velocity is now obtained by (4). Equation (2) remains the same.

$$v_i^{k+1} = \omega v_i^k + D_{best_i} + GD_{best} - 2D_i^k \tag{4}$$

All terms of (4) present the same meaning of (1), as this is a result of considering c_1, c_2, r_1 and r_2 equals to one in the previous equation. This tends to make the method more deterministic and less random. Moreover, the only variable that needs to be tuned is ω.

The particles' initialization is not random in this method, it is now initialized in pre-defined positions. In [7], these positions are not defined, but the number of particles is determined as three. To avoid errors due to the non-limitation of the velocity, the paper proposes a maximum step of 0.035. The rest of the method is equal to PSO.

In DPSO, the method becomes hybrid, that is, the modified PSO is associated with other technique that, in this case, is the P&O. The DPSO is only used in cases of PSC in the mode known as global and the P&O is used when PSC do not occurs, in the mode known as local. The determination of the occurrence of the PSC is realized after the particles' initialization as defined in [7]. If the PSC is evidenced, the DPSO is applied; otherwise, the P&O is used. Furthermore, after the convergence of the global mode, the local mode is activated, and it makes small corrections in the duty cycle for

small variations in irradiance that do not characterize the restart of the DPSO. Although the benefits of the utilization of these two modes, the method proposed in [13] is inefficient to some patterns of PSC, which may result in loss of power if the P&O is used in these conditions.

Despite all modifications, the null velocity, for the particle that presents the best power in the first iteration, persists. The paper suggests the realization of a small change in its position but does not determine how this must be carried out. Furthermore, a new issue in relation to (4) and due to the multiple peaks during the PSC was evidenced. If one particle is in the intermediate position between its best value and the best global value, its velocity will tend to zero, as represented in (5).

$$D = \frac{D_{best_i} + GD_{best}}{2} \tag{5}$$

If this value of D is replaced in (4), the velocity will be influenced only by its last value and ω. This results in the decreasing of velocity and the stuck of the particle.

IV. The Effects of a Variable Inertia Weight

In [14], the relation between the limitation of the velocity, the inertia weight and the performance of the PSO is defined. A small value of maximum velocity results in a slow tracking and in the restriction of the global search, which helps the tracking of a local maximum. In contrast, a large value results in a great global search ability, limited only by the value of ω.

Despite the effects of the maximum velocity value, this do not modify the method performance if its value is not reached. However, the value of ω directly interferes in the velocities of all particles in all iterations.

Consequently, for a better performance of the method, both in the global and local search, the study of ω becomes more attractive. Therefore, some PSO variations use a variable inertia weight. In general, it is modified to present a higher value at initialization, for better performance in the global search, and a lower value close to convergence, to prioritize the local search.

In [8], the PSO logic is not changed but the influence of a linear variation in ω as demonstrated in (6) is analyzed.

$$\omega_k = \omega_{max} - \frac{k}{k_{max}}(\omega_{max} - \omega_{min}) \tag{6}$$

In (6), k represents the iteration number, k_{max} represents the maximum iteration number in which ω changes, ω_k represents the ω value for a certain k and ω_{max} and ω_{min} represents the maximum and minimum values of the coefficient, respectively. In [10], a non-linear variation of ω is proposed, as described in (7).

$$\omega_k = \omega_{max}.exp\left(\frac{-k}{k_{max}}\right) \tag{7}$$

The main difference between the two equations is the need for the lower limitation of ω in (7), because the mathematical structure does not limit it, which occurs in (6).

V. Computational Corrections

As already mentioned in the previous sections, the PSO and its variations may present particular and common errors. In order to compare the methods and to obtain their characteristics, computational corrections were realized.

In all methods, wrong power values persist, because of either the sampling time adopted or the large duty cycle variations. Moreover, the simulation software, PSIM®, may present this error in the method initialization, not caused by the blocking of the particle that presents the best power, but related to the initial transient of the converter in simulation. This error in the power acquisition was adjusted in the simulations because it disables the technique in some conditions. Errors related to the randomness or to the non-delimitation of the maximum velocity were not corrected because they are inherent to the methods' logic and a change to them would result in a modification to the original proposal.

To correct the wrong power value, the particle that presents its best position as the swarm best one, is analyzed. If this particle is close to the swarm best position, and if its power is less than 90% of the swarm best power, the error is evidenced because the power change is substantial for a small position change.

In order to obtain a new swarm best value, the best power and position of the corrected particle becomes the present values and the swarm best values are nullified. So, at the end of the iteration, a new global value is obtained. The proximity of the analyzed particle and the best position is defined as the duty cycle ± 1%.

In Fig. 2(a), an example of this correction applied to PSO in a given pattern of PSC is demonstrated. When analyzing the figure, it is clear how important the correction is, because if it was not applied, the wrong power value would result in a large power fluctuation. This results from the particle's tendency to move towards the individual and global best positions, which do not correspond to the power previously obtained.

In DPSO, the procedure applied to correct the null velocity of the particle that presents its position equal to (5) is similar to the one applied to the wrong power measurement. In this condition, if the particle is close to the medium point between its best position and the swarm best one, and distant at least 0.035, which is the maximum velocity, from the swarm best position, a variable is incremented. In the third increment, the lock is evidenced. In order to make this particle to contribute to the method, its best position and power are updated to the current values. The distance from the medium point was considered as a ± 1% increment of the duty cycle.

Figure 2(b) shows the correction applied to DPSO in a given pattern of PSC. If this correction were not performed, power losses would occur every acquisition cycle of the stuck particle's power, at steady state. Therefore, the need of this correction is evident.

VI. SIMULATIONS

The simulations were carried out in PSIM® and the system schematic is shown in Fig. 3. As demonstrated in Fig. 3, the "Solar Module (physical model)" tool was used to simulate the modules. The chosen module was the Kyocera® KS10 and its parameters can be observed in TABLE I. The comparison between the model and the real module in Standard Test Conditions (STC – irradiance of 1000 W/m² and temperature of 25 °C) is presented in TABLE II.

In Fig. 3, the by-pass diodes were characterized by its ideal behavior and 3.3 nF capacitors were placed in parallel with the modules to enable the simulation of a string with modules submitted to different irradiance values in PSIM®. In TABLE III. , the Buck-Boost converter and simulation parameters are demonstrated.

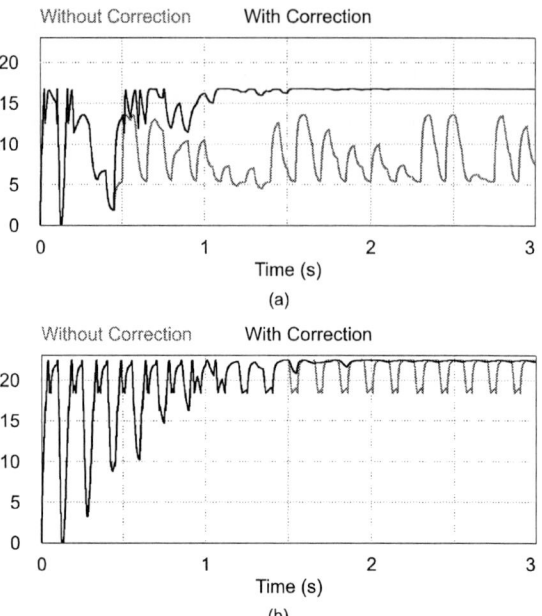

Fig. 2 Power curves with and without computational corrections: (a) PSO, (b) DPSO.

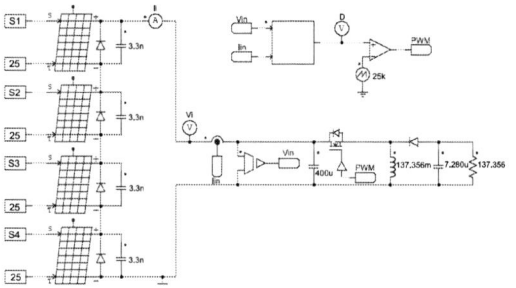

Fig. 3 Schematic used in simulation.

TABLE I.   MODEL DATA.

Parameter	Value
Number of Cells (N_s)	36
Standard Light Intensity (S_0)	1000
Reference Temperature (T_{ref})	25
Series Resistance (R_s)	0.02
Shunt Resistance (R_{sh})	1000
Short-Circuit Current (I_{sc0})	0.63
Saturation Current (I_{s0})	9.995e^{-7}
Band-gap Energy (E_g)	1.12
Ideality Factor (A)	1.75
Temperature Coefficient (C_t)	0.00027
Light Intensity Coefficient (K_s)	0

TABLE II. COMPARISON BETWEEN THE PHYSICAL MODEL AND THE KS10 DATASHEET UNDER STC (1000 W/M² AND 25 °C).

Parameter	KS10	Physical Model
Open-Circuit Voltage	21.7 V	21.643 V
Short-Circuit Current	0.63 A	0.629 A
Maximum Power	10 W	9.927 W
MPP Voltage	17.4 V	17.287 V
MPP Current	0.58 A	0.574 A

TABLE III. SIMULATION DATA.

Parameter	Value
Resistor	137.3562 Ω
Inductor	137.3562 mH
Capacitor	7.2804 µF
Frequency	25 kHz
Bus capacitor	400 µF
Total time	5 s
Time step	1.5 µs
Sampling time	0.05 s

To implement the technique's logic, the "C Block" tool, which allows the structure programming in C language, was used. The input signals were the string current and voltage outputs and the controller's output signal, the PWM signal that drives the Buck-Boost converter.

To obtain better results, the computational corrections were applied to all simulated techniques, which included the PSO, DPSO, PSO with linear and exponential variation of ω, with both k_{max} equal to 10 and 18. In order to carry out the comparison between the different methods, some considerations were made. The number of particles adopted, in all simulations, was three. The particles were not initialized randomly but in defined positions (0.25, 0.5 and 0.75) and the search space was restricted between 0.1 and 0.9. Furthermore, the variables' randomness was realized with the function *rand*. Therefore, they presented the same random value for a specific iteration in all simulations. For example, r_1 presented the same value for a certain particle in the fourth iteration in all simulated conditions. As the problem of the stuck best particle in the first iteration is inherent to all techniques, this particle's variation suggested in [7] was not implemented.

The inertia coefficient adopted in the velocity equation was 0.4, as defined in [7]. For the techniques that apply the variation of this parameter, the ω_{max} was 0.8 and ω_{min} was 0.4. For c_1 and c_2, the chosen values were 0.8 and 1.2, respectively. This way, the particles tend to move more towards the swarm best position than to their best individual.

In relation to DPSO, the velocity was limited in 0.035 as suggested in [7]. In this method, the PSC test was not used and the local mode was used only after convergence and with the step of 0.005.

In reference [8], ω, c_1 and c_2 changes linearly. With the purpose to analyze the benefits of a variable ω, the variation of c_1 and c_2 were not considered. Furthermore, this technique is hybrid, using P&O, but this fact was not considered either.

To perform the comparison between the evaluated techniques, four different patterns of PSC were considered. These patterns are demonstrated in Fig. 4. In all simulations, the parameters obtained were the average power, the number of computational corrections performed and the time to converge, that was considered when the difference between the particles' positions was less than 0.01.

VII. RESULTS AND DISCUSSION

The average power obtained during the simulation time is a parameter that indicates the tracking efficiency, the time required to converge and the power oscillations before the convergence. Thus, this parameter can be used to compare the technique's efficiency in all conditions. The percentage ratio between the average power and the maximum power available is represented in the bar graph of Fig. 5.

As it can be seen in Fig. 5, the efficiency of each technique vary considerably in the performed tests. Consequently, the parameters implicit in the average power must be analyzed. In relation to the convergence time, the DPSO presented the best result in Condition 3 and the PSO presented the best result in the other patterns. Concerning the tracking efficiency, none of the techniques tracked the global maximum in all conditions, but the PSO, DPSO and the techniques that vary linearly and exponentially with 18 iterations, tracked the GMPP under three conditions.

As the PSO presented the lowest power oscillation during the tracking period and better convergence times, this method presented the highest average power in most conditions. The only pattern in which the technique tracked a local maximum was in Condition 4.

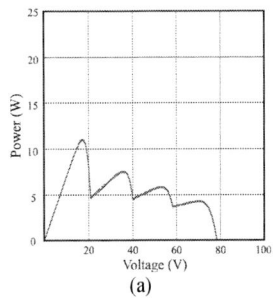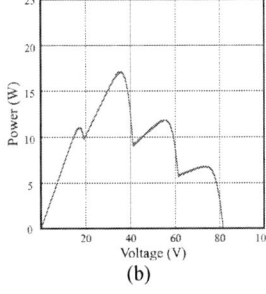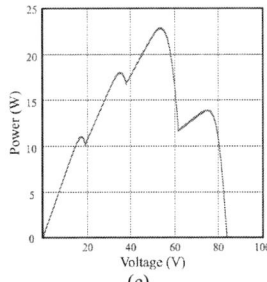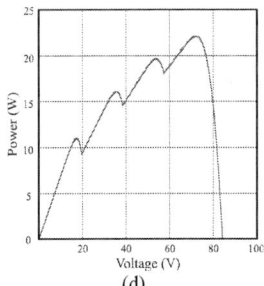

(a)   (b)   (c)   (d)

Fig. 4 Irradiance patterns evaluated: (a) peak to the left of the curve (Condition 1), (b) peak to the left in the central part of the curve (Condition 2), (c) peak to the right in the central part of the curve (Condition 3), (d) peak to the right of the curve (Condition 4).

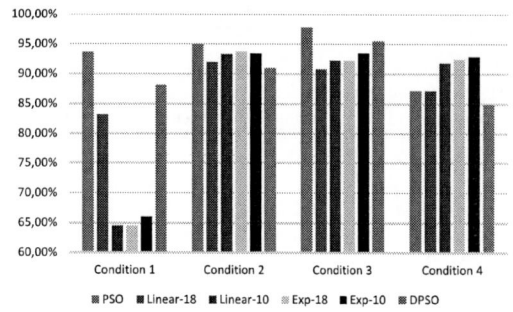

Fig. 5 Efficiency of the techniques in relation to each irradiance pattern.

TABLE IV.   EFFICIENCY OF THE METHODS WITH AND WITHOUT CORRECTIONS.

Method / Condition	Without Correction	With Correction
Linear-10 / Condition 3	83,85%	92,23%
Linear-18 / Condition 1	84,14%	83,19%
Linear-18 / Condition 4	71,01%	87,17%
Exp-10 / Condition 3	88,74%	93,50%
Exp-18 / Condition 3	84,19%	92,18%
DPSO / Condition 1	68,03%	88,11%
DPSO / Condition 2	81,17%	91,02%

Regarding the techniques that vary ω, it was demonstrated that, when the adopted number of maximum iteration is 18, the tracking becomes more efficient, but the time to converge becomes larger, which results in the decrease of the average power, considering techniques that adopt k_{max} equal to 10. Furthermore, the methods based on the exponential reduction results in a shorter convergence time. Because of the large initial velocities, these four were the only methods to present premature convergence.

The DPSO showed good convergence results, having tracked a local maximum only in Condition 4. The velocity limitation and the use of P&O demonstrated excellent solutions for the premature convergence. However, the maximum velocity value must be determined for each system, because using the value proposed by [7], this technique required more time to converge than PSO, which could be less if the velocity was limited in a higher value.

Finally, the computational corrections demonstrated to be determinant in the performance of the techniques. TABLE IV shows the efficiency of some techniques with and without correction in the situations that they were used.

Only for the technique in which the ω varies linearly in 18 iterations in Condition 1, the corrections imply loss of power. In the other conditions of TABLE IV it is evident the performance improvement. In the condition that the power decreased, the two simulations tracked the GMPP, but the corrections resulted in the increase of power oscillations during the tracking period, which decreased the average value.

VIII. CONCLUSION

In this paper, the analysis of PSO and some techniques that are based on this method applied to MPPT was performed. Due to the analytical and computational simplicity, these methods have proven to be excellent solutions for MPPT applied to PV systems under PSC.

Despite the good efficiency of these methods, issues such as the initial locking of the higher power particle, premature convergence, randomness of the techniques, wrong power values obtained and locking of the particles in DPSO, results in power loss. For the correction of these issues, the methodology of [7] and the proposed computational corrections were used.

In relation to the proposed computational corrections, the increase of the methods' efficiency became evident and these proved to be essential for a better performance of the techniques in most partial shading conditions.

ACKNOLEDGEMENT

The authors would like to acknowledge INERGE, UFSJ and Brazilian research funding agencies CAPES, CNPq, and FAPEMIG for the support to this work.

REFERENCES

[1] P. Guerriero, and S. Daliento, "Toward a Hot Spot Free PV Module," IEEE Journal of Photovoltaics, vol.9, pp. 796-802, May 2019.

[2] K. Ishaque, and Z. Salam, "A review of maximum power point tracking techinques of PV system for uniform insolation and partial shading condition," Renewable and Sustainable Energy Reviews, vol. 19, pp. 475-488, March 2013.

[3] L. Bouselham, M. Hajji, B. Hajji, and H. Bouali, "A new MPPT-based ANN for photovoltaic system under partial shading conditions," Energy Procedia, vol. 111, pp. 924-933, March 2017.

[4] S. Farajdadian, and S. Hosseini, "Design of an optimal fuzzy controller to obtain maximum power in solar power generation system", Solar Energy, vol. 182, pp. 161-178, April 2019.

[5] S. Allahabadi, H. Iman-Eini, S. Farhangi, "Neural network based maximum power point tracking technique for PV arrays in Mobile Applications", 2019 10th International Power Electronics, Drive Systems and Technologies Conference (PEDSTC), February 2019.

[6] R. Eberhart, and J. Kennedy, "A new optimizer using particle swarm theory", MHS'95. Proceedings of the sixth international: symposium on micro machine and human science, pp. 39-43, October 1995.

[7] K. Ishaque, and Z. Salam, "A deterministic particle swarm optimization maximum power point tracker for photovoltaic system under partial shading condition," IEEE Transactions on Industrial Electronics, pp. 3195-206, August 2013.

[8] K. Sundareswaran, V. Vignesh Kumar, and S. Palani, "Application of a combined particle swarm optimization and perturb and observe method for MPPT in PV systems under partial shading conditions," Renewable Energy, vol. 75, pp. 308-317, March 2015.

[9] M. Merchaoui, A. Sakly, and M. Mimouni, "Improved fast particle swarm optimization based PV MPPT," 9th Internacional Renewable Energy Congress (IREC), May 2018.

[10] M. Mao, L. Zhou, Z. Yang, Q. Zhang, C. Zheng, B. Xie, Y. Wan, "A hybrid intelligent GMPPT algorithm for partial shading PV system," Control Engineering Practice, vol. 83, pp. 108-115, February 2019.

[11] J. Ahmed, and Z. Salam, "A critical evaluation on maximum power point tracking methods for partial shading in PV systems," Renewable and Sustainable Energy Reviews, vol. 47, pp. 933-953, July 2015.

[12] R. Koad, A. Zobaa, and A. El-Shahat, "A Novel MPPT Algorithm Based on Particle Swarm Optimization for Photovoltaic Systems," IEEE Transactions on Sustainable Energy, vol. 8, pp. 468-476, April 2017.

[13] T. Nguyen, and K. Low, "A global maximum power point tracking scheme employing DIRECT search algorithm for photovoltaic systems," IEEE Transactions on Industrial Electronics, vol. 57, pp. 3456-3467, October 2010.

[14] Y. Shi, and R. Eberhart, "Parameter Selection in Particle Swarm Optimization," EP 98 Proceedings of the 7th international conference on evolutionary programming VII, pp-591-600, March 1998.

978-1-7281-4181-7/19 $31.00 © 2019 IEEE

Evaluation of Techniques to Reduce the Effects of Partial Shading on Photovoltaic Arrays

André Augusto Rodrigues, Paula dos Santos Vicente, Fernando Lessa Tofoli and Eduardo Moreira Vicente
Department of Electrical Engineering
Federal University of São João del-Rei - UFSJ
São João del-Rei, Brazil
rodriguesandre97@hotmail.com, paulasantos@ufsj.edu.br, fernandolessa@ufsj.edu.br, eduardomoreira@ufsj.edu.br

Abstract—**Shading is a harmful effect that decreases the power output of photovoltaic arrays. It can be observed that only a single shaded module is able to drastically reduce the energy production of the photovoltaic array. In addition, when shading reaches a larger number of modules, it becomes even more necessary to use techniques to reduce this effect, because in a series association the less efficient module limits the string current. This paper presents an analysis of six techniques that aim to reduce energy losses in photovoltaic panels operating under partial shading conditions. The techniques are evaluated through computer simulations, with the aid of PSIM® software. Therefore, a comparison is made between these techniques, defining which is the most suitable for each shading profile.**

Keywords—partial shading, reconfiguration, by-pass diodes, photovoltaic modules.

I. INTRODUCTION

In photovoltaic (PV) systems, to supply the power, voltage and current of most applications, series and/or parallel associations of PV modules are necessary. However, the associated modules may operate under different irradiance conditions due to the effect of shading, which may be partial or total. Shading, in general, implies large losses of energy, especially in the case of series connected modules with different irradiance profiles, in which the less efficient module will limit the current of the series string [1]. The main effects caused by this phenomenon are listed below:

- Hot spots appear on the panel, which can lead to PV cell losses. These effects appear due to the fact that shaded cells behave as loads, draining current from the system [2];

- Modification of the PV module operating point and, consequently, reduction of the panel output power [3].

In literature, different techniques that aim to reduce the effects of partial shading are mentioned [4] – [10]. The first method uses by-pass diodes [4], which are connected in antiparallel with the PV modules. This avoids the flow of reverse currents in the module, when they are shaded. However, this technique does not guarantee the full utilization of the available energy of the PV system. The use of by-pass diodes results in the occurrence of multiple peaks in the PxV (power versus voltage) curve of the PV array, which may modify its maximum power point (MPP) [11].

In [5], each module of the PV panel is connected to a DC/DC converter. This strategy aims to operate each module at its MPP, producing good results, because each module operates at its optimal point. As a disadvantage, there is the high cost of these systems, due to the use of several converters.

Usually, the connections between the modules of a PV array are kept fixed. However, some authors [6] – [10] suggest that photovoltaic modules may be rearranged in real time, in order to extract the maximum available energy of the PV panel.

In [6], the objective is to make the sum of the irradiances as similar as possible, for each string of the PV array in the TCT (Total Cross Tied) configuration. Reference [7], adopts the strategy of grouping shaded modules for SP (Series Parallel) arrays. On the other hand, both techniques present a complex switching structure to modify the PV array in its different configurations.

The authors of [7] propose that the modules classified as shaded should be eliminated from the PV system. With the exclusion, it is possible to have the parallel association of series strings with different voltage levels, limiting the energy production of the PV array. Reference [8] suggest the use of DC/DC converters in series with the strings of the PV array, to solve the problem of [7].

The irradiance equalization method, mainly used in TCT arrays, is applied in [9] for PV systems in the SP configuration. In this technique, the PV system must consist of a fixed and a dynamic matrix, which must be associated to the fixed matrix according to its shading level. Nevertheless, the PV modules of the dynamic matrix must be always illuminated.

In [10], a reconfiguration technique is proposed for SP arrays, where all modules can be rearranged, and the reconfiguration aims to group the shaded modules in the smallest possible number of strings. Despite its benefits, this technique does not promote energy gains when there are no possible reconfigurations.

These strategies present advantages and disadvantages for different operating conditions under the influence of partial shading. As a result, it is essential to evaluate them from the point of view of energy production, complexity and applicability in real PV systems, to define which solution is the most interesting for each shading profile. Therefore, the proposed work aims to compare the main aspects of six different methods to reduce the effects of partial shading, evaluating its power output through the analysis and simulation results in PSIM® software.

II. RECONFIGURATION TECHNIQUES FOR PARTIALLY SHADED PV MODULES

For the evaluation of reconfiguration techniques, a PV array in SP configuration have been developed. This panel is composed of six modules, arranged in an array of three rows and two columns (3x2 order). The base system, implemented in PSIM®, is shown in Fig. 1. The array dimension of Fig. 1 was chosen in order to allow a higher number of reconfiguration alternatives for the proposed techniques.

978-1-7281-4181-7/19 $31.00 © 2019 IEEE

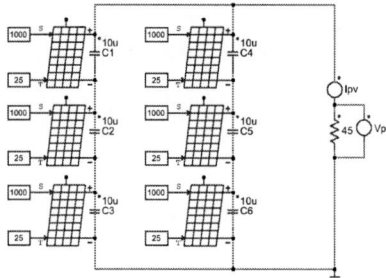

Fig. 1. PV panel implemented in PSIM®.

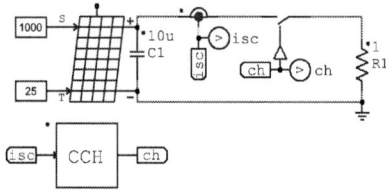

Fig. 2. Circuit used to determine the PV module condition.

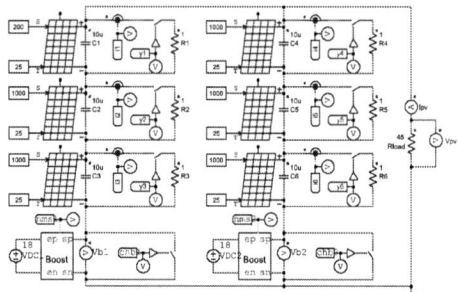

Fig. 3. DC-DC converter associated in series with each string of the PV array.

TABLE I. Parameters of Kyocera KS10-T module.

Parameter	Quantity
Number of Cells	36
Reference Irradiance	1000 W/m²
Reference Temperature	25 °C
Series Resistance	0.02 Ω
Shunt Resistance	1000 Ω
Short-circuit Current	0.63 A
Saturation Current	99.95 μA
Band gap Energy	1.12 eV
Temperature Coefficient	27 mV/°C

The capacitors associated in parallel with each module in Fig. 1 are necessary due to PSIM®' simulation errors when the modules operate with different levels of irradiance. For the simulated system to be as close as possible to a real system, the parameters of the modules were adjusted according to Kyocera's KS10-T module, as seen in TABLE I.

A. Technique 1: By-pass Diodes

If a photovoltaic module is partially shaded, hot spots may appear, due to the fact that a shaded PV cell behaves as a load, when reverse-biased, draining current from the PV array, which can lead to its rupture. To avoid such condition, by-pass diodes are connected in parallel with PV module output, which provides an alternative path for the current and limit the power dissipation in the shaded cells [2].

B. Technique 2: Removal of Shaded PV Modules

As mentioned previously, a shaded cell or module has its photogenerated current reduced due to incident irradiance decrease. Considering that modules connected in series have the string current limited by the less efficient module current, the exclusion of these modules from the PV array becomes a feasible solution [1].

To perform the removal of a PV module, it is necessary to evaluate if it is shaded or not. The circuit presented in Fig. 2 performs this task, based on the module short-circuit current. From TABLE I the short-circuit current in the Standard Test Condition (STC, irradiance at 1000 W/m² and temperature of 25 °C) is 0.63 A. Based on [7], it was specified that if the measured current is 20% less than the reference, i.e. less than 0.504 A, the module will be considered shaded and must be removed from the system.

In this technique, each module of the array of Fig. 1 has a circuit identical to that shown in Fig. 2 associated in parallel with its output. This circuit also performs the elimination function, simply by keeping the switch in the closed state after measuring the short-circuit current and the module is identified as shaded. The one-ohm resistor in series with the switch was used due to inconsistent values of the PV output current in PSIM® simulation, during short-circuit condition.

Note that in this type of system, removing shaded PV modules results in strings with distinct voltage levels associated in parallel. This will force the panel output voltage to be the same as the string with the smallest voltage, thus decreasing the output power of the panel. In addition, in the 3x2 configuration, three modules of the same string should never be eliminated, as this would short-circuit the load terminals permanently.

C. Technique 3: Removal of Shaded PV Modules with Voltage Equalization

In order to solve the problem of strings with distinct voltage levels associated in parallel, a DC/DC converter may be associated in series with each string of the PV module, as shown in Fig. 3. This converter will be used to maintain the voltage level of each string in the same value [8].

The procedure to remove the shaded modules is the same as the previous technique. The difference in this system is the presence of a Boost converter in series with each string (Fig. 3). There may be three distinct exclusion conditions for the modules of each string: only one module, two modules or three modules removed. This system relies on the use of an external voltage source in the correction of the string voltage of the PV array.

D. Technique 4: Irradiance Equalization on SP Arrays

In TCT arrays, it is common to apply reconfiguration techniques using a set of extra modules. The photovoltaic matrix is composed of a fixed and a dynamic part. A similar method is presented in [9], to series-parallel (SP) arrays, where it is proposed that auxiliary modules must be connected to the fixed part in the occurrence of shading. In this method, the modules were not removed from the array; instead, by-pass diodes were connected in parallel.

978-1-7281-4181-7/19 $31.00 © 2019 IEEE

For the circuit of Fig. 1, it was decided to create an auxiliary matrix composed of two extra modules, and the following reconfiguration strategy was adopted: the auxiliary modules may be connected with the two strings of the array, since it has shaded modules. If the level of shading of one string is greater than the other, the string that has the highest number of shaded modules will receive more illuminated modules. If the level of shading is the same, both strings will receive the same amount of extra modules.

E. Technique 5: Array Reconfiguration

When a photovoltaic module is shaded, it has its photogenerated current reduced, decreasing the current supplied to the load and, if it is in a serial connection, it will limit the string current, resulting in high power losses [1]. If there is more than one shaded module on the PV panel, they can be combined in a single serial connection, or in a limited number of serial connections, in order to reduce the system energy losses. Reference [10] addresses these issues, where all PV modules of the array can be reconfigured in order to minimize shading effects.

For the reconfiguration system implemented in PSIM®, the verification of the shaded modules is done by measuring the short-circuit current, with the aid of the circuit of Fig. 2. In the proposed system (3x2), the reconfiguration is viable only when two or three modules are shaded in distinct strings; in the other conditions, the reconfiguration will not improve the output power [10].

F. Technique 6: Array Reconfiguration and Removal of Shaded PV Modules with Voltage Equalization

When the array reconfiguration is not possible, Technique 5 is not capable to improve power output. To address this drawback, Technique 6 is proposed. This method combine the shaded modules and remove them from the array, using a Boost converter to maintain the voltage level. This technique is similar to Technique 3, but instead of using two DC-DC converters when two or three modules are shaded, using reconfiguration, only one converter is necessary, This method is an evolution of Techniques 3 and 5, combining the actions of these two techniques, and is similar to the strategy adopted in [8], however, the exclusion action is added.

III. SIMULATION RESULTS

The simulation results will be presented for the six proposed techniques, when the circuit of Fig. 1 operates in different shading profiles.

In order to highlight the power output of each technique, the evaluated circuits do not have Maximum Power Point Tracking (MPPT) systems. The load connected to the PV array is equal to 45 Ω. This value was specified based on the Maximum Power Point (MPP) of the KS10-T module, as well as the PV array dimension. It is worth mentioning that the operating temperature was kept constant at 25 °C for all simulation profiles.

To evaluate system performance for each technique, the power output of the 3x2 PV array, Fig. 1, is presented in Fig. 4 when all modules operate in the STC. Note that the maximum output power that can be generated is 60 W, since each module can generate a maximum of 10 W in this condition.

A. Case 01: One shaded PV module

Let us consider the shading profile shown in Fig. 5. Only one module is shaded, highlighted in red, with an irradiance level of 200 W/m², while all others are in STC.

Figure 6 shows the simulation results for the proposed techniques in the condition of Fig. 5. From the analysis of Fig. 6.a, it can be seen that the power output of the 3x2 array is severely reduced, near 60%, in the presence of only one shaded module, dropping from 60 W to 25 W. Figure 6.b, shows that the use of by-pass diodes (Technique 1) did not represent any benefit to the panel for this shading profile, keeping the output power at the same level as that shown in Fig. 6.a.

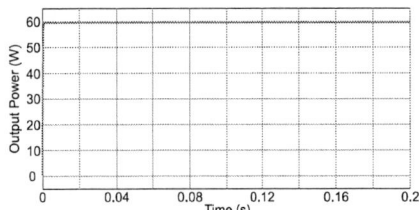

Fig. 4. Maximum output power of the 3x2 PV array in STC.

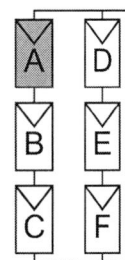

Fig. 5. Shading profile of Case 01.

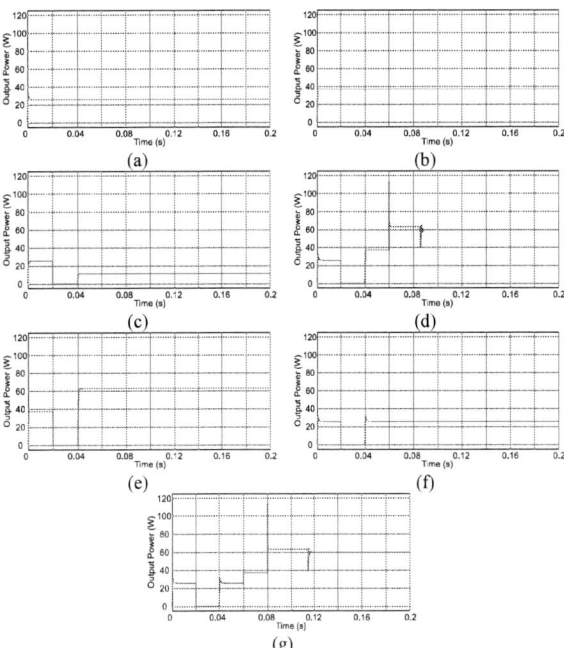

Fig. 6. Simulation results for Case 01: (a) base system; (b) by-pass diodes (Technique 1); (c) removal of shaded PV modules (Technique 2); (d) removal of shaded PV modules with voltage equalization (Technique 3); (e) irradiance equalization on SP arrays (Technique 4); (f) array reconfiguration (Technique 5); (g) Array reconfiguration and removal of shaded PV modules with voltage equalization (Technique 6).

The curve of Fig. 6.c refers to the Removal of Shaded PV Modules (Technique 2). Until 20 ms, the array is maintained in its default configuration, and the output power is the same shown in Fig. 6.a and Fig.6.b, near 25 W. During the interval between 20 and 40 ms, all modules have their outputs short-circuited, in order to evaluate its irradiance level and consequently, the system power drops to zero. The definition of the PV module condition, illuminated or shaded, is given in the same way described in Section II. After 40 ms, the system remove the shaded module from the array and reconnect the others. Note that for this shading profile, the removal of the shaded module causes a voltage unbalance between the two series strings. The array output power has increased from 25 W to almost 37 W, increasing the efficiency of the PV system.

In Fig. 6.d the output power for the Removal of Shaded PV Modules with Voltage Equalization (Technique 3) is shown. This curve is identical to that of Fig. 6c until 60 ms, which means that the operation of the circuits are the same up to this point. Subsequently, the system connects a Boost converter in series with the string that had an excluded module, in order to replace the shaded module. It can be seen that, after a brief oscillation, the output power become stable around 60 W, which is the nominal value.

The analysis of Fig. 6.e reveals that the circuit responsible for Irradiance Equalization on SP Arrays (Technique 4) caused a great increase of system power after the identification of the shaded modules, occurred after 40 ms. Until this time interval, the system operates as Technique 2, but there is no exclusion of the shaded modules due to the use of by-pass diodes, which perform a similar function. The power improvement was caused by the association of two auxiliary modules in the same string, because of the reconfiguration rules adopted. The auxiliary modules are always considered illuminated, but if shade comes to occur in these modules, there would be no increase on output power.

The power output of the Array Reconfiguration (Technique 5) is presented in Fig. 6.f. This method does not remove the shaded modules from the PV array. Between 20 and 40 ms, the outputs of the modules are short-circuited to define their operating state. After 40 ms, the system rearranges the shaded modules in the smallest number of strings. Since there is only one shaded module, the circuit returns to the previous configuration, producing the same output power.

The circuit responsible for the Array Reconfiguration and Removal of Shaded PV Modules with Voltage Equalization (Technique 6) has its simulation result shown in Fig. 6.g. This curve is the same as in Fig. 6.f up to 60 ms. From 60 to 80 ms, the circuit remove the shaded module. Then, the step-up converter is placed in series with the string that had the module excluded, obtaining an output power of 60 W.

B. Case 02: Two Shaded Modules

In the shading profiles of Fig. 7, two shaded modules, highlighted in red, are in the same string, Fig. 7.a, and in different columns of the PV array, Fig. 7.b. The irradiance of the shaded modules is 200 W/m², while the non-shaded modules are at 1000 W/m². The simulation results, from now on, will be presented in the form of tables, due to paper page restrictions and to simplify the results analysis. In

TABLE II, the simulation results for the shading profiles of Fig. 7 are shown.

For the shading profile of Fig. 7.a, with two shaded modules in the same string, the best methods were: Removal of Shaded PV Modules with Voltage Equalization (Technique 3), Irradiance Equalization on SP Arrays (Technique 4), Array Reconfiguration and Removal of Shaded PV Modules with Voltage Equalization (Technique 6). These methods presented similar results, maintaining the power output above 53 W, two times greater than the power output of the base system (26,01 W). The By-pass Diodes (Technique 1) and the Array Reconfiguration (Technique 5) did not produce any improvement in the power output of the base system. The method of Removal of Shaded PV Modules (Technique 2) had the lowest performance among the other techniques in this condition, around 10 W, due to the voltage unbalance caused by the removal of the shaded modules.

For the shading profile of Fig. 7.b, it can be seen that it causes large power losses in the PV array when compared to the nominal case, 60 W, and to the profile of Fig. 7.a, 26.01 W, producing only 3.14 W. The best techniques for this condition are the methods that uses a Boost converter for voltage equalization, in addition to the shaded modules removal. In both cases, the output power is above 55 W.

It is worth mentioning that the step-up converters that are part of the Techniques 3 and 6 were designed to deliver the same power level as the modules removed from the PV array, operating at STC.

C. Case 03: Three Shaded Modules

Two different types of shading, with three shaded modules, are presented in Fig. 8. The shaded modules are in the same string in Fig. 8.a, and in different strings in Fig 8.b. The output powers for this case are shown in TABLE III.

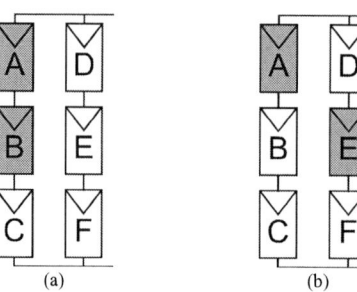

Fig. 7. Shading profiles of Case 02, two shaded modules: (a) in the same string, (b) in different strings.

TABLE II. Simulation results of Case 02.

Techniques	Output Power (W)	
	Shading profile of Fig. 7.a	Shading profile of Fig. 7.b
Base System	26,01 W	3,14 W
Technique 1	26,01 W	28,96 W
Technique 2	09,96 W	28,60 W
Technique 3	59,97 W	59,95 W
Technique 4	53,25 W	28,95 W
Technique 5	26,01 W	26,01 W
Technique 6	55,86 W	55,60 W

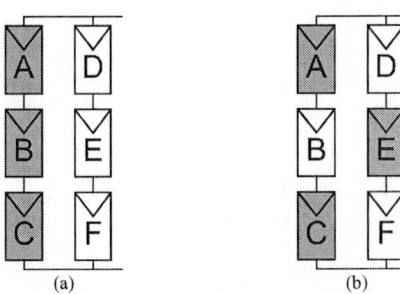

Fig. 8. Shading profiles of Case 03, three shaded modules: (a) in the same string, (b) in different strings.

TABLE III.   Simulation results of Case 03.

Techniques	Output Power (W)	
	Shading profile of Fig. 8.a	Shading profile of Fig. 8.b
Base System	25,50 W	2,88 W
Technique 1	25,50 W	23,66 W
Technique 2	26,36 W	11,29 W
Technique 3	54,04 W	53,45 W
Technique 4	37,20 W	37,20 W
Technique 5	25,50 W	2,88 W
Technique 6	53,24 W	51,30 W

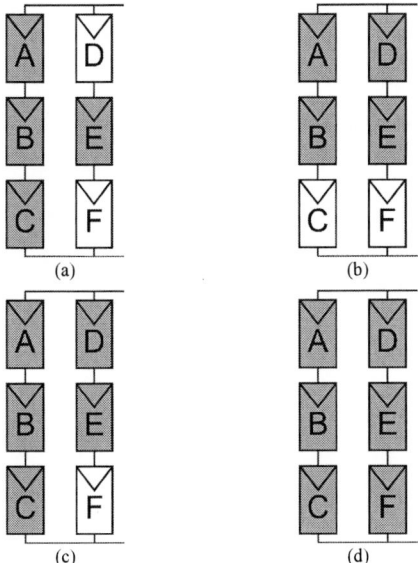

Fig. 9. Shading profiles of Case 04: (a) four shaded modules, different number in each string, (b) four shaded modules, same number in each string, (c) five shaded modules, (d) all shaded modules.

TABLE IV.   Simulation results of Case 04.

Techniques	Output Power (W)			
	Shading profile of Fig. 9.a	Shading profile of Fig. 9.b	Shading profile of Fig. 9.c	Shading profile of Fig. 9.d
Base System	2,86 W	2,83 W	2,85 W	2,85 W
Technique 1	9,56 W	9,66 W	9,10 W	2,86 W
Technique 2	9,37 W	9,24 W	8,80 W	2,86 W
Technique 3	53,34 W	54,01 W	52,61 W	50,69 W
Technique 4	23,64 W	23,18 W	17,83 W	8,40 W
Technique 5	2,86 W	2,86 W	2,85 W	2,85 W
Technique 6	55,31 W	55,21 W	52,12 W	51,43 W

For the type of shade of Fig. 8.a, with three modules shaded in the same string, Techniques 3 and 6 produce the highest power output, in spite of using the step-up converter with an external voltage source. The systems that perform only reconfiguration and exclusion, Techniques 2 and 5, respectively, are not capable to increase power output. The first one because the exclusion of the three modules would short-circuit the terminals of the load, and the second due to the fact that the PV array is already in the best possible configuration. For the shading profile of Fig. 8.b, the methods that presented the best performance, despite a large number of actions, were the techniques that remove the shaded modules and equalize the string voltage, Techniques 3 and 6.

D. Case 04: Four, Five and Six Shaded Modules.

In Figure 9, it is possible to observe the shading profiles characterized by four, five and six shaded modules (irradiance of 200 W/m²). From the data of TABLE IV, it is worth mentioning that Techniques 3 and 6 are the most viable alternatives for these profiles. Such behavior is related to the high level of shade in these profiles, requiring methods that exclude the shaded modules, which limit the currents of both strings.

Considering the results with four shaded modules in different positions, the performance of all techniques were practically the same, as expected, since this shading profile affects both strings, thus limiting their currents.

It is also interesting to note, from the data shown in TABLE IV, that when shading reaches five or six modules, only Techniques 3 and 6 are able to increase considerably the output power of the PV array, as a result of the external voltage sources. Techniques 1, 4 and 5 did not present good results, as they do not have mechanisms for the exclusion of the shaded modules. Besides, Technique 2, which has only the exclusion system, is capable to remove the shaded modules in the arrays of Fig. 9.a, Fig. 9.b and Fig. 9.c, but only in the strings with two shaded modules, since the load would be short-circuited if the three shaded modules were eliminated from the array. This method could not be used in shading profile of Fig. 9.d, as it would need a DC/DC converter to perform voltage equalization of the series string, allowing the exclusion of all modules identified as shaded.

IV. CONCLUSION

The evaluation of six techniques to enhance the power output of PV arrays, operating in partial shading conditions, have been presented. It is worth mentioning that the position of the shade in the PV array is important, since a single shaded module can considerably reduce the energy production of the PV array. Therefore, shaded modules grouped in the same string of the array will result in a higher power output than shaded modules in different strings.

One of the methods that are recommended to shading profiles with few shaded modules is the Array Reconfiguration (Technique 5), which performs the reconfiguration of the PV array, keeping its size by grouping the shaded modules into the fewest possible series strings.

For the shading profiles that were characterized by a large number of shaded modules, the most suitable methods were the ones that had the proposition of excluding the shaded modules from the array and equalize the string voltage through a Boost converter, Techniques 3 and 6. However, these techniques use an external voltage source to power the Boost converter, which is designed to replace the excluded modules.

The system with the addition of by-pass diodes in parallel with the modules (Technique 1) and the exclusion of the shaded modules (Technique 2) are interesting for profiles with a small number of shaded modules arranged in the same array. In these methods, the alternative path provided by the by-pass diodes, as well as the exclusion of the partially shaded modules of the array, results in a voltage unbalance between the strings.

The irradiance equalization method (Technique 4) was effective in partial shading conditions in which the shaded modules were connected in series. However, its potential is limited, since the auxiliary modules are considered always illuminated.

As shown through this work, several techniques can be employed to maximize the energy production of photovoltaic panels operating under partial shading conditions. Techniques 1, 2 and 5 have proved to be more efficient in systems with fewer shaded modules, while Techniques 3, 4 and 6 are suitable for systems with a higher number of shaded modules, depending on external voltage sources and auxiliary modules to raise the system power. It is also worth mentioning that the use of these systems are essential, since only one shaded module can drastically reduce the energy production of the whole PV array.

ACKNOLEDGEMENT

The authors would like to acknowledge INERGE, UFSJ and Brazilian research funding agencies CAPES, CNPq, and FAPEMIG for the support to this work.

REFERENCES

[1] P. dos Santos, E. M. Vicente and E. R. Ribeiro, "Relationship between the Shading Position and the Output Power of a Photovoltaic Panel," Brazilian Power Electronics Conference - COBEP, 2011, pp. 676–681.

[2] K. A. Kim and P.T. Krein, "Photovoltaic Hot Spot Analysis for Cells with Various Reverse-Bias Characteristics through Electrical and Thermal Simulation," IEEE 14th Workshop on Control and Modeling for Power Electronics - COMPEL, 2013, pp. 1-8.

[3] M. Alonso-Garcia, J. Ruiz, and F. Chenlo, "Experimental study of mismatch and shading effects in the I-V characteristic of a photovoltaic module," Solar Energy Materials Solar Cells, vol. 90, no. 3, pp. 329–340, Feb. 2006.

[4] T. Kudoh, F. Sugawara, and H. Tsushima, "A Study of Electrical Properties When Low-Loss, Self-Biased Channel Diodes are Used as Bypass Diodes for Photovoltaic Panels", IEEE 6th Global Conference on Consumer Electronics, 2017.

[5] Quan Li and Wolfs P., "A Review of the Single Phase Photovoltaic Module Integrated Converter Topologies with Three Different DC Link Configurations," IEEE Transaction on Power Electronics, 2008, pp. 1320-33.

[6] K.S. Parlak, "A New Reconfiguration Method for PV Array System," 39th Annual Conference of the IEEE Industrial Electronics Society – IECON, 2013, pp. 1478-1483.

[7] B. Patnaik et al., "Reconfiguration Strategy for Optimization of Solar Photovoltaic Array under Non-uniform Illumination Conditions," 37th IEEE Photovoltaic Specialists Conference - PVSC, 2011. pp. 1859 – 1864.

[8] M. Alahmad et al. "An Adaptive Utility Interactive Photovoltaic System Based on a Flexible Switch Matrix to Optimize Performance in Real-time," Solar Energy, vol.86, no. 3, pp. 951-963, March 2013.

[9] Hak-Gyun Jeong et al. "Development Of Active Module Considering The Shadow Influence of Photovoltaic System," IEEE Vehicle Power and Propulsion Conference - VPPC, 2012, pp. 1362 - 1365.

[10] P. Vicente, T. Pimenta, E. Ribeiro. "Photovoltaic Array Reconfiguration Strategy for Maximization of Energy Production", International Journal of Photoenergy, v. 2015, p. 1-11, 2015.

[11] T. Yetayew, T. Jyothsna, G. Kusuma. "Evaluation of Bay-Pass Diode and DMPPT under Partial Shade Condition of Photovoltaic Systems". 7th International Conference on Power Systems, 2017.

Analysis of Neutral-Point Voltage Balancing in Three-Ports Active Neutral-Point-Clamped Converter

Kaio Vinicius Vilerá, Cassiano Rech

Power Electronics and Control Research Group
Federal University of Santa Maria
Santa Maria-RS, Brazil
kaiovilera@gmail.com, rech.cassiano@gmail.com

Silvio Antonio Teston

Department of Physical Facilities
Federal University of Fronteira Sul
Chapecó-SC, Brazil
silvioteston@gmail.com

Abstract—This paper investigates the neutral-point (NP) voltage balancing problem from the perspective of the space vector modulation for different operating conditions of the recently proposed Three-ports Active Neutral-Point-Clamped (ANPC-3P) topology, which can integrate one energy storage system per inverter leg. Some scenarios to reduce the low-frequency voltage ripple in the NP are analyzed and the effects on the NP current due to the integration are reported. Theoretical propositions were validated by simulation results.

Keywords—Active neutral-point-clamped converter, energy storage system, integrated secondary dc port, space vector modulation, neutral point balancing.

I. INTRODUCTION

CLIMATE change is a topic of concern and the claim to reduce the use of fossil fuel is reinforced by many global efforts. The need of "going green" has encouraged the increasing penetration of renewable energy sources (RES) in power systems. However, one major disadvantage of using renewable energy resources such as solar and wind is that they are not able to dispatch power on demand since they do not have stored energy. Their stochastic input variation causes the power output to vary. Therefore, these RES are called non-dispatchable sources. Energy Storage Systems (ESS) can be integrated to non-dispatchable RES, partially decoupling energy generation from demand and offering many benefits such as load leveling, peak shaving and frequency control [1].

There are different ways of connecting the ESS to the power system. The two most common connections use a dedicated converter for the ESS. On the other hand, it is also possible to connect the ESS directly to the grid-connected inverter, reducing the devices part count since the inverter circuitry is also used for ESS integration. Just a few topologies have the appropriate degree of freedom to allow for this sort

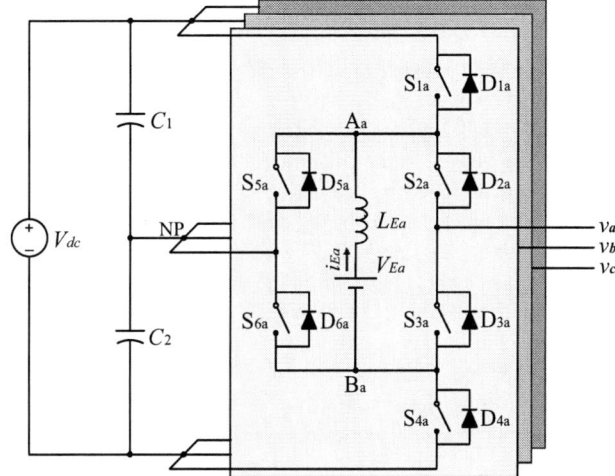

Fig. 1. Three-phase ANPC-3P inverter with one independent ESS per leg.

of connection. The Z-source inverter is an example [2]–[4]. According to [5], an ANPC based topology can be used to integrate one ESS per inverter leg with independent state of charge control as shown in Fig. 1. This topology has a common main dc port, and a secondary birectional dc port and an ac port per inverter leg, so that it can be called Three-Ports Active Neutral-Point Clamped (ANPC-3P) converter. The power flow of each secondary dc port is controlled independently by using the redundant switching states of this topology.

An important issue when dealing with neutral-point-clamped based topologies is the neutral-point (NP) voltage balancing problem. There have been many efforts to mitigate this problem, aiming the elimination of low-frequency oscillation in the NP and reduction of the dc-link capacitance [6]–[10]. Although this issue has been widely reported in literature, it has not yet been systematically analyzed for the recently proposed ANPC-3P topology. Therefore, this paper discusses the theoretical and practical limitations of NP balancing for

---

This study was financed in part by the Coordenação de Aperfeiçoamento de Pessoal de Nível Superior - Brasil (CAPES/PROEX) - Finance Code 001. The authors thank INCT-GD and CNPq (processes 465640/2014-1, 306317/2015-0 and 427987/2018-0), CAPES (process 23038.000776/2017-54) and FAPERGS (17/2551-0000517-1) for the financial support.

978-1-7281-4181-7/19 $31.00 © 2019 IEEE

TABLE I
SWITCHING STATES FOR THE ANPC-3P CONVERTER

State	Semiconductor Devices						Voltage v_x	Voltage v_{ABx}
	S_{1x}	S_{2x}	S_{3x}	S_{4x}	S_{5x}	S_{6x}		
P	1	1	0	0	0	1	$V_{dc}/2$	$V_{dc}/2$
0U4	0	1	1	0	1	0	0	0
0U3	0	1	0	0	1	1	0	0
0U1	0	1	0	1	1	0	0	$V_{dc}/2$
0UL	0	1	1	0	1	1	0	0
0L1	1	0	1	0	0	1	0	$V_{dc}/2$
0L3	0	0	1	0	1	1	0	0
0L4	0	1	1	0	0	1	0	0
N	0	0	1	1	1	0	$-V_{dc}/2$	$V_{dc}/2$

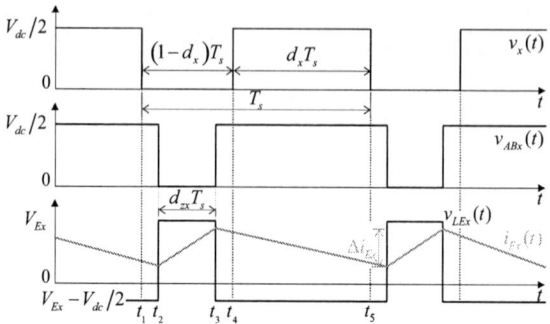

Fig. 2. Voltage and current waveforms for an ANPC-3P leg, detailing the modulation process of a secondary dc port.

different ANPC-3P operating conditions from the perspective of the space vector pulse width modulation (SV PWM). Based on this approach, the impact over the NP voltage balance due to the integration of ESS to the topology is investigated.

II. DESCRIPTION OF THE TOPOLOGY

A simplified schematic of the three-phase ANPC-3P converter is shown in Fig. 1, which is composed of six switches with anti-parallel diodes per leg, a secondary bidirectional dc port per leg and a common symmetric dc bus. The extra degree of freedom provided by the active clamping switches allows certain voltage levels to be applied across terminals A and B, permitting ESS integration into the referred terminals [5]. The ESS are represented as voltage sources in the secondary ports since many of them have this characteristic. Inductors are connected in series with them, allowing the control of the ESS charge and discharge.

Table I shows the switching states for this topology, where the subscript x is the phase index ($x = a, b, c$). Two voltage levels can be imposed on v_{ABx}, $+V_{dc}/2$ and 0, and three levels at the ac output, $+V_{dc}/2$, 0, and $-V_{dc}/2$. There are seven switching states that generate zero level at the ac port. Five of them also generate zero at the secondary dc port. However, only in state 0UL the inner switches are always turned on. Therefore it has a defined switching path, while for the other 4 states the conduction path depends upon the values of the currents. Thus, in this study, the switching states 0U3, 0U4, 0L3 and 0L4 are not considered, simplifying the analysis. On the other hand, the switching states 0U1 and 0L1 generate $V_{dc}/2$ at the ESS port while maintaining zero level at the ac output. When applying $+V_{dc}/2$ or $-V_{dc}/2$ at the output, there is no freedom to choose the v_{ABx} voltage level since there are no redundant switching states for P and N. Therefore, states 0UL, 0U1 and 0L1 are fundamental because it is by adjusting the time duration of these states that it is possible to control the time in which zero level is applied in v_{ABx} during a switching cycle, allowing the power flow control of the ESS.

A. Controllability of i_{Ex}

To control the current i_{Ex}, the average value of the voltage across the inductor within a switching period T_s must be zero. In the example depicted in Fig. 2, the voltage at the ac port

is 0 from t_1 to t_4 and $V_{dc}/2$ from t_4 to t_5. Since there is no redundancy for the P and N switching states, 0UL, 0U1 and 0L1 must be carefully selected during the zero level interval to achieve the volt-second balance in the filter inductor. Thus, the time interval, $d_{zx}T_s$, in which zero level is applied on v_{ABx} can be calculated from

$$d_{zx} = 1 - m_{E_x} \tag{1}$$

where

$$m_{E_x} = \frac{V_{Ex}}{V_{dc}/2} \tag{2}$$

B. AC output voltage space vector and their impact on NP voltage

The space vector diagram for the ANPC-3P is shown in Fig. 3. Since the SV diagram considers only the ac output voltages, it is the same as for the NPC and ANPC case. For example, the switching vector PN0 indicates that phase a is connected to the positive rail (P), phase b to the negative rail (N) and phase c to the NP of the dc link, using 0UL, 0U1 or 0L1, producing line voltages $v_{ab} = V_{dc}$ and $v_{bc} = v_{ca} = -V_{dc}/2$. Thus, the symbol "0" represents the switching states 0UL, 0L1 and 0U1, e.g., P0N can be extended into P(0UL)N, P(0U1)N and P(0L1)N.

For the ANPC topology, not all vectors affect the NP balance. For example, large vectors do not affect the NP voltage because there is no connection to the NP point. Zero vectors do not affect also. However, small and medium vectors affect because they connect one of the current phases to the NP. Therefore, in these cases, the NP potential is dependent on loading conditions. Small vectors that do not change the sign of the phase current connected to the NP are called positive small vectors, or otherwise negative small vectors. Since small vectors come in pairs and they generate the same ac output line-to-line voltages while affecting the NP voltage in an opposite way, their relative duration are usually chosen to compensate for the NP unbalance caused by medium vectors [9], [10]. Table II shows the phase current injected into the NP depending on the applied switching state, which is represented in this case by i_o.

978-1-7281-4181-7/19 $31.00 © 2019 IEEE

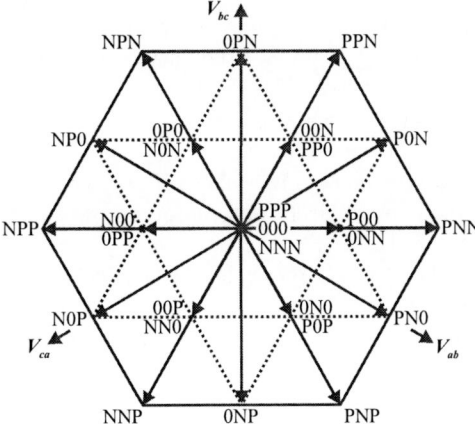

Fig. 3. AC output voltage space vector for the ANPC-3P. "0" represents the switching states 0UL, 0L1 and 0U1.

TABLE II
CONTRIBUTION OF THE AC OUTPUT CURRENTS TO i_{NP} FOR DIFFERENT SWITCHING VECTORS

Positive Small Vectors	i_o	Negative Small Vectors	i_o	Medium Vectors	i_o
0NN	i_a	P00	$-i_a$	P0N	i_b
PP0	i_c	00N	$-i_c$	0PN	i_a
N0N	i_b	0P0	$-i_b$	NP0	i_c
0PP	i_a	N00	$-i_a$	N0P	i_b
NN0	i_c	00P	$-i_c$	0NP	i_a
P0P	i_b	0N0	$-i_b$	PN0	i_c

For the ANPC-3P, the idea is the same regarding the impact on NP voltage due to the ac ports currents, but there are also other currents that might be connected to the NP, which are the ESS currents. Hence, depending on the ESS operation, i.e., whether they are charging, discharging or in floating mode, zero and large vectors might also affect the NP balance. Thus, the NP current for a certain switching state α is given by

$$i_\alpha = i_o + k_a i_{Ea} + k_b i_{Eb} + k_c i_{Ec} \tag{3}$$

where k_x are defined as ESS NP current factors, which depend on the switching state of each inverter leg.

Fig. 4 shows an example in which it is possible to evaluate the NP current composition for the switching state P(0UL)N considering the ESS as current sources. One can observe from Tables I and III that the NP current will be $i_{\mathrm{P(0UL)N}} = i_b + i_{Ea} - i_{Ec}$.

III. ANALYSIS OF THE NEUTRAL POINT CURRENT AND OPERATIONAL LIMITS

The SV diagram has a circular symmetry and, therefore, it is sufficient to present only part of the analysis, as shown in Fig. 5, with θ varying from $0°$ to $60°$, similarly to [9]. The shaded triangular region, shown in Fig. 5, has its vertices

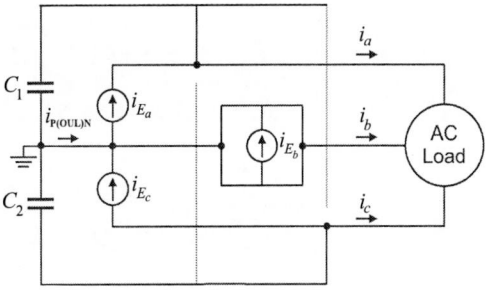

Fig. 4. NP current analysis example for the switching state P(0UL)N.

TABLE III
ESS NP CURRENT FACTORS ACCORDING TO THE SWITCHING STATE OF THE PHASES

Switching state	k_x
P	1
0U1	-1
0UL	0
0L1	1
N	-1

composed by the vectors V_L, V_M and V_{S0}. Consequently, voltage reference vector V_{REF} is synthesized as:

$$V_{\mathrm{REF}} = d_{S0} V_{S0} + d_M V_M + d_L V_L \tag{4}$$

where d_{S0} is the duty-cycle of V_{S0}, d_M is the duty-cycle of V_M, d_L is the duty-cycle of V_L and $d_{S0} + d_M + d_L = 1$. For this sector, the duty-cycles are given by [9] in terms of the modulation index, m, defined as the ratio between the amplitude of the ac port phase voltages and the maximum amplitude of the ac port undistorted sinusoidal phase voltage:

$$d_{S0} = 2 - m\left(\sqrt{3}\cos(\theta) + \sin(\theta)\right) \tag{5}$$

$$d_M = 2m\sin(\theta) \tag{6}$$

$$d_L = -1 + m\left(\sqrt{3}\cos(\theta) - \sin(\theta)\right) \tag{7}$$

Due to the integration of one ESS per inverter leg, there is an additional challenge which is to adjust the relative duration of the redundant switching states to control the ESS currents, i_{Ea}, i_{Eb} and i_{Ec}. The relative duration of positive and negative

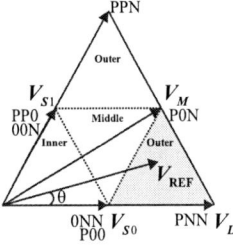

Fig. 5. V_{REF} located in the shaded outer triangle.

978-1-7281-4181-7/19 $31.00 © 2019 IEEE

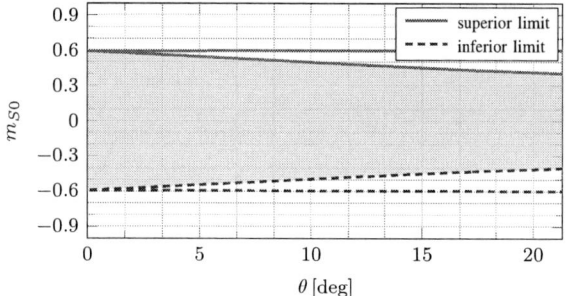

Fig. 6. Limits of m_{S0} for $m = 0.8$ and $m_E = 0.875$.

small vectors, d_{0NN} and d_{P00}, was defined in [9] by the current modulation index $m_{S0} \in [-1, 1]$, such that

$$d_{S0} = d_{0NN} + d_{P00} \tag{8}$$

$$d_{0NN} = (1 + m_{S0}) \frac{d_{S0}}{2} \tag{9}$$

$$d_{P00} = (1 - m_{S0}) \frac{d_{S0}}{2} \tag{10}$$

In addition, a time interval defined by d_{zx}, corresponding to the application interval of 0UL on v_{ABx}, is required to control the ESS currents. Thus, the switching states must be adequately selected to control the ESS currents. In this outer triangle, all the following conditions must be met

$$0 < d_{za} \leq d_{0NN} \tag{11}$$

$$0 < d_{zb} \leq d_{P00} + d_M \tag{12}$$

$$0 < d_{zc} \leq d_{P00} \tag{13}$$

Considering a symmetric topology ($m_{E_a} = m_{E_b} = m_{E_c} = m_E$), so that $d_{za} = d_{zb} = d_{zc} = d_z$, and substituting (2), (9), (10) in (11)-(13), it is possible to determine the limits of m_{S0} as follows

$$Y_{d_z} - 1 \leq m_{S0} \leq 1 - Y_{d_z} \tag{14}$$

where Y_{d_z} is the term related to d_z such that

$$Y_{d_z} = 2 \frac{(1 - m_E)}{d_{S0}} \tag{15}$$

Consequently, the superior and inferior limits of m_{S0} depend on m_E, m and θ. An example is shown in Fig. 6 for the case in which $m = 0.8$ and $m_E = 0.875$. The range of θ depends on m. The grayed area within the limits represents the possible solutions for m_{S0}, which can be used for NP voltage balance.

Since during a sampling period different vectors are used, the resulting NP current is given by the sum of the relative duration in which each switching state is applied times their corresponding current, which are shown in Tables II and III. So, the average NP current for this triangle is given by

$$i_{NP} = d_M i_b + m_{S0} d_{S0} i_a + d_E i_E \tag{16}$$

where

$$d_E = \sum_{x=a,b,c} (d_{Px} + d_{0L1x} - d_{Nx} - d_{0U1x}) \tag{17}$$

d_E is a factor that shows how the combination of the states affects the NP current due to the ESS currents and it can assume positive or negative values. It considers the current factors shown in Table III and the overall duration of each state that affects the NP current.

The middle triangle shown in Fig. 5 is composed of two small vectors and one medium vector. This region is more favorable to control the ESS currents because all vectors can impose 0UL in at least one leg and since there are two small vectors, there are more redundant switching states. Following the same idea as presented before, once the vectors that generate the reference vector are known, the corresponding duty cycles can be calculated [9]. To control all ESS currents, the switching states must be selected considering the following conditions

$$0 < d_{za} \leq d_{0NN} + d_{00N} \tag{18}$$

$$0 < d_{zb} \leq d_{P00} + d_{00N} + d_{P0N} \tag{19}$$

$$0 < d_{zc} \leq d_{P00} + d_{PP0} \tag{20}$$

The relative duration of the positive and negative vectors that compose V_{S0} and V_{S1} are determined respectively by m_{S0} and m_{S1}. Due to the integration of the ESS, the above conditions also impose constraints on m_{S0} and m_{S1}. Considering the same simplifying hypothesis as before, the limits are given by

$$Y_{d_z} - \left(\frac{1 - m_{S1}}{1 + m_{S1}}\right) Y_{d_{S1}} - 1 \leq m_{S0} \leq 1 + Y_{d_{S1}} - Y_{d_z} \tag{21}$$

where $Y_{d_{S1}}$ is the term related to d_{S1} such that

$$Y_{d_{S1}} = \frac{d_{S1}}{d_{S0}}(1 + m_{S1}) \tag{22}$$

In this case, the choice of m_{S1} affects the limits of m_{S0} and vice-versa, therefore, for each θ there is an area of permitted values for m_{S0} and m_{S1} in which it is possible to control the ESS currents, as shown in Fig. 7, with $m = 0.8$ and $m_E = 0.875$. The area between the two limiting surfaces represents the possible values for m_{S0} and m_{S1}. The average NP current for the referred middle triangle is given by

$$i_{NP} = d_M i_b + m_{S0} d_{S0} i_a + m_{S1} d_{S1} i_c + d_E i_E \tag{23}$$

When the reference vector is within the inner triangle shown in Fig. 5, the output voltage is synthesized only by small and zero vectors. Therefore, this region is the most favorable to control the ESS currents and to balance the NP voltage. The operation conditions for this region are determined by

$$0 < d_{za} \leq d_{0NN} + d_{00N} + d_{000} \tag{24}$$

$$0 < d_{zb} \leq d_{P00} + d_{00N} + d_{000} \tag{25}$$

$$0 < d_{zc} \leq d_{P00} + d_{PP0} + d_{000} \tag{26}$$

978-1-7281-4181-7/19 $31.00 © 2019 IEEE

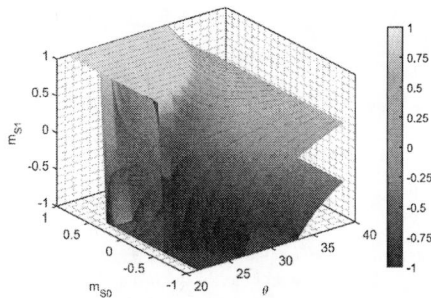

Fig. 7. m_{S0} and m_{S1} limits are represented by the surfaces for $m = 0.8$ and $m_E = 0.875$.

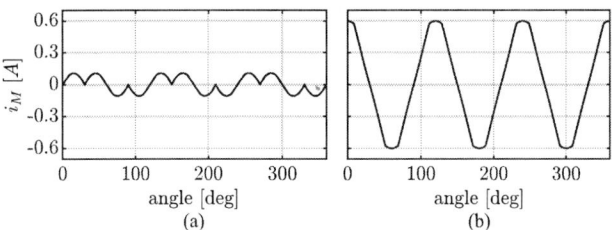

Fig. 8. NP current produced by medium vectors for ac output port currents of amplitude 1 and $m = 0.8$. (a) DPF = 1; (b) DPF = 0.

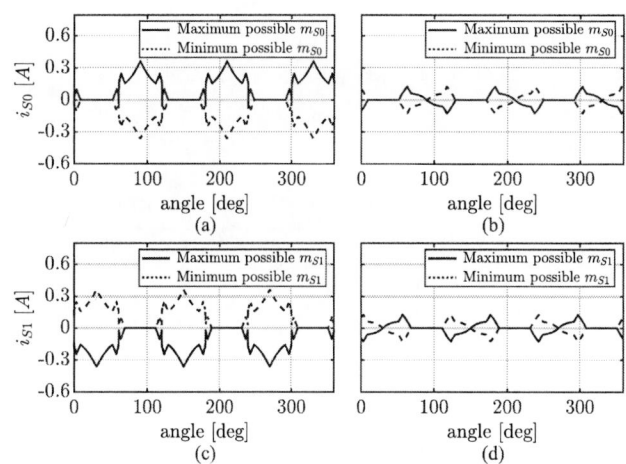

Fig. 9. Current limits that can be generated by small vectors S_0 and S_1 considering the current modulation indexes constraints and ac output port currents of amplitude one, $m = 0.8$ and $m_E = 0.875$. In (a) and (c) DPF = 1; (b) and (d) DPF = 0.

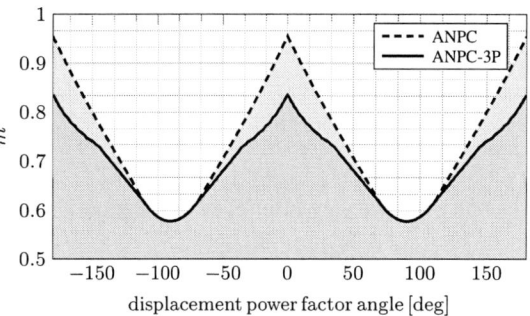

Fig. 10. Modulation index limit in which the low-frequency current ripple in the NP can be suppressed. The ANPC-3P is operating in floating mode with $m_E = 0.875$.

In this case, m_0 determines the zero vectors relative duration such that $d_{000} = (1 + m_0)d_0/2$ and $d_{NNN} + d_{PPP} = (1 - m_0)d_0/2$. Thus, the following constraints must be considered

$$Y_{d_z} - \left(\frac{1 - m_{S1}}{1 + m_{S1}}\right)Y_{d_{S1}} - Y_{d_0} - 1 \leq m_{S0} \leq 1 + Y_{d_{S1}} + Y_{d_0} - Y_{d_z} \tag{27}$$

where Y_{d_0} is the term related to d_0 such that

$$Y_{d_0} = \frac{d_0}{d_{S0}}(m_0 + 1) \tag{28}$$

It is similar to the middle triangle except that this region also has the term (28), dependent on d_0, which increases its limits. The average NP current for this triangle is given by

$$i_{NP} = m_{S0}d_{S0}i_a + m_{S1}d_{S1}i_c + d_E i_E \tag{29}$$

IV. LOW-FREQUENCY CURRENT RIPPLE IN THE NP

The NP current of each sampling instant can be determined by a mapping between the duty cycles of the switching states and the NP current that they produce in terms of the ac port currents and the secondary dc port currents, similar to [9]. The need of controlling the ESS currents impose constraints on all the current modulation indexes as shown in (14), (21) and (27), hindering the control authority over the NP current modulation. The average NP current has a term dependent on the ESS currents as shown in (16), (23) and (29). If the ESS are in floating mode, i_{Ex} equals to zero, these terms can not be used to reduce the LF ripple current flow into the NP, hindering the capacity of suppressing the LF voltage ripple produced. Thus, the ESS currents might be beneficial to balancing the NP voltage since they add extra terms which might be significant depending on ESS currents' magnitude.

Therefore, this section is focused on analyzing the ESS operating in floating mode. Similarly to the ANPC case, the NP current consists only of non-controllable current due to the application of medium vectors, and controllable currents due to the application of small vectors. The non-controllable terms of the NP current produced by medium vectors is shown in Fig. 8 for different displacement power factors (DPF), being higher for low DPF. The controllable terms of the NP current, shown in Fig. 9, are produced by S_1 and S_0 respectively and they can take any value between the limits depending on the control inputs m_{S0} and m_{S1}. The limits of m_{S0} and m_{S1} are reduced due to the ESS integration. As shown in equation (15), a lower m_E leads to more restrictive limits since Y_{d_z} becomes more significant.

Fig. 10 presents two modulation regions in which it is possible to suppress the low-frequency ripple. The first, limited by a dashed line, is of an ANPC inverter, and the second, is of the ANPC-3P operating in floating mode with $m_E = 0.875$. Since the first case has no ESS integration, there is no concern about the control of the ESS currents and the limits of the

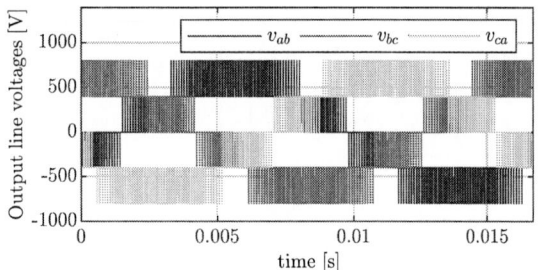

Fig. 11. AC output voltages v_{ab}, v_{bc} and v_{ca} with $m = 0.8$, $m_E = 0.875$, $V_{dc} = 800$V and switching frequency $f_{sw} = 10.26$kHz.

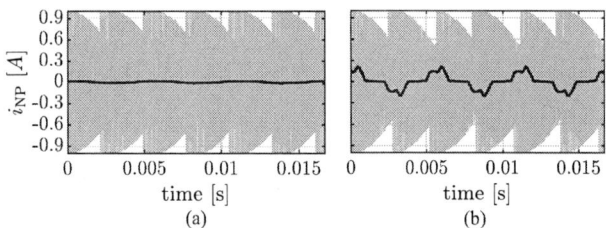

Fig. 12. NP currents for $m_E = 0.875$, $f_{sw} = 10.26$kHz, $\phi = 30°$ and floating mode operation. The black curves represent the moving average of i_{NP}. (a) $m = 0.7$; (b) $m = 0.8$.

NP current modulation indexes are broader. These limits are found by comparing total controllable and non-controllable NP currents defined in every θ for each modulation index.

V. SIMULATION RESULTS

The simulations results were obtained by using the software PSIM and the definition of the switching states per switching period to reach the objectives of synthesizing the ac output voltages, controlling the ESS currents and minimizing the low-frequency voltage ripple in the NP was done by using MATLAB. Fig. 11 shows the ac output voltages v_{ab}, v_{bc} and v_{ca} for $m = 0.8$.

Fig. 12 shows the moving average of i_{NP} for different modulation indexes. As shown in Fig. 10 the maximum modulation index for the displacement power factor angle $\phi = 30°$ in which the low-frequency current ripple in the NP can be suppressed is approximately 0.74. Thus, the moving average of i_{NP} is approximately zero in (a), differently from (b). This result can also be understood by analyzing Fig. 13, in which the controllable term of both small vector are summed and compared with the non-controllable term due to the medium vector. For $m = 0.7$ the non-controllable current can be completely canceled. However, it is not true for $m = 0.8$ since the non-controllable current exceeds the controllable limits in some points. In fact, when the modulation index becomes bigger, the relative duration of the vectors are changed such that the non-controllable term becomes more significant and the controllable terms more restrictive.

VI. CONCLUSION

This paper investigated the NP voltage balancing from the perspective of the SVM method for the recently presented

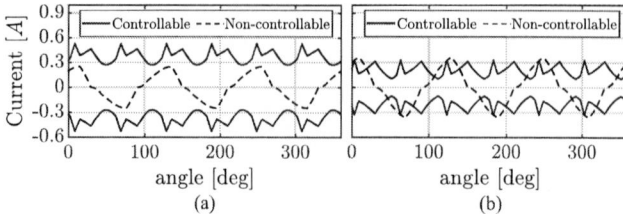

Fig. 13. Maximum and minimum controllable current that can be generated by vectors S_0 and S_1 and non-controllable current due to medium vector for $\phi = 30°$ and $m_E = 0.875$. (a) $m = 0.7$; (b) $m = 0.8$.

ANPC-3P VSI, showing that the ESS integration imposed certain constraints on the freedom to apply some switching states, which reduced the authority over the NP current modulation. When the ESS was operating in floating mode the ANPC-3P VSI required a lower amplitude modulation index to be able to suppress the low-frequency voltage ripple in the NP when compared to the ANPC VSI. Other scenarios in which the ESS is charging or discharging will be deeper discussed in future work. As reported in this paper, the ESS currents also have an impact on the NP current, which might be significant depending on ESS currents' magnitude.

REFERENCES

[1] S. Vazquez, S. M. Lukic, E. Galvan, L. G. Franquelo, and J. M. Carrasco, "Energy storage systems for transport and grid applications," *IEEE Transactions on Industrial Electronics*, vol. 57, no. 12, pp. 3881–3895, 2010.

[2] J. G. Cintron-Rivera, Y. Li, S. Jiang, and F. Z. Peng, "Quasi-Z-Source inverter with energy storage for photovoltaic power generation systems," *IEEE Applied Power Electronics Conference and Exposition (APEC)*, pp. 401–406, 2011.

[3] B. Ge, H. Abu-Rub, F. Z. Peng, Q. Lei, A. T. De Almeida, F. J. Ferreira, D. Sun, and Y. Liu, "An energy-stored quasi-Z-source inverter for application to photovoltaic power system," *IEEE Transactions on Industrial Electronics*, vol. 60, no. 10, pp. 4468–4481, 2013.

[4] S. Hu, Z. Liang, D. Fan, and X. He, "Hybrid ultracapacitor-battery energy storage system based on quasi-Z-source topology and enhanced frequency dividing coordinated control for EV," *IEEE Transactions on Power Electronics*, vol. 31, no. 11, pp. 7598–7610, 2016.

[5] S. A. Teston, M. Mezaroba, and C. Rech, "ANPC Inverter with Integrated Secondary Bidirectional DC Port for ESS Connection," *IEEE Transactions on Industry Applications*, vol. XX, no. X, pp. 1–1, 2019.

[6] S. Busquets-Monge, J. Bordonau, D. Boroyevich, and S. Somavilla, "The nearest three virtual space vector PWM - A modulation for the comprehensive neutral-point balancing in the three-level NPC inverter," *IEEE Power Electronics Letters*, vol. 2, no. 1, pp. 11–15, 2004.

[7] X. Wang, T. Huo, X. Nian, and X. Chu, "Research on balance of neutral-point potential in three level neutral point clamped inverter," *IECON 2017 - 43rd Annual Conference of the IEEE Industrial Electronics Society*, pp. 8329–8333, 2017.

[8] X. Zhou and S. Lu, "A simple zero-sequence voltage injection method to balance the neutral-point potential for Three-level NPC inverters," *IEEE Applied Power Electronics Conference and Exposition (APEC)*, pp. 2471–2475, 2018.

[9] N. Celanovic and D. Boroyevich, "A comprehensive study of neutral-point voltage balancing problem in three-level neutral-point-clamped voltage source PWM inverters," *IEEE Transactions on Power Electronics*, vol. 15, no. 2, pp. 242–249, 2000.

[10] D. H. Lee, S. R. Lee, and F. C. Lee, "An analysis of midpoint balance for the neutral-point-clamped three-level VSI," *PESC Record - IEEE Annual Power Electronics Specialists Conference*, vol. 1, pp. 193–199, 1998.

978-1-7281-4181-7/19 $31.00 © 2019 IEEE

Direct Torque Control Scheme for a Nine-Phase Induction Motor With Reduced Current Harmonic

Gilielson F. da Paz[1], Isaac S. de Freitas[2], Victor F. M. B. Melo[2], Alexandre G. F. da Silva[1]

[1]Post-Graduate Program in Electrical Engineering - PPgEE - UFPB

[2]Federal University of Paraíba, Paraíba, Brazil

Email: gilielson@cear.ufpb.br, isaacfreitas@cear.ufpb.br, victor@cear.ufpb.br, alexandre.silva@cear.ufpb.br

Abstract—This paper discusses a modified Direct Torque Control (DTC) method in order to reduce phase current harmonic distortion in a nine-phase asymmetrical induction motor drive system. Since the classical DTC is a direct adaptation from the three-phase case, it does not take into account the presence of the non-producing torque current components obtained from the Space Vector Decomposition (SVD) technique. Thus, in classical DTC these non-producing torque components assume high amplitudes, causing distortion in phase currents. In this way, it is shown that modified DTC minimizes the undesired components, reducing significantly the distortion in phase currents while maintaining the fast dynamic response of classical DTC.

Index Terms—Asymmetrical nine-phase machines, direct torque control (DTC), harmonic distortion reduction.

Nomenclature

- R_s, R_r - Resistances of stator and rotor, respectively.
- L_m - Mutual inductance.
- L_{ls}, L_{lr} - Stator and rotor leakage inductances, respectively.
- L_s, L_r - Self-inductances of stator and rotor, respectively.
- $V_{s\alpha\beta}$, V_{xy}, V_{uv} - Stator voltages in the $\alpha - \beta$, $x - y$ and $u - v$ planes, respectively.
- $\lambda_{s\alpha\beta}$, λ_{sxy}, λ_{suv} - Stator fluxes in the $\alpha - \beta$, $x - y$ and $u - v$ planes, respectively.
- $i_{s\alpha\beta}$, i_{xy}, i_{uv} - Stator currents in the $\alpha - \beta$, $x - y$ and $u - v$ planes, respectively.
- $\lambda_{r\alpha\beta}$, $i_{r\alpha\beta}$ - Rotor flux and current in $\alpha - \beta$ plane.
- p - Number of pole pairs.
- T_e - Electromagnetic torque.
- ω_r - Rotor angular speed.

I. Introduction

The good features of multiphase machines and drives have been recognized over the years. Some of the reasons to use these systems are better fault tolerance, reduction of per-switch current rating, improved noise characteristics and decrease of the torque ripple without increasing the number of poles [1], [2].

Multiphase machines are highly used in variable-speed applications [2], [3]. They are employed in ship propulsion, locomotive traction, hybrid electric vehicles, battery chargers, elevators, "more-electric" aircraft and high-power industrial applications. In this context, nine-phase machines recently started to have some applications and drive systems investigated [4]–[8]. Fault tolerant strategies for nine-phase permanent magnet machines were developed in [9], [10].

In terms of machine control, the most common type of control used in drive systems is the Field Oriented Control (FOC), which usually employs Proportional-Integral (PI) current controllers and Pulse Width Modulation (PWM) strategies. However, particular attention has been given to the Direct Torque Control (DTC) technique in the last years due to its simplicity and fast response to reference changes, also avoiding the use of current controllers and modulators [11], [12].

In DTC for three-phase drives, voltage vectors based on converter switching states are mapped in an $\alpha - \beta$ plane by means of Clarke's Transformation. On the other hand, by measuring stator currents and voltages, it is possible to estimate electromagnetic torque produced by the machine and amplitude and position of the rotating stator flux in the stationary reference frame in the $\alpha - \beta$ plane also by means of Clarke's Transformation. Band hysteresis limits for torque and flux are defined and depending on the flux position and on the estimated values of torque and flux, a switching voltage vector is selected in order to keep torque and flux within the pre-defined band limits. So, it is usual to elaborate a table with the switching voltage vector that must be applied for each flux position and for each value of torque and flux. For this reason, this control technique is also called Look-up Table DTC (LUT-DTC) or Switching-Table DTC (ST-DTC) [11], [12].

Considering multiphase machines, it is known that the original phase variables (fluxes, voltages and currents) may be transformed into mutually orthogonal domains (or planes) according to the Space Vector Decomposition (SVD) technique. In particular for nine-phase systems, the original nine-phase variables are transformed into four mutually orthogonal two-dimensional domains and one single-dimensional domain by means of Clarke's Transformation.

In this context, DTC has been employed especially for five-[13]–[15] and six-phase [16]–[18] drive systems. In those papers, it is shown that when DTC is employed only considering the $\alpha - \beta$ plane, high distortion appears in the original phase currents. It happens because the selected switching vectors makes the current components in the planes other than $\alpha - \beta$ assume high values. It is known that the current components that produce torque are the $\alpha - \beta$ components in a sinusoidal distributed winding. The current components present in the other planes do not produce torque and only generate losses and harmonic distortion in the original machine currents. In

978-1-7281-4181-7/19 $31.00 © 2019 IEEE

this way, it is interesting to keep these components with null value. However, choosing the switching vectors only taking into account the $\alpha-\beta$ plane does not guarantee that the average current value within a switching period is null in the other planes.

In this way, [18] and [16] discuss a DTC technique for a six-phase permanent magnet motor and for a six-phase induction motor, respectively, that mitigate the presence of currents in the second plane, named $x-y$ plane. In those papers, a performance comparison in terms of torque ripple and current harmonic distortion was performed, comparing the scenario in which the voltage vector selection took into account the $\alpha-\beta$ plane only (this technique is usually called classical DTC) versus the scenario in which the vector selection took into account the minimization of current in the $x-y$ plane (named by the papers' authors as Modified DTC). It was proved that, besides keeping flux and torque within the band limits, Modified DTC also resulted in average null voltage in the $x-y$ plane. This comparison showed that the machine currents presented significant reduction in harmonic distortion in the second scenario. For convenience, [16] named the first scenario as DTC1 and the second scenario as DTC2.

In this context, and following the same nomenclature as [16], this paper discusses DTC1 and DTC2 for an asymmetrical (angle between the three three-phase groups that composed the machine winding is of 20°) nine-phase drive system, which have not been discussed in the literature. As done in [16], [18], the objective is performing a comparative study between the two in order to validate the expected reduction in the current harmonic distortion in DTC2, as observed for six-phase drive systems.

II. NINE-PHASE MOTOR AND INVERTER MODEL

In this paper, a nine-phase induction machine with sinusoidally distributed windings and with three isolated neutrals is studied. Also, the conventional nine-leg converter topology is used, as illustrated in Fig. 1. The machine windings consist of three three-phase groups, here named group 1, composed of phases 1, 4 and 7, group 2, composed of phase 2, 5 and 8, and group 3, composed of phases 3, 6 and 9. Using SVD, the obtained planes are $\alpha-\beta$, $x-y$, $u-v$, $z-w$ and o. Components $z-w$ and o are naturally null due to the neutrals configuration, remaining $\alpha-\beta$, $x-y$ and $u-v$. The Clarke's power-invariant transformation matrix $[T_9]$ is given by:

$$\sqrt{\frac{2}{9}}\begin{bmatrix} 1 & -\frac{1}{2} & -\frac{1}{2} & c(\theta) & -c(2\theta) & -c(4\theta) & c(2\theta) & -c(\theta) & c(4\theta) \\ 0 & \frac{\sqrt{3}}{2} & -\frac{\sqrt{3}}{2} & s(\theta) & s(2\theta) & -s(4\theta) & s(2\theta) & s(\theta) & -s(4\theta) \\ 1 & -\frac{1}{2} & -\frac{1}{2} & -c(4\theta) & c(\theta) & -c(2\theta) & -c(\theta) & c(4\theta) & c(2\theta) \\ 0 & -\frac{\sqrt{3}}{2} & \frac{\sqrt{3}}{2} & s(4\theta) & -s(\theta) & -s(2\theta) & -s(\theta) & s(4\theta) & -s(2\theta) \\ 1 & -\frac{1}{2} & -\frac{1}{2} & -c(2\theta) & -c(4\theta) & c(\theta) & c(4\theta) & c(2\theta) & -c(\theta) \\ 0 & \frac{\sqrt{3}}{2} & -\frac{\sqrt{3}}{2} & s(2\theta) & -s(4\theta) & s(\theta) & -s(4\theta) & s(2\theta) & s(\theta) \\ 1 & 1 & 1 & \frac{1}{2} & \frac{1}{2} & \frac{1}{2} & -\frac{1}{2} & -\frac{1}{2} & -\frac{1}{2} \\ 0 & 0 & 0 & \frac{\sqrt{3}}{2} & \frac{\sqrt{3}}{2} & \frac{\sqrt{3}}{2} & \frac{\sqrt{3}}{2} & \frac{\sqrt{3}}{2} & \frac{\sqrt{3}}{2} \\ \frac{\sqrt{2}}{2} & \frac{\sqrt{2}}{2} & \frac{\sqrt{2}}{2} & -\frac{\sqrt{2}}{2} & -\frac{\sqrt{2}}{2} & -\frac{\sqrt{2}}{2} & \frac{\sqrt{2}}{2} & \frac{\sqrt{2}}{2} & \frac{\sqrt{2}}{2} \end{bmatrix} \tag{1}$$

where s and c denote sine and cosine, respectively, and $\theta = \frac{\pi}{9}$ is the phase shift angle. Using $[T_9]$ in the machine eletrical

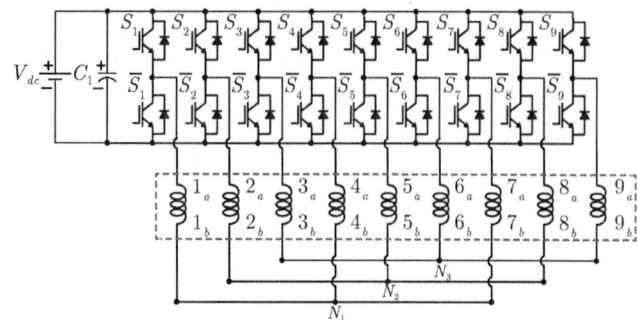

Fig. 1. Studied configuration.

variables equations of the stator and rotor, the machine model in the stationary reference frame is given by:

$$V_{s\alpha\beta} = R_s i_{s\alpha\beta} + \frac{d\lambda_{s\alpha\beta}}{dt} \tag{2}$$

$$V_{sxy} = R_s i_{sxy} + \frac{d\lambda_{sxy}}{dt} \tag{3}$$

$$V_{suv} = R_s i_{suv} + \frac{d\lambda_{suv}}{dt} \tag{4}$$

$$0 = R_r i_{r\alpha} + \frac{d\lambda_{r\alpha}}{dt} - \omega_r \lambda_{r\beta} \tag{5}$$

$$0 = R_r i_{r\beta} + \frac{d\lambda_{r\beta}}{dt} + \omega_r \lambda_{r\alpha} \tag{6}$$

$$\lambda_{s\alpha\beta} = L_s i_{s\alpha\beta} + L_m i_{r\alpha\beta} \tag{7}$$

$$\lambda_{sxy} = L_{ls} i_{sxy} \tag{8}$$

$$\lambda_{suv} = L_{ls} i_{suv} \tag{9}$$

$$\lambda_{r\alpha\beta} = L_m i_{s\alpha\beta} + L_r i_{r\alpha\beta} \tag{10}$$

$$T_e = pL_m(i_{s\beta} i_{r\alpha} - i_{s\alpha} i_{r\beta}) \tag{11}$$

where $L_s = L_{ls} + L_m$. The voltage source inverter of 1 consists of a common DC source and 18 semiconductor switches. Each inverter leg is formed by two switches S_k and $\bar{S}_k$, where $k = (1, 2, 3, 4, 5, 6, 7, 8, 9)$ which can admit two different switching states, being $S_k = 0$ when the switch is open and $S_k = 1$ while conducting, where $\bar{S}_k$ indicates that the switch has the command signal complementary to S_k. For a nine-leg inverter, there are $2^9 = 512$ combinations of possible switching states. Each switching state is defined by a binary sequence formed by the switches $(S_1, ...S_9)$ that generates a corresponding voltage vector V_j, where j is the jth switching combination. Using (1), voltage vectors can be mapped on planes $\alpha-\beta$, $x-y$, $u-v$, $z-w$, and o. The voltage vectors in the $\alpha-\beta$ plane are illustrated in Fig. 2(a).

III. DIRECT TORQUE CONTROL STRATEGY

The classical DTC utilizes only the vectors with the highest amplitudes in the $\alpha-\beta$ plane. Fig. 2(b) shows these vectors. The plane is divided in 18 sectors in which the stator flux vector $\lambda_{s\alpha\beta}$ may be located. Hysteresis band limits (Hf for flux and Ht for torque) determine if flux and torque values must increase or decrease. Two variables, $d\lambda_s$ and dT_e, assume values -1, 0 or 1 such that:

978-1-7281-4181-7/19 $31.00 © 2019 IEEE

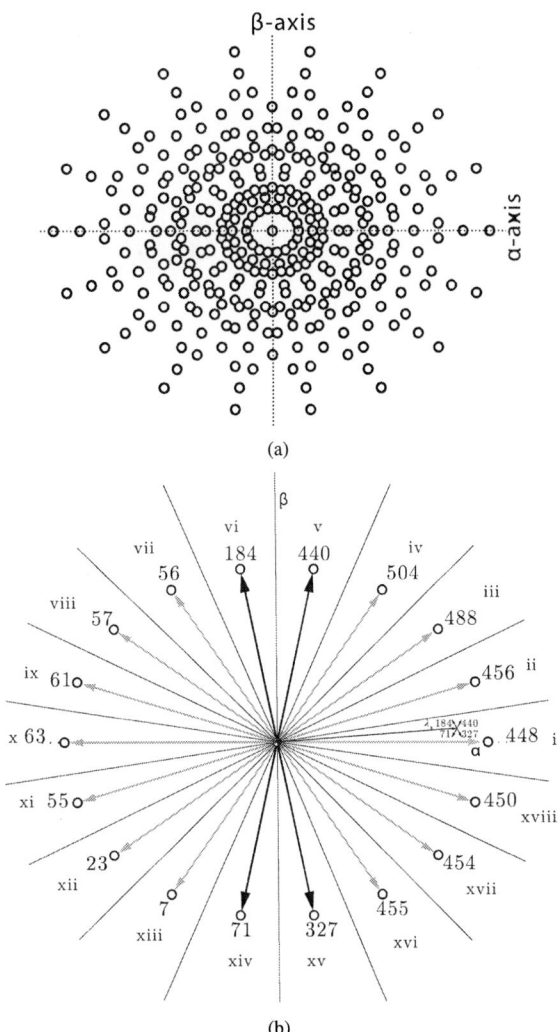

(a)

(b)

Fig. 2. Switching vectors in $\alpha - \beta$ plane. (a) Available switching vectors. (b) Used switching vectors in DTC1.

- if $T_e > T_e^* + Ht$,
  $dT_e = -1$
- else if $T_e < T_e^* - Ht$,
  $dT_e = 1$
- else if $T_e > T_e^* - Ht$ and $T_e < T_e^* + Ht$,
  $dT_e = 0$

where T_e^* is the reference torque value provided by the speed controller, as will be discussed in section IV.

And for flux:

- if $\lambda_s > \lambda_s^* + Hf$,
  $d\lambda_s = 0$
- else if $\lambda_s < \lambda_s^* - Hf$,
  $d\lambda_s = 1$

where λ_s^* is the reference flux value.

Note that torque comparator variable dT_e may assume three different values, making it a three-level comparator. On the other hand, the flux comparator variable $d\lambda_s$ may assume two

different values, making it a two-level comparator. It is worth noting that λ_s is the amplitude of flux rotating vector $\lambda_{s\alpha\beta}$.

As an example, consider that the stator flux $\lambda_{s\alpha\beta}$ is located in sector i. If torque and flux must increase ($dT_e = 1$ and $d\lambda_s = 1$), vector 440 is chosen. Else, if torque must increase but flux must decrease ($dT_e = 1$ and $d\lambda_s = 0$), vector 184 is chosen. Else, if torque must decrease and flux must increase ($dT_e = -1$ and $d\lambda_s = 1$), vector 327 is chosen. Else, if flux and torque must decrease ($dT_e = -1$ and $d\lambda_s = 0$), vector 71 is chosen. When $dT_e = 0$, it is interesting to keep torque within the band limits. In this case, a null vector is applied. As another example, if stator flux $\lambda_{s\alpha\beta}$ is in sector ii, a vector must be chosen among 184, 56, 327, 455 and null in order to maintain torque and flux values within the band limits. The same process must be applied for all sectors.

However, DTC1, which is a direct adaptation from the three-phase case, does not take into account the components $x - y$ and $u - v$. In this case, the selected switching vectors in $\alpha - \beta$ plane corresponds to vectors with high amplitudes in the other planes. Since the current components in these other planes are limited only by the stator resistance and the leakage inductance, which usually has small value, these current components may present high values, distorting the original phase currents.

So, in order to reduce $x - y$ or $u - v$ components, the modified DTC (DTC2) is employed. Considering $u - v$ plane, two semicircles are created, as illustrated in Fig. 3(c). Fluxes in $\alpha - \beta$ and $u - v$ planes must be estimated by measuring stator voltages and currents. As discussed for DTC1, say flux $\lambda_{s\alpha\beta}$ is located in sector i, and the hysteresis limits determine that the flux and torque values must increase ($dT_e = 1$ and $d\lambda_s = 1$). In this case, vector 440 must be applied. But, as seen in Fig. 3(a), vector 248 may also be applied, having similar effects on flux and torque. In addition, in plane $u - v$, vectors 440 and 248 are in opposite directions. So, estimating flux in $u - v$ plane (λ_{suv}), if λ_{suv} is located in one of the sectors covered by the blue semicircle in Fig. 3(c) (sectors i, ii, iii, iv, v, vi, xvi, $xvii$, $xviii$), vector 248 must be applied. Since vector 248 is opposed to flux λ_{suv}, this flux is then reduced. Similarly, if λ_{suv} is located in one of the sectors covered by the red semicircle (sectors vii, $viii$, ix, x, xi, xii, $xiii$, xiv, xv), vector 440 must be applied, reducing λ_{suv}. On the other hand, if λ_{suv} is located in one of the sectors convered by the blue semicircle, vector 248 must be applied in order to reduce λ_{suv}. This flux reduction corresponds directly to a reduction in $u - v$ current components, minimizing, at last, stator currents distortion. This is how method DTC2 operates. Naturally the semicircles are always redefined depending on the values of dT_e and $d\lambda_s$ and on the position of flux $\lambda_{s\alpha\beta}$.

Note that similar proceedings can be done to reduce flux in $x - y$ plane (λ_{sxy}) [see Fig. 3(b)]. However, working on minimizing λ_{suv} shows a more significant decrease in stator current distortion, as will be shown in section V.

978-1-7281-4181-7/19 $31.00 © 2019 IEEE

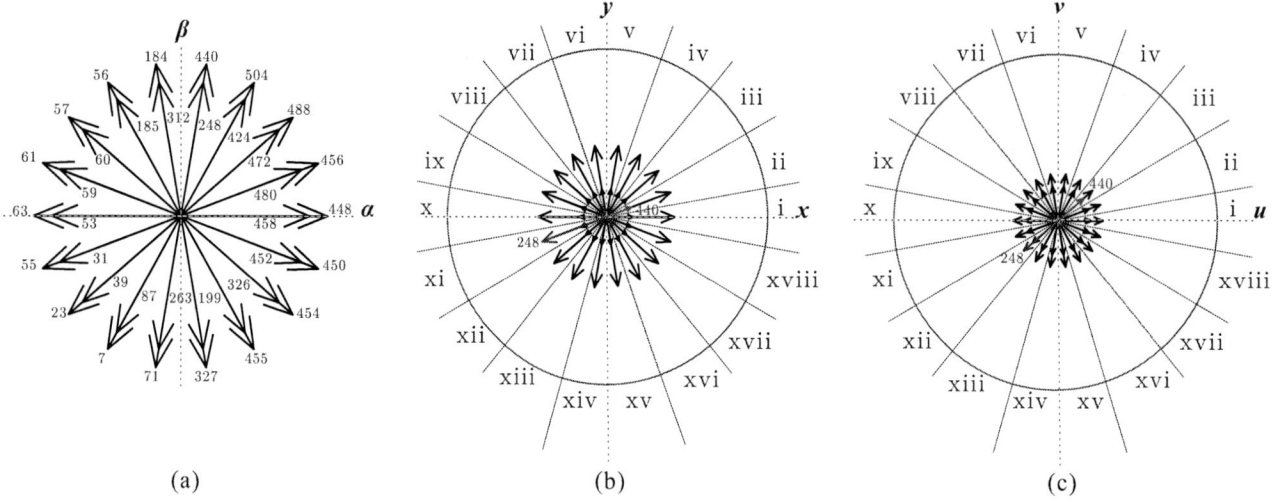

Fig. 3. Used switching vectors in DTC2. (a) $\alpha - \beta$ plane. (b) $x - y$ plane. (c) $u - v$ plane.

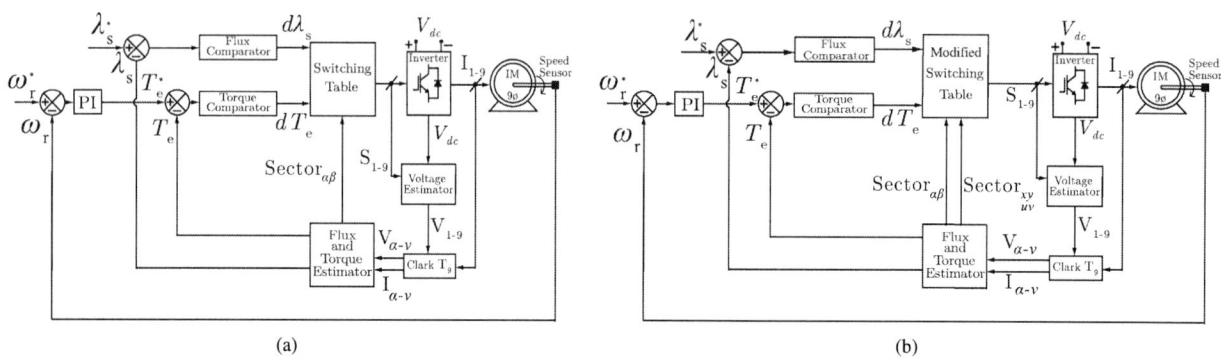

Fig. 4. Control system block diagram. (a) DTC1. (b) DTC2.

IV. CONTROL SYSTEM

The control system block diagram for DTC1 is illustrated in Fig. 4(a). The reference value of the stator flux λ_s^* is compared to the estimated value λ_s in order to define the value of $d\lambda_s$, as explained before. On the other hand, a conventional Proportional-Integral (PI) rotor speed controller defines the value of machine reference electromagnetic torque T_e^*. This value is compared to the estimated value T_e, defining dT_e. Based on $d\lambda_s$ and dT_e, and on the position of the stator flux $\lambda_{s\alpha\beta}$, a voltage vector is chosen from the switching table. Torque and flux are estimated from the measured stator voltages and currents. For the stator flux estimation in the $\alpha - \beta$, $x - y$ and $u - v$ plane, pure integrators with (2), (3) and (4) respectively, are used. The eletromagnetic torque is estimated using (11). The phase voltages are estimated from the switching combination of the binary value states from S_k and the DC Link voltage (V_{dc}) applied to the inverter.

The control system block diagram for method DTC2 is very similar to that of DTC1, but the switching table is modified since the selected switching vector depends on the position of

flux λ_{sxy} or λ_{suv} as well. This block diagram is illustrated in Fig. 4(b).

V. SIMULATION RESULTS

The nine-phase induction machine parameters used in digital simulations are shown in Table I. Simulations were performed using $Ht = 0.2$, $Hf = 0.01$, $\lambda_s^* = 0.988$ Wb and switching frequency of 10 kHz. Fig. 5 illustrates the stator currents of

TABLE I
NINE-PHASE INDUCTION MACHINE PARAMETERS

Parameters	Values
Stator resistance	5.3 Ω
Rotor resistance	2.0 Ω
Stator leakage inductance	24 mH
Rotor leakage inductance	11 mH
Stator-rotor mutual inductance	520 mH
Rated stator flux	0.988 Wb
Rated torque	7 Nm
Frequency	60 Hz
Number of poles	2

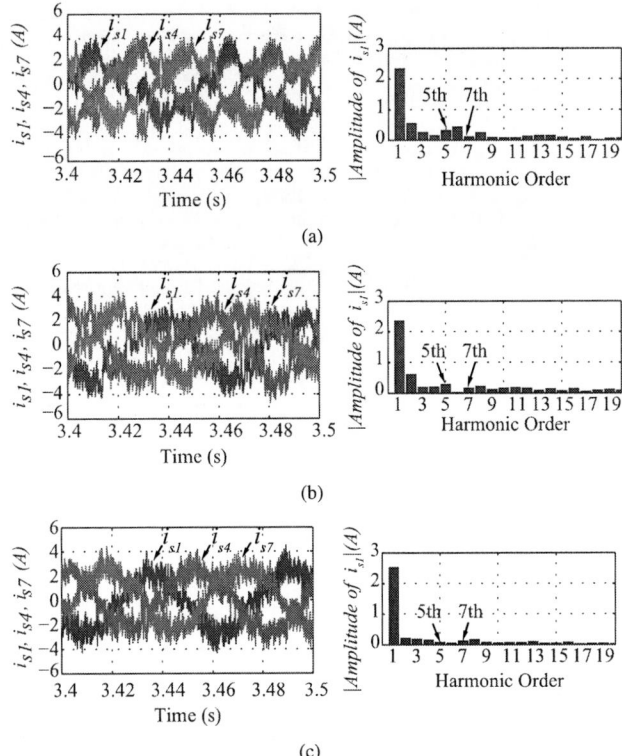

Fig. 5. Currents and frequency spectrum. (a) DTC1. (b) DTC2 minimizing $x-y$. (c) DTC2 minimizing $u-v$.

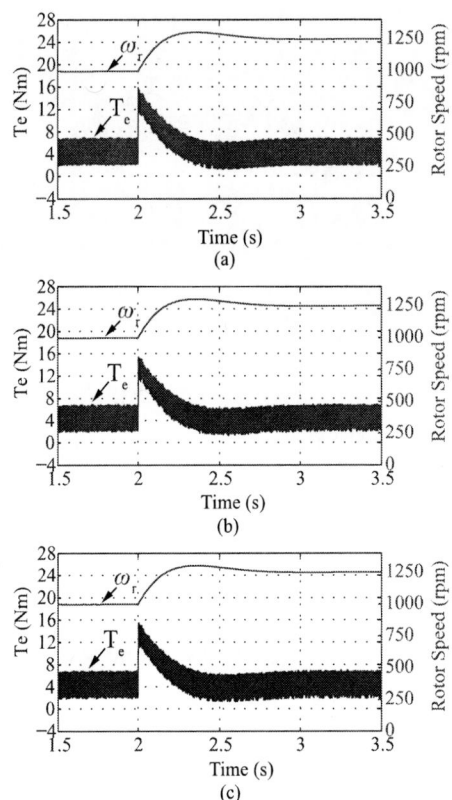

Fig. 6. Dynamic behavior with step in speed reference. (a) DTC1. (b) DTC2 minimizing $x-y$. (c) DTC2 minimizing $u-v$.

one of the three-phase groups with current frequency spectrum for methods DTC1 (Fig. 5(a)), DTC2 minimizing $x-y$ current components (Fig. 5(b)) and DTC2 minimizing $u-v$ current components (Fig. 5(c)). The mechanical torque applied in this case was of 4 Nm. Note that phase currents present high harmonic distortion with DTC1. The phase current total harmonic distortion (THD) obtained with this method was of 54.53%. As explained before, this method does not take into account the current and flux components in other planes than $\alpha-\beta$ plane. Thus, the vector selection only aims to keep flux and torque values within the predetermined hysteresis limits.

However, method DTC2 aims to minimize current and flux in the other planes while keeping flux and torque values within the limits. As can be seen in Fig. 5(b), by choosing to decrease $x-y$ components, the sixth harmonic is significantly reduced. The phase current THD for this case was of 51.39 %. However, by choosing to reduce $u-v$ components, significant harmonic reduction is provided in phase currents, and the phase current THD for this case was of 36.29 %.

Additionally, in order to evaluate the dynamic response and to validate the control system, two tests were performed. First test consisted in applying a step in reference speed from 1000 rpm to 1250 rpm, keeping the same mechanical torque of 4 Nm. Curves of speed and electromagnetic torque are illustrated in Figs. 6(a), 6(b) and 6(c) for DTC1, DTC2 minimizing $x-y$ and DTC2 minimizing $u-v$, respectively. It is possible to

observe that dynamic behavior of DTC2 is essentialy the same as DTC1.

The second test consisted in applying a step in the mechanical torque from 0 to 4 Nm, keeping same reference speed of 1000 rpm. Speed and torque curves for this case are illustrated in Figs. 7(a), 7(b) and 7(c) for DTC1, DTC2 minimizing $x-y$ and DTC2 minimizing $u-v$, respectively. Note that method DTC2 presents just as good performance as DTC1. Thus, from both tests, it is inferred that the dynamic response of the drive is not affected by employing DTC2 instead of DTC1. Note that the torque ripple observed by employing DTC2 is similar to that obtained with DTC1 because the objective of this strategy is the reduction of phase current harmonics only.

VI. CONCLUSION

A modified DTC method for reducing current harmonic distortion in phase currents, discussed in literature for six-phase drive systems, was employed in this paper for an asymmetrical nine-phase induction motor drive system. From SVD, there are the producing torque current components $\alpha-\beta$ and the non-producing torque components ($x-y$ and $u-v$), which generate power losses and phase current distortion. By applying modified DTC, it is possible to minimize either $x-y$ components or $u-v$ components. If the method is applied to minimize $x-y$ components, phase current THD was of

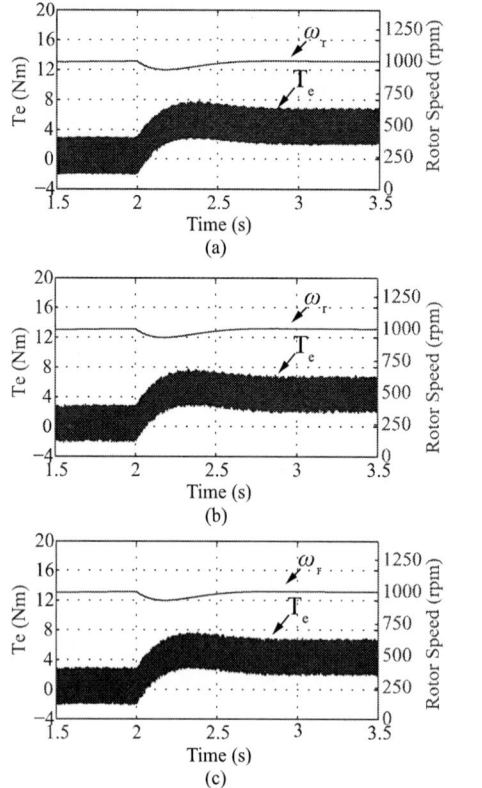

Fig. 7. Dynamic behavior with step in mechanical torque. (a) DTC1. (b) DTC2 minimizing $x - y$. (b) DTC2 minimizing $u - v$.

51.39 % against 54.53 % of classical DTC. If the method is applied to minimize $u - v$ components, phase current THD was of 36.29 %, showing the validity of the proposed method. Besides, it was also shown that the dynamic performance of the system is not affected by using modified DTC instead of classical DTC.

ACKNOWLEDGMENT

The would like to thank in part by the Coordenação de Aperfeiçoamento de Pessoal de Nível Superior - Brasil (CAPES) and the Conselho Nacional de Desenvolvimento Científico e Tecnológico CNPq for the financial support.

REFERENCES

[1] E. Levi, R. Bojoi, F. Profumo, H. Toliyat, and S. Williamson, "Multiphase induction motor drives – a technology status review," *IET Electric Power Applications*, vol. 1, no. 4, p. 489, 2007.

[2] E. Levi, "Multiphase Electric Machines for Variable-Speed Applications," *IEEE Transactions on Industrial Electronics*, no. 5, pp. 1893–1909, may 2008.

[3] ——, "Advances in Converter Control and Innovative Exploitation of Additional Degrees of Freedom for Multiphase Machines," *IEEE Transactions on Industrial Electronics*, vol. PP, no. 99, pp. 1–1, 2015.

[4] M. M. Wogari and O. Ojo, "Nine-phase interior permanent magnet motor for electric vehicle drive," in *2011 IEEE Power and Energy Society General Meeting*. IEEE, jul 2011, pp. 1–8.

[5] E. Jung, H. Yoo, S.-K. Sul, H.-S. Choi, and Y.-Y. Choi, "A Nine-Phase Permanent-Magnet Motor Drive System for an Ultrahigh-Speed Elevator," *IEEE Transactions on Industry Applications*, vol. 48, no. 3, pp. 987–995, may 2012.

[6] I. Subotic, N. Bodo, E. Levi, and M. Jones, "Onboard Integrated Battery Charger for EVs Using an Asymmetrical Nine-Phase Machine," *IEEE Transactions on Industrial Electronics*, vol. 62, no. 5, pp. 3285–3295, may 2015.

[7] V. F. M. B. Melo, C. B. Jacobina, E. R. Braga-Filho, R. M. Cavalcanti, I. S. de Freitas, R. S. Macena, and G. A. d. A. Carlos, "AC-DC-AC nine-phase machine drive system based on H-bridges and three-leg converters," in *IECON 2014 - 40th Annual Conference of the IEEE Industrial Electronics Society*. IEEE, oct 2014, pp. 378–384.

[8] V. F. M. B. Melo, C. B. Jacobina, and N. B. de Freitas, "Open-end nine-phase machine conversion systems," *IEEE Transactions on Industry Applications*, vol. 53, no. 3, pp. 2329–2341, May 2017.

[9] M. Ruba and D. Fodorean, "Analysis of Fault-Tolerant Multiphase Power Converter for a Nine-Phase Permanent Magnet Synchronous Machine," *IEEE Transactions on Industry Applications*, vol. 48, no. 6, pp. 2092–2101, nov 2012.

[10] F. Li, W. Hua, M. Cheng, and G. Zhang, "Analysis of Fault Tolerant Control for a Nine-Phase Flux-Switching Permanent Magnet Machine," *IEEE Transactions on Magnetics*, vol. 50, no. 11, pp. 1–4, nov 2014.

[11] D. Casadei, F. Profumo, G. Serra, and A. Tani, "FOC and DTC: two viable schemes for induction motors torque control," *IEEE Transactions on Power Electronics*, vol. 17, no. 5, pp. 779–787, Sep. 2002.

[12] G. S. Buja and M. P. Kazmierkowski, "Direct torque control of PWM inverter-fed ac motors - a survey," *IEEE Transactions on Industrial Electronics*, vol. 51, no. 4, pp. 744–757, Aug 2004.

[13] M. Bermudez, I. Gonzalez-Prieto, F. Barrero, H. Guzman, X. Kestelyn, and M. J. Duran, "An experimental assessment of open-phase fault-tolerant virtual-vector-based direct torque control in five-phase induction motor drives," *IEEE Transactions on Power Electronics*, vol. 33, no. 3, pp. 2774–2784, March 2018.

[14] M. Bermudez, I. Gonzalez-Prieto, F. Barrero, H. Guzman, M. J. Duran, and X. Kestelyn, "Open-phase fault-tolerant direct torque control technique for five-phase induction motor drives," *IEEE Transactions on Industrial Electronics*, vol. 64, no. 2, pp. 902–911, Feb 2017.

[15] Y. N. Tatte and M. V. Aware, "Torque ripple and harmonic current reduction in a three-level inverter-fed direct-torque-controlled five-phase induction motor," *IEEE Transactions on Industrial Electronics*, vol. 64, no. 7, pp. 5265–5275, July 2017.

[16] J. K. Pandit, M. V. Aware, R. V. Nemade, and E. Levi, "Direct torque control scheme for a six-phase induction motor with reduced torque ripple," *IEEE Transactions on Power Electronics*, vol. 32, no. 9, pp. 7118–7129, Sep. 2017.

[17] J. K. Pandit, M. V. Aware, R. Nemade, and Y. Tatte, "Simplified implementation of synthetic vectors for dtc of asymmetric six-phase induction motor drives," *IEEE Transactions on Industry Applications*, vol. 54, no. 3, pp. 2306–2318, May 2018.

[18] K. D. Hoang, Y. Ren, Z. Zhu, and M. Foster, "Modified switching-table strategy for reduction of current harmonics in direct torque controlled dual-three-phase permanent magnet synchronous machine drives," *IET Electric Power Applications*, vol. 9, no. 1, pp. 10–19, 2015.

Single-Phase AC–DC–AC Five-level X-type Current Source Converter

Louelson A. Costa;
Post-Graduate Program in Electrical Engineering
Federal University of Campina Grande
Campina Grande, Brazil
louelson@gmail.com

Montiê A. Vitorino; and Maurício B. R. Corrêa
Department of Electrical Engineering
Federal University of Campina Grande
Campina Grande, Brazil
vitorino@dee.ufcg.edu.br; mbrcorrea@dee.ufcg.edu.br

Abstract—In this paper a new single-phase AC-DC-AC five-level X-type current source converter is proposed. The use of the X-type topology reduces the conduction losses of the converter, whilst the five-level configuration may reduce the total harmonic distortion under lower frequencies, and allows the increase of the power capability of the converter. The novelty of this paper resides on the multilevel configuration being applied to a X-type current source converter. The operation and sine-triangle pulse-width modulation of the proposed converter are presented, as they are validated by simulation results. Besides that, experimental results for the rectifier side (AC-DC) are presented.

Index Terms—ac-dc-ac, current source converter, dc-link, multilevel, x-type.

I. INTRODUCTION

Single-phase back-to-back converters are often used as uninterrupted power supplies (UPSs), universal power filters (UPFs) or voltage regulators (VRs), and mostly are voltage source converters (VSCs). The current source converters (CSCs), the dual family of the VSCs, are not as used as the VSCs due to two main reasons: high conduction losses in the switches and in the DC-link inductors; and the low power density of the inductors

Despite those drawbacks, the CSCs have some inherent characteristics that are advantageous when compared to the VSCs, such as: natural protection against short-circuit through the DC-link; boost inverter or buck rectifier capability due to the current modulation; and low $\mathrm{d}v/\mathrm{d}t$ in the AC side of the converter [1]. Also, SiC MOSFETs in series with Schottky diodes can achieve low conduction losses while providing the reverse blocking capability needed for CSCs [2], thus mitigating one of the main drawbacks the CSCs. These characteristics show that the CSCs are a viable alternative to the VSCs.

In order to further reduce the conduction losses, the CSCs topologies, when able, can be modified to have a cross-section [8]. This cross-section configuration is named X-type configuration. This configuration reduces the number of conducting switches per switching cycle, thus reducing the conduction losses.

Besides that, the multilevel configuration allows to achieve total harmonic distortion (THD) levels as good as the conventional solution, even under lower frequencies when compared

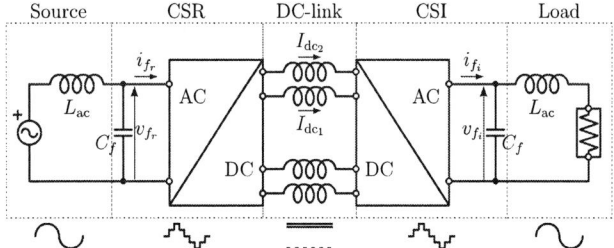

Fig. 1: Single-phase AC-DC-AC five-level CSC supplying a resistive load.

to the three-level solutions. Also, the multilevel topology distributes the power stress over the switches [5], [6]. The combination of the multilevel AC-DC-AC multilevel topology with the X-type CSC results in the multilevel XCSC (m-XCSC). This paper proposes a new single-phase five-level X-type AC-DC-AC current source converter resulting in a back-to-back converter with natural short-circuit protection and high quality AC current waveform.

The paper is divided as following: Section II addresses the back-to-back five-level XCSC topology; Section III explain its sine-triangle pulse-width modulation (SPWM) technique; Section IV presents the control an synchronization technique; Section V presents the simulation and experimental results; and the Conclusion summarize what as developed in this paper.

II. OPERATION AND SWITCHING STATES

The original single-phase back-to-back (b2b) five-level current source converter is shown in Fig. 2a, which has 16 switches and 4 inductors. Considering the asymmetrical five-level CSC, proposed in [7], and connecting two of them in the b2b configuration, the asymmetrical topology can be achieved, as shown in Fig. 2b. It can be seen that the asymmetrical topology has 12 switches and 2 inductors. However, a series connection of two pairs of reverse blocking switches in a CSC can be converted into a cross-section [8], resulting in the X-type topology.

The single-phase b2b five-level XCSC topology is shown in Fig. 2c. It has 12 switches and 2 inductors, less components than the conventional topology and same number of compo-

978-1-7281-4181-7/19 $31.00 © 2019 IEEE

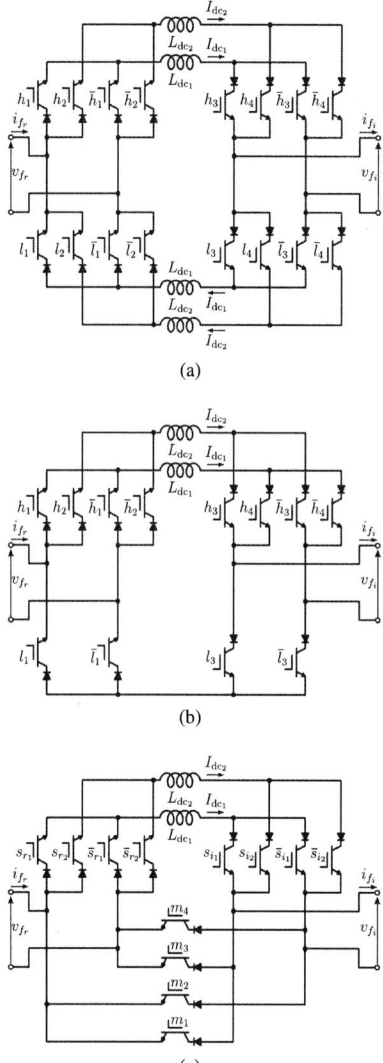

Fig. 2: Single-phase AC-DC-AC five-level CSC topologies: (a) Conventional; (b) Asymmetric [7]; and (c) Proposed X-type.

nents as the asymmetrical topology. However, by using the cross-section configuration, the number of switches conducting by switching cycle is reduced, thus reducing the conduction losses of the converter. Respectively, the conventional, asymmetric and X-type topologies have 8, 6 and 5 switches conducting per switching cycle. Thus, the proposed single-phase b2b five-level XCSC topology has a conduction losses reduction of 37.5%.

Regarding the number of levels, the conventional XCSC has one DC-link inductor and a three-level current; the five-level XCSC (Fig. 2c) has two DC-link inductors and a five-level current; and so on. The number of levels is given by $m = 2n + 1$, being n the number of DC-link inductors of the converter. In this paper, only the five-level configuration is studied, but the development presented in this paper can be expanded to the seven-level, nine-level, etc., topologies.

TABLE I: Half of the switching states of the five-level XCSC, as a function of I_{dc}

s_{r_1}	s_{r_2}	m_k	s_{i_1}	s_{i_2}	i_{f_r}	i_{f_i}	State
0	0	1	0	0	-1	-1	1
0	0	1	0	1	-1	-1/2	2
0	0	1	1	0	-1	-1/2	3
0	0	1	1	1	-1	0	4
0	0	2	0	0	-1	0	5
0	0	2	0	1	-1	+1/2	6
0	0	2	1	0	-1	+1/2	7
0	0	2	1	1	-1	+1	8
0	0	3	0	0	0	-1	9
0	0	3	0	1	0	-1/2	10
0	0	3	1	0	0	-1/2	11
0	0	3	1	1	0	0	12
0	0	4	0	0	0	0	13
0	0	4	0	1	0	+1/2	14
0	0	4	1	0	0	+1/2	15
0	0	4	1	1	0	+1	16
0	1	1	0	0	-1/2	-1	17
0	1	1	0	1	-1/2	0	18
0	1	1	1	0	-1/2	0	19
0	1	1	1	1	-1/2	0	20
0	1	2	0	0	-1/2	0	21
0	1	2	0	1	-1/2	+1/2	22
0	1	2	1	0	-1/2	+1/2	23
0	1	2	1	1	-1/2	+1	24
0	1	3	0	0	+1/2	-1	25
0	1	3	0	1	+1/2	-1/2	26
0	1	3	1	0	+1/2	-1/2	27
0	1	3	1	1	+1/2	0	28
0	1	4	0	0	+1/2	0	29
0	1	4	0	1	+1/2	+1/2	30
0	1	4	1	0	+1/2	+1/2	31
0	1	4	1	1	+1/2	+1	32

In order to find the switching states/vectors of the converter, one has to follow the two basic rules for current source converters switching: there must always be a conducting path for the DC-link current; and there can not be a short-circuit to the AC-link capacitors. Thus, it can be written that:

$$\sum_{k=1}^{2} s_{r_k} + \bar{s}_{r_k} = 1, \quad \sum_{k=1}^{4} m_k = 1, \quad \sum_{k=1}^{2} s_{i_k} + \bar{s}_{i_k} = 1, \quad (1)$$

where s_{r_k} refers to the switches of the rectifier side, m_k refers to the middle switches, and s_{i_k} refers to the switches of the inverter side, as shown in Fig. 2c. It is worth noting that the sub-index "r" refers to the rectifier side (or grid side), whilst the sub-index "i" refers to the inverter side (or load side).

The mean value of the input and output currents are written as

$$I_{f_r} = (s_{r_1} + s_{r_2})I_{dc}/2 - (m_1 + m_2)I_{dc}, \quad I_{f_i} = (s_{i_1} + s_{i_2})I_{dc}/2 - (m_1 + m_3)I_{dc}, \quad (2)$$

where $s_{r_1}, s_{r_2}, m_1, m_2, m_3, s_{i_1}, s_{i_2}$ can be either "1", when the switch is conducting (or closed), or "0", when the switch is not conducing (or open). Thus, Table I can be filled with some of the states/vectors of the converter. It can be noticed that the multilevel currents ($+I_{dc}, +I_{dc}/2, 0, -I_{dc}/2, -I_{dc}$) can be achieved.

The number of states (complexity) of the b2b m-XCSC is: 16 states for the three-level converter; 64 states for the five-

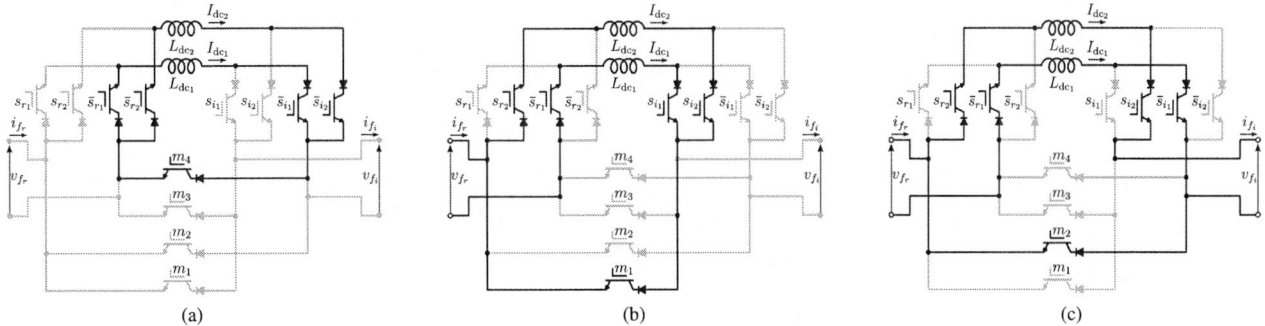

Fig. 3: Single-phase AC-DC-AC five-level XCSC switching states, according to Table I: (a) State 13; (b) State 20; and (c) State 22.

level converter; 256 states for the seven-level converter; and so on. For the conventional b2b m-CSC, the number of states are: 16 states for the three-level converter; 256 states for the five-level converter; 4096 states for the seven-level converter; an so on. Thus, it can be noticed that the m-XCSC can achieve the same number of levels as the conventional topology using less vectors (reduced complexity). Some switching states are illustrated in Fig. 3

III. SINE-TRIANGLE PULSE-WIDTH MODULATION

The sine-triangle pulse-width modulation is based on a combination of the three-level XCSC modulation technique [8] with the modulation technique of the asymmetric multilevel CSR (am-CSR) [9]. The top half of the m-XCSC looks like a conventional m-CSC, and the bottom half is a cross-section of the XCSCs.

For the cross-section correct commutation, at the bottom of Fig. 2c, it only requires input information about the polarity of both reference currents ($i_{f_r}^*$, $i_{f_i}^*$). By using the cross-section logic [8], the middle switches can be properly gated. In this case, there is no comparison with the triangular carriers, as the polarity detector is enough for this logic.

For the switches at the top of Fig. 2c (s_{r_k}, $\bar{s}_{r_k}$, s_{i_k}, $\bar{s}_{i_k}$), one may use the asymmetric SPWM technique presented in [9]. Having two DC-link inductors, two triangular carriers are used, delayed 180° from each other. By using a XOR logical port, the main and the complementary switch (free-wheeling switch) will alternate depending on the polarity of the reference current. The XOR logical port will have the following inputs: the absolute value of the reference current; and the polarity of the reference current. In that way, the XOR logical port will alternate the main and the complementary switch.

Combining those strategies in a single SPWM, the block diagram shown in Fig. 4 is built. It shows the gating block diagram for the switches of the rectifier side (top), middle switches (middle), and inverter side (bottom). This SPWM block diagram is for the five-level topology, but it can be expanded for the seven-level, nine-level, etc., by adding more triangular carries delayed by $360°/n$ from each other, and by adding more XOR logical gate blocks.

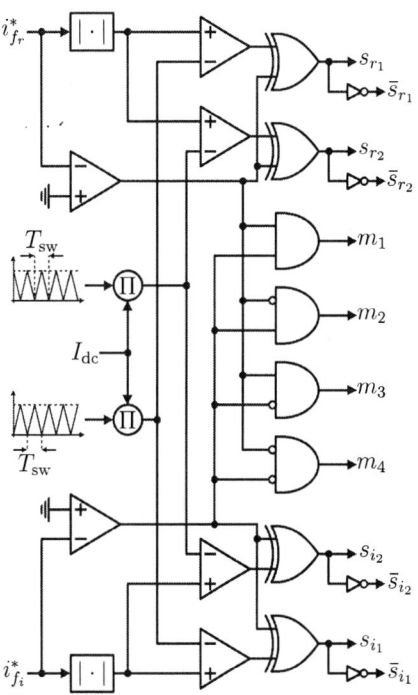

Fig. 4: Single-phase AC-DC-AC five-level XCSC SPWM block diagram.

IV. CONTROL AND SYNCHRONIZATION

The complete back-to-back system is shown in Fig. 6, having: grid, as a voltage source and its line inductance; input AC filter capacitor; single-phase AC-DC-AC multilevel XCSC; output AC filter capacitor; and RL load.

This converter could be controlled by a conventional back-to-back current source control: the rectifier (input) controls de DC-link current; and the inverter (output) controls the voltage of the load. But as a single-phase system, the DC-link presents a twice-the-line frequency oscillation, which causes the DC-link inductor to be bulky.

However, in [4] is proposed a synchronization technique in order to reduce the DC-link oscillation of back-to-back current source converters. Starting from the input and output voltage

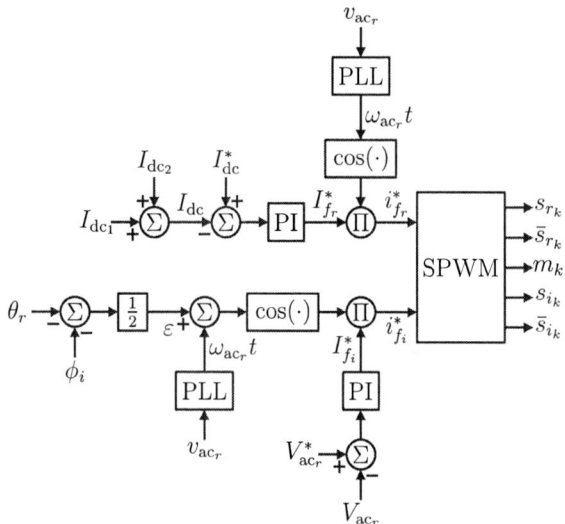

Fig. 5: Single-phase AC-DC-AC five-level XCSC control block diagram, including phase controlled synchronization technique.

and current

$$
\begin{aligned}
v_{\mathrm{ac}_r}(t) &= V_{\mathrm{ac}_r}\cos(\omega_r t), \\
i_{f_r}^*(t) &= I_{f_r}^*\cos(\omega_r t - \theta_r), \\
i_{f_i}^*(t) &= I_{f_i}^*\cos(\omega_i t + \varepsilon), \\
v_{f_i}(t) &= V_{f_i}\cos(\omega_i t + \varepsilon + \phi_i),
\end{aligned} \tag{3}
$$

the synchronization technique controls the phase angle of the output current, which is synchronized with the input current. As shown in Fig. 5, the synchronization technique inputs are the input displacement angle θ_r (calculated by measuring the input AC voltage and current), and the output power factor angle ϕ_i (calculated be measuring the output AC voltage and current).

Besides the input-output currents synchronization technique, the rectifier side is controlled by a conventional strategy: PI controller for the DC-link current; and phase-locked loop (PLL) [10] for grid synchronization. As for the inverter side, it directly controls the output current peak value, in order to control the output voltage, and its phase angle, in order to control the DC-link oscillation. The complete control block diagram is shown in Fig. 5.

Assuming that both input and output frequencies are the same, the DC-link oscillation is given by

$$
L_{\mathrm{dc}} \approx \frac{S_{\mathrm{ac}_i}}{\Delta i_{\mathrm{dc}}\omega_{\mathrm{ac}}I_{\mathrm{dc}}}k_\varepsilon, \tag{4}
$$

being L_{dc} the DC-link inductance, S_{ac_i} the apparent power of the load, Δi_{dc} the DC-link targeted oscillation, ω_{ac} the input and output frequency, and I_{dc} is the DC-link current. The variable k_ε is a function of the input and output angles, and of the efficiency of the converter:

$$
\begin{aligned}
k_\varepsilon &= \sqrt{k_\phi^2 - 2k_\phi\cos(\theta_r + 2\varepsilon + \phi_i) + 1}, \\
k_\phi &= \frac{\cos(\phi_i) + P_t/S_{\mathrm{ac}_i}}{\cos(\theta_r)},
\end{aligned} \tag{5}
$$

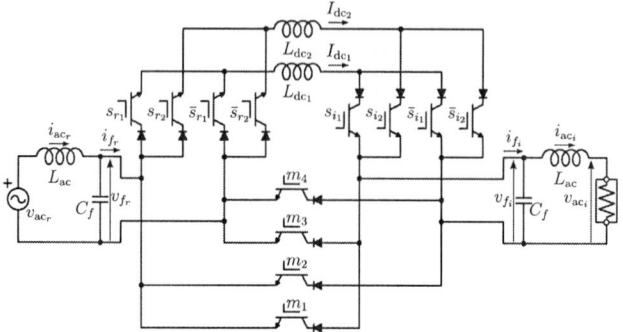

Fig. 6: Single-phase AC-DC-AC five-level XCSC test scenario.

being ϕ_i the power factor of the load, θ_r the input displacement angle, caused by the input AC filter, and P_t is the total power losses of the converter. Thus the output phase angle to achieve the best DC-link reduction is given by

$$
\varepsilon = -(\theta_r + \phi_i)/2. \tag{6}
$$

which is dynamically calculated by means the input AC voltage and current (v_{ac_r} and i_{f_r}), and the output AC voltage and current (v_{f_i} and i_{f_i}).

By controlling the output phase angle ε, one may reduce the DC-link current oscillation by up to 50% [4]. Otherwise, it would be necessary to increase the DC-link inductance, which would result in higher losses, higher costs and lower power density.

For the AC-link capacitors, as shown in [4], they can be limited to be 0.2 pu, thus:

$$
C_f \leq \frac{S_{\mathrm{ac}}}{5\omega_{\mathrm{ac}}V_{\mathrm{ac}}^2}, \tag{7}
$$

being C_f the AC filter capacitor, S_{ac} the apparent power processed by the AC-bus, ω_{ac} the AC-link frequency, and V_{ac} the AC-bus nominal voltage.

It is worth noting that the proposed topology would work all the same without the synchronization technique. The technique is applied in order to have a reduced DC-link oscillation.

V. Simulation and Experimental Results

Simulations results for the single-phase AC-DC-AC five-level XCSC and experimental results for the rectifier side of the converter were taken in order to validate the presented theory for the proposed converter. The test scenario has a voltage source supplying a resistive load using the proposed converter, as shown in Fig. 6.

Table II lists the test scenario parameters: AC filter capacitance C_f; AC frequency f_{ac}; switching frequency f_{sw}; reference DC-link current I_{dc}^*; AC filter inductance L_{ac}; DC-link inductance L_{dc}; modulation index m_i; load power per phase P_{load}; simulation time step T_s; AC voltage source V_{ac}. The simulation results were taken using MATLAB®/Simulink®, using the SimPowerSystems™ library.

978-1-7281-4181-7/19 $31.00 © 2019 IEEE

TABLE II: Test setup parameters

Parameter	Value
C_f	10 μF
f_{ac}	60 Hz
f_{sw}	10 kHz
I_{dc}^*	4 A
L_{ac}	2 mH
L_{dc}	10 mH
P_{load}	250 W
T_s	1 μs
V_{ac}	110 V

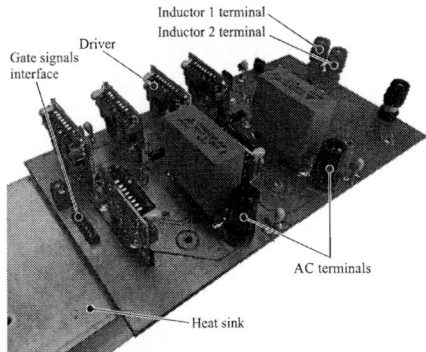

Fig. 7: Single-phase AC-DC five-level CSC prototype.

The five-level CSC prototype is shown in Fig. 7. The prototype was built using IGBTs IRG4PC40U from International Rectifier®, in series with diodes RHRP1560 from Fairchild Semiconductors® in order to achieve the reverse blocking capability. The converter was controlled using a digital signal processor (DSP) TMS320F28335 from Texas Instruments®. The overlap time (dead time) of the converter was implemented in the DSP.

The simulation results for the conventional topology are shown in Fig. 8, and the simulation results for the proposed X-type topology are shown in Fig. 9. It can be seen, in both cases, that the input (rectifier) voltage and current are synchronized, achieving a power factor close to unit. Also, the DC-link current is controlled at 4 A and presents a low oscillation, due to the DC-link control allied to the synchronization technique. The output current has its phase controlled in order to achieve the reduced DC-link oscillation, and the output voltage is controlled at 110 V (150 V peak). The control technique used is the same for both topologies, the only difference between the simulation results are the SPWM strategies, which depends on the topology. A small DC-link distortion is noticed during the current zero-crossing, this is expected as around the zero-crossing region the phase angle calculation might presents some error, reflecting in the DC-link oscillation reduction control.

The experimental results for the rectifier side of the five-level topology are shown in Fig. 10. It can be seen that the input voltage and current are synchronized, achieving a power factor close to unit. Also, the DC-link current is controlled at 4 A, however it presents a twice-line-frequency DC-link oscillation as only the rectifier stage is operating in this result.

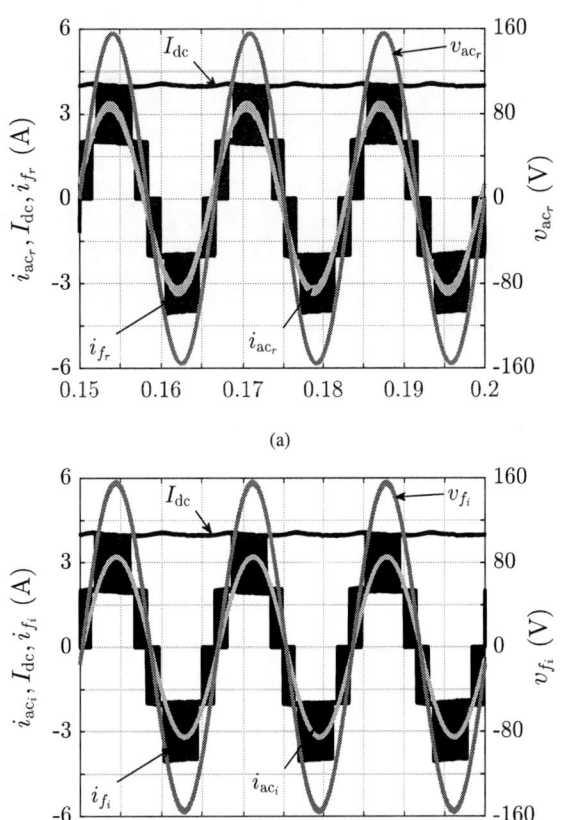

Fig. 8: Simulation results for the conventional single-phase AC-DC-AC five-level CSC: (a) Rectifier, input or grid side; and (b) Inverter, output or load side.

TABLE III: Conventional and proposed X-type AC-DC-AC five-level CSC comparative analysis

Power (pu)		1.0	0.75	0.5
η (%)	Conventional	92.21	89.95	88.98
	Proposed X-type	94.31	92.43	89.25
Input THD (%)	Conventional	2.65	2.75	4.74
	Proposed X-type	2.85	3.07	4.91

Even though being an asymmetric topology, the five-level waveform can be clearly noticed.

The proposed X-type topology have 12 switches and 2 inductors, whilst the conventional solution have 16 switches and 4 inductors. Thus, the proposed solution can operate in the same way as the conventional topology does, using less components. Regarding efficiency (η) and input THD, Table III shows the simulation results for three operation points: 100%, 75% and 50% of nominal power.

As shown in Table III, the proposed X-type topology has an overall better efficiency. On the other hand, the proposed X-type has a slightly worse input THD when compared to the conventional solution. Those results shows that the proposed X-type topology can achieve results as good as the

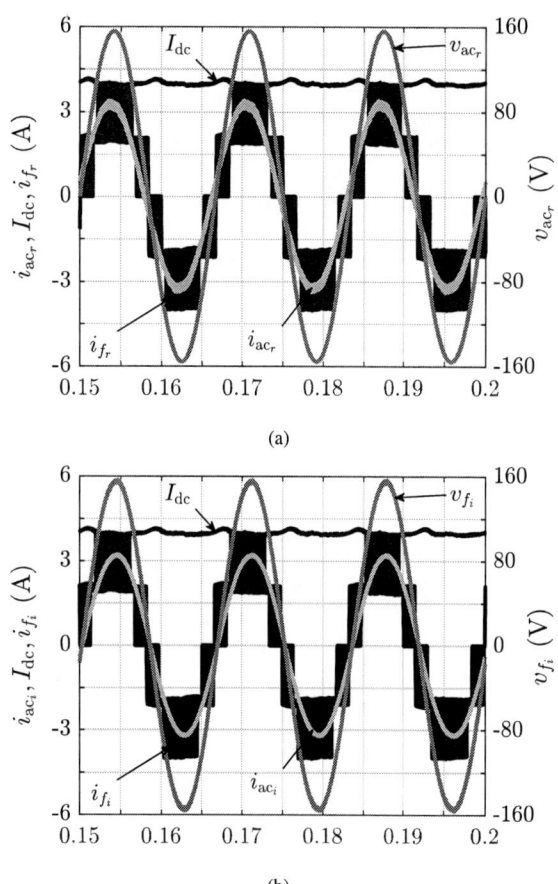

(a)

(b)

Fig. 9: Simulation results for the proposed single-phase AC-DC-AC five-level XCSC: (a) Rectifier, input or grid side; and (b) Inverter, output or load side.

conventional one, even though is has less components, thus improving efficiency and reducing cost.

VI. CONCLUSION

This paper proposed a single-phase AC-DC-AC five-level X-type current source converter. The proposed converter combines the asymmetric m-CSC with the XCSC, achieving the same number of levels and functionality as the conventional AC-DC-AC five-level topology. However, the proposed topology has a reduced number of switches, conducting switches per switching cycle, number of inductors and reduced complexity. The m-XCSC achieved a THD slightly larger than the conventional topology and a higher overall efficiency, but having 25% less number of switches and 50% less number of inductors. The converter operation and its equations were deduced, showing that it can generate a five-level input and output currents. The sine-triangle pulse-width modulation was presented, which is the combination of the asymmetric m-CSC and XCSC cross-section modulations. The theoretical approach was validated by simulation and experimental results, showing that the proposed topology can achieve a performance as good as the conventional topology.

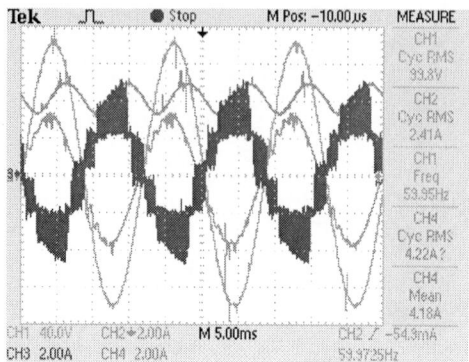

Fig. 10: Experimental results for the rectifier, input or grid side of the five-level XCSC without DC-link synchronization technique. Where CH1 (40 V/div.) is v_{ac_r}; CH2 (2 A/div.) is i_{ac_r}; CH3 (2 A/div.) is $-i_{f_r}$; CH4 (2 A/div.) is I_{dc}; and time (5 ms/div.).

ACKNOWLEDGMENT

The authors gratefully acknowledge the financial support in part by the Post-Graduate Program in Electrical Engineering (PPgEE) of the UFCG, by the National Council for Scientific and Technological Development (CNPq), and by the Coordination of Improvement of Higher Education Personnel (CAPES).

REFERENCES

[1] M. Salo and H. Tuusa, "vector-controlled pwm current-source-inverter-fed induction motor drive with a new stator current control method," *IEEE Transactions on Industrial Electronics*, vol. 52, no. 2, pp. 523–531, April 2005.

[2] L. K. Ries, T. B. Soeiro, M. S. Ortmann, and M. L. Heldwein, "Analysis of carrier-based pwm patterns for a three-phase five-level bidirectional buck +boost-type rectifier," *IEEE Transactions on Power Electronics*, vol. 32, no. 8, pp. 6005–6017, Aug 2017.

[3] M. A. Vitorino, L. F. S. Alves, R. Wang, and M. B. de Rossiter Corrêa, "Low-frequency power decoupling in single-phase applications: A comprehensive overview," *IEEE Transactions on Power Electronics*, vol. 32, no. 4, pp. 2892–2912, April 2017.

[4] L. A. L. de Azevedo Cavalcanti Costa, M. A. Vitorino, and M. B. de Rossiter Corrêa, "Improved single-phase ac-dc-ac current source converter with reduced dc-link oscillation," *IEEE Transactions on Industry Applications*, vol. 54, no. 3, pp. 2506–2516, May 2018.

[5] K. Gnanasambandam, A. K. Rathore, A. Edpuganti, D. Srinivasan, and J. Rodriguez, "Current-fed multilevel converters: An overview of circuit topologies, modulation techniques, and applications," *IEEE Transactions on Power Electronics*, vol. 32, no. 5, pp. 3382–3401, May 2017.

[6] P. Cossutta, M. P. Aguirre, A. Cao, S. Raffo, and M. I. Valla, "Single-stage fuel cell to grid interface with multilevel current-source inverters," *IEEE Transactions on Industrial Electronics*, vol. 62, no. 8, pp. 5256–5264, Aug 2015.

[7] N. Vazquez, H. Lopez, C. Hernandez, E. Vazquez, R. Osorio, and J. Arau, "A different multilevel current-source inverter," *IEEE Transactions on Industrial Electronics*, vol. 57, no. 8, pp. 2623–2632, Aug 2010.

[8] L. A. Costa, M. B. R. Corrêa, M. A. Vitorino, G. G. dos Santos, and D. A. Fernandes, "A family of single-phase current source converters with double outputs," in *2016 IEEE Applied Power Electronics Conference and Exposition (APEC)*, March 2016, pp. 1032–1039.

[9] M. A. Vitorino, L. A. Costa, M. B. R. Corrêa, and C. B. Jacobina, "Multilevel asymmetric single-phase current source rectifiers," in *2016 IEEE Energy Conversion Congress and Exposition (ECCE)*, Sep. 2016, pp. 1–7.

[10] V. Kaura and V. Blasko, "Operation of a phase locked loop system under distorted utility conditions," *IEEE Transactions on Industry Applications*, vol. 33, no. 1, pp. 58–63, Jan 1997.

978-1-7281-4181-7/19 $31.00 © 2019 IEEE

Application of a SHE-PWM modulation for a low switching frequency motor drive with harmonic investigation using the DTFT

Marcos Paulo Brito Gomes, Marina Hassen de Souza, Gabriel Vilkn Ramos,
Alex-Sander Amável Luiz, Marcelo Martins Stopa,
Electrical Engineering Department - CEFET-MG, Belo Horizonte, Brazil.
marcospaulobritogomes@ieee.org, marinahassens@gmail.com, gabrielvilkn@gmail.com,
alex@cefetmg.br, marcelo@cefetmg.br.

Abstract – **This paper presents a study related to an electric drive composed by an induction machine fed by a NPC (neutral point clamped) frequency inverter, extending also to ANPC (active neutral point clamped) topologies, and IGBTs commanded by a SHE-PWM (selective harmonic elimination pulse width modulation) logic pattern. This work also addresses a concern related to the number of dv/dts reproduced in a fundamental period of a waveform, which is related to the semiconductors switching frequency. These dv/dts cause degradations to electric motors and its mitigation contributes minimizing problems on windings, insulations and bearings erosion. A SHE-PWM pattern with very low switching frequency is presented with simulations and experimental results mitigating the 5th and 7th harmonics for the entire drive voltage index modulation range. These harmonics have high amplitude and low frequency, causing high drive losses, and can be reduced by a modulation with semiconductors switching in a frequency equivalent to 360Hz, with high quality and symmetrical voltage and currents waveforms. SHE-PWM was compared with POD-PWM at the same switching rate, through simulation and harmonic content investigated by the DTFT (Discrete Time Fourier Transform).**

Keywords — SHE-PWM modulation, POD-PWM modulation, selective harmonic elimination, dv/dts reduction, NPC, motor drives, Discrete Time Fourier Transform, DTFT.

I. INTRODUCTION

Power converters and frequency inverters provide the operation of induction motors in the industry in a wide range of speed, torque and power. Equipment such as pressurization systems, overhead cranes, conveyor belts, conveyor systems, blasting systems, excavators and loading systems, traction systems, require frequency inverter controls. Assuming that the life of these equipment depends on the duty cycle, environment and power quality, optimized inverters with harmonic elimination, low switching frequency and reduced dv/dt incidence can bring greater durability to these equipment.

Power Quality concerns bring investigation of voltage and current harmonic distortion rates, and *THD* is also regulated by several standards. The IEEE 1459-2010 standard demonstrates the importance of including harmonic content to account for the operating power factor of an electronic device. And since harmonics decrease the power factor, they interfere with the active and reactive energy balance of these systems. The IEEE 519-2014 standard defines examples of

measurement and sampling of harmonic signals, since sampling interferes with harmonic content identification. In this standard, there are definitions of limit metrics for THD for voltages ($THDv$) and currents ($THDi$) individually, as well as the need to mitigate specific harmonic order, when they are more prejudicial for equipment and loads, especially those of high amplitude and low order, [1-4].

When designing power converters and electric drive systems, power semiconductors are chosen, among other parameters, according to their power, operating current, operating voltage, blocking inverse voltage, and thermal dissipation capacity. The switching frequency is related to the semiconductor thermal capacity, conduction and blocking times and this is interconnected with the converter harmonic filter parameters. Switching frequency is typically established for m_f (frequency modulation index) greater than 21, possibly around $80 < m_f < 200$, when semiconductor switch from $f_s = 5kHz$ to $f_s = 12kHz$, with SV-PWM, SPWM, PD-PWM and POD-PWM and others high frequency modulation patterns. Currently with the advent of wide band-gap (WBG) semiconductor materials and high electron mobility transistors (HEMTs), such as silicon carbide (SiC) and gallium nitride (GaN), power electronics can be hard switched at $m_f > 1700$ or higher, when semiconductor can switch above $f_s = 100kHz$, reaching $f_s = 200kHz$.

The ability to switch the semiconductors at high frequency is beneficial as it enables harmonic filtering, and this minimizes heating problems, audible noise, vibrations, hysteresis losses and eddy currents in ferromagnetic materials, electromagnetic emission, interferences in surrounding control circuits, and increases switching losses on inverter semiconductors, concerns for power electronics applied to high efficiency equipment, [5-12,19,20]. However, dv/dts are mainly responsible for bearing degradation by electrostatic discharges, increasing on heating and insulation damage, [19,20]. These problems result in electric machine premature loss, as a result, high number of voltage transitions, dv/dts, are greater concern than $THDi$ and $THDv$. What is observed is that medium and high voltage/power drives still use converters with low switching frequency, $m_f = 21$, when semiconductors operate at frequencies close to $f_s = 1kHz$. Low order harmonics are responsible for most drive problems, vibrations and pulsating torques principally noticeable at low speeds, [5,13]. The 5th, 7th and 11th harmonics can be mitigated by a SHE-PWM pattern for the entire m_i (amplitude modulation index range), at low switching ($f_s = 360Hz$) with low dv/dts incidence, [14-18].

978-1-7281-4181-7/19 $31.00 © 2019 IEEE

II. Motor drive Description

Fig. 1 presents the schematic and the topology of the three-phase and three-level NPC inverter (neutral point clamped by diodes) used to obtain experimental results at the laboratory and whose parameters were used in the simulations. It consists of an WEG W22 Premium induction motor drive with power rated at $15cv$ (approximately $11kW$), fed by three-phase and three-level NPC inverter. The inverter has 16 Semikron IGBTs, which 6 were used as diodes only, A 12-pulse rectifier bridge was also used, linked to a transformer with secondary windings in delta-wye. It is an adopted open loop drive without feedback control. The main objective of this experiment setup is investigate the effects of the POD-PWM and SHE-PWM patterns on the drive, in terms of synthesized voltage and quality of current waveform, harmonic distortions rates THDi and THDv, waveform symmetry, as well as the behavior of vibration, audible noise and heating.

Table 1 lists the WEG W22 Premium 160M 15cv induction motor nominal parameters arranged in the motor nameplate data grouped to the equivalent circuit parameters obtained by testing. These data are of great importance for comparing the simulated results with the experimental results for the two modulation patterns POD-PWM and SHE-PWM. Concerns about harmonic content and dv/dts are mainly related to medium to high voltage power drives. However, the use of a lower power drive is favorable for predicting operation under higher power conditions. From the point of view of the harmonic content of $THDv$ voltage, number of commutations and the incidence of dv/dts are independent of the power rating. This analysis includes the possibility of predicting the current symmetry patterns for higher power drives.

TABLE I.   WEG W22 Pemium 15cv Motor parameters

Parameters - MIT - W22 Premium WEG 160M 15cv			
Nominal Power	11kW(15cv)	Rs	429.0531 mΩ
Nominal Voltage	220/380 Vrms	Rr'	236.8330 mΩ
Nominal Current	38.4/22.2 Arms	Lls	1.96467 mH
Nominal Frequency	60 Hz	Llr'	1.96467 mH
Nominal Speed	1765 rpm	Lm	51.3310 mH
Nominal Efficiency	92.7%	J	0.1843 kgm²
Pole Numbers	4	b	0.00219 Nms
Insulation Class	F	-	-
Temp. of Insul. Class	80K (80°C)	-	-
Regime Duty	S1	-	-

III. The dtft – discrete time Fourier transform

The field of power electronics has a clear interface with signal processing when it comes to DSPs data processing. The same signal processor that is used to generate the pulse pattern for semiconductors can be used to gather drive data. In this case, the DSP sampling effects must be considered. Previous works investigated classical PWM patterns, such as comparison of reference voltages v_a^*, v_b^*, and v_c^* with triangular and sawtooth carrier waveforms, with natural and regular sampling. Often to support the harmonic and spectral analysis mathematical theories were used to calculate the coefficients of the Fourier's trigonometric series coefficients analytically, including analysis by Bessel's series and Jacobi-Anger's expansion, [21]. The DTFT (Discrete Time Fourier Transform), currently widely explored, transforms a signal $f(t)$ in time domain into the frequency domain $F(\omega)$, and can be described in its analysis and synthesis equation eq. (1) e eq. (2).

$$F(\omega) = \mathcal{F}\{f(t)\} = \int_{-\infty}^{\infty} f(t)e^{-i\omega t}\, dt \qquad (1)$$

$$f(t) = \mathcal{F}^{-1}\{F(\omega)\} = \frac{1}{2\pi}\int_{-\infty}^{\infty} F(\omega)e^{i\omega t}\, d\omega \qquad (2)$$

In practical applications, in signal processing and digital systems, the transformation needs to be discretized and implemented in DSPs, assuming the form shown in eq. (3) and eq. (4), [22-24]. The values contained in $F[k]$ are complex numbers, and the amplitude of the individual harmonics can be obtained through the module of those complex numbers, according to eq. (5). These values, $h[k]$, were accounted in this paper to compare spectral content between POD-PWM and SHE-PWM. A DSP sampling frequency of $12kHz$ was considered, totaling $N = 200$ sampling points per fundamental period, $T_1 = 1/60Hz$. The windowing effects could be explored, such as the Hamming, Henning, Blackman and Kaiser windows, but for simplicity were used rectangular windows $L = 200$, [22-24].

$$F[k] = \sum_{n=1}^{N} f[n]\, e^{-i\frac{2\pi}{N}(n-1)(k-1)}, \{k \in \mathbb{Z} | 1 \le k \le N\} \quad (3)$$

$$f[n] = \frac{1}{N}\sum_{k=1}^{N} F[k]\, e^{i\frac{2\pi}{N}(n-1)(k-1)}, \{n \in \mathbb{Z} | 1 \le n \le N\} \quad (4)$$

$$h[k] = \frac{2}{N}\,\big|\,F[k]\,\big| \qquad (5)$$

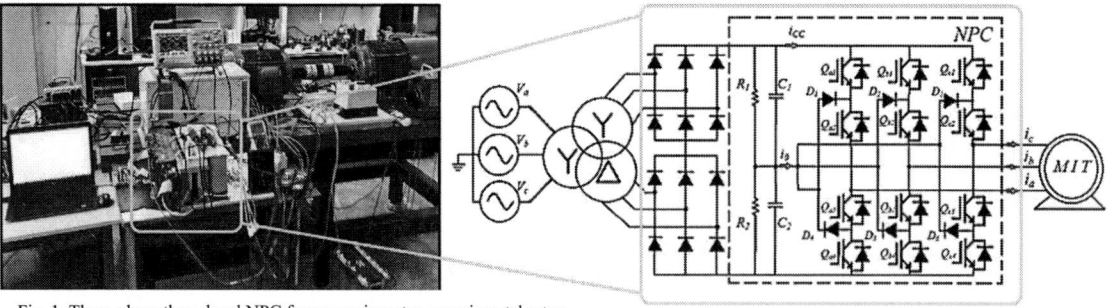

Fig. 1. Three-phase three-level NPC frequency inverter, experimental setup.

IV. THE SHE-PWM - SELECTIVE HARMONIC MODULATION PATTERN

The SHE-PWM pulse pattern composed by selective harmonic elimination has half of a century of research, [1,2]. Although it is a modulation that deviates from the conventional strategies, that consist in comparison between modulation and sawtooth or triangular carriers, and easy to implement for multilevel converters, SHE-PWM has been widely accepted for having a half-period and quarter-period symmetry pulse pattern, highly recommended for low switching frequency applications, [25-30].

The switching angles for selective harmonic elimination are pre-calculated as they consist in transcendental equations solutions. These types of equations are commonly solved by iterative methods, such as Newton Raphson's method., When it comes to index M, for SHE-PWM, it refers to the number of commutations within a quarter of a period of the phase voltage waveform, symmetrical for the remainder of the period. As demonstrated in previous works, this modulation is able to mitigate $(M - 1)$ harmonics for the entire voltage modulation index range. Thus SHE-PWM for M=3 can be selected to mitigate two harmonics, normally chosen the low order ones, the 5th and 7th harmonics, with an equivalent switching frequency of $f_s = 360Hz$, $(m_f = 6)$. SHE-PWM for M=5 in turn can be selected to mitigate the 5th, 7th and 11th harmonics, with an equivalent switching frequency of $f_s = 600Hz$ $(m_f = 10)$. SHE-PWM for M = 4 and other patterns for even values of M were not considered for convenience because these modulations lose energy pulses at the center of the waveform semi-period. For odd values of M, the last harmonic eliminated can be calculated by $(3M - 2)$, selecting low order harmonics to mitigate.

The following steps of eq. (6) to eq. (12) below represent the process of calculating the selective harmonic elimination angles $\overline{a_k}$ for each modulation index value m_i. The order of the matrix in eq. (6) is the same as the M index. Conditional values in vectors $f(\overline{a_k})$ should be given less than a maximum accepted error $\varepsilon < 10^{-4}$, for example. The Jacobian matrix in eq. (9) and eq. (10) has the partial derivatives and therefore is also composed of trigonometric equations.

$$\begin{bmatrix} \cos(\alpha_1) & -\cos(\alpha_2) + & \dots & (-1)^{M+1}\cos(\alpha_M) - \frac{\pi}{4}m_i \\ \cos(5\alpha_1) & -\cos(5\alpha_2) + & \dots & (-1)^{M+1}\cos(5\alpha_M) \\ \vdots & \vdots & & \vdots \\ \cos(n_i\alpha_1) & -\cos(n_i\alpha_2) + & \dots & (-1)^{M+1}\cos(n_i\alpha_M) \end{bmatrix} = \begin{bmatrix} f_{1k} \\ f_{2k} \\ \vdots \\ f_{Mk} \end{bmatrix} \quad (6)$$

$$\overline{a_k} = [a_{1k} \quad a_{2k} \quad \dots \quad a_{Mk}]^T \quad (7)$$

$$f(\overline{a_k}) = [f_{1k} \quad f_{2k} \quad \dots \quad f_{Mk}]^T \quad (8)$$

$$J(\overline{a_k}) = \frac{\partial f(\overline{a})}{\partial \overline{a}}\bigg|_{\overline{a_k}} = \begin{bmatrix} \frac{\partial f_1}{\partial a_1} & \frac{\partial f_1}{\partial a_2} & \dots & \frac{\partial f_1}{\partial a_M} \\ \frac{\partial f_2}{\partial a_1} & \frac{\partial f_2}{\partial a_2} & & \frac{\partial f_2}{\partial a_M} \\ \vdots & \vdots & \ddots & \vdots \\ \frac{\partial f_M}{\partial a_1} & \frac{\partial f_M}{\partial a_1} & \dots & \frac{\partial f_M}{\partial a_M} \end{bmatrix} \quad (9)$$

$$J(\overline{a_k}) = -\begin{bmatrix} \sin(a_1) & -\sin(a_2) & \dots & +(-1)^{M+1}\sin(a_M) \\ 5\sin(5a_1) & -5\sin(5a_2) & \dots & +(-1)^{M+1}5\sin(5a_M) \\ \vdots & \vdots & \ddots & \vdots \\ n_i\sin(n_ia_1) & -n_i\sin(n_ia_2) & \dots & +(-1)^{M+1}n_i\sin(n_ia_M) \end{bmatrix} \quad (10)$$

$$f(\overline{a_k}) = [0 \quad 0 \quad \dots 0]^T \quad (11)$$

$$\overline{a_k} = [a_{1f} \quad a_{2f} \quad \dots \quad a_{Mf}]^T \quad (12)$$

According to an initial value is set for the selective elimination angles $\overline{a_1}$ when performing the first iteration $k = 1$. The values contained in $\overline{a_1}$ are then used to evaluate eq. (6) from which comes the value of $f(\overline{a_1})$. If at least one of in $f(\overline{a_1})$ is greater than the value set for the error ε, the Jacobian matrix $J(\overline{a_1})$ is then calculated. The angles are then updated and will be used for the second interaction $k = 2$ as $\overline{a_2} = -J^{-1}(\overline{a_1}).f(\overline{a_1})$, this is done successively according to the equations eq. (13) to eq. (15).

$$\overline{a}_{k+1} = \overline{a}_k + \Delta\overline{a}_k \quad (13)$$

$$\Delta\overline{a}_k = -J^{-1}(\overline{a}_k).f(\overline{a}_k) \quad (14)$$

$$\overline{a}_{k+1} = \overline{a}_k - J^{-1}(\overline{a}_k).f(\overline{a}_k) \quad (15)$$

The calculation process ends at the last iteration k_f, when the values contained in $f(\overline{a}_{k_f})$ are smaller than the error, ε. At this time the values contained in the vector $\overline{a}_{k_f - 1}$ are the answer to the selected m_i. Selective harmonic angles calculated for SHE-PWM M=3 and SHE-PWM M=5 are shown in the Fig. 2 and Fig. 3 respectively.

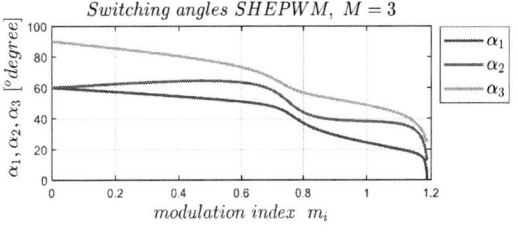

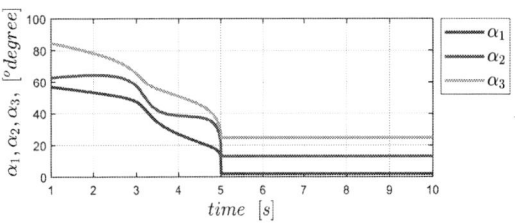

Fig. 2. Switching angles for SHE-PWM, M=3.
a) Angles *versus mi* and b) Angles *versus* time.

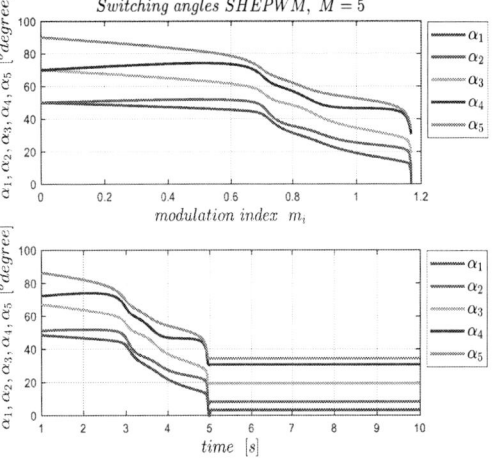

Fig. 3. Switching angles for SHE-PWM, M=5.
a) Angles *versus mi* and b) Angles *versus* time.

V. POD-PWM AND SHE-PWM PATTERNS OPERATING AT LOW FREQUENCY MODULATION INDEX

This section presents the experimental results for phase voltage V_{a0}, line voltages V_{ab} and current waveforms I_a and I_b, obtained from the experimental set up shown in Fig. 1. The intent of this paper is to compare the patterns of POD-PWM (phase opposition disposition pulse width modulation) and SHE-PWM. To establish comparison parameters one premise is to obtain switch frequency equivalence. This is done at the same time that a low switching level is to be achieved using the SHE-PWM of M = 3 and POD-PWM of $m_f = 6$, thus establishing a total of 6 transitions or pulsations per fundamental period. The two modulation patterns will be compared through their harmonic content for the entire modulation index m_i, using the DTFT (Discrete-Time-Fourier-Transform) tool, as explained in eq.(1) to eq.(5).

A. POD-PWM $m_f = 6$ modulation pattern

A major concern for establishing low commutation modulation patterns is waveform symmetry, both semi-period symmetry and a quarter of a period symmetry. Low switching levels imply a low frequency modulation index, $m_f < 21$. For modulations composed by comparison between modulating voltage and carrier with $m_f > 21$ this concern does not exist. However for $m_f < 21$ it is necessary to worry about the waveform symmetry and one of the implications of this concern is the choice for triangular carriers, instead of sawtooth, such that the carriers are in phase opposition disposition, composing the POD-PWM illustrated in Fig. 4, for $m_i = 1,0$ and $m_f = 6$. Analyzing Fig. 4, the perception of symmetry is clear for V_{a0} and V_{ab} waveforms.

The DTFT tool demonstrates that for POD-PWM, the total harmonic distortion for V_a is $THDv = 53.43\%$, for V_{ab} is $THDv = 44.45\%$. As demonstrated in recent papers, without phase opposition PD-PWM, the corresponding values are, for V_a $THDv = 47.57\%$ and for V_{ab} $THDv = 35.29\%$ both are smaller, [1,2]. However as discussed earlier, considering current harmonic distortions, POD-PWM becomes a more interesting modulation than PD-PWM. POD-PWM semi-period and quarter of a period symmetry allows voltage synthetization with minimization of neutral voltage displacement, which decreases common-mode current circulation in practical terms. For POD-PWM, considering induction motor equivalent circuit parameters as shown in Table 1, the drive without load torque results in harmonic current distortion equals to $THDi = 69.73\%$, for nominal load operation this value becomes $THDi = 11.02\%$, because of the fundamental current growth. For PD-PWM respective values are higher, $THDi = 114.40\%$ without load torque and $THDi = 16.12\%$ at nominal load operation, [1,2]. Fig. 4 and Fig. 5 demonstrate the harmonic content up to 50th for all modulation index range m_i, for POD-PWM modulation phase voltage V_{a0} line voltage V_{ab}.

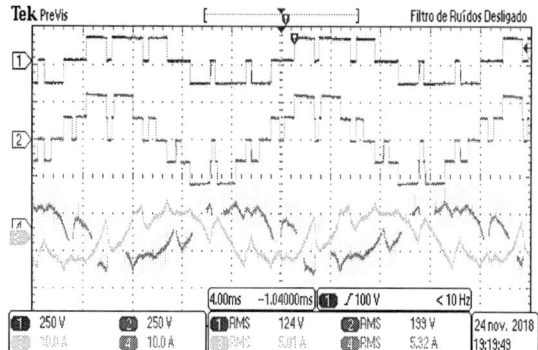

Fig. 4. POD-PWM, $mi = 1.0$ e $mf = 6$. V_{a0}, V_{ab}, I_a, I_b waveforms.

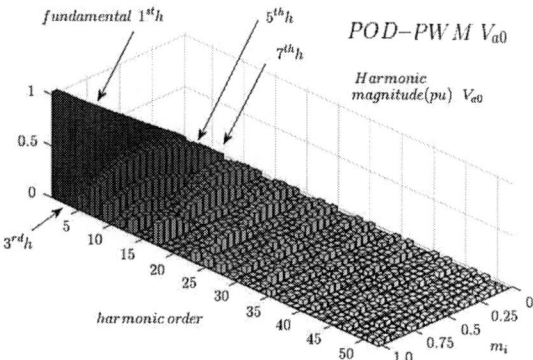

Fig. 5. Harmonic content up to 50th for all modulation index range, for POD-PWM modulation phase voltage V$_{a0}$.

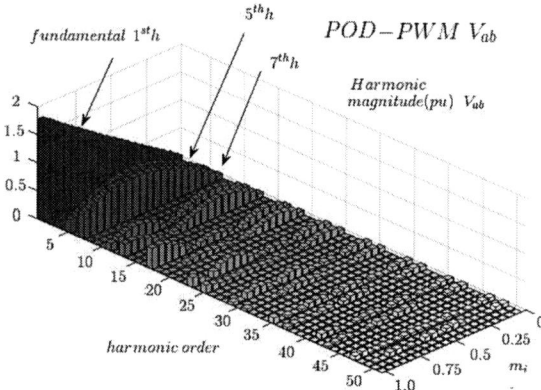

Fig. 6. Harmonic content up to 50th for all modulation index range, for POD-PWM modulation line voltage V$_{ab}$.

Analyzing Fig. 5 it is possible to notice that the fundamental component value of synthesized phase voltage V_a follows the amplitude modulation index value m_i. As expected, the harmonics are located in bands around the converter switching frequency and its multiples. For a POD-PWM with $m_f = 6$, the 5th and 7th harmonics are the most notable for the entire range of amplitude modulation index m_i. Fig. 6 shows thas the 3rd harmonic and its multiple are eliminated for line voltages V_{ab}. In the next section, the SHE-PWM modulation pattern is analyzed for M=3, the harmonic optimization will be discussed based on the similar switching frequency ($f_s = 360Hz$ or $m_f = 6$). The 5th and 7th harmonic will be mitigated by SHE-PWM.

978-1-7281-4181-7/19 $31.00 © 2019 IEEE

B. SHE-PWM M=3 modulation pattern

Experimental results for the NPC inverter commanded by a SHE-PWM M = 3 and $m_i = 1,0$ modulation pattern is illustrated in Fig. 7. As observed for the POD-PWM of $m_f = 3$, the SHE-PWM M = 3 modulation symmetry is clear for both semi-period and quarter of a period. A visual comparison between Fig. 4 and Fig. 7 shows that the selective harmonic elimination modulation becomes more interesting in terms of current total harmonic distortion, In Fig. 7 it is possible to notice a sine wave approximation in the current waveforms. As discussed in the previous topics, this pulse pattern mitigates the 5th and 7th harmonics for the entire modulation index range m_i, for any drive excursion, respecting the criteria of Fig. 2, assuming that the over modulation region has not been explored in this paper.

DTFT tool simulation results are presented in Fig. 8 and it corroborates the quality of those currents waveforms presented in Fig. 7. Analyzing the phase voltage V_{a0} it is also possible to notice the presence of the 3rd harmonic in the modulation, but this harmonic is eliminated in line voltages V_{ab}, as expected for multiple orders of three. Thus SHE-PWM M = 3 synthesizes a range free of low order harmonics for the entire drive excitation in terms of amplitude modulation index, as illustrated in Fig 9. For SHE-PWM M=3 $THDv = 47.75\%$ for V_{a0}, $THDv = 32.57\%$ for V_{ab}, $THDi = 29.43\%$ without load torque and $THDi = 4.10\%$, for the drive operating at rated load torque, being the modulation with lowest harmonic distortion for all cases of voltage and current [1,2].

Fig. 7 and Fig. 8 demonstrate the harmonic content up to 50th for all modulation index range, for SHE-PWM modulation phase voltage V_{a0} line voltage V_{ab}. The presence of higher $THDv$ for low m_i values motivates many works to use high frequency modulations during drive startup, and SHE-PWM for steady state and $m_i = 1,0$, with lower switching rate. The low switching rate is certainly beneficial for medium and high power and voltage drives, usually limited in $f_s = 1kHz$. Further minimizing switching when eliminating low energy pulses near zero crossing in V_{ab}.

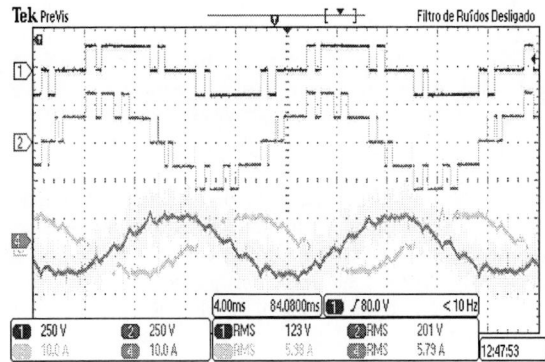

Fig. 7. SHE-PWM, $M = 3$ e $mi = 1.0$.
V_{a0}, V_{ab}, I_a, I_b waveforms.

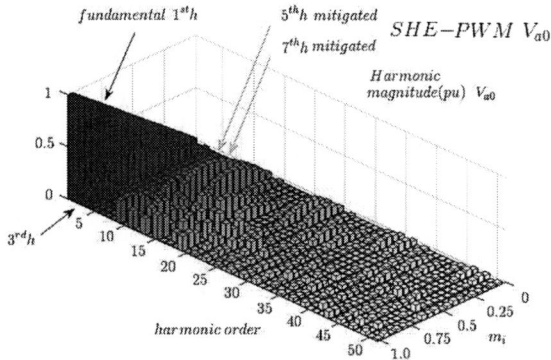

Fig. 8. Harmonic content up to 50th for all modulation index range, for SHE-PWM modulation phase voltage V_{a0}.

The 5th and 7th harmonic are mitigated by SHE-PWM technique in phase voltages V_{a0}, it implies in a harmonic spectrum with a range free of low order harmonics for line voltages V_{ab}, since the 3rd harmonic is also eliminated, as demonstrated in Fig. 8. Here, improved quality of phase and line voltage synthetization also implies an improvement in the current waveform of induction motors, configuring symmetry, and low harmonic distortion rate, with an even better result at nominal torque.

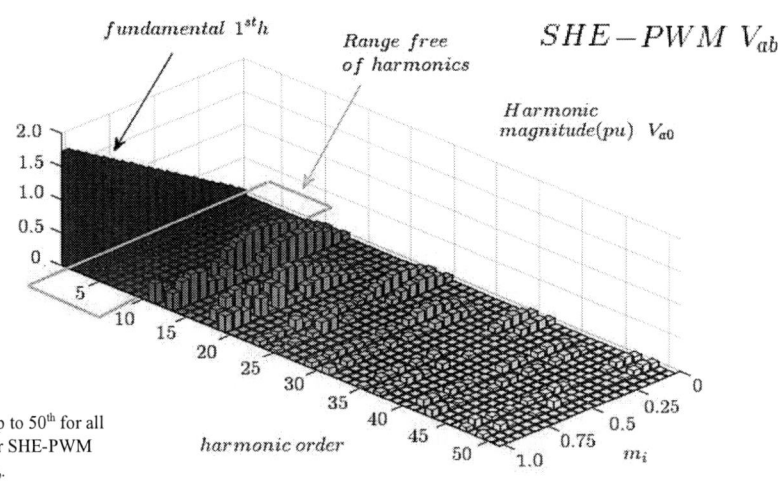

Fig. 9. Harmonic content up to 50th for all modulation index range, for SHE-PWM modulation line voltage V_{ab}.

CONCLUSION

This paper contextualizes the concerns for modulating multilevel power converters applied to medium and high power drives. These types of drives require converter switching frequency limitation, however they require good quality of voltage supply waveform, in terms of harmonic content and dv/dt incidence in electric machines. In this paper, two types of modulation strategies were analyzed and used to control gate drive switching for a three-level, three-phase NPC inverter. For this analysis, one criterion was to establish the same pulsation level per phase voltage waveform. POD-PWM with frequency modulation index equal to $m_f = 6$ (with triangular carriers and adapted for low switching frequency with second carrier in phase opposition disposition) and SHE-PWM M=3. Both modulations have a switching frequency equivalent to $f_s = 360\ Hz$ ($m_f = 6$), when each of the 12 IGBTs on NPC topology switch only six times per synthesized voltage period, which means a very low switching frequency compared to conventional patterns. Both modulations POD-PWM $m_f = 6$ and SHE-PWM M=3, have semi-period symmetry and a quarter of a period symmetry, however SHE-PWM M = 3 has the advantage of eliminating the 5th and 7th harmonics which have low order and high amplitude, for all drive amplitude modulation index m_i excursion. The elimination of these harmonics is beneficial for minimizing drive problems such as heating problems, audible noise, vibrations, hysteresis losses and eddy currents in ferromagnetic materials, electromagnetic emission, interferences in surrounding circuits, and increases switching losses on inverter semiconductors. On the other hand, the low switching frequency decreases the incidence of dv/dt in the electric machine and minimizes bearing degradation and insulation issues. In order to compare the two modulations a DTFT (Discrete Time Fourier Transform) tool was developed and applied to analyze the synthesized spectral content. As a result SHE-PWM M=3 modulation pattern has the lowest total harmonic distortion rate for phase voltages line voltages and currents. The 5th and 7th harmonic are mitigated by this technique, configuring a harmonic spectrum with a range free of low order harmonics, for the entire modulation index range.

ACKNOWLEDGEMENT

The authors would like to acknowledge the CEFET/MG, CAPES, CNPq and FAPEMIG for their financial support.

REFERENCES

[1] M. P. B. Gomes, A.-S. A. Luiz, M. M. Stopa, G. V. Ramos, I. A. Pires. "Assessment of a NPC frequency inverter with low switching frequency modulation for a high speed rating operation of an induction motor".2019 IEEE Applied Power Electronics Conference and Exposition (APEC) (pp. 625-632). IEEE. 2019.

[2] M. P. B. Gomes, A.-S. A. Luiz, M. M. Stopa. "Assessment of operation performance of power converters with low frequency PWM drive issues and field weakening". 2018 13th IEEE International Conference on Industry Applications (INDUSCON) (pp. 591-598). IEEE, 2018.

[3] P. W. Hammond, "A New Approach to Enhance Power Quality for Medium Voltage AC Drives", IEEE Transactions on Industry Applications, Vol. 33, NO. 1, January/February1997, pp 202-208.
B. Wu, "High-Power Converters and AC Drives", IEEE Press, John Wiley, Piscataway, NJ, 2006.

[4] A. G. Tôrres, "Study and characterization of magnetic losses in induction motors". Doctoral Thesis nº 038. Federal University of Minas Gerais. Belo Horizonte, 2004.

[5] N. Mehboob, "Hysteresis Properties of Soft Magnetic Materials". Wien: Autor Edition, v. Doktorin der Naturwissenschaften Dissertation, 2012.

[6] R. P. Machado, "Measures of the classic Película Effect in copper conductors - Proposal of a new model". Ed of the author, v. Master's Dissertation, Coritiba: UFPA, 2007.

[7] A. W. Kelley, S. W. Edwards, J. P. Rhode, M. E. Baran, "Transformer Derating for Harmonic Currents: A Wide-Nand Measurement Approach for Energized Transformers". USA: IEEE, v. vol35, NO.6, November/1999.

[8] M. H. Shin, D. S. Hyun, "Speed Sensorless Stator Flux Oriented Control of Induction Machine in the Field Weakening Region". IEEE, Seoul, Korea, 2001.

[9] H. A. Fattah, M. H. Ismael, A. Bahgat, "Fuzzy Supervisory Control of Field Oriented Controlled AC Drives". Procedings of American Control Conference, AACC. Anchorage, AK, 2002.

[10] J. Kim. J. Jung, K. Nam. "Dual-Inverter Control Strategy for High-Speed Operations of EV Induction Motors". IEEE Transactions on Industry Electronics, Vol. 51. NO. 2, April, 2004.

[11] S. Lim, K. Nam, "Loss-minimising control scheme for induction motors". IEE Proc-Electr. Power Appl, Vol. 151, NO. 4, July, 2004.

[12] G. K. Dubey. "Fundamentals of Electrical Drives". 2nd Edition. ed. Pangbourne, U.K.: Alpha Science International Ltd., 2001.

[13] A. A. Radionov, V. R. Gasiyarov, A. S. Maklakov. "Hybrid PWM on the basis of SVPWM and SHEPWM for VSI as part of 3L-BtB-NPC converter". IECON 2017 43rd Annual Conference of the IEEE Industrial Electronics Society. IEEE, 2017.

[14] H. Akagi, "A New Neutral-Point-Clamped PWM Inverter". IEEE Transactions on Industry Applications, v. IA-17,NO. 05, 1981.

[15] A.-S. A. Luiz and B. J. Cardoso. "Voltages and Sine Currents in Industrial Medium Voltage Converters". Doctoral thesis, Fedral University of Minas Gerais. UFMG, 2009.

[16] C.G. Hu; G. Holmes, W.X. Shen, X.B. Yu, Q.J. Wang, F.L. Luo, "Neutral-point potential balancing control strategy of three-level active NPC inverter based on SHEPWM," IET Power Electronics, Vol.10, no.14, pp. 1755-4535, November 2017.

[17] R. H. Souza; J. L. L .Magalhães, G. R. Silva, A.-S. A. Luiz; Stopa M. M. "Analysis of modulation techniques for an alternative five-level NPC converter used on adjustable-speed drives". Brazilian Power Electronics Conference (COBEP). IEEE, 2017.

[18] A.-S. A. Luiz and B. J. Cardoso. "Minimum Reactive Power Filter Design for High Power Three-level Converters", IECON 2008, Thirtyfourth Anuual Conference of the IEEE Industrial Electronics Society, Orlando, November, 2008.

[19] D. Ma, C. Hu, Q. Ye, et al. "Effect on harmonic performance for three-level ANPC converter with small change in SHEPWM switch angles". Conference on Industrial Electronics and Applications. IEEE, 2016:881-885. IEEE, 2016.

[20] G. J. Su, L. Tang, Z. Wu. Extend Constant-Torque and Constant-Power Speed Range Control of Permanent Magnet Machine Using a Current Source Inverter. IEEE, Knoxville,Tennessee, USA, 2009.

[21] T. A. Lipo, D. G. Holmes. "Pulse Width Modulation for Power Converters Principles and Practice". Piscataway, NJ, USA: John Wiley & Sons, 2003.

[22] A. V. Oppenheim and R. W. Schafer. Discrete-time signal processing. Pearson Education, 2014.

[23] D. Sundararajan. Fourier Analysis, A Signal Processing Approach. Springer, 2018.

[24] G. Konstantinou et al. Interleaved operation of three-level neutral point clamped converter legs and reduction of circulating currents under SHE-PWM. IEEE Transactions on Industrial Electronics, v. 63, n. 6, p. 3323-3332, 2016..

[25] A. Perez Basante, et al. A Universal Formulation for Multilevel Selective-Harmonic-Eliminated PWM With Half-Wave Symmetry. IEEE Transactions on Power Electronics, v. 34, n. 1, p. 943-957, 2018.

[26] A. A. Radionov, V. R. Gasiyarov, and A. S. Maklakov. "Hybrid PWM on the basis of SVPWM and SHEPWM for VSI as part of 3L-BtB-NPC converter." IECON 2017-43rd Annual Conference of the IEEE Industrial Electronics Society. IEEE, 2017

[27] A. M. Omara, and G. Moschopoulos. "Implementation of SHE-PWM Technique for Parallel Voltage Source Inverters Employed in Uninterruptible Power Supplies." 2018 IEEE International Telecommunications Energy Conference (INTELEC). IEEE, 2018.

[28] C. Hu, et al. Neutral-point potential balancing control strategy of three-level active NPC inverter based on SHEPWM. IET Power Electronics, v. 10, n. 14, p. 1943-1950, 2017.

[29] M. Sharifzadeh, H. Vahedi, and K. Al-Haddad. "New constraint in SHE-PWM for single-phase inverter applications." IEEE Transactions on Industry Applications 54.5 (2018): 4554-4562.

[30] A. M. Omara, M. K, El-Nemr, and G. Moschopoulos. "Parallel operation of voltage source inverters using she-pwm based on genetic algorithm." 2017 Nineteenth International Middle East Power Systems Conference (MEPCON). IEEE, 2017.

A Resonant-Switched-Capacitor Step-Down DC–DC Converter in CCM Operation as an LED Driver

Fábio de P. Neres
Mechatronis Research Group
Federal Institute of Ceará
Sobral, Brazil
fabioo.paiva2@gmail.com

Antonia F. da Rocha
Mechatronis Research Group
Federal University of Ceará
Fortaleza, Brazil
antonia.mec@gmail.com

Rodrigo L. dos Santos
Mechatronis Research Group
Federal University of Ceará
Fortaleza, Brazil
linharodrigo@alu.ufc.br

Pedro S. Almeida
Modern Lighting Research Group
Federal University of Juiz de Fora
Juiz de Fora, Brazil
pedro.almeida@ufjf.edu.br

Fernando L. M. Antunes
Group of Power Processing and Control
Federal University of Ceará
Fortaleza, Brazil
fantunes@dee.ufc.br

Edilson M. Sá Jr.
Mechatronis Research Group
Federal Institute of Ceará
Sobral, Brazil
edilson.mineiro@gmail.com

Abstract—This paper proposes a analysis of a LED driver. The driver consists of a resonant switched capacitor converter designed to operate in CCM and fed from nanogrid. This mode of operation allows the reduction of current peaks and, consequently, reduction of conduction losses. There is ZVS in the turn-off of the switches due to the full charge and discharge of the switched capacitor. Therefore, switching losses are also reduced. The inductor operates at twice the switching frequency, contributing to the reduction of the magnetic elements. In addition, this converter has the advantage of stabilizing the power of the LED by adjusting the switching frequency, without the needing sensors and control circuit, which reduces the cost of the circuit. In order to validate the comparative analysis, a 10W prototype was designed in a laboratory, operating with a switching frequency of approximately 58 KHz, and presenting an overall efficiency of 92.8% in nominal conditions.

Index Terms—Resonant Switched Capacitor (RSC) Converter, LED Driver, CCM, ZVS

I. INTRODUCTION

In recent years the growth of distributed generation in the electricity system can be observed due to the increasing share of renewable energy sources in the energy system. In this scenario, DC nanogrids has being highlighted for its efficiency, reliability and safety [1], [2]. Nanogrids are DC power systems that connect distributed generation, consumers, storage elements and controllable DC loads [2]–[4]. Taking into consideration that light emitting diodes (LEDs) are DC-fed loads, it is attractive to direct integrate them into DC nanogrids.

LEDs should become the main source of artificial light in decades to come because of their advantages compared to other lighting sources such as low maintenance, high luminous efficacy, compact size, long lifetime and no harmful effects to the environment [5], [6]. In order to ensure the constant current

supply, LED arrays need drivers that combine features such as high performance, low cost, small volume, high power density and a long lifetime [6], [7].

Among the driver topologies for LEDs, the switched capacitor (SC) converters feature small volume, lightweight and high power density in DC-DC power conversion, The absence of magnetic devices provides a reduced size of the LEDs drivers [8], [9]. For these reasons, several structures have been proposed in the literature using SC converters for LED driving [10]–[12]. However, these circuits generally have high dissipative losses caused by current peaks to which the components of the circuit are submitted and also they have a high EMI emission [14]–[17], they add a small inductor in series with the switched capacitor, which allows them to operate under soft switching conditions (ZVS / ZCS zero voltage / zero current switching), thus reducing the switching losses of the circuit and the EMI emission [13]. Therefore, the RSC converters present attractive features, and can be applied to drive power LEDs [10], [17]–[19].

RSC converters are capable of switching in ZCS operating in discontinuous conduction mode (DCM), which make it possible to increase the switching frequency, and reduce the volume of the magnetic elements, thus reducing the size of the circuit. However, this approach does not reduce the switching losses related to the intrinsic capacitances of the MOSFET switches, when compared to the ZVS technique [20]. In [21], the operation in CCM was proposed, even though there was no ZCS. However, on account of the lower current variation in the inductor, the conduction and on magnetic losses are lower. The converter presented ZVS when turning-off the MOSFETs because of the complete charging and discharging of the switched capacitor. This represents two thirds of eliminated conduction losses [22]. Thus, the converter in CCM presented

978-1-7281-4181-7/19 $31.00 © 2019 IEEE

a higher efficiency than in DCM.

In [19] an RSC step-down converter for powering LEDs was introduced. This converter uses an inductor operating in DCM in series with the switched capacitor to reduce switching losses. In addition, the output current can be stabilized without the need for current sensors and dimerization is achieved by modulating the switching frequency without the need for a closed loop control.

A topology of an RSC DC-DC step-down converter shown in [19] is adapted in this paper to the CCM continuous conduction mode, fed by a nanogrid. The topology analyzed operating in CCM allows the reduction of the current peaks reducing the losses by switching and consequently increase the topology efficiency. The inductor operates at twice the switching frequency. Therefore, even when operating in CCM the volume of the magnetic can be reduced. Because of the full charge and discharge characteristic of the switched capacitor, the switches turn off at zero voltage (ZVS). The operation of the circuit proposed in CCM, as well as its equation are presented in this paper. In order to validate the analysis, a prototype of 10W was built with the same design parameters of [19]. Thus, a comparative analysis was performed between the operation in CCM and DCM.

II. RSC STEP-DOWN LED DRIVER IN CCM

The basic circuit of the RSC DC-DC converter used in this work is shown in Fig. 1. The converter consists of two controlled switches (S_1 e S_2); two fast diodes (D_1 e D_2); one resonant inductor L_r and one resonant switched capacitor C_s; one output filter capacitor C_o and the LED that behaves as a load. This RSC converter works in continuous conduction mode (CCM).

The capacitor C_s is fully charged and discharged in a switching period of the switches. The energy stored in C_s is transferred to the C_o filter and to power LEDs. The resonant inductor L_r provides operation in continuous conduction mode with twice the switching frequency, where the output has a current source characteristic and allows the complete loading and unloading of C_s.

The following considerations are made to simplify the qualitative and quantitative analysis of RSC converter in CCM:

(a) All the components are considered ideal. (b) The switches S_1 and S_2 operate in a complementary way with duty cycle of 0.5. (c) Filter capacitor C_o is large enough to ensure the voltage source characteristic imposed by LEDs.

A. Principle of Operation

The converter can be analyzed in four operations stages, which are shown in Fig. 2. Main waveforms of the circuit can be analyzed through the Fig. 3.

Stage 1 (t_0 – t_2): At instant t = t_0, the C_s capacitor is discharged and the current in the L_r inductor is the minimum (I_{min}). The switch S_2 is turned ON and the diode D_2 is forward biased. Thus, time instant t = t_1, the input source Vin transfers energy to the inductor L_r, capacitor C_s and for the load (C_o and LEDs). Therefore, C_s and L_r are charged. Time instant t_1, the inductor L_r reaches the maximum current value $i_L(t_1) = I_{max}$, thus, the voltage in the capacitor C_s is V_{in}–V_o. From this instant, L_r transferring power to the capacitor and to the load. The capacitor C_s continues charging until it reaches the voltage V_{in}, at the end of step, at t = t_2.

Stage 2 (t_2 – t_3): With capacitor C_s charged, diode D_1 goes into conduction. Thus, the inductor continues to transfer energy to a load through diodes D_1 and D_2. In the final stage, at $t = t_3$, the current in L_r is I_{min} and S_2 turned OFF. During this step, the voltage on S_2 is zero. Thus, the switched turned-off under ZVS condition.

Stage 3 (t_3 – t_5): The switched S_1 is turned ON, the diode D_1 continues in conduction. Thus, the capacitor C_s transfers energy to the load. Therefore, in this step C_s is discharged. The inductor behaves in the same way as in one stage. At $t = t_4$ reaches the maximum current value, $i_L(t4) = I_{max}$, thus, the

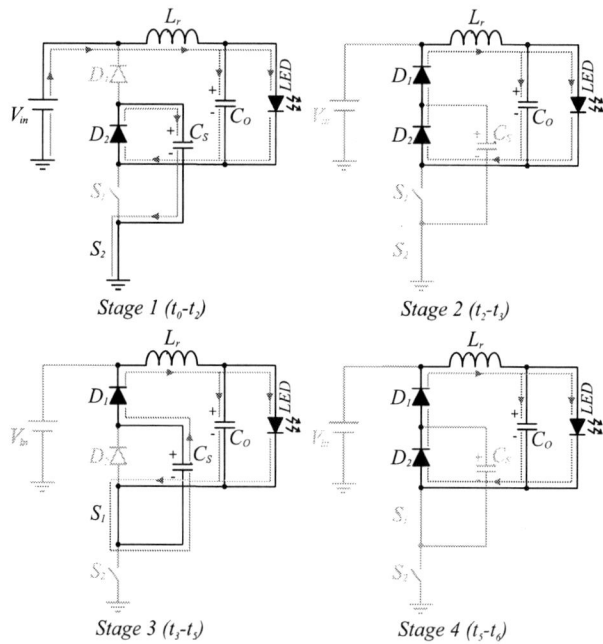

Fig. 2. Operating stages of RSC converter

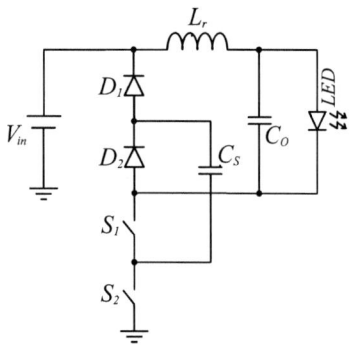

Fig. 1. Analyzed topology proposed

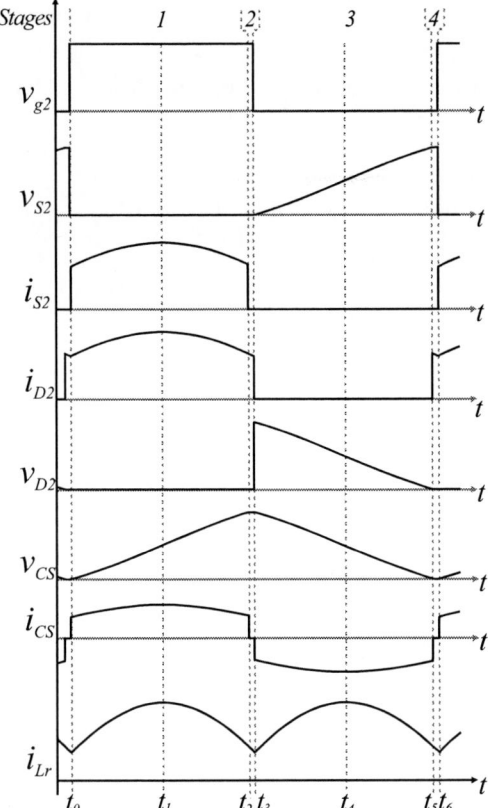

Fig. 3. Main theoretical waveforms of the RSC Converter in CCM

voltage in the capacitor C_s is V_o. When this stage finishes, at $t = t_5$, the capacitor C_s this discharged. Consequently, the voltage at S_1 is zero.

Stage 4 ($t_5 - t_6$): This stage is similar to the second stage. The inductor L_r continues to transfer energy to a load through the diodes D_1 and D_2. When this stage finishes, switched S_1 is turned off under ZVS condition.

B. Load Characteristics

The LED can be represented in simplified form by na ideal diode in series with a source of constant voltage and a resistance, as shown by Fig. 4.

From the electric model, we obtain the mathematical model of the LED. Since the LED is the circuit load, the voltage between the anode and cathode terminals of this device is referred to as the output voltage.

$$V_o = V_{LED} + R_{LED} \cdot I_o \tag{1}$$

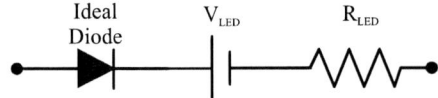

Fig. 4. Simplified electrical model of the LED

Thus, the output power of the converter can be determined by (2).

$$P_o = V_o \cdot I_o \tag{2}$$

C. Calculation of Switched Capacitor

The energy stored in the switched capacitor C_s during the operation period is given in (3).

$$E_c = \frac{1}{2} \cdot C_s \cdot V_{in}^2 \tag{3}$$

As demonstrated by the steps of operation, all the energy stored in the switched capacitor C_s is transferred to the load during half of the switching period of the controlled switches. Therefore, the average power transferred to the output, can be determined by (4). Since T_s defined by $1/f_s$ and f_s is the switching frequency of the converter.

$$P_o = \frac{E_c}{T_s/2} = C_s \cdot V_{in}^2 \cdot \eta \cdot f_s \tag{4}$$

The expression (4) shows that the power of the load depends on the capacitance and the input voltage. For V_{in} variations, the LED power can be stabilized by changing the switching frequency. In addition, if there is a decrease in the voltage of the LEDs, the power of the load tends to remain constant, by raising the current of the LEDs. Of course, for design purposes, the nominal value of the LED current must be respected. Another factor that is demonstrated by this expression that the power dimerization of LEDs is possible. In addition, the output power can be stabilized in open loop.

Isolating C_s (4), gives the SC capacitance as in (5).

$$C_s = \frac{P_o}{V_{in}^2 \cdot f_s \cdot \eta} \tag{5}$$

D. Resonant Inductor

Fig. (5) shows the simplified circuit equivalent to the first step of operation of the converter. The source V_e is the difference between the input voltage and the output voltage V_{in}-V_o. From the circuit $L_r C_s$ of Fig. (5), expressions (6) and (7) are obtained.

$$v_C(t) - V_e = I_{L0} \cdot Z_o \cdot sen(\omega_o t) + (V_{C0} - V_e) \cdot \cos(\omega_o t) \tag{6}$$

$$i_L(t) = I_{L0} \cdot \cos(\omega_o t) - \left(\frac{V_{C0} - V_e}{Z_o}\right) \cdot sen(\omega_o t) \tag{7}$$

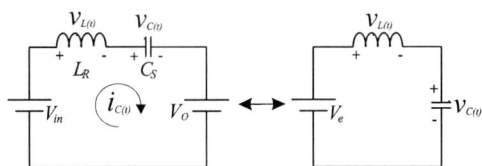

Fig. 5. Simplified circuit of Stage 1

Where is the impedance of the circuit defined by (8) and ω_o is the resonance angular frequency of the circuit defined by (9).

$$Z_o = \sqrt{\frac{L_r}{C_s}} \tag{8}$$

$$\omega_o = \frac{1}{\sqrt{L_r C_s}} \tag{9}$$

In to purpose of simplifying the expressions, it is considered:

$$\omega_o \cdot t = \theta \tag{10}$$

Normalizing (6) and (7) as function of the following factors: V_e = Voltage normalization fator and $V_e/_{Z_o}$ = current normalization factor, obtains:

$$v_{CN}(t) - V_{eN} = I_{L0N} \cdot \text{sen}\theta + (V_{C0N} - V_e) \cdot \cos\theta \tag{11}$$

$$i_{LN}(t) = I_{L0N} \cdot \cos\theta - (V_{C0N} - V_{eN}) \cdot \text{sen}\theta \tag{12}$$

Being (11) and (12) a system, eliminating the θ, (13) is obtained.

$$i_{LN}(t)^2 + (v_{CN}(t) - V_{eN})^2 = I_{L0N}^2 + (V_{C0N} - V_{eN})^2 \tag{13}$$

The voltage across capacitor C_s at instant t_0 is zero ($V_{C0} = 0$). At instant t_1 this voltage equals the difference of the converter input voltage with the converter output voltage ($v_C(t_1) = V_{in} - V_o = V_e$). Applying this condition in (13), the normalized current of the inductor L_r is obtained for the instant t_1 is given by (14).

$$i_{LN}(t_1) = \sqrt{I_{L0N}^2 + 1} \tag{14}$$

Removing the current normalization factor from equation (14) and knowing that at instant t_0 the current in the inductor is the minimum $i_{L0} = I_{min}$ and at instant t_1 this current is the maximum ($i_L(t_1) = I_{max}$), obtain the expression. That determines L_r is defined by (15).

$$L_r = \frac{C_s \cdot V_e^2}{I_{max}^2 - I_{min}^2} \tag{15}$$

The current ripple in the inductor can be defined as:

$$\Delta i_L = I_{max} - I_{min} \tag{16}$$

Considering that in the continuous conduction mode there is a low current ripple, the following approximation can be applied:

$$I_{max} + I_{min} = 2I_o \tag{17}$$

Applying (16), (17) in (15), can be obtained L_r as a function of the current ripple designed for the inductor in (18).

$$L_r = \frac{C_s \cdot (V_{in} - V_o)^2}{2 \cdot I_o \cdot \Delta i_L} \tag{18}$$

III. Design Considerations

In order to evaluate the RSC DC-DC step-down converter in CCM, a 10 W prototype was implemented. Fig. 6 present complete circuit of the implemented RSC DC-DC LED driver. This work used the same PCB layout, components and type of LED, and altered only inductor to obtain a fair comparison with the results of the experiments of this article with the RSC converter in DCM [19]. Tabel I shows the components used in this work.

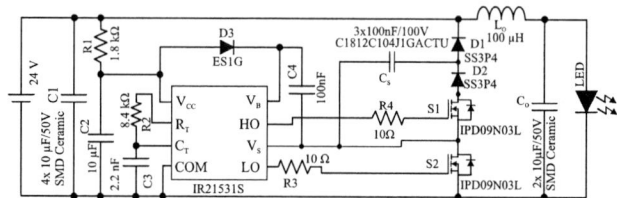

Fig. 6. Complete circuit of implemented LED driver

To emulated a DC-DC nanogrids, a 24 V regulated voltage source has used to power the converter. The RSC converter has been designed to supply a chip on board (COB) LED, where has a series resistance $R_{LED} = 3.726\ \Omega$, forward voltage $V_{LED} = 8.552$ V and rated current of 855 mA.

The capacitor C_s can be calculated form (5), as $C_s = 299.33$ nF, being $C_s = 300$ nF chosen as a commercial value. Inductor L_r can be defined by (18), being $L_r = 103.2\mu H$, considering a current ripple of 30%. The output capacitor C_o is designed to designed to operate at high frequencies according to the guidelines given in [23].

The capacitor C_s is a multilayer ceramic type (C0G C1812C104J1GACTU). This capacitor model has desirable characteristics, such as: low capacitance change with respect to temperature variation and no capacitance decay with time [24]. Capacitive filters were used in the input and output of the circuit for reducing the ripple current. In the input was used an equivalent capacitance of $40\mu F$, consisting of four multilayer ceramic capacitor of $10\mu F$. In the output was used two multilayer ceramic capacitor of $10\mu F$ connected in parallel.

Integrated circuit (IC) IR21531S has been used to drive the switches S_1 and S_2 with switching frequency of 58 kHz.

Table I
Prototype Components

Component	Value	Model
Switched Capacitor (Cs)	3x100nF/100V	C0GC1812C104J1GACTU *KEMET*
Inductor (Lo)	105μH	*CNF*-15/13,5/6-63-IP6
Output Filter Capacitor (Co)	2x10μF/100V	22201C106MAT2A
MOSFETs ($S1,S2$)	-	IPD09N03L *INFINEON*
Diodes ($D1,D2$)	-	SS3P4 *VISHAY*

Dimerization has performed using a variable resistor for a switching frequency of 26–58 kHz. The circuit has assembled on a plate with metal core that also has the function of dissipating the heat generated by the components, avoiding the use of external heatsinks.

IV. EXPERIMENTAL RESULTS

Fig. 7 presents waveforms of input voltage and input current and also the output voltage and current through the LEDs. The average values of the voltage and current input are 24 V and 314.7 mA, respectively. On the other hand, the output voltage is 10 V and the output current is 743.7 mA.

Fig. 8 represents the voltage and current waveforms regarding C_s and L_r, respectively. It can be seen, that C_s is fully charged and discharge over one switching period, the maximum voltage across it is equal input voltage, furthermore the switch turns off under ZVS condition. Besides that, it demonstrates the in CCM mode inductor with twice the switching frequency. Fig. 9 shows the voltage across and the current through switch S_1. It can be seen that the switch is turned off under and ZVS condition, with consequent reduction of switching losses and increase of converter efficiency. Due to the symmetry of the circuit, the waveforms on switch S_2 are similar.

Fig. 10 represents the linear behavior of the output power as a function of the switching frequency. To obtain the plot, the switching frequency was varied to cause the output power to vary. The switching frequency has been reduced from 58 kHz to 26 kHz causing that the output nominal power to be reduced

from 100% to 45%. If the output power is further reduced, the switching frequency is supposed to assume values within the hearing range, and consequently the output power has been limited to 45% in this case.

In order to measure the topology efficiency, calibrated power analyzer model PA4000 manufactured by Tektronix was employed, being obtained an overall efficiency of 92.80% for the rated load condition. Fig. 11 represents the curves of the

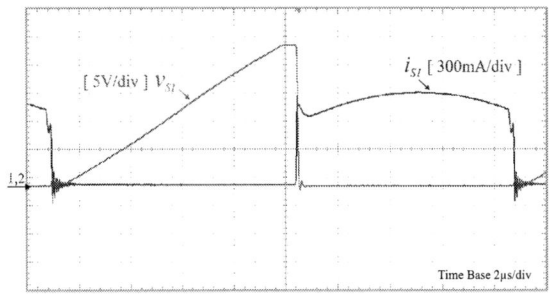

Fig. 9.   Voltage in S1 (CH1) and Current in S1(CH2).

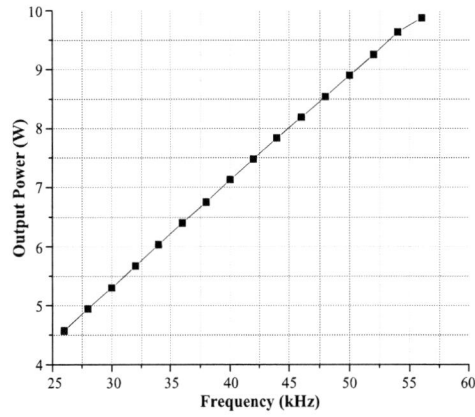

Fig. 10.   Output power variation as a function of the switching frequency

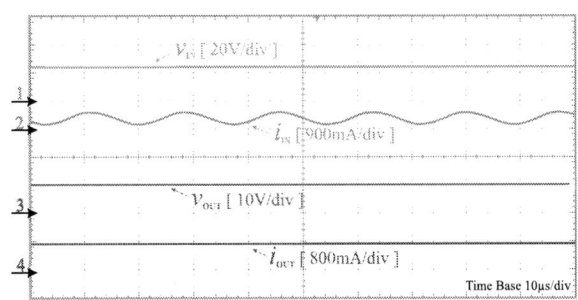

Fig. 7.   Input voltage (CH1) and Input current (CH2), Output voltage (CH3) and Output current (CH4).

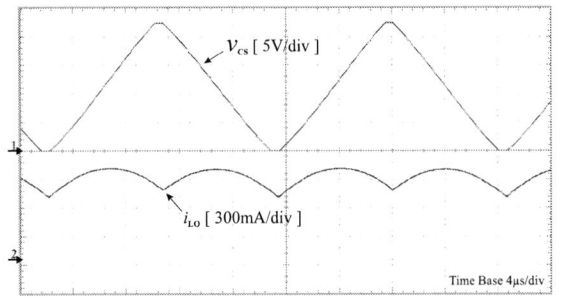

Fig. 8.   Voltage in switched capacitor (CH1) and Inductor current (CH2).

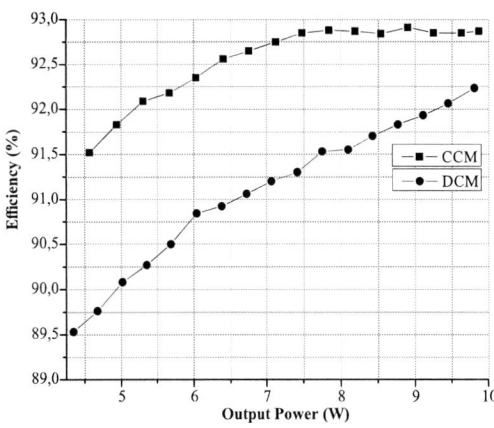

Fig. 11.   Experimental graph of the efficiency x output power with the converter operating in CCM and DCM.

RSC converter efficiency under CCM and DCM operation as a function of the output power. It can be seen that overall efficiency of the topology under CCM operation is higher than 91.5% over the entire load range. Besides, the efficiency of the RSC converter in CCM is higher than the efficiency of the RSC converter in DCM over the entire load range. It is important to mention, that the switches S_1 and S_2 are switched off under ZVS condition over the entire load range. Therefore, it can be stated that changing the mode of operation converter of the DCM to CCM achieves higher efficiency.

V. CONCLUSION

This work proposes a analysis of an RSC DC-DC step down converter in CCM to drive power LEDs fed from a nanogrid. An analysis of the topology demonstrated a structure in the CCM preserves the main characteristics of the topology in DCM, such as: the energy delivered to the load is determined by the switched capacitor, the output power stabilization without the need for current/voltage sensors, the L_r inductor allows the complete charge and discharge of the C_s capacitor, the switchs turns off under ZVS condition and compact size of the converter. The RSC topology in CCM showed na overall yield of 92.8% for the rated load condition. In addition, the overall efficiency of the CCM topology is higher than the DCM across over the entire load range. Future work comprises implementing several LED arrays at the output of the topology, ensuring the power equalization between the arrays and conserving the high efficiency.

ACKNOWLEDGMENT

The authors express their special thanks to Federal University of Ceara, Federal Institute of Ceará (IFCE) - campus Sobral, where the converter prototype was implemented and the financial support provided by the Brazilian government through PROINFRA/IFCE EDITAL N° 01/2017-PRPI, CNPq proc 432647/2016-3 and FUNCAP proc BP3-0139-00380.01.00/18.

REFERENCES

[1] W. Wu, H. Wang, Y. Liu, M. Huang, and F. Blaabjerg, "A Dual-Buck-Boost AC/DC Converter for DC Nanogrid with Three Terminal Outputs," IEEE Trans. Ind. Electron., vol. 64, no. 1, pp. 295–299, 2017.

[2] M. Rafiee Sandgani and S. Sirouspour, "Energy Management in a Network of Grid-Connected Microgrids/Nanogrids Using Compromise Programming," IEEE Trans. Smart Grid, vol. 9, no. 3, pp. 2180–2191, 2018.

[3] B. Nordman, K. Christensen, and A. Meier, "Think Globally, Distribute Power Locally: The Promise of Nanogrids," Computer (Long. Beach. Calif.), vol. 45, no. 9, pp. 89–91, 2012.

[4] D. Burmester, R. Rayudu, W. Seah, and D. Akinyele, "A review of nanogrid topologies and technologies," Renew. Sustain. Energy Rev., vol. 67, pp. 760–775, 2017.

[5] S. W. Lee and H. L. Do, "A Single-Switch AC-DC LED Driver Based on a Boost-Flyback PFC Converter with Lossless Snubber," IEEE Trans. Power Electron. , vol. 32, no. 2, pp. 1375–1384, 2017.

[6] Y. Wang, J. M. Alonso, and X. Ruan, "A Review of LED Drivers and Related Technologies," IEEE Trans. Ind. Electron. , vol. 64, no. 7, pp. 5754–5765, 2017.

[7] E. M. Sá Jr., "Estudo de Estruturas de Reatores Eletrônicos para LEDs de Iluminação," p. 25, 2010.

[8] A. Ioinovici, "Switched-capacitor power electronics circuits," IEEE Circuits Syst. Mag. , vol. 1, no. 3, pp. 37–42, 2001.

[9] On-Cheong Mak, Yue-Chung Wong, and A. Ioinovici, "Step-up DC power supply based on a switched-capacitor circuit," IEEE Trans. Ind. Electron. , vol. 42, no. 1, pp. 90–97, 2002.

[10] R. P. Coutinho, K. C. A. De Souza, F. L. M. Antunes, and E. Mineiro Sá, "Three-Phase Resonant Switched Capacitor LED Driver with Low Flicker," IEEE Trans. Ind. Electron., vol. 64, no. 7, pp. 5828–5837, 2017.

[11] E. Eloi Dos Santos Filho, P. H. A. Miranda, E. M. Sá, and F. L. M. Antunes, "A LED driver with switched capacitor," IEEE Trans. Ind. Appl., vol. 50, no. 5, pp. 3046–3054, 2014.

[12] E. Eloi, E. M. Sá, R. L. Dos Santos, P. A. Miranda, and F. L. M. Antunes, "Single stage switched capacitor LED driver with high power factor and reduced current ripple," Conf. Proc. - IEEE Appl. Power Electron. Conf. Expo. – APEC, vol. 2015–May, no. May, pp. 906–912, 2015.

[13] G. Martinez and J. M. Alonso, "A review on switched capacitor converters with high power density for OLED lamp driving," IEEE Ind. Appl. Soc. - 51st Annu. Meet. IAS 2015, Conf. Rec., pp. 1–8, 2015.

[14] K. W. E. Cheng, "New generation of switched capacitor converters," no. c, pp. 1529–1535, 2002.

[15] S. Li, W. Xie, and K. M. Smedley, "A Family of an Automatic Interleaved Dickson Switched-Capacitor Converter and Its ZVS Resonant Configuration," IEEE Trans. Ind. Electron., vol. 66, no. 1, pp. 255–264, 2019.

[16] S. Li, Y. Zheng, and K. M. Smedley, "A Family of Step-Up Resonant Switched-Capacitor Converter with a Continuously Adjustable Conversion Ratio," IEEE Trans. Power Electron. , vol. 34, no. 1, pp. 378–390, 2019.

[17] A. F. Da Rocha, E. M. Sá, C. E. A. Silva, G. E. Martins, and E. R. Marques, "A step-up switched-capacitor converter for LEDs applied to photovoltaic systems," IEEE Int. Symp. Ind. Electron., vol. 2015–Septe, pp. 1159–1165, 2015.

[18] M. Martins, M. S. Perdigao, A. M. S. Mendes, R. A. Pinto, and J. M. Alonso, "Analysis, design, and experimentation of a dimmable resonant-switched-capacitor LED driver with variable inductor control," IEEE Trans. Power Electron. , vol. 32, no. 4, pp. 3051–3062, 2017.

[19] J. S. Ferreira, P. S. Almeida, M. B. M. Morais, G. E. Martins, and E. M. Sá, "A step-down converter with Resonant Switching Capacitor applied in nanogrids to drive power LEDs," 2017 IEEE 8th Int. Symp. Power Electron. Distrib. Gener. Syst. PEDG 2017, 2017.

[20] R. W. Erickson and D. Maksimovic, "Resonant converters," in Fundamentals of Power Electronics, 2nd ed. Norwell, MA, USA: Kluwer, 2001, ch. 19, pp. 705–759

[21] . F. Rocha, E. R. Marques, C. E. A. Silva, and E. M. Sá, "A step-up converter with switched capacitor using a small inductor in CCM to drive power LEDs," 2015 IEEE 13th Brazilian Power Electron. Conf. 1st South. Power Electron. Conf. COBEP/SPEC 2016, vol. 3, pp. 15–19, 2015.

[22] M. K. Kazimierczuk, Pulse-Width Modulated DC-DC Power Converters. Chichester, West Sussex: John Wiley & Sons, Ltd., 2008.

[23] Cirrus Logic, "AN376 single stage output ripple current and the effect on load current in a LED driver," AN376REV2, Jul. 2013.

[24] X. R. Dielectric and V. D. C. C. Grade, "Packaging C-Spec Ordering Options Table," vol. 29606, no. 864, pp. 1–22, 2015.

High-Gain Bidirectional DC-DC Converter for Battery Charging in DC Nanogrid of Residential Prossumer

1st Fernando Queiroz
Department of Electrical Engineering
Federal University of Ceará
Fortaleza/CE, Brazil
fernando.ufc@gmail.com

2st Paulo Praça
Department of Electrical Engineering
Federal University of Ceará
Fortaleza/CE, Brazil
paulopp@dee.ufc.br

3st Alisson Freitas
Department of Electrical Engineering
Federal Rural University of the Semi-Arid
Mossoro/RN, Brazil
alisson.freitas@ufersa.edu.br

4st Fernando Antunes
Department of Electrical Engineering
Federal University of Ceará
Fortaleza/CE, Brazil
fantunes@dee.ufc.br

Abstract— **This paper presents a bidirectional DC-DC converter with high gain voltage, based on three state switching cell, to charge a battery bank with 96 Vdc operating voltage connected to a 380 Vdc nanogrid bus. This converter can be applied to a DC current nanogrid for residential prosumers and the battery bank can be used to power an electric vehicle, as well as provide power to the nanogrid DC bus. The converter can operate in buck or boost modes. The proposed control aims to guarantee both the bidirectionality of the power flow of the converter and the stability of the reference in steady state. For this, a cascade digital control system is designed through of the average current mode control technique, whose effectiveness of the control implementation and the bidirectionality of the converter are proven and presented by simulation.**

Keywords— *Distributed generation, battery bank, DC nanogrid, decentralized system, energy storage.*

I. INTRODUCTION

With the technological advancement of power electronics, the direct current distribution has become a reality. This exchange of alternating current by direct current has been carried out over the years, but now it is perceived that a new phase has begun, due essentially to two factors: the growth of urban centers and the expansion of renewable energy sources.

One of the focus areas for solving the problems faced by energy generation policies is the microgrid. The microgrid is mainly focused on residential and commercial applications, as these sectors account for about 50% of energy consumption at national and global levels [1],[2].

In residences, when the installed power of the distributed generation plants is less than 25 kW, the term nanogrid can also be used in [3], [4]. Fig. 1 shows the nanogrid schematic used as reference for this paper, where the bidirectional converter used in [5] is connected to the battery bank of an electric vehicle and to the DC bus of 380 Vdc. This nanogrid presents different levels of voltage in its buses, the functionality of which can be realized in two ways, namely grid connected mode or islanded mode.

During normal operation, the nanogrid operates in the mode connected to the utility network and in case of failure, it disconnects from the common coupling point and starts operating in the isolated mode or island mode.

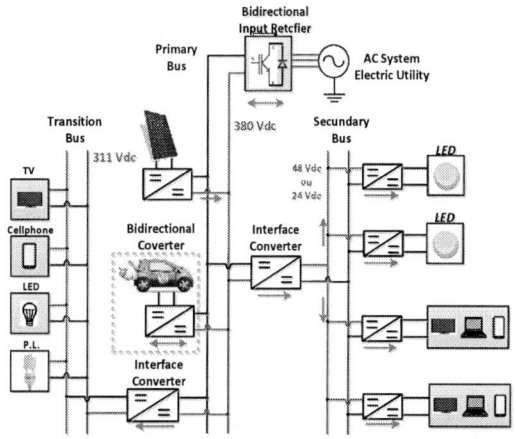

Fig. 1. Nanogrid diagram used as reference, especially the bidirectional DC-DC converter.

Many studies, such as [5], [6] and [7], address control methodologies for each mode of operation of the converter, which is a disadvantage because it makes control more complex and causes greater difficulty in experimental implementation. Thus, in this work a bidirectional converter is proposed that presents a unique control strategy in both directions. The characteristics presented by the proposed control strategy are:

- Ensure stability for both modes of operation;

- Ensure the bidirectionality of the converter without the need to change any parameter of said circuit, regardless of the power flow being processed by the converter;

- Allow the voltage ripples to be reduced on the batteries, guaranteeing longer battery life;

- Ensures that there is no improper switching between operating modes by the presence of voltage noises;

- Indirectly, through the voltage controller, the current reference is dependent on the DC bus voltage, thus avoiding communication between the nanogrid voltage sources;

The results presented throughout this work show the effectiveness of the proposed converter, thus the ability of the

978-1-7281-4181-7/19 $31.00 © 2019 IEEE

control system to guarantee the bidirectionality and the stability of the process.

II. PROPOSED BIDIRECTIONAL DC-DC CONVERTER

The proposed converter is shown in Fig. 2. The topology adopted considers the high input current of the converter and the power processed. This converter consists of the voltage V_1 formed by a bank of batteries of 96 Vdc, and by the source V_2, formed by the 380 Vdc bus, as well as by an energy storage inductor L_b, the transformer Tr, which is composed of the primary windings N_{P1} and N_{P2} and the secondary windings N_{S1} and N_{S2}. The inverter has six switches S_1, S_2, S_3, S_4, S_5 and S_6 and three capacitors C_1, C_2 and C_3, and can operate in buck or boost modes.

The operating principle of the converter in boost mode is dependent on the control signals of the S_1 e S_2, analyzed in continuous conduction mode (CCM), and considering the duty cycle of the signals on the switches greater than 0.5, (overlapping mode), more specifically 0.62. In this mode, the operation of the converter is divided into four stages, during the transfer of power from source V_1 to V_2, i.e., from batteries to the nanogrid DC bus.

In the buck operation mode, power is transferred from the DC bus (V_2) to the battery bank (V_1), considering the CCM and the duty cycle is less than 0.5 (non-overlapping mode), more specifically 0.38. In the buck mode, the switches. In buck mode, the switches S_3, S_4, S_5 and S_6 are controlled, allowing the intrinsic diodes DI_1 and DI_2, In the buck mode, the switches S_3, S_4, S_5 and S_6 are controlled, allowing the intrinsic diodes DI_1 and DI_2, respectively, to correspond to switches S_1 and S_2, to operate passively, thus causing bidirectional power flow. In this scenario, the system operates as a battery charger when the inductor current direction reverses. For this mode of operation, the converter is also divided into four stages.

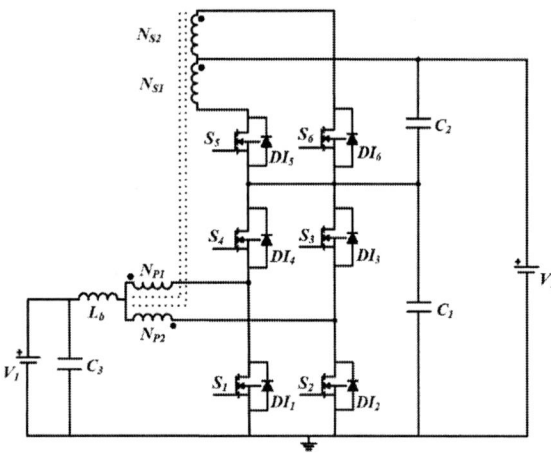

Fig. 2 – Bidirectional high-gain converter proposed.

A. Boost Mode Operation Stages

The main voltage and current waveforms on the proposed converter components in boost mode are shown in Fig. 3, where D_2 is the cyclic ratio or duty cycle, and T_S is the switching period of the switches

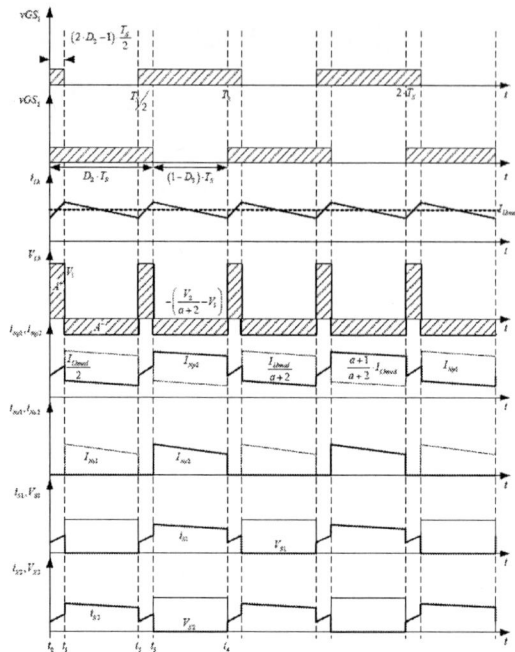

Fig. 3 – Main converter waveforms operating in boost mode.

In this mode the drive operation is divided into four stages, as described below.

First Stage ($t_o \leq t \leq t_1$): In this first stage the switches S_1 and S_2 enter driving, but the switches S_3, S_4, S_5 and S_6 remain locked. In this way the source V_1 supplies power to the inductor L_b through the current I_{Lb}, where I_{Lb} is divided among I_{Np1} and I_{Np2}, which respectively travel through the windings N_{p1} and N_{p2}, passing through the switches S_1 and S_2, the currents I_{S1} and I_{S2}, respectively. In the proposed converter, N_{p1} and N_{p2} have the same number of turns, causing the induced voltage in the autotransformer to equal zero. In this stage the current I_{Lb} increases linearly, storing energy in the inductor L_b and not being transferred to the load, which is supplied by the capacitors C_1 e C_2.

Second stage ($t_1 \leq t \leq t_2$): At this stage, the switch S_1 is locked and the switch S_2 remains on. The diodes of the switches S_4 and S_6 are directly polarized, providing that the source V_2 and the capacitors C_1 and C_2 receive energy from the source V_1 and the inductor L_b, causing the current on it to decrease linearly. Thus, it is assumed that the voltage on the switch S_1 is equal to the voltage on the capacitor C_1.

Third Stage ($t_2 \leq t \leq t_3$): At this point, the switch S_2 remains in the drive and the switch S_1 goes into conduction, making this stage look like the first one. In this way, the switches will be connected, and the energy will be stored in the inductor L_b, where the load is fed by the capacitors C_1 and C_2, which were loaded in the previous stage.

Fourth Stage ($t_3 \leq t \leq t_4$): During this period, the switch S_2 is blocked and the switch S_1 remains conducting, causing the voltage on the switch S_2 to be the same as that of the capacitor C_1, in view that the diodes DI_3 and DI_5 will be directly polarized. In this stage, the energy stored in the inductor L_b, during the third stage is transferred to the capacitors C_1 and C_2, as well as to the load.

The static gain of the converter in the boost mode is determined by means of the average voltage on the inductor and is given by (1).

$$G_{boost} = \frac{V_2}{V_1} = \frac{a+2}{2(1-D_{boost})} = 3.954 \qquad (1)$$

Where:

G_{boost} – Static gain in boost mode
D_{boost} – Duty cycle in boost mode
a – Transformation ratio of transformer

B. Buck Mode Operation Stages

Fig. 4 shows the main waveforms of the voltage and current on the components of the proposed converter in buck mode, where D_1 is the duty cycle.

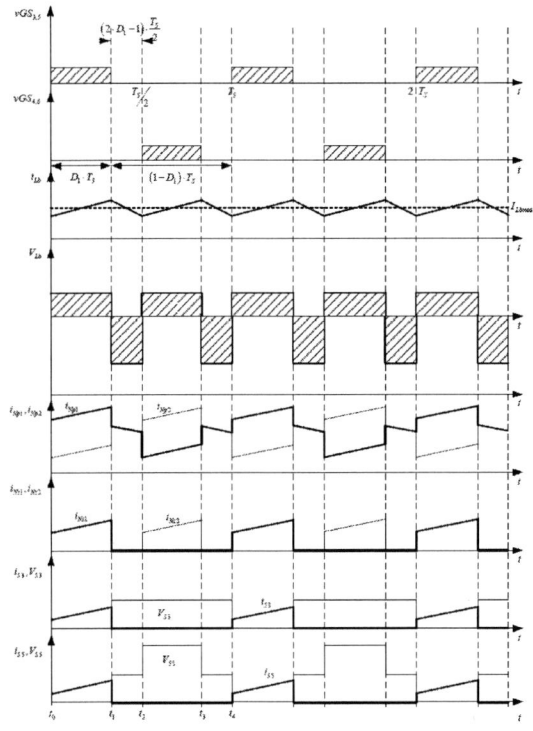

Fig. 4 – Main converter waveforms operating in buck mode.

For this mode of operation, the drive is also divided into four stages, as described below.:

First Stage ($t_o \leq t \leq t_1$): Initially the switches S_3 and S_5 receive the command of the control signal and begin to drive, while the switches S_4 and S_6 remain locked. At this time the diode D_1, of the switch S_1, is reverse polarized, so the winding N_{S1} induces the winding N_{p1}, causing a transfer of energy to inductor L_b, charging the batteries through the current I_{Lb}. The end of this stage is given when the switches S_3 and S_5 are locked.

Second Stage ($t_1 \leq t \leq t_2$): At that stage the switches S_3 and S_5 are blocked by means of the control signal, while the switches S_4 and S_6 remain locked as well as the diode DI_2 intrinsic to the switch S_2 is directly polarized and starts to drive. Source V_1, which represents the batteries, is isolated from source V_2, which represents the DC bus, so only

capacitor C_3 supplies source V_1. It is also assumed that the inductor L_b is discharger linearly, transferring energy to V_1. At this moment the current I_{Lb} enters free wheel through the diodes DI_1 and DI_2, through the windings of the transformer and returning by the inductor.

Third Stage ($t_2 \leq t \leq t_3$): At time t_2 the switch S_4 and the switch $S6$ begin to drive, remaining thus until the end of this stage. However, the switches S_3 and S_5 remain locked and the diode DI_2 is directly polarized. In this stage, the source V_2 transfers energy to the source V_1, via the switches S_4 and S_6, causing the winding N_{S2} to induce the winding N_{p2}, generating a transfer of energy to the input inductor L_b and, consequently, charging the batteries through the current I_{Lb}. It should be mentioned that this stage is similar to the first.

Fourth Stage ($t_3 \leq t \leq t_4$): At the beginning of this period, the switches S_4 and S_6 receive the blocking signal and remain in this state throughout this stage, where the switches S_3 and S_5 remain locked. The energy stored in the input inductor is transferred to the source V_1, causing the diodes DI_1 and DI_2 to operate in freewheeling.

The static gain of the converter in the buck mode is determined by means of the average voltage on the inductor and is given by the expression (2).

$$G_{buck} = \frac{V_1}{V_2} = \frac{2 \cdot D_{buck}}{a+2} = 0.253 \qquad (2)$$

III. Proposed System design

A. Converter Specifications

$V_1 = 96\,Vdc$ (Battery voltage)
$V_2 = 380\,Vdc$ (DC Bus Voltage)
$P_0 = 2000\,W$ (Power output)
$R_0 = 72.2\,\Omega$ (Rated load resistance)
$f_S = 20000\,Hz$ (Switching frequency)
$a = 1$ (Transformation ratio)

B. Inductor Parameter

Considering the first stage of operation, the undulation of the inductor current is given by (3).

$$\Delta I_{Lb} = \frac{V_2(2 \cdot D_{boost} - 1)(1 - D_{boost})}{(a+2) \cdot L_b \cdot f_s} \qquad (3)$$

Where f_s is the switching frequency that is equal to 20k Hz.

Parametrizing equation (3), we have expression (4), through which the graph of Fig. 5, can be traced, varying the duty cycle (D_{boost}) from 0.5 to 1.0, and obtain the maximum point as a function of the cyclic ratio.

$$\overline{\Delta I_{Lb}} = \frac{\Delta I_{Lb}(a+2) \cdot L_b \cdot f_s}{V_2} \qquad (4)$$

$$\overline{\Delta I_{Lb}} = (2 \cdot D_{boost} - 1)(1 - D_{boost}) \qquad (5)$$

A Fig. 5 shows the graph showing the point of the maximum current ripple where the duty cycle is equal to 0.75

and the parameter β_{boost} (which represents the normalized current ripple in the inductor L_b) is equal to 0.125.

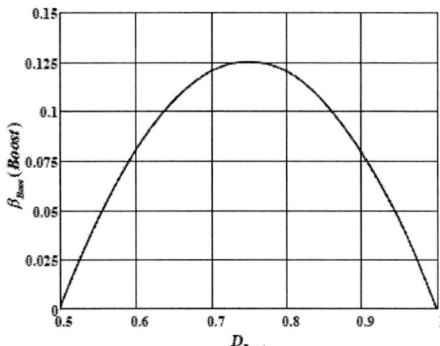

Fig. 5 – Normalized inductor current graph.

Thus, the inductance of the inductor L_b can be determined by expressions (6).

$$L_b = \frac{V_2(2 \cdot D_{boost} - 1)(1 - D_{boost})}{(a + 2) \cdot \Delta I_{Lb} \cdot f_s} \tag{6}$$

Therefore, using the maximum current ripple, shown through Fig. 5, it is assumed that the inductor of the inductor and its respective value is given according to equation (7).

$$L_b = \frac{1}{8} \cdot \frac{V_2}{(a + 2) \cdot \Delta I_{Lb} \cdot f_s} = 364 \, \mu H \tag{7}$$

C. Transformer Design

It is worth mentioning that the high frequency transformer must be designed according to the amount of power processed (P_P), which was determined by equation (8), where P_2 is the output power.

$$P_P = \frac{P_2}{2} \cdot \frac{2a + 1}{2 + a} = 1042 \, W \tag{8}$$

D. Capacitors Design

The capacitance of the capacitors of the proposed converter can be determined by equations (9) and (10).

$$C_1 = C_3 \geq \frac{P_2 \cdot (1 - D_{boost})}{2(a + 2) \cdot \Delta V_2 \cdot f_s \cdot V_1} \tag{9}$$

$$C_1 = C_3 = 8.6 \, \mu F$$

$$C_2 \geq \frac{P_2 \cdot (1 - D_{boost})}{(a + 2) \cdot \Delta V_2 \cdot f_s \cdot V_1} = 17.3 \, \mu F \tag{10}$$

Where ΔV_2 is the maximum total voltage ripple.

It is worth mentioning that, due to the ease of being in the commercial market, it was decided to use the three capacitors in the value 47 μF, with $Rse = 25 \, m\Omega$.

E. Equivalent Converter Model

A Fig. 6 shows the equivalent schematic with modeling variables of the bidirectional converter implemented using the small signal PWM switch model, applied through the technique presented in [8].

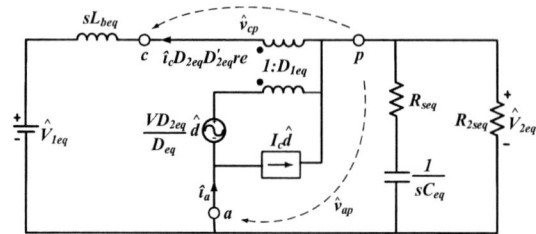

Fig. 6 - Equivalent boost converter model using the PWM switch's CA model.

The Table 1 shows the parameters of the bidirectional converter considering the new switching period, frequency and duty cycle based on the equivalent signal. These results are fundamental to obtain control of the equivalent and bidirectional converter.

TABLE I.   EQUIVANTES CONVERTER PARAMETERS.

Parameter	Equation	Calculated Value
Duty Cycle	D	0.621
Duty Cycle Equivalent	$D_{eq} = 2D - 1$	0.242
Complement of the equivalent Duty Cycle	$D'_{eq} = 1 - D_{eq}$	0.758
Frequency of switching equivalent	$f_{seq} = 2 \cdot f_S$	40k Hz
Voltage of input equivalent	$V_{1eq} = V_1$	96 Vdc
Equivalent Output Voltage	$V_{2eq} = V_{1eq}\left(\frac{1}{1 - D_{eq}}\right)$	126.66 Vdc
Ratio between V2 and V2eq	$Rv = \frac{V_2}{V_{2eq}}$	3
Inductor equivalent	$L_{beq} = L_b$	364 µH
Capacitor equivalent	$C_{eq} = Rv^2 \cdot C_1$	405 µF
Resistance series equivalent	$R_{seq} = \frac{R_{se}}{Rv^2}$	2.8 mΩ
Resistance of load equivalent	$R_{2eq} = \frac{R_2}{Rv^2}$	8.022 Ω

Developing and deducing the expressions of the equivalent circuit of the converter through conventional circuit modeling techniques and performing the manipulation mathematically, the transfer functions for current loop and voltage are obtained. It is necessary to inform that this work only aims at the development and analysis of the proposed converter project. getting the control system for a later approach.

978-1-7281-4181-7/19 $31.00 © 2019 IEEE

IV. CONTROL SYSTEM DESIGN

The control strategy to be applied must predict the bidirectionality of the current in the inductor of the converter in the two modes of operation (buck or boost). For this, the average current mode control is used. In this way, the control strategy was designed to include both modes of operation. Fig. 7 shows the block diagram representing the applied control strategy.

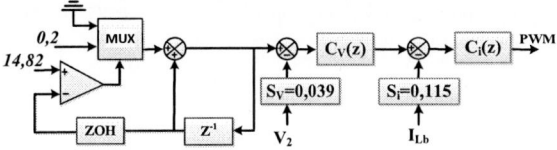

Fig. 7 – Block diagram of the control circuit through the technique by the average current mode.

In the schematic of the control system of Fig. 7, the blocks represented by the components $C_v(z)$ and $C_i(z)$, respectively contain the terms of the transfer functions of the calculated voltage and current grid compensators, which are already discretized, in view of the implementation of digital control.

The block diagram shows that the control uses the cascade technique. The components $S_v = 0.039$ and $S_i = 0.115$ represent, respectively, the gains of the sample measurement sensor of the output voltage and the sample measurement sensor of the current in the inductor of the converter.

In the control block diagram, we have a multiplexer (*MUX*), unit delay operator (z^{-1}), a zero order insurer (*ZOH*), a comparator and two constants, in the value of 0.2 and 14.82. These components constitute a ramp soft-starter model, which has the purpose of allowing the soft starting of the converter. Thus, initially, the value of the voltage reference passes through an integrator and the result is compared with the value of 14.82, which represents 380 Vdc. If they are not equal, the reference value is increased by 0.2, thus guaranteeing the ramp response in the voltage reference until the desired value is reached.

V. SIMULATION RESULTS

The computational simulations were implemented with the purpose of validating the proposed system and were performed through the software PSIM. The results of the simulations were analyzed in steady state and the value of the positive current was adopted as the indication of the direction of charge of the batteries, i.e., the converter operating in buck mode, and the current with negative value indicating the converter in the mode boost.

The system was simulated with the nanogrid scenario disconnected from the utility's DA distribution network, i.e. operating in isolated or islanded mode. Thus, a variation of $\pm 10\%$ of the nominal voltage level of the DC bus was applied, thus seeking to simulate disturbances in a range of voltage rise and sink.

The Fig. 8 shows the voltage in the DC bus, where it is observed that the voltage level varies from 342 Vdc to 418 Vdc, simulating the presence of overvoltage and undervoltage in the bus, but maintaining the average voltage around 380 Vdc. The Fig. 9 shows the behavior of voltage and current through the batteries.

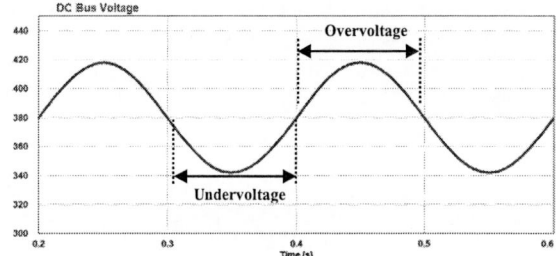

Fig. 8 – Variation of the voltage in the DC bus simulating the insertion of Overvoltage and undervoltage disturbances.

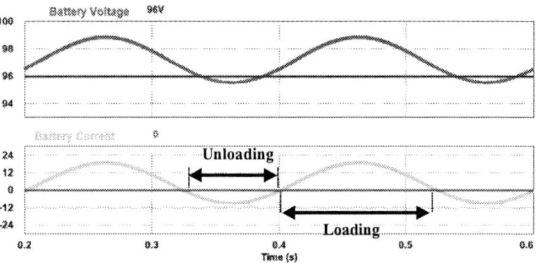

Fig. 9 – Voltage and current in the batteries by varying the voltage level of the DC bus.

It is observed in Fig. 9 that during the period of overvoltage in the DC bus the current is positive and continues to increase as the bus voltage increases, showing that the converter operates in buck mode, soon the batteries are charging. In the undervoltage period on the DC bus, it is noted that the current is negative because the converter starts to operate in boost mode and tends to supply the DC bus through the battery power, causing them to discharge. It is observed that the system presents a longer period with the current of the batteries in the positive value. This shows that the converter operates mostly in buck mode, i.e. with the batteries charging, which is the most desired situation for the proposed system.

In Fig. 10 the voltage and current waveforms are shown in the inductor L_b, where the maximum voltage in the inductor is the same input voltage V_1, i.e., the voltage of the batteries, which is 96 Vdc. The mean current in L_b is the same as the input current I_1. The current value is negative only to indicate that the drive is operating in boost mode, as agreed.

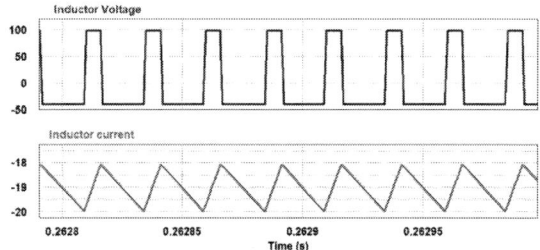

Fig. 10 – Waveforms of the voltage and current in the inductor in boost mode

In Fig. 11 the waveforms of the currents through the switches S_1 and S_2 are shown during the boost mode. It will be seen that during the duty cycle the switches S_1 and S_2

conduct at the same time and then only one of the switches remains in conduction while another is locked.

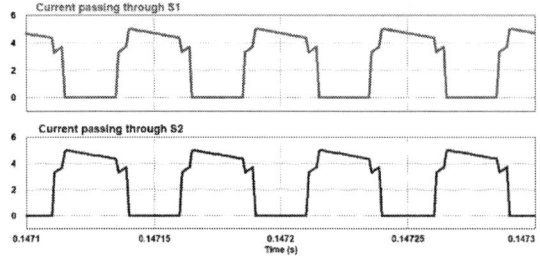

Fig. 11 - Currents waveforms through switches S_1 and S_2.

In Fig. 12 the waveforms of the currents that flow through the intrinsic diodes DI_1 and DI_2 of switches S_1 and S_2, are shown in buck mode.

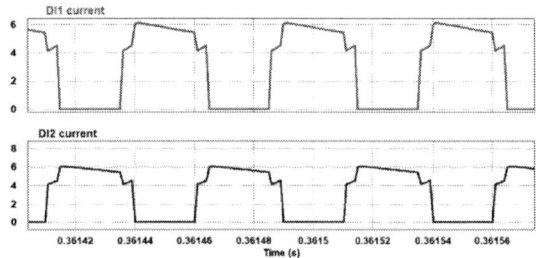

Fig. 12 – Current on the intrinsic diodes of switches S_1 and S_2.

In Fig. 13 the voltage and current waveforms are shown in the filter inductor L_b, where the maximum voltage (in modulus) on the inductor is the same input voltage V_1, i.e., the voltage of the batteries, which is 96 Vdc. The average current in L_b is the same as the input current I_1. It is observed that current value is positive, characterizing that the converter is operating in buck mode, as it was agreed.

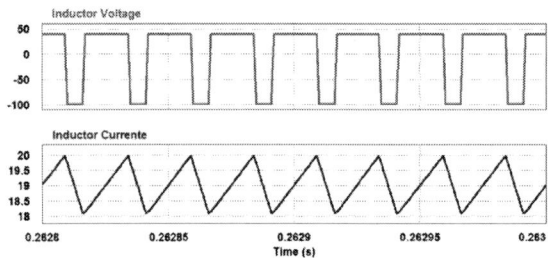

Fig. 13 – Waveforms of the voltage and current in the inductor with the converter operating in buck mode.

VI. CONCLUSION

This work presents a study of the implementation of the control system of a bidirectional DC-DC converter with high voltage gain, based on the three-state switching cell.

The main components of the bidirectional converter and the transfer functions obtained were calculated. The control system was designed on the average current method using PI type controllers with filter. The implementation of the digital control system was done through discretization of the system.

By means of simulations it was possible to validate the converter operation, together with the control system, since the results achieved were compatible with the theory conceived.

The converter operated in both operating modes, i.e., in buck mode and boost mode, showing the efficiency of power flow bidirectionality as well as steady state system stability, proving that it can be perfectly used in the DC nanogrid used as reference.

The next stage of this work is the application of the proposed control strategy on a prototype of the bidirectional DC-DC converter implemented in the laboratory, seeking to prove the theoretical results through experimental application.

ACKNOWLEDGMENT

The authors thank the Energy and Control Processing Group (GPEC), as well as the Post-Graduation Program in Electrical Engineering of the Federal University of Ceará (UFC).

REFERENCES

[1] Energy Information and Administration, "International energy outlook 2013," tech. rep., U.S. Energy Information and Administration (EIA), 2013.

[2] Empresa de Pesquisa Energética (EPE), "Balanço Energético Nacional 2015: Ano base 2014," tech. rep., Ministério de Minas e Energia, Rio de Janeiro, Brasil, 2015.

[3] Boroyevich, D. Future Electronic Power Distribution Systems - A Contemplative View. Maio 2010, Brasov: IEEE, pp. 1369–1380. Disponível em: <http://ieeexplore.ieee.org/document/5510477/>, acesso em 27 de janeiro de 2018.

[4] Schonberger, J. et al., "Dc-bus Signaling: A Distributed Control Strategy for Hybrid Renewable Nanogrid," IEEE Transactions on Industrial Electronics, vol. 53, no. 5, pp. 1453-1460, 2006.

[5] Marques, D. D. Conversor bidirecional CC-CC de alto ganho para aplicação em sistemas autônomos de geração de energia elétrica. XIX Congresso Brasileiro de Automática (CBA), 2012, pp. 1886-1893.

[6] Melo, R. R., Conversor CC-CC bidirecional aplicado a supercapacitores para veículos elétricos, Dissertação (mestrado), Universidade Federal do Ceará, 2014.

[7] Alves, D. B. S., Conversor Boost de alto ganho baseado na versão bidirecional da célula de comutação de três estados, Dissertação (mestrado), Universidade Federal do Ceará, 2014.

[8] Vorperian, V. Simplified Analysis of PWM Converters Using Model of PWM Switch Part I: Continuous Conduction Mode Aerospace And Electronic Systems, IEEE Transactions on Power Systems, Vol.26, No. 2, pp. 490-496, 1990.

Concepts and Case Study of Mismatch Losses in Photovoltaic Modules

Elson Yoiti Sakô
FEEC
University of Campinas (Unicamp)
Campinas, Brazil
elsonyoiti8@gmail.com

João Lucas de Souza Silva
FEEC
University of Campinas (Unicamp)
Campinas, Brazil
jlucas.souzasilva@gmail.com

Daniel de Bastos Mesquita
FEEC
University of Campinas (Unicamp)
Campinas, Brazil
mesquita.b.daniel@gmail.com

Rafael Espino Campos
FEEC
University of Campinas (Unicamp)
Campinas, Brazil
respinoc.154@gmail.com

Hugo Soeiro Moreira
FEEC
University of Campinas (Unicamp)
Campinas, Brazil
moreirahugos@gmail.com

Marcelo Gradella Villalva
FEEC
University of Campinas (Unicamp)
Campinas, Brazil
villalva@unicamp.br

Abstract—**With the diffusion of the implantation of photovoltaic systems, it became necessary to research to increase efficiency and improve the payback of the investment. In this way, losses in photovoltaic systems are an obstacle, and among the several losses, losses by mismatch can be highlighted. Thus, the paper proposes to collect in literature and classify the causes and to point out the solutions for mismatch losses in a PV system, and posteriorly a case study on a PV system comparing the PV modules with dirt and clean. An I-V curve tracer was used to perform measurements with better accuracy. As a result, the mismatch losses happened in the photovoltaic system in the test and had an amplification when the photovoltaic modules were soiling, in both cases, limited the photovoltaic system of reach maximum power. Therefore, it has been shown that mismatch losses can occur in several ways, and can considerably influence the performance of a photovoltaic system.**

Index Terms—**Renewable energy, photovoltaic, architecture, power optimizer for photovoltaic systems, POPS**

I. Introduction

Due to the diffusion of the implantation of photovoltaic (PV) systems, mainly due to the sustainability question [1] and the search for diversification of energy matrices in many countries, it has become essential researches that attempt to understand and solve problems related to generation losses of PV energy. Some factors such as environmental, installation, component manufacturing, cost factors and other miscellaneous factors [2] can influence the performance of PV systems and cause losses.

One of the factors miscellaneous is mismatch losses. Mismatch losses arise due to different characteristics in the PV module of the same string [3], leading to electrical incompatibility. Such losses can be induced by several factors that cause the unbalance in the characteristics between modules connected in the same string, that can be temporary and/or throughout the life of the PV module.

Several are the papers that investigate causes or solutions that may provide mismatch losses in PV systems [3]–[9]. However, it is interesting to gather and classify the different causes and to point out the solutions that exist in the literature. In this way, the problems caused by mismatch losses can be understood and reduced.

Thus, this paper aims to collect in literature and classify the causes, and the solutions of mismatch losses in a PV system, posteriorly present a case study on a PV system comparing the PV modules with dirt and clean. So, it is possible to verify the presence of the mismatch in two situations for the same PV system. Therefore, the following items can be highlighted as a scientific contribution:

- The paper ranked the causes of losses by mismatch effect;
- The paper listed solutions for the mismatch losses;
- The paper presented a case study on a PV system under dirty and clean conditions, showing the mismatch losses in practice.

The paper follows with section II presenting the mismatch losses; agents generatives of mismatch losses, that is, how it is caused; beyond solutions. Then section III presents the case study of mismatch losses in practice, closing with the final considerations.

II. Mismatch Losses in Photovoltaic System

A photovoltaic system consists of a set of modules connected in series and parallel. These modules consist of a set of cells connected in series, parallel or in SPS (series - parallel - series). Such combinations provide the flexibility to achieve desirable voltage, current and power values.

Because of the connection between cells and between modules, it is important that they have the same characteristics. However, even photovoltaic modules that are from the same manufacturer, same batch and with the same classifications are not identical. With cells and photovoltaic modules connected to each other, this non-ideality among system components causes energy losses that are known as mismatch.

Fig. 1 shows an example of mismatch losses. Three PV modules with theoretical power of 190 W each, are inter-

978-1-7281-4181-7/19 $31.00 © 2019 IEEE

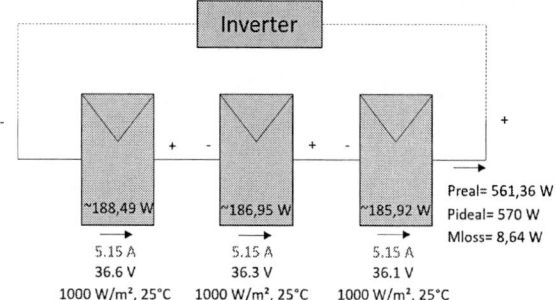

Fig. 1. Example of mismath loss in PV system

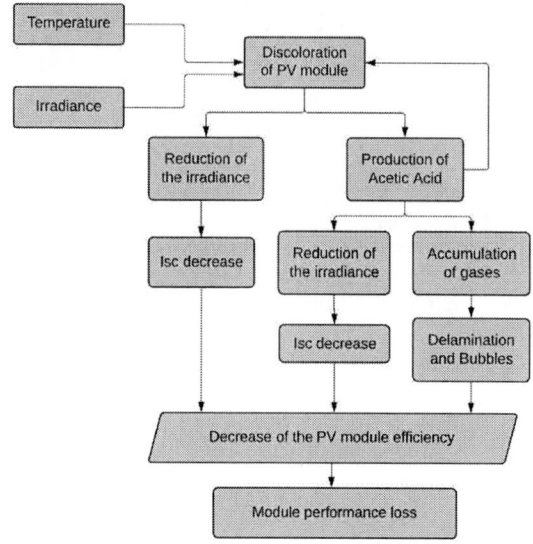

Fig. 2. Fluxogram of the chemical effect that arises over the PV module's mismatch. Figure remade from [10]

connected in series. Due to the manufacturing issue of the modules, one of the PV modules has 5.15 A of current limiting the others to the same current. The voltage, although close, is also different, resulting in the total power of 561.36 W, when it was to be ideally 570 W when exposed in STC condition (1000 W/m^2 and 25 °C).

A. Agents generative of Mismatch Losses

According to [4], there is a close loop between the mismatching and the aging of PV modules, i.e, as long as the modules are subjected to certain conditions that lead to deteriorating itself, more the lifetime is reduced and consequently, increasing the mismatch between the modules that compose a string. That way we can divide the mismatch losses into two parts. There are manufacturing mismatch losses and operating mismatch losses which are the intensification of manufacturing mismatch by other factors. The causes/intensification of the mismatch can be attributed to four groups: chemical, electrical, mechanical and natural. Within these groups are the phenomena listed below and their particularities:

1) Chemical effects: On the chemical field, the most important cause that arises the efficiency of a PV module is the encapsulant discoloration. This degradation happens due to high temperatures and the exposure of high levels of irradiance, resulting in a production of acetic acid over the encapsulant, turning the surface into a brown aspect [10]. This effect is responsible to reduce the irradiance absorbed by the cells due to the variation of the transmittance of the light and, as a consequence, a reduction of the power generated [11]. As an instructive way, a fluxogram involving the consequences of the PV module exposure to the mentioned natural conditions is shown in Fig 2.

2) Partial Shading Conditions: Another crucial aspect that increases the mismatch between PV modules includes natural and electrical conditions, the partial shading conditions (PSC) [12], [13]. This event occurs due to the fact that objects in the nearby can create shadows in the system, being responsible to make the modules operate under different conditions. In this case, modules under higher irradiance will develop a higher current than modules under a lower one. In this case, the modules under lower irradiance limit the generation of the other modules that are not under the shading effect.

3) Cracked cells: Since the PV modules leave the manufactures until being installed, there is a whole process that demands careful with the structure of the modules, due to the risk of mechanical impact which can lead on a crack in the silicon wafer. The PV modules are exposed to vibrations in the process of transportation from the manufacturer until the place where it will be installed. In the field, the modules are always exposed to weather conditions such as wind, rain, and snow that cause thermal stress and can cause cracked cells. Cracked cells can cause hot spots and non-generating areas that reduce the efficiency of module generation as a whole and accelerate its degradation process. It is not always possible to identify cracks and areas without generation only with a visual inspection, and it is necessary to perform electroluminescence (EL) tests to detect such breakdowns.

4) Soiling losses: Soiling losses happen in different forms and intensities that vary according to the region in which the PV system is installed. Generally, soiling losses happen by dust, bird droppings, pollen, sea salt, among others [14]–[16]. When suffering from soiling, the PV system undergoes shading, limiting the passage of the irradiance, causing a decrease in the current of the PV module that limits the other modules to which it is connected, increasing the mismatch losses that accelerate the degradation of the module. Therefore, it is always necessary to keep the PV modules clean. The cleaning of PV modules has different periodicity according to each region, and it is necessary to study the environment in which the PV system will be exposed.

978-1-7281-4181-7/19 $31.00 © 2019 IEEE

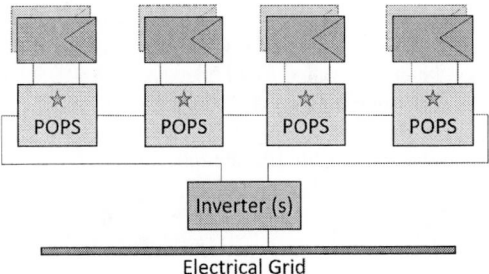

Fig. 3. Architecture with POPS

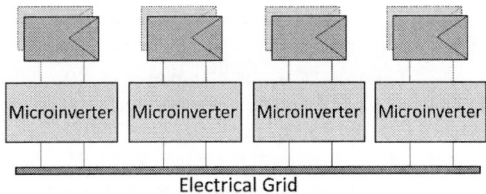

Fig. 4. Architecture with Microinverters

5) Solder degradation: Thermal cycling from external climatic conditions causes constant expansion and contraction of the components of the PV modules. Such a phenomenon is harmful to weld bonds that become more prone to cracks and cracks [4]. These defects are responsible for increasing the series resistance of the modules and tend to increase over time, causing the appearance of hot spots. The poor quality of welding during the manufacturing process increases the probability of the occurrence of this phenomenon [17].

B. Solutions for Mismatch Losses

There are several solutions that seek to minimize mismatch losses. The solutions can be at the level of the architecture of the PV system, the level of the PV module, and at the level of the user.

Considering at the architecture level the main solutions to minimize mismatch are power optimizers [18] and microinverters [19], especially when operating using a differential power processing (DPP) technique that uses DC-DC converter connection in parallel to the PV module.

In the case of the architecture with power optimizers, seen in Fig. 3, losses by mismatch are eliminated, because the PV modules are not directly connected since there is a DC-DC converter at the output of each PV module. The DC-DC converters are interconnected in series or parallel and thereafter the power is delivered to a string inverter for DC-AC conversion.

In the microinverter architecture, seen in Fig. 4, the DC-AC conversion exists in the microinverter itself. It is important to mention that when the architecture with power optimizer or microinverter uses more than one PV module per optimizer or microinverter, mismatch losses are reduced, however, they are not completely extinguished.

Another alternative in architecture level when do not have power optimizers or microinverters is to use different module arrangement techniques to minimize mismatch [20]. This option has the disadvantage of hindering the project of the PV system, because in some places it may be complex to design specific arrangements to avoid mismatch. As well, this technique does not eliminate the mismatch, it only reduces a small percentage.

As for the alternatives at the PV module level, they are considered elements that already come from the factory in the PV module. One can cite the use of more diodes than conventional [21], thus decreasing mismatch during shading of cells; the use of DC-DC converters at the cell level [22], [23]; care in the manufacturing process and high-quality control, mainly obeying IEC 61215.

For the level of the user, it is always necessary to clean the PV modules [14], guaranteeing the reduction in mismatch losses that increase with soiling, and it is important to always monitor the PV system, verifying possible defects if there is a reduction in the generation of energy.

III. CASE STUDY

In order to verify the mismatch effect between PV modules, a case was proposed to analyze the efficiency of a PV power plant composed by twelve PV modules connected in series as it is shown in Fig 5. The PV power plant was installed in Campinas (Brazil) in the year of 2017, so, it is expected that the parameters that contribute to the mismatch and the aging of the equipment be evidenced.

Fig. 5. PV power plant of the case study and how each module will be recognized on the following sessions

A. Metodology

The investigation of the mismatch effect in the PV power plant started with three different cases, focusing on the analysis of the soiling caused by rain that fell over the PV modules

978-1-7281-4181-7/19 $31.00 © 2019 IEEE

one month ago. The data of each case was acquired by use of equipment in the solar market, the I-Ve curve tracer HT Italia 400w. This equipment is capable to provide the most important characteristics of a PV module or a string, such as Isc (short-circuit current), Voc (open circuit voltage), MPP (maximum power point) and also, convert these data to the STC (Standard Test Condition), which is essential for proper comparisons between data obtained under different natural conditions of operation temperature and irradiance. It is also convenient to report the data presented in the STP-190's datasheet: short-circuit current (5.62 A), open circuit voltage (45.2 V) and maximum power point (190 W), with all of the values under STC.

The first test developed was based on cleaning half part of the power plant while the other half was kept dirty, in order to investigate the behavior of each string according to the cleanliness of the PV modules.

On the second test, six modules were compared one by one under two different conditions: with dirt over the surface of the PV modules and after being cleaned. It is expected that the efficiency of each photovoltaic module increases after the effect of the dirt is eliminated.

Finally, in the third test, all of the PV modules were compared for the purpose to discover the highest difference between the modules with higher and lower efficiency under the STC.

B. Discussion and Results

The data collected in the field from the I-Ve Curve tracer was treated with the software TopView, providing I-V curves under STC according to IEC 60891, which establishes proper procedures to translate curves with Operational Conditions (OPC) to the STC.

In the first test, two strings composed of six PV modules were compared in order to confirm the existence of the mismatch between PV modules due to the influence of soiling in the surface. In this case, string one was arbitrarily chosen to be composed with modules 1,2,3,4,5,6 and string two, by the modules 7,8,9,10,11,12 according to the Fig 5. The I-V and P-V curves of the strings, under OPC, were plotted on Fig 6. As it is shown, the curves don't match totally, so this indicates that the soiling found on the surface of string two does influence on its efficiency.

Under an irradiance of 711 W/m^2 and operational temperature of 43.5°C, the string one developed a power of 577.1W, while string two under the same irradiance and temperature, a power of 604.4W. As it was expected, a difference of 27.3W was noticed under almost the same conditions. Under STC and the approach provided by the software, if the strings were considered as a set of identical modules, string one would generate power of 159.23W, while string 2, 166.75W.

In order to develop a numerical comparison, the following equation will be the support for the calculation of the mismatch coefficient.

$$Mismatch = \frac{P_{Nominal} - P_{STC}}{P_{Nominal}} \quad (1)$$

For each occasion, string one showed a mismatch of 16.19% while string two, a mismatch of 12.24%. It is also important to report that the evaluation of the data under STC wasn't possible due to the limitation of the software, but it could be emphasized the mismatch of 27W of the modules operating almost in the same natural conditions.

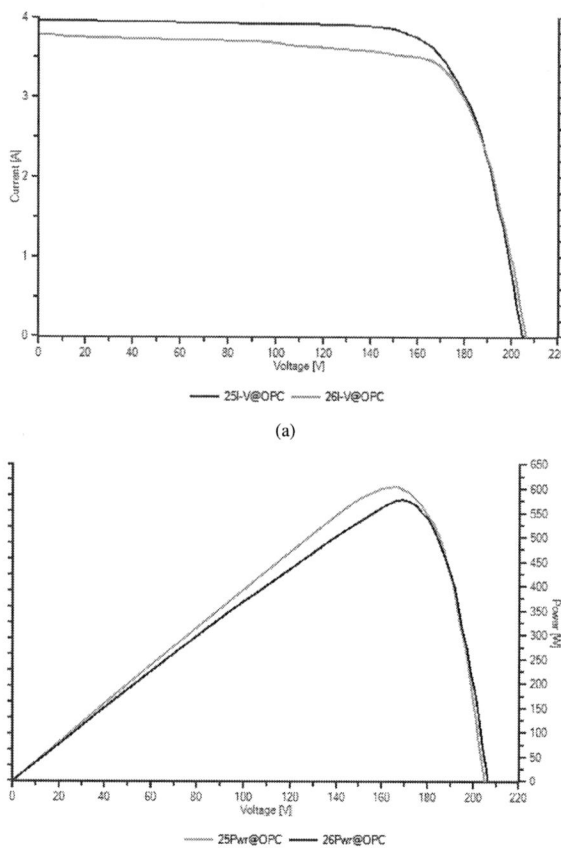

(a)

(b)

Fig. 6. (a) I-V and (b) P-V curve from string 1 and string 2 in OPC

In the second test, the modules that compose string one were compared under two different situations: with soiling generated due to the rain and after being cleaned. The results collected in the field, one by one, of each module under the two mentioned conditions are shown in table I.

According to the data collected in test 2, it is possible to explore individually the six modules under test. The most significant data recorded is the difference between the MPP of module 2. The reason for the discrepancy of 12.09W is the geometry of the soiling on the surface of the module shown in Fig. 7, which shades a solar cell enough to reduce the current that flows over the module. As the shade reduces the MPP, the profile of the I-V curve also changes, generating a curve with two "knees", i.e. with two local MPP, as it is shown in Fig 8.

TABLE I
DATA COLLECTED FROM THE MODULES 1 TO 6 AND NORMALIZED TO STC UNDER DIRTY AND CLEAN CONDITIONS

PV Module	Dirty @STC			Clean @ STC		
	Isc (A)	Voc (V)	MPP (W)	Isc (A)	Voc (V)	MPP (W)
1	5.27	41.36	160.9	5.49	41.72	169.29
2	5.18	41.55	158.75	5.61	41.85	170.84
3	5.18	41.43	159.05	5.5	41.84	167.22
4	5.03	41.08	152.9	5.27	41.55	161.13
5	5.15	41.26	156.54	5.43	41.61	164.18
6	5.13	41.02	155.32	5.38	41.27	161.38

Fig. 7. PV module with dirt caused by animal feces

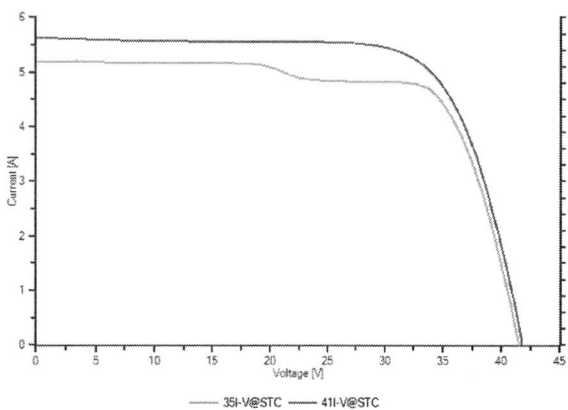

(a)

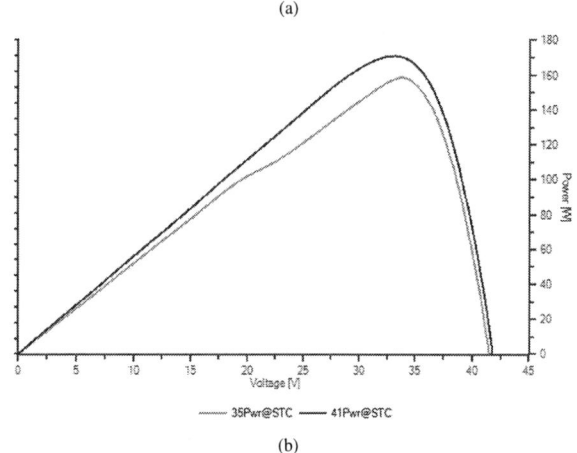

(b)

Fig. 8. (a) I-V and (b) P-V curve from PV module 2 before and after being cleaned in STC

Also about the module 2, the mismatch coefficient under dirt conditions presents a mismatch of 16.44% while, after cleaned, a coefficient of 10.08%. On the other hand, module 6 doesn't seem to be affected to the dirt, due to a mismatch of 18.25% when dirty and 15.06% after cleaned. This aspect indicates that the mismatch found in this module may be attributed to cracked cells, which can only be seen through an electroluminescence test, or due to soldering losses.

Finally, the last test provided some new data from the remaining modules (7,8,9,12) that were cleaned at the beginning of the first test, is shown in the Table II. It is important to point that modules 10 and 11 had some issues during the collect data procedures, and they won't be considered on further individual comparisons on table II.

TABLE II
DATA COLLECTED FROM THE MODULES 7,8,9 AND 12, NORMALIZED TO STC AFTER BEING CLEANED.

PV Module	Isc (A)	Voc (V)	MPP (W)
7	5.48	41.65	167.03
8	5.47	41.78	166.64
9	5.44	41.43	165.35
12	5.44	41.26	163.53

Gathering all the data collected throughout the tests, this stage of the paper will provide a whole comparison between all the modules, even under soiling conditions or after being cleaned. It is also important to mention that, as Equation 1 demands a max power generate under STC, all data also are going to be compared as well.

The module with the highest coefficient of mismatch is the module 4, under soiling conditions, presenting a mismatch of 19.5%, meanwhile, the same module after the cleansing, presented a mismatch of 15.19%, revealing the importance of keeping the PV modules clean. On the other hand, the module with lower mismatch is module 2, presenting a mismatch of 10.08%. Both modules and the nominal dataset's I-V curve were plotted on the same graph on Fig 9.

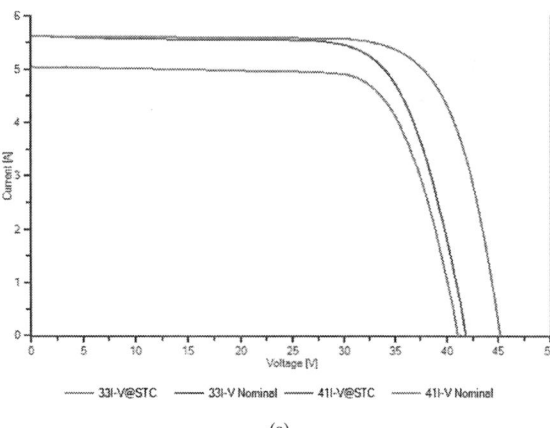

(a)

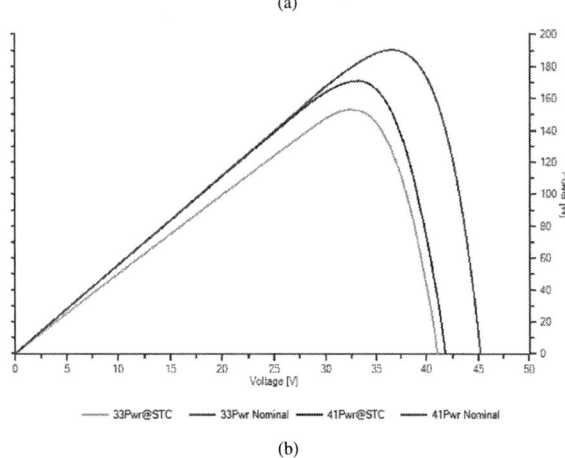

(b)

Fig. 9. (a) I-V and (b) P-V curve at STC of modules 2 and 4, and the nominal behaviour in STC

IV. CONCLUSION

The paper presented concepts of mismatch losses listed causes and pointed out existing solutions in the literature. Subsequently, a case study with a PV system was done. In the first test, half of the plant was cleaned and another part was dirty, comparing both parts, obtaining the power of 604.4W and 577.1W, respectively. Besides that, the module with the highest coefficient of mismatch was the module 4, under soiling conditions, presenting a mismatch coefficient of 19.5%, meanwhile, the same module after the cleansing presented a mismatch coefficient of 15.19%. Thus, it is shown that the mismatch losses cause considerable losses in a photovoltaic system, and that is something that deserves the attention of the photovoltaic designer, seeking to maximize efficiency and decrease the payback period of the investment.

ACKNOWLEDGMENT

This work was supported by the agencies CNPq, CAPES, FAPESP (2016/08645-9) and ANEEL (CPFL-PA3032).

REFERENCES

[1] K. S. Reddy, G. S. Goutham, L. K. Panwar and R. Kumar,"Environmental sustainability analysis of solar photovoltaic (SPV) systems," *2015 Inter.*

Conf. on Comp., Communication and Control (IC4), Indore, 2015, pp. 1-7.

[2] M. M. Fouad, L. A. Shihata, and E. S. I. Morgan, "An integrated review of factors influencing the performance of photovoltaic panels," *Renew. Sustain. Energy Rev.*, vol. 80, no. July 2016, pp. 1499–1511, 2017.

[3] S. Shirzadi, H. Hizam, and N. I. A. Wahab, "Mismatch losses minimization in photovoltaic arrays by arranging modules applying a genetic algorithm," *Sol. Energy*, vol. 108, pp. 467–478, 2014.

[4] P. Manganiello, M. Balato, and M. Vitelli, "A Survey on Mismatching and Aging of PV Modules: The Closed Loop," *IEEE Trans. on Indus. Electron.*, vol. 62, pp. 7276–7286, 2015.

[5] L.L. Bucciarelli Jr., "Power loss in photovoltaic arrays due to mismatch in cell characteristics," *Sol. Energy*, vol. 23, 1979, 277–288.

[6] T. S. Wurster and M. B. Schubert, "Mismatch loss in photovoltaic systems," *Sol. Energy*, vol. 105, pp. 505–511, 2014.

[7] A. Chouder, S. Silvestre, "Analysis model of mismatch power losses in PV systems," *J. Sol. Energy Eng.*, vol. 131, 2009.

[8] S. MacAlpine, C. Deline, R. Erickson, and M. Brandemuehl, "Module mismatch loss and recoverable power in unshaded PV installations," Conf. Rec. IEEE Photovolt. Spec. Conf., pp. 1388–1392, 2012.

[9] D. Picault, B. Raison, S. Bacha, J. De La Casa, J. Aguilera, "Forecasting photovoltaic array power production subject to mismatch losses," *Sol. Energy 84*, vol. 84, 2010, 1301–1309.

[10] A. Sinha, O. S. Sastry and R. Gupta. "Nondestructive characterization of encapsulant discoloration effects in crystalline-silicon PV modules," *Solar Energy Materials and Solar Cells,*, vol. 155, pp. 234–242, 2016.

[11] S. Chattopadhyay, R. Dubey, V. Kuthanazhi, J. J. John, C. S. Solanki, A. Kottantharayil, B. M. Arora, K. L. Narasimhan, V. Kuber, J. Vasi, A. Kumar, and O. S. Sastry, "Visual Degradation in Field-Aged Crystalline Silicon PV Modules in India and Correlation With Electrical Degradation," *IEEE J. Photovoltaics,* vol. 4, no. 6, pp. 1470–1476, 2014.

[12] J. C. Teo, R. H. G. Tan, V. H. Mok, V. K. Ramachandaramurthy, and C. Tan, "Impact of partial shading on the P-V characteristics and the maximum power of a photovoltaic string," *Energies*, vol. 11, no. 7, 2018.

[13] H. S. Moreira, J. L. S. Silva, D. B. Mesquita, B. H. Kikumoto de Paula, and M. G. Villalva, "Experimental Analysis of Photovoltaic Modeling for Partially Shading Conditions," *IEEE/IAS International Conference on Industry Applications*, pp. 242–248, 2018.

[14] J. L. S. Silva, M. M. Cavalcante, H. S. Moreira, D. B. M. Delgado, and M. G. Villalva, "Automated Cleaning System for Photovoltaic Panels," *13th IEEE/IAS Inter. Conf. on Indus. Appl.*, São Paulo-SP, 2018.

[15] A. Gheitasi, A. Almaliky, and N. Albaqawi, "Development of an Automatic Cleaning System for PV Plants," *IEEE PES Asia-Pacific Power and Energy Engineering Conf.*, pp. 7–10, 2015.

[16] H. Becker, W. Vaassen, and W. Herrmann, "Reduced output of solar generators due to pollution," *Proc. of the 14th EU PV Conf.*, 1997.

[17] D. L. King, M. A. Quintana, J. A. Kratochvil, D. E. Ellibee, and B. R. Hansen, "Photovoltaic module performance and durability following long-term field exposure," Progr. Photovolt., Res. Appl., vol. 8, no. 2, pp. 241–256, Mar./Apr. 2000.

[18] J. L. S. Silva, H. S. Moreira, D. B. Mesquita, M. G. Villalva, "Analysis of Power Optimizers in Photovoltaic Power Plant," *13th IEEE/IAS Inter. Conf. on Indus. Appl.*, 2018.

[19] S. Qin, C. B. Barth and R. C. N. Pilawa-Podgurski, "Enhancing Microinverter Energy Capture With Submodule Differential Power Processing," in IEEE Transactions on Power Electronics, vol. 31, no. 5, pp. 3575-3585, May 2016.

[20] A. A. Mansur, Md. R. Amin and K. K. Islam, "Performance Comparison of Mismatch Power Loss Minimization Techniques in Series-Parallel PV Array Configurations," in Energies, vol. 12, no. 874, pp. 1-21, March 2019.

[21] B. B. Pannebakker, A. C. de Waal and W. G. J. H. M. van Sark, "Photovoltaics in the shade: one bypass diode per solar cell revisited," in Progress in Photovoltaic Res. Appl., vol. 25, no., pp. 836-849, May 2017.

[22] C. Olalla, D. Clement, M. Rodriguez, and D. Maksimovic, "Architectures and control of submodule integrated dc-dc converters for photovoltaic applications," *IEEE Trans. Power Electron.*, vol. 28, no. 6, pp. 2980–2997, 2013.

[23] O. Khan and W. Xiao, "An Efficient Modeling Technique to Simulate and Control Submodule-Integrated PV System for Single-Phase Grid Connection," *IEEE Trans. on Sust. Energy*, vol. 7, no. 1, pp. 96-107, Jan. 2016.

NOVEL BUCK-BOOST PFC CONVERTER WITH THREE-STATE SWITCHING CELL

Douglas Carvalho Morais[1], Falcondes José Mendes de Seixas[1], Luís de Oro Arenas[1], Lucas Carvalho Souza[1], Luciano de Souza da Costa e Silva[2]

[1]São Paulo State University (UNESP), School of Engineering, Ilha Solteira – São Paulo – Brazil
[2]Federal Institute of Goiás (IFG) – Jatai – Goiás – Brazil
e-mail: morais.sccp.carvalho@gmail.com, falcondes.seixas@unesp.br, ldeoroa@gmail.com, lucas.souza@ifg.edu.br, lucianocosta_@hotmail.com

Abstract – **In this paper, a buck-boost converter with power factor correction is proposed, using a three-state switching cell named of B configuration, operating in continuous conduction mode. The type B cell insertion on the buck-boost converter keeps conventional structure statistic characteristics and brings as advantage, stress current division in all semiconductors by half inductor current. Moreover, the ripple frequency on converter energy storage elements will be twice switching frequency value, making easier the design of input and output passive filters and reducing the converter final volume and weight. Preliminary simulation results prove the converter effectiveness on power factor correction, keeping output voltage stable in a specified operation point, with a total harmonic distortion of input current below 1.5%.**

Key-words – *buck-boost converter, three-state switching cell, power factor correction, total harmonic distortion, continuous conduction mode.*

I. INTRODUCTION

Nowadays, in the input stage, electronic equipment that need AC-DC conversion make use of a rectifier bridge with diodes and/or thyristors. However, if there is no one second stage to make Power Factor Correction (PFC), the non-linear elements switched in the grid frequency, will cause a distortion on the input current waveform regarding voltage, making small the Power Factor (PF).

Amid this issue, over the last few years solutions were proposed focusing in reduce the Total Harmonic Distortion (THD) of input current which such devices impose to the power grid. Such solutions can be divided into two great categories, passive and active. Active solutions implement DC-DC converters switched at high frequency, and they are preferable regard to passive solutions because [1]:

➢ Weight and volume reduced by increasing of switching frequency;

➢ High efficiency and power density;

➢ Output voltage regulation possibility.

The inherent active solutions advantages can be get directly using converters as buck, boost or buck-boost. Such topologies use two-state switching cells, known too as canonic cells [2]. Due to wide use of these topologies in power factor correction, it is common the proposition of techniques that reduce required stages, or that dividing the current stress in all semiconductors. In this way, high frequency rectification or half-controlled, propitiate the inclusion of the bridge rectifier conventional to the high frequency DC-DC converter stage, thus, this technique is known too as bridgeless [3]. The low frequency bridge rectifier suppression, provides to equipment more reliability,

since less stage are required. Moreover, conduction losses are decreased, because the semiconductors number used in the high frequency stages operation is less regarding classic converter structure.

Interleaving is another technique widely known in the literature [4]. This technique allows the stress current division in all semiconductors, which attenuation is proportional to the number of intercalated converter cells. Moreover, interleaving technique reduce structure weight and volume, once the passive elements ripple frequency is equivalent to the switching frequency multiplied by the number of interleaved converter cells.

Proposed in [5] the Three-State Switching Cell (3SSC) concept comes up. This technique proposes the change of the conventional canonic cell containing one passive and one active semiconductor, by one configuration containing two passive, two active semiconductors and one autotransformer [5-8]. This technique allows stress current division in all semiconductors by half of inductor current. Moreover, ripple frequency of the passive elements is equivalent to twice switching frequency stipulated. Therefore, the design of input and output filters are facilitated and occurs the converter weight and volume reduction.

Due to the three-state switching cell advantages, in this paper is proposed a buck-boost converter using cell B configuration for a new application as PFC circuit. The operation occurs in Continuous Conduction Mode (CCM), and uses the average current control. In section II, converter operation principle and design of passive elements are detailed. In section III, 3SSC buck-boost converter small-signals model is presented, beyond main transfer functions. Lastly, in section IV, simulation results are presented in order to prove the converter effectiveness in power factor correction.

II. 3SSC BUCK-BOOST CONVERTER OPERATION PRINCIPLE WITH CELL B

The three-state switching cell B configuration proposed in [5], can be seen in Fig. 1a. Bilateral inversion is presented in Fig. 1b. Regardless of configuration used, always between 'a' and 'b' points it must have a voltage source or capacitive branch connected, already the 'c' point must contain the connection with a current source or inductive branch. The 3SSC buck-boost converter can be seen in Fig. 1c. In order to realize a converter qualitative analysis, it is assumed that:

➢ The turns ratio of the central tap autotransformer is unitary;

➢ Passive and active components are ideals;

➢ The gate signals in S_2 are 180° delayed regard to S_1;

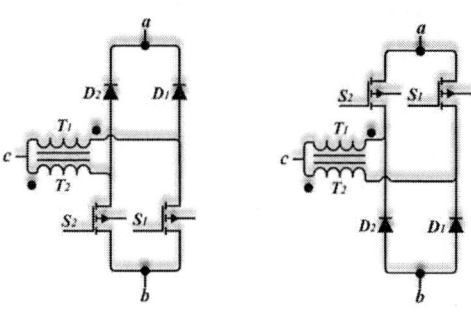
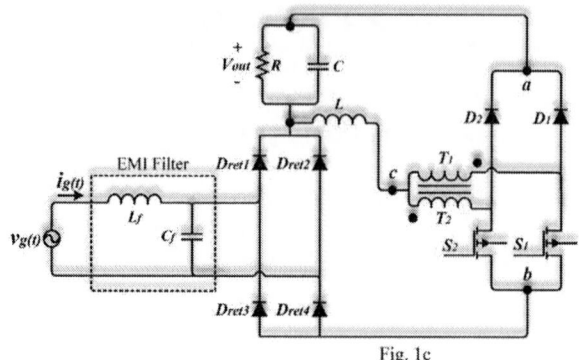

Fig. 1a   Fig. 1b   Fig. 1c

Figure 1a: Cell B; Figure 1b: Bilateral inversion of cell B; Figure 1c: Proposed PFC buck-boost converter with cell B.

➤ The switching frequency is constant and modulation is by pulse width.

Besides that, due to the magnetic effect provided by autotransformer unitary turns ratio, windings voltages are considered equal, as well as the currents, that is:

$$v_{T1(t)} = v_{T2(t)} \qquad (1)$$

$$i_{T1(t)} = i_{T2(t)} \qquad (2)$$

The buck-boost converter presents different operation stages, depending on the duty cycle used. In this case, for duty cycle lower than 0.5, then the converter operates as step-down, buck characteristic. In the other hand, for duty cycle higher than 0.5, then the converter operates as step-up, boost characteristic. Thus, the following analysis can be divided for $d < 0.5$ and $d > 0.5$, respectively [8].

A. Operation Stages for d < 0.5 (Buck Mode).

As step-down, there is four operation stages whose semiconductor states are presented in Table I.

TABLE I: Semiconductor gate signal for $d < 0.5$.

Semiconductor	Δt_1	Δt_2	Δt_3	Δt_4
S_1	ON	OFF	OFF	OFF
S_2	OFF	OFF	ON	OFF
D_1	OFF	ON	ON	ON
D_2	ON	ON	OFF	ON

As can be seen in Table I, for operation with $d < 0.5$ not occurs the overlapping gate signals in the active semiconductors, thus, is usual to call this operation as Non-Overlapping Mode (NOM).

First Stage [Δt_1]: According Table I, only S_1 is turned on, turning-off diode D_1. In the other branch, D_2 is turn-on, since there is no gate signal in S_2. The circuit relating to this stage can be seen in Fig. 2a. It is important to make it clear that for operation stage analysis, the input voltage is replaced by an ideal rectified voltage source.

By analyzing the circuit shown in Fig. 2a, it is possible through the Kirchhoff law, to obtain the inductor voltage and the capacitor current. Thus, for the inductor voltage loop, it is found:

$$v_{L(t)} - v_{T1(t)} - vg_{ret(t)} = 0 \qquad (3)$$

$$v_{L(t)} + v_{T2(t)} + vc_{(t)} = 0 \qquad (4)$$

Replacing (1) into (3) and (4), and putting in evidence inductor voltage, it is obtained:

$$v_{L(t)} = L \frac{d}{dt} i_{L(t)} = \frac{1}{2} \left(vg_{ret(t)} - vc_{(t)} \right) \qquad (5)$$

Similarly, the output capacitor current will be:

$$i_{o(t)} = i_{R(t)} + i_{c(t)} \qquad (6)$$

As shown in Fig. 2a and considering (2), the output current is equivalent to half inductor current. Thus, replacing in (6) and putting in evidence the capacitor current, is obtained:

$$i_{c(t)} = C \frac{d}{dt} v_{c(t)} = \frac{1}{2} i_{L(t)} - i_{R(t)} \qquad (7)$$

Second Stage [Δt_2]: In the second stage, there is no gate signal sent to active semiconductors, thus, both diodes turn-on, disconnecting the input voltage source from circuit remaining. In this way, the second stage circuit can be seen in Fig. 2b.

As in the first stage, inductor voltage and capacitor current can be obtained applying Kirchhoff law. Thus, in the inductor voltage loop, it is obtained:

$$v_{L(t)} - v_{T1(t)} + vc_{(t)} = 0 \qquad (8)$$

$$v_{L(t)} + v_{T2(t)} + vc_{(t)} = 0 \qquad (9)$$

So, for (8) and (9) can be true simultaneously, in both winding voltages must be zero, characterizing a short-circuit, thus:

$$v_{L(t)} = L \frac{d}{dt} i_{L(t)} = -vc_{(t)} \qquad (10)$$

Again, the capacitor current for second stage is given by (6). However, the output current $i_{o(t)}$ in this case is equivalent to total inductor current $i_{L(t)}$, thus, it is obtained:

$$i_{c(t)} = C \frac{d}{dt} v_{c(t)} = i_{L(t)} - i_{R(t)} \qquad (11)$$

Third Stage [Δt_3]: The third stage, though presents S_2 and D_1 turned-on, in mathematical terms, is equal to first stage, so (3) to (7) can be used.

Fourth Stage [Δt_4]: The fourth stage is identical to the second one, being valid from (8) to (11).

B. Operation Stage for d > 0.5 (Boost Mode)

As step-up, there is fourth operation stages whose semiconductor states are presented in Table II. As can be seen, occurs overlapping gate signals sent to active semiconductors, so, in an opposite way to last case, is usual to call this operation as Overlapping Mode (OM).

978-1-7281-4181-7/19 $31.00 © 2019 IEEE

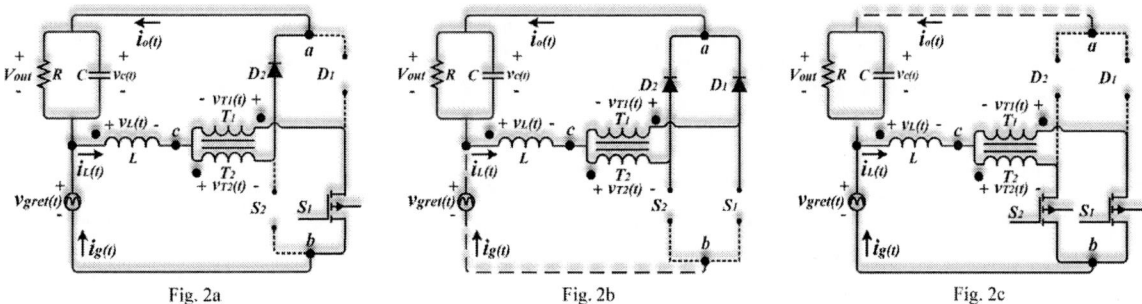

Fig. 2a   Fig. 2b   Fig. 2c

Figure 2a: First operation stage in NOM and second stage in OM; Figure 2b: Second operation stage in NOM; Figure 2c: First operation stage in OM

TABLE II: Semiconductor gate signal for $d > 0.5$.

Semiconductor	Δt_1	Δt_2	Δt_3	Δt_4
S_1	ON	ON	ON	OFF
S_2	ON	OFF	ON	ON
D_1	OFF	OFF	OFF	ON
D_2	OFF	ON	OFF	OFF

First Stage [Δt_1]: In the first stage, both active semiconductors are turn on, so that diodes D_1 and D_2 remain turn-off. In this case, output capacitor supplies the energy required by load. The circuit relating to first stage can be seen in Fig. 2c. Using Kirchhoff's law, the inductor voltage loop is given by:

$$v_{L(t)} - v_{T1(t)} - vg_{ret(t)} = 0 \qquad (12)$$

$$v_{L(t)} + v_{T2(t)} - vg_{ret(t)} = 0 \qquad (13)$$

Since the autotransformer windings voltages are necessarily zero, constituting a short circuit, the inductor current can be obtained as follows:

$$v_{L(t)} = L\frac{d}{dt}i_{L(t)} = vg_{ret(t)} \qquad (14)$$

In this stage, the output current is zero, so the capacitor current is given by:

$$i_{c(t)} = C\frac{d}{dt}v_{c(t)} = -i_{R(t)} \qquad (15)$$

Second Stage [Δt_2]: The second operation stage in OM is identical to first stage in NOM and can be seen in Fig. 2a. In this way, it is valid again (3) to (7).

Third Stage [Δt_3]: The third stage is identical to first one in OM, being (12) to (15) are also valid.

Fourth Stage [Δt_4]: The fourth stage is complementary to the second one, so the same circuit shown in Fig. 2a is used.

C. Voltage Static Gain and Average Inductor Current

Regardless the duty cycle value used, the pulse width sent to any active semiconductors is given by:

$$\Delta t_1 = d \cdot T_s \qquad (16)$$

Being T_s the switching period. Added, the time for first and second stages must be equal to half switching period:

$$\Delta t_1 + \Delta t_2 = \frac{T_s}{2} \rightarrow \Delta t_2 = T_s \cdot \left(\frac{1}{2} - d\right) = d' \cdot T_s \qquad (17)$$

Being $d' = (1/2 - d)$. Considering that inductor current and capacitor voltage are constant in a switching period, (5), (7), (10) and (11), can be written as:

$$\Delta_{IL(1)} = \frac{D \cdot T_s}{2L}(vg - V_c) \qquad (18)$$

$$\Delta_{VC(1)} = \frac{D \cdot T_s}{C}\left(\frac{1}{2}I_L - \frac{V_c}{R}\right) \qquad (19)$$

$$\Delta_{IL(2)} = \frac{-V_c}{L}D' \cdot T_s \qquad (20)$$

$$\Delta_{VC(2)} = \frac{D' \cdot T_s}{C}\left(I_L - \frac{V_c}{R}\right) \qquad (21)$$

A similar procedure can be taken for $d > 0.5$. The voltage static gain can be found considering that the inductor current variation in a switching period is zero, that is:

$$\Delta_{IL(1)} + \Delta_{IL(2)} = 0 \qquad (22)$$

Regardless operation mode used, NOM or OM, it can be shown that the static gain of average voltage is given by:

$$\frac{DT_s}{2L}(vg - V_c) - \frac{V_c}{L}D'T_s \rightarrow \frac{V_c}{V_g} = G_M = \frac{D}{1-D} \qquad (23)$$

Similarly, the average current of capacitor is zero in a switching period, thus:

$$\frac{DT_s}{C}\left(\frac{1}{2}IL - \frac{V_c}{R}\right) + \frac{D'T_s}{C}\left(IL - \frac{V_c}{R}\right) \rightarrow I_L = \frac{V_c}{R \cdot D'} \qquad (24)$$

Both results presented in (23) and (24), coincide with the values found for buck-boost converter structure with canonical cell [9].

D. Inductance and Capacitance of the Buck-Boost Converter with Cell B

Due to rectified sine aspect of the input source, inductance calculation must take into account this input voltage variation at its most critical point, thus, sine wave peak. Considering (18), the current can be written as:

$$\Delta_{IL(1)} = \frac{T_s}{2L} \cdot \left(vg_{ret(t)} - V_c\right) \cdot d_{(t)} \qquad (25)$$

Being,

$$vg_{ret(t)} = V_m \cdot |\sin(\omega t)| \qquad (26)$$

And, replacing (26) in (23),

$$d_{(t)} = \frac{1}{1 + M \cdot |\sin(\omega t)|} \qquad (27)$$

In (26), V_m represents the peak of the input voltage and ωt the angular frequency of power grid. In (27), M represents the inverse of the minimum voltage gain, given by:

$$M = \frac{V_m}{V_c} \qquad (28)$$

Replacing (26) and (27) into (25), the inductor current variation parameterized, is given by:

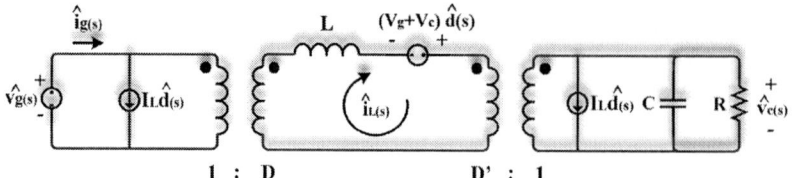

Figure 3: Small signal's model of the proposed buck-boost converter with cell B.

$$\Delta_{IL(1)}{}^* = \frac{M \cdot |\sin(\omega t)| - 1}{1 + M \cdot |\sin(\omega t)|} \tag{29}$$

Being,

$$\Delta_{IL(1)}{}^* = \frac{2L \cdot \Delta_{IL(1)}}{V_c \cdot T_s} \tag{30}$$

To find the critical point, or maximum point, it's enough simply derived (29) and equal zero, in this way:

$$\cos(\omega t) \cdot (V_m + M \cdot V_c) = 0 \tag{31}$$

Thus, the only possible value in the range of 0 to π that make the left side of (31) null is $\omega t = \pi/2$. Replacing this value into (29) and solving as L function, is obtained the NOM inductance given by:

$$L_{NOM} = \frac{Vc \cdot T_s}{2\Delta_{IL}} \left(\frac{M-1}{M+1}\right) \tag{32}$$

Similarly, the OM inductance is given by:

$$L_{OM} = \frac{V_m \cdot T_s}{\Delta_{IL}} \left[\frac{\sin(\alpha)}{1 + M \cdot \sin(\alpha)} - \frac{1}{2} \cdot \sin(\alpha)\right] \tag{33}$$

Being α a M dependent constant, given by:

$$\alpha = arcsen\left(\frac{\sqrt{2}-1}{M}\right) \tag{34}$$

The output capacitance of the 3SSC buck-boost converter, can be calculated with based on the output power P_{out}, line frequency f_{line}, and too:

$$C = \frac{P_{out}}{4\pi \cdot f_{line} \cdot V_c \cdot \Delta V_c} \tag{35}$$

In the next section, will be presented the 3SSC buck-boost converter small-signals model, beyond the main Transfer Functions (TF's).

III. SMALL-SIGNALS MODEL FOR THE BUCK-BOOST CONVERTER WITH CELL B

For the small-signals model construction, it's used the system perturbation and linearization method. A more detailed method explanation can be found in [9].

Firstly, are expressed the inductor average voltage, capacitor average current, and the input average current, all calculated for one switching period, thus:

$$L\frac{d}{dt}i_{L(t)} \approx \left(\bar{v}_{g(t)} + \bar{v}_{c(t)}\right) \cdot d_{(t)} - \bar{v}_{c(t)} \tag{36}$$

$$C\frac{d}{dt}v_{c(t)} \approx \bar{\iota}_{L(t)} \cdot d'_{(t)} - \frac{\bar{v}_{c(t)}}{R} \tag{37}$$

$$i_{g(t)} \approx \bar{\iota}_{L(t)} \cdot d_{(t)} \tag{38}$$

Perturbation and Linearization: Input voltage source and duty cycle are considered system inputs and can be written as

the sum of their steady state values, added to small signals variations represented in the Laplace domain, that is:

$$v_{g(s)} = V_g + \hat{v}_{g(s)} \quad e \quad d_{(s)} = D + \hat{d}_{(s)} \tag{39}$$

Input signals perturbations affect the other circuit state variables, being these: inductor current; capacitor voltage; and input current. Replacing in (36) to (38) the perturbations described in (39), the referred functions can be rewritten as follows:

$$L\frac{d}{dt}\hat{\iota}_{L(s)} = \left(V_g + V_c\right) \cdot \hat{d}_{(s)} + D\hat{v}_{g(s)} - D'\hat{v}_{c(s)} \tag{40}$$

$$C\frac{d}{dt}\hat{v}_{c(s)} = D' \cdot \hat{\iota}_{L(s)} - I_L \cdot \hat{d}_{(s)} - \frac{\hat{v}_{c(s)}}{R} \tag{41}$$

$$\hat{\iota}_{g(s)} = D \cdot \hat{\iota}_{L(s)} + I_L \cdot \hat{d}_{(s)} \tag{42}$$

The terms of (40) to (42) which are multiplied by the duty cycle small signal's variation are considered as controlled sources. The other terms can be combined into ideal transformers with turns ratio given by average duty cycle. Moreover, the derivatives in (40) and (41) are considered as the inductor voltage and the capacitor current, respectively. Thus, (40) to (42) correspond to an electric circuit, which can be seen in Fig. 3.

From the small signals model, it is possible to obtain any circuit transfer function through the superposition principle. Considering the current average values control, it's necessary to find a relation between current to be controlled and duty cycle, and between output voltage and the current desired, that is:

$$G_{igd(s)} = \frac{\hat{\iota}_{g(s)}}{\hat{d}_{(s)}} \quad and \quad G_{vi(s)} = \frac{\hat{v}_{c(s)}}{\hat{\iota}_{g(s)}} \tag{43}$$

In the buck-boost converter case with cell B proposed in this paper, it is desired that the input current follows a rectified sine-wave reference. Thus, the transfer function must relate input current with duty cycle. One way to obtain such transfer function is to first find the relationship between inductor current and duty cycle, which is given in (44):

$$\hat{\iota}_{L(s)} = G_{iLd_{cc}} \cdot \left(\frac{1 + \frac{s}{\omega zid}}{1 + \frac{s}{Q \cdot \omega pid} + \frac{s^2}{\omega pid^2}}\right) \cdot \hat{d}_{(s)} \tag{44}$$

Being:

$$G_{iLd_{cc}} = \frac{2V_c + V_g}{R \cdot (D')^2} \tag{45}$$

$$\omega zid = \frac{2V_c + V_g}{\left(V_c + V_g\right) \cdot R \cdot C} \tag{46}$$

$$Q = R \cdot D' \cdot \sqrt{C/L} \tag{47}$$

$$\omega pid = D'/\sqrt{L \cdot C} \qquad (48)$$

In (45) the DC gain is obtained. In (46) and (48) the zero and pole frequency system are represented, respectively. In (47) the second-order system quality factor is obtained. Replacing (44) into (42) and manipulating some terms, the transfer function that relates input current and duty cycle is obtained and given by (49):

$$G_{igd(s)} = G_{igdCC} \cdot \left(\frac{1 + \dfrac{s}{\omega zid_1} + \dfrac{s^2}{\omega zid_2}}{1 + \dfrac{s}{Q \cdot \omega pid} + \dfrac{s^2}{\omega pid^2}} \right) \qquad (49)$$

Being:

$$\omega zid_1 = \frac{Q \cdot \omega pid \cdot \omega zid}{I_L \cdot \omega zid + Q \cdot G_{iLd_{cc}} \cdot \omega pid} \qquad (50)$$

$$\omega zid_2 = \frac{I_L}{\omega pid^2} \qquad (51)$$

$$G_{igdCC} = I_L + \frac{2V_c + V_g}{R \cdot (D')^2} D \qquad (52)$$

In (49) the relationship between input current and duty cycle is presented, however, it is necessary to relate the same input current also with the output voltage. A way to get such transfer function $G_{vi(s)}$ is firstly finding a relation between output voltage and duty cycle, which is given in (53):

$$G_{vd(s)} = G_{vdCC} \cdot \left(\frac{1 - \dfrac{s}{\omega zvi}}{1 + \dfrac{s}{Q \cdot \omega pid} + \dfrac{s^2}{\omega pid^2}} \right) \qquad (53)$$

Being:

$$G_{vdCC} = \frac{V_g + V_c}{D'} \qquad (54)$$

$$\omega zvi = \frac{(V_c + V_g)}{L \cdot I_L} D' \qquad (55)$$

The direct relationship between output voltage and input current is obtained by division of (53) by (49), that is:

$$G_{vi(s)} = \frac{\hat{v}_{c(s)}}{\hat{d}_{(s)}} \Big/ \frac{\hat{i}_{g(s)}}{\hat{d}_{(s)}} = \frac{\hat{v}_{c(s)}}{\hat{i}_{g(s)}} \qquad (56)$$

Finally, the transfer function $G_{vi(s)}$ can be described as follow:

$$G_{vi(s)} = G_{viCC} \cdot \left(\frac{1 - \dfrac{s}{\omega zvi}}{1 + \dfrac{s}{\omega zid_1} + \dfrac{s^2}{\omega zid_2}} \right) \qquad (57)$$

Being:

$$G_{viCC} = \frac{(V_c + V_g) \cdot R \cdot D'}{I_L \cdot R \cdot (D')^2 + 2V_c + V_g} \qquad (58)$$

It is valid mentioning the zero present in the right half-plane in (53) and (57), traditional buck-boost converter MCC characteristic. The controllers design should be based on (49) and (57) and follow the methodology detailed in [9].

It is also important to say that the same transfer functions are found independently of the duty cycle used, that is, the same controller is used for NOM and OM.

IV. SIMULATION RESULTS

In Table III, the main parameters of the 3SSC buck-boost converter are presented.

TABLE III: Parameters for design.

Description	Symbology	Values
Input voltage (RMS)	Vg(t)	220V/60Hz
Output voltage (d < 0.5)	Vout	100V
Output voltage (d > 0.5)	Vout	400V
Output power	Pout	1000W
Switching frequency	fs	50kHz

In NOM, that is, step-down (buck mode), the input and output voltage and input current are shown in Fig. 4a. Similar result in OM as step-up (boost mode) is also shown in Fig. 4a. As noted, in both operation modes, the input current is very close to a sinusoidal waveform, while the output voltage remains at the operation point specified in Table III. When $d < 0.5$, a power factor of 0.9994 with a THD of 1.45% were obtained. At $d > 0.5$, THD was 1.39% with a PF of 0.9999. In both cases, the 3SSC buck-boost PFC converter meets the IEC 61000-3-2 standard [10].

In Fig. 4b the non-overlapping mode can be seen. In this case, the inductor current does not flow simultaneously through the S_1 and S_2 semiconductors in none instant of time, thus, as shown in Fig. 2b, inside of the switching period, there is a moment that the input voltage is disconnected from the circuit.

In a complementary way, it is possible see also on Fig. 4b for OM, that the gate signals sent to S_1 and S_2 semiconductors have pulse width duration larger than 50%, therefore, overlapping mode occurs. However, differently of the NOM, in this case, the load and the output capacitor are disconnected from the main circuit, being the load, fed by output capacitor, as shown in Fig. 2c.

Moreover, as expected by mathematical analysis, in the NOM or OM operation, the current that flow through S_1 or S_2 semiconductor is half of the inductor current. Also, as can be seen in Fig. 4b, for both cases, the ripple frequency in the inductor is twice switching frequency value stipulated. This is possible because, the inductor charge and discharge two times in a switching period. It is worth noting that the same result is found for the capacitor ripple frequency.

It is also important to note that, as power factor correction, the 3SSC buck-boost converter operating as step-down, will transit between NOM and OM operations mode. For example, considering the output voltage shown in Table III for NOM as operation point. Always that input voltage is smaller than output voltage, the converter needs to operate as boost for ensure that the operation point specified is maintained. This feature, make 3SSC buck-boost converter more reliable against input voltage perturbations, and also guarantee stability for the load operation.

V. CONCLUSIONS

In this paper, a buck-boost converter with power factor correction using three-state switching cell type B was proposed. The average current control technique avoids the zero in right half-plane present on transfer function from input current to output voltage. Moreover, but also allows that input current THD to be minimized, reducing in both operation mode to value smaller than 1.5%.

978-1-7281-4181-7/19 $31.00 © 2019 IEEE

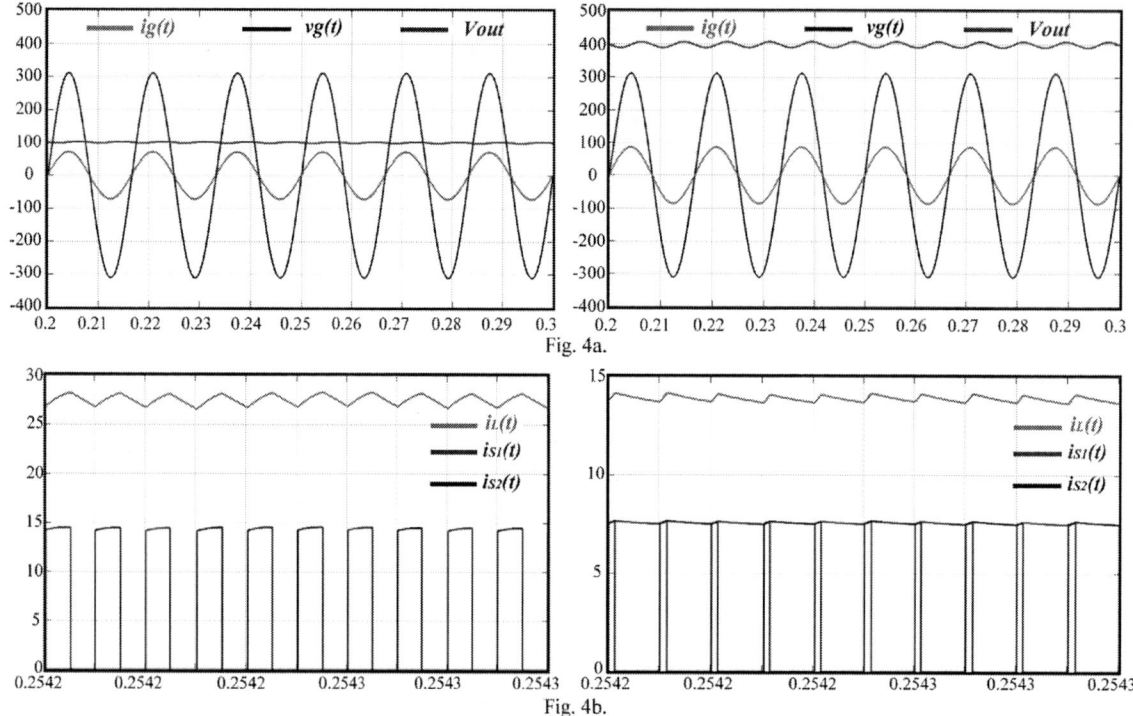

Fig. 4a.

Fig. 4b.

Figure 4a: Input and output voltage and input current (x10) for NOM and OM operation; Figure 4b: Current in both semiconductors and inductor current for NOM and OM operation.

The three-state switching cell B simulated on buck-boost PFC converter provided: the inductor current division, decreasing by half the value that flows in all semiconductors; passive elements increase ripple twice times, allowing weight and volume reduction on the input and output filters.

Finally, a prototype is under construction in order to obtain experimental results to prove buck-boost PFC converter effectiveness with three-state switching cell B.

ACKNOWLEDGMENTS

This study was financed in party by the Coordenação de Aperfeiçoamento de Pessoal de Nível Superior - Brasil (CAPES) - Finance Code 001.

REFERENCES

[1] On Semiconductor ON, "Power Factor Correction (PFC) Handbook", *Choosing the Right Power Factor Controller Solution*. Rev. 5, Apr-2014.

[2] E. E. Landsman, "A unifying derivation of switching DC-DC converter topologies", *1979 IEEE Power Electronics Specialist Conference*, San Diego, CA, USA, 1979, pp.239-243.

[3] B. Zhao, A. Abramovitz and K. Smedley, "Family of Bridgeless Buck-Boost PFC Rectifiers," in *IEEE Transactions on Power Electronics*, vol. 30, no. 12, pp. 6524-6527, Dec. 2015.

[4] L. S. C. e Silva, F. J. M. de Seixas and M. A. G. de Brito, "Bridgeless interleaved boost PFC converter with variable duty cycle control," *XI Brazilian Power Electronics Conference*, Praiamar, 2011, pp. 397-402.

[5] Bascopé, G. V. T.; Barbi, I. "Generation of Family of Non-Isolated DC-DC PWM Converters Using New Three-State Switching Cell". Power Electronics Specialists Conference, 2000. PESC 00. 2000 IEEE 31st Annual, Volume: 2, 18-23 June 2000. p. 858 – 863.

[6] J. P. R. Balestero, F. L. Tofoli, G. V. Torrico-Bascopé and F. J. M. de Seixas, "A DC–DC Converter Based on the Three-State Switching Cell for High Current and Voltage Step-Down Applications," in *IEEE Transactions on Power Electronics*, vol. 28, no. 1, pp. 398-407, Jan. 2013.

[7] J. P. R. Balestero, F. L. Tofoli, R. C. Fernandes, G. V. Torrico-Bascope and F. J. Mendes de Seixas, "Power Factor Correction Boost Converter Based on the Three-State Switching Cell," in *IEEE Transactions on Industrial Electronics*, vol. 59, no. 3, pp. 1565-1577, March 2012.

[8] P. H. Feretti and F. L. Tofoli, "A DC-DC buck-boost converter based on the three-state switching cell," *2017 Brazilian Power Electronics Conference (COBEP)*, Juiz de Fora, 2017, pp. 1-6.

[9] R. W. Erickson, D. Maksimovic, *Fundamentals of Power Electronics*, 2 ed., Springer, Nova York, 2001.

[10] INTERNATIONAL ELECTROTECHNICAL COMMISSION: IEC61000-3-2, Electromagnetic compatibility (EMC) – Part 3-2: Limits – Limits for harmonic current emissions (equipment input current ≤ 16A per phase).

Control Strategy for Multifunctional PV Converter

Luiz Henrique Meneghetti[1], Edivan Laercio Carvalho[2], Emerson Giovani Carati[1],
Jean Patric Costa[1], Carlos Marcelo Oliveira Stein[1], Zeno Luiz Iensen Nadal[3], Rafael Cardoso[1]

[1]*Universidade Tecnológica Federal do Paraná (UTFPR), Pato Branco, Brazil*
[2]*Federal University of Santa Maria - UFSM, Santa Maria, Brazil*
[3]*Copel Distribuição S.A., Curitiba, Brazil*

E-mails: meneghetti.luiz@outlook.com, e.carvalho@ieee.org, emerson@utfpr.edu.br,
jpcosta@utfpr.edu.br, cmstein@utfpr.edu.br, zeno.nadal@copel.com, rcardoso@utfpr.edu.br

Abstract—This paper proposes a multifunctional converter applied to distributed generation with an energy storage system. The main functionalities include the possibility to have programmable power dispatch to the utility grid and trade of energy between the Battery Energy System (BESS) and utility grid, accordingly to the specific interest of an Energy Management System (EMS). In order to allow these functionalities, modifications are proposed in the bidirectional buck-boost converter that is responsible for the BESS power processing. With that, it is also proposed the sharing of the DC bus voltage controller between the bidirectional DC-DC converter and the DC-AC converter. To validate the proposed system, real-time hardware-in-the-loop (HIL) simulations are presented. In the simulations, real data of solar irradiance is considered. Moreover, a model of a lithium-ion battery bank is also incorporated. The proposed system guarantees programmable dispatch with low power errors. In addition, the system is able to operate with and without BESS at the same time it ensures smooth transitions between the operation modes.

Index Terms—Battery charger, Bidirectional converters, Multifunctional PV converter, Hybrid generation systems.

I. INTRODUCTION

Hybrid generation/distribution energy systems are systems that have two or more generation sources and loads. Such systems can operate with coordinated elements to allow better power utilization and programmed power dispatch to the utility grid [1], [2].

A simple example of these systems is the multifunctional photovoltaic converters (MPVC), or hybrid inverters, which in addition to the PV generation, use batteries to allow smoothing of the generation profiles [3] - [6]. This association allows the modification of the PV inverters functionalities related to the utility grid. In this case, different objectives may be desired, including cost optimization [1], [7], peak shaving [8] - [9], uninterrupted power supply [3], [5] and power dispatch control [4].

In grid-tied inverters, it is also possible to give support for local loads and allow energy exchanges between the PV system in addition to the BESS and the utility grid. This degree of freedom allows multifunctional converters to charge/discharge batteries according to specific interests, such as power management and efficient use of the renewable resources [7].

The coordination of the different sources, loads and AC grid is divided into different hierarchical levels, which extend from the basic control loops of the interface converters (such as PV panels boost control, DC-DC bidirectional converter associated to the batteries and DC-AC bidirectional converter associated to the grid interface) to levels related to the power management and supervisory control [1] - [2].

The different control levels aim to guarantee the stability of the system and the best use of energy. So, control strategies shall be designed according to the different operating modes of the PV inverter, including grid-tied mode or off-grid mode [3], [7] - [8].

On the other hand, the reliability of the PV inverters is strongly affected by the generation conditions, and consequently, by the control strategy of the interface converters [10] - [11]. In these types of system, energy exchanges are commonly performed by means of an intermediate DC bus. Therefore, the converters control loops must guarantee the system power balance, as well as the voltage regulation on the DC bus [7].

In [10] and [13], the output dc-ac converter controls the current injected into the AC grid and also regulates the voltage of the intermediate DC bus. This allows the PV converter to operate in MPPT mode, aided by BESS for possible programmed power dispatch. This is the most common mode of operation of PV converters because it is derived from the strategies used in converters without energy storage, as described in [11].

A possible strategy to allow programmable power dispatch is to compensate the power directly in the DC bus using the BESS DC-DC bidirectional converter current. However, when using a current compensation loop for the bidirectional converter, possible power errors can be observed at the point of common coupling (PCC) with the electrical grid.

In order to simplify the control strategy and the DC bus regulation, in this paper, it is proposed that the DC bus voltage control is performed by the bidirectional dc-dc converter. When the state of charge (SoC) of the batteries is outside a pre-set limit, the dc-ac converter takes the responsability for controlling the DC bus voltage. In this case, the BESS can be charged by PV and the utility grid.

The main contribution of this paper is that the DC bus voltage controller is shared in both operation modes, that is, when the BESS or the DC-AC converter controls the DC bus. To achieve that, it is inserted an output inductor in the conventional buck-boost DC-DC converter applied to the BESS. Therefore, transients in the DC bus voltage during the change in operating mode are smoothed. When the dc-ac converter controls the DC bus voltage, a conventional cascade loop is used, as discussed in [11]. When the bidirectional dc-dc converter controls the DC bus voltage, a similar cascade control loop is proposed. Thus, it is possible to attenuate power errors and smooth the transients responses on the DC bus voltage due to the current control on the dc bus.

In the conventional structure, in which the dc-ac converter controls V_{DC}, the DC-DC bidirectional converter reference current is typically generated using the PV power generation and the desired power dispatch [12]. Therefore, the efficiency of the converters related to the PV, the BESS, and the dc-ac interfacing, interferes on the bidirectional converter current reference and contributes to power errors on PCC. On the other hand, when performing the DC bus voltage control from the bidirectional converter, the respective current reference is automatically generated by the control loop. Thus, there is no dependence on the efficiency of the bidirectional converter and the converter related to the PV.

II. System Description

The system consists of a single-phase bidirectional DC-AC converter, a boost converter and a non-isolated bidirectional DC-DC converter that provides power processing to the BESS, as shown in Fig.1, which represent the proposed structure.

The DC-AC converter consists of a single phase-full bridge structure with an LCL filter. The design methodology for the LCL filter can be found in [14]. Possibles methods to design the R_f resistor used to damp the LCL resonance can be found in [15].

The PV power processing uses a DC-DC boost converter, that is implemented with the maximum power point tracking (MPPT) Perturb and Observe (P&O) algorithm [16].

The bidirectional DC-DC converter is used to supply power to a 96 V/40 Ah lithium-ion battery. The low voltage and current ripple discussed in [17] are considered and implemented in this paper using the bidirectional buck-boost topology with the insertion of L_1.

III. Control Strategy

The proposed structure considers that the EMS provides the active and reactive power dispatch references, P^* and Q^*, respectively. Also, the EMS is responsible for monitoring the state of charge (SoC) and decides when the battery bank is charged or discharged. The SoC estimation considers the ampere-hour integral method [18]. The EMS is not implemented in this paper and is considered that it operates accordingly to the energy price, according to other papers found in the literature, i. e. [7]. Figure 2 shows the control strategy diagram, in witch s_d and s_i represent the two switches positions according to the V_{DC} controller.

Two operation modes are considered:

- $20\% \leq \text{SoC} \leq 80\%$

In this case, the EMS provides P^* accordingly to the energy price. So, the battery bank is used to smooth the power fluctuations on the dispatched power due to the stochastic characteristic of the PV generation (P_{PV}) and ensure a constant power injection into the utility grid. In this scenario, the PV plant operates at MPPT, the DC-DC bidirectional converter controls the DC bus voltage (V_{DC}) through I_{L1} current, and the DC-AC converter controls de current injected into the utility grid (i_{inv}).

- $20\% > \text{SoC} > 80\%$

In the case that the SoC > 80% or SoC < 20% the battery bank is disabled to not compromise the battery lifetime. In both cases, the DC bus voltage control loop is changed and the DC-AC converter controls V_{DC}. Considering that the PV plant operates at MPPT, $P^* = P_{PV}$ and the PV power generation is injected into the utility grid. To recharge the batteries, the EMS provides the desired SoC and calculates I_{L1}^* from the desired charging current (I_{L2}^*). The power for charging the battery storage system is supplied by PV power generation and, if necessary, it is drained from the utility grid.

The following subsections describe the control structure depicted in Fig. 2.

A. Frequency Locked Loop (FLL)

The synchronization between the DC-AC converter and the utility grid is done by a standard Second Order Generalized Integrator Frequency Locked Loop (SOGI-FLL). The SOGI-FLL decomposes the voltage at the point of common coupling (V_{PCC}) into components V_α and V_β. These components are used in the current reference generator (i_{inv}^*) of the DC-AC converter. Also, this structure provides the estimated grid frequency $\hat{\omega}$, that is used in the DC-AC current controller. Detail about SOGI-FLL design can be found in [19].

B. Reference Current Generation (i_{inv}^*)

The reference current generation and control of the DC-AC converter are based on the methodology found in [11] and applied in [20]. So, the current reference, synchronized with the the utility grid is given by

$$i_{inv}^* = 2 \frac{V_\alpha(P^* - P_{DC}) - V_\beta Q^*}{V_\alpha^2 + V_\beta^2}, \qquad (1)$$

where, P_{DC} is the power variation due the oscillations of the voltage V_{DC}.

C. Output DC-AC Converter Current Control (i_{inv})

Since the proposed system considers the stationary reference frame $\alpha\beta$, it is necessary an AC controller. It is adopted the ideal proportional-resonant (PR) controller, given by

$$G_{PR}(s) = K_{Pr} + \frac{2K_{Ir}s}{s^2 + \omega^2}. \qquad (2)$$

This structure provides an infinite gain at the resonance frequency ω [11]. So, for a sinusoidal reference current with frequency $\hat{\omega} = \omega$, $G_{PR}(s)$ is capable to track i_{inv}^* with zero error.

978-1-7281-4181-7/19 $31.00 © 2019 IEEE

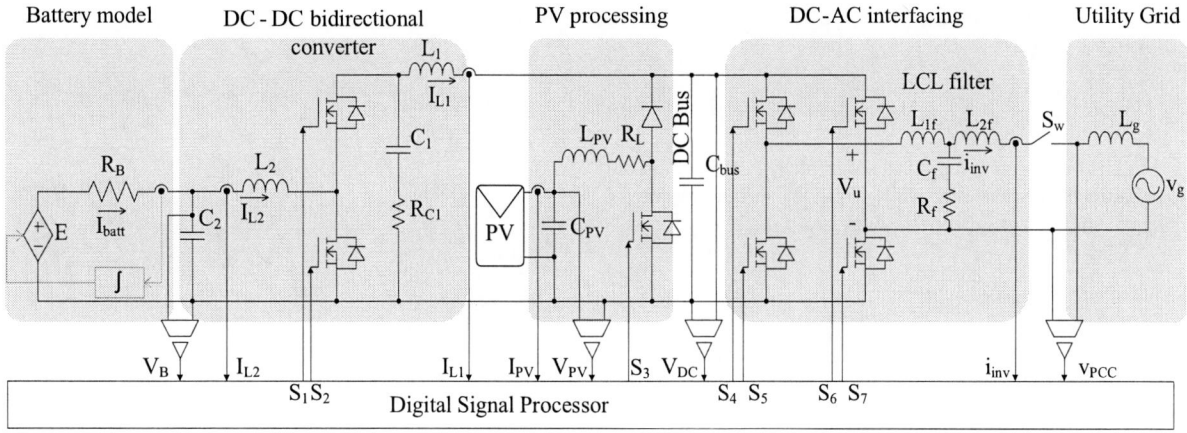

Fig. 1. Diagram of the proposed system.

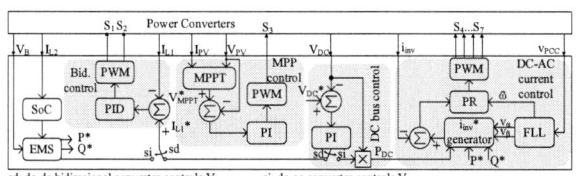

sd: dc-dc bidirecional converter controls V_{DC} si: dc-ac converter controls V_{DC}

Fig. 2. Control strategy diagram.

The design of the PR controller considers the transfer function of the LCL filter given by

$$\frac{i_{inv}(s)}{V_u(s)} = \frac{C_f R_f s + 1}{L_{1f} L_{2f} C_f s^3 + C_f R_f (L_{1f} + L_{2f}) s^2 + (L_{1f} + L_{2f}) s}. \tag{3}$$

D. PV Processing Control

According to [21], the linearized small-signal model of the PV processing plant is

$$\frac{v_{\tilde{P}V}(s)}{\tilde{d}(s)} = \frac{-V_{DC} \frac{r_{PV}}{r_{PV} - R_L}}{\frac{s^2}{\omega_n^2} + \frac{2\xi_n s}{\omega_n} + 1}, \tag{4}$$

where, $\tilde{d}(s)$ is a duty cycle perturbation, r_{PV} is the PV dynamic resistance and R_L is the series resistance of the inductor L_{PV}.

Other parameters of the transfer function are

$$\xi_n = \left(\frac{R_L}{L_{PV}} - \frac{1}{r_{PV} C_{PV}} \right), \tag{5}$$

$$\omega_n = \sqrt{\frac{1}{L_{PV} C_{PV}} \left(1 - \frac{R_L}{r_{PV}} \right)} \tag{6}$$

and

$$r_{PV} = -R_S - \frac{v_t}{i_{sat}} e^{-\frac{V_{PV} + I_{PV} R_S}{v_t}}, \tag{7}$$

where, i_{sat} is the saturation current of the internal PV diode, v_t is the thermal voltage, R_S is the equivalent series resistance

of the PV, and I_{PV} and V_{PV} are the actual current and voltage of the PV array. To simplify the model, it is considered that the PV plant operates at the MPPT with nominal irradiance and temperature. Then, $R_S = 0.0189 \ \Omega$, $i_{sat} = 9.21 \times 10^{-9}$ A, $V_{PV} = 114.3$ V and $I_{PV} = 9.21$ A, according with the adopted PV model [22].

E. DC Bus Voltage Control

When $20\% \leq \text{SoC} \leq 80\%$, the BESS controls V_{DC} by means of I_{L1}. This strategy is adopted to allow the sharing of the V_{DC} controller between the bidirectional DC-DC converter and the DC-AC converter. So, it is necessary to obtain two transfer functions: $\tilde{I}_{L1}(s)/\tilde{d}(s)$ and $V_{DC}(s)/I_{L1}(s)$, where $\tilde{d}(s)$ is the deviation around the nominal duty cycle D of the DC-DC bidirectional converter. The modeling of the DC-DC bidirectional converter considers the methodology discussed in [23]. The insertion of the inductor L_1 in the buck-boost converter provides symmetry so that the model for both charge and discharge modes of the battery bank becomes identical. Equation (8) describes the current transfer function, for both charging and discharging modes, considering the methodology of [23] and the parameters presented in Table I.

$$\frac{\tilde{I}_{L1}(s)}{\tilde{d}(s)} = \frac{-(s + 3.4231 \times 10^9)(s + 2.6546 \times 10^9)(s + 1.2158 \times 10^8)}{(s + 2.267 \times 10^5)(s + 389.1)(s^2 + 1640s + 4.766 \times 10^8)} \tag{8}$$

Considering I_{L1} as a current source, it is possible to obtain the $V_{DC}(s)/I_{L1}(s)$ transfer function given by

$$\frac{V_{DC}(s)}{I_{L1}(s)} = \frac{R_o}{R_o C_{bus} s + 1}, \tag{9}$$

where R_o represents the resistance considering the nominal power and voltage on the DC bus.

When $20\% > \text{SoC} > 80\%$, V_{DC} is controlled by means of i_{inv}. Then, analogously to the previous case, some transfer functions are required to compose the control loop diagram. The current loop is composed by the current controller (Eq.

2) and the DC-AC converter plant (Eq. 3). The DC bus plant, in this case, is given by

$$\frac{v_{\tilde{DC}}(s)}{i_{\tilde{inv}}(s)} = -\frac{\frac{V_{DC}}{V_g}R_o}{R_o C_{bus} s - 1}. \tag{10}$$

Fig. 3(a) shows the control diagram of V_{DC} using the DC-DC bidirectional converter while Fig. 3(b) shows the control diagram of V_{DC} using the DC-AC converter.

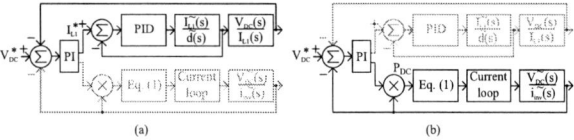

(a)   (b)

Fig. 3. V_{DC} control diagram using (a) DC-DC bidirectional converter and (b) DC-AC converter.

The insertion of the L_1 inductor makes possible to share the external V_{DC} control loop. Furthermore, the I_{L1} current loop is used to charge the battery bank when SoC $< 20\%$. The use of the same controller for distinct functions considerably reduces the complexity of the control system.

F. Battery Charge Control

When SoC $< 20\%$, it is intended to charge the BESS according to the desirable final SoC and charge current (I_{L2}^*). These parameters may be variable over time and are provided by the EMS. The charging power to the batteries can be originated by either the PV panels or the grid, or both, simultaneously. A simple relationship between input and output power of the bidirectional DC-DC converter is used to take advantage of the existing I_{L1} control loop and to promote the battery bank charge. Eq. (11) gives the reference current I_{L1}^* for this case.

$$I_{L1}^* = I_{L2}^* \frac{V_B}{\eta V_{DC}}, \tag{11}$$

where η is the DC-DC bidirectional converter efficiency. In this case, the system is not able to track the dispatched power P^*, making P^* an initial condition on the V_{DC} controller.

IV. RESULTS AND DISCUSSIONS

To validate the proposed system, HIL simulations were performed. Table I presents the parameters of the system and Table III shows the controllers used in the real-time simulation. The power converters (Fig. 1) operates in real-time on a Typhoon HIL 620+ platform. All the controllers and the FLL (Fig. 2) were discretized by the Tustin method and also operate in real-time on TMS320F28335 DSP. The lithium-ion battery and the PV array models are available on Typhoon HIL 602+. Besides, the real-time simulation considers real irradiance data that were obtained experimentally.

As the main result, Fig. 4 illustrate the PV power (P_{PV}), according to real irradiance data applied to the PV array model. Also, it depicts the reference power (P^*), the BESS power injected into the DC bus (P_{BESS}) and the power

TABLE I
MAIN PARAMETERS OF THE SYSTEM.

Utility grid		
Phase Voltage	V_g	$127\ V_{RMS}$
Frequency	f_g	$60\ Hz$
Equivalent inductance	L_g	$1\ mH$
DC-AC coupling filter		
Filter inductors	$L_{1f} - L_{2f}$	$370\ \mu H - 155\ \mu H$
Filter capacitor	C_f	$4.7\ \mu F$
Damping resistor	R_f	$2\ \Omega$
PV power processing		
Filter inductor	L_{PV}	$1.3\ mH$
Filter capacitor	C_{PV}	$220\ \mu F$
Nominal power	P_{max}	$1.05\ kW$
DC bus		
Voltage	V_{DC}	$400\ V$
DC bus capacitance	C_{bus}	$2\ mF$
DC-DC bidirectional converter		
Nominal battery voltage	V_B	$96\ V$
Series resistance	R_B	$0.155\ \Omega$
Filter inductors	$L_1 - L_2$	$1.0\ mH - 1.3\ mH$
Filter capacitors	$C_1 - C_2$	$2.2\ \mu F - 22\ \mu F$
Filter resistor	R_{C1}	$2\ \Omega$
Common parameters for all converters		
Sampling frequency	f_s	$20\ kHz$
Switching frequency	f_{sw}	$20\ kHz$
System nominal power	P	$1.05\ kW$

TABLE II
CONTROL PARAMETERS.

Control loop of	Cutoff frequencies	Controller
V_{PV}	≈ 8100 rad/s	$-0.1 - \frac{14.572}{s}$
i_{inv}	≈ 4200 rad/s	$4.2 + \frac{2(3000s)}{s^2 + \omega^2}$
I_{L1}	≈ 2600 rad/s	$\frac{(s+21.5481)(s+1.9121)}{s(s+6369)}$
V_{DC} by I_{L1}	≈ 125 rad/s	$0.15 + \frac{1}{s}$
V_{DC} by i_{inv}	≈ 377 rad/s	

injected into the utility grid (P_{inv}). Fig. 4 is subdivided into 7 regions ($R_1, ..., R_7$) which will be detailed below.

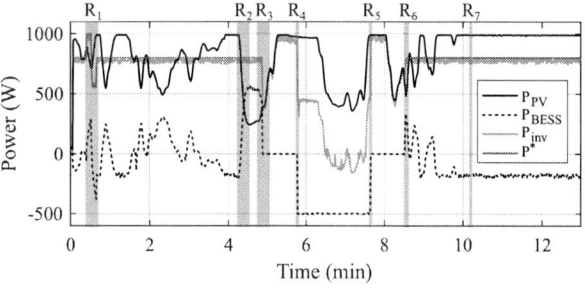

Fig. 4. Power waveforms of the system.

Initially, $20\% <$ SoC $< 80\%$ and $P^* = 800$ W, so the DC-DC bidirectional converter controls V_{DC}. In region R_1, P^* was changed from 800 W to 1000 W, then from 1000 W to 600 W and, finally, from 600 W to 800 W. It is possible to note that the P_{inv} tracks P^* independently of P_{PV}. The difference between P^* and P_{PV} is provided by P_{BESS}. Fig. 5 shows the waveforms in the region R_1. It is possible to note that the lacking or excessive power in the DC bus is compensated by I_{L1}. Moreover, the reference power changes in the form of a ramp, avoiding abrupt transients in the DC bus and the utility grid.

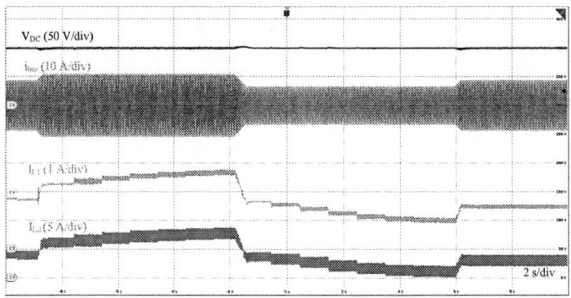

Fig. 5. Waveforms in the region R_1.

Fig. 6 shows detail in the region R_2 that describe the change between charge and discharge of the BESS when $P^* = 800$ W and the DC-DC bidirectional converter controls V_{DC}. It is possible to verify that I_{L1} compensates the power in the DC bus and, consequently, guarantees a constant amplitude of the current flow into the grid.

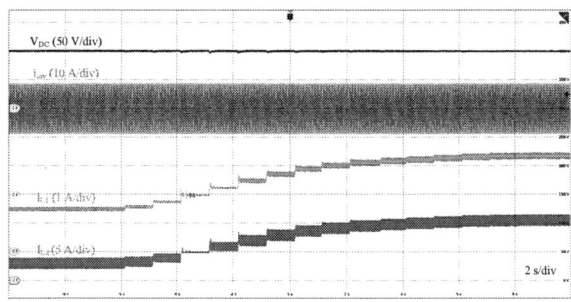

Fig. 6. Waveforms in the region R_2.

In Fig. 7 the effects of the reduced state of charge of the batteries are verified in the region R_3. When SoC < 20%, a power ramp is applied in P^* until it reaches P_{PV}. When $P^* \approx P_{PV}$, the V_{DC} control loop is changed and the DC-AC converter controls the DC bus voltage. This strategy avoids abrupt transients in V_{DC} and, consequently, in P_{inv}.

After changing the V_{DC} control loop, the batteries remain out of operation and the constant power dispatch can not be guaranteed. In this case, the maximum instantaneous power is injected into the utility grid, so, P_{inv} tracks P_{PV}. Since the V_{PCC} is defined by the utility grid, power fluctuations in P_{inv} are a result of the i_{inv} fluctuations, that can be seen in detail in Fig. 7. In Fig. 4 it is possible to note that $P_{inv} \approx P_{PV}$ until region R_4. In region R_4, it is set $I_{L2}^* = 5$ A to charge the batteries. In this case, the DC-AC converter controls V_{DC}.

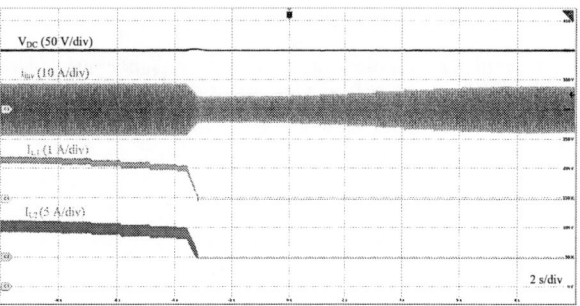

Fig. 7. Waveforms in the region R_3.

Fig. 8a shows the instant when the BESS starts the charging process. In Fig. 4 is showed that the charge power is provided by P_{inv} when P_{PV} is not sufficient.

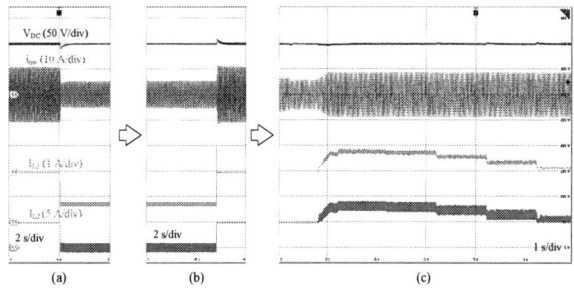

Fig. 8. Waveforms in the region (a) R_4, (b) R_5 and (c) R_6.

The I_{L2} current remains constant from region R_4 to R_5. In region R_5 (Fig. 8b) the charging process is completed and $P_{BESS} = 0$ until region R_6. In region R_6 (Fig. 8c) occurs the transition between the V_{DC} control loops and the power dispatch can be guarantee. Again, the DC-DC bidirectional converter controls V_{DC}.

To verify the total harmonic distortion (THD) of i_{inv} in region R_7, it was set $P^* = 800$ W. Fig. 9 depicts the waveform details of i_{inv}, V_{PCC}, V_{DC} and I_{L1}, besides the first 40 harmonics of i_{inv}. The THD is 2.5% and complies with the IEEE-1547 standard.

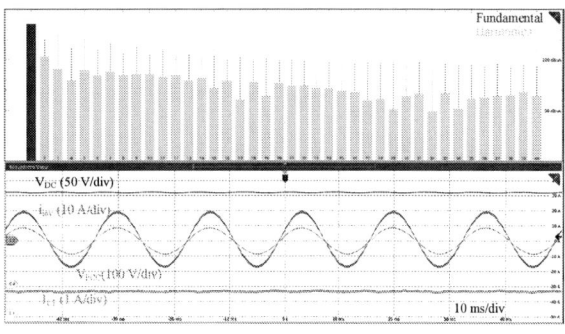

Fig. 9. Waveforms in the region R_7.

V. CONCLUSIONS

This paper proposed the DC bus voltage control loop sharing for the charge and discharge of the batteries in a multifunctional photovoltaic converter, that includes the PV generation, AC grid, and BESS. The DC bus voltage control loop is shared between the bidirectional DC-DC converter and DC-AC converter.

It was shown tests in order to illustrate the system capacity to guarantee the constant power dispatch to the utility grid and power-sharing between the converters when considered the control system hierarchy.

When the BESS has low SoC, the system supplies the maximum power from the PV to the utility grid. Moreover, the system is capable to charge the battery bank by means of the PV and utility grid, simultaneously.

The DC bus voltage control loop sharing simplifies the implementation of the controllers and guarantee smooth transients in the DC bus voltage during the change of the converters that control the bus voltage. Consequently, transients in the power injected in the PCC are also smoothed.

ACKNOWLEDGMENT

This work was funded by the Research and Development project PD 2866-0468/2017, granted by the Brazilian Electricity Regulatory Agency (ANEEL) and Companhia Paranaense de Energia (COPEL). The authors also would like to thank to FINEP, SETI, CNPq, Fundação Araucária, CAPES and UTFPR for additional funding.

REFERENCES

[1] B. Papari, C. S. Edrigton, I. Bhattacharya, G. Radman, "Effective Energy Management of Hybrid Microgrids with Storage Devices," IEEE Transaction on Smart Grid, vol. 10, no. 1, Jan. 2019.

[2] T. Dragicevic, J. M. Guerrero, J. C. Vaquez, D. Skrlec, "Supervisory Control of an Adaptive-Droop Regulated DC Microgrid With Battery Management Capability," IEEE Transaction on Power Electronics, vol. 29, no. 2, pp. 695-706, Feb. 2014.

[3] S. Pranith, S. Kumar, B. Singh, T. S. Bhatti, "Multimode operation of PV-battery system with renewable intermittency smoothening and enhanced power quality," IET Renewable Power Generation, vol. 13, no. 6, pp. 4-29, Sep. 2019.

[4] M. Brenna, F. Foiadelli, M. Longo and D. Zaninelli, "Energy Storage Control for Dispatching Photovoltaic Power," in IEEE Transactions on Smart Grid, vol. 9, no. 4, pp. 2419-2428, July 2018.

[5] N. Saxena, I. Hussain, B. Singh and A. L. Vyas, "Implementation of a Grid-Integrated PV-Battery System for Residential and Electrical Vehicle Applications," in IEEE Transactions on Industrial Electronics, vol. 65, no. 8, pp. 6592-6601, Aug. 2018.

[6] M. R. Sandgani, S. Sirouspour, "Energy Management in a Network of Grid-Connected Microgrids/Nanogrids Using Compromise Programming," IEEE Transaction on Smart Grid, vol. 9, no. 3, pp. 2180-2191, May 2018.

[7] L. V. Belinaso. C. D. Schwertner, L. Michels, "Price-Based Power Management of off-grid Photovoltaic Systems with Centralised with" IET Renewable Power Generation, vol. 10 no. 8, Jul. 2016.

[8] Y. Yang, H. Li, A. Aichhorn, J. Zheng, M. Greenleaf, "Sizing Strategy of Distributed Battery Storage System With High Penetration of Photovoltaic for Voltage Regulation and Peak Load Shaving," IEEE Transaction on Smart Grids, vol. 5, no. 2, pp. 982 - 991, Sep. 2015.

[9] K. Mahmud, M. J. Hossain, G. E. Town, "Peak-Load Reduction by Coordinated Response of Photovoltaics, Battery Storage, and Electric Vehicles," IEEE Access, vol. 6, pp. 29353 - 29365, May 2018.

[10] A. Sangwongwanich, G. Angenendt, S. Zurmühlen, Y. Yang, D. Sera, D. U. Sauer, F. Blaabjerg, "Enhancing PV Inverter Reliability With Battery System Control Strategy" CPSS Transactions on Power Electronics and Applications, vol. 3, no. 2, pp. 93-101, jun. 2018.

[11] R. Teodorescu, M. Liserre, P. Rodrígues, "Grid Converters for Photovoltaic a and nd Wind Power Systems," John Wiley & Sons, New Jersey, 2011.

[12] L.H. Meneghetti, E. L. Carvalho, E. G. Carati, J. P. da Costa, C. M. de Oliveira Stein and R. Cardoso, "Energy Storage System for Programmable Dispatch of Photovoltaic Generation," European Conference on Power Electronics and Applications, Genova, 2019.

[13] N. Mahmud, A. Zahedi and A. Mahmud, "A Cooperative Operation of Novel PV Inverter Control Scheme and Storage Energy Management System Based on ANFIS for Voltage Regulation of Grid-Tied PV System," in IEEE Transactions on Industrial Informatics, vol. 13, no. 5, pp. 2657-2668, Oct. 2017.

[14] F. Li, X. Zhang, H. Zhu, H. Li and C. Yu, "An LCL-LC Filter for Grid-Connected Converter: Topology, Parameter, and Analysis," in IEEE Transactions on Power Electronics, vol. 30, no. 9, pp. 5067-5077, Sept. 2015.

[15] W. Wu, Y. Liu, Y. He, H. S. Chung, M. Liserre and F. Blaabjerg, "Damping Methods for Resonances Caused by LCL-Filter-Based Current-Controlled Grid-Tied Power Inverters: An Overview," in IEEE Transactions on Industrial Electronics, vol. 64, no. 9, pp. 7402-7413, Sept. 2017.

[16] D. Sera, L. Mathe, T. Kerekes, S. V. Spataru and R. Teodorescu, "On the Perturb-and-Observe and Incremental Conductance MPPT Methods for PV Systems," in IEEE Journal of Photovoltaics, vol. 3, no. 3, pp. 1070-1078, July 2013.

[17] E. L. Carvalho, E. G. Carati, J. P. da Costa, C. M. de Oliveira Stein and R. Cardoso, "Analysis, design and implementation of an isolated full-bridge converter for battery charging," 2017 Brazilian Power Electronics Conference (COBEP), Juiz de Fora, 2017, pp. 1-6.

[18] R. Xiong, J. Cao, Q. Yu, H. He and F. Sun, "Critical Review on the Battery State of Charge Estimation Methods for Electric Vehicles," in IEEE Access, vol. 6, pp. 1832-1843, 2018.

[19] S. Golestan, J. M. Guerrero, J. C. Vasquez, A. M. Abusorrah and Y. Al-Turki, "Modeling, Tuning, and Performance Comparison of Second-Order-Generalized-Integrator-Based FLLs," in IEEE Transactions on Power Electronics, vol. 33, no. 12, pp. 10229-10239, Dec. 2018.

[20] R. A. Liston, E. G. Carati, R. Cardoso, J. P. da Costa and C. Marcelo de Oliveira Stein, "A Robust Design of Active Damping With a Current Estimator for Single-Phase Grid-Tied Inverters," in IEEE Transactions on Industry Applications, vol. 54, no. 5, pp. 4672-4681, Sept.-Oct. 2018.

[21] M. Sokolov, "Small-Signal Modelling of Maximum Power Point Tracking for Photovoltaic Systems," thesis, Imperial College London, 2013.

[22] Canadian Solar, "MAXPOWER CS6U - 340 | 345 | 350 High Efficiency Poly Module," Photovoltaic module datasheet, Apr. 2018.

[23] E. L. Carvalho, E. G. Carati, J. P. da Costa, C. M. de Oliveira Stein and R. Cardoso, "Small-Signal Modeling and Analysis of an Isolated Bidirectional Battery Charger," 2018 International Conference on Industry Applications (INDUSCON), São Paulo, 2018, pp. 439-445.

978-1-7281-4181-7/19 $31.00 © 2019 IEEE

Photovoltaic Boost Converter Control Operating in the MPPT and LPPT Modes

Cássia C. C. dos Santos, Cassiano F. Moraes, Jean P. da Costa, Carlos M. O. Stein,
Emerson G. Carati and Rafael Cardoso
PPGEE - Postgraduate Program in Electrical Engineering
UTFPR - Federal University of Technology - Paraná
Pato Branco, Brazil
cassiasantos@alunos.utfpr.edu.br, cassianofmoraes@gmail.com, jpcosta@utfpr.edu.br, cmstein@utfpr.edu.br,
emerson@utfpr.edu.br, rcardoso@utfpr.edu.br

Abstract—**Aiming to generate the maximum power available in photovoltaic systems, maximum power point tracking (MPPT) algorithms are employed. However, with the increased penetration of photovoltaic systems in the grid, it is also necessary that these systems operate with limited power during voltage sags and overvoltage conditions. This paper presents a control method applied to a photovoltaic boost converter to provide a transition between the MPPT and limited power point tracking (LPPT) algorithms. This transition can help the photovoltaic inverter when is necessary to inject less energy to regulate the grid and interrupt the power injection during critical faults. The proposed control method was tested in a boost converter, using a power supply that emulates a photovoltaic module.**
Index Terms—**Photovoltaic, boost, MPPT, LPPT**

I. INTRODUCTION

Photovoltaic (PV) systems generate a direct current (DC) through the irradiation of sunlight. This current is processed by an inverter, which converts it in alternating current (AC) to be applied to AC loads or inject power into the grid. The amount of power generated by the photovoltaic systems depends mainly of solar irradiance level and temperature in the photovoltaic panel. For each level of irradiance and temperature there is a maximum power point (MPP) corresponding which can be extracted for the panel. Tracking the MPP of photovoltaic panel is an essential part of photovoltaic system. For this tracking are used Maximum Power Point Tracking (MPPT) algorithms [1]–[4].

In Brazil with the regulation of net metering, realized in 2012, for the normative resolution nº 482/2012 from Agência Nacional de Energia Elétrica (ANEEL), a high insertion of distributed generation has been observed, mainly through photovoltaic solar generation systems. However, as solar energy is non-dispatchable and its insertion is not considered in the design of the electric distribution system, its high penetration can impact negatively the electric system, causing instability, alteration of voltage profile and rise of losses [5].

When the active power generated by photovoltaic system is greater than the active power demanded by the local loads, the excedent power is injected to the electric grid, raises the voltage of the feeder, causing overvoltage. This problem can be mitigated adjusting the tap of the transformers, increasing the conductors size to reduce the line impedances and store

the power surplus for later use. Nevertheless, these techniques have in common high costs. One alternative is to reduce the active power generation of the photovoltaic system, which can be achieved by forcing the converter to leave the MPP and work in a limited power point (LPP) [6].

In Bellinaso, et al. [7], a cascade control loop is proposed to improve the dynamic performance of photovoltaic system when it operates in the MPPT and the limited power point tracking (LPPT). The control loop proposed is composed of an adaptive voltage controller and a non-linear current controller. However, this work does not approach the LPPT technique and how the transition of the MPPT to the LPPT is performed.

In Lyu, Xu, Zhao, & Wong [8], is proposed to operate the photovoltaic system out of the MPPT in order to the system would have a power backup which can be used later in the system frequency regulation. The strategy to remove the photovoltaic system of the MPP consists in the use of a table containing a voltage value for each temperature and irradiance value, where the voltage value is chosen so that the system has a constant power backup for different irradiance and temperature values. This strategy makes the photovoltaic system always operate out of the MPP, not allowing the operation of the system at other points of the P-V curve. It only operates at the points where the voltage value has been previously set in the table.

In Tafti, Maswood, Konstantinou, Pou, & Blaabjerg [9], a control technique that works both MPPT and out of the MPPT is proposed, this operation makes the constant power generation (CPG) and meets the requirements imposed by the grid. The algorithm is based on the P-V curve operating on both the right and the left side of the curve. It is important to note that the increase of the voltage on the right side of the MPP can lead the PV system to instability if applied to a single stage photovoltaic plant. For the transition between modes of operation, the values of voltage increment and period of these increments are based on a hysteresis band controller.

In Jeon, Lee, Kim & Park [10], is proposed to minimize power stress on photovoltaic Differential Power Processing (DPP) with a LPPT method. This method maintains optimal operating conditions while reducing the total power loss and converter stress in a flyback converter. Other LPPT techniques

978-1-7281-4181-7/19 $31.00 © 2019 IEEE

proposed can be found in [11]–[13]. These techniques are based on Perturb and Observe (P&O) algorithm, however, such papers do not validate experimentally the results obtained.

Thus, this paper aims to present a control system applied to a DC-DC boost converter, which in addition to tracking the MPP by P&O algorithm makes the transition to the LPPT. Thus, it can limit the injection of power in the grid and limit the voltage rise of the DC bus, helping the inverter in the occur of the overvoltages and momentary faults in the power grid. The algorithm developed is also capable to leads the power sent to the DC bus to zero during the occurrence of critical faults.

The control system developed is simple to implement, because it does not demand an additional control loop. It only uses a voltage control loop, and it is not necessary to employ an extra sensor, just using the current and voltage sensors, which are already used for the P&O algorithm. The developed control strategy was validated experimentally with a DC power supply, which emulates the photovoltaic module, a static DC-DC boost converter, a DSP and a resistive load.

II. PROPOSED CONTROL METHOD

The proposed control method, in addition to work at the maximum power point, makes the transition to a limited power point. The limited power value is imposed by external control.

The LPPT algorithm developed was based on the conventional MPPT P&O algorithm. But, differently from the P&O algorithm, that aims to find the MPP by increasing and decreasing the panel voltage (V_{pv}), the LPPT tracks the LPP increasing the panel voltage, as shown in Fig. 1.

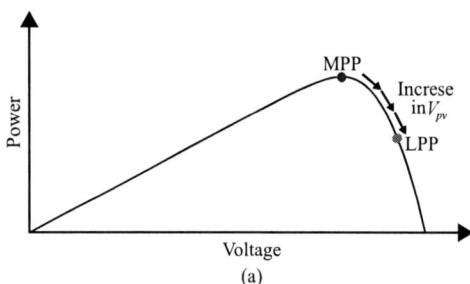

Fig. 1. Power-voltage curve of the photovoltaic panel.

It is observed in the graph shown in Fig. 1 when V_{pv} increases, there is a move change the operating point to the right side of the MPP, decreasing the power value of the panel to the requested LPP.

When no power limitation is required, the external control sends a command to the control of the DC-DC converter warning that it must be with the MPPT algorithm activated, otherwise the LPPT algorithm is activated causing the value of the power generated by the panel to be equal to the value of the P_{ref}, requested by the external control. The flow chart of the proposed algorithm is shown in Fig. 2.

It can be seen from Fig. 2 that if the P_{ref} is lower than the panel power P_{pv}, the reference voltage (V_{ref}) is increased until

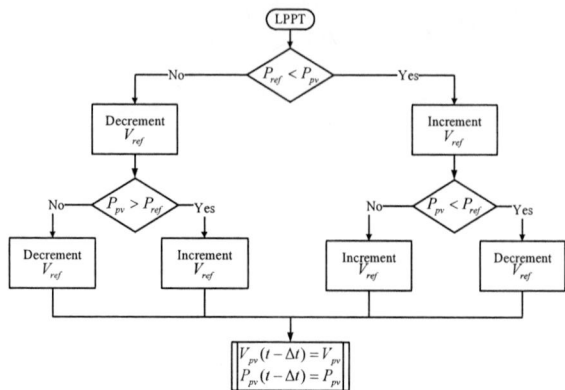

Fig. 2. LPPT flow chart.

a voltage value corresponding to a value of power lower than the P_{ref}. In order to the power value quickly reach the desired value, the voltage increment is multiplied in each iteration. When the power of the panel becomes smaller than P_{ref}, the voltage increment is divided in each iteration, for lessen the oscillation and the algorithm begins to operate with a logic similar to the conventional P&O algorithm, as shown in Fig. 3.

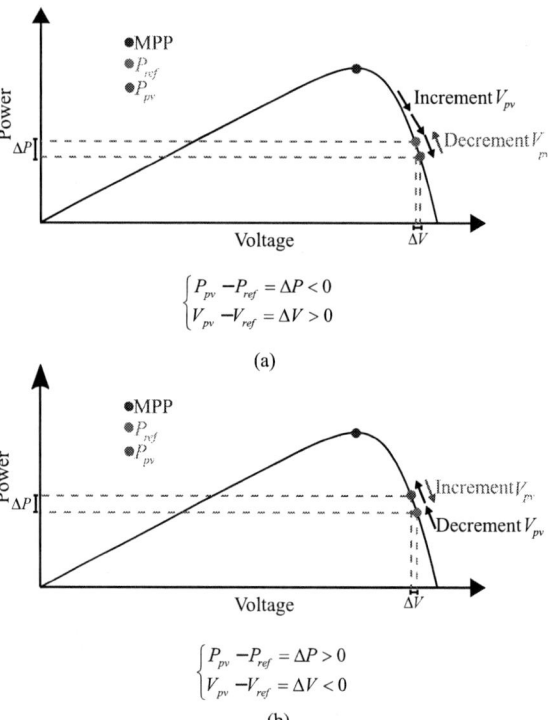

Fig. 3. LPPT algorithm operation for a reference power smaller than MPP.

If there is a negative power change ($\Delta P < 0$) and positive voltage change ($\Delta V > 0$), it means that the power value of the panel is lower than the desired reference power value, so the voltage is decremented as shown in Fig. 3(a). If the power variation ($\Delta P > 0$) is positive and the negative voltage change ($\Delta V < 0$), it means that P_{pv} is higher to P_{ref}, so the voltage is increased, as shown in 3(b). Within this logic the value of the voltage increment is decreased so that the oscillation around the reference point is reduced. If P_{ref} changes to a new reference power (P_{ref_new}), but remains lower than the MPP of the panel and is higher than the previous P_{ref}, the P_{pv} will be lower than the power then the voltage is decremented until P_{pv} reaches a value greater than the P_{ref}. At this point, the algorithm also works with logic similar to the conventional P&O, as shown in Fig. 4.

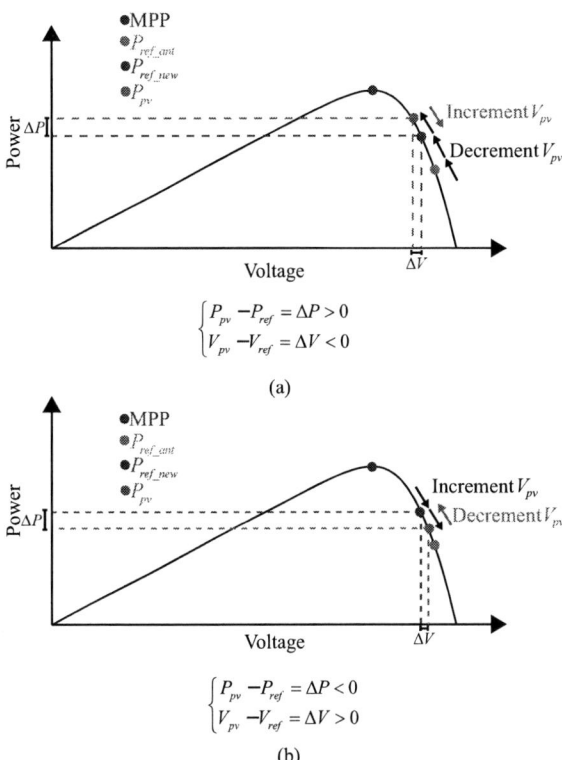

$$\begin{cases} P_{pv} - P_{ref} = \Delta P > 0 \\ V_{pv} - V_{ref} = \Delta V < 0 \end{cases}$$

(a)

$$\begin{cases} P_{pv} - P_{ref} = \Delta P < 0 \\ V_{pv} - V_{ref} = \Delta V > 0 \end{cases}$$

(b)

Fig. 4. LPPT algorithm operation for a new reference power.

If there are a positive power and negative voltage variation around the new reference power, it means that the panel power value is lower than the desired reference power value then the voltage is increased, as shown in Fig. 4(a). If the power variation is negative and the voltage is positive, it means that the power value of the panel is lower than the desired power value, so the voltage is decremented, as shown in Fig. 4(b).

III. SIMULATION RESULTS

To test the developed algorithm, a simulation was performed using the software PSIM. This software has a power element that represents the physical model of a photovoltaic panel, and it can be configured with the characteristics of the desired photovoltaic panel. Fig. 5 shows the circuit implemented on PSIM.

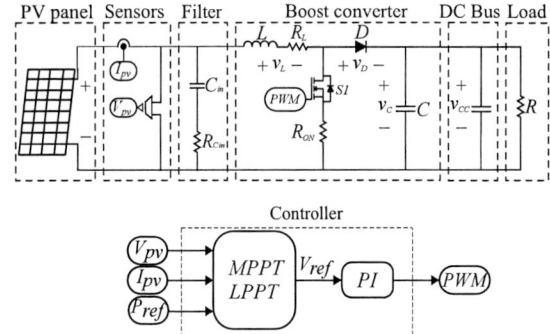

Fig. 5. Boost converter implemented in PSIM.

The source of the circuit shown in Fig. 5 is a PV panel, where two sensors measured the current I_{pv} and voltage V_{pv} of PV panel, in parallel of source there is a filter composed to the input capacitor C_{in} and the input capacitor resistance $R_{C_{in}}$, the boost converter is composed by an inductance L, the inductance resistance R_L, a semiconductor switch $S1$, a diode D and capacitance C. In parallel with the boost converter, there are the bus capacitance v_{cc} and a resistive load R.

The photovoltaic panel considered was the MaxPower CS6U-340P from Canadian Solar. Table I presents the parameters of the PV panel, under standard test conditions (STC), irradiance of 1000 W/m² with an air mass of 1.5 (AM 1.5) spectrum and cell temperature of 25ºC. Table II presents the values of the components for the circuit presented in Fig. 5 .

TABLE I
ELECTRICAL DATA — STC

Parameter	Simbol	Value
Nominal Max. Power	P_{max}	340 W
Optimal Operating Voltage	V_{mp}	37,8 V
Optimal Operation Current	I_{mp}	9,02 A
Open Circuit Voltage	V_{oc}	47,10 V
Short Circuit Current	I_{sc}	10,1 A

TABLE II
BOOST PARAMETERS

Parameter	Value	Parameter	Value
C_{in}	1100 μF	$R_{C_{in}}$	0,1 Ω
L	439 μH	R_L	19 $m\ \Omega$
C	235 μF	C_{bar}	3 $m\ \Omega$
v_D	0,7 V	R_{on}	3,3 $m\ \Omega$

To test if the proposed control alternates between MPPT and LPPT satisfactorily, a hypothetical situation was used, where P_{ref} is changed four times over a period of 9 s. The P_{ref} was started at 340 W, at 3 s it was set at 250 W, at 6 s it was set at 300 W and finally at 8 s it was set again at 340 W. From these conditions the behavior of the power, voltage and current of the PV panel are shown respectively in Fig. 6, Fig. 7 and Fig. 8.

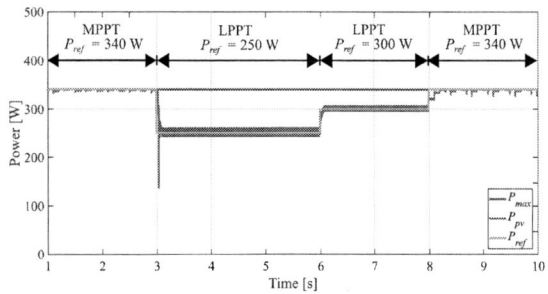

Fig. 6. Simulation result to photovoltaic panel power.

Fig. 6 shows that when the P_{ref} (in yellow) is equal than the maximum power (P_{max}) that can be extracted from the panel, under STC conditions (in blue), in the period from 0 to 2 s, the waveform of the P_{pv} (in orange) is superimposed on the maximum power waveform. This means that at that time the maximum power of the 340 W panel is being extracted, that is the MPPT has been activated. When the P_{ref} is lower than the P_{max}, the power extracted from the panel is no longer the maximum, but is equal to the limited power value of 250 W from 3 to 6 s and 300 W from 6 to 8 s. This means that the LPPT has been activated. In 8 s the P_{ref} returns to be equal than the P_{max}, then the MPPT was activated again.

Fig. 6 also shows that the increment period changed in each condition. In the MPPT was considered a period of 1 s and in the LPPT a period of 0,01 s. The periods were different in each condition so that each condition was better met, in the MPPT a longer time was chosen for each increment, thus it slowed down and consequently had better stability and fewer oscillations. Since the purpose of the LPPT is to go as fast as possible to the reference power since it is only activated when there are problems in the grid, a shorter period was chosen to make the transition faster to the LPPT.

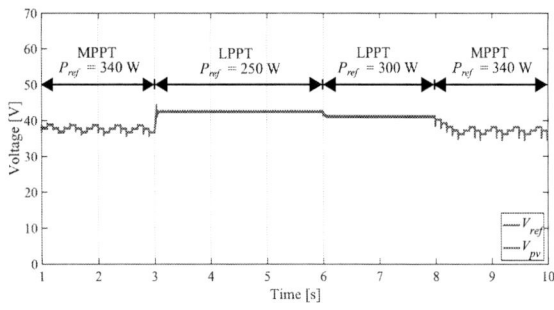

Fig. 7. Simulation result of the photovoltaic panel voltage.

Fig. 7 shows that the panel voltage V_{pv} (in blue) followed the reference voltage V_{ref} generated by the proposed algorithm (in orange). From 0 to 3 s can be observed the characteristic behavior of the algorithm P&O. In 3 s the P_{ref} was changed to 250 W and consequently the V_{ref} was changed, the LPPT took 0,07 s for the V_{pv} to reach the V_{ref}. At 6 s the P_{ref} was changed to 300 W and consequently, V_{pv} took 0,06 s to reach V_{ref}. In the second transition the reference voltage was reached faster because it hears a smaller variation of P_{ref}. In 8 s the P_{ref} returned to a value greater than or equal to P_{max} then the behavior of V_{pv} and V_{ref} returned to the characteristic behavior of P&O.

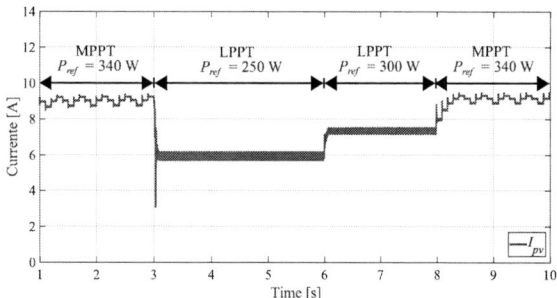

Fig. 8. Simulation result of the photovoltaic panel current.

In Fig. 8 is shown that the current of the photovoltaic panel I_{pv} has the opposite behavior of the P_{pv} when the power is limited to 250 W the voltage is increased since the current is decremented in the same proportion of the voltage. When the power remains limited, but is greater than the previous one the, current is increased.

For this test the initial voltage increment was 1,0 V, in each iteration this increment was multiplied per 1,2 and when the P_{pv} became smaller than P_{ref} the increment was divided per 2,0. This value was chose empirically.

IV. EXPERIMENTAL EVALUATION

For the experimental evaluation, the developed control was implemented in a TMS320F28377S from Texas Instruments. This DSP was used to control the boost converter implemented and a TopCon TC.GSS.20.600.4WR.S.HMI from Regatron was employed to emulate the PV module. The setup in which the control was tested is shown in Fig. 9. The SAS Control shown in Fig. 9 is the software of the TopCon, it allows to configure a curve of the photovoltaic panel. In the experimental evaluation, P_{ref} was set externally by a keyboard.

As with simulation tests, to test if the proposed control alternates between MPPT and LPPT satisfactorily, it was realized a test considered that P_{ref} is changed four times over a period of 40 s. The P_{ref} was started at 340 W, after set at 250 W, 300 W and finally, it was set again at 340 W. Fig. 10 shows when the algorithm is with MPPT activated, with P_{ref} equal to 340 W the panel voltage (in blue) oscillates around 37,2 V and the current (in purple) oscillates around 9,0 A, resulting

Fig. 9. Experimental setup.

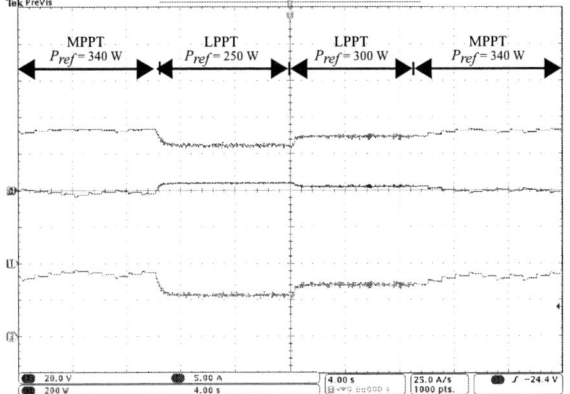

Fig. 10. Experimental result with three diferent values to P_{ref} 340 W, 250 W and 300 W, where P_{pv} in red, V_{pv} in blue and I_{pv} in purple.

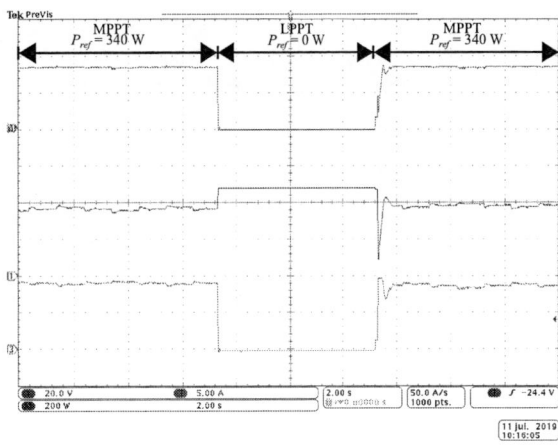

Fig. 11. Experimental result with two P_{ref}, 340 W and 0 W, where P_{pv} in red, V_{pv} in blue and I_{pv} in purple.

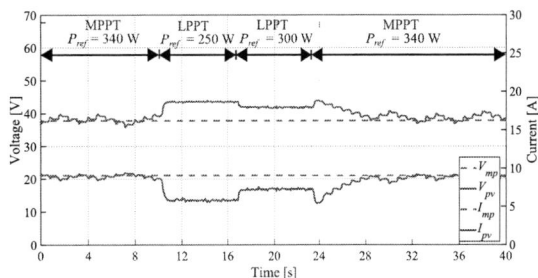

Fig. 12. Voltage and current measured with the power source software.

in a power (in red) of 334,8 W. When the LPPT is activated with a power P_{ref} of 250 W the panel voltage is increased to 43,6 V and the current decremented to 5,6 A, resulting in a limited power generation of 243,8 W. Even with the LPPT activated the P_{ref} is increased to 300 W, so the panel voltage is decremented to 41,8 V and the current is increased to 7,0 A, increasing the value of the power generated in approximately 294,4 W. When the MPPT is activated again, the voltage is decremented up to 37,2 V and the current is increased to 9,0 A until it returns the maximum power point.

Another test was done with a proposed algorithm is interrupt power injection, P_{pv} in 0 W, when there is a grid fault, for this test P_{ref} was set to 0 W. Fig. 11 shows the result obtained.

It is observed in Fig. 11 witch initially the P_{ref} equal to 340 W, so the MPPT mode is activated, the panel voltage (in blue) oscillates around 37,2 V and the current (in purple) oscillates around 9,0 A, resulting in a power (in red) of 334,8 W. When the LPPT is activated with P_{ref} of 0 W, immediately the panel voltage is increased to 47,1 V and the current decremented to 0 A, resulting in a limited power generation of 0 W. After P_{ref} is set to 340 W, so the MPPT mode is activated again.

The SAS control software allows saving the voltage and current data of the panel, Fig. 12 shows the voltage (continuous

line blue) and current (continuous line purple) obtained with the SAS control software and shows that when MPPT is activated, the voltage and current oscillating around the V_{mp} (dashed line blue) and the maximum current (I_{mp}) (dashed line purple) of the photovoltaic panel. The mean values of the panel voltage and current were 37,2 V and 9,0 A, respectively, close to the V_{mp} values of 37,8 V and I_{mp} of 9,02 A.

When the LPPT is activated with P_{ref} of 250 W the voltage is increased and the current is decremented until they reach the values of voltage and current that correspond to the desired power. The mean values of voltage and current were 43,7 V and 5,6 A respectively. Still with the LPPT activated, but with a P_{ref} of 300 W the voltage is decremented and the current increased until reaching the values of voltage and current corresponding to the desired power. The mean values of voltage and current were 42,0 V and 7,0 A respectively.

Fig. 13 shows that when the MPPT is activated the generated power (continuous line red) was 335,8 W near the value of the maximum power of 340W (dashed line red). When LPPT is activated with a P_{ref} of 250 W the power generated was 243,2 W. Continuing with the LPPT activated and with a P_{ref} of 300 W the generated power was of 294,3 W.

Fig. 14 and Fig. 15 show voltage, current and the power behavior of the panel obtained by SAS Control to the second

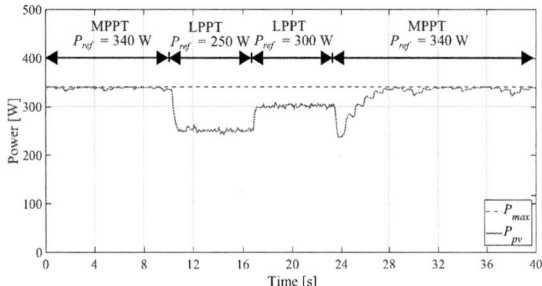

Fig. 13. Power measured with the power source software.

test. When the MPPT was enabled, the voltage and current oscillated around the V_{mp} and I_{mp}, and the power oscillated around P_{max}. When the LPPT was activated to 0 W, the voltage reached 47,1 V, current reached 0 A and the power reached 0 W.

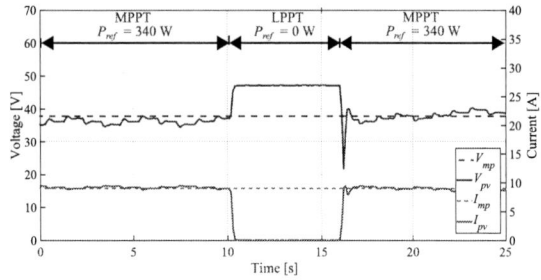

Fig. 14. Voltage and current measured with the power source software for the second test.

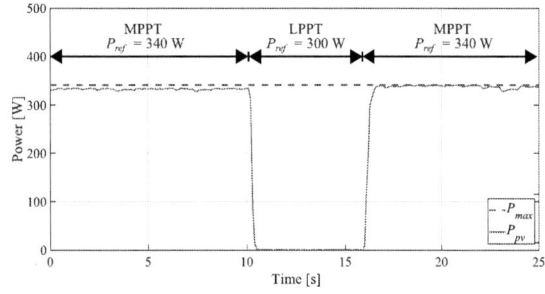

Fig. 15. Power measured with the power source software for the second test.

V. CONCLUSION

A control method applied to a boost converter in addition to tracking the maximum power point of the photovoltaic system tracking a limiting power point was proposed. The limitation of power helping the photovoltaic inverter to regulate the grid and during faults.

The proposed method was based on the operation of the P&O algorithm, but instead of only tracking the maximum

power point, it tracks a limited power point requested by an external control. When this limitation is not requested, the control works as a conventional MPPT algorithm.

The efficacy of the control was tested by simulations in the PSIM software and experimentally using data from a photovoltaic module of 340 Wp. The simulation and experimental results showed that the proposed control method was able to tracking the maximum power value of the panel when the MPPT mode was activated and tracking the limited power value when the LPPT mode was activated. Thus, from the results obtained by simulation and experimental evaluation, it is concluded that the proposed control method was validated.

ACKNOWLEDGEMENT

This work was funded by the Research and Development project PD 2866-0468/2017, granted by the Brazilian Electricity Regulatory Agency (ANEEL) and Companhia Paranaense de Energia (COPEL). The authors also would like to thank to FINEP, SETI, CNPq, Fundação Araucária, CAPES and UTFPR for additional funding.

REFERENCES

[1] M. G. Villalva, J. R. Gazoli, and E. Ruppert Filho, "Comprehensive approach to modeling and simulation of photovoltaic arrays," *IEEE Transactions on power electronics*, vol. 24, no. 5, pp. 1198–1208, 2009.

[2] T. Esram and P. L. Chapman, "Comparison of photovoltaic array maximum power point tracking techniques," *IEEE Transactions on energy conversion*, vol. 22, no. 2, pp. 439–449, 2007.

[3] D. Sera, L. Mathe, T. Kerekes, S. V. Spataru, and R. Teodorescu, "On the perturb-and-observe and incremental conductance mppt methods for pv systems," *IEEE journal of photovoltaics*, vol. 3, no. 3, pp. 1070–1078, 2013.

[4] M. A. G. De Brito, L. Galotto, L. P. Sampaio, G. d. A. e Melo, and C. A. Canesin, "Evaluation of the main mppt techniques for photovoltaic applications," *IEEE transactions on industrial electronics*, vol. 60, no. 3, pp. 1156–1167, 2012.

[5] M. T. Tolmasquim, "Energia renovável: hidráulica, biomassa, eólica, solar, oceânica," *Rio de Janeiro: EPE*, 2016.

[6] R. Tonkoski, L. A. Lopes, and T. H. El-Fouly, "Coordinated active power curtailment of grid connected pv inverters for overvoltage prevention," *IEEE Transactions on Sustainable Energy*, vol. 2, no. 2, pp. 139–147, 2010.

[7] L. V. Bellinaso, H. H. Figueira, M. F. Basquera, R. P. Vieira, H. A. Gründling, and L. Michels, "Cascade control with adaptive voltage controller applied to photovoltaic boost converters," *IEEE Transactions on Industry Applications*, vol. 55, no. 2, pp. 1903–1912, 2018.

[8] X. Lyu, Z. Xu, J. Zhao, and K. P. Wong, "Advanced frequency support strategy of photovoltaic system considering changing working conditions," *IET Generation, Transmission & Distribution*, vol. 12, no. 2, pp. 363–370, 2017.

[9] H. D. Tafti, A. I. Maswood, G. Konstantinou, J. Pou, and F. Blaabjerg, "A general constant power generation algorithm for photovoltaic systems," *IEEE Transactions on Power Electronics*, vol. 33, pp. 4088–4101, May 2018.

[10] Y. Jeon, H. Lee, K. A. Kim, and J. Park, "Least power point tracking method for photovoltaic differential power processing systems," *IEEE Transactions on Power Electronics*, vol. 32, pp. 1941–1951, March 2017.

[11] M. K. K. Reddy and V. Sarkar, "Lppt control of a photovoltaic system against sudden drop of irradiance," in *2017 6th International Conference on Computer Applications In Electrical Engineering-Recent Advances (CERA)*, pp. 562–567, Oct 2017.

[12] P. B. S. Kiran, K. Manjunath, and V. Sarkar, "Limited power control of a single-stage grid connected photovoltaic system," in *2015 Annual IEEE India Conference (INDICON)*, pp. 1–6, Dec 2015.

[13] A. Urtasun, P. Sanchis, and L. Marroyo, "Limiting the power generated by a photovoltaic system," in *10th International Multi-Conferences on Systems, Signals Devices 2013 (SSD13)*, pp. 1–6, March 2013.

978-1-7281-4181-7/19 $31.00 © 2019 IEEE

Predictive Control for a Half-Controlled Boost Rectifier

Gleice Mylena da S. Rodrigues*, Nady Rocha*, Edison Roberto C. da Silva* and Filipe V. Rocha*

*Dept. of Electrical Engineering
Federal University of Paraiba, Joao Pessoa, Paraiba
Email: [gleice.rodrigues, nadyrocha, edison.roberto, filipe.rocha]@cear.ufpb.br

Abstract—**This paper presents a Finite Control Set Model Predictive Control (FCS-MPC) method applied to a Half-Controlled Boost Rectifier (HCBR) that can be used in variable-speed permanent magnet generator. Performance comparison of the system using conventional controls and predictive control methods, in order to regulate the input currents, is accomplished. Results show that the predictive control presents better performance than hysteresis and linear control with pulse width modulation.**

Index Terms—**model predictive control, half-controlled boost rectifier, current control.**

I. INTRODUCTION

The growing energy demand and incentives to parallel connection of renewable energy systems stimulate the rise of electric power generation from small wind energy conversion systems (WECS) based on permanent magnet generators [1], [2]. The need for a rectifier stage in these applications encourages the search for technologies that guarantee the performance of the energy conversion system. Commonly used three-phase circuits include the uncontrolled diode rectifier and the full controlled three-phase rectifier, discussed in [3]–[5].

In the context of balance between good performance and low cost in low power applications, the Half-Controlled Boost Rectifier (HCBR) appears as an advantageous alternative in comparison with those mentioned above, presenting better performance related to the Total Harmonic Distortion (THD) than the diode rectifier and greater simplicity and economy than the fully controlled rectifier [6]. In [7], [8] the half-controlled AC-DC converter is indicated as a good interface between a variable speed permanent magnet generator and a constant output voltage, regarding to losses, reduced number of switches and efficiency.

To improve the performance of converter systems, current control techniques have been research objectives and are constantly reported in the literature. Among the conventional techniques used are: the linear method with Pulse Width Modulation (PWM) and the non-linear method, such as hysteresis control, documented in [9]–[12]. Besides these, another technique that has required research efforts along with the development of power electronics and microprocessors is the Finite Control Set Model Predictive Control (FCS-MPC).

The predictive control technique in power converters consists of predicting future behavior from the finite switching states of a converter and selecting appropriate actions that attenuate the error of the controlled variable, calculated by

the cost function [13]. In the control applied to the current prediction, the state that minimizes the predefined cost function is selected, in order to optimize the current behavior. The applicability of this method has been verified in power converters and demonstrated as an effective alternative to conventional control techniques, as presented in [14]–[17].

This work presents the performance of the FCS-MPC compared to the linear control method with PWM and hysteresis control applied to the half-controlled boost rectifier. The comparative analysis is performed through simulation results and verifies the behavior of the system in relation to the total harmonic distortion of the current and power losses. In addition, experimental results of the PWM and predictive control are presented in order to validate the simulated results.

II. CONVERTER OPERATION

The circuit of the HCBR is shown in Fig. 1. For simplification purposes the machine is expressed by a RLE model and the rectifier structure uses three diodes and three insulated-gate bipolar transistors (IGBTs). In this way, it has the advantage that all switches are connected to the same point, simplifying the control system. However, theoretical analysis previously developed by [7] demonstrate that it presents as a disadvantage the possibility of modulating only the positive half-cycle, resulting in higher harmonic content. This characteristic comes from the possible operation stages of the topology for each phase currents sector shown in Fig. 2.

The six sectors seen in Fig. 2 can be divided into two conditions. The first condition with two positive and one negative phase currents (sectors 1, 3 and 5) and the second condition with two negative currents and one positive (sectors 2, 4 and 6).

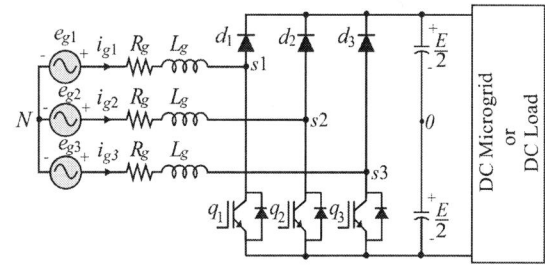

Fig. 1. Half-controlled boost rectifier.

978-1-7281-4181-7/19 $31.00 © 2019 IEEE

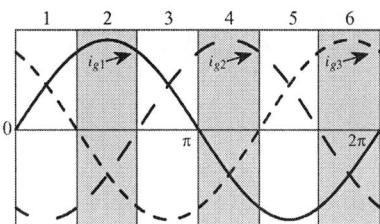

Fig. 2. Phase currents sectors.

By sector 1 analysis, it can be seen that the currents i_{g1} and i_{g3} are positive, while i_{g2} is negative, thus, four of the eight combinations of switch states can be applied, which results in the four operation stages of Fig. 3. The same analysis applies to sectors 3 and 5.

Similarly, analyzing sector 2 in Fig. 2, it is noted that there are only two possible stages of operation, Fig. 4, for the condition with two negatives currents, i_{g2} and i_{g3}, and one positive, i_{g1}, which is also true for sectors 4 and 6. This implies the impossibility of obtaining a modulated negative half cycle of the phase current. Moreover, as mentioned by [7], the combination of states with q_1, q_2 and q_3 equals to zero (blocked switch) cannot be applied to the converter due to the diodes d_1, d_2 and d_3.

III. CONVERTER MODEL

The model of the half-controlled boost rectifier can be described by (1), with the machine expressed by a RLE model:

$$e_{gk} = R_g i_{gk} + L_g \frac{di_{gk}}{dt} + v_{gk} \tag{1}$$

where R_g and L_g are the filter resistances and inductances, respectively, e_{gk} is the back electromotive force and v_{gk} is the voltage generated by the converter ($k = 1, 2, 3$).

The voltage v_{gk} is defined from the pole voltages (v_{k0}), voltage between the intermediate point of each converter leg and the intermediate point of the DC-link, by the relations:

$$v_{gk} = v_{k0} - v_{N0} \tag{2}$$

$$v_{k0} = (1 - 2q_k)\frac{E}{2} \tag{3}$$

$$v_{N0} = \frac{1}{3}(v_{10} + v_{20} + v_{30}) \tag{4}$$

where E is the DC-link voltage and v_{N0} is the voltage between the neutral point N of the AC mains and the midpoint 0 on the DC-link. The pole voltages depend on the switching state of the converter, determined as open for q_k equals to 0 and closed for q_k equals to 1.

From (2)-(4) it is possible to obtain the voltages generated by the converter in function of the switching states, as follows:

$$v_{g1} = \frac{1}{3}\left(-2q_1 + q_2 + q_3\right)E \tag{5}$$

$$v_{g2} = \frac{1}{3}\left(q_1 - 2q_2 + q_3\right)E \tag{6}$$

$$v_{g3} = \frac{1}{3}\left(q_1 + q_2 - 2q_3\right)E \tag{7}$$

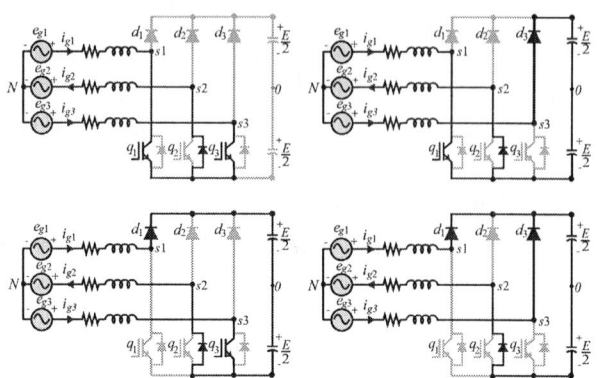

Fig. 3. Operation stages of the topology for the sector 1.

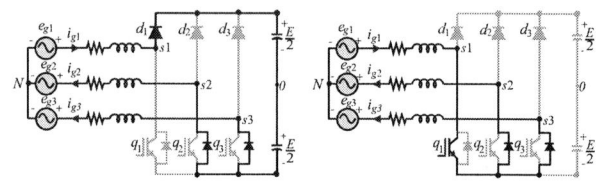

Fig. 4. Operation stages of the topology for the sector 2.

A. Discret Model

For a sampling time T_s, the predicted current can be calculated using the discrete time equation, approaching the derivative with the backward Euler difference method:

$$\frac{di_g}{dt} \approx \frac{i_g(n) - i_g(n - 1)}{T_s} \tag{8}$$

Replacing (8) into (1) and rearranging the terms:

$$i_{gk}(n) = \frac{L_g i_{gk}(n - 1) - v_{gk}(n)T_s + e_{gk}(n)T_s}{R_g T_s + L_g} \tag{9}$$

Shifting (9) one step forward, the equation that calculates the future current is obtained as follows:

$$i_{gk}(n+1) = \frac{L_g i_{gk}(n) - v_{gk}(n+1)T_s + e_{gk}(n+1)T_s}{R_g T_s + L_g} \tag{10}$$

The voltage e_{gk}, for a small sampling time T_s, does not change significantly, then it can be assumed that $e_{gk}(n+1) = e_{gk}(n)$.

B. Predictive Control Method

The predictive control method, shown in diagram form in Fig. 5, uses the prediction of the current error at the beginning of each sampling period, from the evaluation of the effect that each of the possible switching states causes in the cost function employed, as a criterion for selecting the future values.

The cost function can be expressed in different ways depending on the desired control criteria. In its simplest form (11), the control technique has variable switching frequency limited by the sampling frequency, as reported by [14].

$$g = \sum_{k=1}^{3}[i_{gk}^* - i_{gk}(n+1)]^2 \tag{11}$$

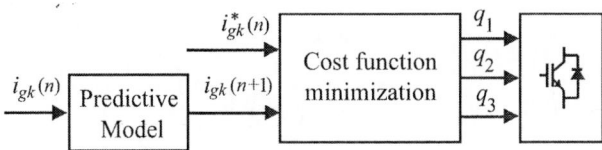

Fig. 5. Predictive control diagram.

where i_{gk}^* is the reference current and $i_{gk}(n+1)$ is the predicted current in each phase.

Steps for implementation of the FCS-MPC for the HCBR are presented in the flowchart of Fig. 6. As displayed, first the phase currents are measured and then the minimization of the cost function is performed. A *for* loop tests all possible switching states in the predicted current calculation given by (10), and the voltage e_{gk} is estimated as (9). The cost function minimum value is stored together with the switching state that generates it. Then, at the end of the loop, the switching state that produces the smallest error is applied.

C. Delay Compensation

When FCS-MPC is implemented, the cost function minimization is performed by evaluating all possible switching states. The whole computation time required causes a delay in implementation, which can introduce errors in the predicted variable, causing it to deviate from the reference [17], [18]. To solve this issue a delay compensation can be performed with the current prediction until instant $n+2$, by time-shifting (10) one step forward.

$$i_{gk}(n+2) = \frac{L_g i_{gk}(n+1) - v_{gk}(n+2)T_s + e_{gk}(n+2)T_s}{R_g T_s + L_g} \quad (12)$$

where $e_{gk}(n+2) = e_{gk}(n+1)$ for a small T_s and $v_{gk}(n+2)$ is the voltage to be applied. Therefore, the voltage $v_{gk}(n+1)$ used in the prediction of the current $i_{gk}(n+1)$ is the voltage selected in the previous sampling time.

IV. SIMULATION RESULTS

Simulations of the HCBR have been performed under three different current control techniques to verify the feasibility of the FCS-MPC compared to conventional control methods. All simulations have used a cascade voltage and current controller, regulating the DC-link voltage from a proportional-integral (PI) controller and generating the current references. Table I shows the parameters used to simulate the circuit.

TABLE I
SIMULATION PARAMETERS

Parameter	Value
Power System P_o	1.6 kW
Phase Voltage e_{gk}	127 V
Filter Inductance L_g	10 mH
Filter Resistance R_g	0.5 Ω
DC-link reference voltage E	400 V

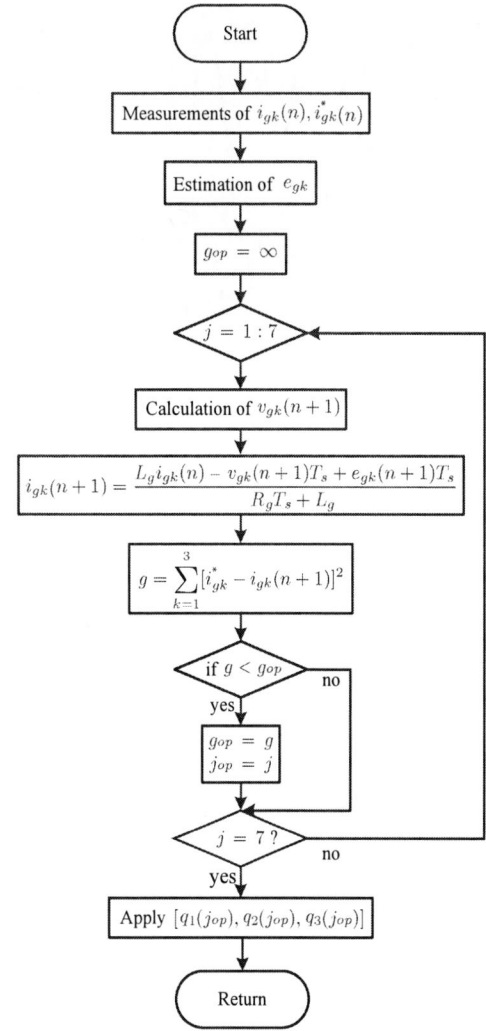

Fig. 6. FCS-MPC flowchart for the half-controlled boost rectifier.

For purpose of comparison with the other control techniques, the predictive control without delay compensation, with voltage estimation e_{gk} and a sampling time T_s of 50 μs, was used. In addition, the parameters applied in the linear control with PWM and hysteresis control techniques were employed to obtain switching frequency comparable to the FCS-MPC, i.e., carrier frequency equal to 3.2 kHz for PWM and hysteresis bandwidth equal to ±0.9 A.

In the linear control model with PWM, the reference and measured currents are compared and the error between them is processed by the PI controller generating the voltage commands for the PWM modulator. The gains used in the PI controller were calculated based on the presented parameters, obtaining a proportional gain (Kp) equal to 1.6 and an integral gain (Ki) equal to 80. The modulation technique applied was the one proposed by [9]. Fig. 7(a) shows the phase voltage (top) and current (bottom) obtained by the circuit simulation using PI with PWM control method, it is noted that the current

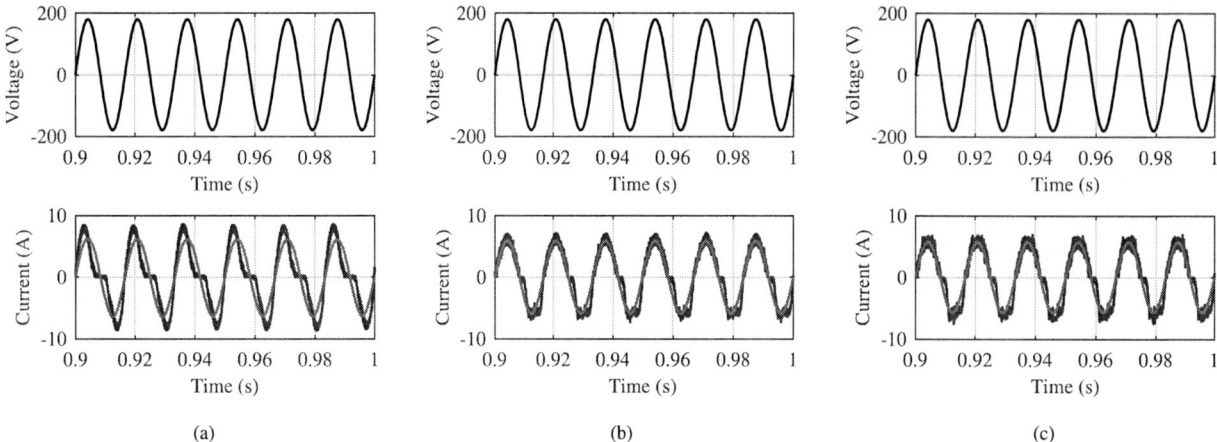

Fig. 7. Simulation results of the generator voltages (top) and generator currents (bottom). (a) PI with PWM control. (b) Hysteresis control. (c) FCS-MPC.

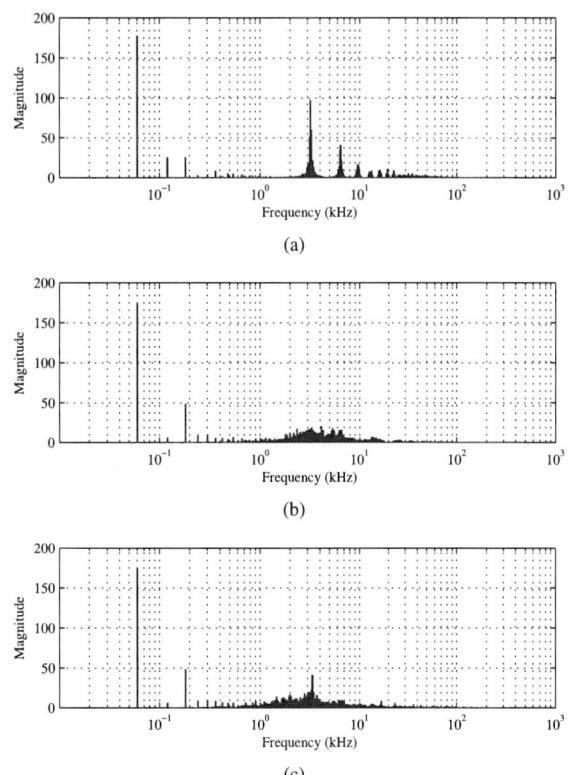

Fig. 8. Simulation results of the pole voltages frequency spectrum. (a) PI with PWM control (b) Hysteresis control. (c) FCS-MPC.

deviates from the reference current and has a THD of 55.92 % due to discontinuities in the transition of the applied states.

When applied the control scheme using hysteresis modulation, the input current of each phase is compared with the respective reference current. Maximum and minimum current limits are set, which establish the hysteresis bandwidth, and the switching operation occurs when these limits are reached, thus,

the input current is controlled around the reference current. Voltage and current obtained for this control scheme are shown in Fig. 7(b). The current appears in phase with the voltage and it traces the reference current. The current THD was 22.55 %.

Fig. 7(c) shows the voltage and current obtained by applying the predictive control method. As shown, the results were very close to those obtained by hysteresis control, including a THD of 23.50 %, and were significantly better than those presented by the PI with PWM modulation.

Despite the good result presented by hysteresis control, the main disadvantage of this method is due to variable switching frequency. Fig. 8 shows the frequency spectrum of the pole voltages for the three control techniques. As shown in Fig. 8(a), the linear control with PWM has a fixed switching frequency determined by the carrier frequency, while the hysteresis control shows a wide frequency range as seen in Fig. 8(b). On the other hand, the switching frequency for the predictive control, shown in Fig. 8(c), is concentrated in 3.2 kHz for an T_s equals to 50 μs.

A power loss analysis performed using the IGBT SKM50GB12T4 thermal module of the PSIM software is presented, as a percentage of the system power, in the bar graph of Fig. 9. It is possible to observe that the converter power losses were below 1.5 % for all control techniques employed, with conduction losses greater than switching losses. The hysteresis control presented the highest power losses percentage, around 1.34 %, while the predictive control presented 1.23 %.

In order to verify the effect of delay compensation on FCS-MPC, simulation was performed. Fig. 10(a) and Fig. 10(b) show the current i_{g1} for cases without delay compensation and with delay compensation, respectively. Through frequency spectrum analysis, it was verified that for the case when the predictive control includes delay compensation the converter generates a higher switching frequency, which was around 4 kHz, and a current THD of 19.66 %.

The performance of the FCS-MPC with delay compensation and a PI controller regulating the DC-link voltage was verified

978-1-7281-4181-7/19 $31.00 © 2019 IEEE

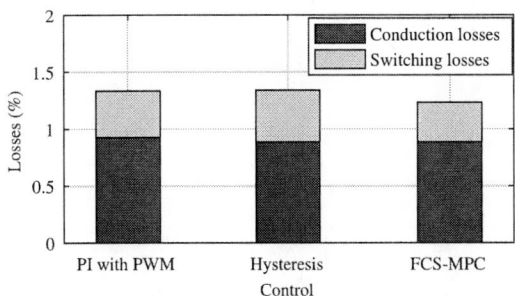

Fig. 9. Power losses for the different control techniques.

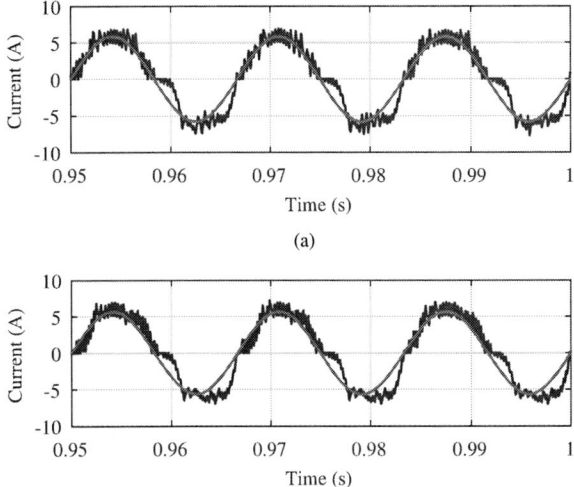

Fig. 10. Simulation results of the phase current i_{g1}. (a) Without compensation delay. (b) With compensation delay.

by applying a load step, varying the system power from 1.6 kW to 2 kW. Fig. 11(a) and Fig. 11(b) show the dynamics of the DC-link and current i_{g1} when the system is stepped, having a recovery time in approximately 0.2 s.

V. EXPERIMENTAL RESULTS

The experimental setup is based on a Digital Signal Processor (DSP) TMS320F28335 with a microcomputer equipped with appropriate plug-in boards and sensors. Results were obtained by oscilloscope Agilent DSO-X 3014A 100MHZ. A three-phase grid with RL filter was used instead of machine and the setup experimental parameters were the same as those used in the simulation. The experimental result was obtained without compensation delay.

Fig. 12(a) illustrates voltage and current at the generator side, e_{g1} and i_{g1}, respectively. It is noted that the current did not present sinusoidal format, with THD of 69.38 % and peak current of 10.4 A. Fig. 13(a) shows frequency spectrum of i_{g1}, where it is observed switching frequency of 2 kHz, such frequency was chosen for comparison with the predictive

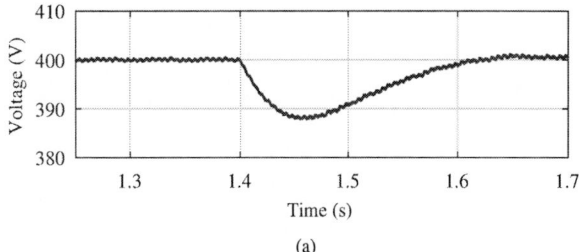

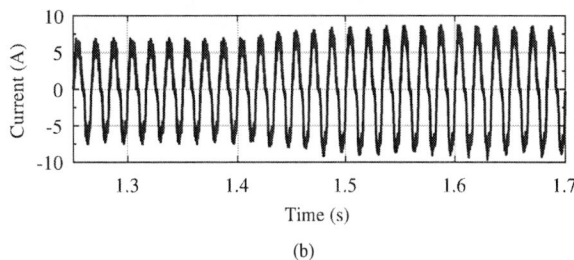

Fig. 11. Simulation results for a step in power from 1.6 kW to 2 kW. (a) DC-link voltage. (b) Current i_{g1}.

control that showed a frequency of approximately 2 kHz for a T_s of 50 μs (see Fig. 12 and Fig. 13).

Fig. 12(b) presents voltage and current for the predictive control. In Fig. 12(b) was observed a best sinusoidal format in comparison with Fig. 12(a) and a smaller THD of 26.22 % and peak current of 7 A. The frequency spectrum of the predictive control is shown in Fig. 13(b), it is noted that the switching frequency was around 2 kHz, which was limited by the digital processing of the FCS-MPC.

From the experimental results, it was concluded that the predictive control has better performance than the PWM control, with better sinusoidal format, less harmonic distortion in the current (THD) and lower peak current values.

VI. CONCLUSION

In this paper a finite control set model predictive control applied to a half-controlled boost rectifier, that can be used in variable-speed permanent magnet generator, was analyzed. Performance comparison of the system using linear pulse-width modulation (PWM) control, hysteresis control and predictive control methods, in order to control the input currents, was accomplished.

Results show that the predictive control presents better performance than the hysteresis and PI with PWM controls. Compared to the linear control with PWM, the harmonic distortion decreased by almost 40 % and compared to hysteresis control, the FCS-MPC presents similar results. However, the hysteresis control has a wide frequency range in rectifier voltage. Simulation and experimental results have been presented to illustrate the correct operation of the FCS-MPC for the half-controlled boost rectifier.

ACKNOWLEDGEMENT

The authors would like to thank in part the Coordenação de Aperfeiçoamento de Pessoal de Nível Superior - Brasil

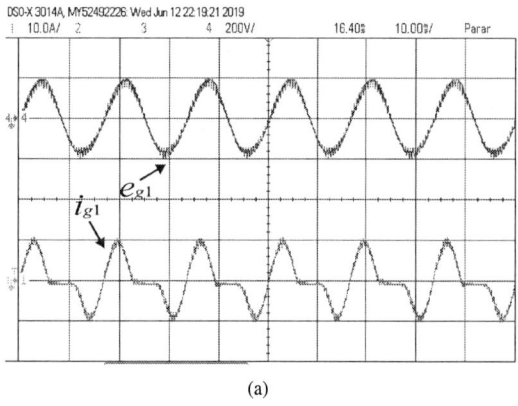 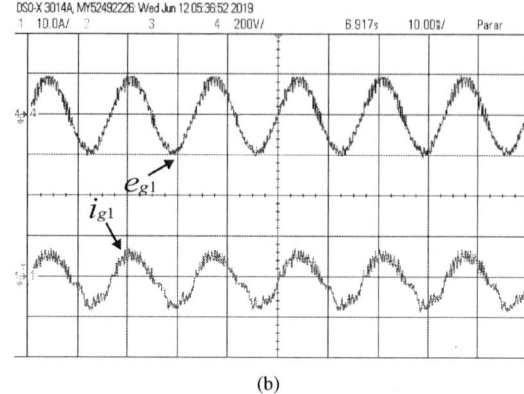

(a)   (b)

Fig. 12. Experimental results of the generator voltages (top) and generator currents (bottom). (a) PWM control. (b) Predictive control with estimated e_{g1}.

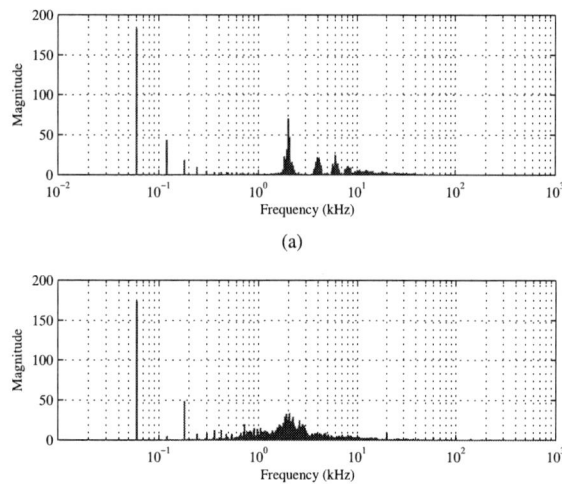

Fig. 13. Experimental results of the pole voltages frequency spectrum. (a) PWM control. (b) Predictive control with estimated e_{g1}.

(CAPES) and the Conselho Nacional de Desenvolvimento Científico e Tecnológico CNPq for the financial support.

REFERENCES

[1] V. Yaramasu, B. Wu, P. C. Sen, S. Kouro, M. Narimani, High-power wind energy conversion systems: State-of-the-art and emerging technologies, Proceedings of the IEEE 103 (5) (2015) 740–788 (May 2015).

[2] A. Uehara, A. Pratap, T. Goya, T. Senjyu, A. Yona, N. Urasaki, T. Funabashi, A coordinated control method to smooth wind power fluctuations of a pmsg-based wecs, IEEE Transactions on Energy Conversion 26 (2) (2011) 550–558 (June 2011).

[3] W. Li, C. Abbey, G. Joós, Control and performance of wind turbine generators based on permanent magnet synchronous machines feeding a diode rectifier, in: 2006 37th IEEE Power Electronics Specialists Conference, 2006, pp. 1–6 (June 2006).

[4] M. M. N. Amin, O. A. Mohammed, Dc-bus voltage control technique for parallel-integrated permanent magnet wind generation systems, IEEE Transactions on Energy Conversion 26 (4) (2011) 1140–1150 (Dec 2011).

[5] Y. Saidi, A. Mezouar, Y. Miloud, M. A. Benmahdjoub, A robust control strategy for three phase voltage t source pwm rectifier connected to a pmsg wind energy conversion system, in: 2018 International Conference

on Electrical Sciences and Technologies in Maghreb (CISTEM), 2018, pp. 1–6 (Oct 2018).

[6] J. Kikuchi, M. D. Manjrekar, T. A. Lipo, Performance improvement of half controlled three phase pwm boost rectifier, in: 30th Annual IEEE Power Electronics Specialists Conference. Record. (Cat. No.99CH36321), Vol. 1, 1999, pp. 319–324 vol.1 (July 1999).

[7] D. Krahenbuhl, C. Zwyssig, J. W. Kolar, Half-controlled boost rectifier for low-power high-speed permanent-magnet generators, IEEE Transactions on Industrial Electronics 58 (11) (2011) 5066–5075 (Nov 2011).

[8] D. S. Oliveira, Jr., M. M. Reis, C. E. A. Silva, L. H. S. Colado Barreto, F. L. M. Antunes, B. L. Soares, A three-phase high-frequency semicontrolled rectifier for pm wecs, IEEE Transactions on Power Electronics 25 (3) (2010) 677–685 (March 2010).

[9] V. Blasko, Analysis of a hybrid pwm based on modified space-vector and triangle-comparison methods, IEEE Transactions on Industry Applications 33 (3) (1997) 756–764 (May 1997).

[10] J. Holtz, Pulsewidth modulation for electronic power conversion, Proceedings of the IEEE 82 (8) (1994) 1194–1214 (Aug 1994).

[11] M. P. Kazmierkowski, L. Malesani, Current control techniques for three-phase voltage-source pwm converters: a survey, IEEE Transactions on Industrial Electronics 45 (5) (1998) 691–703 (Oct 1998).

[12] E. R. C. da Silva, E. C. dos Santos, B. Jacobina, Pulsewidth modulation strategies, IEEE Industrial Electronics Magazine 5 (2) (2011) 37–45 (June 2011).

[13] P. Cortes, S. Kouro, B. La Rocca, R. Vargas, J. Rodriguez, J. I. Leon, S. Vazquez, L. G. Franquelo, Guidelines for weighting factors design in model predictive control of power converters and drives, in: 2009 IEEE International Conference on Industrial Technology, 2009, pp. 1–7 (Feb 2009).

[14] J. Rodriguez, J. Pontt, C. Silva, P. Correa, P. Lezana, P. Cortes, U. Ammann, Predictive current control of a voltage source inverter, IEEE Transactions on Industrial Electronics 54 (1) (2007) 495–503 (Feb 2007).

[15] P. Cortes, L. Vattuone, J. Rodriguez, A comparative study of predictive current control for three-phase voltage source inverters based on switching frequency and current error, in: Proceedings of the 2011 14th European Conference on Power Electronics and Applications, 2011, pp. 1–8 (Aug 2011).

[16] J. Rodriguez, M. P. Kazmierkowski, J. R. Espinoza, P. Zanchetta, H. Abu-Rub, H. A. Young, C. A. Rojas, State of the art of finite control set model predictive control in power electronics, IEEE Transactions on Industrial Informatics 9 (2) (2013) 1003–1016 (May 2013).

[17] P. Cortes, J. Rodriguez, P. Antoniewicz, M. Kazmierkowski, Direct power control of an afe using predictive control, IEEE Transactions on Power Electronics 23 (5) (2008) 2516–2523 (Sep. 2008).

[18] T. Jin, X. Shen, T. Su, R. C. C. Flesch, Model predictive voltage control based on finite control set with computation time delay compensation for pv systems, IEEE Transactions on Energy Conversion 34 (1) (2019) 330–338 (March 2019).

978-1-7281-4181-7/19 $31.00 © 2019 IEEE

Lifetime Investigation of DC-link Capacitors in Multiple Slim Drives System

Shili Huang[1], Haoran Wang[2], Dinesh Kumar[3], Guorong Zhu[1], Huai Wang[2]

1 School of Automation, Wuhan University of Technology, Wuhan, China
2 Department of Energy Technology, Aalborg University, Aalborg, Denmark
3 Danfoss Drives A/S, Global Research and Development Centre, Graasten, Denmark
shili.huang@whut.edu.cn, hao@et.aau.dk, dinesh@danfoss.com, zhgr_55@whut.edu.cn, hwa@et.aau.dk

Abstract—This paper studies the reliability of DC-Link capacitor in multiple slim drives system. In order to obtain the electro-thermal stress of capacitor in individual drive simply and accurately, a time-efficient equivalent model and its analytical model for multiple slim drives system is proposed. Based on the proposed equivalent circuit model and the existing lifetime prediction method, the impact of drive numbers on the reliability of DC-link capacitor in multiple slim drives is analyzed. The results serve as a guideline for capacitor sizing in multiple slim drives system from reliability aspect.

Index Terms—DC-Link capacitor, multiple slim drives, lifetime, adjustable speed drive

I. INTRODUCTION

Adjustable Speed Drive (ASD) with diode rectifier is commonly used in residential, commercial and industrial applications [1], [2]. As an important part of ASD, diode rectifier is energy saving and low cost, but may increase the current harmonics of system, then reduce the power quality and finally cause the distribution network stability issues [3]. To solve the current harmonics problem caused by diode rectifier, a number of harmonic suppression techniques have been developed in many applications, such as passive filters [4], multi-pulse transformer based rectifiers [6] and active harmonic filtering techniques [6], [7], however, these filtering techniques may increase the cost and volume of overall system as well as the complexity of system control. In recent years, slim film capacitor is applied in DC-link filtering, due to the smaller size, higher power quality and potentially higher reliability compared with standard *LC* filtering configuration [8]. The filtering capability of the slim drive is relatively small, which results in the voltage fluctuation appears at the DC link, and finally causes reliability issues of DC-link capacitor [9]. As a result, the reliability evaluation of DC-link capacitors in slim ASD system is of great importance.

Reliability of DC-link capacitors in single ASD have been discussed in literatures [1], [10] and [11]. [1] studied the impact of grid voltage amplitude and phase unbalanced conditions on the reliability of DC-link capacitor in standard ASD system. The reliability evaluation of DC-link capacitor in scalable ASD system power rating is investigated in [1]

and [11]. It concludes that the standard ASD faces reliability challenge in grid voltage unbalanced conditions and higher power rating. The reliability comparison of DC-link capacitor between standard ASD and slim ASD is studied in [1] and [11], and shows that DC-link capacitor in slim ASD is more reliable than that in standard ASD when the grid conditions and power rating are the same. However, in commercial and industrial applications, many slim ASDs are connected in parallel at the Point of Common Coupling (PCC) instead of single drive [2], [12]. The grid voltage conditions and the impedance of ASDs will change accordingly compared with single ASD, so as to vary the electro-thermal stress of the DC-link capacitors in individual drive. The estimated results in single drive are not applicable to multiple drives system anymore.

In order to study the reliability of DC-link capacitors in multiple slim drives system, the electro-thermal stress of capacitor in multiple drives should be obtained through mathematical model, simulation or experiment. However, it is hardly and time-consuming to extend the method of obtaining the capacitor electro-thermal stress from single drive to multiple drives by using simulation or experiment. Moreover, it needs to explore how the drive numbers influence the electro-thermal stress as well as the reliability of DC-link capacitor.

This paper investigates the reliability of DC-link capacitors in multiple slim drives system. Firstly, a time-efficient equivalent model of multiple slim drives is proposed to acquire the electro-thermal stress, which can be implemented in both mathematical model and simulation. Then, based on the proposed equivalent model and the existing lifetime prediction method, the lifetime of DC-link capacitor in multiple slim drives is evaluated. Finally, the impact of drive numbers on DC-link capacitor lifetime is analyzed from the point of view of impedance characteristics. The paper is organized as follows. In section II, a time-efficient equivalent model is proposed and the DC-link capacitor current calculation model is presented. In section III, the lifetime estimation procedure considering lifetime prediction model parameter variations during the evaluation process is presented. Section IV presents the case studies and scalability analysis. Conclusions are followed in section V.

This work was supported by the project of National Natural Science Foundation of China (Grant No: 51777146), and by the Excellent Dissertation Cultivation Funds of Wuhan University of Technology (2018-YS-067).

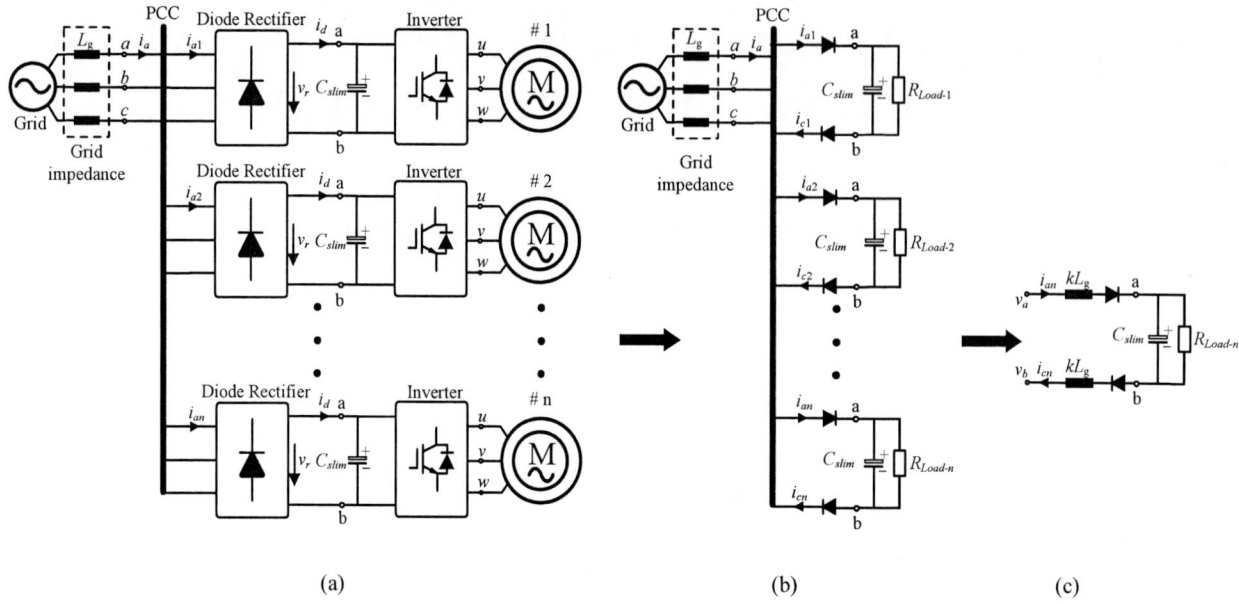

Fig. 1. Block diagram of (a) multiple slim drives configuration, (b) simplified model of the multiple slim drives system, (c) a time-efficient equivalent impedance model of the multiple slim drives system.

II. CURRENT STRESS OF DC-LINK CAPACITOR IN MULTIPLE SLIM DRIVES

A. Proposed Time-efficient Equivalent Model for Multiple Slim Drives

The configuration of multiple slim drives system is shown in Fig. 1 (a). The single slim ASD is implemented with a film capacitor. The front-end of ASD is a three-phase diode rectifier and the rear-end is consist of an inverter. When the three-phase diode rectifier is working on the conduction period, two of input voltage sources are connected to the DC-link and the grid impedance becomes $2L_g$. The load is modeled as a resistor (i.e., R_{Load}), as a result, the single slim ASD system is modeled as a RC circuit. The impedance of single slim ASD from the grid side can be obtained as:

$$Z_{\text{in}} = \frac{1 - 2\omega^2 L_g C_{\text{slim}} + j\omega \frac{2L_g}{R_{\text{Load}}}}{j\omega C_{\text{slim}} + \frac{1}{R_{\text{Load}}}} \tag{1}$$

where the resonant frequency (f_0) and the damping factor (ξ) of a single slim ASD are shown in (2). It can be seen from (1) and (2) that the impedance and harmonic distribution of slim ASD system is determined by grid impedance and slim drive parameters.

$$\begin{cases} f_0 = \dfrac{1}{2\pi\sqrt{2L_g C_{\text{slim}}}} \\ \xi = \dfrac{1}{2R_{\text{Load}}}\sqrt{\dfrac{2L_g}{C_{\text{slim}}}} \end{cases} \tag{2}$$

For multiple slim drives, assuming the interconnection impedance between drives is negligible, the multiple slim drives can be simplified in Fig. 1 (b). The voltage drop of line inductance L_g can be obtained as:

$$V_{\text{L}_g} = j\omega L_g(i_{\text{a1}} + i_{\text{a2}} + \ldots + i_{\text{an}}) \tag{3}$$

The reliability analysis in this paper focuses on the heavy load condition, therefore, the DC-link current in following analysis is based on the continuous current mode. The current of each slim ASD can be expressed as:

$$\begin{cases} i_{\text{a1}} = k_1 i_{\text{a1}} \\ i_{\text{a2}} = k_2 i_{\text{a1}} \\ \quad\vdots \\ i_{\text{an}} = k_n i_{\text{a1}} \end{cases} \tag{4}$$

where k_1, k_2 ... k_n are the scaling coefficients between i_{a1} and i_{a1} ... i_{an}. Substituting (4) into (3), the voltage drop of line inductance L_g is acquired as k times of that in a single slim ASD:

$$\begin{aligned} V_{\text{L}_g} &= j\omega L_g(i_{\text{a1}} + i_{\text{a2}} + \ldots + i_{\text{an}}) \\ &= j\omega(k_1 + k_2 + \ldots + k_n)L_g i_{\text{a1}} \\ &= j\omega(kL_g)i_{\text{a1}} \end{aligned} \tag{5}$$

where $k = k_1 + k_2 + \ldots + k_n$. Therefore, the time-efficient equivalent model is obtained in Fig. 1 (c). Based on the time-efficient equivalent model, the impedance of n numbers of drives system from the grid side is:

$$Z_{\text{in}-\text{n}} = \frac{1 - 2\omega^2 k L_g C_{\text{slim}} + j\omega \frac{2kL_g}{R_{\text{Load}-\text{n}}}}{j\omega C_{\text{slim}} + \frac{1}{R_{\text{Load}-\text{n}}}} \tag{6}$$

The resonant frequency ($f_{0-\text{n}}$) and the damping factor (ξ_{n}) of

978-1-7281-4181-7/19 $31.00 © 2019 IEEE

multiple slim ASDs system are obtained as:

$$\begin{cases} f_{0-n} = \dfrac{1}{2\pi\sqrt{2kL_{\mathrm{g}}C_{\mathrm{slim}}}} \\[2mm] \xi_{\mathrm{n}} = \dfrac{1}{2R_{\mathrm{Load-n}}}\sqrt{\dfrac{2kL_{\mathrm{g}}}{C_{\mathrm{slim}}}} \end{cases} \quad (7)$$

Comparing the equivalent impedance and resonance frequency between single and multiple slim ASDs system, it can be seen that more slim ASDs connected in parallel at PCC will increase the equivalent inductance and decrease the resonance frequency of ASDs system, which finally change the harmonic current distribution of DC-link capacitor.

B. Analytical Model for DC-link Capacitor Current Stress under Continuous Mode

Based on the proposed time-efficient equivalent model, DC-link capacitor current stress in multiple slim drives can be obtained by the analytical model. The three-phase voltage of balanced grid can generally be expressed in the time domain as:

$$\begin{cases} v_{\mathrm{a}} = V_{\mathrm{m}}\sin\theta \\ v_{\mathrm{b}} = V_{\mathrm{m}}\sin(\theta - \frac{2\pi}{3}) \\ v_{\mathrm{c}} = V_{\mathrm{m}}\sin(\theta - \frac{4\pi}{3}) \end{cases} \quad (8)$$

where v_{a}, v_{b} and v_{c} are instantaneous value of grid phase voltage, V_{m} is the amplitude of grid phase voltage, θ equals to ωt. The instantaneous dc side voltage v_{r} can be expressed in terms of the voltage rectifier switching functions S_{a}, S_{b} and S_{c}.

$$v_{\mathrm{r}} = S_{\mathrm{a}}v_{\mathrm{a}} + S_{\mathrm{b}}v_{\mathrm{b}} + S_{\mathrm{c}}v_{\mathrm{c}} \quad (9)$$

Because the analysis is based on the continues current mode, the rectifier switching function S_{a} can be expressed by the Fourier series in (10) [13]. Accordingly, S_{b} and S_{c} can be obtained by replacing θ in (10) by (θ-2π/3) and (θ-4π/3), respectively.

$$S_{\mathrm{a}} = \sum_{q=1,5,7,\ldots}^{\infty} \frac{\sqrt{3}}{\pi}\cdot\frac{(-1)^{l+1}}{q}$$
$$\times\{\sin qu\cos q\theta - (1+\cos qu)\sin q\theta\} \quad (10)$$

where $q=6l\pm1$ (l=0,1,2, ...,q>0), u is an overlap angle. The overlap angle of multiple slim drives can be obtained based on the time-efficient equivalent model in Fig. 1 (c):

$$u = \arccos(1 - \frac{2\omega kL_{\mathrm{g}}I_{\mathrm{d}}}{\sqrt{3}V_{\mathrm{m}}}) \quad (11)$$

where I_{d} is the dc component of dc current. Substituting (8) and (10) into (9), the dc voltage v_{r} is given as:

$$v_{\mathrm{r}} = V_{\mathrm{r}} + v_{\mathrm{rm}}$$
$$= V_{\mathrm{r}} + \sum_{m=6,12,18,\ldots}^{\infty}(A_{\mathrm{dm}}\cos m\theta + B_{\mathrm{dm}}\sin m\theta) \quad (12)$$

where

$$V_{\mathrm{r}} = \frac{3\sqrt{3}V_{\mathrm{m}}}{2\pi}\{1 + \cos u\}$$

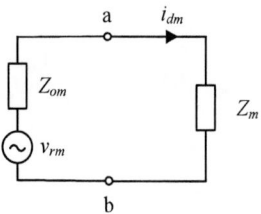

Fig. 2. Equivalent circuit for obtaining harmonic components of dc current.

$$A_{\mathrm{dm}} = \frac{3\sqrt{3}V_{\mathrm{m}}(-1)^p}{2\pi}\{\frac{1+\cos(m+1)u}{m+1}$$
$$- \frac{1+\cos(m-1)u}{m-1}\}$$

$$B_{\mathrm{dm}} = \frac{3\sqrt{3}V_{\mathrm{m}}(-1)^p}{2\pi}\{\frac{\sin(m+1)u}{m+1} - \frac{\sin(m-1)u}{m-1}\}$$

$m=6p$ (p=1,2,3, ...)

The dc current can be obtained by applying the dc voltage and the impedance at the corresponding harmonics frequencies. Based on the time-efficient equivalent model in Fig. 1 (c), the equivalent impedance circuit for obtaining harmonic components of dc current is shown in Fig. 2, where Z_{m} is m th impedance of the load as viewed from terminals a and b ; Z_{om} is m th impedance of the ac side as viewed from terminals a and b; v_{rm} is m th component of dc voltage v_{r}. From Fig. 1 (c), Z_{m} is given by:

$$Z_{\mathrm{m}} = \frac{R_{\mathrm{Load-n}}}{1 + jm\omega C_{\mathrm{slim}}R_{\mathrm{Load-n}}} \quad (13)$$

The impedance of ac side Z_{om} is affected by the overlap angle u, Z_{om} is given by [13]:

$$Z_{\mathrm{om}} = j(2 - \frac{3u}{2\pi})mk\omega L_{\mathrm{g}} \quad (14)$$

The m th harmonic current component of dc current is:

$$i_{\mathrm{dm}} = \frac{v_{\mathrm{rm}}}{Z_{\mathrm{m}} + Z_{\mathrm{om}}} \quad (15)$$

Equation (12), (13), (14) and (15) give:

$$i_{\mathrm{d}} = I_{\mathrm{d}} + \sum i_{\mathrm{dm}}$$
$$= I_{\mathrm{d}} + \sum_{m=6,12,18,\ldots}^{\infty}\sqrt{2}I_{\mathrm{dm}}\cos(m\theta - \beta_{\mathrm{m}}) \quad (16)$$

where

$$I_{\mathrm{d}} = \frac{V_{\mathrm{r}}}{R_{\mathrm{load-n}}}, I_{\mathrm{dm}} = \frac{\sqrt{(A_{\mathrm{dm}}^2 + B_{\mathrm{dm}}^2)/2}}{|Z_{\mathrm{m}} + Z_{\mathrm{om}}|}$$

$$\beta_{\mathrm{m}} = \arctan\frac{B_{\mathrm{dm}}}{A_{\mathrm{dm}}} + \arctan\frac{Im(Z_{\mathrm{m}} + Z_{\mathrm{om}})}{Re(Z_{\mathrm{m}} + Z_{\mathrm{om}})}$$

978-1-7281-4181-7/19 $31.00 © 2019 IEEE

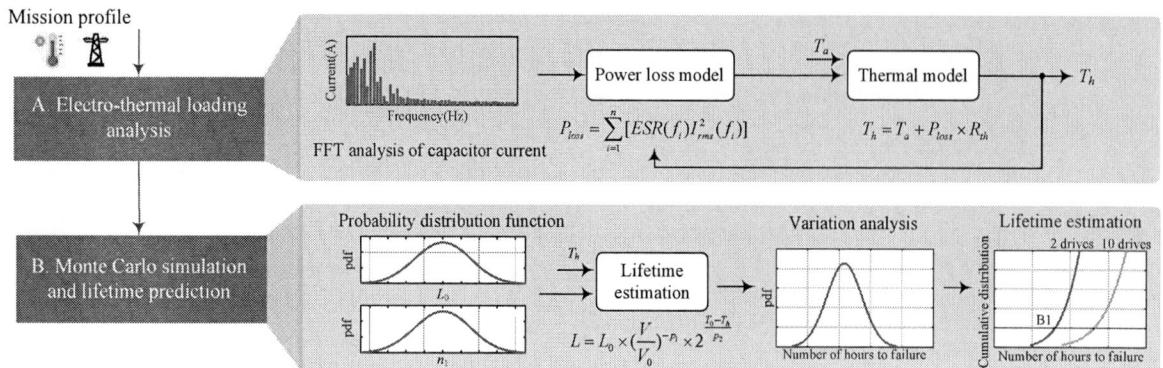

Fig. 3. Lifetime estimation procedure of DC-link capacitor.

TABLE I
SPECIFICATIONS OF THE MULTIPLE SLIM ASDs SYSTEM AND THE DC-LINK CAPACITOR CONFIGURATION.

ASDs system specification		Slim capacitor filter (C_{slim})	
Rated power (kW)	7.5	Physical configurations	1100V/30μF film capacitors
Grid frequency (Hz)	50	Part number	TDK. B32778G0306
Grid phase RMS voltage (V)	230	ESR of single capacitor	14 mΩ @100Hz
DC-link voltage (V)@7.5kW balanced grid voltage	535	Thermal resistance (R_{th})	13 °C/W
Ambient temperature (°C)	60	Rated load lifetime	100000 hours @70 °C and rated ripple current
Grid impedance (L_g)	130μH		

III. RELIABILITY ANALYSIS OF DC-LINK CAPACITOR IN ASDs

The lifetime estimation procedure of DC-link capacitor is shown in Fig. 3, the lifetime model parameter variations are considered. The procedure includes electro-thermal loading analysis and Monte Carlo simulation based lifetime prediction. The loading conditions (mission profile) is applied as the input and the output is the lifetime prediction of DC-link capacitors with a certain confidence level (e.g., 90%) [1].

A. Electro-Thermal Stress Analysis

Based on the simulation (mathematical model or circuit model based simulation), or experiment, the electrical stress (voltage and current) of the capacitor can be obtained. The hot-spot temperature of the capacitor can be estimated by capacitor current spectrum and Equivalent Series Resistance (ESR). ESR is determined by harmonic frequency and hot-spot temperature. The hot-spot temperature can be calculated by thermal model [14]:

$$T_{\text{h}} = T_{\text{a}} + \sum_{i=1}^{n} R_{\text{th}} \times [ESR(f_{\text{i}}) \times I_{\text{rms}}^2(f_{\text{i}})] \quad (17)$$

where T_{h} is the hot-spot temperature and T_{a} is the ambient temperature. R_{th} is the thermal resistance of capacitor between the hot-spot and ambient. $ESR(f_{\text{i}})$ is the ESR at frequency f_{i}, $I_{\text{rms}}(f_{\text{i}})$ is the Root Mean Square (RMS) value of the harmonic current at frequency f_{i}.

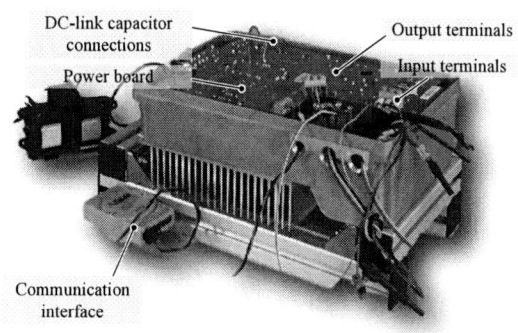

Fig. 4. Experimental prototype of the motor drive with slim capacitor DC-link.

B. Monte Carlo Analysis Based Lifetime Prediction

The lifetime prediction model used in this paper is [14]:

$$L = L_0 \times (\frac{V}{V_0})^{-p_1} \times 2^{\frac{T_0 - T_{\text{h}}}{p_2}} \quad (18)$$

where L_0 is the rated lifetime at rated voltage V_0 and temperature T_0, V and T_{h} are the voltage and hot-spot temperature under operation condition. p_1 is from 7 to 9.4 for film capacitor, p_2 is a coefficient, which equals to 10 [14].

The use of the lifetime model results in a fixed lifetime prediction, which is far from reality since the lifetime model parameters experience a variation. Therefore, Monte Carlo simulation is applied, the temperature-related lifetime con-

978-1-7281-4181-7/19 $31.00 © 2019 IEEE

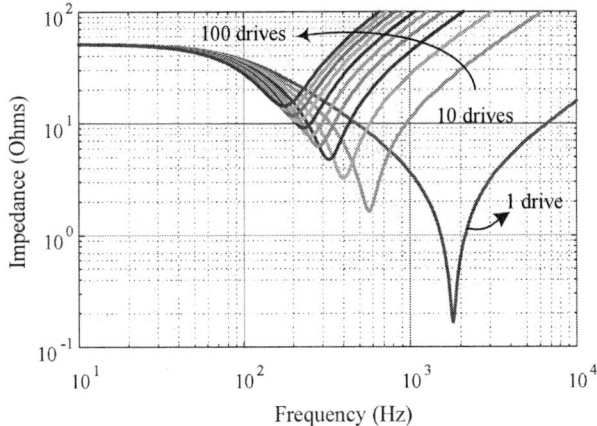

Fig. 5. Impedance characteristics of multiple slim drives with scalable numbers under balanced grid condition.

stants L_0, p_2 and the temperature tolerance-related parameter T_h are experience a variation of 5% with Normal distribution [11].

IV. SCALABILITY ANALYSIS OF CAPACITOR LIFETIME IN MULTIPLE SLIM DRIVES SYSTEM

The specifications of multiple slim ASDs system is shown in Table I. The experimental prototype is shown in Fig. 4. The rated power of multiple slim ASD is 7.5 kW, however, drives operate at partial load conditions in most applications, as a result, 5 kW loading condition is considered in this study. Assuming the power of each slim ASD are the same, then the dc side voltage and current of each slim ASD are the same. Therefore, according to (5), the ac side current and the voltage drop of grid impedance L_g can be acquired as n (the number of ASDs connected in parallel) times of that in single slim ASD:

$$V'_{L_g} = j\omega L_g(ni_{a1}) = j\omega(nL_g)i_{a1} \tag{19}$$

Accordingly, the impedance of multiple slim ASDs system from the grid side is:

$$Z'_{in-n} = \frac{1 - 2\omega^2 nL_g C_{slim} + j\omega \frac{2nL_g}{R_{Load-n}}}{j\omega C_{slim} + \frac{1}{R_{Load-n}}} \tag{20}$$

where the resonant frequency (f_{0-n}) and the damping factor (ξ_n) of multiple slim ASDs system is:

$$\begin{cases} f'_{0-n} = \dfrac{1}{2\pi\sqrt{2nL_g C_{slim}}} \\ \xi'_n = \dfrac{1}{2R_{Load}}\sqrt{\dfrac{2nL_g}{C_{slim}}} \end{cases} \tag{21}$$

The impedance characteristics of multiple slim ASDs is shown in Fig. 5, it can be seen that the impedance at resonance frequency increases with the increases of drive numbers. The resonance frequency of single slim drive is about 1800 Hz, which will result in the capacitor current component at 1800

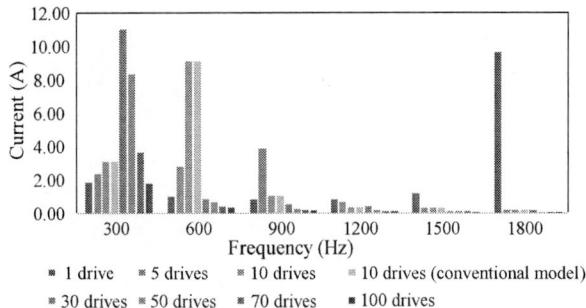

Fig. 6. DC-link capacitor current spectrum of scalable drives in multiple slim ASDs system under balanced grid voltage condition.

Hz is the largest. Similarly, harmonic current distribution at different frequencies can be explained in this way.

Based on the time-efficient equivalent model, the DC-link capacitor current spectrum in multiple slim ASDs system under balanced grid voltage condition can be obtained and shown in Fig. 6. As the ESR of the selected film capacitor is extremely tiny and changes slightly when the frequency is greater then 1800 Hz, the current component that higher then 1800 Hz is negligible. It can be seen from the current spectrum that: a) the capacitor current stress obtained by time-efficient equivalent model is the same as by conventional model with numbers of ASDs, but the complexity is reduced significantly; b) the harmonic current component of capacitor in scalable drives under balanced grid condition is mainly 300 Hz, 600 Hz and 900 Hz; c) with more drives connected in parallel, the maximum component of capacitor current appear at 1800 Hz, 900 Hz, 600 Hz and 300 Hz, respectively.

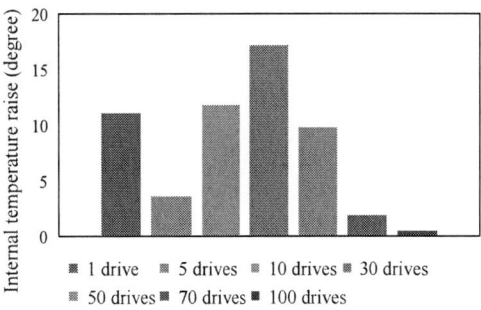

Fig. 7. Estimated hot-spot temperature of the capacitor with scalable drives under balanced grid condition.

Applying lifetime estimation procedure, the estimated hot-spot temperature raise is shown in Fig. 7. The probability density function and B1 lifetime of the capacitors from 1 to 100 drives under balanced grid condition are shown in Fig. 8 (a), (b), respectively. It can be seen that the B1 lifetime in 30 slim drives is the shortest (4.3 years). The accumulated failure at 5 years lifetime of multiple slim drives system is shown in Fig. 9, it can be seen that: a) the accumulated failure at 5

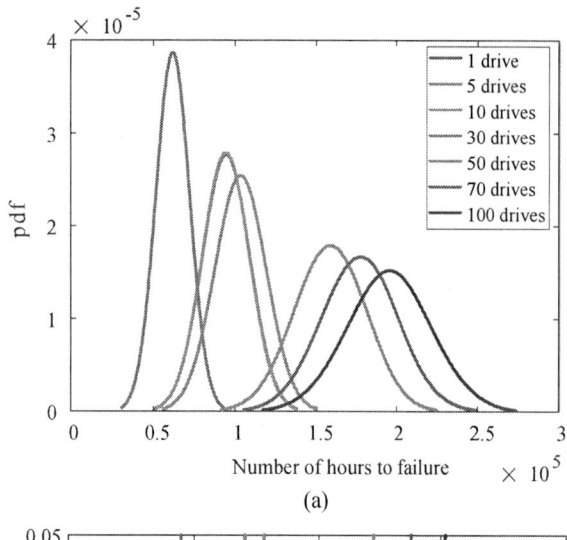

(a)

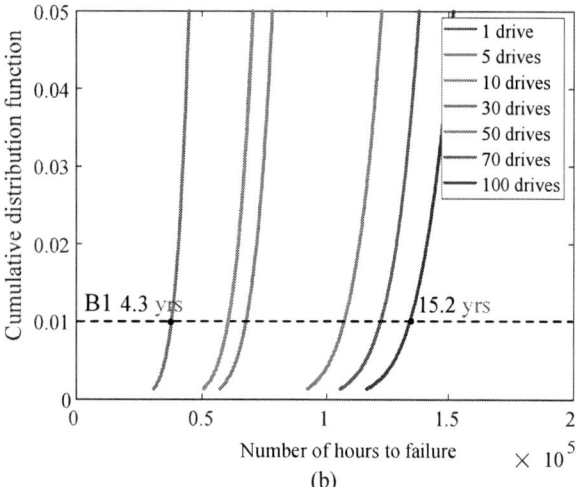

(b)

Fig. 8. Probability density function (pdf) and B1 lifetime of DC-link capacitor in scalable multiple slim ASDs system under balanced grid voltage condition.

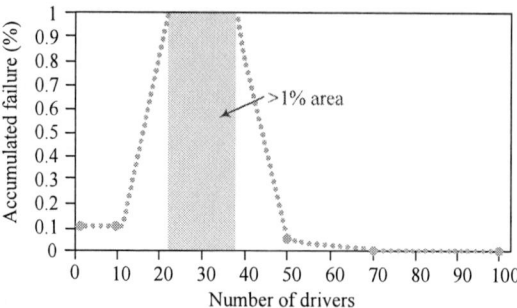

Fig. 9. Accumulated failure at 5 years due to the DC-link capacitor wear out with different number of drives.

the case study with the presented specifications, the lifetime prediction of DC-link capacitor in multiple slim ASDs system are studied. The case study shows that:

1) the lifetime of capacitor in 30 slim drives is the shortest, which because the electro-thermal stress of DC-link capacitor in 30 drives system is the largest;

2) the accumulated failure at 5 years lifetime of multiple slim drives is less than 1% except 30 drives system.

REFERENCES

[1] H. Wang, P. Davari, H. Wang, and D. Kumar, "Lifetime Estimation of DC-Link Capacitors in Adjustable Speed Drives Under Grid Voltage Unbalances," *IEEE Trans. on power Electro.*, vol. 34, no. 5, pp. 4064-4078, May. 2019.

[2] F. Zare, H. Soltani, D. Kumar and P. Davari, "Harmonic Emissions of Three-Phase Diode Rectifiers in Distribution Networks," *IEEE Access.*, vol. 5, pp. 2819-2833, 2017.

[3] B. K. Bose, "Power Electronics and Motor Drives Recent Progress and Perspective," *IEEE Trans. Ind. Electron.*, vol. 56, no. 2, pp. 581-588, Feb. 2009.

[4] J. C. Das, "Passive filters - potentialities and limitations," *IEEE Trans. Ind. Electron.*, vol. 40, no. 1, pp. 232-241, Jan. 2004.

[5] F. Meng, W. Yang, Y. Zhu, and L. Gao, "Load Adaptability of Active Harmonic Reduction for 12-Pulse Diode Bridge Rectifier With Active Interphase Reactor," *IEEE Trans. on power Electro.*, vol. 30, no. 12, pp. 7170-7180, Dec. 2015.

[6] H. Wang and H. Wang, "An active capacitor with self-power and internal feedback control signals," *in Proc. IEEE ECCE Conf*, pp. 3484-3488, 2017.

[7] H. Akagi, "Active Harmonic Filters," *in Proc. IEEE*, vol. 93, no. 12, pp. 2128-2141, Dec 2005.

[8] M. Hinkkanen and J. Luomi, "Induction Motor Drives Equipped With Diode Rectifier and Small DC-Link Capacitance," *IEEE Trans. Ind. Electron.*, vol. 55, no. 1, pp. 312-320, Jan. 2008.

[9] H. Saren, O. Pyrhonen, K. Rauma and O. Laakkonen, "Overmodulation in Voltage Source Inverter with Small DC-link Capacitor," *in Proc. IEEE-PESC Conf*, pp. 892-898, 2005.

[10] H. Wang, P. Davari, D. Kumar and F. Zare, "The impact of grid unbalances on the reliability of DC-link capacitors in a motor drive," *in Proc. IEEE ECCE Conf*, pp. 4345-4350, 2017.

[11] H. Wang, P. Davari, H. Wang, and D. Kumar, "Lifetime benchmarking of two DC-link passive filtering configurations in adjustable speed drives," *in Proc. IEEE APEC Conf*, pp. 228-233, 2018.

[12] H. M. Delpino and D. Kumar, "Line harmonics on systems using reduced DC-link capacitors," *in Proc. IEEE IECON Conf*, pp. 961-966, 2013.

[13] M. Sakui, H. Fujita and M. Shioya, "A method for calculating harmonic currents of a three-phase bridge uncontrolled rectifier with DC filter," *IEEE Trans. Ind. Electron.*, vol. 36, no. 3, pp. 434-440, Aug. 1989.

[14] H. Wang and F. Blaabjerg, "Reliability of Capacitors for DC-Link Applications in Power Electronic Converters—An Overview," *IEEE Trans. Ind. Appl.*, vol. 50, no. 5, pp. 3569-3578, Sep. 2014.

years lifetime of multiple slim drives is less than 1% except 30 drives system, which because the electro-thermal stress of DC-link capacitor in 30 drives system is the largest; b) the accumulated failure at 5 years lifetime of multiple slim drives decreases with the number of drives increases from 50 to 100.

It should be noted that the lifetime prediction in the case study is based on the given drives parameters, grid conditions and ambient temperature. The specific lifetime prediction should be based on the multiple slim drives specifications, grid conditions and ambient temperature. The results can be used for the proper selection of multiple slim drives parameters from the reliability aspect.

V. CONCLUSION

This paper quantitatively investigates the reliability of DC-link capacitors in multiple slim ASDs system. A time-efficient equivalent model to simplify the analysis is proposed and validated. Based on the lifetime prediction procedure and

Direct Power Control with Space Vector Modulation Applied for the Brushless DC Motor

Henrique de Toledo
Instrumentation, Automation and Robotics Engineering
Federal University of ABC
Santo André, Brazil
henrique.toledo@aluno.ufabc.edu.br

José L. Azcue
Instrumentation, Automation and Robotics Engineering
Federal University of ABC
Santo André, Brazil
jose.azcue@ufabc.edu.br

Abstract—Several control strategies have been developed for the brushless dc motor, two of them are the direct torque control and the direct power control. In this paper, the direct power control strategy has been improved by replacing the bang-bang power controller with a space vector pulse width modulation technique, providing a faster speed response and faster power transfer. A different method for estimating the power is also developed to provide easier implementation. To analyze its performance, the developed strategy was compared to the direct power control with bang-bang controller and the direct torque control. Speed, torque and power were measured and analyzed with simulations in the Matlab/Simulink environment.

Keywords—Electrical drive, speed control, torque control, power control, Brushless DC (BLDC).

I. INTRODUCTION

There is a rise in the demand for more effective, reliable, noiseless and durable electric motors, especially in applications such as vehicles, airing, manufacturing and aeromodelling. Many studies have aimed to improve the capacity to convert electrical power into mechanical power, as well as different drive strategies [1]. The brushless DC motor (BLDC) is known for presenting favorable characteristics, but an electronic circuit must be designed in order to sequentially drive its coils. The BLDC motor stands out in servo applications for its good characteristics, such as

good speed and torque relation, good dynamic response and high efficiency [2].

A six-step inverter circuit is used to control the commutation in each one of the motor coils. This allows the implementation of many different closed loop control techniques. One of the main control strategies for sinusoidal AC motors that has been adapted for the BLDC is the direct torque control (DTC), since it presents a fast torque response and is simple to be implemented [3]. The application of a space vector pulse width modulation (SVPWM) allows a better use of the DC voltage source [4] and when applied to the DTC strategy, reduces disturbances in the torque and stator flux linkages [5]. Fig 1 exemplifies a DTC block diagram for BLDC motors

The direct power control (DPC) strategy also presents good results when applied to the BLDC motor, specially regarding speed and power tracking, while also providing the main DTC advantages [6]. This paper develops a different DPC strategy, that applies a space vector modulation replacing the traditional bang-bang (hysteresis) controller and analyzes it by comparing its performance with the DPC with bang-bang controller and the DTC. A new approach to torque and power estimation that does not require the use of look up tables was also developed to allow easier implementation of the strategies.

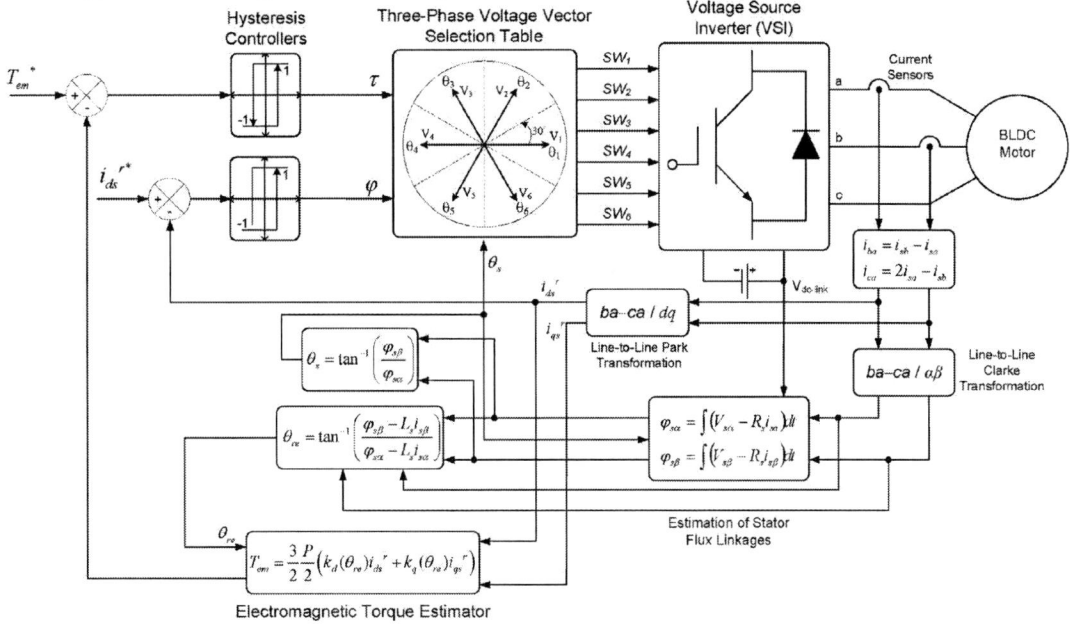

Fig. 1: Direct Torque Control for Brushless DC Motor [2]

II. Brushless DC Motor Operation

The BLDC motors have three phase stator windings and pairs of permanent magnet poles in the rotor. The stator winding distribution is built in a way that the back electromotive force (back-emf) generated has a trapezoidal shape. Because of such characteristics, each phase can be represented as a resistance, a inductance and a trapezoidal back-emf.

The construction is very similar to a permanent magnet synchronous motor. The main difference being the wave shape of the back-emfs. Despite being considered a DC motor, the BLDC operates with three-phase alternating currents.

By properly controlling the current direction in each winding the interaction between stator and rotor electromagnetic fields drives the rotor. To properly control the motor, it is always necessary to know the rotor position, which can be measured directly or estimated. A six step inverter circuit is fed by a DC source and drives the motor. The basic scheme is presented in Fig 2.

III. Modelling

In order to properly model the motor, it is assumed that the three phases have identical resistances, inductances and back-emfs, the motor is not saturated and iron losses are negligible. The Kirchhoff's law can be applied to the equivalent circuit and the following equations are obtained:

$$V_a = R_a i_a + (L_a - M)\frac{di_a}{dt} + E_a \tag{1}$$

$$V_b = R_b i_b + (L_b - M)\frac{di_b}{dt} + E_b \tag{2}$$

$$V_c = R_c i_c + (L_c - M)\frac{di_c}{dt} + E_c \tag{3}$$

where V_x are the phase voltages, R_x are the stator phase resistances, i_x are the phase currents, L_x are the phase self-inductances, M is the mutual inductance between phases, and E_x are the phase back-emfs that can be expressed as:

$$E_a = K f_a(\theta_{re})\omega \tag{4}$$

$$E_b = K f_b(\theta_{re})\omega \tag{5}$$

$$E_c = K f_c(\theta_{re})\omega \tag{6}$$

where K is a proportionality constant, ω is the rotor speed, θ_{re} is the rotor electrical angle and $f_x(\theta_{re})$ are trapezoidal functions of unitary amplitude that depend on the angle value, with $x = \{a, b, c\}$ [7].

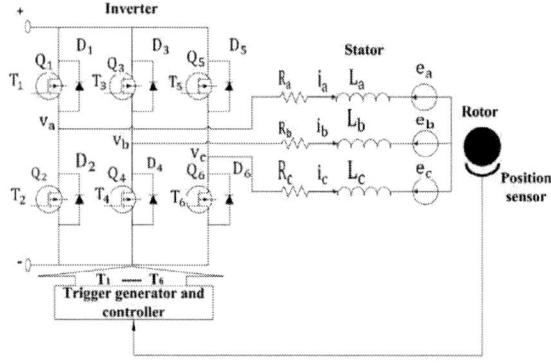

Fig. 2: Brushless DC Motor controlled by six-step inverter [8]

From the expressions, it is observed that the back-emfs amplitudes are proportional to rotor speed, therefore it is possible to define functions that have the same trapezoidal shape but do not vary its amplitude with speed by the following expressions:

$$k_a = \frac{E_a}{\omega} \tag{7}$$

$$k_b = \frac{E_b}{\omega} \tag{8}$$

$$k_c = \frac{E_c}{\omega} \tag{9}$$

where k_x are the phase back-emfs constants with trapezoidal shape.

From Newton's law it is possible do extract a equation that describes the dynamic behavior of the system:

$$\frac{d}{dt}\omega = \frac{T_{em} - T_l - B\omega}{J} \tag{10}$$

where T_{em} is the electromagnetic torque, T_l is the load torque, J is the system's inertia and B is the viscous friction coefficient.

The variables presented in the a, b and c phases can also be represented in two axes systems. With the alpha-beta transform it is possible to represent the phase quantities in two axes that are fixed in the stator, and can be implement by the following equations:

$$X_\alpha = \frac{1}{3}(2X_a - X_b - X_c) \tag{11}$$

$$X_\beta = \frac{1}{\sqrt{3}}(X_b - X_c). \tag{12}$$

With the dq transform it is possible to represent the same quantities in two axes that are fixed in the rotor, and can be implement by the following equations:

$$X_d = \frac{\sqrt{3}}{2}(cos(\theta_{re})X_a + cos(\theta_{re} - \frac{2\pi}{3})X_b + cos(\theta_{re} + \frac{2\pi}{3})X_c) \tag{13}$$

$$X_q = \frac{\sqrt{3}}{2}(-sin(\theta_{re})X_a - sin(\theta_{re} - \frac{2\pi}{3})X_b - sin(\theta_{re} + \frac{2\pi}{3})X_c). \tag{14}$$

IV. Estimators

In order to close the control loops, it is necessary to obtain feedback, which can not always be done by direct measurement with sensors. A good alternative is to measure the phase voltages and currents, calculate its values in dq and alpha-beta references and to estimate all the other necessary quantities, which is called a sensorless drive [9].

The stator flux linkages can be calculated in the alpha-beta reference by the following expressions:

$$\varphi_\alpha = \int(V_\alpha - Ri_\alpha)dt \tag{15}$$

$$\varphi_\beta = \int(V_\beta - Ri_\beta)dt. \tag{16}$$

The stator flux angle can be expressed as:

$$\theta_s = tan^{-1}(\frac{\varphi_\beta}{\varphi_\alpha}) \tag{17}$$

and electrical rotor angle is:

$$\theta_{re} = tan^{-1}(\frac{\varphi_\beta - Li_\beta}{\varphi_\alpha - Li_\alpha}). \tag{18}$$

Most of the methods used to obtain torque and power estimations are based on the use of look up tables to provide correct values accordingly to rotor position, but a different approach is proposed in this paper in which an empirical expression is used, providing the estimation of the necessary values. The constants are obtained offline and the steps required to obtain are:

- With open circuit windings, rotate the rotor at a constant speed greater than zero.
- Measure the phase back-emfs.

- Divide the back-emfs by the rotor speed, obtaining the phase back-emfs constants.
- Apply the dq transform to obtain the phase back-emfs in the dq reference frame.
- Graphically observe the behavior of k_d and k_q in the y axis and θ_{re} in the x axis when the speed is constant.
- Measure the amplitudes, frequency and offset of the plotted curves.

It has been verified empirically that for the motor model described previously, the plotted curves can be estimated by the following expressions:

$$k_d = A_1 sin(2\pi B(\theta_{re} + \pi)) \tag{19}$$

$$k_q = -|A_2 sin(2\pi B(\theta_{re} + \pi))| + C \tag{20}$$

where A_1, A_2, B and C are empirically obtained. The constants can be measured directly from the obtained graphs, since A_1 and A_2 are the amplitudes, B is the frequency and C is an offset. The expressions allow the estimation of k_d and k_q for any given θ_{re}, therefore eliminating the need of look up tables and providing an alternative to traditional estimation methods. The electromagnetic torque is given by:

$$T_{em} = \frac{3P}{4}(k_q(\theta_{re})i_q + k_d(\theta_{re})i_d) \tag{21}$$

where P is the number of rotor poles [3].

V. DIRECT POWER CONTROL PRINCIPLES

In a closed loop speed control system based on DPC, the speed error is calculated by subtracting the feedback speed from the reference speed. A controller (typically PI or PID) acts on the speed error and provides a reference power in the output.

The power error is calculated by subtracting the estimated power from the reference power. Then a hysteresis controller acts on the error and provides a binary output (b_p) that defines if the power should be increased ($b_p = 1$) or decreased ($b_p = 0$). This way, the power should always remain within the hysteresis band. Fig 3 summarizes the hysteresis based DPC strategy. The application of the voltage vectors is summarized in Fig. 4 and Table 1 defines which vectors should be applied for each sector.

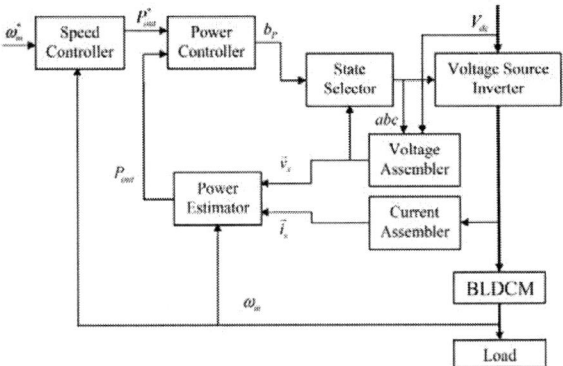

Fig. 3: Block Diagram Of The Direct Power Control Strategy [6]

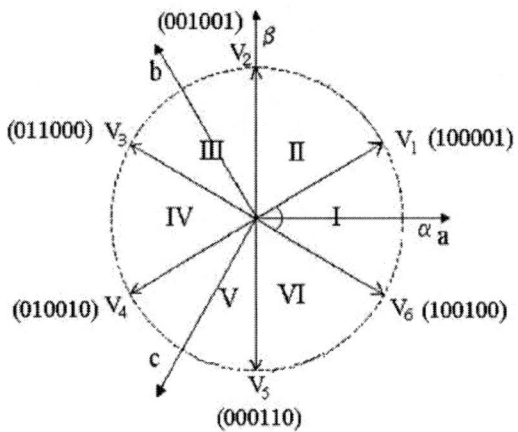

Fig. 4: Voltage Space Vectors for Brushless DC Motor [6]

Table 1: Voltage Vectors Selection

b_p	Sector I	Sector II	Sector III	Sector IV	Sector V	Sector VI
1	V_2	V_3	V_4	V_5	V_6	V_1
0	V_5	V_6	V_1	V_2	V_3	V_4

The state selector table defines which switches should close or open based on the hysteresis signal and the flux sector. The combination of the six switch signals defines the voltage vectors (V_1 to V_6) that can be applied. The flux sector is divided in six sectors (Sector I to Sector VI) to determine which vector should be applied accordingly to the hysteresis signal.

VI. SPACE VECTOR PULSE WIDTH MODULATION PRINCIPLES

As a alternative to the hysteresis based power control loop, a space vector pulse width modulation (SVPWM) based control loop strategy was developed and implemented. The method relies on alternating between two voltage vectors in order to create a desired average voltage vector with a constant switching frequency [10]. It is commonly applied in vector control strategies such as DTC variations.

In the proposed scheme, a PI controller acts on the power error providing a reference voltage for the q-axis and the reference voltage for the d-axis is kept zero. Then a transformation is applied to convert a vector referred in the dq to the alpha-beta reference frame. The gate signals for the inverter will then be generated by alternating two voltage vectors in a way that the average voltage vector corresponds to the alpha and beta voltages desired [11]. The combination of signals that can be applied for a three-phase conduction defines six voltage vectors (S1 to S6). Fig. 5 exemplifies the creation of an average vector and Fig. 6 shows a generic application of the SVPWM technique.

978-1-7281-4181-7/19 $31.00 © 2019 IEEE

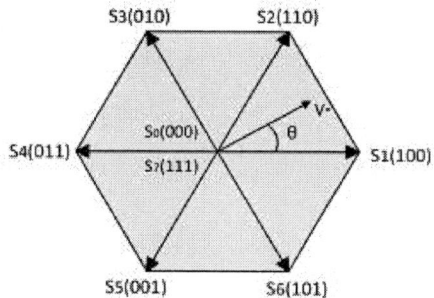

Fig. 5: Block Diagram Of The SVPWM Technique [5]

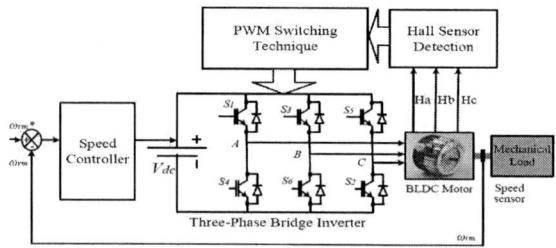

Fig. 6: Application Of The SVPWM for Brushless DC Motor [10]

VII. SIMULATION

To properly validate the proposed strategy, the environment Matlab/Simulink was used to apply the developed model and test the control strategy. Three strategies were simulated, tested, analyzed and compared. The simulated strategies are:

- Direct Torque Control with SVPWM [5],
- Direct Power Control with hysteresis control[6],
- Direct Power Control with SVPWM.

Fig. 7 shows the simulation of the algorithm developed for the DPC with SVPWM

The simulated motor is specified by Table 2.

Table 2: Simulated Motor Parameters

Number of poles (P)	2
Rated Speed (ω_{rated})	1500 rpm
DC Voltage (V_{dc})	300 V
Phase Inductance (L-M)	13 mH
Phase Resistance (R)	0.4 Ω
Viscous Friction Coefficient (B)	0.002 Nm.s/rad
Inertia (J)	0.004 kg.m²

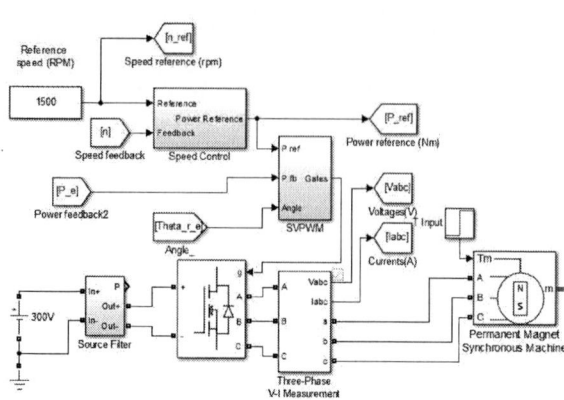

Fig. 7: Simulation Of The Direct Power Control With SVPWM in Simulink

VIII. RESULTS

For easier implementation, it is necessary to obtain the necessary parameters for estimating torque and power in the simulations. The first step necessary is to obtain the empirical values for k_d and k_q for the simulated motor.

The expressions (19) and (20) are verified in Fig. 8 and values graphically found are $A_1 = 0.012318 \frac{Vs}{rad}$, $A_2 = 0.0625 \frac{Vs}{rad}$, $B = 75Hz$ and $C = 0.4666 \frac{Vs}{rad}$. These constants are applied for proper estimation in the following methods.

In order to validate the estimation method proposed, tests were performed comparing the actual and estimated torque in Fig. 9, as well as the actual and estimated power in Fig. 10.

For the performance tests, the reference speed was 1500 rpm with a constant 3 Nm load. The objective is to evaluate the strategies at similar conditions and analyze the speed response. Fig. 11, 12 and 13 respectively show speed, torque and power for all the simulated methods.

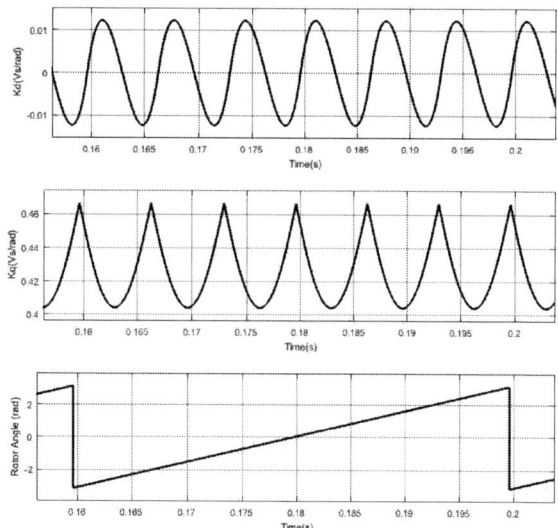

Fig. 8: Back-emfs Constants For Constant Speed

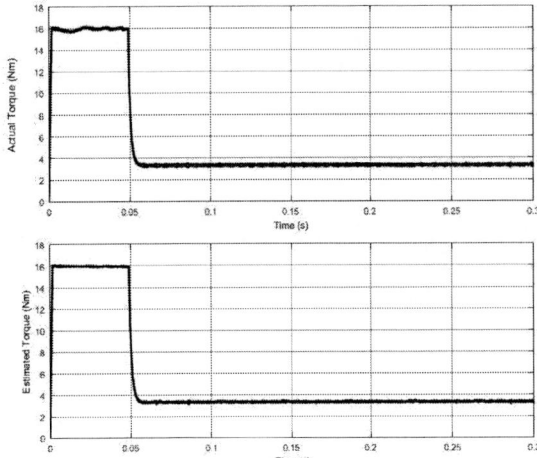

Fig. 9: Analysis Of The Estimated Torque

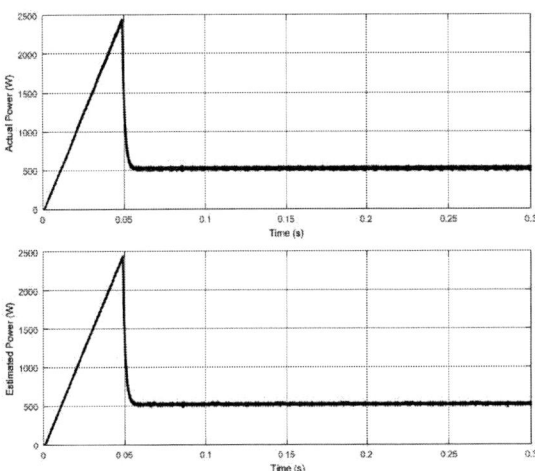

Fig. 10: Analysis Of The Estimated Power

The estimated torque error is less than 0.1 Nm and the estimated power error is less than 20 W. The actual torque and actual power present more oscillations than the estimated because the controller is acting on the estimated values. The estimated torque and power are both sufficiently close to the actual values, which means that different control strategies can be applied using the estimated values as feedback for the controllers.

It is noticeable that the DPC with SVPWM has almost the same rising time as the DTC, which means that these methods have a similar performance regarding speed tracking, the only difference being the non-constant acceleration in the transient for the DPC.

The speed graph of the DPC with SVPWM is very similar to the DPC with hysteresis, but has a faster rising time, which means that the application of SVPWM improved the speed response. All three tested strategies present negligible steady state error (less than 1 rpm) and no ripple.

From the torque graphs it is observed that three strategies present a small torque ripple.

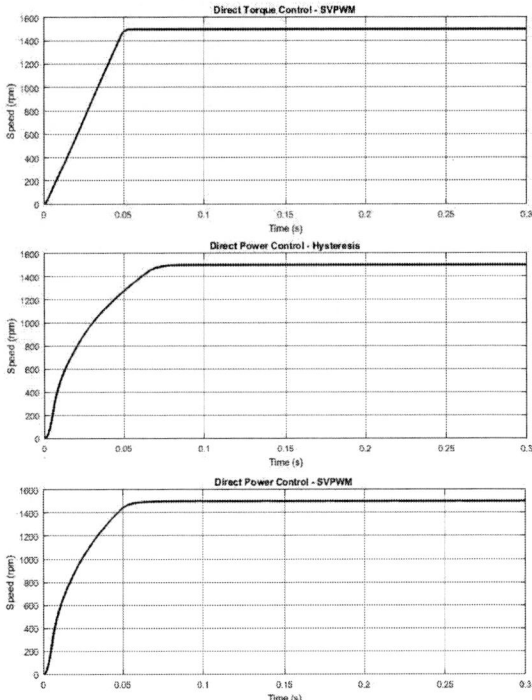

Fig. 11: Analysis Of The Speed Tracking Performance

The main differences in the Torque response are that the DTC presents a constant torque while in transient due to the controller output saturation while the DPC presents a high peak torque that quickly drops causing the variable acceleration already observed in the speed graph. The application of SVPWM reduced slightly the torque ripple.

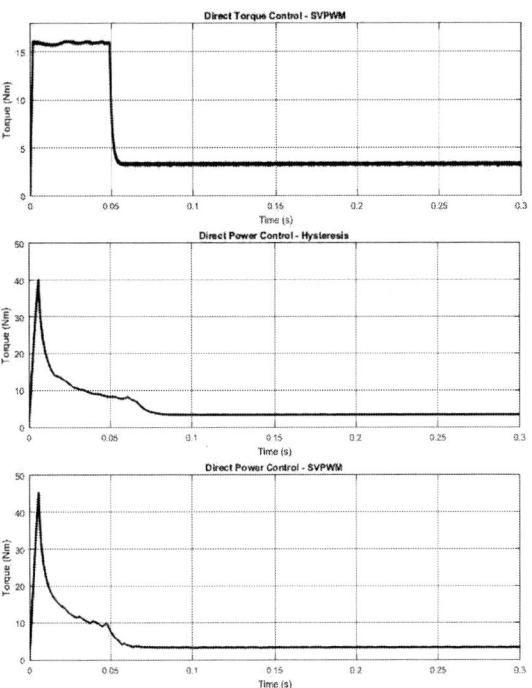

Fig. 11: Analysis Of The Torque Performance

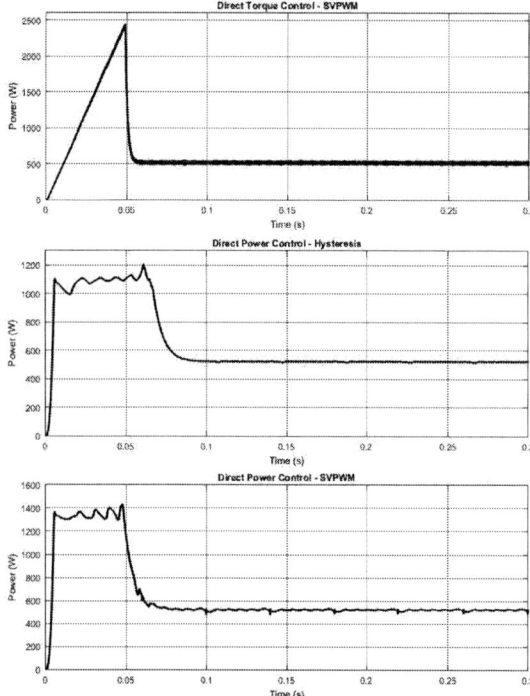

Fig. 12: Analysis Of The Power Performance

From the power graphs it is possible to see that the DTC with SVPWM is slower when delivering output power, increasing it linearly in the transient, since the torque constant and speed also increases linearly, until the system reaches steady state, while the DPC with hysteresis and DPC with SVPWM both quickly deliver power to the output. The maximum power consumption can be limited by the controllers output saturation, so that the electrical system is protected from overloading and voltage drops can be avoided.

IX. Conclusions

The possibilities of replacing the bang-bang hysteresis controller with space vector pulse width modulation in the Direct Power Control strategy have been explored. The DPC with SVPWM is an improvement when compared to the previously developed DPC technique for BLDC motors, specially regarding speed response and torque ripple. The tests show that the performance can be considered as good as the DTC, but with some advantages, such as being easier to implement, having a higher peak torque and having a limited power consumption defined by the controllers output saturation. A deeper evaluation of the saturation effects is left for future works.

The estimation method proposed presented good results and is a good alternative to avoid implementing look up tables. The method could also be applied for many of the control strategies previously developed in the literature that require the estimation of torque or power.

The application of the space vector modulation to the DPC strategy could be a good alternative for applications in which a very fast power transfer is required.

References

[1] P. C. Krause, O. Wasynozuk, and S. D. Sudhoff, "Analysis of Electric Machinery and Drive Systems", IEEE Press, 2002.

[2] M. Cunkas, and O. Aydogdu, "Realization of fuzzy logic Controlled brushless DC motor drives using Matlab/Simulink", Mathematical and Computational Applications, vol. 15, (2), pp.218-229, 2010.

[3] S.B. Ozturk, and H.A. Toliyat, "Direct torque and indirect flux control of brushless DC motor", IEEE/ASME Trans. on Mechatronics, vol. 16, no. 2, pp. 352-358, April 2011.

[4] N. Mohan, Advanced Electric Drives, MNPERE, 2001.

[5] M. T. Lingamsetti, T. T. Sreenu, "Control of Brushless DC Motor with Direct Torque and Indirect Flux using SVPWM Technique", India Journal Of Science and Technology, 2015.

[6] H. Moghbeli, A.H. Niasar, and M. B. Shahrbabak, "Direct Power Control of Brushless DC Motor Drive", Journal of Power Electronics & Power Systems, 2014.

[7] T. Mathew, C. A. Sam, "Modeling and Closed Loop Control of BLDC Motor Using Single Current Sensor", International Journal of Advanced Research in Electrical, Electronics and Instrumentation Engineering, 2013.

[8] N. T. T. Nga, N. T. P. Chi, and N. H. Quang, "Study on Controlling Brushless DC Motor in Current Control Loop Using DC Link Current", American Journal Of Engineering Research, 2018.

[9] J. C. Gamazo-Real, E. Vázquez-Sánchez, and J. Gómes-Gil, "Position and Speed Control of Brushless DC Motors Using Sensorless Techniques and Application Trends", Sensors - Open Access Journals, 2010.

[10] F. R. Yasien, and R. A. Mahmood, "Design New Control System for Brushless DC Motor Using SVPWM", International Journal of Applied Engineering, 2018.

[11] S. Mahammadsoaib,and P. Sajid, "Vector controlled PMSM drive using SVPWM technique – A MATLAB / Simulink Implementation", International Conference on Electrical, Electronics, Signals, Communication and Optimization, 2015.

Modeling and control of a Forward DC-DC converter for Battery Voltage Balancing

Pábulo F. Ciarnoscki, Daniel J. Pagano, *Member, IEEE*, Gabriel M. da Silva

Department of Automation and Systems
Federal University of Santa Catarina, Florianópolis-SC, Brazil
pabulo.felipe@posgrad.ufsc.br, daniel.pagano@ufsc.br, g.manoel@ufsc.br

Abstract—This paper addresses the modeling and control of a novel Forward dc-dc converter with multiplexer for active voltage cell balancing of a bank of Lithium-Ion batteries. The modeling of the converter is developed using a state-space averaging model. The control system is based on a cascade feedback linearization approach for the forward converter and an event-driven control based on the voltage of the cells. Simulation results in time-domain allow to evaluate the performance of the proposed voltage balancing system control.

Index Terms—Active voltage balancing, Forward dc-dc converter, Cascade feedback linearization control, Event-driven control.

I. INTRODUCTION

Nowadays, the use of battery banks [1] [2] has become increasingly necessary in energy electrical distribution systems were renewable energy resources, such as solar and wind power, are penetrating more than ever. These renewable energy resources are intermittent in nature, i.e., the power changes over the time under the influence of the weather fluctuations. This intermittence may produce a strong effect on power system stability over the voltage and frequency variables. In this sense, Battery Energy Storage Systems (BESS) are indispensable to obtain successful in the integration of renewable energy sources into grid power systems.

Electric mobility is another issue where the batteries are at the core of the applications. Second life batteries applications from the spent car batteries that often retain enough energy constitutes a suited recycling use extending the batteries operational lifetime. In these applications where high power is required, Ion-lithium batteries are the most used due to their high efficiency and energy density [3].

In contrast, Li-ion cells require precise control of their conditions of use. Operations outside the limits of voltage, current and temperature of the cell can cause shortening of the service life, bank heating and even explosions [2]. In this way a dedicated circuit is required to control the charge, discharge and the state of the cells. This circuit is known as Battery Management System (BMS) [4].

In a battery bank, the total capacity is affected by the state of each individual cell [2]. Unbalance of the cells in a bank is the main cause of the bank's life and capacity decrease. Unbalance

This work was supported in part by CNPq/BRAZIL under Project 302229/2018-3.

is caused by variations in cell fabrication, variations in internal impedance, and thermal differences in the bank [5].

Conventional topologies of BMSs estimate the state of charge (SoC) of each cell and use resistors to dissipate energy from cells with higher SoC, thus trying to keep all cells in the bank balanced [6]. This method is energetically inefficient, in addition to not being able to balance cells with low SoC.

However there are methods for active balancing, which is intended to transfer load cells with increased energy for cells with less energy. Different topologies are used for this purpose, the most common of which are those that use controlled switches and DC-DC converters [7] [8] [9]. The active methods offer better energy efficiency when compared to the passive method.

The remainder of this paper is organized as follows. Section I introduces the description of the proposed active cell voltage balancing system. Section II addresses the balancing system control, introducing a control algorithm to the voltage balancing of a battery bank compose by three Lithium-ion cells. Sections III and IV are dedicated to system modeling and proposed control system design, respectively. Section V describes the experiment to be performed in this paper and summarizes the assessment of the simulation results. Finally, Section VI presents the main conclusions.

II. ACTIVE CELL VOLTAGE BALANCING SYSTEM

Figure 1 shows the proposed circuit of the cell voltage balancing system. The purpose of this is to withdraw power from the entire battery bank and inject it into the cell with lower voltage. For this, an isolated converter topology with close-loop control is required. The most common isolated topologies are Flyback and Forward. In this work the topology chosen is a Two-Switch Forward Converter. The Forward Converter, when compared to the Flyback Converter, is more energy efficient. In the Flyback converter, the power is temporarily stored in the transformer before being transferred to the secondary. In the Forward converter the energy is transferred directly from the primary to the secondary. The transformer used in the Forward converter is designed to be ideal, with magnetizing current equal to 0 [10].

The batteries are represented by a simplified RC series model. The capacitors represents the capacity of the battery, the voltage on it being the state of charge of the battery and the resistor represents the cell series resistance. The voltage

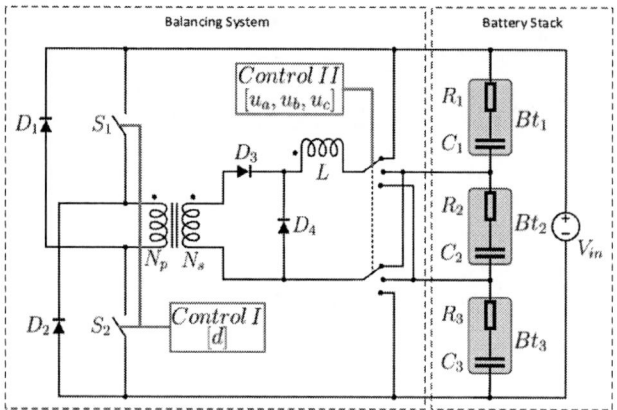

Fig. 1. Voltage balancing system based on a Forward converter for a three cell battery stack.

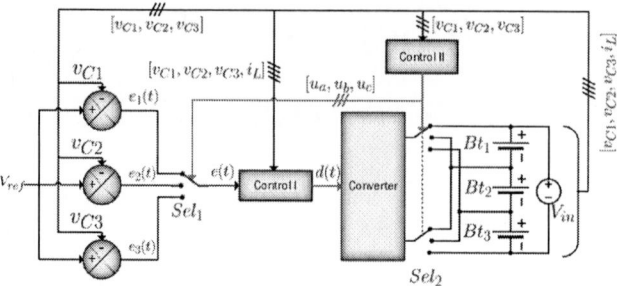

Fig. 2. Proposed control balancing system.

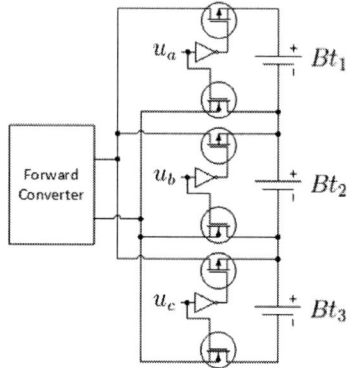

Fig. 3. Circuit based on Mosfet switches for switch Sel_2.

under the capacitor is an internal state of the cell, not being able to measure it. In this work it is considered that the voltage is measurable. In a real application this state would need to be estimated. The voltage source V_{in} represents the battery bank's charger at constant voltage mode, with value equal to the nominal battery bank voltage.

This system is capable of balancing three cells. For banks with the largest number of cells the system can be easily increased or a balancer block can be used for every three cells.

The proposed control is divided into two loops. The general scheme of control can be seen in Fig. 2. The *Control Loop II* is composed by event-driven controller. *Controller II* is responsible for coupling the output of the converter in the cell with lower voltage, through the switch Sel_2. This switch can be realized by a MOSFET arrangement as shown in Fig. 3. Although the number of MOSFETs employed in this circuit be greater than the Forward converter with multiple outputs, the proposed circuit has the advantage of reducing number of inductors and the smaller size of the transformer compared to the multiple output Forward Converter since there is only one secondary coil in the proposed circuit. *Control Loop I* is based on two control schemes: (i) the first consists in a feedback linearization scheme applied on the inductor current dynamics and (ii) the second one in a cascade control composed by an inner current proportional control loop and a voltage outer PI control loop. This control is responsible for controlling the cell's voltage error. The input of the controller is connected to the error of the cell selected under the same control action of the *Control II*, through the switch Sel_1. This switch is implemented by software in the control law. When the first battery (Bt_1) is connected to the converter the voltage error of this same cell is connected to the input of the controller, and so respectively to the other cells.

III. System Modelling

Assuming that only one cell will be connected at to the converter each time, the circuit can be reduced to simplify the analysis as shown in Fig. 4. The drive operates by a fixed

frequency PWM signal applied simultaneously to the S_1 and S_2 switches. The duty cycle of the drive is given by $d = \frac{t_{on}}{T_s}$

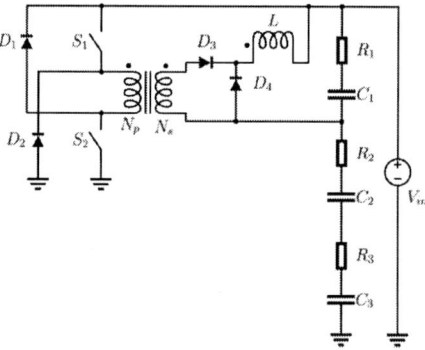

Fig. 4. Simplify schematic diagram used for modeling one cell.

where t_{on} and $T_s = \frac{1}{f_s}$ stand for the time when the switch is on and switching period, respectively.

The circuit have two topological states, the first occurring between 0 and dTs and the second from dTs to Ts.

A. First topological state $[0, dT_s]$

In the first topological state the switches S_1 and S_2 are on, and the voltage V_{in} is apply to the primary of the transformer being the secondary voltage given by $\alpha = \frac{Np}{Ns}$ where N_p and N_s are the number of turns of the primary and the secondary, respectively. The voltage on the secondary forcing the diode

978-1-7281-4181-7/19 $31.00 © 2019 IEEE

D_3 to conduit and the diode D_4 to block. The instantaneous circuit diagram is shown in Fig. 5.

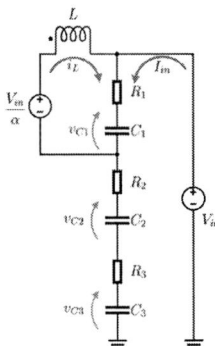

Fig. 5. Circuit diagram of first state for one cell model.

Defining the states of the system $\mathbf{x}(t)$ as

$$\mathbf{x}(t) = \begin{bmatrix} i_L & v_{C1} & v_{C2} & v_{C3} \end{bmatrix}^T, \tag{1}$$

and applying the Kirchhoff's laws to the circuit depicted in Fig. 5, for the first topological state, we have

$$L\frac{di_L}{dt} = \frac{V_{in}}{\alpha} - i_L R_1 - v_{C1} - \left(\frac{-i_L R_1 - v_{C1} - v_{C2} - v_{C3} + V_{in}}{R_t}\right) R_1, \tag{2}$$

$$C_1\frac{dv_{C1}}{dt} = i_L + \frac{-i_L R_1 - v_{C1} - v_{C2} - v_{C3} + V_{in}}{R_t}, \tag{3}$$

$$C_2\frac{dv_{C2}}{dt} = \frac{-i_L R_1 - v_{C1} - v_{C2} - v_{C3} + V_{in}}{R_t}, \tag{4}$$

$$C_3\frac{dv_{C3}}{dt} = \frac{-i_L R_1 - v_{C1} - v_{C2} - v_{C3} + V_{in}}{R_t}. \tag{5}$$

Where $R_t = R_1 + R_2 + R_3$.

B. Second topological state $[dT_s, T_s]$

In the second topological state we have S_1 and S_2 switches in off, due to the magnetizing inductance of the primary coil, the diode D_1 and D_2 are conducting and $-V_{in}$ appears on the primary and $\frac{-V_{in}}{\alpha}$ on the secondary, blocking the diode D_3. Due to the current flowing in the inductor L, the diode D_4 goes into conduction. The instantaneous circuit diagram is shown in Fig. 6.

Applying the Kirchhoff's laws to the circuit presented in Fig. 6, for the second topological state, we have

$$L\frac{di_L}{dt} = -i_L R_1 - v_{C1} - \left(\frac{-i_L R_1 - v_{C1} - v_{C2} - v_{C3} + V_{in}}{R_t}\right) R_1, \tag{6}$$

for the state i_L. For the states v_{C1}, v_{C2} and v_{C3} the equations are equal to (3), (4) and (5), respectively.

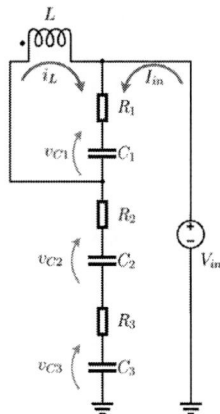

Fig. 6. Circuit diagram of second state for one cell model.

C. State Space Averaging Model

The averaging model for one cell connected system is obtain by multiplying (2) by d and adding with (6) multiplied by $1-d$ as given bellow:

$$L\frac{d\langle i_L \rangle}{dt} = \frac{-R_1(R_2 + R_3)i_L - (R_2 + R_3)v_{C1}}{R_t} \\ \frac{R_1 v_{C2} + R_1 v_{C3} + R_1 V_{in}}{R_t} + \frac{V_{in}}{\alpha} d, \tag{7}$$

and

$$C_1\frac{d\langle v_{C1}\rangle}{dt} = \frac{(R_2 + R_3)i_L - v_{C1} - v_{C2} - v_{C3} + V_{in}}{R_t}, \tag{8}$$

$$C_2\frac{d\langle v_{C2}\rangle}{dt} = \frac{-R_1 i_L - v_{C1} - v_{C2} - v_{C2} + V_{in}}{R_t}, \tag{9}$$

$$C_3\frac{d\langle v_{C3}\rangle}{dt} = \frac{-R_1 i_L - v_{C1} - v_{C2} - v_{C3} + V_{in}}{R_t}. \tag{10}$$

The equilibrium point of (7-10) is given by

$$\begin{aligned} i_L &= 0 \\ v_{C1} &= \frac{V_{in}}{\alpha} d \\ v_{C2} &= V_{in} - v_{C1} - v_{C3} \\ v_{C3} &= V_{in} - v_{C1} - v_{C2} \end{aligned} \tag{11}$$

It can be observed that the output voltage of the inverter (v_{C1}) in steady-state is equal to the output voltage of the standard model of the Forward converter [11]. The voltage v_{C1} is strictly defined, but the voltages v_{C2} and v_{C3} are indeterminated, they fulfil the constraint determined by V_{in}. In this way, a circuit that connects the converter to each cell is necessary, so that all voltages can be balanced.

D. Matrix model

From (7), (8), (9) and (10) and by assuming control $u(t) = d$, we can write the state equations as

$$\dot{\mathbf{x}}(t) = \mathbf{A}\mathbf{x}(t) + \mathbf{B}u(t) + \mathbf{E} \tag{12}$$

978-1-7281-4181-7/19 $31.00 © 2019 IEEE

being for the first coupled cell (a), $\mathbf{A}$, $\mathbf{B}$, $\mathbf{E}$ given by

$$\mathbf{A_a} = \begin{bmatrix} \frac{-R_1(R_2+R_3)}{LR_t} & \frac{-R_2-R_3}{LR_t} & \frac{R_1}{LR_t} & \frac{R_1}{LR_t} \\ \frac{R_2+R_3}{C_1R_t} & \frac{-1}{C_1R_t} & \frac{-1}{C_1R_t} & \frac{-1}{C_1R_t} \\ \frac{-R_1}{C_2R_t} & \frac{-1}{C_2R_t} & \frac{-1}{C_2R_t} & \frac{-1}{C_2R_t} \\ \frac{-R_1}{C_3R_t} & \frac{-1}{C_3R_t} & \frac{-1}{C_3R_t} & \frac{-1}{C_3R_t} \end{bmatrix}, \quad (13)$$

$$\mathbf{B} = \begin{bmatrix} \frac{V_{in}}{L\alpha} & 0 & 0 & 0 \end{bmatrix}^T, \quad (14)$$

$$\mathbf{E_a} = \begin{bmatrix} \frac{R_1V_{in}}{LR_t} & \frac{V_{in}}{C_1R_t} & \frac{V_{in}}{C_2R_t} & \frac{V_{in}}{C_3R_t} \end{bmatrix}^T. \quad (15)$$

For the other coupled cells we can find the model in the same way as before. For the second coupled cell (b) we have

$$\mathbf{A_b} = \begin{bmatrix} \frac{-R_2(R_1+R_3)}{LR_t} & \frac{-R_1-R_3}{LR_t} & \frac{R_2}{LR_t} & \frac{R_2}{LR_t} \\ \frac{-R_2}{C_1R_t} & \frac{-1}{C_1R_t} & \frac{-1}{C_1R_t} & \frac{-1}{C_1R_t} \\ \frac{R_1+R_3}{C_2R_t} & \frac{-1}{C_2R_t} & \frac{-1}{C_2R_t} & \frac{-1}{C_2R_t} \\ \frac{-R_2}{C_3R_t} & \frac{-1}{C_3R_t} & \frac{-1}{C_3R_t} & \frac{-1}{C_3R_t} \end{bmatrix}, \quad (16)$$

$$\mathbf{E_b} = \begin{bmatrix} \frac{R_2V_{in}}{LR_t} & \frac{V_{in}}{C_1R_t} & \frac{V_{in}}{C_2R_t} & \frac{V_{in}}{C_3R_t} \end{bmatrix}^T \quad (17)$$

and for the third coupled cell (c)

$$\mathbf{A_c} = \begin{bmatrix} \frac{-R_3(R_1+R_2)}{LR_t} & \frac{-R_1-R_2}{LR_t} & \frac{R_3}{LR_t} & \frac{R_3}{LR_t} \\ \frac{-R_3}{C_1R_t} & \frac{-1}{C_1R_t} & \frac{-1}{C_1R_t} & \frac{-1}{C_1R_t} \\ \frac{-R_3}{C_2R_t} & \frac{-1}{C_2R_t} & \frac{-1}{C_2R_t} & \frac{-1}{C_2R_t} \\ \frac{R_1+R_2}{C_3R_t} & \frac{-1}{C_3R_t} & \frac{-1}{C_3R_t} & \frac{-1}{C_3R_t} \end{bmatrix}, \quad (18)$$

$$\mathbf{E_c} = \begin{bmatrix} \frac{R_3V_{in}}{LR_t} & \frac{V_{in}}{C_1R_t} & \frac{V_{in}}{C_2R_t} & \frac{V_{in}}{C_3R_t} \end{bmatrix}^T. \quad (19)$$

The matrix $\mathbf{B}$ does not change, being the same for any of the coupled cells.

IV. CONTROL DESIGN

A. Control loop I Design

In order to design the *Control Loop I* we used the Feedback Linearization Control (FLC) technique, similar to that presented in [12]. For the design of this control loop, the first cell is suppose to be the only cell connected to the converter. Due to the similarity of dynamics this control is extended to the other cells.

For this loop it is desired that the control of i_L be decoupled from the other state variables. Looking at Eq. 7, we want the current equation to be of the form:

$$\frac{d\langle i_L \rangle}{dt} = -\frac{R_1(R_2+R_3)}{R_tL}i_L + u(t), \quad (20)$$

where $u(t)$ is the uncoupled control input. This way we can define that

$$u(t) = \frac{-(R_2+R_3)v_{C1} + R_1(v_{C2}+v_{C3}-V_{in})}{R_tL} \\ + \frac{V_{in}}{\alpha L}d(t). \quad (21)$$

Now, from (21) we can obtain $d(t)$

$$d(t) = f_{FLC}(v_{C1}, v_{C2}, v_{C3}) + k_{FLC}u(t), \quad (22)$$

being

$$f_{FLC}(v_{C1}, v_{C2}, v_{C3}) = \\ \frac{((R_2+R_3)v_{C1} - R_1v_{C2} - R_1v_{C3} + R_1V_{in})\alpha}{R_tV_{in}} \quad (23)$$

and

$$k_{FLC} = \frac{\alpha L}{V_{in}}. \quad (24)$$

For the i_L current control, a proportional controller was chosen to limit the maximum current peak. For the control of voltages v_{C1}, v_{C2} and v_{C3} a proportional-integral control was chosen in order to make the final error value equal to zero. The final form of control is showed in the Fig. 7. The battery parameter

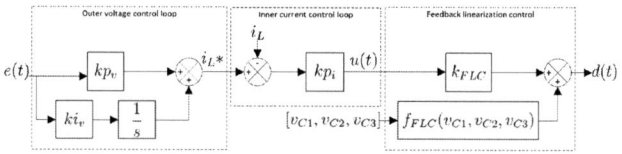

Fig. 7. Cascade PI Control Loop and FLC diagram.

values used for the control loop calculations and simulation were obtained from the Lithium-ion battery model INR18650-25R [13]. The main parameters can be seen in Table I.

TABLE I
PARAMETER VALUES OF INR18650-25R LITHIUM-ION CELL

Parameter	Value	Unit
Nominal Voltage	3.6	V
Max. Voltage	4.2	V
Min. Voltage	2.5	V
Internal Impedance	18	mΩ
Nominal Capacity	2500	mAh

With the parameters of the cell it is possible to estimate the RC model of the battery. Internal resistance can be approximated by the value of internal impedance. The capacitance can be estimated from the nominal battery capacity, being approximately 5294.11[F]. In order to simulate differences between cells, R_1 were defined as equal to the estimated parameter, R_2 was estimated with a 10% increase in nominal value and R_3 was estimated to be the nominal value plus 20%. C_1 were defined as equal to the estimated parameter, C_2 was estimated to be a decrease of 10% of nominal value and C_3 as a decrease of 20% of nominal value. The other parameters were defined empirically. The Table II shows the values of the parameters used.

For the definition of the gain values, it was set $kp_v = 1$. With the need for a subtle integrative action, it was possible to define $ki_v = 10^{-5}$. Finally, kp_i was defined as $kp_i = 20$, so that the peak of current i_L was not greater than $200mA$.

TABLE II
SYSTEM PARAMETER VALUES.

Element	Value	Unit
L	200	mH
R_1	18	mΩ
R_2	19.8	mΩ
R_3	21.6	mΩ
C_1	5294.11	F
C_2	4764.69	F
C_3	4235.28	F
α	1	-

B. Control loop II Design

The second control loop is based on an event-driven control law. This event-driven control has the objective to select through the outputs $U_s = [u_a, u_b, u_c]$ the cell with the lowest voltage. For example $v_{C1} < v_{C2}$ and $v_{C1} < v_{C3}$ we have that

$$U_s = [1, 0, 0]. \tag{25}$$

For $v_{C2} < v_{C1}$ and $v_{C2} < v_{C3}$ the switch sequence is given by

$$U_s = [0, 1, 0]. \tag{26}$$

For $v_{C3} < v_{C1}$ and $v_{C3} < v_{C1}$ we have that

$$U_s = [0, 0, 1]. \tag{27}$$

As an operating condition constraint, only one output can be triggered at the same time. In order to make it, the discrete variables v_i, $i = 1, 2, 3$, $v_i \in \{0, 1\}$ are defined as a function of the voltage capacitor errors

$$
\begin{aligned}
v_1 &= \frac{1}{2}(1 - sign(v_{C2} - v_{C1})), \\
v_2 &= \frac{1}{2}(1 - sign(v_{C3} - v_{C2})), \\
v_3 &= \frac{1}{2}(1 - sign(v_{C1} - v_{C3})).
\end{aligned} \tag{28}
$$

A logical combinatory was designed from these variables allow to determine the outputs $[u_a, u_b, u_c]$ as

$$
\begin{aligned}
u_a &= v_1 v_2, \\
u_b &= (1 - v_1)v_3, \\
u_c &= (1 - v_2)(1 - v_3).
\end{aligned} \tag{29}
$$

This event-driven control is syncronized in the control loop II by using a periodic time-event signal of $100s$.

V. SIMULATION RESULTS

The closed-loop system was simulated using the values of Table II. Initial values are defined as $i_L = 0A$, $v_{C1} = 3.7V$, $v_{C2} = 3.6V$, $v_{C3} = 3.5V$.

Fig. 8 shows the signal variables v_{C1}, v_{C2} and v_{C3} starting from different voltage conditions (unbalanced system). Initially the system starts injecting current into cell 3, which is the cell with the lowest voltage. Due to the restriction imposed by V_{in}, the voltages v_{C2} and v_{C3} begin to decrease.

Approximately at 3400 s the voltages v_{C2} and v_{C3} are equal and at this point the system begins to slide on the line $v_{C1} =$

v_{C3}, increasing the voltage of the two cells until the three cells reach equal values. In this case, the used reference was $V_{ref} = \frac{V_{in}}{3}$.

Sliding actions, like those observed in sliding mode control systems, can be observed on the u_b and u_c signals. Initially with v_{C3} voltage being the lowest, the u_c output remains active until it equals v_{C2}, at this time u_b and u_c outputs begin to oscillate in order to v_{C1} and v_{C3} smoothly reach the reference. The output u_b remains off during this period because the voltage v_{C2} does not become, at any moment, less than v_{C1} and v_{C3}. The slide action can be observed in Fig. 9. After stabilization of the voltages, at 50000 s, the bank charging is simulated with the reference voltage tending to 4.2V. At this point, it is possible to observe that due to the differences between the cells parameters, even after being equalized, the voltages start to diverge and the current of i_L begins to increase to occur balance. The equilibrium occurs approximately at 120000 s. After this time, the bank discharging is simulated.

VI. CONCLUSION

This paper addressed the modeling and control of a Forward DC-DC converter for battery voltage balancing. A novel control based on two control loops was realized. The first control loop employ a feedback linearization control with a control cascade based on an inner current proportional control loop and a voltage outer PI control loop. The second control loop is based on a event-driven control strategy derived from the state of the cell battery voltages. Both control schemes ensure robust performance and stability for uncertainties in the cell battery parameters.

The event-driven control together with the multiplexer synthetized by the selection switch Sel_2 proved to be an interesting alternative when compared to systems equipped with a converter with several outputs or with a converter for each cell, schemes already widely described in the literature [14] [15] [16]. The proposed voltage balancing system has the advantage of employ a lower number of components making the balancing system more compact. The selection of the Forward topology stands out due to its characteristic type of output current source and its high efficiency [10]. The current source type output causes less current oscillations on the cell preventing premature bank degradation due to balancing [17] [18].

Finally, we consider in this paper that the internal states of the cells were directly accessible, which does not happen in a real application. Further work may address (i) the develop of a state observer and (ii) the stability analysis and dynamical response of the system operating with this observer. Another future task will be to optimize the circuit's electrical components and assemble a prototype to validate the simulated results.

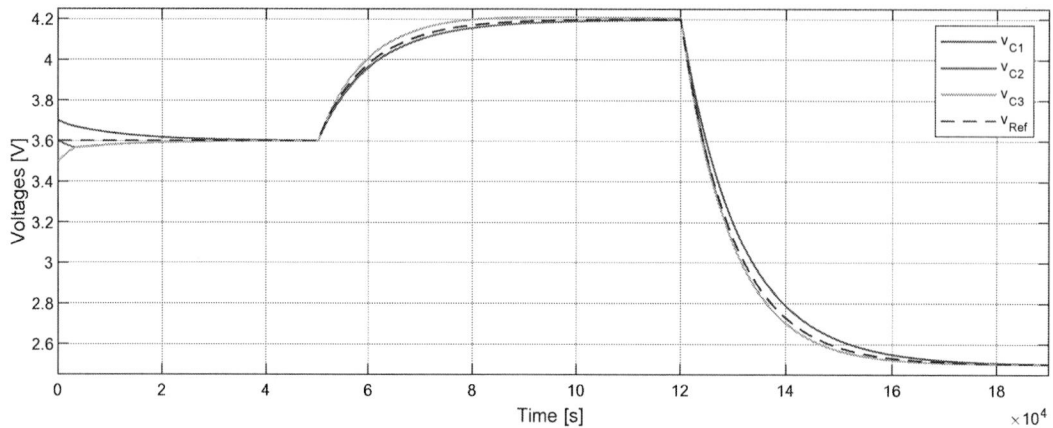

Fig. 8. Closed-loop system time-responses with the proposed controller, starting from unbalanced initial conditions showing the voltage capacitors.

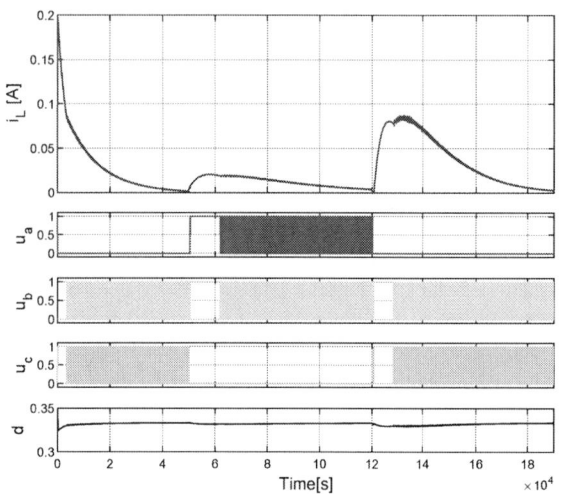

Fig. 9. Closed-loop system time-responses with the proposed controller, starting from unbalanced initial conditions, showing the current signal i_L and the control variables u_a, u_b and u_c.

REFERENCES

[1] T. Morstyn, A. Savkin, B. Hredzak, and V. Agelidis, "Multi-agent sliding mode control for state of charge balancing between battery energy storage systems distributed in a dc microgrid," *IEEE Transactions on Smart Grid*, 2017.

[2] S. Narayanaswamy, M. Kauer, S. Steinhorst, M. Lukasiewycz, and S. Chakraborty, "Modular active charge balancing for scalable battery packs," *IEEE Transactions on Very Large Scale Integration (VLSI) Systems*, vol. 25, no. 3, pp. 974–987, 2017.

[3] A. Pröbstl, S. Park, S. Narayanaswamy, S. Steinhorst, and S. Chakraborty, "Soh-aware active cell balancing strategy for high power battery packs," in *2018 Design, Automation Test in Europe Conference Exhibition (DATE)*, March 2018, pp. 431–436.

[4] D. Xu, L. Wang, and J. Yang, "Research on li-ion battery management system," in *Electrical and Control Engineering (ICECE), 2010 International Conference on*. IEEE, 2010, pp. 4106–4109.

[5] W. Bentley, "Cell balancing considerations for lithium-ion battery systems," in *Battery Conference on Applications and Advances, 1997., Twelfth Annual*. IEEE, 1997, pp. 223–226.

[6] D. Andrea, *Battery management systems for large lithium ion battery packs*. Artech house, 2010.

[7] M. Daowd, N. Omar, P. Van Den Bossche, and J. Van Mierlo, "A review of passive and active battery balancing based on matlab/simulink," *J. Int. Rev. Electr. Eng*, vol. 6, pp. 2974–2989, 2011.

[8] N. H. Kutkut and D. M. Divan, "Dynamic equalization techniques for series battery stacks," in *Telecommunications Energy Conference, 1996. INTELEC'96., 18th International*. IEEE, 1996, pp. 514–521.

[9] K. Zhi-Guo, Z. Chun-Bo, L. Ren-Gui, and C. Shu-Kang, "Comparison and evaluation of charge equalization technique for series connected batteries," in *Power Electronics Specialists Conference, 2006. PESC'06. 37th IEEE*. IEEE, 2006, pp. 1–6.

[10] M. A. K. A. Biabani, "Simulation, mathematical calculation and comparison of power factor and efficiency for forward, fly back and proposed forward-flyback converter," in *2016 International Conference on Electrical, Electronics, and Optimization Techniques (ICEEOT)*. IEEE, 2016, pp. 1583–1589.

[11] B. Singh and G. D. Chaturvedi, "Comparative performance of isolated forward and flyback ac-dc converters for low power applications," in *2008 Joint International Conference on Power System Technology and IEEE Power India Conference*. IEEE, 2008, pp. 1–6.

[12] L. Callegaro, M. Ciobotaru, D. J. Pagano, and J. E. Fletcher, "Feedback linearization control in photovoltaic module integrated converters," *IEEE Transactions on Power Electronics*, vol. 34, no. 7, pp. 6876–6889, 2018.

[13] S. Samsung, "Specification of product," South Korea, Tech. Rep, Tech. Rep., 2013. [Online]. Available: http://dalincom.ru/datasheet/SAMSUNG%20INR18650-25R.pdf

[14] Y.-N. Chang, Y.-S. Shen, H.-L. Cheng, and S.-Y. Chan, "The optimized capacity for lithium battery balance charging/discharging strategy," in *2014 IEEE 23rd International Symposium on Industrial Electronics (ISIE)*. IEEE, 2014, pp. 1842–1847.

[15] M. Evzelman, M. M. U. Rehman, K. Hathaway, R. Zane, D. Costinett, and D. Maksimovic, "Active balancing system for electric vehicles with incorporated low-voltage bus," *IEEE Transactions on Power Electronics*, vol. 31, no. 11, pp. 7887–7895, 2015.

[16] L. Maharjan, S. Inoue, H. Akagi, and J. Asakura, "State-of-charge (soc)-balancing control of a battery energy storage system based on a cascade pwm converter," *IEEE Transactions on Power Electronics*, vol. 24, no. 6, pp. 1628–1636, 2009.

[17] K. Smith, Y. Shi, and S. Santhanagopalan, "Degradation mechanisms and lifetime prediction for lithium-ion batteries—a control perspective," in *American Control Conference (ACC)*, 2015, pp. 728–730.

[18] G. Kaneko, S. Inoue, K. Taniguchi, T. Hirota, Y. Kamiya, Y. Daisho, and S. Inami, "Analysis of degradation mechanism of lithium iron phosphate battery," *World Electric Vehicle Journal*, vol. 6, no. 3, pp. 555–561, 2013.

978-1-7281-4181-7/19 $31.00 © 2019 IEEE

Energy Yield Assessment Methodology for Photovoltaic Microinverters

Andrii Chub
Power Electronics Group
Tallinn University of Technology
Tallinn, Estonia
andrii.chub@taltech.ee

Kosenko Roman
Ubik Solutions OÜ
Tallinn, Estonia
roman@ubiksolutions.eu

Oleksandr Korkh
Power Electronics Group
Tallinn University of Technology
Tallinn, Estonia
oleksandr.korkh@taltech.ee

Dmitri Vinnikov
Power Electronics Group
Tallinn University of Technology
Tallinn, Estonia
dmitri.vinnikov@taltech.ee

Samir Kouro
Electronics Engineering Department
Universidad Técnica Federico Santa María
Valparaiso, Chile
samir.kouro@usm.cl

Abstract—**Two-stage photovoltaic (PV) microinverters with a buck-boost front-end converter can provide unparalleled performance under shading conditions. However, their efficiency curve is not monotonic across the input voltage range. As a result, it is hard to predict energy production under known weather conditions. The proposed methodology utilizes interpolation of microinverter efficiency based on the experimental efficiency look-up table. Modeling of the maximum power point of a PV module based on the Lambert W function could be combined with the interpolated efficiency to estimate energy produced by a microinverter. The proposed methodology requires only two synchronized datasets, solar irradiance and ambient temperature, as the input variables.**

Keywords—*microinverter, photovoltaic, solar energy.*

I. INTRODUCTION

Many residential PV systems suffer from shading of PV modules [1]-[3]. Conventional string PV inverters have limited capability to withstand the shading [1]. The application of PV power optimizers ensures maximum energy harvest by a PV string inverter, but their applicability is limited by a minimum number of PV modules in a string [4]. Contrary to that, microinverters can integrate individual PV modules into the grid. Hence, they avoid the single point of failure and provide rapid module-level shutdown. Additionally, they avoid fire hazard from dc arcing, provide a possibility for PV module health monitoring and dust accumulation detection, flexible PV system design, etc. [1].

PV microinverters are a universal tool for designing small residential PV systems with superior scalability and reliability [5]. They are usually implemented either as single- or two-stage converters [6]. The former results in low cost and limited regulation range, while the latter provide improved input voltage regulation range at the cost of complexity. A two-stage PV microinverter (Fig. 1) contains a front-end dc-dc converter, which results in good decoupling of 100/120 Hz ripple along with the possibility to withstand

partial and opaque shading. The latter feature requires the application of a buck-boost galvanically isolated dc-dc converter [7]. Recently, several converters from that family were justified for PV applications [8]-[10]. They can include either separate buck and boost switching cells [7], [8], or a single integrated buck-boost switching cell [9], [10].

Usually, the buck-boost voltage regulation utilizes three possible operation modes: buck, boost, and pass-through. The latter is the most efficient mode when a converter does not regulate voltage and operates as a dc transformer. If the dc link voltage is fixed, the pass-through mode could be adjusted to the vicinity of the most probable maximum power point (MPP) as shown in Fig. 2. In this case, the buck mode is used during start-up and MPP tracking or in case of operation at low ambient temperatures. The boost mode is used at high ambient temperatures, when MPP voltage of a PV module is reduced, or in case of shading to optimize the MPP tracking. Also, the boost mode is used under partial shading when a global maximum power point appears at a lower voltage defined by the number of bypass diodes [2].

The main issue with the buck-boost front-end converters is their nonmonotonic efficiency dependence on the input voltage as the efficiency is the highest in the pass-through mode. As a result of relatively large efficiency variations, neither peak nor average weighted efficiency values can be used to predict energy production by a microinverter. Hence, a special methodology is proposed in this paper to estimate energy production by a two-stage microinverter with buck-boost front-end dc-dc converter. It requires, first, to measure the efficiency of a microinverter in a target range of voltage and power; second, perform surface fit of the experimental efficiency; third, identify modeling parameters for a target PV module based on the single-diode PV cell model and the Lambert function; finally, an annual experimental profile of solar irradiance and ambient temperature has to be applied to define annual energy yield. These steps are performed in sections II and III.

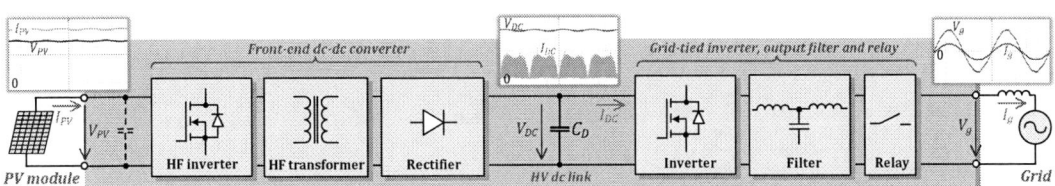

Fig. 1 Generalized structure of a two-stage PV microinverter.

978-1-7281-4181-7/19 $31.00 © 2019 IEEE

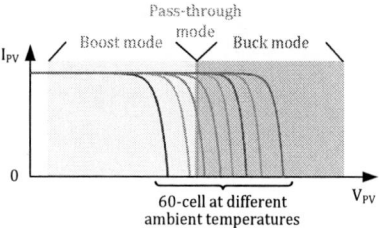

Fig. 2 Utilization of operating modes of a front-end galvanically isolated buck-boost converter.

II. CASE STUDY MICROINVERTER AND EXPERIMENTAL RESULTS

The case study microinverter was described in details in [11]. It is used merely as a case study system. Its topology is shown in Fig. 3. The front-end dc-dc converter implements the boost mode using PWM shoot-through duty cycle (D_{ST}) control by the symmetrical overlap of active states. In the pass-through mode, it operates as the series resonant converter with very high efficiency. It utilizes phase-shifted modulation (PSM) with angle φ to implement voltage buck using the integrated series resonant tank. Its experimental control variables are shown in Fig. 4a.

The efficiency of the experimental prototype was measured in the input voltage range of 25 V to 45 V and from 10% to 100% of the rated power of 300 W. Over 50 reference points were acquired using precision power analyzer Yokogawa WT1800. Next, the experimental efficiency surfaces were obtained in MATLAB using the surface fitting tool and thin-plate spline for interpolation. The contour plots of the experimental data interpolation shown in Fig. 4b follow tightly the experimental data, which results in the coefficient of determination (R^2) equal to unity. These data will be used further for annual performance prediction.

It could be appreciated from Fig. 4 that the fixed dc link voltage control results in a limited region of high efficiency where the front-end dc-dc converter operates in or close to the pass-through mode.

TABLE I. PROTOTYPE PARAMETERS AND COMPONENTS

Parameter or component	Symbol or Block		Type or Value
Transformer	TX	n	6
		L_{lk}	25 μH
		L_m	1000 μH
qZS inductor	L_{qZS}	L_{lkqZS}	0.5 μH
		L_{mqZS}	11 μH
qZS capacitors	C_{qZS1}, C_{qZS2}		12 × 2.2 μF (X7R/100V)
Resonant capacitors	C_1, C_2		43 nF (foil)
Dc link capacitor	C_D		150 μF (Al electrolytic)
LCL filter	L_{f1}		2.6 mH
	L_{f2}		1.8 mH
	C_f		0.47 μF
Dc link voltage	V_{DC}		400 V
Grid	V_g		230 V / 50 Hz
Switching frequency	S_{qZS}		210 kHz
	$S_1...S_4$		105 kHz
	S_5, S_6		50 Hz
	S_7, S_8		20 kHz
LV semiconductors	$S_{qZS}, S_1...S_4$		Infineon BSC035N10NS5
HV semiconductors	D_1, D_2		Wolfspeed C3D02060E
	S_5, S_6		Infineon IPB60R099C7
	S_7, S_8		ROHM SCT2120AFC

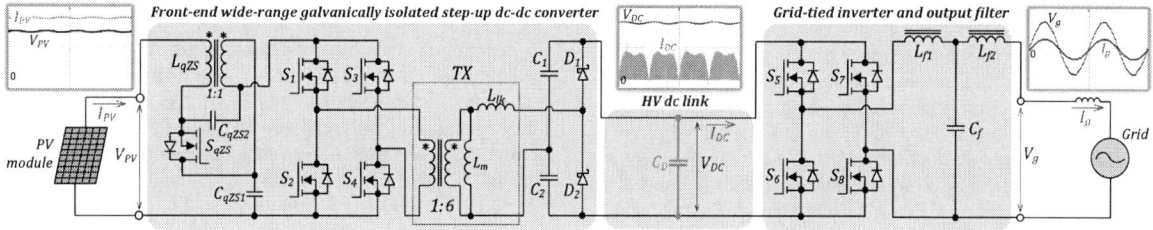

Fig. 3 The topology of the case-study shade-tolerant microinverter [11].

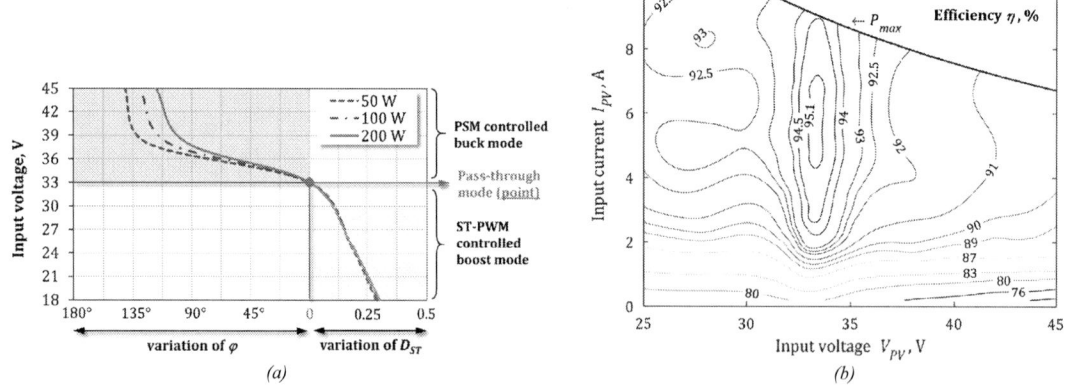

Fig. 4 Control variables of the front-end converter (a) and approximation of the experimental efficiency (b) of the microinverter [11].

III. ESTIMATION OF ANNUAL PV ENERGY PRODUCTION

The interpolations of the experimental efficiency (Fig. 4b) can be used for estimation of the annual energy production and energy loss. The efficiency includes microinverter self-powering from the input PV power. The developed methodology is explained in Fig. 5. It utilizes 1-year mission profiles of the solar irradiance and the ambient temperature for Arizona (Ar.) [12] with 1-minute resolution.

First, the ambient temperature (T_a) is translated into the cell temperature (T_c) using a simple linear model with a low-pass filter to account for thermal dynamics of a PV module:

$$T_c = \left[T_a + \frac{T_{NOCT} - T_{a(NOCT)}}{S_{NOCT}} \cdot S \right] \cdot G_{LPF}(s), \quad (1)$$

where the nomenclature corresponds to Table II. G_{LPF} is the low pass filter transfer function in s-domain:

$$G_{LPF}(s) = \frac{1}{1 + s \cdot \tau}. \quad (2)$$

Second, the PV module has to be simulated using the typical single-diode model of a PV cell (Fig. 6) that requires five parameters as described in [13]-[15]:

$$v_t = \frac{T_c \cdot k_b}{q}, \quad (3)$$

$$I_0 = I_{0(STC)} \cdot \left(\frac{T_c}{T_{STC}} \right)^3 \cdot e^{\left(\frac{q \cdot E_g}{a \cdot k_b} \right) \left(\frac{1}{T_{STC}} - \frac{1}{T_c} \right)}, \quad (4)$$

$$R_s = R_{s(STC)} \cdot \left(1 + k_{rs} \left(T_c - T_{STC} \right) \right), \quad (5)$$

$$I_{Ph} = I_{SC(STC)} \cdot \frac{S}{S_{STC}} \cdot \frac{R_s + R_{sh}}{R_{sh}}, \quad (6)$$

$$w = W_0 \left(\frac{I_{ph} \cdot e^1}{I_0} \right), \quad (7)$$

$$V_{MPP(id)} = Ns \cdot a \cdot v_t \cdot (w - 1), \quad (8)$$

$$I_{MPP(id)} = I_0 \cdot (w - 1) \cdot e^{(w-1)}, \quad (9)$$

$$I_{MPP} = I_{MPP(id)} - \frac{V_{MPP(id)}}{R_{sh}}, \quad (10)$$

$$V_{MPP} = V_{MPP(id)} - I_{MPP} \cdot R_s, \quad (11)$$

$$P_{MPP} = V_{MPP} \cdot I_{MPP}. \quad (12)$$

The given model differs from the conventional one in the approach of simulating series resistance R_s. The experimental results show the positive thermal dependence of R_s [17]. The PV module Jinko Solar JKM300M-60B used in modeling is based on the monocrystalline silicon technology and contains 60 PV cells. For this type of modules, thermal coefficient of R_s (k_{rs}) was defined experimentally in [17].

Third, it is possible to define MPP power and voltage for any solar irradiance and cell temperature. Considering interpolated efficiency of the microinverter η, and assuming $V_{PV} = V_{MPP}$ and $P_{PV} = P_{MPP}$, it is possible to define the

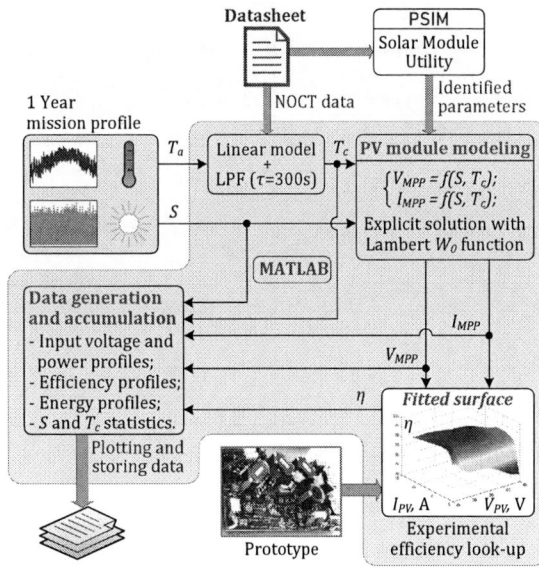

Fig. 5 Flow diagram of the annual energy production estimation.

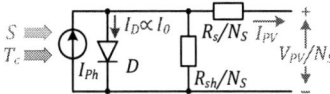

Fig. 6 Single-diode model of a PV cell.

microinverter output power P_g and, consequently, energy production E during an arbitrary time interval t_x:

$$P_g(t) = V_{MPP}(t) \cdot I_{MPP}(t) \cdot \eta \left(V_{MPP}(t); I_{MPP}(t) \right), \quad (13)$$

$$E = \int_0^{t_x} P_g(t) dt. \quad (14)$$

TABLE II. NOMENCLATURE

k_b	Boltzmann constant	v_t	Thermal voltage
q	Elementary charge	R_s	Series resistance
a	Ideality factor	$R_{s(STC)}$	R_s in STC conditions
E_g	Bandgap energy	k_{rs}	R_s thermal coefficient
T_a	Ambient temperature	R_{sh}	Shunt resistor
T_c	Cell temperature	$I_{0(STC)}$	I_0 in STC conditions
S	Solar irradiance	I_0	Saturation current
T_{STC}	T_c in STC conditions	I_{Ph}	Photocurrent
S_{STC}	S in STC conditions	N_S	Number of cells
$T_{a(NOCT)}$	T_a in NOCT conditions	η	Microinv. efficiency
S_{NOCT}	Solar irradiance in NOCT condition	P_g	Microinverter output power
T_{NOCT}	Nominal operating cell temperature (NOCT)	$I_{SC(STC)}$	Short-circuit current of PV module in STC conditions
$V_{MPP(id)}$	MPP voltage of an ideal PV module	τ	Thermal time constant of PV module
$I_{MPP(id)}$	MPP current of ideal PV module	$W_0(\cdot)$	Principle branch of Lambert W function
I_{PV}	Microinv. input current	V_{MPP}	Module MPP voltage
V_{PV}	Microinv. input voltage	I_{MPP}	Module MPP current
P_{PV}	Microinv. input power	P_{MPP}	Module MPP power

TABLE III.   PV MODULE PARAMETERS

Model parameters		Reference data and constants	
Parameter	*Value*	*Parameter*	*Value*
S_{STC}	1000 W/m²	$V_{MPP(STC)}$	32.6 V
T_{STC}	25 °C	$I_{MPP(STC)}$	9.21 A
S_{NOCT}	800 W/m²	$V_{OC(STC)}$	40.1 V
T_{NOCT}	45 °C	$I_{SC(STC)}$	9.72 A
$T_{a(NOCT)}$	20 °C	$V_{MPP(NOCT)}$	30.6 V
a [17]	1.1	$I_{MPP(NOCT)}$	7.32 A
E_g [16]	1.12 eV	$V_{OC(NOCT)}$	37.0 V
$I_{0(STC)}$	5.39·10⁻¹⁰ A	$I_{SC(NOCT)}$	8.01 A
R_{sh}	750 Ω	*Efficiency*	17.98%
$R_{s(STC)}$	0.228 Ω	N_s	60
k_{rs} [17]	0.356 %	k_b	1.38·10⁻²³ J/K
τ [18]	300 s	q	1.602·10⁻¹⁹ C

Table III presents parameters of the solar module obtained from the datasheet, PSIM Solar module utility, or literature (for fundamental and empiric constants).

The considered geographic location can be characterized as mostly sunny, hot, and dry place. Distribution of solar irradiance has a higher expectation around 1000 W/m² as shown in Fig. 7. As a result of high solar irradiance and ambient temperature, the PV module will experience relatively high thermal stress in these conditions: mostly between 10 °C and 75 °C, as shown in Fig. 8.

Next, annual mission profiles of solar irradiance and cell temperature are translated into annual profiles of the MPP voltage and power using the model of the PV module described above. Probability distribution functions of these physical quantities are shown in Figs. 9 and 10. High cell temperature resulting from the high solar irradiance causes the PV module to operate out of the range of high efficiency of the case study microinverters.

Annual mission profiles of the MPP voltage and power can be applied to the efficiency look-up block containing interpolations of the experimental efficiency with thin-plate splines. The probability distribution of the efficiency of the microinverter strongly depends on the geographic location. In the considered conditions corresponding to the state of Arizona, high average MPP power and relatively low MPP voltage result in the microinverter efficiency being distributed mostly in the narrow range between 92 % and 93.5 %, as follows from Fig. 11. The values of efficiency can occasionally reach as low as 77 % and as high as 95.2 %. The resulting average efficiency equals 91.4 % calculated for one year and not considering the nights. The relatively low efficiency could be explained with consideration of microinverter self-powering from the input PV port, which is overlooked often in similar studies.

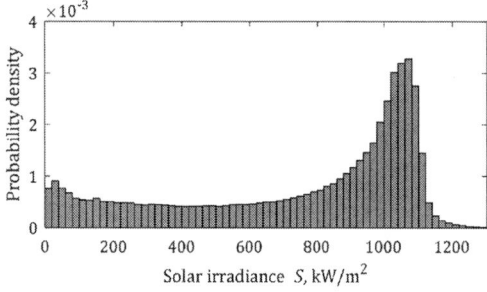

Fig. 7 Probability density distribution of the solar irradiances.

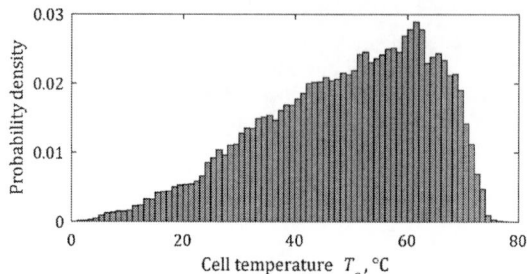

Fig. 8 Probability density distribution of the cell temperatures.

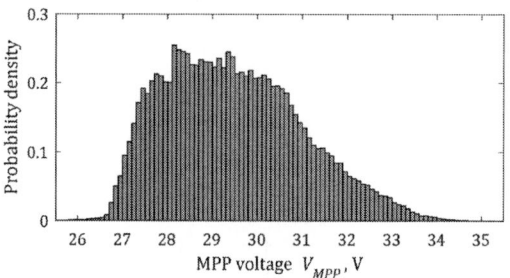

Fig. 9 Probability density distribution of the MPP voltages.

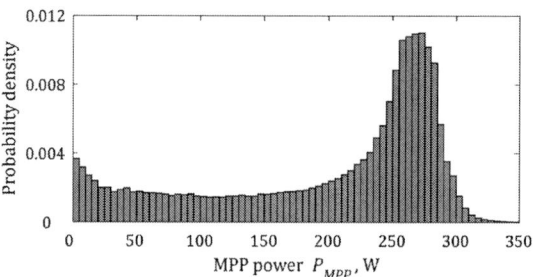

Fig. 10 Probability density distribution of the MPP powers.

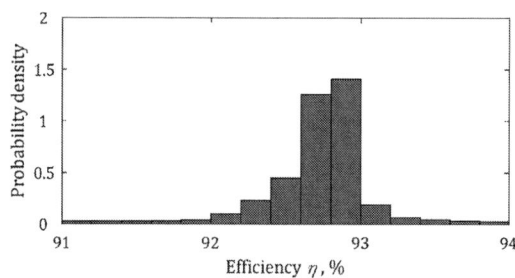

Fig. 11 Probability density distribution of the microinverter efficiencies.

Average values of the annual microinverter efficiency are not enough to reach a conclusion regarding annual energy production as the efficiency values have to be weighted with the input power values. In the considered conditions, the case study PV module can deliver the energy of 861.4 kW·h per year. In this case, the energy loss of 63.4 kW·h per year was predicted. This value corresponds to the relative energy loss of 7.4 %, i.e., the average of 92.6 % from the available PV power was injected into the grid. The estimated value of the relative annual energy yield is by 1.2 % higher than the value of the average converter efficiency estimated at the level of 91.4 %. This estimate proves the assumption that none of the peak efficiency, CEC or EU weighted efficiencies, or average estimated annual efficiency values can predict correctly the energy production. Hence, the need for a new methodology proposed in this paper is justified.

IV. SUMMARY AND CONCLUSIONS

Experimental study of the two-stage PV microinverter with buck-boost front-end dc-dc converter and stable intermediate dc link was performed. The study shows that the microinverter efficiency experience relatively large variation around the point of the optimal efficiency where the front-end dc-dc converter operated in the pass-through mode. Range of the most probable maximum power points of the case study PV module coincides with the range of considerable efficiency variations. Hence, it is important to take into account this effect when estimating annual energy productions by the microinverter.

The paper proposes a simple methodology for quantifying annual energy production by a microinverter based on experimental mission profiles of the solar irradiance and ambient temperatures, PV module parameters identified from a datasheet using PSIM Solar module utility, and experimental efficiency interpolation based on measured values. The methodology is using a single-diode five-parameter model of a PV cell enhanced with a thermally dependent model of the series resistor. The model of a PV module utilizes the Lambert W function to predict instantaneous voltage and current in the maximum power point. Assuming that the microinverter performs perfect tracking of the maximum power point, the calculated values of the current and voltage could be used to define its efficiency from interpolation of the experimental data. Hence, the instantaneous power injected into the grid can be calculated. Integrating this power, energy production could be found for the required period.

The proposed methodology was applied to the case study microinverter fed from Si 60-cell PV module. The microinverter efficiency varies between 77 % and 95 % over the modeled year. The case study PV module is found to be capable of generating 861.4 kW·h per year in the considered Arizona climatic conditions. Then, the case study microinverter will deliver 798 kW·h into the distribution grid per year. Hence, the annual energy loss of 7.4% is expected in the considered conditions.

ACKNOWLEDGMENT

This research was supported in part by the Estonian Centre of Excellence in Zero Energy and Resource Efficient Smart Buildings and Districts, ZEBE, grant 2014-2020.4.01.15-0016 funded by the European Regional Development Fund, in part by SERC Chile (CONICYT/FONDAP/15110019), and in part by AC3E (CONICYT/BASAL/FB0008).

Authors gratefully acknowledge the Department of Energy Technology of Aalborg University, Denmark, for providing access to the high-resolution experimental data of solar irradiances and corresponding ambient temperatures.

REFERENCES

[1] S. Kouro, J. I. Leon, D. Vinnikov and L. G. Franquelo, "Grid-Connected Photovoltaic Systems: An Overview of Recent Research and Emerging PV Converter Technology," *IEEE Ind. Electron. Mag.*, vol. 9, no. 1, pp. 47-61, March 2015.

[2] K. Sinapis et al., "A comprehensive study on partial shading response of c-Si modules and yield modeling of string inverter and module level power electronics," *Solar Energy*, vol. 135, pp. 731–741, 2016.

[3] K. Sinapis, G. Litjens, M. van den Donker, W. Folkerts, and W. van Sark, "Outdoor characterization and comparison of string and MLPE under clear and partially shaded conditions," *Energy Science & Engineering*, vol. 3, no. 6, pp. 510–519, Oct. 2015.

[4] M. Kasper, D. Bortis and J. W. Kolar, "Classification and Comparative Evaluation of PV Panel-Integrated DC–DC Converter Concepts," *IEEE Trans. Power Electron.*, vol. 29, no. 5, pp. 2511-2526, May 2014.

[5] R. Hasan, S. Mekhilef, M. Seyedmahmoudian, and B. Horan, "Grid-connected isolated PV microinverters: A review," *Renewable and Sustainable Energy Reviews*, vol. 67, pp. 1065–1080, Jan. 2017.

[6] K. Alluhaybi, I. Batarseh, H. Hu and X. Chen, "Comprehensive Review and Comparison of Single-Phase Grid-Tied Photovoltaic Microinverters," *IEEE Journal of Emerging and Selected Topics in Power Electronics*, in press. DOI: 10.1109/JESTPE.2019.2900413

[7] C. Yao, X. Ruan, X. Wang and C. K. Tse, "Isolated Buck–Boost DC/DC Converters Suitable for Wide Input-Voltage Range," *IEEE Trans. Power Electron.*, vol. 26, no. 9, pp. 2599-2613, Sept. 2011.

[8] T. LaBella, W. Yu, J. S. Lai, M. Senesky and D. Anderson, "A Bidirectional-Switch-Based Wide-Input Range High-Efficiency Isolated Resonant Converter for Photovoltaic Applications," *IEEE Trans. Power Electron.*, vol. 29, no. 7, pp. 3473-3484, July 2014.

[9] D. Vinnikov, A. Chub, E. Liivik and I. Roasto, "High-Performance Quasi-Z-Source Series Resonant DC–DC Converter for Photovoltaic Module-Level Power Electronics Applications," *IEEE Trans. Power Electron.*, vol. 32, no. 5, pp. 3634-3650, May 2017.

[10] A. Chub, D. Vinnikov, R. Kosenko and E. Liivik, "Wide Input Voltage Range Photovoltaic Microconverter With Reconfigurable Buck–Boost Switching Stage," *IEEE Trans. Ind. Electron.*, vol. 64, no. 7, pp. 5974-5983, July 2017.

[11] D. Vinnikov, A. Chub, E. Liivik, R. Kosenko and O. Korkh, "Solar Optiverter—A Novel Hybrid Approach to the Photovoltaic Module Level Power Electronics," *IEEE Trans. Ind. Electron.*, vol. 66, no. 5, pp. 3869-3880, May 2019.

[12] A. Sangwongwanich, Y. Yang, D. Sera and F. Blaabjerg, "Lifetime Evaluation of Grid-Connected PV Inverters Considering Panel Degradation Rates and Installation Sites," *IEEE Trans. Power Electron.*, vol. 33, no. 2, pp. 1225-1236, Feb. 2018.

[13] G. Farivar, B. Asaei and S. Mehrnami, "An Analytical Solution for Tracking Photovoltaic Module MPP," *IEEE J. Photovoltaics*, vol. 3, no. 3, pp. 1053-1061, July 2013.

[14] A. Kuperman, "Comments on "An Analytical Solution for Tracking Photovoltaic Module MPP"," *IEEE J. Photovoltaics*, vol. 4, no. 2, pp. 734-735, March 2014.

[15] S. Lineykin, M. Averbukh, and A. Kuperman, "An improved approach to extract the single-diode equivalent circuit parameters of a photovoltaic cell/panel," *Renewable and Sustainable Energy Reviews*, vol. 30, pp. 282–289, Feb. 2014.

[16] W. De Soto, S. A. Klein, and W. A. Beckman, "Improvement and validation of a model for photovoltaic array performance," *Solar Energy*, vol. 80, no. 1, pp. 78–88, Jan. 2006.

[17] K. Lee, "Improving the PV Module Single-Diode Model Accuracy with Temperature Dependence of the Series Resistance," *Proc. PVSC'2017*, Washington, DC, 2017, pp. 1526-1530.

[18] S. Armstrong and W. G. Hurley, "A thermal model for photovoltaic panels under varying atmospheric conditions," *Applied Thermal Engineering*, vol. 30, no. 11–12, pp. 1488–1495, Aug. 2010.

Direct Power Control Strategy to Enhance the Dynamic Behavior of DFIG during Voltage Dip

1st Fernando Lino
UFABC/CECS
Universidade Federal do ABC
Santo André - SP, Brazil
flino@ifsp.edu.br

2nd Rogério Vani Jacomini
IFSP
Instituto Federal de São Paulo
Hortolândia - SP, Brazil
jacomini@ifsp.edu.br

3rd Alfeu J. Sguarezi Filho, Senior Member
UFABC/CECS
Universidade Federal do ABC
Santo André - SP, Brazil
alfeu.sguarezi@ufabc.edu.br

Abstract—In this paper is proposed a direct power control strategy based on a rotor voltage equation and that does not ignore the flux transient and the stator voltage that will reduce peaks rotor current during the voltage dips. When there occurs a voltage sag at the stator of DFIG is produced high transients current in the stator and rotor that may damage the rotor-side converter. Given that the new grid code rules state that a wind turbine generator must remain connected to the grid, while still maintaining the supply voltage with the supply of additional reactive power during voltage drops, makes the power converter vulnerable to voltage drops due to high spike currents. To reduce peaks stator and rotor current, the dynamics of the active and reactive stator power is improve from the control of rotor voltages using the stator-voltage-oriented reference frame. To verify the good performance and validate the control strategy propose, simulation results are presented.

Index Terms—Doubly fed induction generation, Stator-voltage-oriented, Direct power control, Voltage Dip, Wind power.

I. INTRODUCTION

In the past, bidirectional power electronic converters (normally a back-to-back converter) protection has been used to prevent the shutdown of wind turbine generators (WTGs) from system against voltage drops that would occur at induction generator power terminals. Thereby, high current transients were produced in the stator and rotor windings due to voltage sags [1]–[3], [6]. These high current transients produces rotor currents and dc-link voltage that exceed the values that the converter can support. Currently, the new grid codes regulatory [4] impose a requirement known as low voltage ride through capability (LVRT), specify that the WTGs must remain connected to the grid when a severe mains voltage drop occurs in one or all phases. In addition, the grid regulatory codes require that the voltage must be proportionally withheld by the provided active power by WTGs and also that the reactive power be maximized to keep the WTG working below their limits during voltage sags [4]. The solution provided by many researchers to meet these requirements proposes LVRT strategies using protection with hardware devices or even using strategies PI control to protect the power electronic converter with the wind turbines that uses the DFIG (doubly-fed induction generator).

A proposed hardware very used solution by the researchers uses a so called crowbars (a set of resistors connected to the rotor windings [1], [3]), bypassing the high rotor currents, limiting in this way, their values. However, the disadvantage of this solution is that when the crowbars are triggered, the DFIG becomes as a squirrel cage induction generator, with the loss of the reactive power, when be required a high amount of reactive power. Another one is related to the extra cost in the overall installation [7].

To avoid the crowbars usage, several control strategies from the rotor converter side have been proposed with the objective to keep the rotor currents and the voltage in the dc-link below the limits supported by the converter during balanced voltage sags. In [8] is proposed a strategy for the control that uses cascade PI controllers for the control of the active and reactive stator power during operation without voltage sag. The rotor currents references are given for external loops power control that use PIs and, when voltage sags occur, the external power control loops get disconnected, and the feedback of the stator currents are the new references. To provide the LVRT capability to control the currents transients schema for the rotor from the converter side, it is proposed in [9] a feed-forward control. In that paper, the stator voltage transients terms are obtained and the feed-forward control is injected into the rotor currents from de converter side controller, according to the transient model of the DFIG. Simulation results shown that the rotor currents transients are reduced during a severe three-phase voltage sag. In [10] is proposed a control scheme that combines PI and hysteresis controllers: the PI controllers regulate the rotor currents under normal operation and the hysteresis controllers govern the rotor currents when the rotor currents or the dc-link voltage exceed their limits. In [5] is proposed a robust controller to ensure the stability of the DFIG under disturbances, particularly under conditions of balanced drop voltage and the results show a fast dynamic behavior; however, it results in undesirable actions in the switching control due the discontinuity problem. The strategy proposed in [7] that uses PI controllers to control the magnetizing current aiming to reduce oscillation in natural component of stator flux during balanced voltage sags, results in high values of rotor currents. It had been found in the experimental results that oscillations have been reduced in the currents of the rotor and stator, although not significantly.

With the generator being subject to voltage sags, this

978-1-7281-4181-7/19 $31.00 © 2019 IEEE

paper proposes a strategy of direct power control for the DFIG that provides a fast dynamic response and also without oscillations in the stator power. Unlike what is found in the technical literature, the rotor dynamic equations are used in the control algorithm. In this case, during voltage sags, are admitted variants in the magnitude of stator flux and stator voltage, being that the control variables are the direct and quadrature components of the rotor voltage. To verify the good performance and to validate the proposed strategy control, simulation results for a 2.2 kW generator in operation under 75% balanced voltage sag are presented, when a comparison is made between the proposed control strategy and the strategy using PI controllers and DPC, both without the consideration of the transient flux and the stator voltage.

II. DOUBLY FED INDUCTION GENERATION MODEL

The DFIG mathematical model by using stator-voltage-oriented (SVO) is used to analyze the dynamical behavior under balanced dips. Therefore the voltage and flux equations of the stator and rotor are written by the following set of equations:

The following set of equations represents the voltage and flux vectors of the stator and rotor :

$$\overline{v}_{s,dq} = jv_{sq} = r_s\overline{i}_{s,dq} + \frac{d\overline{\psi}_{s,dq}}{dt} + j\omega_1\overline{\psi}_{s,dq} \quad (1)$$

$$\overline{v}_{r,dq} = r_r\overline{i}_{r,dq} + \frac{d\overline{\psi}_{r,dq}}{dt} + j\omega_2\overline{\psi}_{r,dq} \quad (2)$$

$$\overline{\psi}_{s,dq} = L_s\overline{i}_{s,dq} + L_m\overline{i}_{r,dq} \quad (3)$$

$$\overline{\psi}_{r,dq} = L_r\overline{i}_{r,dq} + L_m\overline{i}_{s,dq} \quad (4)$$

where: r is the resistance; L is the self inductance; L_m is the magnetizing inductance; σ is the total leakage factor; ω is the angular frequency; v, i, ψ is the voltage, current and flux components, respectively. The subscripts: s, r are stator and rotor; d, q are direct- and quadrature-axis expressed in the synchronous reference frame; $1, 2$ are indices associated to the synchronous and slip speeds.

A. Effects of abrupt changes in the three-phase stator voltages

Preliminarily it is necessary to show how the d- and q-axis stator flux components react to abrupt changes in the stator voltages, when the DFIG is initially operating normally in the steady-state. To study stator dynamics, with the rotor open ($i_r = 0$), we substitute equation (3) into equation (1) getting:

$$\frac{d\psi_{sd}}{dt} = -\frac{1}{\tau_s}\psi_{sd} + \omega_1\psi_{sq} \quad (5)$$

$$\frac{d\psi_{sq}}{dt} = -\omega_1\psi_{sd} - \frac{1}{\tau_s}\psi_{sq} + v_{sq} \quad (6)$$

where $\tau_s = L_s/r_s$ is the stator time constant.

The analytical solution for an abrupt change in Δv_{sq}, assuming that initial conditions of the fluxes and their derivatives be equal to zero, is given by the pair:

$$\psi_{sd} \approx \frac{\Delta v_{sq}}{\omega_1}\left[1 - e^{-t/\tau_s}\cos(\omega_1 t)\right] \quad (7)$$

$$\psi_{sq} \approx \frac{\Delta v_{sq}}{\omega_1} e^{-t/\tau_s}\sin(\omega_1 t) \quad (8)$$

The two flux components present an oscillation of high amplitude at the beginning of transient state and, in the steady-state, it decrease to zero, in accordance with the time constant τ_s of the exponential envelope, as it is showed in Figure 1, to a decrease of $v_{sq} = 0.85$ pu at $t = 1$ s and with a subsequent increase of $v_{sq} = 0.85$ pu at $t = 1.2$ s. The DFIG nominal data used are in the Appendix.

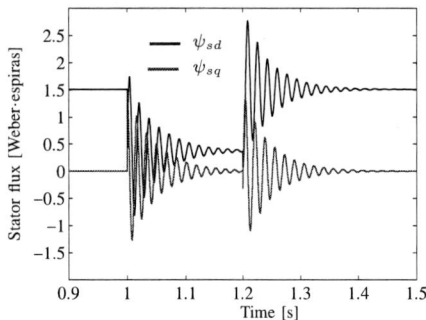

Fig. 1. Stator flux components during voltage dip.

Because of these large amplitudes that occur in the d- and q-axis stator flux components during three-phase stator voltages abrupt changes, it is not convenient to ignore it, i.e, it is required to consider in the machine model the changes in ψ_{sd} and ψ_{sq}.

III. DIRECT POWER CONTROL STRATEGY

In order to get improvement at the dynamic response of stator power is proposed direct power control. The control strategy is based on the rotor dynamic equation, where the direct and quadrature components of the rotor voltage is the controlled variables. In this equation, the stator voltage and the magnitude of stator flux change their values during the voltage dip.

- **Stator reactive and active power**

The instantaneous active and reactive power are computed in terms of the current and voltage space vector as:

$$P_s = \frac{3}{2}\Re e(\overline{V}_{s,dq} \cdot \overline{i}_{s,dq}^*) \quad (9)$$

$$Q_s = \frac{3}{2}\Im m(\overline{V}_{s,dq} \cdot \overline{i}_{s,dq}^*) \quad (10)$$

where (*) means the complex conjugate.

According to equations (3) and (4), the stator current can be computed as:

$$\overline{i}_{s,dq} = \frac{\overline{\psi}_{s,dq}}{\sigma L_s} - \frac{k_r}{\sigma L_s}\overline{\psi}_{r,dq} \quad (11)$$

where $\sigma = (L_sL_r - L_m^2)/L_sL_r$ is the total leakage factor.

By the substitution of equations (1) and (11) into equations (9) and (10), it follows that:

$$P_s = -k_\sigma v_{sq}\left(\psi_{rq} - \frac{\psi_{sq}}{k_r}\right) \quad (12)$$

$$Q_s = -k_\sigma v_{sq}\left(\psi_{rd} - \frac{\psi_{sd}}{k_r}\right) \qquad (13)$$

where $k_s = L_m/L_s$ and $k_r = L_m/L_r$ are the coupling factors of the stator and rotor, respectively, and $k_\sigma = 1.5 k_r/\sigma L_s$.

- *Rotor voltage equations*

According to equations (3) and (4), the rotor current equations in the direct- and quadrature-axis can be expressed as:

$$\overline{i}_{r,dq} = \frac{\overline{\psi}_{r,dq}}{\sigma L_r} - \frac{k_s}{\sigma L_r}\overline{\psi}_{s,dq} \qquad (14)$$

Substituting into equation (2), the rotor voltage equations in the direct- and quadrature-axis can be expressed as

$$v_{rd} = \frac{d\psi_{rd}}{dt} - \omega_2\psi_{rq} + \frac{r_r}{\sigma L_r}\psi_{rd} - \frac{r_r k_s}{\sigma L_r}\psi_{sd} \qquad (15)$$

$$v_{rq} = \frac{d\psi_{rq}}{dt} + \omega_2\psi_{rd} + \frac{r_r}{\sigma L_r}\psi_{rq} - \frac{r_r k_s}{\sigma L_r}\psi_{sq} \qquad (16)$$

Rewriting the equations (12) and (13):

$$\psi_{rq} = \frac{\psi_{sq}}{k_r} - \frac{P_s}{k_\sigma v_{sq}} \qquad (17)$$

$$\psi_{rd} = \frac{\psi_{sd}}{k_r} - \frac{Q_s}{k_\sigma v_{sq}} \qquad (18)$$

and by substituting them into equations (15) and (16) yields:

$$v_{rd} = -\frac{1}{k_\sigma}\frac{d(Q_s/v_{sq})}{dt} + \frac{1}{k_r}\frac{d\psi_{sd}}{dt} + \frac{\omega_2}{k_\sigma v_{sq}}P_s$$
$$- \frac{k_1}{v_{sq}}Q_s + k_2\psi_{sd} - \frac{\omega_2}{k_r}\psi_{sq} \qquad (19)$$

$$v_{rq} = -\frac{1}{k_\sigma}\frac{d(P_s/v_{sq})}{dt} + \frac{1}{k_r}\frac{d\psi_{sq}}{dt} - \frac{k_1}{v_{sq}}P_s$$
$$- \frac{\omega_2}{k_\sigma v_{sq}}Q_s + \frac{\omega_2}{k_r}\psi_{sd} + k_2\psi_{sq} \qquad (20)$$

where $k_1 = r_r/(\sigma L_r k_\sigma)$ and $k_2 = r_r(1 - k_s k_r)/(\sigma L_r k_r)$.

As it was previously written, when voltage dips occur, the instantaneous value stator flux components are not constants, as can be noted in Figure 1, which means the derivatives of the components $d\psi_{sd}/dt \neq 0$ and $d\psi_{sq}/dt \neq 0$. As the instantaneous value of the q- axis stator voltage is also not constant, the derivatives are $d(Q_s/v_{sq})/dt \neq (1/v_{sq})dQ_s/dt$ and $d(P_s/v_{sq})/dt \neq (1/v_{sq})dP_s/dt$. Then, by the application of the chain differentiation rule, these derivatives can be expanded as follows:

$$\frac{d(Q_s/v_{sq})}{dt} = \frac{\partial(Q_s/v_{sq})}{\partial Q_s}\frac{dQ_s}{dt} + \frac{\partial(Q_s/v_{sq})}{\partial v_{sq}}\frac{dv_{sq}}{dt}$$
$$= \frac{1}{v_{sq}}\frac{dQ_s}{dt} - \frac{Q_s}{v_{sq}^2}\frac{dv_{sq}}{dt} \qquad (21)$$

$$\frac{d(P_s/v_{sq})}{dt} = \frac{\partial(P_s/v_{sq})}{\partial P_s}\frac{dP_s}{dt} + \frac{\partial(P_s/v_{sq})}{\partial v_{sq}}\frac{dv_{sq}}{dt}$$
$$= \frac{1}{v_{sq}}\frac{dP_s}{dt} - \frac{P_s}{v_{sq}^2}\frac{dv_{sq}}{dt} \qquad (22)$$

Substituting these equations into the rotor voltage equations (19) and (20), we obtain:

$$v_{rd} = \underbrace{-\frac{1}{k_\sigma v_{sq}}\frac{dQ_s}{dt} + \frac{\omega_2}{k_\sigma v_{sq}}P_s - \frac{k_1}{v_{sq}}Q_s + k_2\psi_{sd}}_{\text{I}}$$
$$+ \underbrace{\frac{Q_s}{k_\sigma v_{sq}^2}\frac{dv_{sq}}{dt} + \frac{1}{k_r}\frac{d\psi_{sd}}{dt} - \frac{\omega_2}{k_r}\psi_{sq}}_{\text{II}} \qquad (23)$$

$$v_{rq} = \underbrace{-\frac{1}{k_\sigma v_{sq}}\frac{dP_s}{dt} - \frac{k_1}{v_{sq}}P_s - \frac{\omega_2}{k_\sigma v_{sq}}Q_s + \frac{\omega_2}{k_r}\psi_{sd}}_{\text{I}}$$
$$- \underbrace{\frac{P_s}{k_\sigma v_{sq}^2}\frac{dv_{sq}}{dt} + \frac{1}{k_r}\frac{d\psi_{sq}}{dt} + k_2\psi_{sq}}_{\text{II}} \qquad (24)$$

where the first four terms of equations (23) and (24) represent the normal operation of DFIG in the grid and the last three terms of these equations represent the influence of the disturbance Δv_{sq} at stator terminals, as it is, for instance, the voltages sag.

A. Control algorithm

The control algorithm is developed by discretizing the components of the rotor equations voltage (23) and (24) for a sampling period T_s, the discrete system is given by the following equations:

$$\overline{v}_r^*(k) = \overline{f}_1(k)\Delta\overline{S}_s(k) + \overline{f}_2(k)\overline{S}_s^*(k) + \overline{f}_3(k)\widehat{\overline{\psi}}_s(k)$$
$$+ \overline{f}_4\Delta\overline{\psi}_s(k) + \overline{f}_5(k)\widehat{\overline{S}}_s(k)\Delta v_{sq}(k) \qquad (25)$$

$$\overline{f}_1(k) = -\frac{1}{k_\sigma T_s}\begin{bmatrix} \frac{1}{\widehat{v}_{sq}(k)} & 0 \\ 0 & \frac{1}{\widehat{v}_{sq}(k)} \end{bmatrix}, \qquad (26)$$

$$\overline{f}_2(k) = \begin{bmatrix} -\frac{k_1}{\widehat{v}_{sq}(k)} & \frac{\widehat{\omega}_2(k)}{k_\sigma \widehat{v}_{sq}(k)} \\ -\frac{\widehat{\omega}_2(k)}{k_\sigma \widehat{v}_{sq}(k)} & -\frac{k_1}{\widehat{v}_{sq}(k)} \end{bmatrix}, \qquad (27)$$

$$\overline{f}_3(k) = \begin{bmatrix} k_2 & -\frac{\widehat{\omega}_2(k)}{k_r} \\ \frac{\widehat{\omega}_2(k)}{k_r} & k_2 \end{bmatrix}, \qquad (28)$$

$$\overline{f}_4 = \begin{bmatrix} G_2 & 0 \\ 0 & G_2 \end{bmatrix}, \qquad (29)$$

and

$$\overline{f}_5(k) = \begin{bmatrix} \frac{G_3}{\widehat{v}_{sq,pu}^2(k)} & 0 \\ 0 & \frac{G_3}{\widehat{v}_{sq,pu}^2(k)} \end{bmatrix} \qquad (30)$$

The controller gains required for control are given by:

$$G_1 = \frac{1}{k_\sigma T_s V_{s,n}}, \qquad (31)$$

$$G_2 = \frac{1}{k_r T_s} \qquad (32)$$

and

$$G_3 = \frac{1}{k_\sigma T_s V_{s,n}^2} \qquad (33)$$

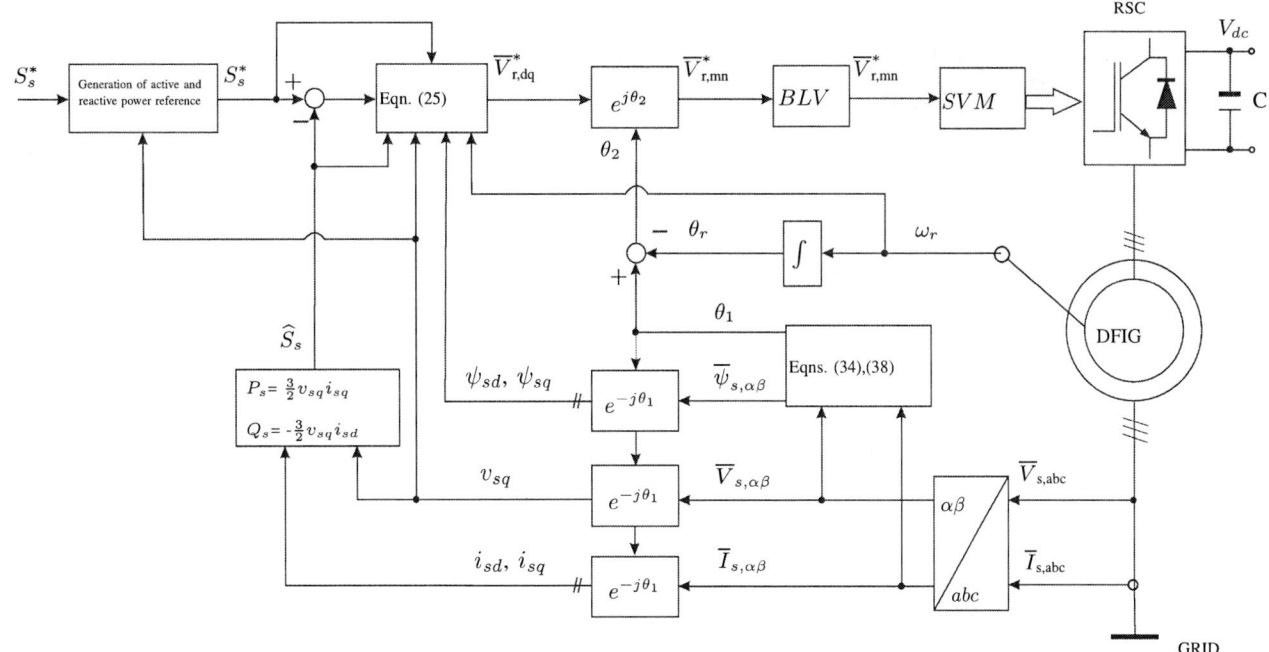

Fig. 2. Control system diagram.

From equation (25), one can analyze that when the generator to operate without disturbance on the grid, the terms that appear $\overline{f}_4$ and $\overline{f}_5(k)$ are null. However, when a disturbance occurs on the grid, the stator voltage and flux vary over time; so the controller must generate a rotor voltage to oppose this variation and so that stator powers follow their respective reference powers, no oscillation and no steady state error.

B. System implementation

In Figure 2 is shown the control system diagram. The stator flux estimation is performed using the following equation:

$$\overline{\psi}_{s,\alpha\beta} = \int_{T_s} (\overline{V}_{s\alpha\beta} - r_s \overline{I}_{s,\alpha\beta}) dt, \qquad (34)$$

If the stator flux, stator voltage and current vectors are known, they are performed to convert the $\alpha\beta$ variables into dq variables, from the following expressions:

$$\overline{\psi}_{s,dq}(k) = e^{-j\theta_1(k)} \overline{\psi}_{s,\alpha\beta}(k) \qquad (35)$$

$$\overline{V}_{s,dq}(k) = e^{-j\theta_1(k)} \overline{V}_{s,\alpha\beta}(k) \qquad (36)$$

$$\overline{I}_{s,dq}(k) = e^{-j\theta_1(k)} \overline{I}_{s,\alpha\beta}(k) \qquad (37)$$

where $\theta_1(k)$ is the stator voltage angle obtained by:

$$\theta_1(k) = \arctan\left(\frac{v_{s\beta}(k)}{v_{s\alpha}(k)}\right) - \frac{\pi}{2}. \qquad (38)$$

The output of the control algorithm generate $v^*_{rd}(k)$ and $v^*_{rq}(k)$, in sequence occurs the change of coordinate system: synchronous reference frame to rotor reference frame, resulting in $v^*_{rm}(k)$ and $v^*_{rn}(k)$. Lastly, they are used to generate the pulse width modulation switching patterns. To maintain the operation of the SVM in the linear region of operation,

the rotor voltage vector $\overline{V}^*_{r,mn}(k)$ is limited by the voltage limiter block (VLB).

IV. SIMULATION RESULTS

The performance of the proposed control strategy is verified by simulation using MATLAB/Simulink. The LVRT test is performed using the German network code [4] to generate the reference reactive power necessary to give the voltage support when occur voltage sags: for each 0.001 p.u. of drop in stator voltage occurs an increase in stator current of 0.02 pu.

A. Simulation Studies

To analyze the performance of the control strategy proposed, the dynamic behavior of the stator powers and stator and rotor currents of the DFIG is compared with two other strategies: (A) PI controllers designed for operating in normal conditions of the grid and (B) DPC conventional, i.e., without considering the dynamics of the stator flux and V_s constant. The parameters of the DFIG and controller gains are shown in the Appendix.

In the control strategy simulated the speed of operation was 1,530 rpm; in the initial instants that the voltage dip occurs, the reference reactive power changes from zero (situation before the occurrence of voltage dip) to 708 VAR in order to satisfy the gird code criteria, and in this depth of the dip was not possible to provide active power because there is no enough current to it, therefore, $P^*_s = 0$.

In Figures 3(b), (c), (f) and (g), high overshoots and oscillations in stator active and reactive power are generated when balanced voltage dip occurs, and consequently, occur high peak currents in the stator and rotor, as shown in Figures 3(d), (e), (h) and (i). In Figure 3(j) and (l) is observed that when the

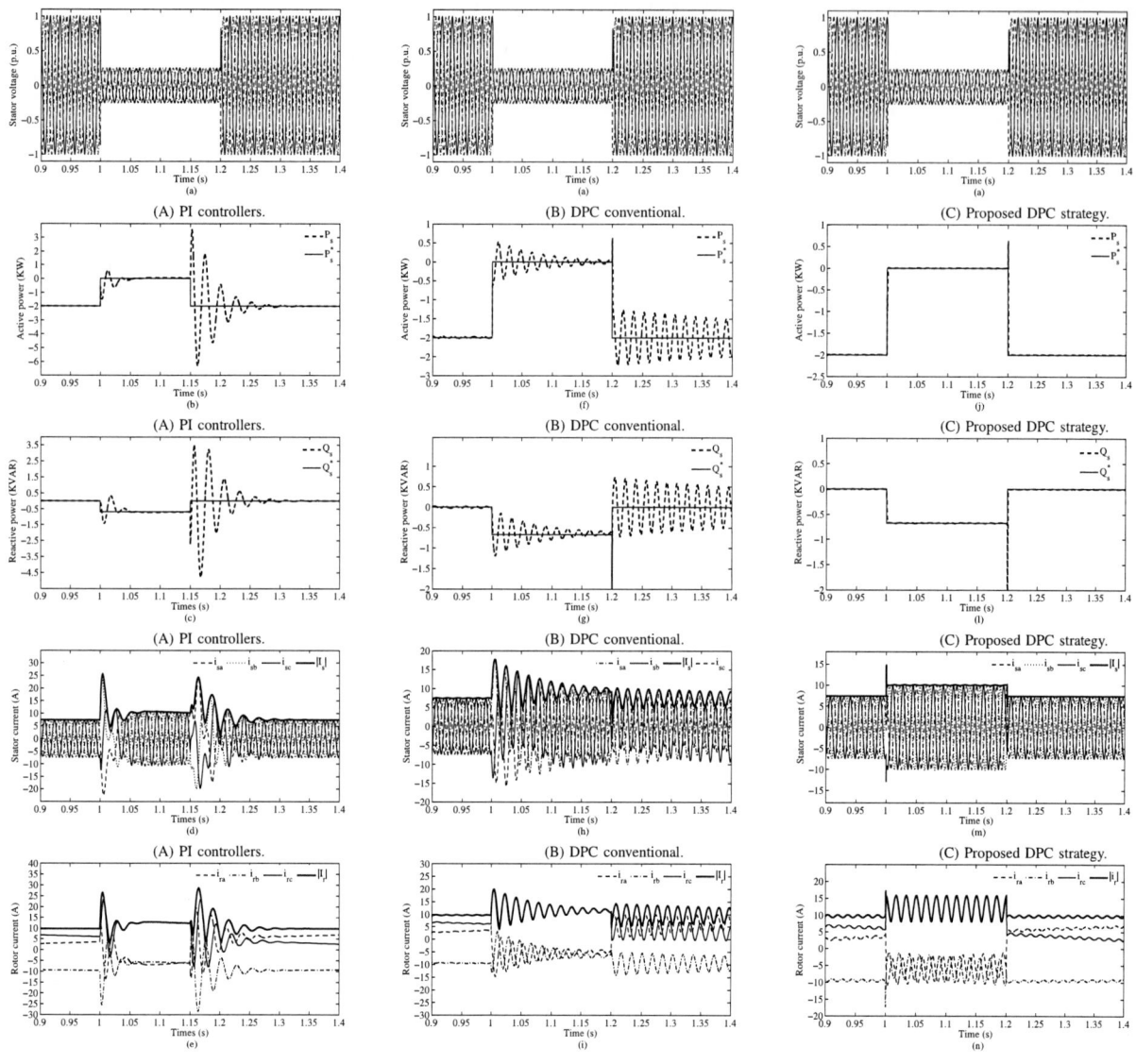

Fig. 3. Balanced voltage sags of $\Delta V_s = 75\%$.

proposed control strategy is used, overshots and oscillations in the stator active and reactive power are eliminated, thereby not appear peak currents, as seen in Figure 3(m) and (n). This comparison aims to show that when considering the dynamics of the stator flux and V_s not constant in the design of the proposed DPC has been possible to control the powers without overshoots and oscillations during voltage dip.

V. CONCLUSION

The dynamic behavior of DFIG during voltage sags is enhanced due to the direct power control strategy presented in this paper. The control strategy is based on rotor voltage dynamic equation where the stator voltage and the magnitude of stator flux change their values during the voltage dip.

Simulation results are presented for a 2.2 kW generator and tested for balanced voltage dip. Comparison is made between the proposed control strategy with the strategy using PI controllers and DPC, the two without consideration of the transient flux and voltage of stator, for testing voltage sag of 75%. In the results, it is verified that the active and reactive power are controlled rapidly without overshoot and oscillation during the voltage sag when using the proposed control strategy; consequently, peaks stator and rotor currents have not appeared, unlike the other two strategy used for the comparison, the cause of the good performance of the proposed strategy is the fact to use the equation of the rotor voltage more precise, with consideration of the flux transient and the stator voltage.

APPENDIX

Parameters of the 2.2 kW DFIG: 2.25 kW, V_s=220 V, f_s=60 Hz, turn ratio=2.73 (N_s/N_r), p=4 poles. R_s=1.2 Ω, R_r=1.24 Ω, L_s=98.14 mH, L_r=98.14 mH, L_m=91.96 mH.

Controller gains values: $G_1 = 0.237$, $G_2 = 5336$, $G_3 = 0.00131$, $k_1 = 0,854$, $k_2 = 13.05$, $k_s = 0.937$, $k_r = 0.937$, $k_\sigma = 117.41$.

REFERENCES

[1] J. Morren, S. W. H. de Haan, "Ridethrough of wind turbines with doubly fed induction generator during a voltage dip," IEEE Trans. on Energy Conversion, vol. 20, pp. 435-441, Sep. 2005.

[2] Dawei Xiang, Li Ran, Peter J. Tavner and Shunchang Yang, "Control of a Doubly Fed Induction Generator in a Wind Turbine During Grid Falt Ride-Through", IEEE Trans. on Energy Conversion, vol. 21, pp. 652-662, Sep. 2006.

[3] J. Lopez, E. Gubia, E. Olea, J. Ruiz and L. Marroyo, "Ride through of wind turbines with doubly fed induction generator under symmetrical voltage dips," IEEE Transactions on Industrial Electronics, vol. 56, pp. 4246-4254, October 2009.

[4] E.ON-NETZ, "Grid Connection Code- extra high voltage", 2006.

[5] J. Costa, H. Pinheiro, T. Degner and G. Arnold, "Robust Controller for DFIGs of Grid-Connected Wind Turbines," IEEE Transactions on Industrial Electronics, vol. 58, pp. 4023-4038, September 2011.

[6] A. Luna, F. K. de Araujo Lima, D. Santos, P. Rodríguez, E. H. Watanabe and S. Arnaltes, "Simplified Modeling of a DFIG for Transient Studies in Wind Power Applications," IEEE Transactions on Industrial Electronics, vol. 58, pp. 9-20, January 2011.

[7] V. Flores Mendes, C. Venicio de Sousa, S. Rocha Silva, B. Cezar Rabelo, and W. Hofmann, "Modeling and Ride-Through Control of Doubly Fed Induction Generators During Symmetrical Voltage Sags," IEEE Transactions on Energy Conversion, vol. 26, pp. 1161-1171, December 2011.

[8] F. Lima, A. Luna, P. Rodriguez, E. Watanabe, and F. Blaabjerg,"Rotor Voltage Dynamics in the Doubly Fed Induction Generator During Grid Faults," IEEE Transactions on Power Electronics, vol. 25, pp. 118-130, January 2010.

[9] J. Liang, W. Qiao, and R. Harley, "Feed-Forward Transient Current Control for Low-Voltage Ride-Through Enhancement of DFIG Wind Turbines" IEEE Transactions on Energy Conversion, vol. 25, pp. 836-843, September 2010.

[10] M. Mohseni, M. Masoum and S. Islam, "Low and high voltage ridethrough of DFIG wind turbines using hybrid current controlled converters," Electric Power Systems Research, vol. 81, pp. 1456-1465, July 2011.

Electrical Simulation of Traction Subway System for Energy Recovery and Energy Saving Studies

Marcio Annibal Pimenta
*Departamento de Engenharia de
Energia e Automação Elétricas
Universidade de São Paulo*
São Paulo, Brazil
marcioapimenta@gmail.com

Wilson Komatsu
*Departamento de Engenharia de
Energia e Automação Elétricas
Universidade de São Paulo*
São Paulo, Brazil
wilsonk@usp.br

Lourenço Matakas Junior
*Departamento de Engenharia de
Energia e Automação Elétricas
Universidade de São Paulo*
São Paulo, Brazil
matakas@usp.br

Abstract – **In the last decades, the energy consumption is the center of several discussions about subway system efficiency due to the risingly cost of energy. Many rail operators have been installing systems to recover energy, including flywheel, super-capacitors and batteries, which can be boarded or stationary. Another way to save energy is the low-cost option of introducing changes in the train operation (e.g. in the trains scheduling, acceleration rate and maximum speed) at the expense of the reduction of the system efficiency by the point of view of the customer service. A simulator is proposed to obtain data to allow comparison between these two options: use energy recovering system or change train operation parameters.**

Keywords – traction simulation, energy recovery, electrical simulation, subway energy consumption.

I. INTRODUCTION

The city of São Paulo, through the XIX and XX centuries, has consolidated its position as one of the main cities in the world's economic scenario. In the mid-1930s it reached its first million inhabitants. In 1967, with 7 million inhabitants, it had a fleet of approximately 493 thousand vehicles, which brought to the limit the downtown road network capacity. The city was characterized by isolated industrial pockets, with a speculative and disordered occupation. In order to contribute to solve the mobility problems of São Paulo, a company called *Metrô de São Paulo* was created in 1968, starting the construction of the first subway line [1]. Today, *Metrô de São Paulo* has 58 stations and 64,7 km of lines, including a fleet of 170 trains, travelling the distance of approximately 1.8 million miles monthly and providing transportation to almost 3 million commuters per day.

To allow the operation of this system, the average consumption of electric energy is 41.2GWh monthly, according to 2019 data. Traction energy is the main responsible for energy spending, corresponding to approximately 69%. Taking into account only one of its lines, the energy dissipated by the braking resistors amounts to 5% of the total traction energy [2], resulting in 2.06GWh, or approximately R\$ 597,400.00 literally burned every month in this line.

Regenerative braking, used in several types of electric vehicles, stores and reuses kinetic energy that would be wasted in decelerating these vehicles, making them more efficient and sustainable. This technology is not new in subway systems. It is used by some countries in Europe, Asia and United States, achieving an energy consumption reduction up to 30 % [3], [4], [5]. Considering this scenario, *Metro de São Paulo* launched in 2014 the *Alternative Energy Sources and Energy Efficiency Policy*, defining basic architecture, systems and rolling stock guidelines for alternative sources of energy and energy efficiency for new lines and modernization of existing ones [6].

Since 2005, once a year, *Metrô* buys electricity at lower rates in the "energy free market". Since then electrical energy cost for *Metro de Sao Paulo* increased around 580%, from R\$50,00 to R\$290,00 per MWh. Fig. 1 shows the energy price evolution along this period. All values are current ones.

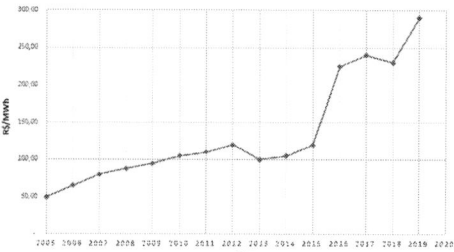

Fig. 1. Evolution of energy cost for *Metrô*.

Therefore, a technique where it is possible to store and reuse this "burned" energy in some way can be attractive, both from an economic, environmental and / or corporate image point of view. This energy can be used in many ways according to company guidelines. For example, be resold to the power grid, used by traction system or by the auxiliary loads, such as stations and tunnel ventilation. The most widely used storage systems in rail and subway systems are batteries, super-capacitors (EDLC) and flywheels, being used individually or together.

These technologies use the "surplus" of energy system to provide it when system requires, converting electrical energy into other forms of energy or "confining" it to use when necessary. For instance, flywheels transform electrical energy into kinetic energy and batteries "confine" electrical energy. The choice of the best technology depends on the response time, weight, volume, operating temperature, charge cycle, power, price, etc [1].

To illustrate the world application scenario of these storage technologies, Table I shows the embedded and Table II the stationary ones [1], where PT is the power installed; S is the energy storage capacity; P is the device weight and V is the system nominal voltage.

This document presents an improved simulation technique suitable for energy saving studies in subway and railway systems. The main novelties of this simulator when compared to previous ones [7] [8] [9] include the addition of storage devices on electrical system, changes on the system modeling and the numerical method to calculate the node voltages of an electrical circuit with time varying parameters. Two energy saving strategies, the "energy storage" and the "speed restriction" are discussed, simulated and compared. The results are compared to field tests results imposing a speed restriction on trains' operation schedule.

978-1-7281-4181-7/19 \$31.00 © 2019 IEEE

TABLE I. BOARDED TECHNOLOGIES

Product	Factory	Type	Main charact.	Application
MITRAC	Bombardier	EDLC	PT: 200 kW S: 1 kWh P: 400 kg	Mannheim Trainway, Germany 2003
Sitras	Siemens	EDLC	PT: 300 kW S: 0.85 kWh P: 820 kg	Austria Trainway 2010
ACR	CAF	EDLC	PT: N/A S: 0.8 kWh P: 800 kg	Seville Saragossa Trainway 2012
STEEM	Alstom	EDLC	PT: N/A S: 0.8 kWh P: 800 kg	Paris Trainway Prototype 2010
Citadis	Alstom and CCM	Flywheel	PT: 325 kW S: 4 kWh P: 1600 kg	Rotterdam Prototype 2004
LRV	Kawasaki	Batterie	PT: 250 kW S: 120 kWh P: 3200 kg	Sapporo Prototype 2007-2008
LFX-300	Kinki Shayro	Batterie	PT: N/A S: 40 kWh P: 3200 kg	Charlotte Prototype 2010
Sitras	Siemens	EDLC + Batterie	PT: 288 + 105 kW S: 0.85 + 18 kWh P: 820 + 826 kg	Lisbon Trainway 2008.

TABLE II. STATIONARY TECHNOLOGIES

Product	Factory	Type	Main charact.	Application
Sitras	Siemens	EDLC	V: 500-750 V PT: 800 kW S: 2.5 kWh	Madrid Subway 2003 and Beijing Subway 2007
NeoGreen	Adetel	EDLC	V: 750 V PT: 300-1000 kW S: 1–4 kWh	Lyon Subway Prototype 2011
Envistore	ABB	EDLC	V: 500-1850 V PT: 750-4500 kW S: 0.8-16.5 kWh	Philadelphia Subway Prototype 2012
Capapost	Meiden-sha	EDLC	V: N/A PT: up to 2000kW S: N/A	Hong Kong Subway 2014
Power Bridge	Piller Power	Flywheel	V: 400-1000 V PT: 1000 kW S: 5 kWh	Rennes Subway 2010
GTR System	Kinetic Traction	Flywheel	V: 570-900 V PT: 200 kW S: 1.5 kWh	New York Subway Prototype 2002
Regen System	Vycon	Flywheel	V: N/A PT: 500 kW S: N/A	Los Angeles Subway 2014
Gigacell BPS	Kawasaki	NiMH	V: 600-1500 V PT: N/A S: 150-400 kWh	Tokyo Subway 2007
B-CHOP	Hitachi	Li-ion	V: 600-1500 V PT: 500–2000 kW S: N/A	Kobe Subway 2007
Intensium	Saft	Li-ion	V: 700 V PT: 900-1500 kW S: 600-400 kWh	Philadelphia Subway Prototype 2012

II. ELECTRIC TRACTION SIMULATION

It is possible to size the power supply components (substations, cables, distance between feeders, etc.) by using a computational traction system simulation in a subway or railway scenario. Besides that, the simulation can be used to expansion and operation planning, since it is possible to insert more trains, new trains schedule, new feeders, stations and increase the length of all system. This convenience can predict the new requirements in terms of number of feeders, required power, conductor's capacity, voltage profile along the line, etc.

This simulator includes three main blocks. The first one, the **Individual Train Simulation** calculates the instantaneous values of position and power consumption for each train based on their individual electrical and mechanical parameters. The second block is the **Traffic Simulation**. It is an extension of the first block with all trains positioned along the track over time, following a predefined trains' schedule. Finally, according to traffic simulation data, the **Electric Simulation** block is performed, returning as output the electric panorama along the track, including current and voltage values at predetermined points and amount of energy consumed, among others.

A. Individual Train Simulation Block

It is basically expected as a result of this step, the instantaneous values of the power consumed or generated (accelerating or braking, respectively) by the individual train and its location at all time instants during its run along the line. The electrical (power consumed or supplied by the train) and mechanical parameters (speed and position) are required for the train to be keep moving as requested by the control system. During vehicle acceleration, a high amount of energy is consumed which mainly depends on acceleration rate value and initial and final velocities. Throughout the time that the train is maintained at constant speed, it absorbs only the necessary power for the composition to oppose friction and anti-movement forces (ramps, curves and velocity). In deceleration period, train generates energy due to regenerative braking. The total amount, as during the acceleration, mainly depends on deceleration rate value and initial and final velocities. Power consumption by auxiliary services (light, air conditioning, etc.) is considered during all the trains journey.

1) Movement Restrictions

As said, among the opposite forces involved in the train movement, three of them can be highlighted. They are associated to ramps, curves and velocity. For this simulation, only the last one is considered. The other two are explained and can be easily included in the simulator.

Ramp restriction value is the proportional share of train weight that opposes to the movement (when ramp is uphill) or helps it (when ramp is downhill). It can be obtained by the ratio of the decomposed vehicle weight parallel to the movement and its total weight. In subway system, this ratio can be replaced by the slope value, r_r, since the angles are small [7], as shown by (1). It is calculated by the ratio of ramp height, Δh, and its length, Δl.

$$r_r = \frac{\Delta h}{\Delta l} \qquad (1)$$

Curve restriction force is given by some conditions. The two main constructive curve characteristics are its superelevation (difference in rails height) and radius. This construction causes centrifugal force, as well as frictional force between wheels and rails, resulting from the train's bogies inscription on the curve, once all its installations are considered rigid. Once the train wheels are installed in a rigid axis and, on a curve, the external rail is longer than the internal one. This restriction force can be calculated by (2), where r_c [N/kN] is the curve restriction force and ρ is its averaged radius [7].

$$r_c = \frac{650}{\rho - 55} \; for \; \rho > 350m$$

$$r_c = \frac{530}{\rho - 55} \; for \; 250 < \rho < 350m \tag{2}$$

$$r_c = \frac{530}{\rho - 3} \; for \; \rho < 250m$$

Velocity restriction force come from the interaction between rails and train wheels and aerodynamic force corresponding to the train friction area with the air. This force can be studied by many forms, one of them is by Davis Formula, which calculate the movement resistance, r_{mpt}, in function of velocity V, with coefficients, A, B and C, pre-established by general train characteristics, as shown in (3) [7]. For the simulation, all train cars were assumed by "motor cars". The resultant comportment of this force versus velocity can be observed in Fig. 2.

$$r_{mpt} = A + B \cdot V + C \cdot V^2 \tag{3}$$

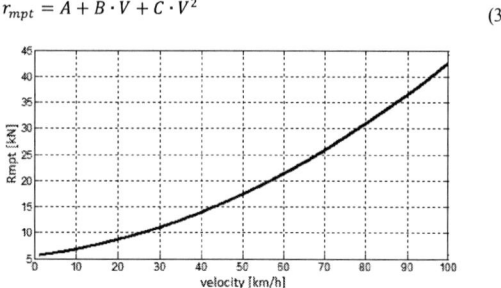

Fig. 2. Movement restriction versus train velocity.

2) Train acceleration and deceleration

The initial train acceleration and deceleration rates are specified on its project. In subway trains, the initial acceleration rate is 1.12 m/s² and the initial deceleration rate is 1.2 m/s². These values are almost constants until a velocity that occurs on traction motors the phenomenon called "field weakening". From this moment on, they can't maintain its torque, but its power. As the motor power is proportional to its mechanical torque multiplied by its angular speed, during the field weakening period, the force that it imposes to train decreases. This behavior is shown in Fig. 3 for acceleration and Fig. 4 to deceleration [1]. After taking into consideration all this forces and rates, finally train position and power are obtained, as shown is Fig. 5.

B. Traffic Simulation Block

Next block is the insertion of all trains in the line, since the operation is done with all trains being driven simultaneously, as shown in Fig. 6. As a result of this step, two matrices are obtained. For the first one each line refers to each corresponding train position on every second. The second matrix, presents the trains power consumption. The number of matrix lines corresponds to how many trains are in the system. The number of columns refers the quantity of necessary seconds to complete the travel. For this document, 8 trains from *Metrô* Line 5-Lilac were simulated.

C. Electrical Simulation Block

After obtaining all trains position and consumed power over time, it is possible to simulate the traction system electric behavior, resulting in the values of power consumed by rectifier substations, voltage values at predefined points along the track (including each train position) and the possibility of predicting the system electrical behavior. Traction system

electrical components (trains, rectifiers, energy storage device, etc.) are modeled as current sources with corresponding parallel resistors, connected to points along track. Trails can be modeled as a time variable resistance between line components (trains, stations or energy storage devices) corresponding to traction cables distance between these points.

1) Train Model

The train is modeled as a power source whose instantaneous value is given by the Traffic Simulation Block plus a parallel resistor. This last one models the effect of the breaking resistor. To ease the calculation of the node voltages by the Modified Nodal Analysis Method – MNA [10], the ideal power sources are substituted by ideal current sources whose instantaneous amplitude is given by the instantaneous values of trains power and voltage. The train voltage and current are iteratively calculated for each time step, as explained in section II-C-3. The voltage initial values are the nominal ones. During acceleration and constant speed, the value of this resistance is very high because the train only "drains" current from the traction circuit. During the deceleration, train "injects" current on system, so the current source signal is inverted. Through this period, there are two situations: firstly, the model is kept maintaining only a current source "injecting" current on system, causing voltage value to increase until the maximum one (defined by project); at this moment, the break resistance is placed in parallel to the current source. Its value is varied in order to keep the voltage on its maximum value, according to (4), where U_{max} is the maximum voltage value of regeneration, P_{el} is the power of the train gives by the Electric Simulation Block and U_n is the node voltage value before the model change (greater than maximum value). If the braking resistor is fully inserted (minimum resistance) and the train voltage continues to increase, the electrical regenerative process is "cutted off" and the mechanical break is applied.

The present proposal to model the trains during breaking period is different from the one commonly adopted in the literature that replaces the current source model by a voltage source and a series conductance during all breaking process [7]. In this proposal, there is no need to change the Modified Nodal Analysis matrices, because the model contains only current sources.

$$R_b = \left| \frac{U_{max}^2 \cdot U_n}{P_{el} \cdot (U_{max} - U_n)} \right| \tag{4}$$

2) Rectifier Model

Rectifier substations can also be modeled by a current source, similar to the procedure adopted for the trains, as shown in Fig. 7a. In this case, the current value I_{se} can be obtained by (5). The resistor value, R_{se}, placed in parallel to this source is equal to the "equivalent internal resistance" value of the rectifier, coming from (6) [7], where P_d is the nominal rectifier power, U_d the nominal rectifier output voltage and U_{d0} the no load voltage.

$$I_{se} = \frac{P_d}{U_{d0} - U_d} \tag{5}$$

$$R_{se} = \frac{(U_{d0} - U_d) \cdot U_d}{P_d} \tag{6}$$

The model discussed above is valid only when rectifier voltage is less or equal to the no load output voltage (rectifier is "turned on"). Due to the fact that diode rectifiers do not allow energy regeneration (it is "turned off") for output voltages higher than the nominal one, the model is switched to a voltage source, with amplitude equal to U_{do} and a series resistor (commonly 1000 Ω) [7], shown in Fig. 7b. With this model, the rectifier diode reverse current is considered, with a small value flowing to it. Other possibility to model this device state is to make I_{se} equals zero and assign a large value to resistance R (10 times more than U_d), as shown is Fig. 7c.

3) DC line modelling and Modified Nodal Analisys

It is known that the trains position is dynamic over time, so the electric traction circuit length between two trains or a train and a rectifier vary, which represent a variable impedance. To solve the complete system (trains, rectifiers, energy storage and DC lines), for every sampling instant, the trains and rectifiers were distributed in their position, according to data obtained by Traffic Simulation, assigning each current source of Fig. 8 to a train or rectifier (or energy storage). In this figure the rail resistances, RR, were moved to the supply rail, R, to ease the evaluation of the circuit by means of Modified Nodal Analysis, without affecting the main nodes voltages ($U1$ to Un). This technique keeps the system fixed (trains position and active power values for trains and storage components) for each sampling time. The final purpose of this simulator is to find the voltage value at each node $U1$ to Un that can be a train, a rectifier or a storage system, by applying iteratively the Modified Nodal Analysis. It is important to emphasize that due to movement of the trains, the role of nodes 1 to n changes for each sampling time. In this paper the electric circuit is iteratively solved for each sampling time.

The use of Modified Nodal Analysis [10] for the circuit shown in Fig. 8 results in (7), where G_n is the sum of all impedances (inverse of resistance) connected to node "n"; G_{xy} is impedance between nodes "x" and "y".

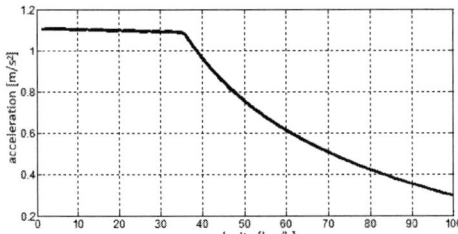

Fig. 3. Acceleration rate versus velocity.

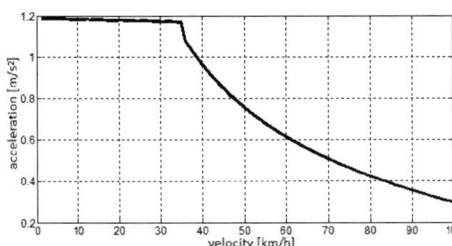

Fig. 4. Deceleration rate versus velocity.

The parameters of (7) varies for every time step. The node voltages are obtained by interactively calculating based on the Preconditioned Conjugate Gradients Method - PCG [11]. MNA was used because the "Classical" Nodal Analysis considers only first Kirchhoff Law (the current one). As in some instants the circuit contains voltage sources, the second Kirchhoff Law has to be involved.

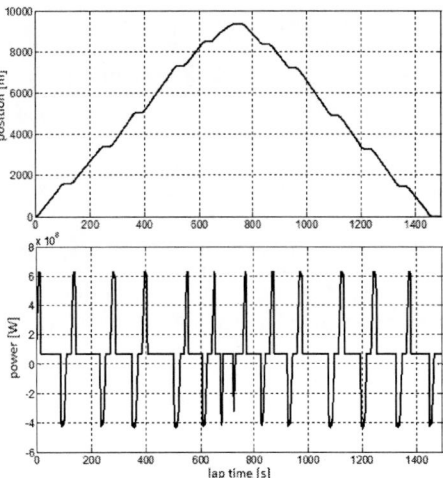

Fig. 5. Train position and power along the track.

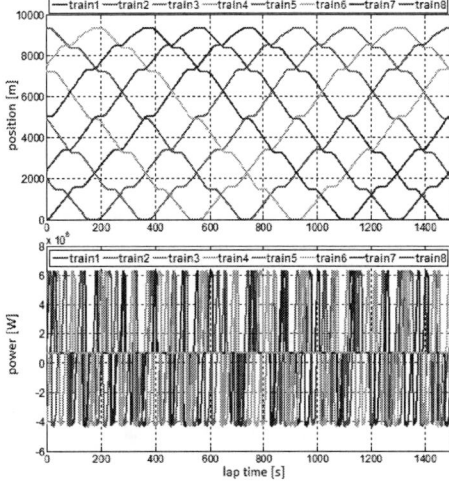

Fig. 6. Line 5-Lilac Traffic Simulation.

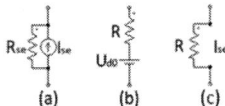

Fig. 7. Rectifier Model when it is on (a) and off (b or c).

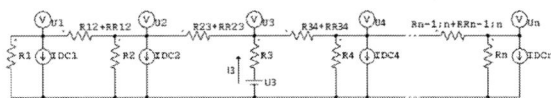

Fig. 8. Circuit example at arbitrary interaction.

4) Software Flowchart

The computational simulation was performed using MatLab software command lines, according to the flowchart shown in Fig. 9. The sample time assumed for this program was one second (the system dynamic is very slow).

First of all, the Individual Train Simulator Block is run, which returns two vectors, one with train position and other with instantaneous train power, as shown in Fig 5. In second, the Traffic Simulation Block is compiled, to position all trains "in the system". For this example, all trains intervals were assumed equal, so the previous vectors were headway shifted for each train, as shown is Fig. 6, forming two matrices (a train position and instantaneous train power with n rows each, where n is number of trains in the system). These steps are made just once, because the results will never change.

Next, with those matrices data, the Electrical Simulation Block is repeated every sampling time until trains complete one entire lap in the system. This iterative process is necessary because the value of the currents associated to each train or energy storage system for the trains are unknown a priori. The inputs for the electrical circuit evaluation are the individual components active power. Currents can be obtained by dividing each value of component's power by their corresponding node voltage (for the first iteration, the nominal voltage value is taken), but the voltages depend on the complete circuit, that's why the iterative process until a specified error (for this project, as mentioned, a Preconditioned Conjugate Gradients Method - PCG [11] is used) . Further iteration is required to change the value of the parallel resistances of each train during braking and of each rectifier when their output currents tends to negative values.

$$
\begin{bmatrix}
G1 & -G12 & 0 & 0 & 0 & 0 & 0 \\
-G12 & G2 & -G23 & 0 & 0 & 0 & 0 \\
0 & -G23 & G3 & -G34 & 0 & 0 & -1 \\
0 & 0 & -G34 & G4 & -G45 & 0 & 0 \\
0 & 0 & 0 & -G45 & \ddots & \vdots & \vdots \\
0 & 0 & 0 & 0 & \cdots & Gn & 0 \\
0 & 0 & -1 & 0 & \cdots & 0 & R3
\end{bmatrix}
*
\begin{bmatrix}
U1 \\ U2 \\ U3 \\ U4 \\ \vdots \\ Un \\ I3
\end{bmatrix}
=
\begin{bmatrix}
-IDC1 \\ -IDC2 \\ 0 \\ -IDC4 \\ \vdots \\ -IDCn \\ U3
\end{bmatrix}
\quad (7)
$$

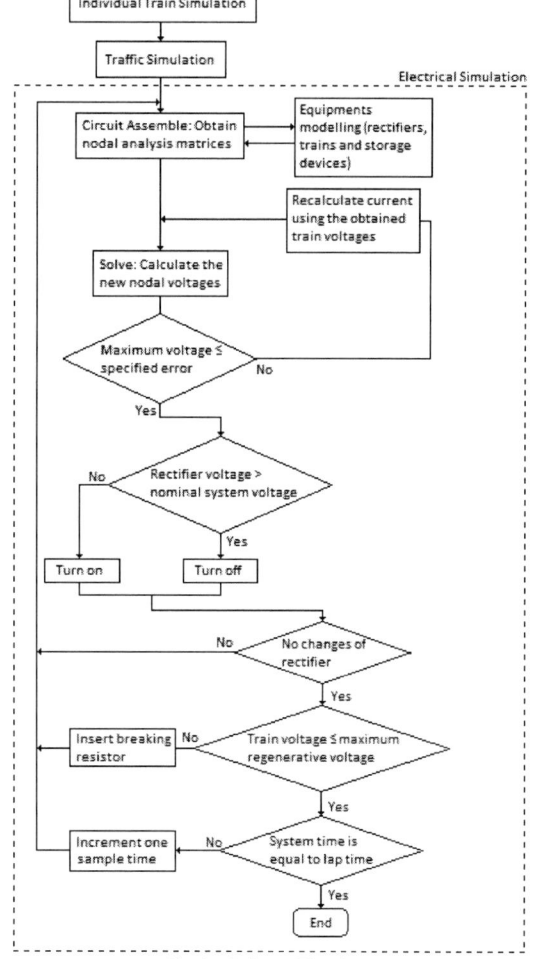

Fig. 9. Software Flowchart.

5) Initial tests- Possibilities of Energy Saving

To give some examples of the Simulator usage some possible scenarios of energy saving can be tested in order to subsidize strategic decisions related to energy saving. These scenarios include not only lower cost solutions, such as lowering maximum train speeds and changing the control strategy for trains braking and acceleration periods, but also the installation of energy storage technologies at track strategic points.

One possibility to locate energy storage device is next to each rectifier. When the rectifier is "switched off" (rated voltage is higher than nominal one), the device stores the additional power generated by the decelerating trains, as shown in Fig. 13. The instantaneous power flow in the storage devices is defined by imposing a device current given by (8).

$$
I_n = \frac{(U_n - U_{d0n})}{R_n} \tag{8}
$$

Where U_n is the node voltage, U_{d0n} is the no load nominal voltage and Rn is chosen to obtain the desired transfer function $\frac{I_n}{U_n}$.

RESULTS

In order to obtain data to verify the energy saving possibilities previously described, two scenarios were simulated.

A. Lowering Maximum Train Speed

This test imposes a lower final velocity, maintaining acceleration rate. This results in considerably smaller consumed power. In Fig. 10 it is possible to verify a real J Fleet [12] train behavior in two moments of acceleration: up to 67 km/h and later 45km/h, operating at Line 1-Blue. The energy increase during the acceleration is much lower in second case (9.28 MWs versus 4.63 MWs, or almost 50% less). The use of simulation program for the same train at the same conditions resulted in 42% less energy consumed, as shown in Fig. 11.

The region of interest in Figs. 10 and 11 is when the velocity grows up, from zero to constant value. For the purpose of comparison, the power and energy comportment only on this area is discussed.

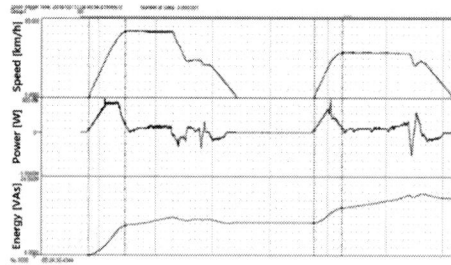

Fig. 10. Real waveforms for a one J Fleet train.

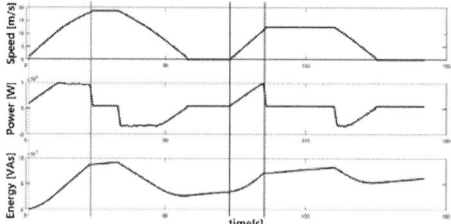

Fig. 11. Simulated waveforms for a one J Fleet train.

978-1-7281-4181-7/19 $31.00 © 2019 IEEE

With the great possibility of energy savings shown above, a greater proportion test was carried out in *Metrô* Line 3-Red, reducing the maximum speed on line sectors where train does not reach commanded speed by line control system, or reaches by a short time, and the proposed speed reduction causes a maximum delay of 4s on this sector. With this in mind, 13 of 34 control system sectors had their maximum speed decreased, generating a total 30s increase in 2,805.5s original route time, or 1.06%. Table III shows the reduction in energy consumption resulting from this change. If the reduction were made on every control system sector, not on 38% of them, the energy consumption economy could be almost 20%, with a lap time increase of 3%.

TABLE III. LINE 3-RED ENERGY CONSUMPTION WITH VELOCITY REDUCTION

	Energy Consumption [kWh]	Specific Consumption [kWh/km]
Original per lap	667.92	15.11
Modified per lap	620.54	14.03
Difference	**47.37**	1.07
	7.09%	
Original per day	400,747.14	
Modified per day	372,321.84	
Difference per day	28,425.30	
Difference per month	**0.85 GWh**	

The same idea was simulated to Line 5-Lilac, but making reductions on every control system sectors. The lap time passed from 1492 to 1634s, or 9.5% more. The energy consumption came from 11,741 to 8,763 MWh, which corresponds to a 25.36% energy saving.

B. Use of Energy Storage Device

Next, simulation data from *Metrô* Line 5-Lilac, with a 7 stations extension, 8 trains circulation, 4 rectifiers and 4 storage devices next to them are presented (going from *Capão Redondo* to *Adolfo Pinheiro* stations- before its newly opened expansion). Calculating data from Fig. 12 and Fig. 13, the potential energy economy with recovery system installed as proposed is about 9.8%. These are only initial results to show the possibilities of use for the proposed simulator. More consistent conclusions about the more adequate choice of energy saving method will require further studies considering storage devices location and the choice of their control strategy.

Considering the results obtained above for the comparison between software and practical results, a whole *Metrô* line can be simulated to check quantitatively the potential of available energy to be stored and reused on system. This result will subsidize the decision to buy or not some energy storage system, as shown in item I, since there will be enough data for a good design, which will translate into savings in its purchase.

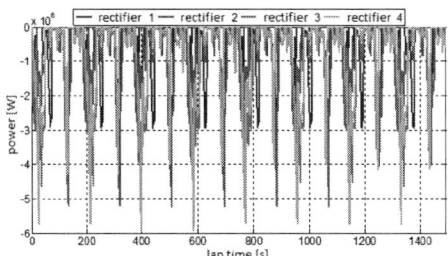

Fig. 12. Power given by rectifiers to system.

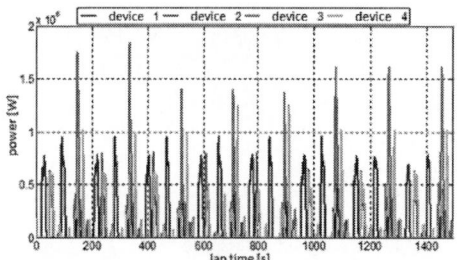

Fig. 13. Power savings locating storage devices next to rectifiers.

CONCLUSION

This paper proposed a simulator for railways and subway systems. To show its adequacy two strategies for energy saving were simulated. To improve the simulated result of 9.8% saving using four energy storage systems, their location and control strategies should be optimized. In any case, cost of this solution will be very high, since it will be necessary to purchase the energy storage system. However, a substantial saving occurs with a small degradation in service level (7% of energy saved in Line 3-Red versus a 1.06% increase in the train lap time), with no additional investment.

REFERENCES

[1] A. F. d. S. Filho, L. C. E. de Queiroz, M. A. Pimenta and M. J. Rosa, *"Uso de simulação para análise de estratégias que visam a economia de energia em sistemas Metroferroviários e tecnologias disponíveis de armazenamento"*, Monografia apresentada à Escola Politécnica da Universidade de São Paulo para obtenção do Título de Especialista em Tecnologia Metroferroviária. São Paulo, 2016.

[2] D. M. de Morais, L. A. da Cruz and R. C. Oliveira, *"Estudo comparativo entre armazenador capacitivo, volante de inércia e inversor"*,. Monografia apresentada à Escola Politécnica da Universidade de São Paulo para obtenção do Título de Especialista em Tecnologia Metroferroviária. São Paulo, 2007.

[3] R. Barrero, X. Tackoen and J. Van Mierlo, "Stationary or onboard energy storage systems for energy consumption reduction in a Metro network," Proceedings of the Institution of Mechanical Engineers, Part F: Journal of Rail and Rapid Transit, 224, 2010.

[4] M. Domínguez, A. P. Cucala, A. Fernández, R. R. Pecharromán and J. Blanquer, "Energy efficiency on train control – design of Metrô ATO driving and impact of energy accumulation devices," 9th World Congress on Railway Research – WCRR, 2011.

[5] M. Chymera, A. Renfrew and M. Barnes, "Analyzing the potential of energy storage on electrified transit systems," 8th World Congress of Railway Research – WCRR, 2008.

[6] Metrô-SP, *"Relatório de Sustentabilidade 2014"*, São Paulo, 2014.

[7] C. L. Pires, *Simulação do sistema de tração elétrica metro-ferroviária*, 370 p., São Paulo, 2006.

[8] Z. Y. Shao, W. S. Chan, J. Allan and B. Mellitt, "A new method of DC power supply modelling for rapid transit railway system simulation", Transactions on the Built Environment vol 6. 1994

[9] Y. Cai, M. R. Irving and S. H. Case, "Iterativa techniques for the solution of complex DC-rail-traction systems including regenerative braking", IEE Proc. Gener. Transm. Distrib., vol 142, No 5, September 1995

[10] L. Q. Orsini and F. A. M. Cipparrone, *Simulação Computacional de Circuitos Elétricos*, 184 p., São Paulo, 2011.

[11] G. Radhika, Support Graph Preconditioners for Sparse Linear Systems. Master of Science submitted to the Office of Graduate Studies of Texas A&M University. USA, 2004.

[12] Metrô-SP, *"Manual de manutenção e operação da Frota J"*. São Paulo, 2009.

Modeling and Simulation of a Stirling-Engine-based Generator Connected to the Grid

Augusto Hayashi
Department of Electrical Engineering
USP
São Paulo, Brasil
augustohayashi@gmail.com

Moacyr Brito
Graduation Program in Electrical
Engineering
UFMS
Mato Grosso do Sul, Brazil
moa.brito@gmail.com

João Onofre
Graduation Program in Electrical
Engineering
UFMS
Mato Grosso do Sul, Brazil
joaonofre@gmail.com

Lourenço Matakas Junior
Department of Electrical Engineering
USP
São Paulo, Brasil
matakas@pea.usp.br

Raymundo Cordero
Graduation Program in Electrical
Engineering
UFMS
Mato Grosso do Sul, Brazil
rcorderog@gmail.com

Luigi Galotto Junior
Graduation in Electrical
Engineering
UFMS
Mato Grosso do Sul, Brazil
lgalotto@gmail.com

Abstract—The Stiling engine converts heat in mechanical energy based on the Stirling cycle, which is the most similar to the Carnot cycle. For that reason, Stirling engine is used in cogeneration process to get electricity from different heat sources. However, few studies were done that integrates the thermodynamic model of the Stirling engine together with the required power electronic systems to connect an electrical generator with a power grid. This paper shows the modeling of the Stirling engine, a DC generator coupled to this engine, the DC-DC converter and the inverter required for connecting the cogeneration system based on Stirling engine to the grid. The synchronization algorithm to connect the inverter to the grid is also explained. The proposed model allows calculating the grid current and voltage, the active and reactive power generated using a Stirling engine as cogeneration system, according to its mechanical and thermodynamic parameters. The Stirling-generator system operates with practically unitary power factor. The aim of this work is to motivate the analysis and use of the Stirling engine in cogeneration.

Keywords—Cogeneration, grid connection, Stirling engine, generation, synchronization.

I. INTRODUCTION

The population growing, the industrial production growing, the use of more transportation systems, the ending of the fossil fuels, the limitation of natural resources and environmental problems motivate the use of renewable energies, in order to supply the increasing global energy demand [1]-[3]. With the development of power electronics techniques, the fabrication of renewable energy conversion systems has increased significantly [3]. These systems can use solar radiation, wind energy, sea wave energy, biomass, geothermal energy and others [3]-[6]. Those renewable energy sources can be used in distributed generation (GD) and being connected to a power grid.

Cogeneration consists in the production of two (or more) types of energy from the same primary energy source. This is one of the best approaches to improve the production of electricity and heat, compared with the process to generate heat and electrical energy separately [7].

The use of Stirling engine is an interesting and widely used alternative in the development of cogeneration systems [3], [8]. This engine is based on the Stirling cycle, which is the most similar thermodynamic process to the Carnot cycle (which has the best conversion efficiency, but it is impossible to be implemented experimentally). Stirling engine receives thermal energy from a heat source (biomass, geothermal, etc.), and convert it to mechanical energy that activate an electric generator. However, few analyses were done about the modeling of the Stirling engine as source of electricity (through cogeneration) connected to a grid.

This paper presents a modeling of a cogeneration system based on Stirling engine, which is connected to the power grid. The thermodynamic model of the Stirling cycle, the DC generator, the power converters and the synchronization algorithm to connect the cogeneration system to the grid were developed in software. The proposed model allows calculating the grid current and voltage, the active and reactive power generated according to the mechanical and electrical parameters of Stirling engine. The control current technique allows the operation of the Stirling-generator system with practically unitary power factor. The model of the Stirling engine, the power converters, the generator and the controllers were developed considering continuous-time systems. It was developed a steady-state model of the Stirling engine. The aim of this work is to motivate the analysis and use of the Stirling engine in cogeneration.

II. STIRLING ENGINE

A. Equations about Heat Transfer

Let consider that a gas is contained in a long tube with length L and constant section area S. Let Q represents the heat that flows to the tube following its axis. Let $T(\mathrm{x}, t)$ be the temperature in the time t and the position x ($x \in [0, L]$) taken in the direction of the heat flow. The quantity of heat ∂Q that flows to the gas during a small time ∂t is [9]:

$$\partial Q = -k.S.\left(\tfrac{\partial T(x,t)}{\partial x}\right)\partial t \qquad (1)$$

978-1-7281-4181-7/19 $31.00 © 2019 IEEE

where k is the thermal conductivity of the gas. Let consider a small volume of the tube, with section area S and thickness Δx. the net quantity of heat, ∂Q, that penetrates that small volume is the heat that enters at the point x less the heat that leaves at the point $x + \Delta x$. Based on (1), this heat can be calculated as follows:

$$\partial Q = k.S.\left(\frac{\partial T(x+\Delta x, t)}{\partial x} - \frac{\partial T(x,t)}{\partial x}\right)\partial t \qquad (2)$$

Besides, the heat that penetrates that small volume must also satisfies the following equation [9]:

$$\partial Q = \rho.c_v.\partial T.S.\Delta x \qquad (3)$$

where ρ and c_v are the density and specific heat of the gas, respectively. For the energy conservation law, (2) and (3) must be equal:

$$\rho.c_v.\partial T.S.\Delta x = k.S.\left(\frac{\partial T(x+\Delta x, t)}{\partial x} - \frac{\partial T(x,t)}{\partial x}\right)\partial t \qquad (4)$$

Therefore:

$$\frac{\partial T}{\partial t} = \left(\frac{k}{\rho.c_v}\right)\left(\frac{\frac{\partial T(x+\Delta x,t)}{\partial x} - \frac{\partial T(x,t)}{\partial x}}{\Delta x}\right) \qquad (5)$$

Calculating the limit of (5) when $\Delta x \to 0$, we get:

$$\frac{\partial T}{\partial t} = \left(\frac{k}{\rho.c_v}\right)\left(\frac{\partial^2 T}{\partial x^2}\right) \qquad (6)$$

According to (6), the speed $\frac{\partial T}{\partial t}$ that a temperature varies in a point depends on $z = \frac{k}{\rho.c_v}$ of the gas. A gas with higher value of z allows the engine operates with higher speed and power [9], i.e., it is better for the implementation of a thermodynamic process. For helium and air, $z = 2.7\times10^{-4}$ m^2/s and $z = 2.7\times10^{-5}$ m^2/s, respectively. Hence, Helium is better than air for the implementation of a thermodynamic process. However, if the engine works at high speed, the heat exchanger works properly, and considering the cost of Helium, many Stirling engines uses air instead of Helium.

B. Modeling of a Stirling-Engine-Based Power Generator

The equations of the Stirling engine used in this paper are based on [9], [10]. Table I shows the parameters of the model. Fig. 1 shows the ideal pression vs volume (PxV) curve of the Stirling cycle. The Stirling process requires a heater, a cooler and a regenerator. Ideally, it is composed by the following four thermodynamic stages [10]:

- 1→2: Isochoric heating: the volume of the working gas keeps constant while it gains thermal energy (heat) from the regenerator. The temperature of the gas rises to T_E.

- 2→3: Isothermal expansion: the gas receives thermal energy from an external heat source. The gas temperature is T_E. This gas expands isothermally, doing work. As a consequence, the pressure of the gas drops.

- 3→4: Isochoric cooling: heat is transferred from the gas to the regenerator, so the temperature drops to T_C. The volume of the gas keeps constant.

- 4→1: Isothermal compression: the volume of the working gas is diminished due to its pressure raises. Heat flows from the gas to an external sink. The temperature of the gas is still T_C.

TABLE I.   PARAMETERS OF STIRLING ENGINE

Name	Symbol	Unit
Engine pressure	P	Pa
Mean pressure of the Stirling cycle	P_{mean}	Pa
Minimum pressure of the Stirling cycle	Pmin	Pa
Swept volume of expansion piston or displacer piston	V_{SE}	m^3
Swept volume of compression piston or power piston	V_{SC}	m^3
Dead volume of expansion space	V_{DE}	m^3
Regenerator volume	V_R	m^3
Dead volume of compression space	V_{DC}	m^3
Expansion space momental volume	V_E	m^3
Compression space momental volume	V_C	m^3
Total momental volume	V	m^3
Total mass of working	M	Kg
Expansion space gas temperature	T_H	K
Compression space gas temperature	T_C	K
Regenerator space gas temperature	T_R	K
Phase angle	dx	Deg
Crank angle	x	
Temperature ratio	t	-
Swept volume ratio	V	-
Dead volume ratio	X	-
Engine speed	n	cycles/s
Expansion energy	W_E	J
Compression energy	W_C	J
Indicated energy	W_i	J
Expansion power	L_E	W
Compression power	L_C	W
Indicated power	L_i	W
Indicated efficiency	η	
Gas constant	R	J/kg.K

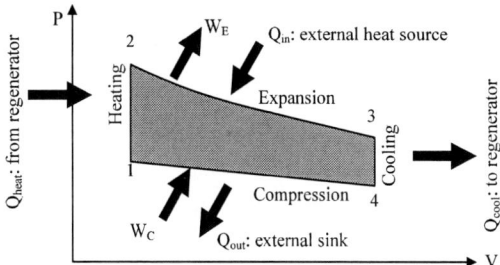

Fig. 1. Thermodynamic Stirling cycle.

There exist three types of Stirling engines: alpha, beta and gamma. In this paper, we will use the Schmidt model of the alpha-type engine, illustrated in Fig. 2, due to it is one of the simplest models. This modeling approach assumes the following statements:

- There is no pressure loss in the heat exchangers.

- The compression and expansion process are isothermal.

- The working gas is ideal gas.

- The regenerative process is perfect.

- The temperatures in the expansion and compression chambers are constant.

- The temperature of the regenerator is the arithmetic mean of the heater and the cooler.

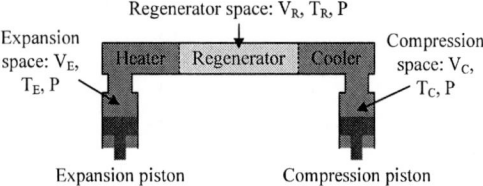

Fig. 2. Alpha-type Stirling engine.

The expansion and compression cylinder volumes are calculated from the crank angle (x) and the phase angle (dx):

$$V_E = \frac{V_{SE}}{2}.(1 - \cos x) + V_{DE} \qquad (7)$$

$$V_C = \frac{V_{SC}}{2}.(1 - \cos(x - dx)) + V_{DC} \qquad (8)$$

The total volume is the sum of the expansion volume, the compression volume and the volume in the regenerator:

$$V = V_E + V_R + V_C \qquad (9)$$

The total mass of working gas is calculated through the equation of ideal gases:

$$P.V = m.R.T \qquad (10)$$

The heater, regenerator and cooler have their own temperature (T_E, T_R, T_C) and volume (V_E, V_R, V_C). On the other hand, it is considered that the pressure (P) is equal in all these parts of the engine. Therefore, equation (10) can be used to estimate the mass in the heater, cooler and regenerator. Thus, the total mass is:

$$m = \frac{P.V_E}{R.T_E} + \frac{P.V_R}{R.T_R} + \frac{P.V_C}{R.T_C} \qquad (11)$$

Let define the variables X_T, X_V, X_{DE}, X_{DC} and X_R as follows:

$$X_T = \frac{T_C}{T_E}; \ X_V = \frac{V_{SC}}{V_{SE}}; \ X_{DE} = \frac{V_{DE}}{V_{SE}}; \ = \frac{V_{DC}}{V_{SE}}; \ X_R = \frac{V_R}{V_{SE}}$$
$$(12)$$

The temperature of the regenerator is:

$$T_R = \frac{T_E + T_C}{2} \qquad (13)$$

Using equations (6) to (13), it is possible to get the equation of the total mass (m) of the working gas:

$$m = \frac{P.V_{SE}}{2.R.T_C}.(S - B.\cos(x - a)) \qquad (14)$$

where:

$$S = X_T + 2.X_T.X_{DE} + \frac{4.X_T.X_R}{1 + X_T} + X_V + 2.X_{DC} \qquad (15)$$

$$B = \sqrt{X_T{}^2 + 2.X_T.X_V.\cos(dx) + X_V{}^2} \qquad (16)$$

$$a = \tan^{-1}\left(\frac{X_V.\sin\theta}{X_T + \cos\theta}\right) \qquad (17)$$

Therefore, the pressure is:

$$P = \frac{2.m.R.T_C}{V_{SE}.(S - B.\cos(x - a))} \qquad (18)$$

$$T_R = \frac{T_E + T_C}{2} \qquad (19)$$

The mean pressure is:

$$P_{mean} = \frac{1}{2\pi}.\oint P.dx = \frac{2.m.R.T_C}{V_{SE}.\sqrt{S^2 - B^2}} \qquad (20)$$

Thus:

$$P = \frac{P_{mean}.\sqrt{S^2 - B^2}}{S - B.\cos(x - a)} = \frac{P_{mean}.\sqrt{1 - c^2}}{1 - c.\cos(x - a)} \qquad (21)$$

where $c = B/S$. When $\cos(x - a) = -1$, the pressure is minimum. The pressure can be expressed as function of the minimum pressure:

$$P_{min} = \frac{2.m.R.T_C}{V_{SE}.(S + B)} \qquad (22)$$

$$P = \frac{P_{min}.(S + B)}{S - B.\cos(x - a)} = \frac{P_{min}.(1 + c)}{1 - c.\cos(x - a)} \qquad (23)$$

With all these previous equations, it is possible to get the expansion energy (W_E), the compression energy (W_C) and indicated energy (W_i) in function of the minimum pressure:

$$W_E = \oint P.dV_E = \frac{P_{min}.V_{SE}.\pi.c.\sin a}{1 + \sqrt{1 - c^2}}.\frac{\sqrt{1 + c}}{\sqrt{1 - c}} \qquad (24)$$

$$W_C = \oint P.dV_C = -\frac{P_{min}.V_{SE}.\pi.c.X_T.\sin a}{1 + \sqrt{1 - c^2}}.\frac{\sqrt{1 + c}}{\sqrt{1 - c}} \qquad (25)$$

$$W_i = W_E + W_C = \frac{P_{min}.V_{SE}.\pi.c.(1 - X_T).\sin a}{1 + \sqrt{1 - c^2}}.\frac{\sqrt{1 + c}}{\sqrt{1 - c}} \qquad (26)$$

Equations (24), (25) and (26) are the energies obtained when the working gas completes one thermodynamic cycle, i.e. when the cycle of Fig. 1 is complete. As a result, the power of the Stirling engine depends on the number of thermodynamic cycles are done per second. Let n be the mechanical speed of the engine (in cycles/second). The expansion, compression and indicated powers are:

$$L_E = W_E.n \qquad (27)$$

$$L_C = W_C.n \qquad (28)$$

$$L_i = W_i . n \qquad (29)$$

The efficiency of the Stirling engine is:

$$\eta = \frac{W_i}{W_E} = 1 - X_T \qquad (30)$$

Considering no thermal losses, W_e also represents the heat that flow to the Stirling engine from a heat source (geothermal, biomass, etc). Thus, (30) represents the efficiency in the conversion of thermal energy to mechanical energy.

The Stirling engine converts thermal energy (heat) into mechanical energy. This mechanical energy is used to move the rotor of an electric machine to generate electric energy. In this work, a DC generator with permanent magnets in its rotor was considered. The steady-state equations of the DC generator are:

$$T_m = k_1 . i_a \qquad (31)$$

$$E_a = k_1 . \omega_m \qquad (32)$$

$$E_a = v_o + r_a . i_a \qquad (33)$$

where T_m is the motor torque, i_a is the armature current of the machine, E_a is the induced voltage, r_a is the armature resistance, and v_o is the voltage at the terminals of the generator. The proposed model of the connection of the Stirling engine-based generator to the grid only considers steady state conditions. In that case, and neglecting mechanical power losses, the mechanical power generated by the Stirling motor, which is the indicated power L_i, is send to the motor. Therefore:

$$L_i = T_m . \omega_m \qquad (34)$$

where ω_m is the mechanical speed of the motor. In the developed model of the electrical generator using Stirling engine, it is considered that the mechanical speed is known. The estimation of the speed would require a transient model of the Stirling engine, which is beyond the objectives of this paper.

III. CONNECTION OF THE STIRLING-ENGINE-BASED POWER GENERATOR TO THE GRID

The power generated by the Stirling engine is injected into the grid with a cascaded Boost DC-DC converter and a current controlled voltage-source inverter. It was used the model of the Stirling-engine-based power generator explained in the previous section. The simplified topology adopted is depicted in Fig. 3 with details regarding the inverter stage and its control system to proper grid connection. The boost topology works with a fixed duty cycle to provide always a V_{BUS} (V_{DC}) voltage greater than the grid peak voltage. It was used the average model for the Boost converter.

A. Grid Synchronization

In distributed generation, the performance of the synchronization algorithm directly affects the power quality, due to it is part of the control loop strategy. For this application, the PLL algorithm is responsible for synchronizing the current to be exactly in phase and with the same frequency as the AC grid voltage, performing active power injection. Fig. 4 shows the diagram of the used PLL-based system.

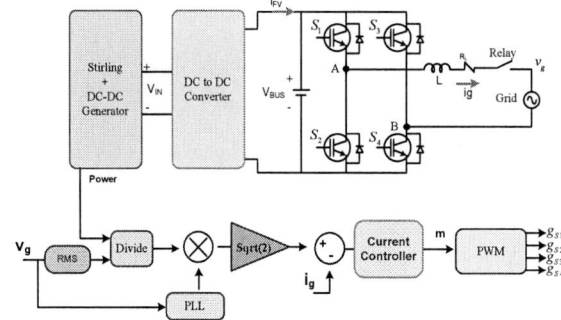

Fig. 3. Simplified topology for grid connection and its control system.

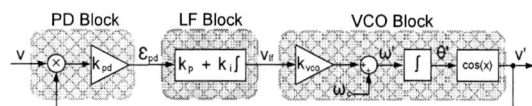

Fig. 4. PLL algorithm for grid synchronization.

The low-pass filter (LF block) is composed by a PI controller and the VCO block is composed by an integrator. The linearized closed loop transfer function of the PLL is defined in (35), considering K_{pd} =1.

$$H(s) = \frac{k_p . s + k_i}{s^2 + k_p . s + k_i} = \frac{2.\xi.\omega_n.s + \omega_n^2}{s^2 + 2.\xi.\omega_n.s + \omega_n^2} \qquad (35)$$

Thus, according to (35) the PLL can be modeled as a 2-order transfer function. Adopting $\xi = 0.9$ and $\omega_n = 37.7$ rad/s, one can found $k_p = 33.9$ and $k_i = 1421.3$.

B. Project of the L-Filter

It was selected a first-order inductive filter (L) due to the low operation power and its simplicity of implementation and control [11]. This inductor acts as a low-pass filter, which attenuates the harmonics created by the inverter, to allow the connection between the converter and the grid. The voltage in the inductor (v_L) is determined by (36).

$$v_L = L . \frac{di_g}{dt} \qquad (36)$$

where i_g is the grid current. The inductance L is selected to minimize the current ripple. The maximum ripple occurs at the grid voltage peak and the voltage across L is the subtraction of this voltage and the DC bus voltage. Let v_{dc} be the value of the voltage at DC bus, v_{gm} be the maximum value of the grid voltage, f_s be the switching frequency, m the modulation index, and Δi_g the desired current ripple. So, the required inductance L is calculated according to (37).

$$L = \frac{m.(v_{dc} - v_{gm})}{f_s . \Delta i_g} \qquad (37)$$

C. Design of the Current Control Loop

From (36) and (37) it is possible to obtain (38) as one can consider the inverter switching as the reduction of the DC bus voltage to a sinusoidal voltage directly proportional to the modulation index m. Equation (38) also considers the inductor resistance (R_L).

$$m. v_{dc} - v_g - R_L. i_g = L. \frac{di_g}{dt} \qquad (38)$$

Considering the small signal analysis and the Laplace Transform, one can derive the current to control transfer function $G_{igm}(s)$ as in (39), considering that the grid voltage is constant in a switching interval.

$$G_{igm}(s) = \frac{Ig(s)}{m(s)} = \frac{V_{DC}}{(s.L+RL)} \qquad (39)$$

A PI plus resonant (PI+Res) controller is adopted to proper inject power to the grid with reduced total harmonic distortion. Equation (40) presents the current controller transfer function [12].

$$C_{PI+Res} = K_p + \frac{K_i}{s} + \frac{K_r.s}{s^2+\omega^2} \qquad (40)$$

Where K_p and K_r are the respective proportional and resonant gains, s is the Laplace operator, ω is the required resonant frequency (377 rad/s) and K_i is the integrator gain. In the project of the gains, it was considered a margin phase of 60° and a cut-off frequency of 2 kHz. The gains are $k_p = 0.06$, $k_i = 475.92$ and $k_r = 950$.

IV. RESULTS

The Stirling-engine-based generator, the converters and the synchronization system were simulated using MATLAB/SIMULINK. The parameters of the simulation are:

- Stirling engine: $n = 295.56$ RPM, $P_{min} = 600000$ Pa (6 bar), $T_C = 50$ °C, $T_H = 850$ °C, $V_{DE} = V_{DC} = 153{\times}10^{-6}$ m³, $V_r = 472.62{\times}10^{-6}$ m³, $V_E = V_C = 472.62{\times}10^{-6}$ m³, $dx = 90°$.

- DC generator: $r_a = 0.893$ Ω, $k_1 = 0{,}616$V.s/rad.

- Boost converter: $L = 4$ mH, $C = 20$ mH, duty cycle $D = 0.65$.

- PLL for the grid synchronization: $k_p = 33.9$, $k_i = 1421.3$.

- L filter: $L = 1.5$ mH;

- Current control loop: $R_L = 0.1$Ω, $k_p = 0.06$, $k_i = 475.92$, $k_r = 950$.

The inverter generates a current in phase with the grid voltage (127 V), as indicated in Table II.

The Stirling engine, the power converters, the DC generator and the controllers were simulated as continuous-time systems.

Fig. 5 shows the simulation diagram of the generator connected to the grid. The indicated power of the Stirling engine is 927.7 W. The DC generator generates 80.9 V. This voltage is send to the boost converter, which gives a mean voltage of 230V. The output voltage of the boost converter is shown in Fig. 6. Fig. 7 shows the grid voltage and the generated current which is in phase with the grid voltage. The active and reactive powers of the grid are 918.2 W and 1.365 VAR, respectively. These values of active and reactive power are obtained by the fact that the generated current is in phase with the grid voltage, as shown in Fig. 8.

TABLE II.   INVERTER MAIN PARAMETERS

Parameter	Value
Nominal Power	1000 W
Vgrid RMS Voltage	127 V
Grid Frequency	60 Hz
Switching Frequency	30 kHz
Current Ripple	10%
L filter	1.5 mH
V_{dc}	250 V

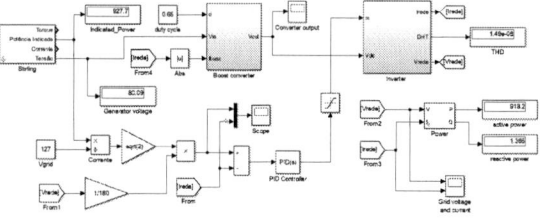

Fig. 5.   Simulation diagram.

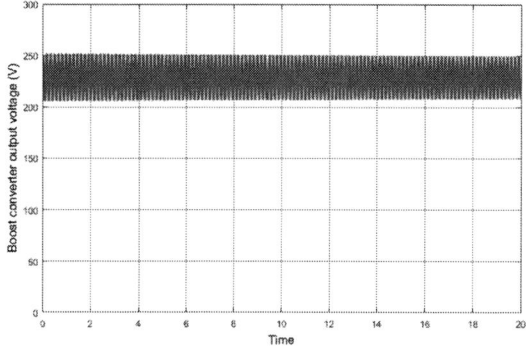

Fig. 6.   Output voltage of the boost inverter.

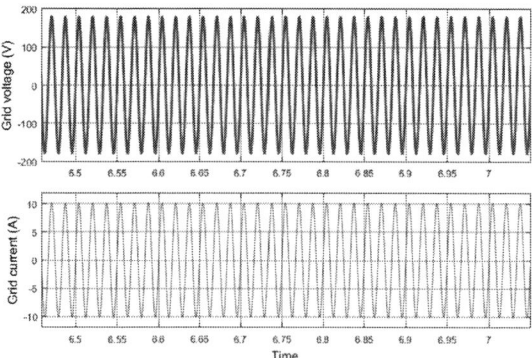

Fig. 7.   Grid voltage and the generated current.

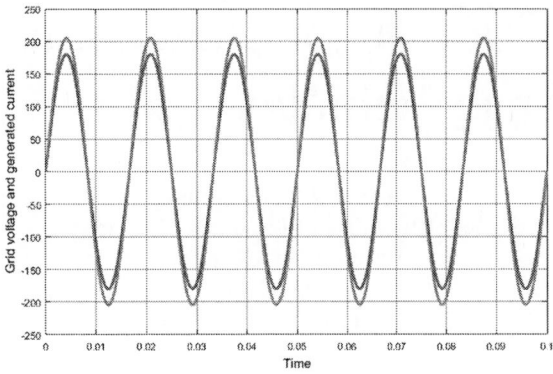

Fig. 8. Grid voltage (red) and the generated current (blue, in scale 20:1).

V. CONCLUSION

This paper shows the modeling and simulation of a DC generator that use an alpha-type Stirling engine, which is connected to the grid. The proposed model allows calculating the grid current and voltage, the active and reactive power generated according to the mechanical and thermodynamic parameters of the Stirling engine. Simulation results show that the Stirling-generator system can give energy to the grid with practically unitary power factor. The aim of this work is to motivate the analysis and use of the Stirling engine in cogeneration. The proposed model shows steady-state results. As future work, transient model of the connection of the Stirling engine to the power grid will be developed.

ACKNOWLEDGMENT

Authors want to thank BATLAB laboratory and Graduation Program in Electrical Engineering, PPGEE-UFMS, for the support to this research.

REFERENCES

[1] K. Yosra, B. Ammar Mohsen, H. A. Hsan. "Modeling and simulation of DSPMSG power generator connected to grid," 16th international conference on Sciences and Techniques of Automatic control & computer engineering, 2015, pp. 656–663.

[2] A. Kulkarni, D. Patil, R. R. Arakerimath, "Experimental Investigation of Heat Sink as a Heat Pipe to Improve Performance of Thermoelectric Generator," International Journal of Scientific Research in Science and Technology, vol. 3, no. 7, pp. 943–948, Oct. 2017 .

[3] K. Yosra, B, H. A. Hsan. "New grid connected hybrid generation system using wind turbine and solar Dish/Stirling engine," 2017 International Conference on Green Energy Conversion Systems (GECS), 2017, pp. 1–7.

[4] A. Viola, M. Trapanense, V. Frazitta, "New solutions for energy supply: sea wave and offshore photovoltaic in Sardinia (Italy)," OCEANS 2018 MTS/IEEE Charleston, 2017, pp. 1–7.

[5] C.R. de Nardin, F.T. Fernandes, F.A. Farret, L.P. Lima, A.J. Longo, M. Godoy Simões, "Surface geothermal energy applied to low cost and low power consumption residential air conditioning," 42nd Annual Conference of the IEEE Industrial Electronics Society - IECON 2016, 2016, pp. 4247–4251.

[6] S. H. Zacchaeus, W. P. Q. Tong, R.T. Naayagi, "Modelling and Simulation of DC microgrid," 2018 IEEE 4th Southern Power Electronics Conference (SPEC), 2018, pp. 1–5.

[7] A. Pietak, M. Meus, S. Nitkiewicz, "The Demand Analysis of the Biogas Plant with Chp System on Substrate, the Aim to Obtain Required Electrical and Thermal Power," Journal of KONES Powertrain and Transport, vol. 19, no. 3, 2012.

[8] P.Surdacki, M. Holuk, K. Bańka, K.Gawkowski, "Investigation of the CHP generation system with the stirling engine," 2017 International Conference on Electromagnetic Devices and Processes in Environment Protection with Seminar Applications of Superconductors (ELMECO & AoS), 2017, pp. 1–4.

[9] J. P. V. Dormael, Theory of Stirling Engine, available in: http://www.robertstirlingengine.com/theory.php

[10] K. Hirata, "SCHMIDT THEORY FOR STIRLING ENGINES", pp. 1–8. Available in: https://www.nmri.go.jp/oldpages/eng/khirata/list/general/schmidt.pdf

[11] Teodorecus, R., Liserre M., Rodrigues P., "Grid converters for photovoltaic and wind power systems", John Willey & Sons, 2011.

[12] M. Liserre, R. Teodorescu, F. Blaabjerg, "Multiple harmonics control for three-phase grid converter systems with the use of PI+RES current controller in a rotating frame," IEEE Transactions on Power Electronics, vol. 21, no. 3, May 2006.

Comparative Strategies of Control for Regenerative Braking in Electric Vehicles

Marina G.S.P. Paredes
School of Electrical and Computer Engineering
University of Campinas
Campinas, SP - Brazil
Email: gabbil@dsce.fee.unicamp.br

José Antenor Pomilio
School of Electrical and Computer Engineering
University of Campinas
Campinas, SP - Brazil
Email: antenor@fee.unicamp.br

Abstract—**This paper approaches an analysis maximum torque control distribution of an electric vehicle for induction machine with traction on the two front wheels. The comparison is based on control strategies between scalar control and direct torque control with space vector modulation (DTC-SVM) during decelerations.**

The strategies focus on critical braking where the combination of regenerative and mechanical braking is necessary to obtain a short distance and recovered energy. The work is simulated on Simulink/Matlab.

Index Terms—**Stored energy, Regenerative torque, Mechanical torque, Induction machine.**

I. INTRODUCTION

Regenerative Braking System (RBS) is a technology to recover energy of electric vehicle (EV) and hybrid electric vehicle (HEV), on braking. The electrical and mechanical braking combination is necessary to obtain braking performance of the vehicle at any circumstance [1], [2].

Modern vehicles make use of technologies such as regenerative braking system (RBS) and Anti-lock braking system (ABS). Regenerative braking system is a technology applicable in both conventional and electric vehicles. The regenerated energy can be stored in batteries, supercapacitors or other devices. ABS allows the wheels of a vehicle to maintain traction while braking, to prevent the wheels from blocking. Hence this way, it helps to avoid uncontrolled skidding [3].

The configuration of the EV case study with traction in two-front-wheels are considered two methods of torque control for driving the induction machine (IM) with space vector modulation (SVM): (i) V/F scalar control and (ii) DTC.

Scalar control is one of the most applied in electric vehicles, but in low frequencies obtain high oscillations and it is difficult to torque control in the driving cycle [4], [5].

Direct Torque Control (DTC-SVM) used by the high power to induction machine reducing oscillations and providing a safety to the passengers, what allows better torque control under decelerations what can be used at electric vehicles [4], [6].

This work develops a simulation tool and proposes control strategies to improve the performance of distributed torque applied to electric vehicles to act during regenerative braking and store energy. The performance analysis considers the capability of the braking system (regenerative plus mechanical)

as well as dynamic modeling of an EV on flat and downhill surfaces [7], [8], [9].

II. STUDY OF VEHICLE MODEL

This study should operate in coordination with the mechanical brake, thus allowing an extension of RBS despite the critical steerability conditions (on flat and downhill surfaces). The vehicle modeling affords an adaptive simulation with both cases of control based on Simulink/Matlab driven by the induction machine with direct torque control (DTC).

The acceleration and braking vehicle performances depend on the available powertrain and vehicle mass. Additionally, it must consider the user safety and comfort [2], [10]. In addition, when a deceleration is required, torque is transferred from the motor and mechanical brakes to the wheels. The braking torque applied by the motor is converted to electric energy and transferred to the storage device (Supercapacitors).

The dynamic simulation of the vehicle model is a part important of study in development research. For this modeling seeks to recovered energy at decelerations during braking, taking into account forces acting on a quarter vehicle is shown in Figure 1, and to Fx denotes the tire braking force at a wheel, the 4-wheels has been represented with fl, fr, rl, and rr (front-left, front-right, rear-left, rear-right), through that will be calculate an estimated mass of a light vehicle.

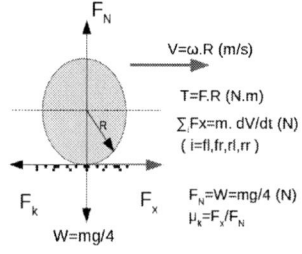

Fig. 1. Wheel longitudinal dynamics

The vehicle total mass used is 1320kg, however, it was reduced to 660kg at simulations for maintaining a deceleration in according with the defined high deceleration above and maximum mechanical torque applied at braking.

978-1-7281-4181-7/19 $31.00 © 2019 IEEE

Table I shows the induction machine model parameters and the torque values, used in the simulation what corresponds to the maximum torque that can be applied to the wheels. As already defined, T_{max}=1.5 T_N. However, to avoid motor overheating during long braking intervals, the motor torque is limited to 240 Nm (T_{Lim}).

The electric power supply is a supercapacitor bank (17F) with initial condition of 380V.

TABLE I
EV PARAMETERS

Induction Machine Power	37kW
Nominal Torque (T_N)	200Nm
IM Maximum Torque (T_{max})	300 Nm
IM Limit Torque (T_{Lim})	240 Nm
Wheel Maximum Torque	750 Nm
Tire Radio (T_R)	0.25 m
Tire rolling resistance coefficient (c_r)	0.015
Rated vertical load	1700 N
Peak longitudinal force at rated load	2500 N
Slip at peak force at rated	10%
Half Vehicle Mass (m)	660 Kg
Path Inclination	0 deg

III. SCALAR CONTROL STRATEGY

To improve scalar control strategy added a closed loop torque control, and the inverter reference is 5% less than the corresponding motor speed applied during the braking by the SP-Control block. This way the induction machine slip is negative, resulting a negative torque. The inverter turns-off at low speeds, avoiding the torque oscillations. Then, Mechanical Torque block applies a friction torque to maintain the same deceleration, as shown in Figure 2.

For the braking stage, the speed reference changes accordingly to the brake pedal and 95% of the motor speed is used as reference.

The torque control block includes two PI regulators. One regulates the speed and produces the torque reference. The second controls the torque loop. It is necessary to know if the required torque is greater than electric motor limit torque. The use of the block "Mechanical Torque" is divided into two parts: (i) "Brake1" and (ii) "Brake2" to apply friction torque according to the torque demand [9], [11].

Figure 3 depicts the flow diagram for two strategies ("Brake1" and "Brake2") to scalar control. The torque developed by the induction motor is directly proportional to the ratio of the applied voltage and the frequency of supply. By varying the voltage and the frequency, but keeping their ratio constant, the torque developed can be kept approximately constant.

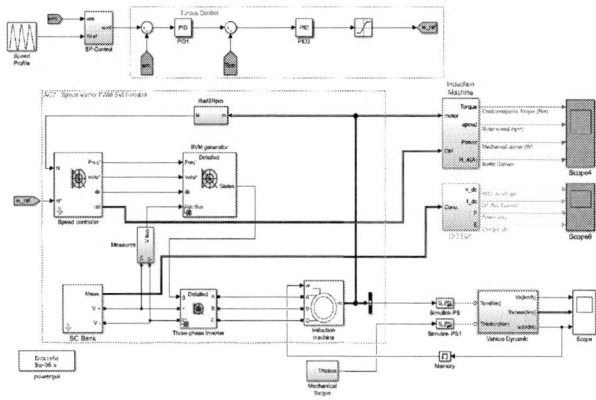

Fig. 2. EV Model System with Scalar Control for driving cycles studies.

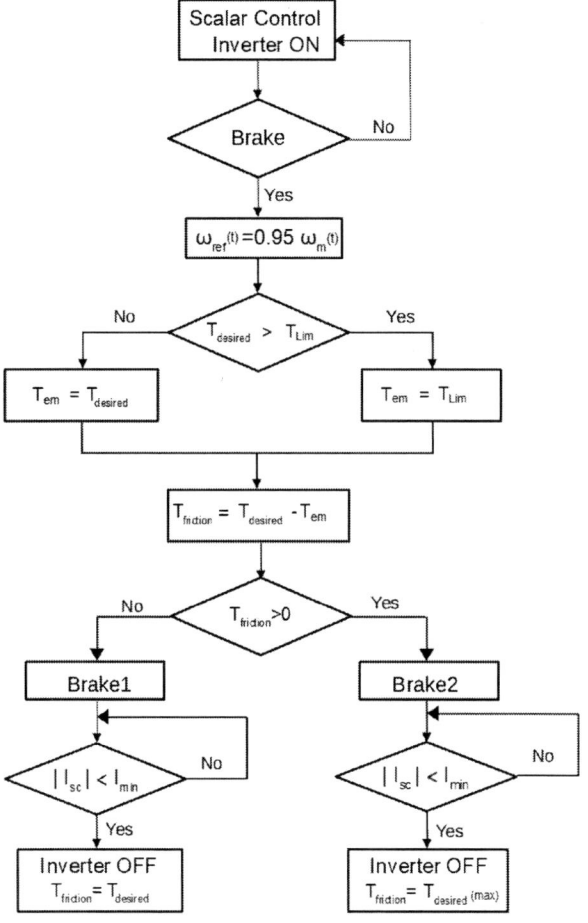

Fig. 3. flow Diagram to Scalar control.

In Brake1 mode the initial friction torque is 0, for a maximum speed reference ω_{ref}. The electrical brake is performed with the desired torque while the current in the SC bank is greater than a minimum value. When the inverter turns-off, the desired torque is applied by the friction brake, maintaining the deceleration and the stability of the vehicle.

In the Brake2 mode the friction torque is greater than zero for a maximum speed reference ω_{ref}. The desired torque is divided between the electrical and the mechanical system. When the regenerative brake losses efficacy, the inverter turns-off and the friction torque assumes the full torque.

IV. DIRECT TORQUE CONTROL STRATEGY

The direct torque control structure is depicted in the "AC4-SVM" red box, shown in the Figure 4. The electrical parameters and the dynamics of the vehicle are the same as for scalar control. Let us consider the intense braking of the vehicle, in which the system uses the motor limit torque of 240 Nm.

Direct torque control (DTC) is based on two PI controllers for torque and flux, which allows better response to the driver's command. It's possible to brake with both, electric and mechanical brakes, in order to achieve maximum torque on the wheels, maintaining the criterion of turning off the electric brake if the motor current is minor than a minimum current [9].

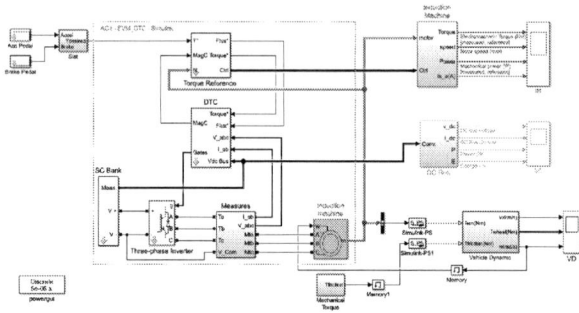

Fig. 4. EV Vehicle Model with DTC

Figure 5 shows the flow diagram of the braking strategies, "Brake1" and "Brake2". The desired torque or torque demand is made by the brake pedal, according to the motor speed vs. torque curve. If the desired torque is greater than the motor limit torque, it is necessary to add the mechanical torque.

Brake1 is performed when the electric brake is enough. The mechanical brake is applied only when the inverter turns-off.

Brake2 uses both, the electric and the mechanical brakes concurrently.

V. RESULTS AND ANALYSIS

A. Maximum Torque Strategy by Scalar Control

Figure 6 shows the comparison of both braking strategies on a dry and flat surface. The Brake1 has a longer braking distance than Brake2, due to the fact that brake demand is lower. The reference speed is applied while the inverter is working.

For the Brake1 mode, the braking torque is variable. This occurs because the scalar strategy defines a speed reference 5% lower than the motor speed. This results a variable slip and, consequently, a variable braking torque. To improve the torque regulation would be necessary to change the speed reference,

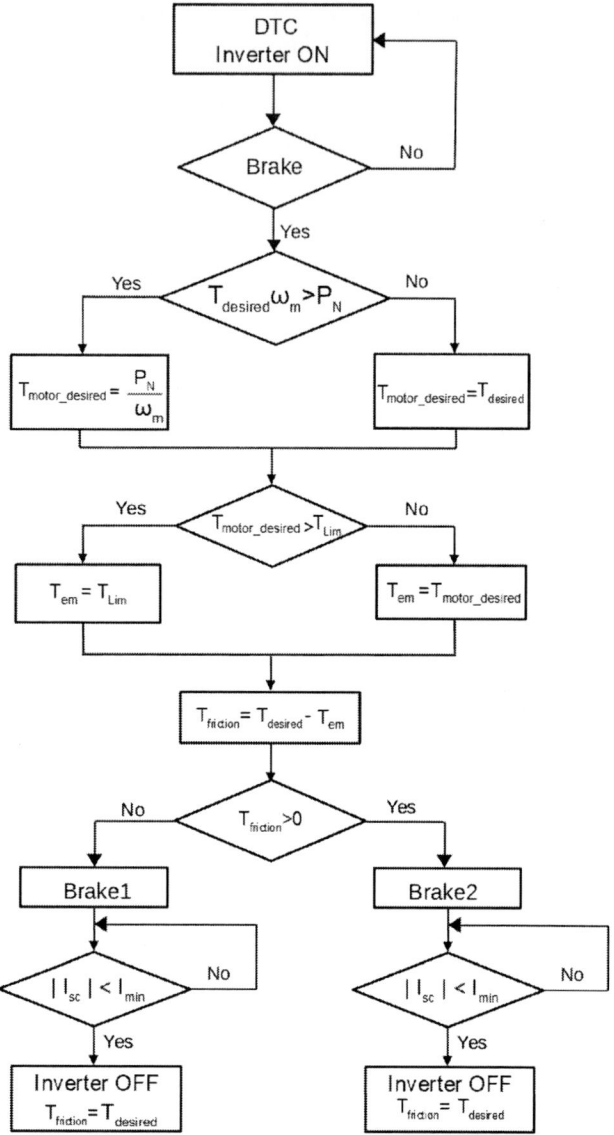

Fig. 5. Direct Control Torque flow Diagram

what doesn't make sense in a simple control strategy as is the scalar control. This will be used in the following DTC strategy.

Table II shows the electrical quantities of the two braking strategies for a flat surface with scalar control. The recovered energy in Brake1 is lower than Brake2 since the initial kinetic energy is lower.

TABLE II
QUANTITIES FOR REGENERATIVE AND MECHANICAL BRAKING

Parameters	Brake1	Brake2
Brake Peak Power (kW)	10.35	19.18
Recovered Energy (kJ)	18.53	32.70

is 14.25m, since the deceleration for Brake2 is greater than Brake1 and it takes a shorter time to stop the vehicle.

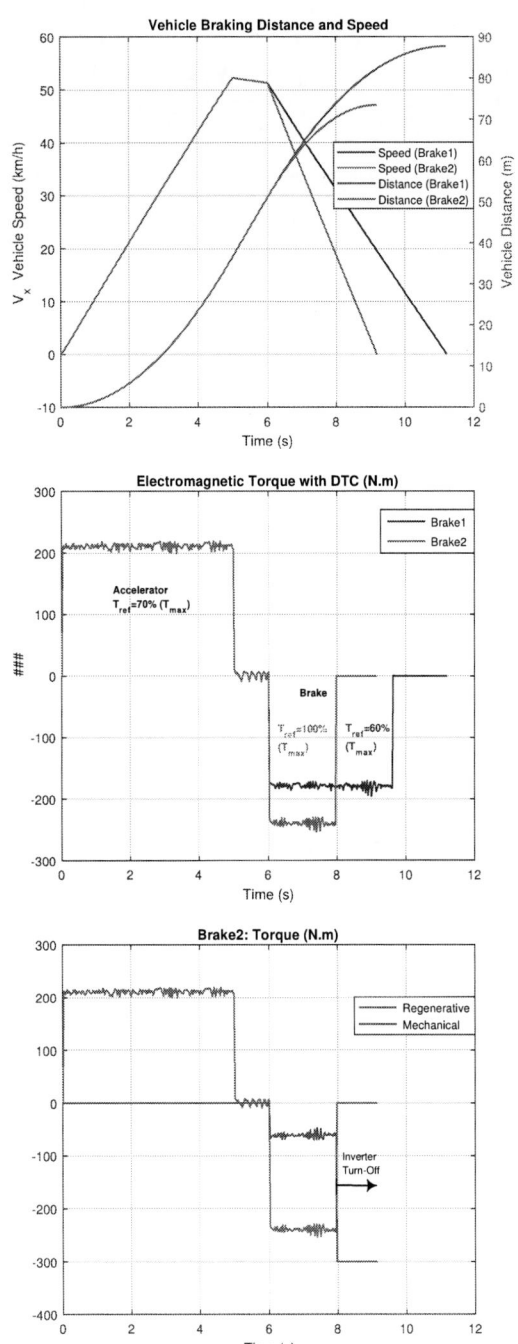

Fig. 7. DTC Simulation on a flat surface.

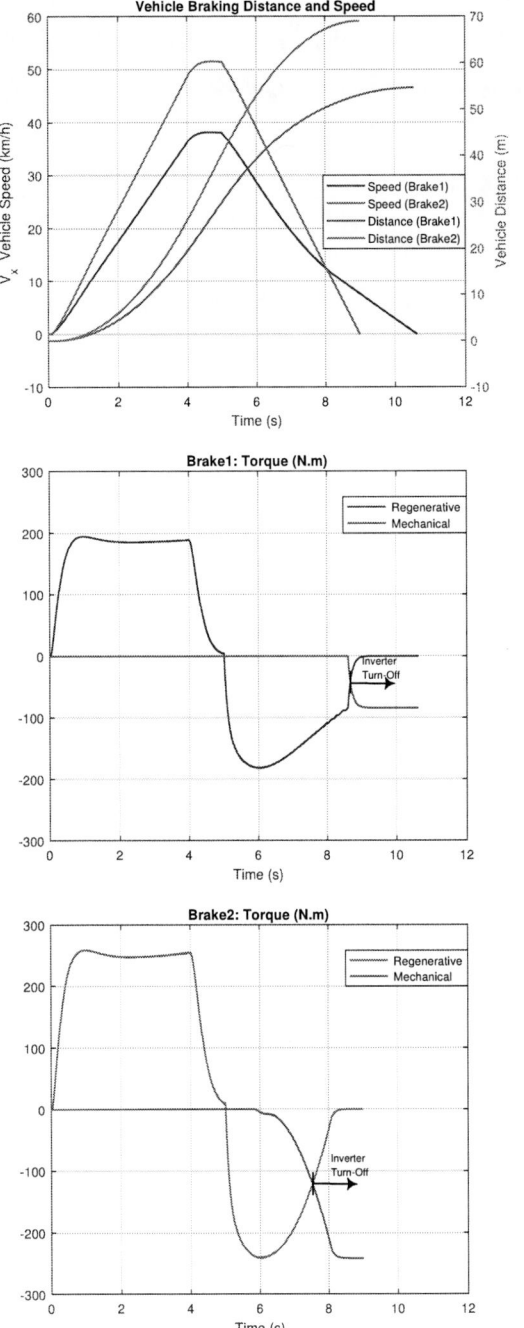

Fig. 6. Simulation Scalar Control on flat surface: Brake1 and Brake 2

B. Direct Torque Control Strategy

The Figure 7 shows simulations when the brake commands 60% of the maximum motor torque (Brake1). If the command corresponds to 100% or more the strategy is Brake2.

For the Brake2, the regenerative energy is lower than Brake1 as the mechanical brake is applied.

The difference braking distance of the vehicle for both cases

The DTC presents a superior behavior compared to the scalar control, due to the precise torque regulation, the following studies consider only DTC systems.

978-1-7281-4181-7/19 $31.00 © 2019 IEEE

TABLE III
QUANTITIES FOR REGENERATIVE AND MECHANICAL BRAKING ON FLAT
SURFACE

Parameters	Brake1	Brake2
Brake Peak Power (kW)	16.95	16.60
Recovered Energy (kJ)	30.12	15.40

C. Comparison between flat/downhill surface by DTC-SVM

This section based on torque strategy describe above with a fixed friction coefficient at 10% slip to dynamic vehicle model. The decelerations are compared and analyzed during two slope angle (0° and -10°).

Figure 8 shows braking during downhill and on the flat surface to compare the traction and braking in both cases, maintaining a fixed coefficient of friction as for a dry surface. In both cases the reference torque changes from an initial braking torque of 60% to 100% of the maximum motor limit torque (Brake 2). The vehicle braking distance obtained between a downhill and on the flat surface is 12m.

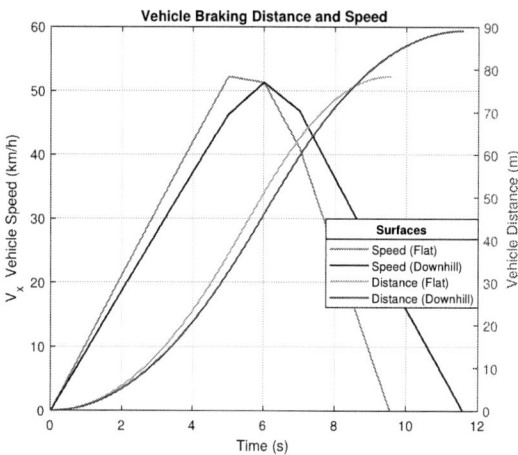

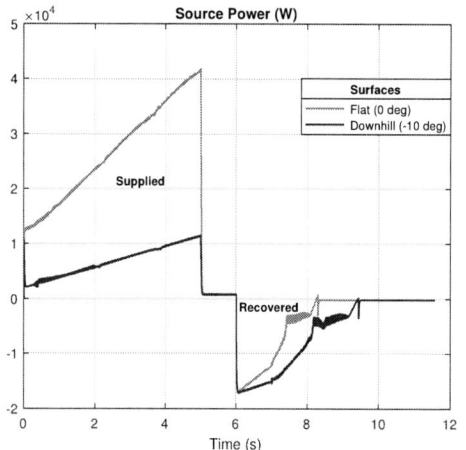

Fig. 8. Electrical Simulation of DTC on flat and downhill surfaces.

Therefore, an adequate vehicle dynamics modeling was required a mechanical torque performance at low frequencies in both cases.

Finally, in downhill surface more energy is recovery since the kinetic energy is summed to the potential energy. For the braking of a light vehicle, it is necessary to have a robust control system with direct torque control, a good option to control the induction motor due to its focus at torque control.

VI. CONCLUSIONS

This work puts forward control strategies for regenerative braking with electric vehicle modeling driven by a single induction machine in a passenger vehicle. Lastly, the proposed dynamic model was improved and incorporate electrical and mechanical control strategies at same simulation. The use of direct torque control (DTC) allowed us to obtain an adequate procedure of braking. Further, the shutdown procedures at low-speed afforded to improve the performance electric (RBS) and mechanical.

REFERENCES

[1] M. G. S. P. Paredes and J. A. Pomilio, "Control system to regenerative and anti-lock braking for electric vehicles," *2018 IEEE Transportation Electrification Conference and Expo (ITEC2018)*, 2018.

[2] P. Gupta, A. Kumar, and S. Deb, "Regenerative Braking Systems (Rbs) (Future of Braking Systems)," *International Journal of Mechanical And Production Engineering*, 2014.

[3] A. Aly, Z. El-shafei, and et al., "An Antilock-Braking Systems (ABS) Control: A Technical Review," *Intelligent Control and Automation*, 2011.

[4] D. B. Hardik D. Desai, "Comparative analysis between scalar control and direct torque control methods for induction motor drives," *International Journal of Current Engineering and Scientific Research (IJCESR)*, 2015.

[5] M. G. S. P. Paredes, J. A. Pomilio, and A. A. Santos, "Combined regenerative and mechanical braking in electric vehicle," *in Proceedings of Brazilian Power Electronics Conference (COBEP)*, 2013.

[6] R. De Oliveira, L. Mattos, and et al., "Regenerative Braking of a Linear Induction Motor used for the Traction of a Maglev Vehicle," *Brazilian Power Electronics Conference (COBEP)*, 2013.

[7] S. Pay and Y. Baghzouz, "Effectiveness of battery-supercapacitor combination in electric vehicles," *IEEE Bologna PowerTech - Conference*, 2015.

[8] K. X. Guoqing Xu, Weimin Li and Z. Song, "An intelligent regenerative braking strategy for electric vehicles," *in Energies 2011, 4, 1461-1477*, 2011.

[9] M. G. S. P. Paredes, "Study of electric vehicle modeling and strategy of torque control for regenerative and anti-lock braking systems," Ph.D. dissertation, School of Electrical and Computer Engineering, State University of Campinas. Brazil, 2018.

[10] A. Santos, "A contribution to the study of friction brakes for application in regenerative braking," Ph.D. dissertation, School of Mechanical Engineering, State University of Campinas. Brazil, 2009.

[11] D. P. M. Aravind Samba Murthy and D. G. Taylor, "Vehicle braking strategies based on regenerative braking boundaries of electric machines," *in Transportation Electrification Conference and Expo (ITEC)*, 2015.

978-1-7281-4181-7/19 $31.00 © 2019 IEEE

Theoretical Solution of the Output Voltage Harmonic Spectra of Dual-Inverter Fed Open-End Winding Loads with Dead Time Effect

Brunno Monteiro Guimarães Ribeiro
Graduate Program in Electrical Engineering (PPGEL)
Federal Center for Tech. Education of Minas Gerais (CEFET-MG)
Belo Horizonte, Brazil
brunnomgr@gmail.com

Marcelo Martins Stopa
Graduate Program in Electrical Engineering (PPGEL)
Federal Center for Tech. Education of Minas Gerais (CEFET-MG)
Belo Horizonte, Brazil
marcelo@des.cefetmg.br

Alex-Sander Amável Luiz
Graduate Program in Electrical Engineering (PPGEL)
Federal Center for Tech. Education of Minas Gerais (CEFET-MG)
Belo Horizonte, Brazil
asal@des.cefetmg.br

Abstract—**Open-end winding electrical machines with one solid-state converter connected to each end of the stator windings are being considered in recent researches for numerous applications including high-power multilevel converters. Due to the absence of the neutral point, this topology creates a path for zero-sequence current, which affects the operation of the machine, increases the losses in the stator windings and raises the voltage of the shaft, leading to potential problems in the bearings. Although many strategies have been proposed to avoid the application of space vectors that generate common-mode voltage across the windings, zero-sequence current still flows due to nonidealities like voltage drops in the solid-state switches and the dead time applied to avoid DC link shoot-through. This paper presents the use of the double Fourier series adapted to extract symbolically the coefficients of the harmonic spectra of open-end winding loads with dual-inverter subjected to dead-time effect. This approach is verified using examples and the predictions are confirmed with the results obtained from Simulink simulations. The influence of dead time in the quality of the synthesized voltage is demonstrated by spectral analysis.**

Keywords — *Dead Time Effect, Double Fourier Series, Dual-Inverter, Harmonic Spectra, Open-End Windings, Theoretical Solutions.*

I. INTRODUCTION

Multilevel converters offer advantages compared to the traditional two-level such as higher voltage capability, lower switching losses, increased power quality and lower dv/dt. These features associated with advancements in semiconductors devices have made them a proven and mature technology in the industry, currently commercialized for a wide range of applications and one of the preferred choice for high voltage and high power applications [1]-[4]. Open-end windings (OEW) electrical machines fed by two sets of three-phase two-level frequency inverters (Fig. 1) have been pointed, since the first researches about the topic [5]-[6], as an alternative to increase drive's capacity while offering other advantages like: elimination of voltage sharing problems, power circuit simplification [7], increased output voltage levels [8], increased reliability [9] and lower switching frequencies [6]. The use of this topology as an alternative to traditional multilevel high-power converters is mentioned by more recent publications [1]-[2].

One of the main challenges regarding the application of the OEW drives is caused by the absence of the neutral point, as it creates a path for zero-sequence currents, which causes the copper losses and cause drive derating [5]. The basic solutions to eliminate this drawback are the use of zero-

sequence reactors or two isolated DC links but due to higher costs, additional losses and space constraints [10] these alternatives might not be possible or feasible. PWM strategies that avoid space vectors that generate common-mode voltages in OEW loads with single DC links were reported in [11]-[12]. Nevertheless, even for this topology, nonidealities such as device voltage drop and dead time (necessary to avoid DC link shoot-through) cause distortions in the output voltages, including zero-sequence voltages. Among these nonidealities, dead time is the one that is expected to cause the major impact [13].

The impact of the dead time on the harmonic spectrum of the output voltage of an inverter might be calculated and assessed in detail by the analytical method in [14]. It is based on unit cells and uses the double Fourier series, which avoids round off errors and physical interferences in the results [15]. Other researches presented also the modelling and the analytical impact of the dead time using different mathematical approaches [16] and for different topologies, including three-level neutral-pointed-clamped inverter [17]. There are also papers exploring dead time issues on OEW drives [18]-[20], but although the presence of zero-sequence voltages have deeper consequences on this topology, the authors have not found any publication dedicated to the study and theoretical determination of their impacts on the output voltage harmonic spectrum of OEW loads.

Aiming to provide a contribution to the analysis of the distortions caused by this nonideality and in order to account for them mathematically in dual-inverter OEW topologies, the authors utilized the dead time effect analysis in [21] to adapt the limits of the integrals that determine the coefficients of the double Fourier series of the phase leg of a voltage source inverter (VSI) [15], for a given pulse-width modulation (PWM) scheme. The output voltage harmonic spectrum of each pole (with respect to the central terminal of the DC input source) are associated to determine the theoretical output voltage harmonic spectrum with and without the dead time effect of each winding of a OEW load fed by two VSIs.

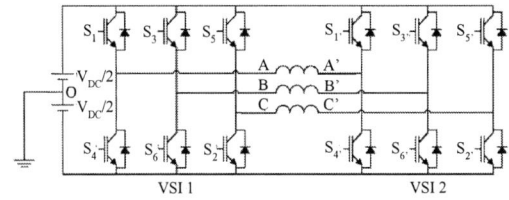

Fig. 1 – Dual-inverter OEW single DC link circuit diagram (adapted from [12]).

This approach has made possible the assessment and study in detail of the impact of zero-sequence harmonics caused by the PWM strategy and by the dead time, offering useful information to compare different modulation schemes, to guide dead time compensation strategies and to establish references to gauge its effectiveness. Simulations results from Simulink are utilized to verify this approach.

This text is organized as follows: Section II presents the modelling of the dead time effect based on the analysis in [21]. Section III presents the theoretical solution of the output voltage harmonic spectra of a VSI phase leg commanded by double-sided natural sampling (DSNS) and the adaptations necessary to account for the dead time and to determine the winding voltages of a dual-inverter OEW topology. Section IV presents the theoretical and simulated spectra with and without dead time effect (based on the section III models) and its analysis. Section V presents the conclusions of this paper followed by references.

II. DEAD TIME MODELLING [21]

In order to mitigate the storage time of the solid-state switches in VSIs, it is necessary to insert a time delay between the on and off commands of complementary switches in the same phase leg, avoiding DC link shoot-through. Besides imposing a minimum pulse width, the dead time distorts the output voltages of the VSI, especially in higher switching frequencies.

To evaluate how the distortions caused by the dead time affect the output voltages the following assumptions are made:

- The reverse storage time of solid-state switches are neglected;
- The switching frequency is much higher than the fundamental frequency of the reference;
- The voltage deviation caused by the dead time in each pulse are uniformly distributed;
- Short pulse dropping does not occur;
- Inverter's output current is nearly sinusoidal;

Considering the assumptions above, the cumulative effect caused by the cyclic time delays (i.e. the dead time effect) can be determined by averaging voltage deviations over each half cycle of the output current. The volt-second area deviation for each pulse is determined by

$$\Delta e = T_d V_{DC} \qquad (1)$$

where T_d is the time delay and V_{DC} is the DC link voltage

and the average voltage deviation over a half cycle of the inverter output is

$$\Delta V = f_c \Delta e = f_c T_d V_{DC} \qquad (2)$$

where f_c is the switching frequency

In addition, the magnitude of the fundamental component is

$$\Delta V_1 = \frac{4}{\pi} \Delta V \qquad (3)$$

Fig. 2 shows how these deviations affect the phase output voltage of a VSI feeding an inductive load with phase angle ϕ (where v_{ref} represents the ideal fundamental output

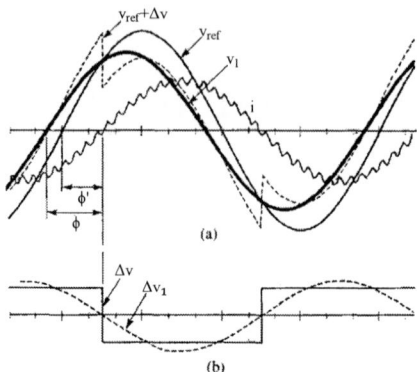

Fig. 2 – a) Dead time impact on VSI fundamental pole voltage; b) Dead time effect represented by a square wave 180° out of phase with load current (adapted from [21]).

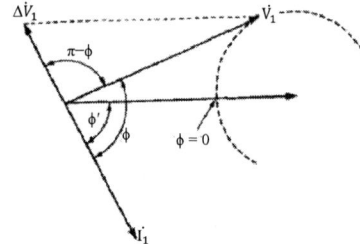

Fig. 3 – Phasor diagram of the of the VSI fundamental pole voltage with dead time effect (adapted from [21]).

voltage and v_1 the output voltage affected by the dead time – modelled as a square-wave 180° out of phase with the current and with magnitude ΔV). The phasor diagram in Fig. 3 describes the same relationships (phasors are highlighted with asterisks). It determines a second-order function by applying trigonometrical identities (4), whose solution is the magnitude of V_1 (5):

$$V_{ref}^2 = \Delta V_1^2 + V_1^2 - \Delta V_1 V_1 \cos(\pi - \phi) \qquad (4)$$

$$V_1 = -\Delta V_1 \cos\phi + \sqrt{V_{ref}^2 - (\Delta V_1 \sin\phi)^2} \qquad (5)$$

Additionally, the output voltage is affected by the harmonics of the square-wave that represents the dead time effect. Equation (6) determine their magnitudes.

$$\Delta V_h = \frac{1}{h} \Delta V_1 \qquad (6)$$

where $h = 2k - 1$ (harmonic order) when $k \geq 2$ and integer

III. THEORETICAL SOLUTION OF THE HARMONIC SPECTRA

A. Double-Sided Natural Sampling

The harmonic components of a VSI output voltage signal can be calculated using theoretical solutions that determine their magnitude and phase accurately and establish a reference to assess different modulation strategies under the same basis [15].

Considering f(x,y) as a time-dependent function determined by the modulation product between the reference and carrier periodic signals, it is possible to calculate its harmonic coefficients using their Fourier series and unit cells

978-1-7281-4181-7/19 $31.00 © 2019 IEEE

(that defines areas in which its modulation product is constant). The coefficients are complex numbers and their magnitudes and phases are determined by an expanded version of this series to a double variable form (7), in which m (carrier index) and n (baseband index) define the frequency of each harmonic $\omega_h = m\omega_c + n\omega_0$ (ω_c and ω_0 are the angular frequency of the carrier and reference, respectively).

$$\overline{C_{mn}} = \frac{1}{2\pi^2} \int_{y_{min}}^{y_{max}} \int_{x_{min}}^{x_{max}} f(x,y) e^{j(mx+ny)} dxdy \quad (7)$$

where $y = \omega_0 t + \theta_0$ and $x = \omega_c t + \theta_c$

Considering a VSI phase leg fed by a V_{DC} DC voltage source and commanded by a double-sided natural sampling PWM technique with a sinusoidal reference, the function and the limits of (7) can be adapted to determine the coefficients of the pole voltage harmonic spectrum (8) [15] (subscript x represents any of the poles of each VSI).

$$\overline{C_{v_{xo_{mn}}}} = \frac{1}{2\pi^2} \int_{-\pi}^{\pi} \int_{-\frac{\pi}{2}(1 + M\cos(y-\theta_r))}^{\frac{\pi}{2}(1 + M\cos(y-\theta_r))} V_{DC} e^{j(mx+ny)} dxdy \quad (8)$$

where M is modulation index; $y = \omega_0 t + \theta_0$; $x = \omega_c t + \theta_c$ and

θ_r is the phase displacement with respect to v_{ao}

B. Double-Sided Natural Sampling with Dead Time Effect

Based on the model presented in Section II, the described effects of the dead time on the output voltage harmonic components may be included in the reference waveform, emulating its effects through the simple modification of the integral limits of (8), which consequently changes the unit cells. Equation (9) shows these modifications to account for the dead time effects on the output voltage of a phase leg commanded by the same double-sided natural sampling technique with a sinusoidal reference and fed by a V_{DC} DC source (subscript x represents any of the poles of each VSI). Comparing to (8) and recurring to (5) and (6) and to Fig. 2 and Fig. 3 it is possible to observe in (9) how the dead time decreases the fundamental component and adds multiple harmonics to pole voltage harmonic spectrum.

$$\overline{C_{v_{xo_{mn}}}} = \frac{1}{2\pi^2} \int_{-\pi}^{\pi} \int_{-\frac{\pi}{2}(1 + V_1 \cos(y-\theta_r) + x_{sum})}^{\frac{\pi}{2}(1 + V_1 \cos(y-\theta_r) + x_{sum})} V_{DC} e^{j(mx+ny)} dxdy \quad (9)$$

where $x_{sum} = \sum_{h=2k-1}^{\infty} \frac{1}{h} \frac{\Delta V_1}{V_{DC}} \sin(h(y - \theta_r) - \cos^{-1}(PF) - \pi)$;

$$V_1 = \sqrt{M^2 - \left(\frac{\Delta V_1}{V_{6\text{-step}}} \operatorname{sen} \phi\right)^2} - \frac{\Delta V_1}{V_{6\text{-step}}} \cos \phi$$

when $k \geq 2$ and integer; $V_{6\text{-step}} = \frac{2\sqrt{6}}{\pi} V_{DC} \frac{\sqrt{2}}{\sqrt{3}} \frac{\sqrt{3}}{2}$ and

ϕ is the angle of the impedance

Equations (8) and (9) are the basic building blocks for the strategy presented in this paper. They are utilized for calculating the harmonic coefficients of each pole voltage without and with dead time effect, respectively. Based on them, and using equations (10)-(14), one can easily determine the winding voltages, zero-sequence voltage and winding voltages with no zero-sequence harmonic coefficients,

respectively (subscripts according to Fig. 1 - subscript x represents any of the poles of each VSI).

$$\overline{C_{v_{aa'_{mn}}}} = \overline{C_{v_{ao_{mn}}}} - \overline{C_{v_{a'o_{mn}}}} \quad (10)$$

$$\overline{C_{v_{bb'_{mn}}}} = \overline{C_{v_{bo_{mn}}}} - \overline{C_{v_{b'o_{mn}}}} \quad (11)$$

$$\overline{C_{v_{cc'_{mn}}}} = \overline{C_{v_{co_{mn}}}} - \overline{C_{v_{c'o_{mn}}}} \quad (12)$$

$$\overline{C_{v_{0_{mn}}}} = \frac{\overline{C_{v_{aa'_{mn}}}} + \overline{C_{v_{bb'_{mn}}}} + \overline{C_{v_{cc'_{mn}}}}}{3} \quad (13)$$

$$\overline{C_{v_{xx'\text{-w/oSeq.0}_{mn}}}} = \overline{C_{v_{xx'_{mn}}}} - \overline{C_{v_{0_{mn}}}} \quad (14)$$

IV. HARMONIC SPECTRA AND ANALYSIS

In order to demonstrate the approach presented above and to analyze the impact of the zero-sequence components on the output voltage harmonic spectra of a dual-inverter feeding an OEW load, a Matlab script was written to symbolic solve the double integrals that determine the harmonic components of each pole voltage (8)-(9). Winding voltages harmonic components are calculated using (10)-(14). The following scenario is analyzed:

- Two three-phase two-level VSIs with ideal IGBT switches with a common DC link feeding a three-phase RL-load from both ends (Fig. 1);

- IGBTs commanded using double-sided natural sampling modulation technique with sinusoidal references phase shifted by 120° between the poles of the same inverter and 180° between the opposite poles of the different inverters. The same triangular carrier is used for both VSIs;

- The main parameters used are modulation index $M_i = 0.9$, DC link voltage $V_{DC} = 600 \text{ V}$, sinusoid reference frequency $f_0 = 60 \text{ Hz}$ and triangular carrier frequency $f_C = 1260 \text{ Hz}$ (carrier ratio $f_C/f_0 = 21$).

To verify the approach of Section III and its symbolic calculations of the harmonic coefficients, simulations using the same parameters above (with Runge-Kutta solver and 5e-7 fixed step) were performed in Simulink and the harmonic spectra extracted using the FFT tool of the SimPowerSystems library. The results are shown in Fig. 4 to Fig. 9, in which the blue bars represent theoretical spectra and the red bars the simulated spectra from 0 to the 60th order (for simplification purposes, only the voltage harmonic spectra related to winding a are presented. They are the same for b and c windings).

Pole voltage harmonic components are shown in Fig.4. It features the reference fundamental component, the carrier fundamental and the equally distributed groups of even (odd) harmonic sideband components around the carrier fundamental odd (even) multiples. Although poles a and a' voltage harmonic components have the same magnitude, they differ from each other by the phases of some components. When subtracting them to extract the winding voltage harmonic components (10), this phase difference caused by the 180º phase shift between the sinusoidal references, leads to the cancellation of the odd carriers and their associated sideband harmonics, but leaves the odd sideband harmonic components around the even carrier groups (Fig. 5). It means

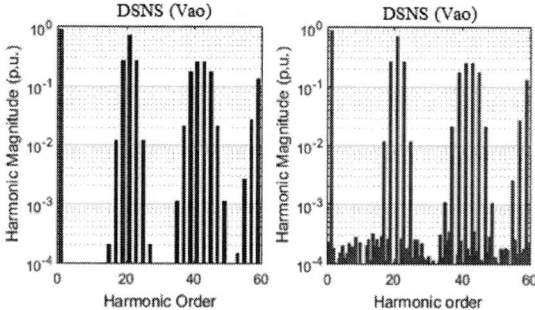

Fig. 4 –Pole voltage harmonic spectrum (theoretical: blue; simulated: red).

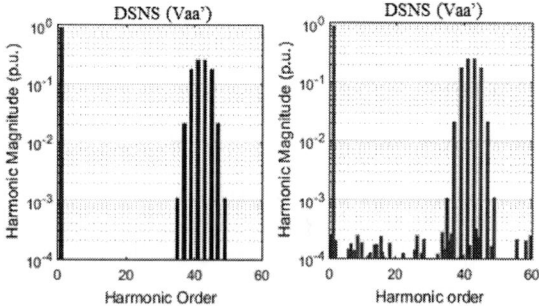

Fig. 5 - Winding voltage harmonic spectrum (theoretical: blue; simulated: red).

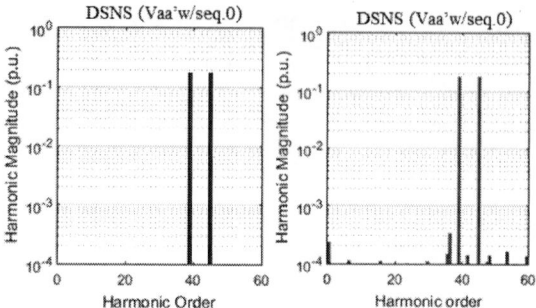

Fig. 6 – Winding zero-sequence voltage harmonic spectrum (theoretical: blue; simulated: red).

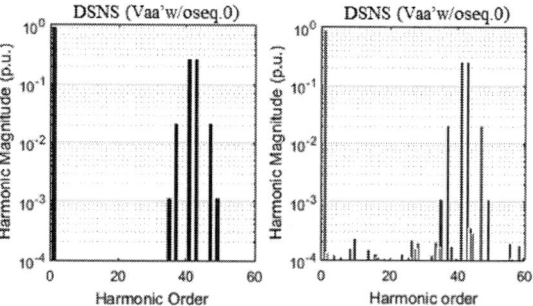

Fig. 7 - Winding voltage without zero-sequence harmonic spectrum (theoretical: blue; simulated: red).

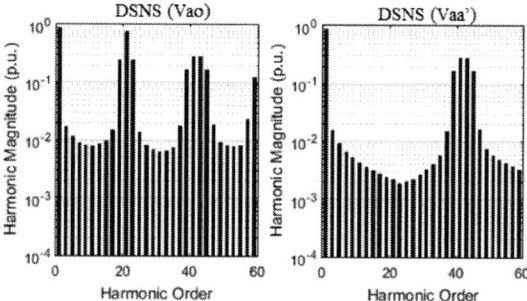

Fig. 8 - Pole voltage and winding voltage harmonic spectra with dead time effect.

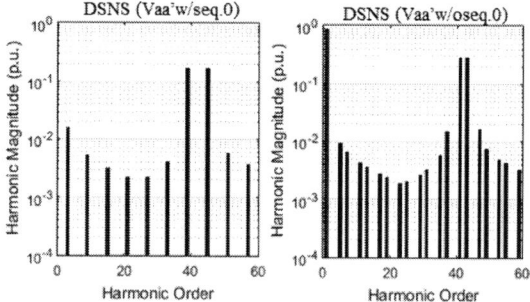

Fig. 9 – Winding zero-sequence voltage and winding voltage without zero sequence harmonic spectra with dead time effect.

that the winding voltage harmonic spectra of a dual-inverter OEW with 180° phase shift has the same profile of the harmonic spectra of the output voltage of a single phase H-bridge VSI.

Comparing the winding voltage of a dual-inverter OEW to the traditional three-phase two-level VSI line voltage, this topology presents less harmonic content (odd carriers and their associated sideband harmonics are absent) and have a higher DC link usage. However, as expected for topologies with common DC link (13), it does not lead to the cancellation of the triplen harmonics (3 ±6h) - in this case 39th and 45th (Fig. 6). As an example, if zero-sequence reactors or independent DC link for each VSIs is adopted, these triplen harmonics (zero-sequence components) are absent in the winding voltage (Fig 7).

If a dead time of 2% of the carrier period and a RL-load with displacement power factor of 0.9 were considered, the pole voltage harmonic spectra would exhibit low-order harmonics that deteriorates its quality, as pointed in (6) and (9). Consequently, it also has effects on the winding voltage harmonic coefficients, which are still absent of the odd carriers and their associated sideband harmonics, but keeps the triplen harmonics (Fig. 8). The additional triplen harmonics, with the remaining 39th and 45th harmonic orders, and the winding voltage harmonic spectra without the zero-sequence components are shown in Fig. 9. Additionally, the expected fundamental voltage magnitude decrease (5) caused by the dead time is observed ($M_i = 0.858$).

V. CONCLUSIONS

This paper presented a theoretical approach to determine the coefficients of the harmonic spectra of a dual-inverter fed OEW load based on the double Fourier series. It was shown how the pole voltages of the VSI interact with each other by cancelling certain harmonic components in the same way as a single-phase H-bridge. It was possible to determine the zero-sequence voltage presence on the winding voltages. Similarly, harmonic spectra of the same topology with dead time effect was studied. It was shown how this nonideality influences the quality of the output voltages, strengthening the common sense that they must be compensated or eliminated, especially in dual-inverter OEW loads, and establishing a basic reference to assess the effectiveness of the compensation methods. The results were verified by simulation that presented similar results.

REFERENCES

[1] H. Abu-Rub, J. Holtz, J. Rodríguez, and G. Baoming, "Medium-voltage multilevel converters - State of the art, challenges and requirements in industrial applications," IEEE Trans. Ind. Electron., vol. 57, no. 8, pp. 2581–2596, Aug. 2010.

[2] S. Kouro, M. Malinowski, K. Gopakumar, J. Pou, L. Franquelo, B. Wu., J. Rodríguez, M. A. Pérez, and J. I. Leon, "Recent advances and industrial applications of multilevel converters," IEEE Trans. Ind. Electron., vol. 57, no. 8, pp. 2553–2580, Aug. 2010.

[3] J. Rodríguez, J-S. Lai, and F. Z. Peng, "Multilevel inverters: A survey of topologies, controls and applications," IEEE Trans. Ind. Electron., vol. 49, no. 4, pp. 724-738, Aug. 2002.

[4] P. K. Steimer and M. D. Manjrekar, "Practical medium voltage converter topologies for high power applications," in Conf. Rec. IEEE 36th IAS Annu. Meeting, Sept. 30–Oct. 4, 2001, pp. 1723-1730.

[5] I. Takahashi and Y. Ohmori, "High-performance direct torque control of an induction motor," IEEE Trans. Ind. Appl., vol.25, no. 2, pp. 257-264, Mar./Apr. 1989.

[6] H. Stemmler and P. Guggenbach, "Configurations of high-power voltage source inverter drives," in Proc. 5th EPE, 1993, vol. 5, pp. 7-14.

[7] K. A. Corzine, S. D. Sudhoff, and C. A. Whitcomb, "Performance characteristics of a cascaded two-level converter," IEEE Trans. Ener. Conv., vol. 13, no. 3, pp. 433-439, Sept. 1999.

[8] E. G. Shivakumar, K. Gopakumar, S. K. Sinha, A. Pittet, and V. T. Ranganathan, "Space vector PWM control of dual-inverter fed open-end winding induction motor drive," in APEC 200116th Annu. IEEE Appl. Power Electron. Conf. and Expo., Mar. 4-8, 2001, pp. 399-405.

[9] B. A. Welchko, T. A. Lipo, T. M. Jahns, S. E. Schulz, "Faul tolerant three-phase AC motor drive topologies: A comparison of features, cost and limitations," IEEE Trans. Power. Electron., vol. 19, no. 4, pp. 1108-1116, Jul. 2004.

[10] T. Kawabata, E. C. Ejiogu, Y. Kawabata, and K. Nishiyama, "New open-winding configurations for high-power inverters," in Proc. ISIE '97 IEEE Intl. Symp. Ind. Electron., Jul. 7-11, 1997, pp. 457-462

[11] M. R. Baiju, K. K. Mohapatra, R. S. Kanchan, and K. Gopakumar, "A dual-two-level inverter scheme with common mode voltage elimination for an indcution motor drive," IEEE Trans. Power. Electron., vol. 19, no.3, pp. 794-805, May. 2004.

[12] B. Zhu, U. R. Prasanna, K. Rajashekara, and H. Kubo, "Comparative study of PWM strategies for three-phase open-end winding induction motor drives," IPEC 2014 Intl. Power Electron. Conf., May 18-21, 2014, pp. 395-402.

[13] A. Somani, R. K. Gupta, K. K. Mohapatra, and N. Mohan, "On the causes of circulating currents in PWM drives with open-end winding AC machines," IEEE Trans. Ind. Electron., vol. 60, no. 9, pp. 3670-3678, Sept. 2013.

[14] C. M. Wu, W-H. Lau, and H. S-H Chung, "Analytical technique for calculating the output harmonics of an H-bridge inverter with dead time," IEEE Trans. Circ. Sys.: Fund. Theory Appl., vol. 46, no. 5, pp. 617-627, May. 1999.

[15] D. G. Holmes and T. A. Lipo, " Pulse width modulation for power converters," New York: Wiley, 2003.

[16] D. C. Moore, M. Odavic, and S. M. Cox, "Dead-time effects on the voltage spectrum of a PWM inverter," IMA Journ. Appl. Mathem., vol. 79, no. 6, pp. 1061-1076, 2014.

[17] I. Dolguntseva, R. Krishna, D. E. Soman, and M. Leijon, "Contour-based dead-time harmonic analysis in a three-level neutral-point-clamped inverter," IEEE Trans. on Ind. Eletron., vol. 62, no. 1, Jan. 2015.

[18] M. Darijevic, M. Jones, and E. Levi, "Analysis of dead-time effects in a five-phase open-end drive with unequal DC link voltages," in Conf. Rec. IEEE 39th Annu. Conf. of the IEEE Ind. Electron. Soc., Nov. 10-13, 2013, pp. 5136-5141.

[19] N. Kalantari and L. A. C. Lopes, "Reduction of dead-time effect on the common mode voltage of an open-end winding machine," IEEE 25th Intl. Symp. Ind. Electron (ISIE), Jun. 8-10, 2016, pp.836-841.

[20] Q. An, J. Liu, Z. Peng, L. Sun, and L. Sun, "Dual-space vector control of open-end winding permanent magnet sychronous motor drive fed by dual-inverter", IEEE Trans. Power. Electron., vol. 31, no. 12, pp. 8329-8342, Dec. 2016.

[21] S-G. Jeong and M-H. Park, "The analysis and compensation of dead-time effects in PWM inverters," IEEE Trans. Ind. Electron., vol. 38, no. 2, pp-108-114, Apr. 1991.

978-1-7281-4181-7/19 $31.00 © 2019 IEEE

BRIDGELESS BUCK-BOOST PFC CONVERTER WITH THREE-STATE SWITCHING CELL

Douglas Carvalho Morais[1], Falcondes José Mendes de Seixas[1], Luís de Oro Arenas[1], Lucas Carvalho Souza[1], Luciano de Souza da Costa e Silva[2]

[1]São Paulo State University (UNESP), School of Engineering, Ilha Solteira – São Paulo – Brazil
[2]Federal Institute of Goiás (IFG) – Jatai – Goiás – Brazil
e-mail: morais.sccp.carvalho@gmail.com, falcondes.seixas@unesp.br, ldeoroa@gmail.com, lucas.souza@ifg.edu.br, lucianocosta_@hotmail.com

Abstract – **In this paper, a bridgeless buck-boost converter with power factor correction is proposed, using a three-state switching cell named of B configuration, operating in continuous conduction mode. The bridgeless technique allows the bridge rectifier inclusion switched on power grid frequency, in the high frequency DC-DC converter stage, reducing the stages number required and increasing the structure reliability. Due to three-state switching cell configuration, there is a current stress division by half of the inductor current in all of circuit semiconductors. Moreover, the ripple frequency on converter energy storage elements is twice switching frequency value, which make easier the design of input and output passive filters and reduces the converter's volume and weight. Preliminary simulation results prove the converter effectiveness on power factor correction, keeping output voltage stable in a specified operation point, with a total harmonic distortion of the input current less than 2%.**

Key-words – *bridgeless buck-boost converter, three-state switching cell, power factor correction, total harmonic distortion, continuous conduction mode.*

I. INTRODUCTION

Nowadays, in the input stage, most of electronic equipment need AC-DC conversion use a rectifier bridge with diodes and/or thyristors. However, if there is no a second stage to do Power Factor Correction (PFC), the non-linear elements switched in the grid frequency, will cause a distortion on the input current waveform regarding input voltage, reducing the Power Factor (PF).

Amid this issue, over the last few years, solutions were proposed focusing in to reduce the Total Harmonic Distortion (THD) of the input current which such equipment imposes on the power grid. These solutions can be divided into two great categories, passive and active. Active solutions implement DC-DC converters switched at high frequency, and they are preferable regard to passive solutions because [1]:

➢ Weight and volume reduced by increasing of switching frequency;

➢ High efficiency and power density;

➢ Output voltage regulation possibility.

The inherent active solutions advantages can be get directly using converters as well as buck, boost or buck-boost. Such topologies make use the two-state switching cell, which is known as canonic cell as well [2]. Due to these topologies wide use in power factor correction, it is common to appear techniques that aim the power electronic processing at higher levels.

Among available techniques for this propose, the high frequency rectification or half-controlled has been employed in several topologies, being them a single/three-phase input

[3-5]. The high frequency rectification propitiates the inclusion of the diode rectifier at low frequency, in the high frequency DC-DC converter stage, this technique is also known as bridgeless [3]. The bridge rectifier suppression switched in low frequency, provides to equipment a larger reliability, since less stages are required. Moreover, conduction losses are decreased, because the semiconductors quantity used in the high-frequency operation stages, is less regarding classic converter structure.

Proposed in [6], the Three-State Switching Cell (3SSC) concept comes up. This technique proposes the change of the conventional canonic cell containing one passive and one active semiconductor, by a configuration containing two passive and, two active semiconductors, and one autotransformer [6-9]. This technique allows current stress division in all semiconductors by half of the inductor current. Moreover, ripple frequency of the passive elements is equivalent to twice switching frequency stipulated. Therefore, the design of input and output filters are facilitated and occurs the converter weight and volume reduction.

Considering the advantages of both techniques, bridgeless and 3SSC, this paper proposes the union of such concepts applied to buck-boost converter for a novel topology operating as PFC stage. The buck-boost converter choice is because that it is able to provide larger autonomy under power grid perturbations. The operation occurs in Continuous Conduction Mode (CCM), and uses the average current control. In section II, converter operation principle and design of passive elements are detailed. In section III, 3SSC bridgeless buck-boost converter small-signals model is presented, beyond main transfer functions. Lastly, in section IV, simulation results are presented in order to prove the converter effectiveness in power factor correction.

II. 3SSC BRIDGELESS BUCK-BOOST CONVERTER OPERATION PRINCIPLE WITH CELL B

The three-state switching cell B configuration proposed in [6], can be seen in Fig. 1a. Bilateral inversion is presented in Fig. 1b. Regardless of the configuration used, always between 'a' and 'b' points it must have a voltage source or capacitive branch connected, already the 'c' point must contain the connection with a current source or inductive branch. The 3SSC bridgeless buck-boost converter is shown in Fig. 1c. In order to realize a converter qualitative analysis, it is assumed that:

➢ The turns ratio of the central tap autotransformer is unitary;

➢ Passive and active components are ideals;

➢ The gate signals are delayed of 180º between S_1 and S_2, as well as the gate signals between S_3 and S_4;

978-1-7281-4181-7/19 $31.00 © 2019 IEEE

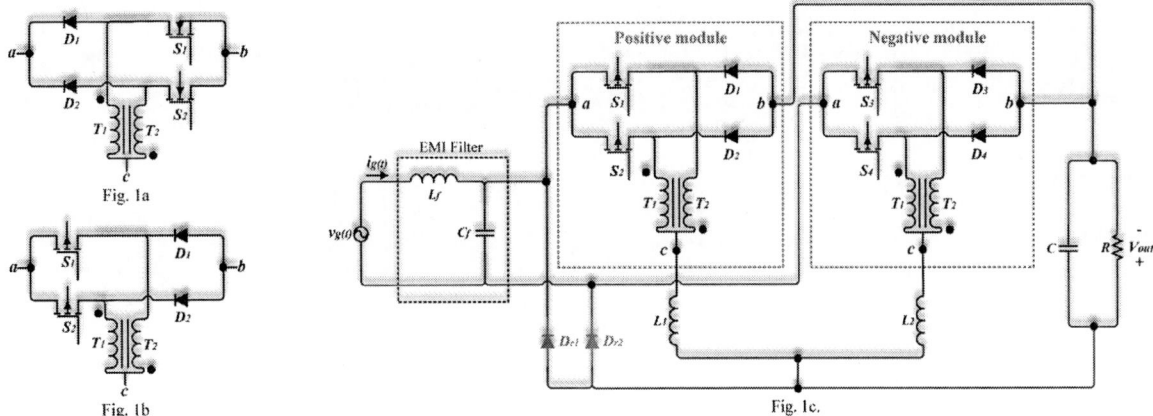

Fig. 1a

Fig. 1b

Fig. 1c.

Figure 1a: Cell B; Figure 1b: Bilateral inversion of cell B; Figure 1c: Proposed PFC bridgeless buck-boost converter with cell B

➤ The switching frequency is constant and modulation is by pulse width.

Besides that, due to the magnetic effect provided by the autotransformer unitary turns ratio, windings voltages are considered equal, as well as the currents, that is:

$$v_{T1(t)} = v_{T2(t)} \qquad (1)$$

$$i_{T1(t)} = i_{T2(t)} \qquad (2)$$

The conduction of any semiconductor presented in Fig. 1c depends on the input voltage semi-cycle, so that: in the positive semi-cycle, the cell B high-frequency switching contained in the positive module occurs, beyond the D_{r2} diode conduction; On the other hand, in the negative semi-cycle, the cell B high-frequency switching contained in the negative module occurs, with the D_{r1} diode conduction. However, it is important to note that the same gate pulses are simultaneously sent to S_1 and S_3, as well as to S_2 and S_4, being the conduction of these, dependent of input voltage semi-cycle. Therefore, it implies that no complexity in the controller project is added, since there is no delay in the gate pulses between modules are required, only between each module semiconductor. Although each cell conduction of bridgeless buck-boost converter depends on the input voltage semi-cycle, the high frequency operation stages of both modules are identical, that is, the converter analysis is independent of input voltage semi-cycle.

Thus, since the bridgeless buck-boost converter switching frequency is much higher than the power grid frequency, a larger stages number in the switching frequency occur in an only semi-cycle in low-frequency. For convenience, considering the high frequency stages that occurs in the input voltage positive semi-cycle, that is, the cell contained in the positive module and the diode D_{r2}, the bridgeless buck-boost converter presents different operation stages, depending on the duty cycle value used. In this case, for duty cycle lower than 0.5, then the converter operates as step-down, buck characteristic. In the other hand, for duty cycle higher than 0.5, then the converter operates as step-up, boost characteristic. Thus, the following analysis can be divided for $d < 0.5$ and $d > 0.5$, respectively [9].

A. Operation Stages for $d < 0.5$ (Buck Mode).

As step-down, there is four operation stages whose semiconductor states are presented in Table I. As can be seen in Table I, for operation with $d < 0.5$ not occurs the overlapping gate signals in the active semiconductors, thus,

is usual to call this operation as Non-Overlapping Mode (NOM).

TABLE I: Semiconductor gate signal for $d < 0.5$.

Semiconductor	Δt_1	Δt_2	Δt_3	Δt_4
S_1	ON	OFF	OFF	OFF
S_2	OFF	OFF	ON	OFF
D_1	OFF	ON	ON	ON
D_2	ON	ON	OFF	ON

First Stage [Δt_1]: According Table I, only S_1 is turned-on, turning-off diode D_1. In the other branch, D_2 turn-on, since there is no gate signal in S_2. The circuit relating to this stage can be seen in Fig. 2a. By analyzing the circuit shown in Fig. 2a, it is possible through the Kirchhoff law, to obtain the inductor voltage and the capacitor current. Thus, for the inductor voltage loop, it is found:

$$v_{L(t)} - v_{T1(t)} - vg_{ret(t)} = 0 \qquad (3)$$

$$v_{L(t)} + v_{T2(t)} + vc_{(t)} = 0 \qquad (4)$$

Replacing (1) into (3) and (4), and putting in evidence inductor voltage, is obtained:

$$v_{L(t)} = L\frac{d}{dt}i_{L(t)} = \frac{1}{2}\left(vg_{ret(t)} - vc_{(t)}\right) \qquad (5)$$

Similarly, the capacitor current will be:

$$i_{o(t)} = i_{R(t)} + i_{c(t)} \qquad (6)$$

As shown in Fig. 2a and considering (2), the output current is equivalent to half inductor current. Thus, replacing in (6) and putting in evidence the capacitor current, is obtained:

$$i_{c(t)} = C\frac{d}{dt}v_{c(t)} = \frac{1}{2}i_{L(t)} - i_{R(t)} \qquad (7)$$

Second Stage [Δt_2]: In the second stage, there is no gate signal sent to active semiconductors, thus, both diodes turn-on, disconnecting the input voltage source from main circuit. In this way, the second stage circuit can be seen in Fig. 2b. The inductor voltage and the capacitor current can be obtained applying Kirchhoff's law. Thus, in the inductor voltage loop, is obtained:

$$v_{L(t)} - v_{T1(t)} + vc_{(t)} = 0 \qquad (8)$$

$$v_{L(t)} + v_{T2(t)} + vc_{(t)} = 0 \qquad (9)$$

So, for (8) and (9) can be true simultaneously, in both winding voltages must be zero, characterizing a short-circuit, thus:

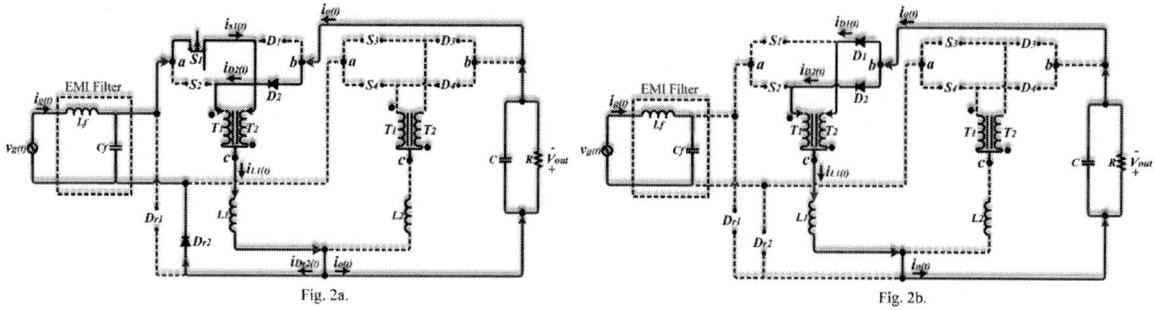

Fig. 2a. Fig. 2b.

Figure 2a: First operation stage in NOM; Figure 2b: Second operation stage in NOM.

$$v_{L(t)} = L \frac{d}{dt} i_{L(t)} = -vc_{(t)} \qquad (10)$$

Again, the capacitor current for second stage is given by (6). However, the output current $i_{o(t)}$ in this case is equivalent to total inductor current $i_{L(t)}$, thus, it is obtained:

$$i_{c(t)} = C \frac{d}{dt} v_{c(t)} = i_{L(t)} - i_{R(t)} \qquad (11)$$

Third Stage [Δt_3]: The third stage, though presents S_2 and D_1 turned-on, is equal to first stage, so (3) to (7) can be used.

Fourth Stage [Δt_4]: The fourth stage is identical to the second one, being valid from (8) to (11).

B. Operation Stage for d > 0.5 (Boost Mode)

As step-up, there is fourth operation stages whose semiconductor states are presented in Table II.

TABLE II: Semiconductor gate signal for $d > 0.5$.

Semiconductor	Δt_1	Δt_2	Δt_3	Δt_4
S_1	ON	ON	ON	OFF
S_2	ON	OFF	ON	ON
D_1	OFF	OFF	OFF	ON
D_2	OFF	ON	OFF	OFF

As can be seen, occurs overlapping gate signals sent to active semiconductors, so, in an opposite way to last case, is usual to call this operation as Overlapping Mode (OM)

First Stage [Δt_1]: In the first stage, both active semiconductors are turn on, so that diodes D_1 and D_2 remain turn-off. In this case, output capacitor supplies the energy required by load. The circuit relating to first stage can be seen in Fig. 3a. Using Kirchhoff's law, the inductor voltage loop is given by:

$$v_{L(t)} - v_{T1(t)} - vg_{ret(t)} = 0 \qquad (12)$$

$$v_{L(t)} + v_{T2(t)} - vg_{ret(t)} = 0 \qquad (13)$$

Since the autotransformer windings voltages are necessarily zero, constituting a short circuit, the inductor current can be obtained as follows:

$$v_{L(t)} = L \frac{d}{dt} i_{L(t)} = vg_{ret(t)} \qquad (14)$$

In this stage, the output current is zero, so the capacitor current is given by:

$$i_{c(t)} = C \frac{d}{dt} v_{c(t)} = -i_{R(t)} \qquad (15)$$

Second Stage [Δt_2]: The second operation stage in OM is identical to first stage in NOM and can be seen in Fig. 2a or Fig. 3b. In this way, it is valid again (3) to (7).

Third Stage [Δt_3]: The third stage is identical to first one in OM, being (12) to (15) are also valid.

Fourth Stage [Δt_4]: The fourth stage is complementary to the second one, so the same circuit shown in Fig. 3b is used.

C. Voltage Static Gain and Average Inductor Current

Regardless the duty cycle value used, the pulse width sent to any active semiconductors is given by:

$$\Delta t_1 = d \cdot T_s \qquad (16)$$

Being T_s the switching period. Adding, the time for first and second stages, it must be equal to half switching period:

$$\Delta t_1 + \Delta t_2 = \frac{T_s}{2} \rightarrow \Delta t_2 = T_s \cdot \left(\frac{1}{2} - d\right) = d' \cdot T_s \qquad (17)$$

Where $d' = (1/2 - d)$. Considering that inductor current and capacitor voltage are constant in a switching period, (5), (7), (10) and (11), can be written as:

$$\Delta_{IL(1)} = \frac{D \cdot T_s}{2L}(vg - V_c) \qquad (18)$$

$$\Delta_{VC(1)} = \frac{D \cdot T_s}{C}\left(\frac{1}{2}I_L - \frac{V_c}{R}\right) \qquad (19)$$

$$\Delta_{IL(2)} = \frac{-V_c}{L}D' \cdot T_s \qquad (20)$$

$$\Delta_{VC(2)} = \frac{D' \cdot T_s}{C}\left(I_L - \frac{V_c}{R}\right) \qquad (21)$$

A similar procedure can be made for $d > 0.5$. The voltage static gain can be found considering that the inductor energy variation in a switching period is zero, that is:

$$\Delta_{IL(1)} + \Delta_{IL(2)} = 0 \qquad (22)$$

Regardless operation mode used, NOM or OM, it can be shown that the static gain of average voltage is given by:

$$\frac{DT_s}{2L}(vg - V_c) - \frac{V_c}{L}D'T_s \rightarrow \frac{V_c}{V_g} = G_M = \frac{D}{1-D} \qquad (23)$$

Similarly, the average energy of capacitor is zero during one switching period, thus:

$$\frac{DT_s}{C}\left(\frac{1}{2}IL - \frac{V_c}{R}\right) + \frac{D'T_s}{C}\left(IL - \frac{V_c}{R}\right) \rightarrow I_L = \frac{V_c}{R \cdot D'} \qquad (24)$$

Both results presented in (23) and (24), coincide with the values found for buck-boost converter structure with canonical cell [10].

D. Inductance and Capacitance of the Bridgeless Buck-Boost Converter with Cell B

Due to rectified sine aspect of the input source, inductance calculation must take into account this input voltage variation

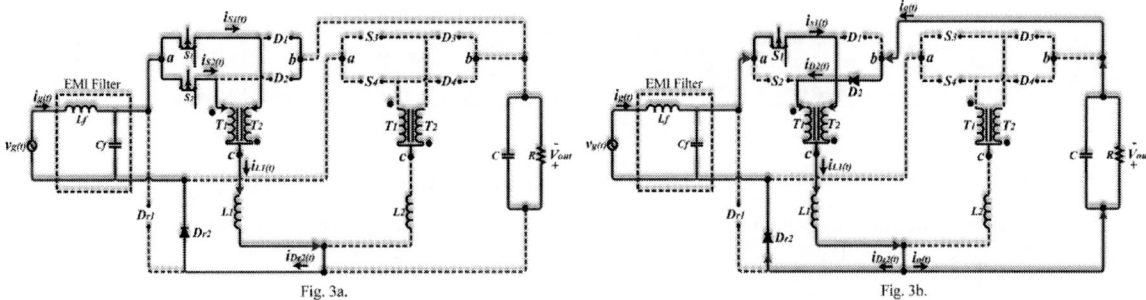

Fig. 3a.   Fig. 3b.

Figure 3a: First operation stage in OM; Figure 3b: Second operation stage in OM.

at its most critical point, thus, sine wave peak. Considering (18), the current can be written as:

$$\Delta_{IL(1)} = \frac{T_s}{2L} \cdot \left(vg_{ret(t)} - V_c\right) \cdot d_{(t)} \qquad (25)$$

Being,

$$vg_{ret(t)} = V_m \cdot |\sin(\omega t)| \qquad (26)$$

And, replacing (26) in (23),

$$d_{(t)} = \frac{1}{1 + M \cdot |\sin(\omega t)|} \qquad (27)$$

In (26), V_m represents the peak of the input voltage and ωt the angular frequency of power grid. In (27), M represents the inverse of the minimum voltage gain, given by:

$$M = \frac{V_m}{V_c} \qquad (28)$$

Replacing (26) and (27) into (25), the inductor current variation parameterized, is given by:

$$\Delta_{IL(1)}^{*} = \frac{M \cdot |\sin(\omega t)| - 1}{1 + M \cdot |\sin(\omega t)|} \qquad (29)$$

Being,

$$\Delta_{IL(1)}^{*} = \frac{2L \cdot \Delta_{IL(1)}}{V_c \cdot T_s} \qquad (30)$$

To find the critical point, or maximum point, it's enough simply derived (29) and equal zero, in this way:

$$\cos(\omega t) \cdot (V_m + M \cdot V_c) = 0 \qquad (31)$$

Thus, the only possible value in the range of 0 to π that make the left side of (31) null is $\omega t = \pi/2$. Replacing this value into (29) and solving as L function, is obtained the NOM inductance given by:

$$L_{NOM} = \frac{Vc \cdot T_s}{2\Delta_{IL}} \left(\frac{M-1}{M+1}\right) \qquad (32)$$

Similarly, the OM inductance is given by:

$$L_{OM} = \frac{V_m \cdot T_s}{\Delta_{IL}} \left[\frac{\sin(\alpha)}{1 + M \cdot \sin(\alpha)} - \frac{1}{2} \cdot \sin(\alpha)\right] \qquad (33)$$

Being α a M dependent constant, given by:

$$\alpha = arcsen\left(\frac{\sqrt{2}-1}{M}\right) \qquad (34)$$

The output capacitance of the 3SSC bridgeless buck-boost converter, can be calculated with based on the output power P_{out}, line frequency f_{line}, and too:

$$C = \frac{P_{out}}{4\pi \cdot f_{line} \cdot V_c \cdot \Delta V_c} \qquad (35)$$

In the next section, will be presented the 3SSC bridgeless buck-boost converter small-signals model, beyond the main Transfer Functions (TF's).

III. SMALL-SIGNALS MODEL FOR THE BRIDGELESS BUCK-BOOST CONVERTER WITH CELL B

For the small-signals model construction, it's used the system perturbation and linearization method. A more detailed method explanation can be found in [10]. Firstly, are expressed the inductor average voltage, capacitor average current, and the input average current, all calculated for one switching period, thus:

$$L\frac{d}{dt}i_{L(t)} \approx \left(\bar{v}_{g(t)} + \bar{v}_{c(t)}\right) \cdot d_{(t)} - \bar{v}_{c(t)} \qquad (36)$$

$$C\frac{d}{dt}v_{c(t)} \approx \bar{\imath}_{L(t)} \cdot d'_{(t)} - \frac{\bar{v}_{c(t)}}{R} \qquad (37)$$

$$i_{g(t)} \approx \bar{\imath}_{L(t)} \cdot d_{(t)} \qquad (38)$$

Perturbation and Linearization: Input voltage source and duty cycle are considered system inputs and can be written as the sum of their steady state values, added to small signals variations represented in the Laplace domain, that is:

$$v_{g(s)} = V_g + \hat{v}_{g(s)} \quad e \quad d_{(s)} = D + \hat{d}_{(s)} \qquad (39)$$

Input signals perturbations affect the other circuit state variables, being these: inductor current; capacitor voltage; and input current. Replacing in (36) to (38) the perturbations described in (39), the referred functions can be rewritten as follows:

$$L\frac{d}{dt}\hat{\imath}_{L(s)} = \left(V_g + V_c\right) \cdot \hat{d}_{(s)} + D\hat{v}_{g(s)} - D'\hat{v}_{c(s)} \qquad (40)$$

$$C\frac{d}{dt}\hat{v}_{c(s)} = D' \cdot \hat{\imath}_{L(s)} - I_L \cdot \hat{d}_{(s)} - \frac{\hat{v}_{c(s)}}{R} \qquad (41)$$

$$\hat{\imath}_{g(s)} = D \cdot \hat{\imath}_{L(s)} + I_L \cdot \hat{d}_{(s)} \qquad (42)$$

The terms of (40) to (42) which are multiplied by the duty cycle small signal's variation are considered as controlled sources. The other terms can be combined into ideal transformers with turns ratio given by average duty cycle. Moreover, the derivatives in (40) and (41) are considered as the inductor voltage and the capacitor current, respectively. Thus, (40) to (42) correspond to an electric circuit, which can be seen in Fig. 4. From the small signals model, it is possible to obtain any circuit transfer function through the superposition principle. In the bridgeless buck-boost converter case with cell B proposed in this paper, it is desired that the input current follows a rectified sine-wave reference.

978-1-7281-4181-7/19 $31.00 © 2019 IEEE

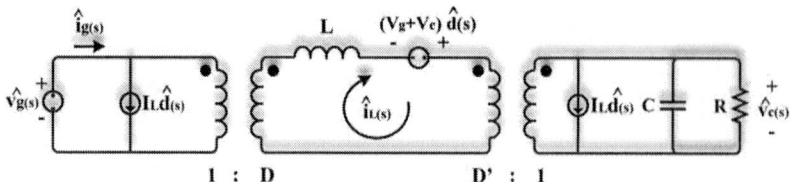

Figure 4: Small signal's model of the proposed bridgeless buck-boost converter with cell B.

Thus, the transfer function must relate input current with duty cycle. One way to obtain such transfer function is to first find the relationship between inductor current and duty cycle, which is given in (43):

$$\hat{i}_{L(s)} = G_{iLd_{cc}} \cdot \left(\frac{1 + \dfrac{s}{\omega zid}}{1 + \dfrac{s}{Q \cdot \omega pid} + \dfrac{s^2}{\omega pid^2}} \right) \cdot \hat{d}_{(s)} \quad (43)$$

Being:

$$G_{iLd_{cc}} = \frac{2V_c + V_g}{R \cdot (D')^2} \quad (44)$$

$$\omega zid = \frac{2V_c + V_g}{(V_c + V_g) \cdot R \cdot C} \quad (45)$$

$$Q = R \cdot D' \cdot \sqrt{C/L} \quad (46)$$

$$\omega pid = D'/\sqrt{L \cdot C} \quad (47)$$

In (44) the DC gain is obtained. In (45) and (47) the zero and pole frequency system are represented, respectively. In (46) the second-order system quality factor is obtained. Replacing (43) into (42) and manipulating some terms, the transfer function that relates input current and duty cycle is obtained and given by:

$$G_{igd(s)} = G_{igdCC} \cdot \left(\frac{1 + \dfrac{s}{\omega zid_1} + \dfrac{s^2}{\omega zid_2}}{1 + \dfrac{s}{Q \cdot \omega pid} + \dfrac{s^2}{\omega pid^2}} \right) \quad (48)$$

Being:

$$G_{igdCC} = I_L + \frac{2V_c + V_g}{R \cdot (D')^2} D \quad (49)$$

$$\omega zid_1 = \frac{Q \cdot \omega pid \cdot \omega zid}{I_L \cdot \omega zid + Q \cdot G_{iLd_{cc}} \cdot \omega pid} \quad (50)$$

$$\omega zid_2 = I_L/\omega pid^2 \quad (51)$$

It is necessary now to relate the same input current also with the output voltage. A way to get such transfer function is firstly finding a relation between output voltage and duty cycle, which is given in (52):

$$G_{vd(s)} = G_{vdCC} \cdot \left(\frac{1 - \dfrac{s}{\omega zvi}}{1 + \dfrac{s}{Q \cdot \omega pid} + \dfrac{s^2}{\omega pid^2}} \right) \quad (52)$$

Being:

$$G_{vdCC} = \frac{V_g + V_c}{D'} \quad (53)$$

$$\omega zvi = \frac{(V_c + V_g)}{L \cdot I_L} D' \quad (54)$$

The direct relationship between output voltage and input current is obtained by division of (52) by (48), which is given by:

$$G_{vi(s)} = G_{viCC} \cdot \left(\frac{1 - \dfrac{s}{\omega zvi}}{1 + \dfrac{s}{\omega zid_1} + \dfrac{s^2}{\omega zid_2}} \right) \quad (55)$$

Being:

$$G_{viCC} = \frac{(V_c + V_g) \cdot R \cdot D'}{I_L \cdot R \cdot (D')^2 + 2V_c + V_g} \quad (56)$$

It is valid mentioning the zero present in the right half-plane in (52) and (55), traditional buck-boost converter MCC characteristic. The controllers design should be based on (48) and (55) and follow the methodology detailed in [9]. It is also important to say that the same transfer functions are found independently of the duty cycle used, that is, the same controller is used for NOM and OM.

IV. SIMULATION RESULTS

In Table III, the main parameters of the 3SSC bridgeless buck-boost converter are presented.

TABLE III: Parameters for design.

Description	Symbology	Values
Input voltage (RMS)	Vg(t)	220V/60Hz
Output voltage (d < 0.5)	Vout	100V
Output voltage (d > 0.5)	Vout	400V
Output power	Pout	1000W
Switching frequency	fs	50kHz

In NOM, that is, step-down (buck mode), the input and output voltage and input current are shown in Fig. 5a. Similar result in OM as step-up (boost mode) is also shown in Fig. 5a. As noted, in both operation modes, the input current is very close to a sinusoidal waveform, while the output voltage remains at the operation point specified in Table III. When d < 0.5, a power factor of 0.9994 with a THD of 1.82% were obtained. At d > 0.5, THD was 1.55% with a PF of 0.9999. In both cases, the 3SSC bridgeless buck-boost PFC converter meets the IEC 61000-3-2 standard [11].

Moreover, as expected by mathematical analysis, in the NOM or OM operation, the current that flow through S_1 or S_2 semiconductor is half of the inductor current. Also, as can be seen in Fig. 5b, for both cases, the ripple frequency in the inductor is twice switching frequency value stipulated. This is possible because, the inductor charge and discharge two times in a switching period. It is worth noting that the same result is found for the capacitor ripple frequency. It is also important to note that, as power factor correction, the 3SSC bridgeless buck-boost converter operating as step-down, will transit between NOM and OM operations mode.

For example, considering the output voltage shown in Table III for NOM as operation point. Always that input voltage is smaller than output voltage, the converter needs to

978-1-7281-4181-7/19 $31.00 © 2019 IEEE   1215

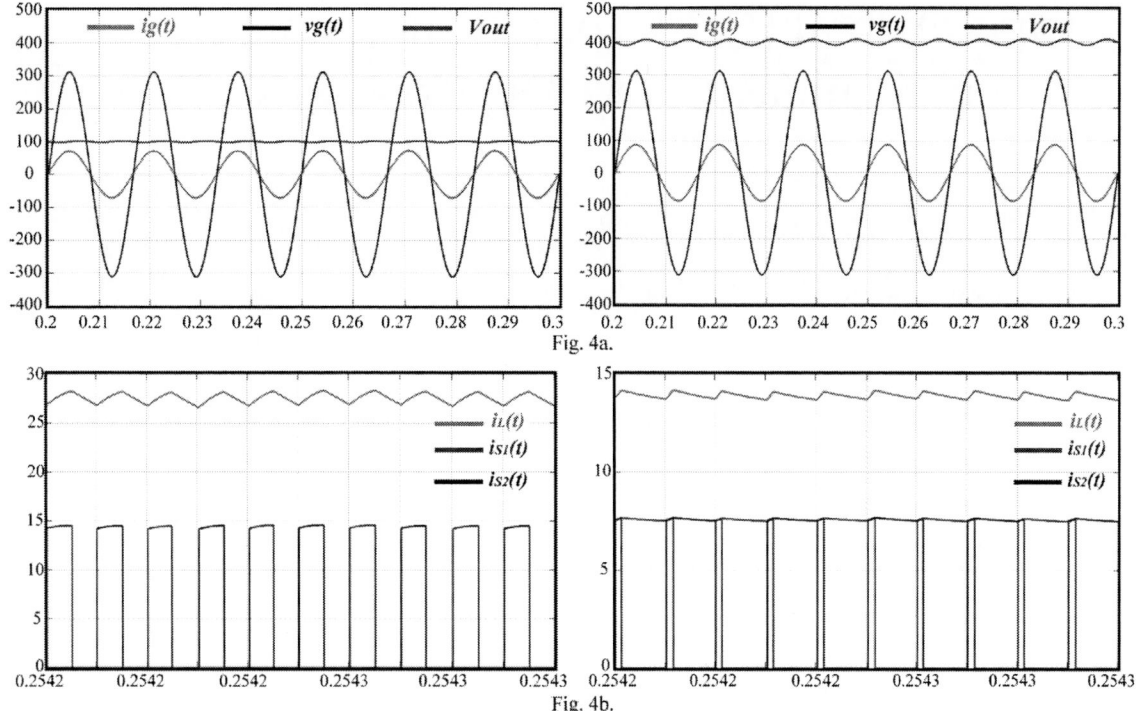

Figure 5a: Input and output voltage and input current (x10) for NOM and OM operation; Figure 5b: Current in both semiconductors and inductor current for NOM and OM operation.

operate as boost for ensure that the operation point specified is maintained. This feature, make 3SSC bridgeless buck-boost converter more reliable against input voltage perturbations, and also guarantee stability for the load operation.

V. CONCLUSION

In this paper, a buck-boost converter with power factor correction using three-state switching cell type B was proposed. The average current control technique avoids the effects by the zero in right half-plane present on transfer function from input current to output voltage. Moreover, it allows that input current THD to be minimized, reducing in both of operation modes to value less than 2%.

The three-state switching cell B simulated on bridgeless buck-boost PFC converter provided: the inductor current division, by half in all of semiconductors elements; passive elements increase the ripple frequency twice times, allowing weight and volume reduction on the input and output filters.

Finally, a prototype is under construction in order to obtain experimental results to prove bridgeless buck-boost PFC converter effectiveness with three-state switching cell B.

ACKNOWLEDGMENTS

This study was financed in party by the Coordenação de Aperfeiçoamento de Pessoal de Nível Superior - Brasil (CAPES) - Finance Code 001.

REFERENCES

[1] On Semiconductor ON, "Power Factor Correction (PFC) Handbook", *Choosing the Right Power Factor Controller Solution*. Rev. 5, Apr-2014.

[2] E. E. Landsman, "A unifying derivation of switching DC-DC converter topologies", *1979 IEEE Power Electronics Specialist Conference*, San Diego, CA, USA, 1979, pp.239-243.

[3] B. Zhao, A. Abramovitz and K. Smedley, "Family of Bridgeless Buck-Boost PFC Rectifiers," in *IEEE Transactions on Power Electronics*, vol. 30, no. 12, pp. 6524-6527, Dec. 2015.

[4] L. S. C. e Silva, F. J. M. de Seixas and M. A. G. de Brito, "Bridgeless interleaved boost PFC converter with variable duty cycle control," *XI Brazilian Power Electronics Conference*, Praiamar, 2011, pp. 397-402.

[5] D. C. Morais, J. M. Falcondes de Seixas, L. C. Souza, L. S. C. e Silva and J. C. P. Júnior, "Three-phase Half-controlled Boost PFC Rectifier with New Variable Duty Cycle Control," *2018 13th IEEE International Conference on Industry Applications (INDUSCON)*, São Paulo, Brazil, 2018, pp. 131-137.

[6] Bascopé, G. V. T.; Barbi, I. "Generation of Family of Non-Isolated DC-DC PWM Converters Using New Three-State Switching Cell". Power Electronics Specialists Conference, 2000. PESC 00. 2000 IEEE 31st Annual, Volume: 2, 18-23 June 2000. p. 858 – 863.

[7] J. P. R. Balestero, F. L. Tofoli, G. V. Torrico-Bascopé and F. J. M. de Seixas, "A DC–DC Converter Based on the Three-State Switching Cell for High Current and Voltage Step-Down Applications," in *IEEE Transactions on Power Electronics*, vol. 28, no. 1, pp. 398-407, Jan. 2013.

[8] J. P. R. Balestero, F. L. Tofoli, R. C. Fernandes, G. V. Torrico-Bascope and F. J. Mendes de Seixas, "Power Factor Correction Boost Converter Based on the Three-State Switching Cell," in *IEEE Transactions on Industrial Electronics*, vol. 59, no. 3, pp. 1565-1577, March 2012.

[9] P. H. Feretti and F. L. Tofoli, "A DC-DC buck-boost converter based on the three-state switching cell," *2017 Brazilian Power Electronics Conference (COBEP)*, Juiz de Fora, 2017, pp. 1-6.

[10] R. W. Erickson, D. Maksimovic, *Fundamentals of Power Electronics*, 2 ed., Springer, Nova York, 2001.

[11] INTERNATIONAL ELECTROTECHNICAL COMMISSION: IEC61000-3-2, Electromagnetic compatibility (EMC) – Part 3-2: Limits – Limits for harmonic current emissions (equipment input current ≤ 16A per phase).

Experimental Results of a Bidirectional Coupled Inductor DC-DC Converter

1st Murilo Brunel da Rosa
Electric Power Processing Group nPEE
Santa Catarina State University UDESC
Joinville - Santa Catarina - Brazil
murilobrunel@gmail.com

2st Menaouar Berrehil El Kattel
Electric Power Processing Group nPEE
Santa Catarina State University UDESC
Joinville - Santa Catarina - Brazil
berrehilelkattel@gmail.com

3nd Robson Mayer
Electric Power Processing Group - nPEE
Santa Catarina State University UDESC
Joinville - Santa Catarina - Brazil
mayerrobson@gmail.com

4rd Sérgio Vidal Garcia Oliveira
Electric Power Processing Group-nPEE
Santa Catarina State University UDESC
Blumenau Regional University FURB
Joinville - Santa Catarina - Brazil
sergio_vidal@ieee.org

Abstract — **In this paper, the theoretical analysis and the experimental results of the bidirectional coupled inductor DC-DC converter are presented. The proposed topology is relatively feasible for low-input-voltage applications for interfacing energy storage elements, such as batteries, and ultracapacitors with the high voltage DC bus in hybrid electric vehicles and electric vehicles. The operation principles, the dc voltage gain, the filters design and the voltage and current stresses are discussed. In order to validate the theoretical study and to prove the converter performance, some experimental results are obtained from a laboratory prototype of 600 W.**

Keywords— Coupled inductor, Step-up DC-DC Converter, Step-down DC-DC Converter, Energy storage.

I. INTRODUCTION

The environmental problems gradually draw more concerns, and electric vehicles (*EVs*) have been developed for future transportation [1]-[2]. Energy storage devices, such as battery banks and ultracapacitors, are needed for the system because of their fast transient response. Therefore, bidirectional DC-DC converters are used to ensure the energy storage device such as battery banks [3]. To reach this goal, power electronics is an enabling technology for the development of electric and hybrid vehicle propulsion systems.

With the advancement of the electronics industry and device technology, several types of power converters with varying degrees of performance are available in the market [4]-[5]. In EVs, the power electronic converters are considered as the key elements that interface their power sources to the drivetrain of the vehicle. The DC-DC converter is used to allow a desired level of dc voltage to be obtained without having to increase the battery bank size. It is illustrated in [6] a bidirectional topology that function as an interface between the DC-link and energy storage system. The structure is used to adapt the voltage levels of the system on board the vehicle or even to recover energy in case of braking.

In this context, this paper provides the theoretical analysis and experimental result of the proposed bidirectional DC-DC converter with coupled inductor

operating in continuous conduction mode (CCM).The main objective of the prototype is to confirm the studied converter capabilities by experimental results. Hence, prototype was built, tested and evaluated to confirm the theoretical analysis.

II. PROPOSED COUPLED INDUCTOR BIDIRECTIONAL DC-DC CONVERTER

In this section, Analysis theoretical and the operating stages for continuous conduction mode (CCM) are done briefly. Next, some simplifying restrictions are imposed without components losses. The converter is in steady state, the variation of the energy stored in the inductors and the energy stored in the capacitors, within switching period, is zero. Therefore, equations (1) and (2) can be given as:

$$\frac{1}{T_s} \int_0^{T_s} v_L(t) \, dt = 0 \tag{1}$$

$$\frac{1}{T_s} \int_0^{T_s} i_C(t) \, dt = 0 \tag{2}$$

Fig. 1 shows the bidirectional coupled inductor DC-DC converter (also known as the bidirectional coupled inductor non-isolated DC-DC converter) which was proposed in [6]. The circuit is composed by three power switches S_1, S_2 and S_3, one coupled inductor L_C, two filter capacitors C_1 and C_2, and DC sources E_1 and E_2. The topology has fewer components and smaller inductor volume compared to the bidirectional cascade, Buck-Boost, Cùk and interleaved bidirectional DC-DC converters [7]-[11], also allowing bidirectional power flow between E_1 and E_2. In general, the operation of proposed converter can be divided into two operating modes, called forward operating mode and the backward operating mode. In forward operating mode, the power flow from E_1 to E_2 is obtained by modulating S_2, keeping S_1 turned-on and S_3 turned-off. In backward operating mode, the power flow from E_2 to E_1 is obtained by modulating S_3, keeping the switches S_2 and S_1 turned-off.

The turns ratio n of the coupled inductor is defined by (3). Where the n_1 is the number of turns on the winding L_1, n_2 is the number of turns on the winding L_2.

978-1-7281-4181-7/19 $31.00 © 2019 IEEE

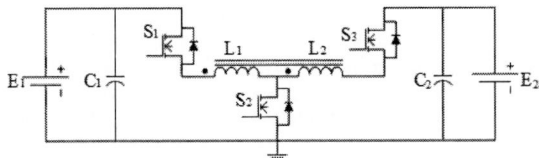

Fig. 1.Topology of the bidirectional coupled inductor DC-DC converter.

$$n = \frac{n_2}{n_1} \qquad or \qquad n = \sqrt{\frac{L_2}{L_1}} \qquad (3)$$

L_1: The inductance value of the winding (L_1)

L_2: The inductance value of the winding (L_2)

A. Forward operating mode (From E_1 to E_2)

In CCM operation, the inductor currents (i_{L1} and i_{L2}) will not reach zero. Based on this, the circuit operation in one switching period (T_S) can be divided into two stages as shown in Fig. 2(a) and 2(b), whereas Fig. 3 illustrates its main idealized waveforms of proposed topology in forward operating mode. The operational mode can be described as follows.

The First Stage [0 → D·Ts]: This stage starts at instant t_0, when the switches S_2 is turned on. The input current (I_1) is flowing through the switches S_1, S_2 and inductor L_1, the current in L_2 is zero and current i_{L1} increases linearly. The voltages on the inductors L_1 and L_2 are E_1 and $E_1 \cdot n_1$, respectively. The load (R_2) receives energy from the output capacitor C_2. The equivalent circuit of first stage is illustrated in Fig. 2(a).

The Second Stage [D·Ts → Ts]: Starts when switch S_2 is turned-off and the internal diode of S_3 is forward biased. The currents i_{L1} and i_{L2} decrease linearly. The voltages across L_1 and L_2 are $(E_1 - E_2)/(1 + n)$ and $(E_1 - E_2) \cdot n/(1 + n)$, respectively. The energy stored in the coupled inductor is transferred to the load (R_2) through L_2 and internal diode of S_3. Furthermore, the output current (I_2) is equal $i_{L2} - i_{C2}$. The equivalent circuit of second stage can be seen in the Fig. 2(b).

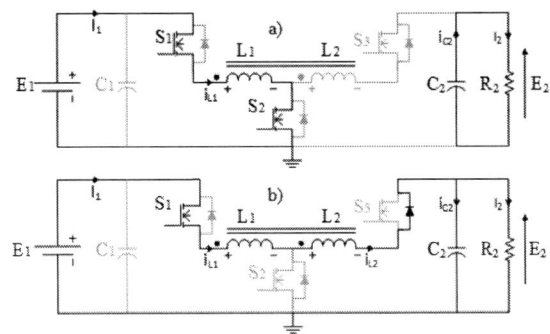

Fig. 2.Topological stages in forward operating mode for CCM.

1. DC Voltage Gain for Forward Operating Mode

The expression for the dc voltage gain for CCM, can be given from the analysis of the average voltage across the inductor winding L_1.

$$V_{L1(Avg)} = \frac{1}{T_S} \left[\int_0^{D \cdot T_S} E_1 \cdot dt + \int_{D \cdot T_S}^{T_S} \left(\frac{E_1 - E_2}{1 + n} \right) \cdot dt \right] = 0 \quad (4)$$

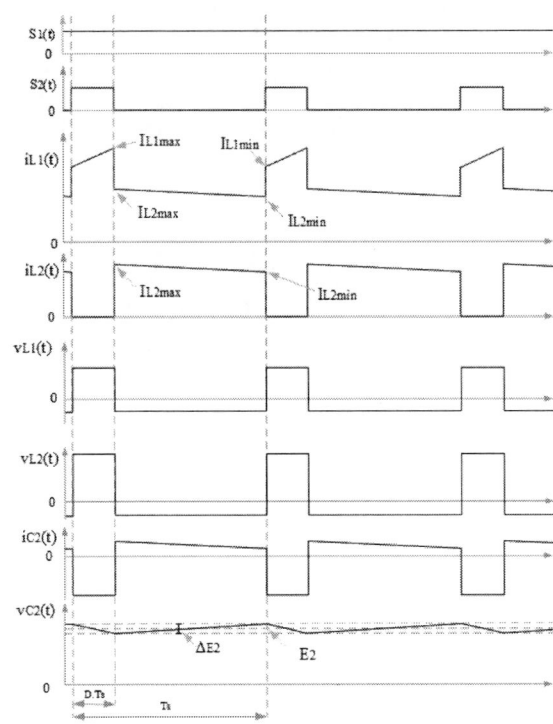

Fig. 3. Idealized current and voltage waveforms in forward operating mode for CCM

Rearranging (4), the dc voltage gain in forward operating mode for CCM is obtained by (5) and, graphically presented and, compared to the conventional Boost, as shown in Fig. 4.

$$G_{E_1 \to E_2} = \frac{E_2}{E_1} = \frac{1 + n \cdot D}{1 - D} \quad (5)$$

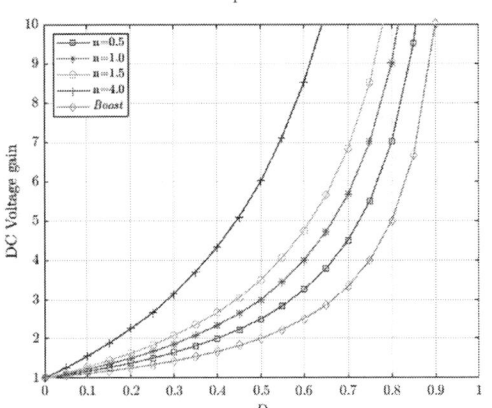

Fig. 4. DC voltage gain in forward operating mode for CCM.

2. Inductor Current Ripple

The current ripple Δi_{L1} in the inductance L_1 can be obtained from the first stage, which is important to determine the inductances values of coupled inductor.

$$V_{L_1} = L_1 \cdot \frac{\Delta I_L}{dt}; \ E_1 = L_1 \cdot \frac{\Delta I_{L_1}}{D \cdot T_S}$$
$$\Delta I_{L_1} = \frac{E_1 \cdot D}{L_1 \cdot f_S} \quad (6)$$

Equation (6) can be used to calculate the value inductances of L_1 and L_2, as shown in (7).

$$L_1 = \frac{E_1 \cdot D}{\Delta I_{L_1} \cdot f_s} \tag{7}$$
$$L_2 = L_1 \cdot n^2$$

3. Voltage Ripple of Output Capacitor

Voltage ripple of the output capacitors (ΔE_2) in terms of the available charge (ΔQ), is expressed by (8).

$$\Delta E_2 = \Delta V_{C_2} = \frac{\Delta Q}{C_2} = \frac{I_{C_2} \cdot \Delta t}{C_2} \tag{8}$$

During the first stage, the capacitor current (I_{C_2}) is equal the output current (I_2); the interval time is equal $D \cdot T_S$, after replacing that in the equation (8), the output voltage ripple is given by (9).

$$\Delta E_2 = \frac{I_2 \cdot D}{f_s \cdot C_2} \tag{9}$$

By means of expression (9), the capacitance is calculated by (10).

$$C_2 = \frac{I_2 \cdot D}{f_s \cdot \Delta E_2} \tag{10}$$

4. Components Stress Calculation for Forward Operating Mode

The RMS current value of capacitor C_2 can be calculated by (11).

$$I_{C_2(RMS)} = \sqrt{\frac{1}{T_S}\left[\int_0^{D \cdot T_S}\left(I_2\right)^2 dt + \int_{D \cdot T_S}^{T_S}\left(\frac{I_2}{1-D}-I_2\right)^2 dt\right]}$$
$$= I_2 \cdot \sqrt{\frac{D}{1-D}} \tag{11}$$

The RMS and average current values through L_1 and L_2 can be expressed by (12) and (13), respectively.

$$I_{L_1(Avg)} = I_1 ; \qquad I_{L_2(Avg)} = I_2 \tag{12}$$

$$I_{L_1(RMS)} = \frac{I_2}{1-D} \cdot \sqrt{(2+n)\cdot n \cdot D + 1}$$
$$I_{L_2(RMS)} = \frac{I_2}{\sqrt{1-D}} \tag{13}$$

From (14) to (16) equations the voltage and current stress on the power switches can be calculate.

$$I_{S_1(Avg)} = I_{L_1(Avg)} ; \qquad I_{S_1(RMS)} = I_{L_1(RMS)} ; \tag{14}$$
$$V_{S1} = 0$$

$$I_{S_2(Avg)} = I_1 - I_2 ; \qquad I_{S_2(RMS)} = \frac{I_1 - I_2}{\sqrt{D}}$$
$$V_{S_2} = \frac{E_1 \cdot n + E_2}{1+n} \tag{15}$$

$$I_{S_3(Avg)} = I_{L_2(Avg)} ; \qquad I_{S_3(RMS)} = I_{L_2(RMS)} ; \tag{16}$$
$$V_{S_3} = E_1 + E_2 \cdot n$$

B. Backward operating mode (From E_2 to E_1)

In backward operating mode, the proposed topology presents two stages for CCM operation in each switching period. The description of the operating principle and idealized waveforms are presented:

The First stage [$0 \to D \cdot T_S$]: In instant $t = 0$, the switch S_3 is turned-on and the coupled inductor windings store energy. The input current (I_2) is flowing through S_1, S_3, L_2 and L_1, the current I_2, i_{L1} and i_{L2} are equals, the voltages across L_1 and L_2 are equal $(E_1 - E_2)/(1 + n)$ and $(E_1 - E_2) \cdot n/(1 + n)$, respectively. The load (R_1) and output capacitor (C_1) receive energy from inductor L_1 through the internal diode of S_1. The equivalent circuit of the first stage is illustrated in Fig. 5(a).

The Second stage [$D \cdot T_S \to T_S$]: In instant $t = D T_S$, the switch S_3 is turned-off and the internal diode of S_2 is forward biased. The current i_{L2} is zero and i_{L1} decrease linearly, the voltage inductor across L_1 and L_2 are equal to E_1 and $E_1 \cdot n_1$, respectively. The energy stored in the coupled inductor is transferred to the load (R_1) through inductor L_1 and internal diode of S_1 and S_2. Additionally, output and input current is equal $i_{L1} - i_{C1}$ and 0, respectively. The equivalent circuit of the second stage is illustrated in Fig. 5(b).

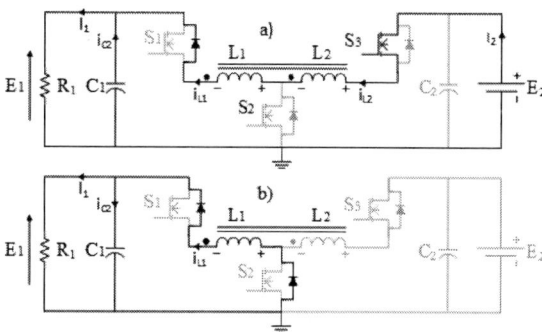

Fig. 5. Topological stages in backward operating mode for CCM.

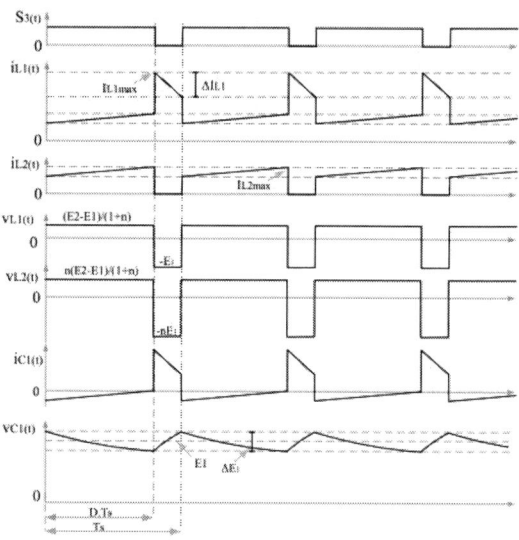

Fig. 6. Idealized waveforms in backward operating mode for CCM.

1. DC Voltage Gain for Backward Operating Mode

Considering the proposed topology is operating in steady state, CCM and backward operating mode, the average voltage value across inductor winding L_2 should be zero for a switching period. Then the dc voltage gain can be given from equation (17).

$$V_{L2(Avg)} = \frac{1}{T_S}\left[\int_0^{D \cdot T_S}\frac{n \cdot (E_2 - E_1)}{1+n} \cdot dt + \int_{D \cdot T_S}^{T_S} -n \cdot E_1 \cdot dt\right] = 0 \tag{17}$$

978-1-7281-4181-7/19 $31.00 © 2019 IEEE

Thus, the dc voltage gain in backward operating mode of CCM is given by (18).

$$G_{E_2 \to E_1} = \frac{E_1}{E_2} = \frac{D}{(1+n) - n \cdot D} \tag{18}$$

The dc voltage gain of the proposed topology in backward operating mode is shown graphically in Fig. 7, and compared to the conventional Buck DC-DC.

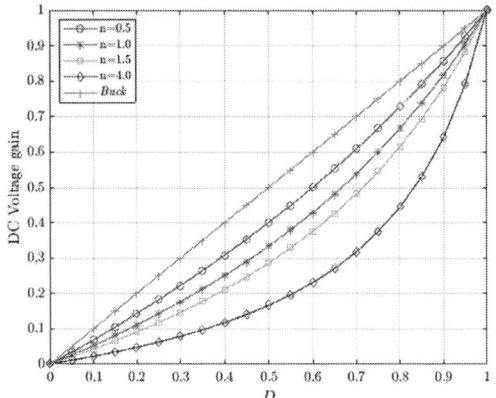

Fig. 7. DC voltage gain in backward operating mode for CCM.

2. Inductor Current Ripple

The inductor current ripple in CCM can be calculated during the second stage. From Fig. 6, the inductor voltage is given by (19).

$$V_{L_2} = L_2 \cdot \frac{\Delta I_{L2}}{dt}; \quad \frac{n \cdot (E_2 - E_1)}{1+n} = L_2 \cdot \frac{\Delta I_{L_2}}{D \cdot T_S} \tag{19}$$

$$\Delta I_{L_2} = \frac{n \cdot (E_2 - E_1) \cdot D}{(1+n) \cdot L_2 \cdot f_s}$$

From (19), the value inductances of L_2 and L_1, are calculated by (20).

$$L_2 = \frac{n \cdot (E_2 - E_1) \cdot D}{(1+n) \cdot \Delta I_{L2} \cdot f_s} \tag{20}$$

$$L_1 = L_2 \cdot n^2$$

3. Voltage Ripple of Output Capacitor

The voltage ripple of output capacitor C_1 caused by charge variation is done by (21).

$$\Delta E_1 = \Delta V_{C_1} = \frac{\Delta Q}{C_1} = \frac{I_{C1} \cdot \Delta t}{C_1} \tag{21}$$

From second stage, the voltage ripple of output capacitor C_1 is defined by (22).

$$\Delta E_1 = \frac{1}{C_1} \frac{I_1 \cdot D - I_2}{f_s} \tag{22}$$

By means of expression (22), the value capacitance is calculated by (23).

$$C_1 = \frac{1}{\Delta E_1} \frac{I_1 \cdot D - I_2}{f_s} \tag{23}$$

4. Components Stress Calculation for Backward Operating Mode

The mathematical expression that describing the RMS value of capacitor current can be given by (24).

$$I_{C_1(RMS)} = \sqrt{\frac{1}{T_S}\left[\int_0^{D \cdot T_S} \left(I_1 - \frac{I_2}{D}\right)^2 dt + \int_{D \cdot T_S}^{T_S} \left(\frac{I_1 - I_2}{1 - D} - I_2\right)^2 dt\right]}$$
$$= (I_2 - I_1 \cdot D) \cdot \sqrt{\frac{1}{D \cdot (1 - D)}} \tag{24}$$

The average and RMS current value flowing through the inductors L_1 and L_2 is defined by (25) and (26), respectively.

$$I_{L_1(Avg)} = I_1; \qquad I_{L_2(Avg)} = I_2 \tag{25}$$

$$I_{L_1(RMS)} = \sqrt{\frac{I_2^2}{D} + \frac{(I_1 - I_2)^2}{1 - D}} \tag{26}$$

$$I_{L_2(RMS)} = \frac{I_2}{\sqrt{D}}$$

The average and RMS values of the switches current are calculated by the flowing equations:

$$I_{S_1(Avg)} = I_{L_1(Avg)}; \qquad I_{S_1(RMS)} = I_{L_1(RMS)}; \\ V_{S1} = 0 \tag{27}$$

$$I_{S_2(Avg)} = I_1 - I_2; \qquad I_{S_2(RMS)} = \frac{I_1 - I_2}{\sqrt{D}} \\ V_{S_2} = E_2 + \frac{n \cdot (E_1 - E_2)}{1+n} \tag{28}$$

$$I_{S_3(Avg)} = I_{L_2(Avg)}; \qquad I_{S_3(RMS)} = I_{L_2(RMS)}; \\ V_{S_3} = E_2 + E_1 \cdot n \tag{29}$$

III. EXPERIMENTAL RESULTS

In order to verify the operation and evaluate the performance of the proposed, a $600\,W$ prototype was designed and it was verified experimentally. The converter specifications are exposed in Table I. A photograph of the implemented prototype is shown in Fig. 8.

TABLE I. MAIN PROTOTYPE SPECIFICATIONS AND PARAMETERS

Parameters	Value
Rated power (P)	600 W
'Source voltage E_1	100 V
Source voltage E_2	300 V
Duty cycle	0.4 (for E_1 to E_2)/ 0.6 (for E_2 to E_1)
Current ripple (ΔI_L)	2.0 A
Turns ratio	1.55
Switching frequency (f_s)	20 kHz

The experimental voltage and current waveforms are obtained from the experimental prototype operating in forward and backward operating mode and are shown in Figs. 9, 10, 11, and 12.

In Fig. 9, the voltages and the windings currents of the coupled inductor L_C are shown for forward operating mode.

Fig. 8.Prototype of proposed topology.

It is observed that the converter presents two operating stages in CCM. Moreover, there is a great similarity between experimental and theoretical current waveforms.

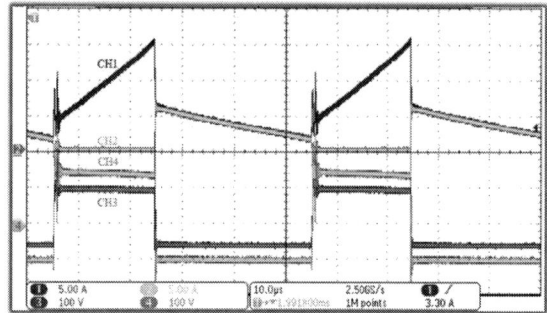

Fig. 9. Voltage and current waveforms through L_1 and L_2 for forward operating mode - Ch1: Inductor current i_{L1}, Ch2: Inductor current i_{L2}, Ch3: Inductor voltage V_{L1}, Ch4: Inductor voltage V_{L2}.

The waveforms shown in Fig 10 were obtained for the following parameters: a duty cycle of 40 %, an input voltage of $100\,V$, an output voltage of $300\,V$, and a power rated $600\,W$. It is noted that at forward operating mode, the input current (I_1) is the same of winding inductor current (i_{L1}). In addition the converter is operating in voltage Step-Up mode.

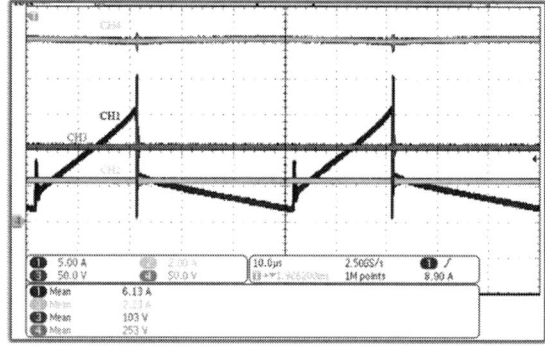

Fig. 10.Voltage and current waveforms in E_1 and E_2 for forward operating mode - Ch1: Input current I_1, Ch2: Output current I_2, Ch3: Input Voltage (E_1), Ch4: Output Voltage (E_2).

In Fig. 11, the voltages and the windings currents of the coupled inductor L_C are shown for backward operating mode. It is observed that the converter is operating in CCM and it presents two stages in each switching period. Moreover, the current waveform is according to the theoretical waveform, shown in Fig. 6.

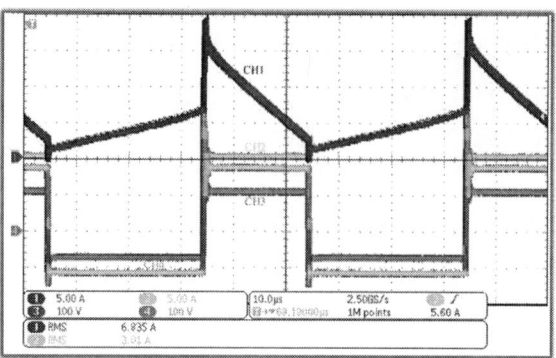

Fig. 11.Voltage and current waveforms through L_1 and L_2 for backward operating mode - Ch1; Inductor current i_{L1}, Ch2: Inductor current i_{L2}, Ch3: Inductor voltage V_{L1}, Ch4: Inductor voltage V_{L2}.

The waveforms shown in Fig 12 were obtained for the following parameters: a duty cycle of 60 %, an input voltage of $300\,V$, an output voltage of $100\,V$, and a power rated $600\,W$. It is noted that at backward operating mode, the input current (I_2) is the same winding inductor current (i_{L2}). Moreover, the converter is operating in voltage Step-Down mode.

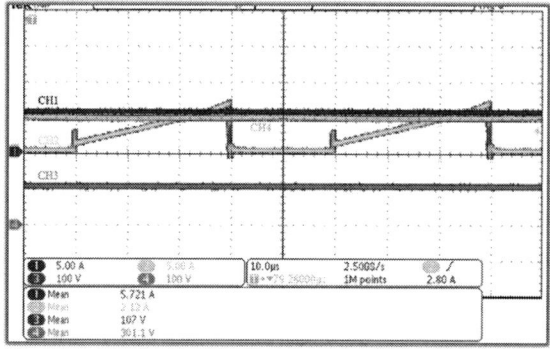

Fig. 12.Voltage and current waveforms in E_1 and E_2 for backward operating mode - Ch1: Output current I_1, Ch2: Input current I_2, Ch3: Output Voltage (E_1), Ch4: Input Voltage (E_2).

IV. CONCLUSIONS

In this paper, a bidirectional coupled inductor DC-DC converter, is introduced. Theoretical analysis of the converter is presented and experimental results are used for its validation. The proposed topology merges the main characteristics of Boost and Buck DC-DC converters, resulting in a converter with small volume, low weight, reduced number of components and, in comparison to the conventional DC-DC converters, it has advantages in terms of dc voltage gain without any isolation, capability of operation in different converter functions, as well as bidirectional operation. Moreover, the converter is competitive and qualified for energy storage systems, such as batteries, and ultracapacitors with the high voltage DC bus in hybrid electric vehicles and electric vehicles.

V. ACKNOWLEDGMENTS

This work was supported in part by the Coordenação de Aperfeiçoamento de Pessoal de Nível Superior – Brasil (CAPES) – Finance Code 001, and in part by the Fundação de Amparo à Pesquisa e Inovação do Estado de Santa Catarina (FAPESC), SC, Brazil. The authors gratefully to UDESC and FURB Universities.

978-1-7281-4181-7/19 $31.00 © 2019 IEEE

VI. REFERENCES

[1] T. Friedli, J. Kolar, "The essence of three-phase PFC rectifier systems", in IEEE 33rd International Telecommunications Energy Conference (INTELEC), Oct. 2011,pp. 1–27.

[2] C. C. Chan, A. Bouscayrol and K. Chen, "Electric, Hybrid, and Fuel-Cell Vehicles: Architectures and Modeling," in IEEE Transactions on Vehicular Technology, vol. 59, no. 2, pp. 589-598, Feb. 2010.

[3] K. Rajashekara, "Present Status and Future Trends in Electric Vehicle Propulsion Technologies," in IEEE Journal of Emerging and Selected Topics in Power Electronics, vol. 1, no. 1, pp. 3-10, March 2013.

[4] K. Tytelmaier, O. Husev, O. Veligorskyi and R. Yershov, "A review of non-isolated bidirectional dc-dc converters for energy storage systems," 2016 II International Young Scientists Forum on Applied Physics and Engineering (YSF), Kharkiv, 2016, pp. 22-28.

[5] K. V. Samarth and M. R. Bachawad, "Comparative analysis of bidirectional converter for V2G application," 2017 International Conference on Energy, Communication, Data Analytics and Soft Computing (ICECDS), Chennai, 2017, pp. 2241-2245.

[6] M. Brunel da Rosa, M. Berrehil El Kattel, R. Mayer, M. D. Possamai and S. V. G. Oliveira, "Analysis and simulation of a novel coupled inductor bidirectional DC-DC converter," 2017 Brazilian Power Electronics Conference (COBEP), Juiz de Fora, 2017, pp. 1-6.

[7] Du, Y.,Zhou, X., Bai, S., Lukic, S., & Huang, A. (2010). Review of non-isolated bi-directional DC-DC converters for plug-in hybrid electric vehicle charge station application at municipal parking decks. 2010 Twenty-Fifth Annual IEEE Applied Power Electronics Conference and Exposition (APEC).

[8] Onar, O. C., Kobayashi, J., &Khaligh, A. (2013). A Fully Directional Universal Power Electronic Interface for EV, HEV, and PHEV Applications. IEEE Transactions on Power Electronics, 28(12), 5489–5498.

[9] Schupbach, R. M., &Balda, J. C. (n.d.). Comparing DC-DC converters for power management in hybrid electric vehicles. IEEE International Electric Machines and Drives Conference, 2003. IEMDC'03.

[10] Lin, C.-C., Wu, G. W., & Yang, L.-S. (2013). Study of a non-isolated bidirectional DC–DC converter. IET Power Electronics, 6(1), 30–37.

[11] Mayer, R., Berrehil El Kattel, M., Possamai, M. D., Bruning, C., & Garcia Oliveira, S. V. (2017). Analysis of a multi-phase interleaved bidirectional DC/DC power converter with coupled inductor. 2017 Brazilian Power Electronics Conference (COBEP).

Design and Test of a SRF-PLL Based Algorithm for Positive-Sequence Synchrophasor Measurements

Gabriel Ubirajara de Carvalho, Gustavo Weber Denardin, Rafael Cardoso, Cassiano Ferro Moraes

Federal University of Technology – Paraná (UTFPR), Pato Branco - PR, Brazil

e-mail: decarvalho.gabriel@gmail.com, gustavo@utfpr.edu.br, rcardoso@utfpr.edu.br, cassianofmoraes@gmail.com

Abstract—**This paper presents the design, analysis and experimental testing of a SRF-PLL based algorithm for positive-sequence synchrophasor measurements in compliance with the IEEE C37.118.1-2011 and C37.118.1a-2014 standards. The proposed approach consists of a three-stage algorithm, being the first one a three-phase demodulation, which detach the positive-sequence from the negative sequence signal in the frequency domain, as well as removes the zero sequence. The second stage is a finite impulse response filter (FIR) that is applied in order to improve the noise and interference rejection. Such class of digital filters was chosen by its characteristics of linear phase and constant group delay. Finally, the last stage is carried out by a synchronous reference frame phase-locked loop featuring magnitude normalization and proportional-integral controller, which estimates amplitude, phase, frequency and rate of change of frequency. The experimental results obtained by a test platform show that steady-state criteria are met.**

Keywords— IEEE C37.118.1, IEEE C37.118.1a-2014, phase-locked-loop (PLL), phasor measurement unit (PMU), power systems, synchrophasor.

I. INTRODUCTION

The growth in the use of distributed energy resources in the electrical system, in addition to the advance of processing capacity and communication of data, has fostered the discussion about smart grids [1]. In this context, the synchronized phasor measurement unit (PMU) is presented as a fundamental component to provide the necessary data from different locations along the system of both transmission and distribution networks [2].

Although there is great interest on implementing synchronized phasor measurement systems (SPMS), budget barriers have prevented their widespread [3]. Therefore, a lot of effort has been put by researchers to propose affordable solutions for PMU. Most of the literature estimation algorithms are based on the Discrete Fourier Transform (DFT) [4]–[6]. This technique has been used since the first symmetrical component distance relay (SCDR) produced in the 1980s, serving as basis for the first PMU at the end of the same decade [7].

The DFT, or its optimized version, the Fast Fourier Transform (FFT), requires reduced computational effort, which motivated its use in the scenario of scarce processing resources at that time. However, this technique has its limitations, especially because of leakage effects that lead to poor performance at off-nominal frequencies. This problem can be more easily solved using a PLL, after a dq-transform and a filter, which is suitable to keep track of phase.

Even though PLLs are widespread knowledge in many engineering areas, inclusively in power systems, it still is an emerging subject for synchrophasor estimation where DFT predominates. Most algorithms in the literature only show simulation results. Due to PLLs being non-linear the real results might significantly differ from the simulated ones, justifying its implementation on hardware and experimental testing for more reliable outcome. Its architecture brings together well-known methods, with the intent to make the design as simple and robust as possible for implementation on a microcontroller and to validate its results under an already available test platform.

This paper presents the design, implementation and analysis of a modular algorithm for positive sequence synchrophasors estimation, which measures amplitude, phase, frequency and rate of change of frequency (ROCOF) measures. It consists of three stages: three-phase demodulation (3PD), pre-filtering (PF) and estimation based on a synchronous reference frame phase-locked loop (SRF-PLL). The goal is to meet the requirements of IEEE standards C37.118.1-2011 and its amendment C37.118.1a-2014 [8], [9].

II. IEEE STANTARDS

In order to define synchrophasor, frequency and ROCOF measurements under all operation conditions, the IEEE standard C37.118.1-2011 introduces methods for evaluation and requirements for compliance [8]. The main concept of the standard is called total vector error (TVE), which is an expression of the difference between the reference phasor and the estimated synchrophasor, as presented in equation (1).

$$TVE = \sqrt{\frac{\left(\hat{X}_r(n)-X_r(n)\right)^2+\left(\hat{X}_i(n)-X_i(n)\right)^2}{\left(X_r(n)\right)^2+\left(X_i(n)\right)^2}} \qquad (1)$$

where $\hat{X}_r(n)$ and $\hat{X}_i(n)$ are the magnitude and phase of the measured synchrophasor, and $X_r(n)$ and $X_i(n)$ are the magnitude and phase of the reference synchrophasor. The standard also defines the maximum allowed TVE as 1%, which means that when the phase error is zero, the maximum magnitude error must be 1%. On the other hand, when the magnitude error is zero, the maximum phase error must be 0,573°, as shown in Fig. 1.

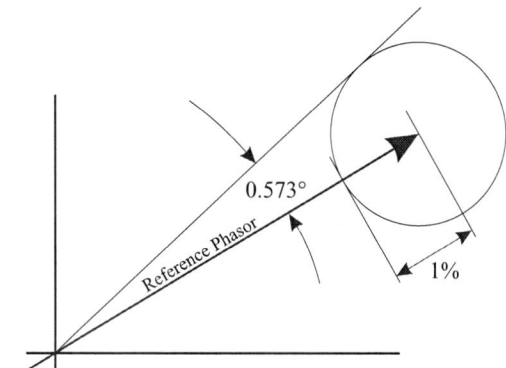

Fig. 1. The TVE criterion of 1% shown at the end of a phasor.

978-1-7281-4181-7/19 $31.00 © 2019 IEEE

Further limits allowances will be presented later in the section V of the paper, in tables used to compare the obtained results with the standards, including the updates brought by the IEEE Std amendment C37.118.1a-2014.

III. PROPOSED ALGORITHM

The processing of the input signal is digital; thus, the variables are in discrete-time domain with sampling interval T. The sampled version of a continuous-time signal $x(t)$ will be presented as $x[n] = x[nT]$, with $n \in \mathbb{Z}$. Subscripts a, b, c refers to the three phases of the input signal, and d, q to the components resulting from dq-transform.

The three-phase demodulation transforms the positive-sequence signal into a low-pass signal, which is a signal having its frequency components centered at 0 Hz. Just to clarify, this calculation of symmetrical components is called demodulation because it mainly brings the signal back to the baseband, around 0 Hz. In closed loop $x_d[n]$ will correspond to the amplitude of the positive-sequence component, and $x_q[n]$ will be brought to zero as the phase-lock is achieved.

In the PF stage the direct axis component $x_d[n]$ is filtered separated from the quadrature axis component $x_q[n]$ resulting in the filtered signals $y_d[n]$ and $y_q[n]$. Also, the use of a PF allows to get a satisfactory interference attenuation, being that only more complex PLL filters are able to deliver a suitable interference rejection rate at a quick response time.

At last, the filtered components follows to the SRF-PLL stage that will be detailed within its design in section IV, giving as result the estimations of amplitude $a_o[n]$, phase $\phi_o[n]$, frequency $\omega_o[n]$ and ROCOF $\alpha_o[n]$.

This modular architecture is proposed with the interest to solve the filtering and tracking problems in a separate manner. This is a simpler approach because it allows to decouple the problems and design one stage at a time individually. Although no algorithm is used for phase compensation due to delays along the loop, specially caused by the PF stage, a calibration process needs to be made to correct some systemic behavior that may occur. Moreover, the attenuation G_{PF} caused by the FIR filtering is compensated right after the filtering stage.

IV. ANALISIS AND DESIGN

In order to present the analysis and design premises, it is important to define the fundamental component of the three-phase input signal as the sum of instantaneous symmetrical components, as shown in equation (2) [10]. Where $a_i[n]$ and $\phi_i[n]$ are the instantaneous amplitudes and angles of phases (a,b,c) or corresponding sequence components (0,1,2), being i = a,b,c,0,1,2. ω_0 is the nominal angular frequency from the electrical system and $T = {2\pi}/{(K\omega_0)}$ is the sampling interval, with an arbitrary integer K greater than 2. For representation simplicity, is considered a narrow-band signal ommiting harmonic interference and noise.

A. Three-Phase Demodulation

The three-phase demodulation is performed through the amplitude invariant Park's Transformation, defined by $x_{dq0}[n] = P[n]x_{abc}[n]$. Note that, for not carrying positive-sequence information, the zero component isn't calculated by excluding the third row of the transformation matrix as shown below:

$$P[n] = \frac{2}{3}\begin{bmatrix} \cos(\phi_a^0[n]) & \cos(\phi_b^0[n]) & \cos(\phi_c^0[n]) \\ -\sin(\phi_a^0[n]) & -\sin(\phi_b^0[n]) & -\sin(\phi_c^0[n]) \end{bmatrix} \quad (3)$$

where $\phi_a^0[n] = \omega_0 nT$, $\phi_b^0[n] = \omega_0 nT - 2\pi/3$ and $\phi_c^0[n] = \omega_0 nT + 2\pi/3$.

This transformation avoids double-frequency oscillations in a balanced situation due to the symmetry of three-phase signals. Another distinguished characteristic provided by this demodulation is that negative-sequence component frequency is shifted to $2\omega_0$, which makes filtration feasible at the next stage even in unbalanced scenarios [11].

B. Pre-filter

Due to phase-locked loops being strictly intolerant to frequency-dependent propagation delays, the linear phase characteristic of FIR filter makes it an appropriate choice for this case. One of the main attributes of FIR is to precisely manage the systems phase response, because it is essentially a taped delay line operating like an accurate phase shifter. A linear phase is achieved by having a symmetrical distribution of the tap-weight coefficients about the FIR filter midpoint [12].

Also, a linear phase FIR filter is known to have a constant group delay, which is a definition used to interpret digitally processed phase response. The group delay is measured in samples and it is invariant to the input frequency. The most common architecture of FIR is the direct-form such as shown in Fig. 2, where $h[n]$ are the Nth tap-weight coefficients, $n \in [0, N-1]$.

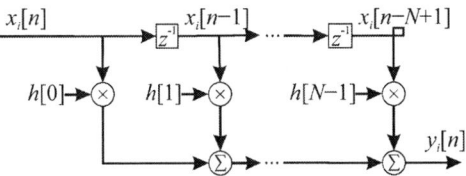

Fig. 2. Direct Nth-order FIR architecture.

The direct-form FIR architecture is an arrangement of shift registers in first-in first-out (FIFO) array configuration, N multipliers, and an accumulator. The tap-weight coefficients correspond to the filter finite duration impulse response.

FIR design is well supported by software tools, giving a stable solution to the desired requirements. The software assisted design begins with a set of specifications, choosing the response type, architecture, filter order, frequency specification such as sampling frequency and cutoff frequency

$$x_{abc}[n] = \begin{bmatrix} x_a[n] \\ x_b[n] \\ x_c[n] \end{bmatrix} = a_0[n]\begin{bmatrix} cos(\omega_0 nT + \phi_0[n]) \\ cos(\omega_0 nT + \phi_0[n]) \\ cos(\omega_0 nT + \phi_0[n]) \end{bmatrix} + a_1[n]\begin{bmatrix} cos(\omega_0 nT + \phi_1[n]) \\ cos\left(\omega_0 nT + \phi_1[n] - \frac{2\pi}{3}\right) \\ cos\left(\omega_0 nT + \phi_1[n] + \frac{2\pi}{3}\right) \end{bmatrix} + a_2[n]\begin{bmatrix} cos(\omega_0 nT + \phi_2[n]) \\ cos\left(\omega_0 nT + \phi_2[n] + \frac{2\pi}{3}\right) \\ cos\left(\omega_0 nT + \phi_2[n] - \frac{2\pi}{3}\right) \end{bmatrix} \quad (2)$$

for the lowpass response, and finally the magnitude specification which in the case of an Equiripple FIR are the passband and stopband attenuations in dB scale.

C. SRF-PLL

The SRF-PLL have a relatively simple structure that makes parameter adjustments easier in comparison with other more complex approaches, and it also presents robust features for digital implementation. Among its features one can mention the ability to track phase ramp with zero steady-state error. However, the magnitude of the input signal strongly influences the loop performance. An alternative to make the loop independent from the magnitude is using normalization by taking the filtered estimated magnitude module $\widehat{U} = \sqrt{y_d^2 + y_q^2}$ and dividing y_q by $\widehat{U}$ [13].

A complete block diagram of the chosen structure is shown in Fig. 3, where the dashed line represents a three-phase connection. It is important to notice that the frequency ω_0 is taken from a path that excludes the proportional action, which is a common strategy to increase accuracy of frequency estimation for a noisy and distorted input signal.

Although PLLs are non-linear systems, a linear analysis is possible by the use of linear approximation. Considering U as the input signal amplitude and $H(s) = k_p + \frac{k_i}{s}$ as the transfer function of the PI controller the resulting characteristic equation of the referred loop is the following [13]:

$$1 + U\frac{H(s)}{s} = 0 \rightarrow s^2 + k_p Us + k_i U = 0 \qquad (4)$$

The gains k_p and k_i can be obtained from the desired poles and zeros location in closed-loop and adjusted in practice due to nonlinear behavior. Nonetheless a trade-off will take place between response speed and steady-state accuracy.

In addition, a curve fit by the linear regression method is applied to the magnitude estimate $a_0[n]$ in equation (5), where G_{PF} is the pre-filter magnitude response considered constant around the frequency of interest for simplicity.

$$a_0 = \left(y_d/G_{PF}\right) \cdot 1.0020549661 - 0.0002139927 \,(5)$$

The final values of $\phi_o[n]$ and $\omega_o[n]$ can be extracted directly from the loop shown in Fig. 3. The final value $\alpha_o[n]$ can be obtained by a straight forward differentiation of $\omega_o[n]$.

V. RESULTS AND DISCUSSION

In this section the obtained experimental results for the proposed algorithm are presented and discussed. The results are gathered from a platform developed, tested and described by [14], [15]. For brief presentation, the test platform was produced in Labview 2010, requiring a Notebook with Labview installed and a data acquisition device (DAQ). The National Instruments USB-6259 [16] was employed as DAQ for signals generation. In the Labview environment, the tests are started by activating the DAQ via a USB communication link. Based on a data stream generated by the platform, the DAQ produces the pulse per second (PPS) required by the PMU and the three-phase signals used to implement the tests defined by the IEEE standards.

The DAQ analog (sinusoidal signal) and digital (PPS signal) outputs are connected to the appropriate analog/digital inputs of the developed PMU running the estimation algorithm. After performing the proposed algorithm in the PMU microcontroller, the estimated data is transmitted back to the Labview application, making possible to compare the produced signals with the measured data.

The development kit used to implement the PMU is the STM32F746G-Discovery [17] from STMicroelectronics featuring an ARM Cortex-M7 microcontroller, with system clock set to 210 MHz (up to 216 MHz). One of the most important reason of its choice is the capability of converting three phase signals simultaneously by master/slave analog-to-digital converter (ADC) configuration and the microcontroller DSP instruction set. The parameters of the algorithm under which the tests were taken are declared in Table I.

TABLE I. ALGORITHM'S DESIGNED PARAMETERS

Parameter	Value
General Specifications	
Nominal grid frequency (Hz)	50
Sampling frequency (kHz)	21
Filter Specifications	
Filter Structure	Direct-Form FIR
Cutoff frequency (Hz)	50
SRF-PLL PI's Specifications	
kp	170
ki	0.38

The nominal grid frequency was chosen to be 50 Hz because it is a multiple of the DAQ clock source that operates at 20 MHz, avoiding undesirable rounding effects originated from non-integer counting at the sampling interval. The Cutoff frequency of the FIR filter was limited to 50 Hz to make the system faster. Lower frequencies would make the

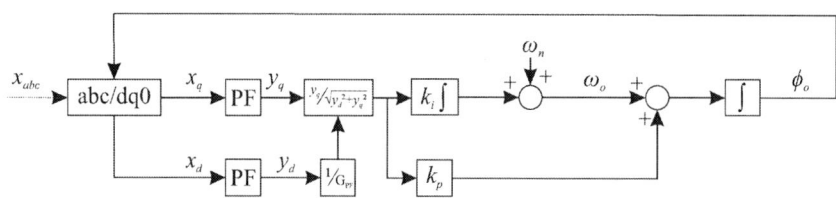

Fig. 3. Three-phase SRF-PLL with magnitude normalization.

system better for interference rejection, but with the drawback of having a slower response.

For all the steady-state tests, the step duration is 2 seconds and the phasor reporting rate is 25 fps. The tables compare the obtained results to the maximum error referenced by the IEEE standard to a M class PMU. Although the results for the P class are not being shown in the tables due to the lack of space, by inspecting Fig. 4-13 it can be verified that requirements are also fulfilled.

A. Magnitude Test

The magnitude test is a variation from 10 to 120 % (0.1 to 1.2 p.u.) in amplitude of the input signal with steps of 0.025 p.u. Table II shows the results. In this test, frequency error (FE) and ROCOF error (RFE) are not applied by IEEE standards.

TABLE II.   MAGNITUDE TEST RESULTS

Test	Magnitude		
Interval	*From 10 to 120 %*		
Error	*Obtained*	*Standard*	*Relative*
TVE (%)	0.31	1	31 %

Fig. 4 and 5 present the TVE and FE for each amplitude percentage, where it is clear that the maximum error occurs at the lowest amplitude value. Considering that normalization strategy is applied to eliminate the input signal magnitude influence over the loop performance, such phenomenon can be attributed to quantization error due to ADC resolution limitation, which is 12 bits.

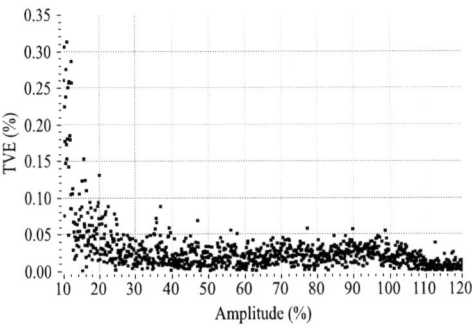

Fig. 4.   TVE for a magnitude interval from 10 to 120 %.

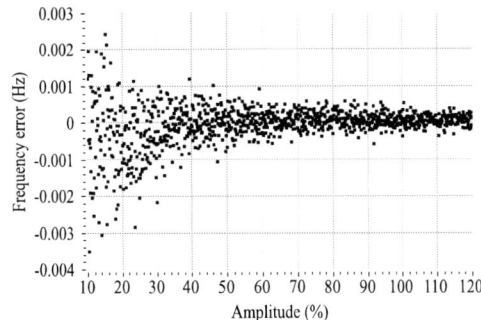

Fig. 5.   Frequency error for a magnitude interval from 10 to 120 %.

B. Frequency Test

Frequency test consists of a variation from 45 to 55 Hz in the input signal frequency with fixed 1 p.u. amplitude and 0°

phase. The frequency step is of 0.2 Hz. Table III shows the results.

TABLE III.   FREQUENCY TEST RESULTS

Test	Frequency		
Interval	*± 5 %*		
Error	*Obtained*	*Standard*	*Relative*
TVE (%)	0.08	1	8 %
FE (mHz)	0.62	5	12 %
RFE (mHz/s)	1	100	0.01 %

Fig. 6 and 7 show the SRF-PLL feature of tracking steady-state phase accurately, resulting in a frequency deviation close to zero.

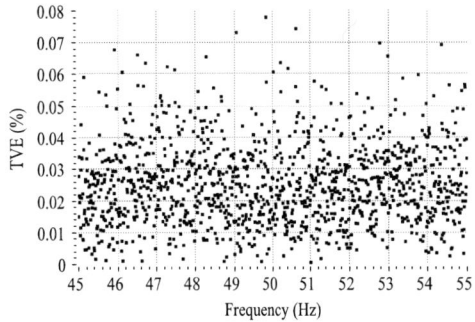

Fig. 6.   TVE for a frequency interval of ± 5 Hz.

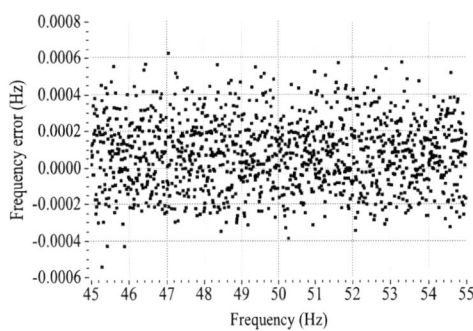

Fig. 7.   Frequency error for a frequency interval of ± 5 Hz.

C. Phase Test

Phase test ranges the input signal phase from -180° to +180° with steps of 10°. Amplitude and frequency are fixed at nominal values. The results are shown in Table IV.

TABLE IV.   PHASE TEST RESULTS

Test	Phase		
Interval	*± 180 °*		
Error	*Obtained*	*Standard*	*Relative*
TVE (%)	0.07	1	7 %

In Fig. 8 and 9 it can be verified that the algorithm tracks the phase properly. However, one can observe disperse points far from the main group of points in Fig. 8. Since this behavior doesn't appears at the frequency results, as seen in Fig. 9, it is probably not a phase tracking error, but a magnitude estimation issue. Thus, despite the algorithm meets the steady-state requirements at this time, adaptive adjustments to PI controller gains may improve the algorithm performance in a future work.

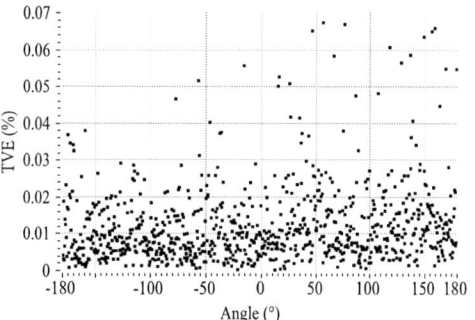

Fig. 8.   TVE for a angle interval of ± 180°.

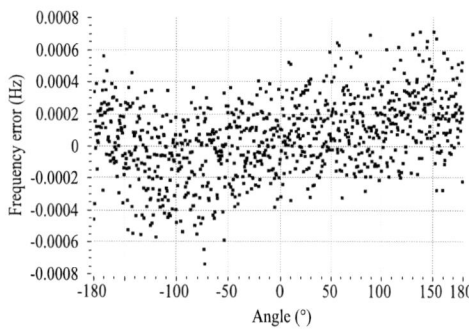

Fig. 9.   Frequency error for a angle interval of ± 180°.

D. Harmonic Distortion Test

The harmonic distortion test is performed from the 2nd to the 50th component with 10% amplitude applied over the fixed input signal of 1 p.u. at 50 Hz. Table V shows the results.

TABLE V.   HARMONIC DISTORTION TEST RESULTS

Test	Harmonic distortion		
Interval	From 2nd to 50th - 10% amplitude		
Error	Obtained	Standard	Relative
TVE (%)	0.05	1	5 %
FE (mHz)	0.62	25	2.5 %

In Fig. 10 and 11 it can be seen the filtering effect provided by the Park's Transformation, resulting in excellent rejection to harmonic distortion, surely reinforced by the PF stage which also reduces unwanted ripples at the measures.

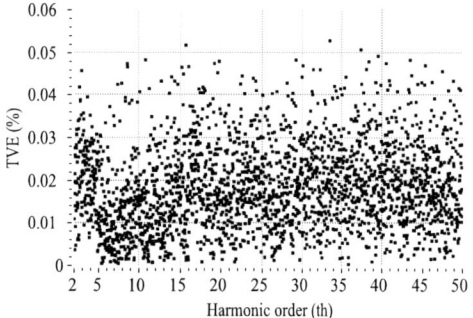

Fig. 10. TVE for a harmonic order interval from 2nd to 50th.

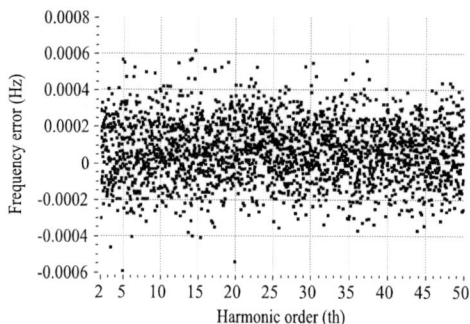

Fig. 11. Frequency error for a harmonic order interval from 2nd to 50th.

E. Out-of-band Interference Test

The out-of-band interference test is realized by adding a sinusoidal component to the input signal, with 10% of the nominal amplitude and frequency going from 2 to 100 Hz (2nd harmonic). Table VI shows the results.

TABLE VI.   OUT-OF-BAND INTERFERENCE TEST RESULTS

Test	Out-of-band interference		
Interval	Until the 2nd harmonic with 10% amplitude		
Error	Obtained	Standard	Relative
TVE (%)	0.07	1.3	5.4 %
FE (mHz)	0.70	10	7 %

Differently from equally simple techniques that requires more complex customization to reject out-of-band interference, the adopted method rejects these frequencies naturally, as Fig. 12 and 13 reveals.

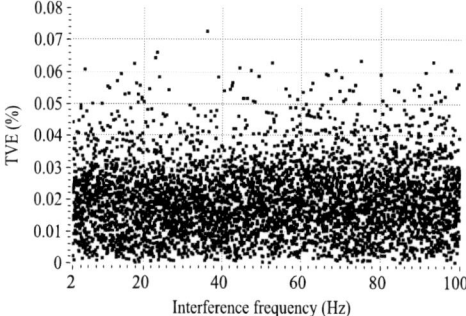

Fig. 12. TVE for a out-of-band interference from 2 to 100 Hz.

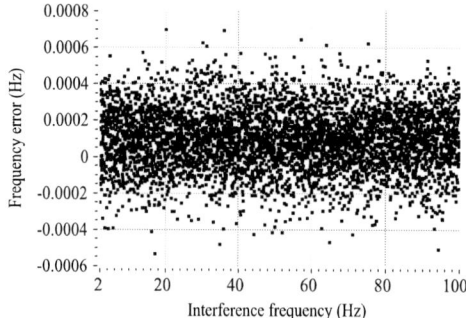

Fig. 13. Frequency error for a out-of-band interference from 2 to 100 Hz.

978-1-7281-4181-7/19 $31.00 © 2019 IEEE

VI. CONCLUSIONS

A steady-state IEEE standard compliant algorithm was presented in this paper along with its experimental results tested under a standardized platform. The algorithm estimates amplitude, phase, frequency and ROCOF, that together forms the positive sequence synchrophasor of three phase signals. Analysis and design of the proposed structure was shown proving to be simple and effective, allowing filtering and tracking systems to be dealt separately thanks for the modular approach adopted.

The filtering stage plays an important role, within the three-phase demodulation, being the Park's Transformation responsible to shift the negative sequence component to the double of the fundamental frequency so the filter can attenuate the ripple making the loop effective both in balanced and unbalanced grid situations. Moreover, it must be emphasized the importance of the linear phase characteristic of the filter, that justified the need of a FIR, otherwise the results would have shown unacceptable errors.

Phase tracking with almost zero steady-state error proved to be true for the SRF-PLL, the same is valid for the frequency and ROCOF measurements made. Also, the amplitude estimation met the requirements with ease too, but contributed to the majority of the TVE, so an improvement may be made, instead of the traditional PI controller it would be interesting to implement an adaptive approach.

Results shows errors far below the maximum limits stipulated by the IEEE standard, emphasizing the excellent rejection to harmonic distortion and out-of-band interference. The continuity of this work is being made to achieve dynamic results as good as the steady-state ones and in the future to compare those results with other algorithms.

ACKNOWLEDGMENT

This study was financed in part by the Coordenação de Aperfeiçoamento de Pessoal de Nível Superior - Brasil (CAPES) - Finance Code 001, Conselho Nacional de Desenvolvimento Científico e Tecnológico (CNPq), Fundação Araucária (FA) and Financiadora de Estudos e Projetos (FINEP). The authors also would like to thank the doctorate student Flavio Lori Grando, from the Federal University of Technology – Paraná (UTFPR), Curitiba – PR, Brazil, for providing the test platform of his own authorship and full support, thereby making this work possible.

REFERENCES

[1] H. Farhangi, "The Path of the Smart Grid," *IEEE Power and Energy Magazine*, vol. 8, no. 1, pp. 18–28, Jan-2010.

[2] F. L. Grando, "Architecture for Development of Synchronized Phasor Measurement Units in the Distribution Level Monitoring," Federal University of Technology - Paraná (UTFPR), 2016.

[3] A. G. Phadke and J. S. Thorp, *Synchronized Phasor Measurements and Their Applications*, 1st ed. Blacksburg: Springer, 2008.

[4] A. Angioni, G. Lipari, M. Pau, F. Ponci, and A. Monti, "A Low Cost PMU to Monitor Distribution

Grids," in *2017 IEEE International Workshop on Applied Measurements for Power Systems (AMPS)*, 2017, p. 6.

[5] F. L. Grando, G. W. Denardin, M. Moreto, and J. D. P. Lopes, "A PMU Prototype for Synchronized Phasor and Frequency Measurements for Smart Grid Applications," in *2015 IEEE 13th Brazilian Power Electronics Conference and 1st Southern Power Electronics Conference (COBEP/SPEC)*, 2015, p. 6.

[6] P. Tosato *et al.*, "A Tuned Lightweight Estimation Algorithm for Low-Cost Phasor Measurement Units," *IEEE Trans. Instrum. Meas.*, vol. 67, no. 5, pp. 1047–1057, 2018.

[7] A. G. Phadke, "Synchronized phasor measurements-a historical overview," in *IEEE/PES Transmission and Distribution Conference and Exhibition*, 2002, pp. 476–479.

[8] I. Power and E. Society, "IEEE Standard for Synchrophasor Measurements for Power Systems," *IEEE Std C37.118.1-2011*, pp. 1–61, 2011.

[9] I. Power and E. Society, "IEEE Standard for Synchrophasor Measurements for Power Systems - Amendment 1: Modification of Selected Performance Requirements," *IEEE Std C37.118.1a-2014 (Amendment to IEEE Std C37.118.1-2011)*, pp. 1–25, 2014.

[10] F. Messina, P. Marchi, L. R. Vega, C. G. Galarza, and H. Laiz, "A Novel Modular Positive-Sequence Synchrophasor Estimation Algorithm for PMUs," *IEEE Trans. Instrum. Meas.*, vol. 66, no. 6, pp. 1164–1175, 2017.

[11] F. M. Gardner, *Phaselock Techniques*, 3rd ed. Palo Alto, California: John Wiley & Sons, Inc., 2005.

[12] F. J. Taylor, *Digital Filters: Principle and Applications with MATLAB*, 1st ed. Hoboken, New Jersey: IEEE Press Series on Digital and Mobile Communication, 2012.

[13] M. Karimi-Ghartemani, *Enhanced Phase-Locked Loop Structures For Power and Energy Applications*. Hoboken, New Jersey: IEEE Press Series on Microelectronic Systems, 2014.

[14] F. L. Grando, A. E. Lazzaretti, G. W. Denardin, M. Moreto, and H. V. Neto, "A Synchrophasor Test Platform for Development and Assessment of Phasor Measurement Units," *IEEE Trans. Ind. Appl.*, vol. 54, no. 4, pp. 3122–3131, 2018.

[15] F. L. Grando, G. W. Denardin, and M. Moreto, "Test Platform for Analysis and Development of Phasor Measurement Units (PMU)," in *12th IEEE International Conference on Industry Applications (INDUSCON)*, 2016, pp. 1–6.

[16] National Instruments, "NI 6259 Device Specifications." pp. 1–26, 2016.

[17] STMicroelectronics, "STM32F769xx Datasheet," no. September. pp. 1–255, 2017.

Economic Analysis of a Peak Shaving System with Diesel Generator

Myrlena R. M. Ferreira
Institute of Electrical Energy
Federal University of Maranhão
São Luís – Maranhão, Brasil
myrlena.raquelly@gmail.com

Luiz A. de S. Ribeiro, IEEE Member
Institute of Electrical Energy
Federal University of Maranhão
São Luís – Maranhão, Brasil
l.a.desouzaribeiro@ieee.org

José G. de Matos, IEEE Member
Institute of Electrical Energy
Federal University of Maranhão
São Luís – Maranhão, Brasil
gomesdematos@ieee.org

Abstract— This paper presents economic analyzes applied to a Peak Shaving System with diesel generator set that is intended to be implemented in a commercial electricity consumer. The consumer considered in the case study is the Federal University of Maranhão, in Brazil. For the elaboration of this study two approaches of this system are made. The first one, called Scenario I, refers to the Peak Shaving System supplying all electricity consumption greater than the contracted power demand. The second approach, named Scenario II, refers to the energy compensation during the period of the day considered as "peak period", in accordance to the tariff structure by which the consumer is charged. The economic analyse considers the value of the kWh produced by the diesel generator set and kWh value purchased from the utility to decide on the economic feasibility of the proposed Peak Shaving System. The conclusion of the analysis is that Scenario II is a viable Peak Shaving solution as it produces economic value for the costumers.

Keywords—peak shaving, generator, rush hour, contracted demand, excess demand.

I. INTRODUCTION

In accordance to the Brazilian regulations for the electric sector, the energy consumers are classified into two groups, considering the voltage level that they are supplied by the utility grid. The Group A consumers are those whose supplied voltage is equal to or greater than 2.3 kV. The other consumers that are supplied with voltages less than 2.3 kV are classified in the Group B. The Group A consumers are charged using a structure tariff classified as binomial and also seasonal time. That is, the prices of electricity vary according to the time of year and to the period of the day. If the period of the year is between May and November, then the tariff corresponding to the dry period is charged, while from December to April the tariff corresponding to the wet period is charged. During the day, there is a price to the electricity if the consumption happens in the period in which the electrical power system of the utility is more loaded, and another price if the consumption happens out of this period. The period of the day when the utility system is more loaded is named is this paper as Peak Hour Time (PHT). On the hand, the period of the day is named of Off-Peak Hour Time (Off-PHT). The PHT comprise a period of three consecutive hours defined by the utility, according to the characteristics of its electrical system. As it was said, PHT falls within the period of greatest consumption. While the O-PHP concerns all other hours of the day. It should be pointed out that PHP and dry periods are characterized by the higher value of electric energy compared to the O-PHP and to the wet period [1].

The demand charged for the consumers of Group A, is the Contracted Demand by the consumer with the utility. However, in case of exceeding an upper 5% or 10% of this demand – depending on the subgroup in which the customer falls according to their level of voltage, he pays for the increased demand. In addition, the difference between the measured and the contracted demand will be charged with the tariff corresponding to approximately three times the tariff charged by the Contracted Demand.

To better understand, the consumer classifieds in Group A are divided into subgroups A1 (voltage level of 230 kV or higher), A2 (voltage level of 88 to 138 kV), A3 (voltage level equal to 69 kV), A3a (voltage level of 30 to 44 kV), A4 (voltage level from 2.3 to 25 kV), and AS (for underground system). For subgroups A1 and A3, the tolerance of excess of demand, related to contracted demand, is 5%, and for the others it is 10%. In addition, for Group A consumers served with voltage less than 30 kV, the tariff structure can still be divided into two types, being the Green hourly rate modality and Blue hourly rate modality. The Green fare modality is characterized by having differentiated tariffs for consumption and only one tariff for demand, that is, it presents a tariff for consumption occurred during the PHP and another for O-PHP However, regardless of the period of the day, the tariff will be the same for demand. On the other hand, the Blue tariff modality has a tariff for PHP and another for O-PHP, both for energy consumption and for electrical power demand [2].

The excess of electricity consumption in periods with higher tariff, and contracted demand being exceeded imply in higher costs to the electrical energy consumers. One must also be observed eventual overloads that this can cause in the transmission and distribution electrical systems. These overloads can generate premature wear in these systems, reducing their quality of supply or even completely damage them. In addition they can cause a disruption of electricity supply to consumers, which is damaging to their facilities and loads [3].

In order to avoid the above mentioned constraints, many consumers are opting by Peak Shaving System (PSS). This type of system is responsible for modeling the energy consumer demand curve. That is, The PSS must supply the consumer's energy demand through another source of electric power other than the utility and thereby mitigate the peak of the consumer demand curve from the point of view of the distribution system.

As it was mentioned above, for the development a PSS is necessary a source of electric energy that replaces the utility (companies that have de public concession for electrical energy distribution in Brazil). In the literature studied, it is possible to distinguish at least three main methods used to develop a PSS. These methods take into consideration the source that will supply electrical energy to the consumer. These main sources are the Diesel Generator Sets, Electrochemical batteries for Energy Arbitrage, and Renewable Energy Sources.

978-1-7281-4181-7/19 $31.00 © 2019 IEEE

PSS using Diesel Generator Sets is the most used method, followed by Energy Arbitrage and lastly Renewable Energy. In the case of the generator, whose primary source is mechanical energy, it must operate in parallel with the utility supplying the loads of the consumer when requested [4]. Energy Arbitrage methods of PSS consists of modulating the demand curve, shifting energy consumption from one period of time to another. In such case, during a period of low-consumption and low-tariff, the consumer must store, for example in a battery bank, the amount of energy sufficient to supply energy to the loads during the periods of high tariffs [5, 6]. While in the renewable energy method the energy source is a renewable energy generation system. Among the most used renewable sources are the photovoltaic energy and the wind energy [7]. And among of them, the photovoltaic energy has been the most used in this type of system [7, 6, 3, 5, 8].

In this work, PSS using diesel generator sets will be studied. For that, in the following sections it will present a case study regarding the consumption of electric energy of a consumer in which economic analyzes are carried out in order to verify how the best to implement this type of system in consumers like those.

II. CASE STUDY

In this work the consumer in question is the Federal University of Maranhão – UFMA, and the period of analysis comprises all the months of the years 2017 and 2018. For this purpose, the data used in this work were got from the electricity bills related to the years 2017 and 2018 and from the mass memory of the consumer meter related to 2017 year. Following, some details about this consumer are shown in TABLE 1. From TABLE 1 one can see that the contracted demand of this consumer is equal to 2500 kW. Therefore, from Fig. 1, it can be seen that the consumer exceeded it in all months in the year 2017. However, the contracted demand for this consumer already reached the limit value allowed by law for a consumer that is served with 13.8 kV rated voltage. Thus, it is not possible for it to contract a higher demand, unless the consumer come to be served with a voltage greater than 13.8 kV, that in this particular case is equal to 69 kV. But, for the consumer be served with 69 kV implies in additional costs. So, in order to receive 69 kV, the consumer would have to pay for the costs of the necessary adaptations in the utility distribution network [1]. This has not yet been done because the consumer does not have the financial resources to do so. Therefore, this solution will not be considered in this work.

TABLE 1. TECHNICAL DETAILS ABOUT THE CONSUMER.

Consumer – UFMA	
Technical details about the consumer	
Voltage group	A4 (2.3 kV – 25 kV)
Phase	Three-phase
Contracted voltage	13.8 kV
Contracted demand	2 500 kW
Horo-seasonal tariff	Green

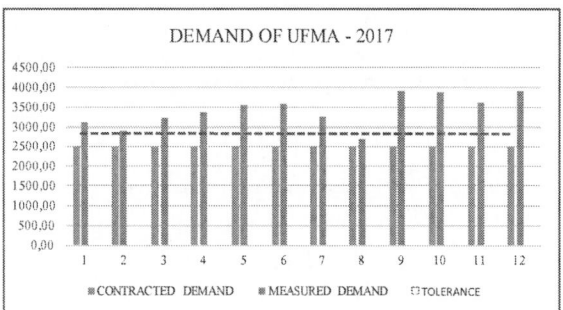

Fig. 1. Monthly record of consumer demand.

In order to better understand the behavior of the loads of this consumer, it is shown in Fig. 2 its demand curve for the day of the highest peak in 2017, whose registration was performed on September 19. The data for plotting this figure was got from the mass memory of the consumer commercial meter. In order to determine the best way to apply a PSS to this consumer profile, the work was carried out contemplating two scenarios. In the Scenario I the PSS compensate all energy consumption above the contracted demand line, and in Scenario II the PSS compensate all energy consumption during the PHT. In this particular consumer location, the PHT is stablished from 18:00 h to 21:00 h every day, disregarding the weekend days and the official national holidays.

For the analysis made for the Scenario I, it can be seen from Fig. 2 that the consumer has exceeded his contracted demand for a period ranging from 8 am to 7 pm, that is, for 11 hours, and that his peak demand was of 3918 kW. Therefore, in designing a PSS for this scenario, one must consider that this consumer needs an alternative energy source that is suitable of delivering that peak of demand over a period of 11 hours. In Scenario II, a peak demand was equal to 2500 kW. So, as the Scenario II is in PHT, in designing a PSS for this scenario, one must consider that this consumer needs an alternative energy source that is suitable for delivering a peak of demand equal to 2000 kW and to function over a period of 3 hours a day.

Considering the above, in the following sections is the analysis performed for this consumer using the Peak Shaving method with diesel generator.

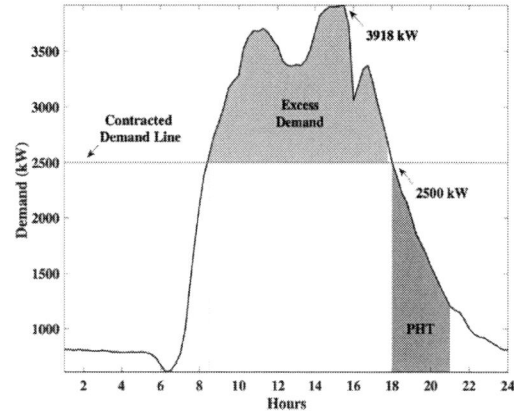

Fig. 2. Demand curve of the highest peak day in 2017.

III. Peak Shaving using the Diesel Generator method

In order to compare the cost of the kWh supplied by the utility with that supplied by the diesel generator set, the tariffs of the concessionaire [9] are shown in TABLE 2. In the composition of the values shown in this table are not being considered the contribution due the public lighting (charged in Brazil), charges due to the excessive reactive power consumes, the tariff flags, and some others non representative taxes.

TABLE 2. TARIFFS OF DEMAND, CONSUMPTION AND EXCESS OF DEMAND.

Tariffs for a Group A4 consumer – Green rate modality		
Demand ($/kW)	5.89	
Consumption ($/kWh)	Peak Hours	0.55
	Off-peak	0.082
Excess of Demand ($/kW)	11.78	

A. Cost of kWh supplied by diesel generator for Scenario I

Scenario I, as quoted in the previous section, refers to the supply of all consumption above the contracted demand line. For this scenario, the rated kVA power of generator (kVA_1)is determined by (1).

$$kVA_1 = \left(\frac{D_{peak} - D_{contracted}}{FP_{generator_{3\phi}}} \right) = 1772.2 \; kVA \qquad (1)$$

Being,

- $D_{peak} \rightarrow$ the highest peak demand recorded during the year: 3917.8 kW;

- $D_{contracted} \rightarrow$ the contracted demand by this consumer: 2500 kW;

- $FP_{generator_{3\phi}} \rightarrow$ the typical rated power factor of three-phase diesel generator: 0.8.

The cost of kWh supplied by the Generator is given by three installments: the cost of diesel per kWh, the depreciation cost of the generator per kWh, and the cost of maintenance per kWh. To find each of these installments the following calculations are required.

The equation (2) [10] is used to calculate the diesel cost per kWh.

$$C_{diesel} = V_{diesel} \times P_{diesel} = 0.3 \times 0.90 = 0.27 \; \$/kWh \qquad (2)$$

Being,

- $C_{diesel} \rightarrow$ the cost of diesel in dollars per kWh;

- $V_{diesel} \rightarrow$ the total diesel consumed in l/kWh. For this work we have used 0.3 l/kWh, which is an approximation commonly used in the literature to determine how many liters of diesel are consumed by the generator to produce 1 kWh when it operates at its nominal power;

- $P_{diesel} \rightarrow$ the diesel price currently in $/liter.

To calculate the generator depreciation cost per kWh was used the equation (3) [10].

$$C_{depreciation} = \frac{\frac{C_{initial}}{G_{life}}}{P_w} = \frac{\frac{710,510.00}{15000}}{1417.8} = 0.0334 \; \$/kWh \qquad (3)$$

Being,

- $C_{depreciation} \rightarrow$ the depreciation cost of the generator in dollars per kWh;

- $C_{initial} \rightarrow$ the initial costs invested in the project. This variable includes the value of the generator, the cost of installation and cost of the materials required for the installation: $ 710,510.00 [11];

- $G_{life} \rightarrow$ the generator useful life: 15000 h [12, 10];

- $P_w \rightarrow$ the power that will be supplied by the generator: P_W = 1417.8 kW.

To calculate the cost of preventive maintenance of the generator per kWh, equation (4) [10] was used.

$$C_{maintenance} = \frac{\frac{C_{man_{prev}}}{h_{work}}}{P_w} = 0.00029369 \; \$/kWh \qquad (4)$$

Being,

- $C_{maintenance} \rightarrow$ the cost of maintenance in dollars per kWh;

- $C_{man_{prev}} \rightarrow$ the cost of preventive maintenance. In this work was used $ 572.85 per semester [13];

- $h_{work} \rightarrow$ the hours worked annually by the generator. In this work was used 11 hours a day, because it is the average time of exceeded contracted demand of this consumer. So, in this calculation it must considered 11 hours a day, 5 days a week, 4 weeks a month, and 12 months a year.

Therefore, the total cost of kWh for Scenario I can be calculated using (5) [10].

$$C_{kWh_{diesel}} = C_{diesel} + C_{depreciation} + C_{maitenance} = 0.3037 \; \$/kWh \qquad (5)$$

B. Economic-financial feasibility analysis of Scenario I

For this analysis, the cost of consumption, demand and excess demand over the year was calculated based on the tariffs shown in TABLE 2. The quantity of kWh and kW over the months was obtained from the electricity bills. And the consumption tariff used was that for the Off-PHT, since in all months the excess of contracted demand occurred outside of the PHT. In addition, as the consumer will have all of his peak consumption met by the generator, the excess demand has also been removed from the total cost to be paid by the consumer. For this, the excess is removed with both the active demand rate and the excess demand rate, because that's how it is charged by the power company. In (2) is the equation used to determine the new total cost after installing SPS.

978-1-7281-4181-7/19 $31.00 © 2019 IEEE

$$C_{total_{new}} = \sum_{m=1}^{12} \Big[C_{t_{utility}}(m) - E_{peak_{utility}}(m)$$
$$+ E_{peak_{GENSET}}(m) \qquad\qquad (6)$$
$$- D_{exc_{da}}(m) - D_{exc_{de}}(m) \Big]$$

Being,

- $C_{total_{new}} \rightarrow$ the new amount paid monthly by the consumer after deploying PSS with GENSET;

- $C_{t_{utility}} \rightarrow$ the total amount paid by the peak consumer monthly to the utility in the year analyzed;

- $E_{peak_{utility}} \rightarrow$ the peah consumption cost calculated through the utility tariff;

- $E_{peak_{GENSET}} \rightarrow$ the peak consumption cost calculated using the GENSET tariff;

- $D_{exc_{da}} \rightarrow$ the cost of the excess of demand calculated through active demand rate;

- $D_{exc_{de}} \rightarrow$ the cost of the excess of demand calculated through excess demand rate;

- $m \rightarrow$ the months 1,2,3..12.

The profit or loss obtained through system deployment can be seen in the analysis of the amount spent by the consumer before and after system deployment. For this, through (3), the result was –$ 148,953.00 whose negative value shows the damage that the system brought to the consumer in Scenario I. Therefore, demonstrating the unfeasibility of implementing SPS with GENSET for supply this amount of consumption in face of a low tariff of the energy concessionaire.

$$E_{annual_I} = \sum_{m=1}^{12} C_{total_{before}}(m)$$
$$- \sum_{m=1}^{12} C_{total_{new}}(m) \qquad (7)$$

Being,

- $E_{annual_I} \rightarrow$ the savings generated after deploying SSP in Scenario I.

- $C_{total_{before}} \rightarrow$ total monthly cost paid before SSP deployment;

- $C_{total_{new}} \rightarrow$ total monthly cost paid after SSP deployment

C. Scenario II

Scenario II, as already mentioned, refers to the supply of all consumption on PHT using the PSS. So, in the following sections it is presented the calculation of the generator rated power for this scenario; the cost per generated kWh by the Diesel Generator is calculated, and the economic-financial analysis of the PSS implementation is performed.

D. Cost of kWh supplied by diesel generator for Scenario II

For Scenario II, the generator rated power (kVA₂) was calculated by (1), changing the equation numerator to the highest peak demand recorded on PHT. As shown in Fig. 2, it was 2500 kW. In this scenario, the generator works for approximately 720 h/year. Thus, the rated power e calculated for the generator would be 3125 kVA.

Similar to Scenario I, the cost of the kWh supplied by the diesel generator was determined by (2) to (5). For a generator with a rated power equal to 2500 kW, whose initial expenses were $ 2,071,057.19, the cost of kWh is 0.3151 $/kWh.

E. Economic-financial feasibility analysis of Scenario II

For this analysis, the cost of the kWh on the PHT obtained based on the utility tariff was compared to the cost of the kWh produced by the diesel generator. It is observed that the peak of demand and the excess of demand were disregarded in the calculations because, as previously mentioned, the demand peaks occur in Off-PHT. The value for the total annual energy cost on PHT with the application of the utility tariff were $ 584,407.73, while $ 315,885.53 was the total cost value of the energy on PHT when the price of kWh produced by diesel generator was used.

Therefore, the total cost of energy during the PHT is lower when the PSS based on diesel generators is used. Thus, from this first simple analysis, it seems that PSS is economically viable to be used in Scenario II. However, other complementary economic analyses are still needed to determine if PSS usage is also advantageous to the consumer if there are other types of available financial investments.

The analyzes performed were the Simple Payback, the Discounted Payback, which and the Net Present Value (NPV) [14, 15].

In order to do these analyzes, it was considered a period of 10 years, which is lower than the useful life of diesel generator. To make the cash flow was considered the consumer's monthly savings after the PSS installation. The annual economy was calculated as the summation of the monthly difference between the value of PHT consumption based on the utility tariff and the value of the PHT consumption based on cost per kWh produced by diesel generator, as shown in (8).

$$E_{annual} = \sum_{m=1}^{12} \Big[V_{PHT_{utility}}(m)$$
$$- V_{PHT_{generator}}(m) \Big] \qquad (8)$$

Being,

- $E_{annual} \rightarrow$ the annual economy in ($);

- $V_{V_{PHT_{utility}}} \rightarrow$ the total monthly cost of consumption in PHT that appears in the consumer's electricity bill, in ($);

- $V_{PHT_{generator}} \rightarrow$ the total monthly cost of consumption in PHT based on the $/kWh produced by the diesel generator, in ($).

Calculations based on (8) were made for the year 2017. In this way, it was found that in 2017 it was saved $ 268,522.19. It is observed that the economy of the year 1 of implementation is the same adopted in each year, because the forecast of the annual economy of the following years proved unfeasible due to the amount of variables that should be considered in the analysis: tariff variation the power utility;

variation in system operation and maintenance expenses; variation in the increase or decrease of consumer charges etc. The Cash Flow of this consumer was generated, which is shown in the columns named Cash Flow in TABLE 3 and in TABLE 4.

- Simple Payback Analysis

The Simple Payback method consists in verifying when the Accumulated Value becomes positive due the Cash Flow that consumer obtains year by year [14]. The time that the Accumulated Value takes to become positive is the time that the project spends to be paid. This period is known as "pay back". In TABLE 3 it is shown how this method is performed. For this purpose, year 0 is the year of the system installation, so the value in cash flow is the investment made in the project and it enters as a negative value. In year 1, and in subsequent years the values are the savings got in that year, calculated as mentioned before. Therefore, these savings enter as a positive value in the Cash Flow. The Accumulated Value in a specific year is equal to the sum of the value of the saving in this year with the Accumulated Value of the immediately previous year. So, as one can be seen from TABLE 3, the project pay back is around 8 years, which is 12 years before the end of the useful life of the generator.

TABLE 3. SIMPLE PAYBACK.

Simple Payback		
Year	Cash Flow	Accumulated Value
0	-$ 2,071,057.19	-$ 2,071,057.19
1	$ 264,625.13	-$ 1,806,432.06
2	$ 264,625.13	-$ 1,541,806.93
3	$ 264,625.13	-$ 1,277,181.80
4	$ 264,625.13	-$ 1,012,556.67
5	$ 264,625.13	-$ 747,931.54
6	$ 264,625.13	-$ 483,306.41
7	$ 264,625.13	-$ 218,681.28
8	$ 264,625.13	$ 45,943.85
9	$ 264,625.13	$ 310,568.98
10	$ 264,625.13	$ 575,194.11
11	$ 264,625.13	$ 839,819.24
12	$ 264,625.13	$ 1,104,444.37
13	$ 264,625.13	$ 1,369,069.49
14	$ 264,625.13	$ 1,633,694.63
15	$ 264,625.13	$ 1,898,319.76
16	$ 264,625.13	$ 2,162,944.89
17	$ 264,625.13	$ 2,427,570.02
18	$ 264,625.13	$ 2,692,195.15
19	$ 264,625.13	$ 2,956,820.28
20	$ 264,625.13	$ 3,221,445.41

- Discounted Payback Analysis

The Discounted Payback method is similar to the Simple Payback. The difference between them is that in the Discounted Payback, its Accumulated Value is calculated based on the Present Value of the Cash Flow, while in the Simple Payback its Accumulated Value is based on only in the Cash Flow values, without any discount applied to them (see TABLE 3) [14]. In

TABLE 4 it is shown how this method is performed. The column of Present Values in this table take into account the Minimum Attractiveness Rate (MAR) discount. The Attractiveness Rate is the rate that represents the minimum that an investor proposes to earn with a specific investment or the maximum that he proposes to pay for a financing. In this context, the Attractiveness Rate is an indication of the opportunity cost, that is, it indicates how much the consumer could profit whether had invested his money in other type of

investment, since he has invested his money in the PSS. To determine the MAR, one can consider the average of the rates of annual profit of several available possibility of investments and add to that the estimated annual rate inflation. In TABLE 5 is shown examples of some possibilities of investments in Brazil and its respective Average Annual Profits (AAP). The average of AAP for these examples is 6.23 %. So, considering an estimated annual inflation rate equal to 4.58%, the MAR obtained was approximately 10%, being this value used in the Discounted Payback Analysis. The present values of the Cash Flow were calculated using (9).

$$PV_t = \frac{CF_t}{(1 + MAR)^t} \tag{9}$$

Being,

- $PV_t \rightarrow$ the Present Value on t year, $t = 0, 1, 2, \dots, 10$.

- $CF_t \rightarrow$ the Cash Flow on the t years;

- MAR $\rightarrow$ the Minimum Attraction Rate, equal 10% (0.1).

TABLE 4. DISCOUNTED PAYBACK.

Discounted Payback			
Year	Cash Flow	Present Value of the Cash Flow	Accumulated Value or Net Present Value
0	-$ 2,071,057.19	-2,071,057.19	-2,071,057.19
1	$ 264,625.13	240,568.3	-1,830,488.89
2	$ 264,625.13	218,698.45	-1,611,790.44
3	$ 264,625.13	198,816.78	-1,412,973.66
4	$ 264,625.13	180,742.52	-1,232,231.14
5	$ 264,625.13	164,311.39	-1,067,919.75
6	$ 264,625.13	149,373.99	-918,545.76
7	$ 264,625.13	135,794.53	-782,751.23
8	$ 264,625.13	123,449.58	-659,301.65
9	$ 264,625.13	112,226.89	-547,074.77
10	$ 264,625.13	102,024.44	-445,050.03
11	$ 264,625.13	92,749.49	-352,300.83
12	$ 264,625.13	84,317.72	-267,983.11
13	$ 264,625.13	76,652.47	-191,330.63
14	$ 264,625.13	69,684.08	-121,646.57
15	$ 264,625.13	63,349.15	-58,297.41
16	$ 264,625.13	57,590.14	-707,28
17	$ 264,625.13	52,354.67	51,647.39
18	$ 264,625.13	47,595.16	99,242.55
19	$ 264,625.13	43,268.32	142,510.87
20	$ 264,625.13	39,334.84	181,845.71

TABLE 5. OTHER INVESTMENT OPTIONS.

Minimum Attractiveness Rate	
Investments	Average Annual Profit
Selic	6.40%
Savings account	4.55%
CDI	6.40%
Ibovespa	7.48%
Average Value	6.23%
Inflation	4.58%
MAR	Average + Inflation = 10.81%

The Accumulated Values or the Net Present Values of Investment (NPV), in last column of

TABLE 4, are calculated by (10).

$$NPV = -I_{initial} + \sum_{t=1}^{n} \frac{CF_t}{(1 + TMA)^t} \tag{10}$$

978-1-7281-4181-7/19 $31.00 © 2019 IEEE

Being,

- $I_{initial} \rightarrow$ the initial investment value;
- $n \rightarrow$ the number of years for which the NPV is being calculated.

For example, for n equal 20, the NPV was equal to $ 181,845.71.

So, as one can be seen from TABLE 4, the project pay back, based on the Discounted Payback Analysis, is around 17 years, which is 3 years before the end of the useful life of the generator. This is an indication of the economic feasibility of the PSS.

IV. CONCLUSION

In this work, an economic analysis about the use of a Peak Shaving System, based on diesel generation, in an electricity consumer with binomial tariff utility in Brazil, was presented. It was simulated the behavior of this type of system in the installation of the consumer from the financial point of view. It was observed that of the use of this system is economically inviable during the periods of the day when the global utility tariff is less than the cost per kWh produced by the diesel generator. This application was studied here as Scenario I. On the hand, during the periods of the day, when the global utility tariff is greater than the cost of the kWh produced by the Peak Shaving System, the study showed that this system is economically viable, and in addition it is competitive and advantageous when compared to other types of available economic investments. These applications were named as Scenario II in this work. Even if there were other investments with a minimum attractiveness rate of around 10% per year, the system during Scenario II proved to be economically attractive.

ACKNOWLEDGMENT

The authors would like to thank the support and motivation provided by the Federal University of Maranhão (UFMA), CEMAR, FAPEMA, CNPq – Brazil, and MEIHAPER CYTED.

References

[1] ANNEL, "Condições gerais de fornecimento de energia elétrica - Resolução Normativa nº 414/2010," Agência Nacional de Energia Elétrica, Brasília, 2017.

[2] Eletrobrás, "Manual de tarifação de energia elétrica," PROCEL, Rio de Janeiro, 2011.

[3] J. Zupancic, E. Lakic, T. Medved and A. Gubina, "Distribution, Advanced Peak Shaving Control Strategies for Battery Storage Operation in Low Voltage," 2017 IEEE Manchester PowerTech, Manchester, UK, 2017.

[4] K. Pandiaraj, B. Fox, D. Morrow, S. Persaud and J. Martin, "Centralised control of diesel gen-sets for [peak shaving and system support," IEE Proceedings -

Generation, Transmission and Distribution, Belfast, UK.

[5] E. Telaretti and L. Dusonchet, "Battery Storage Systems for Peak Load Shaving Applications - Part I," in *Environment and Electrical Engineering* , Florença, Itália, 2016.

[6] B. Wang, M. Zarghami and M. Vaziri, "Energy management and peak-shaving in grid-connected photovoltaic systems integrated with battery storage," in *North American Power Symposium (NAPS)*, Denver, 2016.

[7] F. Braam, R. Hollinger, M. Engesser, S. Muller, R. Kohrs and C. Wittwer, "Peak Shaving with Photovoltaic-Battery Systems," IEEE PES Innovative Smart Grid Technologies, Europe, Istanbul, Turkey, 2014.

[8] E. Telaretti and L. Dusonchet, "Battery Storage Systems for Peak Load Shaving Applications - Part II," in *Environment and Electrical Engineering*, Florença, Itália, 2016.

[9] C. E. d. Maranhão, "Agência Web," CEMAR, 28 August 2018. [Online]. Available: http://www.cemar116.com.br/informacoes/cobranca-de-tarifas-alta-tensao. [Accessed 10 October 2018].

[10] A. Nocera, G. Gomes and V. Pereira, "Análise da viabilidade técnica e financeira da implantação do gerador a diesel no horário de ponta em um hospital de Curitiba," DEPARTAMENTO ACADÊMICO DE ELETROTÉCNICA, Curitiba, 2015.

[11] Cummins, "Engenharia de Aplicações - Manual de aplicações para Grupos Geradores arrefecidos a água," Power Generation, Guarulhos, 2016.

[12] H. d. S. Mota, "Análise técnico econômica de unidades geradoras de energia distribuída," Autarquia Associada à Universidade de São Paulo, São Paulo, 2011.

[13] U. F. d. S. Catarina, "Hospital Universitário Universidade Federal de Santa Catarina," 08 December 2017. [Online]. Available: http://www.hu.ufsc.br/setores/contratos/wp-content/uploads/sites/33/2018/05/Termo-de-Contrato-nº-297.2016-Powercom.pdf. [Accessed 16 November 2018].

[14] R. Bordeaux-Rêgo, G. Paulo, I. Spritzer and L. Zotes, Viabilidade econômico-financeira de projetos, Rio de Janeiro: FGV, 2010.

[15] I. Camargo, Noções Básicas de Engenharia Econômica - Aplicações ao Setor Elétrico, Brasília: Itamarati Ltda, 1998.

Thyristor Triggering, Static and Dynamic Characteristics

Ricardo Alves do Prado
Application Engineering
SEMIKRON Semicondutores Ltda
Carapicuíba, Brazil
ricardo.prado@semikron.com

Ulrich Nicolai
Application Engineering
SEMIKRON Elektronik GmbH & Co.
Nuremberg, Germany
ulrich.nicolai@semikron.com

Abstract—**The purpose of this document is to study the thyristor semiconductor switch through its static, dynamic and triggering parameters, focusing on how they affect and are affected by the variables found in real-life applications.**

Keywords—thyristor, SCR, trigger, latching, gate current.

I. INTRODUCTION

The thyristor semiconductor switch was first introduced by General Electric in 1957 based in Bell Laboratories' discoveries during the previous year about P-N-P-N transistor technologies [3].

Since this first thyristor with 25A of current capability and 300V blocking voltage, its capabilities have greatly improved; modern thyristors can easily reach thousands of amperes and volts and are available in many different structures like fast thyristors and the GTO (Gate Turn-Off thyristor).

This work will cover only the SCR (Silicon Controlled Rectifier) type, also known as Phase-control thyristor, a pretty consolidated and mature technology, which even today can be tricky to master, as it requires specific control and protection circuit behaviors to extract the better it has to offer.

II. STRUCTURE AND WORKING PRINCIPLE

A. Structure and basic characteristics

SCRs are composed by four doped silicon layers, P-N-P-N, being possible to describe them as two BJTs (Bipolar Junction Transistor), one P-N-P and the other an N-P-N, connected by base and collector.

The first P-N junction from the Anode is the responsible for the reverse blocking capability while the second P-N structure gives the thyristor the ability to block forward voltage.

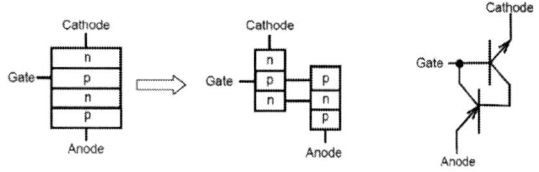

Fig. 1. Silicon structure of a thyristor [1].

When a permissible forward voltage is applied to the device, polarizing anode to cathode, the N-P junction will not allow any current conduction more than the direct leakage current (I_{DD}), however, if an specific electric signal is applied in the second P layer, a relatively small current will flow from gate to cathode, triggering the thyristor.

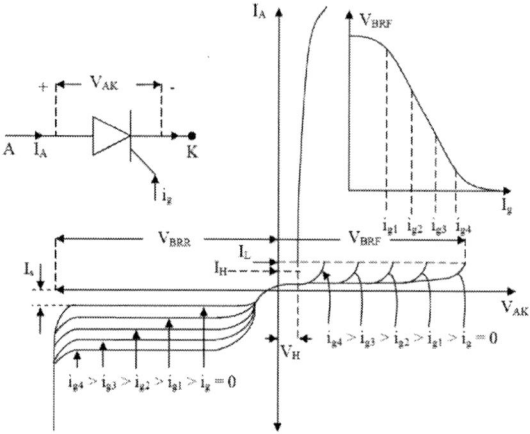

Fig. 2. Voltage-current characteristic of a thyristor [3].

The graphic above shows the mains voltage-current (V_{AK} – I_A) characteristic of a SCR, it is possible to observe that the higher the gate current (I_G), the lower is the forward voltage breakdown value (V_{BRF}).

Gate trigger signal must be sufficient high to reduce V_{BRF} to a tiny value, enough to enable latching in high conduction angles, the standard value used by semiconductor manufacturers to test gate-triggering parameters like V_{GT}, I_{GT}, V_{GD} and I_{GD} (gate trigger and gate non-trigger voltage and current, respectively) is 6V [2].

In power conversion applications, thyristors are not operated at "low" gate current or reverse conducting modes, as these can easily be destructive for them.

B. Latching and gate feedback

Different from other semiconductor switches, once fired, SCRs cannot be put out of conduction, at least not until main current falls under I_H (holding current), however, this also means that they do not require continuous gate signal to keep conducting.

The latching process takes five steps to be completed and it is shown in Fig. 3.

978-1-7281-4181-7/19 $31.00 © 2019 IEEE

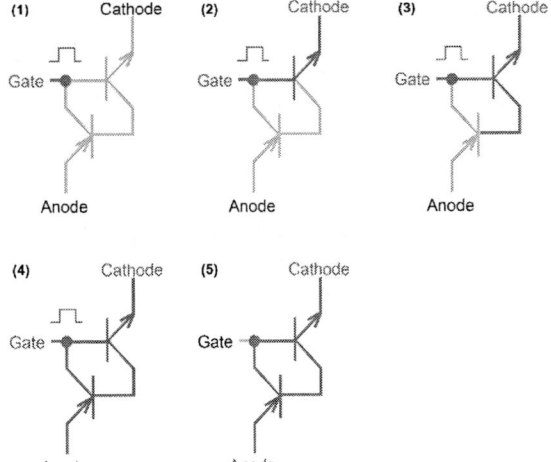

Fig. 3. Thyristor firing sequence, light grey paths not conducting current, red paths conducting.

With the device in forward blocking state, the following actions and reactions take place during latching process:

(1) Gate signal is provided by the driver

(2) Gate current acts as a base current injection for the N-P-N transistor, latching it.

(3) N-P-N transistor begins conduction, dragging current from P-N-P base.

(4) P-N-P transistor begins conduction, feeding N-P-N base.

(5) Current feedback from P-N-P emitter to N-P-N base is complete, gate signal can be removed.

III. STATIC AND THERMAL CHARACTERISTICS

A. Static parameters

The first parameter to be taken into account when designing a power converter is the blocking voltage, in this case there are two types, V_{RRM} (repetitive peak reverse voltage) and V_{DRM} (repetitive peak off-state voltage); the former is the reverse voltage blocking capability while the latter is the forward one.

These parameters are specified from $T_{vj\ min}$ to $T_{vj\ max}$, and are linked to I_{DD} and I_{RD} (direct reverse current), also known as leakage current.

The higher the temperature and the voltage, the higher will be I_{DD} and I_{RD}, these are used to determine the device's voltage class.

Another static characteristic should also be taken into account when dimensioning thyristors, the I_{TRMS}; this is the maximum allowed RMS current for the device, being defined generally by mechanical limitations of the semiconductor, e.g. wire ropes in stud type devices, bus bars in insulated modules, etc.

Thyristors are also defined by another important static parameter, the I_{TSM} (Surge on-state current).

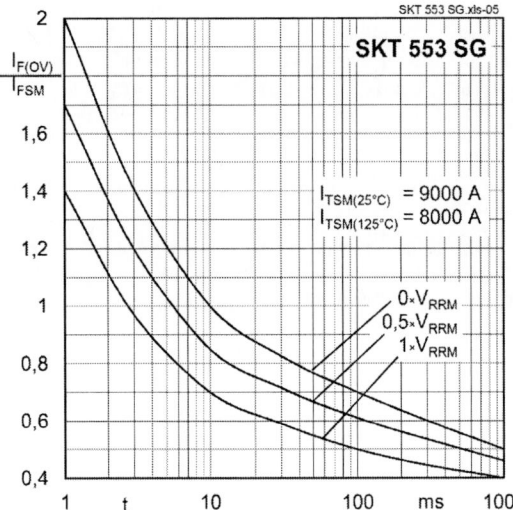

Fig. 4. Overload capability as function of time at different off-state voltages [5].

The I_{TSM} is the maximum peak current during a half-sine wave which the diode can withstand, this value is generally given for 50 and 60Hz (10ms / 8,33ms pulse), for junction temperature of 25°C (ambient temperature) and $T_{j\ MAX}$, with or without reapplied reverse voltage.

Through I_{TSM} value is possible to get i²t value, which is used to size circuit breakers and ultrafast fuses to protect the semiconductor from surge currents generated by short-circuits, triggering mismatch, etc.

It is important to have in mind that these values are given for non-recurrent surge currents, they should not be used for load cycles or continuous operation due to the fact that I_{TSM} rises junction temperature to perilous levels, up to 1300°C [6].

The surge will lead to a fast thermal expansion, causing mechanical stress over the silicon crystalline structure and in most cases damaging it, even if the device survives the surge, this specific region will be now already damaged.

B. The thermal calculus

It is essential in high power applications to know whether the chosen device is able to safely deliver the desired power and load cycle without suffer from overheating.

Nevertheless, prior to temperature calculations, the power losses should be known, thyristor power losses can be summarized by the following equation:

$$P_{THY} = P_{Off} + P_{Cond} + P_{Turn-on} + P_{Turn-off} \quad (1)$$

Being P_{Off} the off-state losses (negligible), P_{Cond} conduction losses, $P_{Turn-on}$ the losses during latching and $P_{Turn-off}$ turn-off losses; for 50-60 Hz and low voltage applications, switching losses are also negligible.

There are different methods to calculate conduction losses in thyristors or diodes (equations will be the same for both), the most used is the one below, as it returns satisfactory values with little time and effort invested.

$$P_{Cond} = V_{T(TO)} \times I_{TAV} + r_T \times I_{RMS}^2 \quad (2)$$

978-1-7281-4181-7/19 $31.00 © 2019 IEEE

Two variables in the equation are project inputs, I_{TAV}, which is the average current passing through the device and I_{TRMS}, its RMS current, the relation between them is known as "form factor", and each topology will have its own form factor (e.g. $\pi/4$ for single-phase rectifiers and static switches).

The other variables are given by the manufacturer in the datasheet, $V_{T(TO)}$ (threshold voltage) and r_T (on-state slope resistance).

With the total power losses, it is time to do the thermal calculations and check whether the device is suitable for the application, this is done through the following equation:

$$T_{vj} = T_{A/W} + P_{THY} \times (R_{TH\ S-A} + R_{TH\ C-S} + R_{TH\ J-C}) \quad (3)$$

The thermal resistances $R_{TH\ J-C}$ and $R_{TH\ C-S}$ are also given in the datasheet, while $R_{TH\ S-A}$ is an information intrinsic to the heatsink, the final calculated junction temperature (T_{vj}) must be lower than $T_{vj\ max}$; in order to avoid stress the device more than necessary, it is advised to use up to 90% of the maximum junction temperature allowed.

For medium voltage applications, turn-off losses become considerable and can be calculated as follows [4]:

$$P_{turn-off} = f \times E_{rr} = f \times \int_{t(I_{TRM=0})}^{t_q} i_T(t) \times v_{AK}(t)\ dt \quad (4)$$

As V_{AK} varies very little during commutation time, it can be considered constant and for the worst case, the peak value should be used; the integral from $t_{(ITRM=0)}$ to t_q of $i_T(t)$ is the own Q_{rr}, so the equation can be rewritten to:

$$P_{turn-off} = 0.5 \times f \times Q_{rr} \times V_{PEAK} \quad (5)$$

Where f is the line frequency, Q_{rr} is the reverse recovery charge of the thyristor (this parameter will be properly explained later) and V_{PEAK} is the peak line-line voltage.

IV. DYNAMIC CHARACTERISTICS

Ideally, commutation to off-state in thyristors are "naturally" done by the circuit, this type of device switch off as soon as anode current drops under I_H, however, in reality it has a reverse recovery charge that works just like in diodes.

As soon as anode current drops under I_H, the thyristor discharge its reverse recovery charge (Q_{rr}) into the circuit, this stored charge value depends on the forward current (I_{TM}), -dI_T/dt (rate of fall for on-state current), junction temperature and the I_{RRM} (reverse current peak) as well.

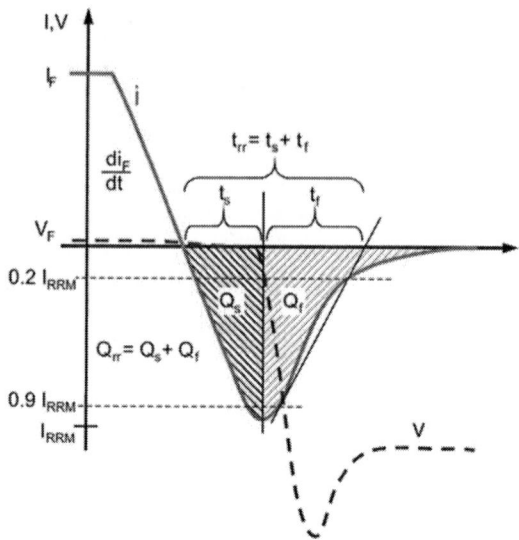

Fig. 5. Reverse recovery characteristic of a SCR [1].

The duration of this process is known as t_{rr} (reverse recovery time), do not mistake it to the t_q, time from $I_{TRM}=0$ to in which forward voltage can be safely reapplied without triggering again the device.

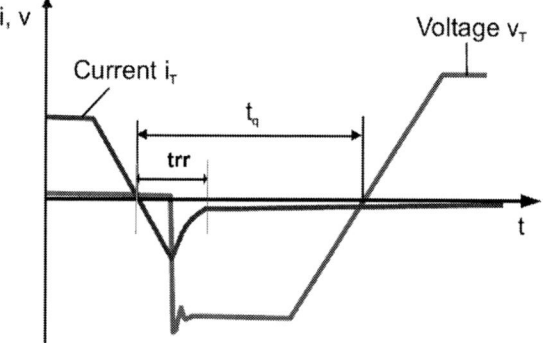

Fig. 6. Differences between t_{rr} and t_q parameters [1].

For phase-control thyristors, t_q is not relevant, while for fast-commutated ones this value ranges from 10 to 100 μs. Circuit commutated turn-off time is directly proportional to junction temperature and dV_D/dt, being inversely proportional to V_{RM} (reverse voltage during turning-off).

The I_{RRM} equation can be found below, being possible to rearrange it to find I_{RRM} or di_T/dt [1].

$$I_{RRM} = \sqrt{2 \times Q_{rr} \times -\frac{dI_T}{dt}} \quad (6)$$

After I_{RRM}, the current recovers quickly, creating an overvoltage due to circuit's inductance, this transient is added to the reverse voltage, and can be calculated through the following equation [2]:

$$V_{L\sigma} = -\sum L_\sigma \times \frac{dI_T}{dt} \quad (7)$$

978-1-7281-4181-7/19 $31.00 © 2019 IEEE   1237

Line inductance can be found by the following equation that considers a 5% line impedance [2].

$$L \geq 0,05 \times \frac{V_V}{2 \times \sqrt{3} \times \pi \times f \times I_V} \qquad (8)$$

Switching-off overvoltages can be dangerous to the device, especially when operating with high inductive loads and electrical connections, to attenuate this effect a RC snubber network can be used to damp the bigger oscillation amplitudes, creating instead lower amplitude oscillations.

The snubber capacitance will be defined by line voltage and the own reverse recovery charge of the used device [1].

$$C = \frac{Q_{rr}}{\sqrt{2} \times V_V} \qquad (9)$$

And the snubber resistor by [1]:

$$R = (1,5 \dots 2) \times \sqrt{\frac{L_S}{C}} \qquad (10)$$

The total power dissipated by the snubber resistor will be [2]:

$$P_{R\ snub} = V_V \times f \times (\sqrt{2} \times 10^{-6} \times Q_{rr} + k_1 \times C \times V_V) \quad (11)$$

Where:

$k_1 = 0$ for diodes in uncontrolled bridge rectifiers

$k_1 = 2*10^{-6}$ for thyristors in controlled one and two pulse center-tapped circuits as well as fully controlled two pulse bridge circuits and AC controllers

$k_1 = 3*10^{-6}$ for thyristors in controlled three and six pulse center-tapped circuits as well as fully controlled two pulse bridge circuits and AC controllers

$k_1 = 4*10^{-6}$ for thyristors and diodes in fully or half-controlled six pulse bridge circuits

V. TRIGGERING

Triggering a thyristor is not a hard task, it is possible to do it with a simple cellphone charger in series with a low value resistor; the true issue is to latch it properly in a way that is possible to harness all the device has to offer.

Gate signal values and shape deeply affects thyristor's capabilities, below there are some trigger parameters that should be taken into account in order to properly analyze and design thyristor triggering:

- I_H: Holding current – Minimum Anode to cathode current to maintain the thyristor in on-state

- I_L: Latching current – Minimum Anode to Cathode current to allow triggering

- V_{GT}: Gate trigger voltage – Minimum gate voltage to trigger the component

- I_{GT}: Gate trigger current – Minimum gate current to trigger the component

- V_{GD}: Gate non-trigger voltage – Maximum gate voltage to not trigger the device

- I_{GD}: Gate non-trigger current – Maximum gate voltage to not trigger de device

- d_{IG} / d_T: Rate of rise of gate current

- t_p: Gate current pulse duration

The relation between the above parameters and other ones are as follows.

TABLE I. RELATION BETWEEN GATE TRIGGERING PARAMETERS

	↑ T_J	↑ V_D	↑ I_G	↑ d_{IG}/dt	↑ t_p
I_H	↓	↓	-	-	-
I_L	↓	↓	↓	↓	↓
V_{GT}	↓	↑	↑	-	-
I_{GT}	↓	-	↓	↓	↓
V_{GD}	↓	-	-	-	-
I_{GD}	↓	↓	-	-	-

a. Page 7 – AN 18-002: Thyristor Triggering and Protection of Diodes and Thyristors, SEMIKRON.

From the table above is possible to observe many relations between these parameters, for instance, junction temperature rise decreases the current and voltage needed to trigger and maintain a thyristor in conduction; this means that as higher the temperature, easier will be to trigger a SCR and maintain it in conduction.

Datasheets generally have information regarding these values, the graph below shows the relation between V_G, I_G and T_{vj}; being the red line the maximum limit for gate voltage and current and the blue line the minimum to trigger the device.

It is recommended to also consider the green line, minimum required to trigger the device in -40°C, to properly size the gate driver.

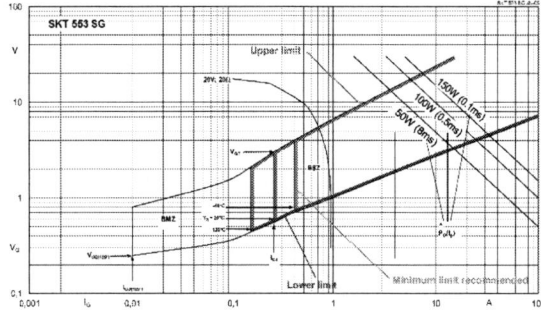

Fig. 7. Triggering parameters for SKT 553 SG capsule thyristor [5].

There are some types of trigger strategies used for thyristors, being each one more adequate for one or another application and with its own positive and negative characteristics.

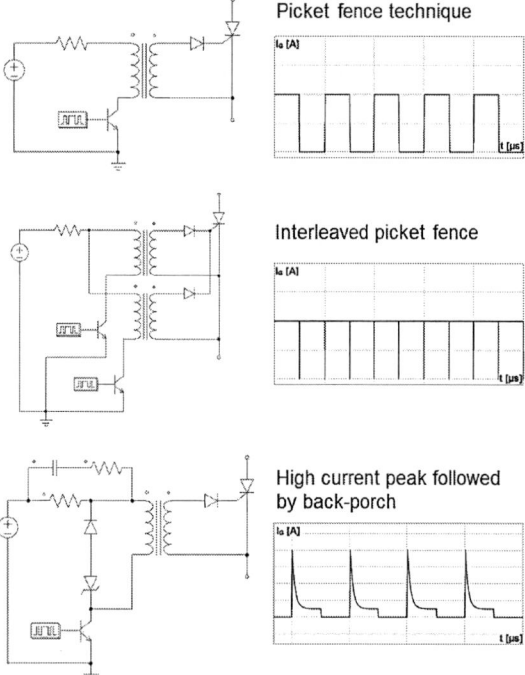

Picket fence technique

Interleaved picket fence

High current peak followed by back-porch

Fig. 8. Thyristor trigger circuits and techniques.

First of all, gate signal pulses must be sufficient long to allow latching, otherwise, gate feedback process (described in chapter 1) will not have been completed, the minimum time recommended is 100μs, in figure 7 is possible to observe that after this time, the increase of pulse length affects needed gate current very little [2].

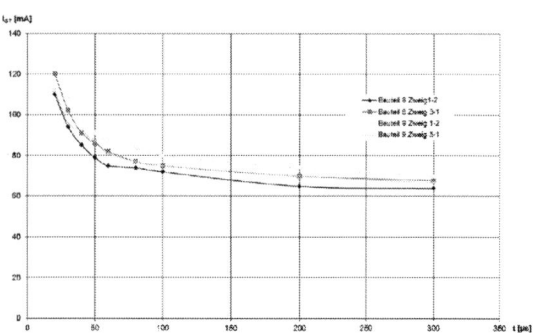

Fig. 9. Gate trigger current vs pulse duration [2].

As previously described, the easiest way to trigger a thyristor is to inject a rectangular wave current into the gate, however, this technique results in poor dI_T/dt capability and dissipates a lot of heat in the gate area due to the all-time conducting wave.

A first improvement to this method is the picket fence technique using 5 to 40 kHz, which can be used to reduce gate and driver power dissipation, however, this method still does not allow great dI_T/dt rates and sometimes can lead to signal stability problems.

Another way to properly trigger an SCR is to use the interleaved picket-fence technique, it allows continuous current injection in the gate with smaller driver components, but still do not solve dI_T/dt.

The technique that allow the thyristor to operate in its maximum capabilities is the injection a high step current (higher than 5 times I_{GT}) from 10 to 20 μs, followed by a minimum 90 μs rectangular or pulsed back-porch wave. The high dI_G/dt solves dI_T/dt capability problem and the pulsed signal reduces unnecessary gate power dissipation [2].

Gate current rising rate must be higher than 1 A/μs for typical applications, for capacitive applications that require much higher dI_T/dt, like magnetizer equipment (where SCRs are used to charge the DC Link), a higher dI_G/dt must be used.

In order to avoid negative feedback from the device to primary side of the gate driver, it is recommended to use a fast diode in the secondary, another interesting tip is to use a high value resistor (between 220 and 2200Ω) and a 10 to 47nF capacitor between gate and cathode (R_{GK} and C_{GK} respectively) to decouple gate signal noise; care must be taken, as this capacitor can reduce gate response to trigger signal [2].

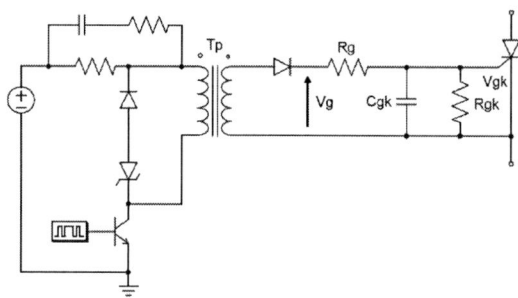

Fig. 10. Possible circuit for thyristor gate driver using high current peak technique and gate signal noise decoupling.

In order to size the gate resistor properly, the following equations can be used:

$$R_G = \frac{V_G - V_{GK}}{I_G} \quad (12)$$

$$P_{RG} = R_G \times I_G{}^2 \times \frac{T_P}{T} \quad (13)$$

The desired gate current (I_G, nominal gate current value during pulse) must be known upfront to use equations 12 and 13, gate current vs gate voltage graphics present in datasheets (e.g. Fig. 7) can be used for this purpose.

And total gate power losses can be calculated as follows:

$$P_G = V_{GK} \times I_G \times \left(\frac{T_P}{T} + \sqrt{\frac{T_P}{T}}\right) \quad (14)$$

Trigger pulses during reverse blocking will increase gate power losses and are not considered in the equations above, as should be avoided at any costs due to unnecessarily overheating the gate area.

The designer should also take great care with the pulse transformers in gate driver, these should be shielded in order to avoid any EMI noise generation and have the lowest inductivity as possible to deliver high dI_G/dt to SCR's gate.

Pulse transformer's inductivity is determined by the core's magnetic permeability and number of winding turns, the more common relations are 1:1, 2:1 and 3:1; in addition to this, it must have an isolation between primary and secondary according to EN50178, of 2,5kV for 400V networks, 4kV for 690V and so on [2].

Besides pulse transformer, the own circuit can add series inductances to gate current path, as stated before, dI_G/dt is reversely proportional to L_{GS} (series gate path inductance), good tips for gate driver board and connections design is to not use more PCB trace length than necessary and twist driver-thyristor cables (gate and auxiliary cathode) to neutralize part of its stray inductance.

Latching can also be achieved other by injecting the right current through the gate, the LTT (Light Triggered Thyristor) for example, can be triggered by light; these components are particularly useful in high EMI environments to avoid gate signal interference.

Nevertheless, undesired triggering may happen by EMI interference, in this situation, latching signal can be transmitted during reverse blocking state, significantly increasing gate power losses.

The additional power losses due to reverse blocking state triggering will force the gate area into a power cycling, a steep temperature increase and decrease during each period, leading to gate solder fatigue or gate crystalline structure damage.

The signal can also just not be enough powerful to properly latch the SCR entirely, overheating the region next to the gate, which will conduct the entire current; anyway, the result of EMI latching will always be the device's damaging and losing of its full capability.

Thyristors are unintended triggered as well by high dV/dt through capacitive displacement, the effects will be the same as for EMI latching, and in order to avoid this a RC snubber network should be applied between anode and cathode, some care must be taken while doing it as the own snubber will create dI_T/dt in the thyristor during latching.

In other cases, triggering parameters can be correct, but main current rise rate (di_T/dt) is too high for the gate driver sets, in this case the triggering will not reach the entire area of the device, with time leading it to failure.

SCR's dI_T/dt capability can be improved by increasing I_G and dI_G/dt accordingly, if even with optimal driver settings the circuit's dI/dt cannot be matched, the rate of rise can be attenuated by using a series inductance with the SCR [2].

$$L_R \geq \frac{V_D}{(di/dt)_{cr}} \tag{16}$$

VI. CONCLUSIONS

Although had been replaced by the MOSFET (Metal-Oxide-Semiconductor Field Effect Transistor) and IGBT (Insulated-Gate Bipolar Transistor) in inverters, DC-DC converters and even in some rectifiers, thyristors are still a widely used semiconductor switch that dominates AC-AC static switches applications and high power controlled rectifiers up to this day due to its robustness, its current, voltage and thermal capabilities.

As stated before, SCRs can be easily used by anyone, but achieve state of the art applying them is a hard task as they require very specific gate signal forms, supporting circuitry and even the right converter's mechanical design to reach their full potential.

Last but not least, it is important to have in mind that latching sensitivity increases with junction temperature, the hotter a thyristor gets, the more likely it is to be undesirably triggered; due to this fact the control over a SCR can be lost in case of overheating, so maximum T_{vj} allowed should always be respected.

REFERENCES

[1] A. Wintrich, U. Nicolai, W. Tursky, T. Reimann, "Application manual power semiconductors", Semikron International GmbH, rev.01, pp. 22-26, 236-249, 2015.

[2] U. Nicolai, "Application note AN 18-002: Thyristor triggering and protection of diodes and thyristors", rev.00, pp. 1-19, February 2018.

[3] Indian Institute of Technology Kharagpur, "Power Semiconductor Devices", rev.02, pp 9.

[4] ABB Switzerland Ltd, "Application note 5SYA 2055-01: Switching losses for phase control and bi-directionally controlled thyristors", rev.01, August 2013.

[5] R. Prado, "SKT 553 SG's datasheet", rev.01, August 2018.

[6] Mitel Semiconductor, "Application note AN4870: Effects of temperature on thyristor performance", rev.1.2, March 1998.

Development of a FPGA-Based Real-Time Simulation System

Yago F. Oliveira, Filipe A. La-Gatta, Rodrigo A. F. Ferreira, Márcio C. B. P. Rodrigues

Group of Power Electronics and Applications
Federal Institute of Education, Science and Technology of Southeast of Minas Gerais
Juiz de Fora, Brazil
marcio.carmo@ifsudestemg.edu.br

Abstract—**This work presents the development of a real-time simulation system based on field-programmable gate array (FPGA). Detailed information regarding the implementation of the proposed real-time simulator (RTS) is presented. A current bidirectional boost converter is considered in order to illustrate the proposed methodology for power converter modeling and implementation of RTS equation solver. The comparison between obtained results and offline commercial simulation is conducted to validate the proposed RTS. The proposed RTS can be used as a low-cost alternative to commercial platforms to perform real-time simulations of power electronic converters.**

Index Terms—**real-time simulation, FPGA, power electronic converter simulation.**

I. INTRODUCTION

Computer simulation of dynamic systems plays an important role on several fields of Engineering [1], [2], being widely applied from teaching to product development. Regarding Power Electronics applications, computer simulation is useful for developing new converter design and control techniques, as well as for optimising existing topologies [3].

In general, computer simulation can be performed using offline software tools or real-time simulation systems [4], [5]. Computer processing speed determines the total time spent on simulation for regular (offline) software tools. Thus, it can be either lower or greater than effectively simulated time of the considered system. On the other hand, regarding a real-time simulation, all internal variables, calculations and outputs must be accurately processed in real-time simulator (RTS) within the same length of time that its physical counterpart would spend, i.e., the time required to compute system solution at a given time-step must be shorter than the actual duration of the time-step [5].

Real-time simulation is more convenient than the regular offline computer simulation for analysis of systems in which different dynamics are present [6]. Power Electronics applications, such as electric vehicles [6]–[10] and renewable energy systems [11], [12] are examples of dynamic systems that have this kind of characteristic. Despite of the advantages of real-time simulation for Power Electronics applications, the relatively high cost of commercial platforms can difficult or restrict the use of this approach for research groups from universities and institutes of technology.

In this context, this paper presents the development of a real-time simulator based on a FPGA (field-programmable gate array) development kit, which can be used as an alternative to commercial platforms to perform real-time simulations of power electronic converters. A thorough description of the RTS is provided, which has been developed by combination of hardware description language and block-oriented design. Also, modeling and simulation approaching adopted to perform simulation of power electronics converters are described. A current bidirectional boost converter is considered in order to illustrate the proposed methodology for power converter modeling and implementation of the RTS equation solver. The comparison of obtained results to offline commercial simulation validates the proposed RTS. The remainder of this paper is organized as follows: Section II describes modeling and simulation methodology considered in this work. The detailed development of proposed RTS is described in Section III. Simulation results and discussion are presented in Section IV, which is followed by paper conclusion and references.

II. SIMULATION OF POWER ELECTRONIC CONVERTERS

The schematic diagram of the bidirectional boost converter (also referred as bidirectional buck/boost converter and DC-DC half-bridge converter in literature) considered in this work is shown in Fig. 1. S_0 and S_1 represent ideal electronic switches, which are complementarily driven.

Inductor and capacitor equivalent series resistances are represented by R_L and R_C, respectively. An ideal voltage source, V_B, is considered as power supply. Load is modeled as an ideal current source, I_O. There are two variables of interest for the power converter simulation: input (inductor) current, I_L, and output voltage, V_O. This approach is similar to the one adopted in [6] for the development of hardware-in-the-loop real-time simulation of an electric vehicle elements.

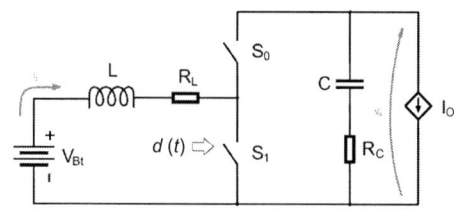

Fig. 1. Current bidirectional boost converter.

978-1-7281-4181-7/19 $31.00 © 2019 IEEE

Since S_0 and S_1 are complementarily driven, a single control signal, $d(t)$, is considered to perform the power converter switching. It allows the definition of two operation stages for the circuit of Fig. 1. This work considers a circuit solution approach based on companion models [13]–[15], where energy-storage elements are replaced for its equivalent circuits, which incorporates the discretization of differential equations. Thus, circuit solving can be performed by means of the analysis and solution of algebraic equations. Fig. 2 shows equivalent circuits for the two operation stages for the power converter under analysis, where Backward Euler discretization method [16] has been considered. Simulation time-step is denoted by Δt. Variables from actual iteration, n, are defined as $I_L(n)$ and $V_O(n)$, while values from the previous iteration are denoted by $I_L(n-1)$ and $V_O(n-1)$.

Duty cycle control signal, $d(n)$, defines which equivalent circuit will be considered for simulation at every time-step, as described in Table I. Thus, it is possible to define the set of equations to be solved in each simulation step. Equations (1) and (2) are related to solution of equivalent circuit of Fig. 2(a). Solution of equivalent circuit fo Fig. 2(b) is performed by solving equations (3) and (4).

$$I_L(n) = \frac{V_{bt}(n) + \frac{L}{\Delta t} I_L(n-1)}{\frac{L}{\Delta t} + R_L} \tag{1}$$

$$V_o(n) = V_c(n-1) - \frac{\Delta t}{C} I_o - I_o R_C \tag{2}$$

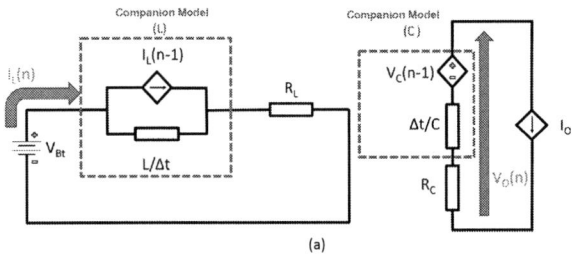

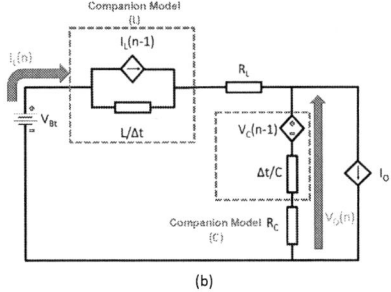

Fig. 2. Equivalent circuits considered for simulation: (a) $d(n) = 1$; (b) $d(n) = 0$.

TABLE I
EQUIVALENT CIRCUIT SELECTION

Duty cycle, $d(n)$	0	1
Switches states	S_0 on and S_1 off	S_0 off and S_1 on
Equivalent circuit	Fig. 2(b)	Fig. 2(a)

$$I_L(n) = \frac{V_{Bt} + \frac{L}{\Delta t} I_L(n-1) - V_c(n-1) + I_O(\frac{\Delta t}{C} + R_C)}{\frac{L}{\Delta t} + R_L + \frac{\Delta t}{C} + R_C} \tag{3}$$

$$V_o(n) = V_c(n-1) - \frac{\Delta t}{C}(I_L(n) - I_O) - R_C(I_L(n) - I_O) \tag{4}$$

III. DESCRIPTION OF THE DEVELOPED FPGA-BASED REAL-TIME SIMULATION SYSTEM

A. ALTERA DE2-115

The FPGA model used in this work is the Intel ALTERA DE2-115 development kit, which has the Cyclone family chip. Its software is the ALTERA Quartus II. For the development of this project, the VHDL language was used in some descriptions It is also used the block diagram design approach to interconnect the elements in its elementary project.

B. Bidirectional Data Bus

In this work a communication interface was used to test the functionality of the developed system, in order to validate the operation of the simulation platform before the external environment and, consequently, the simulation loop. Therefore, creating a communication bus is interesting in this regard.

The bus function is to intermediate the connection between the FPGA board and other components of the simulation set. For the correct control of this plant, it is necessary to receive two inputs with eight bits (values of DC voltage source and load from circuit of Fig. 1), for the calculation and the externalization of two 8-bit outputs ($I_L(n)$ and $V_O(n)$). So, the main purpose is that this communication bus provide a 16-bit bidirectional data flow.

The communication interface of the DE2-115 with the external environment was then made by a bidirectional data bus. This data bus allows the definition of inputs of the 16-bit width provides 256 different values for each variable (8 bits per variable).

1) Simulated Bus: At the programming interface of the FPGA, Quartus II, using the language VHDL, it was made the design of the operation of a tri-state buffer, which makes possible the implementation of the bidirectional port.

After the design in VHDL, this project was compiled and compressed, according to Fig. 3. The buffer shown in Fig. 3 has the function of controlling the communication direction of the bus, since it has three different output states: "0", "1", and "Z" (high impedance). The latter has the purpose of disconnecting the FPGA for writing, thus enabling it for reading on the bus. All of these states are controlled by the OE (output enable) input in the project.

The OE signal is the digital signal that, coming externally, determines the communication direction of the FPGA. Thus

978-1-7281-4181-7/19 $31.00 © 2019 IEEE

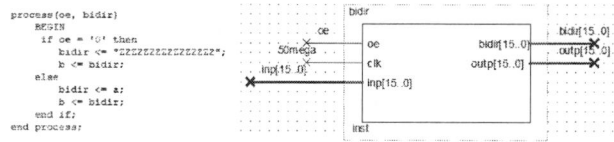

```
process(oe, bidir)
  BEGIN
    if oe = '0' then
      bidir <= "ZZZZZZZZZZZZZZZZ";
      b <= bidir;
    else
      bidir <= a;
      b <= bidir;
    end if;
end process;
```

Fig. 3. Structure of System Bus.

when in low logic signal, the bus is free for writing, that is, the 16 bits are inserted in the bus. However, when OE is at high logic level, the bus becomes available for reading, in which the parallel reading of the 16 external bits occurs. The output port has the function of receiving the value of the "bidir" bus when OE is at the low logic level. The input port has the function of inserting its value in the bus when OE is at high logic level. The "process" structure in VHDL means processes are interpreted line by line, and therefore, within this design structure resembles sequential programming languages. However, processes in this file are executed in parallel.

C. Math Operations

The next step after passing through the bus is performing the mathematical operations with the input vectors. When observing the electronic circuit and its mathematical modeling, it is verified that it has constant elements with values very close to zero, and also values much higher than number of bits of the bus allows to represent.

Thus, to make some operation between these data of disparate values, it is necessary to convert them to the floating-point structure According to the IEEE-754 standard [17].

1) Megafunctions: To perform many functions in this project, special blocks were used. In the Quartus II block simulation environment, it is possible to create blocks of "Megafunctions", which are blocks parameterized and optimized by ALTERA, whose purpose is to perform specific functions frequently used.These blocks are obtained and configured in Quartus by the *MegaWizard Plug-in Manager* wizard.

The block of Megafunction used for data conversion is the element that performs the function of integer-floating-point converter.

According to the function documentation, this operation occurs in 6 cycles of clock. The conversion function is also used in the reverse direction, when the equations are calculated and the vectors need to be transformed to decimal values to exit the block of operations.

Another logical sequence element available in the design is the synchronous counter, for the block of mathematical operations. This counter has the function of executing a count based on the input clock, by means of successive divisions in the initial frequency. The use of this device in this description is to be the delayed clock to save the calculated variables in the memory blocks (flip-flops).

2) Setting up the Equations: The equations referring to the converter model were then formed according to the floating point operations blocks. Both input variables, I_o and V_{Bt}, are part of equations (1)-(4).

All mathematic operations blocks were inserted in another block, called "fpop" (floating point operations). Therefore, within the fpop block, the four equations referring to the converter circuit were added. The equations of the models were generated by follow-up models. Since simulation is based on iterative solution of equations, previous iteration values, which in this case are $V_c(n-1)$ and $I_L(n-1)$, are saved at the end of each operation, on a vector type-D flip-flop. Fig. 4 shows this latch unit, which stores the previous value of simulation math operations. It is driven with the rising edge by the signal of clock responsible for bringing the new vectors, thus synchronizing the entire operation.

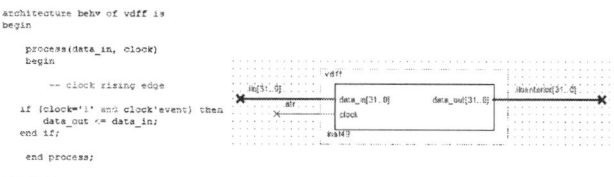

```
architecture behv of vdff is
begin

  process(data_in, clock)
  begin

    -- clock rising edge

    if (clock='1' and clock'event) then
      data_out <= data_in;
    end if;

  end process;

end behv;
```

Fig. 4. 16-bit type-D flip-flop.

For each one of the variables to be expressed by the simulation, there are two possible equations to be solved, and because it is a switched circuit, it depends on the level of the signal $d(n)$. This way, a VHDL multiplexer circuit was implemented, as shown Fig. 5. This block performs the selection of the active equation at every simulation time-step (i.e. the equation that will be solved). This selection is performed by the $d(n)$ signal, which defines the logic level of the SEL input. The multiplexer block is located prior to the flip-flop that stores the previous iterative variable. It was also compressed to a block and saved in the diagram.

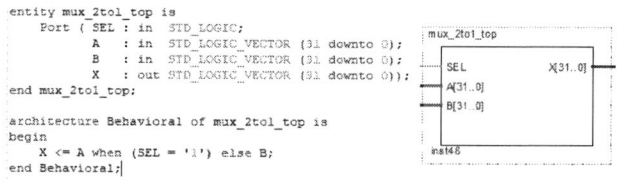

```
entity mux_2to1_top is
    Port ( SEL : in  STD_LOGIC;
           A   : in  STD_LOGIC_VECTOR (31 downto 0);
           B   : in  STD_LOGIC_VECTOR (31 downto 0);
           X   : out STD_LOGIC_VECTOR (31 downto 0));
end mux_2to1_top;

architecture Behavioral of mux_2to1_top is
begin
    X <= A when (SEL = '1') else B;
end Behavioral;
```

Fig. 5. Mux Block.

3) Finishing the System: The four modeled equations are then implemented block by block in the system. As an example, Fig. 6 shows the implementation of equation (4). These four operations are executed in parallel during each simulation step and, as already mentioned, two variables are sent to RTS output. This entire operating environment was then compiled and compressed to a single block, as shown in Fig. 7. This procedure brings more organization to the development environment and makes it more intuitive.

D. Data Input and Output

In order to be certify the correct functioning of the simulation system and to validate the process, it is necessary to ensure

978-1-7281-4181-7/19 $31.00 © 2019 IEEE

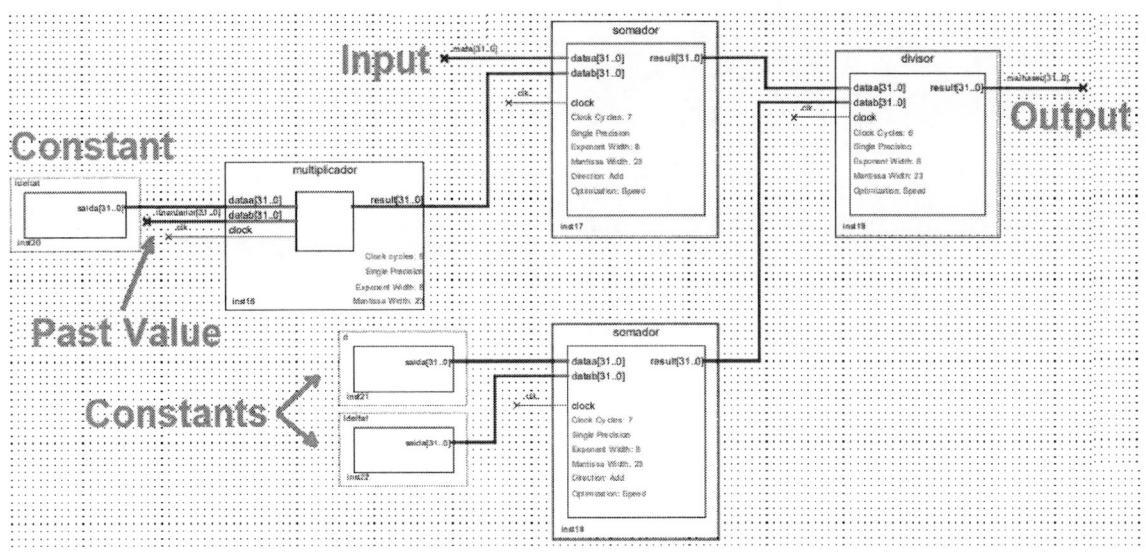

Fig. 6. Block diagram of FPGA implementation of equation (4).

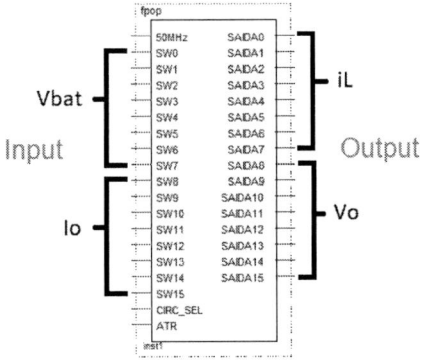

Fig. 7. Operations Block.

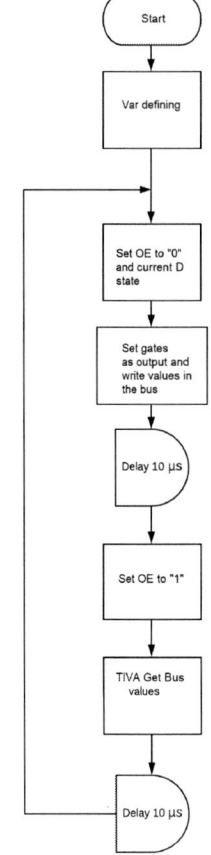

Fig. 8. Flowchart of microcontroller interface operations.

that there is a fidelity of the results and also a synchronization of the operation with the loop. The output of floating point operations block has the width of 16 bits, so the first 8 bits are for the variable I_L, and the second byte for the variable V_O. Simulated values are stored in output flip-flop and sent to bidirectional data bus. A microcontroller is connected to the data bus in order to read output variables, as well as, provide input values and duty cycle signal. These operations have been set according flowchart of Fig. 8. The microcontroller chosen was the Tiva C Series TM4C123GXL due to laboratory availability. The OE signal is controlled by the TIVA, so this is the device that determines the direction of the bus in communication during the procedure. This microcontroller operates as a front-end of FPGA equation solver, which allows either analog-to-digital or digital-to-analog conversions. Fig. 9 shows a photo of the developed RTS prototype.

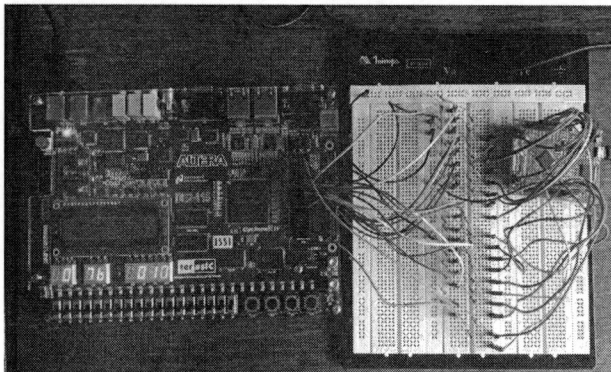

Fig. 9. Real-time simulator prototype.

IV. RESULTS AND DISCUSSION

In order to evaluate the performance of the developed real-time simulator, its simulation results have been compared to the ones obtained with an commercial offline simulation tool.

The comparison among simulation results for I_L and V_O of power converter of Fig. 1 obtained in the developed RTS and in offline commercial simulation tool is presented here in order to validate the proposed real-time simulation system. Reference results have been obtained using MATLAB software and parameters considered in the simulations are listed in Table II. Switching frequency has been set at 5 kHz and simulation time-step has been defined as 20 μs. Regarding the developed RTS, time-step is set by time-delays from flowchart of Fig. 8.

TABLE II
SIMULATION PARAMETERS.

Parameter	Value
V_{Bt}	48 V
C	10 mF
L	1 mH
R_L	1 Ω
R_C	0.1 Ω

Simulation results for input current I_L and output voltage V_O considering two different fixed duty cycles (50% and 75%) are presented in Fig. 10 and Fig. 11. In both cases, it is possible to notice that waveforms obtained from the developed real-time simulator matches the ones from MATLAB offline simulation.

However, it is also possible to observe that waveforms associated to RTS present lower resolution than offline simulation results. There are two factors that impact on this characteristic. First one is related to the 8-bit representation of output variables. The second factor is related to rounding errors associated to the conversion of single-precision floating-point values to decimal numbers, which is performed in the "Math Operations" section of FPGA design. These two issues will be addressed in future work in order to improve RTS

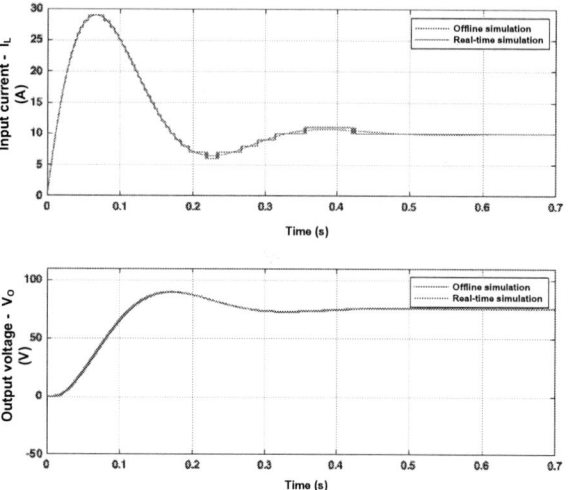

Fig. 10. Comparison among real-time and offline simulation results for fixed duty cycle (50%)

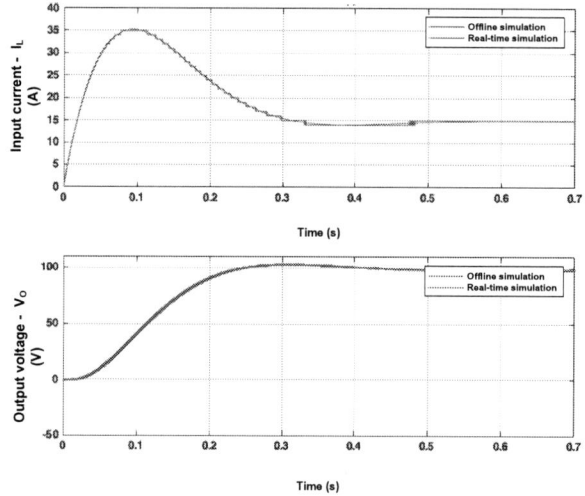

Fig. 11. Comparison among real-time and offline simulation results for fixed duty cycle (75%)

results fidelity. In spite of that, obtained results validates the proposed real-time simulation system.

V. CONCLUSION

The development of FPGA-base real-time simulation system has been presented in this paper. A thorough description of the proposed RTS, which can be used as an alternative to commercial platforms to perform real-time simulations of power electronic converters, has been provided.

A current bidirectional boost converter has been considered in order to illustrate the proposed methodology for power converter modeling and implementation of the RTS equation solver. The comparison of obtained results to offline commercial simulation validated the proposed RTS.

Future work will include the FPGA implementation of ADC and DAC and automation of the conversion from circuit schematics to FPGA blocks.

If requested, FPGA project files used in the RTS development can be shared by authors.

ACKNOWLEDGEMENT

Authors would like to thank CNPq and IF Sudeste MG for the financial support to this work.

REFERENCES

[1] Larry B Rainey, Andreas Tolk, et al. *Modeling and simulation support for system of systems engineering applications.* Wiley Online Library, 2015.

[2] Muhammad H Rashid. *SPICE for power electronics and electric power.* CRC press, 2016.

[3] Francesco Vasca and Luigi Iannelli. *Dynamics and control of switched electronic systems: advanced perspectives for modeling, simulation and control of power converters.* Springer, 2012.

[4] T. Benoit, S. Moten, and Y. Lemmens. Real-time physics-based simulation of mechanisms and systems. In *2017 IEEE/ACM 21st International Symposium on Distributed Simulation and Real Time Applications (DS-RT)*, pages 1–2, Oct 2017.

[5] J. Belanger. The what, where and why of real-time simulation. *IEEE PES General Meeting*, 1(4):37–49, 2010.

[6] J. F. B. Araújo, H. M. Pedrosa, M. C. B. P. Rodrigues, and P. G. Barbosa. Real-time "hardware-in-the-loop" simulation of components of an electric vehicle powertrain: Modeling and implementation. In *2016 12th IEEE International Conference on Industry Applications (INDUSCON)*, pages 1–7. IEEE, 2016.

[7] B. Lu, X. Wu, H. Figueroa, and A. Monti. A low-cost real-time hardware-in-the-loop testing approach of power electronics controls. *IEEE Transactions on Industrial Electronics*, 54(2):919–931, April 2007.

[8] T. Kokenyesi and I. Varjasi. Comparison of real-time simulation methods for power electronic applications. In *4th International Youth Conference on Energy (IYCE)*, 2013.

[9] S. Kermani, R. Trigui, S. Delprat, B. Jeanneret, and T. M. Guerra. PHIL implementation of energy management optimization for a parallel hev on a predefined route. *IEEE Transactions on Vehicular Technology*, 60(3):782–792, March 2011.

[10] H. Hu, G. Xu, and Y. Zhu. Hardware-in-the-loop simulation of electric vehicle powertrain system. In *2009 Asia-Pacific Power and Energy Engineering Conference*, pages 1–5, March 2009.

[11] R. A. F. Ferreira, M. B. C. P. Rodrigues, and P. G. Barbosa. Analysis of an adaptive voltage control of dc microgrids using chil real time simulation. In *2017 Brazilian Power Electronics Conference (COBEP)*, pages 1–8, Nov 2017.

[12] Dalmo C Silva Júnior, Janaína G Oliveira, Pedro M de Almeida, and Cecilia Boström. Control of a multi-functional inverter in an ac microgrid–real-time simulation with control hardware in the loop. *Electric Power Systems Research*, 172:201–212, 2019.

[13] Steven Herbst and Antoine Levitt. Companion models for basic nonlinear and transient devices. 2008.

[14] Christophe Batard, Frédéric Poitiers, Christophe Millet, and Nicolas Ginot. Simulation of power converters using matlab-simulink. *MATLAB–A Fundamental Tool for Scientific Computing and Engineering Applications*, 1:5772, 2012.

[15] Vitor Fernão Pires and Jose Fernando A Silva. Teaching nonlinear modeling, simulation, and control of electronic power converters using matlab/simulink. *IEEE Transactions on Education*, 45(3):253–261, 2002.

[16] S. Buso and P. Mattavelli. *Digital Control in Power Electronics.* MorganClaypool, 2006.

[17] IEEE standard for floating-point arithmetic. *IEEE Std 754-2008*, pages 1–70, Aug 2008.

978-1-7281-4181-7/19 $31.00 © 2019 IEEE

Power Electronics Lab: Converging Knowledge and Technologies

José Antenor Pomilio
School of Electrical and Computer Engineering
University of Campinas
Campinas, Brazil
antenor@fee.unicamp.br

Abstract—**This paper presents some of the experimental activities of the Power Electronics discipline of the School of Electrical and Computer Engineering of the University of Campinas, in Brazil. Simulation is a powerful tool to help the analysis of real circuits since the simulation and experimental results have a strong correlation. The course proposes seven experimental setups, covering the four basic converters: AC-DC, AC-AC, DC-DC and DC-AC. Together with the topologies analysis, typical applications are explored: DC and AC motor speed control, temperature control, UPS, etc. This scenario implies a close collaboration among other disciplines, as control, analog electronics, electric machines, etc. Two of the experiments are discussed to shown the basis of the course, integrating simulation and practical circuits and applications.**

Keywords—Power Electronics Education, Inverters, DC-DC converters, Simulation, Uninterruptible power supply

I. Introduction

The concept that Power Electronics is a field of activity to which the most varied areas of Electrical Engineering contribute is widespread [1-3]. It can be said that in the current context, the Power Electronics development has a strong interface with Electronics, Electric Energy Conversion and Control areas.

As advances occur in each of these fields, the Power Electronics benefits and incorporates such knowledge and solutions. This occurs since the beginning, with the development of thyristors, followed by the evolution of power transistors and nowadays with the development of new materials, such as silicon carbide. Similarly, the constant evolution of dynamic modeling and system control techniques is of great importance for the realization of the power converters and the consequent fulfillment of the requirements of each particular application. The same can be said about the evolution of magnetic materials, as well as modern electric machines.

Possibly the greatest merit of Power Electronics is its ability to perform the processing or conditioning of electrical power with minimal losses. This aspect is essential for the development of many applications; additionally such characteristic is in perfect alignment with modern concepts as energy efficiency and environmental preservation.

In turn, the evolution of Power Electronics raises new challenges not only in these related fields, but also in the regulatory area. The great interest in issues such as harmonic distortions in AC networks or the problems of electromagnetic compatibility are closely related to the operation of the power converters that are at the interface between the equipment and the power grids.

In this broad context, alongside its specific content, the Power Electronics discipline has additional challenges: to realize the connections and the integration of technologically adjacent areas; to apply basic concepts of circuits, signals and systems (such as the equivalence between time domain and frequency domain analyzes); bring to the discussion the pertinence (or obsolescence) of standards and concepts; introduce typical industrial applications, as electric motors speed control, temperature control etc.

Section II presents the conceptual view of the Power Electronics discipline in its theory and experimentation aspects. The role and importance of using simulation tools and the necessary critical view of numerical results are discussed. Section III provides an example of the buck converter, including dynamic response aspects and closed-loop stability analysis. Section IV provides another example of a single phase inverter with output voltage control, also exploring time domain and frequency domain analysis.

II. Power Electronics at Unicamp – Theory and Experimentation

The undergraduate course in Electrical Engineering at the University of Campinas is of the general type, which means that students have a broad vision of the different areas of activity, including electronics, power systems, telecommunications, control, and computing. A deepening in specific areas is possible at the end of the course, when a set of optional credits enables a thematic specialization.

The Power Electronics discipline of the School of Electrical and Computer Engineering (FFEC) at the University of Campinas (UNICAMP), has a 60 hour workload that includes theoretical and experimental activities performed in the laboratory.

The discipline Principles of Electronics is prerequisite to Power Electronics. Although the students have a relatively great freedom to choose when to make the discipline, the suggestion is that it be performed after studying Principles of Control and Principles of Energy Conversion. In these terms, there is a greater freedom to use, even in a very basic way, notions of control by feedback, design of controllers, characteristics of magnetic elements and electric machines, internal properties of semiconductor devices, etc.

The objective is to familiarize the student with different topologies of power electronic converters, identifying typical industrial applications. The objective is also to develop the ability to interpret circuits and waveforms, allowing the analysis of the converters. AC-DC converters (rectifiers), AC-AC (AC regulators and cycle-converters); DC-DC

978-1-7281-4181-7/19 $31.00 © 2019 IEEE

(switching power supplies and choppers), DC-AC (inverters) are studied. Each of these structures considers the most indicated power semiconductor device: diode, thyristor, MOSFET or IGBT. In the experiments, sensors, dedicated ICs and control strategies are also studied. The experiments are organized in such a way as to request from the student the analysis of the phenomena observed in circuits and in the loads.

Laboratory activities are carried out by teams of up to three students and each class holds up to 30 students. The experimental activities, 12 sessions of two hours, are essentially circuit analysis, although there are some items of design. Circuits are available for testing, as well as support equipment (sources, digital oscilloscopes, meters, loads, etc.).

A. Analysis x Design

An important question of defining practical activities in Power Electronics regards the emphasis in design or analysis oriented activities.

As well known to designers and professors of power electronics, the design and construction of converters presents major challenges, many of which are difficult to anticipate by means of simulations. One can cite, mainly, aspects of electromagnetic compatibility that depend heavily on the lay-out of the circuits and the selection of components. The design and construction of magnetic elements is also an item that requires a lot of attention, as well as the thermal design.

Considering the time limitation of the discipline and previous experiences that had an emphasis on the project, in which the great difficulty or impossibility of finalizing the assemblies was verified, it migrated to the current structure that emphasizes the circuit analysis. The importance of designing power converters is emphasized during the course. In some institutions the power electronics design is realized in a second discipline [4,5]. In our case, as this is the unique discipline, some students opt by developing their graduation project in this subject, experiencing the additional challenges of building a prototype.

Some design topics were maintained, not directly in the topologies, but in complementary aspects, such as the scaling of controllers and snubber circuits.

B. The Experiments

The theme of the experiment is in sync with the theoretical classes. Prior to the experiment, a simulation exercise is performed in PSpice. Such a simulation is always very close to the actual experience and the expected results are very close to some of the experimental results. It's available to the students a free PSpice version, which includes all necessary devices and analysis tools. Any other simulator with detailed devices modeling can be used. One important aspect is the availability of complete power switches models for thyristors, TRIAC, MOSFETs and diodes.

Such a link between simulation and practical verification is important for students to mature the concept that every simulation has to be validated. And what validates the simulation, in the last instance, is the realization of an experiment.

There're seven experiments. Five of them are executed in two sessions, while the remaining two experiments are completed in one session. The experiments and the main topics are:

- Diode rectifier I: Single-phase topology; Capacitive and inductive DC filter; AC current (Harmonics, power factor).

- Diode rectifier II: three-phase topology; Capacitive and inductive DC filter, AC current (harmonics, power factor, inrush current); commutation.

- Controlled rectifier I: Half controlled rectifier using thyristors; phase control using TCA785; resistive load behavior and static characteristic.

- Controlled rectifier II: Fully controlled rectifier; DC motor driver; speed control with PI compensator; response to load variations.

- AC-AC converter: Converter with TRIAC and command by whole cycles using integrated circuit CA3059; temperature control with PI regulator; characterization of different temperature sensors.

- Power devices for switching applications I: diodes; comparative analysis of Schottky, fast and slow diodes; turn-on and turn-off dynamic transients.

- Power devices for switching applications II: transistors; comparative analysis of bipolar, MOSFET and IGBT; drivers; turn-on and turn-off commutation.

- DC/DC conversion with MOSFET I: application in switched-mode buck power supply; Pulse Width Modulation using LM3054.

- DC/DC conversion with MOSFET II: Switching losses and turn-off snubber dimensioning; converter efficiency; static characteristic in continuous and discontinuous operation modes; output voltage regulation with PI controller.

- DC/AC conversion with IGBT I: single-phase inverter; three level PWM modulation with analog circuitry; LC output filter and resistive load; spectral analysis.

- DC/AC conversion with IGBT II: different waveforms synthesis; LC filter characterization; sinusoidal output and no linear load; closed loop operation with proportional regulator.

- DC/AC conversion III: three-phase inverter with digital controller; saturation of inductive load by frequency reduction; induction motor drive with scalar V/f control.

C. About the simulations

Unless syntax errors occur, a numerical algorithm will always give a numerical result. It is important that students develop sensitivity and anticipation about what to expect from a simulation.

For example, if the simulated circuit has AC sources of hundreds of Volts, and the loads are of the order of hundreds of Ohms, the expected currents are of the order of Amperes. That is, the simple inspection of the topology should give general indications of the results.

When the results of the numerical calculation distance themselves from this expectation, attention must be paid to the circuit to be simulated, as well as to the programming itself.

In the first case, it is a matter of deeply analyzing the circuit to refine the understanding of its operation, with the objective of accepting or not the simulation result as possible. In the second case, the question is whether the programming performed matches the circuit to be simulated. In both cases there is a reflection on the result, not only the acceptance of a numerical solution that cannot be related to the real system.

The different levels of model complexity allow using the simulator according to the focus of the experiment [6]. When the internal behavior of the power switch isn't relevant, a simple switch model can be adopted, or even a functional block to represent part of the circuit.

The simulations seek to explore different simulator features, mainly transient analysis, frequency analysis (AC sweep) model parameters changes and importing models.

While the time domain results can immediately be recognized during the experimental activities, the frequency domain analysis is used in controller sizing and in the identification of possible dynamic instabilities. The changes of model parameters allow expanding the limited set of available devices included in the student PSpice version [7].

In the sequence some experiments are presented, emphasizing the connection between simulation and experimental results.

III. BUCK CONVERTER

This experiment presents the step-down converter with a MOSFET. The PWM modulation and the features of the IC LM3524/25 are verified. The differences between the continuous and discontinuous conduction modes are tested and the effect of turn-off snubber is verified. It's discussed the dynamic behavior of the converter, as well as investigated the controller design and the converter instabilities.

Concerning the dynamic modeling, one of the simulation tasks is to obtain the Bode diagrams for the output voltage regulation using a PI compensator. Two cases are analyzing, varying the proportional gain. One situation is stable and the other, with reduced proportional gain, is unstable. From the Bode diagram the instability should occur at approximately 400 Hz. Figure 1 shows the equivalent circuit used for the AC analysis, while the Bode diagrams are in figure 2.

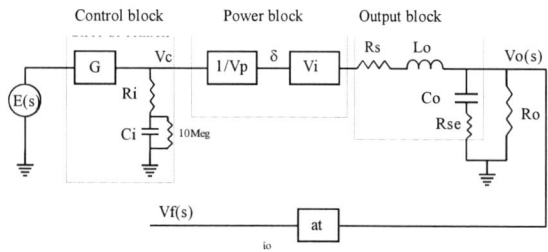

Fig. 1. Equivalent buck circuit for dynamic studies (open loop response at continuous conduction mode).

The experimental results allow verifying the stable operation. Due to the small phase margin (Fig. 2), the response to a load step is poorly damped, as shown in Fig. 3. The control signal is the input of the PWM modulator, what means it's the controller output. Applying the load, the output voltage decreases, making the controller to increase its output in order to widen the duty-cycle. Due to the PI regulator, there's an overshoot, followed by an oscillation, as could be anticipated by the low phase margin verified by simulation.

Also predicted by simulation, the unstable situation is reached reducing the proportional gain (there's a trimpot in the circuit to vary the proportional gain).

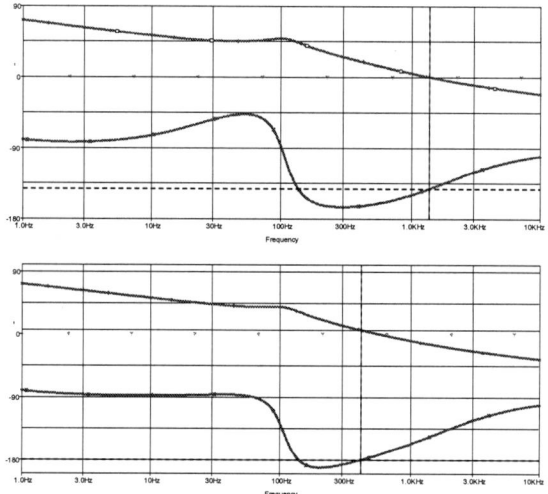

Fig. 2. Bode diagrams: Open loop gain (red) and phase (blue). Above, Stable system, Ri=50kΩ. Bellow, Unstable system, Ri=10kΩ. Horiz: 1 Hz to 10 kHz (log scale).

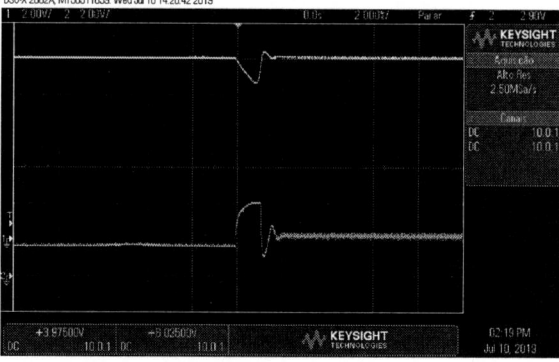

Fig. 3. Output voltage regulation after a load step. Output voltage (yellow) 2 V/div.; and control signal (green) 2 V/div. Horiz.: 2 ms/div.

Figure 4 shows the transition from the stable to unstable operation. While the converter operates in the linear region, the initial instability frequency is approximately 400 Hz, as predicted by the simulation. After becoming unstable, the controller swing between extreme values and the oscillatory frequency decreases, since the circuit non linearity defines a limit cycle and the oscillation.

978-1-7281-4181-7/19 $31.00 © 2019 IEEE   1249

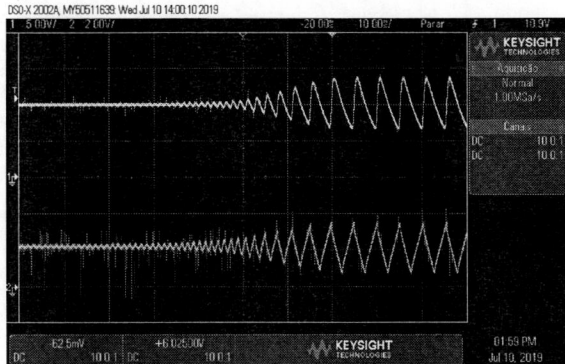

Fig. 4. Induced instability by changing the controller proportional gain. Output voltage (yellow), 5 V/div.; and control signal (green) 2 V/div. Horiz.: 10 ms/div.

Figure 5 shows a detail of the MOSFET turn-off, including the instantaneous power, obtained by the multiplication of drain current (measured on a 1 Ω resistor in series with the drain) and drain to source voltage. Using this result and the MOSFET datasheet the students design a RCD turn-off snubber and insert it in the circuit. The same figure shows the effect of the snubber, drastically reducing the power loss.

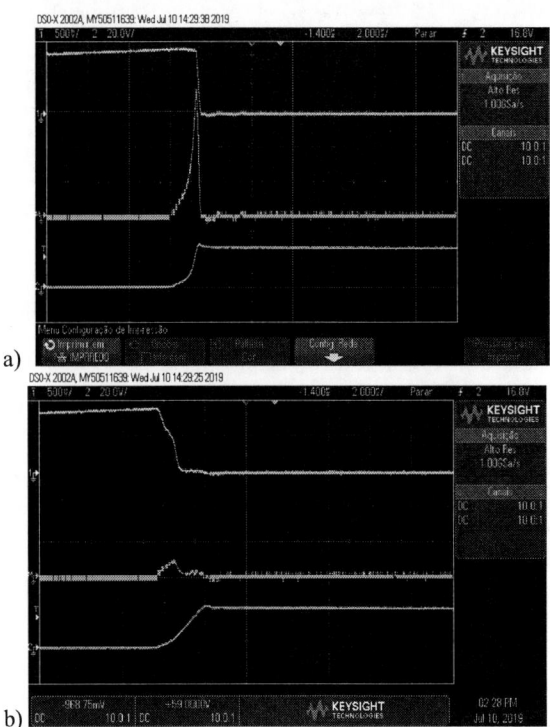

Fig. 5. MOSFET turn-off commutation: Drain current (yellow) 0.5 A/div., Drain to Source voltage (green) 20 V/div., Dissipated power (cyan) 10 W/div. a) Without snubber. b) With RCD snubber. Horiz.: 2 μs/div.

Figure 6 shows the complete circuit. The current measurements are done on the 1 Ω resistors. The transistor driver has an isolated supply (Vgg). A switch allows operating in open or in closed loop.

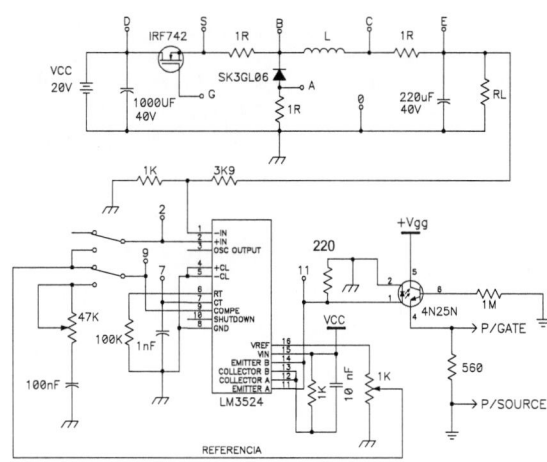

Fig. 6. Experimental circuit.

IV. SINGLE-PHASE INVERTER

The experiment context is the production a sinusoidal voltage, like in an Uninterruptable Power Supply (UPS). In this case, the typical loads are computers which AC power front end is the classical diode single-phase rectifier with capacitive filter (already studied in the first laboratory session). The quality of the sinusoidal output can be measured according to the UPS [5] standards, focusing the harmonic distortion.

The single-phase inverter has a fully analogic PWM circuit. The idea is to verify the generation of the main waveforms: the carrier, the reference and the PWM pulses. It's a three level modulator, using a single triangular carrier, the reference and its inverse for producing the PWM pulses, as illustrated in Fig. 7. At the beginning, the inverter operates in open loop. In the sequence, the output voltage (sinusoidal) is feedback, using a transformer. Fig. 8 shows the block diagram.

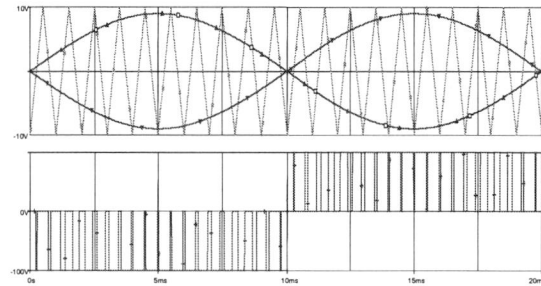

Fig. 7. Conceptual three level PWM signals. Above: triangular carrier and modulating signals. Bellow: resulting inverter output signal.

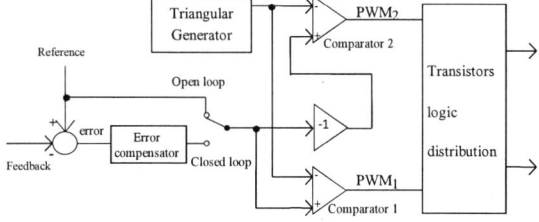

Fig. 8. Analog circuit for commanding and controlling the inverter.

In the simulation exercise, the frequency response of the LC output filter indicates the resonance around 1kHz, without load. Applying a resistive load, the resonance is damped, maintaining the responses at low frequency and the attenuation at the switching frequency, as shown in figure 9.

Also in simulation, but in transient analysis, it's verified the impact of a no linear load on the output voltage waveform. For simplicity and simulation celerity, a two-level PWM signal is obtained conceptually (without the inverter).

With a linear (resistive) load, the voltage at the filter output is sinusoidal. Applying as load a single phase rectifier with capacitive filter, the voltage AC is strongly distorted, as shown in figure 11. The current signal shows the switching component as well as an oscillation around 1 kHz, induced by the undamped filter behavior. The voltage flat top is imposed by the DC capacitor during the recharging intervals.

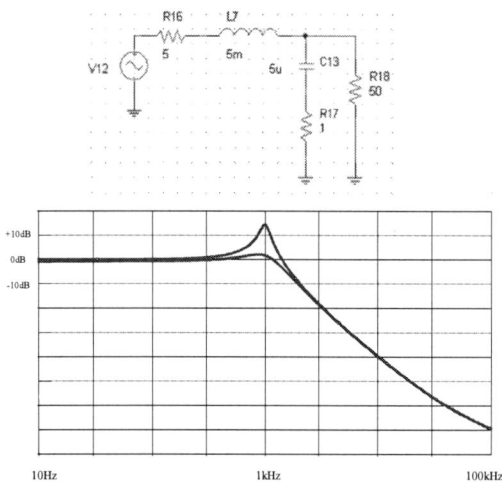

Fig. 9. Simulation results: Circuit (above) and LC filter attenuation (bellow): Undamped (red), damped (blue). Horiz: 10 Hz to 100 kHz (log scale). Vert.: 10 dB/div.

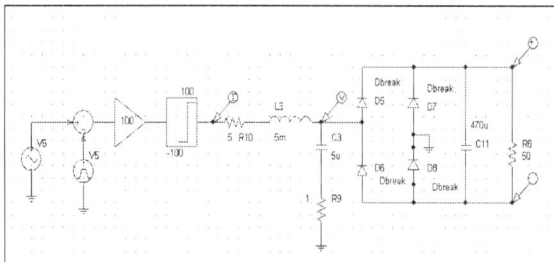

Fig. 10. PWM signal feeding a LC filter with no linear load.

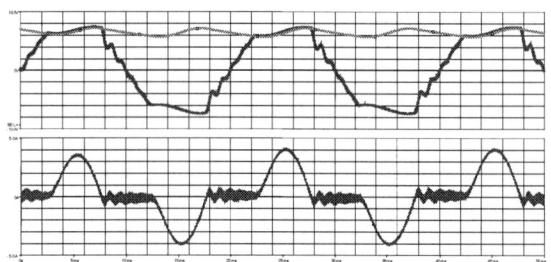

Fig. 11. Simulation results: DC voltage (green); AC voltage (red), 20 V/div. and AC current (blue), 1 A/div. Horiz.: 2 ms/div.

Figure 12 shows the correspondent experimental waveforms. It's evident the similarities between simulation and experimentation. In spite of the experimental setup uses a three-level modulation, there're comparable characteristics: the voltage flat top, the switching frequency components at the current. The oscillation due to the LC filter is attenuated by the circuit losses and can be identified more clearly using the oscilloscope FFT function.

In order to improve the output waveform, and get compliance with the UPS standard [7] it's necessary to operate in closed loop. A switch in the test board (see Fig. 8) commutes from open to closed loop.

In the DC-AC converter, since the reference is a sinusoidal, it's not possible to track the reference with zero error using a simple compensator (P, PI or PID). This could be done with more complex controllers, as a proportional resonant or no linear controller.

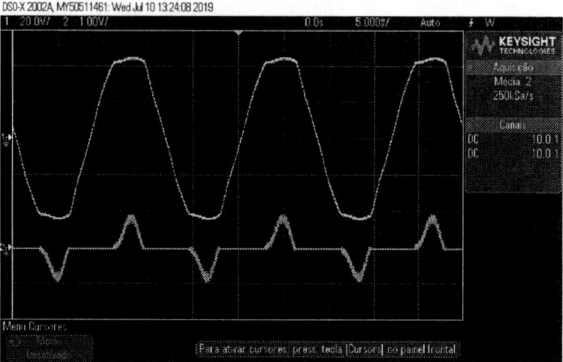

Fig. 12. Experimental waveforms with no linear load: LC filter output voltage (yellow) 20 V/div.; and load current (green) 1 A/div. Horiz.: 5 ms/div.

As the experiment focus is to verify the controller ability to minimize the distortion, the compensator is the simplest one, what means, a proportional gain, enough to reduce the output distortion in order to comply with the standard.

Figure 13 shows the output waveforms in closed loop. The waveform improvement is evident. The internal reference to the PWM modulator is no more a sinusoidal signal, as can be seen in figure 14. The voltage imposed by the inverter has to produce a voltage drop on the passive filter inductor to compensate the current harmonics, resulting an almost sinusoidal voltage at the load terminals.

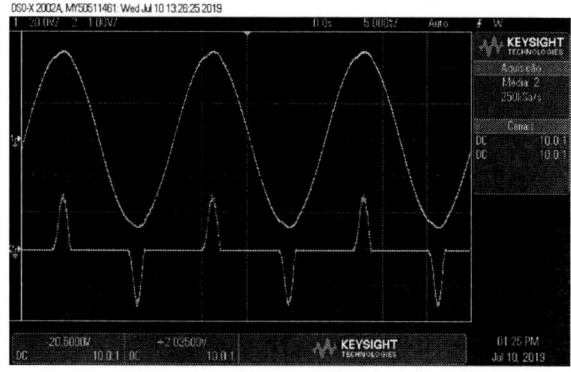

Fig. 13. Closed loop experimental waveforms with no linear load: LC filter output voltage (yellow) 20 V/div.; and load current (green) 1 A/div. Horiz.: 5 ms/div..

978-1-7281-4181-7/19 $31.00 © 2019 IEEE

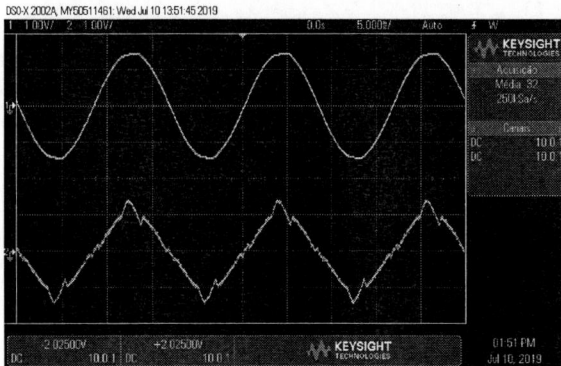

Fig. 14. Internal reference in closed loop: Output voltage (yellow) 1 V/div.; and internal reference (green) 1 V/div., after the controller. Horiz.: 5 ms/div.

Figure 15 shows the electronic circuit for generating the PWM commands and the voltage controller.

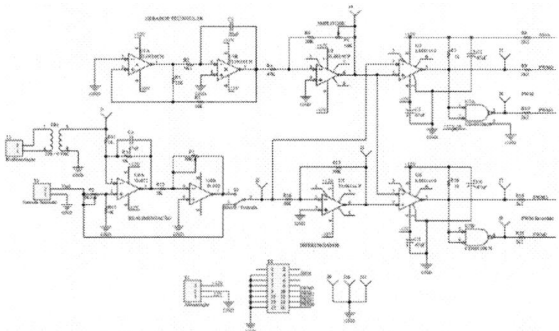

Fig. 15. Electronic circuit with reference, controller, carrier and PWM comparators.

V. CONCLUSIONS

Since its beginning the Power Electronics has been characterized by a strong integration of several areas of Electrical Engineering. In addition to its own content, its association especially with the areas of electric power conversion, control of systems and processes, analog and digital electronics, transforms the discipline into an excellent laboratory for the completion of the training of the Electrical Engineers.

The joint and synergistic use of simulation tools and laboratory practices allow giving consistency to the use of simulators and at the same time enable the development of previous analyzes to be verified in the experimental sessions.

The set of experiments developed at FEEC-UNICAMP allows a broad and general view of the different power converters while, at the same time, explores some typical industrial applications using dedicated integrated circuits. Being the unique discipline on Power Electronics in the EE curriculum, it's a challenge to cover all the content in a single 60-hour course, including theory and experimental applications

All developed material is available on the Internet, including experiment guides and experimental circuits at https://www.fee.unicamp.br/dse/antenor/ee833-eletronica-de-potencia-graduacao .

ACKNOWLEDGMENT

The author would like to thank the FEEC technical support team, especially Nestor de Oliveira, João Paulo Gomes and Bruno Battistella.

REFERENCES

[1] M. Rashid, Ed., "Power Electronics Handbook", Elsevier, 2017.

[2] N. Mohan, T. Undeland and W. Robbins, "Power Electronics: Converters, Applications, and Design", Willey, 2016.

[3] M. Kazmierkowski, F. Blaabjerg, R. Krishnan, "Control in Power Electronics", Academic Press, 2002.

[4] V. Fernão Pires ; A. J. Pires ; O. P. Dias, "Self-learning as a tool for teaching power electronics", *5th IEEE International Conference on E-Learning in Industrial Electronics (ICELIE)*, 2011.

[5] S. Bonho, R. Pizzio, F. A. B. Batista, C. A. Petry, "Teaching power electronics with engineering interdisciplinary projects", *13th Brazilian Power Electronics Conference and 1st IEEE Southern Power Electronics Conference (COBEP/SPEC)*, Fortaleza, Brazil, 2015.

[6] M. Darwish; C. Marouchos, "Simulation levels in teaching power electronics", *48th International Universities' Power Engineering Conference (UPEC)*, 2013.

[7] S. Munk-Nielsen, "Experience with Spice teaching power electronics", *13th European Conference on Power Electronics and Applications*, 2009.

[8] IEC 62040-3, Uninterruptible power systems (UPS) – Part 3: Method of specifying the performance and test requirements

Integration of Solar Photovoltaic (PV) Systems with CCM Inverters into VCM Droop-Controlled Islanded AC Microgrids

Marcus E. T. Souza Jr., Fernando C. Melo, Ernane A. A. Coelho, Luiz C. G. de Freitas

Núcleo de Pesquisa em Eletrônica de Potência (NUPEP) - Faculdade de Engenharia Elétrica (FEELT)
Universidade Federal de Uberlândia (UFU)
Uberlândia, Brasil
marcus11jr@hotmail.com, fernando_cmelo@outlook.com, ernane@ufu.br, lcgfreitas@yahoo.com.br

Abstract—Nowadays, photovoltaic (PV) systems and microgrids are two of the main technologies in electrical engineering researches. Both are important to realize different types of energy management, feeding islanded loads with high reliability, efficiency, social benefits, low cost and environmental impact. This paper assess the possibility of connecting a PV system to a droop-controlled islanded AC microgrid without modified advanced controls, additional components or the use of batteries in a way that only simple standard commercial equipment are used. Two inverters operating with Voltage Control Mode (VCM) employing the droop control method to form an islanded AC microgrid is approached. These inverters function as grid forming converters giving voltage references in order to integrate them with a conventional solar PV system in Current Control Mode (CCM), consolidated technologies which are easily found on the market, since they are used for grid-tie. The proposed complete system is analysed for a typical day, with real measured variations of irradiance and temperature. The computational results of the VCM and CCM inverters integrated is exposed, showing its feasibility and its limitations.

Keywords—*Current Controlled Mode (CCM), Droop Control Method, Inverter, Microgrids, Solar Photovoltaic (PV) Systems, Voltage Controlled-Mode (VCM).*

I. INTRODUCTION

Two technologies have been receiving a great attention from researchers, companies, utilities and users of the power system nowadays. The first one is the solar photovoltaic (PV) systems, which are composed by power production devices that converts sunlight directly into electricity and have multiple social, economic and environmental benefits. The second one is the advent of microgrids, a new structure of the power system that contains local electricity generation, distribution and consumption, and can work in islanded or grid-connected modes. The most common control methods used in both systems are distinct. Solar PV systems are usually current control mode (CCM) technologies, while voltage control mode (VCM) units compose many of the microgrid concepts. This can be seen as a problem, since designers can think that they are incompatible for isolated arrangements and may need modifications for operating in a single assemble. This paper shows exactly the opposite. Although there are some constraints, the two may work together without any changes in their typical control systems or components.

The growing use in all countries [1] (Fig. 1) and the economies of scale are making solar PV systems one of the lowest-cost sources of power generation in the world, as a recent report has shown [2] and illustrated in Fig. 2. In 2014,

This study was financed in part by the Coordenação de Aperfeiçoamento de Pessoal de Nível Superior Brasil (CAPES) - Finance Code 001, and the authors would like to thank the Universidade Federal de Uberlândia (UFU), Post-Graduate Program in Electrical Engineering – FEELT, to CNPq and to FAPEMIG.

solar PV was already competitive with fossil fuel sources. By 2020, together with onshore wind it will be less expensive than any of them, without financial assistance. Solar PV has many advantages among other distributed generation (DG) devices. Besides the falling prices, it does not pollute and do not emit greenhouse gases when in operation, having a very low environmental impact compared to the other technologies.

Since the primary source is sunlight, it is not only a renewable energy, but also an inexhaustible supply of electricity while there is sun. It does not rely on any raw material or human efforts for electricity production after the installation. More than that, solar PV systems are one of the easiest generation technologies to install. It is simple to transport because of its small size and is a modular equipment, ensuring different construction configurations. For these reasons, solar PV systems are clearly a prominent power supply for isolated systems like islanded microgrids. However, they are intermittent and variable sources. That is, in normal situations, their production stops in some moments (in this case, when the sunlight do not reach the PV modules or at night) or varies in time because of changes in the sun's position in relationship to it, by passing clouds and other types of shading (affecting irradiance and temperature). This means that PV inverters, devices that transform the PV module generated DC to AC voltage (the one used in typical loads), rely on Maximum Power Point Tracking (MPPT) techniques and are CCM or grid-feeding converters [3]. These devices operate like ideal current sources that must unavoidably be connected and synchronized to any kind of energized grid that provides them voltage references.

There is not a single and conclusive definition for microgrids, but they can be understood as a local and unique power system structure integrating DG supplying loads in their vicinities, being controlled together to operate in islanded mode or in parallel to the grid [4]. For isolated communities and remote areas, where there is no access to the grid, like islands, rural areas and difficult-to-reach populations, or even for those entities that want to work separated from the grid, islanded AC microgrids are an ideal solution. They bring all the benefits of the traditional grid and DG together by creating

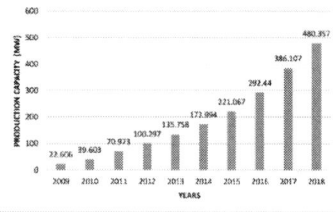

Fig. 1. Global Solar PV Systems production capacity (2009-2018) [1].

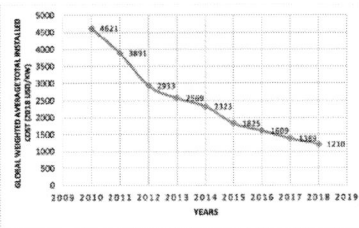

Fig. 2. Global weighted average total installed costs for Solar PV Systems (2010-2018) [2].

a reliable, affordable and sustainable architecture for delivering electricity. There are different approaches for the control and operation of AC microgrids. Some of them use central controllers with communication [5]. Nonetheless one of the most popular microgrids philosophies do not depend on a crucial equipment and have less or no communication lines. This concept is called hierarchical control with droop method [6].

The droop control method is used for the parallelism of inverters in isolated situations (without grid or external references). Using only local power measurements and the droop equations, one can generate amplitude and frequency references in converters for the correct power sharing among them and loads in islanded systems without communication [7]. This type of operation can only happen if the DG is dispatchable: they must deliver energy all the time. Thus, using this proposal for AC microgrids, converters operate as ideal voltage sources and are defined as VCM or grid-forming converters [3].

Thus, this paper is divided in the following way. In section II, the thematic of connecting a traditional PV system to an islanded AC microgrid controlled by droop is discussed and a literature review is made. Section III is dedicated to the theoretical and design descriptions of the CCM PV inverters and the droop control used to realize the parallelism of VCM inverters. In section IV, the test methodology for simulating the connection of both systems is explained and the computational results are presented. Finally, a conclusion is given.

II. DISCUSSIONS AND REVIEW OF PV SYSTEMS CONNECTED TO ISLANDED AC MICROGRIDS

As can be seen, solar PV systems and islanded AC microgrids with droop control are good approaches for the production and use of electricity in isolated areas. However, since they operate in different control modes, the connection of both looks like a difficult task. Some authors have presented solutions for this apparent impasse. In [5], solar PV systems are considered as PQ inverter control type technologies, like the concepts of CCM and grid-feeding. The dispatchable DG converters are voltage source inverters (VSI) and work with the droop control. The same problem appears here, however this solution uses central controllers and communication, as mentioned above, in a master-slave structure, where the VSI units are masters and the PQ inverter control units are slaves. The authors of [8] present a reverse droop control, where the VCM units work with the regular droop method and the CCM units use this modified control. Beyond the problem of having to develop a new system for the CCM systems, the reverse droop, which work inversely to the regular droop by controlling active and reactive powers in function of, respectively, frequency and voltage, is a very sensitive control

in practice. If there are any small variations in the measured voltage (even by measurement imprecisions), for example, a great amount of reactive power can be extracted from the CCM unit. In [9], a complex system called smooth switching droop control (SSDC) is presented. As the name indicates, the units will change their modes of operation from VCM to CCM or vice-versa. Two difficulties arise here. The first one is again a complex modification of regular systems by transforming their controls. The second one is the dependence on energy storage systems (ESS).

Another system is proposed in [10] without modifying the inner control loops. Here an autonomous active power control is used to coordinate ESS, PV systems, and loads in an islanded AC microgrid. Once more, there is a great dependence on ESS. More than that, a master-slave control and bus-signaling method form the base for this approach. A master ESS regulates the frequency bus, while PV systems and loads manage their respective production and consumption by the measured frequency. In [11], the PV systems have their controls modified, as well as their structure, since they are formed as hybrid units with ESS. Here, PV systems are modified to operate as voltage source units following a multi-segment adaptive power/frequency characteristic curve with multi-loop controllers. Again, a hybrid PV/ESS unit is proposed in [12] for microgrids use, but with an adaptive droop control. In [13], a robust control for microgrids with PV systems and wind turbines is discussed. This time, a central controller is applied for regulating the power references for DG units and the droop control for ESS.

Although all these solutions represent great advances in the connection of solar PV systems to islanded AC microgrids, two concerns arise: they invariably have complex modified controls and components and, for many of them, battery use is mandatory. There are still 840 million people in the world without electricity access [14]. Global warming is now around 1°C above pre-industrial levels and can be 1.5°C above between 2030 and 2052 due to human activities [15], putting at risk the fulfillment of the Paris agreement with great negative consequences for human race and nature. These two points show that society and the environment demand fast responses in terms of sustainable energy production, which justifies the adoption of devices that are already consolidated. Solar PV grid-tie inverters are technologies easy to be found in the market with relatively low prices and environmental impacts with great socio-economic benefits, while the use of new, exclusive and specific alternatives can be more expensive, harder to produce and slower to be achieved. In the other hand, eliminating the use of batteries can be a great success, because they represent certain disadvantages. Some of these are: the relative short lifetime (between approximately 1 to 5 years for low capacity batteries), the difficulty in discharge/charging the batteries correctly and the control of state-of-charge (SoC) limits, the large energy cost of manufacture, the maintenance, the high costs per unit, the low relative abundance of materials for the state-of-the-art technologies, the risks of chemical toxicity and explosions and the hard recycling process [16]-[19].

With these difficulties, this paper assess the feasibility of connecting a conventional solar PV system with a CCM type grid-tie inverter to a droop-controlled islanded AC microgrid (with VCM units) without modifications in any of their control systems or components and not depending necessarily on batteries, as illustrated in Fig. 3. This system depends only in

978-1-7281-4181-7/19 $31.00 © 2019 IEEE

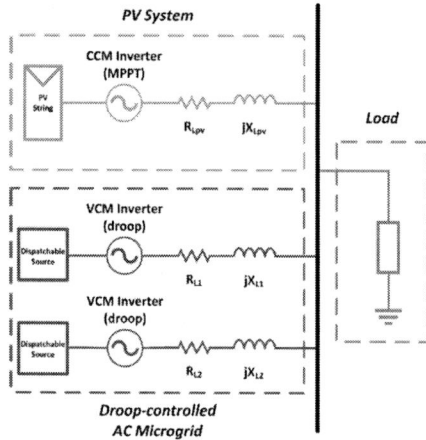

Fig. 3. Complete system with a CCM PV system connected to a droop-controlled AC microgrid with two VCM units and a common load.

only local control and measurements without communication lines, having high reliability. The restraints will be discussed, but it can be recognized that a good planning is a simple form to solve the problem of power sharing. The construction of a microgrid for an isolated community or company will need a good planning because of all the difficulties and inherent requirements that surround them. The smaller size of a microgrid is a benefit for its development and to the optimization of power generation and consumption. The power production will be calculated to supply all the demand and follow its curve. There will be almost all the time a power equilibrium. Any variability in this scenario will happen only concerning the solar PV system, since droop-controlled units will adapt its consumption within their limits. The most probable situation is the excess of energy, since loads can be diminished when the sum of the power produced by the solar PV system and the droop-controlled units is higher than the demand. Although this is a rare case, since usually the demand curve coincides in time with the variation of the PV power curve (working hours, abundant activities, intense use of cooking equipment, use of air conditioning devices and fans, etc., happens typically during the day), some simple ideas can contour this problem. Since a microgrid is composed by an infinitely smaller population (and thus a small load) than a traditional distribution system, they can be guided to conciliate the consumption of electricity with the PV production. Another option is to insert loads that are automatically connected when there is an excess of power generated by the PV system and VCM units. In the worst scenario, if the planning is done correctly, batteries can be inserted in the consumer side to absorb the surplus power.

III. SOLAR PV SYSTEMS AND ISLANDED AC MICROGRIDS

This section presents the main theories about solar PV systems with grid-tie inverters and droop-controlled islanded AC microgrids. Beyond that, examples of design for both of them are exposed. The final designs are used in sequence in computer simulations to verify the possibility of integrating the two systems together. All of the units are single-phase devices with RMS voltage of 127 V and frequency of 60 Hz.

A. Solar PV System with CCM Inverter

The traditional solar PV system for grid connection is composed of PV modules and an inverter (Fig. 4). The first

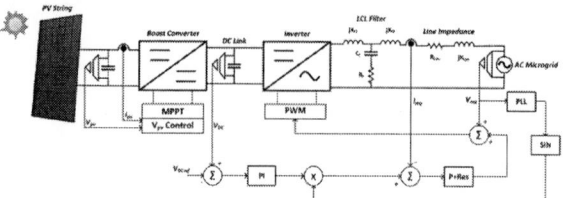

Fig. 4. Complete solar PV system diagram.

ones are formed by PV cells inside a module that converts sunlight directly into electricity. They can be associated in series and in parallel to reach higher voltages and currents, respectively [20]. Here, six PV modules constructed with polycrystalline silicon solar cells are connected in series and their values are shown in Table I.

The output power of PV systems is DC, but the grid and the commonly used loads work with AC power. To make them compatible, a DC-AC converter is normally combined with the PV modules. Recently, the most used technology of converters for PV systems is transformerless and, therefore, it is more efficient and less expensive, bulky and heavy [21]. There are many transformerless PV inverter topologies to improve the performance of the PV system [22]. However, in this paper, a simple transformerless Insulated Gate Bipolar Transistor (IGBT) Full-bridge single-phase topology was adopted. For generating the AC output power, the unipolar sinusoidal pulse-width modulation (PWM) was implemented with 10 kHz of switching frequency. A single-phase LCL filter was used and calculated as proposed in [23]. Its parameters are shown in Table II.

TABLE I. PV MODULES PARAMETERS

Test Conditions	Parameters	Values
STC (1000 W/m² and 25 °C)	Individual Power	330 W
	Total Power	1980 W
	Individual V_{mp}	37.2 V
	Total V_{mp}	223.2 V
	Individual I_{mp}	8.88 A
	Total I_{mp}	8.88 A
	Individual V_{oc}	45.5 V
	Total V_{oc}	273 V
	Individual I_{sc}	9.45 A
	Total I_{sc}	9.45 A

TABLE II. LCL FILTER PARAMETERS

Parameters	Values
Inverter-side Inductor (L_{f1})	4.309 mH
Grid-side Inductor (L_{f2})	93.348 µH
Capacitor (C_f)	16.281 µF
Damping Resistor (R_f)	3.186 Ω

At the inverter input, it is inserted a capacitor to form the DC link with a calculated value of 1818.598 µF and a voltage of 380 V. The external voltage-loop that controls the voltage in the DC link is composed of a Proportional-Integral (PI) with the following transfer function (1), where k_p=1 and T_i=0.01 s:

$$G_{PI}(s) = k_p \left(1 + \frac{1}{sT_i}\right) \qquad (1)$$

For adequate operation, the PV inverter must be synchronized with the microgrid (it depends on external voltage, frequency and phase references) since it is a CCM unit. To do this task, a control technique used to synchronize two systems in frequency and in phase called Second Order Generalized Integrator Phase-Locked Loop (SOGI-PLL) [24] was chosen. This synchronized signal with the resulting value of the PI are then passed through the internal current loop. In this case, the non-ideal Proportional-Resonant (P+Res) controller, given by (2), is the method applied, because it can eliminate the steady-state error in stationary reference frame with an almost infinite gain in the microgrid frequency [25]. The non-ideal P+Res is generally used to take the small frequency variations in account. This is an advantage when connecting the PV inverter to the microgrid because, as it will be seen, its units have some deviations in frequency. The designed values for the P+Res controller are k_i=500, k_p=50, ω_c=5 rd/s and ω_0=2.π.60=377 rd/s.

$$G_{P+Res}(s) = k_p + \frac{2k_i\omega_c s}{s^2 + 2\omega_c s + \omega_0} \qquad (2)$$

The PV modules connected in series (string), may not have high enough voltage for the DC link to impose current to the output of the inverter. To overcome this frequent problem, an interface step-up DC-DC converter (boost) between the PV string and the inverter is used. The parameters of the boost converter are given in Table III.

TABLE III. PV BOOST CONVERTER PARAMETERS

Parameters	Values
Switching Frequency (f_{sw})	20 kHz
Inductor (L_{boost})	5.18536 mH
Input Voltage (V_{mp})	223.2 V
Output Voltage (V_{DC})	380 V
Current Ripple	10%
Duty Cycle (D)	0.4126

The control of the boost converter is made through a PI, also defined by (1), with constant values of k_p=0.1 and T_i=1 s. A MPPT algorithm gives the reference for this controller. This technique makes the PV system work in its maximum power all the time, since the irradiance and temperature variations can force it to operate with a lower value. The implemented MPPT was the classic Perturb and Observe (P&O), where the DC-DC converter duty cycle is constantly disturbed providing voltage variations until the system have its operating point at its maximum [26].

B. Droop-Controlled AC Microgrid with VCM Units

An AC microgrid is basically composed of small-scale DG units, connection lines for the distribution of power and loads. In the hierarchical concept [6], the microgrid have four control levels. The first one is made of the inner control loops, which regulates current and voltage in the internal operation of the DG converters. Since these are inherent elements of these devices, they are defined as a zero level in the hierarchical microgrid control. The primary control, in its turn, defines the islanded operation of the microgrid. The secondary and tertiary levels work, respectively, at the restoration of voltage and frequency and at the synchronization and grid connection of the microgrid.

For this paper, nevertheless, only the zero and the primary

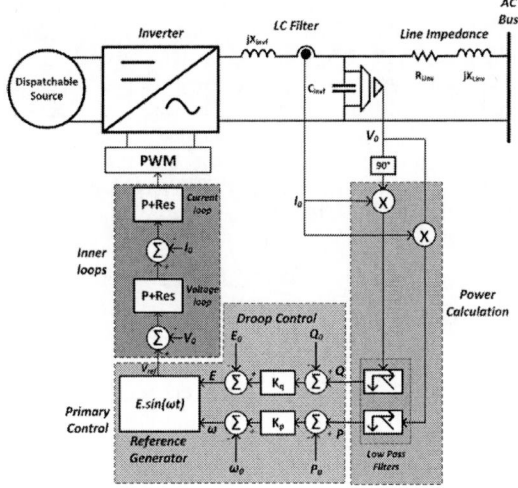

Fig. 5. Droop-Controlled inverter.

levels are approached, where the droop control method is directly applied for the adequate power sharing among the microgrid converters and loads in the islanded operation. The complete system for each inverter is shown in Fig. 5. The zero level is composed of two control loops: an internal one for the current and an external one for the voltage. Both are non-ideal P+Res controllers identical to the one used in the PV system (CCM) through equation (2). For making the tests, two droop-controlled inverters are simulated to represent the islanded AC microgrid and their internal P+Res controllers values are: k_i=100, k_p=2 and ω_c=10 rd/s for the current loop and k_i=10, k_p=0.1 and ω_c=10 rd/s for the voltage loop.

The droop control method allows the appropriate parallelism of inverters by emulating isolated synchronous machines [7]. Applying the Pxω and QxV droop curves illustrated in Fig. 6 and their respective equations (3) and (4), the inverters will share active and reactive powers with the load by creating specific voltage references of amplitude E and frequency ω by only measuring local output powers P and Q from each inverter. It is for this reason that microgrid converters are VCM or grid-forming units. Reference values for the droop equations can be defined as conventional grid values of amplitude E_0 and frequency ω_0 and reference powers P_0 and Q_0. It is possible to note that if the inverter is producing exactly these reference powers, then the frequency and magnitude will be the ones defined by their respective reference values in the droop equations.

$$\omega = \omega_0 - k_p(P - P_0) \qquad (3)$$

$$E = E_0 - k_q(Q - Q_0) \qquad (4)$$

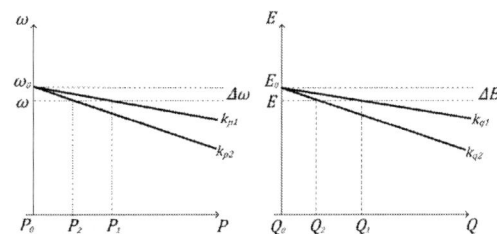

Fig. 6. Pxω and QxV droop curves for 2 inverters.

One of the disadvantages of the droop control method is the voltage amplitude and frequency deviations. Nonetheless, this can be simplified by choosing small droop coefficients k_p and k_q. In this way, even if there is a great power variation, there will be only small amplitude and frequency changes. If the two inverters have different active P_{max1} and P_{max2} and reactive Q_{max1} and Q_{max2} power ratings, their respective droop coefficients, k_{p1}, k_{p2}, k_{q1} and k_{q2}, can be properly chosen and balanced through the following equations [27]:

$$k_{p_1} \cdot P_{max1} = k_{p_2} \cdot P_{max2} \qquad (5)$$

$$k_{q_1} \cdot Q_{max1} = k_{q_2} \cdot Q_{max2} \qquad (6)$$

The use of the droop control method has multiple advantages. The first one is that it can realize parallelism of the inverters and share power with good results among them by calculating only active P and reactive Q output powers through each one of their respective local measured voltages and currents. The second advantage is the lack of any communication among the converters. It is also a very simple technique to be executed with acceptable responses. All the parameters applied to the two converters for the simulation tests are seen in Table IV. It is important to notice that for the correct operation of the droop control, the line impedances must be predominantly inductive ($X>R$). In this paper, their values were properly defined and are presented in Table IV for each converter.

TABLE IV. DROOP-CONTROLLED INVERTERS PARAMETERS

Parameters	Inverter 1	Inverter 2
Line Impedance (Z_{Linv})	0.1+j0.377 Ω	0.2+j0.565 Ω
Pxω Droop Coefficient (k_p)	0.0003 rd/s/W	0.0002 rd/s/W
QxV Droop Coefficient (k_q)	0.0003 V/VAr	0.00015 V/VAr
Maximum Output Apparent Power (S_{inv})	2280 VA	3720 VA
Voltage Reference (E_0)	180 V	180 V
Frequency Reference (ω_0)	377 rd/s	377 rd/s
Active Power Reference/Rating (P_0)	2000 W	3000 W
Reactive Power Reference/Rating (Q_0)	1100 VAr	2200 VAr

IV. SIMULATION RESULTS AND TESTS METHODOLOGY

A. Tests Methodology

To investigate the insertion of the PV system into the islanded AC microgrid, the tests were carried out considering real values of irradiance and PV module temperature that were measured through a weather station at NUPEP laboratory. Two days were considered: the first one (June 29, 2019), when there were low irradiance and temperature variations and the second day (June 25, 2019), when the opposite happens, with fast and great changes. These irradiance and temperature values are illustrated in Fig. 7. Both were inserted in the simulation software. The data is saved every 5 minutes. For the tests, only 5 hours were considered: between 1 PM and 6 PM. In this way, for simulation purposes, the emulated curve is equivalent to the one measured.

Two different tests were made showing all the 5 hours for each day with the PV system integrated to the microgrid since the beginning. The common load is 5 kW (resistive).

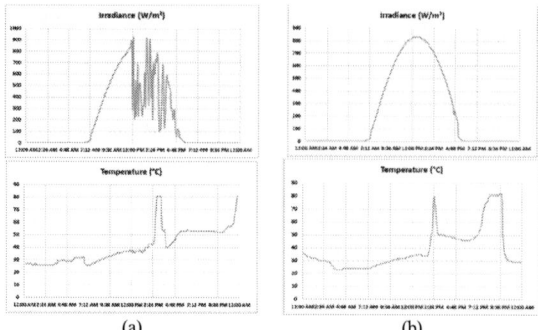

(a)   (b)

Fig. 7. Measured irradiance (W/m²) and PV module temperature (°C) curves: (a) June 25, 2019 and (b) June 29, 2019.

B. Simulation Results

The droop-controlled VCM units adjust their respective active powers as the CCM inverter operates following the variations of the PV system, always supplying the load with its 5 kW. These active powers results can be seen in Figs. 8 and 10. The RMS voltages, shown in Figs. 9 and 11, are lower than 127 V because of the line voltage drops, while the frequencies, also illustrated in Figs. 9 and 11, are stabilized in 60 Hz. Since the PV system follows the voltage magnitude and the frequency values of the microgrid, the correct connection is verified with the VCM units providing voltage references. In this way, the PV system work as plug-and-play element. The only disadvantage presented in the results is that the CCM inverter extracts a lower power than its maximum, but close to it, due to the VCM units limiting its operation, always prioritizing the power balance. The solution is already under development and will be presented in future works.

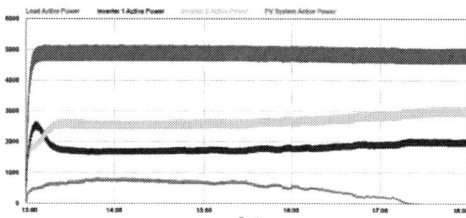

Fig. 8. Active powers - June 29, 2019.

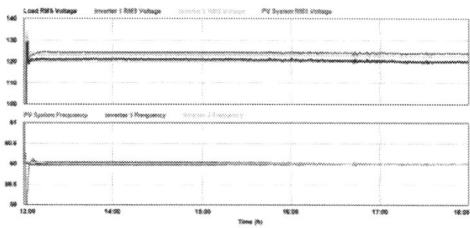

Fig. 9. RMS voltages and frequencies - June 29, 2019.

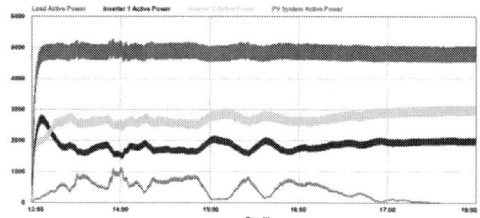

Fig. 10. Active powers - June 25, 2019.

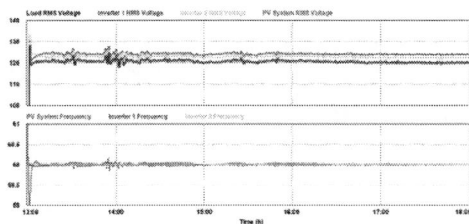

Fig. 11. RMS voltages and frequencies - June 25, 2019.

V. CONCLUSION

This paper evaluated the connection of a conventional solar PV system to an islanded AC microgrid. If a correct planning is made, the integration of a PV CCM inverter (traditionally applied for grid connection) to VCM units controlled by the droop control method proved to be feasible, with the PV system working as a simple plug-and-play device in the microgrid. The paper analysed through simulation results the impact of this incorporation of the PV system with real irradiance and temperature measured values. It was shown that the droop-controlled converters managed, automatically and without any communication, their respective output powers, diminishing proportionally with the power produced by the PV system to supply the load. In future works, this study will be advanced and assessed with different types of load, such as non-linear ones, with the insertion of batteries and with a more flexible management system, where the PV CCM inverter will produce its maximum power in connection with the droop-controlled VCM units.

REFERENCES

[1] IRENA, Renewable Capacity Statistics 2019, March 2019. [Online]. Available: https://www.irena.org/publications/2019/Mar/Renewable-Capacity-Statistics-2019.

[2] IRENA, Renewable Power Generation Costs in 2018, May 2019. [Online]. Available: https://www.irena.org/publications/2019/May/Renewable-power-generation-costs-in-2018.

[3] J. Rocabert, A. Luna, F. Blaabjerg, and P. Rodríguez, "Control of Power Converters in AC Microgrids," IEEE Transactions on Power Electronics, vol. 27, no. 11, pp. 4734-4749, Nov. 2012.

[4] R. H. Lasseter, "MicroGrids," 2002 IEEE Power Engineering Society Winter Meeting. Conference Proceedings (Cat. No.02CH37309), New York, NY, USA, 2002, pp. 305-308 vol.1.

[5] J. A. P. Lopes, C. L. Moreira, and A. G. Madureira, "Defining control strategies for MicroGrids islanded operation," in IEEE Transactions on Power Systems, vol. 21, no. 2, pp. 916-924, May 2006.

[6] J. M. Guerrero, J. C. Vasquez, J. Matas, L. G. de Vicuna, and M. Castilla, "Hierarchical Control of Droop-Controlled AC and DC Microgrids—A General Approach Toward Standardization," in IEEE Transactions on Industrial Electronics, vol. 58, no. 1, pp. 158-172, Jan. 2011.

[7] E. A. A. Coelho, P. C. Cortizo, and P. F. D. Garcia, "Small-signal stability for parallel-connected inverters in stand-alone AC supply systems," in IEEE Transactions on Industry Applications, vol. 38, no. 2, pp. 533-542, Mar.-Apr. 2002.

[8] Dan Wu, Fen Tang, J. C. Vasquez, and J. M. Guerrero, "Control and analysis of droop and reverse droop controllers for distributed generations," 2014 IEEE 11th International Multi-Conference on Systems, Signals & Devices (SSD14), Barcelona, 2014, pp. 1-5.

[9] D. Wu, F. Tang, T. Dragicevic, J. C. Vasquez, and J. M. Guerrero, "A Control Architecture to Coordinate Renewable Energy Sources and Energy Storage Systems in Islanded Microgrids," in IEEE Transactions on Smart Grid, vol. 6, no. 3, pp. 1156-1166, May 2015.

[10] D. Wu, F. Tang, T. Dragicevic, J. C. Vasquez, and J. M. Guerrero, "Autonomous Active Power Control for Islanded AC Microgrids With Photovoltaic Generation and Energy Storage System," in IEEE Transactions on Energy Conversion, vol. 29, no. 4, pp. 882-892, Dec. 2014.

[11] H. Mahmood, D. Michaelson, and J. Jiang, "Decentralized Power Management of a PV/Battery Hybrid Unit in a Droop-Controlled Islanded Microgrid," in IEEE Transactions on Power Electronics, vol. 30, no. 12, pp. 7215-7229, Dec. 2015.

[12] H. Mahmood, D. Michaelson and J. Jiang, "A Power Management Strategy for PV/Battery Hybrid Systems in Islanded Microgrids," in IEEE Journal of Emerging and Selected Topics in Power Electronics, vol. 2, no. 4, pp. 870-882, Dec. 2014.

[13] M. J. Hossain, H. R. Pota, M. A. Mahmud and M. Aldeen, "Robust Control for Power Sharing in Microgrids With Low-Inertia Wind and PV Generators," in IEEE Transactions on Sustainable Energy, vol. 6, no. 3, pp. 1067-1077, Jul. 2015.

[14] IEA, IRENA, UNSD, WB, WHO, Tracking SDG 7: The Energy Progress Report 2019, 2019. [Online]. Available: https://trackingsdg7.esmap.org

[15] IPCC, Global Warming of 1.5°C - Summary for Policymakers, October 2018. [Online]. Available: https://www.ipcc.ch/sr15/

[16] H. Beltran, J. Barahona, R. Vidal, J. C. Alfonso, C. Ariño and E. Pérez, "Ageing of Different Types of Batteries when Enabling a PV Power Plant to Enter Electricity Markets," IECON 2016 - 42nd Annual Conference of the IEEE Industrial Electronics Society, Florence, 2016, pp. 1986-1991.

[17] D. Larcher and J.-M. Tarascon, "Towards Greener and More Sustainable Batteries for Electrical Energy Storage," Nature Chemistry, vol. 7, no. 1, pp. 19-29, 2015.

[18] A. Jossen, J. Garche and D. U. Sauer, "Operation Conditions of Batteries in PV Applications," Solar Energy, vol. 76, pp. 759-769, 2004.

[19] J. Sun et al., "Toxicity, a Serious Concern of Thermal Runaway from Commercial Li-ion Battery," Nano Energy, vol. 27, pp. 313-319, Sep. 2016.

[20] M. G. Villalva, J. R. Gazoli and E. R. Filho, "Comprehensive Approach to Modeling and Simulation of Photovoltaic Arrays," in IEEE Transactions on Power Electronics, vol. 24, no. 5, pp. 1198-1208, May 2009.

[21] T. Kerekes, R. Teodorescu and U. Borup, "Transformerless Photovoltaic Inverters Connected to the Grid," APEC 07 - Twenty-Second Annual IEEE Applied Power Electronics Conference and Exposition, Anaheim, CA, USA, 2007, pp. 1733-1737.

[22] Remus Teodorescu; Marco Liserre; Pedro Rodriguez, "Photovoltaic Inverter Structures," in Grid Converters for Photovoltaic and Wind Power Systems, , IEEE, 2007, pp.

[23] A. Reznik, M. G. Simões, A. Al-Durra and S. M. Muyeen, "LCL Filter Design and Performance Analysis for Grid-Interconnected Systems," in IEEE Transactions on Industry Applications, vol. 50, no. 2, pp. 1225-1232, Mar.-Apr. 2014.

[24] M. Ciobotaru, R. Teodorescu and F. Blaabjerg, "A new single-phase PLL structure based on second order generalized integrator," 2006 37th IEEE Power Electronics Specialists Conference, Jeju, 2006, pp. 1-6.

[25] D. N. Zmood, D. G. Holmes, "Stationary frame current regulation of PWM inverters with zero steady-state error," IEEE Transactions on Power Electronics, vol. 18, no. 3, pp. 814-822, May 2003.

[26] T. Esram and P. L. Chapman, "Comparison of Photovoltaic Array Maximum Power Point Tracking Techniques," in IEEE Transactions on Energy Conversion, vol. 22, no. 2, pp. 439-449, Jun. 2007.

[27] A. Tuladhar, H. Jin, T. Unger and K. Mauch, "Parallel operation of single phase inverters with no control interconnections," in Proc. IEEE APEC'97, vol. 1, 1997, pp. 94–100.

Closed-Form Solutions for Core and Winding Losses Calculation in Single-Phase Boost PFC Rectifiers

Marcos José Jacoboski, André de Bastiani Lange
Marcelo Lobo Heldwein, *Senior Member, IEEE*

Abstract — This work proposes closed-form solutions for core and winding losses for the boost inductor of a single-phase boost PFC converter built in a iron powder toroid core. The core losses expressions are derived based on the analytical loop separation procedure according to the iGSE method. The winding losses are computed based on a set of equations for the approximated Fourier series of the boost inductor current waveform, which is derived in this work. The overall procedure to obtain the Fourier series coefficients are demonstrated, which is used to estimate the winding losses related to the ac resistance for each current harmonic.

Index Terms — iGSE, PFC, Boost inductor, Core losses, Winding Losses, Improved Generalized Steinmetz Equation, Closed-Form Solutions, Fourier Series.

I. INTRODUCTION

Single-phase power factor corrected rectifiers (PFCs) are applied in a number of power supply applications due to equipment standards that limit the harmonic contents of a power supply ac-side currents. Modern PFCs achieve high power conversion efficiency and power density [1-2]. In addition, a fair dc-bus voltage regulation is enabled by active control. Cost, volume and weight are key merit figures for such devices and are steadily reducing for a given rated power level. Thus, the optimized design of PFCs is typically pursued and many research efforts have been made in the recent years in order for this trend to go on. Power semiconductors, cooling system, dc-link capacitors, PFC inductors and EMC filter components, give the main portions of cost and dimensions in a typical PFC. In this context, the relative portion of costs and physical dimensions of PFCs related to their magnetic components is known to be high. Thus, the design procedures to specify and construct such components are of high interest, where mathematical models are typically the basis upon which the design algorithms are built.

The design of a power inductor typically requires a good estimate of the power losses within its magnetic core and windings. There are several methods to compute these, ranging from very simplified analytic models to very detailed finite element method (FEM) analysis with the precision being typically akin to the computation efforts of those methods. Design optimization routines are often based on running the physical model many times in order to analyze the influence of geometry and materials in addition to voltage, currents and frequency variations. It is desired that the routines run as light as possible in computers without losing adequate precision. FEM simulations are burdensome and time consuming and are typically used as a last step in the design routine that is taken to analyze the detailed construction with high precision. Thus, analytical models are preferred in order

to speed up the time required to evaluate a given design within the optimization algorithm.

The required loss models for a typical PFC inductor are the core and winding losses. A well known core loss model to non-sinusoidal excitation waveforms is the Improved Generalized Steinmetz Equation (iGSE) [3], which considers the magnetic flux density and its history in a model that uses the Steinmetz equation parameters often given by magnetic core manufacturers. This method, as originally proposed, requires a piecewise linear vector of the flux density values over time and that this vector is operated in order to split the loss contributions of major and minor loops in the core B × H curve. This is achieved by applying a loop separation algorithm that takes a reasonable computational effort and the according time. Thus, an analytic model that is able to achieve a similar precision but that uses simpler straightforward equations would be preferable.

A number of winding losses models are based on Dowell's models [4]. These lead to a good estimate of the ac resistance for a given winding configuration. The resulting ac resistance values are multiplied by the rms squared values of the existing current harmonic components and summed to evaluate the overall winding losses. Thus, the harmonic spectra of an inductor current also needs to be found. Typical procedures include the use of Fast Fourier Transform (FFT) algorithms applied to numeric simulation results, the double Fourier Transform applied to the theoretic current waveforms [5], among other methods. The FFT requires that a simulation is performed for each design specification and consumes reasonable computation time. The double Fourier Transform gives the series coefficients, but these sometimes need to be post processed to obtain the rms values. In addition, a fair amount of effort is typically needed with this method due to relatively complex integral computations.

This work proposes closed-form solutions for the core and winding losses for the boost inductor of a single-phase boost PFC converter built in an iron powder toroid core.

The core losses expressions are derived based on the analytical loop separation procedure according to the iGSE method, but that result in closed-form expressions. The winding losses are computed based on a set of equations for the approximated Fourier series of the PFC inductor current waveform that are derived here.

II. CORE LOSSES MODEL

The core losses estimation of the boost inductor are derived based on the iGSE method [3], which accurately estimates the core losses for non-sinusoidal excitations using only the original Steinmetz parameters that are typically available from the core datasheet.

978-1-7281-4181-7/19 $31.00 © 2019 IEEE

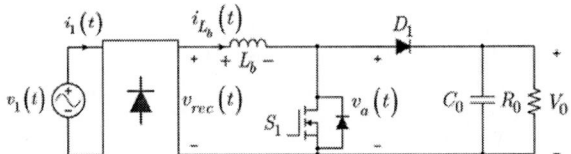

Fig 1: Boost PFC rectifier.

This method consists in the identification and further calculation of the loss contribution of all loops in the B×H curve of the core material using a computational routine that identifies, separates and calculate the contributions of all loops using the flux density function in the inductor core.

The numerical approach, as demonstrated in [6], is often very slow compared to the analytical solutions, because the flux density waveform must be pre-computed in a vector that defines a piecewise linear function to be used as an input to the numerical loop identification routine.

On the other hand, analytical solutions are typically much faster because they use limited and basic information about the converter, such as nominal voltage, current, frequencies and inductance, and require no pre-computed or simulated results.

The analytical calculation of the core losses for the boost PFC rectifier is based on a previous work [6], in which the procedure to obtain the local Steinmetz coefficients and the mathematics that enables the estimation of the core losses based in the iGSE method for a single-phase Bridgeless PFC rectifier using four integrals were demonstrated. Those integrals are referred to the major and minor loops of the rising and falling portions of the flux density waveform.

As will be demonstrated, the difference between the analytical solutions for the core loss of single-phase bridgeless PFC rectifiers and the standard boost PFC rectifiers, given in Fig 1, is only related to the major loop loss.

Using the same approach as in [6], the rising and falling portions of the considered flux density waveform is seen in Fig 2.

The next step is the identification of the number of minor loop levels. To identify how many minor loop levels exists in the flux density waveform, a closer look at a single switching period is shown in Fig 3.

As can be seen, there is no second level minor loops in the flux density waveform of the boost inductor, since there is no minor loop inside a first order minor loop.

Therefore, the loss contributions of the major loop in the kth switching period are given by

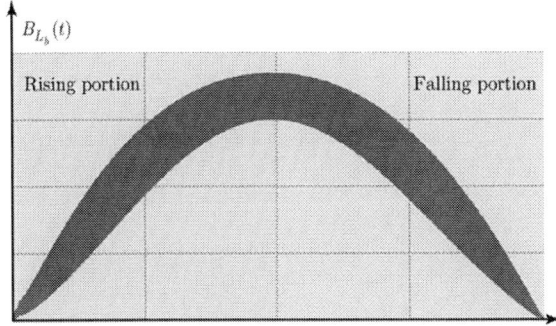

Fig 2: Rising and falling portions of the flux density waveform.

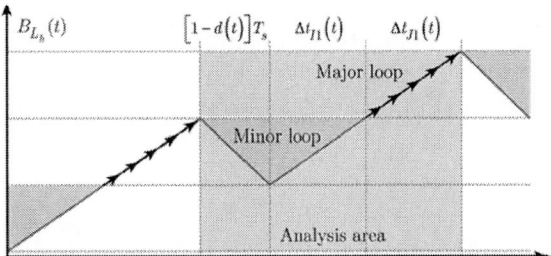

Fig 3: Details of the minor and major loop contributions in one switching period in the rising portion.

$$P_{j1}(t) = \frac{k_i \Delta B_J^{\,\beta-\alpha}}{T_p/2} \frac{|v_1(t)|^\alpha}{(NA_e)^\alpha} \Delta t_{J1}(t), \ \forall \begin{cases} t = t_0 + kT_s \\ t \le \dfrac{T_p}{4} \end{cases}. \tag{1}$$

with the major loop magnetic flux density excursion computed with

$$\Delta B_J = \frac{V_0 T_s V_{1p} - T_s V_{1p}^{\,2} + 2L_b V_0 I_{1p}}{2V_0 A_e N}. \tag{2}$$

The major loop flux density excursion is the maximum excursion found in the flux density waveform, including its high frequency content.

The contribution of the major loop in the rising portion can be calculated summing up all the contributions in all switching periods within the rising portion. Thus,

$$P_{J1} = \sum_{t=t_0,t_0+T_s,\dots}^{T_p/4} \left[P_{j1}(t) \right]. \tag{3}$$

Since the premise of this approach is that the switching frequency is much larger than the fundamental frequency ($T_s \ll T_p$), (3) becomes

$$P_{J1} = \lim_{T_s/T_p \to 0} \left\{ \sum_{t=t_0,t_0+T_s,\dots}^{T_p/4} 2k_i \Delta B_J^{\,\beta-\alpha} \frac{|v_1(t)|^\alpha}{(NA_e)^\alpha} \frac{\Delta t_{J1}(t)}{T_s} \frac{T_s}{T_p} \right\} \tag{4}$$

that can be rewritten as

$$P_{J1} = \lim_{T_s/T_p \to 0} \left\{ \sum_{t/T_p = t_0/T_p,\ t_0/T_p + T_s/T_p,\dots}^{1/4} 2k_i \Delta B_J^{\,\beta-\alpha} \frac{\dfrac{\left| v_1\!\left(\frac{t}{T_p} T_p\right) \right|^\alpha}{(NA_e)^\alpha}}{\dfrac{\Delta t_{J1}\!\left(\frac{t}{T_p} T_p\right)}{T_s}} \frac{T_s}{T_p} \right\}. \tag{5}$$

Defining an auxiliary variable as

$$\gamma = \frac{t}{T_p} \to \Delta\gamma = \frac{T_s}{T_p}, \tag{6}$$

and replacing it into (5) leads to

$$P_{J1} = \lim_{\Delta\gamma \to 0} \left\{ \sum_{\gamma = t_0/T_p,\ t_0/T_p + \Delta\gamma,\dots}^{1/4} 2k_i \Delta B_J^{\,\beta-\alpha} \frac{\left| v_1(\gamma T_p) \right|^\alpha}{(NA_e)^\alpha} \frac{\Delta t_{J1}(\gamma T_p)}{T_s} \Delta\gamma \right\}. \tag{7}$$

Equation (7) is a Riemann Sum [7], which defines an integral. Thus,

$$P_{J1} = \frac{2k_i \Delta B_J^{\,\beta-\alpha}}{(NA_e)^\alpha} \int_{t_0/T_p}^{1/4} \left| v_1(\gamma T_p) \right|^\alpha \frac{\Delta t_{J1}(\gamma T_p)}{T_s} d\gamma. \tag{8}$$

Using the same approach, the contributions of the minor loops in the rising portion are given by

978-1-7281-4181-7/19 $31.00 © 2019 IEEE

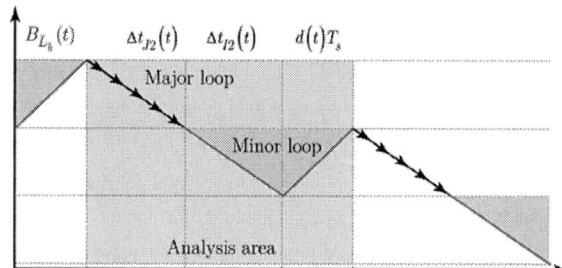

Fig 4: Details of the minor and major loop contributions in one switching period in the falling portion.

$$P_{I1} = \frac{2k_i}{\left(NA_e\right)^\alpha} \int\limits_{t_0/T_p}^{1/4} \Delta B_{I1}^{\beta-\alpha}\left(\gamma T_p\right) \left\{ \begin{array}{l} \left|V_0 - v_1(\gamma T_p)\right|^\alpha \left[1 - d\left(\gamma T_p\right)\right] \\ + \left|v_1(\gamma T_p)\right|^\alpha \dfrac{\Delta t_{I1}\left(\gamma T_p\right)}{T_s} \end{array} \right\} d\gamma . \quad (9)$$

The minor and major loop contributions in the falling portion are depicted in Fig 4.

The contributions of the major and minor loops in the falling portion are calculated using

$$P_{J2} = \frac{2k_i \Delta B_J^{\beta-\alpha}}{\left(NA_e\right)^\alpha} \int\limits_{1/4}^{1/2 - t_0/T_p} \left|V_0 - v_1\left(\gamma T_p\right)\right|^\alpha \frac{\Delta t_{J2}\left(\gamma T_p\right)}{T_s} d\gamma \quad (10)$$

$$P_{I2} = \frac{2k_i}{\left(NA_e\right)^\alpha} \int\limits_{1/4}^{1/2 - t_0/T_p} \Delta B_{I2}^{\beta-\alpha}\left(\gamma T_p\right) \left\{ \begin{array}{l} \left|V_0 - v_1(\gamma T_p)\right|^\alpha \dfrac{\Delta t_{I2}\left(\gamma T_p\right)}{T_s} \\ + \left|v_1(\gamma T_p)\right|^\alpha d\left(\gamma T_p\right) \end{array} \right\} d\gamma . \quad (11)$$

The time intervals functions for both major and minor loops, in the rising and falling portions, are defined as

$$\Delta t_{J1}(t) = \left\{ d(t) - \frac{\left[V_0 - v_1(t)\right]\left[1 - d(t)\right]}{v_1(t)} \right\} T_s, \ \forall \left\{ \begin{array}{l} t = t_0 + kT_s \\ t \le \dfrac{T_p}{4} \end{array} \right. \quad (12)$$

$$\Delta t_{I1}(t) = \frac{\left[V_0 - v_1(t)\right]\left[1 - d(t)\right]}{v_1(t)} T_s, \ \forall \left\{ \begin{array}{l} t = t_0 + kT_s \\ t \le \dfrac{T_p}{4} \end{array} \right. \quad (13)$$

$$\Delta t_{J2}(t) = \left\{ 1 - d(t) - \frac{v_1(t)d(t)}{V_0 - v_1(t)} \right\} T_s, \ \forall \left\{ \begin{array}{l} t = \dfrac{T_p}{4} + kT_s \\ t \le \dfrac{T_p}{2} - t_0 \end{array} \right. \quad (14)$$

$$\Delta t_{I2}(t) = \frac{v_1(t)d(t)}{V_0 - v_1(t)} T_s, \ \forall \left\{ \begin{array}{l} t = \dfrac{T_p}{4} + kT_s \\ t \le \dfrac{T_p}{2} - t_0 \end{array} \right. , \quad (15)$$

and the minor loops flux density excursions functions, in the rising and falling portions, are defined as

$$\Delta B_{I1}(t) = \frac{\left[V_0 - v_1(t)\right]\left[1 - d(t)\right]}{NA_e} T_s, \ \forall \left\{ \begin{array}{l} t = t_0 + kT_s \\ t \le \dfrac{T_p}{4} \end{array} \right. \quad (16)$$

$$\Delta B_{I2}(t) = \frac{v_1(t)d(t)}{NA_e} T_s, \ \forall \left\{ \begin{array}{l} t = \dfrac{T_p}{4} + kT_s \\ t \le \dfrac{T_p}{2} - t_0 \end{array} \right. . \quad (17)$$

The initial time of integration t_0 is given by

$$t_0 = \frac{T_p}{\pi} \text{atan}\left(\frac{\sqrt{4\pi^2 L_b^2 I_{1p}^2 + T_p^2 V_{1p}^2} - T_p V_{1p}}{2\pi L_b I_{1p}} \right). \quad (18)$$

and the details about why start the in integration at t_0 and stop at $T_p/4 - t_0$ are discussed in [6].

For the duty cycle function, a complete function is needed to account for the low frequency voltage drop in the boost inductor. This is found with

$$d(t) = \frac{V_0 - \left[V_{1p}\sin\left(\dfrac{2\pi}{T_p}t\right) - \dfrac{2\pi}{T_p}L_b I_{1p}\cos\left(\dfrac{2\pi}{T_p}t\right)\right]}{V_0}, \ \forall \left\{ \begin{array}{l} t = t_0 + kT_s \\ t \le \dfrac{T_p}{2} - t_0 \end{array} \right. . \quad (19)$$

The grid voltage function is a sinusoidal waveform that is defined by

$$v_1(t) = V_{1p}\sin\left(\frac{2\pi}{T_p}t\right). \quad (20)$$

The total core loss for a single-phase boost PFC rectifier based on the loss expressions regarding the rising and falling portions can be finally computed with

$$P_{core} = P_{rise} + P_{fall} = P_{J1} + P_{I1} + P_{J2} + P_{I2}. \quad (21)$$

III. Winding Losses Model

The amplitude of each harmonic that composes the boost inductor current waveform is needed to analytically calculate the winding losses.

A simple solution for the Fourier Series is demonstrated in the following. This solution is based on the same premise used in the calculation of the core losses, i.e., $T_s \ll T_p$.

The goal is to decompose the switched voltage, defined as $v_a(t)$, which is done by assuming that this voltage waveform is formed by the sum of simple orthogonal switched voltage functions, as seen in Fig 5.

The complex exponential Fourier Series coefficients related to the $v_a(t)$ is given by

$$C(n,m) = \frac{2V_0}{T_p} \left\{ \frac{e^{-jn\frac{4\pi}{T_p}T_s\left[\frac{1-d(mT_s)}{2}\right]} - e^{-jn\frac{4\pi}{T_p}T_s\left[\frac{1-d(mT_s)}{2}\right]}}{-jn\frac{4\pi}{T_p}} \right\} e^{-jn\frac{4\pi}{T_p}mT_s}, \ \left\{ \begin{array}{l} n \in [1,\infty] \\ m \in \left[m_0, \dfrac{T_p}{2T_s} - m_0\right] \end{array} \right. \quad (22)$$

that is the mth orthogonal switched function exponential Fourier coefficient for $t = mT_s$. The complete ac portion of the switched waveform $v_a(t)$ is evaluated by varying m from m_0 to $T_p/2T_s - m_0$, with

$$m_0 = ceil\left(\frac{t_0}{T_s}\right) = \min\left\{ x \in \mathbb{Z} \left| x \ge \frac{t_0}{T_s} \right. \right\}. \quad (23)$$

The ac part of the switched voltage $v_a(t)$ can accordingly be written as

$$v_a^{AC}(t) = \sum_{m=m_0}^{\frac{T_p}{2T_s}-m_0} \sum_{n=1}^{\infty} 2|C(n,m)| \cos\left(n\frac{4\pi}{T_p}t + \arg\left[C(n,m)\right] \right). \quad (24)$$

The ac part of the rectified grid voltage can also be written as a Fourier series given by

$$v_{rec}^{AC}(t) = \sum_{n=1}^{\infty} 2|D(n)| \cos\left(n\frac{4\pi}{T_p}t + \arg\left[D(n)\right] \right), \quad (25)$$

where

$$D(n) = \frac{2V_{1p}}{\pi\left(1 - 4n^2\right)}. \quad (26)$$

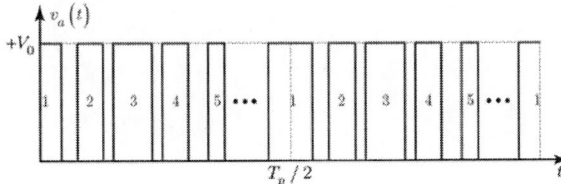

Fig 5: Switched waveform $v_a(t)$.

The ac voltage across the boost inductor can be found subtracting the ac parts of the grid and switched voltages. This results in

$$v_{L_b}^{AC}(t) = v_{rec}^{AC}(t) - v_a^{AC}(t), \qquad (27)$$

that can be rewritten as

$$v_{L_b}^{AC}(t) = \left\{ \begin{array}{c} \sum_{n=1}^{\infty} 2\left|D(n)\right|\cos\left(n\frac{4\pi}{T_p}t + \arg\left[D(n)\right]\right) \\ \frac{T_p}{2T_s}-m_0 \\ -\sum_{m=m_0}^{\infty}\sum_{n=1}^{\infty} 2\left|C(n,m)\right|\cos\left(n\frac{4\pi}{T_p}t + \arg\left[C(n,m)\right]\right) \end{array} \right\}. \qquad (28)$$

$$v_{L_b}^{AC}(t) = \sum_{m=m_0}^{\frac{T_p}{2T_s}-m_0}\sum_{n=1}^{\infty} 2\left|\left[\frac{T_p}{2T_s}-m_0\right]^{-1}D(n)-C(n,m)\right|\cos\left(n\frac{4\pi}{T_p}t + \arg\left[\frac{2T_s}{T_p}D(n)-C(n,m)\right]\right). \qquad (29)$$

Defining an auxiliary variable as

$$M(n,m) \triangleq \left[\frac{T_p}{2T_s}-m_0\right]^{-1}D(n)-C(n,m), \qquad (30)$$

the boost inductor current can be expressed as

$$i_{L_b}(t) = \frac{2I_{1p}}{\pi} + \sum_{m=m_0}^{\frac{T_p}{2T_s}-m_0}\sum_{n=1}^{\infty}\frac{2T_p}{n4\pi L_b}\left|M(n,m)\right|\cos\left(n\frac{4\pi}{T_p}t + \arg\left[M(n,m)\right]-\frac{\pi}{2}\right). \qquad (31)$$

The amplitude of each harmonic related to the boost inductor current can be finally expressed as

$$A_{i_{L_b}}(n) = \left|\sum_{m=m_0}^{\frac{T_p}{2T_s}-m_0}\frac{T_p}{2n\pi L_b}M(n,m)\right|, \quad n=1,2,\dots \qquad (32)$$

IV. EXPERIMENTAL RESULTS

The approximated Fourier series obtained in (31) is plotted as a numerical result in Fig 6. Since the original premise is maintained ($T_s \ll T_p$), the series will converge to the original switched waveform.

The spectrum related to the waveform in Fig 6 is shown in Fig 7. This result is the direct plot of the expression (32). The input parameters needed in the

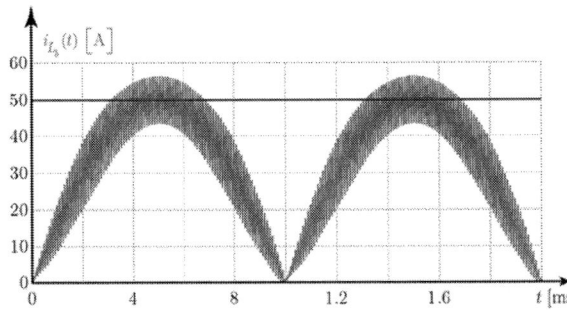

Fig 6: Constructed waveform with the derived Fourier series.

Table 1: Electrical parameters used in the Fourier series example.

Description	Parameter	Value
Grid Frequency	T_p^{-1}	50 Hz
Switching Frequency	T_s^{-1}	10 kHz
Grid peak voltage	V_{1p}	311 V
Grid peak current (50Hz)	I_{1p}	50 A
Boost Inductance	L_b	500 μH
DC bus voltage	V_0	400 V

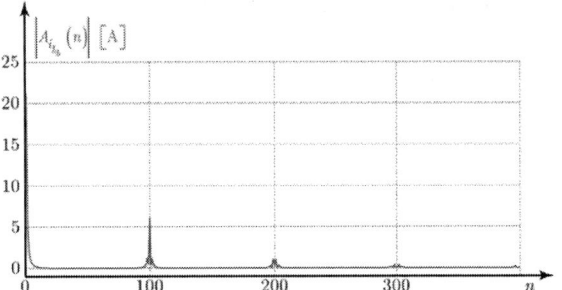

Fig 7: Spectrum of the inductor current Fourier series.

Fourier series are given in Table 1.

To validate the core and winding losses calculation procedure, a 1 kW single-phase boost PFC rectifier was employed. The PFC converter was tested operating at 600 W and 1 kW, and the experimental results containing the theoretical calculations of the core losses along with the temperature rises for both cases are summarized in Table 2. The winding losses and temperature rise are calculated using the same approach described in [6].

The experimental results for 600 W and 1 kW are shown in Fig 8 and Fig 9, respectively, meanwhile the boost PFC rectifier prototype is show in Fig 10.

Table 2: Theoretical calculations and experimental results.

Description	Parameter	Value (600 W)	Value (1 kW)
Core part number	-	KAM168060A	
Grid peak voltage	V_{1p}	$220\sqrt{2}$ V	
Dc output voltage	V_0	380 V	
Switching Frequency	T_s^{-1}	140 kHz	
Core cross sectional area	A_e	1.5 cm^2	
Core magnetic volume	V_e	16.014 cm^3	
Number of turns	N	72	
Measured inductance	L_b	559 μH @ 140 kHz	
Grid reference peak current	I_{1p}	3.9 A	6.5 A
Steinmetz coefficients	α, β, k_c	1.33, 1.99, 10.1 W/m^3	
Copper losses (calculated)	P_{copper}	1.233 W	2.039 W
Core losses (closed-form)	P_{core}^{CF}	0.692 W	0.698 W
Core losses (routine)	P_{core}^{RT}	0.696 W	0.698 W
Calculated temperature rise	ΔT_{calc}	27.5 °C	36.7 °C
Experimental temperature rise	ΔT_{meas}	29.2 °C	37.1 °C

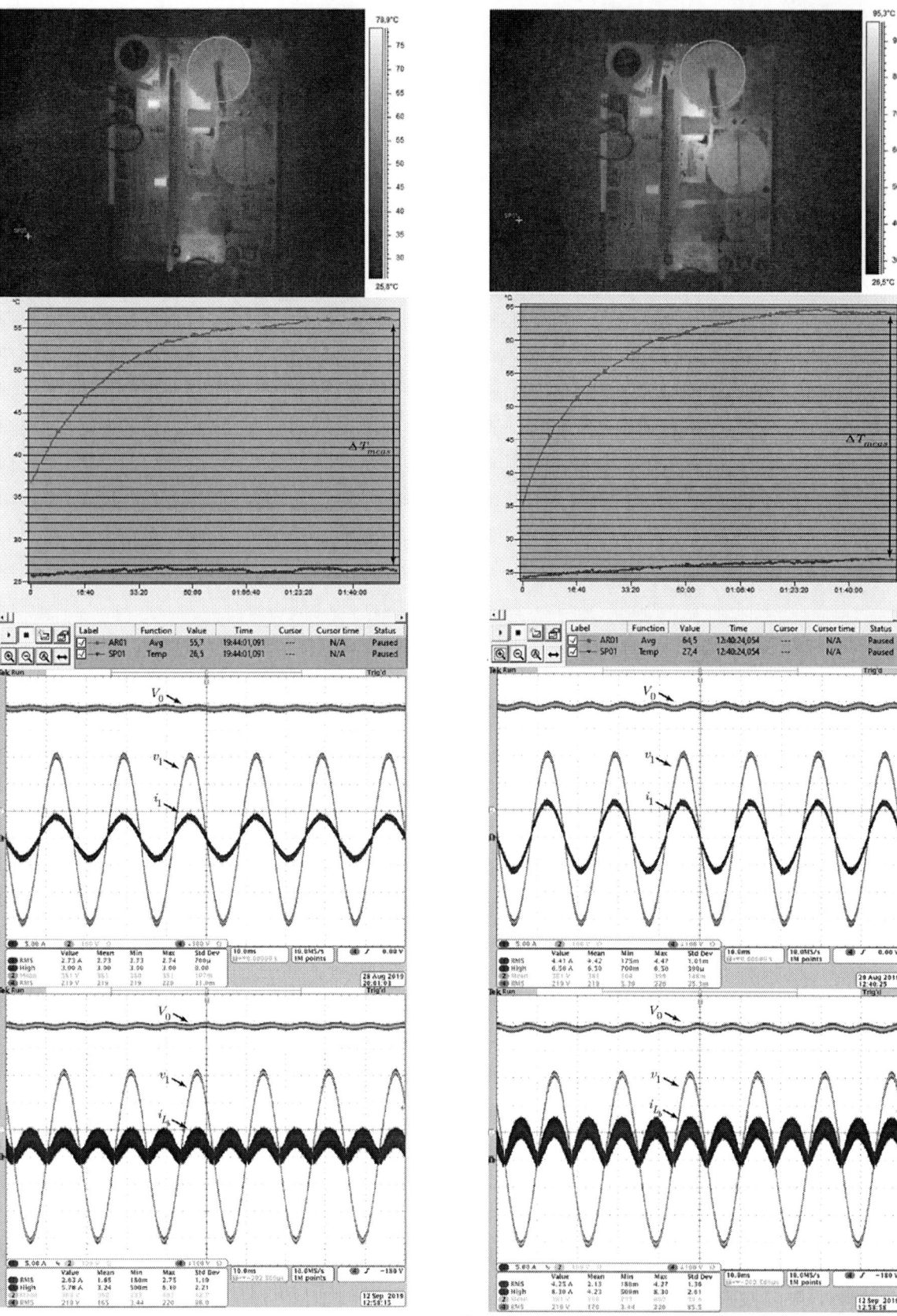

Fig 8: Experimental results for 600 W. Thermal images, temperature rising over time (ΔT_{meas}), dc-link voltage (V_0), grid voltage (v_1), grid current (i_1) and boost inductor current (i_{L_b}).

Fig 9: Experimental results for 1 kW. Thermal images, temperature rising over time (ΔT_{meas}), dc-link voltage (V_0), grid voltage (v_1), grid current (i_1) and boost inductor current (i_{L_b}).

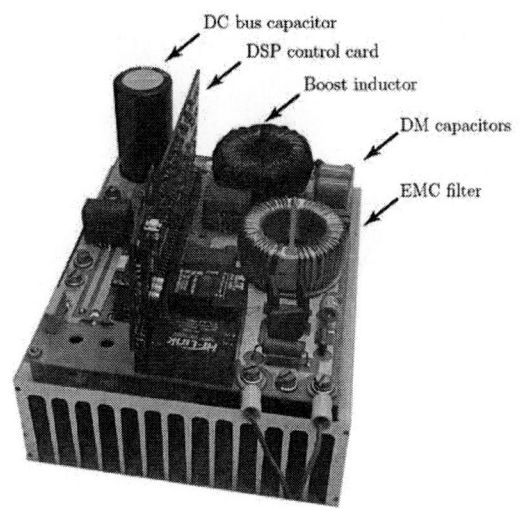

Fig 10: 1kW Boost PFC rectifier prototype.

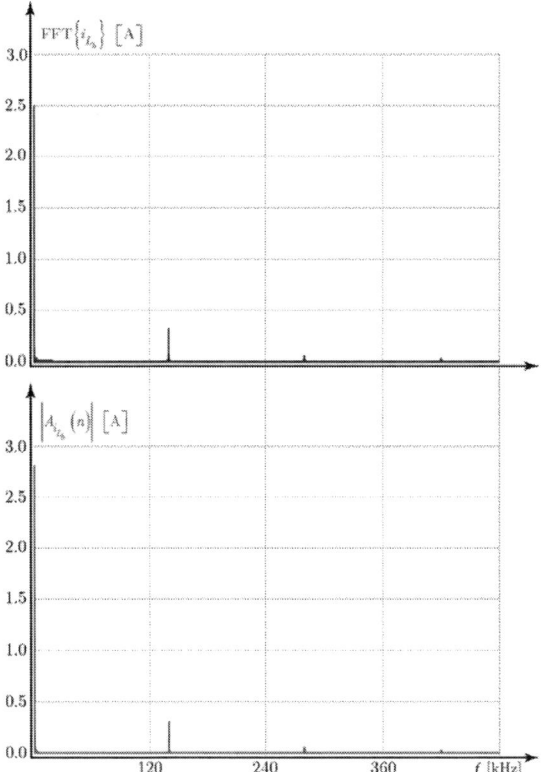

Fig 11: Experimental (blue) and analytical (red) boost inductor current spectrum.

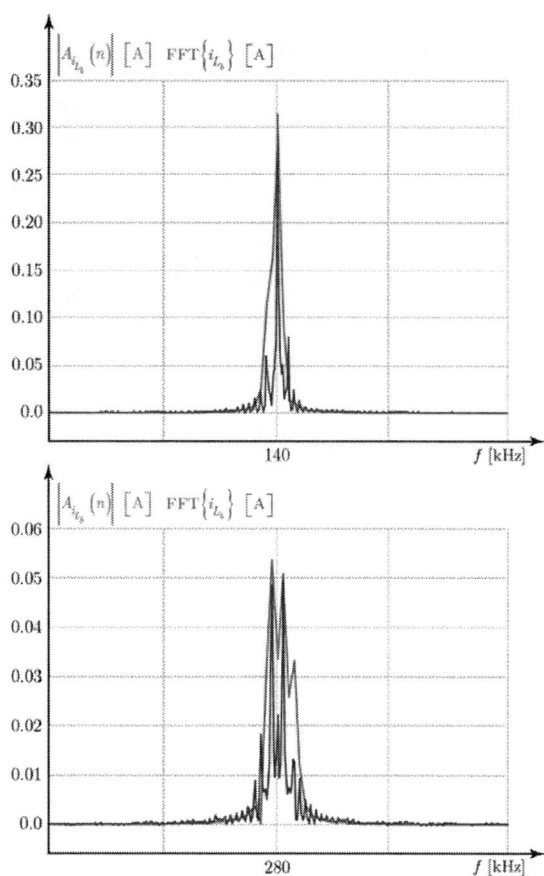

Fig 12: Details of the experimental (blue) and analytical (red) boost inductor current spectrum around 140 kHz and 280 kHz.

each current harmonic derived in this paper, the total winding loss can be calculated more precisely considering the ac resistance related to each harmonic component.

Regarding the analytical equations derived based on the iGSE method, it can be seen that there is a good agreement with respect to the numerical routine, which was validated in [6], showing that the set of equations can be useful in optimization routines, in which an significant reduction of computational time can be achieved.

The comparison between analytical and experimental spectrum for the converter operating at 1 kW is shown in Fig 11 and Fig 12, where a good agreement is observed.

V. CONCLUSIONS

In this work a new analytical approach was used to derive the approximated Fourier series for the boost inductor current in a single-phase PFC rectifier. The premise for the Fourier series convergence is that the switching frequency must be much larger than the fundamental frequency, which can be easily achieved in low power applications.

Using the analytical expression for the amplitude of

REFERENCES

[1] J. W. Kolar and T. Friedli, "The Essence of Three-Phase PFC Rectifier Systems - Part I," *IEEE Trans. Power Electron.*, vol. 28, no. 1, pp. 1–179, 2013.

[2] J. W. Kolar and T. Friedli, "The Essence of Three-Phase PFC Rectifier Systems - Part II," *IEEE Trans. Power Electron.*, vol. 28, no. 1, pp. 1–179, 2013.

[3] K. Venkatachalani, C. R. Sullivan, T. Abdallah, and H. Tacca, "Accurate Prediction of Ferrite Core Loss with Nonsinusoidal Waveforms Using Only Steinmetz Parameters," *2002 IEEE Work. Comput. Power Electron.*, pp. 36–41, 2002.

[4] M. Kazimierczuk, *High-Frequency Magnetic Components.* 2013.

[5] D. G. Holmes and T. A. Lipo, *Pulse Width Modulation for Power Converters.* 2003.

[6] M. J. Jacoboski, A. D. B. Lange, and M. L. Heldwein, "Closed-Form Solution for Core Loss Calculation in Single-Phase Bridgeless PFC Rectifiers Based on the iGSE Method," *IEEE Trans. Power Electron.*, vol. 33, no. 6, pp. 4599–4604, 2017.

Development of a Hybrid PV-Thermoelectric System

Rafael Magalhães Nóbrega de Araújo
Department of Computer Engineering
and Automation (DCA)
Federal University of Rio Grande do Norte (UFRN)
Natal-RN, Brasil
email: ramanoar@gmail.com

Hugo Álisson Alves da Costa
Department of Computer Engineering
and Automation (DCA)
Federal University of Rio Grande do Norte (UFRN)
Natal-RN, Brasil
email: hugo.aac1@gmail.com

João T. de Carvalho Neto
Federal Institute Education,
Science and Technology of Rio Grande do Norte (IFRN)
Natal-RN, Brazil
email: joaoteixeira@dca.ufrn.br

Alexandre Magnus F. Guimarães
School of Sciences and Technology (ECT)
Federal University of Rio Grande do Norte (UFRN)
Natal-RN, Brazil
E-mail: alexandremagnus@ect.ufrn.br

Andrés Ortiz Salazar
Department of Computer Engineering
and Automation (DCA)
Federal University of Rio Grande do Norte (UFRN)
Natal-RN, Brasil
email: andres@dca.ufrn.br

Abstract—**The solar photovoltaic energy is one of the main renewable energy alternatives for countries to achieve a greater diversity in their energy matrix. It benefits from high levels of incident solar irradiance. However, the increase in temperature that comes with it, allied to other environmental variables lowers its overall yield. One cooling option is to use thermoelectric, or Peltier cells. This paper aims to study a hybrid system that seeks an increase of a photovoltaic module yield through its active cooling by a thermoelectric cell. Research strategies include evaluation of controlled temperature, number of thermoelectric cells for a given photovoltaic module, use of maximum power tracking algorithms together with control of the thermoelectric cells power supply to achieve an optimum point of operation. This study uses mathematical models, and simulations to show that the use of thermoelectric cells to cool off the photovoltaic module can increase its yield by up to 7% in the cases studied.**

Index Terms—**Solar energy, Thermoelectric, Peltier, PVs.**

I. Introduction

Throughout the years, several countries have considered solar photovoltaic energy (PV) as an option for their electric energy supply. The vast daily availability of energy from the sun, the low maintenance and handling requirements, and the technological advancement coupled with clean energy incentive programs are some of the reasons behind its growing adoption by countries concerned with diversifying their energy matrix.

The authors would like to thank the Brazilian agencies CAPES and CNPq, which support this work.

Two factors are important in the design and evaluation of PV power generation systems: solar irradiance, and operating temperature. The former is directly related to the amount of energy available for conversion, and the latter to the level of energy losses of the system. The temperature is always linked to the level of solar irradiance, but not limited by it. These two factors are key to determine the energy efficiency of the system, as will be detailed throughout this paper. Due to Earth's curvature, countries located near the equator enjoy a more robust solar irradiance level, so they have a higher power generation potential. However, the increase in irradiance comes with an increase in the temperature of the PV cells, which in turn leads to limited generation potential.

PV cells are subject to the maximum theoretical efficiency of 33.7% for single-junction semiconductors, defined as the Schockley-Queisser limit [1]. However, due to various cell specificities, the construction method, and weather conditions, the practical efficiency is found to be around 26.7% for monocrystalline cells, and 22.3% for polycrystalline cells when operated under standard test conditions in a lab [2].

Several studies were found in the literature that examined the efficiency improvement of PV systems through their cooling, and the increase in energy efficiency through hybrid systems of electric power generation. Some examples include [3] which compared the efficiency between PV systems, PV/Water e PV/Thermoelectric/Water, [4] evaluated the challenges related to hybrid generation PV/Thermoelectric,

978-1-7281-4181-7/19 $31.00 © 2019 IEEE

[5] analyzed the performance of a PV/Thermoelectric hybrid generator according to different climatic conditions, referring to European cities, [6] studied the best compromise between the generation and consumption of the cooling system for a water-cooled PV system, [7] studied performance improvement in a water-cooled PV system on its front surface in arid environments, [8] studied the optimization of a PV system cooled by thermoelectric cells applied to a concentrated solar generator.

The main objective of this work is the evaluation of a hybrid PV-thermoelectric system to obtain a greater energy yield when compared to a regular PV system. The increase in yield will come from the decrease in the operating temperature of the PV module, which will be achieved by the heat transfer provided by the thermoelectric cell.

II. MATHEMATICAL MODELLING

A. PV Electrical Model

The seven parameter model was chosen to represent the PV cell working principles [9]. This model uses two diode saturation currents that represent the premature recombination of the free electrons in the semiconductor, and takes into account a degree of disorder in the manufacturing process, which, in turn, is responsible for a non-uniform junction between the two semiconductor layers. The equivalent circuit is shown in Figure 1. The parameters for this circuit are not readily available in commercial module datasheets.

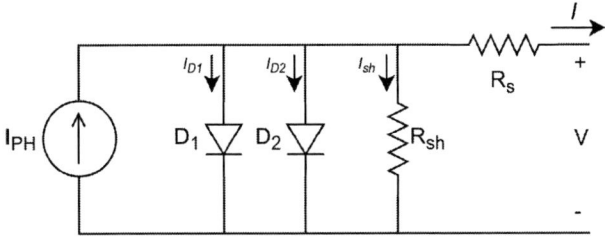

Fig. 1. PV cell equivalent circuit

Therefore, it is necessary to apply a methodology to derive these values from the available data [10]. The equation that governs the model used can be described by:

$$I = -I_{s_1} \exp\left(\frac{V + R_S I}{a_1 N_S V_t} - 1\right) - I_{s_2} \exp\left(\frac{V + R_S I}{a_2 N_S V_t} - 1\right) + I_{PH} - \frac{V + R_S I}{R_{SH}} \quad (1)$$

Where, I, I_{PH}, I_{s_1}, and I_{s_2} are the current available at the terminals, the maximum generated current, and the reverse saturation current of diode 1 and 2, respectively. R_S and R_{SH} are the series and parallel resistors of the equivalent circuit respectively, N_S is the number of cells in series, and a_1 and a_2 are the ideality factors of diodes 1 and 2, respectively. V is the available voltage at the terminals. V_t is the thermal voltage of the module, and is given by kT/q, where k is the Boltzmann constant (1.38×10^{-23} J/K), T is the temperature of the module and q is the electric charge constant (1.602×10^{-19} C).

Through several equivalences, (2) to (5) were derived to support the methodology:

$$I_{s_1} = \frac{V_{oc}(I_{SC} - I_m) - V_m I_{SC}}{V_{oc}[X_{m_1} + K X_{m_2}] - V_m[X_{oc_1} + K X_{oc_2}]} \quad (2)$$

$$I_{s_2} = \left(\frac{T^{2/5}}{3.77}\right) I_{s_1} \quad (3)$$

$$I_{PH} = \frac{V_{oc} I_m + I_{S_1}[V_{oc}(X_{m_1} + K X_{m_2}) - V_m[X_{oc_1} + K X_{oc_2}]}{V_{oc}} \quad (4)$$

$$R_{SH} = \frac{V_m + I_m R_s}{[I_{PH} - I_m - I_{s_1}(X_{m_1} - 1) - I_{s_2}(X_{m_2} - 1)]} \quad (5)$$

Where, V_{oc} and V_m are the open-circuit voltage and voltage at maximum power point, respectively. I_{sc} is the short circuit current, I_m is the current at maximum power point and K is a ratio between I_{s_1} and I_{s_2}. T is the PV cell temperature.

X_m and X_{oc} are variables used to clarify the equations and are of the form given by (6) and (7).

$$X_{oc_{1,2}} = \exp\left(\frac{V_{oc}}{a_{1,2} N_S V_t}\right) \quad (6)$$

$$X_{m_{1,2}} = \exp\left(\frac{V_m + R_S I_m}{a_{1,2} N_S V_t}\right) \quad (7)$$

As it can be seen, several components depend on the value of R_S, although its value is not immediately available.

To address this issue the algorithm displayed in figure 2 was used.

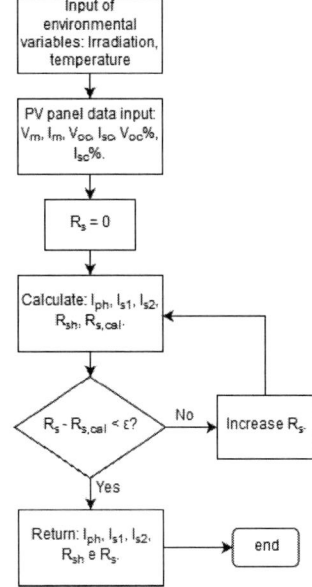

Fig. 2. Iterative algorithm for calculating the equivalent circuit parameters of a PV cell.

$R_{S,cal}$ is given by:

$$R_{S,cal} = \frac{V_m}{I_m} - \frac{1}{\left[\frac{I_{S_1}}{a_1 N_S V_{t_1}} X_{m_1} + \frac{I_{S_2}}{a_2 N_S V_{t_2}} X_{m_2} + \frac{1}{R_{SH}}\right]} \quad (8)$$

978-1-7281-4181-7/19 $31.00 © 2019 IEEE

The methodology as is presented here is used when the PV module data is given for STC conditions. In order to correct for different environmental variables, a set of equations can be used.

$$R_{SH} = R_{SH_{stc}} \frac{G_{stc}}{G} \qquad (9)$$

$$I_{s1} = \frac{I_{PH_{stc}} + K_I(T - T_{stc})}{e^{\frac{V_{oc_{stc}} + K_V \Delta T}{a_1 V_T}} - 1} \qquad (10)$$

$$I_{PH} = \frac{G}{G_{stc}}[I_{PH_{stc}} + K_I(T - T_{stc})] \qquad (11)$$

In which G is the solar irradiance in W/m^2, K_V and K_I are the rates of decrease of voltage and increase of current in $\%/°C$. The subscript *stc* refers to the Standard Test Conditions of $1000\ W/m^2$ irradiance, 25 °C operating temperature and 1.5 air mass.

B. Thermal Analysis

As discussed in the introduction, the operating temperature is a key factor in PV power generation. To calculate the effects of thermoelectric cells over the module's temperature, the thermal behavior needs to be studied. In order to facilitate this, a thermal model of a solar cell is represented as its electric analogue.

The effect of temperature on a PV module can be evaluated through the interaction between its components. A typical commercial polycrystalline PV module can be broken down into [11]: Tempered glass, anti-reflective layer, PV cell, EVA layer, metal contacts, and rear layer (Tedlar).

Each composing layer of a PV module may be thermally represented by a pair of thermal resistance and capacitance. This thermal representation takes into account only the conductive heat transfer between the components.

The thermal resistance defines the measure of material resistance to the passage of heat flow through itself. It is modeled by (12).

$$R_{\Phi_{cd}} = \frac{t}{k \cdot A_s} \qquad (12)$$

where, t is the material thickness (m), k is the module's thermal conductivity $(W/m \cdot K)$, and A_s is the PV module's surface area (m^2).

Thermal capacitance represents the ability of the material to absorb and retain heat. It is modeled by (13).

$$C_{\Phi_{cd}} = \rho \cdot c \cdot A_s \cdot t \qquad (13)$$

where ρ is the material's density (kg/m^3) and c is the material specific heat (J/K).

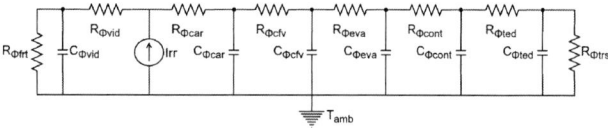

Fig. 3. Equivalent thermal circuit of a PV cell.

The subscripts *frt, vid, car, cfv, eva, cont, ted, trs*, in the figure 3 refer to, respectively: the module's front, the glass, anti-reflexive layer, PV cell, ethylene-vinyl acetate, electric contacts, tedlar, module's back. The current source Irr represents the heat generated through solar irradiance, and the equipotential T_{amb} represents the ambient temperature.

I_{rr} is given by (14) [12]:

$$I_{rr} = \alpha \cdot A_s \cdot G \qquad (14)$$

where α is the solar module absorptance, normally between 70% and 80% and G is the Solar irradiance normal to the PV module (W/m^2).

The equivalent thermal circuit of the PV module shown represents the thermal behavior of the module adjusted for the local environmental conditions. That is, the wind speed and the intensity of the sun will affect the operating temperature, and thus the maximum possible energy to be generated by the module. The wind is factored in $R_{\Omega_{frt}}$ and $R_{\Omega_{trs}}$, which represent the thermal conductance due to convective heat transfers.

Irradiative heat transfer was not taking into consideration, since its effect is negligible [11].

C. Thermoelectric Cell

The thermoelectric cell is capable of transferring heat between its sides when electric power is supplied to its terminals. Alternatively, the presence of a temperature difference between their sides causes an electrical potential difference between their terminals. In this work, the thermoelectric cell functions as a cooling device, usually found in the literature as Thermoelectric Cooler - TEC. Thus the interest here is to understand how the cell behaves thermally when it is supplied with electrical power [13].

From the first law of thermodynamics, one can derive the expressions for the energy balance in each side of the thermoelectric cell. One side is designated as the one which absorbs the heat and transfers it to the side that emits it. Therefore:

$$q_a = \frac{\Delta T}{R_{\Phi_m}} + \alpha_m \cdot T_a \cdot I - \frac{I^2 \cdot R_m}{2} \qquad (15)$$

$$q_e = \frac{\Delta T}{R_{\Phi_m}} + \alpha_m \cdot T_e \cdot I + \frac{I^2 \cdot R_m}{2} \qquad (16)$$

Where: q_a is the heat absorbed by the cell, q_e is the heat emitted by the cell, ΔT is the temperature difference between cell sides, R_{Φ_m} is the thermal resistance of the semiconductor pairs in the direction of heat flow, α_m is the Seebeck coefficient of the cell, T_a is the temperature on the heat-absorbing side, T_e is the temperature on the heat-emitting side, I is the current supplied to the cell, and R_m is the equivalent electrical resistance of the cell.

The subscript m refers to equivalent variables for a complete cell since each of these variables are usually obtained for each thermoelectric pair that makes up the cell as a whole.

Based on (15) and (16) the authors of [13] proposed a thermal equivalent circuit adapted to the electric one shown on figure 4. Where: V_Φ is the variable voltage depending on T_a and the supply current I, R_Φ is the equivalent thermal resistance of the cell, C_Φ is the equivalent thermal capacitance of the cell, P_e is the electric power supplied to the cell. The voltage V_Φ is stated in (17).

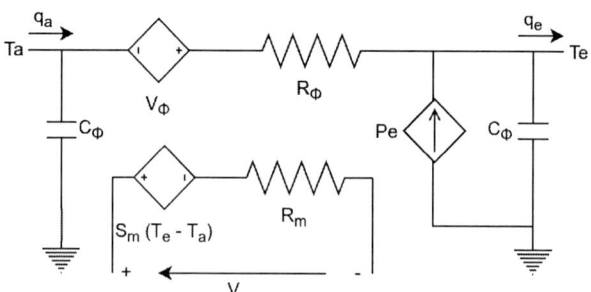

Fig. 4. Equivalent circuit for the thermoelectric cell.

$$V_\Phi = \left(\alpha_m \cdot T_a - \frac{I \cdot R_m}{2}\right) \cdot I \cdot R_\Phi \qquad (17)$$

The values for the components of the equivalent thermal circuit can be drawn indirectly from the information normally available in the datasheet of a thermoelectric cell. The relationships of (18) to (20) were used to define the value of the components used in the simulations.

$$R_m = \frac{V_{max}}{I_{max}} \cdot \frac{(T_h - \Delta T_{max})}{T_h} \qquad (18)$$

$$R_{\Phi_m} = \frac{\Delta T_{max}}{I_{max} \cdot V_{max}} \cdot \frac{2T_h}{(T_h - \Delta T_{max})} \qquad (19)$$

$$\alpha_m = \frac{V_{max}}{T_h} \qquad (20)$$

All information required to solve (18), (19), (20) is taken directly from the datasheet. The meaning of this information is as follows: ΔT_{max}: greatest temperature differential between the sides that the cell can provide according to a certain level of T_h. I_{max}: current supplying the ΔT_{max} at a given temperature T_h. V_{max}: potential difference in the electrical terminals of the cell when it is supplied with I_{max}.

Thermoelectric cell data sheets usually provides two sets of information. Each for a specific value of T_h.

D. Complete System

A commercial PV module was chosen for the simulations due to ease of access to information and because its cells follow the structure previously presented here. The specific data for this module is detailed in the table I.

TABLE I
GCL P6-60 PV MODULE DATA

GCL P6-60	
Silicon type	Polycrystalline
Cell number	60
Efficiency (%)	16
Maximum Power (W)	250
Voltage at MPP (V)	37.71
Current at MPP (A)	8.29
Cell dimensions (mm)	156 x 156

The thermoelectric cell modeled in this work was the cell TEC1-12705, from Heibei I.T., which is physically available for experimental tests in the future. Its data is shown in table II.

TABLE II
THERMOELECTRIC CELL DATA

Hot side temperature (°C)	25 °C	50 °C
Q_{max} (W)	43	49
ΔT_{max} (°C)	66	75
I_{max} (A)	5.3	5.3
V_{max} (V)	14.2	16.2

Where Q_{max} is the maximum power consumed.

The chosen MPPT algorithm was the Incremental Conductance. A Buck converter was used since the expected output voltage is lower than the voltage at maximum power point of the PV module.

The results found in this paper were obtained through computer simulation of the complete interconnected system, as shown in figure 5. It can be observed two circuits interconnected representing the thermal and electrical behaviors of the system. Highlighted in red is the thermal circuit of the module and the thermoelectric cell. From this circuit, it is possible to extract the temperature in the PV cell based on the irradiance that reaches the system and the performance of the Peltier cell. The effect of this temperature is taken into account

in the electric power circuit, highlighted in blue. With these two inputs, irradiance and temperature, it is possible to calculate the power generated by the PV system, deduce the energy expenditure with the thermoelectric cell and thus evaluate the efficiency of the system as a whole.

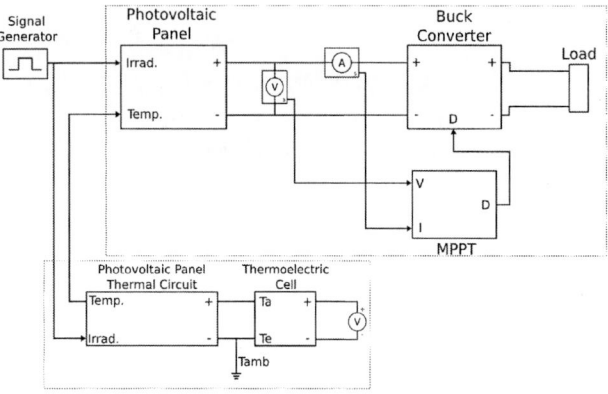

Fig. 5. Complete system schematic.

Tables III, IV and V present the values of each of the components used in the simulations.

The simulations were conducted taking into account specific values of wind speed, and arbitrary values of solar irradiance, the latter to simulate different shading conditions of the system.

III. RESULTS

Preliminary results suggest a net positive balance in the system performance under conditions where the environmental variables contribute to a high operating temperature coupled with regulated power supplied to the thermoelectric cells.

Several simulations were performed for the following cases:

- Winds of 1.0 m/s and 7.5 m/s;
- Solar irradiance values of 1000 W/m^2 and 200 W/m^2;
- Voltage supplied to thermoelectric cell: 1 V and 100 mV;
- Influence of a thermoelectric cell on 1 or 2 areas of PV cell (PVC);
- PV module with and without cooling.

These parameters were chosen to investigate how the system behaves at the boundary conditions imposed by the project. The net yield was calculated as:

$$\eta_\% = \frac{P_{g_c} - P_{te}}{P_{g_r}} \qquad (21)$$

In which $\eta_\%$ is the percentage yield, P_{g_c} is the power generated by the module under cooling, P_{te} is the power consumed by the thermoelectric cells and P_{g_r} is the power generated by the regular module, without cooling.

The results were obtained after 10 minutes of simulation time.

As can be seen in tables VI and VII, the cases with conditions for higher temperature (low wind, 1 m/s) and low power supply for the thermoelectric cell (power supply at 100 mV) show a positive net yield (greater than 100%), that is, once deduced the power consumed by the TEC, the remainder is greater than the power generated by the same module when operated without the cooling.

In order to verify this situation, the best case scenario (1.0 m/s, 1000 W/m^2) was checked against additional power supply levels to the thermoelectric cells. Three new voltage levels were supplied to the cells - 300 mV, 500 mV and 700 mV. The idea was to verify the evolution of the percentage yield with increasing power expenditure.

TABLE III
PV Cell layer components.

	Glass	ARC	Cell	EVA	Elec. Contacts	Tedlar
R_Φ (Ω)	0.0685	1.28×10^{-7}	6.24×10^{-5}	0.0587	1.73×10^{-6}	0.0205
C_Φ (F)	109.5120	0.0040	8.6373	24.4139	0.5914	3.6504

TABLE IV
Convective resistances, R_Φ.

Wind Speed	R_{frt}	R_{trs}
1.0 m/s	4.2794 Ω	15.7114 Ω
7.5 m/s	1.2012 Ω	15.7114 Ω

TABLE V
TE Cell components.

	TEC 12705
R_m (Ω)	2.3469
R_Φ (m K W^{-1})	2.2754
C_Φ (F)	5.68

Through figure 6 it can be verified the importance of controlling the power supplied to the thermoelectric cells. The power to temperature drop ratio is not linear, therefore care must be taken when running experimental scenarios.

It is worth noting that the commercial module used in the study has 60 PV cells. Therefore, in the tests in which the thermoelectric cell acted on one PV cell area (table VI), the power consumed is that of 60 thermoelectric cells. Alternatively, for the cases where the thermoelectric cell acts on two PV cell areas (table VII), the power is consumed by 30 thermoelectric cells.

IV. Conclusion

The system here presented evaluated the feasibility of a hybrid photovoltaic-thermoelectric energy generator, in which the photovoltaic module has increased power output when properly cooled by a group of thermoelectric cells. The models and equivalent electric circuits were presented together with simulation parameters.

Results showed that the use of thermoelectric cells to cool off the PV cell can increase its yield by up to 7% in the studied cases, thus proving the applicability of this technique. The simulations made evident shortcomings with the proposals, since it is not feasible across the whole spectrum of possible environmental conditions.

As next steps, experimental validation can be performed to verify the simulated results. Further research can focus on a few different aspects, such as: more refined control technique, increase in the number of simulation cases, evaluation of thermoelectric cells of

TABLE VI
Percentage Yield ($\eta_\%$) - one peltier cell to one PV cell

Cell Supply	1 V	1 V	100 mV	100 mV
Wind - Area\Irradiance	1000	200	1000	200
1.0 m/s - 1 PVC	95.30%	60.52%	107.72%	100.75%
7.5 m/s - 1 PVC	92.58%	59.72%	103.11%	100.59%

TABLE VII
Percentage Yield ($\eta_\%$) - one peltier cell to two PV cells

Cell Suply	1 V	1 V	100 mV	100 mV
Wind - Area\Irradiance	1000	200	1000	200
1.0 m/s - 2 PVC	98.45%	78.22%	105.91%	99.55%
7.5 m/s - 2 PVC	96.38%	78.04%	102.26%	100.04%

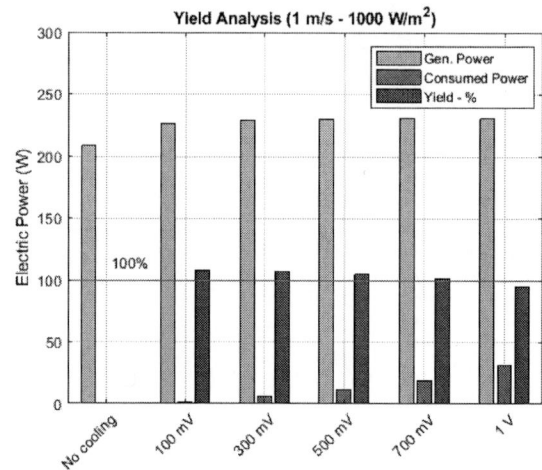

Fig. 6. Further yield investigation.

higher figure of merit, hybrid MPPT algorithms to better control the PV module energy generation while also controlling the power supply to the thermoelectric cells.

Acknowledgment

This work was suported by Coordination for the Improvement of Higher Education Personnel (CAPES) and National Council for Scientific and Technological Development (CNPq).

References

[1] Sven Rühle. Tabulated values of the shockley-queisser limit for single junction solar cells. *Solar Energy*, 130:139 – 147, 2016.

[2] Simon Phillips and Werner Warmuth. Photovoltaics report. Technical report, Fraunhofer Institute for Solar Energy Systems ISE, 2019.

[3] D. Yang and H. Yin. Energy conversion efficiency of a novel hybrid solar system for photovoltaic, thermoelectric, and heat utilization. *IEEE Transactions on Energy Conversion*, 26(2):662–670, June 2011.

[4] D. Narducci and B. Lorenzi. Challenges and perspectives in tandem thermoelectric-photovoltaic solar energy conversion. *IEEE Transactions on Nanotechnology*, 15(3):348–355, May 2016.

[5] F. Attivissimo, A. Di Nisio, A. M. L. Lanzolla, and M. Paul. Feasibility of a photovoltaic-thermoelectric generator: Performance analysis and simulation results. *IEEE Transactions on Instrumentation and Measurement*, 64(5):1158–1169, May 2015.

[6] A. D'Angola, R. Zaffina, D. Enescu, P. Di Leo, G. V. Fracastoro, and F. Spertino. Best compromise of net power gain in a cooled photovoltaic system. In *2016 51st International Universities Power Engineering Conference (UPEC)*, pages 1–6, Sep. 2016.

[7] K.A. Moharram, M.S. Abd-Elhady, H.A. Kandil, and H. El-Sherif. Enhancing the performance of photovoltaic panels by water cooling. *Ain Shams Engineering Journal*, 4(4):869 – 877, 2013.

[8] Hamidreza Najafi and Keith Woodbury. Optimization of a cooling system based on peltier effect for photovoltaic cells. *Solar Energy*, 91:152–160, 05 2013.

[9] Abdelghani Harrag and Sabir Messalti. Three, five and seven pv model parameters extraction using pso. *Energy Procedia*, 119:767 – 774, 2017. International Conference on Technologies and Materials for Renewable Energy, Environment and Sustainability, TMREES17, 21-24 April 2017, Beirut Lebanon.

978-1-7281-4181-7/19 $31.00 © 2019 IEEE

[10] Amin Yahya Khotbehsara and Ali Shahhoseini. A fast and accurate five parameters double-diode model of photovoltaic modules. *2017 Iranian Conference on Electrical Engineering (ICEE)*, pages 265–270, 2017.

[11] S. Armstrong and W.G. Hurley. A thermal model for photovoltaic panels under varying atmospheric conditions. *Applied Thermal Engineering*, 30(11):1488 – 1495, 2010.

[12] Y. Lian, Y. Kao, and H. Tsai. Photo-electro-thermal model for commercial photovoltaic modules. In *2017 International Conference on Applied System Innovation (ICASI)*, pages 158–161, May 2017.

[13] S. B. Lineykin and S. Ben-Yaakov. Pspice-compatible equivalent circuit of thermoelectric cooler. In *2005 IEEE 36th Power Electronics Specialists Conference*, pages 608–612, June 2005.

Experimental Workbench: a tools to help to teaching the techniques of drives of electric machines

1st Victor Camargo Reis
Faculty Electrical Engineering
Federal University of Juiz de Fora
Juiz de Fora-MG, Brazil
reis.victor@engenharia.ufjf.br

2nd Carlos Henrique Silva de Vasconcelos
Department of Electrical and Electronics
Federal Center for Technological Education of Minas Gerais
Leopoldina - MG, Brazil
vasconcelos@cefetmg.br

Abstract—The rapidly growing technological advancements for the electric machines driven systems have made teaching techniques are obsolete. This advancements have introduced difficulties in the teaching process. Aiming to improve the quality of the education, this paper aims to present an experimental workbench for assist in laboratory classes, facilitating the absorption of the content. Will be shown the development and validation of an experimental workbench set up to help teaching. The development of experimental workbench will be divided into three parts: the Power Circuit, the Control Circuit and Precharge Circuit. To validation the experimental workbench as a theaching tool, will be presented the theory of scalar control (V/f), comparing the results from simulation with the experimental. Showing that the experimental workbench as a good tool to improve the quality of the education.

Index Terms—Experimental Workbench, Electral Machines, Electral Drives, Scalar Control, Precharge,

I. INTRODUCTION

In the field of engineering education, the major changes resulting from the evolution of technological, provide more and more knowledge deep into the world from which we are part, providing great transformations. Education, in this way, is also directly influenced, needing to accompany the main advents of the area [2].

In the area of the electrical machines was not different, it has undergone significant changes in the course of the years. However, the main changes did not occur directly electrical machines, but rather, in the systems of drive and control. More specifically in the evolution of the semiconductor power keys. This evolution enabled the alternating current machines to operate in systems with variable speed with the same performance as the continuous current machines, in addition to allowing the implementation of various control and actuations techniques [10] [5].

Several studies have been published about, mainly, when the electric machines are operating as engine and generator [3] [4] [7] . Thus, this growing evolution introduced a difficulty in the visualization and understanding of control techniques by students, often due to its complexity.

Because of the interdisciplinary character of the subject, the control techniques as a whole are complex and have been difficult for most of the students. For facilitating the absorption of content, it is necessary to develop a workbench

that makes visualization and application of theory feasible in practice, facilitating the understanding, being flexible, allowing the study of various techniques, besides, it should enable the study of the control of various types of machines. The most of the workbenchs found on the market dont have flexible to change the control techniques. A type of workbench that presents this flexibility is the DSpace, but it has a high cost when compared to the proposed workbench.

This paper, gives an integrated treatment of the entire subject and present the experimental workbench developed for studies of the techniques of control and drives of electrical machines. Initially displays the configuration of the workbench and its possibilities of use. Only for the validation of the workbench as a tool for aid to education, will be submitted to the technique of scalar control (V/f), where will be presented the simulated and experimental.

II. DESIGN AND CONSTRUCTION

The field of electronic control of electrical machines is a very wide field, and can be applied to various machines. Among them the three-phase induction machine (TIM), synchronous machine and the cascade doubly-fed induction machine (CDFIM). For each type of machine there is a wide-range of control techniques, among them the Scalar Control (V/f), Vector Control and Direct Torque Control (DTC).

In order to guarantee a good absorption of the content by the students, the workbench must guarantee the flexibility of towards several types of machines and their control techniques. In addition, it must be mounted in such a way that allows the machines to operate as engine and generator. Another important application of this workbench is to be used in area-related searches.

The development was divided into three parts: Power Circuit, Control Circuit and Precharge Circuit.

A. Power Circuit

The power circuit must guarantee the flexibility to operate motors drives systems and generators. For this, it was assembled with two frequency inverters (SKS-75F-B6CI-40-V12, Semikron) interconnected in the back-to-back configuration, where one inverter is connected to the motor and the other

978-1-7281-4181-7/19 $31.00 © 2019 IEEE   1271

to the grid, ensuring power flow to both sides. This setting is shown in Figure 1.

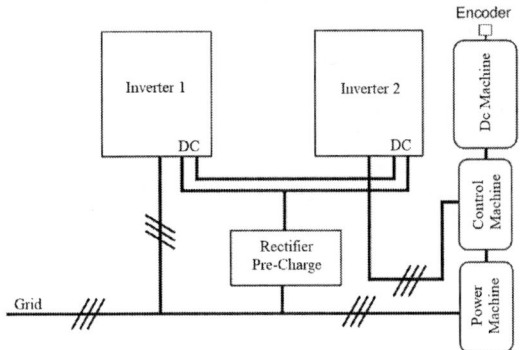

Fig. 1. Power Circuit Configuration.

Among the drive options, CDFIM has the most complex configuration [11]. This machine is composed of two cascaded wound rotor induction machines (DFIM). The Control Machine is fed through the inverter and the Power Machine directly from the grid. For the control of only one induction machine, it is only necessary to drive inverter number 2, connected to it. To control the synchronous machine, the induction motor must be swapped-out. For example, to demonstrate another application, such as the control of a wind turbine that uses a DFIM as generator, just feed the rotor circuit through the inverter. (Control Machine supply) and connect the stator to the mains (Power Machine supply). The direct current (DC) machine is used to simulate the rotation of the blades due to the passage of the wind. The machine bench is shown in Figure 2.

Fig. 2. Machine Workbench (right to left): Encoder, DC Machine and two DFIM (Control Machine and Power Machine).

In order to change from one configuration to another, without having to modify the connections in the power circuit, an electrical control was installed for the system drive. This system consists of a programmable logic module (Logo! 230Rc) and five contactors as shown in Figures 3 and 4.

The power circuit is controlled by four pushbuttons, one normally closed (NC) and three normally open (NO). For this, a command logic has been introduced in accordance with shown below:

- **Pushbutton 01 (NC):** shut down the power circuit;
- **Pushbutton 02 (NA):** closes K5, performs precharge, after 3 seconds closes K2 and opens K5, energizing only the drive inverter 2;

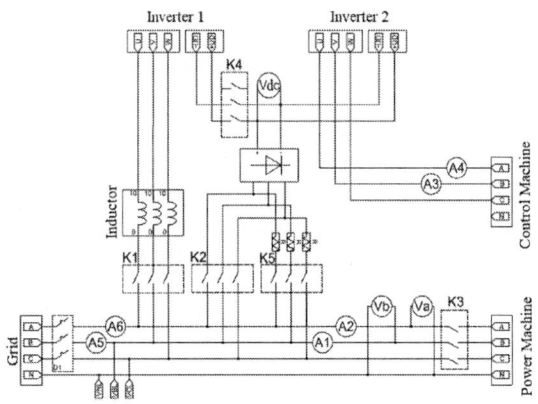

Fig. 3. Control and measurement diagram.

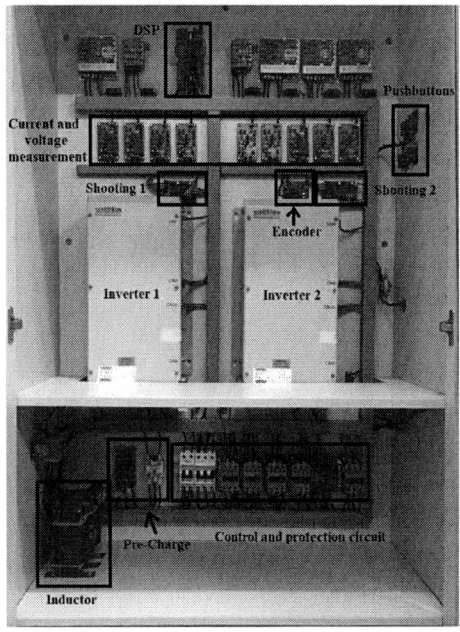

Fig. 4. The Workbench.

- **Pushbutton 05 (NA):** closes K4 and then K5, preloads, after 3 seconds closes K2 and opens K5, energizing the both inverters;
- **Pushbutton 04 (NA):** closes K1, enabling the power circuit of the inverter 1. This button only acts if the busbar is loaded.

The K3 contactor is driven by the Digital Signals Processor (DSP), it is responsible for energizing the power circuit of the Power Machine. This enables you to power up the second inverter separately, keeping the first drive de-energized, and when required, energizes both.

In the Figures 3 and 4 also is show the precharge circuit, consisting of three resistors (100, 20 W) and a three-phase rectifier (SKD30/08A1, SEMIKRON). In addition, the figures show an inductor (5/10 mH) in series with the inverter 1. This

978-1-7281-4181-7/19 $31.00 © 2019 IEEE   1272

inductor has the function of attenuate switching harmonics and reduce the current ripple [9].

B. Control Circuit

The control circuit is designed to control both inverters. The first drive has the function of controlling the power flow between the network and the DC bus. The second one controls the flow between the DC bus and the machine to be driven.

The acquisition of currents and voltages is achieved using six current and three voltage measurement plates were installed, as shown in Figures 3 and 4. The current measurement boards being distributed as follows: two for power machines (A1, A2), two for control machines (A3, A4) and two to measure grid currents (A5, A6). And the voltage measurement boards distributed in: one for the DC bus (Vdc) and two for the measure grid voltages (Va, Vb). In addition, one board was installed for speed measurement and two for conditioning the trip signals coming from the DSP for the inverters. The current and voltage measurement boards were developed in the Power Electronics Laboratory of COPPE/UFRJ.

The DSP (TMS320F28335) of Texas Instruments handles the implementation and processing of the control techniques in the two inverters simultaneously, in which it allows the development and usage of different control techniques through a code of proper instructions, making the bench flexible.

C. Precharge Circuit

A problem with system powerup is the high current during the charging of the DC bus capacitors, which may damage the rectifier components. In order for the current to remain within the tolerable range of the components is inserted in a pre-load circuit, which consists of resistors in series with the power supply and a contactor for the by-pass. The precharge circuit is shown in Figures 1 and 3 [8] [6].

III. DYNAMIC MODEL OF THE INDUCTION MACHINE AND THE CONTROL SYSTEM

In order to understand the operation of the experimental workbench, initially the dynamic model of the induction machine will be shown. The Scalar Control technique will be presented consecutively to demonstrate the flexibility of the workbench.

A. Dynamic Model of the Induction Machine

Considering the sinusoidal magnetic flux density, symmetrical three-phase windings, disregarding magnetic saturation and temperature effects, the mathematical model of IM can be considered linear. The IM model in a generic reference (superscript "g") can be represented as follows by:

$$\bar{u}_s^g = R_s \bar{i}_s^g + \frac{d\bar{\psi}_s^g}{dt} + j\omega^g \bar{\psi}_s^g \tag{1}$$

$$\bar{u}_r^g = R_r \bar{i}_r^g + \frac{d\bar{\psi}_r^g}{dt} + j(\omega^g - \omega_r)\bar{\psi}_r^g \tag{2}$$

$$\bar{\psi}_s^g = L_s \bar{i}_s^g + M \bar{i}_r^g \tag{3}$$

$$\bar{\psi}_r^g = M \bar{i}_s^g + L_r \bar{i}_r^g \tag{4}$$

And the mechanical dynamics of the machine and given by:

$$T(t) - T_L(t) = J\frac{d\omega_m}{dt} + B\omega_m \tag{5}$$

The electromagnetic conjugate ($T(t)$) of TIM and given by:

$$T(t) = -\frac{3}{2}\frac{P}{2}M(\bar{i}_s^g \times \bar{i}_r^g) \tag{6}$$

Where the subscript s e r refers to the stator and the rotor respectively, u the voltage, i the current, ψ the flow, P the number of poles, ω_m the mechanical speed, T_L the the load conjugate, R, L and M the resistance, own inductance and mutual inductance, respectively.

The mechanical speed of TIM is given by:

$$\omega_m = (1 - s)\omega_s \tag{7}$$

Where s is the machine slip and ω_s the synchronous speed.

$$\omega_s = \frac{4\pi f}{P} \tag{8}$$

B. Scalar Control (V/f)

The scalar control technique consists of maintaining a constant relationship between voltage and supply frequency (constant magnetic flux), avoiding the saturation of the magnetic flux and, consequently, affecting the conjugate available in the machine. This control technique is widely used in drives that do not require high precision, as well as being a control technique that is easy to implement and low cost.

The operating principle of the scalar control is explained in Equation 1. Disregarding the voltage drop in the resistance and reactance of dispersion of the stator we can write:

$$\bar{\psi}_s^g = \frac{\bar{u}_s^g}{\omega^g} \tag{9}$$

This consideration is only valid if the voltage drops above are negligible and while meeting the condition below:

$$\|(R_s + j\omega_s L_s)\| << \omega_s M \tag{10}$$

Rewriting Equation 6 as a function of flow and currents, we can express the electromagnetic conjugate by:

$$T(t) = \frac{3}{2}\frac{P}{2}(\psi_s \times \bar{i}_s^g) \tag{11}$$

Considering that the current depends on the load and the load will be constant, it is observed that maintaining the relationship between the voltage and constant frequency, the conjugate will also be constant. Figure 5 shows the voltage curve by frequency.

The V/f ratio is kept constant up to the nominal frequency. Above this point a saturation is introduced in the voltage component, requiring the motor to operate in the field weakening region, as shown in Figure 6. Without this saturation the motor would operate with an voltage overload , in other words, it

978-1-7281-4181-7/19 $31.00 © 2019 IEEE   1273

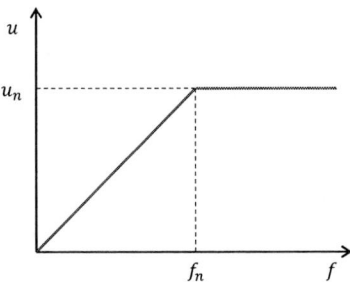

Fig. 5. Frequency x voltage curve.

would be applied a voltage above the nominal one, causing damage to the motor.

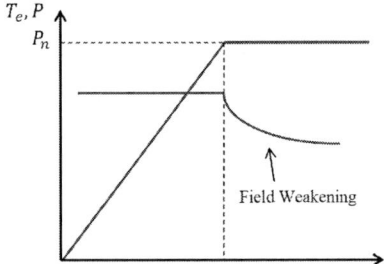

Fig. 6. Power curve and electromagnetic conjugate.

It is observed that in operation at low frequencies, mainly in low power engines, the condition of Equation 10 is not met, reducing the maximum conjugate for low frequencies. So that the motor delivers nominal conjugate at low frequency, a compensation is entered in the V/f ratio, decreasing the effects of the voltage drop in the stator resistance and reactance. This compensation is shown in Figure7. It is observed that the compensation occurred only for frequencies below half the nominal.

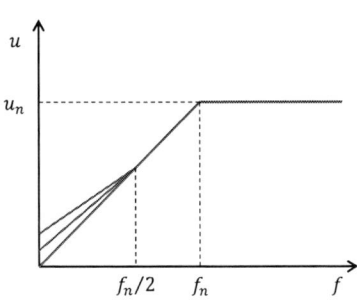

Fig. 7. Frequency x voltage curve with the compensation.

Figure 8 shows the block diagram of the open loop scalar control.

In this configuration, the mechanical speed of the machine varies according to the load due to slipping. To avoid this problem is inserted a speed control loop in series with the

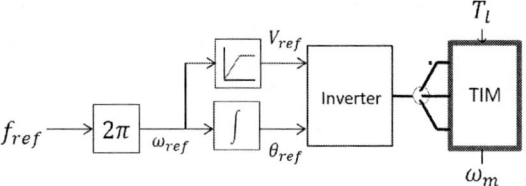

Fig. 8. Block diagram of the open loop V/f control.

scalar control. Where, by applying the error between the setpoint speed $(\omega_m^\star)$ and the measured speed (ω_m) to a PI controller, the slip value (ω_{sl}) is determined. Adding the slip to the speed the electrical control of the rotor (ω_r), you have the speed reference (ω_{ref}) for the control [1]. The speed control loop is shown in Figure 9.

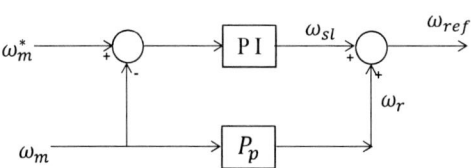

Fig. 9. Closed loop speed control.

IV. SIMULATION AND EXPERIMENTAL RESULTS

For the proof of scalar technique shown in section III-B, a system has been simulated consisting of a three-phase frequency inverter with a switching frequency of 12 kHz, a precharge circuit and a 2 HP induction machine. To prevent abrupt accelerations and decelerations on the machine, an acceleration ramp has been introduced in the speed reference. Table I shows the parameters of the induction machine used in the simulation.

TABLE I
PARAMETERS OF THE INDUCTION MACHINE.

Power	$2HP$	$1.472W$	—
Voltage and Current	$220V$	$9A$	—
Rated Seed	$178Rad/s$	$1.700RPM$	—
R_s	4.08Ω	L_s	$315.4mH$
R_s	4.87Ω	L_r	$323.5mH$
M	$305mH$	Class	B
Inertia	$0.018Kg.m^2$	Friction	$0.002N.m$

A. Precharge Results

To show the behavior and effect of the precharge circuit, was simulated the energization of the system without your application, since performing this test on the fisical bench could cause damage in the circuit. Later, with the presence of the circuit of precharge, the test was carried out experimentally, so that it would be possible to analyze the effects of preload. Figure 10 shows the currents during system energization without the precharge circuit. It is observed that there is a current peak in the order of 300 A, much higher than that supported by the rectifier.

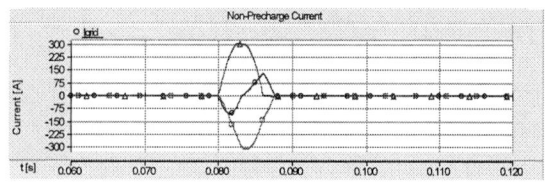

Fig. 10. Precharge (simulation): Mains current during system energisation without precharge.

To verify the operation of the precharge circuit, the precharge was initiated by pressing the pushbutton 02, explained in section II-A. Figures 11 and 12 show the voltage and current curves during the preload process, respectively.

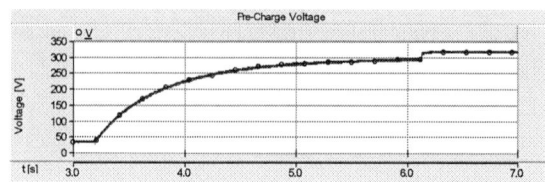

Fig. 11. Precharge (experimental): DC bus voltage.

In Figure 11, it is observed a slow load of the DC bus and after 3 seconds, with the removal of the preload resistors, the voltage reaches its steady state value.

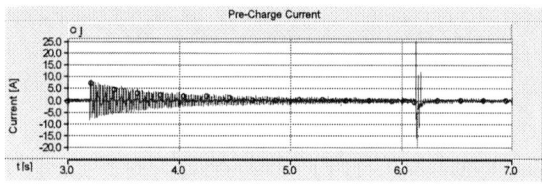

Fig. 12. Precharge (experimental): Grid current during precharge.

In Figure 12 it is observed that the current during the preload suffered a reduction of approximately ten times, thus within the tolerance range of the components.

B. Scalar Control Simulations Results

Initially, the machine was started with a frequency of 70 Hz applied to the acceleration ramp to the control presented in Figure 8. Figure 13 shows the speed curve of the machine during starting with the acceleration ramp. A linear and smooth acceleration is observed.

Figure 14 shows the voltage reference to be applied to the machine during startup. Comparing Figures 13 and 14, it is observed that the V/f ratio is constant before and after 3.5 seconds, when the frequency reaches the value of 30 Hz. However, a small difference can also be observed in the angle between of the straight lines. This is due to the insertion of the compensation in order to reduce the effects of the voltage

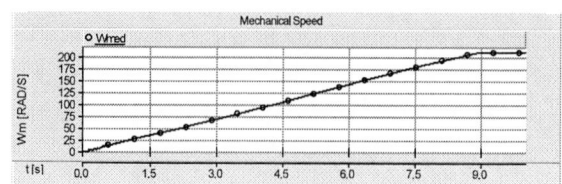

Fig. 13. V/f control (simulated): Machine start with acceleration ramp.

drop on the stator resistance, as shown in section III-B. After this point, the compensation is removed, causing the V/f ratio to return to regular slope.

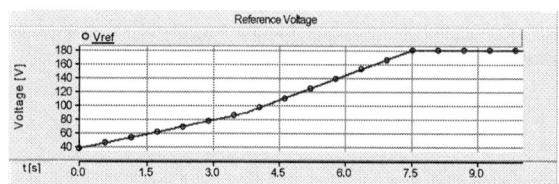

Fig. 14. V/f control (simulated): Curve of the voltage ramp applied to the machine.

It can be pointed out in Figure 14 that from 7.5 seconds, when the frequency reaches the value of 60 Hz, a saturation was introduced in the value of the voltage to be applied to the machine, placing it to operate in the region of field weakening. While in Figure 13 the speed continues up to rise to the reference value.

To demonstrate the behavior of the control in the face of speed variations, the machine was initially started with a frequency of 40 Hz and during the interval of 10 to 15 seconds the reference was set at 60 Hz. Figure 15 shows the behavior of the machine in relation to the speed variation. There is an error between the reference speed and the measurement, this occurs due to the slipping of the machine.

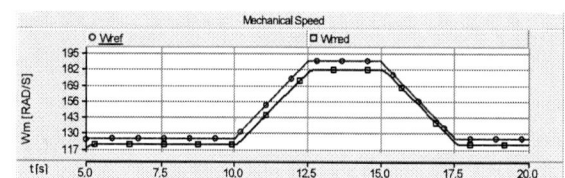

Fig. 15. V/f control (simulated): Variation of the mechanical speed in the open loop.

To compensate for this speed error in regime, the speed control loop presented in Figure 9 was added to the V/f control. In Figure 16, a reference speed of 100 rad/s was applied and during the interval of 10 to 15 seconds the reference was adjusted to 130 rad/s. It is noticed that the speed mesh eliminated the speed error.

C. Experimental Results of Scalar Control

To aid absorption and understanding of the technique of scalar control, the system presented in section IV-B was

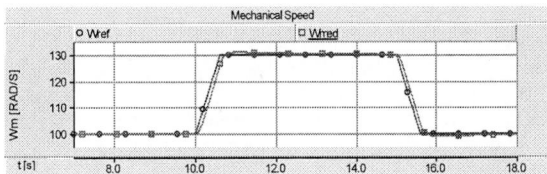

Fig. 16. V/f control (simulated): Variation of the mechanical speed in the closed loop.

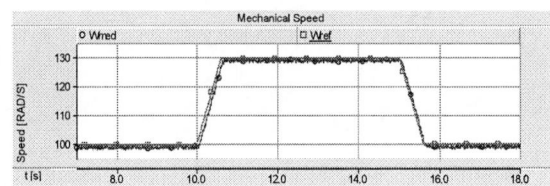

Fig. 20. V/f control (experimental): Variation of the mechanical speed in the closed loop.

submitted under the same conditions on the experimental bench.

Figures 13 and 14 were experimentally reproduced and are shown in Figures 17 and 18, respectively. It is observed that the results of the simulations reproduced the experimental results.

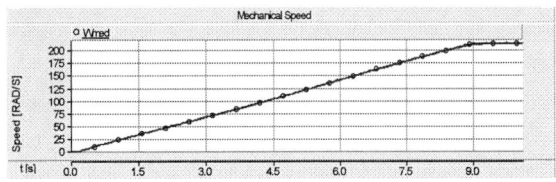

Fig. 17. V/f control (experimental): Machine start with acceleration ramp.

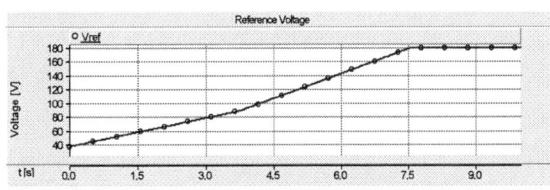

Fig. 18. V/f control (experimental): Curve of the voltage ramp applied to the machine.

Figure 19 shows the dynamics of the system in relation to speed variation. It is observed that the dynamics of the experimental system was similar to the simulated one. The same occurred with Figure 20.

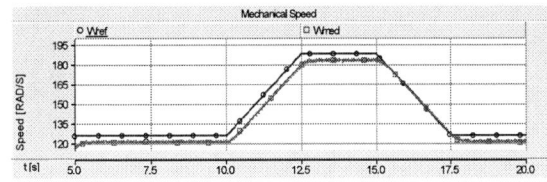

Fig. 19. V/f control (experimental): Variation of the mechanical speed in the open loop.

V. CONCLUSIONS

This work has shown the need for new means to better absorb content arising from technological developments. The use of simulation tools in conjunction with experimental attestation provides a better understanding of the content.

It can also be observed that the results obtained experimentally prove the theory studied and presented, as well were

equivalent to those acquired in simulations, showing that the workbench is a good tool for the improvement of teaching.

It is observed that the acquisition of data between the workbench and the computer is an important tool for the absorption of content.

Therefore, the inclusion of the experimental part in the learning process helps to absorb the theory of control and drive techniques of electrical machines.

ACKNOWLEDGMENT

The authors appreciate the support given by Federal Center for Technological Education of Minas Gerais for the development of this work.

REFERENCES

[1] B. K. Bose, "Modern Power Electronics", Prentice Hall, 2002.
[2] J. E. Froyd, P. C. Wankat and K. A. Smith, "Five major shifts in 100 years of engineering education", Proceedings of the IEEE 100 (SPL CONTENT): pp. 1344-1360, 2012.
[3] B. Hopfensperger, D. Atkinson and R. Lakin, "Stator-flux-oriented control of a doubly-fed induction machine with and without position encoder", Electric Power Applications, IEE Proceedings on 147(4): pp. 241-250, 2000.
[4] J. Hu, J. Zhu and D. Dorrell, "A new control method of cascaded brushless doubly fed induction generators using direct power control", Energy Conversion, IEEE Transactions on 29(3): pp. 771-779, 2014.
[5] W. Leonhard, "Control of Electrical Drives", Springer-verlag, 2011.
[6] N. Mohan, T. M. Undeland and W. P. Robbins, "Power Electronics: Converters, Applications, and Design", 3 edn, Wiley, 2003.
[7] J. M. Peña and E. V. Díaz, "Implementation of v/f scalar control for speed regulation of a three-phase induction motor", IEEE ANDESCON, pp. 1-4, 2016.
[8] H. Sepahvand, M. Khazraei, K. A. Corzine and M. Ferdowsi, "Start-up procedure and switching loss reduction for a single-phase flying capacitor active rectifier", IEEE Transactions on Industrial Electronics on 60(9): pp. 3699-3710., 2013
[9] L. T. F. Soares, C. M. Pimenta, S. I. S. Junior and S. R. Silva, "Modelagem e controle de um conversor back-to-back para aplicação em geração de energia eólica", IV Brazilian Symposium of Electronic Systems (SBSE), 2012.
[10] P. Vas, "Sensorless Vector and Direct Torque Control", Oxford University Press, 1998.
[11] C. H. S. Vasconcelos, A. C. Ferreira and R. M. Stephan, "Stator flux orientation control of the cascaded doubly fed induction machine", 2015 IEEE 24th International Symposium on Industrial Electronics (ISIE), pp. 548-553, 2015.

Predicting the Life Time of Power Semiconductor Modules

Clovis N. L. Gajo
Product Management
SEMIKRON Semicondutores Ltda.
São Paulo, Brazil
clovis.gajo@semikron.com

Arendt Wintrich
Application Management
SEMIKRON International GmbH
Nuremberg, Germany
arendt.wintrich@semikron.com

Paul Drexhage
Application Management
SEMIKRON, Inc.
Hudson, USA
paul.drexhage@semikron.com

Abstract—Methods to estimate the lifetime of Power Semiconductor Modules, the typical aging factors, how they act, and some of the new improvement technologies that lead to higher lifetime.

Keywords—Semiconductor, wear, aging, cycling, lifetime, Power, Module

I. INTRODUCTION

Power semiconductors are a key component on current and future applications; they are present on Wind Energy, Solar Energy, Energy Storage Systems, Motor Drives, Electric Vehicles, and even on small home appliances like washing machines and air conditioning. It is without surprise that their lifetime and reliability are a major factor for all those applications. Here the main causes of power semiconductor modules aging are indicated, with emphasis on mechanical stress, and ways to predict the end of the lifetime are summarized and explained.

II. PRESENT POWER MODULE STRUCTURE

Standard module technology have chips soldered on an insulant DBC (Direct Bonded Copper) which is also soldered to a copper baseplate, with interconnections between chips, DBC and terminals by wire bonds – aluminum wire ultrasonically welded to the metal-coated surface ("metallization") of the semiconductor chip.

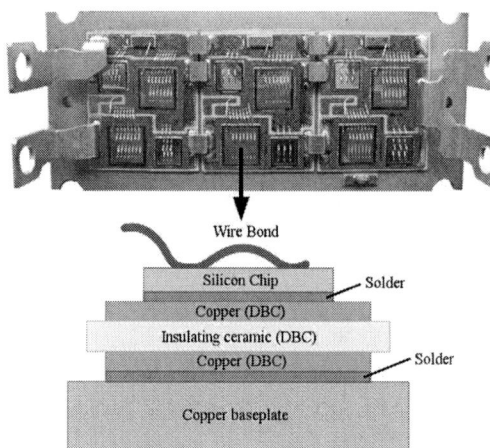

Fig. 1. Typical module structure.

The materials show on Fig. 1 are the most typical on standard modules, although not mandatory.

- Aluminum wires are wedge-bonded by applying force and ultrasonic power.

- For higher current levels, multiple wires are paralleled.

- Usually each wire is bonded on more than one point to the chip.

The assembly with solder layers and interconnections by wire bonding is easy to produce, flexible, easy to automate and cost effective. It has some disadvantages, which limits its lifetime on applications subject to high power/thermal cycling, mainly due to different thermal expansion.

Each material expands as temperature increases and contracts when it decreases, although at different rates because:

- Each material has its own coefficient of thermal expansion (CTE).

- There are temperature gradients across each part, implying on induced mechanical stress even with materials with identical CTE.

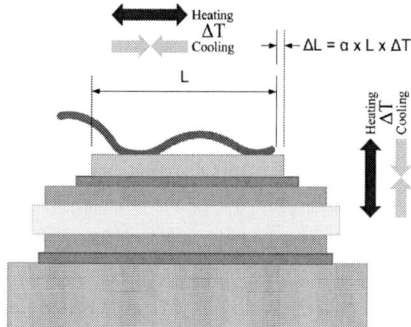

Fig. 2. Thermal expansion

Expansion occurs in all directions, and is usually more significant on the larger dimension of any part or in the dimension were the part is connected to other parts/materials.

III. WEAR-OUT MECHANISMS

Aging/wear-out occurs even on modules properly designed, produced and operated within their datasheet limits, mostly due to the factors below:

- Wire bond fatigue.

- Solder Layers degradation.

- Metallization Reconstruction.

- Corrosion of both chip metallization and bond wires by humidity, electric field and temperature.

Failures can occur at any time of the module lifetime. Aging/wear-out are mostly responsible for the increase of failure probability at module End-of-Life, as show on the bathtub curve:

978-1-7281-4181-7/19 $31.00 © 2019 IEEE

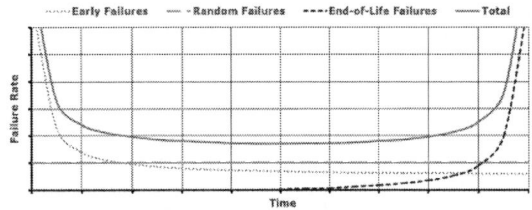

Fig. 3. Bathtub curve – Failure rate vs. Time

A. Wire bond fatigue

Heel crack - Partial or complete interruption of the wire bond close to the bent point were it connects to the chip (near the wire bond "foot"). This can lead to high conduction resistance or even a complete interruption of the current flow.

Fig. 4. Heel crack - SEM (Scanning Electron Microscope) image showing a huge crack on the closest wire bond. The photo also shows the area were another wire bond was connected before lifting-off. [1]

Lift off - Complete break of the ultrasonic weld bonding that connect the wire bond "foot" to the surface metallization.

- Cracks start to occur in the welded joint owing to different CTE also due the force applied by the wire expansion and spring force.

- Wire spring force pulls the connection away from surface, effectively lifting it off the chip.

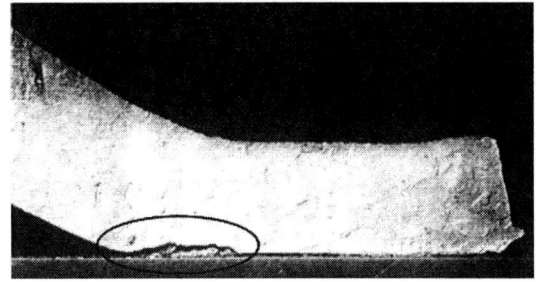

Fig. 5. Lift off – bonding between wire bond and chip metallization is cracked and will lead to a complete disconnection of the wire. [3]

The degraded connection due to wire bond cracks or disconnection from the chip leads to higher resistance, higher power losses, higher temperatures, higher expansion/contraction cycles, higher mechanical stress which will speed-up the connection degradation, closing a vicious circle that will cause chip destruction by overheating or by electrical arc (when last wire bond lift off).

B. Solder Layer Degradation

Brittleness and cracking in the solder connecting chips to DBC substrate and DBC substrate to baseplate.

Main reason are the different CTEs of chip and solder and also between copper and solder.

Cracking may occur within the alloys (i.e. Cu5Sn6) formed between the solder and the copper.

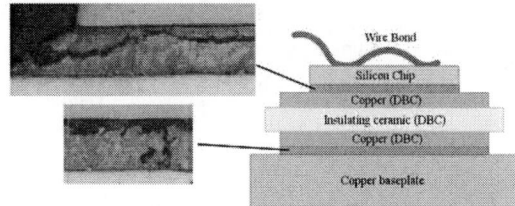

Fig. 6. Cracks on chip to DBC and DBC to baseplate solders.

Cracked solders layers generate voids inside the solder or its alloys, reducing thermal and electrical conduction.

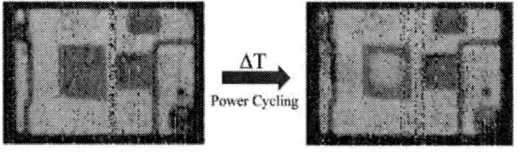

Fig. 7. Voids generated in chip solder after power cycling, similar to those that appear on modules close to end-of-life. They are show as lighter areas on the right side SAM (Scanning Acoustic Microscope) image above.

A cracked solder layer increases both electrical and thermal resistance of the chips, leading to higher resistance, higher power losses, higher temperatures, higher expansion/contraction cycles, higher mechanical stress which will speed-up the degradation, closing a vicious circle that will cause chip destruction by overheating.

C. Metallization Reconstruction

"Reconstruction" is the extrusion of the aluminum grains within the surface metallization on the chip that appears as dark areas on chip surface. The average temperature and the differing CTEs of metallization and silicon are factors that influence it.

Fig. 8. Darker areas on chips due to metallization reconstruction on IGBT (left) and on diode (right). [2]

The reconstruction increases sheet resistance in the chip and manifests itself as a higher V_{CE} for a given current. It reduces the effective number of IGBT cells on the chip, thereby reducing the effective conduction area and leading to hot spots and potential melting.

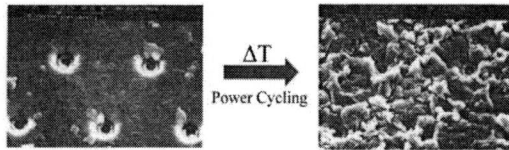

Fig. 9. Metallization reconstruction - SEM image showing the surface of an IGBT before and after millions of power cycles. [4]

The higher resistance leads to a reduction of the conduction area, higher power losses, higher temperatures, higher expansion/contraction cycles, higher mechanical stress which will speed-up the degradation, closing a vicious circle that will cause chip destruction by overheating.

IV. THE LESIT STUDY

Modern lifetime estimate curves are based on an intensive study funded by the Swiss government at the Swiss Federal Institute of Technology (ETH).

LESIT (Leistungselektronik, Systemtechnik, Informationstechnologie) study (1991-1998)

- Survey of large number of power semiconductor modules from European and Japanese manufacturers.

- Subjected modules to power cycling until failure occurred.

- Established the effect of chip junction temperature (T_j) variation on lifetime.

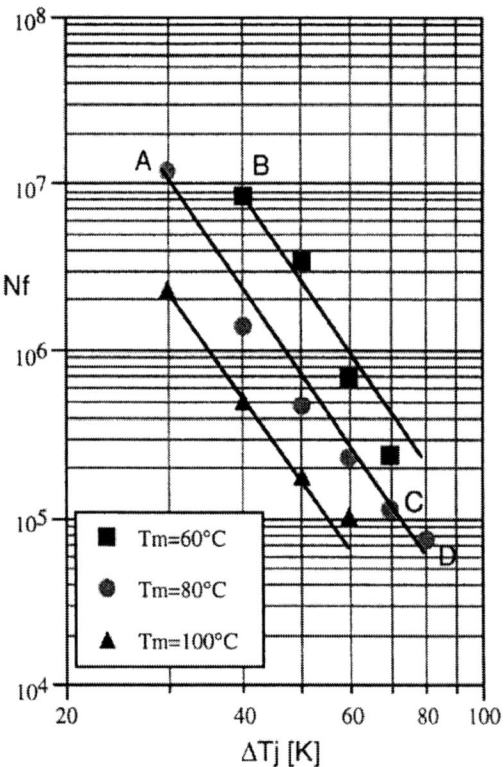

Fig. 10. Number of cycles to failure N_f as function of ΔT_j with T_{medium} as parameter. [3]

$$N_f = A \times \Delta T_j{}^\alpha \times e^{\left(\frac{Q}{R \times T_m}\right)}$$

(1)
[3]

N_f = number of cycles to failure

- "cycle": change in temperature from $T_{j,min}$ to $T_{j,max}$ to $T_{j,min}$

- "failure": 5% increase of the V_{ce}

$\Delta T_j = T_{j,max} - T_{j,min}$ (K)

$T_m = T_{j,min} + \Delta T_j / 2$ (K)

A = 640 (Technology scaling factor)

α = -5

Q = 7.8 x 10^4 (J/mol)

R = 8.314 (J/mol*K) $\rightarrow$ Gas constant

In the LESIT formula above, "A" is the Technology Scaling Factor, "$\Delta T_j{}^\alpha$" came from Coffin-Manson law and the remaining from Arrhenius equation.

V. PACKAGING IMPROVEMENTS

Improvements since the LESIT study (1997-2009)

- Production technology (soldering, wire bonding)

- New chip generations (chip thickness)

- Bond wire geometry optimization

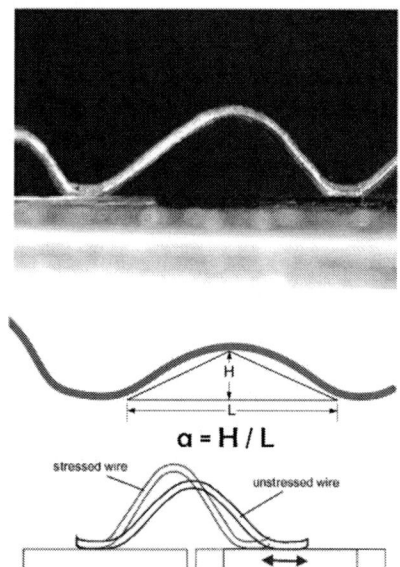

Fig. 11. Bond Wire Geometry Optimization

Bond wire geometry optimization: the shape of bond wire affects fatigue characteristics during thermal cycling.

- Bond wire aspect ratio, $\alpha = H / L$

- Increasing aspect ratio increases power cycling capability up to a point.

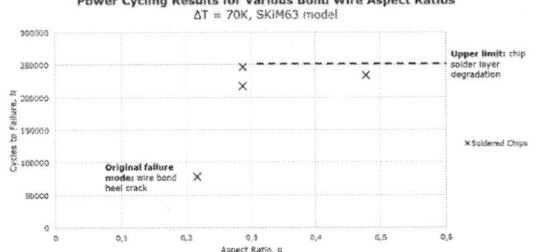

Fig. 12. Power Cycling results vs. Bond Wire aspect ratio "a".

With the optimized values (higher "a" aspect ratio), the failure mode is no longer on wire bonds, but on solder layer.

VI. MODERN POWER CYCLING CURVES

Technology improvements have increased the lifetime since the publication of the LESIT study. The improvements occurred in materials, construction, and manufacturing methods.

Improved lifetime curves were developed by scaling the original LESIT curves.

- Two curves published in 2009: "Advanced" and "Standard".

- Power cycling were performed only with high ΔT_j (70K, 110K) to reduce test time.

- The curves are the same for baseplate and baseplate-less technologies.

  - Power cycle test time on the order of seconds mainly triggers the failure mechanism "bond wire failure" and/or "chip solder fatigue"

  - Baseplate solder fatigue results from longer power cycles (minutes) or passive temperature cycles (minutes)

The 2009 revision of the Lifetime equation considered additional factors that influences the number of cycles to failure (N_f), with $t_{pulse} > 2$ s

$$N_f = \underbrace{A}_{\substack{\text{Technology} \\ \text{scaling factor}}} \times \underbrace{\Delta T_j^{-5.039}}_{\substack{\text{Coffin-Manson} \\ \text{law}}} \times \underbrace{e^{\left(\frac{E_a}{k_B \times T_{jm}}\right)}}_{\substack{\text{Arrhenius} \\ \text{equation}}} \qquad (2)$$

A = Technology factor

- Standard → A = 648000

- Advanced → A = 3846000

ΔT_j = Temperature swing = $T_{j,max} - T_{j,min}$ (K)

E_a = Activation Energy = 9.891×10^{-20} (J)

k_B = Boltzman constant = 1.38×10^{-23} (J/K)

T_{jm} = Medium junction temperature = $T_{j,min} + \Delta T_j / 2$ (K)

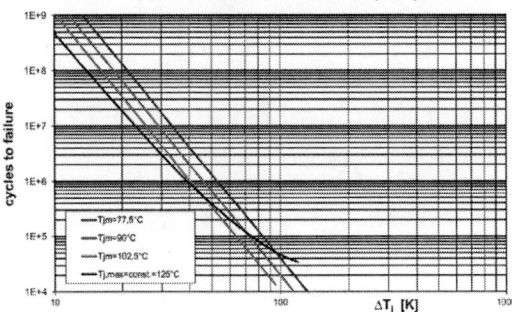

Fig. 13. Standard IGBT modules - N_f (power cycles to failure) as a function of the temperature cycling amplitude ΔT_j and the mean temperature T_{jm}. [5]

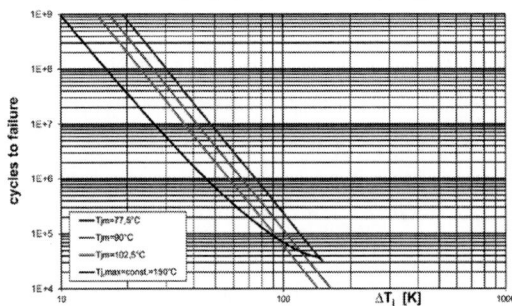

Fig. 14. N_f (power cycles to failure) as a function of the temperature cycling amplitude ΔT_j and the mean temperature T_{jm} for Advanced IGBT modules with improved bond wires or improved die attach (e.g. by sintering). [5]

VII. THE EVOLUTION OF THE LIFETIME MODEL

The original LESIT lifetime equation from 1997 (1) considers only the temperature swing (ΔT_j) and medium temperature (T_m). It does not considers factors like the impact of pulse duration, bond wire aspect ratio and chip thickness.

New lifetime prediction models considering additional influencing factors were proposed, similar to the one that was proposed during CIPS 2008 conference [6] for IGBT modules:

$$N_f = \underbrace{A}_{\substack{\text{Technology} \\ \text{scaling factor}}} \cdot \underbrace{\Delta Tj^{-4.416}}_{\substack{\text{Coffin-} \\ \text{Manson law}}} \cdot \underbrace{e^{\left(\frac{1285}{T_{j,min}+273}\right)}}_{\substack{\text{Arrhenius} \\ \text{equation}}} \cdot \underbrace{t_{on}^{-0.463}}_{\substack{\text{Pulse} \\ \text{duration}}} \cdot \underbrace{I_B^{-0.716}}_{\substack{\text{Current} \\ \text{per} \\ \text{bond}}} \cdot \underbrace{V_C^{-0.761}}_{\substack{\text{Voltage} \\ \text{class}}} \cdot \underbrace{D^{-0.5}}_{\substack{\text{Bond} \\ \text{wire} \\ \text{diameter}}} \qquad (3)$$

A = Technology factor

- Standard → A = 2.03×10^{14}

- Advanced (IGBT4) → A = 9.34×10^{14}

ΔT_j = Temperature swing = $45\ldots150$ (K)

$T_{j,min}$ = Min. chip Temperature = $20\ldots120$ (°C)

T_{on} = Pulse duration = $1\ldots15$ (s)

I_B = Current per bond "foot" = $3\ldots23$ (A)

V_C = Voltage class (rel. to Chip thick.) = $6\ldots33$ (V/100)

D = Bond wire diameter = $75\ldots100$ (µm)

Another model was proposed on PCIM 2013 conference that takes in account the user of sintered instead of solder contacts and optimized Al wire bonds (e.g.: SKiM63) [7]:

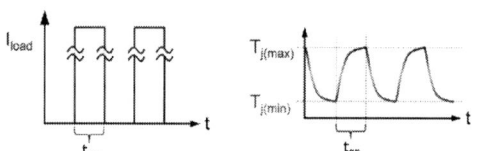

$$n_f = A \cdot \Delta T_j^{\alpha} \cdot ar^{\beta_1 \cdot \Delta T_j + \beta_0} \cdot \left(\frac{C + t_{on}^{\gamma}}{C + 1} \right) \cdot \exp\left(\frac{E_a}{k_B \cdot T_{jm}} \right) \cdot f \quad (4)$$

A = 3.4368 x 10^{14}

α = -4.923   (64 K ≤ ΔT_j ≤ 113 K)

ar = Wire bond aspect ratio

β_1 = -9.012 x 10^{-3} (0.19 ≤ ar ≤ 0.42)

β_0 = 1.942   (0.19 ≤ ar ≤ 0.42)

C = 1.434   (0.07 s ≤ t_{on} ≤ 63 s)

γ = -1.208   (0.07 s ≤ t_{on} ≤ 63 s)

E_a = 0.06606   (eV, 32.5°C ≤ T_{jm} ≤ 122°C)

k_B = 1.38 x 10^{-23} (J/K, Boltzman constant)

f = 0.6204   (for 1200 V diodes)

This model has the following differences:

- Considers the impact of bond wire geometry, cycle duration and chip thickness.

- The impact of medium temperature was reduced.

- It was validated through tests on 1200V IGBTs and diodes in single sided sintered modules with optimized bond wire aspect ratio (i.e. SKiM63 and SKiM93).

LESIT and other older studies do not fully acknowledge the pulse duration (heating time and t_{on}) influence and their results were a mix of various pulse times on the order of seconds. Both "Standard" and "Advanced" curves are valid for pulses with t_{on} of at least 2 s.

The CIPS 2008 and SKiM63 models do include a factor for accounting the t_{on} influence on lifetime.

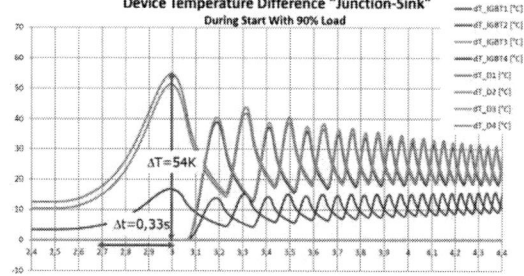

Fig. 15. Pulse duration t_{on} and T_j response.

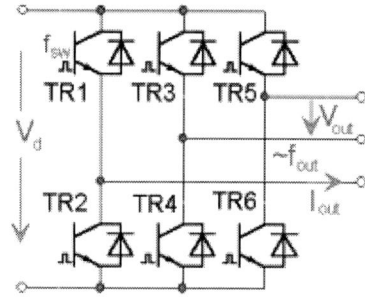

Fig. 16. Resulting ΔT_j-s for a 0.33 s power pulse.

Shorter pulses lead to higher number of cycles to fail as temperature gradients do not have time to fully develop across the chip. With short pulse times like 0.33 s the temperature profile and stress does not develop fully in the critical interconnections between bond wires and chip or chip and DBC solder.

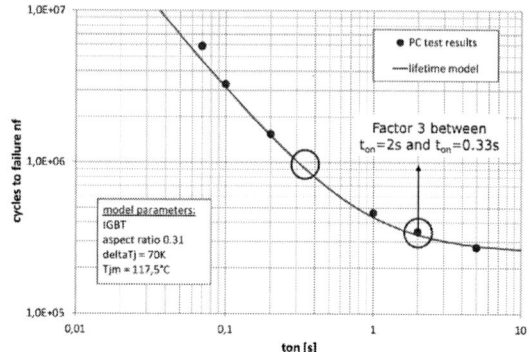

Fig. 17. N_f dependency to pulse time.

Another point correlated to the above is that small temperature swings (e.g. ΔT < 20 K) can be neglected, either when the low temperature swing is due to too short pulses or to other causes (e.g. low power).

As a guidance for design it can be said that for most applications, when temperature cycles have small amplitude, their effect on product lifetime is not significant.

VIII. ESTIMATING THE LIFETIME

As the models and their graphs can be used to estimate the lifetime of the modules, the main concern is to properly choose the best available model and input the correct parameters of the application on it.

Data from the module came directly from datasheets or by contacting the manufacturer, as some data, like bond wire specifications, chip disposition and wire bond quantity per chip usually is not included by default.

The temperatures are estimated through calculus, simulations, or measured on real parts. Some manufacturers have simulation programs that can be used to properly estimate $T_{j,min}$, ΔT_j and T_m.

For a quick and simple example, the thermal calculations for a 3Ø IGBT motor drive are performed through a manufacturer simulator:

Fig. 18. Power circuit for a 71 kW, 3Ø motor drive

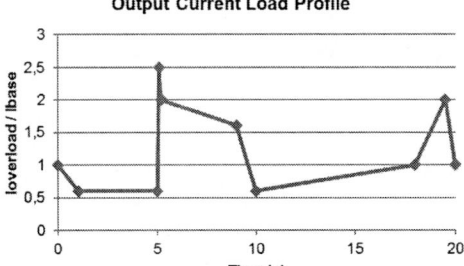

Fig. 19. Data used in the simulation. Image at top is the load cycle.

V_{in}	750 V		
V_{out}	480 V	f_{out}	60 Hz
I_{out}	100 A	Cos_{phi}	0.85
P_{out}	71.00 kW	f_{sw}	5.0 kHz
Overload Factor	2.00	Overload Duration	5 s
$f_{min\ out}$	60.00 Hz	$V_{min\ out}$	480 V

Simulated device: SEMiX453GB12E4s modules (IGBT4, soldered, with copper baseplate, 3 units in the circuit).

Resulting graph from SemiSel simulation:

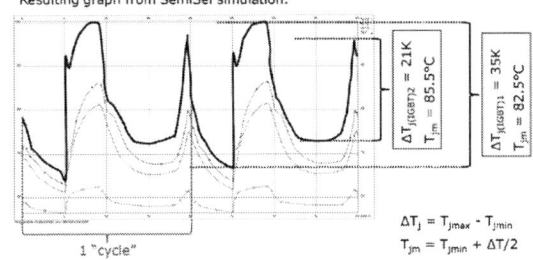

Fig. 20. Simulation results.

The thicker curve is from IGBT temperature excursions and will be analyzed. The diode (thinner line) analysis is similar, although as the $\Delta T_{j(diode)}$ is much lower it can be ignored.

Using (2) for Advanced IGBT modules (2009 revision):

$\Delta T_{j(IGBT)1} = 35$ K, $T_{jm} = 82.5°C \rightarrow 3.60 \times 10^7$ cycles

$\Delta T_{j(IGBT)2} = 21$ K, $T_{jm} = 85.5°C \rightarrow 3.99 \times 10^8$ cycles

Effective number of cycles:

$1 / N_{effective} = 1 / 3.60 \times 10^7 + 1 / 3.99 \times 10^8$

$N_{effective} = 3.3 \times 10^7$ cycles.

As load cycle is of 20 s (see Fig. 19.) and considering continuous operation:

Operation years = $3.3 \times 10^7 \times 20 / 60 / 60 / 24 / 365.25$

Operation years = 20,9 years.

IX. CONCLUSION

It must be taken into account that other failure causes must be considered for Power Semiconductors besides mechanical stress by thermal cycling. In any case, it is the major aging agent for many applications, like motor drivers for elevators, electric vehicles and electric tools, and is always a good starting point for any lifetime estimation, switching or complementing with others when estimation show high values, indicating that other wear causes can be so or even more significant.

REFERENCES

[1] Amro & Lutz. "Power Cycling with High Temperature Swing of Discrete Components based on Different Technologies." PESC 2004. pp. 2593-2598.

[2] Ciappa, M. "Selected failure mechanisms of modern power modules." Microelectronics Reliability 42, 2002. pp. 653-667.

[3] Held et al. "Fast Power Cycling Test for IGBT Modules in Traction Applications." PEDS, 1997. pp. 425-430.

[4] Hamidi et al. "Increased Lifetime of Wire Bond Connections for IGBT Power Modules." IEEE Applied Power Electronic Conference and Exhibition (APEC), Anaheim, 2001.

[5] A. Wintrich, U. Nicolai, W. Tursky, T. Reimann, "Application Manual Power Semiconductors", 2nd edition, ISLE Verlag 2015, ISBN 978-3-938843-83-3

[6] Bayerer, R.; Herrmann, T.; Licht, T.; Lutz, J.; Feller, M.: "Model for Power Cycling Lifetime of IGBT Modules – Various Factors Influencing Lifetime", CIPS 2008, Conference Proceedings.

[7] Scheuermann & Schmidt. "A New Lifetime Model for Advanced Power Modules with Sintered Chips and Optimized Al Wire Bonds." PCIM 2013. pp. 810-817.

Workbench for Monitoring and Operation of Electric Machines

1st Guilherme Afonso Pillon de C. A. Pessoa
Department of Electrical Engineering (DEE)
Federal University of Rio Grande do Norte (UFRN)
Natal-RN, Brazil
guilhermepillon@hotmail.com

2nd Evandro Ailson de Freitas Nunes
Federal Institute of Bahia (IFBA)
Federal University of Rio Grande do Norte (UFRN)
Natal-RN, Brazil
evandro.nunes@ifba.edu.br

3rd Werbet Luiz Almeida da Silva
Department of Computer Engineering and Automation (DCA)
Federal University of Rio Grande do Norte (UFRN)
Natal-RN, Brazil
werbethluizz@hotmail.com

4th Ricardo Ferreira Pinheiro
Department of Computer Engineering and Automation (DCA)
Federal University of Rio Grande do Norte (UFRN)
Natal-RN, Brazil
ricpinh@gmail.com

5th Andres Ortiz Salazar
Department of Computer Engineering and Automation (DCA)
Federal University of Rio Grande do Norte (UFRN)
Natal-RN, Brazil
andres@dca.ufrn.br

Abstract—In this paper, a generic platform to improve the test, evaluation and research regarding to electric machine is developed. The proposed system can measure experimental data by means of electric sensors and acquire these data by means a data acquisition device connected to a microcomputer which can be employed to record, compare, estimate and chart the result in real-time. Also, the system allows closed-loop control strategies implementation to a huge variety of DC and AC electric machines. Experimental results obtained from case studies are presented to validate the platform applicability and robustness.

Index Terms—electric machines, supervisory systems, real time evaluating.

I. INTRODUCTION

Nowadays, electric machines are present in almost every residential and industrial appliances. Regarding to industrial sector, it is estimated that these devices are responsible for 65% of the total energy consumption [1]. In the last years, the use of electric machines experimented an accentuated increase due to their applicability in promising trends such as hybrid electric vehicles [2], [3] and [4], distributed energy resources (e.g., wind energy conversion systems) [5] and flywheel energy storage systems [6]. All these applications have the electric machines testing as standard practice since the machine fabrication, during maintenance and even in electrical engineering teaching [7].

Traditionally, electric machine testing measurements are performed by commercial instruments (e.g, oscilloscopes and multimeters), writing down the data for further calculations. According to [8], in case of some large projects, recording

measurements generate a big volume of data and labor, which necessarily require computer use. In order to overcome this testing limitation, several studies have proposed testing systems aiming accuracy of data acquisition. In [9], the authors propose an educational platform to perform transformers and electrical machine testing. The authors performs open-circuit and short-circuit experiments with transformers and evaluated synchronous motor V type characteristic curve. Even though the results are interesting, the main drawback of the work is the platform limitation to the referred experiments.

Over the years, electric machine testing aimed to avoid using oscilloscope in laboratory experiments and to provide friendly user interface. In [10], authors propose an automated system for frequency analysis in LabVIEW-based programming language called virtual instrument, which performs frequency response tests on different electric machines. On the other hand, author in [11] proposes a system that allows a system platform to perform evaluation of transformers, DC machines, three-phase synchronous machines, single phase induction motors and three-phase induction motors. The main feature of this work is the recording and data plotting in real-time. However, there is no possibility to perform any control strategy by adopting the previously mentioned systems.

Thereby, the main contribution of this paper is to provide a generic platform to perform not only traditional testing evaluation for general applications, but also to allow control strategy implementation aiming research purposes. Two case studies are studied to validate the approach. The first one consists of a traditional electric machine testing and the second

978-1-7281-4181-7/19 $31.00 © 2019 IEEE

one consists on the use of the proposed platform to give support to a new wind power conversion system proposed by [12].

II. MEASURING, ACQUISITION AND SUPERVISORY SYSTEM

The built platform can be divided into three specific stages aiming the complete system variables evaluation. The first one consists in measuring voltage and current variables in DC and AC systems. The second one is responsible for acquiring mechanical speed variable. These stages are performed by means of transducers. The last one employs acquisition data and supervisory system to perform estimations, record waveform signals and their features and chart real-time results. The entire block diagram is depicted at Fig. 1 and the stages are explained as follows.

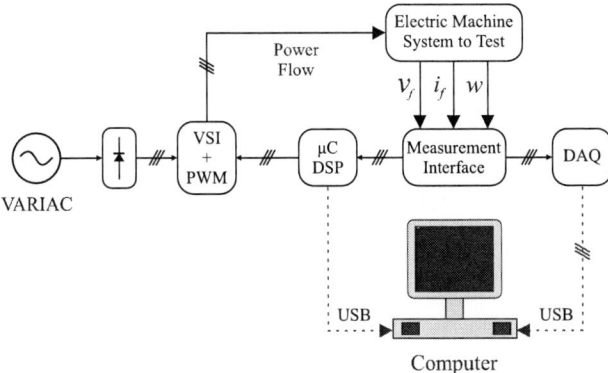

Figure 1. Architecture of the proposed monitoring system.

A. Voltage and Current Transducers

Electric machine can be defined as direct current or alternate current machines (DC or AC, respectively). The proposed platform has electronic devices to deal with individual machine features. This stage is responsible for converting power electrical variables into suitable low amplitude signals to be quantized and sampled. The conditioning signals system schematic diagram is depicted in Fig. 2. The electrical current conversion is performed by means of LEM LA25-NP transducers while voltage conversion employs a LEM LV20-P. These transducers works with good relative error rates below 1%. Additional reducer transformer is needed to provided an output voltage signal determined by equation 1:

$$V_{out} = 2.5 V_{in}\left(\frac{30 + POT_G}{250 + R_{eq}}\right) + \frac{15POT_O}{POT_O + 10K}, \quad (1)$$

wherein POT_G and POT_O represent the gain and offset adjustment potentiometers, respectively, whose resistance is in Ω. V_{out} and V_{in} represent the small signal output voltage and the measured voltage of interest, respectively. R_{eq} represents the equivalent resistance of current limiting resistors association in the input. The current transducer circuit is designed similarly to the circuit shown in Fig. 2 with a slight difference. In

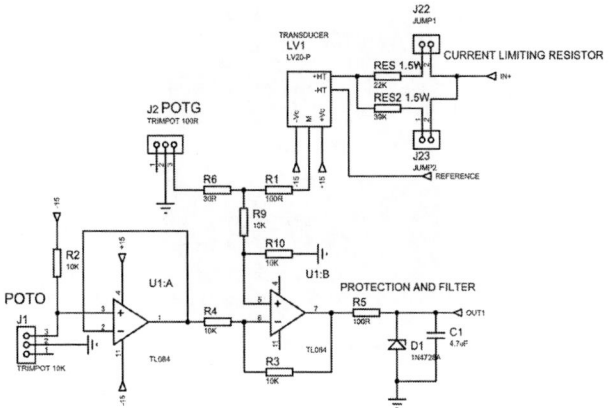

Figure 2. Signal conditioning system schematic diagram.

measuring electric current, there is no need to employ current limiter resistors on the transducer input. This feature leads to another output voltage signal, described by equation 2:

$$V_{out} = 10^{-3}I_p(30 + POT_G) + \frac{15POT_O}{POT_O + 10K}. \quad (2)$$

wherein I_p is the current of interest. The final result of the sensoring and acquisition interface is shown in Fig. 3. The signal conditioning board is used to convert 0-3 V levels generated by the DSP to the IGBT driver rated voltage range that is 15 up to -7.5 V. In addition to generating the complement of the signal, to control the complementary power switch.

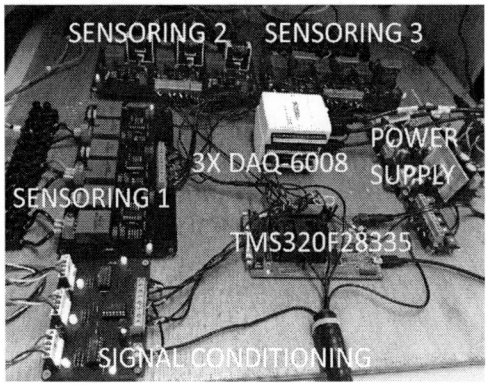

Figure 3. Printed Circuit Board of Measuring and Conditioning Signals.

B. Speed Transducers

Optical encoders with sixty orifices are employed for angular speed monitoring. This device was designed on SketchUp 3D software and printed by means of PLA filament, as shown in Fig. 4. The encoder is positioned close to the electric machine shaft that is under monitoring. In addition, an optical barrier sensor must be tangentially fixed to the encoder.

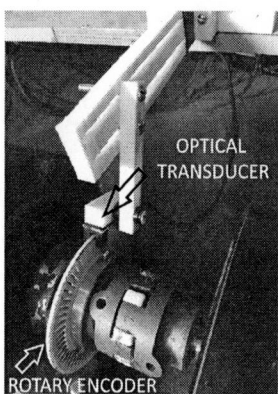

Figure 4. Installed Speed Transducer

Equation 3 is adopted to obtain angular rotation speed:

$$speed_{rot/s} = \frac{nPulse}{\Delta t(s) * 60},\qquad(3)$$

wherein $\Delta t(s)$ represents the sampling window, $nPulse$ is the acumulated sum of intercepted orifices during a $\Delta t(s)$ and $speed_{rot/s}$ is the rotational speed in rotations per second.

C. Acquisition and Supervisory System

The acquisition and supervisory system were built by means of three National Instruments digital acquisition devices DAQ-6008 and a LabView-based programming language. Three DAQ parallel operation allow a real-time visualization of up to twenty signals, among which nine of them are used for measuring AC, DC or pulsed voltage signals. Another nine inputs are used to measure AC, DC or pulsed current signals and two for acquiring angular speeds.

Three operation interfaces were developed in LabView and are depicted in Figs 5, 6 and 7. Fig 5 shows the interface responsible for signal configuration and routing. This interface is useful for calibrating signals obtained from the transducers stage. In this screen, user may define gain and offset values and the kind of measurement, choosing between AC or DC.

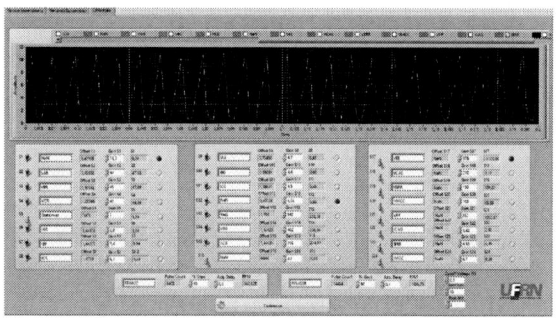

Figure 5. Interface for routing and sensoring calibration.

The second interface depicted in Fig. 6 shows all the interest variables emplying charts. It is possible to monitor every signal from the experimental setup under evaluation. In addition, circular gauges shows frequency, average values and RMS

information. The calculated variables are shown in the third screen, e.g., voltage and current phasors, power triangle and slipping.

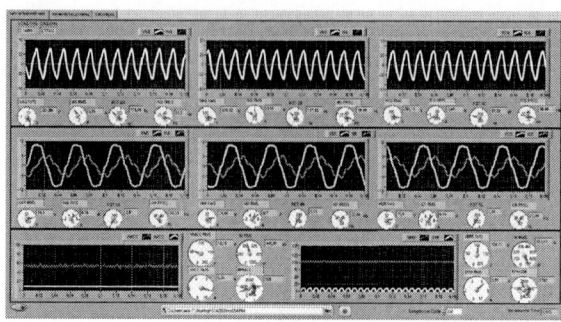

Figure 6. Interface for variable real-time evaluation.

Instantaneous power are obtained by applying clark transform and calculated by means of MatLab scripts that solve the matricial equation. The powers triangle and the three-phase phasors are depicted in Fig. 7.

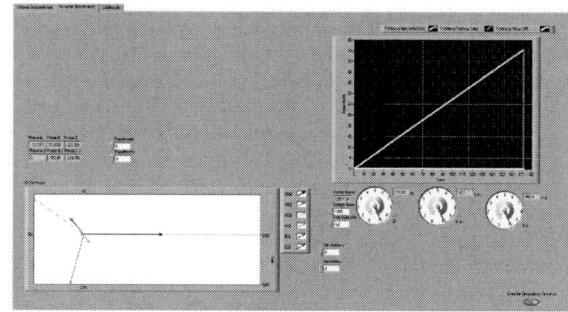

Figure 7. Interface for calculated variables evaluation.

Direct measurement of voltage and current variables is not always viable (e.g., measuring phase current in a delta-connected machine). So, current and voltage transform matrices were implemented on the supervisor, allowing the user to select which kind of physical variable is desired to work with. From nodal and mesh analysis, considering an arbitrary three-phase delta-connected load, the current and voltage transform were defined. As described by equation 4 and 5.

$$\begin{bmatrix} V_{AN} \\ V_{BN} \\ V_{CN} \end{bmatrix} = \begin{bmatrix} 1 & -1 & 0 \\ 0 & 1 & -1 \\ -1 & 0 & 1 \end{bmatrix}^{-1} \cdot \begin{bmatrix} V_{AB} \\ V_{BC} \\ V_{CA} \end{bmatrix}\qquad(4)$$

$$\begin{bmatrix} I_{AN} \\ I_{BN} \\ I_{CN} \end{bmatrix} = \begin{bmatrix} 1 & 0 & -1 \\ -1 & 1 & 0 \\ 0 & -1 & 1 \end{bmatrix}^{-1} \cdot \begin{bmatrix} I_{AB} \\ I_{BC} \\ I_{CA} \end{bmatrix}\qquad(5)$$

For the implementation of these matrix calculation, the usage of MatLab scripts embedded in the supervisory has become crucial, mainly for the easy implementation of the pseudo-inverse transformations.

978-1-7281-4181-7/19 $31.00 © 2019 IEEE

III. CASE STUDIES

In order to validate the proposed platform, two case studies were carried out. The experimental setup developed by [12] is adopted in this work to perform traditional machine testing and regulation process, employing Electromagnetic Frequency Regulator (REF), as follows.

A. System Description

According to [12] the platform consists of a DC motor responsible for emulating the mechanical behavior (e.g, a wind hub). Its shaft is connected to the REF asynchronous rotor by means of a mechanical coupling, both supported by a mechanical bearing. The synchronous rotor consists of a squirrel cage (short-circuited) rotor and the speed output of the REF that drives the synchronous machine. The block diagram is presented in Fig. 8. The absorbed currents by REF are measured and sent to the referred acquisition system and to a DSP (Digital Signal Processing) TMS320f28335, of Texas Instruments, which is responsible for all prototype control routines. The system operation is provided by means of a Voltage Source Inverter (VSI). To validate the developed system, no-load and blocked rotor tests were conducted. As well as power flow and inertia tests of the REF rotor. The implemented test platform is depicted in Fig. 9.

From a voltage/frequency control, embedded at the TMS320f28335 platform, the currents injected into the REF coils are controlled. Utilizing a three-phase inverter, with has a L connection filter of 0.5 mH per phase. Fig. 10 shows the schematic diagram of system connections, wherein blocks marked with the prefix (V) represent voltage sensors. (A) represents current sensors and (Wn) represent speed sensors. Gains applied to the sampled signals were defined from measurements using the YOKOGAWA Energy Analyzer CW500.

Uncontrolled three-phase rectifiers independently feeds VSI DC link, DC motor and electromagnetic field of the synchronous generator, sourcer that were simplified by VCC_0, VCC_1 and VCC_2. The star-connected synchronous generator has its windings associated with a bank of resistive loads of 75

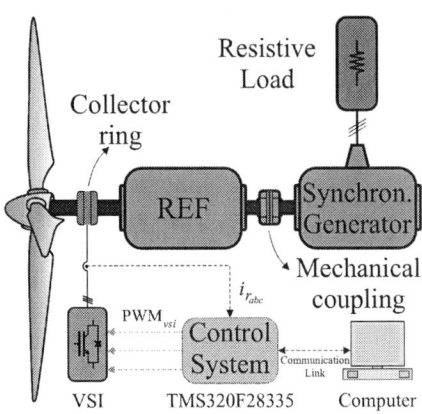

Figure 8. System block diagram.

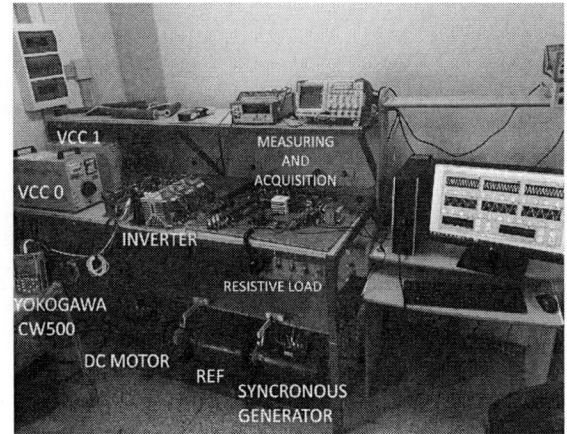

Figure 9. Developed platform apllied to REF system.

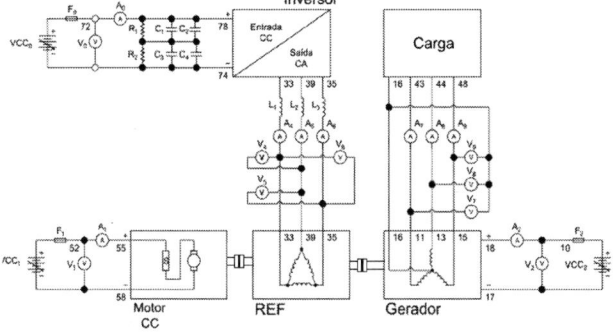

Figure 10. Measuring blocks applied to REF system.

Ω per phase. W_{n2} is the generator rotational speed, obtained using the encoder setup.

B. No-Load and Blocked Rotor Tests

A consolidated strategy used test for machine parameters identification that allows to define magnetization and dispersion reactances is the combination of no-load and blocked rotor tests. In order to simplify the machine operation during blocked rotor test, the REF rotary stator was kept locked, and three-phase currents were injected from VSI. The synchronous rotor was set free to rotate without any load connected in its shaft. For no-load test, 13 tests were performed employing linear incremental voltage levels to the REF, until reaching its rated voltage. In blocked rotor test, incremental current levels were applied to the machine until reaching is rated value of 6 A, totaling 8 tests. From the measurements, Clarke voltages and currents are calculated in real-time and to chart the data, Matlab was used to concatenate the results exported from the supervisory. Fig. 11 depicts Clark RMS values obtained from (a) and (b) at blocked rotor and (c) and (d) at no load tests. From Clarke voltage and currents, the real power, reactive power and apparent power, are calculated in real-time. MatLab was used to concatenate the results exported from the supervisory system. The Fig. 12 depicts power values obtained at blocked rotor (a) and (b) at no-load tests.

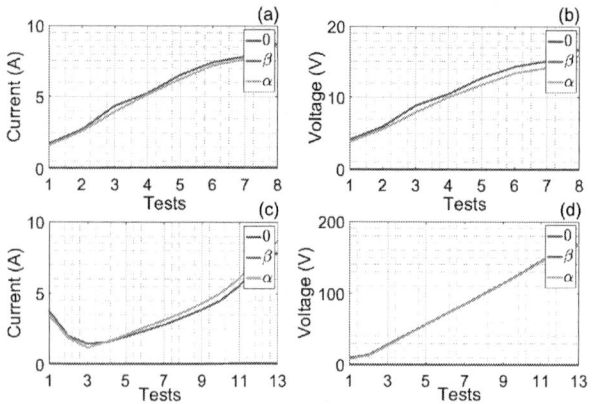

Figure 11. Clarke RMS values.

In no-load tests, the machine rotor does not have a mechanical load, but its own rotor weight leading to very low real power values. In this way, apparent power is predominantly reactive. For this reason, the curves are almost superimposed on Fig. 12 (b).

C. REF Rotor Inertia Test

Rotor inertia test starts from applying rated voltage to the machine. After variables stabilization, the energy supplier is turned off, in order to observe rotor slowdown as a function of time. From supervisory, we could evaluate the waveforms involved obtain essential parameters to determine the machine inertia. Fig. 13 (a) depicts the rotor angular speed decay in response to power shutdown. It also show three-phase voltages and currents. Since the system is balanced, the voltages and currents are very similar. In Fig. 13 (b), the moment of shutdown is shown. The injected current departs to zero at the moment in which signal is ceased, but the voltage decays exponentially due to the storaged energy in the winding coils.

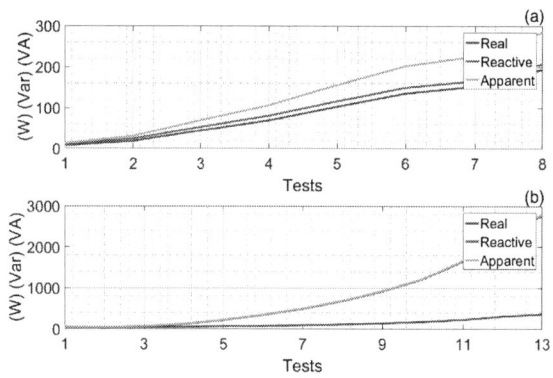

Figure 12. Powers for blocked rotor and no load tests.

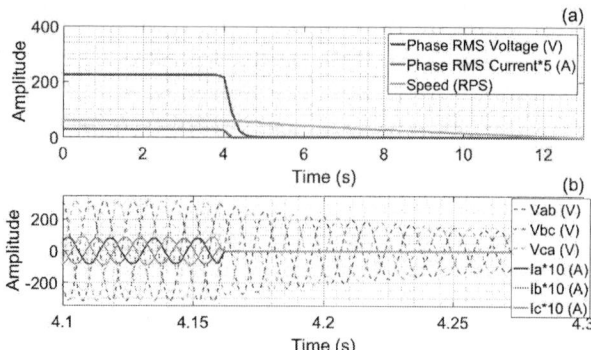

Figure 13. Rotor speed decay in response to power shutdown.

D. Power Flow Test

During power flow tests the REF stator was kept free and mechanically coupled to the DC motor shaft. The REF coils are supplied with three-phase currents from the VSI. The ratio of stator frequency of rotation and the frequency injected by the inverter results in a rotational speed in the REF rotor, that is mechanically coupled to the synchronous generator rotor. The Fig. 14 depicts the voltages and currents consumed by the DC motor (VMCC and IMCC, respectively) at 839.2 RPM and the DC link (VBRR and IBRR, respectively) as a time function window. In the legend, average values at the test are displayed.

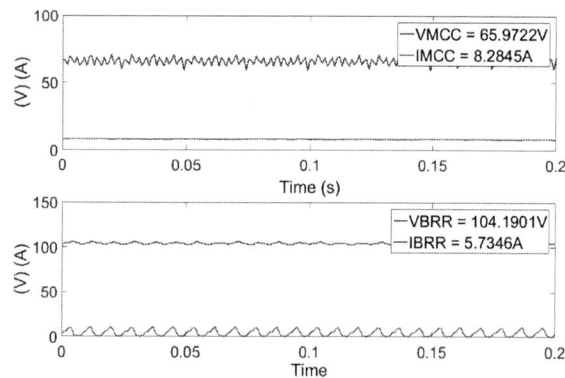

Figure 14. DC motor and DC link voltage and current signals.

At the output of the synchronous generator with current at a frequency of 59.082 Hz, all three-phase variables are monitored and stored, allowing the calculation of powers, through the Clarke transform. The power factor has come very close to unity, resulting in 0.99604 since the load is purely restive, powering a three-phase restive star load, with 75 Ω per phase. The Fig. 15 (a) and (b) depicts three-phase voltages and currents generated at the output of the synchronous motor. The Fig. 15 (c), (d) and (e) depicts the voltage and current per phase. In the legend, RMS values at the test are displayed. In Fig. 16 the instantaneous output powers of the synchronous generator are shown. As the power factor is very close to unity, the active power is almost equals to apparent.

978-1-7281-4181-7/19 $31.00 © 2019 IEEE

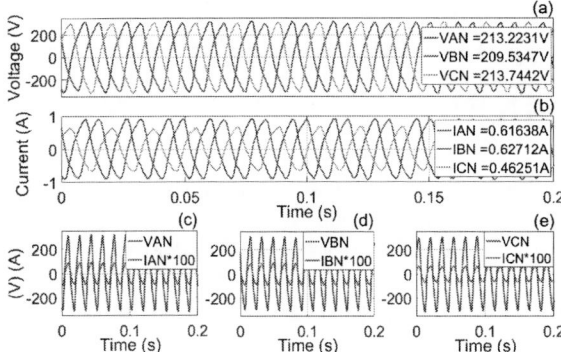

Figure 15. Synchronous generated voltages and currents

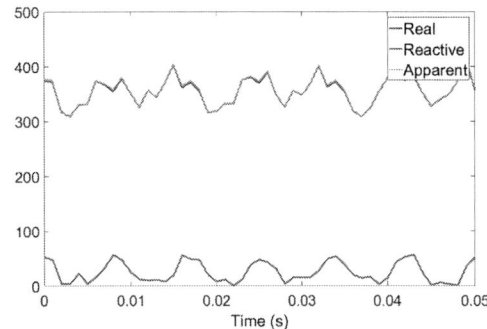

Figure 16. Instantaneous output powers of synchronous generator

All three-phase inverter output variables are monitored. With a power factor of 0.8398, considering the injected current of frequency equal to 17.08 Hz. In the legend, RMS values at the test are displayed. Fig.17 (a) and (b) depicts three-phase voltages and currents injected to REF coils. The Fig. 17 (c), (d) and (e) depicts the voltage and current per phase. In the legend, RMS values at the test are displayed.

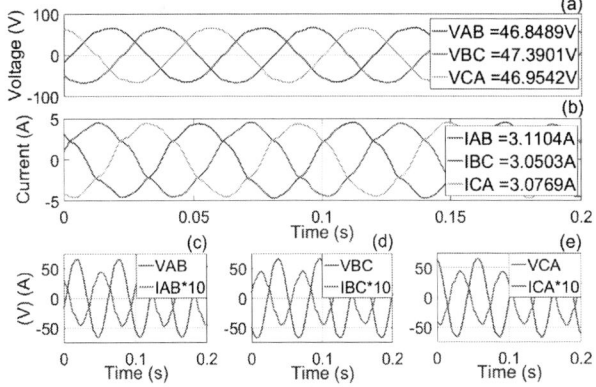

Figure 17. REF injected voltages and currents

CONCLUSIONS

The versatility of the platform becomes great, as well as being able to acquire the most diverse information (currents

AC/DC, voltages AC/DC, Speed). The associating with Matlab resources, has permitted any type of calculations. Such as instant power, positioning of spinning vectors calculations and Δ/Y transformations. In addition, the benefits of monitoring the state of the system operation in real time showed good performance, when testing traditional machines and new prototypes such as the REF. As an improvement to the system, it is expected to acquire a more advanced data acquisition platform than the DAQ-6008, in order to use only one module for sampling and quantization of the 18 signals, with higher sampling frequency. The developed supervisory system has been used in several research projects in the Control and Power Systems Laboratory of the Control and Automation Department, at the Federal University of Rio Grande do Norte.

IV. ACKNOWLEDGEMENT

The authors thank both PRPGI/IFBA, Federal Institute of Bahia and UFRN, Federal University of Rio Grande do Norte by the financial support. This study was financed in part by the Coordenação de Aperfeiçoamento de Pessoal de Nível Superior - Brasil (CAPES) - Finance Code 001.

REFERENCES

[1] S. Mojlish, N. Erdogan, D. Levine, and A. Davoudi, "Review of hardware platforms for real-time simulation of electric machines," *IEEE Transactions on Transportation Electrification*, vol. 3, no. 1, pp. 130–146, March 2017.

[2] V. Ivanov, D. Savitski, and B. Shyrokau, "A survey of traction control and antilock braking systems of full electric vehicles with individually controlled electric motors," *IEEE Transactions on Vehicular Technology*, vol. 64, no. 9, pp. 3878–3896, Sep. 2015.

[3] W. Xu, J. Zhu, Y. Guo, S. Wang, Y. Wang, and Z. Shi, "Survey on electrical machines in electrical vehicles," in *2009 International Conference on Applied Superconductivity and Electromagnetic Devices*, Sep. 2009, pp. 167–170.

[4] Z. Q. Zhu and D. Howe, "Electrical machines and drives for electric, hybrid, and fuel cell vehicles," *Proceedings of the IEEE*, vol. 95, no. 4, pp. 746–765, April 2007.

[5] M. Malinowski, A. Milczarek, R. Kot, Z. Goryca, and J. T. Szuster, "Optimized energy-conversion systems for small wind turbines: Renewable energy sources in modern distributed power generation systems," *IEEE Power Electronics Magazine*, vol. 2, no. 3, pp. 16–30, Sep. 2015.

[6] D. Gerada, A. Mebarki, N. L. Brown, C. Gerada, A. Cavagnino, and A. Boglietti, "High-speed electrical machines: Technologies, trends, and developments," *IEEE Transactions on Industrial Electronics*, vol. 61, no. 6, pp. 2946–2959, June 2014.

[7] E. V. Lyubimov, S. P. Gladyshev, D. A. Istselemov, and N. A. Belyaev, "Universal software-hardware measurement complex for testing wide range of synchronous machines," in *2016 IEEE International Conference on Electro Information Technology (EIT)*, May 2016, pp. 0252–0257.

[8] M. Enache, A. Campeanu, S. Enache, and I. Vlad, "Testing stand for low-power electrical machines," in *2018 International Conference on Applied and Theoretical Electricity (ICATE)*, Oct 2018, pp. 1–6.

[9] A. Keyhani and S. Hao, "Microcomputer-aided data acquisition system for laboratory testing of transformers and electrical machines," *IEEE Transactions on Power Systems*, vol. 3, no. 3, pp. 1328–1334, Aug 1988.

[10] F. S. Sellschopp and M. A. Arjona L., "An automated system for frequency response analysis with application to an undergraduate laboratory of electrical machines," *IEEE Transactions on Education*, vol. 47, no. 1, pp. 57–64, Feb 2004.

[11] Y. C. Chin, "Development of educational platform for experiments of electric machines," in *2014 International Symposium on Computer, Consumer and Control*, June 2014, pp. 593–596.

[12] P. Vitor Silva, R. Ferreira Pinheiro, A. Ortiz Salazar, L. Pereira do Santos Junior, and C. Cesar de Azevedo, "A proposal for a new wind turbine topology using an electromagnetic frequency regulator," *IEEE Latin America Transactions*, vol. 13, no. 4, pp. 989–997, April 2015.

DC-link voltages balance method for single-phase NPC inverter operating with reactive power compensation

Marco Fajardo
Grupo de Sistemas Industriales de Electrónica de Potencia
Universidad Simón Bolívar
Caracas, Venezuela
mfajardom@est.ups.edu.ec

Julio Viola, José Manuel Aller, Flavio Quizhpi
Carrera de Ingeniería Eléctrica
Universidad Politécnica Salesiana
Cuenca, Ecuador
jviola@ups.edu.ec; jaller@ups.edu.ec; fquizhpi@ups.edu.ec

Abstract—**Neutral point clamped (NPC) are one of the most used topologies for inverters in photovoltaic (PV) applications due to several operative advantages. The same hardware used for DC/AC conversion can be exploited to add reactive power compensation capabilities. In absence of PV generation (i.e. night hours) the DC link voltage have to be controlled by injecting or extracting active power to or from the DC link capacitors. While this technique is widely used in active power filters (APF) based on H-bridge inverters, the use of NPC inverters imposes a restriction due to potential imbalances in the voltages of DC-link capacitors, produced by DC components in the current of nonlinear loads being compensated. In this paper an analysis of DC currents effect is presented and an enabling method to allow reactive power compensation with single-phase NPC inverters is proposed and tested.**

Index Terms—**active power filter, neutral point clamped, grid connected inverter, reactive power compensation, single-phase systems**

I. INTRODUCTION

There are several works focused in the operation of single-phase grid connected inverters with the capability of reactive power compensation. In [1] authors propose a current control to regulate active and reactive power with a single-phase inverter connected to a microgrid. In [2] a proportional-resonant controller, and in [3], a *dq* transformation, are used to extract maximum active power from a PV array and to inject specified reactive power to the utility. In [4] a bidirectional boost/buck DC-DC converter is inserted before the inverter, and reactive power injection to the grid is also controlled. Most of these works, however, consider only the H-bridge topology. The three-level NPC topology is widely used in three- and single-phase PV applications due to many advantages over other inverters, such as two-level half-bridge or H-bridge. First NPC inverters use the "I" topology, but in last years the "T" topology is gaining acceptation among designers since the introduction of power modules from several manufacturers [5]. Both topologies use a DC-link split in two (ideally equal) voltages [6]. In photovoltaic applications, the voltage generated by PV arrays is typically applied to both halves of the DC-link allowing for an almost fixed voltage. For applications of reactive power compensation the DC-link voltage is controlled by an inner active power or current loop, so that the voltage level is between preestablished values that allows the correct operation of the inverter. In [7] an example of a three-phase NPC inverter used as reactive power compensator is presented. This technique, however, fails in obtaining a voltage balance between both halves of DC-link

when asymmetrical loads are connected in parallel with the single-phase NPC inverter. As it will be shown, even a small quantity of DC component in the load current, leads to a permanent unbalance of the two halves of DC-link.

In this work the extended operation of NPC grid connected inverters is analyzed so that the compensation of current harmonics is made possible even when DC component is present in the nonlinear load current. The analysis considers the operation under two different scenarios of asymmetric loads and when no power is being generated by the PV panels (i.e. night hours).

The paper is structured as follows: Section II describes fundamental concepts of reactive power compensation by using APF. In section III the different effects of asymmetrical loads on H-bridge and NPC inverters is addressed, while in section IV the DC-link voltages balance method is presented. Finally, in section V, the behavior of NPC inverter operating as APF is presented, and the proposed balance method is tested by simulations.

II. REACTIVE POWER COMPENSATION

When an inverter is used as a reactive power compensator, whether it be an active power filter or a grid connected inverter operating with extended capabilities, the voltage in the DC-link capacitors is controlled by controlling the active power taken from the grid or from the PV arrays if it is available. Once the desired voltage level is reached, the control loop just take the active power required to compensate the inverter losses. Figure 1 depicts a scheme of a power inverter acting as reactive power compensator when nonlinear loads are connected to the grid. Ideally all harmonics contained in the load current are canceled in the point of common coupling (PCC) and since active power to or from the APF can be carried only at fundamental frequency, just reactive power is involved in the APF branch.

III. ASYMMETRIC LOADS

In some cases, besides of generating current harmonics, nonlinear loads can have DC current component, mainly due to the presence of half-wave rectifiers (less common nowadays), or asymmetric operation of full-wave rectifiers (sometimes due to malfunction or damage in one or more diode) typically found in the input stage of most electronic appliances (e.g. power supply units, battery chargers, LED lamps, etc). The DC current component will circulate through

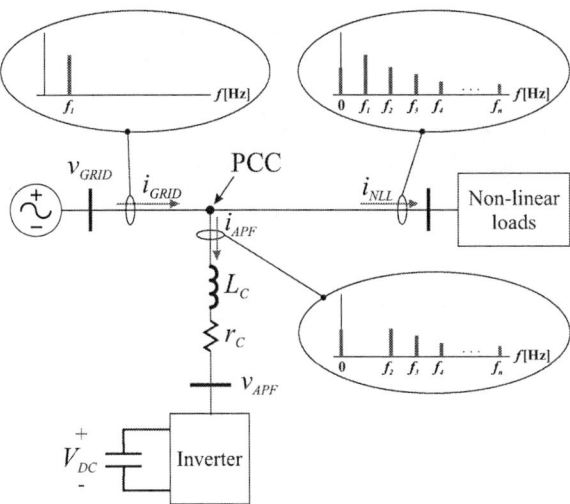

Fig. 1: Inverter operating as active power filter.

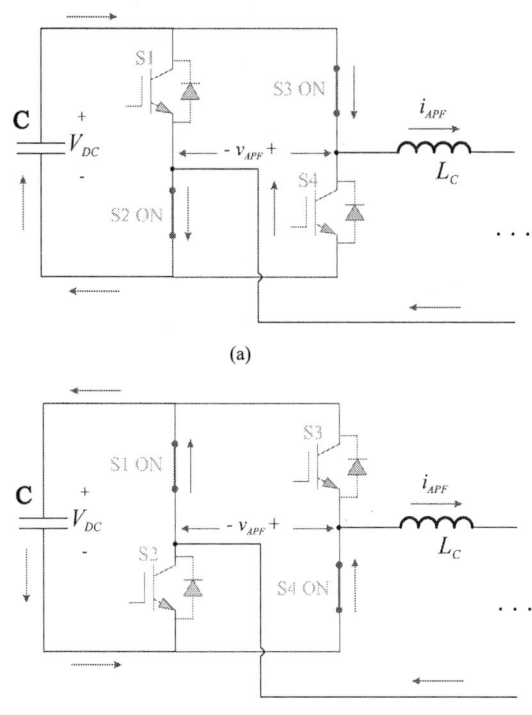

Fig. 2: DC current effect in an active power filter based on H-bridge inverter, a) during $\bar{v}_{APF} > 0$, and b) during $\bar{v}_{APF} < 0$.

the AC system generating saturation and overheating in distribution transformers. If the APF current control loop includes the compensation of this DC current component, then in the PCC an equivalent DC current with opposite sign have to be injected by the inverter. Those APFs based on H-bridge inverters can deal with this situation since the DC current will circulate through the DC-link capacitor in one direction when the inverter output voltage is positive, and in the opposite direction when the inverter output voltage is negative. This situation is depicted in Figs.2(a) and (b). As a consequence capacitor voltage will have a 60Hz voltage ripple, but its average voltage can be can controlled to a pre specified reference value.

While the H-bridge has this self-balancing capability, for an APF based on an NPC inverter the effect of a DC current is different. In Fig. 3(a) and (b) a scheme of an NPC inverter operating as APF is shown. In this type of converter the positive voltages are synthesized by using the voltage coming from the upper capacitor and, conversely, the negative voltages are obtained from the voltage of the lower capacitor. Considering this, when a DC current is injected from the inverter it produces a sustained discharge of the upper capacitor and a charge of the lower capacitor. When the voltage of the upper capacitor lowers enough, the inverter loses control over the current through the coupling inductor and the entire operation of the inverter is lost.

In general, an APF based on an NPC converter cannot compensate DC currents, and a method to overcome this situation is proposed in the next section.

IV. DC-LINK VOLTAGES BALANCE METHOD

The proposed balance method considers that the inner loop is based on a predictive current control [8]. This method requires to know the value of the coupling inductor to determine what voltage have to be synthesized at the inverter output to obtain the desired current.

Taking as reference Fig. 1, the equation for predictive control of inductor current can be obtained as:

$$v_{GRID} - \bar{v}_{APF} = L_C \frac{di_{APF}}{dt} + i_{APF} r_C \qquad (1)$$

- v_{GRID} is the grid voltage

- $\bar{v}_{APF}$ is the mean value of inverter output voltage
- i_{APF} is the compensation current
- L_C is inductance value of coupling inductor
- r_C is the intrinsic resistance of coupling inductor

In most cases r_C is small and can be neglected. A discrete time approximation can be used for the current derivative:

$$L_C \frac{di_{APF}}{dt} \approx L_C \frac{i_{APF}(k+1) - i_{APF}(k)}{T_s} \qquad (2)$$

being T_s the data sampling time used in the discrete implementation. From the previous equation an expression for inverter voltage can be obtained so that a desired current value can be met:

$$\bar{v}_{APF} = v_{GRID} + \frac{L_C}{T_s}(i_{APF}(k+1) - i_{APF}(k)) \qquad (3)$$

Considering that in the next control cycle (k+1) the i_{APF} must have the desired value of current reference $i_{APF\,ref}$ is introducing, resulting in:

$$\bar{v}_{APF} = v_{GRID} + \frac{L_C}{T_s}(i_{APF\,ref}(k) - i_{APF}(k)) \qquad (4)$$

Figure 4 shows the implementation of this predictive scheme, where besides the inner current loop, the outer voltage loop and current reference calculation are depicted.

Controller PI$_1$ is in charge of regulate the total DC-link voltage ($V_{DC1} + V_{DC2}$). The $V_{DC\,ref}$ is supposed constant in this analysis, but in PV applications typically it depends on the MPPT algorithm. In that case the variation of $V_{DC\,ref}$ is slow and has no influence in the performance of the inner current loop. Grid current reference is obtained as the product of grid voltage and the output of PI$_1$ to ensure

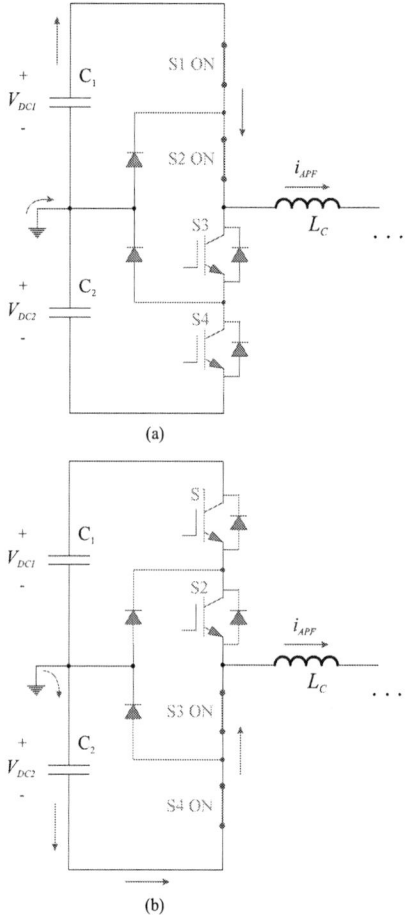

(a)

(b)

Fig. 3: DC current effect in an NPC inverter operating as active power filter, a) during $\bar{v}_{APF} > 0$, and b) during $\bar{v}_{APF} < 0$.

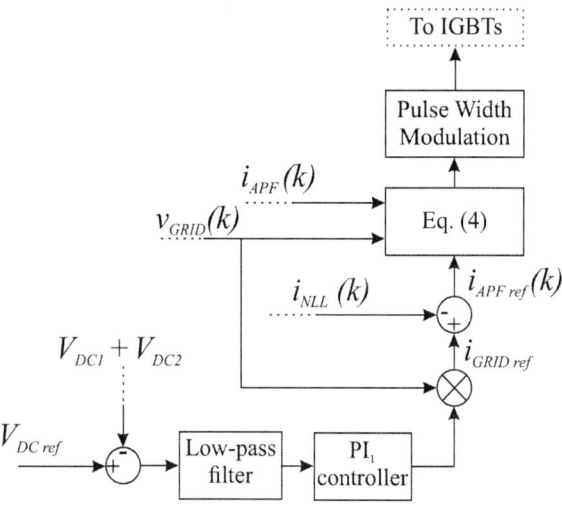

Fig. 4: Block diagram for a predictive current control.

sinusoidal waveform and unit power factor. The load current is subtracted from $i_{GRID\,ref}$, to obtain the value of $i_{APF\,ref}$ used in (4).

A. Load current DC component

When a DC component is present in the load current this component is injected with opposite sign to the predictive control expression described in (4), generating the unbalance in the voltages of the DC-link capacitors. This DC component can be removed by computing the mean value of current with a moving window and subtracting this value to the reference current. There is, however, another way by which DC components can be injected in the current loop. The output of the PI_1 controller typically has ripple at fundamental frequency (50Hz or 60Hz) due to the existence of ripple in DC voltages. Even when measured DC voltage is passed through a low-pass filter, a component of fundamental frequency still is present in the PI's output. When this output is multiplied by the line voltage a product of two sines at fundamental frequency results. The identity for product of sines is:

$$\sin(a)\sin(a) = \frac{1}{2}(1 - \cos(2a)) \tag{5}$$

which leads to the creation of DC component and second harmonic component (100Hz or 120Hz). This DC component is also injected to the predictive expression generating again unbalance in the DC-link voltages.

B. Balance method based on PI action

To overcome the detrimental effect that DC component has on the balance of NPC inverter DC-link voltages, the action of a PI controller is proposed. The input to the controller is the difference between voltages V_{DC1} and V_{DC2}. The controller action is to add or subtract a DC component to the current reference sent to the predictive current control described in (4). In Fig. 5 the block diagram of the modified control loop is shown, where in red color is the controller PI_2. The proposed balance method does not alter the settings of the original control loop and just add the action of controller PI_2. When voltage error is positive ($V_{DC1} > V_{DC2}$) the PI_2 controller will add a positive DC component to the current reference, which will lead to a decrease in V_{DC1}, and an increase in V_{DC2}. After a certain time the voltage difference is reduced to zero. The same is true for the case of negative error ($V_{DC1} < V_{DC2}$), when the PI_2 will add a negative DC component to the current reference.

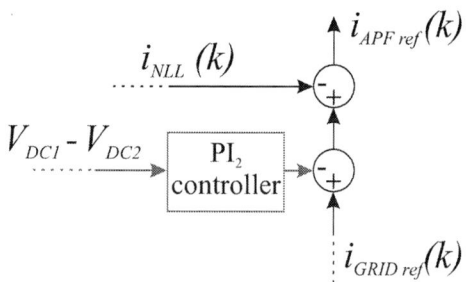

Fig. 5: PI controller to obtain DC-link voltages balance.

As it will be shown in the next section, the final effect of this compensation method is that the grid current will have a DC component, but the harmonics present in the current of non-linear load can be compensated. In general, the proposed method enables the operation of the single-phase NPC inverter as APF.

V. SIMULATION OF THE PROPOSED METHOD

Proposed methods are simulated with MATLAB/Simulink® to analyze the performance of the NPC inverter when it used as an APF and considering that the non-linear load current contains DC component. Since no power is injected from PV panels, the DC-link voltages are entirely determined by active power taken from the grid, which is controlled by the PI_1 controller in the voltage loop. Inverter settings for these tests are: $V_{DC\,ref} = 500$V, $L_C = 10$mH, $C_1 = C_2 = 3300\mu$F, and $fs = 10$kHz. The non-linear load connected to the grid is simulated as a full-wave rectifier with a filter capacitor with C=2200μF, and a resistor with R=50Ω connected in the rectifier's output. Two different scenarios of load asymmetry are simulated:

- (1) Diodes in each branch of the rectifier bridge having different internal parameters
- (2) Diode with open-circuit fault in one branch of the rectifier bridge

In Fig. 6 the load and grid currents for Scenario 1 are shown, where a small asymmetry in load current can be observed. The inverter compensates the harmonics and provides a grid current greatly improved with a lower THD. The DC component present in the load current is canceled by the inverter action but this leads to a permanent drift in the DC-link capacitor voltages as shown in Fig. 7. The sum of both voltages remains constant (500V in this tests) since the PI_1 controller in the voltage outer loop assures the voltage value is controlled, but the permanent drift in individual voltages will lead to a current control loss in the inverter.

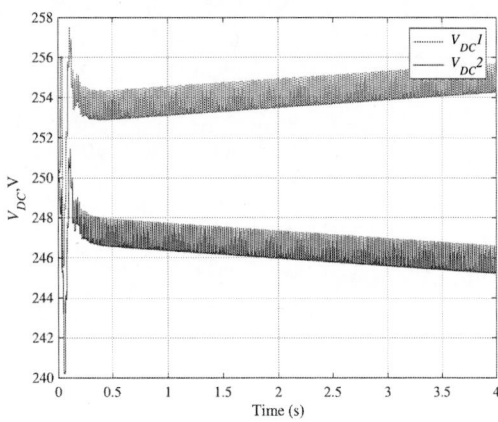

Fig. 7: Drift in DC-link capacitors voltages for non-linear load behaving as described in Scenario 1.

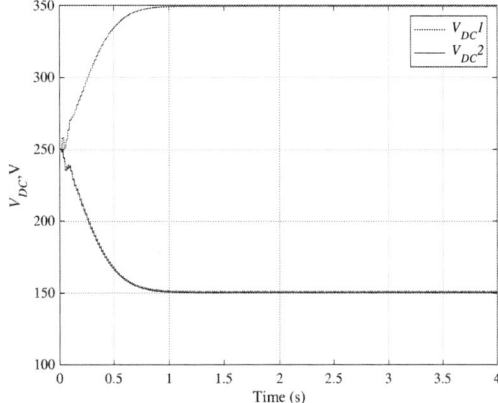

Fig. 8: Drift in DC-link capacitors voltages for non-linear load behaving as described in Scenario 2.

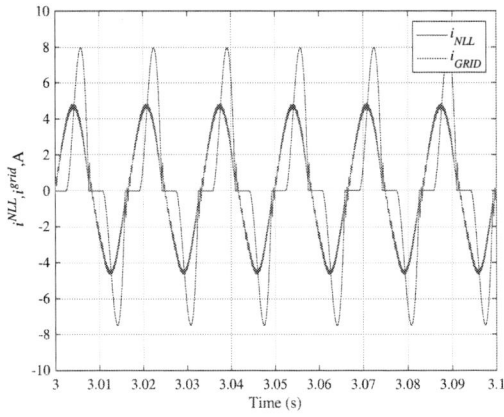

Fig. 6: Grid current evolution for non-linear load behaving as described in Scenario 1.

In a similar manner, if an open-circuit fault is simulated in one of the diodes of the full-wave rectifier, the load current DC component is even larger than the obtained in Scenario 1. The compensation action of the inverter leads to a faster unbalance in the DC-link voltages as it is shown in Fig. 8. After less than one second, one of the capacitors is discharged enough for the inverter to lose control.

Both previous scenarios are tested again but now applying the proposed balance method. In Fig. 9 load and grid currents

are shown. Due to the effect of balance method a DC component appears in grid current, nevertheless the THD of current (not considering the DC component) is reduced from 53% in the load to 7.5% in the grid. In Fig. 10 the evolution of DC-link voltages is shown. An intentional unbalance in the initial capacitor voltages is set to analyze the action of the balance method.

When a diode open-circuit fault is simulated, the resulting load current is a half-wave with THD=84%. The grid current exhibits a DC component as a result of the compensation method, but its THD is 9.4%. Both current are shown in Fig. 11. Evolution of DC voltage for this scenario is shown in Fig. 12.

Figure 13 shows a phase comparison between grid current and voltage, where the displacement power factor is $PF_D \approx 0.99$.

In Fig. 14 evolution of DC-link voltage is shown when a diode open-circuit fault is simulated at t=1s and the balance method is activated at t=1.5s.

Finally, a test is run to assess the method's robustness when a difference in capacitance values due to manufacturing tolerance is simulated. In this case a $\pm5\%$ variation over the nominal value is applied resulting in $C_1 = 3135\mu$F and

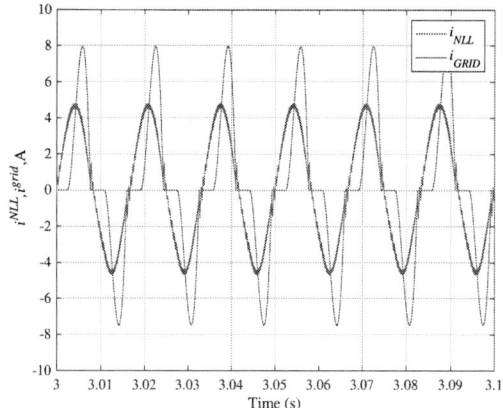

Fig. 9: Grid current evolution for non-linear load behaving as described in Scenario 1.

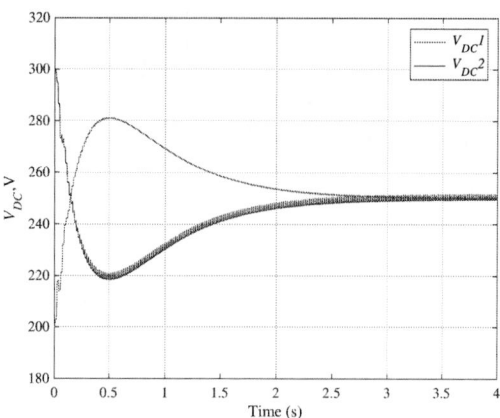

Fig. 12: Evolution of DC-link capacitors voltages for non-linear load behaving as described in Scenario 2.

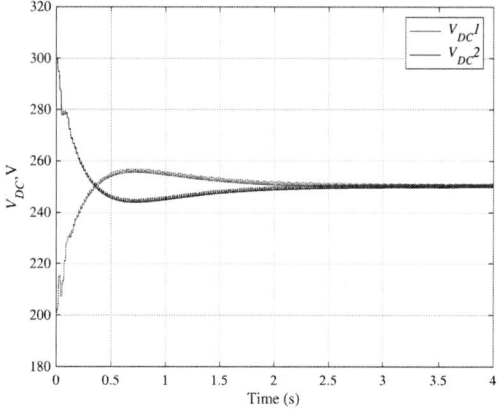

Fig. 10: Evolution of DC-link capacitors voltages for non-linear load behaving as described in Scenario 1.

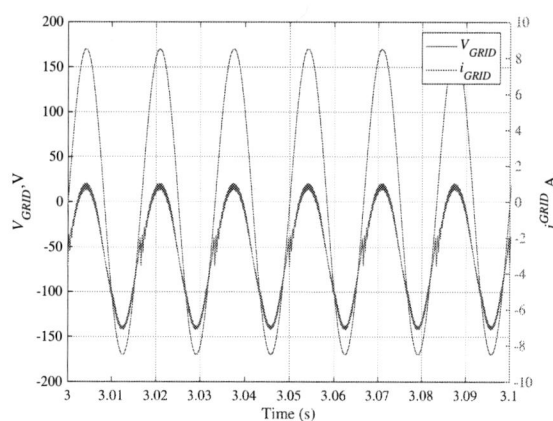

Fig. 13: Comparison of grid current and grid voltage for Scenario 2.

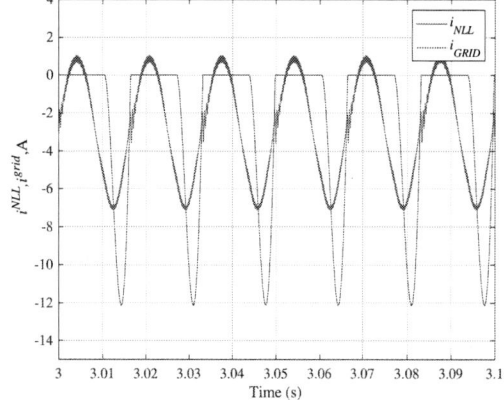

Fig. 11: Grid current evolution for non-linear load behaving as described in Scenario 2.

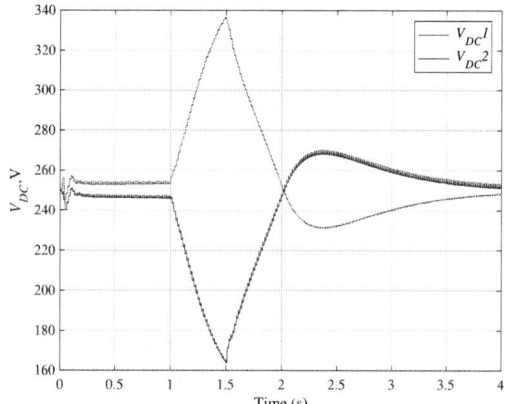

Fig. 14: Evolution of DC-link capacitors voltages when a diode open-circuit fault is simulated.

$C_2 = 3465\mu F$. In addition an inductance value of 11mH is used in the simulations but the L_C value in (4) is maintained as 10mH. The test is run under the same conditions of diode fault shown in Fig. 14, resulting in the voltages evolution shown in Fig. 15. After both transients, capacitors voltages tend to balance.

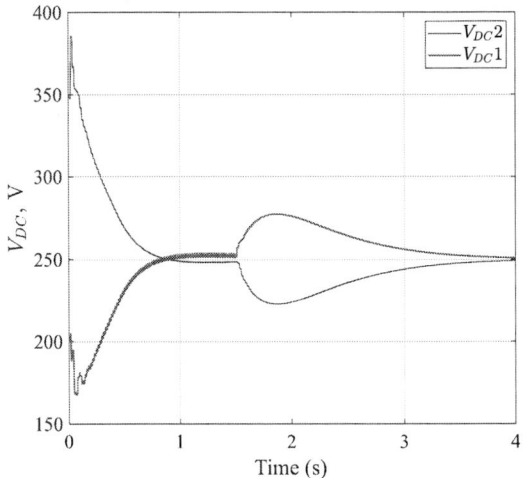

Fig. 15: Evolution of DC-link capacitors voltages when a diode open-circuit fault is simulated and different capacitance values are considered.

VI. CONCLUSION

The operation of a NPC single-phase grid connected inverter capable of reactive power compensation is analyzed when no power from the PV arrays is available. The detrimental effect in the balance of the DC-link voltages produced by the DC component in the non-linear load current is studied. A method based on a PI controller is proposed to assure balance of the DC-link voltages. The controller does not affect the previously existent predictive control current loop. The method is tested with simulations, showing that stable operation is possible even when asymmetrical loads are connected to the grid. Future work include experimental tests using the platform reported in [9].

ACKNOWLEDGMENT

The authors want to thank to Universidad Politecnica Salesiana in Cuenca, Ecuador, and Universidad Simon Bolivar in Caracas, Venezuela, for the institutional and financial support to this work.

REFERENCES

[1] S. Dasgupta, S. K. Sahoo, and S. K. Panda, "Single-phase inverter control techniques for interfacing renewable energy sources with microgridpart i: Parallel-connected inverter topology with active and reactive power flow control along with grid current shaping," *IEEE Transactions on Power Electronics*, vol. 26, pp. 717–731, March 2011.

[2] S. Agrawal, P. C. Sekhar, and S. Mishra, "Control of single-phase grid connected pv power plant for real as well as reactive power feeding," in *2013 IEEE International Conference on Control Applications (CCA)*, pp. 667–672, Aug 2013.

[3] F. E. Aamri, H. Maker, A. Mouhsen, and M. Harmouchi, "A new strategy to control the active and reactive power for single phase grid-connected pv inverter," in *2015 3rd International Renewable and Sustainable Energy Conference (IRSEC)*, pp. 1–6, Dec 2015.

[4] G. Min, K. Lee, J. Ha, and M. H. Kim, "Design and control of single-phase grid-connected photovoltaic microinverter with reactive power support capability," in *2018 International Power Electronics Conference (IPEC-Niigata 2018 -ECCE Asia)*, pp. 2500–2504, May 2018.

[5] V. Jerinic, K. Lenz, and R. Hinken, "IGBT power module in three-level neutral point clamped type 2 (NPC2, T-NPC, mixed voltage) topology in short circuit modes," in *PCIM Europe 2016; International Exhibition and Conference for Power Electronics, Intelligent Motion, Renewable Energy and Energy Management*, pp. 1–8, May 2016.

[6] H. Shin, K. Lee, J. Choi, S. Seo, and J. Lee, "Power loss comparison with different PWM methods for 3L-NPC inverter and 3L-T type inverter," in *2014 International Power Electronics and Application Conference and Exposition*, pp. 1322–1327, Nov 2014.

[7] C. A. Weishaupt, L. A. Morn, J. R. Espinoza, J. W. Dixon, and G. Joos, "A reactive power compensator topology based on multilevel single-phase npc converters," in *2010 IEEE International Conference on Industrial Technology*, pp. 1339–1344, March 2010.

[8] C. R. D. Osório, G. S. da Silva, J. C. Giacomini, and C. Rech, "Comparative analysis of predictive current control techniques applied to single-phase grid-connected inverters," in *2017 Brazilian Power Electronics Conference (COBEP)*, pp. 1–6, Nov 2017.

[9] J. Viola, J. Restrepo, F. Quizhpi, M. I. Gimenez, J. Aller, V. Guzman, and A. Bueno, "A Flexible Hardware Platform for Applications in Power Electronics Research and Education," pp. 226–232, IEEE, Nov. 2014.

New Five-Level Flying Capacitor Inverter Fed by a Boost-Flyback DC-DC Voltage Source

Antônio Venâncio de M. Lacerda Filho[*], André E. L. da Costa[†], Edison R. C. da Silva[*†]
Ronnan de B. Cardoso[*], Darlan A. Fernandes[*]
[*]Post-Graduate Program in Electrical Engineering – PPGEE – LOSE – CEAR
[*]Federal University of Paraíba – UFPB, João Pessoa - PB, Brazil
[†]Post-Graduate Program in Electrical Engineering – PPGEE – COPELE – LEIAM
[†]Federal University of Campina Grande, Campina Grande - PB, Brazil
e-mail: antonio.lacerda@cear.ufpb.br, andre.lucenadacosta@gmail.com, ercdasilva@gmail.com
ronnan.cardoso@cear.ufpb.br, darlan@cear.ufpb.br

Abstract—This paper reports a new five-level pulse-width modulation (PWM) that includes a flying capacitor supplied by a boost/flyback DC-DC converter that also provides the DC-bus voltage. There is a significant reduction in the number of power devices and capacitors required to implement a conventional five-level flying capacitor inverter. The topology is free from complex PWM control to regulate the capacitor voltage and is capable of operating under any power factor condition. Simulated and experimental results validate the converter principle.

Index Terms—Multi-level inverter, hybrid converter, active NPC inverter, flying capacitor inverter, DC-DC converter-fed inverter.

I. INTRODUCTION

The Full-Bridge Cascade (FBC) [1], [2], the Neutral Point Clamped (NPC) [3], [4], and the Floating Capacitor (FC) [5], [6] are considered as conventional multilevel topologies. The most basic FBC is constituted by two cascaded full-bridges (eight switches) individually fed by isolated sources providing five levels (5L); its three-phase connection is in Y. Although three level inverters are very often employed, the motivation of using five levels is to reduce the switch voltage stress and their ability to generate voltage waveform with less harmonic distortion at reduced switching frequency. Both conventional 5L-NPC and 5L-FC inverters need eight main switches. In case of 5L-NPC twelve (12) additional clamping diodes are required, since cascaded diodes are used to block the higher voltage; four cascaded voltage sources (or capacitors dividing the DC-link voltage) of equal value form the DC-link voltage and provide nodes to which the 5L inverter output can be connected. In contrast, the 5L-FC requires ten (10) additional clamping capacitors that need to be regulated; this increases cost and control complexity of the inverter.

Other well known converter topologies are: the ANPC in which the NPC clamping diodes are replaced by active switches enabling loss-balancing among the switches [7]; the second version of the NPC in which the neutral point is connected via a bidirectional switch [4], and nowadays also known as T connection; and the hybrid ANPC-FC that reduces fluctuations of the neutral point DC voltage and maintains balanced blocking voltages across each device at turn-off [8].

In special, many topologies can be derived from the general structure of hybrid ANPC-FC inverter [9], [10]. One of these is the five-level active NPC converter (5L-ANPC), shown in Fig. 1(a) [11]. This topology combines the flexibility of the multilevel floating capacitor converter with the robustness of ANPC converters with only eight switches and three capacitors (two of them dividing the DC-link voltage). By regulating the value of the floating capacitor at $V_i/2$ (the DC-link voltage is $2V_i$) allows for four levels operation [12]; the fifth one (zero level) is achieved by using the ANPC principle. Other arrangements and further reduction in the number of components have been introduced such as the eight-switch 5L-ANPC (8S-5L-ANPC) [13] and the six-switch (6S-5L-ANPC) [14]. The 6S-5L-ANPC topology is shown in Fig. 1(b). The reduction in the number of components is evident. However, the 5L-ANPC is more indicated for high power factor applications [14]. Regulation of the floating capacitor voltage of these topologies is obtained with the help of an adequate Pulse-Width Modulation (PWM) strategy. Avoiding its control complexity requires the use of a fixed floating source.

Different voltage step-up DC-DC converter topologies can be used as inverter sources [15]–[17]. Applied examples have been shown in [18] and [19] for the 3L-NPC inverter. Also, a high gain boost/flyback DC converter has been combined with a FC inverter to produce a four level output [20]. Note that in this last case, two isolated voltage outputs are needed: one step-up for feeding the DC-link; and step-down for regulating the voltage value of the floating capacitor.

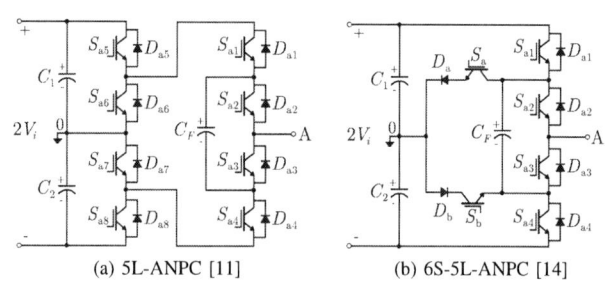

(a) 5L-ANPC [11]   (b) 6S-5L-ANPC [14]

Fig. 1. Five-level topologies

978-1-7281-4181-7/19 $31.00 © 2019 IEEE

This paper proposes an alternative five-level hybrid TFC inverter (5L-TFC) that requires a fixed flying capacitor voltage. For proper operation it is combined with a step-up voltage gain DC-DC boost-flyback converter with dual output using single power stage. Regulation of the flying capacitor is easily obtained without any complicated PWM strategy.

The paper is organized as follows: Section II describes operating principles of the proposed five-level combined converter; Section III discusses its modulation strategy and provides design considerations for the topology; Section IV gives the simulation verification; Section V presents the experimental results; Section VI presents the comparison between 5L-TFC and 6S-5L-ANPC and Section VII gives the conclusion.

II. THE PROPOSED FIVE-LEVEL COMBINED CONVERTER

Consider the five-level hybrid TFC topology presented in Fig. 2, in which the pole voltage is connected to the neutral point 0 via the bidirectional switch S_{a5}–S_{a6}, as in the figure.

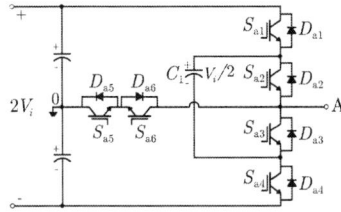

Fig. 2. Proposed hybrid 5L-TFC inverter topology.

Possible states of operation are indicated in Table I. When the flying capacitor voltage v_{C1} is chosen to be $V_i/2$ then five-level operation is obtained at the inverter output.

TABLE I
STATE OF OPERATION.

State	S_{a1}	S_{a2}	S_{a3}	S_{a4}	S_{a5}	S_{a6}	v_{A0}	v_{C1}
1	1	1	0	0	0	0	V_i	V_i
2	1	0	1	0	0	0	V_i-v_{C1}	$V_i/2$
3	0	0	0	0	1	1	0	0
4	0	1	0	1	0	0	-V_i+v_{C1}	-$V_i/2$
5	0	0	1	1	0	0	-V_i	-V_i

The use of a DC-DC boost-flyback converter (BFC) for feeding the flying capacitor results in the converter proposed in Fig. 3. Note that one more switch is added, S_1; switch S will not be taken into account since the transformer primary circuit can be considered as part of the circuit providing the DC-bus voltage.

The boost-flyback part of the proposed converter was conceived from the combination of a rearranged buck-boost converter [21] and a flyback converter. As shown in Fig. 3, such arrangement evidences the utilization of voltage $v_o = V_i + v_{C2}$ as DC-link voltage while providing the neutral point connection. This part is constituted of two diodes, D and D_1, two switches, S and S_1, a high-frequency transformer and two output capacitors, C_1 and C_2. The switch connected in anti-parallel with diode D_1 helps clamping the floating capacitor voltage at the desired value.

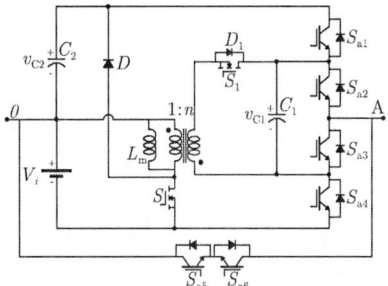

Fig. 3. Proposed combined boost/flyback/TFC converter.

A. The DC-DC Boost-Flyback Converter

The operation frequency of the boost/flyback converter (BFC) is higher than that of the power inverter itself. When considered in separate the BFC has two modes of operation: one when switch S is *on* and the other when it is *off*. Then for each operation state of the inverter, indicated in Table I, two modes of the DC-DC converter do occur. Then, there are twenty modes of operation for the combined converter. However, only ten of them, corresponding to the positive load current half-cycle, are presented in Fig. 4.

Before considering these ten modes of the combined converter, a description of the two modes of BFC will be given. One of them, Mode A, is shown in figures 4(a), 4(c), 4(e), 4(g) and 4(i); the other one, Mode B, is shown in figures 4(b), 4(d), 4(f), 4(h) and 4(j). In the following description, the coupled inductor is modeled as an ideal transformer, which consists of a turn ratio of $1/n$ and a magnetizing inductor L_m. It is considered that its operation is in the continuous mode (CCM). In addition, all components are considered to be ideal. The typical waveforms of the DC-DC converter are illustrated in Fig. 5, for CCM operation.

• Mode A $[0 < t < DT_s]$

At instant t_0 switch S is turned *on* and the magnetizing inductance is fed from the voltage source, that is, $v_{Lm} = V_i$. During this time interval diodes D_1 and D are reverse-biased and switch S_1 is *off*. The magnetizing current i_{Lm} increases (see Fig. 5) and reaches its peak value at instant DT_s. Same for current i_S. Capacitor C_1 maintains the voltage obtained in precedent mode of the combined converter. Capacitor C_2 together with the voltage source V_i supplies energy to the DC-link. Equations for Mode A, are:

$$\frac{di_{Lm}}{dt} = \frac{V_i}{L_m}; \quad \frac{dv_{C1}}{dt} = \frac{i_a}{C_1}; \quad \frac{dv_{C2}}{dt} = \frac{-i_a}{C_2} \quad (1)$$

In these equations i_a is the phase current. It will be seen that in certain modes the load current does not flow through the capacitor so that $i_a = 0$ in (1).

• Mode B $[DT_s < t < T_s]$

At instant DT_s switch S is turned *off*. Part of the energy stored in the magnetizing inductance L_m, at the end of the first mode, is transferred to capacitor C_2 via diode D. At the same time S is turned *off*, S_1 is turned *on*. Conduction of either D_1 or S_1 keeps the voltage across capacitor C_1 equal to that of the

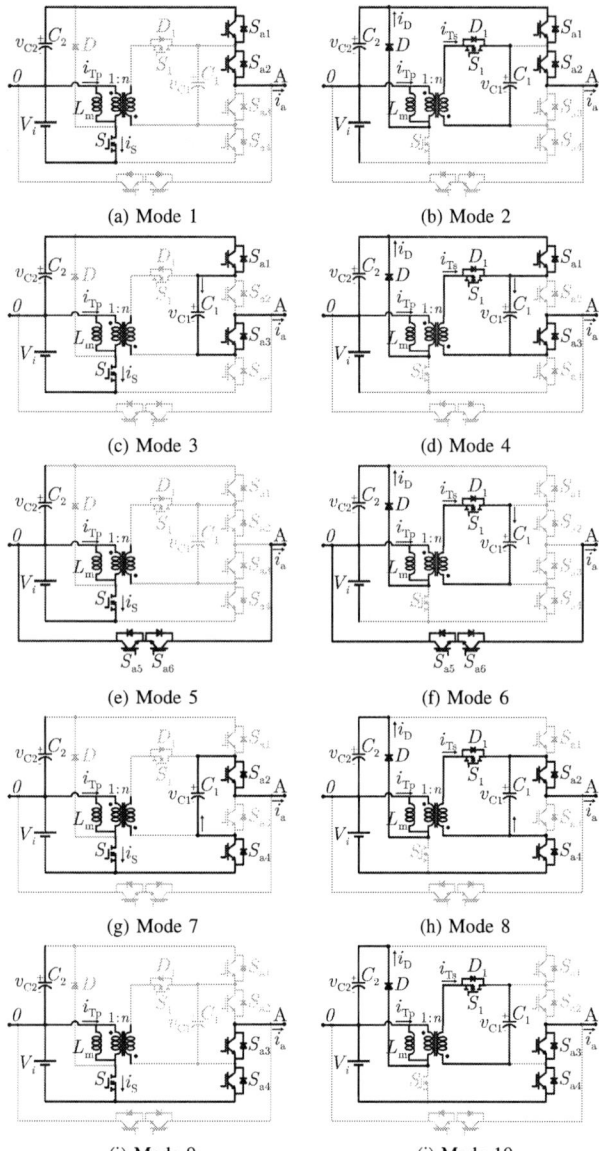

(a) Mode 1   (b) Mode 2

(c) Mode 3   (d) Mode 4

(e) Mode 5   (f) Mode 6

(g) Mode 7   (h) Mode 8

(i) Mode 9   (j) Mode 10

Fig. 4. Modes of operation.

transformer secondary voltage. Conduction of these devices depends on the inverter modes; in four of them capacitor C_1 is charged from the energy stored in the transformer. However, in other four ones the load current i_a plays an important role in both charge of capacitor C_1 and conduction of S_1. Equations for this interval are:

$$v_{Lm} = -V_{C2} = -V_i \qquad \frac{di_{Lm}}{dt} = \frac{-V_i}{L_m}$$

$$\frac{dv_{C1}}{dt} = \frac{\frac{i_{Lm}}{n} - i_a}{C_1} \qquad (2)$$

$$\frac{dv_{C2}}{dt} = \frac{i_{Lm} - i_a}{C_2}$$

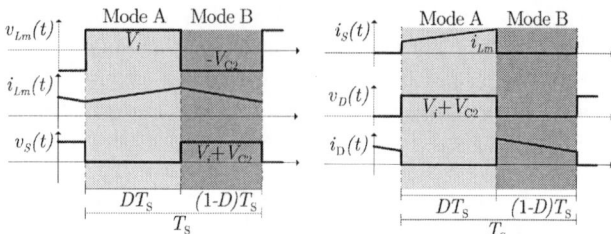

Fig. 5. Typical waveforms of the DC-DC converter

There are modes in which the load current does not flow through the capacitor so that $i_a = 0$ in (2). It should be noticed that mainly in this mode the secondary current i_{Ts} can be either positive (D_1 conducts) or negative (S_1 conducts), depending on the current imposed by the inverter.

The static gain is obtained through the energy balance in the magnetizing inductance with the help of Fig. 5, in which diode D_1 conducts the transformer secondary current. The gain corresponding to equations in (3) for winding ratio $n = 0.5$ is depicted in Fig. 6.

$$\frac{V_{C2}}{V_i} = \frac{D}{(1-D)} ; \quad \frac{V_{C1}}{V_i} = \frac{D}{n(1-D)} ; \quad \frac{V_0}{V_i} = \frac{1}{(1-D)} \quad (3)$$

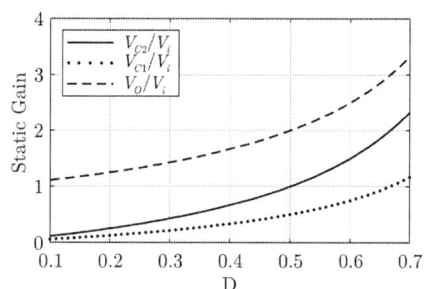

Fig. 6. Converter static gains.

For $D = 0.5$, the DC-link voltage is $2V_i$ and the flying capacitor voltage is $V_i/2$. These conditions are maintained for $D = 0.5$ even when winding resistances r_{e1} and r_{e2} in the coupled inductor, on-state equivalent resistances r_{e1} and r_{e2} of paralleled diodes D_1/S_1 and D_2/S_2, and on-state resistance $r_{DS(on)}$ of the power switch are considered.

B. The Five-Level Combined Converter

It has been shown above that operation of the DC-DC converter with $D = 0.5$ allows conditions so that the inverter can deliver five-level voltage at the phase output. Operation modes of the combined converter will be now described.

Mode 1 – In this mode S, S_{a1} and S_{a2} are conducting (S_{a3} and S_{a4} are *off*) while the magnetizing inductance absorbs energy from the voltage source. As in Mode A, diodes D_1 and D are reverse-biased and switch S_1 is *off*. The voltage applied to the phase output is $v_{A0} = +V_i$.

Mode 2 – S_{a1} and S_{a2} continue conducting but S is turned *off* and diode D starts conducting thus transferring energy from the magnetizing inductance to capacitor C_2. Diode D_1 makes

978-1-7281-4181-7/19 $31.00 © 2019 IEEE

possible to transfer part of the energy stored in magnetizing inductance into capacitor C_1. As in Mode B, switch S_1 is turned *on* in synchronism with S opening just to guarantee the voltage across capacitor C_1 remains around nV_i. The phase output is also $v_{A0} = +V_i$.

Mode 3 – During this stage S_{a2} is turned *off*, S_{a1} continues conducting and S_{a3} is turned *on*. Switch S is conducting so that the transformer stores energy and the behavior of the DC-DC converter is the same as in *Mode 1*. The voltage applied to the phase output is around $v_{A0} = +V_i - V_i/2 = +V_i/2$, since v_{C1} can suffers from variation in the previous mode.

Mode 4 – In this mode, S_{a2} continues to be *off* and S_{a3} is *on*, as in *Mode 3*, but S is turned *off*. Since the voltage across capacitor C_1 is intended to remain equal to nV_i then part of the load current flowing through S_{a1} and S_{a3} flows via S_1. The phase output is near $+V_i/2$.

Mode 5 – In this mode, except S_{a5} and S_{a6} all switches of the inverter are *off* and the voltage at the output is zero. Behavior of the DC-DC converter is the same as in *Mode 1*.

Mode 6 – The inverter behavior is the same as in *Mode 5* and the boost/flyback behavior is the same as in *Mode 2*.

Mode 7 – During this interval, S_{a1} and S_{a3} are *off* while S_{a2} and S_{a4} are *on*. As in *Mode 3*, the transformer is energized via switch S. The voltage applied to the phase output is $v_{A0} = -V_i + V_i/2 = -V_i/2$.

Mode 8 – As in *Mode 7*, S_{a1} and S_{a3} are *off* while S_{a2} and S_{a4} are *on* during this interval with the difference that S is turned *off*. As in *Mode 4*, the voltage across capacitor C_1 is intended to remain equal to nV_i; then part of the load current flowing through S_{a4} and S_{a2} flows via D_1. Phase output is also $v_{A0} = -V_i/2$.

Mode 9 – During this interval, S_{a1} and S_{a2} are *off* while S_{a3} and S_{a4} are *on*. As in *Mode 1*, S is conducting and the magnetizing current increases. Diodes D_1 and D are reverse-biased and switch S_1 is *off*. The voltage applied to the phase output is $-V_i$.

Mode 10 – During this stage, the inverter operates as in *Mode 9*, that is, S_{a1} and S_{a2} are *off*, S_{a3} and S_{a4} are *on*, S is *off*. Either diode D_1 is or switch S_1 is turned *on*, are able to conduct. Phase output voltage is also $-V_i$.

C. Design Methodology

The value of D is chosen from Fig. 6, the load current and magnetizing current are calculated from power specification. Inductances and capacitances are calculated from respective voltage ripple and current specifications. For instance, the flying capacitance can be calculated as a function of the peak current and the voltage variation from [14].

$$C_1 = \frac{I_{pk}}{2\,\Delta V_{C1}\,f_s\,m} \qquad (4)$$

in which I_{pk} is the peak current and m is the modulation index. Semiconductors are specified from *rms* and average values. In order to avoid current spikes in D_1/S_1, a small inductance should be connected in series with the transformer secondary winding.

III. Modulation Strategy

The pulse pattern, as shown in Fig. 7, can be obtained from a carrier-based PWM strategy. Four triangular carriers are level-shifted in phase disposition (PD PWM). Intersection of the modulating signal with each of the carriers from the top to the bottom generates pulses to S_{a3}, S_{a1}, S_{a5}, S_{a6}, S_{a4}, S_{a2}. It is worth mentioning that switches S_{a1} is complementary to S_{a4} and S_{a2} to S_{a3}, except when S_{a5} and/or S_{a6} are conducting; in this case they are *off*. The BFC uses the conventional ramp modulation with $D = 0.5$.

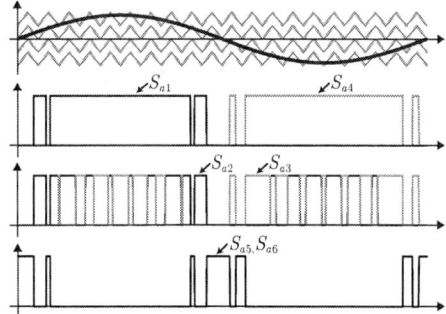

Fig. 7. PWM strategy and pulse pattern.

IV. Simulation Results

The topology of Fig. 3 was simulated with the help of PSIM for voltages $V_{C1} = 75$ V and a DC-link voltage $V_0 = 300$ V, from a DC voltage source, V_i, of 150 Volts. A duty cycle $D = 0.5$ was employed in the DC-DC converter what allows continuous conduction mode of operation. In order to reduce the transformer volume, the converter was operated at 40 kHz. A R-L circuit composed the load: $R_{Load} = 16\,\Omega$; $L_{Load} = 7$ mH. Design specifications and parameters are given in Table II and Table III.

TABLE II
DESIGN SPECIFICATION.

Specification	Symbol	Value
Ouput power of output 1	P_{01}	220 W
Inverter ouput power	P_0	450 W
Current ripple in L_m	Δi_{Lm}	$0.1 i_{Lm}$
Maximal voltage ripple in C_1	ΔV_{C1}	$0.017 V_{C1}$
Maximal voltage ripple in C_2	ΔV_{C2}	$0.004 V_{C2}$
DC converter modulation frequency	f_s	40 kHz
Duty-cycle	D	0.5
Inverter modulation frequency	f_0	10 kHz

TABLE III
CONVERTER PARAMETERS.

Specification	Symbol	Value
Winding ratio	n	0.5
Magnetizing inductance	L_m	4.3 mH
Floating capacitance	C_1	2200 μF
DC-link capacitance	C_2	2200 μF
Load resistance	R_{Load}	16 Ω
Load inductance	L_{Load}	7 mH

978-1-7281-4181-7/19 $31.00 © 2019 IEEE

From Fig. 8(a), it is observed that the output voltage has five levels and that the load current is satisfactory. Current total harmonic distortion (THD) and voltage weighted THD (WTHD) are shown in Fig. 8(b) and 8(c) respectively: they decrease with the modulation index and increase with reduction of the flying capacitance (not shown). Other results not shown are: losses decreases with power increase; adequate operation is achieved for any power factor; voltages V_{C1} and V_0 reach the expected values of 75 V and 300 V, respectively, from an input voltage of 150 V.

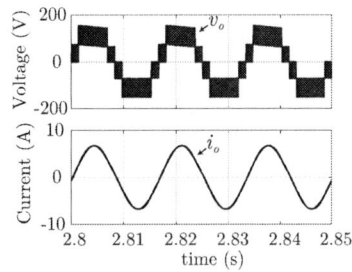

(a) Output voltage and load current.

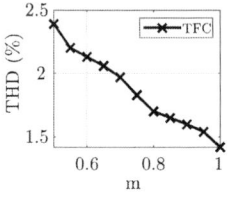

(b) THD vs. modulation index.

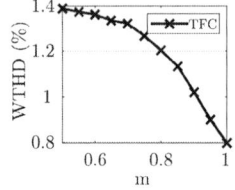

(c) WTHD vs. modulation index.

Fig. 8. Simulation results.

V. EXPERIMENTAL RESULTS

The proposed single-phase five-level combined converter was tested using the experimental setup of Fig. 9, for 450 W. An input voltage of 150 V originated a DC-link voltage of 300 V and a floating capacitor voltage of 75 V, as desired. This is shown in Fig. 10. Command signals of switches S_{a1}, S_{a2}, and S_{a5} are shown in Fig. 11. As expected, five levels have been obtained at the output voltage, as shown in Fig. 12. As it can be seen in Fig. 13, the voltage stress of devices are asymmetrical but at most equal to half of the DC-link.

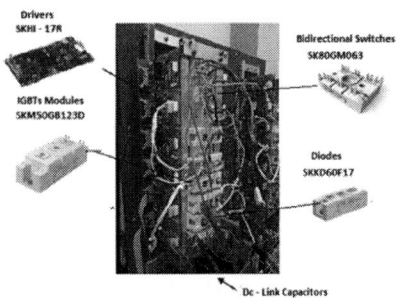

Fig. 9. Experimental setup.

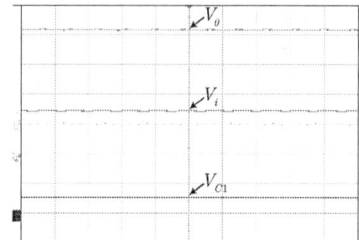

Fig. 10. Experimental results: DC-link voltage (top), DC-source voltage (middle); floating source voltage, 100 V/div.

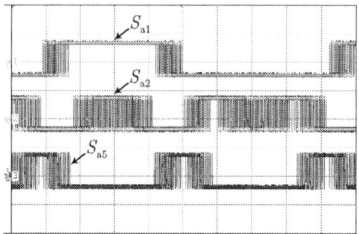

Fig. 11. Command signals for switches S_{a1} (top), S_{a2} (middle) and S_{a5} (bottom), 20 V/div.

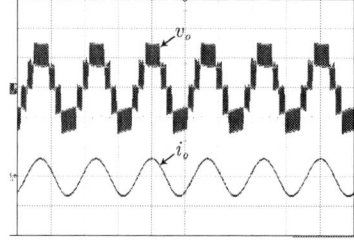

Fig. 12. Experimental results: Output voltage (100 V/div.); load current (20 A/div.).

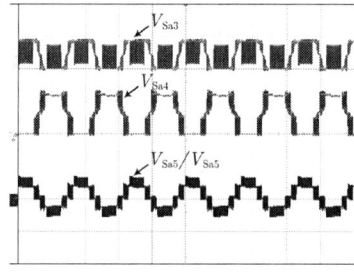

Fig. 13. Voltage stress of devices: from the top to the bottom: S_{a3} (100 V/div.); S_{a4} (200 V/div.); S_{a5}/S_{a6} (200 V/div.).

VI. COMPARISON BETWEEN PROPOSED 5L-TFC AND 6S-5L-ANPC

Compared to the conventional 5L-ANPC inverters, the 6S-5L-ANPC reduces two active switches and has lower conduction loss [14]. For comparison with 5L-TFC, the 6S-5L-ANPC DC-bus will be considered to be fed by the transformer primary side circuit. The comparison between 5L-TFC and 6S-5L-ANPC in Fig. 14 shows that both topologies have equivalent losses even if 5L-TFC includes an additional switch

978-1-7281-4181-7/19 $31.00 © 2019 IEEE

S_1. 5L-TFC can operate under any power factor with a conventional open-loop PWM strategy, while 6S-5L-ANPC needs a specific modulation strategy. Also, the voltage stress of devices is similar, as shown in Table IV, so that they can use same common devices.

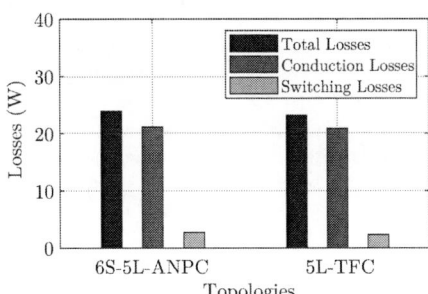

Fig. 14. Losses comparison.

TABLE IV
VOLTAGE STRESS OF DEVICES.

Device	6S-5L-ANPC	5L-TFC
S_{a1}/D_{a1}	$0.75V_i$	$0.75V_i$
S_{a2}/D_{a2}	$0.25V_i$	$0.25V_i$
S_{a3}/D_{a3}	$0.25V_i$	$0.25V_i$
S_{a4}/D_{a4}	$0.75V_i$	$0.75V_i$
S_{a5}/D_{a5} S_{a6}/D_{a6}	-	V_i
S_a/D_a	V_i	-
S_b/D_b	V_i	-

Disadvantages of the 5L-TFC are: difference between the secondary and floating capacitor voltages provokes spikes in D_1-S_1 current although they can be reduced by proper transformer design; modes 4 and 8 cause an increase in the transformer secondary current; additional switch S_1 and transformer increase the converter cost.

VII. CONCLUSIONS

This paper investigated a hybrid 5-level TFC inverter fed by an step-up DC-DC converter. The DC-DC converter is able to generate three output voltages DC, one of type boost, other of type buck-boost and third one of type flyback, of which the output voltage depends on the transformer winding ratio. The converter employs only one winding and a single switch. Two of these outputs were used, one to feed the DC-link voltage and the other to regulate the 5-level inverter floating capacitor voltage. Analysis of operation and explanation of the PWM technique have been included. Simulation results are presented and verified by experimental results. Different from the counterpart 6S-5L-ANPC, the proposed converter is able to operate under any power factor, while maintaining equivalent losses even having an extra switch and a high frequency transformer.

ACKNOWLEDGMENT

Authors acknowledge the Conselho Nacional de Desenvolvimento Científico e Tecnológico (CNPq), from Brazil, for funding this investigation

REFERENCES

[1] R. H. Baker and L. H. Bannister, "Electric Power Converter," Feb. 18 1975, United States Patent 3,867,643.

[2] M. Marchesoni, M. Mazzucchelli, and S. Tenconi, "A non conventional power converter for plasma stabilization," in *Proc. of PESC'88*, 1988, pp. 212–219.

[3] G. S. Zinoviev and N. N. Lopatkin, "Evolution of multilevel voltage source inverters," in *9th International Conference on Actual Problems of Electronic Instrument Engineering*, vol. 1, 2008, pp. 125–136.

[4] A. Nabae, I. Takahashi, and H. Akagi, "A new neutral-point-clamped PWM inverter," *IEEE Transactions on Industry Applications*, no. 5, pp. 518–523, 1981.

[5] H. Sugimoto, "New PWM control method for a three-level inverter," 1982, 80260 (Japan).

[6] T. Meynard and H. Foch, "Multi-level conversion: high voltage choppers and voltage-source inverters," in *PESC'92 Record. 23rd Annual IEEE Power Electronics Specialists Conference*, 1992, pp. 397–403.

[7] T. Bruckner and S. Bernet, "Loss balancing in three-level voltage source inverters applying active NPC switches," in *32nd Annual Power Electronics Specialists Conference (PESC)*, vol. 2, 2001, pp. 1135–1140.

[8] D.-Y. Lee, I. Choy, and D.-S. Hyun, "A new pwm DC/DC converter with isolated dual output using single power stage," in *IEEE Industry Applications Conference. 36th IAS Annual Meeting*, vol. 3, 2001, pp. 1889–1895.

[9] F. Z. Peng, "A generalized multilevel inverter topology with self voltage balancing," *IEEE Transactions on industry applications*, vol. 37, no. 2, pp. 611–618, 2001.

[10] X. Yuan, "Derivation of voltage source multilevel converter topologies," *IEEE Transactions on Industrial Electronics*, vol. 64, no. 2, pp. 966–976, 2016.

[11] P. Barbosa, P. Steimer, J. Steinke, M. Winkelnkemper, and N. Celanovic, "Active-neutral-point-clamped (ANPC) multilevel converter technology," in *European Conference on Power Electronics and Applications*, 2005.

[12] X. Kou, K. A. Corzine, and Y. L. Familiant, "Full binary combination schema for floating voltage source multi-level inverters," in *IEEE Industry Applications Conference. 37th IAS Annual Meeting*, vol. 4, 2002, pp. 2398–2404.

[13] T. B. Soeiro, R. Carballo, J. Moia, G. O. Garcia, and M. L. Heldwein, "Three-phase five-level active-neutral-point-clamped converters for medium voltage applications," in *Brazilian Power Electronics Conference*, 2013, pp. 85–91.

[14] H. Wang, L. Kou, Y.-F. Liu, and P. C. Sen, "A new six-switch five-level active neutral point clamped inverter for pv applications," *IEEE Transactions on Power Electronics*, vol. 32, no. 9, pp. 6700–6715, 2016.

[15] T.-J. Liang and K. Tseng, "Analysis of integrated boost-flyback step-up converter," *IEE Proceedings-Electric Power Applications*, vol. 152, no. 2, pp. 217–225, 2005.

[16] J. R. Dreher, F. Marangoni, J. L. Ortiz, M. L. d. S. Martins, H. T. Camara, and L. D. Flora, "High step-up voltage gain integrated DC/DC converters," in *IEEE International Symposium on Power Electronics for Distributed Generation Systems (PEDG)*, 2012, pp. 125–132.

[17] J. C. Giacomini, P. F. Costa, A. Andrade, L. Schuch, and M. L. Martins, "Desenvolvimento de um conversor CC–CC boost-forward integrado para aplicações com elevado ganho de tensão," *Eletrônica de Potência*, vol. 22, no. 2, pp. 206–214, 2017.

[18] A. Nami, F. Zare, A. Ghosh, and F. Blaabjerg, "Multi-output DC–DC converters based on diode-clamped converters configuration: topology and control strategy," *IET power electronics*, vol. 3, no. 2, pp. 197–208, 2010.

[19] J. C. Rosas-Caro, J. M. Ramírez, and P. M. García-Vite, "Novel DC–DC multilevel boost converter," in *IEEE Power Electronics Specialists Conference*, 2008, pp. 2146–2151.

[20] A. S. Andrade and E. R. da Silva, "DC-link control of a three-level NPC inverter fed by shaded photovoltaic system," in *13th Brazilian Power Electronics Conference and 1st Southern Power Electronics Conference (COBEP/SPEC)*, 2015, pp. 1–5.

[21] F. Z. Peng, L. M. Tolbert, and F. Khan, "Power electronics' circuit topology-the basic switching cells," in *IEEE Workshop Power Electronics Education*, 2005, pp. 52–57.

Integrated Zeta Inverter Applied in a Single-Phase Grid-Connected Photovoltaic System

Leonardo Poltronieri Sampaio
Departament of Electrical Engineering
Federal University of Techonology –
UTFPR-CP
Cornélio Procópio-PR, Brazil
sampaio@utfpr.edu.br

Sérgio Augusto Oliveira da Silva
Departament of Electrical Engineering
Federal University of Techonology –
UTFPR-CP
Cornélio Procópio-PR, Brazil
augus@utfpr.edu.br

Paulo Júnior Silva Costa
Departament of Electrical Engineering
Federal University of Techonology –
UTFPR-CP
Cornélio Procópio-PR, Brazil
paulojcosta@utfpr.edu.br

Abstract— This paper presents a novel integrated topology for step-up the photovoltaic array voltage, as well as to inject all the extracted PV array energy into the single-phase AC utility grid. The topology is based on the Zeta converter, named as Zeta Inverter, and its output current is controlled using hysteresis current control. The maximum power point tracking is performed employing the perturb and observe algorithm. Besides the obtaining low total harmonic distortion in the grid-injected current, the proposed system is proposed to replace the traditional double stage photovoltaic system. This paper presents the converter operation, the control technique, and the main simulation results in order to demonstrate the feasibility of the proposed system.

Keywords—Photovoltaic System, Integrated Converter, Zeta Converter, Power Converters

I. INTRODUCTION

Taking into account the technology evolution, as well as the power conversion, emerges the necessity to deal with different kinds of electrical energy sources for several conditions and applications. In this context, power electronics has been contributing significantly to the power conversion, developing and producing even more efficient products with a high-power factor, reduced losses, as well as in agreement with the required by the commercial and industrial sectors [1]–[4].

Commonly, there are situations in which the electrical energy source does not meet with the required by some equipment and/or system. For instance, the photovoltaic (PV) module and the fuel cell (FC) supplies energy in DC system, thus, they require power conversion stages to injected an AC current into the grid [5]–[7]. In addition, typically, the effective voltage for various residential/commercial applications is higher than the average output voltage of the PV and FC alternative sources [8], [9].

Usually, to overcome this limitation, an association of a step-up converter followed by a voltage source inverter (VSI) is employed [6], [10], [11]. This traditional double stage conversion is widely used due to the simplicity control designing and implementation, since each power conversion can be treated separately and the VSI presents a stable dynamic behavior [6], [12]–[14]. On the other hand, the aforementioned system can present reduced efficiency, great number of components, as well as high costs. One alternative could be the use of a single integrated topology, which is able to step-up the DC voltage and produces an AC output current with reduced number of components, increasing the overall system efficiency and reducing the costs [15]–[17].

In [15] has been presented a three-phase buck-boost inverter which can operate both grid-connected and in islanded mode. In [18] and [19] have been presented the integrated Boost DC/AC, which is deployed using two bidirectional Boost converters connected in parallel to the load or grid. Therefore, researches have proposed the development of new topologies suitable for photovoltaic applications, which are able to step-up and simultaneously convert to AC the input DC voltage[15], [17], [20]–[22].

Therefore, this paper proposes a single-phase integrated topology based on Zeta converter, named as Zeta Inverter. The Zeta inverter can step-up the output PV array voltage and inject into the utility grid AC current with reduced total harmonic distortion (THD). To perform the maximum power point tracking (MPPT) it is employed the well-known Perturb and Observe algorithm. The Zeta inverter output current is controlled by using the hysteresis current control.

This paper is organized as follows: Section 2 presents the main description of the PV system involving the Zeta Inverter. The P&O technique, the hysteresis control, the phased locked loop (PLL) system, as well as the voltage unbalance control used to control the Zeta Inverter are presented in Section 3. Simulation results are presented in Section 4,while the conclusions of this study are presented in Section 5.

II. DESCRIPTION OF THE PHOTOVOLTAIC SYSTEM

Fig. 1 presents the integrated Zeta Inverter employed to inject the extracted PV array energy into the utility grid. It comprises a PV array and an Integrated Zeta Inverter. The output current of the Zeta Inverter is controlled using hysteresis current control while the MPPT is performed by adopting the P&O-based MPPT.

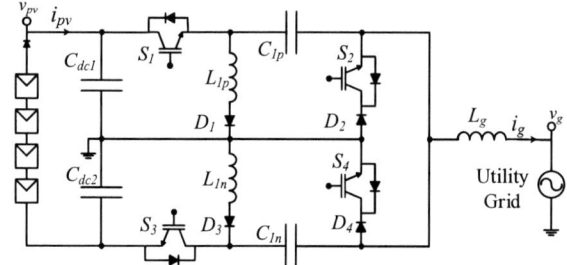

Fig. 1 Electrical circuit of the Integrated Zeta Inverter applied in a single-phase grid-tied Photovoltaic System.

A. Mathematical Model of the Adopted PV Cell

The PV module shown in Fig. 1 can be modelled by means of an equivalent electric circuit of a PV cell. The adopted model of the PV cell is depicted in Fig. 2 [23], where I_{ph} is the photocurrent; i_{Dpv} is the diode current; i_{pv_c} represents the output current of the PV cell; and v_{pv_c} is the output voltage of the PV cell. In addition, the series and parallel resistances are represented by R_s and R_p, respectively.

978-1-7281-4181-7/19 $31.00 © 2019 IEEE

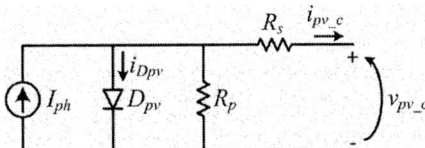

Fig. 2. PV cell equivalent circuit.

The i_{pv_c} and I_{ph} quantities can be computed, respectively, by (1) and (2), while the reverse saturation current I_r, the reverse saturation current in STC (Standard Test Condition) ($I_{r(STC)}$), the PV module voltage (v_{pv_m}), and the open circuit voltage of the PV system (V_{oc_m}) can be determined from (3), (4), (5) and (6), respectively.

$$i_{pv\_c} = I_{ph} - I_r \left[e^{q\frac{v_{pv\_c}+i_{pv}R_s}{\eta kT}} - 1 \right] - \frac{v_{pv\_c}+i_{pv}R_s}{R_p} \quad (1)$$

$$I_{ph} = [I_{SC} + \alpha(T - T_r)]\frac{G}{1000} \quad (2)$$

$$I_r = I_{r(STC)} \left(\frac{T}{T_r}\right)^3 e^{\left[\frac{qE_g}{\eta k}\left(\frac{1}{T_r}-\frac{1}{T}\right)\right]} \quad (3)$$

$$I_{r(STC)} = \frac{I_{SC} - \frac{V_{oc\_c}}{R_p}}{e^{\frac{qV_{oc\_c}}{\eta kT_r}} - 1} \quad (4)$$

$$v_{pv\_m} = v_{pv\_c}N_s \quad (5)$$

$$V_{oc\_m} = V_{oc\_c}N_s \quad (6)$$

where q is the electron charge; η is the ideality factor of the p-n junction; k is the Boltzmann constant; T_r is the nominal temperature in Kelvin (298 K); I_{SC} is the short-circuit current in STC ($T_r = 298$ K and G = 1000 W/m²); α is the temperature coefficient; E_g is the band gap energy (1.1 eV); V_{oc_c} is the open-circuit voltage of the photovoltaic cell; and N_s is the number of photovoltaic cells.

The group of equations from (1) to (6) may be solved using a numerical approach. In this paper, the Newton-Raphson algorithm is employed to calculate the PV cell current i_{pv_c} for a certain operation point, as follows:

$$i_{pv_{n+1}} = i_{pv_n} - \frac{f\left(i_{pv_n}\right)}{f'\left(i_{pv_n}\right)} \quad (7)$$

$$f\left(i_{pv_n}\right) = I_{ph} - i_{pv_n} - I_r \left[e^{q\frac{v_{pv\_c}+i_{pv_n}R_s}{\eta kT}} - 1 \right] \\ - \frac{v_{pv\_c}+i_{pv_n}R_s}{R_p} \quad (8)$$

$$f'\left(i_{pv_n}\right) = -1 - I_r \left(e^{q\frac{v_{pv\_c}+i_{pv_n}R_s}{\eta kT}} \right)\frac{qR_s}{\eta kT} - \frac{R_s}{R_p} \quad (9)$$

where n is the n-th iteration of the algorithm, and $f'\left(i_{pv\_c_n}\right)$ is the derivative of the current i_{pv_c} corresponding to iteration n. Therefore, (8) and (9) are used to determine the current of the photovoltaic cell for each iteration.

Finally, from the set of equations described in this section, the voltage (v_{pv}) and current (i_{pv}), resulting from the conception of a certain series and/or parallel PV arrangement can be determined.

B. Operation of the Zeta Inverter

The proposed converter is controlled by a hysteresis control strategy, in which there is the imposition of the injected current into the grid. The magnitude of the injected current is determined by means of the PV array voltage controller.

The operating principle of the proposed converter is based on the hysteresis control, in which during the positive half-cycle of the grid voltage the switches S_1 and S_2 are commuted, while during the negative half cycle are commuted the switches S_3 and S_4.

1) Positive half cycle:

a) Interval t_a: switch S_1 is turned on while S_2 is turned off, energizing the inductors L_{1p} and L_g.

b) Interval t_b: S_1 is turned off and S_2 is turned on, desenergizing the inductors L_{1p} and L_g. At the end of the interval t_b, the current through the inductor L_{1p} reaches zero and the diode D_1 is turned off.

c) Interval t_c: inductor L_g remains discharging until the end of the switching period.

2) Negative half cycle:

a) Interval $t_{a'}$: switch S_3 is turned on while S_4 is turned off, energizing the inductors L_{1n} and L_g.

b) Interval $t_{b'}$: S_3 is turned off and S_4 is turned on, deenergizing the inductors L_{1n} and L_g. At the end of the interval $t_{b'}$, the current through the inductor L_{1n} reaches zero and the diode D_2 is turned off.

c) Interval $t_{c'}$: The inductor L_g remains discharging until the end of the switching period.

Fig. 3 shows the diagram control of the Zeta Inverter switches' gates, whereas Fig. 4 shows the equivalent electrical circuit for each time interval previously described.

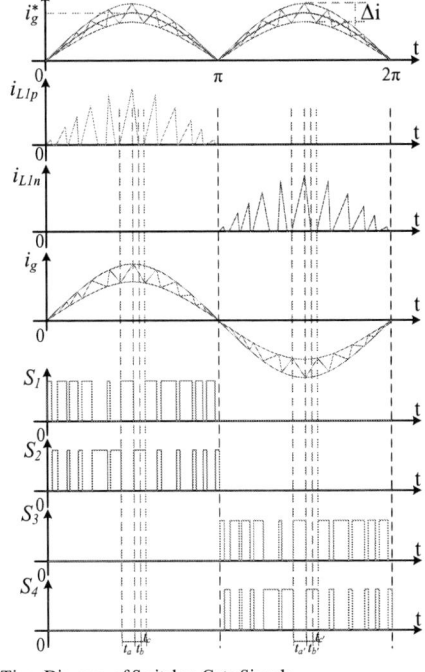

Fig. 3. Time Diagram of Switches Gate Signals.

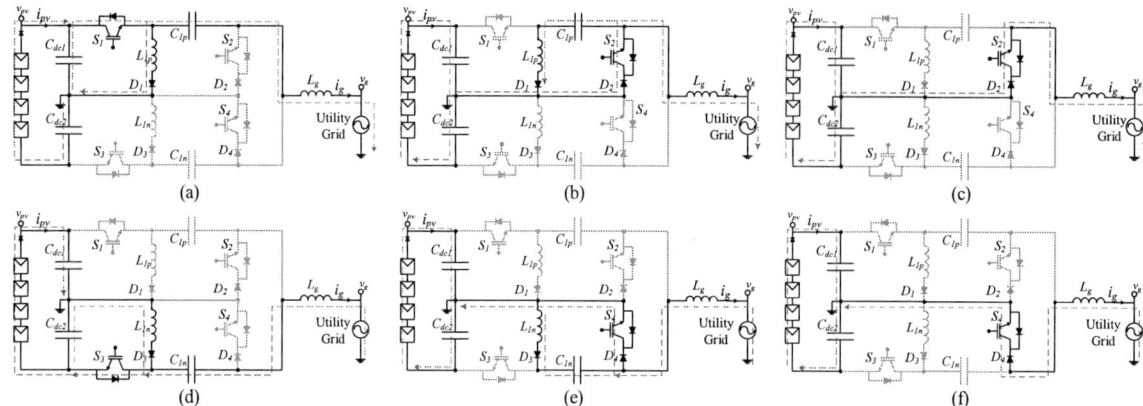

Fig. 4. Equivalent circuit of Zeta Inverter during the time intervals: (a) t_a, (b) t_b, (c) t_c, (d) $t_{a'}$, (e) $t_{b'}$, (f) $t_{c'}$.

According to the electrical circuit, the ideal converter static gain can be determined as:

$$G(\omega t) = \frac{v_g(\omega t)}{V_{pv}} = \frac{D_a(\omega t)^2 + D_a(\omega t)D_b(\omega t)}{2D_b(\omega t)} \quad (10)$$

From (10), the static gain related to the peak amplitude of the grid voltage can be expressed as follows:

$$G_M = \frac{V_g}{V_{pv}} = \frac{D_A{}^2 + D_A D_B}{2D_B} \quad (11)$$

where v_g is the sinusoidal grid voltage; V_g is the peak amplitude of v_g; V_{pv} is the average PV array voltage; D_a and D_b represent the respective subinterval lengths of t_a and t_b; and D_A and D_B represent, respectively, the subinterval lengths of t_a and t_b related to the peak amplitude of V_g.

The quantities D_A and D_B can be found as

$$D_A = \frac{-\frac{V_{pv}}{2} + \sqrt{\left(\frac{D_B V_{pv}}{2}\right)^2 + 2V_{pv}D_B V_g}}{V_{pv}} \quad (12)$$

$$D_B = \frac{4I_g L_1 f_s}{V_{pv}} \quad (13)$$

where f_s is the switching frequency and I_g is the peak current of the output inductance L_g.

III. Control Techniques

This section presents the P&O-based MPPT technique, the hysteresis current control, the DC-bus voltage balance control and the PLL system adopted in this paper to control the grid-injected current.

A. P&O MPPT Algorithm

The P&O method is a well-know MPPT algorithm in which its main principle is based on the increment or decrement of the controlled variable (voltage, current or duty cycle) [24]. This method is also known as the Hill-Climbing method, which its operation principle is to find the maximum power point of the PV array characteristic curve ($P_{pv} \times v_{pv}$). In this paper, the MPPT algorithm based on P&O is used to determine the voltage reference (v_{pv}^*) employed in the PV array voltage (v_{pv}) control. The flowchart of the P&O MPPT algorithm is presented in Fig. 5.

B. PV Arry Voltage Control

By means of the PV array voltage controller, the current reference (I_g^*) used in the hysteresis current control is obtained. Disregarding the power losses, the power absorbed or supplied by the PV array dc-bus can be expressed as:

$$v_{Pv}i_{Cdc} = \frac{V_g I_g}{2} \quad (14)$$

From (14), the transfer function of the PV array voltage control is given by:

$$G_{v_{pv}}(s) = \frac{\hat{v}_{pv}(s)}{\hat{i}_{cdc}(s)} = \frac{V_g}{2V_{pv}C_{dc}s} \quad (15)$$

where i_{Cdc} is the current that flows through the capacitors C_{dc1} and C_{dc2}; and C_{dc} is the equivalent PV array dc-bus capacitance.

C. Hysteresis Current Control

From the P&O MPPT technique, the value of the voltage reference is determined and employed to control the PV array voltage. Therefore, by means of a proportional-integral (PI) controller the PV array voltage is controlled, generating the peak magnitude of the current reference (I_g^*) to be injected into the grid. In addition, a PLL system is used to estimate the utility voltage phase-angle (θ), which is used to calculate the coordinate $\sin(\theta)$, which is employed to determine the output current reference for the hysteresis current control (i_g^*). Fig. 6 illustrated the block diagram of the hysteresis current control adopted for the Zeta Inverter.

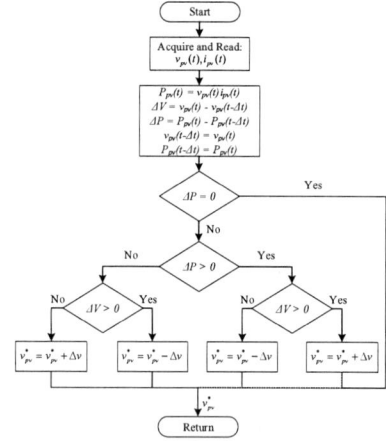

Fig. 5. Flowchart of the P&O-based MPPT algorithm.

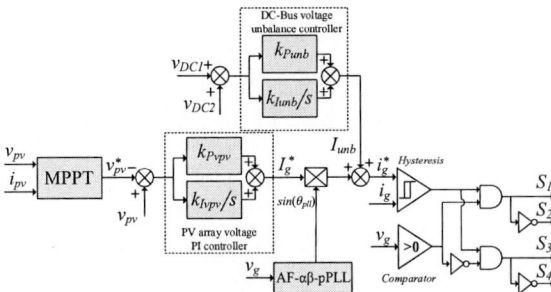

Fig. 6. Block diagram of the hysteresis current control.

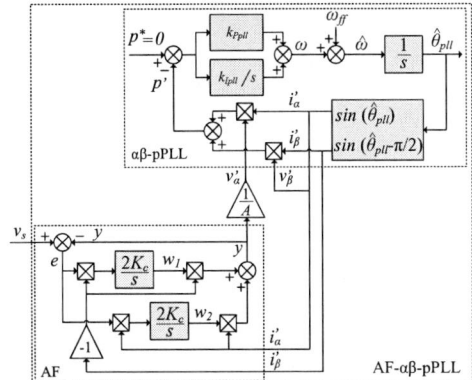

Fig. 7. Block diagram of single-phase AF-αβ-pPLL scheme.

D. Unbalance DC-bus Voltage Control

Once the presented topology requires a split capacitor configuration in the DC-bus, voltage unbalances can occur between the v_{DC1} and v_{DC2} leading the system to operate under undesired conditions [25]. Therefore, to overcome this problem, a PI controller must be used to control the voltage balancing between the DC-bus capacitors. Thus, as shown in Fig. 6, the output signal of the PI controller (I_{unb}) is used to compose the current reference of the hysteresis current control.

E. PLL System

Fig. 7 presents the single-phase PLL algorithm used in the Zeta Inverter, where its description is detailed in [26]. In the adopted PLL, a non-autonomous adaptive filter (AF) operates in conjunction with the $\alpha\beta$-pPLL in order to improve the PLL performance when the utility voltage has disturbances like, for instance, voltage harmonics. Moreover, the complete procedures used to determine the proportional (k_{Ppll}) and integral (k_{Ipll}) of the PLL controller gains are also presented in [26].

IV. SIMULATION RESULTS

The proposed integrated Zeta Inverter was simulated using the software MATLAB & Simulink. In this paper it is adopted four series-connected PV modules with total PV array power of 980 Wp. Table I presents the parameters of the PV module employed in this paper, whereas Table II summarizes the common parameters used for the Zeta Inverter. Table III presents the main controller parameters used in the system implementation.

Fig. 8 presents the PV array power (P_{pv}), voltage (v_{pv}) and current (i_{pv}), as well as the injected current into the grid (i_g) for the system operating at MPP. As can be noted, the solar

irradiance varies from 500 to 750 W/m² and after that from 750 to 1000 W/m² and the MPP is achieved in all situations. It can also be noticed that the PV array voltage and current present low oscillations during the solar irradiance changes.

TABLE I.   STANDARD TEST CONDITIONS (STC) OF THE SOLARWORLD SUNMODULE PLUS SW 245 PV MODULE

Maximum PV Power	P_{MAX} = 245 W
MPP voltage	V_{MPP} = 30.8 V
MPP current	I_{MPP} = 7.96 A
Open-circuit voltage	V_{OC} = 37.5 V
Short-circuit current	I_{SC} = 8.49 A

TABLE II.   ZETA INVERTER PARAMETERS

Nominal utility rms voltage	V_g = 127 V
Nominal utility frequency	f_g = 60 Hz
Zeta Inverter nominal power	S_{inv} = 1000 VA
Nominal input DC-bus voltage	V_{pv} = 123 V
Input DC-Bus capacitance	$C_{dc1} = C_{dc2}$ = 4 mF
Input inductive filter	$L_{Ip} = L_{In}$ = 50 µH
Output inductive filter	L_g = 6 mH
Coupling capacitances	$C_{Ip} = C_{In}$ = 40 µF

TABLE III.   CONTROLLERS' PARAMETERS

Hysteresis current limits	Δi = 0.1 A
PV Array voltage PI controller gains	k_{Pvpv} = 0.04; k_{Ivpv} = 0.3
Unbalance voltage controller gains	k_{Punb} = 0.08; k_{Iunb} = 0.01
PLL PI controller gains	k_{Ppll} = 424.3; k_{Ipll} = 32234
Adaptive filter step size parameter (AF-pPLL)	μ = 0.007
Sampling time (AF-pPLL)	T = 16.66 µs
Adaptive filter gain (AF-pPLL)	K_c = 420
Cut off frequency (SRF Controller) (2nd order LPF Butterworth filter)	f_c = 30 Hz
P&O sampling time	$T_{S_{MPPT}}$ = 1 s
P&O voltage step size	Δv = 1 V

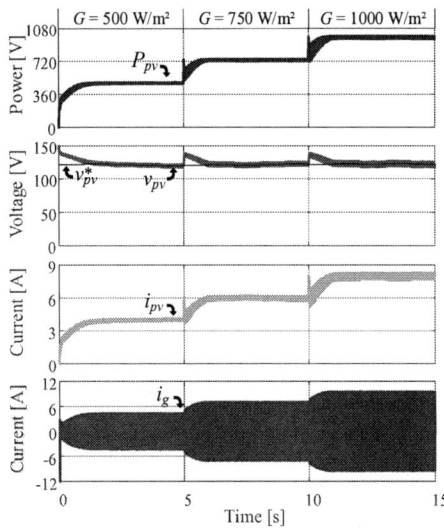

Fig. 8. PV array power (P_{pv}), voltage (v_{pv}) and current (i_{pv}) and grid current (i_g) during transients of solar irradiance using the P&O MPPT algorithm.

978-1-7281-4181-7/19 $31.00 © 2019 IEEE

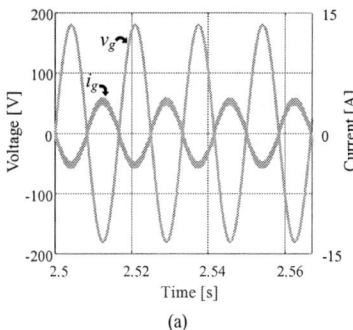 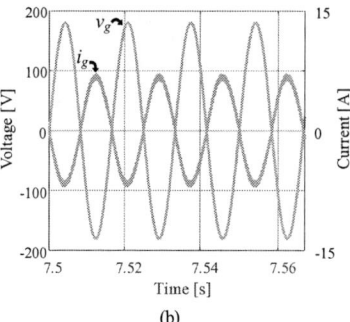 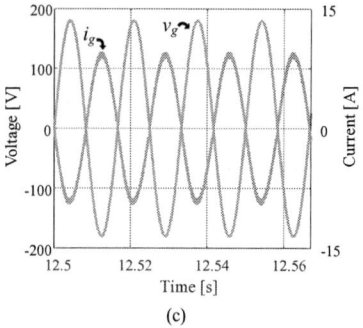

(a)   (b)   (c)

Fig. 9. Grid voltage (v_g) and injected grid current (v_g) considering three different solar irradiance values: (a) $G = 500$ W/m²; (b) $G = 750$ W/m²; and (c) $G = 1000$ W/m².

Fig. 9 shows the utility grid voltage (v_g) and the grid-injected current (i_g) considering the solar irradiance changes presented in Fig. 8. As can be seen, for all the three conditions of solar irradiance the injected grid current presented sinusoidal waveforms with THD lower than 4.5 %, complying to the IEEE and IEC standards [27], [28]. Table IV presents the THD of the injected grid current for the PV system operating at MPP for the three evaluated solar irradiation values.

TABLE IV.   THD OF THE INJECTED CURRENT INTO THE GRID

Interval (s)	Solar Irradiance (W/m²)	THDig (%)
0 − 5	500	4.7
5 −10	750	2.6
10 − 15	1000	2.2

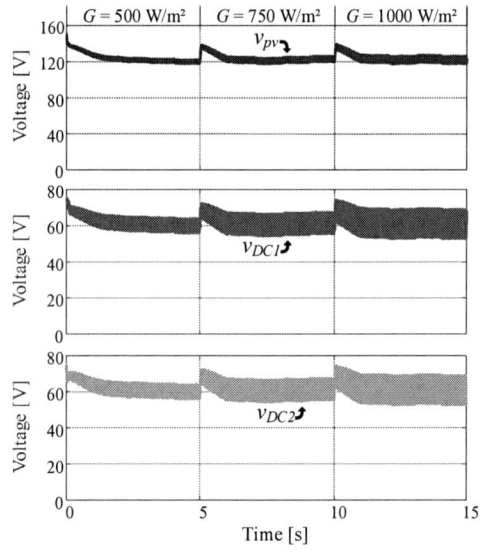

Fig. 10. PV array voltage (v_{pv}) and input capacitors' voltages (v_{DC1} and v_{DC2}) considering changes in the solar irradiance.

The PV array voltage (v_{pv}) and the voltages across the input capacitors (v_{DC1} and v_{DC2}) are shown in Fig. 10. It can be noted that the voltage v_{DC1} presents very similar behavior and magnitude compared to the voltage v_{DC2}, due to the use of the voltage unbalance controller, guarantying the safe converter operation, mainly, when the solar irradiance variations occur.

V. CONCLUSION

This paper presented a proposal of a novel inverter topology based on the Zeta converter, named as Zeta Inverter. The inverter topology was evaluated in a single-phase grid-tied PV system, where boost DC PV array voltage, as well as injection into the grid sinusoidal current with low THD were achieved.

The Zeta Inverter current was controlled employing hysteresis control, which presented low THD for different values of solar irradiance, being in compliance with the IEEE and IEC standards.

The presented control was able to regulate the PV array voltage, as well as the balance voltage of the input capacitors used to determine the current reference. In addition, the P&O technique was used to perform the MPPT, which achieved the MPP for all the solar irradiance evaluated cases.

Finally, by means of simulation results, the Zeta Inverter topology presented a good performance, as well as demonstrated be a suitable topology to replace the traditional double stage system composed of a step-up converter associated with a voltage source inverter.

ACKNOWLEDGMENT

The authors gratefully acknowledge the financial support received from CNPq (Process 400837/2016-1).

REFERENCES

[1] A. Bindra and T. Keim, "Exciting Advances in Power Electronics: APEC 2018 Divulges Latest Advances in Magnetics, Wide-Bandgap Devices, Vehicle Batteries, 3-D Packaging, and More," *IEEE Power Electron. Mag.*, vol. 5, no. 2, pp. 49–55, 2018.

[2] F. Blaabjerg, Y. Yang, K. Ma, and X. Wang, "Power electronics - the key technology for renewable energy system integration," in *2015 International Conference on Renewable Energy Research and Applications (ICRERA)*, 2015, pp. 1618–1626.

[3] D. L. R. Vidor, E. Rosa, N. Rigo, and J. R. Pinheiro, "A future trend for converters: Switched and linear together," in *2017 IEEE 8th International Symposium on Power Electronics for Distributed Generation Systems (PEDG)*, 2017, pp. 1–7.

[4] B. K. Bose, "The past, present, and future of power electronics [Guest Introduction]," *IEEE Ind. Electron. Mag.*, vol. 3, no. 2, pp. 7-11,14, 2009.

[5] M. Forouzesh, Y. P. Siwakoti, S. A. Gorji, F. Blaabjerg, and B. Lehman, "Step-Up DC–DC Converters: A Comprehensive Review of Voltage-Boosting Techniques, Topologies, and Applications," *IEEE Trans. Power Electron.*, vol. 32, no. 12, pp. 9143–9178, 2017.

[6] M. H. de Freitas Takami, S. A. Oliveira da Silva, and L. P. Sampaio, "Dynamic performance comparison involving grid-connected PV systems operating with active power-line conditioning and subjected to sudden solar irradiation changes," *IET Renew. Power Gener.*, vol. 13,

978-1-7281-4181-7/19 $31.00 © 2019 IEEE   1305

no. 4, pp. 587–597, 2019.

[7] F. L. Tofoli, D. d. C. Pereira, W. Josias de Paula, and D. d. S. Oliveira Júnior, "Survey on non-isolated high-voltage step-up dc–dc topologies based on the boost converter," *IET Power Electron.*, vol. 8, no. 10, pp. 2044–2057, 2015.

[8] R. Mechouma, B. Azoui, and M. Chaabane, "Three-phase grid connected inverter for photovoltaic systems, a review," in *2012 First International Conference on Renewable Energies and Vehicular Technology*, 2012, pp. 37–42.

[9] J. C. S. de Morais, R. Gules, J. L. S. de Morais, and L. G. Fernandes, "Transformerless DC-DC converter with high voltage gain based on a switched-inductor structure applied to photovoltaic systems," in *2017 Brazilian Power Electronics Conference (COBEP)*, 2017, pp. 1–6.

[10] S. A. O. da Silva, L. P. Sampaio, F. M. de Oliveira, and F. R. Durand, "Feed-forward DC-bus control loop applied to a single-phase grid-connected PV system operating with PSO-based MPPT technique and active power-line conditioning," *IET Renew. Power Gener.*, vol. 11, no. 1, pp. 183–193, 2017.

[11] H. Xiang, Y. Yan, and H. Jiang, "A two-stage PV grid-connected inverter with optimized anti-islanding protection method," in *2009 International Conference on Sustainable Power Generation and Supply*, 2009, pp. 1–4.

[12] N. Altin, S. Ozdemir, H. Komurcugil, I. Sefa, and S. Biricik, "Two-stage grid-connected inverter for PV systems," in *2018 IEEE 12th International Conference on Compatibility, Power Electronics and Power Engineering (CPE-POWERENG 2018)*, 2018, pp. 1–6.

[13] S. F. P. Dias and M. Stefanello, "Control of weighted average current of Voltage sourced Inverters with LCL-filter for grid connected applications," in *2015 IEEE 13th Brazilian Power Electronics Conference and 1st Southern Power Electronics Conference (COBEP/SPEC)*, 2015, pp. 1–5.

[14] B. A. Angélico, S. A. O. da Silva, and L. B. G. Campanhol, "Proportional–integral/proportional–integral-derivative   tuning procedure of a single-phase shunt active power filter using Bode diagram," *IET Power Electron.*, vol. 7, no. 10, pp. 2647–2659, 2014.

[15] M. A. G. de Brito, L. P. Sampaio, G. De Azevedo Melo, and C. A. Canesin, "Three-phase tri-state buck-boost integrated inverter for solar applications," *IET Renew. Power Gener.*, vol. 9, no. 6, pp. 557–565, 2015.

[16] H. G. Cabral *et al.*, "A PV module-integrated inverter using a Ćuk converter in DCM with island detection scheme," in *2015 IEEE 13th Brazilian Power Electronics Conference and 1st Southern Power Electronics Conference (COBEP/SPEC)*, 2015, pp. 1–6.

[17] L. Galotto, M. A. G. De Brito, L. P. Sampaio, and C. A. Canesin, "Integrated single-stage converters with Tri-state modulation suitable for photovoltaic systems," in *COBEP 2011 - 11th Brazilian Power Electronics Conference*, 2011.

[18] R. O. Caceres and I. Barbi, "A boost DC-AC converter: analysis, design, and experimentation," *IEEE Trans. Power Electron.*, vol. 14, no. 1, pp. 134–141, Jan. 1999.

[19] P. Sanchis, A. Ursaea, E. Gubia, and L. Marroyo, "Boost DC-AC inverter: a new control strategy," *IEEE Trans. Power Electron.*, vol. 20, no. 2, pp. 343–353, 2005.

[20] L. G. Junior, M. A. G. De Brito, L. P. Sampaio, and C. A. Canesin, "Evaluation of integrated inverter topologies for low power PV systems," in *3rd International Conference on Clean Electrical Power: Renewable Energy Resources Impact, ICCEP 2011*, 2011.

[21] F. C. Melo, L. S. Garcia, L. C. de Freitas, E. A. A. Coelho, V. J. Farias, and L. C. G. de Freitas, "Proposal of a Photovoltaic AC-Module With a Single-Stage Transformerless Grid-Connected Boost Microinverter," *IEEE Trans. Ind. Electron.*, vol. 65, no. 3, pp. 2289–2301, 2018.

[22] D. B. Bizarro, R. B. Godoy, L. Galotto, and J. O. P. Pinto, "New topology and predictive control for photovoltaic application with inductive power decoupling capability," in *2015 IEEE 13th Brazilian Power Electronics Conference and 1st Southern Power Electronics Conference (COBEP/SPEC)*, 2015, pp. 1–6.

[23] M. M. C. D. C. Martins, "Modelo de Arranjo Fotovoltaico Destinado a Análise em Eletrônica de Potência via Simulação," *Eletrônica Potência - SOBRAEP (in Port.*, vol. 13, no. 3, pp. 141–146, 2008.

[24] M. A. G. De Brito, L. Galotto, L. P. Sampaio, G. De Azevedo Melo, and C. A. Canesin, "Evaluation of the main MPPT techniques for photovoltaic applications," *IEEE Trans. Ind. Electron.*, vol. 60, no. 3, 2013.

[25] L. B. G. Campanhol, S. A. O. Da Silva, A. A. De Oliveira, and V. D. Bacon, "Dynamic Performance Improvement of a Grid-Tied PV System Using a Feed-Forward Control Loop Acting on the NPC Inverter Currents," *IEEE Trans. Ind. Electron.*, vol. 64, no. 3, pp. 2092–2101, 2017.

[26] V. D. Bacon, S. A. O. da Silva, L. B. G. Campanhol, and B. A. Angélico, "Stability analysis and performance evaluation of a single-phase phase-locked loop algorithm using a non-autonomous adaptive filter," *IET Power Electron.*, vol. 7, no. 8, pp. 2081–2092, 2014.

[27] IEEE Std 1547-2018, "IEEE Standard for Interconnection and Interoperability of Distributed Energy Resources with Associated Electric Power Systems Interfaces Sponsored by the IEEE Standard for Interconnection and Interoperability of Distributed Energy Resources with Associate," 2018.

[28] IEC, "IEC 61727- Photovoltaic (PV) Systems - Characteristics of the Utility Interface," 2004.

A Novel Phase-Locked Loop with Positive and Negative Sequence Detection Capability

Jean M. L. Fonseca*, Francisco Kleber A. Lima*, Carlos Gustavo C. Branco*, Renato G. Araujo*
*Federal University of Ceará
Fortaleza, Brazil
Emails: jlobodaf@alu.ufc.br, klima@dee.ufc.br, gustavo@dee.ufc.br, renato.g.a@hotmail.com

Abstract—This paper proposes the expansion of two phase-locked loop structures to detect positive and negative sequence components. The first structure is a positive sequence detector only and the second one is a modification of the first one where an adaptive filter is added to increase filtering capability in scenarios with interharmonic distortion. Both proposed structures were implemented in the hardware-in-the-loop (HIL) dSPACE 1103 platform together with two well known structures, the DSOGI-FLL and the DDSRF-PLL. Steady-state and dynamic response of each structure were analyzed in scenarios with either harmonic distortion only or harmonic and interharmonic distortion. Results showed that the proposed structures are highly accurate, specifically under harmonic distortion, however the structure with adaptive filter is slower. In addition, results showed that the structures are at least as precise and fast as the DSOGI-FLL and DDSRF-PLL. Lastly, under interharmonic distortion, it is also found that the PLL with adaptive filter has an overall better performance.

Index Terms—Phase-locked loop, Harmonics, Interharmonics, Negative Sequence Detector, dSPACE.

I. INTRODUCTION

Grid-connected power converters have a burgeoning role in the modern power grid. They already are a vital part of distributed energy resources (DERs), highlighting, for instance, their use in Photovoltaic Generation (PV) and Wind Power Generation (WPG) technologies like the doubly-fed induction generator (DFIG) [1]. These power electronics-based equipment, which have experienced a downfall in cost, also play an important role in solving common power quality problems with structures such as the static synchronous compensator (STATCOM) [2] and active filters [3] becoming popular solutions even at the utility distribution level.

In order to operate correctly, power converters need specific information about some of the grid electric quantities, namely: frequency, currents, voltages and phase angle [4]. This information regarding the state of the grid is employed in control structures, where determined objectives are well defined. Examples of objectives are: voltage regulation and power factor correction (reactive power compensation) for STATCOMs [2]; actively preventing undesirable non-fundamental currents into the grid for active filters [3].

The authors would like to thank the National Council for Scientific and Technological Development (CNPq) and Coordination for the Improvement of Higher Level Personnel (CAPES) for funding this research.

There are several solutions for the online monitoring of these variables, however, the most widespread alternatives are based on phase-locked loops (PLLs) [5], [6]. PLLs are supposed to obtain this information in a fast and precise manner. There are several PLL structures in the academic literature, nevertheless, structures like the SRF-PLL [7], DDSRF-PLL [8] and DSOGI-FLL [9] deserve more attention because of their simplicity and exactness.

In this paper however, two different structures will be analyzed [10]–[12]. The first structure was proposed in [10] and is a robust positive sequence detector, that is, it estimates phase angle, amplitude and fundamental frequency of the positive sequence component of the voltage/current. The second structure is a modification of the first structure where a adaptive filter (AF) is added in order to enhance filtering capability in scenarios with interharmonic distortion [11]–[13].

This paper proposes to expand both structures so that they are not only able to detect positive sequence components, but negative sequence components as well. Negative sequence components arise from grid voltage unbalance caused by unbalanced loads and/or grid faults [14]. Detecting negative sequence components are important to indicate the severity of the voltage unbalance [14] but they can also be used in others control algorithms [15] that will allow the power converter to stay connected while still providing (a more limited) functionality, thus increasing low voltage right through (LVRT) and abiding the ever demanding grid code regulations.

This paper is organized as follows: Section II provides the mathematical modeling of the proposed structures based on the original ones presented in [10]–[12]; Section III provides steady-state and dynamic experimental results obtained in scenarios involving harmonic and interharmonic distortion; Section IV concludes the paper with final remarks regarding the proposed structures.

II. MODELING THE PROPOSED STRUCTURES

The two proposed structures operate very similarly, as such, the basic operation of the first structure will be explained thoroughly. After this, the second structure will be presented and the differences will be clearly pointed out.

978-1-7281-4181-7/19 $31.00 © 2019 IEEE

A. Basic operation (sequence extraction)

The complete structure of the proposed positive and negative sequence detector are illustrated in Figure 1. The proposed structures are capable of estimating the frequency ($\hat{f}_1$), phase angle and amplitude of both positive and negative sequence components of a generic input signal $u_{\alpha\beta}(t)$, defined in the stationary frame $\alpha\beta$. It does so by projecting $u_{\alpha\beta}(t)$ onto two complex subspaces defined as:

$$
\begin{cases}
e^{-(2\pi\hat{f}_1 t)} = cos(2\pi\hat{f}_1 t) - jsin(2\pi\hat{f}_1 t) \\
e^{-(2\pi\hat{f}_1 t - \pi/2)} = cos(2\pi\hat{f}_1 t - \pi/2) - jsin(2\pi\hat{f}_1 t - \pi/2)
\end{cases},
$$
(1)

for the positive sequence components, and:

$$
\begin{cases}
e^{-(2\pi\hat{f}_1 t)} = cos(2\pi\hat{f}_1 t) - jsin(2\pi\hat{f}_1 t) \\
e^{-(2\pi\hat{f}_1 t + \pi/2)} = cos(2\pi\hat{f}_1 t + \pi/2) - jsin(2\pi\hat{f}_1 t + \pi/2)
\end{cases},
$$
(2)

for the negative sequence components. It is important to highlight that to obtain these subspaces, the input signal fundamental frequency should be properly estimated, that is, $\hat{f}_1 = f_1$. The projection, also called inner-product in [10], can be defined as:

$$
\begin{cases}
g_{\alpha\beta}^{+}(t) = \begin{bmatrix} g_\alpha^+(t) \\ g_\beta^+(t) \end{bmatrix} = \begin{bmatrix} \int_{t-T_1}^{t} u_\alpha(t)e^{-j(2\pi\hat{f}_1 t)}dt \\ \int_{t-T_1}^{t} u_\beta(t)e^{-j(2\pi\hat{f}_1 t - \pi/2)}dt \end{bmatrix} \\
g_{\alpha\beta}^{-}(t) = \begin{bmatrix} g_\alpha^-(t) \\ g_\beta^-(t) \end{bmatrix} = \begin{bmatrix} \int_{t-T_1}^{t} u_\alpha(t)e^{-j(2\pi\hat{f}_1 t)}dt \\ \int_{t-T_1}^{t} u_\beta(t)e^{-j(2\pi\hat{f}_1 t + \pi/2)}dt \end{bmatrix}
\end{cases},
$$
(3)

where $T_1 = 1/\hat{f}_1$. To calculate the inner-product $g_{\alpha\beta}^+(t)$ and $g_{\alpha\beta}^-(t)$ are separated in real and imaginary parts:

$$
\begin{bmatrix} Re\{g_{\alpha\beta}^+(t)\} \\ Im\{g_{\alpha\beta}^+(t)\} \\ Re\{g_{\alpha\beta}^-(t)\} \\ Im\{g_{\alpha\beta}^-(t)\} \end{bmatrix} = \begin{bmatrix} Re\{g_\alpha^+(t)\} + Re\{g_\beta^+(t)\} \\ Im\{g_\alpha^+(t)\} + Im\{g_\beta^+(t)\} \\ Re\{g_\alpha^-(t)\} + Re\{g_\beta^-(t)\} \\ Im\{g_\alpha^-(t)\} + Im\{g_\beta^-(t)\} \end{bmatrix}.
$$
(4)

From (1), (2) and (3), one can rewrite the calculation of the real and imaginary parts as shown in Equations (5) and (6).

$$
\begin{cases}
Re\{g_\alpha^+(t)\} = \int_{t-T_1}^{t} u_\alpha(t)cos(2\pi\hat{f}_1 t)dt \\
Re\{g_\beta^+(t)\} = \int_{t-T_1}^{t} u_\beta(t)cos(2\pi\hat{f}_1 t - \pi/2)dt = \\
\qquad \int_{t-T_1}^{t} u_\beta(t)sin(2\pi\hat{f}_1 t)dt \\
Re\{g_\alpha^-(t)\} = \int_{t-T_1}^{t} u_\alpha(t)cos(2\pi\hat{f}_1 t)dt \\
Re\{g_\beta^-(t)\} = \int_{t-T_1}^{t} u_\beta(t)cos(2\pi\hat{f}_1 t + \pi/2)dt = \\
\qquad -\int_{t-T_1}^{t} u_\beta(t)sin(2\pi\hat{f}_1 t)dt
\end{cases}.
$$
(5)

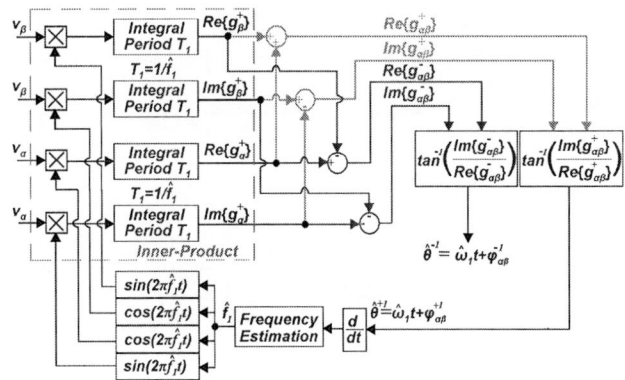

Fig. 1. Proposed positive and negative sequence detector.

$$
\begin{cases}
Im\{g_\alpha^+(t)\} = -\int_{t-T_1}^{t} u_\alpha(t)sin(2\pi\hat{f}_1 t)dt \\
Im\{g_\beta^+(t)\} = -\int_{t-T_1}^{t} u_\beta(t)sin(2\pi\hat{f}_1 t - \pi/2)dt = \\
\qquad \int_{t-T_1}^{t} u_\beta(t)cos(2\pi\hat{f}_1 t)dt \\
Im\{g_\alpha^-(t)\} = -\int_{t-T_1}^{t} u_\alpha(t)sin(2\pi\hat{f}_1 t)dt \\
Im\{g_\beta^-(t)\} = -\int_{t-T_1}^{t} u_\beta(t)sin(2\pi\hat{f}_1 t + \pi/2)dt = \\
\qquad -\int_{t-T_1}^{t} u_\beta(t)cos(2\pi\hat{f}_1 t)dt
\end{cases}.
$$
(6)

From (5), it can be concluded that $Re(g_\alpha^+(t)) = Re(g_\alpha^-(t))$ and $Re(g_\beta^+(t)) = -Re(g_\beta^-(t))$, while from (6), it can be concluded that $Im(g_\alpha^+(t)) = Im(g_\alpha^-(t))$ and $Im(g_\beta^+(t)) = -Im(g_\beta^-(t))$. This is an important conclusion since by using the same output values from the original inner-products defined in [10] (used only as a positive sequence detector), it is possible to obtain both the real and imaginary parts of the negative sequence.

By summing up the real and imaginary parts, as defined in (4), the phase angle is obtained by:

$$
\begin{cases}
\hat{\theta}_{\alpha\beta}^{+1}(t) = tan^{-1}\left(\dfrac{Im\{g_{\alpha\beta}^+(t)\}}{Re\{g_{\alpha\beta}^+(t)\}} \right) = \left(2\pi\hat{f}_1 t + \hat{\phi}_{\alpha\beta}^{+1}\right) \\
\hat{\theta}_{\alpha\beta}^{-1}(t) = tan^{-1}\left(\dfrac{Im\{g_{\alpha\beta}^-(t)\}}{Re\{g_{\alpha\beta}^-(t)\}} \right) = \left(2\pi\hat{f}_1 t + \hat{\phi}_{\alpha\beta}^{-1}\right)
\end{cases},
$$
(7)

where $\hat{\phi}_{\alpha\beta}^{+1}$ and $\hat{\phi}_{\alpha\beta}^{-1}$ are the phase shift angle of the positive and negative sequence of the input signal.

B. Frequency Estimation

The frequency estimation remain unchanged compared to the structures presented in [10]–[12], which is calculated based solely on the positive sequence phase angle. The structure is presented in the Figure 2.

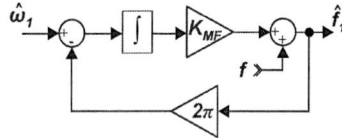

Fig. 2. Frequency estimation structure (Low-Pass Filter).

If the estimated frequency is different than the input signal frequency ($\hat{f}_1 \neq f_1$), there will be a frequency error that can be defined as $\Delta f_1 = \hat{f}_1 - f_1$. Thus, the real positive sequence phase angle is $\hat{\theta}_{\alpha\beta}^{+1} = 2\pi \hat{f}_1 t + 2\pi \Delta f_1 + \phi_{\alpha\beta}^{+1}$. By differentiating both sides of the expression and isolating the term Δf_1, one can find:

$$\Delta f_1 = \frac{\dfrac{\Delta\hat{\theta}_{\alpha\beta}^{+1}}{dt} - 2\pi\hat{f}_1}{2\pi}. \tag{8}$$

From (8), it is possible to calculate the correct frequency by integrating the right side over Δt:

$$\hat{f}_1(t) = K_{MF} \int \left(\frac{1}{2\pi} \frac{d\theta_{\alpha\beta}^{+1}}{dt} - \hat{f}_1(t) \right) dt + f, \tag{9}$$

where K_{MF} is a gain in rad^{-1}, and f is the initial nominal frequency estimation (in Hz). The transfer function is given by:

$$\frac{\hat{f}_1(s)}{\hat{\omega}_1(s)} = \frac{K_{MF}}{s + 2\pi K_{MF}}. \tag{10}$$

The structure acts as a low-pass filter where the input $\hat{\omega}_1$ is the derivative of the phase angle $\hat{\theta}_{\alpha\beta}^{+1}$. K_{MF} is the filter setting gain. The estimated frequency also employed as an argument of the sine and cosine functions used in the inner-product calculation.

The amplitude of the positive and negative sequence components are calculated as:

$$\begin{cases} \hat{V}_{\alpha\beta}^{+1} = Re\{g_{\alpha\beta}^{+}(t)\}cos(\hat{\theta}_{\alpha\beta}^{+1})(t) + Im\{g_{\alpha\beta}^{+}(t)\}sin(\hat{\theta}_{\alpha\beta}^{+1})(t) \\ \hat{V}_{\alpha\beta}^{-1} = Re\{g_{\alpha\beta}^{-}(t)\}cos(\hat{\theta}_{\alpha\beta}^{-1})(t) + Im\{g_{\alpha\beta}^{-}(t)\}sin(\hat{\theta}_{\alpha\beta}^{-1})(t) \end{cases} \tag{11}$$

Lastly, the positive and negative sequence components AC signals are given by:

$$\begin{cases} y_{\alpha}^{+1}(t) = \hat{V}_{\alpha\beta}^{+1}(t)cos(\hat{\theta}_{\alpha\beta}^{+1})(t) \\ y_{\beta}^{+1}(t) = \hat{V}_{\alpha\beta}^{+1}(t)cos(\hat{\theta}_{\alpha\beta}^{+1}(t) - \pi/2) \\ y_{\alpha}^{-1}(t) = \hat{V}_{\alpha\beta}^{-1}(t)cos(\hat{\theta}_{\alpha\beta}^{-1})(t) \\ y_{\beta}^{-1}(t) = \hat{V}_{\alpha\beta}^{-1}(t)cos(\hat{\theta}_{\alpha\beta}^{-1}(t) + \pi/2) \end{cases} \tag{12}$$

C. Structure with adaptive filter

The PLL proposed in [10] has only one single parameter to be tuned, K_{MF}. In [11], [12], it was noted that the original structure has a poor performance (even for low values of K_{MF}) under scenarios with interharmonic distortion. As such, by adding an AF, shown in Figure 4, the filtering capability under these scenarios could be enhanced. The complete structure of the proposed positive and negative sequence detector

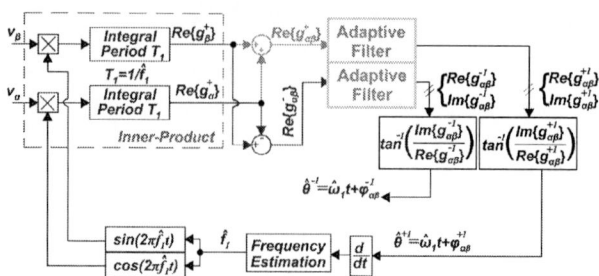

Fig. 3. Proposed positive and negative sequence detector with AF.

with AF are illustrated in Figure 3. The structure of the AF may be seen in [12] and observed as follows.

The AF output signals are defined as $\hat{x}_\alpha$ and $\hat{x}_\beta$, where $\hat{x}_\alpha = Re\{g_{\alpha\beta}^{+1}(t)\}$ and $\hat{x}_\beta = Im\{g_{\alpha\beta}^{+1}(t)\}$. Their transfer functions are expressed in (13) and (14). The bode diagram of the AF can be found in Figure 5.

$$\frac{\hat{x}_\alpha}{x_\alpha}(s) = \frac{K_{FA}(s+1)}{s^2 + s(K_{FA}+1) + K_{FA} + \hat{\omega}_1^2}, \tag{13}$$

$$\frac{\hat{x}_\beta}{x_\alpha}(s) = \frac{K_{FA}\hat{\omega}_1}{s^2 + s(K_{FA}+1) + K_{FA} + \hat{\omega}_1^2}. \tag{14}$$

It can be observed by the frequency response that the output signals are delayed by 90° at the central frequency (which is provided by the frequency estimation structure). This signal is equivalent to the imaginary part of the inner-product of the original structure, expressed in Equation (6). As such, in this modified structure there is no need to calculate the imaginary part as it is provided by the AF, thus reducing the computational burden (there is a higher demand if calculated using the inner-product). Furthermore, the bode diagram shows a high attenuation of interharmonic, highlighting its filtering capability. The modeling of the AF structure follows the original one, therefore Equations (7)-(12) are still used.

III. EXPERIMENTAL RESULTS

Experimental results were carried out in order to assess the performance of the proposed structures. The assessment is made by comparing the PLLs with the DDSRF-PLL and the DSOGI-FLL, two very popular structures. The parameters

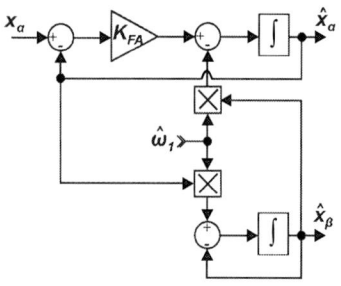

Fig. 4. Adaptive filter employed.

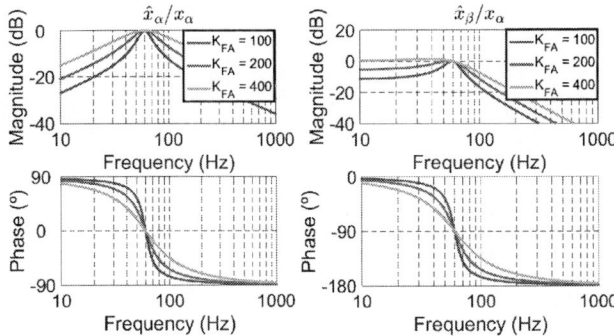

Fig. 5. Frequency response of the AF.

TABLE I
MEASURED THD AND RMSE (HARMONIC DISTORTION SCENARIO)

Analyzed Signal	Synchronization Algorithm			
	DDSRF-PLL	DSOGI-FLL	Prop.	Prop. w/ AF
THD V^{+1}	1.2%	0.58%	0.06%	0.06%
THD V^{-1}	6.18%	5.71%	0.04%	0.04%
RMSE V^{+1}	3.8%	0.65%	0.22%	0.22%
RMSE V^{-1}	7.5%	6%	0.4%	0.4%

TABLE II
MEASURED TWD AND RMSE (INTERHARMONIC DISTORTION SCENARIO)

Analyzed Signal	Synchronization Algorithm			
	DDSRF-PLL	DSOGI-FLL	Prop.	Prop. w/ AF
TWD V^{+1}	3.29%	3.00%	9.02%	2.82%
TWD V^{-1}	11.57%	11.07%	12.36%	3.33%
RMSE V^{+1}	5%	3.1%	9%	2.9%
RMSE V^{-1}	12.8%	13.2%	12.2%	3.5%

TABLE III
SETTLING TIMES (HARMONIC DISTORTION SCENARIO)

Analyzed Sequence	Synchronization Algorithm			
	DDSRF-PLL	DSOGI-FLL	Prop.	Prop. w/ AF
Positive	>200 ms	160 ms	60 ms	170 ms
Negative	190 ms	180 ms	50 ms	160 ms

TABLE IV
SETTLING TIMES (INTERHARMONIC DISTORTION SCENARIO)

Analyzed Sequence	Synchronization Algorithm			
	DDSRF-PLL	DSOGI-FLL	Prop.	Prop. w/ AF
Positive	>200 ms	160 ms	50 ms	150 ms
Negative	150 ms	150 ms	50 ms	160 ms

used in the DSOGI-FLL are $k_{sogi} = 0{,}25$ and $\Gamma = 25$ while the parameters used in the DDSRF-PLL are $K_p = 46$, $K_i = 2116$ and low-pass filter cut-off frequency $\omega_c = 2\pi 6$. Both were chosen favoring filtering capability using guidelines provided in [8], [9]. The parameters used for the proposed structures are $K_{FA} = 100$ and $K_{MF} = 9$, as used in [11], [12].

All the structures are implemented in MATLAB/Simulink in order to be used in the dSPACE 1103 platform. A signal generator is employed to create the signals and by using the D/A channels, the output of each structure is captured in real time by oscilloscopes. The operating frequency is 8 kHz.

A. Steady-state scenarios

Two steady-state scenarios are considered for the experimental results, one with harmonic distortion and another with both harmonic and interharmonic distortion.

1) Harmonic distortion: The input signal for this scenario is the following: v_{abc}: $V^{+1} = 1.0$ p.u., $V^{-1} = 0.2$ p.u, $V^{-2} = 0.1$ p.u., $V^{z3} = 0.2$ p.u., $V^{-5} = 0.2$ p.u. e $V^{+7} = 0.2$ p.u.

Figure 6(a) shows the input signal as well as the positive and negative sequence component estimated by each of structure. Figure 7(a) presents the root mean square error (RMSE) defined as the rms value of the difference between the actual positive/negative sequence component and the ones estimated by each structure. Table I provides the total harmonic distortion (THD) of the output signal of each structure as well as the RMSE for both positive and negative sequence components.

It can be seen from Table I that both proposed structures (with and without AF) reject the harmonic distortion almost completely. As such their filtering capability is greater by a large margin. Their RMSE are also much lower than the other structures.

2) Interharmonic distortion: The input signal for this scenario is the same as the one with harmonic distortion however, it is also added the interharmonic components 0.1 p.u. at 30 Hz and 0.1 p.u. at 90 Hz.

Figure 6(b) shows the estimated positive and negative sequence components. The error is shown in Figure 7(b). Since interharmonic distortion is present, the total wave distortion (TWD) is calculated instead of the THD (as it only consider components multiple of the fundamental frequency). The

TWD information is provided along with the measured RMSE in Table II.

Table II shows that the proposed PLL without AF has low filtering capability in this scenario. When the AF is added, the performance is greatly improved, specially considering the negative sequence component. The measured RMSE suggests that the proposed PLL with AF is the most precise one.

B. Dynamic response

Two dynamic response scenarios are considered, one with only harmonic distortion and, again, another with both harmonic and interharmonic distortion.

1) Harmonic distortion: The harmonic distortion present in the system is the same as the steady-state scenario, however, at a certain point, the positive and negative sequence components change from $V^{+1} = 1$ p.u. to $V^{+1} = 0.8$ p.u. and $V^{-1} = 0$ p.u. to $V^{-1} = 0.4$ p.u., with phase angle jump of $(\theta^{+1}) = +60°$ and $(\theta^{-1}) = -45°$. Furthermore, the fundamental frequency goes from 60 Hz to 66 Hz. Figure 8 shows the RMSE response.

It can be seen from Figure 8 that the proposed PLL has a much faster response (approximately 50 ms) in comparison to the others. The structure with AF has a settling time that is close to the DSOGI-FLL's, however, as shown in the steady-

978-1-7281-4181-7/19 $31.00 © 2019 IEEE

1310

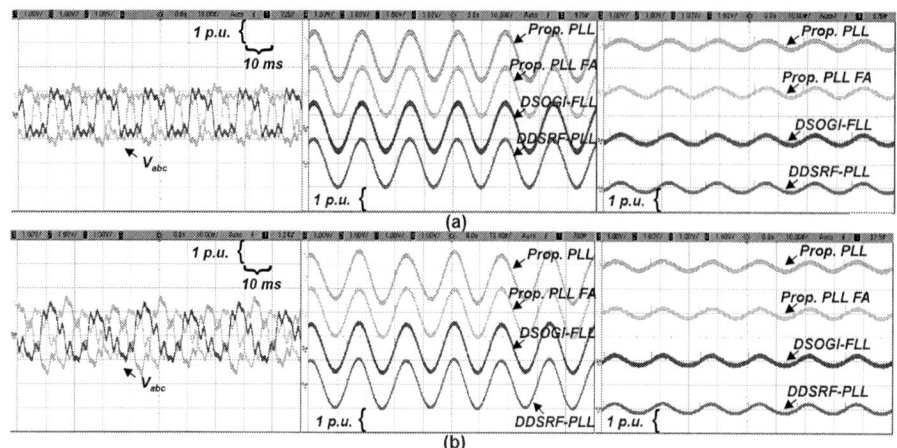

Fig. 6. Steady-state response of the proposed PLL (yellow), with AF (green), DSOGI-FLL (purple) and DDSRF-PLL (pink). Plots from left to right: input signals, positive sequence component estimation and negative sequence component estimation; (a) harmonic scenario (b) interharmonic scenario.

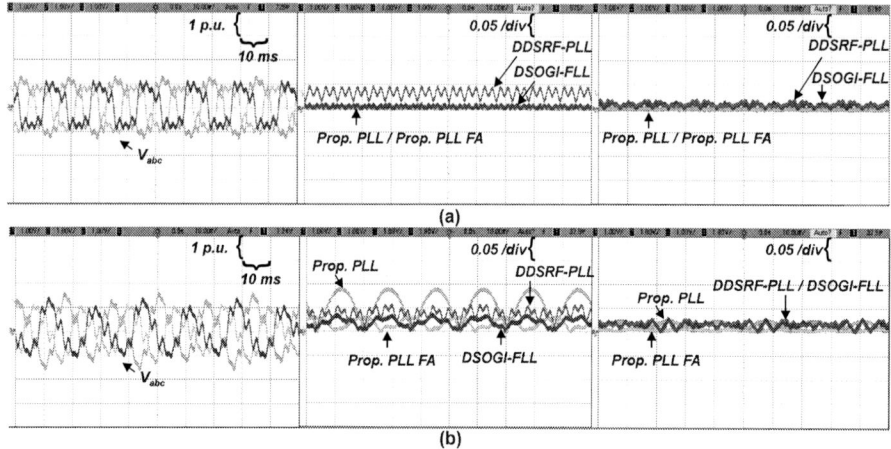

Fig. 7. Steady-state response of the proposed PLL (yellow), with AF (green), DSOGI-FLL (purple) and DDSRF-PLL (pink). Plots from left to right: input signals, positive sequence component RMSE and negative sequence component RMSE; (a) harmonic scenario (b) interharmonic scenario.

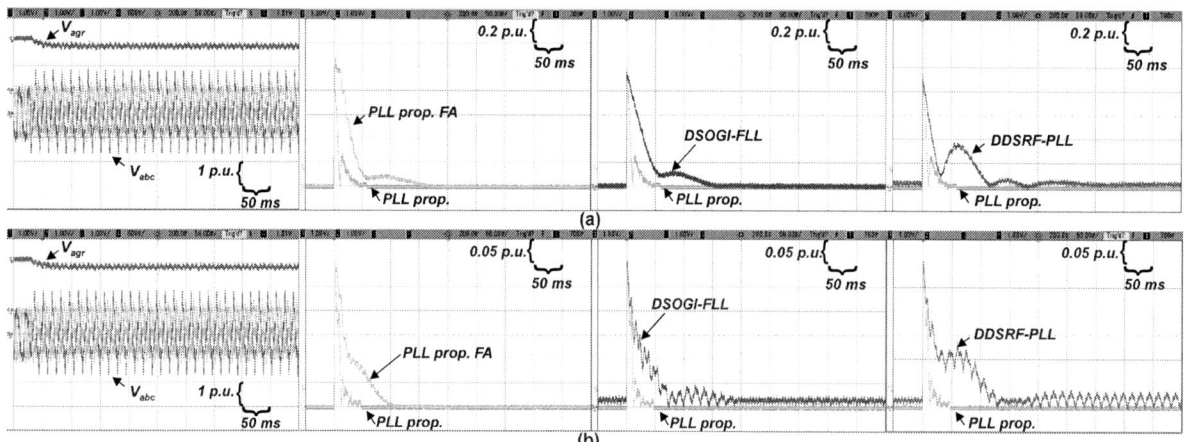

Fig. 8. Dynamic response of the proposed PLL (yellow), with AF (green), DSOGI-FLL (purple) and DDSRF-PLL (pink). Plots from left to right: input signals with harmonic distortion; (a) positive sequence component RMSE; (b) negative sequence component error.

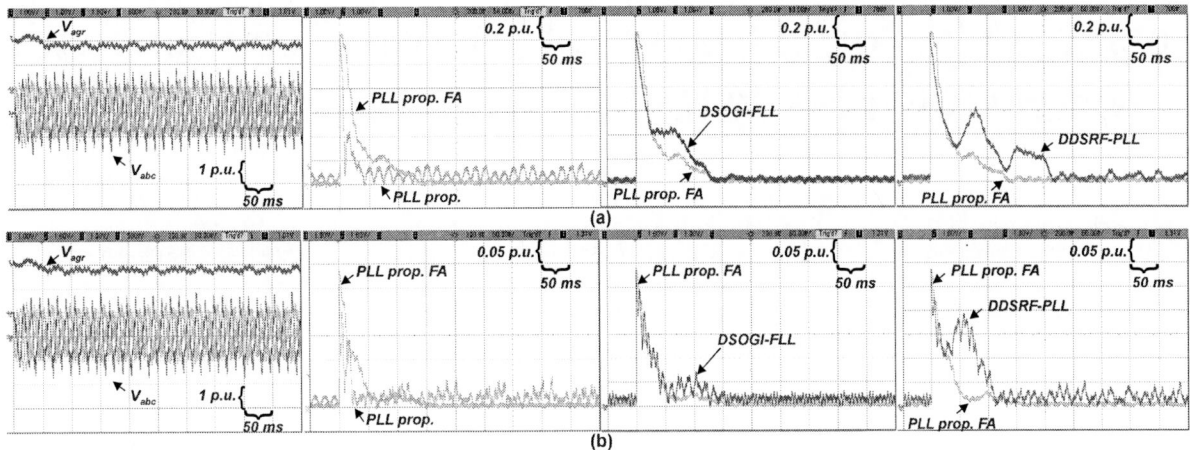

Fig. 9. Dynamic response of the proposed PLL (yellow), with AF (green), DSOGI-FLL (purple) and DDSRF-PLL (pink). Plots from left to right: input signals with interharmonic distortion; (a) positive sequence component RMSE; (b) negative sequence component error.

state scenario, the proposed PLLs are more precise. Details regarding the settling time can be found in Table III.

2) Interharmonic distortion: The interharmonic distortion present in the system is the same as the steady-state scenario, however, the same changes promoted in the dynamic response harmonic distortion scenario are considered here. Figure 9 shows the RMSE response.

The dynamic response for the interharmonic distortion showed in Figure 9 makes clear that the structure with AF has a similar response to the DSOGI-FLL with respect to settling time, however, as shown previously, the proposed structure is more precise. Furthermore, although much faster, the original structure lacks the precision offered by the addition of an AF. Table IV details the settling time of each structure.

IV. CONCLUSIONS

The steady-state scenario with harmonic distortion showed that the proposed PLLs are more precise in every aspect when compared to the DDSRF-PLL and DSOGI-FLL. It can also be pointed out that, in this scenario, the proposed structures have nearly equal results. In the case of interharmonic distortion, the use of the AF is justified as without it, the response is much worse. Furthermore, in the interharmonic distortion scenario, it is shown that the structure with AF has a higher precision than the DDSRF-PLL and the DSOGI-FLL for both positive and negative sequence component estimation.

Dynamic response scenarios showed that the PLL without AF is much faster and its use is recommended when only harmonic distortion is present. If interharmonic components are present in the input signal, the proposed PLL with AF is recommended as without it the precision is worsened.

REFERENCES

[1] F. Blaabjerg, Z. Chen, and S. B. Kjaer, "Power electronics as efficient interface in dispersed power generation systems," *IEEE transactions on power electronics*, vol. 19, no. 5, pp. 1184–1194, 2004.

[2] N. G. Hingorani, L. Gyugyi, and M. El-Hawary, *Understanding FACTS: concepts and technology of flexible AC transmission systems.* IEEE press New York, 2000, vol. 1.

[3] H. Akagi, "Active harmonic filters," *Proceedings of the IEEE*, vol. 93, no. 12, pp. 2128–2141, 2005.

[4] F. Blaabjerg, R. Teodorescu, M. Liserre, and A. V. Timbus, "Overview of control and grid synchronization for distributed power generation systems," *IEEE Transactions on Industrial Electronics*, vol. 53, no. 5, 2006.

[5] S. Golestan, J. M. Guerrero, and J. C. Vasquez, "Three-phase plls: A review of recent advances," *IEEE Transactions on Power Electronics*, vol. 32, no. 3, pp. 1894–1907, 2017.

[6] N. Jaalam, N. A. Rahim, A. H. A. Bakar, C. Tan, and A. M. A. Haidar, "A comprehensive review of synchronization methods for grid-connected converters of renewable energy source," *Renewable and Sustainable Energy Reviews*, vol. 59, pp. 1471–1481, 2016.

[7] S.-K. Chung, "A phase tracking system for three phase utility interface inverters," *IEEE Transactions on Power electronics*, vol. 15, no. 3, pp. 431–438, 2000.

[8] P. Rodríguez, J. Pou, J. Bergas, J. I. Candela, R. P. Burgos, and D. Boroyevich, "Decoupled double synchronous reference frame pll for power converters control," *IEEE Transactions on Power Electronics*, vol. 22, no. 2, pp. 584–592, 2007.

[9] P. Rodriguez, A. Luna, R. S. Munoz-Aguilar, I. Etxeberria-Otadui, R. Teodorescu, and F. Blaabjerg, "A stationary reference frame grid synchronization system for three-phase grid-connected power converters under adverse grid conditions," *IEEE transactions on power electronics*, vol. 27, no. 1, pp. 99–112, 2012.

[10] J. A. M. Neto, L. Lovisolo, B. W. França, and M. Aredes, "Robust positive-sequence detector algorithm," in *2009 35th Annual Conference of IEEE Industrial Electronics.* IEEE, 2009, pp. 788–793.

[11] F. K. A. Lima, R. G. Araujo, F. L. Tofoli, and C. G. C. Branco, "A phase-locked loop algorithm for single-phase systems with inherent disturbance rejection," *IEEE Transactions on Industrial Electronics*, 2019.

[12] F. K. de Araújo Lima, R. G. Araújo, F. L. Tofoli, and C. G. C. Branco, "A three-phase phase-locked loop algorithm with immunity to distorted signals employing an adaptive filter," *Electric Power Systems Research*, vol. 170, pp. 116–127, 2019.

[13] J. M. L. Fonseca, S. S. Queiroz, S. R. Lima, W. da Silva Lima, R. G. Almeida, F. K. A. Lima, and C. G. C. Branco, "Performance analysis of synchronization algorithms for grid-connected power converters under sub and inter-harmonics distortion," in *2018 IEEE Applied Power Electronics Conference and Exposition (APEC).* IEEE, 2018, pp. 2952–2958.

[14] M. H. Bollen, *Understanding power quality problems.* IEEE press, 2000.

[15] P. Rodriguez, A. V. Timbus, R. Teodorescu, M. Liserre, and F. Blaabjerg, "Flexible active power control of distributed power generation systems during grid faults," *IEEE Transactions on Industrial Electronics*, vol. 54, no. 5, pp. 2583–2592, 2007.

Performance analysis of alternatives switching commands for Three-level Boost Rectifier with Hysteresis Current Control.

Gabriel Vilkn Ramos
Electrical Engineering departament
CEFET-MG
Belo Horizonte, Brazil
gabrielvilkn@gmail.com

Alex-Sander Amável Luiz
Electrical Engineering departament
CEFET-MG
Belo Horizonte, Brazil
asal@des.cefetmg.br

Marcelo Martins Stopa
Electrical Engineering departament
CEFET-MG
Belo Horizonte, Brazil
marcelo@des.cefetmg.br

Marcos Paulo Brito Gomes
Electrical Engineering departament
UFMG
Belo Horizonte, Brazil
marcospaulobritogomes@ieee.org

Ramon Henriques de Souza
Electrical Enginneering departarment
CEFET-MG
Belo Horizonte, Brazil
ramonhs@cefetmg.br

Daniel Franco Leal
Electrical Enginneering departarment
CEFET-MG
Belo Horizonte, Brazil
danielfl.bhz@gmail.com

Abstract— **This work presents and analyzes converter switching strategies in order to achieve higher efficiency and better current waveforms of a hysteresis current control for a high power and low cost three-phase and three-level active AC-DC rectifier. The rectifier consists of an alternative converter composed by two active semiconductors switches and four power diodes per phase. It implies in a good voltage and current quality and lower maintenance and acquisition costs. One advantage of this topology is that it allows the use of a large variety of command techniques with different switching strategies. The present work will focus on the study and evaluation of alternative pulse commands of the hysteresis current controller. The study is conducted through simulation and experiments. The results obtained reinforce and confirm the proposed ideas.**

Keywords— *Power rectifier; Three-Level Converter; Switching analysis; Hysteresis current control.*

I. INTRODUCTION

Rectifiers convert voltages and currents from sinusoidal form to continuous, making their circuits extremely used in several processes of AC-DC energy conversion. There are many topologies of rectifiers already developed and established. Among them, it can be found the uncontrolled and controlled structures in single and polyphase versions. Additionally, multipulse rectifiers can be obtained through a three phase six pulses bridges, improving passively the system input power factor with an appropriate design [1].

Controlled rectifiers are extremely important because they can present sinusoidal current, better voltage regulation and power factor operation control. Traditionally they are based on simpler structures with two level voltages in their input that results in a high dv/dt when switching semiconductor devices. This can be a source of problems such as grid voltage degradation, electromagnetic interference EMI and higher stress of power switches and reactive elements (transmission lines, cables, transformers, etc.) [2].

In this context exists a demand for studying of three levels rectifiers thanks to the benefits of higher voltage levels, smaller ripples, a better total harmonic distortion - THD, higher range of variation of power factor and bidirectional energy flow [3].

Three level active rectifiers appeared with the advent of the frequency inverters due the necessity of more levels in input voltage, better current waveforms, and lower harmonic distortion [4]. Besides, the ability of dealing with high power

and medium voltage loads with smaller semiconductor ratings permits them to be employed as AFE in huge plants and applications in industries [5].

The topology of the controlled rectifier of this work was proposed by Y. Zhao [6] and it is presented in Figure 1. It shows to be a good alternative in AC-DC conversion since the three-level standard imposes a lower dv/dt to the grid, provides a better wave form, when compared to a two-level rectifier which is the most common controlled ac-dc converter. In addition, the semiconductor devices need to withstand only half of the DC bus voltage reducing the switching and conduction losses and costs [7].

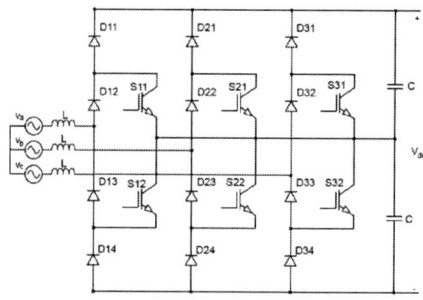

Fig. 1. RNPC Three Level Boost Rectifier.

The main disadvantage of this structure is its unidirectional power flow when compared to other VSI and NPC topologies [7]. However, it presents some good characteristics of the three level NPC rectifiers, using half number of active switches and four power diodes, once more reducing the manufacturing cost. Additionally, the topology does not generate short-circuit in the DC bus, showing flexibility of switching command then allowing different switching strategies. [8]

The two power switches in each leg of the converter are responsible for the positive and negative half-cycle operation and traditionally in the literature are switched simultaneously, it was presented in [6] with hysteresis control and, in [9], with SHE-PWM modulation.

The present work will focus on the study and evaluation of alternative pulse commands to the hysteresis current modulation and compared with [8] that propose alternative pulse commands with PWM technique, it tends to improve in

978-1-7281-4181-7/19 $31.00 © 2019 IEEE

quality of the current waveforms and the efficiency of the converter.

II. HYSTERESIS CURRENT CONTROL

The hysteresis current modulation is a technique that establishes a band for the current variation, between a maximum and minimum error around a reference current. It is one of the simplest and most direct techniques to command a converter. Besides, according to [10], the advantages of hysteresis band controller over traditional PWM techniques among others, are direct machine torque control of an AC drive, converter current limitation and protection in short circuit and fault conditions it is also more natural choice of switching strategy for active rectifiers, STATCOM, SVG, etc.

Its operation is based on measuring current in converter phases and comparing them with references current through hysteresis comparators. These comparators verify if the measured current errors reach the limits established by the hysteresis band. When the current errors reach the maximum or minimum current limits the comparator turns switches either on or off to keep the current between the established limits according to figure 2.

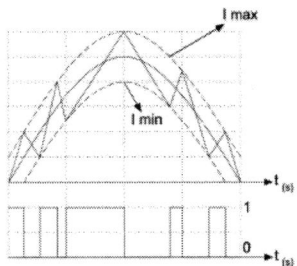

Fig.2. Hysteresis band switching scheme.

This modulation technique generates commands with variable switching frequency, and due to that occurs a scattering in harmonic spectrum, making difficult the filtering process. However, the individual harmonic amplitude is in general lower than all that found with any PWM command strategies.

Figure 3 presents an alternative diagram for hysteresis control.

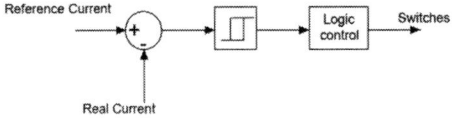

Fig.3. Hysteresis Current Controller for one phase.

Figure 4 shows an implementation for active rectifier control. In this figure, a hysteresis reference current can be obtained by an instantaneous input voltage sample, multiplied by the output signal of the DC voltage PI controller. The error between the reference currents Ia*, Ib* and Ic* and the converter instantaneous currents, Ia, Ib and Ic, is submitted to hysteresis comparator generating pulse to the power switches due to preset limits. The DC voltage PI controller is responsible for DC bus voltage regulation and stabilization.

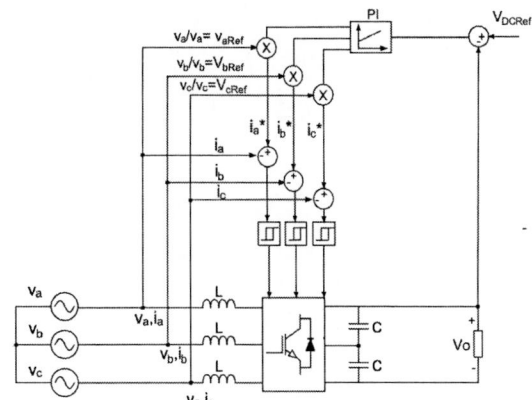

Fig. 4. Hysteresis current control applied a rectifier converter.

III. SWITCHING STRATEGIES

One advantage of the Three Level rectifier topology proposed here it's the possibility of setting different pulse commands for semiconductor devices. It is particularly important since it can affect both converter thermal losses and harmonic current distortion [8].

Traditionally for these converters the switches responsible for the positive and negative half-cycle are switched simultaneously with the same pulse command as can be seen in the works [6] and [9]. The figure 5 shows the logical implementation of the classical switching, and the figure 6 shows the pulses pattern for the switches S11 and S12, responsible for the current flow in the positive and negative half-cycle respectively. VRef is the reference phase voltage sampled as in fig 4.

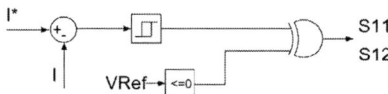

Fig. 5. Logic control circuit for classical implementation.

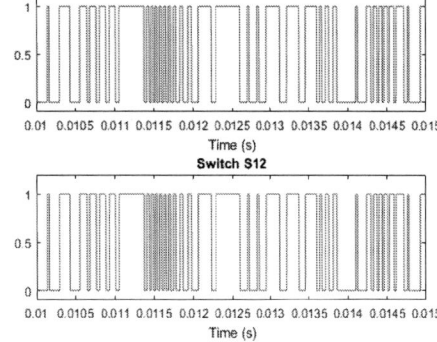

Fig.6. Switching commands for classical switching.

However, there are other possibilities beyond this classical technique [8] which implies in four different command schemes for this topology, the classical one and other three strategies.

The first possible different command scheme for this converter consists of controlling both switches with complementary pulse signals, which means one in opposition to the other. The figure 7 shows the logical implementation for the complementary switching, and the figure 8 the pulses for S11 and S12 responsible for the positive and negative half-cycle respectively.

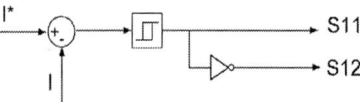

Fig.7. Logic control circuit for complementary implementation.

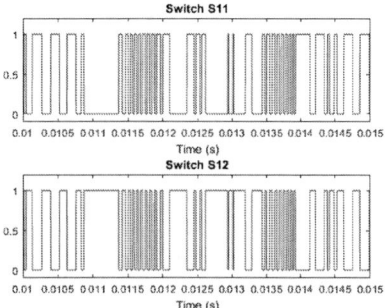

Fig.8. Switching commands for complementary switching.

Other command scheme for this converter can be seen in the Figures 9 and 10. Here when the instantaneous reference voltage VRef is in the positive half-cycle the upper device switches while the lower switch is kept steadily on. When the polarity of the phase voltage is negative, the bottom device switches during this the negative half-cycle while the upper switch is kept Steadily on [8].

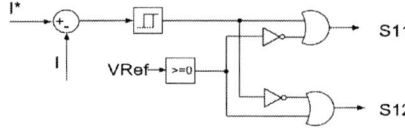

Fig.9 Logic control circuit for Steadily on implementation.

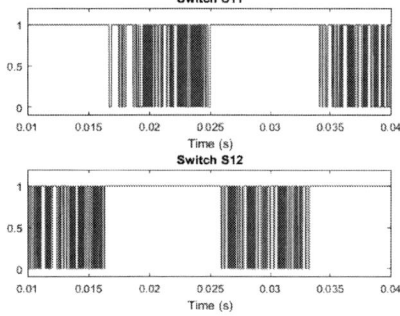

Fig.10. Switching commands for Steadily on switching.

The last command scheme for this converter consists of commuting the supper switch in the positive half-cycle while the lower switch remains steadily off. The opposite must be performed during negative half-cycle keeping the upper device Steadily off while the bottom device switches with complementary signal [8].

The figure 11 shows the logical implementation and the figure 12 the pulses for S11 and S12 responsible for the positive and negative half-cycle respectively.

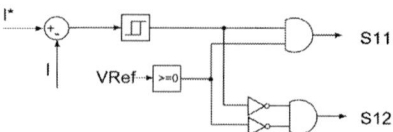

Fig.11. Logic control circuit for Steadily on implementation.

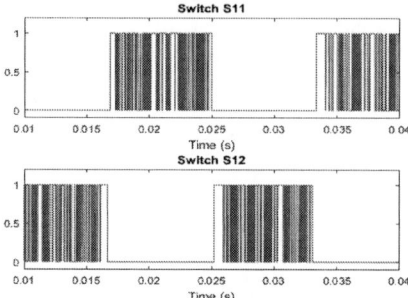

Fig.12. Switching commands for Steadily on switching.

IV. SIMULATION RESULTS

In order to compare the switching strategy performances, simulations of the power rectifier are accomplished by using the SIMULINK® software with the following specifications: Line-to-line grid voltage of 220 V and 60Hz, three input inductances of 0.9 mH, two DC link capacitors of 470 uF, a DC load resistance of 75 ohms with a DC bus voltage of 380V and a hysteresis band of 10%.

The system was controlled with hysteresis current band control according to that shown in figure 4. The figure 13 and 14 present the simulation results for the current waveforms for the four types of switching.

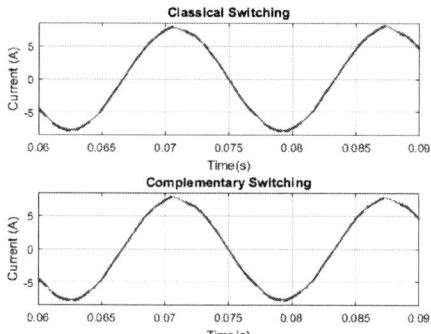

Fig.13. Currents wave forms for classical and complementary switching.

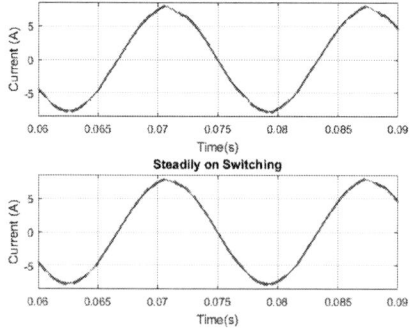

Fig.14.Currents wave forms for Steadily on and off switching.

Comparing the switching strategies performance of the rectifier the figures 15, 16, 17 and 18 shows the current spectrum for each case of current waveform show in figures 13 and 14. It was analyzed the system performance and so the harmonics.

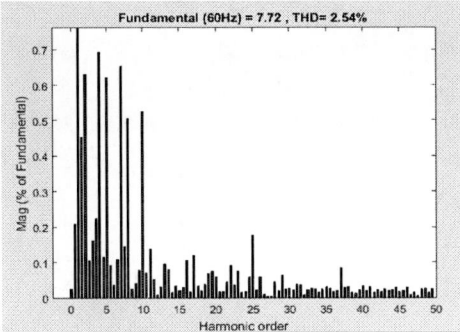

Fig.15. Current Spectrum for classical switching.

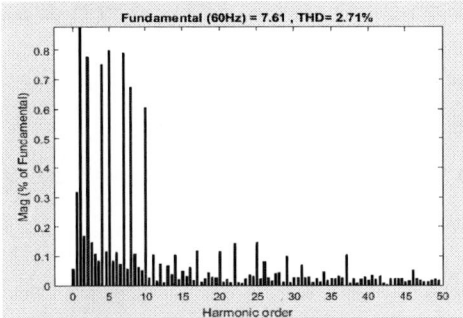

Fig.16. Current Spectrum for complementary switching.

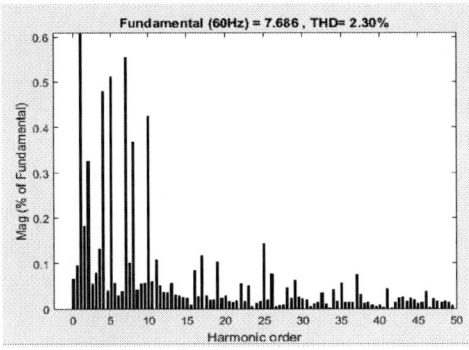

Fig.17. Current Spectrum for Steadily on switching.

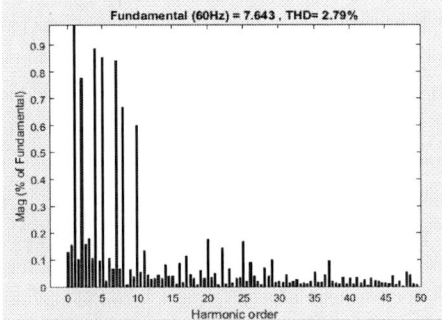

Fig.18. Current Spectrum for Steadily off switching.

Analyzing the currents spectrum for each switching strategies, the total harmonic distortion shows that the steadily on switching strategy generates the best current waveforms in harmonics content. Observing the current spectrum figures, the complementary and Steadily off switching shows the worst current waveform. The classical switching even shows a better performance than complementary and Steadily off switching.

Comparing all of the current spectrums its possible see more difference in low harmonic content due the switching mode, where the opening of the switching cause degradation in current waveforms since the current cannot flow reversed to the grid in this rectifier topology. This fact it's confirmed in the lower THD levels and harmonic amplitudes by steadily on switching strategy.

In the same way figures 19 and 20 shows the waveforms of the line to line voltage response for the rectifier for each switching strategy.

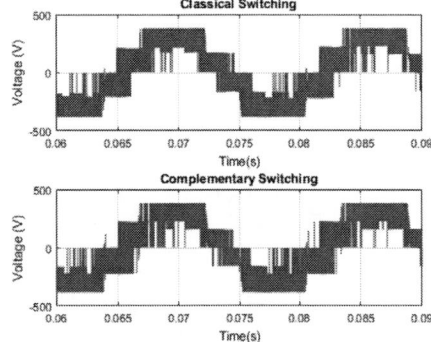

Fig.19. input voltage waveforms for classical and complementary switching.

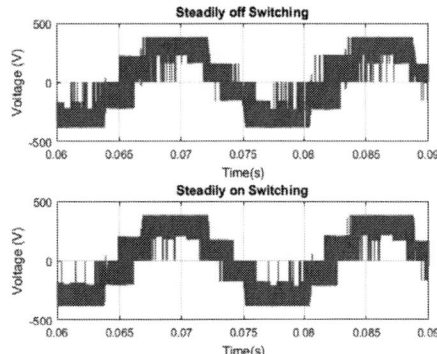

Fig 20. Input voltage waveforms for Steadily on and off switching.

The figures 21,22,23 and 24 shows the voltage spectrum for all switching strategies. This figures again shows that the results in classical and steadily on switching shows the best results in THD levels, mainly comparing all of the voltage spectrum in low frequency region.

Again, the complementary and steadily off switching pattern show the worst results, since the current cannot flow reversed to the grid. In this case, the converter loses the zero-voltage reference of the DC bus in the opening of switch and it is influenced by the voltage associated to energy stored in the input inductor.

978-1-7281-4181-7/19 $31.00 © 2019 IEEE   1316

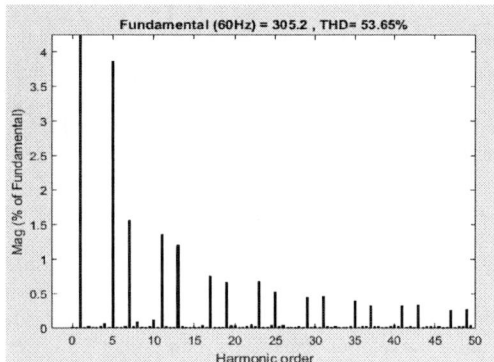

Fig.21.Voltage Spectrum for classical switching.

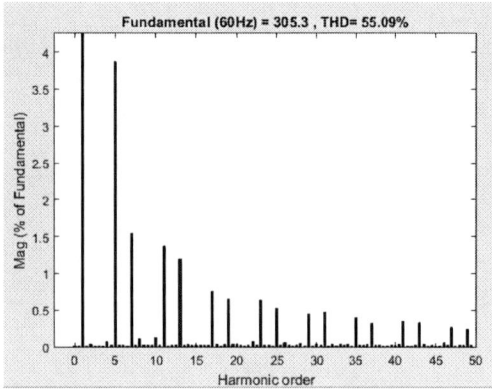

Fig.22. Voltage Spectrum for complementary switching.

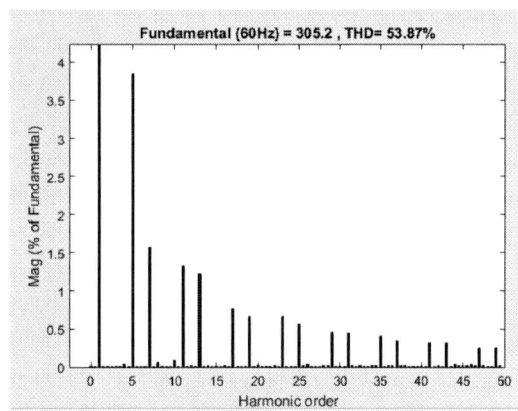

Fig.23. Voltage Spectrum for Steadily on switching.

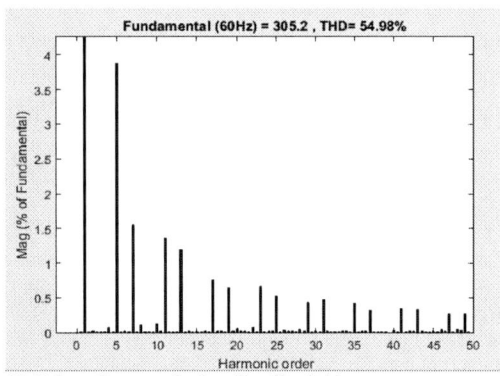

Fig.24. Voltage Spectrum for Steadily off switching.

V. PRATICAL RESULTS

A small prototype of the rectifier was built and it can be seen in the figure 25. The initial parameters are in given by following the specifications: Line-to-line grid voltage of 15V and 60Hz, three input inductances of 0.9 mH, two capacitors in DC link of 470uF and a DC load resistance of 75 ohms.

Fig. 25. Practical prototype of the rectifier.

Figures 26 and 27 show the results obtained with the prototype for classical switching strategy presented in last section.

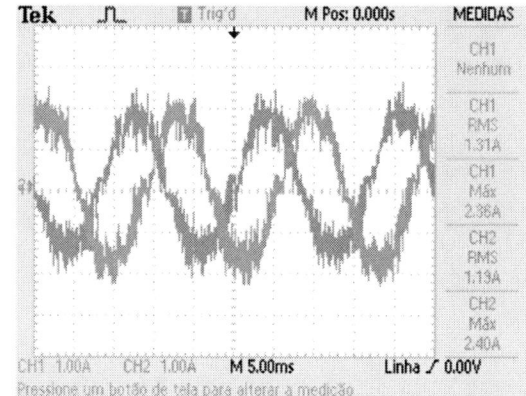

Fig. 26. Currents wave forms for classical switching.

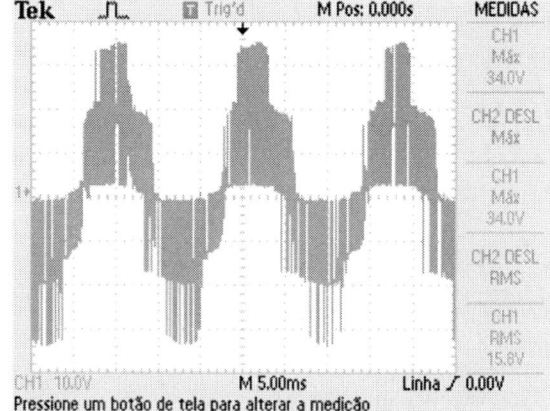

Fig. 27. Input voltage waveforms for classical switching.

Figures 28 and 29 show the results obtained with the prototype for Steadily on switching strategy presented in last section.

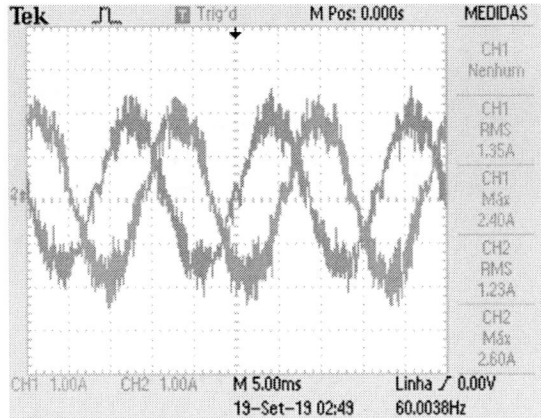

Fig.28. Currents wave forms for Steadily on switching.

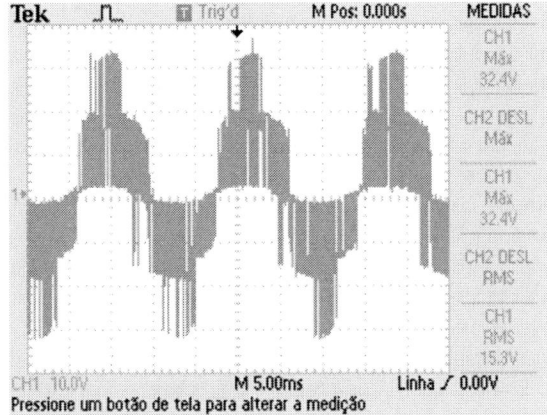

Fig.29. Input voltage waveforms for Steadily on switching.

The table 1 show the results of THD of input current and voltage of the prototype for classical and Steadily on switching strategy presented in last section.

TABLE I. THD OF INPUT CURRENT AND VOLTAGE FOR CLASSICAL AND STEADILY ON SWITCHING STRATEGIES.

Switching Strategy	Current THD	Voltage THD
Classical switching	22.5%	80.69%
Steadily on switching	20.26%	72.82%

VI. CONCLUSION

The paper presented an assessment of switching strategies for a three-level boost rectifier controlled with hysteresis current control. This technique establishes a band for the current variation, between a maximum and minimum error around a reference current. As it is well known it is simplest and most direct techniques to command a converter.

The converter topology studied here presents various advantages over other structures. Some of them are its lower dv/dt applied to the grid and the semiconductor device, the switches need to withstand only half of the DC bus voltage. The converter also permits to design four different switching patterns: classical, complementary, steadily on and steadily off strategies that were presented, discussed and evaluated in this work. It is possible to verify differences in the THD values of these four strategies.

The Steadily On switching strategy presented the lower THD. The Steadily Off switching strategy showed the higher THD. Due to the lower current THD this paper point out the Steadily On switching pattern as the best strategy to commute de switches. Oppositely since the current cannot flow reversed to the grid the converter loses the zero-voltage reference of the DC bus with the opening of switch.

Practical results presented confirms the simulation ones and the measured THD levels verify the steadily on as the best switching mode comparing with the classical switching present in the works [6] and [9].

REFERENCES

[1] MOHAN, N.; UNDERLAND, T.M.; ROBBINS, W. P., Power Electronics: Converters, applications, and Design.3/e. John Wiley & Sons,2003.

[2] RASHID MUHAMMAD H; Power Electronics Handbook; Academic press, 2001.

[3] Portilho, T.B; Luiz, A-S. A.; Stopa M.M A low Cost Three-Level High Power Rectifier for active Power Factor Compensation. In: 18th International Symposium on Electromagnetic Fields in Mechatronics, Electrical and Electronic Engineering, 2017, Lodz.

[4] A. A. Luiz and B. de Jesus Cardoso Filho, "A new design of selective harmonic elimination for adjustable speed operation of AC motors in mining industry," *2017 IEEE Applied Power Electronics Conference and Exposition (APEC)*, Tampa, FL, 2017, pp. 607-614.

[5] A. A. Luiz and B. de Jesus Cardoso Filho, "Improving power quality in mining industries with a three-level active front end," *2015 IEEE Industry Applications Society Annual Meeting*, Addison, TX, 2015, pp. 1-9.

[6] Y. Zhao, Y. Li, T. A. Lipo, "Force commutated three level boost type rectifier," IEEE Transactions on Industry Applications, 1995.

[7] A. A. Luiz and B. J. Cardoso Filho, "Sinusoidal voltages and currents in high power converters," *2008 34th Annual Conference of IEEE Industrial Electronics*, Orlando, FL, 2008, pp. 3315-3320.

[8] Portilho, T. B; Luiz, A-S. A.; Stopa M.M Performance and Efficiency Analysis of Switching Commands For Three-Level Boost Rectifier. In: 18th International Symposium on Electromagnetic Fields in Mechatronics, Electrical and Electronic Engineering, 2017, Lodz.

[9] K. O'Brien, R. Teichmann, S. Bernet, "Active rectifier for medium voltage drive systems," Applied Power Electronics Conference and Exposition, 2001.

[10] Leal, D. F.; Luiz, A-S. A.; Ribeiro, E.K.B.; Souza, R. H.; Stopa M.M, Hysteresis Band Current Control Applied to a Five Level NPC Type G Converter. In: 13th IEEE International Conference on Industry Applications, 2018.

High Power Factor Rectifier Using One Cycle Control Strategy

Amanda Thayla S Monteiro
Graduate Program in Electrical Engineering
Federal University of Piaui
Teresina, Piaui
amathayla@hotmail.com

Prof° Dr° Rafael Rocha Matias
Graduate Program in Electrical Engineering
Federal University of Piaui
Teresina, Piaui
rafaelrocha@ufpi.edu.br

Jailson Leite Silva
Graduate Program in Electrical Engineering
Federal University of Piaui
Teresina, Piaui
jailson_ls@hotmail.com

Jose Antonio dos Santos Neto
Graduate Program in Electrical Engineering
Federal University of Piaui
Teresina, Piaui
nsantos93@hotmail.com

Abstract—Due to the considerable increase in the insertion of electronic components to Electric Power Systems (EPSs), the present article analyzes a method to eliminate undesired currents originated by nonlinear loads connected to the grid and further performs the Power Factor (PF) correction on single-phase rectifier loads. The proposed solution uses the topology of the DC-DC boost converter working in the Boundary Conduction Mode (BCM) as a Power Factor Correction (PFC) circuit and implemented using the One Cycle Control (OCC) strategy, which proposes high efficiency during the switching period. The simulations yielded interesting results regarding the proposed control method, with the results showing the increase of the PF and reduction of the source current Total Harmonic Distortion (THD), benefiting the power quality of the system.

Index Terms—Retifiers, One Cycle Control, THD, Power Factor, Power Quality.

I. Introduction

Since the twentieth century, the generation, transmission and distribution stages of electric power systems utilize alternating current. This system is subjected to several undesired factors such as the use of non-linear loads, atmospheric discharges, short circuits and saturation of magnetic cores [1]. These factors promote the separation of currents and voltages sinusoidal profiles, leading to numerous risks such as: heating in motors and transformers, faults in capacitors and in electronic equipment, electronic interference and noise [2].

In AC systems it is a known fact that only the fundamental components transfer energy to the load, thus, the components that present phase shift regarding the grid voltage or present a different frequency do not transfer energy to the load, contributing to the total power losses in the system [3], which causes the load to display a low Power Factor (PF).

The PF is one of the most relevant parameters of the distribution system, it is an indicative that expresses the efficiency with which the energy is being consumed, so it is possible to measure the quality of the power supplied by the distribution systems and the power consumed by the users [4]. Hence, a

low PF can cause malfunction of devices connected to the grid and it also limits the maximum active power collected from the grid.

One of the major causes of low PF are non-linear loads such as single-phase rectifiers. These circuits are present in numerous fundamental applications such as motor drives, lighting circuits, power amplification circuits, power supplies for computers and televisions [5]. Hence, it is necessary to search for solutions to correct the waveform of the grid current, to maintain the sinusoidal profile of the AC system parameters [6].

A commonly used solution for PF correction is the use of Power Factor Correction (PFC) circuits, such as the one displayed in Fig. 1. In basic terms, it can be implemented by inserting a control stage between a rectifier and a capacitive filter, forcing the input current to have the same waveform as the input voltage.

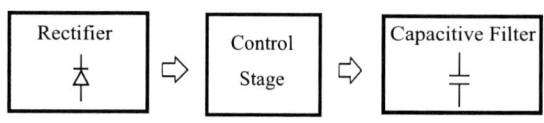

Fig. 1. PFC circuit schematics.

In theory, all DC-DC static converters can be used as high PF pre-regulators. These circuits, very similar to switch-mode sources, generally perform the conversion by applying pulsed DC voltage to an inductor or transformer at a certain frequency/period, causing the current flow to store magnetic energy, which is then used by the system output [7].

By adjusting the duty cycle, the voltage at the output can be changed, or preferably, kept constant, through a proper control method, even if changes in load and current occur. Among

978-1-7281-4181-7/19 $31.00 © 2019 IEEE

the topologies of static converters for applications in PFC circuits, the boost converter that received major attention, due to the fact that it presents several advantages. Such advantages are: simplified control, since the current naturally follows the input voltage, the converter requires only a voltage regulation loop; reduction of switching losses due to the lack of reverse recovery on the diode; and zero current switching [8].

The converter can be implemented in three modes of conduction: Continuous Conduction Mode (CCM), Discontinuous Conduction Mode (DCM) and Boundary Conduction Mode (BCM). By setting a constant conduction time for the switch, the input current peaks are lead to naturally follow a sinusoidal envelope. The off time is variable, which means that the operating frequency is not fixed, varying accordingly with input and output conditions.

Boost converter applications in the CCM presented problems of reverse diode recovery and high switching losses, characteristic of conventional continuous mode. One way to suppress such losses is to change the boost converter to DCM operation. According to [9] this mode of operation is quite effective and unitary PF was achieved by implementing an open loop control strategy.

However, this mode of operation has some drawbacks, such as the fact that it generates high RMS currents through the switches, which increases the conduction losses on the semiconductors.

One alternative that aims to reduce the current through the switches while keeping a high PF is to implement the BCM operation. The reduction of the input RMS current occurs because the transistor enters the conduction state when the current reaches zero. Hence, the system creates the conditions to turn-off the diode and the transistor conducts under zero current, not presenting diode reverse recovery issues. Therefore, there is no zero current time interval and the RMS input current is lower than in the Discontinuous Mode, increasing its efficiency [10].

A boost converter can be operated in several ways, regarding the control method. The most traditional techniques presented in the literature are: peak current control, hysteresis control and average current control, which uses analog multipliers to generate a reference current. The first two strategies produce distortions in the input current, while the latter requires a search on the non-linearity of the converter for the controller design.

In this scenario, the One Cycle Control (OCC) method appears as an alternative to the previously mentioned strategies, for implementation in PFC circuits. This method has the following advantages: rejection of disturbances in the power source within one switching cycle; the average value of the controlled variable follows the dynamic reference within a switching cycle and the controller corrects the switching errors, also within one switching cycle.

In addition, there is no steady-state error or dynamic error between the control reference and the average value of the controlled variable [11]. The work by [10] implemented an OCC control loop in a DC-DC boost converter in BCM mode, with DC input, presenting a satisfactory response and achieving the dynamic control from transient state to steady state within one cycle. This hence confirms its easy implementation, since it only uses analog devices and does not need a current sensor for the converter operation and functionality.

From the mathematical model of the boost converter and its OCC control loops for BCM operation, it is possible to design a converter prototype able to attain output voltage regulation for load variations, resulting in high PF and reduced harmonic distortion of the AC input current. The following section shows the theoretical basis explanation for the OCC implementation in the boost converter, the implementation of the control in a circuit with a single-phase rectifier and the results obtained in the PSIM® software.

II. OCC WORKING PRINCIPLES FOR THE BOOST CONVERTER OPERATING IN BCM

Considering initially a pure DC voltage at the input, the following observations can be outlined. The input voltage, v_S, "detects" an equivalent resistance at the input terminals of the boost converter, which is represented by R_i. In the OCC control technique, the converter operates as a resistance emulator, imposing a linear relationship R_i between the input voltage v_S and input current i_i. In this control strategy the converter's characteristic equation is used as control law.

To operate in BCM mode, i_i is replaced by v_S/R_i. Thus the OCC strategy is implemented with electronic elements, starting from the principle that the control law will be the characteristic equation of the converter.

A. OCC for BCM operation

To operate in BCM, the OCC strategy must generate a variable frequency and the converter must operate with constant conduction time (t_{on}). In other words, to synthesize an OCC strategy in BCM, it is necessary to find the control law that generates the conduction time t_{on}.

In the boost converter, during the magnetization period, the switch is conducting and the inductor begins to store input voltage energy, the current in the inductor starts to increase, so from (1), it is possible to estimate a value of the variation of the magnetizing current , considering $v_L = V_s$, since the inductor is in parallel with the input voltage source.

$$\Delta i_{L_{on}} = \frac{v_L}{L} t_{on} = \frac{V_s}{L} \frac{D}{f_{sw}} \tag{1}$$

In Fig. 2 it is shown the waveform of the converter operating in BCM and it is possible to express the average current in the inductor as seen in (2):

$$I_{Lm} = \frac{\Delta I_{L_{on}}}{2} \tag{2}$$

Substituting (1) into (2), yields (3):

$$I_{Lm} = \frac{\Delta I_{L_{on}}}{2} = \frac{v_S}{2L f_{sw}} \left(1 - \frac{v_S}{V_o}\right) \tag{3}$$

The average current can be rewritten as $I_{Lm} = (v_S/R_i)$. The switching frequency is calculated as seen in:

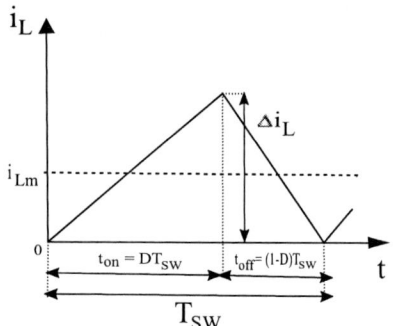

Fig. 2. Waveform of the current in the inductor of a boost operating in BCM.

$$f_{sw} = \frac{R_i}{2L}\left(1 - \frac{v_S}{V_o}\right) \tag{4}$$

Rewriting (4), one can find (5):

$$\frac{V_o}{R_i} = (V_o - v_S)\frac{T_{sw}}{2L} \tag{5}$$

Substituting $D = 1 - v_S/V_o$ and the relation given by (4) in $D = t_{on}/Tsw$, t_{on} can be expressed by (6):

$$t_{on} = DT_{sw} = \frac{D}{f_{sw}} = \frac{2L}{R_i} \tag{6}$$

Finally, the product of (6) and the output voltage V_o results in (7):

$$\frac{V_o}{R_i} = V_o\frac{t_{on}}{2L} \tag{7}$$

Equations (5) and (7) are essential to carry out the control strategy. The value of V_o/R_i represents a virtual current that is used as a parameter of the OCC strategy. This value is determined using a PI controller, which is shown latter.

B. OCC implementation

The objective is to synthesize a circuit to perform OCC based on (5) and (7). The general scheme of the OCC method is shown in Fig. 3. The "T_{sw} Processor" block has the function of determining the duration of a switching period. On the other hand, the "Processor of t_{on}" block determines the switch conducting time. The variable v_{sw} in the output of the "T_{sw} Processor" represents the implementation of the control law represented in (5). Similarly, the v_{on} variable in the output of the "t_{on} Processor" represents the processing of the control law seen in (7).

The PI controller, R_c, determines the voltage value that is compared with the output v_{sw} and v_{on} to generate the time required to regulate the output voltage of the converter to the reference voltage V_{ref}. The value of the output of R_c, v_m, represents the virtual current seen in the control law equations. The cutoff frequency of the controller must be 10% or less the value of the switching frequency.

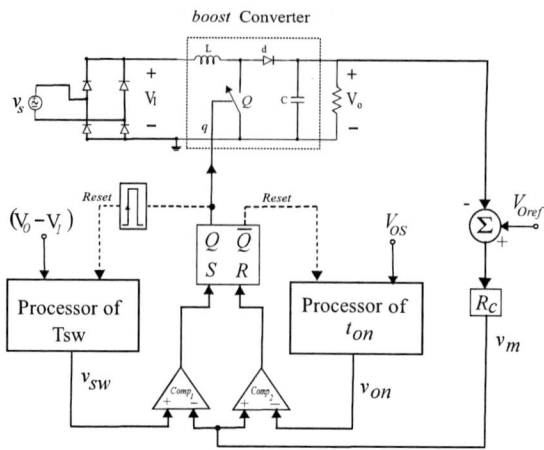

Fig. 3. Generalized scheme for OCC control operating in BCM.

The comparator 1 determines the intersection between the signals v_{sw} and v_m, as well as the end of the switching period. This intersection is at the instant of equality seen in (5). The comparator 2 determines the intersection between v_{on} and v_m, and determines when the end of t_{on} and the beginning of t_{off} occurs, representing the equality in the control law seen in (7).

A signal conditioning is performed on V_o and v_s to adjust them to the electronic circuit. The conditioning is carried out by gain K_v, decreasing the amplitude of the signals. Thus, this gain can be seen in (8):

$$K_v = \frac{V_{OS}}{V_o} = \frac{v_{IS}}{v_i} \tag{8}$$

here K_v has the interval of $0 < K_v \le 1$, for the purpose of decreasing the signal's amplitude.

C. OCC Integrator in BCM

The control laws for the OCC, (5) and (7), must be implemented electronically by means of a simple resettable integrator, as shown in Fig. 4. The integrator is responsible for determining the required time interval.

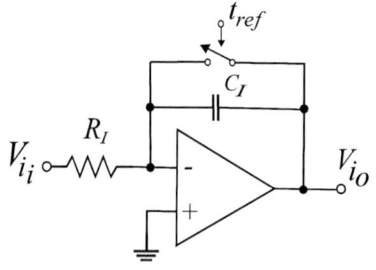

Fig. 4. Integration circuit used in the OCC strategy.

978-1-7281-4181-7/19 $31.00 © 2019 IEEE

Considering that V_{i_i} and V_{i_o} can be constant in a switching period, the transfer function of the integrator shown in Fig. 4 is calculated by (9) as follows:

$$V_{i_o}(t) = -\frac{1}{R_I C_I} \int_0^{t_{ref}} V_{i_i}(t) d\tau$$
$$V_{i_o}(t) = -V_{i_i}(t) \frac{t_{ref}}{R_I C_I} \tag{9}$$

To operate in BCM, the OCC strategy uses a modulated current, which despite representing a virtual current, has its value generated at the output of the regulator R_c, producing the signal v_m. This value is compared to the output of the integrators to generate t_{on} and T_{sw}. After performing an analysis to determine T_{sw}, equaling the common terms in (5), (9) and using them in (8), the relation represented by (10) can be attained:

$$V_{i_o}(t) = \frac{V_{OS}}{R_I}$$
$$-V_{i_i}(t) \frac{t_{ref}}{R_I C_I} = (V_{OS} - v_{IS}) \frac{T_{sw}}{2L} \tag{10}$$

Assuming that $-V_{i_i} = (V_{OS} - v_{IS})$ and $t_{ref} = T_{sw}$, these terms can be plugged in (10), thus resulting in (11):

$$(V_{OS} - v_{IS}) \frac{T_{sw}}{R_I C_I} = (V_{OS} - v_{IS}) \frac{T_{sw}}{2L} \tag{11}$$
$$R_I C_I = 2L$$

The analysis to determine t_{on} is carried out in the same way as T_{sw}. Equating the terms of (7) and (9), it is possible to find (12):

$$V_{i_o}(t) = \frac{V_{OS}}{R_I}$$
$$-V_{i_i}(t) \frac{t_{ref}}{R_I C_I} = V_{OS} \frac{t_{on}}{2L} \tag{12}$$

It can be noticed that $-Vi_i = V_{OS}$ and $t_{ref} = t_{on}$. Substituting these terms into (12) results in (13):

$$V_{OS} \frac{t_{on}}{R_I C_I} = V_{OS} \frac{t_{on}}{2L} \tag{13}$$
$$R_I C_I = 2L$$

In (11) and (13), the term $R_I C_I$ determines the time constant of the integrator. Thus, the integrator achieves a faster response if the variables of the converter obey the relation given by (14):

$$R_I C_I \geq 2L \tag{14}$$

The signals v_{sw}, v_{on} and v_m of the circuit of Fig. 3 are calculated by (15), (16) and (17), respectively:

$$v_m = \frac{V_{OS}}{R_I} = R_c[V_{OS_{ref}} - V_{OS}] \tag{15}$$

$$v_{sw} = \frac{V_{OS-v_{IS}}}{R_I C_I} \int_0^{T_{sw}} d\tau = \frac{V_{OS} - v_{IS}}{R_I C_I} T_{sw} \tag{16}$$

$$v_{on} = \frac{V_{OS}}{R_I C_I} \int_0^{t_{on}} d\tau = \frac{V_{OS}}{R_I C_I} t_{on} \tag{17}$$

The term T_{sw} is determined by the intersection of the signal v_{sw} with v_m. On the other hand, t_{on} is determined by the intersection of the v_{on} signal with v_m. The waveforms of the signals v_m, v_{sw} and v_{on} are shown in Fig. 5.

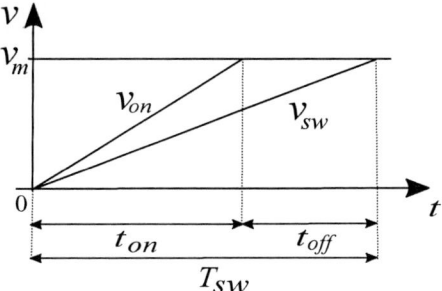

Fig. 5. Waveforms of the output signal of the integrators.

The slope of the signals is determined by the value of V_{OS} for v_{on} and the value of $(V_{OS} - v_{IS})$ for v_{sw}.

III. RESULTS OBTAINED BY THE IMPLEMENTATION OF THE OCC CONTROL STRATEGY

The operation logic of the OCC control follows the proposal that, in steady-state, the average voltage across the inductance is zero, and the output voltage, V_o, is equal to the average voltage on the diode. The voltage across the diode, however, will vary between practically zero (when the component is conducting) and the supply voltage, V_s. Its average value at each cycle must be equal to the output voltage, obtained by integrating V_o.

The integrated signal is compared to the reference and until it reaches its value, the switch remains on (voltage V_s applied on the diode). When both voltages have equal value, the capacitor of the integrator circuit is discharged and the comparator changes its state, turning off the transistor until the beginning of the next cycle, which is determined by the clock according to Fig. 3.

The implemented simulation is composed of a non-linear load connected to a boost converter fed by a single-phase AC system. The parameters used are shown in Table I, the capacitor of the boost converter, C, had an initial voltage of 100 V, defining the system's reference voltage value, and a unit gain K_v.

Fig. 6 shows the control signals v_m, v_{on} and v_{sw}, processed by the OCC control circuit to generate the t_{on} and T_{sw} values, allowing the boost converter to operate in BCM, curves obtained from the simulations in the PSIM®.

978-1-7281-4181-7/19 $31.00 © 2019 IEEE

TABLE I
VALUES USED FOR THE SIMULATION.

VARIABLE	VALUE
Maximum output voltage variation, ΔV_o	3 %
Input Voltage, $v_s(t)$	$24\sin(wt)$
Output Voltage, V_o	100 V
Maximum output power, P_o	350 W
Single Phase Frequency, f	60 Hz
Inductor, L	55 μH
Capacitor, C	1.24 mF

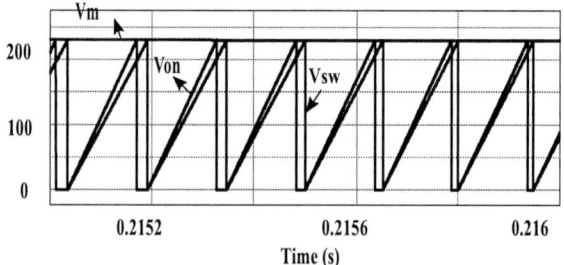

Fig. 6. Waveforms of the output signals of the implemented system integrators.

Fig. 7 shows the waveform of the current flowing from the source, which acts as a BCM mode typical sinusoidal envelope [10].

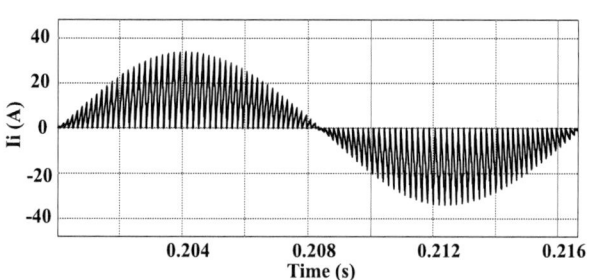

Fig. 7. AC source current waveform of the implemented system.

The peak amplitude of the input current was greater than the source voltage peak value, as seen in Fig. 8. The waveforms are in phase and synchronized with each other. The calculation tool embedded in PSIM® measured THD = 55.58% for the input current and FP = 0.8756, when comparing the source's voltage and current.

In Fig. 9 the inductor current waveform can be seen. By analyzing it, one can observe the characteristic envelope of the single-phase full-wave rectifiers, with the boost converter switching in BCM operation and the current reaching zero at the end of each of the inductor's magnetization period.

The source current harmonic spectrum after the inclusion of the boost converter with OCC control is shown in Fig. 10. From its analysis, one can note that only a small amount of low frequency harmonic components is present.

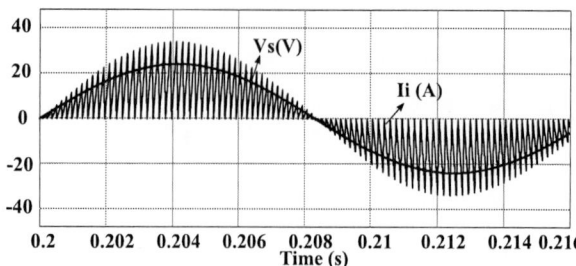

Fig. 8. Current and AC input voltage waveforms of the implemented system.

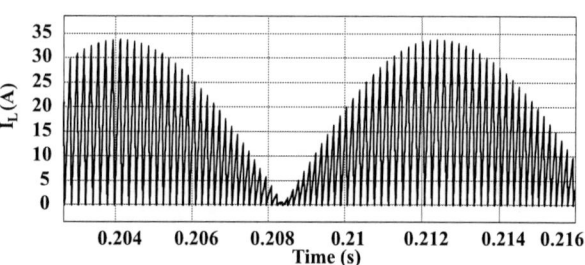

Fig. 9. Waveform of the current flowing through the inductor in BCM operation.

Some regulatory organizations related to electric power, such as the Institute of Electrical and Electronics Engineers (IEEE) and the International Electrotechnical Commission (IEC), have set up parameters and limitations for harmonics generation in order to control their propagation. The international established standards are the American IEEE 519 and the European IEC 61000 3-2. Hence, the data collected from the simulations is compared to recommended values established by the aforementioned standards.

Table II shows a comparison between the collected data and THD levels recommended by the IEEE-519 standard.

TABLE II
COMPARISON BETWEEN THE THD LEVELS RECOMMENDED BY THE IEEE-519 AND THE VALUES MEASURED IN THE BOOST CONVERTER CONTROLLED USING THE OCC METHOD.

MAXIMUM HARMONIC CURRENT DISTORTION IN PERCENTAGE OF I_L (FUNDAMENTAL CURRENT)							
DISTORTION	Order of odd harmonics						
	h=3	h=5	h=7	h=9	h=11	h=13	THD
RECOMMENDED	12%	12%	12%	12%	5.5%	5.5%	15%
MEASUREMENT	0.079%	0.018%	0.0098%	0.031%	0.019%	0.03%	55.58%

Table III shows the IEC 61000 3-2 standard and its tolerances compared to the THD levels obtained using simulation data:

Among the measured harmonic components, none exceeded the individual limits established by IEEE and IEC. However, the general THD level is high due to a large number of high frequency harmonic components, which, despite having

978-1-7281-4181-7/19 $31.00 © 2019 IEEE

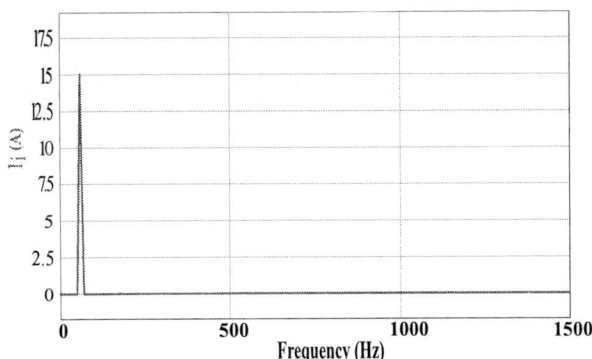

Fig. 10. Harmonic spectrum of the AC source current

TABLE III
COMPARISON BETWEEN THE THD LEVELS RECOMMENDED
BY IEC 61000 3-2 AND THE VALUES MEASURED IN THE BOOST
CONVERTER CONTROLLED USING THE OCC METHOD.

| DISTORTION | HARMONIC CURRENT DISTORTION | | | | | |
| | Order of odd harmonics | | | | | |
	h=3	h=5	h=7	h=9	h=11	h=13
RECOMMENDED	2.3	1.14	0.77	0.4	0.33	0.21
MEASUREMENT	0.012	2.73×10^{-3}	1.49×10^{-3}	4.6×10^{-3}	2.83×10^{-3}	4.29×10^{-3}

a low magnitude interfere with the THD computation. Current distortions when approaching zero also contribute to the high value of THD.

Fig. 11 shows the output voltage waveform, displaying the expected shape and good voltage regulation.

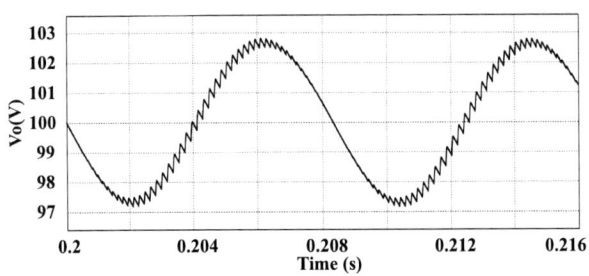

Fig. 11. Output voltage waveform of the boost converter.

IV. CONCLUSIONS

The boost converter operating in BCM mode proved to be an interesting topology because it did not dissipate the energy on the transistor's switching instant, reducing the switching losses and consequently maximizing its efficiency. The operation of the converter using the OCC control strategy is greatly beneficial due to the converter being fully controlled by analog electronics and does not requiring a current sensor for the control loop, resulting in a simplified structure that only requires the direct use the output and input voltage values.

Based on the spectra acquired during the simulation, it is possible to infer that the OCC control was effective in suppressing undesired currents related to individual harmonic components up to the 13th order, since all the individual values of magnitude measured were below the established standards.

In spite of this, the control was not able to reach the limits established by IEEE-519 with respect to Total Harmonic Distortion, due to the presence of harmonic components in high frequency and the crossover distortion present in the current waveform, which directly interfere with THD calculations.

Thus, a better parameter adjustment and the insertion of a filter to contain the high frequency harmonics are required in order to better control similar AC systems. During the simulations it was observed that any variation in reference value, input voltage or load affects the transistor's conducting time interval. However, the average voltage on the diode is always kept equal to the value determined by the reference.

REFERENCES

[1] R. R. Matias. Compensadores Estáticos de Potência para Sistemas Trifásicos. Tese (Doutorado) — Universidade Federal de Campina Grande, 2012.
[2] J. Stones and A. Collinson Power quality. Power Engineering Journal, v. 15, n. 2, p. 58–64, 2001.
[3] M. C. L. Souto. Conversor Boost PFC com controle digital. Trabalho de Conclusão de Curso —Universidade do Estado de Santa Catarina, 2015.
[4] L. Roggia. Estudo, Controle e Implementação do Conversor Boost PFC Operando em Modo de Condução Mista. Dissertação (Mestrado)—Universidade Federal de Santa Maria,2009.
[5] M. Crestani. Com uma terceira portaria, o novo fator de potência já vale em Abril. [S.l.]: Ano XXII, 1994.
[6] A. P. Monteiro. Estratégias de controle aplicadas a um filtro ativo de potência paralelo para a supressão de componentes harmônicas. Trabalho de Conclusão de Curso— Universidade Federal do Piauí, 2017.
[7] J. A. Pomilio. Conversores ca-cc - Retificadores. 2014. Este texto foi elaborado em função da disciplina "Eletrônica de Potência", ministrada nos cursos de pós-graduação em Engenharia Elétrica na Faculdade de Engenharia Elétrica e de Computação da Universidade Estadual de Campinas. Disponível em: ¡http://www.dsce.fee.unicamp.br/ antenor/pdffiles/eltpot/cap3.pdf/¿. Acesso em: 01 maio 2018.
[8] J. A. A. Dias. Conversores Monofásico-Trifásicos com Otimização de Perdas, Tolerância à Falha e Comparação Multicritério. Tese (Doutorado) — Universidade Federal de Campina Grande, 2010.
[9] J. Lazar and S. Cuk. Open loop control of a unity power factor, discontinuous conduction mode boost rectifier. Power Electronics Group - IEEE, Caltech, Pasadena, CA, p. 671–677, 1995.
[10] M. A. Vitorino, et al. Design of boost converter operating in crm controlled by occ. IEEE, n. 6810, p. 440–447, 2013.
[11] K. M. Smedley and S. Cuk. One cycle control of switching converters. Power Electronics, IEEE Transactions, n. 10(6), p. 625–633, 1995.

A comparison among different Finite Control Set approaches and Convex Control Set Model-based Predictive Control applied in a Three-Phase Inverter with RL load

Arthur G. Bartsch
Electrotechnology Department
Federal Institute of Santa Catarina
Jaraguá do Sul, SC, Brazil
arthur.bartsch@ifsc.edu.br

Dayse M. Cavalcanti, Mariana S. M. Cavalca, Ademir Nied
Electrical Engineering Department
Santa Catarina State University
Joinville, SC, Brazil
daysemaranhao@gmail.com, mariana.cavalca@udesc.br, ademir.nied@udes.br

Abstract—**This paper presents a comparative study between two model-based predictive control (MPC) methods applied in a three-phase power inverter in terms of design and performance: finite control set (FCS-MPC) and convex control set (CCS-MPC). Moreover, different FCS-MPC strategies are tested. In order to analyze their advantages and disadvantages, it is shown a brief explanation about the inverter modeling and the control algorithms implementation. Note that the same parameters are used in the controllers numerical analyses, including unitary prediction horizon and an analytical solution for FCS-MPC. Furthermore, a different cost function for FCS-MPC is proposed in order to remove steady state error.**

Index Terms—**Three-phase inverter, FCS-MPC, CCS-MPC**

I. INTRODUCTION

The improvement in energy conservation techniques and energy use seeks the comfort of modern society by promoting a conscious technological advance. In this context, power electronics appears as the science of control and conversion of electric power, i.e., the control of the flow of energy between electrical systems [1]. However, the performance of the energy conversion is typically subject to a control algorithm, main topic of this article.

There is a wide range of controllers, varying in complexity, robustness and adaptability, that provides accurate and stable control looping responses. Proportional-Integrative-Derivative (PID) controller family is hugely applied in processes since it is intuitive and simple to implement, also, adjustments can make it acceptable for almost any case. Usually, PI controllers are employed in power electronics equipments, but researches show that other controllers, such as Model-based Predictive Control (MPC), can stand out in specific situations [2]–[4].

MPC appoints to a range of control methods based on the idea of choosing the control actions as the solution to an optimal control problem with a receding horizon strategy [4]. As the name suggests, it relies on the mathematical model capacity of representing the system behaviour, and repeat the calculation process at each iteration to analyze the future reference path.

Two interesting strategies in the proposed study area are the Finite Control Set Model-based Predictive Control (FCS-MPC) and the Convex Control Set or Continuous Control Set Model-based Predictive Control (CCS-MPC), since they consider the discrete nature of power converters [5], [6]. In short, FCS-MPC has a finite number of control actions and it can treat commutation nonlinearities, but has a variable operation frequency. CCS-MPC has a infinite number of solutions and it implies the modelling of the system considering the duty cycle as the control action.

For power electronics applications, in the current technology state, it is not possible to directly treat constraints using CCS-MPC while it is possible with FCS-MPC [4]. In other hand, conventional FCS-MPC has problems with steady-state error and high current ripple [7]. These features supported the need to search for other FCS-MPC algorithms to compare with conventional FCS-MPC and with CCS-MPC.

The main objective of this paper is to compare the performance between different FCS-MPC approaches and CCS-MPC when determining the behavior of a three-phase inverter in order to obtain a desired output, while reviewing basic concepts of the proposed device and techniques. Being organized as follows: Section II describes the case study; Section III describes the controller methods; Section IV presents the simulation results; and Section V provides conclusions.

II. SYSTEM ANALYSIS

An inverter is an electronic device that converts direct current (DC) to alternating current (AC), with power handling capacity depending on its design. For this analysis, it is considered an ideal three-phase inverter, which means there is no energy loss on its elements due to intrinsic factors of components, and steady state operation.

A. Continuous time model

The system configuration is presented in Fig. 1. Batteries or ultracapacitors may be used as V_{dc}.

978-1-7281-4181-7/19 $31.00 © 2019 IEEE

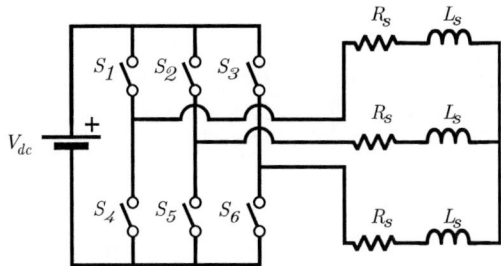

Fig. 1. Circuit diagram of a three-phase inverter with RL load.

The phase voltages that feed the load can be described as

$$
\begin{bmatrix} v_{as}(t) \\ v_{bs}(t) \\ v_{cs}(t) \end{bmatrix} = \begin{bmatrix} \dfrac{2}{3} & -\dfrac{1}{3} & -\dfrac{1}{3} \\ -\dfrac{1}{3} & \dfrac{2}{3} & -\dfrac{1}{3} \\ -\dfrac{1}{3} & -\dfrac{1}{3} & \dfrac{2}{3} \end{bmatrix} \begin{bmatrix} S_1(t) \\ S_2(t) \\ S_3(t) \end{bmatrix} V_{dc} \quad (1)
$$

where S_1, S_2 and S_3 are the command signals for the inverter legs upper switches and V_{dc} is the voltage of DC-link.

The RL load can be expressed in dq coordinates, with reference frame $\theta_e = \omega_e t$. Therefore [8], [9]

$$
L_{qs}\frac{di_{qs}(t)}{dt} = v_{qs}(t) - R_s i_{qs}(t) - L_{ds}\omega_e i_{ds}(t) \quad (2)
$$

$$
L_{ds}\frac{di_{ds}(t)}{dt} = v_{ds}(t) - R_s i_{ds}(t) + L_{qs}\omega_e i_{qs}(t) \quad (3)
$$

where

$$
L_{qs} = L_{ds} = L_{ls} + \frac{3}{2}L_{ms},
$$

$$
\begin{bmatrix} \varphi_{qs}(t) \\ \varphi_{ds}(t) \\ f_{0s}(t) \end{bmatrix} = T_{qd}(\theta_e) \begin{bmatrix} \varphi_{as}(t) \\ \varphi_{bs}(t) \\ \varphi_{cs}(t) \end{bmatrix}
$$

where φ means voltage v or current i and

$$
T_{qd}(\theta_e) = \frac{2}{3} \begin{bmatrix} \cos(\theta_e) & \cos\left(\theta_e - \dfrac{2\pi}{3}\right) & \cos\left(\theta_e + \dfrac{2\pi}{3}\right) \\ \sin(\theta_e) & \sin\left(\theta_e - \dfrac{2\pi}{3}\right) & \sin\left(\theta_e + \dfrac{2\pi}{3}\right) \\ \dfrac{1}{2} & \dfrac{1}{2} & \dfrac{1}{2} \end{bmatrix}
$$

B. Discrete time model

By simplicity and assuming a sufficiently high sampling period T_s, Euler's Forward discretization may be applied to describe the sampling phenomena in the inverter. Thus, the RL load discrete dynamical behaviour is described as

$$
i_{qs}[k+1] = \left(1 - \frac{R_s T_s}{L_{qs}}\right) i_{qs}[k] - \frac{L_{ds}\omega_e T_s i_{ds}[k]}{L_{qs}} + \frac{T_s v_{qs}[k]}{L_{qs}}
$$

$$
i_{ds}[k+1] = \left(1 - \frac{R_s T_s}{L_{ds}}\right) i_{ds}[k] + \frac{L_{qs}\omega_e T_s i_{qs}[k]}{L_{ds}} + \frac{T_s v_{ds}[k]}{L_{ds}}
$$
$$(4)$$

III. PREDICTIVE CONTROL FUNDAMENTALS

Model-based predictive control can easily handle multivariable cases, operational constraints and nonlinearities, being indicated for increased performance and higher efficiency demands. Its implementation can be complex, but the basic formulation just requires: the prediction model, a cost function, and obtaining the control law [4].

The prediction horizon N represents how far one wishes to predict the behavior of the system and the control horizon M aims to minimize the deviation from the reference trajectory. The cost function J penalizes the deviation of the predicted control outputs from the reference trajectory, being described as a performance measurement tool for optimal control systems.

A. Finite Control Set

Finite control set is a class of MPC used to treat processes with event-based characteristics, being commonly applied in the power electronics field. [1], [2], [9] explain in details its operating principle and have implementation examples. Algorithm 1 presents a simplified FCS-MPC algorithm.

Algorithm 1 Algorithm for conventional FCS-MPC

1: Read currents $i_{qs}[k]$ and $i_{ds}[k]$
2: **for** $n \leftarrow 1 : N_{vv}$ **do**
3:     Determine $v_{qs,n}[k]$ and $v_{ds,n}[k]$ associated to state n
4:     Calculate $i_{qs,n}[k+1]$ and $i_{ds,n}[k+1]$ due to $v_{qs,n}[k]$ and $v_{ds,n}[k]$
5:     Determine J using (5)
6:     **if** $J_n[k] \leq J^\star$ **then**
7:         $J^\star \leftarrow J_n[k]$
8:         $v_{qs}^\star \leftarrow v_{qs,n}[k]$
9:         $v_{ds}^\star \leftarrow v_{ds,n}[k]$
10:    **end if**
11: **end for**
12: Apply $v_{qs}^\star$ and $v_{ds}^\star$
13: Wait for next sampling period

Essentially, the optimization problem is simplified to the prediction of the system behavior only for the possible switching combinations as the control input is chosen to be one of the available N_{vv} voltage vectors. For unitary prediction horizon, the cost function often employed in the conventional FCS-MPC is

$$
J = (i_{qs,r}[k+1] - i_{qs}[k+1])^2 + (i_{ds,r}[k+1] - i_{ds}[k+1])^2 \quad (5)
$$

where $i_{qs,r}$ and $i_{ds,r}$ are the reference for i_{qs} and i_{ds}, respectively. Also, it is important to note that $i_{qs}[k+1]$ and $i_{ds}[k+1]$ are the predicted values for these currents given the conditions in the instant k.

Since FCS-MPC requires a large amount of calculation, it is not interesting to work with long horizons once it increases the computation complexity. So, provided that the horizon is small, the algorithm is easy and intuitive to implement as it consists of evaluating the cost function online for every

978-1-7281-4181-7/19 $31.00 © 2019 IEEE

possible sequence [10]. Note that there is no need for a modulator, but the switching frequency can vary according to the load.

B. Finite Control Set with integrator

One of the major problems with conventional finite control set is its steady state error, which is usually masked by the noise in the signals [7]. As most of the controllers uses a cascade structure, the external loop compensates the error of the internal loop augmenting the reference. But, for applications that use a single loop, this FCS-MPC characteristic could be undesired. Therefore, a change in the cost function is proposed, which yields

$$J_w =(i_{qs,r}[k+1] - i_{qs}[k+1])^2 + (i_{ds,r}[k+1] - i_{ds}[k+1])^2$$
$$\rho_w(w_{iqs}^2[k+1] + w_{ids}^2[k+1]) + \rho_u(v_{qs}^2[k] + v_{ds}^2[k]) \tag{6}$$

where w_{iqs} and w_{ids} are, respectively, the accumulated errors for i_{qs} and i_{ds}, ρ_u is a penalization for control action spent and ρ_w is a penalization for accumulated error growth. The accumulated errors are calculated with

$$w_{iqs}[k+1] = w_{iqs}[k] + i_{qs,r}[k+1] - i_{qs}[k+1] \tag{7}$$
$$w_{iqs}[k+1] = w_{iqs}[k] + i_{ds,r}[k+1] - i_{ds}[k+1] \tag{8}$$

The algorithm to solve the optimization problem is the same of the conventional case (presented in Algorithm 1). But instead of solving (5), it is necessary to solve (6). Also, before solving (6), it is needed to calculate (7) and (8).

C. Finite Control Set with greater resolution

In order to reduce the noise in the FCS-MPC controlled variables, it is possible to combine the concept of FCS-MPC with some type of modulation. The simplest one is to establish virtual voltage vectors, other than the real voltage vectors, and to synthesize them using pulse-width modulation (PWM) or space-vector modulation (SVM) [11].

The only change in the main algorithm is the vectors evaluated. To be effective, this approach will require more vectors than the usual eight ones. Therefore, the computational time will increase. The same cost function with the integrator can be used, but it is also possible to include a soft constraint in the cost function [12], using

$$J_{w,c} = J_w + \rho_c \left(\frac{i_{ds}[k+1]}{i_{ds,\max}} \right)^2 \mu(i_{ds}[k+1] - i_{ds,\max}) \tag{9}$$

being ρ_c a massive penalization factor, $i_{ds,\max}$ the maximum value of $i_{ds}[k+1]$ current and $\mu(i_{ds}[k+1] - i_{ds,\max})$ a step function given by

$$\mu(i_{ds}[k+1] - i_{ds,\max}) = \begin{cases} 1 & \text{if} \quad i_{ds}[k+1] > i_{ds,\max} \\ 0 & \text{if} \quad i_{ds}[k+1] \le i_{ds,\max}. \end{cases} \tag{10}$$

As this controller has better resolution, the soft constraint is often respected, differently of the previous FCS-MPC. Moreover, this strategy has higher switching frequency due to the modulation.

D. Convex Control Set

Convex control set can be seen as the traditional MPC techniques, such as Dynamic Matrix Control (DMC) and Generalized Predictive Control (GPC). There are several researches involving this sort of controllers in the power electronics area, highlighting [5], [13], [14].

Basically, it synthesizes a new vector by adjusting the application duration of two adjacent voltage vectors, in that way, the optimization problem is solved by setting the derivative of the cost function to zero [4]. Algorithm 2 presents a simplified algorithm of CCS-MPC.

Algorithm 2 Algorithm for conventional CCS-MPC

1: Read currents $i_{qs}[k]$ and $i_{ds}[k]$
2: Calculate the accumulated errors $w_{iqs}[k]$ and $w_{ids}[k]$ using (7) and (8)
3: Calculate the voltages $v_{qs}^\star$ and $v_{ds}^\star$ using (12) and (13)
4: Saturate $v_{qs}^\star$ and $v_{ds}^\star$ if necessary
5: Send these voltages as references for a modulator
6: Wait for next sampling period

For the specific case of unitary prediction horizon, and a three-phase RL load, assuming that the modulation can provide any v_{qs} and v_{ds} (unconstrained case), the solution for the problem

$$\min_{v_{qs}[k], v_{ds}[k]} J_w \tag{11}$$
$$\text{subject to (4)}$$

is given by

$$v_{qs}^\star[k] = K_{iqs,r} i_{qs,r}[k+1] + K_{iqs,q} i_{qs}[k] + K_{ids,q} i_{ds}[k] + K_{iqs,w} w_{iqs} \tag{12}$$
$$v_{ds}^\star[k] = K_{ids,r} i_{ds,r}[k+1] + K_{iqs,d} i_{qs}[k] + K_{ids,d} i_{ds}[k] + K_{ids,w} w_{ids} \tag{13}$$

with

$$K_{iqs,r} = \frac{(L_{qs}\rho_w + L_{qs})T_s}{(\rho_w + 1)T_s^2 + L_{qs}^2 \rho_u}$$

$$K_{iqs,q} = \frac{(R_s\rho_w + R_s)T_s^2 + (-L_{qs}\rho_w - L_{qs})T_s}{(\rho_w + 1)T_s^2 + L_{qs}^2 \rho_u}$$

$$K_{ids,q} = \frac{(L_{ds}\rho_w + L_{ds})T_s^2 \omega_e}{(\rho_w + 1)T_s^2 + L_{qs}^2 \rho_u}$$

$$K_{iqs,w} = \frac{L_{qs}\rho_w T_s}{(\rho_w + 1)T_s^2 + L_{qs}^2 \rho_u}$$

$$K_{ids,r} = \frac{(L_{ds}\rho_w + L_{ds})T_s}{(\rho_w + 1)T_s^2 + L_{ds}^2 \rho_u}$$

$$K_{iqs,d} = -\frac{(L_{ds}\rho_w + L_{ds})T_s^2 \omega_e}{(\rho_w + 1)T_s^2 + L_{ds}^2 \rho_u}$$

$$K_{ids,d} = \frac{(R_s\rho_w + R_s)T_s^2 + (-L_{ds}\rho_w - L_{ds})T_s}{(\rho_w + 1)T_s^2 + L_{ds}^2 \rho_u}$$

978-1-7281-4181-7/19 $31.00 © 2019 IEEE

$$K_{ids,w} = \frac{L_{ds}\rho_w T_s}{(\rho_w + 1)T_s^2 + L_{ds}^2 \rho_u}$$

As Equations (12) and (13) could result any real number, the CCS-MPC treats the process with the entire number of real voltage signals that are sent to the modulator as control action possibilities, fixing the switching frequency and being able to work with long horizons [4].

IV. METHODOLOGY, RESULTS AND DISCUSSION

The simulation results were obtained using JuliaPro 1.0.3.1 Language [15]. The following procedures were used to achieve the results: develop the algorithm for each studied controller; consider the use of the same cost function, sampling time, switching frequency f_{sw} and other parameters to obtain a fair comparison; analyze the results both in dq and abc coordinate reference systems; verify if the dynamical responses have steady state error, ripple and noise; verify if the constraint treatment was effective; and make a comparative analysis.

Table I presents all parameters used in the simulation, including control, modulation (when applied) and inverter. The total simulated points were 250000. The minimum and fixed simulation step size was 1.0 μs.

Figures 2-6 present i_{as} and v_{as} dynamics for conventional FSC-MPC, FCS-MPC with integrator, FCS-MPC with integrator and modulation (using $N_{vv} = 16$), FCS-MPC with integrator and modulation (using $N_{vv} = 16$) and constraint treatment ($i_{ds,\max} = 0.5$ A) and, finally, CCS-MPC with integrator, respectively.

Figures 7-11 present i_{qs} and i_{ds} dynamics for conventional FSC-MPC, FCS-MPC with integrator, FCS-MPC with integrator and modulation (using $N_{vv} = 16$), FCS-MPC with integrator and modulation (using $N_{vv} = 16$) and constraint treatment ($i_{ds,\max} = 0.5$ A) and, finally, CCS-MPC with integrator, respectively.

In Figures 2 and 7 it is clear that conventional FCS-MPC does not achieve the current reference, presenting steady-steady error. Also, the small switching frequency, about 1/6 of sampling frequency [10], [12], increases the current ripple. However, this characteristic reduces audible noise and the commutation losses in the inverter switches [16].

In Figures 3 and 8, the change in the cost function allows the closed loop system to achieve the reference and reduce even more the switching frequency. This frequency reduction

is associated with the control action penalization while the reference tracking is related to the accumulated error penalization. This proposed strategy is most suitable for the cases in that lower switching frequencies are fit for.

As seen in the other Figures, the modulated FCS-MPC and the CCS-MPC were capable to reduce significantly the current ripple, due to the modulation used (PWM in this case). However, even in the FCS-MPC case that i_{ds} constraint

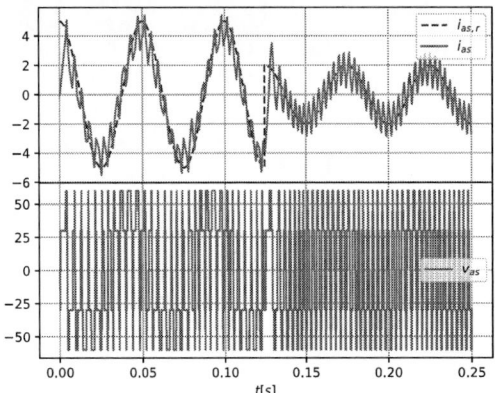

Fig. 2. i_{as} and v_{as} for conventional FCS-MPC.

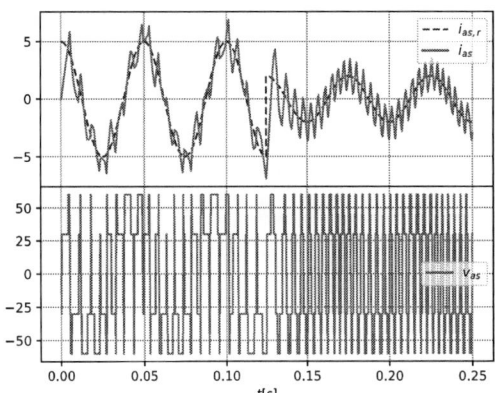

Fig. 3. i_{as} and v_{as} for FCS-MPC with integrator.

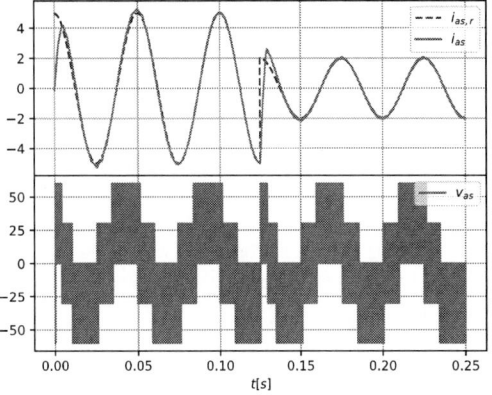

Fig. 4. i_{as} and v_{as} for FCS-MPC with better resolution.

TABLE I
SIMULATION PARAMETERS.

Parameter	Value	Parameter	Value
N	1	M	1
R_s	5.0 Ω	ω_e	125.66 rad/s
L_{ls}	2.5 mH	L_{ms}	16.4 mH
ρ_w	0.01	ρ_u	13.62 μ(A/V)^2
T_s	0.1 ms	f_{sw}	10 kHz
V_{dc}	90 V	ρ_c	$10 \cdot 10^{10}$
$i_{ds,\max}$	0.5 A	N_{vv}	8 or 16

978-1-7281-4181-7/19 $31.00 © 2019 IEEE

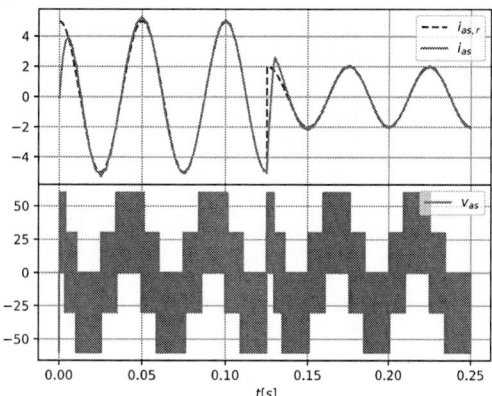

Fig. 5. i_{as} and v_{as} for FCS-MPC with better resolution and constraint treatment.

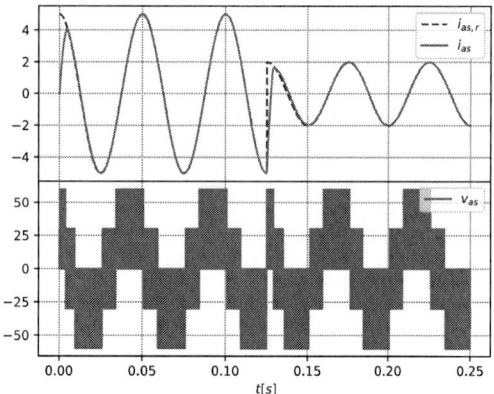

Fig. 6. i_{as} and v_{as} for CCS-MPC with integrator.

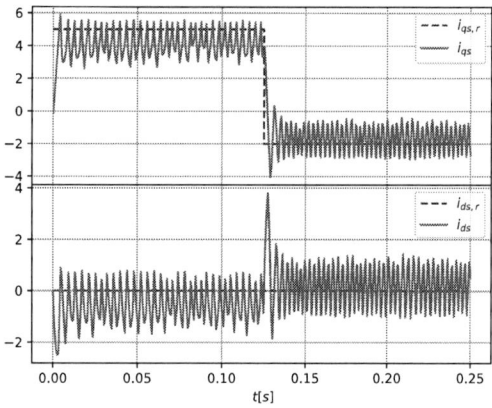

Fig. 7. i_{qs} and i_{ds} for conventional FCS-MPC.

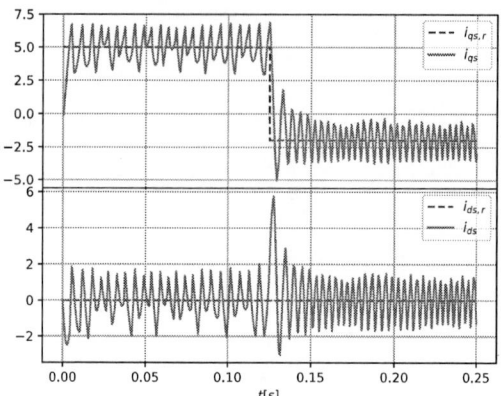

Fig. 8. i_{qs} and i_{ds} for FCS-MPC with integrator.

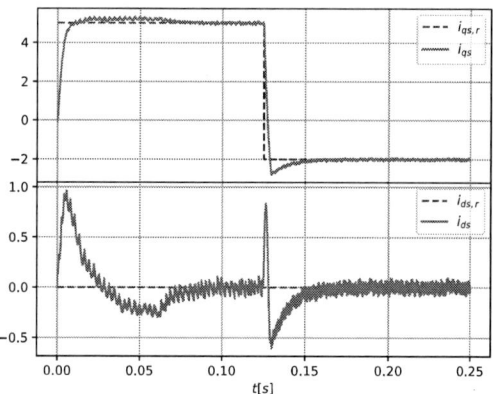

Fig. 9. i_{qs} and i_{ds} for FCS-MPC with better resolution.

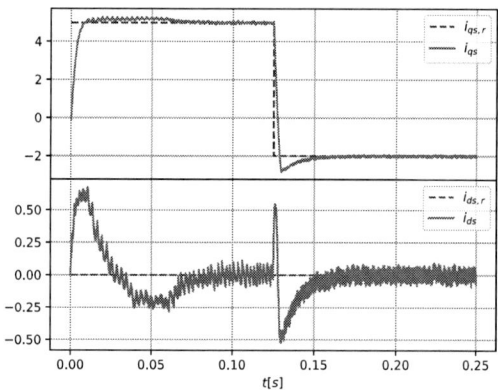

Fig. 10. i_{qs} and i_{ds} for FCS-MPC with better resolution and constraint treatment.

was treated in the cost function, the CCS-MPC accomplished this requirement with better results, due to its capability to synthesize any voltage vector.

Experimental results from other works imply that this study case with $T_s = 0.1$ ms could be implemented in conventional DSPs. [1] studied the application of MPC in a VSI with sampling time equal to $100\mu s$ and $20\mu s$. [4] compared the use of GPC and FCS-MPC for handling power converters at sampling times of, respectively, $83.33\mu s$ and from $50\mu s$ to $25\mu s$. [6] suggested both CCS-MPC and FCS-MPC for induction motors with sampling time equal to $0.1ms$. [9] applied FCS-MPC for handling a 3L-NPC and a cascaded H-bridge at sampling times of, respectively, $100\mu s$ and $200\mu s$. [12] proposed FCS-MPC for a PMSM drive system with

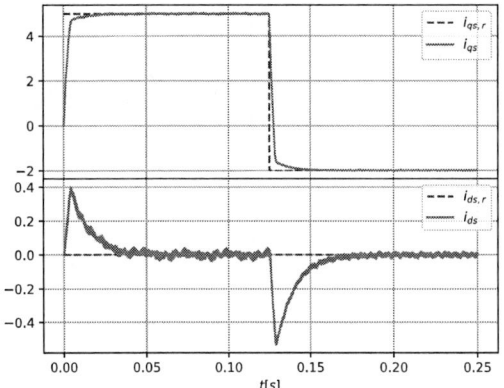

Fig. 11. i_{qs} and i_{ds} for for CCS-MPC with integrator.

sampling time equal to $100\mu s$.

V. CONCLUSION

This work presented a comparison among different strategies for FCS-MPC implementations and a CCS-MPC algorithm. All controllers were applied in a three-phase inverter feeding a RL load. Characteristics as ripple, dynamic response, steady state error, constraint treatment and switching frequency were observed.

In practically all tested cases, CCS-MPC algorithm presented better characteristics, i.e., smallest ripple, fast dynamics and null steady state error. Besides it does not directly treat the system constraints, it was possible to accomplish this requirement with only the cost function tuning. However, these characteristics were mainly provided by the high switching frequency due to the modulation, what could amplify some audible noise and increase switching losses.

In relation to FCS-MPC, using the modulated FCS-MPC, even with only 16 virtual vectors, a similar dynamic response was achieved. With this approach it is also possible to treat systems constraints, besides some over-signal tolerance. However, it spent twice computational cost than other FCS-MPC algorithms. The FCS-MPC with integrator has a significantly better response than conventional FCS-MPC, particularly in terms of steady steady error and current ripple frequency, even with a lower switching frequency and besides a higher ripple amplitude.

Therefore, for the studied case and similar, FCS-MPC with integrator is recommended over conventional FCS-MPC, even with slightly higher computational cost. For applications that need lower ripple, CCS-MPC approach is recommended, even if it cannot formally treat system constraints due to small sampling period requirement. It was seen that it is possible to deal with the constraints adjusting the cost function tuning, what could be sufficient in a large amount of cases.

ACKNOWLEDGMENT

This work was supported by the Santa Catarina State University (UDESC), Federal Institute of Santa Catarina (IFSC), Programa de Bolsas Universitárias de Santa Catarina (UNIEDU/FUMDES), Fundação de Amparo à Pesquisa e Inovação do Estado de Santa Catarina (FAPESC) and Coordenação de Aperfeiçoamento de Pessoal de Nível Superior (CAPES).

REFERENCES

[1] J. Rodriguez, J. Pontt, C. A. Silva, P. Correa, P. Lezana, P. Cortes, and U. Ammann, "Predictive current control of a voltage source inverter," *IEEE Transactions on Industrial Electronics*, vol. 54, no. 1, pp. 495–503, feb 2007.

[2] S. Vazquez, J. I. Leon, L. G. Franquelo, J. Rodriguez, H. A. Young, A. Marquez, and P. Zanchetta, "Model predictive control: A review of its applications in power electronics," *IEEE Industrial Electronics Magazine*, vol. 8, no. 1, pp. 16–31, mar 2014.

[3] F. Niu, B. Wang, A. S. Babel, K. Li, and E. G. Strangas, "Comparative evaluation of direct torque control strategies for permanent magnet synchronous machines," *IEEE Transactions on Power Electronics*, vol. 31, no. 2, pp. 1408–1424, feb 2016.

[4] C. Bordons and C. Montero, "Basic principles of MPC for power converters: Bridging the gap between theory and practice," *IEEE Industrial Electronics Magazine*, vol. 9, no. 3, pp. 31–43, sep 2015.

[5] M. Preindl, "Robust control invariant sets and lyapunov-based MPC for IPM synchronous motor drives," *IEEE Transactions on Industrial Electronics*, vol. 63, no. 6, pp. 3925–3933, jun 2016.

[6] A. A. Ahmed, B. K. Koh, and Y. I. Lee, "A comparison of finite control set and continuous control set model predictive control schemes for speed control of induction motors," *IEEE Transactions on Industrial Informatics*, vol. 14, no. 4, pp. 1334–1346, April 2018.

[7] P. Lezana, R. Aguilera, and D. Quevedo, "Steady-state issues with finite control set model predictive control," in *2009 35th Annual Conference of IEEE Industrial Electronics*, 2009, pp. 1776–1781.

[8] P. C. Krause, O. Wasynczuk, S. D. Sudhoff, and S. Pekarek, *Analysis of Electric Machinery and Drive Systems*. Wiley-IEEE Press, 2013.

[9] J. Rodriguez, M. P. Kazmierkowski, J. R. Espinoza, P. Zanchetta, H. Abu-Rub, H. A. Young, and C. A. Rojas, "State of the art of finite control set model predictive control in power electronics," *IEEE Transactions on Industrial Informatics*, vol. 9, no. 2, pp. 1003–1016, may 2013.

[10] T. Geyer, G. Papafotiou, and M. Morari, "Model predictive direct torque control—part i: Concept, algorithm, and analysis," *IEEE Transactions on Industrial Electronics*, vol. 56, no. 6, pp. 1894–1905, jun 2009.

[11] Y. Wang, X. Wang, W. Xie, F. Wang, M. Dou, R. M. Kennel, R. D. Lorenz, and D. Gerling, "Deadbeat model-predictive torque control with discrete space-vector modulation for pmsm drives," *IEEE Transactions on Industrial Electronics*, vol. 64, no. 5, pp. 3537–3547, 2017.

[12] M. Preindl and S. Bolognani, "Model predictive direct torque control with finite control set for PMSM drive systems, part 2: Field weakening operation," *IEEE Transactions on Industrial Informatics*, vol. 9, no. 2, pp. 648–657, may 2013.

[13] A. Linder and R. Kennel, "Model predictive control for electrical drives," in *2005 IEEE 36th Power Electronics Specialists Conference*. IEEE, 2005, pp. 1793–1799.

[14] M. Preindl, E. Schaltz, and P. Thogersen, "Switching frequency reduction using model predictive direct current control for high-power voltage source inverters," *IEEE Transactions on Industrial Electronics*, vol. 58, no. 7, pp. 2826–2835, jul 2011.

[15] J. Bezanson, A. Edelman, S. Karpinski, and V. B. Shah, "Julia: A fresh approach to numerical computing," *SIAM review*, vol. 59, no. 1, pp. 65–98, 2017. [Online]. Available: https://doi.org/10.1137/141000671

[16] M. Preindl and S. Bolognani, "Model predictive direct speed control with finite control set of PMSM drive systems," *IEEE Transactions on Power Electronics*, vol. 28, no. 2, pp. 1007–1015, feb 2013.

Design and Analysis of Output Filter with Long Lifetime E-Cap for AC-DC LED Driver

Andressa da S. Fernandes
Mechatronis Research Group
Federal University of Ceará
Sobral, Brazil
andressafernandes06.af@gmail.com

Felipe C. do Nascimento
Mechatronis Research Group
Federal Institute of Ceará
Sobral, Brazil
lipecn01@gmail.com

Antonia F. da Rocha
Mechatronis Research Group
Federal University of Ceará
Fortaleza, Brazil
antonia.mec@gmail.com

Rodrigo L. dos Santos
Mechatronis Research Group
Federal University of Ceará
Fortaleza, Brazil
linharodrigo@alu.ufc.br

Fernando L. M. Antunes
Group of Power Processing and Control
Federal University of Ceará
Fortaleza, Brazil
fantunes@dee.ufc.br

Edilson M. Sá Jr.
Mechatronis Research Group
Federal Institute of Ceará
Sobral, Brazil
edilson.mineiro@gmail.com

Abstract—**This paper proposes a design and analysis of output filter with long lifetime electrolytic capacitor for a single stage resonant converter based on a switched capacitor for LED driver. A study about conventional capacitors and long life capacitors is presented. Furthermore, the parameters that degrade the electrolytic capacitors lifetime and the effect of current ripple and of LED intrinsic series resistance on the output filter capacitor C_o are discussed. Experimental results are presented in order to validate the methodology proposed for the output filter sizing.**

Index Terms—**Current ripple, E-Cap, filter capacitor, long-life**

I. INTRODUCTION

Light emitting diodes (LEDs) have been increasingly used in various lighting applications [1], due to advantages such as low-cost, low-volume and low maintenance cost [2]–[4]. Besides have a luminous efficiency of over 200 lm/W and have a service lifetime exceeding 50 000 h [5], [6]. The LEDs require a constant current drive so that their luminous flux is constant [4]. Due to the existence of an AC-DC stage in LED drivers, the flickering frequency is typically rated at twice the AC mains frequency, which generates a variation of the luminous flux output of the LED, also known as flicker. A flicker modulation high can lead to human health problems, such as the risks of epilepsy, cause, visual disturbances, visual fatigue, vertigo and panic attacks [7], [8].

The recommendation IEEE Std 1789:2015 limits the percent flicker occurs at 100 Hz or 120 Hz, as it must be restricted to 8% and 9.6%, respectively [9]. The percent flicker can be approximated to half the LED current ripple, resulting in a current ripple percentage of 19.2% for a low risk level. The LED drivers, shall allow the reduction of the output current ripple and have a useful life equal to or greater than the LEDs [10].

The LED drivers can be classified into two-stage, integrated-stage and single-stage [11]. Fig. 1(a) shows the structure of the two-stage driver, which consists of power factor correction

(PFC) stage and a power control (PC) stage. This driver structure requires a greater number of components, which can result in a lower efficiency [5], [12], [13]. Fig. 1(b) shows a integrated LED driver. This method combines topologies of the PFC stage and the PC stage under shared switches. However, the peaks current in this switches can raise, which can reduce the driver efficiency [12]. Fig. 1(c) shows a single stage driver, which consists of one stage for PFC and to PC (to limit the LED current). In single driver the efficiency should be higher, because the energy is processed once, unlike two stage driver and integrated drivers [5], [11].

The electrolytic capacitor (E-Cap) can be used as an output filter to reduce low frequency current ripple in single-stage drivers. Several authors claim that the use of E-Cap conventional in drivers for LEDs can reduce the drivers lifetime expectancy, because this capacitor technology has a useful

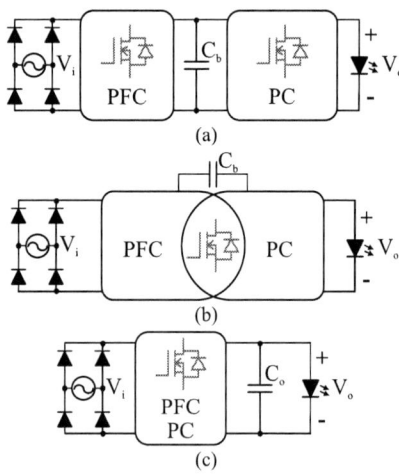

Fig. 1. Structure drivers for LEDs. (a) Two-stage, (b) Integrated-stage, (c) single-stage.

978-1-7281-4181-7/19 $31.00 © 2019 IEEE

life less than the LEDs lifetime [14]–[22]. However, new electrolyte technologies are being used in electrolytic capacitors. So, allow to obtain a stable behavior of the dissipation factor and a linear increase of the temperature in the capacitor, enabling a longer E-cap lifetime [23]–[25].

In this context, this paper presents a design and analysis of output filter with long lifetime electrolytic capacitor for a single stage resonant converter based on a switched capacitor for LED driver. In addition it is analyzed the influence of current ripple and of the LED intrinsic series resistance in output filter capacitance. In addition, an analysis of the life expectancy of long-term electrolytic capacitors is evaluated.

II. Calculation of the Output Filter Capacitor

Initially, the output filter capacitor value must be determined to allow the choice of capacitor technology to be used. Fig. 2 shows a single-stage AC-DC LED driver used for the design and analysis of output filter with long lifetime electrolytic capacitor in this paper.

The input voltage $v_i(t)$ is defined by (1), and for a unity power factor and a low harmonic distortion, the input current $i_i(t)$, as (2), must follow the input voltage waveform. Where V_m is the amplitude of the input voltage, I_m is the amplitude of the input current and ω is the line angular frequency.

$$v_i(t) = V_m sen(\omega t) \tag{1}$$

$$i_i(t) = I_m sen(\omega t) \tag{2}$$

The instantaneous input power can be obtained by (3).

$$p_i(t) = v_i(t) \cdot i_i(t) = V_m I_m sen^2(\omega t) \tag{3}$$

Substituting a trigonometric identities (4) in (3), the instantaneous input power can be rewriten as (5).

$$\sin^2 a = \frac{1}{2} - \frac{1}{2} \cdot \cos(2a) \tag{4}$$

$$p_i(t) = V_m I_m \left(\frac{1}{2} - \frac{1}{2} \cos 2(\omega t) \right) \tag{5}$$

The output power can be assumed as (6), since LEDs present voltage source characteristic, thus the output voltage (V_o) can be considered constant. Besides that, the converter does not store energy and has constant efficiency.

$$p_o(t) = V_o \cdot i_o(t) = \eta \cdot p_i(t) \tag{6}$$

Applying (5) in (6), the output current $i_o(t)$, can be obtained by (7)

$$i_o(t) = \frac{\eta V_m I_m}{V_o} \left(\frac{1}{2} - \frac{1}{2} \cos(2\omega t) \right) \tag{7}$$

The output capacitor Co is high enough to keep the LED on. Thus, electric model can be simplified by a LED intrinsic series resistance (R_{LED}) and a LED forward voltage (V_{LED}) [31]. In this condition, electric LED model can be considered linear, which allows the application of the superposition theorem.

From (7), can be seen that converter output current is composed of a DC component $I_{o.DC}$ and a AC component $i_{o.CA}(t)$, defined by (8) and (9), respectively. Fig. 3(a) shows a simplified circuit with both components of the output current.

$$I_{o.DC} = \frac{\eta V_m I_m}{2V_o} \tag{8}$$

$$i_{o.CA}(t) = -\frac{\eta V_m I_m}{2V_o} \cos(2\omega t) \tag{9}$$

In this work only the AC component is analyzed. Since it is current ripple through LED causes variation of luminous flux in time (flicker). Therefore, Fig. 3(a) can be simplified for the Fig. 3(b) for analysis. From (9) and (8), rms value of the output current $I_{o.CA}$ can be defined by (10).

$$I_{o.AC} = \frac{\eta V_m I_m}{2V_o} \cdot \frac{\sqrt{2}}{2} = I_{o.DC} \frac{\sqrt{2}}{2} \tag{10}$$

From the analysis of the circuit shown in Fig. 3(b), the rms value of AC LED current component and rms value of current through C_o are defined as (11) and (12), respectively.

$$I_{ac.LED} = I_{o.DC} \frac{\sqrt{2}}{2} \cdot \frac{1}{\sqrt{1 + (2\omega C_o \cdot R_{LED})^2}} \tag{11}$$

$$I_{ac.Co} = \frac{I_{o.DC} \sqrt{2(\omega C_o R_{LED})^2}}{\sqrt{1 + (2\omega C_o R_{LED})^2}} \tag{12}$$

The output current ripple Δi_o can be obtained by (13), as illustrated in Fig. 4. Where $\Delta i_{o\%}$ is the current ripple percentage.

$$\Delta i_o = I_{o.DC} \cdot \Delta i_{o\%} \tag{13}$$

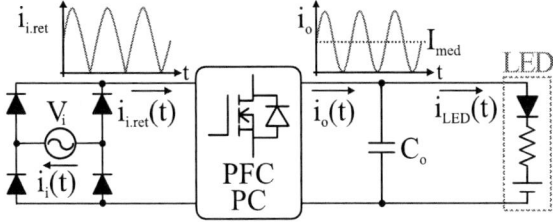

Fig. 2. Output current analysis in single-stage AC-DC LED driver.

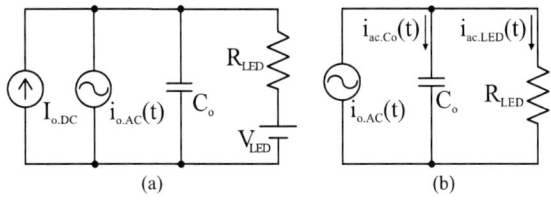

Fig. 3. Simplified circuit for output current analysis. (a) DC and AC components; (b) AC component.

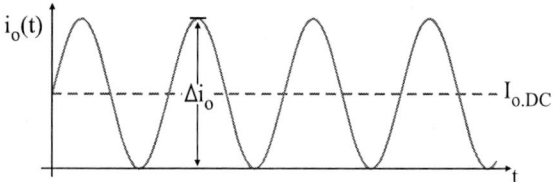

Fig. 4. Output current waveform.

Considering that the output current ripple occurs twice line frequency , because C_o is high, the rms value of the alternating component of the current in the LED as a function of the variation of the output current expressed in (13) can be obtained by (14).

$$I_{ac.LED} = \frac{\sqrt{2}}{4} I_{o.DC} \cdot \Delta i_{o\%} \qquad (14)$$

Substituting (11) in (14), (15) gives the output filter capacitor C_o.

$$C_o = \frac{\sqrt{4 - \Delta i_{o\%}^2}}{2\omega \cdot \Delta i_{o\%} \cdot R_{LED}} \qquad (15)$$

Considering only a twice line frequency ripple, the percent flicker in LED is theoretically equal to half of current ripple percentage $\Delta i_{o\%}$. Thus, to limit the possible adverse biological effects of flicker in general illumination [9], it is recommended a $\Delta i_{o\%}$ up to 19.2% for a 120 Hz current ripple. Can be seen in (15), output filter capacitor is inversely proportional to LED intrinsic series resistance R_{LED}, therefore C_o is reduced to a high R_{LED}.

III. CAPACITOR TECHNOLOGY

Since LEDs have a long lifespan (> 50000), the capacitor technology must be selected to enable the driver lifetime to be compatible with the life of the LEDs. Currently, the multilayer ceramic capacitors (MLCC), the film capacitors (FC) and E-Caps are the most used technologies in drivers for LEDs [26]. MLCCs have a long life expectancy, may be equivalent to or longer than the useful life of the LEDs. Besides that, the MLCCs have low equivalent series resistance R_{esr} values, being lower than FCs and E-Caps. However, the commercial capacitance values are reduced ($< 2.2\ \mu F$) for higher voltages (> 250 V), which requires the use of several capacitors in parallel and raises the cost of the driver.

FCs also have a high life expectancy, may be equivalent to or longer than the useful life of the LEDs. The FCs support higher voltages and commercially there are capacitances of the order of tens of microfarads. Besides that, FCs have low equivalent series resistance R_{esr} values, being superior to the MLCC and inferior to the E-Cap. However, this technology is more bulky, when compared to MLCCs and E-Caps. In applications where volume is a limiting factor, its use is impracticable or some method must be used to reduce the value of its capacitance. Several authors have proposed methods to reduce the output filter capacitor value [6], [10], [27], [28], but additional circuits are required to cancel the current ripple, converter in order to absorb the low frequency ripple, current regulators, circuit for quasi-constant power control or complex techniques, which can increase cost and reduce driver performance.

Conventional E-Caps are designed with a minimum amount of electrolyte, so any change in temperature can affect the thermal stability of the capacitor reducing its life. Currently, new E-Caps were developed, which are produced with an additional amount of electrolyte, besides the addition of acids/bases and chemicals, allowing greater thermal stability of the electrolyte, which increases its expectancy useful life [23]. Also known as long life E-Caps, these capacitors support higher temperature levels and higher current ripple levels. Thus, the useful life of these new E-Caps should be evaluated, because this technology of capacitors has a low volume in addition to supporting high levels of tension for high values of capacitances. Therefore, it is an attractive solution to avoid the use of additional methods to reduce output filter capacitance values and to enable its use in applications where volume and cost are a limiting factor.

A. E-Cap Lifetime Estimation

The end of the life of an E-Cap is defined by the change of its parameters, such as: Current ripple is another parameter that degradation the lifetime. A high current ripple generates an increase in the internal temperature of the capacitor, due to the higher dissipation factor. Therefore, a current ripple induce self-heating of the capacitor, which considerably reduces the lifetime [29].

The increase in temperature in the capacitor, above the maximum levels supported, promotes a process of evaporation and deterioration of the electrolyte. The reduction of the volume of the electrolyte causes a decrease of the effective surface area, decreasing capacitance and increasing equivalent series resistance which propitiates reducing the expected lifetime of the capacitor [30].

The parameters such as: temperature in the capacitor core, current ripple in the capacitor, dissipation factor between the core and the surface of the capacitor must be take into account for estimation the expected lifetime of the E-cap.

The expected lifetime of the electrolytic capacitors can be determined by law of Arrhenius [31], [32], as described in (16). Where L_{exp} the expected useful life, L_o the manufacturer's lifetime, T_{max} the maximum temperature supported by the capacitor and t_a at ambient temperature in the application.

$$L_{exp} = L_o \cdot 2^{\frac{T_{max} - T_a}{10}} \qquad (16)$$

The area of total capacitor dissipation A_t is another factor needed to estimate the useful life of the electrolytic capacitor. A_t can be obtained by (17), considering a heat dissipation area. Since d the diameter of the capacitor surface and h the height of the capacitor.

$$A_t = \frac{\pi}{4} d(d + 4h) \qquad (17)$$

The temperature variation on capacitor external area, due to the application of a current ripple, is obtained by (18). Since

β is a factor that determine the radiated constant heat of the capacitor.

$$\Delta T_{\text{sup}.ripple} = \frac{I_{ac.Co}{}^2 \cdot R_{esr}}{\beta \cdot A_t} \qquad (18)$$

From the temperature variation of the capacitor external area is obtained a temperature variation in the capacitor core, as described by (19). Where α the temperature difference factor between the core and the surface of the capacitor.

$$\Delta T_{nuc.ripple} = \alpha \cdot \Delta T_{\text{sup}.ripple} \qquad (19)$$

The temperature variation in the capacitor core to the nominal ripple current can be defined by (20).

$$\Delta T_{nuc.nom} = \frac{\Delta T_{nuc.ripple}}{\left(\frac{I_{ac.Co}}{I_{Co.nom}}\right)^2} \qquad (20)$$

Thus, the temperature rise factor in the capacitor core related with current ripple applied is obtained as (21) and the temperature rise factor in the capacitor core related with nominal current ripple is given by (22).

$$A_{ripple} = 10 - 0,25 \cdot \Delta T_{nuc.ripple} \qquad (21)$$

$$A_{nom} = 10 - 0,25 \cdot \Delta T_{nuc.nom} \qquad (22)$$

The useful life estimate of the electrolytic capacitor as a function of the temperature variation in the capacitor core can be obtained by (23), considering the factors presented.

$$L_{temp} = L_o \cdot 2^{\frac{T_{\max} - T_a}{10}} \cdot 2^{\left(\frac{\Delta T_{nuc.nom}}{A_{nom}} - \frac{\Delta T_{nuc.ripple}}{A_{ripple}}\right)} \qquad (23)$$

IV. DESIGN CONSIDERATIONS

The output filter capacitor C_o was used in a single stage resonant converter based on a switched capacitor for LED driver. This driver was designed to provide an output voltage V_o and a LED current I_{LED} of 385 V and 230 mA, respectively. The output filter capacitor C_o is designed considering that the current ripple percentage is 19.2% of the output average current. The parameters used for the design specifications of the output filter capacitor are presented in Table I and calculated from (15).

However, to compensate for the 20% reduction in capacitance at the end-of-life of the E-cap and still ensure the current ripple percentage designed, a capacitance 1.2 times higher than the value calculated in (15) is required. Thus, a 68 μF commercial value long life use capacitor of the KXF series was selected to be implemented in the converter. The expected lifetime of the E-cap is estimated by (23). It is considered a working temperature of up to 75°C and the parameters of the selected capacitor are described in Table II.

TABLE I
DESIGN PARAMETERS

Parameter	Symbol	Value
LED intrinsic series resistance	R_{LED}	270.2 Ω
LED forward voltage	V_{LED}	323.4 V
LED rated current	I_{LED}	230 mA
LED current ripple percentage	$\Delta I_{LED\%}$	19.2%
Line frequency	f_r	60 Hz

TABLE II
PARAMETERS FROM THE DATASHEET OF THE 68 μF E-CAP

Case size	Diameter(mm)	18
	Length (mm)	31.5
	Ripple current (A)	0.52
	Tanδ (120Hz)	0.24
	Nominal lifetime at 105°C	20000

V. RESULTS AND DISCUSSIONS

Fig. 5 presents waveforms of input voltage and input current, and also the output voltage and current through the LEDs in the nominal load condition, where the power factor is 0.993 and the current total harmonic distortion (THD) is 8.85%. Furthermore, the rms values of the input voltage and current are 222.04 V and 426.87 mA, respectively. The average output voltage is 389.5 V and the average LED current is 237.7 mA and an output power of 90.31 W.

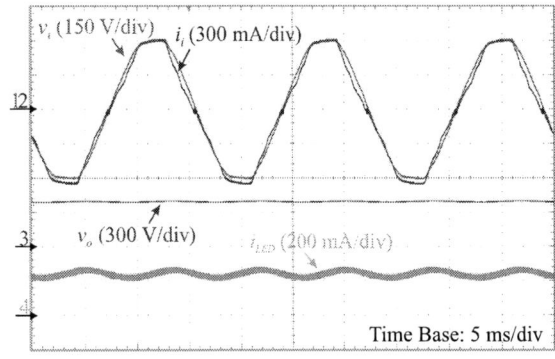

Fig. 5. Input voltage (v_i) and input current (i_i); voltage (v_o) and current (i_{LED}) waveforms in the LEDs.

Fig. 6 presents waveforms of the LED current i_{LED}, the current output capacitor $i_{ac.Co}$ and the output current i_o. The LED current ripple experimental was a value of 40 mA, i.e., 16.83% of the average LED current. As expected, the current ripple percentage value is lower than that considered in project. Afterall, an output capacitance value greater than calculated was used to compensate a reduction in capacitance at the end of life the capacitors. The experimental rms value of the capacitor current was 169 mA, which validates equation (12), since it obtained a theoretical rms value of current of 162.21 mA. Under working conditions that the output capacitor was designed in this paper, a expected lifetime of over 100 000 h was obtained as given by (23). Therefore, a C_o lifetime compatible with LED lifetime. According to analysis of section II, i_o is composed by the sum of the components i_{LED} and $i_{ac.Co}$, as can be seen in the experimental result (Fig. 6).

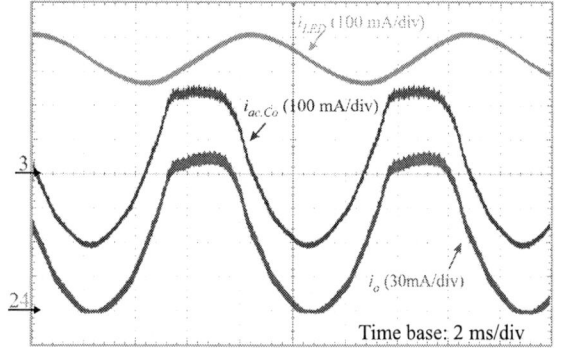

Fig. 6. LED current i_{LED}, current output capacitor $i_{ac.Co}$ and output current i_o

Fig. 7 shows the recommended operating area defined according to the percent flicker and ripple frequency [9]. The flicker of LEDs was obtained through a BPW21R photodiode and according to the methodology presented in [33]. This device has a spectral sensitivity that approaches the sensitivity curve of the human eye [34]. The measured percent flicker was of 8.62%. This value is in accordance with the limits recommended for IEEE Std 1789-2015.

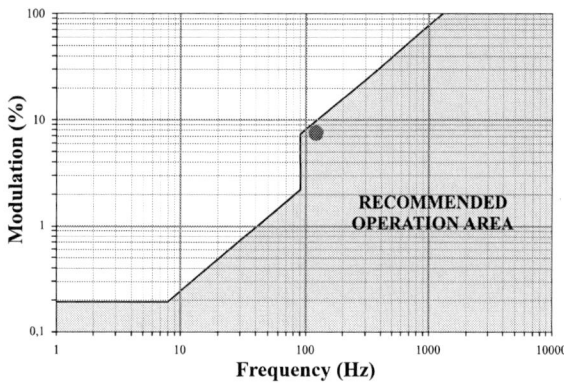

Fig. 7. Recommended operating area in terms of percent flicker as a function of the ripple frequency [9]

VI. CONCLUSION

This paper presented a design and analysis of the output filter with long lifetime electrolytic capacitor for single-stage LED driver. A discussion of conventional capacitors and long lifetime electrolytic capacitors was also presented.

To limit the possible adverse biological effects of flicker in general illumination, it is recommended a $\Delta i_{o\%}$ up to 19.2% for a 120 Hz current ripple. Therefore, from the dimensioning methodology proposed to output filter, R_{LED} is the main parameter that reduces the output capacitor, because R_{LED}

is inversely proportional to C_o. The output filter was a 68 μF capacitor KXF series.

The current ripple percentage and the percent flicker were 16.83% and 8.62%, respectively. This value is in accordance with the limits recommended for IEEE Std 1789-2015. Furthermore, the experimental result of the rms value of the capacitor current also conforms to the theoretical value. Therefore, the experimental results validate the proposed methodology for dimensioning the output filter capacitance.

The Life expectancy of the long-life E-cap was of over 100 000 h, under the working conditions presented, equivalent to the lifespan of the LEDs. Therefore, long life E-cap are no longer the limiting factor in LED drivers.

ACKNOWLEDGMENT

The authors express their special thanks to Federal University of Ceará, Federal Institute of Ceará (IFCE) - campus Sobral, where the converter prototype was implemented and the financial support provided by the Brazilian government through PROINFRA/IFCE EDITAL Nº 01/2017-PRPI, CNPq proc 432647/2016-3 and FUNCAP proc BP3-0139-00380.01.00/18.

REFERENCES

[1] J. W. Kim, J. M. Choe, and J. S. J. Lai, "Nonisolated single-switch two-channel LED driver with simple lossless snubber and low-voltage stress," IEEE Trans. Power Electron., vol. 33, no. 5, pp. 4306–4316, 2018.

[2] K. I. Hwu and W. Z. Jiang, "Nonisolated Two-Phase Interleaved LED Driver With Capacitive Current Sharing," IEEE Trans. Power Electron., vol. 33, no. 3, pp. 2295–2306, Mar. 2018.

[3] H. Khalilian, H. Farzanehfard, E. Adib, and M. Esteki, "Analysis of a New Single-Stage Soft-Switching Power-Factor-Correction LED Driver With Low DC-Bus Voltage," IEEE Trans. Ind. Electron., vol. 65, no. 5, pp. 3858–3865, May 2018.

[4] M. G. Kim, "High-Performance Current-Mode-Controller Design of Buck LED Driver with Slope Compensation," IEEE Trans. Power Electron., vol. 33, no. 1, pp. 641–649, 2018.

[5] M. Arias, I. Castro, D. G. Lamar, A. Vazquez, and J. Sebastian, "Optimized Design of a High Input-Voltage-Ripple-Rejection Converter for LED Lighting," IEEE Trans. Power Electron., vol. 33, no. 6, pp. 5192–5205, Jun. 2018.

[6] H. Valipour, G. Rezazadeh, and M. R. Zolghadri, "Flicker-Free Electrolytic Capacitor-Less Universal Input Offline LED Driver With PFC," IEEE Trans. Power Electron., vol. 31, no. 9, pp. 6553–6561, Sep. 2016.

[7] A. Wilkins, J. Veitch, and B. Lehman, "LED lighting flicker and potential health concerns: IEEE standard PAR1789 update," 2010 IEEE Energy Convers. Congr. Expo. ECCE 2010 - Proc., pp. 171–178, 2010.

[8] L. Peretto, E. Pivello, R. Tinarelli, and A. E. Emanuel, "Theoretical analysis of the physiologic mechanism of luminous variation in eye-brain system," IEEE Trans. Instrum. Meas., vol. 56, no. 1, pp. 164–170, 2007.

[9] IEEE Power Electronics Society, IEEE Recommended Practices for Modulating Current in High-Brightness LEDs for Mitigating Health Risks to Viewers. 2015.

[10] H. Kim, S. Member, M. C. Choi, S. Kim, and S. Member, "An AC - DC LED Driver with a Two Parallel Inverted Buck Topology for Reducing the Light Flicker in Lighting Applications to Low - Risk Levels," vol. 8993, no. 99, pp. 3879–3891, 2016.

[11] Y. Wang, J. M. Alonso, and X. Ruan, "A Review of LED Drivers and Related Technologies," IEEE Trans. Ind. Electron., vol. 64, no. 7, pp. 5754–5765, 2017.

[12] S. Zhang, Y. Qiu, Y. Wang, X. Liu, J. Marcos Alonso, and D. Xu, "A high-power-factor integrated-stage AC-DC LED driver based on flyback-class e converter," 2017 IEEE Ind. Appl. Soc. Annu. Meet. IAS 2017, vol. 2017-Janua, pp. 1–5, 2017.

[13] S. Lee and H. Do, "Boost-Integrated Two-Switch Forward AC–DC LED Driver With High Power Factor and Ripple-Free Output Inductor Current," IEEE Trans. Ind. Electron., vol. 64, no. 7, pp. 5789–5796, Jul. 2017.

[14] X. Liu, Q. Yang, Q. Zhou, J. Xu, and G. Zhou, "Single-Stage Single-Switch Four-Output Resonant LED Driver with High Power Factor and Passive Current Balancing," IEEE Trans. Power Electron., vol. 32, no. 6, pp. 4566–4576, 2017.

[15] J. Baek and S.-Y. Chae, "Single-Stage Buck-Derived LED Driver With Improved Efficiency and Power Factor Using Current Path Control Switches," IEEE Trans. Ind. Electron., vol. 64, no. 10, pp. 1–1, 2017.

[16] Z. P. Da Fonseca, A. J. Perin, E. A. Junior, and C. B. Nascimento, "Single-Stage High Power Factor Converters Requiring Low DC-Link Capacitance to Drive Power LEDs," IEEE Trans. Ind. Electron., vol. 64, no. 5, pp. 3557–3567, 2017.

[17] Y. Ni, S. Pervaiz, S. Member, M. Chen, and K. K. Afridi, "Capacitor Energy Buffers Through Capacitance Ratio Optimization," vol. 32, no. 8, pp. 6363–6380, 2017.

[18] J. M. Alonso, M. S. Perdigão, M. A. Dalla Costa, G. Martínez, and R. Osorio, "Analysis and Experiments on a Single-Inductor Half-Bridge LED Driver With Magnetic Control," IEEE Trans. Power Electron., vol. 32, no. 12, pp. 9179–9190, 2017.

[19] H.-A. Ahn, S.-K. Hong, and O.-K. Kwon, "A Highly Accurate Current LED Lamp Driver With Removal of Low-Frequency Flicker Using Average Current Control Method," IEEE Trans. Power Electron., vol. 33, no. 10, pp. 8741–8753, Oct. 2018.

[20] J. Lam, N. El-Taweel, and M. Abbasi, "An Output-Current-Dependent DC-Link Energy Regulation Scheme for a Family of Soft-Switched AC/DC Offline LED Drivers Without Electrolytic Capacitors," IEEE Trans. Ind. Electron., vol. 64, no. 7, pp. 5838–5850, 2017.

[21] G. M. Soares, P. S. Almeida, J. M. Alonso, and H. A. C. Braga, "Capacitance minimization in offline LED drivers using an active-ripple-compensation technique," IEEE Trans. Power Electron., vol. 32, no. 4, pp. 3022–3033, 2017.

[22] B. Sun, X. Fan, L. Li, H. Ye, W. Van Driel, and G. Zhang, "A Reliability Prediction for Integrated LED Lamp with Electrolytic Capacitor-Free Driver," IEEE Trans. Components, Packag. Manuf. Technol., vol. 7, no. 7, pp. 1081–1088, 2017.

[23] Virshay, "Capacitors -Aluminum in Power Supplies," Appl. Note, pp. 1–22.

[24] Cornell Dubilier, "Aluminum Electrolytic Capacitor Application Guide," Technology, no. 864, pp. 1–22, 2011.

[25] L. Eliasson, "Aluminium Electrolytic Capacitors Designed for Long Operational Life," Cart. Int. 2012 Proc., pp. 1–6, 2012.

[26] P. Y. Huang, H. Nagasaki, and T. Shimizu, "Capacitor Characteristics Measurement Setup by Using B-H Analyzer in Power Converters," IEEE Trans. Ind. Appl., vol. 54, no. 2, pp. 1602–1613, 2018.

[27] Y. Gao, L. Li, and P. K. T. Mok, "An AC Input Switching-Converter-Free LED Driver with Low-Frequency-Flicker Reduction," IEEE J. Solid-State Circuits, vol. 52, no. 5, pp. 1424–1434, 2017.

[28] P. Fang and Y. F. Liu, "Energy Channeling LED Driver Technology to Achieve Flicker-Free Operation with True Single Stage Power Factor Correction," IEEE Trans. Power Electron., vol. 32, no. 5, pp. 3892–3907, 2017.

[29] Vishay, "Engineering Solutions Aluminum Capacitors in Power Supplies," p. 1–18.

[30] CDM Cornell Dubilier, "Aluminum Electrolytic Capacitor Application Guide," Technology, pp.1–22.

[31] S. G. Parler Jr, "Deriving Life Multipliers for Electrolytic Capacitors," IEEE Power Electron. Soc. Newsl., vol. 16, no. 1, pp. 11–12, 2004.

[32] Reliability data handbook – Universal model for reliability prediction of electronics components, PCBs and equipment, IEC TR 62380, pp. 1–90, 2004.

[33] M. B. M. de Morais, D. M. Rufino, P. S. B. Ferreira, R. L. dos Santos and E. M. Sá Jr, "A Comparative Study the Percent Flicker and Photometric Measurement in Three-Phase and Single-Phase Drivers,"2017 Brazilian Power Electronics Conference (COBEP), Juiz de Fora, 2017, pp. 1-5.

[34] Vishay Semiconductors, "Silicon Photodiode BPW21R," 2011.

Hybrid Model of Electric Vehicle

1st Débora de Souza Martins
Federal University of Espírito Santo
UFES
Vitória, Brazil
debora.martins@aluno.ufes.br

2nd Danilo P. e Silva
Federal University of Espírito Santo
UFES
Vitória, Brazil
danilo.silva@ifes.edu.br

3rd Jussara Farias Fardin
Federal University of Espírito Santo
UFES
Vitória, Brazil
jussara.fardin@ufes.br

Abstract—With the increase of Electric Vehicles (EVs) and the use of energy from renewable sources emerge the necessity to integrate both systems, such as Vehicle to Grid (V2G). This integration involves the evaluation of continuous and discrete variables. Therefore, systems can be modeled as a hybrid system. This paper presents an EV hybrid model on a V2G system characterized by a Charging Station (CS) powered by wind and photovoltaic (PV) sources. The model is based on a Energy Storage System (ESS) equivalent circuit, Discrete Hybrid Automata system (DHA) and Mixed Logical Dynamics (MLD).

Index Terms—Electrical Vehicle, Energy Storage Systems, Smart Grid, Energy System, Mixed Logical Dynamical System.

I. Introduction

Electric mobility is expanding at a rapid pace around the world. According to the International Energy Agency (IEA), in 2018, the global electric car fleet exceeded 5.1 million, up 2 million from the previous year. China remains the world's largest Electric Vehicle (EV) market, followed by Europe and the United States [1]. This number is also expected to increase in coming years. In 2016, the European Environmental Agency (EEA) has set targets to expand the number of EV in the European continent by replacing all vehicles with EV in European cities until 2050 [2].

It is further expected a large expansion in the EV market. The increasing of using this technology will generate considerable changes in Energy System (ES) network, such as, the boost in ES's load caused by EV charging and the possibility of utilize the energy stored in vehicles to avoid interruptions. Thus, EV can be used as Energy Storage System (ESS) to backup power for higher priority loads, but it can also affect the operation of the energy system in terms of stability and with reliability [3].

The concept of Vehicle-to-Grid (V2G) comes to formulate resolutions of network overcharged problems with EV integration on the energy system. The objective of V2G is to analyze the best moment to connect the EV to charge or supply the energy to network. If the load demand on a ES and the energy price are low the EV should be charged, on the other hand, if the load system is high, it is possible to connect EVs to supply the energy network [4]. However, the EV should not access to ES freely and unmanageable. It is necessary to control its dynamic to understand the behavior of EV and ES connection.

Besides the development in EV and ES integration, there are fosters investment and studies about Renewable Energy Sources (RES) and Smart Grids (SG). This area of research is being explored because of a large increase in electrical energy consumption despite concerns about environmental problems [5]. The SG is a flexible architecture for the management of energy resources that enables end-users to actively participate in energy market [6].

In this context, studies are realized about V2G in the SG system. It is presented, in [6], a simulation of EV and SG interaction in a bidirectional power flux way, which means Grid to Vehicle(G2V) and Vehicle to Grid (V2G) modes. In the G2V mode, EV behaves as an electrical load; in the V2G mode is acting like injecting source. It was evaluated the power flux and some conditions that enable EV to receive or supply power to the electrical system. In the same context, in [7], it is proposed a real-life military application of vehicle-to-grid and vehicle-to-vehicle (V2G-V2V)-based Microgrid (MG) system, which can be defined as integrated energy system within the SG definition [8] [9]. Moreover, in [10], it is analyzed the possibility of extending the life cycle of EV plug-in batteries to an MG system.

Implementing a V2G-G2V system requires analysis of continuous and discrete variables. By continuous variables analysis, it is possible to define discrete components to make decisions. For example, to define to connect or not the EV to the ES or to establish a system as V2G or G2V, which are discrete variables, it is important to verify continuous variables such as vehicle battery state of charge, electrical current level and the quantity of energy generated in the electrical system. Systems that contain continuous and discrete components can be defined as Hybrid Systems [11].

Hybrid modeling is widely used in literature for application of Model Predictive Control (MPC) or optimization problems such as Mixed Integer Linear Programming (MILP). V2G and/or G2V systems are also being explored as optimization problems such as done in [12], where it discusses V2G implementation to frequency regulation at MG system using optimization approach by MILP system. Moreover, in [13], there is an optimal Energy Management Strategy (EMS) to optimize the electrical system with photovoltaic (PV) generation and an EV hybrid system in Smart Building with a V2G concept.

Hybrid systems encompass many different mathematical formulations, as Mixed Logical Dynamics (MLD), Piecewise Affine System (PWA) and Discrete Hybrid Automata (DHA). These subclasses of a hybrid system are equivalent under some

978-1-7281-4181-7/19 $31.00 © 2019 IEEE

considerations [11]. For this reason, it is possible to apply theoretical control properties through tools for conversion between subclasses [14].

This paper presents an EV hybrid model on a V2G system characterized by a Charging Station (CS) powered by wind and PV sources. Firstly, the model is defined as a DHA system and then it is transformed into MLD description.

The scenario defined to model's elaboration consists of a CS able to send information to EV that is roading in a delimited area. This information corresponds to the level of power generation from renewable sources and it is sent to the vehicle each 5 seconds. The output decision also depends on the generation level of energy on the CS.

The delimited area is around the CS and if the vehicle is on the furthest point from CS, inside the area, the time required for EV to reach the station is 190 seconds, or 3 minutes and 10 seconds. Disconnecting EV is also not instantaneous and takes 35 seconds to complete. These time delays are considered to approximate model to real EV behavior.

The output model consists of binary decision and continuous variables. The second type is monitored mainly to determine accuracy in the calculations performed by the EV battery circuit model. The input system is defined by electrical current, connection status between EV and the charging station, State of Charge (SoC) and Voltage drop at the previous time. The model is developed in a scenario which there is a communication between CS and the EV. So it is possible to process information that comes from CS status.

The paper is structured as follows: Section II presents the battery circuit model and methodology used to construct the hybrid model, including the definition of classes DHA, PWA, and MLD. Section III presents the result of two model tests. The first test verifies the calculation of continuous system output: terminal voltage. The second test verifies if logical operations are being done correctly. Finally, section IV presents the conclusions and discussions about the model and discussions to improve.

II. ELECTRIC VEHICLE MODEL

The first step to develop the electrical vehicle DHA model is to define an equivalent circuit to represent and calculate its electrical dynamic. This is shown in Section II-A. Then, circuit elements are defined in Section II-B based on EV battery type characteristics. From the equivalent circuit, it is possible to deduce state-space equations of the system and define the DHA model. At this point, it is also important to define how discrete variables will be changed by the model rules defined in Section II-C.

A. Equivalent Circuit

There are different equivalent circuits to represent a battery system [15]. Thévènin equivalent circuit, shown in Fig.1, is the most used [16] and it will be used in this paper, where R_0 represents internal losses; R_1 and C_1 represent a transient relaxation effect on the battery terminal current. The voltage drop in R_1C_1 branch is V_{RC}; SoC is battery State of Charge,

and Q_R is the battery capacity. These circuit parameters are defined in Section II-B. Continuous system output is terminal voltage: V_T; and continuous system input is the current i_L. For this model, there are two possibilities of the current input. The electrical current may come from the charging station PV and wind generators or it may come from the vehicle circuit while it is inroad. The choice of which current to use for the input is explained in Section II-C.

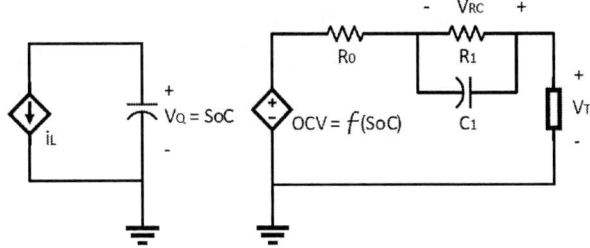

Fig. 1. Equivalent Circuit

To improve the analysis and compose accuracy results, more RC branches can be added to the circuit system. However, it is a trade-off between accuracy and complexity to implement the model [16]. This paper considers one RC branch to prioritizing adequacy for model construction in hybrid systems.

The open-circuit-voltage (OCV) has a nonlinear relationship with SoC. It depends on the type of material of the battery and the ambient temperature. However, in Lithium batteries normally used in EV, the curve representing the relationship between $OCV - SoC$ is flat. The voltage varies little with the loading and unloading of the system [16].

In the present modeling, it will be considered the range of values where the $OCV - SoC$ curve can be considered as linear. In (1) is shown the linear equation which uses parameters β_1 and β_0. Their definition are presented in Section II-B.

$$V_{OCV} = \beta_1(SoC) + \beta_0 \tag{1}$$

The discrete-time state-space model is used in the elaboration of the hybrid system model and it is deduced from the battery circuit shown in Fig.1. It is shown in (2), where T_S is sampling time:

$$\begin{cases} \begin{bmatrix} SoC[k+1] \\ V_{RC}[k+1] \end{bmatrix} = \begin{bmatrix} 1 & 0 \\ 0 & e^{-\frac{T_S}{R_1C_1}} \end{bmatrix} \begin{bmatrix} SoC[k] \\ V_{RC}[k] \end{bmatrix} + \begin{bmatrix} \frac{T_S}{Q_R} \\ R_1\left(1 - e^{-\frac{T_S}{R_1C_1}}\right) \end{bmatrix} i_L \\ V_T[k] = \begin{bmatrix} \beta_1 & 1 \end{bmatrix} \begin{bmatrix} SoC[k] \\ V_{RC}[k] \end{bmatrix} + R_0 i_L[k] + \beta_0 \end{cases} \tag{2}$$

B. Parameters Definition

In the present paper the $OCV - SoC$ relation is considered linear. Thus, considering (1), their parameters values are $\beta_1 = 0.0107$ and $\beta_0 = 3.0950$. These values were obtained by linearization of the $OCV - SoC$ curve from INR18650-20R battery at 25°C [17]. Based on the same $OCV - SoC$ curve and on the Low Current OCV test it is obtained circuit parameters: $R_0 = 0.0710\Omega$; $R_1 = 0.0222\Omega$ and $C_1 = 1201.4F$

978-1-7281-4181-7/19 $31.00 © 2019 IEEE

C. Hybrid Model

The subclass of DHA hybrid systems is the result of the relationship between Switched Affine Cystem (SAS) and a Finite State Machine (FSM). Therefore, a DHA system encompasses mathematically continuous variables from SAS and decision discrete variables from FSM [14]. This relationship is possible due to connection elements: Event Generator (EG) and Mode Selector (MS).

The EG extracts logic signals from the continuous variables and its decisions generate a state transition in the FSM. Besides, the MS combines all logic variables to select a dynamic continuous mode in SAS. The DHA model generalizes many models oriented to computational tools for hybrid systems and it also represents the starting point for solution of complex analysis and problems of hybrid systems syntheses. However, DHA verification and simulation procedures have a high computational cost [14].

The MLD is a modeling structure that allows the description of several systems classes, such as continuous and discrete inputs systems, DHA systems, models with qualitative outputs and PWA systems. it is based on matrix mathematical expressions and it has a lower computational cost to be run. Thus, from the DHA model system, for example, it is possible to transform the model into an MLD system. The transformation of the first-principles description of the hybrid systems in the MLD form requires the application of a set of rules [18]. However, it is presented in [19] a tool to calculate matrices that describe an MLD system computationally, which these are adequate to optimize, control and simulate dynamic systems: the HYbrid System DEscription Language (HYSDEL).

HYSDEL software can convert nonlinearities into linear functions through propositional logics. It can extract the MLD formulation from the matrix form and simulate the model. In this simulation, it uses mixed-integer programming.

From the representation in state variables of the battery circuit, a hybrid model is defined to EV. Therefore, it is considered that the electric vehicle will road in a specific route and its battery will be charged in a charging station whose energy comes from wind and PV sources. Continuous input variables of the system at instant k are shown in (3).

$$
\begin{aligned}
&\text{Continuous Input} \\
&SoC[k-1] \rightarrow & \text{State of Charge} \\
&V_{RC}[k-1] \rightarrow & \text{Voltage Drop in RC branch} \\
&I_{bat}[k] \rightarrow & \text{Battery EV Current} \\
&I_{PV}[k] \rightarrow & \text{PV generatior current} \\
&I_{Wind}[k] \rightarrow & \text{Wind generator current} \\
&E_{PV}[k] \rightarrow & \text{PV energy generator} \\
&E_{Wind}[k] \rightarrow & \text{Wind energy generator}
\end{aligned} \tag{3}
$$

There is also a discrete input u_1 defined as EV and charger connection status. If $u_1 = 1$, the EV is connected to the charger.

A range of 7 auxiliary binary variables $\delta = [\delta_1 \ \delta_2 \ \delta_3 \ \delta_4 \ \delta_5 \ \delta_6 \ \delta_7]$ defines logical characteristics for the current system from continuous variables. This range is the DHA Event Generator (EG).

δ_1 and δ_3 are flags to indicate if SoC is higher than 70% and if it is already 100%. HYSDEL do not process equal in logical variables. Thus, boundary conditions are defined as the model. δ_2 denotes in the current direction. If it is greater than zero, the EV is being charged. This is analyzed when the EV is connected to the charger. In operation disconnected to the charger, EV internal circuit current may have positive and negative values that do not indicate whether it is being charged or not. From δ_5 until δ_6 there are logical operations between the binary input and δ_1 to δ_4. E_{Max} is a parameter that indicates the maximum power generation capacity of power sources.

In (4), it is defined some security factors based on the necessity of minimizing EV's battery degradation and also to be able to charge EV at the charging station for a safe time. These security factors are defined in δ_1 and δ_4. The first one is based on EV should be parked if its battery has an SoC of 50% or less [20]. Considering that EV can be far away from charging station, for safety, it is defined that if SoC is less than 70%, EV can be charged. it is also defined, in δ_4, that charging station can supply energy to EV if generation is at least 80% of maximum capacity.

$$
EG : \begin{cases}
[\delta_1 = 1] \leftrightarrow [SoC[k] \geq 70\%] \\
[\delta_2 = 1] \leftrightarrow [I_{PV}[k] + I_{Wind} \geq 0] \\
[\delta_3 = 1] \leftrightarrow [SoC[k] \geq 100\%] \\
[\delta_4 = 1] \leftrightarrow [E_{PV} + E_{Wind} \geq 0.8 \times E_{Max}] \\
[\delta_5 = 1] \leftrightarrow [u_1[k] \wedge \neg\delta_2 \wedge \neg\delta_3 \wedge \delta_4] \\
[\delta_6 = 1] \leftrightarrow [\neg u_1[k] \wedge \neg\delta_1 \wedge \neg\delta_3 \wedge \delta_4]
\end{cases} \tag{4}
$$

The continuous auxiliary variable of this model is represented by $z = i_l$ and it corresponds to the electric current used as input to the model. The value of this auxiliary variable depends on EV and charging station connection status. If EV is connected to CS, the current will come from the power sources; If it is not connected, the current will correspond to the one when EV is roading.

Binary outputs of the model indicate if the EV can connect or should be disconnect from the CS: $y_1[k] = 1 \rightarrow \delta_5$ Keep EV connected and $y_2[k] = 1 \rightarrow \neg\delta_6$ Keep EV disconnected.

Continuous output are terminal battery voltage V_T, state of charge SoC and V_{RC} at present instant. To calculate system states, defined in (2), it is important to verify previous states. Thus, to the model, states are also output variables to be sent to the input system in the next instant. Also, the time delay defined to change EV connection status at each iteration is calculated out of the MLD model.

There are two possible situations for EV: connected or disconnected from the battery charger. These situations are defined by the Mode Selector (MD) of the DHA model which is shown in 5.

$$
MS : m_p(k) = \begin{cases}
1 & \text{if } u_1 \\
2 & \text{if } \neg u_1
\end{cases} \tag{5}
$$

From MS, system states and outputs are calculated. For example, to calculate SoC as is shown in Switched Affine

System (SAS) represented in (6). In this case, I_1 and I_2 represent, respectively, PV and wind generators current circuit and the electrical current when EV is on road. Similarly, it is possible to calculate the drop RC branch V_{RC} state and the terminal voltage output V_T.

$$SAS : SoC[k] = \begin{cases} SoC[k-1] + I_1 \frac{T_S}{Q_R} & \text{if } u_1[k] = 1 \\ SoC[k-1] + I_2 \frac{T_S}{Q_R} & \text{if } u_1[k] = 0 \end{cases}$$

(6)

Through the HYSDEL software, DHA operations are transformed into MLD systems, as is shown in (7).

$$SoC[k] = SoC[k-1] + u_1[k].I_1 \frac{T_S}{Q_R} + \neg u_1[k].I_1 \frac{T_S}{Q_R} \quad (7)$$

Bounded to $E_2\delta(t) \leq E_1 u(t) + E_4 x(t) + E_5$, where E_1, E_2, E_3, E_4 and E_5 are matrices which lines dimension is 56 and columns dimension is between 1 and 16. They will be omitted in this article because of space limitation.

III. Modeling Results

The model verification procedure consists of two steps. The first one is to verify if the MLD model calculates correctly the output system and the second test consists of analyzing the discrete outputs of the system.

A. Continuous Output Analysis

In order to evaluate the continuous output variable of the model, the test scenario is the EV not connected to the CS. Thus, the input current is from the internal circuit of the EV. Two different current input profiles are used: Federal Urban Driving Schedule (FUDS) and Highway Driving Schedule (US06). With these databases, in [17], it is presented real experiments with the INR 18650-20R battery at different temperatures and initial SoC, to get the respective terminal voltage values at each instant. These values will be used to compare the terminal voltage results obtained with the model.

The battery model is the same used to determine the circuit parameters and the factors in $OCV - SoC$ linear approximation. Thus, the present test is intended to verify that the MLD model is correctly calculating the continuous output. Because of this, the database used for testes is obtained for the same temperature and battery type that the one used to parameters definition in [17]. If other data is used, it is necessary to adjust battery circuit parameters or to construct an estimator for them.

Therefore, define parameters and to validate results, data is obtained at 25°C and initial battery SoC at 80%. The continuous output variable is terminal voltage and it will be compared to the real measured terminal voltage from FUDS and US06 database. The current profiles used for checking are shown in Fig.2.

Figures 3 and 4 show the terminal voltage values measured in the battery, according to the current profiles US06 and FUDS, respectively. Also, it is possible to verify terminal voltage values calculated by model.

From the analysis of the graphs, it is possible to observe that calculated values vary as the real terminal voltage. It confirms

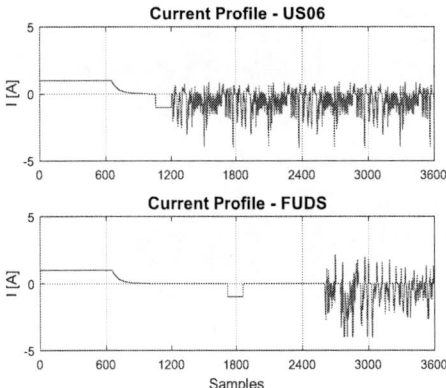

Fig. 2. Current Profiles

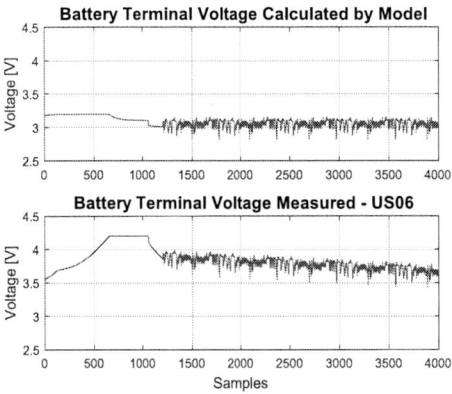

Fig. 3. Terminal Voltage Measured and Calculated - US06 Database

that the calculation result by the MLD model is obtained as expected. However, it is also verified that calculated voltage varies around a different value that real voltage. This may be caused by battery circuit parameters varying by the temperature, load state, or variation of current being applied. Since circuit parameters are not updated in this model, this error can

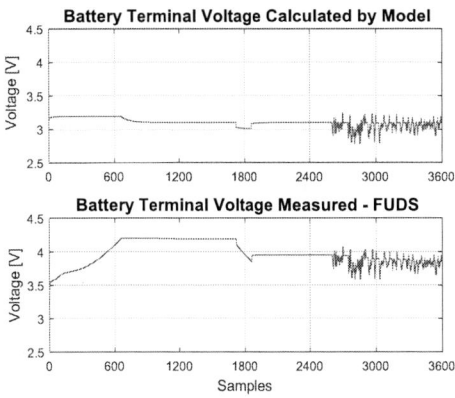

Fig. 4. Terminal Voltage Measured and Calculated - FUDS Database

be propagated.

B. Discrete Output Analysis

To evaluate the DHA model, it will be analyzed EV model decision-making responses based on system inputs and states. The discrete output is based on the EV model states and information from wind and PV generation on the charger. Test scenarios and their results are presented in III-B2 and III-B3.

1) Wind and PV battery charger parameters: To this paper, it is considered empirically profiles for PV generation and wind generation. The maximum value of all energy generated in this case is $12kWh$. Thus the value from which the charger can supply power to the vehicle is $0.8 \times 12 = 9.6kWh$, as is defined in logical operations in Equation (4).

2) First Test: In the first situation, EV is moving on the delimited area and it is not connected to CS. The EV SoC become less than the limit defined to enable EV moving without prejudice its battery, as is defined in δ_1 in (4). Also, the power generation on the charger becomes higher than $9.6kWh$ after $Time = 1285s$. Thus, the vehicle should be connected to the charger which means that output decision to keep EV disconnect remains to zero: $y_2 = 0$.

It is also considered 3 minutes and 10seconds delay, or 38 samples with 5 seconds time sampling, to start the connection. This delay finishes at $Time = 1475s$ and it represents the time required for the vehicle to travel from the furthest point to the charger, inside the bounded area.

After time delay, status connection will be changed from zero to 1: $u_1 = 1$ and also output decision to keep EV connected will be changed to 1: $y_1 = 1$. The Fig.5 shows outputs y_1, y_2 and also EV connection to charger status.

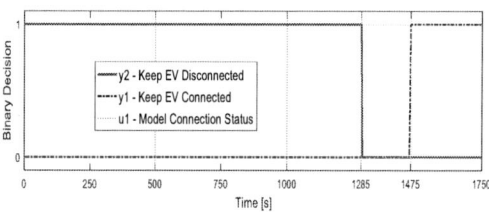

Fig. 5. Model binary decision and connection status between EV and the charger.

It is also expected that EV SoC starts to increase because of the charging at CS. In Fig.6 it is possible to verify that SoC also starts to increase at $Time = 1475s$, as expected.

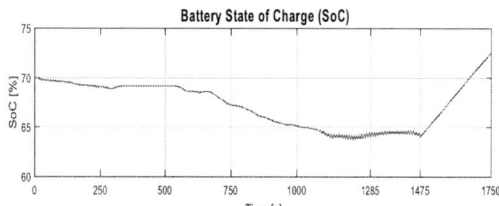

Fig. 6. EV Battery State of Charge.

3) Second Test: In the second situation, the EV is already connected to the charging station and it is being charged until $SoC = 100\%$. The wind and PV generation adopted in this situation is higher than the threshold that allows the charger to send power to EV ($9.6kWh$).

It is possible to observe that SoC is increasing linearly from 75% to 100% in Fig.7. The linearity is observed because of $OCV - SoC$ relation.

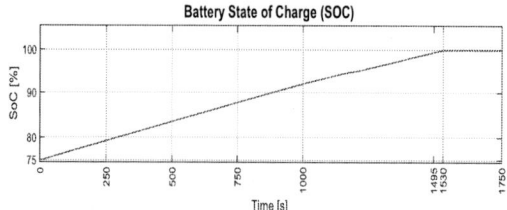

Fig. 7. EV Battery State of Charge

Once the SoC reaches 100%, at $Time = 1495s$ the model output indicates that the vehicle should not remain connected. The output y_1 which means to keep EV connected to the charger changes from 1 to 0, as it is shown in Fig.8. After that, a 35-second delay, or 7 samples with the 5 seconds time sampling, is considered for the vehicle to be disconnected.

After the delay, at $Time = 1530s$ output y_2 which means to keep EV disconnected changes from 0 to 1. Also, EV connection status changes at the same time from 1 to 0, as it is shown in Fig.8.

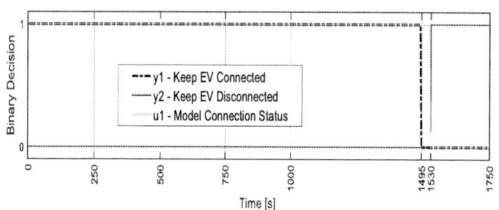

Fig. 8. Model binary decision and connection status between EV and the charger.

4) Third Test: In the third test, EV is connected to the charger in the beginning and its SoC is not maximum. However, the wind and PV power generation become less than $9.6kWh$ at $Time = 390s$. Thus, the expected decision of the model is to disconnect EV from the charger even if the battery is not charged.

After disconnection, in this case, it was adopted that the vehicle would remain stationary. Therefore, the input current has been switched to zero and the SoC remains constant. The Fig.9 shows SoC values during the test. It is possible to observe that SoC stops to increase after the delay at $Time = 425s$.

Model decision and connection status are shown in Fig.10. It is possible to observe the decision to disconnect EV from the charger at $Time = 390s$ because of the less generation. After the delay, at $Time = 425s$, it is possible to observe the change of output decision to keep EV disconnected and EV status connection.

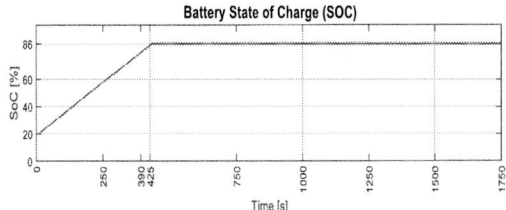

Fig. 9. EV Battery State of Charge.

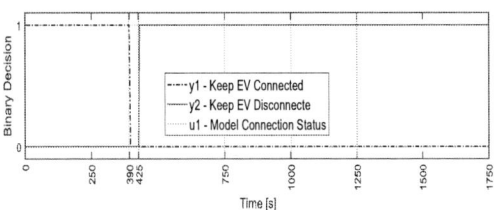

Fig. 10. Model binary decision and connection status between EV and the charger.

IV. CONCLUSION

To evaluate the advantages and limitations of the MLD model of an EV presented in this paper, two tests were done with the available database. The first one is an analysis of the continuous variables and it shows that the model correctly computes the variation of terminal voltage values. However, there is still an error between measured and calculated values. It may be due to factors such as circuit parameters variation which may lead to $OCV - SoC$ modification. The online estimation of circuit parameters and states can be an alternative for improvement of the presented model. Thus, in addition to increasing accuracy concerning data variation, the model can be generalized more easily for other battery types.

In the analysis of decision output, it is concluded that the model responses are obtained as it is expected. From this good operation, it is possible to increase the complexity of the EV model by inserting more possibilities for analysis and response to situations that are not addressed in the present paper. For example, consider that EV can provide energy to some element of the Charging Station, like an Energy Storage System, depending on EV SoC. This kind of structure makes possible to apply control techniques as predictive control, for example.

V. ACKNOWLEDGMENT

Authors thank the Coordenação de Aperfeiçoamento de Pessoal de Nível Superior (CAPES) and the Fundação de Amparo à Pesquisa e Inovação do Espírito Santo (FAPES) for research founding. Also, the Center for Advanced Life Cycle (CALCE) Battery Research Group for database available on the Internet.

REFERENCES

[1] Iea - global ev outlook 2019. www.iea.org/publications/reports/globalevoutlook2019, May 2019.

[2] Electric vehicles in europe. https://www.eea.europa.eu/publications/electric-vehicles-in-europe, Sep 2016.

[3] Evangelos Karfopoulos, P Papadopoulos, Spyros Skarvelis-Kazakos, Iñaki Unda, Liana Cipcigan, Nikos Hatziargyriou, and Nick Jenkins. Introducing electric vehicles in the microgrids concept. *2011, IEEE 16th International Conference on Intelligent System Application to Power Systems*, pages 1–6, 09 2011.

[4] Yimin Zhou. Vehicle to grid technology: A review. 07 2015.

[5] ENERDATA Intelligence and Consulting. Global energy trends. https://www.enerdata.net, Nov 2018.

[6] I. Sami, Z. Ullah, K. Salman, I. Hussain, S. M. Ali, B. Khan, C. A. Mehmood, and U. Farid. A bidirectional interactive electric vehicles operation modes: Vehicle-to-grid (v2g) and grid-to-vehicle (g2v) variations within smart grid. In *2019 International Conference on Engineering and Emerging Technologies (ICEET)*, pages 1–6, Feb 2019.

[7] M. A. Masrur, A. G. Skowronska, J. Hancock, S. W. Kolhoff, D. Z. McGrew, J. C. Vandiver, and J. Gatherer. Military-based vehicle-to-grid and vehicle-to-vehicle microgrid—system architecture and implementation. *IEEE Transactions on Transportation Electrification*, 4(1):157–171, March 2018.

[8] M. Pipattanasomporn, H. Feroze, and S. Rahman. Multi-agent systems in a distributed smart grid: Design and implementation. In *2009 IEEE/PES Power Systems Conference and Exposition*, pages 1–8, March 2009.

[9] C. Marnay, S. Chatzivasileiadis, C. Abbey, R. Iravani, G. Joos, P. Lombardi, P. Mancarella, and J. von Appen. Microgrid evolution roadmap. In *2015 International Symposium on Smart Electric Distribution Systems and Technologies (EDST)*, pages 139–144, Sep. 2015.

[10] S. Beer, T. Gomez, D. Dallinger, I. Momber, C. Marnay, M. Stadler, and J. Lai. An economic analysis of used electric vehicle batteries integrated into commercial building microgrids. *IEEE Transactions on Smart Grid*, 3(1):517–525, March 2012.

[11] B. D.& Bemporad A. Heemels, W. & Schutter. Equivalence of hybrid dynamical models. *Elsevier Automatica*, 37(7):1085–1091, 2001.

[12] M. Endo and K. Tanaka. Evaluation of storage capacity of electric vehicles for vehicle to grid considering driver's perspective. In *2018 IEEE International Conference on Environment and Electrical Engineering and 2018 IEEE Industrial and Commercial Power Systems Europe (EEEIC / I CPS Europe)*, pages 1–5, June 2018.

[13] H. Turker and I. Colak. Multiobjective optimization of grid-photovoltaic- electric vehicle hybrid system in smart building with vehicle-to-grid (v2g) concept. In *2018 7th International Conference on Renewable Energy Research and Applications (ICRERA)*, pages 1477–1482, Oct 2018.

[14] D. P. e Silva, M. D. Queiroz, J. F. Fardin, J. L. F. Sales, and M. T. D. Orlando. Hybrid modeling of energy storage system and electrical loads in a pilot-microgrid. In *2018 13th IEEE International Conference on Industry Applications (INDUSCON)*, pages 433–438, Nov 2018.

[15] Xiaosong Hu, Shengbo Li, and Huei Peng. A comparative study of equivalent circuit models for li-ion batteries. *Journal of Power Sources*, 198:359–367, 01 2012.

[16] G. S. Misyris, T. Tengnér, A. G. Marinopoulos, D. I. Doukas, and D. P. Labridis. Battery energy storage systems modeling for online applications. In *2017 IEEE Manchester PowerTech*, pages 1–6, June 2017.

[17] Fangdan Zheng, Yinjiao Xing, Jiuchun Jiang, Bingxiang Sun, Jonghoon Kim, and Michael Pecht. Influence of different open circuit voltage tests on state of charge online estimation for lithium-ion batteries. *Applied Energy*, 183, 12 2016.

[18] A. Bemporad, G. Ferrari-Trecate, and M. Morari. Observability and controllability of piecewise affine and hybrid systems. *IEEE Transactions on Automatic Control*, 45(10):1864–1876, Oct 2000.

[19] D. Mignone, A. Bemporad, and M. Morari. A framework for control, fault detection, state estimation, and verification of hybrid systems. In *Proceedings of the 1999 American Control Conference (Cat. No. 99CH36251)*, volume 1, pages 134–138 vol.1, June 1999.

[20] Markus Eider and Andreas Berl. Dynamic ev battery health recommendations. 06 2018.

Static Transfer Switch Applied to Single-Phase Uninterruptible Power Supply

Marcus Vinícius Maia Rodrigues
IFSP-Instituto Federal de São Paulo
Avaré, Brasil
marcus.rodrigues@ifsp.edu.br

Newton da Silva
UEL-Univ. Estadual de Londrina
Londrina, Brasil
newton.silva@uel.br

Flávio Alessandro Serrão Gonçalves
UNESP-Univ. Estadual Paulista
Sorocaba, Brasil
flavio.as.goncalves@unesp.br

Abstract — This paper presents a Static Transfer Switch (STS) applied to Single-Phase Uninterruptible Power Supply. The performance of two STS topologies was evaluated, one traditional composed of thyristors and the other formed by the association of IGBTs and diodes. In addition, two approaches for detection of disturbances in the grid voltage waveform are explored: the first method is based on the dq0 transformation and the second one is performed through a PLL (Phase Locked Loop) algorithm. Simulation and experimental results were performed to evaluate the performance of topologies and methods for disturbances detection.

Keywords— Static Transfer Switch, IGBT, UPS, Passive Standby

I. INTRODUCTION

Despite the efforts accomplished to reduce the disturbances affecting the quality of electricity, the available energy in electrical grid may not present acceptable quality for some applications. With the development of the electronics industry, it is increasingly common for electronic equipment to be present in fundamental sectors of society, such as, telephone exchanges, hospitals, factories and bank branches, where a given electronic system cannot have its power supply interrupted. No matter how good the power supply system for these consumers, there is still a possibility of disturbances in the electrical grid. These problems are considered as Power Quality (PQ) phenomena [1].

The grid disturbances can be imperceptible for final consumers and may slowly damage electronic components of the products. Thus, equipment has its useful life reduced due to issues quality of the power supplied, and users, who are not aware of this factor, consider product with inferior quality [1].

The use of Uninterruptible Power Supply (UPS) has emerged as a way of solving these problems, being quite efficient for protection of loads against disturbances, that is, transients voltage sag/swell, or short interruptions in power supply.

According to the International Electrotechnical Commission (IEC 62040-3 Ed. 2.0 b: 2011), three types of UPS topologies are defined, one of which is known as Passive Standby, as shown in Fig. 1 [2].

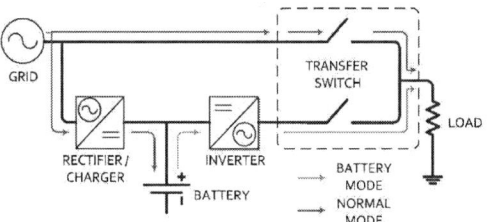

Fig. 1. Block diagram of UPS Passive Standby.

For Passive Standby topology, there are two modes of operation: normal mode and battery mode. In normal mode, AC power (grid) feeds directly load and in battery mode a voltage source inverter transfers power to load while there is energy stored in battery pack. The transfer switch is responsible for transferring power to the load according to operating mode, considering conception of two input ports and one output port, as shown in Fig. 1.

The performance of an UPS is defined by a series of characteristics related to quality of energy delivered to load. One of the most important is the total load transfer time (*tt*), basically represents the amount of time required by the commutation of transfer switches to exchange power source that feeds the load (electric grid or inverter). For a good performance it is desired that total transfer time to be smallest possible, allowing load in question keeps less exposed to oscillations of source that is suffering disturbance. Thus, ensuring greater reliability of power system [1-3].

The transfer switch performance is investigated in [4]. This work defines that total load transfer time (*tt*) corresponds to the sum of detection time (*td*) and transfer time (*tf*). The detection time is defined by the time interval between the occurrence and the detection of the disturbance. The transfer time is defined by the time interval demanded between the disturbance detection and the exchange of the power supply that feeds the load. Thus, reducing the transfer time causes the load to be less exposed to the source being disturbed.

It is common to use power semiconductor devices for implementing the transfer switch, thus being called the Static Transfer Switch (STS). The STS has been used to replace the mechanical/electromechanical switches, due to various advantages such as: it allows a high number of switching operations, it operates without arc formation (suitable for explosive environments), it has fast switching time, it is more robust and do not generate audible noise [5].

Considering the bidirectional operation in current and voltage required for STS, a pair of thyristors type SCR (Silicon Controlled Rectifier) or TRIAC (Triode for Alternating Current) are regularly used in STS implementation, according to power rate. However, STS based on thyristors presents some drawbacks related to total load transfer time [6-11].

The STS commutates between operation modes when a disturbance occurs in any power source that is connecting the load. Usually, instantaneous voltages of the power sources are monitored and used as the triggering factor for the STS commutation, for example, whenever the peak voltage value exceeds nominal ranges defined as reference.

Two common control strategies of STS are the break-before-make (BBM) and the make-before-break (MBB) approaches. The BBM controller transfers the load to the alternative power feeder after preferred feeder current has ceased, whereas the MBB controller initiates the transfer to the alternative source before the current on the corresponding switch in the preferred source has become zero. A clear advantage of the MBB over the BBM approach is its inherently faster transfer time. MBB method, however, is prone to errors. Under adverse circumstances, MBB controlled STS may cause an undesirable connection between the preferred and the alternative feeders. As a result, a high amplitude current surge and cause damages [6].

An application of a UPS considering the integration of photovoltaic energy is presented in [12]. Basically, it consists of a Standby UPS, if the preferred power supply does not supply power to the loads, a battery-powered inverter is used to continue supplying power to the loads. The battery is charged with energy from PV and the transfer switch is controlled by an intelligent relay.

The transfer time from grid mode to backup power mode is very important to evaluate the performance of the UPS system [13, 14]. There are works that also show the importance of STS in other applications than UPS. In [15] the introspective review of the AC and DC microgrid (MG) systems with renewable based distributed generation (DG) units is presented. This work shows that one of major challenges in MG operations is the protection system to properly respond to both modes of operations (grid-connected mode and islanded mode). When a disturbance occurs beyond acceptable range, the switching device (for example STS) is activated to isolate the faulted section by sending the signal to the appropriate switch. To avoid damage equipment and maintain system stability, the protective switching device should respond in the least possible time in any abnormal conditions i.e. 'the speed of response' is important parameter.

In [16] investigates mitigation of unbalance issues in secondary system caused by electric vehicles charging on split-phase. Through adding phase reconfiguration devices (each phase reconfiguration device requires a set of 4 static transfer switch) in the secondary system, it is possible to minimize neutral current at the transformer and reducing the systems unbalance.

As [17] presents a study related to the implementation of a static transfer switch applied in an uninterrupted power source, responsible for connecting the load to the AC grid or to a battery storage system. However two different methods for disturbances detection in grid voltage waveform are explored now. Moreover, this work presents comparisons between the conventional topology of STS based on thyristors and a new topology formed by association of IGBTs and diodes.

II. STATIC TRANSFER SWITCH

The STS enables very fast transfer from one power source to an alternative power source, providing adequate power supply for load in face of various power quality problems affecting electrical grid, such as, voltage sags and voltage swells, interruptions.

The electrical circuit in which the STS is inserted is shown in Fig. 2. There are two power sources in the circuit, one preferred source (grid) and one alternative source (inverter). In normal operation, the load is supplied by preferred source (grid). When a disturbance occurs, load is transferred to alternative source. After the disturbance, the load is transferred to the grid again.

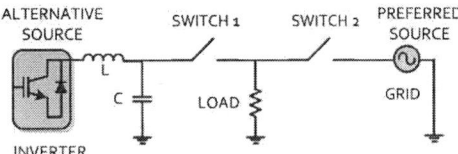

Fig. 2. Simplified electrical schematic of the static transfer switch.

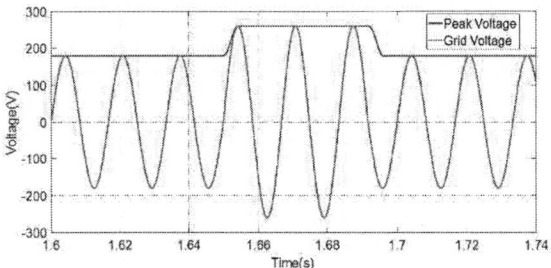

Fig. 3. Tracking of voltage peak value from voltage sources.

A. Control strategy for disturbance detection

Two approaches for detection of disturbances in the grid voltage waveform to control the STS are explored: the first method is based on the dq frame system and the second one is performed through a PLL (Phase Locked Loop) algorithm, which performs the detection of the angle of phase of a voltage waveform.

In both methods there is the tracking of voltage peak value from sources, the occurrence of disturbance is characterized through reduction of voltage peak value (sag) or through increase (swell), as shown by Fig. 3.

B. Disturbance detection strategy based on dq0 transformation

The block diagram for disturbance detection strategy based on dq0 transformation is shown in Fig. 4. The strategy is naturally suitable for three-phase systems. However, it can be applied in single-phase systems, as is the case of the STS discussed in this article, considering the creation of some hypothetical waveforms.

First, each voltage waveform sampled serves as input to the "ABC HYPOTHETICAL" block, where three waveforms are generated (Va, Vb and Vc). The three waveforms are ideally phase-shifted of 120° each other and have same amplitude. Thus, the vector sum of the three waveforms is zero, ie without homopolar component. The simplified transformation matrix regarding three-phase abc system to orthogonal dq synchronous system is given by (1).

$$\begin{bmatrix} Vd \\ Vq \end{bmatrix} = \frac{2}{3} \begin{bmatrix} \cos\theta & \cos\left(\theta - \frac{2\pi}{3}\right) & \cos\left(\theta + \frac{2\pi}{3}\right) \\ \sin\theta & \sin\left(\theta - \frac{2\pi}{3}\right) & \sin\left(\theta + \frac{2\pi}{3}\right) \end{bmatrix} \begin{bmatrix} Va \\ Vb \\ Vc \end{bmatrix} \quad (1)$$

After the three ideally lagged signals are transformed into components d and q, the square root of the sums of squares of these components is obtained. The obtained value is then subtracted from a reference of value 1 and obtained its value in module, according to (2).

$$Vs = \left| 1 - \sqrt{Vd^2 + Vq^2} \right| \quad (2)$$

The result of the module is applied to a hysteresis function and finally the output of that function is the disturbance detection signal.

The disturbance detection signal has logical level high when a disturbance occurs and low when it does not. Thus, the disturbance detection signal can be used as input signal for triggering static transfer switch logic.

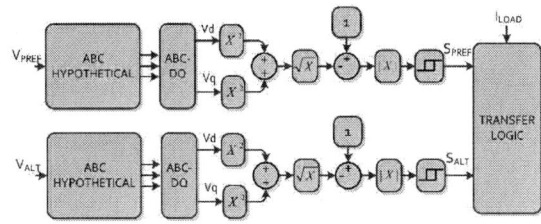

Fig. 4. Block diagram for strategy based on coordinate system dq.

C. Disturbance detection strategy based on PLL

The strategy is based on algorithm for positive sequence detection using PLL proposed by [18]. Fig. 5 shows block diagram for the strategy, where sampled voltage waveforms serves as input to the PLL algorithms, which are used to determine the phase angle of the voltage waveforms.

A sinusoidal waveform is produced in phase with input waveform and with amplitude normalized to 1 by using estimated angle. The scalar product of two signals in phase is made, which results in a waveform whose mean value multiplied by two is equal to peak value of input waveform fundamental component. The peak value is then subtracted from unit value which corresponds to 1 pu and module result of this subtraction is subjected to a hysteresis function.

Finally, the output of the hysteresis function is the disturbance detection signal. The signal has logical level high when a disturbance occurs and low level when it does not.

D. Hysteresis Function

The disturbance strategies methods employ a hysteresis function in their algorithms. The flowchart and block diagram of the Hysteresis function are shown in Fig. 6. Basically, the operating principle of hysteresis function is based on thresholds of input signal X. When X exceeds 0.1 pu, output X2 presents high logic level (logic level 1) and when it is below 0.04 pu, X2 presents low logic level. However, case the input is between 0.1 and 0.04 pu, output signal will depend on previous condition of itself.

E. Description of static transfer switch operation criteria

It is necessary to detect disturbances in voltage waveforms of preferred and alternative source to control the STS. The disturbance detection consists in monitoring peak values of voltages waveforms provided by power sources using one of presented strategies. So, when voltage of one of power source increases or decreases its peak value, it means that the source is going through a certain type of disturbance.

After the disturbance detection has been performed by one of the strategies, transfer logic is applied determining the configuration of the command signals to be used in the static switch. The load current waveform is used in the transfer logic to define the operating criterion.

The STS transfers power from preferred power source or alternative power source to the load according with transfer logic and eventual disturbance detection.

The S_{PREF} and S_{ALT} parameters have a Boolean domain. The first parameter (S_{PREF}) indicates the occurrence of disturbance detection at the preferred source ($S_{PREF} = 1$), or its non-occurrence ($S_{PREF} = 0$). Similarly, the S_{ALT} parameter indicates disturbance detection at the alternate source.

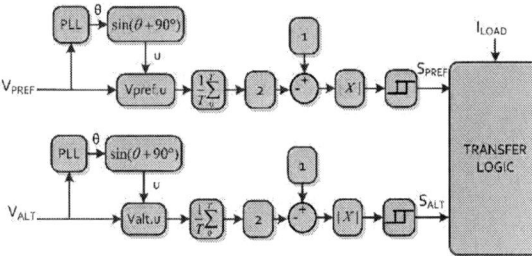

Fig. 5. Block diagram for strategy based on PLL amplitude detection

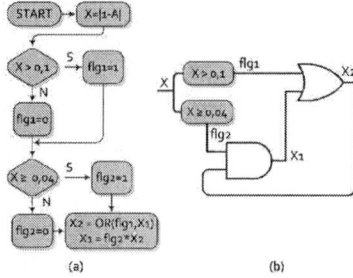

Fig. 6. (a) Flowchart of the Hysteresis function. (b) Block diagram of the Hysteresis function.

The transfer logic complies with following criteria:

- If there is no disturbance detection at preferred power source ($S_{PREF} = 0$), the static switch connected to preferred power source will provide power for load;

- If there is disturbance detection only at preferred power source ($S_{PREF} = 1$ and $S_{ALT} = 0$), load is transferred to alternative power source;

- If disturbances are detected in both power sources ($S_{PREF} = 1$ and $S_{ALT} = 1$), load should continue to be powered by preferred power source.

In the prototype, signals (V_{PREF}, V_{ALT} and I_{LOAD}) are acquired by analog-to-digital converter of DSC and converted to per unit values (pu). These signals serve as input for disturbance strategies and for voltage peak values calculation, as can be seen in the flowchart of Fig. 7.

III. TOPOLOGIES OF STS

A. Topology with Thyristors

It is very common in the literature, the use of thyristors, especially the SCR or TRIAC in STS implementation [6-11].

However, these two power devices behave as a retention switch, once fired and when conducting current cannot be blocked. The only way to block a thyristor is by reducing value of the anode current to a value below holding current (minimum anode current that holds thyristor in conduction). This feature makes thyristor a slow device in turn-off condition, because even after gate signal is removed, the thyristor remains turned on.

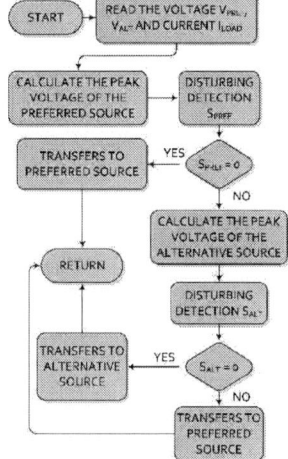

Fig. 7. Flowchart of the STS operation criteria

978-1-7281-4181-7/19 $31.00 © 2019 IEEE

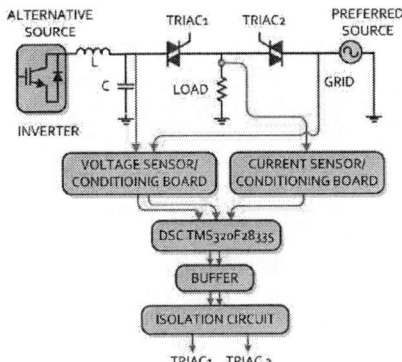

Fig. 8. STS electrical schematic with thyristor.

Due to waiting time for thyristor circulating current to become less than holding current required for the turning off, the performance of the STS with thyristors is impaired during transfer of the load power between the sources.

It is important to note that in this topology, when there is a disturbance detection in one power source, load supply is transferred to other power source only at zero crossing of load current. Thus, there is no short circuit between power sources.

Fig. 8 illustrates an electrical schematic of static transfer switch implemented with two TRIACs model TIC226D. This power device supports 8 A (RMS), 70 A (peak value), maximum blocking voltage of 400 V and typically 20 mA current in its gate terminal for drive.

In electrical schematic there are current and voltage sensors, signals from sensors go through a conditioning circuit to adjust voltage levels for A/D converter inputs of TMS320F28335 DSC used in control system. The TRIACs are driven by an isolated driver which, in Fig. 8, is represented by an isolation circuit. The buffer is used because the DSC has low output current capacity at its terminals.

B. Topology with IGBTs and Diodes

In order to alleviate the drawbacks of STS based in thyristor topology, it is possible to construct a four-quadrant power switch, controlled in both commutation situations (turn-off to turn-on and vice versa) using IGBTs and diodes [19].

Fig. 9 illustrates two arranges of power switches to configure (four quadrants/ bidirectional in current) power switch. Two IGBTs can be arranged using common emitter configuration with two diodes (Fig. 9 (a)), or used in common collector configuration with two diodes (Fig. 9 (b)). Each power transistor is responsible for conducting one direction of the load current, called (IGBT$_F$) and (IGBT$_R$) according with current flow direction, respectively.

The transference of the load from preferred power source to alternative power source occurs through four steps, as is shown in Fig. 10. The bidirectional switch operates in four quadrants as shown in Fig. 10. The common emitter topology was used because it results in simpler command circuitry as only one power supply is required to drive the pair of IGBT. The positive direction of current is indicated by I$_{LOAD}$>0. In case of current flow to be reverse, I$_{LOAD}$<0, the process is analogous to Fig. 9 with the same switching logic.

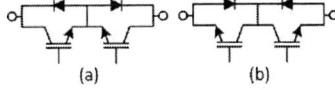

Fig. 9. Bidirectional switches topologies using IGBTs and diodes: (a) common emitter configuration, (b) common collector configuration

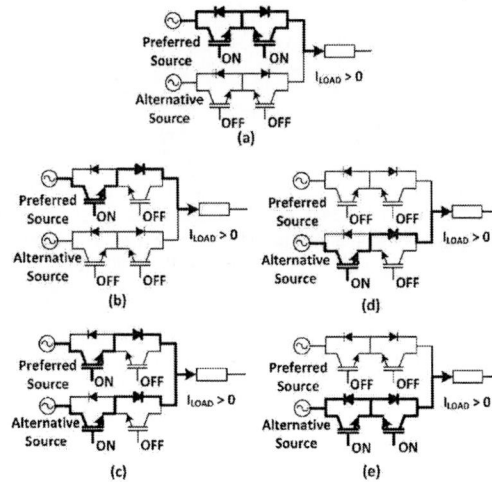

Fig. 10. Illustration of STS topology and its four steps of operation. (a) Load connected by the preferred source, (b) 1st Step, (c) 2nd Step, (d) 3rd Step and (e) 4th Step.

1st Step: The IGBT$_F$ from STS branch connected in preferred source is turned-off (blocked) while the load current is flowing in its opposite direction (I$_{LOAD}$>0). In this way, the current direction cannot be reversed.

2nd Step: The IGBT$_F$ from STS branch connected in alternative source is turned-on. Fig. 10 (c) illustrates the load current flowing from both the power sources (preferred and alternative). However, the operating situation depends on instantaneous voltage value of both sources.

For instance, if at the time of commutation the preferred source voltage is higher than the voltage from alternative source, diode of the STS direct branch connected to alternative source (which should conduct) will be reverse-biased, and hence the load current will flow only from preferred source. On the other hand, if instantaneous voltage of preferred source is lower than the voltage from alternative source, diode of the STS direct branch connected to preferred source is blocked and the IGBT$_F$ from STS branch connected in alternative source instantaneously assumes the integral load current.

It is important to note that in this stage the load current flows unidirectionally. Thus, there is no possibility of the current flowing between the sources, causing a short circuit.

3rd Step: The IGBT$_F$ from STS branch connected in preferred source is turned-off, the load is fed only by alternative power source.

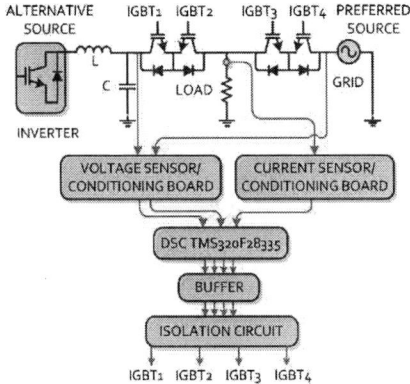

Fig. 11. STS electrical schematic with IGBTs and diode.

978-1-7281-4181-7/19 $31.00 © 2019 IEEE   1346

4th Step: The IGBT$_R$ from STS branch connected in preferred source is turned-on. It is possible to re-establish the bi-directional switch characteristic, so that the load current can circulate in both directions.

The duration of steps 1 and 4 is not critical because the transistors are not conducting and will not conduct current instantaneously. These steps can occur as fast as possible. The steps 2 and 3 are considered critical stages, once there is IGBTs commutations to conduct or to block the load current. This fact must be considered according to the characteristics of the IGBTs used in topology [7].

The Fig. 11 shows electrical schematic of static transfer switch implemented with bidirectional switch topology based on IGBTs with common emitter configuration. For prototype evaluation the STS was implemented using IGBT model IRGB15B60KD. The power device supports 15 A (RMS value) and maximum blocking voltage of 600 V.

IV. RESULTS

A. Performance of disturbance detection strategies

In order to evaluate the performance of disturbance detection strategies, evaluations were accomplished considering voltage disturbances related with sag and swell for several cases. For a more complete analysis the disturbances were performed in different regions of voltage waveform (peak, intermediate value and crossover), and exposed in electrical grid interruption situations. Table I shows detection times (td) considering the cases evaluated.

It is possible to conclude from Table I that the detection time, *td*, depends on severity and instant that disturbance occurs. Considering the same disturbance severity, in both methods, the disturbances performed at the peak and at the intermediate points were detected faster than ones in the crossover voltage.

The strategy based on dq coordinate detected all disturbances faster than the PLL based strategy. Therefore, this algorithm was the most efficient and it was adopted for STS experimental evaluation.

TABLE I.   STS DISTURBANCE DETECTION TIMES

STS Disturbance Detection Times			
Voltage disturbances		*Methods for disturbances detection*	
		Dq coordinate system	*PLL algorithm*
Peak point	Sag (70%)	0.1 ms	1.3 ms
	Sag (30%)	0.1 ms	3.2 ms
	Swell (70%)	0.1 ms	4.5 ms
	Swell (30%)	0.1 ms	6.7 ms
	Short interruption	0.1 ms	1.0 ms
Crossoover point	Sag (70%)	1.3 ms	3.2 ms
	Sag (30%)	1.8 ms	4.9 ms
	Swell (70%)	1.0 ms	5.5 ms
	Swell (30%)	1.6 ms	6.9 ms
	Short interruption	1.3 ms	2.8 ms
Intermediate point	Sag (70%)	0.1 ms	1.6 ms
	Sag (30%)	0.1 ms	3.2 ms
	Swell (70%)	0.1 ms	4.5 ms
	Swell (30%)	0.1 ms	6.7 ms
	Short interruption	0.1 ms	1.1 ms

B. Performance static transfer switches topologies

In order to evaluate the performance of STS based on thyristors, sag and swell voltage disturbances were made in the region of voltage crossover, with a magnitude of 20% source peak voltage. The choice was made because it is a difficult case for disturbance detection, since it does not generate a very large change in voltage amplitude and, as mentioned earlier, is the worst instant to disturbance detection. The STS performance is also tested considering different power factors (PF) values, as is shown in a Table II.

It is possible to notice a reasonable difference in the detection times (*td*) presented in Tables I and II, due to the different severities of the applied disturbances.

As can be seen, transfer time (*tf*) is dependent on the load power factor. As is well known, the load is transferred to the other source only at the current zero crossing load, so that there is no short circuit between the sources.

It is possible to conclude that transfer time (*tf*) increases as the load power factor decreases, consequently, STS performance worsens with power factor decreases.

Similarly, in order to evaluate the performance of the STS IGBT topology, the same tests performed for the thyristor topology were performed, as is shown in a Table III. However, since this topology does not depend on load current to be transferred from one source to another, no different power factors load tests were performed.

It is important to note that in IGBT based STS topology, the transfer time (*tf*) in the two disturbance situations was 0.266 ms. This is because there are 4 steps for bidirectional switch and a processor frequency of 15 kHz (66.66 us), so 66.66 us x 4 equals 0.266 ms.

Experimental results concerning the two STS topologies under different voltage conditions (voltage sag, voltage swell and power grid interruption) and considering load with unit power factor (R = 100 Ω) are shown in Figs 12 and 13. The grid voltage waveform is located at the top, while the load voltage waveform at the bottom.

Since in the experimental results the load is purely resistive, during the transition from step 1 to step 2, no current flows through the load for a few microseconds, causing the notch in the voltage load waveform.

TABLE II.   STS PERFORMANCE WITH THYRISTOR TOPOLOGY

STS performance with thyristor topology				
Voltage disturbances		*td*	*tf*	*tt*
PF = 1	Sag (20%)	2.4 ms	6.0 ms	8.4 ms
	Swell (20%)	2.3 ms	6.1 ms	8.4 ms
PF = 0.9	Sag (20%)	2.4 ms	7.3 ms	9.7 ms
	Swell (20%)	2.3 ms	7.4 ms	9.7 ms
PF = 0.8	Sag (20%)	2.4 ms	7.8 ms	10.2 ms
	Swell (20%)	2.3 ms	7.9 ms	10.2 ms
PF = 0.7	Sag (20%)	2.4 ms	8.1 ms	10.5 ms
	Swell (20%)	2.3 ms	8.2 ms	10.5 ms

TABLE III.   STS PERFORMANCE WITH IGBTs AND DIODES TOPOLOGY

STS performance with IGBTs and diodes topology			
Voltage disturbances	*td*	*tf*	*tt*
Voltage Sag (20%)	2,4 ms	0,266 ms	2,666 ms
Voltage Swell (20%)	2,3 ms	0,266 ms	2,566 ms

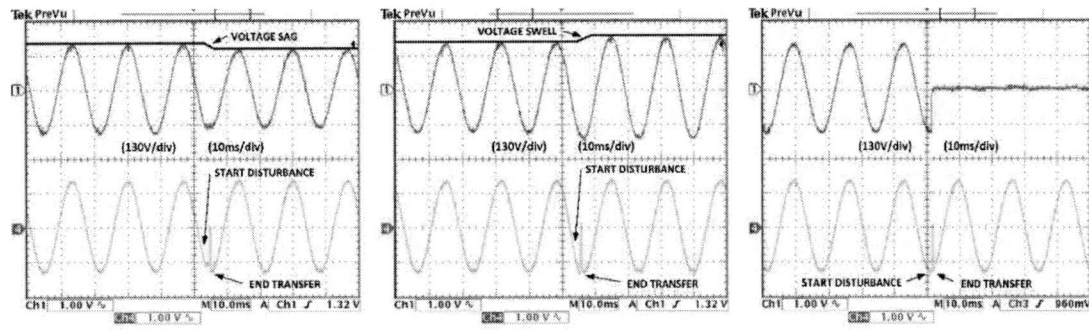

Fig. 12. Experimental results using STS based on IGBT with Diodes.

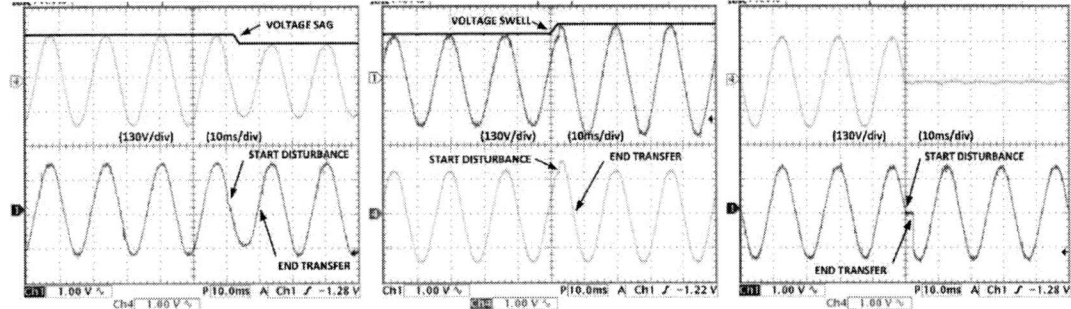

Fig. 13. Experimental results using STS based on Thyristor.

V. CONCLUSIONS

The operation of the STS for protection of sensitive loads against voltage disturbance has been analyzed in this paper. Two topologies of STS were presented. The topology of IGBTs and diodes presented a shorter transfer time than the traditional one, composed of thyristors. The reduction time was significant, contributing to the load, during the transfer of its power is not without power for a long period.

Two methods for detecting disturbance were also evaluated. The method for detecting the peak voltage value of the sources using the dq coordinate was the most efficient, presenting a shorter time in relation to the PLL method.

ACKNOWLEDGMENT

This Project was financially supported by Grant #2016/08645-9 and Grant #2018/24331-0, São Paulo Research Foundation (FAPESP).

REFERENCES

[1] A. L. B. Ferreira, UPS de 5KVA tipo Passive Stand-by com integração de painéis solares. Master's Thesis. Universidade Estadual de Londrina, Londrina, Brasil, 2009.

[2] IEC 62040-3. Uninterruptible power systems (UPS) – Part 3: Method of specifying the performance and test requirements. International Electrotechnical Commission. Ed. 2.0. 2011

[3] A. L. B. Ferreira, C. H. G. Treviso, L. C. Mathias, M. V. M. Rodrigues and W. R. B. M. Nunes. No-Break de 3kVA para aplicações residenciais com interface para painéis solares e controle microprocessado. Semina: Ciências Exatas e Tecnológicas, Londrina, vol. 35, n. 1. 2014, pp. 39-54.

[4] M. N. Moschakis, N. D. Hatziargyriou, "A detailed model for a thyristor-based static transfer switch", IEEE Transactions on Power Delivery, v. 18, nº 4, pp. 1442 – 1449, Outubro 2003.

[5] N. Verma, K. Gupta and S. Mahapatra. Implementation of Solid State Relays For Power System Protection. International Journal of Scientific & Technology Research, vol 4, June 2015, pp 65-70.

[6] J. Yao, A. Abramovitz, Y. Wang, H. Weng and J. Zhao. Safe-triggering-region control scheme for suppressing cross current in static transfer switch. Electric Power Systems Research 125. Aug, 2015, pp. 245–253.

[7] M. U. Cuma, Digital signal processor based implementation of custom power device controllers. PhD Thesis, Institute of Natural and Applied Sciences. Çukurova University, 2010.

[8] H. Mokhtari, S. B. Dewan and M. Reza Iravani, Performance evaluation of thyristor based static transfer switch, IEEE Transactions on Power Power Delivery, vol. 15. 2000, pp. 960–966.

[9] G.Bertuzzi, U.S.Cinti, E.Cevenini, A.Nalbone, Static transfer switch (STS): application solutions .Correct use of the STS in systems providing maximum power reliability, in:Telecom. Energy Conference, 2007, INTELEC, 2007, pp.587–594.

[10] Ambra Sannino, Static transfer switch: analysis of switching conditions and actual transfer time, in: Power Engineering Society Winter Meeting, 2001, IEEE, vol.1, 2001, pp. 120–125.

[11] M. R. Javed, T. Mahmmood and M. A. Choudhry. Performance analysis of static transfer switch using MATLAB/Simulink. Power Generation System and Renewable Energy Technologies (PGSRET), June 2015.

[12] I. Daut, M. Irwanto and S. Hardi. Photovoltaic Powered Uninterruptible Power Supply Using Smart Relay. The 4rd International Power Engineering and Optimization Conference (PEOCO2010), Shah Alam, Selangor, Malaysia. 23-24 June 2010.

[13] M. A. Abusara and J. M. Guerrero. Line-Interactive UPS for Microgrids. IEEE Transactions on Industrial Electronics. Vol.: 61 , Issue: 3, March, 2014.

[14] M. Aamir, K. A. Kalwar and S. Mekhilef. Review: Uninterruptible Power Supply (UPS) system. Renewable and Sustainable Energy Reviews 58. pp 1395 – 1410, 2016.

[15] J. J. Justo, F. Mwasilu, J. Lee and J-W. Jung. AC-microgrids versus DC-microgrids with distributed energy resources: A review. Renewable and Sustainable Energy Reviews 24. 2013, pp. 387–405.

[16] M. K. Gray and W. G. Morsi. Economic assessment of phase reconfiguration to mitigate theunbalance due to plug-in electric vehicles charging. Electric Power Systems Research 140. 2016, pp. 329–336.

[17] M. V. M. Rodrigues and N. da Silva. Controlador multimalhas para inversor monofásico e controle de chave estática de transferência aplicados em fontes ininterruptas de energia. Revista Eletrônica de Potência, vol 21. Fev. 2016, pp. 52-62.

[18] M. S. de Pádua. Digital techniques for power grid synchronization, with applications in distributed generation. Master's Thesis. Universidade Estadual de Campinas, Campinas, Brazil, 2006.

[19] M. P. Kazmierkowski, F. Blaabjerg, R. Krishnan, Control in power electronics, Academic Press, 1st Edition, 2003.

Analysis of Rectifiers for RF Energy Harvesting Aiming Low Power Sensing Applications

Humberto Pereira da Paz
Engineering, Modeling and Applied
Social Sciences Center (CECS)
Universidade Federal do ABC
São Paulo, Brazil
humberto.paz@ufabc.edu.br

Vinícius Santana da Silva
Engineering, Modeling and Applied
Social Sciences Center (CECS)
Universidade Federal do ABC
São Paulo, Brazil
vinicius.santana@ufabc.edu.br

Eduardo Vicente Valdés Cambero
Engineering, Modeling and Applied
Social Sciences Center (CECS)
Universidade Federal do ABC
São Paulo, Brazil
eduardo.valdes@ufabc.edu.br

Humberto Xavier de Araújo
Department of PPGMCS
Universidade Federal do Tocantins
Palmas, Tocantins, Brazil
hxaraujo@uft.edu.br

Ivan Roberto Santana Casella
Engineering, Modeling and Applied
Social Sciences Center (CECS)
Universidade Federal do ABC
São Paulo, Brazil
ivan.casella@ufabc.edu.br

Carlos Eduardo Capovilla
Engineering, Modeling and Applied
Social Sciences Center (CECS)
Universidade Federal do ABC
São Paulo, Brazil
carlos.capovilla@ufabc.edu.br

Abstract—Seeking to increase the autonomy of wireless sensing devices, this article analyzes different RF rectifier topologies for ambient energy harvesting applications at 2.45 GHz, focusing on scenarios where the input received power level of the RF rectifier is between -10 dBm and -20 dBm. For this propose, an equivalent diode RF model is taken into account for better describing the diode effect on the efficiency of each rectifier topology, presenting the impedance curve of four commercial diodes, and the respective conducting thresholds, without loss of generality. Furthermore, a computational rectifier optimization methodology is given, discussing the relation of the load impedance with the resulting circuit efficiency.

Index Terms—Rectifier, energy harvesting, radio frequency, diode model.

I. Introduction

The development of wireless sensors is advancing according to the creation of new device integration technologies, aiming the enhancement of remote sensing systems, and limiting factors as the device lifetime have been on focus. Sensing, processing, and transmitting data of physical, chemical, and biological processes are normally limited by the device mobility, which is generally powered by cables. Even in the case of sensing wireless networks, powered exclusively by batteries, it is also necessary to manage battery recharges periodically. Due to this factor, dedicated power sources based on energy harvesting from the environment have received much attention in the past recent years, as a manner to decrease the system restrictions [1]–[3], and radio frequency (RF) energy harvesting is considered promising due to its uninterrupted power availability.

The research project is funded by the Brazilian National Council for Scientific and Technological Development (CNPq), and the Coordination for the Improvement of Higher Education Personnel (CAPES).

The process of energy harvesting from RF sources has been proposed from many different aspects, such as available power density [4], [5], maximum transmitting distance [1], [2], wideband signal [6], and others. The recent advances in this field are increasing the importance of the RF rectifiers development, the device responsible for converting the RF signal to a DC level, principally for physically isolated and independent systems, dedicated to communication networks with power consumption in the order of micro-watts.

The design of an RF rectifier is commonly developed for elevated power levels at its input, P_{in}, over -10 dBm, considering the ISM (Industrial, Scientific, and Medical) at 2.45 GHz [2]–[4], [7]–[9], but a lower number of works develop energy harvesting for P_{in} levels below -10 dBm [5], [6], [10], [11]. Wi-Fi routers, for example, are limited by a maximum EIRP (Effective Isotropic Radiated Power) of 20 dBm and 18 dBm for the IEEE 802.11b and IEEE 802.11g protocols, respectively. Nevertheless, due to propagation losses, to obtain P_{in} levels over -10 dBm at the receiver, the system needs to be located extremely close to the transmitter [1]. This factor highlights the necessity to elevate the efficiency of RF rectifiers for P_{in} levels in the order of tens of micro-watts. Additionally, the efficiency increase is also correlated to an increment of the maximum operating distance of the harvesting system.

Even though the development goal of a rectifier is to optimize its efficiency, it is necessary to have a viable project design to be put into production, making use of established communication devices as sources for the energy harvesting, without the addition of high gain dedicated amplifiers to increase the transmitted power level.

In this context, a study of the rectifier topologies that present the most expressive efficiency responses under the desired power levels (-10 dBm and -20 dBm) is presented in this

978-1-7281-4181-7/19 $31.00 © 2019 IEEE

work, filling in this way, a gap in the literature. Besides, this study includes a method for the diode evaluation according to its impedance, since it is dependent on the frequency and P_{in} level. So, the rectifier optimization process is shown as a contribution to the development of any dedicated RF energy harvesting rectifier.

Besides this introduction, this article is organized in the following order. In section II, it is described the diode RF model, and four established commercial diodes are compared, based on their impedance behavior. In section III, the rectifier topologies are described, the design of each topology is optimized for P_{in} levels equal to -20 dBm and -10 dBm. In section IV, a conclusion is presented.

II. DIODE RF MODEL

In RF applications, the parasitic effects related to the diode package have a significant impact over the component losses [6], and the reactive part of its impedance becomes mostly capacitive when reversely polarized. Therefore, due to the component non-linearities summed up with the parasitic effects, it becomes a vital factor to describe the diode using an equivalent model for RF small signal analysis. As shown in Fig. 1, it can be done by adding a parallel parasitic capacitance, C_p, and a series parasitic inductance, L_s, to the equivalent conventional diode model, according to [12], [13].

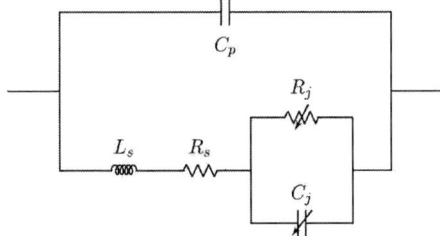

Fig. 1: Equivalent small signal RF model of the diode with the package parasitic components.

In Fig. 1, R_s is the series resistance, which is constant. The junction resistance and capacitance, R_j and C_j, respectively, are dependent of the voltage over the diode, V_d, according to (1) and (2) [12], [13], where V_j is the junction voltage, C_{jo} is the zero voltage capacitance, m is the grading coefficient, I_s is the saturation current, n is the ideality factor, and T is the temperature in Kelvin.

$$R_j = \frac{8.331 \times 10^{-5}\, nT}{I_s + I_d} \tag{1}$$

$$C_j = \frac{C_{j0}}{\left(1 - \frac{V_d}{V_j}\right)^m} \tag{2}$$

Analyzing (1) and (2), it can be noticed that only R_j is not directly related to V_d, so, considering that (3) represents the current I_d, for V_t as the thermal voltage, and substituting (3) in (1) (applying a Taylor expansion over the exponential term

and truncating it at the third-order resulting term), the derived equation, (4), represents the direct relation of R_j with V_d.

$$I_d = I_s \left[e^{\frac{V_d}{n V_t}} - 1 \right] \tag{3}$$

$$R_j = 8.33 \times 10^{-5} nT I_s^{-1} \left[1 - V_d + \frac{V_d^2}{2} \right]^{\frac{1}{n V_t}} \tag{4}$$

The R_j increases according to $|V_d|$, for negative V_d values, due to its negative first-order term, (4). Therefore, the diode impedance Z_d equation is presented in (5).

$$Z_d = \left[\frac{1}{((jwC_j + \frac{1}{R_j})^{-1} + R_s + jwL_s)^{-1}} + jwC_p \right]^{-1} \tag{5}$$

The Table I presents the commercial diodes SMS7630, SMS7621, HSM282 and HSM286 capable of operating at 2.45 GHz, according to the manufactures Skyworks and Broadcom.

TABLE I: Diode parameters for impedance analysis.

Diode	SMS7630	SMS7621	HSM282	HSM286
I_s (uA)	5	0.04	0.022	0.05
R_s (Ohm)	20	12	6	6
C_{jo} (pF)	0.14	0.1	0.7	0.18
V_j (V)	0.34	0.51	0.65	0.65
M	0.4	0.35	0.5	0.5
n	1.05	1.05	1.08	1.08
B_v (V)	2	3	15	7
Package	079LF	079LF	SOT-23	SOT-23
C_p (pF)	0.16	0.16	0.08	0.08
L_s(nF)	0.8	0.8	2	2

The parameter B_v, shown in Table I, represents the maximum reverse voltage over the diode, when $V_d < -B_v$ the diode efficiency decays significantly due to an avalanche breakdown effect. However, this factor can be put aside in this analysis, considering the low P_{in} level design goal of this study.

Developing the impedance module, $|Z_d|$, for each diode at 300 K and 2.45 GHz, it becomes possible to observe the conduction threshold voltage, V_l (the point where $|Z_d|$ reaches its lowest value when the diode is directly polarized), and the $|Z_d|$ behavior when the diode is reverse and direct polarized. In Fig. 2, although the SMS7630 does not present the most elevated impedance, when reverse polarized, it reaches the conducting threshold for a voltage level significantly lower than the other diodes, V_l approximately 0.15 V. So, even though the diode HSM282 presents lower $|Z_d|$ values for positive voltages lower than 0.15 V, it presents an impedance more than two times lower than the others diodes when reverse polarized. This factor is responsible for decreasing rectifiers efficiency severely due to a reverse current flow over the diode when it should be ideally working as an open circuit. Therefore, due to the lower threshold and the relatively high reverse polarized impedance, the SMS7630 presents the most

978-1-7281-4181-7/19 $31.00 © 2019 IEEE

practical characteristics for an RF rectifier diode, according to Fig. 2.

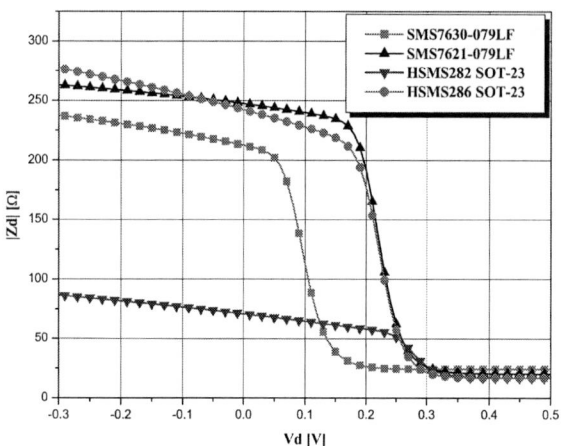

Fig. 2: Z_d module as a function of the diode voltage.

III. RECTIFIER TOPOLOGIES ANALYSIS

Each RF rectifier topology has its optimum point of operation based on the P_{in} level, signal frequency and load resistance, R_l. By exploiting it, the rectifier efficiency is analyzed in this work for the series [10], [12], shunt [7], [14], doubler [6], [8], and symmetric [4], [11] topologies.

The rectifier input is considered to be connected to the output of an antenna, for the modeling proposes, and this antenna is described by its equivalent Thévenin circuit model using a voltage source, v_{IN}, and its series impedance, Z_s. The rectifiers are composed by a set of impedance matching network Z_{line}, the diode D, an output capacitor C_o, and the R_l.

The series rectifier is shown in Fig. 3. It can be verified that its diode conducts, ideally, during the positive half-cycle of the input signal. To ensure that this characteristic is kept in an RF rectifier, it is necessary to add a DC short-circuit at the anode side of the diode. It becomes necessary to eliminate the DC voltage to avoid the diode to be reversely polarized during both input signal half-cycles, which is a problem that can surge when using antenna models similar to patchs and monopoles that present $Z_s = \infty$ for DC. The DC voltage at the diode anode works as a shifter in the diode impedance curve, presented in the last section, turning the threshold higher according to its intensity. In counterpart, for the shunt rectifier, presented in Fig. 4, it is necessary to block the DC current that can flow from the circuit output back to the antenna, for the cases that the antenna presents $Z_s = 0$ intrinsically to its design.

In the case of the symmetric rectifier shown in Fig. 5, which is formed by the diodes, D_1 and D_2, the input matching impedances, Z_{line1} and Z_{line2}, and the output capacitors, C_{o1} and C_{o2}, it is possible to compare it to a two-branch series rectifier working in opposite half-cycles. In the same way, the doubler rectifier, presented in Fig. 6 that is composed by the

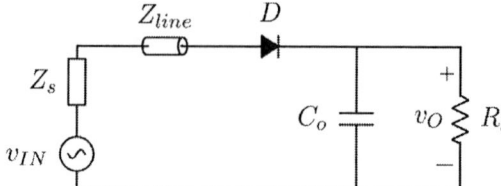

Fig. 3: Series rectifier.

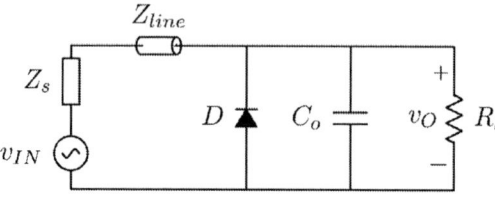

Fig. 4: Shunt rectifier.

diodes, D_1 and D_2, can be described by the junction of a series and a parallel rectifier, but with the capacitor C_1 responsible for increasing the current flux over D_1 during its conduction period.

It is important to note that only for the doubler rectifier, it is not necessary to add a short-circuit at its input. In these last two topologies, both diodes are working at different half-cycles, and this factor increases the rectification period of the signal. However, this increase does not represent a real enhancement of the output power over R_l. This factor can be demonstrated by simulating the rectifier topologies and presenting their efficiencies according to P_{in}, using the Harmonic Balance (HB) simulation and following the efficiency calculus, η_{RF-DC}, shown in (6) with the DC component of v_O, V_o.

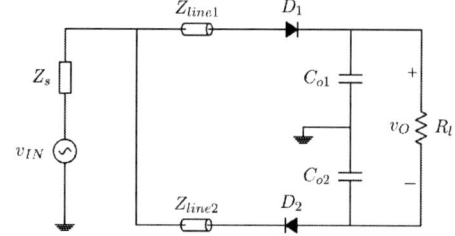

Fig. 5: Symmetric rectifier.

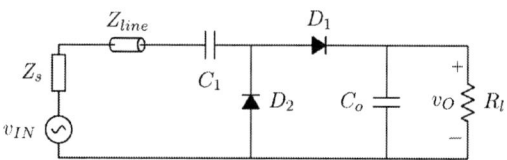

Fig. 6: Doubler rectifier.

$$\eta_{RF-DC} = \frac{V_o^2}{P_{in} R_l} \qquad (6)$$

978-1-7281-4181-7/19 $31.00 © 2019 IEEE   1351

Aiming to analyze these topologies for low P_{in} levels, the four circuits are optimized using the software ADS (Advanced Design System) for -20 dBm and for -10 dBm, using the diode SMS7630-079LF and keeping the output capacitor equal to 100 pF to guarantee an inexpressive ripple at the output, and in the case of the doubler rectifier $C_1 = 100$ pF. To extract the maximum efficiency of each topology, the input impedance of the rectifiers are kept matched to 50 Ω. To reach this input impedance value, each Z_{line} is formed by an ideal LC network, and using the LSSP (Large Signal S-Parameters) in conjunction with the HB simulations Z_{line} and R_l are optimized for each specific circuit, reaching the maximum efficiency value at each desired P_{in} level.

In Fig. 7 and Fig. 8, it is possible to see the final efficiency values obtained through simulation. It is essential to clarify that the most elevated efficiency level reached for each rectifier can be obtained for a P_{in} level higher than the designed. This behavior is expected to happen, since the percentage of power dissipated over the diode decreases with the increase of P_{in}, assuming that the input impedance is kept relatively matched around the design P_{in}.

Therefore, in Fig. 7, for P_{in} equal to -20 dBm, the series and the shunt rectifiers present similar efficiency values, with a slightly higher value for the series rectifier, and the same can be noticed for the symmetric rectifier when comparing it to the doubler rectifier. However, the one diode topologies present more meaningful results, almost 10% over the two diode topologies, for -20 dBm. This behavior is carried out for the results presented in Fig. 8, where the circuits are optimized for -10 dBm, but with a lower efficiency difference (around 5%).

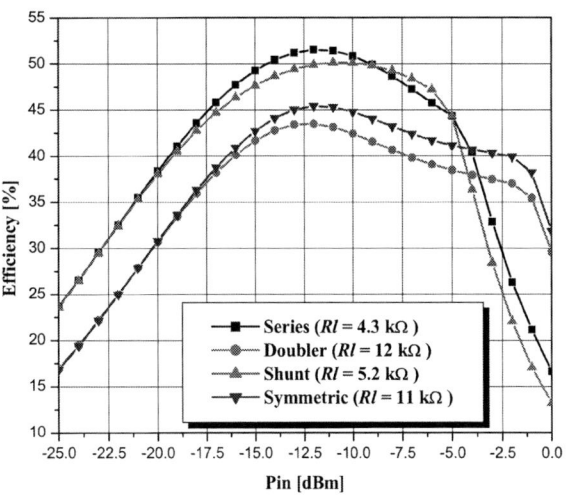

Fig. 7: Rectifiers efficiency when optimized for -20 dBm and at 2.45 GHz.

As each rectifier is optimized for its maximum efficiency point, a factor that defined the R_l value in Fig. 7 and Fig. 8, the doubler and the symmetric rectifiers present an optimized R_l value twice as high the series and shunt rectifiers. Although

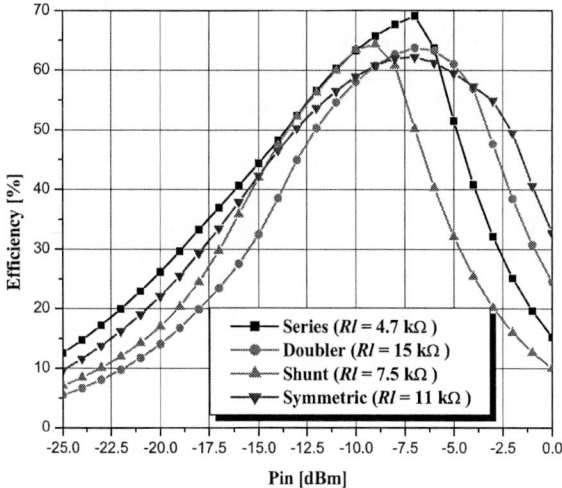

Fig. 8: Rectifiers efficiency when optimized for -10 dBm and at 2.45 GHz.

a higher R_l tends to result on a higher V_o level, it is also essential to understand the efficiency behavior of the rectifiers according to R_l for the cutting edge P_{in} level analyzed in this work, -20 dBm, making possible to compare their efficiencies for the same R_l value. So, in Fig. 9, the efficiency of the rectifiers for -20 dBm at 2.45 GHz is presented varying the R_l value, demonstrating that the series rectifier is the most efficient for low R_l values.

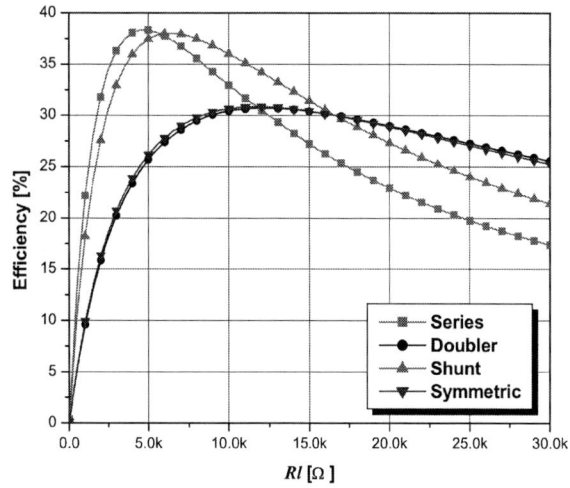

Fig. 9: Rectifiers efficiency according to the load impedance for -20 dBm at 2.45 GHz.

IV. CONCLUSION

This work analyzed RF rectifiers for energy harvesting applications at 2.45 GHz, emphasizing important aspects as the diode RF model, circuit topology, and efficiency according to the input power level. Four different diodes were analyzed, and their respective impedances and conducting thresholds were discussed using the given manufacture parameters and the package parasitic effects, demonstrating that the diode SMS7630 reaches the conducting threshold for a lower voltage when compared to the SMS7621, HSM828 and HSM286, making it a viable option for RF rectifiers design. Furthermore, a performance analysis of four rectifier topologies based on low input power levels, -20 dBm and -10 dBm, described that the series and the shunt rectifier topologies are the most efficient, and that the series rectifier presents the most elevated efficiency for low output load values at -20 dBm.

REFERENCES

[1] S. Kim, R. Vyas, J. Bito, K. Niotaki, A. Collado, A. Georgiadis, and M. M. Tentzeris, "Ambient rf energy-harvesting technologies for self-sustainable standalone wireless sensor platforms," *Proceedings of the IEEE*, vol. 102, no. 11, pp. 1649–1666, Nov. 2014.

[2] C.-H. Lin, C.-W. Chiu, and J.-Y. Gong, "A wearable rectenna to harvest low-power rf energy for wireless healthcare applications," Oct. 2018, pp. 1–5.

[3] E. L. Chuma, L. d. l. T. Rodríguez, Y. Iano, L. L. B. Roger, and M. Sanchez-Soriano, "Compact rectenna based on a fractal geometry with a high conversion energy efficiency per area," *IET Microwaves, Antennas Propagation*, vol. 12, no. 2, pp. 173–178, Feb. 2018.

[4] H. Takhedmit, B. Merabet, L. Cirio, B. Allard, F. Costa, C. Vollaire, and O. Picon, "A 2.45-GHz dual-diode RF-to-DC rectifier for rectenna applications," in *The 40th European Microwave Conference*, Sep. 2010, pp. 37–40.

[5] Y.-S. Chen and C.-W. Chiu, "Theoretical limits of rectifying efficiency for low-power wireless power transfer," *International Journal of RF and Microwave Computer-Aided Engineering*, vol. 28, no. 4, p. e21218. [Online]. Available: https://onlinelibrary.wiley.com/doi/abs/10.1002/mmce.21218

[6] M. ur Rehman, W. Ahmad, and W. T. Khan, "Highly efficient dual band 2.45/5.85 GHz rectifier for RF energy harvesting applications in ISM band," in *2017 IEEE Asia Pacific Microwave Conference (APMC)*, Nov. 2017, pp. 150–153.

[7] Q. Zhang, J. Ou, Z. Wu, and H. Tan, "Novel microwave rectifier optimizing method and its application in rectenna designs," *IEEE Access*, vol. 6, pp. 53 557–53 565, Sep. 2018.

[8] Y. Shi, J. Jing, Y. Fan, L. Yang, and M. Wang, "Design of a novel compact and efficient rectenna for wifi energy harvesting," *Progress In Electromagnetics Research C*, vol. 83, pp. 57–70, Jan. 2018.

[9] D. Lin, C. Yu, C. Lin, and Y. Lee, "Dual band rectenna with one rectifier," in *2017 International Symposium on Electronics and Smart Devices (ISESD)*, Oct. 2017, pp. 268–272.

[10] V. Palazzi, C. Kalialakis, F. Alimenti, P. Mezzanotte, L. Roselli, A. Collado, and A. Georgiadis, "Design of a ultra-compact low-power rectenna in paper substrate for energy harvesting in the wi-fi band," in *2016 IEEE Wireless Power Transfer Conference (WPTC)*, May. 2016, pp. 1–4.

[11] M. Zeng, A. S. Andrenko, X. Liu, B. Zhu, Z. Li, and H. Z. Tan, "Differential topology rectifier design for ambient wireless energy harvesting," in *2016 IEEE International Conference on RFID Technology and Applications (RFID-TA)*, Sep. 2016, pp. 97–101.

[12] X. Gu, S. Hemour, L. Guo, and K. Wu, "Integrated cooperative ambient power harvester collecting ubiquitous radio frequency and kinetic energy," *IEEE Transactions on Microwave Theory and Techniques*, vol. 66, no. 9, pp. 4178–4190, Sep. 2018.

[13] S. Shieh and M. Kamarei, "Transient input impedance modeling of rectifiers for RF energy harvesting applications," *IEEE Transactions on Circuits and Systems II: Express Briefs*, vol. 65, no. 3, pp. 311–315, Mar. 2018.

[14] H. Sun, Y. Guo, M. He, and Z. Zhong, "Design of a high-efficiency 2.45-GHz rectenna for low-input-power energy harvesting," *IEEE Antennas and Wireless Propagation Letters*, vol. 11, pp. 929–932, Aug. 2012.

A Simplified Analysis of Buck-Type Interleaved DC-DC Converter for Battery Chargers Application

1st Menaouar Berrehil El Kattel
Electric Power Processing Group - nPEE
Santa Catarina State University - UDESC
Joinville - Santa Catarina - Brazil
berrehilelkattel@gmail.com

2nd Robson Mayer
Electric Power Processing Group - nPEE
Santa Catarina State University -UDESC
Joinville - Santa Catarina - Brazil
mayerrobson@gmail.com

3rd Maicon Douglas Possamai
Electric Power Processing Group-nPEE
Santa Catarina State University - UDESC
Joinville - Santa Catarina - Brazil
possamai.maicon@gmail.com

4rd Sérgio Vidal Garcia Oliveira
Electric Power Processing Group - nPEE
Santa Catarina State University - UDESC
Blumenau Regional University - FURB
Joinville - Santa Catarina - Brazil
sergio_vidal@ieee.org

Abstract—**This paper presents the analysis of a new buck-type interleaved dc-dc non-isolated converter based on the classic interleaved topologies of the buck converter operating in the continuous conduction mode. A simple comparative analysis between the proposed converter and two dc-dc buck converters is conducted. For this purpose, a conventional buck converter, an interleaved buck converter and the proposed converter were compared. In this paper the operation principles, the dc voltage gain, the filters design and the voltage and current stresses are discussed. The proposed converter could replace the conventional dc-dc step-down voltage converter stage with poor power quality and low energy density, such as used in high power ac-dc battery chargers, dc sources among others. In order to validate the theoretical study and to prove the converter performance, some experimental results, obtained from a laboratory prototype of 4 kW, are presented.**

Keywords—Buck Converter, Interleaved DC-DC Converter, Battery Charger.

I. INTRODUCTION

Many industrial applications in the world have been modernized and adapted to the new market requirements and regulatory standards, especially in relation to the electricity consumption of the grid.

Figure 1 illustrates a typical industrial application which requires the power rectification from the grid in AC current to DC current.

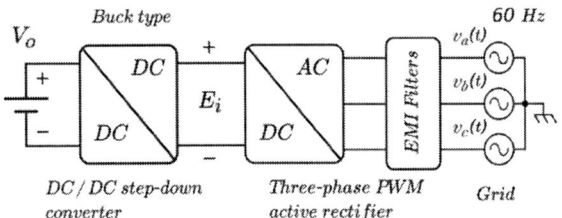

Fig. 1.Configuration of a two-stage non-insolated AC-DC battery charger.

This three-phase active rectifier system, which is mainly used at high power (> 20 kW) to improve the quality of the

drained power energy, requires an input controlled rectifier with power factor correction and, sometimes, Electromagnetic Interference (EMI) Filters, being the interface with the AC power grid and a secondary DC-DC stage, in charge of the adequacy of energy, to the load requirements [1].

Figure 1 shows the typical system for charging high-power batteries, widely used in Uninterruptible Power Supply (UPS), telecommunications, battery electric vehicles and energy storage systems, by means the batteries or supercapacitors [2-6]. However, the load represented by the battery (V_o) can also be an induction furnace, welding machine, DC motor, and among other applications that require the control and the high quality of power supply.

In the AC-DC stage, several topologies of the three-phase PWM rectifier can be applied to attend a wide power range, draining sinusoidal currents from the grid and providing a stabilized output voltage for the second stage [6]-[13]. The Buck-type converter topologies are widely employed in this second stage, providing a controlled output voltage lower than the input voltage, in addition to load current control, especially in applications requiring low current and voltage ripples [14]. To attend the requirements of high efficiency, high power density and modularity, interleaved topologies have been largely applied and developed for high power processing [15].

The Buck converter interleaving technique is further well-known in the industry in applications such as low volume battery chargers, offering a more efficient and simple load current control [16]-[23]. Buck-type interleaved converters are important devices for system weight and volume reduction, providing average output voltage lower than input voltage, as well as, controlled low-ripple current. Thus, for applications that comprise the power range from tens to hundreds of kilowatts, the interleaving technique is fundamental. In this work, a new interleaved converter topology is introduced and compared to the more classic solutions, which are the conventional Buck converter and the Two-phase interleaved buck converter.

II. CONVENTIONAL CONVERTER BUCK STRUCTURES

This section summarizes the most conventional Buck topologies, which are widely used in power electronics with the main purpose of reducing the supply voltage.

978-1-7281-4181-7/19 $31.00 © 2019 IEEE

A. Conventional Buck DC-DC Converter

The conventional Buck DC-DC converter is used to reduce input voltage (E_i) level and for application that required the output of converter is characterized as the current source, according to the illustrating in figure 2 [14]. The energy flow between the input and converter output is controlled by means adjusting the duty cycle (D), which it is the division between the conduction time (t_{on}) and the switching period (T_S) of the switch. This constant adjustment is done in the duty cycle which controls the output voltage (V_o) and the load current, maintaining a constant output power.

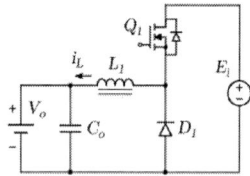

Fig. 2.Conventional DC-DC Buck Converter.

The main idealized waveforms in the continuous conduction mode (CCM) are illustrated in Figure 3. The conventional Pulse Width Modulation (PWM) technique is utilized for switch and power flow controls. The following mathematical analyses will be based on theoretical waveforms.

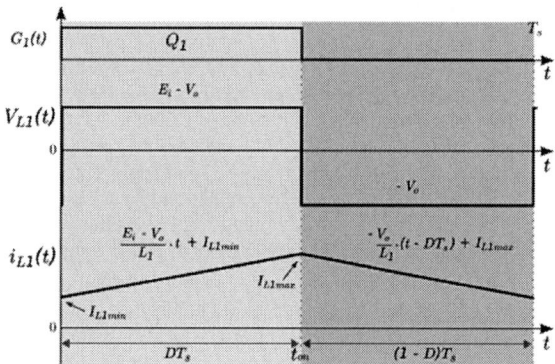

Fig. 3.Voltage and current idealized waveforms of conventional Buck converter.

At the steady state the dc voltage gain of the conventional Buck converter can be obtained considering null the average inductors voltage and it is presented in (1) considering the CCM operation.

$$\frac{V_0}{E_i} = D \tag{1}$$

The inductor (L_1) is designed and can be defined from the analysis of the first stage, where it accumulates energy from the input source. The mathematical expression for describing the inductor ripple current is given by (2).

$$\Delta I_L = I_{L\,max} - I_{L\,min} = \frac{V_0\left(1-D\right)}{L_1 \cdot f_S} \tag{2}$$

The maximum and minimum inductor current can be defined by (3) and (4).

$$I_{L\,max} = I_{L(avg)} + \frac{\Delta I_L}{2} \tag{3}$$

$$I_{L\,min} = I_{L(avg)} - \frac{\Delta I_L}{2} \tag{4}$$

The output filter capacitance (C_o) is also defined as a function of voltage and current ripple flowing through it. The voltage and current capacitor behavior is represented by the following expressions:

$$\Delta v_C = \frac{1}{C} \cdot \int_0^t i_C(t)\, dt\,; \qquad i_C(t) = C \cdot \frac{dv_C}{dt} \tag{5}$$

The output voltage ripple can be obtained by (6).

$$\Delta V_0 = \frac{V_0 \cdot \left(1-D\right)}{8\,L_1 \cdot C_0 \cdot f_S^{\,2}} = \frac{\Delta i_L}{8\,C_0 \cdot f_S} \tag{6}$$

The average and RMS values of the current flowing through the inductor L_1 is defined by (7).

$$I_{L(avg)} = I_0\,; \qquad I_{L(RMS)} = I_0 \cdot \sqrt{1 + \frac{1}{12}\cdot\left(\frac{\Delta I_L}{I_0}\right)^2} \tag{7}$$

Where the average load current (I_0) is obtained by P_o and V_o.

The RMS current of output capacitor is given by (8).

$$I_{C_0(RMS)} = \frac{V_0 \cdot \left(1-D\right)}{\sqrt{12}\cdot L_1 \cdot f_S} = \frac{\Delta I_L}{2\,\sqrt{3}} \tag{8}$$

The average and RMS value of the switch current is given by (9).

$$I_{Q(avg)} = I_0 \cdot D\,; \qquad I_{Q(RMS)} = I_0 \cdot \sqrt{D} \tag{9}$$

The average and RMS values of the diode current is defined by (10).

$$I_{D(avg)} = I_0 \cdot \left(1-D\right)\,; \qquad I_{D(RMS)} = I_0 \cdot \sqrt{1-D} \tag{10}$$

B. Interleaved Buck DC-DC Converter

When larger quantity of energy is require to be transferred to the load, strategies such as parallelism and interleaving of converters are widely used [17], considering that the interleaved converter appears as an advantageous solution in several applications to solves some of the limitations, such as presented by conventional Buck converter, which are low efficiency, high conduction losses, high switching losses and low power density.

The interleaving technique provides efficient solutions, such as, the division of load current between the converter phases, thereby reducing losses in the semiconductors, as well as, the reduction of the volume of passive elements [16].Therefore, the two-phase intercalated Buck DC-DC converter, shown in Figure 4, is widely used, especially in energy storage systems using batteries [4], [5], [18], [19].

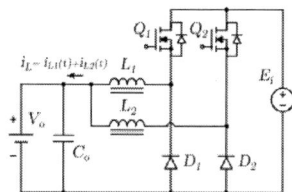

Fig. 4.Two-phase interleaved buck DC-DC converter.

In the interleaved Buck converter of Figure 4, the switches Q_1 and Q_2 are controlled by PWM and driven with the phase-shift angle of 180 degrees. The complementary switching function can produces the voltage and current ripple cancelation in output converter, since $i_{L1}(t) + i_{L2}(t)$ yield the current $i_L(t)$.

Figure 5 shows the currents in L_1 and L_2 for $D = 0.5$.

978-1-7281-4181-7/19 $31.00 © 2019 IEEE

Therefore, the converter has two distinct operating regions, when $D > 0.5$ and $D < 0.5$. In the condition $D = 0.5$, the theoretical ripple of voltage and output current are null.

The mathematical analysis is developed based on waveforms witch correspond to the operation with a duty cycle of $D > 0.5$.

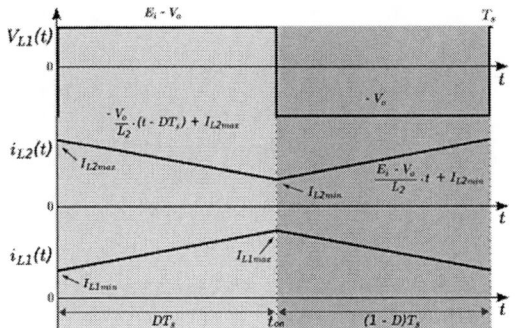

Fig. 5. Voltage and current idealized waveforms of Two-phase interleaved buck DC-DC converter, when D=0.5.

The expression of the dc voltage gain in CCM is the same as the conventional Buck DC-DC converter, according to (1).

Considering the first stage of operation, the equivalent current ripple is defined from the voltage across inductor by $v_L = L \cdot (\Delta i_L / \Delta t)$. Therefore, the equation of output current ripple Δi_L can be expressed as:

$$\Delta I_L = \frac{V_0 \left(1 - 2D\right)}{L_1 \cdot f_S} \tag{11}$$

The output filter capacitance is given by (12).

$$\Delta V_0 = \frac{V_0 \cdot \left(1 - 2D\right)}{16 L_1 \cdot C_0 \cdot f_S^2} = \frac{\Delta i_L}{16 C_0 \cdot f_S} \tag{12}$$

The average and RMS values of current flowing in each inductor (L_1 and L_2) are defined by (13).

$$I_{L(avg)} = \frac{I_0}{2}; \qquad I_{L(RMS)} = \frac{I_0}{2} \cdot \sqrt{1 + \frac{4}{3} \cdot \left(\frac{\Delta I_L}{I_0}\right)^2} \tag{13}$$

The RMS value of current flowing through the filter capacitor C_o is given by (14).

$$I_{C_0(RMS)} = \frac{V_0 \cdot \left(1 - 2D\right)}{\sqrt{12} \cdot L_1 \cdot f_S} = \frac{\Delta I_L}{2\sqrt{3}} \tag{14}$$

The average and RMS values of the switches current (Q_1, Q_2) are expressed by (15).

$$I_{Q(avg)} = \frac{I_0 \cdot D}{2}; \qquad I_{Q(RMS)} = \frac{I_0 \cdot \sqrt{D}}{2} \tag{15}$$

The average and RMS values of the diodes (D_1, D_2) currents are defined by (16).

$$I_{D(avg)} = I_0 \cdot \frac{\left(1 - D\right)}{2}; \qquad I_{D(RMS)} = I_0 \cdot \frac{\sqrt{1 - D}}{2} \tag{16}$$

The obtained expressions indicate the necessity of the output filter capacitor with half value of the conventional Buck converter, as well as, the smaller RMS current in the inductors and switches, by means of this, it is possible to conclude that the interleaved topology is better for applications requiring the higher power processing levels.

III. PROPOSED STRUCTURE

In this section, the proposed topology is presented as well as, it's operating principle and idealized mathematical analyses, are derived.

A. Proposed Interleaved Buck DC-DC Converter

As previously noted, the most commonly used way to reduce the volume of input and output filters is increasing the frequency of the inductor current ripple and the output voltage ripple. This converter consists of two-phase intercalated, where each phase has two sets of switches connected in parallel. By employing the interleaved modulation, it is possible to achieve the operating frequency of filters is four times higher than the switching frequency. The structure has four active switches Q_1, Q_2, Q_3 and Q_4 with their respective diodes D_1, D_2, D_3 and D_4 for the free-wheeling stage of the current, an autotransformer T_a and an output filter composed of the inductor L_1 and the capacitor C_0. The autotransformer enables the inductor current ripple to be reduced due to the mutual effect and the use of the interleaving technique.

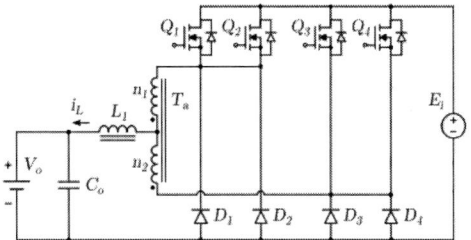

Fig. 6. Proposed Interleaved Buck DC-DC Converter.

The topology switches are commanded by means of PWM modulation and shifted by 90°, as shown in Figure 7. Similar to that of the conventional interleaved buck converter, the currents flow through the windings n_1 and n_2 produce the equivalent output current $i_{L1}(t) = i_{n1}(t) + i_{n2}(t)$. Therefore, the converter also has two distinct operating regions, when $D > 0.25$ and $D < 0.25$. In the condition $D = 0.25$, the theoretical output ripples are completely canceled. The converter analysis is approached based on the operation stages and idealized waveforms shown in Figure 7, considering the operation region for $D > 0.25$.

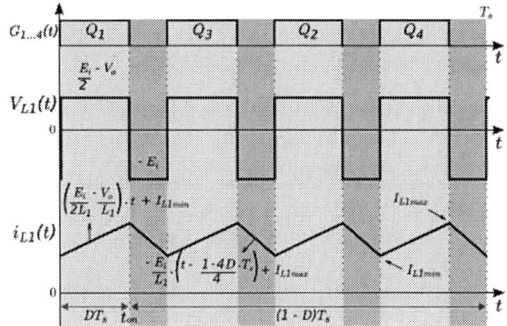

Fig. 7. Voltage and current idealized waveforms of proposed interleaved Buck DC-DC Converter.

Considering the proposed topology operates in steady state, the dc voltage gain is expressed by (17).

$$\frac{V_0}{E_i} = 2D \tag{17}$$

The mathematical expression that describes the output inductor current ripple (L_l) is given by (18).

$$\Delta I_L = I_{L\,max} - I_{L\,min} = \frac{V_0 \cdot (1 - 4D)}{4\,L_1 \cdot f_S} \qquad (18)$$

The capacitor output voltage ripple Co is obtained by means of (5) and given by (19).

$$\Delta V_0 = \frac{V_0 \cdot (1 - 4D)}{128\,L_1 \cdot C_0 \cdot f_S^2} = \frac{\Delta I_L}{32\,C_0 \cdot f_S} \qquad (19)$$

The average and RMS values of the current flowing through the output inductor (L_l) is defined by (20).

$$I_{L(avg)} = I_0; \qquad I_{L(RMS)} = I_0 \cdot \sqrt{1 + \frac{1}{3} \cdot \left(\frac{\Delta I_L}{2\,I_0}\right)^2} \qquad (20)$$

The average and RMS values of the current flowing through the autotransformer windings n_l and n_2 is given by (21).

$$I_{L(avg)} = \frac{I_0}{2}; \qquad I_{L(RMS)} = \frac{I_0}{2} \cdot \sqrt{1 + \frac{1}{3} \cdot \left(\frac{2\,\Delta I_L}{2\,I_0}\right)^2} \qquad (21)$$

The RMS value of the current flowing through the output filter capacitor is defined by (22).

$$I_{C_0(RMS)} = \frac{V_0 \cdot (1 - 4D)}{16\sqrt{2} \cdot L_1 \cdot f_S} = \frac{\Delta I_L}{4\sqrt{2}} \qquad (22)$$

The average and RMS values of currents flowing through the switches (Q_1, Q_2, Q_3 and Q_4) are expressed by (23).

$$I_{Q(avg)} = \frac{I_0 \cdot D}{2}; \qquad I_{Q(RMS)} = \frac{I_0 \cdot \sqrt{D}}{2} \qquad (23)$$

The average and RMS values of currents flowing through the diodes (D_1, D_2, D_3 and D_4) are defined by (24).

$$I_{D(avg)} = I_0 \cdot \frac{(1 - 2D)}{4}; \qquad I_{D(RMS)} = I_0 \cdot \frac{\sqrt{1 - 2D}}{4} \qquad (24)$$

As shown, it is verified that the proposed converter has smaller current ripple and less current stresses in switches and the output capacitor, compared with the traditional topologies presented previously. The dc voltage gain of the proposed converter is shown in Figure 8 compared to the other converters. This theoretical curve illustrates the output voltage behavior by means of varying the duty cycle, allowing the same voltage gain to be obtained for smaller values of D. It can be seen that the dc voltage gain for the proposed converter is higher than those of the other buck DC-DC converters.

IV. EXPERIMENTAL RESULTS

In this section, a proposed converter design example is presented, where a laboratory prototype was implemented and tested to verify the principle of the proposition with the specifications listed in Table I. Figure 9 shows the prototype of the proposed converter that was implemented in the laboratory. The voltage and current waveforms at the input source and the load are shown in Figure 10 for rated power.

The dc voltage gain of the proposed converter, expression (17), and the current ripple frequency over L_1 can be verified in the waveforms of Figure 11.

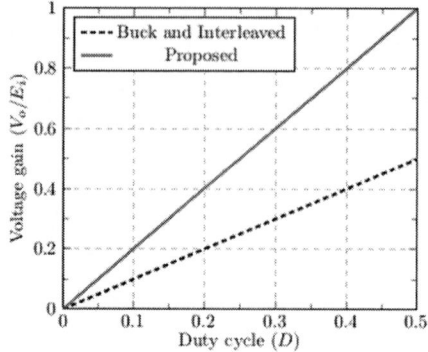

Fig. 8. Theoretical dc voltage gain curves for different values of duty cycle.

TABLE I. PARAMETERS OF THE EXPERIMENTAL PROTOTYPE

Parameters	Value
Rated power (P_o)	4.8 kW
Output Voltage (V_o)	200 V
Input voltage (E_i)	550 V
Output voltage ripple (ΔV_o)	0.1 V
Current ripple (ΔI_L)	2.0 A
Turns ratio ($n_l : n_2$)	18:18
Switching frequency (f_s)	20 kHz

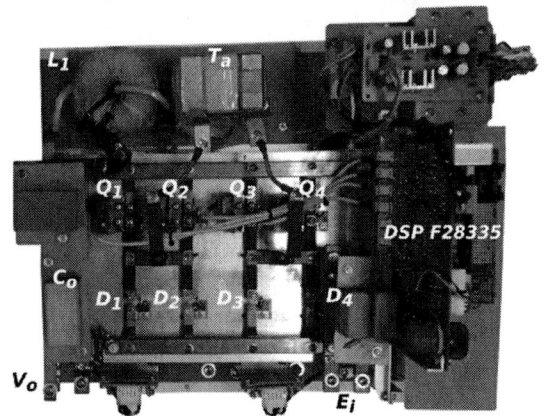

Fig. 9. Prototype of proposed interleaved buck DC-DC converter.

As shown, the output current ripple and the output voltage ripple are four times higher than the switching frequency ($4f_s$). These advantages allow the volume of passive elements to be reduced.

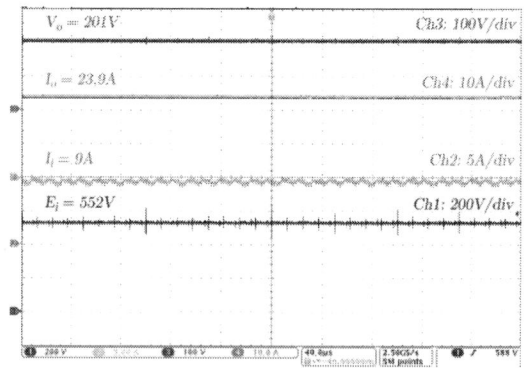

Fig. 10. Voltage and current waveforms at the source and the load - Ch1; Input voltage E_i, Ch2: Source current (I_i), Ch3: Output voltage (V_o), Ch4: Output current (I_o).

978-1-7281-4181-7/19 $31.00 © 2019 IEEE

Figures 12 and 13 show the current and maximum voltage through switch Q_1 and the free-wheel diode D_1, respectively.

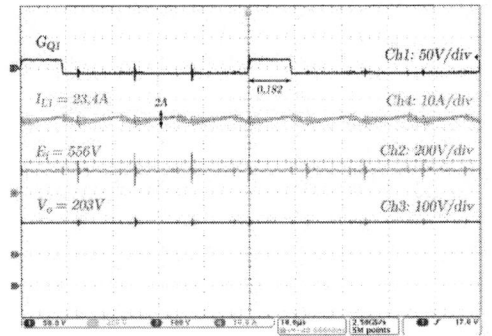

Fig. 11. Voltage and current waveforms - Ch1: Command signal of Q_1 ($D = 0.182$), Ch2: Input voltage (E_i), Ch3: Output voltage (V_o), Ch4: Inductor current I_{L1}.

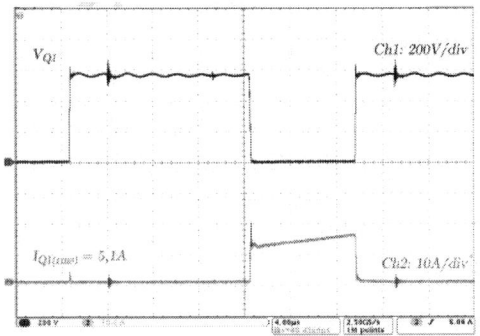

Fig. 12. Voltage and current waveforms over switch Q_1 - Ch1: Switch voltage, Ch2: Switch instantaneous current.

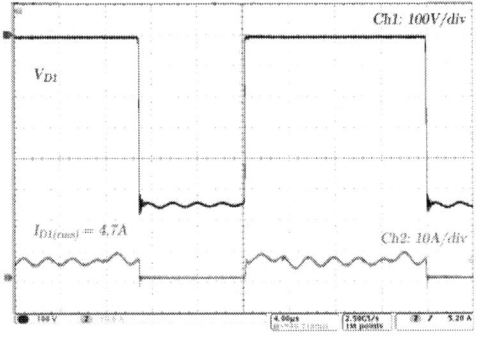

Fig. 13. Voltage and current waveforms over diode D_1 - Ch1: Diode voltage, Ch2: Diode instantaneous current.

Figure 14 shows the voltage and current on the output filter capacitor. In Figure 15, the electrical quantities such as voltage and current at the input and output converter are measured with the Tektronix power analyzer model PA4000. In the rated load, the instantaneous efficiency (*97.19 %*) and the instantaneous powers are also measured.

V. CONCLUSIONS

In this paper, a new interleaved Buck-type converter has been proposed for the application and optimization of battery recharge systems or DC storage sources that require high output currents. This paper has presented a comparative analysis between the proposed and two widely used converters, being the conventional Buck converter and the

Two-phase Interleaved Buck. The experimental results demonstrated that the output inductor volume can be reduced by four times and the capacitor filter by two times compared to the previously described converters. In addition, the voltage and output current ripple is four times higher than switching frequency.

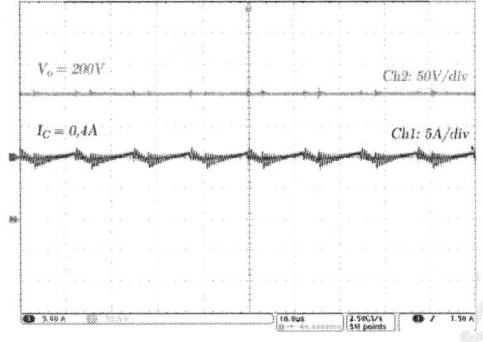

Fig. 14. Voltage and current waveforms through the filter capacitor C_o - Ch1: instantaneous current and Ch2: output voltage

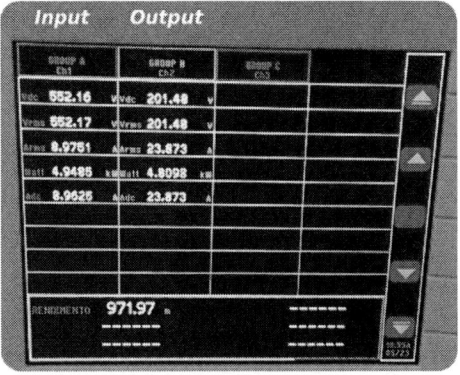

Fig. 15. Photo of the power analyzer of the input/output electrical quantities along to the efficiency

VI. ACKNOWLEDGMENTS

This work was supported in part by the Coordenação de Aperfeiçoamento de Pessoal de Nível Superior – Brasil (CAPES) - Finance Code 001, and in part by the Fundação de Amparo à Pesquisa e Inovação do Estado de Santa Catarina (FAPESC), SC, Brazil. The authors gratefully to UDESC and FURB Universities.

VII. REFERENCES

[1] T. Friedli, J. Kolar, "The essence of three-phase PFC rectifier systems", in IEEE 33rd International Telecommunications Energy Conference (INTELEC), Oct. 2011,pp. 1–27.

[2] K-M. Yoo, K.-D. Kim, J.-Y. Lee, "Single-and three- phase PHEV onboard battery charger using small link capacitor", IEEE Transactions on Industrial Electronics, vol. 60, no. 8, pp. 3136–3144, Aug. 2013.

[3] R. Hou, A. Emadi, "A primary full-integrated active filter auxiliary power module in electrified vehicles with single-phase onboard chargers", IEEE Transactions on Power Electronics, vol. 32, no. 11, pp. 8393–8405, Nov. 2017.

[4] C-Y. Oh, D-H. Kim, D-G. Woo, W-Y. Sung, Y-S. Kim, B-K. Lee, "A high-efficient nonisolated single- stage on-board battery charger for electric vehicles", IEEE Transactions on Power Electronics, vol. 28, no. 12, pp. 5746–5757, Dec. 2013.

[5] J. Bae, B. L. Nguyen, H. Cha, "A novel battery formation equipment

978-1-7281-4181-7/19 $31.00 © 2019 IEEE

using two-stage differential buck converter", in Transportation Electrification Asia- Pacific (ITEC Asia-Pacific), 2016 IEEE Conference and Expo, pp. 740–744, IEEE, June 2016.

[6] C. Shi, A. Khaligh, "A Two-Stage Three-Phase Integrated Charger for Electric Vehicles With Dual Cascaded Control Strategy", IEEE Journal of Emerging and Selected Topics in Power Electronics, vol. 6, no. 2, pp. 898–909, June 2018.

[7] K. Inazuma, H. Utsugi, K. Ohishi, H. Haga, "High- power-factor single-phase diode rectifier driven by repetitively controlled IPM motor", IEEE Transactions on Industrial Electronics, vol. 60, no. 10, pp. 4427– 4437, Oct. 2013.

[8] T-H. Liu, Y. Chen, P-H. Yi, J-L. Chen, "Integrated battery charger with power factor correction for electric-propulsion systems", IET Electric Power Applications, vol. 9, no. 3, pp. 229–238, March 2015.

[9] J. Doval-Gandoy, C. M. Penalver, "Dynamic and steady state analysis of a three phase buck rectifier",IEEE Transactions on power electronics, vol. 15, no. 6, pp. 953–959, Nov. 2000.

[10] M. Chomat, L. Schreier, "Power factor correction in voltage source inverter under unbalanced three-phase voltage supply conditions", in Industrial Electronics, ISIE'03. 2003 IEEE International Symposium on, vol. 1, June 2003, pp. 264–269.

[11] M. L. Heldwein, S. A. Mussa, I. Barbi, "Three-phase multilevel PWM rectifiers based on conventional bidirectional converters", IEEE Transactions on Power Electronics, vol. 25, no. 3, pp. 545–549, March 2010.

[12] T. Nussbaumer, J. W. Kolar, "Comparison of 3-phase wide output voltage range PWM rectifiers", IEEE Transactions on Industrial Electronics, vol. 54, no. 6, pp. 3422–3425, Dec. 2007.

[13] L. K. Ries, T. B. Soeiro, M. S. Ortmann, M. L. Heldwein, "Analysis of Carrier-Based PWM Patterns for a Three-Phase Five-Level bidirectional Buck-Boost-Type Rectifier", IEEE Transactions on Power Electronics, vol. 32, no. 8, pp. 6005–6017, Aug. 2017.

[14] Y.-C. Chuang, "High-efficiency ZCS buck converter for rechargeable batteries", IEEE Transactions on Industrial Electronics, vol. 57, no. 7, pp. 2463–2472, July 2010.

[15] H. Choi, "Interleaved boundary conduction mode (BCM) buck power factor correction (PFC) converter", IEEE Transactions on Power Electronics, vol. 28, no. 6, pp. 2629–2634, June 2013.

[16] R. Mayer, A. Péres, S. V. G. Oliveira, "Conversor cc-cc multifásico bidirecional em corrente não isolado aplicado a sistemas elétricos de tração de veículos elétricos e híbridos", Eletrônica de Potência - SOBRAEP, vol. 20, no. 3, pp. 311–321, Jun./Ago. 2015.

[17] L. Tan, N. Zhu, B. Wu, "An Integrated Inductor for Eliminating Circulating Current of Parallel Three-Level DC–DC Converter-Based EV Fast Charger", IEEE Transactions on Industrial Electronics, vol. 63, no. 3, pp. 1362–1371, March 2016.

[18] A. C. Schittler, D. Pappis, A. Campos, M. A. Dalla Costa, J. M. Alonso, "Interleaved buck converter applied to high-power HID lamps supply: Design, modeling and control", IEEE Transactions on Industry Applications, vol. 49, no. 4, pp. 1844–1853, July/Aug. 2013.

[19] I-O. Lee, S-Y. Cho, G-W. Moon, "Interleaved buck converter having low switching losses and improved step-down conversion ratio", IEEE Transactions on Power Electronics, vol. 27, no. 8, pp. 3664–3675, Aug. 2012.

[20] M. Esteki, B. Poorali, E. Adib, H. Farzanehfard, "Interleaved Buck Converter With Continuous Input Current, Extremely Low Output Current Ripple, Low Switching Losses, and Improved Step-Down Conversion Ratio.", IEEE Transactions on Industrial Electronics, vol. 62, no. 8, pp. 4769–4776, Aug. 2015.

[21] YQiu, M. Xu, K. Yao, J. Sun, F. C. Lee, "Multifrequency small-signal model for buck and multiphase buck converters", IEEE transactions on power electronics, vol. 21, no. 5, pp. 1185–1192, Sept. 2006.

[22] Y-M. Chen, S-Y. Tseng, C-T. Tsai, T-F. Wu, "Interleaved buck converters with a single-capacitor turn-off snubber", IEEE Transactions on Aerospace and Electronic Systems, vol. 40, no. 3, pp. 954–967, July 2004.

[23] G. V. T. Bascopé, I. Barbi, "Generation of a family of non-isolated DC-DC PWM converters using new three-state switching cells", in Power Electronics Specialists Conference, 2000. PESC 00. 2000 IEEE 31st Annual, vol. 2, pp. 858–863, IEEE, June 2000.

978-1-7281-4181-7/19 $31.00 © 2019 IEEE

PERFORMANCE OF A MULTI-GROUNDED DISTRIBUTION NETWORK WITH A FOUR-WIRE THREE-PHASE POWER CONDITIONER

Samuel N. Duarte, Bruno C. Souza, Pedro M. Almeida, Leandro R. Araújo, Pedro G. Barbosa

Federal University of Juiz de Fora, Graduate Program in Electrical Engineering, Juiz de Fora–MG, Brazil

Email: samuel.neves@engenharia.ufjf.br, bruno.cortes@engenharia.ufjf.br,
pedro.machado@ufjf.edu.br, leandro.araujo@ufjf.edu.br, pedro.gomes@ufjf.edu.br

Abstract - **This paper presents a simplified analysis of a multi-grounded distribution network compensated by a four-wire three-phase power conditioner. The effect of the neutral conductors and grounding resistances on the voltage unbalance factor are demonstrated, in steady-state, performing digital simulations using the OpenDSS package. Then, a power conditioner is connected to the distribution network to compensate for the currents drained by an unbalanced load. The control loops used to control the AC currents and the DC voltage of the power conditioner are also presented. Results of time-domain simulations show that, although the unbalance can not be fully mitigated, the power conditioner can reduce the voltage unbalance and neutral voltages of the distribution network.**

Keywords – **Distribution Network, Power Conditioner, Voltage Unbalance.**

I. INTRODUCTION

In recent years, due to the concepts of smart grids, the number of studies involving applications of power electronics converters in distribution networks increased considerably [1–3]. Moreover, since these modern networks feed a large number of non-linear equipment and operate with unbalanced voltages and currents [4, 5], the power quality indexes along the network feeders must be in accordance with the recommendations of standards [6, 7].

Regarding the distribution network modeling, the neutral conductor is frequently neglected in the papers presented in the literature. The detailed representation of this wire is very important when dealing with four-wire three-phase systems [8–10] since several equipments are connected between one phase and the neutral wires, forcing a non-zero neutral current flowing through the feeders. Thus, if the neutral conductor is not solidly-grounded, the feeder will present neutral voltages [10–13] and the simulation results may not properly represent the behavior of the distribution network.

Different topologies of power electronic converters have being connected to modern distribution networks to perform ancillary services, such as harmonic compensation, power factor correction, among others [14]. However, time domain studies involving the operation of power electronic converters on four-wire three-phase multi-grounded systems are not much discussed in the literature. In fact, the majority of the digital simulation studies in this field is done in the frequency domain (e.g. power flow programs, among others). Notwithstanding, in some cases it is necessary to represent both neutral conduc-

tor and grounding impedances [10, 12, 13].

Therefore, this work presents analysis of a four-wire three-phase multi-grounded distribution network. This electric network configuration is one of the most complexes and it is suitable to verify the arise of voltage unbalance and neutral voltage in the system. Firstly, analysis in steady-stage are presented in order to demonstrate the importance of representing the neutral conductors and the grounding resistances. Then, control loops for a power conditioner, parallel-connected with a load of the system, are presented to control the AC currents and DC voltage of the converter. Results of digital simulations of the power conditioner used to compensate for the unbalanced currents drained by the load are presented in order to verify the impact of its operation on the electric network.

II. SYSTEM DESCRIPTION

Fig. 1 shows the topology of the four-wire three-phase multi-grounded distribution network. The resistance R_f and the inductance L_f are the parameters of the feeders 1 and 2, while the resistance R_L and the inductance L_L are the parameters of the loads 1 and 2. The loads 1 and 2 are grounded through the resistances R_{g1} and R_{g2}, respectively. Thus, the voltages $v_{pcc,n}^{B1}$ and $v_{pcc,n}^{B2}$ are the neutral-to-ground voltages at the bus 1 and bus 2, respectively. The superscripts $(^{B1})$ and $(^{B2})$ were used to refer to the bus 1 and bus 2, respectively. The voltage v_{sn}, across the grounding resistance R_g, is the neutral-to-ground voltage at the secondary of the transformer, whose phase voltages are v_{sa}, v_{sb} and v_{sc}. The parameter L_g is the inductance of the reactor connected to the neutral of the transformer.

A. Steady-state Analysis

This section presents steady-state analysis for the distribution network of Fig. 1. This network was modeled in the OpenDSS program in order to perform a power flow analysis. The parameters of the electric network, feeders and loads are given in Tables III, IV and V (Appendix), respectively.

The aim here is to analyze the sensitivity of the distribution network due to the variation of the grounding resistances. Thus, the values of the grounding resistance of the transformer (R_g) and the grounding resistances of the loads (R_{g1} and R_{g2}) were varied according to 18 multipliers. That is, these grounding resistances were multiplied by 0.00001 until 5000, resulting in 18 grounding configurations.

Tables I and II show the voltage unbalance factors (VUFs) at bus 1 and bus 2, respectively, for each grounding configu-

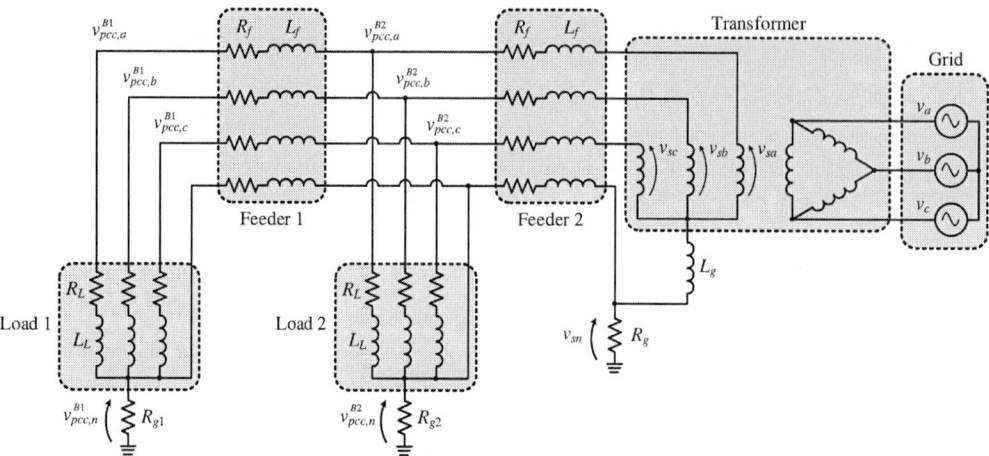

Fig. 1: Topology of the distribution network.

ration. The grounding configurations 1 and 18 represent the extremes cases. That is, solidly grounded and isolated neutral, respectively. The VUF can be calculated by the recommendation of NEMA or IEC [6, 7]. Usually, international standards set the limit for the VUF to 2 % [15]. In the aforementioned tables, the percentage of the neutral voltage is calculated by the relation between the RMS neutral voltage and the average value ($\bar{V}$) of the RMS three-phases voltages [10]. Note that in both tables there are some buses with unbalance factor above and below the 2 % limit.

It can be seen that bus 1 presents higher VUFs than bus 2. This may occur due to the fact that bus 1 is located at the end of the feeder. Moreover, both tables show that the calculated VUFs are quite different and are not capable to consider the neutral voltage appropriately. Although it is not the aim of this paper, a standardization should be proposed, since the use of more than one VUF may lead to contradictory conclusions.

Fig. 2 shows the phase-to-earth voltage at bus 1 for each grounding configuration. It can be noted that the relation between the phase-to-earth voltage and the grounding resistances is non-linear and that an abrupt voltage variation occurs between the grounding configurations 9 and 12. This figure also indicates that there are more than one value for the grounding resistances that will result in the same phase voltage.

Fig. 3 shows the neutral voltages of the distribution network for different grounding configurations. Since the system is radial and both loads have the same parameters, the neutral voltage profiles for each bus are quite similar. The lower neutral voltage occurs across the grounding resistance of the transformer, while the higher occurs at bus 1. The curves shown in Fig. 3 also confirms the non-linear relation between the system voltages and the values of the grounding resistances.

In order to reduce the voltage unbalance and neutral voltages of the distribution network of Fig. 1, the next section will present the control loops [16, 17] of a power conditioner that will be connected to the bus 1.

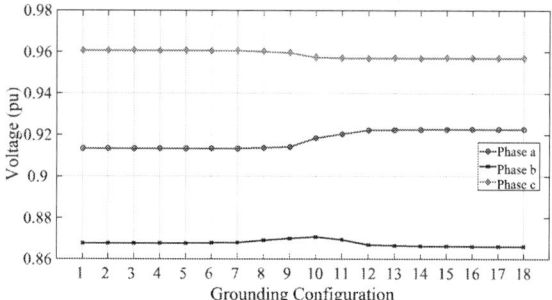

Fig. 2: Phase-to-earth voltage profile for the bus 1

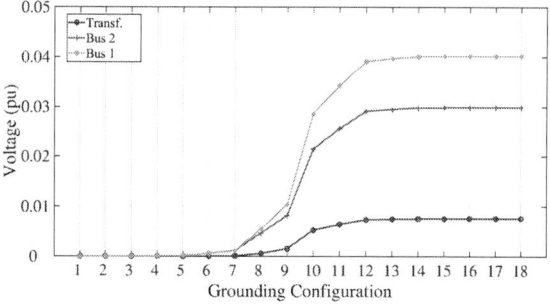

Fig. 3: Neutral-to-ground voltage profile for each bus

III. THE POWER CONDITIONER

Fig. 4 shows the topology of the four-wire three-phase power conditioner connected to an equivalent distribution network through a first-order passive filter. The parameters R_i and L_i represent the resistance and inductance of the interface filter. The equivalent DC capacitor of the power conditioner is modeled by the parameter C_{eq}. The equivalent electric network of Fig. 4 represent the distribution network of Fig. 1.

Besides regulating its DC bus voltage, the power conditioner will compensate for the unbalanced currents drained by the load connected at bus 1. However, the positive and negative-sequence components of the measured voltages and currents must be separated before be sent to the converter con-

978-1-7281-4181-7/19 $31.00 © 2019 IEEE

TABLE I
Bus 1 VUF for different groundings

Conf.	Multiplier	VUFs [%]			
		NEMA	$(V_n/\bar{V})$	IEC (V_2/V_1)	IEC (V_0/V_1)
1	0.00001	5.122	0	4.040	3.202
2	0.00005	5.122	0.001	4.040	3.202
3	0.0001	5.121	0.002	4.040	3.202
4	0.0005	5.121	0.012	4.040	3.201
5	0.001	5.119	0.024	4.040	3.200
6	0.005	5.110	0.121	4.038	3.190
7	0.01	5.099	0.242	4.036	3.179
8	0.05	5.013	1.202	4.025	3.085
9	0.1	4.916	2.254	4.019	2.973
10	0.5	4.889	6.248	4.061	2.488
11	1	5.049	7.498	4.102	2.371
12	5	5.290	8.546	4.154	2.345
13	10	5.328	8.666	4.161	2.351
14	50	5.360	8.759	4.168	2.357
15	100	5.364	8.770	4.168	2.358
16	500	5.367	8.780	4.169	2.359
17	1000	5.367	8.781	4.169	2.359
18	5000	5.368	8.782	4.169	2.359

TABLE II
Bus 2 VUF for different groundings

Conf.	Multiplier	VUFs [%]			
		NEMA	$(V_n/\bar{V})$	IEC (V_2/V_1)	IEC (V_0/V_1)
1	0.00001	3.407	0	2.637	2.115
2	0.00005	3.407	0.001	2.637	2.115
3	0.0001	3.407	0.003	2.637	2.115
4	0.0005	3.406	0.013	2.636	2.114
5	0.001	3.406	0.025	2.636	2.114
6	0.005	3.400	0.123	2.635	2.107
7	0.01	3.392	0.241	2.634	2.100
8	0.05	3.338	0.996	2.628	2.040
9	0.1	3.277	1.731	2.625	1.970
10	0.5	3.225	4.569	2.651	1.660
11	1	3.321	5.451	2.676	1.583
12	5	3.472	6.184	2.708	1.563
13	10	3.496	6.267	2.713	1.567
14	50	3.516	6.331	2.716	1.570
15	100	3.518	6.339	2.717	1.571
16	500	3.520	6.345	2.717	1.571
17	1000	3.520	6.346	2.717	1.571
18	5000	3.521	6.346	2.717	1.571

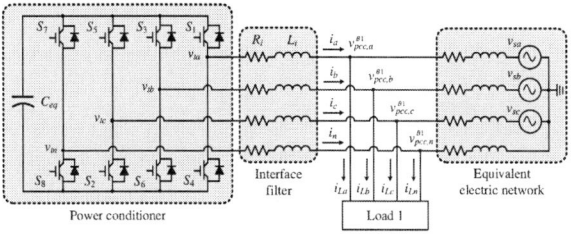

Fig. 4: Topology of the power conditioner connected to an equivalent distribution network.

trollers, as will be presented in the next section.

A. The Separation of the Positive and Negative Sequences

The positive and negative-sequence voltages at PCC are separated by using the algorithm shown in Fig. 5. After be

transformed to the $\alpha\beta$ coordinates through the Clark transformation, the measured voltages feed two SOGI circuits whose outputs are combined to separate the positive and negative sequences. The positive and negative-sequence voltages are then transformed to the dq frame, where $d\rho/dt = \omega$ is the angular frequency tracked by the Phase-Locked Loop based on the Synchronous Reference Frame (SRF-PLL). The subscripts $(_1)$ and $(_2)$ indicate the positive and negative sequences, respectively. Fig. 6 shows the internal structure of the SOGI circuit.

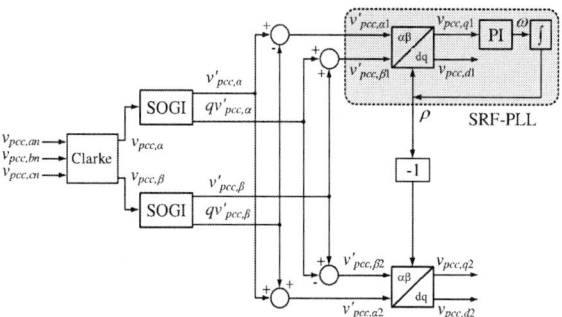

Fig. 5: Block diagram for the algorithm used to separate the positive and negative-sequence voltages.

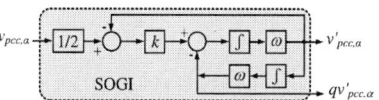

Fig. 6: Block diagram for the SOGI circuit.

The separation of the positive and negative-sequence currents are performed by using notch filters instead of the SOGI circuit, since the currents of the power conditioner present a faster dynamic. After the power conditioner and load currents be measured and transformed to the dq frame, their oscillating parcels are eliminated though notch filters tuned at $\omega_o = 2\omega$.

B. The Positive and Negative-sequence Current Control Loops

Fig. 7 shows the block diagram for the direct and quadrature axes positive-sequence current control loops [16, 17]. In this figure, $K_{i1}(s)$ is a proportional-integral (PI) controller, $R_{eq} = (R_i + R_{igbt})$, R_{igbt} is the resistance of the IGBTs, m_{d1} and m_{q1} are the modulation indexes used to control the currents I_{d1} and I_{q1} of the power conditioner, respectively, while V_{dc} is the DC bus voltage and $V_{t,d1}$ and $V_{t,q1}$ are the direct and quadrature axes terminal voltages of the power conditioner.

It is also possible to build negative-sequence current control loops for the power conditioner, similar to those shown in Fig. 7. However, in this case the synchronous reference frame should rotate on the clockwise direction. Furthermore, the cross-coupling between the negative-sequence currents have opposite signs compared to those shown in Fig. 7.

C. The Zero-sequence Current Control Loop

The use of the synchronous reference frame allows the implementation of PI controllers which makes the design of the

978-1-7281-4181-7/19 $31.00 © 2019 IEEE

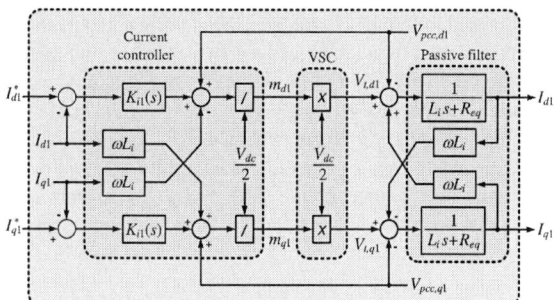

Fig. 7: Block diagram for the direct and quadrature axes positive-sequence current control loops.

controller gains easier. However, for the case of zero-sequence quantities, it is not possible directly to use the synchronous reference frame transformation. In order to overcome this problem, in [17] it is presented a control strategy that allows the use of PI controllers to control the zero-sequence current. Thus, a transport delay buffer will be used to generate the 90-degrees shift-delay current $i_{\beta 0}$ for the zero-sequence current $i_0 = i_{\alpha 0}$.

Fig. 8 shows the block diagram for the transport delay buffer, where $T = (1/f)$ is the fundamental period of the zero-sequence current. Thus, the synchronous reference frame transformation can then be applied in the currents $i_{\alpha 0}$ and $i_{\beta 0}$ in order to obtain and control the zero-sequence currents I_{d0} and I_{q0}, as shown in the simplified block diagram for the zero-sequence current control loop of Fig. 9. The modulation indexes m_{d0} and m_{q0} are used to calculate the index m_0, that is added to the three-phase modulation indexes m_a, m_b and m_c [17].

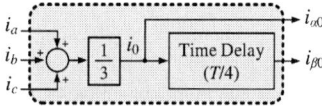

Fig. 8: Block diagram for the transport delay buffer used to generate the zero-sequence quadrature currents.

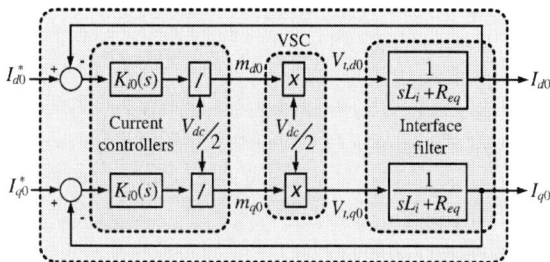

Fig. 9: Simplified block diagram for the direct and quadrature axes zero-sequence current control loops.

D. The DC voltage control

Fig. 10 shows the block diagram for the DC voltage control loop of the power conditioner, where the symbols (˜) and (¯) are used to indicate the small signal and steady-state variables [17]. The block $K_{vdc}(s)$ represents the PI controller, while $\bar{V}_{pcc,d1}$ is the steady-state direct-axis positive-sequence

voltage at PCC. Since the time constant of the DC voltage controller is designed to be greater than the positive-sequence current controller of the power conditioner, the current closed-loop transfer function can be considered as an unitary gain.

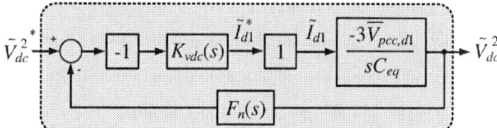

Fig. 10: Block diagram for the DC voltage control loop.

Since the DC bus voltage starts oscillating with twice the grid frequency due to the unbalances at PCC, a notch filter $F_n(s)$ tuned at $\omega_o = 2\omega$ and with an unitary quality factor is used in the feed-back of the control loop of Fig. 10 in order to ensure the correct operation of the DC voltage controller [17].

IV. RESULTS OF DIGITAL SIMULATIONS

The power conditioner of Fig. 4, connected to the distribution network of Fig. 1, was modeled in the PSIM program in order to demonstrate the impact of its operation on the distribution network. Besides regulating its DC bus voltage, the power conditioner will compensate for the negative and zero-sequence currents of the load 1 of the distribution network of Fig. 1. Tables VI and VII, in the Appendix, show the parameters of the power conditioner and its controllers, respectively. The IGBTs of the power conditioner are switched with a sinusoidal PWM strategy.

Fig. 11 (a) and Fig. 11 (b) show the waveforms of the AC terminal currents and the DC bus voltage of the power conditioner, respectively. Fig. 11 (b) shows also the output signal of the notch filter used to eliminate the oscillation in the DC bus voltage.

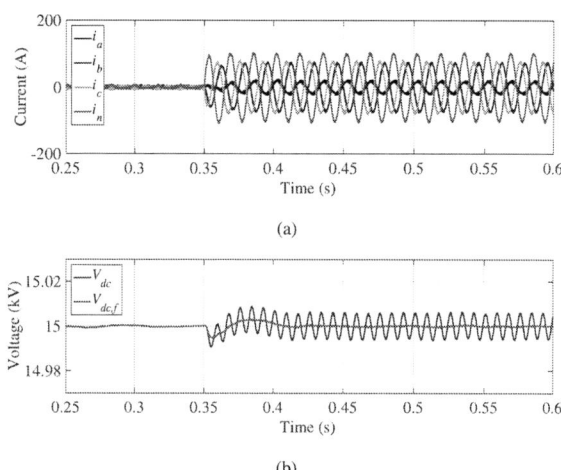

(a)

(b)

Fig. 11: Waveforms of the power conditioner: (a) AC terminal currents and (b) DC bus voltage.

Fig. 12 (a) and Fig. 12 (b) show the RMS values and the negative-sequence components of the voltages at bus 1, respectively. On the other hand, Fig. 13 (a) and Fig. 13 (b) show the RMS values and the negative-sequence components of the voltages at bus 2, respectively.

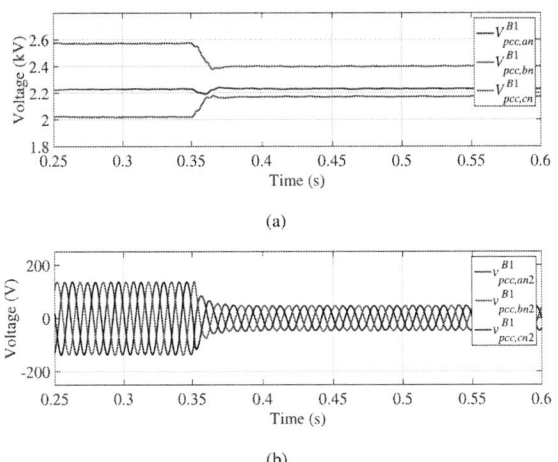

(a)

(b)

Fig. 12: Voltage waveforms of the bus 1: (a) RMS values and (b) negative-sequence components.

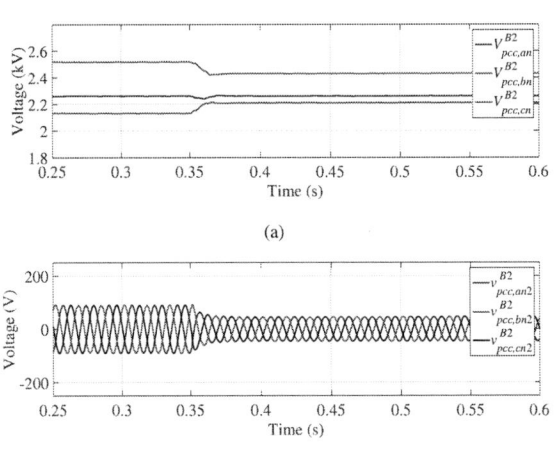

(a)

(b)

Fig. 13: Voltage waveforms of the bus 2: (a) RMS values and (b) negative-sequence components.

Fig. 14 (a) and Fig. 14 (b) show the zero-sequence components and the neutral voltages at the bus 1, bus 2 and at the secondary of the transformer.

Before t = 0.35 s the voltages across the loads 1 and 2 are unbalanced, as shown in Fig. 12 (a) and Fig. 13 (a), respectively. At t = 0.35 s the power conditioner starts synthesizing currents in order to compensate for the unbalanced currents of the load 1, as shown in Fig. 11 (a). Fig. 11 (b) shows the DC bus voltage starts oscillating due to the unbalanced currents synthesized by the power conditioner at the AC side. It can be noted in Fig. 12 (b), Fig. 13 (b), Fig. 14 (a) and Fig. 14 (b) that the neutral voltages of the system and the negative and zero-sequence voltages at bus 1 and bus 2 are reduced due to the power conditioner operation.

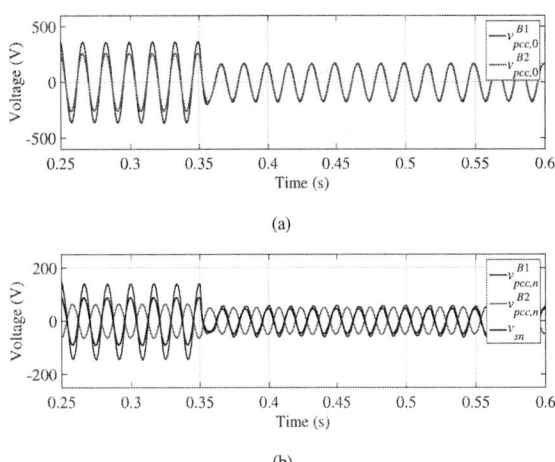

(a)

(b)

Fig. 14: Voltage waveforms of the system: (a) zero-sequence and (b) neutral.

V. CONCLUSIONS

This paper presented studies in a four-wire three-phase multi-grounded distribution network. Firstly, steady-state analysis were performed in order to demonstrate the importance of the neutral conductor and grounding representations. It was shown that the electric network presents different voltage profiles depending on the values of the grounding resistances. Moreover, the voltage unbalance factors can exceed the recommendations of standards. Then, a power conditioner was connected to a bus of the distribution network in order to compensate for the unbalanced currents drained by the load connected at PCC. The control loops of the power conditioner were also presented. Results of digital simulations were presented to demonstrate the performance of the distribution network when the power conditioner compensates for the unbalanced load currents. Although the unbalance in the system voltages can not be fully eliminated, the results shown that the power conditioner can mitigate part of the voltage unbalance and neutral voltages of the distribution network.

ACKNOWLEDGMENT

This study was financed in part by the Coordenação de Aperfeiçoamento de Pessoal de Nível Superior - Brasil (CAPES) - Finance Code 001, the National Council for Scientific and Technological Development (CNPq), the State Funding Agency of Minas Gerais (FAPEMIG) and the National Institute for Electric Energy (INERGE).

APPENDIX

Parameters of the distribution network.

REFERENCES

[1] K. Schneider, B. Mather, B. Pal, C.-W. Ten, G. Shirek, H. Zhu, J. Fuller, J. Pereira, L. Ochoa, L. De Araujo *et al.*, "Analytic considerations and design basis for the IEEE distribution test feeders," *IEEE Transactions on Power Systems*, vol. 33, no. 3, pp. 3181–3188, 2017.

TABLE III
Network parameters

Parameter	Value
Line RMS voltage (V_s)	4.16 kV
Fundamental frequency (f_s)	60 Hz
Current limiter reactor (L_g)	0.8 mH
Grounding resistance (R_g)	3 Ω
Grounding resistance of load 1 (R_{g1})	7 Ω
Grounding resistance of load 2 (R_{g2})	7 Ω

TABLE IV
Impedances of the feeders

Phase	Value
"a"	$0.1900 + j0.8671$ Ω
"b"	$0.1900 + j0.8671$ Ω
"c"	$0.1900 + j0.8671$ Ω
"n"	$0.2500 + j0.8671$ Ω

TABLE V
Impedances of the loads

Phase	Value
"a"	$17.22 + j9.74$ Ω
"b"	$11.15 + j8.12$ Ω
"c"	$38.24 + j21.93$ Ω

TABLE VI
Parameters of the power conditioner

Parameter	Value
Equivalent capacitance (C_{eq})	3 mF
DC voltage (V_{dc}^*)	15 kV
Inductance of the interface filter (L_i)	24 mH
Resistance of the interface filter (R_i)	90 mΩ
Resistance of the IGBTs (R_{igbt})	13 mΩ
Switching frequency (f_{sw})	9 kHz

TABLE VII
Controller gains of the power conditioner

Controller	Gain	Value
K_{i1}, K_{i2} and K_{i0}	Proportional	48 V/A
	Integral	206 V/(A s)
K_{vdc}	Proportional	39.2 μA/V^2
	Integral	1.3 mA/(V^2 s)

[2] R. Yan and T. K. Saha, "Analysis of unbalanced distribution lines with mutual coupling across different voltage levels and the corresponding impact on network voltage," *IET Generation, Transmission & Distribution*, vol. 9, no. 13, pp. 1727–1737, 2015.

[3] N. A. do Amaral Filho, L. R. de Araujo, D. R. R. Penido, and F. de Alcântara Vieira, "Impacts of the representation of mutual coupling between feeders in distribution systems," *International Journal of Electrical Power & Energy Systems*, vol. 105, pp. 17–27, 2019.

[4] A. Von Jouanne and B. Banerjee, "Assessment of voltage unbalance," *IEEE transactions on power delivery*, vol. 16, no. 4, pp. 782–790, 2001.

[5] L. R. de Araujo, D. R. R. Penido, J. L. R. Pereira, and S. Carneiro, "Voltage security assessment on unbalanced multiphase distribution systems," *IEEE Transactions on*

Power Systems, vol. 30, no. 6, pp. 3201–3208, 2014.

[6] N. S. P. M. 1-2016, "Motors and Generators," 2016.

[7] IEC, "61000-3-13 - electromagnetic compatibility (EMC) - limits - assessment of emission limits for the connection of unbalanced installations to MV, HV and EHV power systems," 2008.

[8] D. R. R. Penido, L. R. de Araujo, S. Carneiro, J. L. R. Pereira, and P. A. N. Garcia, "Three-phase power flow based on four-conductor current injection method for unbalanced distribution networks," *IEEE Transactions on Power Systems*, vol. 23, no. 2, pp. 494–503, 2008.

[9] R. M. Ciric, A. P. Feltrin, and L. F. Ochoa, "Power flow in four-wire distribution networks - general approach," *IEEE Transactions on Power Systems*, vol. 18, no. 4, pp. 1283–1290, 2003.

[10] B. Cortes, L. R. Araújo, and D. R. Penido, "A study of grounding in a multiphase unbalanced distribution network with neutral conductor: An analysis on the IEEE NEV test feeder," in *2018 Brazilian Symposium on Electrical Systems (SBSE)*. IEEE, 2018, pp. 1–6.

[11] T.-H. Chen and W.-C. Yang, "Analysis of multi-grounded four-wire distribution systems considering the neutral grounding," *IEEE Transactions on Power Delivery*, vol. 16, no. 4, pp. 710–717, 2001.

[12] E. R. Collins and J. Jiang, "Analysis of elevated neutral-to-earth voltage in distribution systems with harmonic distortion," *IEEE transactions on power delivery*, vol. 24, no. 3, pp. 1696–1702, 2009.

[13] L. Moreno-Díaz, E. Romero-Ramos, A. Gómez-Expósito, E. Cordero-Herrera, J. R. Rivero, and J. S. Cifuentes, "Accuracy of electrical feeder models for distribution systems analysis," in *2018 International Conference on Smart Energy Systems and Technologies (SEST)*. IEEE, 2018, pp. 1–6.

[14] J. C. da Cunha, W. de Oliveira Rossi, A. L. Batschauer, and M. Mezaroba, "A simple control scheme to a voltage regulator based in a current controlled STATCOM," in *2013 Brazilian Power Electronics Conference*. IEEE, 2013, pp. 1212–1218.

[15] H. Markiewicz and A. Klajn, "Voltage disturbances standard EN 50160 - voltage characteristics in public distribution systems," *Wroclaw University of Technology*, vol. 21, pp. 215–224, 2004.

[16] S. N. Duarte, G. A. Fogli, P. M. Almeida, and P. G. Barbosa, "Energization and de-energization strategies of a distribution static synchronous compensator," *Brazilian Power Electronics Journal*, vol. 23, no. 1, mar 2018.

[17] S. N. Duarte, F. T. Ghetti, P. M. de Almeida, and P. G. Barbosa, "Zero-sequence voltage compensation of a distribution network through a four-wire modular multilevel static synchronous compensator," *International Journal of Electrical Power & Energy Systems*, vol. 109, pp. 57–72, 2019.

A new strategy of modulation based on Space Vector Modulation and Annotated Paraconsistent Logic for a three-phase converter

1st Antonio Carlos Duarte Ricciotti
Department of Electrical Engineering
UNIR
Porto Velho, Brazil
acdricciotti@unir.br

2nd João Inácio da Silva Filho
Lab. of Applied Paraconsistent Logic
UNISANTA
Santos, Brazil
inacio@unisanta.br

3rd Raphael A. Bispo de Oliveira
Lab. of Applied Paraconsistent Logic
UNISANTA
Santos, Brazil
adamelk@gmail.com

4th Viviane B. da S. Duarte Ricciotti
Department of Electrical Engineering
UNIR
Porto Velho, Brazil
viviane@unir.br

5th Hyghor Miranda Côrtes
Lab. of Applied Paraconsistent Logic
UNISANTA
Santos, Brazil
hyghorcortes@gmail.com

6th André Miguel Nicolini
GEPOC-UFSM
UFSM
Santa Maria, Brazil
andrenicoliniee@gmail.com

Abstract—In the last decades, the popularization of distributed generation systems has been increasing exponentially. That has led to the development of new modulation strategies aimed at improving the commutation of switches in power converters. Among the desired improvements is the best choice of the sequence of switching to obtain an excellent performance of the converter, as well as the mitigation of unwanted harmonics. Space Vector Modulation (SVM) is one of the most intensively studied strategies. This work presents a new approach of modulation, the paraconsistent space vector modulation (PSVM), that is based on SVM, but uses the a non classical Paraconsistent Annotated Logic (PAL) to define witch vector will be applied. The PAL theoretical basis allows the treatment of contradictory signs, and the main characteristic is the tolerance of contradiction in its foundations. As a result, it was possible to prove the validation of the technique and to provide subsidies for its improvement.

Index Terms—Space Vector Modulation, Modulation of Converters, Paraconsistent Annotated Logic, Inverters, Power Electronics.

I. INTRODUCTION

The growth of the popularization of distributed generation systems has as its primary foundation the use of renewable energy sources, for example, sources from biomass, solar, wind, among others, and have as one of the main characteristics the low impact on the environment [1]. Other factors also drive this growth are related to the maturation of the technologies associated with the converters. New topologies of converters dedicated to the integration of renewable sources to the electric grid, for instance, Highly Efficient and Reliable Inverter Concept (HERIC) that interconnects the converter to the network without the need of a transformer. New generations of semiconductor devices that are more efficient, support large currents, higher switching frequency, among other characteristics, for example, metal-oxide-semiconductor (MOSFET) field diodes and transistors, diodes, and IGBTs

of silicon carbide (SIC). Advances in digital signal processors (DSPs) that allow the control and supervision of power converters [2] and [3], and consequently the expansion of distributed generation systems is intrinsically linked with the technologies used in power converters, which allow flexible and efficient interconnections between different renewable generation plants, e.g., wind generation, and the electric power system. Therefore, with the advancement of the new devices, they require modulation techniques that improve the efficiency of the converters, reducing the losses in the switches [4]. The correct use of the modulation techniques makes the converters efficient means of obtaining high voltage levels [5], by attenuating the total harmonic distortion (THD) rate at the output [6]. [7] and [8] used the selective elimination of harmonics in pulse width modulation (PWM) to improve voltage THD. The delta modulation proposed by [9] presented attenuation of low order harmonics in the output voltages and reduced number of switching for high amplitude modulation indices. The modulation by sinusoidal pulse width (SPWM) presents constant switching frequency and distances the harmonic components of the output voltages of the fundamental frequency inverter [10] and [11]. The space vector modulation (SVM) techniques applied in inverters allow the reduction of the number of switch changes, decrease the harmonic content of the output voltage, increasing the amplitude modulation index of the inverter, and they can implement in microprocessors [12]. For example, [13] propose a control strategy for wind turbines based on synchronous generators of permanent magnets operating with parallel converters. Using a discontinuous SVM modulation modified by a PWM carrier and a mean two-phase three-phase rectifier model, it was possible to implement a current control in coordinates *dqo*, and thereby reduce the circulating currents in the parallel converters.

978-1-7281-4181-7/19 $31.00 © 2019 IEEE

II. SPACE VECTOR MODULATION

The technique that uses spatial vectors is SVM modulation and consists of selecting the switching vectors which usually are the three closest vectors to the reference, calculating the duration of the vectors (the duty cycles) and determining a sequence of vectors [3]. Fig. 1 shows an example of three-phase two-level inverter where can possible see three blocks, the inverter, the filter, and the connection to the grid. Note that the converter using six switches S_a, S_b, and S_c, that correspond the superior devices, whereas $\bar{S}_a$, $\bar{S}_b$, and $\bar{S}_c$ correspond the inferior devices. This topology allows for generating the modulation desired.

Fig. 1. Two-level Inverter. Adapted [4]

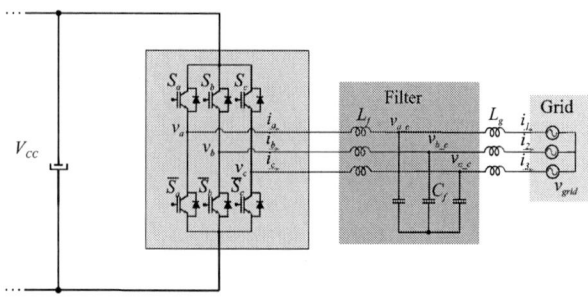

Fig. 2 shows the simplified diagram a two-levels inverter for the understanding of the process of the SVM. where the devices are substituted by six switches, that named of S_a, S_b, S_c, $\bar{S}_a$, $\bar{S}_b$, and $\bar{S}_c$. The switches have two states (ON and OFF) and are used to compose the space vectors modulation. The vectors produced corresponding the eight conditions possible of implementation [14].

Fig. 2. Simplified two-levels inverter model.

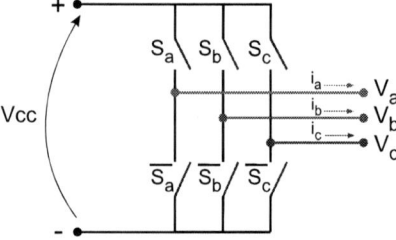

The correct combination of the switching states produces the output two-level phase voltage inverter. This combination sequence is called the switching state.

The states for a two-level inverter are arranged in vectors ($V_0 - V_7$). Six of them ($V_1 - V_6$) are located at the vertices of the hexagon shown in Fig. 3, and the others (V_0 and V_7) are in its center. The hexagon is divided in six sectors ($S_1 - S_6$) and the reference voltage (V^*) is produced by the voltages V_a and V_b.

The dwell times or time necessary for providing the reference voltage based in each vector is using to calculate the voltages V_a and V_b to generate the V^* at that particular point of time. For sector 1, S_1, of 0 to $\pi/3$, for example, V^* can be generated with V0, V1, and V2, as shown in Fig. 3.

In terms of the duration time, V^* is given by:

$$V^* * T_s = V_1 * T_1 + V_2 * T_2 + V_0 * T_0 \tag{1}$$

where T_s is the sampling time, and T_1, T_2, and T_0 are the time periods for which V_1, V_2, and V_0 are applied.

The SVM strategy consists in to approximate the reference voltage to the value of the line voltage output, employing the combination of the switching states corresponding to the space vectors. Thus, To obtain a high efficient index, the analysis must identify very accurately the location of the vectors and the reference vector V^* in the figure represented in the hexagon of Fig. 3.

Fig. 3. Diagram Space Vector. Adapted [14].

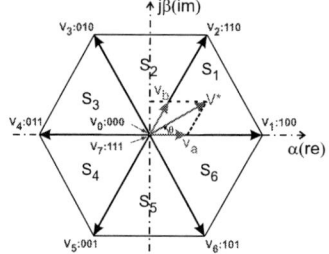

III. PARACONSISTENT LOGIC

Paraconsistent Logic (PL) belongs to the class of non-classical logics, whose structures differ from classical binary logic, opposing the law of non-contradiction. Due to this, the PL can be adapted to act as theoretical support for the algorithms that make up the sequence detector system optimized for the commutation of the inverter. The main characteristic of Paraconsistent Logic is the derogation from the principle of non-contradiction, the pillar of classical logic since it proposes to admit contradiction in its fundamentals without trivialization. In summary, one can consider that the following theoretical principles support the Paraconsistent Logic: - Given a theory (deductive) **T**, based on logic **L**, it is said to be consistent, if among its theorems there are no such that, one is the denial of the other; on the contrary, **T** is called inconsistent. The theory **T** is trivial if all the sentences (closed formulas) of its language are theorems; if this does not occur, **T** is non-trivial. If **L** is one of the standard logics, like the classical one, the **T** theory is trivial if and only if it is inconsistent [15].

A. Paraconsistent Annotated Logic

PL can be studied through an annotation concept in which it can have an associated reticulated. With the structuring of an annotated logic, the PL is capable of handling inconsistent information where the annotation constants are represented at the vertices of its reticulated which interpretation will connotations of extreme logic states propositions. In this work, a particular form of Paraconsistent Annotated Logic (PAL) is used with two values (μ, λ) to form the annotation. With the annotated paraconsistent logic annotation of two values (PAL2v) may obtain a representation of how the notes or evidence express knowledge of a proposition $P_{(\mu,\lambda)}$, using a grid formed by ordered pairs, such that: $\tau = \{(\mu, \lambda) \mid \mu, \lambda \in [0,1] \subset \Re\}$. The lattice four vertices associated with the paraconsistent annotated logic with two values (PAL2v) represented logic conditions, and present the evidence degrees that make the propositions $P_{(\mu,\lambda)}$, as can be seen in Fig. 4(a). Note that a proposition can have four maximum states set values (False, True, Inconsistent, and Indeterminate). On the other hand, the Unitary Square in the Cartesian Plane (USCP), defined by a linear transformation of the Hasse's lattice, is exposed in Fig. 4(b).

Fig. 4. (a) Finite lattice of Hasse; (b) Unitary Square in the Cartesian Plane. Adapted [15]

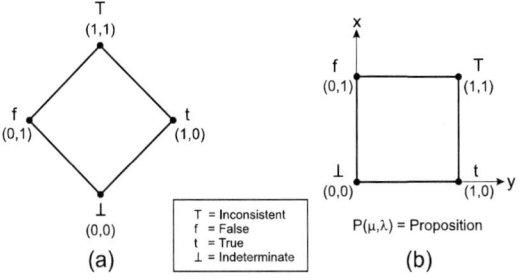

B. Interpretation of Paraconsistent Annotated Logic of two values

For a better representation of an annotation, and also for the practical use of the lattice PALv2 in the treatment of uncertainties, some algebraic interpretations involving a Unitary Square in the Cartesian Plane and the representative lattice of PAL2v are made. The transformations between USCP and PAL2v is defined through three steps: scale change (transformation one - Tf_1), rotation (transformation two - Tf_2), and translation (transformation three - Tf_3). By making the composition of the three steps that generated the transformations ($Tf_3 \ \theta \ Tf_2 \ \theta \ Tf_1$) to obtain the linear transformation represented by 2,and nomenclature of PAL2v given by 3:

$$Tf_{(X,Y)} = (x - y, x + y - 1) \tag{2}$$

$$Tf_{(\mu,\lambda)} = (\mu - \lambda, \mu + \lambda - 1) \tag{3}$$

where, in the nomenclature of PAL2v, X is equal a μ and it is the degree of favorable evidence, and Y is equal λ and it is the degree of unfavorable evidence.

The Degree of Certainty (D_C) is the relationship between the degree of favorable evidence and degree of unfavorable evidence, and it is obtained by:

$$D_C = \mu - \lambda \tag{4}$$

On the other hand, the Degree of Contradiction (D_{ct}) is the obtained by:

$$D_{ct} = \mu + \lambda - 1 \tag{5}$$

Fig. 5(a) display the Unitary Square in the Cartesian Plane and Fig. 5(b) shows the PAL2v lattice. Note that the USCP situated the proposition between the degree of favorable evidence (μ) and the degree of unfavorable evidence (λ) and the intervals are [(0..1),(0..1)]. Until this moment, there are divided four regions through limits of the degree of Certainty and degree of contradiction and for a proposition only there are four logic states (Truth, of Falsehood, of Inconsistency or of Indefinite) [16] and [17].

Fig. 5. (a) Unitary Square in the Cartesian Plane; (b) PAL2v Reticulated. Adapted [15]

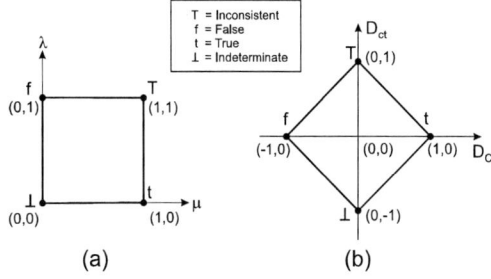

Based on a specialist, one can obtain twelve logical states; for this, one must create limits for the areas of certainty, contradiction, truth, and falsity. The limits are the value of certainty control superior (V_{CCS}), value of uncertainty control superior (V_{CtCS}), value of certainty control inferior (V_{CCI}), and value of uncertainty control inferior (V_{CtCI}). Fig 6 shows the twelve logical states for a specialist with 50 % of knowledge, for example. Note that now its have the four states extreme (True, False, Indeterminate, and Inconsistent) and eight states no-extreme (Indeterminate tending to false - $\perp \to F$, Indeterminate tending to true - $\perp \to V$, Inconsistent tending to false - $T \to F$, Inconsistent tending to true - $T \to V$, Quasi-true tending to Inconsistent - $Q_V \to T$, Quasi-false tending to Inconsistent - $Q_F \to T$, Quasi-true tending to Indeterminate - $Q_V \to \perp$, Quasi-false tending to Indeterminate - $Q_F \to \perp$)). It is possible to verify that certainty control limit or the uncertainty control limit can change of areas the extremes states.

C. Paraconsistent Analyzer Node

The method for the treatment of uncertainties using An algorithm presents Paraconsistent Annotated Logic denominated Paraconsistent Analysis Node (PAN). The PAN is capable of receiving evidence and supplying a certainty value accompanied by its interval of certainty ($\varphi(\pm)$). Therefore, it is the analysis that receives evidence degrees in their inputs and supplies two values; one that represents the Real Certainty Degree D_{CR} in the Interval of Certainty $\varphi(\pm)$. The D_{CR}

978-1-7281-4181-7/19 $31.00 © 2019 IEEE

Fig. 6. Twelve Logical States. Adapted [15]

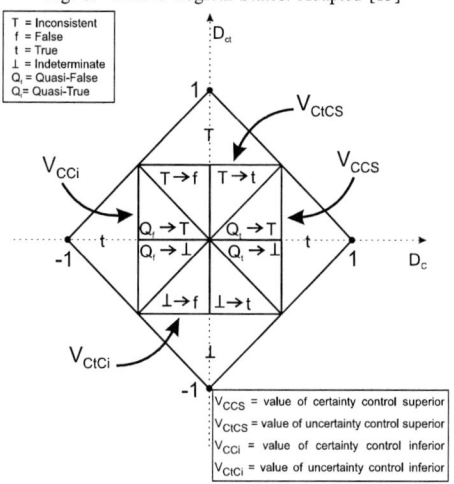

is considered a value of the Degree of Certainty DC free of contradiction effects and given by:

For $D_C > 0$,

$$D_{CR} = 1 - \sqrt{(1 - |D_C|)^2 + D_{ct}{}^2} \qquad (6)$$

and for $D_C < 0$,

$$D_{CR} = \sqrt{(1 - |D_C|)^2 + D_{ct}{}^2} - 1 \qquad (7)$$

The Interval of Certainty $\varphi(\pm)$ is calculate by:

$$\varphi = (1 - |D_{ct}| \qquad (8)$$

If D_{ct} is positive, the signal of Interval of Certainty is positive ($\varphi = +\varphi$), otherwise, the Interval of Certainty is negative ($\varphi = -\varphi$)

D. Algorithm of the Paraconsistent Analyzer Node

With the considerations presented here, it can be computed values using the obtained equations and build a system of paraconsistent analysis capable of offering a satisfactory answer derived from information collected from uncertain knowledge database. The algorithm of paraconsistent analysis (PAL2v) is described below:

1. Enter with the input values
μ */ favorable evidence Degree $0 \leq \mu \leq 1$
λ */ unfavorable evidence Degree $0 \leq \lambda \leq 1$
2. Calculate the Contradiction Degree
$D_{ct} = \mu + \lambda) - 1$
3. Calculate the Interval of Certainty
$\varphi = 1 - |D_{ct}|$
4. Determine the output signal
If $\varphi \leq 0.25$ Then do $_1 = 0.5$ and $S_2 = \varphi$:
Indefinite and go to the item 10
else go to the next step
5. Calculate the Certainty Degree
$D_C = \mu - \lambda$

6. Calculate the distance D
$D = \sqrt{(1 - |D_C|)^2 + D_{ct}{}^2}$
7. Calculate the real Certainty Degree
If $D_C > 0$ then $D_{Cr} = (1 - D)$
If $D_C < 0$ then $D_{Cr} = (D - 1)$
8. Determine the signaling of the Interval of Certainty
If $\mu + \lambda > 1$ then Signal positive $\varphi(\pm) = \varphi(+)$
If $\mu + \lambda < 1$ then Signal negative $\varphi(\pm) = \varphi(-)$
If $\mu + \lambda = 1$ then Signal zero $\varphi(\pm) = \varphi(0)$
9. Calculate the real Evidence Degree
$\mu_{Er} = \dfrac{D_{Cr} + 1}{2}$
10. Present the outputs
Do $S_1 = \mu_{Er}$ and $S_2 = \varphi(\pm)$
11. End

Fig.7 shows the symbolic representation of a PAN, where the PAN has two inputs; favorable Evidence Degree μ and unfavorable Evidence Degree λ of the regarding analyzed Proposition $P(\mu, \lambda)$ and two output signals of results; the Real Certainty Degree (μ_{Er}), and the Interval of Certainty symbolized by $\varphi(\pm)$.

Fig. 7. Paraconsistent Analyzer Node. Adapted [15]

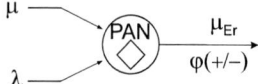

IV. PARACONSISTENT VECTOR MODULATION

The Paraconsistent Vector Modulation (PSVM) uses the vector modulation techniques and the fundamentals of the Paraconsistent Annotated Logic to optimize the switching pattern. The paraconsistent algorithm that promotes this strategy of modulation is presented below.

A. Algorithm of Modeling and Extraction of Degree of Evidence

To extract the degree of evidence of an entry in the paraconsistent annotated logic is necessary to model the input and normalize the values of the degree of evidence in the interval [0,1]. Similar to the activation functions of an artificial neural network, the modeling can be, for example, a ramp function, that below a minimum value the degree of evidence equal 0, above a maximum amount the output value of the degree of evidence is equal 1 and a linear variation occur between the maximum and minimum values. Note that others function can be adopted since they depend on the type of signal to be modeled. Thus, the algorithm of the ramp is described below follows:

1. Determine the maximum limit value (V_{MaxL}) to form the Discourse Universe.
2. Enter the minimum limit value (V_{MinL}).
3. Display the value of measured magnitude (V_X).

978-1-7281-4181-7/19 $31.00 © 2019 IEEE

4. Determine the degree of evidence favorable through the equations:

$$\mu \begin{cases} \dfrac{V_X - V_{MminL}}{V_{MaxL} - V_{MinL}} & if \ V_X \in [V_{MinL}, V_{MaxL}] \\ 1 & if \ V_X \geq V_{MaxL} \\ 0 & if \ V_X \leq V_{MinL} \end{cases} \quad (9)$$

5. Determine the degree of unfavorable evidence by completing the degree of evidence favorable

$$\lambda = (1 - \mu) \quad (10)$$

B. Algorithm of Paraconsistent Vector Modulation

The algorithm of PSVM starts with the construction of a paraconsistent evidence-detector system. The evidence-detection module uses the transformation alpha-beta coordinates, to obtain the degrees of evidence that are based on the paraconsistent logic fundamentals, two sources of information are using, and the algorithm makes a comparison between this sources.

The first source of information to be considered by the algorithm is the identification of the sector in which the Reference Vector (V^*) is inserted. The second source of data is the identification of the triangle within the hexagon with the determination of the Nearest Three Vectors (NTV). With this information, the standardized results in the form of degrees of evidence are presented through the paraconsistent algorithm in the PAN that will provide the result of the evidence analysis.

The values of the degrees of evidence obtained by the paraconsistent algorithms are compared for getting the cyclic switching ratio, which defines the optimized switching pattern.

The first source of information is obtained from the sectors that make up the hexagon and the location of the reference vector. For example, the reference vector in the triangle in sector S1 showed in Fig. 8. Its location is compared to the degrees of lag between the switching states V_1 (100) and V_2 (110). The state V_1 corresponding to zero degrees of deviation, wherein the analysis of PAL2v it means the value of the degree of evidence null ($\mu = 0$). The state V_2 corresponding to 60 degrees of deviation, wherein the analysis of PAL2v means the value of the degree of evidence maximum ($\mu = 1$). The second source of information is the limit reference voltage. The PAN returns the twelve paraconsistent logic states that is possible calculated the reference voltage module vector ($ModV^*$), reference voltage vector angle (θ), reference voltage optimized (V_1^*), and the three-phase reference voltage (V_{al}, V_{bl}, and V_{cl}). Finally, the dwell times are calculated, and the switches signal are generating.

C. Results of Paraconsistent Vector Modulation

With the PAL2v algorithms interconnected the vector modulation techniques, a paraconsistent vector modulation system was created for application in a two-level inverter. The method called Para-ModVetorLPA2v was written in LabView and shipped in the FPGA of the NI sb-RIO-9607 Controller with a NI-9683 (General Purpose Inverter Controller RIO.Mezzanine

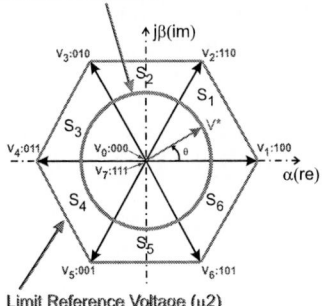

Fig. 8. Sources of Information for Determining Degrees of Evidence.

Card) board. The CompactRIO NI sb-RIO 9607 single-board RIO consists of a controller with 667 MHz dual-core CPU, Zynq-7020 FPGA, 512 MB DRAM, 512 MB storage. In the Hardware-in-the-loop (Typhoon-HIL 402), a three-phase Inverter was implemented with DC bus voltage of 512 VDC and 30 kW. The three-phase resistive load applied at the conversion output. The Fig. 9 shows this setup developed with HIL-402, NI sb-RIO 9607, and NI-9683.

Fig. 9. Setup System (HIL-402, NI sbRIO-9607, and NI-9683).

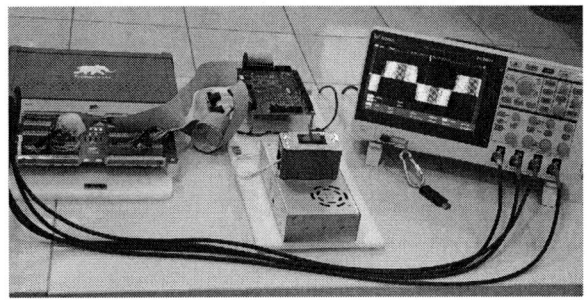

Fig. 10 shows the waveform of line-to-line voltage output from line "ab" at the inverter implemented in HIL-402 with PAL2v algorithm.

Fig. 11 shows the waveform of phase current output from phase "a" at the inverter implemented in HIL-402 with PAL2v algorithm. Phase current in output inverter is the 120.4 A (RMS). THD present in signal is 18.57 %.

Fig. 12 shows the waveform of phase current output from phase "a" at the inverter implemented in HIL-402 with SVPWM algorithm. THD present in signal is 23.48 %.

When compared to Fig. 11 and Fig. 12 it is possible to notice the good performance that the paraconsistent vector modulation presents.

D. Conclusions

In this paper was presented a method of vector modulation structured in the fundamentals of the Paraconsistent Logic to effect switching sequence optimization for Inverters. This

978-1-7281-4181-7/19 $31.00 © 2019 IEEE

type of approach is innovative and brings advantages over the techniques known to date because the hybrid algorithms allow greater flexibility in the detection of evidence related to the location of the reference vector and its amplitude. With this detection being expressed through degrees of evidence ranging from 0 to 1, it was possible to use the PAL2v Algorithms that can adapt logical treatments to contradictory signals. In this way, we obtain results that attenuate the effects of the harmonics producing a better optimization of the sequence of switches. From these fundamentals, a logical structure was implemented, composed of interconnected algorithms, thus forming a Paraconsistent Vector Modulation System called Para-ModVetorLPA2v. To verify the operation and to validate the algorithms of the System, several simulations were made with the LabView Software, and the results were very promising, for future work will be used the algorithms developed in this research to use this type of Paraconsistent Vector approach in modulation for other kinds of Inverters.

Fig. 10. Result of line voltage output from line "ab" at the inverter implemented in HIL-402 with PAL2v algorithm.

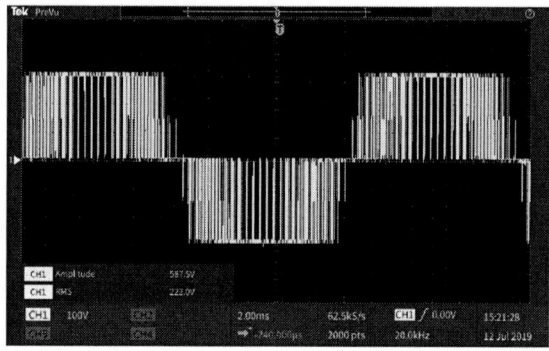

Fig. 11. Result of phase current output from phase "a" at the inverter implemented in HIL-402 with PAL2v algorithm.

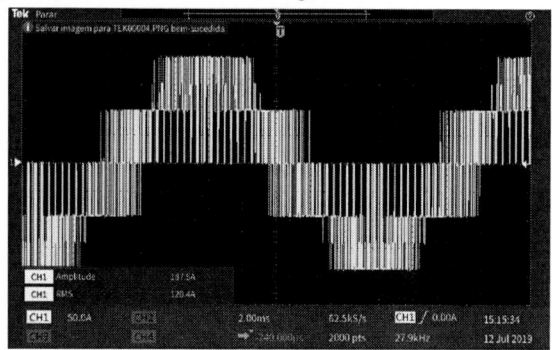

Fig. 12. Result of phase current output from phase "a" at the inverter implemented in HIL-402 with SVM.

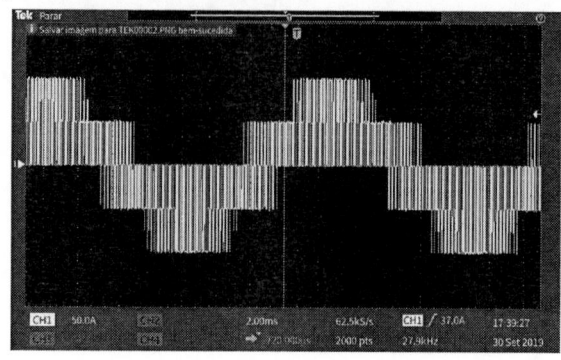

REFERENCES

[1] R. Teodorescu, M. Liserre, and P. Rodrigues, *Grid Converters for Photovoltaic and Wind Power Systems*. United Kingdom: John Wiley & Sons, Ltd., 2011.

[2] L. G. Franquelo, J. I. Leon, and E. Dominguez, "Recent advances in high-power industrial applications," *IEEE Transactions on Industrial Electronics*, 2010.

[3] S. Kouro, M. Malinowski, K. Gopakumar, J. Pou, L. G. Franquelo, B. Wu, J. Rodriguez, M. A. Prez, and J. I. Leon, "Recent advances and industrial applications of multilevel converters," *IEEE TRANSACTIONS ON INDUSTRIAL ELECTRONICS*, vol. 57, pp. 2553–2580, 08 2010.

[4] A. C. D. Ricciotti, "Modulação descontínua para conversores com pernas em paralelo magneticamente acopladas," 2017.

[5] K. Ilves, A. Antonopoulos, S. Norrga, and H. Nee, "A new modulation method for the modular multilevel converter allowing fundamental switching frequency," *IEEE Transactions on Power Electronics*, vol. 27, pp. 3482–3494, Aug 2012.

[6] G. Falahi, W. Yu, and A. Q. Huang, "Thd minimization of modular multilevel converter with unequal dc values," in *2014 IEEE Energy Conversion Congress and Exposition (ECCE)*, pp. 2153–2158, Sep. 2014.

[7] H. S. Patel and R. G. Hoft, "Generalized techniques of harmonic elimination and voltage control in thyristor inverters: Part i–harmonic elimination," *IEEE Transactions on Industry Applications*, vol. IA-9, pp. 310–317, May 1973.

[8] P. N. Enjeti, P. D. Ziogas, and J. F. Lindsay, "Programmed pwm techniques to eliminate harmonics: a critical evaluation," *IEEE Transactions on Industry Applications*, vol. 26, pp. 302–316, March 1990.

[9] P. D. Ziogas, "The delta modulation technique in static pwm inverters," *IEEE Transactions on Industry Applications*, vol. IA-17, pp. 199–204, March 1981.

[10] S. R. Bowes and Yen-Shin Lai, "The relationship between space-vector modulation and regular-sampled pwm," *IEEE Transactions on Industrial Electronics*, vol. 44, pp. 670–679, Oct 1997.

[11] M. A. Boost and P. D. Ziogas, "State-of-the-art carrier pwm techniques: a critical evaluation," *IEEE Transactions on Industry Applications*, vol. 24, pp. 271–280, March 1988.

[12] H. Pinheiro, F. Botteron, C. Rech, L. Schuch, R. F. Camargo, H. L. Hey, H. A. Grundlin, and J. R. Pinheiro, "Modulação space vector para inversores aalimentados em tensão: Uma abordagem unificada," *Controle e automação Sociedade Brasileira de Automática*, vol. 16, pp. 13 – 24, 03 2005.

[13] Z. Xu, R. Li, H. Zhu, D. Xu, and C. H. Zhang, "Control of parallel multiple converters for direct-drive permanent-magnet wind power generation systems," vol. 27, no. 3, pp. 1259–1270, 2012.

[14] D. G. Holmes and T. A. Lipo, *Pulse Width Modulation for Power Converters: Principles and Practice*. NY, 1 ed., 10 2003.

[15] J. I. da Silva Filho, G. Lambert-Torres, and J. M. Abe, *Uncertainty Treatment Using Paraconsistent Logic - Introducing Paraconsistent Artificial Neural Networks*, vol. 211 of *Frontiers in Artificial Intelligence and Applications*. IOS Press, 2010.

[16] J. i. D. Silva, A. Rocco, M. C. Mario, and L. F. Ferrara, "Annotated paraconsistent logic applied to an expert system dedicated for supporting in an electric power transmission systems re-stablish," in *2006 IEEE PES Power Systems Conference and Exposition*, pp. 2212–2220, Oct 2006.

[17] J. I. Da Silva Filho and A. Rocco, "Power systems outage possibilities analysis by paraconsistent logic," in *2008 IEEE Power and Energy Society General Meeting - Conversion and Delivery of Electrical Energy in the 21st Century*, pp. 1–6, July 2008.

Techniques of Solar Irradiance Estimation from Datasheet Information of Photovoltaic Panels

1st Itaiara F. Carvalho
Department of Electric Engineering (DEE)
Universidade Federal de Campina Grande (UFCG)
Campina Grande, Brazil
itaiara.carvalho@ee.ufcg.edu.br

2nd Maurício B. R. Correa
Department of Electric Engineering (DEE)
Universidade Federal de Campina Grande (UFCG)
Campina Grande, Brazil
mbrcorrea@dee.ufcg.edu.br

Abstract—Due to the increasing proportion of photovoltaic production in the electric generation mix, it is important to know the capacity of photovoltaic generation to estimate the efficiency of the solar power plant in any condition and thus to predict the point where the maximum power is converted. The process of converting solar energy into electrical energy is directly influenced by the level of irradiance and temperature of the photovoltaic module. In this way, equipment and tools that enable the acquisition of such data contribute directly to the knowledge of the efficiency of the photovoltaic system. While the acquisition of temperature values in photovoltaic panels is easy to implement, the same does not occur for irradiance values, where the costs of the sensors and their calibration are some of the problems faced. With the objective of exploring and comparing irradiance estimation techniques, the present work presents four solar radiation estimation algorithms, based on PV panel datasheet.

Index Terms—Photovoltaic Energy, Irradiance Estimator, Irradiance.

I. INTRODUCTION

A growing participation of photovoltaic systems in the world energy matrix is becoming increasingly evident. This is due not only to the already intrinsic advantages of renewable sources, but also due to the low environmental impact presented, as well as being a viable alternative to be applied in rural areas and difficult to access.

The estimation of real-time photovoltaic (PV) generation potential assists in the development of self-consumption strategies, [1], [2], local control of energy systems, [3], [4] and creation of datasets that are widely used to infer local behavior of photovoltaic power generation.

Environmental factors such as temperature and solar irradiance have a significant effect on the output power of PV array. To increase the overall efficiency of a PV plant, it is an important task to force the PV array to function in its maximum power point (MPP) condition. Although the MPP is unique for a uniform and constant climatic condition, it will be altered by variations in the solar irradiance and the PV array temperature. Due to the high correlation between these environmental factors and the generation of energy, it is important that there be tools for measuring or predicting these climatic factors. While the acquisition of temperature values in the PV array is easy to implement, for the irradiance the costs

of the sensors and their calibration are some of the problems faced.

The pyranometer is a device commonly used to measure solar irradiance, however, there are some negatives that hinder its acquisition and use, such as its high cost, and the fact that sensors can return measurements located at a specific point does not correspond to the actual average irradiance of the PV plant.

In monitored PV installations, measured DC voltage, current and temperature can be used for the elaboration of algorithms that return estimates of irradiance and output power. The technique of estimation of irradiance, besides eliminating extra costs, avoids the complexity of installation and maintenance of irradiance sensors. In scenarios with large PV plants, the irradiance estimation technique eliminates the mentioned problems related to the use of sensors and is of great importance for purposes such as monitoring, detection of faults in one or more panels of the plant, thus contributing to the increased efficiency of the PV plant.

In this paper four irradiance estimation techniques are explored and compared. The estimators are based on the following PV models: Simplified Model of the Photovoltaic Panel [5], represented by a current source, an anti-parallel diode and the series resistance (the presence of the shunt resistance is ignored); One-Diode Five-Parameter Model [6], in which the shunt resistance is considered; and also by a Modified Model [7] in which the anti-parallel diode is replaced by an external control current source.

The paper is organized as follows: Section II introduces the Simplified Model of the Photovoltaic Panel. Section III introduces the One-Diode Five-Parameter Model. Section IV describes the four methods choosed to estimate the irradiance. Section V shows and discusses the main results. Section VI draws the main conclusions.

II. SIMPLIFIED MODEL OF THE PHOTVOLTAIC PANEL

The simplified model of a photovoltaic panel presented here, Fig. 1, is composed of a current source whose output is directly proportional to the light irradiating the module, in parallel with a diode modeled by Shockley's exponential characteristic curve [5], [8].

978-1-7281-4181-7/19 $31.00 © 2019 IEEE

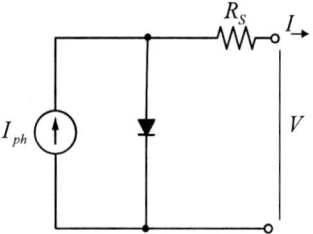

Fig. 1. Simplified Model of the Photovoltaic Panel.

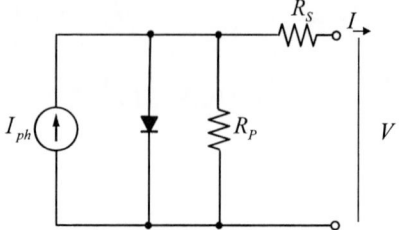

Fig. 2. Five parameter circuit model of a PV cell.

In this model the dependence of the temperature of the photocurrent I_{ph} and the saturation current of the diode is considered. A series resistor R_s was included, but shunt resistance was not considered [9]. The model represented can then be described by (1) [10]:

$$V(I, G, T_c) =$$
$$= V_T(T_c) ln \left[\frac{I}{I_{ph}(G, T_c)} + \left(1 - \frac{I}{I_{ph}(G, T_c)}\right) e^{\frac{V_{oc}(G, T_c)}{V_T}} \right]$$
$$- R_s I, \tag{1}$$

where I is the output current and V is the output voltage of the photovoltaic module; G is the solar radiation intensity incident on the PV panel surface; T_c the PV cell temperature; I_{ph} and V_{oc} are respectively the, photocurrent and the open circuit voltage of the module; V_T is the thermal voltage of the diode of the equivalent circuit; and R_s is the series resistance.

Equation (1) shows that four parameters are required to characterize the model: I_{ph}, V_{oc}, V_T, and R_s. Most of the time these parameters are not clearly available in the datasheet, however, there are several ways of inferring them (analytically or numerically) using the information provided by manufacturers [11] [12] or extracting them from experimental results [13] . In this work the parameters: I_{ph}, V_{oc}, and V_T were determined using the following equations [10]:

$$I_{ph}(G, T_c) \cong I_{sc}(G, T_c) = I_{sc0} \frac{G}{G_0} [1 + \alpha(T_c - T_0)] \tag{2}$$

$$V_{oc}(G, T_c) = V_{oc0} [1 + \beta(T_c - T_0)] + V_{T0} ln \left(\frac{G}{G_0}\right) \tag{3}$$

$$V_T(T_c) = V_{T0} \frac{T_c}{T_0}, \tag{4}$$

where I_{sc0} and V_{oc0} are the short-circuit current and the open-circuit voltage, α and β its respective temperature coefficients at the Standard Test Conditions (STC) and V_{T0} is the equivalent thermal voltage in the same conditions.

III. ONE-DIODE FIVE-PARAMETER MODEL

The one-diode five-parameter model [14], Fig. 2, describes the relationship between the DC voltage (V), current (I),

solar irradiance (G), and cell temperature (T_c) for a PV panel composed of N_p cells in parallel and N_s in series as [15]:

$$f(V, I, T_c, G) = 0 = I_{ph}(T_c, G)N_p + -\frac{V + R_s I \frac{N_s}{N_p}}{R_p G \frac{N_s}{N_p}} - I$$
$$+ -I_0(T_c)N_p \left[e^{\left(q \frac{V + R_s I \frac{N_s}{N_p}}{n k T_c N_s}\right)} - 1 \right], \tag{5}$$

where I_{ph} is the irradiance current; I_0 is the cell reverse saturation current; q is the electron charge ($q = 1,602$ x 10^{-19} C); n is the diode ideality factor; k is the Boltzmann constant ($k = 1.3806503$ x 10^{-23} J/K); T_c is the PV cell temperature; and R_s and R_p are respectively the series resistance and the shunt resistance.

The five parameters: n, R_s, I_{ph}, I_0 e R_p depend on the reference parameters at STC ($T_0 = 25\ °C$ and $G_0 = 1000$ W/m^2) :

$$n = n_{ref} \tag{6}$$

$$R_s = R_{s,ref} \tag{7}$$

$$I_{ph} = \frac{G}{G_0} [I_{ph,ref} + \alpha(T_c - T_0)] \tag{8}$$

$$I_0 = I_{0,ref} \left[\frac{T_c}{T_0}\right]^3 e^{\left[\frac{E_{g,ref}}{k T_0} - \frac{E_g}{k T_c}\right]} \tag{9}$$

$$R_p = \frac{G_0}{G} R_{p,ref}, \tag{10}$$

$$E_g = 1.17 - 4.73 \times 10^{-4} \frac{T_c^2}{T_c + 636} \tag{11}$$

where G corresponds to the solar radiation, while G_0 is the solar irradiation at STC; T_c is the temperature of the cell, while T_0 is the temperature at STC; $E_{g,ref}$ is the band gap energy (eV) at STC; and α corresponds to the temperature coefficient of the short-circuit current at STC.

The five unknown parameters at STC are: n_{ref}, $R_{s,ref}$, $I_{ph,ref}$, $I_{0,ref}$, and $R_{p,ref}$.

978-1-7281-4181-7/19 $31.00 © 2019 IEEE

The five-parameter model makes use of G_0, T_0, α, as well as other information available in the datasheet, such as: open circuit voltage (V_{oc0}), short circuit current (I_{sc0}), voltage (V_{M0}) and current (I_{M0}) at maximum power (MPP) this procedure is detailed in [16], where the five-parameter model is reduced to a two-parameter model by improving the efficiency of the solution algorithm.

IV. ESTIMATORS OF SOLAR IRRADIANCE

A. Irradiance Estimator: The Simplified Four-Parameter Model

The simplified four-parameter model (Section II) can be easily inverted and an estimation of solar irradiance incident on the PV cells can be performed inverting (1) and considering that $e^{\frac{V_{oc}(G,T_c)}{V_T(T_c)}}$ and $V_T \cong V_{T0}$

$$
\begin{aligned}
G_{est}(T_c, V, I) &= \\
&= \frac{G_0 I}{I_{sc0}\left[1 + \alpha(T_c - T_0)\right]} + G_0 e^{\frac{V + R_s I - V_{oc}[1 + \beta(T_c - T_0)]}{V_{T0}}},
\end{aligned} \quad (12)
$$

where V_T is the thermal voltage while V_{T0} is the thermal voltage at STC.

B. Irradiance Estimator: The Five-Parameter Model

The Five-Parameter Model (Section III) describes the relationship between voltage (V), current (I), solar irradiance (G) and cell temperature (T_c) for a PV panel [15]. Replacing the equations (6) - (11) into (5) and solving for G the irradiance estimator is obtained (13)

$$
\begin{aligned}
&G_{est} = \\
&= \frac{I + I_{0,ref} N_p \left[\frac{T_c}{T_0}\right]^3 e^{\left(\frac{E_{g,ref}}{kT_0} - \frac{E_g}{kT_c}\right)} \left[e^{q \frac{V + R_{s,ref} I \frac{N_s}{N_p}}{nkT_c N_s}} - 1\right]}{\frac{1}{G_0}\left[N_p(I_{ph,ref} + \alpha(T_c - T_0)) - \frac{V + R_{s,ref} I \frac{N_s}{N_p}}{R_{p,ref} \frac{N_s}{N_p}}\right]}
\end{aligned} \quad (13)
$$

where N_p is the number of cells in series and N_s is the number of string in parallel.

C. Irradiance Estimator: Model Independent of PV Cell Temperature

In this section is presented an irradiance estimator whose equation is independent of the temperature of the photovoltaic cell [17], [18].

The currents and voltages measured in the PV array are subject to variations due to climatic conditions such as temperature and irradiance, where $DeltaI$ e $DeltaV$ are [19]:

$$
\Delta I = \alpha \left(\frac{G}{G_0}\right)(T_c - T_0) + \left(\frac{G}{G_0} - 1\right) I_{sc0} \quad (14)
$$

$$
\Delta V = -\beta (T_c - T_0) - R_s \Delta I \quad (15)
$$

The irradiance can then be estimated by solving simultaneously (14) and (15)

$$
G_{est} = G_0 \left(\frac{I_{sc0} + \Delta I}{I_{sc0} - \frac{\alpha}{\beta}(R_s \Delta I + \Delta V)}\right) \quad (16)
$$

Changes in the current and voltage of the PV array due to variations in the climatic conditions in relation to its STC values can also be calculated as [19]:

$$
\Delta I = I - I_{M0} \quad (17)
$$

$$
\Delta V = V - V_{M0} \quad (18)
$$

where I and V are real-time measurements of the current and voltage of the PV panel, which are the most likely values of MPP under real environmental conditions. In addition, I_{M0} and V_{M0} are the maximum current and voltage values of the photovoltaic module at STC, found in the datasheet.

D. Irradiance Estimator: Modified Circuit Model

This estimator makes use of a modified circuit introduced in [7] that replaces the anti–parallel diode by an external control current source. The generalized model that relates the voltage of the PV array (V) and the current (I) is given by [20]:

$$
I = I'_{ph} - I'_0 \left[exp\left(\frac{q(V + IR'_s)}{N_s nkT_c}\right) - 1\right] - \frac{V + IR'_s}{R'_p} \quad (19)
$$

The values I'_{ph}, I'_0, R'_s e R'_p are related to their cell–specific counterparts as follows: $I'_{ph} := N_p I_{ph}$; $I'_0 := N_p I_0$; $R'_s = \frac{N_s}{N_p} R_s$ e $R'_p := \frac{N_s}{N_p}$; where I_{ph} is the irradiance current; I_0 is the cell reverse saturation current; and R_s and R_p are respectively the series resistance and the shunt resistance. The parameters depend on G, T_c and certain reference values at STC, that is, G_0, T_0, $I_{ph,ref}$, $I_{0,ref}$, $R_{p,ref}$, $R_{s,ref}$ e n_{ref}, as described by (6)-(10).

Assuming that I, V and T_c are available for measurement, the estimation of the irradiance G_{est} follows from algebraic manipulations of the characteristic $I - V$ (19) plus (6) - (10). In addition, considering the dependence of the shunt resistance (R_p) on the irradiance ie $R_p = R_{p,ref}$,

$$
G_{est} = G_0 \frac{I + I'_{0,ref}\left(\frac{T_c}{T_0}\right)^3 e^{K_1} K_2 + \frac{V + IR'_s}{R'_p}}{I'_{ph,ref} + \alpha(T_c - T_0)} \quad (20)
$$

$$
K_1 = \frac{E_{g,ref}}{kT_0} - \frac{E_g}{kT_c} \quad (21)
$$

$$
K_2 = e^{\frac{q(V + IR'_s)}{N_s nkT_c}} - 1. \quad (22)
$$

V. RESULTS

A measurement and data acquisition system was installed in the Laboratory of Industrial Electronics and Drive of Machines (LEIAM) of the Universidade Federal de Campina Grande (UFCG) to perform tests.

The tests were performed using a 130 W 36 cells polycrystalline photovoltaic module manufactured by Kyocera (model KC130TM) and the *SolarModule LED-Flasher*, which can emit flashes with intensities of: 1000 W, 800 W, 600 W, 400 W, and 200 W. The flash duration time during power measurement is given in milliseconds (ms), whose default value is 200 ms. After the measurement is performed, the curve diagram $I - V$ is displayed and information such as power at the point of maximum power (MPP), short circuit current, open circuit voltage, current and voltage at the point of maximum power, irradiance power, temperature characteristics, and other characteristics are displayed on the screen.

All tests were performed with the same intensity of flash and temperature (301.15 K). For all tests, the flash intensity of 600 W was chosen, which returned a measured value of 615 W/m^2. The voltage and current values measured with the *SolarModule LED-Flasher* equipment were also the same for all the tested estimators.

A. Irradiance Estimator: The Simplified Four-Parameter Model

Considering the procedures described for the irradiance estimation using the Simplified Four-Parameter Model, (12) was solved utilizing the software MatLab ®, Fig. 3.

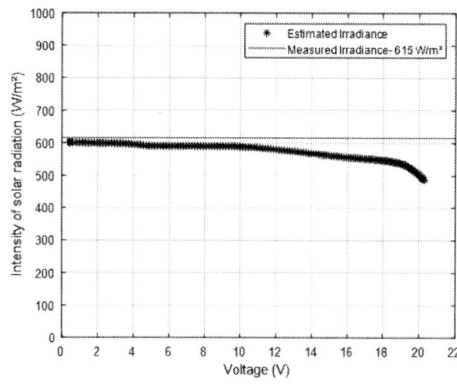

Fig. 3. Irradiance estimator: Simplified four parameter model.

In Fig. (3) it can be seen that (12) allows a good estimation of the incident solar radiation, except when the voltage approaches its open circuit value ($\approx 21\ V$), in this case the estimated irradiance gives results that differ substantially from the reference irradiance. This mismatch is due to the assumption $V_T \cong V_{T0}$.

B. Irradiance Estimator: The Five-Parameter Model

In order to experimentally verify the equation (13) the system used in the previous section, composed of a PV panel

(KYOCERA KC130TM) and the SolarModule LED-Flasher was considered.

Using the MatLab ®software and applying the found parameters [16] for the PV array (KYOCERA KC130TM) in (13) the irradiance estimations are obtained, Fig. 4.

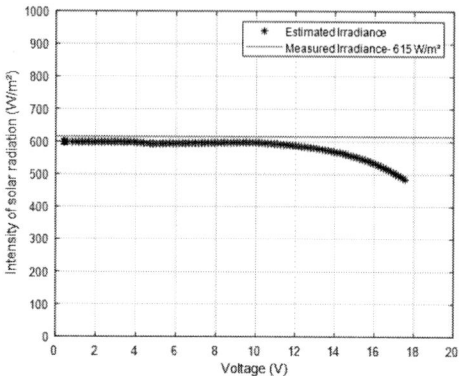

Fig. 4. Irradiance estimator: The five-parameter model.

The estimator based on the five parameter model (13) allows a good irradiance estimate up to 12 $V \sim 14\ V$, especially from 16 V it is possible to notice an increase of the estimated irradiance value. Thus, the estimator reliability is concentrated up to the possible maximum power voltage values.

C. Irradiance Estimator: Model Independent of PV Cell Temperature

Considering (16) which is independent of the panel temperature to estimate the irradiance, the irradiance estimation is performed, Fig 5.

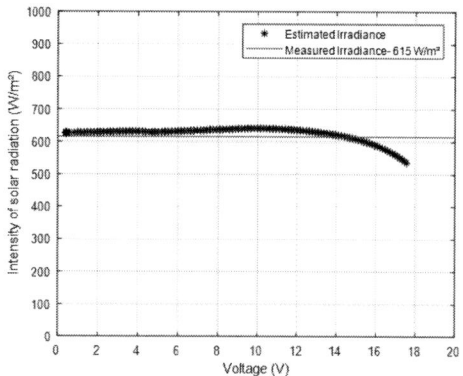

Fig. 5. Irradiance estimator: The five-parameter model.

Equation (16) generates a good estimation of solar radiation, especially up to the MPP condition, since in (17) and (18), I and V must be real-time measurements of the PV module current and voltage, which are the most likely MPP values under real environmental conditions.

D. Irradiance Estimator: Modified Circuit Model

The last estimator, based on the modified circuit model, was tested using (20) - (22), whose Fig. 6 illustrates the behavior of the estimator against the measured irradiance.

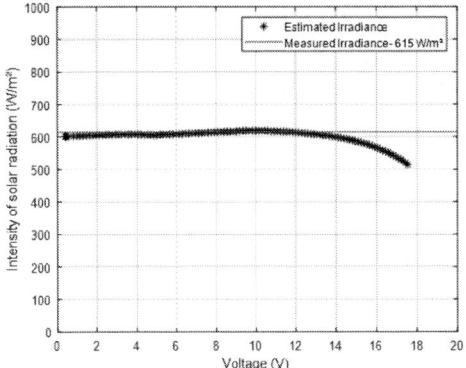

Fig. 6. Irradiance estimator: The five-parameter model.

This estimator offers the best estimate of irradiance when compared to the other three estimators, in the range up to $\approx 14\ V$. When the voltage values are increasing, the estimator obtains undesired values (as well as the other estimators), however the estimated irradiance was closer to the measured irradiance value than in the other estimators.

VI. CONCLUSION

In this paper it was possible to analyze four different estimators [8], [15], [18], [20] that provide irradiance values from datasheet information. The first irradiance estimator, *Irradiance Estimator: The Simplified Four-Parameter Model*, provided acceptable irradiance values compared to the actual value (measured irradiance). The advantage of this estimator lies in the simplicity of the model, which leads to fast processing algorithms. The second irradiance estimator, *Irradiance Estimator: The Five-Parameter Model*, different from the previous estimator, it considers the shunt resistance effect. For this estimator the estimated irradiance values also had a acceptable approximation compared to the measured values, however when the voltage approaches its open circuit value, the obtained irradiance values also deviate from the expected values. The last two estimators: *Irradiance Estimator: Model Independent of PV Cell Temperature* and *Irradiance Estimator: Modified Circuit Model* presented the best estimates among the four estimators. The *Irradiance Estimator: Model Independent of PV Cell Temperature* has the advantage of having a model that is independent of the PV cell temperature, thus dispensing with the use of temperature sensors. This estimator presented the best irradiance values around the MPP condition. Finally, the *Irradiance Estimator: Modified Circuit Model* also presents satisfactory irradiance estimates. Irradiance estimators are of great importance for purposes such as monitoring and detection of faults, in this way the four estimators analyzed here are presented as good alternatives for estimation of irradiance.

REFERENCES

[1] F. Sossan, A. M. Kosek, S. Martinenas, M. Marinelli and H. Bindner, "Scheduling of domestic water heater power demand for maximizing PV self-consumption using model predictive control," IEEE PES ISGT Europe 2013, Lyngby, 2013, pp. 1-5.

[2] R. Luthander, D. Widén, J.Palm, "Photovoltaic selfconsumption in buildings: A review," Appl. Energy, vol 142, pp. 80–94, March 2015.

[3] M. Hossain, T. Saha, N. Mithulananthan, H. Pota, "Robust control strategy for PV system integration in distribution systems," Appl. Energy, vol 99, pp. 355–362, November 2012.

[4] M. Nick, R. Cherkaoui and M. Paolone, "Stochastic day-ahead optimal scheduling of Active Distribution Networks with dispersed energy storage and renewable resources," 2014 IEEE Conference on Technologies for Sustainability (SusTech), Portland, OR, 2014, pp. 91-96.

[5] L. Cristaldi, M. Faifer, F.Ponci, M. Rossi, "Monitoring of a PV system: the role of the panel model," in Proc. Workshop on Applied Measurement for Power Systems, Aachen, Germany, 2011, pp.90–95.

[6] W. De Soto , S. Klein, and W. Beckman, "Improvement and validation of a model for photovoltaic array performance," Solar Energy, vol. 80, no. 1, pp. 78–88, 2006.

[7] H. Tian, F. Mancilla-David, K. Ellis, E. Muljadi, and P. Jenkins, "A cell-to-module-to-array detailed model for photovoltaic panels," Solar Energy, vol. 86, no. 9, pp. 2695 – 2706, 2012.

[8] L. Cristaldi, M. Faifer, M. Rossi and S. Toscani, "MPPT definition and validation: A new model-based approach," 2012 IEEE International Instrumentation and Measurement Technology Conference Proceedings, Graz, 2012, pp. 594-599.

[9] G. Walker, "Evaluating MPPT converter topologies using a MATLAB PV model", Journal of Electrical & Electronics Engineering, Australia, IEAust, vol.21, No. 1, 2001, pp.49-56.

[10] . Brofferio, L. Cristaldi, F. Della Torre and M. Rossi, "An in-hand model of photovoltaic modules and/or strings for numerical simulation of renewable-energy electric power systems," 2010 IEEE Workshop on Environmental Energy and Structural Monitoring Systems, Taranto, 2010, pp. 14-18.

[11] R. Khezzar, M. Zereg and A. Khezzar, "Comparative study of mathematical methods for parameters calculation of current-voltage characteristic of photovoltaic module," 2009 International Conference on Electrical and Electronics Engineering - ELECO 2009, Bursa, 2009, pp. I-24-I-28.

[12] E. Matagne, R. Chenni and R. El Bachtiri, "A photovoltaic cell model based on nominal data only," 2007 International Conference on Power Engineering, Energy and Electrical Drives, Setubal, Portugal, 2007, pp. 562-565.

[13] F. Adamo, F. Attivissimo and M. Spadavecchia, "A tool for Photovoltaic panels modeling and testing," 2010 IEEE Instrumentation & Measurement Technology Conference Proceedings, Austin, TX, 2010, pp. 1463-1466

[14] W. De Soto , S. Klein, and W. Beckman, "Improvement and validation of a model for photovoltaic array performance," Solar Energy, vol. 80, no. 1, pp. 78–88, 2006.

[15] E. Scolari, F. Sossan and M. Paolone, "Photovoltaic-Model-Based Solar Irradiance Estimators: Performance Comparison and Application to Maximum Power Forecasting," in IEEE Transactions on Sustainable Energy, vol. 9, no. 1, pp. 35-44, Jan. 2018.

[16] A. Laudani, F. Mancilla-David, F. Riganti-Fulginei, and A. Salvini, "Reduced-form of the photovoltaic five-parameter model for efficient computation of parameters," Solar Energy, vol. 97, pp. 122–127, 2013.

[17] A. Chikh and A. Chandra, "An Optimal Maximum Power Point Tracking Algorithm for PV Systems With Climatic Parameters Estimation," in IEEE Transactions on Sustainable Energy, vol. 6, no. 2, pp. 644-652, April 2015.

[18] E. Moshksar and T. Ghanbari, "Real-time estimation of solar irradiance and module temperature from maximum power point condition," in IET Science, Measurement & Technology, vol. 12, no. 6, pp. 807-815, 9 2018.

[19] A. Chikh and A. Chandra, "Adaptive neuro-fuzzy based solar cell model," in IET Renewable Power Generation, vol. 8, no. 6, pp. 679-686, August 2014.

[20] A. Laudani, F. R. Fulginei, A. Salvini, M. Carrasco and F. Mancilla-David, "A fast and effective procedure for sensing solar irradiance in photovoltaic arrays," 2016 IEEE 16th International Conference on Environment and Electrical Engineering (EEEIC), Florence, 2016, pp. 1-4.

Extra Reactive Power Analysis on a Distribution Grid with High Integration of PV Generation

Daniel C. Pompermayer, Caroline Marin, Mariana A. Mendes, Luann G. O. Queiroz, Luiz G. R. Tonini,
Murillo C. Vargas, and Oureste E. Batista
Postgraduate Program in Electrical Engineering, PPGEE
Federal University of Espírito Santo, UFES
Vitória, Brazil
daniel.pompermayer@ufes.br, caroline.m.lima@aluno.ufes.br, mariana.a.mendes@aluno.ufes.br,
luann.queiroz@aluno.ufes.br,luiz.tonini@aluno.ufes.br, murillo.vargas@aluno.ufes.br, oureste.batista@ufes.br

Abstract—The Federal University of Espírito Santo (UFES) has acquired a 5.441 kWp solar photovoltaic distributed generator, which will be installed over the roofs of the buildings of the Goiabeiras Campus in the city of Vitória, Espírito Santo, Brazil. Knowing that a big amount of active power will be injected in the Goiabeiras Campus' feeder, while the consumed reactive power will remain the same, the influence of the solar plant over the power factor has been studied. This paper employed data from a meteorological automatic station located in the campus and metering record data to estimate the power factor. Results allow to state that a significant decrease in power factor may be observed in hour intervals with big generation. Nevertheless, results show that possible charges due to extra reactive power are not significant.

Index Terms—Electric power systems, energy management, energy quality, photovoltaic systems, power factor.

I. Introduction

Worldwide power generation is concentrated in large-scale power plants connected to high voltage networks. Transmission lines carry energy over long distances to downstream substations which distribute electrical energy through radial distribution feeders, supplying final consumers [1].

In the last decades, there has been a rise of the so called distributed generation. Distributed Generation (DG) is a new electrical energy generation paradigm in which small-scale power plants connected to the radial distribution feeders are employed side-by-side with large-scale and far power plants. It is the reason of a growing use of renewable technologies and an expansion of the idea of decentralized production, increasing its importance and applicability [2].

Due to the good index of solar irradiance in Brazil throughout the year [3], the use of the solar energy to generate electrical energy through photovoltaic (PV) panels is the preferred DG source among Brazilians. In July 2019, about 90% of the DG plants were based on PV technology [4]. It represents 86.9% of the DG capacity installed in the country [4].

Despite the many advantages [5], the high insertion of PV DG in distribution feeders can cause changes in its voltage profile, in its level of network current, in its power flow and in its fault location and level [6]–[11]. Likewise, an increase in the injected active energy, while the consumed reactive energy remains unchanged, may cause a variation in the power factor of DG facilities [12].

Federal University of Espírito Santo (UFES) contracted on February 11th, 2019 the installation of 5.441 kWp of PV generation systems on grid at the Goiabeiras Campus, in Vitória, Espírito Santo, Brazil [13]. The university PV plant will supply approximately 70% of the electrical energy consumption of the Goiabeiras Campus.

The solar micro-generators that will be installed in the various buildings of the campus will make the university feeder a network with high insertion of DG, suffering the aforementioned effects. Among the expected effects, it is a significant change in the metered power factor due to the high injection of active power.

The general objective of this paper is to analyze the financial impacts of the changes in the power factor of UFES Goiabeiras Campus due to the solar PV plants.

This paper is divided in four sections: this introduction; an extra reactive power analysis section where the expected injected active power is estimated and related to historical consumption data, the new power factor is obtained and the financial effects of the power factor changes are modeled; a results section where the found impacts are presented and discussed; and a conclusion section pointing the study conclusions and its recommendations to future works.

II. Extra Reactive Power Analysis

Since the metered power factor is a ratio of the metered active electrical power and the metered apparent electrical power, as a mean to evaluate how it can be influenced by the DG of UFES' Goiabeiras Campus, the generated electrical energy must be estimated. It has been done employing the meteorological data from an automatic station located in the campus.

The analysis employs a daily temporal window methodology consisting of grouping data by hour. It is done in order to evaluate how the power factor varies within a single day. Differences between the days of a given month and days of another month are only considered as data variation.

978-1-7281-4181-7/19 $31.00 © 2019 IEEE

The new power factor estimation relies on also knowing the consumed active and reactive electrical energy. The following subsections describe the methodology employed for the estimation of the generated electrical energy and for the estimation of the power factor.

A. Estimation of the Generated Electrical Energy

The electrical energy generated by a photovoltaic generator can be estimated from the solar radiation on the surface of the equipment and is given by (1).

$$E_{PV}(kWh) = E_{sol}(kWh/m^2) \times A(m^2) \times \eta_{PV} \times \eta_{PCU} \quad (1)$$

where E_{PV} is the electrical energy generated by the PV generator in an hour, E_{sol} is the incident solar energy within one hour interval, A is the area of the PV panel, η_{PV} is the efficiency of the generator, and η_{PCU} is the efficiency of the power conditioning unit [14].

The National Institute of Meteorology (INMET), an agency of the Ministry of Agriculture, Livestock and Supply of Brazil, holds the Vitoria-A612 automatic station at the Goiabeiras Campus of UFES [15]. Among its available data, the station provides the solar radiation incident within one hour intervals, which may be used to estimate the electrical energy generation at Goiabeiras Campus.

Since the solar radiation data from Vitoria-A612 station is given in kJ/m^2, (1) must be modified by

$$E_{PV}(kWh) = \frac{E_{sol}(kJ/m^2)}{3600} \times A(m^2) \times \eta_{PV} \times \eta_{PCU} \quad (2)$$

The efficiency η_{PV} of a real PV panel varies due to its surface temperature. Such an effect is quantified by the Power Temperature Coefficient (PTC), which is a percentage loss in the efficiency of a PV panel for every degree Celsius of temperature above the standard one [14]. In order to include such an efficiency loss, (2) must be modified by

$$E_{PV}(kWh) = \frac{E_{sol}(kJ/m^2)}{3600} \times A(m^2) \times \eta_{PCU} \times \eta_{PV} \times$$
$$\left\{ 1 + \frac{PTC \times [T_{surface}(^\circ C) - T_{std}(^\circ C)]}{100} \right\} \quad (3)$$

Fig. 1 presents a box plot illustrating the solar radiation data from Vitória-A612 station. The presented data set is grouped by hour and contains from 697 to 705 different values of solar radiation to each one of the 24 hours of the day. Data was collected from February 22nd, 2015 to March 24th, 2017.

UFES' solar plant will be composed by 10.660 JA Solar JAM60S01-315/PR PV panels of 315 Wp and 6.720 JA Solar JAM60S01-310/PR PV panels of 310 Wp controlled by 55 Goodwe GW30KLV-MT inverters of 30 kVA and 82 Goodwe GW35KLV-MT inverters of 36 kVA. The efficiency (η) of each equipment is shown in Table I.

Despite having two different nominal powers, both models of PV panels have a surface area of 1.6628 m² and a PTC of

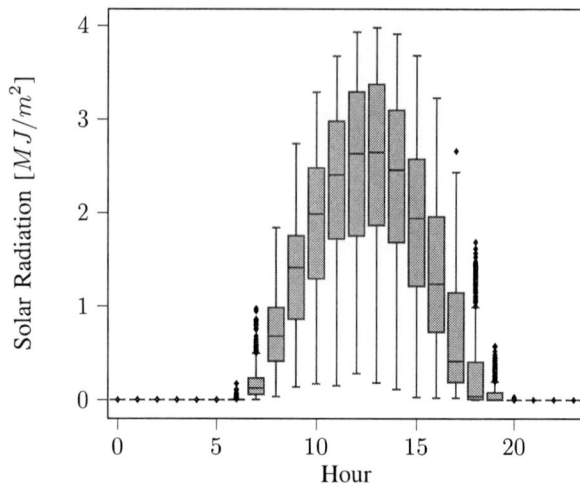

Fig. 1. Solar radiation data from Vitoria-A612 automatic station

TABLE I
EFFICIENCY OF EMPLOYED EQUIPMENT

Equipment	Efficiency (η)
310 Wp Panel	18.9%
315 Wp Panel	18.6%
30 kVA Inverter	98.3%
36 kVA Inverter	98.5%

$-0.36\%/^\circ C$, and both were tested under a operation temperature of 25°C.

Employing an operation temperature of 48.5°C [16] and data from Fig. 1 and from Table I in (3), it's been estimated what would be the generated electrical energy in the observation period. Each hourly estimated generated electrical energy was up limited to the maximum value of electrical energy which must be delivered by each inverter, which is numerically equal to its rated power.

Fig. 2 presents the estimated generated electrical energy. In the figure, the data set is grouped by hour, having as many values per hour as the radiation data set.

Fig. 2 shows that the energy generation is manly concentrated in the interval between 8h and 16h. Within this interval, one may observe a high variation in the generated energy, demonstrated by an inter-quartile range minimum of 573 kWh at 8h and an inter-quartile range maximum of 1.535 kWh at 12h. Such a variation may be explained by the strong relation between the solar energy generation and the weather conditions and seasons.

In order to verify the validity of the estimated generated electrical energy obtained from (3), one may compare the estimated productivity of the employed PV system with the productivity presented by the official data. According to the National Brazilian Electricity Regulatory Agency (ANEEL) [17], the productivity of the PV systems in the state of Espírito Santo is of the order of 1.450 kWh/kWp/year.

Equation (4) presents the estimation of the productivity from

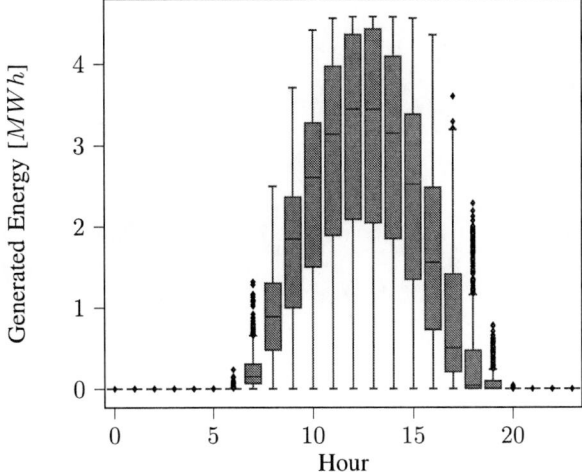

Fig. 2. Estimated generated electrical energy

the electrical energy obtained from (3).

$$Pr\left(\frac{kWh}{kWp \cdot year}\right) = \qquad (4)$$

$$= \frac{\sum\limits_{i=1}^{n} E_{PV_i}}{n_{310} \times 0.310(Wp) + n_{315} \times 0.315(Wp)}$$

where n is the number of hours in the estimation, n_{310} is the number of employed 310 Wp PV panels and n_{315} is the number of employed 315 Wp PV panels.

The productivity obtained by the application of the data in Fig. 2 related to the year of 2016 into (4) is of the order of 1.420 kWh/Kwp/year, i.e. 2% smaller than the official value. The found value is then reasonable and the energy estimation is valid.

B. Estimation of the Power Factor

In Brazil, medium voltage fed facilities must keep a high power factor. The national regulation states that a charge will be applied whenever the demand for reactive power is beyond the limits [18]. In the state of Espírito Santo, the limits are:

- between 0h and 6h: any value of inductive power factor or capacitive power factor greater than 0.92;
- between 6h and 0h: any value of capacitive power factor or inductive power factor greater than 0.92;

Equation (5) presents the power factor calculation from the generated and consumed energy.

$$pf = \frac{E_C - E_{PV}}{\sqrt{(E_C - E_{PV})^2 + (E_{RI} - E_{RC})^2}} \qquad (5)$$

where E_C is the consumed energy, E_{PV} is the generated energy, E_{RI} is the consumed reactive energy and E_{RC} is the furnished reactive energy.

UFES' administration made available a metering record having the consumption data from February 22nd, 2015 to March 24th, 2017. Fig. 3 presents data of active and reactive energy consumption grouped by hour. From 760 to 763 values of energy is available to each one of the 24 hours of the day.

A 300 kVAr capacitor bank was recently installed in Goiabeiras Campus and has also been considered in the calculations.

C. Estimation of the Financial Impacts

The effects of a possible low power factor occurrence are more visible when its financial charges are presented. According to the resolutions in force [18], the applied charge for the extra reactive power is given by (6).

$$E_{RE} = \sum_{T=1}^{n}\left[EEAM_T \times \left(\frac{f_R}{f_T} - 1\right)\right] \times VR_{ERE} \qquad (6)$$

where E_{RE} is the applied charge for extra reactive power, $EEAM_T$ is the metered active energy in an hour interval, f_R is the reference power factor (0.92), f_T is the metered power factor and V_{RE} is the extra reactive energy fee.

The regulations in force in Brazil do not state how the generated energy must be treated in the computation of the extra reactive power demand. Nevertheless, the local energy distributor do not apply any charges when the facility is injecting power to the network. On the other hand, the extra reactive power is charged when the facility is supplied by the grid.

Using the distributive property of the product in (6), it is possible to find the charges within an hour interval.

$$E_{RE_T} = EEAM_T \times \left(\frac{f_R}{f_T} - 1\right) \times VR_{ERE} \qquad (7)$$

An even better understanding would be achieved by presenting the low power factor charges as a percentage of the economy due to the solar plant. Equation (8) presents the computation of the charges percentage within an hour interval.

$$E_{RE_T}(\%) = \frac{EEAM_T \times \left(\frac{f_R}{f_T} - 1\right) \times VR_{ERE}}{E_{PV} \times (TE + TUSD)} \qquad (8)$$

where TE is the active energy fee and $TUSD$ is the distribution system utilisation fee.

In Brazil, the applied electricity fees depends on each facilities' service voltage and, for those which are served in a medium voltage level, whether the consumption is registered in the rush hours. Rush hours occurs from 18h to 21h in the business days.

In the state of Espírito Santo, the current electricity fees are defined by the Homologation Resolution 2.432 of August 7th, 2018 [19]. Table II presents the fees applied to the facilities, UFES included, which are served in a voltage level from 30 kV to 44 kV.

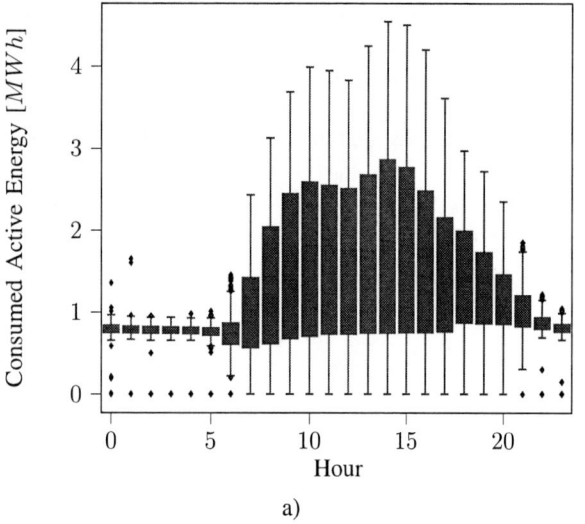
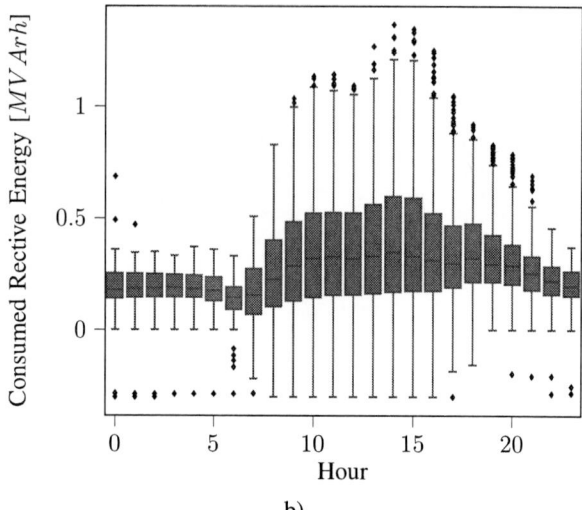

a)   b)

Fig. 3. Consumed active (a) and reactive (b) electrical energy

TABLE II
CURRENT ELECTRICITY RATES IN THE STATE OF ESPÍRITO SANTO

| Rate | Value (R$/kWh) | |
	Conventional	Rush
TE	0.28876	0.46341
TUSD	0.08153	1.29819
VR_{ERE}	0.30331	

TABLE III
UNDER REGULATORY POWER FACTOR OCCURRENCES

Hour	New PF	Old PF
0	0.41%	0.41%
1	0.27%	0.27%
2	0.27%	0.27%
3	0.27%	0.27%
4	0.14%	0.14%
5	0.14%	0.14%
6	0.55%	0.55%
7	10.70%	0.41%
8	35.25%	0.55%
9	49.93%	0.41%
10	46.36%	0.27%
11	39.37%	0.27%
12	34.71%	0.14%
13	36.49%	0.14%
14	39.64%	0.14%
15	42.39%	0.14%
16	38.13%	0.14%
17	27.98%	0.27%
18	14.68%	0.14%
19	10.43%	0.00%
20	0.14%	0.14%
21	0.14%	0.14%
22	0.27%	0.27%
23	0.27%	0.27%

III. RESULTS AND DISCUSSION

Consumption data and estimated generation data were applied in (5) in order to generate from 728 to 729 values of estimated power factor for each one of the 24 hours of the day. Fig. 4 presents two box plots containing the absolute value of the new estimated power factor and the old power factor obtained from the consumption data set.

Fig. 4a) shows that the old power factor was a well-behaved data with an inter-quartile range varying from 0.017 to 0.030, which means that 50% of the power factor data is located within ±3% of the median value. Nevertheless, Fig. 4b) shows a significant increase in the data variation within the interval from 7h to 17h. The inter-quartile range within varies from 0.02 to 0.30.

Table III presents the percentage of occurrences of power factor less than the regulatory power factor of 0.92 with and without the PV plant operation. The values in the table means that, if the employed radiation and consumption patterns occur again after putting the PV plant into operation, a power factor smaller than the regulatory one will occur with the given percentage.

Applying the estimated values in (8), the charges due to extra reactive power was found as a percentage of the estimated economy in the same hourly interval. Fig. 5 presents a cumulative histogram where X-axis presents the percentage

of occurrences of percentage charges smaller than the charges on Y-axis.

Fig. 5 shows that, despite the significant reduction in power factor, 90% of the hour intervals presented charges smaller than 15% of the estimated economy in the same interval. This behavior is not unexpected since the extra reactive power charges are proportional to the metered energy which tends to be small in the intervals with a big estimated generation.

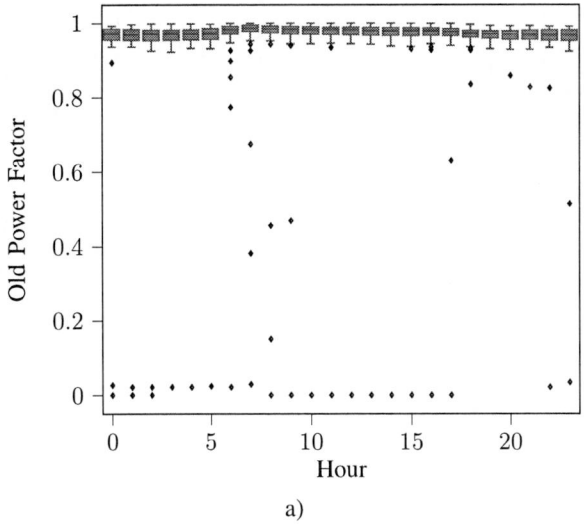

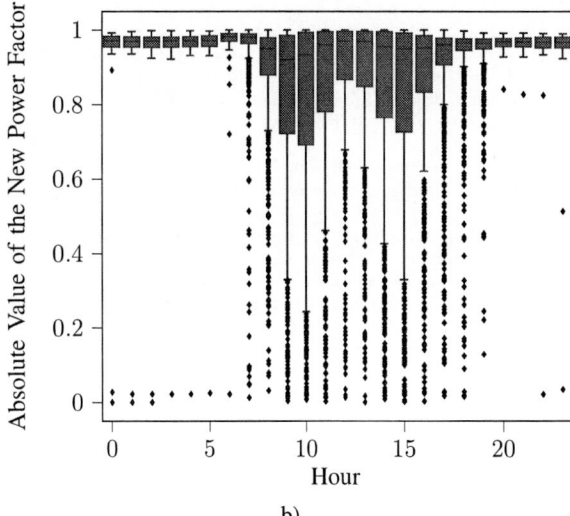

Fig. 4. Old (a) and new (b) power factor

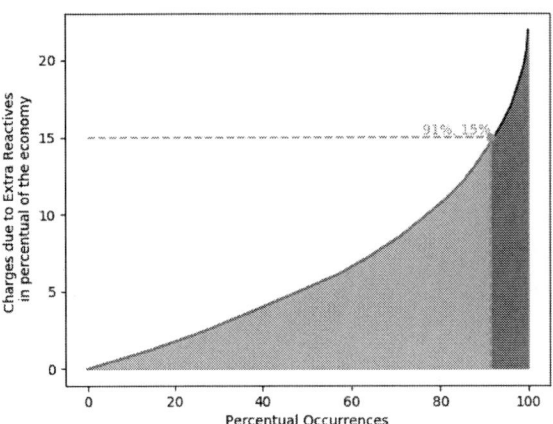

Fig. 5. Cumulative occurrences of percentage charges

IV. CONCLUSION

This paper investigated the influence of the PV generator plant over the power factor of UFES' Goiabeiras Campus.

The study employed data from a meteorological automatic station located in Goiabeiras Campus to estimate the generated electrical energy and a metering record consumption data to evaluate what would be the campus power factor with the solar plant in operation.

Results allow to state that if the radiation and consumption patterns of February 22nd, 2015 to March 24th, 2017 occur again after putting the PV into operation, the campus power factor can decrease significantly in moments of energy generation. Nevertheless, since the extra reactive power charges is proportional to the metered energy, which tends to be small

in the intervals of big generation, the charges showed to be smaller than 15% in 90% of the occurrences. None of the estimated proportional charges showed to be greater than 25%.

Future works may consider to employ stochastic methods to better predict the future behavior. The radiation pattern, as well as the consumption pattern, may be represented as a random variable and the power factor, as well as the proportional charges, may be modeled by a density probability function.

UFES' Goiabeiras Campus has its own distribution electrical feeder. This paper considered the power factor only in the main meter. It's also important to observe how the DG affects the power factor all over the feeder and to evaluate whether it is interesting to generate some reactive power in the PV inverters.

REFERENCES

[1] F. Martín-Martínez, A. Sánchez-Miralles, M. Rivier, and C. F. Calvillo, "Centralized vs distributed generation. A model to assess the relevance of some thermal and electric factors. Application to the Spanish case study," *Energy*, vol. 134, pp. 850–863, Sep. 2017.

[2] S. Montoya-Bueno, J. I. Muñoz-Hernández, and J. Contreras, "Uncertainty management of renewable distributed generation," *Journal of Cleaner Production*, vol. 138, pp. 103–118, Dec. 2016.

[3] P. G. V. Sampaio and M. O. A. González, "Photovoltaic solar energy: Conceptual framework," *Renewable and Sustainable Energy Reviews*, vol. 74, pp. 590–601, Jul. 2017.

[4] Agência Nacional de Energia Elétrica (ANEEL), "Consumer units with distributed generation [Unidades consumidoras com Geração Distribuída]," 2019. [Online]. Available: http://www2.aneel.gov.br/scg/gd/GD_Fonte.asp

[5] G. Pepermans, J. Driesen, D. Haeseldonckx, R. Belmans, and W. D'haeseleer, "Distributed generation: Definition, benefits and issues," *Energy Policy*, vol. 33, no. 6, pp. 787–798, Apr. 2005.

[6] R. A. Walling, R. Saint, R. C. Dugan, J. Burke, and L. A. Kojovic, "Summary of distributed resources impact on power delivery systems," *IEEE Transactions on Power Delivery*, vol. 23, no. 3, pp. 1636–1644, Jun. 2008.

[7] P. Barker and R. D. Mello, "Determining the impact of distributed generation on power systems. Part I. Radial distribution systems," in *2000 Power Engineering Society Summer Meeting (Cat. No.00CH37134)*. IEEE, Jul. 2000, pp. 1645–1656.

[8] K. Balamurugan, D. Srinivasan, and T. Reindl, "Impact of distributed generation on power distribution systems," *Energy Procedia*, vol. 25, pp. 93–100, 2012.

[9] M. A. Mendes, "Análise dos Impactos da Alta Inserção de Geração Distribuída Fotovoltaica na Proteção de Sobrecorrente Temporizada Análise dos Impactos da Alta Inserção de Geração Distribuída Fotovoltaica na Proteção de Sobrecorrente Temporizada," *UFES*, p. 92, 2018.

[10] M. A. Mendes, M. C. Vargas, O. E. Batista, and D. S. L. Simonetti, "A review on the methods for mitigate the impacts of photovoltaic distributed generation in power systems protection," in *2018 Simposio Brasileiro de Sistemas Eletricos (SBSE)*. IEEE, May 2018.

[11] M. C. Vargas, M. A. Mendes, and O. E. Batista, "Faults Location Variability in Power Distribution Networks with High PV Penetration Level," in *2018 13th IEEE International Conference on Industry Applications (INDUSCON)*, Jan. 2019, pp. 459–466.

[12] F. L. Albuquerque, G. G. Caixeta, A. J. Morais, and S. B. Silva, "Análise da Curva de Carga em Prédios Públicos com Sistemas Fotovoltaicos Conectados à Rede dotados de Compensação de Potência Reativa," in *2012 Simpósio Brasileiro de Sistemas Elétricos (SBSE)*, Mar. 2012.

[13] Universidade Federal do Espírito Santo (UFES), "Processo nº 23068.044202/2018-11: Instalação de sistema de geração de energia elétrica solar captada por placas fotovoltaicas a serem instaladas nas coberturas das edificações existente." [Online]. Available: https://protocolo.ufes.br/{#}/documentos/2299313/

[14] R. Singh and R. Banerjee, "Estimation of rooftop solar photovoltaic potential of a city," *Solar Energy*, vol. 115, pp. 589–602, May 2015.

[15] National Institute of Meteorology [Instituto Nacional de Meteorologia], "Automatic Stations [Estações Automáticas]," 2019. [Online]. Available: http://www.inmet.gov.br/portal/index.php?r=estacoes/estacoesAutomaticas

[16] G. M. Villalva, *Energia Solar Fotovoltaica: conceitos e aplicações*. São Paulo: Érica, 2012.

[17] Agência Nacional de Energia Elétrica (ANEEL), "Technical Note 0056/2017-SRD/ANEEL [Nota Técnica 0056/2017-SRD/ANEEL]," p. 26, 2017. [Online]. Available: http://www.aneel.gov.br/documents/656827/15234696/Nota+T%C3%A9cnica_0056_PROJE%C3%87%C3%95ES+GD+2017/38cad9ae-71f6-8788-0429-d097409a0ba9

[18] ——, "Normative Resolution 414 of November 9th, 2010 [Resolução Normativa 414 de 9 de novembro de 2010]," p. 148, 2010. [Online]. Available: http://www2.aneel.gov.br/cedoc/ren2010414comp.pdf

[19] ——, "Homologation Resolution 2432 of August 7th, 2018 [Resolução Homologatória de 7 de agosto de 2018]," p. 11, 2018. [Online]. Available: http://www2.aneel.gov.br/cedoc/reh20182432ti.pdf

Alternative FCS-MPC concepts for cascade free motor speed control

Filipe Fernandes, José de Oliveira, Ademir Nied
Electrical Engineering Department
Santa Catarina State University
Joinville, SC, Brazil
filipe02.fernandes@gmail.com, jose.oliveira@udesc.br, ademir.nied@udesc.br

Arthur G. Bartsch
Electrotechnology Department
Federal Institute of Santa Catarina
Jaraguá do Sul, SC, Brazil
arthur.bartsch@ifsc.edu.br

Abstract—This paper provides alternative FCS-MPC algorithms for cascade-free motor speed control and compares with conventional FCS-MPC. The proposed algorithms uses the concept of virtual vectors combined with a modulation to reduce the ripple caused by conventional FCS-MPC. The difference in the proposed algorithms is related with the search of the virtual vectors in order to reduce computational cost, memory use or processing. Simulation results are provide showing the benefits of the proposed algorithms.

Index Terms—Cascade free, speed control, FCS-MPC.

I. Introduction

Model-based predictive control (MPC) is a set of control techniques increasingly used in power converters and drivers that has gained attention in the research community [1]. Even though MPC feature high computational burden [2], it has a flexible control scheme that can easily deal with multivariable case, system constraints, and nonlinearities [3].

Powerful microprocessors that perform large amount of calculations in a reduced time are available with a reduced cost, what makes the MPC's computational burden a minor problem to some applications [4].

The fundamentals of MPC consist in using the modeled process to predict its future behavior within a predefined time. An optimization problem that includes the control objectives, the predicted variables, and the possible constraints of the system is solved, evaluating the control actions to be applied [5].

A very attractive strategy mainly for drivers and power converters is the finite-control-set model predictive control (FCS-MPC) [6], [7]. Since this kind of systems can have a finite set of inputs [8], as the switches' on or off states [9].

In each sample time, many future values of the controlled variable are predicted with the equations of the modeled system and the current system conditions, each value corresponding to a input in the finite control set. Then each value is evaluated through a cost function and the input that generate the minor cost is chosen and applied to the system, leading its way to the desired reference [10].

In the conventional FCS-MPC the switch states are used to make the prediction, what leads to a non constant switching frequency and a low resolution in the control action, important issues for power electronics converters [11]. The low resolution, can cause a non zero steady state error and a noisy current

dynamic [12]. The inaccuracy in the parameters can also lead to a non zero steady-state error [13], [14], as usually FCS-MPC does not include an integral action.

To avoid this drawbacks the FCS-MPC can be adapted with the pulse width modulation (PWM) technique, which has a fixed frequency, and the control actions are the duty cycle instead of the switch states [15]. The PWM improves the current performance in plants with a low inductance, as the DC motor. With a low inductance the current is allowed to have abrupt changes, for non constant and reduced switching frequency as the FCS-MPC one. In this situations the PWM strategy delivers a less noisy dynamic due to higher switching frequency.

Some strategies to this adaptation are proposed by [16] in order to improve properties in steady state of the controlled variables. These concepts were applied to a nonlinear first order fluid flow process. Therefore, it was possible to ally the good properties of FCS-MPC, as easy constrain treatment and nonlinearity treatment, with smaller ripples in the controlled variables. If a convex control set MPC (CCS-MPC) was used, for the sampling time needed, the soft constraint treatment can not be imposed.

Therefore, this paper intends to apply the strategies of FCS-MPC proposed by [16] to a second order system and analyze its behavior. For this, will be used the DC-motor, a second order system commonly used as benchmark and often fed by power converters. Also, the original algorithms were change for including integral action and, consequently, rejecting load disturbance. This change was performed using a disturbance estimator [17]. Another contribution is the some improvements over the original algorithms proposed by Bartsch *et al* [16], and detailed better in the proper section. Moreover, the proposed strategy is a direct-speed control, also known as cascade-free control, since it deals directly with speed and current at same scheme [18], [19].

Section II presents the DC-motor modeling for further use in the simulation and prediction. The conventional and the proposed FCS-MPC algorithms theoretical background are exposed in Section III. The methodology, simulation results and further discussions are presented in Section IV. Finally, the conclusion is presented in Section V

978-1-7281-4181-7/19 $31.00 © 2019 IEEE

Algorithm 1 Algorithm for conventional FCS-MPC

1: Measure $\omega(k)$
2: Set $J_{\min} \leftarrow \infty$
3: **for** $i \leftarrow 0$ to n_s **do**
4:　　Calculate $\omega(k+1)$ due to u_i applied in k
5:　　Evaluate the cost functional J
6:　　**if** If $J < J_{\min}$ **then**
7:　　　　$J_{\min} \leftarrow J$
8:　　　　$u_{i_{\min}} \leftarrow u_i$
9:　　**end if**
10: **end for**
11: Set $u_k(k) = u_k^\star(k) \leftarrow u_{i_{\min}}$
12: Wait for next sampling period

II. MOTOR MODEL

According to [20] the voltage in the DC-motor armature can be modeled by a series circuit. Which leads in discrete time the following equations:

$$i_a(k+1) = \left(1 - \frac{R_a \Delta t}{L_a}\right) i_a(k) - \frac{K_a \Delta t}{L_a}\omega(k) + \frac{\Delta t}{L_a}u_a(k) \tag{1}$$

$$\omega(k+1) = \left(1 - \frac{B \Delta t}{J}\right)\omega(k) + \frac{K_t \Delta t}{J}i_a(k) - \frac{\Delta t}{J}\tau_L(k). \tag{2}$$

With these equations is possible to see that u_a has no direct influence in ω. It's seen that u_a changes i_a, which in the next iteration will changes ω, this behavior characterize this system as a second order system. This phenomena is consequence of the adopted discretization and it is valid for small sampling periods Δt.

III. PROPOSED CONTROL THEORETICAL BACKGROUND

As said before the FCS-MPC will be focused. In the first moment, the conventional FCS-MPC algorithms will be explained, and the improvements proposed by [16], will be shown in the other subsections. The improvements involving dynamic set will not be analyzed here, giving more space to a deeper study of the other techniques proposed.

A. Classic FCS-MPC

As described in [4], the idea behind the FCS-MPC consists in testing how a certain range of inputs will influence the output and apply the input value that best fits the desired reference. This idea can be easily seen in the Algorithm 1.

One problem of the classic FCS-MPC is that its necessary to evaluate all values of u_i in order to find the ideal u_k. For that [16] propose a dynamic search.

A dynamic set range consists in change the evaluated control action set $\mathbb{U}_k$ according to process current conditions. Since is not reasonable to test all u_k possible values, is possible to change, the $\mathbb{U}_k$ that will be tested. This strategy is an extra control design resource and can be chosen by the designer [16].

Algorithm 2 Algorithm for FSDS

1: Define $n_s \gg N_s$
2: $i_0 \leftarrow i_{\min} - N_s$
3: **if** $i_0 < 0$ **then**
4:　　$i_0 \leftarrow 0$
5: **end if**
6: $i_f \leftarrow i_{\min} + N_s$
7: **if** $i_f > n_s$ **then**
　　$i_f \leftarrow n_s$
8: **end if**
9: **for** $i \leftarrow i_0$ to i_f **do**
10:　　Predict ω due to u_i
11:　　Evaluate the cost functional J
12:　　**if** If $J < J_{\min}$ **then**
13:　　　　$J_{\min} \leftarrow J$
14:　　　　$i_{\min} \leftarrow i$
15:　　**end if**
16: **end for**
17: Set $u_k(k) = u_k^\star(k) \leftarrow u_{i_{\min}}$
18: Wait for next sampling period

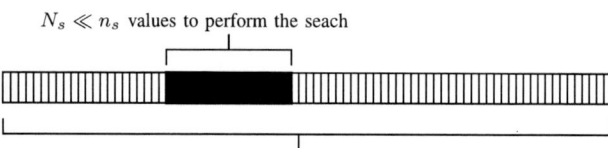

Fig. 1. FSDS algorithm concept

B. Fixed set with dynamic search (FSDS)

The first method proposed by [16] consists in using a search horizon N_s. Given a large number of possible states n_s. A search of size N_s is made in the neighborhood of the current control action. According to the Algorithm 2 (other parts, involving measurement *et cetera*, are omitted).

The number of evaluations, in the worst case, is $2N_s + 1$ searches, with the approach presented before. High N_s values imply higher computational loads for the algorithm. Fig. 1 helps to understand the FSDS algorithm concept: there is a main set with a large amount (n_s) of possibilities while the search for the optimal value is performed in only $N_S \ll n_s$ cases.

C. Fixed set with dynamic subset (FSDSs)

The second method proposed by [16] is to use a fixed values set with larger steps. Besides, there is a second set, called as dynamic subset. A search is made in the larger set than a second search is made in the smaller subset, around the value found in the first search. This approach is useful to reduce the oscillations in the control action and, consequently, in the output. A possible algorithm to proceed with this strategy is shown in the Algorithm 3.

Note that the subsets can be implemented recursively, without the need of a vector storage. But even with the storage

Algorithm 3 Algorithm for FSDSs
___
1: Define $n_s > n_{s,b}$
2: $du \leftarrow (u_1 - u_0)/n_{s,b}$ //*defines the small search length*
3: Proceed the first search (in u_i) //*conventional FCS algorithm*
4: $u_{\min} \leftarrow u(i_{\min} - 1)$
5: $u_{\max} \leftarrow u(i_{\min} + 1)$
6: $u_j = u_{\min}$
7: **while** $u_j < u_{\max}$ **do**
8:     Predict ω due to u_j
9:     Evaluate the cost functional J
10:    **if** If $J < J_{\min}$ **then**
11:       $J_{\min} \leftarrow J$
12:       $u_{i_{\min}} \leftarrow u_j$
13:    **end if**
14:    $u_{j+1} \leftarrow u_j + du$
15:    $j \leftarrow j + 1$
16: **end while**
17: Set $u_k(k) = u_k^\star(k) \leftarrow u_{i_{\min}}$
18: Wait for next sampling period
___

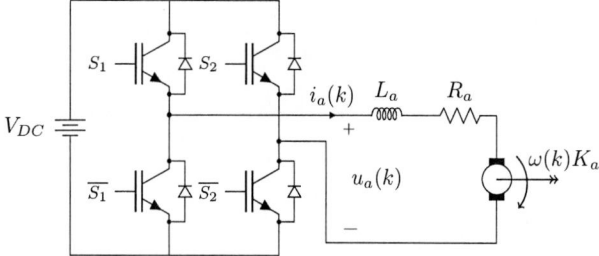

Fig. 3. Circuit model and four-quadrant converter

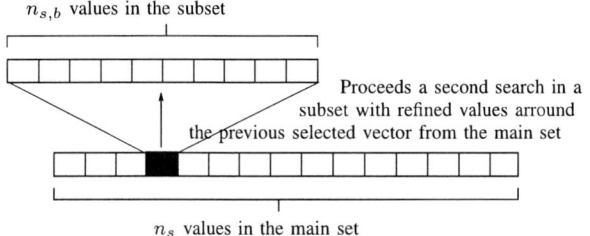

Fig. 2. FSDSs algorithm concept

in the memory, this procedure reduces the total amount of processor storage in relation to FSDS algorithm. Also, to avoid vector superposition, it is possible to define $du \leftarrow (u_{k,\max} - u_{k,\min})/(n_{s,b} + 1)$ and define $du_0 \leftarrow u_{k,\min} + du$. These procedures would improve the voltage resolution and an improvement in relation to the original algorithm presented in [16].

Fig. 2 helps to understand the FSDSs algorithm concept: there is a main set with a small amount (n_s) of possibilities, even though this amount is greater than the conventional algorithm. After the algorithm choose a value in the main set, a second search is performed in a related subset (with $n_{s,b}$ size) for refining the optimal solution value.

It is important to note that in this approach the search horizon N_s is equal to the desired number of states n_s.

IV. CONTROL DESIGN AND SIMULATION RESULTS

The evaluated DC-Motor has the parameters K_a = 0.3091 Vs, R = 16.525 Ω, L = 65.5 mH, J = 20.2108 μKgm^2 and B_m = 0.622171 Nms. The PWM dead time nonlinearity is neglected. The established sampling time T_s = 2 ms is small enough for a effective control. The simulation step was set to Δt = 1 μs. The PWM

frequency operation is 10 kHz. It is assumed that a four quadrant converter feeds the DC-motor. Fig. 3 shows the model circuit with the four-quadrant converter.

The tests were performed in 1.0 s of total simulation time. The transitions were made with ramps instead of steps to avoid confusions in the torque estimation. For that the reference through time was set as follows: the reference angular speed goes from zero to 20.0 rad/s with a ramp of 33.3 ms, then stay in 20.0 rad/s until 3.3 s, then goes from 20.0 rad/s to 40.0 rad/s again with a ramp of 33.3 ms, then stay in 40.0 rad/s until the end of the simulation. In 0.5 s a load of 71.4 mNm was connected to the rotor.

A random noise of 20 mA was added to the current measurement to a better proximity to the real processes.

The cost function employed was:

$$J(\omega(k)) = (\omega^*(k) - \omega(k))^2 \tag{3}$$

were $\omega^*(k)$ is the reference angular speed in the instant k.

The load disturbance is supposed unknown by the controller. For that [17] uses a torque estimator. Based on that, it was develop a disturbance estimator defined as

$$\hat{\tau}_L(k) = \hat{\tau}_L(k - 1) + K_{est}T_s(\omega^*(k) - \omega(k)) \tag{4}$$

where K_{est} is the estimation gain. In this work, K_{est} = 0.2 Nm. Note that any speed error changes the estimator value, including references changes. Therefore, ramp speed references are preferable for the estimator giving better estimations.

All the simulations were made in a CPU with an Intel®Core™i3-2350M @ 2.30 GHz processor, 4.00GB of RAM, and Windows 10 Home 64 Bits. The software used for the simulations and figures plot was developed in a script code, in Python language (version 3.7.0).

For further results comparison, the conventional FCS-MPC was simulated. The angular speed can be seen in Fig. 4, the current in Fig. 5 and the control action in Fig. 6.

In the conventional FCS-MPC, the possible control actions are the switches states (on or off), that leads, with the four quadrant converter, to three states -30 V, 0 V, 30 V, as can be seen in Fig. 6. This control action has a low resolution and a variable frequency (limited to half of the sampling frequency) what generates a current and angular speed ripple, as the angular speed oscillates around the reference but never reach it.

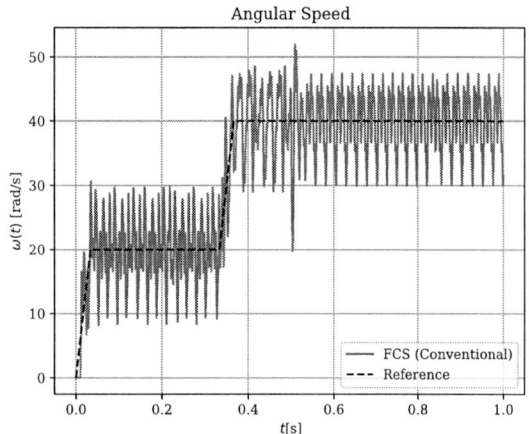

Fig. 4. Angular speed dynamic response

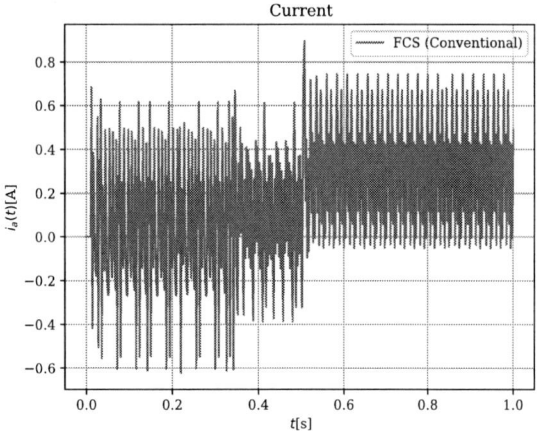

Fig. 5. Current dynamic response

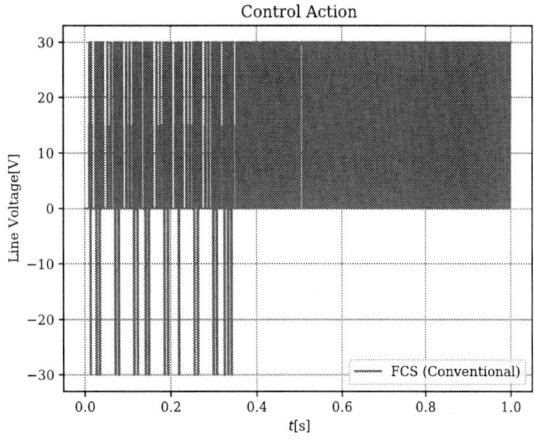

Fig. 6. Duty Cycle dynamic

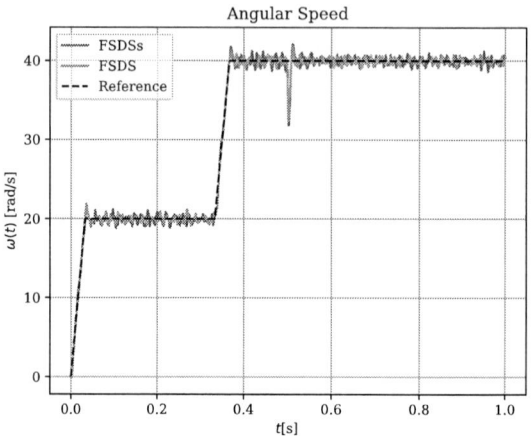

Fig. 7. Angular speed dynamic response

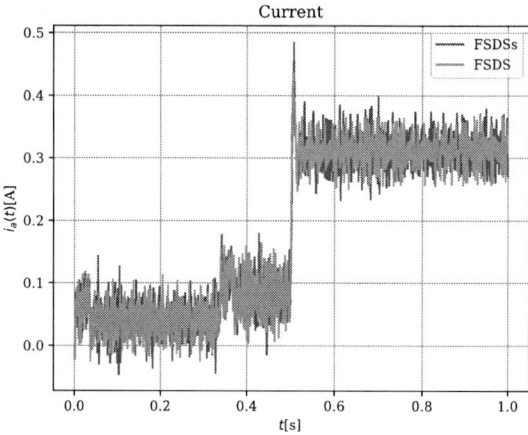

Fig. 8. Current dynamic response

Fig. 7, Fig. 8, and Fig. 9 presents the results for both strategies, FSDS and FSDSs. With a prediction horizon of $N = 2$ steps ahead, as this is a second order process, two steps are the minimum that can be used. For the FSDS was used $n_s = 48$ and $N_s = 12$, for the FSDSs was used $n_s = 4$ and $n_{s,b} = 8$. With these values, a fair comparison of the strategies computational burden can be made as both make 12 searches in each sample time, with 48 total voltages possibilities.

In both strategies, FSDS and FSDSs, the angular speed follows the transition ramps and presents a small ripple when the reference stabilizes if compared with conventional FCS-MPC. However, the mean speed value is the same of the reference, since there is a load torque estimation in each case (see Fig. 10). In the speed references transitions, there are some estimation errors, since the estimator estimates the acceleration torque instead the load torque. As smooth the ramp transition is, lower is this error.

In steady state, the angular speed and current dynamic

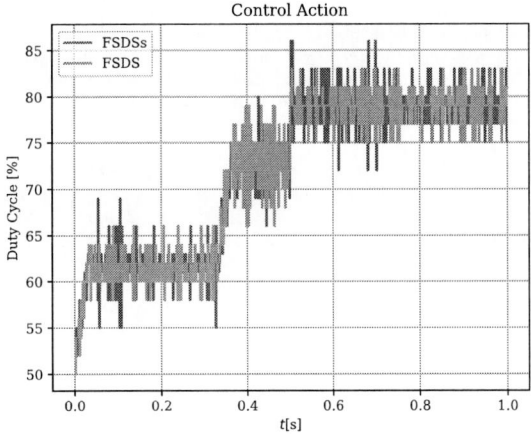

Fig. 9. Duty Cycle dynamic

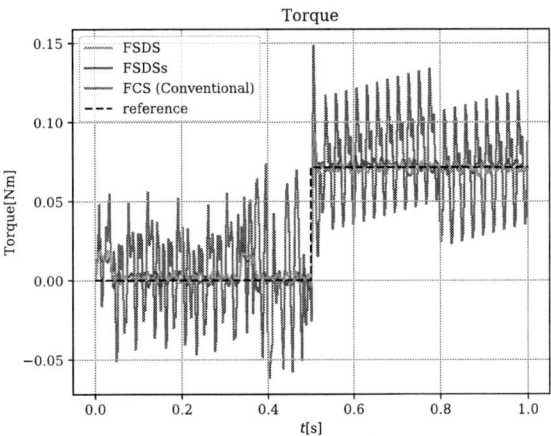

Fig. 10. Estimated torque dynamic

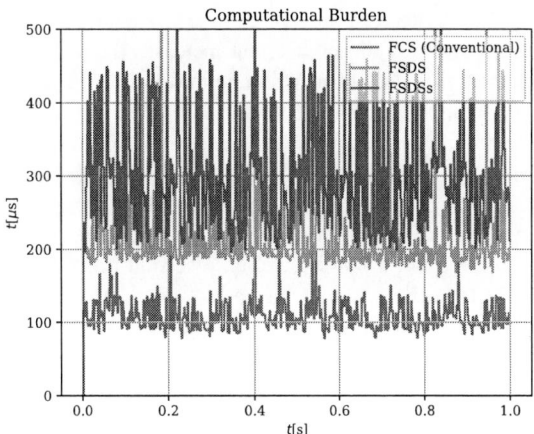

Fig. 11. Computational burden through simulation time

are noisy, because of the current inaccuracy measurement (there is an extra random noise in the measurement), PWM commutation, and the control action low resolution. Thus, with a more accurate current measurement and a bigger resolution the steady error tends to be more behaved.

At 0.5 s a load is connected to the motor, as shown in Fig. 10, in reaction the angular speed drops and the current rises. As the disturbance estimator reacts to that, the control action rises leading the angular speed back to the reference. This happens with an oscillation, that is caused by the estimator and, for bad estimator design, it could lead to instability.

The current rises to a level when the reference is in 20 rad/s and rises again when the reference goes to 40 rad/s. As the dynamic shows the current is sensitive to the load torque change, it react as a step when a load step is applied.

The FSDS and FSDSs control had a resolution of 48 positions. This made both strategies to have a really similar behavior and maintaining the angular speed close to the

reference. As the FSDS make searches around the previous control action for a big step of reference it would take a long time to reach the reference, said that FSDS would be ideal for process that don't have abrupt changes in the reference. In this case the FSDSs strategy would be better.

The estimated torque, as shown in Fig. 10, in general followed the load torque. At the moments of reference change the torque estimation felt a disturbance, but after them it came back to the right estimation. And the load step was followed, was wanted.

For comparison of the strategies the computational burden was measured. As the time to process varies according to the software and computer used and can be influenced by the system interruptions and processes, will be made a qualitative analysis of the average process time, to avoid misinterpretations of the results. The computational burden is showed in Fig. 11, as can be seen the conventional FCS-MPC strategy has the faster processing, as it has a low resolution, the number of states to be evaluated are low leading to a low time to process. Both, FSDS and FSDSs had a similar computational burden, with FSDS having a slightly faster result, that because FSDSs have to do two different searches meanwhile FSDS make just one kind of search.

Can be seen that both strategies have a better performance than the conventional FCS control, with a comparable computational burden.

V. CONCLUSION

This paper explored different algorithms using the finite control set concept. The proposed algorithms use modulation which reduces the ripple in the different variables, specially in the motor speed. The FSDS strategy needs a large memory storage, since it uses a vector with a large number of possible control actions, but uses a small search horizon. This characteristic provides slower dynamic answer than the FSDSs strategy in closed loop. In other hand, the FSDSs strategy performs two searches. The first is in large range with low resolution and

978-1-7281-4181-7/19 $31.00 © 2019 IEEE

the second is in a small range, neighbor of the first choice. The second search can be programmed recursively, which reduces significantly the storage, but increases the processing. As expected, in the performed simulations the FSDS strategy had lower computational burden than FSDSs strategy.

As contributions, this paper proposed some changes in the algorithms in order to make them faster. Also, a disturbance estimator was proposed, since the FCS-MPC strategy does not have integral action natively. Finally, this paper tested these algorithms in a power electronic second order process, differently than the original concept of use these techniques for non event-based processes. The algorithms have proved itself, since both were capable to reduce significantly the speed and current ripple, with a acceptable computational burden increase. Between the algorithms, FSDS had a slightly faster performance by its simpler implementation. And, using a ramp reference, its slower dynamics was not perceived. However, this algorithm needs more storage, and it could be not suitable for every processor for practical applications. Therefore, for this kind of process and for process with abrupt reference changes, the FSDSs is the best strategy, since it can deliver a faster control action change and can be implemented with lower storage.

ACKNOWLEDGMENT

The authors thank Santa Catarina State University and Federal Institute of Santa Catarina, the "Programa de Bolsas de Iniciação Científica da UDESC" (PROBIC) and the "Programa de Bolsas Universitárias de Santa Catarina" (UNIEDU/FUMDES) for the finnancial support.

REFERENCES

[1] X. Shi, J. Zhu, L. Li, and Y. Qu, "Model predictive control of pwm ac/dc converters for bi-directional power flow control in microgrids," in *2015 Australasian Universities Power Engineering Conference (AUPEC)*, Sep. 2015, pp. 1–4.

[2] F. An, W. Song, B. Yu, and K. Yang, "Model predictive control with power self-balancing of the output parallel dab dc–dc converters in power electronic traction transformer," *IEEE Journal of Emerging and Selected Topics in Power Electronics*, vol. 6, no. 4, pp. 1806–1818, Dec 2018.

[3] J. Hu, J. Zhu, G. Lei, G. Platt, and D. G. Dorrell, "Multi-objective model-predictive control for high-power converters," *IEEE Transactions on Energy Conversion*, vol. 28, no. 3, pp. 652–663, Sep. 2013.

[4] S. Vazquez, J. I. Leon, L. G. Franquelo, J. Rodriguez, H. A. Young, A. Marquez, and P. Zanchetta, "Model predictive control: A review of its applications in power electronics," *IEEE Industrial Electronics Magazine*, vol. 8, no. 1, pp. 16–31, mar 2014.

[5] H. A. Young, M. A. Perez, J. Rodriguez, and H. Abu-Rub, "Assessing finite-control-set model predictive control: A comparison with a linear current controller in two-level voltage source inverters," *IEEE Industrial Electronics Magazine*, vol. 8, no. 1, pp. 44–52, mar 2014.

[6] M. Narimani, Bin Wu, V. Yaramasu, Zhongyuan Cheng, and N. R. Zargari, "Finite control-set model predictive control (fcs-mpc) of nested neutral point-clamped (nnpc) converter," *IEEE Transactions on Power Electronics*, vol. 30, no. 12, pp. 7262–7269, Dec 2015.

[7] J. Rodriguez, M. P. Kazmierkowski, J. R. Espinoza, P. Zanchetta, H. Abu-Rub, H. A. Young, and C. A. Rojas, "State of the art of finite control set model predictive control in power electronics," *IEEE Transactions on Industrial Informatics*, vol. 9, no. 2, pp. 1003–1016, may 2013.

[8] R. Heydari, T. Dragicevic, and F. Blaabjerg, "Coordinated operation of vscs controlled by mpc and cascaded linear controllers in power electronic based ac microgrid," in *2018 IEEE 19th Workshop on Control and Modeling for Power Electronics (COMPEL)*, June 2018, pp. 1–4.

[9] Y. Feng, P. Kou, L. Yu, and H. Feng, "Time switching finite-control-set model predictive control for surface-mounted permanent magnet synchronous motor," in *2018 37th Chinese Control Conference (CCC)*, July 2018, pp. 3514–3519.

[10] R. P. Aguilera, P. Lezana, and D. E. Quevedo, "Finite-control-set model predictive control with improved steady-state performance," *IEEE Transactions on Industrial Informatics*, vol. 9, no. 2, pp. 658–667, may 2013.

[11] C. Bordons and C. Montero, "Basic principles of MPC for power converters: Bridging the gap between theory and practice," *IEEE Industrial Electronics Magazine*, vol. 9, no. 3, pp. 31–43, sep 2015.

[12] P. Lezana, R. Aguilera, and D. Quevedo, "Steady-state issues with finite control set model predictive control," in *2009 35th Annual Conference of IEEE Industrial Electronics*, Nov 2009, pp. 1776–1781.

[13] R. Mendez, D. Sbarbaro, and J. Espinoza, "High dynamic and static performance fcs-mpc strategy for static power converters," in *2016 IEEE Energy Conversion Congress and Exposition (ECCE)*, Sep. 2016, pp. 1–7.

[14] S. Jeong and S. Song, "Improvement of predictive current control performance using online parameter estimation in phase controlled rectifier," *IEEE Transactions on Power Electronics*, vol. 22, no. 5, pp. 1820–1825, Sep. 2007.

[15] M. Gendrin, J. Gauthier, and X. Lin-Shi, "A finite control set model predictive control based hybrid pwm technique," in *2015 IEEE International Conference on Industrial Technology (ICIT)*, March 2015, pp. 2218–2223.

[16] A. G. Bartsch, R. T. Preuss, M. S. M. Cavalca, and J. A. de Oliveira, "Finite control set model-based predictive control applied to a non-event based process," *IEEE Transactions on Industrial Eletronics*, vol. 61, no. 10, pp. 5249–5258, 2018.

[17] M. Preindl and E. Schaltz, "Load torque compensator for model predictive direct current control in high power pmsm drive," in *2010 IEEE International Symposium on Industrial Electronics*. IEEE, 2010, pp. 1347 – 1352.

[18] M. Preindl and S. Bolognani, "Model predictive direct speed control with finite control set of PMSM drive systems," *IEEE Transactions on Power Electronics*, vol. 28, no. 2, pp. 1007–1015, feb 2013.

[19] E. Fuentes, D. Kalise, J. Rodriguez, and R. M. Kennel, "Cascade-free predictive speed control for electrical drives," *IEEE Transactions on Industrial Electronics*, vol. 61, no. 5, pp. 2176–2184, may 2014.

[20] E. Hernández-Márquez, C. A. Avila-Rea, R. Silva-Ortigoza, J. R. García-Sánchez, M. Antonio-Cruz, H. Taud, and M. Marcelino-Aranda, "Modeling and simulation of a dc motor fed by a full-bridge buck inverter," in *2017 International Conference on Mechatronics, Electronics and Automotive Engineering (ICMEAE)*, Nov 2017, pp. 98–103.

978-1-7281-4181-7/19 $31.00 © 2019 IEEE

Comparative Study of the Power Sharing Techniques for Microgrids in Autonomous Mode

Gustavo M. S. Azevedo, Erik C. Oliveira, Marcelo C. Cavalcanti and Leonardo R. Limongi

Department of Electrical Engineering - Federal University of Pernambuco, Recife - PE, Brazil

e-mail: gustavo.msazevedo@ufpe.br, erik.cruz.oliveira@gmail.com, marcelo.ccavalcanti@ufpe.br and leonardo.limongi@ufpe.br

Abstract—This work brings a comparative study of the main techniques of power sharing for microgrids operating in the autonomous mode. The techniques considered here are the droop control for inductive and resistive lines and the derivative voltage droop. A set of figures of merit is defined in order to perform a helpful and reliable comparison. Also, it is considered a simple three-phase microgrid with two converters and loads where is evaluated the influence of the microgrid line impedances and the power sharing techniques. The simulation results bring valuable insights regarding the steady state performance of the evaluated power sharing techniques.

Index Terms—distributed generation, microgrids, power sharing, droop control

I. Introduction

Generally, the microgrids operates connected to the main electrical grid. In this connected mode, the deficit (or excess) in generated power can be supplied (or absorbed) by the main grid. The microgrids are also able to operate disconnected from the main grid, i.e., in an autonomous mode. In such mode, the generated power must match the demanded load. Therefore, the main challenge is properly share the demanded load power among the distributed power sources of the microgrid. In this sense, many power sharing techniques have been developed [1].

Load power sharing through the parallel operation of power converters has been studied for a long time. First, it was applied in uninterruptible power supply (UPS) to increase its power capability and reliability. After that, it has been extensively applied in distributed generation systems and microgrids [2].

Basically, these power sharing controls can be grouped in two categories: with communication and without communication [3], [4]. The typical example of communication based power sharing control is the master-slave control, where the master converter sends the set point of power for each slave converter through a low band-width communication link [5]. The main disadvantage of this control is its dependence of the communication link.

On the other hand, the communication free techniques use the information (e.g. voltage and current) available in the output of each converter to perform the power sharing. In this category of control, the most common technique is the droop control, which consists in emulating the behavior of large power generators [6]. However, the droop control presents power sharing inaccuracy and, for this reason, many adaptations have been proposed. Many of these adaptations, along with other power sharing techniques, are discussed in the review presented in [1] and [3].

Nevertheless, none of these reported works have been provided a deep performance comparison of the power sharing through well-defined criterions and under the same base. Therefore, this work proposes four figures of merit, which are used to compare some of the most popular power sharing techniques. The selected techniques are the droop control for inductive and resistive lines and the derivative voltage droop.

II. Power Sharing Control Techniques

A. Inductive Droop Control - LDC

The conventional droop control uses the same principle of large power generators in the electrical power system, i.e., the converters are controlled in such a way to present active power-frequency (P-ω) and reactive power-voltage (Q-V) characteristics similar to a synchronous machine operating in steady state [6]. If the impedance connecting the sources are purely inductive, the system frequency is more dependent of the active power while the voltages are more dependent of the reactive power [7]. Therefore, the droop regulation of frequency and voltage through the active and reactive powers is [8]:

$$\omega_i = \omega_0 - m_i P_i \tag{1}$$
$$V_i = V_0 - n_i Q_i, \tag{2}$$

where ω_i and V_i are the reference of frequency and voltage amplitude of the *i-th* converter; ω_0 and V_0 are the frequency and amplitude of the voltage without load; m_i and n_i are the droop coefficients of frequency and amplitude of the converter i, respectively; P_i and Q_i are the average active and reactive power delivered by the converter i. Note that the P-ω and Q-V characteristics imposed by (1) and (2) have been obtained supposing an inductive line impedance, then this technique will be called inductive droop control (LDC).

If the converters have different power ratings, the power sharing should be proportional to their rated power capacity [3]. This is done by choosing the proper droop coefficients in order to ensure the following relation:

$$m_1 S_{R1} = m_2 S_{R2} = \ldots = m_i S_{Ri} \tag{3}$$
$$n_1 S_{R1} = n_2 S_{R2} = \ldots = n_i S_{Ri}, \tag{4}$$

where S_{Ri} is the rated power capacity of the converter i. However, this is not all, since m_i and n_i have impact upon microgrid voltage and frequency regulation and its stability, so they must be carefully and appropriately designed [3].

978-1-7281-4181-7/19 $31.00 © 2019 IEEE

B. Resistive Droop Control - RDC

If the connection impedances were purely resistive, the power characteristics changes, i.e., the system frequency is more dependent of the reactive power while the voltages are more dependent of the active power [9]. Thus, it is suggested to use the rules:

$$\omega_i = \omega_0 + m_i Q_i \qquad (5)$$
$$V_i = V_0 - n_i P_i, \qquad (6)$$

This technique will be called resistive droop control (RDC), since it is derived of the power characteristics in the system with resistive lines.

The relation given by (3) and (4) is still applicable in the RDC.

C. Voltage Derivative Droop Control - VDDC

The conventional droop control for predominantly inductive lines is able to guarantee the active power sharing among the converters in a manner proportional to the nominal capacity of each converter. However, the same does not occur with the reactive power sharing that presents deviations from the desired value. Such reactive power error is due to the impedance asymmetry of the cables between the converters and the loads. Thus, the converter that is closer to the load tends to delivered more reactive power to this load. To mitigate this effect, [10] proposed an adaptation of the LDC where the voltage derivative is used instead of the voltage. This technique will be called voltage derivative droop control (VDDC).

The VDDC also uses (1) to perform the active power sharing. On the other hand, the reactive power sharing is governed by [10]:

$$\dot{V}_i = \dot{V}_0 - n_{Di}(Q_i - Q_{0i}) \qquad (7)$$
$$V_i = V_0 + \int_t \dot{V}_i \, d\tau, \qquad (8)$$

where $\dot{V}_i$ represents the time rate of change of V_i, n_{Di} is the $\dot{V}_i$ droop coefficient, $\dot{V}_0$ is the nominal $\dot{V}_i$ which is set to $0\,V/s$, and Q_{0i} is the reactive power set point at the nominal $\dot{V}_i$. In fact, [10] uses Q_{0i} to provide a voltage restoration given by

$$\frac{d}{dt}Q_{0i} = K_{res}Q_{Ri}(\dot{V}_0 - \dot{V}_i), \qquad (9)$$

where K_{res} is the $\dot{V}_i$ restoration gain and Q_{Ri} is the rated reactive power capacity of the converter i.

According to [10], it is necessary to ensure the relation

$$n_{D1}Q_{R1} = n_{D2}Q_{R2} = \ldots = n_{Di}Q_{Ri} \qquad (10)$$

to have a power sharing proportional to the converters rated power. Besides, the restoration gain K_{res} must be the same for all converters.

III. FIGURES OF MERIT

The performance of the power sharing control techniques is evaluated through four figures of merit, which are defined and explained in following. Such figures of merit are evaluated when the system is operating in steady state.

A. Active Power Error

The active power that each converter must deliver is proportional to its rated capacity compared with the rated capacity of the other converts. Neglecting the losses and considering a microgrid with N converters and M loads, the active power delivered by the converter i should by

$$P_i^* = \frac{S_{R_i}}{\sum_{i=1}^{N} S_{R_i}} \sum_{j=1}^{M} P_{L_j}, \qquad (11)$$

where S_{R_i} is the rated capacity of the converter i and P_{L_j} is the active power of the load j.

If P_i^* is the desired power, the active power error, for the converter i, can be obtained by

$$EP_i\% = \frac{P_i - P_i^*}{P_i^*} 100\%. \qquad (12)$$

where P_i is the active power delivered by the converter.

B. Reactive Power Error

In the same way, the reactive power that each converter must deliver is proportional to its rated capacity compared with the other ones. Thus, the reactive power injected by the converter i should by

$$Q_i^* = \frac{S_{R_i}}{\sum_{i=1}^{N} S_{R_i}} \sum_{j=1}^{M} Q_{L_j}, \qquad (13)$$

where Q_{L_j} is the reactive power of the load j.

Therefore, the reactive power error, for the converter i, can be obtained by

$$EQ_i\% = \frac{Q_i - Q_i^*}{Q_i^*} 100\%. \qquad (14)$$

where Q_i is the reactive power delivered by the converter.

C. Voltage Regulation

The voltage regulation of the converter i indicates how far its output RMS voltage, V_i, is from of the rated value, V_R. The voltage regulation ΔV_i is given by

$$\Delta V_i\% = \frac{V_i - V_R}{V_R} 100\%. \qquad (15)$$

In case of three-phase unbalanced voltages, it is considered the phase with the highest voltage regulation.

D. Frequency Regulation

The frequency regulation of the microgrid indicates how far its voltage frequency, f, is from of the rated value, f_R. The frequency regulation Δf is given by

$$\Delta f\% = \frac{f - f_R}{f_R} 100\%. \qquad (16)$$

Note that, all the microgrid converters operating at the same frequency when the microgrid is in steady state. Therefore, just one frequency regulation should be assigned to the entire microgrid.

978-1-7281-4181-7/19 $31.00 © 2019 IEEE

IV. Comparison Among the Power Sharing Control Techniques

The comparison is performed through simulations carry out in the MATLAB/Simulink. A simple and hypothetical two converters three-phase microgrid, operating in autonomous mode, is considered in such way the effect of their parameter variation on the power sharing techniques could be evaluated. The electrical diagram of such microgrid is shown in Fig. 1. It is considered that the rated power capacity of Converter #1 is twice as the Converter #2. Z_1 and Z_2 represents the cables impedances that connect the converters to a concentrated load. The averaged power components P_i and Q_i are obtained through first-order low-pass filters with cut-off frequency of $12\,Hz$. For simplicity, the converters are modeled as a voltage source with a LC filter ($L = 1\,mH$ and $C = 4.7\,\mu F$).

Three kind of loads, with different characteristics, are connected one-by-one in order to evaluate their effect on the microgrid: L_1 is purely resistive with $10\,kW$; L_2 is purely inductive with $5\,kVAr$; and L_3 is resistive-inductive with $5\,kW + j5\,kVAr$. L_1 is connected at $t = 5\,s$, L_2 at $t = 15\,s$, L_3 at $t = 25\,s$ and all the three loads are disconnected at $t = 35\,s$.

Since the line impedances plays an important role under the power sharing behavior, each power sharing control technique is analyzed for five different scenarios concerning the Z_1 and Z_2 characteristics. The scenarios considered are:

Case 1– Inductive line with the same impedance;
Case 2– Resistive line with the same impedance;
Case 3– Resistive-Inductive line with the same impedance;
Case 4– Inductive line with different impedance;
Case 5– Resistive line with different impedance.

The results and analysis for each case is presented in following sections.

A. Case 1 - Inductive line with the same impedance

In this case, it is considered an inductive line with the same impedance, $Z_1 = Z_2 = j0.55\,\Omega$. Although it is expected a predominantly resistive line in low voltage systems, this hypothetical case is used to comparison purpose. Thus, these values of Z_1 and Z_2 are numerically equal to the resistance of an $100\,m$, $4\,mm^2$ typical cable.

Each power sharing technique is simulated under the same conditions. Their time responses are shown in Fig. 2, where it could be see the power components demanded by the load and injected by the converters. Besides, it is also show the

frequency reference and the output voltage RMS value (of the phase a) of each converter.

For the LDC, it could be observed that the Converter #1 delivers approximately the twice active power of the Converter #2 and they operate at the same frequency in steady state. The maximum active power error occurs when the three loads are on (between $25\,s$ and $35\,s$) reaching 1.6% and the frequency regulation of -0.32%, as shown in the Table I. On the other hand, the reactive power does not have the desired proportion: the Converter #1 delivers less reactive power than the desired (an error of -7.9%) and the Converter #2 delivers 43.5% more reactive power than the desired. The maximum voltage regulation reaches -3.2%, as shown in the Table I.

This behavior completely changes in the RDC technique with inductive line impedance. The active power error becomes very high and different (see Table I) because now its depends on the voltage (instead of the frequency as in the LDC). This is a considerable drawback since the active power is provided by some primary power source connected to the converter DC side and it may result in an undesired consumption of this source. The reactive power sharing is better compared to LDC and both converters have the same error (-8.6%).

The VDDC have presented an interesting result given rise to the same results of the LDC, as shown in the Table I. Thus, the reactive power sharing enhancement claimed by their authors is not reached. The difference is only the transient behavior, as can be seen in the Fig. 2. In fact, a detailed analysis of [10] reveals that the relation between $\Delta V_i = V_i - V_o$ and Q_i is given by the transference function

$$\Delta V_i(s) = \frac{V_i(s) - V_o(s)}{Q_i(s)} = \frac{-n_{Di}}{s + n_{Di}\,K_{res}\,Q_{Ri}}. \quad (17)$$

Therefore, their proposal just adds a first-order dynamic under V_i, with a time constant of $1/(n_{Di}\,K_{res}\,Q_{Ri})$, and the effective droop coefficient for the reactive power (in steady state) is $1/(K_{res}\,Q_{Ri})$.

B. Case 2 - Resistive line with the same impedance

In this case, it is considered a resistive line with the same impedance, $Z_1 = Z_2 = 0.55\,\Omega$. This value corresponds to the resistance of an $100\,m$, $4\,mm^2$ typical cable. It worth to note that Z_1 has the same impedance module for all the study cases, although its characteristic changes. The same is valid for Z_2.

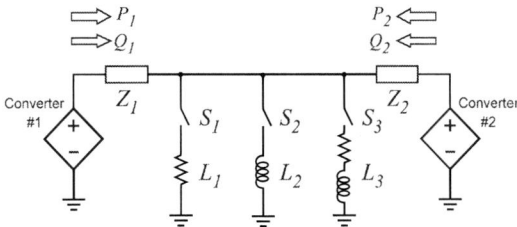

Fig. 1. One-line diagram of the three-phase microgrid considered in the comparison.

TABLE I
COMPARISON OF THE POWER SHARING TECHNIQUES FOR THE CASE 1.

Case	Parameter	Power Sharing Technique		
		LDC	RDC	VDDC
1	EP_1	1,6%	-14,7%	1,6%
	EP_2	1,6%	33,9%	1,6%
	EQ_1	-7,9%	8,6%	-7,9%
	EQ_2	43,5%	8,6%	43,4%
	ΔV_1	-2,1%	-3,1%	-2,1%
	ΔV_2	-3,2%	-4,8%	-3,2%
	Δf	-0,32%	0,20%	-0,31%

The time response results of each power sharing technique are shown in Fig. 3 and a summary of the steady state results are provided in the Table II.

For the LDC, although the active power sharing have a significant increase in its error (now is 5.3%), it keep very above the error of the reactive power (64, 2%), as can see in Table II. The reactive power sharing also got worse in this resistive line case.

The RDC has its reactive power sharing improved related to the Case 1 (8.6% to 3.6% of error). However, the active power sharing error keeps very high (33.7%). Thus, although many have recommended to use the RDC in resistive lines, these results show that it is not a good choice, since the active power error remains high.

The VDDC also presents the same results of the LDC, as shown in the Table II. The difference is only the transient behaviour, as can be seen in the Fig. 3.

C. Case 3 - Resistive-Inductive line with the same impedance

In this case, it is considered a resistive and inductive line with the same impedance, $Z_1 = Z_2 = 0.39 + j0.39\,\Omega$. This impedance has the same module of that one used in the previous cases.

The time response results of each power sharing technique are shown in Fig. 4 and a summary of the steady state results are provided in the Table III.

Observing the Table III and comparing with the previous cases, it could be noted that both LDC, RDC and VDDC techniques have presented an intermediated perform between the Case 1 and Case 2. The VDDC also presents the same results of the LDC.

D. Case 4 - Inductive line with different impedance

Typically, the distances among converters and loads are different in a microgrid, then the connection line impedances also will be different. Therefore, in order to take such characteristic into account, it is considered an inductive line where the length of one of circuit is half. Thus, it was selected $Z_1 = j0.55\,\Omega$ and $Z_2 = Z_1/2$.

The time response results of each power sharing technique are shown in Fig. 5 and a summary of the steady state results are provided in the Table IV.

The LDC and the VDDC have the same steady state performance and they still present the lower active power error compared with the RDC.

Comparing the reactive power error with the Case 1, one can note that it has increased. This is due to the fact that the load is now closer to Converter #2, then it has more impact over the output voltage of the Converter #2 (decreasing it) and, according to (2), the reactive power increases. On the other hand, the Converter #1 injects less reactive power than the necessary because the load is far from it.

E. Case 5 - Resistive line with different impedance

Similar to the previous case, cable lines with different lengths is considered, nevertheless with a resistive characteristic. Thus, it was selected $Z_1 = 0.55\,\Omega$ and $Z_2 = Z_1/2$.

The time response results of each power sharing technique are shown in Fig. 6 and a summary of the steady state results are provided in the Table V.

Again, the LDC and the VDDC present the lowest active power error, while the reactive power error is the highest considering all the five cases.

TABLE II
COMPARISON OF THE POWER SHARING TECHNIQUES FOR THE CASE 2.

Case	Parameter	Power Sharing Technique		
		LDC	RDC	VDDC
2	EP_1	5,3%	-9,2%	5,3%
	EP_2	5,3%	33,7%	5,3%
	EQ_1	-26,4%	3,6%	-26,3%
	EQ_2	64,2%	3,6%	63,6%
	ΔV_1	-1,6%	-3,2%	-1,6%
	ΔV_2	-3,6%	-4,7%	-3,6%
	Δf	-0,32%	0,18%	-0,32%

TABLE III
COMPARISON OF THE POWER SHARING TECHNIQUES FOR THE CASE 3.

Case	Parameter	Power Sharing Technique		
		LDC	RDC	VDDC
3	EP_1	4,2%	-12,9%	4,2%
	EP_2	4,2%	37,8%	4,2%
	EQ_1	-15,8%	7,3%	-17,7%
	EQ_2	58,4%	7,3%	58,1%
	ΔV_1	-1,8%	-3,0%	-1,8%
	ΔV_2	-3,4%	-4,8%	-3,5%
	Δf	-0,31%	0,19%	-0,31%

TABLE IV
COMPARISON OF THE POWER SHARING TECHNIQUES FOR THE CASE 4.

Case	Parameter	Power Sharing Technique		
		LDC	RDC	VDDC
4	EP_1	1,6%	-17,9%	1,6%
	EP_2	1,6%	40,3%	1,6%
	EQ_1	-13,6%	7,9%	-13,7%
	EQ_2	52,8%	7,9%	52,8%
	ΔV_1	-2,0%	-3,0%	-2,0%
	ΔV_2	-3,5%	-5,1%	-3,5%
	Δf	-0,31%	0,20%	-0,32%

TABLE V
COMPARISON OF THE POWER SHARING TECHNIQUES FOR THE CASE 5.

Case	Parameter	Power Sharing Technique		
		LDC	RDC	VDDC
5	EP_1	4,7%	-15,1%	4,7%
	EP_2	4,7%	43,4%	4,7%
	EQ_1	-34,4%	3,6%	-34,4%
	EQ_2	80,2%	3,6%	80,0%
	ΔV_1	-1,5%	-3,0%	-1,5%
	ΔV_2	-4,0%	-5,1%	-4,0%
	Δf	-0,32%	0,19%	-0,32%

978-1-7281-4181-7/19 $31.00 © 2019 IEEE

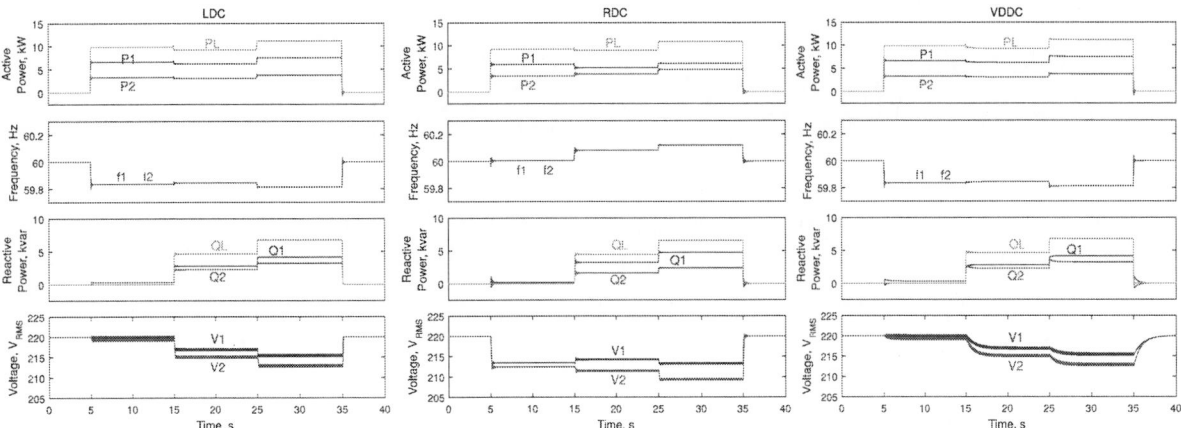

Fig. 2. Simulation results of the power sharing techniques for a microgrid with inductive line impedance of the same value (Case 1).

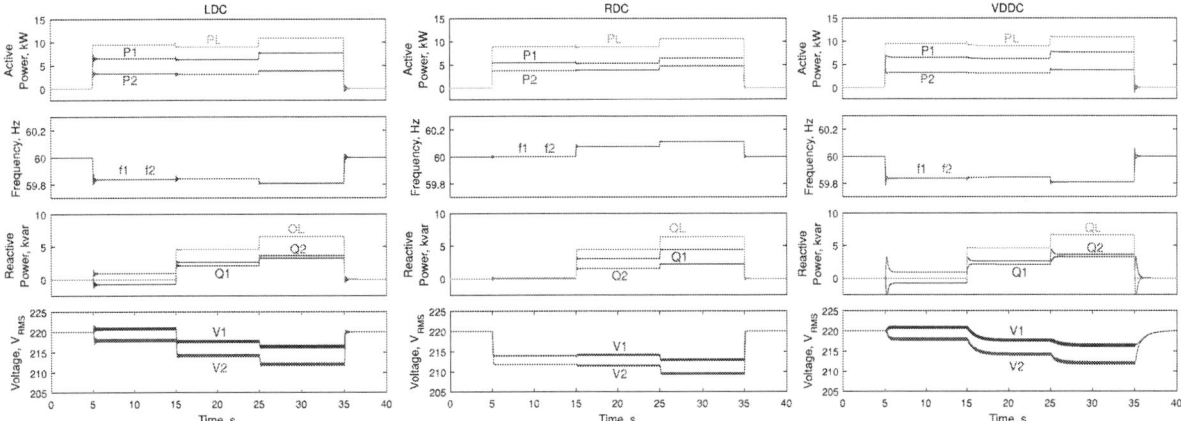

Fig. 3. Simulation results of the power sharing techniques for a microgrid with resistive line impedance of the same value (Case 2).

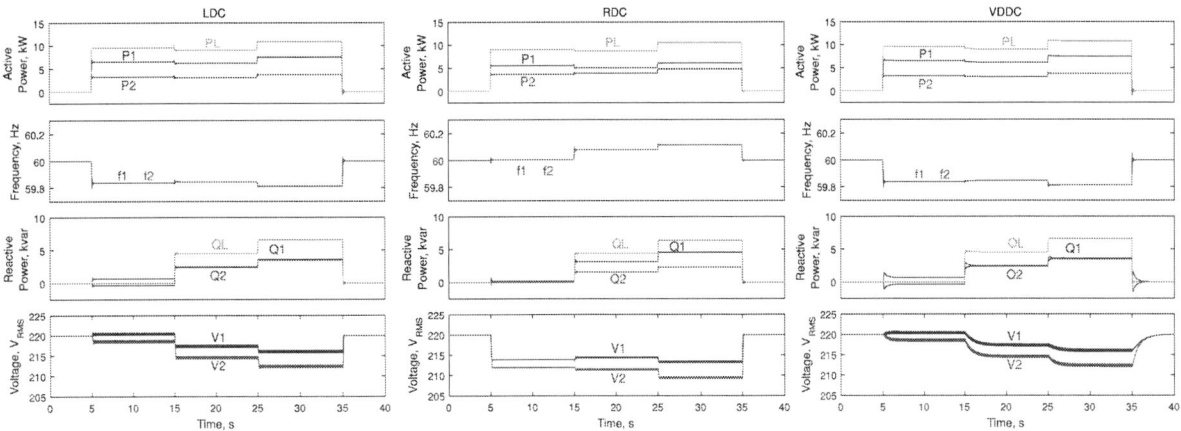

Fig. 4. Simulation results of the power sharing techniques for a microgrid with resistive and inductive line impedance of the same value (Case 3).

V. CONCLUSIONS

In this work, it is proposed four figures of merit to evaluate the steady state performance of the power sharing techniques in microgrids operating in autonomous mode. Besides, a simple simulation setup is recommended to use these figures of merit. Thus, the well-know LDC and RDC techniques are evaluated along with the VDDC, which supposedly has a improved reactive power sharing.

It is observed that the LDC presents small active power sharing errors for any kind of line, although it is better for inductive lines. However, its reactive power error is high. The RDC presents an inverse behavior, i.e., small reactive

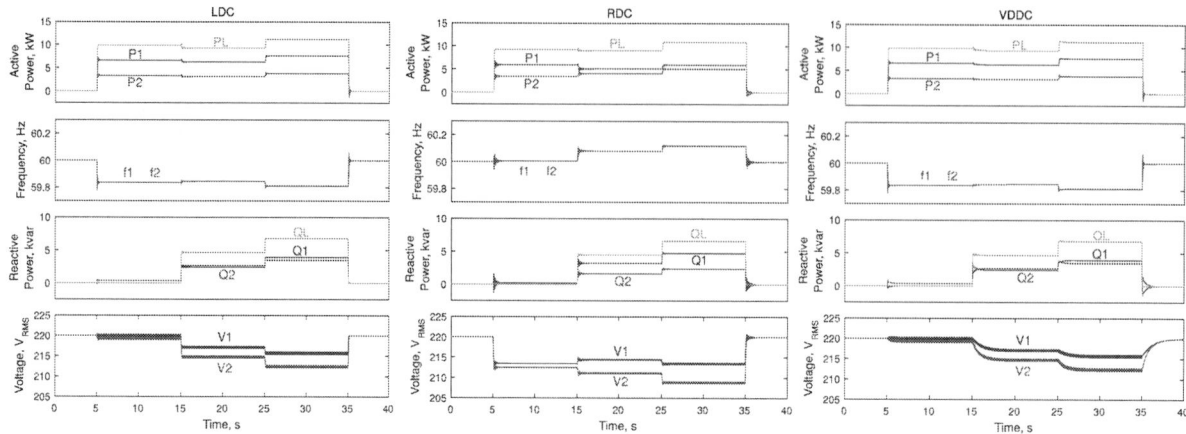

Fig. 5. Simulation results of the power sharing techniques for a microgrid with inductive line impedance of different values (Case 4).

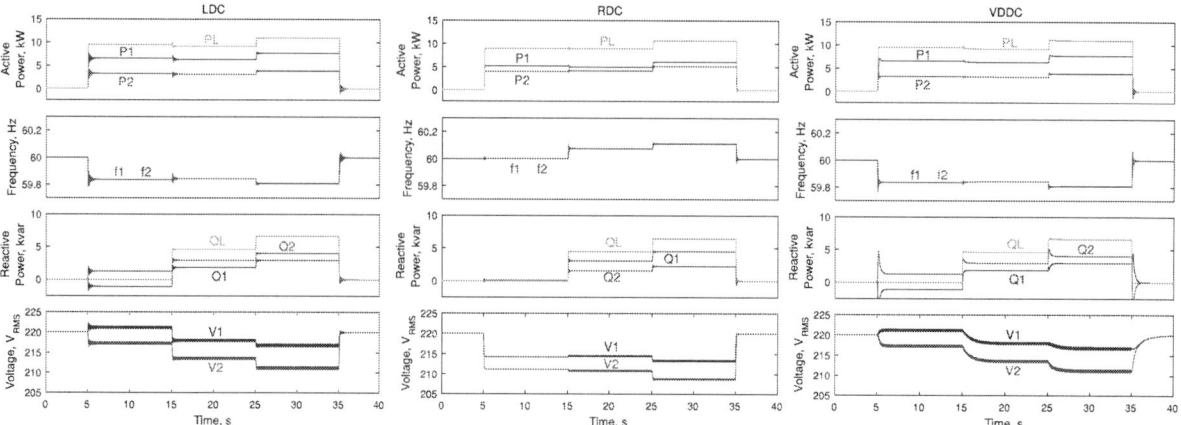

Fig. 6. Simulation results of the power sharing techniques for a microgrid with resistive line impedance of different values (Case 5).

power error and high active power error. Therefore, it is recommended to use the LDC, even in the resistive lines, to keep the control over the converter primary power source. Even in the case of resistive lines, the RDC presents high error of active power sharing, which is a major drawback.

It is observed that the VDDC does not improve the reactive power sharing as stated. The VDDC steady state behavior is equal to the LDC if the product $K_{res} Q_{Ri}$ is equal to the droop coefficient of the LDC (n_i). Thus, the difference is just the transient behavior.

ACKNOWLEDGMENT

This work was supported by INCT-GD, with funding of CNPq (465640/2014-1), CAPES (23038.000776/2017-54) and FAPERGS (17/2551-0000517-1) and also supported by FACEPE (APQ-0903-3.04/14).

REFERENCES

[1] Y. Han, H. Li, P. Shen, E. A. A. Coelho and J. M. Guerrero, "Review of Active and Reactive Power Sharing Strategies in Hierarchical Controlled Microgrids," IEEE Trans. on Power Electr., vol. 32, no. 3, pp. 2427–2451, 2017.

[2] G. M. S. Azevedo, M. C. Cavalcanti, F. A. S. Neves, P. Rodriguez and J. Rocabert, "Performance improvement of the droop control for single-phase inverters" 2011 IEEE International Symposium on Industrial Electronics (ISIE), pp. 1465–1470, 2011.

[3] H. Han, X. Hou, J. Yang, J. Wu, M. Su and J. M. Guerrero, "Review of Power Sharing Control Strategies for Islanding Operation of AC Microgrids," IEEE Trans. on Smart Grid, vol. 7, pp. 200–215, 2016.

[4] G. M. S. Azevedo, J. P. Arruda, M. C. Cavalcanti, and L. R. Limongi, "Fault detection system to trigger the microgrid disconnection procedures," 13th Brazilian Power Electronics Conference and 1st Southern Power Electronics Conference (COBEP/SPEC), pp. 1–6, 2015.

[5] G. M. S. Azevedo, M. C. Cavalcanti, F. A. S. Neves, L. R. Limongi and F. Bradaschia, "A control of microgrid power converter with smooth transient response during the change of connection mode," 2013 Brazilian Power Electronics Conference (COBEP), pp. 1008–1015, 2013.

[6] J. M. Guerrero, L. G. Vicuna, J. Matas, M. Castilla and J. Miret, "Output impedance design of parallel-connected UPS inverters with wireless load-sharing control," IEEE Trans. on Indus. Electr., vol. 52, pp. 1126–1135, 2005.

[7] K. De Brabandere, B. Bolsens, J. Van den Keybus, A. Woyte, J. Driesen and R. Belmans, "A Voltage and Frequency Droop Control Method for Parallel Inverters," IEEE Trans. on Power Electr., vol. 22, pp. 1107-1115, 2007.

[8] T. D. Cardoso, G. M. S. Azevedo, M. C. Cavalcanti, F. A. S. Neves, L. R. Limongi, "Implementation of droop control with enhanced power calculator for power sharing on a single-phase microgrid," 2017 Brazilian Power Electronics Conference (COBEP), pp. 1–6, 2017.

[9] J. M. Guerrero, J. Matas, L. Garcia de Vicuna, M. Castilla and J. Miret, "Decentralized Control for Parallel Operation of Distributed Generation Inverters Using Resistive Output Impedance," IEEE Trans. on Indus. Electr., vol. 54, pp. 994-1004, 2007.

[10] Chia-Tse Lee, Chia-Chi Chu and Po-Tai Cheng, "A New Droop Control Method for the Autonomous Operation of Distributed Energy Resource Interface Converters," IEEE Trans. on Power Electr., vol. 28, pp.14, 2013.

978-1-7281-4181-7/19 $31.00 © 2019 IEEE

Hybrid One-Cycle Control Strategy for Buck+Boost PFC Battery Charger

Aluísio Alves de Melo Bento
Department of Electric Engineering and Telecommunication
State University of Rio de Janeiro
Rio de Janeiro, Brazil
aluisio.uerj@gmail.com

Raphael Perci Santiago
Department of Electric Engineering and Telecommunication
State University of Rio de Janeiro
Rio de Janeiro, Brazil
rpercieng@gmail.com

Abstract— **Electric vehicles emerge as a growing market. The utilization of electric vehicles will progressively create the need of high performance battery chargers. The buck+boost PFC rectifier comes out as an effective and efficient plug-in battery charger. Several works report application of this topology supported by digital controllers. These controllers normally employ input and output voltage sensors to perform input current reference and the operation mode decision, i.e. buck or boost modes, with continuous comparison between input and output voltages. On the other hand, analogical control solutions requires, besides the use of input and output voltage sensors, the use of analog multiplier to perform input current template and mode managing. This work proposes an analog control strategy based on the One-Cycle Control technique. Features of such technique includes the elimination of grid and output voltage sensors, elimination of analog multiplier and any current control loop design, spontaneous and smooth transition between buck and boost operation modes, and simple low cost circuitry. This paper carry out comparisons concerning the proposed and the existent control strategies, including cost, robustness and performance. Simulation results are carried out to a first validation of the proposal.**

Keywords—one-cycle control technique, power factor correction, analogical control, battery charger, hybrid converter

I. INTRODUCTION

Very recently, the technology of electric vehicles has been shown to be very attractive and it has a great potential to become the new target of the market. The growing usage of electric vehicles creates a pertinent concern about how to support this new market with high quality energy and high performance battery charging systems.

The charging process is strongly depended of high quality converter that must allow step-up and step-down conversions. An alternative is the basic buck-boost dc-dc converter. However, the main disadvantage of this converter is the huge inductor volume and current/voltage stresses, which results in lower efficiency when compared to the boost or boost converters. The buck+boost based PFC rectifier is shown in Fig.1 and, besides sharing the inductor, differs of individual parts by an additional diode during buck mode and switch S1, which has the function to provide voltage drop during boost mode. Features of this topology are simplicity, high power density and efficiency, low input current THD and high power factor while accomplishes international standards [1]-[3]. High frequency insulation and soft-switching solution are proposed [4] to this topology but with six or more switches. Bidirectional dc-dc step-up/down converters are reported for power flow management between dc-link and batteries [5][6].

Digital control realization dominates the control solution scenario for buck+boost topology [1][2],[4]-[6]. Features of digital controllers include flexibility, accuracy, low noise level and a large number of software tools besides progressive cost decreasing. However, the quasi totality of digital control strategies for buck+boost PFC rectifier employ input and output voltage sensors in addition to a periodic mathematical routs execution. This requires expertise related to the programing and mounting procedures. On the other hand, analog controllers are low cost and fast response solutions, besides being very effective. Although have less or none flexibility, are suitable for simple control strategies implementations [7]. An analog controller was proposed in the decade of 1993 for buck+boost PFC rectifier [3]. There, the necessity of input and output voltage sensors, analogic multiplier and variable switching frequency operation make it a disadvantageous choice for converter topology.

Among analog control techniques, the One-Cycle Control (OCC) technique is suitable for PCF rectifiers and active power filters [8][9]. Such technique eliminates explicit current controller, a fast control loop, thus eliminating analog multiplier and grid voltage sensor. Hybrid OCC strategies for a single-phase boost PFC to operates between the DCM and CCM modes, with smooth transition was reported in [11], and a Hybrid OCC PWM strategy was applied for three-phase PFC allowing operation between continuous and discontinuous PWM [12]. However, OCC strategy to support buck+boost converter has not yet been mentioned in the literature.

This work presents an analog OCC strategy alternative to digital, and analog, control strategies employed in the buck+boost converter. Features of the proposed control strategy include the elimination of grid voltage sensor, elimination of the analog multiplier, spontaneous and smooth transition between buck and boost operation modes, and simple and low cost circuitry. Simulation essays confirm the theory of proposal.

II. BASICS

This section supports the understanding of the principle of operation and synthetizes the proposed control strategy, developed in section III. First, the fundamentals of buck and

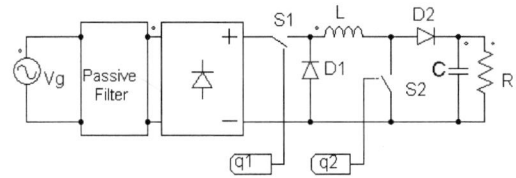

Fig. 1. Buck+boost PFC rectifier.

Identify applicable funding agency here. If none, delete this text box.

boost converters are shown. Then, a brief introduction of the OCC technique principle is done and, finally, the synthesis of OCC strategies for both buck and boost based PFC rectifiers are performed.

A. Basic DC-DC Converters Operation Limitations

It is well known that the principal limitations of boost and buck converters, shown in Fig.2, are dictated by the input and output voltages. For boost converter, the instantaneous output voltage must to be greater than or equal to the input one. Otherwise, for the buck converter, the instantaneous output voltage must be smaller than or equal to the input one.

For the boost-based PFC rectifiers, the output voltage must be higher than or equal to the instantaneous input voltage. This characteristic prohibits it employment for battery charging once battery voltage are usually lower than the grid peak voltage. An advantage of boost PFC is the continuous input current, which reduces the size of the input low pass filter.

For the buck-based PFC rectifier, the dc output voltage must be lower than or equal to the instantaneous input voltage, becoming suitable for battery charging. Moreover, the inductor, current is continuous and reduces the size of the output LC low pass filter. However, in the extremities of grid voltage cycle the converter cannot operate once the input voltage is lower than the output voltage. A important drawback of buck operation is the discontinuous input current and imposes an input filter employment.

B. The One-Cycle Control Technique

The OCC technique is a suitable solution for PFC and active power filter, and has been applied for several power conditioning systems [10]. The OCC technique principle, illustrated in Fig. 3(a), does not use grid voltage sensor nor any current template or analogue/digital multipliers. On the other hand, grid current sensor signal is directly applied to the PWM comparator input, where the other input is feed by a variable amplitude carrier signal. Therefore, it is not necessary to have any concern with the current controller design, which results in a scheme with increased robustness and a reduced number of devices. For instance, the classical control scheme is shown in Fig. 3(b). It is evident the different complexity in terms of circuit assembly for analog realization and very simple in digital implementations.

Based on Fig. 3(a), the output of R_{dc}, the dc-link regulator, controls the sawtooth carrier amplitude V_m to provide sinusoidal grid current. There, V_{Os} and V^*_{Os} are the sensed output voltage and the respective reference signals; the grid current sensor output is a voltage signal given by $R_s i_g$ where, R_s is the sensor output resistance.

The principle of the OCC technique consists in emulating an equivalent resistance R_e at the PFC rectifier input. The result is a grid power factor close to unity:

$$R_e = \frac{v_g}{i_g} = \frac{V_g}{I_g}. \tag{1}$$

The controller synthesis is based on the converter control law, given by the quasi-steady-state transfer function $f(d)$, valid inside a switching period [13], and represented here by

$$R_e i_g = f(d)V_O. \tag{2}$$

By introducing the current sensor resistance R_s in (2) results,

$$R_s i_g = f(d)\frac{R_s V_O}{R_e}. \tag{3}$$

For dc-dc unidirectional static converters, the function $f(d)$ is a unipolar sawtooth carrier. Thus, the right side of (3) is realized by a carrier with amplitude equal to V_m as follow:

$$R_s i_g = f(d)V_m. \tag{4}$$

Considering, the modulation index m defined as

$$m = \frac{V_g}{V_O}, \tag{5}$$

in (4) the following condition is established

$$I_g = mV_m/R_s. \tag{6}$$

This states that the input current magnitude is imposed by carrier amplitude V_m adjustment. For a closed loop, as in Fig. 3(a), it is done through R_{dc} regulator. The grid current is guaranteed by spontaneous convergence of the OCC technique, which is similar to the voltage follower behavior and convergence speed is strongly dependent on the boost inductor value L [14].

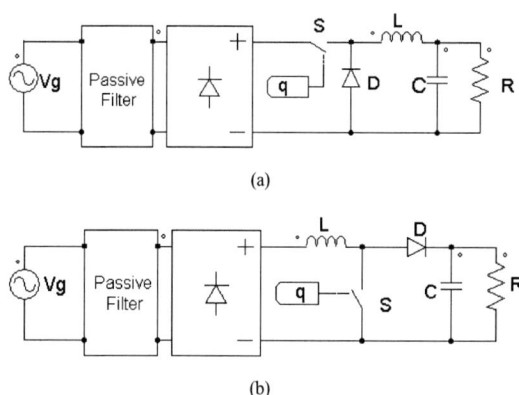

Fig. 2. Based on basic dc-dc converters PFC rectifiers. (a) Basic buck PFC. (b) Basic boost PFC rectifier.

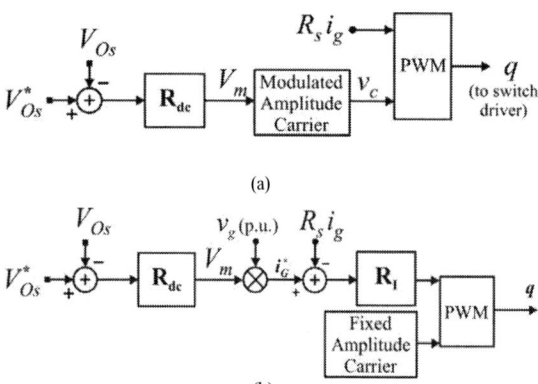

Fig. 3 The principle of the OCC technique (a) and classical control technique (b).

C. OCC Control Law for the Boost-based PFC Rectifier

The synthesis of OCC strategies takes place on the quasi-steady-state transfer function, which considers the solution into a switching period [1]. The quasi-steady-state transfer function for the boost converter, Fig. 2(b), is given by the following equation:

$$V_O = \frac{v_{in}}{1-d}, \tag{7}$$

where $v_{in} = |v_{grid}|$, being

$$v_{grid} = V_g \sin(\omega_g t), \tag{8}$$

$$f_g = \omega_g / 2\pi, \tag{9}$$

ω_g is the angular grid frequency and f_g is the grid frequency.

Finally, d is the duty cycle, defined by

$$d = \frac{t_{on}}{T} = t_{on} f, \tag{10}$$

being t_{on} the switch conduction time interval, T the switching period and f the switching frequency respectively. Considering $f \gg f_g$, the entities V_O and v_{in} in (7) can be considered constant into a switching period T.

By imposing unity power factor for the PFC input (7), it can be rewritten as

$$R_e i_{in} = (1-d)V_O, \tag{11}$$

Where, i_{in} is the average inductor current into a switching period T. Rearranging (11), the boost based PFC rectifier control law is given by

$$i_{in} = (1-d)\frac{V_O}{R_e}, \tag{12}$$

D. OCC Control Law of the Buck-based PFC Rectifier

The quasi-steady-state transfer function for the buck converter, Fig. 2(b), is given by

$$V_O = v_{in} d, \tag{13}$$

Similarly to the boost PFC, the buck based PFC rectifier control law is

$$i_{in} d = \frac{V_O}{R_e}. \tag{14}$$

The relationship between the inductor and input (switch) average currents is given by

$$i_{in} = i_L d. \tag{15}$$

Substituting (15) in (14) results in the buck based PFC rectifier control law as a function of inductor current,

$$i_L d^2 = \frac{V_O}{R_e}. \tag{16}$$

In order to integrate the control of both boost and buck PFC rectifiers, the respective control laws should be based on the inductor current.

E. Analog Realizations

The analog realization of boost control law (6) is carried out with resettable integrator and ordinary devices as showed in Fig. 4(a) and mathematically represented by

$$i_L = \frac{V_O}{R_e} - \frac{V_O}{R_e} \frac{1}{T} \int_0^{t_{on}} dt \tag{17}$$

The analog realization of buck control law (16) is carried out with two resettable integrators and ordinary devices as showed in Fig. 4(b) and mathematically represented by

$$\frac{2}{T} \int_0^{t_{on}} \left[\int_0^{t_{on}} i_L dt \right] dt = \frac{V_O}{R_e}. \tag{18}$$

As stated before, (17) and (18) consider that the average inductor current i_L does not vary into a switching period.

III. THE PROPOSAL

The proposed OCC hybrid controller is a consequence of the fact that the converters operate in mutually exclusive regions. The simplified scheme for PFC battery charger, composed by the power circuit and proposed OCC strategy, is shown in Fig. 5. There, the boost mode is achieved by maintaining S1 in on-state and S2 in high frequency PWM mode of operation. Note that, in this mode, the diode D1 does not operate due to reverse polarization. Moreover, the buck mode is achieved by maintaining S2 in off-state and S1 in high frequency PWM mode of operation. As explained next, the proposed OCC strategy operates with spontaneous and smooth transition between the boost and buck modes.

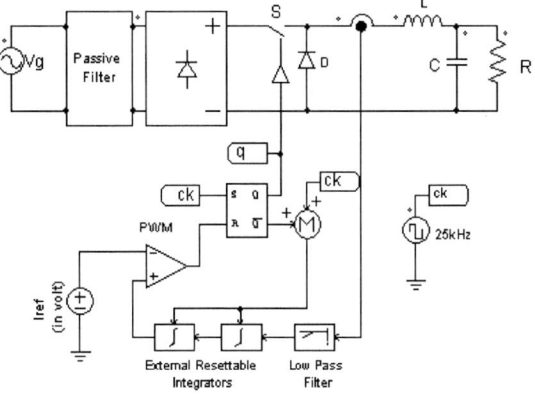

(a)

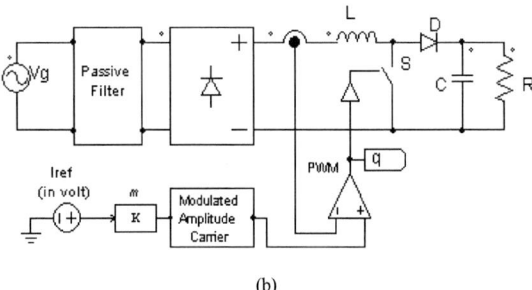

(b)

Fig. 4. Analog realization of OCC technique applied to (a) buck based PFC rectifiers and (b) boost based PFC rectifiers.

For boost converter, the PWM acts only for input voltage lower than the battery voltage. After this point, the current sensor signal (inductor current sensor) becomes bigger than the carrier amplitude and S2 remains open. Meanwhile, in control buck part, the output signal of the integrators path reaches the constant reference signal, as predicted in (14). Thus the buck switch, before in on-state, starts to commutate in switching frequency.

Starting at the left end of the grid voltage cycle, that is, when the input voltage is lower than the output one, the boost mode takes place. The right-hand side of (12), a modulated amplitude carrier, is conventionally implemented by a ramp generator through a resettable integrator plus one added. However, better results are obtained with modulated

amplitude triangular carrier. There, the amplitude dictates the magnitude of the inductor current, which is the input current in boost mode. After the input voltage reaches the output voltage, the inductor current crosses the carrier and the switch S2 remains on off-state. At the same time, the double integration of the inductor current in the buck mode control law (14) starts to reach the constant V_O/R_e and the switch S1 starts the PWM switching. It must be noticed that, in boost mode operation, the double integrator output does not reach the constant V_O/R_e. Thus, S1 remains in on-state. Finally, one can note that the scheme in Fig. 5 does not employ input voltage sensor, explicit current controller neither analog multiplier.

IV. DESIGN AND SIMULATION RESULTS

The design and simulation of the system are based the PSpice scheme shown in Fig.6. The input filter, composed by two inductances, a capacitance and a damp resistor, was adapted from [1] and [15].

The operation of control core circuit is explained in parts. The double resettable integrator must match its gains based on (18) and the hardware gains given by $1/(R_5C_1R_{10}C_2)$. The modulated amplitude carrier is defined by its gain and (17), $1/(R_{15}C_3)$ and the switching frequency. The switching frequency is defined by the clock generator time constant $R_{16}C_4$ and the reset pulse width by the time constant $R_{17}C_5$. The resistors of differential amplifiers are equal to 220 kΩ. Table I summarizes design and device specifications.

Simulation results are depicted in Fig. 8, from top to bottom: instantaneous and average grid current; inductor current, carrier and S2 command signals; and grid current magnitude reference, output of double resettable integrator, and S1 command signals. Zoomed curves of PWM blocks

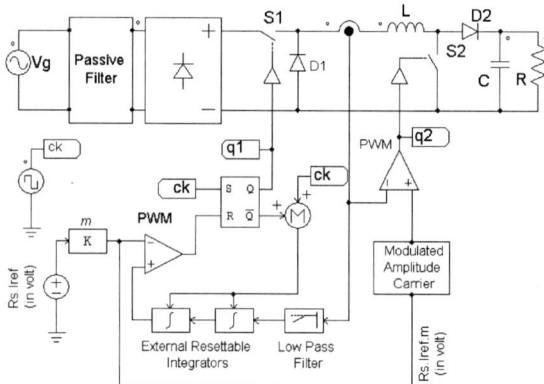

Fig. 5. Block diagram of power and control (proposed) circuits.

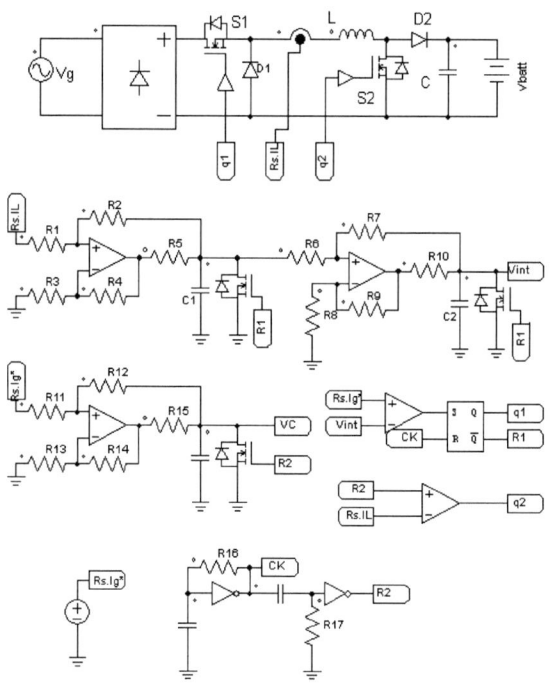

Fig. 6. Schematic diagram of the OCC strategy Control Core used for simulation essays (PSpice).

TABLE I. DESIGN SPECIFICATIONS AND DEVICES

Design Variable	Specification
Grid voltage (rms)	226 V
Input power	1600 W
Batter voltage	160 V
Switching frequency, f	25 kHz
Inductor, L	750 μH
Output capacitor, C	1000 μF/250V
R1 to R4; R11 to R14; and R6 to R9	220 kΩ
R5; R10 / C1; C2	1 kΩ / 20 nF
R15 / C3	1 kΩ / 40 nF
R16 / C4	470 Ω / 15 nF
R17 / C5	270 Ω / 1 nF
Operational amplifiers	AD823A
Voltage comparators	LM311
Signal MOSFETs Q3; Q4; and Q5	2N7000
Power MOSFET Q1	IRF
Power MOSFET Q2	IRF

978-1-7281-4181-7/19 $31.00 © 2019 IEEE

are shown in Fig. 8. The top frame presets boost PWM input-output signals, i.e., the carrier, inductor current and switch S2 command. In addition, the bottom frame is the buck PWM input-output signals, i.e., the grid current magnitude reference, double resettable integrator output, and switch S1 command signals. It can notice the spontaneous locking of both switches states when out of respective modes. Finally, grid voltage and current are shown in Fig.9. And current THD of 1.09 % and PF of 0,999 were obtained.

Some drawbacks are carried out during simulation tests. Phase delay during boost mode may cause some current oscillation in the mode transition. It has been observed that connecting a low pass filter before the double integrator and a direct connection of the inductor current to the boost PWM comparator, minimizes this oscillation. In addition, it was observed that the inductor choice strongly affects the performance of controller.

CONCLUSIONS

The work proposed a control strategy for a plug-in battery charger derived from OCC technique. The objective is to present an alternative to digital control strategies. The synthesis of the proposed control strategy was supported by an introduction of converter and the OCC technique particularities. The scheme to produce the proposed strategy was composed by ordinary low cost devices. Advantages of such strategy include no need of grid voltage sensor, analog multiplier, simplicity, robustness and low cost. Simulation tests showed up the effectiveness of proposed control strategy, which presented no critical points during the synthesis and simulation procedures. One characteristic of the solution is the easy extension to higher power operation. During the operation and design of this system, it should be taken special attention for some details, such as, the stability and parameter tuning, as for instance, the Inductor and Low pass filter.

REFERENCES

[1] F. J. B. Brito Jr., S. Daher, R. P. T. Bascopé, B. B. Chaves, D. P. Damasceno and B. R. de Almeida,"Design and Control for a Single-Phase-Stage Buck_Boost-Type PWM PFC Rectifier," Proc. of 13th IEEE International Conference on Industry Applications (INDUSCON). 2018, pp.530-536.

[2] O. Chang-Yeol, K. Dong-Hee, W. Dong-Gyun, S. Won-Yong,K. Yun-Sung and L. Byoung-Kuk,"High-Efficient Nonisolated Single-Stage On-Board Battery Charger for Electric Vehicles", In IEEE Transactions on Power Electronics, Vol. 28, No. 12, Dec. 2013.

[3] M. C. Ghanem, K. Al-Haddad, and G. Roy,"A new single phase buckboost converter with unity power factor," In Proc. 1993 IEEE Ind. Appl. Soc. Annu. Meet., vol. 2, Oct. 1993, pp. 785–792.

[4] Y. Kwang-Min, K. Kyung-Dong, and L. Jun-Young, "Single- and Three-Phase PHEV Onboard Battery Charger Using Small Link Capacitor", In IEEE Transactions on Power Electronics, Vol. 60, No. 8, Aug 2013, pp. 3136- 3144

[5] S. Waffler and J. W. Kolar, "A novel low-loss modulation strategy for high-power bidirectional buck+boost converters," IEEE Trans. Power Electron., vol. 24, no. 6, pp. 1589–1599, June 2009.

[6] M. Anun, M. Ordonez, I. Galiano and G. Oggier, "Bidirectional Power Flow with Constant Power Load in Electric Vehicles: A Non-linear Strategy for Buck+Boost Cascade Converters", In IEEE Applied Power Electronics Conference and Exposition – APEC. 2014.

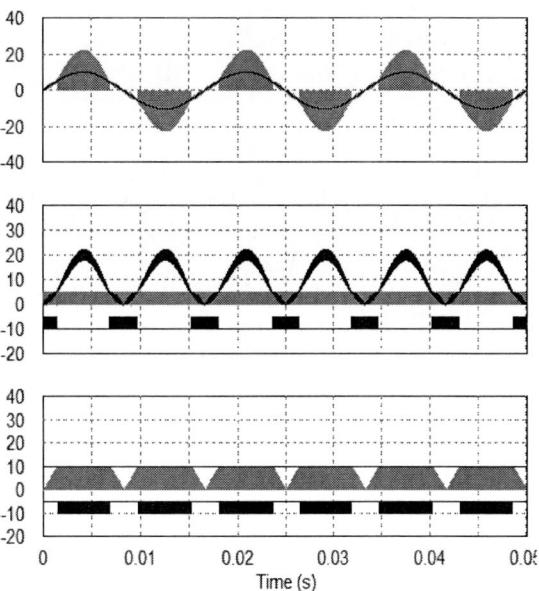

Fig. 7. Simulation results with scheme in Fig. 5. From up to down: instantaneous and average input current; input and output signal of boost PWM block; input and output signal of buck PWM block.

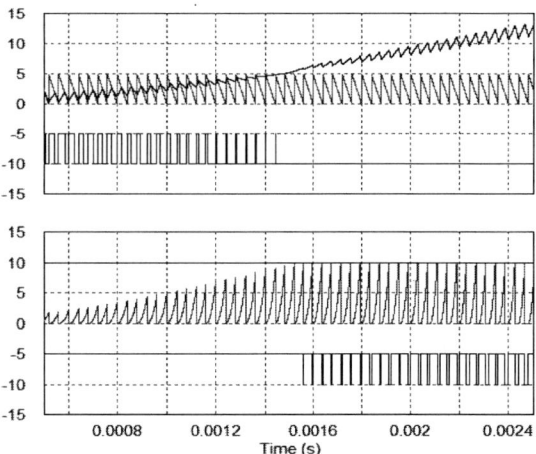

Fig. 8. Zoomed view of PWM signals which highlights the smoothly transition between both operation modes. Top curves: inductor current, triangular carrier and S2 command. Bottom curves: Reference current, integrator output and S1 command

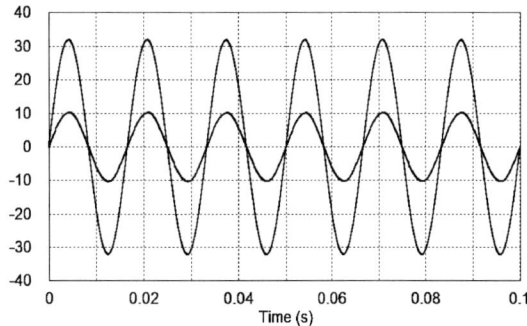

Fig. 9. Grid voltage (large curve, 100 V/div.) and Grid current (smaller curve, 10 A/div.).

[7] S. Ben-Yaakov, I. Zeltser, A. Katz, G. Golembo, "The best of two worlds: a mixed-mode front-end controller IC," In Power Electronics Technology, PET´06, 2006

[8] K. M. Smedley; C. Slobodan, "One cycle control of switching converters," In IEEE Transactions on Power Electronics, Vol. 10, No.8, Nov. 1995, pp. 625-633.

[9] B. Hua, Z .Zhengming, M, Shuo, et al.,"Comparison of the three PWM strategies – SPWM, SVPWM and one-cycle control". Proc. of IEEE Power Electronics and Drive System, 2003, pp.1313-1316.

[10] T. Jin, K. M. Smedley, "Operation of unified constant-frequency integration controlled three-phase active power filter with unbalanced load". Proc. of IEEE APEC 2003, vol. 1, pp. 148-153.

[11] A. A. M. Bento, E. R. C da Silva,"Hybrid One-Cycle Controller for Boost PFC Rectifier", In IEEE Trans. on Ind. Appl., Vol. 45, Issue 1, pp. 268-277. 2009.

[12] A. A. M. Bento, A. Lock, E. R. C. da Silva, D. A. Fernandes, "Hybrid one-cycle control technique for three-phase power factor control," In IET Power Electronics. Vol. 11 , Issue 3, pp. 484- 490. 2018.

[13] J. Rajagopalan, F.C. Lee, P. A. Nora,"Generalized Technique for Derivation of Average Current Mode Control Laws for Power Factor Correction without Input Voltage Sensing," In IEEE Transaction on Power Electronics, vol. 14, pp. 663-672, July 1999.

[14] K. M. Smedley, "Tricks of the Trade: Poincare Stability Analysis of Switching Converters with Nonlinear Control," In IEEE Power Electronic Society Newsletters, vol. 14, no. 1, January, 2002. pp. 5-6.

[15] R. W. Erickson, "Optimal single resistors damping of input filters," in Proc. 14th Annual Appl. Power Electron. Conf. Expo., vol. 2, Mar. 1999, pp. 1073–1079.

Trapezoidal Current Mode for Bidirectional High Step Ratio Modular Multilevel dc-dc Converter

C. Pineda, J. Pereda, S. Neira, P. Bravo
Department of Electrical Engineering
Pontificia Universidad Católica de Chile
Santiago, Chile
cnpineda@uc.cl

J. Rodríguez, C. García
Faculty of Engineering
Universidad Andres Bello
Santiago, Chile
jose.rodriguez@unab.cl

Abstract—High voltage direct current (HVDC) transmission systems and low voltage direct current (LVDC) networks are becoming popular due to their advantages and present feasibility, mainly pushed by the advance in power electronics. However, one of the main drawbacks of dc networks is the lack of a dc-dc converter that performs as the ac transformer in terms of efficiency, reliability and high-step voltage ratio, specially in medium and high voltage. Modular multilevel converters are an attractive alternative because they can manage medium and high dc voltages with standard semiconductor devices, and they can achieve high efficiency with soft switching modulation, such as resonant, trapezoidal or triangular currents usually used in double active bridges (DAB). This paper proposes a novel generalized trapezoidal current mode applied to high step ratio Modular Multilevel dc-dc Converters. The proposed modulation increases the efficiency and achieves bidirectional control of the power, soft-switching and a natural balance of the voltage in the capacitors. The simulation results show the bidirectional operation and the capacitor voltage balance of the converter under different operating conditions with high efficiency (98.2%).

Index Terms—dc-dc conversion, modular multilevel converter, high step ratio conversion, trapezoidal current mode, triangular current mode, soft switching.

I. Introduction

Direct current was displaced by alternating current time ago, at the beginning of the electric power system, and the main reason was the ac transformer. However, the development of power electronics had enabled the return of dc power systems. DC has important advantages over alternating current in the transmission of energy, such as lower power losses, less and smaller cables and absence of reactive power nor peak voltages. Additionally, dc power is increasingly used in renewable energies such as photovoltaic solar systems and wind generators, battery and fuel cells energy storage systems, and in several loads such as computers, modern home appliances, electric vehicles and data centers. However, the penetration of dc power networks will take time because it requires displacing the actual ac infrastructure and there is still a competitiveness gap between the dc conversion technology and the mature ac transformer, especially for high voltage dc-dc converters.

Currently, the most promising technology for HVDC and MVDC systems is the modular multilevel converter (MMC)

[1], but it is mainly used to interconnect ac and dc systems, working as an inverter or rectifier. On the other hand, the development of medium and low voltage direct current (MVDC-LVDC) networks is also of great interest, so the dc-dc converter that achieves a suitable interconnection between HVDC, MVDC and LVDC systems will enable the massive penetration of dc networks. This converter should provide high efficiency, bidirectional power flow, voltage regulation and fault blocking capability [2], [3].

Several modular multilevel converters have been proposed for dc-dc conversion, but the front-to-front (FTF-MMC) topologies stand out. These topologies can be separated in hard switching modulation (HS-MMC), and soft switching modulation (SS-MMC). The hard switching topology uses two MMCs connected through a medium-frequency transformer and generate multilevel ac voltages but it requires capacitor voltage balance control and its hard-switching operation reduce efficiency [4]–[10]. To solve this two drawbacks, soft switching topologies have been proposed using resonant tanks (R-MMC) [11]–[13] or triangular currents mode (TCM-MMC) [14], achieving natural balance of the capacitor voltages and soft switching.

The resonant converter (R-MMC) achieves soft switching but it limits the voltage regulation and generates high peak currents, causing high conduction losses and low efficiency despite its soft-switching operation. On the other hand, trapezoidal current mode (TCM) scheme applied to dual active bridge (DAB) dc-dc converters [15], [16] achieves bidirectional control of the power with higher efficiency, maintaining zero-current-switching (ZCS) operation without resonance. Therefore, trapezoidal current mode is the best option for MMC dc-dc converter and it was recently proposed using a triangular current mode [14], which is a specific case of trapezoidal current mode where the soft switching is achieved optimally.

This paper presents a generalized trapezoidal current mode (TCM) for high step ratio dc-dc MMC based on [14], which reduces the peak current in the ac link over triangular operation. The proposal keeps the advantages of the original work; achieves bidirectional power flow through a simple control law, ZCS operation and natural balance of the voltage in the capacitors. The high step ratio dc-dc MMC topology

978-1-7281-4181-7/19 $31.00 © 2019 IEEE

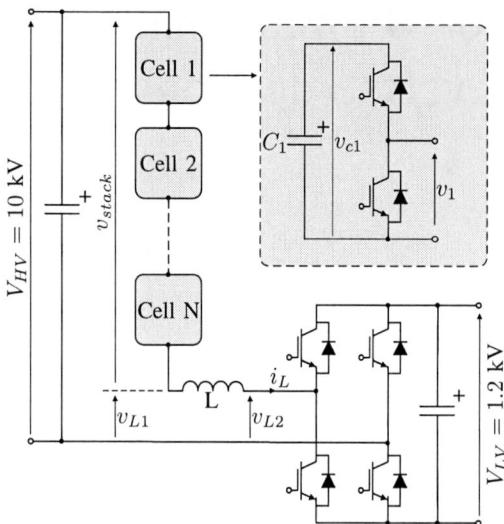

Fig. 1. Topology of the high step ratio dc-dc MMC using half-bridges in cells.

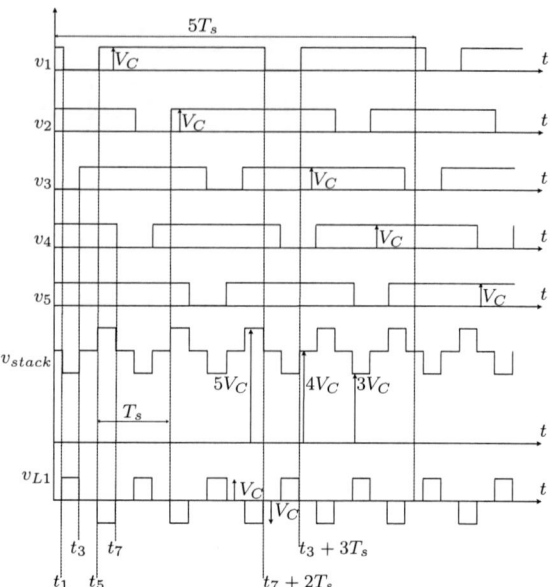

Fig. 2. Operation waves in high voltage side for N=5 cells and j=2.

is presented in Section II, the proposed modulation of the converter is described in Section III and the dynamic balance of power is analysed in Section IV. Finally, the simulation results are shown in Section V.

II. TOPOLOGY

The topology of the high step ratio dc-dc MMC is formed by a stack of N half-bridge cells wich support the high-voltage side (V_{HV}), while the low-voltage side (V_{LV}) is connected by an active full-bridge converter, as is shown in Fig. 1. Both subsystems are connected through an inductor L and the bidirectional power transfer is managed by controlling the current flow on it. The converter achieves high-step ratio without the need for a transformer, however it could be added to provide galvanic isolation. Furthermore, filters are placed at both sides to decrease voltage and currents oscillations, but they are not considered in the following work for simplicity.

The current through the inductor is controlled by the voltages v_{L1} and v_{L2} following (1). The voltage v_{L1} is the difference between the voltage V_{HV} and the total voltage in the stack v_{stack}, given by the sum of the half-bridges output voltages v_i. Thus, v_{stack} is controlled by the switching states of each cell. On the other hand, v_{L2} is determined by the full-bridge switching state and the low voltage V_{LV}.

$$L\frac{di_L}{dt} = v_{L1} - v_{L2} \tag{1}$$

Therefore, the current across the inductance L is controlled by the difference of the voltages v_{L1} and v_{L2} at each side of it in order to control the power flow between both ports.

III. TRAPEZOIDAL MODULATION

To generate a trapezoidal current in the inductor L, the voltages v_{L1} and v_{L2} need to be controlled to generate a three-level voltage in both terminals of the inductor. The full-bridge works with unipolar modulation, generating three levels in v_{L2}, but the stack requires a special modulation to get the needed three levels at v_{L1}.

In order to achieve this, the cells are modulated with the same waveform using a phase shift of T_S to increase the resultant frequency v_{L1} and to keep the balance among the cells [14]. The output voltage of the first half-bridge cell v_1 is defined as a periodical signal of period $N \cdot T_s$ specified by the transition times as is shown in Fig. 2. Therefore, a time-controlled voltage v_{stack} of three-levels centered on $(N-1)V_C$, amplitude V_C and period T_s is generated in the stack if the voltage in the cells are phase shifted by T_s and all the capacitor voltages in the cells are balanced and equal to V_C, as is determined in (2).

$$v_{stack}(t) = \begin{cases} & 0 < t < t_1 \\ (N-1)V_C & t_3 < t < t_5 \\ & t_7 < t < T_s \\ (N-2)V_C & t_1 < t < t_3 \\ N \cdot V_C & t_5 < t < t_7 \end{cases} \tag{2}$$

Thus, v_{L1} is a three-level voltage with the same parameters of v_{stack} but centred on zero if (3) holds.

$$V_C = \frac{V_{HV}}{(N-1)} \tag{3}$$

The presented modulation waveforms for each cell are illustrated in Fig. 2, taking $N = 5$ as an example. Successful operation relies on all cells being healthy. However, in case of failure it is possible to bypass the faulty cell and adjust the V_C to the new number of cells.

Using the stack modulation previously presented, a trapezoidal current through the inductor could be generated by managing the duty cycles (D_1, D_2, D_3, D_4) of the voltages v_{L1}, v_{L2} and two additional phase-shift variables Φ_1 and Φ_2 (Figs.3(a)-(b)). The resultant waveform is defined in (4), considering the conditions $0 < D_1, D_3, (\Phi_1 + D_2), (\Phi_2 + D_4) < 0.5$. This type of modulation allows to manage the same level of power with different trapezoidal current waveforms, thus the extra degrees of freedom (Φ_1, Φ_2) can be used to minimize current peaks and conduction losses.

$$
i_L(t) = \begin{cases}
0 & \text{if } t \in (0, t_1), (t_4, t_5) \\
& \quad \vee \ (t_8, T_s) \\[6pt]
\frac{V_C}{L}(t - t_1) & \text{if } t \in (t_1, t_2) \\[6pt]
i(t_2) - \frac{V_{LV} - V_C}{L}(t - t_2) & \text{if } t \in (t_2, t_3) \\[6pt]
i(t_3) - \frac{V_{LV}}{L}(t - t_3) & \text{if } t \in (t_3, t_4) \\[6pt]
-\frac{V_C}{L}(t - t_4) & \text{if } t \in (t_5, t_6) \\[6pt]
i(t_6) + \frac{V_{LV} - V_C}{L}(t - t_6) & \text{if } t \in (t_6, t_7) \\[6pt]
i(t_7) + \frac{V_{LV}}{L}(t - t_7) & \text{if } t \in (t_7, t_8)
\end{cases}
\tag{4}
$$

The trapezoidal current mode shows smaller current peaks compared with the asymmetric triangular current mode (ATCM) for the same power level. Additionally, there is no restriction for the voltage ratio $r_V = (V_C/V_{LV})$ as the current goes to zero during (t_3, t_4) with $V_{LV} \sim V_C$. However, the ATCM presents two more soft switching transitions compared with the trapezoidal case. This comparative behaviour is shown in Figs. 3(a)-(c) for a power flow from high-voltage side to low-voltage side and in Figs. 3(b)-(d) for the opposite power flow direction. A further analysis shows that the ATCM is a specific case of the trapezoidal modulation with $D_1 = \Phi_1 + D_2$ and $D_3 = \Phi_2 + D_4$, as long as the voltage ratio ensures that the current return to zero. Thus, when systems show constant voltages with $r_V < 1$ the triangular current mode is suitable, on the other hand, the trapezoidal modulation is suitable for systems with variable r_V, even allowing $r_V = 1$.

Condition in (5) ensures ZCS for half of the commutations in the proposed trapezoidal current mode. The peak currents in the inductor (6) depends on $\{\Phi_1, \Phi_2\}$, enabling the optimization of the current shape for a given power transfer reference based on these parameters.

$$
D_{i+1} = r_V \cdot D_i \qquad i = 1, 3
\tag{5}
$$

$$
\begin{aligned}
i_L^{p1} &= i_L(t_2) = \frac{V_C}{f_s L}\Phi_1 \\[6pt]
i_L^{p2} &= i_L(t_6) = -\frac{V_C}{f_s L}\Phi_2
\end{aligned}
\tag{6}
$$

IV. POWER MODELING

To modeling exchange of power between the high-voltage port, the low-voltage port and the cell stack, the area covered for the inductor current in the first (A_{Ltp}^1) and second (A_{Ltp}^2) half cycles of the switching cycle are defined in (7) and illustrated in Fig. 3 (a). These areas can be defined as function (f_{tp}) of the phase shift Φ_i and the cycle D_i if the ZCS condition is achieved.

$$
\begin{aligned}
A_{Ltp}^1 &= f_{tp}(\Phi_1, D_1) = \int_0^{\frac{T_s}{2}} i_L \cdot dt \\[6pt]
A_{Ltp}^2 &= -f_{tp}(\Phi_2, D_3) = \int_{\frac{T_s}{2}}^{T_s} i_L \cdot dt
\end{aligned}
\tag{7}
$$

$$
f_{tp}(\Phi, D) = \frac{V_c \cdot D \cdot T}{2fL}\left(2\Phi - D\left(1 - r_V\right)\right)
$$

The average power delivered by the high-voltage port at one switching cycle P_{HV} is defined by the sum of the areas A_{Ltp}^1 and A_{Ltp}^2 due to the inductor current waveform is the same as in the high-voltage port if filters are not considered (8).

$$
P_{HV} = V_{HV} \cdot \left(\frac{A_{Ltp}^1 + A_{Ltp}^2}{T_s}\right)
\tag{8}
$$

The previous analysis supposes that the mean voltage in the cell capacitors remains balanced during their operation. In order to found the voltage capacitor balance condition, the inductor current area between times t_3 and t_4 (A_{Ltg}^1) and between times t_7 and t_8 (A_{Ltg}^2) are defined in (9) and illustrated in Fig. 3 (a). Also, these areas can be defined as function (f_{tg}) of the phase shift Φ_i and the duty cycle D_i if the ZCS condition is achieved.

$$
\begin{aligned}
A_{Ltg}^1 &= f_{tg}(\Phi_1, D_1) = \int_{t_3}^{t_4} i_L \cdot dt \\[6pt]
A_{Ltg}^2 &= -f_{tg}(\Phi_2, D_3) = \int_{t_7}^{t_8} i_L \cdot dt
\end{aligned}
\tag{9}
$$

$$
f_{tg}(\Phi, D) = \frac{V_{LV} \cdot T}{2fL}\left(\Phi - D\left(1 - r_V\right)\right)^2
$$

Considering the trapezoidal current mode and the proposed modulation, the voltage in each cell capacitor can be found integrating the current flowing through it. Then, the capacitor voltage at the end of the modulation time window $N \cdot T_s$ can

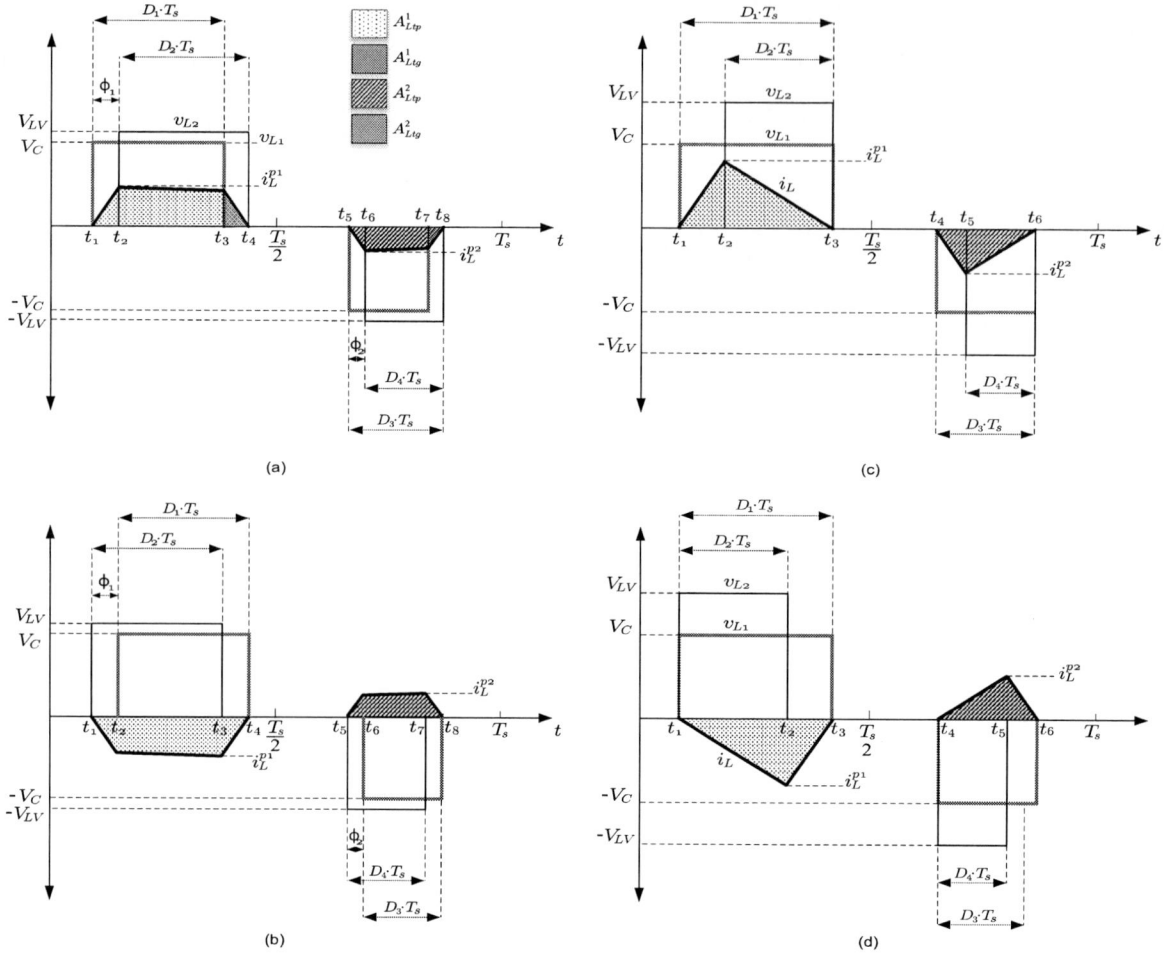

Fig. 3. Current and voltage waveforms for converter under trapezoidal and triangular modulations: (a) power flow from high-voltage side to low voltage side for trapezoidal modulation. (b) power flow from low-voltage side to high-voltage side for trapezoidal modulation. (c) power flow from high-voltage side to low voltage side for triangular modulation. (d) power flow from low-voltage side to high-voltage side for triangular modulation.

be expressed using the areas A_{Ltp}^1, A_{Ltp}^2, A_{Ltg}^1 and A_{Ltg}^2 as is shown in (10).

$$v_{ci}(t_0 + N \cdot T_s) = v_{ci}(t_0) +$$
$$\frac{A_{Ltp}^1 \cdot (N-2) + A_{Ltg}^1 + A_{Ltp}^2 \cdot N - A_{Ltg}^2}{C_i}$$
$$(10)$$

Therefore, the steady state voltage balance condition in the stack can be found in (11), imposing that the voltage in the capacitor remains constant at the end of the modulation window $N \cdot T_s$.

$$A_{Ltp}^1 \cdot (N-2) + A_{Ltg}^1 + A_{Ltp}^2 \cdot N - A_{Ltg}^2 = 0 \quad (11)$$

Thus, solving (8) and (11) is possible to control the power transfer between high-voltage and low-voltage ports, keeping the voltage balance in the capacitors. These equations has a set of solutions for Φ_1, Φ_2, D_1 and D_3, which can be optimized

considering peak current and losses. A similar analysis can be performed for negative power flow inverting the phase shift of voltages v_{L1} and v_{L2} (Φ_1 and Φ_2) as is illustrated in Fig 3 (b). Therefore, these relationships allow full bidirectional control of the power flow maintaining the ZCS and voltage balance in the converter.

V. RESULTS

This section presents the PLECS simulation in a 1MW full-scale converter. The parameters of the full-scale proposed converter are listed in Table I and the semiconductor devices were modelled using the loss model provided by the manufacturer. Different capacitance values around 120 mF are considered for each capacitor in the cells to evaluate the natural balance in real conditions (capacitance tolerance of 20%).

Fig. 4 (a) shows the operation of the converter at 500 kW with the low-voltage side as load, illustrating the voltages

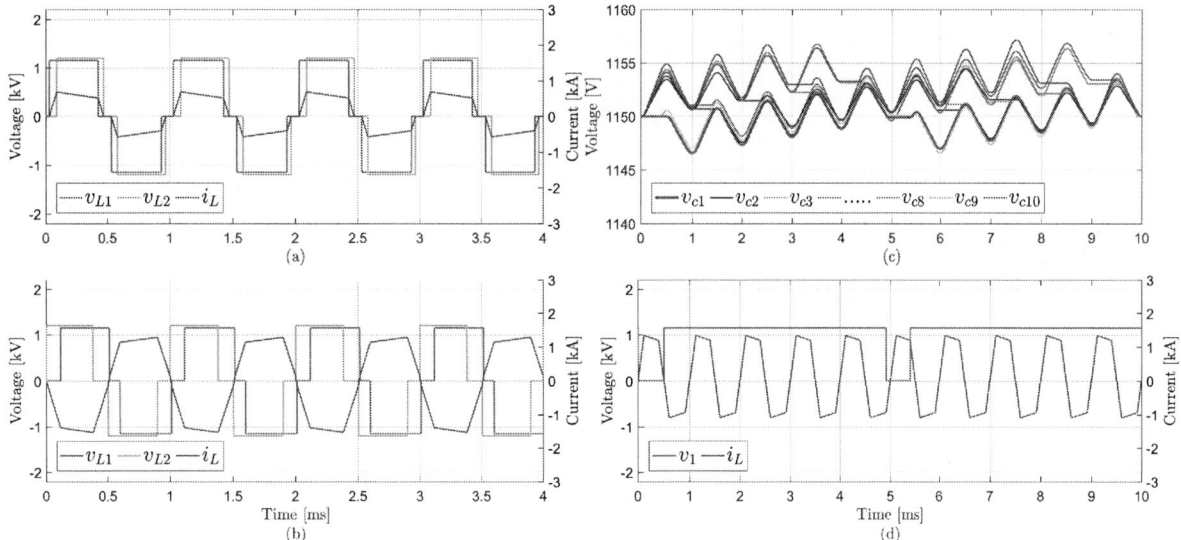

Fig. 4. Simulation waveforms. (a) Voltage and current from high voltage port to low voltage port at 500 kW. (b) Voltage and current from low voltage port to high voltage port at 1 MW. (c) Waveforms of capacitor cells voltages at 1MW. (d) Voltage and current of one cell at 1 MW.

and current of the inductor in four switching periods. The bidirectional capability is illustrated in Fig. 4 (b), showing the waveform of the converter at 1 MW of power from low voltage port to high voltage port in the same time window.

The voltages of each cell capacitor are shown in Fig. 4 (c) for one modulation time window, verifying the correct power balance of the converter and the natural balance of the cell capacitors, even with different capacitance values. The ripple voltage in the capacitors are limited to 1% by design.

The voltage and cell current for one cell are shown in Fig. 4 (d), verifying the ZCS in this side of the converter. The voltage v_{c1} and v_1 represents the voltage of the capacitor and the output voltage of cell 1 respectively. Thus, the charge and discharge process of the capacitor v_{c1} can be observed in Fig. 4 (c) and (d). Also, the condition of balance presented in (11) can be deducted from Fig. 4 (d), looking the number of cycles of positive current over negative current in the cell while it is connected ($v_1 = V_C$).

The overall simulated efficiency is 98.2% at full power. The following power losses were taken into account: (i) IGBT conduction losses; (ii) IGBT turn-on and turn-off switching losses; (iii) diode reverse recovery losses; and (v) diode conduction losses. The power losses in the cells stack at full power represents 80.03% of total losses and are composed mostly by conduction losses (94.6%) due to the ZCS operation. The power losses in the low-voltage full-bridge at full power represents the other 19.97% of total losses and are distributed more evenly between conduction (70.4%) and switching losses (29.6%). The peak current is 1.6 times the average current at 1 MW.

Both efficiency and peak current achieve better results with the proposed modulation than those previously presented

TABLE I
PARAMETERS OF SIMULATION

Description	Parameter	Value
Maximum Power	P_{tmax}	1 MW
HV side Voltage	V_{HV}	10.35 kV
LV side Voltage	V_{LV}	1.2 kV
Number of cells	N	10
Inductor inductance	L	100 μH
Capacitance of cell capacitors	C_i	120 mF $\pm$ 20%
Switching frequency of the full-bridge	f_s	1 kHz
IGBTs ABB 5SNA3600E17000	V / I	1.7 kV / 3.6 kA

using a resonant operation for a similarly designed converter [11]. The triangular current mode [14] is a specific operation point of the trapezoidal current mode (TCM), where the soft switching is achieved twice times the trapezoidal one, but the peak current is higher. Therefore, the mode of operation (trapezoidal or triangular) can be selected in real time to optimize the overall efficiency of the converter according to the operation point, choosing to reduce the switching losses (triangular) or the conduction losses (trapezoidal).

VI. CONCLUSIONS

A generalized trapezoidal current mode (TCM) for dc-dc MMC has been proposed and validated, achieving bidirectional power control, ZCS and natural balance of the voltage in capacitor cells. A simple and general condition law is proposed to modulate the converter. Theoretical analysis has been verified with full-scale simulations, showing higher efficiency (98.2%) and lower peak current compared to previous reports of this topology using resonant operation and triangular current mode. However, triangular current mode, a specific operation point of the proposed TCM, minimizes the switching losses because it achieves the double of soft switching than general TCM. Finally, the converter using general TCM has a high potential to be a suitable solution for high step ratio dc-dc conversion.

ACKNOWLEDGMENT

The authors gratefully acknowledge the financial support provided by CONICYT through FONDECYT 1171142, SERC Chile (CONICYT/FONDAP/15110019), and the UC Energy Research Center of Pontificia Universidad Católica de Chile.

REFERENCES

[1] A. Lesnicar and R. Marquardt, "An innovative modular multilevel converter topology suitable for a wide power range," in *2003 IEEE Bologna Power Tech Conference Proceedings,*, vol. 3, pp. 6–pp, IEEE, 2003.

[2] S. P. Engel, M. Stieneker, N. Soltau, S. Rabiee, H. Stagge, and R. W. De Doncker, "Comparison of the modular multilevel dc converter and the dual-active bridge converter for power conversion in hvdc and mvdc grids," *IEEE Transactions on Power Electronics*, vol. 30, no. 1, pp. 124–137, 2014.

[3] J. D. Páez, D. Frey, J. Maneiro, S. Bacha, and P. Dworakowski, "Overview of dc–dc converters dedicated to hvdc grids," *IEEE Transactions on Power Delivery*, vol. 34, no. 1, pp. 119–128, 2018.

[4] N. Denniston, A. M. Massoud, S. Ahmed, and P. N. Enjeti, "Multiple-module high-gain high-voltage dc–dc transformers for offshore wind energy systems," *IEEE Transactions on Industrial Electronics*, vol. 58, no. 5, pp. 1877–1886, 2010.

[5] T. Lüth, M. M. Merlin, T. C. Green, F. Hassan, and C. D. Barker, "High-frequency operation of a dc/ac/dc system for hvdc applications," *IEEE Transactions on Power Electronics*, vol. 29, no. 8, pp. 4107–4115, 2013.

[6] J. A. Ferreira, "The multilevel modular dc converter," *IEEE Transactions on Power Electronics*, vol. 28, no. 10, pp. 4460–4465, 2013.

[7] K. Filsoof and P. W. Lehn, "A bidirectional modular multilevel dc–dc converter of triangular structure," *IEEE Transactions on Power Electronics*, vol. 30, no. 1, pp. 54–64, 2014.

[8] I. Gowaid, G. Adam, A. M. Massoud, S. Ahmed, D. Holliday, and B. Williams, "Quasi two-level operation of modular multilevel converter for use in a high-power dc transformer with dc fault isolation capability," *IEEE Transactions on Power Electronics*, vol. 30, no. 1, pp. 108–123, 2014.

[9] R. Vidal-Albalate, J. Barahona, D. Soto-Sanchez, E. Belenguer, R. S. Peña, R. Blasco-Gimenez, and H. Z. de la Parra, "A modular multilevel dc-dc converter for hvdc grids," in *IECON 2016-42nd Annual Conference of the IEEE Industrial Electronics Society*, pp. 3141–3146, IEEE, 2016.

[10] S. Cui, N. Soltau, and R. W. De Doncker, "A high step-up ratio soft-switching dc–dc converter for interconnection of mvdc and hvdc grids," *IEEE Transactions on Power Electronics*, vol. 33, no. 4, pp. 2986–3001, 2017.

[11] X. Zhang, T. C. Green, and A. Junyent-Ferré, "A new resonant modular multilevel step-down dc–dc converter with inherent-balancing," *IEEE Transactions on Power Electronics*, vol. 30, no. 1, pp. 78–88, 2014.

[12] X. Xiang, X. Zhang, G. P. Chaffey, and T. C. Green, "An isolated resonant mode converter with flexible modulation and variety of configurations for mvdc application," *IEEE Transactions on Power Delivery*, vol. 33, no. 1, pp. 508–519, 2017.

[13] X. Zhang, M. Tian, X. Xiang, J. Pereda, T. C. Green, and X. Yang, "Large step ratio input-series-output-parallel chain-link dc-dc converter," *IEEE Transactions on Power Electronics*, 2018.

[14] C. Pineda, J. Pereda, X. Zhang, and F. Rojas, "Triangular current mode for high step ratio modular multilevel dc-dc converter," in *2018 IEEE Energy Conversion Congress and Exposition (ECCE)*, pp. 5185–5189, IEEE, 2018.

[15] G. Ortiz, H. Uemura, D. Bortis, J. W. Kolar, and O. Apeldoorn, "Modeling of soft-switching losses of igbts in high-power high-efficiency dual-active-bridge dc/dc converters," *IEEE Transactions on Electron Devices*, vol. 60, no. 2, pp. 587–597, 2012.

[16] F. Jauch and J. Biela, "Generalized modeling and optimization of a bidirectional dual active bridge dc-dc converter including frequency variation," *IEEJ Journal of Industry Applications*, vol. 4, no. 5, pp. 593–601, 2015.

Homothetic Method to Compute Winding Losses in the Design of Power Inductors

André Furlan[1,2], Alvaro Morentin[1], Guillaume Fontes[1], Guillaume Delamare[1], Marcelo L. Heldwein[2], Thierry Meynard[1,3]

[1] *Power Design Technologies, Toulouse, France*
[2] *Power Electronics Institute (INEP), Federal University of Santa Catarina (UFSC), Florianópolis, Brazil*
[3] *LAPLACE, Université de Toulouse, CNRS, INPT, UPS, France*
furlan@inep.ufsc.br, {alvaro.morentin, guillaume.fontes, guillaume.delamare}@powerdesign.tech,
heldweing@inep.ufsc.br, thierry.meynard@laplace.univ-tlse.fr

Abstract—**This work proposes an improvement for the consideration of ac winding losses in the design process of inductors to be employed in power converters, where the geometrical parameters of the inductors are not known before starting the design. The presented method consists of the pre-computation of a homothetic multivariate regular grid surface by finite element simulations, from which the winding resistance can be obtained by direct multivariate linear interpolation. Special attention is given to Dowell's analytical formulations. These are often used in the design but typically with little details given on how the final expressions are derived. The results with the new method are shown to be more accurate than the analytical calculations over a wide range of design points and, thus, are a useful design tool.**

Index Terms—**Winding resistance, finite element analysis, magnetic components, air gap**

I. INTRODUCTION

Dowell's equations [1] are often used to calculate the equivalent ac resistance (R_{AC}) of inductor windings, taking skin and proximity effects into account. These equations are derived through a 1D approximation to calculate the window magnetic field under certain simplifying hypotheses. However, inductors are generally gapped, and this 1D approximation is not applicable since the fringing field of the air gap cannot be described in a 1D manner [2]. Gapped inductors can be accurately analyzed using the finite element method (FEM) with the computation efforts and CPU time that this represents. When a fast computation of the R_{AC} is needed, to avoid a computationally intensive finite element simulation, some works applied tuned equations based in Dowell's methods [3] [4]. In such works, additional coefficients are introduced through a data fitting process against FEM results. The derived relation coefficients depend on the inductor geometrical parameters. Consequently, this approach can be used only for specific geometries. Some analytical methods for gapped inductors are presented in [2] [5] [6]. Nevertheless, a final validation with a FEM simulation is indispensable.

Fig. 1 describes a typical power inductor design procedure. After a fast design, using the area product approach ($AwAc$) [7], with the geometric parameters, the winding loss can be evaluated through different methods. Following this evaluation, the calculated and projected power losses are compared.

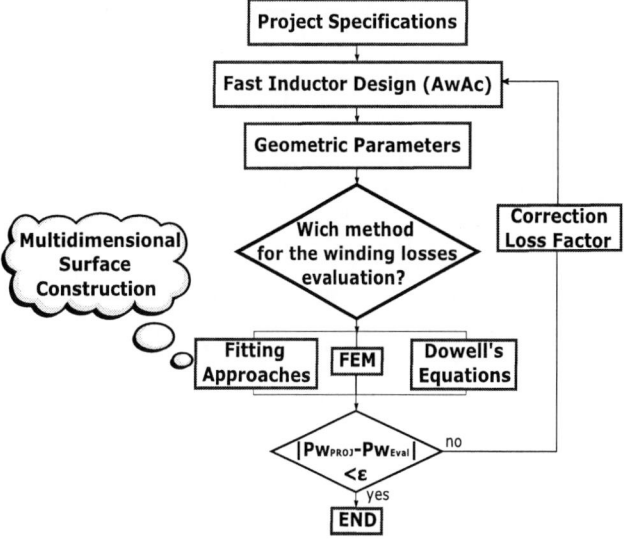

Fig. 1. Inductor design procedure.

If the error is greater than the tolerance, the design process is repeated once again, using a correction loss factor.

In this work, a general approach based on the interpolation of a surface that is pre-computed using 2D FEM is proposed to obtain the R_{AC} of a generic E-core inductor in a reduced CPU time. Thus, generating a process that is adequate for use in iterative and optimization design tools. The homothetic properties of these structures as well as relations between different conductivities are explored to reduce the number of variables. Generating this surface requires a high computational effort. Nevertheless, it can be used in the design of an arbitrary number of inductors after being computed offline.

First, the topology of the studied inductor is presented, and Dowell's equations are derived with two different approaches: neglecting the gap in the calculation of the orthogonal magnetic field (H_E) and considering the gap in the evaluation of H_E.

After describing the formulation, the homothetic and conductivity relations are studied to reduce the number of vari-

978-1-7281-4181-7/19 $31.00 © 2019 IEEE

ables of the problem. Different values of inductor geometric parameters are combined, and many FEM simulations have been run to establish a reduced multidimensional FEM surface. R_{AC} is then evaluated through direct multivariate linear interpolation from the FEM results. The resulting values are shown to match FEM simulations more closely than analytical formulations.

II. CHARACTERISTICS OF THE FOIL INDUCTOR

Fig. 2 describes the cross section of the right-side core-winding window of an E-core inductor that is built with foil conductors using air gaps in the center and outer legs.

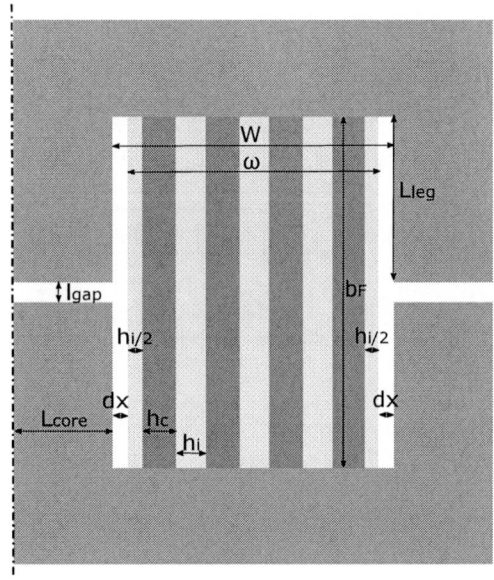

Fig. 2. Cross section of the right-side core-winding window of an E-core inductor.

The following parameters are defined:
- f : excitation frequency [Hz];
- N: number of turns;
- L_{core}: core leg width [m];
- W: core window width [m];
- w: winding window width [m];
- L_{leg}: core leg height [m];
- b_F: core window total height [m];
- h_c: foil conductor thickness [m];
- h_i: insulation thickness [m];
- l_{gap}: air gap length [m];
- dx: distance between coil insulation and vertical core legs ($dx = (W - w)/2$) [m];
- σ: foil conductivity [S/m];
- μ_{cr}: core relative permeability;
- k_w: winding fill factor. Ratio between the winding section area and the winding window area without spacing ($k_w = N \cdot h_c / w$).

III. CLASSICAL ANALYTICAL FORMULATION

Two different approaches for Dowell's equations are analyzed in this section. The first approach is widely used in previous works: it was developed for ungapped cores, resulting in the classical Dowell's equation. The second approach considers the gap in the expression of the orthogonal field (H_E). Usually, little details are given about the formulation and its application.

Classical methods to compute R_{AC} are based in the orthogonality existing between skin effect and proximity effect in conductors. This makes possible to decouple the two effects and simplifies the analysis [8]. In the next subsections, these effects will be described in terms of losses for a sinusoidal excitation at frequency f in order to calculate R_{AC}.

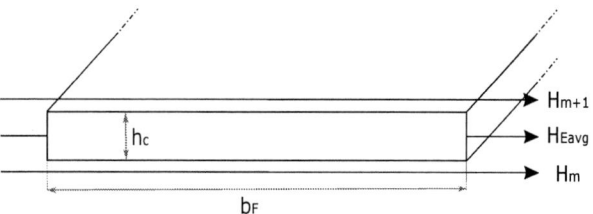

Fig. 3. Cross section of a foil conductor that is influenced by an external magnetic field (H_E)

A. Skin Effect

The skin-effect losses for a single winding layer (including DC losses) per unit length is calculated as [2]

$$P_{Skin} = F_{Skin} \cdot r_{DC} \cdot \widehat{I}^2 \tag{1}$$

where $\widehat{I}$ is the peak current flowing through the coil, Δ is the foil conductor thickness normalized to the skin depth (h_c/δ), δ is the skin depth ($1/\sqrt{\pi \mu_0 \sigma f}$), μ_0 is the vacuum permeability, r_{DC} is the single layer DC resistance ($1/(\sigma b_F h_c)$), and F_{Skin} is the factor that describes the rise of the conductor resistance due to the skin effect (2).

$$F_{Skin} = \frac{\Delta}{4} \frac{\sinh \Delta + \sin \Delta}{\cosh \Delta - \cos \Delta} \tag{2}$$

B. Proximity Effect

The losses due to the proximity effect for a single winding layer are described as [2]

$$P_{Prox} = F_{Prox} \cdot r_{DC} \cdot H_{Eavg}^2 \tag{3}$$

where F_{Prox} is a factor that describes the amount of winding losses due to the proximity effect

$$F_{Prox} = b_F^2 \Delta \frac{\sinh \Delta - \sin \Delta}{\cosh \Delta + \cos \Delta} \tag{4}$$

and H_{Eavg} is the average value of the two magnetic fields of the two sides of the coil, illustrated in Fig. 3 and described by the following equation

$$H_{Eavg} = \frac{H_m + H_{m-1}}{2} \tag{5}$$

where m is the number of the layer, and

$$H_m = \frac{m\widehat{I} - h_{cg}l_g}{b_F} \tag{6}$$

978-1-7281-4181-7/19 $31.00 © 2019 IEEE   1409

where h_{cg} is the magnetic field at the center gap. Usually, this variable is not included in the Dowell's method demonstration, which would lead to a better understanding of Dowell's equations.

With (5) and (6), H_{Eavg}^2 becomes

$$H_{Eavg}^2 = \left(\frac{(2m-1)\widehat{I}}{2b_F}\right)^2 - \frac{(2m-1)\widehat{I}h_{cg}l_g}{b_F} + \left(\frac{h_{cg}l_g}{b_F}\right)^2 \tag{7}$$

C. Total Losses

The total proximity losses per unit length for N turns is

$$P_{ProxTot} = F_{Prox} \cdot r_{DC} \cdot \sum_{m=1}^{N} H_{Eavg}^2 \tag{8}$$

where

$$\sum_{m=1}^{N} H_{Eavg}^2 =$$
$$\sum_{m=1}^{N} \left[\left(\frac{(2m-1)\widehat{I}}{2b_F}\right)^2 - \frac{(2m-1)\widehat{I}h_{cg}l_g}{b_F} + \left(\frac{h_{cg}l_g}{b_F}\right)^2 \right]$$
$$= \frac{(4N^3-N)\widehat{I}^2}{12b_F^2} - \frac{h_{cg}l_g N^2 \widehat{I}}{b_F^2} + \frac{h_{cg}^2 l_g^2 N}{b_F^2} \tag{9}$$

For an ungapped core $l_g = 0$. Thus,

$$\sum_{m=1}^{N} H_{Eavg}^2 = \frac{(4N^3-N)\widehat{I}^2}{12b_F^2} \tag{10}$$

and

$$\sum_{m=1}^{N} H_{Eavg}^2 = \frac{(N^3-N)\widehat{I}^2}{12b_F^2} \tag{11}$$

for $h_{cg} = N\widehat{I}/2l_g$.

The total losses per unit length including proximity and skin effects can be found with

$$P_{Tot} = \sum_{m=1}^{N} (P_{Prox} + P_{Skin}) \tag{12}$$

1) Ungapped Core:
For ungapped cores (12) becomes

$$P_{Tot} = R_{DC}\widehat{I}^2\frac{\Delta}{4}\left(\frac{\sinh\Delta + \sin\Delta}{\cosh\Delta - \cos\Delta} + \frac{4N^2-1}{3}\frac{\sinh\Delta - \sin\Delta}{\cosh\Delta + \cos\Delta}\right) \tag{13}$$

where $R_{DC} = Nr_{DC}$. Thus, it follows that,

$$R_{AC} = R_{DC}\frac{\Delta}{2}\left(\frac{\sinh\Delta + \sin\Delta}{\cosh\Delta - \cos\Delta} + \frac{4N^2-1}{3}\frac{\sinh\Delta - \sin\Delta}{\cosh\Delta + \cos\Delta}\right) \tag{14}$$

Equation (14) presented in [2] is equivalent to classic Dowell's formulation presented in several works, in particular [9]. Indeed, by applying (15) into (14), the classic Dowell's equation (16) is obtained.

$$\frac{\sinh\Delta + \sin\Delta}{\cosh\Delta - \cos\Delta} = 2\frac{\sinh 2\Delta + \sin 2\Delta}{\cosh 2\Delta - \cos 2\Delta} - \frac{\sinh\Delta - \sin\Delta}{\cosh\Delta + \cos\Delta} \tag{15}$$

$$R_{AC} = R_{DC}\Delta\left(\frac{\sinh 2\Delta + \sin 2\Delta}{\cosh 2\Delta - \cos 2\Delta} + \frac{2(N^2-1)}{3}\frac{\sinh\Delta - \sin\Delta}{\cosh\Delta + \cos\Delta}\right) \tag{16}$$

2) Gapped Core:
For the gapped core shown in Fig. 2, (12) becomes

$$P_{Tot} = R_{DC}\widehat{I}^2\frac{\Delta}{4}\left(\frac{\sinh\Delta + \sin\Delta}{\cosh\Delta - \cos\Delta} + \frac{N^2-1}{3}\frac{\sinh\Delta - \sin\Delta}{\cosh\Delta + \cos\Delta}\right) \tag{17}$$

the above expression gives,

$$R_{AC} = R_{DC}\frac{\Delta}{2}\left(\frac{\sinh\Delta + \sin\Delta}{\cosh\Delta - \cos\Delta} + \frac{N^2-1}{3}\frac{\sinh\Delta - \sin\Delta}{\cosh\Delta + \cos\Delta}\right) \tag{18}$$

Equation (18) differs from the classical approach (16), and it is found in the literature much less frequently, in particular (18) is found in [3].

IV. MULTI-DIMENSIONAL SURFACE DESIGN

The construction of a regular homothetic multi-dimensional surface is proposed in this study. This surface can be used in the Fitting Approaches described in Fig. 1, allowing a simple multivariate interpolation to evaluate R_{AC} for generic gapped E-core inductors. In the multivariate problem, linear multidimensional interpolation is an attractive choice due to its computation speed and simplicity.

Throughout this section, the homothetic properties of the proposed surface, as well as relations between different conductivities, are explored to reduce the number of variables. Finally, the surface construction is described.

A. Homothetic

A possible approach to reduce the number of variables is to consider fixed shapes of magnetic cores with dimensions that vary according to a homothetic law [10].

According to Dowell's equations, for two inductors with foil conductor thicknesses h_{c1} and h_{c2}, to have the same R_{AC}/R_{DC}, we need to have $\Delta_1 = \Delta_2$, i.e. $h_{c1}/\delta_1 = h_{c2}/\delta_2$. For homothetic inductors, $N_1 = N_2$ and $k_{w1} = k_{w2}$, thus,

$$w_1/\delta_1 = w_2/\delta_2 \tag{19}$$

Therefore, instead of working directly with the frequency in our design of experiments, we can use the foil conductor thickness normalized to the skin depth (w/δ). The geometric compensation is contained inside this relation. For the same foil materials ($\sigma_1 = \sigma_2$) applying the relation (19) and the skin depth equation leads to

$$f_2 = f_1 \cdot K_{g12}^2 \qquad (20)$$

where $K_{g12} = w_1/w_2$ is the geometric factor of two homothetic inductors.

All geometric variables must follow the geometric factor to take advantage of the homothetic property, except the w/δ ratio, where the geometric factor is already implicit. To have a homothetic surface, we can set a geometric parameter and change the other parameters in function of their relations with it.

B. Conductivity Relations

Another important advantage of working with the w/δ ratio is that the same surface, constructed by fixing a winding material, can be used for different coil conductivities. The conductive factor is also implicit in the thickness of the foil conductor normalized to the skin depth. For two identical inductors with only different winding material, the relation (19) becomes:

$$\delta_2 = \delta_1 \implies f_2 = f_1 K_{\sigma 12} \qquad (21)$$

where $K_{\sigma 12} = \sigma_1/\sigma_2$ is the conductivity factor of two identical inductors with different foil conductivities, σ_1 and σ_2, respectively.

C. Surface Construction

The multivariate regular grid surface implies that all possible combinations between the different variables are covered. To generate this surface, the value of the fixed parameter needs to be chosen. Upper and lower limits for the remaining variables are defined. The discretization of the resulting intervals and a FEM simulation for each of the possible combinations are done in the sequence. The resulting total number of points on the surface depends on the number of different discrete values of each variable.

The flowchart illustrated in Fig. 4 describes the construction of the proposed homothetic surface. Subsequently, each item of this flowchart will be discussed.

1) Variables of the Problem : In order to construct a regular grid surface, the relative permeability of the core is fixed to 2100, and aluminum foil conductors are used for the inductor winding. Eight variables result, namely: f, N, W, w, L_{core}, L_{leg}, l_{gap}, and k_w.

2) Fixed-value Parameter : The variable W was chosen to be fixed. Then the following function represents the multidimensional surface

$$R_{AC}/R_{DC} = F_s\left(N, k_w, \frac{w}{\delta}, \frac{w}{W}, \frac{L_{core}}{W}, \frac{L_{leg}}{W}, \frac{l_{gap}}{W}\right) \quad (22)$$

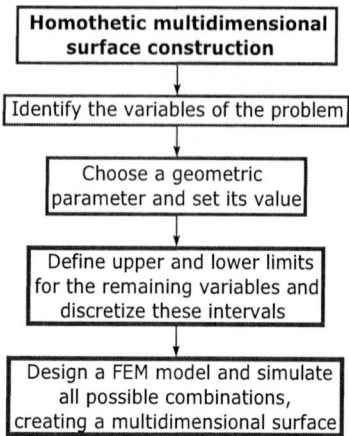

Fig. 4. Proposed Homothetic Surface.

It can be noted that all variables of (22) are dimensionless, which is a typical characteristic of a homothetic problem.

In terms of simulation, W was fixed to a small value (0.05 m). Although the surface is constructed with a fixed W, due to the homothetic property, as we are working with geometric normalized variables, this surface can be used to interpolate any value of W. A reduced maximum frequency for larger topologies is identified by considering (20), which is not a problem for the design since higher frequencies tend to reduce magnetics size.

3) Discretization of Parameters: The next step is to determine the range of the ratios between the remaining geometric variables and the fixed-value parameter (w/W, L_{core}/W, L_{leg}/W and l_{gap}/W). To define upper and lower limits for the core variables L_{core}/W and L_{leg}/W, the E cores present in [12] were taken as reference. The minimum and the maximum air gap normalized by the core window select for our test surface were $W/200$, and $W/2$, respectively. The winding window (w) was set between 50% and 99.8% of W. Finally, the winding fill factor was set between 0.2 and 0.8, and the number of turns was limited to 10, to reduce the calculation time to construct the surface test. The discretization of the variables intervals is described below:

$$
\begin{aligned}
[N] &= [1, 5, 10] \\
[L_{core}/W] &= [0.4, 0.8, 1.2] \\
[w/W] &= [0.5, 0.7, 0.8, 0.98, 0.99, 0.998] \\
[L_{leg}/W] &= [1.0, 1.5, 2.0, 2.5, 3.0] \\
[l_{gap}/W] &= [0.005, 0.05, 0.5] \\
[k_w] &= [0.2, 0.4, 0.6, 0.8] \\
[w/\delta] &= [0.1, 10, 25, 100, 250, 500]
\end{aligned}
\qquad (23)
$$

The total number of FEM simulations to construct a regular grid surface with the above variables intervals is: $range(N) \cdot range(\frac{L_{core}}{W}) \cdot range(\frac{w}{W}) \cdot range(\frac{L_{leg}}{W}) \cdot range(\frac{l_{gap}}{W}) \cdot range(k_w) \cdot range(\frac{w}{\delta}) = 19440$.

4) FEM Design: The 2D FEM simulations have been carried out with the FEMM software [13]. The harmonic magnetodynamic formulation is used to calculate the equivalent

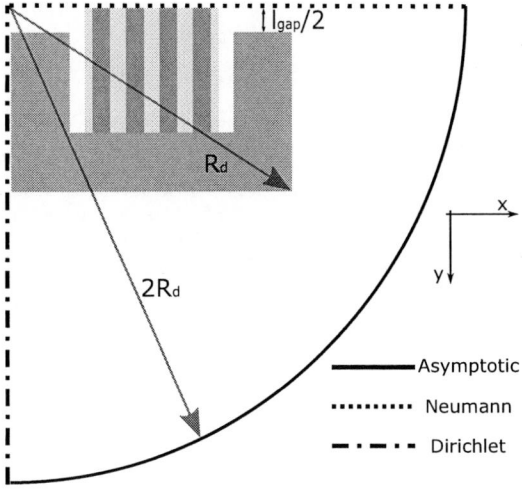

Fig. 5. Reduced FEM Domain

resistance of the windings. Details of its formulation are found in [14].

Due to the symmetry of the problem, only one-quarter of the domain is modeled as shown in Fig. 5. The application of Asymptotic Boundary Conditions (ABC) is proposed for the FEMM model, which is a way to approximate an open boundary, other than truncation, as described in [15]. Classical Neumann and Dirichlet boundary conditions are applied for the symmetric axes of the model, x and y, respectively.

Mesh refinement is essential to increase the accuracy of finite element calculations. The mesh refinement must be compatible with the skin depth (δ) of the winding. Usually, 1.5 to 2 elements per skin depth are required for a correct solution [14]. In this work, the conductors were divided into layers of δ-thickness from the surface layer up to a depth of 3δ. Assuming the maximum size of each triangular element is equal to $\delta/2$ in these layers, and outside of these layers the finite element mesh automatically generated by FEMM, as shown in Fig. 6, a sufficient level of refinement is achieved leading to consistent results.

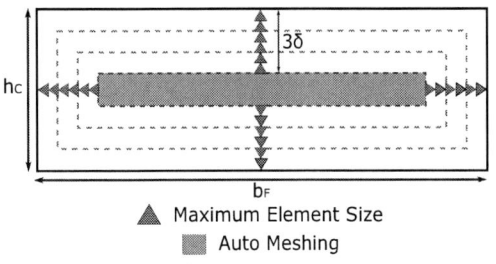

Fig. 6. Mesh Refinement Criteria

V. RESULTS

In order to test the accuracy of the new method, the R_{AC}/R_{DC} values of 1000 randomly-chosen design points are obtained from the constructed surface by direct multivariate linear interpolation.

The interpolation is implemented using the SciPy library, an open source scientific library for Python used for scientific and technical computing [11]. Particularly, the interpolation is done with the *scipy.interpolate.RegularGridInterpolator* class, an interpolation method on a regular grid in arbitrary dimensions. A linear interpolation method has been chosen and will be called to interpolate each required design.

The random points are created between the upper and lower limits of the variables of the problem described in (23). For the width of the core window (W), fixed in our FEM simulation, we explore the homothetic property with random values between 0.005 and 1 m, spanning a very wide range of applications. The random variables are described in (24).

$$
\begin{aligned}
N_{Rand} &= int(random(1,10)) \\
L_{core}/W_{Rand} &= random(0.4,1.2) \\
w/W_{Rand} &= random(0.5,0.998) \\
L_{leg}/W_{Rand} &= random(1,3) \\
l_{gap}/W_{Rand} &= random(0.005,0.5) \\
k_{wRand} &= random(0.2,0.8) \\
w/\delta_{Rand} &= random(1,500) \\
W_{Rand} &= random(0.005,1)
\end{aligned}
\tag{24}
$$

Formulations (18), (16) and the proposed interpolation method are compared in Fig. 7 relative to FEM simulations. The errors are respectively defined as

$$
ERR_{ACtest\%} = 100 \times \left(\frac{R_{ACtest} - R_{ACFEM}}{R_{ACFEM}} \right)
\tag{25}
$$

where R_{ACtest} can be the R_{AC} calculated with (16), (18) or by the interpolation method.

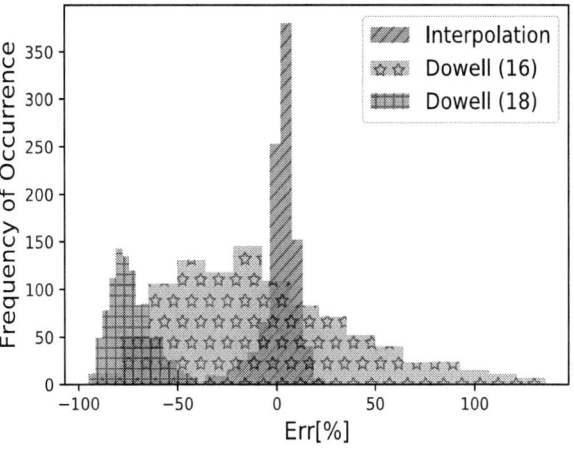

Fig. 7. R_{AC} errors: Dowell (16), Dowell (18) and Interpolation.

The results with the proposed method are more accurate than analytical calculations as shown in Fig. 7. The mean absolute error applying (16) is lower than the mean absolute error of (18). Otherwise, with (16), the data values are more spread out from the mean. The comparison is more evident in Tab. I, with the evaluation of the mean and the standard deviation (Std) of $|ERR_{ACtest\%}|$, maximum and minimum values of $ERR_{ACtest\%}$ and variance of $ERR_{ACtest\%}$.

As shown in Tab. I, the estimated absolute error (Mean $\pm$ Std) in this application of the proposed method is 6.7 ± 7.0, a lower value comparing to 37.3 ± 25.8 (16) and 73.7 ± 10.6 (18). Although the estimated absolute error is lower with the Dowell classic formulation than with the air gap formulation, its variance is much higher when compared to the other approaches. What was expected by looking at the Fig. 7, where the Dowell (16) distribution is more spread out than the Dowell (18) distribution. Therefore, a correction factor would be more easily applied to (18) than to (16).

TABLE I
MEAN , STANDARD DEVIATION (STD), MIN AND MAX OF ERRORS AND VARIANCE

Method	$|ERR_{AC\%}|$ Mean	$|ERR_{AC\%}|$ Std	$ERR_{AC\%}$ Min	$ERR_{AC\%}$ Max	$ERR_{AC\%}$ Var
Dowell(16)	37.3	25.8	-93.3	135.8	2011
Dowell(18)	73.7	10.6	-95.2	-40.1	113
Interpolation	6.7	7.0	-58.1	29.1	92

VI. CONCLUSIONS

This work proposed an improvement for taking into account ac winding losses in the inductor design process, where the geometrical parameters are not defined before starting the design.

First, special attention was paid to Dowell's formulations. Several works use the Dowell's method applying (18) or (16) to calculate inductors the equivalent resistance of the windings, but, usually, no more details are given about the equations. However, it was shown that these equations give different results. Generally, both formulations can have significant errors compared to FEM simulation.

A new homothetic method was proposed for E-core inductors, improving R_{AC} calculations. The presented method consists of the construction of a homothetic multivariate regular grid FEM surface, from which R_{AC} is obtained through direct multivariate linear interpolation. In the surface construction, the geometric variable W (width of the core window) was set, but due to the homothetic properties, the proposed surface can be used to interpolate any value of W. The same surface can also be used for winding materials with different conductivities.

The results with the proposed method typically are more accurate than analytical calculations. The accuracy of the new approach could be further improved by increasing the discretization of the problem variables, at the cost of more computational effort. However, this computation can be performed offline once and used a posteriori for designing an arbitrary number of different inductors.

REFERENCES

[1] P. L. Dowell, "Effects of eddy currents in transformer windings," in Proceedings of the Institution of Electrical Engineers, vol. 113, no. 8, pp. 1387-1394, August 1966.

[2] J. Muehlethaler, Modeling and multi-objective optimization of inductive power components. Zurich, Switzerland: Eidgenössische Technische Hochschule Zürich, 2012.

[3] F. A. Holgun, R. Asensi, R. Prieto and J. A. Cobos, "Simple analytical approach for the calculation of winding resistance in gapped magnetic components," 2014 IEEE Applied Power Electronics Conference and Exposition - APEC 2014, Fort Worth, TX, 2014, pp. 2609-2614.

[4] Xi Nan and C. R. Sullivan, "An improved calculation of proximity-effect loss in high-frequency windings of round conductors," IEEE 34th Annual Conference on Power Electronics Specialist, 2003. PESC '03., Acapulco, Mexico, 2003, pp. 853-860 vol.2.

[5] A. Van den Bossche and V. Valchev, "Eddy current losses and inductance of gapped foil inductors," IEEE 2002 28th Annual Conference of the Industrial Electronics Society. IECON 02, Sevilla, 2002, pp. 1190-1195 vol.2.

[6] P. Wallmeier, N. Frohleke and H. Grotstollen, "Improved analytical modeling of conductive losses in gapped high-frequency inductors," Conference Record of 1998 IEEE Industry Applications Conference. Thirty-Third IAS Annual Meeting (Cat. No.98CH36242), St. Louis, MO, USA, 1998, pp. 913-920 vol.2.

[7] T. Meynard, Analysis and Design of Multicell DC/DC Converters Using Vectorized Models. Hoboken, NJ: Wiley, 2015, ch. 7.

[8] J. A. Ferreira, "Improved analytical modeling of conductive losses in magnetic components," in IEEE Transactions on Power Electronics, vol. 9, no. 1, pp. 127-131, Jan. 1994.

[9] Z. Shen, Z. Li, L. Jin and H. Wang, "An AC resistance optimization method applicable for inductor and transformer windings with full layers and partial layers," 2017 IEEE Applied Power Electronics Conference and Exposition (APEC), Tampa, FL, 2017, pp. 2542-2548.

[10] F. Forest, E. Laboure, T. Meynard and M. Arab, "Analytic Design Method Based on Homothetic Shape of Magnetic Cores for High-Frequency Transformers," in IEEE Transactions on Power Electronics, vol. 22, no. 5, pp. 2070-2080, Sept. 2007.

[11] (2019, Apr.) SciPy. [Online]. Available:https://scipy.org

[12] (2019, Apr.) Soft Ferrites and Accessories - Data Handbook 2013 [Online]. Available: https://www.ferroxcube.com

[13] D. Meeker, Finite Element Method Magnetics, Version 4.2 (12Jan 20162006 Build).

[14] N. Ida and J. P. A. Bastos, Electromagnetics and Calculation of Fields, 2nd ed. NewYork, NY: Springer, 1997, p. 432.

[15] D. Meeker, Finite element method magnetics, Version 4.2, Users Manual, October 2015.

Evaluation of Power Quality Impacts due to Photovoltaic Penetration in Distribution Grids via Time-Domain Simulation

Henrique Pires Corrêa[†], Flávio Henrique Teles Vieira[†]

[†]Escola de Engenharia Elétrica, Mecânica e de Computação, UFG, Goiânia 74605-010, Brazil

Abstract—In this paper, the problem of estimating the maximum penetration of distributed photovoltaic power in a distribution grid for mantaining acceptable power quality is considered. We evaluate power quality supplied by the grid in terms of root mean square value and total harmonic distortion of bus voltages, which are two important parameters taken into account by the utility in management of grid supply quality. A simple simulation model is proposed for evaluating these parameters as functions of the photovoltaic penetration factor. The model is used in a simulation of an hypotethical distribution feeder in order to validate the proposed method of estimating penetration factor limits.

Index Terms—Photovoltaics, distributed generation, penetration factor, power quality.

I. INTRODUCTION

A parameter of great interest in the evaluation of distributed photovoltaic (PV) generation implemented in power grids is the penetration factor. This parameter is related to the percentage of total load in a grid that is supplied by distributed photovoltaic generators directly connected to it. Considering that disturbances in power quality may arise from intense power injection by PV sources, it is desirable to estimate maximum acceptable bounds for the penetration factor. Although power quality standards exist for distribution grids, determining the maximum penetration factor is a complex task due to its dependence on distributed generation and grid topologies. Hence, simulation is an attractive way of solving the problem for engineering practicioners, since a general analytical solution would be exceedingly complex and require data not usually available to the power utility [1].

In this sense, we propose a simple simulation model that enables evaluation of power quality indexes for arbitrary grids as functions of photovoltaic penetration and topology of the generators. More precisely, we focus on the evaluation of voltage root mean square values and total harmonic distortion, which are often considered of great importance by utilies. The proposed method is based on the injection of current waveforms in the grid, which are previously obtained via simulation of a standalone photovoltaic system.

The method is validated by carrying out simulations of multiple photovoltaic penetration scenarios in a hypothetical low-voltage distribution feeder. The simulations contemplate

This work was funded in part by CAPES under Finance Code 001.

variable PV generation both in terms of penetration factor absolute value and distributed generation topology. The proposed model has the advantage of enabling simultaneous estimation of root mean square voltage levels and harmonic distortion in grid buses, which allows fast assessment of penetration factor influence with respect to both parameters. Moreover, the method is valid for arbitrary inverter harmonic spectra.

This work is organized as follows. In Section II, we briefly discuss power quality parameters, related norms and recently adopted definitions for the penetration factor. In Section III, we present the proposed simulation method of injecting previously sampled AC currents in the power grid buses. In Section IV, a study case consisting in the application of the proposed method to a hypothetical grid is presented. In section V, obtained simulation results are given. In Section VI, we present conclusions concerning the results obtained in this paper.

II. POWER QUALITY AND PENETRATION FACTOR

From the standpoint of power quality supplied by the distribution grid, the parameters usually considered as of greatest importance are voltage total harmonic distortion (THD) and root mean square value (RMS). Although quality disturbances arise from the injection of undesired currents in the system, voltage is a better power quality indicator because the propagation of harmonic currents is dependent on local grid characteristics (i.e. installed equipment and line impedances). In fact, the IEEE-519 standard takes this dependence into account by establishing harmonic current injection limits as functions of the short-circuit capacity of the bus [2].

The expansion of PV source integration to distribution grids motivate the establishment of limits for installed photovoltaic power, since this is crucial for mantaining power quality in accessible levels. Existing studies concerning this problem agree in establishing a main parameter, usually denoted as distributed generation penetration factor (PF). However, there is no consensus in regards to the precise definition of this parameter. Some adopted definitions in the literature are:

- Ratio between peak PV power and feeder capacity;
- Ratio between peak PV power and transformer rating;
- Ratio bebtween peak PV power and maximum demand.

Regardless of the adopted definition, it is clear that power quality problems become more significant with the growth of PF. In this sense, we shall use simulation to establish PF limits for an hypothetical low-voltage feeder. Usually, such lines are fed by a single transformer whose nominal rating is equal to the peak demand of connected loads. Hence, we opt for defining PF as the ratio between peak PV power and the nominal rating of the transformer associated to the feeder, given by:

$$PF = \frac{P_{photovoltaic}(kW)}{S_{transformer}(kVA)} \quad (1)$$

It should be noted that usually, up to a certain value of PF, distributed generation is beneficial to grid voltage levels. In fact, the RMS values of voltage at the end of feeders tend to be lower and may be close to the lower acceptable limit. The injection of PV power in the feeder reduces current flow toward the loads, contributing to voltage regulation.

The maximum PF for a given harmonic distortion limit is dependent on the type of inverter used to connect PV sources to the grid. For example, the use of PWM inverters is not prone to generating severe harmonic problems. In fact, these converters transform the DC input into AC by means of high-frequency switching, which provides an output spectrum with high-order harmonics that is easily filtered [3]. Statistical studies have been carried out [4] regarding the distribution of inverter types in grid-connected PV souces. The results showed that an equivalent inverter spectrum with significant presence of low-order harmonics must be assumed in PV studies. Hence, integration of high-harmonic injection equipment (such as square-wave inverters) to the grid is significant.

The objective of our analysis is establishing a general procedure for assessing power quality indexes (voltage RMS and THD) as functions of photovoltaic penetration. In what follows, we will take as references the following standard quality limits for low-voltage grids:

- Voltage level: $202V \leq V_{LL} \leq 231V$, according to [5];
- Harmonic distortion: $THD_V \leq 5\%$, according to [2].

III. SIMULATION METHOD FOR PF EVALUATION

We propose a simulation method that consists of injecting adequately scaled current waveforms in the grid, which correspond to the previously simulated current waveforms for a load being fed by a PV system. In what follows, we first suppose the current in one of the inverter phases $i_a(t)$ as given and describe the proposed injection method. We then proceed to presenting the PV system simulation used to capture the inverter output current waveform.

Given the current $i_a(t)$, we multiply it by a scaling factor $K \ll 1$, obtaining a new waveform $i'_a(t)$ with small RMS value. Hence, a certain PV penetration PF can be simulated by injecting in the grid a current $n \cdot i'_a(t)$, where n is an adequately chosen integer. If K is sufficiently small, incrementing n allows the simulation of an approximately continuous variation of the penetration factor. In order to evaluate three-phase harmonic injection, corresponding waveforms for phases B and C can be generated by applying time delays to $i'_a(t)$. In this sense, phase balance is assumed. In Figure 1, the general method is illustrated by means of a flowchart.

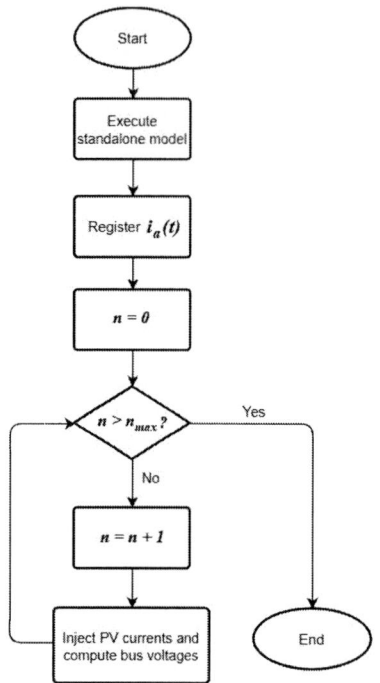

Figure 1: Flowchart of proposed method.

We now describe the simulation used in order to acquire the $i_a(t)$ waveform. We first implemented, in the *Simulink* platform provided by MATLAB, a real-time PV panel model as described in [6]. This model is convenient because it allows for the assessment of electrical transients in the panel output. Furthermore, the complete PV system is implemented by using *SimPowerSystems* functional blocks that emulate square-wave single-phase inverter and buck converter with maximum power point tracking via Perturb & Observe algorithm [7]. A resistive load is then connected to inverter output and the simulation is executed. The AC current fed to the load is captured and stored as $i_a(t)$. The implemented system and adopted values are given in Figure 2 and Table I, respectively. The latter were chosen in order to restrict current ripple for full duty cycle under 100 mA and achieve filtering of high-order DC-DC switching harmonics, with a rated 1 kW output power.

We observe that the resistor R in Figure 2 actually represents the 1 kW load interfaced by means of a square-wave inverter with 220 V nominal output voltage. The actual full configuration of inverter and three-phase unit power factor load is not shown due to space constraints.

Table I: Standalone PV system filter values

Parameter	Value
L	10 mH
C	1 mF
C_1	1 mF
Load Power	1 kW
DC-DC Switching Frequency	40 kHz

We opted for using a square-wave inverter because there

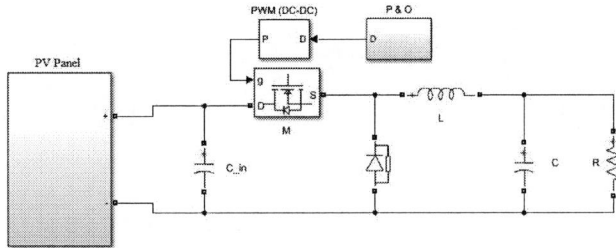

Figure 2: Standalone PV system model.

are, on average, significant lower-harmonic injections by PV equipment [3]. Furthermore, from an engineering standpoint, this may be considered a worst-case evaluation: if adequate THD levels are guaranteed for square-wave inverters, any other grid scenario will present lower harmonic distortion.

In Figure 3, we depict the method used to inject the captured current waveform in the power grid. The controlled current sources are driven by $i_a(t)$ and delayed versions of this signal, in order to simulate balanced three-phase power injection. A high-valued resistance R_{aux} is used in order to prevent convergence problems due to current injection in inductive elements of the grid model.

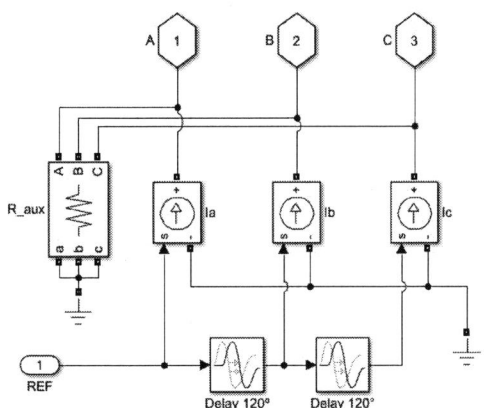

Figure 3: Model for injecting current in the power grid.

IV. DISTRIBUTED GENERATION AND FEEDER MODEL

The use of low-voltage distribution by means of lines deriving from a transformer secondary is frequently employed for small loads, including residential consumers. Hence, we shall consider this scenario for evaluating the impact of PV penetration.

Impedance of the feeders is essential for evaluating power quality problems. In an ideal situation, there should be no impedance between PV source and the low-voltage bus to which it is connected. If this was the case, injected current flow would not generate voltage drops along the feeder, thus not distorting the bus voltage. Clearly, impedance is significant in low-voltage feeders and its magnitude is of great influence in power quality [4]. In fact, high impedances imply that reverse

power flows may cause significant overvoltage problems and that harmonic injection may distort the bus voltage severely.

We considered the hypothetical system represented in Figure 4. It consists of a three-phase feeder supplied by a $13.8kV : 380V$ transformer with nominal rating of $25kVA$ and low-voltage tap setting of 3%. Data regarding transformer impedance as a function of nominal rating [5] suggests adopting a single-phase impedance of $Z_t = 0.1165 + j0.0751\Omega$ at a frequency of $60Hz$.

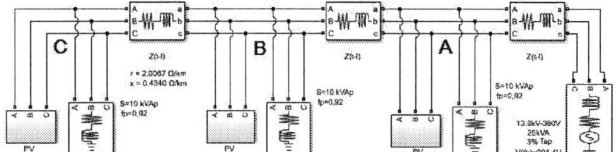

Figure 4: Secondary feeder model.

We suppose the feeder supports a bulk load equal to the transformer nominal power. Analogously to previous works [8], we model the load distribution along the feeder by equally spaced points with equal power consumption. In this particular problem, we suppose three points with demands equal to $8.33kVA$, as given in Figure 4. The bulk load implies a current $I \approx 38A$ drawn from the transformer.

Given the estimated supplied current, we consider the feeder as being dimensioned for a capacity of $60A$. The average impedance of copper conductors with this capacity is $z = 2.0067 + j0.4340\Omega/km$ [5]. For each feeder segment between load fractions, we consider a distance of $200m$. For the segment between transformer and first load fraction, we assume a smaller $50m$ distance, which is justified by the fact that transformer are usually installed close to one of the loads. The adopted distances give the following impedances for load-load and transformer-load feeder segments, respectively: $Z_{(l-l)} = 0.4013 + j0.0868\Omega$ and $Z_{(t-l)} = 0.1003 + j0.021\Omega$. All pertinent impedance values are given in Table II.

Table II: Feeder impedance values.

Impedance	Value (Ω)
Z_t	$0.1165 + j0.0751$
$Z_{(t-l)}$	$0.1003 + j0.021$
$Z_{(l-l)}$	$0.4013 + j0.0868$

We assume the primary distribution grid may be approximated by an ideal Thévenin voltage source, given that the transformation ratio significantly reduces impedance due to the primary network, as seen from the transformer secondary.

In Figure 4, the PV blocks correspond to the current injection model given in Figure 3.

V. SIMULATION RESULTS

We used the proposed current injection model and simulation procedure in order to evaluate variations of voltage RMS and THD values as functions of photovoltaic penetration. More precisely, we evaluated all bus voltages in Figure 4 for the following scenarios:

- PV generation only at the initial feeder section;
- PV generation only at the middle feeder section;
- PV generation only the the last feeder section;
- Equal PV generation in all buses.

Given that different values of upstream feeder impedance are seen from each load bus, it is expected that different power quality impacts may arise in terms of PV source location. Hence, varying the distribution of sources along the feeder is important in order to assess disturbances in the grid.

The results obtained in the simulations are presented in Figures 5 to 12. These plots show the variation of voltage RMS values and total harmonic distortion for all buses (A, B and C in Figure 4). The vertical lines show the minimum PF value for which the power quality standard is violated.

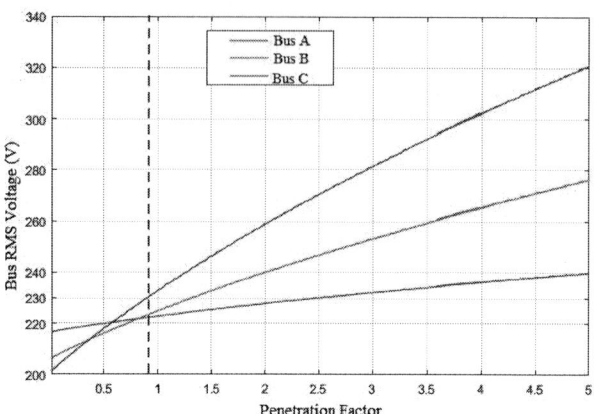

Figure 7: Phase-neutral voltages *versus* PF (Generation in C).

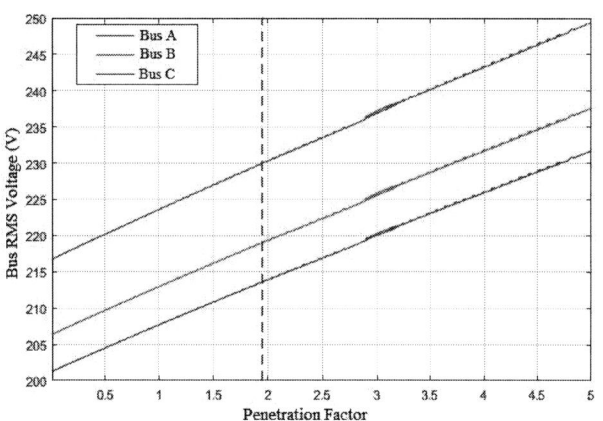

Figure 5: Phase-neutral voltages *versus* PF (Generation in A).

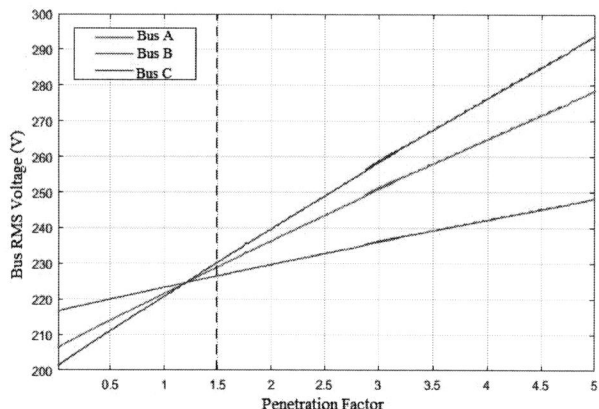

Figure 8: Phase-neutral voltages *versus* PF (Equal generation).

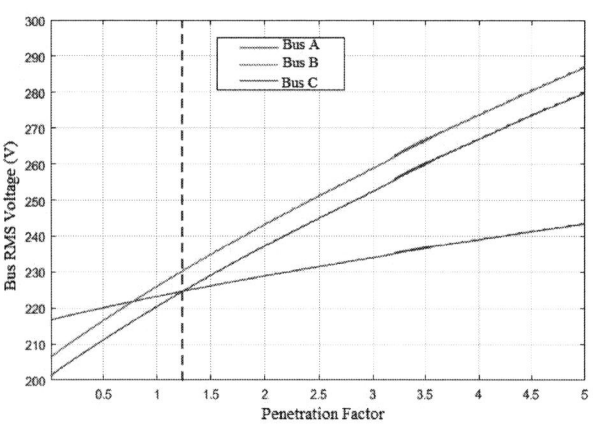

Figure 6: Phase-neutral voltages *versus* PF (Generation in B).

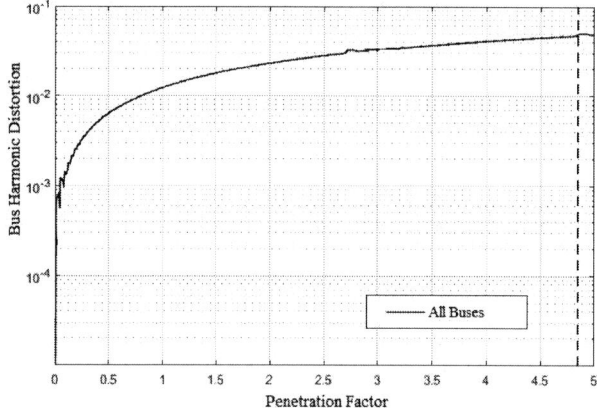

Figure 9: Harmonic distortion *versus* PF (Generation in A).

For all simulated scenarios, a penetration factor $PF < 2$ caused violation of at least one of the considered power quality standards. Furthermore, the overvoltage limit was in all cases violated for a smaller PF in comparison to the harmonic distortion limit. Hence, in the hypothetical system, photovoltaic penetration is limited by overvoltage problems.

Although the violation of THD limits only happened for relatively high PF values, harmonic injection should not be neglected. Considering that distortion effects are able to propagate in the network, widespread voltage distortion may

978-1-7281-4181-7/19 $31.00 © 2019 IEEE   1417

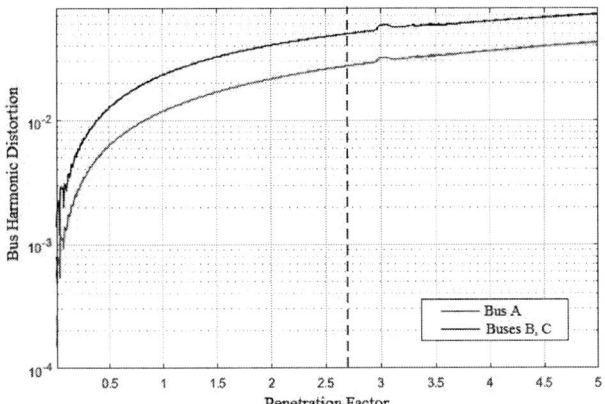

Figure 10: Harmonic distortion *versus* PF (Generation in B).

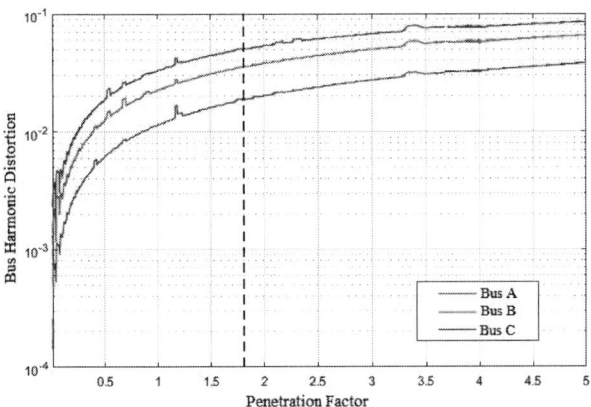

Figure 11: Harmonic distortion *versus* PF (Generation in C).

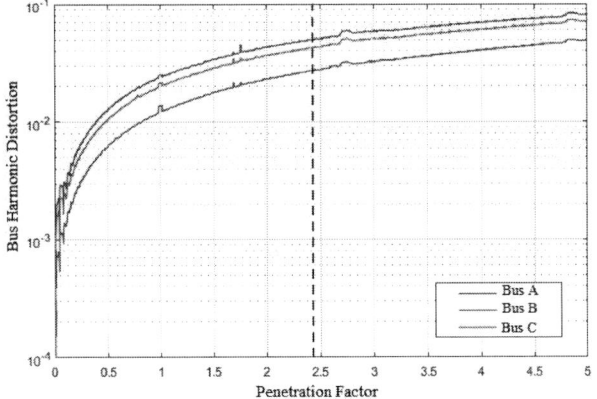

Figure 12: Harmonic distortion *versus* PF (Equal generation).

contribute to higher harmonic indexes in adjacent higher-voltage networks, whose THD limits are stricter.

The results show that power quality problems for a given PF were greater for photovoltaic generation positioned farther away from the source transformer. This is a reasonable result, since lower impedance between the PV source and transformer allow for smaller distortion of voltage waveforms along the grid. High levels of PF are tolerable for generation concentrated in bus A due to its small distance to the transformer.

The obtained PF limits for each scenario are given in Table III. We define these limits as the maximum PF values for which no power quality parameter (voltage or harmonic distortion) violates the adopted standards. The third column of this table gives results obtained in [8], where a similar analysis is carried out with data pertaining to a real $400V$ low-voltage network supplied by a $630kVA$ transformer. It is notable that, in spite of different nominal voltage and bulk power values, the obtained PF limits were similar.

Table III: Photovoltaic Penetration Factor Limits

Generation Bus(es)	Maximum PF	Maximum PF in [8]
A	190%	205%
B	125%	130%
C	92%	81%
ABC	150%	–

We now discuss the similarity between results obtained in our simulation with a hypothetical network and those given in [8]. The dimensioning of feeder conductors is executed according to current capacity safety margins and maximum voltage drop. Considering that the real system of [8] is properly designed, the voltage drop in cable impedances will be similar to those obtained in the hypothetical system, whose conductors were dimensioned according to the previously mentioned criteria. Finally, the similar PF limits show that the chosen feeder lenghts are representative of a real system.

VI. CONCLUSION

A simple circuit model for evaluating grid impacts due to photovoltaic penetration was proposed and validated by means of computer simulation and comparison with previous works. In the adopted study case, the proposed model was used in order to obtain plots of power quality parameters as functions of the penetration factor, which allow the evaluation of PF limits. The obtained results suggest that the model is consistent with distributed generation theory and can be used as a simple tool for evaluating PF limits for arbitrary grid and PV generation topologies.

REFERENCES

[1] R. A. Shayani, Método para a determinação do limite de penetração da geração distribuída fotovoltaica em redes radiais de distribuição, Ph.D. thesis, Universidade de Brasília (2010).

[2] IEEE. IEEE Recommended Practices and Requirements for Harmonic Control in Electric Power System, 1991.

[3] N. Mohan, T. Undeland, Power electronics: converters, applications, and design. Wiley India, 2007. ISBN 9788126510900.

[4] A. A. Latheef, Harmonic impact of photovoltaic inverter systems on low and medium voltage distribution systems, Master's thesis, University of Wollongong (2006).

[5] ANEEL. Normative Resolution nº 482, 2012.

[6] M. G. Villalva, J.R. Gazoli, E. R. Filho. Comprehensive approach to modeling and simulation of photovoltaic arrays. IEEE Transactions on Power Electronics, 2009.

[7] M. Elgendy, B. Zahawi, D. Atkinson, Assessment of perturb and observe mppt algorithm implementation techniques for pv pumping applications, IEEE Transactions on Sustainable Energy.

[8] R. A. Kordkheili, B. Bak-Jensen, J. R-Pillai, P. Mahat, Determining maximum photovoltaic penetration in a distribution grid considering grid operation limits, in: 2014 IEEE PES General Meeting, 2014.

Control Strategy and Power Management for Multifunctional Inverters with BESS and Reactive Power Compensation

Luiz Henrique Meneghetti[1], Edivan Laercio Carvalho[2], Gustavo B. K. Schmidt[1], Emerson Giovani Carati[1], Jean Patric da Costa[1], Carlos Marcelo de Oliveira Stein[1], Zeno Luiz Iensen Nadal[3], Rafael Cardoso[1]

[1]*Universidade Tecnológica Federal do Paraná (UTFPR), Campus Pato Branco, Brazil*
[2]*Federal University of Santa Maria - UFSM, Santa Maria, Brazil*
[3]*Copel Distribuição S.A., Curitiba, Brazil*

E-mails: meneghetti.luiz@outlook.com, e.carvalho@ieee.org, gustavoschmidt@alunos.utfpr.edu.br, emerson@utfpr.edu.br, jpcosta@utfpr.edu.br, cmstein@utfpr.edu.br, zeno.nadal@copel.com, rcardoso@utfpr.edu.br

Abstract—Due to the non-controllable characteristics of the photovoltaic (PV) generation, like the solar radiance intermittence, technical challenges arise for the electrical power system, when considered a high penetration of PV generation in the utility grid. One possible solution for these problems is to integrate a Battery Energy Storage System (BESS) with the generation systems. Herewith, it is possible to consider the distributed generator similarly as controllable sources, allowing many operating modes for the PV systems, including reactive power compensation. In this way, this paper proposes a low-level power management strategy for the compensation of active and reactive power to local loads. It aims to guarantee an adequate power factor to the installation reducing cost associated with the energy bill. Besides that, real scenarios are considered, in which the limits of the state of charge (SOC) cannot be exceeded. In this case, the system can promote the BESS recharge only by the PV or by the PV plus the utility grid. To evaluate the proposed structure, PSIM software simulations are presented. It is considered the physical model of a PV array, which uses real data obtained experimentally.

Index Terms—Active var compensation, Multifunctional inverters, Photovoltaic systems, Reactive power, Battery energy storage systems.

I. INTRODUCTION

The increase in distributed generation in low voltage distribution systems has introduced a series of challenges in the power quality field. Especially with high inductive loads and in areas with weak grids, where the low power quality can cause a lot of undesirable consequences in the distribution systems, including the rise in the reactive power consumption and overload in the distribution grid [1], [2].

Commonly, grid-tie PV inverters operate injecting active power into the ac grid. But intermittency problems and the integration of a non-controllable source to the electrical distribution system can limit the application of PV inverters to maintain the power quality [1] - [3].

In this context, the multifunctional PV inverter attained attention in recent years, due to the potential benefits for power quality improvement, when associating a BESS to the PV generator in a single device. As a consequence, it is granted to the multifunctional inverter the operation with different objectives, being possible to realize programmable power dispatch and power peak shaving. It also permits islanded operation [4] - [8].

This degree of freedom can be used in multifunctional inverters to provide power compensation (active and reactive) when supplying local loads. Consequently, it is possible to improve the power quality at the point of common coupling (PCC), even with the presence of inductive local loads.

In [9] it is described a study of a three-phase system with a PV-BESS for the reactive power compensation in a medium voltage grid. However, it is considered that the reactive power reference is given by the system operator, that is, not automatically generated locally. In [6] it is presented a control strategy for the integration of a PV-BESS set to a single-phase electrical grid applied to residences and electrical vehicles. The intent is to supply continuous power to the grid with reactive compensation. But, the method employed is not shown clearly. In [10], the BESS is used for PV power peak shaving and to reduce the cost of the electrical bill by supplying the local load using a PV-BESS set, without the grid support. However, [10] considers only active power compensation.

This paper proposes a PV distributed generation system composed by a three-phase inverter and a BESS, aiming to supply distinct local loads and to compensate reactive power. The system considers the battery SOC and different conditions of power generation. As the main contribution it is proposed a power management done locally in the inverter. The system structure is illustrated in Fig. 1. Depending on the SOC, three different operation modes are evaluated, as described below.

Mode 1 ($20\% \leq SOC \leq 90\%$): The priority of the PV inverter is to compensate reactive power to avoid extra energy bill charges due to excessive reactive power demand. The output current is limited according to the inverter specifications and in case of excessive active power generation, the battery is charged absorbing this energy.

Mode 2 ($SOC < 20\%$): For this case, the reactive power

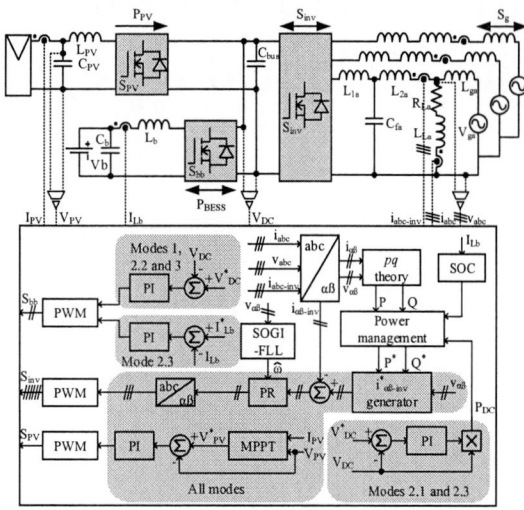

Fig. 1. Representation of the proposed system.

compensation is disabled to avoid power transients due to the photovoltaic intermittence. The batteries can be recharged in two ways: only using the PV or using the PV in addition to the utility grid.

Mode 3 (SOC>90%): The inverter injects the rated power into the electrical grid with the aid of the BESS. In this mode, the reactive power compensation is disabled and the interest is to take benefits from the active power injected into the utility grid due to benefits of the energy price.

II. SYSTEM DESCRIPTION

The system consists of a 5.44 kVA three-phase full-bridge inverter, connected to the grid using an LCL filter. The methodology for the filter design can be found in [11] - [12]. The PV power generation is processed through a Boost dc-dc converter. For the BESS energy processing, it is utilized one buck-boost dc-dc bidirectional converter. The photovoltaic array consists of 16 panels that amounts 5440 W of power [13]. The battery bank consists of a lithium-ion array of 240 V/100 Ah.

The control structure considers the stationary reference frame $\alpha\beta$. The active and reactive power calculation uses the pq theory [14], and are given by

$$P = v_\alpha i_\alpha + v_\beta i_\beta \qquad (1)$$

and

$$Q = v_\beta i_\alpha - v_\alpha i_\beta, \qquad (2)$$

where, v_α, v_β, i_α and i_β are the voltage and current components in the load $R_L + j\omega_g L_L$. With that, it is possible to define the active and reactive power demand from the load and allow the compensation by the frequency inverter using the currents

$i_{\alpha-inv}$ and $i_{\beta-inv}$. Two proportional-resonant (PR) controllers are used to control $i_{\alpha-inv}$ and $i_{\beta-inv}$. The power references that the inverter needs to track are obtained according to the operation mode. The operation modes of the system are changed according to the SOC of the storage system.

A. Operation Modes

On each operation mode, the PV system operates at the maximum power point using the Perturb and Observe (P&O) algorithm [15]. The algorithm provides the voltage reference of the PV converter (V_{PV}^*) at the maximum power point.

A proportional-integral (PI) controller is applied to the voltage control V_{PV}^*. The SOC is evaluated by means of the ampere-hour counting method [16]. A Second Order Generalized Integrator-Frequency-Locked-Loop (SOGI-FLL) is in charge to estimate the frequency of the utility grid [3], [17]. Moreover, according to [3], the inverter reference currents are given by

$$\begin{bmatrix} i_{\alpha-inv}^* \\ i_{\beta-inv}^* \end{bmatrix} = \frac{1}{v_\alpha^2 + v_\beta^2} \begin{bmatrix} v_\alpha & v_\beta \\ v_\beta & -v_\alpha \end{bmatrix} \begin{bmatrix} -P^* \\ -Q^* \end{bmatrix}, \quad (3)$$

where, P^* e Q^* are the active and reactive power references of the inverter, according to the pq theory and the proposed power management. Fig. 2 illustrates the flowchart of power management, which has three operation modes that will be detailed in what follows.

1) Mode 1 (20%≤SOC≤90%): This operation mode is used to compensate the reactive and active power of the local load. This takes advantage of energy pricing and avoids extra charges due to the possible inadequate power factor. To achieve that, both powers demanded by the local load are evaluated through the pq theory. Considering S the total apparent power demanded by the load and S_{max} the maximum rated apparent power of the inverter, three possible scenarios are presented:

- $S \leq S_{max}$− it is implemented the compensation of the active and reactive power according to the references calculated by the pq theory. Thus, the utility grid does not supply either active nor reactive power. Therefore $P^* = P$ and $Q^* = Q$.

- $S > S_{max}$− in this case, P, Q or both are greater than S_{max}. So, if:

  $Q > S_{max}$ the active power reference becomes null and it is injected the maximum reactive power into the load. The utility grid supplies active power to the load, while the power factor at the PCC is the highest possible. Therefore $P^* = 0$ and $Q^* = S_{max}$. On the other hand, if

  $Q < S_{max}$ the active power demanded by the load surpasses the inverter limit. Therefore, the inverter supplies the calculated reactive power and the active power is limited to not exceed S_{max}. In this case, the power factor at the PCC is compensated by the inverter. Thus $P^* = \sqrt{S_{max}^2 - Q^{*2}}$ and $Q^* = Q$.

In this operation mode, the voltage in the dc-bus is controlled by the bidirectional converter using a PI controller.

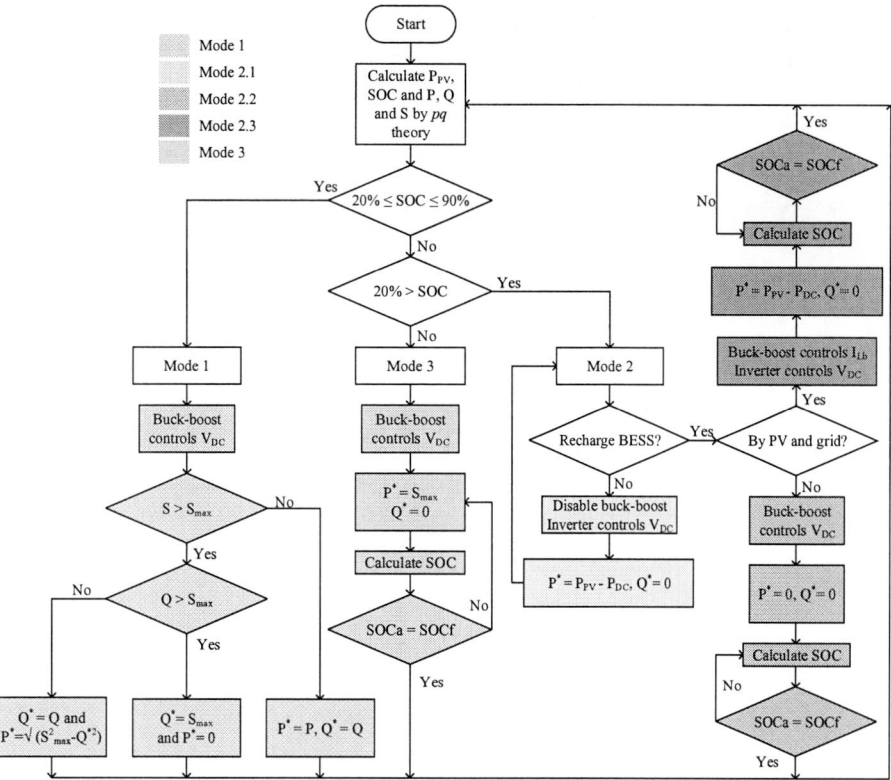

Fig. 2. Flowchart for the proposed management system, representing the structure for each mode and sub-mode of operation.

2) Mode 2 (SOC<20%): This mode is enabled when the batteries have low SOC to avoid damage to them. The mode 2 is divided into three sub-modes:

- Mode 2.1– The buck-boost bidirectional converter is disabled and there are neither charge nor discharge in the storage system. The power injected into the PCC is the maximum instantaneous power available (ideally P_{PV}). The support to reactive power is disabled due to the intermittence characteristic of the PV generation, avoiding random reactive power transients at the PCC. Due to the disabled state of the storage system, the inverter takes charge controlling V_{DC} through P_{DC}, according to [3]. Therefore $P^* = P_{PV} - P_{DC}$ and $Q^* = 0$. The definition to start charging the storage system exits this operation mode.

- Mode 2.2– This operation mode is an option to recharge the BESS. The voltage at the dc-bus V_{DC} is controlled by the buck-boost bidirectional converter and both power references of the inverter become null. Thus, the inverter does not supply power into the PCC and the power generated by the PV system is directed to the storage system through the dc-bus. Therefore, $P^* = 0$ and $Q^* = 0$. When the actual SOC (SOCa) of the BESS

reaches the desired final SOC (SOCf) the power manager exits this mode.

- Mode 2.3– In this operation mode, both PV the utility grid are responsible to charge the BESS. In this mode, the battery charging current I^*_{Lb} is defined and it is tracked by the buck-boost bidirectional converter using a PI controller. In this case, the dc-bus voltage is controlled by the inverter. If the power necessary to charge the BESS exceeds the PV generation, the utility grid supplies the deficit. If the power from the PV generation surpasses the power necessary to charge the BESS, the power in excess is injected into the PCC. Hence, $P^* = P_{PV} - P_{DC}$ and $Q^* = 0$. When the SOCa reaches the SOCf the power manager exits this mode.

3) Mode 3: This operation mode is used in the case of a high SOC and there is the necessity to reduce it. To achieve that, it is injected the maximum power into the PCC to take advantage from the sale of active power to the utility grid. Reactive power compensation is disabled. The dc-bus voltage is controlled by the buck-boost bidirectional converter. Thus, $P^* = S_{max}$ and $Q^* = 0$.

978-1-7281-4181-7/19 $31.00 © 2019 IEEE

III. RESULTS AND DISCUSSIONS

To analyze the proposed system, PSIM software simulations were used. Table I shows the parameters used in the simulations. To simulate the PV array, the PV physical model available in the software was used. Besides, real irradiance and temperature data obtained experimentally were also used. It is important to mention that the irradiance data was compacted to allow the simulation. Thus, the power fluctuation from the PV system becomes more abrupt than in real systems.

TABLE I
MAIN PARAMETERS OF THE SYSTEM.

Utility grid		
Line-to-line voltage	V_g	220 V_{RMS}
Frequency	f_g	60 Hz
Equivalent inductance	L_g	500 μH
LCL filter (per phase)		
Filter inductors	$L_1 - L_2$	155 μH - 50 μH
Filter capacitor	C_f	13 μF
Boost converter and PV		
Filter inductor	L_{PV}	500 μH
Filter capacitor	C_{PV}	220 μF
Nominal power	P_{PV}	5.44 kW
DC bus		
Voltage	V_{DC}	500 V
DC bus capacitance	C_{bus}	3 mF
Buck-boost converter and BESS		
Nominal battery voltage	V_b	240 V
Filter inductors	L_b	2.0 mH
Filter capacitors	C_b	22 μF
Common parameters for all converters		
Switching frequency	f_{sw}	20 kHz
System nominal power	S_{max}	5.44 kVA

Initially, Fig. 3 presents simulation results for mode 1. In this case, three possible scenarios are analyzed, as described below.

During 0.2s $\leq$ t <0.32s, since the power demanded by the load is inferior to the power available in the inverter, only the inverter supplies the load. The power profile of the load is illustrated in Fig. 3(a). Hence, there is no charge from the electrical company due to energy consumption, neither charge due to reactive excess. Fig. 3(b) illustrates the PV intermittent power compensated by the BESS.

During 0.32s $\leq$ t < 0.66s additional load is inserted in such a way that the apparent power drained by the loads becomes greater than the available power provided by the inverter, while the reactive power demanded by the load is inferior to the capability of the inverter. In this case, it is possible to verify that there is no reactive power being drained from the utility grid (Fig. 3(c)). In this case, the inverter provides all the reactive power to the load. The active power available in the inverter is injected into the PCC, and the load consumes the minimum active power possible from the utility grid (Fig. 3(d)).

During 0.66s $\leq$ t < 1s the apparent power of the load is higher than the available power of the inverter. So, to minimize the reactive power flow from the grid, the inverter injects the maximum possible reactive power into the PCC. It reduces the

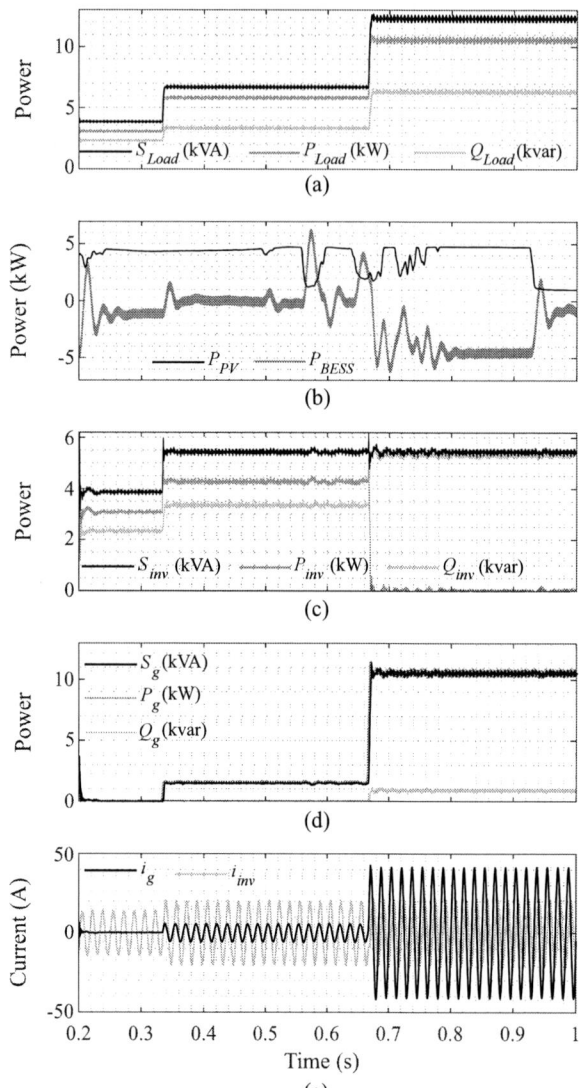

Fig. 3. Waveforms of the system operating in mode 1: (a) local load apparent (S_{Load}), active (P_{Load}) and reactive (Q_{Load}) powers; (b) PV generation power (P_{PV}) and BESS power (P_{BESS}); (c) inverter apparent (S_{inv}), active (P_{inv}) and reactive (Q_{inv}) powers; (d) apparent (S_g), active (P_g) and reactive (Q_g) powers of the utility grid and (e) grid current of phase A (i_g) and inverter current of phase A (i_{inv}).

energy bill charge due to excessive reactive consumption. In Fig. 3(e) it is possible to observe the inverter output waveform of phase A (i_{inv}) and the utility grid (i_g).

Fig. 4 illustrates the waveforms of the system during the transition between modes 1 and 2.2. During 0.2s $\leq$ t < 0.3s the system operates at mode 1. Fig. 4 (a) shows that there is no active power being drained from the utility grid. It is due to the generated power P_{PV} be higher than the inverter power

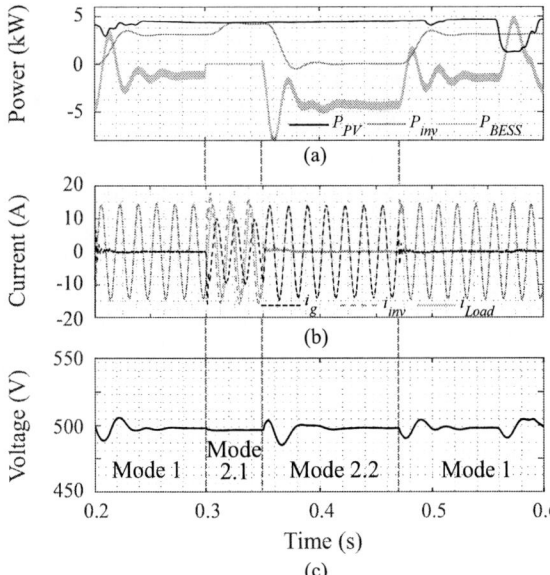

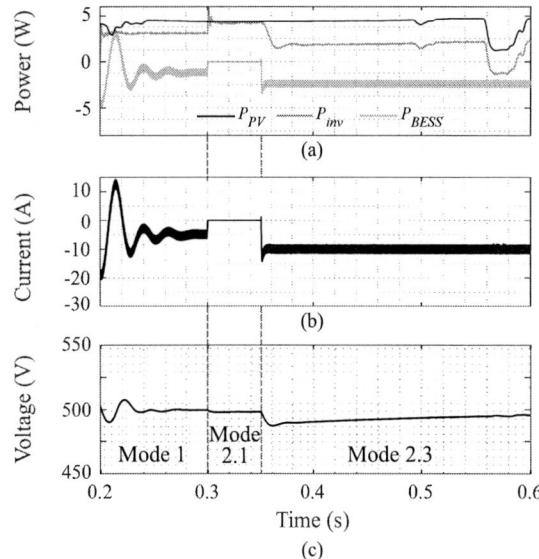

Fig. 4. Waveforms of the system on the operation modes 2.1 and 2.2: (a) PV generation power (P_{PV}), BESS power (P_{BESS}) and inverter active power (P_{inv}); (b) utility grid current of phase A (i_g), inverter current of phase A (i_{inv}) and local load current of phase A (i_{Load}) and (c) DC bus voltage (V_{DC}).

Fig. 5. Waveforms of the system on the operation modes 2.1 and 2.3: (a) PV generation power (P_{PV}), BESS power (P_{BESS}) and inverter active power (P_{inv}); (b) BESS charge current (I_{Lb}) and (c) DC bus voltage(V_{DC}).

P_{inv}. This difference in power is used to charge the BESS through P_{BESS}. Fig. 4(b) corroborates this behavior since the grid current i_g is null.

During $0.3s \leq t < 0.35s$ the BESS is disabled to prove the proposed strategy. The power from the BESS becomes null and the power injected into the PCC by the inverter follows the profile from the PV generation. During $0.35s \leq t < 0.47s$ the power supplied by the inverter into the dc-bus is null and all the power generated by the PV panels is utilized to charge the battery bank. Finally, during $0.35s \leq t < 0.6s$, the BESS is in operation again and the load is supplied totally by the inverter. In Fig. 4(c) it is illustrated the voltage waveforms of the dc-bus to verify the respective oscillation due to the power oscillations and the changes of the converters that control it.

Fig. 5 depicts the waveforms of the system during the transition of modes 1 and 2.3. During $0.2s \leq t < 0.3s$ the system operates at mode 1. During $0.3s \leq t < 0.35s$ the BESS is disabled to prove the proposed strategy. The BESS is recharged by the utility grid and or/PV panels during $0.35s \leq t < 0.6s$. So, the system operates at mode 2.3. In this mode, it is possible to set the charging current of the storage system. The charging current was set to 10 A, as illustrated in Fig. 5(b). It can be observed that the inverter power is positive when the power of the loads and storage system is less than the power generated by the panels, as illustrated in Fig. 5(a). In this case, the exceeding power is injected into the PCC. When the PV power generation is inferior to the power of the loads and batteries, the inverter demands power from the utility grid and P_{inv} becomes negative. Fig. 5(c) presents the

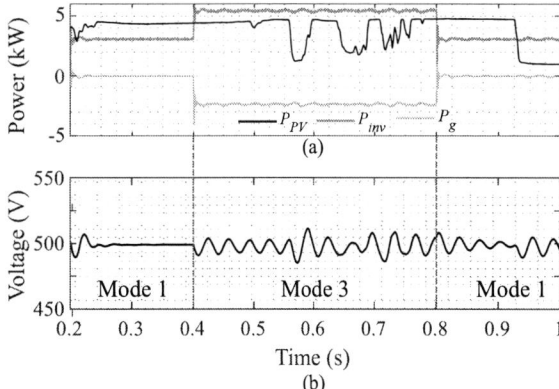

Fig. 6. Waveforms of the system on the operation mode 3: (a) PV generation power (P_{PV}), utility grid power (P_g) and inverter active power (P_{inv}) and (b) DC bus voltage (V_{DC}).

dc-bus voltage waveform.

Fig. 6 illustrates the transition between the operation modes 1 and 3. During $0.2s \leq t < 0.4s$ the system operates at mode 1. At $t = 0.4s$ it is considered that the SOC of the storage system is greater than the established limit (90%). So, during $0.4s \leq t < 0.8s$, the inverter injects into the PCC the maximum active power possible, that is $P^* = S_{max}$ and $Q^* = 0$. At $t < 0.8s$, it is considered that the storage system reaches the desired final SOCf, and system returns to mode 1. In Fig. 6(c) it is possible to verify the dc-bus voltage waveform.

978-1-7281-4181-7/19 $31.00 © 2019 IEEE   1424

IV. Conclusions

This paper presented a structure for the control and management of a multifunctional grid-tie inverter. The system can supply the load with active and reactive power. It aims to reduce the energy bill reducing the active power consumption from the grid at the same tie reduces the reactive power demand from the grid. The system can adequate itself to different demands of active and reactive power as the load requires. Moreover, there are predicted scenarios that the local load exceeds the maximum rated power of the inverter. In this case, the utility grid provides energy to the load.

In real situations, it is necessary to guarantee the operation of the BESS withing the SOC limits to avoid battery damage. In this context, the proposed system has some functionalities. When the SOC is below the minimum limit, the BESS is disabled and the energy generated by the PV panels is injected into the PCC. With that, it is possible to reduce the active power consumption from the grid. To recharge the BESS, the system has two options: one is to recharge the batteries using only PV generation. This scenario is attractive for the case when there are small local loads connected at the PCC. Another possibility is to recharge the BESS using the PV generation and the utility grid, so the rate of charge of the BESS becomes adjustable. This scenario becomes attractive for periods of lower energy prices. Besides that, when the power necessary to the charge is lower than the PV generation, the exceeding power is injected into the PCC, also reducing the active power consumption from the grid. Finally, if the BESS exceeds the upper SOC limit, the inverter can inject a constant power into the utility grid, despite the intermittent PV generation.

Acknowledgment

This work was funded by the Research and Development project PD 2866-0468/2017, granted by the Brazilian Electricity Regulatory Agency (ANEEL) and Companhia Paranaense de Energia (COPEL). The authors also would like to thank to FINEP, SETI, CNPq, Fundação Araucária, CAPES and UTFPR for additional funding.

References

[1] Y. Liu, S. Rau, C. Wu and W. Lee, "Improvement of Power Quality by Using Advanced Reactive Power Compensation," in IEEE Transactions on Industry Applications, vol. 54, no. 1, pp. 18-24, Jan.-Feb. 2018.

[2] L. Wang, C. Lam and M. Wong, "Multifunctional Hybrid Structure of SVC and Capacitive Grid-Connected Inverter (SVC//CGCI) for Active Power Injection and Nonactive Power Compensation," in IEEE Transactions on Industrial Electronics, vol. 66, no. 3, pp. 1660-1670, March 2019.

[3] R. Teodorescu, M. Liserre, P. Rodrígues. "Grid Converters for Photovoltaic a and nd Wind Power Systems," John Wiley & Sons, New Jersey, 2011.

[4] L. V. Belinaso. C. D. Schwertner, L. Michels, "Price-Based Power Management of off-grid Photovoltaic Systems with Centralised with" IET Renewable Power Generation, vol. 10 no. 8, Jul. 2016.

[5] M. Brenna, F. Foiadelli, M. Longo and D. Zaninelli, "Energy Storage Control for Dispatching Photovoltaic Power," in IEEE Transactions on Smart Grid, vol. 9, no. 4, pp. 2419-2428, July 2018.

[6] N. Saxena, I. Hussain, B. Singh and A. L. Vyas, "Implementation of a Grid-Integrated PV-Battery System for Residential and Electrical Vehicle Applications," in IEEE Transactions on Industrial Electronics, vol. 65, no. 8, pp. 6592-6601, Aug. 2018.

[7] S. Pranith, S. Kumar, B. Singh, T. S. Bhatti, "Multimode operation of PV-battery system with renewable intermittency smoothening and enhanced power quality," IET Renewable Power Generation, vol. 13, no. 6, pp. 4-29, Sep. 2019.

[8] Y. Yang, H. Li, A. Aichhorn, J. Zheng, M. Greenleaf, "Sizing Strategy of Distributed Battery Storage System With High Penetration of Photovoltaic for Voltage Regulation and Peak Load Shaving," IEEE Transaction on Smart Grids, vol. 5, no. 2, pp. 982 - 991, Sep. 2015.

[9] T. D. C. Busarello, A. Mortezaei, A. S. Bubshait and M. G. Simões, "Three-phase battery storage system with transformerless cascaded multilevel inverter for distribution grid applications," in IET Renewable Power Generation, vol. 11, no. 6, pp. 742-749, 10 5 2017.

[10] V. T. Tran, M. R. Islam, D. Sutanto and K. M. Muttaqi, "Mitigation of Solar PV Intermittency Using Ramp-Rate Control of Energy Buffer Unit," in IEEE Transactions on Energy Conversion, vol. 34, no. 1, pp. 435-445, March 2019.

[11] F. Li, X. Zhang, H. Zhu, H. Li and C. Yu, "An LCL-LC Filter for Grid-Connected Converter: Topology, Parameter, and Analysis," in IEEE Transactions on Power Electronics, vol. 30, no. 9, pp. 5067-5077, Sept. 2015.

[12] W. Wu, Y. Liu, Y. He, H. S. Chung, M. Liserre and F. Blaabjerg, "Damping Methods for Resonances Caused by LCL-Filter-Based Current-Controlled Grid-Tied Power Inverters: An Overview," in IEEE Transactions on Industrial Electronics, vol. 64, no. 9, pp. 7402-7413, Sept. 2017.

[13] Canadian Solar, "MAXPOWER CS6U - 340 | 345 | 350 High Efficiency Poly Module," Photovoltaic module datasheet, Apr. 2018.

[14] E. H. Watanabe, M. Aredes and H. Akagi, "The p-q Theory for Active Filter Control: Some Problems and Solution," in Journal of Control, Automation and Electrical Systems, vol. 15, no. 1, pp. 78-84, March 2004.

[15] D. Sera, L. Mathe, T. Kerekes, S. V. Spataru and R. Teodorescu, "On the Perturb-and-Observe and Incremental Conductance MPPT Methods for PV Systems," in IEEE Journal of Photovoltaics, vol. 3, no. 3, pp. 1070-1078, July 2013.

[16] R. Xiong, J. Cao, Q. Yu, H. He and F. Sun, "Critical Review on the Battery State of Charge Estimation Methods for Electric Vehicles," in IEEE Access, vol. 6, pp. 1832-1843, 2018.

[17] S. Golestan, J. M. Guerrero, J. C. Vasquez, A. M. Abusorrah and Y. Al-Turki, "Modeling, Tuning, and Performance Comparison of Second-Order-Generalized-Integrator-Based FLLs," in IEEE Transactions on Power Electronics, vol. 33, no. 12, pp. 10229-10239, Dec. 2018.

Sizing of Supercapacitor and BESS for peak shaving applications

Kaique Ferreira
Electrical Engineering Department
Federal University of Espírito Santo
Vitória, Brazil
kaiquee.f@hotmail.com

Walbermark Marques dos Santos
Electrical Engineering Department
Federal University of Espírito Santo
Vitória, Brazil
walbermark.santos@ufes.br

Augusto César Rueda Medina
Electrical Engineering Department
Federal University of Espírito Santo
Vitória, Brazil
augusto.rueda@ufes.br

Abstract - **In the last decades, the growth of electric energy demand made the study and application of peak shaving systems grown rapidly. One of the most crucial parts of these systems is the Energy Storage System (ESS) sizing. Numerous methods have already been proposed for sizing, forecasting and ESS charge/discharge planning. However, optimization sizing techniques aiming economic benefits have not been explored yet for Battery Energy Storage Systems (BESS) associated with Supercapacitors. This paper proposes a sizing methodology to determine the quantity of BESS and Supercapacitor from a load profile to obtain maximum financial savings. The code uses genetic algorithm to find the best ESS combination and calculates its respective operating points to determine which has the greatest economic potential. The outputs are the BESS and Supercapacitor values (in kWh) that the consumer should acquire in order to obtain the highest rate of return. Showing that, depending on the analyzed load profile, the association of different ESS technologies can lead to a higher economic benefit than using only one.**

Keywords—Peak Shaving, BESS, Supercapacitor, Genetic Algorithm.

I. INTRODUCTION

Over the years, there is a natural increase in the electric power demand caused by the increase of end users [1]. This increase can lead to problems such as instability, power failure and voltage fluctuation if the grid does not support the required level of demand [1].

The most immediate solution to avoid this scenario would be increasing the generating capacity and the number of transmission and distribution lines, which results in substantial investments. An alternative is to apply demand management methodologies, highlighting in this context, the peak shaving methodology.

"Peak shaving" can be defined as a set of techniques that aims to reduce power demand at the highest consumption periods to lower the demand charges of the electricity bill or delay in the expansion of investments that the power utility must do by not supporting the power demand during these periods.

There are several examples where diesel generators are used to supply part of power demand during the abovementioned periods [1] [2]. However, in addition to the high Operation and Maintenance (O&M) costs of diesel generators [1], concerns about emission of greenhouse gases to the atmosphere led to an increase in the utilization of renewable energy sources [3].

Due the intermittent nature of renewable energy sources such as photovoltaic (PV) and wind, the use of these sources directly connected to the power utility bus can lead to voltage fluctuations and frequency instability [1] [4] [5]. It is a common practice to use renewable energy sources tied to ESS.

The lowering of ESS prices, the increase in the electric energy demand prices and peak shaving interest [6] makes the use of only ESS for peak shaving a viable option. An example is utilize ESS to accumulate energy in periods where the prices are lower to utilize it when the prices are higher, as shown in [6] [7] [8].

One of the key steps for starting a peak shaving project using only ESS is sizing the ESS that will be used. Once it is not correctly sized, it can lead to reduction of profits that the project could have during the project's lifetime, and it can also lead to cost increase and rise of system losses [1] as well as early battery aging [6]. Once well sized, it can lead to a more even power consumption profile, reducing the need for power utility bus upgrades [6] as well as economic benefits due to the reduction of power demand.

Several sizing methods have already been proposed. As per Linear Programing to minimize energy costs and find the best financial return (without the full analysis of the consumer profile) [9]; Linear programming using a dual-simplex algorithm [6]; Graphical or numerical demonstration of the objective function for the possible input values, such as load profile, energy fees and battery parameters, utilizing dynamic programming to the operating strategy [2].

However, a topic that has not been yet explored is the sizing of multiple ESS types working alongside, since all of the aforementioned papers size for one type of ESS at a time.

Thus, this work presents a methodology for sizing the best combination of BESS and Supercapacitors in order to obtain the maximum financial return possible from an input consumer profile.

There are two main components in the electricity bill of most commercial and industrial consumers: the consumption charge [$/kWh] and demand charge, which represents the maximum power demand registered in the period (usually per month) [$/kW] [10]. The implemented strategy aims to store electrical energy in ESS and unload it at times where the power demand exceeds a pre-stablished value to decrease the maximum power demand registered by the power utility per month. It is noteworthy that there are studies which focus on other improvements or where the energy bill is done in other methods, varying the strategy adopted [6] [7] [11]. Thus, it is shown that, depending on the consumer profile, the highest financial return can be found in the combination of different ESS, which this paper summarized by BESS and supercapacitors.

To improve the proposed idea presentation, the sections of this paper are divided as follows: Section II describes how the peak shaving processes works; Section III explains how

978-1-7281-4181-7/19 $31.00 © 2019 IEEE

the methodology presented obtains the quantity of BESS and supercapacitors; Section IV performs the methodology described in section III for two consumers with different load profiles, but the same amount of total energy; Conclusions are presented in Section IV.

II. PEAK SHAVING

A. Peak shaving utilization Example

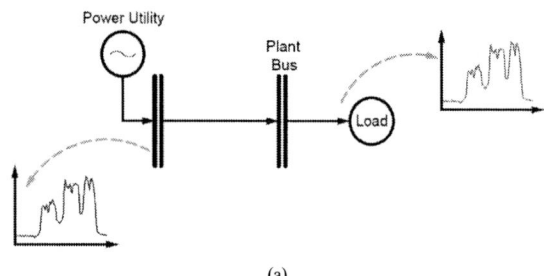

(a)

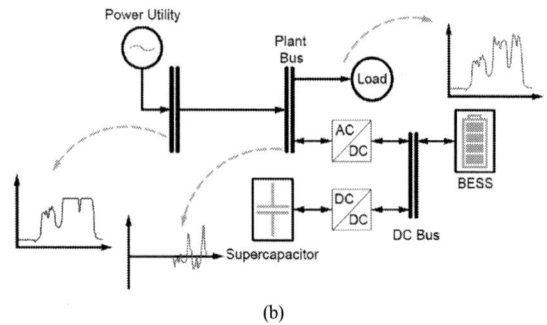

(b)

Fig. 1. Example of Peak Shaving using BESS and Supercapacitor.

Figure 1(a) shows a load profile of a hypothetical consumer without using any peak shaving techniques for one day load. It can be observed that the consumption profile registered in the power utility is the same as the consumer profile.

In Figure 1(b) it is shown, to the same consumer, how his load profile could have been registered by the power utility in the same period of time if peak shaving techniques were used. In this example, BESS and supercapacitors are used for peak shaving, taking into account their respective rates of charge, discharge and self-discharge.

Thus, the peak shaving technique can change the consumer's demand profile seen by the power utility, and the peak energy demand are no longer coming from the utility bus, but it is supplied by the ESS.

B. ESS model

The circuit model chosen in this paper to represent the BESS and Supercapacitor's circuit is represented by Figure 2 [12]. The BESS and supercapacitor differences considered in this paper are the presented model values. Therefore the Charge/Discharge/Self-discharge rates are different. It is noteworthy that this circuit is a simplified model. There are other variables that may influence the State of Charge (SoC) of the ESS, such as Temperature, State of Health (SoH) and utilization time [6] [13].

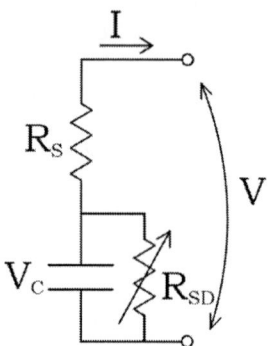

Fig. 2. Energy Storage System Circuit Model

In order to calculate to SoC during the program execution, the Coulomb Counting method is used, which calculates SoC by integrating the current over time [14].

C. Peak Shaving examples around the world

Peak Shaving projects has already been implemented in several different countries, using a wide range of energy capacities and different types of ESS technologies. Table I illustrates some examples of peak shaving applications around the world [1].

TABLE I. PEAK SHAVING EXAMPLES AROUND THE WORLD

Project name	ESS	Energy capacity [MWh]	Location [Country]
Smarter Network Storage	BESS	10	England
Gwansei Gakuin University	Vanadium Redox Flow	5	Japan
Prudent Energy Inc	Vanadium Redox Flow	3,6	EUA
Soma, Kapan (IHI)	BESS	2,8	Japan
Wailea, HI	BESS	1	EUA
Zurich battery energy storage system	BESS	0,5	Switzerland

III. SIZING METODOLOGY

The search for the best set of ESS and points of operation is based on the Bellman's optimization principle, where regardless of the initial state and condition, subsequent decisions must be optimal for this initial condition [15]. Once each set of ESS contains an optimal set of operation points, the program only has to find the best operation points to each set of ESS and analyze which has the highest financial return.

Before determining the amount of BESS and supercapacitors that the analyzed profile should use, an assessment of how these components would work during the project execution was made. For this, the program simulates the charging and discharging conditions of the ESS during a year of project execution, assuming that there will be no major changes in the load profile in the coming years [9].

The algorithm is divided into three parts: New Consumption; Threshold determination and ESS Determination.

A. New Consumption

For illustrative purpose, it will be explained how the program works using only one ESS technology.

978-1-7281-4181-7/19 $31.00 © 2019 IEEE

The program starts with the ESS fully charged. The algorithm changes the SoC of the ESS according to its charge/discharge/idle requirement, causing the new demand registered in the power utility to be changed, as already shown in Figure 1(b).

To stablish the need for charging/discharging/idle abovementioned, a relationship is made between the current consumption and a pre-stablished threshold, called demand threshold (thr). If the demand is higher than thr, the ESS will discharge the required power (amount of energy that exceeded the thr) in the grid bus. Otherwise, the ESS may be in idle (when it is fully charged) or in charging state (when the ESS is not fully charged).

When the ESS need to be charged, the new demand registered by the power utility is given by Equation (1).

$$Dem = Dem_{original} + ESS_{charge} + ESS_{Autodischarge} \quad (1)$$

Equation (2) describes how the new demand is calculated when it is necessary for the ESS to discharge into the grid bus.

$$Dem = Dem_{original} - ESS_{discharge} \quad (2)$$

If it is not necessary to charge or discharge the ESS, the new demand registered only needs to supply the demand required by the consumer and maintain the SoC of the ESS in order to supply the self-discharge as described in Equation (3). It is worth noting that in the current paper, losses related to power conversion, SOH and O&M are not considered.

$$Dem = Dem_{original} + ESS_{Autodischarge} \quad (3)$$

The possible ESS states (Charge/Discharge/Idle) are shown in Figure 3.

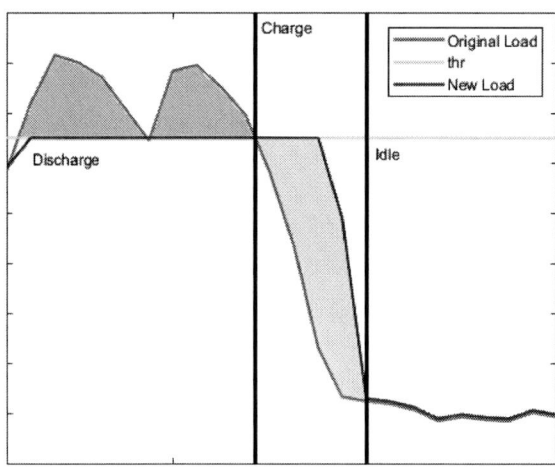

Fig. 3. Possible ESS states.

The program inputs to calculate the new demand registered by the power utility are: Consumer load profile (historical data averaged in 15 minutes resolution)/ ESS charge/discharge rates [kWh]; ESS power capacity [kWh]; Self-discharge rate [% p.d.] and demand threshold.

Figure 4 shows an example of the algorithm's output (the new demand represented by the blue chart) for one day analysis. This new demand profile is calculated from the previous load profile and the load variation of the ESS, following the Equations (1), (2) and (3).

It can be observed that the maximum demand registered by the power utility in the new profile demand is lower than the registered in the original load, decreasing the demand charges, which can represent 50% of the total amount paid in the electricity bill [2]. Once having the new demand profile, the new energy bill can be calculated, and thus evaluate if the project is economically feasible.

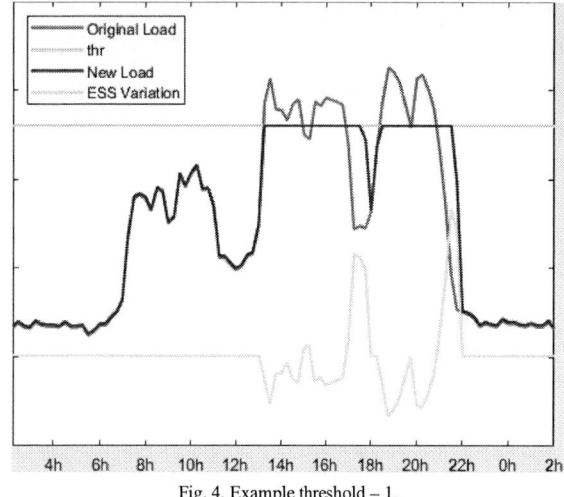

Fig. 4. Example threshold – 1.

B. Threshold determination

The demand threshold (thr) value determination is not trivial, some authors do not specify how it was determined and/or use it as input to the program [6] [9].

In order to illustrate how poor thr determination can lead to high demand values, the same input data from the example of Figure 4 were used to generate the results shown in Figure 5. The only difference being the thr. After comparing the results, it is evident that there was an increase in the maximum value registered in the new demand for the second presented case.

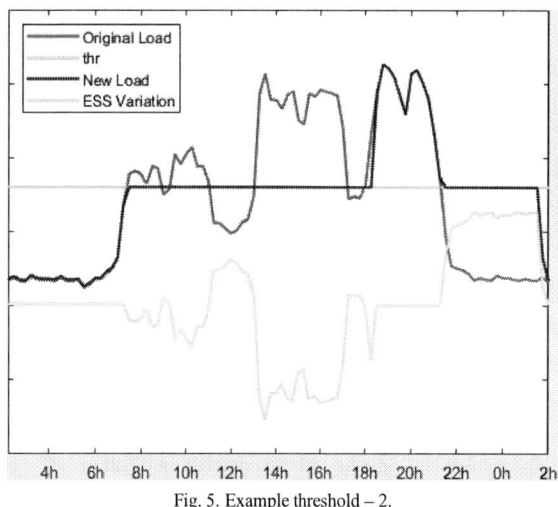

Fig. 5. Example threshold – 2.

In this paper, the lowest value of thr associated with the lowest peak of demand is considered as the optimal threshold, as shown in Figure 6.

There are other ways to determine the thr, as a relationship between the area above the thr and the energy capacity of the selected ESS [2] [10], but it does not take into account the battery charge/discharge rate and time between peaks, which means that the ESS may not be used to its maximum potential.

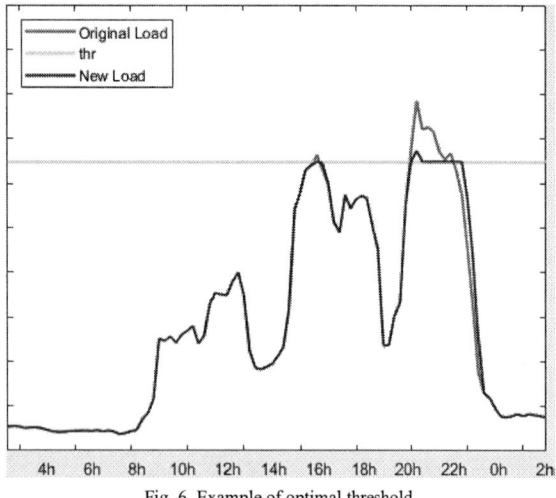

Fig. 6. Example of optimal threshold.

The highest demand registered in the new demand profile is not necessarily the thr, since this value is limited to the ESS discharge and charge rates, as already demonstrated in [9].

The algorithm that searches for the best value of thr is shown below:

```
Step = 0.001*max(Original_Profile);
Thr = max(Original_Profile);
Best_thr = Thr;
Precision = 5;
For i=1:1:Precision
    Exit =1;
    While exit ~= 0
        Thr = Thr – Step;
        [New_Profile]=Function _ New_Profile(Thr);
        If max(New_Profile) <=
                max(Function _New_Profile(Best_thr))
            Best_thr = thr;
        Else
            Exit = 0;
        End
    End
    Thr = Best_thr;
    Step =Step/10;
End
```

C. ESS Determination

The BESS and supercapacitor selection criteria is based on the highest financial return over the project's lifetime.

One of the program constraints for the ESS selection is the maximum investment, which represents the maximum value spent on the set of BESS and supercapacitors. Thus, the program has an input value of investment limit.

In order to obtain a better discrete combination value of BESS and supercapacitors, a genetic algorithm containing the abovementioned investment restrictions was used.

The genetic algorithm performs the calculations to determine the thr on each set of individuals, making each set

of BESS and supercapacitors to operate at their full potential, as described in section B. Thus, in each reproduction there are sets of ESS, and for each set of ESS there is an output that represents the final economy. The set of ESS that has the greatest economy is selected. Figure 7 illustrates how the final program works.

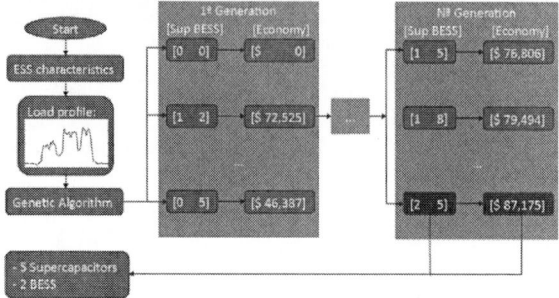

Fig. 7. Example of the proposed sizing methodology.

IV. SIMULATION

Two different hypothetical consumer's profiles were analyzed: Consumer A and Consumer B. In order to make a fair analyzes, both profiles are from the same year and have the same total annual energy consumption.

For illustrative purpose, Figure 8 demonstrates how the demand profile is during a business day on each consumer.

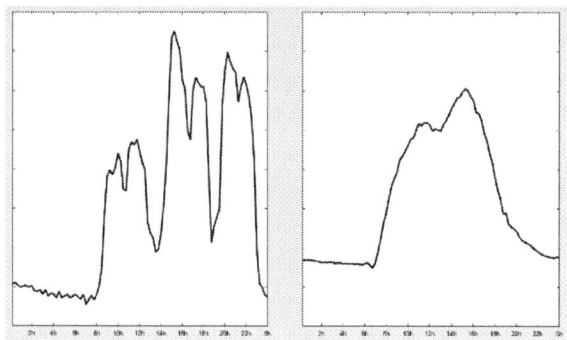

Fig. 8. Demand profile – Consumer A and Consumer B.

The genetic algorithm searched for the best set of BESS and supercapacitors for both consumers, as well as their respective thr.

The simulation was done considering an entire year of consumption in order to take into account the climatic variations caused by each season of the year.

The thr value changes every month, since the electricity bill considered in this paper charges consumption and demand monthly.

Table 2 shows the electrical characteristics of the ESS used in the simulation. It is worth noting that the simulation does not consider variations in temperature; charge and discharge rates are considered constant regardless the ESS SoC, being only limited to the total energy capacity [kWh].

TABLE II. ESS CHARACTERISTICS

	Energy Capacity [kWh]	Discharge \ Charge Rate [%\15min]	Auto Discharge [% p.d.]
Supercapacitor	3.55	33.8%	~0.2
BESS	22.272	25%	~0.01

Table 3 presents the results for the abovementioned consumers. The Payback time, Final Economy and the set of BESS and supercapacitors are shown.

TABLE III. PROGRAM RESULTS

	Supercapacitor [kWh]	BESS [kWh]	Payback [year]	Final Economy [$]
A	17.75	44.544	4.9516	88,708.00
B	0	44.544	7.5333	29,118.00

V. CONCLUSIONS

Two consumers were presented with the same total electricity demand, but different load profiles.

The results of the simulation shows that Consumer A would have the highest financial return using a combination of BESS and supercapacitors, while Consumer B should only utilize BESS in order to make a larger profit. One of the likely explanations for these different set of ESS is that Consumers A has a greater demand variation compared to Consumer's B. Therefore, it is found that the need for usage of BESS and/or supercapacitor in order to obtain the highest economic return is intrinsically connected to the consumer's load profile. Concluding that depending on the load profile, the combined usage of different ESS technologies can lead to greater savings than using only one type.

ACKNOWLEDGMENT

The authors thank the Federal University of Espírito Santo (UFES) for infrastructure and the Coordination for the Improvement of Higher Education Personnel (CAPES) for the financial support.

REFERENCES

[1] Uddin, Moslem, et al. "A review on peak load shaving strategies." *Renewable and Sustainable Energy Reviews* 82 (2018): 3323-3332.

[2] Oudalov, Alexandre, Rachid Cherkaoui, and Antoine Beguin. "Sizing and optimal operation of battery energy storage system for peak shaving application." *2007 IEEE Lausanne Power Tech*. IEEE, 2007.

[3] Teitzel, B. C., M. H. Haque, and R. Inwood. "Energy storage for rooftop solar photovoltaic systems to reduce peak demand." *8th International Conference on Electrical and Computer Engineering*. IEEE, 2014.

[4] Cheng, Chun-Tian, et al. "Short-term peak shaving operation for multiple power grids with pumped storage power plants." *International Journal of Electrical Power & Energy Systems* 67 (2015): 570-581.

[5] Shivashankar, S., et al. "Mitigating methods of power fluctuation of photovoltaic (PV) sources–A review." *Renewable and Sustainable Energy Reviews* 59 (2016): 1170-1184.

[6] Martins, Rodrigo, et al. "Optimal component sizing for peak shaving in battery energy storage system for industrial applications." *Energies* 11.8 (2018): 2048.

[7] Madhuranga, A. G. N., and I. A. Premaratne. "Energy storage battery bank system to reduce peak demand for domestic consumers." *2017 International Conference on Computer, Communications and Electronics (Comptelix)*. IEEE, 2017.

[8] Dabbagh, Mehiar, et al. "Shaving data center power demand peaks through energy storage and workload shifting control." *IEEE Transactions on Cloud Computing* (2017).

[9] Fisher, Michael, Jay Whitacre, and Jay Apt. "A Simple Metric for Predicting Revenue from Electric Peak-Shaving and Optimal Battery Sizing." *Energy Technology* 6.4 (2018): 649-657.

[10] Chua, Kein Huat, Yun Seng Lim, and Stella Morris. "Energy storage system for peak shaving." *International Journal of Energy Sector Management* 10.1 (2016): 3-18.

[11] Yang, Ye, et al. "Sizing strategy of distributed battery storage system with high penetration of photovoltaic for voltage regulation and peak load shaving." *IEEE Transactions on Smart Grid* 5.2 (2013): 982-991.

[12] Niu, Jianjun, Brian E. Conway, and Wendy G. Pell. "Comparative studies of self-discharge by potential decay and float-current measurements at C double-layer capacitor and battery electrodes." *Journal of power sources* 135.1-2 (2004): 332-343.

[13] Xing, Yinjiao, et al. "State of charge estimation of lithium-ion batteries using the open-circuit voltage at various ambient temperatures." *Applied Energy* 113 (2014): 106-115.

[14] G. S. Misyris, T. Tengnér, A. G. Marinopoulos, D. I. Doukas, D. P. Labridis, "Battery energy storage systems modeling for online applications", *IEEE Manchester PowerTech*, pp. 1-6, 2017.

[15] Zhang, Hua-Guang, et al. "An overview of research on adaptive dynamic programming." *Acta Automatica Sinica* 39.4 (2013): 303-311

Two-Stage SEPIC-Buck Topology for Neighborhood Electric Vehicle Charger

Rafael H. Eckstein
Power Electronic Institute (INEP)
Federal University of Santa Catarina (UFSC)
Florianópolis, Brazil
rafael.eckstein@ifsc.edu.br

Telles B. Lazzarin
Power Electronic Institute (INEP)
Federal University of Santa Catarina (UFSC)
Florianópolis, Brazil
telles@inep.ufsc.br

Gierri Waltrich
Power Electronic Institute (INEP)
Federal University of Santa Catarina (UFSC)
Florianópolis, Brazil
gierri@gmail.com

Abstract—**It is no exaggeration to consider electric vehicles a reality in our current society. Comparing to combustion vehicles they have better efficiency and monetary profitability by distance traveled. However, the necessary infrastructure for the proper functioning of EVs, as batteries, individual chargers, charge stations, etc., are still precarious in several places. This paper introduces a new interleaved model for a low voltage battery pack charger for neighborhood electric vehicles (NEV). The system has two power stages, a SEPIC PFC rectifier and a DC-DC Buck converter, which process 1.5 kW per cell a total of 4.5 kW. The proposed system is adequate to lower power single-phase systems, which the battery pack voltage is lower than the grid.**

Index Terms—**Neighborhood electric vehicle, SEPIC PFC Rectifier, DC-DC Buck Converter, 3 Level Battery Charger.**

I. INTRODUCTION

The interest in electric vehicles (EVs) has been steadily increased in the last years due to better overall system efficiency and the higher cost of fossil fuel compared to electricity [1], [2]. A counterpoint on electric vehicles has always been the low autonomy, as the battery pack (BP) could not provide power for more than a few dozens of kilometers. While this is still a disadvantage, studies have been shown an improvement in the power density of the batteries.

Neighborhood Electric Vehicles (NEV) present a clever choice for situations where fuel consumption is more important than high autonomy. Several locations require vehicles for short-distance travel, such as shopping malls, business conglomerates, golf courses, universities, resorts and so on. In these zones when the ride is complete, the vehicle should be moved back to its exit point or remains at the end point. In both cases, the venture may provide locations to charge the BBs while waiting for the next trip.

The NEVs are usually smaller in size compared to an EV and therefore have a smaller battery pack. The battery voltage levels range from 40 to 90 V, and the capacity of the mounted battery is generally around 100 Ah, for this capacity a 1.5 kW charger is already sufficient [3]. However, charging can

be faster by increasing the electric current supplied to the BP. To meet the desire of brazilians entrepreneurs and end users, a battery charger that can charge up the BP in 3 different levels 1.5, 3.0 and 4.5 kW (slow, moderate and fast charge, respectively) is proposed in this paper.

For these values of power, the disturbances in the mains rise and becomes necessary to operate with unit power factor [3]. The system should also provide a battery current with low ripple during the whole charge cycle. Generally, the EV chargers consist of two stages, a PFC Boost converter, with a diode bridge operating as a rectifier and an isolated DC-DC converter, to control the battery charge [4]. Fig. 1 presents a very common topology for EV chagers, a PFC Boost rectifier and an isolated Full-Bridge converter. A disadvantage of this is the low efficiency of the Full-Bridge when operating between a high bus, generated by the PFC boost converter, and a low voltage battery pack, as is the case with NEV.

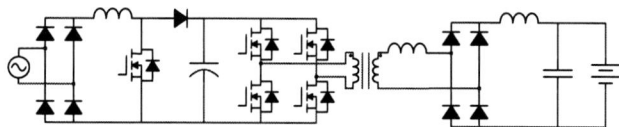

Fig. 1. A usual topology for electric vehicle charger.

A solution to this problem is the use of resonant rectifiers, because compared to the traditional, the phase shift Full Bridge converter presents improvement in efficiency and total volume [5]. The disadvantage is the difficult implementation of the system in order to present different load levels. Another way to solve the system efficiency for a low voltage BP is the presence of a rectifier that reduces the input voltage, different from the PFC Boost converter. In order to have a robust, low-cost and efficiency electric vehicle charger, this work presents a system composed of a interleaved three level PFC SEPIC rectifier and a DC-DC Buck converter (Fig. 2).

978-1-7281-4181-7/19 $31.00 © 2019 IEEE

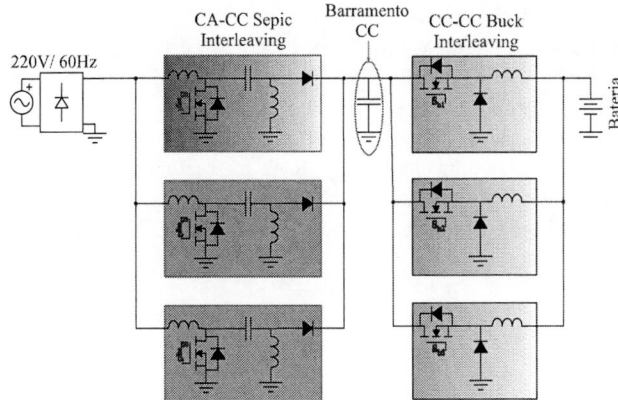

Fig. 2. Proposed system.

II. PROPOSED SYSTEM AND CONTROL STRATEGY

This section presents the operation steps of only one module (Fig. 3) of the proposed system, the description of each power stage, the control strategy used and finally analyzes the interleaved technique used.

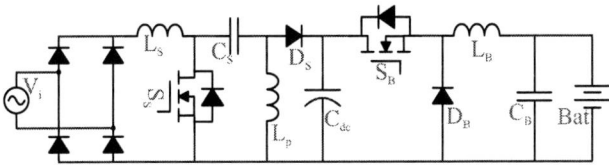

Fig. 3. Proposed system - one module.

A. Proposed System

The proposed system consists of a single-phase diode bridge, a SEPIC converter (first stage), a dc-link and a DC-DC Buck converter (second stage) connected to the battery terminals. Each power stage has two operational steps. Fig. 4 shows the operational steps of the SEPIC converter in continuous conduction mode:

- Step 1 - During the course of this step the switch S_s is conducting, and the diode D_s, is blocked. The currents in the inductors L_s and L_p increase linearly and the fictional load R_o is fed by the capacitor C_{dc}.
- Step 2 - The second step starts the moment the switch S_s is blocked. At this moment, the diode D_s conducts the current, and the energy stored in each inductor is transferred to the output. The L_s and L_p currents decrease linearly. This step ends when switch S_s is triggered again, returning to the first step.

The operatinal steps of the Buck converter are shown in Fig. 5 and are described below. Fig. 6 shows the relationship between the duty cycle and the state of charge (SOC) of the battery.

- Step 1 - The switch, S_B, is conducting and the source current flows through the inductor. The power delivered

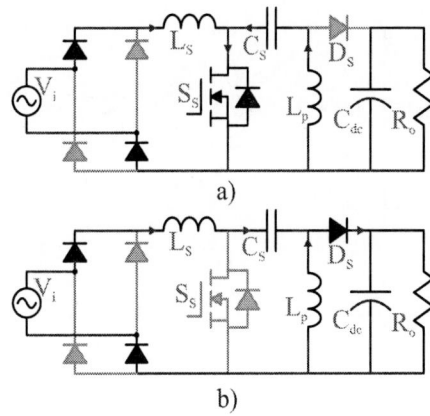

Fig. 4. a) First step - SEPIC rectifier. b) Second step - SEPIC rectifier.

to the load comes from the input source (DC bus). The current in the inductor increases linearly.

- Step 2 - When the switch is open the energy stored in the inductor is transferred to load through the free wheel diode, D_B. In this step the current of the inductor decreases linearly.

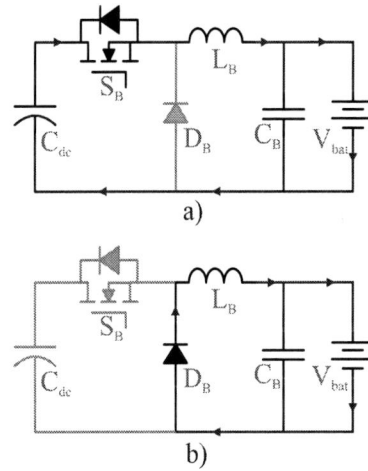

Fig. 5. a) First step - Buck converter. b) Second step - Buck converter.

The choice of the converters for each power stage was made because the topologies present some positives aspects, such as:

- SEPIC: (i) Operate with static gain less than unitary (buck type); (ii) Present current input, wich facilitates PFC systems; (iii) It has only one switch, which does not require an isolated driver; (iv) Enables the galvanic isolation of the system by substituting the inductor Lp by a transformer;
- Buck: (i) Presents a current characteristic in the output; (ii) Reduces input voltage (NEV battery packs have low voltage level); (iii) It has only one active switch; (iv) Capability to interleaved operation to reduce the output current ripple;

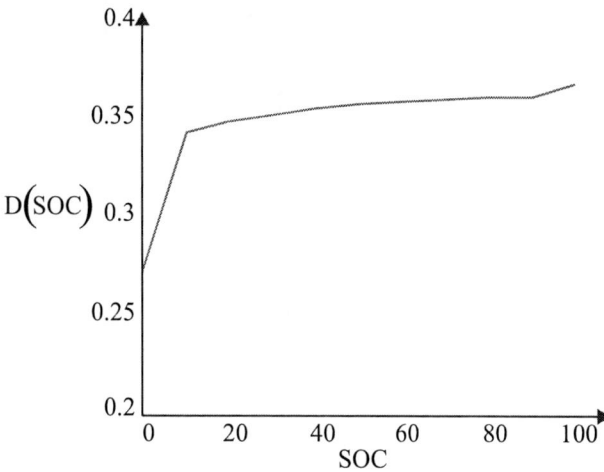

Fig. 6. relationship between the duty cycle and the state of charge (SOC) of the battery.

In comparison to the classical electric vehicle charger structures present in the literature, the system proposed has a lower number of components, a lower number of active semiconductors, and a lower DC bus voltage, which allows an optimum operation of the converters during the whole voltage range of the electric grid and the battery.

B. Control Strategy

Both converters presents two PI controllers. The SEPIC Rectifier controls the voltage on the DC - bus and the input current, operating at a power factor close to the unit. While the CC-CC Buck converter performs CC-CV (Constant current - constant voltage) control for proper battery charging.

The SEPIC control uses a voltage sensor to obtain a sine waveform signal equal to the mains. The modulus of this signal is multiplied with the voltage compensator output signal and is compared to the signal obtained by an inductor current sensor, L_s. The result error is compensated by the controller and the final signal is used for the PWM modulation of the switch. Fig. 7 presents the SEPIC rectifier control block diagram.

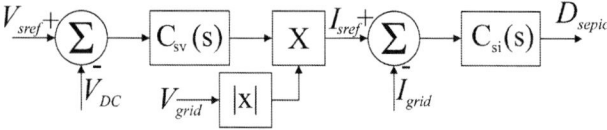

Fig. 7. SEPIC rectifier control block diagram.

For the Buck converter control, assuming the battery is discharged, with minimum voltage at the terminals, a constant reference (V_{bref}) of value equal to the maximum bank voltage is applied to the voltage loop. Considering that a battery pack has high inertia, the voltage variation at the terminals is considerably slow in relation to a switching period or even from the mains.

Thus, the error generated by the difference of the reference voltage and the measurement at the terminals, after being

limited by the maximum value of the battery charge current, becomes the reference for Buck converter current control. This stage is known as constant current (CC) and lasts until the battery terminals voltage reaches the reference value.

In the second stage (CV) the external loop controls the converter, keeping the voltage level at the terminals constant while the injected current is reduced until reaching a value of 5% rated when the charge is finished. Fig. 8 presents the block diagram of the control technique employed.

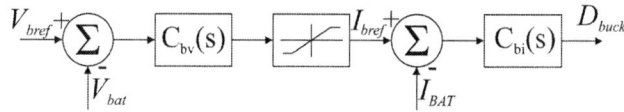

Fig. 8. Buck converter control block diagram.

C. Interleaved Strategy

The system operates with three modules in interleaved configuration, with each module processing 1.5 KW, so the maximum load power is 4.5 kW. In addition to allowing a higher value of processed power, a interleaved technique, which type of module switch is triggered by a lagged signal of $2\pi/n$ (where n is the number of modules operating), reduces the ripple of the battery current, which is one of the critical parameter to prolong the battery life.

III. SIMULATION RESULTS

The proposed system was simulated for charging the battery pack of the Li, an electric vehicle from Mobilis, a company located in Florianópolis, Brazil. The battery pack parameters are shown in Table I.

A dynamic simulation of the system was performed to show the behavior of the voltages and currents of the electric grid and the battery. Fig. 9 presents the voltage levels of the mains, the DC bus and the BP. Fig. 10 a) and 10 b) shows the waveforms of voltage and current at the battery terminals, respectively. When the battery voltage reaches 55V, the control system that previously controlled the current starts to control the voltage until the current reaches 5% at rated current finishing the charge cycle. The ripples of the charge and inductors currents are presented in Fig. 10 c) and Fig. 10 d), respectivaly. The Fig. 11 shows the grid voltage and current, which are in phase, presenting a power factor very close to the unit.

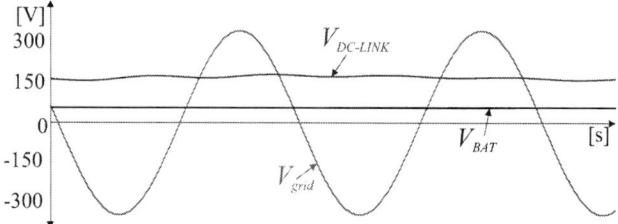

Fig. 9. Relation between mains, dc bus and battery pack voltage.

978-1-7281-4181-7/19 $31.00 © 2019 IEEE

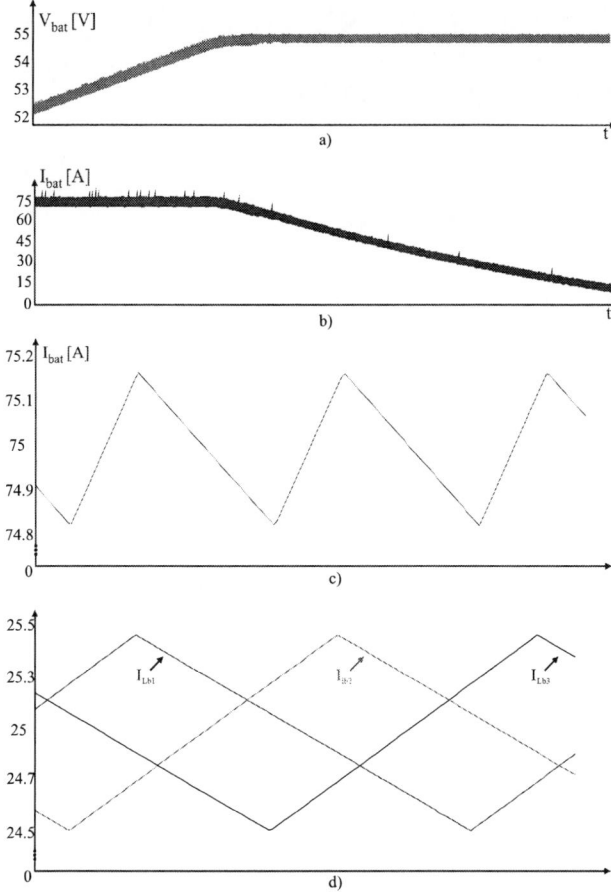

Fig. 10. Simulations results: a) Battery pack voltage during one cycle. b) Charge current during one cycle. c) Charge current ripple. d) Current ripple of the Buck converter inductors.

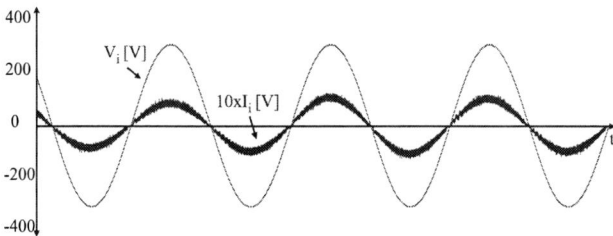

Fig. 11. Waveforms of the voltage and current of the mains.

TABLE I
PARAMETERS OF THE PROPOSED SYSTEM

Parameter	Value	Parameter	Value
Input Voltage (rms)	220 V	Maximum Output Voltage	55 V
DC Voltage	150 V	Nominal Output Current (p/ Cell)	25 A
Nominal Power (p/ Cell)	1.5 kW	Buck Switching Frequency	100 kHz
SEPIC Switching Frequency	70 kHz	Number of Cells	3

IV. CONCLUSION

This paper presents a study of an electronic system capable of charging a neighborhood electric vehicle, which presents a lower voltage BP, using a SEPIC rectifier and a DC-DC Buck converter. The system has some technical advantages such as the low number of switches, low dc-link voltage, good duty cycle ratio, and the possibility of galvanic isolated operation. The results of the first simulations were satisfactory, battery current ripple is small and the current and voltage of the electric mains are in phase with each other.

ACKNOWLEDGMENT

The authors would like to thank CAPES for their contribution to this work in the form of a fellowship awarded to Telles Brunelli Lazzarin, Dr.[program: POS-DOC -Pesquisa Pós-doutoral no Exterior/Process n° 88881.119841/2016-01].

REFERENCES

[1] Z. Li and H. Wang, "Comparative analysis of high step-down ratio isolated DC/DC topologies in PEV applications," in 2016 IEEE Applied Power Electronics Conference and Exposition (APEC), 2016, pp. 1329–1335.

[2] D. Wei, F Darie, and H. Wang, "Neighborhood-level collaborative fair charging scheme for electric vehicles," in 2014 IEEE Innovative Smart Grid Technologies Conference (ISGT), 2014, pp. 1-5.

[3] C. S. Lee, J. B. Jeong, B. H. Lee, and J. Hur, "Study on 1.5 kW battery chargers for neighborhood electric vehicles," in 2011 IEEE Vehicle Power and Propulsion Conference, 2011, pp. 1–4.

[4] M. M. U. Alam, W. Eberle, and F. Musavi, "A hybrid resonant bridgeless AC-DC power factor correction converter for off-road and neighborhood electric vehicle battery charging," in 2014 IEEE Applied Power Electronics Conference and Exposition - APEC 2014, 2014, pp. 1641–1647.

[5] I. O. Lee, "Hybrid DC-DC Converter With Phase-Shift or Frequency Modulation for NEV Battery Charger," IEEE Transactions on Industrial Electronics, vol. 63, no. 2, pp. 884–893, Feb. 2016.

Modeling Battery Energy Storage System operating in DC microgrid with DAB converter

Rafael dos Santos[1] *, Flávio A. S. Gonçalves[2] *, José A. Olimpio Filho[3] *, Fernando P. Marafão[4] *, Eric Gil[5]†

*UNESP - Instituto de Ciência e Tecnologia de Sorocaba
Av. Três de Março, 511-Aparecidinha, Sorocaba-SP, Brazil 18087-180
†Emerson Process Management AB
Hammarby Allé 150, SE-120 66 Stockholm-Sweden
[1]rafaelsantosrdsrds@gmail.com, [2]flavio.as.goncalves@unesp.br, [3]jose.olimpio@unesp.br,
[4]fernando.marafao@unesp.br, [5]eng.ericgil@gmail.com

Abstract—**This paper presents a computational model for studying the energy flow at the integration of a battery bank (BB) to a DC microgrid, through the use of the Dual Active Bridge (DAB) converter. The one time constant model was adopted for the battery, with its parameters obtained through exponential regression by the Least Squares method, through its hysteresis curve between terminal voltage and load level percentage. The curve was synthesized by data collected from a LiFeYPO4 reference battery. The DAB converter was controlled by the Phase-Shift modulation technique. The converter power management occurs through a state machine, which defines the conditions of energy flow between the microgrid and the energy storage system. This paper has educational aims for the study of energy flow at the integration of BB into DC microgrids.**

Index Terms—**DAB, State Machines, Battery Banks, DC Microgrid.**

I. INTRODUCTION

The field of electricity generation based on renewable energy sources (RES) has concentrated many research and development efforts around the world. These researches has been motivated by the promotion of sustainable energy policies, for example through credit incentives, tax subsidies, etc. [1], [2]. The technologies based on RES have increasingly offered attractive solutions for the integration of distributed energy resources such as: Electric vehicles, energy storage tecnologies and microgrids [3], [4]. In particular, microgrids have been considered as an attractive and efficient solution for applications in industrial parks, commercial, rural and residential groupings, besides scenarios where there are distributed renewable resources [4], [5]. However, one of the most challenging problems in these structures comes from the RES power delivery intermittence, which makes necessary the use of energy storage elements to compensate this effect. The main energy storage devices are the bank of batteries (BB), supercapacitors, flywheels, among others. However, batteries are a good alternative for storage systems, due to their high energy density and capacity for long-term energy supply [6]. The interconnection between the batteries and the DC microgrid requires an interface based on power electronics with bi-directional power flow capability. There are several topologies that may be employed for this purpose, such as the buck-boost converters [7], full-bridge push-pull converters

[8], and different Dual Active Bridge (DAB) configurations [9]. The DAB converters are preferred due to high voltage gain, high efficiency and galvanic insulation [10], [11].

Despite the massification of energy storage devices into DC microgrids, there are currently no consolidated standards that address the design of this type of microgrid, including the energy storage structure with batteries [12]. Although there are some regulations, such as the sections on safety standards in the National Electric Code American battery systems [13], the IEEE-1679-2010 standard for characterization and evaluation of lithium ion batteries in stationary applications [14], and the IEEE-1375-1998 standard for battery protection [15], there are still difficulties related to lack of practical knowledge about security issues, DC installation compatible products and experience in DC microgrids projects [12]. This context creates opportunities for the development of standards in the area, which are created on discussions based on studies, simulations, assessments and applications in this field. In this sense, this work aims to propose a simulation environment to explore different behaviors of BB integrated to DC microgrid bus, through a DAB power converter, in order to evaluate and analyze the main effects of this type of operation. The power management by the converter will occur through a state machine [16], which defines the conditions of energy flow between the microgrid and the energy storage system.

This article is organized as follows:

- Section II: Presentation of the DC microgrid.
- Section III: Presentation of the BB modeling method.
- Section IV: Description of BB and microgrid integration.
- Section V: Presentation of state machine model.
- Section VI: Simulations results.
- Section VII: Brief presentation of droop control strategy as an possible improvement for the control scheme.
- Section VIII: Discussions.
- Section IX: Main conclusions and application possibilities.

II. DC MICROGRID

A microgrid involves a low voltage electrical grid (LVD-C/LVAC) or medium voltage electrical grid (MVDC/MVAC), which can inject or receive energy from a main-grid. The arrangement of distributed generators systems such as: internal combustion engines, gas turbines, micro-turbines, photovoltaics, fuel cells and wind power, energy storage systems and a cluster of linear loads (RL), can form a microgrid. Microgrids has benefits related to control possibilities that grants grid reliability and power quality requirement, besides the ease of integration with distributed power sources and higher energy gains [6], [17], [18]. In Fig. 1, it is shown the microgrid adopted in this paper, which has the following characteristics: a) Presence of a bi-directional Li-ion BB, sharing the DC bus with a gas turbine, a photovoltaic system, DFIG wind turbine system and linear loads (RL) b) A DC bus voltage capacity of 1.2 kV with 50 kW of power available.

III. BB MODELING METHODOLOGY

A. Types of Battery Models

Batteries can be modeled according to the level of details that one want to take into account. Therefore, there are three types of models: Physical models, empirical models and abstract models. The physical models describe the behavior of the battery with a high degree of precision, since it takes into consideration the materials that composes the battery, its physical-chemical, thermal, and other dynamics phenomena associated with electric charges. However, this complete description can cause processing overhead and a greater need for memory on the part of computational systems during simulations. Empirical models treat the battery as an enclosed block, with its parameters determined experimentally, without having a physical meaning. These parameters determine the correlation between the variables of input and output of the battery. It also makes the model not so faithful to a real battery, for being too simple. The abstract models represent the battery as a equivalent electric circuit. The advantage of this form of modeling is that the complex electrochemical phenomena are represented by electrical components, providing greater fidelity to the behavior of the battery, with a relatively simple computational simulation [19]. There are several types of possible circuits depending on the behavior that one want to model. In this work, the one-time constant model was used to represent the battery and its charging and discharging dynamics, as can be seen in Fig. 2:

V_{oc}: Voltage measured at battery terminals, when in open-circuit operation.

V_{bat}: Voltage measured at the battery terminals while conducting.

R_s: Resistance that represents the conduction losses due to the internal characteristics of the battery.

R_p e C_p: Used to simulate load and battery discharge dynamics.

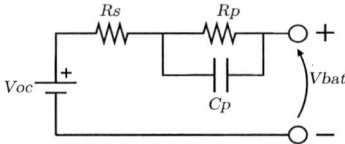

Fig. 2. One time constant abstract model used to model the battery.

B. Real battery reference for the BB modeling

In order to implement the battery model, it was used the OPR48-1250-65k battery parameters according to data supplied by the manufacturer [20]. The BB was modeled as a serial association of 5 batteries with the characteristics described in Tab. I. The objective was to achieve a nominal voltage of 240 V_{cc}, since the operation of this battery association occurs with a DC bus of 1200 V, by a transformer with a voltage gain of 5 (see Fig. 7 and Tab. V).

TABLE I
DATA FROM OPR48-1250-65K BATTERY USED FOR BB MODELING [20].

Parameters	Value
Nominal Voltage	48 V_{cc}
Capacity	65000 Wh (1354 Ah)
Max. Voltage	56.2 V_{cc}
Cutoff Voltage	46.4 V_{cc}
Total charge	4.87 10^6 C

C. Obtaining the V_{oc} by the Least Squares method with exponencial regression

The variable V_{oc} depends on battery state of charge (SOC). The variable SOC is defined according to (1):

$$ SOC = \frac{Q_a}{Q_t} = SOC_0 - \frac{1}{C} \int_0^t I_b(t)dt \qquad (1) $$

Q_a: Battery charge, Q_{total}: Total battery charge.
SOC_0: Initial battery charge, C: Battery capacity [C].
$I_b(t)$: Battery current.

Although there is no exact relationship between the variable SOC and V_{oc} [21], the latter can be deduced as a function of its logarithmic relationship with the variable SOC, according to a curve whose contours usually resemble the Fig. 3. Thus, SOC values were determined for different values of V_{oc} as can be seen in the Tab. II. Therefore, a data disposition with similar characteristics to the Fig. 3 was obtained. The main objective was to model a typical behavior between V_{oc}

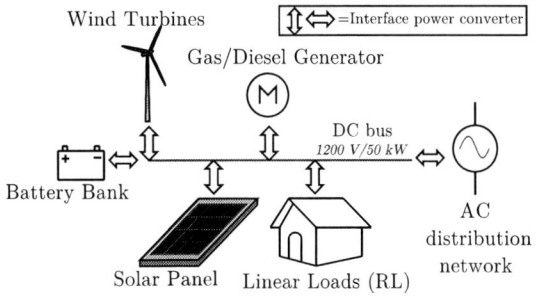

Fig. 1. Adopted DC microgrid.

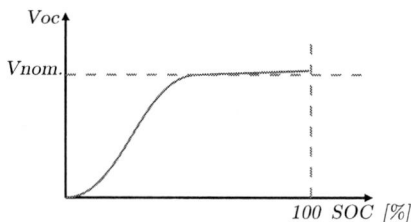

Fig. 3. Correlation between V_{oc} and SOC.

and SOC, considering a generalized analysis without taking specific details of battery manufacturing technologies.

Therefore, using the Least Squares method to fit a curve with the data presented in the Tab. II, the expression (2), which describes the behavior of the variable V_{oc}, was determined.

$$V_{oc} = 116,68 + 29,19\ln(SOC) \quad (2)$$

Substituting (1) into (2) one obtain (3):

$$V_{oc} = 116,68 + 29,19\ln(SOC_0 - \frac{1}{C}\int_0^t I_b(t)dt) \quad (3)$$

D. Determination of R_s, R_p and C_p

The resistance R_s was neglected for the simulation model development. The R_p and the capacitor C_p were determined by consecutive simulations, in order to obtain sufficiently satisfactory battery charge and discharge curves. Thus R_p was defined as the value of 0.03 Ω and C_p with a capacitance of 35.000 F with an initial voltage of -41 V in order to simulate the discharge behavior when the battery is fully charged.

IV. INTEGRATION OF BB WITH DC MICROGRID

A. DAB Converter

In order do provide the integration of BB with DC microgrid, a DC-DC DAB converter was adopted, which is represented in Fig. 4. The DAB converter consists of two H bridges, where the primary bridge of high voltage (HV) is connected to the DC current bus and the low voltage bridge (LV) is connected to the BB. The interconnection of these bridges is made through a high frequency transformer (T) and the inductance (L) represents its magnetic flux dispersion effect. Transformer (T) also performs the galvanic isolation between the bridges. During the process of BB charging, the power flow comes from the HV to the LV bridge, and

TABLE II
VALUES OF SOC [%] ASSOCIATED WITH DIFFERENT BATTERY VOLTAGE LEVELS IN OPEN CIRCUIT.

SOC [%]	V_{oc} [V]	SOC [%]	V_{oc} [V]	SOC[%]	V_{oc} [V]
0.01	0	10	100	20	200
30	232	40	234	45	236
50	239	55-80	240	85	245
90	250	95	255	100	281

during the discharge process the opposite situation occurs. Further details on the operation steps of the DAB converter are described in [22].

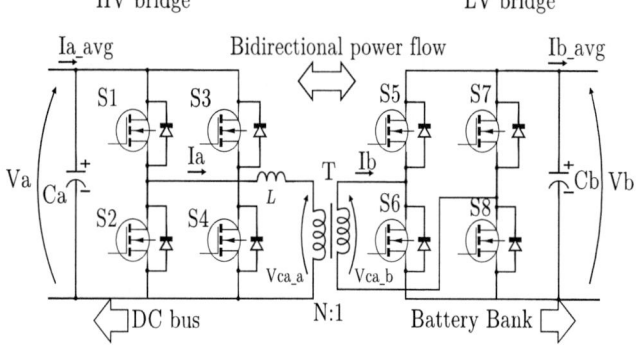

Fig. 4. DAB converter integrating the BB and DC bus.

B. The Phase-Shift modulation technique

According to [23], [24] and [25] the control of power transfer (P_t) in DAB converter can be accomplished by the phase-shift modulation strategy by (4). The lag (ϕ) is given by the expression (5) and is regulated by the current I_b determined by (6). The current I_b is controlled by the duty cycle D, given by (7)

$$P_t = \frac{V_a V_b N\phi(\pi - |\phi|)}{2L\pi^2 f_s}, -\pi < \phi < \pi \quad (4)$$

$$\phi = \frac{\pi}{2}[1 - \sqrt{1 - \frac{8LI_b fs}{NV_a}}] \quad (5)$$

$$I_b = \frac{V_a N(D - D^2)}{2Lf_s} \quad (6)$$

$$D = \frac{1}{2}[1 - \sqrt{1 - \frac{8L|P_t|fs}{NV_a}}] \quad (7)$$

Where:

V_a: HV Bridge voltage, V_b: LV Bridge voltage

N: Transformation ratio of T, L: Dispersion inductor of T.

f_s: Switching frequency of DAB converter.

ϕ: Angular lag between the voltages V_a and V_b in radians.

If the angular lag between V_a and V_b given by the variable ϕ is negative, power transfer occurs from the LV bridge to the HV bridge. Otherwise if ϕ is positive, one have the power transfer from HV bridge to the LV bridge.

V. MICROGRID BEHAVIOR MODEL WITH STATE MACHINE

A. Controlling the DAB Converter in PSIM

The control of the DAB converter by the Phase-Shift modulation technique was executed with the C-block in the PSIM® software (Fig. 5). A square wave generator operating at 20 kHz was used to drive the converter switches. The input $Vbar$ and $Vbat$ represent the DC bus and BB voltage

978-1-7281-4181-7/19 $31.00 © 2019 IEEE

respectively. The outputs S_1, $S1_{bar}$, S_2, $S2_{bar}$ are responsible for the control of the converter switches. From (4) a value of ϕ equal to $\frac{\pi}{4}$ rad. was used to transfer the energy for charging the BB and $\frac{-\pi}{4}$ rad. to deliver the energy from battery to the DC bus, in order to approximately have the maximum bidirectional power transfer between BB and DC bus considering the parameters given in Tab.V.

B. State machine implementation

The algorithm implemented in C-Block detects the rising and falling edge of the reference square wave source signal (Fig. 5). From this detection, a command signal is used to represent the time related to the lag of $\pm\frac{\pi}{4}$ rad. between the voltages V_a and V_b. After the generation of the control waves to the converter, the implemented code contains an algorithm that ensures that the battery has the load conditions required to be used in the DC microgrid. A Moore state machine contains the states that determine whether the battery must be charged (receive current), discharged (provide current) or be in hold state, without supplying or receiving current. These states are defined and identified according to the voltage level measured at the battery and DC bus, in accordance with the tables III and IV. Tab. III shows the different outputs of the Moore machine for each state of the DC bus and BB. The transition between states is performed from the voltage level of the battery and the DC bus. The "hold" state occurs when the voltage level in the bus is defined as "medium" and there is the possibility of

TABLE III
STATE DIAGRAM FOR THE OPERATION OF THE DAB CONVERTER.

■ *Hold* ■ *Charge* ■ *Discharge*			
State	V_{bar}		
V_{bat}	Low/Low	Low/Medium	Low/High
	Medium/Low	Medium/Medium	Medium/High
	High/Low	High/Medium	High/High

TABLE IV
VOLTAGE CONDITIONS TO REPRESENT DIFFERENT BATTERY STATES.

V	Value	V	Value
V_{bat} high	>265 V	V_{bar} high	>1250 V
V_{bat} medium	[235,265] V	V_{bar} medium	[1150,1250] V
V_{bat} low	<235 V	V_{bar} low	<1150 V

TABLE V
PARAMETERS USED FOR SIMULATION.

Parameters	Value	Parameter	Value
V_a	1200 V_{cc}	N	5
V_b	240 V_{cc}	ω	$4\pi * 10^4 \frac{rad}{s}$
P	50 kW	L	135 μH
f_s	20 kHz	C_{1min}	47 μF
ϕ	45°	C_{2min}	390 μF
D	0.5	$L_{ca,bat}$	1 mH

a variation in its voltage, if it provides or receives energy from the batteries. The "discharge" state occurs when the batteries provide power, if they have medium or high voltage, and the bus has low voltage. The "charge" state occurs when the bus has a high voltage, and the battery has medium or low voltage.

VI. SIMULATIONS AND RESULTS

A. Microgrid simulation in PSIM®

The circuit shown in Fig. 7 was simulated using the PSIM® software and represents the DAB converter circuit with the BB coupled to the right and the DC bus represented by a triangular source with an offset voltage of 1100 V and peak-to-peak variation of 200 V, representing the intermittent supply of energy from RES in the microgrid. These values were chosen so that the converter operates in its 3 conditions: Charge, discharge and hold. The parameters of the simulated electrical system can be found on Tab. V. The converter project was based on [24] and [25], although L_{ca} and L_{bat} where determined empirically in order to have better current behavior.

B. BB Model

The BB model (Fig. 6) was built by implementing the mathematical approximation obtained in (3) and the data on Tab. I. The values in the limiter represent the actual maximum and minimum voltage limits of the battery.

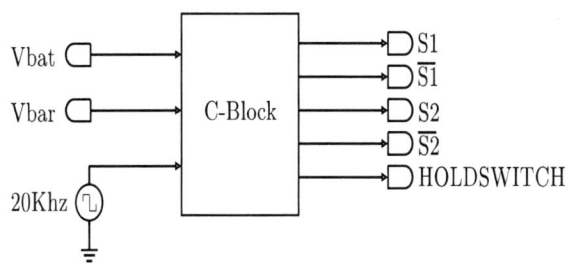

Fig. 5. C-Block and control circuit scheme in PSIM.

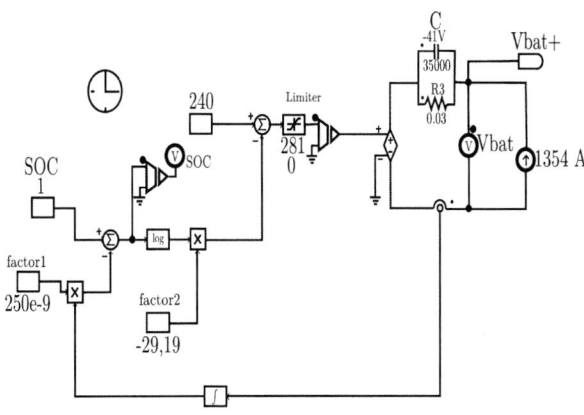

Fig. 6. BB model implemented in PSIM®.

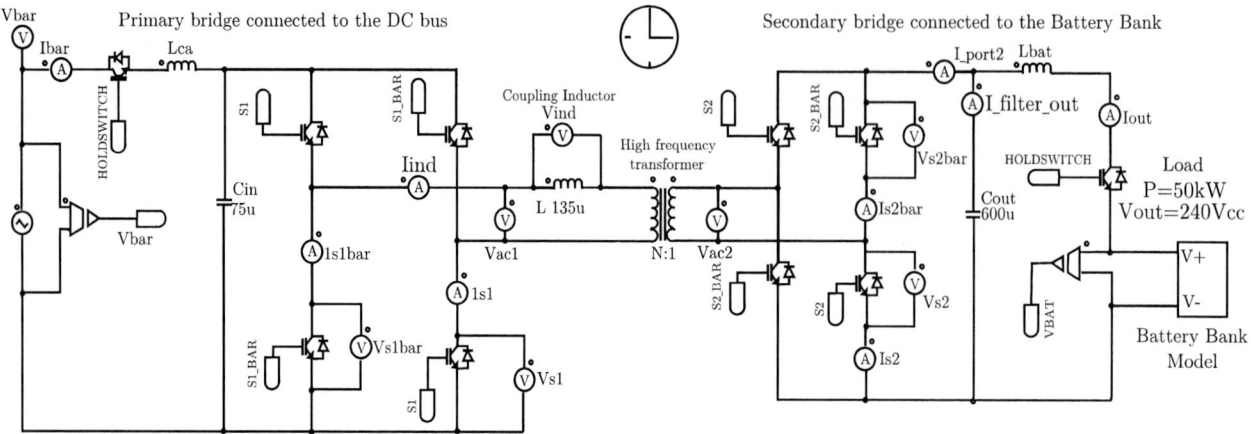

Fig. 7. DAB converter implemented in PSIM.

C. BB charge and discharge curves

Figures 8 and 9 show simulation results considering charge, discharge and hold operations, represented by stages 1 to 3. In stage 1, defined by the period of 1 s to 5 s, it is represented the initial charging of the BB so that it can be used by the microgrid, with positive currents and voltage. In stage 2, defined by the period of 5 s to 6 s, the battery was in the discharge condition, with negative currents. In stage 3, defined by the period of 7 s to 9 s, the circuit did not demand energy and did not show any energy surplus to load the BB that remained in hold. After the third stage, these operations are repeated according to the different load, battery voltage,

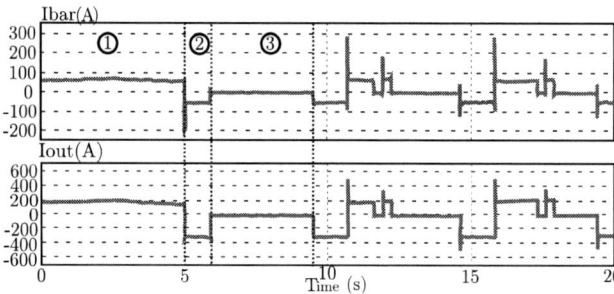

Fig. 8. DC bus and BB current.

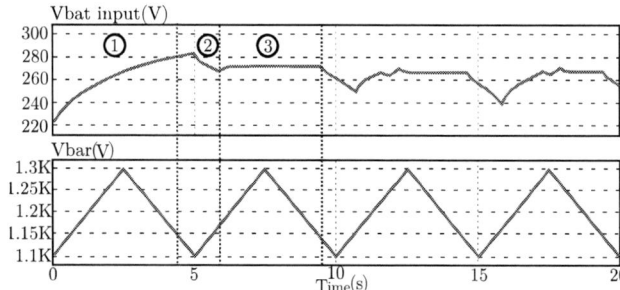

Fig. 9. BB and DC bus voltages.

and bus voltage situations. Due to the presence of the switches operating at a switching frequency of 20 Khz, the simulations were done with Step Time 5E-006 and 20 s of duration.

VII. CONTROL IMPROVEMENT FOR POWER MANAGEMENT

The proposed topology presented in this paper can be improved using droop control, as explored in [26] and [27]. This control approach can be used to implement an autonomous distributed control. Droop control is utilized for battery as shown in Fig. 10. This strategy promotes a smooth transition between battery charging and discharging operation (trajectory from point A to point B). Regardless of system operating condition, if the bus voltage reaches a predetermined value, the battery will discharge or recharge at its maximum values, fully utilizing the battery power rating. Battery will run into standby mode if the SOC is out of limit.

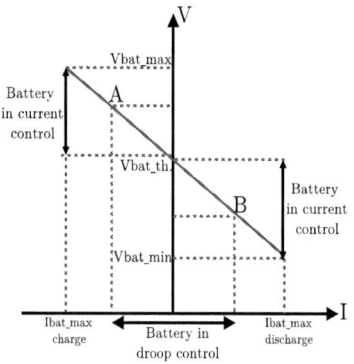

Fig. 10. Example of droop curve for battery charging and discharging behavior as exposed in [26].

VIII. DISCUSSIONS

Initially, the values used for the battery model characterized their operation over a period of one hour. In this way, the value of the capacitance C and the integrator time constant $\int$ (Fig. 6) was divided by 3600. Thus, the battery started to have an

978-1-7281-4181-7/19 $31.00 © 2019 IEEE

average capacity in Ampere-seconds, with a more convenient dynamics for joint simulation with high-frequency power switches. Therefore, changing the value of the integration factor $\int$ in the battery model can be performed to achieve higher or lower battery charge and discharge times. This can be done to evaluate battery behavior for shorter or longer simulation periods, depending on what is more convenient. Finally, according to the figures 8 and 9 the proposed model sucessfully represented the operations of battery charging and discharging operations, as well as situations in which no action should be performed.

IX. Conclusion

This paper presented the modeling of an energy storage system (1200 V - 50 kW DC) for integration into a DC microgrid, with the educational aim of allowing the study of energy flow and some effects of integration of distributed micro generators based on RES. The energy storage system was composed of a lithium-ion battery bank and a DAB converter to perform the integration to the DC bus. Power flow control was implemented with the DAB converter, through phase shift modulation, and power management between the storage system and the DC bus via a Moore finite state machine.

X. Acknoledgements

This project was financially supported by Grants #2016/08645-9 and #2018/24331-0, São Paulo Research Foundation (FAPESP).

References

[1] D. E. Olivares, A. Mehrizi-Sani, A. H. Etemadi, C. A. Cañizares, R. Iravani, M. Kazerani, A. H. Hajimiragha, O. Gomis-Bellmunt, M. Saeedifard, R. Palma-Behnke *et al.*, "Trends in microgrid control," *IEEE Transactions on smart grid*, vol. 5, no. 4, pp. 1905–1919, 2014.

[2] M. Soshinskaya, W. H. Crijns-Graus, J. M. Guerrero, and J. C. Vasquez, "Microgrids: Experiences, barriers and success factors," *Renewable and Sustainable Energy Reviews*, vol. 40, pp. 659–672, 2014.

[3] A. Lidula, NWA Rajapakse, "Voltage balancing and synchronization of microgrids with highly unbalanced loads," *Renewable and Sustainable Energy Reviews*, vol. 31, pp. 907–920, 2014.

[4] R. Palma-Behnke, C. Benavides, F. Lanas, B. Severino, L. Reyes, J. Llanos, and D. Sáez, "A microgrid energy management system based on the rolling horizon strategy," *IEEE Transactions on smart grid*, vol. 4, no. 2, pp. 996–1006, 2013.

[5] R. Bhoyar and S. Bharatkar, "Potential of microsources, renewable energy sources and application of microgrids in rural areas of maharashtra state india," *Energy Procedia*, vol. 14, 2012.

[6] S. Sinha, A. K. Sinha, and P. Bajpai, "Solar pv fed standalone dc microgrid with hybrid energy storage system," in *2017 6th International Conference on Computer Applications In Electrical Engineering-Recent Advances (CERA)*. IEEE, 2017, pp. 31–36.

[7] A. K. Verma, B. Singh, and D. Shahani, "Grid to vehicle and vehicle to grid energy transfer using single-phase half bridge boost ac-dc converter and bidirectional dc-dc converter," *International Journal of Engineering, Science and Technology*, vol. 4, no. 1, pp. 46–54, 2012.

[8] G. Chen, Y.-S. Lee, D. Xu, and Y. Wang, "A novel soft-switching and low-conduction-loss bidirectional dc-dc converter," in *Proceedings IPEMC 2000. Third International Power Electronics and Motion Control Conference (IEEE Cat. No. 00EX435)*, vol. 3. IEEE, 2000, pp. 1166–1171.

[9] A. R. Alonso, J. Sebastian, D. G. Lamar, M. M. Hernando, and A. Vazquez, "An overall study of a dual active bridge for bidirectional dc/dc conversion," in *2010 IEEE Energy Conversion Congress and Exposition*. IEEE, 2010, pp. 1129–1135.

[10] R. Naayagi, A. J. Forsyth, and R. Shuttleworth, "High-power bidirectional dc–dc converter for aerospace applications," *IEEE Transactions on Power Electronics*, vol. 27, no. 11, pp. 4366–4379, 2012.

[11] P. F. Costa, P. H. Löbler, A. Toebe, L. Roggia, and L. Schuch, "Modelagem e controle do conversor dab aplicado à carga de baterias."

[12] T. R. de Oliveira, *Microrredes em corrente contínua*, 2016, acessado em 16-Abril-2019. [Online]. Available: https://www.troliveira.com/l/distcc

[13] S. McCluer, *DC Systems and Battery Safety Evolution through Codes and Government Regulations*, 2019, acessado em 16-Abril-2019. [Online]. Available: https://www.ieeepes.org/presentations/gm2014/2756.pdf

[14] IEEE, *IEEE Guide for the Characterization and Evaluation of Lithium-Based Batteries in Stationary Applications*, 2017, acessado em 16-Abril-2019. [Online]. Available: https://standards.ieee.org/standard/1679_1-2017.html

[15] ——, *IEEE 1375-1998 - Guide for Protection of Stationary Battery Systems*, 1998, acessado em 16-Abril-2019. [Online]. Available: https://standards.ieee.org/standard/1375-1998.html

[16] Y. Han, X. Meng, G. Zhang, Q. Li, and W. Chen, "An energy management system based on hierarchical control and state machine for the pemfc-battery hybrid tramway," in *2017 IEEE Transportation Electrification Conference and Expo, Asia-Pacific (ITEC Asia-Pacific)*. IEEE, 2017, pp. 1–5.

[17] H. Qian, J. Zhang, J.-S. Lai, and W. Yu, "A high-efficiency grid-tie battery energy storage system," *IEEE transactions on power electronics*, vol. 26, no. 3, pp. 886–896, 2010.

[18] M. S. Ballal, K. V. Bhadane, R. M. Moharil, and H. M. Suryawanshi, "A control and protection model for the distributed generation and energy storage systems in microgrids," *Journal of Power Electronics*, vol. 16, no. 2, pp. 748–759, 2016.

[19] F. Saidani, F. X. Hutter, R.-G. Scurtu, W. Braunwarth, and J. N. Burghartz, "Lithium-ion battery models: a comparative study and a model-based powerline communication," *Advances in Radio Science*, vol. 15, pp. 83–91, 2017.

[20] *48V LiFeYPO4 Battery (Lithium Iron Yttrium Phosphate)-datasheet*, OMNIPOWER. [Online]. Available: https://store.sinetech.co.za/wp-content/uploads/2018/02/OPR48V-OmniPower-LiFeYPO4-Battery.pdf

[21] Z. Yu, R. Huai, and L. Xiao, "State-of-charge estimation for lithium-ion batteries using a kalman filter based on local linearization," *Energies*, vol. 8, no. 8, pp. 7854–7873, 2015.

[22] W. Silva, "Estudo e implementação de um conversor bidirecional como interface na regulação de tensão em barramento cc e carregamento de baterias em um sistema nanorrede," *Programa de Pós-Graduação em Engenharia Elétrica–UFMG. Belo Horizonte, Setembro de*, 2013.

[23] F. Krismer, "Modeling and optimization of bidirectional dual active bridge dc-dc converter topologies," Ph.D. dissertation, ETH Zurich, 2010.

[24] H. R. Mamede *et al.*, "Interligação de conversores dab para aplicação em transformadores de estado sólido," 2016.

[25] R. Piveta, "Otimização do rendimento do conversor dab aplicado ao transformador eletrônico," Master's thesis, Universidade Federal de Santa Maria, Rio Grande do Sul, Brasil, 2015.

[26] Q. Ye, R. Mo, and H. Li, "Impedance modeling and verification of a dual active bridge (dab) dc/dc converter enabled dc microgrid in freedm system," in *2016 IEEE 8th International Power Electronics and Motion Control Conference (IPEMC-ECCE Asia)*. IEEE, 2016, pp. 2875–2879.

[27] X. Yu, X. She, X. Zhou, and A. Q. Huang, "Power management for dc microgrid enabled by solid-state transformer," *IEEE Transactions on Smart Grid*, vol. 5, no. 2, pp. 954–965, 2013.

Multi-modular scalable DC-AC power converter for current injection to the grid based on predictive voltage control

S. Toledo and M. Rivera
Dept. of Electrical Engineering
Univerisdad de Talca
Curicó, Chile
{stoledo,mrivera}@utalca.cl

E. Maqueda, M. Ayala, J. Pacher, C. Romero
and R. Gregor
Engineering Faculty
Universidad Nacional de Asunción
Luque, Paraguay
{emaqueda,mayala,jpacher,cromero,rgregor}@ing.una.py

T. Dragicevic
Dept. of Energy Technology
Aalborg University
Aalborg, Denmark
tdr@et.aau.dk

P. Wheeler
Dept. of Electrical and Electronic Engineering
University of Nottingham
Nottingham, UK
pat.wheeler@nottingham.ac.uk

Abstract—Increasing world electrical energy demand make necessary to develop new efficient and reliable schemes for power injection to the existing grid under distributed generation frames. In this work a novel DC-AC control scheme is proposed, which is based on one hand, on several 2L-VSI working together as a multi-modular converter in a parallel switching connection topology in the power converter stage and, in the other hand, a control strategy using two loops, an internal based on predictive voltage control and another external applying a PR current control, achieving THD levels lower than 1% for the injected current, accomplishing with the international standards for these kind of applications.

Index Terms—Current injection, distributed generation systems, multi-modular VSI, parallel switches, predictive voltage control.

I. INTRODUCTION

IN the last decades, different renewable energies, such as: micro-hydro, solar, wind farms, etc [1]–[3], have been studied in order to use them in isolated or grid-on electric systems. In every case, power converters are often used as a fixed power source [4]–[6]. These converters have to secured a controlled active and reactive power flux, reduce the harmonic content in currents and voltages, reduce the resonance effect and secure a good grid synchronization, in case of distributed systems. To accomplish these requirements, different control techniques have been studied, such as proportional integral (PI) control [7], proportional resonant (PR) control [8], predictive control [9], etc, and in many cases, they are combined

The authors express their gratitude to Consejo Nacional de Ciencia y Tecnología de Paraguay (CONACYT), for the support and financing through Project PINV15-0584, to CONICYT of Chile through the FONDE-CYT Regular Project 1160690, Project MEC 80150056 and CONICYT-PFCHA/Doctorado Nacional/2019-21192003.

with modulation techniques, such as: pulse width modulation (PWM) [10], space vector modulation (SVM) [11]–[13], etc. However, the majority of power converters are commonly used as a fixed power source, which complicates the scalability in terms of energy conversion system, mainly in applications where an extension of the system, based on renewable energies, is planned in order to respond efficiently to an increase in demand associated with a variation in load requirements.

This paper presents the design of the architecture of a multi-modular power converter to obtain scalable power, where each module consists in a three-phase voltage source inverter (VSI), with a novel control scheme based on two close loops, which consist in an internal predictive voltage control and an outer PR current control. This feature is interesting in applications where power converters with reduced volume, high energy density and a good availability are required.

This paper is organized as follows: the proposed current control scheme is shown in Section II. In Section III, the model predictive control (MPC) applied as a voltage control is presented. Section IV describes the reference voltage generator stage. Section V depicted some experimental results for parallel connected switches. The grid connection process are described in Section VI. Obtained simulation results are shown in Section VII regarding transient and steady-state performance. Finally, conclusions are presented in the last section.

II. PROPOSED CURRENT CONTROL SCHEME

As it is shown in Fig. 1, the proposed control scheme consists of several two level VSI (2L-VSI) modules working together, where every switch is connected in parallel with the corresponding switches on other VSI dividing the circulating

978-1-7281-4181-7/19 $31.00 © 2019 IEEE

currents by *"n"* (where *n* represents the number of modules) allowing, for example, the increase of power capability of the whole system, or to decrease the charge for every leg allowing the use of smaller semiconductors for the same rated power. In this case, the switching signal is the same for every module, that is, the same commutation pattern is applied in every VSI connected in parallel such as the commutation occurs at the same time achieving a multi-tracking configuration for the circulating current. The mentioned configuration will be treated in a more detailed way in Section V and some experimental results using parallel switches will be depicted in terms of the circulating currents behavior in order to validate the proposed scheme.

The DC source comes from a 2.4 kW solar photo-voltaic (PV) system at the input of the power converter and a LC filter at the output being the capacitor voltage (v_o) the output signal that must be controlled in order to achieve an interconnection to the grid. A bypass switch is needed to connect the generation system with the grid in the common coupling point (PCC) once both stages were synchronized. The control consists of an inner predictive voltage control (PVC) stage with a fast transient response and an outer PR current control plus a feed-forward control to generate the reference for the voltage control stage. The feed-forward control comes from a voltage reference generator that takes the phase of the grid voltage (v_g) from a phase-locked loop (PLL) generating a synchronized signal with the grid as the desired output of the converter stage. At the beginning, the desired injected grid currents (i_g) must to be set to zero as well as v_g and v_o should be the same, then the PR output will be zero at the starting point. Once both voltages have been synchronized, the bypass connects the generation stage with the grid and the current can be controlled by setting the desired current at the input of the PR controller. The PVC selects the optimal switching combination, in every sampling time, to control the output capacitor voltage that generates the desired current injection. In the next section the implemented PVC technique is depicted.

III. PREDICTIVE VOLTAGE CONTROL

MPC strategy consists on predicting the future behaviour of the analyzed system for each possible input state based on dynamic equations which define it, and picking the best state which fulfills the desired output. In this way, the first step to apply this control is to obtain the system model.

A. Power Converter Model

The scheme of a basic module of the converter is shown in Fig. 2, which is composed of a DC source, a 2L-VSI and an output LC filter.

The 2L-VSI is one of the most used topologies in the literature [13]–[15]. In this case, it is possible to define a switching function as $S_x \in \{1, 0\}$ where x denotes each phase ($x = a, b, c$). Given that each leg is conformed by two switches that should be activated in a complementary form to avoid short-circuits between the terminals of the DC source v_{dc}, the valid switching states are the following:

$$
\begin{aligned}
S_a &= \begin{cases} 1, & \text{if } S_1 = 1 \text{ and } S_4 = 0, \\ 0, & \text{if } S_1 = 0 \text{ and } S_4 = 1, \end{cases} \\
S_b &= \begin{cases} 1, & \text{if } S_2 = 1 \text{ and } S_5 = 0, \\ 0, & \text{if } S_2 = 0 \text{ and } S_5 = 1, \end{cases} \\
S_c &= \begin{cases} 1, & \text{if } S_3 = 1 \text{ and } S_6 = 0, \\ 0, & \text{if } S_3 = 0 \text{ and } S_6 = 1. \end{cases}
\end{aligned} \tag{1}
$$

Only eight switching states are valid for the 2L-VSI. Likewise, it is possible to synthetize the voltages at the points a, b and c regarding the point N (v_{aN}, v_{bN} y v_{cN}) depending on the switches of the 2L-VSI and v_{dc} as:

$$
v_{aN} = S_a v_{dc}, \quad v_{bN} = S_b, v_{dc}, \quad v_{cN} = S_c v_{dc}. \tag{2}
$$

In order to determine the effective voltages applied for each phase (i.e., from a, b, y c to n), the common mode voltage v_{nN} must be subtracted from the equation (2). It is possible to determine v_{nN} in a simple way by applying Kirchhoff voltage law as follows:

$$
v_{nN} = \frac{v_{aN} + v_{bN} + v_{cN}}{3}. \tag{3}
$$

So, the effective phase voltage is given by:

$$
v_{an} = v_{aN} - v_{nN}, \quad v_{bn} = v_{bN} - v_{nN} \text{ and } v_{cn} = v_{cN} - v_{nN}. \tag{4}
$$

By applying Clarke's transformation [16] to each valid state of the 2L-VSI module, it is possible to represent the space vector graph which is depicted on Fig. 3 showing the eight valid voltage vectors in the $\alpha - \beta$ plane.

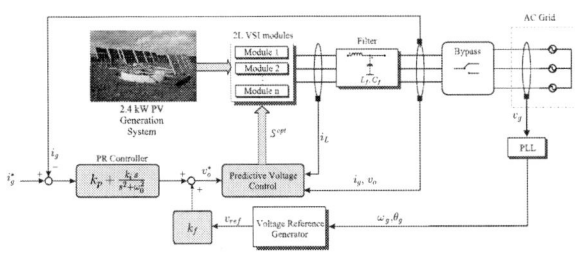

Fig. 1. Proposed control scheme.

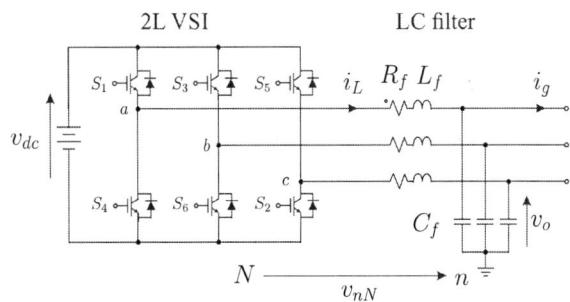

Fig. 2. Conversion module topology.

978-1-7281-4181-7/19 $31.00 © 2019 IEEE

B. Output Filter Model

As can be seen in Fig. 2, the 2L-VSI module is connected to a load through an LC output filter. Each converter leg possess an inductor L_f and its corresponding leakage resistance R_f and a capacitor C_f. By considering the inductor currents i_L and the capacitor voltage v_o as state variables of this second order system and assuming that the parameters are equal for the three legs, the system dynamic in $\alpha - \beta$ plane is given by:

$$L_f \frac{di_{L\alpha\beta}}{dt} = v_{\alpha\beta} - v_{o\alpha\beta} - R_f i_{L\alpha\beta}, \qquad (5)$$

where $v_{\alpha\beta}$ corresponds to one of the valid vectors shown in Fig. 3. On the other hand, the dynamic behaviour of the capacitor voltage is defined as follows:

$$C_f \frac{dv_{o\alpha\beta}}{dt} = i_{L\alpha\beta} - i_{g\alpha\beta}. \qquad (6)$$

In this way, the state space representation of the system is:

$$\frac{d\mathbf{x}}{dt} = \mathbf{Ax} + \mathbf{Bu}, \qquad (7)$$

where

$$\mathbf{x} = \begin{bmatrix} i_{L\alpha\beta} \\ v_{o\alpha\beta} \end{bmatrix}, \quad \mathbf{A} = \begin{bmatrix} -\frac{R_f}{L_f} & -\frac{1}{L_f} \\ \frac{1}{C_f} & 0 \end{bmatrix},$$

$$\mathbf{u} = \begin{bmatrix} v_{\alpha\beta} \\ i_{g\alpha\beta} \end{bmatrix} \text{ y } \mathbf{B} = \begin{bmatrix} \frac{1}{L_f} & 0 \\ 0 & -\frac{1}{C_f} \end{bmatrix}. \qquad (8)$$

The previous equations define completely the continuous model for the LC filter by considering the output voltages of the 2L-VSI $v_{\alpha\beta}$ and the load currents $i_{g\alpha\beta}$ as inputs.

C. Discrete Time Model of LC Filter

The discrete model of the system is given by:

$$\mathbf{x}(k+1) = \mathbf{A}_d\mathbf{x}(k) + \mathbf{B}_d\mathbf{u}(k), \qquad (9)$$

being $\mathbf{A}_d = e^{\mathbf{A}T_s}$, $\mathbf{B}_d = \int_0^{T_s} e^{\mathbf{A}(T_s - \tau)}\mathbf{B}d\tau$ and T_s corresponding to the sampling time. Thus, by using a discrete model that describes the system it is possible to predict the states of v_o e i_L as follows:

$$i_L(k+1) = a_{11}i_L(k) + a_{12}v_o(k) + b_{11}v(k) + b_{12}i_g(k), \qquad (10)$$

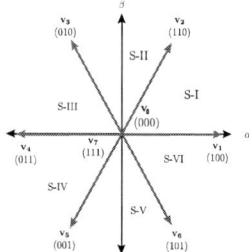

Fig. 3. Valid vectors $\alpha - \beta$ plane.

$$v_o(k+1) = a_{21}i_L(k) + a_{22}v_o(k) + b_{21}v(k) + b_{22}i_g(k), \qquad (11)$$

siendo

$$\mathbf{A}_d = \begin{bmatrix} a_{11} & a_{12} \\ a_{21} & a_{22} \end{bmatrix}, \mathbf{B}_d = \begin{bmatrix} b_{11} & b_{12} \\ b_{21} & b_{22} \end{bmatrix}, \qquad (12)$$

where k is the actual instant time and $k + 1$ indicates the beginning of the next sample instant. Through the previous equations it is possible to predict the inductor currents values and the voltage output to implement the MPC.

D. Predictive Voltage Control

The basic operating principle of the algorithm is the following: first, at the beginning of the sampling instant, new measurements of v_o, i_L and i_g are obtained. These measurements define the starter point from which the algorithm predicts the future trajectory of the state space variables by considering (10) and (11), for each feasible voltage vector. Every predicted value is evaluated with a pre-designed cost function (CF), and the vector with the lowest CF is applied to the 2L-VSI switches. The presented technique, in this section, is based on the proposed control in [17], [18]. The selected cost function is the following:

$$g = (v_{o\alpha}^* - v_{o\alpha})^2 + (v_{o\beta}^* - v_{o\beta})^2 + \lambda_d g_{der}, \qquad (13)$$

being $v_{o\alpha}^*$ and $v_{o\beta}^*$ the desired voltages on the $\alpha - \beta$ plane and defining:

$$g_{der} = \left(C_f \omega_{ref} v_{o\beta}^* - i_{L\alpha} + i_{g\alpha}\right)^2 + \left(C_f \omega_{ref} v_{o\alpha}^* - i_{L\beta} + i_{g\beta}\right)^2. \qquad (14)$$

The term g_{der} is introduced with the purpose of correcting the incapacity of the classic predictive control of tracking the capacitor voltage derivative resulting in a high total harmonic distortion (THD), creating an ideal regulator which controls the voltage capacitor and its derivative effectively. The effect of the derivative term is controlled with a weighting factor λ_d.

This strategy is based on the evaluation, at every sampling instant, of the cost function g for all the valid vectors and to apply the vector which minimizes the CF in the next sampling instant achieving a desired voltage tracking.

IV. OUTER CURRENT CONTROL LOOP FOR REFERENCE VOLTAGE GENERATION

As mentioned above, the proposed scheme is based on the implementation of two control loops, an internal loop corresponding to the predictive voltage control and an external loop that generates the voltage reference to inject a controlled current into the grid. In Fig. 1 the external control loop is shown where the difference between the injected current i_g and the desired current i_g^* is calculated to be applied as the input of a PR controller [19] which output is added to a feedforward control signal v_{ref} which is used at system startup to synchronize the output voltage of the conversion stage v_o with the grid voltage v_g so as to be connected through a bypass. This generated reference voltage is the input to

978-1-7281-4181-7/19 $31.00 © 2019 IEEE

the voltage predictive control block to calculate the optimum vector to be applied. The voltage v_{ref} is generated from the measurement of the phase of the grid θ obtained from the PLL and the amplitude of the signal of the grid. The value of i_g^* is determined from the current amplitude and phase according to the desired active and reactive power.

Once the desired currents generate the desired voltages and the predictive voltage controller selects the vector to be applied, the same commutation signals are applied to every module generating several paths for the current to be splitted from the total flow decreasing the thermal stress and providing an increased power management capability. In order to validate the idea of using the same commutation signal to control parallel switches to split the circulating current, in next section experimental tests of the proposed parallel topology are depicted.

V. EXPERIMENTAL TEST OF THE PROPOSED PARALLEL SWITCHES TOPOLOGY

The proposed topology implies the application of the same switching signal to every module of the converters hoping that the current splits to every path and the reaction to contingencies is based on the capability of the system to manage the additional current in case of broken switch for example. In this section a brief study of the behavior of parallel switches is presented.

In order to verify the parallel operation of several modules of the VSI converter, an experimental test bench has been designed. The test circuit can be seen in Fig. 4. The circuit shows the modular interconnection of the first branch of three identical VSI converters connected in parallel. Each branch is composed of two switches in parallel and in the same way each switch of the converter branch is parallel to each other. The circuit is connected at the input to a DC voltage source and at the output to a resistive load. By means of this design, an analysis is carried out that consists of controlling the switches through the same signal for each VSI converter where the behavior of the current delivered by the first branch of each converter is studied to show how the flow is splitted and how it behaves under faults in any switch.

Fig. 5 shows the setup test implemented in order to evaluate the operation of several modules of the VSI converters in parallel connection. The figure shows the electronic boards of each branch of the converters, in modular form, the DC input source and an oscilloscope. The main components, used in the implementation of the VSI converter, are: **a)** for the switch the SiC MOSFET power semiconductor CAS120M12 from the

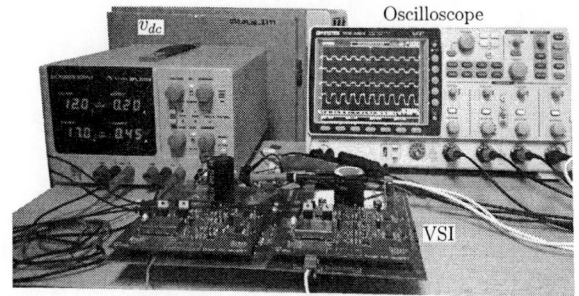

Fig. 5. Experimental test setup.

manufacturer CREE, **b)** the IR2110 driver for the control of the trip signal of the switches, and **c)** fiber optic transceivers HFBR-0500Z for signal reception of programmable control devices. In the circuit the measurements of the output current of each branch of the converters were made, represented in Fig. 4 by $i_{L(1)}$, $i_{L(2)}$ e $i_{L(3)}$ respectively. The sum of these currents is represented by i_L and it corresponds in turn to the current that passes through the resistance R.

The waveforms of the currents $i_{L(1)}$, $i_{L(2)}$, $i_{L(3)}$ e i_L can be seen in Fig. 6. It is observed that initially each current of each branch of the converters are very similar, 2.5 A pk-pk respectively, while the total current i_L displayed by the channel 4 of the oscilloscope showing a value of 7.5 A pk-pk (i.e. three times the current in every branch as is expected), being an expected behavior. Further on, it is possible to observe the moment of fault over the path of the current $i_{L(3)}$, corresponding to the extraction of the third VSI converter of the system. At that instant, the magnitude of the current $i_{L(3)}$ is evenly distributed in the currents $i_{L(1)}$ and $i_{L(2)}$ as shown in Fig. 6, and also, for the transient, the current i_L over the load R does not present disturbances.

These results allow to confirm the feasibility of using parallel switching configuration with the same commutation signal achieving a multi-tracking operation that can manage

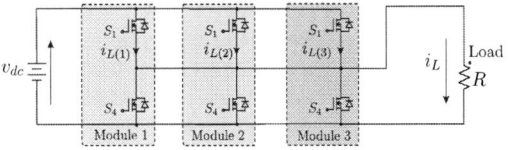

Fig. 4. Tests circuit diagram.

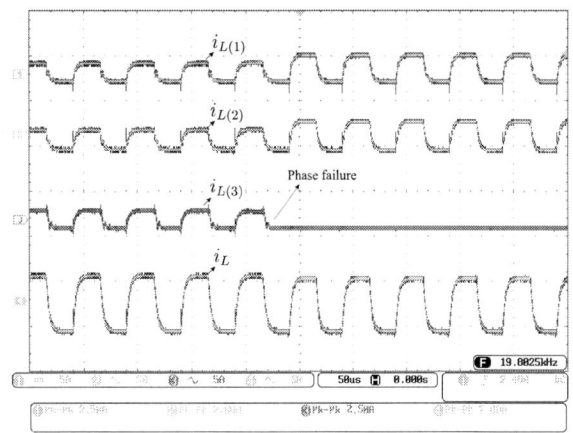

Fig. 6. Switching waveform of the current of three branches of the VSI in parallel.

978-1-7281-4181-7/19 $31.00 © 2019 IEEE   1444

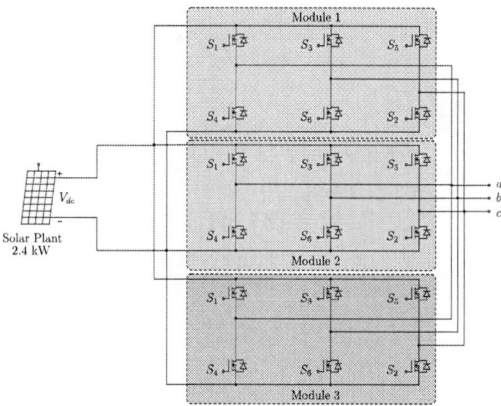

Fig. 7. multi-modular DC-AC converter scheme.

TABLE I
SIMULATION PARAMETERS.

Parameter	Simulation parameters		
	Symbol	Value	Unit
Grid voltage	v_g	311	V_p
Grid frequency	f_g	50	Hz
DC link voltage	v_{dc}	700	V
Leakage inductor resistance	R_f	10	mΩ
Filter inductance	L_f	2.4	mH
Filter capacitance	C_f	24	μF
Sampling time	T_s	25	μs
Sampling frequency	f_s	40	kHz
Derivative control weighting	λ_d	0.2	
PR proportional control constant	k_p	10	
PR integral control constant	k_i	1500	

contingencies and even increase the power supplied capability of the whole system adding more modules depending on the demanded energy.

Fig. 7 shows the multi-modular configuration for three parallel VSIs that will be used on the final implementation of the proposed grid connected generation system.

VI. GRID CONNECTION PROCESS

The grid connection process is performed as follows:

1) The predictive voltage control is used to generate an output voltage (v_o) synchronized with the grid voltage (v_g), that is, the voltage of the capacitor is synchronized with the grid in order to be connected on the CCP.

2) Next, once the voltages are equal, the bypass is used to connect the systems with the reference current (i_g^*) set to zero.

3) Finally, the feed-forward signal can be set to zero and all the reference voltage comes from the PR controller, then it is possible to select the desired injected current (i_g).

Based on the above described procedure, it is possible to evaluate the proposal performing some simulations using MATLAB/Simulink environment.

VII. SIMULATION RESULTS OF THE PROPOSED GRID CONNECTED GENERATION SYSTEM

The multi-modular grid connected generation system consists on three modules (i.e. as shown in Fig. 7) under the control scheme depicted on Fig.1 and with the parameters of Table I. In Fig. 8 the interconnection process and the behavior under steps of desired currents are shown. At the beginning, the injected current is zero because the generation system is not connected to the grid and the output voltage (v_o) is still in synchronization process regarding the grid voltage (v_g). This process takes around 0.14 seconds. After 0.2 seconds, the bypass connects the output capacitors with the grid but i_g is still set to zero generating a short transient during the connection instant. Then the system is ready to inject the

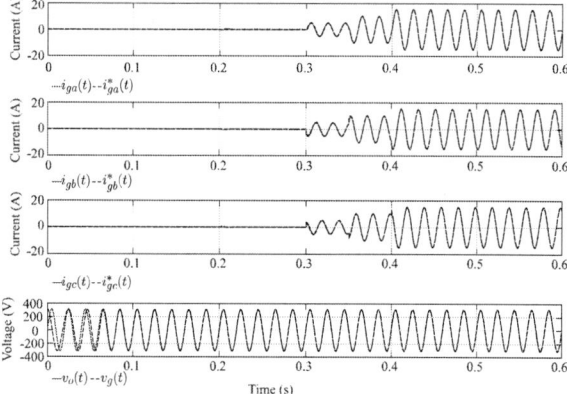

Fig. 8. Injected current and output voltage of the proposed control scheme.

desired current (at this point it is possible to set the feed-forward control to zero gradually changing k_f from 1 to 0, being careful not to generate high current pikes). At 0.3 seconds the desired current i_g^* is set to 5 amps, then at 0.35 seconds is set to 10 amps and finally, at 0.4 seconds is set to 15 amps. During this current steps it is possible to appreciate a good tracking of the reference. The THD oscillates from 0.33 to 0.9 % in the injected current. The tracking error during the input step, applied at 0.4 seconds, is shown in Fig. 9 where the transient stage can be measured to be around 30 milliseconds approximately. That is due to the use of the PR controller in the outer loop. In this case it is possible to use the inner predictive voltage control since this technique presents a fast transient response compared with the PR controller.

Besides the current tracking test, an under failure test has been performed and the results are reflected in Fig. 10. In this Figure it is possible to appreciate the behavior of the multi-modular converter at the moment of a failure occurs. As it can be seen, the circulation current is homogeneously distributed between every switch (some minimal differences can be found in the experimental implementation due to the parametric variation between different Sic-MOSFET devices). i_{L1}, i_{L2} and i_{L3} correspond to the circulating current through modules 1, 2 and 3, respectively. i_L corresponds to the total current supplied by the converter to the output filter and finally,

978-1-7281-4181-7/19 $31.00 © 2019 IEEE

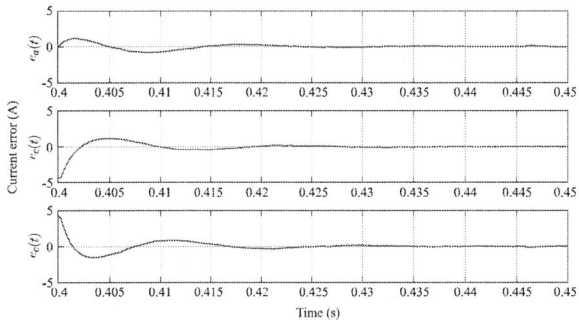

Fig. 9. Current error under input steps.

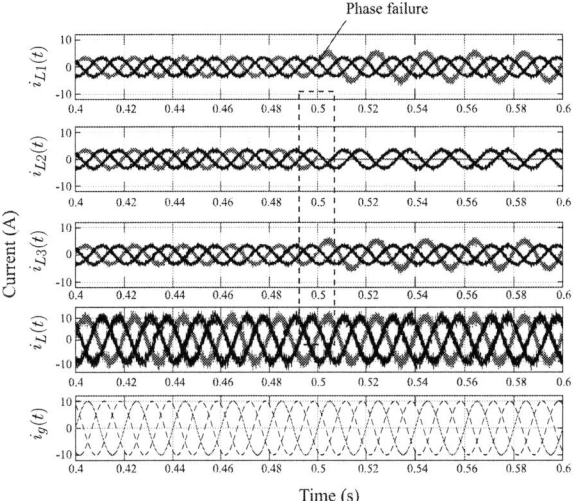

Fig. 10. Current behavior under a phase failure.

i_g is the total injected current. At 0.5 seconds, a failure in the module 2 is simulated, it is seen that the path was broken and i_{L2} goes to zero. In order to hold the injected current i_g, the current splits between i_{L1} and i_{L3} such a way that the effect of this disconnection is negligible. These results are consistent with those observed in the experimental tests of the section V. Concerning the voltage, it presents a THD around 0.77 % that is a good value, this due to the derivative control term added to the classic PVC.

VIII. CONCLUSION

The main contribution of this paper lies on the proposal of a novel injected current control strategy based on two control loops with a multi-modular VSI converter topology. The multi-modular topology provides more energy availability due to the possibility of splitting the current between several paths at the moment that a failure occurs. Moreover, this topology provides in turn scalability in the implementation due to the parallel topology shows a good performance sharing the same commutation signal which means it is feasible to connect the necessary number of modules in order to provide and manage the desired power. The control strategy shows a good performance for current tracking, error signals and THD

reduction, accomplishing with all the international standards related to this kind of application.

REFERENCES

[1] W. Xiao, M. S. El Moursi, O. Khan, and D. Infield, "Review of grid-tied converter topologies used in photovoltaic systems," *IET Renewable Power Generation*, vol. 10, no. 10, pp. 1543–1551, 2016.

[2] Ö. Göksu, R. Teodorescu, C. L. Bak, F. Iov, and P. C. Kjær, "Instability of wind turbine converters during current injection to low voltage grid faults and pll frequency based stability solution," *IEEE Trans. Power Systems*, vol. 29, no. 4, pp. 1683–1691, 2014.

[3] S. Toledo, M. Rivera, and J. L. Elizondo, "Overview of wind energy conversion systems development, technologies and power electronics research trends," in *2016 IEEE International Conference on Automatica (ICA-ACCA)*, Oct 2016, pp. 1–6.

[4] A. Lesnicar and R. Marquardt, "An innovative modular multilevel converter topology suitable for a wide power range," in *2003 IEEE Bologna Power Tech Conference Proceedings,*, vol. 3. IEEE, 2003, pp. 1–6.

[5] M. Hagiwara and H. Akagi, "Control and experiment of pulsewidth-modulated modular multilevel converters," *IEEE Trans. Power Electron.*, vol. 24, no. 7, pp. 1737–1746, 2009.

[6] A. Antonopoulos, L. Angquist, and H.-P. Nee, "On dynamics and voltage control of the modular multilevel converter," in *2009 13th European Conference on Power Electronics and Applications*. IEEE, 2009, pp. 1–10.

[7] V. Valouch, M. Bejvl, P. Šimek, and J. Škramlík, "Power control of grid-connected converters under unbalanced voltage conditions," *IEEE Trans. Ind. Electron.*, vol. 62, no. 7, pp. 4241–4248, 2014.

[8] D. Pérez-Estévez, J. Doval-Gandoy, A. G. Yepes, O. Lopez, and F. Baneira, "Enhanced resonant current controller for grid-connected converters with lcl filter," *IEEE Trans. Power Electron.*, vol. 33, no. 5, pp. 3765–3778, 2017.

[9] D. Caballero, S. Toledo, M. Rivera, E. Maqueda, F. Gavilan, C. Romero, and R. Gregor, "Predictive voltage control using matrix converter for a stand-alone wind energy based microgrid," in *2018 IEEE International Conference on Automation/XXIII Congress of the Chilean Association of Automatic Control (ICA-ACCA)*. IEEE, 2018, pp. 1–6.

[10] J. A. Riveros, J. Prieto, M. Rivera, S. Toledo, and R. Gregor, "A generalised multifrequency pwm strategy for dual three-phase voltage source converters," *Energies*, vol. 12, no. 7, p. 1398, 2019.

[11] M. Rivera, S. Toledo, C. Baier, L. Tarisciotti, P. Wheeler, and S. Verne, "Indirect predictive control techniques for a matrix converter operating at fixed switching frequency," in *2017 IEEE International Symposium on Predictive Control of Electrical Drives and Power Electronics (PRE-CEDE)*. IEEE, 2017, pp. 13–18.

[12] O. Gonzalez, M. Ayala, J. Doval-Gandoy, J. Rodas, R. Gregor, and M. Rivera, "Predictive-fixed switching current control strategy applied to six-phase induction machine," *Energies*, vol. 12, no. 12, p. 2294, 2019.

[13] F. Gavilan, D. Caballero, S. Toledo, E. Maqueda, R. Gregor, J. Rodas, M. Rivera, and I. Araujo-Vargas, "Predictive power control strategy for a grid-connected 2l-vsi with fixed switching frequency," in *Power, Electronics and Computing (ROPEC), 2016 IEEE International Autumn Meeting on*. IEEE, Nov 2016, pp. 1–6.

[14] S. Wang, C. Li, C. Che, and D. Xu, "Direct torque control for 2l-vsi pmsm using switching instant table," *IEEE Trans. Ind. Electron.*, vol. 65, no. 12, pp. 9410–9420, Dec 2018.

[15] Q. Li and D. Jiang, "Dc-link current analysis of three-phase 2l-vsi considering ac current ripple," *IET Power Electronics*, vol. 11, no. 1, pp. 202–211, Nov 2017.

[16] L. Zhan, Y. Liu, and Y. Liu, "A clarke transformation-based dft phasor and frequency algorithm for wide frequency range," *IEEE Trans. Smart Grid*, vol. 9, no. 1, pp. 67–77, Jan 2018.

[17] T. Dragičević, "Model predictive control of power converters for robust and fast operation of ac microgrids," *IEEE Trans. Power Electron.*, vol. 33, no. 7, pp. 6304–6317, Jul 2018.

[18] T. Dragicevic, C. Zheng, J. Rodriguez, and F. Blaabjerg, "Robust quasi-predictive control of lcl-filtered grid converters," *IEEE Trans. Power Electron.*, pp. 1–1, 2019.

[19] S. A. Richter and R. W. De Doncker, "Digital proportional-resonant (pr) control with anti-windup applied to a voltage-source inverter," in *Proceedings of the 2011 14th European Conference on Power Electronics and Applications*, Aug 2011, pp. 1–10.

978-1-7281-4181-7/19 $31.00 © 2019 IEEE

FSC-MPC Current Control of a 5-level Half-Bridge/ANPC Hybrid Three-phase Inverter

Jailson Leite Silva
Department of Electrical Engineering
Federal University of Piauí
Teresina, Brazil
jailson_ls@hotmail.com

Rafael Rocha Matias
Department of Electrical Engineering
Federal University of Piauí
Teresina, Brazil
rafaelrocha@ufpi.edu.br

Max Dannyel de Carvalho Alves
Department of Electrical Engineering
Federal University of Piauí
Teresina, Brazil
maxdannyel@hotmail.com

Ranoyca Nayana Alencar Leão e Silva
Department of Electrical Engineering
University of International Integration
of the Afro-Brazilian Lusophony
Redenção, Brazil
ranoyca@unilab.edu.br

Amanda Thayla Silva Monteiro
Department of Electrical Engineering
Federal University of Piauí
Teresina, Brazil
amathayla@hotmail.com

Kristian Pessoa dos Santos
Eletrotechnique Department
Federal Institute of Piauí
Parnaíba, Brazil
kristianpessoa@*ifpi*.edu.br

Abstract—**Multilevel converters are well-established in literature as topologies used in applications aiming to reduce current and voltage Total Harmonic Distortion (THD). This work proposes a current control loop for a 5-level three-phase inverter using Finite Control Set - Model Predictive Control (FCS-MPC). The adopted scheme allows the current control of a system composed of the inverter connected to a RL load, so that the phase currents of the load follow sinusoidal references whose amplitude can be adjusted. The optimal switching sequence is found by the minimization of a cost function that is calculated for the available voltage vectors. The used control method is straightforward, and the constraints can be adjusted in a simple manner during the project. The used method was able to maintain line voltage THD ratios below other techniques such as SHE-PWM, PD-PWM and CS-PWM.**

Keywords—*Harmonic distortion, multilevel inverter, predictive control*

I. Introduction

Power converters are devices widely used in the industry and represent considerable economic interest nowadays. DC-to-AC power converters can be used in several modern applications, such as variable speed AC drives, inductive heating, auxiliary power supplies and uninterruptible power supplies [1], [2]. In literature, several inverter control techniques can be found, such as PWM, SVPWM, deadbeat type control, hysteresis control, among others. Each of these techniques has advantages and disadvantages which must be taken into account for each application [3]–[7].

One form of nonlinear control that is currently being the aim of many research efforts in power electronic applications is known as the MPC. This control technique was limited to processes with slow dynamics, such as those involving chemical reactions, since the required calculation demanded a considerable amount of time. With the development of faster microcontrollers and DSPs in recent years, the benefits of MPC could be applied to the power electronics industry, since now complex calculations can be performed on-line in a matter of a few microseconds, while employing relatively inexpensive hardware.

MPC can be applied in multivariable systems using a very simple approach and presents a fast dynamic response. It also includes nonlinearities and constraints to be incorporated into the control law intuitively [8].

The MPC technique uses information from the mathematical model of the system to predict future values of the variable to be controlled and to change the control signal as needed. The system treated in this work consists of a 5-level three-phase inverter with wye-connected RL load, which follows a current reference by means of a Finite Control Set Predictive Control (FSC-MPC) loop. The system was simulated using the Simulink®, environment from the MATLAB®, software.

II. Control Strategy Elements

A. Characteristics of Model Predictive Control

There are two main groups of MPC techniques that have been applied to power electronics. The first one is known as Continuous Control Set MPC and contains a modulator which generates the switching states after receiving the continuous output of the predictive controller. The second one is the Finite Control Set MPC (FCS-MPC), which utilizes the limited number of switching states of the power converter in order to solve the optimization problem.

This approach adopts a discrete model to predict the system output for each possible switching sequence up to the chosen prediction horizon and the best switching operation is determined by the minimization of a predefined cost function. The switching operation that yields the minimum cost function value is applied in the next sampling instant. This technique has the advantage of being straightforward, applying the control signal directly to the power converter and eliminating the need of a modulator.

The concept of current control by means of predictive control is derived from that and based on the fact that only a finite number of possible of current values can be generated by the static converter, due to the inverter voltage output being defined by the number of electronic switches and their combinations [10]–[12].

The aim of the current control scheme is to minimize the error between the measured currents and the reference values. This requirement can be written in the form of a cost function, as mentioned before. The cost function g in (1) is expressed in orthogonal coordinates and measures the error between the references and the predicted currents [7], [10]. Where $i_\alpha^p(k+1)$ and i_β^p k+1 are, respectively, the real and imaginary parts of the vector of predicted load currents for a given voltage vector, which is obtained using the load model, using

978-1-7281-4181-7/19 $31.00 © 2019 IEEE

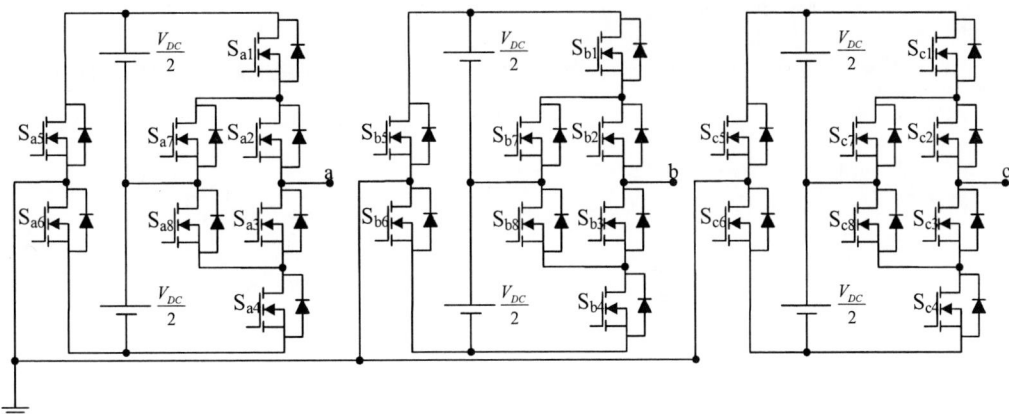

Fig. 1 Topology of the Half-bridge/ANPC Three-phase Inverter [9], [13].

a prediction horizon of one sampling instant. The reference currents $i_{\alpha}^{*}(k+1)$ and $i_{\beta}^{*}(k+1)$ are, respectively, the real and imaginary parts of the reference current vector i*(k+1). It is assumed that this reference current does not vary significantly during a sampling interval, so the relation i*(k+1) = i*(k) will be considered as valid during the calculations.

$$g = \left| i_{\alpha}^{*}(k+1) - i_{\alpha}^{p}(k+1) \right| + \left| i_{\beta}^{*}(k+1) - i_{\beta}^{p}(k+1) \right| \quad (1)$$

B. 5-Level Inverter

A new topology of a multilevel inverter derived from the Half-bridge and Active Neutral Point Clamped (ANPC) topologies, depicted in Fig. 1 was developed by [9]. Among the advantages of the topology is the use of a single power bus for each phase and flexibility for the implementation of high-performance modulation techniques [9], [13]. In addition, this topology can be used in large motor drive systems.

The great advantage of this inverter topology is that it has the positive characteristics of the ANPC, as it will inherit the best loss distribution among its semiconductors. On the other hand, its Half-bridge features allow 5 levels for the phase voltage output [9], [13]. There are some restrictions in the control of this topology due to the circuit characteristics. Considering x = {a,b,c}, the semiconductor switches S_{x5} and S_{x6} cannot be turned on nor off at the same time, S_{x2} and S_{x3} cannot be turned off at the same instant and finally, S_{x4} cannot conduct simultaneously with S_{x8}, the same being applied for S_{x1} and S_{x7}.

The inverter phase voltage can assume the values $+V_{DC}$, $+V_{DC}/2$, 0, $-V_{DC}/2$ and $-V_{DC}$. The output voltage for each phase, generated by each switching state S_x, can be described by Table 1 with x = {a,b,c}. There are many other possible switching states to generate the 5 voltage levels, but only 5 where chosen in order to reduce computational burden.

The inverter has multiple output voltage vectors, i.e., combinations of phase voltages according to switching states, according to (2):

$$v = \frac{2}{3}\left(v_{a} + av_{b} + a^{2}v_{c}\right) \quad (2)$$

Table 1 – Switching states for each phase of the inverter

S_x	S_{x1}	S_{x2}	S_{x3}	S_{x4}	S_{x5}	S_{x6}	S_{x7}	S_{x8}	v_x
+2	1	1	0	0	0	1	0	0	$+V_{DC}$
+1	0	1	0	0	1	0	1	0	$+V_{DC}/2$
0	0	0	1	1	1	0	0	0	0
-1	0	0	1	0	0	1	0	1	$-V_{DC}/2$
-2	0	0	1	1	0	1	0	0	$-V_{DC}$

Using the switching state vector **S** given in (3), (2) can be rewritten as shown in (4):

$$\mathbf{S} = \frac{2}{3}\left(S_{a} + aS_{b} + a^{2}S_{c}\right) \quad (3)$$

$$\mathbf{v} = \mathbf{S}\frac{V_{DC}}{2} \quad (4)$$

There are 125 possible voltage vectors found using (4), but only 61 were used to implement the system, since the others represent repeated vectors [9], [13]. The complete output voltage vector map can be viewed in Fig. 2.

Generally, induction motors or RL loads are connected to the output of an inverter, so an RL load was used in the system. The load dynamics can be described by (5) considering load resistance by R and inductance by L.

$$\mathbf{v} = R\mathbf{i} + L\frac{d\mathbf{i}}{dt} + \mathbf{e} \quad (5)$$

Where current vector **i** and back-emf vector (**e**) are presented in (6) and (7), respectively.

$$\mathbf{i} = \frac{2}{3}\left(i_{a} + ai_{b} + a^{2}i_{c}\right) \quad (6)$$

$$\mathbf{e} = \frac{2}{3}\left(e_{a} + ae_{b} + a^{2}e_{c}\right) \quad (7)$$

The block diagram in Fig. 3 shows the basic concepts behind the FSC-MPC technique used for current control.

978-1-7281-4181-7/19 $31.00 © 2019 IEEE

Basically, the present time load current measurement is inserted in the predictive model and, between the 61 possible future values of the load current, the one closest to the present time current reference is chosen. The switching sequence that generated the optimal load current value appears in form of the variables Sa, Sb and Sc in the block diagram. This switching sequence then is applied to the inverter and the cycle is performed again for the next sampling time.

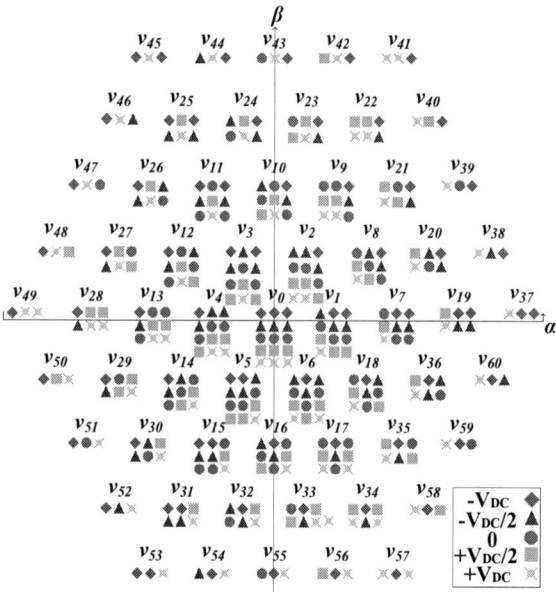

Fig. 2. Voltage vector map [9], [13].

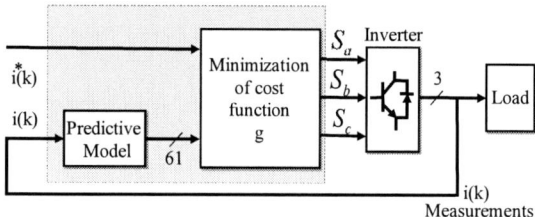

Fig. 3. Block diagram of the system.

C. Discretization

The content of (3) must be discretized in order to calculate the load current using a sampling T_s. The discrete-time model is used to predict the future value of the load current from voltages and currents measured at the k-th sampling instant. Considering that the load can be modelled as a first order system, the discrete time model can be obtained by a simple approximation of the derivative [7].

The derivative di/dt is replaced by Euler Forward Approximation where the derivative is approximated as presented in (8):

$$\frac{d\mathbf{i}}{dt} \approx \frac{\mathbf{i}(k+1) - \mathbf{i}(k)}{T_s} \qquad (8)$$

This result is inserted in (3) in order to obtain the prediction of the future charge current at the instant k + 1, for each of the 61 values of the voltage vector $\mathbf{v}(k)$ generated by the inverter. The resulting relation is presented by (9):

$$\mathbf{i}^p(k+1) = \left(1 - \frac{RT_s}{L}\right)\mathbf{i}(k) + \frac{T_s}{L}\left(\mathbf{v}(k) - \hat{\mathbf{e}}(k)\right) \qquad (9)$$

The term $\hat{\mathbf{e}}(k)$ denotes the estimated back-emf and p index means predicted variables. The back-emf vector necessary to each iteration, can be calculated considering load voltage and current measurements using (10):

$$\hat{\mathbf{e}}(k-1) = \mathbf{v}(k-1) - \frac{L}{T_s}\mathbf{i}(k) - \left(R - \frac{L}{T_s}\right)\mathbf{i}(k-1) \qquad (10)$$

The necessary term $\hat{\mathbf{e}}(k)$ can be found using the consideration $\hat{\mathbf{e}}(k) = \hat{\mathbf{e}}(k-1)$, since the sampling time is small. This result is inserted in (7) in order to obtain the prediction of the future load current at instant k + 1, for each of the 61 values of the voltage vector $\mathbf{v}(k)$.

The complete implementation process and its steps are summarized in the flow diagram in Fig. 4, that elaborates on the block diagram of Fig. 3. First the load currents are measured and used to evaluate the back-emf vector, which is necessary in later calculations.

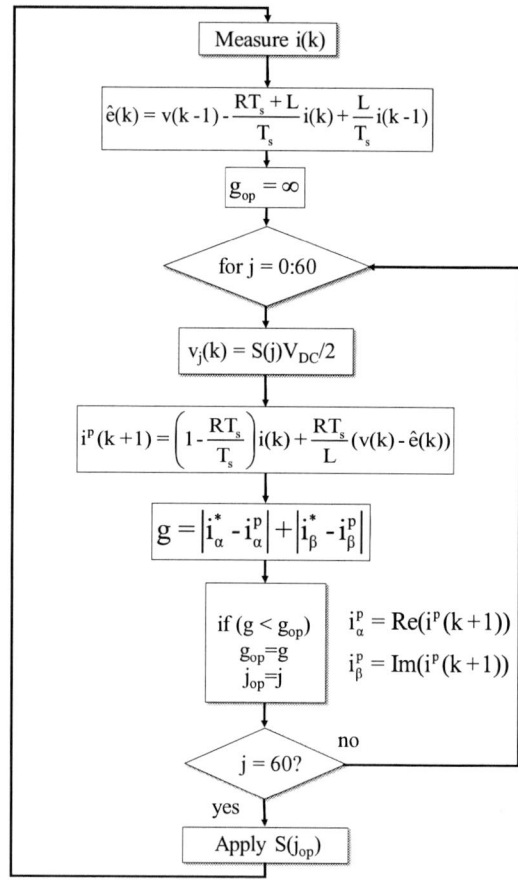

Fig. 4. Flow diagram of the control algorithm.

Afterwards, the initial optimal value for the cost function g_{op} is set to infinity, so that the optimization process can be initialized. The cost function is then calculated for all 61 voltage vectors and the one that generates the current closest to the required reference is chosen to be applied in the next sampling interval, i.e., there is a delay. It is important to point

978-1-7281-4181-7/19 $31.00 © 2019 IEEE   1449

out that the prediction horizon is of one step, meaning that every control action is determined one sampling interval in advance.

III. RESULTS

In the performed simulation the reference current used was a sinusoidal signal with $i^*_{peak} = 16.12$ A for each phase and the sampling time $T_s = 50\,\mu s$ was chosen, as seen in Table 2. The sampling time was chosen with base in previous works presented in literature and simulation tests [7]. The resulting current waveforms using a calculation step of $T_c = 1\,\mu s$ can be observed in Fig. 5.

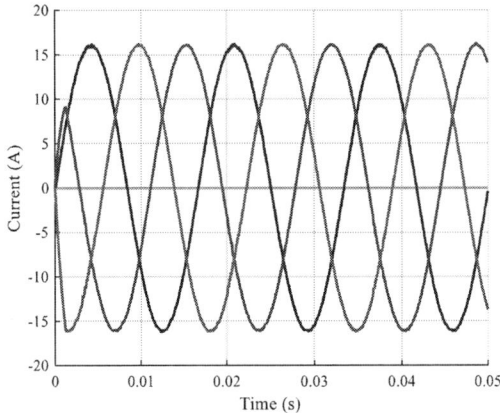

Fig. 5. Load currents i_a, i_b and i_c.

Table 2 – Specifications of the performed simulation

	Parameter	Value
V_{DC}	DC link voltage	340 V
L	Load inductance	20.02 mH
R	Load Resistance	17.71 Ω
i^*_{peak}	Reference current	16.12 A
T_s	Sampling Time	50μs
f_s	Switching Frequency	6 kHz

The output power of the converter is 7.5 kVA to a RL load, presenting a fundamental phase voltage value of $V_a = 225.81$ V and line voltage of $V_{ab} = 379.90$ V. The switching frequency is an average between the individual measured frequencies of all electronic switches and is found as being approximately $f_s = 6$ kHz. The phase voltage output containing all frequency components can be observed in Fig. 6 and the line voltage output in Fig.7. It can be seen in Fig. 6 that sometimes the imposed voltage vectors not change and therefore there is no commutation in the semiconductor switches of the analyzed phase, since the optimal vector not change since the last sampling time.

As the control loop main target is the load current, this variable presents a sinusoidal waveform, which is translated in a very small harmonic content. To reach the reference values of load current, the output phase voltage is varied accordingly whenever necessary, meaning that switching sequence is not fixed as in open loop PWM techniques. This may interfere in the voltage waveform symmetry, but the system still maintains the output voltage THD in reasonable levels without connecting a filter before the load.

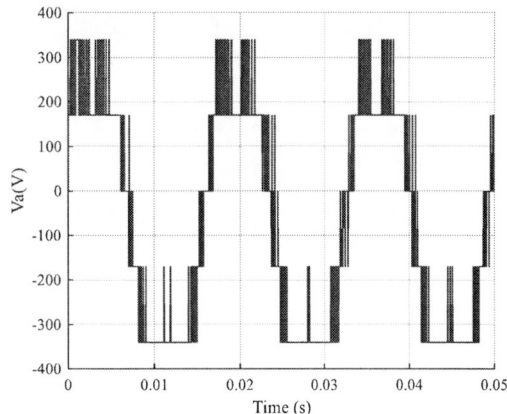

Fig. 6. Phase voltage output V_a (t).

Using the parameters of Table 2, the THD found it was equals to 0.4% for the load current whereas for the phase voltage it was equals to 24.57% and finally, for the line voltage it was equals to 5.95%, considering up to the 51st harmonic [5]. This result demonstrates the effectiveness of the method to reduce the harmonic components of the system.

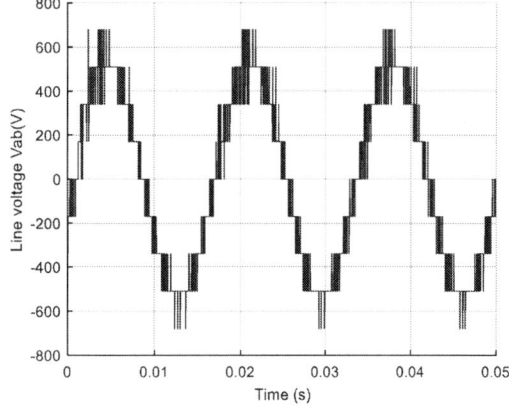

Fig. 7. Line voltage output V_{ab}(t).

Fig. 8 and Fig. 9 shown the behavior of the output current THD i_a (%) and THD V_a (%), respectively, with varying sampling time T_s. The harmonic distortion for i_a is clearly directly proportional to the sampling time. For V_a this relation is apparently different for some data points, but the general trend is that the load voltage presents lower THD (%) with shorter sampling times. This is linked to the $i^*(k+1) = i^*(k)$ assumption, since with shorter T_s this approximation is closer to reality. Similarly, the ripples in the current waveform are reduced with the decrease in sampling time. Using the same parameters of Table 2, the resulting load currents for a reference current step change of $i_{peak1}^* = 13$ A to $i_{peak2}^* = 17$ A are shown in Fig. 10. The initial reference $i_{peak1}^* = 13$ A is used prior to the instant t = 0.017 s, after which it is increased to $i_{peak2}^* = 17$ A. In addition, Fig.10 shows that the load currents follow the desired reference value.

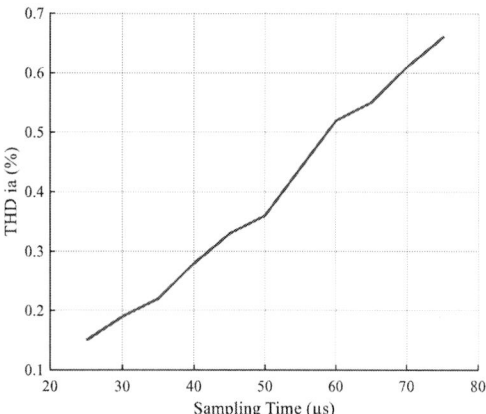

Fig. 8. THD (%) of output current $i_a(t)$ versus sampling time.

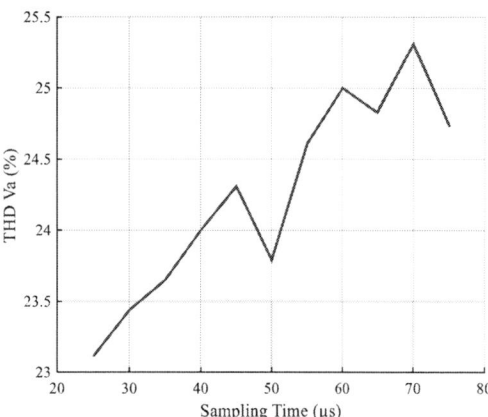

Fig. 9. THD (%) of phase voltage output $V_a(t)$ versus sampling time.

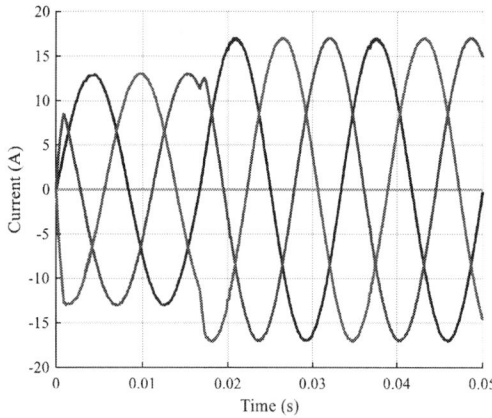

Fig.10 Load currents $i_a(t)$, $i_b(t)$ and $i_c(t)$ for the reference step.

The system performance using FSC-MPC was compared with 3 different modulation techniques using data from [14]. Fig. 11 and Fig. 12 show the THD (%) and WTHD (%), respectively, for the line voltage V_{ab} using SHE-PWM, PD-PWM, CSV-PWM [14]. The FSC-MPC method presented lower THD and WTHD compared to the others, which once again demonstrates its feasibility. To develop this comparison, the magnitude of the output voltage fundamental component

used to calculate the reference current for the FSC-MPC algorithm was similar to the other cited methods.

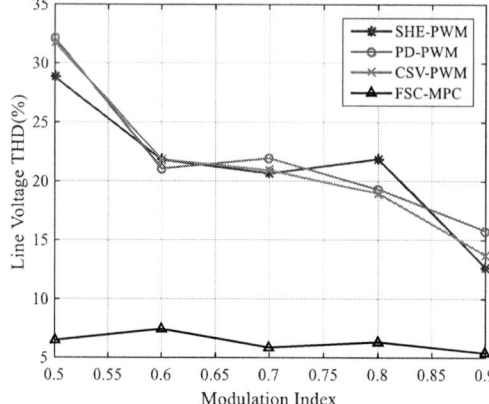

Fig. 11 THD (%) for $V_{ab}(t)$ versus modulation index.

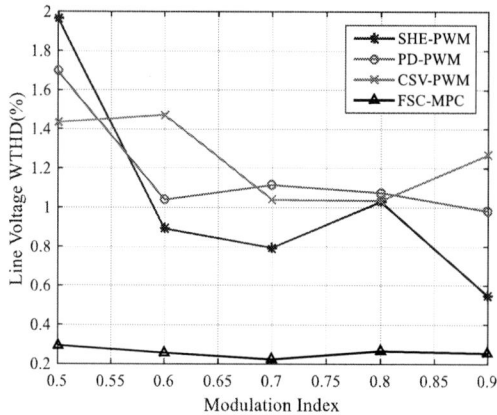

Fig. 12 WTHD (%) for $V_{ab}(t)$ versus modulation index.

IV. CONCLUSION

This paper presented a FSC-MPC current control method for a 5-level Half-bridge/ANPC hybrid inverter. The control system at hand was able to track the imposed current reference, hence, proving its viability. It was also demonstrated that the control was able to maintain the reference tracking for distinct current magnitudes, by being submitted to a reference step change and quickly reaching the requested values. The system displayed smaller THD and WTHD ratios in comparison with other methods, such as SHE-PWM, PD-PWM and CSV-PWM. The simplicity of the project and easy modification of constraints turns this control technique appealing to power electronics applications. As a disadvantage of this method, it employs a closed loop, which means that current sensors are necessary to its implementation. As future work, an experimental implementation will be developed to validate the propose control technique. In addition, a deeper mathematical analysis of the cost function can be developed so that current ripples are reduced even when larger sampling times are used.

ACKNOWLEDGMENTS

This work was achieved with the financial support of CAPES (National Council for the Improvement of Higher

Education) through FAPEPI (Foundation for Research Support of Piauí).

REFERENCES

[1] J. Rodríguez, J. S. Lai, and F. Z. Peng, "Multilevel inverters: A survey of topologies, controls, and applications," *IEEE Trans. Ind. Electron.*, 2002.

[2] S. Vazquez, J. Rodriguez, M. Rivera, L. G. Franquelo, and M. Norambuena, "Model Predictive Control for Power Converters and Drives: Advances and Trends," *IEEE Trans. Ind. Electron.*, 2017.

[3] I. Barbi, *Eletrônica de Potência*, 3rd ed. Florianópolis: EDUFSC, 2000.

[4] D. W. Hart, *Eletrônica de Potência - Análise e Projetos de Circuitos*. Porto Alegre: MacGraw-Hill, 2012.

[5] N. Mohan, T. M. Undeland, and W. P. Robbins, *Power Electronics: Converters, Applications, and Design*. John Wiley & Sons, 2003.

[6] M. H. Rashid, *Power Electronics: Circuits, Devices and Applications*. São Paulo: Makron Books, 1999.

[7] J. Rodriguez and P. Cortes, *Predictive Control of Power Converters and Electrical Drives*. 2012.

[8] M. Morari and J. H. Lee, "Model predictive control: Past, present and future," in *Computers and Chemical Engineering*, 1999.

[9] R. N. A. L. Silva, L. H. S. C. Barreto, P. P. Praça, D. S. O. JR., M. L. Heldwein, and S. A. Mussa, "Conversor híbrido simétrico de cinco níveis baseado nas topologias half-bridge/anpc," *Rev. Eletrônica Potência - SOBRAEP*, vol. 17, no. 3, pp. 623–631, 2012.

[10] E. F. Camacho and C. Bordons, *Model Predictive Control*, 2nd ed. London: Springer-Verlag, 2007.

[11] S. Kouro, P. Cortes, R. Vargas, U. Ammann, and J. Rodriguez, "Model Predictive Control:A Simple and Powerful Method to Control Power Converters," *IEEE Trans. Ind. Electron.*, 2009.

[12] V. Yaramasu, M. Rivera, B. Wu, and J. Rodriguez, "Model predictive current control of two-level four-leg inverters- Part i: Concept, algorithm, and simulation analysis," *IEEE Trans. Power Electronics.*, 2013.

[13] R. N. A. L. Silva, "Inversor multinível híbrido simétrico trifásico de cinco níveis baseado nas topologias half-bridge e ANPC," UFC, 2013.

[14] M. D. de C. Alves, "Modulação SHE-PWM Aplicada em um Inversor Multinível Híbrido Simétrico Trifásico de Cinco Níveis Baseado nas Topologias Half-Bridge e ANPC," UFPI, 2018.

Modeling and Simulation of a Stirling Engine in SCILAB

Raymundo Cordero
Graduation Program in Electrical
Engineering
UFMS
Mato Grosso do Sul, Brazil
rcorderog@gmail.com

Thyago Estrabis
Graduation Program in Electrical
Engineering
UFMS
Mato Grosso do Sul, Brazil
thyago.estrabis@gmail.com

João Onofre
Graduation Program in Electrical
Engineering
UFMS
Mato Grosso do Sul, Brazil
joaonofre@gmail.com

Felipe Alexandre Monteiro
Graduation Program in Electrical
Engineering
UFMS
Mato Grosso do Sul, Brazil
felipealexandremonteiro3@gmail.com

Augusto Hayashi
Department of Electrical Engineering
USP
São Paulo, Brazil
augustohayashi@gmail.com

Abstract—**Stirling cycle is the most similar to the Carnot cycle, which theoretically is the most efficient thermodynamic cycle. For that reason, Stirling engine is a powerful alternative in cogeneration, i.e., to produce electricity from different heat sources. The Stirling cycle is complicated to model, due to it has many equations. This paper shows the model of a Stirling engine developed in the free and open source SCILAB. The objective of this paper is to be a didactic tutorial to develop the model of the Stirling engine to be used as primary machine for the simulation of electrical generators used in cogeneration. The modeling using free software allows many students and researches to use and edit the model. The developed model gives information about pressure, volume, generated power and efficiency. Simulation results shows that the operative model done in SCILAB can be used in the analysis of Stirling engine.**

Keywords—Cogeneration, free and open software, Stirling engine, modeling.

I. Introduction

Nowadays, the increasing of the energy consumption, the ending of oil, coil and other fossil fuels and the conservation of the environment motivates the study and development of new energy source based on renewable energies [1], [2];

Cogeneration is an effective alternative in the generation of heat and electricity using a single primary energy source [3]. In many cases, a heat source (biomass, geothermal, solar heat, etc.) is available. In these cases, it is possible to use the heat source in a thermodynamic cycle in order to generate mechanical energy. This mechanical energy is used to move an electrical generator which produces electricity.

Many thermodynamic cycles were proposed in literature, for different applications [4]. Brayton cycle is used in gas turbines and electric power generation. Internal combustion engines (such as fuel vehicles) are based on Otto or Diesel cycles. Rankine cycle is also used in power generation plants. Theoretically, Carnot cycle is the process which has the highest efficiency in the generation of mechanical energy from a heat source. However, this cycle cannot be implemented experimentally.

Stirling cycle is a special case. The Stirling cycle can use air as working gas, which is an advantage because air it is free and, different to water steam (used in Ranking cycle), air does not produce problems in the mechanical parts. Stirling cycle has the most similar to the Carnot cycle (in terms of heat to mechanical energy conversion efficiency) which can be experimentally implemented [4], [5].

The modeling of the Stirling cycle is difficult due to it is composed by many non-linear equations. Its modeling is important to get information about the thermodynamic conditions and the energy available to generate electricity.

On the other hand, free and open software allows students and researches doing simulations in many topics due to its availability and resources [6]-[8]. Between different options, SCILAB is a free and open source numerical software which has an application, XCOS, which allows doing simulations using block diagram, which is an advantage when complex system must be modeled [9]. Few open source numerical softwares have a SIMULINK-like simulation interface. On the other hand, many tutorials can be found in order to learn how to modelate dynamic systems with SCILAB/XCOS. For that reason, SCILAB will be used in this paper.

This paper presents a modeling of the Stirling engine developed in SCILAB/XCOS. The objective of this work is to motivate the use of free software in engineering, especially in cogeneration. Although SCILAB does not have all the mathematical tools that MATLAB has, it can be used to model nonlinear systems and power electronics circuits [8], [9]. The developed model gives information about pressure, volume, generated power and efficiency. Simulation results shows that the operative model done in SCILAB can be used in the analysis of Stirling engine. The block diagram can be easily coupled with models of electrical generators modelled in SCILAB. The developed model of the Stirling engine is done considering steady-state condition. The dynamic of electrical systems are much faster than the dynamics of mechanical systems. For that reason, a steady state model of the Stirling engine can be used.

978-1-7281-4181-7/19 $31.00 © 2019 IEEE

II. Modeling of a Stirling Engine

The modeling of the Stirling engine used in this paper is based on [10], [11]. This model consider an adiabatic analysis, i.e., no heat transfer occurs between system and the environment. This also applies that the sum of the work developed by the engine with the internal energy is zero. The ideal adiabatic analyze, as in Fig. 1, implies maximum efficiency. Table 1 shows the parameters of the model. The ideal pressure vs volume (PxV) thermodynamic curve is illustrated in Fig. 1.

The Stirling cycle is composed by 4 cycles: Isochoric heating (1→2), isothermal expansion (2→3), isochoric cooling (3→4) and isothermal compression (4→1). The difference between Stirling and Carnot Cycle, in theory, is that the compression and expansion stages (1→2 and 3→4) in Stirling cycle are isochoric (done in constant volume), while in Carnot cycle, these stage are adiabatic (no heat flow). The stages of the Stirling cycle described as follows:

- 1→2: Isochoric heating: the working gas gains energy from a regenerator while its volume keeps constant.

- 2→3: Isothermal expansion: the working gas receives energy from the external thermal source, increasing its volume. The gas makes a work, and its pressure drops.

- 3→4: Isochoric cooling: the working gas gives energy to the regenerator, while its volume keeps constant.

- 4→1: Isothermal compression: the volume of the working gas drops due to its pressure raises. Heat flows from the gas to an external sink.

There exist three types of Stirling engines: alpha, beta and gamma. In this paper, it is used the gamma-type Stirling engine, illustrated in Fig. 2. The following conditions are assumed:

- There is no pressure loss in the heat exchangers.

- The compression and expansion process are isothermal.

- The working gas is ideal gas.

- The regenerative process is perfect.

- The temperatures in the expansion and compression chambers are constant.

- The temperature of the regenerator is the arithmetic mean of the heater and the cooler.

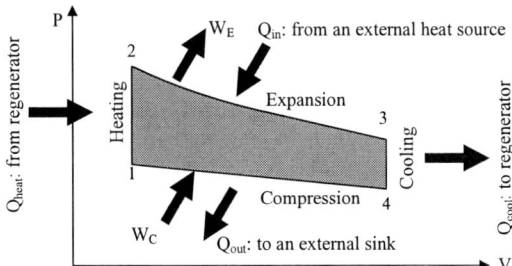

Fig. 1. Scheme of the Stirling cycle.

TABLE I. Parameters of Stirling Engine

Name	Symbol	Unit
Engine pressure	P	Pa
Swept volume of expansion piston or displacer piston	V_{SE}	m³
Swept volume of compression piston or power piston	V_{SC}	m³
Dead volume of expansion space	V_{DE}	m³
Regenerator volume	V_R	m³
Dead volume of compression space	V_{DC}	m³
Expansion space momental volume	V_E	m³
Compression space momental volume	V_C	m³
Total momental volume	V	m³
Total mass of working	M	Kg
Expansion space gas temperature	T_H	K
Compression space gas temperature	T_C	K
Regenerator space gas temperature	T_R	K
Phase angle	dx	Deg
Crank angle	x	
Temperature ratio	t	-
Swept volume ratio	v	-
Dead volume ratio	X	-
Engine speed	n	Hz
Expansion energy	W_E	J
Compression energy	W_C	J
Indicated energy	W_i	J
Expansion power	L_E	W
Compression power	L_C	W
Indicated power	L_i	W
Indicated efficiency	η	
Generator efficiency	$\eta_{generator}$	
Electrical power given by the generator	$P_{generator}$	W
Thermal power (Heat given to the Stirling engine per second)	Q_s	W
Mechanical output power	P_{mec}	W
Gas constant	R	J/kg.K
Mean pressure	P_{mean}	P

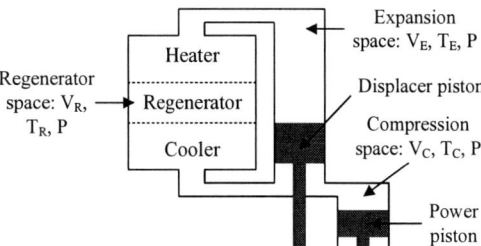

Fig. 2. Gamma-type Stirling engine.

The expansion and compression cylinder volumes are:

$$V_E = \frac{V_{SE}}{2}.(1 - \cos x) + V_{DE} \tag{1}$$

$$V_C = \frac{V_{SE}}{2}.(1 - \cos x) + \frac{V_{SC}}{2}.(1 - \cos(x - dx)) + V_{DC} \tag{2}$$

The crack angle (x), in rad, can be defined as the integral of the mechanical speed.

The volume in the heater, cooler and regenerator change in time. However, considering that the inner volume of the pistons are constants, then the total momental volume are the sum of the expansion, compression and regeneration volumes:

$$V = V_E + V_R + V_C \qquad (3)$$

The expansion, compression and regeneration regions have their own temperatures and volumes. However, it is possible to consider that the pressure in these regions is equal. Assuming that the process gas is ideal, it is possible to use the ideal gas formula to get the masses:

$$P.V = m.R.T \qquad (4)$$

The temperature, volume and pressure in the expansion, regeneration and compression regions are (T_E, V_E, P), (T_R, V_R, P) and (T_E, V_C, P), respectively. The temperature in the regenerator can be modeled as the average of the temperatures in the heater and cooler:

$$T_R = \frac{T_E + T_C}{2} \qquad (5)$$

Thus, based on (4), the mass in the expansion, compression and regeneration regions are:

$$m_E = \frac{P.V_E}{R.T_E} \qquad (6)$$

$$m_R = \frac{P.V_R}{R.T_R} \qquad (7)$$

$$m_C = \frac{P.V_C}{R.T_C} \qquad (8)$$

Based on the mass conservation law, the total mass is:

$$m = m_E + m_R + m_C \qquad (9)$$

Using equations (1) to (9), the total mass (m) of the working gas is

$$m = \frac{P.V_{SE}}{2.R.T_C}.(S - B.\cos(x - a)) \qquad (10)$$

where:

$$t = \frac{T_C}{T_E}; \qquad (11)$$

$$v = \frac{V_{SC}}{V_{SE}}; \ X_{DE} = \frac{V_{DE}}{V_{SE}}; \ = \frac{V_{DC}}{V_{SE}}; \ X_R = \frac{V_R}{V_{SE}} \qquad (12)$$

$$S = \left(1 + 2.X_{DE} + \frac{4.X_R}{1+t}\right).t + v + 2.X_{DC} + 1 \qquad (13)$$

$$B = \sqrt{(t-1)^2 + 2.(t-1).v.\cos(dx) + v^2} \qquad (16)$$

$$a = \tan^{-1}\left(\frac{v.\sin(dx)}{t + \cos(dx) + 1}\right) \qquad (17)$$

Let define the variable c as follows:

$$c = \frac{B}{S} \qquad (18)$$

According to [10], [11] the engine pressure (P) depends on the following formula:

$$P = \frac{P_{mean}.\sqrt{1-c^2}}{1 - c.\cos(x-a)} \qquad (21)$$

The expansion energy (W_E), the compression energy (W_C) and the indicated energy (W_i) can be expressed in function of the mean pressure (P_{mean}):

$$W_E = \oint P.dV_E = \frac{P_{MEAN}.V_{SE}.\pi.c.\sin a}{1 + \sqrt{1-c^2}} \qquad (22)$$

$$W_C = \oint P.dV_C = -\frac{P_{MEAN}.V_{SE}.\pi.c.t.\sin a}{1 + \sqrt{1-c^2}} \qquad (23)$$

$$W_i = W_E + W_C = \frac{P_{MEAN}.V_{SE}.\pi.c.(1-t).\sin a}{1 + \sqrt{1-c^2}} \qquad (24)$$

The indicated energy is the available energy of the Stirling process. It is equal to the inner area in the diagram of Fig. 1. All these energies are calculated considering one thermodynamic cycle (see Fig. 1). In consequence, the power generated by the Stirling engine depends on the mechanical speed, which defines the number of thermodynamic cycles done per second Let n be the mechanical speed of the engine (in cycles/second). The expansion, compression and indicated powers are:

$$L_E = W_E.n \qquad (25)$$

$$L_C = W_C.n \qquad (26)$$

$$L_i = W_i.n \qquad (27)$$

The efficiency of the Stirling engine is:

$$\eta = \frac{W_i}{W_E} = 1 - t \qquad (28)$$

The efficiency shown in (28) is the ratio between the indicate energy and expansion energy. It is also the relationship between the thermal power (heat transferred to the Stirling engine per second, Q_s) with the mechanical power produced by the engine (P_{mec}), as indicated in (29):

$$\eta = \frac{P_{mec}}{Q_s} \qquad (29)$$

Let $\eta_{generator}$ be the efficiency of an electrical generator, i.e., the relationship between the mechanical input power and the electrical output power. Thus, (30) describes the relationship between the primary mechanical power that the generator receives with the electrical power produced by the generator.

$$\eta.\eta_{generator} = \frac{P_{mec}}{Q_s}.\frac{P_{generator}}{P_{mec}} \qquad (30)$$

Therefore, the product of these two efficiencies is the relationship between the electrical power produced by the generator with the thermal power given to the Stirling engine.

$$\frac{P_{generator}}{Q_s} = \eta.\eta_{generator} \qquad (31)$$

978-1-7281-4181-7/19 $31.00 © 2019 IEEE

Therefore:

$$P_{generator} = \eta \cdot \eta_{generator} \cdot Q_s \qquad (31)$$

Equation (32) indicates the electrical power generated by the Stirling-generator system according to the heat given to the engine.

All these equations were developed considering steady state conditions. Generally, transient behavior is not considered.

III. SIMULATION DIAGRAM IN SCILAB

Fig. 3 shows the full representation of Gamma-type Stirling engine in SCILAB. It was used the application XCOS to implement the equations described in the previous section. Information about the use of XCOS can be found in [9].

The schemes of each sub-blocks that composes the model of the Stirling engine are shown in Figs. 4 to 8. Fig. 4 represents the implementation, in block diagram, of (13), (18), (16), in order to get the variables s, c and B. Equations (17) and (21) were implemented as illustrated in Figs, 5 and 6, respectively, in order to get the variables a and P. The estimation of the momental volume, through equations (1) to (3), was modelled as shown in Fig. 7. The estimation of the expansion power, compression power, indicated power and indicated efficiency, according to (25), (26), (27) and (28), was implemented as shown in Fig. 8.

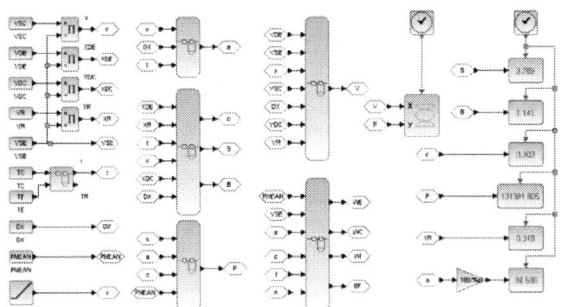

Fig. 3. Stirling engine simulation diagram in SCILAB.

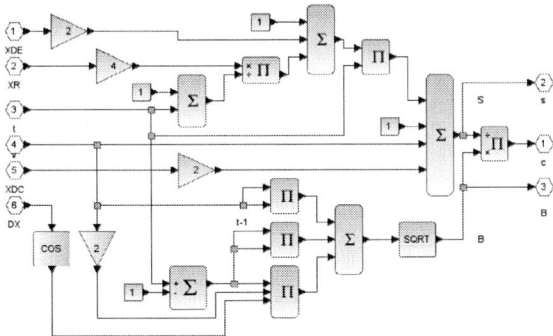

Fig. 4. Estimation of variables "s", "c" and "B".

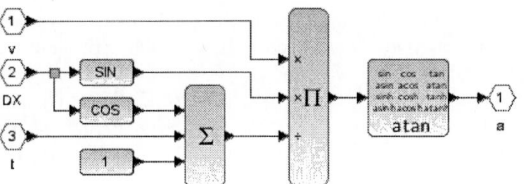

Fig. 5. Estimation of the variable "a".

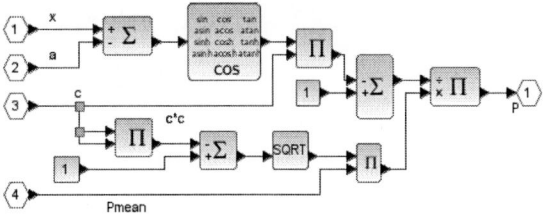

Fig. 6. Estimation of the engine total pressure (P).

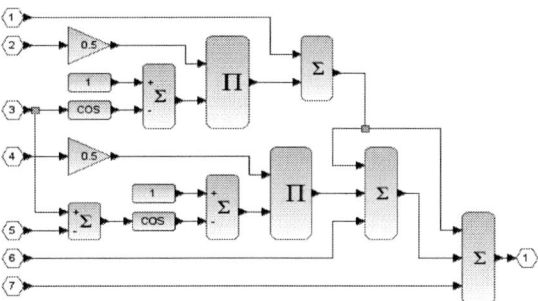

Fig. 7. Estimation of momental volume (V).

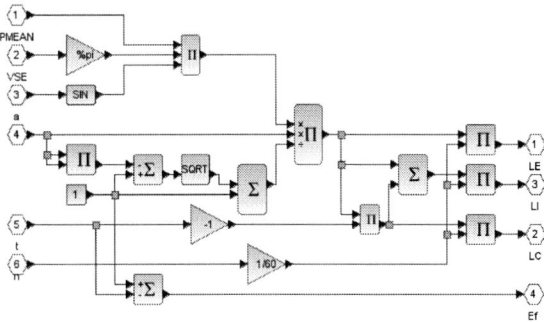

Fig. 8. Estimation of expansion power, compression power, indicated power and indicated efficiency.

IV. RESULTS

Table II shows the parameters of Stirling engine used to perform the simulation. The pressure vs volume (PxV) diagram is shown in Fig. 9. The simulation result is some different to the ideal cycle in Fig. 1, but is it similar to the results shown in [11]. The pressure goes to 75 kPa to almost 140 kPa.

The block diagram done in SCILAB gives the values of the formulas explained previously. Table III shows the obtained values of *a*, *t*, L_E, L_C, L_i, and the efficiency. The values in Table III were verified using a code done in SCILAB that implements the equations of the Stirling model explained previously. The SCILAB code is compatible with MATLAB.

TABLE II. SIMULATION PARAMETERS

Parameter	Value
T_E	400 °C
T_C	30 °C
n	2000 RPM
V_{SE}	$0.628*10^{-6}$ m³
V_{SC}	$0.628*10^{-6}$ m³
V_{DE}	$0.2*10^{-6}$ m³
V_{DC}	$0.2*10^{-6}$ m³
V_R	$0.2*10^{-6}$ m³
P_{mean}	101300 Pa
dx	90 °

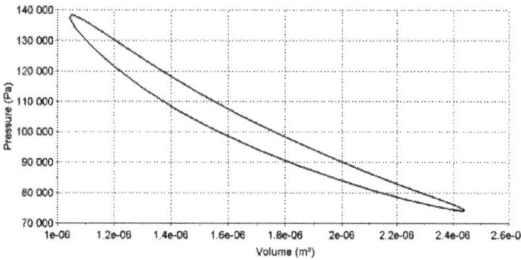

Fig. 9. Curve *PxV* obtained in the simulation.

The result obtained in Fig. 9 is similar to the obtained in [11]. However, it is different to the *PxV* curve shown in Fig. 1. Fig. 1 is a theoretical thermodynamic curve considering an ideal adiabatic conditions (impossible to achieve), where the flux of heat between the mass and the environment is zero. Besides, it is impossible to create, experimentally, a mechanical system that allows the cooling or heating in a perfectly constant volume. That means, an perfect isochoric system is impossible to create (but it is possible to create systems similar to the isochoric process). Note that the main difference between Figs. 1 and 9 is the size of the isochoric curves (1→2, 3→4). For those reasons, Figs. 1 and 9 are different. However, according to [11], Fig. 9 is more similar to the experimental Stirling process.

V. CONCLUSION

This paper shows the didactic modeling and simulation of a Stirling engine in SCILAB/XCOS. The block diagram estimates all the parameters required to analyze the Stirling engine. This block can be used to integrate the Stirling machine with other blocks such a CC or AC generator. Working with block diagram is more didactic and useful than developing a textual code of the model. This paper also tries to encourage researches to develop models about power electronics and electrical systems in SCILAB, in order to make tests about cogeneration using open source softwares.

TABLE III. SIMULATION RESULTS

Parameter	Value
a	34.59°
t	0.459
L_E	0.586 W
L_C	-0.264 W
L_i	0.322 W
Efficiency	55 %

ACKNOWLEDGMENT

Authors want to thank BATLAB laboratory and Graduation Program in Electrical Engineering, PPGEE-UFMS, for the support to this research.

REFERENCES

[1] K. Yosra, B. Ammar Mohsen, H. A. Hsan. "Modeling and simulation of DSPMSG power generator connected to grid," 16th international conference on Sciences and Techniques of Automatic control & computer engineering, 2015, pp. 656–663.

[2] A. Kulkarni, D. Patil, R. R. Arakerimath, "Experimental Investigation of Heat Sink as a Heat Pipe to Improve Performance of Thermoelectric Generator," International Journal of Scientific Research in Science and Technology, vol. 3, no. 7, pp. 943–948, Oct. 2017 .

[3] A. Pietak, M. Meus, S. Nitkiewicz, "The Demand Analysis of the Biogas Plant with Chp System on Substrate, the Aim to Obtain Required Electrical and Thermal Power," Journal of KONES Powertrain and Transport, vol. 19, no. 3, 2012.

[4] T.K. Ghosh and M.A. Prelas, Energy Resources and Systems: Volume 1: Fundamentals and Non-Renewable Resources, pp. 89–140. Springer Science + Business Media B.V. 2009.

[5] P.Surdacki, M. Holuk, K. Bańka, K.Gawkowski, "Investigation of the CHP generation system with the stirling engine," 2017 International Conference on Electromagnetic Devices and Processes in Environment Protection with Seminar Applications of Superconductors (ELMECO & AoS), 2017, pp. 1–4.

[6] A. M. Ibrahim, "Free Alternatives to Matlab for Undergraduate Electronics Engineering Curricula ," 32nd Annual Conference on IEEE Industrial Electronics, 2006, pp. 4598–4603.

[7] B. Kanoj, M. Arjun Sanu and A. B. Raju, "Steady state analysis of autonomous wind energy conversion system for irrigation purpose employing induction machine using Python — A free and open source software," 2013 IEEE Global Humanitarian Technology Conference: South Asia Satellite, 2013, pp. 292–297.

[8] B. P. Meneses, M. T. S. Quixada, J. C. T. Campos and L. L. N. dos Reis, "Simulation of converters using Scilab/Scicoslab," XI Brazilian Power Electronics Conference, 2011, pp. 1032–1036.

[9] SCILAB Web Site, [Online] Available at: <www.scilab.org>

[10] J. P. V. Dormael, Theory of Stirling Engine, available in: http://www.robertstirlingengine.com/theory.php

[11] K. Hirata, "SCHMIDT THEORY FOR STIRLING ENGINES", pp. 1–8. Available in: https://www.nmri.go.jp/oldpages/eng/khirata/list/general/schmidt.pdf

A Low Cost Photovoltaic Panel Emulator

Aluísio Alves de Melo Bento
Department of Electric Engineering and Telecommunication
State University of Rio de Janeiro
Rio de Janeiro, Brazil
aluisio.uerj@gmail.com

Raphael Perci Santiago
Department of Electric Engineering and Telecommunication
State University of Rio de Janeiro
Rio de Janeiro, Brazil
rpercieng@gmail.com

Abstract— **Photovoltaic panels are a low efficiency energy source. These require a relative large area in order to generate a specified output power and operate subject to uncontrollable environmental conditions. Photovoltaic panel emulators offer several important advantages in the context of testing MPPT techniques and static converter controller performance besides contributing to educational tool. This paper proposes a photovoltaic emulator whose the main characteristics are low electric inertia, flexibility, simplicity, low cost and robustness while maintaining a good approximation of the solar I-V profile. The design procedure is detailed and experimental results are carried out to confirm the feasibility of proposal.**

Keywords—photovoltaic cell, PV panel emulator, operational amplifier, analog controller

I. INTRODUCTION

Solar energy has been shown a large market increase due to continued development of photovoltaic (PV) technology over the past decade due to technological improvement, better awareness, rising electricity bills and reduction of system cost [1-2]. The international effort to substitute polluted sources of energy by renewables also plays a fundamental role in this change of mindset.

However, there are several problems related with the real time experimentation with solar photovoltaic, such as: (i) need of large amount of area to capture solar irradiation, (ii) expensive and complex experimental setup, (iii) need of specific levels of solar irradiation and climate to realize suitable experimentation. This motivates the development of equipment with equivalent behavior characteristics as real PV panels under light irradiation. Photovoltaic panel emulators (PVE) have been presented in a variety of models along the last years [3]. The main task for a SPE is to reproduce the I-V and P-V curve of a practical PV panel. In this sense, it should mimic the behavior of a real PV module with reasonable accuracy and make possible PV research at indoor environment through electronic circuits. An additional advantage of those circuits is the reduced size combined with great design flexibility.

In technical literature, converter based emulators have been broadly investigated with ordinary dc-dc converter topologies [3]-[16]. SEPIC and other dc-dc converters based emulators were shown in [3]. In [11], a mathematical model that represents the equivalent circuit of a PV cell, making it possible to emulate the PV behavior under atmospheric variations. The authors in [15] shows an interesting motivation for a PVE, where they proposed the use in prototype smart grid for studying the behavior of systems having grid-connected PV system with the aid of embedded microelectronics.

There are also solutions based on sophisticated control strategies, such as in [12][13], which accurately describe the dependence of the modules on solar irradiance and cell temperature. A low cost emulator of an integration of PV array and a dc-dc boost converter was presented in [21]. Digital control techniques were also used based on FPGA, ARM (Advanced RISC Machines) processors and DSP controllers to achieve better performance in terms of the operating conditions of the module under certain climate conditions and shading [17]-[19].

The solution presented in [16] uses a D.C programmable power supply for testing P&O MPPT algorithm. A great advantage of solutions applying D.C sources is the fact that those avoids the bandwidth problems related with electronics PV emulators [16]. In [12], real-time simulation through a computer controlled D.C power supply leads to a control strategy similar to the one used today for commercial companies that commercializes PVE. Commercial companies offer PVE mainly based on controllable dc voltage sources. However, the cost and complexity are some easily verified drawbacks of digital based PVE.

On the other hand, PV emulators is an alternative to these drawbacks [3][4][7][8][14]. Analog solution normally uses the diode based approximation techniques as a model to reproduce the intrinsic characteristics of a single cell that composes the basis of a photovoltaic module [20]-[25]. This method successfully reproduces exact I-V and P-V curves, being a reliable and accurate. The analog approach also eliminates the need of a software or lookup table to build the PV characteristics curves, being suitable to a low cost circuit solution that can be used, for example, as a test plataform for PV inverters [14].

Analog solutions for PVE have been considered since long time. In [7] and [9] analog PVE solutions were proposed. Although have good performance, disadvantage include high number of device (discrete, operational amplifiers, analog multipliers) and high power losses. A hybrid solution combined static converter (buck), analog circuit and digital processor as a control strategy for a high performance PVE was proposed in [8]. A cost guide solution with decreasing of power dissipation was presented in [23]. However, a poor performance was reported in this paper. An analog PVE composed by operational amplifiers, analog switches and discrete devices was presented in [4]. This solution utilizes PID controller but doesn't emulate temperature phenomenon.

This paper deals with the development of an analog PVE which minimizes the assembly device parts and the power dissipation level while produces a good performance. The proposed circuit accomplishes fast response of linear series regulator and satisfactory. Besides, this solution emulates real time temperature and light variations and parasitic parallel resistance on PV panel. The characteristic I-V of the PV panel is obtained from a real low power matrix four-diode. A simple voltage sweep load circuit is suggested and

978-1-7281-4181-7/19 $31.00 © 2019 IEEE

implemented in order to support the proposed experimental essays. Experimental results are carried out to validate theory and confirm the feasibility of proposal.

II. BASICS

A basic review of PV cell and analog concepts is carried out in order to support the comprehension and synthesis of the proposal.

The energy feed by PV panels are in dc voltage-current format and depends on the incident light, temperature, (and power flow to the load) the connected load. Figure 1 shown the behavior of a standard PV panel under temperature variation, Fig.1(a), and incident light intensity, Fig.1(b). This strongly nonlinear behavior imposes a technique that is capable to control those changes, since it greatly affects the performance of the system. Techniques for tracking the maximum power point (MPPT) define an important area in power electronic and control theory.

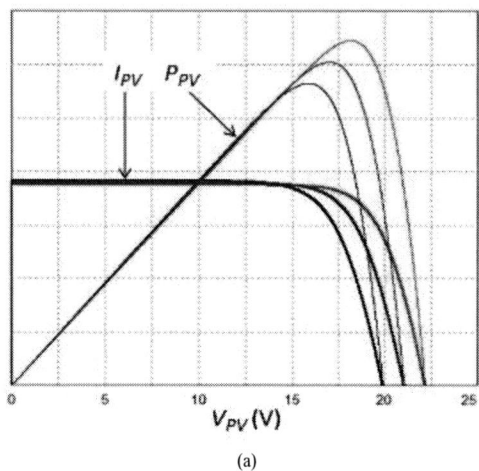

(a)

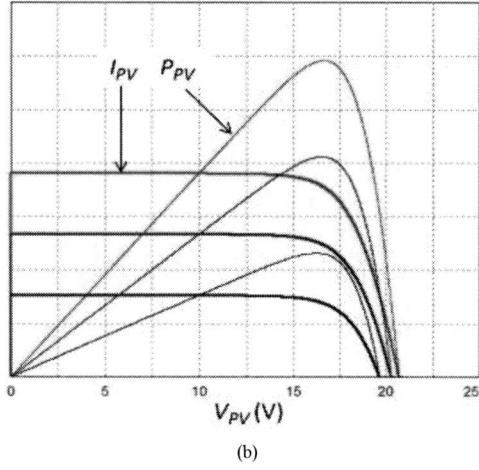

(b)

Fig.1. PV panel illustrative characteristics. Current (1 A/div.) and power (10 W/div.) as a function of voltage from PV module. Temperature standard level T = 30 °C. (a) Temperature parameter: 10 °C (black), 25 °C (blue) e 40 °C (red). Irradiation standard level of parameterization S=1000 W/m². (b) Light Intensity parameterization S: 1000 W/m² (red), 700 W/m² (blue) and 400 W/m² (black).

B. Differential Amplifier with Series Feedback

The proposed analog PVE is basically composed by three structures whose main characteristic is: an input differential voltage controls am output series voltage (or current). This in order to realize a (i) modulated amplitude sawtooth carrier ; (ii) a controlled current diode and (iii) an output floating voltage amplifier. An operational amplifier based circuit that performs these functions is presented in Fig. 3. It employs series feedback instead parallel one.

Based on Fig. 3, where the operational amplifier input impedance is considered as infinity, the current of R_1 and R_2 can be given by

$$I_{R1} = \frac{V_1 - V_3}{R_1} = -\frac{V_{x2} - V_3}{R_2}. \tag{1}$$

Analogously, the current of R_3 and R_4 is given by

$$I_{R3} = \frac{V_2 - V_3}{R_3} = -\frac{V_{x1} - V_3}{R_4}. \tag{2}$$

If it is considering the relationship

$$V_{Rx} = V_{x1} - V_{x2} \tag{3}$$

in (1) and (2), then, handling these, one obtains

$$V_{Rx} = \frac{R_2}{R_1}V_1 - \frac{R_4}{R_3}V_2 + V_3\left(\frac{R_4}{R_3} - \frac{R_2}{R_1}\right). \tag{4}$$

Considering $R_4/R_3 = R_2/R_1$ in (4), it is simplified to

$$V_{Rx} = \frac{R_2}{R_1}\left(V_1 - V_2\right). \tag{5}$$

Thus, a simple relationship between the input differential voltage and I_x is obtained.

$$I_x = \frac{V_{Rx}}{R_x} = \frac{R_2}{R_1 R_x}\left(V_1 - V_2\right) \tag{6}$$

Finally, considering $R_2 \gg R_x$, as depict in Fig. 3, the output current can be assumed equal to I_x.

$$I_x = \frac{R_2}{R_1 R_x}\left(V_1 - V_2\right). \tag{7}$$

C. Modulated Amplitude Sawtooth Carrier Generator Based on Differential Amplifier with Series Feedback

The circuit shown in Fig. 4(a) is a modulated amplitude sawtooth carrier. There, the capacitor receives current controlled by the input voltage (V_1-V_2) and a clock signal

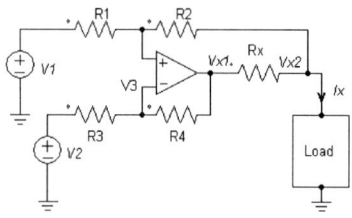

Fig. 3. Differential voltage amplifier with series feedback.

978-1-7281-4181-7/19 $31.00 © 2019 IEEE

from the TTL generator reset, i.e., discharge the capacitor. This circuit was designed to have high speed and low r_{DS} of the signal MOSFET, and can be used to support PVE tests with a not shown voltage load sweep.

D. Current Source Controlled by Voltage Based on Differential Amplifier with Series Feedback

The circuit in Fig. 4(b) feed a current obtained by the difference of input voltages. This is suitable to emulate the diode current, once it is the difference between the photovoltaic and panel output current. One may attempt to the fact that the photovoltaic and panel output currents are processed by voltage. This circuit allows current scale down with an advantageous solution explained in next section.

E. Output Floating Voltage Amplifier Based on Differential Amplifier with Series Feedback

The circuit shown in Fig. 4(c) allows the production of a controlled voltage in series with a power MOSFET. Then, it is possibly to feedback current trough a source resistor or active current sensor, and control the resistor voltage drop accurately. Also, to avoid oscillation caused by the voltage control loop, a RC low pass filter is connected between the operational amplifier and the gate terminal.

III. THE PROPOSAL

The proposed PVE synthesis is a natural application of the circuits presented in the previous section.

A. Block Diagram

The block diagram of the PVE is illustrated in Fig. 5. There, the scheme is functionally divided in three parts: the diode current scale down circuit; the low power diode model; and the diode voltage scale up. The objective in this scheme is to obtain the I-V parameters directly from a real diode.

The PVE output voltage applied to the load produces the PVE current Ipv, which serves as part of the feedback control mechanism, in order to scale down circuit. Due to the output sensor impedance, this signal is in voltage form, R_sI_{pv}. The second input of scale down circuit is fed by a voltage level equivalent to the photovoltaic current, i.e., the current produced by the light incidence intensity, I_s. The differential voltage, $k_1(R_SI_S - R_SI_{PV})$ is produced in the resistor R_5,. Thus, the current applied in the diode model is given by $R_SI_S - R_SI_{PV})/R_5$. The choice of R_5, k_1 and R_S dictates the model diode current scale down. A very low current (tens of milliamps) driven through the signal diode is used to extract the emulated PV panel characteristic, is based on four diode model, as depicted in Fig. 6.

A resistance is placed in parallel to diode set. Additionally, an electrically isolated and thermally coupled resistor is used to emulate the temperature phenomenon. Also, diffusion and junction capacitances are intrinsically computed in the model. Other must to consider that a vast work is presented in order to choose different diodes and obtain different curves. Once the current of the model is established and the diode voltage is produced. Then, this voltage is scaled up to produces the PVE output voltage. Instantaneous V-I characteristic and low power model are the principal predicted features of the proposed PVE.

B. Schematic Diagram

The schematic diagram is shown in Fig. 6. Next, equations to calculate devices are developed.

The voltage that represents the current in the diode model is given by

$$V_{Idm} = R_5I_{pm} \tag{8}$$

The current sensor output resistance R_s defines the voltage feedback signal.

$$V_{Ipv} = R_sI_{pv} \tag{9}$$

Also, it is convenient to employ this value to the resistance associate to the photovoltaic current.

$$V_{Is} = R_sI_s \tag{10}$$

The transfer function of the scale down circuit is given by

$$V_{R5} = \frac{R_2}{R_1}\left(V_{Is} - V_{Ipv}\right). \tag{11}$$

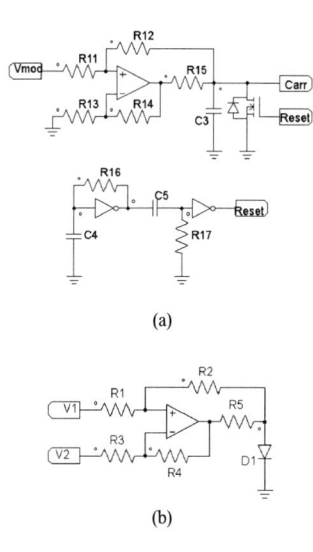

(a)

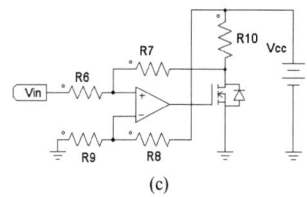

(b)

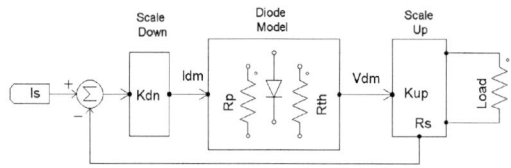

(c)

Fig. 4. Circuits based on the differential amplifier with series feedback. (a) Modulated Amplitude Sawtooth Carrier Generator. (b) Current Source Controlled by Voltage. (c) Output Floating Voltage Amplifier.

Fig. 5. Proposed PVE block diagram.

978-1-7281-4181-7/19 $31.00 © 2019 IEEE   1460

Considering (8), (9) and (10) in (4), results in

$$I_{dm} = \frac{R_s R_2}{R_5 R_1}\left(I_s - I_{pv}\right) \qquad (12)$$

Defining a scale down gain, k_{dn}, as

$$k_{dn} = \frac{R_s R_2}{R_5 R_1}, \qquad (13)$$

the diode model current is given by

$$I_{dm} = k_{dn}\left(I_s - I_{pv}\right). \qquad (14)$$

The maximum diode model current occurs for $I_{pv}=0$.

$$I_{dm} = k_{dn} I_s. \qquad (15)$$

This expression is used to choice the signal diodes to compose the model.

The voltage scale up gain, k_{up}, is defined by

$$V_{pv} = k_{up} V_{dm}. \qquad (16)$$

Where, V_{dm} the diode model voltage. Thus, for open circuit condition, the diode model voltage is defined as

$$V_{oc} = k_{up} V_{dm,max}. \qquad (17)$$

Considering the transfer of scale up in Fig. 6, k_{up} is given by

$$k_{up} = \frac{R_7}{R_6} \qquad (18)$$

The I_s imposition can be made by mechanical as potentiometer or by an external controlled voltage from an analog or digital structure. The parallel resistance, i.e., the model diode, variation can be produced by mechanical way or supported by voltage controlled resistor with MOSFET. Finally, thermal effects are produced by a thermal coupled resistor polarized, by auxiliary circuit source, in such way to increase the temperature of the model diodes. The coupling, beside a lock positioning, is increased by thermal paste with high thermal conductivity.

IV. EXPERIMENTAL RESULTS

The project of the proposed system starts by the diode choice based on its I-V characteristics. Then, it is taken account the PV panel output open circuit voltage and the short-circuit current. For the prototype developed here, the diode used was the 1N4148, whose maximum current is 300 mA. The diode model current is fixed in 20 mA to avoid self-warm-up. Meanwhile a thermal coupled resistor R_{th} with controlled current produces model temperate variations.

An initial guest of V_{oc} is 12 V and a short circuit current equal to 3 A. Operational amplifiers are the AD823N. This established, the equations presented in previous section are used to evaluate passive devices. In order to minimize losses, R_s is chosen equal to 0.1 Ω. Calculus using (8)-(18) results in: $V_{Is} = 0.3$ V; R7 = R9 = 820 kΩ; R6 = R8 = 100 kΩ; R1= R3 = R2 = R4 = 220 kΩ; R5 = 15 Ω; R10 = 470 Ω. The thermal resistor was dimensioned to produce 1 W for maximum temperature (not measured).

The results in Fig. 7 illustrate the output variables. The sweep voltage is clamped by the panel open circuit voltage. And Fig. 8 shows the effect produced in the model diodes by

the thermal resistor. Based on [26], the sweep time was chosen, equal to 25 ms.

CONCLUSIONS

Dealing with the development of an analog PVE, this paper presented the theory to understand the proposal, a design tutorial and experimental results. The performance and feasibility of the proposed PV panel emulator was verified by experimental results, and it was able to achieve characteristics that are typical of PV modules, such as low electric inertia, flexibility, simplicity, and robustness, were confirmed. The simplicity of proposed circuit suggests itself the low cost while maintaining a good approximation of the solar I-V profile.

REFERENCES

[1] A. Luque Handbook of photovoltaic science and engineering. New York:Wiley; 2003.

[2] A. Reinders, A. Dijk, E. Wiemken, W. Turkenburg, "Technical and Economical analysis of grid-connected PV systems by means of simulation," Proc. of Photovoltaic. Research Applications, pp.71-82, 1999.

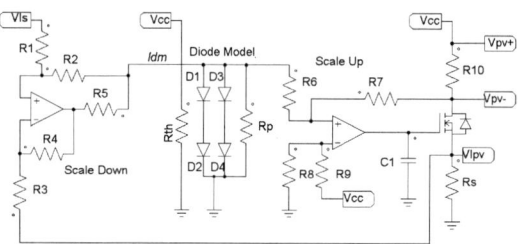

Fig. 6. Proposed PVE schematic diagram.

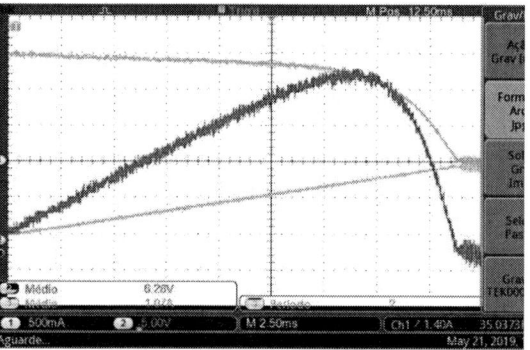

Fig. 7. PVE response. Output voltage (cyan); output current (orange) and power (red).

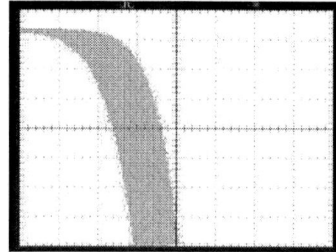

Fig. 8. Thermal response in the model diodes after the connection of thermal resistor.

[3] R. J. Prasanth, et al. "Analysis on solar PV emulators: A review," Renewable and Sustainable Energy Reviews, vol. 81, Jan 2018, pp. 149-160, 2018.

[4] A. A. Brown, N. Gammon, R. Altuwayjiri, S. Rahil, A. Ali, Samira. "An Analogue Computation based Photovoltaic Emulator for realistic Inverter Testing, " Proc. of 3rd International Conference on Energy Engineering Faculty of Energy engineering – Egypt. Dec 28-30, 2015.

[5] L. DDC, Q. Nguyen. "A photovoltaic panel emulator using a buck-boost DC/DC converter and a low cost micro-controller, " Sol Energy, vol. 86(5), pp. 1477–1484, 2012.

[6] Z. Zhou, P. M. Holland, P. Igic. "MPPT algorithm test on a photovoltaic emulating system constructed by a DC power supply and an indoor solar panel, " Energy Convers Manag, vol. 85, pp. 460–469, 2014.

[7] P. E. Marenholtz, "Programmable Solar Array Simulator," IEEE Transactions on Aerospace and Electronic Systems, vol. AES-2, no. 6, pp. 104–107, 1966.

[8] Vachtsevanos, G., & Kalaitzakis, K. "A Hybrid Photovoltaic Simulator for Utility Interactive Studies," IEEE Transactions on Energy Conversion, vol. EC-2, pp. 227–231, 1987.

[9] I. Moussa,A. Khedher, A. Bouallegue, "Design of a Low-Cost PV Emulator Applied for PVECS," Electronics, Vol. 8, pp. 232, 2019.

[10] S. Rutu, R. Ankur. "Voltage controlled buck converter based photovoltaic emulator," Proc. of the international conference on electrical, electronics, signals, communication and optimization, EESCO, pp. 1– 4, 2015.

[11] L. P. Sampaio,S. A. O. Silva. "Robust control applied to a photovoltaic array emulator using buck converter," Proc. of the international conference on renewable energies and power quality, 2015.

[12] K. Hoffman. "Real-time simulation of photovoltaic modules," Sol Energy, vol. 56, 1996.

[13] M. Cirrincione,M. C. Di Piazza, M. Pucci, G. Vitale, "Real-time simulation of photovoltaic arrays by growing neural gas controlled DC-DC converter," Proc. of the IEEE power electronics specialists conference, 2008.

[14] J. Chavarria, D. Biel, F. Guinjoan Francese, A. Poveda, F. Masana, E. Alarcon. "Low cost photovoltaic array emulator design for the test of PV grid-connected Inverters," IEEE, pp.1–6, 2014.

[15] Y. Amirthagunaraj, F. M. Fazil, M. Sawsan, P. J. Binduhewa, J. B. Ekanayake, K. M. Liyanage, A.

Atputharajah, K. Samarakoon. "Photovoltaic emulator for smart grid test facility," Proc. of the Peradeniya Univ. International Research Sessions Sri Lanka, vol. 18, pp. 156, 2014.

[16] Z. Zhou, P.M Holland, P. Igic. "MPPT algorithm test on a photovoltaic emulating system constructed by a DC power supply and an indoor solar panel," Energy Convers Manag, vol. 85, pp. 460–469, 2014.

[17] G. M. Tornez-Xavier, F. Gomez-Castaneda , C. Moreno, Flores-Nava LM. "FGPA development and implementation of a solar panel emulator," Proc. of the 10th international conference on electrical engineering, computing science and automatic control. CCE, pp. 467–472, 2013.

[18] D. C. Lu Dylan,N. Q. Nguyen Quang. "A photovoltaic panel emulator using a buck-boost DC/DC converter and a low cost micro-controller," Sol Energy, vol. 86, pp. 1477–1484, 2012.

[19] A. A. Castillo, C. J. Vazquez, O. Jaime, C. Roberto, G. J. Sandoval, C. Adrian. "A high accuracy photovoltaic emulator system using arm processors," Sol Energy, vol. 120, pp. 389–398, 2015.

[20] O. Nagayoshi. "Novel PV array/module IV curve simulator circuit," Proc. of the photovoltaic specialists conference, pp.1535–1538, 2002.

[21] O.M. Midtgard. "A simple photovoltaic simulator for testing of power electronics," Proc. of the European conference on power electronics and applications. Aalborg, pp. 1–10, 2007.

[22] D. Schofield, M.P. Foster, D. Stone. "Low-cost solar emulator for evaluation of maximum power point tracking methods," Electron Lett., pp. 3-47, 2011.

[23] S. Ahmed, A. K. Hassan. "Design and implementation of a low cost solar panel emulator," Proc. of the 42 nd IEEE photovoltaic specialist conference (PVSC), 2015.

[24] K. Wang, L. Yongdong, R. Jianye, S. Min. "Design and implementation of a solar array simulator," Proc. of the international conference on electrical machines and systems (ICEMS), pp. 2633–2636, 2008.

[25] C. Hayrettin, I. Damla, P. K. Sener. "A new numerical solution approach for the real-time modeling of photovoltaic panels," Proc. of the Asia-Pacific power and energy engineering conference, 2012.

[26] A. Sanaullah, H. A. Khan, "Design and implementation of a low cost Solar Panel emulator," Proc. IEEE 42nd Photovoltaic Specialist Conference (PVSC), 2015.

Combining Model-Based and Extremum Seeking Control for Fast Tracking the Maximum Power Point of Lundell Alternator

Gabriel Sales Lins Rodrigues
Departamento de Engenharia Elétrica
Universidade Federal de Campina
Grande
Campina Grande, Brazil
gabriel.rodrigues@ee.ufcg.edu.br

Alexandre Cunha Oliveira
Departamento de Engenharia Elétrica
Universidade Federal de Campina
Grande
Campina Grande, Brazil
aco@dee.ufcg.edu.br

Antonio Marcus Nogueira Lima
Departamento de Engenharia Elétrica
Universidade Federal de Campina
Grande
Campina Grande, Brazil
amnlima@dee.ufcg.edu.br

Abstract—Avoiding constructive changes in the Lundell alternator to avert manufacturing drawbacks and space limitations, the short-term alternative to increase power generation capabilities is to provide new degrees-of-freedom to the machine control system and take advantage of them. In order to achieve this, replacing only the passive rectifier with a semi-controlled switched-mode rectifier, a novel maximum power point tracking control strategy for a Lundell alternator is proposed. Using a detailed simulation model to act as the real alternator, it is shown that a model-based approach to maximum power point tracking is fast but may be inaccurate due to control project simplifications. It is then proposed a combination of the model-based tracking strategy with the Extremum Seeking Control real-time optimization technique – which by itself is accurate but slow – and both accurate and fast maximum power point tracking capabilities are observed, while still maintaining a simple control project. Simulation results are presented to demonstrate the control strategy feasibility and how it compares to the other studies in the subject with regard of power generated and convergence time.

Index Terms—automotive, alternator, extremum seeking control, mppt, rectifier

I. INTRODUCTION

The demand for electric power in conventional vehicles has been growing. Besides the addition of new functionalities, this is due to many devices, once driven directly by the engine, now being switched for electrically powered ones [1], [2]. Since most of the automotive fleet is still made of internal combustion engine (ICE) powered vehicles and the transition between conventional vehicles to electrical ones is still going to take some time – especially in less developed countries – there is a need for short-term solutions to supply the increasing power demand.

The generation systems of conventional ICE-powered vehicles are based on the claw-pole alternator (also known as the Lundell alternator), which is driven by a system of pulley and belt connected to the ICE. The Lundell alternator has been used for so long due to its high power density [2], robustness and low cost manufacturing [3]. On the other hand, it has an efficiency problem, with some alternators operating under

50 % [4]. Another issue is that the alternator is more efficient in lower speeds, close to idle, when there is less energy available from the primary source; and when the speed increases, the efficiency decreases [4].

While there has been some studies suggesting to remove the Lundell alternator and replace with a different type of machine, this solution would demand substantial changes in the vehicle production chain. Others have proposed some constructive changes in it, such as using permanent magnets between rotor poles to increase efficiency and power outputs [2] but it faces the same manufacturing issue. In order to avoid that, a less invasive approach is to explore the alternator controllability limitation.

In conventional systems, the alternator is connected to a three-phase diode rectifier unable to be used for control purposes. The only degree-of-freedom available to control the alternator is by means of its field circuit current, which is switched on and off to meet the battery and electrical loads voltage requirement. Because of that, it has been discussed the possibility to replace the diode rectifier for a controlled rectifier to provide new degrees-of-freedom to the system.

The most successful attempt in doing so came in [1], when the concept of "load matching", developed in [5], was used to track an alternator maximum power point (MPP) across a wide range of operating points by using a three-phase semi-controlled boost rectifier. While their control technique had an upside of having a computational low cost, needing simple experimentally obtained parameters from the machine, and fast convergence, there was some downsides. It was a static model-based approach, relying on a simplified mathematical model, which would give some approximate, but not exact answer. As it was based in machine parameters obtained experimentally, this would also contribute to diverge from the optimal point. Besides that, it could not account for thermal variations, machine aging process and, if one intended to scale up, it would be face issues with manufacturing process variations, since it is unfeasible to characterize each and every alternator produced.

978-1-7281-4181-7/19 $31.00 © 2019 IEEE

Aiming at overcoming the limits of the strategy developed by [1], [6] – and subsequently [7], using a multivariate approach – proposed a model-free alternator maximum power point tracking (MPPT) technique using the same switched-mode rectifier (SMR) topology and the same load matching concept. Their strategy was based on a combination of sliding-mode and extremum seeking control (ESC). ESC, a real-time non-model optimization technique that was very popular around 1950, had lost its relevance due to stability uncertainty. Due to formal proof of its stability given in [8], it came back to be used in engineering applications. Although the simulation results showed it delivered on accuracy, it also showed it could be slow at some situations, with convergence times being in the tenths of seconds, which can be critical, depending on the application.

In this work, a hybrid MPPT technique for Lundell alternators is proposed. It combines the fast response of the strategy developed in [1] with ESC capability to perform real-time optimization in order to reach the actual MPP accurately. Although the control approach may be different, it is also based in the load matching concept and uses the same semi-controlled rectifier topology. To show the deficiencies of the strategy developed by [1], besides validating the hybrid strategy and comparing it with different approaches, a detailed simulation model of the alternator and rectifier was used, including factors like magnetic saturation and non-ideal components.

The second section of this paper presents the Lundell alternator mathematical model. In the third section, it is explained the load matching concept. In the fourth section, it is discussed how to use load matching to track the alternator MPP. The fifth section presents the simulation used in this study. The sixth section discusses the problems with the model-based approach to track the alternator MPP and then the seventh section presents the ESC in order to, in the eighth section, show the combined strategy proposed in this paper.

II. LUNDELL ALTERNATOR

The Lundell Alternator is a salient-pole synchronous machine [3]. However, as it is pointed out in [9], the saliency ratio is close to unit and, therefore, it is reasonable to treat the alternator as a round rotor machine. It can be seen in Fig. 1 a simplified electric circuit of a Y-connected alternator. The field current regulator is responsible to control the field current, i_f, in order to keep the DC bus voltage, v_{dc}, at a given setpoint. This is done by switching on and off the circuit input voltage. A characteristic of these alternators is a long field circuit time constant [1].

The armature circuit may be modeled as a set of three-phase sinusoidal back emf voltages (e_a, e_b, e_c); three inductors (l_a, l_b, l_c) of equal inductance (l_s), which may be constant or, in order to represent the saturation effect, variable; and three resistors (r_a, r_b, r_c) of equal resistance (r_s). The back emfs amplitude, e_s, may be expressed as (1) [1], [10].

$$e_s = m_f \, \omega \, i_f \qquad (1)$$

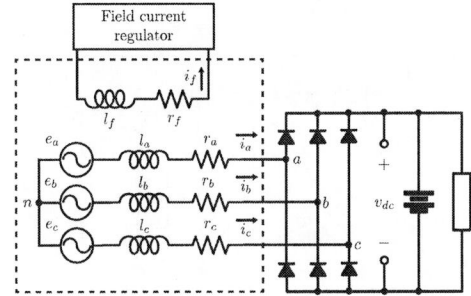

Fig. 1. Y-connected Lundell alternator circuit in conventional automotive systems.

where ω is the electrical frequency, related to the rotor mechanical frequency, ω_r, by the number of pole-pairs, p, as all synchronous machines ($\omega = p\omega_r$); and m_f is the mutual inductance between field and stator circuits. This mutual inductance, just like the phase inductances l_s, may be simplified as a constant or described as a function to replicate the saturation effect. This is all summarized in (2).

$$
\begin{aligned}
v_a(t) &= e_s \sin\left(\omega t\right) - \left(r_s i_a + l_s \frac{di_a}{dt} + i_a \frac{dl_s}{dt}\right) \\
v_b(t) &= e_s \sin\left(\omega t + \frac{2\pi}{3}\right) - \left(r_s i_b + l_s \frac{di_b}{dt} + i_b \frac{dl_s}{dt}\right) \quad (2) \\
v_c(t) &= e_s \sin\left(\omega t - \frac{2\pi}{3}\right) - \left(r_s i_c + l_s \frac{di_c}{dt} + i_c \frac{dl_s}{dt}\right)
\end{aligned}
$$

where v_a, v_b, v_c are the terminal voltages, and i_a, i_b, i_c are the phase currents. If l_s is assumed to be constant, it is possible to simplify the derivative terms in (2).

Due to the substantial power difference, the alternator's torque influence over the ICE is minimal and, therefore, alternator mechanical model may be simplified as a speed source.

III. LOAD MATCHING CONCEPT

In [5] it was shown that, for a three-phase Y-connected source coupled to a three-phase diode bridge and a constant-voltage load v_o (Fig. 2), the current i_o delivered to the load is given by (3).

$$i_o = \frac{3}{\pi} \frac{\sqrt{e_s^2 - \frac{16}{\pi^2}\left(\frac{v_o}{2} + v_d\right)^2}}{\omega l_s} \qquad (3)$$

where v_d is the diode on-voltage. Consequently, the output power p_o is given by (4).

$$p_o = \frac{3v_o}{\pi} \frac{\sqrt{e_s^2 - \frac{16}{\pi^2}\left(\frac{v_o}{2} + v_d\right)^2}}{\omega l_s} \qquad (4)$$

The circuit in Fig. 2 does not have resistive components in any of the phases. The authors point out that it is possible to take the phase resistances into account in the load current expression at the expense of simplicity. By doing that, (3) becomes a much more complex expression. It also does not account for magnetic saturation, thermal variations, and

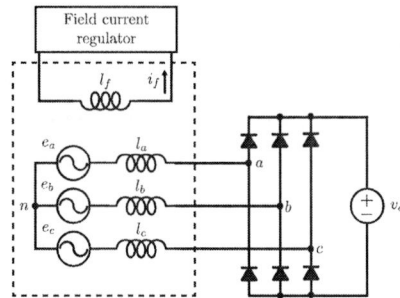

Fig. 2. Y-connected machine-like circuit coupled to a three-phase diode bridge and a constant-voltage load.

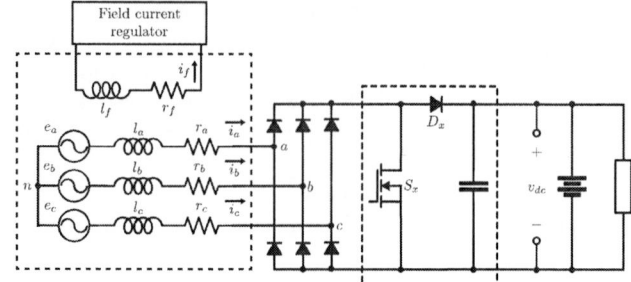

Fig. 4. Concept switched-mode rectifier to achieve load matching.

other non-idealities. Therefore, both (3) and (4) are simplified expressions and, consequently, any prediction made by using them introduces the error of not accounting for those effects. This may be an issue depending on the necessary accuracy level of the given application.

A set of curves was plotted, which may be seen in Fig. 3, using (4). It is clear from those curves that there is an optimal value of v_o that maximizes the power output. This is called "load matching" due to the fact that the load is being adjusted to match the source characteristics [1]. By taking (4), ignoring the diodes on-voltage, deriving it with respect to v_o and setting the result to zero, the optimal output voltage, $v_o^{\max}$, that maximizes the power output is given by (5).

$$v_o^{\max} = \frac{\pi e_s}{2\sqrt{2}} = \frac{\pi\,(m_f\,\omega\,i_f)}{2\sqrt{2}} \tag{5}$$

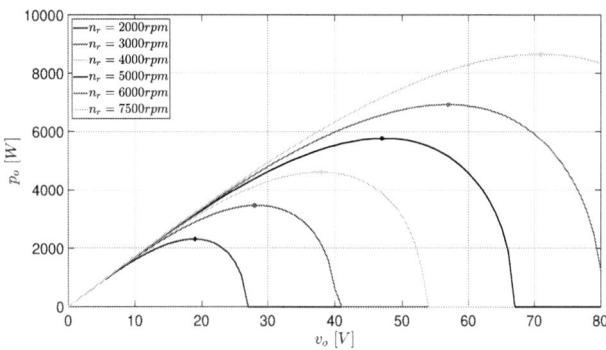

Fig. 3. Model-based alternator output power plots indexed by v_o for different rotor speeds (n_r) and maximum field current.

IV. ALTERNATOR MAXIMUM POWER POINT TRACKING

In order to track the alternator MPP in real applications, the conventional three-phase diode bridge has to be replaced by an active topology that provides some new degrees-of-freedom. The SMR topology shown in Fig. 4 was used in [1] to explain how this is possible.

By controlling S_x, switching it on and off in high frequency using a PWM signal, the voltage over the switch, v_x, goes from zero (switch on) to v_{dc} (switch off), considering both S_x and D_x are ideal switches. Therefore, by controlling the switch

duty ratio, the average value of v_x is going to be something between zero and v_{dc}. Voltage v_x, which is also the voltage seen in the rectifier output, is equivalent to v_o from Fig. 2. This means that, if v_{dc} is guaranteed to be higher than a given operating point optimal v_o, it is possible track the alternator MPP.

A. Model-Based Tracking

As it was not asserted until now, disregarding any negative effects it may cause due to saturation, and according to (4), the higher e_s is, the higher p_o is. In other words, higher frequencies (which also means, higher speeds) and higher field currents are beneficial to maximize power output[1]. Although speed is not a variable that can be controlled for this purpose, i_f may be maximized in order to achieve more power.

Based on (5), which approximates the optimal v_o, one may, based on the concept SMR operation principles, determine an analytical expression to calculate the optimal duty cycle for all operating conditions. As it was said before, the average value of v_x, $\langle v_x \rangle$, is a fraction of v_{dc} depending on the switch duty cycle d_{smr} according to (6).

$$\langle v_x \rangle = (1 - d_{\text{smr}})\,v_{dc} \tag{6}$$

By replacing v_x in (6) with (5) and then isolating d_{smr}, it is determined the approximate duty cycle which will give the optimal voltage.

$$d_{\text{smr}}^{\max} = 1 - \frac{\pi}{2\sqrt{2}} \frac{m_f\,\omega\,i_f}{v_{dc}} \tag{7}$$

The downside of using (7) alone to determine the duty cycle is the lack of accuracy. The mathematical analysis that led to (7) is quite simplified in terms of having no winding resistance, no saturation effect, no thermal variations, ideal switches. Besides that, there is inaccuracy in the experimental data needed, such as the determination of parameter m_f. Incorporating all of these non-idealizations to the model would make it way more complex and may even require sensors that aren't present in a vehicle. Another option would be to conduct an extensive experimental step going through several

---

[1]Even though the internal machine reactance follows an inverse relation with regard to the output power, and the higher the frequency, the higher the reactance, this contribution is minimal when compared to the back emf. Therefore, the power balance with a higher speed is still positive

operating situations to try and map the model behavior by using a neural network or some data-driven model. Whichever case one chooses to pick, it does not scale up well because it is unfeasible to do it for every alternator being produced. This way, if tests are conducted for only a few, the other ones are still subject to manufacturing variation. Even if this was not true and all of the past steps were completely viable, it would still not account for environment exposure and device aging and, after some time, the control would be deviating from optimal.

V. SIMULATION MODEL

In order to show the inaccuracy when compared to a more detailed system representation, a simulation model was built using MATLAB®/Simulink® and its Simscape™ toolbox. The simulation was created similar to [3] and the alternator characteristics were extracted from the same study. The alternator is a Bosch LiX 180 A, 2.4 kW and, differently from what was used so far in this work, it has a delta-connected stator winding. Its crankshaft to alternator pulley ratio is 1:2.5, which gives a speed range of operation of 2000–7500 rpm, since a typical light duty vehicle has an ICE speed range of 1000–3000 rpm. The maximum field current is around 5 A. This simulation includes resistance in both field and stator circuits, stator winding resistance variation due to thermal effects, rotor-stator coupling saturation, phase inductance saturation, field circuit dynamics, non-ideal switches – piecewise linear diode models, as in [3], and MOSFETs modeled by the Shichman-Hodges equation – and sensors.

First of, the m_f parameter is treated as function of i_f, given by a fitted curve displayed in Fig. 5. The same goes for the inductance l_s that can be seen in Fig. 6. The field circuit resistance , r_f, is set to 1.9 Ω at 20 °C. For the stator windings, knowing that their typical operating temperature is in the range of 120 °C–180 °C [11], the resistance is modeled considering the thermal characteristics for copper, and is given by (8).

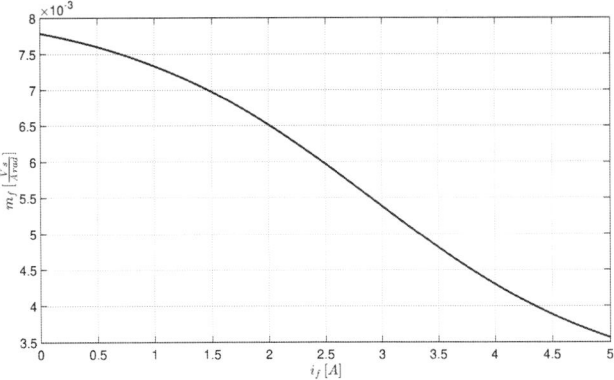

Fig. 5.   Altenator rotor-stator mutual inductance function.

$$r_s = 0.03 \left(1 + 0.0068 \left(T - 20\right)\right) \tag{8}$$

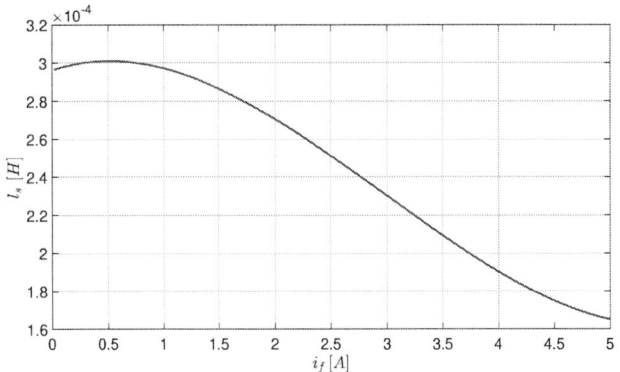

Fig. 6.   Altenator stator inductance function.

where T is the winding temperature in degree Celsius; 0.03 Ω is the resistance at a reference temperature of 20 °C; and 0.0068 °C^{-1} is the copper resistivity temperature coefficient. The alternator mechanical model is simplified as a speed source for the reasons previously stated.

When applying the model-based load matching MPPT to this machine, it is important to make adjustments regarding the stator connection type. As all the mathematical development was conducted considering a Y-connected stator and e_s was the magnitude of its phase voltage, in order to a delta-connected stator be compliant with it, there has to be a scaling process to determine what would be the equivalent phase voltage if it was connected in Y. Therefore, when using (7), the m_f parameter is divided by $\sqrt{3}$. Also, as it is suggested in [1], since i_f will be maximized, it is considered constant at $i_f^{\max}$ for (7), including the evaluation of the m_f function. These simplifications are used for model-based optimal duty cycle calculation only, and not for the model simulation itself.

VI. DISCREPANCIES OF THE MODEL-BASED MPPT STRATEGY

Using the simulation model to do a parameter sweep in order to trace more realistic power curves and then comparing them to the model-based results, the inconsistency is visible (Fig. 7). Both the power expectations and, more importantly, the correspondent voltage MPP are deviating from the more realistic simulation model by, respectively, over 1 kW and 8 V, considering the case of $n_r = 7500$ rpm. It is also important to point out that these results were obtained while considering maximum field current. If that condition changes (as in i_f dynamical behavior), since the model-based approach is static, the current variation would not be considered, the saturation effects would be visible, and the results could differ even more.

In short terms of operation, thermal variations can also increase the discrepancies, while aging will do that but in the long term. Summarizing, any parameter variation that is not accounted for, contributes to the decrease in tracking capability and power output. As it is too complex or even unfeasible to add all factors, a complimentary approach of a model-free optimization is proposed to improve MPPT effectiveness.

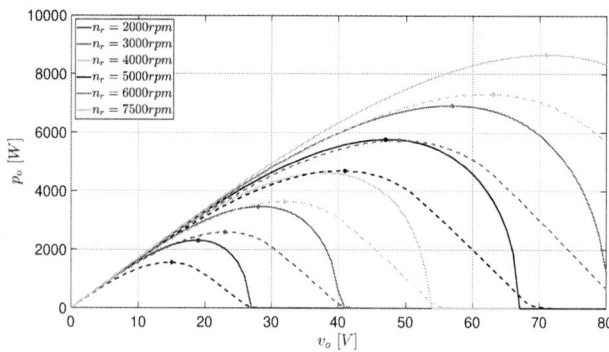

Fig. 7. Comparison between simulation power and model-based curves. The solid lines are the model-based results, while the dashed ones are from simulation.

VII. EXTREMUM SEEKING CONTROL

Extremum Seeking Control is a real-time model-free optimization strategy that was popular between 1950-1960 but was left aside due to lack of formal stability proof. Only in 2000, in [8], the authors were able to develop that stability proof and brought back some attention to the topic.

ESC is used to maximize (or minimize) in real-time a given objective function related to a physical problem. Being a model-free technique makes it useful for complex or non-linear systems. Originally proposed as a continuous-time strategy, it can also be adapted to discrete-time, paying attention to sampling times and system dynamics. The typical scheme of this control system is displayed in Fig. 8.

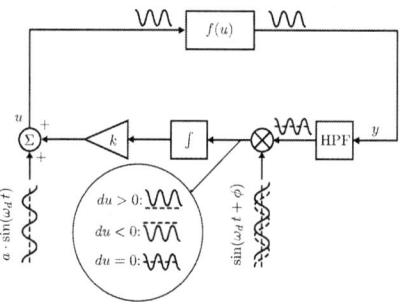

Fig. 8. Typical Extremum Seeking Control scheme with sinusoidal disturbance.

The scheme shown in Fig. 8 works in a series of steps. A given disturbance is added to the physical system input u. In this case, the disturbance is a sinusoidal with amplitude a and frequency ω_d. The sinusoidal disturbance generates a sinusoidal response in the system output. This sinusoidal effect in the system output may be filtered out, usually with a high-pass filter (HPF). Then, the filtered signal is multiplied with another sine wave of the same frequency of the original. This may be given a phase ϕ to account for a system delay. The result of this multiplication will indicate in which direction u should vary in order to maximize or minimize the system output. At the end, this multiplication result is integrated, possibly with a gain, and added to the current value of u.

ESC may also be understood as a formal and more versatile version of other widely used extremum seeking algorithms, such as Perturb and Observe. Also, it is not necessary for ESC to use sinusoidal disturbances. They only need to have a zero mean and positive mean when squared.

An ESC adaptation, combined with sliding-mode control, was presented in [6], and a multivariate version was proposed in [7], to track automotive alternators MPP based on the load matching concept. The SMR structure was used to do so. Although the results showed that the system was able to track the optimal power point, it also showed a response time that could be considered slow for certain applications, as the strategy took around half a second to converge in one of the examples [7]. Therefore, the strategy offered a trade-off, gaining in accuracy and losing in speed.

VIII. HYBRID ALTERNATOR MPPT TECHNIQUE

The combination of model-based and model-free strategies, as shown in [12], can be quite useful to bring the best of both approaches: the speed of model-based and the accuracy of model-free. Therefore, it is proposed a hybrid alternator MPPT strategy, combining the model-based approach developed in [1] and an ESC strategy. These combinations may be compared to a proportional-integral (PI) controller, where the proportional controller can't take the system exactly to the setpoint, but gets it close and fast; while the integral controller is too slow to take the whole task, but can guarantee zero steady state error.

The proposed scheme is displayed in Fig. 9. Although the concept boost-like SMR used before was useful to explain the strategy working principle, it has some disadvantages [1]. It is preferred then to use the semi-controlled three-phase SMR and command the switches all together, switching them on and off at the same time. This will give the circuit an equivalent behavior while using less switches and minimizing losses. This is the same circuit used in [1].

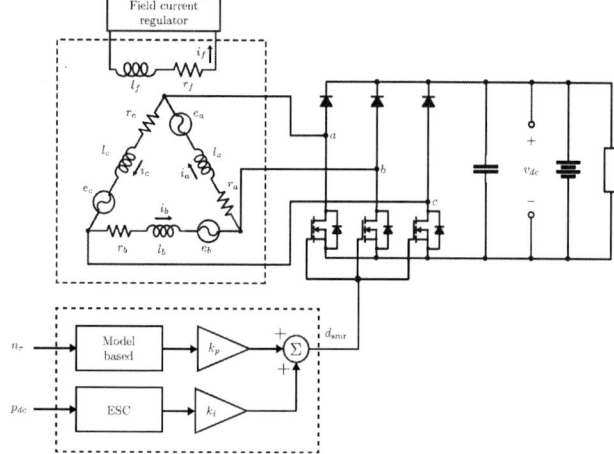

Fig. 9. Proposed hybrid MPPT scheme to fast track the alternator MPP.

The adopted switching frequency was $10\,\text{kHz}$, while the system sampling frequency was $100\,\text{kHz}$. It was enough to produce good results with a digital version of the ESC. The filter designed for the ESC was a passband filter, instead of the typical high-pass filter, so it could filter out both the machine natural frequency, DC components and the switching frequency. Two digital filters, a 4th order low-pass filter and a 3rd order high-pass filter, composed the passband filter.

One important detail is that there are gains, k_p and k_i, in the output of both controllers so it is possible to control the level of influence of each one. In this case, the values for k_p and k_i were, respectively, 1.0 and 0.5.

To showcase the strategy performance, steps were applied to the rotor speed, both increasing and decreasing it, and the results were compared to the model-based strategy (equivalent to $k_p = 1$ and $k_i = 0$) for accuracy and speed. Fig. 10 shows how the hybrid technique is able to follow the maximum power point references and achieves approximately the same convergence times as the model-based strategy in a rotor speed step up from 5500 rpm to 6500 rpm. The same goes for the rotor speed step down from 6500 rpm to 5500 rpm displayed in Fig. 11. The convergence times are around 10 ms, which is a great improvement from the works in [6] and [7].

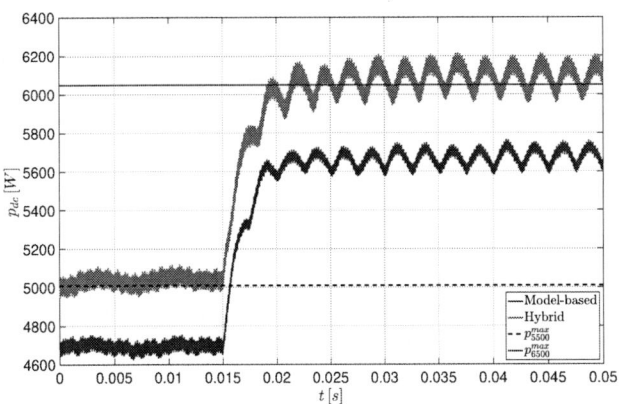

Fig. 10. MPPT strategies comparison in a speed step from 5500 rpm to 6500 rpm.

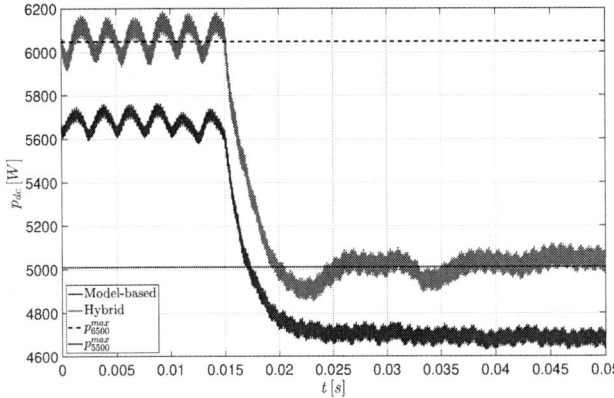

Fig. 11. MPPT strategies comparison in a speed step from 6500 rpm to 5500 rpm.

IX. CONCLUSIONS

In this paper, it was shown that the model-based approach to track the alternator MPP, although very fast, is full of inaccuracies that deviate the result from the optimal point. Considering the alternator in study, In terms of power expectation and optimal voltage, the error was, respectively, up to 1 kW and 8 V. Based on the same load matching concept and using the same SMR structure in the attempt to get fast responses but more accurately, it was proposed a combination of the model-based load matching strategy and ESC, a model-free optimization strategy. Considering [1], the proposed technique is easily implemented by changing only the control law, while the tuning steps of the additional ESC controller are also simple. The results obtained with the new technique show that both goals were reached, as the convergence times were very similar to the model-based approach – around 10 ms – while reaching the expected maximum power point, what resulted in a power increase of about 6–7 % for the evaluated scenarios.

ACKNOWLEDGMENT

The authors gratefully acknowledge the Graduate Program in Electrical Engineering – PPgEE/COPELE and CNPq for their financial support.

REFERENCES

[1] D. J. Perreault and V. Caliskan, "Automotive power generation and control," *IEEE Transactions on Power Electronics*, vol. 19, no. 3, pp. 618–630, 2004.

[2] L. Tutelea, D. Ursu, I. Boldea, and S. Agarlita, "IPM claw-pole alternator system for more vehicle braking energy recuperation," *Journal of Electrical Engineering*, vol. 12, no. 4, pp. 1–10, 2012.

[3] D. Sarafianos, R. A. Mcmahon, T. J. Flack, and S. Pickering, "Characterisation and Modelling of Automotive Lundell Alternators," no. June, pp. 928–933, 2015.

[4] R. Ivankovic, J. Cros, M. T. Kakhki, C. A. Martins, and P. Viarouge, "Power Electronic Solutions to Improve the Performance of Lundell Automotive Alternators," in *New Advances in Vehicular Technology and Automotive Engineering*. Quebec City, Canadá: INTECH Open Access Publisher, 2012, ch. 6, pp. 169–190.

[5] V. Caliskan, D. J. Perreault, T. M. Jahns, and J. G. Kassakian, "Analysis of Three-Phase Rectifiers with Constant-Voltage Loads," *IEEE Transactions on Circuits and Systems I: Fundamental Theory and Applications*, vol. 50, no. 9, pp. 1220–1226, 2003.

[6] S. F. Toloue and M. Moallem, "A Multi-surface Sliding-mode Extremum Seeking Controller for Alternator Maximum Power Point Tracking," in *Conf. 41st IEEE/IECON*. Yokohama: IEEE, 2015.

[7] S. F. Tolue and M. Moallem, "Multivariable Sliding-mode Extremum Seeking Control with Application to Alternator Maximum Power Point Tracking," in *Conf. 42nd IEEE/IECON*. Florence, Italy: IEEE, 2016.

[8] M. Krstić and H. H. Wang, "Stability of extremum seeking feedback for general nonlinear dynamic systems," *Automatica*, vol. 36, no. 4, pp. 595–601, 2000.

[9] Y. Tamto, "Determination des Paramètres d'une machine à Griffes. Application au domaine automobile." Ph.D. thesis, Institut National Polytechnique de Grenoble - INPG, 2011.

[10] D. Stoia, M. Cernat, and R. Rabinovici, "An electromagnetic model for the lundell alternator with switched-mode rectifier," in *IEEE Convention of Electrical and Electronics Engineers in Israel*. Eilat, Israel: IEEE, 2008.

[11] S. C. Tang, T. A. Keim, and D. J. Perreault, "Thermal modeling of Lundell alternators," *IEEE Transactions on Energy Conversion*, vol. 20, no. 1, pp. 25–36, 2005.

[12] L. V. Hartmann, M. A. Vitorino, M. B. d. R. Corrêa, and A. M. N. Lima, "Combining model-based and heuristic techniques for fast tracking the maximum-power point of photovoltaic systems," *IEEE Transactions on Power Electronics*, vol. 28, no. 6, pp. 2875–2885, 2013.

978-1-7281-4181-7/19 $31.00 © 2019 IEEE

Z-Source Inverter for Photovoltaic Microgeneration

Felipe A. F. Almeida
IFSP - Instituto Federal de Educação,
Ciência e Tecnologia de São Paulo
Boituva, Brasil
felipe.almeida@ifsp.edu.br

Felipe Guerra
Control and Automation Engineering
UNESP-Univ. Estadual Paulista
Sorocaba, Brasil

Flávio Alessandro Serrão Gonçalves
Control and Automation Engineering
UNESP-Univ. Estadual Paulista
Sorocaba, Brasil
flavio.as.goncalves@unesp.br

Abstract—The number of grid-connected photovoltaic systems has been growing in the world as a sustainable alternative for power generation. This paper presents a study of Z-source inverter applied to a low voltage residential photovoltaic microgeneration system. The ZSI topology is evaluated through the expressions related to its equivalent electrical circuits, with the differential of considering asymmetry and accounting for the ohmic losses due to the equivalent series resistances in the inductors and capacitors of the impedance network. In addition to the mathematical model of the ZSI converter, it is also presented a design methodology for the selection of impedance network elements. Finally, computational simulations considering the Simple Boost modulation technique are developed to evaluate the performance and effectiveness of the proposed system.

Keywords—Z-source Inverter, Grid-tie Microgeneration, Design of Impedance Network.

I. INTRODUCTION

The microgeneration power systems can be connected to the grid (on-grid) or isolated (stand-alone). Stand-alone systems usually use energy storage elements such as batteries to supply the intermittency caused by renewable sources in generation of electric power to the loads. [1]

A photovoltaic system connected to the grid basically consists of the main elements: photovoltaic module (PV), DC/AC converter and load / grid. This system can be connected to the grid in two ways: directly to the grid, where the energy is quickly drained to the system and used by the nearest consumers or through the point of common coupling (PCC), where the electric energy generated is consumed by the building itself, and only surplus is supplied to the grid. [2]

One of the disadvantages of photovoltaic systems is low efficiency and relatively high installation cost. In order to make the system more attractive there is a need to extract its maximum power. However, the photovoltaic cell presents great sensitivity to solar irradiance and temperature, modifying the maximum power point (MPP), this characteristic can be overcome through an algorithm of Maximum Power Point Tracking (MPPT) implemented in the interconnection stage of photovoltaic modules. [1]

The distributed generation based on renewable energies, using as primary sources photovoltaic cells and/or fuel cells, presents a natural characteristic of supplying only continuous voltage. Thus, distributed generation systems connected to AC grid requires a structure that acts as interface between the DC and AC bus, called the DC-AC converter, or just the inverter. [4]

In the last decades, many DC-AC converter topologies have been proposed in the literature, including the Z-Source Inverter (ZSI). [5] The ZSI was originally proposed to overcome certain limitations of conventional Voltage Source Inverter (VSI) converters, as follows:

• Voltage step-down characteristic, the input stage voltage must be higher than the output voltage, so usually a DC-DC converter is required to aid in voltage rise,

• Simultaneous conduction of the switches in same leg of inverter is not allowed, therefore time intervals should be added in control signals (dead time), which may generate undesirable distortion of the current waveform.

The ZSI converter is based on an impedance network formed by a set of inductors and capacitors. The configuration allows a new operational state called by shoot-through zero state (ST), which can be added to provide an output voltage step up behavior, without causing damages in semiconductor switches, as there is in regular null state. These characteristics can represent an advantageous solution for grid-connected photovoltaic systems, especially considering microgeneration at range of few kW. Once it provides a single conversion stage capable of controlling the photovoltaic module in the MPP and the current in the PAC, allowing the reduction of the volume and number of switches used and consequently the cost reduction and increased efficiency system. [5-7]

On the other hand, it is important to note that according with the application and the power level involved the regular solution may become more advantageous in terms of efficiency and cost than ZSI. [5-8]

In this context, this article presents a study of the ZSI converter topology applied to a low voltage residential photovoltaic microgeneration system. The ZSI topology is evaluated through expressions related to its equivalent electrical circuits, with the differential of considering ohmic losses due to equivalent series resistances – ESR, in the inductors and capacitors of the impedance network. In addition to the mathematical model of the ZSI converter, it is also presented a design methodology for the selection of impedance network elements. Finally, computational simulations are developed to evaluate the performance and effectiveness of the proposed system.

II. Z-SOURCE INVERTER (ZSI)

A. Principle of Operation

The original Z-source deals with the use of a single impedance network (consisting of an array of inductors L1 and L2 and capacitors C1 and C2 in the form of X) to create a new operational state, which can be added to provide an output voltage step up behavior. The configuration allows the elimination of an additional power stage, usually a DC-DC converter, responsible for stepping up the bus voltage in conventional converters (VSI or CSI), according to Figure 1 (a). [9]

Figure 1 (b) shows a structure of a ZSI converter fed by a DC power source and supplying energy for a three-phase three-wire load. The VSI converter has eight switching states, six active-function states where the input DC voltage is supplied at load and two null-function states where the load terminals are shorted (0 V). The additional state available in the ZSI converter is called a zero state (ST) shoot-through, it represents the concurrent conduction of the lower and upper switches in one or more inverter arms. The feature is prohibited in the regular VSI converter, because the short in DC source would damage the switches. However, the state ST allows ZSI to provide a voltage step-up and step-down characteristic. [4-7]

B. Mathematical Modeling

Several converter modeling techniques have appeared in the literature, including the Current Injected, Circuit Averaging and State Space Averaging methods. The state-space description of system dynamics is one of the modern control theories, and State Space Averaging (SSA) makes use of this description to develop averaged small signal equations of PWM-switched converters. Thus, a PWM-modulated converter that can be represented by equivalent linear circuits and its operational stages, can be modeled using SSA method. [11].

There are several works in the literature that perform ZSI modeling using SSA. [9]-[11-21] However, approaches differ mainly with respect to the equivalent circuit, state space model variable definitions, control variables, and impedance network considerations and simplifications. Considering different approaches to model the ZSI converter, this paper considers an impedance network with ohmic losses in the passive elements according to the equivalent circuit of Figure 2 (a). The circuit has two different operational states, controlled by switches S1 and S2. When S1 is ON and S2 is OFF, the converter operates in Non-Shoot-Through state (NST) during time interval T_1. When S1 is blocked (OFF) and S2 is conducting (ON), the converter operates in Shoot-Through state (ST) during the time interval T_0.

In order to evolve in SSA representation, inductor currents and capacitor voltages were established as state variables and listed as output variables. In addition, input variables were adopted as voltage and current at the input stage. Thus, it was possible to write the vectors x (t), u (t) and y (t), according to (1).

$$x(t) = \begin{bmatrix} i_{L1}(t) \\ i_{L2}(t) \\ v_{C1}(t) \\ v_{C2}(t) \end{bmatrix} u(t) = \begin{bmatrix} V_0(t) \\ i_i(t) \end{bmatrix} e\, y(t) = \begin{bmatrix} v_{C1}(t) \\ i_{L1}(t) \end{bmatrix} \quad (1)$$

Inverter can be represented as a short circuit of input terminals (link DC) while ZSI operates in ST state, according with circuit shown in Figure 2 (b). Thus, considering ZSI operating in ST state during T_0 time interval, it is possible to infer state space matrices, as presented in (2).

$$A_1 = \begin{bmatrix} -(R_{L1}+R_{C1}) & 0 & 1 & 0 \\ 0 & -(R_{L2}+R_{C2}) & 0 & 1 \\ -1 & 0 & 0 & 0 \\ 0 & -1 & 0 & 0 \end{bmatrix}$$

$$B_1 = \begin{bmatrix} 0 & 0 \\ 0 & 0 \\ 0 & 0 \\ 0 & 0 \end{bmatrix} C_1 = \begin{bmatrix} 0 & 0 & 1 & 0 \\ 1 & 0 & 0 & 0 \end{bmatrix} E_1 = \begin{bmatrix} 0 & 0 \\ 0 & 0 \end{bmatrix} \quad (2)$$

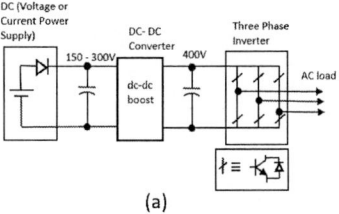

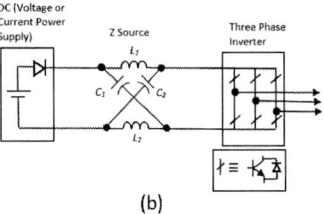

Fig. 1. (a) Two-stage conventional topology. (b) Three-phase ZSI converter powered by a fuel cell. [5]

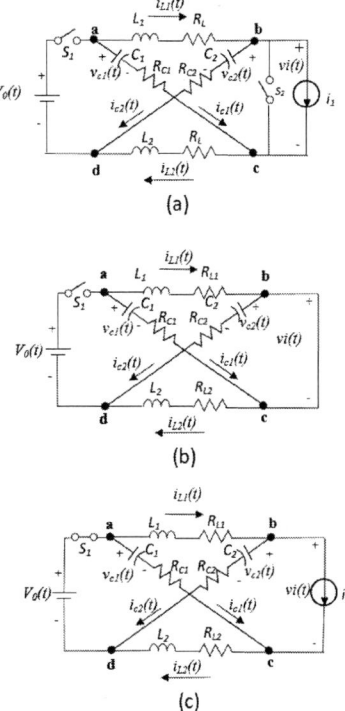

Fig. 2. Equivalent circuits: (a) ZSI converter (b), ZSI in ST state and (c) ZSI in non-shoot through state (NST).

On the other hand, while ZSI operates in non-shoot-through state (NST) the converter operates like as a regular VSI, according with circuit shown in Figure 2 (c).

Thereby, ZSI can be modeled as a current source (with zero value for traditional null states and nonzero for the other states). Considering ZSI converter operating in NST state during interval T_1, state space matrices can be obtained as written in (3). Where T is defined as a switching period ($T = T_0 + T_1$) and $D = T_0/T$ is defined as duty cycle of shoot-through.

978-1-7281-4181-7/19 $31.00 © 2019 IEEE   1470

$$A_2 = \begin{bmatrix} -(R_{L1}+R_{C2}) & 0 & 0 & -1 \\ 0 & -(R_{L2}+R_{C1}) & -1 & 0 \\ 0 & 1 & 0 & 0 \\ 1 & 0 & 0 & 0 \end{bmatrix}$$

$$B_2 = \begin{bmatrix} 1 & RC_2 \\ 1 & RC_1 \\ 0 & -1 \\ 0 & -1 \end{bmatrix} \quad C_2 = \begin{bmatrix} 0 & 0 & 1 & 0 \\ 1 & 0 & 0 & 0 \end{bmatrix} \quad E_2 = \begin{bmatrix} 0 & 0 \\ 0 & 0 \end{bmatrix}$$

(3)

The state space model of ZSI converter can be derived using SSA approach and matrices (2) and (3), as presented in (4).

$$A = \begin{bmatrix} (D-1)(R_{L1}+R_{C2})-D(R_{L1}+R_{C1}) & 0 & D & D-1 \\ 0 & (D-1)(R_{L2}+R_{C1})-D(R_{L2}+R_{C2}) & D-1 & D \\ -D & 1-D & 0 & 0 \\ 1-D & -D & 0 & 0 \end{bmatrix}$$

$$B = \begin{bmatrix} 1-D & RC_2 \cdot (1-D) \\ 1-D & RC_1 \cdot (1-D) \\ 0 & D-1 \\ 0 & D-1 \end{bmatrix} \quad C = \begin{bmatrix} 0 & 0 & 1 & 0 \\ 1 & 0 & 0 & 0 \end{bmatrix} \quad E = \begin{bmatrix} 0 & 0 \\ 0 & 0 \end{bmatrix} \quad K = \begin{bmatrix} L_1 & 0 & 0 & 0 \\ 0 & L_2 & 0 & 0 \\ 0 & 0 & L_1 & 0 \\ 0 & 0 & 0 & L_2 \end{bmatrix}$$

(4)

The steady state conditions are presented in (5), considering inverter DC link current as I_I and input DC source voltage as V_0.

$$X = (D-1) \begin{bmatrix} \dfrac{I_I}{(2D-1)} \\[2mm] \dfrac{I_I}{(2D-1)} \\[2mm] \dfrac{(2D-1) \cdot V_0 + (1-D) \cdot R_{L2} \cdot I_I + I_I \cdot D \cdot (R_{C1}+R_{C2}+R_{L1})}{(2D-1)^2} \\[2mm] \dfrac{(2D-1) \cdot V_0 + (1-D) \cdot R_{L1} \cdot I_I + I_I \cdot D \cdot (R_{C1}+R_{C2}+R_{L2})}{(2D-1)^2} \end{bmatrix}$$

(5)

In order to operate the Z-source converter in continuous current conduction mode, some operating situations should be avoided by properly designing the passive elements of the impedance network [21].

In addition, regarding ZSI model development, it is important to note that there is a difference when considering ZSI with a voltage source at its input or a PV module at its input. Assuming the use of a PV module at the input stage requires a coupling capacitor between the PV module and the ZSI loop input diode. As the PV module can be modeled as a current source in parallel to the coupling capacitor, the equivalent circuit at the input of the ZSI loop would be a current source parallel to a capacitor, changing the state space system representation [11].

C. Design of Impedance Source Network

In general, two methodologies for carrying out the impedance network design are found in the literature, differentiated by the way of considering the system in voltage mode or in current mode [11-22]. The impedance source network design methodology described in [21] considers the influence of the load, such as power factor in a voltage mode approach. On the other hand, the design methodology of [22] uses a current mode approach and the following parameters: DC input voltage V_0, AC peak voltage at the inverter output $\hat{v}_{ac}$, input source power P_{cc} and voltage ripple factors k_v and current k_i.

The relationship between the voltage boost factor (B) and shoot through duty cycle (D), resultant of the ST state, can be calculated by (6).

$$B = \frac{T_1}{T_1 - T_0} = \frac{1}{1 - 2D}$$

(6)

Using simple boost pulse width modulation strategy, the converter output peak voltage can be determined by (7) as a function of D and V_0.

$$\hat{v}_{ac} = \frac{1-D}{1-2D} \cdot V_0$$

(7)

Thus, the capacitance and inductance specifications of impedance source network can be calculated by (8) and (9), respectively.

$$C = \frac{P_{cc}(2\hat{v}_{ac} - V_0)}{4\hat{v}_{ac}k_v f_s V_0 (4\hat{v}_{ac} - V_0)}$$

(8)

$$L = \frac{\hat{v}_{ac}V_0(2\hat{v}_{ac} - V_0)}{P_{cc}k_i f_s(4\hat{v}_{ac} - V_0)}$$

(9)

Table 1 presents a summary of parameters adopted as design input data and results obtained for impedance source network elements.

The equivalent series resistances (ESR) were adopted considering capacitor datasheets information and estimated winding resistance values from physical designs of the inductors.

TABLE I. SUMMARY OF INPUT DATA AND RESULTS

Parameters	Values	Results	Values
P_{cc}	6395 W	$L_1 = L_2$	3.6 mH
V_{ac}	179.6 V	$C_1 = C_2$	1 mF
f_s	4.98 kHz	k_v	0.01
V_0	72 V	k_i	0.05
$r_{c1} = r_{c2}$	0.025 Ω	$r_{l1} = r_{l2}$	0.01 Ω

III. EXPLORATION OF PARAMETRIC VARIATION

As shown in (10) and (11), the specification of capacitance and inductance depend on several parameters (phase peak voltage $\hat{v}_{ac}$, the output voltage of the photovoltaic module V_0, of the switching frequency f_s, the voltage ripple indices k_v and current k_i e of the power of the P_{cc} system).

Thus, it is important to determine how the variations in parameters influence capacitance and inductance values determined by design methodology, and consequently modify the operating characteristics of the ZSI.

Figure 3 shows the capacitance and inductance values obtained by design methodology as a function of voltage and current ripples, respectively, and considering different switching frequencies (f_s) [1.26 kHz, 5 kHz and 12.6 kHz] and ($k_i = 0.05, k_v = 0.01, \hat{v}_{ac} = 127\sqrt{2}$ V, $V_0 = 72$ V).

A. Variation of photovoltaic module power

The maximum power supplied by the photovoltaic module is influenced by ambient conditions such as irradiation and temperature.

Figure 4 (a) shows the influence of the PV module rated power on capacitance and inductance design. The curves show that the inductance value has a decreasing nonlinear behavior with power increasing, while the capacitance value increases linearly with power increasing.

978-1-7281-4181-7/19 $31.00 © 2019 IEEE

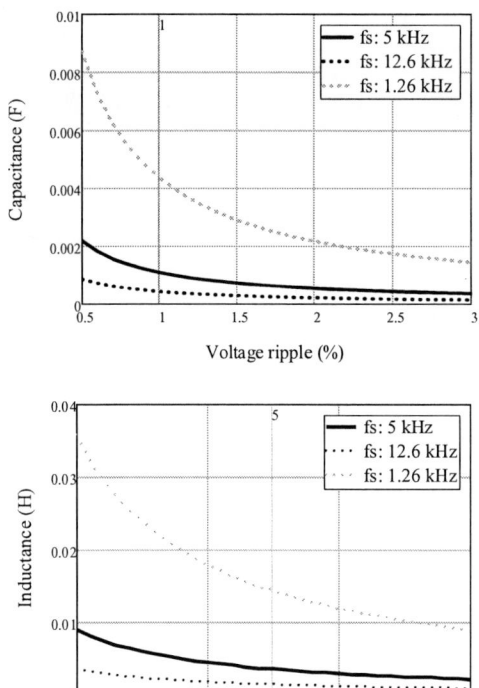

In addition, voltage and current ripples change as a function of the power supplied by the photovoltaic system. Therefore, investigation of the compliance of the ripples with the initial design specifications for a power range is necessary.

Figure 4 (b) presents curves demonstrating the influence on ripples as a function of the PV module, considering the operating values shown in Table I. In this case it can be concluded that as the power increases, the voltage ripple also increases while the current ripple decreases. Furthermore, if the maximum allowable voltage and current ripples would be 1,5% and 10%, respectively, the curves in Figure 4 (b) indicate that the system would meet the limits in the power range of 3 kW to 9.6 kW. Operations outside this power range would cause the system to operate with voltage and current ripples greater than maximum allowable values.

The rated power of the module was chosen to be 6kW under nominal solar irradiance and temperature conditions. In case of variations in solar irradiance, for example, the energy supplied by the module may decrease. The minimum allowable solar irradiance for the project is related to the lower limit of the allowable power range, specifically 3 kW, because for values below the current inductor ripple is greater than 10%.

From the analysis performed, it can be concluded that to ensure voltage ripple compliance the rated PV power must be considered. On the other hand, the current ripple has the opposite behavior. Thus, to ensure a current ripple lower than the maximum allowed, it is necessary to consider the maximum power of PV (9.6 kW). Therefore, it is important to be aware of the power range that the project can meet.

Fig. 3. Capacitance and Inductance as a function of ripple (voltage and current) for different switching frequencies.

B. Voltage variation of the photovoltaic module

The voltage delivered by the photovoltaic module as well as the power may vary depending on ambient conditions such as solar irradiance and temperature. In these situations, the MPPT algorithms must ensure that the PV operates at the point of maximum power with a given operating point.

Figure 5 shows the PV output voltage influences in the designed values of capacitance and inductance, consequently, theirs size and weight, keeping the agreement with ripple boundaries. As can be seen, as lower the PV output voltage, as higher the capacitance and the inductance required. Therefore, it is necessary to choose a certain value of the output voltage of the PV. One of the selection criteria may be to determine the maximum gain provided by the ZSI converter impedance network to meet the design voltage value of the DC bus.

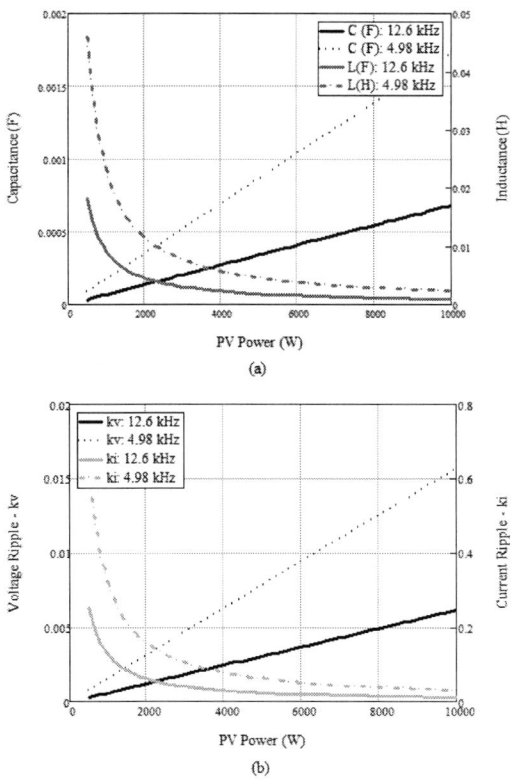

Fig. 4. Design of the impedance network of the ZSI converter (a) PV power and (b) voltage and current ripple.

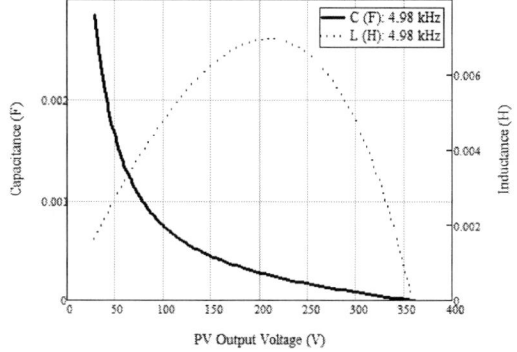

Fig. 5. Variation of PV voltage. ZSI converter impedance network design.

Considering the capacitor voltage set to 400 V and the inverter operating in the linear region on 127 Vrms three-phase system, Figure 6 shows the relationship between the shoot-through duty cycle with respect to the PV output voltage, disregarding non-idealities.

The lower the voltage available for the PV, the greater shoot-through ratio required to meet the design voltage in the capacitor. According to this analysis, the higher the allowable gain, the greater the range of voltage amplitude in the PV allowed to meet the capacitor design voltage. From this point of view, being more flexible to deal with variations due to solar irradiation and temperature.

Two situations of voltage output values from the PV module and their impacts on voltage and current ripples are presented in Figure 7. The first situation considers PV voltage at 216V leading to a shoot-through ratio of 0.315, while the second considers 72V and a shoot-through ratio of 0.41.

In both cases at the design point the established ripple limits are met. The voltage ripple curve has the behavior described by a decay curve, while the current ripple curve has shape of quadratic curve. Thus, it is possible to observe that as the PV output voltage increases the voltage ripple decreases keeping within the limits.

However, the same behavior does not occur in current ripple. For the case of lower PV voltage and higher shoot-through ratio there is a greater variation of the resulting ripple in relation to the case of higher PV voltage, leaving the established limit region. While in the case of the higher PV voltage the curve remains within the boundary compliance region.

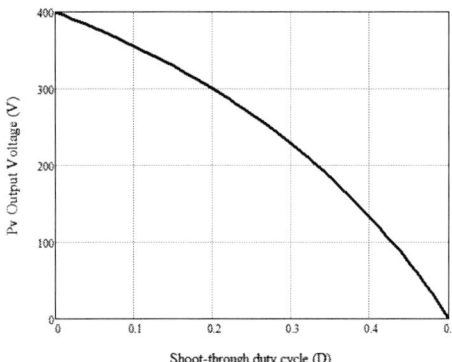

Fig. 6. PV output voltage as s function of shoot-through duty cycle

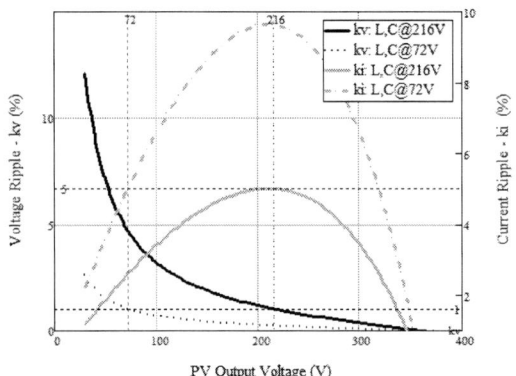

Fig. 7. Variation of PV voltage. ZSI converter impedance network design.

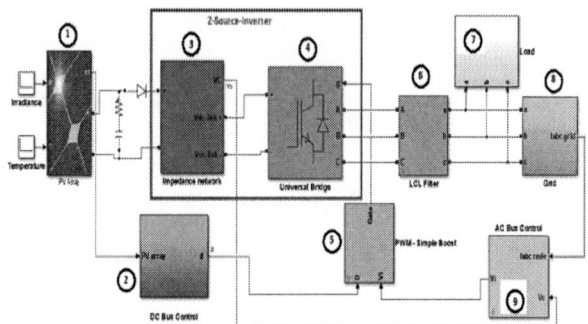

Fig. 8. Photovoltaic system model connected to the grid via a ZSI converter.

Therefore, using design configurations considering lower PV voltages and higher use of shoot-through enables the system to operate with a wider PV voltage range. Nevertheless, current ripple is affected and may limit operation with respect to meeting ripple limits and maintaining the required current conduction mode.

IV. RESULTS AND DISCUSSIONS

A. Simulink system modeling

In order to evaluate system operation, a complete microgeneration system was developed using models in the Simulink, as is shown in Figure 8. The system consists in: (1) Photovoltaic module; (2) DC bus controller with P&O MPPT algorithm; (3) Z-Source Impedance Network, (4) Three-phase Inverter; (5) PWM modulation - Simple Boost; (6) LCL Filter; (7) Three-phase load; (8) Electrical grid; (9) AC bus controller.

B. Variations of ambient conditions

The simulation analyzes system's ability to maintain operation at the maximum available power regardless of environmental conditions while injects power into the AC grid and controls the DC link voltage. Maximum power point tracking was performed using the PO (Perturb & Observe) algorithm.

Two situations are explored without AC load connected in bus, therefore, the generated power from PV modules is supplied to the grid. Initially the irradiation is maintained fixed at 1000 W / m² and the temperature is changed every 2 seconds to 25 °C, 45 °C and 10 °C. Subsequently, temperature is fixed at 25ºC and the solar irradiation is changed to 500, 800 and 100 W/m.

An overview of simulation results from microgeneration system operating under different temperature and irradiation conditions are shown in Figure 9. The results demonstrate that the control system can regulate the DC link voltage at the set value of 400V, and properly tracking the power supplied by the PV module.

Figure 9 (b) shows a detail of the simulation results considering the condition of 1000 W / m² and 45 ° C. It is possible to see the power injection to the grid provided by the ZSI-based microgeneration system, considering the current waveform with 20 A amplitude and 180 ° delay angle in relation to the voltage.

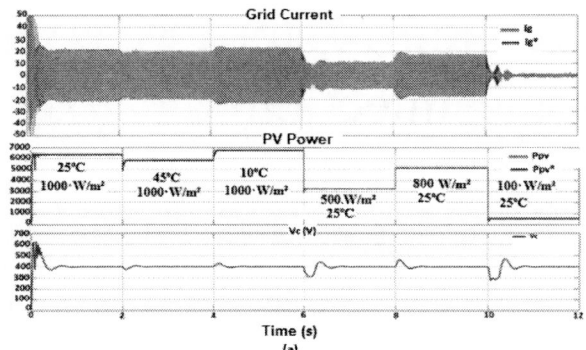

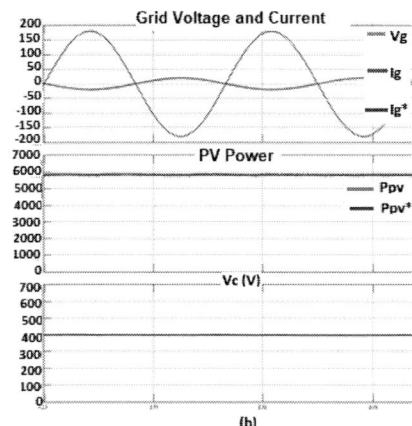

Fig. 9. (a) Grid current (Ig), PV power (Ppv) and capacitor voltage (Vc) for variations in temperature and irradiation (b) 1000 W / m² and 45°C

V. CONCLUSIONS

A Z-source inverter-based microgeneration system is presented and modeled using state space averaging approach.

Analyzes and discussions involving the specification and design of the impedance network components are carried out considering different situations (switching frequency, PV module power and PV module voltage) in order to ensure voltage and current ripple compliance.

Simulation results considering the grid connected microgeneration system are presented demonstrating its effectiveness. The control system has demonstrated the ability to adequately track the maximum power in photovoltaic modules considering different environmental conditions.

ACKNOWLEDGMENT

This Project was financially supported by Grant #2016/08645-9 and Grant #2018/24331-0, São Paulo Research Foundation (FAPESP).

REFERENCES

[1] M. G. Villalva, "Energia Solar fotovoltaica: conceitos e aplicações", 1 ed. São Paulo: Érica 2012

[2] H. Abu-Rub, M. Malinowski, K. Al-Haddad, "Power electronics for renewable energy systems" in Transportation and Industrial Applications, Hoboken, NJ, USA: Wiley, Jul. 2014.

[3] S. Z. Mohammad Noor, A. M. Omar, N. N. Mahzan and I. R. Ibrahim, "A review of single-phase single stage inverter topologies for photovoltaic system," IEEE 4th Control and System Graduate Research Colloquium, 2013, pp. 69-74

[4] S. Chaudhary and S. N. Singh, "Single Phase Grid Interactive Solar Photovoltaic Inverters: A Review," 2018 National Power Engineering Conference (NPEC), Madurai, 2018, pp. 1-6

[5] F. Z. Peng, "Z-source inverter", IEEE Trans. Ind. Appl., vol. 39, no. 2, pp. 504-510, 2003.

[6] Y. Huang, M. Shen, F. Z. Peng and J. Wang, " Z Source Inverter for Residential Photovoltaic Systems," in IEEE Transactions on Power Electronics, vol. 21, no. 6, pp. 1776-1782, Nov 2006

[7] F. Blaabjerg, C. Zhe and S. B. Kjaer, "Power electronics as efficient interface in dispersed power generation systems", IEEE Trans. Power Electron., vol. 19, no. 5, pp. 1184-1194, 2004.

[8] D.I. Brandão, F.A.S. Gonçalves, F.M. Marafão, H.K.M. Paredes. "Decoupled Reference Generator for Shunt Active Filters Using the Conservative Power Theory". Journal of Control, Automation and Electrical Systems, v. 24, p. 522-534, 2013.

[9] M. Shen, A. Joseph, J. Wang, F. Z. Peng and D. J. Adams, "Comparison of traditional inverters and Z-source inverter for fuel cell vehicles", IEEE Trans. Power Electron., vol. 22, no. 4, pp. 1453-1463, 2007.

[10] R. Erickson, Fundamentals of power electronics, Kluwer Academic Publisher, 2001.M.

[11] D. Sun et al., "Modeling, Impedance Design, and Efficiency Analysis of Quasi- Z Source Module in Cascaded Multilevel Photovoltaic Power System," in IEEE Transactions on Industrial Electronics, vol. 61, no. 11, pp. 6108-6117, Nov. 2014.

[12] P. C. Loh, D. M. Vilathgamuwa, C. J. Gajanayake, Y. R. Lim and C. W. Teo, "Transient Modeling and Analysis of Pulse-Width Modulated Z-Source Inverter," in IEEE Transactions on Power Electronics, vol. 22, no. 2, pp. 498-507, March 2007.

[13] C. J. Gajanayake, D. M. Vilathgamuwa and P. C. Loh, "Development of a Comprehensive Model and a Multiloop Controller for Z-Source Inverter DG Systems," in IEEE Transactions on Industrial Electronics, vol. 54, no. 4, pp. 2352-2359, Aug. 2007.

[14] J. Jokipii and T. Suntio, "Dynamic characteristics of three-phase Z-source-based photovoltaic inverter with asymmetric impedance network," 2015 9th International Conference on Power Electronics and ECCE Asia (ICPE-ECCE Asia), Seoul, 2015, pp. 1976-1983.

[15] C. S. Thelukuntla and M. Veerachary, "Analysis of single-phase asymmetric Z-source inverter," 2015 International Conference on Energy, Power and Environment: Towards Sustainable Growth (ICEPE), Shillong, 2015, pp. 1-6.

[16] Jingbo Liu, Jiangang Hu and Longya Xu, "A modified space vector PWM for Z-source inverter-modeling and design," 2005 International Conference on Electrical Machines and Systems, Nanjing, 2005, pp. 1242-1247, Vol. 2.

[17] J. Liu, J. Hu and L. Xu, "Dynamic Modeling and Analysis of Z Source Converter—Derivation of AC Small Signal Model and Design-Oriented Analysis," in IEEE Transactions on Power Electronics, vol. 22, no. 5, pp. 1786-1796, Sept. 2007.

[18] M. Shen, Q. Tang and F. Z. Peng, "Modeling and Controller Design of the Z-Source Inverter with Inductive Load," 2007 IEEE Power Electronics Specialists Conference, Orlando, FL, 2007, pp. 1804-1809.

[19] Cheng Ruqi, Zhao Gengshen, Guo Tianyong, Zhao Ergang, Zhao yao and Qi chao, "Modeling and state feedback control of Z-source inverter," Power Electronics and Intelligent Transportation System (PEITS), 2009 2nd International Conference on, Shenzhen, 2009, pp. 125-129.

[20] V. P. N and M. K. Kazimierczuk, "Small-Signal Modeling of Open-Loop PWM Z-Source Converter by Circuit-Averaging Technique," in IEEE Transactions on Power Electronics, vol. 28, no. 3, pp. 1286-1296, March 2013.

[21] S. Rajakaruna and L. Jayawickrama, "Steady-State Analysis and Designing Impedance Network of Z-Source Inverters," in IEEE Transactions on Industrial Electronics, vol. 57, no. 7, pp. 2483-2491, July 2010.

[22] R. Santana, "Inversor de fonte de impedância para aplicações em fontes de energia renovável," Lisboa, 2013.

IFOC for a Nine-phase Induction Motor Drive with Current Harmonic Injection

Alexandre G. F. da Silva[1], Isaac S. de Freitas[1], Member, IEEE,
Simplício A. da Silva[1], Victor F. M. B. Melo[1], Member, IEEE
and Gilielson F. da Paz[1], Member, IEEE
[1]Federal University of Paraíba (UFPB) - João Pessoa - PB - Brazil
e-mails: [alexandre.silva, isaacfreitas, sarnaud, victor, gilielson]@cear.ufpb.br

Abstract—**The multiphase machines present a lot of advantages when compared to the traditional three-phase machines, as example, for the same power, they operate with lower bus voltage and lower current per phase, present higher fault tolerance, and more specifically when it comes to concentrated windings machines, the harmonic current injection may be used to increase the torque production. Taking advantage of this last feature, in this work is developed the IFOC (indirect field oriented control) with current harmonic injection for a concentrated windings nine-phase induction motor drive, what confers the variable-speed machine operation with flux waveform optimization, reducing its peak value, giving better use to the iron core and allowing power density improvement.**

Index Terms—**IFOC, nine-phase induction machine, current harmonic injection**

I. INTRODUCTION

Traditionally, the three-phase induction machines dominate the industrial environment, since the power distribution grid is also three-phase. However, because the power electronics devices development, and the variable-speed AC drives, the phase number of power supply is not a delimiting factor for the use of multiphase machine anymore [1].

When it is compared to the three-phase machine, the multiphase machine introduces a number of advantages, as less torque pulsation and fault tolerance, because it can still working even when open-circuit or short-circuit occurs, in one or more phases [2].

Another multiphase machine advantage is the smaller current per phase than a three-phase machine with same power [2], what becomes it even more interesting in high power applications, because it inflicts less stress in the semiconductor devices, allowing the use of less robust components, and reducing the losses.

As shown in [3], the current harmonic injection in multiphase machines induces torque increment, and the non sinusoidal flux application reduce the flux peak value, promoting better exploitation from the iron core, so, multiphase machines use less iron to produce the same mechanical power than three-phase machines, reducing their weight and volume, becoming more attractive in electrical vehicle applications.

The multiphase machine drive is similar to the three-phase machine, and, usually, are divided in field-oriented control (FOC), direct torque control (DTC) and model predictive control (MPC) [2].

Multiphase machines DTC implementation examples can be found in [4] – [6], and multiphase machine MPC implementation examples can be found in [7] – [9], however, DTC has the disadvantage of the switching states number grows exponentially with the number phase increase, making its implementation complicated, while MPC has a high computational cost.

The field-oriented controller has already been richly explored in three-phase machines application [10] – [17], and it has already been implemented in five-phase machines [18] – [20]. The indirect field-oriented controller (IFOC) is one of the most used drives, for its simple implementation and do not require flux or torque calculator, being less dependent of the machine parameters [21].

In the literature is possible find IFOC implementation for nine-phase machine drive [22], but not with current harmonics injection. In this scene, an indirect field-oriented control (IFOC) with current harmonic injection for a nine-phase induction machine is proposed for a variable-speed operation with flux waveform optimization and power density improvement.

II. MATHEMATICAL DEVELOPMENT

The nine-phase machine mathematical model is already introduced in literature [23] – [25], and in the dq domain, with a generic reference frame, is given by (1) – (6).

$$\hat{\boldsymbol{\lambda}}_{sdqh}^{gh} = l_{sh}\hat{\boldsymbol{i}}_{sdqh}^{gh} + l_{mh}\hat{\boldsymbol{i}}_{rdqh}^{gh} \tag{1}$$

$$\hat{\boldsymbol{\lambda}}_{rdqh}^{gh} = l_{rh}\hat{\boldsymbol{i}}_{rdqh}^{gh} + l_{mh}\hat{\boldsymbol{i}}_{sdqh}^{gh} \tag{2}$$

$$\hat{\boldsymbol{v}}_{sdqh}^{gh} = r_s\hat{\boldsymbol{i}}_{sdqh}^{gh} + j\omega_{gh}\hat{\boldsymbol{\lambda}}_{sdqh}^{gh}\frac{d\hat{\boldsymbol{\lambda}}_{sdqh}^{gh}}{dt} \tag{3}$$

$$0 = r_{rh}\hat{\boldsymbol{i}}_{rdqh}^{gh} + j\left(\omega_{gh} - h\omega_r\right)\hat{\boldsymbol{\lambda}}_{rdqh}^{gh} + \frac{d\hat{\boldsymbol{\lambda}}_{rdqh}^{gh}}{dt} \tag{4}$$

$$c_{eh} = phl_{mh}\left(i_{sdh}^{gh}i_{rqh}^{gh} - i_{rdh}^{gh}i_{sqh}^{gh}\right) \tag{5}$$

$$C_e = \sum_{h=1,3,5,\ldots}^{H} c_{eh} \tag{6}$$

In (1) – (6), the subscript h refers to the harmonic order. $\hat{\boldsymbol{\lambda}}_{sdqh}^{gh}$, $\hat{\boldsymbol{\lambda}}_{rdqh}^{gh}$, $\hat{\boldsymbol{i}}_{sdqh}^{gh}$, $\hat{\boldsymbol{i}}_{rdqh}^{gh}$ and $\hat{\boldsymbol{v}}_{sdqh}^{gh}$ are, respectively, the stator flux, the rotor flux, the stator current, the rotor current and the stator voltage. They are rotating vectors represented

978-1-7281-4181-7/19 $31.00 © 2019 IEEE

in orthogonal complexes plans (one plan for each harmonic) whose axis is a generic rotating vector gh. The machine variables have a direct component (indicated by the subscript d) and a quadrature component (indicated by the subscript q).

The machine parameters l_{sh}, l_{mh}, l_{rh}, r_s, r_{rh} and p are, respectively, the stator self inductance, the mutual inductance, the rotor self inductance, the stator resistance, the rotor resistance and the number of pole pairs. ω_{gh} is the reference frame angular speed, ω_r is the rotor angular speed, and C_e is the electromagnetic torque.

The harmonics until the 7th order are considered ($h = 1, 3, 5$ and 7), so, the Clarke-Park transformation confers, to the machine variables, representation in 4 orthogonal plans ($dq1$, $dq3$, $dq5$ and $dq7$).

This way, the field oriented control can be implemented for each plan, separately, using similar techniques to the applied in three-phase induction machine, as in [10] and [11], moving the reference used in (1) - (6) from the generic reference frame to other reference frames to get the decoupling between torque control and flux control.

In Fig. 1 is possible to see some vectors which could be used as reference frame, as the rotor flux fundamental component vector, the rotor flux third harmonic component vector and the stationary reference.

From (2) and (4), it achieves (7), where $\tau_{rh} = l_{rh}/r_{rh}$.

$$\frac{l_{mh}}{\tau_{rh}}\hat{i}_{sdqh}^{gh} = \frac{1}{\tau_{rh}}\hat{\lambda}_{rdqh}^{gh} + \frac{d}{dt}\hat{\lambda}_{rdqh}^{gh} + j\left(\omega_{gh} - h\omega_r\right)\hat{\lambda}_{rdqh}^{gh} \tag{7}$$

Using the rotor flux vector of each harmonic component as reference frame, the flux vector is resumed to its real component and therefore

$$\lambda_{rdh}^{bh} = \lambda_{rh} \quad \lambda_{rqh}^{bh} = 0 \tag{8}$$

$$\frac{l_{mh}}{\tau_{rh}}\dot{i}_{sdh}^{bh} = \frac{1}{\tau_{rh}}\lambda_{rh} + \frac{d}{dt}\lambda_{rh} \tag{9}$$

Considering the steady state condition induction machine, the derivative term in (9) is nullified, and the stator current d component reference can be determined by (10), which relates it to the reference rotor flux.

$$i_{sdh}^{bh*} = \frac{1}{l_{mh}}\lambda_{rh}^* \tag{10}$$

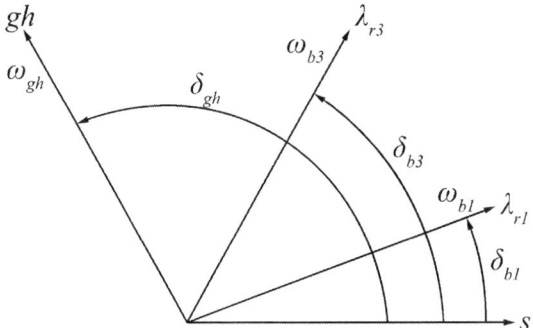

Fig. 1. Example of reference frames.

Meanwhile, in the rotor flux reference frame, (5) can be simplified in (11), and the stator current q component reference is given by (12), which relates it to the electromagnetic torque reference, characterizing the decoupling between the flux control and torque control.

$$c_{eh} = ph\frac{l_{mh}}{l_{rh}}\lambda_{rh}i_{sqh}^{bh} \tag{11}$$

$$i_{sqh}^{bh*} = \frac{l_{rh}}{phl_{mh}}\frac{c_{eh}^*}{\lambda_{rh}^*} \tag{12}$$

In order that the current control be SISO, it is required change to the stationary reference frame, but for this, it is necessary calculate the vector phase for each rotor flux component.

The vector phase of rotor flux fundamental component can be estimated by the imaginary part of (7), according to what is shown in (13) and (14). The vectors phases of rotor flux harmonics components are determined as multiples of vector phase of rotor flux fundamental component.

$$\omega_{br}^* = \frac{l_{m1}}{\tau_{r1}}\frac{i_{sq1}^{b1*}}{\lambda_{r1}^*} \tag{13}$$

$$\delta_{b1}^* = \int_0^t \omega_{br}^*\left(\tau\right)d\tau + \int_0^t \omega_r\left(\tau\right)d\tau \tag{14}$$

$$\delta_{b3}^* = 3\delta_{b1}^* \quad \delta_{b5}^* = 5\delta_{b1}^* \quad \delta_{b7}^* = 7\delta_{b1}^* \tag{15}$$

With (1) - (4) in the stationary reference frame, it is possible combine and rearrange them to come to (16).

$$\hat{v}_{sdqh}^s = \left(r_s + \frac{r_{rh}l_{mh}^2}{l_{rh}^2}\right)\hat{i}_{sdqh}^s + \left(l_{sh} - \frac{l_{mh}^2}{l_{rh}}\right)\frac{d}{dt}\hat{i}_{sdqh}^s$$
$$+ \frac{l_{mh}}{l_{rh}}\left(jh\omega_r - \frac{r_{rh}}{l_{rh}}\right)\hat{\lambda}_{rdqh}^s \tag{16}$$

Considering the term $\frac{l_{mh}}{l_{rh}}\left(jh\omega_r - \frac{r_{rh}}{l_{rh}}\right)\hat{\lambda}_{rdqh}^s$ as a perturbation, (16) can be used to design resonant PI controllers, in order to track the sinusoidal references.

Lastly, (17), which describes the machine mechanical performance (F_m and J_m are the friction coefficient and the moment of inertia, respectively), can be used to design the shaft speed PI controller. The development established in this paper results in the graph shown in Fig. 2.

$$\frac{d\omega_r}{dt} = -\frac{F_m}{J_m}\omega_r + \frac{p}{J_m}\left(C_e - C_m\right) \tag{17}$$

III. Simulation Results

The drive represented in Fig. 2 diagram was simulated with the software Matlab. It was considered a concentrated windings symmetric 40° phase-shifted nine-phase induction machine, 50 Hz. The machine parameters are presented in Table I. To allow the 3rd current harmonic injection, the machine is connected to the inverter with isolated neutral, according to Fig. 3.

As shown in Fig. 2, the rotor flux reference, as well as the electromagnetic torque reference, can be partitioned between

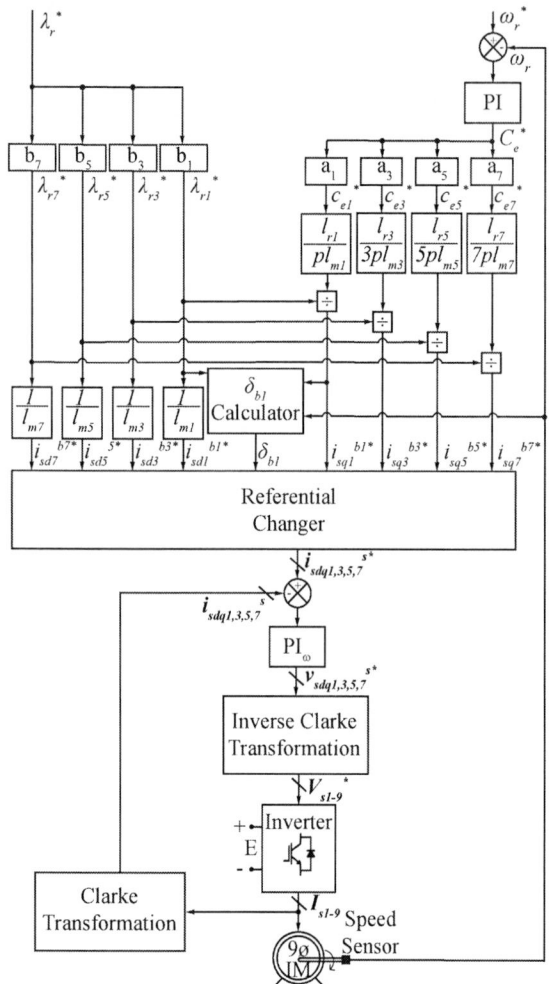

Fig. 2. IFOC for a nine-phase machine drive with harmonic injection diagram.

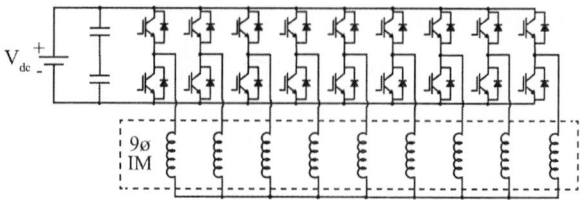

Fig. 3. Machine connection.

TABLE I
MACHINE PARAMETERS.

Parameter	Value	Parameter	Value
l_{s1}	192.1 mH	l_{r1}	1.341 μH
l_{s3}	21.0 mH	l_{r3}	1.374 μH
l_{s5}	7.6 mH	l_{r5}	1.434 μH
l_{s7}	4.3 mH	l_{r7}	1.508 μH
l_{m1}	331.0 μH	r_{r1}	6.3 $\mu\Omega$
l_{m3}	99.0 μH	r_{r3}	27.4 $\mu\Omega$
l_{m5}	47.2 μH	r_{r5}	65.4 $\mu\Omega$
l_{m7}	23.0 μH	r_{r7}	112.8 $\mu\Omega$
r_s	1.8 Ω	p	2
J_m	0.00257 Kg.m^2	F_m	0.00324 N.m/rad/s

the harmonic components, what can be made in different proportions.

Using harmonic injection, the flux waveform can takes a non sinusoidal waveform. This way, to confer better use to the iron core, the constants b_h were chosen as $b_1 = 1$, $b_3 = -0.16$, $b_5 = 0.016$, ensuring a trapezoidal waveform and reducing the

flux peak value, what allows the machine operates with higher mechanical loads, outside the saturation.

As shown in Table I, the magnetization inductance decreases as the harmonic order increase, therefor, although the flux 7th harmonic component has little influence in the flux waveform, it leads to meaning increase in the peak current value, so, the injection of 7th harmonic current component should be avoided, setting $b_7 = 0$.

Similarly, the constants a_h were chosen in a way to keeps the proportions between the harmonic components, at the same time that respects the relationship established in (6), with $a_1 = 0.8503$, $a_3 = 0.1361$, $a_5 = 0.0136$ and $a_7 = 0$.

In the simulation, the machine starts, accelerating until 0,25 s and holds constant speed until 0,5 s, when it slows down until 0,75 s, when it reaches the speed of 0 rad/s. From this point, the machine accelerates with reverse rotation until 1 s, keeping constant speed from then on.

In Fig. 4 – 5, Fig. 6 – 7 and Fig. 8 – 9, the steady state rotor flux and stator current in a phase are shown, for just the fundamental current application (1H), fundamental with 3rd current harmonic injection (1H3H) and fundamental with 3rd and 5th current harmonic injection (1H3H5H), respectively.

As expected, with the higher order harmonic injection, the flux acquires a more trapezoidal waveform, what allows to increase the effective flux, with no iron saturation. In Table II, the rotor flux peak value and the stator current peak value are related for each situation.

Using 3rd and 5th current harmonic injection, in Fig. 10 is shown the motor shaft speed, and in Fig. 11 – 14, the contribution of each harmonic current component can be observed in the dq plans.

IV. CONCLUSION

In this work, for a nine-phase induction machine, the decoupling between torque control and flux control was secured for all the dq plans, and the controllers tracked the speed and the currents, both fundamental and harmonics components, to their respectively references, therefore, it was successful in the nine-phase induction machine IFOC with harmonic injection design.

In addition, it was demonstrated that, the 3rd and 5th harmonic current injection can reduce the flux peak value up to 16 %, what allows the machine operates with higher mechanical loads with no iron saturation.

978-1-7281-4181-7/19 $31.00 © 2019 IEEE

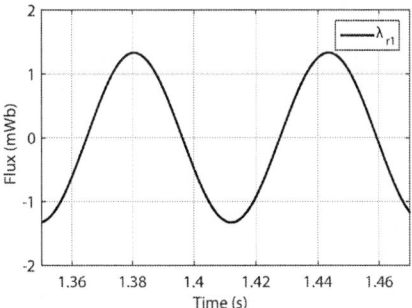

Fig. 4. Rotor flux with no harmonic injection.

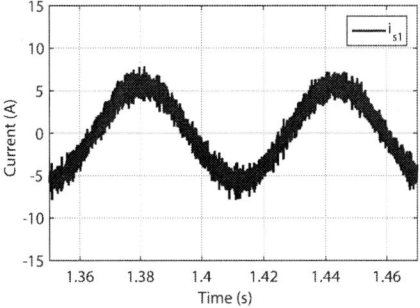

Fig. 5. Stator current with no harmonic injection.

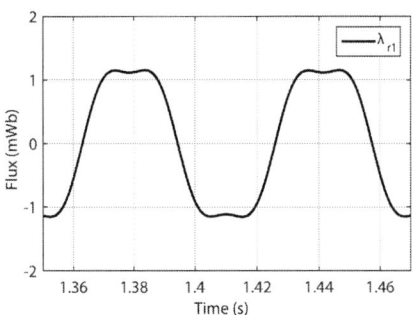

Fig. 6. Rotor flux with 3rd harmonic injection.

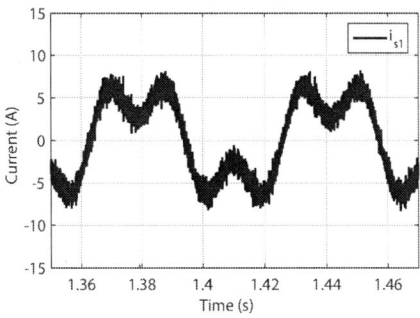

Fig. 7. Stator current with 3rd harmonic injection.

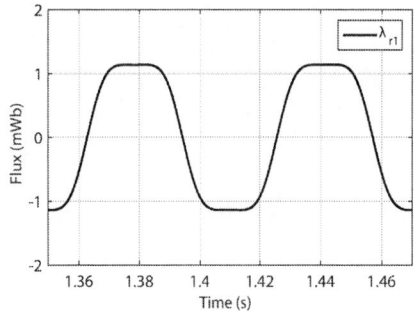

Fig. 8. Rotor flux with 3rd and 5th harmonic injection.

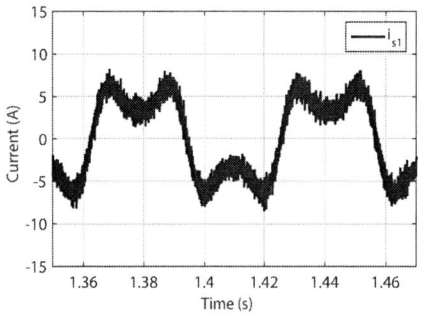

Fig. 9. Stator current with 3rd and 5th harmonic injection.

TABLE II
ROTOR FLUX AND STATOR CURRENT PEAK VALUE.

situation	λ_{rpk}	$\frac{\lambda_{rpk}}{\lambda_{rpk(1H)}}$	i_{spk}	$\frac{i_{spk}}{i_{spk(1H)}}$
1H	1.3293 mWb	1.0000	7.8553 A	1.0000
1H3H	1.1537 mWb	0.8679	8.4778 A	1.0792
1H3H5H	1.1387 mWb	0.8566	8.2384 A	1.0488

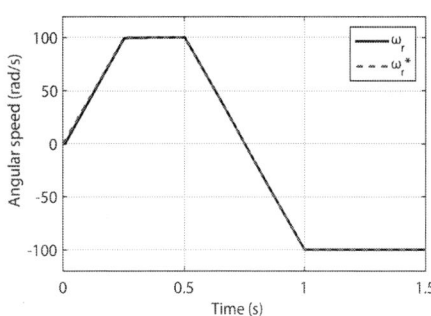

Fig. 10. Shaft angular speed.

ACKNOWLEDGMENT

The would like to thank in part by the Coordenação de Aperfeiçoamento de Pessoal de Nível Superior - Brasil (CAPES) and the Conselho Nacional de Desenvolvimento Científico e Tecnológico CNPq for the financial support.

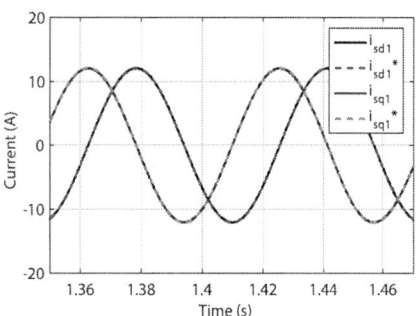

Fig. 11. Stator current $dq1$ components.

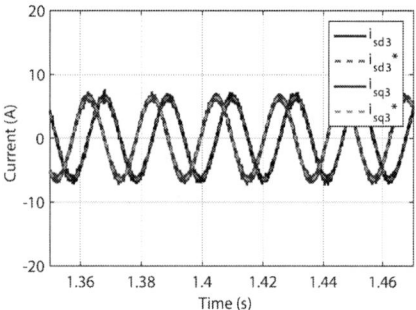

Fig. 12. Stator current $dq3$ components.

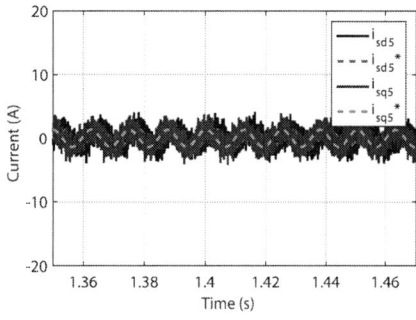

Fig. 13. Stator current $dq5$ components.

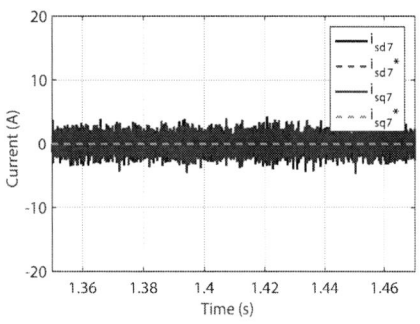

Fig. 14. Stator current $dq7$ components.

REFERENCES

[1] E. Levi, "Multiphase electric machines for variable-speed applications" IEEE Transaction Industry Electronics., vol. 55, n. 5, pp. 1893–1909, May 2008.

[2] Z. Liu, Y. Li, Z. Zheng, "A review of drive techniques for multiphase machines," CES Transactions on Electrical Machines and Systems., vol. 2, n.2, pp. 243–251, Jun 2018.

[3] R. O. C. Lyra, T. A. Lipo, "Torque density improvement in a six-phase induction motor with third harmonic current injection," IEEE Transactions on Industry Applications, vol. 38, n. 5, pp. 1351–1360, Sep/Oct 2002.

[4] H. A. Toliyat, H. Shu, "A novel direct torque control (DTC) method for five-phase induction machines," APEC 2000. Fifteenth Annual IEEE Applied Power Electronics Conference and Exposition (Cat. No.00CH37058), New Orleans, LA, USA, Feb 2000.

[5] L. Parsa, H. A. Toliyat, "Sensorless direct torque control of five-phase interior permanent magnet motor drives," IEEE Transactions on Industry Applications, vol. 43, n. 4, pp. 952–959, Jul 2007.

[6] U. Mahanta, D. Patinaik, B. P. Panigrahi, A. K. Panda, "Dynamic modeling and simulation of SVM-DTC of five phase induction Mmotor," 2015 International Conference on Energy, Power and Environment: Towards Sustainable Growth (ICEPE), Shillong, India, Jun 2015.

[7] R. Gregor el al. "Predictive-space vector PWM current control method for asymmetrical dual three-phase induction motor drives," IET Electric Power Applications, vol. 4, n. 1, pp. 26–34, Jan 2010.

[8] F. Barrero et al, "An enhanced predictive current control method for asymmetrical six-phase motor drives," IEEE Transactions on Industrial Electronics, vol. 58, n. 8, pp. 3242–3252, Aug 2011.

[9] J. A. Rivero, F. Barrero, E. Levi, M. J. Durán, S. Toral, M. Jones, "Variable-speed five-phase induction motor drive based on predictive torque control," IEEE Transactions on Industrial Electronics, V. 60, n. 8, pp. 2957–2968, Aug 2013.

[10] R. W. De Doncker, D. W. Novotny, "The universal field oriented controller," IEEE Transactions on Industry Applications, vol. 30, n. 1, pp. 92–100, Jan/Fev 1994.

[11] X. Xu, R. De doncker, D. W. Novotny "A stator flux oriented induction machine drive," , PESC '88 Record., 19th Annual IEEE Power Electronics Specialists Conference, pp. 870–876, Kyoto, Japan, Apr 1988.

[12] G. Wang, Y. Yu, R. Yang, W. Chen, D. Xu, "A robust speed controller for speed sensorless fild-oriented controlled induction motor drives," 2008 IEEE Vehicle Power and Propulsion Conference, Harbin, China, Sept 2008.

[13] P. Shrawane, "Indirect field - oriented control of induction motor," 12th IEEE International Power Electronics Congress, San Luis Potosi, Mexico, Aug 2010.

[14] A. Kumar, T. Ramesh, "Direct field oriented control of induction motor drive," 2015 Second International Conference on Advances in Computing and Communication Engineering, Dehradun, India, May 2015.

[15] Y. Liu, G. Tau, H. Wang, F. Blaarbjeg "Analysis of indirect rotor field oriented control-based induction machine performance under inaccurate field-oriented condition," IECON 2017 - 43rd Annual Conference of the IEEE Industrial Electronics Society, Beijing, China, Nov 2017.

[16] Y. Wang, Z. Zhang, F. Gao, "A comparative investigation of SDMPC, FOC, and Bang-bang control for induction motor drives," 2018 IEEE International Power Electronics and Application Conference and Exposition (PEAC), Shenzhen, China, Nov 2018.

[17] B. K. Nisahd, R. Sharma, "Induction motor control using modified indirect field oriented control," 2018 8th IEEE India International Conference on Power Electronics (IICPE), Jaipur, India, Dec 2018.

[18] H. Xu, H. A. Toliyat, L. J. Petersen, "Rotor field oriented control of five-phase induction motor with the combined fundamental and third harmonic currents," APEC 2001. Sixteenth Annual IEEE Applied Power Electronics Conference and Exposition (Cat. No.01CH37181), Anaheim, CA, USA, Mar 2001.

[19] F. Wilczyński, M. Morawiec, P. Strankowski, J. Guziński, A. Lewicki, "Sensorless field oriented control of five phase induction motor with third harmonic injection," 2017 11th IEEE International Conference on Compatibility, Power Electronics and Power Engineering (CPE-POWERENG)," Cadiz, Spain, Apr 2017.

[20] P. Zhu, M. Qiao, Y. Wei, Y. Xia, "Research on five-phase induction motor system control with third harmonic current injection," The Journal of Engineering, vol. 2017, n. 13, pp. 2559–2563, 2017.

[21] B. Bose, "Power Electronics and AC drives," Englewood Cliffs, N Prentice-Hall, 1986.

[22] M. Sowmiya, G. Renukadevi, K. Rajambal, "IFOC of a nine phase induction motor drive," International Journal of Engineering Science and Innovative Technology (IJESIT), vol. 2, n. 3, pp. 72–78, May, 2013.

[23] M. Ajaz, A. B. Anwar, A. K. Tanwar, B. Vashisht, "Simulation and mathematical modeling of ninePhase induction motor drive," 2015 Annual IEEE India Conference (INDICON), New Dheli, India, Dec 2015.

[24] R. R. Bastos, L. A. R. Silva, R. M. Valle, B. J. C. Filho, "Modelagem de uma máquina de indução de nove fases," Anais do XX Congresso Brasileiro de Automática, Belo Horizonte, MG, Brazil, Sep 2014.

[25] G. B. Vanderley et al. "Modelagem e acionamento de uma máquina de indução eneafásica com injeção de harmônico para aumento de conjugado eletromagnético," O XXII Congresso Brasileiro de Automática, João Pessoa, PB, Brazil, Sept 2018.

A Comparison of a Discrete-Time PI and an Indirect MPC Current Controllers for a Single-Phase Grid-Connected Inverter Operating with Distorted Grid and Significant Computation Feedback Delay

Sergio Pires Pimentel[*†], Oleksandr Husev[†‡], Dmitri Vinnikov[†], Serhii Stepenko[†‡], Lauri Kutt[†], and Jose Rodriguez[§]

[*]School of Electrical, Mechanical and Computer Engineering, Federal University of Goias (UFG), Goiania, Brazil
[†]Department of Electrical Power Engineering and Mechatronics
Tallinn University of Technology (TalTech), Tallinn, Estonia
Email: sergio.pimentel@ieee.org, oleksandr.husev@ieee.org, dmitri.vinnikov@ieee.org
[‡]Chernihiv National University of Technology, Chernihiv, Ukraine
[§]Universidad Andrés Bello (UNAB), Santiago, Chile

Abstract—In this paper, one technical gap related to the indirect model predictive control (MPC) is discussed. This technical gap corresponds to the performance of such a controller to deal with distorted grid voltages and a significant computation delay through the feedback regulation loop. A basic case study involving a traditional 3-level h-bridge and a first-order output filter was considered in order to magnify the effect of time horizon prediction and voltage over the regulated grid current waveform. A comparison related to the covered MPC controller (in two different forms) and a classical discrete-time proportional-integral (PI) controller is performed by means of the achieved total harmonic distortion (THD) values. Simulation results from a model built on PSIM/Powersim are presented and discussed. It can be concluded that the MPC controller offers a wider operating range with stable conditions and lower measured values of THD from regulated grid current. And such kind of special feature becomes even more evident if a significant computation delay at the feedback control loop must be overtaken.

Index Terms—Predictive Control, Discrete-time Systems, Power Electronics, Industrial Applications, Distributed Power Generation.

I. INTRODUCTION

Different structures of the control systems have been proposed for grid-connected inverters. The linear current controllers such as Proportional-Integral (PI) based, Proportional-Resonant (PR) based or Repetitive Controller (RC) based have a wide usage. It can be underlined that the main problems concern stability of the controller at a significant computational delay and a limited range of harmonic compensation. For instance, each of harmonics requires an additional control loop (in the case of a PR controller).

The achieved progress observed through last two decades on new industrial electronic devices, digital microntrollers and dedicated digital signal processors (especially in terms of time running, memory capacity, and feasibility of coding in blocks or high-level code language) can be associated with

a more intense interest for new discrete-time feedback control methods in Power Electronics and other industrial applications.

Among those observed improvements, a feedback system based on model predictive control (MPC) has been established as a more simple and robust control method in terms of perturbation rejection. It provides better results on simulation and experimental evaluations compared to other conventional methodologies (for instance, linear and PR controllers) [1]–[4]. However, it is natural that there are some technical issues regarding MPC control method that are not still so high explored or solved. One of these technical issues can be associated with the receding horizon and the cost function composition in the case of grid distorted or low-price fast running microprocessors.

In parallel, it cannot be ignored how many grid-connected PV systems have been installed over the last years, especially in terms ac low-voltage single-phase power systems. There are a lot of well-known features regarding their dynamics, benefits and drawbacks related to feedback operation control of power converters and its interactions with grid-utility [5]–[7]. Renewable energy sources are increasingly important in the actual electrical system. European Commission set up a target to raise the share of renewable energy to 20% before 2020 [8].

Considering both scientific and industrial contexts, this paper is devoted to the indirect MPC study for an inverter connected to the distorted grid based on the complex control system. This control system consists of a combination of a Field-Programmable Gate Array (FPGA) and a Digital Signal Processing (DSP) controller, which in turn leads to an additional computational delay for data exchange and needs a specific approach in the MPC application.

An additional current controller based on discrete-time PI form was considered as reference for comparing performances from both current controller. And for evaluating the perfor-

978-1-7281-4181-7/19 $31.00 © 2019 IEEE

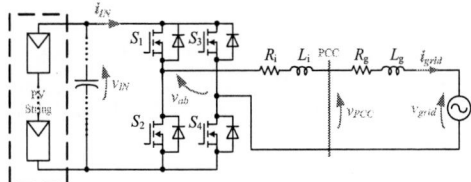

Fig. 1. PV grid-connected system based on a single-phase h-bridge inverter.

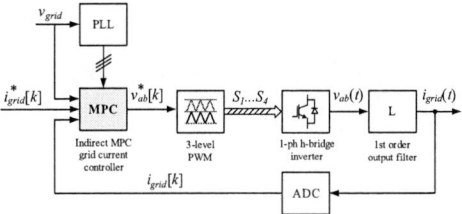

Fig. 2. Control diagram of the indirect model-predictive grid current controller (MPC) adapted from [9].

mance of such different settings, the value of THD (Total Harmonic Distortion) observed on grid current was considered as the key parameter for such kind of comparison. The main aim is simple: to quantify the impact of each grid current controller in terms of the achieved value of THD.

Next sections provide a description from the considered case of study and both methodologies for grid current controlling, followed by a discussion of results and conclusions.

II. Case of Study

In order to magnify the effect of time horizon prediction and voltage over the regulated grid current waveform, a traditional (and simple) inverter topology was applied. The case of study covered by this work is presented in Fig. 1.

As expected, the case of study consists of a classical single-phase h-bridge three-level inverter fed by a PV string and connected to a low voltage (distribution level) grid-utility through a first order filter here composed by the terms R_i and L_i. In addition, a first order L filter was considered instead of a second order LCL filter for reducing the impact of grid voltage distortions into grid current through any resonance condition. Other specification values from system shown in Fig. 1 are described in Tab. I. This work dealt with an European electrical application and, for this reason, different values to nominal grid frequency and voltage from the Brazilian standards have been considered.

TABLE I
Considered Specifications for the Case of Study

Dc input voltage v_{IN} (nominal, STC)	400 V
Filter resistance R_i	0.05 Ω
Filter inductance L_i	2.60 mH
Estimated grid resistance R_g	0.01 μΩ
Estimated grid resistance L_g	400 μH
Grid-utility nominal voltage (L1-L0)	230 V
Grid-utility fundamental frequency	50 Hz
Switching frequency (devices) f_{PWM}	65 kHz
Sampling frequency $1/T_s$	7.6 kHz

Power semiconductors devices from Fig. 1 have a switching pattern defined by a (once again) traditional modulation strategy based on the Pulse Width Modulation (PWM) with 3 levels of range. Basically, voltage deviations (or small voltage steps) are added to inverter voltage v_{ab} according to the grid voltage available at PCC (Point of Common Coupling) v_{PCC} and reference for grid current i_{grid}^*. Such voltage deviations

are responsible for modulation the inverter output voltage in order to inject a specific amount of power through the L filter.

III. Grid Current Control

A. Indirect MPC controller

In order to regulate the injection of power into the grid-current, it was considered a current control based on the indirect MPC controller proposed by [9]. Its original structure was adapted in order to become more related with the case study shown in Fig. 1. Therefore, the current control loop presented in Fig. 2 was used by this work.

Based on the conclusions described in [10], this work considered the cost function $J[p,d]$ described in (1) for composing the indirect continuous-control set MPC (CCS-MPC) grid current control method.

$$J[p,d] = \left\| i_{grid}^*[p] - i_{pred}[p,d] \right\| \quad (p \geq k) \quad (1)$$

The function $J[j,d]$ also quantizes estimations of different grid current errors. And each of them can be associated with a possible scenario d, which it is expected to occur at instant pT_s and result on a predictive grid current value $i_{pred}[p,d]$.

Based on the optimization procedure (minimum value) applied to function $J[p,d]$ a possible suitable scenario d is selected. By it, the indirect MPC controller from Fig. 2 provides a specific reference to inverter voltage v_{ab}^* (from a finite range of voltage deviations) which is considered by the modulation strategy during elaboration of switching pattern. The composition of inverter reference voltage v_{ab}^* can be described then by (2).

$$v_{ab}^*[p,d] = v_{grid}[p] + \Delta v_{ab}^*[p,d] \quad (p \geq k) \quad (2)$$

This works considered only two ranges of voltage deviations: one with 9 parcels ($d = 9$); and another one with 21 parcels ($d = 21$). Tab. II details the voltage deviations for the case with 9 parcels ($d = 9$) and a maximum voltage mismatch between PCC voltage and output inverter equal to 40 V [9]. This maximum value corresponds to 10% of the input dc voltage applied to inverter topology.

Same as it can be observed during the quantization on a discretization process, a higher number of parcels corresponds to a lower voltage gap among consecutive deviation states. In terms of a feedback current regulation by a voltage manipulation, a higher number of parcels (or deviations) on voltage offers a more sensitive grid current control capability. By this reason, only two ranges of voltage deviations have been checked through this work. Additional details regarding the composition of reference voltage v_{ab}^* and the associated cost function composition are described in [9].

TABLE II
SCENARIOS AND VOLTAGE DEVIATIONS FOR COST FUNCTION J

Scenario	Evaluation band	$\Delta v_{ab}^*[p,d]$	$\Delta v_{ab,pu}^*[p,d]$
$d=1$	-100.0%	-40.0 V	-10.0 pu
$d=2$	-75.0%	-30.0 V	-7.5 pu
$d=3$	-50.0%	-20.0 V	-5.0 pu
$d=4$	-25.0%	-10.0 V	-2.5 pu
$d=5$	null	0.0 V	0.0 pu
$d=6$	$+25.0\%$	$+10.0$ V	$+2.5$ pu
$d=7$	$+50.0\%$	$+20.0$ V	$+5.0$ pu
$d=8$	$+75.0\%$	$+30.0$ V	$+7.5$ pu
$d=9$	$+100.0\%$	$+40.0$ V	$+10.0$ pu

B. Discrete-time PI controller

In terms of comparison, it was considered a simple and well-known discrete-time PI controller which form on z-domain is shown in (3) [11]. It is associated with a Bilinear/Tustin equivalent form of PI controller.

$$ PI(z) = \left[\frac{K_p}{2} + \frac{K_p}{T_i} \cdot \frac{T_s(z+1)}{4(z-1)} \right] \cdot Z\left\{ e^{-\tau s} \right\} \qquad (3) $$

The term τ does not correspond to a feature of system (plant). In fact, it corresponds to an accumulated time-transport delay (dominant time constant) through the regulated control loop. The term T_s corresponds to the applied sampling period.

The here applied PI controller was designed (K_p and T_i specification) through a frequency domain methodology and based on the following parameters: phase margin equals to 65 deg; cut-off frequency equal to 3.6 kHz; sampling frequency equal to 6.5 kHz (as in Tab. I); and a time-transport delay equals to five sampling periods T_s ($\tau = 5T_s$).

The considered diagram control associated with the discrete-time PI controller is equivalent to that one presented in Fig. 2 except for the replacement of the indirect MPC grid current controller by the PI controller described in (3).

IV. RESULTS

The comparison between both grid current controllers in terms of the selected case of study was evaluated by a simulation model built on PSIM/Powersim [12].

A. Computational Feedback Delay

During the digital implementation of grid current control loop it was observed an overall computational feedback time-delay. And this feature matches with one of main motivations for this work: a significant computational delay application.

It was observed that the overall computational feedback time-delay could be split into two different delays: a first

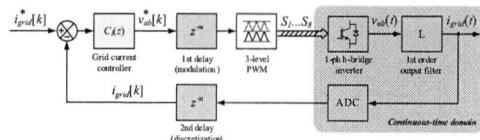

Fig. 3. Modeling (green blocks) of overall computational feedback delay observed during the digital implementation of grid current control loop.

time-delay associated with the modulation strategy; and a second one related to the discretization process of a monitored continuous-time variable signal. Fig. 3 presents the adopted modeling for both time-delays observed through feedback grid current control loop (green blocks). In terms of a simplification procedure, the first and second delays (respectively, z^{-m} and z^{-n}) were considered as integer multiple values of the T_s.

In this work, the selection of values to m and n for representing the additional computational delays for data exchange during the feedback loop was based on two points: any sampled signal takes up to three sampling periods for being allowed to be read ($n = 3$); and an second delay of two sampling periods was associated with PWM modulation ($m = 2$). It means that the total computational delay considered by this work is equal to five sampling periods ($m+n = 5$). As the implemented MPC and PI routines involve only simple commands, it can be assumed that their codes did not affect the overall computational delay because they take a neglected part of sampling period for being fully concluded.

It must be mentioned that both time-delays shown in Fig. 3 are intrinsic to this kind of application and they are not caused by any of two here compared grid current controllers (yellow block in Fig. 3). Actually, both grid controllers covered by this work must be able to deal with current regulation under such kind of drawbacks. Obviously, it is not a simple task. For instance, the here indirect MPC control investigated dealt with feedback grid current samples delayed by a factor of 3 (in discrete time-domain) for providing a future (or predicted) suitable voltage reference by a factor of 2 (in discrete time-domain). Same happened to PI controller verification but without any prediction scenario, a simple time-transport delay τ was considered as it is shown in (3).

B. Distorted Grid Voltage

A grid-utility with a typical harmonic distortion on voltage was considered. It corresponds to a low level of harmonic distortion and it attends fully the electrical policies and standards in Europe (IEC Std. 61727), in Brazil (ANEEL RN685/2015) and from USA (IEEE Std. 1547). The composition of considered distorted grid voltage is detailed in Tab. III, offering a THD value of 3.09%.

C. Transient Analysis

The starting up behavior of grid-connected system for both grid current controllers was checked considering different levels of power injected into the grid-utility. This methodology was motivated by the characteristic from grid-connected

TABLE III
COMPOSITION OF DISTORTED GRID VOLTAGE

Component	Frequency	Magnitude	Phase
Fundamental	50 Hz	325 V	0 deg
5th harmonic	250 Hz	10 V	0 deg

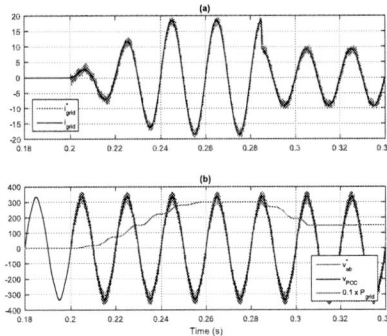

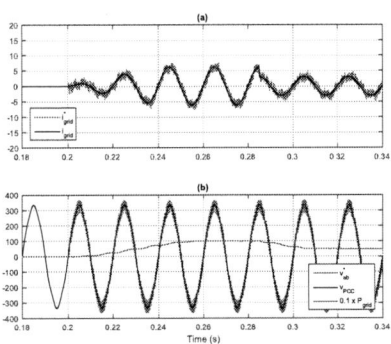

Fig. 4. PI controller: grid current signals (top) and voltage at PCC (bottom) during a startup from 0 to 3.0 kW at $t = 0.2$ s, followed by a step-down of 50% on the reference current at $t = 0.285$ s.

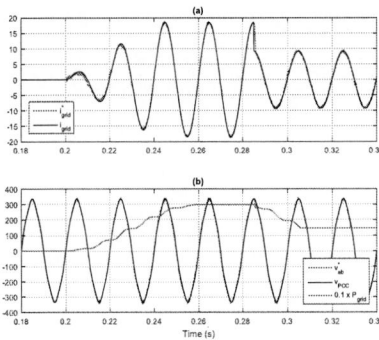

Fig. 5. PI controller: grid current signals (top) and voltage at PCC (bottom) during a startup from 0 to 1.0 kW at $t = 0.2$ s, followed by a step-down of 50% on the reference current at $t = 0.285$ s.

inverters to offer better results closed to their nominal power level. In this work, the considered nominal power level was approximately 3.0 kW.

Figs. 4 and 5 present two different power levels in the case of grid current regulation by a discrete-time PI controller. It can be observed how the PI controller is not so effective in terms of matching the grid current waveform with its reference signal. It is also related to the h-bridge topology from the inverter and the first-order output filter. Even so, it is possible to observe a stable operation during and after the startup from the system.

A few moments later, a step-down response is required at $t = 0.285$ s (according to simulated time). Grid current reference signal was suddenly reduced to half of its steady-state value. Fig. 4 and 5 show that PI controller is also suitable

Fig. 6. MPC controller ($d = 9$): grid current signals (top) and voltage at PCC (bottom) during a startup from 0 to 3.0 kW at $t = 0.2$ s, followed by a step-down of 50% on the reference current at $t = 0.285$ s.

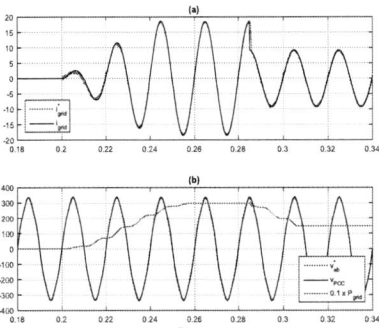

Fig. 7. MPC controller ($d = 21$): grid current signals (top) and voltage at PCC (bottom) during a startup from 0 to 3.0 kW at $t = 0.2$ s, followed by a step-down of 50% on the reference current at $t = 0.285$ s.

for recovering the stable operation point, unless the ripple turned to become a little bit higher.

Figs. 6, 7, 8, and 9 are related to the performance of an indirect MPC controller considering the same situation previously observed with PI controller. As it was mentioned in section III, there are two ranges of voltage deviations covered by this work: 9 and 21 parcels (voltage scenarios). That is the reason for duplicating the number of graphics associated with the PI controller.

From a simple comparison among Figs. 4, 6, and 7, it is possible to notice the impact of MPC current controller in terms of matching the grid current waveform with its reference signals and reducing the current ripple injected into the grid-utility. Such kind of improvement is visible right after the connection of inverter to grid-utility at $t = 0.2$ s and it persists through all the shown period of time.

As a consequence, a more stable active power injection into the grid-utility was observed by using the MPC controllers for the case from 0 to 3.0 kW. Same conclusion can be addressed to a lower power level (from 0 to 1.0 kW), as it is presented in Figs. 6 and 9. Unless it must be highlighted

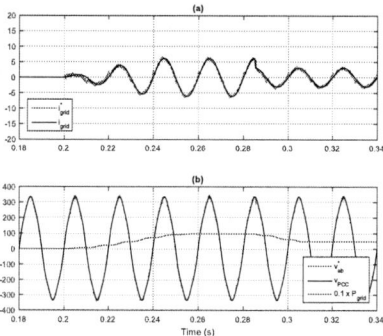

Fig. 8. MPC controller ($d = 9$): grid current signals (top) and voltage at PCC (bottom) during a startup from 0 to 1.0 kW at $t = 0.2$ s, followed by a step-down of 50% on the reference current at $t = 0.285$ s.

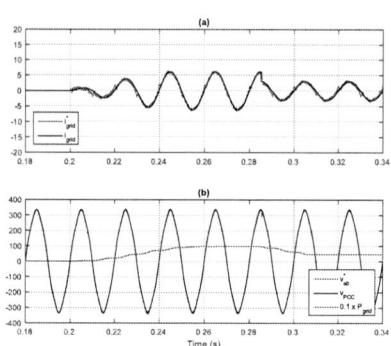

Fig. 9. MPC controller ($d = 21$): grid current signals (top) and voltage at PCC (bottom) during a startup from 0 to 1.0 kW at $t = 0.2$ s, followed by a step-down of 50% on the reference current at $t = 0.285$ s.

that an operation condition under a lower power level affect the fitting of regulated grid current with its reference signal, especially on the range of zero-crossing values. As such issue is more visible with MPC controller than with PI controller, it suggests an inaccuracy of the internal dynamic model (model-prediction) to properly deal with the 5th harmonic distortion on the PCC voltage. Some variations on the internal model were tested but they did not provide better results. Even being based on a simulation model, it suggests another investigation of zero-crossing region and the deadtime limitation of power semiconductors with MPC controllers.

Finally, as it was expected, the MPC controller based on 21 voltage deviations (Figs. 7 and 9) provided a reference voltage v_{ab}^* with less voltage gaps related to the PCC voltage than the MPC controller based on 9 voltage deviations (Figs. 6 and 8). In terms of a real digital implementation, such feature could be even more explored for reducing the amount of switching losses over each fundamental period.

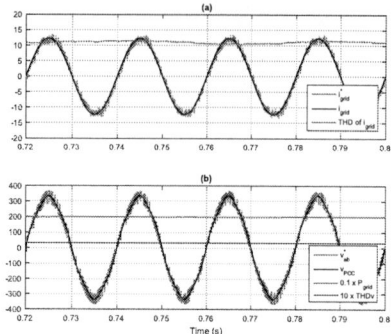

Fig. 10. PI controller: current (top) and voltage (bottom) waveforms during a stable 2.0 kW operation point (grid current with THD = 11.2 %).

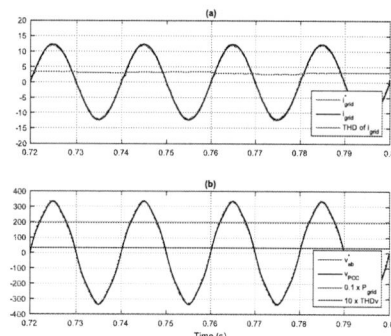

Fig. 11. MPC controller ($d = 21$): current (top) and voltage (bottom) waveforms during a stable 2.0 kW operation point (THDi = 3.4 %).

D. Steady-State Analysis

This work proposes a comparison of discrete-time PI current controller and an indirect MPC current controller. Such comparison can be quantified in terms of the capability from both current controllers to provide a regulated grid current with a low THD value. A fair comparison was performed considering the case of a power sharing between inverter and grid-utility at the same level for all here evaluated grid current controllers. The chosen power level was 2.0 kW, which corresponds to an operation condition less than nominal power (best THD values) and higher than too low power levels (worst THD values).

Figs. 10 and 11 present some results during a steady-state operation around 2.0 kW. From them, it is possible to verify that the MPC current controller based on 21 voltage deviations provided the lowest THD value. Moreover, such conclusion stills valid for an operation on wider range of power levels as it can be estimated from the achieved THD values described in Tab. IV.

It is valid to reinforce that the achieved THD values provided the MPC current controllers and presented in Tab. IV were obtained by a fixed maximum voltage mismatch between

978-1-7281-4181-7/19 $31.00 © 2019 IEEE

TABLE IV
ACHIEVED THD$_i$ VALUES BY GRID CURRENT CONTROLLERS

Power Level	PI	MPC ($d = 9$)	MPC ($d = 21$)
0.5 kW	45.8%	17.2%	16.3%
1.0 kW	23.9%	8.1%	6.7%
2.0 kW	11.2%	3.7%	3.4%
3.0 kW	7.3%	2.3%	2.2%

PCC voltage and output inverter equal to 40 V.

In order to check the dependency of indirect MPC controller to this maximum voltage, several tests were performed. For instance, considering a maximum voltage equal to 10 V (75% lower than it was considered initially) and a power level at 2.0 kW, both MPC controllers were able to provide a grid current with a THD value equal to: 3.2% ($d = 9$); and 3.0% ($d = 21$). Since these values were similar to those presented in Tab. IV for the same power level, it can be concluded that a reduction on voltage deviation would not decrease the THD value proportionally. Based on this specific test and others, it was possible to conclude two main facts: the MPC controller with highest number of voltage deviations would provide the lowest THD value; and the reducing of maximum voltage mismatch between PCC voltage and output inverter does not provide a significant improvement on the THD value at the same proportion.

Other possible conclusion based on the achieved results is related to the cost-benefit trade-off for increasing the voltage deviations on MPC controllers. In fact, an experimental implementation of higher amount of voltage deviations would require longer time-consumption routines and a quantitative analysis of such issue is outside the scope of this work. However, a qualitative analysis in terms of performance and based on the achieved results shown in Tab. IV can be performed. Such qualitative analysis indicates that even a wider gap on the number of voltage deviations (from $d = 9$ to 21) would not reduce significantly (or proportionally) the THD value.

V. CONCLUSION

In this paper, an indirect MPC current controller was compared to a discrete-time PI controller in order to regulate the grid current provided by a grid-connected single-phase inverter. Distorted grid and significant computational delay on the feedback loop was also considered as additional tasks to both current controllers. Such tasks are totally in accordance with the nowadays grid-utility (a more and more presence of nonlinear loads) and the wide availability of low-price fast-running microprocessors for digital implementations. This last feature can associate this work with a lot of low-energy industrial applications.

The performances from both current controllers were quantified in terms of the achieved THD values on regulated grid current. Based on the discussed simulation results, it can be concluded that MPC controller, in comparison with discrete-time PI controller, offer a wider operating range of stable conditions with lower values of THD from regulated grid current. And such special feature becomes even more visible if a high computation delay must be overcame, as it was expected by this work.

The analysis of obtained results do also suggest some additional investigations related to the indirect MPC controller regarding the internal model accuracy for distorted grid applications and also the power losses during an experimental digital implementation.

ACKNOWLEDGMENT

Co-authors O. Husev and D. Vinnikov was supported by the Estonian Research Council (grant PUT1443) and by the Estonian Centre of Excellence in Zero Energy and Resource Efficient Smart Buildings and Districts (ZEBE), grant 2014-2020.4.01.15-0016 funded by the European Regional Development Fund. Besides, co-author S. Stepenko was supported by the Estonian Research Council (grant MOBJD126) and co-author J. Rodriguez acknowledges the support of Conicyt through projects FB 0008 and 1170167.

REFERENCES

[1] J. Rodriguez and P. Cortes, *Predictive Control of Power Converters and Electrical Drives*. Wiley-IEEE Press, 2012.

[2] T. Geyer, *Model Predictive Control of High Power Converters and Industrial Drives*. John Wiley & Sons Inc, 2016.

[3] S. Vazquez, R. P. Aguilera, P. Acuna, J. Pou, J. I. Leon, L. G. Franquelo, and V. G. Agelidis, "Model predictive control for single-phase NPC converters based on optimal switching sequences," *IEEE Transactions on Industrial Electronics*, vol. 63, no. 12, pp. 7533–7541, dec 2016.

[4] L. R. A. Pinto, S. P. Pimentel, E. G. Marra, B. Alvarenga, T. M. Cesar, and C. K. D. Lima, "Proposal of model predictive control (mpc) method for a three-phase four-leg inverter applied in a distributed generation system," in *2017 Brazilian Power Electronics Conference (COBEP)*, Nov 2017, pp. 1–6.

[5] S. P. Pimentel, R. M. M. Martinez, and J. A. Pomilio, "Single-phase distributed generation system based on asymmetrical cascaded multilevel inverter," in *2009 Brazilian Power Electronics Conference*. IEEE, sep 2009.

[6] O. Husev, C. Roncero-Clemente, E. Makovenko, S. P. Pimentel, D. Vinnikov, and J. Martins, "Optimization and implementation of the proportional-resonant controller for grid-connected inverter with significant computation delay," *IEEE Transactions on Industrial Electronics*, pp. 1–1, 2019.

[7] S. P. Pimentel, O. Husev, D. Vinnikov, C. Roncero-Clemente, E. Makovenko, and S. Stepenko, "Voltage control tuning of a single-phase grid-connected 3l qzs-based inverter for pv application," in *2018 IEEE 38th International Conference on Electronics and Nanotechnology (ELNANO)*, Apr. 2018.

[8] "Renewables 2018 global status report," Renewable Energy Policy Network for the 21st Century, Paris, France, Tech. Rep., Jun. 2018.

[9] S. P. Pimentel, O. Husev, D. Vinnikov, C. Roncero-Clemente, and S. Stepenko, "An indirect model predictive current control (ccs-mpc) for grid-connected single-phase three-level npc quasi-z-source pv inverter," in *2018 IEEE 59th International Scientific Conference on Power and Electrical Engineering of Riga Technical University (RTUCON)*, Nov 2018, pp. 1–6.

[10] S. P. Pimentel, O. Husev, D. Vinnikov, and S. Stepenko, "Analysis of cost function composition based on the horizon time prediction of an indirect mpc current control in single-phase grid-connected pv inverters," in *2019 IEEE 60th International Scientific Conference on Power and Electrical Engineering of Riga Technical University (RTUCON)*, Oct. 2019.

[11] M. S. Fadali and A. Visioli, *Digital Control Engineering*. Elsevier Science Publishing Co Inc, 2012.

[12] Powersim, "Psim electronic simulation software," Jul. 2019. [Online]. Available: https://powersimtech.com/

978-1-7281-4181-7/19 $31.00 © 2019 IEEE

A NON-ISOLATED DC-DC BOOST CONVERTER WITH THREE-STATE SWITCHING CELL

Lucas Carvalho Souza[2], Falcondes José Mendes de Seixas[1], Luís de Oro Arenas[1], Douglas Carvalho Morais[1], Luciano de Souza da Costa e Silva[2]

[1]São Paulo State University (UNESP), School of Engineering, Ilha Solteira – São Paulo – Brazil
[2]Federal Institute of Goiás (IFG) – Jataí – Goiás – Brazil
e-mail: lucas.souza@ifg.edu.br, falcondes.seixas@unesp.br, ldeoroa@gmail.com, morais.sccp.carvalho@gmail.com,
lucianocosta_@hotmail.com

Abstract— This paper presents a boost converter based on the three-state switching cell (3SSC), which has been widely used in the design of several topologies of electronic power converters. The structure is part of a family of non-isolated DC-DC converters that has been reported previously in the literature, however, detailed modeling and analysis have not been presented before. The topology consists of type-A cell, one of the five variants of 3SSC. This structure has the advantages associated to the use of 3SSC, such as reduction of the current stress in semiconductors and distribution of losses on them. A qualitative analysis in continuous conduction mode is performed, the control pulses of the main switches being non-overlapping, the operational stages are analyzed and a design procedure is obtained. The modeling of the topology is performed, obtaining the small-signal model of the converter. Experimental results are presented and duly discussed to validate the theoretical assumptions.

Keywords— *boost converter, continuous conduction mode, converter modeling, DC-DC converters, three-state switching cell*

I. INTRODUCTION

The growing search for technological advances coupled with the increase in energy consumption has directly affected the research and development of semiconductor power components and new topologies of power converters. In this perspective, the continuous improvement of the electronic power converters, for the most diverse applications, is characterized by the requirements of lower volume, lower weight and lower cost of production. In addition to these requirements, it is also expected that these converters will have the highest possible efficiency in a wide power range, in order to attend the economic restrictions and the negative environmental effects associated to the losses during the electrical energy processing [1- 2].

Because of the challenge of finding new structures capable of attending the demands and requirements presented, non-isolated DC-DC converters based on the three-state switching cell (3SSC) were proposed for high power and high current applications [3]. Five distinct switching cells named A, B, C, D and E have been presented as different configurations that can be used to replace the two-state switching cell of the non-isolated classical converters.

Basically, the three-state cell structure consists of two controlled switches, two diodes, and an autotransformer with unit turns ratio. As noted in the literature [4-5], because of this structural feature, the use of 3SSC allows the switches to process only part of the energy transferred to the load, while the diodes process the remainder energy. In this way, there is a reduction of the current stress through each main switch, evenly dividing the losses between all of semiconductors, with the current division between the branches is guaranteed by the presence of the autotransformer with unit turns ratio. Since the losses are distributed over a larger number of components, the thermal dissipation design of the converters using the 3SSC is also facilitated. Another advantage attributed to 3SSC and its application in power electronics converters, there is the reduction of weight and volume of the energy storage elements, because they are designed to operate under twice the switching frequency.

According to the analysis performed in [6], which presents a review of several converter structures based on 3SSC, the most DC-DC converters applications employ the type-B cell, mainly because the fact that applying topologies 3SSC buck, boost, buck-boost, Ćuk, SEPIC, and Zeta present the same static gains over the entire duty cycle range of classical DC-DC topologies. In this way, the objective of this paper is to present the 3SSC boost converter using the type-A cell, proposed and described in [3], as shown in Fig. 1. The type-A cell originates from the classical isolated converter known in the literature as a voltage-fed push-pull converter, which contrasts with the traditional current-fed boost topologies, resulting in a voltage-fed boost converter.

On the other hand, the dynamic response of DC-DC converters is gaining more interest with the advent of digital and portable applications of low voltage power electronics [7]. Several studies have been focused to the design of power converters with fast dynamic performance, however, it could be a challenging task when it comes to boost power converters, since these traditionally have a slow dynamic response due to the presence of a Right Half Plane Zero (RHPZ) in its transfer function. In contrast, the 3SSC boost converter using the type-A cell has a transfer function with a minimum phase characteristic, which gives it a faster dynamic response and simpler controller design [8].

The maximum duty cycle of the 3SSC boost converter using type-A cell must be kept less than 0.5 for each transistor, preventing the possibility of simultaneous conduction of the transistors and consequent destruction of the structure. Thus, as will be shown in section II, the voltage gain of the converter in Continuous Conduction Mode (CCM) is limited to a maximum of twice the input voltage. However, even with this restriction, the proposed converter becomes interesting for applications with a high power processing, a size smaller than traditional structures, and that do not require high voltage gain.

978-1-7281-4181-7/19 $31.00 © 2019 IEEE

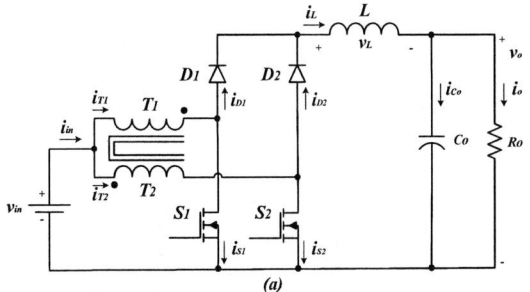

Fig. 1. Non-isolated DC-DC boost converter using the 3SSC type-A cell.

II. THE 3SSC BOOST: STATIC ANALYSIS

In this section, to explain the principle of operation, the 3SSC boost converter is analyzed in steady state. The topology of the 3SSC boost converter, using type-A cell, consists of an autotransformer with windings (T_1, T_2), two controlled switches (S_1, S_2), two diodes (D_1, D_2), an inductor L and the load R_o, connected in parallel with the output capacitor C_o. This topology has a functional limitation from the command standpoint of the switches, i.e. the converter operates only at a duty cycle less than 0.5.

A. Principle of Operation: Ideal Devices

In this paper, the converter is analyzed in continuous conduction mode, and the switch control is performed by fixed pulse width modulation, with a 180-degree delay between the command signals. The duty cycle is less than 0.5, thus characterizing the operation in non-overlapping mode, which is a behavior associated with 3SSC [4]. All converter components are considered ideal. During one switching period, the structure presents four operation stages, as shown in Fig. 2. The main theoretical waveforms are shown in Fig. 3, where each of the variables is defined as follows:

V_{GS1}, V_{GS2} – Command signals applied to switches S_1 and S_2, respectively;

$i_{in}(t)$ –   input current;

$i_{S1}(t)$ – S_1 switch current;

$i_{D1}(t)$ – D_1 diode current;

$i_L(t)$ – L inductor current,

$i_o(t)$ – R_o load current;

$i_{Co}(t)$ – C_o capacitor current;

$v_{S1}(t)$ – S_1 switch voltage;

$v_{D1}(t)$ – D_1 diode voltage;

$v_L(t)$ – L inductor voltage.

In Table I the semiconductor states are summarized for each converter operation stages.

TABLE I.   SWITCH STATES DURING A 3SSC BOOST CONVERTER SWITCHING PERIOD OPERATING IN NON-OVERLAPPING MODE AND CCM.

Semiconductor	$t_0<t<t_1$	$t_1<t<t_2$	$t_2<t<t_3$	$t_3<t<t_S$
S_1	ON	OFF	OFF	OFF
S_2	OFF	OFF	ON	OFF
D_1	OFF	ON	ON	ON
D_2	ON	ON	OFF	ON

First stage ($t_0 <t <t_1$) [Fig. 2a]: initially, the switch S_1 is turned-on, while the switch S_2 is turned-off. Since the turns

ratio of the autotransformer is unitary, the current i_{in} is divided equally between these windings. Half of the current i_{in} flows through T_1 and diode D_2 and another through T_2 and switch S_1. In this stage, the current i_L increases linearly and the inductor L stores energy. Considering that the autotransformer windings have the same impedance, the voltages on the windings T_1 and T_2 are equal to the voltage V_{in}. The voltage v_L, the current i_{in}, as well as the current i_{Co} during this stage are expressed by expressions (1), (2) and (3) respectively.

$$v_{L(t)} = L\frac{di_{L(t)}}{dt} = 2v_{in(t)} - v_{o(t)} \tag{1}$$

$$i_{in(t)} = 2i_{L(t)} \tag{2}$$

$$i_{Co(t)} = C\frac{dv_{o(t)}}{dt} = i_{L(t)} - i_{o(t)} \tag{3}$$

Second stage ($t_1 <t <t_2$) [Fig. 2b]: switch S_1 is turned-off, while switch S_2 remains turned-off. Diode D_1 is forward-biased and D_2 remains turned-off. The current flowing through T_1 and T_2, due to the effect of the autotransformer, causes a zero magnetic flux in the core. The voltage on the inductor L is reverse, and the current on it decreases linearly, transferring the stored energy and V_{in} energy to the load. The voltage v_L, the current i_{in}, as well as the current i_{Co} during this stage are expressed by expressions (4), (5) and (6), respectively.

$$v_{L(t)} = L\frac{di_{L(t)}}{dt} = v_{in(t)} - v_{o(t)} \tag{4}$$

$$i_{in(t)} = i_{L(t)} \tag{5}$$

$$i_{Co(t)} = C\frac{dv_{o(t)}}{dt} = i_{L(t)} - i_{o(t)} \tag{6}$$

Third stage ($t_2 <t <t_3$) [Fig. 2c]: due to the circuit symmetry, this stage is similar to the first, although switch S_2 is turned-on and S_1 remains turned-off. The diode D_1 continues turned-on and D_2 turned-off.

Fourth stage ($t_3 <t <t_s$) [Fig. 2d]: this stage is identical to the second stage, where the current through the inductor L flows through the diodes D_1, D_2 and the autotransformer windings.

B. Static Gain

The relation of the input voltage and the output voltage, called the static gain, is obtained by using differential equations (1) and (4), considering that the average voltage across the inductor must be equal to zero during one switching period. From this assumption, the equation (7) is obtained.

$$G_v = \frac{V_o}{V_{in}} = 1 + 2D \tag{7}$$

In Fig. 4, the static gain behavior. Observing this figure, it can be seen that the static gain of the converter as a function of the duty cycle is linear. In addition, it is verified that the maximum voltage gain of the topology is restricted to twice the input voltage, since if $D>0.5$ the transistors S_1 and S_2 conduct simultaneously, v_{T1} and v_{T2} would be null and v_{in} would be short-circuited.

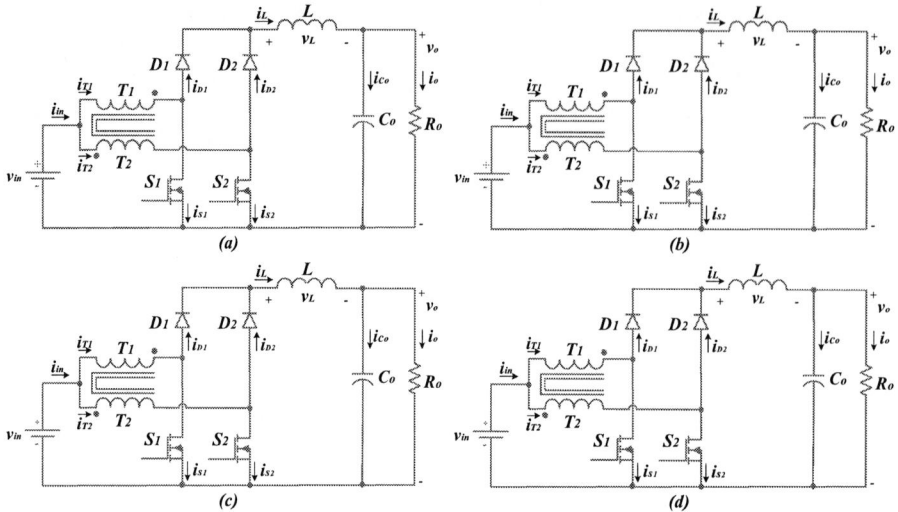

Fig. 2. Operation stages of the 3SSC boost converter in NOM-CCM. *(a) First stage. (b) Second stage. (c) Third stage. (d) Fourth stage.*

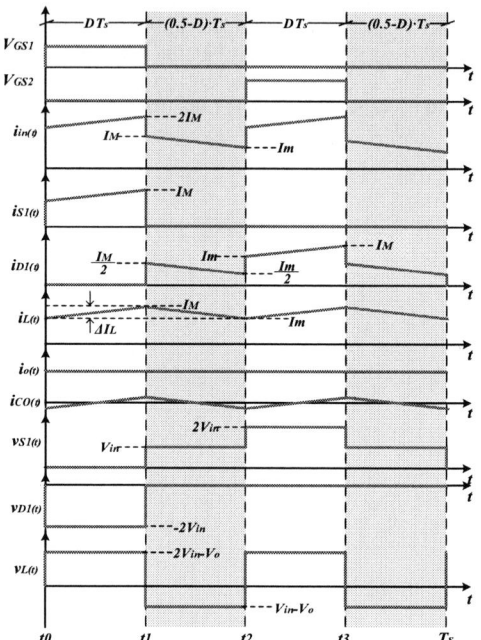

Fig. 3. Main theoretical waveforms for NOM-CCM.

C. Indutor Design

The current ripple in the inductor ΔI_L is equal in the first and second stages. Then, by analyzing the current i_L and using expressions (1), (4) and (7), the current ripple through the inductor L is given in (8).

$$\Delta I_L = \frac{(1-2D)\cdot D\cdot V_o\cdot T_S}{(1+2D)\cdot L} \qquad (8)$$

The inductance value can be determined as (9).

$$L = \frac{(1-2D)\cdot D\cdot V_o\cdot T_S}{(1+2D)\cdot \Delta I_L} \qquad (9)$$

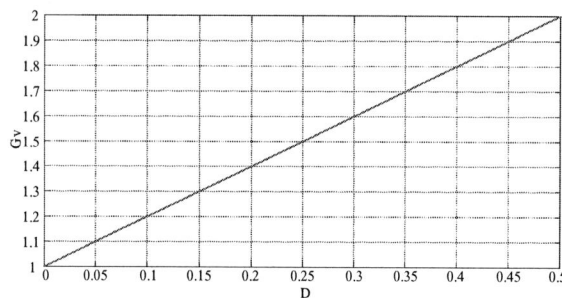

Fig. 4. Static gain as a function of the duty cycle in the continuous conduction mode.

D. Output Capacitor Design

The capacitance value of the output capacitor can be calculated from the electrical charge variation during one switching period, by the expression (10), where ΔV_o is the voltage ripple desired.

$$C \geq \frac{1}{16}\cdot\frac{(1-2D)\cdot D\cdot V_{in}\cdot T_s^2}{L\cdot\Delta V_O} \qquad (10)$$

III. SMALL SIGNAL MODEL OF BOOST CONVERTER WITH TYPE-A CELL

The modeling of the converter is performed by using the model of small-signal exposed in [9]. It is considered the converter operates under ideal conditions, without resistive losses in its elements and without the presence of inductance and parasitic capacitances in its electronic components.

Under steady-state conditions, waveforms and equations of voltages and currents of the DC-DC 3SSC boost converter were determined in section II using the basic principles of electrical circuit analysis. The average voltage of the inductor and the average currents of the capacitor are, in general, non-linear functions of the signals present in the converter, constituting, therefore, a set of non-linear differential equations. In order to obtain a simplified model, a small signal model is usually constructed, linearizing around an operating point, in which the harmonics of the modulation frequency are neglected.

978-1-7281-4181-7/19 $31.00 © 2019 IEEE

A. Average Waveforms of the Indutor, Capacitor and Input Current

As can be seen in Fig 2, the voltage at the inductor, the current at the input, as well as the current at the capacitor, contains dc components, besides the ripple at the switching frequency and its harmonics. The ripple at the switching frequency is very small in magnitude when compared to the dc components. Thus, according to [9], it is possible to approximate small undulations by replacing the waveforms with their mean low-frequency values, obtaining:

$$L\frac{d\bar{i}_{L(t)}}{dt} \approx \bar{v}_{in(t)}(1+2\bar{d}_{(t)}) - \bar{v}_{o(t)} \tag{11}$$

$$\bar{i}_{in(t)} \approx \bar{i}_{L(t)}(1+2\bar{d}_{(t)}) \tag{12}$$

$$C\frac{d\bar{v}_{o(t)}}{dt} \approx \bar{i}_{L(t)} - \frac{\bar{v}_{o(t)}}{R_o} \tag{13}$$

The terms with bar in equations (11) to (13) represent the mean value of the variable in a switching period. Therefore, these are the calculated averaged equations describing the variations dc and ac of low frequency of the converter.

B. Perturbation and Linearization

Equations (11) to (13) are not linear because they involve the multiplication of quantities varying in time. Thus, theses expressions can be linearized through the insertion of a small signal disturbance, so the variables are composed of an average value and a small signal disturbance, as follows:

$$\bar{i}_{L(t)} = I_L + \hat{i}_{L(t)} \tag{14}$$

$$\bar{v}_{in(t)} = V_{in} + \hat{v}_{in(t)} \tag{15}$$

$$\bar{v}_{o(t)} = V_o + \hat{v}_{o(t)} \tag{16}$$

$$\bar{d}_{(t)} = D + \hat{d}_{(t)} \tag{17}$$

By substituting (14), (15), (16) and (17) in (11), (12) and (13), and disregarding the constant and second-order terms containing the product of small quantities, is obtained:

$$L\frac{d\hat{i}_{L(t)}}{dt} = 2V_{in}\hat{d}_{(t)} + \hat{v}_{in(t)}(1+2D) - \hat{v}_{o(t)} \tag{18}$$

$$\hat{i}_{in(t)} = 2I_L\hat{d}_{(t)} + \hat{i}_{L(t)}(1+2D) \tag{19}$$

$$c\frac{d\hat{v}_{o(t)}}{dt} = \hat{i}_{L(t)} - \frac{\hat{v}_{o(t)}}{R_o} \tag{20}$$

C. Construction of the Small-Sgnal Equivalent Circuit Model

Equations (18), (19) and (20) are the small signal ac expressions of the three-state switching type-A cell boost converter, which can be seen in Fig. 5. Equation (18) results from Kirchhoff law, the small signal model of the inductor current. Equation (20) describes the small signal model currents flowing to the node connected to the capacitor. Expression (19) represents the current ac of the small signal model, which flows in the converter through the input voltage source. In such equations, terms multiplied by the variation to small signals of the duty cycle are considered as controlled sources, since the duty cycle is a control input. The other terms can be combined in ideal transformers with a turns ratio given as a function of the duty cycle.

D. Converter Transfer Function

The equivalent circuit of Fig. 5 can be represented in the Laplace domain and solved using conventional linear circuit analysis techniques to find the converter transfer functions, input and output impedances, among other desired relationships. The transfer function G_{vd} is found by setting the input voltage variations to zero, and then solving the equivalent circuit model $\hat{v}_{o(s)}$ for the function $\hat{d}_{(s)}$:

$$G_{vd(s)} = \frac{\hat{v}_{o(s)}}{\hat{d}_{(s)}}\Bigg|_{\hat{v}_{in(s)}=0} \tag{21}$$

$$G_{vd(s)} = \frac{2V_{in}}{s^2 C_o L + s\dfrac{L}{R_o} + 1} \tag{22}$$

Equation (22) is the transfer function which describes how the control input variations $\hat{d}_{(s)}$ influences the output voltage $\hat{v}_{o(s)}$. Analyzing this equation is verified that this converter has a minimum phase characteristic, that is, there is no presence of zeros in the right half plane, which allows the design of a control project with high bandwidth. Furthermore, as can be seen in Fig. 1, the converter structure does not have an input inductor, which results in a voltage fed circuit with a considerable ripple in the input current. However, due to the fact that this converter has no zero in the right half-plane, a faster dynamic response is achieved than the current-fed converters. It is important to highlight that the transfer function of the 3SSC boost converter is similar to the Buck converter, when they operate in CCM.

It should be noted that controllers for non-minimal-phase systems are more difficult to design because of the gain increases in a conventional controller, the closed-loop poles are attracted by the zeros in the right half-plane. Because of this feature, it is often difficult to design high bandwidth controllers, and the transient response may not be as fast as desired. In addition, since the poles tend to be destabilized, the phase margin is often limited, which makes the system more

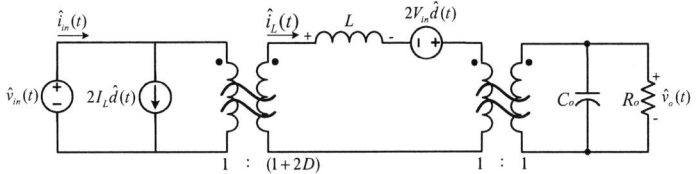

Fig. 5. 3SSC boost converter with type-A cell small-signal ac equivalent circuit.

978-1-7281-4181-7/19 $31.00 © 2019 IEEE

sensitive to computational or controller delays [10]. Therefore, due to the minimum phase characteristic of the converter in question, the design of the control technique is less complex when compared to the structures that present zeros in the right half-plane in their transfer function.

IV. EXPERIMENTAL RESULTS

To validate the theoretical assumptions, an experimental prototype of the DC-DC converter was assembled according to the specifications and main parameters presented in Table II. A photograph of the experimental prototype is shown in Fig. 6.

Fig. 7 shows control signals of switches S_1 and S_2 and the current i_{D2} and the voltage v_{S1}. The gate pulses on the switches show the converter operates in NOM, with D=0.33. According to Fig. 7, the maximum voltage v_{S1} is approximately 360 V. When the switch S_1 is conducting, the current i_{S1} is equivalent to the current i_{D2} at this time, being approximately 2 A peak.

Fig. 8 shows the voltage V_{in} and the voltage V_o, together with the current i_L and the current i_{D2}. These results demonstrate the characteristic of the converter operating as a voltage elevator, with V_o in approximately 300 V, conforming to the gain required. Moreover, it can be verified by the current i_L, in approximately 2 A_{avg}, that the converter is operating in

TABLE II. DESIGN SPECIFICATIONS

Parameter	Value/Specification
Input (V_{in})	180 V
Output (V_o)	300 V
Output power (P_o)	600 W
Switching freq. (f_s)	25 kHz
Ripple current through the inductor L (ΔI_L)	$0.3 \cdot I_o$
Ripple voltage across the capacitor C_o (ΔV_o)	$0.1 \cdot V_o$
Inductor L	$L=1.3mH$, core NEE-42/21/15-IP12 by Thornton, $N = 68$ turns – 5 x AWG26
Capacitor C_o	$C_o = 82\eta F$ / 1.1 kV, metal polyester film capacitor
Autotransformer (T)	$N_p/N_s=1/1$, core NEE-65/33/39-IP12 by Thornton, $N = 7$ turns – 5 x AWG26
Switch S_1-S_2	IGBT module, $I_C = 43$ A, $V_{CES} = 1200$ V, by SEMIKRON
Diode D_1-D_2	$V_R = 1500$ V, $I_F = 20$ A, $V_F = 1.7$ V, by ON Semiconductor

Fig. 6. Experimental prototype of the 3SSC boost converter with type-A cell.

CCM. It is also important to note that the current ripple frequency is twice of the switching frequency, in 50 kHz.

Fig. 9 represents the voltage v_{D2}, current i_L, and currents i_{D1} and i_{D2}. The maximum reverse voltage on the diodes is twice the voltage V_{in}, in approximately -360 V, which is in accordance with the theoretical analysis. Looking at the results, it can be seen that the diodes D_1 and D_2 conduct simultaneously when the current i_L decreases linearly, that is, the moment when the switches S_1 and S_2 are not conducting. When the diodes are turned-on simultaneously, the current through each of these semiconductors is half of the current i_L that flows through the L inductor, in approximately 1 A, thus the division in the semiconductor current stress is verified.

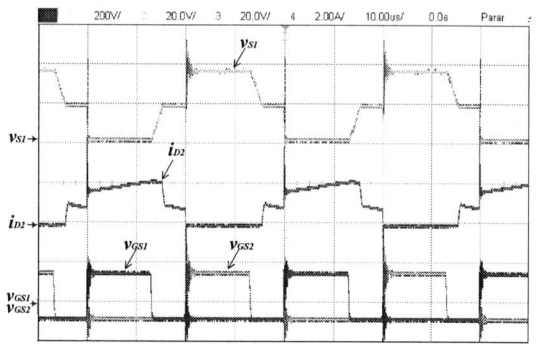

Fig. 7. V_{GS1} (20 V/div); V_{GS2} (20 V/div); i_{D2} (2 A/div); v_{S1} (200 V/div); time: 10 μs/div.

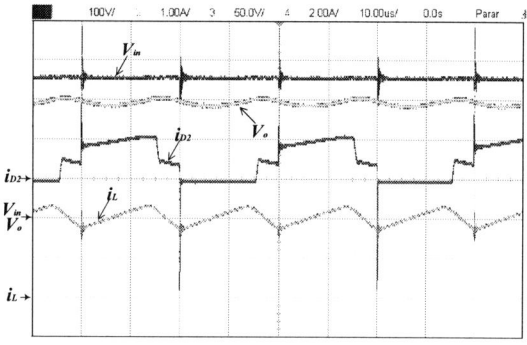

Fig. 8. V_{in} (50 V/div); V_o (100 V/div); i_L (1 A/div); i_{D2} (2 A/div); time: 10 μs/div.

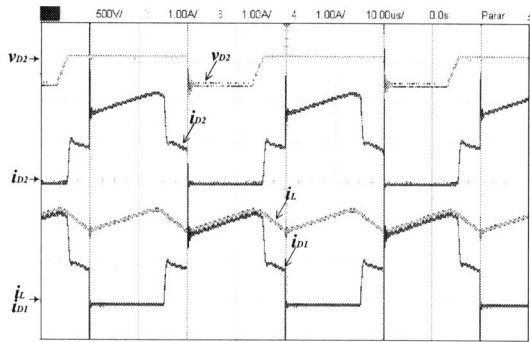

Fig. 9. v_{D2} (500 V/div); i_L (1 A/div); i_{D1} (1 A/div); i_{D2} (1 A/div); time: 10 μs/div.

V. CONCLUSION

This paper presented the non-isolated DC-DC 3SSC boost converter using 3SSC boost converter with type-A cell. According to the theoretical analysis and experimental results show the advantages of the application of 3SSC incorporated in the presented topology, for example, the current is divided among the semiconductors, reducing the current stresses, with the distribution of the losses between the elements. In addition, only a portion of the energy from the input source flows through the active switches, while the remaining part is transferred directly to the load without being processed by those switches, i.e., that energy is delivered to the load via the diodes and windings of the autotransformer.

The converter transfer function has a minimum phase characteristic, i.e. there are no zeros in the right half-plane. Thus, when compared to the traditional boost structures that have RHPZ in its transfer function, the 3SSC boost converter, with a suitable controller design, can present fast dynamic response, giving greater simplicity in the design of a control project.

Finally, the structure becomes very attractive solution in power electronics applications that do not require high voltage gains and require structures with small dimensions and high-power density.

ACKNOWLEDGMENT

This study was financed in party by the Coordenação de Aperfeiçoamento de Pessoal de Nível Superior - Brasil (CAPES) - Finance Code 001, and with the support of the Instituto Federal de Educação, Ciência e Tecnologia de Goiás, Campus Jataí.

REFERENCES

[1] J. W. Kolar, U. Drofenik, J. Biela, M. L. Heldwein, H. Ertl, T. Friedli, and S. D. Round, "PWM Converter Power Density Barriers," IEEE Power Conversion Conference, Nagoya, 2007.

[2] J. W. Kolar, J. Biela, S. Waffler, T. Friedli, U. Badstuebner, "Performance Trends and Limitations of Power Electronic Systems," 2010 6th International Conference on Integrated Power Electronics Systems, Nuremberg, 2011.

[3] Bascopé, G. V. T.; Barbi, I. "Generation of Family of Non-Isolated DC-DC PWM Converters Using New Three-State Switching Cell". Power Electronics Specialists Conference, 2000. PESC 00. 2000 IEEE 31st Annual, Volume: 2, 18-23 June 2000. p. 858 – 863.

[4] J. P. R. Balestero, F. L. Tofoli, G. V. Torrico-Bascopé and F. J. M. de Seixas, "A DC–DC Converter Based on the Three-State Switching Cell for High Current and Voltage Step-Down Applications," in *IEEE Transactions on Power Electronics*, vol. 28, no. 1, pp. 398-407, Jan. 2013.

[5] G. C. Silveira, F. L. Tofoli, L. D. S. Bezerra and R. P. Torrico-Bascopé, "Analysis and Small-Signal Modeling of a Noinsolated High Voltage Step-Up DC-DC Boost Converter," 2015 IEEE 13th Brazilian Power Electronics Conference and 1st Southern Power Electronics Conference (COBEP/SPEC), Fortaleza, 2015.

[6] F. L.Tofoli, D. de A. Tavares and J. I. de A. Saldanha, "Survey on Topologies Based on The Three-State and Multi-State Switching Cells," in IET Power Electronics, vol. 12, no. 5, pp. 967-982, May 2019.

[7] K. Hwu, Y. Yau, "A KY boost converter", in *IEEE Transactions on Power Electronics*, 2010, vol. 25 no.11, pp. 2699–2703.

[8] M. Soltani, A. Mostaan, Y. P. Siwakoti, P. Davari, F. Blaabjerg, "Family of Step-Up DC/DC Converters With Fast Dynamic Response For Low Power Applications," in IET Power Electronics, vol. 9, no. 14, pp. 2665-2673, Nov. 2016.

[9] R. W. Erickson, D. Maksimovic, *Fundamentals of Power Electronics*, 2 ed., Springer, Nova York, 2001.

[10] M. Forouzesh, Y. P. Siwakoti, S. A. Gorji, and B. Lehman, "Step-Up DC-DC Converters: A Comprehensive Review of Voltage-Boosting Techniques, Topologies, and Applications," in *IEEE Transactions on Power Electronics*, vol. 32, no. 12, pp. 9143-9178, Dec. 2017.

Input Voltage Range Extension Methods in the Series-Resonant DC-DC Converters

Andrii Chub
Power Electronics Group
Tallinn University of Technology
Tallinn, Estonia
andrii.chub@taltech.ee

Dmitri Vinnikov
Power Electronics Group
Tallinn University of Technology
Tallinn, Estonia
dmitri.vinnikov@taltech.ee

Jih-Sheng Lai
Future Energy Electronics Center
Virginia Tech
Blacksburg, VA, USA
laijs@vt.edu

Abstract—**The series resonant dc-dc converters are mature technology used mostly in applications requiring constant dc voltage gain. They were recently enhanced by the application of the boost rectifiers. This approach showed new opportunities for input voltage regulation range extension. The given paper summarizes known operating modes and shows a way of significant input voltage range extension through the utilization of hybrid full-bridge switching cell. Two converter topologies are introduced in this paper. Possible operating modes are identified and summarized.**

Keywords—dc-dc converter, dc microgrid, residential systems, renewable energy.

I. INTRODUCTION

The greatest challenge for humanity is matching the growing demand for energy needed for economic development with pollution prevention required for environmental sustainability [1]. Economical way of electrification is envisioned as one of the most effective measures of global sustainability [2]. The end use of energy in electric form was long overlooked in favor of appliances powered on-site from natural gas, coal, etc. [3]. However, the electrification is held back by due to lack of long-term policy support [4]. Meanwhile, numerous studies demonstrate that electrification can significantly reduce the carbon footprint of buildings [5], transportation [6]-[8], industry [9]-[11], etc.

Microgrids are the backbone technology for the future electrification as they ensure higher resiliency and energy efficiency at the system level [12]. The dc microgrids were justified as an advantageous replacement for the ac systems [13]. The dc microgrid suits well for renewable electric energy sources, lighting, electric drives, and many appliances that have intermediate dc link, as well as for the integration of energy storages that are dc due to their electrochemical nature [14]. Deployment of fully-functional dc microgrids requires wide adoption of power electronic converters [15]. Recent advancements in power electronic technology enable penetration of low-voltage (LV) dc microgrids in small-scale systems [16]. This is a relatively new technology that needs standardization of voltage levels and, consequently, standard equipment for technology deployment [17].

This paper considers residential LV dc microgrids, similar to the one presented in Fig. 1. As shown, dc distribution and storage of renewable energy are associated with the use of different dc-dc converters. Their standardization is possible through the development of topologies with a wide input voltage range, i.e., capable of operation with different energy sources and storages [18]. However, the input voltage range extension is associated with moderate efficiency across the extended input voltage range [19]. Promising technology for the wide input voltage range applications is galvanically isolated buck-boost dc-dc

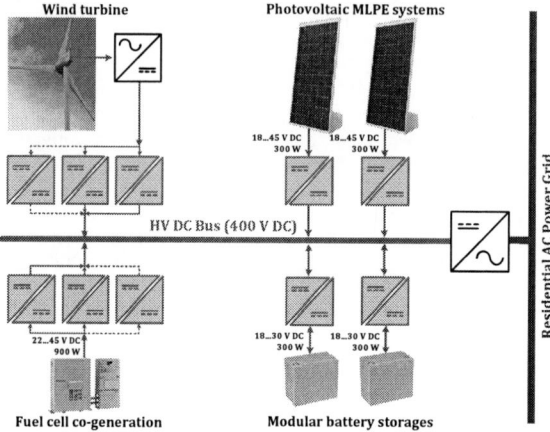

Fig. 1 Example of residential dc microgrid utilizing modular dc-dc converters [18].

converters (IBBCs) [20]. These topologies could be realized by controlling either a single switching cell, like the impedance-source IBBC [21] or by controlling both switching cells connected to a two-winding isolating transformer [22]. The latter usually utilize the LV side cell for performing the voltage step-down and high-voltage (HV) side cell for the voltage step-up [20], [22].

Recently, it was demonstrated that IBBC topologies could be synthesized with the implementation of a bridgeless power factor correction (PFC) rectifier circuit at the HV side of an IBBC. This approach was demonstrated for phase-shifted voltage-fed topologies [23] as well as SRC-based topologies [24], [25]. The latter demonstrate higher efficiency mostly due to lower circulating current in case of the discontinuous resonant current mode (DrCM) in SRC, i.e., when a resonant tank is implemented with the quality factor (Q) lower than unity in any conditions and, consequently, the resonant current drops to zero when the applied voltage is set to zero. The SRC is capable of the input voltage step-down at the constant switching frequency [26] in the DrCM and, thus, is advantageous for synthesis of the IBBC topologies.

The focus of this study is on the class of the full-bridge IBBC topologies based on the DrCM SRC with boost rectifier utilizing leakage inductance for voltage step-up. This class of IBBCs includes several existing topologies with the following boost cells: voltage multipliers with four-quadrant switch [24]; bridgeless boost PFC rectifier [27]; symmetrical and asymmetrical boost voltage doubler rectifiers (VDRs) with single-pulse control [28]-[30], boost VDR with double-pulse control [31]; simplified single-switch boost VDR [32].

978-1-7281-4181-7/19 $31.00 © 2019 IEEE

The given paper aims to show how numerous control methods could be combined in a single IBBC based on the DrCM SRC. Previously, most of these control methods were demonstrated for different converters but never in a single converter simultaneously. The next section introduces two converter topologies capable of integration of different control methods to enable unparalleled input voltage range. The third section describes how these converters can perform voltage step-down, while the fourth section introduces different control methods for voltage step-up. The fifth section explains the utilization of the control methods over the extended input voltage regulation range and provides useful references to the theoretical basis for the methods published before. Finally, conclusions are drawn.

II. IBBC CONVERTER FAMILY BASED ON THE DrCM SRC TOPOLOGY

The basic bidirectional topology (Fig. 2a) is very similar to conventional SRC. The main difference is in the implementation of the full-bridge switching cells that contain series capacitor and thus are capable of operating in different modes, i.e., as a full-bridge cell or as an asymmetrical half-bridge cell [33]. In the given case, the input full-bridge cell ($S_1...S_4$) is considered LV side, and the output full-bridge cell ($Q_1...Q_4$) is on the HV side as the isolating transformer TX steps the voltage up with turns ratio of 1:n. Moreover, the transformer features embedded leakage inductance L_{lk} used as the part of the resonant tank $C_2 - L_{lk} - C_3$. The resonant frequency could be calculated as follows:

$$f_r = \frac{1}{2 \cdot \pi \cdot \sqrt{L_{lk} \cdot C_r}}, \tag{1}$$

where the equivalent resonant capacitance C_r equals

$$C_r = \frac{C_2 \cdot C_3}{C_2 + C_3 \cdot n^2}. \tag{2}$$

To achieve soft-switching in this mode designated as dc transformer (DCX) mode, the magnetizing inductance has to be rated for preselected dead-time, as explained in [34]. For the same reason, the switching frequency ($f_{SW} = 1/T_{SW}$) should be 5-10% lower than the resonant frequency.

A unidirectional embodiment of the topology from Fig. 2a is shown in Fig. 2b, where designators of HV switches are kept for compatibility with further descriptions of the control methods. It features fewer switches at the output and can operate only in a limited number of modes comparing to that of the topology in Fig. 2a. However, it is still capable of covering the same input voltage range. The normalized dc voltage gain of the converter G will be used to define the input voltage range in relative terms:

$$G = \frac{V_{OUT}}{n \cdot V_{IN}}. \tag{3}$$

III. DCX AND VOLTAGE STEP-DOWN MODES

A. DCX Modes

In the DCX modes, the given IBBC operates most efficiently with full soft-switching in semiconductors as the current of the magnetizing inductor L_m recharges parasitic capacitances of the switches during dead-time. Hence, the active semiconductor components operate with a duty cycle of 0.5 less a short dead-time needed for soft-switching. Considering that each switching cell can operate as a full-bridge (FB) or as a half-bridge (HB) cell, there are four possible DCX modes:

- DCX-FB-FB (Fig. 3a): both switching cells operate in the FB configuration, which results in $G = 1$.
- DCX-HB-HB (Fig. 3b): both switching cells operate in the HB configuration, which results in $G = 1$.
- DCX-HB-FB (Fig. 3c): the input side cell is configured as the HB, while the output is the FB, which causes gain reduction down to $G = 0.5$.
- DCX-FB-HB (Fig. 3d): the input side cell is configured as the FB, while the output is the HB, which results in the higher gain of $G = 2$.

In Fig. 3 and all further figures, dashed lines correspond to switches that perform synchronous rectification and, thus, could be turned off. Yellow boxes mark the dead-times. All further discussions regarding implementation of the voltage step-up or step-down consider that any other mode is based on one of the DCX modes and converges to this DCX mode when the control variable minimizes the regulation effort.

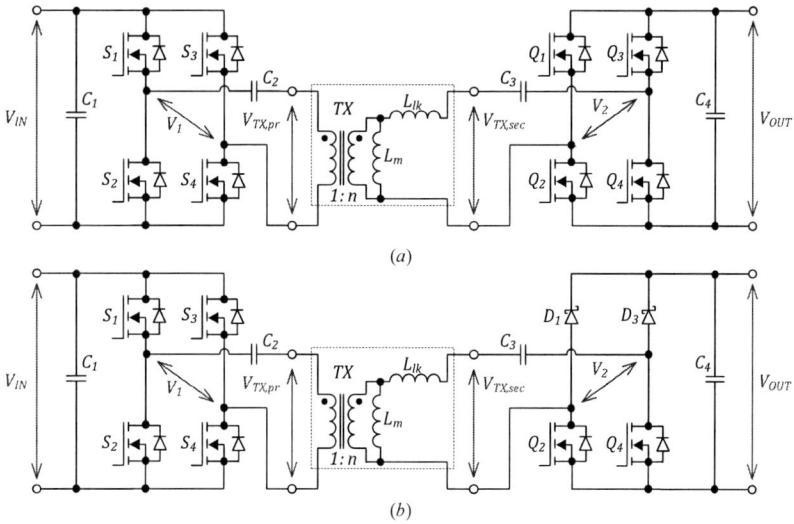

(a)

(b)

Fig. 2 Family of IBBCs based on the DrCM SRC topology: (a) bidirectional and (b) unidirectional implementation.

978-1-7281-4181-7/19 $31.00 © 2019 IEEE   1494

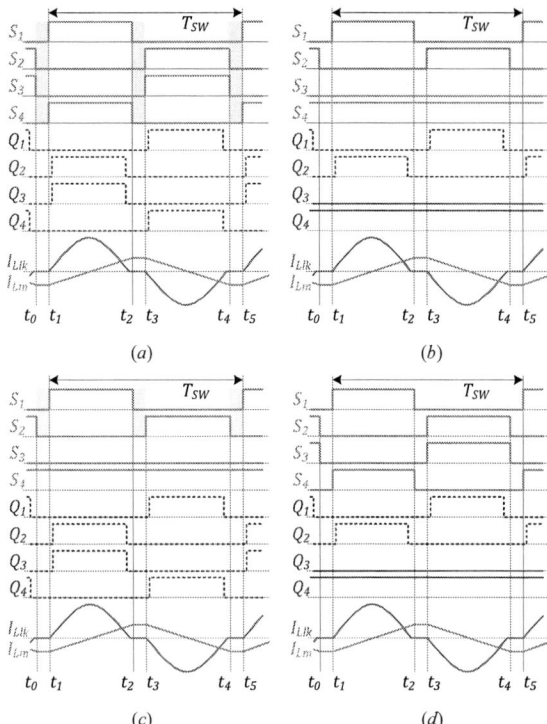

(a)

(b)

(c)

(d)

Fig. 3 Modulation of the IBBC in the (a) DCX-FB-FB, (b) DCX-HB-HB, (c) DCX-HB-FB, and (d) DCX-FB-HB modes.

The following subsections present several possibilities of voltage-step-down implementation trough modulation of the input side switching cell, while the output side cell acts as a full-bridge rectifier, either synchronous or passive, as an example.

B. Implementation of the Voltage Step-Down with Half-Bridge Input Side Inverter

The input side switching cell steps down the input voltage naturally when configured as the HB. However, the resonant tank could be used to step down the input voltage even further using pulse-width modulation (PWM). HB configuration can be easily achieved by turning on permanently any switch while turning off the adjacent switch in the same leg of the inverter [33].

SD-HB-PWM mode (Fig. 4a): the most typical approach is to apply the PWM to the input side HB. The resonant

current deflects from the sinusoidal shape due to DrCM design of the resonant tank. The dc voltage gain dependence on the duty cycle D_{SD} is nonlinear, and also depends on the resonant tank quality factor, i.e., on the operating power [26].

SD-HB-APWM mode (Fig. 4b): Another possibility to implement the voltage step-down using HB input side cell is the application of the asymmetrical PWM (APWM) in the DrCM SRC [35]. One half-wave of the resonant current is always sinusoidal in this control method, which improved soft-switching performance. Moreover, proper design of the magnetizing inductance enables soft-switching without any auxiliary circuit in a wide range of voltage and power [36].

C. Implementation of the Voltage Step-Down with Full-Bridge Input Side Inverter

The FB configuration of the input side switching cell provides several implementation possibilities of the input voltage step-down.

SD-FB-PSM mode (Fig. 5a): The simplest approach is in the application of the phase-shifted modulation (PSM) the FB cell in the given IBBC [26]. In this mode, the converter features soft-switching as long as it maintains DrCM operation [37]. Operation at the switching frequency near the resonant frequency ensures the most comprehensive soft-switching range in this mode [38].

SD-FB-HPSM mode (Fig. 5b): The previous method features relatively low efficiency at light load. This issue can be resolved with the application of the hybrid PSM along with the DrCM as proposed in [39]. This modulation could improve light load efficiency and minimizes parasitic oscillations in the given IBBC.

SD-FB-APWM mode (Fig. 5c): The APWM shows high performance with the HB in the previous subsection. Hence,

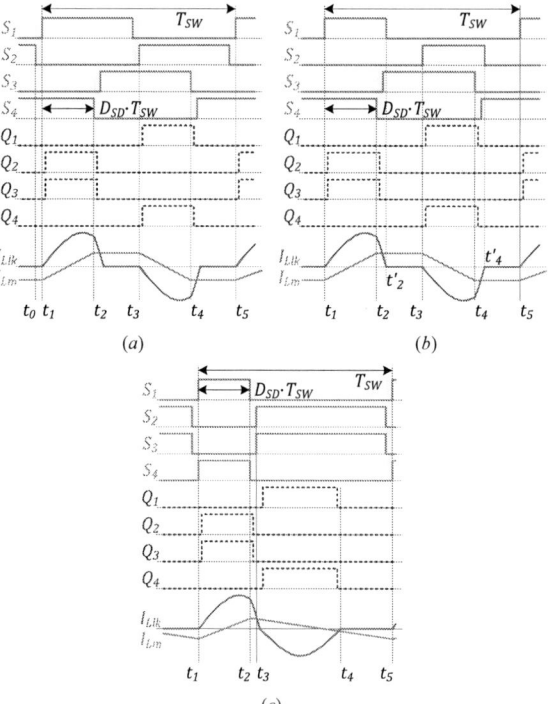

(a)

(b)

(c)

Fig. 5 Modulation of the IBBC in the (a) SD-FB-PSM, (b) SD-FB-HPSM, and (c) SD-FB-APWM modes with a synchronous FBR at the output.

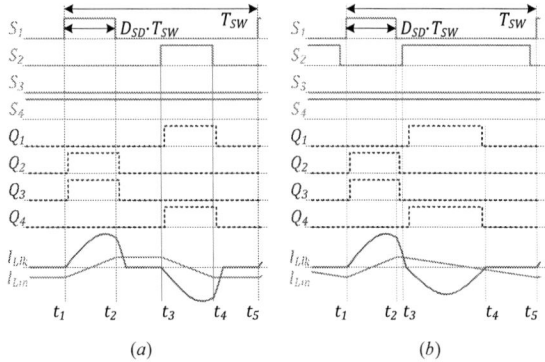

(a)

(b)

Fig. 4 Modulation of the IBBC in the (a) SD-HB-PWM and (b) SD-HB-APWM modes and considering a synchronous FBR at the output.

978-1-7281-4181-7/19 $31.00 © 2019 IEEE

it could be suggested to verify its performance in case of the FB configuration of the input switching cell. This operating mode was not reported for the FB circuits before.

IV. VOLTAGE STEP-UP MODES

A. Implementation of the Voltage Step-Up with Full-Bridge Output Side Rectifier

SU-BFBR-I mode (Fig. 6a): top (Q_1, Q_3) or bottom (Q_2, Q_4) switches perform voltage step-up by overlapping control signals through their phase and duty cycle control, while the other pair of switches is inactive or performs synchronous rectification [27]. In the given case, the bottom switches short-circuit the output side winding of the transformer TX when it is fed with voltage from the input side. This forces its leakage inductance L_{lk} to increase stored energy. Hence, this inductance acts as the boost inductor.

SU-BFBR-II mode (Fig. 6b): this mode is similar to the previous. Here, one leg of the output side switching cell (Q_1, Q_2 or Q_3, Q_4) can perform voltage step-up through PSM relative to the input side switches. This method is derived here from an idea in [40] that was not applied before to the SRC topology. The switches Q_3 and Q_4 apply voltage to the leakage inductance, causing an increase of its stored energy.

In both cases, the leakage inductance increases stored energy during time intervals [t_0; t_1], [t_2; t_3], and [t_4; t_5]. Then the stored energy is released to the output capacitors. The longer the energy storage interval of $D_{SU} \cdot T_{SW}$, the higher dc voltage gain can be achieved by the IBBC [27].

B. Implementation of the Voltage Step-Up with Half-Bridge Output Side Rectifier

Half-bridge configuration of the output side switching cell can be regarded as the boost VDR. The given IBBC is capable of reconfiguration into asymmetrical VDR, also known as Greinacher VDR. The reconfiguration can be performed by permanently turning on any switch (Q_4 in this case) and turning off the adjacent switch in the same leg of the inverter (Q_3). There are two existing control methods applicable to the given boost asymmetrical VDR:

SU-BVDR-1P mode (Fig. 7a): single pulse PWM is applied to the switches Q_1 and Q_2. When any of them is turned on, high voltage is applied to the leakage inductance to increase its stored energy during $D_{SU} \cdot T_{SW}$ [28]-[30]. This is similar to the previous methods for the boost FB rectifiers. However, the dc voltage gain of the VDR based circuits is inherently higher for the same D_{SU}.

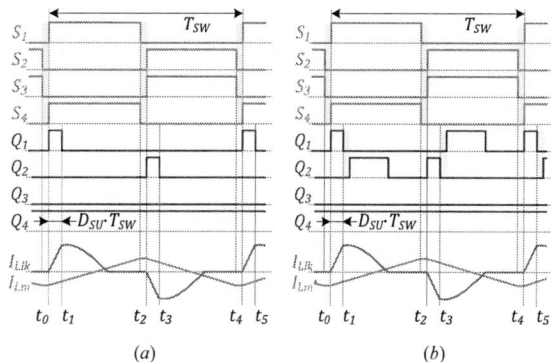

Fig. 7 Modulation of the IBBC in the (a) SU-BVDR-1P and (b) SU-BVDR-2P modes implemented with the FB inverter at the input.

SU-BVDR-2P mode (Fig. 7b): the single-pulse PWM suffers from relatively high conduction losses. Thus double-pulse PWM was proposed in [31]. It takes advantage of the implementation of the synchronous rectification to lower the conduction losses in the output side semiconductors. In every other way, it is similar to the previous modulation.

SU-1SwBVDR mode (Fig. 8): the previous two modulations suffer from hard turn-on of the switches Q_1 and Q_2 at the instants t_0, t_2, and t_4. These switching losses were shown to contribute significantly to the reduction of the converter efficiency, as reported in [32]. The same work suggests the application of APWM with only one hard turn-on during the switching period instead of two. Significant reduction of the rectifier switching losses was demonstrated for low power applications. On the other side, this control mode requires higher values of D_{SU} than that in the two previous modes.

V. GENERALIZATION, DISCUSSION, AND FUTURE RESEARCH

The previous sections overview numerous existing control methods applicable to the proposed IBBC that is based on the DrCM SRC implemented with hybrid full-bridge switching cells. This section aims to summarize and generalize their application in the given IBBC. Arrangement of the different control modes is shown in Fig. 9. It is worth mentioning that each step-down control mode can be implemented with FB rectifier (FBR) or the VDR rectifier at the output, either synchronous or passive. Similarly, any voltage step-up mode can be implemented with either a FB inverter (FBI) or a HB inverter (HBI) at the input side that

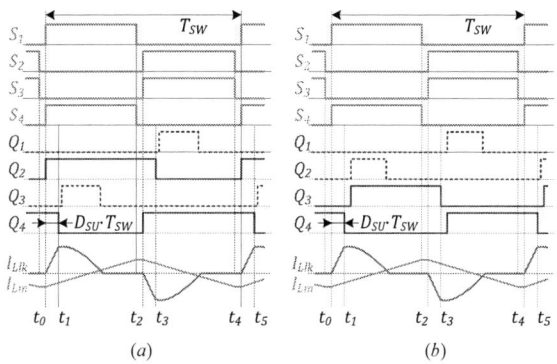

Fig. 6 Modulation of the IBBC in the SU-BFBR-I (a) and SU-BFBR-II (b) modes implemented with the FB inverter at the input.

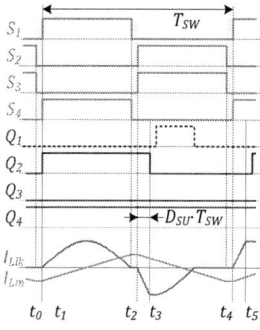

Fig. 8 Modulation of the IBBC in the SU-1SwBVDR mode implemented with a FB inverter at the input.

978-1-7281-4181-7/19 $31.00 © 2019 IEEE

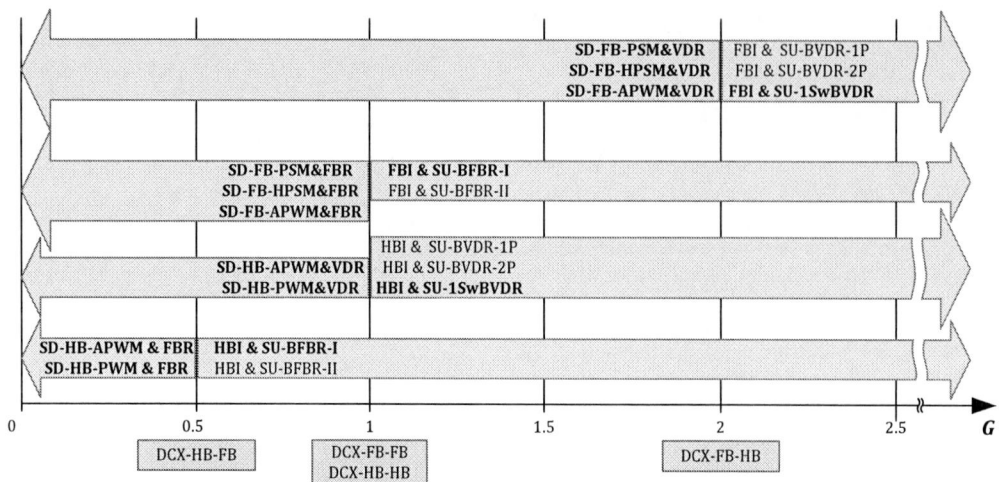

Fig. 9 Arrangement of the operating modes across the wide range of the dc voltage gain of the considered IBBC (bold font designates modes available in the unidirectional version of the given IBBC).

operates with the switch duty cycle of close to 0.5, i.e., in the same way as in the corresponding DCX mode.

It could be appreciated from Fig. 9 that each mode starts from a corresponding DCX mode. The arrow direction designates efficiency decreasing as losses increase when the IBBC is moving further from the starting DCX mode. Also, $G = 1$ could be achieved with two different DCX modes.

Most of the control methods overviewed in this paper were presented previously in some form. The contribution of this paper is in the introduction of the IBBC that is based on the DrCM SRC implemented with hybrid full-bridge switching cells and in the generalization of its control flexibility. It is worth mentioning that these control modes were never combined in any existing SRC-based IBBC. As a result, the input voltage range can be significantly extended. For example, the input voltage range can be as wide as 1:16 if the voltage regulation in any mode is limited to twofold from the corresponding DCX mode to avoid efficiency penalties: in this way, the minimum gain value is $G = 0.25$ (half of $G = 0.5$ in DCX-HB-FB mode), while the maximum gain then equals $G = 4$ (twice $G = 2$ in DCX-FB-HB).

This paper avoids presenting detailed operating principle for the modes included as they were previously published as parts of other research and thus repeating those yields no novelty nor it can be done in a concise way. References are summarized in Table I for a curious reader. The table omits references to DCX modes as they can be regarded as a subset of the step-up and step-down voltage regulation modes.

The modes listed in the Table I were all but one presented previously and their feasibility is proven beforehand. The SD-FB-APWM mode was synthesized and suggested in this work. Its experimental feasibility and performance metrics are to be tested in the future. It follows from Fig. 9 that up to seven operating modes could cover the same input voltage operating range. Selection of the proper mode concerning the operating power and dc voltage gain is the main focus of future research. For example, for $0.5 \leq G \leq 1$, an arbitrary gain value can be obtained by stepping down from $G = 1$ using one out of the three available step-down modes, or it can be stepped up from $G = 0.5$ using one out of the two available step-up modes.

TABLE I. OPERATION PRINCIPLE OF THE IBBC MODES

Mode	Figure	Reference
SD-HB-PWM	Fig. 4a	[26]
SD-HB-APWM	Fig. 4b	[35], [36]
SD-FB-PSM	Fig. 5a	[26], [37], [38]
SD-FB-HPSM	Fig. 5b	[39]
SD-FB-APWM	Fig. 5c	Derived here from [36]
SU-BFBR-I	Fig. 6a	[27]
SU-BFBR-II	Fig. 6b	Derived here from [40]
SU-BVDR-1P	Fig. 7a	[28] - [30]
SU-BVDR-2P	Fig. 7b	[31]
SU-1SwBVDR	Fig. 8	[32]

Special attention has to be paid to limitations of the available modes. Many of them feature soft-switching that imposes limitations on converter parameters, like magnetizing inductance value. It is possible that those requirements would contradict each other when the converter is designed and some modes would not be feasible within control set of ultra-wide range SRC-based IBBC.

VI. CONCLUSIONS

The main contribution of this paper is in the integration of modified SRC topology with a resonant tank designed to operate in DrCM and numerous existing control modes enabling step-up and step-down voltage regulation. Most of these operating modes are known from the literature. However, none of the SRC-based IBBCs was demonstrated with more than three control modes at once, which resulted in a narrow input voltage range. This work generalizes the implementation principles of ultra-wide range SRC-based IBBC that requires no additional auxiliary components.

Future research will focus on a detailed analysis of compatibilities and contradictions between the existing control modes as well as the selection of the best performing control mode for each dc gain value or range.

ACKNOWLEDGMENT

This research was supported in part by the Estonian Research Council grant PSG206, and in part by the Estonian Centre of Excellence in Zero Energy and Resource Efficient Smart Buildings and Districts, ZEBE, grant 2014-2020.4.01.15-0016 funded by the European Regional Development Fund.

References

[1] B. Elleuch, F. Bouhamed, M. Elloussaief, and M. Jaghbir, "Environmental sustainability and pollution prevention," *Environmental Science and Pollution Research*, vol. 25, no. 19, pp. 18223–18225, Jul. 2018.

[2] T. Moore, "Electrification and global sustainability," *EPRI Journal*, vol. 23, no. 1, p. 42-49, Jan.-Feb. 1998.

[3] K. Dennis, "Environmentally Beneficial Electrification: Electricity as the End-Use Option," *The Electricity Journal*, vol. 28, no. 9, pp. 100–112, Nov. 2015.

[4] M. Wei et al., "Deep carbon reductions in California require electrification and integration across economic sectors," *Environmental Research Letters*, vol. 8, no. 1, p. 14038, Mar. 2013.

[5] J. Van Roy, B. Verbruggen, and J. Driesen, "Ideas for Tomorrow: New Tools for Integrated Building and District Modeling," *IEEE Power and Energy Mag.*, vol. 11, no. 5, pp. 75–81, Sep. 2013.

[6] O. Y. Edelenbosch, A. F. Hof, B. Nykvist, B. Girod, and D. P. van Vuuren, "Transport electrification: the effect of recent battery cost reduction on future emission scenarios," *Climatic Change*, vol. 151, no. 2, pp. 95–108, Sep. 2018.

[7] M. Weiss, P. Dekker, A. Moro, H. Scholz, and M. K. Patel, "On the electrification of road transportation – A review of the environmental, economic, and social performance of electric two-wheelers," *Transportation Research Part D: Transport and Environment*, vol. 41, pp. 348–366, Dec. 2015.

[8] D. McCollum, V. Krey, P. Kolp, Y. Nagai, and K. Riahi, "Transport electrification: A key element for energy system transformation and climate stabilization," *Climatic Change*, vol. 123, no. 3–4, pp. 651–664, Oct. 2013.

[9] Z. J. Schiffer and K. Manthiram, "Electrification and Decarbonization of the Chemical Industry," *Joule*, vol. 1, no. 1, pp. 10–14, Sep. 2017.

[10] S. Lechtenböhmer, L. J. Nilsson, M. Åhman, and C. Schneider, "Decarbonising the energy intensive basic materials industry through electrification – Implications for future EU electricity demand," *Energy*, vol. 115, pp. 1623–1631, Nov. 2016.

[11] P. Lombardi, P. Komarnicki, R. Zhu and M. Liserre, "Flexibility options identification within Net Zero Energy Factories," *Proc. IEEE PowerTech 2019*, Milan, Italy, June 23-27, 2019, 6 pp.

[12] A. Gholami, F. Aminifar and M. Shahidehpour, "Front Lines Against the Darkness: Enhancing the Resilience of the Electricity Grid Through Microgrid Facilities," *IEEE Electrification Magazine*, vol. 4, no. 1, pp. 18-24, March 2016.

[13] J. J. Justo, F. Mwasilu, J. Lee, and J.-W. Jung, "AC-microgrids versus DC-microgrids with distributed energy resources: A review," *Renew. Sust. Energy Reviews*, vol. 24, pp. 387–405, Aug. 2013.

[14] L. E. Zubieta, "Are Microgrids the Future of Energy?: DC Microgrids from Concept to Demonstration to Deployment," *IEEE Electrification Magazine*, vol. 4, no. 2, pp. 37-44, June 2016.

[15] N. Kirby, "Current Trends in dc: Voltage-Source Converters," *IEEE Power and Energy Magazine*, vol. 17, no. 3, pp. 32-37, May 2019.

[16] T. Dragicevic, J. C. Vasquez, J. M. Guerrero and D. Skrlec, "Advanced LVDC Electrical Power Architectures and Microgrids: A step toward a new generation of power distribution networks.," *IEEE Electrification Magazine*, vol. 2, no. 1, pp. 54-65, March 2014.

[17] E. Rodriguez-Diaz, F. Chen, J. C. Vasquez, J. M. Guerrero, R. Burgos and D. Boroyevich, "Voltage-Level Selection of Future Two-Level LVDC Distribution Grids: A Compromise Between Grid Compatibiliy, Safety, and Efficiency," *IEEE Electrification Magazine*, vol. 4, no. 2, pp. 20-28, June 2016.

[18] D. Vinnikov, A. Chub, R. Kosenko, and E. Liivik, "Versatile Power Electronic Building Block for Residential DC Microgrids," *Proc. SPEEDAM'2018*, June 20-22, 2018, Amalfi, Italy, pp. 735-742.

[19] A. Chub, D. Vinnikov, E. Liivik and T. Jalakas, "Multiphase Quasi-Z-Source DC–DC Converters for Residential Distributed Generation Systems," *IEEE Trans. Ind. Electron.*, vol. 65, no. 10, pp. 8361-8371, Oct. 2018.

[20] C. Yao, X. Ruan, X. Wang and C. K. Tse, "Isolated Buck–Boost DC/DC Converters Suitable for Wide Input-Voltage Range," *IEEE Trans. Power Electron.*, vol. 26, no. 9, pp. 2599-2613, Sept. 2011.

[21] A. Chub, D. Vinnikov, R. Kosenko and E. Liivik, "Wide Input Voltage Range Photovoltaic Microconverter With Reconfigurable Buck–Boost Switching Stage," *IEEE Trans. Ind. Electron.*, vol. 64, no. 7, pp. 5974-5983, July 2017.

[22] D. Vinnikov, A. Chub, R. Kosenko, J. Zakis and E. Liivik, "Comparison of Performance of Phase-Shift and Asymmetrical Pulsewidth Modulation Techniques for the Novel Galvanically Isolated Buck–Boost DC-DC Converter for Photovoltaic Applications," *IEEE J. Emerg. Sel. Top. Power Electron.*, vol. 5, no. 2, pp. 624-637, June 2017.

[23] H. Wu, Y. Lu, T. Mu and Y. Xing, "A Family of Soft-Switching DC–DC Converters Based on a Phase-Shift-Controlled Active Boost Rectifier," *IEEE Trans. Power Electron.*, vol. 30, no. 2, pp. 657-667, Feb. 2015.

[24] T. LaBella, W. Yu, J. Lai, M. Senesky and D. Anderson, "A Bidirectional-Switch-Based Wide-Input Range High-Efficiency Isolated Resonant Converter for Photovoltaic Applications," *IEEE Trans. Power Electron.*, vol. 29, no. 7, pp. 3473-3484, July 2014.

[25] H. Wu, T. Xia, X. Zhan and Y. Xing, "High step-up isolated resonant converter with voltage quadrupler rectifier and dual-phase-shift control," *IET Power Electron.*, vol. 8, no. 12, pp. 2462-2470, 12 2015.

[26] J.-P. Vandelac and P. D. Ziogas, "A DC to DC PWM series resonant converter operated at resonant frequency," *IEEE Trans. Ind. Electron.*, vol. 35, no. 3, pp. 451-460, Aug. 1988.

[27] X. Zhao, L. Zhang, R. Born and J. Lai, "A High-Efficiency Hybrid Resonant Converter With Wide-Input Regulation for Photovoltaic Applications," *IEEE Trans. Ind. Electron.*, vol. 64, no. 5, pp. 3684-3695, May 2017.

[28] S. Son, O. A. Montes, A. Junyent-Ferré and M. Kim, "High Step-Up Resonant DC/DC Converter With Balanced Capacitor Voltage for Distributed Generation Systems," *IEEE Trans. Power Electron.*, vol. 34, no. 5, pp. 4375-4387, May 2019.

[29] S. Kim, B. Kim, B. Kwon and M. Kim, "An Active Voltage-Doubler Rectifier Based Hybrid Resonant DC/DC Converter for Wide-Input-Range Thermoelectric Power Generation," *IEEE Trans. Power Electron.*, vol. 33, no. 11, pp. 9470-9481, Nov. 2018.

[30] B. Kim, S. Kim, D. Huh, J. Choi and M. Kim, "Hybrid resonant half-bridge DC/DC converter with wide input voltage range," *Proc. APEC'2018*, San Antonio, TX, 2018, pp. 1876-1881.

[31] X. Zhao, C. Chen and J. Lai, "A High-Efficiency Active-Boost-Rectifier-Based Converter With a Novel Double-Pulse Duty Cycle Modulation for PV to DC Microgrid Applications," *IEEE Trans. Power Electron.*, vol. 34, no. 8, pp. 7462-7473, Aug. 2019.

[32] J. Kim, M. Park, J. Han, M. Lee and J. Lai, "PWM Resonant Converter with Asymmetric Modulation for ZVS Active Voltage Doubler Rectifier and Forced Half Resonance in PV Application," *IEEE Trans. Power Electron.* doi: 10.1109/TPEL.2019.2914016

[33] D. Vinnikov, A. Chub, O. Korkh and M. Malinowski, "Fault-Tolerant Bidirectional Series Resonant DC-DC Converter with Minimum Number of Components," *Proc. IEEE ECCE'2019*, Baltimore (MD), USA, Sep. 29 - Oct 3, 2019, 5 pp

[34] D. Vinnikov, A. Chub, E. Liivik and I. Roasto, "High-Performance Quasi-Z-Source Series Resonant DC–DC Converter for Photovoltaic Module-Level Power Electronics Applications," *IEEE Trans. Power Electron.*, vol. 32, no. 5, pp. 3634-3650, May 2017.

[35] G. Moschopoulos and P. Jain, "A series-resonant DC/DC converter with asymmetrical PWM and synchronous rectification," *Proc. PESC'2000*, Galway, Ireland, 2000, pp. 1522-1527 vol.3.

[36] K. Ali, P. Das and S. K. Panda, "Analysis and Design of APWM Half-Bridge Series Resonant Converter With Magnetizing Current Assisted ZVS," *IEEE Trans. Ind. Electron.*, vol. 64, no. 3, pp. 1993-2003, March 2017.

[37] Y. V. Singh, K. Viswanathan, R. Naik, J. A. Sabate and R. Lai, "Analysis and control of phase-shifted series resonant converter operating in discontinuous mode," *Proc. APEC'2013*, Long Beach, CA, 2013, pp. 2092-2097.

[38] E. Kim and B. Kwon, "Zero-Voltage- and Zero-Current-Switching Full-Bridge Converter With Secondary Resonance," *IEEE Trans. Ind. Electron.*, vol. 57, no. 3, pp. 1017-1025, March 2010.

[39] M. Pahlevani, S. Pan and P. Jain, "A Hybrid Phase-Shift Modulation Technique for DC/DC Converters With a Wide Range of Operating Conditions," *IEEE Trans. Ind. Electron.*, vol. 63, no. 12, pp. 7498-7510, Dec. 2016.

[40] H. Wu, J. Zhang, X. Qin, T. Mu and Y. Xing, "Secondary-Side-Regulated Soft-Switching Full-Bridge Three-Port Converter Based on Bridgeless Boost Rectifier and Bidirectional Converter for Multiple Energy Interface," *IEEE Trans. Power Electron.*, vol. 31, no. 7, pp. 4847-4860, July 2016.

Extended Kalman Filter based Speed Estimation for the Control of PMSG

Mukhtiar Singh

Electrical Engineering Deptt.
Delhi Technological University, Delhi, India
Email: smukhtiar_79@yahoo.co.in

Abstract—This paper deals with the speed and position estimation algorithm for the sensorless control of wind energy conversion system (WECS) having permanent magnet synchronous generator (PMSG) as electromechanical converter. Here, the main purpose is to get rid of the sensors and their issues related to maintenance, connection, tearing and wearing etc. A thorough mathematical modelling of PMSG has been presented. Moreover, Extended Kalman Filter (EKF) based observer has been proposed for the speed and position estimation of PMSG. The proposed PMSG based WECS has been modelled and simulated using the different blocksets of MATLAB/*SIMPOWERSYSTEM*. An extensive simulation study has been conducted and various results have been discussed and analyzed to validate the EKF based sensorless control of PMSG driven wind turbine.

Keywords— PMSG, Wind Turbine, Sensorless Control, Kalman Filter, Extended Kalman Filter

I. INTRODUCTION

Wind energy can be harvested in several ways. Stall or pitch-controlled turbines with fixed speed operation were very common in earliest commercial wind farms. With the advancement in power electronics, now a days variable speed wind energy conversion systems are being installed all over world. Most of these systems generally consist of gear box to run the generator on high speed in order to reduce number of poles, which further results in lowering the size, weight and inertia of the generator. But the use of gear box can reduces the efficiency up to 10%, depending on design and gear ratio. The gear box also requires regular maintenance due to tearing and wearing resulting into an extra cost and reduced reliability. Thus, the WECS having gear box are not suitable for off shore or remote/isolated power generation [1].

For the elimination of gearbox, the design of generator needs to be modified to make it suitable for low speed operation with speed range in between 15-75 rpm. It is very difficult to have low speed Asynchronous generators as it will necessitate to have small pole pitch with a large air-gap. This kind of generators are practically unsuitable as they require to have large amount of reactive power support. Therefore, for directly driven low speed wind turbines, the synchronous generators are more preferable [2]. In synchronous generator, the rotor needs to be excited separately in order to generate sufficient magnetic field. The excitation system consists of carbon brushes and slip rings, which results in additional losses in field windings and requires regular maintenance of brushes/slip rings. Moreover, the use of multi-poles increases the size and weight of generator due to bulky field windings, which is highly undesirable as it introduces additional cost of transportation and needs strong tower structure. With advancement in magnetic material, the permanent magnet machines can now proved to be alternative choice. In PMSG, the magnetic field is provided by the permanent magnets mounted on rotor shaft itself. This eliminates the excitation system and its side effects related to rotor winding loss, tearing-wearing and maintenance of slip rings/brushes etc. The use of advance magnetic material can also increases the power density and compactness of generator. Thus, PMSG are more suitable for wind power generation, especially for directly driven wind turbines.

The main advantage of directly driven variable speed WECS is their ability to keep on changing their operating speed according to variable wind velocity so that maximum power could be generated under all operating conditions. For this purpose exact position and speed of rotor is required as input to speed controller. Generally an encoder or Techo-generator is used for this purpose. But the inclusion of these types of sensors generally suffers from following drawbacks.

* The speed or position sensors may suffer under hostile operating conditions related to weather, temperature variation, moisture, corrosive gasses and vibrations in generator.

* Problems of noise present in electrical signals which are used as input to speed and position sensors.

* Chances of mechanical damage to sensor, connecting cables or connectors.

* Inclusion of sensors also increases cost and complexity of the overall system, especially in case of small rating turbine where the main emphasis is given not only to reduce the no. of components but cost as well.

Therefore, a sensor less control strategy is required to exhibit maximum power point tracking (MPPT). The sensorless control MPPT can be performed by direct or indirect methods. In direct method, the speed of generator is estimated from generator terminal voltage and currents. There are lot of methods based on Motional EMF, model reference adaptive system (MRAS), variation of Flux-linkage, and Kalman Filter

978-1-7281-4181-7/19 $31.00 © 2019 IEEE

etc. However, it is very difficult to estimate the precise rotor position & speed especially at low amplitude of voltage and current at lower speed. Further, the estimation may also suffer due to the simplified modelling of machine which is based on on several guesstimates, and neglection of parameter variation [3]-[19]. Further the accurate prediction of position and speed is almost impossible in case of directly driven WECS, where the generator rotor speed is very low ranging from 15-75 rpm, depending on wind velocity. Therefore, an alternative method based on EKF has been proposed to estimate speed and position. The observer based EKF method offers a powerful computation-intensive alternative. Basically, it is a full-order observer for recursive state estimation of nonlinear dynamic systems (in the presence of noise).

II. System Configuration

The proposed system mainly comprises PMSG based directly driven WECS where two back to back connected fully controllable converters have been used for grid integration and PMSG control. The complete block diagram of proposed system is shown in Fig. 1.

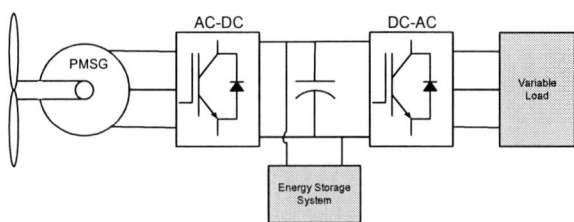

Fig. 1. System Block Diagram

A. Control Description

The comprehensive control diagram of thesystem is shown in Fig. 2. Here, the generator side converter is controlled to perform MPPT. For this purpose, the generator terminal voltage and currents are sensed. The clarke's transformation is applied to convert these quantities in to two phase quantities (stationary refrence frame). These transformed machine terminal voltages and currents are applied to EKF based observer in order to estimate the rotor speed and position. The estimated speed is further applied to torque speed characteristics of the generator in order to perform MPPT.

The grid side converter is primarily used for grid synchronization and DC-Link voltage regulation purpose. The output of DC voltage regulator gives the reference active current component. This refrence current is further multipled with unity vector template driven from the grid voltages using phase lock loop (PLL). The complete control diagram of the system is shown below:

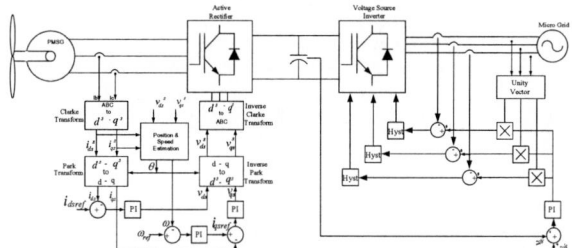

Fig. 2. Control Diagram of Grid connected WECS

III. Control Design

The system design and control is primarily focused on the speed and position estimation of PMSG. Therfore, the control system is mainly devided in the two categories: Mathematical modelling of PMSG and EKF based speed estimator design.

A. Modelling of PMSG

The PMSG is found to have inherent advantages of reduced copper loss, lower inertia, high power density and lower maintenance requirements due to the absence of rotor field winding and its associated excitation system. Here, it is pertinent to mention that the equivalent circuit of machine remains same as of a wound rotor except that the resultant field current I_f must be taken as constant or the flux linkage $\psi_f = L_m I'_f$ = constant. The equivalent transient circuits of PMSG in synchronously rotating frame for quadrature and direct axis are shown in Fig. 3 and Fig. 4, respectively with the assumption that the machine do not have any damper winding. So finally in case of PMSG analysis the wound rotor equivalent circuit can be considered by ignoring the dotted portion of its equivalent circuit [20].

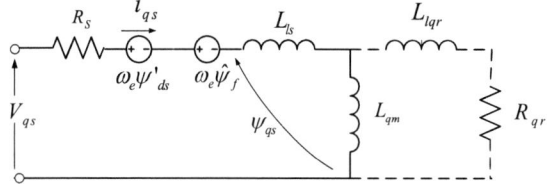

Fig.(3) q-axis Equivalent Model of PMSG

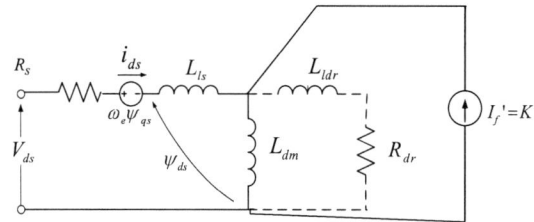

Fig.(4) d-axis Equivalent Model of PMSG

From the d-q equivalent model of PMSG the circuit equations are as follows:

978-1-7281-4181-7/19 $31.00 © 2019 IEEE

$$V_{qs} = R_s i_{qs} + \omega_e \psi'_{ds} + \omega_e \hat{\psi}_f + \frac{d}{dt}\psi_{qs} \qquad (1)$$

$$V_{ds} = R_s i_{ds} + \omega_e \psi_{ds} + \frac{d}{dt}\psi_{ds} \qquad (2)$$

Where

$$\hat{\psi}_f = L_{dm} I'_f \qquad (3)$$

$$\psi'_{ds} = (L_{dm} + L_{ls})i_{ds} = L_{ds}i_{ds} \qquad (4)$$

$$\psi_{ds} = \psi'_{ds} + \hat{\psi}_f \qquad (5)$$

$$\psi_{qs} = (L_{qm} + L_{ls})i_{qs} = L_{qs}i_{qs} \qquad (6)$$

And the torque equation is given as

$$T_e = \frac{3}{2}\left(\frac{P}{2}\right)(\psi_{ds}i_{qs} - \psi_{qs}i_{ds}) \qquad (7)$$

On substituting equations (3) - (6) in (1), (2) and (7) and simplifying, we will get

$$\frac{di_{qs}}{dt} = \frac{\omega_b}{X_{qs}}\left[V_{qs} - R_s i_{qs} - X_{ds}\frac{\omega_e}{\omega_b}i_{ds} - \frac{\omega_e}{\omega_b}V_f\right] \qquad (8)$$

$$\frac{di_{ds}}{dt} = \frac{\omega_b}{X_{ds}}\left[V_{ds} - R_s i_{ds} - X_{qs}\frac{\omega_e}{\omega_b}i_{qs}\right] \qquad (9)$$

$$T_e = \frac{3P}{4\omega_b}\left[(V_f + F'_{ds})i_{qs} - F_{qs}i_{ds}\right] \qquad (10)$$

Where $V_f = \omega_b \hat{\psi}_f$, $X_{ds} = \omega_b L_{ds}$, $X_{qs} = \omega_b L_{qs}$,

, $F'_{ds} = \omega_b \psi'_{ds}$, $F_{qs} = \omega_b \psi_{qs}$, and $\omega_b =$ base frequency.

The time derivative terms in equations (1) and (2) may be neglected under steady-state operating conditions of machine, and accordingly they may be re-written as

$$V_{qs} = R_s i_{qs} + \omega_e(\psi_f + L_{ds}i_{ds}) \qquad (11)$$

$$= R_s i_{qs} + V_f + X_{ds}i_{ds}$$

$$V_{ds} = R_s i_{ds} - X_{qs}i_{qs} \qquad (12)$$

The information of speed can be obtained from rotating flux vector and the flux vector can be estimated from generator terminal voltages and currents by using following equations [36].

$$\psi_{ds}^{\ s} = \int(v_{ds}^{\ s} + R_s i_{ds}^{\ s}).dt \qquad (13)$$

$$\psi_{qs}^{\ s} = \int(v_{qs}^{\ s} + R_s i_{qs}^{\ s}).dt \qquad (14)$$

$$\hat{\psi}_s = \sqrt{\psi_{ds}^{\ s^2} + \psi_{qs}^{\ s^2}} \qquad (15)$$

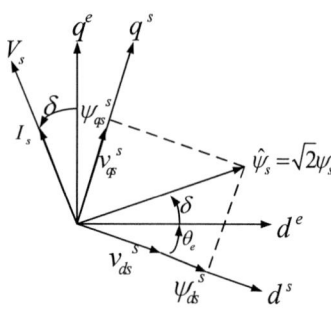

Fig. (5) Phasor Diagram based on d-q axis

$$\cos(\theta_e + \delta) = \frac{\psi_{ds}^{\ s}}{\hat{\psi}_s} \qquad (16)$$

$$\sin(\theta_e + \delta) = \frac{\psi_{qs}^{\ s}}{\hat{\psi}_s} \qquad (17)$$

$$\omega_e = \frac{d\theta_e}{dt} = \frac{(v_{qs}^{\ s} + i_{qs}^{\ s})\psi_{ds}^{\ s} - (v_{ds}^{\ s} + i_{ds}^{\ s})\psi_{qs}^{\ s}}{\hat{\psi}_s^{\ 2}} \qquad (18)$$

Here, $i_{ds}^{\ s}$, $i_{qs}^{\ s}$, $v_{ds}^{\ s}$, and $v_{qs}^{\ s}$ are the stationary reference frame (Clarke-transform) currents and voltages on d^s-q^s axis.

B. Extended Kalman FilterDesign

Among the few Observer-based position and speed estimation methods of PMSM suggested in literature, the Extended Kalman Filter method offers a powerful computation-intensive alternative. Being a full-order observer involving noise, it can be very helpful in estimating the state of nonlinear dynamic systems in real time using powerful DSP. The machine terminal voltages (V_{abc}) and currents (I_{abc}) are converted to stationary ref. from to calculate currents $i_{ds}^{\ s}$, $i_{qs}^{\ s}$ and voltages $v_{ds}^{\ s}$, $v_{qs}^{\ s}$. These values are further given as input to EKF based observer in order to predict the rotor position and speed. Here for computation easiness $L_{qs} = L_{ds}$ is a valid assumption especially in case of surface mounted magnets. The voltage equations comprising stator flux linkages terms in stationary reference frame are written as

$$V_{ds}^{\ s} = R_s i_{ds}^{\ s} - \frac{d}{dt}\psi_{ds}^{\ s} \tag{19}$$

$$V_{qs}^{\ s} = R_s i_{qs}^{\ s} + \frac{d}{dt}\psi_{qs}^{\ s} \tag{20}$$

$$\psi_{ds}^{\ s} = L_s i_{ds}^{\ s} + \hat{\psi}_f \cos\theta_e \tag{21}$$

$$\psi_{qs}^{\ s} = L_s i_{qs}^{\ s} + \hat{\psi}_f \sin\theta_e \tag{22}$$

Where L_s= Synchronous Inductance of the machine. Now on differentiating equations (21) and (22) and then substituting in equations (19) and (20), respectively, we get

$$V_{ds}^{\ s} = R_s i_{ds}^{\ s} + L_s \frac{d}{dt}i_{ds}^{\ s} - \hat{\psi}_f \omega_e \sin\theta_e \tag{23}$$

$$V_{qs}^{\ s} = R_s i_{qs}^{\ s} + L_s \frac{d}{dt}i_{qs}^{\ s} + \hat{\psi}_f \omega_e \cos\theta_e \tag{24}$$

$$\frac{d\theta_e}{dt} = N_P.\omega_r$$

These equations contain the states $i_{ds}^{\ s}$, $i_{qs}^{\ s}$, ω_r, and θ_e, out of which the last two states are need to be determined. The machine dynamic model in state-variable form may be expressed as:

$$\frac{dX}{dt} = A.x + B.U \tag{25}$$

$$Y = CX \tag{26}$$

Where X=[$i_{ds}^{\ s}$ $i_{qs}^{\ s}$ ω_r θ_e]^T is the state vector, U=[$v_{ds}^{\ s}$ $v_{qs}^{\ s}$]^T is the input vector, Y=[$i_{ds}^{\ s}$ $i_{qs}^{\ s}$]^T is the output vector, where the matrices A, B and C are as follows:

$$A = \begin{bmatrix} -\dfrac{R_s}{L_s} & 0 & \dfrac{\hat{\psi}_f.N_P}{L_S}\sin\theta_e & 0 \\ 0 & -\dfrac{R_s}{L_s} & -\dfrac{\hat{\psi}_f.N_P}{L_S}\cos\theta_e & 0 \\ 0 & 0 & 0 & 0 \\ 0 & 0 & N_P & 0 \end{bmatrix} \tag{27}$$

$$B = \begin{bmatrix} \dfrac{1}{L_s} & 0 \\ 0 & \dfrac{1}{L_s} \\ 0 & 0 \\ 0 & 0 \end{bmatrix} \tag{28}$$

and

$$C = \begin{bmatrix} 1 & 0 & 0 & 0 \\ 0 & 1 & 0 & 0 \end{bmatrix} \tag{29}$$

Here, the non-linear model of machine as per equations (25) and (26) may be approximated in its discrete form as
$$x_k = (A.T + I).x_{k-1} + B.T.u_{k-1} \tag{30}$$

$$y_k = C.x_k \tag{31}$$

Where I and T represents the unity matrix and sampling time interval, respectively. Now the matrix at K_{th} interval can be written as

$$A_k = \begin{bmatrix} 1 - T.\dfrac{R_s}{L_s} & 0 & T.\dfrac{\hat{\psi}_f.N_P}{L_S}\sin\theta_e & 0 \\ 0 & 1 - T.\dfrac{R_s}{L_s} & -T.\dfrac{\hat{\psi}_f.N_P}{L_S}\cos\theta_e & 0 \\ 0 & 0 & 1 & 0 \\ 0 & 0 & T.N_P & 1 \end{bmatrix}_k \tag{32}$$

Now the prediction of state x_k can be determined in two steps:
1. Prediction/Time updating step:

$$x_k = f(x_{k-1}, U_k, 0), \tag{33}$$

$$P_k^- = A_k.P_{k-1}.A_k^T + W_k.Q_{k-1}.W_k^T \tag{34}$$

2. Correction/Measurement updating step:

$$K_k = P_k^-.C_{k-1}^T (C_k.P_k^- C_k^T + V_k.R.V_k^T)^{-1} \tag{35}$$

$$x_k = x_k^- + K_k(y_k - C(x_k^-, 0) \tag{36}$$

$$P_k = (1 - K_k C_k)P^- \tag{37}$$

Where x_{k-1} and x_k^- are the posteriori and priori state estimation at step k-1; and k, respectively. Here, P_k and P_k^- stands for posteriori and priori state prediction error covariance matrix at step k; W_k, V_k are jacobian matrix, which are further reduced to unity matrices. Q_k and R_k are constants. A block diagram representation of Extended Kalman Filter is shown below.

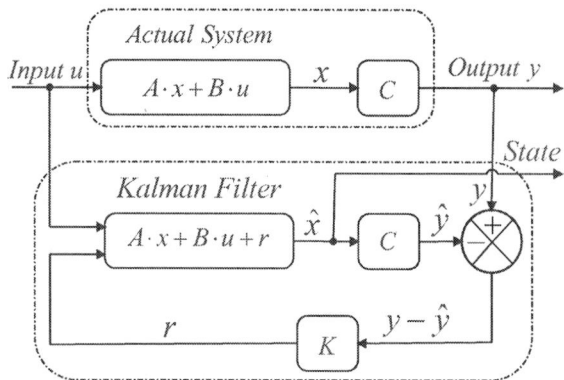

Fig. (6) Block Diagram of Extended Kalman Filter

IV. RESULT ANALYSIS AND DISCUSSION

The above discussed EKF based position and speed estimating algorithm has been designed and simulated in MATLAB. The detailed simulation fallouts for the sensor less control of PMSG are shown in Fig. 7. The variable torque according to variable wind velocity is applied to the PMSG and then the speed of generator is regulated in order to track the maximum power point. The Fig.7 (a) shows the applied variable torque. In Fig. 7(b), the profile of actual speed, reference speed, and estimated speed have been presented while the actual and estimated rotor positions are shown in Fig. 7(c). The error between actual and estimated speed is shown in Fig. 7(d), which is almost zero for different speeds starting from very low speed of 10 rad/sec. to a maximum speed of 30rad/sec. The pattern of 3–phase PMSG current waveforms are presented in Fig. 7(e). Here, the variable speed control and operation is confirmed by the variable frequency and amplitude of these currents waveforms. Fig. 7(f), presents the generated power under different operating speeds.

Initially, the applied torque is 25 Nm; at this torque the reference speed is 10 rad/sec and the controller generates the speed control command to follow the reference speed. The corresponding rotor position, current and power are as shown in Fig. (7). At t=0.5sec, the applied input torque starts increasing and get settled at 30 Nm, in response to increased applied torque, the PMSG speed also increases, resulting in to the increased value of generator current and power. At t=1sec, the input torque is again increased to 45 Nm, which results into increased speed of 30rad/sec., the peak currents of 35 A and power of 1050 Watts. The applied torque starts decreasing at t=1.5 sec. and finally, sets to a new value of 35 Nm. As a response, the decrease in generator speed, current and power can be observed from their respective traces. At t=2.5sec., the applied torque is again decreased to the value of 25Nm and the systems behaves just as in the beginning. From the traces of generator speed, it is quite apparent that the estimator is able to predict the generator speed and position accurately under varying operating conditions, especially at lower speed region, which is the desired goal as the directly driven PMSG are primarily expected to operate at lower speeds.

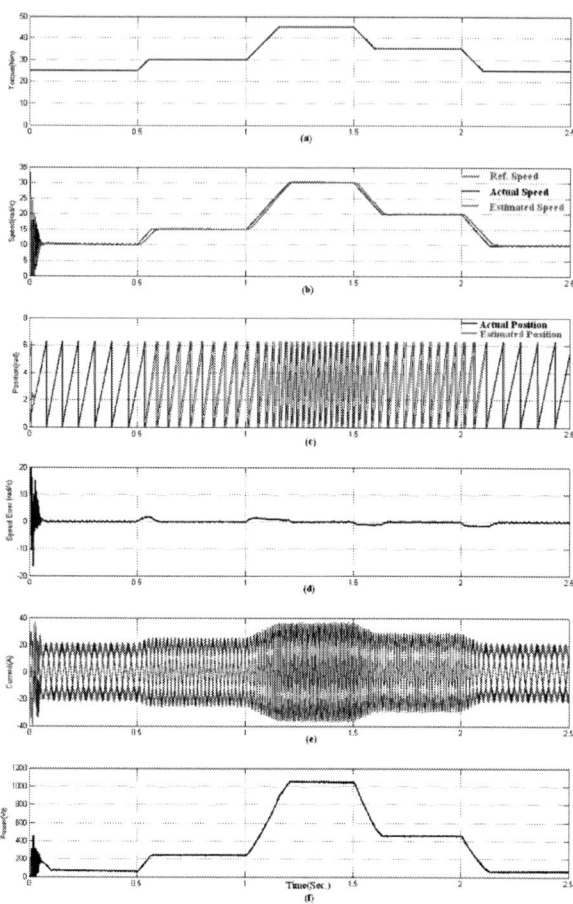

Fig. 7. Simulation Results for (a) Input Torque, (b) Traces of Actual, Ref., and Estimated Speed, (c) Actual and estimated position, (d) Error between Actual and Estimated Speed, (e) PMSG current, and (f) Power at DC-Link

V. CONCLUSION

In this work, EKF based observer is developed to estimate the speed and position of PMSG. The proposed algorithm is able to estimate the speed and position within vast operating range with almost zero error under steady state and dynamic conditions of operatin. Moreover, the system was able to work precisely under low speed operating region, which is the major requirement of directly driven wind turbines. A systematic mathematical model of PMSG is presented to design the EKF speed and position estimator. The proposed algorithm very precisely predicts the position and speed, which are further utilized to control the speed of PMSG in order to extract maximum power under different operating conditions. The proposed system is simulated in MATLAB/SimPowerSystem environment and the effectiveness of proposed algorithm is demonstrated through detailed simulation results.

References

[1] Westlake A. J. G., Bumby J.R., Spooner E.,"Damping the power-angle oscillations of a permanent magnet synchronous generator with

978-1-7281-4181-7/19 $31.00 © 2019 IEEE

particular reference to wind turbine applications". *IEE Proceedings, Electr. Power Appl.*, Vol 143, No 3, May, 1996.

[2] Binder A., Schneider T.,"Permanent magnet synchronous generators for regenerative energy conversion– a survey". European Conference on Power Electronics and Applications, EPE 2005, 11-14 Sept., Dresden, Germany, 2005.

[3] T. Senjyu, E. Billy, A.Yona, N. Urasaki "Maximum Wind Power Capture by Sensorless Rotor Position and Wind Velocity Estimation from Flux Linkage and Sliding Observer," *International Journal of Emerging Electric Power Systems,* The Berkeley Electronic Press, Vol. 8,Issue 2, Article3, 2007.

[4] K. Tan, S. Islam, " Optimum Control Strategies in Energy Conversion of PMSG Wind Turbine System Without Mechanical Sensors" *IEEE Trans. Energy Conversion*, vol. 19, no. 2, June 2004

[5] M. Abdelrahem, C. M. Hackl and R. Kennel, "Finite Position Set-Phase Locked Loop for Sensorless Control of Direct-Driven Permanent-Magnet Synchronous Generators," in *IEEE Transactions on Power Electronics*, vol. 33, no. 4, pp. 3097-3105, April 2018.

[6] Y. Zhao, C. Wei, Z. Zhang and W. Qiao, "A Review on Position/Speed Sensorless Control for Permanent-Magnet Synchronous Machine-Based Wind Energy Conversion Systems," in *IEEE Journal of Emerging and Selected Topics in Power Electronics*, vol. 1, no. 4, pp. 203-216, Dec. 2013.

[7] R. I. Putri, M. Pujiantara, A. Priyadi, T. Ise and M. H. Purnomo, "Maximum power extraction improvement using sensorless controller based on adaptive perturb and observe algorithm for PMSG wind turbine application," in *IET Electric Power Applications*, vol. 12, no. 4, pp. 455-462, 4 2018.

[8] B. Terzic and M. Jadric, "Design and implementation of the extended Kalman filter for the speed and rotor position estimation of brushless DC motor," *IEEE Trans. Ind. Electron.*, vol. 48, no. 6, pp. 1065–1073, Dec. 2001.

[9] S. Bolognani, L. Tubiana, and M. Zigliotto, "Extended Kalman filter tuning in sensorless PMSM drives," *IEEE Trans. Ind. Appl.*, vol. 39, no. 6, pp. 1741–1747, Nov./Dec. 2003.

[10] J. Hu and B. Wu, "New integration algorithms for estimating motor flux over a wide speed range," *IEEE Trans. Power Electron.*, vol. 13, no. 5, pp. 969–977, Sep. 1998.

[11] Z. Zhang, Y. Zhao, W. Qiao and L. Qu, "A Space-Vector-Modulated Sensorless Direct-Torque Control for Direct-Drive PMSG Wind Turbines," in IEEE Transactions on Industry Applications, vol. 50, no. 4, pp. 2331-2341, July-Aug. 2014.

[12] Z. Chen, M. Tomita, S. Doki, and S. Okuma, "An extended electromotive force model for sensorless control of interior permanent-magnet synchronous motors," *IEEE Trans. Ind. Electron.*, vol. 50, no. 2, pp. 288–295, Apr. 2003.

[13] T. Mahmoud, Z. Y. Dong and J. Ma, "Integrated optimal active and reactive power control scheme for grid connected permanent magnet

synchronous generator wind turbines," in *IET Electric Power Applications*, vol. 12, no. 4, pp. 474-485, 4 2018.

[14] M. Singh and A. Chandra, "Application of Adaptive Network-Based Fuzzy Inference System for Sensorless Control of PMSG-Based Wind Turbine With Nonlinear-Load-Compensation Capabilities," in *IEEE Transactions on Power Electronics*, vol. 26, no. 1, pp. 165-175, Jan. 2011.

[15] S. Morimoto, K. Kawamoto, and Y. Takeda, "Position and speed sensorles control for IPMSM based on estimation of position error," *Electr. Eng. Jpn.*, vol. 144, no. 2, pp. 43–52, 2003.

[16] M. Singh and A. Chandra, "Comparative study of sliding mode and ANFIS based observers for speed & position sensor-less control of variable speed PMSG," *CCECE 2010*, Calgary, AB, 2010, pp. 1-4.

[17] Paul P. Acarnley, John F. Watson, "Review of Position-Sensorless Operation ofBrushless Permanent-Magnet Machines," *IEEE Trans. Ind. Electron.*, vol.53, no. 2, April 2006

[18] L. Ying and N. Ertugrul, "A novel, robust DSP-based indirect rotor position estimation for permanent magnet AC motors without rotor saliency," *IEEE Trans. Power Electron.*, vol. 18, no. 2, pp. 539–546, Mar. 2003.

[19] M. G. Simoes, B. K. Bose, and R. J. Spiegel, "Fuzzy logic based intelligent control of a variable speed cage machine wind generation system,"*IEEE Trans. Power Electron.*, vol. 12, no. 1, pp. 87–95, Jan. 1997.

[20] B. K. Bose, Modern Power Electronics and AC Drives, Prentice Hall 2002.

Mukhtiar Singh received the B.Tech. and M.Tech. degrees in electrical engineering from National Institute of Technology (Erstwhile REC Kurukshetra), Kurukshetra, India, in 1999 and 2001, respectively, and the Ph.D. degree from Ecole de Technologie Superieure, University of Quebec, Montreal, QC, Canada. He is presently working as full Professor in the Department of Electrical Engineering, Delhi Technological University (DTU), New Delhi, India.

Prof. Singh is recipient of the IEEE student scholarship for one of the best paper in 34th IEEE Conference on Industrial Electronics, IECON-2008, held at Orlando, FL, USA. He is also the winner of Researchers Merit Scholarship of the University of Quebec, Montreal, Canada for the three years consecutively, 2008, 2009, and 2010.

AUTHOR INDEX

Abrantes-Ferreira, Armando J. G. 1010
Afonso, Leonardo Carlos............445
Aguiar, Maria Júlia Rosa 826
Albuquerque, Lorena 838
Albuquerque, Lorena Lorraine Oliveira............1040
Aller, José Manuel 1289
Almeida, Andrei De O. 25
Almeida, Antonio D. D. 868
Almeida, Felipe A. F. 1469
Almeida, Pedro M. 19, 585, 1360
Almeida, Pedro S............730, 1052, 1118
Almeida, Pedro Santos............370, 579
Alonso, Augusto Matheus Dos Santos 269, 445
Altuna, José A. Torrico 952
Alves, Fábio 880
Alves, Marcos Henrique Da Silva 160
Alves, Max Dannyel De Carvalho 1447
Amaral, Fernando Venâncio............1004
Amorim, W. C. S. 790
Andersen, Romero 838
Andersen, Romero Leandro............1040
Andrade, Allan V. S. 621
Andrade, António M. S. S. 856
Andrade, Lucas Mendonça 1082
Andrea, Cristiano Quevedo............998
Antunes, Fernando............1124
Antunes, Fernando L. M. 1118, 1331
Araújo, Leandro R. 19, 1360
Araújo, Lucas 435
Araujo, Lucas S. 597
Araujo, Renato G. 1307
Archetti, João A. G. 742
Aredes, Mauricio............184
Aredes, Maurício............880
Arenas, Luís De Oro 1136, 1211, 1487
Attuati, Gabriel 1064
Ayala, M. 1441
Azcue, José L. 347, 1166
Azeredo, Lucas F. S. 898
Azevedo, Gustavo M. S. 1390
Bacheti, Gabriel G. 772
Balbino, Anderson José 886
Barbi, Ivo............917
Barbosa, Paula Stael S. 1028
Barbosa, Pedro G. 19, 25, 263, 585, 910, 1360
Barbosa, Pedro Gomes 13, 423
Barbosa, Samanta Gadelha............591, 680
Barbosa, Victor E. S. 472, 993
Barcelos, R. P. 969
Barcelos, Silvangela L. 400
Barcelos, Silvangela L. S. L. 1034
Barreto, Luiz Henrique Silva Colado 680
Barros, Luciano Sales 748
Barros, Tárcio André Dos S. 686

Bartsch, Arthur G. 923, 1325, 1384
Bartsch, Camila R. S. 923
Batista, Edson Antonio 998
Batista, Oureste E............898, 1378
Batista, Oureste Elias 1
Batschauer, A. L. 220
Batschauer, Alessandro L............724
Batschauer, Alessandro Luiz 208
Bekoski, Kleber Chan 245
Belchior, Fernando Nunes 531
Bellinaso, Lucas Vizzotto............406, 940
Bento, Aluísio Alves De Melo 1396, 1458
Bernardeli, Victor Regis 305
Bertoldi, B. 417
Bessa, Isaías V............172
Bessa, Iury V. 172, 1076
Bettiol, Arlan Luiz 917
Bisogno, Fábio E. 233
Blaabjerg, Frede 1
Boaventura, W. C. 615
Bonaldi, Erik Leandro 85
Bonaldo, Jakson Paulo 269
Bonifacio, Joao............55
Borchardt, Mauricio 376
Borges, Victor Luiz Flor 917
Borges-Da-Silva, Luiz Eduardo 85
Borin, Lucas C. 202, 519
Bradaschia, F. 220, 716
Braga, Alexandre Viana............49
Braga, Henrique Antônio Carvalho............370, 579
Braga, Mateus Freitas 423
Branco, Carlos Gustavo C. 1307
Brandao, Danilo I. 364, 466, 597
Brandao, Danilo Iglesias 445
Brasil, Thiago Americano Do 934
Bravo, P. 1402
Bressan, Marcos Vinicius 724
Brinker, Tobias 975
Brito, Moacyr 1195
Brito, Moacyr A. G. 501
Bueno, Alexandre Galvão 311
Bueno, Mateus De F. 376
Busarello, Tiago Davi Curi 125
Buso, Simone............143
Buticchi, Giampaolo............257
Cabrera, Ana 507
Cacau, Ronny Glauber De Almeida............886
Cagnini, Paulo R. 993
Caicedo, Jorge 880
Caldeira, Carolina A. 820, 868
Caldognetto, Tommaso 143
Calil, Henrique A. M. 92
Caliman, João Olímpio 844
Callegari, João M. S. 478

AUTHOR INDEX

Cambero, Eduardo Vicente Valdés 627, 1349
Campos, Rafael De Farias 299
Campos, Rafael Espino 1130
Capovilla, Carlos E. 952
Capovilla, Carlos Eduardo 627, 1349
Carati, Emerson G. 472, 993, 1148
Carati, Emerson Giovani......................1142, 1420
Cárdenas, Roberto 61
Cardoso, Rafael 472, 993, 1142, 1148, 1223, 1420
Cardoso, Ronnan De B. 1295
Carralero, Leandro L. O. 778, 964
Carvalho, Edivan Laercio 1142, 1420
Carvalho, Henrique T. M. 288
Carvalho, Itaiara F. 1372
Casella, Ivan Roberto Santana.....................627, 1349
Cavalca, Mariana S. M. 1325
Cavalcante, Levy R. 958
Cavalcanti, Dayse M. 1325
Cavalcanti, Marcelo C. 1390
Cavilha, Pedro P. 692
Chacon, Igor V. 668
Chub, Andrii 1178, 1493
Ciarnoscki, Pábulo F. 1172
Coelho, Ernane A. A. 98, 288, 1253
Coelho, Roberto F. 251
Coelho, Roberto Francisco 451, 639, 651
Colque, Juan C. 347
Conceição, Claudio Alvares 1004
Conde, Eliomar R.D. 73
Conde, Manoel 411
Cordero, Raymundo 1195, 1453
Cornelius, Richard G. 461, 543
Corrêa, Douglas Rosa 281
Corrêa, Henrique Pires 1414
Correa, Maurício B. R. 1106, 1372
Corrêa, Rafael Di Lorenzo 49
Côrtes, Hyghor Miranda 1366
Cortizo, P. C. 178
Costa, Fabiano F. 778, 964
Costa, Jean Patric 1142
Costa, Louelson A. 1106
Costa, Paulo Júnior Silva 1301
Costabeber, Alessandro 257
Cupertino, A. F. 573, 615, 790
Cupertino, Allan F. 466, 478, 850
Da Câmara, Raphael A. 904
Da Costa, André E. L. 1295
Da Costa, Hugo Álisson Alves....................1265
Da Costa, Jean P. 472, 1148
Da Costa, Jean Patric 993, 1420
Da Fonseca, Brayan Sobral 946
Da Luz, Caio Meira Amaral................................239
Da Paz, Gililelson F. 820, 1100, 1475
Da Paz, Humberto Pereira 627, 1349

Da Rocha, Antonia F. 1118, 1331
Da Rocha, Lorrana F. 1058
Da Rosa, Murilo Brunel................................ 1217
Da Silva, Alexandre G. F. 1100, 1475
Da Silva, Bruno Alves Sousa 680
Da Silva, Carlos Henrique 531
Da Silva, Edison R. C. 1295
Da Silva, Edison Roberto C. 1022, 1154
Da Silva, Eduardo Luiz Santos........................ 382
Da Silva, Gabriel M..................................... 1172
Da Silva, Guilherme S. 461, 543
Da Silva, João Lucas 674
Da Silva, Luiz Eduardo Borges 531
Da Silva, Mauricio M. 856
Da Silva, Newton 125, 1343
Da Silva, Rogerio Luiz 917
Da Silva, Sérgio Augusto Oliveira.............. 484, 1301
Da Silva, Simplicio A. 1475
Da Silva, Vinícius Santana........................ 627, 1349
Da Silva, Werbet Luiz Almeida........................ 1283
Da Silveira, D. B. 615
Da Siveira, Augusto W. Fleury Veloso 981
Dacol, Rodrigo Patrício................................ 341
Dai, Hang 67
Dalmolin, Thiago Brezolin............................. 406
Damasceno, Debora Pereira 591
Dantas, Marcos.. 838
De Aguiar, Manoel L.................................. 108
De Aguiar, Manuel Luis................................ 892
De Albuquerque, Vinicius M. 730, 1052
De Almeida, Antonio D. D. 820
De Almeida, Bruno Ricardo 591, 680
De Almeida, Pedro M..............25, 263, 910, 1058
De Almeida, Rogério Gaspar......................... 987
De Almeida, Thales E. P. 149
De Alvarenga, Bernardo Pinheiro 1070
De Andrade, Darizon Alves 305
De Andrade, Guilherme Giglio 934
De Andrade, Jessika Melo 451
De Andrade, Khristian M. 149
De Araújo, Humberto Xavier 627, 1349
De Araújo, Rafael Magalhães Nóbrega 1265
De Barros, R. C. 615
De Bessa, Isaías Valente.............................. 190
De Bessa, Iury Valente................................ 190
De Brito, Moacyr A. G. 513, 832
De Camargo, Robinson Figueiredo..................... 1064
De Campos, Luis F. F. 923
De Carvalho, Gabriel Ubirajara 1223
De Carvalho, Marenice M. 1076
De Castro, Allan G. 108
De Castro, Allan Gregori 335, 892
De Castro, Marcelo 910
De Figueiredo, Hugo F. M. 814, 874

AUTHOR INDEX

De França, Pedro Henrique Thiesen 633
De Freitas, Isaac S. 1100, 1475
De Freitas, Isaac Soares 1040
De Freitas, Luiz C. G. 288, 1253
De Freitas, Marcos Antonio Arantes 305
De Freitas, Tiara Rodrigues Smarssaro 844
De Lacerda, Macklyster Lãnucy Scherre Stofel 455
De Lima, Gustavo B. 288
De Lucena, Lucas Fabrício M. 1022
De Matos, José G. 555, 1229
De Matos, José Gomes 275, 400
De Medeiros, Luiz Otávio Campos 49
De Morais, Aniel Silva 79, 281
De Morais, Fernanda Carnielutti 929
De Novaes, Yales R. 376
De Novaes, Yales Rômulo 214, 784
De Oliveira, André L. P. 478
De Oliveira, Azauri A. 484
De Oliveira, Carlos M. R. 108, 335
De Oliveira, Carlos Matheus R. 892
De Oliveira, Cássio Alves 981
De Oliveira, Fellipe André Lucena 1040
De Oliveira, Janaína G. 567, 742, 910, 1028
De Oliveira, José 299, 923, 1384
De Oliveira, Júlio Cesar Lopes 495
De Oliveira, Leonardo W. 567, 742
De Oliveira, Pedro Andrade..85
De Oliveira, Raphael A. Bispo 1366
De Oliveira, Rodrigues 1046
De Paula, Geyverson T. 149, 335
De Sá, Marcos Victor Dantas 1040
De Seixas, Falcondes José Mendes 1136, 1211, 1487
De Sousa, Gean J. M. 802
De Sousa, R. O. 790
De Souza, H. E. P. 716
De Souza, Igor D. N. 263, 910
De Souza, Marina Hassen..1112
De Souza, Ramon Henriques.....................................1313
De Toledo, Henrique 1166
De Vasconcelos, Carlos Henrique Silva.................1271
Delamare, Guillaume...1408
Denardin, Gustavo W. 814, 874
Denardin, Gustavo Weber....................................245, 1223
Dias, Bruno H. 567
Dias, Mateus P. 766
Dias, Robson F. Da S. 1010
Dias, Robson F. S. 1034
Dicler, Felipe Novaes Francis 184
Diesel, José Adriano Damacena 633
Dill, Gustavo Kaefer 490
Dohler, Jessica S. 1028, 1058
Dos Reis, Mauro Sandro..934
Dos Santos, Ângelo Marcílio M................................862
Dos Santos, Cássia C. C. 1148

Dos Santos, Kristian Pessoa.............................561, 1447
Dos Santos, Rafael 1435
Dos Santos, Rodrigo L. 1118, 1331
Dos Santos, Stefan T. C. A. 108
Dos Santos, Tadeu F. 668
Dos Santos, Walbermark Marques........455, 844, 1426
Dragicevic, T. 1441
Drexhage, Paul... 1277
Duarte, Jorge L. 43
Duarte, Samuel N. 19, 585, 1360
Duarte, Samuel Neves 423
Eckstein, Rafael H. 1431
El Kattel, Menaouar Berrehil..................... 1217, 1354
Enomoto, Kelly 411
Escobar, Gerardo 376
Espindula, Carla J. 898
Estrabis, Thyago.. 1453
Estrabis, Thyago Vasconcelos 998
Ewerling, Marcos Vinícius Mosconi 358
Fabricio, Edgard L. L. 868
Faceroli, Silvana T... 704
Faistel, Tiago M. K. 856
Fajardo, Marco... 1289
Fardin, Jussara Farias 1337
Farias, João. V. M. 850
Fauth, Leon 975
Feltrin, Valesca Bettim 406
Feretti, Paulo Henrique 166
Fernandes, Andressa Da S....................................... 1331
Fernandes, Darlan A. 1295
Fernandes, Darlan Alexandria................................. 987
Fernandes, Filipe 1384
Fernandes, Marcelo C. 263
Fernandes, Marcelo De C. 730
Fernandes, Nicolas T. D. 597
Fernandes, Samir W. 609
Fernández-Ramírez, Luis M.................................... 904
Ferreira, Andre 1028
Ferreira, Andre A.. 1058
Ferreira, André Augusto 826
Ferreira, Julio Cesar 934
Ferreira, Kaique 1426
Ferreira, Myrlena R. M. 1229
Ferreira, Rodrigo A. F. 704, 1241
Ferst, Matheus K... 814, 874
Fiamoncini, Lucas 651
Filho, A. J. Sguarezi 73
Filho, Alfeu J. Sguarezi 778, 1183
Filho, Antonio Venâncio De M. Lacerda 1295
Filho, Braz De J. Cardoso............................ 160, 1004
Filho, Braz J. Cardoso 597
Filho, Ernesto Ruppert 686
Filho, Hermínio Miguel Oliveira 561
Filho, João Inácio Da Silva 1366

AUTHOR INDEX

Filho, José A. Olimpio 1435
Filho, José De Arimatéia Olímpio......269
Filho, Joselino Santana......85
Filho, Marcos José De Moraes......981
Flores, Mendes......1046
Fogli, Gabriel A. 263, 364, 910
Fonseca, Jean M. L. 1307
Font, Carlos Henrique Illa 358
Fontes, Guillaume 1408
Foster, João G. L. 131
Foster, Joao Gabriel Luppi 85
França, Bruno 880
Fraytag, Jeferson 525
Freitas, Alisson 1124
Freitas, Isaac 838
Freitas, Isaac S. 868
Freitas, Luiz C. G. 98
Friebe, Jens 975
Furlan, André......1408
Furtado, Pablo C. De S. 13
Gabbi, Thieli 698
Gajo, Clovis N. L. 1277
Galea, Michael......257
Galotto, Luigi 832
Galottojunior, Luigi......1195
Gama, Felipe O. S. 668
García, C. 1402
García, Raymundo Cordero......998
García-Triviño, Pablo 904
Gehrke, Camila S. 820
Gil, Eric......1435
Gobbato, Cassio 245
Godoy, Ruben Barros 832
Gomes, Hugo M. C......778
Gomes, Lucas Do N. 1010
Gomes, Luciano Coutinho......305, 981
Gomes, Marcos Paulo Brito 1112, 1313
Gonçalves, Flávio A. S. 1435
Gonçalves, Flávio Alessandro Serrão......1343, 1469
Gonzatti, Robson B. 131
Gonzatti, Robson Bauwelz 85, 531
Grabovski, E. F. C. 417, 1016
Grassi, Márcio Afonso Soleira 998
Gregor, R. 1441
Griffiths, Alison......227
Grigoletto, Felipe Bovolini 929
Gründling, Hilton A......549
Gruner, Víctor Ferreira 651
Guamán, Alex......507
Guazzelli, Paulo R. U. 108, 335
Guazzelli, Paulo Roberto U. 892
Guerra, Felipe 1469
Guerreiro, Joel F. 766
Guerreiro, Joel Filipe......736, 754

Guerrero, Josep M. 686
Guillardi, Hildo 766
Guimarães, Alexandre Magnus F. 1265
Guimarães, Bruno P. B. 531
Guo, Yanjie 104
Ha, Hengxu 227
Harikumaran, Jayakrishnan 257
Hartmann, Lucas V......820
Hayashi, Augusto 1195, 1453
Hayashi, Pedro......411
He, Zhengyou......441
Heerdt, Joselito Anastácio 341, 633
Heldwein, M. L. 417, 1016
Heldwein, Marcelo L......692, 802, 1408
Heldwein, Marcelo Lobo......525, 710, 1259
Henning, Luiz Fernando...... 495
Herrera, Felipe 61
Hoch, Henrique J. 856
Honório, Dalton De Araújo 680
Huang, Shili 1160
Husev, Oleksandr 1481
Idalgo, Gabriel F. 952
Ikhide, Monday 227
Imai, Fernando 214
Inácio, Cleber Onofre 1004
Jacobina, Cursino Brandão 1022
Jacoboski, Marcos José 1259
Jacomini, Rogério Vani 1183
Jahns, Thomas M. 67
Junior, Ademir Da S. T. 910
Júnior, Ademir S. T. 263
Junior, Dalmo C. Silva 1058
Júnior, Demercil De Souza Oliveira 561, 680
Junior, Emerson Abreu Bastos 239
Junior, Florindo A. C. 1076
Junior, Florindo A. C. Ayres 172, 190
Junior, Guilherme Penha Da Silva 748
Júnior, Hildo Guillardi 736, 754
Junior, Ildenor Davi S. 862
Junior, Josemar Alves Dos Santos 981
Junior, Lourenço Matakas 1189
Junior, Luigi G. 513
Junior, Nery De Oliveira...... 49
Júnior, Paulo R. M. 850
Junior, Sergio L. B. 692
Junior, Sérgio Luiz Sambugari 125
Kawahara, Yoshihiro 137
Kennel, Ralph 55
Khera, Fatma A. 196
Kirsten, André Luís......382, 525
Klumpner, Christian 196
Koch, Gustavo G. 202, 519
Koleff, Lucas...... 411, 429, 435
Komatsu, Wilson 411, 429, 435, 1189

AUTHOR INDEX

Korkh, Oleksandr 1178
Kouro, Samir......................................1178
Kraemer, Rodrigo A. S.............................472
Kumar, Dinesh 1160
Kutt, Lauri.......................................1481
Lafay, Jean M. S. 993
La-Gatta, Filipe A. 1241
Lago, Lucas F. R. 704
Lai, Jih-Sheng 1493
Lambert, Gustavo 784
Lambert-Torres, Germano 85, 131
Lange, A. B. 417
Lange, André De Bastiani 1259
Laurindo, Bruno 880
Lazzari, Thiago 698
Lazzarin, T. B....................................969
Lazzarin, Telles B. 388, 394, 796, 1431
Lazzarin, Telles Brunelli 358, 451, 639, 886
Leal, Daniel Franco...............................1313
Lee, Woongkul.....................................67
Leguizamón, Daniel M. Barrera 294
Leite, Valter J. S................................609
Lemes, Lucas José.................................305
Lessa, Tofoli.....................................1046
Li, Fang..353
Li, Shufan 353
Lima, Antonio Marcus Nogueira 1463
Lima, Francisco Kleber A. 1307
Lima, Gustavo B. 98
Lima, Mateus L. 742
Limongi, Leonardo R...............................1390
Lin, Siqi...975
Lino, Fernando....................................73, 1183
Liu, Qing...143
Liu, Yeran..441
Liu, Zhimeng......................................104
Lopes, Juliano De Pelegrini.......................245
López, Diego A. Bautista 294
López, Fabián R. Jiménez 294
Lucas, Kevin E. 31, 172, 190, 1076
Luiz, Alex-Sander Amável 1112, 1206, 1313
Lumertz, Mateus M. 335
Lumertz, Mateus Moro 892
Macedo, Elienai O.................................1034
Macedo, Tatiana Saviato...........................455
Machado, Sebastián De J. Manrique.................484
Madawala, Udaya K. 441, 663
Mai, Leonardo S. 796
Mai, Ruikun.......................................441
Maia, Amanda C. 461, 543
Maqueda, E. 1441
Marafão, Fernando P...............................1435
Marafão, Fernando Pinhabel 269, 445
Marcilio, Wagner 155

Marin, Caroline 1378
Marques, Luciana 429
Marques, Rúbio Campos 579
Marques, Vanessa Da Costa 987
Marra, Enes Goncalves 1070
Martins, Débora De Souza 1337
Martins, Denizar C. 251
Martins, Denizar Cruz 455, 651
Martins, Eduardo 155
Martins, Mário L. Da S. 856
Matakas, Lourenço.................411, 429, 435, 1195
Matias, Rafael Rocha 1319, 1447
Mattavelli, Paolo 766
Mattos, Everson 519
Mayer, Robson.................................1217, 1354
Medeiros, Lucas Taylan P. 862, 958
Medeiros, Renan L. P.............31, 172, 190, 1076
Medina, Augusto César Rueda 1426
Melo, Fernando C. 288, 1253
Melo, Victor F. M. B. 1100, 1475
Mendes, Mariana A.............................898, 1378
Mendes, Mariana Altoé............................1
Mendes, Victor F. 364, 466, 478
Mendes, Victor Flores 537
Mendonça, D. C. 573, 790
Mendonça, Gabriel A. 850
Mendonça, Lucas S. 233
Meneghetti, Luiz H................................993
Meneghetti, Luiz Henrique....................1142, 1420
Mengatto, Alisson 633
Mertens, Thomas M. 37
Mesquita, Daniel De Bastos 1130
Meynard, Thierry..................................1408
Mezaroba, M. 220
Mezaroba, Marcello 208, 603
Michels, Leandro..............................406, 940
Miranda, Luis Colque 710
Mollica, Denis 85, 131
Montagner, Vinícius F. 202, 519, 549
Montecinos, Werner Jara 710
Monteiro, Amanda Thayla S 1319
Monteiro, Amanda Thayla Silva 1447
Monteiro, Felipe Alexandre........................1453
Monteiro, José R. B. A. 108, 335, 484
Monteiro, Jose Roberto B. A. 892
Monteiro, Marina V. C. 1028
Moraes, Caio G. Da S. 388, 394, 692
Moraes, Cassiano F. 1148
Moraes, Cassiano Ferro 245, 1223
Morais, Douglas Carvalho.............1136, 1211, 1487
Morais, L. M. F...............................178, 790
Moreira, Adson B. 958
Moreira, Adson Bezerra 862
Moreira, Ana Carolina 495

AUTHOR INDEX

Moreira, Hugo Soeiro 1130
Moreira, Marcos 155
Morentin, Alvaro 1408
Mota, Denisia De V. 958
Mussa, S. A. 1016
Mussa, Samir A. 796
Mussa, Samir Ahmad 710
Musse, Bernardo F. 742
Nadal, Zeno L. I. 993
Nadal, Zeno Luiz Iensen 1142, 1420
Naidon, Thiago C. 233
Narusue, Yoshiaki 137
Nascimento, Felipe C. Do 1331
Nascimento, Saulo O. 730
Neira, S. 1402
Neres, Fábio De P. 1118
Neto, Antônio O. Costa 98
Neto, Augusto Nery De Lima 317
Neto, João Amin Moor 934
Neto, João T. De Carvalho 1265
Neto, Jose Antonio Dos Santos 1319
Neto, Pedro Jose Dos Santos 686
Neto, R. C. 716
Neves, F. A. S. 220, 716
Neves, Francisco De Assis Dos Santos 208
Neves, Marcello 880
Neves, Vitor G. 772
Nicolai, Ulrich 1235
Nicolini, André Miguel 1366
Nied, Ademir 299, 923, 1325, 1384
Nolasco, Rafael César 537
Norambuena, Margarita 929
Nunes, Evandro Ailson De Freitas 1283
Ogoulola, Christel Enock Ghislain 49
Oliveira, Alexandre Cunha 1463
Oliveira, Demercil De S. 591, 904
Oliveira, Erik C. 1390
Oliveira, Fellipe 838
Oliveira, Hércules A. 555
Oliveira, Hércules Araújo 275
Oliveira, Isabel R. H. 609
Oliveira, Janaina G. 730, 1052, 1058
Oliveira, Matheus H. M. Zanchetta 1004
Oliveira, Sérgio Vidal Garcia 1217, 1354
Oliveira, Tatiane Martins 79, 239
Oliveira, Yago F. 1241
Omine, Leandro T. 501
Onofre, João 1195, 1453
Orige, Mateus C. 376
Oshiro, Marcos Roberto 832
Osório, Caio R. D. 202, 519, 549
Ota, João I. Y. 766
Ota, João Inácio Yutaka 736
Pacas, Mario 507

Pacheco, Antonio L. S. 692
Pacher, J. 1441
Pagano, Daniel J. 31, 802, 1172
Pagano, Daniel Juan 382
Paredes, Carlos 710
Paredes, Helmo Kelis Morales 269
Paredes, Marina G. S. P. 1201
Parreiras, Thiago Morais 160
Pastuch, Luiz Fernando Martins 639
Pelizari, Ademir 952
Pellini, Eduardo 411, 429, 435
Pereda, J. 1402
Pereira, Dênis De Castro 370, 579
Pereira, H. A. 573, 615, 790
Pereira, Heverton A. 466, 478, 850
Pereira, Luana K. Melgaço 674
Pereira, Rondineli R. 131, 531
Pereira, Rondineli Rodrigues 85
Pereira, Thiago A. 251
Pessoa, Guilherme A. P. De C. A. 668
Pessoa, Guilherme Afonso Pillon De C. A. 1283
Piccoli, Daniel Cesar 495
Pimenta, Marcio Annibal 1189
Pimentel, Sergio Pires 1070, 1481
Pineda, C. 1402
Pinheiro, Guilherme G. 531
Pinheiro, Humberto 929
Pinheiro, Lilian V. 742
Pinheiro, Lucas De Paula Assunção 555
Pinheiro, Ricardo Ferreira 1283
Pinheiro, Vinícius Marcos 981
Piontkewicz, Rodrigo Jose 495
Pomilio, José A. 766
Pomilio, José Antenor 311, 317, 645,
657, 736, 754, 1201, 1247
Pompermayer, Daniel C. 1378
Possamai, Carlos Eduardo 917
Possamai, Maicon Douglas 1354
Pozo, Isaac 507
Pozo, Marcelo 507
Pozo, Nataly 507
Praça, Paulo 1124
Praça, Paulo P. 904
Praça, Paulo Peixoto 561
Prado, Ricardo Alves Do 1235
Puma, J. L. Azcue 73
Qiu, Hao 137
Queiroz, Eliabe Duarte 645, 657
Queiroz, Fernando 1124
Queiroz, Luann G. O 1378
Quizhpi, Flavio 1289
Rabelo, Everton Bernard Figueiredo 370
Rael, Victor 429
Ramos, Gabriel Vilkn 1112, 1313
Ramos, Marcos Leonardo 49

AUTHOR INDEX

Raposo, Rafael Fernandes 233
Rech, Cassiano 603, 724, 940, 1094
Regina, Bruno De Almeida................................826
Rego, Rosana C. B...120
Reis, Victor Camargo 1271
Rendón, Manuel A. 730, 1052
Resende, Ênio C. 288
Rezek, Angelo José Junqueira 49
Ribeiro, Brunno Monteiro Guimarães.................1206
Ribeiro, Enio Roberto...166
Ribeiro, Luiz A. De S. 275, 400, 555, 1229
Ribeiro, Luiz Felipe Corrêa De Sá Santos 184
Ribeiro, Paulo F. 531
Ricciotti, Antonio Carlos Duarte 1366
Ricciotti, Viviane B. Da S. Duarte 1366
Riedemann, Javier 710
Rios, Lara Ana Rodarte 79, 239
Rios, Nicolas Parma..808
Rivera, M...1441
Rivera, Marco 8, 61
Riveros, José A. 8, 61
Rocha, Filipe V. 1154
Rocha, Nady 868, 987, 1022, 1154
Rocha-Osorio, C. M. 73
Rodrigues, André Augusto 1088
Rodrigues, Andressa De Melo.............................760
Rodrigues, Gabriel Sales Lins............................1463
Rodrigues, Gleice M. S. 868
Rodrigues, Gleice Mylena Da S.........................1154
Rodrigues, Gleice Mylena Da Silva 987
Rodrigues, José Carlos Grilo 49
Rodrigues, Márcio C. B. P............. 704, 1052, 1241
Rodrigues, Marcus Vinícius Maia 1343
Rodríguez, J..1402
Rodriguez, Jose 929, 1481
Roe, Maurice G. L..43
Roig, Mateo D. G...................................388, 394
Rojas, Jeimy C. Sanabria 294
Rolim, Luís G. B..1010
Rolim, Luis Guilherme Barbosa 184
Roman, Kosenko 1178
Romero, C..1441
Rospirski, Alexandra..796
Ruiz, Flávia P...958
Ruiz-Caballero, Domingo.....................................710
Ruppert, Ernesto...347
Sá, Edilson M. 1118, 1331
Saavedra, Osvaldo Ronald 275, 555
Sacco, Francesco 411
Saccol, Gabriel Avila 940
Sakô, Elson Yoiti 1130
Sakurai, Takayasu 137
Salazar, Andrés O. 668
Salazar, Andres Ortiz 1265, 1283

Salvador, Marcos Antonio 639
Salvadori, Fabiano 820
Sampaio, Leonardo Poltronieri 1301
Sant'Ana, Wilson C. 85, 131
Santana, Luan S. 778
Santana, R. A. S. 178
Santana, Wilson Cesar 531
Santiago, Raphael Perci 1396, 1458
Santisteban, José Andrés 946
Santos, Hugo E. 149
Santos, João 155
Sarlioglu, Bulent 67
Sarrias-Mena, Raúl 904
Scalcon, Filipe 698
Scalcon, Filipe P. 549
Schardong, Charles 940
Scherer, Lucas Giuliani 1064
Schlickmann, Henrique R.....................................820
Schmidt, Gustavo B. K.......................................1420
Schmidt, Luiz H. T. 802
Schmitz, Lenon 651
Schowantz, G. S. 969
Schuetz, Dimas Alã 929
Schulter, Wolfgang 952
Seixas, P. F. 178
Seleme, S. I. 573
Seleme, Seleme I. 674, 772
Silva, Andre Felicio De Sousa 1070
Silva, Danilo P. E 1337
Silva, Gabriel S. Barbara Da S. E 964
Silva, Jailson Leite...............................1319, 1447
Silva, João Lucas De Souza 1130
Silva, Laylla Fernandes..760
Silva, Leonardo P. S. 862, 958
Silva, Luciano De Souza Da Costa E..................760,
    1136, 1211, 1487
Silva, R. P. 615
Silva, Rafael M. 364, 772
Silva, Ranoyca Nayana Alencar
Leão E 1447
Silva, Renata C. 772
Silva, Sidelmo Magalhães 1004
Silva, Vinicius Zimmermann 49
Silva, Walquíria Do N. 567
Silveira, Joao Pedro C. 686
Simões, Marcelo Godoy 269
Simonetti, Domingos 844
Simonetti, Domingos S. L. 329
Singh, Mukhtiar 1499
Soares, Ana L. 98
Soares, Débora M. 92
Soares, Emerson L. 1022
Soares, Guilherme Marcio.........370, 423, 579, 808
Soares, Marcus Vieira 784
Solís-Chaves, J. S...73

AUTHOR INDEX

Sousa, Clodualdo V. 364, 772
Sousa, Silas M. 609
Sousa, Thiago William Pires 490
Souza, Bruno C. 19, 585, 1360
Souza, Géssica C. De A. 668
Souza, Lucas Carvalho 760, 1136, 1211, 1487
Souza, Marcus E. T. 1253
Spiazzi, Giorgio 114, 143
Stangler, E. V. 220, 716
Stangler, Eduardo Vasconcelos............................208
Stefanelo, Márcio 698
Stein, Carlos M. 993
Stein, Carlos M. O. 472, 1148
Stein, Carlos Marcelo De Oliveira 1420
Stein, Carlos Marcelo Oliveira...........................1142
Stepenko, Serhii 1481
Stephan, Richard M. 37, 92, 621
Stopa, Marcelo M. 850
Stopa, Marcelo Martins 1112, 1206, 1313
Suárez, José H. 778
Tahim, André P. N. 964
Takamiya, Makoto 137
Tao, Chengxuan 104, 353
Tavares, Pedro Laguardia.....................................579
Tedeschi, Elisabetta 445
Teixeira, Estêvão Coelho......................................808
Teixeira, Rodrigo De A. 668
Teixeira, Vanessa S. C. 958
Teixeira, Vanessa Siqueira De C..........................862
Tennakoon, Sarath..227
Teodorescu, R. 573
Teston, Silvio Antonio.................................603, 1094
Tian, Haonan 323
Tibola, Gabriel 43
Tofoli, Fernando Lessa 79, 166, 239, 281,
     370, 579, 1082, 1088
Toledo, S. 1441
Tonini, Luiz G. R. 898, 1378
Toniolo, Francesco 143
Torres, Bruno Silva..85
Torres, Renato A. 67
Torres, Vitor C. S. 1052
Tripathi, Anshuman 323
Vaca, David A..1076
Vaisambhayana, Sriram 323
Valentim, Gustavo 429
Vargas, Murillo C. 898, 1378
Vargas, Murillo Cobe 1
Vargas, Rodrigo Z...347
Vasquez, Juan Carlos 686
Vaz, Jerson Rogério Pinheiro 275
Vedovatte, Marcos V. A. 513
Veras, Caio Kerson O. 680
Veras, Leonilson Dos Santos 555
Viajante, Ghunter Paulo 305

Vicente, Eduardo Moreira.....................239, 1082, 1088
Vicente, Paula Dos Santos............................1082, 1088
Vieira, Flávio Henrique Teles..............................1414
Vieira, Rodrigo P..549, 698
Vilela, Wellington M. 149
Vilerá, Kaio Vinicius 603, 1094
Vilkn, P. H. J. 178
Villalva, Marcelo Gradella 1130
Vinnikov, Dmitri 1178, 1481, 1493
Viola, Julio 1289
Vitorino, Montiê A. 1106
Volpato, Cesar 698
Waltrich, Gierri..........................341, 692, 1431
Wang, Haoran 1160
Wang, Huai 1160
Wang, Lei 663
Wang, Lifang 104, 353
Wang, Liye 353
Wang, Qi 8
Watanabe, Edson H. 1034
Wermuth, Matheos C. 461, 543
Wheeler, P. 8, 61, 257, 1441
Wheeler, Pat W. 196
Wintrich, Arendt 1277
Wong, Man-Chung...663
Xavier, Lucas S. 466, 478
Yang, Yongheng 1
Yin, Chengliang 104
Yuan, Xibo 329
Zacarias, Joice D. S. 466
Zambon, Mário..435
Zhang, Ya 43
Zhu, Guorong 1160
Zimann, F. J. 220
Zimann, Felipe Joel 208
Zucuni, Jordan 929